Organic Chemistry

Organic

Chemistry

JONATHAN CLAYDEN

University of Manchester

NICK GREEVES

University of Liverpool

STUART WARREN

University of Cambridge

PETER WOTHERS

University of Cambridge

OXFORD

UNIVERSITY PRESS

OXFORD

UNIVERSITY PRESS

Great Clarendon Street, Oxford OX2 6DP

Oxford University Press is a department of the University of Oxford
It furthers the University's objective of excellence in research, scholarship,
and education by publishing worldwide in

Oxford New York

Auckland Bangkok Buenos Aires
Cape Town Chennai Dar es Salaam Delhi Hong Kong Istanbul
Karachi Kuala Lumpur Madrid Melbourne Mexico City Mumbai
Nairobi São Paulo Singapore Taipei Tokyo Toronto

A catalogue record for this book is available from the British Library

Library of Congress Cataloging in Publication Data
(Data available)

ISBN 0 19 850346 6

Typeset by Wyvern 21 Ltd, Bristol
Printed and bound in Italy by
Giunti Industrie Grafiche, Florence

Preface

Students of Organic Chemistry are not hard-pressed to find a general text to support their learning during the first years at University. The shelves of a University bookshop will usually offer a choice of at least half a dozen – all entitled Organic Chemistry; all with substantially more than 1000 pages. Closer inspection of these titles quickly disappoints expectations of variety. Almost without exception, general Organic Chemistry texts have been written to accompany traditional American sophomore courses, and their rather precisely defined requirements. This has left the authors of these books little scope for reinvigorating their presentation of chemistry with new ideas.

We wanted to write a book whose structure develops around the development of ideas rather than on the sequential presentation of facts. We believe that students benefit most of all from a book which leads from familiar concepts to unfamiliar ones, not just encouraging them to *know* but to *understand* and to understand *why*. We were spurred on by the nature of the best modern University chemistry courses, which themselves follow this pattern: this is after all how science itself develops. We also knew that if we did this we could, from the start, relate the chemistry we were talking about to the two most important sorts of chemistry that exist – the chemistry that is known as life, and the chemistry as practised by chemists solving real problems in laboratories.

We aimed at an approach which would make sense to and appeal to today's students. But all of this meant taking the axe to the roots of some long-standing textbook traditions. The best way to find out how something works is to take it apart and put it back together again, so we started with the tools for constructing chemical ideas. Chemists present chemistry in terms of structural diagrams, and for this reason our first chemical chapter, Chapter 2, is not about bonding or about alkanes, but about drawing structures well—the handwriting of chemistry. We want students to appreciate that the structures they write have demonstrable reality so we introduce spectroscopy in Chapter 3 as our means of communication with the molecular world. It is only as far as spectroscopy reveals it that the chemistry described in the book can be shown to be true.

All practising chemists protect themselves from being crushed by the vastness of organic chemistry by moulding it and ordering it with curly arrows. Without curly arrows, chemistry is chaos, and impossible to learn. Curly arrows unify chemistry, and are essential to the solution of problems. They allow known chemistry to be presented mechanism by mechanism and unknown chemistry to be predicted. We devote most of Chapter 5 to the technique of writing curly arrow mechanisms, and throughout the rest of the book we continually stress the use of curly arrows as the chemist's most important tool outside the laboratory.

Curly arrows work because reactions are orbital-controlled, and we introduce orbitals in Chapters 4 as a way of explaining the structural reality we see with spectroscopy and in Chapter 6 onwards as a way of explaining the mechanistic reality we express in terms of curly arrows. By calling on curly arrows and ordering chemistry according to mechanism we allow ourselves to discuss mechanistically (and orbitally) simple reactions (addition to C=O, for example) before more complex and involved ones such as S_N1 and S_N2. Complexity follows in its own time, but we have deliberately omitted detailed discussion of obscure reactions of little value, or of variants of reactions which lie a simple step of mechanistic logic from our main story: some of these are explored in the problems at the end of each chapter. We have similarly aimed to avoid exhuming principles and rules (from those of Le Chatelier through Markovnikov, Saytseff, Least Motion, and the like) to explain things which are better understood in terms of fundamental thermodynamic or mechanistic concepts.

As four authors, we have found ourselves in general agreement about our precepts of how to present chemistry in an understandable, learnable, interesting and appealing way. But, as is guaranteed among any group of chemists, we have not always found ourselves in agreement about the details. We have presented chemistry as something whose essence is truth, of provable veracity, but which is embellished with opinions and suggestions to which not all chemists subscribe. We aim to avoid dogma, and promote the healthy weighing up of evidence, and on occasion we are content to leave readers to draw their own conclusions.

We worked separately on different chapters, meeting three or four times a year during the four years of writing to discuss our progress and examine critically each others' work. Some chapters received important revision in the hands of a second author, and jarring changes of style or overlaps or omissions of material were eliminated by thorough revision of each chapter by a second (or third) author. Some external reviewers also made very helpful comments. We believe that the chapters that you now read retain some of their inherent individuality, but remain unified by having adapted together to the same environment.

The driving force behind the slow but steady nurturing process which brought each chapter to maturity was our editor, Michael Rodgers, without whose gently insistent encouragement we would have lost hope of ever seeing the book complete well before Chapter 53. Our families and friends forebore early mornings, late nights and lost weekends in front of Macintoshes and piles of proofs. Our research groups put up with our absences and preoccupations. We would like to thank and express our appreciation to all of these.

Science is important not just to scientists, but to society. Our aim has been to write a book which itself takes a scientific standpoint – 'one foot inside the boundary of the known, the other just outside'[1]—and encourages the reader to do the same.

Manchester	JPC
Liverpool	NG
Cambridge	SW
Cambridge	PDW

1. McEvedy, C. *The Penguin Atlas of Ancient History*, Penguin Books 1967.

Outline

Contents

Julian Hutcheo

Julian Hutcheon

What is organic chemistry?

1

Organic chemistry and you

You are already a highly skilled organic chemist. As you read these words, your eyes are using an organic compound (retinal) to convert visible light into nerve impulses. When you picked up this book, your muscles were doing chemical reactions on sugars to give you the energy you needed. As you understand, gaps between your brain cells are being bridged by simple organic molecules (neuro-transmitter amines) so that nerve impulses can be passed around your brain. And you did all that without consciously thinking about it. You do not yet understand these processes in your mind as well as you can carry them out in your brain and body. You are not alone there. No organic chemist, however brilliant, understands the detailed chemical working of the human mind or body very well.

We, the authors, include ourselves in this generalization, but we are going to show you in this book what enormous strides have been taken in the understanding of organic chemistry since the science came into being in the early years of the nineteenth century. Organic chemistry began as a tentative attempt to understand the chemistry of life. It has grown into the confident basis of vast multinational industries that feed, clothe, and cure millions of people without their even being aware of the role of chemistry in their lives. Chemists cooperate with physicists and mathematicians to understand how molecules behave and with biologists to understand how molecules determine life processes. The development of these ideas is already a revelation at the beginning of the twenty-first century, but is far from complete. We aim not to give you the measurements of the skeleton of a dead science but to equip you to understand the conflicting demands of an adolescent one.

Like all sciences, chemistry has a unique place in our pattern of understanding of the universe. It is the science of molecules. But organic chemistry is something more. It literally creates itself as it grows. Of course we need to study the molecules of nature both because they are interesting in their own right and because their functions are important to our lives. Organic chemistry often studies life by making new molecules that give information not available from the molecules actually present in living things.

This creation of new molecules has given us new materials such as plastics, new dyes to colour our clothes, new perfumes to wear, new drugs to cure diseases. Some people think that these activities are unnatural and their products dangerous or unwholesome. But these new molecules are built by humans from other molecules found on earth using the skills inherent in our natural brains. Birds build nests; man makes houses. Which is unnatural? To the organic chemist this is a meaningless distinction. There are toxic compounds and nutritious ones, stable compounds and reactive ones—but there is only one type of chemistry: it goes on both inside our brains and bodies and also in our flasks and reactors, born from the ideas in our minds and the skill in our hands. We are not going to set ourselves up as moral judges in any way. We believe it is right to try and understand the world about us as best we can and to use that understanding creatively. This is what we want to share with you.

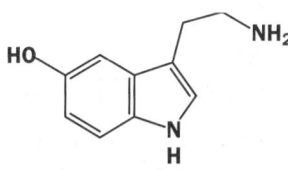

11-*cis*-retinal
absorbs light when we see

serotonin
human neurotransmitter

> We are going to give you structures of organic compounds in this chapter—otherwise it would be rather dull. If you do not understand the diagrams, do not worry. Explanation is on its way.

Organic compounds

Organic chemistry started as the chemistry of life, when that was thought to be different from the chemistry in the laboratory. Then it became the chemistry of carbon compounds, especially those found in coal. Now it is both. It is the chemistry of the compounds of carbon along with other elements such as are found in living things and elsewhere.

■
You will be able to read towards the end of the book (Chapters 49–51) about the extraordinary chemistry that allows life to exist but this is known only from a modern cooperation between chemists and biologists.

The organic compounds available to us today are those present in living things and those formed over millions of years from dead things. In earlier times, the organic compounds known from nature were those in the 'essential oils' that could be distilled from plants and the alkaloids that could be extracted from crushed plants with acid. Menthol is a famous example of a flavouring compound from the essential oil of spearmint and *cis*-jasmone an example of a perfume distilled from jasmine flowers.

menthol   *cis*-jasmone   quinine

Even in the sixteenth century one alkaloid was famous—quinine was extracted from the bark of the South American cinchona tree and used to treat fevers, especially malaria. The Jesuits who did this work (the remedy was known as 'Jesuit's bark') did not of course know what the structure of quinine was, but now we do.

The main reservoir of chemicals available to the nineteenth century chemists was coal. Distillation of coal to give gas for lighting and heating (mainly hydrogen and carbon monoxide) also gave a brown tar rich in aromatic compounds such as benzene, pyridine, phenol, aniline, and thiophene.

benzene   pyridine   phenol   aniline   thiophene

Phenol was used by Lister as an antiseptic in surgery and aniline became the basis for the dyestuffs industry. It was this that really started the search for new organic compounds made by chemists rather than by nature. A dyestuff of this kind—still available—is Bismarck Brown, which should tell you that much of this early work was done in Germany.

Bismarck Brown Y

■
You can read about polymers and plastics in Chapter 52 and about fine chemicals throughout the book.

CH_3——$(CH_2)_{\overline{n}}$——CH_3

n = an enormous number
length of molecule is $n + 2$
carbon atoms

CH_3——$(CH_2)_{\overline{n}}$——CH_2——CH_3

n = an enormous number
length of molecule is $n + 3$
carbon atoms

In the twentieth century oil overtook coal as the main source of bulk organic compounds so that simple hydrocarbons like methane (CH_4, 'natural gas') and propane ($CH_3CH_2CH_3$, 'calor gas') became available for fuel. At the same time chemists began the search for new molecules from new sources such as fungi, corals, and bacteria and two organic chemical industries developed in parallel—'bulk' and 'fine' chemicals. Bulk chemicals like paints and plastics are usually based on simple molecules produced in multitonne quantities while fine chemicals such as drugs, perfumes, and flavouring materials are produced in smaller quantities but much more profitably.

At the time of writing there were about 16 million organic compounds known. How many more are possible? There is no limit (except the number of atoms in the universe). Imagine you've just made the longest hydrocarbon ever made—you just have to add another carbon atom and you've made another. This process can go on with any type of compound *ad infinitum*.

But these millions of compounds are not just a long list of linear hydrocarbons; they embrace all kinds of molecules with amazingly varied properties. In this chapter we offer a selection.

What do they *look* like? They may be crystalline solids, oils, waxes, plastics, elastics, mobile or volatile liquids, or gases. Familiar ones include white crystalline sugar, a cheap natural compound isolated from plants as hard white crystals when pure, and petrol, a mixture of colourless, volatile, flammable hydrocarbons. Isooctane is a typical example and gives its name to the octane rating of petrol.

The compounds need not lack colour. Indeed we can soon dream up a rainbow of organic compounds covering the whole spectrum, not to mention black and brown. In this table we have avoided dyestuffs and have chosen compounds as varied in structure as possible.

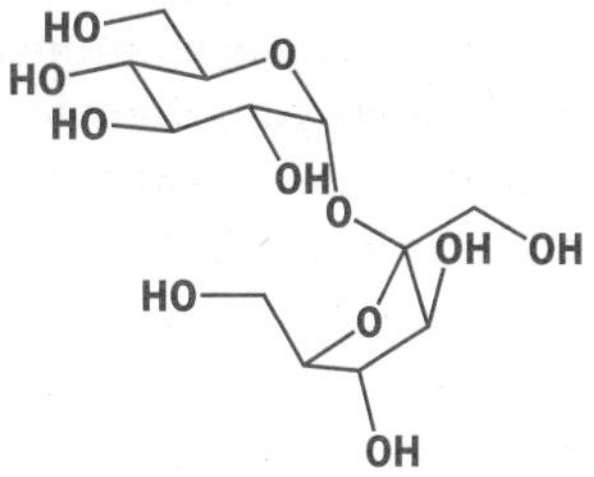

sucrose – ordinary sugar
isolated from sugar cane
or sugar beet
white crystalline solid

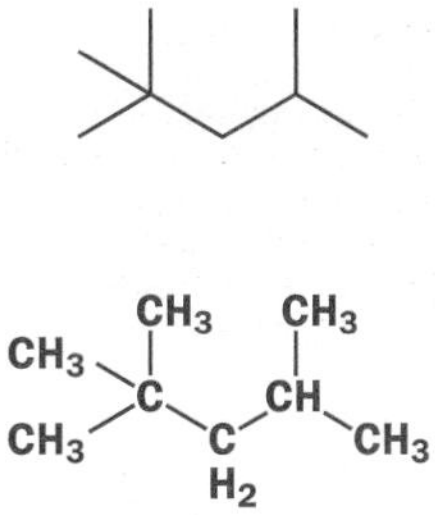

isooctane (2,2,5-trimethylpentane)
a major constituent of petrol
volatile inflammable liquid

Colour	Description	Compound	Structure
red	dark red hexagonal plates	3′-methoxybenzocycloheptatriene-2′-one	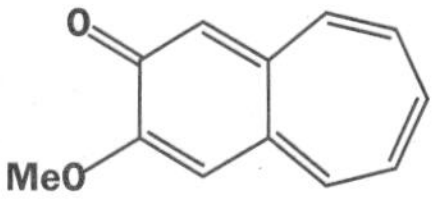
orange	amber needles	dichloro dicyano quinone (DDQ)	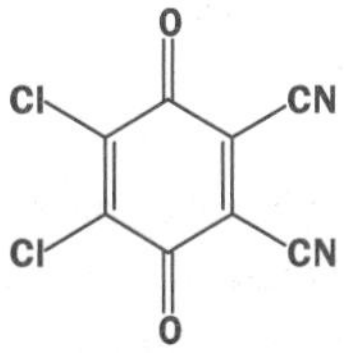
yellow	toxic yellow explosive gas	diazomethane	
green	green prisms with a steel-blue lustre	9-nitroso julolidine	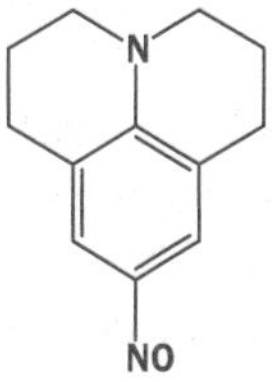
blue	deep blue liquid with a peppery smell	azulene	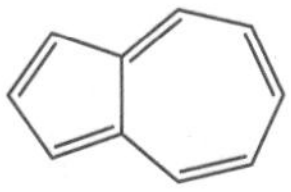
purple	deep blue gas condensing to a purple solid	nitroso trifluoromethane	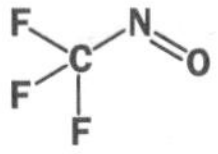

Colour is not the only characteristic by which we recognize compounds. All too often it is their odour that lets us know they are around. There are some quite foul organic compounds too; the smell of the skunk is a mixture of two thiols—sulfur compounds containing SH groups.

skunk spray contains:

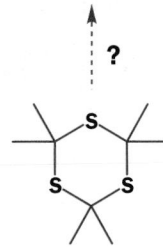

thioacetone

?

trithioacetone;
Freiburg was evacuated
because of a smell from
the distillation this compound

propane
dithiol

4-methyl-4-
sulfanylpentan-
2-one

two candidates for
the worst smell in the world

no-one wants to find the winner!

the divine smell
of the black truffle
comes from this compound

damascenone - the smell of roses

But perhaps the worst aroma was that which caused the evacuation of the city of Freiburg in 1889. Attempts to make thioacetone by the cracking of trithioacetone gave rise to 'an offensive smell which spread rapidly over a great area of the town causing fainting, vomiting and a panic evacuation 'the laboratory work was abandoned'.

It was perhaps foolhardy for workers at an Esso research station to repeat the experiment of cracking trithioacetone south of Oxford in 1967. Let them take up the story. 'Recently we found ourselves with an odour problem beyond our worst expectations. During early experiments, a stopper jumped from a bottle of residues, and, although replaced at once, resulted in an immediate complaint of nausea and sickness from colleagues working in a building two hundred yards away. Two of our chemists who had done no more than investigate the cracking of minute amounts of trithioacetone found themselves the object of hostile stares in a restaurant and suffered the humiliation of having a waitress spray the area around them with a deodorant. The odours defied the expected effects of dilution since workers in the laboratory did not find the odours intolerable … and genuinely denied responsibility since they were working in closed systems. To convince them otherwise, they were dispersed with other observers around the laboratory, at distances up to a quarter of a mile, and one drop of either acetone *gem*-dithiol or the mother liquors from crude trithioacetone crystallisations were placed on a watch glass in a fume cupboard. The odour was detected downwind in seconds.'

There are two candidates for this dreadful smell—propane dithiol (called acetone *gem*-dithiol above) or 4-methyl-4-sulfanylpentan-2-one. It is unlikely that anyone else will be brave enough to resolve the controversy.

Nasty smells have their uses. The natural gas piped to our homes contains small amounts of deliberately added sulfur compounds such as *tert*-butyl thiol $(CH_3)_3CSH$. When we say small, we mean *very small*—humans can detect one part in 50 000 000 000 parts of natural gas.

Other compounds have delightful odours. To redeem the honour of sulfur compounds we must cite the truffle which pigs can smell through a metre of soil and whose taste and smell is so delightful that truffles cost more than their weight in gold. Damascenones are responsible for the smell of roses. If you smell one drop you will be disappointed, as it smells rather like turpentine or camphor, but next morning you and the clothes you were wearing will smell powerfully of roses. Just like the compounds from trithioacetone, this smell develops on dilution.

Humans are not the only creatures with a sense of smell. We can find mates using our eyes alone (though smell does play a part) but insects cannot do this. They are small in a crowded world and they find others of their own species and the opposite sex by smell. Most insects produce volatile compounds that can be picked up by a potential mate in incredibly weak concentrations. Only 1.5 mg of serricornin, the sex pheromone of the cigarette beetle, could be isolated from 65 000 female beetles—so there isn't much in each beetle. Nevertheless, the slightest whiff of it causes the males to gather and attempt frenzied copulation.

The sex pheromone of the Japanese beetle, also given off by the females, has been made by chemists. As little as 5 μg (micrograms, note!) was more effective than four virgin females in attracting the males.

serricornin

the sex pheromone of the cigarette beetle
Lasioderma serricorne

japonilure

the sex pheromone of the Japanese beetle
Popilia japonica

The pheromone of the gypsy moth, disparlure, was identified from a few μg isolated from the moths and only 10 μg of synthetic material. As little as 2×10^{-12} g is active as a lure for the males in field tests. The three pheromones we have mentioned are available commercially for the specific trapping of these destructive insect pests.

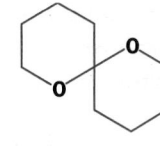

disparlure

the sex pheromone of the Gypsy moth
Portheria dispar

olean

sex pheromone of the olive fly
Bacrocera oleae

this mirror image isomer
attracts the males

this mirror image isomer
attracts the females

Don't suppose that the females always do all the work; both male and female olive flies produce pheromones that attract the other sex. The remarkable thing is that one mirror image of the molecule attracts the males while the other attracts the females!

What about taste? Take the grapefruit. The main flavour comes from another sulfur compound and human beings can detect 2×10^{-5} parts per billion of this compound. This is an almost unimaginably small amount equal to 10^{-4} mg per tonne or a drop, not in a bucket, but in a good-sized lake. Why evolution should have left us abnormally sensitive to grapefruit, we leave you to imagine.

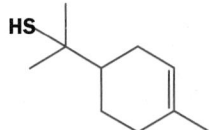

flavouring principle of grapefruit

For a nasty taste, we should mention 'bittering agents', put into dangerous household substances like toilet cleaner to stop children eating them by accident. Notice that this complex organic compound is actually a salt—it has positively charged nitrogen and negatively charged oxygen atoms—and this makes it soluble in water.

bitrex
denatonium benzoate
benzyldiethyl[(2,6-xylylcarbamoyl)methyl]ammonium benzoate

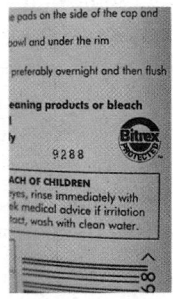

Other organic compounds have strange effects on humans. Various 'drugs' such as alcohol and cocaine are taken in various ways to make people temporarily happy. They have their dangers. Too much alcohol leads to a lot of misery and any cocaine at all may make you a slave for life.

Again, let's not forget other creatures. Cats seem to be able to go to sleep at any time and recently a compound was isolated from the cerebrospinal fluid of cats that makes them, or rats, or humans go off to sleep quickly. It is a surprisingly simple compound.

This compound and disparlure are both derivatives of fatty acids, molecules that feature in many of the food problems people are so interested in now (and rightly so). Fatty acids in the diet are a popular preoccupation and the good and bad qualities of saturates, monounsaturates, and polyunsaturates are continually in the news. This too is organic chemistry. One of the latest molecules to be recognized as an anticancer agent in our diet is CLA (conjugated linoleic acid) in dairy products.

CH_3 OH

alcohol
(ethanol)

CH_3 CO_2Me

cocaine
- an addictive alkaloid

a sleep-inducing fatty acid derivative
cis-9,10-octadecenoamide

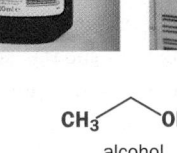

CLA (Conjugated Linoleic Acid)
cis-9-*trans*-11 conjugated linoleic acid
dietary anticancer agent

Another fashionable molecule is resveratrole, which may be responsible for the beneficial effects of red wine in preventing heart disease. It is a quite different organic compound with two benzene rings and you can read about it in Chapter 51.

For our third edible molecule we choose vitamin C. This is an essential factor in our diets—indeed, that is why it is called a vitamin. The disease scurvy, a degeneration of soft tissues, particularly in the mouth, from which sailors on long voyages like those of Columbus suffered, results if we don't have vitamin C. It also is a universal antioxidant, scavenging for rogue free radicals and so protecting us against cancer. Some people think an extra large intake protects us against the common cold, but this is not yet proved.

resveratrole from the skins of grapes
is this the compound in red wine
which helps to prevent heart disease?

> Vitamin C (ascorbic acid) is a vitamin for primates, guinea-pigs, and fruit bats, but other mammals can make it for themselves.

vitamin C (ascorbic acid)

Organic chemistry and industry

Vitamin C is manufactured on a huge scale by Roche, a Swiss company. All over the world there are chemistry-based companies making organic molecules on scales varying from a few kilograms to thousands of tonnes per year. This is good news for students of organic chemistry; there are lots of jobs around and it is an international job market. The scale of some of these operations of organic chemistry is almost incredible. The petrochemicals industry processes (and we use the products!) over 10 million litres of crude oil every day. Much of this is just burnt in vehicles as petrol or diesel, but some of it is purified or converted into organic compounds for use in the rest of the chemical industry. Multinational companies with thousands of employees such as Esso (Exxon) and Shell dominate this sector.

Some simple compounds are made both from oil and from plants. The ethanol used as a starting material to make other compounds in industry is largely made by the catalytic hydration of ethylene from oil. But ethanol is also used as a fuel, particularly in Brazil where it is made by fermentation of sugar cane wastes. This fuel uses a waste product, saves on oil imports, and has improved the quality of the air in the very large Brazilian cities, Rio de Janeiro and São Paulo.

monomers for polymer
manufacture

styrene

Plastics and polymers take much of the production of the petrochemical industry in the form of monomers such as styrene, acrylates, and vinyl chloride. The products of this enormous industry are everything made of plastic including solid plastics for household goods and furniture, fibres for clothes (24 million tonnes per annum), elastic polymers for car tyres, light bubble-filled polymers for packing, and so on. Companies such as BASF, Dupont, Amoco, Monsanto, Laporte, Hoechst, and ICI are leaders here. Worldwide polymer production approaches 100 million tonnes per annum and PVC manufacture alone employs over 50 000 people to make over 20 million tonnes per annum.

acrylates

The washing-up bowl is plastic too but the detergent you put in it belongs to another branch of the chemical industry—companies like Unilever (Britain) or Procter and Gamble (USA) which produce soap, detergent, cleaners, bleaches, polishes, and all the many essentials for the modern home. These products may be lemon and lavender scented but they too mostly come from the oil industry. Nowadays, most products of this kind tell us, after a fashion, what is in them. Try this example—a well known brand of shaving gel along with the list of contents on the container:

vinyl chloride

Does any of this make any sense?

It doesn't all make sense to us, but here is a possible interpretation. We certainly hope the book will set you on the path of understanding the sense (and the nonsense!) of this sort of thing.

Ingredients
aqua, palmitic acid, triethanolamine, glycereth-26, isopentane, oleamide-DEA, oleth-2, stearic acid, isobutane, PEG-14M, parfum, allantoin, hydroxyethyl-cellulose, hydroxypropyl-cellulose, PEG-150 distearate, CI 42053, CI 47005

Ingredient	Chemical meaning	Purpose
aqua	water	solvent
palmitic acid	$CH_3(CH_2)_{14}CO_2H$	acid, emulsifier
triethanolamine	$N(CH_2CH_2OH)_3$	base
glycereth-26	glyceryl$(OCH_2CH_2)_{26}OH$	surfactant
isopentane	$(CH_3)_2CHCH_2CH_3$	propellant
oleamide-DEA	$CH_3(CH_2)_7CH=CH(CH_2)_7CONEt_2$	
oleth-2	Oleyl$(OCH_2CH_2)_2OH$	surfactant
stearic acid	$CH_3(CH_2)_{16}CO_2H$	acid, emulsifier
isobutane	$(CH_3)_2CHCH_3$	propellant
PEG-14M	polyoxyethylene glycol ester	surfactant
parfum	perfume	
allantoin	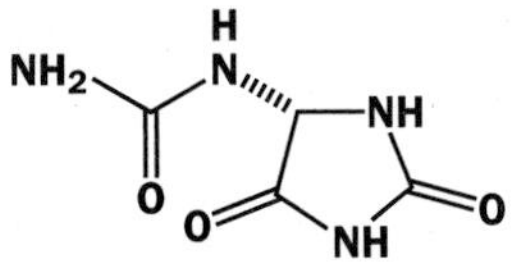 allantoin	promotes healing in case you cut yourself while shaving
hydroxyethyl-cellulose	cellulose fibre from wood pulp with $-OCH_2CH_2OH$ groups added	gives body
hydroxypropyl-cellulose	cellulose fibre from wood pulp with $-OCH_2CH(OH)CH_3$ groups added	gives body
PEG-150 distearate	polyoxyethylene glycol diester	surfactant
CI 42053	Fast Green FCF (see box)	green dye
CI 47005	Quinoline Yellow (see box)	yellow dye

The structures of two dyes

Fast Green FCF and Quinoline Yellow are colours permitted to be used in foods and cosmetics and have the structures shown here. Quinoline Yellow is a mixture of isomeric sulfonic acids in the two rings shown.

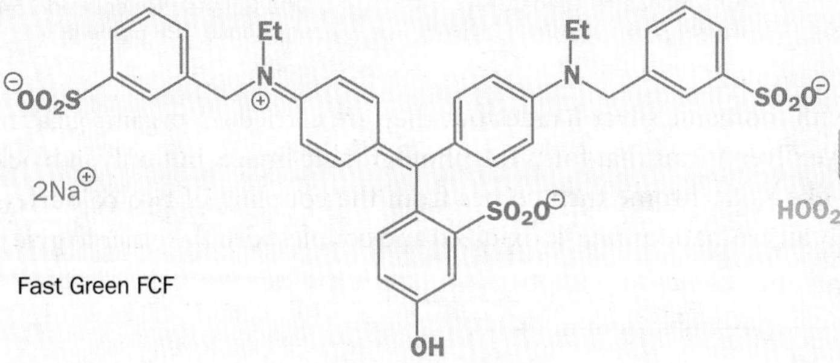

Fast Green FCF

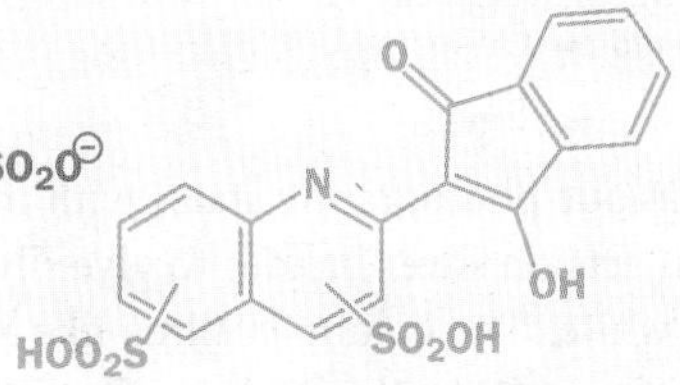

Quinoline Yellow

The particular acids, bases, surfactants, and so on are chosen to blend together in a smooth emulsion when propelled from the can. The result should feel, smell, and look attractive and a greenish colour is considered clean and antiseptic by the customer. What the can actually says is this: 'Superior lubricants within the gel prepare the skin for an exceptionally close, comfortable and effective shave. It contains added moisturisers to help protect the skin from razor burn. Lightly fragranced.'

Another oil-derived class of organic chemical business includes adhesives, sealants, coatings, and so on, with companies like Ciba–Geigy, Dow, Monsanto, and Laporte in the lead. Nowadays aircraft

Superglue bonds things together when this small molecule joins up with hundreds of its fellows in a polymerization reaction

■ The formation of polymers is discussed in Chapter 52.

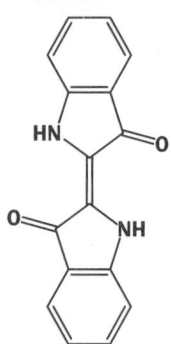

are glued together with epoxy-resins and you can glue almost anything with 'Superglue', a polymer of methyl cyanoacrylate.

There is a big market for intense colours for dyeing cloth, colouring plastic and paper, painting walls, and so on. This is the dyestuffs and pigments industry and leaders here are companies like ICI and Akzo Nobel. ICI have a large stake in this aspect of the business, their paints turnover alone being £2 003 000 000 in 1995.

The most famous dyestuff is probably indigo, an ancient dye that used to be isolated from plants but is now made chemically. It is the colour of blue jeans. More modern dyestuffs can be represented by ICI's benzodifuranones, which give fashionable red colours to synthetic fabrics like polyesters.

We see one type of pigment around us all the time in the form of the colours on plastic bags. Among the best compounds for these are the metal complexes called phthalocyanines. Changing the metal (Cu and Fe are popular) at the centre and the halogens round the edge of these molecules changes the colour but blues and green predominate. The metal atom is not necessary for intense pigment colours—one new class of intense 'high performance' pigments in the orange–red range are the DPP (1,4-diketopyrrolo[3,4-c]pyrroles) series developed by Ciba–Geigy. Pigment Red 254 is used in paints and plastics.

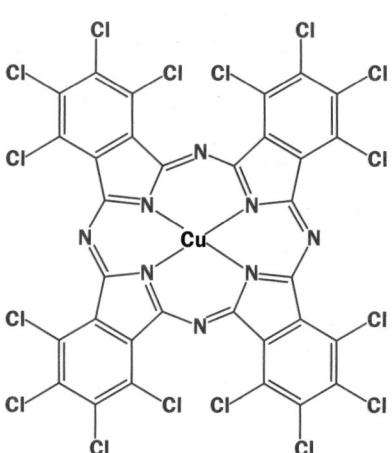

indigo
the colour of blue jeans

ICI's Dispersol benzodifuranone
red dyes for polyester

ICI's Monastral Green GNA
a good green for plastic objects

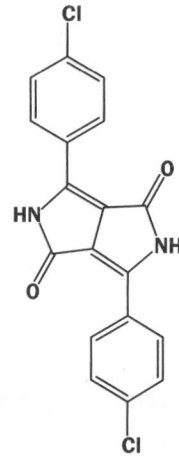

Ciba Geigy's Pigment Red 254
an intense DPP pigment

■ You can read in Chapter 7 why some compounds are coloured and others not.

Colour photography starts with inorganic silver halides but they are carried on organic gelatin. Light acts on silver halides to give silver atoms that form the photographic image, but only in black and white. The colour in films like Kodachrome then comes from the coupling of two colourless organic compounds. One, usually an aromatic amine, is oxidized and couples with the other to give a coloured compound.

colourless aromatic amine

light, silver
photographic developer

magenta pigment from two colourless compounds

colourless cyclic amide

cis-jasmone
the main compound
in jasmine perfume

That brings us to flavours and fragrances. Companies like International Flavours and Fragrances (USA) or Givaudan–Roure (Swiss) produce very big ranges of fine chemicals for the perfume, cosmetic, and food industries. Many of these will come from oil but others come from plant sources. A typical perfume will contain 5–10% fragrances in an ethanol/water (about 90:10) mixture. So the perfumery industry needs a very large amount of ethanol and, you might think, not much perfumery material. In fact, important fragrances like jasmine are produced on a >10 000 tonnes per annum scale. The cost of a pure perfume ingredient like *cis*-jasmone, the main ingredient of jasmine, may be several hundred pounds, dollars, or euros per gram.

The world of perfumery

Perfume chemists use extraordinary language to describe their achievements: 'Paco Rabanne pour homme was created to reproduce the effect of a summer walk in the open air among the hills of Provence: the smell of herbs, rosemary and thyme, and sparkling freshness with cool sea breezes mingling with warm soft Alpine air. To achieve the required effect, the perfumer blended herbaceous oils with woody accords and the synthetic aroma chemical dimethylheptanol which has a penetrating but indefinable freshness associated with open air or freshly washed linen'. (J. Ayres, *Chemistry and Industry*, 1988, 579)

Chemists produce synthetic flavourings such as 'smoky bacon' and even 'chocolate'. Meaty flavours come from simple heterocycles such as alkyl pyrazines (present in coffee as well as roast meat) and furonol, originally found in pineapples. Compounds such as corylone and maltol give caramel and meaty flavours. Mixtures of these and other synthetic compounds can be 'tuned' to taste like many roasted foods from fresh bread to coffee and barbecued meat.

an alkyl pyrazine
from coffee and
roast meat

furonol
roast meat

corylone
caramel
roasted taste

maltol
E-636 for cakes
and biscuits

Some flavouring compounds are also perfumes and may also be used as an intermediate in making other compounds. Two such large-scale flavouring compounds are vanillin (vanilla flavour as in ice cream) and menthol (mint flavour) both manufactured on a large scale and with many uses.

vanillin
found in vanilla pods;
manufactured
on a large scale

menthol
extracted from mint;
25% of the world's supply
manufactured

Food chemistry includes much larger-scale items than flavours. Sweeteners such as sugar itself are isolated from plants on an enormous scale. Sugar's structure appeared a few pages back. Other sweeteners such as saccharin (discovered in 1879!) and aspartame (1965) are made on a sizeable scale. Aspartame is a compound of two of the natural amino acids present in all living things and is made by Monsanto on a large scale (over 10 000 tonnes per annum).

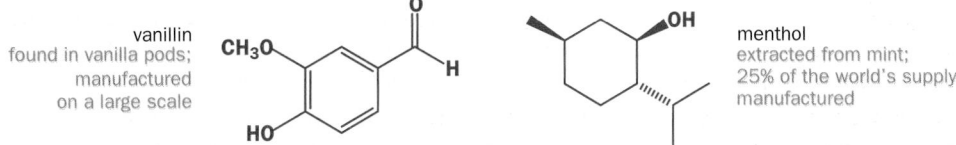

aspartame ('NutraSweet')
200 × sweeter than sugar

is made from
two amino acids –

aspartic
acid

methyl ester of
phenylalanine

The pharmaceutical businesses produce drugs and medicinal products of many kinds. One of the great revolutions of modern life has been the expectation that humans will survive diseases because of a treatment designed to deal specifically with that disease. The most successful drug ever is ranitidine (Zantac), the Glaxo–Wellcome ulcer treatment, and one of the fastest-growing is Pfizer's sildenafil (Viagra). 'Success' refers both to human health and to profit!

You will know people (probably older men) who are 'on β-blockers'. These are compounds designed to block the effects of adrenaline (epinephrine) on the heart and hence to prevent heart disease. One of the best is Zeneca's tenormin. Preventing high blood pressure also prevents heart disease and certain specific enzyme inhibitors (called 'ACE-inhibitors') such as Squibb's captopril work in this way. These are drugs that imitate substances naturally present in the body.

The treatment of infectious diseases relies on antibiotics such as the penicillins to prevent bacteria from multiplying. One of the most successful of these is Smith Kline Beecham's amoxycillin. The four-membered ring at the heart of the molecule is the 'β-lactam'.

Glaxo-Wellcome's ranitidine
the most successful drug to date
world wide sales peaked >£1,000,000,000 per annum

Pfizer's sildenafil (Viagra)
three million satisfied customers in 1998

Zeneca's tenormin
cardioselective β-blocker
for treatment and prevention
of heart disease

Squibb's captopril
specific enzyme inhibitor
for treatment and
prevention of hypertension

SmithKline Beecham's amoxycillin
β-lactam antibiotic
for treatment of bacterial infections

We cannot maintain our present high density of population in the developed world, nor deal with malnutrition in the developing world unless we preserve our food supply from attacks by insects and fungi and from competition by weeds. The world market for agrochemicals is over £10 000 000 000 per annum divided roughly equally between herbicides, fungicides, and insecticides.

At the moment we hold our own by the use of agrochemicals: companies such as Rhône-Poulenc, Zeneca, BASF, Schering–Plough, and Dow produce compounds of remarkable and specific activity. The most famous modern insecticides are modelled on the natural pyrethrins, stabilized against degradation by sunlight by chemical modification (see coloured portions of decamethrin) and targeted to specific insects on specific crops in cooperation with biologists. Decamethrin has a safety factor of >10 000 for mustard beetles over mammals, can be applied at only 10 grams per hectare (about one level tablespoon per football pitch), and leaves no significant environmental residue.

a natural pyrethin
from *pyrethrum* - daisy-like flowers from East Africa

decamethrin
a modified pyrethrin - more active and stable in sunlight

As you learn more chemistry, you will appreciate how remarkable it is that Nature should produce three-membered rings and that chemists should use them in bulk compounds to be sprayed on crops in fields. Even more remarkable in some ways is the new generation of fungicides based on a five-membered ring containing three nitrogen atoms—the triazole ring. These compounds inhibit an enzyme present in fungi but not in plants or animals.

One fungus (potato blight) caused the Irish potato famine of the nineteenth century and the various blights, blotches, rots, rusts, smuts, and mildews can overwhelm any crop in a short time. Especially now that so much is grown in Western Europe in winter, fungal diseases are a real threat.

benomyl
a fungicide which controls
many plant diseases

propiconazole
a *triazole* fungicide

You will have noticed that some of these companies have fingers in many pies. These companies, or groups as they should be called, are the real giants of organic chemistry. Rhône–Poulenc, the French group which includes pharmaceuticals (Rhône–Poulenc–Rorer), animal health, agrochemicals, chemicals, fibres, and polymers, had sales of about 90 billion French Francs in 1996. Dow, the US group which includes chemicals, plastics, hydrocarbons, and other bulk chemicals, had sales of about 20 billion US dollars in 1996.

Organic chemistry and the periodic table

All the compounds we have shown you are built up on hydrocarbon (carbon and hydrogen) skeletons. Most have oxygen and/or nitrogen as well; some have sulfur and some phosphorus. These are the main elements of organic chemistry but another way the science has developed is an exploration of (some would say take-over bid for) the rest of the periodic table. Some of our compounds also had fluorine, sodium, copper, chlorine, and bromine. The organic chemistry of silicon, boron, lithium, the halogens (F, Cl, Br, and I), tin, copper, and palladium has been particularly well studied and these elements commonly form part of organic reagents used in the laboratory. They will crop up throughout this book. These 'lesser' elements appear in many important reagents, which are used in organic chemical laboratories all over the world. Butyllithium, trimethylsilyl chloride, tributyltin hydride, and dimethylcopper lithium are good examples.

The halogens also appear in many life-saving drugs. The recently discovered antiviral compounds, such as fialuridine (which contains both F and I, as well as N and O), are essential for the fight against HIV and AIDS. They are modelled on natural compounds from nucleic acids. The naturally occurring cytotoxic (antitumour) agent halomon, extracted from red algae, contains Br and Cl.

BuLi
butyllithium

Me_3SiCl
trimethylsilyl chloride

Bu_3SnH
tributyltin hydride

Me_2CuLi
dimethylcopper lithium

halomon
naturally occurring
antitumour agent

fialuridine
antiviral compound

Another definition of organic chemistry would use the periodic table. The key elements in organic chemistry are of course C, H, N, and O, but also important are the halogens (F, Cl. Br, I), p-block elements such as Si, S, and P, metals such as Li, Pd, Cu, and Hg, and many more. We can construct an organic chemist's periodic table with the most important elements emphasized:

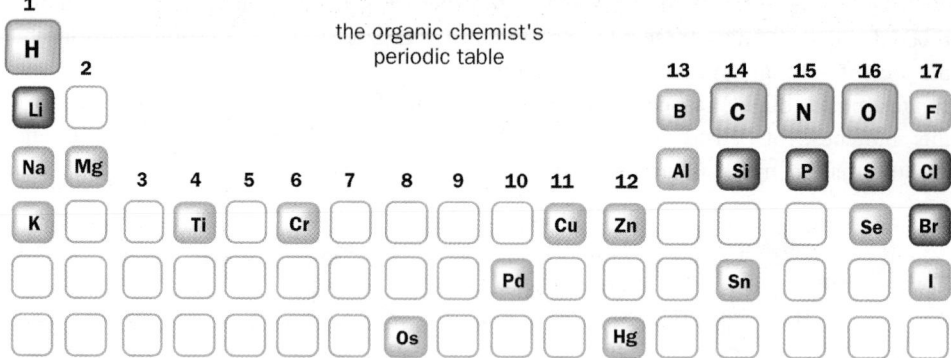

the organic chemist's periodic table

> You will certainly know something about the periodic table from your previous studies of inorganic chemistry. A basic knowledge of the groups, which elements are metals, and roughly where the elements in our table appear will be helpful to you.

So where does inorganic chemistry end and organic chemistry begin? Would you say that the antiviral compound foscarnet was organic? It is a compound of carbon with the formula CPO_5Na_3 but is has no C–H bonds. And what about the important reagent tetrakis triphenylphosphine palladium? It has lots of hydrocarbon—twelve benzene rings in fact—but the benzene rings are all joined to phosphorus atoms that are arranged in a square around the central palladium atom, so the molecule is held together by C–P and P–Pd bonds, not by a hydrocarbon skeleton. Although it has the very organic-looking formula $C_{72}H_{60}P_4Pd$, many people would say it is inorganic. But is it?

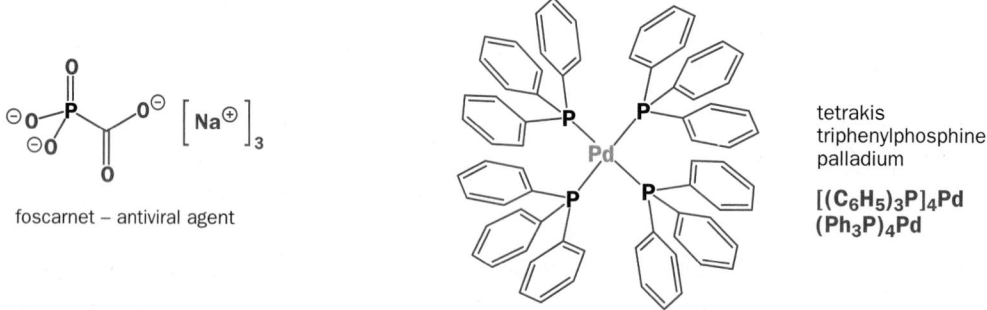

foscarnet – antiviral agent

tetrakis triphenylphosphine palladium

$[(C_6H_5)_3P]_4Pd$
$(Ph_3P)_4Pd$

The answer is that we don't know and we don't care. It is important these days to realize that strict boundaries between traditional disciplines are undesirable and meaningless. Chemistry continues across the old boundaries between organic chemistry and inorganic chemistry on the one side and organic chemistry and biochemistry on the other. Be glad that the boundaries are indistinct as that means the chemistry is all the richer. This lovely molecule $(Ph_3P)_4Pd$ belongs to *chemistry*.

Organic chemistry and this book

We have told you about organic chemistry's history, the types of compounds it concerns itself with, the things it makes, and the elements it uses. Organic chemistry today is the study of the structure and reactions of compounds in nature of compounds, in the fossil reserves such as coal and oil, and of those compounds that can be made from them. These compounds will usually be constructed with a hydrocarbon framework but will also often have atoms such as O, N, S, P, Si, B, halogens, and metals attached to them. Organic chemistry is used in the making of plastics, paints, dyestuffs, clothes, foodstuffs, human and veterinary medicines, agrochemicals, and many other things. Now we can summarize all of these in a different way.

- ● **The main components of organic chemistry as a discipline are these**
  - ● **Structure determination**—how to find out the structures of new compounds even if they are available only in invisibly small amounts
  - ● **Theoretical organic chemistry**—how to understand those structures in terms of atoms and the electrons that bind them together
  - ● **Reaction mechanisms**—how to find out how these molecules react with each other and how to predict their reactions
  - ● **Synthesis**—how to design new molecules—and then make them
  - ● **Biological chemistry**—how to find out what Nature does and how the structures of biologically active molecules are related to what they do

This book is about all these things. It tells you about the structures of organic molecules and the reasons behind them. It tells you about the shapes of those molecules and how the shape relates to their function, especially in the context of biology. It tells you how those structures and shapes are discovered. It tells you about the reactions the molecules undergo and, more importantly, how and why they behave in the way they do. It tells you about nature and about industry. It tells you how molecules are made and how you too can think about making molecules.

We said 'it tells' in that last paragraph. Maybe we should have said 'we tell' because we want to speak to you through our words so that you can see how we think about organic chemistry and to encourage you to develop your own ideas. We expect you to notice that four people have written this book and that they don't all think or write in the same way. That is as it should be. Organic chemistry is too big and important a subject to be restricted by dogmatic rules. Different chemists think in different ways about many aspects of organic chemistry and in many cases it is not yet possible to be sure who is right.

We may refer to the history of chemistry from time to time but we are usually going to tell you about organic chemistry as it is now. We will develop the ideas slowly, from simple and fundamental ones using small molecules to complex ideas and large molecules. We promise one thing. We are not going to pull the wool over your eyes by making things artificially simple and avoiding the awkward questions. We aim to be honest and share both our delight in good complete explanations and our puzzlement at inadequate ones. So how are we going to do this? The book starts with a series of chapters on the structures and reactions of simple molecules. You will meet the way structures are determined and the theory that explains those structures. It is vital that you realize that theory is used to explain what is known by experiment and only then to predict what is unknown. You will meet mechanisms—the dynamic language used by chemists to talk about reactions—and of course some reactions.

The book starts with an introductory section of four chapters:

1   What is organic chemistry?

2   Organic structures

3   Determining organic structures

4   Structure of molecules

In Chapter 2 you will look at the way in which we are going to present diagrams of molecules on the printed page. Organic chemistry is a visual, three-dimensional subject and the way you draw molecules shows how you think about them. We want you too to draw molecules in the best way available now. It is just as easy to draw them well as to draw them in an old-fashioned inaccurate way.

Then in Chapter 3, before we come to the theory of molecular structure, we shall introduce you to the experimental techniques of finding out about molecular structure. This means studying the interactions between molecules and radiation by **spectroscopy**—using the whole electromagnetic spectrum from X-rays to radio waves. Only then, in Chapter 4, will we go behind the scenes and look at the theories of why atoms combine in the ways they do. Experiment comes before theory. The spectroscopic methods of Chapter 3 will still be telling the truth in a hundred years time, but the theories of Chapter 4 will look quite dated by then.

We could have titled those three chapters:

2   What shapes do organic molecules have?

3   How do we know they have those shapes?

4   Why do they have those shapes?

You need to have a grasp of the answers to these three questions before you start the study of organic reactions. That is exactly what happens next. We introduce organic reaction mechanisms in Chapter 5. Any kind of chemistry studies **reactions**—the transformations of molecules into other molecules. The dynamic process by which this happens is called **mechanism** and is the language of organic chemistry. We want you to start learning and using this language straight away so in Chapter 6 we apply it to one important class of reaction. This section is:

5   Organic reactions

6   Nucleophilic addition to the carbonyl group

Chapter 6 reveals how we are going to subdivide organic chemistry. We shall use a mechanistic classification rather than a structural classification and explain one type of reaction rather than one type of compound in each chapter. In the rest of the book most of the chapters describe types of reaction in a mechanistic way. Here is a selection.

9   Using organometallic reagents to make C–C bonds

17   Nucleophilic substitution at saturated carbon

20   Electrophilic addition to alkenes

22   Electrophilic aromatic substitution

29   Conjugate Michael addition of enolates

39   Radicals

Interspersed with these chapters are others on physical aspects, organic synthesis, stereochemistry, structural determination, and biological chemistry as all these topics are important parts of organic chemistry.

'Connections' section

Chemistry is not a linear subject! It is impossible simply to start at the beginning and work through to the end, introducing one new topic at a time, because chemistry is a network of interconnecting ideas. But, unfortunately, a book is, by nature, a beginning-to-end sort of thing. We have arranged the chapters in a progression of difficulty as far as is possible, but to help you find your way around we have included at the beginning of each chapter a 'Connections' section. This tells you three things divided among three columns:

(a) what you should be familiar with before reading the chapter—in other words, which previous chapters relate directly to the material within the chapter ('Building on' column)

(b) a guide to what you will find within the chapter ('Arriving at' column)

(c) which chapters later in the book fill out and expand the material in the chapter ('Looking forward to' column)

The first time you read a chapter, you should really make sure you have read any chapter mentioned under (a). When you become more familiar with the book you will find that the links highlighted in (a) and (c) will help you see how chemistry interconnects with itself.

Boxes and margin notes

The other things you should look out for are the margin notes and boxes. There are four sorts, and they have all appeared at least once in this chapter.

> ● **Heading**
>
> **The most important looks like this. Anything in this sort of box is very important—a key concept or a summary. It's the sort of thing you would do well to hold in your mind as you read or to note down as you learn.**

Heading

Boxes like this will contain additional examples, amusing background information, and similar interesting, but inessential, material. The first time you read a chapter, you might want to miss out this sort of box, and only read them later on to flesh out some of the main themes of the chapter.

▶ Sometimes the main text of the book needs clarification or expansion, and this sort of margin note will contain such little extras to help you understand difficult points. It will also remind you of things from elsewhere in the book that illuminate what is being discussed. You would do well to read these notes the first time you read the chapter, though later, as the ideas become more familiar, you might choose to skip them.

■ This sort of margin note will mainly contain cross-references to other parts of the book as a further aid to navigation. You will find an example on p. 20.

End-of-chapter problems

You can't learn organic chemistry—there's just too much of it. You can learn trivial things like the names of compounds but that doesn't help you understand the principles behind the subject. You have to understand the principles because the only way to tackle organic chemistry is to learn to work it out. That is why we have provided end-of-chapter problems. They are to help you discover if you have understood the material presented in each chapter. In general, the 10–15 problems at the end of each chapter start easy and get more difficult. They come in two sorts. The first, generally shorter and easier, allow you to revise the material in that chapter. The second asks you to extend your understanding of the material into areas not covered by the chapter. In the later chapters this second sort will probably revise material from previous chapters.

If a chapter is about a certain type of organic reaction, say elimination reactions (Chapter 19), the chapter itself will describe the various ways ('mechanisms') by which the reaction can occur and it will give definitive examples of each mechanism. In Chapter 19 there are three mechanisms and about 65 examples altogether. You might think that this is rather a lot but there are in fact millions of examples known of these three mechanisms and Chapter 19 only scrapes the surface. Even if you totally comprehended the chapter at a first reading, you could not be confident of your understanding about elimination reactions. There are 13 end-of-chapter problems for Chapter 19. The first three ask you to interpret reactions given but not explained in the chapter. This checks that you can use the ideas in familiar situations. The next few problems develop specific ideas from the chapter concerned with why one compound does one reaction while a similar one behaves quite differently.

Finally there are some more challenging problems asking you to extend the ideas to unfamiliar molecules.

The end-of-chapter problems should set you on your way but they are not the end of the journey to understanding. You are probably reading this text as part of a university course and you should find out what kind of examination problems your university uses and practise them too. Your tutor will be able to advise you on suitable problems for each stage of your development.

The solutions manual

The problems would be of little use to you if you could not check your answers. For the maximum benefit, you need to tackle some or all of the problems as soon as you have finished each chapter without looking at the answers. Then you need to compare your suggestions with ours. You can do this with the solutions manual (*Organic Chemistry: Solutions Manual*, Oxford University Press, 2000). Each problem is discussed in some detail. The purpose of the problem is first stated or explained. Then, if the problem is a simple one, the answer is given. If the problem is more complex, a discussion of possible answers follows with some comments on the value of each. There may be a reference to the source of the problem so that you can read further if you wish.

Colour

You will already have noticed something unusual about this book: almost all of the chemical structures are shown in red. This is quite intentional: emphatic red underlines the message that structures are more important than words in organic chemistry. But sometimes small parts of structures are in other colours: here are two examples from p. 11, where we were talking about organic compounds containing elements other than C and H.

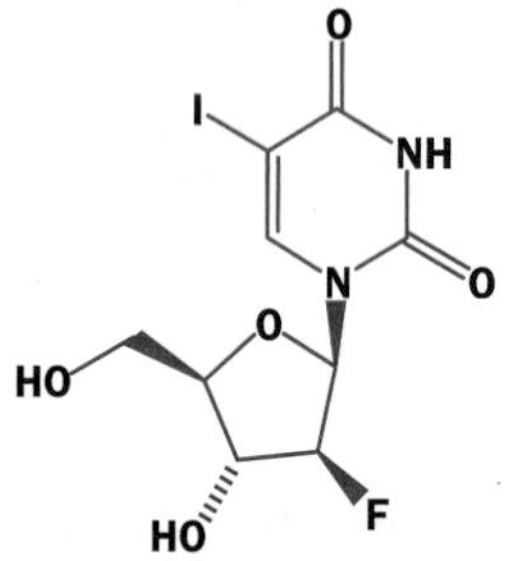

fialuridine
antiviral compound

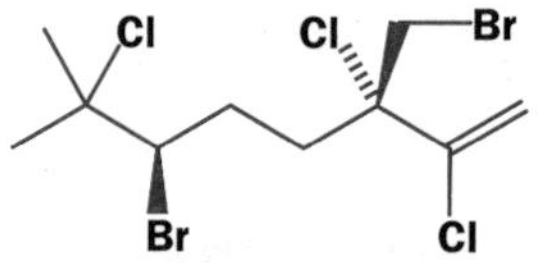

halomon
naturally occurring antitumour agent

Why are the atom labels black? Because we wanted them to stand out from the rest of the molecule. In general you will see black used to highlight important details of a molecule—they may be the groups taking part in a reaction, or something that has changed as a result of the reaction, as in these examples from Chapters 9 and 12.

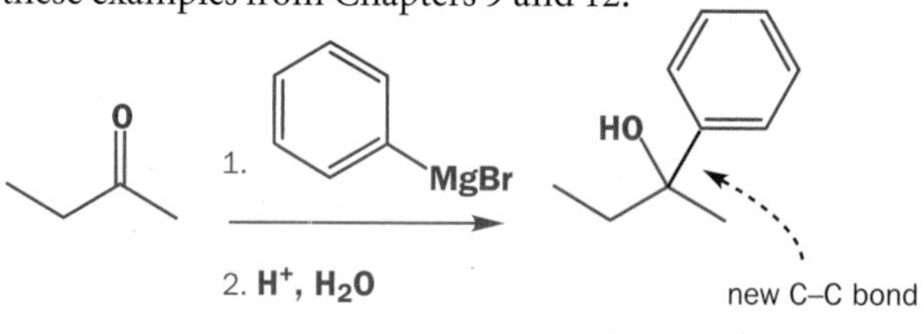

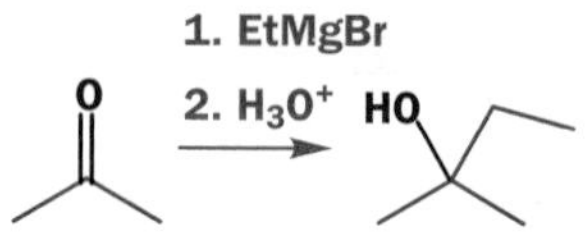

new C–C bond

We shall often use black to emphasize 'curly arrows', devices that show the movement of electrons, and whose use you will learn about in Chapter 5. Here is an example from Chapter 10: notice black also helps the '+' and '–' charges to stand out.

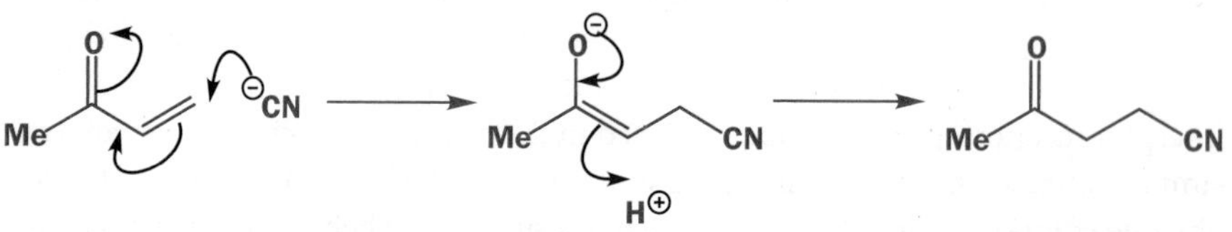

Occasionally, we shall use other colours such as green, or even orange, yellow, or brown, to high-light points of secondary importance. This example is part of a reaction taken from Chapter 19: we want to show that a molecule of water (H_2O) is formed. The green atoms show where the water comes from. Notice black curly arrows and a new black bond.

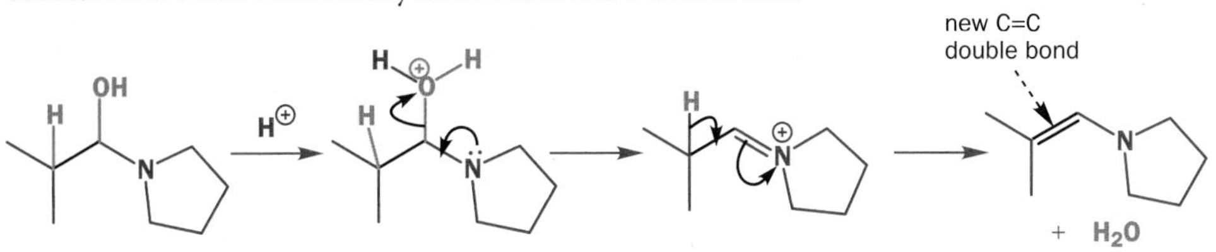

Other colours come in when things get more complicated—in this Chapter 24 example, we want to show a reaction happening at the black group in the presence of the yellow H (which, as you will see in Chapter 9, also reacts) and also in the presence of the green 'protecting' groups, one of the topics of Chapter 24.

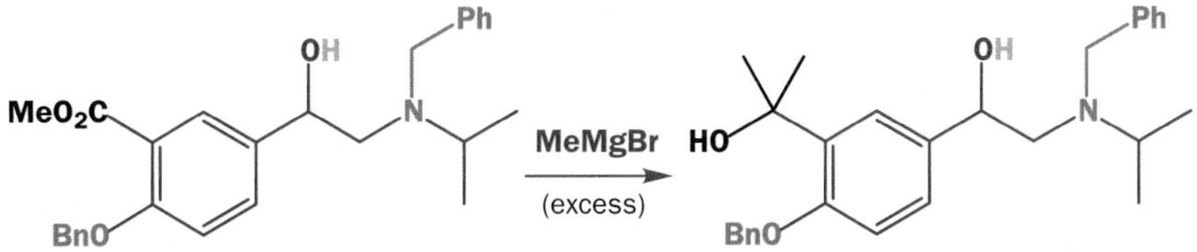

And, in Chapter 16, colour helps us highlight the difference between carbon atoms carrying four different groups and those with only three different groups. The message is: if you see something in a colour other than red, take special note—the colour is there for a reason.

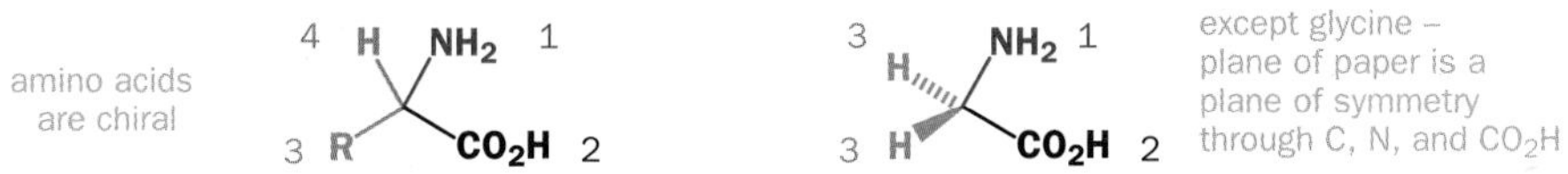

That is all we shall say in the way of introduction. On the next page the real chemistry starts, and our intention is to help you to learn real chemistry, and to enjoy it.

Organic structures

2

Connections

Building on:

- This chapter does not depend on Chapter 1

Leading to:

- The diagrams used in the rest of the book
- Why we use these particular diagrams
- How organic chemists name molecules in writing and in speech
- What is the skeleton of an organic molecule
- What is a functional group
- Some abbreviations used by all organic chemists
- Drawing organic molecules realistically in an easily understood style

Looking forward to:

- Ascertaining molecular structure spectroscopically ch3
- What determines a molecule's structure ch4

There are over 100 elements in the periodic table. Many molecules contain well over 100 atoms—palytoxin, for example (a naturally occurring compound with potential anticancer activity) contains 129 carbon atoms, 221 hydrogen atoms, 54 oxygen atoms, and 3 nitrogen atoms. It's easy to see how chemical structures can display enormous variety, providing enough molecules to build even the most complicated living creatures. But how can we understand what seems like a recipe for confusion? Faced with the collection of atoms we call a molecule, how can we make sense of what we see? This chapter will teach you how to interpret organic structures. It will also teach you how to draw organic molecules in a way that conveys all the necessary information and none of the superfluous.

Palytoxin

Palytoxin was isolated in 1971 in Hawaii from *Limu make o Hane* ('deadly seaweed of Hana') which had been used to poison spear points. It is one of the most toxic compounds known requiring only about 0.15 microgram per kilogram for death by injection. The complicated structure was determined a few years later.

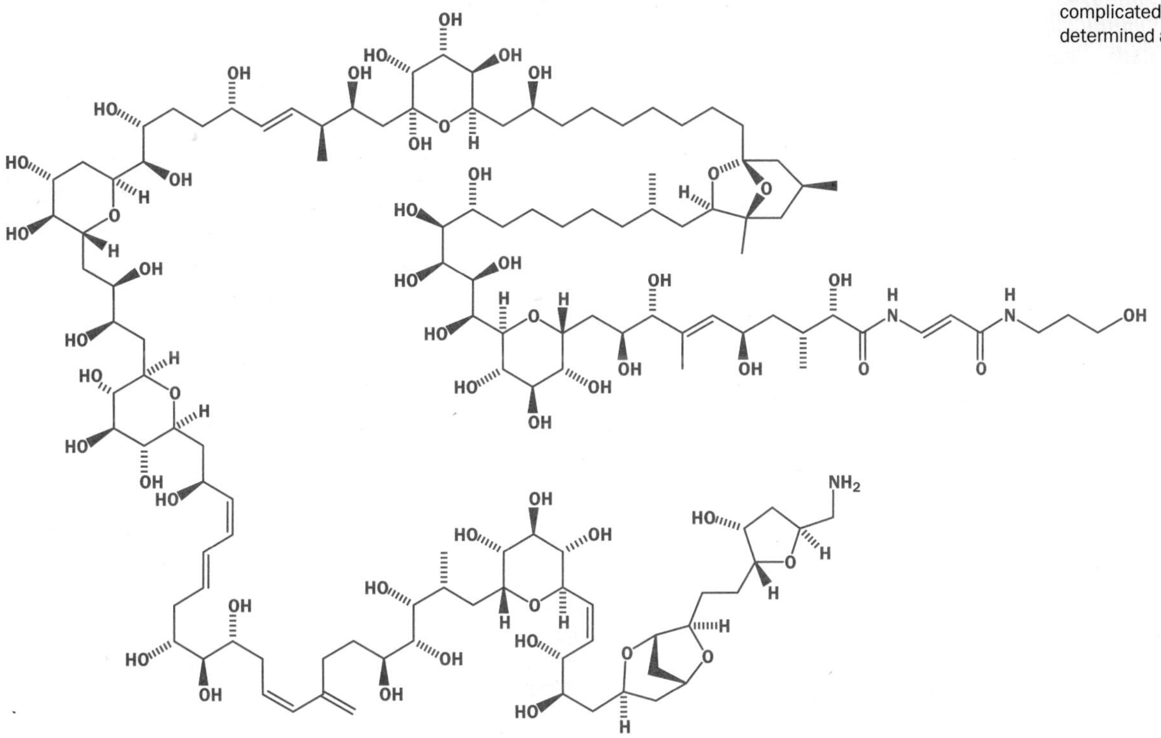

Hydrocarbon frameworks and functional groups

As we explained in Chapter 1, organic chemistry is the study of compounds that contain carbon. Nearly all organic compounds also contain hydrogen; most also contain oxygen, nitrogen, or other elements. Organic chemistry concerns itself with the way in which these atoms are bonded together into stable molecular structures, and the way in which these structures change in the course of chemical reactions.

Some molecular structures are shown below. These molecules are all amino acids, the constituents of proteins. Look at the number of carbon atoms in each molecule and the way they are bonded together. Even within this small class of molecules there's great variety—*glycine* and *alanine* have only two or three carbon atoms; *phenylalanine* has nine.

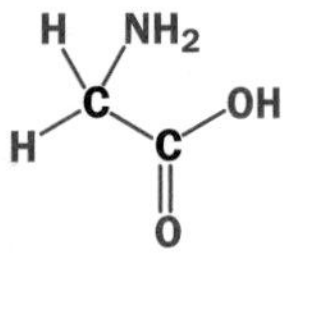

glycine 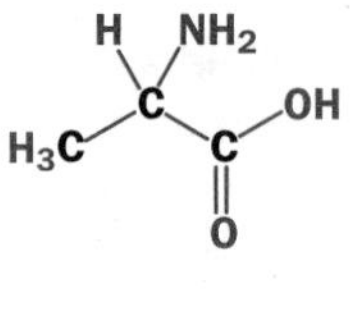   alanine 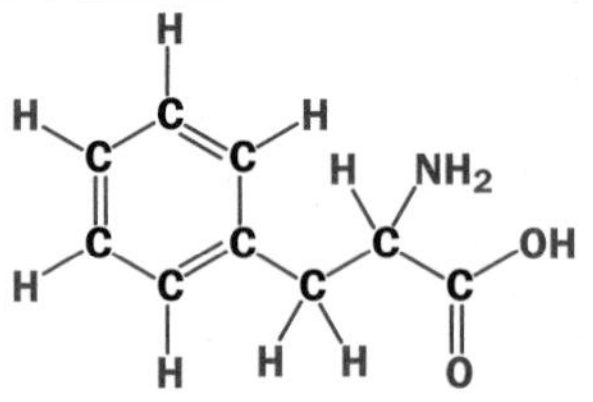   phenylalanine

Lysine has a chain of atoms; *tryptophan* has rings.

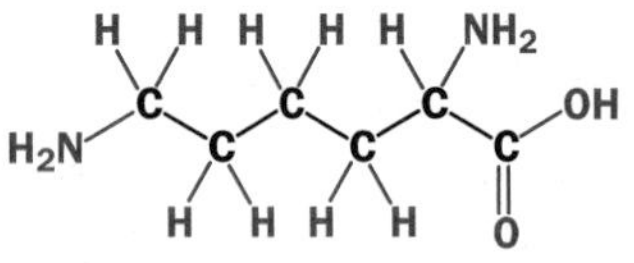

lysine 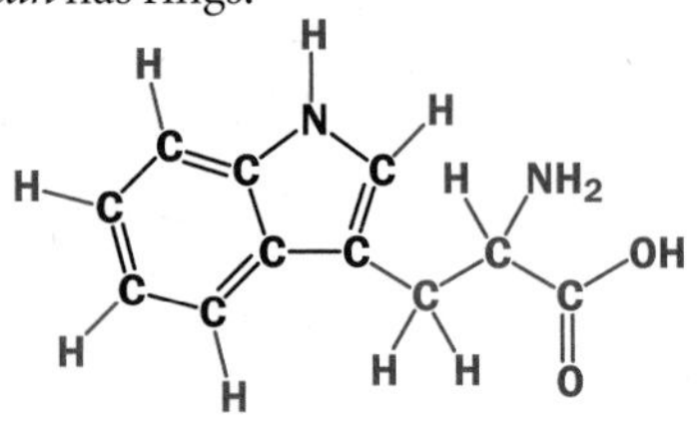   tryptophan

In *methionine* the atoms are arranged in a single chain; in *leucine* the chain is branched. In *proline*, the chain bends back on itself to form a ring.

> ■
> We shall return to amino acids as examples several times in this chapter, but we shall leave detailed discussions about their chemistry till Chapters 24 and 49, when we look at the way in which they polymerize to form peptides and proteins.

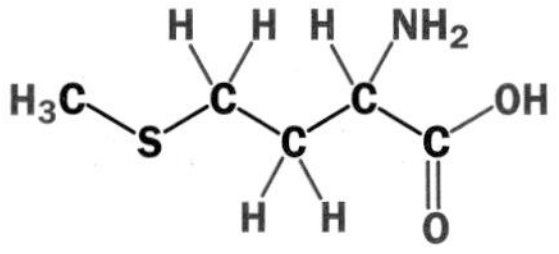

methionine 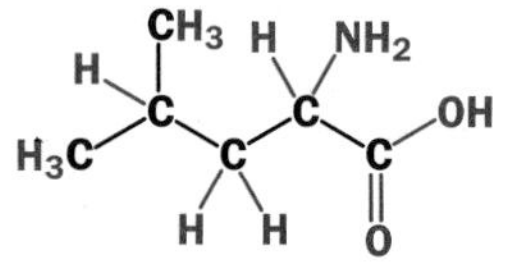   leucine 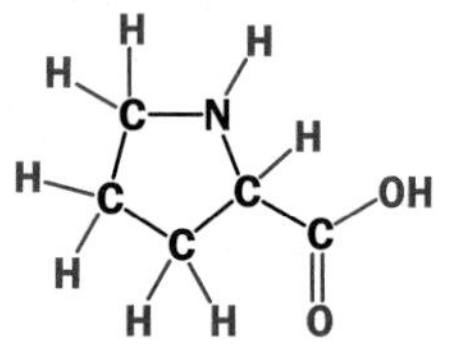   proline

Yet all of these molecules have similar properties—they are all soluble in water, they are all both acidic and basic (amphoteric), they can all be joined with other amino acids to form proteins. This is because the chemistry of organic molecules depends much less on the number or the arrangement of carbon or hydrogen atoms than on the other types of atoms (O, N, S, P, Si...) in the molecule. We call parts of molecules containing small collections of these other atoms **functional groups**, simply because they are groups of atoms that determine the way the molecule works. All amino acids contain two functional groups: an amino (NH_2 or NH) group and a carboxylic acid (CO_2H) group (some contain other functional groups as well).

● The **functional groups** determine the way the molecule works both chemically and biologically.

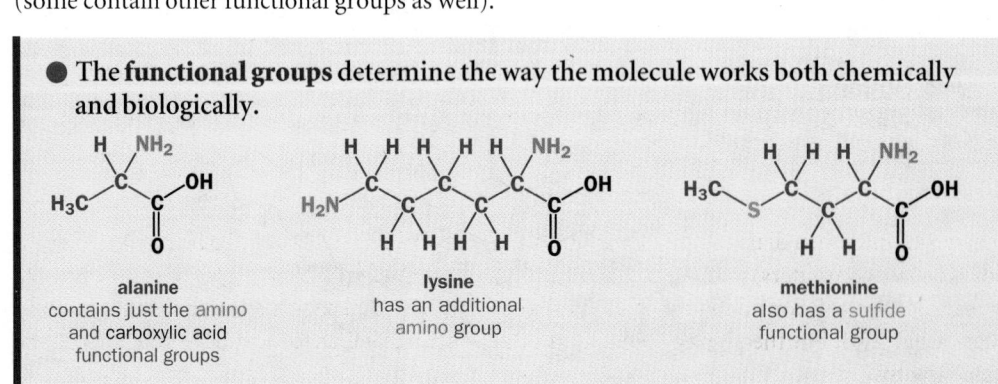

alanine
contains just the amino and carboxylic acid functional groups

lysine
has an additional amino group

methionine
also has a sulfide functional group

That isn't to say the carbon atoms aren't important; they just play quite a different role from those of the oxygen, nitrogen, and other atoms they are attached to. We can consider the chains and rings of carbon atoms we find in molecules as their skeletons, which support the functional groups and allow them to take part in chemical interactions, much as your skeleton supports your internal organs so they can interact with one another and work properly.

● The **hydrocarbon framework** is made up of chains and rings of carbon atoms, and it acts as a support for the functional groups.

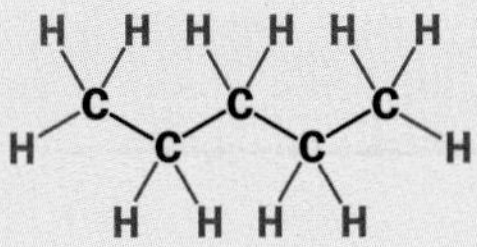

a chain 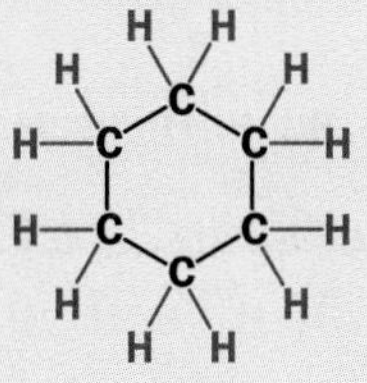   a ring 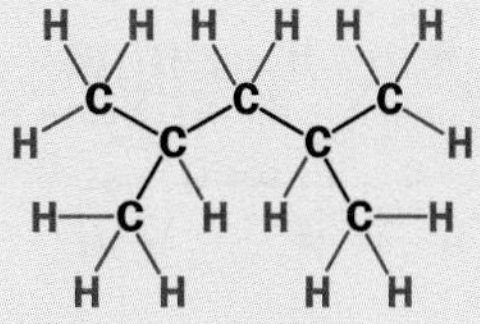   a branched chain

Organic skeletons

Organic molecules left to decompose for millions of years in the absence of light and oxygen become literally carbon skeletons—crude oil, for example, is a mixture of molecules consisting of nothing but carbon and hydrogen, while coal consists of little else but carbon. Although the molecules in coal and oil differ widely in chemical structure, they have one thing in common: no functional groups! Many are very unreactive: about the only chemical reaction they can take part in is combustion, which, in comparison to most reactions that take place in chemical laboratories or in living systems, is an extremely violent process. In Chapter 5 we will start to look at the way that functional groups direct the chemical reactions of a molecule.

We will see later how the interpretation of organic structures as hydrocarbon frameworks supporting functional groups helps us to understand and rationalize the reactions of organic molecules. It also helps us to devise simple, clear ways of representing molecules on paper. You saw in Chapter 1 how we represented molecules on paper, and in the next section we shall teach you ways to draw (and ways not to draw) molecules—the handwriting of chemistry. *This section is extremely important,* because it will teach you how to communicate chemistry, clearly and simply, throughout your life as a chemist.

Drawing molecules

Be realistic

Below is another organic structure—again, you may be familiar with the molecule it represents; it is a fatty acid commonly called linoleic acid.

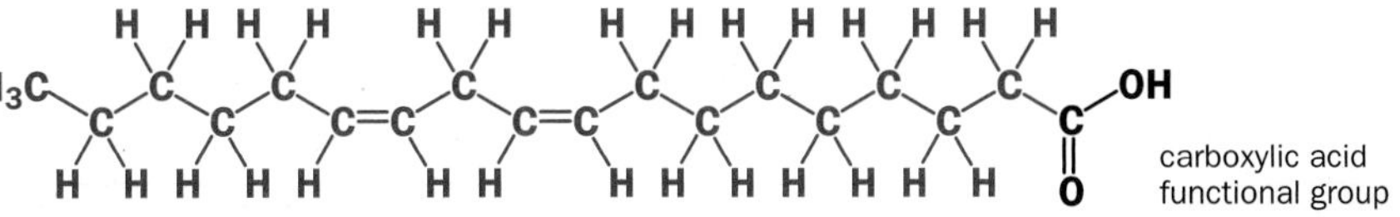

linoleic acid

carboxylic acid functional group

We could also depict linoleic acid as

CH₃CH₂CH₂CH₂CH₂CH=CHCH₂CH=CHCH₂CH₂CH₂CH₂CH₂CH₂CO₂H

linoleic acid

or as

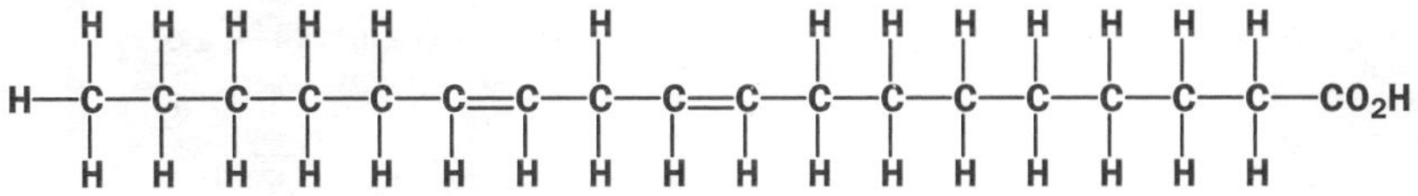

linoleic acid

You may well have seen diagrams like these last two in older books—they used to be easy to print (in the days before computers) because all the atoms were in a line and all the angles were 90°. But are they realistic? We will consider ways of determining the shapes and structures of molecules in more detail in Chapter 3, but the picture below shows the structure of linoleic acid determined by X-ray crystallography.

Three fatty acid molecules and one glycerol molecule combine to form the fats that store energy in our bodies and are used to construct the membranes around our cells. This particular fatty acid, linoleic acid, cannot be manufactured in the human body, and is an essential part of a healthy diet found, for example, in sunflower oil.

Fatty acids differ in the length of their chains of carbon atoms, yet they have very similar chemical properties because they all contain the carboxylic acid functional group. We shall come back to fatty acids in Chapter 49.

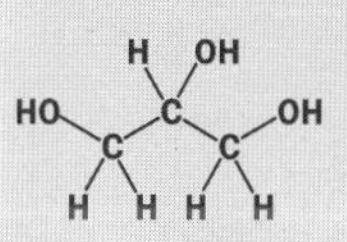

glycerol

► X-ray crystallography discovers the structures of molecules by observing the way X-rays bounce off atoms in crystalline solids. It gives clear diagrams with the atoms marked a circles and the bonds as rods joining them together.

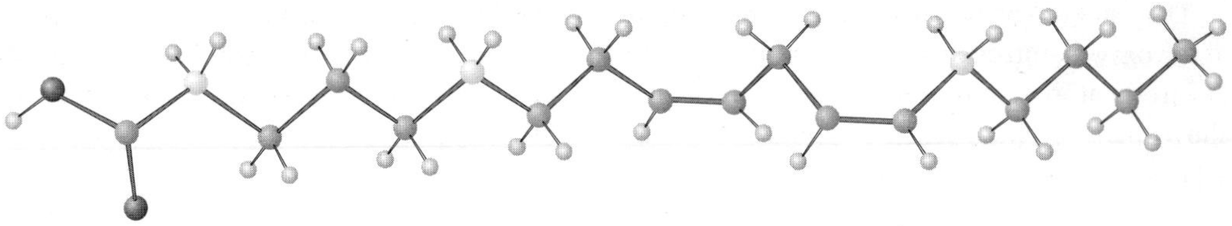

You can see that the chain of carbon atoms is not linear, but a zig-zag. Although our diagram is just a two-dimensional representation of this three-dimensional structure, it seems reasonable to draw it as a zig-zag too.

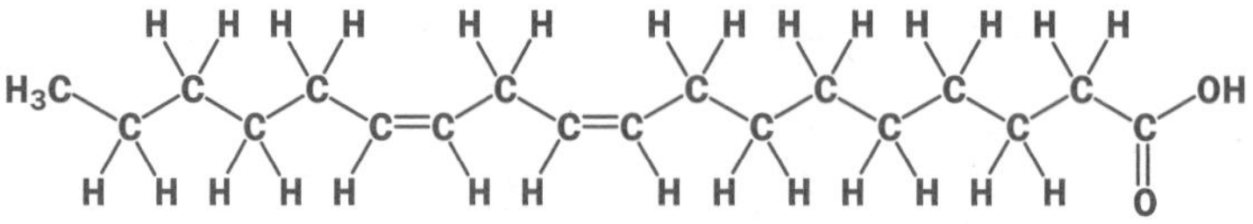

linoleic acid

This gives us our first guideline for drawing organic structures.

> ● **Guideline 1**
> **Draw chains of atoms as zig-zags**

Realism of course has its limits—the X-ray structure shows that the linoleic acid molecule is in fact slightly bent in the vicinity of the double bonds; we have taken the liberty of drawing it as a 'straight zig-zag'. Similarly, close inspection of crystal structures like this reveals that the angle of the zig-zag is about 109° when the carbon atom is not part of a double bond and 120° when it is. The 109° angle is the 'tetrahedral angle', the angle between two vertices of a tetrahedron when viewed from its centre. In Chapter 4 we shall look at why carbon atoms take up this particular arrangement of bonds. Our realistic drawing is a projection of a three-dimensional structure onto flat paper so we have to compromise.

Be economical

When we draw organic structures we try to be as realistic as we can be without putting in superfluous detail. Look at these three pictures.

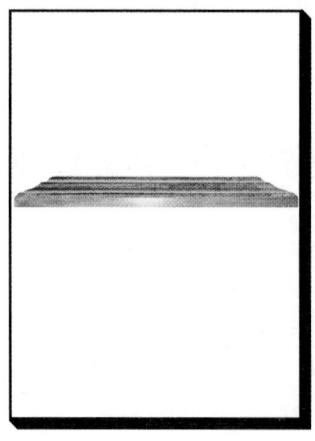

1   **2**   **3**

(1) is immediately recognizable as Leonardo da Vinci's Mona Lisa. You may not recognize (2)—it's also Leonardo da Vinci's Mona Lisa—this time viewed from above. The frame is very ornate, but the picture tells us as much about the painting as our rejected linear and 90° angle diagrams did about

our fatty acid. They're both correct—in their way—but sadly useless. What we need when we draw molecules is the equivalent of (3). It gets across the idea of the original, and includes all the detail necessary for us to recognize what it's a picture of, and leaves out the rest. And it was quick to draw—this picture was drawn in less than 10 minutes: we haven't got time to produce great works of art!

Because functional groups are the key to the chemistry of molecules, clear diagrams must emphasize the functional groups, and let the hydrocarbon framework fade into the background. Compare the diagrams below:

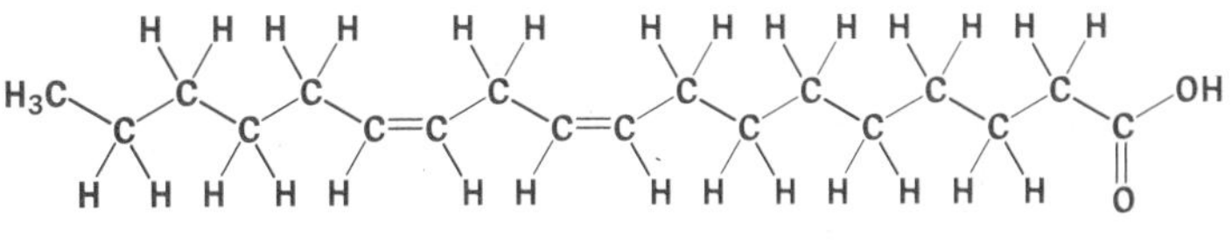

linoleic acid

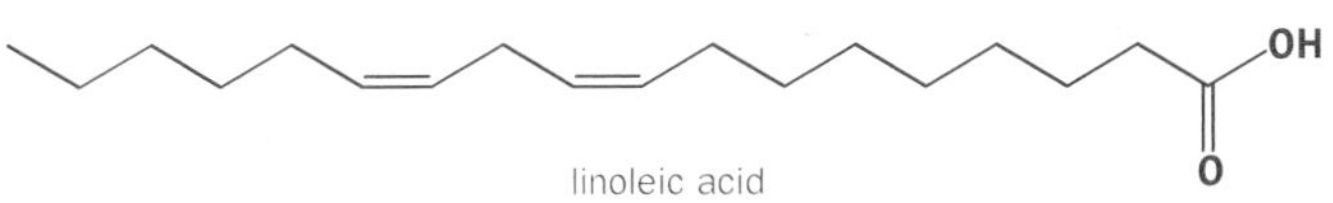

linoleic acid

The second structure is the way that most organic chemists would draw linoleic acid. Notice how the important carboxylic acid functional group stands out clearly and is no longer cluttered by all those Cs and Hs. The zig-zag pattern of the chain is much clearer too. And this structure is much quicker to draw than any of the previous ones!

To get this diagram from the one above we've done two things. Firstly, we've got rid of all the hydrogen atoms attached to carbon atoms, along with the bonds joining them to the carbon atoms. Even without drawing the hydrogen atoms we know they're there—we assume that any carbon atom that doesn't appear to have its potential for four bonds satisfied is also attached to the appropriate number of hydrogen atoms. Secondly, we've rubbed out all the Cs representing carbon atoms. We're left with a zig-zag line, and we assume that every kink in the line represents a carbon atom, as does the end of the line.

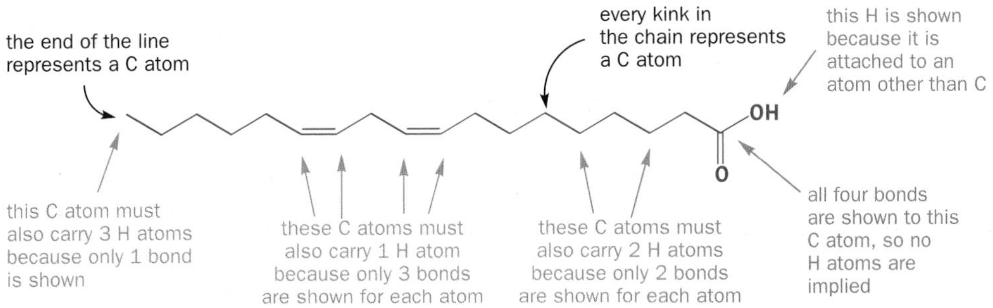

the end of the line represents a C atom

every kink in the chain represents a C atom

this H is shown because it is attached to an atom other than C

this C atom must also carry 3 H atoms because only 1 bond is shown

these C atoms must also carry 1 H atom because only 3 bonds are shown for each atom

these C atoms must also carry 2 H atoms because only 2 bonds are shown for each atom

all four bonds are shown to this C atom, so no H atoms are implied

We can turn these two simplifications into two more guidelines for drawing organic structures.

● **Guideline 2**

Miss out the Hs attached to carbon atoms, along with the C–H bonds (unless there is a good reason not to)

● **Guideline 3**

Miss out the capital Cs representing carbon atoms (unless there is a good reason not to)

Be clear

Try drawing some of the amino acids represented on p. 20 in a similar way, using the three guidelines. The bond angles at tetrahedral carbon atoms are about 109°. Make them look about 109° projected on to a plane! (120° is a good compromise, and it makes the drawings look neat.)

Start with leucine — earlier we drew it as the structure to the right. Get a piece of paper and do it now; then see how your drawing compares with our suggestions.

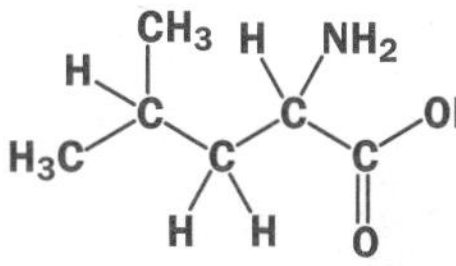

▶

What is 'a good reason not to'? One is if the C or H is part of a functional group. Another is if the C or H needs to be highlighted in some way, for example, because it's taking part in a reaction. Don't be too rigid about these guidelines: they're not rules. Better is just to learn by example (you'll find plenty in this book): if it helps clarify, put it in; if it clutters and confuses, leave it out. One thing you must remember, though: if you write a carbon atom as a letter C then you *must* add all the H atoms too. If you don't want to draw all the Hs, don't write C for carbon.

leucine

It doesn't matter which way up you've drawn it, but your diagram should look something like one of these structures below.

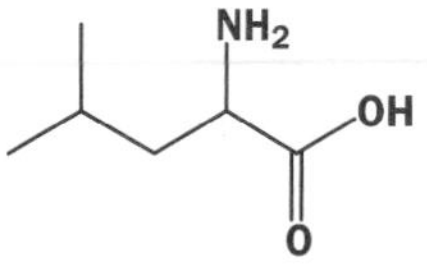

The guidelines we gave were only guidelines, not rules, and it certainly does not matter which way round you draw the molecule. The aim is to keep the functional groups clear, and let the skeleton fade into the background. That's why the last two structures are all right—the carbon atom shown as 'C' is part of a functional group (the carboxyl group) so it can stand out.

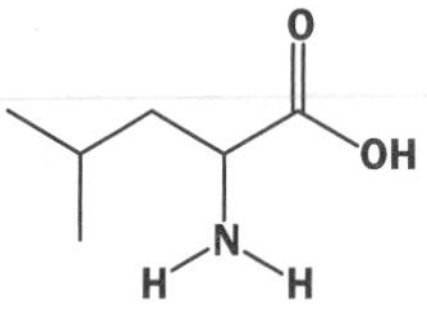

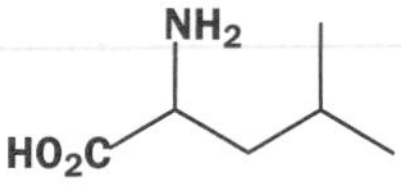

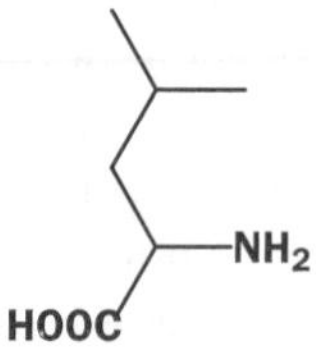

Now turn back to p. 20 and try redrawing the some of the other eight structures there using the guidelines. Don't look at our suggestions below until you've done them! Then compare your drawings with our suggestions.

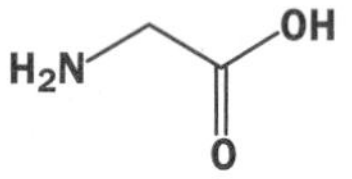

Remember that these are only suggestions, but we hope you'll agree that this style of diagram looks much less cluttered and makes the functional groups much clearer than the diagrams on p. 20. Moreover, they still bear significant resemblance to the 'real thing'—compare these crystal structures of lysine and tryptophan with the structures shown above, for example.

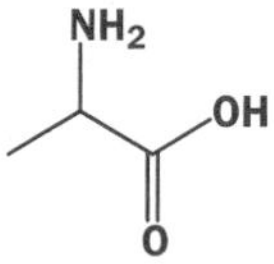

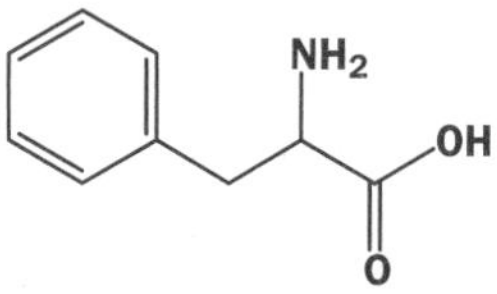

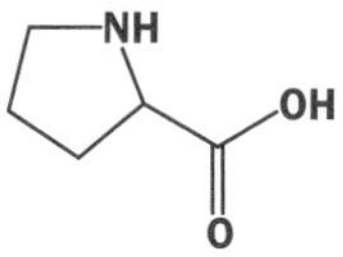

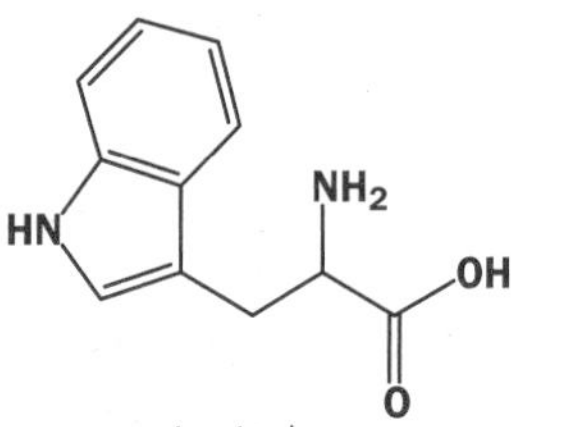

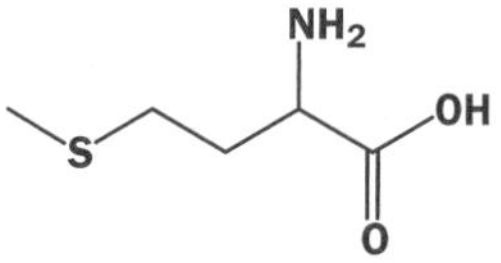

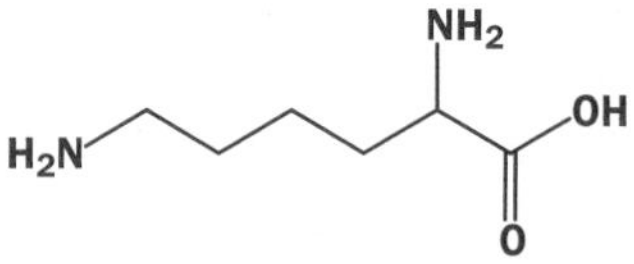

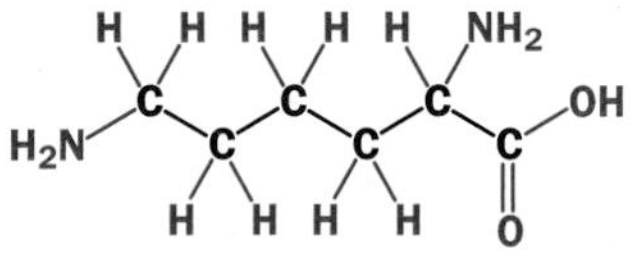

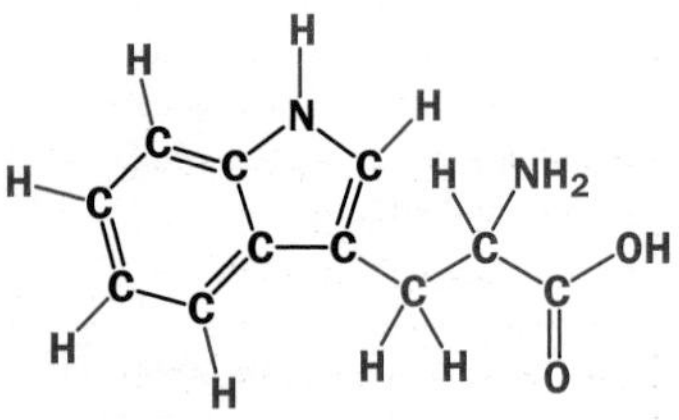

Structural diagrams can be modified to suit the occasion

You'll probably find that you want to draw the same molecule in different ways on different occasions to emphasize different points. Let's carry on using leucine as an example. We mentioned before that an amino acid can act as an acid or as a base. When it acts as an acid, a base (for example, hydroxide, OH^-) removes H^+ from the carboxylic acid group in a reaction we can represent as

The product of this reaction has a negative charge on an oxygen atom. We have put it in a circle to make it clearer, and we suggest you do the same when you draw charges: +'s and −'s are easily mislaid. We shall discuss this type of reaction, the way in which reactions are drawn, and what the 'curly arrows' in the diagram mean in Chapter 5. But for now, notice that we drew out the CO_2H as the fragment on the left because we wanted to show how the O–H bond was broken when the base attacked. We modified our diagram to suit our own purposes.

■
Not all chemists put circles round their plus and minus charges—it's a matter of personal choice.

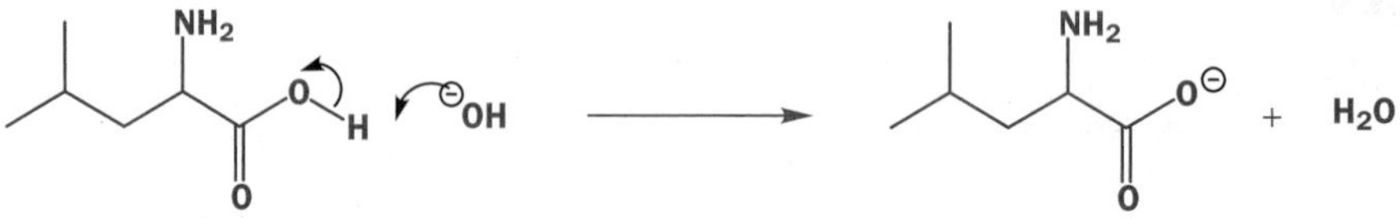

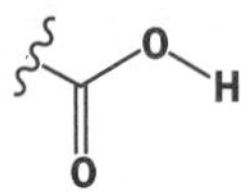

■
The wiggly line is a graphical way of indicating an incomplete structure: it shows where we have mentally 'snapped off' the CO_2H group from the rest of the molecule.

When leucine acts as a base, the amino (NH_2) group is involved. The nitrogen atom attaches itself to a proton, forming a new bond using its *lone pair*.

We can represent this reaction as

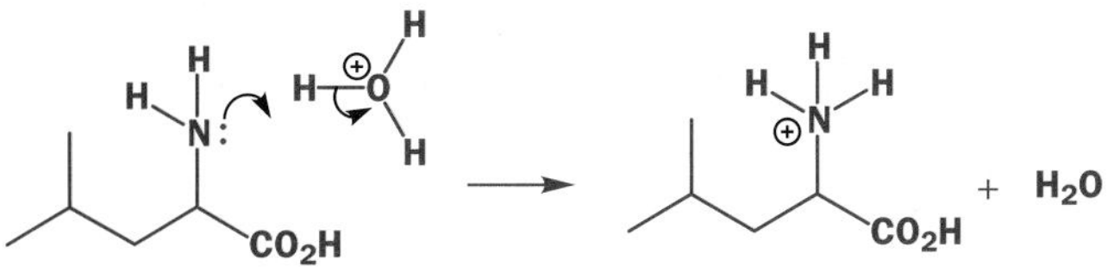

Notice how we drew the lone pair at this time because we wanted to show how it was involved in the reaction. The oxygen atoms of the carboxylic acid groups also have lone pairs but we didn't draw them in because they weren't relevant to what we were talking about. Neither did we feel it was necessary to draw CO_2H in full this time because none of the atoms or bonds in the carboxylic acid functional group was involved in the reaction.

Structural diagrams can show three-dimensional information on a two-dimensional page

Of course, all the structures we have been drawing give only an idea of the real structure of the molecules. For example, the carbon atom between the NH_2 group and the CO_2H group of leucine has a tetrahedral arrangement of atoms around it, a fact which we have so far completely ignored.

We might want to emphasize this fact by drawing in the hydrogen atom we missed out at this point as in structure 1 (in the right-hand margin). We can then show that one of the groups attached to this carbon atom comes towards us, out of the plane of the paper, and the other one goes away from us, into the paper. There are several ways of doing this. In structure 2, the bold, wedged bond suggests a perspective view of a bond coming towards you, while the hashed bond suggests a bond fading away from you. The other two 'normal' bonds are in the plane of the paper.

Alternatively we could miss out the hydrogen atom and draw something a bit neater though slightly less realistic as structure 3. We can assume the missing hydrogen atom is behind the plane of the paper, because that is where the 'missing' vertex of the tetrahedron of atoms attached to the carbon atom lies. These conventions allow us to give an idea of the three-dimensional shape (stereochemistry) of any organic molecule—you have already seen them in use in the diagram of the structure of palytoxin at the beginning of this chapter.

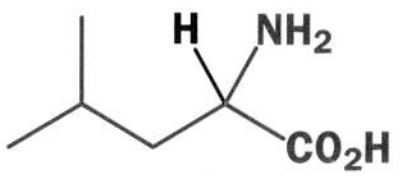

> ● **Reminder**
>
> Organic structures should be drawn to be *realistic, economical, clear.*
>
> We gave three guidelines to help you achieve this when you draw structures:
>
> • Guideline 1: Draw chains of atoms as zig-zags
>
> • Guideline 2: Miss out the Hs attached to carbon atoms, along with the C–H bonds
>
> • Guideline 3: Miss out the capital Cs representing carbon atoms

The guidelines we have given and conventions we have illustrated in this section have grown up over decades. They are used by organic chemists because they work! We guarantee to follow them for the rest of the book—try to follow them yourself whenever you draw an organic structure. Before you ever draw a capital C or a capital H again, ask yourself whether it's really necessary!

Now that we have considered how to draw structures, we can return to some of the structural types that we find in organic molecules. Firstly, we'll talk about hydrocarbon frameworks, then about functional groups.

■ A lone pair is a pair of electrons that is not involved in a chemical bond. We shall discuss lone pairs in detail in Chapter 4. Again, don't worry about what the curly arrows in this diagram mean—we will cover them in detail in Chapter 5.

1

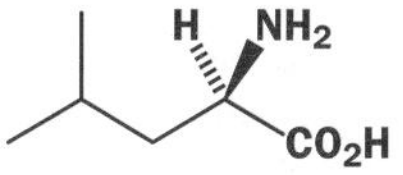

2

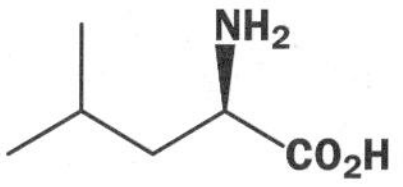

3

▶ When you draw diagrams like these to indicate the three-dimensional shape of the molecule, try to keep the hydrocarbon framework in the plane of the paper and allow functional groups and other branches to project forwards out of the paper or backwards into it.

■ We shall look in more detail at the shapes of molecules—their *stereochemistry*—in Chapter 16.

Hydrocarbon frameworks

Carbon as an element is unique in the variety of structures it can form. It is unusual because it forms strong, stable bonds to the majority of elements in the periodic table, including itself. It is this ability to form bonds to itself that leads to the variety of organic structures that exist, and indeed to the possibility of life existing at all. Carbon may make up only 0.2% of the earth's crust, but it certainly deserves a whole branch of chemistry all to itself.

Chains

The simplest class of hydrocarbon frameworks contains just chains of atoms. The fatty acids we met earlier have hydrocarbon frameworks made of zig-zag chains of atoms, for example. Polythene is a polymer whose hydrocarbon framework consists entirely of chains of carbon atoms.

a section of the structure of polythene

At the other end of the spectrum of complexity is this antibiotic, extracted from a fungus in 1995 and aptly named linearmycin as it has a long linear chain. The chain of this antibiotic is so long that we have to wrap it round two corners just to get it on the page.

We haven't drawn whether the CH_3 groups and OH groups are in front of or behind the plane of the paper, because (at the time of writing this book) the stereochemistry of linearmycin is unknown.

linearmycin

> Notice we've drawn in four groups as CH_3—we did this because we didn't want them to get overlooked in such a large structure. They are the only tiny branches off this long winding trunk.

Names for carbon chains

It is often convenient to refer to a chain of carbon atoms by a name indicating its length. You have probably met some of these names before in the names of the simplest organic molecules, the alkanes. There are also commonly used abbreviations for these names: these can be very useful in both writing about chemistry and in drawing chemical structures, as we shall see shortly.

> The names for shorter chains (which you must learn) exist for historical reasons; for chains of 5 or more carbon atoms, the systematic names are based on Greek number names.

Names and abbreviations for carbon chains

Number of carbon atoms in chain	Name of group	Formula[†]	Abbreviation	Name of alkane (= chain + H)
1	methyl	$-CH_3$	Me	methane
2	ethyl	$-CH_2CH_3$	Et	ethane
3	propyl	$-CH_2CH_2CH_3$	Pr	propane
4	butyl	$-(CH_2)_3CH_3$	Bu	butane
5	pentyl	$-(CH_2)_4CH_3$	—[‡]	pentane
6	hexyl	$-(CH_2)_5CH_3$	—[‡]	hexane
7	heptyl	$-(CH_2)_6CH_3$	—[‡]	heptane
8	octyl	$-(CH_2)_7CH_3$	—[‡]	octane
9	nonyl	$-(CH_2)_8CH_3$	—[‡]	nonane
10	decyl	$-(CH_2)_9CH_3$	—[‡]	decane

[†] This representation is not recommended.
[‡] Names for longer chains are not commonly abbreviated.

Organic elements

You may notice that the abbreviations for the names of carbon chains look very much like the symbols for chemical elements: this is deliberate, and these symbols are sometimes called 'organic elements'. They can be used in chemical structures just like element symbols. It is often convenient to use the 'organic element' symbols for short carbon chains for tidiness. Here are some examples. Structure 1 to the right shows how we drew the structure of the amino acid methionine on p. 24. The stick representing the methyl group attached to the sulfur atom does, however, look a little odd. Most chemists would draw methionine as structure 2, with 'Me' representing the CH_3 (methyl) group. Tetraethyllead used to be added to petrol to prevent engines 'knocking', until it was shown to be a health hazard. Its structure (as you might easily guess from the name) is shown as item 3. But it's much easier to write as $PbEt_4$ or Et_4Pb

Remember that these symbols (and names) can only be used for terminal chains of atoms. We couldn't abbreviate the structure of lysine from

to

NOT CORRECT

for example, because Bu represents

and not

1 methionine

2 methionine

3 tetraethyllead

4

Before leaving carbon chains, we must mention one other very useful organic element symbol, R. R in a structure can mean *anything*—it's a sort of wild card. For example, structure 4 would indicate any amino acid, where R = H is glycine, R = Me is alanine… As we've mentioned before, and you will see later, the reactivity of organic molecules is so dependent on their functional groups that the rest of the molecule can be irrelevant. In these cases, we can choose just to call it R.

Carbon rings

Rings of atoms are also common in organic structures. You may have heard the famous story of Auguste Kekulé first realizing that benzene has a ring structure when he dreamed of snakes biting their own tails. You have met benzene rings in phenylalanine and aspirin. Paracetamol also has a structure based on a benzene ring.

benzene

phenylalanine   paracetamol   aspirin

When a benzene ring is attached to a molecule by only *one* of its carbon atoms (as in phenylalanine, but not paracetamol or aspirin), we can call it a 'phenyl' group and give it the organic element symbol Ph.

the phenyl group, Ph

is equivalent to

Kekulé's snake dream inspired this figure that appeared in a spoof edition of the German chemical Journal, *Berichte der Deutschen Chemischen Gesellschaft* in 1886

Benzene has a ring structure

In 1865, August Kekulé presented a paper at the Academie des Sciences in Paris suggesting a cyclic structure for benzene, the inspiration for which he ascribed to a dream. However, was Kekulé the first to suggest that benzene was cyclic? Some believe not, and credit an Austrian schoolteacher, Josef Loschmidt with the first depiction of cyclic benzene structures. In 1861, 4 years before Kekulé's dream, Loschmidt published a book in which he represented benzene as a set of rings. It is not certain whether Loschmidt or Kekulé—or even a Scot named Archibald Couper—got it right first.

► Of course, Ar = argon too, but so few argon compounds exist that there is never any confusion.

PhOH = ![OH on benzene ring] phenol

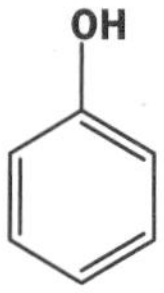

![muscone structure]

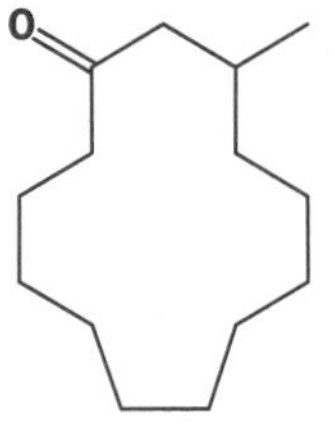

muscone

Any compound containing a benzene ring, or a related (Chapter 7) ring system is known as 'aromatic', and another useful organic element symbol related to Ph is Ar (for 'aryl'). While Ph always means C_6H_5, Ar can mean any *substituted* phenyl ring, in other words, phenyl with any number of the hydrogen atoms replaced by other groups.

For example, while PhOH always means phenol, ArOH could mean phenol, 2,4,6-trichlorophenol (the antiseptic TCP), paracetamol or aspirin (among many other substituted phenols). Like R, the 'wild card' alkyl group, Ar is a 'wild card' *aryl* group.

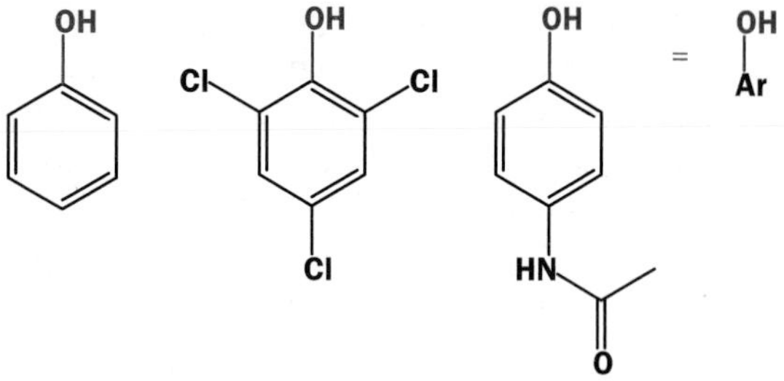

The compound known as muscone has only relatively recently been made in the lab. It is the pungent aroma that makes up the base-note of musk fragrances. Before chemists had determined its structure and devised a laboratory synthesis the only source of musk was the musk deer, now rare for this very reason. Muscone's skeleton is a 13-membered ring of carbon atoms.

![phenol, 2,4,6-trichlorophenol, paracetamol structures, = Ar-OH]

phenol   2,4,6-trichlorophenol   paracetamol

The steroid hormones have several (usually four) rings fused together. These are testosterone and oestradiol, the important human male and female sex hormones.

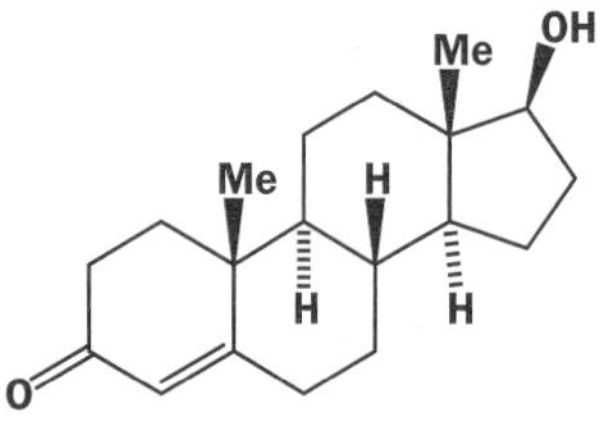

![testosterone structure]

testosterone

![oestradiol structure]

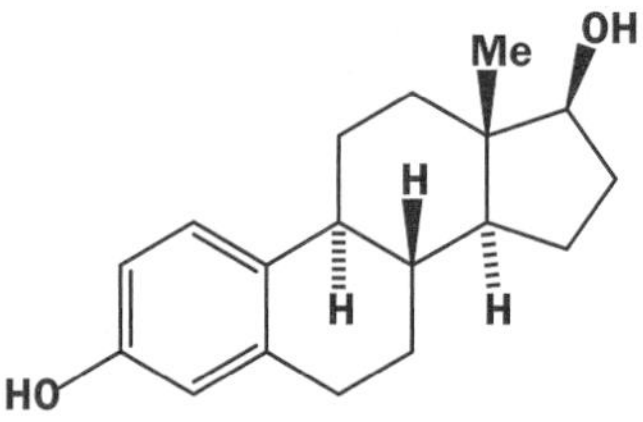

oestradiol

Some ring structures are much more complicated. The potent poison strychnine is a tangle of interconnecting rings.

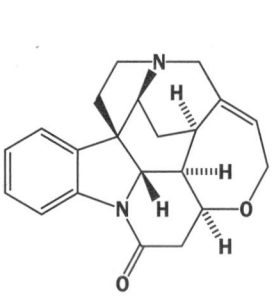

strychnine

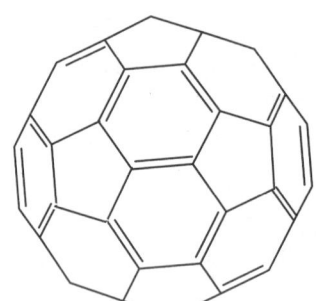

Buckminsterfullerene

Buckminsterfullerene

Buckminsterfullerene is named after the American inventor and architect Richard Buckminster Fuller, who designed the structures known as 'geodesic domes'.

One of the most elegant ring structures is shown above and is known as Buckminsterfullerene. It consists solely of 60 carbon atoms in rings that curve back on themselves to form a football-shaped cage. Count the number of bonds at any junction and you will see they add up to four so no hydrogens need be added. This compound is C_{60}. Note that you can't see all the atoms as some are behind the sphere.

Rings of carbon atoms are given names starting with 'cyclo', followed by the name for the carbon chain with the same number of carbon atoms.

To the right, structure 1 shows chrysanthemic acid, part of the naturally occurring pesticides called pyrethrins (an example appears in Chapter 1), which contains a cyclopropane ring. Propane has three carbon atoms. Cyclopropane is a three-membered ring. Grandisol (structure 2), an insect pheromone used by male boll weevils to attract females, has a structure based on a cyclobutane ring. Butane has four carbon atoms. Cyclobutane is a four-membered ring. Cyclamate (structure 3), formerly used as an artificial sweetener, contains a cyclohexane ring. Hexane has six carbon atoms. Cyclohexane is a six-membered ring.

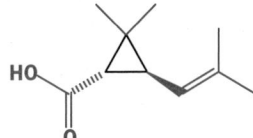

1   chrysanthemic acid

2   grandisol

3   cyclamate

Branches

Hydrocarbon frameworks rarely consist of single rings or chains, but are often branched. Rings, chains, and branches are all combined in structures like that of the marine toxin palytoxin that we met at the beginning of the chapter, polystyrene, a polymer made of six-membered rings dangling from linear carbon chains, or of β-carotene, the compound that makes carrots orange.

part of the structure of polystyrene

β-carotene

Just like some short straight carbon chains, some short branched carbon chains are given names and organic element symbols. The most common is the isopropyl group. Lithium diisopropylamide (also called LDA) is a strong base commonly used in organic synthesis.

the isopropyl group
i-Pr

is equivalent to LiN*i*-Pr$_2$

lithium di**isopropyl**amide (LDA)

Notice how the 'propyl' part of 'isopropyl' still indicates three carbon atoms; they are just joined together in a different way—in other words, as an *isomer* of the straight chain propyl group. Sometimes, to avoid confusion, the straight chain alkyl groups are called '*n*-alkyl' (for example, *n*-Pr, *n*-Bu)—*n* for 'normal'—to distinguish them from their branched counterparts.

Iproniazid is an antidepressant drug with *i*-Pr in both structure and name.

is equivalent to

iproniazid

● *Isomers* are molecules with the same kinds and numbers of atoms joined up in different ways *n*-propanol, *n*-PrOH, and isopropanol, *i*-PrOH, are isomeric alcohols. Isomers need not have the same functional groups—these compounds are all isomers of C$_4$H$_8$O.

■
'Isopropyl' may be abbreviated to *i*-Pr, *i*Pr, or Pr*^{i}*. We will use the first in this book, but you may see the others used elsewhere.

the isobutyl group
i-Bu

The isobutyl (*i*-Bu) group is a CH_2 group joined to an *i*-Pr group. It is *i*-PrCH$_2$–
Two isobutyl groups are present in the reducing agent diisobutyl aluminium hydride (DIBAL).

diisobutyl aluminium hydride (DIBAL)
is equivalent to **HAl*i*-Bu$_2$**

The painkiller ibuprofen (marketed as Nurofen®) contains an isobutyl group.

CO$_2$H

Ibuprofen

Notice how the invented name ibuprofen is a medley of 'ibu' (from i-Bu for isobutyl) + 'pro' (for propyl, the three-carbon unit shown in gold) + 'fen' (for the phenyl ring). We will talk about the way in which compounds are named later in this chapter.

the *sec*-butyl group
s-Bu

There are two more isomers of the butyl group, both of which have common names and abbreviations. The *sec*-butyl group (*s*-butyl or *s*-Bu) has a methyl and an ethyl group joined to the same carbon atom. It appears in an organolithium compound, *sec*-butyl lithium, used to introduce lithium atoms into organic molecules.

Li

is equivalent to **s-BuLi**

the *tert*-butyl group
t-Bu

The *tert*-butyl group (*t*-butyl or *t*-Bu) group has three methyl groups joined to the same carbon atom. Two *t*-Bu groups are found in BHT ('butylated hydroxy toluene'), an antioxidant added to some processed foods.

OH

BHT

is equivalent to

OH
t-Bu   *t*-Bu

Me

BHT

● **Primary, secondary, and tertiary**

The prefixes *sec* and *tert* are really short for secondary and tertiary, terms that refer to the carbon atom that attaches these groups to the rest of the molecular structure.

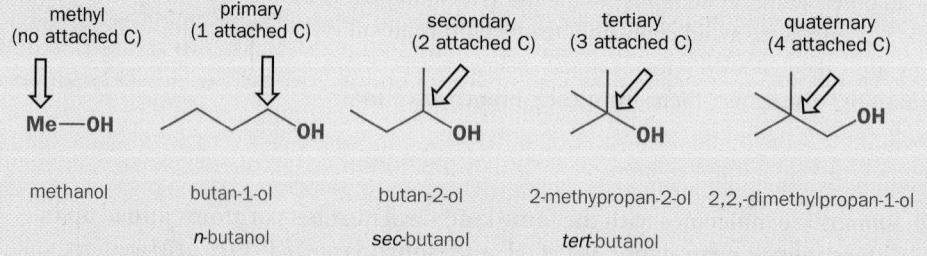

methyl (no attached C)	primary (1 attached C)	secondary (2 attached C)	tertiary (3 attached C)	quaternary (4 attached C)
Me—OH	OH	OH	OH	OH
methanol	butan-1-ol	butan-2-ol	2-methypropan-2-ol	2,2,-dimethylpropan-1-ol
	n-butanol	*sec*-butanol	*tert*-butanol	

A primary carbon atom is attached to only one other C atom, a secondary to two other C atoms, and so on. This means there are five types of carbon atom.

These names for bits of hydrocarbon framework are more than just useful ways of writing or talking about chemistry. They tell us something fundamental about the molecule and we shall use them when we describe reactions.

This quick architectural tour of some of the molecular edifices built by nature and by man serves just as an introduction to some of the hydrocarbon frameworks you will meet in the rest of this chapter and of this book. Yet, fortunately for us, however complicated the hydrocarbon framework might be, it serves only as a support for the functional groups. And, by and large, a functional group in one molecule behaves in much the same way as it does in another molecule. What we now need to do, and we start in the next section, is to introduce you to some functional groups, and to explain why it is that their attributes are the key to understanding organic chemistry.

Functional groups

If you take ethane gas (CH_3CH_3, or EtH, or even ⌒ , though a single line like this doesn't look much like a chemical structure) and bubble it through acids, bases, oxidizing agents, reducing agents—in fact almost any chemical you can think of—it will remain unchanged. Just about the only thing you can do with it is burn it. Yet ethanol (CH_3CH_2OH, or ⌒OH, or preferably EtOH) not only burns, it reacts with acids, bases, and oxidizing agents.

The difference between ethanol and ethane is the functional group—the OH or hydroxyl group. We know that these chemical properties (being able to react with acids, bases, and oxidizing agents) are properties of the hydroxyl group and not just of ethanol because other compounds containing OH groups (in other words, other alcohols) have similar properties, whatever their hydrocarbon frameworks.

Your understanding of functional groups will be the key to your understanding of organic chemistry. We shall therefore now go on to meet some of the most important functional groups. We won't say much about the properties of each group; that will come in Chapter 5 and later. Your task at this stage is to learn to recognize them when they appear in structures, so make sure you learn their names. The classes of compound associated with some functional groups also have names: for example, compounds containing the hydroxyl group are known as alcohols. Learn these names too as they are more important than the systematic names of individual compounds. We've told you a few snippets of information about each group to help you to get to know something of the group's character.

Ethanol

The reaction of ethanol with oxidizing agents makes vinegar from wine and sober people from drunk ones. In both cases, the oxidizing agent is oxygen from the air, catalysed by an enzyme in a living system. The oxidation of ethanol by microorganisms that grow in wine left open to the air leads to acetic acid (ethanoic acid) while the oxidation of ethanol by the liver gives acetaldehyde (ethanal).

acetaldehyde   liver   ethanol   micro-organism   acetic acid

Human metabolism and oxidation

The human metabolism makes use of the oxidation of alcohols to render harmless other toxic compounds containing the OH group. For example, lactic acid, produced in muscles during intense activity, is oxidized by an enzyme called lactate dehydrogenase to the metabolically useful compound pyruvic acid.

lactic acid O_2 lactate dehydrogenase   pyruvic acid

Alkanes contain no functional groups

The alkanes are the simplest class of organic molecules because they contain no functional groups. They are extremely unreactive, and therefore rather boring as far as the organic chemist is concerned. However, their unreactivity can be a bonus, and alkanes such as pentane and hexane are often used as solvents, especially for purification of organic compounds. Just about the only thing alkanes will do is burn—methane, propane, and butane are all used as domestic fuels, and petrol is a mixture of alkanes containing largely isooctane.

pentane   hexane   isooctane

Alkenes (sometimes called olefins) contain C=C double bonds

It may seem strange to classify a type of bond as a functional group, but you will see later that C=C double bonds impart reactivity to an organic molecule just as functional groups consisting of, say, oxygen or nitrogen atoms do. Some of the compounds produced by plants and used by perfumers are alkenes (see Chapter 1). For example, pinene has a smell evocative of pine forests, while limonene smells of citrus fruits.

α-pinene

limonene

You've already met the orange pigment β-carotene. Eleven C=C double bonds make up most of its structure. Coloured organic compounds often contain chains of C=C double bonds like this. In Chapter 7 you will find out why this is so.

β-carotene

Alkynes contain C≡C triple bonds

Just like C=C double bonds, C≡C triple bonds have a special type of reactivity associated with them, so it's useful to call a C≡C triple bond a functional group. Alkynes are **linear** so we draw them with four carbon atoms in a straight line. Alkynes are not as widespread in nature as alkenes, but one fascinating class of compounds containing C≡C triple bonds is a group of antitumour agents discovered during the 1980s. Calicheamicin is a member of this group. The high reactivity of this combination of functional groups enables calicheamicin to attack DNA and prevent cancer cells from proliferating. For the first time we have drawn a molecule in three dimensions, with two bonds crossing one another—can you see the shape?

calicheamicin
(R = a string of sugar molecules)

Alcohols (R–OH) contain a hydroxyl (OH) group

We've already talked about the hydroxyl group in ethanol and other alcohols. Carbohydrates are peppered with hydroxyl groups; sucrose has eight of them for example (a more three-dimensional picture of the sucrose molecule appears in Chapter 1).

Molecules containing hydroxyl groups are often soluble in water, and living things often attach sugar groups, containing hydroxyl groups, to otherwise insoluble organic compounds to keep them in solution in the cell. Calicheamicin, a molecule we have just mentioned, contains a string of sugars for just this reason. The liver carries out its task of detoxifying unwanted organic compounds by repeatedly hydroxylating them until they are water-soluble, and they are then excreted in the bile or urine.

sucrose

■ Remember that R can mean any alkyl group.

■ If we want a structure to contain more than one 'R', we give the R's numbers and call them R^1, R^2... Thus R^1–O–R^2 means an ether with two different unspecified alkyl groups. (*Not* R_1, R_2..., which would mean $1 \times R$, $2 \times R$...)

diethyl ether
"ether"   THF

Ethers (R^1–O–R^2) contain an alkoxy group (–OR)

The name **ether** refers to any compound that has two alkyl groups linked through an oxygen atom. 'Ether' is also used as an everyday name for diethyl ether, Et_2O. You might compare this use of the word 'ether' with the common use of the word 'alcohol' to mean ethanol. Diethyl ether is a highly flammable solvent that boils at only 35 °C. It used to be used as an anaesthetic. Tetrahydrofuran (THF) is another commonly used solvent and is a cyclic ether.

Brevetoxin B is a fascinating naturally occurring compound that was synthesized in the laboratory in 1995. It is packed with ether functional groups in ring sizes from 6 to 8.

■ Another common laboratory solvent is called 'petroleum ether'. Don't confuse this with diethyl ether! Petroleum ether is in fact not an ether, but a mixture of alkanes. 'Ether', according to the *Oxford English Dictionary*, means 'clear sky, upper region beyond the clouds', and hence used to be used for anything light, airy, and volatile.

Brevetoxin B

Brevetoxin B is one of a family of polyethers found in a sea creature (a dinoflagellate *Gymnodinium breve*, hence the name) which sometimes multiplies at an amazing rate and creates 'red tides' around the coasts of the Gulf of Mexico. Fish die in shoals and so do people if they eat the shellfish that have eaten the red tide. The brevetoxins are the killers. The many ether oxygen atoms interfere with sodium ion (Na^+) metabolism.

brevetoxin

Amines ($R-NH_2$) contain the amino (NH_2) group

We met the amino group when we were discussing the amino acids: we mentioned that it was this group that gave these compounds their basic properties. Amines often have powerful fishy smells: the smell of putrescine is particularly foul. It is formed as meat decays. Many neurologically active compounds are also amines: amphetamine is a notorious stimulant.

putrescine

amphetamine

Nitro compounds ($R-NO_2$) contain the nitro group (NO_2)

The nitro group (NO_2) is often incorrectly drawn with five bonds to nitrogen which you will see in Chapter 4, is impossible. Make sure you draw it correctly when you need to draw it out in detail. If you write just NO_2 you are all right!

Several nitro groups in one molecule can make it quite unstable and even explosive. Three nitro groups give the most famous explosive of all, TNT (trinitrotoluene), its kick.

the nitro group

nitrogen cannot have five bonds!

incorrect structure for the nitro group

TNT

nitrazepam

However, functional groups refuse to be stereotyped. Nitrazepam also contains a nitro group, but this compound is marketed as Mogadon®, the sleeping pill.

Alkyl halides (fluorides R–F, chlorides R–Cl, bromides R–Br, or iodides R–I) contain the fluoro, chloro, bromo, or iodo groups

These four functional groups have similar properties—though alkyl iodides are the most reactive and alkyl fluorides the least. PVC (polyvinyl chloride) is one of the most widely used polymers—it has a chloro group on every other carbon atom along a linear hydrocarbon framework. Methyl iodide (MeI), on the other hand, is a dangerous carcinogen, since it reacts with DNA and can cause mutations in the genetic code.

a section of the structure of PVC

■ These compounds are also known as haloalkanes (fluoroalkanes, chloroalkanes, bromoalkanes or iodoalkanes).

▶ Because alkyl halides have similar properties, chemists use yet another 'wild card' organic element, X, as a convenient substitute for Cl, Br, or I (sometimes F). So R–X is any alkyl halide.

Aldehydes (R–CHO) and ketones (R^1–CO–R^2) contain the carbonyl group C=O

Aldehydes can be formed by oxidizing alcohols—in fact the liver detoxifies ethanol in the bloodstream by oxidizing it first to acetaldehyde (ethanal, CH_3CHO). Acetaldehyde in the blood is the cause of hangovers. Aldehydes often have pleasant smells—2-methylundecanal is a key component of the fragrance of Chanel No 5™, and 'raspberry ketone' is the major component of the flavour and smell of raspberries.

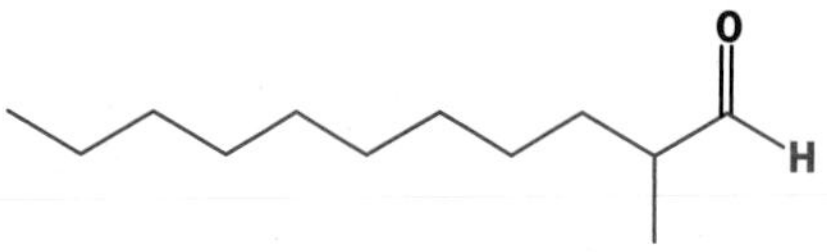

2-methylundecanal

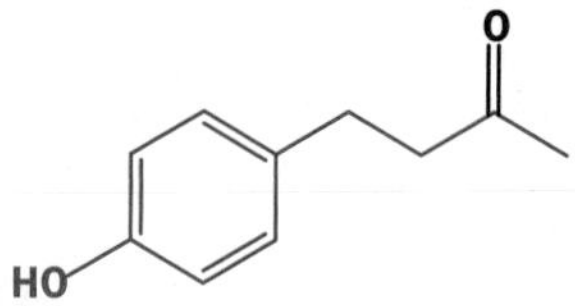

"raspberry ketone"

■
–CHO represents:

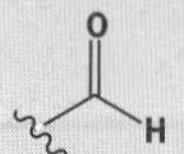

When we write aldehydes as R–CHO, we have no choice but to write in the C and H (because they're part of the functional group)—one important instance where you should ignore Guideline 3 for drawing structures. Another point: always write R–CHO and never R–COH, which looks too much like an alcohol.

Carboxylic acids (R–CO_2H) contain the carboxyl group CO_2H

As their name implies, compounds containing the carboxylic acid (CO_2H) group can react with bases, losing a proton to form carboxylate salts. Edible carboxylic acids have sharp flavours and several are found in fruits—citric, malic, and tartaric acids are found in lemons, apples, and grapes, respectively.

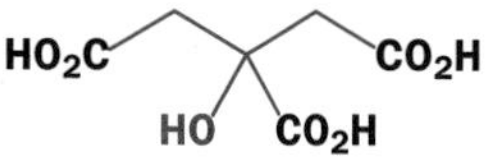

citric acid

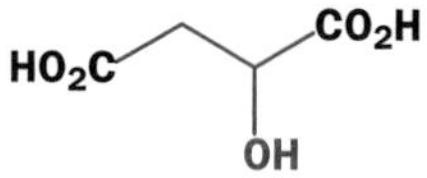

malic acid

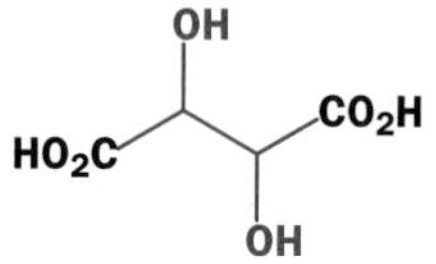

tartaric acid

Esters (R^1–CO_2R^2) contain a carboxyl group with an extra alkyl group (CO_2R)

Fats are esters; in fact they contain three ester groups. They are formed in the body by condensing glycerol, a compound with three hydroxyl groups, with three fatty acid molecules.

Other, more volatile esters, have pleasant, fruity smells and flavours. These three are components of the flavours of bananas, rum, and apples:

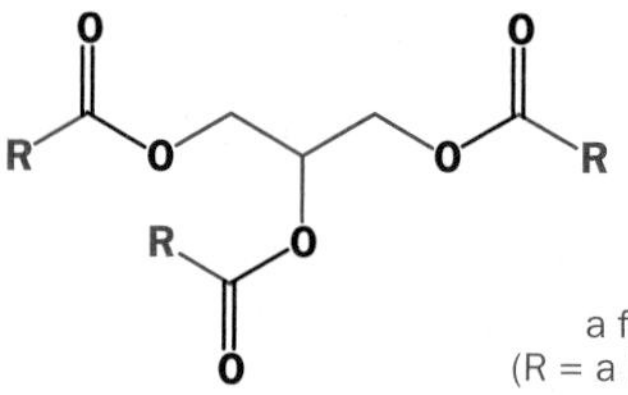

a fat molecule
(R = a long alkyl chain)

■
The terms 'saturated fats' and 'unsaturated fats' are familiar—they refer to whether the R groups are saturated (no C=C double bonds) or unsaturated (contains C=C double bonds)—see the box on p. 32. Fats containing R groups with several double bonds (for example, those that are esters formed from linoleic acid, which we met at the beginning of this chapter) are known as 'polyunsaturated'.

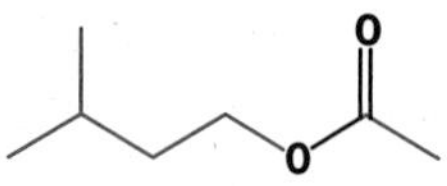

isopentyl acetate
(bananas)

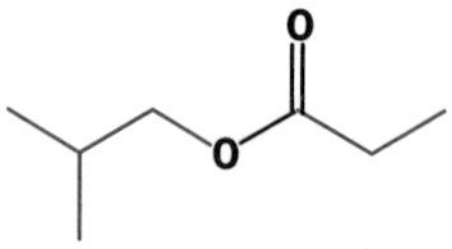

isobutyl propionate
(rum)

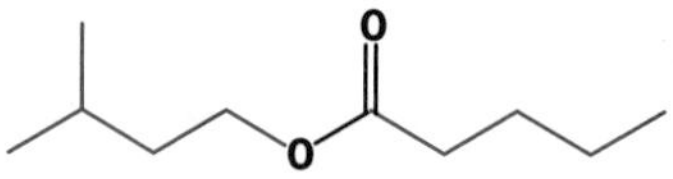

isopentyl valerate
(apples)

Amides (R–$CONH_2$, R^1–$CONHR^2$, or R^1–$CONR^2R^3$)

Proteins are amides: they are formed when the carboxylic acid group of one amino acid condenses with the amino group of another to form an amide linkage (also known as a peptide bond). One protein molecule can contain hundreds of amide bonds. Aspartame, the artificial sweetener marketed as NutraSweet®, on the other hand contains just two amino acids, aspartic acid and phenylalanine, joined through one amide bond. Paracetamol is also an amide.

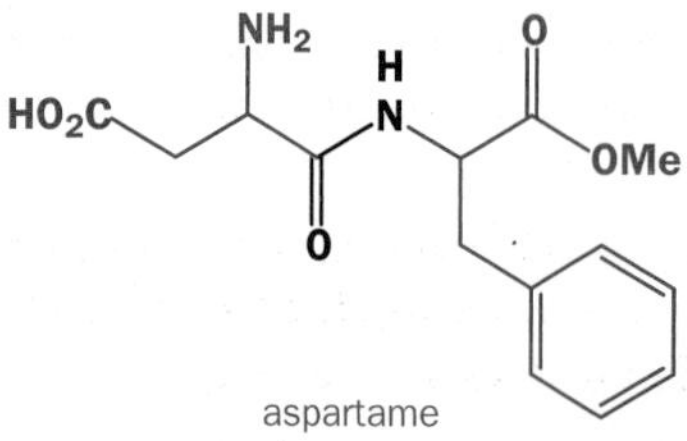

aspartame

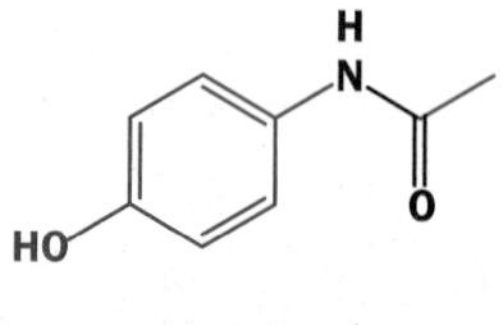

paracetamol

Nitriles or cyanides (R–CN) contain the cyano group –C≡N

Nitrile groups can be introduced into molecules by reacting potassium cyanide with alkyl halides. The organic nitrile group has quite different properties associated with lethal inorganic cyanide: Laetrile, for example, is extracted from apricot kernels, and was once developed as an anticancer drug. It was later proposed that the name be spelt 'liar-trial' since the results of the clinical trials on laetrile turned out to have been falsified!

laetrile

Acyl chlorides (acid chlorides) (R–COCl)

Acyl chlorides are reactive compounds used to make esters and amides. They are derivatives of carboxylic acids with the –OH replaced by –Cl, and are too reactive to be found in nature.

acetyl chloride

Acetals

Acetals are compounds with two single bonded oxygen atoms attached to the same carbon atom. Many sugars are acetals, as is laetrile which you have just met.

an acetal   sucrose   laetrile

Carbon atoms carrying functional groups can be classified by oxidation level

All functional groups are different, but some are more different than others. For example, the structures of a carboxylic acid, an ester, and an amide are all very similar: in each case the carbon atom carrying the functional group is bonded to two **heteroatoms**, one of the bonds being a double bond. You will see in Chapter 12 that this similarity in structure is mirrored in the reactions of these three types of compounds, and in the ways in which they can be interconverted. Carboxylic acids, esters, and amides can be changed one into another by reaction with simple reagents such as water, alcohols, or amines plus appropriate catalysts. To change them into aldehydes or alcohols requires a different type or reagent, a reducing agent (a reagent which adds hydrogen atoms). We say that the carbon atoms carrying functional groups that can be interconverted without the need for reducing agents (or oxidizing agents) have the same oxidation level—in this case, we call it the 'carboxylic acid oxidation level'.

> ▶ **A heteroatom is an atom that is not C or H**
> You've seen that a functional group is essentially any deviation from an alkane structure, either because the molecule has fewer hydrogen atoms than an alkane (alkenes, alkynes) or because it contains a collection of atoms that are not C and not H. There is a useful term for these 'different' atoms: heteroatoms. A **heteroatom** is any atom in an organic molecule other than C or H.

● **The carboxylic acid oxidation level**

carboxylic acids   esters   amides   nitriles   acyl chlorides

In fact, amides can quite easily be converted into nitriles just by dehydration (removal of water), so we must give nitrile carbon atoms the same oxidation level as carboxylic acids, esters, and amides. Maybe you're beginning to see the structural similarity between these four functional groups that you could have used to assign their oxidation level? In all four cases, the carbon atom has *three* bonds to heteroatoms, and only one to C or H. It doesn't matter how many heteroatoms there are, just how many bonds to them. Having noticed this, we can also assign both carbon atoms in 'CFC-113', one of the environmentally unfriendly aerosol propellants/refrigerants that have caused damage to the earth's ozone layer, to the carboxylic acid oxidation level.

Aldehydes and ketones contain a carbon atom with *two* bonds to heteroatoms; they are at the 'aldehyde oxidation level'. The common laboratory solvent dichloromethane also has two bonds to heteroatoms, so it too contains a carbon atom at the aldehyde oxidation level, as do acetals.

> ▶ Don't confuse oxidation **level** with oxidation **state**. In all of these compounds, carbon is in oxidation state +4.

"CFC-113"

● **The aldehyde oxidation level**

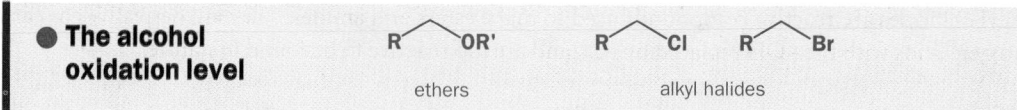

aldehydes   ketones   acetals   dichloromethane

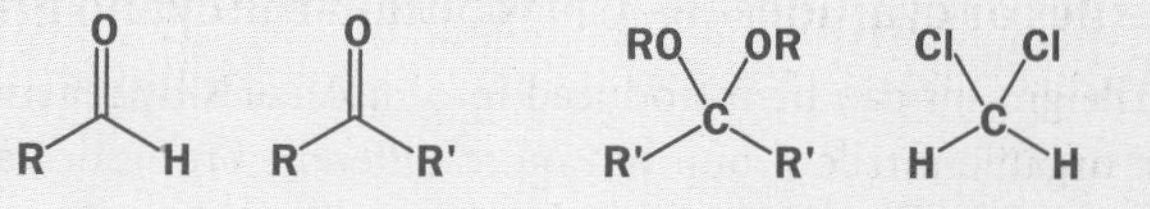

Alcohols, ethers, and alkyl halides have a carbon atom with only *one* single bond to a heteroatom. We assign these the 'alcohol oxidation level', and they are all easily made from alcohols without oxidation or reduction.

● **The alcohol oxidation level**

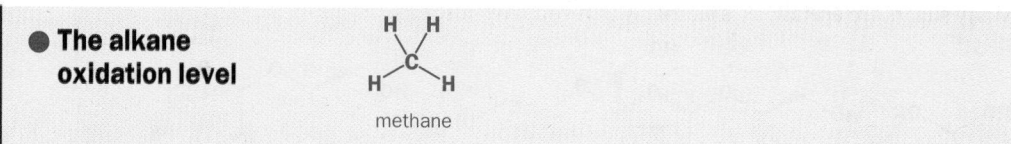

ethers   alkyl halides

We must include simple alkanes, which have no bonds to heteroatoms, as an 'alkane oxidation level'.

● **The alkane oxidation level**

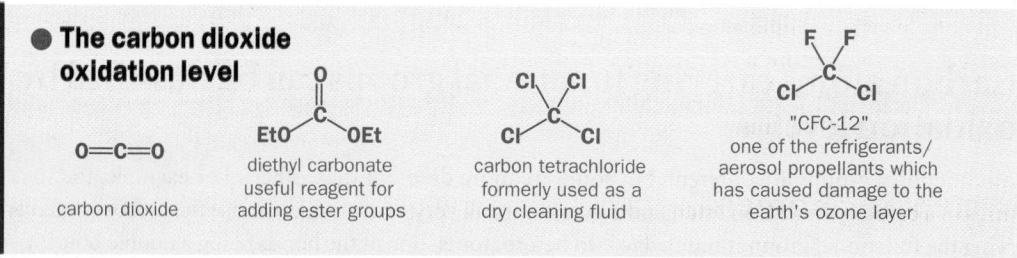

methane

The small class of compounds that have a carbon atom with four bonds to heteroatoms is related to CO_2 and best described as at the carbon dioxide oxidation level.

● **The carbon dioxide oxidation level**

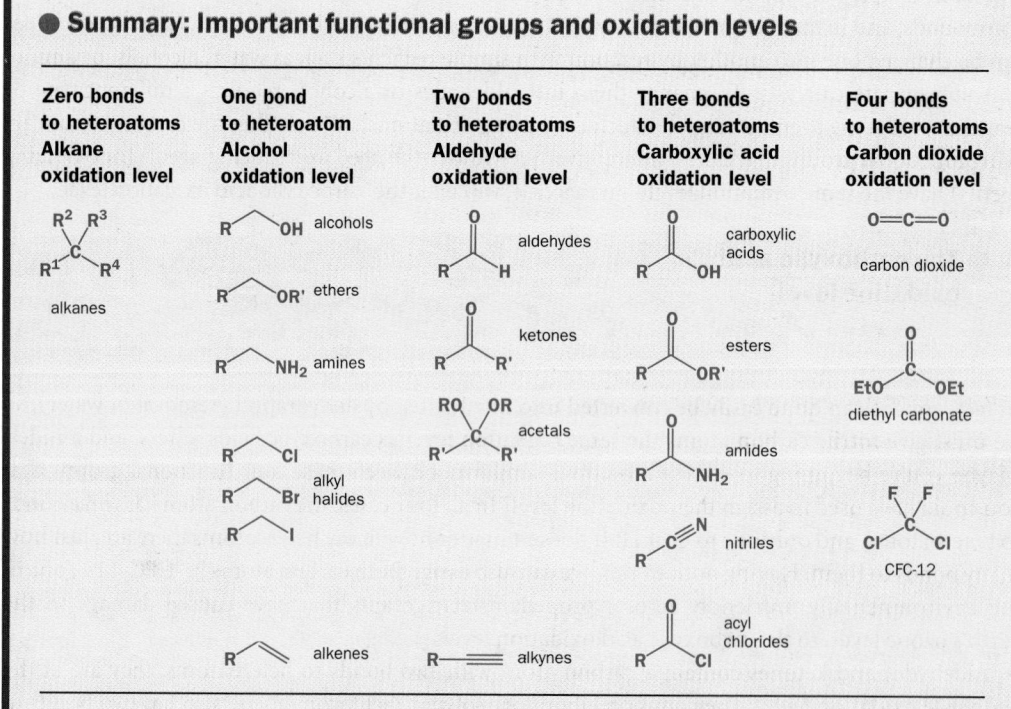

carbon dioxide

diethyl carbonate
useful reagent for
adding ester groups

carbon tetrachloride
formerly used as a
dry cleaning fluid

"CFC-12"
one of the refrigerants/
aerosol propellants which
has caused damage to the
earth's ozone layer

● **Summary: Important functional groups and oxidation levels**

Zero bonds to heteroatoms Alkane oxidation level	One bond to heteroatom Alcohol oxidation level	Two bonds to heteroatoms Aldehyde oxidation level	Three bonds to heteroatoms Carboxylic acid oxidation level	Four bonds to heteroatoms Carbon dioxide oxidation level
alkanes	alcohols	aldehydes	carboxylic acids	carbon dioxide
	ethers	ketones	esters	diethyl carbonate
	amines	acetals	amides	CFC-12
	alkyl halides		nitriles	
			acyl chlorides	
	alkenes	alkynes		

Alkenes and alkynes obviously don't fit easily into these categories as they have no bonds to heteroatoms. Alkenes can be made from alcohols by dehydration without any oxidation or reduction so it seems sensible to put them in the alcohol column. Similarly, alkynes and aldehydes are related by hydration/dehydration without oxidation or reduction.

Naming compounds

So far, we have talked a lot about compounds by name. Many of the names we've used (palytoxin, muscone, brevetoxin…) are simple names given to complicated molecules without regard for the actual structure or function of the molecule—these three names, for example, are all derived from the name of the organism from which the compound was first extracted. They are known as **trivial names**, not because they are unimportant, but because they are used in everyday scientific conversation.

Names like this are fine for familiar compounds that are widely used and referred to by chemists, biologists, doctors, nurses, perfumers alike. But there are over 16 million known organic compounds. They can't all have simple names, and no one would remember them if they did. For this reason, the IUPAC (International Union of Pure and Applied Chemistry) have developed **systematic nomenclature**, a set of rules that allows any compound to be given a unique name that can be deduced directly from its chemical structure. Conversely, a chemical structure can be deduced from its systematic name.

The problem with systematic names is that they tend to be grotesquely unpronounceable for anything but the most simple molecules. In everyday speech and writing, chemists therefore do tend to disregard them, and use a mixture of systematic and trivial names. Nonetheless, it's important to know how the rules work. We shall look next at systematic nomenclature, before going on to look at the real language of chemistry.

Systematic nomenclature

There isn't space here to explain all the rules for giving systematic names for compounds—they fill several desperately dull volumes, and there's no point knowing them anyway since computers will do the naming for you. What we will do is to explain the principles underlying systematic nomenclature. You should understand these principles, because they provide the basis for the names used by chemists for the vast majority of compounds that do not have their own trivial names.

Systematic names can be divided into three parts: one describes the hydrocarbon framework; one describes the functional groups; and one indicates where the functional groups are attached to the skeleton.

You have already met the names for some simple fragments of hydrocarbon framework (methyl, ethyl, propyl…). Adding a hydrogen atom to these alkyl fragments and changing -yl to -ane makes the alkanes and their names. You should hardly need reminding of their structures:

Names for the hydrocarbon framework

one carbon	methane	CH_4		
two carbons	ethane	CH_3—CH_3		
three carbons	propane	 CH_3‿CH_3	cyclopropane	△

Names for the hydrocarbon framework (continued)

four carbons	butane	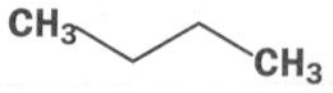 CH_3 CH_3	cyclobutane	
five carbons	pentane	 CH_3 CH_3	cyclopentane	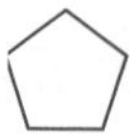
six carbons	hexane	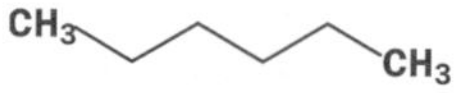 CH_3 CH_3	cyclohexane	
seven carbons	heptane	 CH_3 CH_3	cycloheptane	
eight carbons	octane	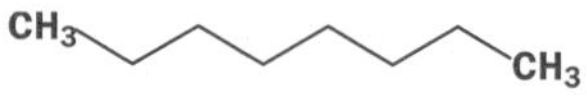 CH_3 CH_3	cyclo-octane	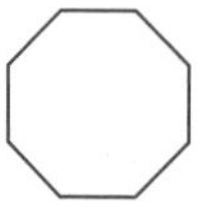
nine carbons	nonane	 CH_3 CH_3	cyclononane	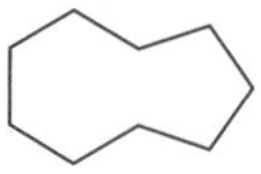
ten carbons	decane	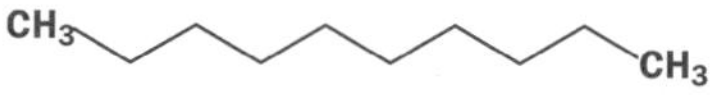 CH_3 CH_3	cyclodecane	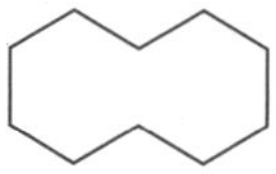

The name of a functional group can be added to the name of a hydrocarbon framework either as a suffix or as a prefix. Some examples follow. It is important to count all of the carbon atoms in the chain, even if one of them is part of a functional group: so pentanenitrile is actually BuCN.

CH_3OH

methanol   ethanal 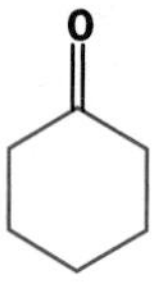   cyclohexanone 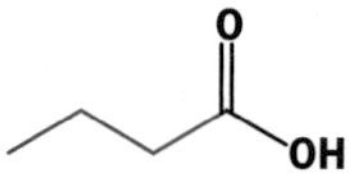   butanoic acid 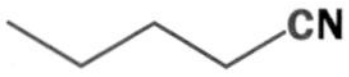   pentanenitrile

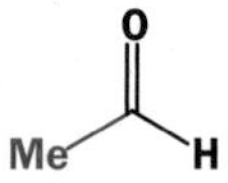

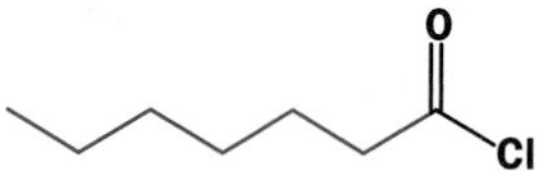

heptanoyl chloride $HC \equiv CH$ ethyne   ethoxyethane CH_3NO_2 nitromethane   propene

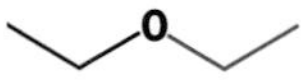

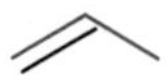

iodobenzene

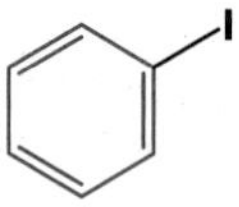

Compounds with functional groups attached to a benzene ring are named in a similar way.

Numbers are used to locate functional groups

Sometimes a number can be included in the name to indicate which carbon atom the functional group is attached to. None of the above list needed a number—check that you can see why not for each one. When numbers are used, the carbon atoms are counted from one end. In most cases, either of two numbers could be used (depending on which end you count from); the one chosen is always the lower of the two. Again, some examples will illustrate this point. Notice again that some functional groups are named by prefixes, some by suffixes, and that the number always goes directly before the functional group name.

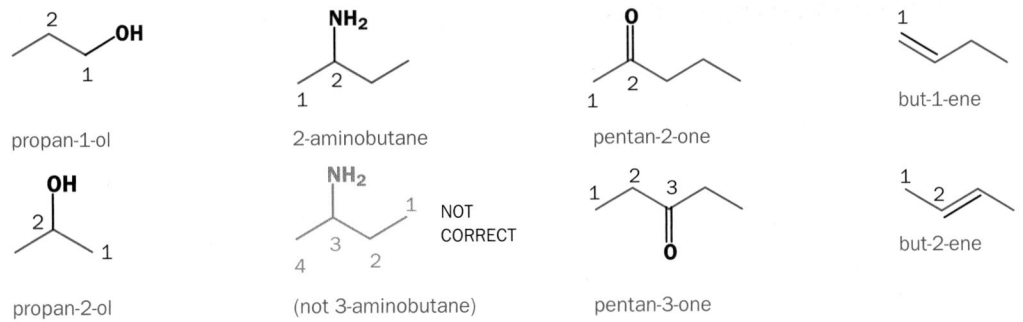

propan-1-ol

2-aminobutane

pentan-2-one

but-1-ene

propan-2-ol

(not 3-aminobutane)

NOT CORRECT

pentan-3-one

but-2-ene

Here are some examples of compounds with more than one functional group.

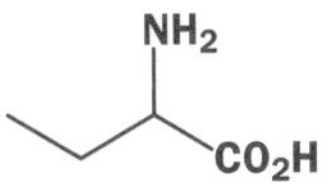

2-aminobutanoic acid 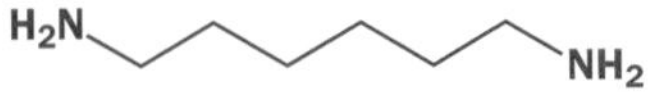   1,6-diaminohexane 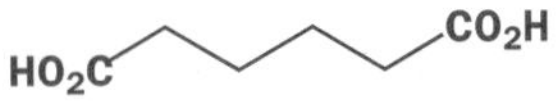   hexanedioic acid   tetrabromomethane   1,1,1-trichloroethane

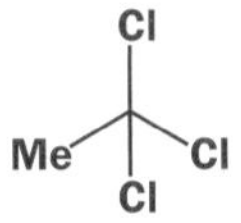

Again, the numbers indicate how far the functional groups are from the end of the carbon chain. Counting must always be from the same end for each functional group. Notice how we use di-, tri-, tetra- if there are more than one of the same functional group.

With cyclic compounds, there isn't an end to the chain, but we can use numbers to show the distance between the two groups—start from the carbon atom carrying one of the functional groups, then count round.

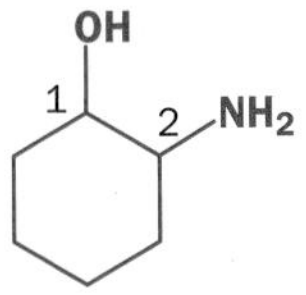

2-aminocyclohexanol

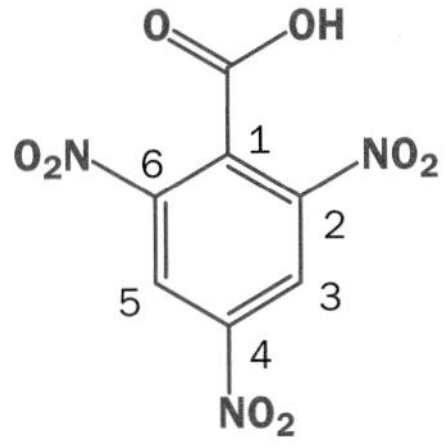

2,4,6-trinitrobenzoic acid

These rules work for hydrocarbon frameworks that are chains or rings, but many skeletons are branched. We can name these by treating the branch as though it were a functional group:

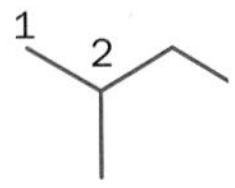

2-methylbutane

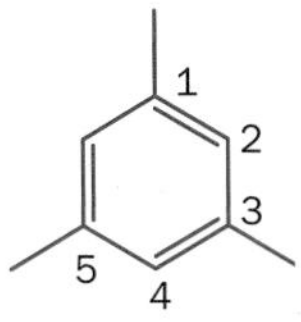

1,3,5-trimethyl benzene

1-butylcyclopropanol

Ortho, meta, and *para*

With substituted benzene rings, an alternative way of identifying the positions of the substituents is to use the terms *ortho, meta,* and *para. Ortho* compounds are 1,2-disubstituted, *meta* compounds are 1,3-disubstituted, and *para* compounds are 1,4-disubstituted. Some examples should make this clear.

▶

ortho, meta, and *para* are often abbreviated to *o, m,* and *p.*

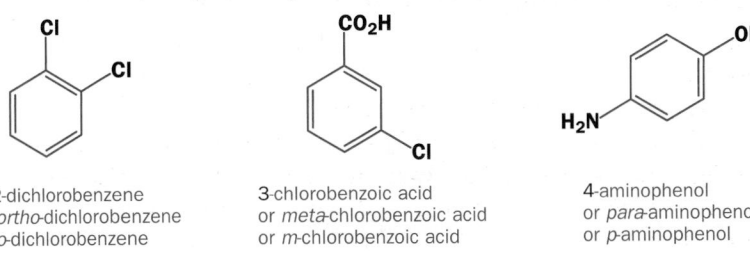

1,2-dichlorobenzene
or *ortho*-dichlorobenzene
or *o*-dichlorobenzene

3-chlorobenzoic acid
or *meta*-chlorobenzoic acid
or *m*-chlorobenzoic acid

4-aminophenol
or *para*-aminophenol
or *p*-aminophenol

► Beware! *Ortho*, *meta*, and *para* are used in chemistry to mean other things too: you may come across orthophosphoric acid, metastable states, and paraformaldehyde—these have nothing to do with the substitution patterns of benzene rings.

The terms *ortho*, *meta*, and *para* are used by chemists because they're easier to remember than numbers, and the words carry with them chemical meaning. 'Ortho' shows that two groups are next to each other on the ring even though the atoms may not happen to be numbered 1 and 2. They are one example of the way in which chemists don't always use systematic nomenclature but revert to more convenient 'trivial' terms. We consider trivial names in the next section.

What do chemists really call compounds?

The point of naming a compound is to be able to communicate with other chemists. Most chemists are happiest communicating chemistry by means of structural diagrams, and structural drawings are far more important than any sort of chemical nomenclature. That's why we explained in detail how to draw structures, but only gave an outline of how to name compounds. Good diagrams are easy to understand, quick to draw, and difficult to misinterpret.

● **Always give a diagram alongside a name unless it really is something very simple, such as ethanol.**

But we do need to be able to communicate by speech and by writing as well. In principle we could do this by using systematic names. In practice, though, the full systematic names of anything but the simplest molecules are far too clumsy for use in everyday chemical speech. There are several alternatives, mostly based on a mixture of trivial and systematic names.

Names for well known and widely used simple compounds

A few simple compounds are called by trivial names not because the systematic names are complicated, but just out of habit. We know them so well that we use their familiar names.

You may have met this compound before (left), and perhaps called it ethanoic acid, its systematic name. But in a chemical laboratory, everyone would refer to this acid as acetic acid, its trivial name. The same is true for all these common substances.

► We haven't asked you to remember any trivial names of molecules yet. But these 10 compounds are so important, you must be able to remember them. Learn them now.

Trivial names like this are often long-lasting, well understood historical names that are less easy to confuse than their systematic counterparts. 'Acetaldehyde' is easier to distinguish from 'ethanol' than is 'ethanal'.

Trivial names also extend to fragments of structures containing functional groups. Acetone, acetaldehyde, and acetic acid all contain the acetyl group (MeCO-, ethanoyl) abbreviated Ac and chemists often use this 'organic element' in writing AcOH for acetic acid or EtOAc for ethyl acetate.

Chemists use special names for four fragments because they have mechanistic as well as structural significance. These are vinyl and allyl; phenyl and benzyl.

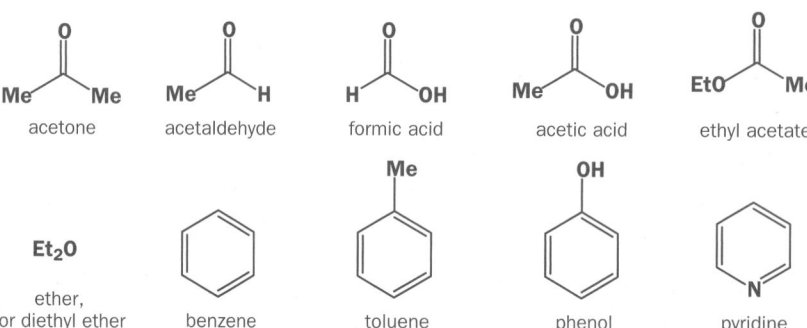

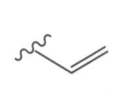

the vinyl group

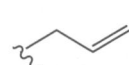

the allyl group

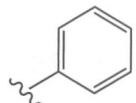

the phenyl group: Ph

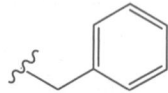

the benzyl group: Bn

Giving the vinyl group a name allows chemists to use simple trivial names for compounds like vinyl chloride, the material that polymerizes to give PVC (poly vinyl chloride) but the importance of the name lies more in the difference in reactivity (Chapter 17) between vinyl and allyl groups.

vinyl chloride

a section of the structure of PVC - Poly Vinyl Chloride

diallyl disulfide

allicin

The allyl group gets its name from garlic (*Allium* sp.), because it makes up part of the structure of the compounds on the right responsible for the taste and smell of garlic.

Allyl and vinyl are different in that the vinyl group is attached directly to a double bonded C=C carbon atom, while the allyl group is attached to a carbon atom *adjacent* to the C=C double bond. The difference is extremely important chemically: allyl compounds are typically quite reactive, while vinyl compounds are fairly unreactive.

For some reason, the allyl and vinyl groups have never acquired organic element symbols, but the benzyl group has and is called Bn. It is again important not to confuse the benzyl group with the phenyl group: the phenyl group is joined through a carbon atom in the ring, while the benzyl group is joined through a carbon atom attached to the ring. Phenyl compounds are typically unreactive but benzyl compounds are often reactive. Phenyl is like vinyl and benzyl is like allyl.

allyl acetate   **vinyl** acetate   **benzyl** acetate   **phenyl** acetate

We shall review all the organic element symbols you have met at the end of the chapter.

Names for more complicated but still well known molecules

Complicated molecules that have been isolated from natural sources are always given trivial names, because in these cases, the systematic names really are impossible!

Strychnine is a famous poison featured in many detective stories and a molecule with a beautiful structure. All chemists refer to it as strychnine as the systematic name is virtually unpronounceable. Two groups of experts at IUPAC and *Chemical Abstracts* also have different ideas on the systematic name for strychnine. Others like this are penicillin, DNA, and folic acid.

But the champion is vitamin B_{12}, a complicated cobalt complex with a three-dimensional structure of great intricacy. No chemist would learn this structure but would look it up in an advanced textbook of organic chemistry. You will find it in such books in the index under vitamin B_{12} and not under its systematic name. We do not even know what its systematic name might be and we are not very interested.

Even fairly simple but important molecules, the amino acids for example, that have systematic names that are relatively easy to understand are normally referred to by their

strychnine, or
(1R,11R,18S,20S,21S,22S)-12-oxa-8.17-
diazaheptacyclo [15.5.0^{1,8}.0^{2,7}.0^{15,20}]
tetracosa-2,4,6,14-tetraene-9-one (IUPAC)
or
4aR-[4aα,5aα,8aR*,15aα,15bα,15cβ]-
2,4a,5,5a,7,8,15,15a,15b,15c-decahydro-
4,6-methano-6H,14H-indolo[3,2,1-*ij*]oxepino
[2,3,4-*de*]pyrrolo[2,3-*h*]quinolone
(*Chemical Abstracts*)

vitamin B_{12}, or....

trivial names which are, with a bit of practice, easy to remember and hard to muddle up. They are given in full in Chapter 49.

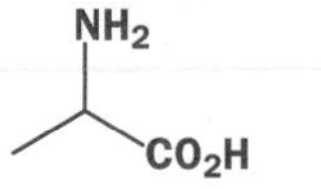

alanine, or
2-aminopropanoic acid

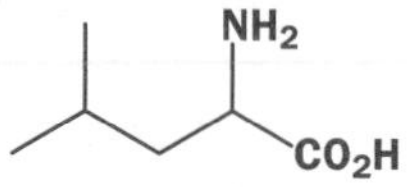

leucine, or
2-amino-4-methylpentanoic acid

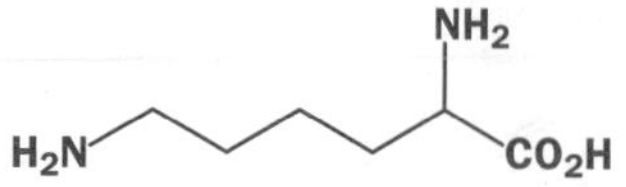

lysine, or
2,6-diaminohexanoic acid

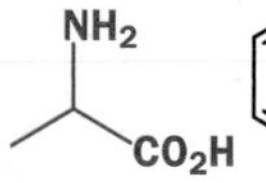

A very flexible way of getting new, simple names for compounds can be to combine a bit of systematic nomenclature with trivial nomenclature.

Alanine is a simple amino acid that occurs in proteins. Add a phenyl group and you have phenylalanine a more complex amino acid also in proteins.

alanine   phenylalanine

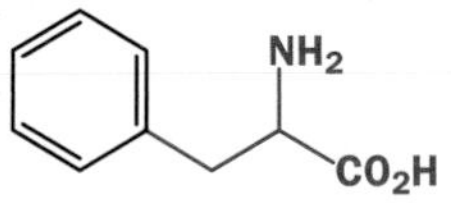

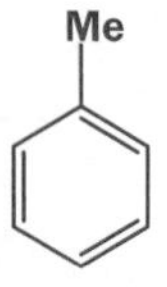

Toluene, the common name for methylbenzene, can be combined (both chemically and in making names for compounds!) with three nitro groups to give the famous explosive trinitrotoluene or TNT.

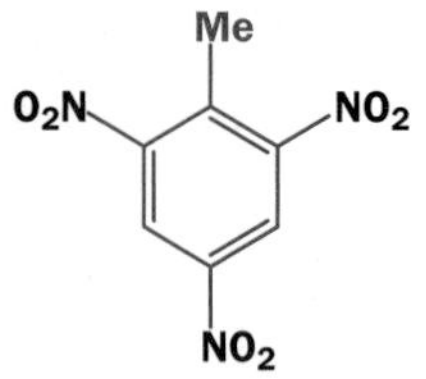

toluene   2,4,6-trinitrotoluene

Compounds named as acronyms

Some compounds are referred to by acronyms, shortened versions of either their systematic or their trivial name. We just saw TNT as an abbreviation for TriNitroToluene but the commoner use for acronyms is to define solvents and reagents in use all the time. Later in the book you will meet these solvents.

▶
The names and structures of these common solvents need learning too.

THF
(TetraHydroFuran)

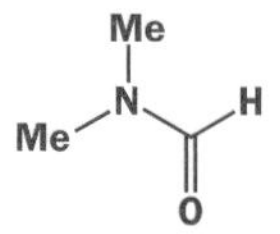

DMF
(DiMethylFormamide)

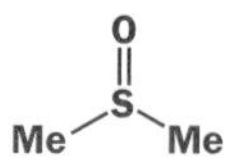

DMSO
(DiMethylSulfOxide)

The following reagents are usually referred to by acronym and their functions will be introduced in other chapters so you do not need to learn them now. You may notice that some acronyms refer to trivial and some to systematic names. There is a glossary of acronyms for solvents, reagents and other compounds on page 1513.

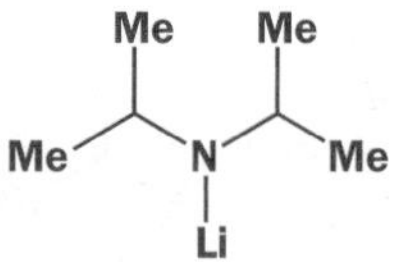

LDA
Lithium Di-isopropylAmide

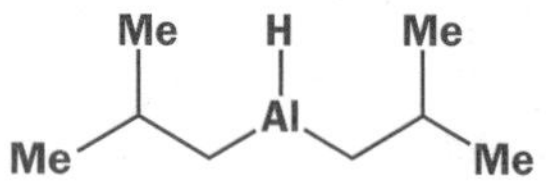

DIBAL
Di-IsoButylALuminium hydride

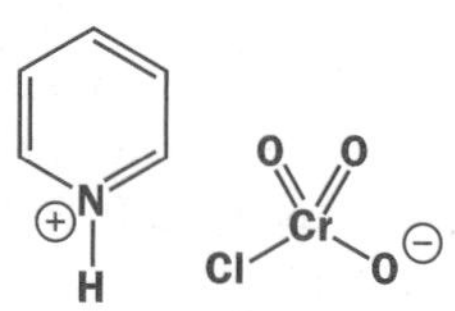

PCC
Pyridinium ChloroChromate

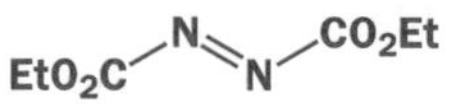

DEAD
DiEthyl Azo-Dicarboxylate

Compounds for which chemists use systematic names

You may be surprised to hear that practising organic chemists use systematic names at all in view of what we have just described, but they do! Systematic names really begin with derivatives of pentane (C_5H_{12}) since the prefix pent- means five, whereas but- does not mean four. Chemists refer to simple derivatives of open chain and cyclic compounds with 5 to about 20 carbon atoms by their systematic names, providing that there is no common name in use. Here are some examples.

cyclopentadiene 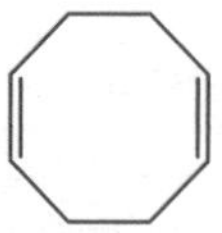   cyclo-octa-1,5-diene 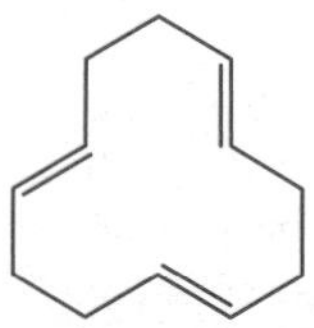   cyclododeca-1,5,9-triene 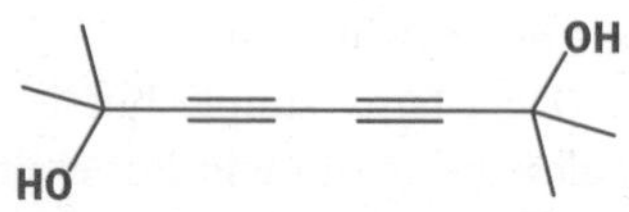   2,7-dimethyl-3,5-octadiyne-2,7-diol

11-bromo-undecanoic acid

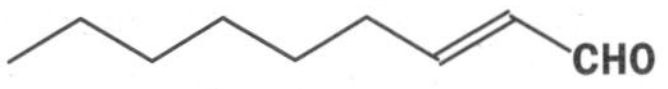

non-2-enal

These names contain a syllable that tells you the framework size: penta- for C_5, octa- for C_8, nona- for C_9, undeca- for C_{11}, and dodeca- for C_{12}. These names are easily worked out from the structures and, what is more important, you get a clear idea of the structure from the name. One of them might make you stop and think a bit (which one?), but the others are clear even when heard without a diagram to look at.

Complicated molecules with no trivial names

When chemists make complex new compounds in the laboratory, they publish them in a chemical journal giving their full systematic names in the experimental account, however long and clumsy those names may be. But in the text of the paper, and while talking in the lab about the compounds they have made, they will just call them 'the amine' or 'the alkene'. Everyone knows which amine or alkene is meant because at some point they remember seeing a chemical structure of the compound. This is the best strategy for talking about almost any molecule: draw a structure, then give the compound a 'tag' name like 'the amine' or 'the acid'. In written chemistry it's often easiest to give every chemical structure a 'tag' number as well.

To illustrate what we mean, let's talk about this compound.

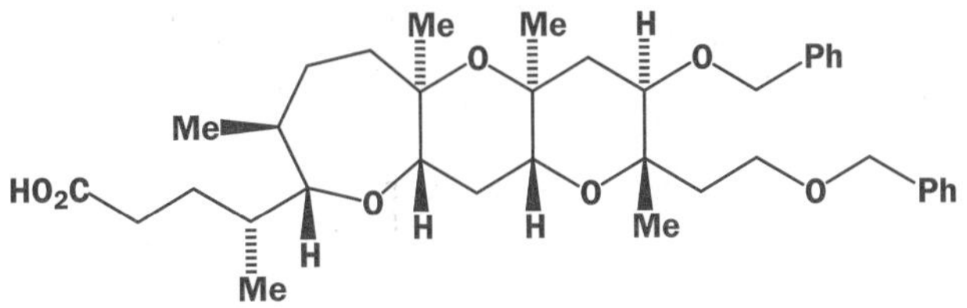

19

This carboxylic acid was made and used as an intermediate when chemists in California made brevetoxin (see p. 33) in 1995. Notice how we can call a complicated molecule 'this acid'—a 'tag' name—because you've seen the structure. It also has a tag number (19), so we can also call it 'compound 19', or 'acid 19', or 'brevetoxin fragment 19'. How much more sensible than trying to work out its systematic name.

How should you name compounds?

So what should you call a compound? It really depends on circumstances, but you won't go far wrong if you follow the example of this book. We shall use the names for compounds that real

---

● **Our advice on chemical names—six points in order of importance**
- Draw a structure first and worry about the name afterwards
- Learn the names of the *functional groups* (ester, nitrile, etc.)
- Learn and use the names of a few simple compounds used by all chemists
- In speech, refer to compounds as 'that acid' (or whatever) while pointing to a diagram
- Grasp the principles of systematic (IUPAC) nomenclature and use it for compounds of medium size
- Keep a notebook to record acronyms, trivial names, structures, etc. that you might need later

chemists use. There's no need to learn all the commonly used names for compounds now, but you should log them in your memory as you come across them. Never allow yourself to pass a compound name by unless you are sure you know what chemical structure it refers to.

We've met a great many molecules in this chapter. Most of them were just there to illustrate points so don't learn their structures! Instead, learn to recognize the names of the functional groups they contain. However, there were 10 names for simple compounds and three for common solvents that we advised you to learn. Cover up the right hand of each column and draw the structures for these 13 compounds.

Important structures to learn

acetone		toluene	
ether or diethyl ether		pyridine	
acetaldehyde		phenol	
formic acid		THF (tetrahydrofuran)	
acetic acid or AcOH		DMF or (dimethylformamide) Me₂NCHO	
benzene			
ethyl acetate or EtOAc		DMSO (dimethylsulfoxide)	

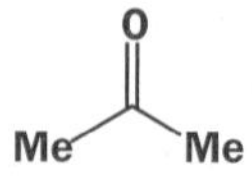

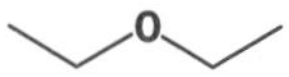

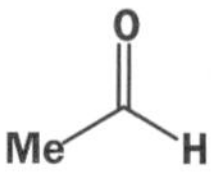

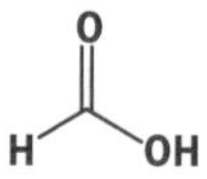

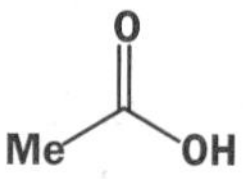

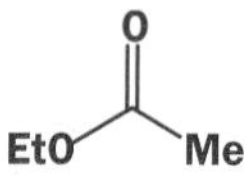

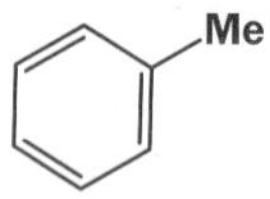

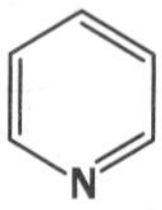

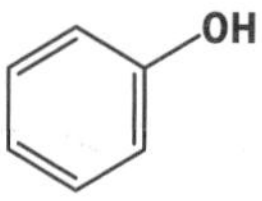

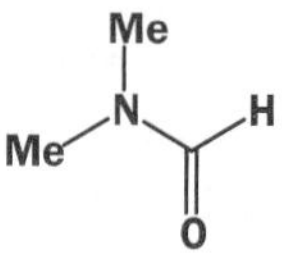

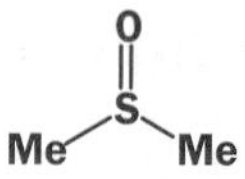

That's all we'll say on the subject of nomenclature—you'll find that as you practise using these names and start hearing other people referring to compounds by name you'll soon pick up the most important ones. But, to reiterate, make sure you never pass a compound name by without being absolutely sure what it refers to—draw a structure to check.

● Review box: Table of fragment names and organic elements

R	alkyl		***t*-Bu**	*tert*-butyl	
Me	methyl	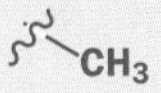 CH$_3$	**Ar**	aryl	any aromatic ring
Et	ethyl	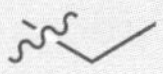	**Ph**	phenyl	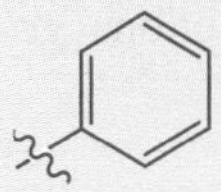
Pr (or *n*-Pr)	propyl		**Bn**	benzyl	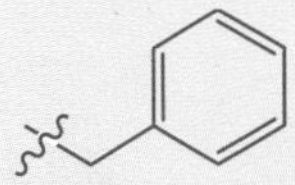
Bu (or *n*-Bu)	butyl		**Ac**	acetyl	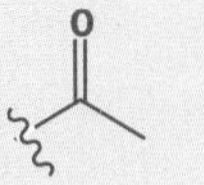
***i*-Pr**	isopropyl	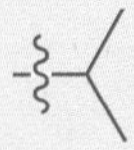		vinyl	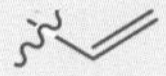
***i*-Bu**	isobutyl	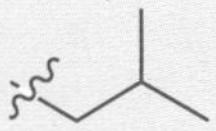		allyl	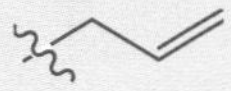
***s*-Bu**	*sec*-butyl	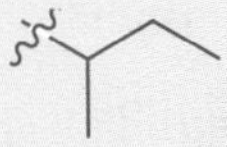	**X**	halide	**F, Cl, Br, or I**

Problems

1. Draw good diagrams of saturated hydrocarbons with seven carbon atoms having (a) linear, (b) branched, and (c) cyclic frameworks. Draw molecules based on each framework having both ketone and carboxylic acid functional groups.

2. Study the structure of brevetoxin on p. 33. Make a list of the different types of functional group (you already know that there are many ethers) and of the numbers of rings of different sizes. Finally study the carbon framework—is it linear, cyclic, or branched?

3. What is wrong with these structures? Suggest better ways of representing these molecules.

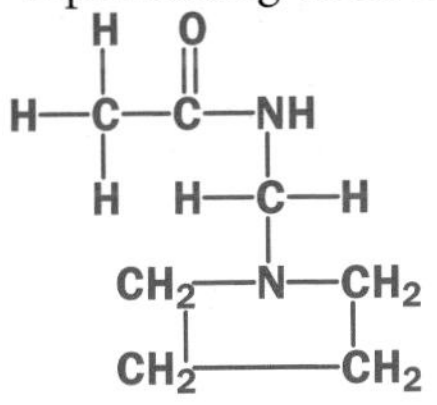

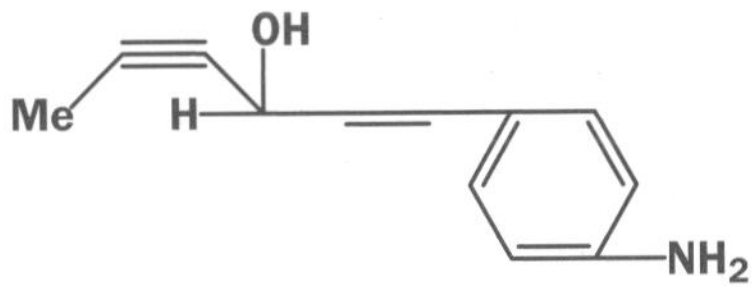

4. Draw structures corresponding to these names. In each case suggest alternative names that might convey the structure more clearly to someone who is listening to you speak.

(**a**) 1,4-di-1(1-dimethylethyl)benzene

(**b**) 3-(prop-2-enyloxy)prop-1-ene

(**c**) cyclohexa-1,3,5-triene

5. Draw one possible structure for each of these molecules, selecting any group of your choice for the 'wild card' substituents.

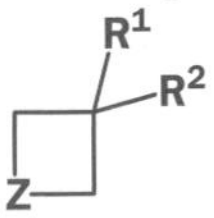

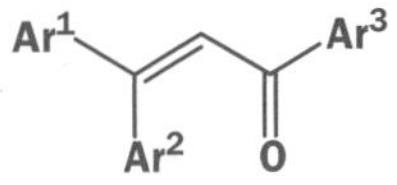

6. Translate these very poor 'diagrams' of molecules into more realistic structures. Try to get the angles about right and, whatever you do, don't include any square coplanar carbon atoms or other bond angles of 90°!

$C_6H_5CH(OH).(CH_2)_4COC_2H_5$

$O(CH_2CH_2)_2O$

$(CH_3O)_2CHCH=CHCH(OMe)_2$

7. Suggest at least six different structures that would fit the formula C_4H_7NO. Make good realistic diagrams of each one and say which functional group(s) are present.

8. Draw and name a structure corresponding to each of these descriptions.

(**a**) An aromatic compound containing one benzene ring with the following substituents: two chlorine atoms having a *para* relationship, a nitro group having an *ortho* relationship to one of the chlorine atoms, and an acetyl group having a *meta* relationship to the nitro group.

(b) An alkyne having a trifluoromethyl substituent at one end and a chain of three carbon atoms at the other with a hydroxyl group on the first atom, an amino group on the second, and the third being a carboxyl group.

9. Draw full structures for these compounds, displaying the hydrocarbon framework clearly and showing all the bonds present in the functional groups. Name the functional groups.

$AcO(CH_2)_3NO_2$

$MeO_2C.CH_2.OCOEt$

$CH_2{=}CH.CO.NH(CH_2)_2CN$

10. Identify the oxidation level of each of the carbon atoms in these structures with some sort of justification.

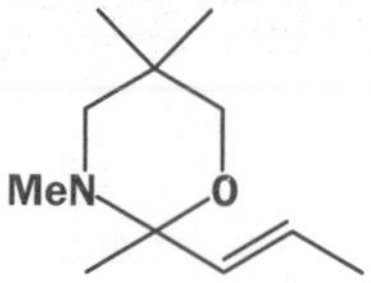

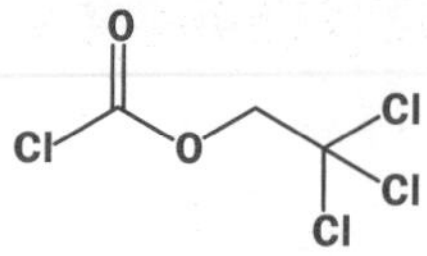

11. If you have not already done so, complete the exercises on pp. 23 (drawing amino acids) and 44 (giving structures for the 10 common compounds and three common solvents).

Determining organic structures

3

Connections

Building on:	**Arriving at:**	**Looking forward to:**
• What sorts of structure organic molecules have ch2	• Determining structure by X-ray crystallography	• ^{1}H NMR spectroscopy ch11
	• Determining structure by mass spectrometry	• Solving unknown structures spectroscopically ch15
	• Determining structure by ^{13}C NMR spectroscopy	
	• Determining structure by infrared spectroscopy	

Introduction

Organic structures can be determined accurately and quickly by spectroscopy

Having urged you, in the last chapter, to draw structures realistically, we now need to answer the question: what is realistic? How do we know what structures molecules actually have? Make no mistake about this important point: *we really do know what shape molecules have.* You wouldn't be far wrong if you said that the single most important development in organic chemistry in modern times is just this certainty, as well as the speed with which we can *be* certain. What has caused this revolution can be stated in a word—**spectroscopy**.

● What is spectroscopy?

Rays or waves interact with molecules:

- X-rays are scattered
- Radio waves make nuclei resonate
- Infrared waves are absorbed

Spectroscopy:

- measures these interactions
- plots charts of absorption
- relates interactions with structure

X-rays give bond lengths and angles. **Nuclear magnetic resonance** tells us about the carbon skeleton of the molecule. **Infrared spectroscopy** tells us about the types of bond in a molecule.

Structure of the chapter

We shall first consider structure determination as a whole and then introduce three different methods:

- Mass spectrometry (to determine mass of molecule and atomic composition)
- Nuclear magnetic resonance (NMR) spectroscopy (to determine carbon skeleton of molecule)
- Infrared spectroscopy (to determine functional groups in molecule)

Of these, NMR is more important than all the rest put together and so we shall return to it in Chapter 11. Then in Chapter 15, after we've discussed a wider range of molecules, there will be a review chapter to bring the ideas together and show you how unknown structures are really determined. If you would like more details of any of the spectroscopic methods we discuss, you should refer to a specialized book.

X-ray is the final appeal

▶

X-ray crystal structures are determined by allowing a sample of a crystalline compound to diffract X-rays. From the resulting diffraction pattern, it is possible to deduce the precise spatial arrangement of the atoms in the molecule—except, usually, the hydrogen atoms, which are too light to diffract the X-rays and whose position must be inferred from the rest of the structure.

In Chapter 2 we suggested you draw saturated carbon chains as zig-zags and not in straight lines with 90° or 180° bond angles. This is because we know they *are* zig-zags. The X-ray crystal structure of the 'straight' chain diacid, hexanedioic acid, is shown below. You can clearly see the zig-zag chain the planar carboxylic acid groups, and even the hydrogen atoms coming towards you and going away from you. It obviously makes sense to draw this molecule *realistically* as in the second drawing.

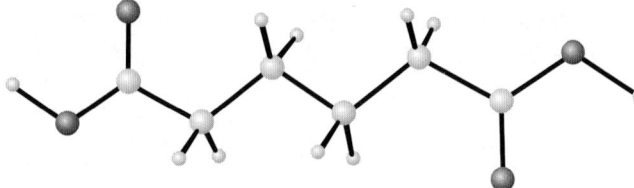

HO_2C—$(CH_2)_4$—CO_2H

hexanedioic acid

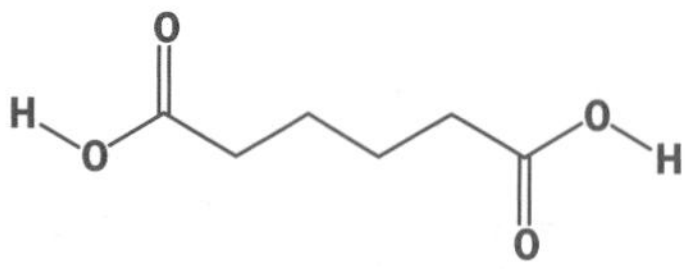

shape of hexanedioic acid

X-ray crystal structure of hexanedioic acid. Data for structure taken from Cambridge Crystallographic Data Centre

▶

Coenzymes are small molecules that work hand-in-hand with enzymes to catalyse a biochemical reaction.

This is one question that X-ray answers better than any other method: what shape does a molecule have? Another important problem it can solve is the structure of a new unknown compound. There are bacteria in oil wells, for example, that use methane as an energy source. It is amazing that bacteria manage to convert methane into anything useful, and, of course, chemists really wanted to know how they did it. Then in 1979 it was found that the bacteria use a coenzyme, given the trivial name 'methoxatin', to oxidize methane to methanol. Methoxatin was a new compound with an unknown structure and could be obtained in only very small amounts. It proved exceptionally difficult to solve the structure by NMR but eventually methoxatin was found by X-ray crystallography to be a polycyclic tricarboxylic acid. This is a more complex molecule than hexanedioic acid but X-ray crystallographers routinely solve much more complex structures than this.

■

If you like systematic names, you can call methoxatin 4,5-dihydro-4,5-dioxo-1*H*-pyrrolo[2,3-*f*]quinoline-2,7,9-tricarboxylic acid. But you may feel, like us, that 'methoxatin' and a diagram or the tag name 'the tricarboxylic acid' are better.

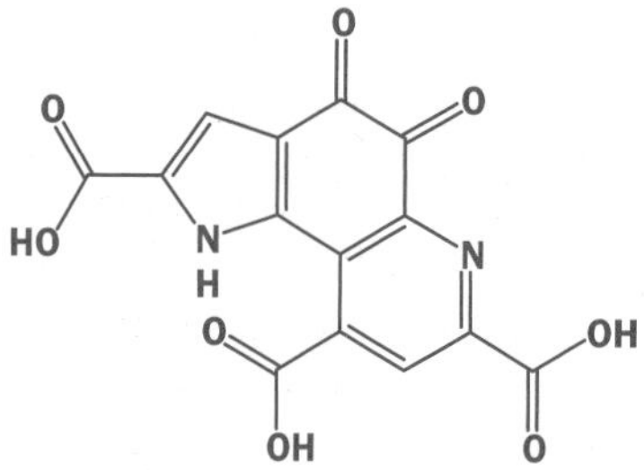

methoxatin

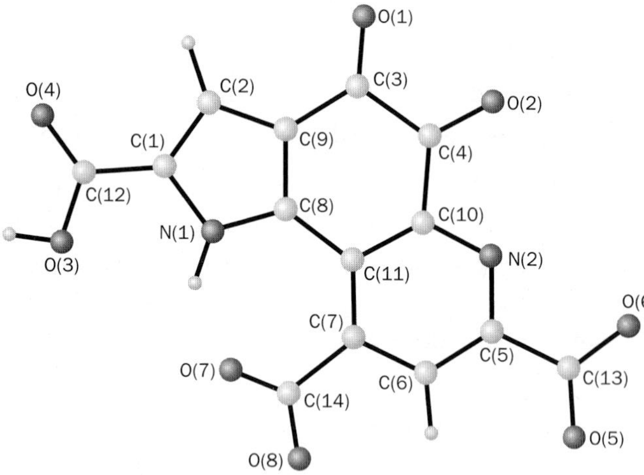

X-ray crystal stucture of methoxatin. Data for the X-ray structure taken from the Cambridge Crystallographic Data Centre

X-ray crystallography has its limitations

If X-ray crystallography is so powerful, why do we bother with other methods? There are two reasons.

- X-ray crystallography works by the scattering of X-rays from electrons and requires crystalline solids. If an organic compound is a liquid or is a solid but does not form good crystals, its structure cannot be determined in this way.

- X-ray crystallography is a science in its own right, a separate discipline from chemistry because it requires specific skills, and a structure determination can take a long time. Modern methods have reduced this time to a matter of hours or less, but nonetheless by contrast a modern NMR machine with a robot attachment can run more than 100 spectra in an overnight run. So we normally use NMR routinely and reserve X-rays for difficult unknown structures and for determining the detailed shape of important molecules.

Outline of structure determination by spectroscopy

Put yourself in these situations.

- Finding an unknown product from a chemical reaction
- Discovering an unknown compound from Nature
- Detecting a suspected food contaminant
- Routinely checking purity during the manufacture of a drug

In all cases except perhaps the second you need a quick and reliable answer. Suppose you are trying to identify the heart drug propranolol, one of the famous 'beta blockers' used to reduce high blood pressure and prevent heart attacks. You would first want to know the molecular weight and atomic composition and this would come from a *mass spectrum*: propranolol has a molecular weight (relative molecular mass) of 259 and the composition $C_{16}H_{21}NO_2$. Next you would need the carbon skeleton—this would come from *NMR*, which would reveal the three fragments shown.

propranolol $C_{16}H_{21}NO_2$

fragments of propranolol from the NMR spectrum

There are many ways in which these fragments could be joined together and at this stage you would have no idea whether the oxygen atoms were present as OH groups or as ethers, whether the nitrogen would be an amine or not, and whether Y and Z might or might not be the same atom, say N. More information comes from the **infrared spectrum**, which highlights the functional groups, and which would show that there is an OH and an NH in the molecule but not functional groups such as CN or NO_2. This still leaves a variety of possible structures, and these could finally be distinguished by another technique, ^{1}H NMR. We are in fact going to avoid using ^{1}H NMR in this chapter, because it is more difficult, but you will learn just how much information can be gained from mass spectra, IR spectra, and ^{13}C NMR spectra.

Now we must go through each of these methods and see how they give the information they do. For this exercise, we will use some compounds you encounter in everyday life, perhaps without realizing it.

■
^{1}H NMR makes an entrance in Chapter 11.

Method and what it does	What it tells us	Type of data provided
Mass spectrum weighs the molecule	molecular weight (relative molecular mass) and composition	259; $C_{16}H_{21}NO_2$
^{13}C NMR reveals all different carbon nuclei	carbon skeleton	no C=O group; ten carbons in aromatic rings; two carbons next to O; three other saturated C atoms
Infrared reveals chemical bonds	functional groups	no C=O group; one OH; one NH

Mass spectrometry

Mass spectrometry weighs the molecule

A mass spectrometer has three basic components: something to volatilize and ionize the molecule into a beam of charged particles; something to focus the beam so that particles of the same mass:charge ratio are separated from all others; and something to detect the particles. All spectrometers in common use operate in a high vacuum and usually use positive ions. Two methods are used to convert neutral molecules into cations: electron impact and chemical ionization.

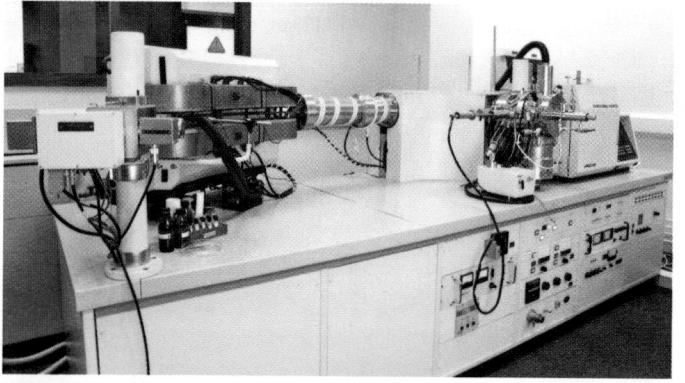

> Mass spectrometry uses a different principle from the other forms of spectroscopy we discuss: what is measured is not absorption of energy but the mass of the molecule or fragments of molecule.

In this picture of a mass spectrometer you can see the large copper-coloured electromagnetic coils used to deflect the charged ions. The sample is introduced into the instrument via the port in the right of the picture where it is also volatilized and ionized. The ions are then deflected by the electromagnets before hitting the ion detector shown in the very left of the picture.

Mass spectrometry by electron impact

In **electron impact (E.I.) mass spectrometry** the molecule is bombarded with highly energetic electrons that knock a weakly bound electron out of the molecule. If you think this is strange, think of throwing bricks at a brick wall: the bricks do not stick to the wall but knock loose bricks off the top of the wall. Losing a single electron leaves behind a radical cation: an unpaired electron and a positive charge. The electron that is lost will be one of relatively high energy (the bricks come from the *top* of the wall), and this will typically be one not involved in bonding, for example, an electron from a lone pair. Thus ammonia gives $NH_3^{+\bullet}$ and a ketone gives $R_2C=O^{+\bullet}$. If the electron beam is not too high in energy, some of these rather unstable radical cations will survive the focusing operation and get to the detector. Normally two focusing operations are used: the beam is bent magnetically and electrostatically to accelerate the cations on their way to the detector and it takes about 20 μs for the cations to get there. But if, as is often the case, the electron beam supplies more than exactly the right amount of energy to knock out the electron, the excess energy is dissipated by fragmentation of the radical cation. Schematically, an unknown molecule first forms the radical cation $M^{+\bullet}$ which then breaks up (fragments) to give a radical $X^\bullet$ and a cation Y^+. Only charged particles (cations in most machines) can be accelerated and focused by the magnetic and electrostatic fields and so the detector records only the molecular ion $M^{+\bullet}$ and positively charged fragments Y^+. Uncharged radicals $X^\bullet$ are not recorded.

loss of one electron
leaves a radical cation

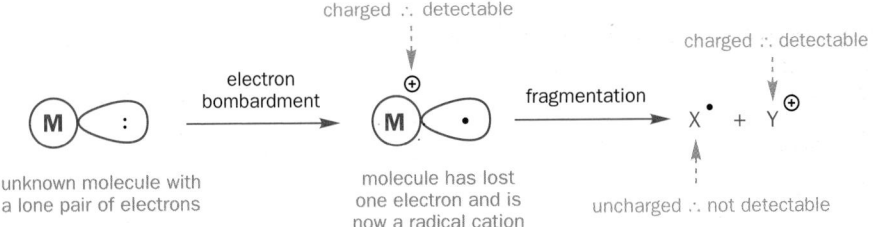

charged ∴ detectable

charged ∴ detectable

electron bombardment

fragmentation

unknown molecule with a lone pair of electrons

molecule has lost one electron and is now a radical cation

uncharged ∴ not detectable

A typical result is the E.I. mass spectrum for the alarm pheromone of the honey bee. The bees check every insect coming into the hive for strangers. If a strange insect (even a bee from another hive) is detected, an alarm pheromone is released and the intruder is attacked. The pheromone is a simple volatile organic molecule having this mass spectrum.

■ Insects communicate by releasing compounds with strong smells (to the insect!). These have to be small volatile molecules, and those used to communicate between members of the same species are called **pheromones**.

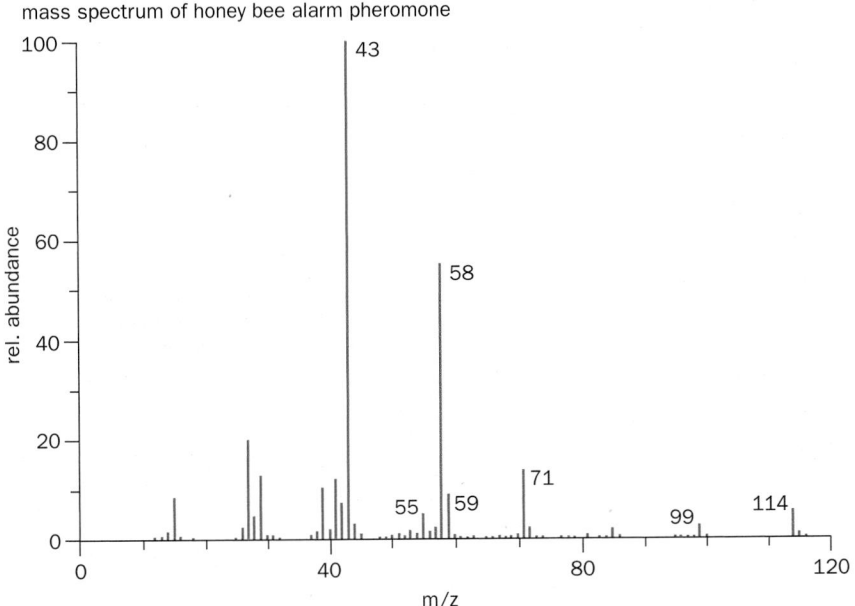

mass spectrum of honey bee alarm pheromone

The strongest peak, at 43 mass units in this case, is assigned an 'abundance' of 100% and called the **base peak**. The abundance of the other peaks is shown relative to the base peak. In this spectrum, there is only one other strong peak (58 at 50%) and the peak of highest mass at 114 (at 5%) is the molecular ion corresponding to a structure $C_7H_{14}O$. The main fragmentation is to a $C_5H_{11}^{\bullet}$ radical (not observed as it isn't charged) and a cation $C_2H_3O^+$, which forms the base peak. The pheromone is the simple ketone heptan-2-one.

▶ Mass spectroscopy requires minute quantities of sample—much less than the amounts needed for the other techniques we will cover. Pheromones are obtainable from insects only on a microgram scale or less.

heptan-2-one
(only one lone pair shown)

electron bombardment

fragmentation

$M^{+\bullet} = 114 = C_7H_{14}O$

$X^{\bullet} = C_5H_{11} = 71$

$Y^{\oplus} = C_2H_3O = 43$

The problem with E.I. is that for many radical cations even 20 μs is too long, and all the molecular ions have decomposed by the time they reach the detector. The fragments produced may be useful in identifying the molecule, but even in the case of the bee alarm pheromone it would obviously be better to get a stronger and more convincing molecular ion as the weak (5%) peak at 114 might also be a fragment or even an impurity.

Mass spectrometry by chemical ionization

In **chemical ionization (C.I.) mass spectrometry** the electron beam is used to ionize a simple molecule such as methane which in turn ionizes our molecule by collision and transfer of a proton. Under electron bombardment, methane loses a bonding electron (it doesn't have any other kind) to give $CH_4^{+\bullet}$ which reacts with an unionized methane molecule to give $CH_3^{\bullet}$ and CH_5^{+}. Before you write in complaining about a mistake, just consider that last structure in a bit more detail. Yes, CH_5^{+} does have a carbon atom with *five* bonds. But it has only eight electrons! These are distributed between five bonds (hence the + charge) and the structure is thought to be trigonal bipyramidal. This structure has not been *determined* as it is too unstable. It is merely proposed from theoretical calculations.

proposed structure of $CH_5^{\oplus}$
the two black bonds share two electrons

This unstable compound is a powerful acid, and can protonate just about any other molecule. When it protonates our sample, a proton has been added rather than an electron removed, so the resulting particles are simple cations, not radical cations, and are generally more stable than the radical cations produced by direct electron impact. So the molecular ion has a better chance of lasting the necessary 20 μs to reach the detector. Note that we now observe $[M + H]^{+}$ (i.e. one more than the molecular mass) rather than M^{+} by this method.

Having more functional groups helps molecular ions to decompose. The aromatic amine 2-phenylethylamine is a brain active amine found in some foods such as chocolate, red wine, and cheese and possibly implicated in migraine. It gives a poor molecular ion by E.I., a base peak with a mass as low as 30 and the only peak at higher mass is a 15% peak at 91. The C.I. mass spectrum on the other hand has a good molecular ion: it is $[M + H]^{+}$ of course. Normally a fragmentation gives one cation and another radical, only the cation being detected. It is relatively unusual for one bond to be able to fragment in either direction, but here it does, which means that both fragments are seen in the spectrum.

the radical cation can fragment in two ways

Mass spectrometry separates isotopes

You will know in theory that most elements naturally exist as mixtures of isotopes. If you didn't believe it, now you will. Chlorine is normally a 3:1 mixture of ^{35}Cl and ^{37}Cl (hence the obviously false relative atomic mass of '35.5' for chlorine) while bromine is an almost 1:1 mixture of ^{79}Br and ^{81}Br (hence the 'average' mass of 80 for bromine). Mass spectrometry separates these isotopes so that you get true not average molecular weights. The molecular ion in the E.I. mass spectrum of the bromo-amide below has two peaks at 213 and 215 of roughly equal intensity. This might just represent the loss of molecular hydrogen from a molecular ion 215, but, when we notice that the first fragment (and base peak) has the same pattern at 171/173, the presence of bromine is a more likely explanation. All the smaller fragments at 155, 92, etc. lack the 1:1 pattern and also therefore lack bromine.

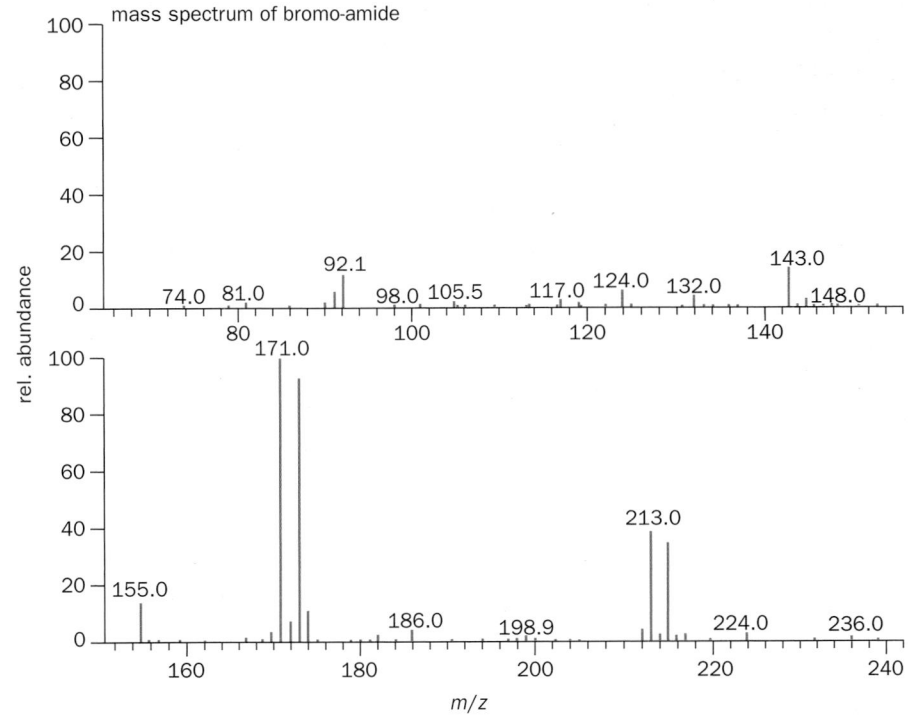

The mass spectrum of chlorobenzene (PhCl, C_6H_5Cl) is very simple. There are two peaks at 112 (100%) and 114 (33%), a peak at 77 (40%), and very little else. The peaks at 112/114 with their 3:1 ratio are the molecular ions, while the fragment at 77 is the phenyl cation (Ph$^+$ or $C_6H_5^+$).

The mass spectrum of DDT is very revealing. This very effective insecticide became notorious as it accumulated in the fat of birds of prey (and humans) and was phased out of use. It can be detected easily

> It's worth remembering that the Ph$^+$ weighs 77: you'll see this mass frequently.

Table 3.1 Summary table of main isotopes for mass spectra

Element	Carbon	Chlorine	Bromine
isotopes	^{12}C, ^{13}C	^{35}Cl, ^{37}Cl	^{79}Br, ^{81}Br
rough ratio	1.1% ^{13}C (90:1)	3:1	1:1

by mass spectrometry because the five chlorine atoms produce a complex molecular ion at 352/354/356/358/360 with ratios of 243:405:270:90:15:1 (the last is too small to see). The peak at 352 contains nothing but ^{35}Cl, the peak at 354 has four atoms of ^{35}Cl and one atom of ^{37}Cl, while the invisible peak at 360 has five ^{37}Cl atoms. The ratios need some working out, but the fragment at 335/337/339 in a ratio 9:6:1 is easier. It shows just two chlorine atoms as the CCl_3 group has been lost as a radical, leaving $C_{13}H_9Cl_2^+$.

> Remember: mass spectroscopy is very good at detecting minute quantities.

Isotopes in DDT

The ratio comes from the 3:1 isotopic ratio like this:

- chance of one ^{35}Cl in the molecule: $\frac{3}{4}$

- chance of one ^{37}Cl in the molecule: $\frac{1}{4}$

If the molecule or fragment contains two chlorine atoms, as does our $C_{13}H_9Cl_2^+$, then

- chance of two ^{35}Cls in the molecule: $\frac{3}{4} \times \frac{3}{4} = \frac{9}{16}$

- chance of one ^{35}Cl and one ^{37}Cl in the molecule: $\left[\frac{3}{4} \times \frac{1}{4}\right] + \left[\frac{1}{4} \times \frac{3}{4}\right] = \frac{6}{16}$

- chance of two ^{37}Cls in the molecule: $\frac{1}{4} \times \frac{1}{4} = \frac{1}{16}$

The ratio of these three fractions is 9:6:1, the ratio of the peaks in the mass spectrum.

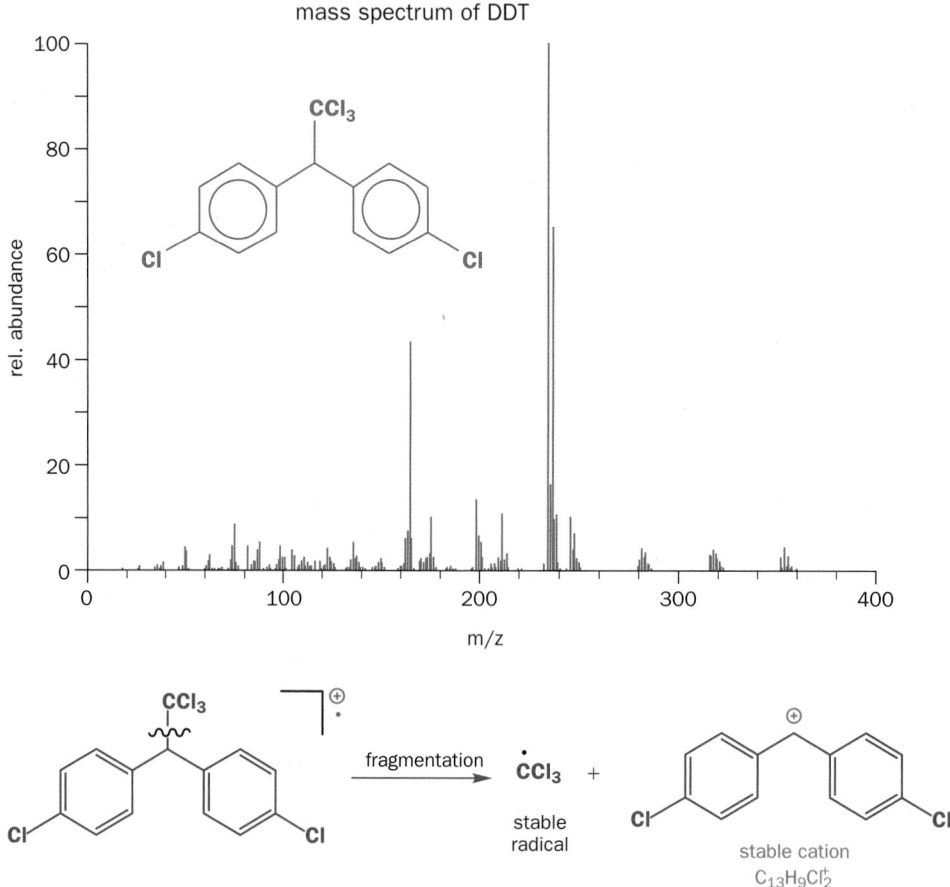

Carbon has a minor but important isotope ^{13}C

Many elements have minor isotopes at below the 1% level and we can ignore these. One important one we cannot ignore is the 1.1% of ^{13}C present in ordinary carbon. The main isotope is ^{12}C and you may recall that ^{14}C is radioactive and used in carbon dating, but its natural abundance is minute. The stable isotope ^{13}C is not radioactive, but it is NMR active as we shall soon see. If you look back at the mass spectra illustrated so far in this chapter, you will see a small peak one mass unit higher than each peak in most of the spectra. This is no instrumental aberration: these are genuine peaks containing ^{13}C instead of ^{12}C. The exact height of these peaks is useful as an indication of the number of carbon atoms in the molecule. If there are n carbon atoms in a molecular ion, then the ratio of M^+ to $[M + 1]^+$ is $100 : (1.1) \times n$.

The electron impact mass spectrum of BHT gives a good example. The molecular ion at 220 has an abundance of 24% and $[M + 1]^+$ at 221 has 4–5% abundance but is difficult to measure as it is so weak. BHT is $C_{15}H_{26}O$ so this should give an $[M + 1]^+$ peak due to ^{13}C of $15 \times 1.1\%$ of M^+, that is, 16.5% of M^+ or $24 \times 16.5 = 4.0\%$ actual abundance. An easier peak to interpret is the base peak at 205 formed by the loss of one of the six identical methyl groups from the *t*-butyl side chains (don't forget what we told you in Chapter 2—all the 'sticks' in these structures are methyl groups and not hydrogen atoms). The base peak (100%) 205 is $[M—Me]^+$ and the ^{13}C peak 206 is 15%, which fits well with $14 \times 1.1\% = 15.4\%$.

BHT

BHT is used to prevent the oxidation of vitamins A and E in foods. It carries the E-number E321. There has been some controversy over its use because it is a cancer suspect agent, but it is used in some 'foods' like chewing gum. BHT stands for 'Butylated HydroxyToluene': you can call it 2,6-di-*t*-butyl-4-methylphenol if you want to, but you may prefer to look at the structure and just call it BHT. You met BHT briefly in Chapter 2 when you were introduced to the tertiary butyl group.

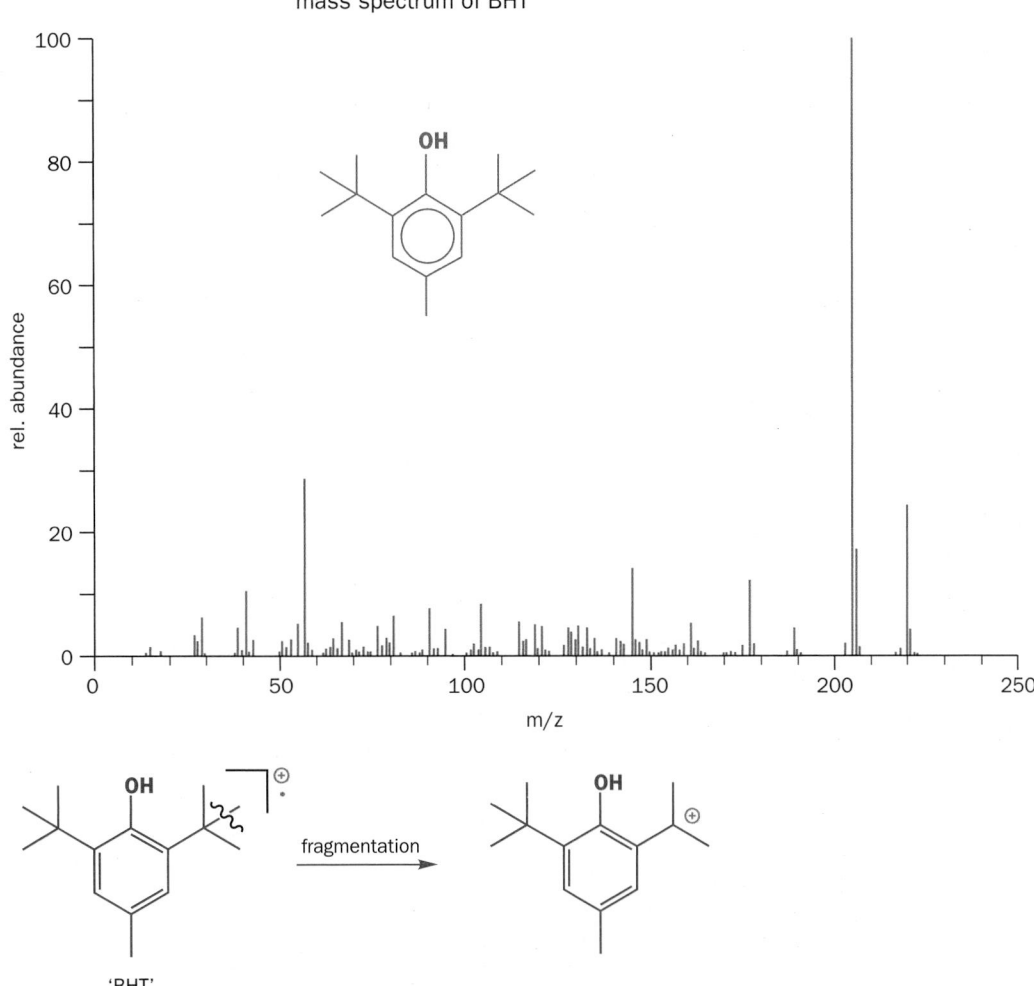

mass spectrum of BHT

Other examples you have seen include the DDT spectrum, where the peaks between the main peaks are ^{13}C peaks: thus 236, 238, and 240 are each 14% of the peak one mass unit less, as this fragment has 13 carbon atoms. If the number of carbons gets very large, so does the ^{13}C peak; eventually it is *more* likely that the molecule contains one ^{13}C than that it doesn't. We can ignore the possibility of two ^{13}C atoms as 1.1% of 1.1% is very small (probability of 1.32×10^{-5}).

Table 3.2 summarizes the abundance of the isotopes in these three elements. Notice that the ratio for chlorine is not exactly 3:1 nor that for bromine exactly 1:1; nevertheless you should use the simpler ratios when examining a mass spectrum. Always look at the heaviest peak first: see whether there is chlorine or bromine in it,

Table 3.2 Abundance of isotopes for carbon, chlorine, and bromine

Element	Major isotope: abundance	Minor isotope: abundance
carbon	^{12}C: 98.9%	^{13}C: 1.1%
chlorine	^{35}Cl: 75.8%	^{37}Cl: 24.2%
bromine	^{79}Br: 50.5%	^{81}Br: 49.5%

and whether the ratio of M$^+$ to [M + 1]$^+$ is about right. If, for example, you have what seems to be M$^+$ at 120 and the peak at 121 is 20% of the supposed M$^+$ at 120, then this cannot be a ^{13}C peak as it would mean that the molecule would have to contain 18 carbon atoms and you cannot fit 18 carbon atoms into a molecular ion of 120. Maybe 121 is the molecular ion.

Atomic composition can be determined by high resolution mass spectrometry

Ordinary mass spectra tell us the molecular weight (MW) of the molecule: we could say that the bee alarm pheromone was MW 114. When we said it was $C_7H_{14}O$ we could not really speak with confidence because 114 could also be many other things such as C_8H_{18} or $C_6H_{10}O_2$ or $C_6H_{14}N_2$. These different atomic compositions for the same molecular weight can nonetheless be distinguished if we know the exact molecular weight, since individual isotopes have non-integral masses (except ^{12}C by definition). Table 3.3 gives these to five decimal places, which is the sort of accuracy you need for meaningful results. Such accurate mass measurements are called **high resolution mass spectrometry**.

For the bee alarm pheromone, the accurate mass turns out to be 114.1039. Table 3.4 compares possible atomic compositions, and the result is conclusive. The exact masses to three places of decimals fit the observed exact mass only for the composition $C_7H_{14}O$. You may not think the fit is very good when you look at the two numbers, but notice the difference in the error expressed as parts per million. One answer stands out from the rest. Note that even two places of decimals would be enough to distinguish these four compositions.

A more important case is that of the three ions at 28: nitrogen, carbon monoxide, and ethylene (ethene, $CH_2{=}CH_2$). Actually mass spectra rarely go down to this low value because some nitrogen is usually injected along with the sample, but the three ions are all significant and it is helpful to see how different they are. Carbon monoxide CO is 27.9949, nitrogen N_2 is 28.0061, and ethylene 28.0313.

> ● In the rest of the book, whenever we state that a molecule has a certain atomic composition, you can assume that it has been determined by high resolution mass spectrometry on the molecular ion.

One thing you may have noticed in Table 3.4 is that there are no entries with just one nitrogen atom. Two nitrogen atoms, yes; one nitrogen no! This is because any complete molecule with *one nitrogen in it has an odd molecular weight*. Look back at the mass spectrum of the compounds giving good molecular ions by C.I. for an example. The nitro compound had M = 127 and the amine M = 121. This is because C, O, and N all have even atomic weights—only H has an odd atomic weight. Nitrogen is the only element from C, O, and N that can form an odd number of bonds (3). Molecules with one nitrogen atom must have an odd number of hydrogen atoms and hence an odd molecular weight. Molecules with only C, H, and O or with even numbers of nitrogen atoms have even molecular weights.

If we are talking about fragments, that is, cations or radicals, the opposite applies. A fragment has, by definition, an unused valency. Look back at the fragments in this section and you will see that this is so. Fragments with C, H, O alone have odd molecular weights, while fragments with one nitrogen atom have even molecular weights.

▶

The reason that exact masses are not integers lies in the slight mass difference between a proton (1.67262×10^{-27} kg) and a neutron (1.67493×10^{-27} kg) and in the fact that electrons have mass (9.10956×10^{-31} kg).

Table 3.3 Exact masses of common elements

Element	Isotope	Atomic weight	Exact mass
hydrogen	^{1}H	1	1.00783
carbon	^{12}C	12	12.00000
carbon	^{13}C	13	13.00335
nitrogen	^{14}N	14	14.00307
oxygen	^{16}O	16	15.99492
fluorine	^{19}F	19	18.99840
phosphorus	^{31}P	31	30.97376
sulfur	^{32}S	32	31.97207
chlorine	^{35}Cl	35	34.96886
chlorine	^{37}Cl	37	36.96590
bromine	^{79}Br	79	78.91835
bromine	^{81}Br	81	80.91635

Table 3.4 Exact mass determination for the bee alarm pheromone

Composition	Calculated M$^+$	Observed M$^+$	Error in p.p.m.
$C_6H_{10}O_2$	114.068075	114.1039	358
$C_6H_{14}N_2$	114.115693	114.1039	118
$C_7H_{14}O$	**114.104457**	**114.1039**	5
C_8H_{18}	114.140844	114.1039	369

▶

This rule holds as long as there are only C, H, N, O, S atoms in the molecule. It doesn't work for molecules with Cl or P atoms for example.

Nuclear magnetic resonance

What does it do?

Nuclear magnetic resonance (NMR) allows us to detect atomic nuclei and say what sort of environment they are in, within their molecule. Clearly, the hydrogen of, say, propanol's hydroxyl group is different from the hydrogens of its carbon skeleton—it can be displaced by sodium metal, for example. NMR (actually ^{1}H, or proton, NMR) can easily distinguish between these two sorts of hydrogens. Moreover, it can also distinguish between all the other different sorts of hydrogen atoms present. Likewise, carbon (or rather ^{13}C) NMR can easily distinguish between the three different carbon atoms. In this chapter we shall look at ^{13}C NMR spectra and then in Chapter 11 we shall look at proton (^{1}H) NMR spectra in detail.

NMR is incredibly versatile: it can even scan living human brains (see picture) but the principle is still the same: being able to detect nuclei (and hence atoms) in different environments. We need first to spend some time explaining the principles of NMR.

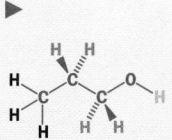

▶

Proton NMR can distinguish between the different coloured hydrogens. Carbon NMR can distinguish between all the carbons.

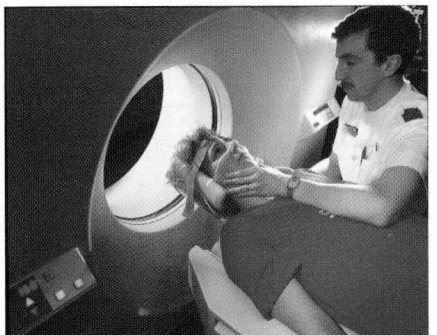

Magnetic Resonance Imaging

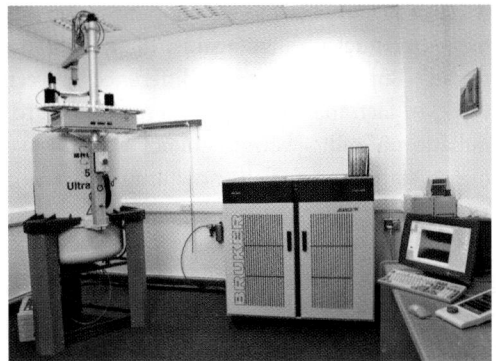

An NMR machine

■

When NMR is used medically it is usually called Magnetic Resonance Imaging (MRI) for fear of frightening patients wary of all things *nuclear*.

NMR uses a strong magnetic field

Imagine for a moment that we were able to 'switch off' the earth's magnetic field. One effect would be to make navigation much harder since all compasses would be useless. They would be free to point in whatever direction they wanted to and, if we turned the needle round, it would simply stay where we left it. However, as soon as we switched the magnetic field back on, they would all point north—their lowest energy state. Now if we wanted to force a needle to point south we would have to use up energy and, of course, as soon as we let go, the needle would return to its lowest energy state, pointing north.

In a similar way, some atomic nuclei act like tiny compass needles and have different energy levels when placed in a magnetic field. The compass needle can rotate through 360° and have an essentially infinite number of different energy levels, all higher in energy than the 'ground state' (pointing north). Fortunately, our atomic nucleus is more restricted—its energy levels are quantized, just like the energy levels of an electron, which you will meet in the next chapter, and there are only certain specific energy levels it can adopt. This is like allowing our compass needle to point, say, only north or south. Some nuclei (including 'normal' carbon-12) do not interact with a magnetic field at all and cannot be observed in an NMR machine. The nuclei we shall be looking at, ^{1}H and ^{13}C, do interact and have just two different energy levels. When we apply a magnetic field to these nuclei, they can either align themselves with it, which would be the lowest energy state, or they can align themselves against the field, which is higher in energy.

Let us return to the compass for a moment. We have already seen that if we could switch off the earth's magnetic field it would be easy to turn the compass needle round. When it is back on we need to push the needle (do work) to displace it from north. If we turned up the earth's magnetic field still more, it would be even harder to displace the compass needle. Exactly how hard it is to turn the compass needle depends on how strong the earth's magnetic field is and also on how well our needle is magnetized—if it is only weakly magnetized, it is much easier to turn it round and, if it isn't magnetized at all, it is free to rotate.

Likewise, with our nucleus in a magnetic field, the difference in energy between the nuclear spin aligned with and against the applied field depends on how strong the magnetic field is, and also on the properties of the nucleus itself. The stronger the magnetic field we put our nucleus in, the greater the energy difference between the two alignments. Now here is an unfortunate thing about NMR: the energy difference between the nuclear spin being aligned with the magnetic field and against it is really *very* small—so small that we need a very, very strong magnetic field to see any difference at all.

◀

This picture shows a typical NMR instrument. The extremely powerful superconducting magnet is shown on the left. This model features a robotic arm to change the samples automatically so many spectra can be run overnight. The large box in the centre of the picture is the radio wave generator and receiver. This is much larger than the computer needed to process the data which simply sits on the bench.

▶

Nuclei that interact with magnetic fields are said to possess **nuclear spin**. The exact number of different energy levels a nucleus can adopt is determined by this nuclear spin, I, of the particular isotope. The nuclear spin I can have various values such as 0, $\frac{1}{2}$, 1, $\frac{3}{2}$ and the number of energy levels is given by $2I + 1$. Some examples are: ^{1}H, $I = \frac{1}{2}$; ^{2}H (= D), $I = 1$; ^{11}B, $I = \frac{5}{2}$; ^{12}C, $I = 0$.

NMR machines contain very strong electromagnets

The earth's magnetic field has a field strength of 2×10^{-5} tesla. A typical magnet used in an NMR machine has a field strength of between 2 and 10 tesla, some 10^5 times stronger than the earth's field. These magnets are dangerous and no metal objects must be taken into the rooms where they are: stories abound of unwitting workmen whose metal toolboxes have become firmly attached to NMR magnets. Even with the immensely powerful magnets used the energy difference is still so small that the nuclei only have a very small preference for the lower energy state. Fortunately, we can just detect this small preference.

NMR also uses radio waves

A ^{1}H or ^{13}C nucleus in a strong magnetic field can have two energy levels. We could do work to make our nucleus align against the field rather than with it (just like turning the compass needle round). But since the energy difference between the two states is so small, we don't need to do much work. In fact, the amount of energy needed to flip the nucleus can be provided by electromagnetic radiation of radio-wave frequency. Radio waves flip the nucleus from the lower energy state to the higher state. The nucleus now wants to return to the lower energy state and, when it does so, the energy comes out again and this (a tiny pulse of radiofrequency electromagnetic radiation) is what we detect.

We can now sum up how an NMR machine works.

1 The sample of the unknown compound is dissolved in a suitable solvent and put in a very strong magnetic field. Any atomic nuclei with a nuclear spin now have different energy levels, the exact number of different energy levels depending on the value of the nuclear spin. For ^{1}H and ^{13}C NMR there are two energy levels

2 The sample is irradiated with a short pulse of radiofrequency energy. This disturbs the equilibrium balance between the two energy levels: some nuclei absorb the energy and are promoted to a higher energy level

3 We then detect the energy given out when the nuclei fall back down to the lower energy level using what is basically a sophisticated radio receiver

4 After lots of computation, the results are displayed in the form of intensity (i.e. number of absorptions) against frequency. Here is an example, which we shall return to in more detail later.

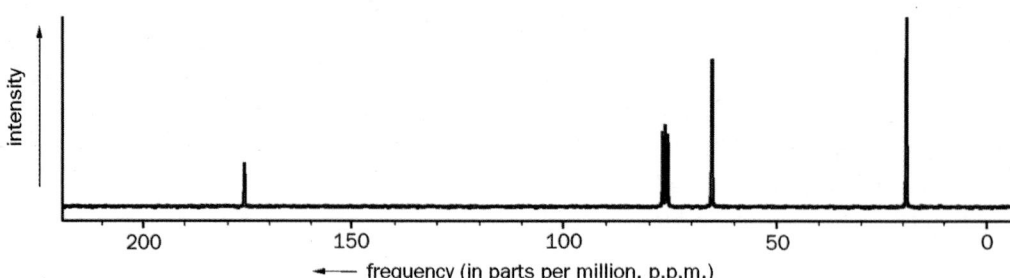

Why do chemically distinct nuclei absorb energy at different frequencies?

In the spectrum you see above, each peak represents a different kind of carbon atom: each one absorbs energy (or **resonates**—hence the term nuclear magnetic *resonance*) at a different frequency. But why should carbon atoms be 'different'? We have told you two factors that affect the energy difference (and therefore the frequency)—the magnetic field strength and what sort of nucleus is being studied. So you might expect all carbon-13 nuclei to resonate at one particular frequency and all protons (^{1}H) to resonate at one (different) frequency. But they don't.

The variation in frequency for different carbon atoms must mean that the energy jump from nucleus-aligned-with to nucleus-aligned-against the applied magnetic field must be different for each type of carbon atom. The reason there are different types of carbon atom is that their nuclei experience a magnetic field that is not quite the same as the magnetic field that we apply. Each nucleus is surrounded by electrons, and in a magnetic field these will set up a tiny electric current. This current will set up its own magnetic field (rather like the magnetic field set up by the electrons of an electric current moving through a coil of wire or solenoid), which will oppose the magnetic field that we apply. The electrons are said to **shield** the nucleus from the external magnetic field. If the electron distribution varies from ^{13}C atom to ^{13}C atom, so does the local magnetic field, and so does the resonating frequency of the ^{13}C nuclei. Now, you will see shortly (in Chapter 5) that a change in electron density at a carbon atom also alters the *chemistry* of that carbon atom. NMR tells us about the chemistry of a molecule as well as about its structure.

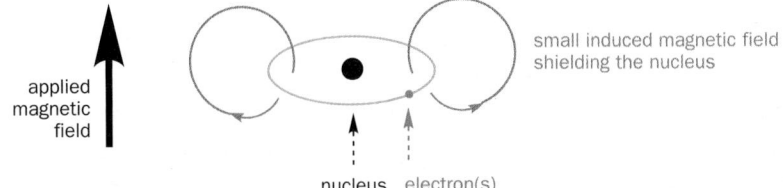

shielding of nuclei from an applied magnetic field by electrons:

applied magnetic field

small induced magnetic field shielding the nucleus

nucleus   electron(s)

> ● Changes in the **distribution of electrons** around a nucleus affect:
>   • the *local magnetic field* that the nucleus experiences
>   • the *frequency* at which the nucleus resonates
>   • the *chemistry* of the molecule at that atom
> This variation in frequency is known as the **chemical shift**. Its symbol is δ.

As an example, consider ethanol (right). The red carbon attached to the OH group will have relatively fewer electrons around it compared to the green carbon since the oxygen atom is more electronegative and draws electrons towards it, away from the carbon atom.

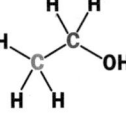

ethanol

■
We have shown all the Cs and Hs here because we want to talk about them.

The magnetic field that this (red) carbon nucleus feels will therefore be slightly greater than that felt by the (green) carbon with more electrons since the red carbon is less shielded from the applied external magnetic field—in other words it is **deshielded**. Since the carbon attached to the oxygen feels a stronger magnetic field, there will be a greater energy difference between the two alignments of its nucleus. The greater the energy difference, the higher the resonant frequency. So for ethanol we would expect the red carbon with the OH group attached to resonate at a higher frequency than the green carbon, and indeed this is exactly what the ^{13}C NMR spectrum shows.

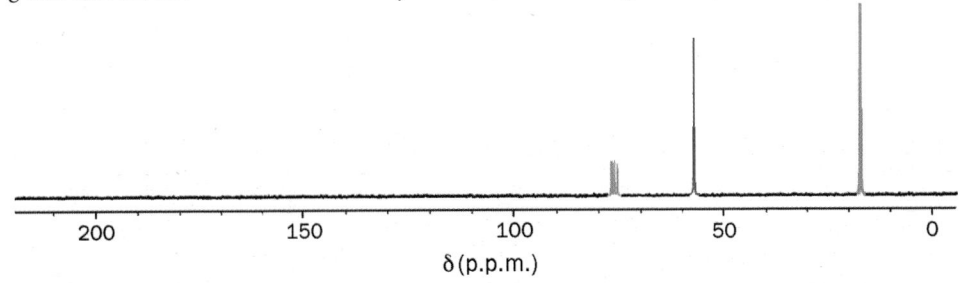

δ (p.p.m.)

▶
The peaks at 77 p.p.m., coloured brown, are those of the usual solvent ($CDCl_3$) and can be ignored for the moment. We shall explain them in Chapter 15.

The chemical shift scale

When you look at an NMR spectrum you will see that the scale does not appear to be in magnetic field units, nor in frequency units, but in 'parts per million' (p.p.m.). There is an excellent reason for this and we need to explain it. The exact frequency at which the nucleus resonates depends on the external applied magnetic field. This means that, if the sample is run on a machine with a different magnetic field, it will resonate at a different frequency. It would make life very difficult if we couldn't say exactly where our signal was, so we say how far it is from some reference sample, as a fraction of the operating frequency of the machine. We know that all protons resonate at approximately the same frequency in a given magnetic field and that the *exact* frequency depends on what sort of chemical environment it is in, which in turn depends on its electrons. This approximate frequency is the operating frequency of the machine and simply depends on the strength of the magnet—the stronger the magnet, the larger the operating frequency. The precise value of the operating frequency is simply the frequency at which a standard reference sample resonates. In everyday use, rather than actually referring to the strength of the magnet in tesla, chemists usually just refer to its operating frequency. A 9.4 T NMR machine is referred to as a 400 MHz spectrometer since that is the frequency in this strength field at which the protons in the reference sample resonate; other nuclei, for example ^{13}C, would resonate at a different frequency, but the strength is arbitrarily quoted in terms of the proton operating frequency.

The reference sample—tetramethylsilane, TMS

H3C, CH3 Si H3C CH3

tetramethylsilane, TMS

▶

Silicon and oxygen have opposite effects on an adjacent carbon atom: silicon shields; oxygen deshields.

Electronegativities: Si: 1.8; C: 2.5; O: 3.5.

The compound we use as a reference sample is usually tetramethylsilane, TMS. This is silane (SiH_4) with each of the hydrogen atoms replaced by methyl groups to give $Si(CH_3)_4$. The four carbon atoms attached to silicon are all equivalent and, because silicon is more electropositive than carbon, are fairly electron-rich (or *shielded*), which means they resonate at a frequency a little less than that of most organic compounds. This is useful because it means our reference sample is not bang in the middle of our spectrum!

The chemical shift, δ, in parts per million (p.p.m.) of a given nucleus in our sample is defined in terms of the resonance frequency as:

$$\delta = \frac{\text{frequency (Hz)} - \text{frequency TMS (Hz)}}{\text{frequency TMS (MHz)}}$$

No matter what the operating frequency (i.e. strength of the magnet) of the NMR machine, the signals in a given sample (e.g. ethanol) will always occur at the same chemical shifts. In ethanol the (red) carbon attached to the OH resonates at 57.8 p.p.m. whilst the (green) carbon of the methyl group resonates at 18.2 p.p.m. Notice that by definition TMS itself resonates at 0 p.p.m. The carbon nuclei in most organic compounds resonate at greater chemical shifts, normally between 0 and 200 p.p.m.

Now, let's return to the sample spectrum you saw on p. 58 and which is reproduced below, and you can see the features we have discussed. This is a 100 MHz spectrum; the horizontal axis is actually frequency but is usually quoted in p.p.m. of the field of the magnet, so each unit is one p.p.m. of 100 MHz, that is, 100 Hz. We can tell immediately from the three peaks at 176.8, 66.0, and 19.9 p.p.m. that there are three different types of carbon atom in the molecule.

■
Again, ignore the brown solvent peaks—they are of no interest to us at the moment. You also need not worry about the fact that the signals have different intensities. This is a consequence of the way the spectrum was recorded.

But we can do better than this: we can also work out what sort of chemical environment the carbon atoms are in. All ^{13}C spectra can be divided into four major regions: saturated carbon atoms (0–50 p.p.m.), saturated carbon atoms next to oxygen (50–100 p.p.m.), unsaturated carbon atoms (100–150 p.p.m.), and unsaturated carbon atoms next to oxygen, i.e. C=O groups (150–200 p.p.m.).

Regions of the ^{13}C NMR spectrum (scale in p.p.m.)			
Unsaturated carbon atoms next to oxygen (C=O)	**Unsaturated carbon atoms (C=C and aromatic carbons)**	**Saturated carbon atoms next to oxygen (CH$_3$O, CH$_2$O, etc.)**	**Saturated carbon atoms (CH$_3$, CH$_2$, CH)**
$\delta = 200\text{--}150$	$\delta = 150\text{--}100$	$\delta = 100\text{--}50$	$\delta = 50\text{--}0$

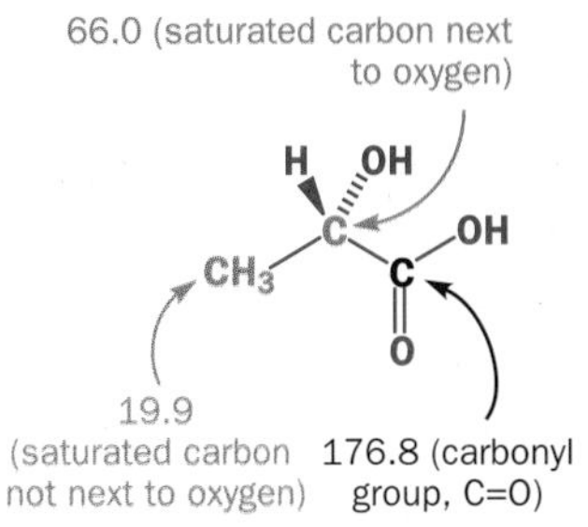

The spectrum you just saw is in fact of lactic acid (2-hydroxypropanoic acid). When you turned the last page, you made some lactic acid from glucose in the muscles of your arm—it is the breakdown product from glucose when you do anaerobic exercise. Each of lactic acid's carbon atoms gives a peak in a different region of the spectrum.

lactic acid (2-hydroxypropanoic acid)

66.0 (saturated carbon next to oxygen)

19.9 (saturated carbon not next to oxygen) 176.8 (carbonyl group, C=O)

Different ways of describing chemical shift

The chemical shift scale runs to the left from zero (where TMS resonates)—i.e. backwards from the usual style. Chemical shift values around zero are obviously small but are confusingly called 'high field' because this is the high magnetic field end of the scale. We suggest you say 'large' or 'small' chemical shift and 'large' or 'small' δ, but 'high' or 'low' field to avoid confusion. Alternatively, use 'upfield' for high field (small δ) and 'downfield' for low field (large δ).

One helpful description we have already used is **shielding**. Each carbon nucleus is surrounded by electrons that shield the nucleus from the applied field. Simple saturated carbon nuclei are the most shielded: they have small chemical shifts (0–50 p.p.m.) and resonate at high field. One electronegative oxygen atom moves the chemical shift downfield into the 50–100 p.p.m. region. The nucleus has become deshielded. Unsaturated carbon atoms experience even less shielding (100–150 p.p.m.) because of the way in which electrons are distributed around the nucleus. If the π bond is to oxygen, then the nucleus is even more deshielded and moves to the largest chemical shifts around 200 p.p.m. The next diagram summarizes these different ways of talking about NMR spectra.

> NMR spectra were originally recorded by varying the applied field. They are now recorded by variation of the frequency of the radio waves and that is done by a pulse of radiation. The terms 'high and low field' are a relic from the days of scanning by field variation.

■ If you are coming back to this chapter after reading Chapter 4 you might like to know that unsaturated C atoms are further deshielded because a π bond has a *nodal plane*. π Bonds have a plane with no electron density in at all, so electrons in π bonds are less efficient at shielding the nucleus than electrons in σ bonds.

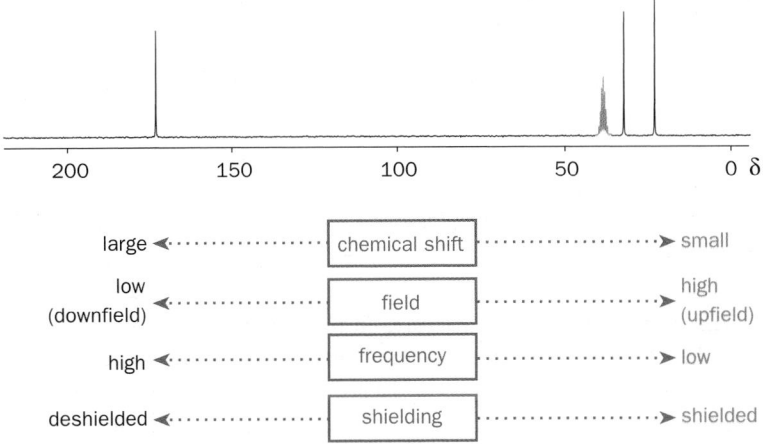

A guided tour of NMR spectra of simple molecules

We shall first look at NMR spectra of a few simple compounds before looking at unknown structures. Our very first compound, hexanedioic acid, has the simple NMR spectrum shown here. The first question is: why only three peaks for six carbon atoms? Because of the symmetry of the molecule, the two carboxylic acids are identical and give one peak at 174.2 p.p.m. By the same token C2 and C5 are identical while C3 and C4 are identical. These are all in the saturated region 0–50 p.p.m. but it is likely that the carbons next to the electron-withdrawing CO$_2$H group are more deshielded than the others. So we assign C2/C5 to the peak at 33.2 p.p.m. and C3/C4 to 24.0 p.p.m.

■ Why isn't this compound called 'hexane-1,6-dioic acid'? Well, carboxylic acids can only be at the end of chains, so no other hexanedioic acids are possible: the 1 and 6 are redundant.

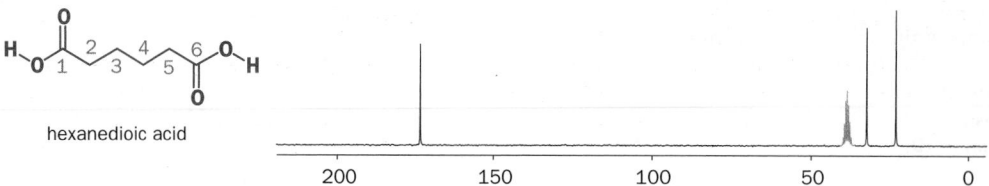

hexanedioic acid

The bee alarm pheromone (heptan-2-one) has no symmetry so all its seven carbon atoms are different. The carbonyl group is easy to identify (208.8 p.p.m., highlighted in red) but the rest are more difficult. Probably the two carbon atoms next to the carbonyl group come at lowest field, while C7 is certainly at highest field (13.9 p.p.m.). It is important that there are the right number of signals at about the right chemical shift. If that is so, we are not worried if we cannot assign each frequency to a precise carbon atom.

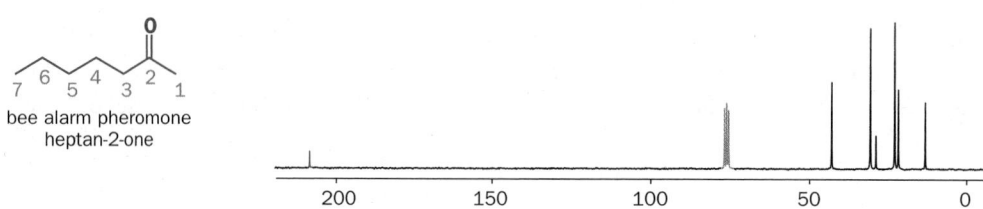

bee alarm pheromone
heptan-2-one

You met BHT on p. 55: its formula is $C_{15}H_{24}O$ and the first surprise in its NMR spectrum is that there are only seven signals for the 15 carbon atoms. There is obviously a lot of symmetry; in fact the molecule has a plane of symmetry vertically as it is drawn here. The very strong signal at $\delta = 30.4$ p.p.m. belongs to the six identical methyl groups on the *t*-butyl groups and the other two signals in the 0–50 p.p.m. range are the methyl group at C4 and the central carbons of the *t*-butyl groups. In the aromatic region there are only four signals as the two halves of the molecule are the same. As with the last example, we are not concerned with exactly which is which; we just check that there are the right number of signals with the right chemical shifts.

OH

Me 'BHT' $C_{15}H_{24}O$

plane of symmetry

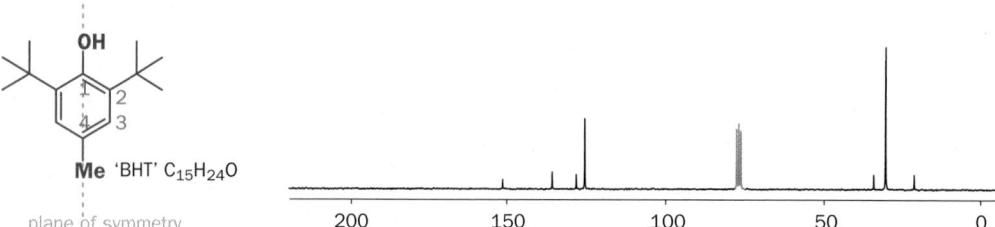

Paracetamol is a familiar painkiller with a simple structure—it too is a phenol but in addition it has an amide on the benzene ring. Its NMR spectrum contains one saturated carbon atom at 24 p.p.m. (the methyl group of the amide side chain), one carbonyl group at 168 p.p.m., and four other peaks at 115, 122, 132, and 153 p.p.m. These are the carbons of the benzene ring. Why four peaks? The two sides of the benzene ring are the same because the NHCO CH3 side chain can rotate rapidly so that C2 and C6 are the same and C3 and C5 are the same. Why is one of these aromatic peaks in the C=O region at 153 p.p.m.? This must be C4 as it is bonded to oxygen, and it just reminds us that carbonyl groups are not the only unsaturated carbon atoms bonded to oxygen (see the chart on p. 61), though it is not as deshielded as the true C=O group at 168 p.p.m.

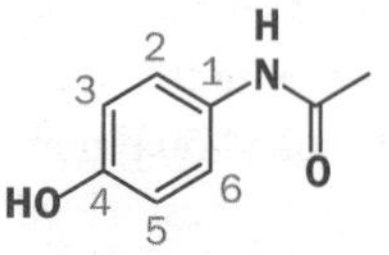

HO

paracetamol

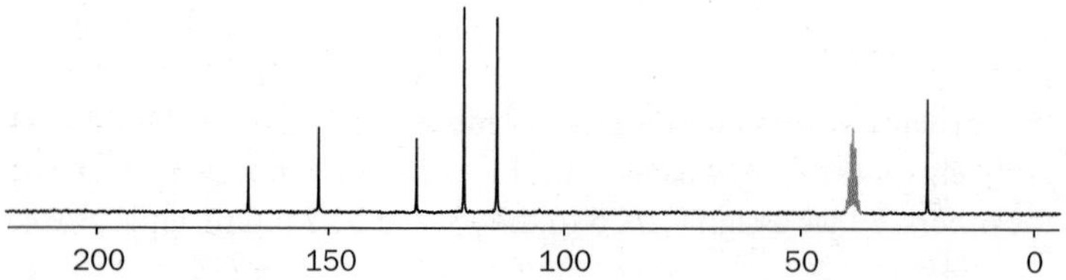

The effects of deshielding within the saturated carbon region

We have mentioned deshielding several times. The reference compound TMS (Me$_4$Si) has very shielded carbon atoms because silicon is more electropositive than carbon. Oxygen moves a saturated carbon atom downfield to larger chemical shifts (50–100 p.p.m.) because it is much more electro*negative* than carbon and so pulls electrons away from a carbon atom by polarizing the C–O bond. In between these extremes was a CO$_2$H group that moved its adjacent carbon down to around 35 p.p.m. These variations in chemical shift within each of the 50 p.p.m. regions of the spectrum are a helpful guide to structure as the principle is simple.

> ● Electro*negative* atoms move adjacent carbon atoms *down*field (to larger δ) by *de*shielding.

For the carbon atom next to the carboxylic acid, the oxygen atoms are, of course, no longer adjacent but one atom further away, so their deshielding effect is not as great.

The reverse is true too: electro*positive* atoms move adjacent carbon atoms upfield by shielding. This is not so important as there are few atoms found in organic molecules that are more electropositive than silicon and so few carbons are more shielded than those in Me$_4$Si. About the only important elements like this are the metals. When a carbon atom is more shielded than those in TMS, it has a negative δ value. There is nothing odd about this—the zero on the NMR scale is an arbitrary point. Table 3.5 shows a selection of chemical shift changes caused to a methyl group by changes in electronegativity.

Table 3.5 Effect of electronegativity on chemical shift

Electronic effect	Electronegativity of atom bonded to carbon	Compound	δ(CH$_3$)	δ(CH$_3$) – 8.4
donation	1.0	CH$_3$–Li	–14	–22.4
↑	2.2	CH$_3$–H	–2.3	–10.7
weak	1.8	CH$_3$–SiMe$_3$	0.0	–8.4
no effect	2.5	CH$_3$–CH$_3$	8.4	0
weak	3.1	CH$_3$–NH$_2$	26.9	18.5
↓	—	CH$_3$–COR	~30	~22
↓	3.5	CH$_3$–OH	50.2	41.8
withdrawal	4.1	CH$_3$–F	75.2	66.8

The last column in Table 3.5 shows the effect that each substituent has when compared to ethane. In ethane there is no electronic effect because the substituent is another methyl group so this column gives an idea of the true shift caused by a substituent. These shifts are roughly additive. Look back at the spectrum of lactic acid on p. 60: the saturated carbons occur at 19.9 and 66.0. The one at 66.0 is next both to an oxygen atom and a carbonyl group so that the combined effect would be about 42 + 22 = 64—not a bad estimate.

We shall look at similar but more detailed correlations in Chapters 11 and 15.

NMR is a powerful tool for solving unknown structures

Simple compounds can be quickly distinguished by NMR. These three alcohols of formula C$_4$H$_{10}$O have quite different NMR spectra.

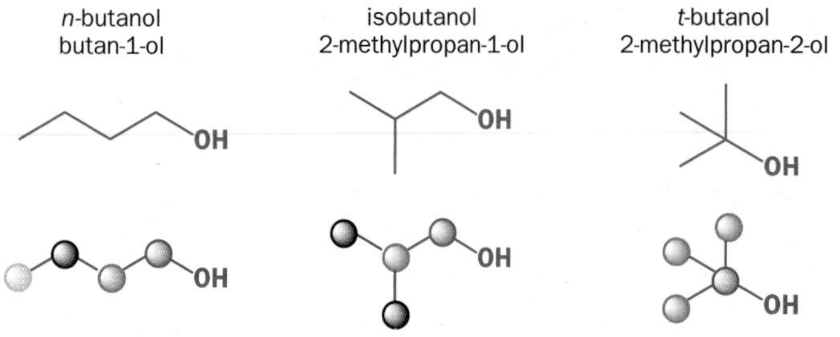

n-butanol butan-1-ol	isobutanol 2-methylpropan-1-ol	*t*-butanol 2-methylpropan-2-ol	δ (p.p.m.)	*n*-butanol	isobutanol	*t*-butanol
				62.9	70.2	69.3
				36.0	32.0	32.7
				20.3	20.4	—
				15.2	—	—

▶

The C atoms have been arbitrarily colour-coded.

■

The meanings of *n*-, iso-, and *t*- were covered in Chapter 2 (p. 29–30).

planes of symmetry

A

B

C

D

E

F

G

■

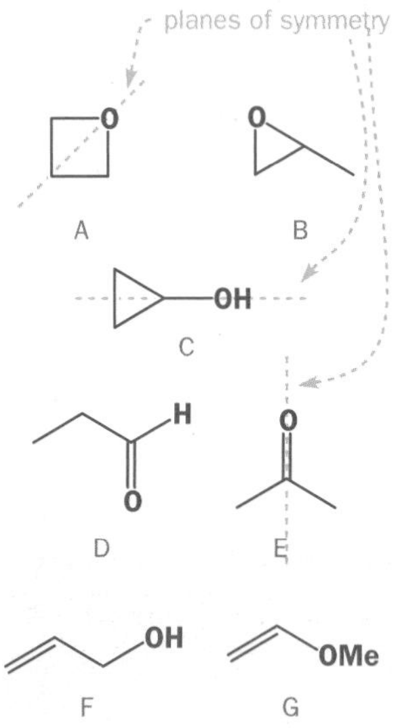

An epoxide is a three-membered cyclic ether.

Each alcohol has a saturated carbon atom next to oxygen, all close together. Then there are carbons next door but one to oxygen: they are back in the 0–50 p.p.m. region but at its low field end—about 30–35 p.p.m.. Notice the similarity of these chemical shifts to those of carbons next to a carbonyl group (Table 3.5 on p. 63). In each case we have C–C–O and the effects are about the same. Two of the alcohols have carbon(s) one further away still at yet smaller chemical shift (further upfield, more shielded) at about 20 p.p.m., but only the *n*-butanol has a more remote carbon still at 15.2. The *number* and the *chemical shift* of the signals identify the molecules very clearly.

A more realistic example would be an unknown molecule of formula C_3H_6O. There are seven reasonable structures, as shown. Simple symmetry can distinguish structures A, C, and E from the rest as these three have only two types of carbon atom. A more detailed inspection of the spectra makes identification easy. The two carbonyl compounds, D and E, each have one peak in the 150–220 p.p.m. region but D has two different saturated carbon atoms while E has only one. The two alkenes, F and G, both have one saturated carbon atom next to oxygen, but F has two normal unsaturated carbon atoms (100–150 p.p.m.) while the enol ether, G, has one normal alkene and one unsaturated carbon joined to oxygen. The three saturated compounds (A–C) present the greatest problem. The epoxide, B, has two different carbon atoms next to oxygen (50–100 p.p.m.) and one normal saturated carbon atom. The remaining two both have one signal in the 0–50 and one in the 50–100 p.p.m. regions. Only proton NMR (Chapter 11) and, to a certain extent, infrared spectroscopy (which we will move on to shortly) will distinguish them reliably.

Here are NMR spectra of three of these molecules. Before looking at the next page see if you can assign them to the structures on the left. Try also to suggest which signals belong to which carbon atoms.

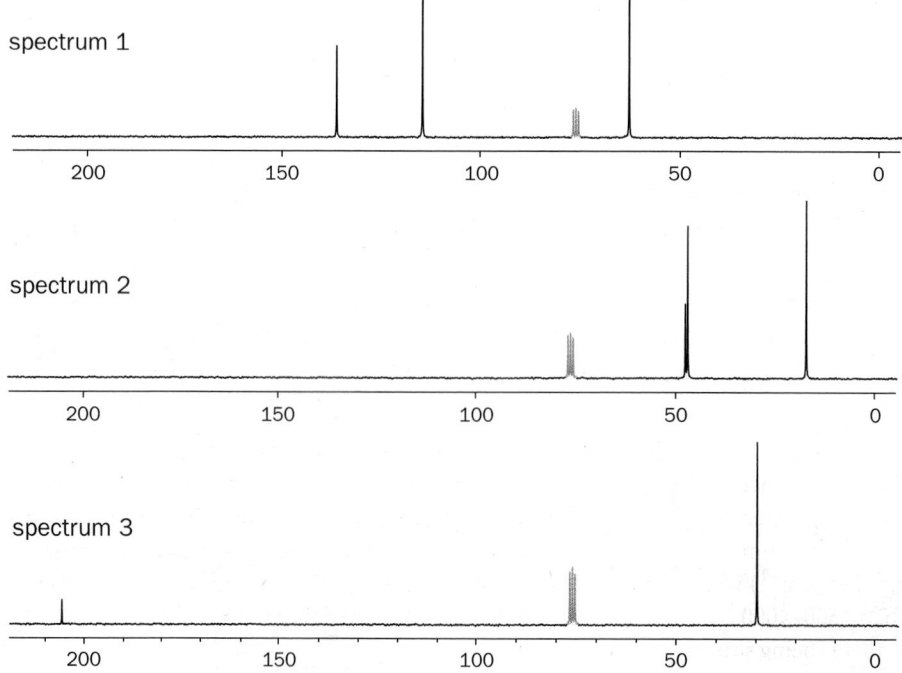

spectrum 1

spectrum 2

spectrum 3

These shouldn't give you too much trouble. The only carbonyl compound with two identical carbons is acetone, Me_2CO (E) so spectrum 3 must be that one. Notice the very low field C=O signal (206.6 p.p.m.) typical of a simple ketone. Spectrum 1 has two unsaturated carbons and a saturated carbon next to oxygen so it must be F or G. In fact it has to be F as both unsaturated carbons are similar (137 and 116 p.p.m.) and neither is next to oxygen (>150 p.p.m., cf. 206.6 in spectrum 3). This leaves spectrum 2, which appears to have no carbon atoms next to oxygen as all chemical shifts are less than 50 p.p.m. No compound fits that description (impossible for C_3H_6O anyway!) and the two signals at 48.0 and 48.2 p.p.m. are suspiciously close to the borderline. They are, of course, next to oxygen and this is compound B.

Infrared spectra

Functional groups are identified by infrared spectra

Some functional groups, for example, C=O or C=C, can be seen in the NMR spectrum because they contain carbon atoms, while the presence of others like OH can be inferred from the chemical shifts of the carbon atoms they are joined to. Others cannot be seen at all. These might include NH_2 and NO_2, as well as variations around a carbonyl group such as COCl, CO_2H, and $CONH_2$. Infrared (IR) spectroscopy provides a way of finding these functional groups because it detects the stretching and bending of bonds rather than any property of the atoms themselves. It is particularly good at detecting the stretching of unsymmetrical bonds of the kind found in functional groups such as OH, C=O, NH_2, and NO_2.

NMR requires electromagnetic waves in the radio-wave region of the spectrum to make nuclei flip from one state to another. The amount of energy needed for stretching and bending individual bonds, while still very small, corresponds to rather shorter wavelengths. These wavelengths lie in the infrared, that is, heat radiation just to the long wavelength side of visible light. When the carbon skeleton of a molecule vibrates, all the bonds stretch and relax in combination and these absorptions are unhelpful. However some bonds stretch essentially independently of the rest of the molecule. This occurs if the bond is either:

- much stronger or weaker than others nearby, or
- between atoms that are much heavier or lighter than their neighbours

Indeed, the relationship between the frequency of the bond vibration, the mass of the atoms, and the strength of the bond is essentially the same as Hooke's law for a simple harmonic oscillator.

$$v = \frac{1}{2\pi c}\sqrt{\frac{f}{\mu}}$$

The equation shows that the frequency of the vibration v is proportional to the (root of) a **force constant** f—more or less the bond strength—and inversely proportional to the (root of) a **reduced mass** μ, that is, the product of the masses of the two atoms forming the bond divided by their sum.

$$\mu = \frac{m_1 m_2}{m_1 + m_2}$$

Stronger bonds vibrate faster and so do lighter atoms. You may at first think that stronger bonds ought to vibrate more slowly, but a moment's reflection will convince you of the truth: which stretches and contracts faster, a tight steel spring or a slack steel spring?

Infrared spectra are simple absorption spectra. The sample is exposed to infrared radiation and the wavelength scanned across the spectrum. Whenever energy corresponding to a specific wavelength is absorbed, the intensity of the radiation reaching a detector momentarily decreases, and this is recorded in the spectrum. Infrared spectra are usually recorded using a frequency measurement called **wavenumber** (cm^{-1}) which is the inverse of the true wavelength λ in centimetres to give convenient numbers (500–4000 cm^{-1}). Higher numbers are to the left of the spectrum because it is really wavelength that is being scanned.

bond vibration in the infrared

■ **Hooke's law** describes the movement of two masses attached to a spring. You may have met it if you have studied physics. You need not be concerned here with its derivation, just the result.

We need to use another equation here:

$$E = h\nu = h\frac{c}{\lambda} \text{ since } \lambda = \frac{c}{\nu}$$

The energy, E, required to excite a bond vibration can be expressed as the inverse of a wavelength λ or as a frequency ν. Wavelength and frequency are just two ways of measuring the same thing. More energy is needed to stretch a strong bond and you can see from this equation that larger E means higher wavenumbers (cm^{-1}) or smaller wavelength (cm).

To run the spectrum, the sample is either dissolved in a solvent such as $CHCl_3$ (chloroform) that has few IR absorptions, pressed into a transparent disc with powdered solid KBr, or ground into an oily slurry called a **mull** with a hydrocarbon oil called 'Nujol'. Solutions in $CHCl_3$ cannot be used for looking at the regions of C–Cl bond stretching nor can Nujol mulls be used for the region of C–H stretching. Neither of these is a great disadvantage, especially as nearly all organic compounds have some C–H bonds anyway.

▶ h is **Planck's constant** and c the velocity of light.

▶ You should always check the way the spectrum was run before making any deductions!

▶ This simple bench-top IR machine was used to record the spectra used in this book. The metal plate holder, used to hold the sample in the bath of the IR beam, can be seen inside. The spectrum is manipulated using the computer on the right.

We shall now examine the relationship between bond stretching and frequency in more detail. Hooke's law told us to expect frequency to depend on both mass and bond strengths, and we can illustrate this double dependence with a series of bonds of various elements to carbon.

Values chiefly affected by mass of atoms: (lighter atom, higher frequency)

C–H	C–D	C–O	C–Cl
$3000\ cm^{-1}$	$2200\ cm^{-1}$	$1100\ cm^{-1}$	$700\ cm^{-1}$

Values chiefly affected by bond strength (stronger bond, higher frequency)

C≡O	C=O	C–O
$2143\ cm^{-1}$	$1715\ cm^{-1}$	$1100\ cm^{-1}$

Just because they were first recorded in this way, infrared spectra have the baseline at the top and peaks going downwards. You might say that they are plotted upside down and back to front. At least you are now accustomed to the horizontal scale running backwards as that happens in NMR spectra too. A new feature is the change in scale at $2000\ cm^{-1}$ so that the right-hand half of the spectrum is more detailed than the left-hand half. A typical spectrum looks like this.

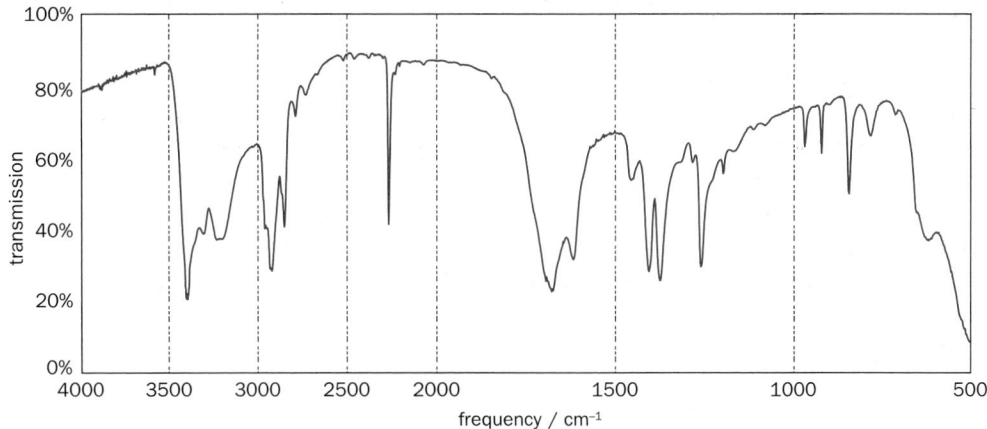

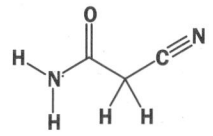

IR spectra are plotted 'upside down' because they record **transmission** (the amount of light reaching the detector) rather than absorbance.

cyanoacetamide

(spectrum taken as a Nujol mull)

There are four important regions of the infrared spectrum

You will see at once that the infrared spectrum contains many lines, particularly at the right-hand (lower frequency) end; hence the larger scale at this end. Many of these lines result from several bonds vibrating together and it is actually the left-hand half of the spectrum that is more useful.

The first region, from about 4000 to about 2500 cm^{-1} is the region for C–H, N–H, and O–H bond stretching. Most of the atoms in an organic molecule (C, N, O, for example) are about the same weight. Hydrogen is an order of magnitude lighter than any of these and so it dominates the stretching frequency by the large effect it has on the reduced mass. The reduced mass of a C–C bond is $(12 \times 12)/(12 + 12)$, i.e. $144/24 = 6.0$. If we change one of these atoms for H, the reduced mass changes to $(12 \times 1)/(12 + 1)$, i.e. $12/13 = 0.92$, but, if we change it instead for F, the reduced mass changes to $(12 \times 19)/(12 + 19)$, i.e. $228/31 = 7.35$. There is a small change when we increase the mass to 19 (F), but an enormous change when we decrease it to 1 (H).

Even the strongest bonds—triple bonds such as C≡C or C≡N—absorb at slightly lower frequencies than bonds to hydrogen: these are in the next region from about 2500 to 2000 cm^{-1}. This and the other two regions of the spectrum follow in logical order of bond strength as the reduced masses are all about the same: double bonds such as C=C and C=O from about 1900–1500 cm^{-1} and single bonds at the right-hand end of the spectrum. These regions are summarized in this chart, which you should memorize.

The concept of reduced mass was introduced on p. 65.

Remember: Hooke's law says that frequency depends on both mass and a force constant (bond strength).

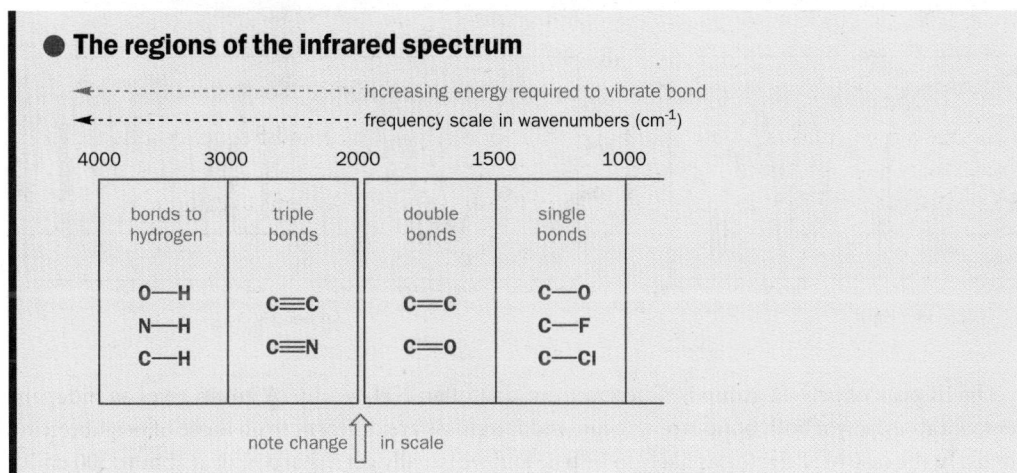

Looking back at the typical spectrum, we see peaks in the X–H region at about 2950 cm^{-1} which are the C–H stretches of the CH$_3$ and CH$_2$ groups. The one rather weak peak in the triple bond region (2270 cm^{-1}) is of course the C≡N group and the strong peak at about 1670 cm^{-1} belongs to the C=O group. We shall explain soon why some IR peaks are stronger than others. The rest of the spectrum is in the single bond region. This region is not normally interpreted in detail but is characteristic of the compound as a whole rather in the way that a fingerprint is characteristic of an individual human being—and, similarly, it cannot be 'interpreted'. It is indeed called the **fingerprint region**. The useful information from this spectrum is the presence of the CN and C=O groups and the exact position of the C=O absorption.

The X–H region distinguishes C–H, N–H, and O–H bonds

The reduced masses of the C–H, N–H, and O–H combinations are all about the same. Any difference between the positions of the IR bands of these bonds must then be due to bond strength. In practice, C–H stretches occur at around 3000 cm^{-1} (though they are of little use as virtually all organic compounds have C–H bonds), N–H stretches occur at about 3300 cm^{-1}, and O–H stretches higher still. We can immediately deduce that the O–H bond is stronger than N–H which is stronger than C–H. IR is a good way to measure such bond strengths.

Table 3.6 IR bands for bonds to hydrogen

Bond	Reduced mass, μ	IR frequency, cm^{-1}	Bond strength, kJ mol^{-1}
C–H	12/13 = 0.92	2900–3200	CH$_4$: 440
N–H	14/15 = 0.93	3300–3400	NH$_3$: 450
O–H	16/17 = 0.94	3500–3600[a]	H$_2$O: 500

[a]When not hydrogen-bonded: see below.

The X–H IR stretches are very different in these four compounds.

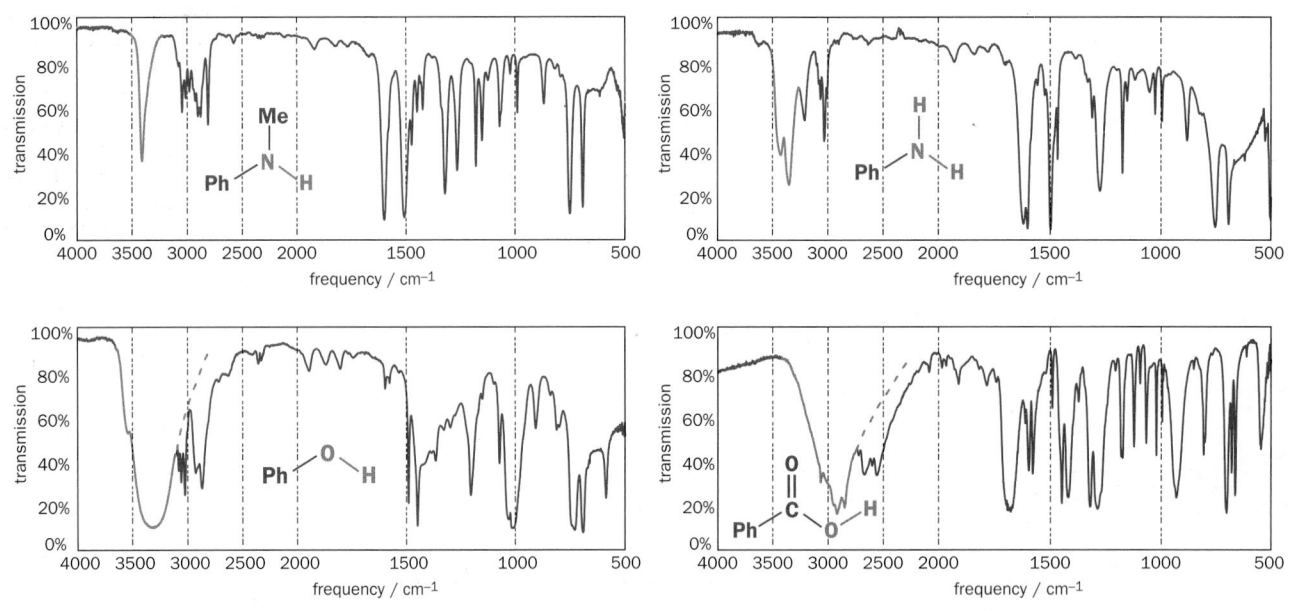

The IR peak of an NH group is different from that of an NH$_2$ group. A group gives an independent vibration only if both bond strength and reduced mass are different from those of neighbouring bonds. In the case of N–H, this is likely to be true and we usually get a sharp peak at about 3300 cm^{-1}, whether the NH group is part of a simple amine (R$_2$NH) or an amide (RCONHR). The NH$_2$ group is also independent of the rest of the molecule, but the two NH bonds inside the NH$_2$ group have identical force constants and reduced masses and so vibrate as a single unit. Two equally strong bands appear, one for the two N–H bonds vibrating in phase (symmetric) and one for the two N–H bonds vibrating in opposition (antisymmetric). The antisymmetric vibration requires more energy and is at slightly higher frequency.

The O–H bands occur at higher frequency, sometimes as a sharp absorption at about 3600 cm^{-1}. More often, you will see a broad absorption at anywhere from 3500 to 2900 cm^{-1} This is because OH groups form strong hydrogen bonds that vary in length and strength. The sharp absorption at 3600 cm^{-1} is the non-hydrogen-bonded OH and the lower the absorption the stronger the H bond.

Alcohols form hydrogen bonds between the hydroxyl oxygen of one molecule and the hydroxyl hydrogen of another. These bonds are variable in length (though they are usually rather longer than normal covalent O–H bonds) and they slightly weaken the true covalent O–H bonds by varying amounts. When a bond varies in length and strength it will have a range of stretching frequencies distributed about a mean value. Alcohols typically give a rounded absorption at about 3300 cm^{-1} (contrast the sharp N–H stretch in the same region). Carboxylic acids (RCO$_2$H) form hydrogen-bonded dimers with two strong H bonds between the carbonyl oxygen atom of one molecule and the acidic hydrogen of the other. These also vary considerably in length and strength and usually give very broad V-shaped absorbances.

hydrogen bonding in an alcohol

the hydrogen-bonded dimer of a carboxylic acid

Good examples are paracetamol and BHT. Paracetamol has a typical sharp peak at 3330 cm^{-1} for the N–H stretch and then a rounded absorption for the hydrogen-bonded O–H stretch from 3300 down to 3000 cm^{-1} in the gap between the N–H and C–H stretches. By contrast, BHT has a sharp absorption at 3600 cm^{-1} as the two large and roughly spherical *t*-butyl groups prevent the normal H bond from forming.

the hydrogen-bonded OH group in paracetamol

Hydrogen bonds are weak bonds formed from electron-rich atoms such as O or N to hydrogen atoms also attached by 'normal' bonds to the same sorts of atoms. In this diagram of a hydrogen bond between two molecules of water, the solid line represents the 'normal' bond and the green dotted line the longer hydrogen bond. The hydrogen atom is about a third of the way along the distance between the two oxygen atoms.

hydrogen bond

The ^{13}C NMR spectra of these two compounds are on page 62.

We can use the N–H and O–H absorptions to rule out an alternative isomeric structure for paracetamol: an ester with an NH$_2$ group instead of an amide with NH and OH. This structure must be wrong as it would give two similar sharp peaks at about 3300 cm^{-1} instead of one sharp and one broad peak actually observed.

alternative and wrong structure for paracetamol

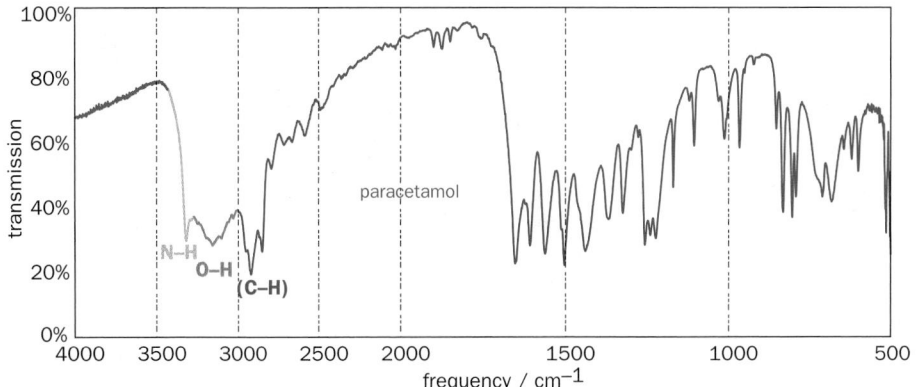

paracetamol

N–H
O–H
(C–H)

BHT

O–H
(C–H)

in BHT H-bonding is prevented by large *t*-butyl groups

You may be confused the first time you see the IR spectrum of a terminal alkyne, R–C≡C–H, because you will see a strongish sharp peak at around 3300 cm^{-1} that looks just like an N–H stretch. The displacement of this peak from the usual C–H stretch at about 3000 cm^{-1} cannot be due to a change in the reduced mass and must be due to a marked increase in bond strength. The alkyne C–H bond is shorter and stronger than alkane C–H bonds.

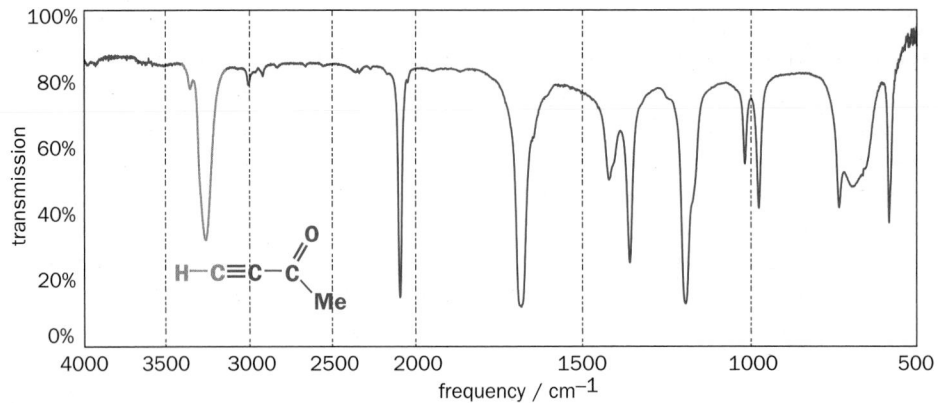

► In Chapter 4, you will see that carbon uses an sp^3 orbital to make a C–H bond in a saturated structure but has to use an sp orbital for a terminal alkyne C–H. This orbital has one-half s character instead of one-quarter s character. The electrons in an s orbital are held closer to the carbon's nucleus than in a p orbital, so the sp orbital makes for a shorter, stronger C–H bond.

■ What are the other peaks in this spectrum?

● **The summary chart shows some typical peak shapes and frequencies for X–H bonds in the region 4000–3000 cm^{-1}.**

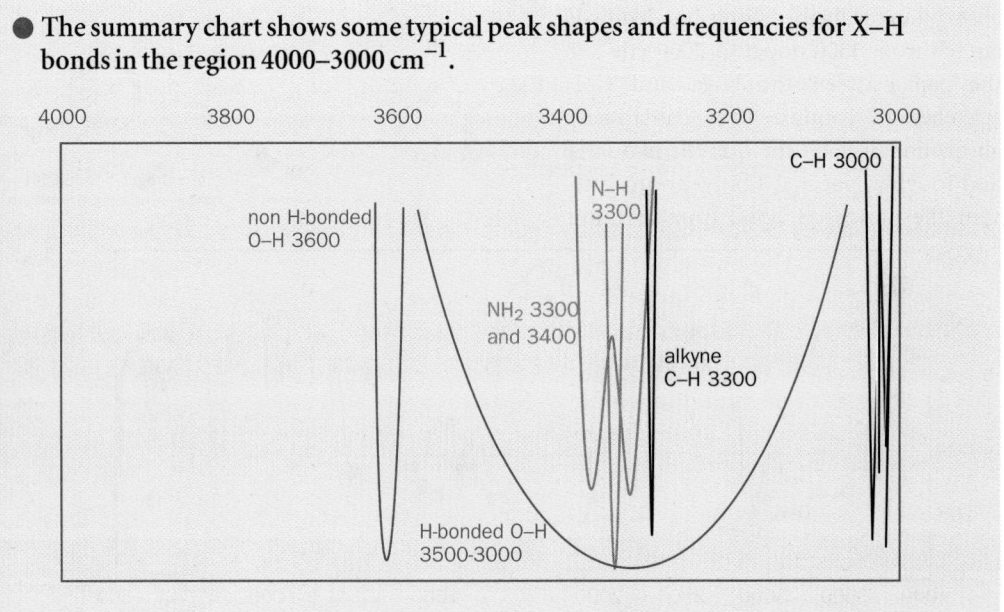

The double bond region is the most important in IR spectra

In the double bond region, there are three important absorptions, those of the carbonyl (C=O), alkene (C=C), and nitro (NO$_2$) groups. All give rise to sharp bands: C=O to one strong (i.e. intense) band anywhere between 1900 and 1500 cm^{-1}; C=C to one weak band at about 1640 cm^{-1}; and NO$_2$ to two strong (intense) bands in the mid-1500s and mid-1300s cm^{-1}. The number of bands is easily dealt with. Just as with OH and NH$_2$, it is a matter of how many identical bonds are present in the same functional group. Carbonyl and alkene clearly have one double bond each. The nitro group at first appears to contain two different groups, N$^+$–O$^-$ and N=O, but delocalization means they are identical and we see absorption for symmetrical and antisymmetrical stretching vibrations. As with NH$_2$, more work is needed for the antisymmetrical vibration which occurs at higher frequency (>1500 plus cm^{-1}).

► Delocalization is covered in Chapter 7; for the moment, just accept that both NO bonds are the same.

delocalization in the nitro group

symmetric
NO_2 stretch (~1350)

antisymmetric
NO_2 stretch (~1550)

The strength of an IR absorption depends on dipole moment

Now what about the variation in strength (i.e. intensity, the amount of energy absorbed)? The strength of an IR absorption varies with the change of dipole moment when the bond is stretched. If the bond is perfectly symmetrical, there is no change in dipole moment and there is no IR absorption. Obviously, the C=C bond is less polar than either C=O or N=O and is weaker in the IR. Indeed it may be absent altogether in a symmetrical alkene. By contrast the carbonyl group is very polar (Chapter 4) and stretching it causes a large change in dipole moment and C=O stretches are usually the strongest peaks in the IR spectrum. You may also have noticed that O–H and N–H stretches are stronger than C–H stretches (even though most organic molecules have many more C–H bonds than O–H or N–H bonds): the reason is the same.

> ▶
> Contrast the term 'strength' applied to absorption and to bonds. A stronger absorption is a *more intense* absorption—i.e. one with a big peak. A strong *bond* on the other hand has a *higher frequency* absorption (other things being equal).

Dipole moments

Dipole moment depends on the variation in distribution of electrons along the bond, and also its length, which is why stretching a bond can change its dipole moment. For bonds between unlike atoms, the larger the difference in electronegativity, the greater the dipole moment, and the more it changes when stretched. For identical atoms (C=C, for example) the dipole moment, and its capacity to change with stretching, is much smaller. Stretching frequencies for symmetrical molecules are measured using Raman spectra. This is an IR-based technique using scattered light that relies on polarizability of bonds. Raman spectra are outside the scope of this book.

This is a good point to remind you of the various deductions we have made so far about IR spectra.

● Absorptions in IR spectra

Position of band depends on →	reduced mass of atoms	**light** atoms give **high** frequency
	bond strength	**strong** bonds give **high** frequency
Strength of band depends on →	change in dipole moment	**large** dipole moment gives **strong** absorption
Width of band depends on →	hydrogen bonding	**strong H** bond gives **wide** peak

We have seen three carbonyl compounds so far in this chapter and they all show peaks in the right region (around 1700 cm^{-1}) even though one is a carboxylic acid, one a ketone, and one an amide. We shall consider the exact positions of the various carbonyl absorptions in Chapter 15 after we have discussed some carbonyl chemistry.

hexanedioic acid
1720 cm^{-1}

heptan-2-one
1710 cm^{-1}

paracetamol
1667 cm^{-1}

The single bond region is used as a molecular fingerprint

The region below 1500 cm^{-1} is where the single bond vibrations occur. Here our hope that individual bonds may vibrate independently of the rest of the molecule is usually doomed to disappointment. The atoms C, N, and O all have about the same atomic weight and C–C, C–N, and C–O single bonds all have about the same strength.

In addition, C–C bonds are likely to be joined to other C–C bonds with virtually identical strength and reduced mass, and they have essentially no dipole moments. The only one of these single bonds of any value is C–O which is polar enough and different enough (Table 3.7) to show up as a strong absorption at about 1100 cm^{-1}. Some other single bonds such as C–Cl (weak and with a large reduced mass) are quite useful at about 700 cm^{-1}. Otherwise the single bond region is usually crowded with hundreds of absorptions from vibrations of all kinds used as a 'fingerprint' characteristic of the molecule but not really open to interpretation.

Among the hundreds of peaks in the fingerprint region, there are some of a quite different kind. Stretching is not the only bond movement that leads to IR absorption. Bending of bonds, particularly C–H and N–H bonds, also leads to quite strong peaks. These are called **deformations**. Bending a bond is easier than stretching it (which is easier, stretching or bending an iron bar?). Consequently, bending absorptions need less energy and come at lower frequencies than stretching absorptions for the same bonds. These bands may not often be useful in identifying molecules, but you will notice them as they are often strong (they are usually stronger than C=C stretches for example) and may wonder what they are.

Finally in this section, we summarize all the useful absorptions in the fingerprint region. Please be cautious in applying these as there are other reasons for bands in these positions.

Table 3.7 Single bonds

Pair of atoms	Reduced mass	Bond strength
C–C	6.0	350 kJ mol^{-1}
C–N	6.5	305 kJ mol^{-1}
C–O	6.9	360 kJ mol^{-1}

Table 3.8 Useful deformations (bending vibrations)

Group	Frequency, cm^{-1}	Strength
CH$_2$	1440–1470	medium
CH$_3$	~1380	medium
NH$_2$	1550–1650	medium

You may not yet understand all the terms in Table 3.9, but you will find it useful to refer back to later.

Table 3.9 Useful absorptions in the fingerprint region

Frequency, cm^{-1}	Strength	Group	Comments
1440–1470	medium	CH$_2$	deformation (present in nujol)
~1380	medium	CH$_3$	deformation (present in nujol)
~1350	strong	NO$_2$	symmetric N=O stretch
1250–1300	strong	P=O	double bond stretch
1310–1350	strong	SO$_2$	antisymmetric S=O stretch
1120–1160	strong	SO$_2$	symmetric S=O stretch
~1100	strong	C–O	single bond stretch
950–1000	strong	C=CH	*trans* alkene (out-of-plane deformation)
~690 *and* ~750	strong	Ar–H	five adjacent Ar–H (out-of-plane)
~750	strong	Ar–H	four adjacent Ar–H (out-of-plane)
~700	strong	C–Cl	single bond stretch

Mass spectra, NMR, and IR combined make quick identification possible

If these methods are each as powerful as we have seen on their own, how much more effective they must be together. We shall finish this chapter with the identification of some simple unknown compounds using all three methods. The first is an industrial emulsifier used to blend solids and liquids into smooth pastes. Its electron impact mass spectrum has peaks at 74 and 72 and a base peak at 58. The two peaks at 74 and 72 cannot be isotopes of bromine or chlorine as the ratio is neither 1:1 nor 3:1. It looks as though 74 might be the molecular ion. However a chemical ionization mass spectrum reveals a molecular ion at 90 (MH^+) and hence the true molecular ion at 89. An odd molecular weight (89) suggests one nitrogen atom, and high resolution mass spectrometry reveals that the formula is $C_4H_{11}NO$.

the electron impact mass spectrum of an industrial emulsifier

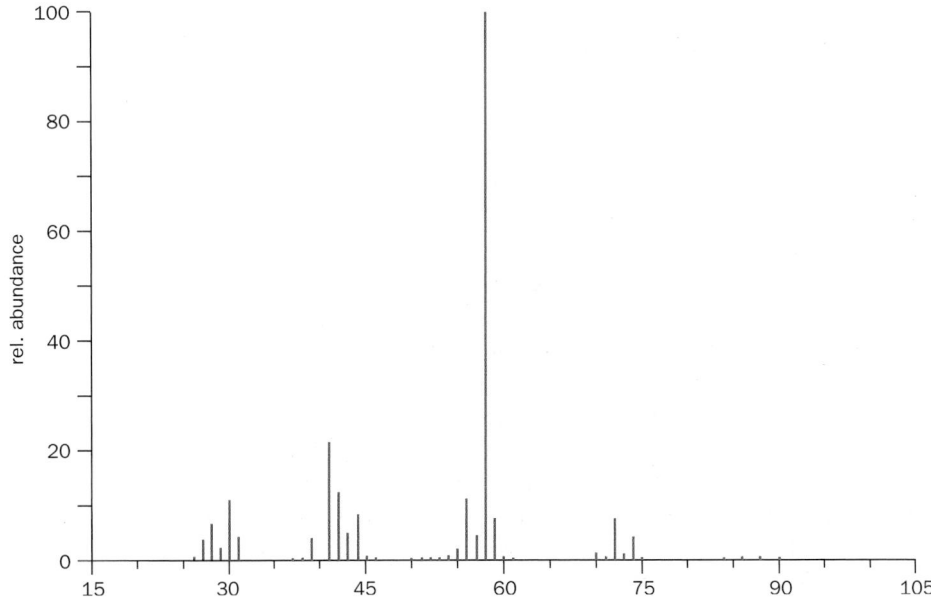

The ^{13}C NMR spectrum has only three peaks so two carbon atoms must be the same. There is one signal for saturated carbon next to oxygen, and two for other saturated carbons, one more downfield than the other. The IR spectrum reveals a broad peak for an OH group with two sharp NH_2 peaks just protruding. If we put this together, we know we have C–OH and C–NH_2. Neither of these carbons can be duplicated (as there is only one O and only one N!) so one of the remaining carbons must be duplicated.

^{13}C NMR spectrum of the emulsifier

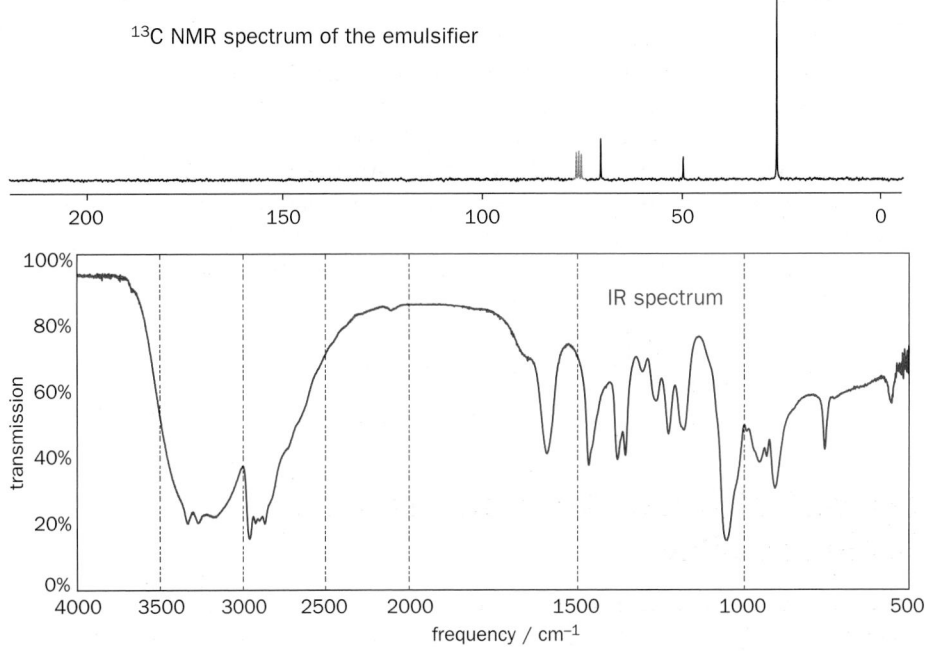

The next stage is one often overlooked. We don't seem to have much information, but try and put the two fragments together, knowing the molecular formula, and there's very little choice. The carbon chain (shown in red) could either be linear or branched and that's it!

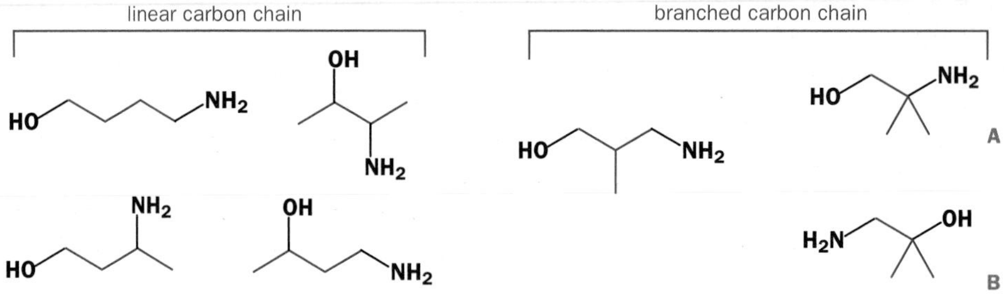

linear carbon chain   branched carbon chain

> By **chain terminating** we mean only attachable to one other atom.

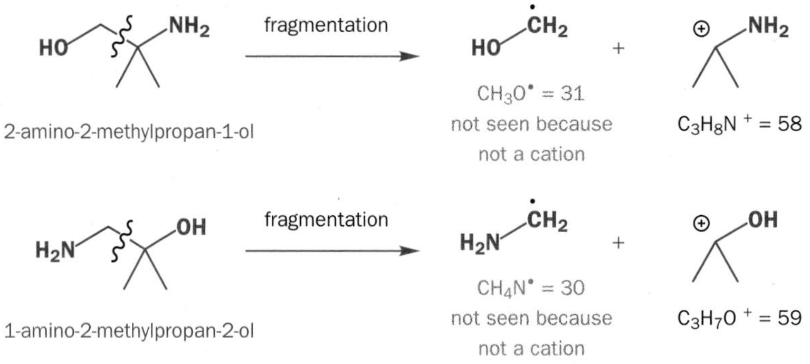

There is no room for double bonds or rings because we need to fit in the eleven hydrogen atoms. We cannot put N or O in the chain because we know from the IR that we have the *chain terminating* groups OH and NH_2. Of the seven possibilities only the last two, A and B, are serious since they alone have two identical carbon atoms (the two methyl groups in each case); all the other structures would have four separate signals in the NMR. How can we choose between these? The base peak in the mass spectrum was at 58 and this fits well with a fragmentation of one structure but not of the other: the wrong structure would give a fragment at 59 and not 58. The industrial emulsifier is 2-amino-2-methylpropan-1-ol.

2-amino-2-methylpropan-1-ol

$CH_3O^{\bullet} = 31$
not seen because not a cation

$C_3H_8N^+ = 58$

1-amino-2-methylpropan-2-ol

$CH_4N^{\bullet} = 30$
not seen because not a cation

$C_3H_7O^+ = 59$

Double bond equivalents help in the search for a structure

The last example was fully saturated but it is usually a help in deducing the structure of an unknown compound if, once you know the atomic composition, you immediately work out how much unsaturation there is. This is usually expressed as 'double bond equivalents'. It may seem obvious to you that, if $C_4H_{11}NO$ has no double bonds, then C_4H_9NO (losing two hydrogen atoms) must have one double bond, C_4H_7NO two double bonds, and so on. Well, it's not quite as simple as that. Some possible structures for these formulae are shown below.

some structures for C_4H_9NO

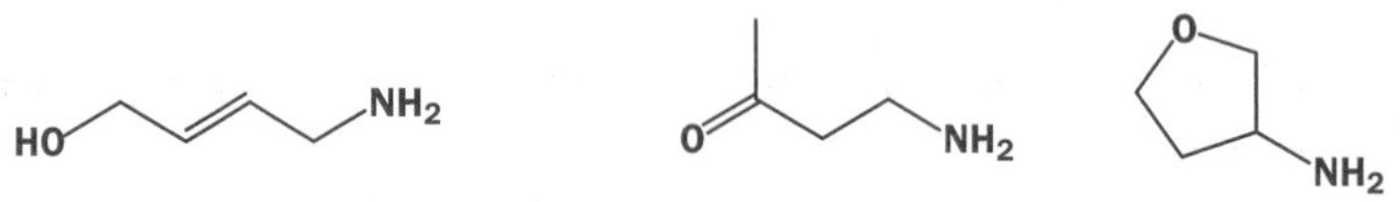

some structures for C_4H_7NO

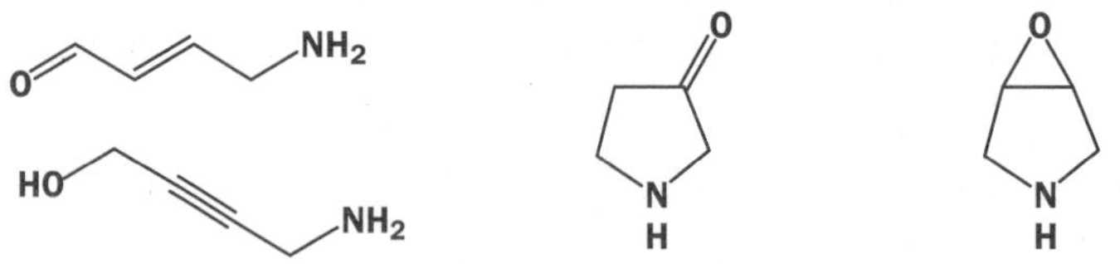

Some of these structures have the right number of double bonds (C=C and C=O), one has a triple bond, and three compounds use rings as an alternative way of 'losing' some hydrogen atoms. Each time you make a ring or a double bond, you have to lose two more hydrogen atoms. So double bonds (of all kinds) and rings are called **Double Bond Equivalents** (**DBEs**).

You can work out how many DBEs there are in a given atomic composition just by making a drawing of one possible structure (all possible structures have the same number of DBEs). Alternatively, you can calculate the DBEs if you wish. A saturated hydrocarbon with n carbon atoms has $(2n + 2)$ hydrogens. Oxygen doesn't make any difference to this: there are the same number of Hs in a saturated ether or alcohol as in a saturated hydrocarbon.

So, for a compound containing C, H, and O only, take the actual number of hydrogen atoms away from $(2n + 2)$ and divide by two. Just to check that it works, for the unsaturated ketone $C_7H_{12}O$ the calculation becomes:

1 Maximum number of H atoms for 7 Cs $\quad\quad 2n + 2 = 16$

2 Subtract the actual number of H atoms (12) $\quad\quad 16 - 12 = 4$

3 Divide by 2 to give the DBEs $\quad\quad 4/2 = 2$

$C_7H_{12}O$ = two DBE

The unsaturated ketone does indeed have an alkene and a carbonyl group. The unsaturated cyclic acid has: $16 - 10 = 6$ divided by $2 = 3$ DBEs and it has one alkene, one C=O and one ring. Correct. The aromatic ether has $16 - 8 = 8$ divided by 2 gives 4 DBEs and it has three double bonds in the ring and the ring itself. Correct again.

Nitrogen makes a difference. Every nitrogen adds *one extra hydrogen* atom because nitrogen can make three bonds. This is one fewer hydrogen to subtract. The formula becomes: subtract actual number of hydrogens from $(2n + 2)$, *add one for each nitrogen atom*, and divide by two. We can try this out too.

saturated hydrocarbon C_7H_{16}

saturated alcohol $C_7H_{16}O$

saturated ether $C_7H_{16}O$

All have $(2n + 2)$ H atoms

$C_7H_{10}O_2$ = three DB

C_7H_8O = four DBE

saturated C_7 compound with nitrogen

$C_7H_{17}N = (2n + 3)$Hs

$C_7H_{15}NO_2$ = one DBE

$C_7H_{13}NO$ = two DBE

C_7H_9NO= four DBE

$C_7H_{10}N_2$ = four DBE

The saturated compound has $(2n + 3)$ Hs instead of $(2n + 2)$. The saturated nitro compound has $(2n + 2) = 16$ less 15 (the actual number of Hs) plus one (the number of nitrogen atoms) = 2. Divide this by 2 and you get 1 DBE, which is the N=O bond. The last compound (we shall meet this later as 'DMAP') has:

1 Maximum number of H atoms for 7 Cs $\quad\quad 2n + 2 = 16$

2 Subtract the actual number of H atoms (10) $\quad\quad 16 - 10 = 6$

3 Add number of nitrogens $\quad\quad 6 + 2 = 8$

4 Divide by 2 to give the DBEs $\quad\quad 8/2 = 4$

There are indeed three double bonds and a ring, making four in all. You would be wise to check that you can do these calculations without much trouble.

If you have other elements too it is simpler just to draw a trial structure and find out how many DBEs there are. You may prefer this method for all compounds as it has the advantage of finding one possible structure before you really start! One good tip is that if you have few hydrogens relative to the number of carbon atoms (and at least four DBEs) then there is probably an aromatic ring in the compound.

▶

Do not confuse this calculation with the observation we made about mass spectra that the molecular weight of a compound containing one nitrogen atom must be odd. This observation and the number of DBEs are, of course, related but they are different calculations made for different purposes.

● **Working out the DBEs for an unknown compound**

1   Calculate the expected number of Hs in the saturated structure

   (a) For C_n there would be: $2n + 2$ Hs if C, H, O only

   (b) For C_nN_m there would be $2n + 2 + m$ Hs

2   Subtract the actual number of Hs and divide by 2. This gives the DBEs

3   If there are other atoms (Cl, B, P, etc.) it is best to draw a trial structure

4   If there are few Hs, e.g. less than the number of Cs, suspect a benzene ring

5   A benzene ring has *four* DBEs (three for the double bonds and one for the ring)

6   A nitro group has *one* DBE only

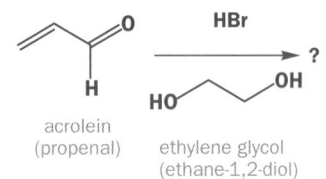

acrolein
(propenal)

ethylene glycol
(ethane-1,2-diol)

An unknown compound from a chemical reaction

Our second example addresses a situation very common in chemistry—working out the structure of a product of a reaction. The situation is this: you have treated propenal (acrolein) with HBr in ethane-1,2-diol (or glycol) as solvent for one hour at room temperature. Distillation of the reaction mixture gives a colourless liquid, compound X. What is it?

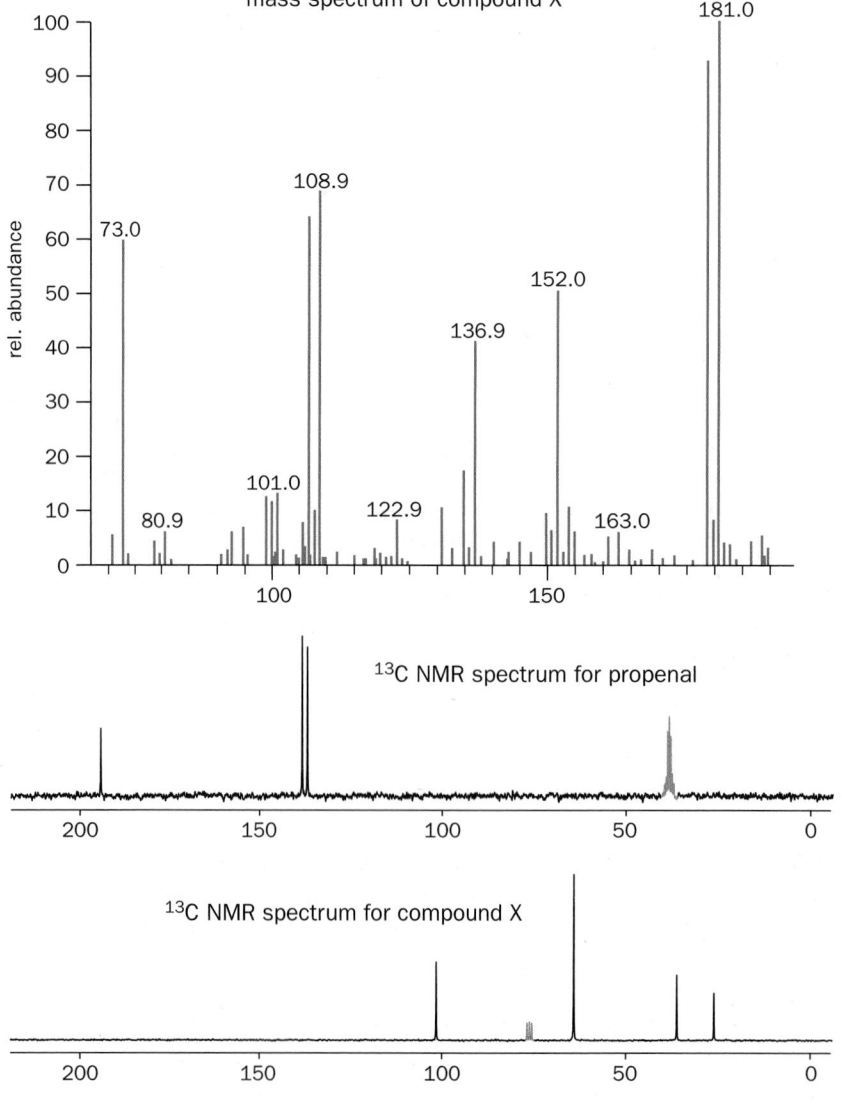

mass spectrum of compound X

^{13}C NMR spectrum for propenal

^{13}C NMR spectrum for compound X

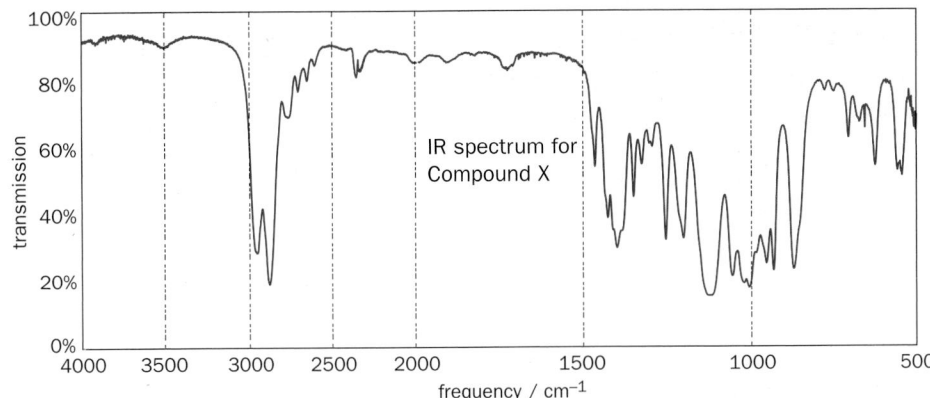

The mass spectrum shows a molecular ion (181) much heavier than that of the starting material, $C_3H_4O = 56$. Indeed it shows two molecular ions at 181/179 typical of a bromo-compound, so it looks as if HBr has added to the aldehyde somehow. High resolution reveals a formula of $C_5H_9BrO_2$ and the five carbon atoms make it look as though the glycol has added in too. If we add everything together we find that the unknown compound is the result of the three reagents added together less one molecule of water. A trial structure reveals *one* DBE.

The next thing is to see what remains of the propenal. The NMR spectrum of $CH_2=CH-CHO$ clearly shows one carbonyl group and two carbons on a double bond. These have all disappeared in the product and for the five carbon atoms we are left with four signals, two saturated, one next to oxygen, and one at 102.6 p.p.m. just creeping into the double bond region. It can't be an alkene as an alkene is impossible with only one carbon atom! The IR spectrum gives us another puzzle—there appear to be no functional groups at all! No OH, no carbonyl, no alkene—what else can we have? The answer is an ether—or rather two ethers as there are two oxygen atoms. Now that we suspect an ether, we can look for the C–O single bond stretch in the IR spectrum and find it at 1128 cm^{-1}. Each ether oxygen must have a carbon atom on each side of it. Two of these could be the same, but where are the rest?

We can solve this problem with a principle you may have guessed at before. If one oxygen atom takes a saturated carbon atom downfield to 50 p.p.m. or more, what could take a carbon downfield to 100 p.p.m. or more? We have established that chemical shifts are roughly additive so two oxygen atoms would just do. This would give us a fragment C–O–C–O–C accounting for three of the five carbon atoms. If you try and join the rest up with this fragment, you will find that you can't do it without a double bond, for example, the structure in the margin.

But we know we haven't got a double bond, (no alkene and no C=O) so the DBE must be a ring. You might feel uncomfortable with rings, but you must get used to them. Five-, six-, and seven-membered rings are very common. In fact, most known organic compounds have rings in them. We could join the skeleton of the present molecule up in many rings of various sizes like this one in the margin.

But this won't do as it would have five different carbon atoms. It is much more likely that the basic skeletons of the organic reagents are preserved, that is, that we have a two-carbon and a three-carbon fragment joined through oxygen atoms. This gives four possibilities.

These are all quite reasonable, though we might prefer the third as it is easier to see how it derives from the reagents. A decision can easily be reached from the base peak in the mass spectrum at 73. This is a fragment corresponding to the five-membered ring and not to the six-membered ring. The product is in fact the third possibility.

Looking forward to Chapters 11 and 14

We have only begun to explore the intricate world of identification of structure by spectroscopy. It is important that you recognize that structures are assigned, not because of some theoretical reason or because a reaction 'ought' to give a certain product, but because of sound evidence from spectra. You have seen three powerful methods—mass spectra, ^{13}C NMR, and IR spectroscopy in this chapter. In Chapter 11 we introduce the most important of all—proton (^{1}H) NMR and, finally, in Chapter 14 we shall take each of these a little further and show how the structures of more complex unknown compounds are really deduced. The last problem we have discussed here is not really solvable without proton NMR and in reality no-one would tackle any structure problem without this most powerful of all techniques. From now on spectroscopic evidence will appear in virtually every chapter. Even if we do not say so explicitly every time a new compound appears, the structure of this compound will in fact have been determined spectroscopically. Chemists make new compounds, and every time they do they **characterize** the compound with a full set of spectra. No scientific journal will accept that a new compound has been made unless a full description of all of these spectra are submitted with the report. Spectroscopy lets the science of organic chemistry advance.

Problems

1. How does the mass spectrum give evidence of isotopes in the compounds of bromine, chlorine, and carbon? Assuming the molecular ion of each of these compounds is of 100% abundance, what peaks (and in what intensity) would appear around that mass number? (a) C_2H_5BrO, (b) C_{60}, (c) C_6H_4BrCl? Give in cases (a) and (c) a possible structure for the compound. What compound is (b)?

2. The ^{13}C NMR spectrum for ethyl benzoate contains these peaks: 17.3, 61.1, 100–150 p.p.m. (four peaks), and 166.8 p.p.m. Which peak belongs to which carbon atom?

ethyl benzoate

3. The thinner used in typists' correction fluids is a single compound, $C_2H_3Cl_3$, having ^{13}C NMR peaks at 45.1 and 95.0. What is its structure? A commercial paint thinner gives two spots on thin layer chromatography and has ^{13}C NMR peaks at 7.0, 27.5, 35.2, 45.3, 95.6, and 206.3. Suggest what compounds might be used to make up this thinner.

4. The 'normal' O–H stretch (i.e. without hydrogen bonding) comes at about 3600 cm^{-1}. What is the reduced mass (μ) for O–H? What happens to the reduced mass when you double the atomic weight of each atom in turn, that is, what is μ for O–D and what is μ for S–H? In fact, both O–D and S–H stretches come at about 2500 cm^{-1}. Why?

5. Three compounds, each having the formula C_3H_5NO, have the IR spectra summarized here. What are their structures? Without ^{13}C NMR data, it may be easier to tackle this problem by first writing down all the possible structures for C_3H_5NO. In what specific ways would ^{13}C NMR data help?

(**a**) One sharp band above 3000 cm^{-1}; one strong band at about 1700 cm^{-1}

(**b**) Two sharp bands above 3000 cm^{-1}; two bands between 1600 and 1700 cm^{-1}

(**c**) One strong broad band above 3000 cm^{-1}; a band at about 2200 cm^{-1}

6. Four compounds having the molecular formula $C_4H_6O_2$ have the IR and ^{13}C NMR spectra given below. How many DBEs are there in $C_4H_6O_2$? What are the structures of the four compounds? You might again find it helpful to draw out some or all possibilities before you start.

(**a**) IR: 1745 cm^{-1}; ^{13}C NMR: 214, 82, 58, and 41 p.p.m.

(**b**) IR: 3300 (broad) cm^{-1}; ^{13}C NMR: 62 and 79 p.p.m.

(**c**) IR: 1770 cm^{-1}; ^{13}C NMR: 178, 86, 40, and 27 p.p.m.

(**d**) IR: 1720 and 1650 (strong) cm^{-1}; ^{13}C NMR: 165, 131, 133, and 54 p.p.m.

7. Three compounds of molecular formula C_4H_8O have the IR and ^{13}C NMR spectra given below. Suggest a structure for each compound, explaining how you make your deductions.

compound A   IR: 1730 cm^{-1}; ^{13}C NMR: 13.3, 15.7, 45.7, and 201.6 p.p.m.

compound B   IR: 3200 (broad) cm^{-1}; ^{13}C NMR: 36.9, 61.3, 117.2, and 134.7 p.p.m.

compound C   IR: no peaks except CH and fingerprint; ^{13}C NMR: 25.8 and 67.9 p.p.m.

Compound A reacts with NaBH$_4$ to give compound D. Compound B reacts with hydrogen gas over a palladium catalyst to give the same compound D. Compound C reacts with neither reagent. Suggest a structure for compound D from the data given and explain the reactions. (*Note.* H$_2$ reduces alkenes to alkanes in the presence of a palladium catalyst.)

compound D   IR: 3200 (broad) cm^{-1}; ^{13}C NMR: 15.2, 20.3, 36.0, and 62.9 p.p.m.

8. The situation is as follows: You have dissolved *t*-BuOH (Me$_3$COH) in MeCN with an acid catalyst, left the solution overnight, and found crystals with the following characteristics there in the morning. What are they?

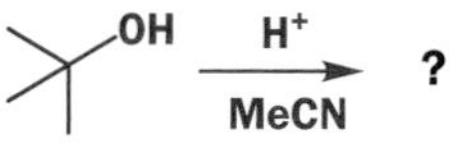

IR: 3435 and 1686 cm^{-1}
^{13}C NMR: 169, 50, 29, and 25 p.p.m.

mass spectrum (%): 115 (7), 100 (10), 64 (5), 60 (21), 59 (17), 58 (100), and 56 (7). (Don't try to assign all of these!)

9. How many isomers of trichlorobenzene are there? The 1,2,3-trichloro isomer is illustrated. Could they be distinguished by ^{13}C NMR?

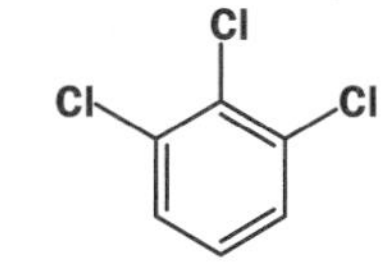

10. How many signals would you expect in the ^{13}C NMR of the following compounds?

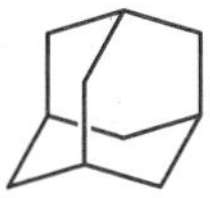

A 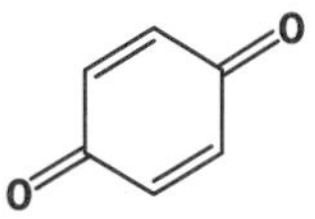   B 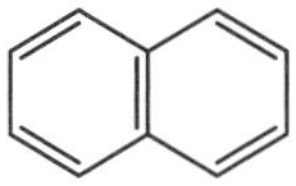   C

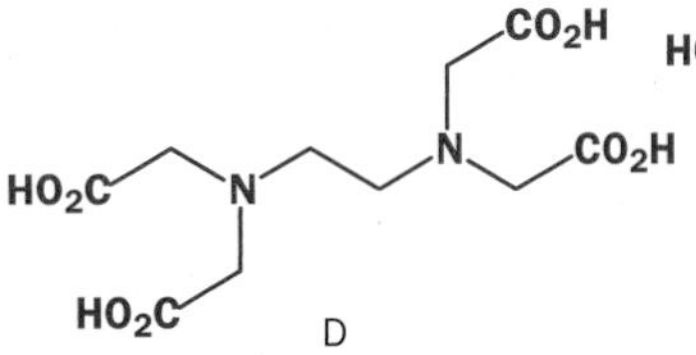

D 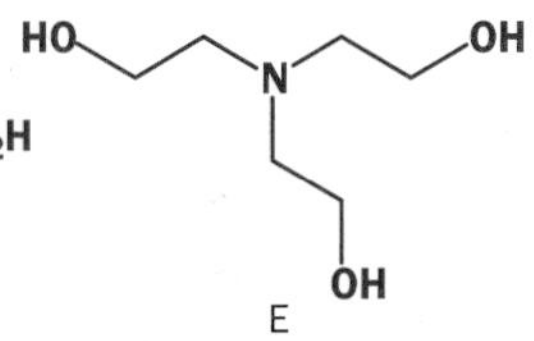   E

11. How would mass spectra help you distinguish these structures?

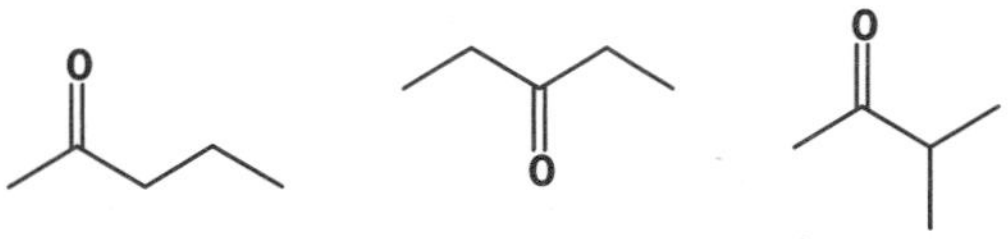

Structure of molecules

4

Note from the authors to all readers

This chapter contains mathematical material that some readers may find daunting. Organic chemistry students come from many different backgrounds since organic chemistry occupies a middle ground between the physical and the biological sciences. We hope that those from a more physical background will enjoy the material as it is. If you are one of those, you should work your way through the entire chapter. If you come from a more biological background, especially if you have done little maths at school, you may lose the essence of the chapter in a struggle to understand the equations. We have therefore picked out the more mathematical parts in boxes and you should abandon these parts (and any others!) if you find them too alien. The general principles behind the chapter—why molecules have the structures they do—are obviously so important that we cannot omit this essential material but you should try to grasp the principles without worrying too much about the equations. The ideas of atomic orbitals overlapping to form bonds, the molecular orbitals that result, and the shapes that these orbitals impose on organic molecules are at least as central for biochemistry as they are for organic chemistry. Please do not be discouraged but enjoy the challenge.

Introduction

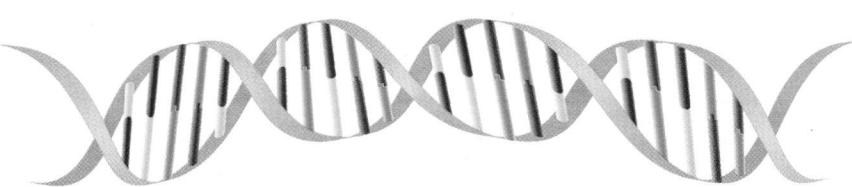

You may recognize the model above as DNA, the molecule that carries the genetic information for all life on earth. It is the exact structure of this compound that determines precisely what a living thing is—be it man or woman, frog, or tree—and even more subtle characteristics such as what colour eyes or hair people have.

What about this model?

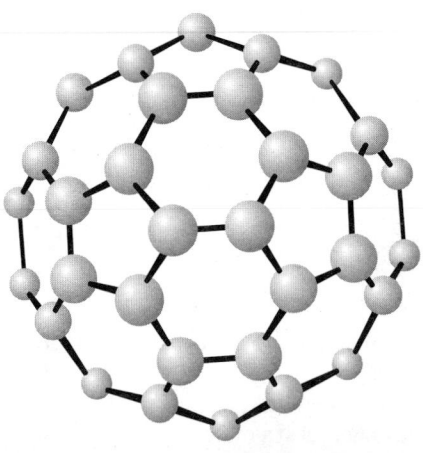

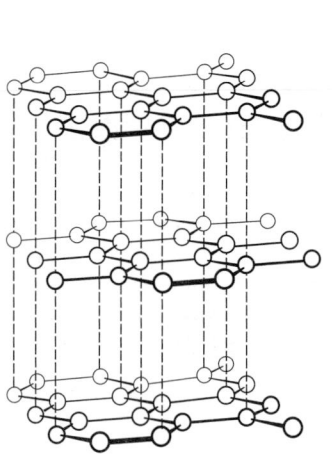

graphite

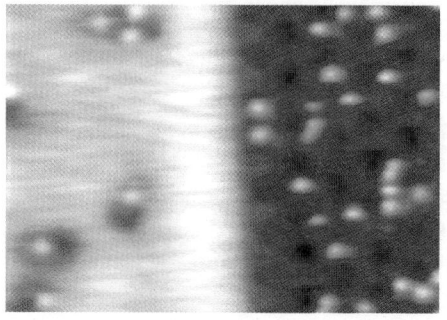

▶
The dark brown blobs in this STM picture recorded at a temperature of 4 K are individual oxygen atoms adsorbed on a silver surface. The light blobs are individual ethylene (ethene) molecules. Ethylene will only adsorb on silver if adjacent to an oxygen atom. This is an atomic scale view of a very important industrial process—the production of ethylene oxide from ethylene and oxygen using a silver catalyst.

▶
The picture on the right is an X-ray structure of a catenane—a molecule consisting of two interlocking rings joined like two links in a chain. The key to the synthesis depends on the self-stacking of the planar structures prior to ring closure.

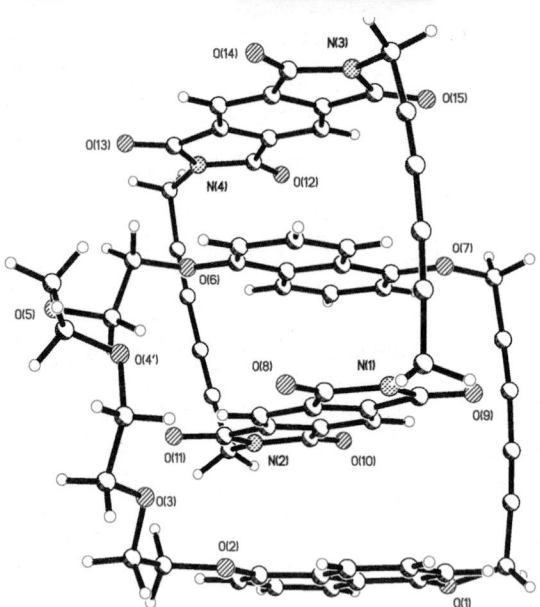

You may also have recognized this molecule as buckminsterfullerene, a form of carbon that received enormous interest in the 1980s and 1990s. The question is, how did you recognize these two compounds? You recognized their *shapes*. All molecules are simply groups of atoms held together by electrons to give a definite three-dimensional shape. What exactly a compound might be is determined not only by the atoms it contains, but also by the arrangement of these atoms in space—the shape of the molecule. Both graphite and buckminsterfullerene are composed of carbon atoms only and yet their properties, both chemical and physical, are completely different.

There are many methods available to chemists and physicists to find out the shapes of molecules. One of the most recent techniques is called **Scanning Tunnelling Microscopy (STM)**, which is the closest we can get to actually 'seeing' the atoms themselves.

Most techniques, for example, X-ray or electron diffraction, reveal the shapes of molecules indirectly.

In Chapter 3 you met some of the spectroscopic methods frequently used by organic chemists to determine the shape of molecules. Spectroscopy would reveal the structure of methane, for example, as tetrahedral—the carbon atom in the centre of a regular tetrahedron with the hydrogen atoms at the corners. In this chapter we are going to discuss *why* compounds adopt the shapes that they do.

This tetrahedral structure seems to be very important—other molecules, both organic and inorganic, are made up of many tetrahedral units. What is the origin of this tetrahedral structure? It could simply arise from four pairs of electrons repelling each other to get as far as possible from each other. That would give a tetrahedron.

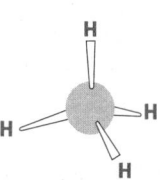

methane is tetrahedral

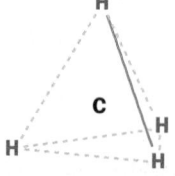

the H atoms form
a tetrahedron

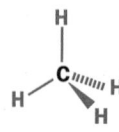

methane is
tetrahedral

This simple method of deducing the structure of molecules is called **Valence Shell Electron Pair Repulsion Theory (VSEPRT)**. It says that all electron pairs, both bonding and nonbonding, in the outer or valence shell of an atom repel each other. This simple approach predicts (more or less) the correct structures for methane, ammonia, and water with four electron pairs arranged tetrahedrally in each case.

VSEPRT seems to work for simple structures but surely there must be more to it than this? Indeed there is. If we really want to understand *why* molecules adopt the shapes they do, we must look at the atoms that make up the molecules and how they combine. By the end of this chapter, you should be able to predict or at least understand the shapes of simple molecules. For example, why are the bond angles in ammonia 107°, while in hydrides of the other elements in the same group as nitrogen, PH_3, AsH_3, and SbH_3, they are all around 90°? Simple VSEPRT would suggest tetrahedral arrangements for each.

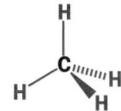

tetrahedral methane
four bonds and no lone pairs

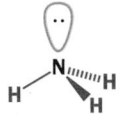

tetrahedral ammonia
three bonds and one lone pair

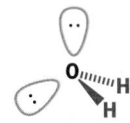

tetrahedral water
two bonds and two lone pairs

Atomic structure

You know already what makes up an atom—protons, neutrons, and electrons. The protons and neutrons make up the central core of an atom—the nucleus—while the electrons form some sort of cloud around it. As chemists, we are concerned with the electrons in atoms and more importantly with the electrons in molecules: chemists need to know how many electrons there are in a system, where they are, and what energy they have. Before we can understand the behaviour of electrons in molecules, we need to look closely at the electronic structure of an atom. Evidence first, theory later.

Atomic emission spectra

Many towns and streets are lit at night by sodium vapour lamps. You will be familiar with their warm yellow-orange glow but have you ever wondered what makes this light orange and not white? The normal light bulbs you use at home have a tungsten filament that is heated white hot. You know that this white light could be split by a prism to reveal the whole spectrum of visible light and that each of the different colours has a different frequency that corresponds to a distinct energy. But where does the orange street light come from? If we put a coloured filter in front of our white light, it would absorb some colours of the spectrum and let other colours through. We could make orange light this way but that is not how the street lights work—they actually generate orange light and orange light only. Inside these lights is sodium metal. When the light is switched on, the sodium metal is slowly vaporized and, as an electric current is passed through the sodium vapour, an orange light is emitted. This is the same colour as the light you get when you do a flame test using a sodium compound.

The point is that only one colour light comes from a sodium lamp and this must have one specific frequency and therefore one energy. It doesn't matter what energy source is used to generate the light, whether it be electricity or a Bunsen burner flame; in each case light of one specific energy is given out. Looking at the orange sodium light through a prism, we see a series of very sharp lines with two particularly bright orange lines at around 600 nm. Other elements produce similar spectra—indeed two elements, rubidium and cesium, were discovered by Robert Bunsen after studying such spectra. They are actually named after the presence of a pair of bright coloured lines in their spectra—cesium from the Latin *caesius* meaning bluish grey and rubidium from the Latin *rubidus* meaning red. Even hydrogen can be made to produce an atomic spectrum and, since a hydrogen atom is the simplest atom of all, we shall look at the atomic spectrum of hydrogen first.

If enough energy is supplied to a hydrogen atom, or any other atom, an electron is eventually knocked completely out of the atom. In the case of hydrogen a single proton is left. This is, of course, the ionization of hydrogen.

$$H\cdot \xrightarrow{\text{energy}} H^{\oplus} + e^{\ominus}$$

H atom   proton   electron

What if we don't quite give the atom enough energy to remove an electron completely? It's not too hard to imagine that, if the energy is not enough to ionize the atom, the electron would be

Quantum mechanics tells us that energy is quantized. Light does not come in a continuous range of energies but is divided up into minute discrete packets (quanta) of different noncontinuous (discrete) energies. The energy of each of these packets is related to the frequency of the light by a simple equation: $E = h\nu$ (E is the energy, ν the frequency of the light, and h is Planck's constant). The packet of light released from sodium atoms has the frequency of orange light and the corresponding energy.

'loosened' in some way—the atom absorbs this energy and the electron moves further away from the nucleus and now needs less energy to remove it completely. The atom is said to be in an **excited state**. This process is a bit like a weight lifter lifting a heavy weight—he can hold it above his head with straight arms (the excited state) but sooner or later he will drop it and the weight will fall to the ground. This is what happens in our excited atom—the electron will fall to its lowest energy, its **ground state**, and the energy put in will come out again. This is the origin of the lines in the atomic spectra not only for hydrogen but for all the elements. The flame or the electric discharge provides the energy to promote an electron to a higher energy level and, when this electron returns to its ground state, this energy is released in the form of light.

Line spectra are composed of many lines of different frequencies, which can only mean that there must be lots of different energy transitions possible, but not just *any* energy transitions. Quantum mechanics says that an electron, like light, cannot have a continuous range of energies, only certain definite energies, which in turn means that only certain energy transitions are possible. This is rather like trying to climb a flight of stairs—you can jump up one, two, five, or even all the steps if you have enough energy but you cannot climb up half or two-thirds of a step. Likewise coming down, you can jump from one step to any other—lots of different combinations are possible *but there is a finite number, depending on the number of steps*. This is why there are so many lines in the atomic spectra— the electron can receive energy to promote it to a higher energy level and it can then fall to any level below and a certain quantity of light will be released.

We want to predict, as far as we can, where all the electrons in different *molecules* are to be found including the ones not involved with bonding. We want to know where the molecule can accommodate *extra* electrons and from where electrons can be removed most easily. Since most molecules contain many electrons, the task is not an easy one. However, the electronic structure of atoms is somewhat easier to understand and we can approximate the electronic structure of *molecules* by considering how the component atoms combine.

The next section is therefore an introduction to the electronic structure of *atoms*—what energies the electrons have and where they may be found. Organic chemists are rarely concerned with atoms themselves but need to understand the electronic structure in atoms before they can understand the electronic structure in molecules. As always, evidence first!

The atomic emission spectrum of hydrogen

The atomic emission spectrum of hydrogen is composed of many lines but these fall into separate sets or series. The first series to be discovered, not surprisingly, were those lines in the visible part of the spectrum. In 1885, a Swiss schoolmaster, Johann Balmer, noticed that the wavelengths, λ, of the lines in this series could be predicted using a mathematical formula. He did not see why; he just saw the relationship. This was the first vital step.

$$\lambda = \text{constant} \times \frac{n^2}{n^2 - 2^2} \quad (n \text{ is an integer greater than 2})$$

As a result of his work, the lines in the visible spectrum are known as the Balmer series. The other series of lines in the atomic emission spectrum of hydrogen were discovered later (the next wasn't discovered until 1908). These series are named after the scientists who discovered them; for example, the series in the ultraviolet region is known as the Lyman series after Theodore Lyman.

Balmer's equation was subsequently refined to give an equation that predicts the frequency, ν, of any of the lines in any part of the hydrogen spectrum rather than just for his series. It turns out that his was not the most fundamental series, just the first to be discovered.

$$\nu = \text{constant} \times \left(\frac{1}{n_1^2} - \frac{1}{n_2^2} \right)$$

Each series can be described by this equation if a particular value is given for n_1 but n_2 is allowed to vary. For the Lyman series, n_1 remains fixed at 1 while n_2 can be 2, 3, 4, and so on. For the Balmer series, n_1 is fixed at 2 while n_2 can be 3, 4, 5, and so on.

Atomic emission spectra are evidence for electronic energy levels

Atomic emission spectra give us our first clue to understanding the electronic energy levels in an atom. Since the lines in the emission spectrum of hydrogen correspond to the electron moving between energy levels and since frequency is proportional to energy, $E = h\nu$, the early equations must represent just the difference between two energy levels. This in turn tells us that the electron's energy levels in an atom must be inversely proportional to the square of an important integer 'n'. This can be expressed by the formula

$$E_n = -\frac{\text{constant}}{n^2}$$

where E_n is the energy of an electron in the nth energy level and n is an integer ≥ 1 known as the **principal quantum number**. Note that, when $n = \infty$, that is, when the electron is no longer associated with the nucleus, its energy is zero. All other energy levels are lower than zero because of the minus sign in the equation. This is consistent with what we know already—we must put energy in to ionize the atom and remove the electron from the nucleus.

Electronic energy levels

In more detail, the constant in this equation can be broken down into a universal constant, the Rydberg constant R_H, which applies to any electron on any atom, and a constant Z which has a particular value for each atom.

$$E_n = -\frac{R_H Z^2}{n^2}$$

The Rydberg constant R_H, is measured in units of energy. For a given atom (i.e. Z is constant) there are many different energy levels possible (each corresponding to a different value of n). Also, as n gets bigger, the energy gets smaller and smaller and approaches zero for large n. The energy gets smaller as the electron gets further away from the nucleus. For electrons in the same energy level but in different atoms, (i.e. keeping n constant but varying Z), the energy of an electron depends on the square of the atomic number. This makes sense too—the more protons in the nucleus, the more tightly the electron is held in the atom.

● The electrons in any atom are grouped in energy levels whose energies are universally proportional to the inverse square of a very important number n. This number is called the **principal quantum number** and it can have only a few integral values ($n = 1, 2, 3\ldots$). The energy levels also depend on the type of atom.

An energy level diagram gives some idea of the relative spacing between these energy levels.

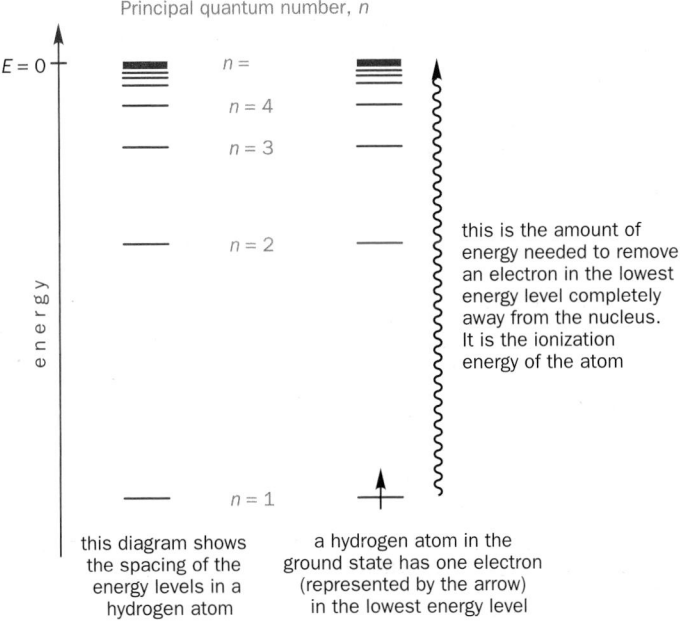

Principal quantum number, n

this is the amount of energy needed to remove an electron in the lowest energy level completely away from the nucleus. It is the ionization energy of the atom

this diagram shows the spacing of the energy levels in a hydrogen atom

a hydrogen atom in the ground state has one electron (represented by the arrow) in the lowest energy level

▶ Notice how the spacing between the energy levels gets closer and closer. This is a consequence of the energy being inversely proportional to the square of the principal quantum number. It tells us that it becomes easier and easier to remove an electron completely from an atom as the electron is located in higher and higher energy levels. As we shall see later, the increasing value of the principal quantum number also correlates with the electron being found (on average) further and further from the nucleus and being easier and easier to remove. This is analogous to a rocket escaping from a planet— the further away it is, the less it experiences the effects of gravity and so the less energy it requires to move still further away. The main difference is that there seems to be no quantization of the different energy levels of the rocket—it appears (to us in our macroscopic world at least) that any energy is possible. In the case of the electron in the atom, only certain values are allowed.

Three quantum numbers come from the Schrödinger equation

There is no doubt about the importance of n, the principal quantum number, but where does it come from? This quantum number and two other quantum numbers come from solving the **Schrödinger equation**. We are not going to go into any details regarding Schrödinger's equation or how to solve it—there are plenty of more specialized texts available if you are interested in more detail.

Solutions to Schrödinger's equation come in the form of **wave functions** (symbol Ψ), which describe the energy and position of the electrons thought of as waves. You might be a little unsettled to find out that we are describing electrons using waves but the same wave–particle duality idea applies to electrons as to light. We regularly think of light in terms of waves with their associated wavelengths and frequencies but light can also be described using the idea of photons—individual little light 'particles'. The same is true of the electron; up to now, you will probably have thought of electrons only as particles but now we will be thinking of them as waves.

It turns out that there is not one specific solution to the Schrödinger equation but many. This is good news because the electron in a hydrogen atom can indeed have a number of different energies. It turns out that each wave function can be defined by three quantum numbers (there is also a fourth quantum number but this is not needed to define the wave function). We have already met the principal quantum number, n. The other two are called the **orbital angular momentum quantum number** (sometimes called the azimuthal quantum number), ℓ, and the **magnetic quantum number, m_ℓ**.

A specific wave function solution is called an **orbital**. The different orbitals define different energies and distributions for the different electrons. The name 'orbital' goes back to earlier theories where the electron was thought to orbit the nucleus in the way that planets orbit the sun. It seems to apply more to an electron seen as a particle, and orbitals of electrons thought of as particles and wave functions of electrons thought of as waves are really two different ways of looking at the same thing. Each different orbital has its own individual quantum numbers, n, ℓ, and m_ℓ.

> Don't worry about the rather fancy names of these quantum numbers; just accept that the three numbers define a given wave function.

Summary of the importance of the quantum numbers

What does each quantum number tell us and what values can it adopt? You have already met the principal quantum number, n, and seen that this is related to the energy of the orbital.

The principal quantum number, n

Different values for n divide orbitals into groups of similar energies called **shells**. Numerical values for n are used in ordinary speech. The first shell ($n = 1$) can contain only two electrons and the atoms H and He have one and two electrons in this first shell, respectively.

The orbital angular momentum quantum number, ℓ

The orbital angular momentum quantum number, ℓ, determines, as you might guess, the angular momentum of the electron as it moves in its orbital. This quantum number tells us the shape of the orbital, spherical or whatever. The values that ℓ can take depend on the value of n: ℓ can have any value from 0 up to $n - 1$: $\ell = 0$, 1, 2, ... , $n - 1$. The different possible values of ℓ are given letters rather than numbers and they are called s, p, d, and f.

value of n	1	2	3	4
possible values of ℓ	0	0, 1	0, 1, 2	0, 1, 2, 3
name	1s	2s, 2p	3s, 3p, 3d	4s, 4p, 4d, 4f

The magnetic quantum number, m_ℓ

The magnetic quantum number, m_ℓ, determines the spatial orientation of the angular momentum. In simple language it determines where the orbitals are in space. Its value depends on the value of ℓ, varying from $-\ell$ to $+\ell$: $m_\ell = \ell, \ell - 1, \ell - 2, ... , -\ell$. The different possible values of m_ℓ are given suffixes on the letters

value of n	1	2	2
value of ℓ	0	0	1
name	1s	2s	2p
possible values of m_ℓ	0	0	+1, 0, −1
name	1s	2s	$2p_x$, $2p_y$, $2p_z$

defining the quantum number ℓ. These letters refer to the direction of the orbitals along the x-, y-, or z-axes. Organic chemists are concerned mostly with s and p orbitals ($\ell = 0$ or 1) so the subdivisions of the d orbitals can be omitted.

Each quantum number gives subdivisions for the one before. There are no subdivisions in the lowest value of each quantum number: and the subdivisions increase in number as each quantum number increases. Now we need to look in more detail at the meanings of the various values of the quantum numbers.

Atomic orbitals

Nomenclature of the orbitals

For a hydrogen atom the energy of the orbital is determined only by the principal quantum number, n, and n can take values 1, 2, 3, and so on. This is the most fundamental division and is stated first in the description of an electron. The electron in a hydrogen atom is called $1s^1$. The 1 gives the value of n: the most important thing in the foremost place. The designation s refers to the value of ℓ. These two together, 1s, define and name the orbital. The superscript 1 tells us that there is one electron in this orbital.

The orbital angular momentum quantum number, ℓ, determines the shape of the orbital. Instead of expressing this as a number, letters are used to label the different shapes of orbitals. s orbitals have $\ell = 0$, and p orbitals have $\ell = 1$.

Using both these quantum numbers we can label orbitals 1s, 2s, 2p, 3s, 3p, 3d, and so on. Notice that, since ℓ can only have integer values up to $n - 1$, we cannot have a 1p or 2d orbital.

> ● **Names of atomic orbitals**
> - The first shell ($n = 1$) has only an s orbital, 1s
> - The second shell ($n = 2$) has s and p orbitals 2s and 2p
> - The third shell ($n = 3$) has s, p, and d orbitals, 3s, 3p, and 3d

One other point to notice is that, for the hydrogen atom (and, technically speaking, any one-electron ion such as He^+ or Li^{2+}), a 2s orbital has exactly the same energy as a 2p orbital and a 3s orbital has the same energy as the 3p and 3d orbitals. Orbitals that have the same energy are described as **degenerate**. In atoms with more than one electron, things get more complicated because of electron–electron repulsion and the energy levels are no longer determined by n alone. In such cases, the 2s and 2p or the 3s, 3p, and 3d orbitals or any other orbitals that share the same principal quantum number are no longer degenerate. In other words, in multielectron atoms, the energy of a given orbital depends not only on the principal quantum number, n, but also in some way on the orbital angular momentum quantum number, ℓ.

Values of the magnetic quantum number, m_ℓ, depend on the value of ℓ. When $\ell = 0$, m_ℓ can only take one value (0); when $\ell = 1$, m_ℓ has three possible values (+1, 0, or −1). There are five possible values of m_ℓ when $\ell = 2$ and seven when $\ell = 3$. In more familiar terms, there is only one sort of s orbital; there are three sorts of p orbitals, five sorts of d orbitals, and seven sorts of f orbitals. All three p orbitals are degenerate as are all five d orbitals and all seven f orbitals (for both single-electron and multielectron atoms). We shall see how to represent these orbitals later.

There is a fourth quantum number

The spin of an electron is the angular momentum of an electron spinning *about its own axis*, although this is a simplified picture. This angular momentum is different from the angular momentum, ℓ, which represents the electron's angular momentum about the nucleus. The magnitude of the electron's spin is constant but it can take two orientations. These are represented using the fourth quantum number, the **spin angular momentum quantum number**, m_s, which can take the value of $+\frac{1}{2}$ or $-\frac{1}{2}$ in any orbital, regardless of the values of n, ℓ, or m_ℓ. Each

s, p, d, f

These letters hark back to the early days of spectroscopy and refer to the appearance of certain lines in atomic emission spectra: 's' for 'sharp', 'p' for 'principal', 'd' for 'diffuse', and 'f' for 'fundamental'. The letters s, p, d, and f matter and you must know them, but you do not need to know what they originally stood for.

Value of ℓ	Name of orbital
0	s
1	p
2	d
3	f

You have already come across another spin—the nuclear spin—which gives rise to NMR. There is an analogous technique, **electron spin resonance**, ESR, which detects unpaired electrons.

orbital can hold a maximum of two electrons and then only when the electrons have different 'spin', that is, they must have different values of m_s, $+\frac{1}{2}$ or $-\frac{1}{2}$. The rule that no more than two electrons may occupy any orbital (and then only if their spins are paired) is known as the **Pauli exclusion principle**.

> ● **Every electron is unique!**
>
> If electrons are in the same atom, they must have a unique combination of the four quantum numbers. Each orbital, designated by three quantum numbers, n, ℓ, and m_ℓ, can contain only two electrons and then only if their spin angular quantum numbers are different.

How the periodic table is constructed

All the quantum numbers for all the electrons with $n = 1$ and 2 can now be shown in a table like the ones earlier in this chapter. Though we have so far been discussing the hydrogen atom, in fact, the H atom never has more than two electrons. Fortunately, the energy levels deduced for H also apply to all the other elements with some minor adjustments. This table would actually give the electronic configuration of neon, Ne.

In this table, the energy goes up from left to right, though all the 2p orbitals are degenerate. To add $n = 3$, one column for the 3s, three columns for the 3p, and five columns for the 3d orbital would be needed. Then all five 3d orbitals would be degenerate.

value of n	1	2	2	2	2
value of ℓ	0	0	1	1	1
name	1s	2s	2p	2p	2p
possible values of m_ℓ	0	0	+1	0	−1
name	1s	2s	$2p_x$	$2p_y$	$2p_z$
possible values of m_s	$+\frac{1}{2}, -\frac{1}{2}$	$+\frac{1}{2}, -\frac{1}{2}$	$+\frac{1}{2}, -\frac{1}{2}$	$+\frac{1}{2}, -\frac{1}{2}$	$+\frac{1}{2}, -\frac{1}{2}$
electrons	$1s^2$	$2s^2$	$2p_x^2$	$2p_y^2$	$2p_z^2$

Another way to show the same thing is by an energy diagram showing how the quantum numbers divide and subdivide.

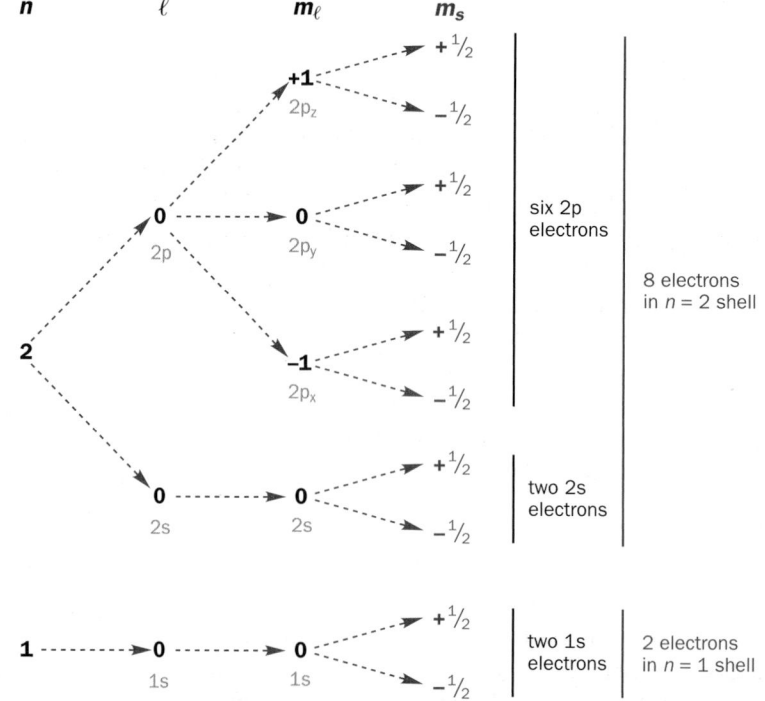

These numbers explain the shape of the periodic table. Each element has one more electron (and one more proton and perhaps more neutrons) than the one before. At first the lowest energy shell (n = 1) is filled. There is only one orbital, 1s, and we can put one or two electrons in it. There are therefore two elements in this block, H and He. Next we must move to the second shell (n = 2), filling 2s first so we start the top of groups 1 and 2 with Li and Be. These occupy the top of the red stack marked 's block' because all the elements in this block have one or two electrons in their outermost s orbital and no electrons in the outermost p orbital. Then we can start on the 2p orbitals. There are three of these so we can put in six electrons and get six elements B, C, N, O, F, and Ne. They occupy the top row of the black p block. Most of the elements we need in this book are in those blocks. Some, Na, K, and Mg for example, are in the s block and others, Si, P, and S for example, are in the second row of the p block.

The layout of the periodic table

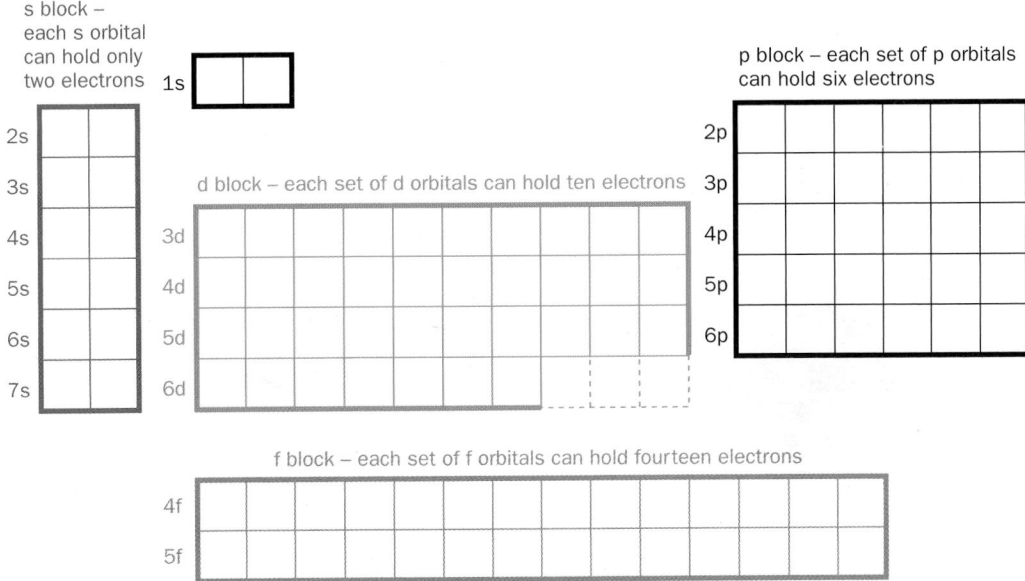

s block –
each s orbital
can hold only
two electrons

1s

2s 3s 4s 5s 6s 7s

d block – each set of d orbitals can hold ten electrons

3d 4d 5d 6d

p block – each set of p orbitals
can hold six electrons

2p 3p 4p 5p 6p

f block – each set of f orbitals can hold fourteen electrons

4f 5f

Other orbitals

Organic chemists are really concerned only with s and p orbitals since most of the elements we deal with are in the second row of the periodic table. Later in the book we shall meet elements in the second row of the p block (Si, P, S) and then we will have to consider their d orbitals, but for now we are not going to bother with these and certainly not with the f orbitals. But you may have noticed that the 4s orbital is filled before the 3d orbitals so you may guess that the 4s orbital must be slightly lower in energy than the 3d orbitals. Systems with many electrons are more complicated because of electron repulsion and hence the energies of their orbitals do not simply depend on n alone.

Graphical representations of orbitals

One problem with wave functions is trying to visualize them: what does a wave function look like? Various graphs of wave functions can be plotted but they are not much help as Ψ itself has no physical meaning. However the *square* of the wave function, Ψ^2, does have a practical interpretation; it is proportional to the probability of finding an electron at a given point. Unfortunately, we can't do

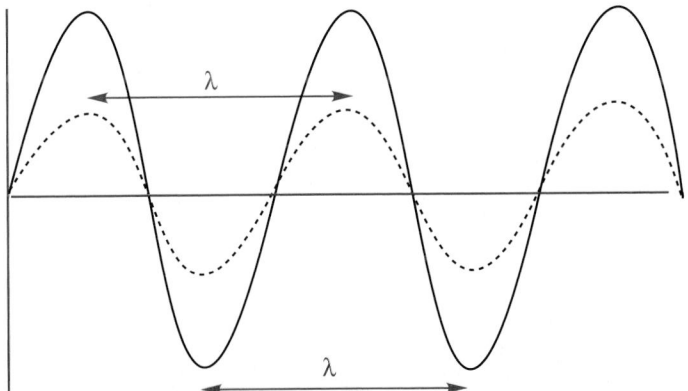

These two waves both have the same wavelength, λ, but the dashed wave is less intense than the other wave. The intensity is proportional to the amplitude squared.

▶

There is some justification for this interpretation that the wave function squared is proportional to the probability of finding an electron. With light waves, for example, while the wavelength provides the colour (more precisely the energy) of the wave, it is the amplitude squared that gives the brightness.

But this is looking at light in terms of waves. In terms of particles, photons, the intensity of light is proportional to the density of photons.

better than probability as we are unable to say exactly where the electron is at any time. This is a consequence of **Heisenberg's uncertainty principle**—we cannot know both the exact position and the exact momentum of an electron simultaneously. Here we know the momentum (energy) of the electron and so its exact position is uncertain.

How do we depict a probability function? One way would be to draw contours connecting regions where there is an equal probability of finding the electron. If Ψ^2 for a 1s orbital is plotted, a three-dimensional plot emerges. Of course, this is a two-dimensional representation of a three-dimensional plot—the contours are really spherical like the different layers of an onion. These circles are rather like the contour lines on a map except that they represent areas of equal probability of finding the electron instead of areas of equal altitude.

Another way to represent the probability is by a **density plot**. Suppose we could see exactly where the electron was at a given time and that we marked the spot. If we looked again a little later, the electron would be in a different place—let us mark this spot too. Eventually, if we marked enough spots, we would end up with a fuzzy picture like those shown for the 1s and 2s orbitals. Now the *density* of the dots is an indication of the probability of finding an electron in a given space—the more densely packed the dots (that is, the darker the area), the greater the probability of finding the electron in this area. This is rather like some maps where different altitudes are indicated by different colours.

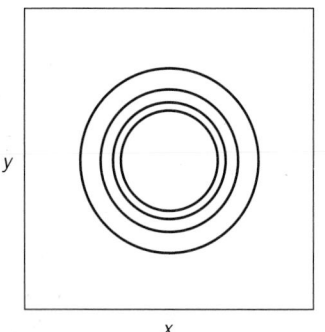

y

x

contour diagram of 1s orbital

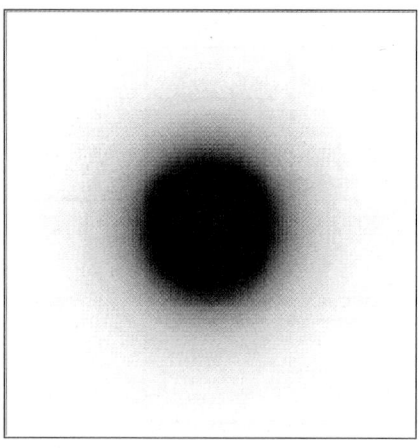

density plot of 1s orbital

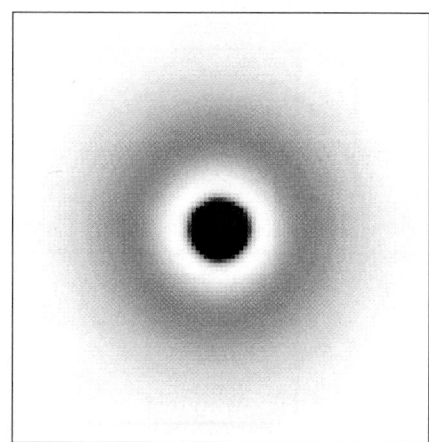

density plot of 2s orbital

The 2s orbital, like the 1s orbital, is spherical. There are two differences between these orbitals. One is that the 2s orbital is bigger so that an electron in a 2s orbital is more likely to be found further away from the nucleus than an electron in a 1s orbital. The other difference between the orbitals is that, within the 2s orbital but not within the 1s orbital, there is a region where there is no electron density at all. Such a region is called a **nodal surface**. In this case there is no electron density at one set radius from the nucleus; hence this is known as a **radial node**. The 2s orbital has one radial node.

Nodes are important for musicians

You can understand these nodal surfaces by thinking in terms of waves. If a violin or other string instrument is plucked, the string vibrates. The ends cannot move since they are fixed to the instrument. The note we hear is mainly due to the string vibrating as shown in the diagram for the first harmonic.

However, there are other vibrations of higher energy known as **harmonics**, which help to give the note its timbre (the different timbres allow us to tell the difference between, say, a flute and a violin playing the same note). The second and third harmonics are also shown.

Each successive harmonic has one extra node—while the first harmonic has no nodes (if you don't count the

end stops), the second harmonic has one and the third has two and so on. These are points where the string does not vibrate at all (you can even 'select out' the second harmonic on a stringed instrument if you gently press halfway along the vibrating string).

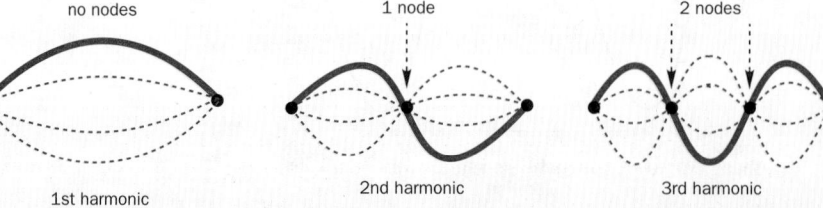

no nodes

1 node

2 nodes

1st harmonic

2nd harmonic

3rd harmonic

● **Shapes of s orbitals**
- The 1s orbital is spherically symmetrical and has no nodes
- The 2s orbital has one radial node and the 3s orbital two radial nodes. They are both spherically symmetrical

What does a Ψ^2 for a 2p orbital look like? The probability density plot is no longer spherically symmetrical. This time the shape is completely different—the orbital now has an orientation in space and it has two lobes. Notice also that there is a region where there is no electron density between the two lobes—another nodal surface. This time the node is a plane in between the two lobes and so it is known as a **nodal plane**. One representation of the 2p orbitals is a three-dimensional plot, which gives a clear idea of the true shape of the orbital.

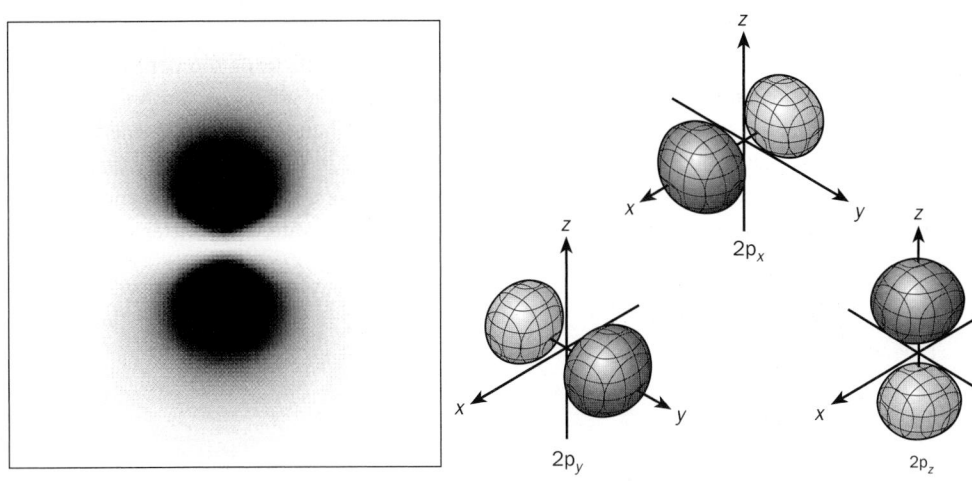

density plot of 2p orbital

three-dimensional plot of the 2p orbitals

Plots of 3p and 4p orbitals are similar—each has a nodal plane and the overall shape outlined in each is the same. However, the 3p orbital also has a radial node and the 4p has two radial nodes and once again the size of the orbital increases as the principal quantum number increases.

All this explains why the shape of an orbital depends on the orbital angular quantum number, ℓ. All s orbitals ($\ell = 0$) are spherical, all p orbitals ($\ell = 1$) are shaped like a figure eight, and d orbitals ($\ell = 2$) are yet another different shape. The problem is that these probability density plots take a long time to draw—organic chemists need a simple easy way to represent orbitals. The contour diagrams were easier to draw but even they were a little tedious. Even simpler still is to draw just one contour within which there is, say, a 90% chance of finding the electron. This means that all s orbitals can be represented by a circle, and all p orbitals by a pair of lobes.

The phase of an orbital

The wave diagrams need further discussion to establish one fine point—the **phase** of an orbital.

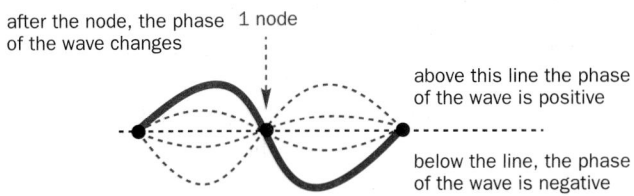

after the node, the phase of the wave changes   1 node

above this line the phase of the wave is positive

below the line, the phase of the wave is negative

Just as an electromagnetic wave, or the wave on a vibrating string, or even an ocean wave possesses different 'phases' (for example, the troughs and peaks of an ocean wave) so too do the atom's wave functions—the orbitals. After each node in an orbital, the phase of the wave function changes. In the

▶ You might have noticed that each orbital in the *n*th energy level has the same total number of nodes, $n - 1$. The total number of nodes is the sum of the numbers of radial nodes and nodal planes. Thus both the 2s and 2p have one node (a radial node in the case of the 2s and a nodal plane in the case of the 2p) while the 3s, 3p, and 3d orbitals each have two nodes (the 3s have two radial nodes, the 3p orbitals each have one radial node and one nodal plane, and the 3d orbitals each have two nodal planes).

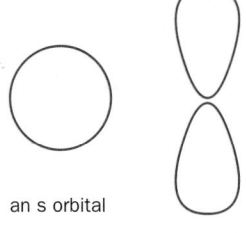

an s orbital

a p orbital

■ Remember that the orbitals are three-dimensional and that these drawings represent a cross-section. A three-dimensional version would look more like a sphere for an s orbital and an old-fashioned hour-glass for a p orbital. Actually, each lobe of a p orbital is much more rounded than the usual representation, but that is not so important.

2p orbital, for example, one lobe is one phase; the other lobe is another phase with the nodal plane in between. In the standing wave above the different phases are labelled positive and negative. The phases of a p orbital could be labelled in the same way (and you may sometimes see this) but, since chemists use positive and negative signs to mean specific charges, this could get confusing. Instead, one half of the p orbital is usually shaded to show that it has a different phase from the other half.

here the different phases of the p orbital are labelled positive and negative – this can be confusing and so is best avoided   here the different phases of the p orbital are shown by shading one half and not the other

The magnetic quantum number, m_ℓ

The magnetic quantum number, m_ℓ, determines the spatial orientation of the orbital's angular momentum and takes the values $-\ell$ to $+\ell$. An s orbital ($\ell = 0$), being spherical, can only have one orientation in space—it does not point in any one direction and hence it only has one value for $m_\ell(0)$. However, a p orbital could point in any direction. For a p orbital ($\ell = 1$) there are three values of m_ℓ: -1, 0, and $+1$. These correspond to the p orbitals aligned along the mutually perpendicular x-, y-, and z-axes. These orbitals, designated p_x, p_y, and p_z, are all degenerate. They differ only in their spatial orientations.

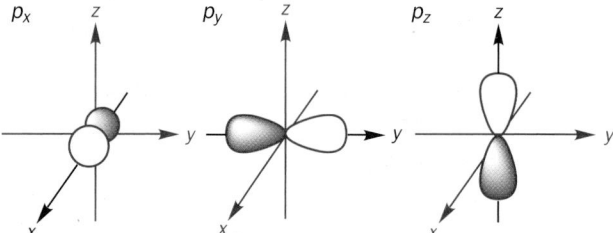

the three degenerate p orbitals are aligned along perpendicular axes

Summary so far

- Electrons in atoms are best described as waves

- All the information about the wave (and hence about the electron) is in the wave function, Ψ, the solution to the Schrödinger equation

- There are many possible solutions to the Schrödinger equation but each wave function (also called an orbital) can be described using three quantum numbers

- The principal quantum number, n, is largely responsible for the energy of the orbital (in one-electron systems, such as the hydrogen atom, it alone determines the energy). It takes integer values 1, 2, 3, 4, and so on, corresponding to the first, second, third, and so on shells of electrons

- The orbital angular momentum quantum number, ℓ, determines the angular momentum that arises from the motion of an electron moving in the orbital. Its value depends on the value of n and it takes integer values 0, . . ., $n-1$ but the orbitals are usually known by letters (s when $\ell = 0$, p when $\ell = 1$, d when $\ell = 2$, and f when $\ell = 3$). Orbitals with different values of ℓ have different shapes—s orbitals are spherical, p orbitals are shaped like a figure of eight

- The magnetic quantum number, m_ℓ, determines the spatial orientation of the orbital. Its value depends on the value of ℓ and it can take the integer values: $-\ell$, . . . 0, . . . $+\ell$. This means that there is only one type of s orbital, three different p orbitals (all mutually perpendicular), five different d orbitals, and seven different f orbitals. The three different p orbitals are all degenerate, that is, they have the same energy (as do the five d orbitals and the seven f orbitals)

- There is also a fourth quantum number, the spin angular momentum quantum number, m_s, which can take values of $+\frac{1}{2}$ or $-\frac{1}{2}$. The spin is not a property of orbitals but of the electrons that we put in the orbitals

- No two electrons in any one atom can have all four quantum numbers the same—this means that each orbital as described by the (first) three quantum numbers can hold a maximum of two electrons and then only if they have opposing spins

- We usually use a shorthand notation to describe an orbital such as 1s or $2p_y$

> ● **These three quantum numbers n, ℓ, and m_ℓ, define an orbital**
>
> The number tells us the principal quantum number, n
>
> | 2s | $3p_x$ | $4p_y$ |
> |---|---|---|
>
> the letter tells us the orbital angular momentum quantum number, ℓ — the subscript letter tells us the magnetic quantum number, m_ℓ

A few points are worth emphasis. Orbitals do not need to have electrons in them—they can be vacant (there doesn't have to be someone standing on a stair for it to exist!). So far we have mainly been talking about the hydrogen atom and this has only one electron. Most of the time this electron is in the 1s orbital (the orbital lowest in energy) but if we give it enough energy we can promote it to a vacant orbital higher in energy, say, for example, the $3p_x$ orbital.

Another point is that the electrons may be found anywhere in an orbital except in a node. In a p orbital containing one electron, this electron may be found on either side but never in the middle. When the orbital contains two electrons, one electron doesn't stay in one half and the other electron in the other half—both electrons could be anywhere (except in the node).

Finally, remember that all these orbitals are superimposed on each other. The 1s orbital is *not* the middle part of the 2s orbital. The 1s and 2s orbitals are separate orbitals in their own rights and each can hold a maximum of two electrons but the 2s orbital does occupy some of the same space as the 1s orbital (and also as the 2p orbitals, come to that). Neon, for example, has ten electrons in total: two will be in the 1s orbital, two in the 2s orbital, and two in each of the 2p orbitals. All these orbitals are superimposed on each other but the pairs of electrons are restricted to their individual orbitals. If we tried to draw all these orbitals, superimposed on each other as they are, in the same diagram the result would be a mess!

Putting electrons in orbitals

Working out where the electrons are in any atom, that is, which orbitals are populated, is easy. We simply put two electrons into the lowest energy orbital and work upwards. This 'building up' of the different atoms by putting electrons in the orbitals until they are full and then filling up the orbital next lowest in energy is known as the **Aufbau principle** (*Aufbau* is German for 'building up'). The first and only electron in the hydrogen atom must go into the 1s orbital. In this sort of diagram the energy levels are represented as horizontal lines stacked roughly in order with the lowest energy at the bottom. Electrons are represented as vertical arrows. Arrows pointing upwards show one spin ($m_s = +\frac{1}{2}$ or $-\frac{1}{2}$) and arrows pointing downwards the other (which is which doesn't matter).

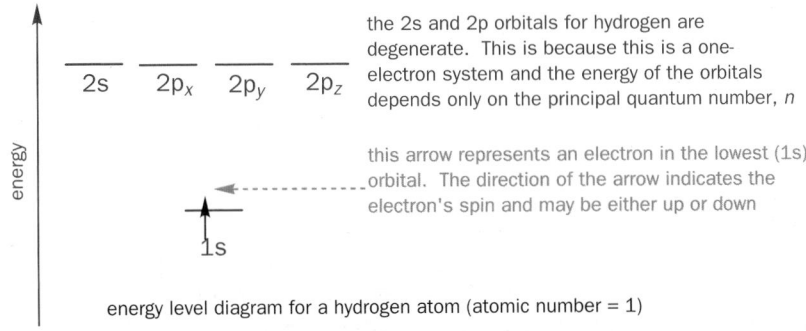

energy

2s $2p_x$ $2p_y$ $2p_z$ the 2s and 2p orbitals for hydrogen are degenerate. This is because this is a one-electron system and the energy of the orbitals depends only on the principal quantum number, n

this arrow represents an electron in the lowest (1s) orbital. The direction of the arrow indicates the electron's spin and may be either up or down

1s

energy level diagram for a hydrogen atom (atomic number = 1)

The helium atom has two electrons and they can both fit into the 1s orbital providing they have opposite spins. The other change to the diagram is that, with two electrons and electron repulsion a factor, the 2s orbital is now lower in energy than the three 2p orbitals, though these three are still degenerate.

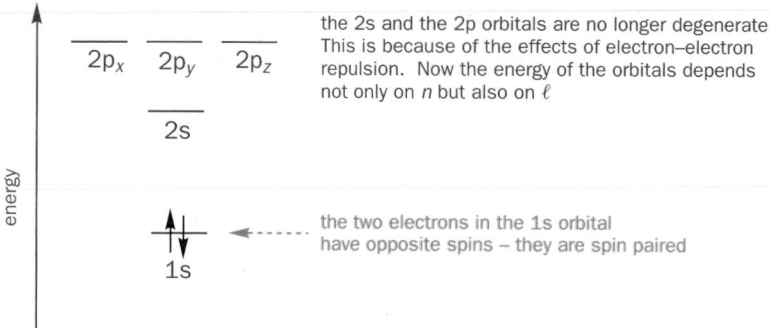

the 2s and the 2p orbitals are no longer degenerate. This is because of the effects of electron–electron repulsion. Now the energy of the orbitals depends not only on n but also on ℓ

the two electrons in the 1s orbital have opposite spins – they are spin paired

energy level diagram for a helium atom (atomic number = 2)

Lithium has one more electron but the 1s orbital is already full. The third electron must go into the next lowest orbital and that is the 2s. In this three-electron system, like that of the two-electron He atom, the three 2p orbitals are higher in energy than the 2s orbital. By the time we come to boron, with five electrons, the 2s is full as well and we must put the last electron into a 2p orbital. It doesn't matter which one; they are degenerate.

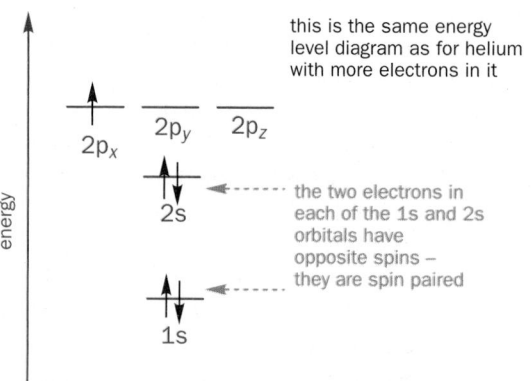

this is the same energy level diagram as for helium with more electrons in it

the two electrons in each of the 1s and 2s orbitals have opposite spins – they are spin paired

energy level diagram for a boron atom (atomic number = 5)

Carbon has one more electron than boron but now there is a bit of a problem—where does the last electron go? It could either be paired with the electron already in one of the p orbitals or it could go into one of the other degenerate p orbitals. It turns out that the system is lower in energy (electron–electron repulsion is minimized) if the electrons are placed in different degenerate orbitals with their spins parallel (that is, both spins $+\frac{1}{2}$ or both $-\frac{1}{2}$). Another way of looking at this is that putting two electrons into the same orbital with their spins paired (that is, one $+\frac{1}{2}$, one $-\frac{1}{2}$) requires some extra amount of energy, sometimes called **pairing energy**.

▶

This is known as **Hund's rule**. An atom adopts the electronic configuration that has the greatest number of unpaired electrons in degenerate orbitals. Whilst this is all a bit theoretical in that isolated atoms are not found very often, the same rule applies for electrons in degenerate orbitals in molecules.

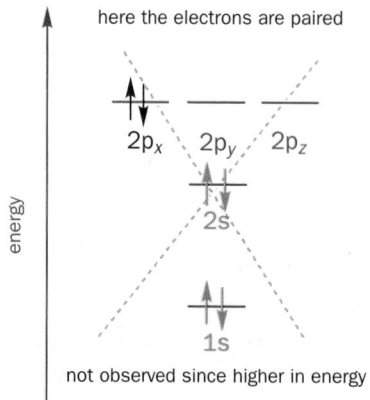

here the electrons are paired

here the electrons are in different degenerate 2p orbitals with their spins parallel

not observed since higher in energy

lower in energy and the one the carbon atom actually adopts

the two possible arrangements for the electrons in a carbon atom

Nitrogen, with one more electron than carbon, has a single electron in each of the 2p orbitals. For oxygen, the new electron pairs up with another already in one of the 2p orbitals. It doesn't enter the 3s orbital (the orbital next lowest in energy) since this is so much higher in energy and to enter the 3s orbital would require more energy than that needed to pair up with a 2p electron.

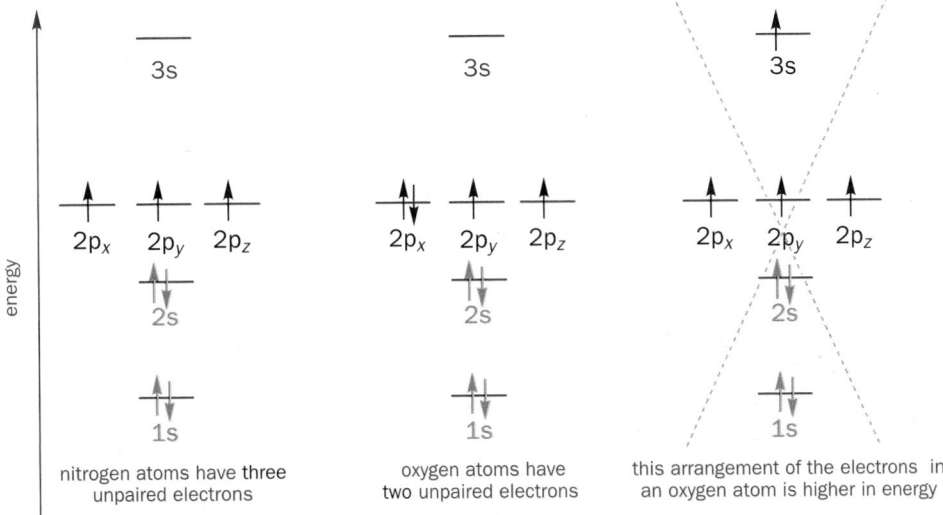

nitrogen atoms have **three** unpaired electrons

oxygen atoms have **two** unpaired electrons

this arrangement of the electrons in an oxygen atom is higher in energy

Molecular orbitals—homonuclear diatomics

So far the discussion has concerned only the shapes and energies of **atomic orbitals (AOs)**. Organic chemists really need to look at the orbitals for whole molecules. One way to construct such **molecular orbitals (MOs)** is to combine the atomic orbitals of the atoms that make up the molecule. This approach is known as the **Linear Combination of Atomic Orbitals (LCAO)**.

Atomic orbitals are wave functions and the different wave functions can be combined together rather in the way waves combine. You may be already familiar with the ideas of combining waves—they can add together constructively (in-phase) or destructively (out-of-phase).

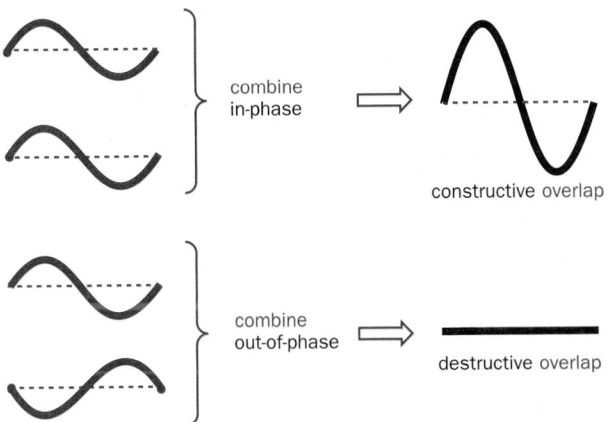

combine in-phase

constructive overlap

combine out-of-phase

destructive overlap

the two ways of combining a simple wave – in-phase and out-of-phase

Atomic orbitals can combine in the same way—in-phase or out-of-phase. Using two 1s orbitals drawn as circles (representing spheres) with dots to mark the nuclei and shading to represent phase, we can combine them in-phase, that is, add them together, or out-of-phase when they cancel each other out in a nodal plane down the centre between the two nuclei. The resulting orbitals belong to both atoms—they are molecular rather than atomic orbitals. As usual, the higher energy orbital is at the top.

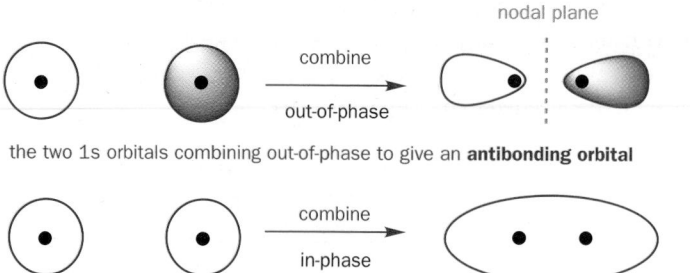

the two 1s orbitals combining out-of-phase to give an **antibonding orbital**

the two 1s orbitals combining in-phase to give a **bonding orbital**

When the two orbitals combine out-of-phase, the resulting molecular orbital has a nodal plane between the two nuclei. This means that if we were to put electrons into this orbital there would be no electron density in between the two nuclei. By contrast, if the molecular orbital from in-phase combination contained electrons, they would be found in between the two nuclei. Two exposed nuclei repel each other as both are positively charged. Any electron density between them helps to bond them together. So the in-phase combination is a **bonding molecular orbital**. As for the electrons themselves, they can now be shared between two nuclei and this lowers their energy relative to the 1s atomic orbital. Electrons in the orbital from the out-of-phase combination do not help bond the two nuclei together; in fact, they hinder the bonding. When this orbital is occupied, the electrons are mainly to be found anywhere *but* between the two nuclei. This means the two nuclei are more exposed to each other and so repel each other. This orbital is known as an **antibonding molecular orbital** and is higher in energy than the 1s orbitals.

The combination of the atomic 1s orbitals to give the two new molecular orbitals is simply shown on an energy level diagram. With one electron in each 1s orbital, two hydrogen atoms combine to give a hydrogen molecule.

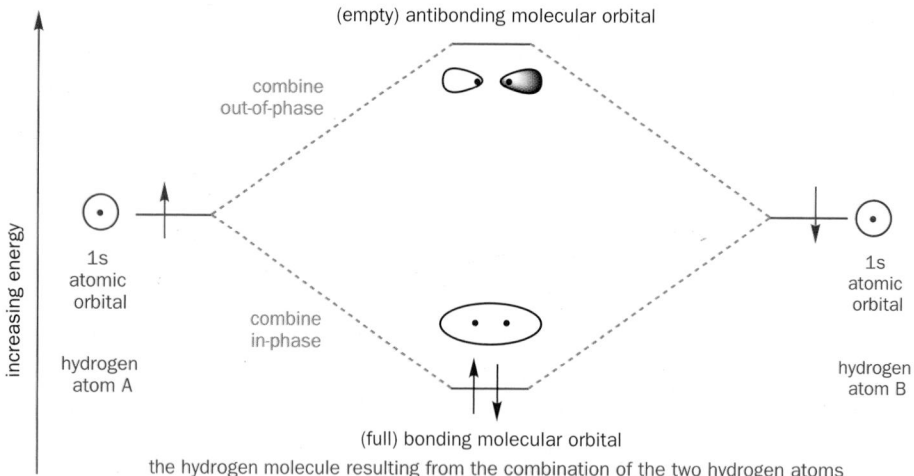

the hydrogen molecule resulting from the combination of the two hydrogen atoms

There are several points to notice about this diagram.

- *Two* atomic orbitals (AOs) combine to give *two* molecular orbitals (MOs)
- By LCAO we add the two AOs to make the bonding orbital and subtract them to make the antibonding orbital
- Since the two atoms are the same, each AO contributes the same amount to the MOs
- The bonding MO is *lower* in energy than the AOs
- The antibonding MO is *higher* in energy than the AOs
- Each hydrogen atom initially had one electron. The spin of these electrons is unimportant

- The two electrons end up in the MO lowest in energy. This is the bonding MO
- Just as with AOs, each MO can hold two electrons *as long as the electrons are spin paired*
- The two electrons between the two nuclei in the bonding MO hold the molecule together—they are the chemical bond
- Since these two electrons are lower in energy in the MO than in the AOs, energy is given out when the atoms combine
- Or, if you prefer, we must put in energy to separate the two atoms again and to break the bond

From now on, we will always represent molecular orbitals in energy order—the highest-energy MO at the top (usually an antibonding MO) and the lowest in energy (usually a bonding MO and the one in which the electrons are most stable) at the bottom. We suggest you do the same.

When we were looking at the electronic configuration of atoms, we simply filled up the atomic orbitals starting from the lowest in energy and worked up. With molecules we do the same: we just fill up the molecular orbitals with however many electrons we have, starting from the lowest in energy and remembering that each orbital can hold two electrons and then only if they are spin paired.

Breaking bonds

If an atom is supplied with energy, an electron can be promoted to a higher energy level and it can then fall back down to its ground state, giving that energy out again. What would happen if an electron were promoted in a hydrogen molecule from the lowest energy level, the bonding MO, to the next lowest energy level, the antibonding MO? Again, an energy level diagram helps.

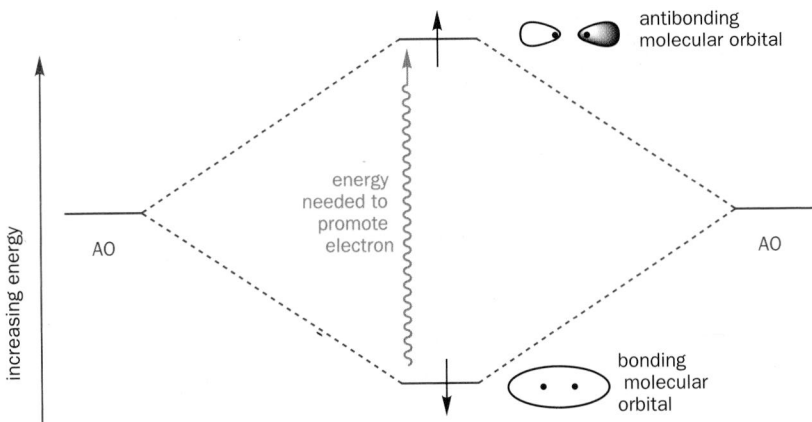

we can supply energy to promote an electron from the bonding MO to the antibonding MO

Now the electron in the antibonding orbital 'cancels out' the bonding of the electron in the bonding orbital. Since there is no overall bonding holding the two atoms together, they can drift apart as two separate atoms with their electrons in 1s atomic orbitals. In other words, promoting an electron from the bonding MO to the antibonding MO breaks the chemical bond. This is difficult to do with hydrogen molecules but easy with, say, bromine molecules. Shining light on Br_2 causes it to break up into bromine atoms.

■
This idea will be developed in Chapters 5 and 6 when we look at bond-breaking steps in organic reaction mechanisms.

Bonding in other elements: helium

A hydrogen molecule is held together by a single chemical bond since the pair of electrons in the bonding orbital constitutes this single bond. What would the MO energy level diagram for He_2 look like? Each helium atom has two electrons ($1s^2$) so now both the bonding MO and the antibonding MO are full. Any bonding due to the electrons in the bonding orbital is cancelled out by the electrons in the antibonding orbital.

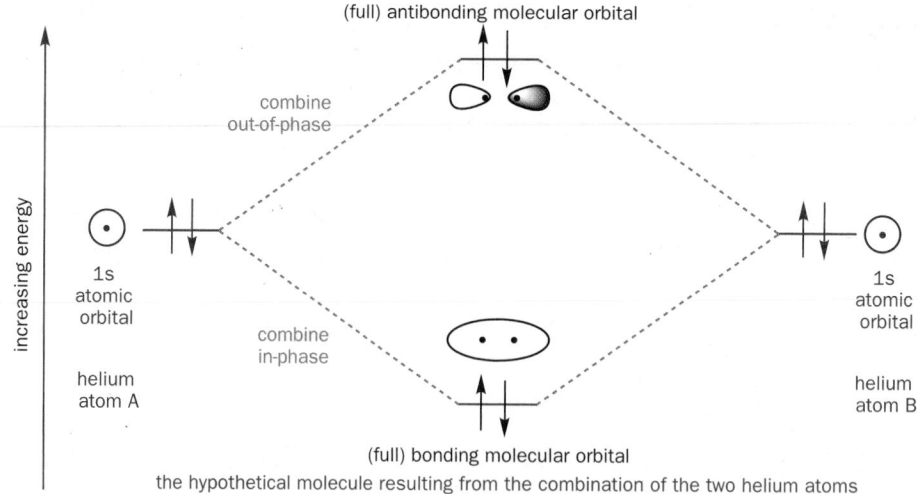

the hypothetical molecule resulting from the combination of the two helium atoms

There is no overall bonding, the two helium atoms are not held together, and He$_2$ does not exist. Only if there are more electrons in bonding MOs than in antibonding MOs will there be any bonding between two atoms. In fact, we define the number of bonds between two atoms as the **bond order** (dividing by two since *two* electrons make up a chemical bond).

$$\text{bond order} = \frac{(\text{no. of electrons in bonding MOs}) - (\text{no. of electrons in antibonding MOs})}{2}$$

Hence the bond orders for H$_2$ and He$_2$ are

$$\text{bond order (H}_2) = \frac{2-0}{2} = 1 \quad \text{i.e. a single bond}$$

$$\text{bond order (He}_2) = \frac{2-2}{2} = 0 \quad \text{i.e. no bond}$$

Bond formation using 2s and 2p atomic orbitals

So far we have been looking at how we can combine the 1s atomic orbitals to give the molecular orbitals of simple molecules. However, just as there are lots of higher, vacant energy levels in atoms, so there are in molecules too. Other atomic orbitals combine to give new molecular orbitals and the 2s and 2p orbitals concern organic chemistry most of all. The 2s AOs combine in exactly the same way as the 1s orbitals do and also give rise to a bonding and an antibonding orbital. With p orbitals as well, there are more possibilities.

Since we are beginning to talk about lots of different MOs, we shall need to label them with a little more thought. When s orbitals combine, the resulting MOs, both bonding and antibonding, are totally symmetrical about the axis joining the two nuclei.

■
Antibonding orbitals are designated with a * e.g. σ*, or π*

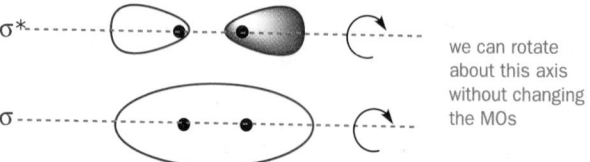

we can rotate
about this axis
without changing
the MOs

both MOs have rotational symmetry about the axis through the two nuclei

When orbitals combine in this end-on overlap to give cylindrically symmetrical MOs, the resulting orbitals are said to possess **sigma (σ) symmetry**. Hence the bonding MO is a **sigma orbital** and electrons in such an orbital give rise to a **sigma bond**. In the hydrogen molecule the two hydrogen atoms are joined by a σ bond.

What MOs result from the combination of two p orbitals? There are three mutually perpendicular p orbitals on each atom. As the two atoms approach each other, these orbitals can combine in two different ways—one p orbital from each atom can overlap end-on, but the other two p orbitals on each atom must combine side-on.

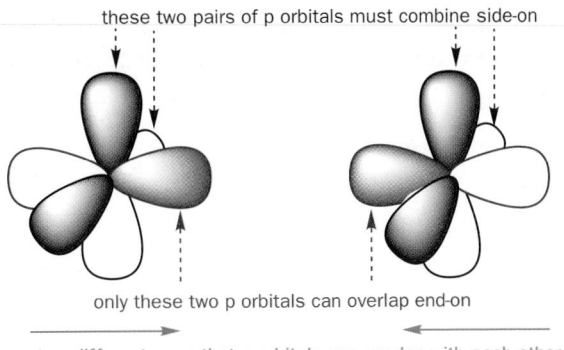

these two pairs of p orbitals must combine side-on

only these two p orbitals can overlap end-on

two different ways that p orbitals can overlap with each other

The end-on overlap (in-phase and out-of-phase) results in a pair of MOs that are cylindrically symmetrical about the inter-nuclear axis—in other words, these combinations have σ symmetry. The two molecular orbitals resulting from the end-on combination of two 2p orbitals are labelled the 2pσ and the 2pσ* MOs.

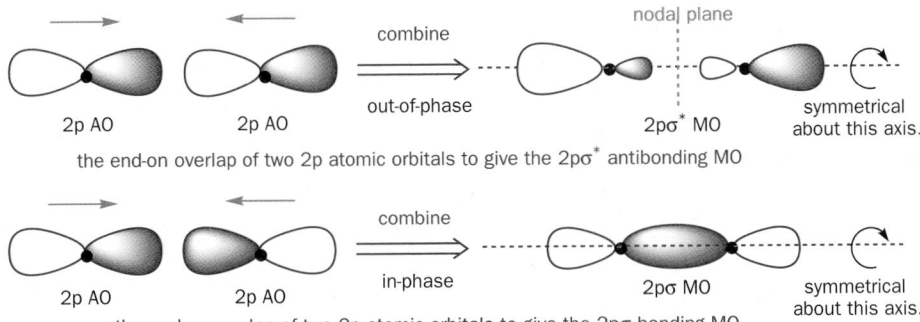

nodal plane

2p AO   2p AO   combine
                              out-of-phase   2pσ* MO   symmetrical about this axis.

the end-on overlap of two 2p atomic orbitals to give the 2pσ* antibonding MO

2p AO   2p AO   combine
                              in-phase   2pσ MO   symmetrical about this axis.

the end-on overlap of two 2p atomic orbitals to give the 2pσ bonding MO

The side-on overlap of two p orbitals forms an MO that is no longer symmetrical about the inter-nuclear axis. If we rotate about this axis, the phase of the orbital changes. The orbital is described as having π **symmetry**—a π **orbital** is formed and the electrons in such an orbital make up a π **bond**. Since there are *two mutually perpendicular pairs* of p orbitals that can combine in this fashion, there are a *pair* of degenerate mutually perpendicular π bonding MOs and a *pair* of degenerate mutually perpendicular π* antibonding MOs.

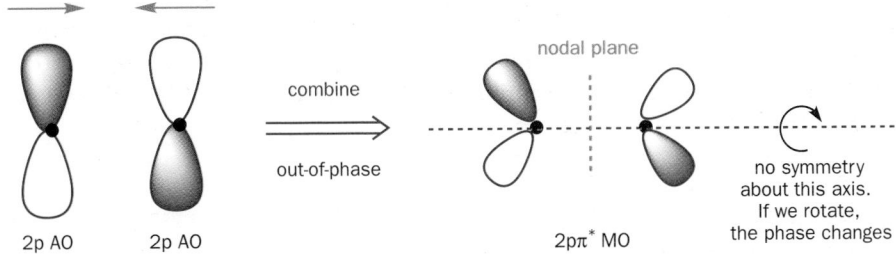

nodal plane

2p AO   2p AO   combine
                              out-of-phase   2pπ* MO   no symmetry about this axis. If we rotate, the phase changes

the side-on overlap of two 2p atomic orbitals to give the 2pπ* antibonding MO

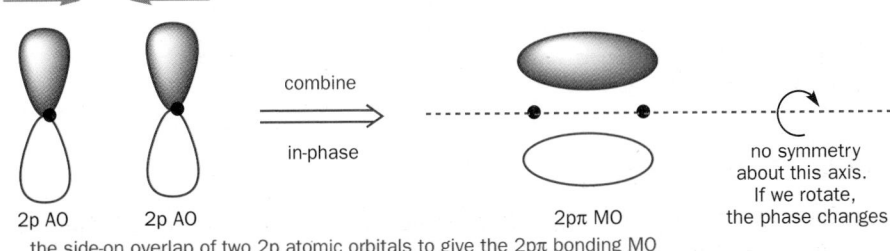

2p AO   2p AO   combine
                              in-phase   2pπ MO   no symmetry about this axis. If we rotate, the phase changes

the side-on overlap of two 2p atomic orbitals to give the 2pπ bonding MO

The two sorts of molecular orbitals arising from the combinations of the p orbitals are not degenerate—more overlap is possible when the AOs overlap end-on than when they overlap side-on. As a result, the pσ orbital is lower in energy than the pπ orbital. We can now draw an energy level diagram to show the combination of the 1s, 2s, and 2p atomic orbitals to form molecular orbitals.

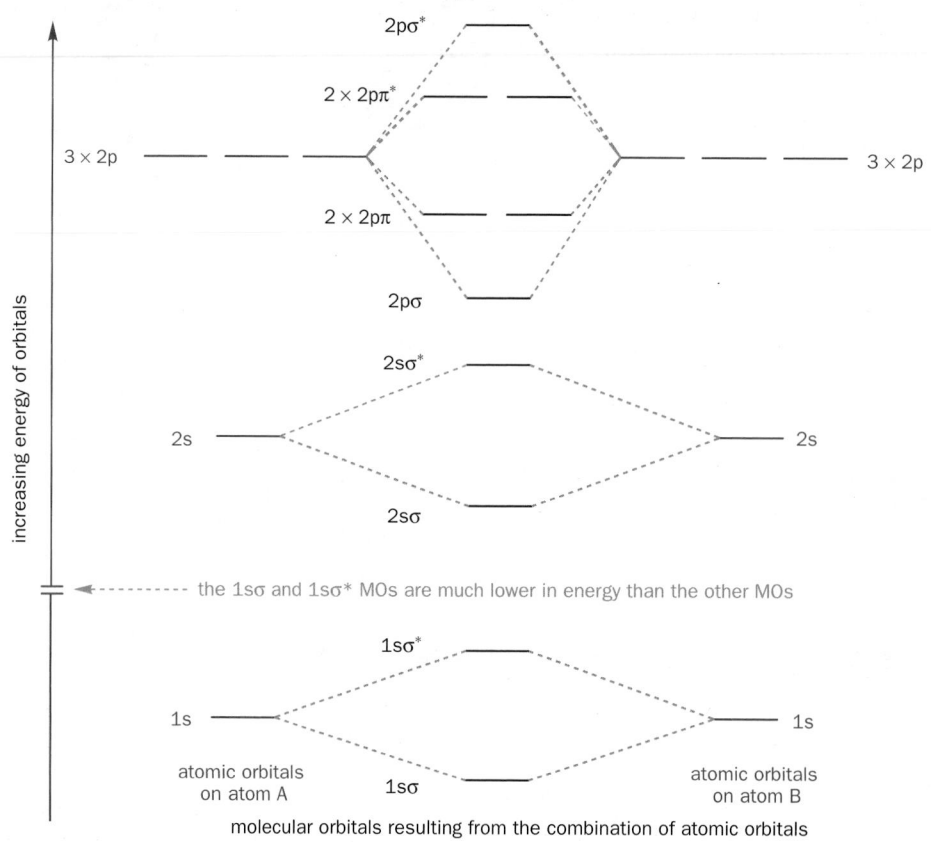

molecular orbitals resulting from the combination of atomic orbitals

Let us now look at a simple diatomic molecule—nitrogen. A nitrogen molecule is composed of two nitrogen atoms, each containing seven electrons in total. We shall omit the 1s electrons because they are so much lower in energy than the electrons in the 2s and 2p AOs and because it makes no difference in terms of bonding since the electrons in the 1sσ* cancel out the bonding due to the electrons in the 1sσ MO. The electrons in the 1s AOs and the 1s MOs are described as core electrons and so, in discussing bonding, we shall consider only the electrons in the outermost shell, in this case the 2s and 2p electrons. This means each nitrogen contributes five bonding electrons and hence the molecular orbitals must contain a total of ten electrons.

The electrons in the σ and σ* MOs formed from the 2s MOs also cancel out—these electrons effectively sit on the atoms, two on each, and form **lone pairs**—nonbonding pairs of electrons that do not contribute to bonding. All the bonding is done with the remaining six electrons. They fit neatly into a σ bond from two of the p orbitals and two π bonds from the other two pairs. Nitrogen has a triple bonded structure.

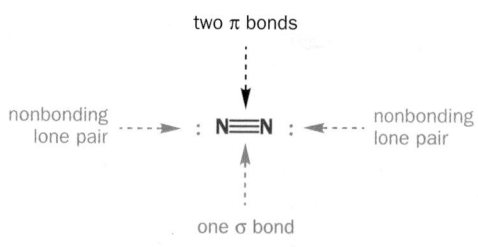

Heteronuclear diatomics

Up to now we have only considered combining two atoms of the same element to form homonuclear diatomic molecules. Now we shall consider what happens when the two atoms are different. First of all, how do the atomic orbitals of different elements differ? They have the same sorts of orbitals 1s, 2s, 2p, etc. and these orbitals will be the same shapes but the orbitals will have different energies. For

Homonuclear and heteronuclear refer to the nature of the atoms in a diatomic molecule. In a **homonuclear molecule** the atoms are the same (such as H_2, N_2, O_2, F_2) while in a **heteronuclear molecule** they are different (as in HF, CO, NO, ICl).

example, removing an electron completely from atoms of carbon, oxygen, or fluorine (that is, ionizing the atoms) requires different amounts of energy. Fluorine requires most energy, carbon least, even though in each case we are removing an electron from the same orbital, the 2p AO. The energies of the 2p orbitals must be lowest in fluorine, low in oxygen, and highest in carbon.

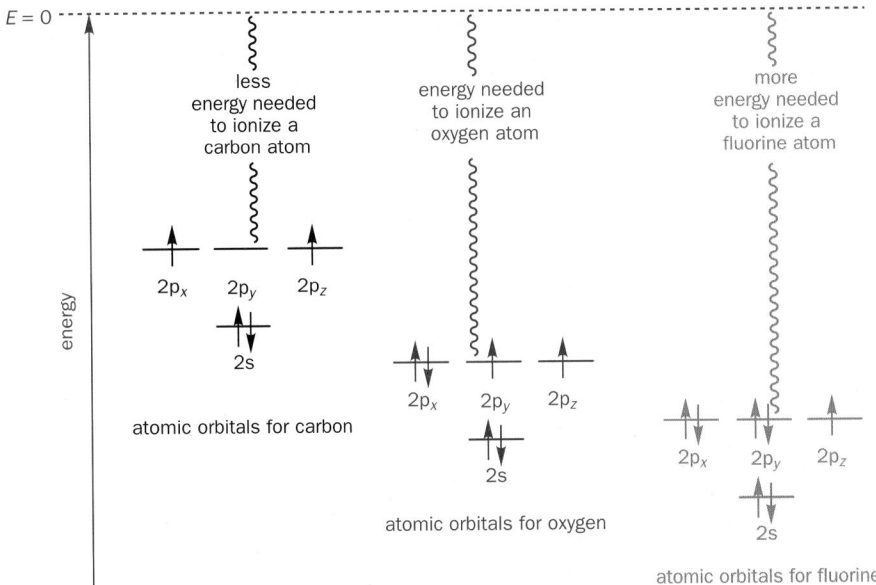

We are talking now about electronegativity. The more electronegative an atom is, the more it attracts electrons. This can be understood in terms of energies of the AOs. The more electronegative an atom is, the lower in energy are its AOs and so any electrons in them are held more tightly. This is a consequence of the increasing nuclear charge going from left to right across the periodic table. As we go from Li across to C and on to N, O, and F, the elements steadily become more electronegative and the AOs lower in energy.

So what happens if two atoms whose atomic orbitals were vastly different in energy, such as Na and F, were to combine? An electron transfers from sodium to fluorine and the product is the ionic salt, sodium fluoride, Na^+F^-.

The important point is that the atomic orbitals are too far apart in energy to combine to form new molecular orbitals and no covalent bond is formed. The ionic bonding in NaF is due simply to the attraction between two oppositely charged ions. When the atomic orbitals have exactly the same energy, they combine to form new molecular orbitals, one with an energy lower than the AOs, the other with an energy higher than the AOs. When the AOs are very different in energy, electrons are transferred from one atom to another and ionic bond-

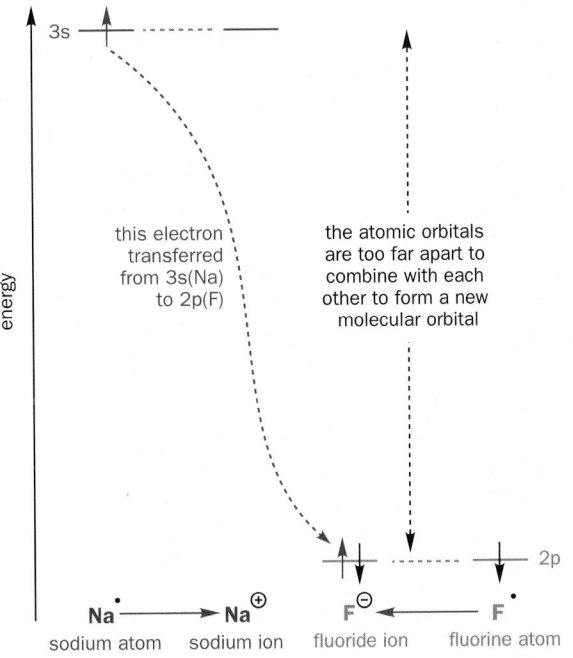

ing results. When the AOs are *slightly* different in energy, they do combine and we need now to look at this situation in more detail.

The AOs combine to form new MOs but they do so unsymmetrically. The more electronegative atom, perhaps O or F, contributes more to the bonding orbital and the less electronegative element (carbon is the one we shall usually be interested in) contributes more to the antibonding orbital. This applies both to σ bonds and to π bonds so here is an idealized case.

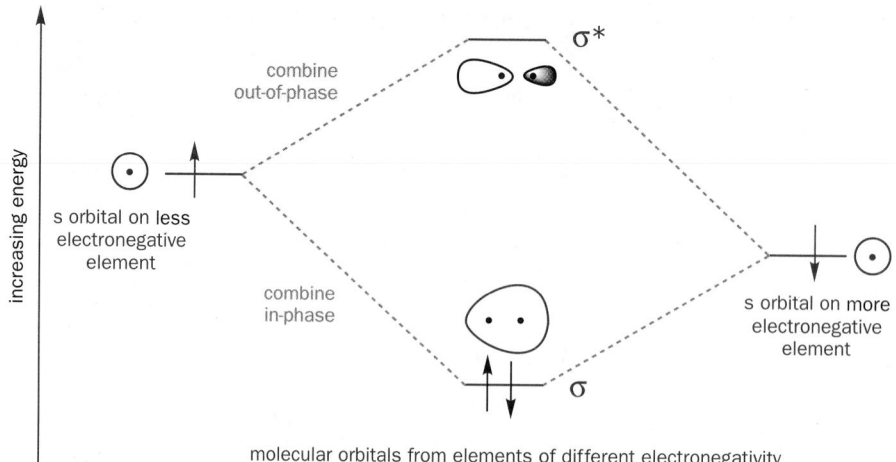

molecular orbitals from elements of different electronegativity

These three different cases where the two combining orbitals differ greatly in energy, only a little, or not at all are summarized below.

Energies of AOs both the same	**AO on atom B is a *little* lower in energy than AO on atom A**	**AO on atom B is a *lot* lower in energy than AO on atom A**
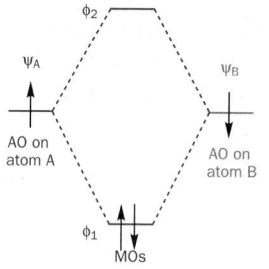	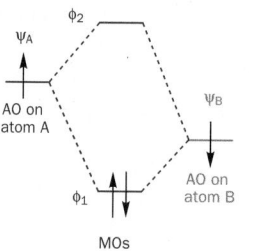	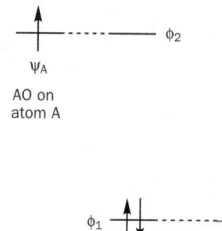
large interaction between AOs	less interaction between AOs	AOs are too far apart in energy to interact
bonding MO much lower in energy than AOs	bonding MO is lowered only by a small amount relative to AO on atom B	the filled orbital on the anion has the same energy as the AO on atom B
antibonding MO is much higher in the energy than the AOs	antibonding MO is raised in energy by only a small amount relative to AO on atom B	the empty orbital on the cation has same energy as the AO on atom A
both AOs contribute equally to the MOs	the AO on B contributes more to the bonding MO and the AO on A	only one AO contributes to each 'MO'
electrons in bonding MO are shared equally between the two atoms	electrons in bonding MO are shared between atoms but are associated more with atom B than A	electrons in the filled orbital are located only on atom B
bond between A and B would classically be described as purely covalent	bond between A and B is covalent but there is also some electrostatic (ionic) attraction between atoms	bond between A and B would classically be described as purely ionic
easiest to break bond into two radicals (homolytic fission). Heterolytic fission of bond is possible and could give either A^+ and B^- or A^- and B^+	easiest to break bond into two ions, A^+ and B^-, although it is also possible to give two radicals	compound already exists as ions A^+ and B^-

> Homolytic and heterolytic refer to the fate of the electrons when a bond is broken. In **homolytic fission** one electron goes to each atom. In **heterolytic fission** both electrons go to the same atom. Thus I_2 easily gives two iodine atoms by homolytic fission ($I_2 \rightarrow 2I^·$) while HI prefers heterolytic fission ($HI \rightarrow H^+ + I^-$). The dot in $I^·$ means a single unpaired electron.

As an example of atomic orbitals of equal and unequal energies combining, let us consider the π bonds resulting from two carbon atoms combining and from a carbon atom combining with an oxygen atom. With the C–C π bond, both p orbitals have the same energy and combine to form a symmetrical π bond. If the bonding MO (π) is occupied, the electrons are shared equally over both carbon atoms. Compare this with the π bond that results from combining an oxygen p AO with a carbon p AO.

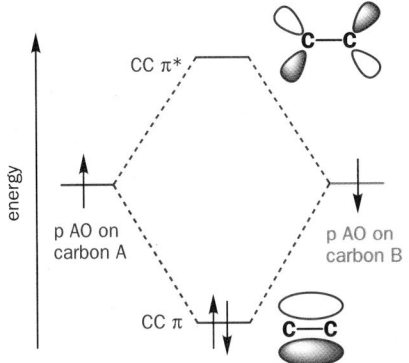

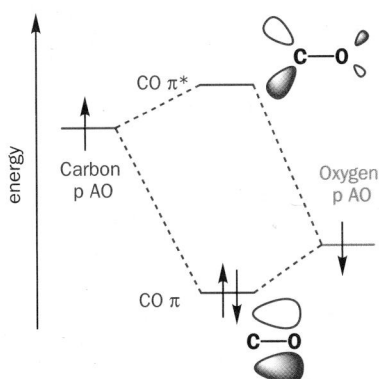

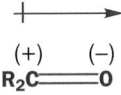

Now the bonding MO (π) is made up with a greater contribution from the oxygen p orbital than from the carbon p orbital. If this MO contained electrons, there would be more electrons around the oxygen atom than around the carbon. This C–O π bond is covalent but there is also some electrostatic contribution to its bond strength. This electrostatic interaction actually makes a C–O double bond much stronger than a C–C double bond (bond strength for C=O, about 725–60 kJ mol^{-1}; for C=C, 600–25 kJ mol^{-1}: compare also a C–O single bond, 350–80 kJ mol^{-1} with a C–C single bond, 340–50 kJ mol^{-1}). Because the electrons in the populated MO (π) are associated more with the oxygen atom than with the carbon, it is easier to break this bond heterolytically with both electrons moving completely on to the oxygen atom than it is to break it homolytically to get a diradical with one electron moving on to the carbon and one on to the oxygen atom. This will be the first chemical reaction we study in detail in Chapters 5 and 6.

(+)   (−)
R$_2$C══════O

Other factors affecting degree of orbital interaction

Having similar energies is not the only criterion for good interaction between two atomic orbitals. It also matters how the orbitals overlap. We have seen that p orbitals overlap better in an end-on fashion (forming a σ bond) than they do side-on (forming a π bond). Another factor is the size of the atomic orbitals. For best overlap, the orbitals should be the same size—a 2p orbital overlaps much better with another 2p orbital than it does with a 3p or 4p orbital.

A third factor is the symmetry of the orbitals—two atomic orbitals must have the appropriate symmetry to combine. Thus a 2p$_x$ orbital cannot combine with a 2p$_y$ or 2p$_z$ orbital since they are all perpendicular to each other (they are **orthogonal**). In one case the two p orbitals have no overlap at all; in the other case any constructive overlap is cancelled out by equal amounts of destructive

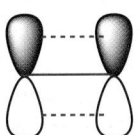

efficient overlap of p orbitals of the same size (same principal quantum number n)

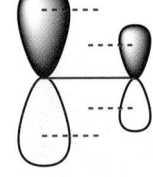

inefficient overlap of p orbitals of different size (different principal quantum numbers n)

p$_z$ and p$_x$

these two p orbitals cannot combine because they are perpendicular to each other

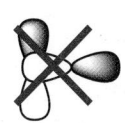

p$_z$ and p$_y$

here any constructive overlap is cancelled out by equal amounts of destructive overlap

p and s (side-on)

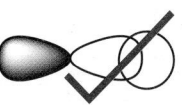

p and s (end-on)

however, s and p orbitals can overlap end-on

overlap. Likewise, an s orbital can overlap with a p orbital only end-on. Sideways overlap leads to equal amounts of bonding and antibonding interactions and no overall gain in energy.

Molecular orbitals of molecules with more than two atoms

We now need to look at ways of combining more than two atoms at a time. For some molecules, such as H_2S and PH_3, that have all bond angles equal to 90°, the bonding should be straightforward—the p orbitals (which are at 90°) on the central atom simply overlap with the 1s orbitals of the hydrogen atoms.

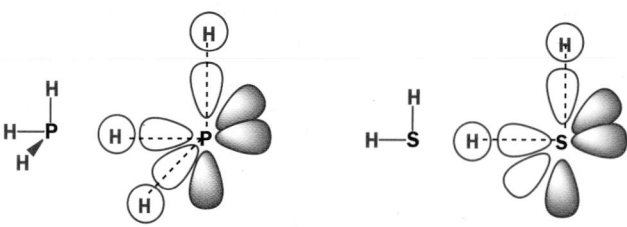

the 90° angles in PH_3 and H_2S come from the overlap of the hydrogen 1s AO with the p AO of the phosphorus or sulfur

But how do we account for the bond angles in water (104°) and ammonia (107°) when the only atomic orbitals are at 90° to each other? All the covalent compounds of elements in the row Li to Ne raise this difficulty. Water (H_2O) and ammonia (NH_3) have angles between their bonds that are roughly tetrahedral and methane (CH_4) is exactly tetrahedral but how can the atomic orbitals combine to rationalize this shape? The carbon atom has electrons only in the first and second shells, and the 1s orbital is too low in energy to contribute to any molecular orbitals, which leaves only the 2s and 2p orbitals. The problem is that the 2p orbitals are at right angles to each other and methane does not have any 90° bonds. (So don't draw any either! Remember Chapter 2.). Let us consider exactly where the atoms are in methane and see if we can combine the AOs in such a way as to make satisfactory molecular orbitals.

Methane has a tetrahedral structure with each C–H bond 109 pm and all the bond angles 109.5°. To simplify things, we shall draw a molecule of methane enclosed in a cube. It is possible to do this since the opposite corners of a cube describe a perfect tetrahedron. The carbon atom is at the centre of the cube and the four hydrogen atoms are at four of the corners.

Now, how can the carbon's 2s and 2p atomic orbitals combine with the four hydrogen 1s atomic orbitals? The carbon's 2s orbital can overlap with all four hydrogen 1s orbitals at once with all the orbitals in the same phase. In more complicated systems like this, it is clearer to use a diagram of the AOs to see what the MO will be like.

Each of the 2p orbitals points to opposite faces of the cube. Once more all four hydrogen 1s orbitals can combine with each p orbital but this time the hydrogen AOs on the opposite faces of the cube must be differently phased.

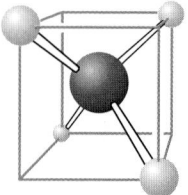

a molecule of methane enclosed in a cube

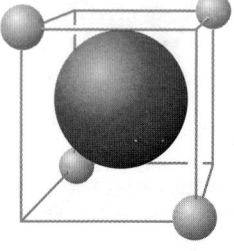

the carbon 2s AO can overlap with all four hydrogen 1s AOs at once

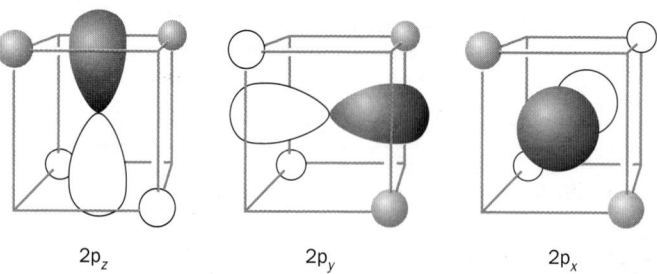

$2p_z$ $2p_y$ $2p_x$

the hydrogen 1s orbitals can overlap with the three 2p orbitals

Again we are not going to draw these three molecular orbitals but you can see from the AO diagrams what they look like. They are degenerate (that is, they have the same energy) and each orbital has one nodal plane (it is easiest to see in the middle diagram passing vertically down the middle of the cube and dividing shaded orbitals on the right from unshaded orbitals on the left). Only the bonding overlap between the AOs is shown but of course there is an antibonding interaction for every bonding interaction, which means there are eight MOs altogether (which is correct since there were eight AOs to start with).

Organic chemists can just about understand this 'correct' MO picture of methane and theoretical chemists are able to construct correct MOs for very much more complex molecules than methane. There is experimental evidence too that these pictures are correct. Other experiments reveal that all four C–H bonds in methane are exactly the same and yet the MOs for methane are not all the same. *There is no contradiction here!* The molecular orbital approach tells us that there is one MO of one kind and three of another but the electrons in them are shared out over all five atoms. No one hydrogen atom has more or less electrons than any other—they are all equivalent. Techniques that tell us the structure of methane do not tell us where bonds are; they simply tell us where the atoms are located in space—*we* draw in bonds connecting atoms together. Certainly the *atoms* form a regular tetrahedron but exactly where the electrons are is a different matter entirely. The classical picture of two atoms held together by a pair of electrons is not necessarily correct—the five atoms in methane are held together by electrons but these are in molecular orbitals, which spread over all the atoms. We are going to need the classical picture when we draw mechanisms. Methane only has one carbon atom—imagine what it would be like with larger compounds that can contain hundreds of carbon atoms! Fortunately, there is another, simpler method we can use to describe bonding that preserves the important points from this theory.

Hybridization of atomic orbitals

For most of organic chemistry, it is helpful to consider the molecule as being made up of atoms held together by bonds consisting of a pair of electrons. When working out the MOs for methane, we used the carbon 2s and all three of the 2p orbitals to combine with the hydrogen 1s orbitals. Each orbital combined with all the hydrogen orbitals equally. Another way to consider the bonding would be to combine the carbon 2s and 2p orbitals first to make four new orbitals. Each of these orbitals would be exactly the same and be composed of one-quarter of the 2s orbital and three-quarters of one of the p orbitals. The new orbitals are called sp³ hybrid orbitals to show the proportions of the AOs in each. This process of mixing is called **hybridization**.

Combining four atomic orbitals on the same atom gives the same total number of hybrid orbitals. Each of these has one-quarter s character and three-quarters p character. The sp³ orbital has a planar node through the nucleus like a p orbital but one lobe is larger than the other because of the extra contribution of the 2s orbital, which adds to one lobe but subtracts from the other.

The four sp³ orbitals on one carbon atom point to the corners of a tetrahedron and methane can be formed by overlapping the large lobe of each sp³ orbital with

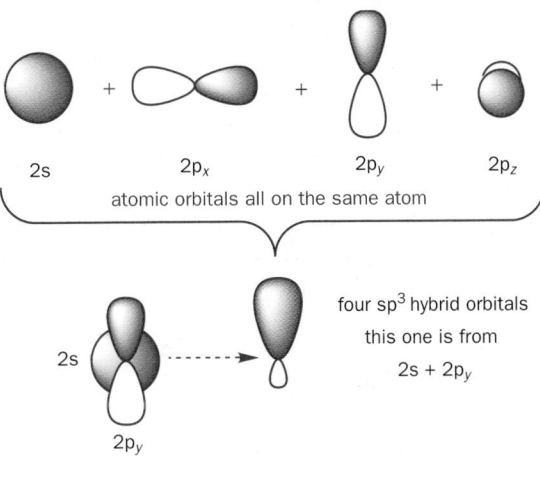

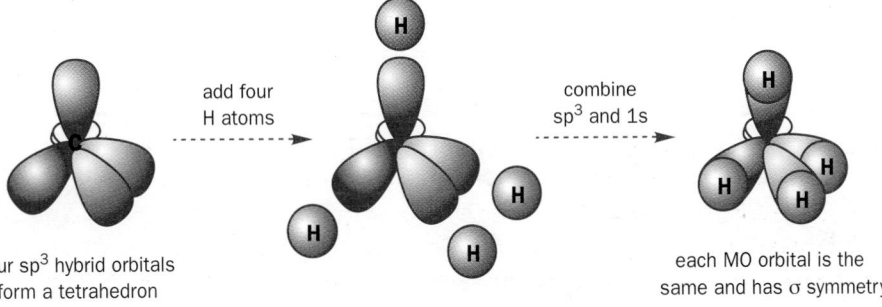

four sp³ hybrid orbitals form a tetrahedron

add four H atoms

combine sp³ and 1s

each MO orbital is the same and has σ symmetry

the 1s orbital of a hydrogen atom. Each overlap forms an MO (2sp³ + 1s) and we can put two electrons in each to form a C–H σ bond. There will of course also be an antibonding MO, σ* (2sp³ – 1s) in each case, but these orbitals are empty.

The great advantage of this method is that it can be used to build up structures of much larger molecules quickly and without having to imagine that the molecule is made up from isolated atoms. So it is easy to work out the structure of ethene (ethylene) the simplest alkene. Ethene is a planar molecule with bond angles close to 120°. Our approach will be to hybridize all the orbitals needed for the C–H framework and see what is left over. In this case we need three bonds from each carbon atom (one to make a C–C bond and two to make C–H bonds).

Therefore we need to combine the 2s orbital on each carbon atom with two p orbitals to make the three bonds. We could hybridize the 2s, $2p_x$, and $2p_y$ orbitals (that is, all the AOs in the plane) to form three equal sp² hybrid atomic orbitals, leaving the $2p_z$ orbital unchanged. These sp² hybrid orbitals will have one-third s character and only two-thirds p character.

The three sp² hybrid atomic orbitals on each carbon atom can overlap with three other orbitals (two hydrogen 1s AOs and one sp² AO from the other carbon) to form three σ MOs. This leaves the two $2p_z$ orbitals, one on each carbon, which combine to form the π MO. The skeleton of the molecule has five σ bonds (one C–C and four C–H) in the plane and the central π bond is formed by two $2p_z$ orbitals above and below the plane.

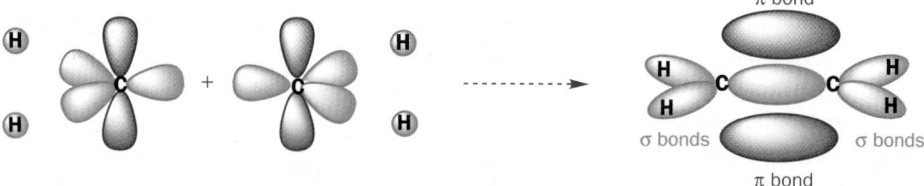

Ethyne (acetylene) has a C–C triple bond. Each carbon bonds to only two other atoms to form a linear CH skeleton. Only the carbon 2s and $2p_x$ have the right symmetry to bind to only two atoms at once so we can hybridize these to form two sp hybrids on each carbon atom leaving the $2p_y$ and $2p_z$ to form π MOs with the 2p orbitals on the other carbon atom. These sp hybrids have 50% each s and p character and form a linear carbon skeleton.

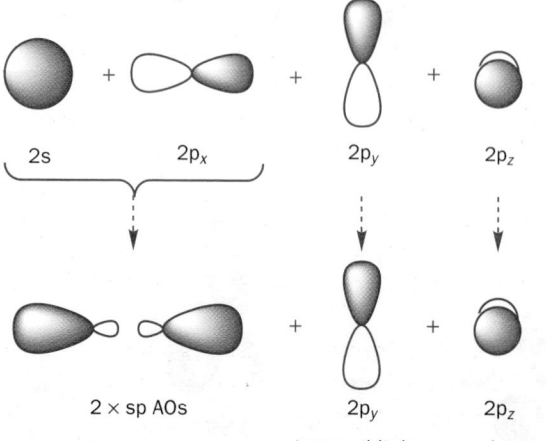

We could then form the MOs as shown below. Each sp hybrid AO overlaps with either a hydrogen 1s AO or with the sp orbital from the other carbon. The two sets of p orbitals combine to give two mutually perpendicular π MOs.

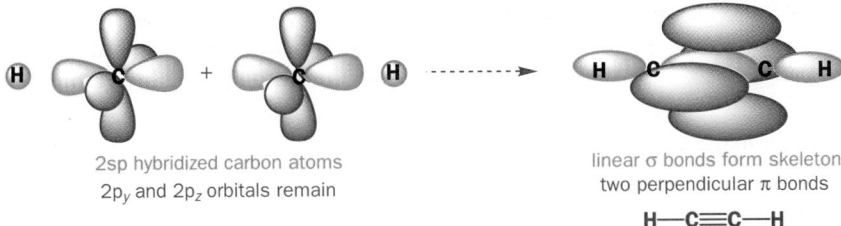

2sp hybridized carbon atoms
$2p_y$ and $2p_z$ orbitals remain

linear σ bonds form skeleton
two perpendicular π bonds

H—C≡C—H

Hydrocarbon skeletons are built up from tetrahedral (sp^3), trigonal planar (sp^2), or linear (sp) hybridized carbon atoms. It is not necessary for you to go through the hybridization process each time you want to work out the shape of a skeleton. In real life molecules are not made from their constituent atoms but from other molecules and it doesn't matter how complicated a molecule might be or where it comes from; it will have an easily predictable shape. All you have to do is count up the single bonds at each carbon atom. If there are two, that carbon atom is linear (sp hybridized), if there are three, that carbon atom is trigonal (sp^2 hybridized), and, if there are four, that carbon atom is tetrahedral (sp^3 hybridized).

This hydrocarbon (hex-5-en-2-yne) has two linear sp carbon atoms (C2 and C3), two trigonal sp^2 carbon atoms (C5 and C6), a tetrahedral sp^3 CH$_2$ group in the middle of the chain (C4), and a tetrahedral sp^3 methyl group (C1) at the end of the chain. We had no need to look at any AOs to deduce this—we needed only to count the bonds.

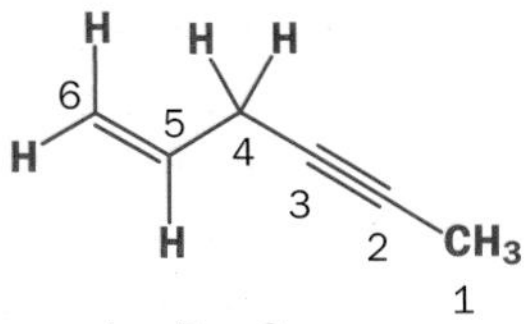

hex-5-en-2-yne

If you had drawn the molecule more professionally as shown in the margin, you would have to check that you counted up to four bonds at each carbon. Of course, if you just look at the double and triple bonds, you will get the right answer without counting single bonds at all. Carbon atoms with no π bonds are tetrahedral (sp^3 hybridized), those with one π bond are trigonal (sp^2 hybridized), and those with two π bonds are linear (sp hybridized). This is essentially the VSEPRT approach with a bit more logic behind it.

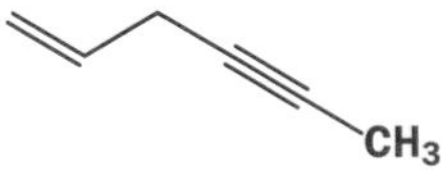

> ● All normal compounds of carbon have eight electrons in the outer shell ($n = 2$) of the carbon atom, all shared in bonds. It doesn't matter where these electrons come from; just fit them into the right MOs on sp, sp^2, or sp^3 atoms.

► Notice that atoms 1–4 are drawn in a straight line. Alkynes are linear—draw them like that!

CH$_3$

We can hybridize any atoms

Hybridization is a property of AOs rather than specifically of carbon and, since all atoms have AOs, we can hybridize any atom. A tetrahedral arrangement of atoms about any central atom can be rationalized by describing the central atom as sp^3 hybridized. The three molecules shown here all have a tetrahedral structure and in each case the central atom can be considered to be sp^3 hybridized.

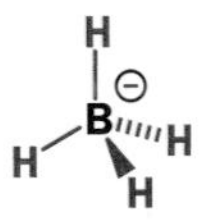

Each of these three molecules has four equivalent σ bonds from the central tetrahedral sp^3 atom, whether this is B, C, or N, and the same total number of bonding electrons—the molecules are said to be **isoelectronic**. These three elements come one after the other in the periodic table so each nucleus has one more proton than the last: B has 5, C has 6, and N has 7. This is why the charge on the central atom varies.

Compounds of the same three elements with only three bonds are more complicated. Borane, BH$_3$, has only three pairs of bonding electrons. The central boron atom bonds to only three other atoms. We can therefore describe it as being sp^2 hybridized with an empty p orbital.

Each of the B–H bonds results from the overlap of an sp^2 orbital with the hydrogen 1s orbital. The

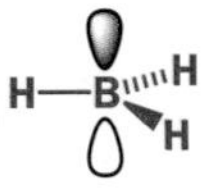

borohydride anion

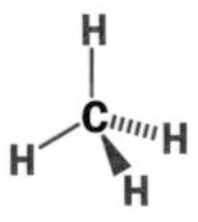

methane

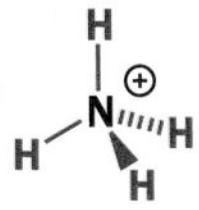

ammonium cation

vacant p orbital   trigonal borane

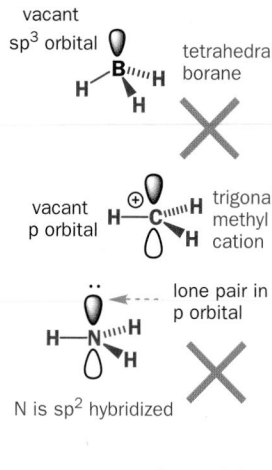

vacant sp³ orbital — tetrahedral borane

vacant p orbital — trigonal methyl cation

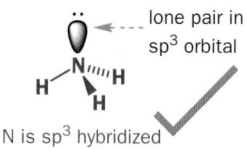

lone pair in p orbital

N is sp² hybridized

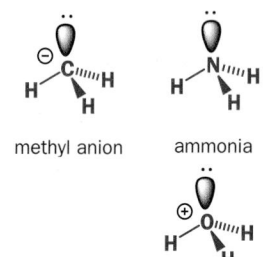

lone pair in sp³ orbital

N is sp³ hybridized

methyl anion   ammonia

hydronium ion

p orbital is not needed and contains no electrons. Do not be tempted by the alternative structure with tetrahedral boron and an empty sp³ orbital. You want to populate the lowest energy orbitals for greatest stability and sp² orbitals with their greater s character are lower in energy than sp³ orbitals. Another way to put this is that, if you have to have an empty orbital, it is better to have it of the highest possible energy since it has no electrons in it and doesn't affect the stability of the molecule.

Borane is isoelectronic with the methyl cation, CH_3^+. All the arguments we have just applied to borane also apply to Me^+ so it too is sp² hybridized with a vacant p orbital. This will be very important when we discuss the reactions of carbocations in Chapter 17.

Now what about ammonia, NH_3? Ammonia is *not* isoelectronic with borane and Me^+! As well as three N–H bonds, each with two electrons, the central nitrogen atom also has a lone pair of electrons. We have two choices: either we could hybridize the nitrogen atom sp² and put the lone pair in the p orbital or we could hybridize the nitrogen sp³ and have the lone pair in an sp³ orbital.

This is the opposite of the situation with borane and Me^+. The extra pair of electrons *does* contribute to the energy of ammonia so it should be in the lower-energy orbital, sp³, rather than pure p. Experimentally the H–N–H bond angles are all 107.3°. Clearly, this is much closer to the 109.5° sp³ angle than to the 120° sp² angle. But the bond angles are not exactly 109.5°, so ammonia cannot be described as pure sp³ hybridized. VSEPRT says the lone pair repels the bonds more than they repel each other. Alternatively, you could say that the orbital containing the lone pair must have slightly more s character while the N–H bonding orbitals must have correspondingly more p character.

The methyl anion, CH_3^-, and hydronium ion, H_3O^+, are both isoelectronic with ammonia so that all share the same pyramidal structure. Each is approximately tetrahedral with a lone pair in an sp³ orbital. These elements follow each other in the periodic table so the change in charge occurs because each nucleus has one more proton than the last. VSEPRT also gives this answer.

Shape of phosphine

Phosphine, PH_3, has bond angles of about 90° and there is no need for hybridization. The three H 1s AOs can overlap with the three 3p orbitals of the phosphorus atom, which leaves the lone pair in the 3s orbital. This 'pure s' lone pair is less energetic and therefore less reactive than the sp³ lone pair in ammonia which explains why ammonia is more basic than phosphine (see Chapter 8). In general atoms from Na to Ar are less likely to be hybridized than those from Li to Ne because the longer bonds mean the substituents are further from the central atom and steric interaction is less. VSEPRT does *not* give this answer.

Double bonds to other elements

The C=O double bond is the most important functional group in organic chemistry. It is present in aldehydes, ketones, acids, esters, amides, and so on. We shall spend Chapters 5–10 discussing its chemistry so it is important that you understand its electronic structure. As in alkenes, the two atoms that make up this double bond are sp² hybridized. The carbon atom uses all three sp² orbitals for overlap with other orbitals to form σ bonds, but the oxygen uses only one for overlap with another orbital (the sp² orbitals on the carbon atom) to form a σ bond. However, the other two sp² orbitals are not vacant—they contain the oxygen's two lone pairs. A p orbital from the carbon and one from the oxygen make up the π bond which also contains two electrons.

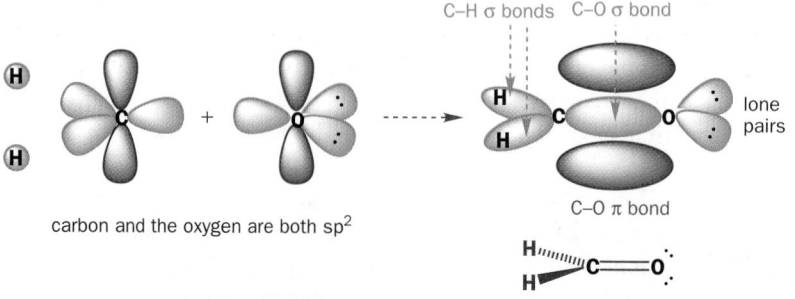

C–H σ bonds   C–O σ bond

lone pairs

C–O π bond

carbon and the oxygen are both sp²

The less important double bonds to nitrogen (imines) are very similar but now there is only one lone pair on nitrogen and a second σ bond to whatever substituent is on the nitrogen atom. Looking down on the planar structures of alkenes, imines, and ketones we see only the ends of the p orbitals but the rest of the structures are clearly related.

Alkenes have a planar trigonal framework of sp² carbon atoms. Each uses one sp² orbital to form a σ bond to the other carbon atom and two sp² orbitals to form σ bonds to the substituents (here the general 'R'). Two carbon p orbitals are used for a C–C π bond. There are no lone pairs of electrons on either carbon atom.

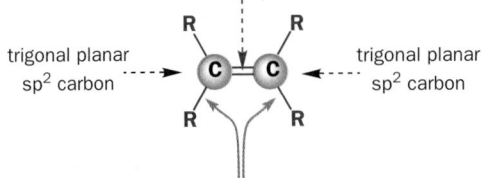

Imines have a planar trigonal framework of an sp² carbon atom and an sp² nitrogen atom. Each uses one sp² orbital to form a σ bond to the other atom and a p orbital to form a π bond to the other atom. The carbon uses two sp² orbitals and the nitrogen one to form σ bonds to the substituents (here the general 'R'). There is one lone pair of electrons on the nitrogen atom.

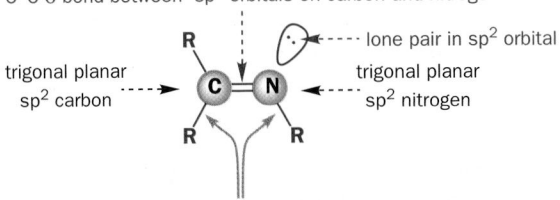

Carbonyl compounds have a planar trigonal framework of an sp² carbon atom and an sp² oxygen atom. Each uses one sp² orbital to form a σ bond to the other atom and a p orbital to form a π bond to the other atom. The carbon uses two sp² orbitals to form σ bonds to the substituents (here the general 'R'). There are two lone pairs of electrons on the oxygen atom.

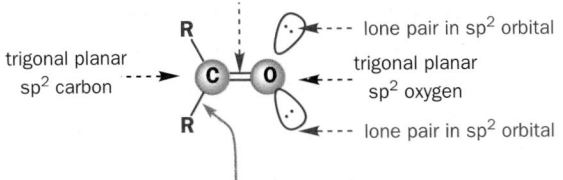

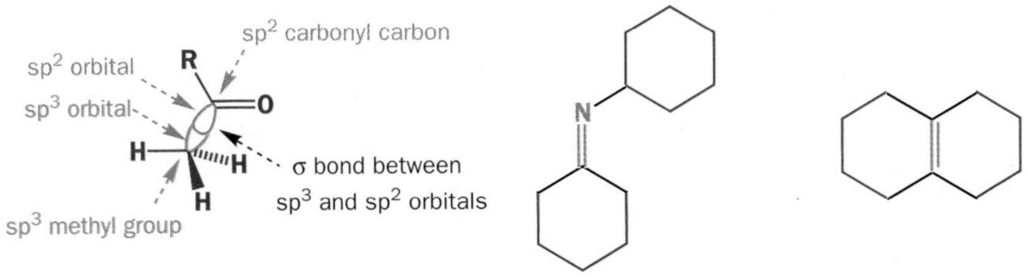

Where 'R' is joined to the double bond through a carbon atom, the nature of R determines which orbital will be used to pair up with the sp² orbital. In all the compounds shown below a saturated carbon atom with four bonds is joined to the double bond. The C–C single bond is a σ bond between an sp^2 orbital on the ketone, imine, or alkene and an sp^3 orbital on the substituent. It doesn't make any difference that the second two compounds contain rings. In all cases the black bond joins a saturated, tetrahedral, sp^3 carbon atom to the double bond and all the black σ bonds are between sp^2 and sp^3 carbons or nitrogens.

All the other combinations are possible—here are just a few. It should be clear by now that σ bonds can form between any sort of orbitals that can point towards each other but that π bonds can form only between p orbitals.

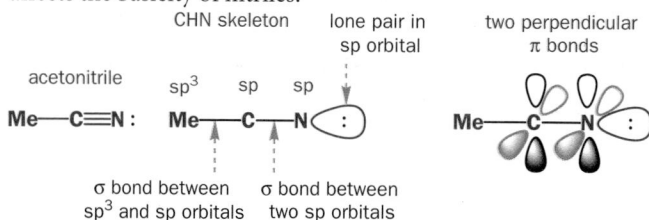

σ bond between   σ bond between   σ bond between
two sp^2 orbitals   1s and sp^2 orbitals   sp and sp^2 orbitals

Triple bonds can be formed between carbon and other elements too. The most important is the CN triple bond present in cyanides or nitriles. Both C and N are sp hybridized in these linear molecules, which leaves the lone pair on nitrogen in an sp orbital too. You will see (Chapter 8) how this affects the basicity of nitriles.

> ● **All normal compounds of nitrogen have eight electrons in the outer shell ($n = 2$) of the nitrogen atom, six shared in bonds and two in a lone pair. All normal compounds of oxygen have eight electrons in the outer shell ($n = 2$) of the oxygen atom, four shared in bonds and four in lone pairs. It doesn't matter where these electrons come from; just fit them into the right MOs on sp, sp^2, or sp^3 atoms.**

Conclusion

We have barely touched the enormous variety of molecules, but it is important that you realize at this point that these simple ideas of structural assembly can be applied to the most complicated molecules known. We shall use AOs and combine them into MOs to solve the structure of very small molecules and to deduce the structures of small parts of much larger molecules. With the additional ideas in Chapter 7 (conjugation) you will be able to grasp the structures of any organic compound. From now on we shall use terms like AO and MO, 2p orbital, sp^2 hybridization, σ bond, energy level, and populated orbital without further explanation. If you are unsure about any of them, refer back to this chapter for an explanation.

Problems

1. In the (notional and best avoided in practice) formation of NaCl from a sodium atom and a chlorine atom, descriptions like this abound in textbooks: 'an electron is transferred from the valency shell of the sodium atom to the valency shell of the chlorine atom'. What is meant, in quantum number terms, by 'valency shell'? Give a complete description in terms of all four quantum numbers of that transferred electron: (**a**) while it is in the sodium atom and (**b**) after it has been transferred to the chlorine atom. Why is the formation of NaCl by this process to be discouraged?

2. What is the electronic structure of these species? You should consult a periodic table before answering.

H$^{\ominus}$ HS$^{\ominus}$ K$^{\oplus}$ Xe

3. What sort of bonds can be formed between s orbitals and p orbitals? Which will provide better overlap, 1s + 2p or 1s + 3p? Which bonds will be stronger, those between hydrogen and C, N, O, and F on the one hand or those between hydrogen and Si, P, S, and Cl on the other? Within the first group, bond strength goes in this order: HF > OH > NH > CH. Why?

4. Though no helium 'molecule' He$_2$ exists, an ion He$_2^+$ does exist. Explain.

5. You may be surprised to know that the molecule CH$_2$, with divalent carbon, can exist. It is of course very unstable but it is known and it can have two different structures. One has an H–C–H bond angle of 180° and the other an angle of 120°. Suggest structures for these species and say which orbitals will be occupied by all bonding and nonbonding electrons. Which structure is likely to be more stable?

6. Construct an MO diagram for the molecule LiH and suggest what type of bond it might have.

7. Deduce the MOs for the oxygen molecule. What is the bond order in oxygen and where are the 2p electrons?

8. Construct MOs for acetylene (ethyne) without hybridization.

9. What is the shape and hybridization of each carbon atom in these molecules?

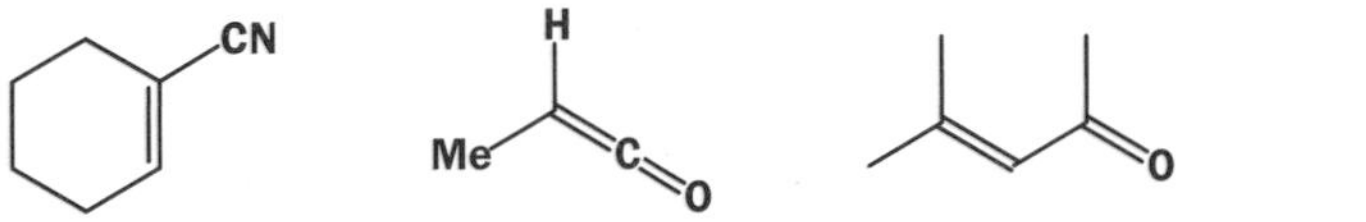

10. Suggest detailed structures for these molecules and predict their shapes. We have deliberately made noncommittal drawings to avoid giving away the answer to the question. Don't use these sorts of drawing in your answer.

CO$_2$, CH$_2$=NCH$_3$, CHF$_3$, CH$_2$=C=CH$_2$, (CH$_2$)$_2$O

Organic reactions

5

Connections

Building on:	Arriving at:	Looking forward to:
• Drawing molecules realistically ch2	• Why molecules generally *don't* react with each other!	• The rest of the chapters in this book
• Ascertaining molecular structure spectroscopically ch3	• Why sometimes molecules *do* react with each other	
• What determines molecular shape and structure ch4	• In chemical reactions electrons move from full to empty orbitals	
	• Molecular shape and structure determine reactivity	
	• Representing the movement of electrons in reactions by curly arrows	

Chemical reactions

Most molecules are at peace with themselves. Bottles of water, or acetone (propanone, $Me_2C=O$), or methyl iodide (iodomethane CH_3I) can be stored for years without any change in the chemical composition of the molecules inside. Yet when we add chemical reagents, say, HCl to water, sodium cyanide (NaCN) to acetone, or sodium hydroxide to methyl iodide, chemical reactions occur. This chapter is an introduction to the reactivity of organic molecules: why they don't and why they do react; how we can understand reactivity in terms of charges and orbitals and the movement of electrons; how we can represent the detailed movement of electrons—the mechanism of the reaction—by a special device called the curly arrow.

To understand organic chemistry you must be familiar with two languages. One, which we have concentrated on so far, is the structure and representation of molecules. The second is the description of the reaction mechanism in terms of curly arrows and that is what we are about to start. The first is static and the second dynamic. The creation of new molecules is the special concern of chemistry and an interest in the mechanism of chemical reactions is the special concern of organic chemistry.

Molecules react because they move. They move internally—we have seen (Chapter 3) how the stretching and bending of bonds can be detected by infrared spectroscopy. Whole molecules move continuously in space, bumping into each other, into the walls of the vessel they are in, and into the solvent if they are in solution. When one bond in a single molecule stretches too much it may break and a chemical reaction occurs. When two molecules bump into each other, they may combine with the formation of a new bond, and a chemical reaction occurs. We are first going to think about collisions between molecules.

Not all collisions between molecules lead to chemical change

All organic molecules have an outer layer of many electrons, which occupy filled orbitals, bonding and nonbonding. Charge–charge repulsion between these electrons ensures that all molecules repel each other. Reaction will occur only if the molecules are given enough energy (the activation energy for the reaction) for the molecules to pass the repulsion and get close enough to each other. If two molecules lack the required activation energy, they will simply collide, each bouncing off the electrons on the surface of the other and exchanging energy as they do so, but remain chemically

> ▶
> The **activation energy**, also called the **energy barrier** for a reaction, is the minimum energy molecules must have if they are to react. A population of a given molecule in solution at room temperature has a range of energies. If the reaction is to occur, some at least must have an energy greater than the activation energy. We shall discuss this concept in more detail in Chapter 13.

unchanged. This is rather like a collision in snooker or pool. Both balls are unchanged afterwards but are moving in different directions at new velocities.

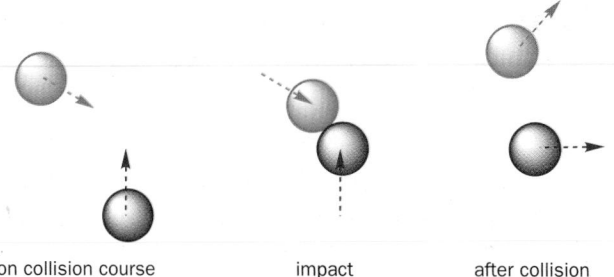

on collision course   impact   after collision

Charge attraction brings molecules together

In addition to this universal repulsive force, there are also important attractive forces between molecules if they are charged. Cations (+) and anions (−) attract each other electrostatically and this may be enough for reaction to occur. When an alkyl chloride, RCl, reacts with sodium iodide, NaI, in acetone (propanone, $Me_2C=O$) solution a precipitate of sodium chloride forms. Sodium ions, Na^+, and chloride ions, Cl^-, ions in solution are attracted by their charges and combine to form a crystalline lattice of alternating cations and anions—the precipitate of crystalline sodium chloride.

> We saw why these atoms form an ionic compound in Chapter 4.

This inorganic style of attraction is rare in organic reactions. A more common cause of organic reactions is attraction between a charged reagent (cation or anion) and an organic compound that has a dipole. An example that we shall explore in this chapter is the reaction between sodium cyanide (a salt, NaCN) and a carbonyl compound such as acetone. Sodium cyanide is made up of sodium cations, Na^+, and cyanide anions, CN^-, in solution. Acetone has a carbonyl group, a $C=O$ double bond, which is polarized because oxygen is more electronegative than carbon. The negative cyanide ion is attracted to the positive end of the carbonyl group dipole.

> We analysed the orbitals of the carbonyl group in Chapter 4 and established that the reason for the polarity is the greater electronegativity of the oxygen atom.

It is not even necessary for the reagent to be charged. Ammonia also reacts with acetone and this time it is the **lone pair** of electrons —a pair of electrons not involved in bonding and concentrated on the nitrogen atom of the uncharged ammonia molecule—that is attracted to the positive end of the carbonyl dipole.

Polarity can arise from σ bonds too. The most electronegative element in the periodic table is fluorine and three fluorine atoms on electropositive boron produce a partially positively charged boron atom by σ bond polarization. The negative end of the acetone dipole (the oxygen atom) is attracted to the boron atom in BF_3.

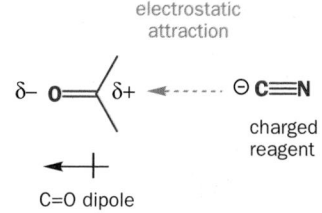

But we have not told you the whole story about BF_3. Boron is in group 3 and thus has only six electrons around it in its trivalent compounds. A molecule of BF_3 is planar with an empty p orbital. This is the reverse of a lone pair. An empty orbital on an atom does *not* repel electron-rich areas of other molecules and so the oxygen atom of acetone is attracted electrostatically to the partial positive charge and one of the lone pairs on oxygen can form a bonding interaction with the empty orbital. We shall develop these ideas in the next section.

So, to summarize, the presence of a dipole in a molecule represents an imbalance in the distribution of the bonding electrons due to polarization of a σ bond or a π bond or to a pair of electrons or an empty orbital localized on one atom. When two molecules with complementary dipoles collide and together have the required activation energy to ensure that the collision is sufficiently energetic to overcome the general electronic repulsion, chemical change or reaction can occur.

Orbital overlap brings molecules together

Other organic reactions take place between completely uncharged molecules with no dipole moments. One of the old 'tests' for unsaturation was to treat the compound with bromine water. If the brown colour disappeared, the molecule was unsaturated. We don't use 'tests' like these any more (spectroscopy means we don't need to) but the reaction is still an important one. A simple symmetrical alkene combines with symmetrical bromine in a simple addition reaction.

The only electrons that might be useful in the kind of attraction we have discussed so far are the lone pair electrons on bromine. But we know from many experiments that electrons flow out of the alkene towards the bromine atom in this reaction—the reverse of what we should expect from electron distribution. The attraction between these molecules is not electrostatic. In fact, we know that reaction occurs because the bromine molecule has an empty orbital available to accept electrons. This is not a localized atomic orbital like that in the BF_3 molecule. It is the antibonding orbital belonging to the Br–Br σ bond: the $\sigma*$ orbital. There is therefore in this case an attractive interaction between a full orbital (the π bond) and an empty orbital (the $\sigma*$ orbital of the Br–Br bond). The molecules are attracted to each other because this one interaction is between an empty and a full orbital and leads to bonding, unlike all the other repulsive interactions between filled orbitals. We shall develop this less obvious attraction as the chapter proceeds.

Most organic reactions involve interactions between full and empty orbitals. Many also involve charge interactions, and some inorganic reactions involve nothing but charge attraction. Whatever the attraction between organic molecules, reactions involve electrons moving from one place to another. We call the details of this process the **mechanism of the reaction** and we need to explain some technical terms before discussing this.

■
Terms such as σ bond, $\sigma*$ orbital, π bond, $\pi*$ orbital, lone pair, atomic and molecular orbital, and bonding and antibonding orbital, are all explained in Chapter 4.

Electron flow is the key to reactivity

The vast majority of organic reactions are polar in nature. That is to say, electrons flow from one molecule to another as the reaction proceeds. The electron donor is called a **nucleophile** (nucleus-loving) while the electron acceptor is called the **electrophile** (electron-loving). These terms come from the idea of charge attraction as a dominating force in reactions. The nucleophile likes nuclei because they are positively charged and the electrophile likes electrons because they are negatively charged. Though we no longer regard reactions as controlled only by charge interactions, these names have stuck.

Examples of reactions where the nucleophile is an anion and the electrophile is a cation and a new bond is formed simply by charge attraction leading to the combination of opposite charges include the reaction of sodium hydroxide with positively charged phosphorus compounds. The new bond between oxygen and phosphorus is formed by the donation of electrons from the nucleophile (hydroxide ion HO^-) to the electrophile (the positively charged phosphorus atom).

▶
Nucleophiles do not really react with the nucleus but with empty electronic orbitals. Even so, electrostatic attraction (and repulsion) may well play a crucial role in determining the course of the process. If a molecule has a positive charge, it is because there are more protons in its nuclei than there are electrons around them.

More often, reaction occurs when electrons are transferred from a lone pair to an empty orbital as in the reaction between an amine and BF_3. The amine is the nucleophile because of the lone pair of electrons on nitrogen and BF_3 is the electrophile because of the empty p orbital on boron.

■ A 'dative covalent bond' is just an ordinary σ bond whose electrons happen to come from one atom. Most bonds are formed by electron donation from one atom to another and a classification that makes it necessary to know the history of the molecule is not useful. Forget 'dative bonds' and stick to σ bonds or π bonds.

The kind of bond formed in these two reactions used to be called a 'dative covalent bond' because both electrons in the bond were donated by the same atom. We no longer classify bonds in this way, but call them σ bonds or π bonds as these are the fundamentally different types of bonds in organic compounds. Most new bonds are formed by donation of both electrons from one atom to another.

These simple charge or orbital interactions may be enough to explain simple inorganic reactions but we shall also be concerned with nucleophiles that supply electrons out of bonds and electrophiles that accept electrons into antibonding orbitals. For the moment accept that polar reactions usually involve electrons flowing *from a nucleophile* and *towards an electrophile*.

● **In reaction mechanisms**
- Nucleophiles donate electrons
- Electrophiles accept electrons

Since we are describing a dynamic process of electron movement from one molecule to another in this last reaction, it is natural to use some sort of arrow to represent the process. Organic chemists use a curved arrow (called a 'curly arrow') to show what is going on. It is a simple and eloquent symbol for chemical reactions.

The curly arrow shows the movement of a pair of electrons from nitrogen into the gap between nitrogen and boron to form a new σ bond between those two atoms. This representation, what it means, and how it can be developed into a language of chemical reactions is our main concern in this chapter.

Orbital overlap controls angle of successful attack

Electrostatic forces provide a generalized attraction between molecules in chemical reactions. In the reaction between chloride anions and sodium cations described above, the way in which these two spherical species approached one another was unimportant because the charges attracted one another from any angle. In most organic reactions the orbitals of the nucleophile and electrophile are directional and so the molecular orbitals of the reacting molecules exert important control. If a new bond is to be formed as the molecules collide, the orbitals of the two species must be correctly aligned in space. In our last example, only if the sp^3 orbital of the lone pair on nitrogen points directly at the empty orbital of the BF_3 can bond formation take place. Other collisions will not lead to reaction. In the first frame a successful collision takes place and a bond can be formed between the orbitals. In the second frame are three examples of unsuccessful collisions where no orbital overlap is possible. There are of course many more unproductive collisions but only one productive collision. Most collisions do not lead to reaction.

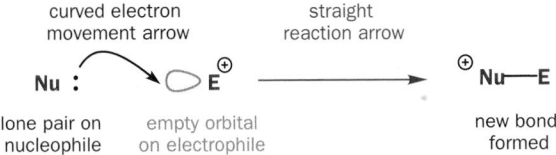

The orbitals must also have about the right amount of energy to interact profitably. Electrons are to be passed from a full to an empty orbital. Full orbitals naturally tend to be of lower energy than empty orbitals—that is after all why they are filled! So when the electrons move into an empty orbital they have to go up in energy and this is part of the activation energy for the reaction. If the energy gap is too big, few molecules will have enough energy to climb it and reaction will be bad. The ideal would be to have a pair of electrons in a filled orbital on the nucleophile and an empty orbital on the electrophile of the same energy. There would be no gap and reaction would be easy. In real life, a small gap is the best we can hope for.

Now we shall discuss a generalized example of a neutral nucleophile, Nu, with a lone pair donating its electrons to a cationic electrophile, E, with an empty orbital. Notice the difference between the curly arrow for electron movement and the straight reaction arrow. Notice also that the nucleophile has given away electrons so it has become positively charged and that the electrophile has accepted electrons so it has become neutral.

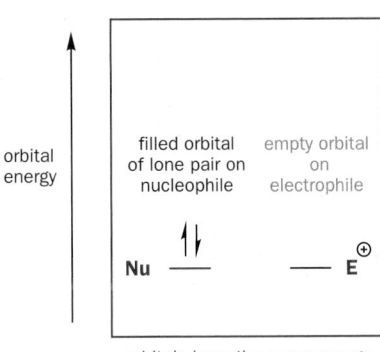

If we look at different possible relative energies for the lone pair orbital and the empty orbital, we might have equal energies, a small gap, or a large gap. Just as in Chapter 4, the horizontal lines represent energy levels, the arrows on them represent electrons, and the vertical scale is energy with high energy at the top and low energy at the bottom.

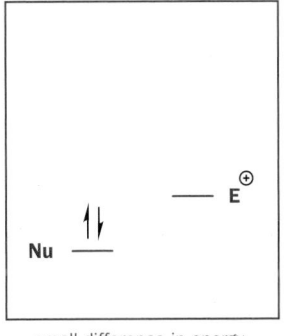

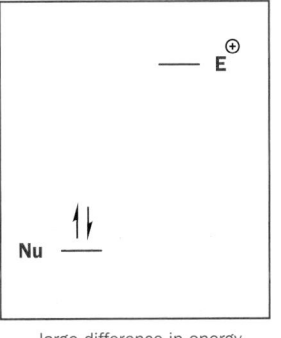

orbitals have the same energy / small difference in energy of filled and empty orbitals / large difference in energy of filled and empty orbitals

At first this picture suggests that the electrons will have to climb up to the empty orbital if it is higher in energy than the filled orbital. This is not quite true because, when atomic orbitals interact, their energies split to produce two new molecular orbitals, one above and one below the old orbitals. This is the basis for the static structure of molecules described in the last chapter and is also the key to reactivity. In these three cases this is what will happen when the orbitals interact (the new molecular orbitals are shown in black between the old atomic orbitals).

■
These diagrams of molecular energy levels combining to form new bonding and antibonding orbitals are almost identical to those we used in Chapter 4 to make molecular orbitals from atomic orbitals.

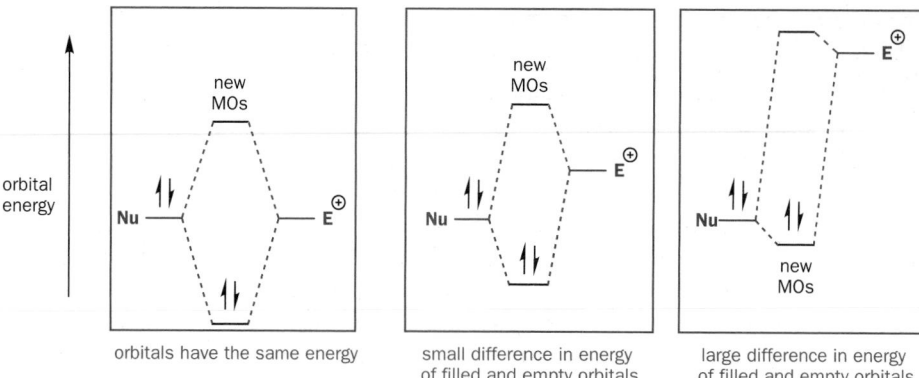

orbitals have the same energy   small difference in energy   large difference in energy
                                   of filled and empty orbitals   of filled and empty orbitals

> ■
> We saw exactly the same response when we combined AOs of different energies to make MOs in Chapter 4.

In each case there is actually a gain in energy when the electrons from the old lone pair drop down into the new stable bonding molecular orbital formed by the combination of the old atomic orbitals. The energy gain is greatest when the two orbitals are the same and least when they are very far apart in energy. The other new MO is higher in energy than either of the old AOs but it does not have to be occupied.

Only the highest-energy occupied orbitals of the nucleophile are likely to be similar in energy to only the lowest unoccupied orbitals of the electrophile. This means that the lower-lying completely filled bonding orbitals of the nucleophile can usually be neglected and only the **highest occupied molecular orbital (HOMO)** of the nucleophile and the **lowest unoccupied molecular orbital (LUMO)** of the electrophile are relevant. These may be of about the same energy and can then interact strongly. Orbital overlap—of both direction and energy—is therefore an important requirement for successful reaction between two organic molecules.

> ● **Molecules repel each other because of their outer coatings of electrons.**
>
> **Molecules attract each other because of:**
> * **attraction of opposite charges**
> * **overlap of high-energy filled orbitals with low-energy empty orbitals**
>
> **For reaction, molecules must approach each other so that they have:**
> * **enough energy to overcome the repulsion**
> * **the right orientation to use any attraction**

We need now to look at which types of molecules are nucleophiles and which types are electrophiles. When you consider the reactivity of any molecule, this is the first question you should ask: is it nucleophilic or electrophilic?

Nucleophiles donate high-energy electrons to electrophiles

Nucleophiles are either negatively charged or neutral species with a pair of electrons in a high energy filled orbital that they can donate to electrophiles. The most common type of nucleophile has a nonbonding lone pair of electrons. Usually these are on a heteroatom such as O, N, S, or P.

water   ammonia   trimethylphosphine   dimethylsulfide

These four neutral molecules, ammonia, water, trimethylphosphine, and dimethylsulfide, all have lone pairs of electrons in sp^3 orbitals and in each case this is the donor or nucleophilic orbital. The group VI atoms (O and S) have two lone pairs of equal energy. These are all nonbonding electrons and therefore higher in energy than any of the bonding electrons.

Anions are often nucleophiles too and these are also usually on heteroatoms such as O, S, or halogen which may have several lone pairs of equal energy. The first diagram for each of our examples shows the basic structure and the second diagram shows all the lone pairs. It is not possible to allocate the negative charge to a particular lone pair as they are the same.

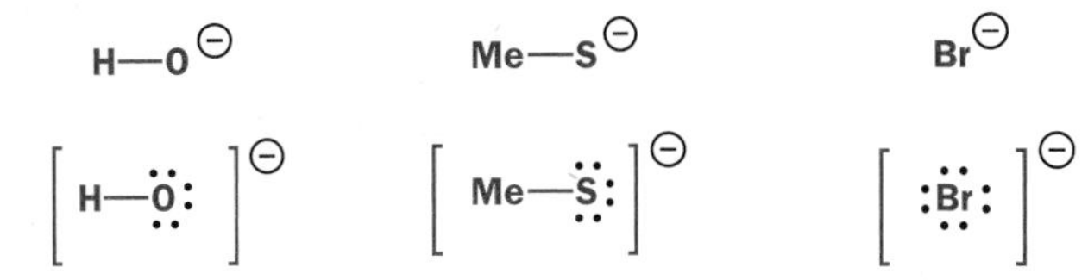

hydroxide   methane thiolate   bromide

There are a few examples of carbon nucleophiles with lone pairs of electrons, the most famous being the cyanide ion. Though linear cyanide has a lone pair on nitrogen and one on carbon, the nucleophilic atom is usually anionic carbon rather than neutral nitrogen as the sp orbital on carbon has a higher energy than that on the more electronegative nitrogen. Most anionic nucleophiles containing carbon have a heteroatom as the nucleophilic atom such as the anion methane thiolate shown above.

■ This point will be important in Chapter 6 as well.

Neutral carbon electrophiles usually have a π bond as the nucleophilic portion of the molecule. When there are no lone pair electrons to supply high-energy nonbonding orbitals, the next best is the lower-energy filled π orbitals rather than the even lower-energy σ bonds. Simple alkenes are

cyanide ion   sp lone pair   sp lone pair

weakly nucleophilic and react with strong electrophiles such as bromine. In Chapter 20 we shall see that the reaction starts by donation of the π electrons from the alkene into the σ* orbital of the bromine molecule (which breaks the Br–Br bond) shown here with a curly arrow. After more steps the dibromoalkane is formed but the molecules are attracted by overlap between the full π orbital and the empty σ* orbital.

It is possible for σ bonds to act as nucleophiles and we shall see later in this chapter that the borohydride anion, BH_4^-, has a nucleophilic B–H bond and can donate those electrons into the π* orbital of a carbonyl compound breaking that bond and even-

tually giving an alcohol as product. The first stage of the reaction has electrons from the B–H single bond of nucleophilic anion BH_4^-, which lacks lone pair electrons or π bonds, as the nucleophile.

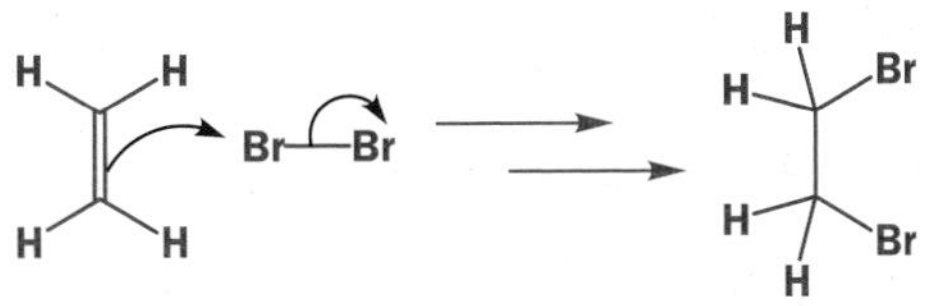

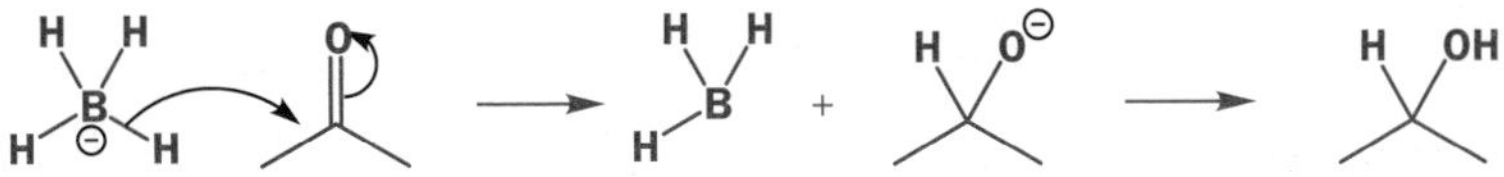

In this section you have seen lone pairs on anions and neutral molecules acting as nucleophiles and, more rarely, π bonds and even σ bonds able to do the same job. In each case the nucleophilic electrons came from the HOMO—the highest occupied molecular orbital—of the molecule. Don't worry if you find the curly arrows strange at the moment. They will soon be familiar. Now we need to look at the other side of the coin—the variety of electrophiles.

Electrophiles have a low-energy vacant orbital

Electrophiles are neutral or positively charged species with an empty atomic orbital (the opposite of a lone pair) or a low-energy antibonding orbital. The simplest electrophile is the proton, H+, a species without any electrons at all and a vacant 1s orbital. It is so reactive that it is hardly ever found and almost any nucleophile will react with it.

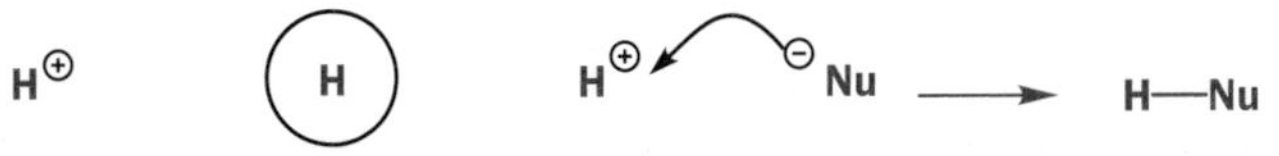

proton   empty 1s orbital   reaction with anionic nucleophile

Each of the nucleophiles we saw in the previous section will react with the proton and we shall look at two of them together. Hydroxide ion combines with a proton to give water. This reaction is governed by charge control. Then water itself reacts with the proton to give H_3O^+, the true acidic species in all aqueous strong acids.

hydroxide as   water as
nucleophile   nucleophile

We normally think of protons as acidic rather than electrophilic but an acid is just a special kind of electrophile. In the same way, Lewis acids such as BF_3 or $AlCl_3$ are electrophiles too. They have empty orbitals that are usually metallic p orbitals. We saw above how BF_3 reacted with Me_3N. In that reaction BF_3 was the electrophile and Me_3N the nucleophile. Lewis acids such as $AlCl_3$ react violently with water and the first step in this process is nucleophilic attack by water on the empty p orbital of the aluminium atom. Eventually alumina (Al_2O_3) is formed.

water as   empty   new
nucleophile p orbital   σ bond

Protic and Lewis acids

Protic acids (also known as Brønsted acids) are electrophiles (like HCl) that can donate protons (H^+) to nucleophiles. They will be discussed in detail in Chapter 8. Lewis acids are also electrophiles but they donate more complicated cations to nucleophiles. They are usually metal halides such as LiCl, BF_3, $AlCl_3$, $SnCl_4$, and $TiCl_4$. We shall meet them in many later chapters, particularly in Chapters 22–8 when we discuss carbon–carbon bond formation.

Few organic compounds have vacant atomic orbitals and most organic electrophiles have low-energy antibonding orbitals. The most important are π* orbitals as they are lower in energy than σ* orbitals and the carbonyl group (C=O) is the most important of these—indeed it is the most important functional group of all. It has a low-energy π* orbital ready to accept electrons and also a partial positive charge on the carbon atom. Previously we said that charge attraction helped nucleophiles to find the carbon atom of the carbonyl group.

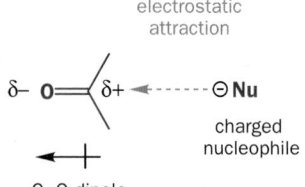

electrostatic attraction

charged nucleophile

C=O dipole

high-energy filled orbitals of the carbonyl group

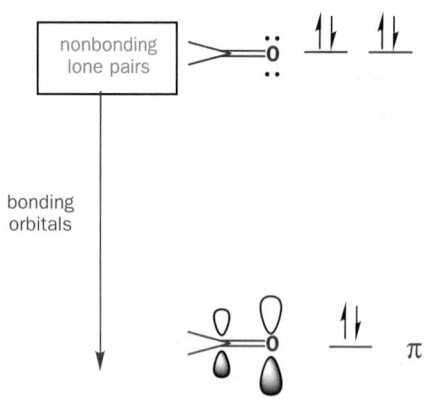

nonbonding lone pairs

bonding orbitals

Charge attraction is important in carbonyl reactions but so are the orbitals involved. Carbonyl compounds have a low-energy bonding π orbital. Carbonyl compounds have a dipole because in this filled orbital the electrons are more on electronegative oxygen than on carbon. The same reason (electronegative oxygen) makes this an exceptionally low-energy orbital and the carbonyl group a very stable structural unit. This orbital is rarely involved in reactions. Going up the energy scale we next have two degenerate (equal in energy) lone pairs in nonbonding orbitals. These are the highest-energy electrons in the molecule (HOMO) and are the ones that react with electrophiles.

When we consider the carbonyl group as an electrophile, we must look at antibonding orbitals too. The only one that concerns us is the relatively low-energy π* orbital of the C=O double bond (the LUMO). This orbital is biased towards the carbon to compensate for the opposite bias in the filled π orbital. How do we know this if there are no electrons in it? Simply because nucleophiles, whether charged or not, attack carbonyl groups at the carbon atom. They get the best overlap with the larger orbital component of the π* orbital.

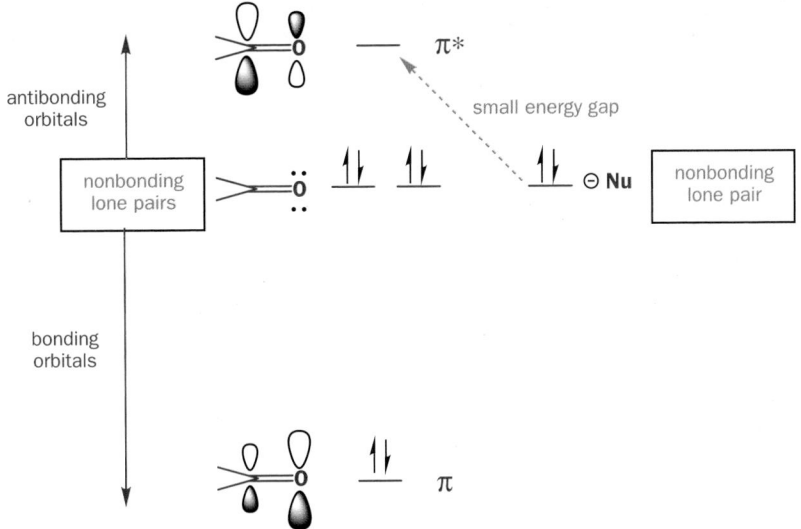

So now we can draw a mechanism for the attack of a nucleophile on the carbonyl group. The lone pair electrons on the nucleophile move into the π^* orbital of the C=O double bond and so break the π bond, though not, of course, the σ bond. Here is that process in curly arrow terms.

The lone pair electrons on oxygen interact better with empty orbitals such as the 1s of the proton and so carbonyl compounds are protonated on oxygen.

The resulting cation is even more electrophilic because of the positive charge but nucleophiles still attack the carbon atom of the carbonyl group because the π^* orbital still has more contribution from carbon. The positive charge is neutralized even though the nucleophile does not attack the positively charged atom.

Even σ bonds can be electrophilic if the atom at one end of them is sufficiently electronegative to pull down the energy of the σ^* orbital. Familiar examples are acids where the acidic hydrogen atom is joined to strongly electronegative oxygen or a halogen thus providing a dipole moment and a relatively low-energy σ^* orbital.

These two diagrams suggest two different ways of looking at the reaction between a base and an acid, but usually both interactions are important. Notice that an acid is just an electrophile that has an electrophilic hydrogen atom and a base is just a nucleophile that acts on a hydrogen atom. This question is explored more in Chapter 8. Bonds between carbon and halogen are also polarized in some cases though the electronegativity difference is sometimes very small.

It is easy to exaggerate the importance of single-bond polarization. The electronegativity difference between H and Cl is 0.9 but that

Quick guide to important electronegativities

		H			
		2.2			
Li		B	C	N	O	F
0.98		2.04	2.55	3.04	3.44	3.98
	Mg	Al	Si	P	S	Cl
	1.31	1.61	1.9	2.19	2.58	3.16
						Br
						2.96
						I
						2.66

These are Pauling electronegativities, calculated by Linus Pauling (1901–94) who won the chemistry Nobel prize in 1954 and the Nobel peace prize in 1983 and from whose ideas most modern concepts of the chemical bond are derived. Born in Portland, Oregon, he worked at 'CalTech' (the California Institute of Technology at Pasadena) and had exceptionally wide-ranging interests in crystallography, inorganic chemistry, protein structure, quantum mechanics, nuclear disarmament, politics, and taking vitamin C to prevent the common cold.

between C and Br only 0.3 while the C–I bond is not polarized at all. When carbon–halogen σ bonds act as electrophiles, polarity hardly matters but a relatively low-energy σ* orbital is vitally important. The bond strength is also important in these reactions too as we shall see.

Some σ bonds are electrophilic even though they have no dipole at all. The halogens such as bromine (Br_2) are examples. Bromine is strongly electrophilic because it has a very weak Br–Br σ bond. Symmetrical bonds have the energies of the σ orbital and the σ* orbital roughly evenly distributed about the nonbonding level. A weak symmetrical σ bond means a small energy gap while a strong symmetrical σ bond means a large energy gap. Bromine is electrophilic but carbon–carbon σ bonds are not. Reverting to the language of Chapter 4, we could say that the hydrocarbon framework is made up of strong C–C bonds with low-energy populated and high-energy unpopulated orbitals, while the functional groups react because they have low LUMOs or high HOMOs.

An example would be the rapid reaction between a sulfide and bromine. No reaction at all occurs between a sulfide and ethane or any other simple C–C σ bond. Lone pair electrons are donated from sulfur into the Br–Br σ* orbital, which makes a new bond between S and Br and breaks the old Br–Br bond.

■
Notice how putting charges in circles (Chapter 2) helps here. There is no problem in distinguishing the charge on sulfur (in a ring) with the plus sign (not in a ring) linking the two products of the reaction.

Summary: interaction between HOMO and LUMO leads to reaction

Organic reactions occur when the HOMO of a nucleophile overlaps with the LUMO of the electrophile to form a new bond. The two electrons in the HOMO slot into the empty LUMO. The reacting species may be initially drawn together by electrostatic interaction of charges or dipoles but this is not necessary. Thus at this simplest of levels *molecular recognition* is required for reaction. The two components of a reaction must be matched in terms of both charge–charge attraction and the energy and orientation of the orbitals involved.

Nucleophiles may donate electrons (in order of preference) from a lone pair, a π bond, or even a σ bond and electrophiles may accept electrons (again in order of preference) into an empty orbital or into the antibonding orbital of a π bond (π* orbital) or even a σ bond (σ* orbital). These antibonding orbitals are of low enough energy to react if the bond is very polarized by a large electronegativity difference between the atoms at its ends or, even for unpolarized bonds, if the bond is weak.

The hydrocarbon framework of organic molecules is unreactive. Functional groups such as NH_2 and OH are nucleophilic because they have nonbonding lone pairs. Carbonyl compounds and alkyl halides are electrophilic functional groups because they have low-energy LUMOs (π* for C=O and σ* for C–X, respectively).

Organic chemists use curly arrows to represent reaction mechanisms

You have seen several examples of curly arrows so far and you may already have a general idea of what they mean. The representation of organic reaction mechanisms by this means is so important that we must now make quite sure that you do indeed understand exactly what is meant by a curly arrow, how to use it, and how to interpret mechanistic diagrams as well as structural diagrams.

A **curly arrow** represents the *actual movement of a pair of electrons* from a filled orbital into an empty orbital. You can think of the curly arrow as representing a pair of electrons thrown, like a climber's grappling hook, across from where he is standing to where he wants to go. In the simplest cases, the result of this movement is to form a bond between a nucleophile and an electrophile. Here are two examples we have already seen in which lone pair electrons are transferred to empty atomic orbitals.

hydroxide ion   empty   new   water as   empty   new
as nucleophile 1s orbital   σ bond   nucleophile   p orbital   σ bond

Note the exact position of the curly arrow as the value of this representation lies in the precision and uniformity of its use. The arrow always starts with its tail on the source of the moving electrons, representing the filled orbital involved in the reaction. The head of the arrow indicates the final destination of the pair of electrons—the new bond between oxygen and hydrogen or oxygen and aluminium in these examples. As we are forming a new bond, the head of the arrow should be drawn to a point on the line between the two atoms.

When the nucleophile attacks an antibonding orbital, such as the weak Br–Br bond we have just been discussing, we shall need two arrows, one to make the new bond and one to break the old.

The bond-making arrow is the same as before but the bond-breaking arrow is new. This arrow shows that the two electrons in the bond move to one end (a bromine atom) and turn it into an anion. This arrow should start in the centre of the bond and its head should rest on the atom (Br in this case) at the end of the bond. Another example would be the attack of a base on the strong acid HBr.

It is not important how much curvature you put into the arrows or whether they are above or below the gaps of the bonds, both on the same side, or on opposite sides so long as they begin and end in the right places. All that matters is that someone who reads your arrows should be able to deduce exactly what is happening in the reaction from your arrows. We could have drawn the ammonia/HBr reaction like this if we had wished.

Charge is conserved in each step of a reaction

In all these examples we have reacted neutral molecules together to form charged species. Because the starting materials had no overall charge, neither must the products. If we start with neutral molecules and make a cation, we must make an anion too. Charge cannot be created or destroyed. If

> ► Some chemists prefer to place this point halfway between the atoms but we consider that the representation is clearer and more informative if the arrowhead is closer to the atom to which the new bond is forming. For these examples the difference is minimal and either method is completely clear but in more complex situations, our method prevents ambiguity as we shall see later. We shall adopt this convention throughout this book.

our starting materials have an overall charge—plus or minus—then the same charge must appear in the products.

starting materials have   products must also have
overall positive charge   overall positive charge

When it is a π bond that is being broken rather than a σ bond, only the π bond is broken and the σ bond should be left in place. This is what commonly happens when an electrophilic carbonyl group is attacked by a nucleophile. Just as in the breaking of a σ bond, start the arrow in the middle of the π bond and end by putting the arrowhead on the more electronegative atom, in this case oxygen rather than carbon.

π bond is broken   C–O σ bond remains

In this case the starting materials had an overall negative charge and this is preserved as the oxyanion in the product. The charge disappears from the hydroxide ion because it is now sharing a pair of electrons with what was the carbonyl carbon atom and a charge appears on what was the carbonyl oxygen atom because it now has both of the electrons in the old π bond.

Electrons can be donated from π bonds and from σ bonds too. The reaction of an alkene with HBr is a simple example of a C–C π bond as nucleophile. The first arrow (on the nucleophile) starts in the middle of the π bond and goes into the gap between one of the carbon atoms and the hydrogen atom of HBr. The second arrow (on the electrophile) takes the electrons out of the H–Br σ bond and puts them on to the bromine atom to make bromide ion. This sort of reaction make us place alkenes among the functional groups as well as as part of the framework of organic molecules.

Notice that it was important to draw the two reagents in the right orientation since both are unsymmetrical and we want our arrow to show which end of the alkene reacts with which end of HBr. If we had drawn them differently we should have had trouble drawing the mechanism. Here is a less satisfactory representation.

If you find yourself making a drawing like this, it is worth having another go to see if you can be clearer. Drawing mechanisms is often rather experimental—try something and see how it looks: if it is unclear, try again. One way to avoid this particular problem is to draw an *atom-specific* curly arrow passing through the atom that reacts. Something like this will do.

This reaction does not, in fact, stop here as the two ions produced (charge conservation requires one cation and one anion in this first reaction) now react with each other to form the product of the reaction. This reaction is pretty obvious as the anion is the nucleophile and the cation, with its empty p orbital, is the electrophile.

The reaction that occurs between the alkene and HBr occurs in two stages—the formation of the ions and their combination. Many reactions are like this and we call the two stages **steps** so that we talk about 'the first step' and 'the second step', and we call the ions **intermediates** because they are formed in one step and disappear in the next. We shall discuss these intermediates in several later chapters (for example, 17 and 19).

When σ bonds act as nucleophiles, the electrons also have to go to one end of the σ bond as they form a new bond to the electrophile. We can return to an earlier example, the reaction of sodium borohydride (NaBH₄) with a carbonyl compound, and complete the mechanism. In this example, one of the atoms (the hydrogen atom) moves away from the rest of the BH₄ anion and becomes bonded to the carbonyl compound. The LUMO of the electrophile is, of course, the π* orbital of the C=O double bond.

The arrow on the nucleophile should again start in the middle of the bond that breaks and show which atom (the black H in this case) is transferred to the electrophile. The second arrow we have seen before. Here again you could use an atom-specific arrow to make it clear that the electrons in the σ bond act as a nucleophile through the hydrogen and not through the boron atom.

This reaction also occurs in two steps and the oxyanion is an intermediate, not a product. The reaction is normally carried out in water and the oxyanion reacts with water by proton transfer.

We shall discuss this reaction, the reduction of carbonyl compounds by NaBH₄, in detail in Chapter 6.

The decomposition of molecules

So far we have described reactions involving the combination of one molecule with another. Many reactions are not like this but involve the spontaneous decomposition of one molecule by itself without any assistance from any other molecule. In these reactions there is no electrophile or nucleophile. The usual style of reaction consists of a weak, often polarized σ bond breaking to give two new molecules or ions. The dissociation of a strong acid HX is a simple example.

In organic chemistry spontaneous dissociation of diazonium salts, compounds containing the N₂⁺ group, occurs very easily because one of the products, nitrogen gas ('dinitrogen') is very stable. It does not much matter what R is (alkyl or aryl); this reaction happens spontaneously at room temperature.

This is not, of course, the end of the reaction as R⁺ is very reactive and we shall see the sort of things it can do in Chapters 17 and 19. More commonly, some sort of catalysis is involved in decomposition reactions. An important example is the decomposition of tertiary alcohols in acid solution. The carbon–oxygen bond of the alcohol does not break by itself but, after the oxygen atom has been protonated by the acid, decomposition occurs.

This two-step mechanism is not finished because the positive ion (one particular example of R⁺) reacts further (Chapter 17). In the decomposition step the positive charge on the oxygen atom as well as the fact that the other product is water helps to break the strong C–O σ bond. In these three examples, the functional group that makes off with the electrons of the old σ bond (X, N_2^+, and OH_2^+) is called the **leaving group**, and we shall be using this term throughout the book. The spontaneous decomposition of molecules is one of the clearest demonstrations that curly arrows mean the movement of two electrons. Chemical reactions are dynamic processes, molecules really do move, and electrons really do leave one atomic or molecular orbital to form another.

These three examples all have the leaving group taking both electrons from the old σ bond. This type of decomposition is sometimes called **heterolytic fission** or simply **heterolysis** and is the most common in organic chemistry. There is another way that a σ bond can break. Rather than a pair of electrons moving to one of the atoms, one electron can go in either direction. This is known as **homolytic fission** as two species of the same charge (neutral) will be formed. It normally occurs when

> Each bromine radical has an unpaired electron in an atomic orbital.

similar or indeed identical atoms are at each end of the σ bond to be broken. Both fragments have an unpaired electron and are known as radicals. This type of reaction occurs when bromine gas is subjected to sunlight.

The weak Br–Br bond breaks to form two bromine **radicals**. This can be represented by two single-headed curly arrows, **fish hooks**, to indicate that only one electron is moving. This is virtually all you will see of this special type of curly arrow until we consider the reactions of radicals in more detail (Chapter 39). When you meet a new reaction you should assume that it is an ionic reaction and use two-electron arrows unless you have a good reason to suppose otherwise.

Curly arrows also show movement of electrons within molecules

So far all the mechanisms we have drawn have used only one or two arrows in each step. In fact, there is no limit to the number of arrows that might be involved and we need to look at some mechanisms

> Don't be alarmed—these mechanisms will all be discussed in full later in the book, this particular one in Chapter 10.

with three arrows. The third arrow in such mechanisms usually represents movement of electrons inside of the reacting molecules. Some pages back we drew out the addition of a nucleophile to a carbonyl compound.

This is a two-arrow mechanism but, if we lengthen the structure of the carbonyl compound by adding a double bond in the right position, we can add the nucleophile to a different position in the molecule by moving electrons within the molecule using a third arrow.

The first arrow from the nucleophile makes a new σ bond and the last breaks the carbonyl π bond. The middle arrow just moves the C–C π bond along the molecule. If you inspect the product you will see that its structure follows precisely from the arrows. The middle arrow starts in the middle of a π bond and ends in the middle of a σ bond. All it does is to move the π electrons along the molecule. It turns the old π bond into a σ bond and the old σ bond into a π bond. We shall discuss this sort of reaction in Chapter 10.

In some mechanisms there is a second step in the mechanism and both are three-arrow processes. Here is the first step in such a mechanism. See if you can understand each arrow before reading the explanation in the next paragraph.

The arrow from the hydroxide ion removes a proton from the molecule making a new O–H bond in a molecule of water. The middle arrow moves the electrons of a C–H bond into a C–C bond making it into a π bond and the third arrow polarizes the carbonyl π bond leaving an oxyanion as the product. Charge is conserved—an anion gives an anion. In fact this 'product' is only an intermediate and the second step also involves three arrows.

Starting from the oxyanion, the first arrow re-forms the carbonyl group, the middle arrow moves a π bond along the molecule, and the third arrow breaks a C–O σ bond releasing hydroxide ion as one of the products of the reaction. We shall meet this sort of reaction in detail later (Chapters 19 and 27).

Mostly for entertainment value we shall end this section with a mechanism involving no fewer than eight arrows. See if you can draw the product of this reaction without looking at the result.

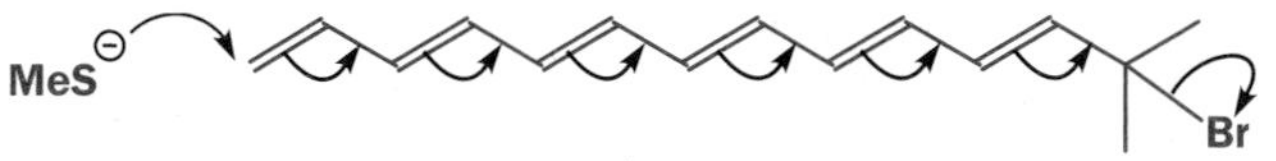

The first arrow forms a new C–S σ bond and the last arrow breaks a C–Br σ bond but all the rest just move π bonds along the molecule. The product is therefore:

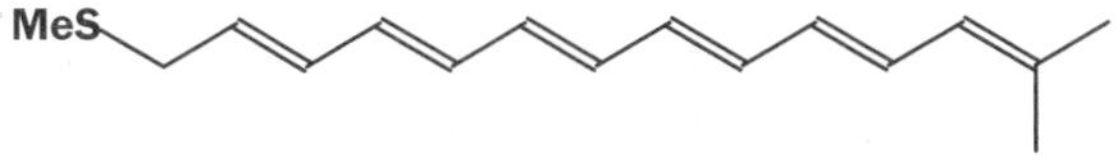

We shall not be discussing this reaction anywhere in the book! We have included it just to convince you that, once you understand the principle of curly arrows, you can understand even very complicated mechanisms quite easily. At this stage we can summarize the things you have learned about interpreting a mechanism drawn by someone else.

Summary: what do curly arrows mean?

- A curly arrow shows the movement of a pair of electrons
- The tail of the arrow shows the source of the electron pair, which will be a filled orbital (HOMO)
  - such as a lone pair or a π bond or a σ bond
- The head of the arrow indicates the ultimate destination of the electron pair which will either be:
  - an electronegative atom that can support a negative charge (a *leaving group*)
  - *or* an empty orbital (LUMO) when a new bond will be formed
  - *or* an antibonding orbital (π* or σ*) when that bond will break
- Overall charge is always conserved in a reaction. Check that your product obeys this rule

Now would be a good time to do Problems 1 and 2 at the end of the chapter, which will give you practice in the interpretation of mechanisms.

Drawing your own mechanisms with curly arrows

Curly arrows must be drawn carefully! The main thing you need to remember is that curly arrows must start where there is a pair of electrons and end somewhere where you can leave a pair of electrons without drawing an absurd structure. That sounds very simple—and it is—but you need some practice to see what it means in detail in different circumstances. Let us look at the implications with a reaction whose products are given: the reaction of triphenylphosphine with methyl iodide.

$$MeI + Ph_3P \longrightarrow Ph_3\overset{\oplus}{P}{-}Me + I^{\ominus}$$

First observe what has happened: a new bond has been formed between the phosphorus atom and the methyl group and the carbon–iodine bond has been broken. Arrows represent movement of electron pairs *not* atoms so the reactants must be drawn within bonding distance before the mechanism can be drawn. This is analogous to the requirement that molecules must collide before they can react. First draw the two molecules so that the atoms that form the new bond (P and C) are near each other and draw out the bonds that are involved (that is, replace 'MeI' with a proper chemical structure).

$$Ph_3P \qquad CH_3{-}I$$

Now ask: which is the electrophile and which the nucleophile (and why)? The phosphorus atom has a lone pair and the carbon atom does not so Ph$_3$P must be the nucleophile and the C–I bond of MeI must be the electrophile. All that remains is to draw the arrows.

$$Ph_3P\colon\;\; CH_3{-}I \longrightarrow Ph_3\overset{\oplus}{P}{-}CH_3 + I^{\ominus}$$

Admittedly, that was quite an easy mechanism to draw but you should still be pleased if you succeeded at your first try.

Warning! Eight electrons is the maximum for B, C, N, or O

We now ought to spell out one thing that we have never stated but rather assumed. Most atoms in organic molecules, if they are not positively charged, have their full complement of electrons (two in the case of hydrogen, eight in the cases of carbon, nitrogen, and oxygen) and so, if you make a new bond to one of those elements, *you must also break an existing bond*. Suppose you just 'added' Ph_3P to MeI in this last example without breaking the C–I bond: what would happen?

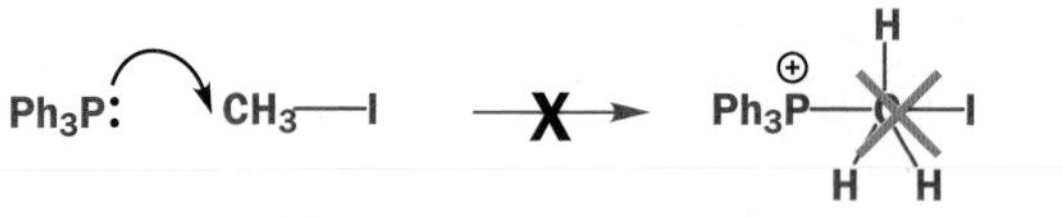

wrong mechanism

impossible structure
carbon has five bonds

This structure must be wrong because carbon cannot have five bonds—if it did it would have ten electrons in the 2s and the 2p orbitals. As there are only four of those (2s, $2p_x$, $2p_y$, and $2p_z$) and they can have only two electrons each, eight electrons is the maximum and that means that four bonds is the maximum.

> ● **If you make a new bond to uncharged H, C, N, or O you must also break one of the existing bonds in the same step.**

There is a nasty trap when a charged atom has its full complement of electrons. Since BH_4^- and NH_4^+ are isoelectronic with methane and have four σ bonds and hence eight electrons, no new bonds can be made to B or N. The following attractive mechanisms are impossible because boron has no lone pair in BH_4^- and nitrogen has no empty orbital in NH_4^+.

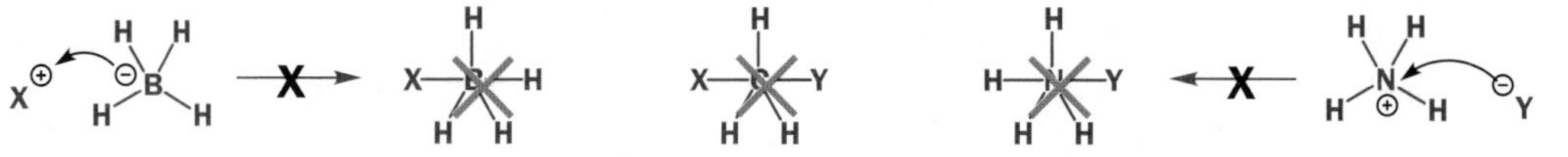

impossible reaction

impossible structure
boron has five bonds

impossible structure
carbon has five bonds

impossible structure
nitrogen has five bonds

impossible reaction

Reactions with BH_4^- always involve the loss of H and a pair of electrons using the BH bond as nucleophile and reactions with NH_4^+ always involve the loss of H without a pair of electrons using the NH bond as electrophile.

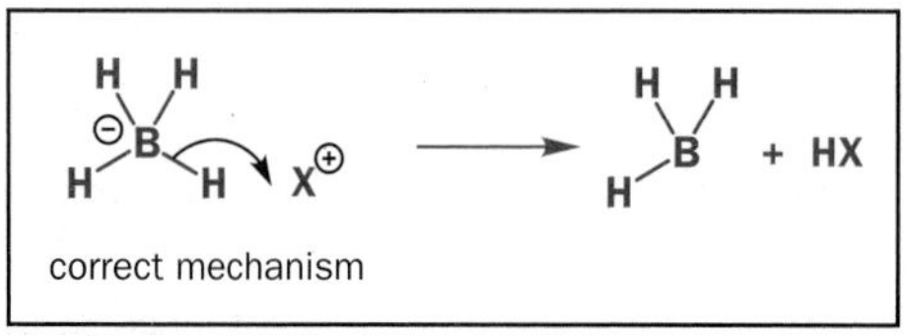

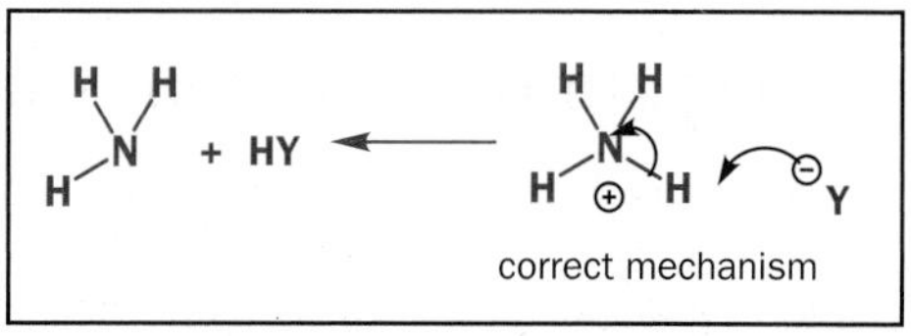

correct mechanism

correct mechanism

Similarly, nucleophiles do not attack species like H_3O^+ at oxygen, even though it is the oxygen atom that carries the positive charge. Reaction occurs at one of the protons, which also neutralizes the positive charge. Or, to put it another way, H_3O^+ is an acid (electrophilic at hydrogen) and not electrophilic at oxygen.

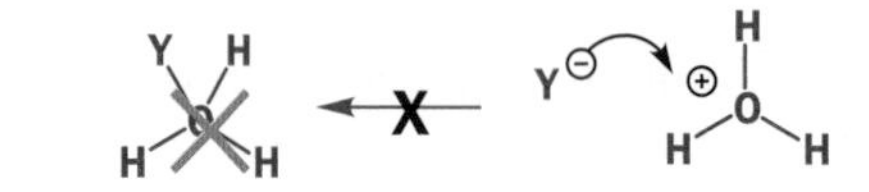

impossible structure
oxygen has four bonds

impossible reaction

correct mechanism

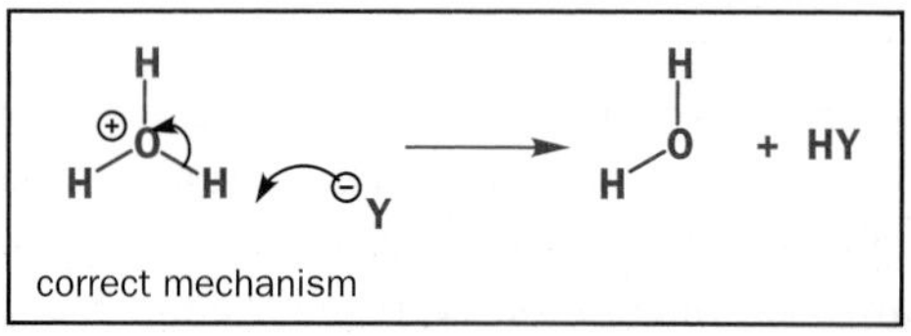

Try a simple example: primary alcohols can be converted into symmetrical ethers in acid solution. Suggest a mechanism for this acid-catalysed conversion of one functional group into another.

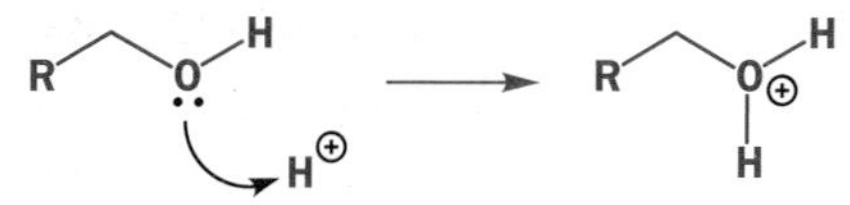

The reaction must start by the protonation of something and the only candidate is the oxygen atom as it alone has lone pair electrons. This gives us a typical oxonium ion with three bonds to oxygen and a full outer shell of eight electrons.

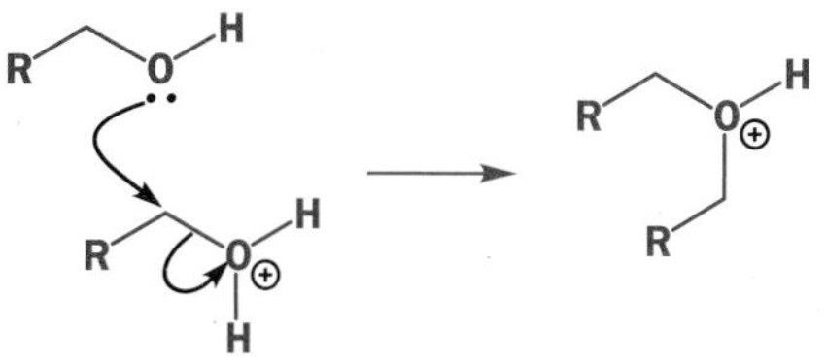

To make the ether a second molecule of alcohol must be added but we must not now be tempted to attack the positively charged oxygen atom with the nucleophilic OH group. The second molecule could attack a proton, but that would just make the same molecules. Instead it must attack at carbon expelling a molecule of water as a leaving group and creating a new oxonium ion.

Finally, the loss of the proton from the new oxonium ion gives the ether. Though this is a three-step mechanism, two of the steps are just proton transfers in acidic solution and the only interesting step is the middle one. Here is the whole mechanism.

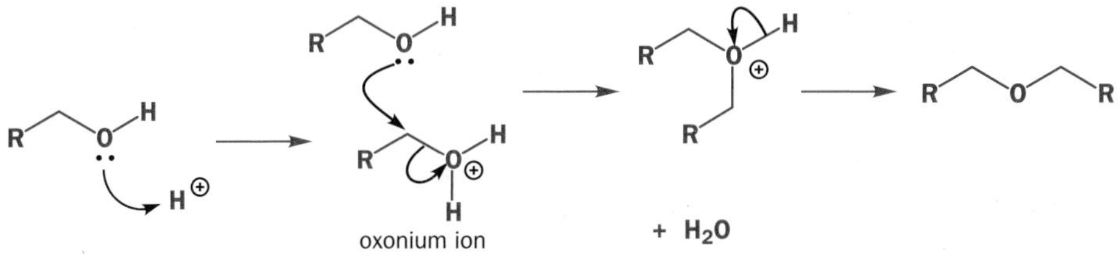

oxonium ion

+ H_2O

Drawing a two-step mechanism: cyanohydrin formation

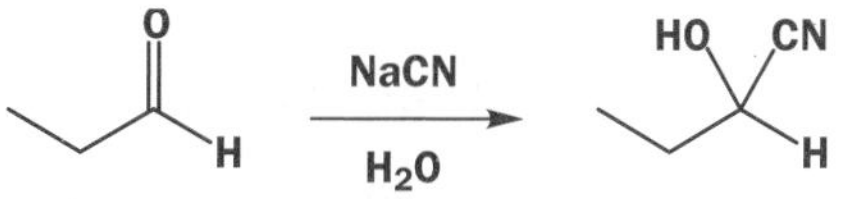

Now what about this slightly more complicated example? Sodium cyanide is added to a simple aldehyde in aqueous solution. The product is a **cyanohydrin** and we shall discuss this chemistry in Chapter 6.

This reaction is presented in a style with which you will become familiar. The organic starting material is written first and then the reagent over the reaction arrow and the solvent under it. We must decide what happens. NaCN is an ionic solid so the true reagent must be cyanide ion. As it is an anion, it must be the nucleophile and the carbonyl group must be the electrophile. Let us try a mechanism.

This is a good mechanism but it doesn't quite produce the product. There must be a second step in which the oxyanion picks up a proton from somewhere. The only source of protons is the solvent, water, so we can write:

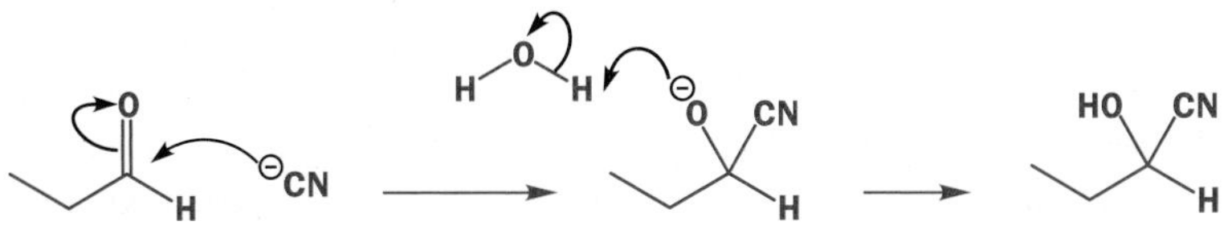

This is the complete mechanism and we can even make a prediction about the reaction conditions from it. The second step needs a proton and water is not a very good proton donor. A weak acid as catalyst would help.

Now for a real test: can you draw a mechanism for this reaction?

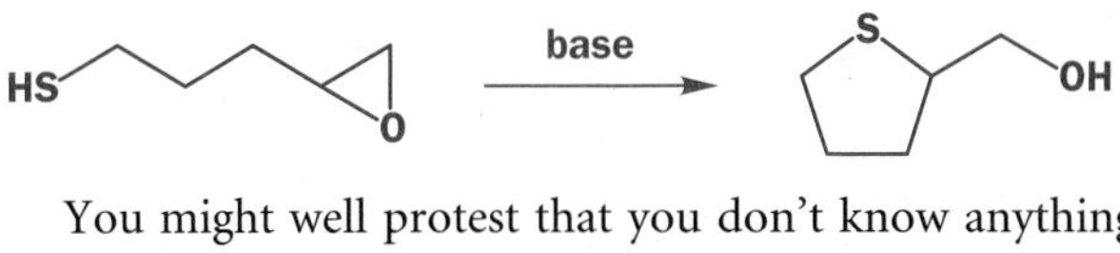

You might well protest that you don't know anything about the chemistry of three-membered rings or of either of the functional groups, SH and cyclic ether. Be that as it may, you can still draw a mechanism for the reaction. It is important that you are prepared to try your hand at mechanisms for new reactions as you can learn a lot this way. Ask first of all: which bonds have been formed and which broken? Clearly the S–H bond has been broken and a new S–C bond formed. The three-membered ring has gone by the cleavage of one of the C–O bonds. The main chain of carbon atoms is unchanged. We might show these ideas in some way such as this.

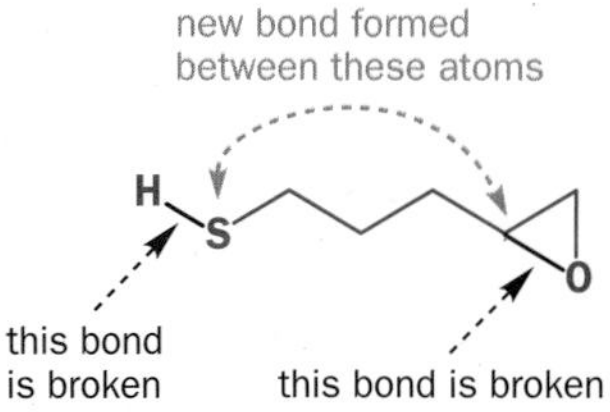

new bond formed
between these atoms

this bond
is broken

this bond is broken

Now you could continue in many ways. You might say 'what breaks the SH bond?' This must be the role of the base as a base removes protons. You might realize that the reaction cannot happen while the sulfur atom is so far away from the three-membered ring (no chance of a collision) and redraw the molecule so that the reaction can happen.

Now draw the mechanism. It is easy once you have done the preparatory thinking. The sulfur anion must be the nucleophile so the C–O bond in the three-membered ring must be the electrophile. Here goes!

That is not quite the product so we must add a proton to the oxyanion. Where can the proton come from? It must be the proton originally removed by the base as there is no other. We can write B for the base and hence BH$^+$ for the base after it has captured a proton.

Your mechanism probably didn't look as neat as the printed version but, if you got it roughly right, you should be proud. This is a three-step mechanism involving chemistry unknown to you and yet you could draw a mechanism for it. Are you using coloured arrows, by the way? We are using black arrows on red diagrams but the only point of that is to make the arrows stand out. We suggest you use any colour for your arrows that contrasts with your normal ink.

Decide on a 'push' or a 'pull' mechanism

In one step of a reaction mechanism electrons flow from a site rich in electrons to an electron-deficient site. When you draw a mechanism you must make sure that the electrons flow in one direction only and neither meet at a point nor diverge from a point. One way to do this is to decide whether the mechanism is 'pushed' by, say, a lone pair or an anion or whether it is 'pulled' by, say, a cation, an empty orbital, or by the breaking of a reactive weak π bond or σ bond. This is not just a device either. Extremely reactive molecules, such as fluorine gas, F_2, react with almost anything—in this case because of the very electrophilic F–F σ bond (low energy F–F σ^* orbital). Reactions of F_2 are 'pulled' by the breaking of the F–F bond. The nearest thing in organic chemistry is probably the reactions of carbon cations such as those formed by the decomposition of diazonium salts.

In the first step the electrons of the σ bond are pulled away by the positive charge and the very stable leaving group, N_2. In the second step lone pair electrons are pulled into the very reactive cation by the nonbonding empty orbital on carbon. Even very weak nucleophiles such as water will react with such cations as a real example shows.

In all our previous examples we have drawn the first arrow from the nucleophile, anion, lone pair, or whatever and pushed the electrons along the chain of arrows. This is a natural thing to do; indeed the skill of drawing mechanisms is sometimes derisively referred to as 'electron pushing', but some mechanisms are more easily understood as 'electron pulling'. In general, if a cation, an acid, or a Lewis acid is a reagent or a catalyst, the reaction is probably *pulled*. If an anion or a base is involved as a reagent, the reaction is probably *pushed*. In any case it isn't so important which approach you adopt as that you should do one or the other and not muddle them up.

A more interesting example of a pull mechanism is the reaction of isoprene (2-methylbutadiene) with HBr. The product is an unsaturated alkyl bromide (a bromoalkene).

What has happened? HBr has clearly added to the diene while one of the double bonds has vanished. However, the remaining double bond, whichever it is, has moved to a new position in the middle of the molecule. So how do we start? HBr is a strong acid so the reaction must begin with the protonation of some atom in the diene by HBr. Which one? If you examine the product you will see that one atom has an extra hydrogen and this must be where protonation occurs.

number of protons on each carbon atom in starting material and product

The only change is at the left-hand end of the molecule where there is an extra proton. We must add the proton of HBr to that atom. The highest-energy orbital at that atom is the rather unreactive alkene π bond so we must use that as the nucleophile, though the electrons are really being pulled out of the π bond by reactive HBr.

It is not necessary to draw in that hydrogen atom in the product of this step. It is, of course, necessary to put the positive charge on the carbon atom in the middle that has lost electrons. Now we can add bromide ion (the other product of the first step) to this cation but not where we have written the plus charge as that will not give us the right product. We must move the remaining double bond along the molecule as we add the bromide ion. This too is a 'pulled' reaction as the unstable plus charge on carbon pulls electrons towards itself.

So this is a two-step reaction and the driving force for the two steps is a strongly acidic electrophile in the first and a strongly electrophilic cation at carbon in the second. Here is the full mechanism.

Now we can summarize the extra points we have made in this section as a series of guidelines.

Extra guidelines for writing your own mechanisms
- Decide on the structure of any ambiguous reagents, for example, salt or a covalent compound?
- Decide which is the nucleophilic and which the electrophilic atom
- Decide whether to think in a *push* or a *pull* manner
- Mark lone pairs on the nucleophilic atom
- Draw the molecule(s) in a spatial arrangement that makes reaction possible
- Curly arrows always move in the same direction. They never meet head on!
- If you make a new bond to H, C, N, or O you must also break one of the existing bonds in the same step
- Draw your arrows in colour to make them stand out
- Mark charges clearly on reactants and intermediates
- Make sure that overall charge is conserved in your mechanism

We have only given you a preliminary trial run as a learner driver of curly arrows in this section. The way forward is practice, practice, practice.

Curly arrows are vital for learning organic chemistry

Curly arrows can be used to explain the interaction between the structure of reactants and products and their reactivity in the vast majority of organic reactions, regardless of their complexity. When used correctly they can even be used to predict possible outcomes of unknown processes and hence to design new synthetic reactions. They are thus a powerful tool for understanding and developing organic chemistry and it is vital that you become proficient in their use. They are the dynamic language of organic reaction mechanisms and they will appear in every chapter of the book from now on.

Another equally important reason for mastering curly arrows now, before you start the systematic study of different types of reactions, is that the vast number of 'different reactions' turn out not to be so different after all. Most organic reactions are ionic; they therefore all involve nucleophiles and electrophiles and two-electron arrows. There are relatively few types of organic electrophiles and nucleophiles and they are involved in all the 'different' reactions. If you understand and can draw mechanisms, the similarity between seemingly unrelated reactions will become immediately apparent and thus the number of distinct reaction types is dramatically reduced.

Drawing curly arrow mechanisms is a bit like riding a bike. Before you've mastered the skill, you keep falling off. Once you've mastered the skill, it seems so straightforward that you wonder how you ever did without it. You still come across busy streets and complex traffic junctions, but the basic skill remains the same.

If you still feel that drawing mechanisms for yourself is difficult, this stage-by-stage guide may help you. Once you've got the idea, you probably won't need to follow it through in detail.

A guide to drawing mechanisms with curly arrows

1   Draw out the reagents as clear structures following the guidelines in Chapter 2. Check that you understand what the reagents and the solvent are under the conditions of the reaction, for example, if the reaction is in a base, will one of the compounds exist as an anion?

2   Inspect the starting materials and the products and assess what has happened in the reaction. What new bonds have been formed? What bonds have been broken? Has anything been added or removed? Have any bonds moved around the molecule?

3   Identify the nucleophilic centres in all the reactant molecules and decide which is the most nucleophilic. Then identify the electrophiles present and again decide which is the most electrophilic

4   If the combination of these two centres appears to lead to the product, draw the reactants, complete with charges, so as to position the nucleophilic and electrophilic centres within bonding distance ensuring that the angle of attack of the nucleophile is more or less consistent with the orbitals involved

5   Draw a curly arrow from the nucleophile to the electrophile. It must start on the filled orbital or negative charge (show this clearly by just touching the bond or charge) and finish on the empty orbital (show this clearly by the position of the head). You may consider a 'push' or a 'pull' mechanism at this stage

6   Consider whether any atom that has been changed now has too many bonds; if so one of them must be broken to avoid a ridiculous structure. Select a bond to break. Draw a curly arrow from the centre of the chosen bond, the filled orbital, and terminate it in a suitable place

7   Write out the structures of the products specified by the curly arrows. Break the bonds that are the sources of the arrows and make those that are the targets. Consider the effect on the charges on individual atoms and check that the overall charge is not changed. Once you have drawn the curly arrows, the structure of the products is already decided and there is no room for any further decisions. Just write what the curly arrows tell you. If the structure is wrong, then the curly arrows were wrong so go back and change them

8   Repeat stages 5–7 as required to produce a stable product

When you have read through all the different types of reaction mechanism, practise drawing them out with and without the help of the book. Complete the exercises at the end of the chapter and then try to devise mechanisms for other reactions that you may know. You now have the tools to draw out in the universal pictorial language of organic chemists virtually all the mechanisms for the reactions you will meet in this book and more besides!

Problems

1. Each of these molecules is electrophilic. Identify the electrophilic atom and draw a mechanism for reaction with a generalized nucleophile Nu⁻, giving the product in each case.

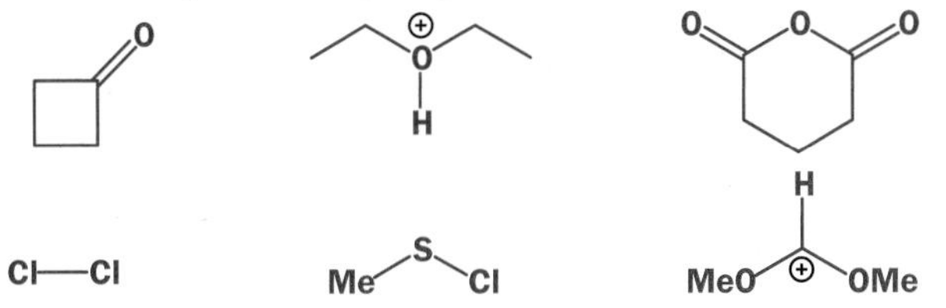

2. Each of these molecules is nucleophilic. Identify the nucleophilic atom and draw a mechanism for reaction with a generalized electrophile E⁺, giving the product in each case.

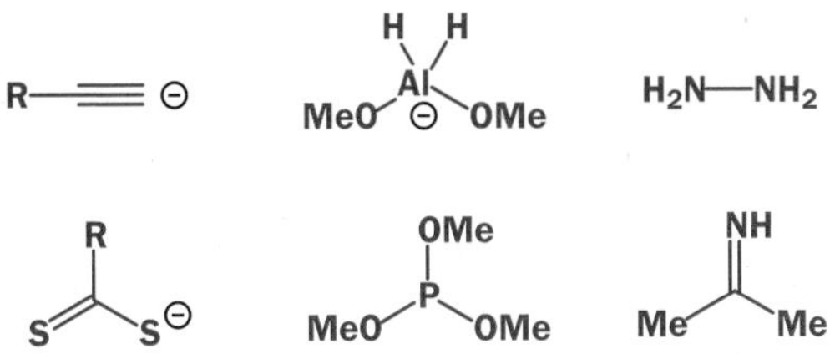

3. Complete these mechanisms by drawing the structure of the products in each case.

(a)  ⟶ ?

(b) 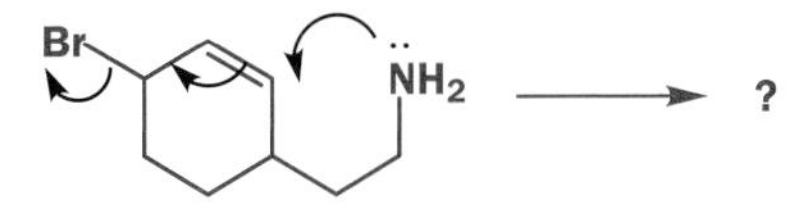 ⟶ ?

4. Each of these electrophiles could react with a nucleophile at (at least) two different atoms. Identify these atoms and draw a mechanism for each reaction together with the products from each.

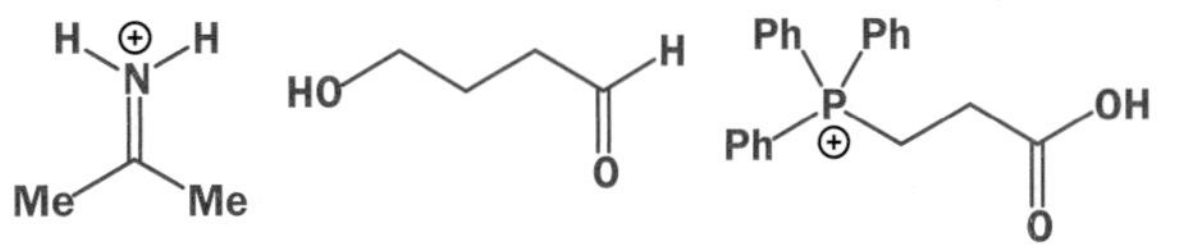

5. Put in the arrows on these structures (which have been drawn with all the atoms in the right places!) to give the products shown.

(a)

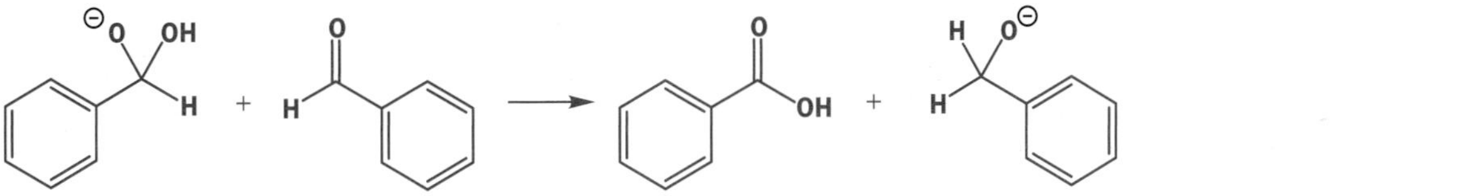

(b)

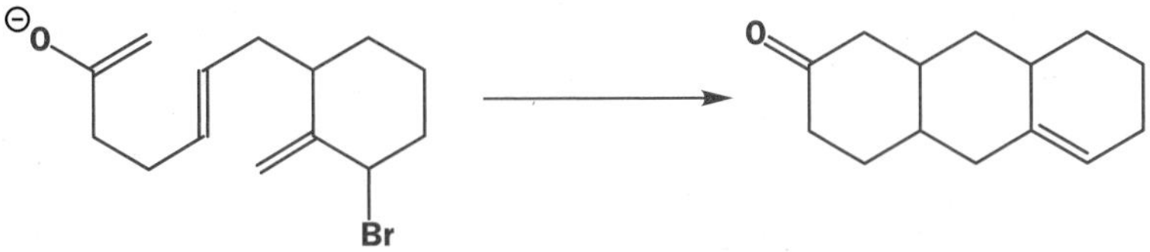

6. Draw mechanisms for these reactions. The starting materials have not necessarily been drawn in a helpful way.

(a)

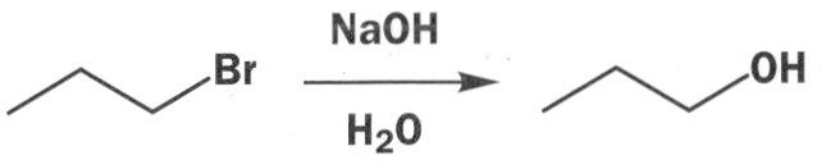

(b)

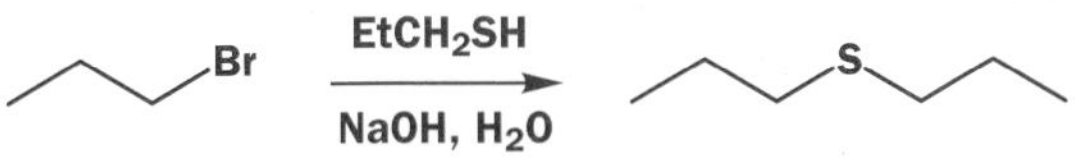

(c)

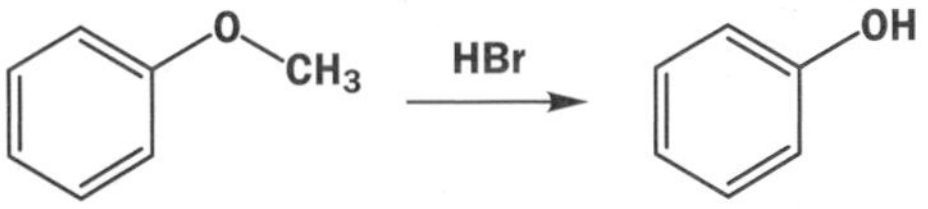

7. Draw a mechanism for this reaction.

$$PhCHBr.CHBr.CO_2H + NaHCO_3 \longrightarrow PhCH=CHBr$$

Hints. First draw good diagrams of the reagents. $NaHCO_3$ is a salt and a weak base—strong enough only to remove which proton? Then work out which bonds are formed and which broken, decide whether to push or pull, and draw the arrows. What are the other products?

Nucleophilic addition to the carbonyl group

6

Connections

Building on:

- Functional groups, including the carbonyl group (C=O) ch2
- Identifying the functional groups in a molecule spectroscopically ch3
- How molecular orbitals explain molecular shapes and functional groups ch4
- How, and why, molecules react together, the involvement of functional groups, and using curly arrows to describe reactions ch5

Arriving at:

- How and why the C=O group reacts with nucleophiles
- Explaining the reactivity of the C=O group using molecular orbitals and curly arrows
- What sorts of molecules can be made by reactions of C=O groups
- How acid or base catalysts improve the reactivity of the C=O group

Looking forward to:

- Additions of organometallic reagents ch9
- C=O groups with an adjacent double bond ch10
- How the C=O group in derivatives of carboxylic acids promotes substitution reactions ch12
- Substitution reactions of the C=O group's oxygen atom ch14

Molecular orbitals explain the reactivity of the carbonyl group

We are now going to leave to one side most of the reactions you met in the last chapter—we will come back to them all again later in the book. In this chapter we are going to concentrate on just one of them—probably the simplest of all organic reactions—the addition of a nucleophile to a carbonyl group. The carbonyl group, as found in aldehydes, ketones, and many other compounds, is without doubt the most important functional group in organic chemistry, and that is another reason why we have chosen it as our first topic for more detailed study.

You met nucleophilic addition to a carbonyl group on p. 114 and 119, where we showed you how cyanide reacts with acetone to give an alcohol. As a reminder, here is the reaction again, with its mechanism.

nucleophilic addition of CN⁻ to the carbonyl group

protonation

> We will frequently use a device like this, showing a reaction scheme with a mechanism for the same reaction looping round underneath. The reagents and conditions next to the arrow across the top will tell you how you might carry out the reaction, and the pathway shown underneath will tell you how it actually works.

The reaction has two steps: nucleophilic addition of cyanide, followed by protonation of the anion. In fact, this is a general feature of all nucleophilic additions to carbonyl groups.

> ● Additions to carbonyl groups generally consist of two mechanistic steps:
> 1 Nucleophilic attack on the carbonyl group
> 2 Protonation of the anion that results

The addition step is more important, and it forms a new C–C σ bond at the expense of the C=O π bond. The protonation step makes the overall reaction addition of HCN across the C=O π bond.

Why does cyanide, in common with many other nucleophiles, attack the carbonyl group? And why does it attack the *carbon* atom of the carbonyl group? To answer these questions we need to look in detail at the structure of carbonyl compounds in general and the orbitals of the C=O group in particular.

The carbonyl double bond, like that found in alkenes (whose bonding we discussed in Chapter 4), consists of two parts: one σ bond and one π bond. The σ bond between the two sp² hybridized atoms—carbon and oxygen—is formed from two sp² orbitals. The other sp² orbitals on carbon form the two σ bonds to the substituents while those on oxygen are filled by the two lone pairs. The sp² hybridization means that the carbonyl group has to be planar, and the angle between the substituents is close to 120°. The diagram illustrates all this for the simplest carbonyl compound, formaldehyde (or methanal, CH₂O). The π bond then results from overlap of the remaining p orbitals—again, you can see this for formaldehyde in the diagram.

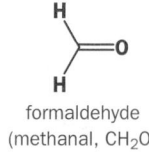

formaldehyde
(methanal, CH₂O)

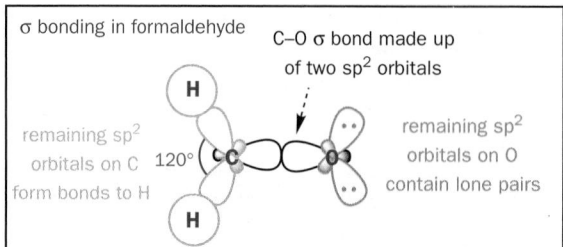

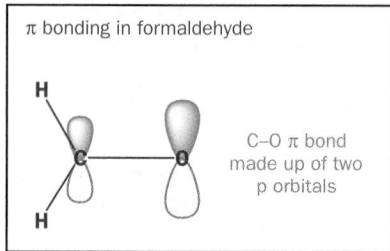

■
You were introduced to the polarization of orbitals in Chapter 4, and we discussed the case of the carbonyl group on p. 103.

Notice that we have drawn the π bond skewed towards oxygen. This is because oxygen is more electronegative than carbon, polarizing the orbital as shown. Conversely, the unfilled π* antibonding orbital is skewed in the opposite direction, with a larger coefficient at the carbon atom. Put all of this together and we get the complete picture of the orbitals of a carbonyl group.

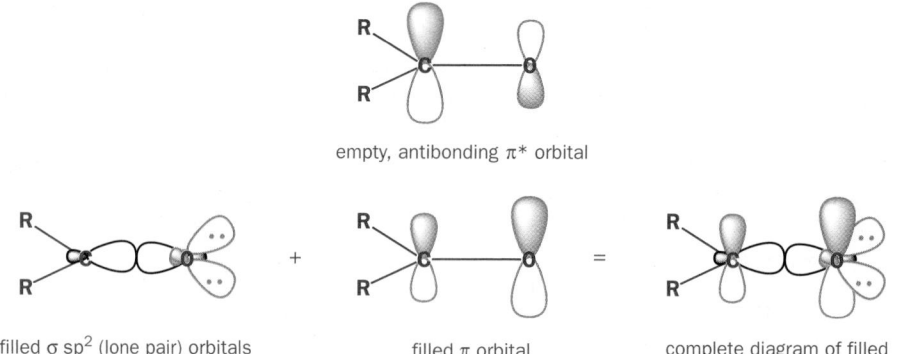

empty, antibonding π* orbital

filled σ sp² (lone pair) orbitals   filled π orbital   complete diagram of filled orbitals of C=O bond

Electronegativities, bond lengths, and bond strengths

Representative bond energies, kJ mol⁻¹		Representative bond lengths, Å		Electronegativity	
C–O 351	C=O 720	C–O 1.43	C=O 1.21	C 2.5	O 3.5

Because there are two types of bonding between C and O, the C=O double bond is rather shorter than a typical C–O single bond, and also over twice as strong—so why is it so reactive? Polarization is the key. The polarized C=O bond gives the carbon atom some degree of positive charge, and this charge attracts negatively charged nucleophiles (like cyanide) and encourages reaction. The polarization of the antibonding π* orbital towards carbon is also important, because, when the carbonyl group reacts with a nucleophile, electrons move from the HOMO of the nucleophile (an sp orbital in this case) into the LUMO of the electrophile—in other words the π* orbital of the C=O bond. The greater coefficient of the π* orbital at carbon means a better HOMO–LUMO interaction, so this is where the nucleophile attacks.

As our nucleophile—which we are representing here as 'Nu⁻'—approaches the carbon atom, the electron pair in its HOMO starts to interact with the LUMO (antibonding π*) to form a new σ bond.

Filling antibonding orbitals breaks bonds and, as the electrons enter the antibonding π^* of the carbonyl group, the π bond is broken, leaving only the C–O σ bond intact. But electrons can't just vanish, and those that were in the π bond move off on to the electronegative oxygen, which ends up with the negative charge that started on the nucleophile. You can see all this happening in the diagram below.

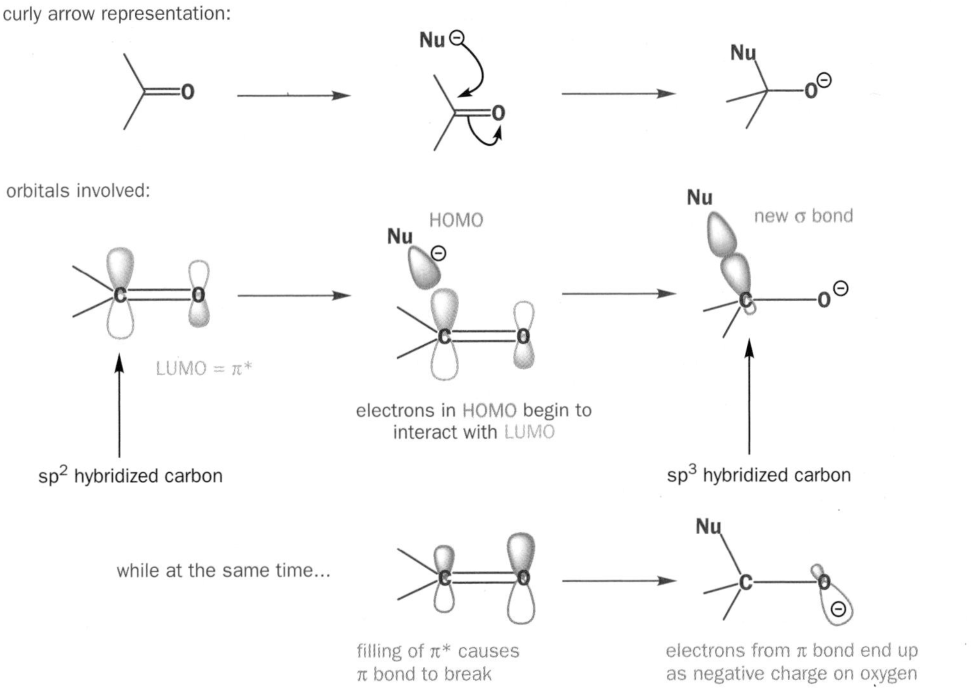

curly arrow representation:

orbitals involved:

HOMO

Nu

LUMO = π^*

sp^2 hybridized carbon

Nu

new σ bond

electrons in HOMO begin to interact with LUMO

sp^3 hybridized carbon

while at the same time...

Nu

filling of π^* causes π bond to break

electrons from π bond end up as negative charge on oxygen

The HOMO of the nucleophile will depend on what the nucleophile is, and we will meet examples in which it is an sp or sp^3 orbital containing a lone pair, or a B–H or metal–carbon σ orbital. We shall shortly discuss cyanide as the nucleophile; cyanide's HOMO is an sp orbital on carbon.

Notice how the trigonal, planar sp^2 hybridized carbon atom of the carbonyl group changes to a tetrahedral, sp^3 hybridized state in the product. For each class of nucleophile you meet in this chapter, we will show you the HOMO–LUMO interaction involved in the addition reaction.

Cyanohydrins from the attack of cyanide on aldehydes and ketones

Now that we've looked at the theory of how a nucleophile attacks a carbonyl group, let's go back to the real reaction with which we started this chapter: cyanohydrin formation from a carbonyl compound and sodium cyanide. Cyanide contains sp hybridized C and N atoms, and its HOMO is an sp orbital on carbon. The reaction is a typical nucleophilic addition reaction to a carbonyl group: the electron pair from the HOMO of the CN⁻ (an sp orbital on carbon) moves into the C=O π^* orbital; the electrons from the C=O π orbital move on to the oxygen atom. The reaction is usually carried out in the presence of acid, which protonates the resulting alkoxide to give the hydroxyl group of the composite functional group known as a cyanohydrin. The reaction works with both ketones and aldehydes, and the mechanism below shows the reaction of a general aldehyde.

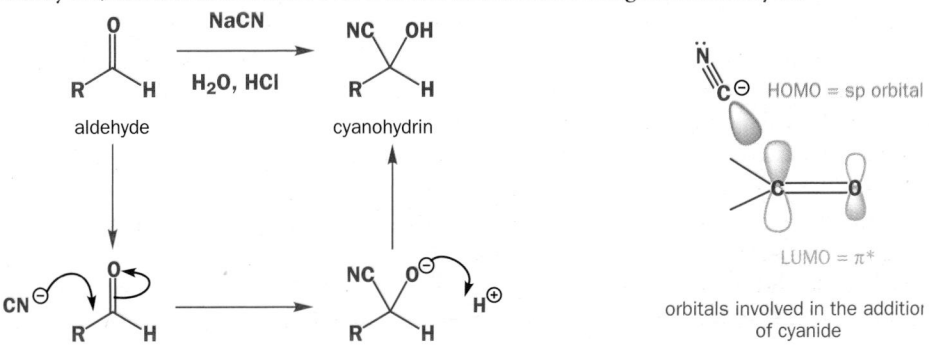

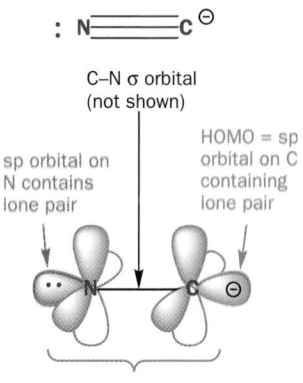

orbitals of the cyanide ion

C–N σ orbital (not shown)

sp orbital on N contains lone pair

HOMO = sp orbital on C containing lone pair

two pairs of p orbitals make two orthogonal πbonds

HOMO = sp orbital

LUMO = π^*

orbitals involved in the addition of cyanide

Cyanohydrins in synthesis

Cyanohydrins are important synthetic intermediates—for example, the cyanohydrin formed from this cyclic amino ketone forms the first step of a synthesis of some medicinal compounds known as 5HT$_3$ agonists, which were designed to reduce nausea in chemotherapy patients. Cyanohydrins are also components of many natural and industrial products, such as the insecticides cypermethrin (marketed as 'Ripcord', 'Barricade', and 'Imperator') and fluvalinate.

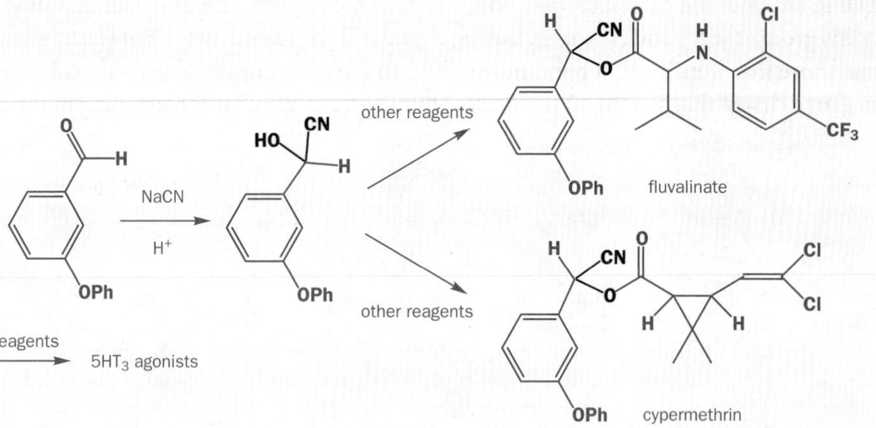

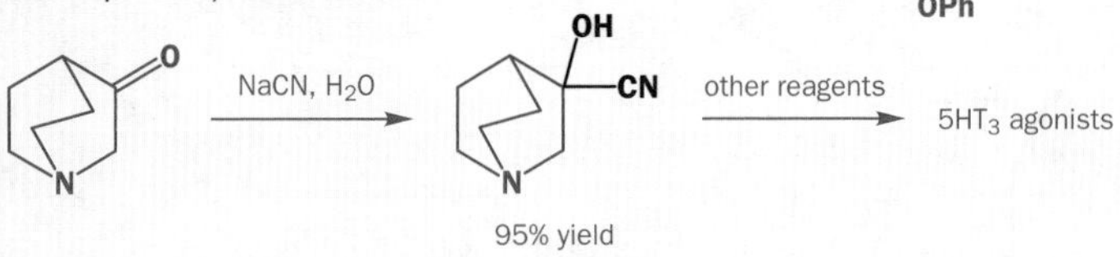

95% yield

> This is because the cyanide is a good *leaving group*—we'll come back to this type of reaction in much more detail in Chapter 12.

Cyanohydrin formation is reversible: just dissolving a cyanohydrin in water can give back the aldehyde or ketone you started with, and aqueous base usually decomposes cyanohydrins completely.

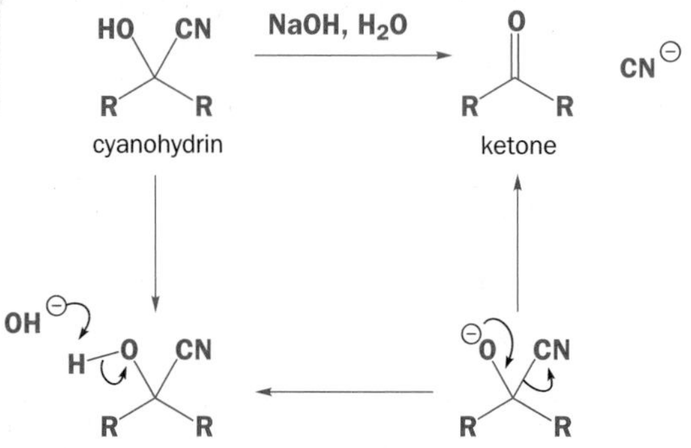

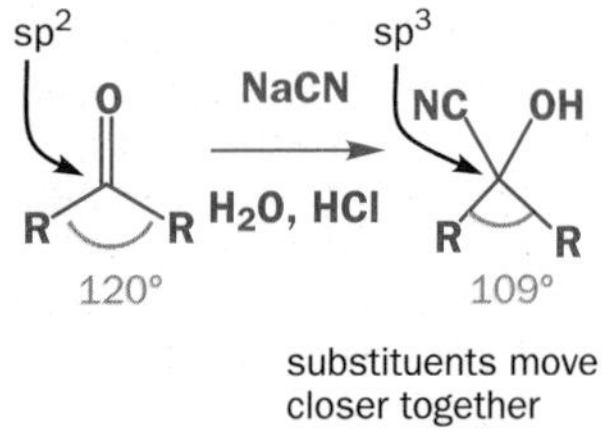

substituents move closer together

Cyanohydrin formation is therefore an equilibrium between starting materials and products, and we can only get good yields if the equilibrium favours the products. The equilibrium is more favourable for aldehyde cyanohydrins than for ketone cyanohydrins, and the reason is the size of the groups attached

Some equilibrium constants

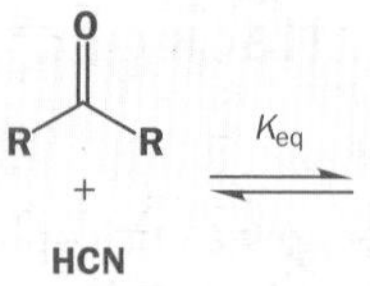

aldehyde or ketone	K_{eq}
PhCHO	212
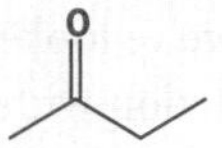 (ketone)	28

Cyanohydrins and cassava

The reversibility of cyanohydrin formation is of more than theoretical interest. In parts of Africa the staple food is cassava. This food contains substantial quantities of the glucoside of acetone cyanohydrin (a glucoside is an acetal derived from glucose). We shall discuss the structure of glucose later in this chapter, but for now, just accept that it stabilizes the cyanohydrin.

The glucoside is not poisonous in itself, but enzymes in the human gut break it down and release HCN. Eventually 50 mg HCN per 100 g of cassava can be released and this

is enough to kill a human being after a meal of unfermented cassava. If the cassava is crushed with water and allowed to stand ('ferment'), enzymes in the cassava will do the same job and then the HCN can be washed out before the cassava is cooked and eaten.

The cassava is now safe to eat but it still contains some glucoside. Some diseases found in eastern Nigeria can be traced to long-term consumption of HCN. Similar glucosides are found in apple pips and the kernels inside the stones of fruit such as peaches and apricots. Some people like eating these, but it is unwise to eat too many at one sitting!

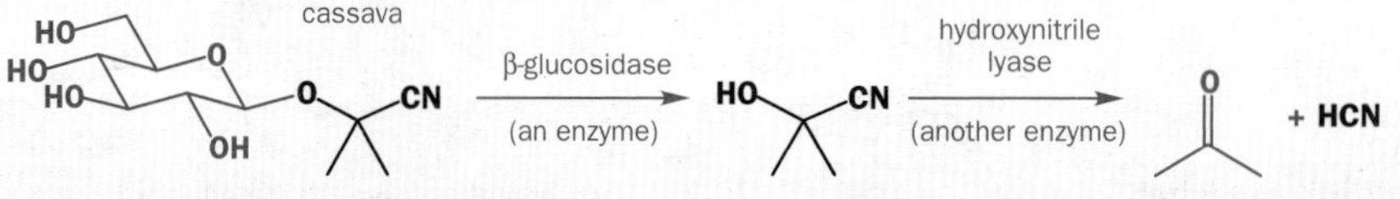

to the carbonyl carbon atom. As the carbonyl carbon atom changes from sp^2 to sp^3, its bond angles change from about 120° to about 109°—in other words, the substituents it carries move closer together. This reduction in bond angle is not a problem for aldehydes, because one of the substituents is just a (very small) hydrogen atom, but for ketones, especially ones that carry larger alkyl groups, this effect can disfavour the addition reaction. Effects that result from the size of substituents and the repulsion between them are called **steric effects**, and we call the repulsive force experienced by large substituents **steric hindrance**.

> Steric hindrance (*not* hinder*ance*) is a consequence of repulsion between the electrons in all the filled orbitals of the alkyl substituents.

The angle of nucleophilic attack on aldehydes and ketones

Having introduced you to the sequence of events that makes up a nucleophilic attack at C=O (interaction of HOMO with LUMO, formation of new σ bond, breakage of π bond), we should now tell you a little more about the *direction* from which the nucleophile approaches the carbonyl group. Not only do nucleophiles always attack carbonyl groups at carbon, but they also always approach from a particular angle. You may at first be surprised by this angle, since nucleophiles attack not from a direction perpendicular to the plane of the carbonyl group but at about 107° to the C=O bond. This approach route is known as the **Bürgi–Dunitz trajectory** after the authors of the elegant crystallographic methods that revealed it. You can think of the angle of attack as the result of a compromise between maximum orbital overlap of the HOMO with π* and minimum repulsion of the HOMO by the electron density in the carbonyl π bond.

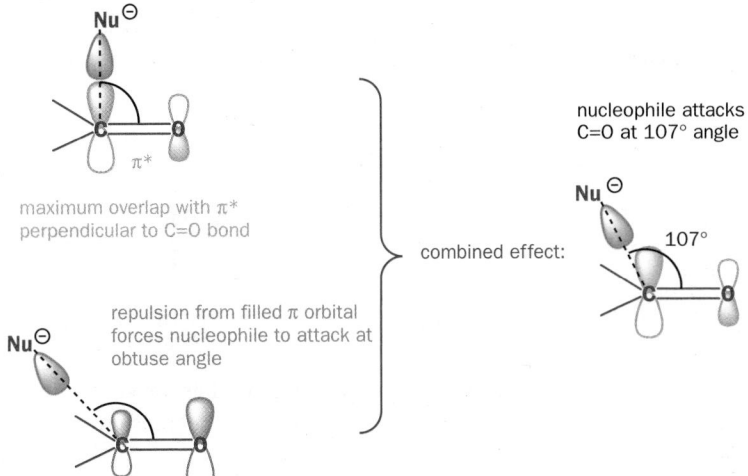

maximum overlap with π* perpendicular to C=O bond

repulsion from filled π orbital forces nucleophile to attack at obtuse angle

combined effect:

nucleophile attacks C=O at 107° angle

> ■ The Bürgi–Dunitz angle
> Bürgi and Dunitz deduced this trajectory by examining crystal structures of compounds containing both a nucleophilic nitrogen atom and an electrophilic carbonyl group. They found that, when the two got close enough to interact, but were not free to undergo reaction, the nitrogen atom always lay on or near the 107° trajectory decribed here. Theoretical calculations later gave the same 107° value for the optimum angle of attack.

Any other portions of the molecule that get in the way of (or, in other words, that cause *steric hindrance* to) the Bürgi–Dunitz trajectory will greatly reduce the rate of addition and this is another reason why aldehydes are more reactive than ketones. The importance of the Bürgi–Dunitz trajectory will become more evident later—particularly in Chapter 34.

> Although we now know precisely from which direction the nucleophile attacks the C=O group, this is not always easy to represent when we draw curly arrows. As long as you bear the Bürgi–Dunitz trajectory in mind, you are quite at liberty to write any of the variants shown here, among others.

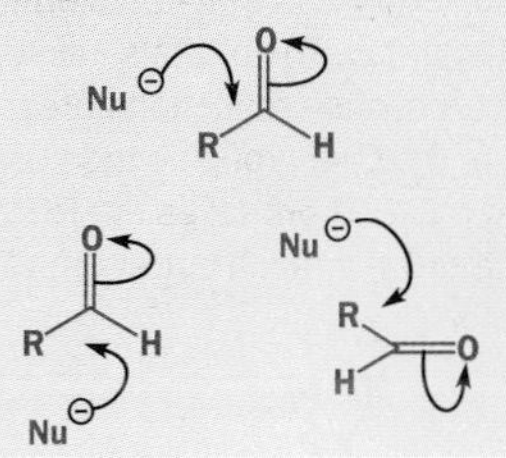

Nucleophilic attack by 'hydride' on aldehydes and ketones

Nucleophilic attack by the hydride ion, H^-, is not a known reaction. This species, which is present in the salt sodium hydride, NaH, is so small and has such a high charge density that it only ever reacts as a base. The reason is that its filled 1s orbital is of an ideal size to interact with the hydrogen

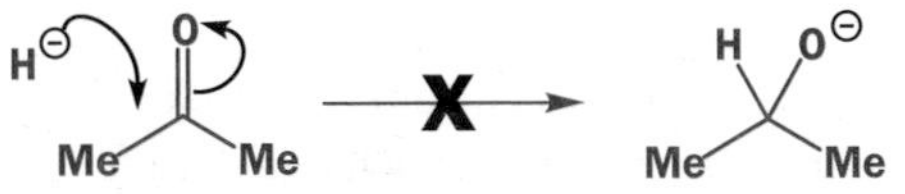

nucleophilic attack by H^-
never happens

H^- *always* reacts as a base

$H_2 + X^{\ominus}$

atom's contribution to the σ* orbital of an H–X bond (X can be any atom), but much too small to interact easily with carbon's more diffuse 2p orbital contribution to the LUMO (π*) of the C=O group.

Nevertheless, adding H⁻ to the carbon atom of a C=O group would be a very useful reaction, as the result would be the formation of an alcohol. This process would involve going down from the aldehyde or ketone oxidation level to the alcohol oxidation level (Chapter 2, pp. 25–36) and would therefore be a reduction. It cannot be done with NaH, but it can be done with some other compounds containing nucleophilic hydrogen atoms.

reduction of a ketone to an alcohol

The most important of these compounds is sodium borohydride, $NaBH_4$. This is a water-soluble salt containing the tetrahedral BH_4^- anion, which is isoelectronic with methane but has a negative charge since boron has one less proton in the nucleus than does carbon.

But beware! The boron's negative charge doesn't mean that there is a lone pair on boron—there isn't. You cannot draw an arrow coming out of this charge to form another bond. If you did, you would get a pentacovalent B(V) compound, which would have 10 electrons in its outer shell. Such a thing is impossible with a first row element as there are only four available orbitals ($1 \times 2s$ and $3 \times 2p$). Instead, since all of the electrons (including that represented by the negative charge) are in B–H σ orbitals, it is from a B–H bond that we must start any arrow to indicate reaction of BH_4^- as a nucleophile. By transferring this pair of electrons we make the boron atom neutral—it is now trivalent with just six electrons.

What happens when we carry out this reaction using a carbonyl compound as the electrophile? The hydrogen atom, together with the pair of electrons from the B–H bond, will be transferred to the carbon atom of the C=O group.

Though no hydride ion, H⁻, is actually involved in the reaction, the transfer of a hydrogen atom with an attached pair of electrons can be regarded as a 'hydride transfer'. You will often see it described this way in books. But be careful not to confuse BH_4^- with the hydride ion itself. To make it quite clear that it is the hydrogen atom that is forming the new bond to C, this reaction may also be helpfully represented with a curly arrow *passing through* the hydrogen atom.

The oxyanion produced in the first step can help stabilize the electron-deficient BH_3 molecule by adding to its empty p orbital. Now we have a tetravalent boron anion again, which could transfer a second hydrogen atom (with its pair of electrons) to another molecule of aldehyde.

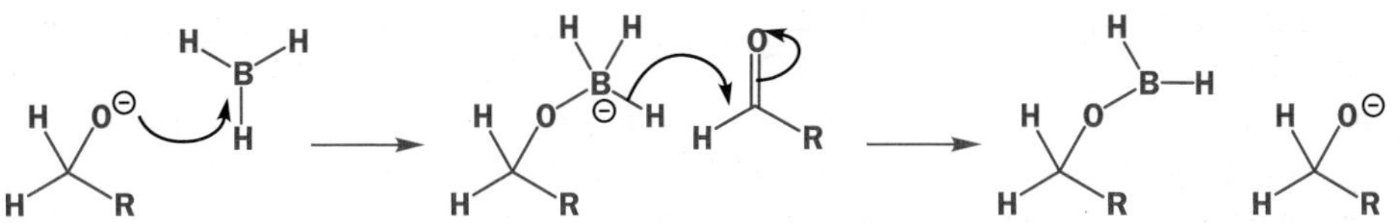

This process can continue so that, in principle, all four hydrogen atoms could be transferred to molecules of aldehyde. In practice the reaction is rarely as efficient as that, but aldehydes and ketones are usually reduced in good yield to the corresponding alcohol by sodium borohydride in water or alcoholic solution. The water or alcohol solvent provides the proton needed to form the alcohol from the alkoxide.

examples of reductions with sodium borohydride

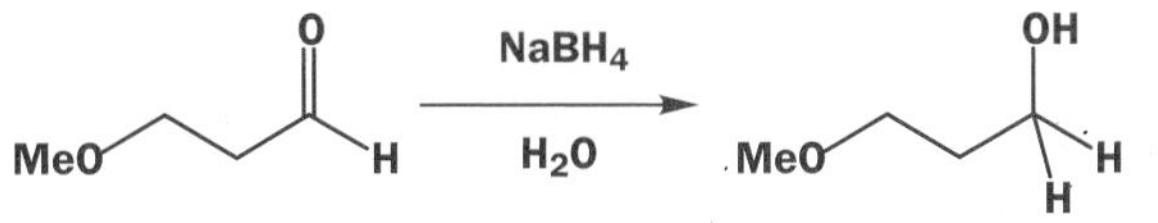

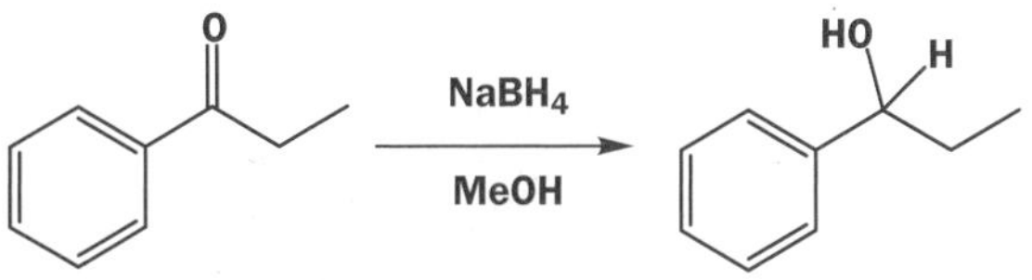

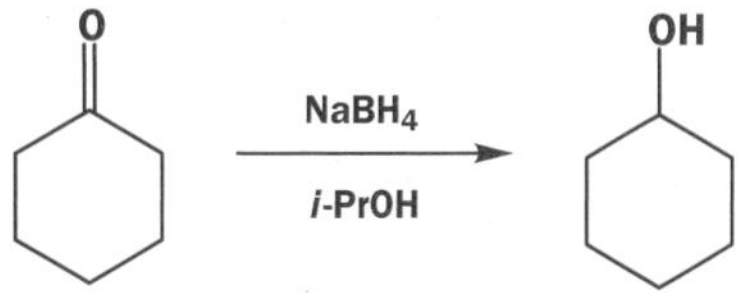

Sodium borohydride is one of the weakest hydride donors available. The fact that it can be used in water is evidence of this as more powerful hydride donors such as lithium aluminium hydride, LiAlH$_4$, react violently with water. Sodium borohydride reacts with both aldehydes and ketones, though the reaction with ketones is slower: for example, benzaldehyde is reduced about 400 times faster than acetophenone in isopropanol.

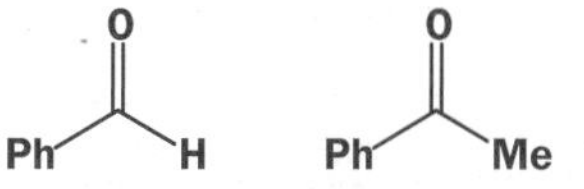

Sodium borohydride does not react at all with less reactive carbonyl compounds such as esters or amides: if a molecule contains both an aldehyde and an ester, only the aldehyde will be reduced.

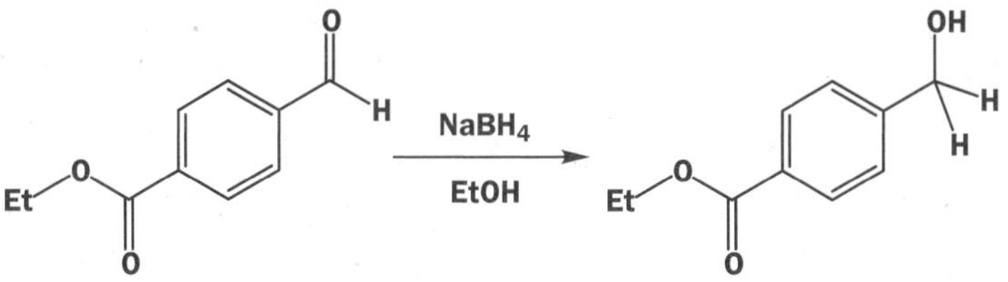

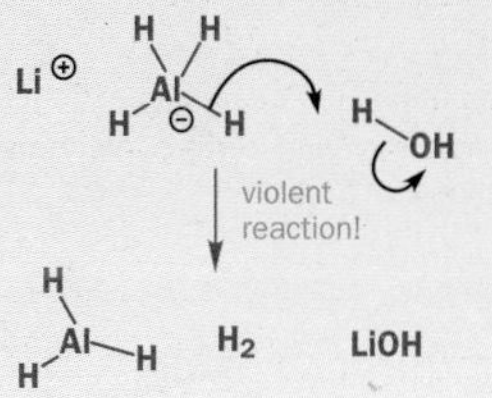

The next two examples illustrate the reduction of aldehydes and ketones in the presence of other reactive functional groups. No reaction occurs at the nitro group in the first case or at the alkyl halide in the second.

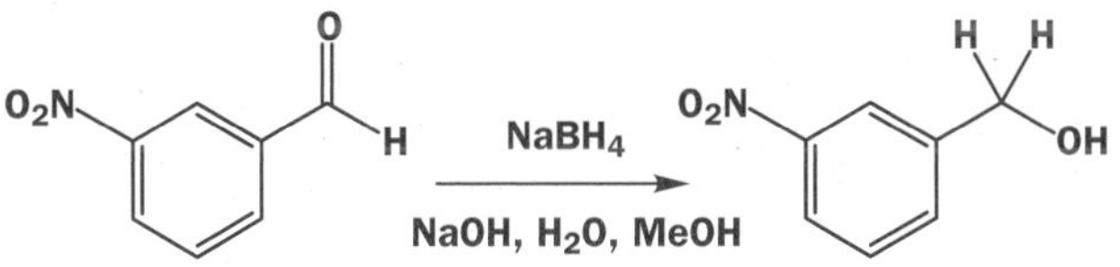

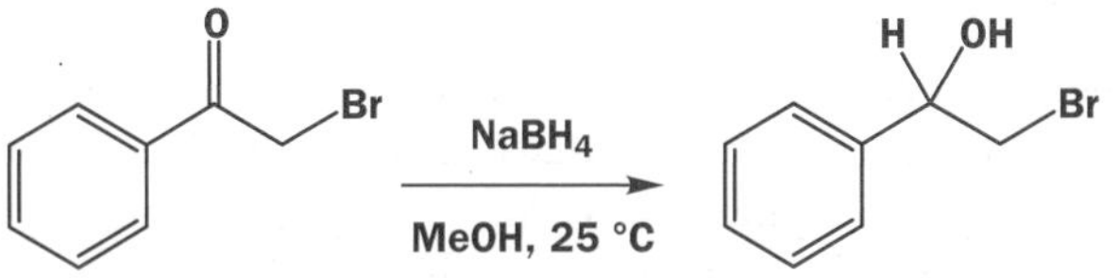

benzaldehyde   acetophenone

▶

Aluminium is more electropositive (more metallic) than boron and is therefore more ready to give up a hydrogen atom (and the associated negative charge), whether to a carbonyl group or to water. Lithium aluminium hydride reacts violently and dangerously with water in an exothermic reaction that produces highly flammable hydrogen.

violent reaction!

H$_2$ LiOH

▶

Organometallic compounds always have a metal–carbon bond.

Electronegativities

C	Li	Mg
2.5	1.0	1.2

■

We explained on p. 102 the origin of the polarization of bonds to electropositive elements.

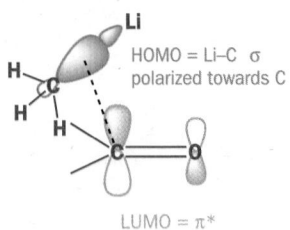

HOMO = Li–C σ polarized towards C

LUMO = π*

orbitals involved in the addition of methyllithium

organometallics are destroyed by water

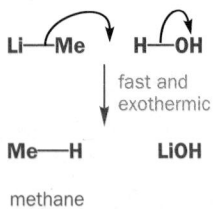

fast and exothermic

Me—H   LiOH

methane

■

Aprotic solvents contain no acidic protons, unlike, say, water or alcohols.

■

'Secondary' and 'tertiary' alcohols are defined on p. 30.

■

Victor Grignard (1871–1935) of the University of Lyon was awarded the Nobel Prize for chemistry in 1912 for his discovery of these reagents.

Addition of organometallic reagents to aldehydes and ketones

The next type of nucleophile we shall consider is the organometallic reagent. Lithium and magnesium are very electropositive metals, and the Li–C or Mg–C bonds in organolithium or organomagnesium reagents are highly polarized towards carbon. They are therefore very powerful nucleophiles, and attack the carbonyl group to give alcohols, forming a new C–C bond. For our first example, we shall take one of the simplest of organolithiums, methyllithium, which is commercially available as a solution in Et_2O, shown here reacting with an aldehyde. The orbital diagram of the addition step shows how the polarization of the C–Li bond means that it is the carbon atom of the nucleophile that attacks the carbon atom of the electrophile and we get a new C–C bond.

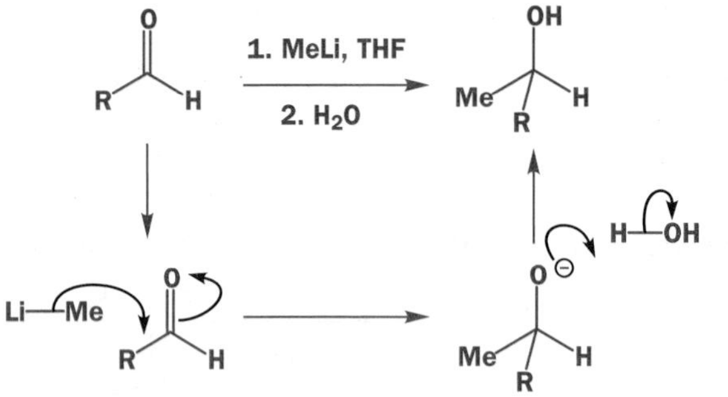

The course of the reaction is much the same as you have seen before, but we need to highlight a few points where this reaction scheme differs from those you have met earlier in the chapter. First of all, notice the legend '1. MeLi, THF; 2. H_2O'. This means that, first, MeLi is added to the aldehyde in a THF solvent. Reaction occurs: MeLi adds to the aldehyde to give an alkoxide. Then (and only then) water is added to protonate the alkoxide. The '2. H_2O' means that water is added in a separate step only when all the MeLi has reacted: it is not present at the start of the reaction as it was in the cyanide reaction and some of the borohydride addition reactions. In fact, water *must not* be present during the addition of MeLi (or of any other organometallic reagent) to a carbonyl group because water destroys organometallics very rapidly by protonating them to give alkanes (organolithiums and organomagnesiums are strong bases as well as powerful nucleophiles). The addition of water, or sometimes dilute acid or ammonium chloride, at the end of the reaction is known as the **work-up**.

Because they are so reactive, organolithiums are usually reacted at low temperature, often –78 °C (the sublimation temperature of solid CO_2), in aprotic solvents such as Et_2O or THF. Organolithiums also react with oxygen, so they have to be handled under a dry, inert atmosphere of nitrogen or argon.

Other common, and commercially available, organolithium reagents include *n*-butyllithium and phenyllithium, and they react with both aldehydes and ketones. Note that addition to an aldehyde gives a secondary alcohol while addition to a ketone gives a tertiary alcohol.

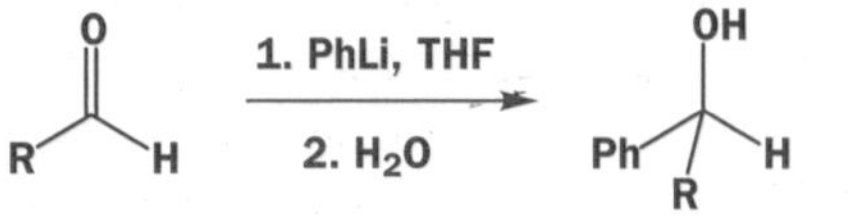

secondary alcohol 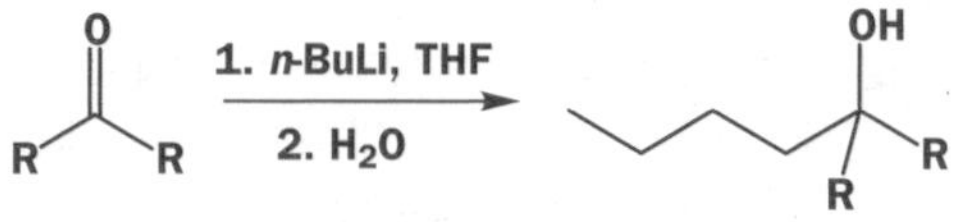   tertiary alcohol

Organomagnesium reagents known as **Grignard reagents** (RMgX) react in a similar way. Some simple Grignard reagents, such as methyl magnesium chloride, MeMgCl, and phenyl magnesium bromide, PhMgBr, are commercially available, and the scheme shows PhMgBr reacting with an aldehyde. The reactions of these two classes of organometallic reagent—organolithiums and Grignard reagents—with carbonyl compounds are among the most important ways of making carbon–carbon bonds, and we will consider them in more detail in Chapter 9.

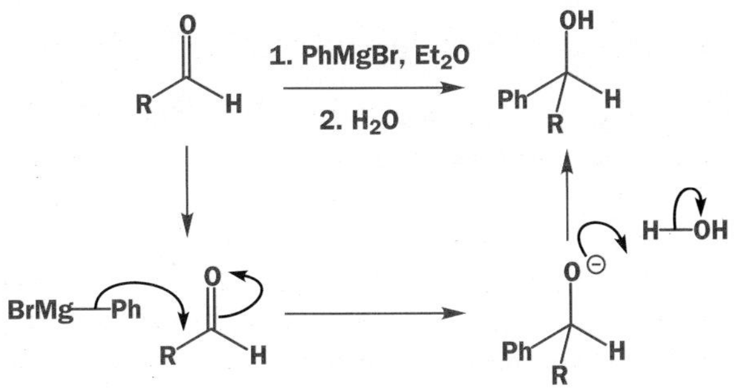

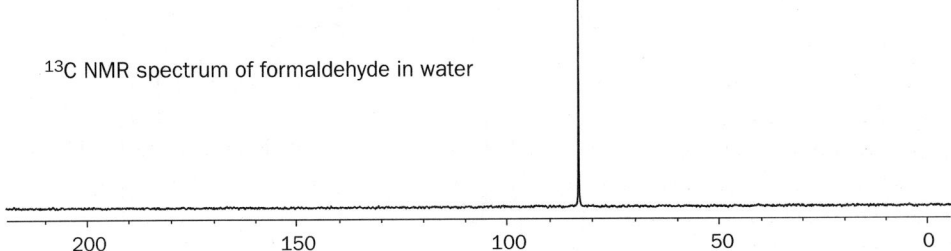

> Grignard reagents are made by reacting alkyl or aryl halides with magnesium 'turnings'.
>
> Mg, ether
> Ph—Br ⟶ Ph—Mg—Br

Addition of water to aldehydes and ketones

Nucleophiles don't have to be highly polarized or negatively charged to react with aldehydes and ketones: neutral ones will as well. How do we know? This ^{13}C NMR spectrum was obtained by dissolving formaldehyde, $H_2C=O$, in water. You will remember from Chapter 3 that the carbon atoms of carbonyl groups give ^{13}C signals typically in the region of 150–200 p.p.m. So where is formaldehyde's carbonyl peak? Instead we have a signal at 83 p.p.m.—where we would expect tetrahedral carbon atoms singly bonded to oxygen to appear.

^{13}C NMR spectrum of formaldehyde in water

200   150   100   50   0

What has happened is that water has added to the carbonyl group to give a compound known as a hydrate or 1,1-diol.

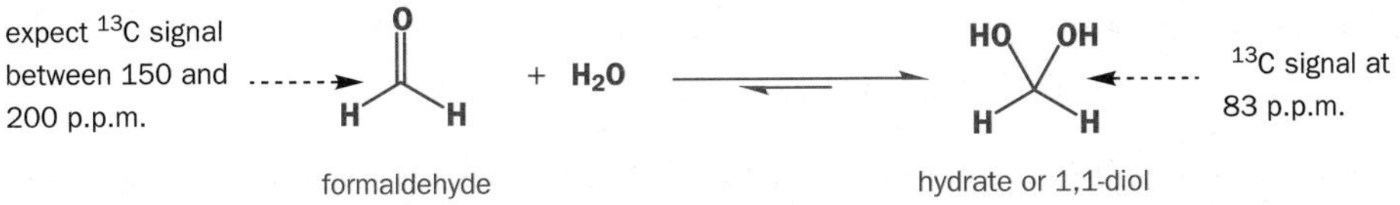

expect ^{13}C signal between 150 and 200 p.p.m. formaldehyde + H_2O ⇌ hydrate or 1,1-diol ^{13}C signal at 83 p.p.m.

This reaction, like the cyanohydrin formation we discussed at the beginning of the chapter, is an equilibrium, and is quite general for aldehydes and ketones. But, as with the cyanohydrins, the position of the equilibrium depends on the structure of the carbonyl compound. Generally, the same steric factors (pp. 138–139) mean that simple aldehydes are hydrated to some extent while simple ketones are not. However special factors can shift the equilibrium towards the hydrated form even for ketones, particularly if the carbonyl compound is reactive or unstable.

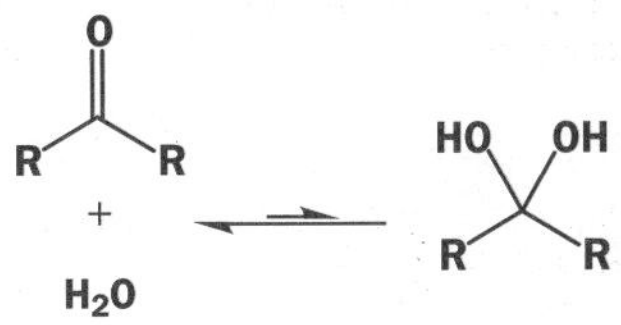

Formaldehyde is an extremely reactive aldehyde as it has no substituents to hinder attack—it is so reactive that it is rather prone to polymerization (Chapter 52). And it is quite happy to move from sp^2 to sp^3 hybridization because there is very little increased steric hindrance between the two hydrogen atoms as the bond angle changes from 120° to 109° (p. 139). This is why our aqueous solution of formaldehyde contains essentially no CH_2O—it is completely hydrated. A mechanism for the hydration reaction is shown below. Notice how a proton has to be transferred from one oxygen atom to the other, mediated by water molecules.

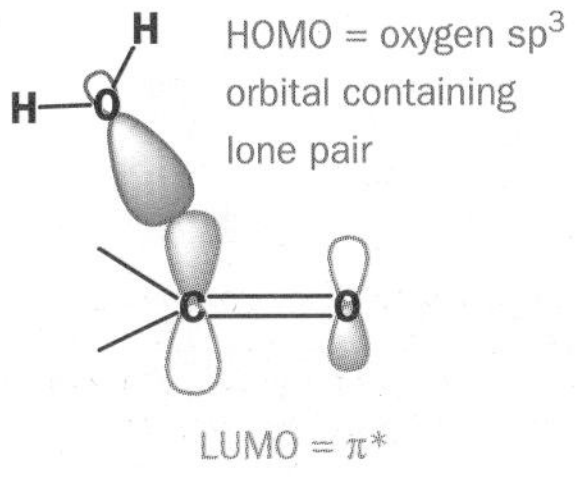

significant concentrations of hydrate are generally formed only from aldehydes

HOMO = oxygen sp^3 orbital containing lone pair

LUMO = π^*

orbitals involved in the addition of water

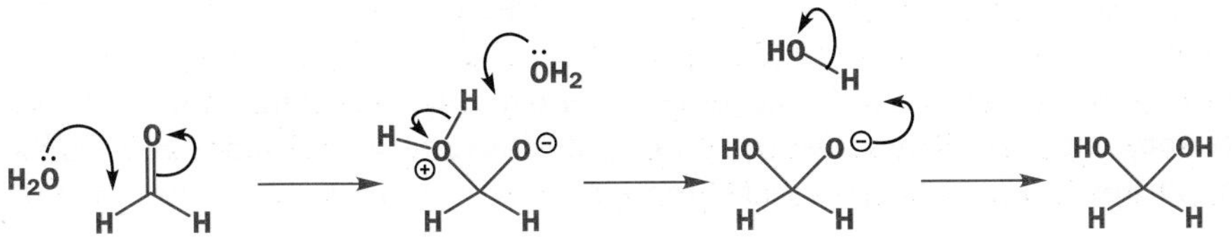

Monomeric formaldehyde

The hydrated nature of formaldehyde poses a problem for chemistry that requires anhydrous conditions such as the organometallic additions we have just been talking about. Fortunately *cracking* (heating to decomposition) the polymeric 'paraformaldehyde' can provide monomeric formaldehyde in anhydrous solution.

polymeric 'paraformaldehyde'

$$HO\!-\!\left[\!-\!O\!-\!\right]_n\!-\!OH \xrightarrow{\Delta} CH_2O$$

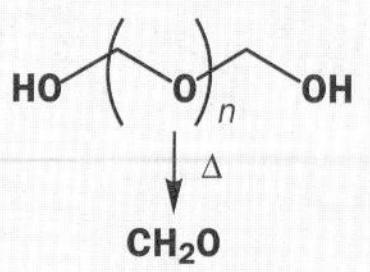

Formaldehyde reacts with water so readily because its substituents are very small: a steric effect. Electronic effects can also favour reaction with nucleophiles—electronegative atoms such as halogens attached to the carbon atoms next to the carbonyl group can increase the extent of hydration according to the number of halogen substituents and their electron-withdrawing power. They increase the polarization of the carbonyl group, which already has a positively polarized carbonyl carbon, and make it even more prone to attack by water. Trichloroacetaldehyde (chloral, Cl_3CHO) is hydrated completely in water, and the product 'chloral hydrate' can be isolated as crystals and is an anaesthetic. You can see this quite clearly in the two IR spectra. The first one is a spectrum of chloral hydrate from a bottle—notice there is no strong absorption between 1700 and 1800 cm^{-1} (where we would expect C=O to appear) and instead we have the tell-tale broad O–H peak at 3400 cm^{-1}. Heating drives off the water, and the second IR spectrum is of the resulting dry chloral: the C=O peak has reappeared at 1770 cm^{-1}, and the O–H peak has gone.

■ Chloral hydrate is the infamous 'knock out drops' of Agatha Christie or the 'Mickey Finn' of Prohibition gangsters.

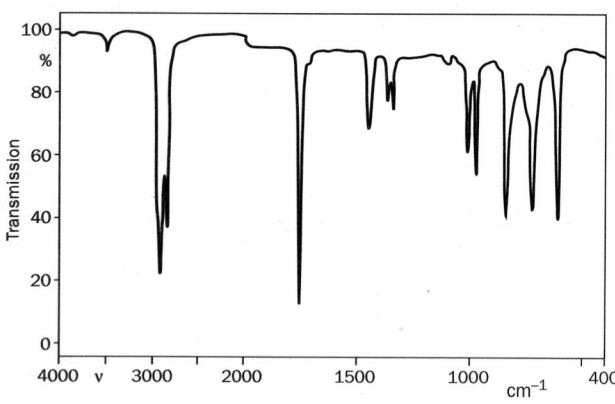

IR spectrum of chloral (nujol)

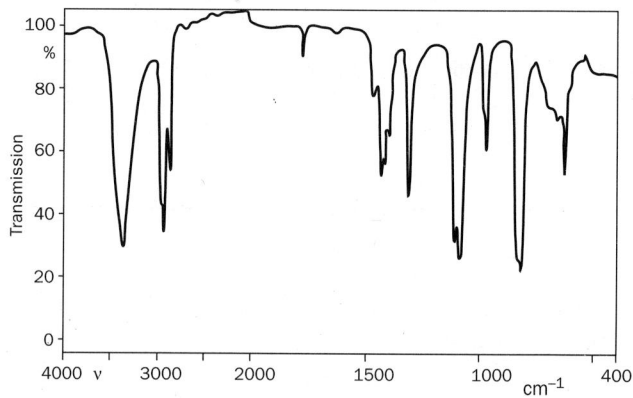

IR spectrum of chloral hydrate (nujol)

The chart shows the extent of hydration (in water) of a small selection of carbonyl compounds: hexafluoroacetone is probably the most hydrated carbonyl compound possible!

$$\underset{R}{\overset{O}{\|}}\!\!\underset{R}{\diagdown}\; +\; H_2O\; \underset{\longleftarrow}{\overset{K}{\longrightarrow}}\; \underset{R}{\overset{HO\;\;OH}{\diagup\diagdown}}\!\underset{R}{}$$

		equilibrium constant K		
acetone		0.001		
acetaldehyde	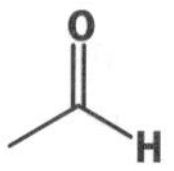	1.06	chloral	2000
formaldehyde		2280	hexafluoroacetone	1 200 000

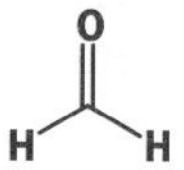

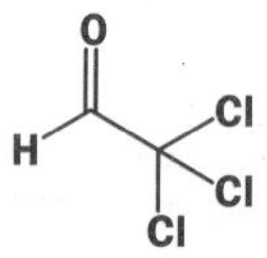

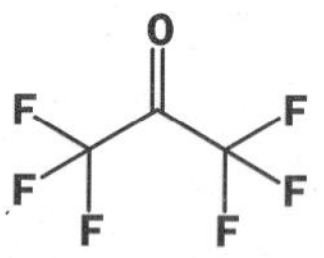

■ The larger the equilibrium constant, the more the equilibrium is to the *right*.

Cyclopropanones—three-membered ring ketones—are also hydrated to a significant extent, but for a different reason. You saw earlier how *acyclic* ketones suffer increased steric hindrance when the bond angle changes from 120° to 109° on moving from sp^2 to sp^3 hybridization. Cyclopropanones (and other small-ring ketones) conversely prefer the small bond angle because their substituents are already confined within a ring. Look at it this way: a three-membered ring is really very strained, with bond angles forced to be 60°. For the sp^2 hybridized ketone this means bending the bonds 60° away from their 'natural' 120°. But for the sp^3 hybridized hydrate the bonds have to be distorted by only 49° (= 109° − 60°). So addition to the C=O group allows some of the strain inherent in the small ring to be released—hydration is favoured, and indeed cyclopropanone and cyclobutanone are very reactive electrophiles.

> ● **The same structural features that favour or disfavour hydrate formation are important in determining the reactivity of carbonyl compounds with other nucleophiles, whether the reactions are reversible or not. Steric hindrance and more alkyl substituents make carbonyl compounds less reactive towards any nucleophile; electron-withdrawing groups and small rings make them more reactive.**

Hemiacetals from reaction of alcohols with aldehydes and ketones

Since water adds to (at least some) carbonyl compounds, it should come as no surprise that alcohols do too. The product of the reaction is known as a **hemiacetal**, because it is halfway to an acetal, a functional group, which you met in Chapter 2 (p. 35) and which will be discussed in detail in Chapter 14. The mechanism follows in the footsteps of hydrate formation: just use ROH instead of HOH.

A proton has to be transferred from one oxygen atom to the other: we have shown ethanol doing this job, with one molecule being protonated and one deprotonated. There is no overall consumption of ethanol in the protonation/deprotonation steps, and the order in which these steps happen is not important. In fact, you could reasonably write them in one step as shown in the margin, without involving the alcohol, and we do this in the next hemiacetal-forming reaction below. As with all these carbonyl group reactions, what is really important is the addition step, not what happens to the protons.

Hemiacetal formation is reversible, and hemiacetals are stabilized by the same special structural features as those of hydrates. However, hemiacetals can also gain stability by being cyclic—when the carbonyl group and the attacking hydroxyl group are part of the same molecule. The reaction is now an **intramolecular** (within the same molecule) addition, as opposed to the **intermolecular** (between two molecules) ones we have considered so far.

> cyclopropanone
>
> sp^2 C wants 120°, but gets 60°
>
> ↕ H₂O
>
> sp^3 C wants 109°, but gets 60°
>
> cyclopropanone hydrate

> hemiacetal   acetal
>
> hemiacetal from ketone (or "hemiketal")   a cyclic hemiacetal (or "lactol")
>
> names for functional groups

> • *Inter*molecular reactions occur between two molecules
> • *Intra*molecular reactions occur within the same molecule

> ■ We shall discuss the reasons why intramolecular reactions are more favourable, and why cyclic hemiacetals and acetals are more stable, in Chapter 14.

aldehyde   → EtOH →   hemiacetal

HÖEt

hydroxyaldehyde   →   cyclic hemiacetal

intramolecular attack of hydroxyl group

Although the cyclic hemiacetal (also called 'lactol') product is more stable, it is still in equilibrium with some of the open-chain hydroxyaldehyde form. Its stability, and how easily it forms, depend on the size of the ring: five- and six-membered rings are free from strain (their bonds are free to adopt 109° or 120° angles—compare the three-membered rings on p. 145), and five- or six-membered hemiacetals are common. Among the most important examples are many sugars. Glucose, for example, is a hydroxyaldehyde that exists mainly as a six-membered cyclic hemiacetal (>99% of glucose is cyclic in solution), while ribose exists as a five-membered cyclic hemiacetal.

▶

The way we have represented some of these molecules may be unfamiliar to you: we have shown **stereochemistry** (whether bonds come out of the paper or into it—the wiggly lines indicate a mixture of both) and, for the cyclic glucose, **conformation** (the actual shape the molecules adopt). These are very important in the sugars: we devote Chapter 16 to stereochemistry and Chapter 18 to conformation.

hydroxyaldehyde   can be drawn as   hydroxyaldehyde   cyclic **glucose**: >99% in this form

hydroxyaldehyde   can be drawn as   hydroxyaldehyde   cyclic **ribose**

Ketones can form hemiacetals

Hydroxyketones also form hemiacetals, but (as you should now expect) they usually do so less readily than hydroxyaldehydes. However, this hydroxyketone must exist solely as the cyclic hemiacetal because it shows no C=O stretch in its IR spectrum. The reason? The starting hydroxyketone is already cyclic, with the hydroxyl group poised to attack the ketone—it can't get away, so cyclization is highly favoured.

Acid and base catalysis of hemiacetal and hydrate formation

▶

A **catalyst** increases the rate of a chemical reaction but emerges from the reaction unchanged.

In Chapter 8 we shall look in detail at acids and bases, but at this point we need to tell you about one of their important roles in chemistry: they act as catalysts for a number of carbonyl addition reactions, among them hemiacetal and hydrate formation. To see why, we need to look back at the mechanisms of hemiacetal formation on p. 145 and hydrate formation on p. 143. Both involve proton-transfer steps like this.

ethanol acting as a base

ethanol acting as an acid

■

We introduced protonation by acid in Chapter 5, pp. 119–121.

In the first proton-transfer step, ethanol acts as a **base**, removing a proton; in the second it acts as an **acid**, donating a proton. Strong acids or strong bases (for example, HCl or NaOH) increase the rate of hemiacetal or hydrate formation because they allow these proton-transfer steps to occur *before* the addition to the carbonyl group.

In acid (dilute HCl, say), this is the mechanism. The first step is now protonation of the carbonyl group's lone pair: the positive charge makes it much more electrophilic so the addition reaction is faster. Notice how the proton added at the beginning is lost again at the end—it really is a catalyst. In acid it is also possible for the hemiacetal to react further with the alcohol to form an acetal, but this need not concern you at present.

hemiacetal formation in acid

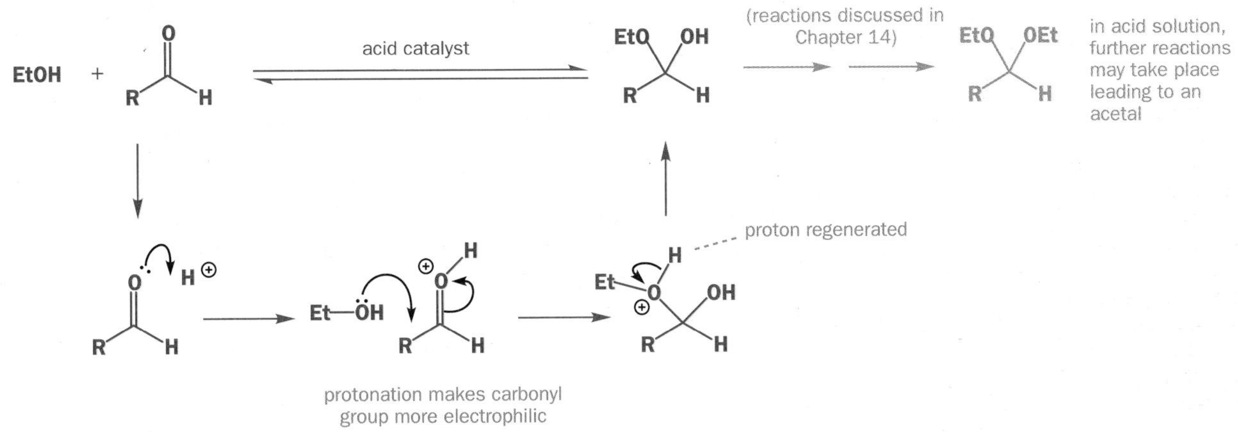

protonation makes carbonyl
group more electrophilic

And this is the mechanism in basic solution. The first step is now deprotonation of the ethanol by hydroxide, which makes the addition reaction faster by making the ethanol more nucleophilic. Again, base (hydroxide) is regenerated in last step, making the overall reaction catalytic in base. The reaction in base always stops with the hemiacetal—acetals never form in base.

hemiacetal formation in base

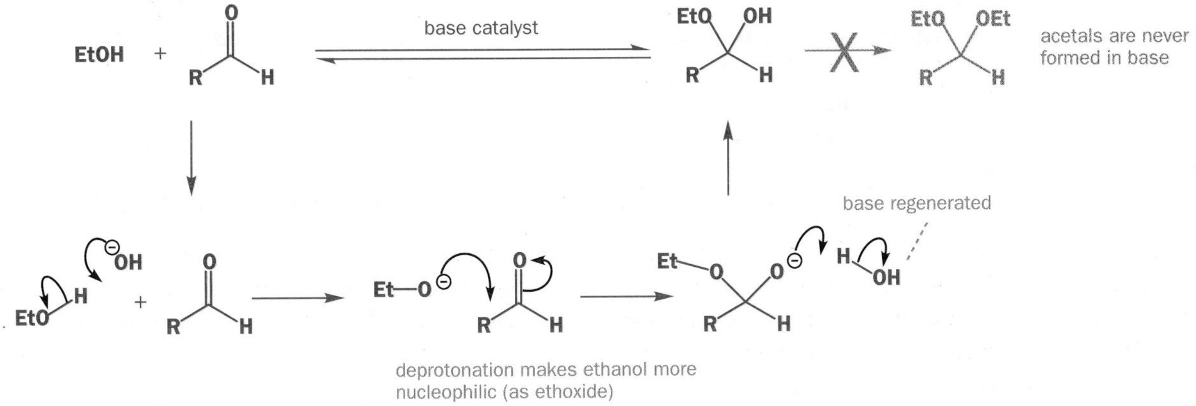

deprotonation makes ethanol more
nucleophilic (as ethoxide)

The final step could equally well involve deprotonation of ethanol to give alkoxide—and alkoxide could equally well do the job of catalysing the reaction. In fact, you will often come across mechanisms with the base represented just as 'B⁻' because it doesn't matter what the base is.

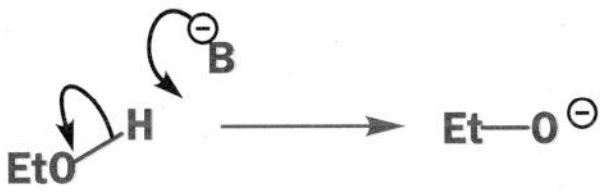

These two mechanisms typify acid- and base-catalysed additions to carbonyl groups and we can summarize the effects of the two catalysts.

● **For nucleophilic additions to carbonyl groups:**
  • **Acid catalysts work by making the carbonyl group more electrophilic**
  • **Base catalysts work by making the nucleophile more nucleophilic**

Bisulfite addition compounds

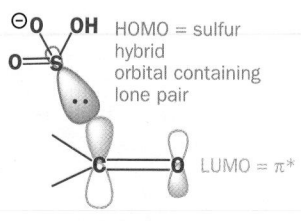

orbitals involved in the addition of bisulfite

> The structure of NaHSO$_3$, sodium bisulfite, is rather curious. It is an oxyanion of a sulfur(IV) compound with a lone pair of electrons—the HOMO—on the sulfur atom, but the charge is formally on the more electronegative oxygen. As a 'third row' element (third row of the periodic table, that is) sulfur can have more that just eight electrons—it's all right to have four or six bonds to S or P, unlike, say, B or N.

The last nucleophile of this chapter, sodium bisulfite, NaHSO$_3$, adds to aldehydes and some ketones to give what is usually known as a **bisulfite addition compound**. The reaction occurs by nucleophilic attack of a lone pair on the carbonyl group, just like the attack of cyanide. This leaves a positively charged sulfur atom but a simple proton transfer leads to the product.

bisulfite addition compound

The products are useful for two reasons. They are usually crystalline and so can be used to purify liquid aldehydes by recrystallization. This is of value only because this reaction, like several you have met in this chapter, is reversible. The bisulfite compounds are made by mixing the aldehyde or ketone with saturated aqueous sodium bisulfite in an ice bath, shaking, and crystallizing. After purification the bisulfite addition compound can be hydrolysed back to the aldehyde in dilute aqueous acid or base.

crystalline solid

The reversibility of the reaction makes bisulfite compounds useful intermediates in the synthesis of other adducts from aldehydes and ketones. For example, one practical method for making cyanohydrins involves bisulfite compounds. The famous practical book 'Vogel' suggests reacting acetone first with sodium bisulfite and then with sodium cyanide to give a good yield (70%) of the cyanohydrin.

70% yield

What is happening here? The bisulfite compound forms first, but only as an intermediate on the route to the cyanohydrin. When the cyanide is added, reversing the formation of the bisulfite compound provides the single proton necessary to to give back the hydroxyl group at the end of the reaction. No dangerous HCN is released (always a hazard when cyanide ions and acid are present together).

Other compounds from cyanohydrins

Cyanohydrins can be converted by simple reactions into hydroxyacids or amino alcohols. Here is one example of each, but you will have to wait until Chapter 12 for the details and the mechanisms of the reactions. Note that one cyanohydrin was made by the simplest method—just NaCN and acid—while the other came from the bisulfite route we have just discussed.

hydroxyacids by hydrolysis of CN in cyanohydrin

amino alcohols by reduction of CN in cyanohydrin

The bisulfite compound of formaldehyde (CH_2O) has special significance. Earlier in this chapter we mentioned the difficulty of working with formaldehyde because it is either an aqueous solution or a dry polymer. One readily available monomeric form is the bisulfite compound. It can be made in water (in which it is soluble) but addition of ethanol (in which it isn't) causes it to crystallize out.

The compound is commercially available and, together with the related zinc salt, is widely used in the textile industry as a reducing agent.

The second reason that bisulfite compounds are useful is that they are soluble in water. Some small (that is, low molecular weight) aldehydes and ketones are water-soluble—acetone is an example. But most larger (more than four or so carbon atoms) aldehydes and ketones are not. This does not usually matter to most chemists as we often want to carry out reactions in organic solvents rather than water. But it can matter to medicinal chemists, who make compounds that need to be compatible with biological systems. And in one case, the solubility of bisulfite adduct in water is literally vital.

Dapsone is an antileprosy drug. It is a very effective one too, especially when used in combination with two other drugs in a 'cocktail' that can be simply drunk as an aqueous solution by patients in tropical countries without any special facilities, even in the open air. But there is a problem! Dapsone is insoluble in water.

The solution is to make a bisulfite compound from it. You may ask how this is possible since dapsone has no aldehyde or ketone—just two amino groups and a sulfone. The trick is to use the formaldehyde bisulfite compound and exchange the OH group for one of the amino groups in dapsone.

dapsone: antileprosy drug; insoluble in water

water-soluble "pro-drug"

Now the compound will dissolve in water and release dapsone inside the patient. The details of this sort of chemistry will come in Chapter 14 when you will meet imines as intermediates. But at this stage we just want you to appreciate that even the relatively simple chemistry in this chapter is useful in synthesis, in commerce, and in medicine.

Problems

1. Draw mechanisms for these reactions.

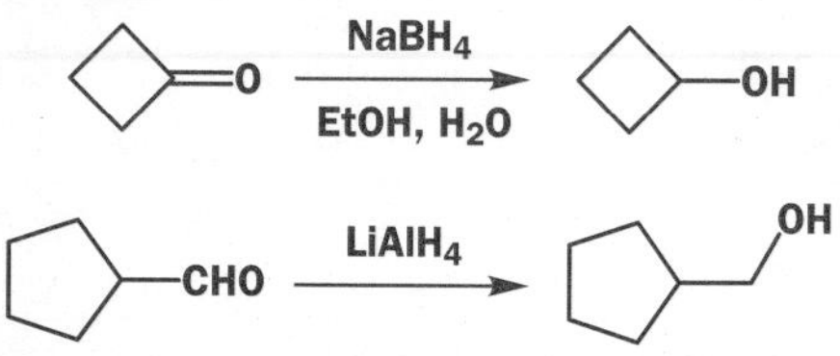

2. Cyclopropanone exists as the hydrate in water but 2-hydroxyethanal does not exist as its hemiacetal. Explain.

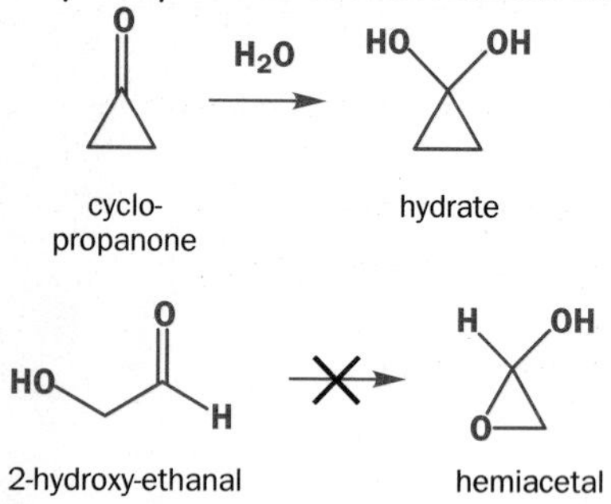

3. One way to make cyanohydrins is illustrated here. Suggest a detailed mechanism for the process.

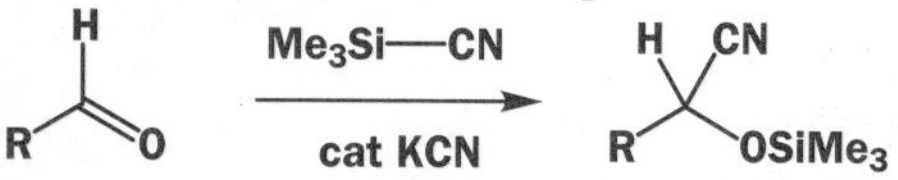

4. There are three possible products from the reduction of this compound with sodium borohydride. What are their structures? How would you distinguish them spectroscopically, assuming you can isolate pure compounds?

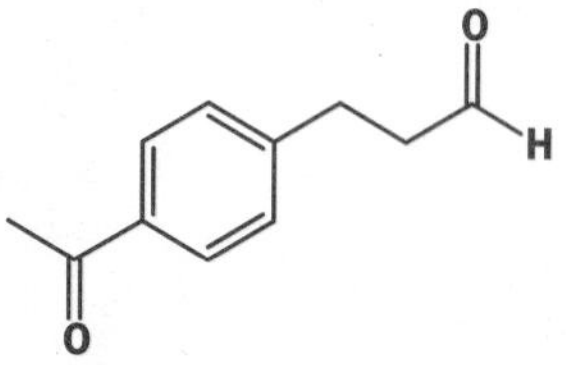

5. The triketone shown here is called 'ninhydrin' and is used for the detection of amino acids. It exists in aqueous solution as a monohydrate. Which of the three ketones is hydrated and why?

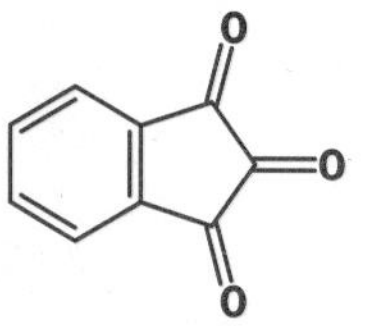

ninhydrin

6. This hydroxyketone shows no peaks in its infrared spectrum between 1600 and 1800 cm^{-1} but it does show a broad absorption at 3000 to 3400 cm^{-1}. In the ^{13}C NMR spectrum, there are no peaks above 150 p.p.m. but there is a peak at 110 p.p.m. Suggest an explanation.

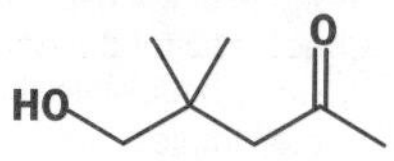

7. Each of these compounds is a hemiacetal and therefore formed from an alcohol and a carbonyl compound. In each case give the structure of these original materials.

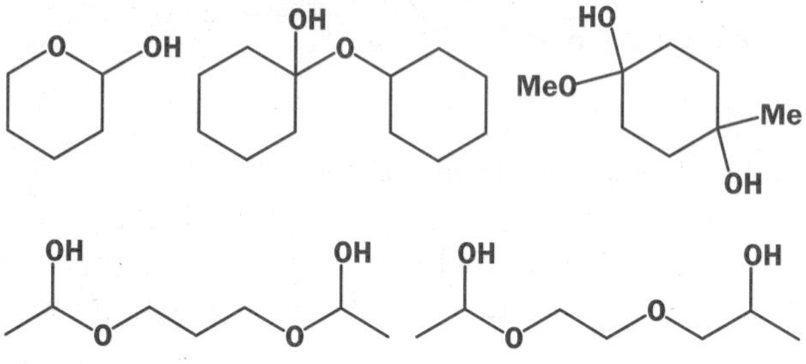

8. Trichloroethanol may be prepared by the direct reduction of chloral hydrate in water with sodium borohydride. Suggest a mechanism for this reaction. (Warning! Sodium borohydride does *not* displace hydroxide from carbon atoms!)

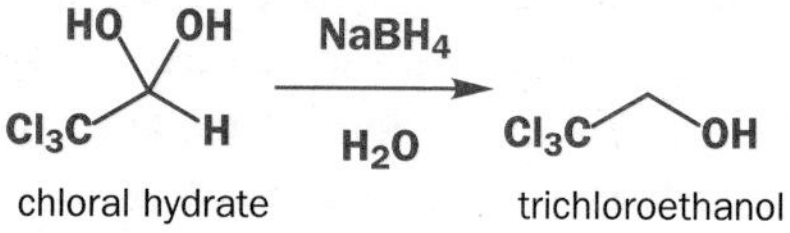

chloral hydrate   trichloroethanol

9. It has not been possible to prepare the adducts from simple aldehydes and HCl. What would be the structure of such compounds, if they could be made, and what would be the mechanism of their formation? Why cannot these compounds in fact be made?

10. What would be the products of these reactions? In each case give a mechanism to justify your predictions.

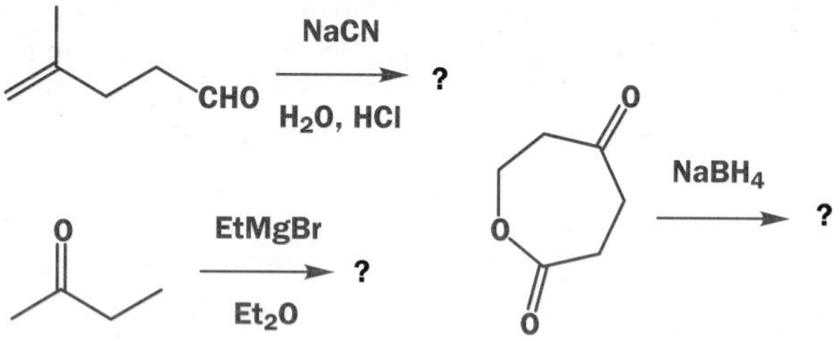

11. The equilibrium constant K_{eq} for formation of the cyanohydrin of cyclopentanone and HCN is 67, while for butan-2-one and HCN it is 28. Explain.

Delocalization and conjugation

7

Connections

Building on:
- Orbitals and bonding ch4
- Representing mechanisms by curly arrows ch5
- Ascertaining molecular structure spectroscopically ch3

Arriving at:
- Interaction between orbitals over many bonds
- Stabilization by the sharing of electrons over molecules
- Where colour comes from
- Molecular shape and structure determine reactivity
- Representing one aspect of structure by curly arrows
- Structure of aromatic compounds

Looking forward to:
- Acidity and basicity ch8
- Conjugate addition and substitution ch10
- Chemistry of aromatic compounds ch22 & ch23
- Enols and enolates ch21, ch25–ch29
- Chemistry of heterocycles ch43 & ch44
- Chemistry of life ch49–ch51

Introduction

As you look around you, you will be aware of many different colours—from the greens and browns outside to the bright blues and reds of the clothes you are wearing. All these colours result from the interaction of light with the pigments in these different things—some frequencies of light are absorbed, others scattered. Inside our eyes, chemical reactions detect these different frequencies and convert them into electrical nerve impulses sent to the brain. All these different pigments have one thing in common—lots of double bonds. For example, the pigment responsible for the red colour in tomatoes, lycopene, is a long-chain polyalkene.

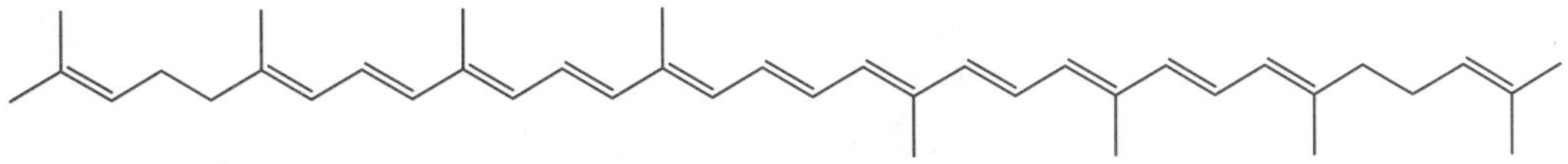

lycopene, the red pigment in tomatoes, rose hips, and other berries

Lycopene contains only carbon and hydrogen while most pigments contain many other elements but nearly all contain double bonds. This chapter is about the properties, such as colour, of molecules that have several double bonds and that depend on the joining up or **conjugation** of these double bonds.

In earlier chapters, we talked about basic carbon skeletons made up of σ bonds. In this chapter we shall see how, in some cases, we can also have a large π framework spread over many atoms and how this dominates the chemistry of such compounds. We shall see how this π framework is responsible for the otherwise unexpected stability of certain cyclic polyunsaturated compounds, including benzene and other aromatic compounds. We shall also see how this framework gives rise to the many colours in our world. To understand such molecules properly, we need to start with the simplest of all unsaturated compounds, ethene.

The structure of ethene (ethylene, $CH_2=CH_2$)

The structure of ethene (ethylene) is well known. It has been determined by electron diffraction and is **planar** (*all* atoms are in the same plane) with the bond lengths and angles shown below. The carbon atoms are roughly trigonal and the C–C bond distance is shorter than that of a C–C σ bond.

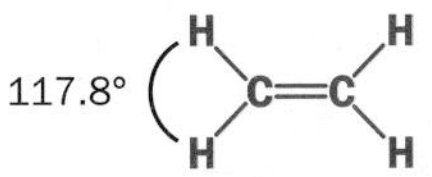

117.8°

C–H bond length 108 pm
C=C bond length 133 pm

▶

Important point. Ethene is not actually formed by bringing together two carbon atoms and four hydrogen atoms: individual carbon atoms do not hybridize their atomic orbitals and then combine. We are simply trying to rationalize the shapes of molecular orbitals. Hybridization and LCAO are tools to help us accomplish this.

We shall use the approach of Chapter 4 (p. 106) and rationalize the shapes of molecular orbitals by combining the atomic orbitals of the atoms involved using the LCAO (Linear Combination of Atomic Orbitals) approach. Hybridizing the atomic orbitals first makes this simpler. We mix the 2s orbital on each carbon atom with two of the three 2p orbitals to give three sp^2 orbitals leaving the third p orbital unchanged. Two of the sp^2 orbitals overlap with the hydrogen 1s orbitals to form molecular orbitals, which will be the C–H σ bonds. The other sp^2 orbital forms the σ C–C bond by overlapping with the sp^2 orbital on the other carbon. The remaining p orbital can overlap with the p orbital on the other carbon to form a molecular orbital that represents the π bond.

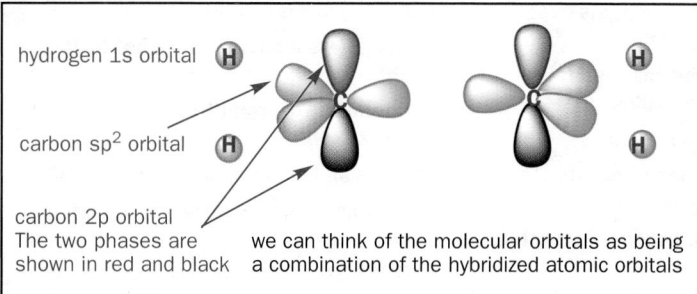

hydrogen 1s orbital

carbon sp^2 orbital

carbon 2p orbital
The two phases are shown in red and black

we can think of the molecular orbitals as being a combination of the hybridized atomic orbitals

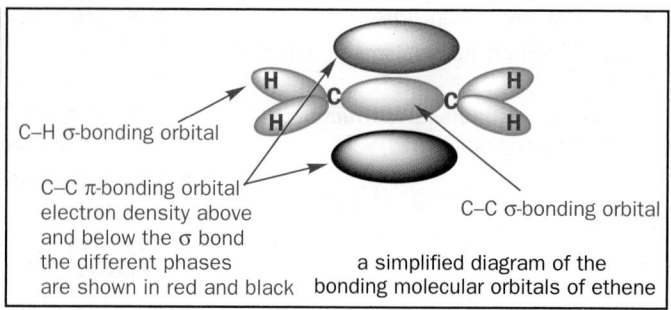

C–H σ-bonding orbital

C–C π-bonding orbital electron density above and below the σ bond the different phases are shown in red and black

C–C σ-bonding orbital

a simplified diagram of the bonding molecular orbitals of ethene

Eth*ene* is chemically more interesting than eth*ane* because of the π bond. In fact, the π bond is the most important feature of ethene. In the words of Chapter 5, the C–C π orbital is the HOMO (Highest Occupied Molecular Orbital) of the alkene, which means that the electrons in it are more available than any others to react with something that wants electrons (an electrophile). Since this orbital is so important, we will look at it more closely.

The π orbital results from combining the two 2p orbitals of the separate carbon atoms. Remember that when we combine *two* atomic orbitals we get *two* molecular orbitals. These result from combining the p orbitals either in-phase or out-of-phase. The in-phase combination accounts for the bonding molecular orbital (π), whilst the out-of-phase combination accounts for the anti-bonding molecular orbital (π*). As we progress to compounds with more than one alkene, so the number of π orbitals will increase but will remain the same as the number of π* orbitals.

α β notation

Theoretical chemists would label the energy of an electron in the p orbital as 'α' and that of an electron in the molecular orbital resulting from the combination of two p orbitals as 'α + β'. (Both α and β are, in fact, negative, which means that α + β is lower in energy than α.) The corresponding energy of an electron in the antibonding orbital is 'α − β'. Whilst α represents the energy an electron would have in an *atomic* orbital, β represents the change in energy when the electron is delocalized over the two carbon atoms. Since the π bond contains two electrons, both in the lowest-energy molecular orbital, the π orbital, the total energy of the electrons is 2α + 2β. If, instead, the two electrons remained in the atomic orbitals, their energy would be just 2α. Therefore the system is 2β lower in energy if the electrons are in the π molecular orbital rather than the atomic orbitals.

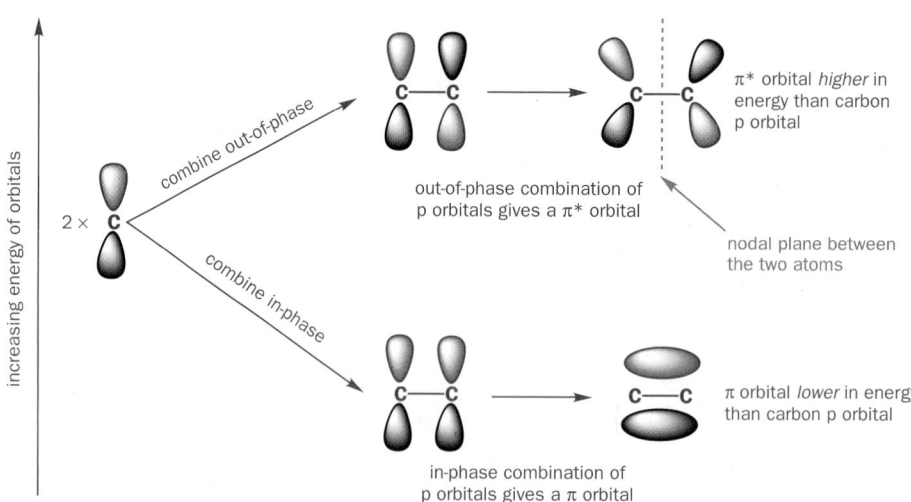

increasing energy of orbitals

$2 \times$

combine out-of-phase

combine in-phase

out-of-phase combination of p orbitals gives a π* orbital

π* orbital *higher* in energy than carbon p orbital

nodal plane between the two atoms

in-phase combination of p orbitals gives a π orbital

π orbital *lower* in energy than carbon p orbital

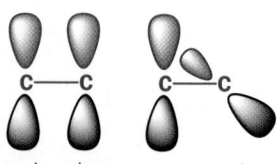

good overlap   poor overlap

the two p orbitals can only overlap if they are parallel

The π bond contains two electrons and, since we fill up the energy level diagram from the lowest-energy orbital upwards, both these electrons go into the bonding molecular orbital. In order to have a strong π bond, the two atomic p orbitals must be able to overlap effectively. This means they must be parallel.

There are two isomers (*cis* and *trans* or *E* and *Z*) of many alkenes

The π bond has electron density both above and below the σ bond as the parallel p orbitals overlap locking the bond rigid. Hence no rotation is possible about a double bond—the π bond must be broken before rotation can occur. One consequence of this locking effect of the double bond is that there are two isomers of a disubstituted alkene. One is called a *cis* or *Z* alkene, the other a *trans* or *E* alkene.

> By contrast, a C–C σ bond has electron density along the line joining the two nuclei and allows free rotation.

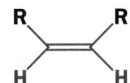

a *trans* or *E* alkene

a *cis* or *Z* alkene

Alkenes resist rotation

Maleic and fumaric acids were known in the nineteenth century to have the same chemical composition and the same functional groups and yet they were different compounds—why remained a mystery. That is, until 1874 when van't Hoff proposed that free rotation about double bonds was restricted. This meant that, whenever each carbon atom of a double bond had two different substituents, isomers would be possible. He proposed the terms *cis* (Latin meaning 'on this side') and *trans* (Latin meaning 'across or on the other side') for the two isomers. The problem was: which isomer was which? On heating, maleic acid readily loses water to become maleic anhydride so this isomer must have both acid groups on the same side of the double bond.

fumaric acid → heat → no change

trans carboxylic acid groups

maleic acid → heat, − H_2O → maleic anhydride

cis carboxylic acid groups

It is possible to interconvert *cis* and *trans* alkenes, but the π bond must be broken first. This requires a considerable amount of energy—around 260 kJ mol^{-1}. One way to break the π bond would be to promote an electron from the π orbital to the π* orbital (from HOMO to LUMO). If this were to happen, there would be one electron in the bonding π orbital and one in the antibonding π* orbital and hence no overall bonding. Electromagnetic radiation of the correct energy could promote the electron from HOMO to LUMO. The correct energy actually corresponds to light in the ultraviolet (UV) region of the spectrum. Thus, shining UV light on an alkene would promote an electron from its bonding π molecular orbital to its antibonding π* molecular orbital, thereby breaking the π bond (but not the σ bond) and allowing rotation to occur.

> Notice that it takes less energy to break a C–C π bond than a C–C σ bond (about 260 kJ mol^{-1} compared to about 350 kJ mol^{-1}). This is because the sideways overlap of the p orbitals to form a π bond is not as effective as the head-on overlap of the orbitals to form a σ bond. Enough energy is available to break the π bond if the alkene is heated to about 500 °C.

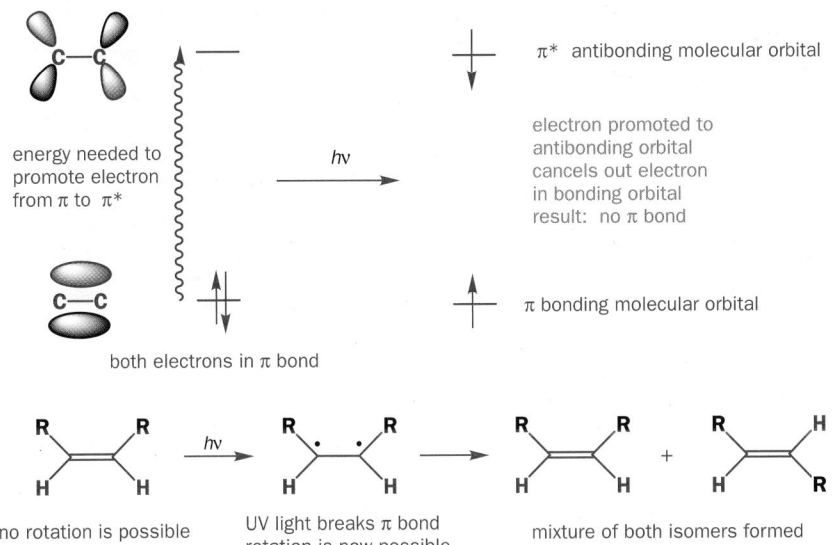

energy needed to promote electron from π to π*

hν

both electrons in π bond

π* antibonding molecular orbital

electron promoted to antibonding orbital cancels out electron in bonding orbital result: no π bond

π bonding molecular orbital

overlap of orbitals is more efficient in a σ bond than in a π bond

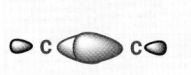

no rotation is possible

hν UV light breaks π bond rotation is now possible

mixture of both isomers formed

Molecules with more than one C–C double bond

Benzene has three strongly interacting double bonds

The rest of this chapter concerns molecules with more than one C–C double bond and what happens to the π orbitals when they interact. To start, we shall take a bit of a jump and look at the structure of benzene. Benzene has been the subject of considerable controversy since its discovery in 1825. It was soon worked out that the formula was C_6H_6, but how were these atoms arranged? Some strange structures were suggested until Kekulé proposed the correct structure in 1865.

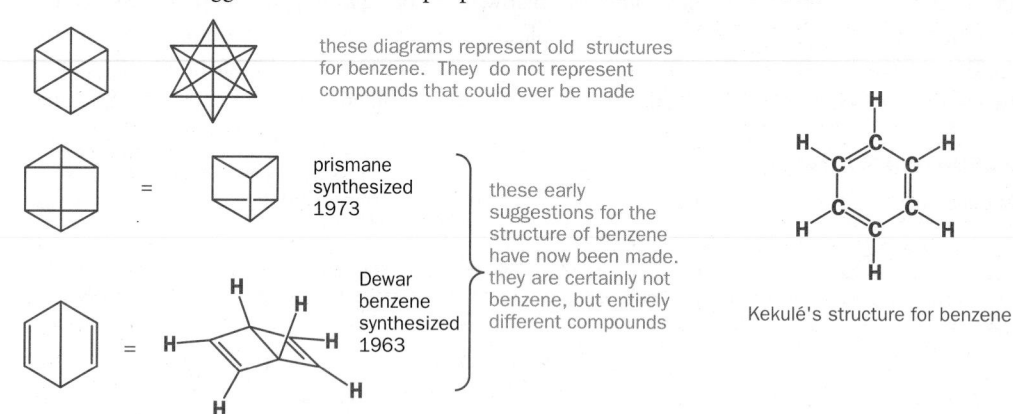

these diagrams represent old structures for benzene. They do not represent compounds that could ever be made

prismane synthesized 1973

Dewar benzene synthesized 1963

these early suggestions for the structure of benzene have now been made. they are certainly not benzene, but entirely different compounds

Kekulé's structure for benzene

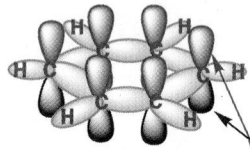

σ bonds shown in green

a single p orbital different phases shown in red and black

Let's look at the molecular orbitals for Kekulé's structure. As in simple alkenes, each of the carbon atoms is sp^2 hybridized leaving the remaining p orbital free.

The σ framework of the benzene ring is like the framework of an alkene. The problem comes with the p orbitals—which pairs do we combine to form the π bonds? There seem to be two possibilities.

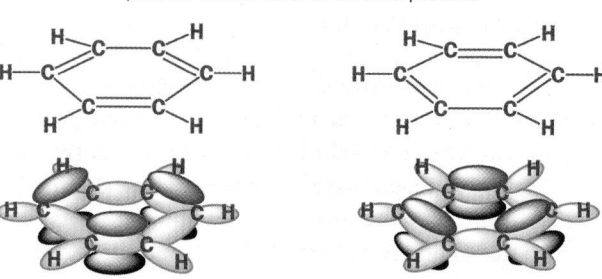

combining different pairs of p orbitals puts the double bonds in different postions

With benzene itself, these two forms are equivalent but, if we had a 1,2- or a 1,3-disubstituted benzene compound, these two forms would be different. A synthesis was designed for these two compounds but it was found that both compounds were identical. This posed a bit of a problem to Kekulé—his structure didn't seem to work after all. His solution was that benzene rapidly equilibrates, or 'resonates' between the two forms to give an averaged structure in between the two.

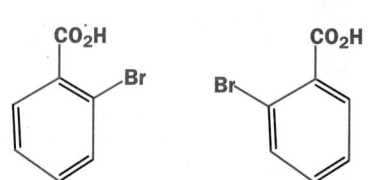

2-bromobenzoic acid   '6'-bromobenzoic acid

if the double bonds were localized then these two compound would be chemically different. (the double bonds are drawn shorter than the single bonds to emphasize the difference)

■

Combining these six atomic p orbitals actually produces *six* molecular orbitals. We shall consider the form of all these orbitals later in the chapter when we discuss benzene and aromaticity more fully.

The molecular orbital answer to this problem, as you may well know, is that all six p orbitals can combine to form (six) new molecular orbitals, one of which (the one lowest in energy) consists of a ring of electron density above and below the plane of the molecule. Benzene *does not resonate* between the two Kekulé structures—the electrons are in molecular orbitals spread equally over all the carbon atoms. However the term 'resonance' is still sometimes used (but not in this book) to describe this mixing of molecular orbitals.

We shall describe the π electrons in benzene as **delocalized**, that is, no longer localized in specific double bonds between two particular carbon atoms but spread out, or delocalized, over all six atoms in the ring. An alternative drawing for benzene shows the π system as a ring and does not put in the double bonds.

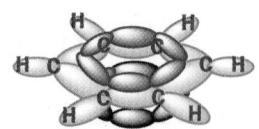
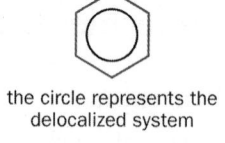

the circle represents the delocalized system

The Kekulé structure is used for mechanisms

This representation does present a slight problem to modern organic chemists, however: it is not possible to draw mechanisms using the delocalized representation of benzene. The curly arrows we use represent two electrons. This means that in order to write sensible mechanisms we still draw benzene as though the double bonds were localized. Keep in mind though that these double bonds are not really localized and it does not matter which way round we draw them.

this representation cannot be used for mechanisms

it does not matter which way round you draw the benzene ring – both drawings give the same result but one uses more arrows to get there

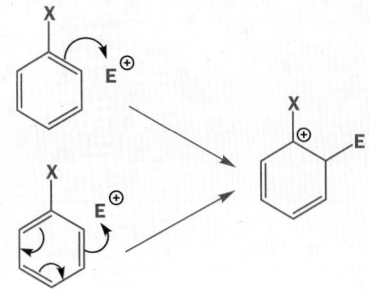

We are saying that *the π electrons are not localized* in alternating double bonds but are actually *spread out over the whole system* in a molecular orbital shaped like a ring (we will look at the shapes of the others later). The electrons are therefore said to be **delocalized**. Theoretical calculations confirm this model, as do experimental observations. Electron diffraction studies show benzene to be a regular, planar hexagon with all the carbon–carbon bond lengths identical (139.5 pm). This bond length is in between that of a carbon–carbon single bond (154.1 pm) and a full carbon–carbon double bond (133.7 pm). A further strong piece of evidence for this ring of electrons is revealed by proton NMR and is discussed on on p. 251.

● **Delocalization terminology**

What words should be used to describe delocalization is a vexed question. Terms such as resonance, mesomerism, conjugation, and delocalization are only a few of the ones you will find in books. You will already have noticed that we don't like 'resonance' because it suggests that the structure vibrates rapidly between localized structures. We shall use conjugation and delocalization: **conjugation** focuses on the sequence of alternating double and single bonds while **delocalization** focuses on the molecular orbitals covering the whole system. Electrons are *delocalized* over the whole of a *conjugated* system.

Noncyclic polyenes

What would the structure be like if the three C–C double bonds were not in a ring as they are in benzene but were instead in a chain. What is the structure of hexatriene? Are the bond lengths still all the same?

cut here →

benzene

hexatriene a 'cut open benzene ring'?

There are two isomers of hexatriene: a *cis* form and a *trans* form. The name refers to the geometry about the central double bond. The two isomers have different chemical and physical properties. Rotation is still possible about the single bonds (although slightly more difficult than around a normal single bond) and there are three different planar conformations possible for each isomer. Keeping the central black double bond the same, we can rotate about each of the green σ bonds in turn. Each row simply shows different ways to draw the same compound.

cis-hexatriene → rotation about single bond → → rotation about single bond →

trans-hexatriene → rotation about single bond → → rotation about single bond →

The structures of both *cis*- and *trans*-hexatriene have been determined by electron diffraction and two important features emerge.

- Both structures are essentially planar (the *cis* form is not quite for steric reasons)

- There are double and single bonds but the central double bond in each case is slightly longer than the end double bonds and the single bonds are slightly shorter than a 'standard' single bond

The most stable structure of *trans*-hexatriene is shown here.

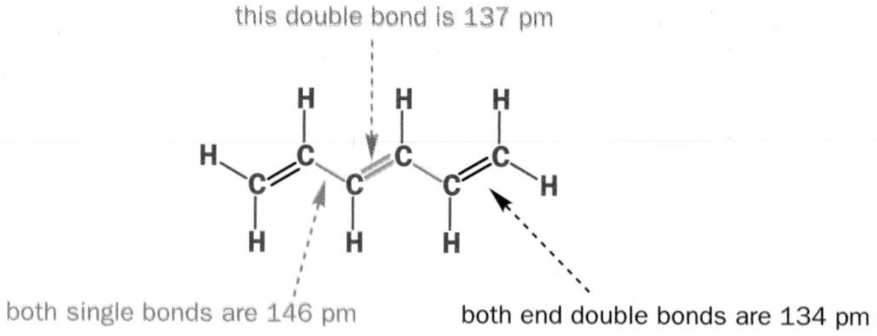

Why is this structure planar and why are the bond lengths different from their 'standard' values? This sounds like the situation with benzene and again the answers lie in the molecular orbitals that can arise from the combination of the six p

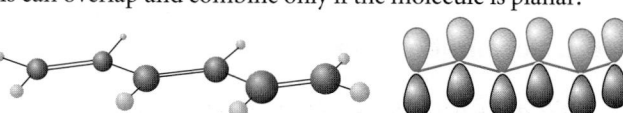

this double bond is 137 pm

both single bonds are 146 pm

both end double bonds are 134 pm

standard values: single bond: 154 pm
double bond: 134 pm

orbitals. Just as in benzene, these orbitals can all combine to give one big molecular orbital over the whole molecule. However, the p orbitals can overlap and combine only if the molecule is planar.

Since the p orbitals on carbons 2 and 3 overlap, there is some partial double bond character in the central σ bond, helping to keep the structure planar. This overlap means that it is slightly harder to rotate this 'formal single bond' than might be expected—it requires about 30 kJ mol^{-1} to rotate it whereas the barrier in propene is only around 3 kJ mol^{-1}.

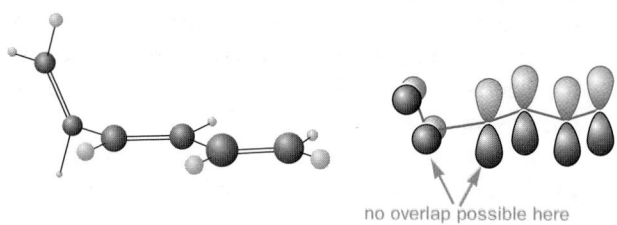

when all the atoms are planar all six p orbitals can overlap

no overlap possible here

if we rotate about a single bond,
one pair of p orbitals can no longer overlap with the others

This explains why the compound adopts a planar structure but, in order to understand why the bond lengths are slightly different from their expected values or even why they are not all the same as in benzene, we must look at *all* the molecular orbitals for hexatriene. Before we can do this, we must first study some simpler systems and address the important question of conjugation seriously.

Conjugation

In benzene and hexatriene every carbon atom is sp^2 hybridized with the remaining p orbital available to overlap with its neighbours. The uninterrupted chain of p orbitals is a consequence of having alternate double and single bonds. When two double bonds are separated by just one single bond, the two double bonds are said to be **conjugated**. Conjugated double bonds have different properties from isolated double bonds, both physically (they are often longer as we have already seen) and chemically (Chapters 10, 23, and 35).

Conjugated systems

In the dictionary, 'conjugated' is defined, among other ways, as 'joined together, especially in pairs' and 'acting or operating as if joined'. This does indeed fit very well with the behaviour of such conjugated double bonds since the properties of a conjugated system are often different from those of the component parts. We are using *conjugation* to describe bonds and *delocalization* to describe electrons.

■ This aspect of structure is called **conformation** and is the subject of Chapter 18.

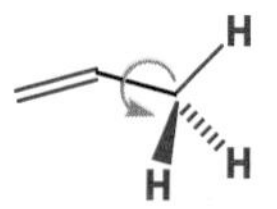

rotation about this bond requires only about 3 kJ mol^{-1}

because of the overlap of the p orbitals on carbons 2 and 3, rotation is now harder and requires more than **30** kJ mol^{-1}

You have already met several conjugated systems: remember lycopene at the start of this chapter and β-carotene in Chapter 3? All eleven double bonds in β-carotene are separated by only one single bond. We again have a long chain in which all the p orbitals can overlap to form molecular orbitals.

β-carotene – all eleven double bonds are conjugated

Another very important highly conjugated compound is chlorophyll. This is the green pigment in plants without which life on earth as we know it could not exist.

the structure of chlorophyll
the ring shown in green
is fully conjugated

It is not necessary to have two carbon–carbon double bonds in order to have a conjugated system—the C–C and C–O double bonds of propenal (acrolein) are also conjugated. The chemistry of such conjugated carbonyl compounds is significantly different from the chemistry of their component parts (Chapter 10).

propenal (acrolein):
here the C-C double bond is
conjugated with an aldehyde group

What is important though is that the double bonds are separated by *one and only one* single bond. Remember the unsaturated fatty acid, linoleic acid, that you met in Chapter 2? Another fatty acid with even more unsaturation is arachidonic acid. None of the four double bonds in this structure are conjugated since in between any two double bonds there is an sp^3 carbon. This means there is no p orbital available to overlap with the ones from the double bonds. The saturated carbon atoms insulate the double bonds from each other.

these four double bonds are not conjugated
– they are all separated by two single bonds

these tetrahedral sp^3 carbons prevent any possible
overlap of the p orbitals in the double bonds

If an atom has two double bonds directly attached to it, that is, there are no single bonds separating them, again no conjugation is possible. The simplest compound with such an arrangement is allene.

If we look at the arrangement of the p orbitals in this system, it is easy to see why no delocalization is possible—the two π bonds are perpendicular to each other.

$H_2C\!=\!C\!=\!CH_2$

allene

central carbon is sp hybridized
end carbons are sp^2 hybridized | end carbons are sp^2 hybridized

the π bonds formed as a result
of the overlap of the p orbitals
must be at right angles to each other

not only are the two π bonds perpendicular,
but the two methylene groups are too

● **Requirements for conjugation**
* Conjugation requires double bonds separated by one single bond
* Separation by two single bonds or no single bonds will not do

The allyl system

The allyl cation

We would not say that two p orbitals are conjugated—they just make up a double bond—so just how many p orbitals do we need before something can be described as conjugated? It should be clear that in butadiene the double bonds are conjugated—here we have four p orbitals.

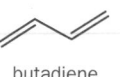

butadiene

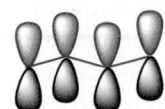

in butadiene four
p orbitals interact

Is it possible to have three p orbitals interacting? How can we get an isolated p orbital–after all, we can't have half a double bond. Let us look for a moment at allyl bromide (prop-2-enyl bromide or 1-bromoprop-2-ene). Carbon 1 in this compound has got four atoms attached to it (a carbon, two hydrogens, and a bromine atom) so it is tetrahedral (or sp³ hybridized).

allyl bromide

Bromine is more electronegative than carbon and so the C–Br bond is polarized towards the bromine. If this bond were to break completely, the bromine would keep both electrons from the C–Br bond to become bromide ion, Br^-, leaving behind an organic cation. The end carbon would now only have three groups attached and so it becomes trigonal (sp² hybridized). This leaves a vacant p orbital that we can combine with the π bond to give a new molecular orbital for the allyl system.

the p orbital has the correct symmetry to combine with
the π bond to form a new molecular orbital for the allyl system

$Br^\ominus$

▶

As more orbitals combine it becomes more difficult to represent the molecular orbitals convincingly. We shall often, from now on, simply use the atomic orbitals to represent the molecular orbitals.

▶ Remember:

1   We are simply combining atomic *orbitals* here—whether or not any of the orbitals contain any electrons is irrelevant. We can simply fill up the resultant molecular orbitals later, starting with the orbitals lowest in energy and working upwards until we have used up any electrons we may have.

2   This method allows us to work out, without too much difficulty, the shapes and energies of the molecular orbitals. The *compound* does not split its molecular orbitals into atomic orbitals and then recombine them into new molecular orbitals; we do.

Rather than trying to combine the p orbital with the π bond, it is easier for us to consider how three p orbitals combine; after all, we thought of the π bond as a combination of two p orbitals. Since we are combining three atomic orbitals (the three 2p orbitals on carbon) we shall get three molecular orbitals. The lowest-energy orbital will have them all combining in-phase. This is a bonding orbital since all the interactions are bonding.

The next orbital requires one node, just as higher-energy atomic orbitals have extra nodes (Chapter 4). The only way to include a node and maintain the symmetry of the system is to put the node through the central atom. This means that when this orbital is occupied there will be no electron density on this central atom. Since there are no interactions between adjacent atomic orbitals (either bonding or antibonding), this is a nonbonding orbital.

The final molecular orbital must have two nodal planes. All the interactions of the atomic orbitals are out-of-phase so the resulting molecular orbital is an antibonding orbital.

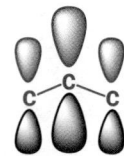

the bonding molecular
orbital of the allyl system, Ψ_1

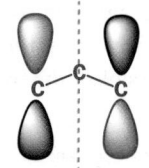

nodal plane
through the middle atom
nonbonding Ψ_2

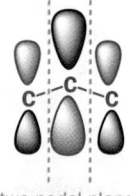

two nodal planes
antibonding Ψ_3

We can summarize all this information in a molecular orbital energy level diagram.

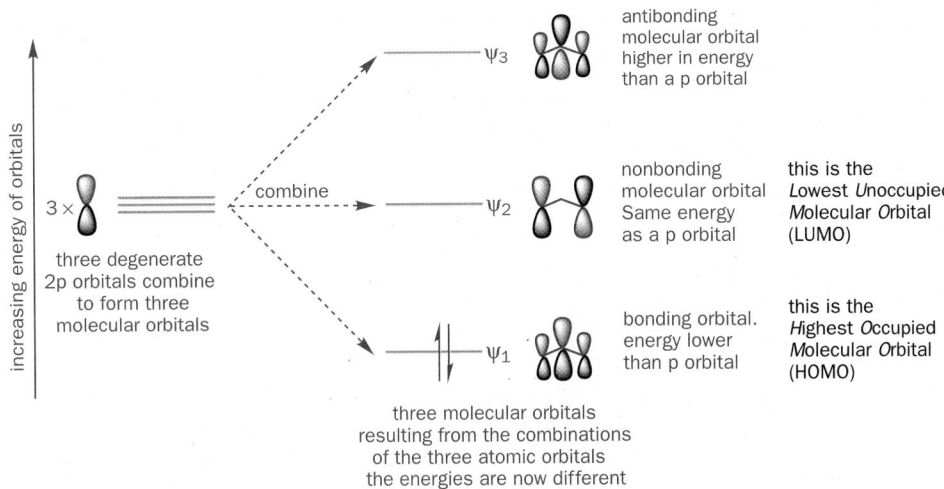

the π molecular orbitals of the allyl system: the allyl cation

The two electrons that were in the π bond now occupy the orbital lowest in energy, the bonding molecular orbital Ψ_1, and now spread over three carbon atoms. The electrons highest in energy and so most reactive are those in the HOMO. However, in this case, since the allyl cation has an overall positive charge, we wouldn't really expect it to act as a nucleophile. Of far more importance is the vacant nonbonding molecular orbital—the LUMO, the nonbonding Ψ_2. It is this orbital that must be attacked if the allyl cation reacts with a nucleophile. From the shape of the orbital, we can see that the incoming electrons will attack the end carbon atoms not the middle one since, if this orbital were full, all electron density in it would be on the end carbon atoms, not the middle one. A different way of looking at this is to see which carbon atoms in the system are most lacking in electron density. The only orbital in this case with any electrons in it is the bonding molecular orbital Ψ_1. From the relative sizes of the coefficients on each atom we can see that the middle carbon has more electron density on it than the end ones; therefore the end carbons must be more positive than the middle one and so a nucleophile would attack the end carbons.

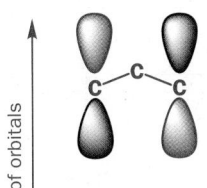

empty ψ_2 nonbonding MO
if it did have electrons in it they would be on the end carbon atoms so nucleophiles attack here

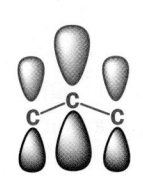

occupied ψ_1 bonding MO
most electron density is on the central carbon which means the end carbons must be more positively charged

> There are also all the molecular orbitals from the σ framework but we do not need to consider these: the occupied σ-bonding molecular orbitals are considerably lower in energy than the molecular orbitals for the π system and the vacant antibonding molecular orbitals for the σ bonds are much higher in energy than the π antibonding molecular orbital.

> The term **coefficient** describes the contribution of an individual atomic orbital to a molecular orbital. It is represented by the size of the lobes on each atom.

Representations of the allyl cation

How can we represent all this information with curly arrows? The simple answer is that we can't. Curly arrows show the *movement* of a pair of electrons. The electrons are not really moving around in this system—they are simply spread over all three carbon atoms with most electron density on the middle carbon. Curly arrows can give us an indication of the equivalence of the two end carbons, showing that the positive charge is shared over these two atoms.

The curly arrows we used in this representation are slightly different from the curly arrows we used (Chapter 5) to represent mechanisms by the forming and breaking of bonds. We still arrive at the second structure by supposing that the curly arrows mean the movement of two electrons so that the right-hand structure results from the 'reaction' shown on the left-hand structure, but these 'reactions' would be the movement of electrons *and nothing more*. In particular, no atoms have moved and no σ bonds have been formed or broken. These two structures are just two different ways of

curly arrows show the positive charge is shared over both the end atoms

drawing the same species. The arrows are **delocalization arrows** and we use them to remind us that our simple fixed-bond structures do not tell the whole truth. To remind us that these are delocalization arrows, we use a different reaction arrow, a single line with arrowheads on each end (↔).

The problem with these structures is that they seem to imply that the positive charge (and the double bond for that matter) is jumping from one end of the molecule to the other. This, as we have seen, is just not so. Another and perhaps better picture uses dotted lines and partial charges. However, as in the representation of benzene with a circle in the middle, we cannot draw mechanisms on this structure. Each of the representations has its value and we shall use both.

> Do not confuse this delocalization arrow with the equilibrium sign. A diagram like this would be wrong:
>
> The equilibrium arrows may be used only if atoms have moved and the species differ by at least a σ bond. Maybe the simplest reaction that could be shown this way would be the protonation of water where a proton moves and an O–H σ bond, shown in black, is formed or broken.

a structure to emphasize the equivalence of both bonds and the sharing of the charge at both ends

● A summary of the allyl cation system

- The two electrons in the π system are spread out over all three carbon atoms with most electron density on the central carbon
- There are no localized double and single bonds—both C–C bonds are identical and in between a double and single bond
- Both end carbons are equivalent
- The positive charge is shared equally over the two end carbons. The LUMO of the molecule shows us that this is the site for attack by a nucleophile

The delocalized allyl cation can be compared to localized carbocations by NMR

In the reaction below, a very strong acid (called 'superacid'—see Chapter 17) protonates the OH group of 3-cyclohexenol, which can then leave as water. The resulting cation is, not surprisingly, unstable and would normally react rapidly with a nucleophile. However, at low temperatures and if there are no nucleophiles present, the cation is relatively stable and it is even possible to record a carbon NMR spectrum (at –80 °C).

FSO_3H-SbF_5

liquid SO_2, –80 °C

$-H_2O$

The NMR spectrum of this allylic cation reveals a plane of symmetry, which confirms that the positive charge is spread over two carbons. The large shift of 224 p.p.m. for these carbons indicates very strong deshielding (that is, lack of electrons) but is nowhere near as large as a localized cation. The middle carbon's shift of 142 p.p.m. is almost typical of a normal double bond indicating that it is neither significantly more nor less electron-rich than normal.

141.9

224.4   224.4

37.1   37.1

17.5

the ^{13}C NMR shifts in p.p.m.
notice the plane of symmetry down the middle

Carbocation ^{13}C shift

This localized carbocation shows an enormous shift of 330 p.p.m. indicating very little shielding of the positively charged carbon atom. Again, due to the instability of this species, the ^{13}C spectrum was recorded at low temperature.

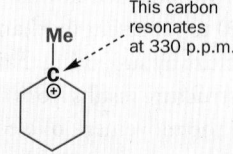

This carbon resonates at 330 p.p.m.

The allyl radical

When we made the allyl cation from allyl bromide, the bromine atom left as bromide ion taking both the electrons from the C–Br bond with it—the C–Br bond broke **heterolytically**. What if the bond broke **homolytically**—that is, carbon and bromine each had one electron? A bromine atom and an allyl radical (remember a radical has an unpaired electron) would be formed. This reaction can be shown using the single-headed fish hook curly arrows from Chapter 5: normal double-headed arrows show the movement of two electrons; single-headed arrows show the movement of one.

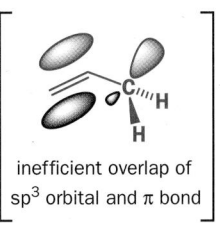

homolytic cleavage of the C–Br bond forms
a bromine atom and the allyl radical

Now the end carbon has a single unpaired electron. What do we do with it? Before the bond broke, the end carbon was tetrahedral (sp^3 hybridized). We might think that the single electron would still be in an sp^3 orbital. However, since an sp^3 orbital cannot overlap efficiently with a π bond, the single electron would then have to be localized on the end carbon atom. If the end carbon atom becomes trigonal (sp^2 hybridized), the single electron could be in a p orbital and this could overlap and combine with the π bond. This would mean that the radical could be spread over the molecule in the same orbital that contained the cation.

inefficient overlap of
sp^3 orbital and π bond

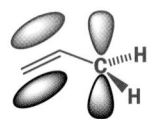

efficient overlap of
p orbital and π bond

So once again we have three p orbitals to combine. This is the same situation as before. We have the same atoms, the same orbitals, and so the same energy levels. In fact, the molecular orbital energy level diagram for this compound is *almost the same* as the one for the allyl cation: the only difference is the number of electrons in the π system. Whereas in the allyl cation π system we only had two electrons, here we have three (two from the π bond plus the single one). Where does this extra electron go? Answer: in the next lowest molecular orbital—the nonbonding molecular orbital.

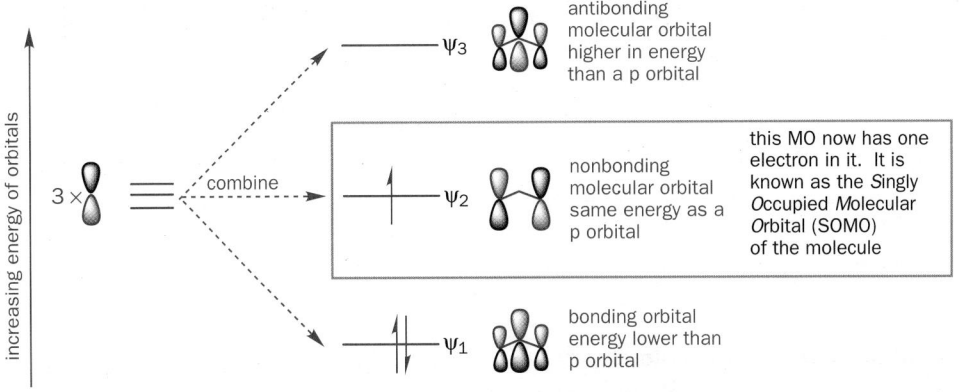

the π molecular orbitals of the allyl system: the allyl radical

The extra electron is in an orbital all by itself. This orbital must be the HOMO of the molecule but is also the LUMO since it still has room for one more electron. It is actually called the **Singly Occupied Molecular Orbital (SOMO)**, for obvious reasons. The shape of this orbital tells us that the single electron is located on the end carbon atoms. This can also be shown using delocalization arrows (again single-headed arrows to show movement of one electron).

the single electron can be on either of the end carbon atoms

The allyl anion

What would have happened if both electrons from the C–Br bond in allyl bromide had stayed behind on the carbon? If we had removed the bromine atom with a metal, magnesium for example (Chapter 9), both electrons would remain leaving an overall negative charge on the allyl system.

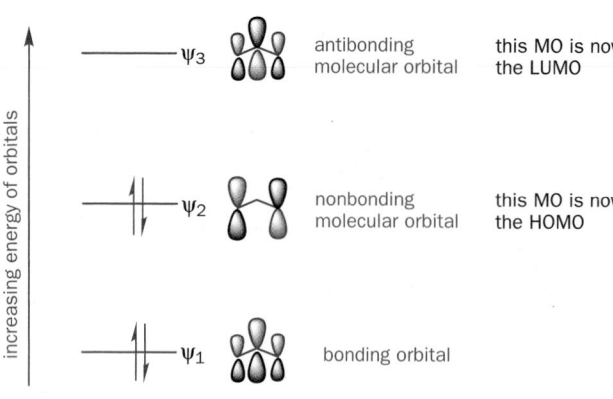

reaction of allyl bromide with a metal gives the allyl anion

ψ₃ antibonding molecular orbital   this MO is now the LUMO

ψ₂ nonbonding molecular orbital   this MO is now the HOMO

ψ₁ bonding orbital

increasing energy of orbitals

the π molecular orbitals of the allyl system: the allyl anion

Again, this system is much more stable if the negative charge can be spread out rather than localized on one end carbon. This can be accomplished only if the negative charge is in a p orbital rather than an sp^3 orbital. The molecular orbital energy level diagram is, of course, unchanged: all we have to do is put the extra electron in the nonbonding orbital. Altogether we now have four electrons in the π system—two from the π bond and two from the negative charge. Both the bonding and the nonbonding orbitals are now fully occupied.

Where is the electron density in the allyl anion π system? The answer is slightly more complicated than that for the allyl cation because now we have two full molecular orbitals and the electron density comes from a sum of both orbitals. This means there is electron density on all three carbon atoms. However, the HOMO for the anion is now the nonbonding molecular orbital. It is this orbital that contains the electrons highest in energy and so most reactive. In this orbital there is no electron density on the middle carbon; it is all on the end carbons. Hence it will be the end carbons that will react with electrophiles. This is conveniently represented by curly arrows.

the curly arrows give a good representation of the HOMO and they show the negative charge concentrated on the end carbon atoms. But the structures suggest localized bonds and charges

these two structures emphasize the equivalence of the bonds and that the charge is spread out

> ● **A summary of the allyl anion system**
> - There are no localized double and single bonds—both C–C bonds are the same and in between a double and single bond
> - Both end carbons are the same
> - The four electrons in the π system are spread out over all three carbon atoms. In the bonding orbital most electron density is on the central carbon but, in the nonbonding orbital, there is electron density only on the end carbons
> - The electrons highest in energy and so most reactive (those in the HOMO) are to be found on the end carbons. Electrophiles will therefore react with the end carbons

Such predictions from a consideration of the molecular orbitals are confirmed both by the reactions of the allyl anion and by its NMR spectrum. It is possible to record a carbon NMR spectrum of the allyl anion directly (for example, as its lithium derivative). The spectrum shows only two signals: the middle carbon at 147 p.p.m. and the two end carbons both at 51 p.p.m.

The central carbon's shift of 147 p.p.m. is almost typical of a normal double bond carbon whilst the end carbons' shift is in between that of a double bond and a saturated carbon bearing a negative charge. Notice also that the central carbon in the allyl cation and the anion have almost identical chemical shifts—142 and 147 p.p.m., respectively. If anything, the *anion* central carbon is more deshielded. Compare this with the spectra for methyllithium and propene itself. Methyllithium shows a single peak at –15 p.p.m. and propene shows three ^{13}C signals as indicated below.

the central carbon
resonates at 147 p.p.m.

both end carbons
resonate at 51 p.p.m.

a localized structure like this would
have a very different spectrum

methine carbon
resonates at 134 p.p.m.

the methyl carbon
resonates at –15 p.p.m.

methylene carbon
resonates at 116 p.p.m.

methyl carbon
resonates at 19.5 p.p.m.

Other allyl-like systems

The carboxylate anion

You may already be familiar with one anion very much like the allyl anion—the carboxylate ion formed on deprotonating a carboxylic acid with a base. In this structure we again have a double bond adjacent to a single bond but here oxygen atoms replace two of the carbon atoms.

a carboxylic acid

a carboxylate anion

X-ray crystallography shows both carbon–oxygen bond lengths in this anion to be the same (136 pm), in between that of a normal carbon–oxygen double bond (123 pm) and single bond (143 pm). The negative charge is spread out equally over the two oxygen atoms.

the electrons are delocalized over the π system

these structures emphasize the equivalence
of the two C–O bonds and that the negative
charge is spread over both oxygen atoms

The molecular orbital energy diagram for the carboxylate anion is the very similar to that of the allyl system. There are just two main differences.

1 The coefficients of the atomic orbitals making up the molecular orbitals will change because oxygen is more electronegative than carbon and so has a greater share of electrons

2 The absolute values of the energy levels will be different from those in the allyl system, again because of the difference in the electronegativities. Compare with the differences between the molecular orbitals for ethene and a carbonyl, p. 103

The nitro group

The nitro group consists of a nitrogen bonded to two oxygen atoms and a carbon (for example, an alkyl group). There are two ways of representing the structure: one using formal charges, the other using a dative bond. Notice in each case that one oxygen is depicted as being doubly bonded, the other singly bonded. Drawing both oxygen atoms doubly bonded is incorrect—*nitrogen cannot have five bonds* since this would represent ten electrons around it and there are not enough orbitals to put them in.

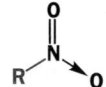

two ways of representing the nitro group
the stucture on the left has formal charges
on the nitrogen and one oxygen, the other
has a dative bond from the nitrogen

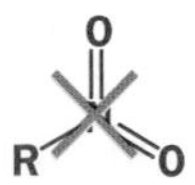

incorrect drawing
of the nitro group
*nitrogen cannot
have five bonds*

▶
Notice that the delocalization over
the nitro group is similar to that
over the carboxylate group. In
fact, the nitro group is
isoelectronic with the carboxylate
group, that is, both systems have
the same number of electrons.

The problem with the two correct drawings is that they do not show the equivalence of the two
N–O bonds. However, we do have an N–O double bond next to an N–O single bond which means
that the negative charge is delocalized over both of the oxygen atoms. This can be shown by curly
arrows.

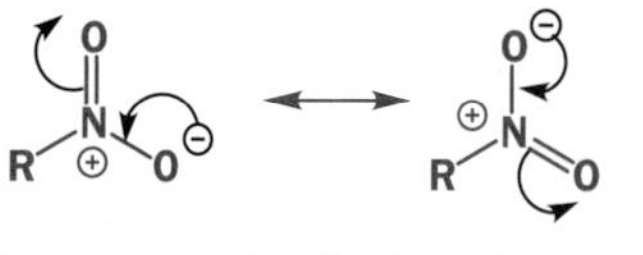

the electrons are delocalized over the π system

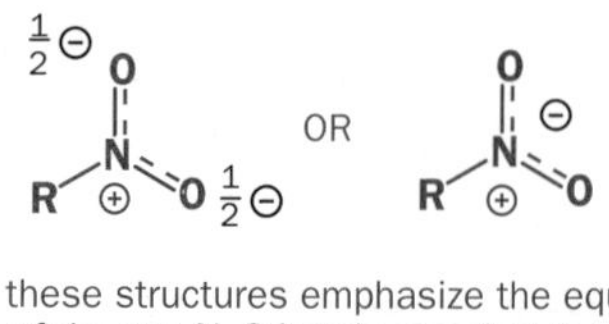

these structures emphasize the equivalence
of the two N–O bonds and that the negative
charge is spread over both oxygen atoms

Just to reiterate, the same molecular orbital energy diagram can be used for the allyl systems and
the carboxylate and nitro groups. Only the absolute energies of the molecular orbitals are different
since different elements with different electronegativities are used in each.

The amide group

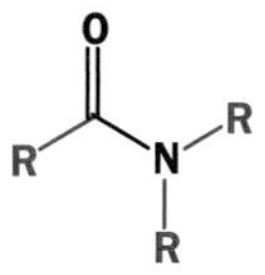

The amide is a very important group in nature since it is the link by which amino acids join together
to form peptides, which make up the proteins in our bodies. The structure of this deceptively simple
group has an unexpected feature, which is responsible for much of the stability of proteins.

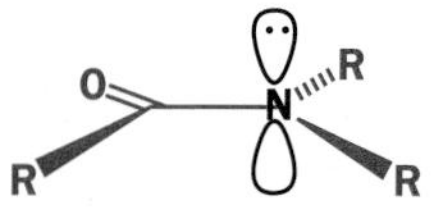

an amide

In the allyl anion, carboxylate, and nitro systems we
had four electrons in the π system spread out over three
atoms. The nitrogen in the amide group also has a pair of
electrons that could conjugate with the π bond of the carbonyl group. Again, for effective overlap with the π bond,
the lone pair of electrons must be in a p orbital. This in
turn means that the nitrogen must be sp² hybridized.

nitrogen is trigonal with
its lone pair in a p orbital

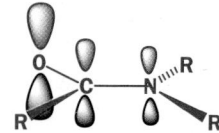

the lowest π orbital of the amide
the same arrangement
of p orbitals as in the allyl system

In the carboxylate ion, a negative charge was shared (equally) between two oxygen atoms. In an
amide there is no charge as such—the lone pair on nitrogen is shared between the nitrogen and the
oxygen. However, since oxygen is more electronegative
than nitrogen, it has more than its fair share of the electrons in this π system. (This is why the p orbital on the
oxygen atom in the lowest bonding orbital shown above is
slightly larger than the p orbital on the nitrogen.) The
delocalization can be shown using curly arrows.

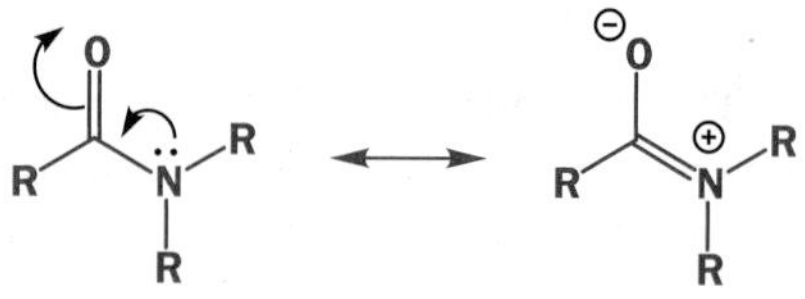

This representation suffers from the usual problems. Curly arrows show the movement of a pair
of electrons. The structure on the left, therefore, suggests that electrons are flowing from the nitrogen
to the oxygen. *This is not true*: the molecular orbital picture tells us that the electrons are unevenly
distributed over the three atoms in the π system with a greater electron density on the oxygen. The
curly arrows show us how to draw an alternative diagram. The structure on the right implies that the
nitrogen's lone pair electrons have moved completely on to the oxygen. Again this is not true; there is
simply more electron density on the oxygen than on the nitrogen. The arrows are useful in that they
help us to depict how the electrons are unevenly shared in the π system.

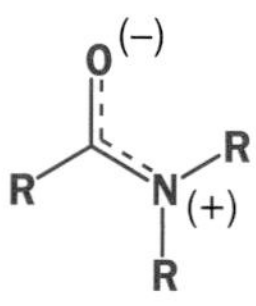

A better representation might be this structure. The charges in brackets indicate substantial,
though not complete, charges, maybe about a half plus or minus charge. However, we cannot draw
mechanisms on this structure and all these representations have their uses.

● Let us summarize these points.

- The amide group is planar—this includes the first carbon atoms of the R groups attached to the carbonyl group and to the nitrogen atom
- The lone pair electrons on nitrogen are delocalized into the carbonyl group
- The C–N bond is strengthened by this interaction—it takes on partial double bond character. This also means that we no longer have free rotation about the C–N bond which we would expect if it were only a single bond
- The oxygen is more electron-rich than the nitrogen. Hence we might expect the oxygen rather than the nitrogen to be the site of electrophilic attack
- The amide group as a whole is made more stable as a result of the delocalization

The amide is a functional group of exceptional importance so we shall look at these points in more detail.

The structure of the amide group

How do we know the amide group is planar? X-ray crystal structures are the simplest answer. Other techniques such as electron diffraction also show that simple (noncrystalline) amides have planar structures. N,N-dimethylformamide (DMF) is an example.

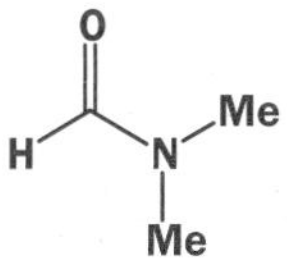

The C–N bond length to the carbonyl group is closer to that of a standard C–N double bond (127 pm) than to that of a single bond (149 pm). This partial double bond character is responsible for the restricted rotation about this C–N bond. We must supply 88 kJ mol^{-1} if we want to rotate the C–N bond in DMF (remember a full C–C double bond takes about 260 kJ mol^{-1}). This amount of energy is not available at room temperature and so, for all intents and purposes, the amide C–N bond is locked at room temperature as if it were a double bond. This is shown in the carbon NMR spectrum of DMF. How many carbon signals would you expect to see? There are three carbon atoms altogether and three signals appear—the two methyl groups on the nitrogen are different. If free rotation were possible about the C–N bond, we would expect to see only two signals. In fact, if we record the spectrum at higher temperatures, we do indeed only see two signals since now there is sufficient energy available to overcome the rotational barrier and allow the two methyl groups to interchange.

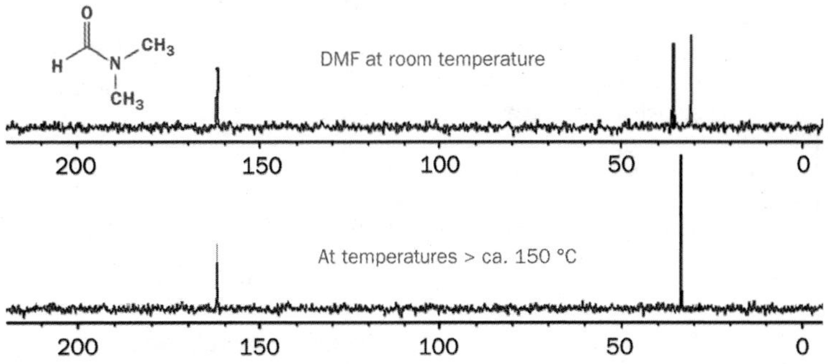

Proteins are composed of many amino acids joined together with amide bonds. The amino group of one can combine with the carboxylic acid group of another to give an amide. This special amide, which results from the combining of two amino acids, is known as a **peptide**—two amino acids join to form a dipeptide; many join to give a polypeptide.

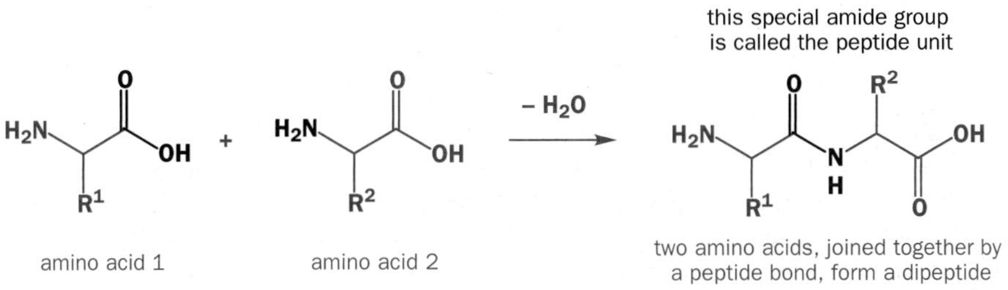

The peptide unit so formed is a planar, rigid structure since there is restricted rotation about the C–N bond. This means that two isomers should be possible—a *cis* and a *trans*.

C=O and N–H are *trans*   C=O and N–H are *cis*

It is found that nearly all the peptide units found in nature are *trans*. This is not surprising since the *cis* form is more crowded (a *trans* disubstituted double bond is lower in energy than a *cis* for the same reason).

Protein shape and activity

This planar, *trans* peptide unit poses serious limitations on the shapes proteins can adopt. Understanding the shapes of proteins is very important—enzymes, for example, are proteins with catalytic properties. Their catalytic function depends on the shape adopted: alter the shape in some way and the enzyme will no longer work.

Reactivity of the amide group

Just as delocalization stabilizes the allyl cation, anion, and radical, so too is the amide group stabilized by the conjugation of the nitrogen's lone pair with the carbonyl group. This, together with the fact that the amine part is such a poor leaving group, makes the amide one of the least reactive carbonyl groups (we shall discuss this in Chapter 12).

Furthermore, the amine part of the amide group is unlike any normal amine group. Most amines are easily protonated. However, since the lone pair on the amide's nitrogen is tied up in the π system, it is less available for protonation or, indeed, reaction with any electrophile. As a result, an amide is preferentially protonated on the oxygen atom but it is difficult to protonate even there (see next chapter, p. 201). Conjugation affects reactivity.

The conjugation of two π bonds

The simplest compound that can have two conjugated π bonds is butadiene. As we would now expect, this is a planar compound that can adopt two different conformations by rotating about the single bond. Rotation is somewhat restricted (around 30 kJ mol^{-1}) but nowhere near as much as in an amide (typically 60–90 kJ mol^{-1}). What do the molecular orbitals for the butadiene π system look like? The lowest-energy molecular orbital will have all the p orbitals combining in-phase. The next lowest will have one node, and then two, and the highest-energy molecular orbital will have three nodes (that is, all the p orbitals will be out-of-phase).

Isomers of butadiene

Butadiene normally refers to 1,3-butadiene. It is also possible to have 1,2-butadiene which is another example of an allene (p. 157).

rotation about this single bond is only slightly restricted

1,2-butadiene
an allene

1,3-butadiene
a conjugated diene

The molecular orbitals of butadiene

Butadiene has two π bonds and so four electrons in the π system. Which molecular orbitals are these electrons in? Since each molecular orbital can hold two electrons, only the two molecular orbitals lowest in energy are filled. Let's have a closer look at these orbitals. In Ψ_1, the lowest-energy bonding orbital, the electrons are spread out over all four carbon atoms (above and below the plane) in one continuous orbital. There is bonding between all the atoms. The other two electrons are in Ψ_2. This orbital has bonding interactions between carbon atoms 1 and 2, and also between 3 and 4 but an *antibonding* interaction between carbons 2 and 3. Overall, in both the occupied π orbitals there are

In Chapter 3 we saw that IR spectroscopy shows carbonyl groups in the region 1600–1800 cm^{-1}. At the higher end of that region the C=O stretches of very reactive acid chlorides and acid anhydrides show that they have full C=O double bonds. At the lower end of that region the C=O stretching frequency of amides comes at about 1660 cm^{-1} showing that they are halfway to being single bonds. These relationships will be explored further in Chapter 15. The conjugation of the nitrogen's lone pair with the carbonyl bond *strengthens* the C–N bond but *weakens* the carbonyl bond. The weaker the bond, the less energy it takes to stretch it and so the lower the IR absorption frequency. Overall the molecule is more stable, as is reflected in the reactivity (or lack of it) of the amide group, Chapter 12.

electrons between carbons 1 and 2 and between 3 and 4, but the antibonding interaction between carbons 2 and 3 in Ψ_2 partially cancels out the bonding interaction in Ψ_1. This explains why all the bonds in butadiene are not the same and why the middle bond is more like a single bond while the end bonds are double bonds. If we look closely at the coefficients on each atom in orbitals Ψ_1 and Ψ_2, it can be seen that the bonding interaction between the central carbon atoms in Ψ_1 is greater than the antibonding one in Ψ_2. Thus butadiene does have some double bond character between carbons 2 and 3, which explains why there is the slight barrier to rotation about this bond.

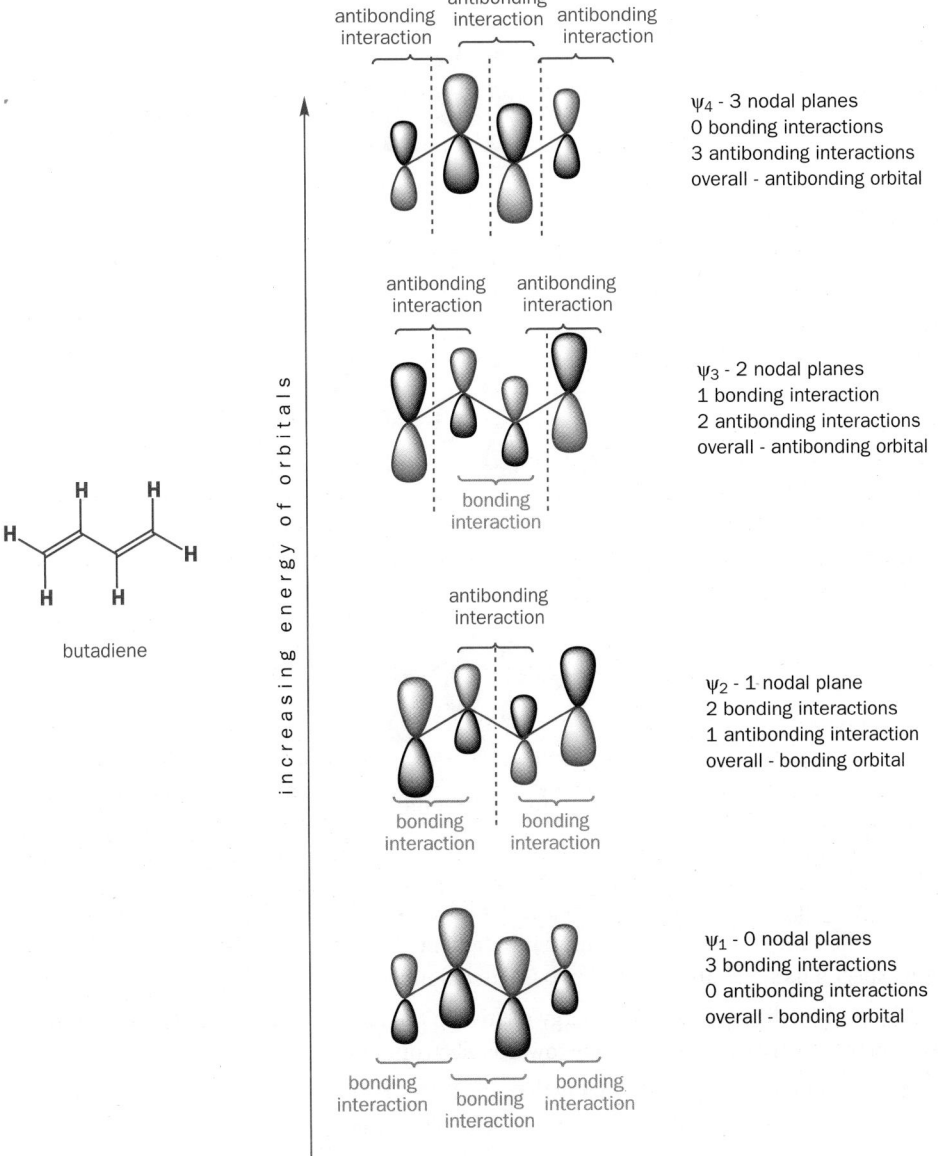

In our glimpse of hexatriene earlier in this chapter we saw a similar effect, which we could now interpret if we looked at all the molecular orbitals for hexatriene. We have three double bonds and two single bonds with slightly restricted rotation. Both butadiene and hexatriene have double bonds and single bond: neither compound has all its C–C bond lengths the same, yet both compounds are conjugated. What is the real evidence for conjugation? How does the conjugation show itself in the properties and reactions of these compounds? To answer these questions, we need to look again at the energy level diagram for butadiene and compare it with that of ethene. A simple way to do this is to make the orbitals of butadiene by combining the orbitals of ethene.

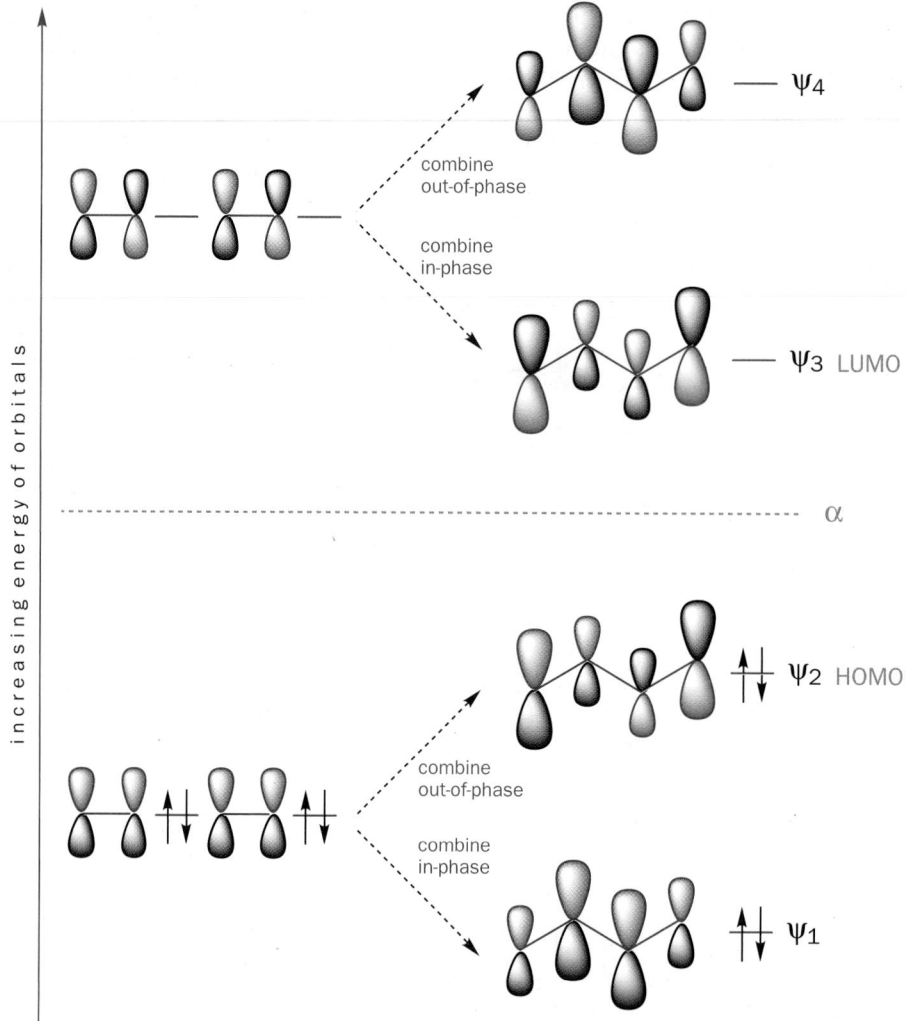

We have drawn the molecular orbital diagram for the π molecular orbitals of butadiene as a result of combining the π molecular orbitals of two ethene molecules. There are some important points to notice here.

- The overall energy of the two bonding butadiene molecular orbitals is lower than that of the two molecular orbitals for ethene. This means that butadiene is more thermodynamically stable than we might expect if its structure were just two isolated double bonds

- The HOMO for butadiene is *higher* in energy relative to the HOMO for ethene. This means butadiene should be *more* reactive than ethene towards electrophiles

- The LUMO for butadiene is *lower* in energy than the LUMO for ethene. Consequently, butadiene would be expected to be *more* reactive towards nucleophiles than ethene

- So whilst butadiene is more stable than two isolated double bonds, *it is also more reactive* (Chapter 20)

Recall that on p. 153 we saw how it was possible to promote an electron from HOMO to LUMO in an isolated double bond using UV light and that this allowed rotation about this bond. In butadiene, however, promoting an electron from the HOMO to the LUMO actually *increases* the electron density between the two central atoms and so stops rotation.

Butadiene model

A simple theoretical model of the butadiene system predicts the energy of the bonding Ψ_1 orbital to be $[\alpha + 1.62\beta]$ and that of bonding orbital Ψ_2 to be $[\alpha + 0.62\beta]$. With both of these orbitals fully occupied, the total energy of the electrons is $[4\alpha + 4.48\beta]$. Remember that the energy of the bonding π molecular orbital for ethene was $[\alpha + \beta]$ (p. 152) so, if we were to have two localized π bonds (each with two electrons), the total energy would be $[4\alpha + 4\beta]$. This theory predicts that butadiene with both double bonds conjugated is lower in energy than it would be with two localized double bonds by 0.48β. Both α and β are negative; hence $[4\alpha + 4.48\beta]$ is *lower* in energy than $[4\alpha + 4\beta]$ by 0.48β.

UV and visible spectra

In Chapter 2 we saw how, if given the right amount of energy, electrons can be promoted from a low-energy atomic orbital to a higher-energy one and how this gives rise to an atomic absorption spectrum. Exactly the same process can occur with molecular orbitals. In fact, we have already seen (p. 153) that UV light can promote an electron from the HOMO to the LUMO in a double bond.

HOMO–LUMO gap

Electrons can be promoted from any filled orbital to any empty orbital. The smallest energy difference between a full and empty molecular orbital is between the HOMO and the LUMO. The smaller this difference, the less energy will be needed to promote an electron from the HOMO to the LUMO: the smaller the amount of energy needed, the longer the wavelength of light needed since $\Delta E = h\nu$. Therefore, an important measurement is the wavelength at which a compound shows maximum absorbance, λ_{max}. A difference of more than about 4 eV (about 7×10^{-19} J) between HOMO and LUMO means that λ_{max} will be in the ultraviolet region (wavelength, λ, < 300 nm). If the energy difference is between about 3 eV (about 4×10^{-19} J) and 1.5 eV (about 3×10^{-19} J) then λ_{max} will be in the visible part of the spectrum.

■

We can get a good estimate of the absolute energies of molecular orbitals from photoelectron spectroscopy and electron transmission spectroscopy (see Chapter 4). Such experiments suggest energies for the HOMO and LUMO of butadiene to be –9.03 and +0.62 eV, respectively, whilst for ethene they are –10.51 and +1.78 eV, respectively.

We have seen above that the energy difference between the HOMO and LUMO for butadiene is less than that for ethene. Therefore we would expect butadiene to absorb light of longer wavelength than ethene (the longer the wavelength the lower the energy, $\Delta E = hc/\lambda$). This is found to be the case: butadiene absorbs at 215 nm compared to 185 nm for ethene. The conjugation in butadiene means it absorbs light of a longer wavelength than ethene. In fact, this is true generally.

● **The more conjugated a compound is, the smaller the energy transition between its HOMO and LUMO and hence the longer the wavelength of light it can absorb. Hence UV–visible spectroscopy can tell us about the conjugation present in a molecule.**

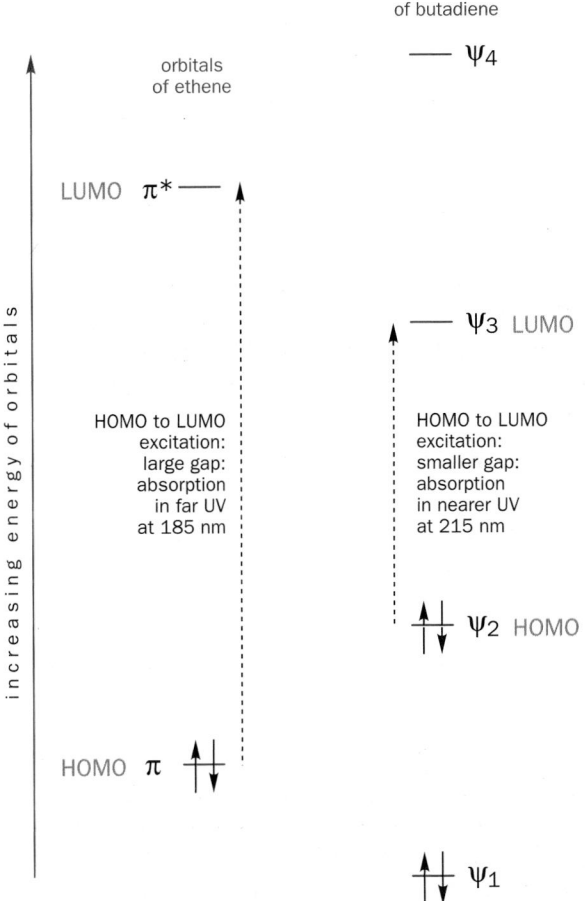

orbitals of butadiene

orbitals of ethene

increasing energy of orbitals

—— ψ_4

LUMO π^* ——

—— ψ_3 LUMO

HOMO to LUMO excitation: large gap: absorption in far UV at 185 nm

HOMO to LUMO excitation: smaller gap: absorption in nearer UV at 215 nm

ψ_2 HOMO

HOMO π

ψ_1

Both ethene and butadiene absorb in the far-UV region of the electromagnetic spectrum (215 nm is just creeping into the UV region) but, if we extend the conjugation further, the gap between HOMO and LUMO will eventually be sufficiently decreased to allow the compound to absorb visible light and hence be coloured. A good example is the red pigment in tomatoes we introduced at the start of the chapter. It has eleven conjugated double bonds (plus two unconjugated) and absorbs light at about 470 nm.

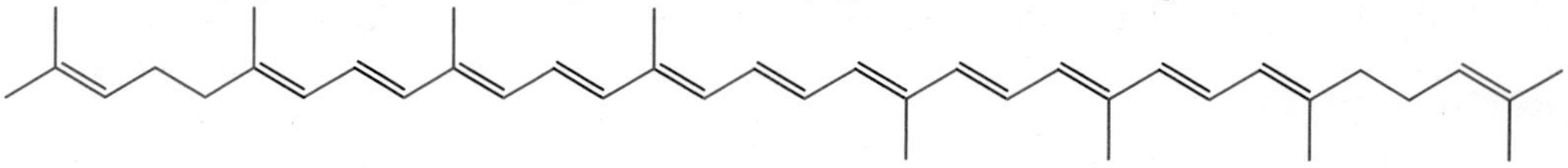

lycopene, the red pigment in tomatoes, rose hips, and other berries

The colour of pigments depends on conjugation

You can see now that it is no coincidence that this compound and the two other highly conjugated compounds we met earlier, chlorophyll and β-carotene, are all highly coloured natural pigments. In fact, all dyes and pigments are highly conjugated compounds.

Natural pigments

The similarities between lycopene and β-carotene are easier to see if the structure of lycopene is twisted. Lycopene is a precursor of carotene so, when a cell makes carotene, it makes lycopene *en route*.

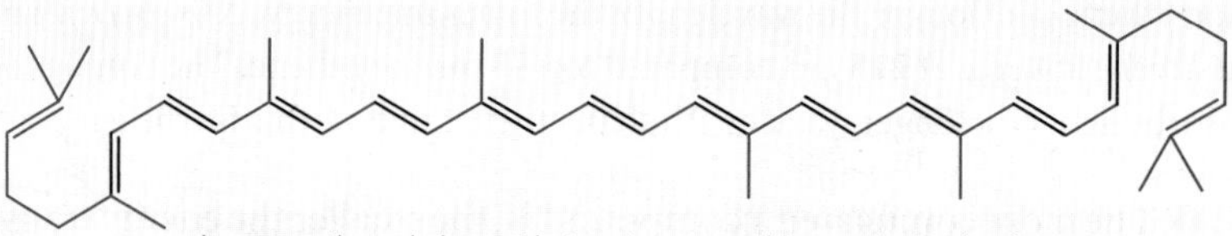

lycopene, the red pigment in tomatoes, rose hips, and other berries

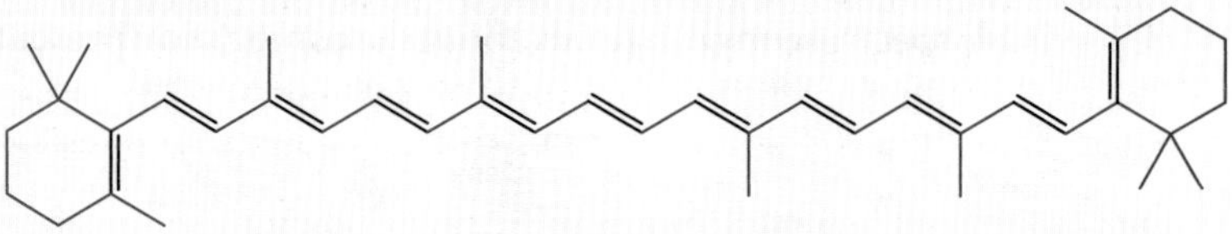

β-carotene, the red pigment in carrots and other vegetables

If a compound absorbs one colour, it is the complementary colour that is transmitted—the red glass of a red light bulb *doesn't* absorb red light; it absorbs everything else letting only red light through. Here is a table of approximate wavelengths for the various colours. The last column gives the approximate length a conjugated chain must be in order to show the colour in question. The number n refers to the number of double bonds in conjugation.

Approximate wavelengths for different colours

Absorbed frequency, nm	Colour absorbed	Colour transmitted	$R(CH=CH)_nR$, $n =$
200–400	ultraviolet	—	< 8
400	violet	yellow-green	8
425	indigo-blue	yellow	9
450	blue	orange	10
490	blue-green	red	11
510	green	purple	
530	yellow-green	violet	
550	yellow	indigo-blue	
590	orange	blue	
640	red	blue-green	
730	purple	green	

Every extra conjugated double bond in a system increases the wavelength of light that is absorbed. If there are fewer than about eight conjugated double bonds, the compound absorbs in the ultraviolet and we don't notice the difference. With more than eight conjugated double bonds, the absorption creeps into the visible and, by the time it reaches 11, the compound is red. If we wanted a blue or green compound, we should need a very large number of conjugated double bonds and such pigments do not usually rely on π bonds alone.

Transitions from bonding to antibonding π orbitals are called π → π* transitions. A much smaller energy gap is available if we use electrons in a nonbonding orbital as the electrons start off much higher in energy and can be promoted to low-lying antibonding π orbitals. We call these transitions n → π*, where the 'n' stands for nonbonding. It is easy to find coloured compounds throughout the whole range of wavelengths by this means. The colour of blue jeans comes from the pigment indigo. The two nitrogen atoms provide the lone pairs that can be excited into the π* orbitals of the rest of the molecule. These are low in energy because of the two carbonyl groups. Yellow light is absorbed by this pigment and indigo-blue light transmitted.

colourless precursor to indigo   indigo: the pigment of blue jeans

Jeans are dyed by immersion in a vat of reduced indigo, which is colourless since the conjugation is interrupted by the central single bond. When the cloth is hung up to dry, the oxygen in the air oxidizes the 'pigment' to indigo and the jeans turn blue. Conjugation is the key to colour.

Many conjugated compounds are yellow because, although they have their λ_{max} in the UV, the broad absorption tails into the visible and the compound weakly absorbs violet light making it pale yellow. An example is this imine with a long conjugated system joining the two aromatic rings together.

The imine is yellow but when it is reduced to the amine, breaking the conjugation in the middle so that the two benzene rings are no longer linked together, the result is a dark orange compound. This is rather surprising because you would normally expect the compound with the longer conjugated system to absorb at longer wavelengths. Check with the table above to see that an orange compound definitely absorbs at longer wavelengths than a yellow compound.

the yellow imine   the orange amine

The answer to this paradox lies in the change of hybridization of the nitrogen atom. In the imine, the nitrogen is trigonal and the lone pair is in an sp^2 orbital in the plane of the conjugated system. No delocalization of the lone pair is possible and the UV absorption comes from a simple π → π* transition. When the imine is reduced, the C–N bond can rotate and the amine can be trigonal too, but with the N–H bond in the plane and the lone pair in a p orbital conjugated with the right-hand benzene ring. The absorption giving the orange colour is an n → π* transition not a π → π* transition. Even delocalization of a lone pair into one benzene ring with a nitro group can give a longer wavelength absorption than a conjugated system of bonding electrons.

Aromaticity

Let us now return to the structure of benzene. Benzene is unusually stable for an alkene and is not normally described as an alkene at all. For example, whereas normal alkenes readily react with

This chemistry is discussed in Chapter 22.

bromine to give dibromoalkane *addition* products, benzene reacts with bromine only with difficulty—it needs a Lewis acid catalyst and then the product is a mono*substituted* benzene and not an addition compound.

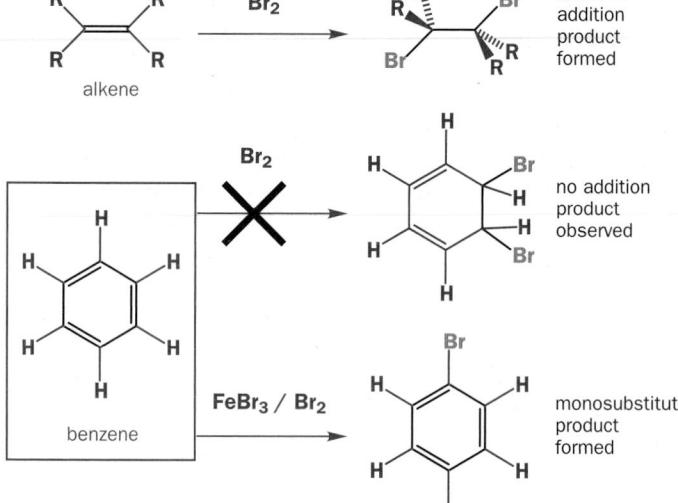

Bromine reacts with benzene in a substitution reaction (a bromine atom replaces a hydrogen atom), *keeping the benzene structure intact.* This ability to retain its ring structure through all sorts of chemical reactions is one of the important differences of benzene compared to alkenes and one that originally helped to define the class of aromatic compounds to which benzene belongs.

Cyclooctatetraene has four double bonds in a ring. What do you think its structure will be?

cyclooctatetraene

You will probably be surprised to find cyclooctatetraene (COT for short), unlike benzene, is *not* planar. Also *none* of the double bonds are conjugated—there are indeed alternate double and single bonds in the structure but conjugation is possible only if the p orbitals of the double bonds can overlap; here they do not. Since there is no conjugation, there are two C–C bond lengths in cyclooctatetraene—146.2 and 133.4 pm—which are typical for single and double C–C bonds. If possible, make a model of cyclooctatetraene for yourself—you will find the compound naturally adopts the shape below. This shape is often called a 'tub'.

Chemically, cyclooctatetraene behaves like an alkene not like benzene. Bromine, for example, does not form a substitution product but an addition product. There is something strange going on here—why is benzene so different from other alkenes and why is cyclooctatetraene so different from benzene? The mystery deepens when we look at what happens when we treat cyclooctatetraene with powerful oxidizing or reducing agents.

If 1,3,5,7-tetramethylcyclooctatetraene is treated at low temperature (−78 °C) with SbF_5/SO_2ClF (strongly oxidizing conditions) a dication is formed. This cation, unlike the neutral compound, is *planar* and all the C–C bond lengths are the same.

neutral compound is tub-shaped   dication is planar

Drawing the dication

The dication still has the same number of atoms as the neutral species only fewer electrons. Where have the electrons been taken from? The π system now has two electrons less. We could draw a structure showing two localized positive charges but this would not be ideal since the charge is spread over the whole ring system.

one structure with localized charges   the charges can be delocalized all round the ring   structure to show equivalence of all the carbon atoms

It is also possible to *add* electrons to cyclooctatetraene by treating it with alkali metals and a di*anion* results. X-ray structures reveal this dianion to be planar, again with all C–C bond lengths the same (140.7 pm). The difference between the anion and cation of cyclooctatetraene on the one hand and cyclooctatetraene on the other is the number of electrons in the π system. The cation has six π electrons, the anion has ten, but neutral cyclooctatetraene has eight.

Substituted benzene compounds, such as the one below with six silicon atoms around the edge, can also react with lithium to give a dianion. This dianion, with eight π electrons, is now no longer planar.

Treatment of benzene itself with the strongly oxidizing SbF₅/SO₂ClF reagent has no effect but it is possible to oxidize substituted derivatives. Hexakis(dimethylamino)-benzene, for example, can be oxidized with iodine. Again, the resulting dication is nonplanar and all the C–C bond lengths are not the same.

Do you see a pattern forming? The important point is not the number of conjugated atoms but the *number of electrons in the π system*. When they have 4 or 8 π electrons, both benzene and cyclooctatetraene adopt nonplanar structures; when they have 6 or 10 π electrons, a planar structure is preferred.

planar neutral compound   nonplanar dianion

planar neutral compound   nonplanar dication

If you made a model of cyclooctatetraene, you might have tried to force it to be flat. If you managed this you probably found that it didn't stay like this for long and that it popped back into the tub shape. The strain in planar COT can be overcome by the molecule adopting the tub conformation. The strain is due to the numbers of atoms and double bonds in the ring—it has nothing to do with the number of electrons. The planar dication and dianion of COT still have this strain. The fact that these ions do adopt planar structures must mean there is some other form of stabilization that outweighs the strain of being planar. This extra stabilization is called **aromaticity**.

Heats of hydrogenation of benzene and cyclooctatetraene

It is possible to reduce unsaturated C=C double bonds using hydrogen gas and a catalyst (usually nickel or palladium) to produce fully saturated alkanes. This process is called hydrogenation and it is exothermic (that is, energy is released) since a thermodynamically more stable product, an alkane, is produced.

Margarine manufacture

This reaction is put to good use in the manufacture of margarines. One of the ingredients in many margarines is hydrogenated vegetable oil. When polyunsaturated fats are hydrogenated they become more solid. This means that, rather than having to pour our margarine on to our toast in the morning, we can spread it. We saw the second acid in this series, linoleic acid, at the start of Chapter 2.

linolenic acid
m.p. –11 °C

linoleic acid
m.p. –5° C

oleic acid
m.p. 16 °C

stearic acid
m.p. 71 °C

melting points (m.p.s) of some common fatty acids

When *cis*-cyclooctene is hydrogenated, 96 kJ mol⁻¹ of energy is released. Cyclooctatetraene releases 410 kJ mol⁻¹ on hydrogenation. This value is approximately four times one double bond's worth, as we might expect. However, whereas the heat of hydrogenation for cyclohexene is 120 kJ mol^{-1}, on hydrogenating benzene, only 208 kJ mol^{-1} is given out, which is much less than the 360 kJ mol^{-1} that we would have predicted. This is shown in the energy level diagram below.

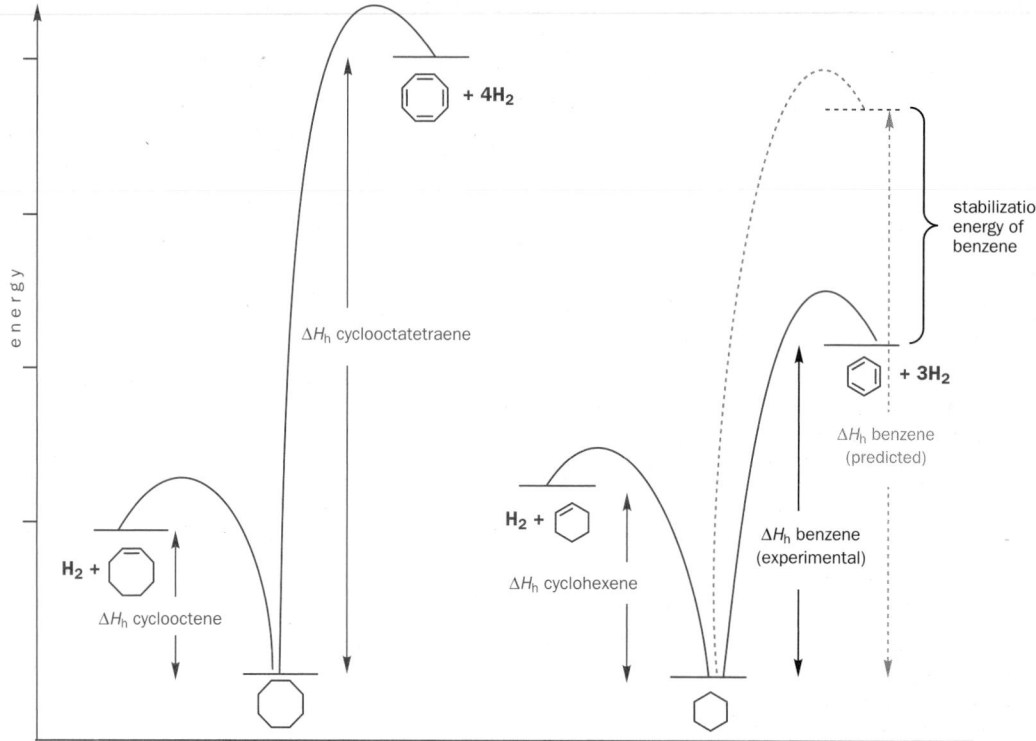

Benzene has six π molecular orbitals

The difference between the amount of energy we expect to get out on hydrogenation (360 kJ mol^{-1}) and what is observed (208 kJ mol^{-1}) is about 150 kJ mol^{-1}. This represents a crude measure of just how extra stable benzene really is relative to what it would be like with three localized double bonds.

In order to understand the origin of this stabilization, we must look at the molecular orbitals. We can think of the π molecular orbitals of benzene as resulting from the combination of the six p orbitals. We have already encountered the molecular orbital lowest in energy with all the orbitals combining in-phase.

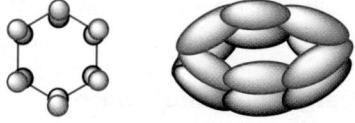

the lowest energy MO for benzene has all the p orbitals combining in-phase

The next lowest molecular orbital will have one nodal plane. How can we divide up the six atoms symmetrically with one nodal plane? There are two ways depending on whether or not the nodal plane passes through a bond or an atom.

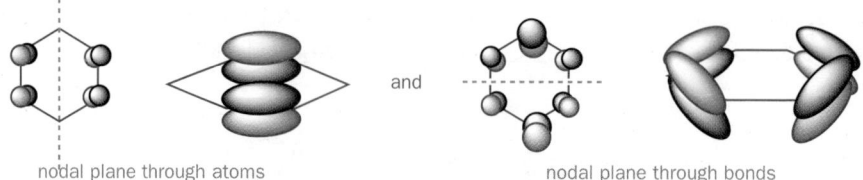

nodal plane through atoms   nodal plane through bonds

there are two ways of symmetrically dividing the six carbon atom
one has a node through two atoms, the other through two C–C bonds

It turns out that these two different molecular orbitals both have exactly the same energy, that is, they are **degenerate**. This isn't obvious from looking at them but, nevertheless, it is so.

The next molecular orbital will have two nodal planes and again there are two ways of arranging these, which lead to two degenerate molecular orbitals.

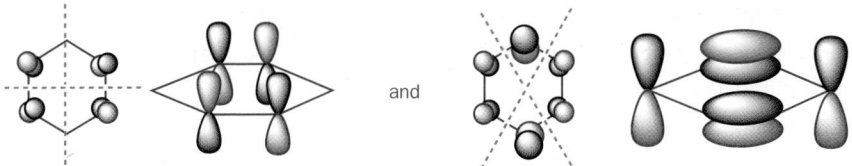

with two nodal planes, there are again two possible molecular orbitals

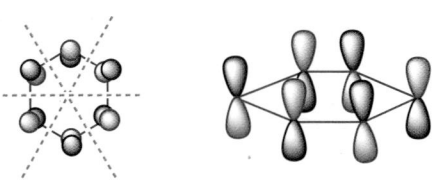

the MO highest in energy has all p orbitals combining out-of-phase

The final molecular orbital will have three nodal planes, which must mean all the p orbitals combining out-of-phase.

These then are the six π molecular orbitals for benzene. We can draw an energy level diagram to represent them.

Benzene model

Whilst the HOMOs for benzene are the degenerate π molecular orbitals (Ψ_2), the next molecular orbital down in energy is not actually the π molecular orbital (Ψ_1). This all-bonding π molecular orbital (Ψ_1) is so stable that *four* σ bonding molecular orbitals actually come in between the π molecular Ψ_1 and Ψ_2 orbitals but are not shown in this molecular orbital energy level diagram. The greatest contribution to stability comes from this lowest-energy π bonding molecular orbital (Ψ_1). This allows bonding interactions between all adjacent atoms. Theory tells us that the energy of this orbital is $\alpha + 2\beta$ whilst that of the degenerate bonding molecular orbitals is $\alpha + \beta$. When all these bonding molecular orbitals are fully occupied, the total energy of the electrons is $6\alpha + 8\beta$, which is 2β lower in energy than we would predict for three localized double bonds. Butadiene had a theoretical stabilization energy of just 0.48β relative to two isolated double bonds so 2β is really quite significant.

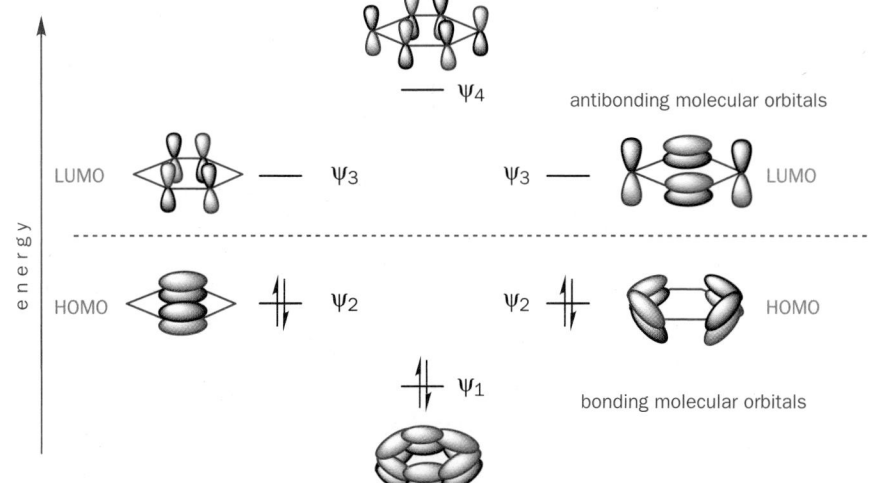

the π molecular orbitals for benzene. The dashed line represents the energy of an isolated p orbital all orbitals below this line are bonding, all above it are antibonding. Benzene has six electrons in its π system so all the bonding MOs are fully occupied

The π molecular orbitals of conjugated cyclic hydrocarbons can be easily predicted

Notice that the layout of the energy levels is a regular hexagon with its apex pointing downwards. It turns out that the energy level diagram for the molecular orbitals resulting from the combination of *any* regular cyclic arrangement of p orbitals can be deduced from the appropriately sided polygon. If we take a regular polygon with one corner pointing downwards and draw a circle round it so that all the corners touch the circle, the energies of the molecular orbitals will be where the corners touch the circle. The circles should be of the same size and the polygons fitted inside the circle. The horizontal diameter represents the energy of a carbon p orbital and so, if any energy levels are on this line, they must be nonbonding. All those below are bonding; all those above antibonding.

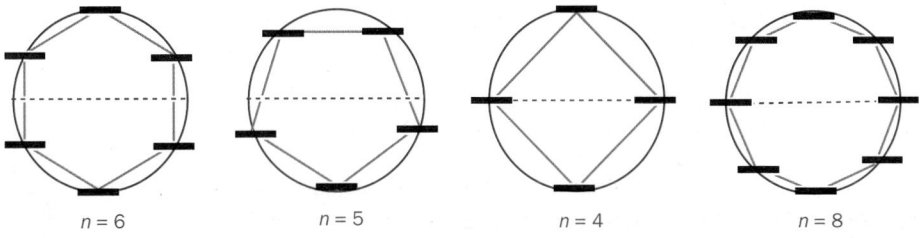

$n = 6$ $n = 5$ $n = 4$ $n = 8$

▬ = energy level of MO relative to energy of p orbital indicated by dashed line

n = number of carbon atoms in ring

Notes on these energy level diagrams:

- This method predicts the energy levels for the molecular orbitals of planar, monocyclic, arrangements of identical atoms (usually all C) only
- The dashed line represents an energy level α and in each case the circle radius is 2β
- There is always one single molecular orbital lower in energy than all the others (at energy α + 2β). This is because there is always one molecular orbital where all the p orbitals combine in-phase
- If there are an even number of atoms, there is also a single molecular orbital highest in energy; otherwise there will be a pair of degenerate molecular orbitals highest in energy
- All the molecular orbitals come in degenerate pairs except the one lowest in energy and, for even-numbered systems, the one highest in energy

Now we can begin to put all the pieces together and make sense of what we know so far. Let us compare the energy level diagrams for benzene and planar cyclooctatetraene. We are not concerned with the actual shapes of the molecular orbitals involved, just the energies of them.

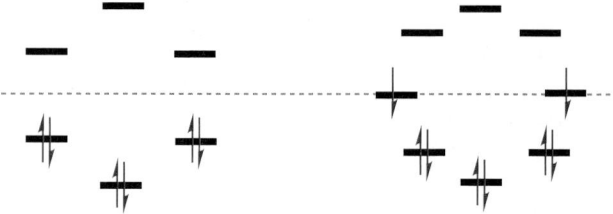

MO level diagram for benzene

MO level diagram for *planar* cyclooctatetraene

Benzene has six π electrons, which means that all its bonding molecular orbitals are fully occupied giving a closed shell structure. COT, on the other hand, has eight electrons. Six of these fill up the bonding molecular orbitals but there are two electrons left. These must go into the degenerate pair of nonbonding orbitals. Hund's rule (Chapter 4) would suggest one in each. Therefore this planar structure for COT would not have the closed shell structure that benzene has—it must either lose or gain two electrons in order to have a closed shell structure with all the electrons in bonding orbitals. This is exactly what we have already seen—both the dianion and dication are planar, allowing delocalization all over the ring, whereas neutral COT adopts a nonplanar tub shape with localized bonds.

Hückel's rule tells us if compounds are aromatic

Using this simple method to work out the energy level diagrams for other rings, we find that there is always a single low-energy bonding orbital (composed of all p orbitals combining in-phase) and then pairs of degenerate orbitals. Since the single orbital will hold two electrons when full and the degenerate pairs four, we shall have a closed shell of electrons in these π orbitals only when they contain 2 + 4n electrons (n is an integer 0, 1, 2, etc.). This is the basis of Hückel's rule.

● **Hückel's rule**

Planar, fully conjugated, monocyclic systems with (4n + 2) π electrons have a closed shell of electrons all in bonding orbitals and are exceptionally stable. Such systems are said to be aromatic.

Analogous systems with 4n π electrons are described as anti-aromatic

That the π system is fully conjugated and planar are important conditions for aromaticity. The next (4n + 2) number after six is ten so we might expect this cyclic alkene to be aromatic.

If this annulene with five *cis* double bonds were planar, each internal angle would be 144°. Since a normal double bond has bond angles of 120°, this would be far from ideal. This compound can be made but it does *not* adopt a planar conformation and therefore is not aromatic even though it has ten π electrons. By contrast, [18]annulene, which is also a (4n + 2) π electron system (n = 4), does adopt a planar conformation and *is* aromatic (as shown by proton

all-*cis*-[10]annulene

[18]-annulene

▶ Of course, this isn't the molecular orbital energy level diagram for real cyclooctatetraene since COT is not planar but tub-shaped.

■ This is not a strict definition of aromaticity. It is actually very difficult to give a concise definition. Hückel's rule is certainly a good guide but also important is the extra stability of the compound (shown, for example, in resistance to changes to its π system) and low reactivity towards electrophiles. Perhaps the best indication as to whether or not a compound is aromatic is the proton NMR spectrum. The protons attached to an aromatic ring are further downfield than would otherwise be expected (Chapter 11).

▶ **Annulenes** (meaning ring alkenes) are compounds with alternating double and single bonds. The number in brackets tells us how many carbon atoms there are in the ring. Using this nomenclature, you could call benzene [6]annulene and cyclooctatetraene [8]annulene—but don't.

NMR). Note the *trans–trans–cis* double bonds: all bond angles can be 120°. [20]annulene presumably could become planar (it isn't quite) but since it is a 4n p electron system rather than a 4n + 2 system, it is not aromatic and the structure shows localized single and double bonds.

The importance of the system being monocyclic is less clear. The problem that often arises is 'exactly how do we count the π electrons?'. Taking a simple example, should we consider naphthalene as two benzene rings joined together or as a ten π electron system?

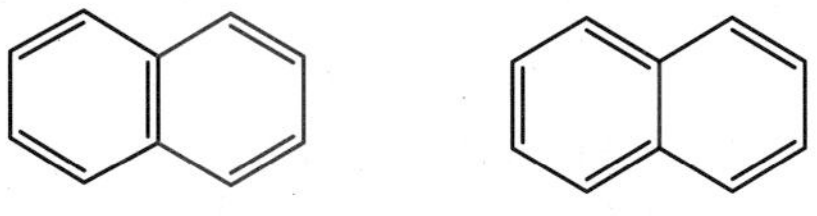

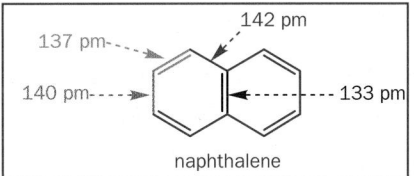

should we count naphthalene as two benzene
rings or one large ring with 10 π electrons?

1,6-Methano[10]annulene is rather like
naphthalene but with the middle bond
replaced by a methylene bridging group.
This compound is almost flat (carbons
1 and 6 are raised slightly out of the
plane) and shows aromatic character.

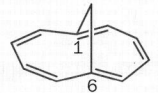

1,6-methano[10]annulene

From its chemistry, it is very clear that naphthalene is aromatic but perhaps a little less so than benzene itself. For example, naphthalene can easily be reduced to tetralin (1,2,3,4-tetrahydronaphthalene) which still contains a benzene ring. Also, in contrast to benzene, all the bond lengths in naphthalene are not the same.

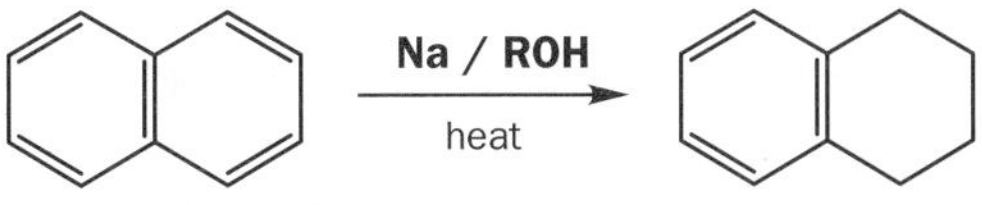

naphthalene → **Na / ROH** heat → tetralin

137 pm 142 pm
140 pm 133 pm
naphthalene

Hückel's rule is very useful and it helps us to predict and understand the aromatic stability of numerous other systems. Cyclopentadiene, for example, has two double bonds that are conjugated but the whole ring is not conjugated since there is a methylene group in the ring. However, this compound is relatively easy to deprotonate (see next chapter, p. 196) to give a very stable anion in which all the bond lengths are the same. How many electrons does this system have? Each of the double bonds contributes two electrons and the negative charge (which must be in a p orbital to complete the conjugation) contributes a further two making six altogether. The energy level diagram shows us that six π electrons completely fill the bonding molecular orbitals thereby giving a stable structure.

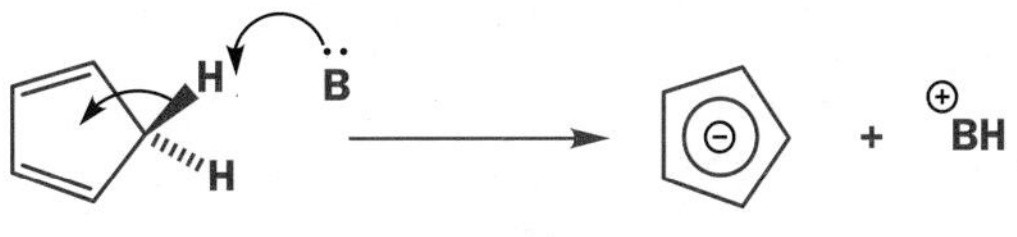

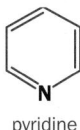

deprotonation of cyclopentadiene gives
the stable cyclopentadienyl anion

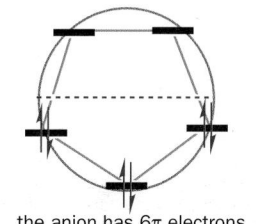

the anion has 6π electrons
completely filling the bonding MOs

Aromatic heterocyclic compounds

So far all the aromatic compounds you have seen have been hydrocarbons. However, most aromatic systems are **heterocyclic**—that is, involving atoms other than carbon and hydrogen. A simple example is pyridine.

In this structure a nitrogen replaces one of the CH groups in benzene. The ring still has three double bonds and thus six π electrons. Consider the structure shown below, pyrrole. This is also aromatic but how can we count six π electrons?

In the cyclopentadiene ring above, there were also two double bonds and on deprotonation one carbon could formally contribute the other two electrons needed for aromaticity. In pyrrole the nitrogen's lone pair can make up the six π electrons needed for the system to be aromatic.

We are really just beginning to scratch the surface of aromatic chemistry. You will meet many aromatic compounds in this book: in Chapter 22 we shall look at the chemistry of benzene and in Chapters 43 and 44 we shall discuss heterocyclic aromatic compounds. We shall finish off this chapter with a few more examples of some common aromatic compounds. In each case the aromatic part of the molecule—which may be one ring or several rings—is outlined in black.

pyridine

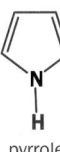

pyrrole

First, a compound released by many cut plants, especially grasses, with a fresh delightful smell usually called 'new mown hay'. Coumarin is also present in some herbs such as lavender. It contains a benzene ring and an α-pyrone fused together.

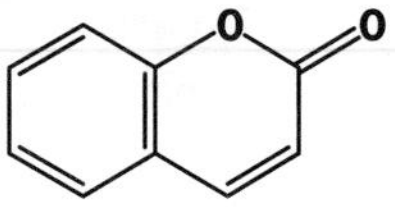

coumarin – the smell of 'new mown hay' also found in lavender

benzene 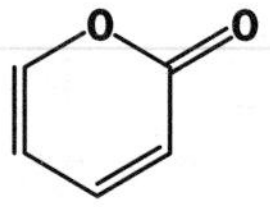   α-pyrone

Next, pirimicarb, a selective insecticide that kills sap-sucking aphid pests but does not affect the useful predators such as ladybirds (ladybugs) that eat them. It contains a pyrimidine ring—a benzene ring with two nitrogen atoms.

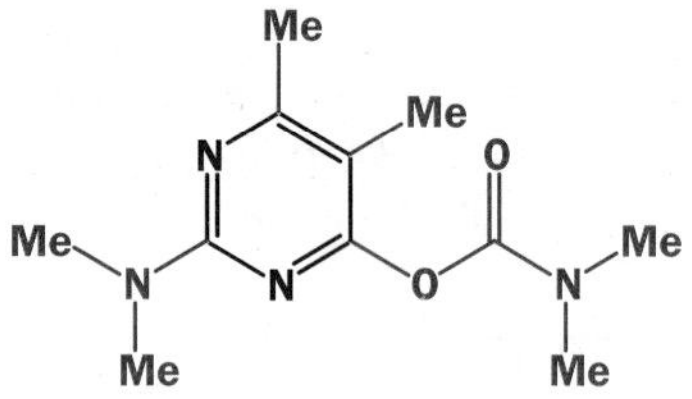

pirimicarb – a selective insecticide which kills aphids but not ladybirds

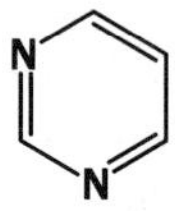

pyrimidine

LSD stands for LySergic acid Diethylamide. It is the hallucinogenic drug 'acid'. When people walk off a building claiming that they can fly, they are probably on acid. It contains an indole ring made up of a benzene ring and a pyrrole ring fused together.

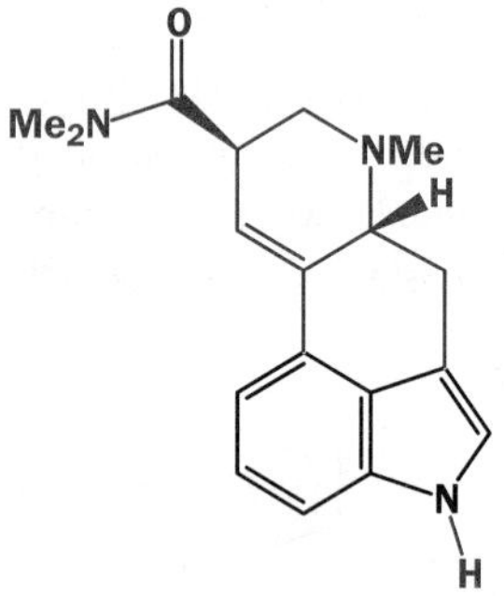

LSD, lysergic acid diethylamide, the infamous 'acid' giving hallucinations and unfounded confidence in flying

benzene 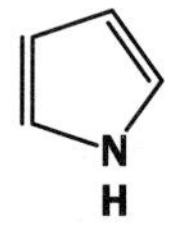   pyrrole

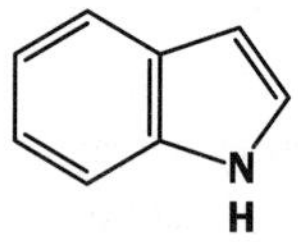

indole

The world's best selling medicine in 1998 was Omeprazole, an antiulcer drug from Astra. It prevents excess acid in the stomach and allows the body to heal ulcers. It contains a pyridine ring and a benzimidazole ring, two aromatic heterocycles.

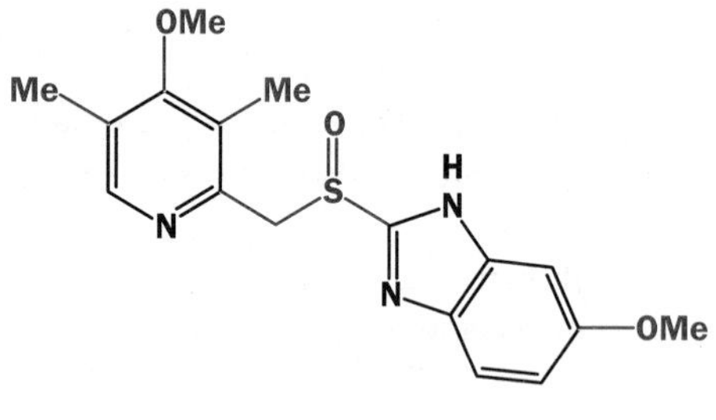

Omeprazole Astra's best selling antiulcer drug

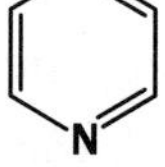

pyridine

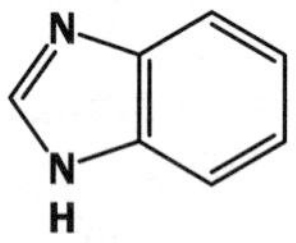

benzimidazole

The drug in the news in 1999 was Viagra, Pfizer's cure for male impotence. In the first three months after its release in 1998, 2.9 million prescriptions were issued for Viagra. It contains a simple benzene ring and a more complex heterocyclic system, which can be divided into two aromatic heterocyclic rings.

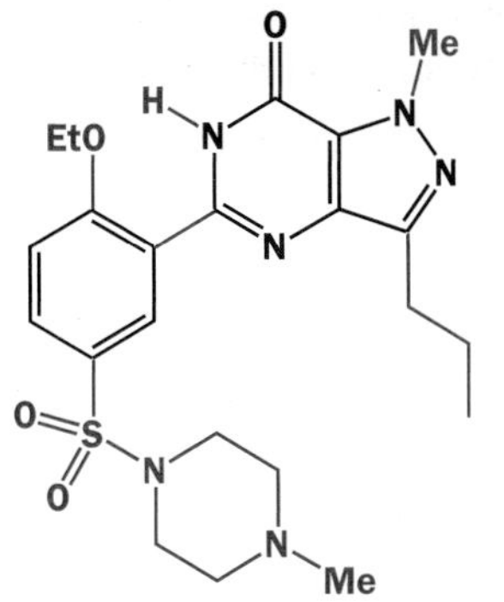

Viagra
Pfizer's new treatment
for male impotence
(male erectile dysfunction)

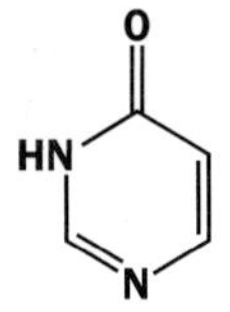

pyrimidone 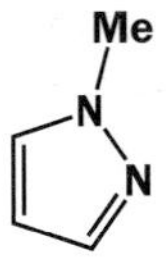   pyrazole

benzene

Finally, the iron compound haem, part of the haemoglobin molecule we use to carry oxygen around in our bloodstream. It contains the aromatic porphyrin ring system with its eighteen electrons arranged in annulene style. Chlorophyll, mentioned earlier in this chapter, has a similar aromatic ring system.

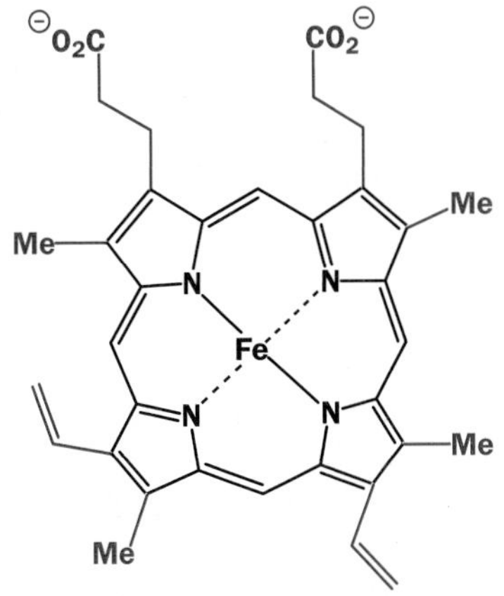

haem – part of the haemoglobin that
transports oxygen in the blood

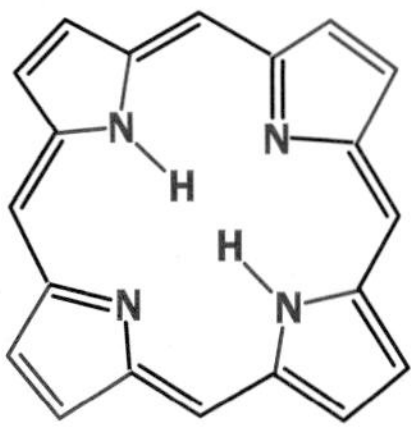

porphyrin or porphin
one 18 π electron ring is shown in black.
Others are possible

Problems

1. Are these molecules conjugated? Explain your answer in any reasonable way.

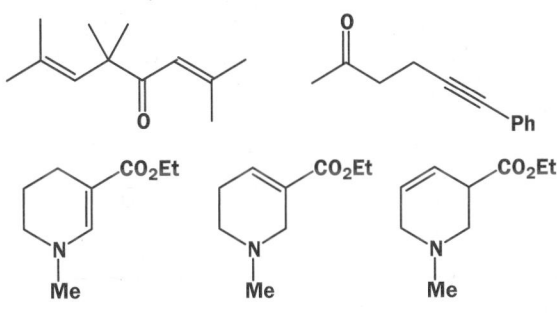

2. Draw a full orbital diagram for all the bonding and antibonding π orbitals in the three-membered cyclic cation shown here. The molecule is obviously very strained. Might it survive by also being aromatic?

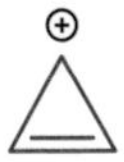

3. How extensive are the conjugated systems in these molecules?

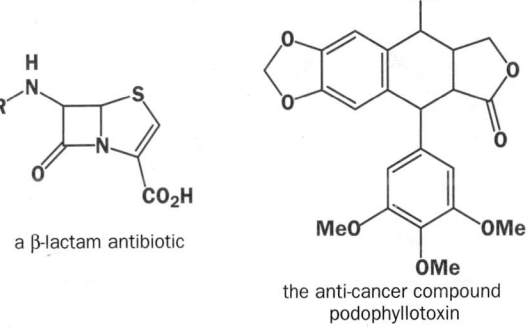

a β-lactam antibiotic

the anti-cancer compound
podophyllotoxin

4. Draw diagrams to illustrate the conjugation present in these molecules. You should draw three types of diagram: (**a**) conjugation arrows to give at least two different ways of representing the molecule joined by the correct 'reaction' arrow; (**b**) a diagram with dotted lines and partial charges (if any) to show the double bond and charge distribution (if any); and (**c**) a diagram of the atomic orbitals that make up the lowest-energy bonding molecular orbital.

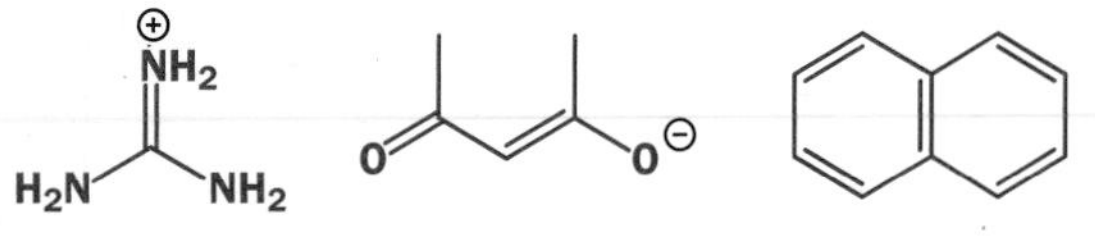

5. Which of these compounds are aromatic? Justify your answer with some electron counting. You are welcome to treat each ring separately or two or more rings together, whichever you prefer.

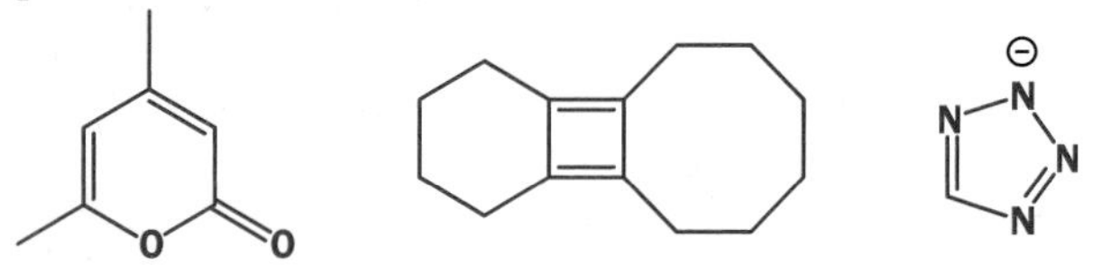

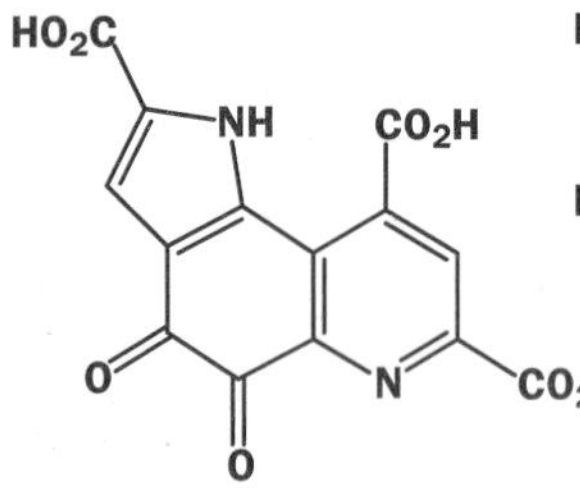

methoxatin: co-enzyme from bacteria living on methane

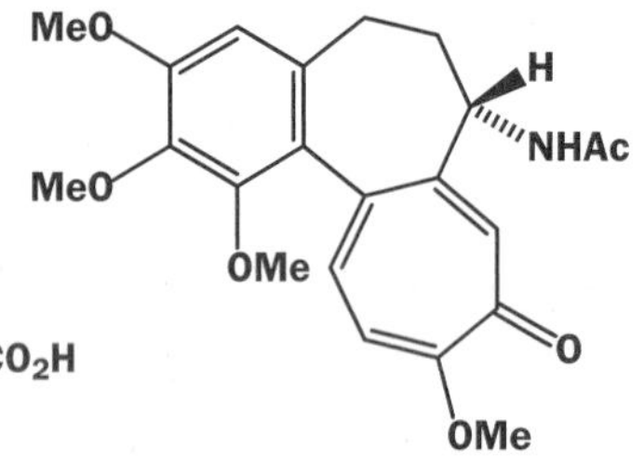

colchicine: compound from Autumn crocus used to treat gout

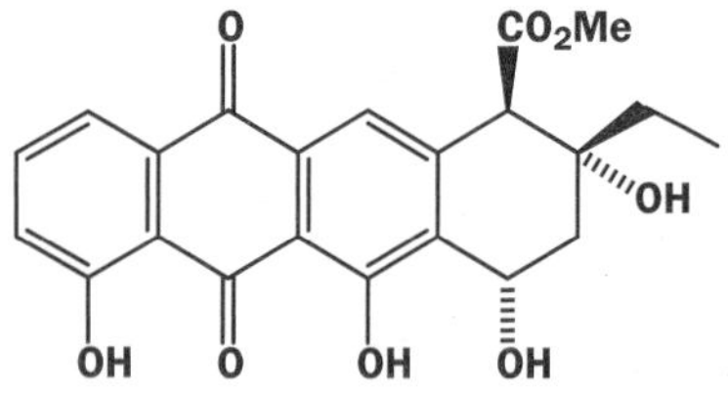

aklavinone: a tetracycline antibiotic

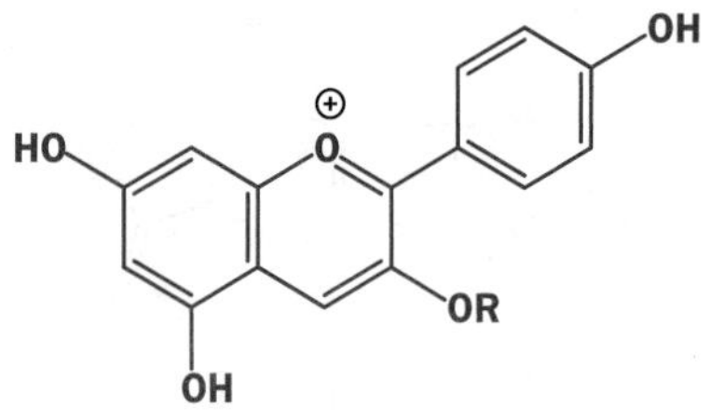

callistephin: natural red flower pigment

6. A number of water-soluble pigments in the green/blue/violet ranges used as food dyes are based on cations of the type shown here. Explain why the general structure shows such long wavelength absorption and suggest why the extra functionality (OH group and sulfonate anions) is put into 'CI food green 4' a compound approved by the EU for use in food under E142.

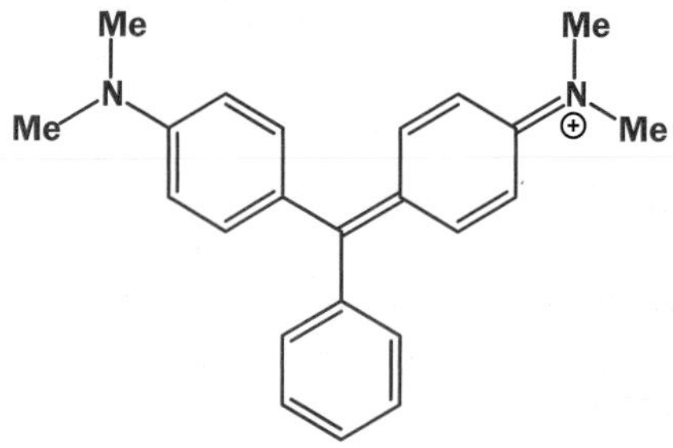

general structure for water soluble food dye in the green/blue/violet range

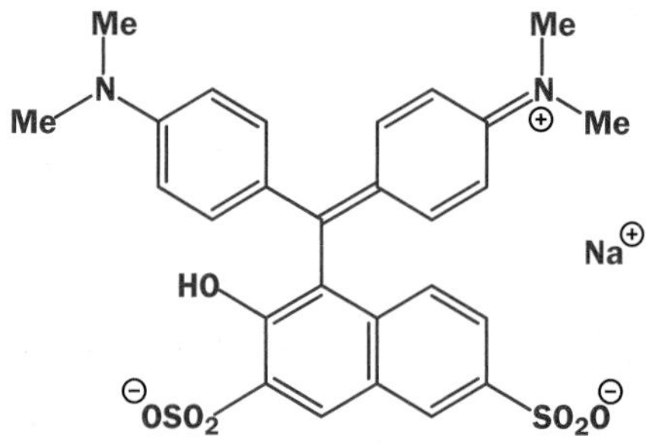

green food dye 'CI food green 4' [E142]

7. Turn to Chapter 1 and look at the structures of the dyes in the shaving foam described on p. 7. Comment on the structures in comparison with those in Problem 6 and suggest where they get their colour from and why they too have extra functional groups. Then turn to the beginning of Chapter 1 (p. 3) and look at the structures of the compounds in the 'spectrum of molecules'. Can you see what kind of absorption leads to each colour? You will want to think about the conjugation in each molecule but you should not expect to correlate structures with wavelengths in any even roughly quantitative way.

8. Go through the list of aromatic compounds at the end of the chapter and see how many electrons there are in the rings taken separately or taken together (if they are fused). Are all the numbers of the $(4n + 2)$ kind?

Acidity, basicity, and pK_a

Note from the authors to all readers

This chapter contains physical data and mathematical material that some readers may find daunting. Organic chemistry students come from many different backgrounds since organic chemistry occupies a middle ground between the physical and the biological sciences. We hope that those from a more physical background will enjoy the material as it is. If you are one of those, you should work your way through the entire chapter. If you come from a more biological background, especially if you have done little maths at school, you may lose the essence of the chapter in a struggle to understand the equations. We have therefore picked out the more mathematical parts in boxes and you should abandon these parts if you find them too alien. We consider the general principles behind the chapter so important that we are not prepared to omit this essential material but you should try to grasp the principles without worrying too much about the equations. The ideas of acidity, basicity, and pK_a values together with an approximate quantitative feel for the strength and weakness of acids and bases are at least as central for biochemistry as they are for organic chemistry. Please do not be discouraged but enjoy the challenge.

Introduction

This chapter is all about acidity, basicity, and pK_a. Acids and bases are obviously important because many organic and biological reactions are catalysed by acids or bases, but what is pK_a and what use is it? pK_a tells us how acidic (or not) a given hydrogen atom in a compound is. This is useful because, if the first step in a reaction is the protonation or deprotonation of one of the reactants, it is obviously necessary to know where the compound would be protonated or deprotonated and what strength acid or base would be needed. It would be futile to use too weak a base to deprotonate a compound but, equally, using a very strong base where a weak one would do would be like trying to crack open a walnut using a sledge hammer—you would succeed but your nut would be totally destroyed in the process.

The aim of this chapter is to help you to understand *why* a given compound has the pK_a that it does. Once you understand the trends involved, you should have a good feel for the pK_a values of commonly encountered compounds and also be able to predict the values for unfamiliar compounds.

Originally, a substance was identified as an acid if it exhibited the properties shown by other acids: a sour taste (the word acid is derived from the Latin *acidus* meaning 'sour') and the abilities to turn blue vegetable dyes red, to dissolve chalk with the evolution of gas, and to react with certain 'bases' to form salts. It seemed that all acids must therefore contain something in common and at the end of the eighteenth century, the French chemist Lavoisier erroneously proclaimed this common agent to be oxygen (indeed, he named oxygen from the Greek *oxus* 'acid' and *gennao* 'I produce'). Later it was realized that some acids, for example, hydrochloric acid, did not contain oxygen and soon hydrogen was identified as the key species. However, not all hydrogen-containing compounds are acidic, and at the end of the nineteenth century it was understood that such compounds are acidic only if they produce hydrogen ions H^+ in aqueous solution—the more acidic the compound, the more hydrogen ions it produces. This was refined once more in 1923 by J.N. Brønsted who proposed simple definitions for acids and bases.

> **● Brønsted definitions of acids and bases**
> - **An acid is a species having a tendency to lose a proton**
> - **A base is a species having a tendency to accept a proton**

Other definitions of acids and bases are useful, the most notable being those of Lewis, also proposed in 1923. However, for this chapter, the Brønsted definition is entirely adequate.

Acidity

An isolated proton is incredibly reactive—formation of H_3O^+ in water

Hydrochloric acid is a strong acid: the free energy $\Delta G°$ for its ionization equilibrium in water is -40 kJ mol^{-1}.

$$\text{HCl (aq)} \rightleftharpoons \text{H}^+ \text{(aq)} + \text{Cl}^- \text{(aq)} \qquad \Delta G°_{298K} = -40 \text{ kJ mol}^{-1}$$

Such a large negative $\Delta G°$ value means that the equilibrium lies well over to the right. In the gas phase, however, things are drastically different and $\Delta G°$ for the ionization is $+1347$ kJ mol^{-1}.

$$\text{HCl (g)} \rightleftharpoons \text{H}^+ \text{(g)} + \text{Cl}^- \text{(g)} \qquad \Delta G°_{298K} = +1347 \text{ kJ mol-1}$$

This $\Delta G°$ value corresponds to 1 molecule of HCl in 10^{240} being dissociated! This means that HCl does not spontaneously ionize in the gas phase—it does not lose protons at all. Why then is HCl such a strong acid in water? The key to this problem is, of course, the water. In the gas phase we would have to form an isolated proton (H^+, hydrogen ion) and chloride ion and this is energetically very unfavourable. In contrast, in aqueous solution the proton is strongly attached to a water molecule to give the very stable **hydronium ion**, H_3O^+, and the ions are no longer isolated but solvated. Even in the gas phase, adding an extra proton to neutral water is highly exothermic.

$$\text{H}_2\text{O (g)} + \text{H}^+ \text{(g)} \longrightarrow \text{H}_3\text{O}^+ \text{(g)} \qquad \Delta H° = -686 \text{ kJ mol}^{-1}$$

In fact, an isolated proton is so reactive that it will even add on to a molecule of methane in the gas phase to give CH_5^+ in a strongly exothermic reaction (you have already encountered this species in mass spectrometry on p. 52). We are therefore extremely unlikely to have a naked proton in the gas phase and certainly *never* in solution. In aqueous solution a proton will be attached to a water molecule to give a hydronium ion, H_3O^+ (sometimes called a hydroxonium ion). This will be solvated just as any other cation (or anion) would be and hydrogen bonding gives rise to such exotic species as $H_9O_4^+$ ($H_3O^+·3H_2O$) shown here.

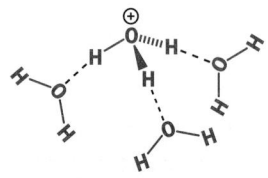

a structure for a solvated hydronium ion in water the dashed bonds represent hydrogen bonds

Every acid has a conjugate base

In water, hydrogen chloride donates a proton to a water molecule to give a hydronium ion and chloride ion, both of which are strongly solvated.

$$\text{HCl (aq)} + \text{H}_2\text{O (l)} \rightleftharpoons \text{H}_3\text{O}^+ \text{ (aq)} + \text{Cl}^- \text{(aq)}$$

In this reaction water is acting as a base, according to our definition above, by accepting a proton from HCl which in turn is acting as an acid by donating a proton. If we consider the reverse reaction (which is admittedly insignificant in this case since the equilibrium lies well over to the right), the chloride ion accepts a proton from the hydronium ion. Now the chloride is acting as a base and the hydronium ion as an acid. The chloride ion is called the **conjugate base** of hydrochloric acid and the hydronium ion, H_3O^+, is the **conjugate acid** of water.

> ● **For any acid and any base**
>
> $$\text{AH} + \text{B} \rightleftharpoons \text{BH}^+ + \text{A}^-$$
>
> where AH is an acid and A^- is its conjugate base and B is a base and BH^+ is its conjugate acid, that is, *every acid has a conjugate base associated with it and every base has a conjugate acid associated with it.*

For example, with ammonia and acetic acid

$$\text{CH}_3\text{COOH} + \text{NH}_3 \rightleftharpoons \text{NH}_4^+ + \text{CH}_3\text{COO}^-$$

the ammonium ion, NH_4^+, is the conjugate acid of the base ammonia, NH_3, and the acetate ion, CH_3COO^-, is the conjugate base of acetic acid, CH_3COOH.

Water can behave as an acid or as a base

If a strong acid is added to water, the water acts as a base and is protonated by the acid to become H_3O^+. If we added a strong base to water, the base would deprotonate the water to give hydroxide ion, OH^-, and here the water would be acting as an acid. Such compounds that can act as either an acid or a base are called **amphoteric**.

With a strong enough acid, we can protonate almost anything and, likewise, with a strong enough base we can deprotonate almost anything. This means that, to a certain degree, all compounds are amphoteric. For example, hydrochloric acid will protonate acetic acid.

In this example acetic acid is acting as a base! Other compounds need acids even stronger than HCl to protonate them. Remember that, in chemical ionization mass spectrometry (p. 52), protonated methane, CH_5^+, was used to protonate whatever sample we put in to the machine in order to give us a cation; CH_5^+ is an incredibly strong acid.

The amino acids you encountered in Chapter 2 are amphoteric. Unlike water, however, these compounds have separate acidic and basic groups built into the same molecule.

When amino acids are dissolved in water, the acidic end protonates the basic end to give a species with both a positive and a negative charge on it. A neutral species that contains both a positive and a negative charge is called a **zwitterion**.

an amino acid zwitterion

How the pH of a solution depends on the concentration of the acid

You are probably already familiar with the pH scale: acidic solutions all have a pH of less than 7—the lower the pH the more acidic the solution; alkaline solutions all have pHs greater than 7—the higher the pH, the more basic the solution. Finally, pH 7 is neither acidic nor alkaline but neutral.

The pH of a solution is only a measure of the acidity of the solution; it tells us nothing about how strong one acid might be relative to another. The pH of a solution of a given acid varies with its concentration: as we dilute the solution, the acidity falls and the pH increases. For example, as we decrease the concentration of HCl in an aqueous solution from 1 to 0.1 to 0.01 to 0.001 mol dm^{-3}, the pH changes from 0 to 1 to 2 to 3.

What a pH meter actually measures is the concentration of hydronium ions in the solution. The scale is a logarithmic one and is defined as

$$pH = -\log[H_3O^+]$$

Our solutions of HCl above therefore have hydronium ion concentrations of $[H_3O^+] = 10^0, 10^{-1}, 10^{-2},$ and 10^{-3} mol dm^{-3} respectively. Since the scale is logarithmic, a pH difference of 1 corresponds to a factor of 10 in hydronium ion concentration, a pH difference of 2 corresponds to a factor of 100, and so on.

The ionization of water

Pure water at 25 °C has a pH of 7.00. This means that the concentration of hydronium ions in water must be 10^{-7} mol dm^{-3} (of course, it is actually the other way round: the hydronium ion concentration in pure water is 10^{-7} mol dm^{-3}; hence its pH is 7.00). Hydronium ions in pure water can arise only from the self-dissociation or autoprotolysis of water.

$$H_2O + H_2O \rightleftharpoons H_3O^+ \text{ (aq)} + OH^- \text{ (aq)}$$

In this reaction, one molecule of water is acting as a base, receiving a proton from the other, which in turn is acting as an acid by donating a proton. From the equation we see that, for every hydronium ion formed, we must also form a hydroxide ion and so in pure water the concentrations of hydroxide and hydronium ions are equal.

$$[H_3O^+] = [OH^-] = 10^{-7} \text{ mol dm}^{-3}$$

The product of these two concentrations is known as the ionization constant of water, K_W (or as the ionic product of water, or maybe sometimes as the autoprotolysis constant, K_{AP})

$$K_W = [H_3O^+][OH^-] = 10^{-14} \text{ mol}^2 \text{ dm}^{-6} \text{ at 25 °C}$$

This is a constant in aqueous solutions, albeit a very, very small one. This means that, if we know the hydronium ion concentration, we also know the hydroxide concentration and vice versa since the product of the two concentrations always equals 10^{-14}.

■
Don't worry—water is still safe to drink despite all this acid and hydroxide in it! This is, of course, because the concentrations of hydronium and hydroxide ions are *very* small (10^{-7} mol dm^{-3} corresponds to about 2 parts per billion). This very low concentration means that there are not enough free hydronium (or hydroxide) ions in water either to do us any harm when we drink it, *or to catalyse chemical reactions.*

For example

It is easy to work out the pH of a 0.1 M solution of sodium hydroxide,

$$[NaOH] = 0.1 \text{ M}$$

and, since the sodium hydroxide is fully ionized,
$[OH^-] = 0.1$ M but $[OH^-] \times [H_3O^+] = 10^{-14}$.

So

$$[H_3O^+] = \frac{10^{-14}}{0.1} = 10^{-13} \text{mol dm}^{-3}$$

$$pH = -\log[H_3O^+] = -\log(10^{-13}) = 13$$

How the pH of a solution also depends on the acid in question

If we measured the pH of an aqueous solution of an organic acid and compared it to an equally concentrated solution of HCl, we would probably find the pHs different. For example, whilst 0.1M HCl has a pH of 1, the same concentration of acetic acid has a pH of 3.7 and is much less acidic. This can only mean that a 0.1M solution of acetic acid contains fewer hydronium ions than a 0.1M solution of HCl.

● Aqueous hydrochloric acid (or any strong acid) has a lower pH than an equal concentration of aqueous acetic acid (or any weak acid) because it is more fully dissociated and thereby produces more hydronium ions.

For hydrochloric acid, the equilibrium lies well over to the right: in effect, HCl is completely dissociated.

$$HCl \text{ (aq)} + H_2O \text{ (l)} \rightleftharpoons H_3O^+ \text{ (aq)} + Cl^- \text{ (aq)}$$

Acetic acid is not fully dissociated—the solution contains both acetic acid and acetate ions.

$$CH_3COOH \text{ (aq)} + H_2O \text{ (l)} \rightleftharpoons H_3O^+ \text{ (aq)} + CH_3COO^- \text{ (aq)}$$

Acids as preservatives

Acetic acid is used as a preservative in many foods, for example, pickles, mayonnaise, bread, and fish products, because it prevents bacteria and fungi growing. However, its fungicidal nature is not due to any lowering of the pH of the foodstuff. In fact, it is the *undissociated* acid that acts as a bactericide and a fungicide in concentrations as low as 0.1–0.3%. Besides, such a low concentration has little effect on the pH of the foodstuff anyway.

Although acetic acid can be added directly to a foodstuff (disguised as E260), it is more common to add vinegar which contains between 10 and 15% acetic acid. This makes the product more 'natural' since it avoids the nasty 'E numbers'. Actually, vinegar has also replaced other acids used as preservatives, such as propionic (propanoic) acid (E280) and its salts (E281, E282, and E283).

The definition of pK_a

Now we need to be clearer about 'strong' and 'weak' acids. In order to measure the strength of an acid relative to water and find out how effective a proton donor it is, we must look at the equilibrium constant for the reaction

$$AH \text{ (aq)} + H_2O \text{ (l)} \rightleftharpoons H_3O^+ \text{ (aq)} + A^- \text{ (aq)}$$

The position of equilibrium is measured by the equilibrium constant for this reaction K_{eq}.

$$K_{eq} = \frac{[H_3O^+][A^-]}{[AH][H_2O]}$$

The concentration of water remains essentially constant (at 55.56 mol dm^{-3}) with dilute solutions of acids wherever the equilibrium may be and a new equilibrium constant, K_a, is defined and called the **acidity constant**.

$$K_a = \frac{[H_3O^+][A^-]}{[AH]}$$

Like pH, this is also expressed in a logarithmic form, pK_a.

> ● p$K_a = -\log K_a$
>
> Because of the minus sign in this definition, the lower the pK_a, the larger the equilibrium constant, K_a, is and hence the stronger the acid. *The pK_a of the acid is the pH where it is exactly half dissociated.* At pHs above the pK_a, the acid HA exists as A$^-$ in water; at pHs below the pK_a, it exists as undissociated HA.

At pHs above the pK_a of the acid, it will also be more soluble in water. Hydrocarbons are insoluble in water—oil floats on water, for example. Unless a compound has some hydrophilic groups in it that can hydrogen bond to the water, it too will be insoluble. Ionic groups considerably increase a compound's solubility and so the ion A$^-$ is much more soluble in water than the undissociated acid HA. In fact water can solvate both cations and anions, unlike some of the solvents you will meet later. This means that we can increase the solubility of a neutral acid in water by increasing the proportion of its conjugate base present. All we need to do is raise the pH.

A simple example is aspirin: whilst the acid itself is not very soluble in water, the sodium salt is much more soluble (soluble aspirin is actually the sodium or calcium salt of 'normal' aspirin).

aspirin
not very soluble in water

the sodium (or calcium) salt of aspirin is more soluble in water

Conversely, if the pH of a solution is lowered, the amount of the acidic form present increases, and the solubility decreases. In the acidic environment of the stomach (around pH 1–2), soluble aspirin will be converted back to the normal acidic form and precipitate out of solution.

▶ How concentrated is water? Very concentrated you may say—but the concentration is limited. We know that one mole of pure water has a mass of 18 g and occupies 18 cm^3. So, in one dm^3, there are $1000/18 = 55.56$ mol. *You cannot get more concentrated water than this* (unless you did something drastic like taking it into a black hole!).

▶ This is how we can work out that the pK_a of the acid is the pH at which it is exactly half dissociated: we can rearrange the equation for K_a to give

$$[H_3O^+] = K_a \times \frac{[AH]}{[A^-]}$$

Taking minus the log of both sides gives us

$$pH = pK_a + \log\left(\frac{[A^-]}{[AH]}\right)$$

If the concentrations of acid AH and its conjugate base A$^-$ are equal, the term in brackets equals 1 and log (1) = 0 and so the pH simply equals the pK_a of the acid.

This means that, if we took a 0.1 M aqueous solution of acetic acid and raised its pH from 3.7 (its natural pH) to 4.76 (the pK_a of acetic acid) using dilute sodium hydroxide solution, the resultant solution would contain equal concentrations of acetic acid and acetate ion.

In the same way, organic bases such as amines can be dissolved by *lowering* the pH. Codeine (7,8-didehydro-4,5-epoxy-3-methoxy-17-methylmorphinan-6-ol) is a commonly used painkiller. Codeine itself is not very soluble in water but it does contain a basic nitrogen atom that can be protonated to give a more soluble salt. It is usually encountered as a phosphate salt. The structure is complex, but that doesn't matter.

neutral codeine
sparingly soluble in water

the conjugate acid is much
more soluble in water

Charged compounds can be separated by acid–base extraction

Adjusting the pH of a solution often provides an easy way to separate compounds. Since weak acids form soluble anions at pHs above their pK_a values, this presents us with an easy method for extracting organic acids from mixtures of other compounds. For example, if we dissolve the mixture of compounds in dichloromethane (which is **immiscible** with water, that is, it will not mix with water but instead forms a separate layer) and 'wash' this solution with aqueous sodium hydroxide, any organic acids present will be converted to their water-soluble salts and dissolve into the water layer. We have extracted the organic acids into the aqueous layer. If we then separate and acidify the aqueous layer, the acid form, being less soluble in water, will precipitate out. If the acid form has a charge and the conjugate base is neutral as with amines, for example, now the cationic acid form will be more soluble in water than the conjugate base.

● **Acid–base extraction**

For a neutral weak organic acid HA

$$HA(aq) + H_2O \rightleftharpoons H_3O^{\oplus}(aq) + A^{\ominus}(aq)$$

- Anionic A⁻ is more soluble *in water* than the neutral acid HA
- Neutral acid HA is more soluble *in organic solvents* than anionic A⁻

For a neutral weak organic base B

$$HB^{\oplus}(aq) + H_2O \rightleftharpoons H_3O^{\oplus}(aq) + B(aq)$$

- The cationic acid HB⁺ is more soluble *in water* than the neutral conjugate base B
- The neutral conjugate base, B is more soluble *in organic solvents* than the cationic acid HB⁺

Separating a mixture of benzoic acid (PhCO$_2$H) and toluene (PhMe) is easy: dissolve the mixture in CH$_2$Cl$_2$, add aqueous NaOH, shake the mixture of solutions, and separate the layers. The CH$_2$Cl$_2$ layer contains all the toluene. The aqueous layer contains the sodium salt of benzoic acid. Addition of HCl to the aqueous layer precipitates the insoluble benzoic acid.

insoluble
in water

insoluble
in water

insoluble
in water

soluble
in water

In the same way, any basic compounds dissolved in an organic layer could be extracted by washing the layer with dilute aqueous acid and recovered by raising the pH, which will precipitate out the less soluble neutral compound.

Whenever you do any extractions or washes in practical experiments, just stop and ask yourself: 'What is happening here? In which layer is my compound and why?' That way you will be less likely to throw the wrong layer (and your precious compound) away!

Benzoic acid preserves soft drinks

Benzoic acid is used as a preservative in foods and soft drinks (E210). Like acetic acid, it is only the acid form that is effective as a bactericide. Consequently, benzoic acid can be used as a preservative only in foodstuffs with a relatively low pH, ideally less than its pK_a of 4.2. This isn't usually a problem: soft drinks, for example, typically have a pH of 2–3. Benzoic acid is often added as the sodium salt (E211), perhaps because this can be added to the recipe as a concentrated solution in water. At the low pH in the final drink, most of the salt will be protonated to give benzoic acid proper, which presumably remains in solution because it is so dilute.

A graphical description of the pK_a of acids and bases

For both cases, adjusting the pH alters the proportions of the acid form and of the conjugate base. The graph plots the concentration of the free acid AH (green curve) and the ionized conjugate base A⁻ (red curve) as percentages of the total concentration as the pH is varied.

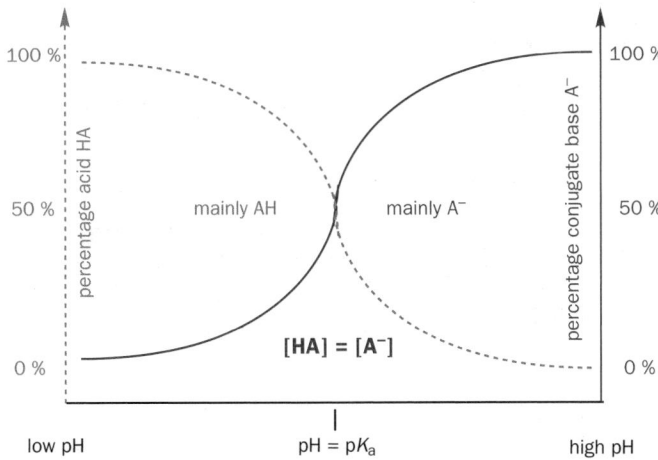

At low pH the compound exists entirely as AH and at high pH entirely as A⁻. At the pK_a the concentration of each species, AH and A⁻, is the same. At pHs near the pK_a the compound exists as a mixture of the two forms.

An acid's pK_a depends on the stability of its conjugate base

HCl is a much stronger acid than acetic acid: the pK_a of HCl is around −7 compared to 4.76 for acetic acid. This tells us that in solution K_a for hydrogen chloride is 10^7 mol dm⁻³ whilst for acetic acid it is only $10^{-4.76} = 1.74 \times 10^{-5}$ mol dm⁻³. Why are the equilibria so different? Why does hydrogen chloride fully dissociate but acetic acid do so only partially?

$$\text{HCl (aq)} + \text{H}_2\text{O (l)} \rightleftharpoons \text{H}_3\text{O}^+ \text{(aq)} + \text{Cl}^- \text{(aq)} \qquad K_a = 10^7$$

$$\text{CH}_3\text{COOH (aq)} + \text{H}_2\text{O (l)} \rightleftharpoons \text{H}_3\text{O}^+ \text{(aq)} + \text{CH}_3\text{COO}^- \text{(aq)} \qquad K_a = 1.74 \times 10^{-5}$$

The answer must have something to do with the conjugate base A⁻ of each acid HA, since this is the only thing that varies from one acid to another. In both the equilibria above, water acts as a base by accepting a proton from the acid. For the hydrochloric acid equilibrium in the reverse direction, the chloride ion is not a strong enough base to deprotonate the hydronium ion. Acetate, on the other hand, is easily protonated by H_3O^+ to give neutral acetic acid, which means that acetate must be a stronger base than chloride ion.

● Acid and conjugate base strength
- The stronger the acid HA, the weaker its conjugate base, A⁻
- The stronger the base A⁻, the weaker its conjugate acid AH

▶

An alternative way of looking at this is that chloride ion is much happier being a chloride ion than acetate is being an acetate ion: *the chloride ion is fundamentally more stable than is the acetate ion.*

▶

Have a close look at Table 8.1 for there are some interesting points to notice.

• Look at the acids themselves—we have neutral, cationic, and even anionic acids
• Notice the range of different elements carrying the negative charge of the conjugate bases—we have iodine, chlorine, oxygen, sulfur, nitrogen, and carbon and many more are possible
• Most importantly, notice the vast range of pK_a values: from around –10 to 50. This corresponds to a difference of 10^{60} in the equilibrium constants and these are by no means the limits. Other compounds or intermediates can have pK_a values even greater or less than these.

For example, hydrogen iodide has a very low pK_a of –10. This means that HI is a strong enough acid to protonate most things. Its conjugate base, iodide ion, is therefore not very basic at all—in fact, we very rarely think of it as a base—it will not deprotonate anything. A very powerful base is methyllithium, MeLi. Here we effectively have CH_3^- (but see Chapter 9), which can accept a proton to become neutral methane, CH_4. Methane is therefore the conjugate acid in this case. Clearly, methane isn't at all acidic—its pK_a is about 48.

Table 8.1 gives a list of compounds and their approximate pK_a values.

Over the next few pages we shall be considering the reasons for these differences in acid strength but we are first going to consider the simple consequences of mixing acids or bases of different strength.

The difference in pK_a values tells us the equilibrium constant between two acids or bases

If we have a mixture of two bases in a pot and we throw in a few protons, where will the protons end up? Clearly, this depends on the relative strengths of the bases—if they are equally strong, then the protons will be shared between them equally. If one base is stronger than the other, then this base will get more than its fair share of protons. If we put into our pot not two bases but one base and an acid, then it's exactly the same as putting in two bases and then adding some protons—the protons end up on the strongest base. Exactly how the protons are shared depends on the difference in strengths of the two bases, which is related to the difference in the pK_as of their conjugate acids.

Table 8.1 The pK_a value of some compounds

Acid	pK_a	Conjugate base
HI	ca. –10	I^-
HCl	ca. –7	Cl^-
H_2SO_4	ca. –3	HSO_4^-
HSO_4^-	2.0	SO_4^{2-}
CH_3COOH	4.8	CH_3COO^-
H_2S	7.0	HS^-
NH_4^+	9.2	NH_3
C_6H_5OH	10.0	$C_6H_5O^-$
CH_3OH	15.5	CH_3O^-
CH_3COCH_3	20.0	$CH_3COCH_2^-$
$CH{\equiv}C-H$	25	$CH{\equiv}C^-$
NH_3	33	NH_2^-
C_6H_6	ca. 43	$C_6H_5^-$
CH_4	ca. 48	CH_3^-

■

That the difference in pK_as gives the log of the equilibrium constant can easily be shown by considering, as an example, the equilibrium for the reaction between hydrogen sulfate and acetate.

$$HSO_4^- + CH_3OO^-(aq) \rightleftharpoons CH_3COOH(aq) + SO_4^{2-}$$

$$K_{eq} = \frac{[SO_4^{2-}][CH_3COOH]}{[HSO_4^-][CH_3COO^-]}$$

The equilibrium constant for this reaction is simply the K_a for the hydrogen sulfate equilibrium divided by the K_a for the acetic acid equilibrium.

$$K_{eq} = \frac{[SO_4^{2-}][CH_3COOH]}{[HSO_4^-][CH_3COO^-]} = \frac{[H_3O^+][SO_4^{2-}]}{[HSO_4^-]} \times \frac{[CH_3COOH]}{[H_3O^+][CH_3COO^-]}$$

$$K_{eq} = K_a(HSO_4^-) \times \frac{1}{K_a(CH_3COOH)} = \frac{10^{-2}}{10^{-4.8}} = 10^{2.8} \approx 600$$

This tells us in our case that, if we mixed sodium hydrogen sulfate and sodium acetate in water, we would end up with mainly sodium sulfate and acetic acid, the equilibrium constant for the reaction above being approximately 600.

> ● In a mixture of two acids or two bases
> • The ratio of K_a values gives us an indication of the equilibrium constant for the reaction between a base and an acid
> • The difference in pK_as gives us the log of the equilibrium constant.

As an example, let us look at a method for acetylating aromatic amines in aqueous solution. This reaction has a special name—the Lumière–Barbier method. We shall consider the acetylation of aniline $PhNH_2$ (a basic aromatic amine) using acetic anhydride. The procedure for this reaction is as follows.

1 Dissolve one equivalent of aniline in water to which one equivalent of hydrochloric acid has been added.

Aniline is not soluble in water to any significant degree. This isn't surprising as aniline is just a hydrophobic hydrocarbon with an amine group. The HCl (pK_a –7) protonates the aniline (pK_a of the conjugate acid of aniline is 4.6) to give the hydrochloride. Now we have a salt that is very soluble in water.

aniline
insoluble in water

anilinium ion
soluble in water

2 Warm to 50°C and add 1.2 equivalents of acetic anhydride followed by 1.2 equivalents of aqueous sodium acetate solution.

The acetic anhydride could be attacked by either the water, the acetate, or by aniline itself. Aniline is much more nucleophilic than the other two nucleophiles but only aniline itself can attack the anhydride: *protonated aniline has no lone pair and is not nucleophilic*. This, then, is the role of the sodium acetate—to act as a base and deprotonate the aniline hydrochloride. The pK_as of the aniline hydrochloride and acetic acid are about the same, around 4.7. An equilibrium will be set up to give some neutral aniline which will then attack the acetic anhydride and form the amide.

aniline attacks π* of acetic anhydride   deprotonation of ammonium ion

acetate is a good leaving group   the product - an amide

3 Cool in ice and filter off crystals of product, acetanilide.

The product is insoluble in water and, because it is an amide, is much less basic than aniline (pK_a of conjugate acid < 0) and so is not protonated to give a water-soluble salt.

More from pK_as: Calculating the pK_a values for water acting as a base and as an acid

The material in this box is quite mathematical and may be skipped if you find it too alien.

How easy is it to protonate or deprotonate water?

All our reactions so far have been in water and it is easy to forget that water itself also competes for protons. If, for example, we have both sulfuric acid H_2SO_4 and hydrochloric acid HCl in aqueous solution, hydrochloric acid with its lower pK_a (–7) will not protonate the conjugate base of sulfuric acid (pK_a –3), hydrogen sulfate HSO_4^-: both acids will protonate water instead. So water is a stronger base than either chloride or hydrogen sulfate ions. In fact, we can work out the pK_a for the protonation of water.

We want to answer the questions: 'How easy is it to protonate water? What strength of acid do we need?'

Look at this simple reaction:

$$H_3O^+(aq) + H_2O(l) \rightleftharpoons H_2O(l) + H_3O^+(aq)$$

Obviously, the equilibrium constant for the equation above will be 1 since both sides of the equation are the same. But we don't use normal equilibrium constants—we use the acidity constant, K_a, which is slightly different. Remember that this is actually the normal equilibrium constant for the reaction multiplied by [H_2O] = 55.56, the 'concentration' of water. This is normally useful in that it cancels out the [H_2O] term in the denominator but not in this case.

Here we have

$$K_a = K_{eq} \times [H_2O] = \frac{[H_2O][H_3O^+]}{[H_2O][H_3O^+]} \times [H_2O] = [H_2O] = 55.56$$

so the pK_a for the protonation of water is: pK_a(H_3O^+) = –log(55.56) = –1.74.

The pK_a also equals the pH when we have equal concentrations of acid and conjugate base. A solution would be quite acidic if exactly half of the number of water molecules present were hydronium ions—its pH would be –1.74.

So with water acting as a base

$$AH(aq) + H_2O(l) \rightleftharpoons H_3O^+(aq) + A^-(aq)$$

It is clear that for any acid with a lower pK_a than –1.74, the equilibrium will lie over to the right.

> **Acids with a lower pKa than –1.74 will *protonate* water completely**

We can also work out the pK_a for water acting as an acid. Now the equilibrium is

$$H_2O(l) + H_2O(l) \rightleftharpoons H_3O^+(aq) + OH^-(aq)$$

Going through the same calculations as before, we find

$$K_a = K_{eq} \times [H_2O] = \frac{[H_3O^+][OH^-]}{[H_2O][H_2O]} \times [H_2O]$$

$$= \frac{K_{AP}}{[H_2O]} = \frac{10^{-14}}{55.56} = 1.80 \times 10^{-16}$$

so the pK_a for the *deprotonation* of water is: pK_a(H_2O) = –log(1.80 × 10^{-16}) = 15.74.

This means that, if we put in water a base whose conjugate acid's pK_a is greater than 15.74, it will simply be protonated by the water and give an equivalent amount of hydroxide ions.

- **Bases B whose conjugate acid HB has a higher pK_a than 15.74 will *deprotonate* water completely**

■ Any sharp-eyed readers may notice an inconsistency in the statement that the pK_a equals the pH when we have equal concentrations of acid and conjugate base. If, when [A^-] = [AH], the pK_a = pH, then the pK_a for water equals the pH when [H_2O] = [H_3O^+]. We *assume* that [H_2O] is constant at 55.56 mol dm^{-3} and so [H_3O^+] must also equal 55.56 mol dm^{-3} and hence pH = pK_a = –log(55.56). This assumption cannot be valid here: rather [H_2O] + [H_3O^+] should equal approximately 55.56 mol dm^{-3}.

> ● The strongest base in aqueous solution is OH⁻ and the strongest acid in aqueous solution is H_3O^+. Remember that:
>
> - Addition of stronger bases than OH⁻ just gives more OH⁻ by the deprotonation of water
> - Addition of stronger acids than H_3O^+ just gives more H_3O^+ by protonation of water
>
> Also remember that:
>
> - The pH of pure water at 25°C is 7.00 (not the pK_a)
> - The pK_a of H_2O is 15.74
> - The pK_a of H_3O^+ is –1.74

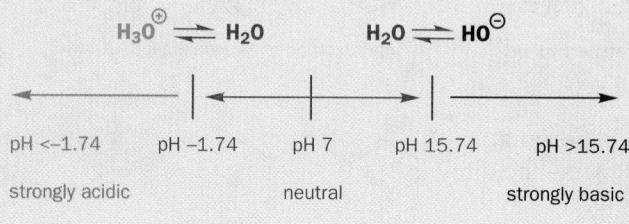

The choice of solvent limits the pK_a range we can use

In water, our effective pK_a range is only –1.74 to 15.74, that is, it is determined by the solvent. This is known as the **levelling effect** of the solvent. This is an important point. It means that, if we want to remove the proton from something with a high pK_a, say 25–30, it would be impossible to do this in water since the strongest base we can use is hydroxide. If we do need a stronger base than OH⁻, we must use a different solvent system.

For example, if we wanted to deprotonate ethyne (acetylene, pK_a 25), then hydroxide (the strongest base we could have in aqueous solution, pK_a 15.7) would establish an equilibrium where only 1 in $10^{9.3}$ ($10^{15.7}/10^{25}$) ethyne molecules were deprotonated. This means about 1 in 2 billion of our ethyne molecules will be deprotonated at any one time. Since, no matter what base we dissolve in water, we will only at best get hydroxide ions, this is the best we could do in water. So, in order to deprotonate ethyne to any appreciable extent, we must use a different solvent that does not have a pK_a less than 25. Conditions often used to do this reaction are sodium amide ($NaNH_2$) in liquid ammonia.

$$CH\equiv C-H + NH_2^{\ominus} \xrightarrow{\quad NH_3 \text{ (l)} \quad} CH\equiv C^{\ominus} + NH_3$$

Using the pK_as of NH_3 (ca. 33) and ethyne (25) we would predict an equilibrium constant for this reaction of 10^8 ($10^{-25}/10^{-33}$)—well over to the right. Amide ions can be used to deprotonate alkynes.

Since we have an upper and a lower limit on the strength of an acid or base that we can use in water, this poses a bit of a problem: How do we know that the pK_a for HCl is greater than that of H_2SO_4 if both completely protonate the water? How do we know that the pK_a of methane is greater than that of ethyne since both the conjugate bases fully deprotonate water? The answer is that we can't simply measure the equilibrium for the reaction in water—we can do this only for pK_as that fall between the pK_a values of water itself. Outside this range, pK_a values are determined in other solvents and the results are extrapolated to give a value for what the pK_a in water might be.

Constructing a pK_a scale

We now want to look at ways to rationalize the different pK_a values for different compounds—we wouldn't want to have to memorize all the values. You will need to get a feel for the pK_a values of

▶

Because the pK_a values for very strong acids and bases are so hard to determine, you will find that they often differ in different texts—sometimes the values are no better than good guesses! However, while the absolute values may differ, the relative values (which is the important thing because we need only a rough guide) are usually consistent.

different compounds and, if you know what factors affect them, it will make it much easier to predict an approximate pK_a value, or at least understand why a given compound has the pK_a value that it does.

A number of factors affect the strength of an acid, AH.

$$\text{AH (solvent)} \rightleftharpoons \text{A}^- \text{ (solvent)} + \text{H}^+ \text{ (solvent)}$$

These include:

1   Intrinsic stability of the conjugate base, anion A$^-$. Stability can arise, for example, by having the negative charge on an electronegative atom or by spreading the charge over other groups. Either way, the more 'stable' the conjugate base, the less basic it will be and so the stronger the acid

2   Bond strength A–H. Clearly, the easier it is to break this bond, the stronger the acid

3   The solvent. The better the solvent is at stabilizing the ions formed, the easier it is for the reaction to occur

● **Acid strength**

- The most important factor in the strength of an acid is the stability of the conjugate base—the more stable the conjugate base, the stronger the acid

- An important factor in the stability of the conjugate base is which element the negative charge is on—the more electronegative the element, the more stable the conjugate base

The negative charge on an electronegative element stabilizes the conjugate base

The pK_a values for second row hydrides CH_4, NH_3, H_2O, and HF are about 48, 33, 16, and 3, respectively. This trend is due to the increasing electronegativities across the period: F$^-$ is much more stable than CH_3^-, because fluorine is much more electronegative than carbon.

Weak A–H bonds make stronger acids

However, on descending group VII (group 17), the pK_a values for HF, HCl, HBr, and HI decrease in the order 3, –7, –9, and –10. Since the electronegativities decrease on descending the group we might expect an increase in pK_as. The decrease observed is actually due to the weakening bond strengths on descending the group and to some extent the way in which the charge can be spread over the increasingly large anions.

Delocalization of the negative charge stabilizes the conjugate base

The acids HClO, $HClO_2$, $HClO_3$, and $HClO_4$ have pK_a values 7.5, 2, –1, and about –10, respectively. In each case the acidic proton is on an oxygen attached to chlorine, that is, *we are removing a proton from the same environment in each case*. Why then is perchloric acid, $HClO_4$, some 17 orders of magnitude stronger in acidity than hypochlorous acid, HClO? Once the proton is removed, we end up with a negative charge on oxygen. For hypochlorous acid, this is localized on the one oxygen. With each successive oxygen, the charge can be more delocalized, and this makes the anion more stable. For example, with perchloric acid, the negative charge can be delocalized over all four oxygen atoms.

> That the charge is spread out over all the oxygen atoms equally is shown by electron diffraction studies: whereas perchloric acid has two types of Cl–O bond, one 163.5 pm and the other three 140.8 pm long, in the perchlorate anion all Cl–O bond lengths are the same, 144 pm, and all O–Cl–O bond angles are 109.5°.

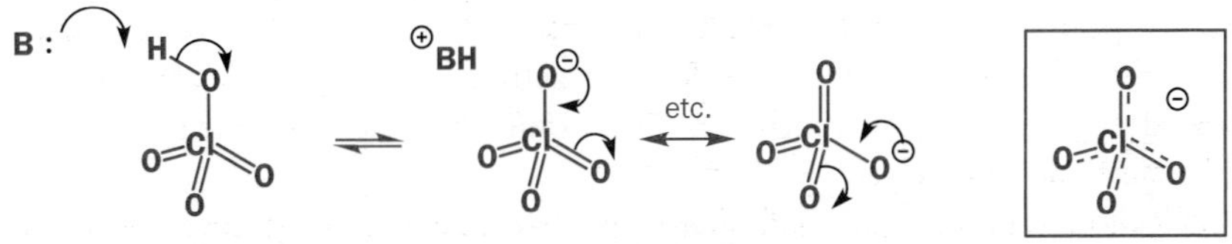

the negative charge on the perchlorate anion can be delocalized equally over all four oxygens

Similar arguments explain the pK_as for other oxygen acids, for example, ethanol (pK_a, 15.9), acetic acid (4.8), and methane sulfonic acid (−1.9). In ethoxide, the negative charge is localized on one oxygen atom, whilst in acetate the charge is delocalized over two oxygens and in methane sulfonate it is spread over three oxygens.

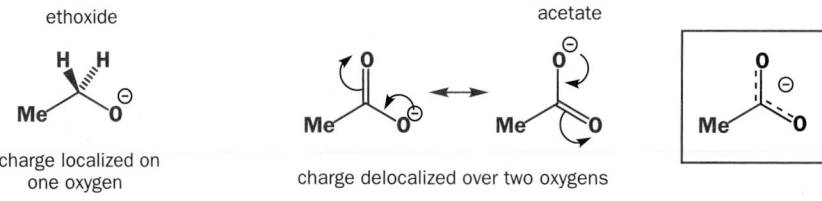

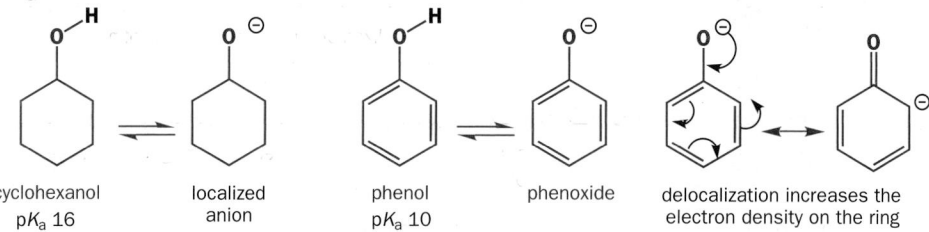

In phenol, PhOH, the OH group is directly attached to a benzene ring. On deprotonation, the negative charge can again be delocalized, not on to other oxygens but this time on to the aromatic ring itself.

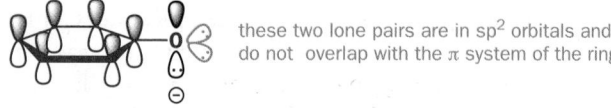

The effect of this is to stabilize the phenoxide anion relative to the conjugate base of cyclohexanol where no delocalization is possible and this is reflected in the pK_as of the two compounds: 10 for phenol but 16 for cyclohexanol.

> ● **Get a feel for pK_as!**
>
> Notice that these oxygen acids have pK_as that conveniently fall in units of 5 (approximately).
>
> | Acid | RSO$_2$OH | RCO$_2$H | ArOH | ROH |
> |---|---|---|---|---|
> | Approx. pK_a | 0 | 5 | 10 | 15 |

The same delocalization of charge can stabilize anions derived from deprotonating carbon acids. These are acids where the proton is removed from carbon rather than oxygen and, in general, they are weaker than oxygen acids because carbon is less electronegative. If the negative charge can be delocalized on to more electronegative atoms such as oxygen or nitrogen, the conjugate base will be stabilized and hence the acid will be stronger.

Table 8.2 shows a selection of carbon acids with their conjugate bases and pK_as. In each case the proton removed is shown in black.

Table 8.2 The conjugate bases and pK_as of some carbon acids

Acid	Conjugate base	pK_a	Comments
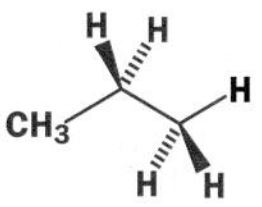	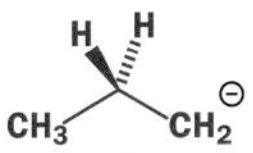	~50	charge is localized on one carbon—difficult since carbon is not very electronegative
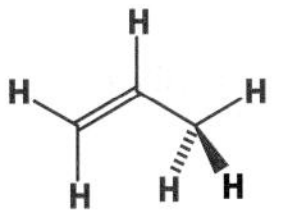	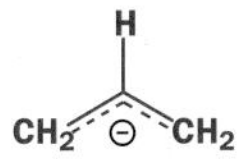	~43	charge is delocalized over π system—better but still not really good
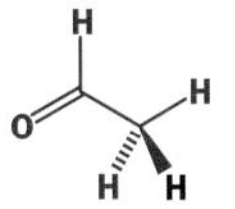	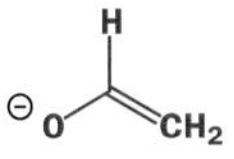	13.5	charge is delocalized over π system but is mainly on the electronegative oxygen—much better
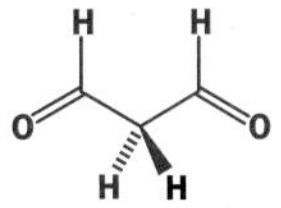	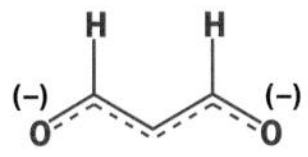	5	charge delocalized over π system but mainly over two oxygens—better still
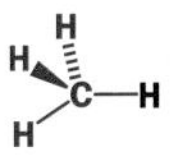	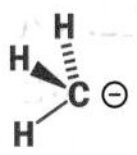	~48	charge is localized on one carbon—again very unsatisfactory
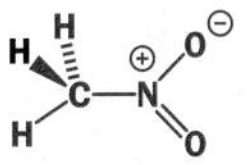	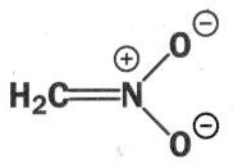	10	charge is delocalized but mainly on oxygens of nitro group
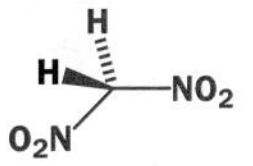	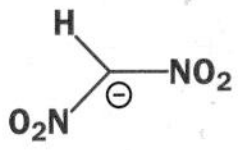	4	charge can be delocalized over two nitro groups—more stable anion
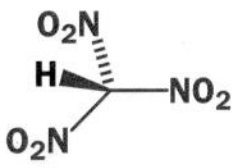	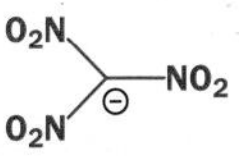	0	charge can be delocalized over three nitro groups—very stable anion

> electron withdrawing groups lowering the pK_as of carbon acids 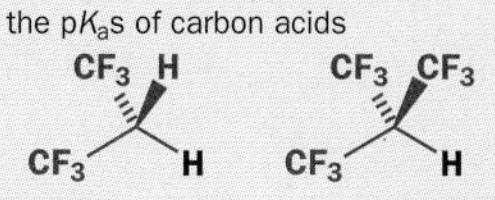
>
>
> pK_a ca. 22   pK_a ca. 10
>
> CF$_3$—H   CCl$_3$—H   CBr$_3$—H
>
> fluoroform   chloroform   bromoform
>
> pK_a 26   pK_a 15   pK_a 9
>
> Study carefully the pK_as for the haloform series, CHX$_3$—they may not do what you think they should! Chloroform is much more acidic than fluoroform even though fluorine is more electronegative (likewise with bromoform and chloroform). The anion CF$_3^-$ must be slightly destabilized because of some backdonation of electrons. The anion from chloroform and bromoform may also be stabilized by some interaction with the d orbitals (there aren't any on fluorine). The conjugate base anion of bromoform is relatively stable—you will meet this again in the bromoform/iodoform reaction (Chapter 21).

It isn't necessary for a group to be conjugated in order to spread the negative charge: any group that withdraws electrons will help to stabilize the conjugate base and therefore increase the strength of the acid. Some examples are shown below for both oxygen and carbon acids.

electron-withdrawing groups lowering the pK_a of carboxylic acids

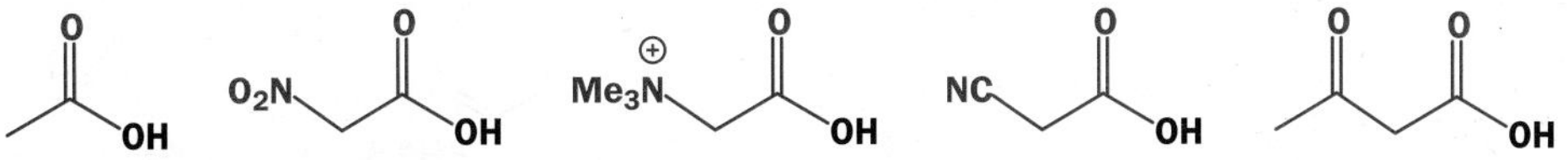

pK_a 4.76   pK_a 1.7   pK_a 1.8   pK_a 2.4   pK_a 3.6

electron-withdrawing groups lowering the pK_a of alcohols

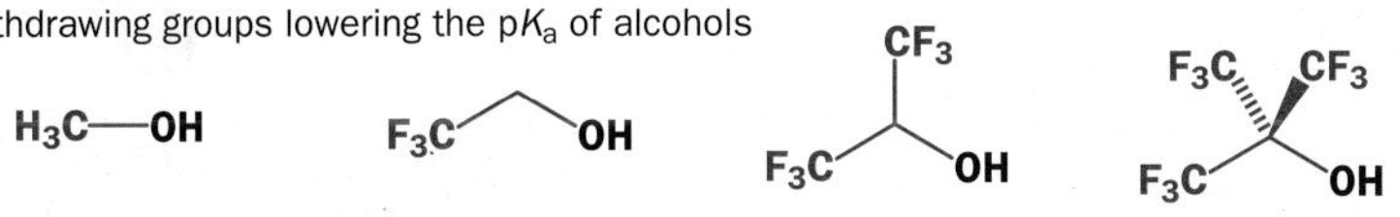

pK_a 15.5   pK_a 12.4   pK_a 9.3   pK_a 5.4

> • Notice how very electron-withdrawing the nitro group is—it lowers the pK_a of acetic acid even more than a quaternary ammonium salt!
>
> • Notice also that the fourth alcohol with three CF$_3$ groups is almost as acidic as acetic acid.

Picric acid is a very acidic phenol

Electron-withdrawing effects on aromatic rings will be covered in more detail in Chapter 22 but for the time being note that electron-withdrawing groups can considerably lower the pK_as of substituted phenols and carboxylic acids, as illustrated by picric acid.

2, 4, 6-Trinitrophenol's more common name, picric acid, reflects the strong acidity of this compound (pK_a 0.7 compared to phenol's 10.0). Picric acid used to be used in the dyeing industry but is little used now because it is also a powerful explosive (compare its structure with that of TNT!).

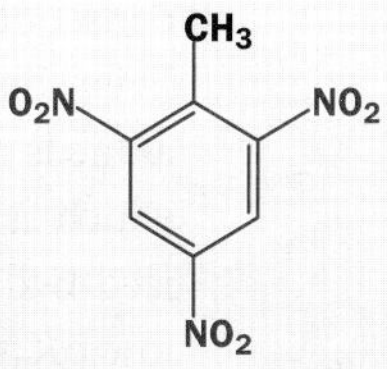

trinitrotoluene, TNT 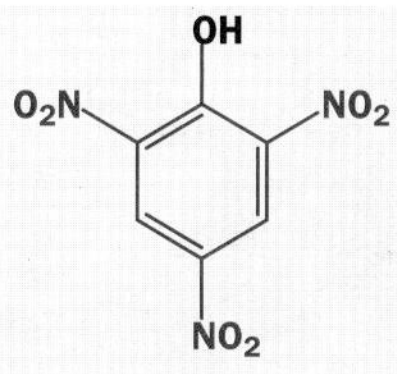   picric acid

Electron withdrawal in these molecules is the result of σ bond polarization from an inductive effect (Chapter 5). The electrons in a σ bond between carbon and a more electronegative element such as N, O, or F will be unevenly distributed with a greater electron density towards the more electronegative atom. This polarization is passed on more and more weakly throughout the carbon skeleton. The three fluorine atoms in CF$_3$H reduce the pK_a to 26 from the 48 of methane, while the nine fluorines in (CF$_3$)$_3$CH reduce the pK_a still further to 10.

Such inductive effects become less significant as the electron-withdrawing group gets further away from the negative charge as is shown by the pK_as for these chlorobutanoic acids: 2-chloro acid is significantly stronger than butanoic acid but by the time the chlorine atom is on C4, there is almost no effect.

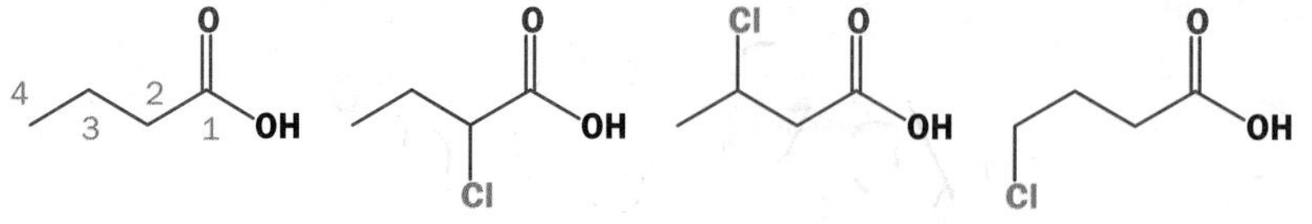

pK_a 4.8   pK_a 2.8   pK_a 4.1   pK_a 4.5

Hybridization can also affect the pK_a

The hybridization of the orbital from which the proton is removed also affects the pK_a. Since s orbitals are held closer to the nucleus than are p orbitals, the electrons in them are lower in energy, that is, more stable. Consequently, the more s character an orbital has, the more tightly held are the electrons in it. This means that electrons in an sp orbital (50% s character) are lower in energy than those in an sp^2 orbital (33% s character), which are, in turn, lower in energy than those in an sp^3 orbital (25% s character). Hence the anions derived from ethane, ethene, and ethyne increase in stability in this order and this is reflected in their pK_as. Cyanide ion, ⁻CN, with an electronegative element as well as an sp hybridized anion, is even more stable and HCN has a pK_a of about 10.

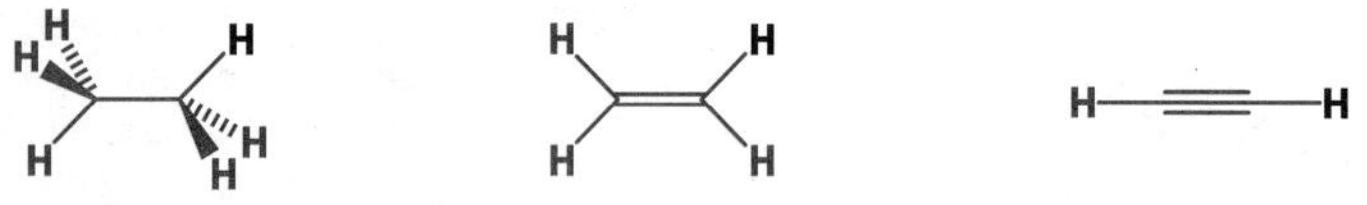

pK_a ca. 50   pK_a ca. 44   pK_a ca. 26

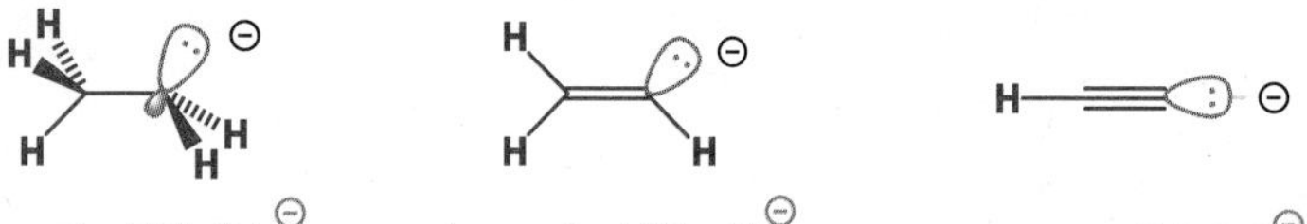

lone pair of CH$_3$CH$_2$⁻ in sp^3 orbital   lone pair of CH$_2$=CH⁻ in sp^2 orbital   lone pair of HC≡C⁻ in sp orbital

More remote hybridization is also important

The more s character an orbital has, the more it holds on to the electrons in it. This makes an sp hybridized carbon less electron-donating than an sp^2 one, which in turn is less electron-donating than an sp^3 carbon. This is reflected in the pK_as of the compounds shown here.

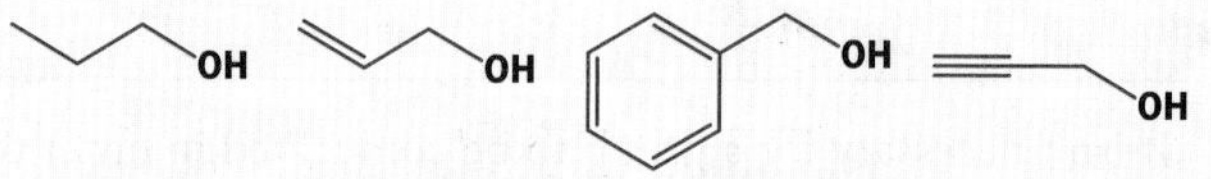

pK_a 16.1   pK_a 15.5   pK_a 15.4   pK_a 13.5

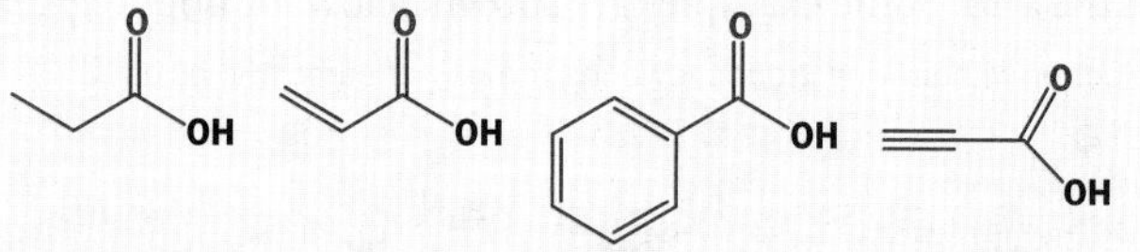

pK_a 4.9   pK_a 4.2   pK_a 4.2   pK_a 1.9

Highly conjugated carbon acids

If we can delocalize the negative charge of a conjugate anion on to oxygen, the anion is more stable and consequently the acid is stronger. Even delocalization on to carbon alone is good if there is enough of it, which is why some highly delocalized hydrocarbons have remarkably low pK_as for hydrocarbons. Look at this series .

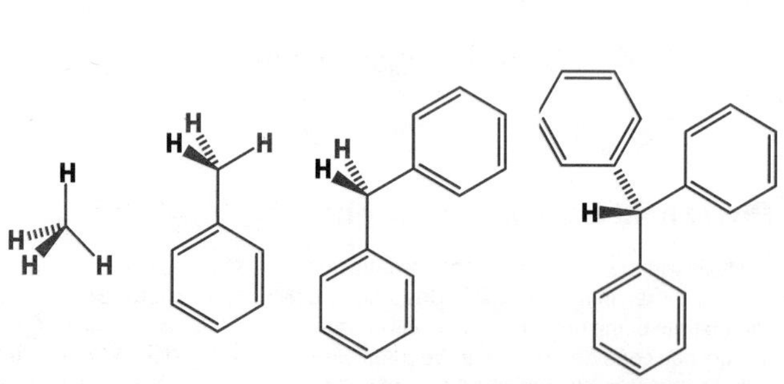

pK_a ca. 48   pK_a ca. 40   pK_a ca. 33   pK_a ca. 32

Increasing the number of phenyl groups decreases the pK_a—this is what we expect, since we can delocalize the charge over all the rings. Notice, however, that each successive phenyl ring has less effect on the pK_a: the first ring lowers the pK_a by 8 units, the second by 7, and the third by only 1 unit. In order to have effective delocalization, the system must be planar (Chapter 7). Three phenyl rings cannot arrange themselves in a plane around one carbon atom because the *ortho*-hydrogens clash with each other (they want to occupy the same space) and the compound actually adopts a propeller shape where each phenyl ring is slightly twisted relative to the next.

the hydrogens in the *ortho* positions try to occupy the same space

each phenyl ring is staggered relative to the next

Even though complete delocalization is not possible, each phenyl ring does lower the pK_a because the sp^2 carbon on the ring is electron-withdrawing. If we force the system to be planar, as in the compounds below, the pK_a is lowered considerably.

fluorene, pK_a 22.8
in the anion, the whole
system is planar

9-phenylfluorene, pK_a 18.5
in the anion, only the two
fused rings can be planar

fluoradene pK_a 11
in the anion, the whole
system is planar

The 'Fmoc' protecting group

Sometimes in organic chemistry, when we are trying to do a reaction on one particular functional group, another group in the molecule may also react with the reagents, often in a way that we do not want. If a compound contains such a vulnerable group, we can 'protect' it by first converting it into a different less reactive group that can easily be converted back to the group that we want later. An example of such a 'protecting group' is the Fmoc group used (for example, as the chloride, X = Cl) to protect amines or alcohols.

The protecting group is removed using a base. This works because of the acidity of the proton in position 9 on the fluorene ring. Removal of that proton causes a breakup of the molecule with the release of the amine at the end.

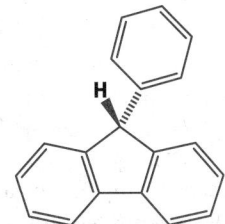

Fmoc-X
Fmoc = 9-**F**luorenyl**m**ethyl**o**xy**c**arbonyl

Fmoc-Cl + NHR$_2$ $\longrightarrow$ Fmoc-NR$_2$ + HCl

protected amine

anion is stabilized by conjugation
but can undergo elimination

9-methylenefluorene

CO$_2$ + NHR$_2$
de-protected
amine

We saw in Chapter 7 how some compounds can become aromatic by gaining or losing electrons. Cyclopentadiene is one such compound, which becomes aromatic on deprotonation. The stability gained in becoming aromatic is reflected in the compound's pK_a.

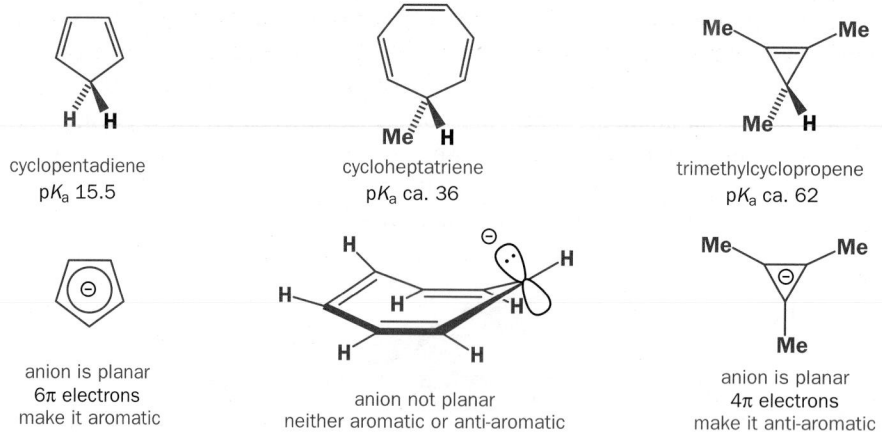

cyclopentadiene
pK_a 15.5

cycloheptatriene
pK_a ca. 36

trimethylcyclopropene
pK_a ca. 62

anion is planar
6π electrons
make it aromatic

anion not planar
neither aromatic or anti-aromatic

anion is planar
4π electrons
make it anti-aromatic

This is by no means as far as we can go. The five cyanide groups stabilize this anion so much that the pK_a for this compound is about −11 and it is considerably more acidic than cyclopentadiene (pK_a 15.5).

Compare the pK_a of cyclopentadiene with that of cycloheptatriene. Whilst the anion of the former has 6 π electrons (which makes it isoelectronic with benzene), the anion of the latter has 8 π electrons. Remember that on p. 176 we saw how 4n π electrons made a compound anti-aromatic? The cycloheptatrienyl anion does have 4n π electrons but it is not anti-aromatic because it isn't planar. However, it certainly isn't aromatic either and its pK_a of around 36 is about the same as that of propene. This contrasts with the cyclopropenyl anion, which must be planar since any three points define a plane. Now the compound is anti-aromatic and this is reflected in the very high pK_a (about 62). Other compounds may become aromatic on losing a proton. We looked at fluorene a few pages back: now you will see that fluorene is acidic because its anion is aromatic (14 π electrons).

● **The more stabilized the conjugate base, A⁻, the stronger is the acid, HA. Ways to stabilize A⁻ include:**
- **Having the charge on an electronegative element**
- **Delocalizing the negative charge over other carbon atoms, or even better, over more electronegative atoms**
- **Spreading out the charge over electron-withdrawing groups by the polarization of σ bonds (inductive)**
- **Having the negative charge in an orbital with more s character**
- **Becoming aromatic**

Electron-donating groups decrease acidity

The anions are also stabilized by solvation. Solvation is reduced by increasing the steric hindrance around the alkoxide.

All of the substituents in the examples above have been electron-withdrawing and have helped to stabilize the negative charge of the conjugate base, thereby making the acid stronger. What effect would electron-donating groups have? As you would expect, these destabilize the conjugate base because, instead of helping to spread out the negative charge, they actually put more in. The most common electron-donating groups encountered in organic chemistry are the alkyl groups. These are weakly electron-releasing (p. 416).

formic (methanoic) acid
pK_a 3.7

acetic (ethanoic) acid
pK_a 4.8

methanol
pK_a 15.5

ethanol
pK_a 16.0

isopropyl alcohol
pK_a 17.1

tert-butyl alcohol
pK_a 19.2

Although to a lesser extent than amides (p. 165), the ester group is also stabilized by conjugation. In this case, the 'ethoxide part' of the ester is electron-releasing. This explains the pK_as shown below.

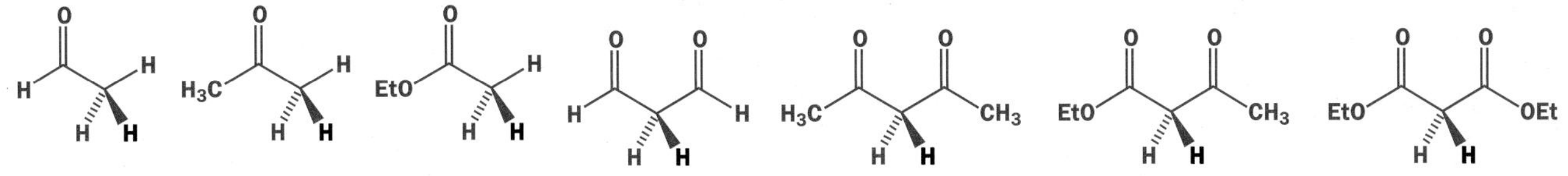

| acetaldehyde (ethanal) pK_a 13.5 | acetone (propanone) pK_a 20 | ethyl acetate (ethyl ethanoate) pK_a 25 | propandial pK_a ca. 5 | acetylacetone (2,4-pentanedione) pK_a 8.9 | ethyl acetoacetate (ethyl 3-oxobutanoate) pK_a 10.6 | diethyl malonate (diethyl propanedioate) pK_a 12.9 |

Nitrogen acids

Oxygen acids and carbon acids are by far the most important examples you will encounter and by now you should have a good understanding of why their pK_a values are what they are. Before we move on to bases, it would be worthwhile to remind you how different nitrogen acids are from oxygen acids, since the conjugate bases of amines are so important. The pK_a of ammonia is much greater than the pK_a of water (about 33 compared with 15.74). This is because oxygen is more electronegative than nitrogen and so can stabilize the negative charge better. A similar trend is reflected in the pK_as of other nitrogen compounds, for example, in the amide group. Whilst the oxygen equivalent of an amide (a carboxylic acid) has a low pK_a, a strong base is needed to deprotonate an amide. Nevertheless, the carbonyl group of an amide does lower the pK_a from that of an amine (about 30) to around 17. It's not surprising, therefore, that the two carbonyl groups in an imide lower the pK_a still further, as in the case of phthalimide. Amines are not acidic, amides are weakly acidic (about the same as alcohols), and imides are definitely acidic (about the same as phenols).

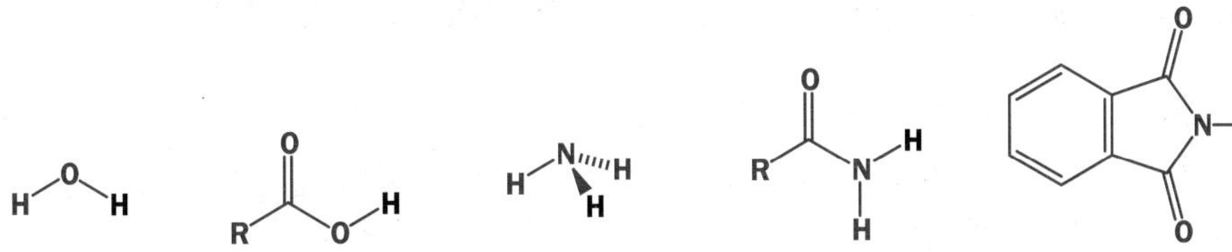

| water pK_a 15.74 | carboxylic acid pK_a ca. 5 | ammonia pK_a ca. 33 | amide pK_a ca. 17 | phthalimide pK_a 8.3 |

The potassium salt of 6-methyl-1,2,3-oxathiazin-4-one 2,2-dioxide known as acesulfame-K is used as an artificial sweetener (trade name Sunett). Here the negative charge is delocalized over both the carbonyl and the sulfone groups.

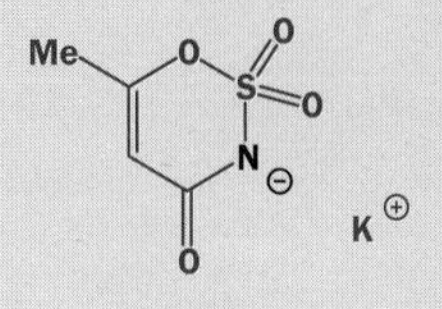

acesulfame-K

Basicity

A **base** is a substance that can accept a proton by donating a pair of electrons. We have already encountered some—for example, ammonia, water, the acetate anion, and the methyl anion. The question we must now ask is: how can we measure a base's strength? To what extent does a base attract a proton? We hope you will realize that we have already addressed this problem by asking the same question from a different viewpoint: to what extent does a protonated base want to keep its proton? For example if we want to know which is the stronger base—formate anion or acetylide anion—we look up the pK_as for their conjugate acids. We find that the pK_a for formic acid (HCO_2H) is 3.7, whilst the pK_a for ethyne (acetylene) is around 25. This means that ethyne is much more reluctant to part with its proton, that is, acetylide is much more basic than formate. This is all very well for anions—we simply look up the pK_a value for the neutral conjugate acid, but what if we want to know the basicity of ammonia? If we look up the pK_a for ammonia we find a value around 33 but this is the value for deprotonating neutral ammonia to give the amide ion, NH_2^-.

▶ **Amides**

Do not get confused by the two uses of the word 'amide' in chemistry. Both the carbonyl compound and the 'ionic' base formed by deprotonating an amine are known as amides. From the context it should be clear which is meant—most of the time chemists (at least organic chemists!) mean the carbonyl compound.

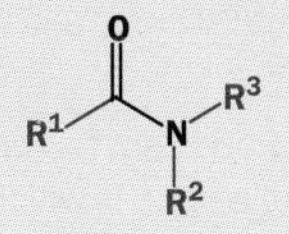

the amide group

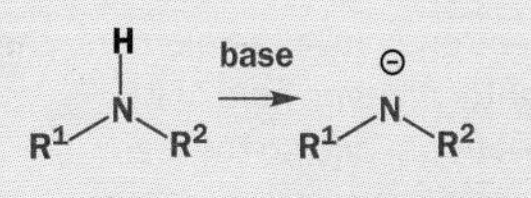

an amine   an amide base
e.g. sodium amide
$NaNH_2$

▶ **Get a feel for pK_as!**

Remember that the pK_a also represents the pH when we have equal concentrations of acid and conjugate base, that is, NH$_3$ and NH$_4^+$ in this case. You know that ammonia is a weak base and that an aqueous solution is alkaline so it should come as no surprise that its pK_a is on the basic side of 7. To be exact, at pH 9.24 an aqueous solution of ammonia contains equal concentrations of ammonium ions and ammonia.

If we want to know the *basicity* of ammonia, we must look up the pK_a of its conjugate acid, the ammonium cation, NH$_4^+$, protonated ammonia. Its pK_a is 9.24 which means that ammonia is a weaker base than hydroxide—the pK_a for water (the conjugate acid of hydroxide) is 15.74 (p. 190). Now we can summarize the states of ammonia at different pH values.

Scales for basicity—pK_B and pK_{aH}

The material in this box is quite mathematical and may be skipped if you find it too alien.

It is often convenient to be able to refer to the basicity of a substance directly. In some texts a different scale is used, pK_B. This is derived from considering how much hydroxide ion a base forms in water rather than how much hydronium ion the conjugate acid forms.

For the pK_B scale:

$$B(aq) + H_2O \rightleftharpoons OH^-(aq) + BH^+(aq)$$

$$K_B = \frac{[OH^-][BH^+]}{[B]}$$

Hence

$$pK_B = -\log(K_B)$$

For the pK_a scale:

$$BH^+(aq) + H_2O \rightleftharpoons H_3O^+(aq) + B(aq)$$

$$K_A = \frac{[H_3O^+][B]}{[BH^+]}$$

Hence

$$pK_A = -\log(K_A)$$

Just as in the acid pK_a scale, the lower the pK_a the stronger the acid, in the basic pK_B scale, the lower the pK_B, the stronger the base. The two scales are related: the product of the equilibrium constants simply equals the ionic product of water.

$$K_B \times K_A = \frac{[OH^-][BH^+]}{[B]} \times \frac{[H_3O^+][B]}{[BH^+]}$$

$$= [OH^-][H_3O^+] = K_W = 10^{-14}$$

that is,

$$pK_A + pK_B = pK_W = 14$$

There is a separate scale for bases, but it seems silly to have two different scales, the basic pK_B and the familiar pK_a, when one will do and so we will stick to pK_a. However, to avoid any misunderstandings that can arise from amphoteric compounds like ammonia, whose pK_a is around 33, we will either say:

• The pK_a of ammonia's conjugate acid is 9.24

or, more concisely,

• The pK_{aH} of ammonia is 9.24 (where pK_{aH} simply means the pK_a of the conjugate acid)

What factors affect how basic a compound is?

This really is the same as the question we were asking about the strength of an acid—the more 'stable' the base, the weaker it is. The more accessible the electrons are, the stronger the base is. Therefore a negatively charged base is more likely to pick up a proton than a neutral one; a compound in which the negative charge is delocalized is going to be less basic than one with a more concentrated, localized charge, and so on. We have seen that carboxylic acids are stronger acids than simple alcohols because the negative charge formed once we have lost a proton is delocalized over two oxygens in the carboxylate but localized on just one oxygen for the alkoxide. In other words, the alkoxide is a stronger base because its electrons are more available to be protonated. Since we have already considered anionic bases, we will now look in more detail at neutral bases.

▶

The most important factor in the strength of a base is which element the lone pair (or negative charge) is on. The more electronegative the element, the tighter it keeps hold of its electrons, and so the less available they are to accept a proton, and the weaker is the base.

There are two main factors that determine the strength of a neutral base: how accessible is the lone pair and to what extent can the resultant positive charge formed be stabilized either by delocalization or by the solvent. The accessibility of the lone pair depends on its energy—it is usually the HOMO of the molecule and so, the higher its energy, the more reactive it is and hence the stronger the base. The lone pair is lowered in energy if it is on a very electronegative element or if it can be delocalized in some manner.

This explains why ammonia is 10^{10} times more basic than water: since oxygen is more electronegative than nitrogen, its lone pair is lower in energy. In other words, the oxygen atom in water wants to keep hold of its electrons more than the nitrogen in ammonia does and is therefore less likely to donate them to a proton. The pK_{aH} for ammonia (that is, the pK_a for ammonium ion) is 9.24 whilst the pK_{aH} for water (the pK_a for hydronium ion) is −1.74. Nitrogen bases are the strongest neutral bases commonly encountered by the organic chemist and so we will pay most attention to these in the discussion that follows.

> **▶ pK_{aH}**
>
> We use pK_{aH} to mean the pK_a of the conjugate acid.

Neutral nitrogen bases

Ammonia is the simplest nitrogen base and has a pK_{aH} of 9.24. Any substituent that increases the electron density on the nitrogen therefore raises the energy of the lone pair thus making it more available for protonation and increasing the basicity of the amine (larger pK_{aH}). Conversely, any substituent that withdraws electron density from the nitrogen makes it less basic (smaller pK_{aH}).

Effects that increase the electron density on nitrogen

We can increase the electron density on nitrogen either by attaching an electron-releasing group or by conjugating the nitrogen with an electron-donating group. The simplest example of an electron-releasing group is an alkyl group (p. 416). If we successively substitute each hydrogen in ammonia by an electron-releasing alkyl group, we should increase the amine's basicity. The pK_{aH} values for various mono-, di-, and trisubstituted amines are shown in Table 8.4.

Points to notice in Table 8.4:

- All the amines have pK_{aH}s greater than that of ammonia (9.24)
- All the primary amines have approximately the same pK_{aH} (about 10.7)
- All the secondary amines have pK_{aH}s that are slightly higher
- Most of the tertiary amines have pK_{aH}s lower than those of the primary amines

Table 8.4 pK_{aH} values for primary, secondary, and tertiary amines

R	pK_{aH} RNH$_2$	pK_{aH} R$_2$NH	pK_{aH} R$_3$N
Me	10.6	10.8	9.8
Et	10.7	11.0	10.8
n-Pr	10.7	11.0	10.3
n-Bu	10.7	11.3	9.9

The first point indicates that our prediction that replacing the hydrogens by electron-releasing alkyl groups would increase basicity was correct. A strange feature though is that, whilst substituting one hydrogen of ammonia increases the basicity by more than a factor of ten (one pK_a unit), substituting two has less effect and in the trisubstituted amine the pK_{aH} is actually *lower*. So far we have only considered one cause of basicity, namely, the availability of the lone pair but the other factor, the stabilization of the resultant positive charge formed on protonation, is also important. Each successive alkyl group does help stabilize the positive charge because it is electron-releasing but there is another stabilizing effect—the solvent. Every hydrogen attached directly to nitrogen will be hydrogen bonded with solvent water and this also helps to stabilize the charge: the more hydrogen bonding, the more stabilization. The observed basicity therefore results from a combination of effects: (1) the increased availability of the lone pair and the stabilization of the resultant positive charge, which increases with successive replacement of hydrogen atoms by alkyl groups; and (2) the stabilization due to solvation, an important part of which is due to hydrogen bonding and this effect decreases with increasing numbers of alkyl groups.

more stabilization of positive charge from alkyl groups →

← more stabilization of positive charge from hydrogen bonding with solvent

> **Gas phase acidity**
>
> If we look at the pK_{aH} values in the gas phase, we can eliminate the hydrogen bonding contribution and we find the basicity increases in the order we expect, that is, tertiary > secondary > primary.

Introducing alkyl groups is the simplest way to increase the electron density on nitrogen but there are other ways. Conjugation with an electron-donating group produces even stronger bases (p. 202) but we could also increase the electron density by using elements such as silicon. Silicon is more

■ Remember that Me_4Si, tetramethyl silane, in carbon NMR resonates at 0.0 p.p.m.? This is another consequence of silicon being more electropositive than carbon—the methyl carbons of TMS are more shielded and so resonate at a smaller chemical shift than other saturated carbons.

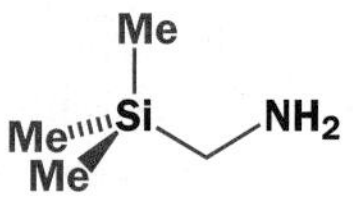

pK_{aH} 11.0

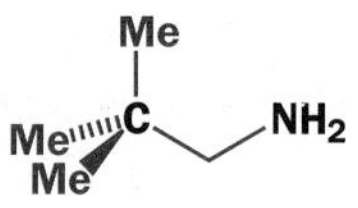

pK_{aH} 10.2

electropositive than carbon, that is, it pushes more electron density on to carbon. This extra donation of electrons also means that the silicon compound has a higher pK_{aH} value than its carbon analogue since the nitrogen's lone pair is higher in energy.

■ Compare this summary with the one for the stabilization of the conjugate base A^- on p. 196. In both cases, we are considering the same factors.

● Effects that decrease the electron density on nitrogen

The lone pair on nitrogen will be *less* available for protonation, and the amine *less* basic, if:

- **The nitrogen atom is attached to an electron-withdrawing group**
- **The lone pair is in an sp or sp^2 hybridized orbital**
- **The lone pair is conjugated with an electron-withdrawing group**
- **The lone pair is involved in maintaining the aromaticity of the molecule**

The pK_{aH}s of some amines in which the nitrogen is attached either directly or indirectly to an electron-withdrawing group are shown below. We should compare these values with typical values of about 11 for simple primary and secondary amines.

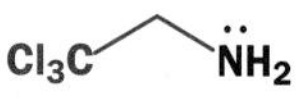

pK_{aH} 5.5

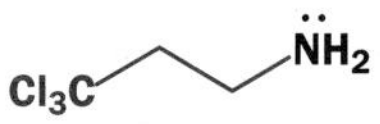

pK_{aH} 9.65

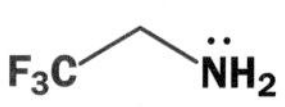

pK_{aH} 5.7

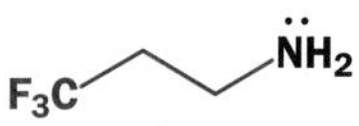

pK_{aH} 8.7

The strongly electron-withdrawing CF_3 and CCl_3 groups have a large effect when they are on the same carbon atom as the NH_2 group but the effect gets much smaller when they are even one atom further away. Inductive effects fall off rapidly with distance.

Hybridization is important

As explained on p. 194, the more s character an orbital has, the more tightly it holds on to its electrons and so the more electron-withdrawing it is. This is nicely illustrated by the series in Table 8.5.

- These effects are purely inductive electron withdrawal. Satisfy yourself there is no conjugation possible
- The last compound's pK_{aH} is very low. This is even less basic than a carboxylate ion.

Table 8.5 pK_{aH}s of unsaturated primary, secondary, and tertiary amines

R	RNH_2	R_2NH	R_3N
$H_3C—CH_2—CH_2—$	10.7	11.0	10.3
$H_2C=CH—CH_2—$	9.5	9.3	8.3
$HC≡C—CH_2—$	8.2	6.1	3.1

If the lone pair itself is in an sp^2 or an sp orbital, it is more tightly held (the orbital is lower in energy) and therefore much harder to protonate. This explains why the lone pair of the nitrile group is not at all basic and needs a strong acid to protonate it.

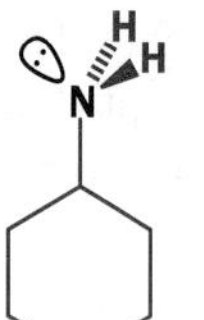

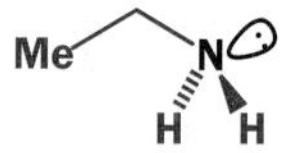

lone pair in sp^3 orbital
pK_{aH} 10.7

lone pair in sp^2 orbital
pK_{aH} 9.2

lone pair in sp^3 orbital
pK_{aH} 10.8

lone pair in sp orbital
pK_{aH} ca. –10

■ Of course, electron-withdrawing groups on the benzene ring will affect the availability of the lone pair. For example, the pK_{aH} of *p*-nitro aniline is only 1.11. This explains why certain aromatic amines (for example, nitroanilines and dibromoanilines) can't be acetylated using the Lumière–Barbier method (p. 188).

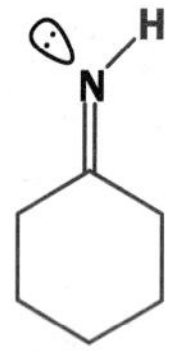

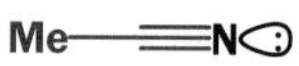

The low pK_{aH} of aniline ($PhNH_2$), 4.6, is partly due to the nitrogen being attached to an sp^2 carbon but also because the lone pair can be delocalized into the benzene ring. In order for the lone pair to be fully conjugated with the benzene ring, the nitrogen would have to be sp^2 hybridized with the lone pair in the p orbital. This would mean that both hydrogens of the NH_2 group would be in the same plane as the benzene ring but this is not found to be the case. Instead, the plane of the NH_2 group is about 40° away from the plane of the ring. That the lone pair is partially conjugated into the ring is shown indirectly by NMR shifts and by the chemical reactions that aniline undergoes. Notice that, when protonated, the positive charge cannot be delocalized over the benzene ring and any stabilization derived from the lone pair in unprotonated aniline being delocalized into the ring is lost.

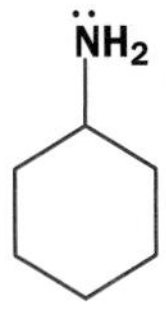

cyclohexylamine
pK_{aH} 10.7

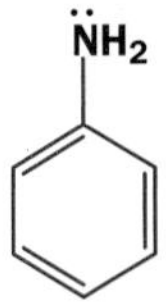

aniline
pK_{aH} 4.6

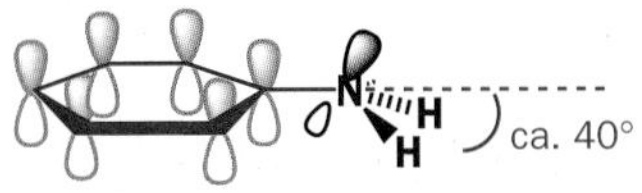

the NH_2 group is about 40° away
from being in the plane of the ring

Amides are weak bases protonated on oxygen

In contrast to aromatic amines, the amide group is completely planar (p. 165) with the nitrogen sp^2 hybridized and its lone pair in the p orbital, thereby enabling it to overlap effectively with the carbonyl group.

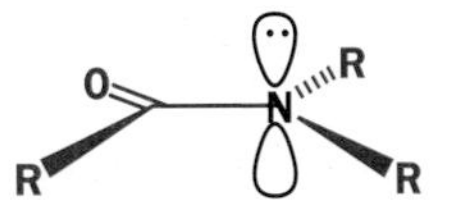

nitrogen is sp^2 hybridized with
its lone pair in a p orbital

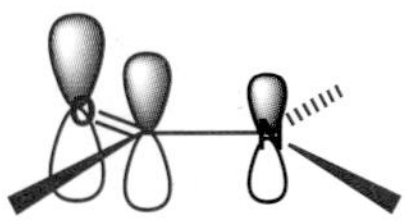

good overlap
with the carbonyl group

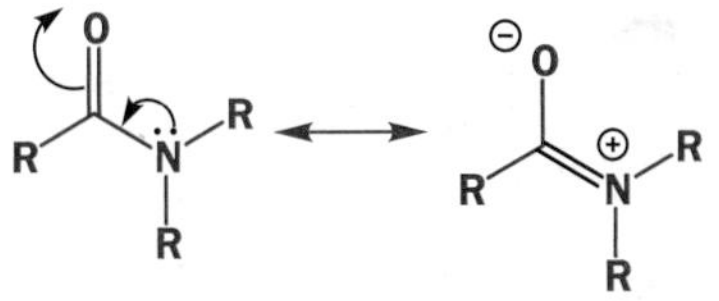

delocalization of nitrogen's
lone pair into π system

This delocalization 'ties up' the lone pair and makes it much less basic: the pK_{aH} for an amide is typically between 0 and −1. Because of the delocalization amides are not protonated on nitrogen.

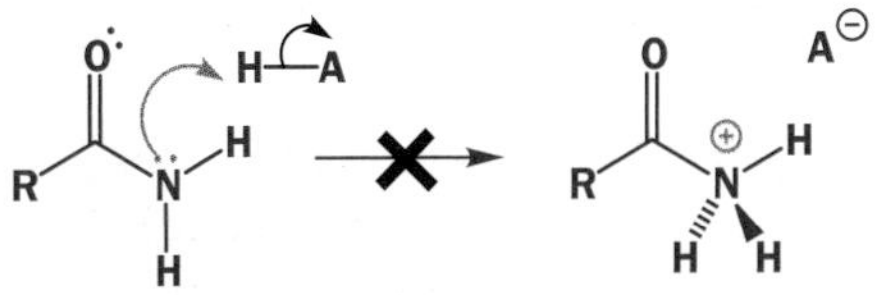

no protonation occurs on the nitrogen atom

Protonation at nitrogen would result in a positive charge on the nitrogen atom. Since this is adjacent to the carbonyl, whose carbon is also electron-deficient, this is energetically unfavourable. Protonation occurs instead on the carbonyl oxygen atom. We can draw the mechanism for this using either a lone pair on oxygen or on nitrogen.

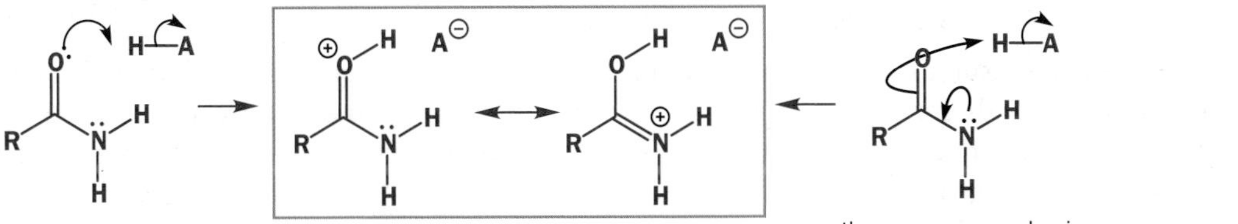

protonation occurs
on the oxygen atom

these structures are just two different
ways to draw the same delocalized cation

these arrows emphasise
the contribution of the
nitrogen's lone pair.

Furthermore, if the amide were protonated at nitrogen, the positive charge could not be delocalized on to the oxygen but would have to stay localized on the nitrogen. In contrast, when the amide is protonated on the oxygen atom, the charge can be delocalized on to the nitrogen atom making the cation much more stable. We can see this if we draw delocalization arrows on the structures in the green box.

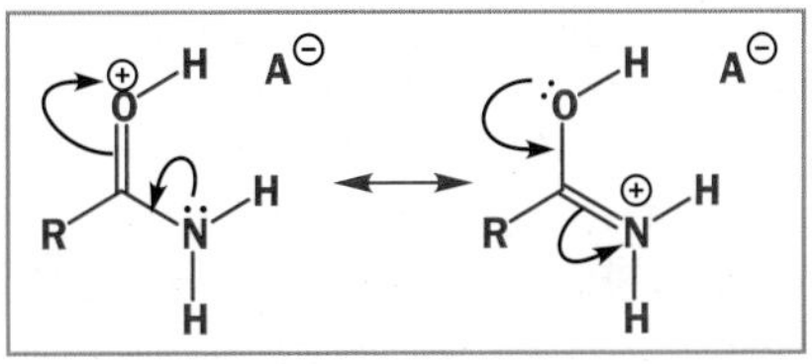

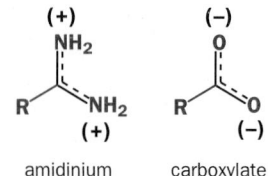

an amidine   pK_{aH} 12.4

Amidines are stronger bases than amides or amines

An amidine is the nitrogen equivalent of an amide—a C=NH group replaces the carbonyl. Amidines are much more basic than amides, the pK_{aH}s of amidines are larger than those of amides by about 13 so there is an enormous factor of 10^{13} in favour of amidines. In fact, they are among the strongest neutral bases.

An amidine has two nitrogen atoms that could be protonated—one is sp^3 hybridized, the other sp^2 hybridized. We might expect the sp^3 nitrogen to be more basic but protonation occurs at the sp^2 nitrogen atom. This happens because we have the same situation as with an amide: only if we protonate on the sp^2 nitrogen can the positive charge be delocalized over both nitrogens. We are using *both* lone pairs when we protonate on the sp^2 nitrogen.

amidinium cation

carboxylate anion

The electron density on the sp^2 nitrogen in an amidine is increased through conjugation with the sp^3 nitrogen. The delocalized amidinium cation has identical C–N bond lengths and a positive charge shared equally between the two nitrogen atoms. It is like a positively charged analogue of the carboxylate ion.

Amidine bases

Two frequently used amidine bases are DBN (1,5-*diaza*bicyclo[3.4.0]*nonene*-5) and DBU (1,8-*diaza*bicyclo[5.4.0]*undecene*-7).

They are easier to make, more stable, and less volatile than simpler amidines.

DBN   DBU

Guanidines are very strong bases

Even more basic is guanidine, pK_{aH} 13.6, nearly as strong a base as NaOH! On protonation, the positive charge can be delocalized over three nitrogen atoms to give a very stable cation. All three nitrogen lone pairs cooperate to donate electrons but protonation occurs, as before, on the sp^2 nitrogen atom.

guanidine: pK_a 13.6

very stable guanidinium cation each (+) is a third of a positive charge

very stable carbonate dianion each (–) is two-thirds of a negative charge

This time the resulting guanidinium ion can be compared to the very stable carbonate dianion. All three C–N bonds are the same length in the guanidinium ion and each nitrogen atom has the same charge (about one-third positive). In the carbonate dianion, all three C–O bonds are the same length and each oxygen atom has the same charge (about *two*-thirds negative as it is a dianion).

Imidazoline is a simple cyclic amidine and its pK_{aH} value is just what we expect, around 11. Imidazole, on the other hand, is less basic (pK_{aH} 7.1) because both nitrogens are attached to an electron-withdrawing sp^2 carbon. However, imidazole, with its two nitrogen atoms, is more basic than pyridine (pK_{aH} 5.2) because pyridine only has one nitrogen on which to stabilize the positive charge.

imidazoline
pK_{aH} 11

imidazole
pK_{aH} 7.1

imidazolium cation

pyridine
pK_{aH} 5.2

pyridinium cation

Both imidazole and pyridine are aromatic—they are flat, cyclic molecules with 6π electrons in the conjugated system (p. 177). Imidazole has one lone pair that is and one that is not involved in the aromaticity (Chapter 43).

this lone pair is in an sp^2 orbital and is not involved with the aromaticity of the ring. Protonation occurs here $--\blacktriangleright$

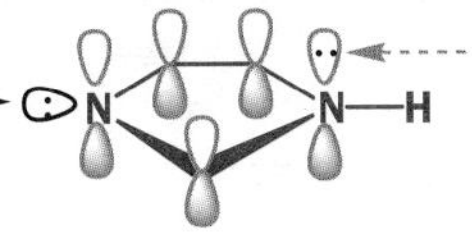

this lone pair is in a p orbital contributing to the 6π electrons in the aromatic ring

the aromaticity of imidazole

Protonation occurs on the nitrogen atom having the sp^2 lone pair because both lone pairs contribute and the resulting delocalized cation is still aromatic. Pyridine is also protonated on its sp^2 lone pair (it is the only one it has!) and the pyridinium ion is also obviously aromatic—it still has three conjugated π bonds in the ring.

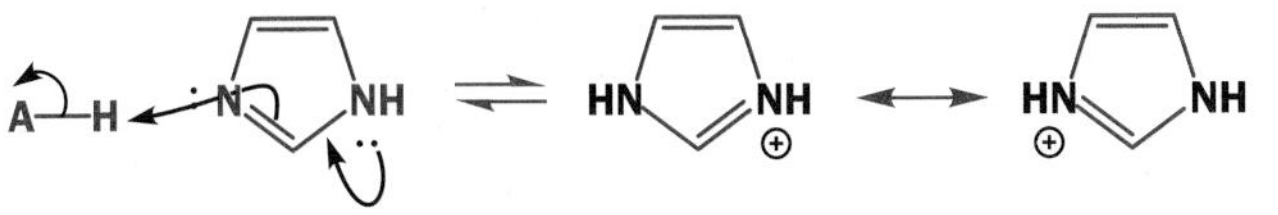

aromatic imidazole

aromatic imidazoium ion

This contrasts to pyrrole in which the lone pair on the only nitrogen atom is needed to complete the six aromatic π electrons and is therefore delocalized around the ring. Protonation, if it occurs at all, occurs on carbon rather than on nitrogen since the cation is then delocalized. But the cation is no longer aromatic (there is a saturated CH_2 group interrupting the conjugation) and so pyrrole is not at all basic (pK_{aH} about -4).

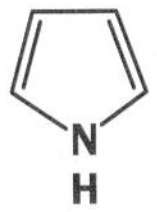

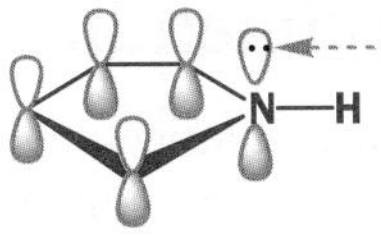

this lone pair is in a p orbital contributing to the 6π electrons in the aromatic ring

pyrrole
pK_{aH} ca. -4 the aromaticity of pyrrole

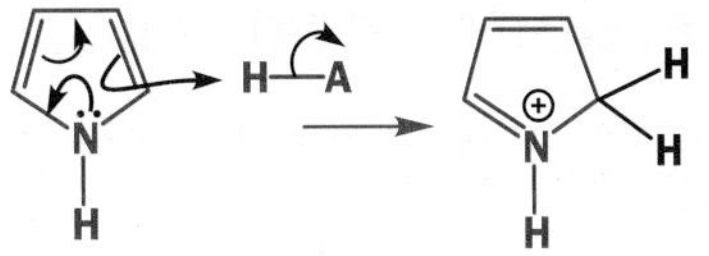

aromatic pyrrole

nonaromatic cation

Neutral oxygen bases

We have already seen that water is a much weaker base than ammonia because oxygen is more electronegative and wants to keep hold of its electrons (p. 199). Oxygen bases in general are so much weaker than their nitrogen analogues that we don't regard them as bases at all. It is still important to know the pK_{aH}s of oxygen compounds because the first step in many acid-catalysed reactions is protonation at an oxygen atom. Table 8.6 gives a selection of pK_{aH}s of oxygen compounds.

Table 8.6 pK_{aH}s of oxygen compounds

Oxygen compound	Oxygen compound (conjugate base A)	Approximate pK_{aH} of oxygen compound (pK_a of acid HA)	Conjugate acid HA of oxygen compound
ketone	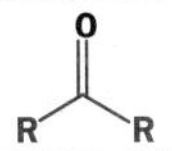	-7	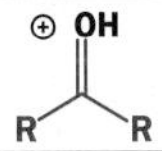
carboxylic acid	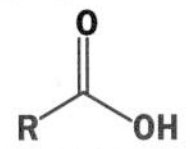	-7	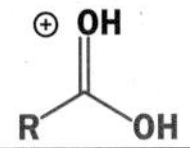
phenol	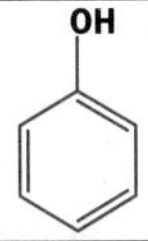	-7	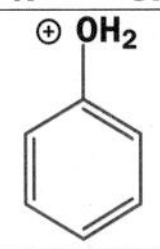
carboxylic ester	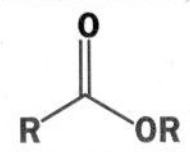	-5	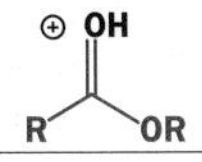

Table 8.6 (continued)

Oxygen compound	Oxygen compound (conjugate base A)	Approximate pK_{aH} of oxygen compound (pK_a of acid HA)	Conjugate acid HA of oxygen compound
alcohol		–4	
ether		–4	
water		–1.74	
amide		–0.5	

All the same factors of electron donation and withdrawal apply to oxygen compounds as well as to nitrogen compounds, but the effects are generally much less pronounced because oxygen is so electronegative. In fact, most oxygen compounds have pK_{aH}s around –7, the notable exception being the amide, which, because of the electron donation from the nitrogen atom, has a pK_{aH} around –0.5 (p. 201). They are all effectively nonbasic and strong acids are needed to protonate them.

pK_a in action—the development of the drug cimetidine

The development of the anti-peptic ulcer drug cimetidine gives a fascinating insight into the important role of pK_a in chemistry. Peptic ulcers are a localized erosion of the mucous membrane, resulting from overproduction of gastric acid in the stomach. One of the compounds that controls the production of the acid is histamine. (Histamine is also responsible for the symptoms of hay fever and allergies.)

cimetidine

histamine

Histamine in this example is an **agonist** in the production of gastric acid. It binds to specific sites in the stomach cells (receptor sites) and triggers the production of gastric acid (mainly HCl).

An **antagonist** works by binding to the same receptors but not stimulating acid secretion itself. This prevents the agonist from binding and stimulating acid production.

Histamine works by binding into a receptor in the stomach lining and stimulating the production of acid. What the developers of cimetidine at SmithKline Beecham wanted was a drug that would bind to these receptors without activating them and thereby prevent histamine from binding but not stimulate acid secretion itself. Unfortunately, the antihistamine drugs successfully used in the treatment of hay fever did not work—a different histamine receptor was involved. Notice that cimetidine and histamine both have an imidazole ring in their structure. This is not coincidence—cimetidine's design was centred around the structure of histamine.

In the body, most histamine exists as a salt, being protonated on the primary amine and the early compounds modelled this. The guanidine analogue was synthesized and tested to see if it had any antagonistic effect (that is, if it could bind in the histamine receptors and prevent histamine binding). It did bind but unfortunately it acted as an *agonist* rather than an *antagonist* and stimulated acid secretion rather than blocking it. Since the guanidine analogue has a pK_{aH} even greater than histamine (about 14.5 compared to about 10), it is effectively all protonated at physiological pH.

the major form of histamine at physiological pH (7.4)

pK_a 10

the guanidine analogue
the extra carbon in the chain was found to increase the efficacy of the drug

pK_a 14.5

The agonistic behaviour of the drug clearly had to be suppressed. The thought occurred to the SmithKline Beecham chemists that perhaps the positive charge made the compound agonistic, and so a polar but much less basic compound was sought. Eventually, they came up with burimamide. The most important change is the replacement of the C=NH in the guanidine compound by C=S. Now instead of a guanidine we have a thiourea which is much less basic. (Remember that amidines, p. 202, are very basic but that amides aren't? The thiourea is like the amide in that the sulfur withdraws electrons from the nitrogens.) The other minor adjustments, increasing the chain length and adding the methyl group on the thiourea, further increased the efficacy.

burimamide

introduction of sulfur decreased the p*K*$_a$ to ca. −1 so now this group is no longer protonated

extra chain length and the methyl group increased the activity still further

The new compound was a fairly good antagonist (that is, bound in the receptors and blocked histamine) but more importantly shown no agonistic behaviour at all. The compound was such a breakthrough that it was given a name, 'burimamide', and even tested in man. Burimamide was good, but unfortunately not good enough—it couldn't be given orally. A rethink was needed and this time attention was focused on the imidazole ring.

positive charge here withdraws electrons and decreases p*K*$_{aH}$ of ring

thiourea too far away from ring to influence p*K*$_{aH}$
alkyl chain is electron-donating and raises p*K*$_{aH}$ of ring

imidazole
p*K*$_{aH}$ 6.8

histamine
p*K*$_{aH}$ of imidazole ring 5.9

burimamide
p*K*$_{aH}$ of imidazole ring 7.25

The p*K*$_{aH}$ of the imidazole ring in burimamide is significantly greater than that in histamine: the longer alkyl group in burimamide is electron-donating and raises the p*K*$_{aH}$ of the ring. In histamine, on the other hand, the positive charge of the protonated amine withdraws electrons and decreases the p*K*$_{aH}$. This means, of course, that there will be a greater proportion of protonated imidazole (imidazolium cation) in burimamide and this might hinder effective binding in the histamine receptor site. So the team set out to lower the p*K*$_{aH}$ of the imidazole ring. It was known that a sulfur occupies just about the same space as a methylene group, –CH$_2$–, but is more electron-withdrawing. Hence 'thiaburimamide' was synthesized.

tautomers of thiaburimamide: p*K*$_{aH}$ of imidazole ring 6.25

It turns out that one tautomer of the imidazole ring binds better than the other (and much better than the protonated form). The introduction of a methyl group on the ring was found to increase the proportion of this tautomer and did indeed improve binding to the histamine receptor, even though the p*K*$_{aH}$ of the ring was raised because of the electron-donating character of the methyl group.

A   B
these two tautomers are in rapid equilibrium
we want tautomer 'A'

A   B
introduction of an electron-releasing group
favoured tautomer 'A'

metiamide: p*K*$_{aH}$ of imidazole ring 6.8

When the drug was invented, the company was called Smith, Kline, and French (SKF) but after a merger with Beechams the company became SmithKline Beecham or SB. SB and Glaxo-Welcome have now also merged to form GlaxoSmithKline (GSK). Things may have changed further by the time you read this book.

■
Tautomers are isomers differing only in the positions of hydrogen atoms and electrons. Otherwise the carbon skeleton is the same. They will be explained in Chapter 21.

The new drug, metiamide, was ten times more effective than burimamide when tested in man. However, there was an unfortunate side-effect: in some patients, the drug caused a decrease in the number of white blood cells, leaving the patient open to infection. This was eventually traced back to the thiourea group. The sulfur had again to be replaced by oxygen, to give a normal urea and, just to see what would happen, by nitrogen to give another guanidine.

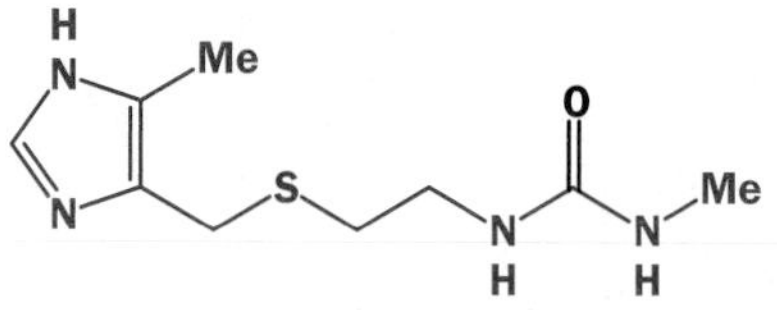

urea analogue of metiamide 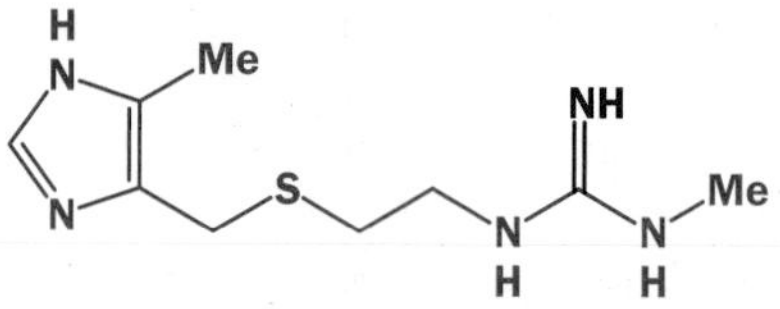   guanidine analogue of metiamide

Neither was as effective as metiamide but the important discovery was that the new guanidine no longer showed the agonistic effects of the earlier guanidine. Of course, the guanidine would also be protonated so we had the same problem we had earlier—how to decrease the pK_{aH} of the guanidine.

A section of this chapter considered the effect of electron-withdrawing groups on pK_{aH} and showed that they reduce the pK_{aH} and make a base less basic. This was the approach now adopted—the introduction of electron-withdrawing groups on to the guanidine to lower its pK_{aH}. Table 8.7 shows the pK_{aH}s of various substituted guanidines.

Table 8.7 pK_{aH}s of substituted guanidines

R	H	Ph	CH$_3$CO	NH$_2$CO	MeO	CN	NO$_2$
pK$_{aH}$	14.5	10.8	8.33	7.9	7.5	–0.4	–0.9

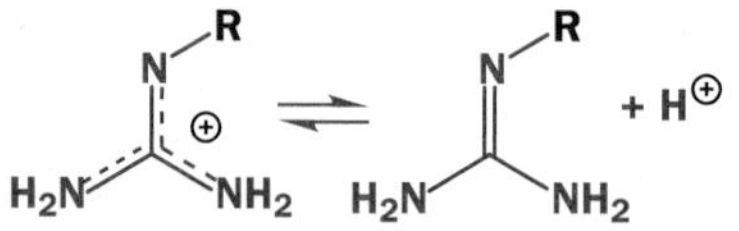

Clearly, the cyano- and nitro-substituted guanidines would not be protonated at all. These were synthesized and found to be just as effective as metiamide but without the nasty side-effects. Of the two, the cyanoguanidine compound was slightly more effective and this was developed and named 'cimetidine'.

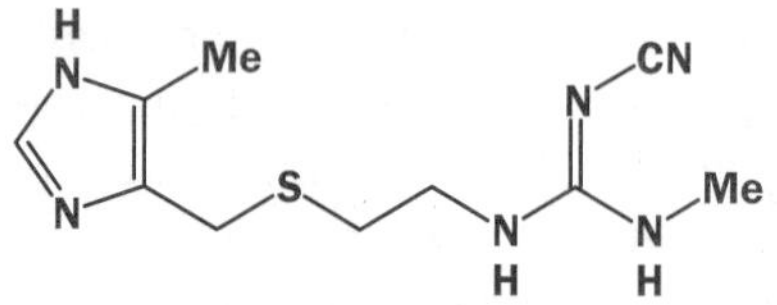

the end result, cimetidine

The development of cimetidine by Smith, Kline, and French from the very start of the project up to its launch on the market took thirteen years. This enormous effort was well rewarded—Tagamet (the trade name of the drug cimetidine) became the best-selling drug in the world and the first to gross more than one billion dollars per annum. Thousands of ulcer patients worldwide no longer had to suffer pain, surgery, or even death. The development of cimetidine followed a rational approach based on physiological and chemical principles and it was for this that one of the scientists involved, Sir James Black, received a share of the 1988 Nobel Prize for Physiology or Medicine. None of this would have been possible without an understanding of pK_as.

Problems

1. If you wanted to separate a mixture of naphthalene, pyridine, and *p*-toluic acid, how would you go about it?

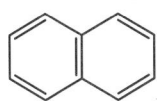 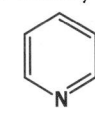 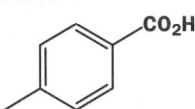

naphthalene   pyridine   *para*-toluic acid

2. In the separation of benzoic acid from toluene we suggested using NaOH solution. How concentrated a solution would be necessary to ensure that the pH was above the pK_a of benzoic acid (pK_a 4.2)? How would you estimate how much solution to use?

3. What species would be present if you were to dissolve this hydroxy-acid in: (a) water at pH 7; (b) aqueous alkali at pH 12; or (c) a concentrated solution of a mineral acid?

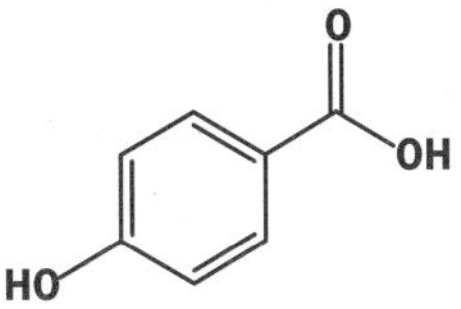

4. What would you expect to be the site of (**a**) protonation and (**b**) deprotonation if the compounds below were treated with an appropriate acid or base? In each case suggest a suitable acid or base for both purposes.

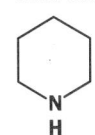 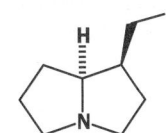

5. Suggest what species would be formed by each of these combinations of reagents. You are advised to use pK_a values to help you and to beware of some cases where 'no change' might be the answer.

(a)

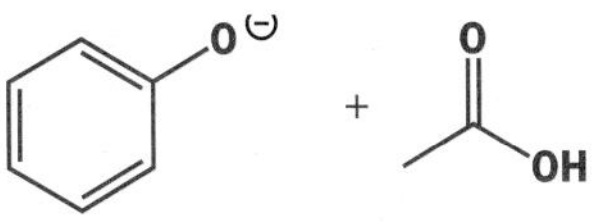

(b) 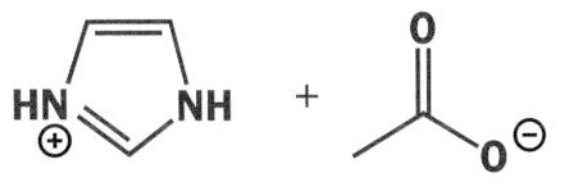 HN   NH +

(c)

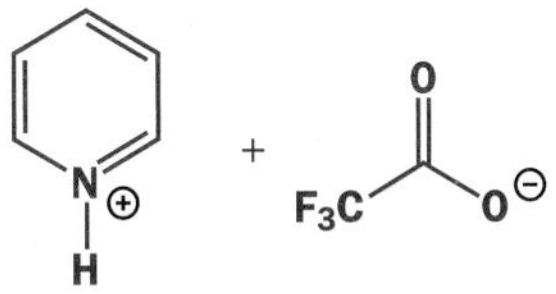

6. What is the relationship between these two molecules? Discuss the structure of the anion that would be formed by the deprotonation of each compound.

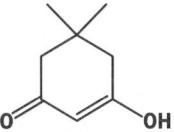

7. What species would be formed by treating this compound with: (**a**) one equivalent; (**b**) two equivalents of NaNH$_2$ in liquid ammonia?

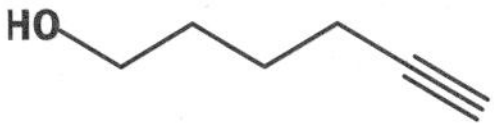

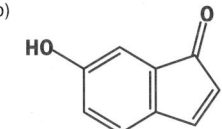

8. The carbon NMR spectra of these compounds could be run in D$_2$O under the conditions shown. Why were these conditions necessary and what spectrum would you expect to observe?

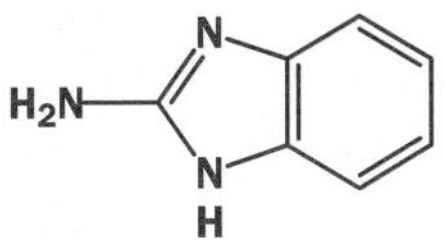

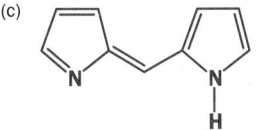

^{13}C spectrum run in DCl/D$_2$O ^{13}C spectrum run in NaOD/D$_2$O

9. The phenols shown here have approximate pK_a values of 4, 7, 9, 10, and 11. Suggest with explanations which pK_a value belongs to which phenol.

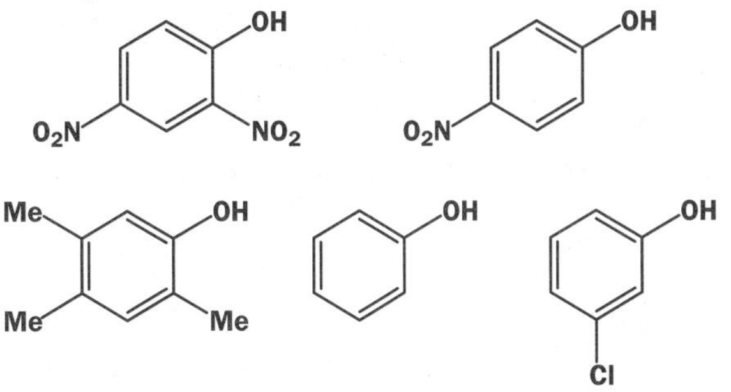

10. Discuss the stabilization of the anions formed by the deprotonation of (**a**) and (**b**) and the cation formed by the protonation of (**c**). Consider delocalization in general and the possibility of aromaticity in particular.

(a) 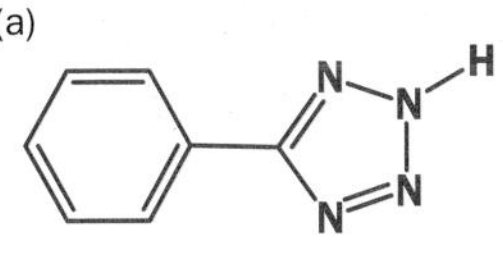   (b)

(c)

11. The pK_a values for the amino acid cysteine are 1.8, 8.3, and 10.8. Assign these pK_a values to the functional groups in cysteine and draw the structure of the molecule in aqueous solution at the following pHs: 1, 5, 9, and 12.

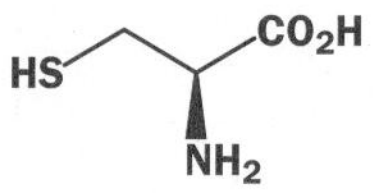

cysteine

12. Explain the variations in the pK_a values for these carbon acids.

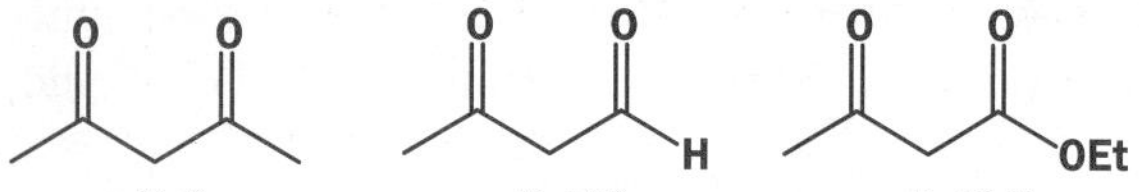

pK_a 9   pK_a 5.9   pK_a 10.7

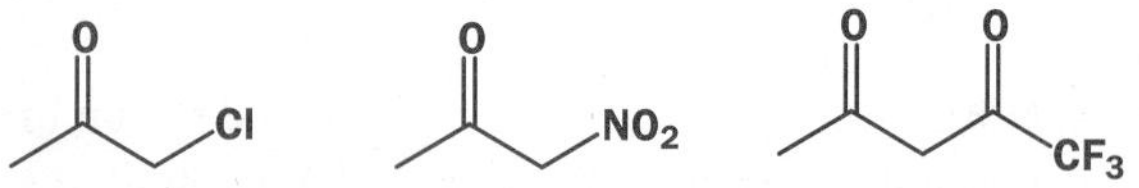

pK_a 16.5   pK_a 5.1   pK_a 4.7

13. Explain the various pK_a values for these derivatives of the naturally occurring amino acid glutamic acid. Say which pK_a belongs to which functional group and explain why they vary in the different derivatives.

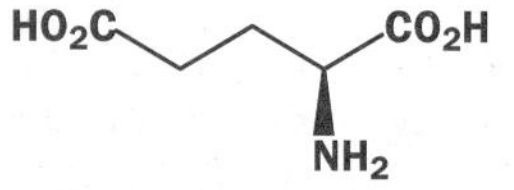

glutamic acid; pK_as 2.19, 4.25, and 9.67

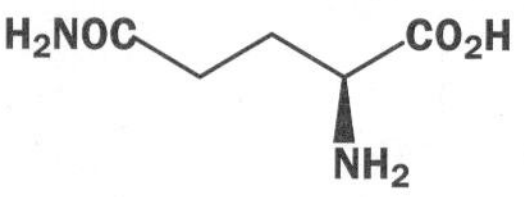

glutamine; pK_as 2.17 and 9.13

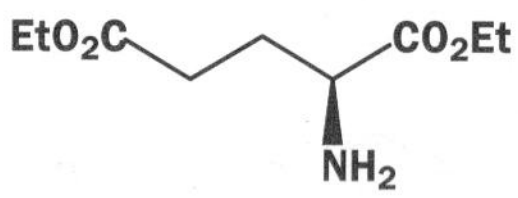

diethyl ester; pK_a 7.04

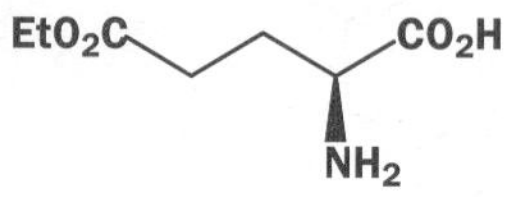

monoethyl ester;
pK_as 2.15 and 9.19

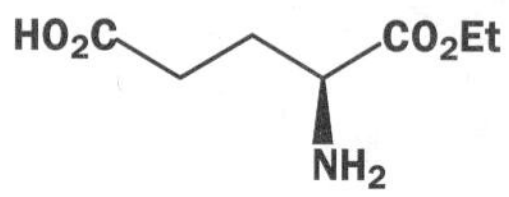

monoethyl ester;
pK_as 3.85 and 7.84

14. Neither of these methods of making pentan-1,4-diol will work. Explain why not—what will happen instead?

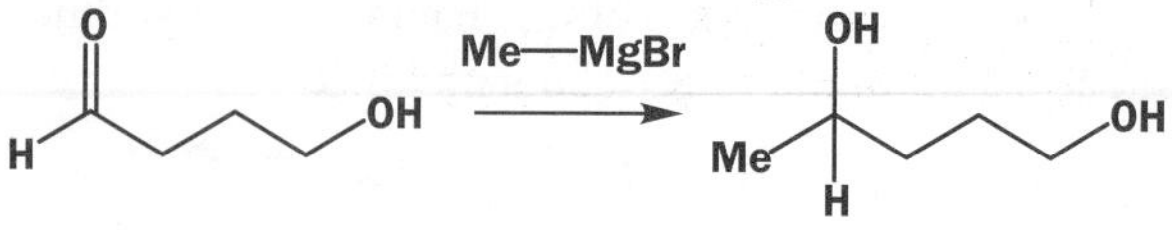

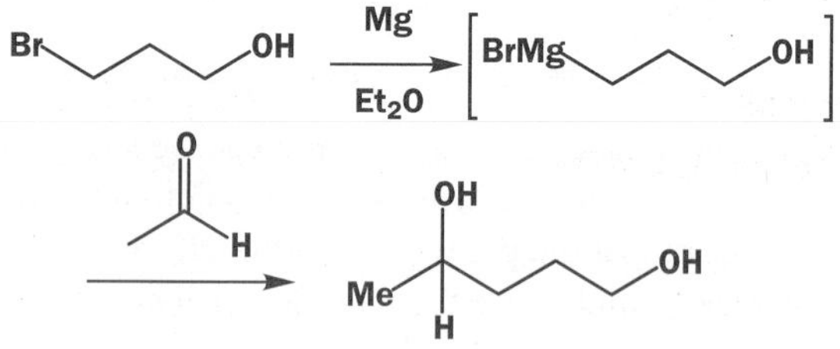

Using organometallic reagents to make C–C bonds

9

Connections

Building on:

- Electronegativity and the polarization of bonds ch4
- Grignard reagents and organolithiums attack carbonyl groups ch6
- C–H deprotonated by very strong bases ch8

Arriving at:

- Organometallics: nucleophilic and often strongly basic
- Making organometallics from halo-compounds
- Making organometallics by deprotonating carbon atoms
- Using organometallics to make new C–C bonds from C=O groups

Looking forward to:

- More about organometallics ch10 & ch48
- More ways to make C–C bonds from C=O groups ch26–ch29
- Synthesis of molecules ch 25 & ch30

Introduction

In Chapters 2–8 we covered basic chemical concepts, which mostly fall under the headings 'structure' (Chapters 2–4 and 7) and 'reactivity' (Chapters 5, 6, and 8). These concepts are the bare bones supporting all of organic chemistry, and now we shall start to put flesh on these bare bones. In Chapters 9–23 we will tell you about the most important classes of organic reaction in more detail.

One of the things organic chemists do, for all sorts of reasons, is to make molecules. And making organic molecules means making C–C bonds. In this chapter we are going to look at one of the most important ways of making C–C bonds: using organometallics, such as organolithiums and Grignard reagents, and carbonyl compounds. We will consider reactions such as these.

■
You met these types of reactions in Chapter 6: in this chapter we will be adding more detail with regard to the nature of the organometallic reagents and what sort of molecules can be made using the reactions.

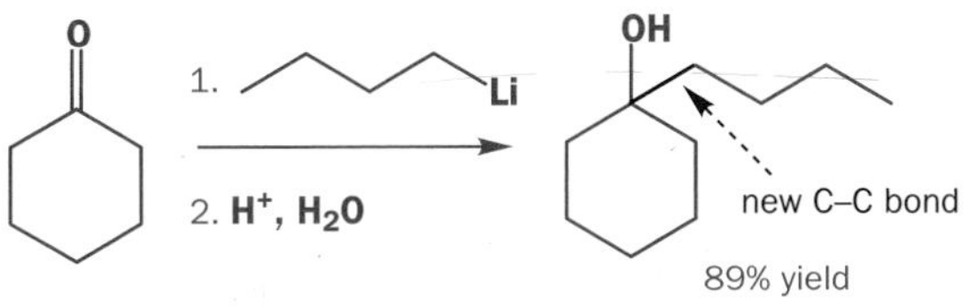

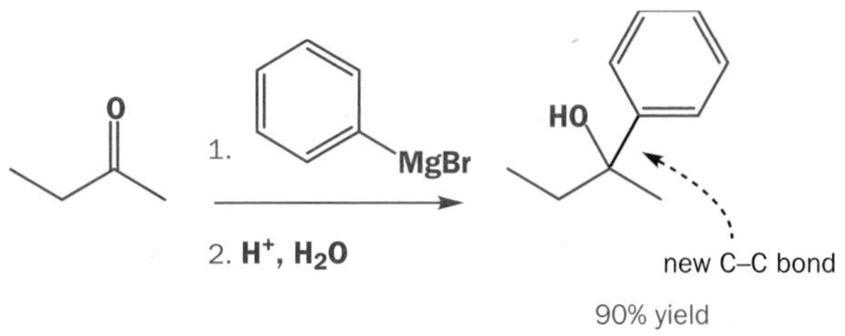

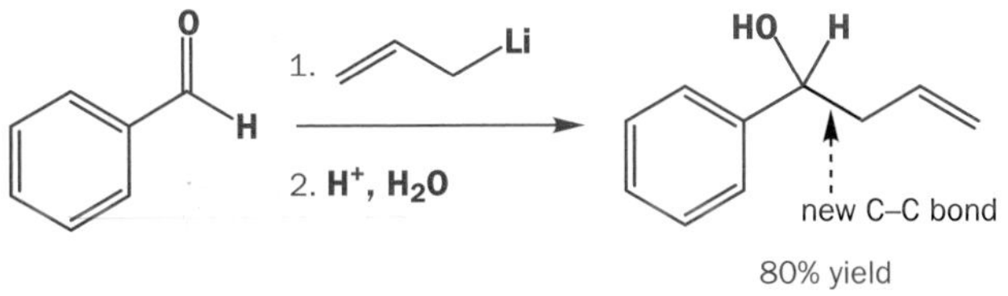

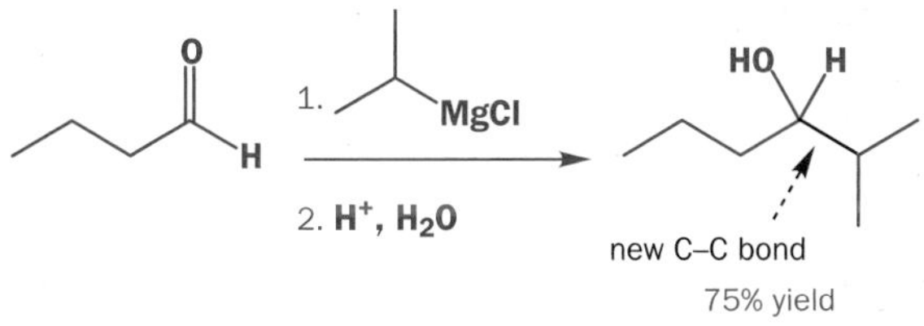

The organometallic reagents act as nucleophiles towards the electrophilic carbonyl group, and this is the first thing we need to discuss: why are organometallics nucleophilic? We then move on to, firstly, how to make organometallics, then to the sort of electrophiles they will react with, and then finally to the sort of molecules we can make with them.

Organometallic compounds contain a carbon–metal bond

The polarity of a covalent bond between two different elements is determined by electronegativity. The more electronegative an element is, the more it attracts the electron density in the bond. So the

How important are organometallics for making C–C bonds?

As an example, let's take a molecule known as 'juvenile hormone'. It is a compound that prevents several species of insects from maturing and can be used as a means of controlling insect pests. Only very small amounts of the naturally occurring compound can be isolated, but it can instead be made in the lab from simple starting materials. At this stage you need not worry about how, but we can tell you that, of the sixteen C–C bonds in the final product, seven were made by reactions of organometallics, many of them the sort of reactions we will describe in this chapter. This is not an isolated example. As further proof, take this important enzyme inhibitor, closely related to arachidonic acid which you met in Chapter 7. It has been made by a succession of C–C bond-forming reactions using organometallics: eight of the twenty C–C bonds in the product were formed using organometallic reactions.

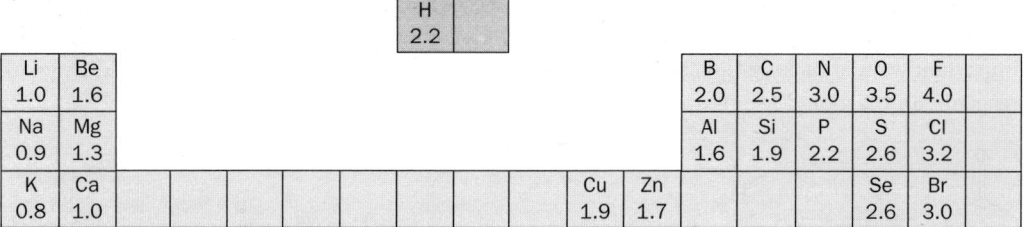

black bonds made by organometallic reactions

Cecropia juvenile hormone

an enzyme inhibitor

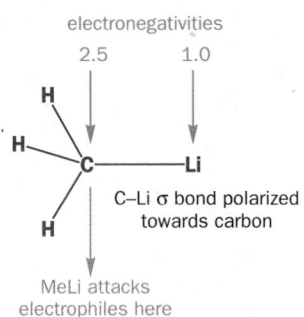

electronegativities
2.5   3.5

C=O π bond polarized towards oxygen

nucleophiles attack here

electronegativities
2.5   1.0

C–Li σ bond polarized towards carbon

MeLi attacks electrophiles here

greater the *difference* between the electronegativities, the greater the difference between the attraction for the bonding electrons, and the more polarized the bond becomes. In the extreme case of complete polarization, the covalent bond ceases to exist and is replaced by electrostatic attraction between ions of opposite charge. We discussed this in Chapter 4 (p. 101), where we considered the extreme cases of bonding in NaF.

When we discussed (in Chapter 6) the electrophilic nature of carbonyl groups we saw that their reactivity is a direct consequence of the polarization of the carbon–oxygen bond towards the more electronegative oxygen, making the carbon a site for nucleophilic attack. In organolithium compounds and Grignard reagents the key bond bond is polarized in the opposite direction—*towards* carbon—making carbon a nucleophilic centre. This is true for most organometallics because, as you can see from this edited version of the periodic table, metals (such as Li, Mg, Na, K, Ca, and Al) all have lower electronegativity than carbon.

Pauling electronegativities of selected elements

						H 2.2											
Li 1.0	Be 1.6											B 2.0	C 2.5	N 3.0	O 3.5	F 4.0	
Na 0.9	Mg 1.3											Al 1.6	Si 1.9	P 2.2	S 2.6	Cl 3.2	
K 0.8	Ca 1.0								Cu 1.9	Zn 1.7					Se 2.6	Br 3.0	

The orbital diagram—the kind you met in Chapter 4—represents the C–Li bond in methyllithium in terms of a sum of the atomic orbitals of carbon and lithium. Remember that, the more

orbital diagram for the C–Li bond of MeLi

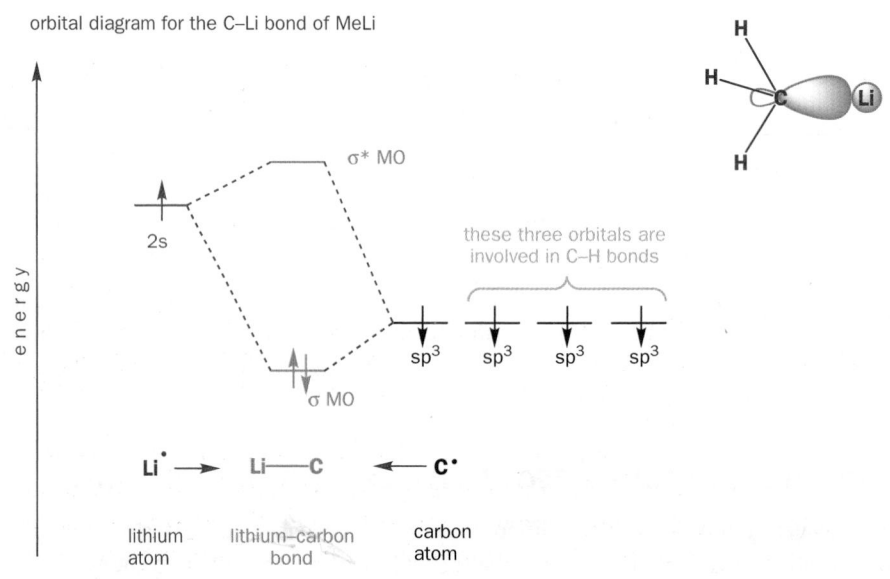

electronegative an atom is, the lower in energy its atomic orbitals are (p. 101). The filled C–Li σ orbital that arises is closer in energy to the carbon's sp³ orbital than to the lithium's 2s orbital, so we can say that the carbon's sp³ orbital makes a greater contribution to the C–Li σ bond and that the C–Li bond has a larger coefficient on carbon. Reactions involving the filled σ orbital will therefore take place at C rather than Li. The same arguments hold for the C–Mg bond of Grignard reagents.

We can also say that, because the carbon's sp³ orbital makes a greater contribution to the C–Li σ bond, the σ bond resembles a filled C sp³ orbital—in other words it *resembles* a lone pair on carbon. This is a useful idea because it allows us to think about the way in which methyllithium reacts—as though it were an ionic compound Me⁻Li⁺—and you may sometimes see MeLi or MeMgCl represented in mechanisms as Me⁻.

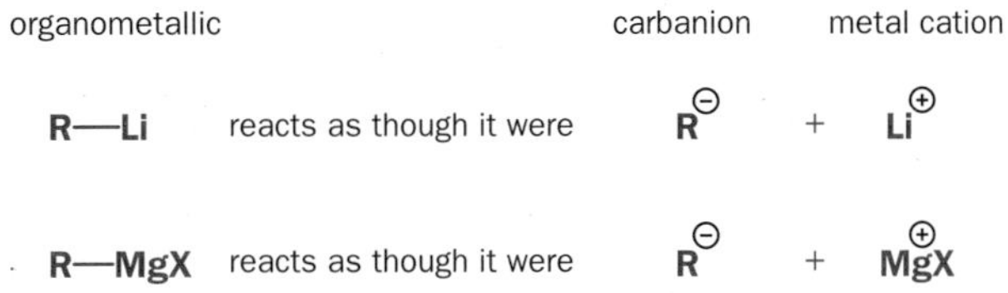

organometallic		carbanion	metal cation
R—Li	reacts as though it were	R⁻	+ Li⁺
R—MgX	reacts as though it were	R⁻	+ MgX⁺

> You have already met cyanide (p. 119), a carbon nucleophile that really does have a lone pair on carbon. Cyanide's lone pair is stabilized by being in a lower-energy sp orbital (rather than sp³) and by having the electronegative nitrogen atom triply bonded to the carbon.

> Carbon atoms that carry a negative charge, for example Me⁻, are known as **carbanions**.

The true structure of organolithiums and Grignard reagents is rather more complicated!

Even though these organometallic compounds are extremely reactive towards water and oxygen, and have to be handled under an atmosphere of nitrogen or argon, a number have been studied by X-ray crystallography in the solid state and by NMR in solution. It turns out that they generally form complex aggregates with two, four, six, or more molecules bonded together, often with solvent molecules. In this book we shall not be concerned with these details, and it will suffice always to represent organometallic compounds as simple monomeric structures.

Making organometallics

How to make Grignard reagents

Grignard reagents are made by reacting magnesium turnings with alkyl halides in ether solvents to form solutions of alkylmagnesium halide. Iodides, bromides, and chlorides can be used, as can both aryl and alkyl halides, though they cannot contain any functional groups that would react with the Grignard reagent once it is formed. Here are some examples.

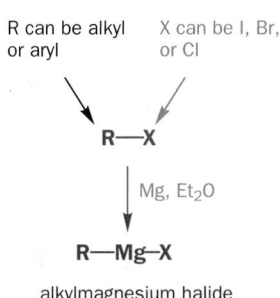

R can be alkyl or aryl   X can be I, Br, or Cl

R—X

↓ Mg, Et₂O

R—Mg—X

alkylmagnesium halide (Grignard reagent)

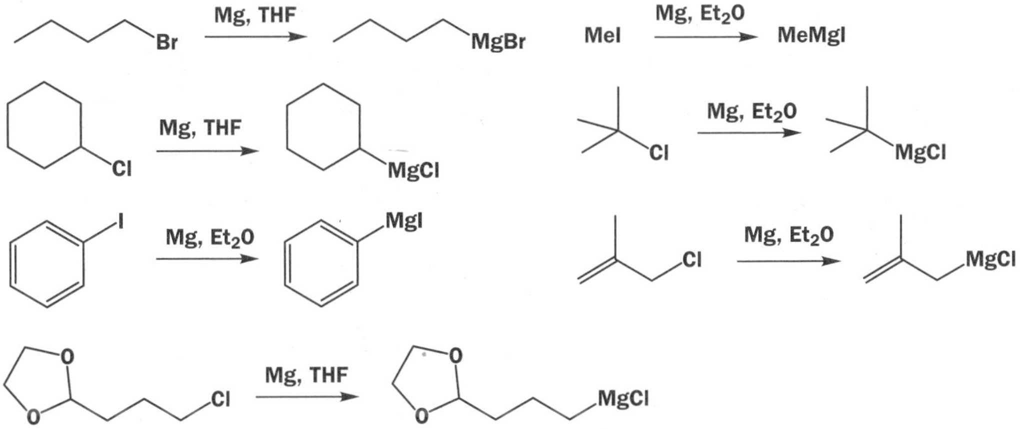

> Diethyl ether (Et₂O) and THF are the most commonly used solvents, but you may also meet others such as dimethoxyethane (DME) and dioxane.
>
> common ether solvents
>
> diethyl ether   THF (tetrahydrofuran)
>
> dioxane   DME (dimethoxyethane)

oxidative insertion

magnesium(0)

magnesium inserts into this bond

magnesium(II)

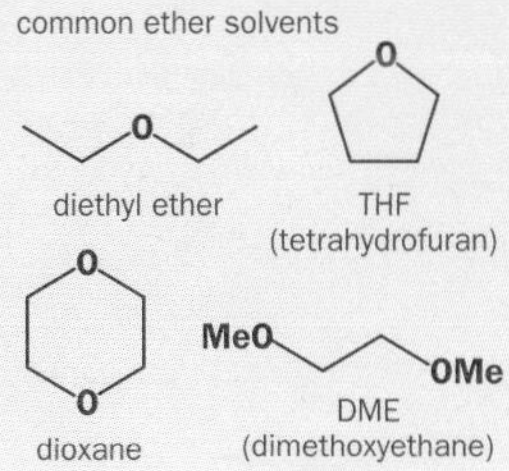

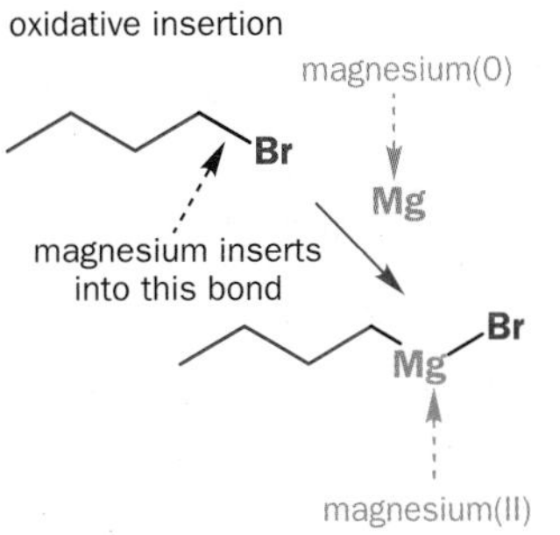

The reaction scheme is easy enough to draw, but what is the mechanism? Overall it involves an *insertion* of magnesium into the new carbon–halogen bond. There is also a change in oxidation state of the magnesium, from Mg(0) to Mg(II). The reaction is therefore known as an **oxidative insertion** or **oxidative addition**, and is a general process for many metals such as Mg, Li (which we meet shortly), Cu, and Zn.

The mechanism of the reaction is not completely understood but a possible (but probably not very accurate) way of writing the mechanism is shown here: the one thing that is certain is that the first interaction is between the metal and the halogen atom.

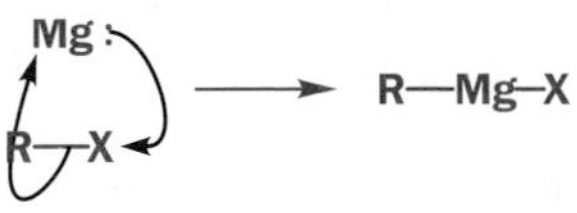

Mg + R—X → R—Mg—X

complex between
Lewis-acidic metal atom
and lone pairs of THF

R can be alkyl   X can be I, Br
or aryl   or Cl

R—X

Li, THF

R—Li   LiX

alkyllithium plus lithium halide

The reaction takes place not in solution but on the surface of the metal, and how easy it is to make a Grignard reagent can depend on the state of the surface—how finely divided the metal is, for example. Magnesium is usually covered by a thin coating of magnesium oxide, and Grignard formation generally requires 'initiation' to allow the metal to come into contact with the alkyl halide. Initiation can be accomplished by adding a small amount of iodine or 1,2-diiodoethane, or by using ultrasound to dislodge the oxide layer. The ether solvent is essential for Grignard formation because (1) ethers (unlike, say, alcohols or dichloromethane) will not react with Grignards and, more importantly, (2) only in ethers are Grignard reagents soluble. In Chapter 5 you saw how triethylamine forms a complex with the Lewis acid BF_3, and much the same happens when an ether meets a metal ion such as magnesium or lithium: the metals are Lewis-acidic because they have empty orbitals (2p in the case of Li and 3p in the case of Mg) that can accept the lone pair of the ether.

How to make organolithium reagents

Organolithium compounds may be made by a similar oxidative insertion reaction from lithium metal and alkyl halides. Each inserting reaction requires two atoms of lithium and generates one equivalent of lithium halide salt. As with Grignard formation, there is really very little limit on the types of organolithium that can be made this way.

Some Grignard and organolithium reagents are commercially available

Most chemists (unless they were working on a very large scale) would not usually make the simpler organolithiums or Grignard reagents by these methods, but would buy them in bottles from chemical companies (who, of course, *do* use these methods). The table lists some of the most important commercially available organolithiums and Grignard reagents.

Commercially available organometallics

methyllithium (MeLi)	methylmagnesium chloride, bromide, and iodide (MeMgX)
n-butyllithium (*n*-BuLi or just BuLi)	ethylmagnesium bromide (EtMgBr)
sec-butyllithium (*sec*-BuLi or *s*-BuLi)	butylmagnesium chloride (BuMgCl)
tert-butyllithium (*tert*-BuLi or *t*-BuLi)	allylmagnesium chloride and bromide
phenyllithium (PhLi)	phenylmagnesium chloride and bromide (PhMgCl or PhMgBr)

Organometallics as bases

Organometallics need to be kept absolutely free of moisture—even moisture in the air will destroy them. The reason is that they react very rapidly and highly exothermically with water to produce alkanes. Anything that can protonate them will do the same thing. If we represent these protonation reactions slightly differently, putting the products on the left and the starting materials (represented, just for effect, as 'carbanions') on the right, you can see that they are acid–base equilibria from the last chapter. The organometallic acts as a base, and is protonated to form its conjugate acid—methane or benzene in these cases.

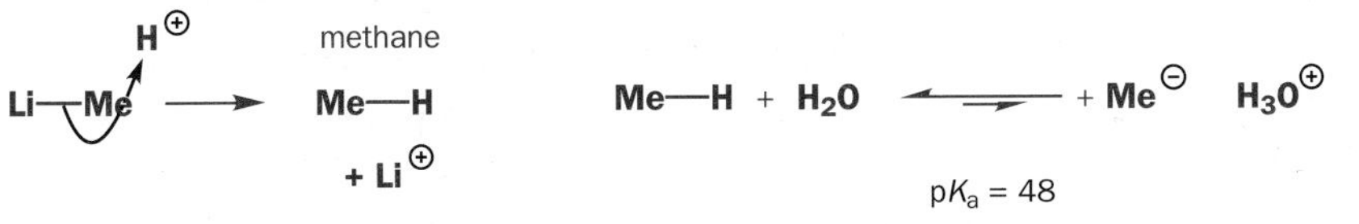

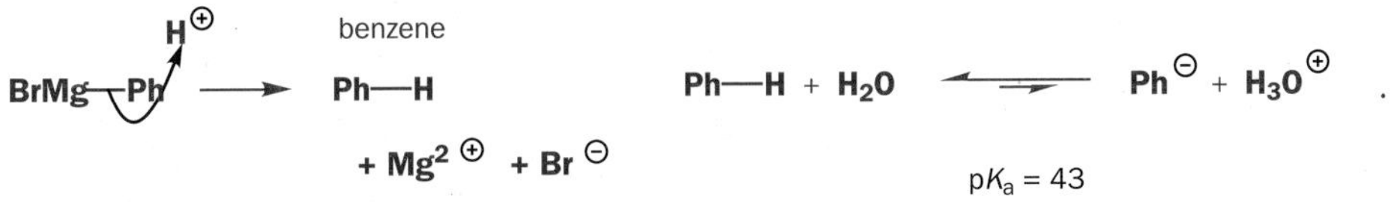

The equilibria lie vastly to the left: the pK_a values indicate that methane and benzene are extremely weak acids and that methyllithium and phenylmagnesium bromide must therefore be extremely strong bases. Some of the most important uses of organolithiums—butyllithium, in particular—are as bases and, because they are so strong, they will deprotonate almost anything. That makes them very useful as reagents for making *other* organolithiums.

Making organometallics by deprotonating alkynes

In Chapter 8 (p. 194) we talked about how hybridization affects acidity. Alkynes, with their C–H bonds formed from sp orbitals, are the most acidic of hydrocarbons, with pK_as of about 25. They can be deprotonated by more basic organometallics such as butyllithium or ethylmagnesium bromide. Alkynes are sufficiently acidic to be deprotonated even by nitrogen bases, and another common way of deprotonating alkynes is to use $NaNH_2$ (sodium amide), obtained by reacting sodium with liquid ammonia. An example of each is shown here: we have chosen to represent the alkynyllithium and alkynylmagnesium halide as organometallics and the alkynyl sodium as an ionic salt. Propyne and acetylene are gases, and can be bubbled through a solution of the base.

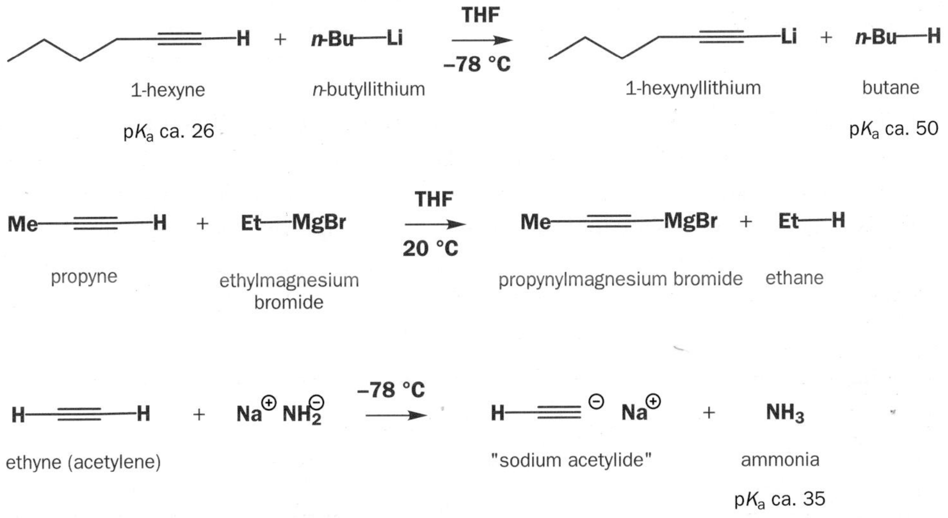

The metal derivatives of alkynes can be added to carbonyl electrophiles as in the following examples. The first (we have reminded you of the mechanism for this) is the initial step of an important synthesis of the antibiotic, erythronolide A, and the second is the penultimate step of a synthesis of the widespread natural product, farnesol.

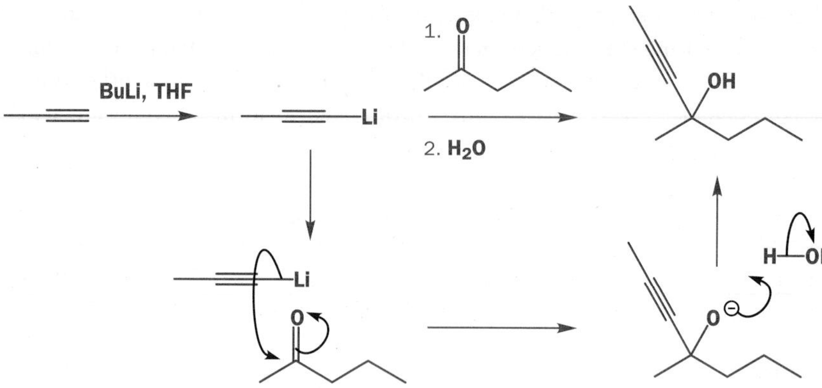

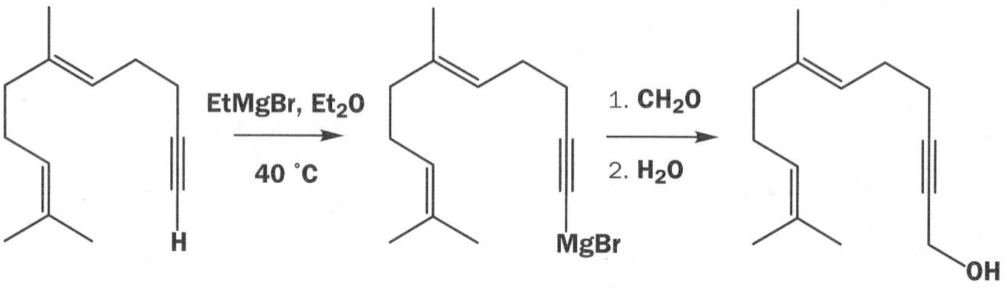

Ethynyloestradiol

The ovulation-inhibiting component of many oral contraceptive pills is a compound known as ethynyloestradiol, and this compound too is made by an alkynyllithium addition to the female sex hormone

oestrone. A range of similar synthetic analogues of hormones containing an ethynyl unit are used in contraceptives and in treatments for disorders of the hormonal system.

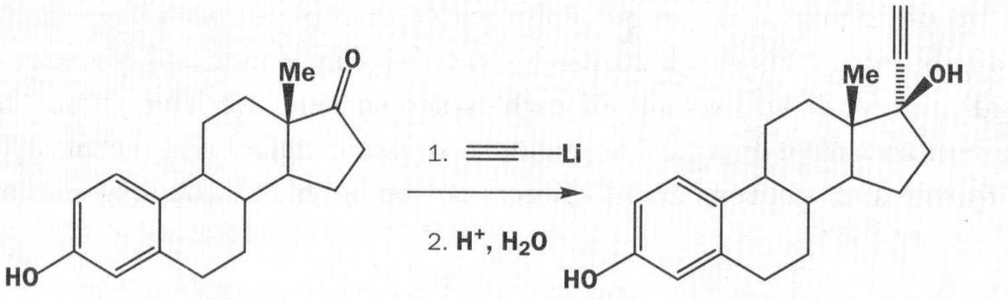

Making organometallics by deprotonating aromatic rings: ortholithiation

Look at the reaction below: in some ways it is quite similar to the ones we have just been discussing. Butyllithium deprotonates an sp^2 hybridized carbon atom to give an aryllithium. It works because the protons attached to sp^2 carbons are more acidic than protons attached to sp^3 carbons (though they are a lot less acidic than alkyne protons).

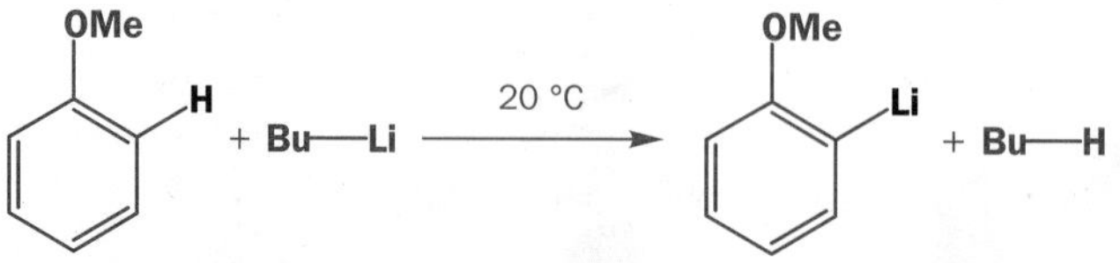

But there is another factor involved as well. There has to be a functional group containing oxygen (sometimes nitrogen) next to the proton to be removed. This functional group 'guides' the butyllithium, so that it attacks the adjacent protons. It does this by forming a complex with the Lewis-acidic lithium atom, much as ether solvents dissolve Grignard reagents by complexing their Lewis-acidic metal ions. This mechanism means that it is only the protons *ortho* to the functional group that can be removed, and the reaction is known as an **ortholithiation**.

■ The terms *ortho*, *meta*, and *para* were defined on p. 39.

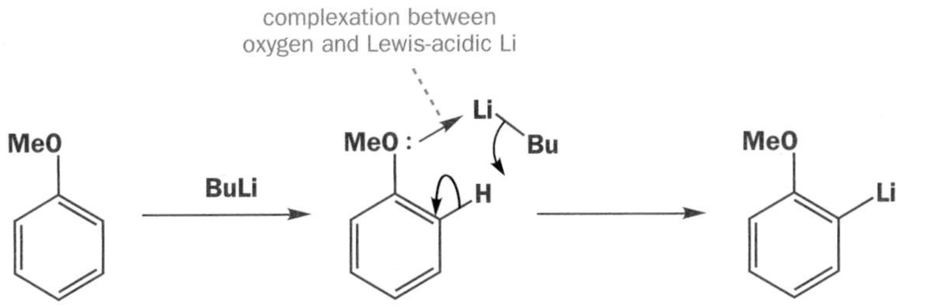

complexation between
oxygen and Lewis-acidic Li

The example below shows an organolithium formed by ortholithiation being used to make a new C–C bond. Here it is a nitrogen atom that directs attack of the butyllithium.

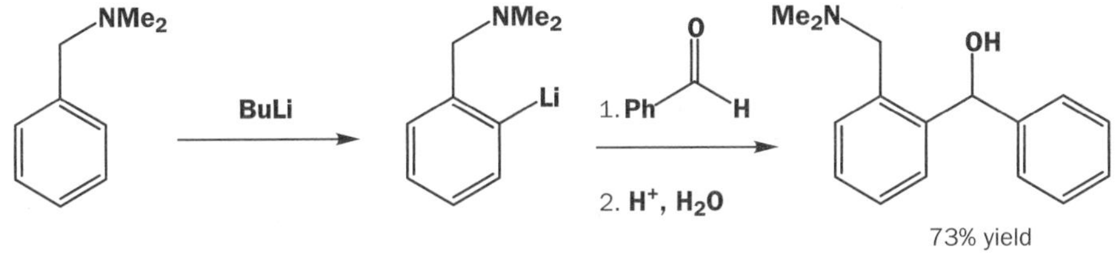

73% yield

Ortholithiation is useful because the starting material does not need to contain a halogen atom. But it is much less general than the other ways we have told you about for making organolithiums, because there are rather tight restrictions on what sorts of groups the aromatic ring must carry.

Fredericamycin

Fredericamycin is a curious aromatic compound extracted in 1981 from the soil bacterium *Streptomyces griseus*. It is a powerful antibiotic and antitumour agent, and its structure is shown below. The first time it was made in the laboratory, in 1988, the chemists in Boston started their synthesis with three consecutive lithiation reactions: two are ortholithiations, and the third is slightly different. You needn't be concerned about the reagents that react with the organolithiums; just look at the lithiation reactions

themselves. In each one, an oxygen atom (colour-coded green) directs a strongly basic reagent to remove a nearby proton (colour-coded black). As it happens, none of the steps uses *n*-BuLi itself, but instead its more reactive cousins, *sec*-BuLi and *tert*-BuLi (see the table on p. 212). The third lithiation step uses a different kind of base, made by deprotonating an amine (pK_a about 35). The yellow proton removed in this third lithiation is more acidic because it is next to an aromatic ring (p. 194).

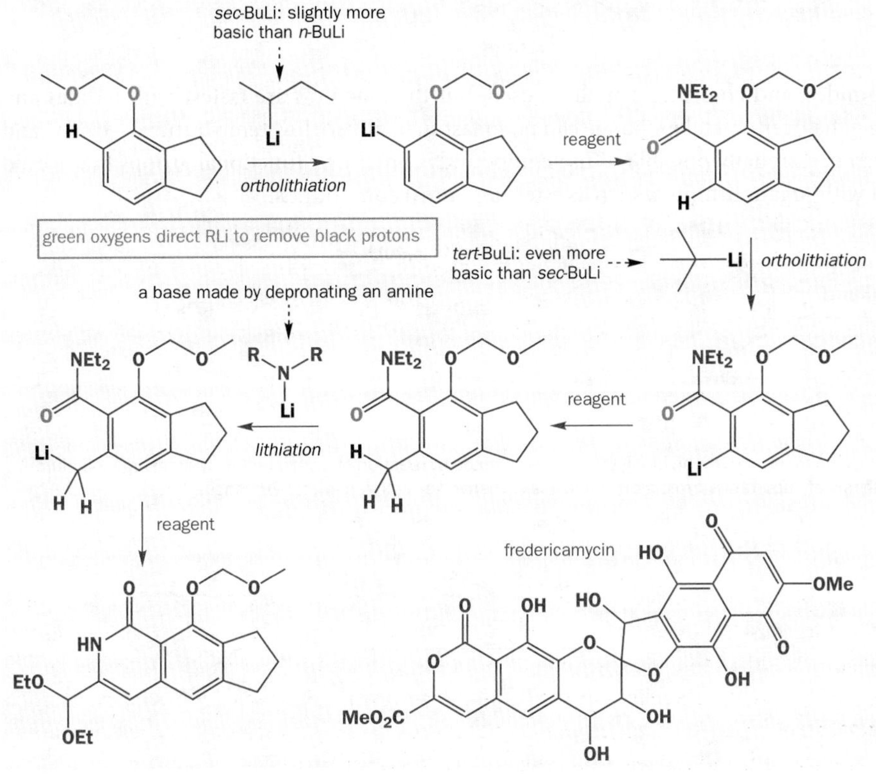

sec-BuLi: slightly more basic than *n*-BuLi

ortholithiation

green oxygens direct RLi to remove black protons

tert-BuLi: even more basic than *sec*-BuLi

ortholithiation

a base made by deprotonating an amine

lithiation

reagent

reagent

fredericamycin

Halogen–metal exchange

Deprotonation is not the only way to use one simple organometallic reagent to generate another more useful one. Organolithiums can also remove halogen atoms from alkyl and aryl halides in a reaction known as **halogen–metal exchange.** Look at this example and you will immediately see why.

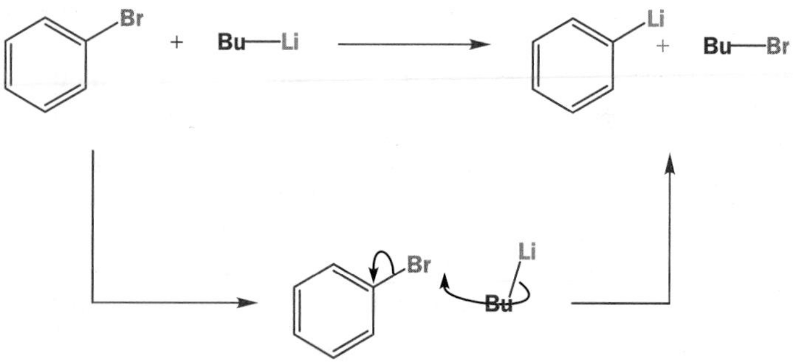

pK_{aH} = ca. 50

Bu—Li can be considered as $Bu^{\ominus}$ $Li^{\oplus}$

Ph—Li can be considered as $Ph^{\ominus}$ $Li^{\oplus}$

pK_{aH} = ca. 43

The bromine and the lithium simply swap places. As with many of these organometallic processes, the mechanism is not altogether clear, but can be represented as a nucleophilic attack on bromine by the butyllithium. But why does the reaction work? The key, again, is pK_a. The reaction works because the organolithium that is formed (phenyllithium, which protonated would give benzene, pK_a about 43) is less basic (more stable) than the organolithium we started with (BuLi, which protonated would give butane, pK_a about 50). The following reactions are also successful halogen–metal exchanges, and in each case the basicity of the organolithium decreases.

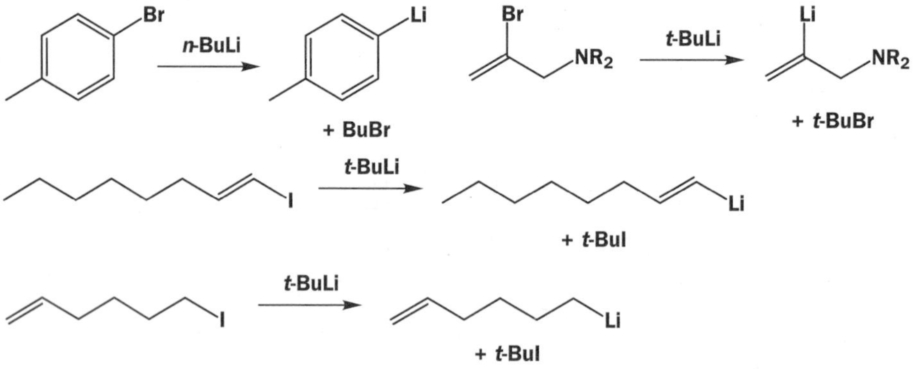

Iodides, bromides, and chlorides can all be used, but the reactions are fastest with iodides and bromides. In fact, halogen–metal exchange can be so fast that, at very low temperature (–100 °C and below), it is even occasionally possible to use compounds containing functional groups that would otherwise react with organolithiums, such as esters and nitro compounds.

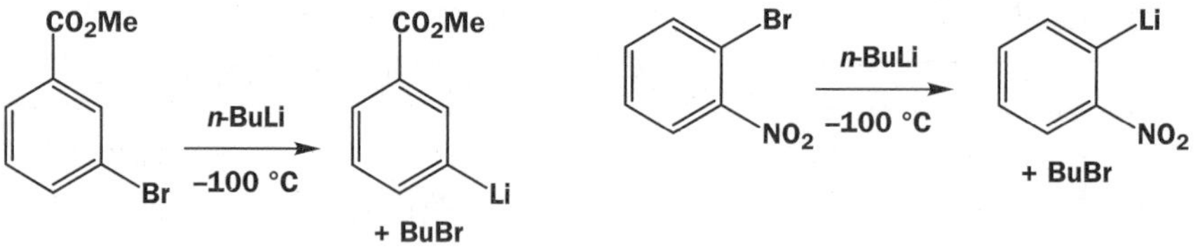

Fenarimol

Fenarimol is a fungicide that works by inhibiting the fungus's biosynthesis of important steroid molecules. It is made by reaction of a diarylketone with an organolithium derived by halogen–metal exchange.

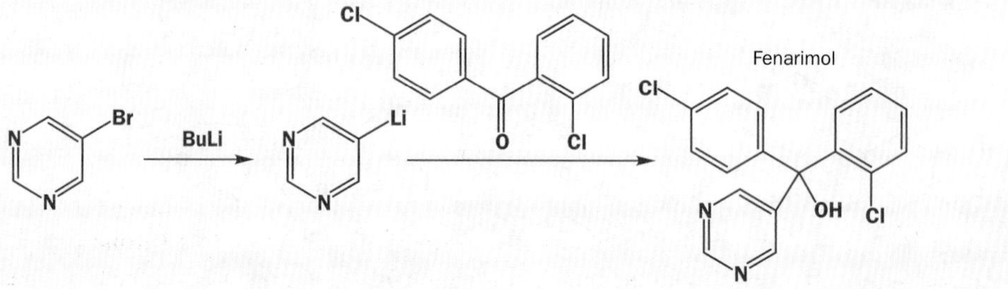

You will see several examples of transmetallation with copper salts in the next chapter.

You met the idea that carbonyl groups, and aromatic rings, acidify adjacent protons in Chapter 8 p. 192.

> ### ▶ *tert*-Butyllithium
>
> Alkyl substituents are slightly electron-donating, so more substituted organolithiums are less stable because the carbon atom is forced to carry even more of a negative charge. Instability reaches a peak with *tert*-butyllithium, which is the most basic of the commonly available organolithium reagents, and so is particularly useful for halogen exchange reactions. (It is so unstable that even in solution it will spontaneously catch fire in contact with air.) Its importance is enhanced by a subtlety in its reactions that we have not yet mentioned: a problem with halogen–metal exchanges is that the two products, an organolithium and an alkyl halide, sometimes react with one another in a substitution (Chapter 17) or elimination (Chapter 19) reaction. This problem is overcome provided two equivalents of *t*-BuLi are used. The first takes part in the halogen–metal exchange, while the second immediately destroys the *t*-butyl bromide produced by the exchange, preventing it from reacting with the organolithium product.
>
>
> Do not be concerned about the mechanism at this stage: we will come back to this sort of reaction in Chapter 19.

Transmetallation

Organolithiums can be converted to other types of organometallic reagents by **transmetallation**— simply treating with the salt of a less electropositive metal. The more electropositive lithium goes into solution as an ionic salt, while the less electropositive metal (magnesium and cerium in these examples) takes over the alkyl group.

But why bother? Well, the high reactivity—and in particular the basicity—of organolithiums, which we have just been extolling, sometimes causes unwanted side-reactions. You saw in Chapter 8 that protons next to carbonyl groups are moderately acidic (pK_a about 20), and because of this organolithiums occasionally act as bases towards carbonyl compounds instead of as nucleophiles. Organoceriums, for example, are rather less basic, and may give higher yields of the nucleophilic addition products than organolithiums or Grignard reagents.

> ### An instance where transmetallation is needed to produce another organometallic, which does act as a base but not as a nucleophile!
>
> Dialkylzincs are stable, distillable liquids that can be made by transmetallating Grignard reagents with zinc bromide. They are much less reactive than organolithium or organomagnesium compounds, but they are still rather basic and react with water to give zinc hydroxides and alkanes. They are used to preserve old books from gradual decomposition due to acid in the paper. The volatile dialkylzinc penetrates the pages thoroughly, where contact with water produces basic hydoxides that neutralize the acid, stopping the deterioration.
>

Acidic protons were a major problem in several syntheses of the anticancer compounds, daunorubicin and adriamycin, which start with a nucleophilic addition to a ketone with a pair of particularly acidic protons. Organolithium and organomagnesium compounds remove these protons rather than add to the carbonyl group, so some Japanese chemists turned to organocerium compounds. They made ethynylcerium dichloride (HC≡CCeCl$_2$) by deprotonating acetylene, and then transmetallating with cerium trichloride. They found that it reacted with the ketone to give an 85% yield of the alcohol they wanted.

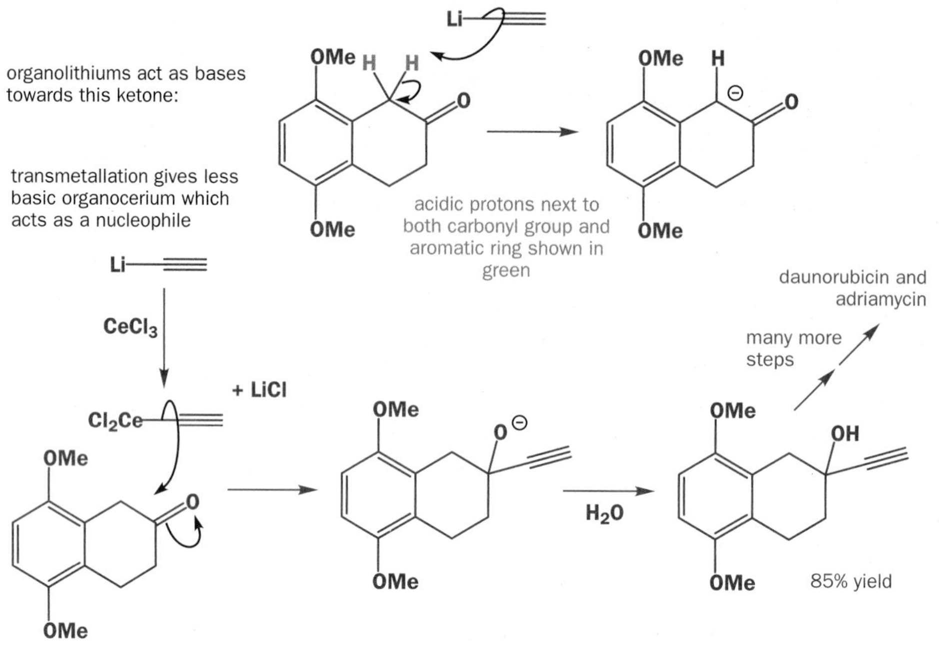
organolithiums act as bases towards this ketone:
transmetallation gives less basic organocerium which acts as a nucleophile
Li
OMe
H
O
acidic protons next to both carbonyl group and aromatic ring shown in green
CeCl3
+ LiCl
Cl2Ce
H2O
OH
many more steps
daunorubicin and adriamycin
85% yield

Using organometallics to make organic molecules

Now that you have met all of the most important ways of making organometallics (summarized here as a reminder), we shall move on to consider how to use them to make molecules: what sorts of electrophiles do they react with and what sorts of products can we expect to get from their reactions? Having told you how you can make other organometallics, we shall really be concerned for the rest of this chapter only with Grignard reagents and organolithiums. In nearly all of the cases we shall talk about, the two classes of organometallics can be used interchangeably.

● Ways of making organometallics

- Oxidative insertion of Mg into alkyl halides
- Oxidative insertion of Li into alkyl halides
- Deprotonation of alkynes
- Ortholithiation of functionalized benzene rings
- Halogen–metal exchange
- Transmetallation

Making carboxylic acids from organometallics and carbon dioxide

Carbon dioxide is a carbonyl compound, and it is an electrophile. It reacts slowly with water, for example, to form the unstable compound carbonic acid—you can think of this as a hydration reaction of a carbonyl group.

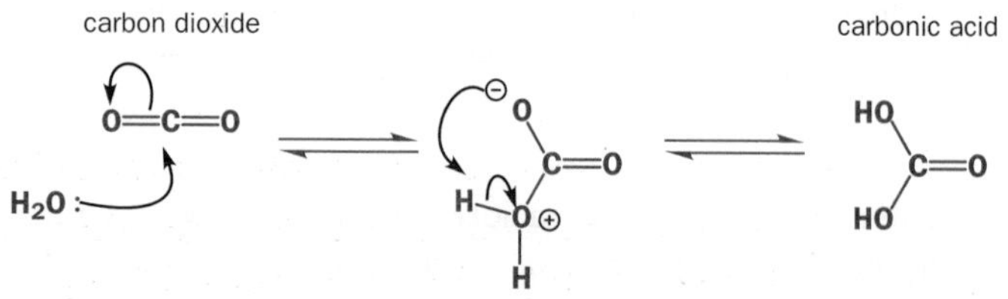
carbon dioxide
carbonic acid
O=C=O
H2O:
C=O
HO
H

Carbon dioxide reacts with organolithiums and Grignard reagents to give carboxylate salts. Protonating the salt with acid gives a carboxylic acid with one more carbon atom than the starting organometallic. The reaction is usually done by adding solid CO_2 to a solution of the organolithium in THF or ether, but it can also be done using a stream of dry CO_2 gas.

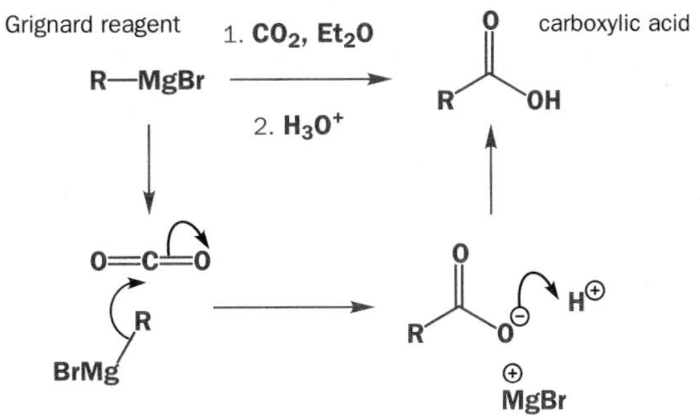

The examples below show the three stages of the reaction: (1) forming the organometallic; (2) reaction with the electrophile (CO_2); and (3) the acidic work-up or **quench**, which protonates the product and destroys any unreacted organometallic left over at the end of the reaction. The three stages of the reaction have to be monitored carefully to make sure that each is finished before the next is begun—in particular it is absolutely essential that there is no water present during either of the first two stages—water must be added only at the end of the reaction, when the organometallic has all been consumed by reaction with the electrophile. You may occasionally see schemes written out without the quenching step included—but it is nonetheless always needed.

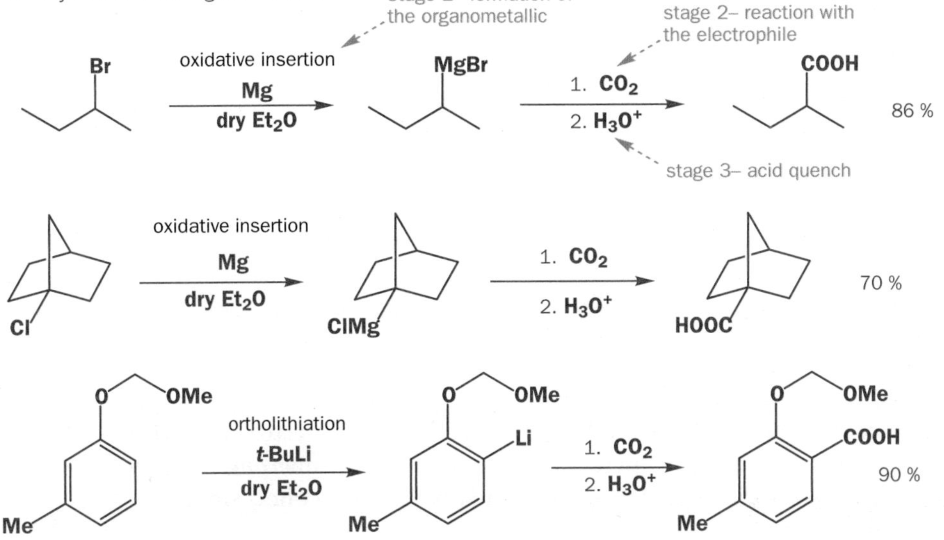

Methicillin synthesis

Methicillin is an important antiobiotic compound because it works even against bacteria that have developed resistance to penicillin, whose structure is quite similar. It can be made from an acid obtained by reaction of carbon dioxide with an organolithium. In this case the organolithium is made by an ortholithiation reaction of a compound with two oxygen atoms that direct removal of the proton in between them.

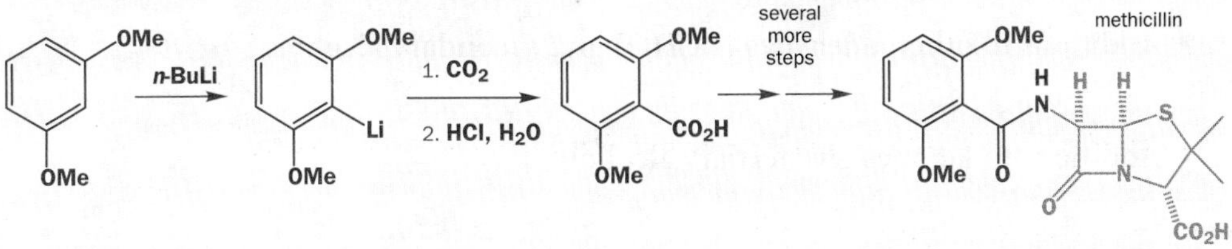

Making primary alcohols from organometallics and formaldehyde

You met formaldehyde, the simplest aldehyde, in Chapter 6, where we discussed the difficuties of using it in anhydrous reactions: it is either hydrated or a polymer (paraformaldehyde, $(CH_2O)_n$) and, in order to get pure, dry formaldehyde, it is necessary to heat ('crack') the polymer to

■
Primary, secondary, and tertiary are
defined on p. 30.

decompose it. But formaldehyde is a remarkably useful reagent for making **primary** alcohols, in other words, alcohols that have just one carbon substituent attached to the hydroxy-bearing C atom. Just as carbon dioxide adds one carbon and makes an acid, fomaldehyde adds one carbon and makes an alcohol.

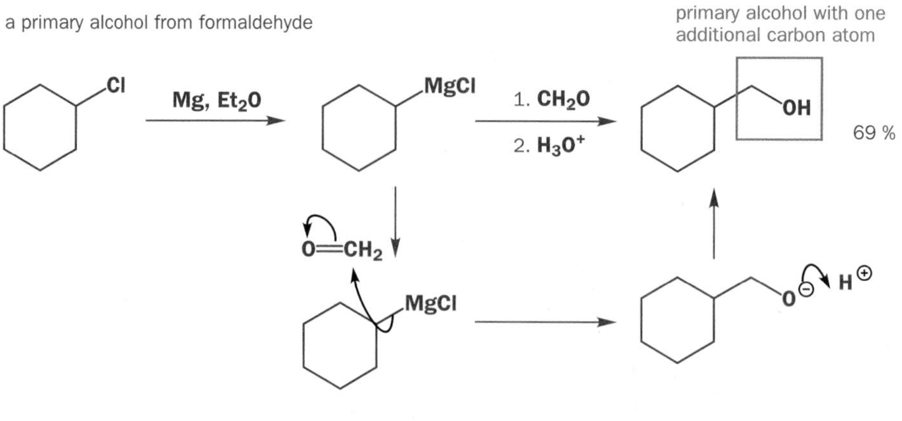

a primary alcohol from formaldehyde

primary alcohol with one
additional carbon atom

In the next examples, formaldehyde makes a primary alcohol from two deprotonated alkynes. The second reaction here (for which we have shown organolithium formation, reaction, and quench simply as a series of three consecutive reagents) forms one of the last steps of the synthesis of *Cecropia* juvenile hormone whose structure you met right at the beginning of the chapter.

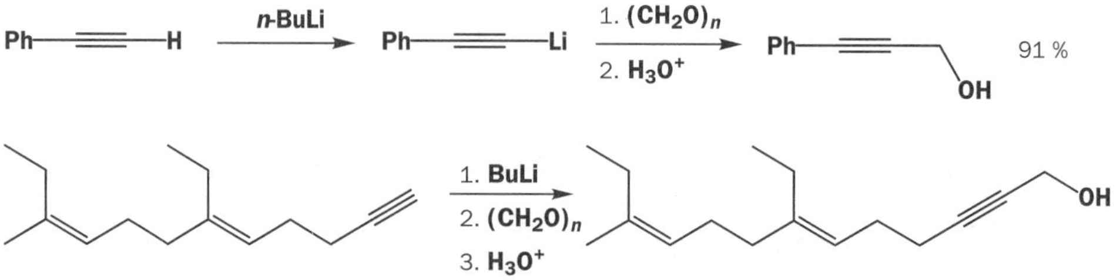

● Something to bear in mind with all organometallic additions to carbonyl
compounds is that the addition takes the oxidation level *down one.* In other
words, if you start with an aldehyde, you end up with an alcohol. More
specifically,

● Additions to CO_2 give carboxylic acids

● Additions to formaldehyde (CH_2O) give primary alcohols

● Additions to other aldehydes (RCHO) give secondary alcohols

● Additions to ketones give tertiary alcohols

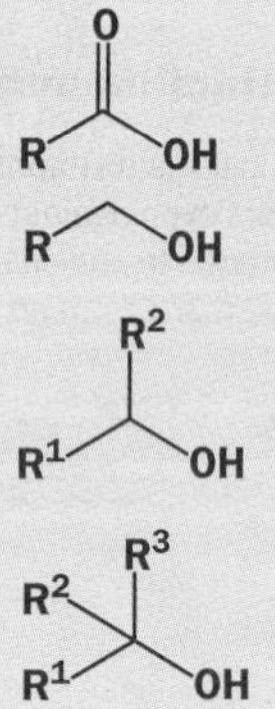

Secondary and tertiary alcohols: which organometallic, which aldehyde, which ketone?

Aldehydes and ketones react with Grignard or organolithium reagents to form secondary and tertiary alcohols, respectively, and some examples are shown with the general schemes here.

secondary alcohols from aldehydes

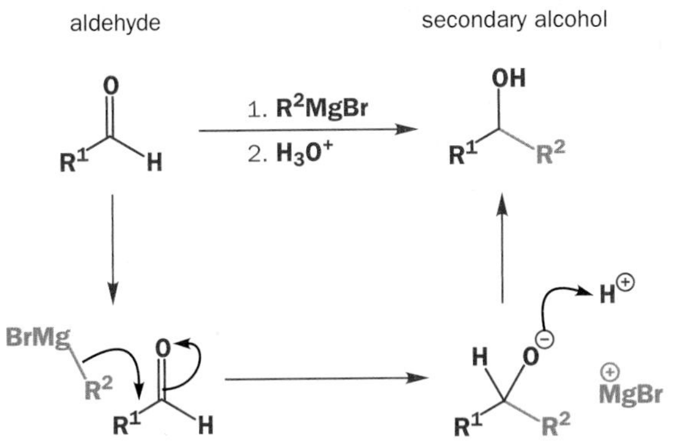

aldehyde   secondary alcohol

tertiary alcohols from ketones

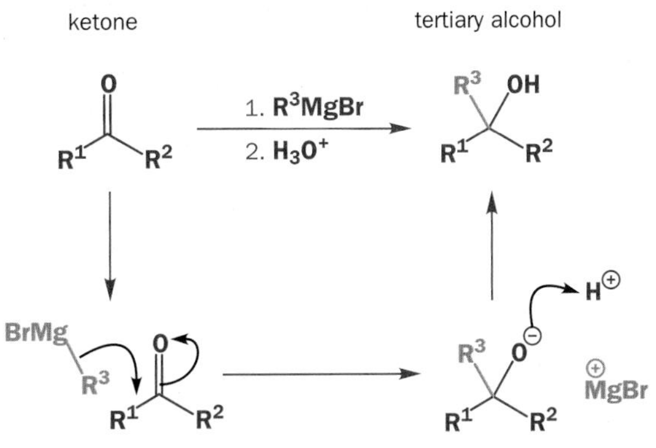

ketone   tertiary alcohol

two examples:

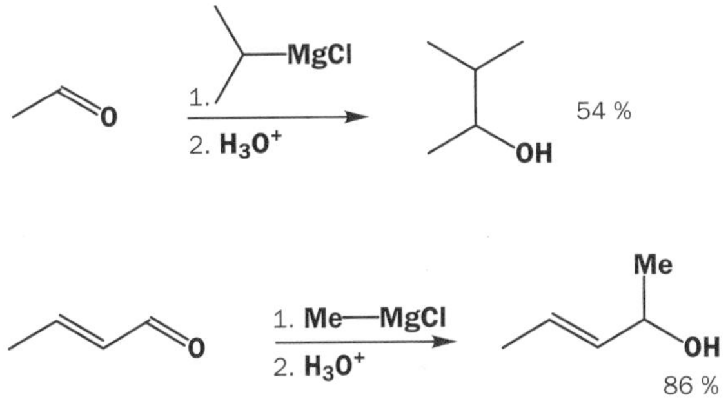

54 %

86 %

two examples:

89 %

81%

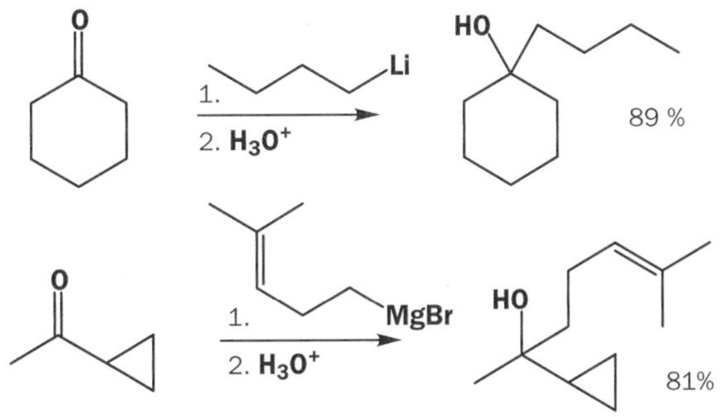

To make any secondary alcohol, however, there is often a choice of two possible routes, depending on which part of the molecule you choose to make the organometallic and which part you choose to make the aldehyde. For example, the first example here shows the synthesis of a secondary alcohol from isopropylmagnesium chloride and acetaldehyde. But it is equally possible to make this same secondary alcohol from isobutyraldehyde and methyllithium or a methylmagnesium halide.

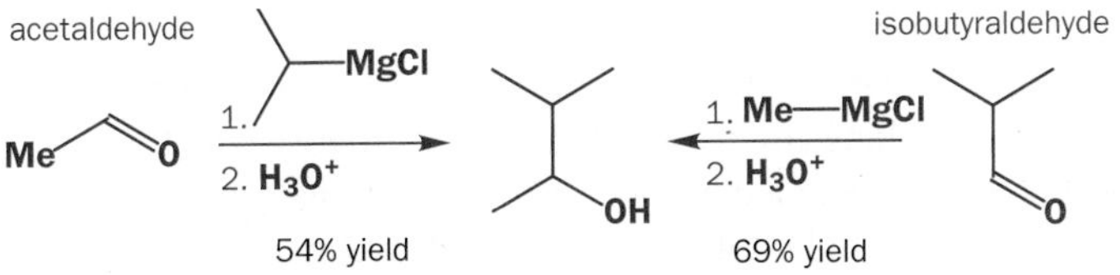

acetaldehyde   isobutyraldehyde

54% yield   69% yield

Indeed, back in 1912, when this alcohol was first described in detail, the chemists who made it chose to start with acetaldehyde, while in 1983, when it was needed as a starting material for a synthesis, it was made from isobutyraldehyde. Which way is better? The 1983 chemists probably chose the isobutyraldehyde route because it gave a better yield. But, if you were making a secondary alcohol for the first time, you might just have to try both in the lab and see which one gave a better yield. Or you might be more concerned about which uses the cheaper, or more readily available, starting materials—this was probably behind the choice of methylmagnesium chloride and the unsaturated aldehyde in the second

Flexibility in the synthesis of alcohol

As an illustration of the flexibility available in making secondary alcohols, one synthesis of bongkrekic acid, a highly toxic compound that inhibits transport across certain membranes in the cell, required both of these (very similar) alcohols. The chemists making the compound at Harvard University chose to make each alcohol from quite different starting materials: an unsaturated aldehyde and an alkyne-containing organolithium in the first instance, and an alkyne-containing aldehyde and vinyl magnesium bromide in the second.

alcohols needed for the synthesis of bongkrekic acid

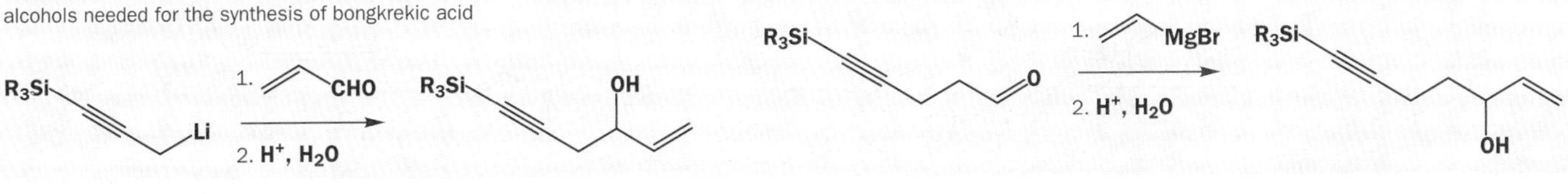

example. Both can be bought commercially, while the alternative route to this secondary alcohol would require a vinyllithium or vinylmagnesium bromide reagent that would have to be made from a vinyl halide, which is itself not commercially available, along with difficult-to-dry acetaldehyde.

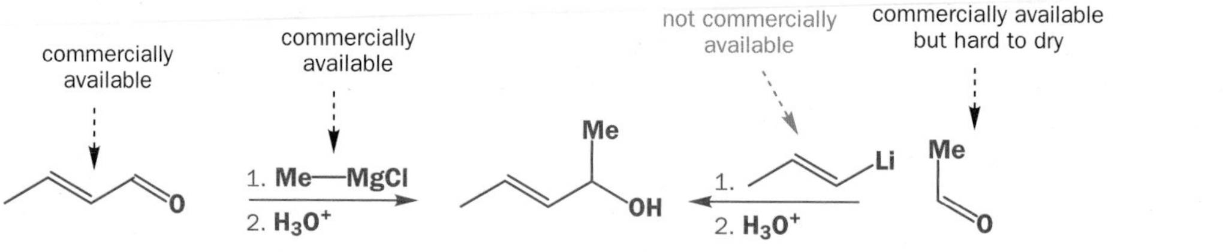

With tertiary alcohols, there is even more choice. The last example in the box is a step in a synthesis of the natural product, nerolidol. But the chemists in Paris who made this tertiary alcohol could in principle have chosen any of these three routes.

three routes to a tertiary alcohol

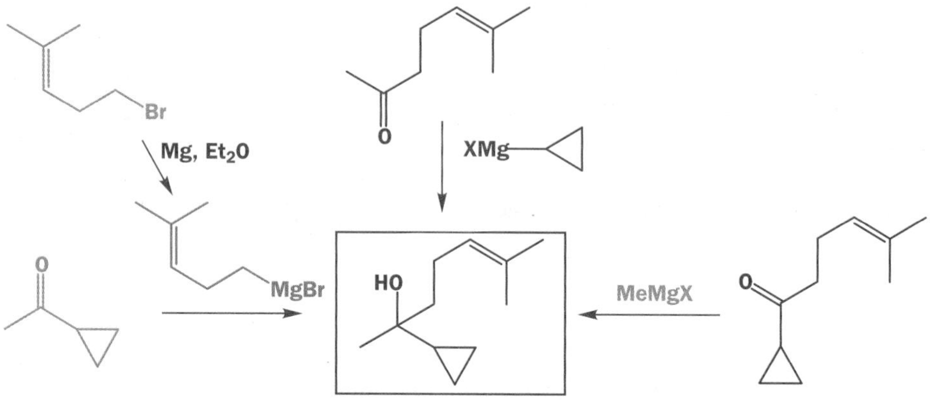

> ■
> Note we have dropped the aqueous quench step from these schemes to avoid cluttering them.

Only the reagents in orange are commercially available, but, as it happens, the green Grignard reagent can be made from an alkyl bromide, which is itself commercially available, making the route on the left the most reasonable.

Now, do not be dismayed! We are not expecting you to remember a chemical catalogue and to know which compounds you can buy and which you can't. All we want you to appreciate at this stage is that there are usually two or three ways of making any given secondary or tertiary alcohol, and you should be able to suggest alternative combinations of aldehyde or ketone and Grignard reagent that will give the same product. You are not expected to be able to assess the relative merits of the different possible routes to a compound. That is a topic we leave for a much later chapter on retrosynthetic analysis, Chapter 30.

Ketones by oxidation of secondary alcohols

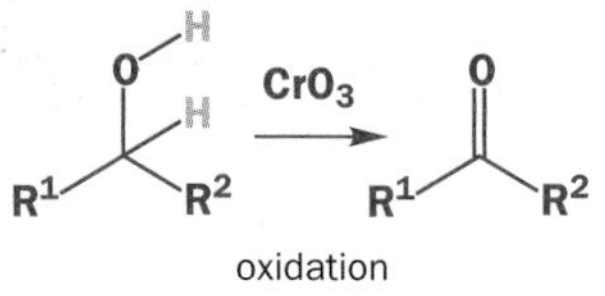

Tertiary alcohols can be made from ketones, and secondary ones from aldehydes, but we should now show you that ketones can be made from secondary alcohols by an oxidation reaction. There are lots of possible reagents, but a common one is an acidic solution of chromium trioxide. We will look in much more detail at oxidation later, when we will discuss the mechanism of the reaction, but for now take it from us that secondary alcohols give ketones on treatment with CrO_3. Note that you can't oxidize tertiary alcohols (without breaking a C–C bond). The link between secondary alcohols and ketones means that the ketones needed for making tertiary alcohols can themselves ultimately be made by addition of organometallics to carbonyl compounds. Here, for example, is a sequence of reactions leading to a compound needed to make the drug viprostol.

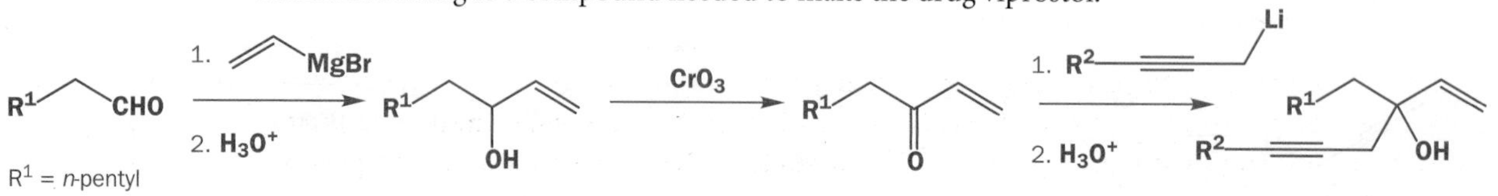

A closer look at some mechanisms

We finish this chapter with some brief words about the mechanism of the addition of organometallics to carbonyl compounds. The problem with this reaction is that no-one really knows precisely what happens during the addition reaction. We know what the organic products are because we can isolate them and look at them using NMR and other spectroscopic techniques. But what happens to the metal atoms during the reaction?

You will have noticed that we always write the addition reaction with the metal atom just falling off the organometallic as it reacts, and then appearing near to the anionic oxygen atom of the product. In other words, we have not been specific about what the metal atom is actually doing during the addition; in fact, we have been deliberately vague so as not to imply anything that may not be true. But there is one thing that is certain about this process, and before we discuss it we need to remind you of something we talked about in Chapter 6: the effect of acid on the addition of nucleophiles to carbonyl groups. We said that acid tends to catalyse addition reactions by protonating the carbonyl group, making it positively charged and therefore more electrophilic.

Now, of course, in our organometallic addition reactions we have no acid (H$^+$) present, because that would destroy the organometallic reagent. But we do have *Lewis-acidic* metal atoms—Li or Mg—and these can play exactly the same role. They can coordinate to the carbonyl's oxygen atom, giving the carbonyl group positive charge and therefore making it more electrophilic. In one possible version of the mechanism, a four-centred mechanism allows coordination of the magnesium to the oxygen while the nucleophilic carbon atom attacks the carbonyl group. The product ends up with a (covalent) Mg–O bond, but this is just another way of writing RO$^-$ MgBr$^+$.

which can also be written as

■ Lewis acids were introduced in Chapter 5 (p. 120).

dotted bond indicates new bond forming

dotted bond indicates old bond breaking

possible transition state for Grignard addition

The four-centred mechanism is quite hard to visualize just with curly arrows: what they are saying is that the O–Mg interaction is forming at the same time as the new C–C bond, and that simultaneously the old C–Mg bond and C=O π bond are breaking. A neat way of representing all of this is to draw what we might see if we took a snapshot of the reaction halfway through, using dotted lines to represent the partially formed or partially broken bonds. It would look something like this, and such a snapshot is known as the **transition state** for the reaction.

▶ The term 'transition state' has, in fact, a more precise definition, which we will introduce in Chapter 13.

An alternative possibility is that two molecules of the Grignard reagent are involved, and that the transition state is a six-membered ring. We are telling you all this not because we want to confuse you but because we want to be honest: there is genuine uncertainty about the mechanism, and this arises because, while it is easy to determine the products of a reaction using spectroscopy, it is much harder to determine mechanisms.

■ We shall devote two chapters entirely to mechanism and how it is studied: Chapters 13 and 41.

six-centred transition state

But, for one type of Grignard reagent, it is certain that the addition proceeds through a six-membered ring. Here is a reaction between an allylic Grignard reagent and a ketone. The product is a tertiary alcohol, but perhaps not the tertiary alcohol you would expect. The Grignard reagent appears to have attached itself via the wrong carbon atom. We can explain this by a six-membered transition state, but one involving only one molecule of Grignard reagent.

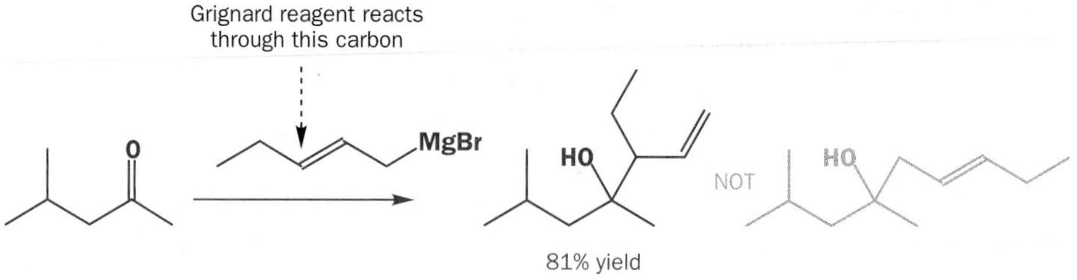

Grignard reagent reacts
through this carbon

81% yield

curly-arrow representation of the mechanism

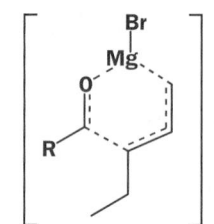

six-centred transition state

Mg, Et₂O

plus

dimer

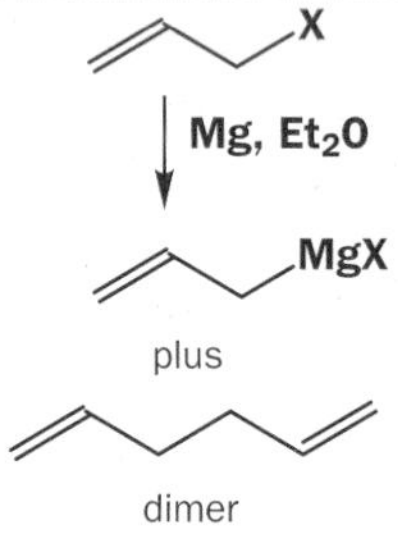

Allylic Grignard reagents are unusual for more than one reason, and it turns out that they are, in fact, quite hard to make in good yield from allyl halides. The problem is that the allyl halide is highly reactive towards the Grignard reagent as it forms, and a major by-product tends to be a dimer. The way round this problem is to make the Grignard reagent actually in the presence of the carbonyl compound. This method works in a number of cases, not just with allylic Grignards, and is often called the Barbier method.

For example, it is a straightforward matter to make these three alcohols, provided the allylic halide, aldehyde, and magnesium are all mixed together in one flask. The Grignard reagent forms, and immediately reacts with the aldehyde, before it has a chance to dimerize. In the second example, notice again that the allylic Grignard reagent must have reacted through a six-membered transition state because the allyl system has 'turned around' in the product.

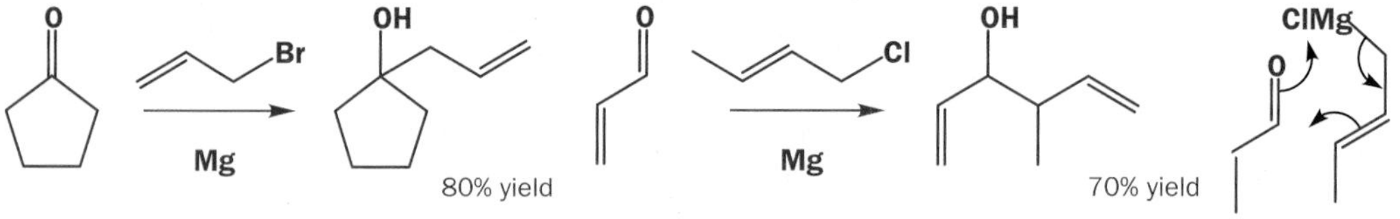

80% yield   70% yield

The last reaction above leads us nicely into the next chapter where we will look at an alternative way for such unsaturated aldehydes to react—by *conjugate addition.*

Problems

1. Propose mechanisms for the first four reactions in the chapter.

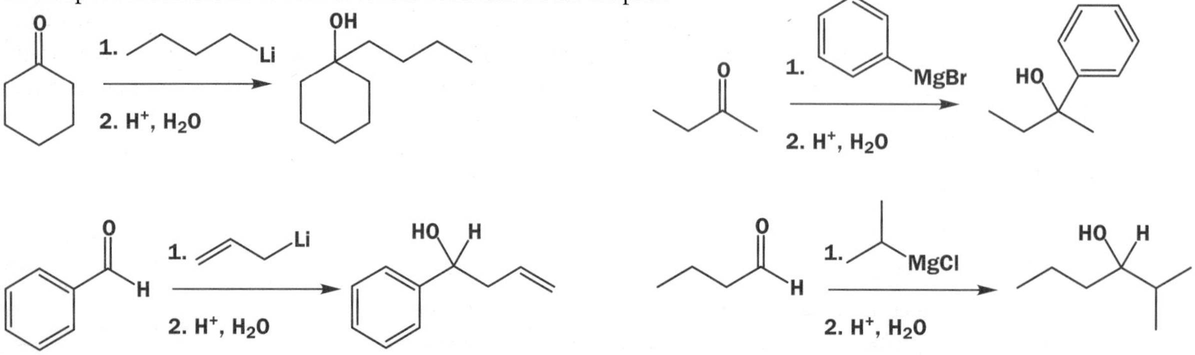

2. When this reaction is carried out with allyl bromide labelled as shown with ^{13}C, the label is found equally distributed between the ends of the allyl system in the product. Explain how this is possible. How would you detect the ^{13}C distribution in the product?

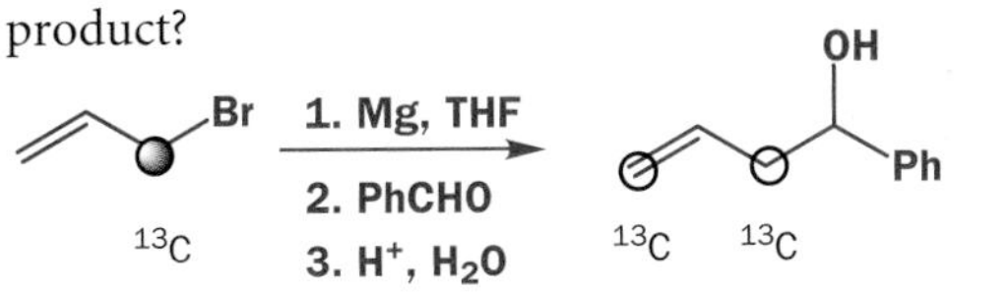

3. What products would be formed in these reactions?

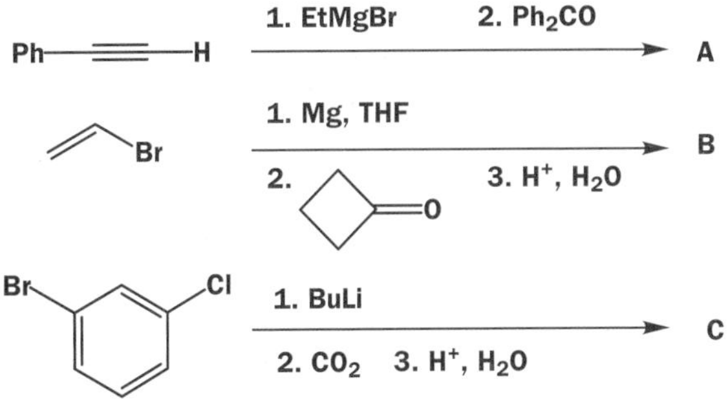

4. Suggest alternative routes to fenarimol—that is, different routes from the one shown in the chapter (p. 216).

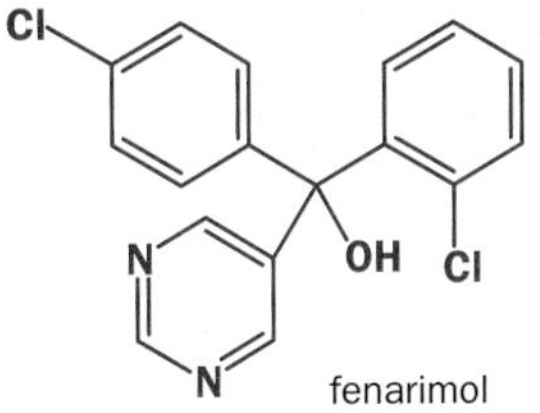

fenarimol

5. The synthesis of the gastric antisecretory drug rioprostil requires this alcohol.

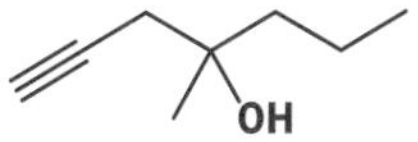

(**a**) Suggest possible syntheses starting from ketones and organometallics and (**b**) suggest possible syntheses of the ketones in part (**a**) from aldehydes and organometallics (don't forget about CrO_3 oxidation!).

6. Suggest two syntheses of the bee pheromone heptan-2-one.

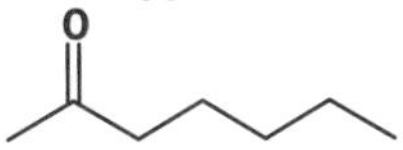

heptan-2-one

7. How could you prepare these compounds using *ortho*-lithiation procedures?

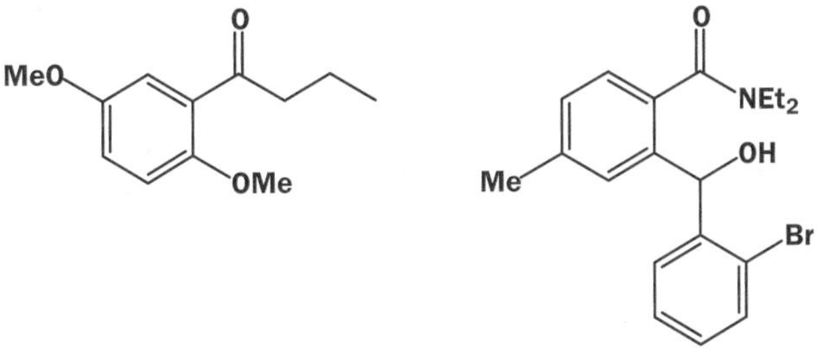

8. Why is it possible to make the lithium derivative A by Br/Li exchange, but not the lithium derivative B?

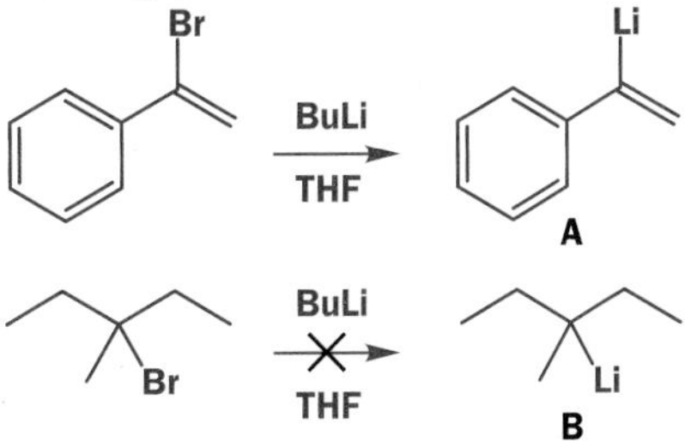

9. Comment on the selectivity (that is, say what else might have happened and why it didn't) shown in this Grignard addition reaction used in the manufacture of an antihistamine drug.

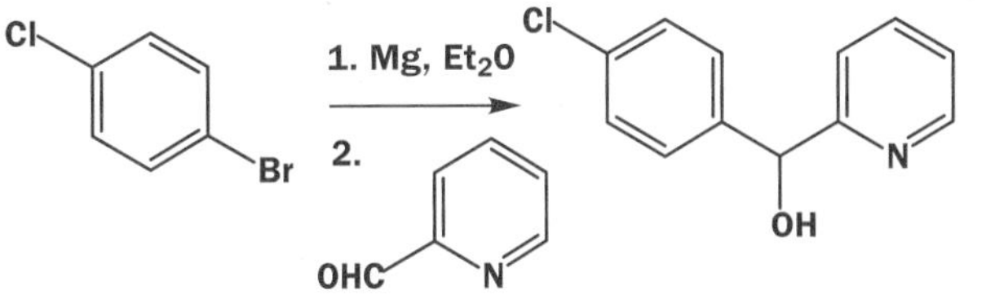

10. The antispasmodic drug biperidin is made by the Grignard addition reaction shown here. What is the structure of the drug? Do not be put off by the apparent complexity of the compounds—the chemistry is the same as that you have seen in this chapter. How would you suggest that the drug procyclidine should be made?

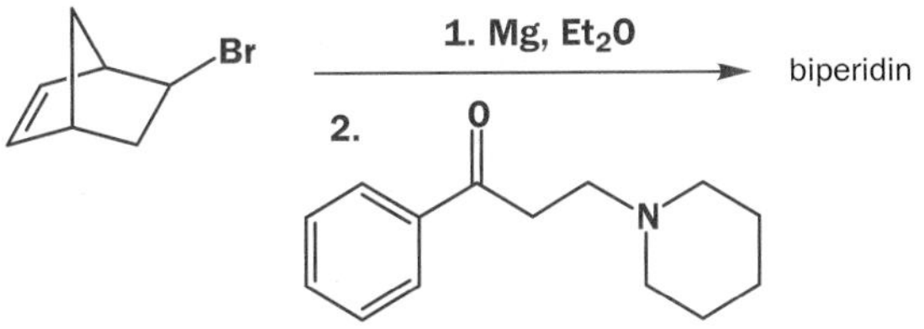

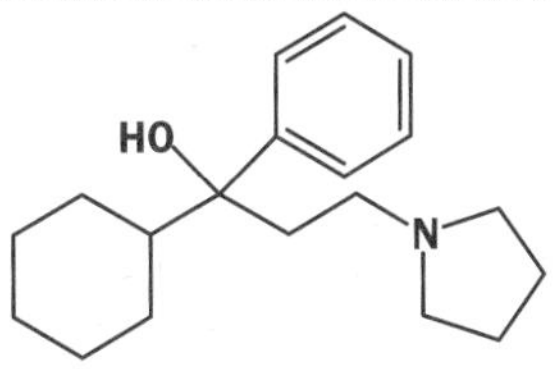

procyclidine

11. Though heterocyclic compounds, such as the nitrogen ring system in this question, are introduced rather later in this book, use your knowledge of Grignard chemistry to draw a mechanism for what happens here. It is important that you prove to yourself that you can draw mechanisms for reactions on compounds that you have never met before.

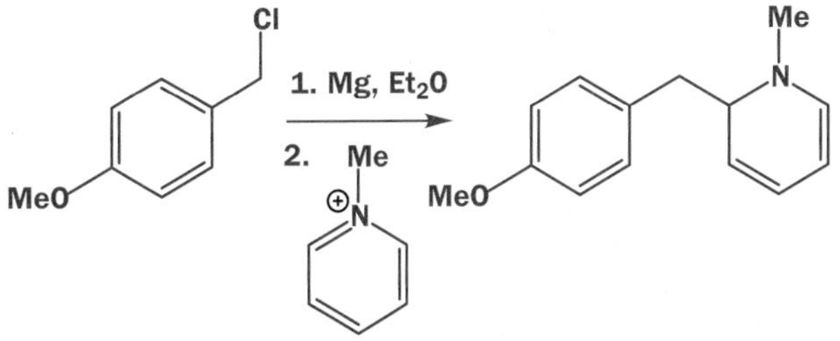

Conjugate addition

Connections

Building on:	Arriving at:	Looking forward to:
● Reactions of C=O groups ch6 & ch9 ● Conjugation ch7	● How conjugation affects reactivity ● What happens to a C=O group when it is conjugated with a C=C bond ● How the C=C double bond becomes electrophilic, and can be attacked by nucleophiles ● Why some sorts of nucleophiles attack C=C while others still attack the C=O group	● Conjugate addition in other electrophilic alkenes ch23 ● Conjugate addition with further types of nucleophiles ch29 ● Alkenes that are *not* conjugated with C=O ch20

Conjugation changes the reactivity of carbonyl groups

To start this chapter, here are four reactions of the same ketone. For each product, the principal absorptions in the IR spectrum are listed. The pair of reactions on the left should come as no surprise to you: nucleophilic addition of cyanide or a Grignard reagent to the ketone produces a product with no C=O peak near 1700 cm^{-1}, but instead an O–H peak at 3600 cm^{-1}. The 2250 cm^{-1} peak is C≡N; C=C is at 1650 cm^{-1}.

■ If you need to review IR spectroscopy, turn back to Chapter 3. Chapter 6 dealt with addition of CN$^-$ to carbonyl compounds, and Chapter 9 with the addition of Grignard reagents.

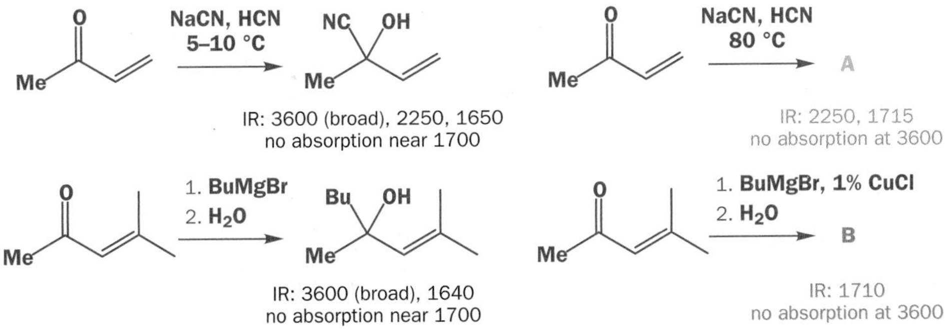

IR: 3600 (broad), 2250, 1650
no absorption near 1700

IR: 2250, 1715
no absorption at 3600

IR: 3600 (broad), 1640
no absorption near 1700

IR: 1710
no absorption at 3600

But what about the reactions on the right? Both products A and B have kept their carbonyl group (IR peak at 1710 cm^{-1}) but have lost the C=C. Yet A, at least, is definitely an addition product because it contains a C≡N peak at 2200 cm^{-1}.

Well, the identities of A and B are revealed here: they are the products of addition, not to the carbonyl group, but to the C=C bond. This type of reaction is called conjugate addition, and is what this chapter is all about. The chapter will also explain how such small differences in reaction conditions (temperature, or the presence of CuCl) manage to change the outcome completely.

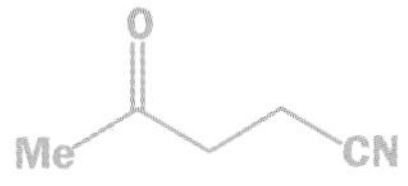

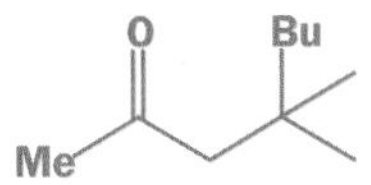

direct addition to the C=O group

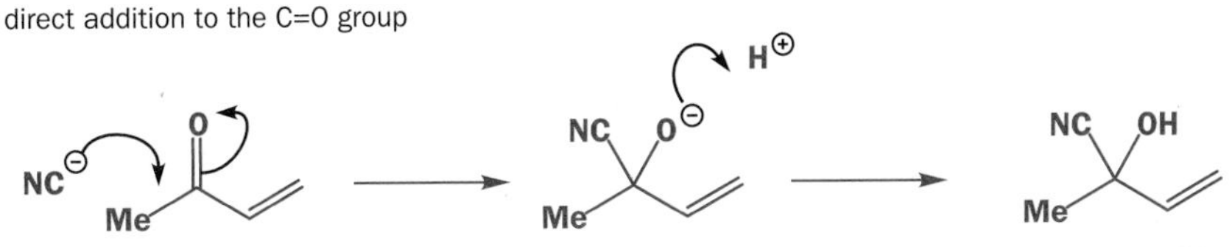

conjugate addition to the C=C double bond

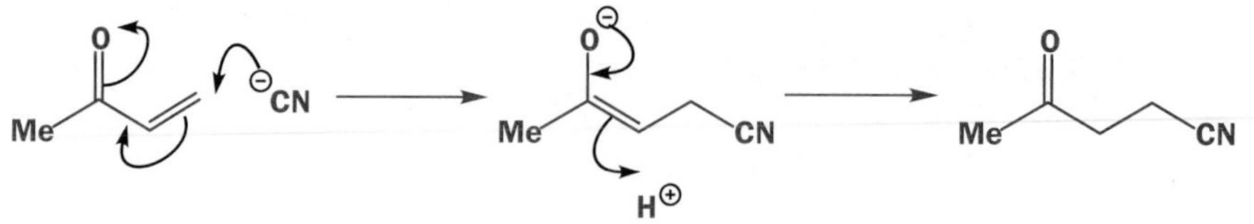

Conjugate addition to the C=C double bond follows a similar course to direct addition to the C=O group, and the mechanisms for both are shown here. Both mechanisms have two steps: addition, followed by protonation. Conjugate additions only occur to C=C double bonds next to C=O groups. They don't occur to C=C bonds that aren't immediately adjacent to C=O (see the box on p. 229 for an example).

Compounds with double bonds adjacent to a C=O group are known as **α,β-unsaturated carbonyl compounds**. Many α,β-unsaturated carbonyl compounds have trivial names, and some are shown here. Some *classes* of α,β-unsaturated carbonyl compounds also have names such as 'enone' or 'enal', made up of 'ene' (for the double bond) + 'one' (for ketone) or 'ene' + 'al' (for aldehyde).

▸ The α and β refer to the distance of the double bond from the C=O group: the α carbon is the one next to C=O (*not* the carbonyl carbon itself), the β carbon is one further down the chain, and so on.

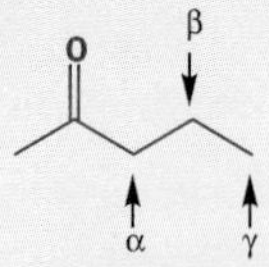

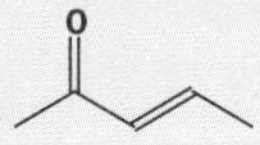

α,β-unsaturated ketone

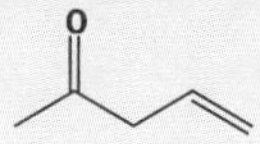

β,γ-unsaturated ketone

an α,β-unsaturated aldehyde (an enal)

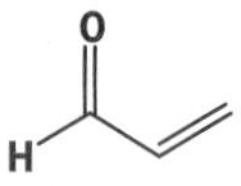

propenal (trivial name = acrolein)

an α,β-unsaturated ketone (an enone)

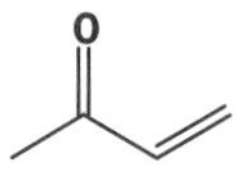

but-3-en-2-one (trivial name = methyl vinyl ketone)

an α,β-unsaturated acid

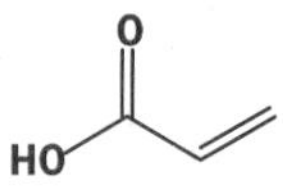

propenoic acid (trivial name = acrylic acid)

an α,β-unsaturated ester

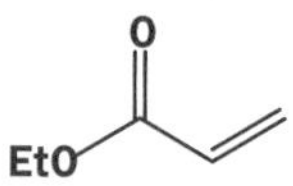

ethyl propenoate (trivial name = ethyl acrylate)

A range of nucleophiles will undergo conjugate additions with α,β-unsaturated carbonyl compounds, and six examples are shown below. Note the range of nucleophiles, and also the range of carbonyl compounds: esters, aldehydes, acids, and ketones.

types of nucleophile which undergo conjugate addition

cyanide **KCN** +

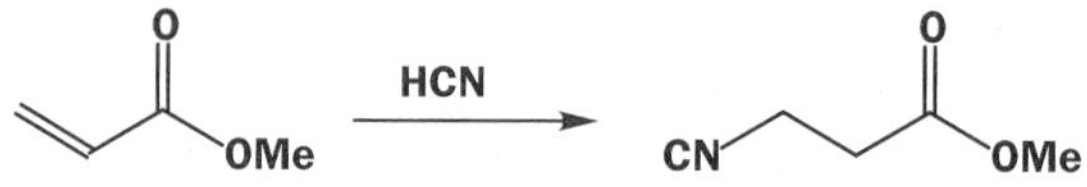

amines **Et_2NH** +

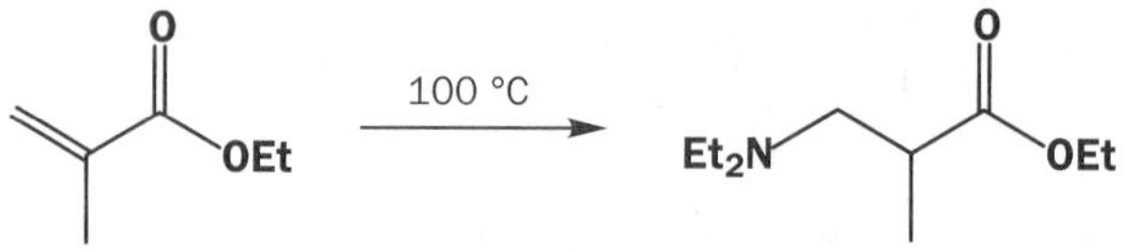

alcohols **MeOH** +

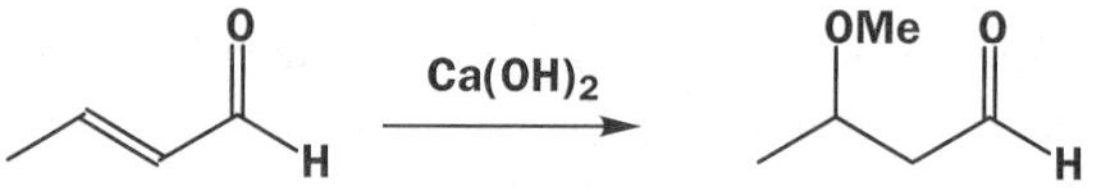

thiols **MeSH** +

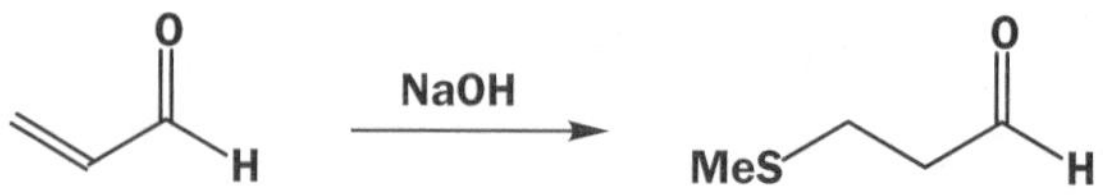

bromide **HBr** +

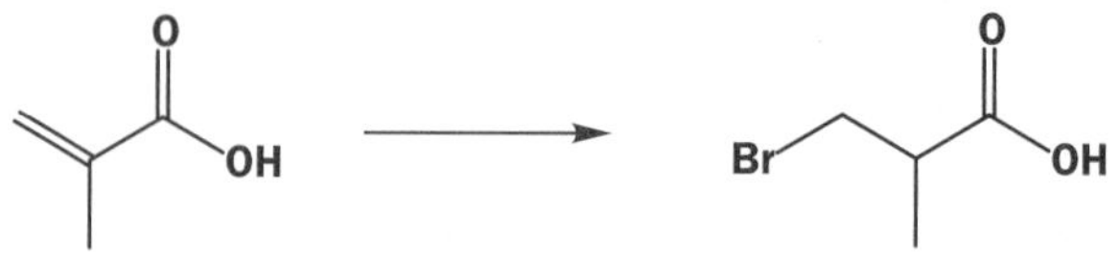

chloride **HCl** +

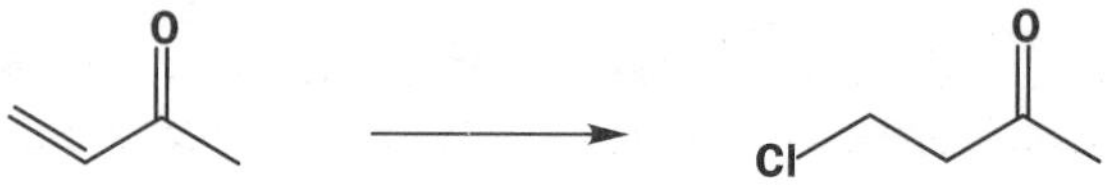

The reason that α,β-unsaturated carbonyl compounds react differently is conjugation, the phenomenon we discussed in Chapter 7. There we introduced you to the idea that bringing two π systems (two C=C bonds, for example, or a C=C bond and a C=O bond) close together leads to a stabilizing interaction. It also leads to modified reactivity, beacause the π bonds no longer react as independent functional groups but as a single, conjugated system.

Termite self-defence and the reactivity of alkenes

Soldier termites of the species *Schedorhinotermes lamanianus* defend their nests by producing this compound, which is very effective at taking part in conjugate addition reactions with thiols (RSH). This makes it highly toxic, since many important biochemicals carry SH groups. The worker termites of the same species—who build the nests—need to be able to avoid being caught in the crossfire, so they are equipped with an enzyme that allows them to reduce compound 1 to compound 2. This still has a double bond, but the double bond is completely unreactive towards nucleophiles because it is not conjugated with a carbonyl group. The workers escape unharmed.

Alkenes conjugated with carbonyl groups are polarized

You haven't met many reactions of alkenes yet: detailed discussion will have to wait till Chapter 20. But we did indicate in Chapter 5 that they react with electrophiles. Here is the example from p. 124: in the addition of HBr to isobutene the alkene acts as a nucleophile and H–Br as the electrophile.

This is quite different to the reactivity of a C=C double bond conjugated with a carbonyl group, which, as you have just seen, reacts with nucleophiles such as cyanide, amines, and alcohols. The conjugated system is different from the sum of the isolated parts, with the C=O group profoundly affecting the reactivity of the C=C double bond. To show why, we can use curly arrows to indicate delocalization of the π electrons over the four atoms in the conjugated system. Both representations are extremes, and the true structure lies somewhere in between, but the polarized structure indicates why the conjugated C=C bond is electrophilic.

curly arrows indicate delocalization of electrons

true electron distribution lies somewhere in between these extremes

▶
You may be asking yourself why we can't show the delocalization by moving the electrons the other way, like this.

Think about electronegativities: O is much more electronegative than C, so it is quite happy to accept electrons, but here we have taken electrons away, leaving it with only six electrons. This structure therefore cannot represent what happens to the electrons in the conjugated system.

● Conjugation makes alkenes electrophilic

- • Isolated C=C double bonds are nucleophilic

- • C=C double bonds conjugated with carbonyl groups are electrophilic

Polarization is detectable spectroscopically

IR spectroscopy provides us with evidence for polarization in C=C bonds conjugated to C=O bonds. An unconjugated ketone C=O absorbs at 1715 cm^{-1} while an unconjugated alkene C=C absorbs (usually rather weakly) at about 1650 cm^{-1}. Bringing these two groups into conjugation in an α,β-unsaturated carbonyl compound leads to two peaks at 1675 and 1615 cm^{-1}, respectively, both quite strong. The lowering of the frequency of both peaks is consistent with a weakening of both π bonds (notice that the polarized structure has only single bonds where the C=O and C=C double bonds were). The increase in the *intensity* of the C=C absorption is consistent with polarization brought about by conjugation with C=O: a conjugated C=C bond has a significantly larger dipole moment than its unconjugated cousins.

The polarization of the C=C bond is also evident in the ^{13}C NMR spectrum, with the signal for the sp^2 carbon atom furthest from the carbonyl group moving downfield relative to an unconjugated alkene to about 140 p.p.m., and the signal for the other double bond carbon atom staying at about 120 p.p.m.

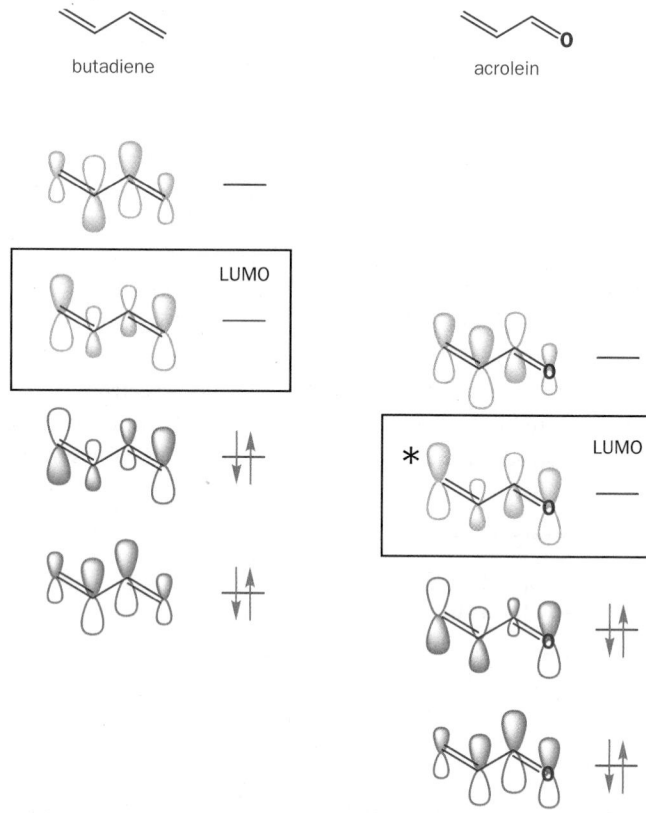

143 p.p.m. 132 p.p.m.

compared with

124 p.p.m. 119 p.p.m.

Molecular orbitals control conjugate additions

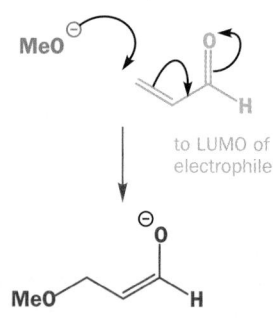

electrons must move from HOMO of nucleophile

MeO⁻

to LUMO of electrophile

We have spectroscopic evidence that a conjugated C=C bond is polarized, and we can explain this with curly arrows, but the actual bond-forming step must involve movement of electrons from the HOMO of the nucleophile to the LUMO of the unsaturated carbonyl compound. The example in the margin has methoxide (MeO⁻) as the nucleophile.

But what does this LUMO look like? It will certainly be more complicated than the π* LUMO of a simple carbonyl group. The nearest thing you have met so far (in Chapter 7) are the orbitals of butadiene (C=C conjugated with C=C), which we can compare with the α,β-unsaturated aldehyde acrolein (C=C conjugated with C=O). The orbitals in the π systems of butadiene and acrolein are shown here. They are different because acrolein's orbitals are perturbed (distorted) by the oxygen atom (Chapter 4). You need not be concerned with exactly how the sizes of the orbitals are worked out, but for the moment just concentrate on the shape of the LUMO, the orbital that will accept electrons when a nucleophile attacks.

butadiene

acrolein

LUMO

LUMO

*

> In acrolein, the HOMO is in fact not the highest filled π orbital you see here, but the lone pairs on oxygen. This is not important here, though, because we are only considering acrolein as an electrophile, so we are only interested in its LUMO.

In the LUMO, the largest coefficient is on the β carbon of the α,β-unsaturated system, shown with an asterisk. And it is here, therefore, that nucleophiles attack. In the reaction you have just seen, the HOMO is the methoxide oxygen's lone pair, so this will be the key orbital interaction that gives rise to the new bond. The second largest coefficient is on the C=O carbon atom, so it's not surprising that some nucleophiles attack here as well—remember the example right at the beginning of the chapter where you saw cyanide attacking either the double bond or the carbonyl group depending on the conditions of the reaction. We shall next look at some conjugate additions with alcohols and amines as nucleophiles, before reconsidering the question of where the nucleophile attacks.

Ammonia and amines undergo conjugate addition

Amines are good nucleophiles for conjugate addition reactions, and give products that we can term **β-amino carbonyl compounds** (the new amino group is β to the carbonyl group). Dimethylamine is a gas at room temperature, and this reaction has to be carried out in a sealed system to give the ketone product.

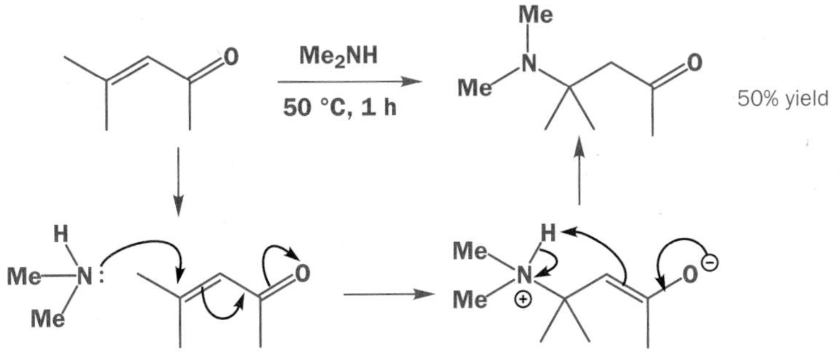

This is the first conjugate addition mechanism we have shown you that involves a neutral nucleophile: as the nitrogen adds it becomes positively charged and therefore needs to lose a proton. We can use this proton to protonate the negatively charged part of the molecule as you have seen happening before. This proton-transfer step can alternatively be carried out by a base: in this addition of butylamine to an α,β-unsaturated ester (ethyl acrylate), the added base (EtO⁻) deprotonates the nitrogen atom once the amine has added. Only a catalytic amount is needed, because it is regenerated in the step that follows.

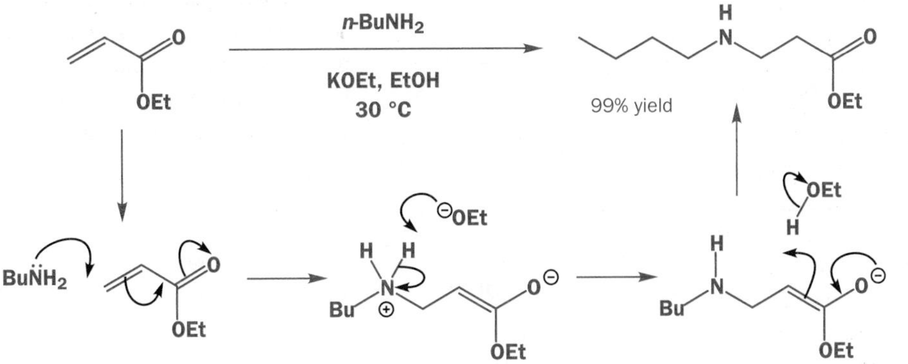

Ammonia itself, the simplest amine, is very volatile (it is a gas at room temperature, but a very water-soluble one, and bottles of 'ammonia' are actually a concentrated aqueous solution of ammonia), and the high temperatures required for conjugate addition to this unsaturated carboxylic acid can only be achieved in a sealed reaction vessel.

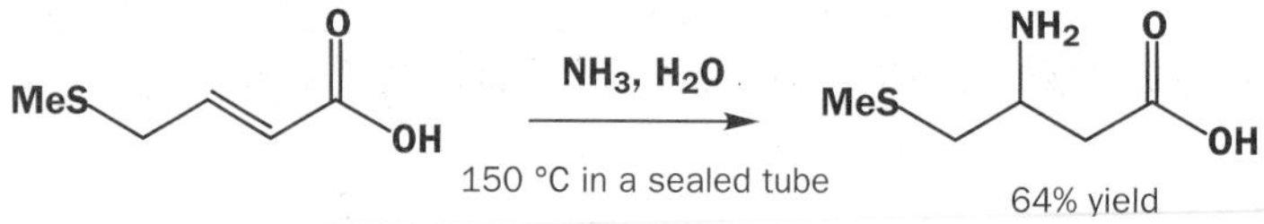

Amines are bases as well as nucleophiles, and in this reaction the first step must be deprotonation of the carboxylic acid: it's the ammonium carboxylate that undergoes the addition reaction. You would not expect a negatively charged carboxylate to be a very good electrophile, and this may well be why ammonia needs 150 °C to react.

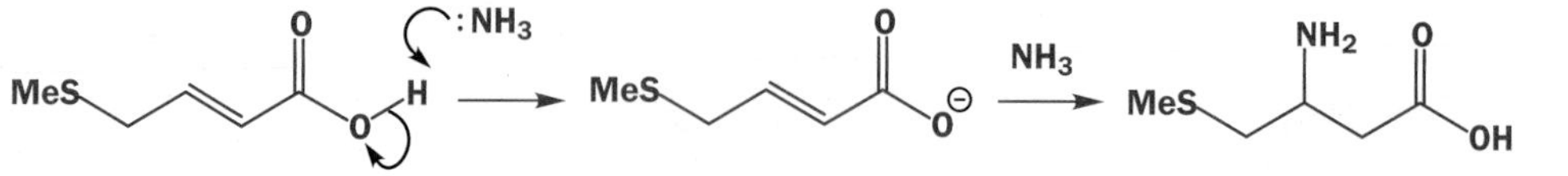

The β-amino carbonyl product of conjugate addition of an amine is still an amine and, provided it has a primary or secondary amino group, it can do a second conjugate addition. For example, methylamine adds successively to two molecules of this unsaturated ester.

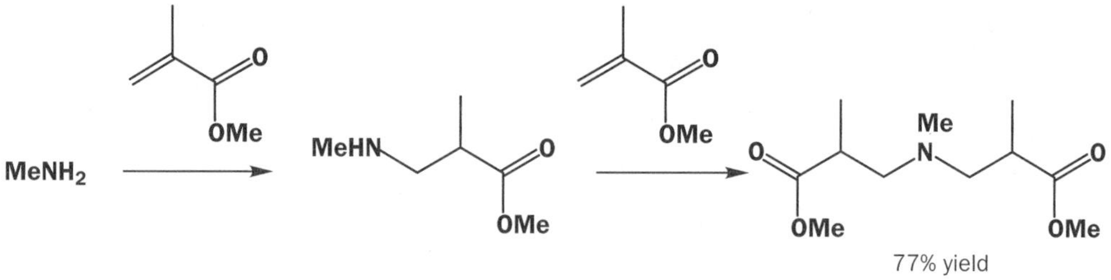

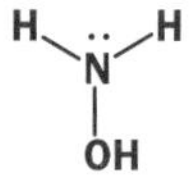

> ► Tertiary amines can't give conjugate addition products because they have no proton to lose.

Two successive conjugate additions can even happen in the same molecule. In the next example, hydroxylamine is the nucleophile. Hydroxylamine is both an amine and an alcohol, but it always reacts at nitrogen because nitrogen (being less electronegative than oxygen) has a higher-energy (more reactive) lone pair. Here it reacts with a cyclic dienone to produce a bicyclic ketone, which we have also drawn in a perspective view to give a better idea of its shape.

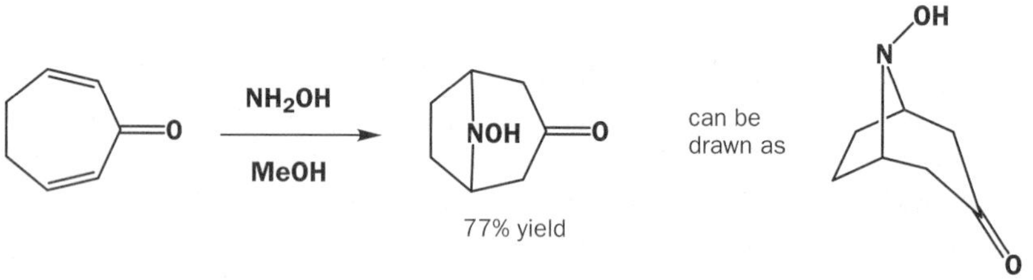

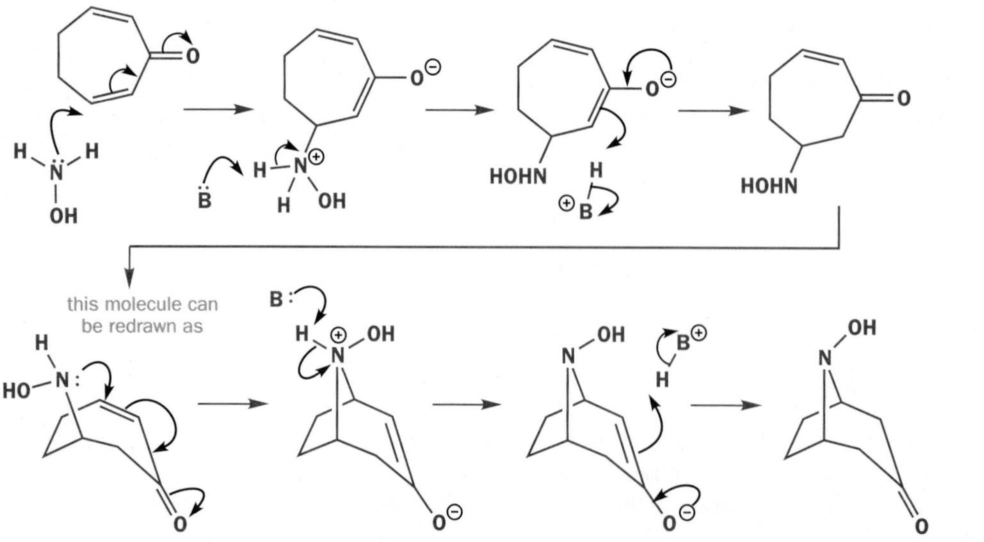

hydroxylamine

The reaction sequence consists of two conjugate addition reactions. The first is intermolecular, and gives the intermediate enone. The second conjugate addition is intramolecular, and turns the molecule into a bicyclic structure. Again, the most important steps are the C–N bond-forming reactions, but there are also several proton transfers that have to occur. We have shown a base 'B:' carrying out these proton transfers: this might be a molecule of hydroxylamine, or it might be a molecule of the solvent, methanol. These details do not matter.

Conjugate addition of alcohols can be catalysed by acid or base

Alcohols undergo conjugate addition only very slowly in the absence of a catalyst: they are not such good nucleophiles as amines for the very reason we have just mentioned in connection with the reactivity of hydroxylamine—oxygen is more electronegative than nitrogen, and so its lone pairs are of lower energy and are therefore less reactive. Alkoxide *anions* are, however, much more nucleophilic. You saw methoxide attacking the orbitals of acrolein above: the reaction in the margin goes at less than 5 °C.

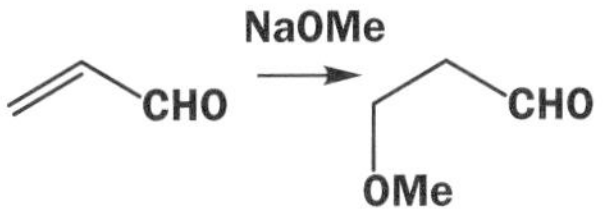

The alkoxide doesn't have to be made first, though, because alcohols dissolved in basic solution are at least partly deprotonated to give alkoxide anions. How much alkoxide is present depends on the pH of the solution and therefore the pK_a of the base (Chapter 8), but even a tiny amount is acceptable because once this has added it will be replaced by more alkoxide in acid–base equilibrium with the alcohol. In this example, allyl alcohol adds to pent-2-enal, catalysed by sodium hydroxide in the presence of a buffer.

■ In Chapter 6 we discussed the role of base and acid catalysts in the direct addition of alcohols to carbonyl compounds to form hemiacetals. The reasoning—that base makes nucleophiles more nucleophilic and acid makes carbonyl groups more electrophilic—is the same here.

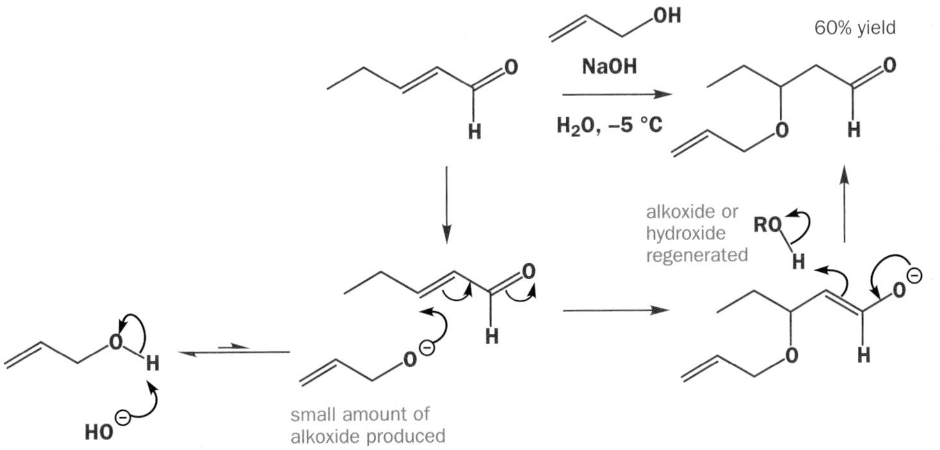

Only a catalytic amount of base is required as the deprotonation of ROH (which can be water or allyl alcohol) in the last step regenerates more alkoxide or hydroxide. It does not matter that sodium hydroxide (pK_{aH} 15.7) is not basic enough to deprotonate an alcohol (pK_a 16–17) completely, since only a small concentration of the reactive alkoxide is necessary for the reaction to proceed.

We can also make rings using alkoxide nucleophiles, and in this example the phenol (hydroxybenzene) is deprotonated by the sodium methoxide base to give a phenoxide anion. Intramolecular attack on the conjugated ketone gives the cyclic product in excellent yield. In this case, the methoxide (pK_{aH} about 16) will deprotonate the phenol (pK_a about 10) completely, and competitive attack by MeO^- acting as a nucleophile is not a problem as intramolecular reactions are usually faster than their intermolecular equivalents.

■ There are some important exceptions to this depending on the size of ring being formed, and some of these are described in Chapter 42.

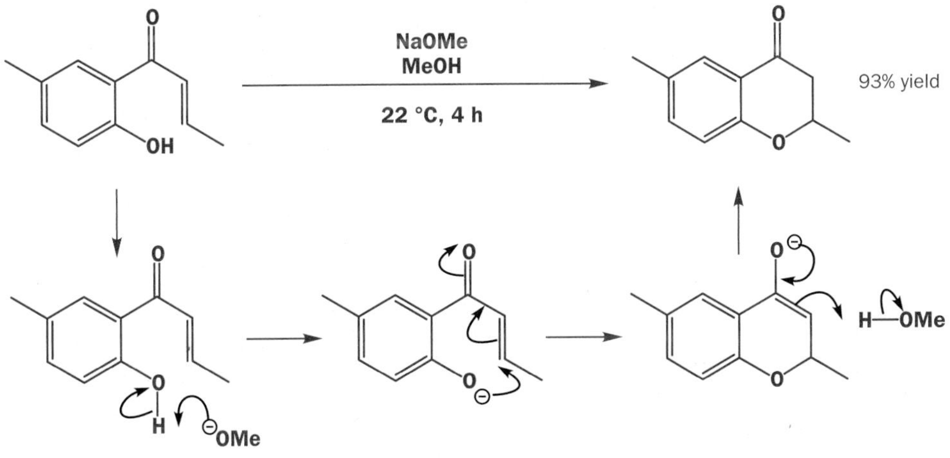

Acid catalysts promote conjugate addition of alcohols to α,β-unsaturated carbonyl compounds by protonating the carbonyl group and making the conjugated system more electrophilic. Methanol adds to this ketone exceptionally well, for example, in the presence of an acid catalyst known as 'Dowex 50'. This is an acidic resin—just about as acidic as sulfuric acid in fact, but completely insoluble, and therefore very easy to remove from the product at the end of the reaction by filtration.

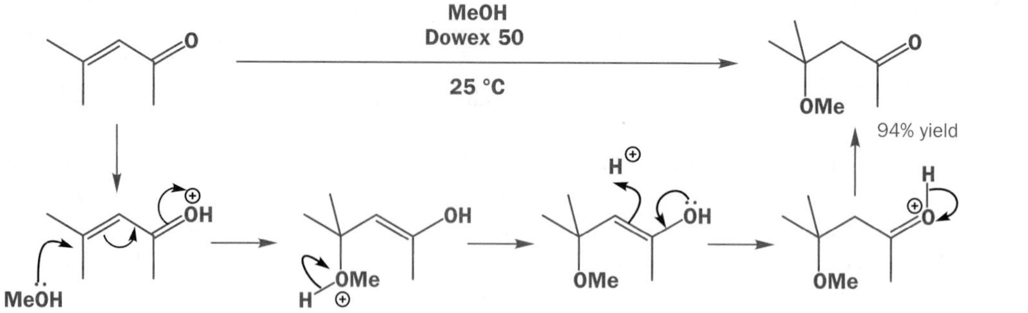

Once the methanol has added to the protonated enone, all that remains is to reorganize the protons in the molecule to give the product. This takes a few steps, but don't be put off by their complexity—as we've said before, the important step is the first one—the conjugate addition.

Conjugate addition or direct addition to the carbonyl group?

We have shown you several examples of conjugate additions using various nucleophiles and α,β-unsaturated carbonyl compounds, but we haven't yet addressed one important question. When do nucleophiles do conjugate addition (also called '1,4-addition') and when do they add directly to the carbonyl group ('1,2-addition')? Several factors are involved—they are summarized here, and we will spend the next section of this chapter discussing them in turn.

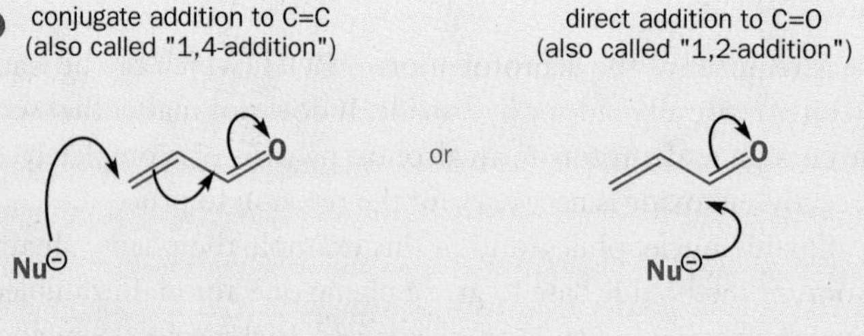

The way that nucleophiles react depends on:
- the conditions of the reaction
- the nature of the α, β-unsaturated carbonyl compound
- the type of nucleophile

Reaction conditions

The very first conjugate addition reaction in this chapter depended on the conditions of the reaction. Treating an enone with cyanide and an acid catalyst at low temperature gives a cyanohydrin by direct attack at C=O, while heating the reaction mixture leads to conjugate addition. What is going on?

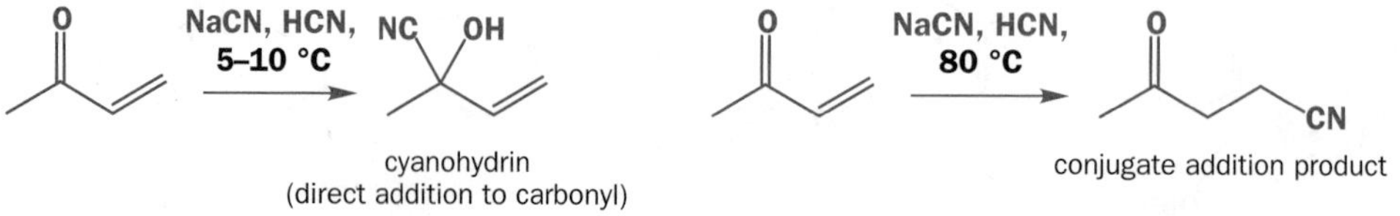

We'll consider the low-temperature reaction first. As you know from Chapter 6, it is quite normal for cyanide to react with a ketone under these conditions to form a cyanohydrin. Direct addition to the carbonyl group turns out to be faster than conjugate addition, so we end up with the cyanohydrin.

conjugate addition product

slow but irreversible

fast but reversible

cyanohydrin

thermodynamic product:
more stable

kinetic product:
forms faster

Now, you also know from Chapter 6 that cyanohydrin formation is reversible. Even if the equilibrium for cyanohydrin formation lies well over to the side of the products, at equilibrium there will still be a small amount of starting enone remaining. Most of the time, this enone will react to form more cyanohydrin and, as it does, some cyanohydrin will decompose back to enone plus cyanide—such is the nature of a dynamic equilibrium. But every now and then—at a much slower rate—the starting enone will undergo a conjugate addition with the cyanide. Now we have a different situation: conjugate addition is essentially an *irreversible* reaction, so once a molecule of enone has been converted to conjugate addition product, its fate is sealed: it cannot go back to enone again. Very slowly, therefore, the amount of conjugate addition product in the mixture will build up. In order for the enone–cyanohydrin equilibrium to be maintained, any enone that is converted to conjugate addition product will have to be replaced by reversion of cyanohydrin to enone plus cyanide. Even at room temperature, we can therefore expect the cyanohydrin to be converted bit by bit to conjugate addition product. This may take a very long time, but reaction rates are faster at higher temperatures, so at 80 °C this process does not take long at all and, after a few hours, the cyanohydrin has all been converted to conjugate addition product.

The contrast between the two products is this: cyanohydrin is formed faster than the conjugate addition product, but the conjugate addition product is the more stable compound.

Typically, kinetic control involves lower temperatures and shorter reaction times, which ensures that only the fastest reaction has the chance to occur. And, typically, thermodynamic control involves higher temperatures and long reaction times to ensure that even the slower reactions have a chance to occur, and all the material is converted to the most stable compound.

● **Kinetic and thermodynamic control**

- The product that forms faster is called the **kinetic product**
- The product that is the more stable is called the **thermodynamic product**
  Similarly,
- Conditions that give rise to the kinetic product are called **kinetic control**
- Conditions that give rise to the thermodynamic product are called **thermodynamic control**

Why is direct addition faster than conjugate addition? Well, although the carbon atom β to the C=O group carries some positive charge, the carbon atom of the carbonyl group carries more, and so electrostatic attraction for the charged nucleophiles will encourage it to attack the carbonyl group directly rather than undergo conjugate addition.

attack is possible at
either site

but electrostatic attraction to
C=O is greater

LUMO

And why is the conjugate addition product the more stable? In the conjugate addition product, we gain a C–C σ bond, losing a C=C π bond, but keeping the C=O π bond. With direct addition, we still gain a C–C bond, but we lose the C=O π bond and keep the C=C π bond. C=O π bonds are stronger than C=C π bonds, so the conjugate addition product is the more stable.

lose C=O π bond
369 kJ mol⁻¹

gain C–C σ bond

gain C–C σ bond

lose C=C π bond
280 kJ mol⁻¹

We will return to kinetic and thermodynamic control in Chapter 13, where we will analyse the rates and energies involved a little more rigorously, but for now here is an example where conjugate addition is ensured by thermodynamic control. Note the temperature!

HCN, KCN
160 °C

75% yield

Structural factors

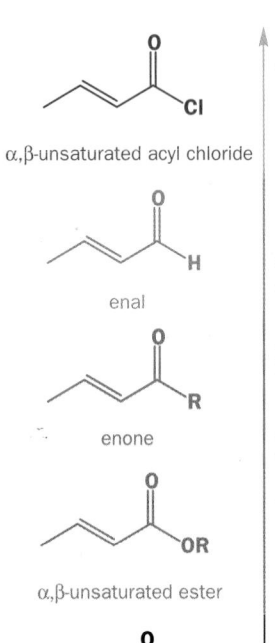

α,β-unsaturated acyl chloride

enal

enone

α,β-unsaturated ester

α,β-unsaturated amide

Not all additions to carbonyl groups are reversible: additions of organometallics, for example, are certainly not. In such cases, the site of nucleophilic attack is determined simply by reactivity: the more reactive the carbonyl group, the more addition to C=O will result. The most reactive carbonyl groups, as you will see in Chapter 12, are those that are not conjugated with O or N (as they are in esters and amides), and particularly reactive are acyl chlorides and aldehydes. In general, the proportion of direct addition to the carbonyl group follows the reactivity sequence in the margin.

Compare the way butyllithium adds to this α,β-unsaturated aldehyde and α,β-unsaturated amide. Both additions are irreversible, and BuLi attacks the reactive carbonyl group of the aldehyde, but prefers conjugate addition to the less reactive amide. Similarly, ammonia reacts with this acyl chloride to give an amide product that derives (for details see Chapter 12) from direct addition to the carbonyl group, while with the ester it undergoes conjugate addition to give an amine.

1. BuLi, –70 °C to +20 °C
2. H₂O

1. BuLi, –70 °C to +20 °C
2. H₂O

NH₃

NH₃

Sodium borohydride is a nucleophile that you have seen reducing simple aldehydes and ketones to alcohols, and it usually reacts with α,β-unsaturated aldehydes in a similar way, giving alcohols by direct addition to the carbonyl group.

NaBH₄, EtOH

97% yield

NaBH₄, EtOH

99% yield

Quite common with ketones, though, is the outcome on the right. The borohydride has reduced not only the carbonyl group but the double bond as well. In fact, it's the double bond that's reduced first in a conjugate addition, followed by addition to the carbonyl group.

This reaction, and how to control reduction of C=O and C=C, will be discussed in more detail in Chapter 24.

For esters and other less reactive carbonyl compounds conjugate addition is the only reaction that occurs.

Steric hindrance also has a role to play: the more substituents there are at the β carbon, the less likely a nucleophile is to attack there. Nonetheless, there are plenty of examples where nucleophiles undergo conjugate addition even to highly substituted carbon atoms.

The concept of steric hindrance was introduced in Chapter 6.

The nature of the nucleophile: hard and soft

Among the best nucleophiles of all at doing conjugate addition are **thiols**, the sulfur analogues of alcohols. In this example, the nucleophile is thiophenol (phenol with the O replaced by S). Remarkably, no acid or base catalyst is needed (as it was with the alcohol additions), and the product is obtained in 94% yield under quite mild reaction conditions.

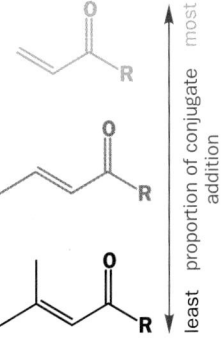

Why are thiols such good nucleophiles for conjugate additions? Well, to explain this, and why they are much less good at direct addition to the C=O group, we need to remind you of some ideas we introduced in Chapter 5. There we said that the attraction between nucleophiles and electrophiles is governed by two related interactions—electrostatic attraction between positive and negative charges and orbital overlap between the HOMO of the nucleophile and the LUMO of the electrophile. Successful reactions usually result from a combination of both, but sometimes reactivity can be dominated by one or the other. The dominant factor, be it electrostatic or orbital control, depends on the nucleophile and electrophile involved. Nucleophiles containing small, electronegative atoms (such as O or Cl) tend to react under predominantly electrostatic control, while nuclophiles containing larger atoms (including the sulfur of thiols, but also P, I, and Se) are predominantly subject to control by orbital overlap. The terms 'hard' and 'soft' have been coined to describe these two types of reagents. **Hard nucleophiles** are typically from the early rows of the periodic table and have higher charge density, while **soft nucleophiles** are from the later rows of the periodic table—they are either uncharged or have larger atoms with higher-energy, more diffuse orbitals.

Table 10.1 divides some nucleophiles into the two categories (plus some that lie in between)—but don't try to learn it! Rather, convince yourself that the properties of each one justify its location in the table. Most of these nucleophiles you have not yet seen in action, and the most important ones at this stage are indicated in **bold type**.

Table 10.1 Hard and soft nucleophiles

Hard nucleophiles	Borderline	Soft nucleophiles
F^-, **OH^-**, **RO^-**, SO_4^{2-}, Cl^-,	N_3^-, **CN^-**	I^-, **RS^-**, RSe^-, S^{2-}
H_2O, **ROH**, ROR', $RCOR'$,	**RNH_2**, **$RR'NH$**,	**RSH**, RSR', R_3P
NH_3, **$RMgBr$**, **RLi**	Br^-	alkenes, aromatic rings

Not only can nucleophiles be classified as hard or soft, but electrophiles can too. For example, H^+ is a very hard electrophile because it is small and charged, while Br_2 is a soft electrophile: its orbitals are diffuse and it is uncharged. You saw Br_2 reacting with an alkene earlier in the chapter, and we explained in Chapter 5 that this reaction happens solely because of orbital interactions: no charges are involved. The carbon atom of a carbonyl group is also a hard electrophile because it carries a partial positive charge due to polarization of the C=O bond. What is important to us is that, in general, hard nucleophiles prefer to react with hard electrophiles, and soft nucleophiles with soft electrophiles. So, for example, water (a hard nucleophile) reacts with aldehydes (hard electrophiles) to form hydrates in a reaction largely controlled by electrostatic attraction. On the other hand, water does not react with bromine (a soft electrophile). Yet bromine reacts with alkenes while water does not. Now this is only a very general principle, and you will find plenty of examples where hard reacts with soft and soft with hard. Nonetheless it is a useful concept, which we shall come back to later in the book.

● **Hard/soft reactivity**
  - Reactions of hard species are dominated by charges and electrostatic effects
  - Reactions of soft species are dominated by orbital effects
  - Hard nucleophiles tend to react well with hard electrophiles
  - Soft nucleophiles tend to react well with soft electrophiles

What has all this to do with the conjugate addition of thiols? Well, an α,β-unsaturated carbonyl compound is unusual in that it has two electrophilic sites, one of which is hard and one of which is soft. The carbonyl group has a high partial charge on the carbonyl carbon and will tend to react with hard nucleophiles, such as organolithium and Grignard reagents, that have a high partial charge on the nucleophilic carbon atom. Conversely, the β carbon of the α,β-unsaturated carbonyl system does not have a high partial positive charge but is the site of the largest coefficient in the LUMO. This makes the β carbon a soft electrophile and likely to react well with soft nucleophiles such as thiols.

● **Hard/soft—direct/conjugate addition**
  - Hard nucleophiles tend to react at the carbonyl carbon (hard) of an enone
  - Soft nucleophiles tend to react at the β-carbon (soft) of an enone and lead to conjugate addition

Anticancer drugs that work by conjugate addition of thiols

helenalin

vernolepin

Drugs to combat cancer act on a range of biochemical pathways, but most commonly on processes that cancerous cells need to use to proliferate rapidly. One class attacks DNA polymerase, an enzyme needed to make the copy of DNA that has to be provided for each new cell. Helenalin and vernolepin are two such drugs, and if you look closely at their structure you should be able to spot two α,β-unsaturated carbonyl groups in each. Biochemistry is just chemistry in very small flasks called cells, and the reaction between DNA polymerase and these drugs is simply a conjugate addition reaction between a thiol (the SH group of one of the enzyme's cysteine residues) and the unsaturated carbonyl groups. The reaction is irreversible, and shuts down completely the function of the enzyme.

Copper(I) salts have a remarkable effect on organometallic reagents

Grignard reagents add directly to the carbonyl group of α,β-unsaturated aldehydes and ketones to give allylic alcohols: you have seen several examples of this, and you can now explain it by saying that the hard Grignard reagent prefers to attack the harder C=O rather than the softer C=C electrophilic centre. Here is a further example—the addition of MeMgBr to a cyclic ketone to give an allylic alcohol, plus, as it happens, some of a diene that arises from this alcohol by loss of water (dehydration). Below this example is the same reaction to which a very small amount (just 0.01 equivalents, that is, 1%) of copper(I) chloride has been added. The effect of the copper is dramatic: it makes the Grignard reagent undergo conjugate addition, with only a trace of the diene.

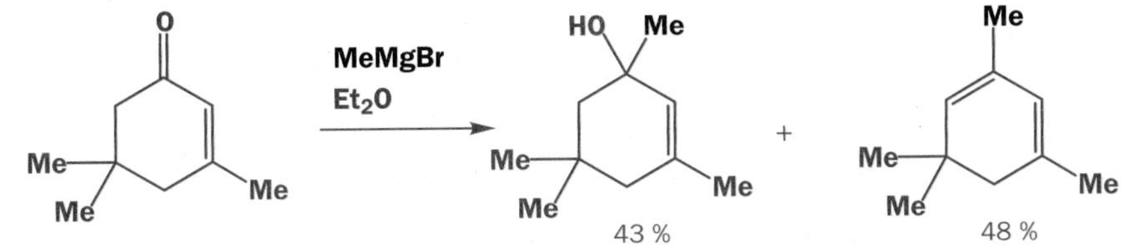

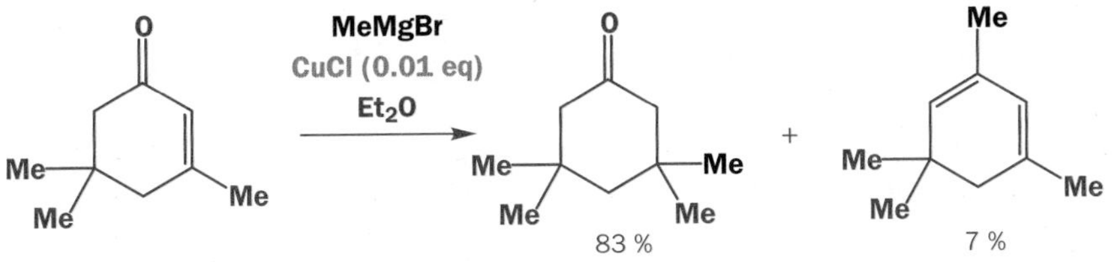

Organocopper reagents undergo conjugate addition

The copper works by transmetallating the Grignard reagent to give an organocopper reagent. Organocoppers are softer than Grignard reagents, and add in a conjugate fashion to the softer C=C double bond. Once the organocopper has added, the copper salt is available to transmetallate some more Grignard, and only a catalytic amount is required.

> Organocoppers are softer than Grignard reagents because copper is less electropositive than magnesium, so the C–Cu bond is less polarized than the C–Mg bond, giving the carbon atom less of a partial negative charge. Electronegativities: Mg, 1.3; Cu, 1.9.

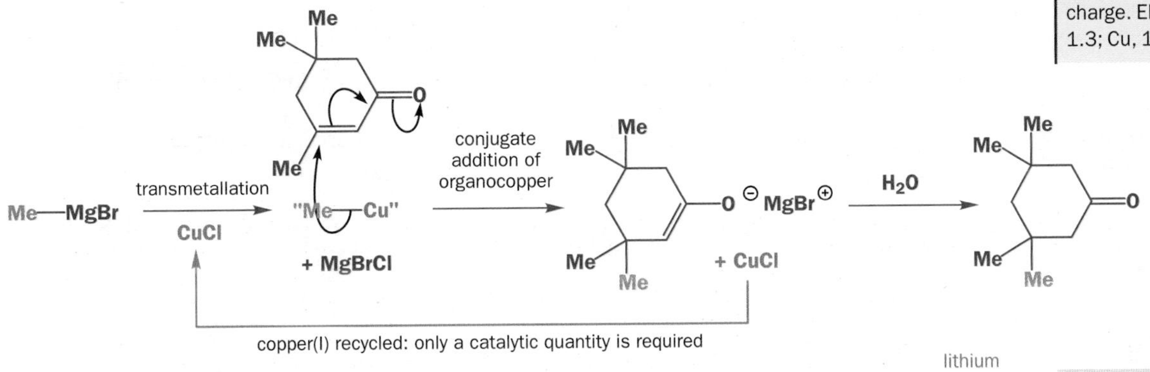

copper(I) recycled: only a catalytic quantity is required

> ■ We discussed transmetallation in Chapter 9.

The organocopper is shown here as 'Me–Cu' because its precise structure is not known. But there are other organocopper reagents that also undergo conjugate addition and that are much better understood. The simplest result from the reaction of two equivalents of organolithium with one equivalent of a copper (I) salt such as CuBr in ether or THF solvent at low temperature. The lithium cuprates (R$_2$CuLi) that are formed are not stable and must be used immediately.

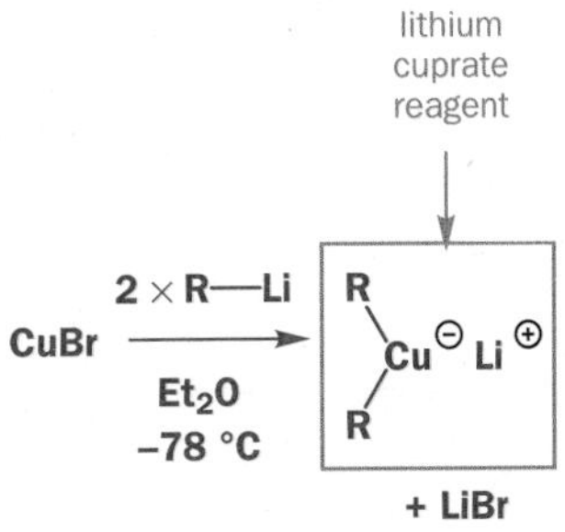

lithium cuprate reagent

> ■ As with the organolithiums that we introduced in Chapter 9, the exact structure of these reagents is more complex than we imply here: they are probably tetramers (four molecules of R$_2$CuLi bound together), but for simplicity we will draw them as monomers.

The addition of lithium cuprates to α,β-unsaturated ketones turns out to be much better if trimethylsilyl chloride is added to the reaction—we will explain what this does shortly, but for the moment here are two examples of lithium cuprate additions.

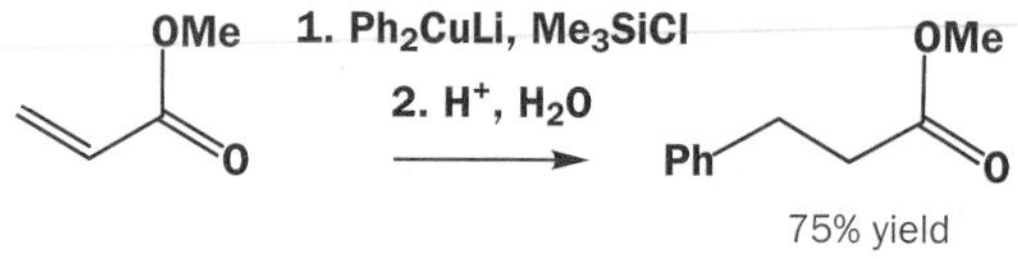

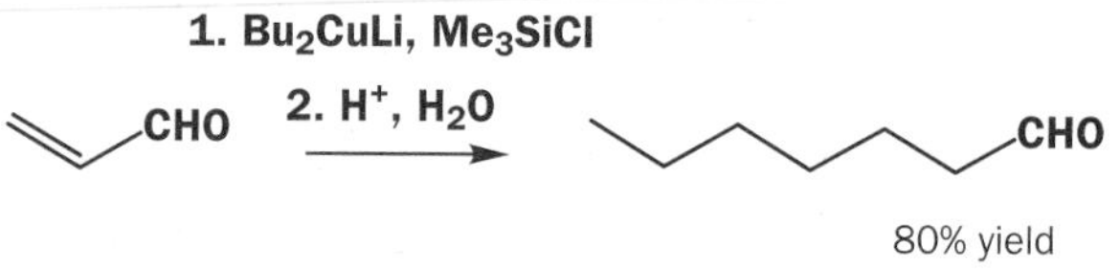

The silicon works by reacting with the negatively charged intermediate in the conjugate addition reaction to give a product that decomposes to the carbonyl compound when water is added at the end of the reaction. Here is a possible mechanism for a reaction between Bu_2CuLi and an α,β-unsaturated ketone in the presence of Me_3SiCl. The first step is familiar to you, but the second is a new reaction. Even so, following what we said in Chapter 5, it should not surprise you: the oxygen is clearly the nucleophile and the silicon the electrophile, and a new bond forms from O to Si as indicated by the arrow. The silicon-containing product is called a **silyl enol ether**, and we will come back to these compounds and their chemistry in more detail in later chapters.

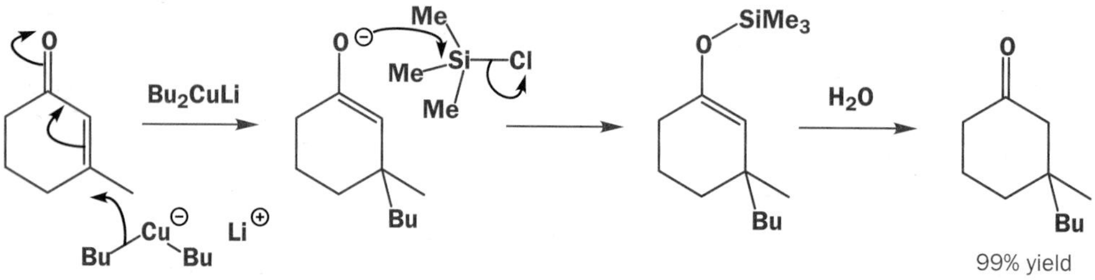

Conclusion

We end with a summary of the factors controlling the two modes of addition to α,β-unsaturated carbonyl compounds, and by noting that conjugate addition will be back again—in Chapters 23 (where we consider electrophilic alkenes conjugated with groups other than C=O) and 29 (where the nucleophiles will be of a different class known as enolates).

● **Summary**

	Conjugate addition favoured by	Direct addition to C=O favoured by
Reaction conditions (for reversible additions):	• thermodynamic control: high temperatures, long reaction times	• kinetic control: low temperatures, short reaction times
Structure of α,β-unsaturated compound:	• unreactive C=O group (amide, ester)	• reactive C=O group group (aldehyde, acyl chloride)
	• unhindered β carbon	• hindered β carbon
Type of nucleophile:	• soft nucleophiles	• hard nucleophiles
Organometallic:	• organocoppers or catalytic Cu(I)	• organolithiums, Grignard reagents

Problems

1. Draw mechanisms for this reaction and explain why this particular product is formed.

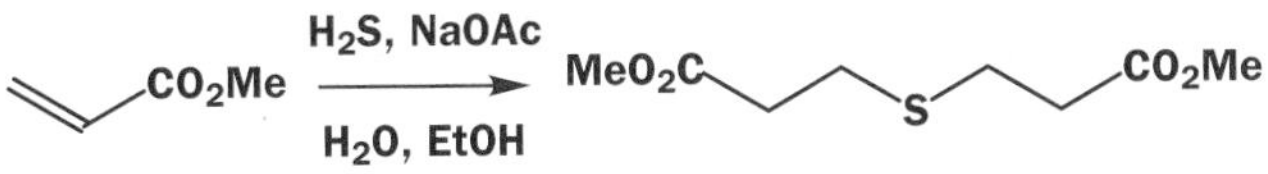

2. Which of the two routes shown here would actually lead to the product? Why?

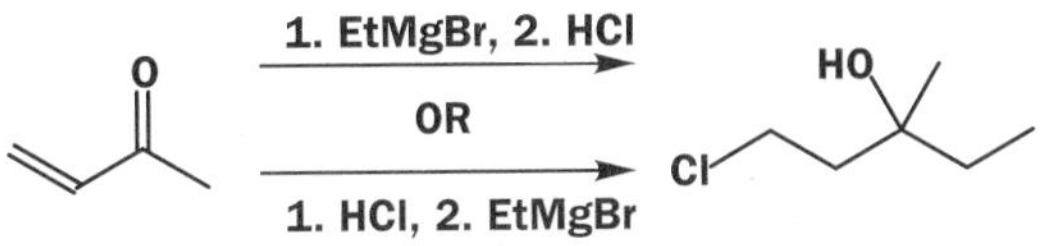

3. Suggest reasons for the different outcomes of the following reactions (your answer must, of course, include a mechanism for each reaction).

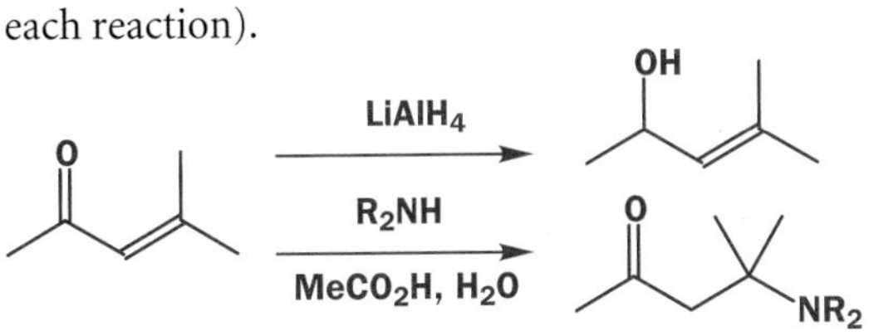

4. Addition of dimethylamine to the unsaturated ester A could give either product B or C. Draw mechanisms for both reactions and show how you would distinguish them spectroscopically.

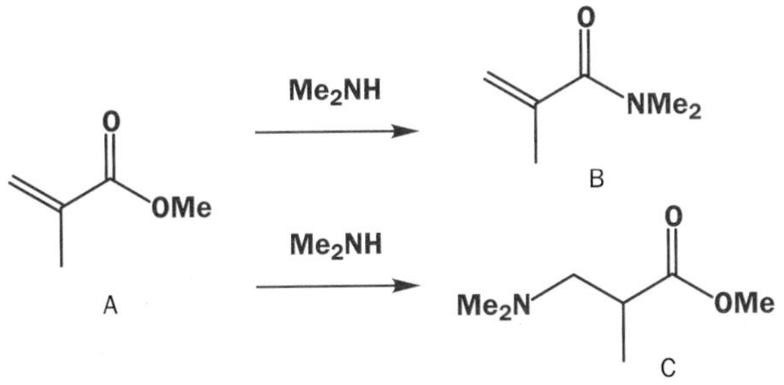

5. Suggest mechanisms for the following reaction.

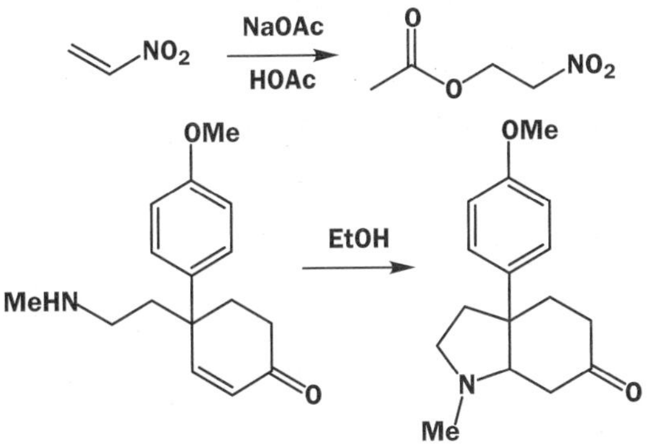

6. Predict the product of these reactions.

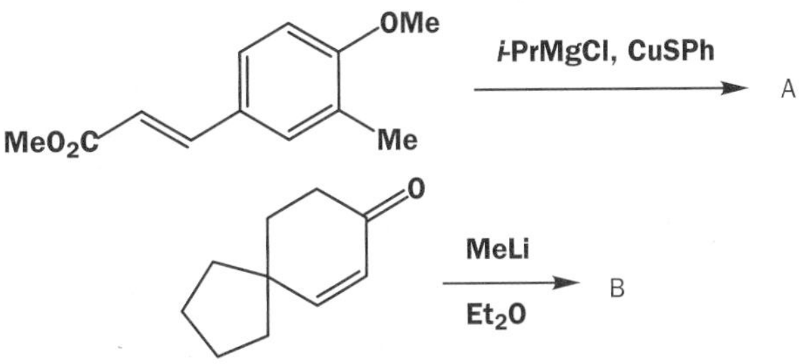

7. Two routes are proposed for the preparation of this amino alcohol. Which do you think is more likely to succeed and why?

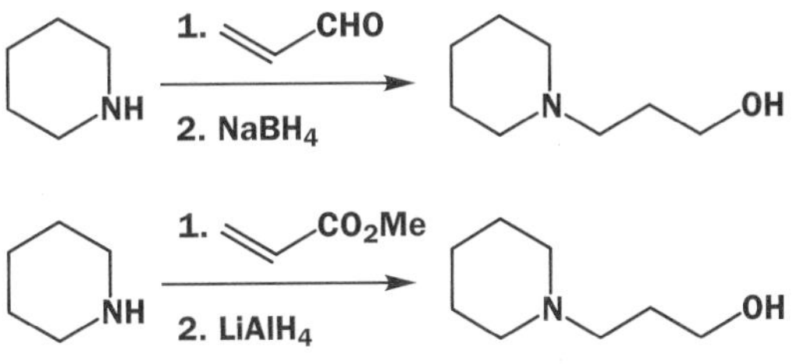

8. How would you prepare these compounds by conjugate addition?

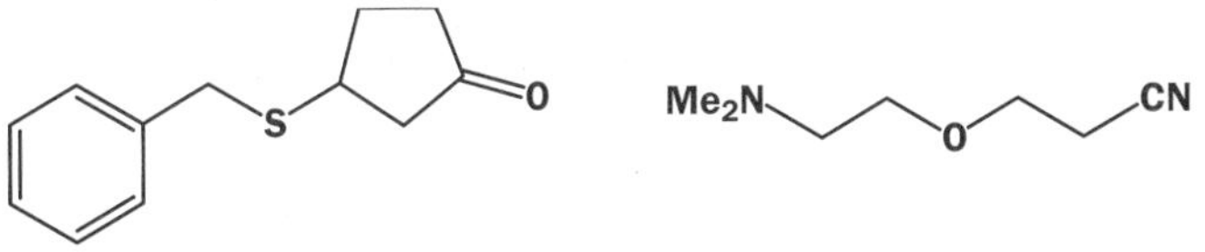

9. How might this compound be made using a conjugate addition as one of the steps? You might find it helpful to consider the preparation of tertiary alcohols as decribed in Chapter 9 and also to refer back to Problem 1 in this chapter.

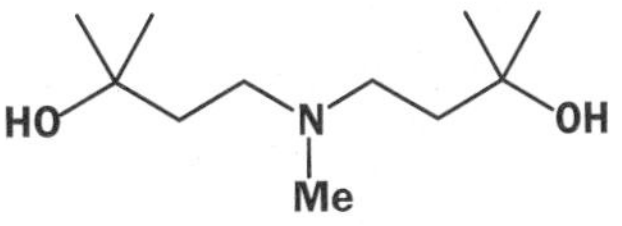

10. When we discussed reduction of cyclopentenone to cyclopentanol, we suggested that conjugate addition of borohydride must occur before direct addition of borohydride; in other words, this scheme must be followed.

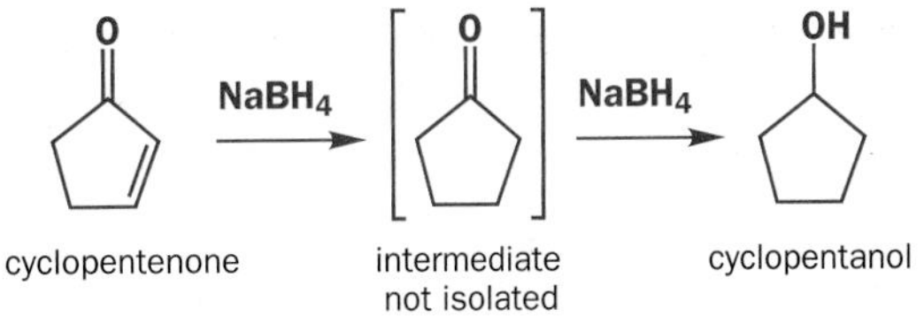

What is the alternative scheme? Why is the scheme shown above definitely correct?

11. Suggest a mechanism for this reaction. Why does conjugate addition occur rather than direct addition?

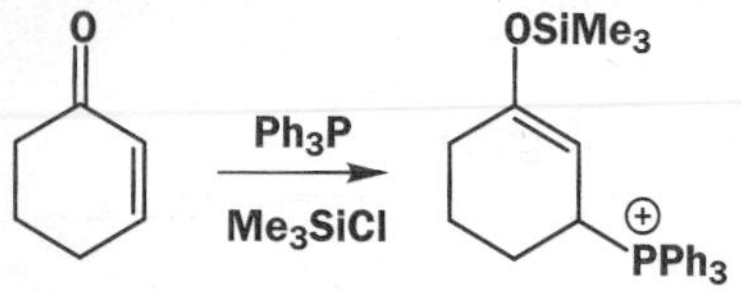

Why is the product shown as a cation? If it is indeed a salt, what is the anion?

12. How, by choice of reagent, would you make this reaction give the direct addition product (route A)? How would you make it give the conjugate addition product (route B)?

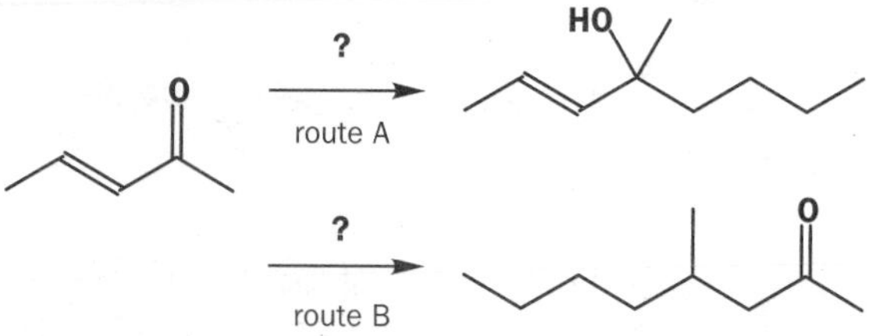

Proton nuclear magnetic resonance

11

Connections

Building on:
- X-ray crystallography, mass spectrometry, ^{13}C NMR and infrared spectroscopy ch3

Arriving at:
- Proton (or ^1H) NMR spectroscopy
- How ^1H NMR compares with ^{13}C NMR
- How 'coupling' in ^1H NMR provides most of the information needed to find the structure of an unknown molecule

Looking forward to:
- Using ^1H NMR with other spectroscopic methods to solve structures rapidly ch15
- Using ^1H NMR to investigate the detailed shape (stereochemistry) of molecules ch32
- ^1H NMR spectroscopy is referred to in most chapters of the book as it is the most important tool for determining structure; you must understand this chapter before reading further

The differences between carbon and proton NMR

We used ^{13}C NMR in Chapter 3 as part of a three-pronged attack on the problem of determining molecular structure. Important though these three prongs are, we were forced to confess at the end of Chapter 3 that we had delayed the most important technique of all—proton (^1H) NMR—until a later chapter because it is more complicated than ^{13}C NMR. This is that delayed chapter and we must now tackle those complications. We hope you will see ^1H NMR for the beautiful and powerful technique that it surely is. The difficulties are worth mastering for this is the chemist's primary weapon in the battle to solve structures.

Proton NMR differs from ^{13}C NMR in a number of ways.

- ^1H is the major isotope of hydrogen (99.985% natural abundance), while ^{13}C is only a minor isotope (1.1%)

- ^1H NMR is quantitative: the area under the peak tells us the number of hydrogen nuclei, while ^{13}C NMR may give strong or weak peaks from the same number of ^{13}C nuclei

- Protons interact magnetically ('couple') to reveal the connectivity of the structure, while ^{13}C is too rare for coupling between ^{13}C nuclei to be seen

- ^1H NMR shifts give a more reliable indication of the local chemistry than that given by ^{13}C spectra

We shall examine each of these points in detail and build up a full understanding of proton NMR spectra. The other spectra remain important, of course.

Proton NMR spectra are recorded in the same way as ^{13}C NMR spectra: radio waves are used to study the energy level differences of nuclei, but this time they are ^1H and not ^{13}C nuclei. Hydrogen nuclei have a nuclear spin

▶ **An instance where ^1H NMR was *not* useful**

In Chapter 3 you met methoxatin. Proton NMR has little to tell us about its structure as it has so few protons (i.e. ^1H atoms) (it is $C_{14}H_6N_2O_8$). Carbon NMR and eventually an X-ray crystal structure gave the answer. There are four OH and NH protons (best seen by IR) and only two C-H protons. The latter protons are the kind that proton NMR reveals best. Fortunately, most compounds have lots more than this.

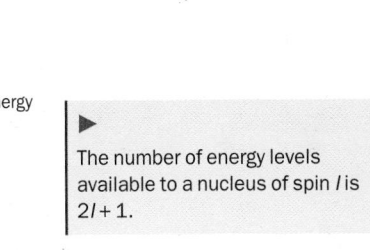

methoxatin

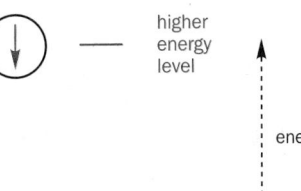

applied magnetic field B_0

nucleus aligned against applied magnetic field — higher energy level

energy

nucleus aligned with applied magnetic field — lower energy level

▶ The number of energy levels available to a nucleus of spin I is $2I + 1$.

of a half and so have two energy levels: they can be aligned either with or against the applied magnetic field.

The spectra look much the same: the scale runs from right to left and the zero point is given by the same reference compound though it is the proton resonance of Me_4Si rather than the carbon resonance that defines the zero point. You will notice at once that the scale is much smaller, ranging over only about 10 p.p.m. instead of the 200 p.p.m. needed for carbon. This is because the variation in the chemical shift is a measure of the shielding of the nucleus by the electrons around it. There is inevitably less change possible in the distribution of two electrons around a hydrogen nucleus than in that of the eight valence electrons around a carbon nucleus. Here is a simple ^1H NMR spectrum.

> ▶
> This 10 p.p.m. scale is not the same as any part of the ^{13}C NMR spectrum. It is at a different frequency altogether.

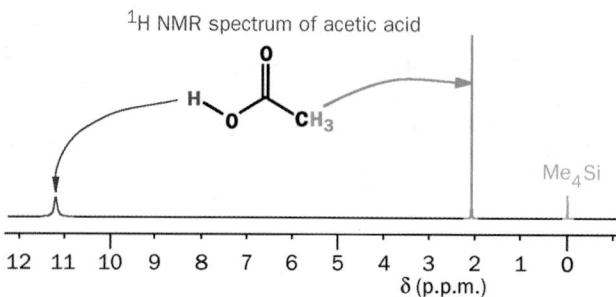

^1H NMR spectrum of acetic acid

Integration tells us the number of hydrogen atoms in each peak

The chemical shift of the twelve hydrogen atoms of the four identical methyl groups in Me_4Si is defined as zero. The methyl group in the acid is next to the carbonyl group and so slightly de-shielded at about δ 2.0 p.p.m. and the acidic proton itself is very deshielded at δ 11.2 p.p.m. The same factor that makes this proton acidic—the O–H bond is polarized towards oxygen—also makes it resonate at low field. So far things are much the same as in carbon NMR. Now for a difference. Notice that the ratio of the peak heights in this spectrum was about 3:1 and that that is also the ratio of the number of protons. In fact, it's not the peak height but the area under the peaks that is exactly proportional to the number of protons. Proton spectra are normally **integrated**, that is, the area under the peaks is computed and recorded as a line with steps corresponding to the area, like this.

> ■
> It is not enough simply to measure the relative heights of the peaks because, as here, some peaks might be broader than others. Hence the area under the peak is measured.

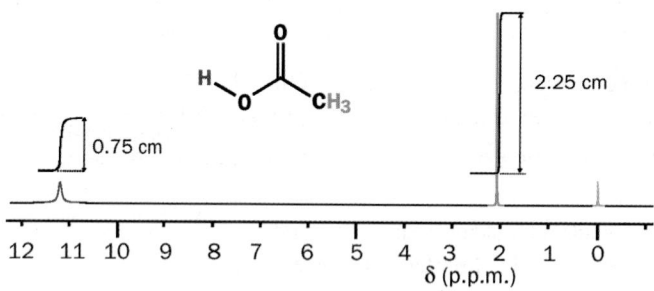

Simply measuring the height of the steps with a ruler gives you the *ratio* of the numbers of protons represented by each peak. Knowing the atomic composition from the mass spectrum, we also know the distribution of protons of various kinds. Here the heights are 0.75 and 2.25 cm, a ratio of about 1:3. The compound is $C_2H_4O_2$ so, since there are 4 H atoms altogether, the peaks must contain $1 \times H$ and $3 \times H$, respectively.

In the spectrum of 1,4-dimethoxybenzene, there are just two signals in the ratio of 3:2. This time the compound is $C_8H_{10}O_2$ so the true ratio must be 6:4. Assigning the spectrum requires the same attention to symmetry as in the case of ^{13}C spectra.

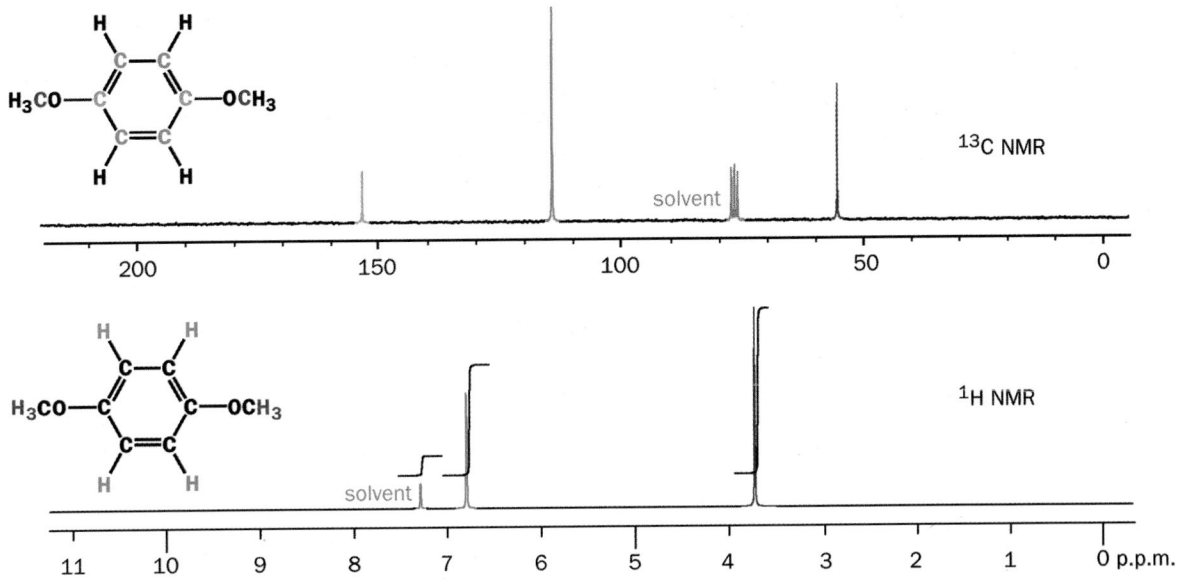

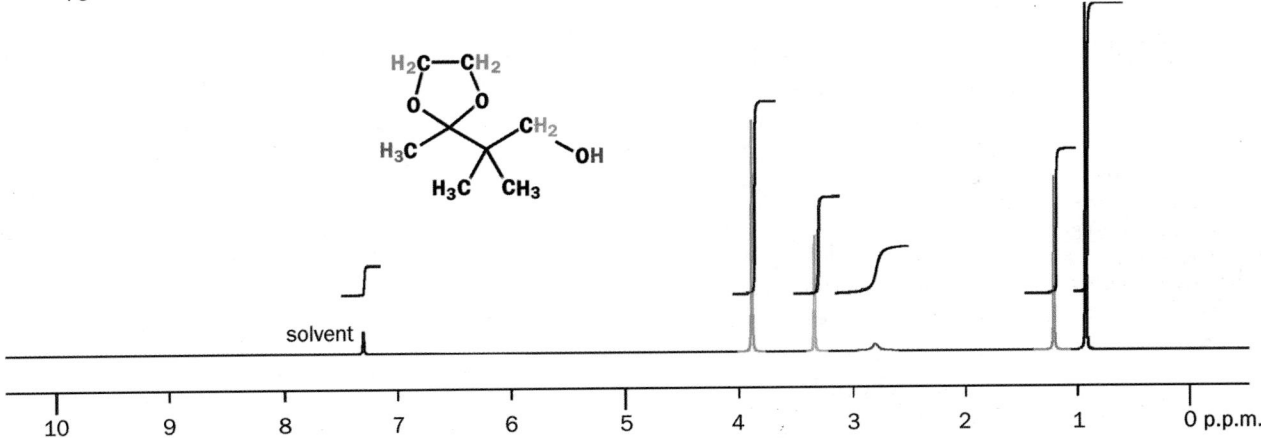

In this next example it is easy to assign the spectrum simply by measuring the steps in the integral. There are two identical methyl groups (CMe_2) having 6 Hs, one methyl group by itself having 3 Hs, the OH proton (1 H), the CH_2 group next to the OH (2 Hs), and finally the CH_2CH_2 group between the oxygen atoms in the ring (4 Hs).

Proton NMR spectra are generally recorded in solution in deuterochloroform ($CDCl_3$)—that is, chloroform with the ^1H replaced by ^2H. The proportionality of the size of the peak to the number of protons tells you why: if you ran a spectrum in $CHCl_3$, you would see a vast peak for all the solvent Hs because there would be much more solvent than the compound you wanted to look at. Using $CDCl_3$ cuts out all extraneous protons.

Regions of the proton NMR spectrum

The integration gives useful—indeed essential—information, but it is much more important to understand the reasons for the exact chemical shift of the different types of proton. In the last example you can see one marked similarity to carbon spectra: protons on saturated carbon atoms next to oxygen are shifted downfield to larger δ values (here 3.3 and 3.9 p.p.m.). The other regions of the proton NMR spectrum are also quite similar in general outline to those of ^{13}C spectra. Here they are.

regions of the proton NMR spectrum   Me$_4$Si

protons on unsaturated carbons next to oxygen: aldehydes	protons on unsaturated carbons: benzene, aromatic hydrocarbons	protons on unsaturated carbons: alkenes	saturated CH$_3$ CH$_2$ CH next to oxygen	saturated CH$_3$ CH$_2$ CH not next to oxygen

10.5   8.5   6.5   4.5   3.0   δ (p.p.m.)   0.0

These regions hold for protons attached to C: protons attached to O or N can come almost anywhere on the spectrum. Even for C–H signals, the regions are approximate and overlap quite a lot. You should use the chart as a basic guide, but you will need a more detailed understanding of proton chemical shifts than you did for ^{13}C chemical shifts. To achieve this understanding, we now need to examine each class of proton in more detail and examine the reasons for particular shifts. It is important that you grasp these reasons. An alternative is to learn all the chemical shifts off by heart (not recommended).

Protons on saturated carbon atoms

Chemical shifts are related to the electronegativity of substituents

▶

In this chapter you will see a lot of numbers—chemical shifts and differences in chemical shifts. We need these to show that the ideas behind ^{1}H NMR are securely based in fact. You do *not* need to learn these numbers. Comprehensive tables can be found at the end of Chapter 15, which we hope you will find useful for reference while you are solving problems. Again, do not attempt to learn the numbers!

We shall start with protons on saturated carbon atoms. If you study Table 11.1 you will see that the protons in a methyl group are shifted more and more as the atom attached to them gets more electronegative.

When we are dealing with simple atoms as substituents, these effects are straightforward and more or less additive. If we go on adding electronegative chlorine atoms to a carbon atom, electron density is progressively removed from it and the carbon nucleus and the hydrogen atoms attached to it are progressively deshielded.

Table 11.1 Effects of electronegativity

Atom	Electronegativity	Compound	^{1}H NMR shift, p.p.m.
Li	1.0	CH$_3$–Li	–1.94
Si	1.9	CH$_3$–SiMe$_3$	0.0
N	3.0	CH$_3$–NH$_2$	2.41
O	3.4	CH$_3$–OH	3.50
F	4.0	CH$_3$–F	4.27

▶

The second two compounds, dichloromethane CH$_2$Cl$_2$ and chloroform CHCl$_3$, are commonly used as solvents and their shifts will become familiar to you if you look at a lot of spectra.

	CH$_3$Cl	CH$_2$Cl$_2$	CHCl$_3$
^{1}H NMR shift, p.p.m.	3.06	5.30	7.27
^{13}C NMR shift, p.p.m.	24.9	54.0	77.2

Proton chemical shifts tell us about chemistry

The truth is that shifts and electronegativity are not perfectly correlated. The key property is indeed electron withdrawal but it is the electron-withdrawing power of the whole substituent in comparison with the carbon and hydrogen atoms in the CH skeleton that matters. Methyl groups joined to the same element, say, nitrogen, may have very different shifts if the substituent is an amino group (CH$_3$–NH$_2$ has δ$_H$ for the CH$_3$ group = 2.41 p.p.m.) or a nitro group (CH$_3$–NO$_2$ has δ$_H$ 4.33 p.p.m.). A nitro group is much more electron-withdrawing than an amino group.

▶

You have seen δ used as a symbol for chemical shift. Now that we have two sorts of chemical shift—in the ^{13}C NMR spectrum and in the ^{1}H NMR spectrum—we need to be able to distinguish them. δ$_H$ means chemical shift in the ^{1}H NMR spectrum, and δ$_C$ chemical shift in the ^{13}C NMR spectrum.

What we need is a quick guide rather than some detailed correlations, and the simplest is this: all functional groups except very electron-withdrawing ones shift methyl groups from 1 p.p.m. (where you find them if they are not attached to a functional group) downfield to about 2 p.p.m. Very electron-withdrawing groups shift methyl groups to about 3 p.p.m.

● **Approximate chemical shifts for methyl groups**

No electron-withdrawing functional groups	Less electron-withdrawing functional groups X	More electron-withdrawing functional groups X
Me at about 1 p.p.m.	MeX at about 2 p.p.m. (i.e. add 1 p.p.m.)	MeX at about 3 p.p.m. (i.e. add 2 p.p.m.)
aromatic rings, alkenes, alkynes	carbonyl groups: acids (CO_2H), esters (CO_2R), ketones (COR), nitriles (CN)	oxygen-based groups: ethers (OR), esters (OCOR)
	amines (NHR)	amides (NHCOR)
	sulfides (SR)	sulfones (SO_2R)

Rather than trying to fit these data to some atomic property, even such a useful one as electronegativity, we should rather see these shifts as a useful measure of the electron-withdrawing power of the group in question. The NMR spectra are telling us about the chemistry. Among the largest shifts possible for a methyl group is that caused by the nitro group, 3.43 p.p.m., at least twice the size of the shift for a carbonyl group. This gives us our first hint of some important chemistry: one nitro group is worth two carbonyl groups when you need electron withdrawal. You have already seen that electron withdrawal and acidity are related (Chapter 8) and in later chapters you will see that we can correlate the anion-stabilizing power of groups like carbonyl, nitro, and sulfone with proton NMR.

Methyl groups give us information about the structure of molecules

It sounds rather unlikely that the humble methyl group could tell us much that is important about molecular structure—but just you wait. We shall look at four simple compounds and their NMR spectra— just the methyl groups, that is. The first two are the acid chlorides on the right.

The first compound shows just one methyl signal containing 9 Hs at δ_H 1.10 p.p.m. This tells us two things. All the protons in each methyl group are the same; and all three methyl groups in the tertiary butyl (*t*-butyl, or Me₃C–) group are the same. This is because rotation about C–C single bonds, both about the CH₃–C bond and about the (CH₃)₃C–C bond, is fast. Though at any one instant the hydrogen atoms in one methyl group, or the methyl groups in the *t*-butyl group, may differ, on average they are the same. The time-averaging process is fast rotation about a σ bond. The second compound shows two 3H signals, one at 1.99 and one at 2.17 p.p.m. Now rotation is slow—indeed the C=C double bond does not rotate at all and so the two methyl groups are different. One is on the same side of the alkene as (or '*cis* to') the –COCl group while the other is on the opposite side (or '*trans*').

The second pair of compounds contain the CHO group. One is a simple aldehyde, the other an amide of formic acid: it is DMF, dimethylformamide. The first has two sorts of methyl group: a 3H signal at δ_H 1.81 p.p.m. for the SMe group and a 6H signal for the CMe₂ group. The two methyl groups in the 6H signal are the same, again because of fast rotation about a C–C σ bond.

The second compound also has two methyl signals, at 2.89 and 2.98 p.p.m., each 3H, and these are the two methyl groups on nitrogen. Restricted rotation about the N–CO bond must be making the two Me groups different. You will remember from Chapter 7 (p. 164) that the N–CO amide bond has considerable double bond character because of conjugation: the lone pair electrons on nitrogen are delocalized into the carbonyl group.

> Rotation about single bonds is generally very fast (you are about to meet an exception); rotation about double bonds is generally very, very slow (it just doesn't happen). This was discussed in Chapter 7.

Chemical shifts of CH₂ groups

Shifts of the same order of magnitude occur for protons on CH₂ groups and the proton on CH groups, but with the added complication that CH₂ groups have *two* other substituents and CH groups *three*. A CH₂ (methylene) group resonates at 1.3 p.p.m., about 0.4 p.p.m. further downfield than a comparable CH₃ group (0.9 p.p.m.), and a CH (methine) group resonates at 1.7 p.p.m., another 0.4 p.p.m. downfield. Replacing each hydrogen atom in the CH₃ group by a carbon atom causes a small downfield shift as carbon is slightly more electronegative (C 2.5 p.p.m.; H 2.2 p.p.m.) than hydrogen and therefore shields less effectively.

> ● **Chemical shifts of protons in CH, CH₂, and CH₃ groups with no nearby electron-withdrawing groups**
>
> | CH group | CH₂ group | CH₃ group |
> |---|---|---|
> | 0.4 p.p.m. downfield ← | 0.4 p.p.m. downfield ← | |
> | 1.7 p.p.m. | 1.3 p.p.m. | 0.9 p.p.m. |

$\delta(CH_2)$ ~3.0 p.p.m.

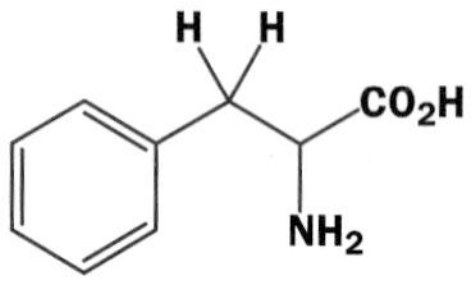

phenylalanine

■ You'll meet this reaction in the next chapter, and we shall discuss protection and protecting groups in Chapter 24. For the moment, just be concerned with the structure of the product.

The benzyl group (PhCH₂–) is very important in organic chemistry. It occurs naturally in the amino acid phenylalanine, which you met in Chapter 2. Phenylalanine has its CH₂ signal at 3.0 p.p.m. and is moved downfield from 1.3 p.p.m. mostly by the benzene ring.

Amino acids are often protected as the 'Cbz' derivatives (Carboxybenzyl) by reaction with an acid chloride. Here is a simple example together with the NMR spectrum of the product. Now the CH₂ group has gone further downfield to 5.1 p.p.m. as it is next to both oxygen and phenyl.

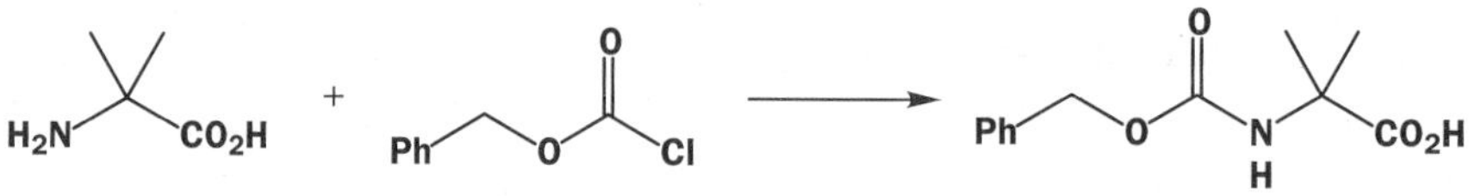

amino acid   "Cbz chloride"   "Cbz protected" amino acid
               (benzyl chloroformate)

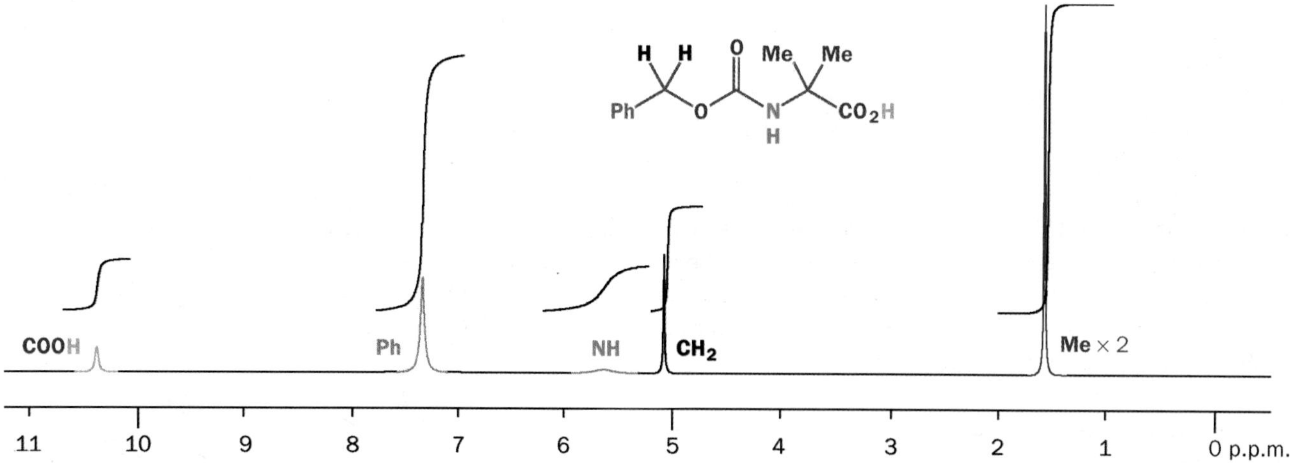

Like double bonds, **cage structures** prevent bond rotation, and can make the two protons of a CH₂ group appear different. There are many flavouring compounds from herbs that have structures like this. In the example here—myrtenal, from the myrtle bush—there is a four-membered ring bridged across a six-membered ring. The CH₂ group on the bridge

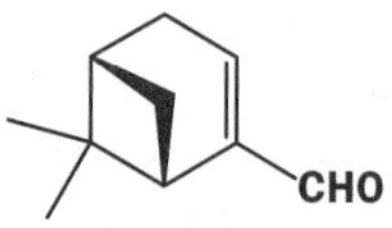

myrtenal

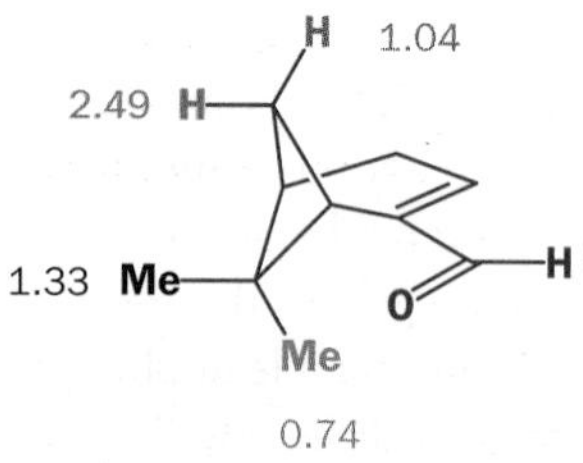

2.49 H   H 1.04

1.33 Me

Me

0.74

has two different hydrogen atoms—one is over a methyl group and the other is over the enal system. No rotation of any bonds in the cage is possible, so these hydrogens are always different and resonate at different frequencies (1.04 and 2.49 p.p.m.). The methyl groups on the other bridge are also different for the same reason.

Chemical shifts of CH groups

A CH group in the middle of a carbon skeleton resonates at about 1.7 p.p.m.—another 0.4 p.p.m. downfield from a CH_2 group. It can have up to three substituents and these will cause further downfield shifts of about the same amount as we have already seen for CH_3 and CH_2 groups. Here are three examples from nature: nicotine, the compound in tobacco that causes the craving (though not the death, which is doled out instead by the carbon monoxide and tars in the smoke), has one hydrogen atom trapped between a simple tertiary amine and an aromatic ring at 3.24 p.p.m. Lactic acid has a CH proton at 4.3 p.p.m. You could estimate this with reasonable accuracy by taking 1.7 (for the CH) and adding 1.0 (for C=O) plus 2.0 (for OH) = 4.7 p.p.m. Vitamin C (ascorbic acid) has two CHs. One at 4.05 p.p.m. is next to an OH group (estimate 1.7 + 2.0 for OH = 3.7 p.p.m.) and one next to a double bond and an oxygen atom at 4.52 p.p.m. (estimate 1.7 + 1 for double bond + 2 for OH = 4.7 p.p.m.).

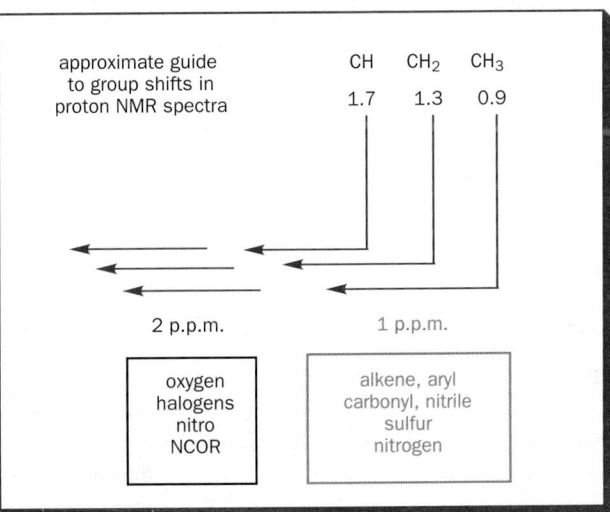

nicotine

methyl ester of lactic acid

vitamin C (ascorbic acid)

An interesting case is the amino acid phenylalanine whose CH_2 group we looked at a moment ago. It also has a CH group between the amino and the carboxylic acid groups. If we record the ^1H NMR spectrum in D_2O, either in basic (NaOD) or acidic (DCl) solutions we see a large shift of that CH group. In basic solution the CH resonates at 3.60 p.p.m. and in acidic solution at 4.35 p.p.m. There is a double effect here: CO_2H and NH_3^+ are both more electron-withdrawing than CO_2^- and NH_2 so both move the CH group downfield.

> D_2O, NaOD, and DCl have to be used in place of their ^1H equivalents to avoid swamping the spectrum with H_2O protons. All acidic protons are replaced by deuterium in the process — more on this later.

phenylalanine

Your simple guide to chemical shifts

We suggest you start with a very simple (and therefore oversimplified) picture, which should be the basis for any further refinements. Start methyl groups at 0.9, methylenes (CH_2) at 1.3, and methines (CH) at 1.7 p.p.m. Any functional group is worth a *one* p.p.m. downfield shift except oxygen and halogen which are worth *two* p.p.m. This diagram summarizes the basic position.

approximate guide to group shifts in proton NMR spectra

CH	CH₂	CH₃
1.7	1.3	0.9

2 p.p.m.

1 p.p.m.

oxygen
halogens
nitro
NCOR

alkene, aryl
carbonyl, nitrile
sulfur
nitrogen

If you want more detailed information, you can refer to the tables in Chapter 15 or better still the more comprehensive tables in any specialized text.

This is a very rough and ready guide and you can make it slightly more accurate by adding subdivisions at 1.5 and 2.5 p.p.m. and including the very electron-withdrawing groups (nitro, ester, fluoride), which shift by 3 p.p.m. This gives us the summary chart on this page, which we suggest you use as a reference.

Summary chart of proton NMR shifts
values to be added to 0.9 for CH_3, 1.3 for CH_2 or 1.7 for CH

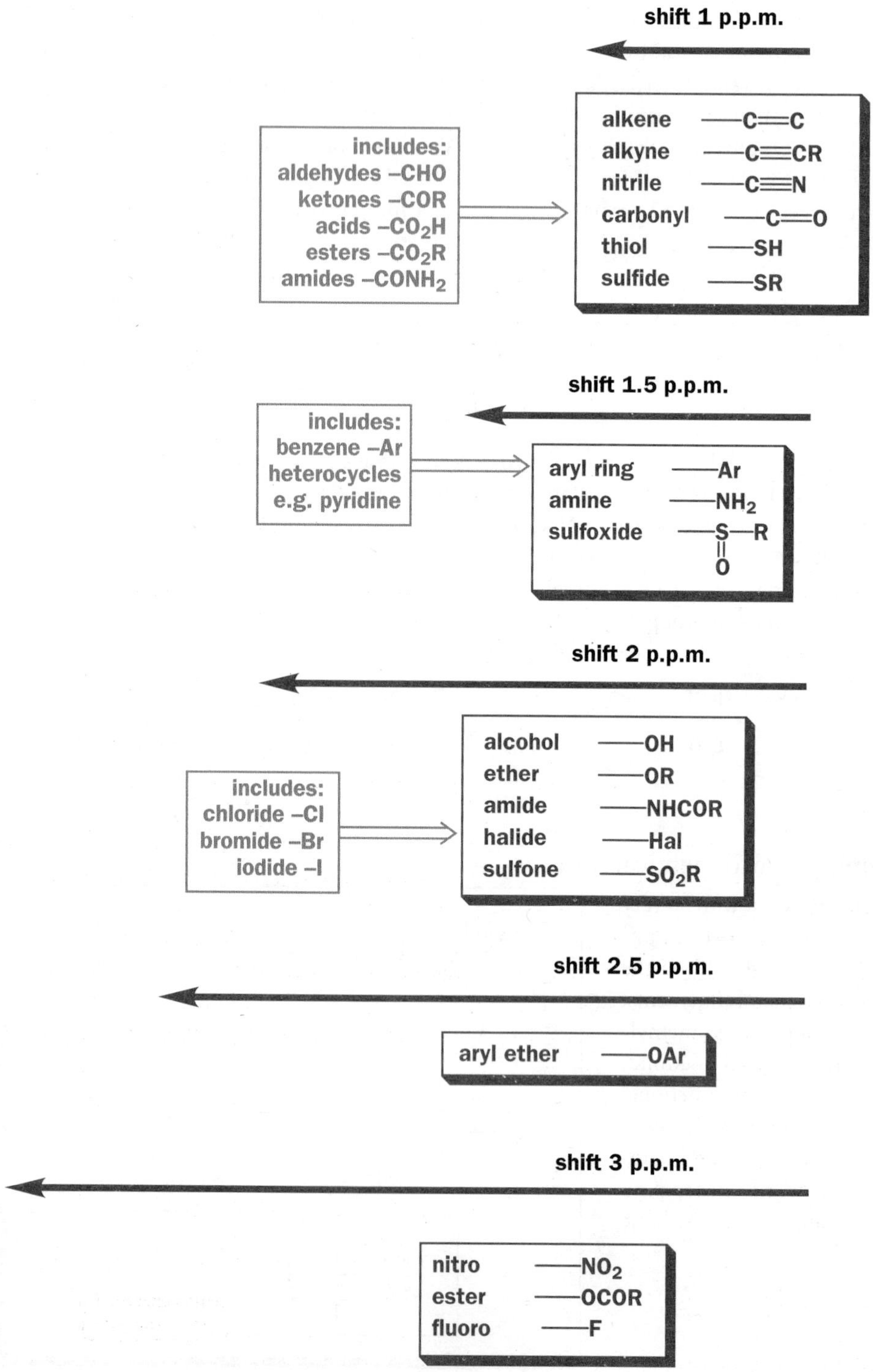

shift 1 p.p.m.

includes:
aldehydes –CHO
ketones –COR
acids –CO_2H
esters –CO_2R
amides –$CONH_2$

alkene	—C=C
alkyne	—C≡CR
nitrile	—C≡N
carbonyl	—C=O
thiol	—SH
sulfide	—SR

shift 1.5 p.p.m.

includes:
benzene –Ar
heterocycles
e.g. pyridine

aryl ring	—Ar
amine	—NH_2
sulfoxide	—S—R (‖O)

shift 2 p.p.m.

includes:
chloride –Cl
bromide –Br
iodide –I

alcohol	—OH
ether	—OR
amide	—NHCOR
halide	—Hal
sulfone	—SO_2R

shift 2.5 p.p.m.

aryl ether	—OAr

shift 3 p.p.m.

nitro	—NO_2
ester	—OCOR
fluoro	—F

Answers deduced from this chart won't be very accurate but will give a good guide. Remember—these shifts are additive. Take a simple example, the ketoester below. There are just three signals and the integration alone distinguishes the two methyl groups from the CH_2 group. One methyl has been shifted from 0.9 p.p.m. by about 1 p.p.m., the other by more than 2 p.p.m. The first must be next to C=O and the second next to oxygen. More precisely, 2.14 p.p.m. is a shift of 1.24 p.p.m. from our standard value (0.9 p.p.m.) for a methyl group, about what we expect for a methyl ketone, while 3.61 p.p.m. is a shift of 2.71 p.p.m., close to the expected 3.0 p.p.m. for an ester joined through the oxygen atom. The CH_2 group is next to an ester and a ketone carbonyl group and so we expect it at 1.3 + 1.0 + 1.0 = 3.3 p.p.m., an accurate estimate, as it happens. We shall return to these estimates when we look at spectra of unknown compounds.

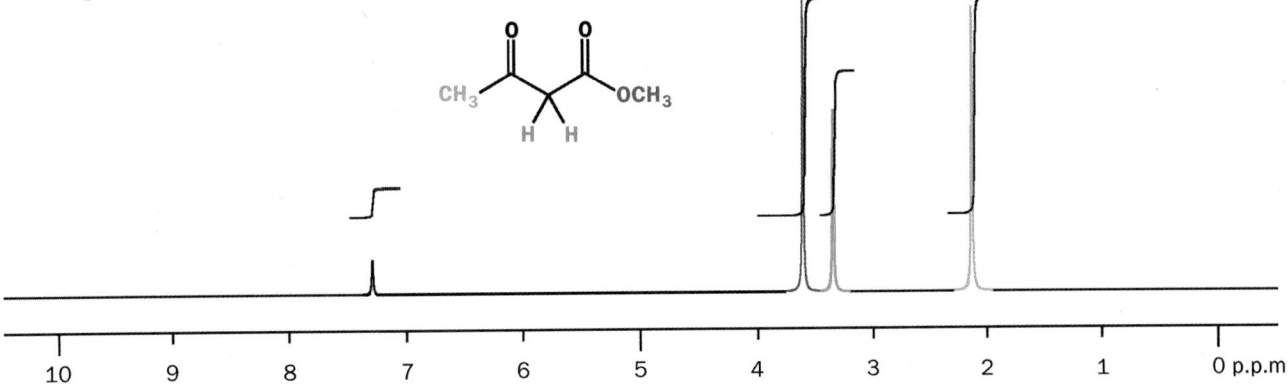

The alkene region and the benzene region

In ^{13}C NMR, one region was enough for both of these, but see how different things are with proton NMR.

The two carbon signals are almost the same (1.3 p.p.m. difference < 1% of the total 200 p.p.m. scale) but the proton signals are very different (1.6 p.p.m. difference = 16% of the 10 p.p.m. scale). There must be a fundamental reason for this.

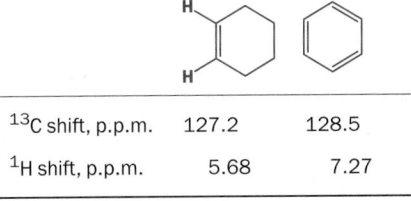

^{13}C shift, p.p.m.	127.2	128.5
^1H shift, p.p.m.	5.68	7.27

The benzene ring current causes large shifts for aromatic protons

A simple alkene has an area of low electron density in the plane of the molecule because the π orbital has a node there, and the carbons and hydrogen nuclei lying in the plane gain no shielding from the π electrons.

The benzene ring looks similar at first sight, and the plane of the molecule is indeed a node for all the π orbitals. However, benzene is 'aromatic'—it has extra stability because the six π electrons fit into three very stable orbitals and are delocalized round the whole ring.

The applied field sets up a ring current in these delocalized electrons that produces a local field rather like the field produced by the electrons around a nucleus. Inside the benzene ring, the induced field opposes the applied field but, outside the ring, it reinforces the applied field. The carbon atoms are in the ring itself and experience neither effect, but the hydrogens are outside the ring, feel a stronger applied field, and appear less shielded.

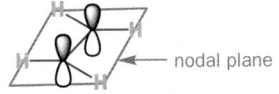
← nodal plane

■
Chapter 7 was devoted to a discussion of aromaticity and delocalization.

▶
Magnetic fields produced by circulating electrons are all around you: electromagnets and solenoids are exactly this.

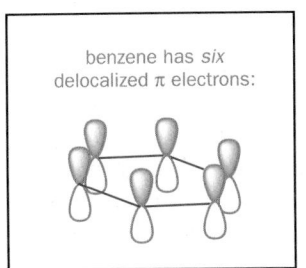

benzene has *six* delocalized π electrons:

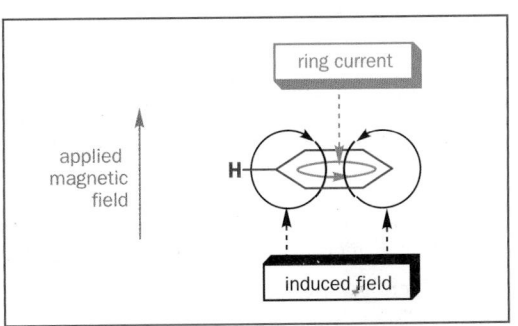

ring current

applied magnetic field

H

induced field

Cyclophanes and annulenes

You may think that it is rather pointless imagining what goes on inside an aromatic ring as we cannot have hydrogen atoms literally *inside* a benzene ring. However, we can get close. Compounds called cyclophanes have loops of saturated carbon atoms attached at both ends to the same benzene rings. You see here a structure for [7]*para*-cyclophane, which has a string of seven CH_2 groups attached to the *para* positions of the same benzene ring. The four protons on the benzene ring itself appear as one line at a normal δ 7.07 p.p.m. The two CH_2 groups joined to the benzene ring (C1) are deshielded by the ring current at δ 2.64 p.p.m. The next two sets of CH_2 groups on C2 and C3 are neither shielded nor deshielded at δ 1.0 p.p.m. The middle CH_2 group in the chain (C4) must be pointing towards the ring in the middle of the π system and is heavily shielded by the ring current at negative δ (–0.6 p.p.m.).

[7]-*para*-cyclophane

With a larger aromatic ring, it *is* possible actually to have hydrogen atoms inside the ring. Compounds are aromatic if they have $4n + 2$ delocalized electrons and this ring with nine double bonds, that is, 18 π electrons, is an example. The hydrogens outside the ring resonate in the aromatic region at rather low field (9.28 p.p.m.) but the hydrogen

atoms inside the ring resonate at an amazing –2.9 p.p.m. showing the strong shielding by the ring current. Such extended aromatic rings are called annulenes: you met them in Chapter 7.

Hs outside the ring δ_H +9.28 p.p.m.

Hs inside the ring δ_H –2.9 p.p.m.

Uneven electron distribution in aromatic rings

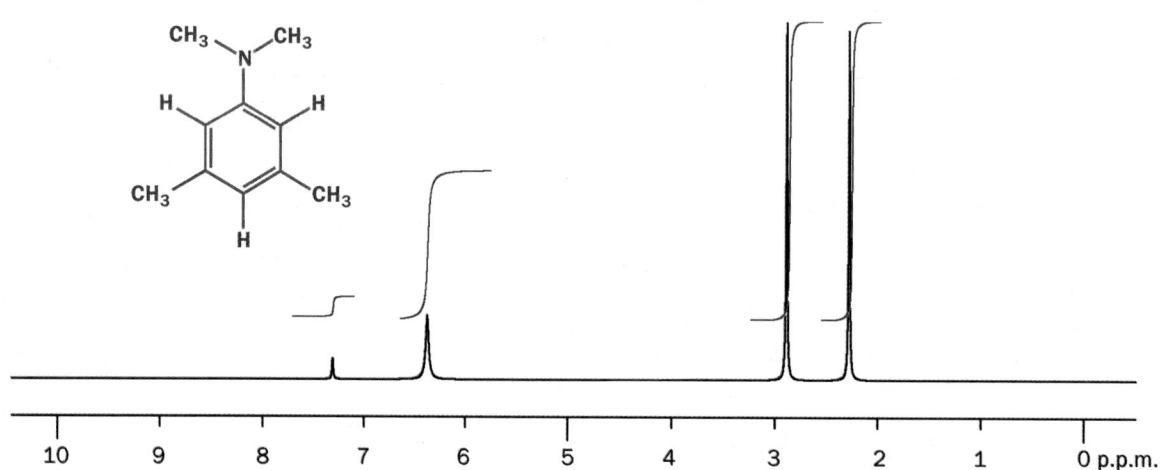

The NMR spectrum of this simple aromatic amine has three peaks in the ratio 1:2:2 which must be 3H:6H:6H. The 6.38 p.p.m. signal clearly belongs to the protons round the benzene ring, but why are they at 6.38 and not at 7.27 p.p.m.? We must also distinguish the two methyl groups at 2.28 p.p.m. from those at 2.89 p.p.m. The chart on p. 250 suggests that these should both be at about 2.4 p.p.m., close enough to 2.28 p.p.m. but not to 2.89 p.p.m. The solution to both these puzzles is the distribution of electrons in the aromatic ring. Nitrogen feeds electrons into the π system making it electron-rich: the ring protons are more shielded and the nitrogen atom becomes positively charged and its methyl groups more deshielded. The peak at 2.89 p.p.m. belongs to the NMe_2 group.

lone pair in
p orbital
on nitrogen

Other groups, such as simple alkyl groups, hardly perturb the aromatic system at all and it is quite common for all five protons in an alkyl benzene to appear as one signal instead of the three we might expect. Here is an example with some nonaromatic protons too: there is another on p. 248—the Cbz-protected amino acid.

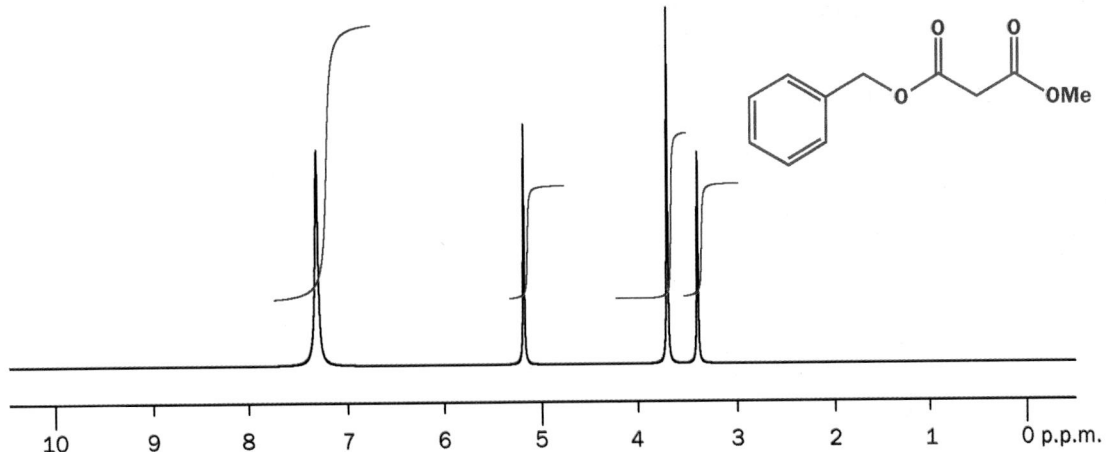

The five protons on the aromatic ring all have the same chemical shift. The OCH_3 group is typical of a methyl ester (the chart on p. 250 gives 3.9 p.p.m.). One CH_2 group is between two carbonyl groups (cf. δ 3.35 p.p.m. for the similar CH_2 group on p. 251). The other is next to an ester and a benzene ring: we calculate 1.3 + 1.5 + 3.0 = 5.8 p.p.m. for that—reasonably close to the observed 5.19 p.p.m.

How electron donation and withdrawal change chemical shifts

We can get an idea of the effect of electron distribution by looking at a series of 1,4-disubstituted benzenes. This pattern makes all the remaining hydrogens in the ring the same. The compounds are listed in order of chemical shift: largest shift (lowest field) first. Benzene itself resonates at 7.27 p.p.m. Conjugation is shown by the usual curly arrows, and inductive effects by a straight arrow by the side of the group. Only one effect and one hydrogen atom are shown; in fact, both groups exert the same effect on all four identical hydrogen atoms.

> Conjugation, as discussed in Chapters 7 and 10, is felt through π bonds, while inductive effects are the effects of electron withdrawal or donation felt simply by polarization of the σ bonds of the molecule. See p. 193.

electron-withdrawing groups

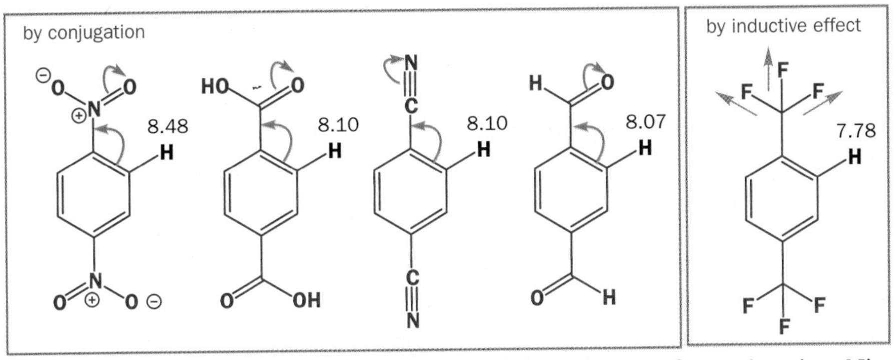

by conjugation

8.48   8.10   8.10   8.07

by inductive effect

7.78

The largest shifts come from groups that withdraw electrons by conjugation. Nitro is the most powerful—this should not surprise you as we saw the same in nonaromatic compounds both in ^{13}C and ^{1}H NMR spectra. Then come the carbonyl groups and nitrile followed by the few groups showing simple inductive withdrawal. CF_3 is an important example of this kind of group—three fluorine atoms combine to exert a powerful effect.

electron-donating and -withdrawing groups

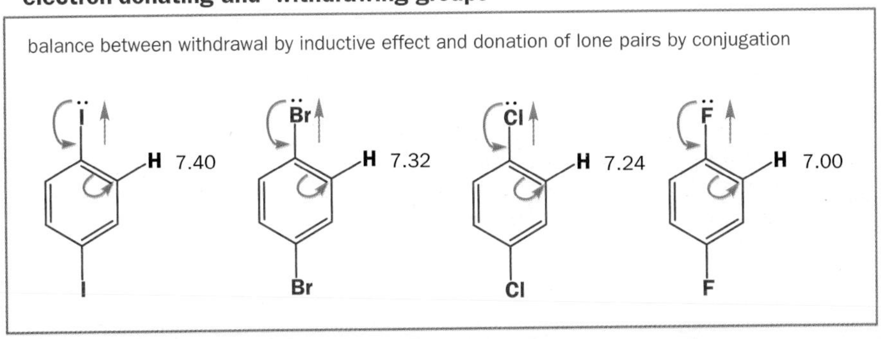

balance between withdrawal by inductive effect and donation of lone pairs by conjugation

H 7.40   Br H 7.32   Cl H 7.24   F H 7.00

In the middle, around the position of benzene itself at δ 7.27 p.p.m., come the halogens whose inductive electron withdrawal and lone pair donation are nearly balanced.

electron-donating groups

This all has very important consequences for the reactivity of differently substituted benzene rings: their reactions will be discussed in Chapter 22.

Proton NMR is, in fact, a better guide to the electron density at carbon than is carbon NMR.

Alkyl groups are weak inductive donators and at the smallest shift we have the groups that, on balance, donate electrons to the ring and increase the shielding at the carbon atoms. Amino is the best of these. So a nitrogen-based functional group (NO_2) is the best electron withdrawer while another (NH_2) is the best electron donor.

As far as the donors with lone pairs are concerned, two factors are important—the size of the lone pairs and the electronegativity of the element. If we look at the four halides (central box above) the lone pairs are in 2p(F), 3p(Cl), 4p(Br), and 5p(I) orbitals. In all cases the orbitals on the benzene ring are 2p so the fluorine orbital is of the right size and the others too large. Even though fluorine is the most electronegative, it is still the best donor.

Now comparing the groups in the first row of the p block elements. F, OH, NH_2, all have lone pairs in 2p orbitals so electronegativity is the only variable. As you would expect, the most electronegative element, F, is now the weakest donor.

Element	Electronegativity	δ_H, p.p.m.	Shift from 7.27
F	4.1	7.00	−0.27
O	3.5	6.59	−0.68
N	3.1	6.35	−0.92

Electron-rich and electron-deficient alkenes

The same sort of thing happens with alkenes. We'll concentrate on cyclohexene so as to make a good comparison with benzene. The six identical protons of benzene resonate at 7.27 p.p.m.; the two identical alkene protons of cyclohexene resonate at 5.68 p.p.m. A conjugating and electron-withdrawing group such as a ketone removes electrons from the double bond as expected—but unequally. The proton nearer the C=O group is only slightly downfield from cyclohexene but the more distant one is over 1 p.p.m. downfield. The curly arrows show the electron distribution, *which we can deduce* from the NMR spectrum.

In Chapter 10 we used ^{13}C NMR to convince you that a carbonyl group polarized a conjugated alkene; we hope you find the ^{1}H NMR data even more convincing. Conjugate addition occurs to those very atoms whose electron deficiency we can measure by proton NMR.

Oxygen as a conjugating electron donor is even more dramatic. It shifts the proton next to it downfield by the inductive effect but pushes the more distant proton upfield again by a whole p.p.m. by donating electrons. The separation between the two protons is nearly two p.p.m.

For both types of substituent, the effects are more marked on the more distant (β) proton. If these shifts reflect the true electron distribution, we can deduce that nucleophiles will attack the electron-deficient site in the nitroalkene, while electrophiles will be attacked by the electron-rich sites in silyl enol ethers and enamines. These are all important reagents and do indeed react as we predict, as you will see in later chapters. Look at the difference—there are nearly 3 p.p.m. between the nitro compound and the enamine!

electron-deficient
nitroalkene — **H** 7.31

electron-rich
silyl enol ether — **H** 4.73

electron-rich
enamine — **H** 4.42

Structural information from the alkene region

Alkene protons on different carbon atoms can obviously be different if the carbon atoms themselves are different and we have just seen examples of that. Alkene protons can also be different if they are on the same carbon atom. All that is necessary is that the substituents at the other end of the double bond should themselves be different. The silyl enol ether and the unsaturated ester below both fit into this category. The protons on the double bond must be different, because each is *cis* to a different group. The third compound is an interesting case: the different shifts of the two protons on the ring prove that the N–Cl bond is at an angle to the C=N bond. If it were in line, the two hydrogens would be identical. The other side of the C=N bond is occupied by a lone pair and the nitrogen atom is trigonal (sp² hybridized).

> DMF is similar: as we saw earlier (p. 247), it has two different methyl groups because of the double bond.

silyl enol ether

OSiMe₃ — 3.78, 3.93 — 1.02 (9H)

unsaturated ester

CO₂Me — **H** 6.10 — 1.95 (3H) — **H** 5.56

chloroimine

7.50 **H** — **H** 7.99

The aldehyde region: unsaturated carbon bonded to oxygen

The aldehyde proton is unique. It is directly attached to a carbonyl group—one of the most electron-withdrawing groups that exists—and is very deshielded, resonating with the largest shifts of any CH protons in the 9–10 p.p.m. region. The examples below are all compounds that we have met before. Two are just simple aldehydes—aromatic and aliphatic. The third is the solvent DMF. Its CHO proton is less deshielded than most—the amide delocalization that feeds electrons into the carbonyl group provides some extra shielding.

> **Aliphatic** is a catch-all term for compounds that are not aromatic.

10.14 — an aromatic aldehyde

9.0 **H** — an alphatic aldehyde

8.01 **H** — DMF

Conjugation with an oxygen atom has much the same effect—formate esters resonate at about 8 p.p.m.—but conjugation with π bonds does not. The simple conjugated aldehyde below and myrtenal both have CHO protons in the normal region (9–10 p.p.m.).

~8.0 — a formate ester

1.99 2.19 — Me **H** 9.95 — **H** 5.88 — 3-methylbut-2-enal

9.43 — myrtenal

Two other types of protons resonate in this region: some aromatic protons and some protons attached to heteroatoms like OH and NH. The first of these will provide our discussion on structural information and the second will be the subject of the section following that discussion.

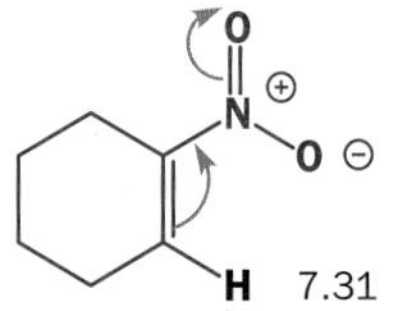

Structural information from the aldehyde region

Protons on double bonds, even very electron-deficient double bonds like those of nitroalkenes, hardly get into the aldehyde region. However, some benzene rings with very electron-withdrawing groups do manage it because of the extra downfield shift of the ring current, so beware of nitrobenzenes as they may have signals in the 8–9 p.p.m. region.

More important molecules with signals in this region are the aromatic heterocycles such as pyridine, which you met in Chapter 7. The NMR shifts clearly show that pyridine is aromatic and we discussed its basicity in Chapter 8. One proton is at 7.1 p.p.m., essentially the same as benzene, but the others are more downfield and one, at C2, is in the aldehyde region. This is not because pyridine is 'more aromatic' than benzene but because nitrogen is more electronegative than carbon. Position C2 is like an aldehyde—a proton attached to sp^2 C bearing a heteroatom—while C4 is electron-deficient by conjugation (the electronegative nitrogen is electron-withdrawing). Isoquinoline is a pyridine and a benzene ring fused together and has a proton even further downfield at 9.1 p.p.m.— this is an imine proton that experiences the ring current of the benzene ring.

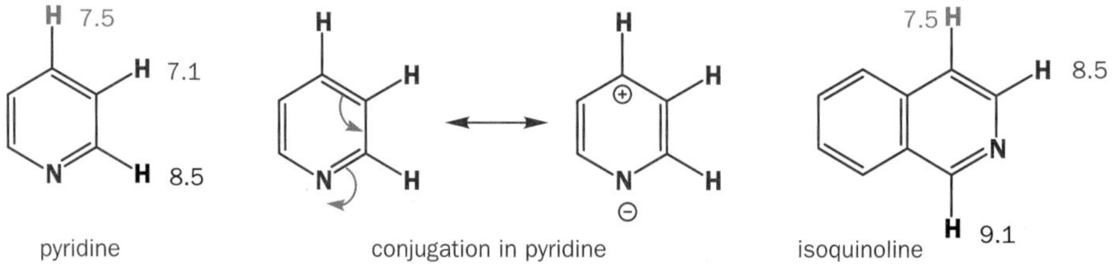

pyridine   conjugation in pyridine   isoquinoline

Protons on heteroatoms are more variable than protons on carbon

Protons directly attached to O, N, or S (or any other heteroatom, but these are the most important) also have signals in the NMR spectrum. We have avoided them so far because the positions of these signals are less reliable and because they are affected by exchange.

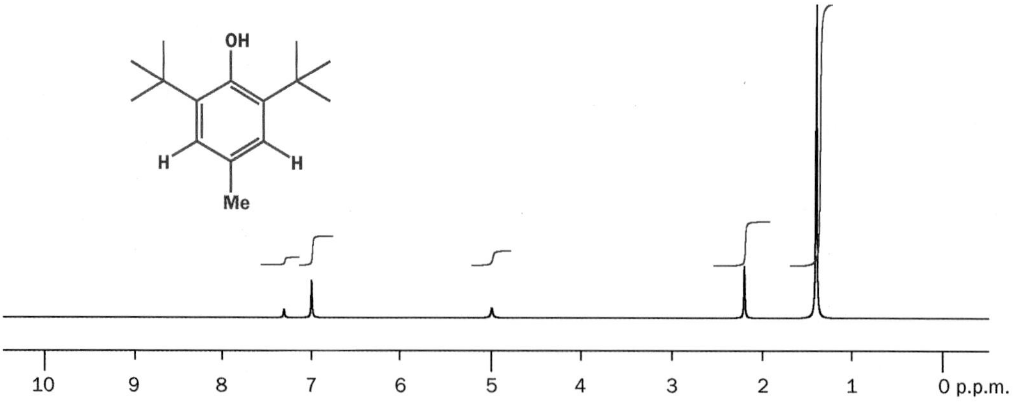

In Chapter 3 we looked at the ^{13}C NMR spectrum of BHT. Its proton NMR is very simple, consisting of just four lines with integrals 2, 1, 3, and 18. The chemical shifts of the *t*-butyl group, the methyl group on the benzene ring, and the two identical aromatic protons should cause you no surprise. What is left, the 1H signal at 5.0 p.p.m., must be the OH. Earlier on in this chapter we saw the spectrum of acetic acid CH_3CO_2H, which showed an OH resonance at 11.2 p.p.m. Simple alcohols such as *t*-butanol have OH signals in $CDCl_3$ (the usual NMR solvent) at around 2 p.p.m. Why such differences?

electron-deficient
nitroalkene

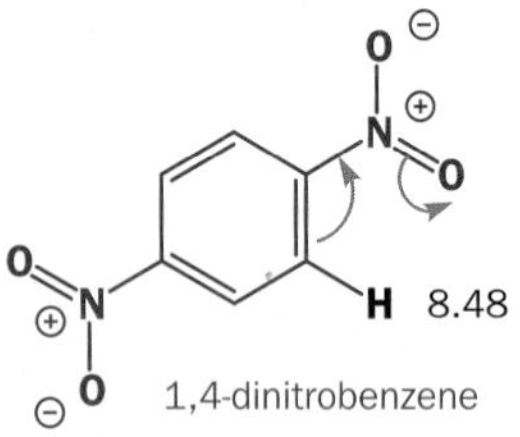

1,4-dinitrobenzene

► Please note that the alternative 'conjugation' shown in the structure below is wrong. The structure with two adjacent double bonds in a six-membered ring is impossible and, in any case, as you saw in Chapter 8, the lone pair electrons on nitrogen are in an sp^2 orbital orthogonal to the p orbitals in the ring. There is no interaction between orthogonal orbitals.

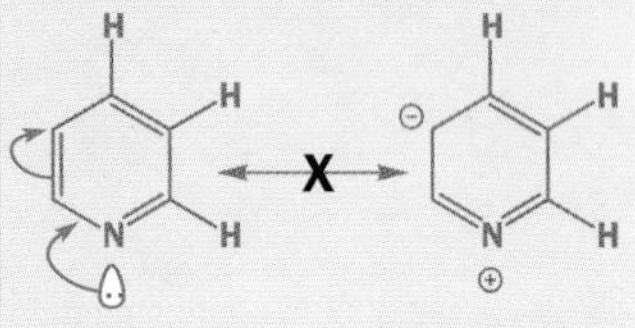

incorrect   impossible
delocalization   structure

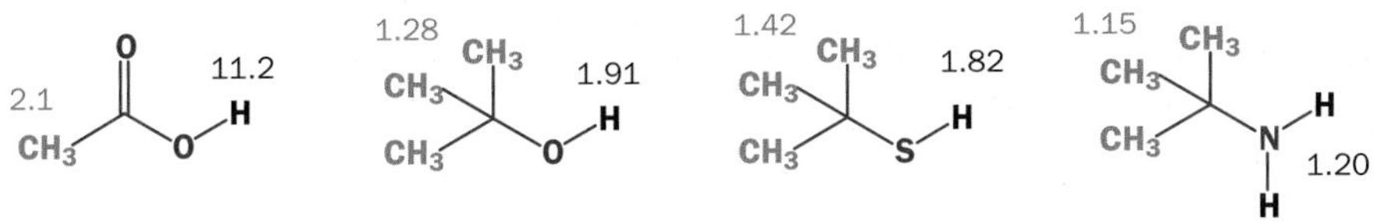

acetic acid

t-BuOH in CDCl$_3$

t-BuSH in CDCl$_3$

t-BuNH$_2$ in CDCl$_3$

This is a matter of acidity. The more acidic a proton is—that is, the more easily it releases H$^+$ (this is the definition of acidity from Chapter 8)—the more the OH bond is polarized towards oxygen. The more the RO–H bond is polarized, the closer we are to free H$^+$, which would have no shielding electrons at all, and so the further the proton goes downfield. The OH chemical shifts and the acidity of the OH group are very roughly related.

Functional group	Alcohol ROH	Phenol ArOH	Carboxylic acid RCO$_2$H
pK_a	16	10	5
δ_H(OH), p.p.m.	2.0	5.0	>10

Thiols (RSH) behave in a similar way to alcohols but are not so deshielded, as you would expect from the smaller electronegativity of sulfur (phenols are all about 5.0 p.p.m., PhSH is at 3.41 p.p.m.). Alkane thiols appear at about 2 p.p.m. and arylthiols at about 4 p.p.m. Amines and amides show a big variation, as you would expect for the variety of functional groups involved, and are summarized below. Amides are slightly acidic, as you saw in Chapter 8, and amide protons resonate at quite low fields. Pyrroles are special—the aromaticity of the ring makes the NH proton unusually acidic and they appear at about 10 p.p.m.

chemical shifts of NH protons

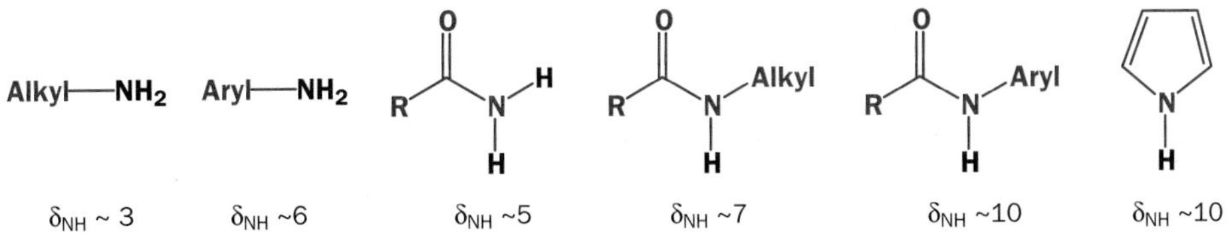

Alkyl—NH$_2$ Aryl—NH$_2$

$\delta_{NH} \sim 3$ $\delta_{NH} \sim 6$ $\delta_{NH} \sim 5$ $\delta_{NH} \sim 7$ $\delta_{NH} \sim 10$ $\delta_{NH} \sim 10$

Exchange of acidic protons is revealed in proton NMR spectra

Compounds with very polar groups often dissolve best in water. NMR spectra are usually run in CDCl$_3$, but heavy water, D$_2$O, is an excellent NMR solvent. Here are some results in that medium.

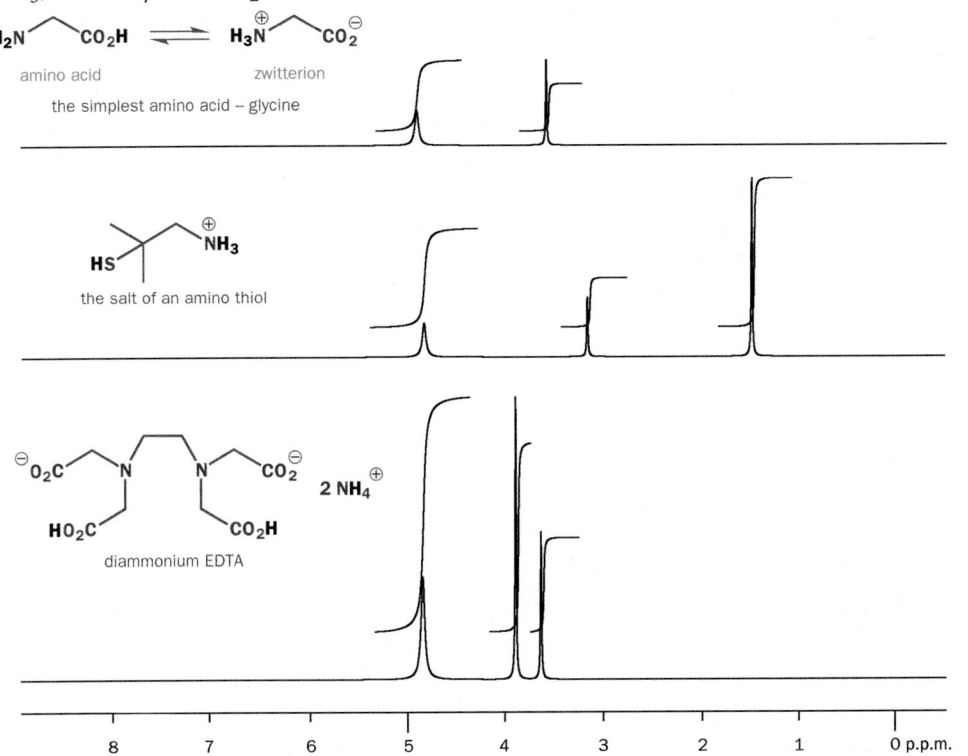

amino acid   zwitterion

the simplest amino acid – glycine

the salt of an amino thiol

diammonium EDTA

► EDTA is ethylenediamine tetraacetic acid, an important complexing agent for metals. This is the salt formed with just two equivalents of ammonia.

Glycine is expected to exist as a zwitterion (Chapter 8, p. 183). It has a 2H signal for the CH_2 between the two functional groups, which would do for either form. The 3H signal at 4.90 p.p.m. might suggest the NH_3^+ group, but wait a moment before making up your mind. The aminothiol salt has the CMe_2 and CH_2 groups about where we would expect them, but the SH and NH_3^+ protons appear as one 4H signal. The double salt of EDTA has several curious features. The two CH_2 groups in the middle are fine, but the other four CH_2 groups all appear identical as do all the protons on both the CO_2H and NH_3^+ groups.

The best clue to why this is so involves the chemical shifts of the OH, NH, and SH protons in these molecules. They are all the same within experimental error: 4.90 p.p.m. for glycine, 4.80 p.p.m. for the aminothiol, and 4.84 p.p.m. for EDTA. They all correspond to the same species: HOD. Exchange between XH (where X = O, N, or S) protons is extremely fast, and the solvent, D_2O, supplies a vast excess of exchangeable *deuteriums*. These immediately replace all the OH, NH, and SH protons in the molecules with D, forming HOD in the process. Recall that we do not see signals for deuterium atoms (that's why deuterated solvents are used). They have their own spectra at a different frequency.

The same sort of exchange between OH or NH protons with each other or with traces of water in the sample means that the OH and NH peaks in most spectra in $CDCl_3$ are rather broader than the peaks for CH protons.

Two questions remain. First, can we tell whether glycine is a zwitterion in water or not? Not really: the spectra fit either or an equilibrium between both. Other evidence leads us to prefer the zwitterion in water. Second, why are all four CH_2CO groups in EDTA the same? This we can answer. As well as the equilibrium exchanging the CO_2H protons with the solvent, there will be an equally fast equilibrium exchanging protons between CO_2H and CO_2D. This makes all four 'arms' of EDTA the same.

You should leave this section with an important chemical principle firmly established in your mind.

● **Proton exchange is fast**

Proton exchange between heteroatoms, particularly O, N, and S, is a *very fast* process in comparison with other chemical reactions, and often leads to averaged peaks in the ^1H NMR spectrum.

You will need this insight as you study organic mechanisms.

Coupling in the proton NMR spectrum

Nearby hydrogen nuclei interact and give multiple peaks

So far proton NMR has been not unlike carbon NMR on a smaller scale. However, we have yet to discuss the real strength of proton NMR, something more important than chemical shifts and something that allows us to look not just at individual atoms but also at the way the C–H skeleton is joined together. This is the result of the interaction between nearby protons known as **coupling**.

An example we could have chosen in the last section is the nucleic acid component, cytosine, which has exchanging NH_2 and NH protons giving a peak for HDO at 4.5 p.p.m. We didn't choose this example because the other two peaks would have puzzled you. Instead of giving just one line each, they give two lines each—doublets as you will learn to call them—and it is time to discuss the origin of this 'coupling'.

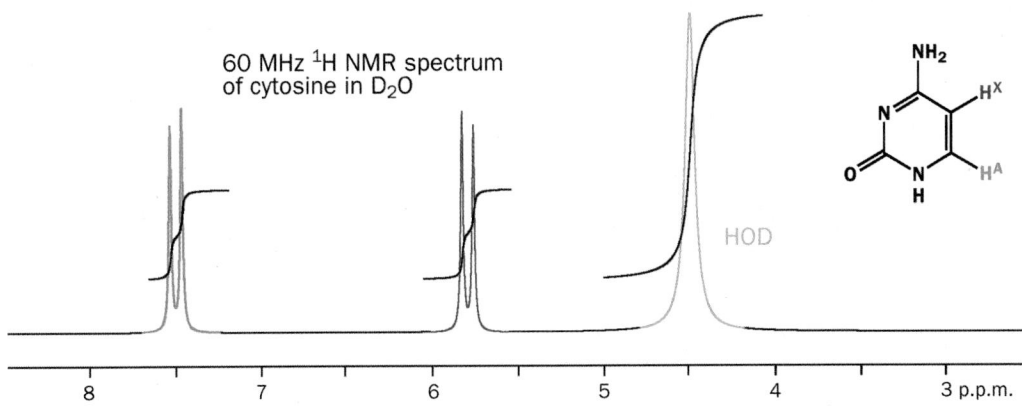

▶

Cytosine is one of the four bases that, in combination with deoxyribose and phosphate, make up DNA. It is a member of the class of heterocycles called **pyrimidines**. We come back to the chemistry of DNA towards the end of this book, in Chapter 49.

You might have expected a spectrum like that of the heterocycle below, which is also a pyrimidine. It too has exchanging NH_2 protons and two protons on the heterocyclic ring. But these two protons give the expected two lines instead of the four lines in the cytosine spectrum. It is easy to assign the spectrum: proton H^A is attached to an aldehyde-like C=N and so comes at lowest field. The proton H^X is *ortho* to two electron-donating NH_2 groups and so comes at high field for an aromatic proton (p. 254). These protons do not couple with each other because they are too far apart. They are separated by five bonds whereas the ring protons in cytosine are separated by just three bonds.

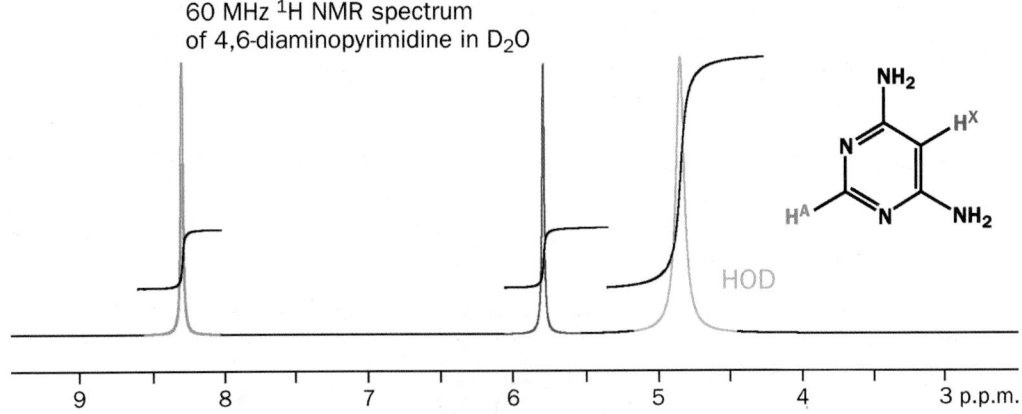

Understanding this phenomenon is so important that we are going to explain it in three different ways—you choose which appeals to you most. Each method offers a different insight.

The pyrimidine spectrum has two single lines (**singlets** we shall call them from now on) because each proton, H^A or H^X, can be aligned either with or against the applied magnetic field. The cytosine spectrum is different because each proton, say, H^A, is near enough to experience the small magnetic field of the other proton H^X as well as the field of the magnet itself. The diagram shows the result.

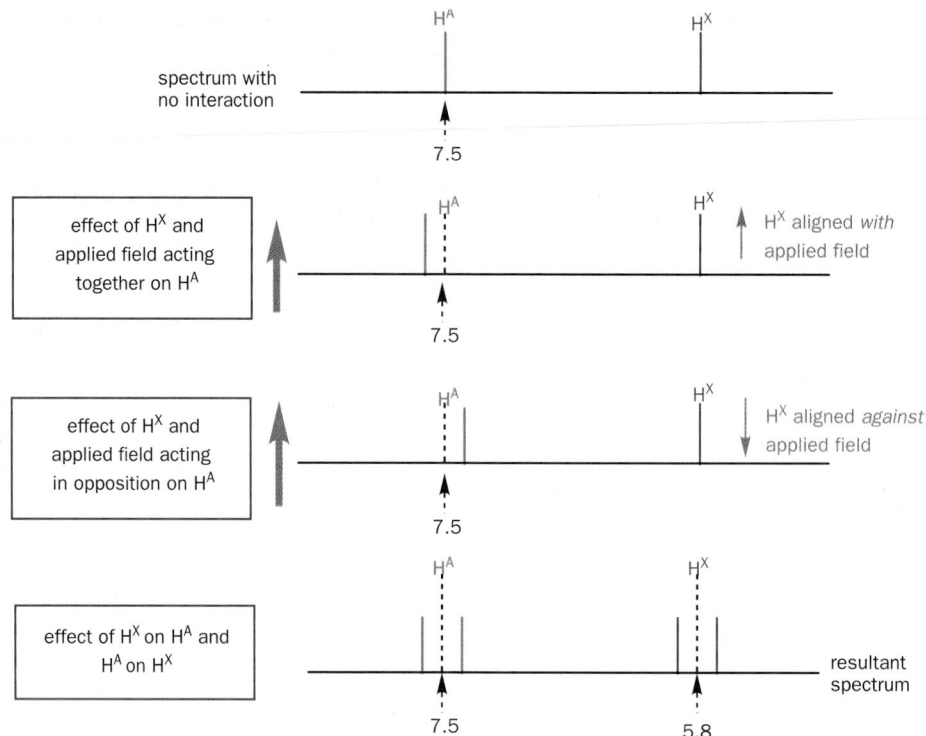

If each proton interacted only with the applied field we would get two singlets. But proton H^A actually experiences two slightly different fields: the applied field *plus* the field of H^X or the applied field *minus* the field of H^X. H^X acts either to increase or to decrease the field experienced by H^A. The position of a resonance depends on the field experienced by the proton so these two situations give rise to two slightly different peaks—a **doublet** as we shall call it. And whatever happens to H^A happens to H^X as well, so the spectrum has two doublets, one for each proton. Each couples with the other. The field of a proton is a very small indeed in comparison with the field of the magnet and the separation between the lines of a doublet is very small. We shall discuss the size of the coupling later (p. 269).

The second explanation takes into account the energy levels of the nucleus. In Chapter 4, when we discussed chemical bonds, we imagined electronic energy levels on neighbouring atoms interacting with each other and splitting to produce new molecular energy levels, some higher in energy and some lower in energy than the original atomic energy levels. When hydrogen *nuclei* are near each other in a molecule, the nuclear energy levels also interact and split and produce new energy levels. If a single hydrogen nucleus interacts with a magnetic field, we have the picture on p. 243 of this chapter: there are *two* energy levels as the nucleus can be aligned with or against the applied magnetic field, there is one energy jump possible, and there is a resonance at one frequency. This you have now seen many times and it can be summarized as shown below.

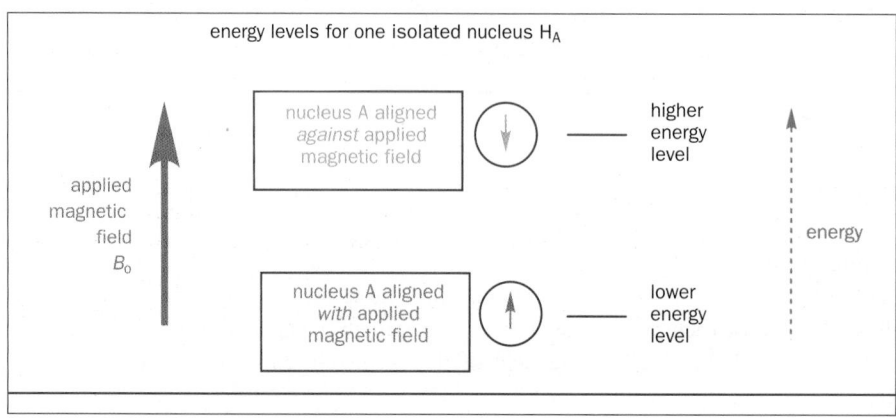

The spectrum of the pyrimidine on p. 259 showed two protons each independently in this situation. Each had two energy levels, each gave a singlet, and there were two lines in the spectrum. But, in the cytosine molecule, each proton has another hydrogen nucleus nearby and there are now *four* energy levels. Each nucleus H^A and H^X can be aligned with or against the applied field. There is one most stable energy level where they are both aligned with the field and one least stable level where they are both aligned against. In between there are two different energy levels in which one nucleus is aligned with the field and one against. Exciting H^A from alignment with to alignment against the applied field can be done in two slightly different ways, shown as A_1 and A_2 on the diagram. The result is two resonances very close together in the spectrum.

energy levels for two interacting nuclei H^A and H^X

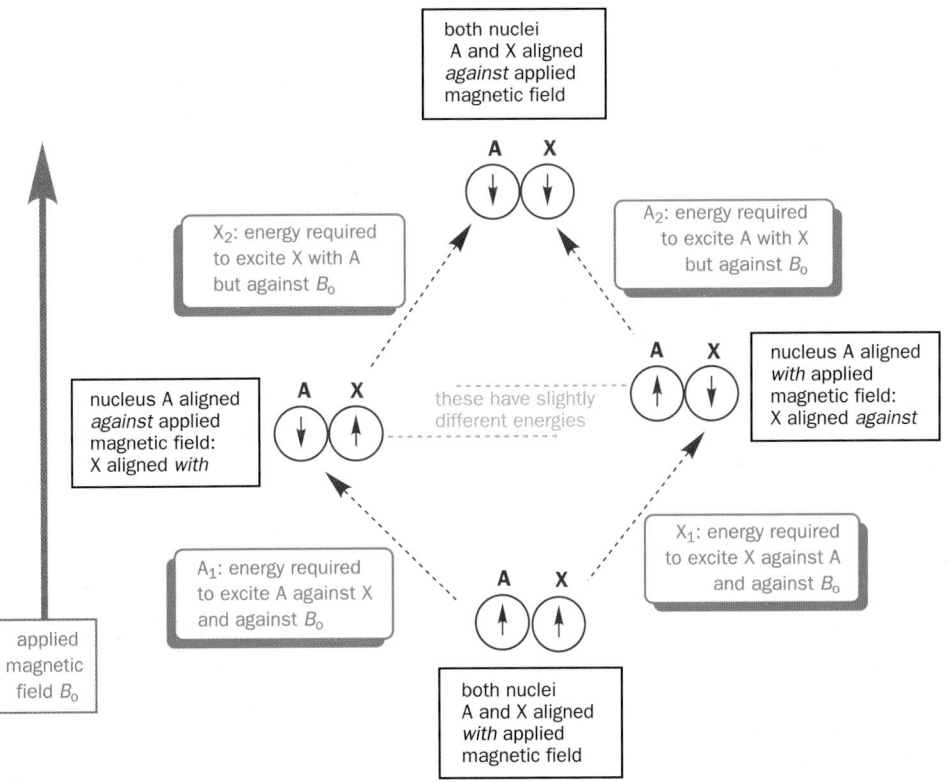

Please notice carefully that we cannot have this discussion about H^A without discussing H^X in the same way. If there are two slightly different energy jumps to excite H^A, there must also be two slightly different energy jumps to excite H^X. The difference between A_1 and A_2 is exactly the same as the difference between X_1 and X_2. Each proton now gives two lines (a doublet) in the NMR spectrum and the splitting of the two doublets is *exactly the same*. We describe this situation as **coupling**. We say 'A and X are coupled' or 'X is coupled to A' (and vice versa, of course). We shall be using this language from now on and so must you.

Now look back at the spectrum of cytosine at the beginning of this section. You can see the two doublets, one for each of the protons on the aromatic ring. Each is split by the same amount (this is easy to check with a ruler) and the separation of the lines is the **coupling constant** and is called J. In this case $J = 4$ Hz. Why do we measure J in hertz and not in p.p.m.? We measure chemical shifts in p.p.m. because we get the same number regardless of the rating of the NMR machine in MHz. We measure J in Hz because we also get the same number regardless of the machine.

▶ **Measuring coupling constants in hertz**

▶ **Measuring coupling constants in hertz**

To measure a coupling constant it is essential to know the rating of the NMR machine in MHz (MegaHertz). This is why you are told that each illustrated spectrum is, say, a '250 MHz ^{1}H NMR spectrum'. To measure the coupling, measure the distance between the lines by ruler or dividers and use the horizontal scale to find out the separation in p.p.m. The conversion is then easy—to turn parts per million of megahertz into hertz you just leave out the million! So 1 p.p.m. on a 300 MHz machine is 300 Hz. On a 90 MHz machine it would be 90 Hz.

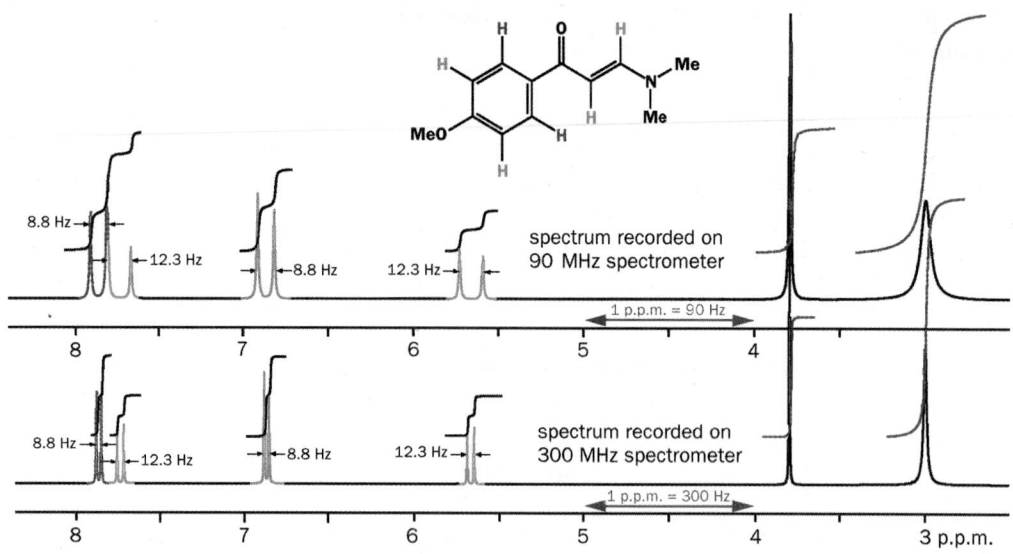

spectrum recorded on 90 MHz spectrometer

1 p.p.m. = 90 Hz

spectrum recorded on 300 MHz spectrometer

1 p.p.m. = 300 Hz

● **Spectra from different machines**

When you change from one machine to another, say, from an 80 MHz to a 500 MHz NMR machine, chemical shifts (δ) stay the same in p.p.m. but coupling constants (J) stay the same in Hz.

Now for the third way to describe coupling. If you look again at what the spectrum would be like without interaction between H^A and H^X you would see this, with the chemical shift of each proton clearly obvious.

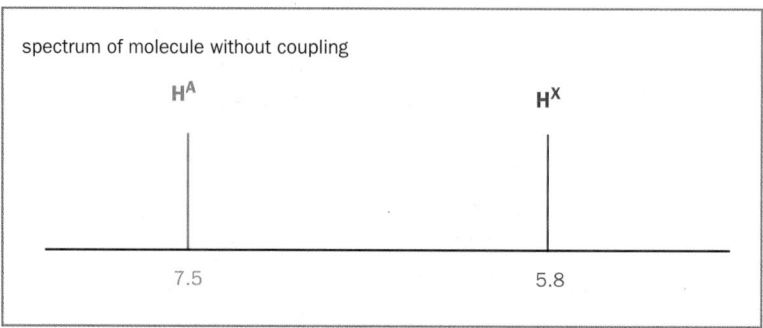

spectrum of molecule without coupling

But you don't see this because each proton couples with the other and splits its signal by an equal amount either side of the true chemical shift.

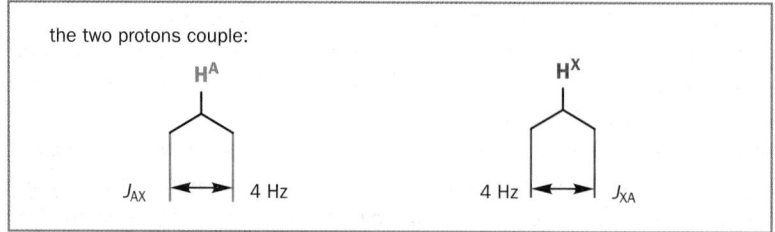

the two protons couple:

The true spectrum has a pair of doublets each split by an identical amount. Note that no line appears at the true chemical shift, but it is easy to measure the chemical shift by taking the midpoint of the doublet.

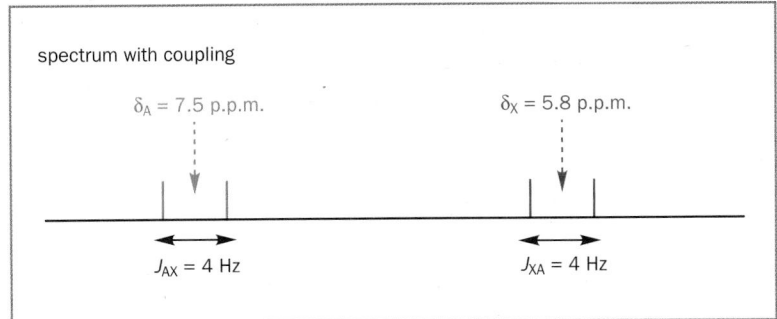

spectrum with coupling

$\delta_A = 7.5$ p.p.m.

$\delta_X = 5.8$ p.p.m.

$J_{AX} = 4$ Hz

$J_{XA} = 4$ Hz

So this spectrum would be described as δ_H 7.5 (1H, d, *J* 4 Hz, H^A) and 5.8 (1H, d, *J* 4 Hz, H^X). The main number gives the chemical shift in p.p.m. and then, in brackets, comes the integration as the number of Hs, the shape of the signal (here 'd' for doublet), the size of coupling constants in Hz, and the assignment, usually related to a diagram. The integration refers to the combined integral of both peaks in the doublet. If the doublet is exactly symmetrical, each peak integrates to half a proton. The combined signal, however complicated, integrates to the right number of protons.

We have described these protons as A and X with a purpose in mind. A spectrum of two equal doublets is called an **AX spectrum**. A is always the proton you are discussing and X is a proton with a very different chemical shift. The alphabet is used as a ruler: nearby protons (on the chemical shift scale—not necessarily nearby in the structure!) are called B, C, etc. and distant ones are called X, Y, etc. You will see the reason for this soon.

If there are more protons involved, the splitting process continues. Here is the NMR spectrum of a famous perfumery compound supposed to have the smell of 'green leaf lilac'. The compound is an acetal with five nearly identical aromatic protons at the normal benzene position (7.2–7.3 p.p.m.) and six protons on two identical OMe groups.

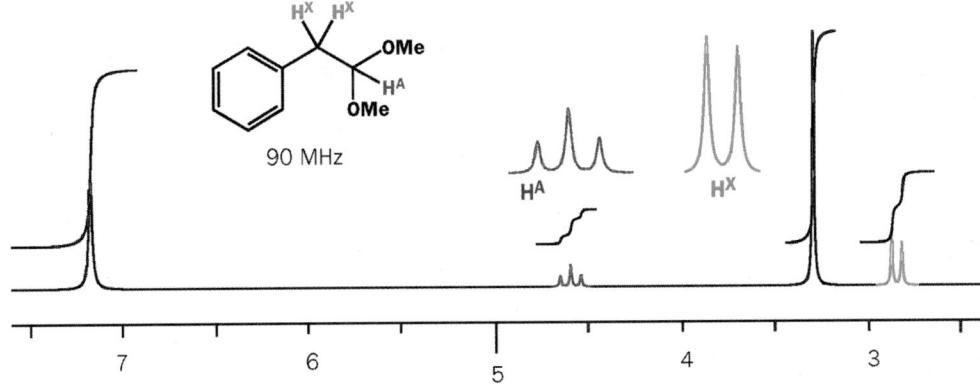

90 MHz

H^A

H^X

It is the remaining three protons that interest us. They appear as a 2H doublet at 2.9 p.p.m. and a 1H *triplet* at 4.6 p.p.m. In NMR talk, **triplet** means three equally spaced lines in the ratio 1:2:1. The triplet arises from the three possible states of the two identical protons in the CH$_2$ group.

If one proton H^A interacts with two protons H^X, it can experience three states of proton H^X. Both protons H^X can be aligned with the magnet or both against. These states will increase or decrease the applied field just as before. But if one proton H^X is aligned with, and one against the applied field, there is no net change to the field experienced by H^A and there are two possibilities for this (see diagram). We therefore see a signal of double intensity for H^A at the correct chemical shift, one signal at higher field and one at lower field. In other words, a 1:2:1 triplet.

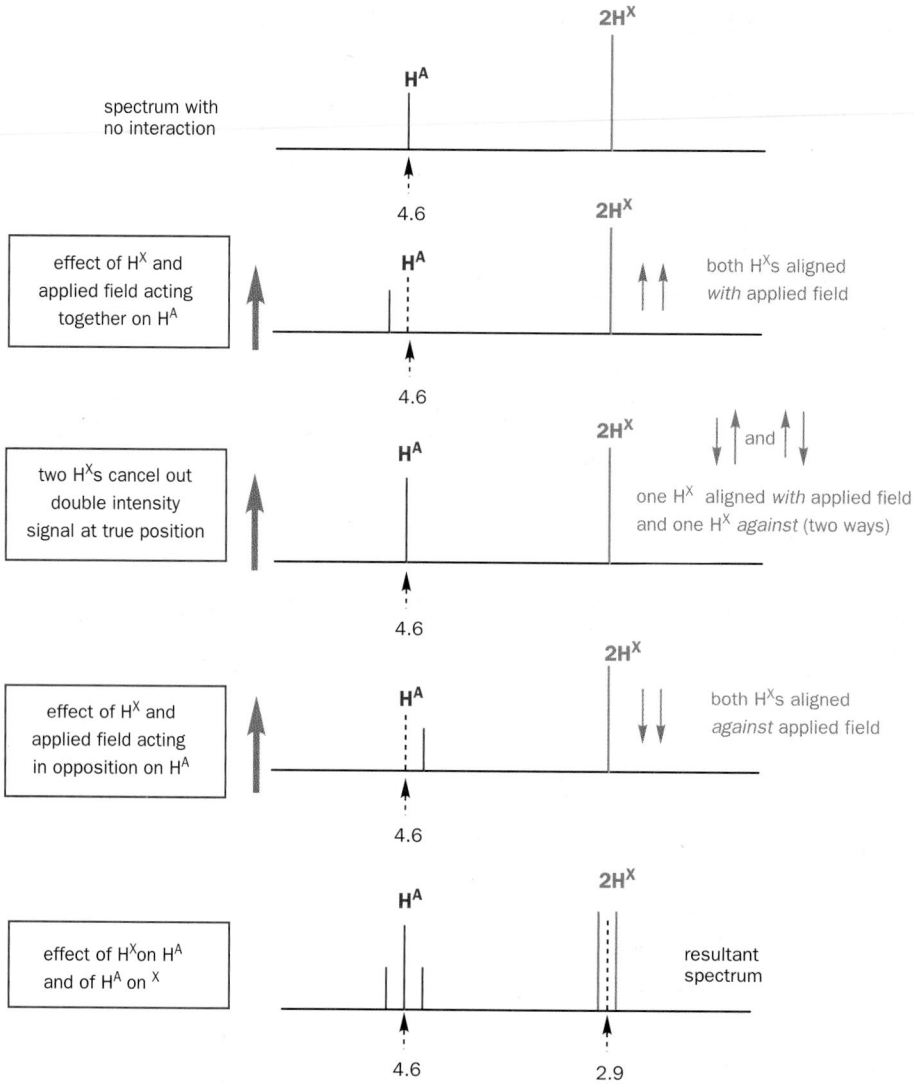

We could look at this result by our other methods too. There is one way in which both nuclei can be aligned with and one way in which both can be aligned against the applied field, but two ways in which they can be aligned one with and one against. Proton H^A interacts with each of these states. The result is a 1:2:1 triplet.

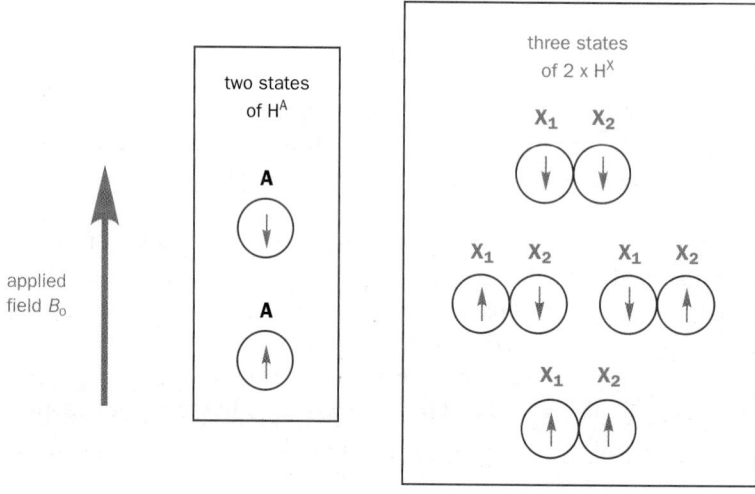

Using our third way to see how the triplet arises, we can look at the splitting as it happens.

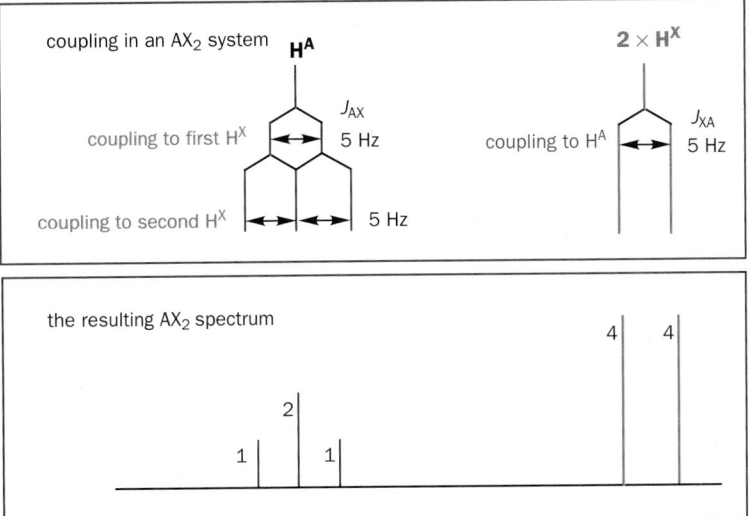

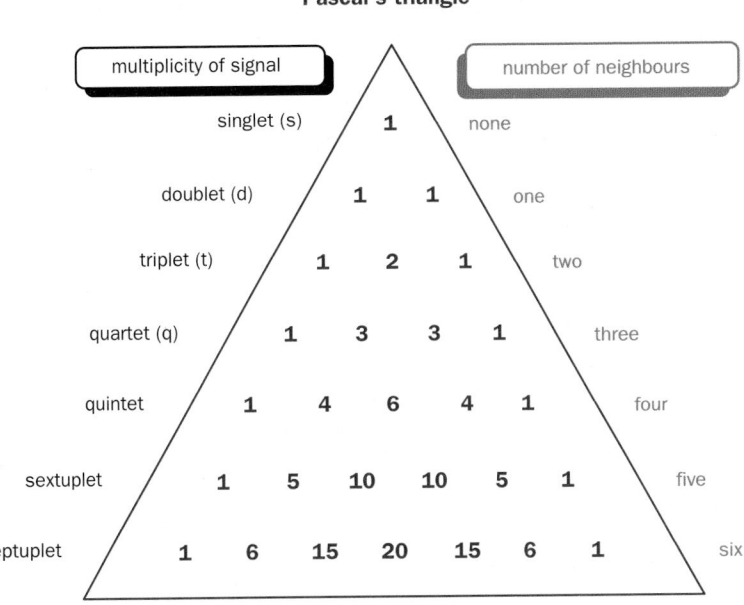

If there are more protons involved, we continue to get more complex systems, but the intensities can all be deduced simply from Pascal's triangle, which gives the coefficients in a binomial expansion. If you are unfamiliar with this simple device, here it is.

Pascal's triangle

multiplicity of signal		number of neighbours

singlet (s)				1				none
doublet (d)			1		1			one
triplet (t)		1		2		1		two
quartet (q)		1	3		3		1	three
quintet	1		4	6		4	1	four
sextuplet	1	5		10	10		5	1	five
septuplet	1	6	15		20		15	6	1	six

> ▶ **Constructing Pascal's triangle**
>
> Put '1' at the top and then add an extra number in each line by adding together the numbers on either side of the new number in the line above. If there is no number on one side, that counts as a zero, so the lines always begin and end with '1'.

You can read off from the triangle what pattern you may expect when a proton is coupled to n equivalent neighbours. There are always $n + 1$ peaks with the intensities shown by the triangle. So far, you've seen 1:1 doublets (line 2 of the triangle) from coupling to 1 proton, and 1:2:1 triplets (line 3) from coupling to 2. You will often meet ethyl groups (CH_3–CH_2X) where the CH_2 group appears as a 1:3:3:1 quartet and the methyl group as a 1:2:1 triplet and isopropyl groups $(CH_3)_2CHX$ where the methyl groups appear as a 6H doublet and the CH group as a septuplet. The outside lines of a septuplet are so weak (1/20th of the middle line) that it is often mistaken for a quintet. Inspection of the integral should put you on the right track.

Here is a simple example, the four-membered cyclic ether oxetane. Its NMR spectrum has a 4H triplet for the two identical CH_2 groups next to oxygen and a 2H quintet for the CH_2 in the middle. Each proton H^X 'sees' four identical neighbours (H^A) and is split equally by them all to give a

▶ **Constructing Pascal's triangle**

Remember, the coupling comes from the *neighbouring* protons: it doesn't matter how many protons form the signal itself (2 for H^X, 4 for H^A)—it's how many are next door (4 next to H^X, 2 next to H^A) that matters. It's *what you see* that counts not *what you are*.

1:4:6:4:1 quintet. Each proton H^A 'sees' two identical neighbours H^X and is split into a 1:2:1 triplet. The combined integral of all the lines in the quintet together is 2 and of all the lines in the triplet is 4.

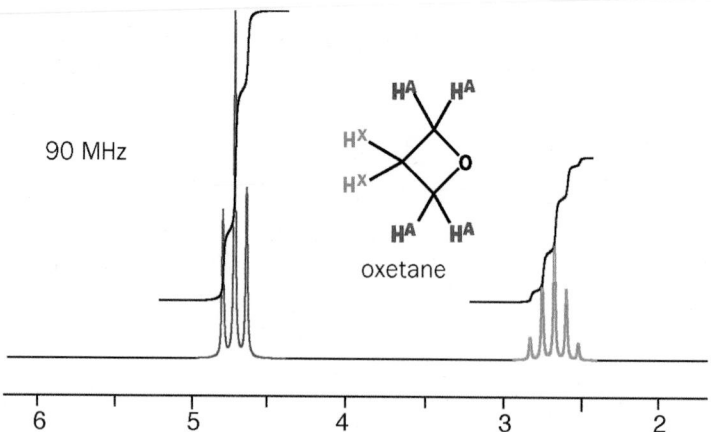

A slightly more complicated example is the diethyl acetal below. It has a simple AX pair of doublets for the two protons on the 'backbone' (red and green) and a typical ethyl group (2H quartet and 3H triplet). An ethyl group is attached to only one substituent through its CH_2 group, so the chemical shift of that CH_2 group tells us what it is joined to. Here the peak at 3.76 p.p.m. can only be an OEt group. There are, of course, two identical CH_2 groups in this molecule.

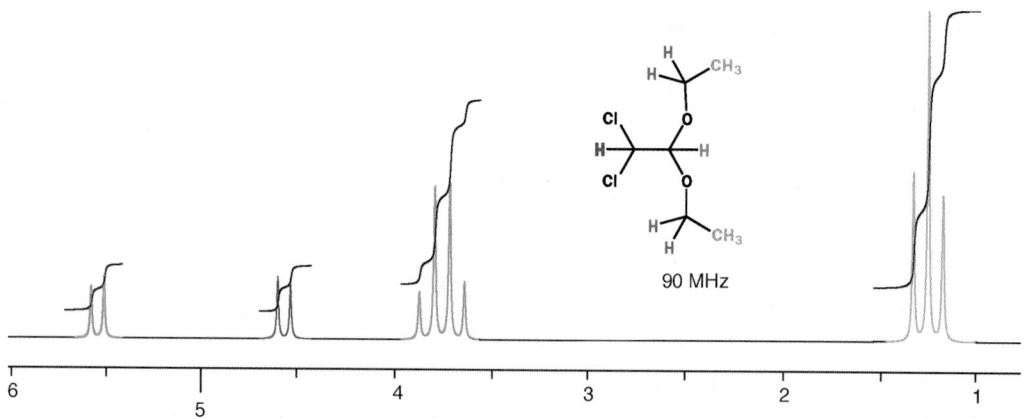

So far, we have seen situations where a proton has several neighbours, but the coupling constants to all the neighbours have been the same. What happens when coupling constants differ? Chrysanthemic acid, the structural heart of the natural pyrethrin insecticides, gives an example of the simplest situation—where a proton has two different neighbours.

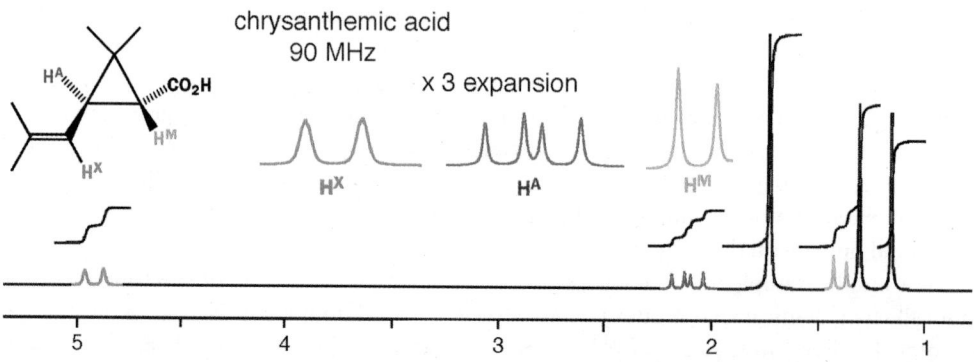

This is an interesting three-membered ring compound produced by pyrethrum flowers (Chapter 1). It has a carboxylic acid, an alkene, and two methyl groups on the three-membered ring. Proton H^A has two neighbours, H^X and H^M. The coupling constant to H^X is 8 Hz, and that to H^M is 5.5 Hz. The splitting pattern looks like this (right).

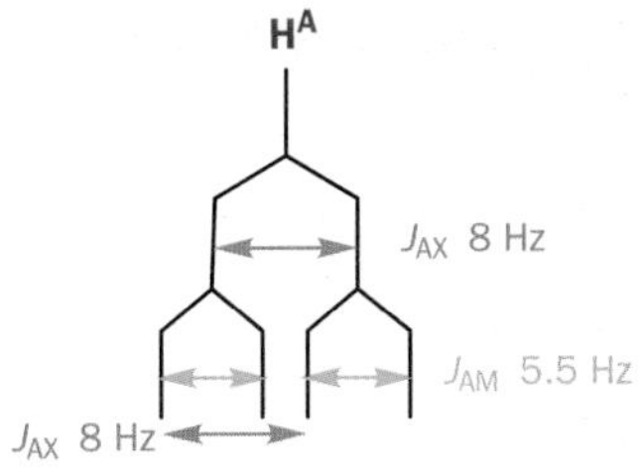

Abbreviations used for style of signal

Abbreviation	Meaning	Comments
s	singlet	might be 'broad'
d	doublet	equal in height
t	triplet	should be 1:2:1
q	quartet	should be 1:3:3:1
dt	double triplet	other combinations too, such as dd, dq, tq...
m	multiplet	avoid if possible but sometimes necessary to describe complicated signals

The result is four lines of equal intensity called a **double doublet** (or sometimes a doublet of doublets), abbreviation dd. The smaller coupling constant can be read off from the separation between lines 1 and 2 or between lines 3 and 4, while the larger coupling constant is between lines 1 and 3 or between lines 2 and 4. You could see this as an imperfect triplet where the second coupling is too small to bring the central lines together: alternatively, look at a triplet as a special case of a double doublet where the two couplings are identical.

▶ Coupling constants to two identical protons must be identical but, if the protons differ, the coupling constants must also be different (though sometimes by only a very small amount).

Coupling is a through bond effect

Neighbouring nuclei might interact through space or through the electrons in the bonds. We know that coupling is in fact a 'through bond effect' because of the way coupling constants vary with the shape of the molecule. The most important case occurs when the protons are at either end of a double bond. If the two hydrogens are *cis*, the coupling constant J is typically about 10 Hz but, if they are *trans*, J is much larger, usually 15–18 Hz. These two chloro acids are good examples.

H atoms distant orbitals parallel   hydrogens are *trans* $J = 15$ Hz   H atoms close orbitals not parallel   hydrogens are *cis* $J = 9$ Hz

If coupling were through space, the nearer *cis* hydrogens would have the larger J. In fact, coupling occurs *through the bonds* and the more perfect parallel alignment of the orbitals in the *trans* compound provides better communication and a larger J.

Coupling is at least as helpful as chemical shift in assigning spectra. When we said that the protons on cyclohexenone had the chemical shifts shown, how did we know? It was coupling that told us the answer. The proton next to the carbonyl group has one neighbour and appears as a doublet with $J = 11$ Hz, just right for a proton on a double bond with a *cis* neighbour. The proton at the other end appears as a double triplet. Inside each triplet the separation of the lines is 4 Hz and the two triplets are 11 Hz apart. This means the following diagramatically.

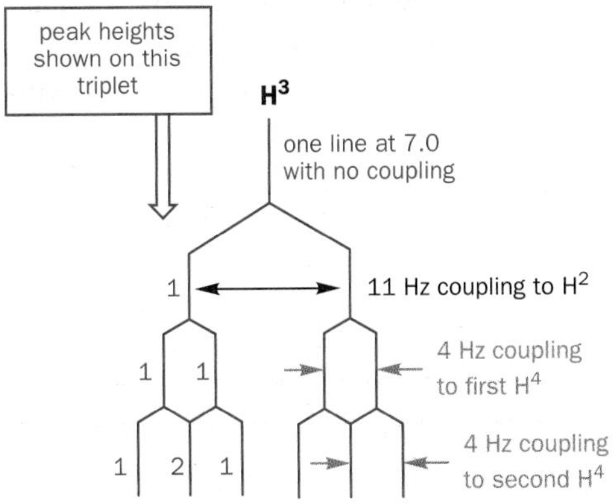

▶ For the same reason—orbital overlap—this *anti* arrangement of substituents is also preferred in chemical reactions such as elimination (Chapter 19) and fragmentation (Chapter 38).

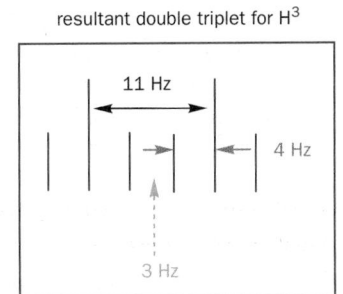

peak heights shown on this triplet

H^3

one line at 7.0 with no coupling

11 Hz coupling to H^2

4 Hz coupling to first H^4

4 Hz coupling to second H^4

resultant double triplet for H^3

11 Hz

4 Hz

3 Hz

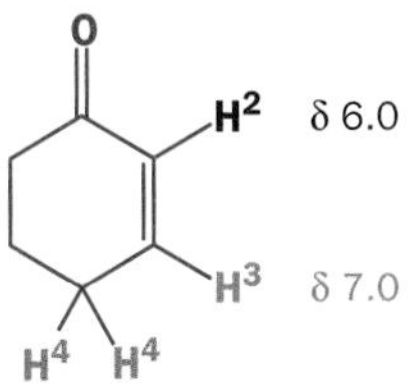

H^2 δ 6.0

H^3 δ 7.0

H^4 H^4

This is what happens when a proton couples to different groups of protons with different coupling constants. Many different coupling patterns are possible, many can be interpreted, but others cannot. However, machines with high field magnets make the interpretation easier. As a demonstration, let us turn back to the bee alarm pheromone that we met in Chapter 3. An old 90 MHz NMR spectrum of this compound looks like this.

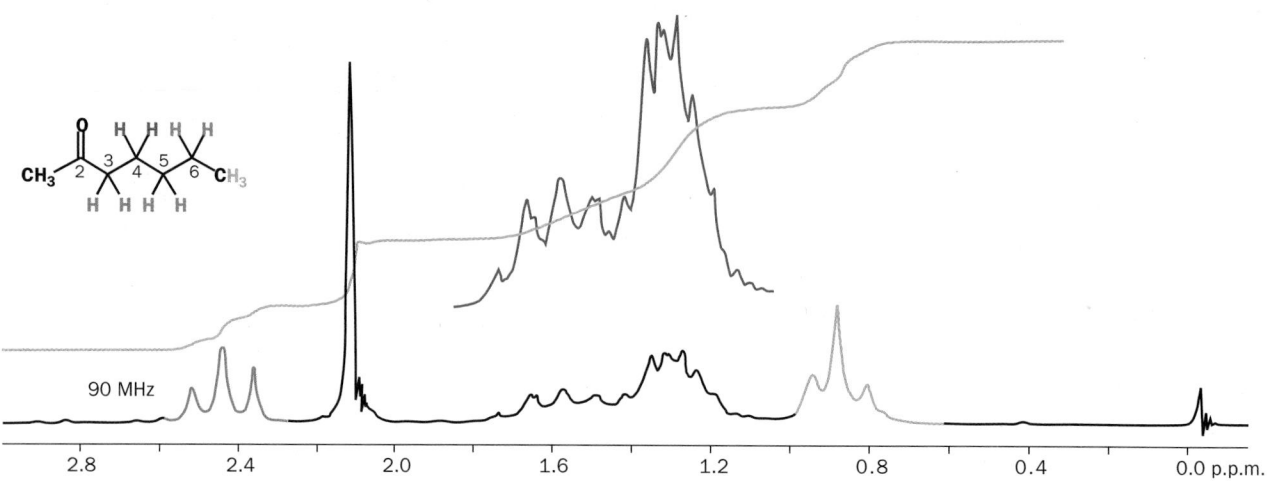

You can see the singlet for the isolated black methyl group and just about make out the triplets for the green CH_2 group next to the ketone (C3) at about 2.5 p.p.m. and for the orange methyl group at 0.9 p.p.m. (C7) though this is rather broad. The rest is frankly a mess. Now see what happens when the spectrum is run on a more modern 500 MHz spectrometer.

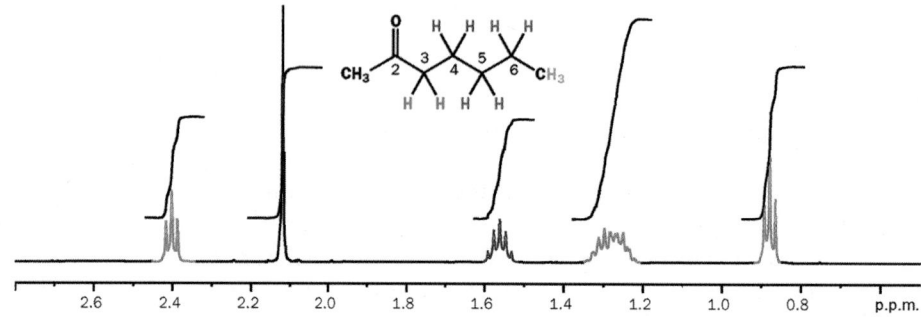

Notice first of all that the chemical shifts have not changed. However, all the peaks have closed up. This is because J stays the same in Hz and the 7 Hz coupling for the methyl group triplet was 7/90 = 0.07 p.p.m. at 90 MHz but is 7/500 = 0.014 p.p.m. at 500 MHz. In the high field spectrum you can easily see the singlet and the two triplets but you can also see a clear quintet for the red CH_2 group at C4, which couples to both neighbouring CH_2 groups with the same J (7 Hz). Only the two CH_2 groups at C5 and C6 are not resolved. However, this does not matter as *we know they are there from the coupling pattern*. The triplet for the orange methyl group at C7 shows that it is next to a CH_2 group, and the quintet for the CH_2 group at C4 shows that it is next to two CH_2 groups. We know about one of them, at C5, so we can infer the other at C6.

Coupling constants depend on three factors

In heptanone all the coupling constants were about the same but in cyclohexenone they were quite different. What determines the size of the coupling constant? There are three factors.

- Through bond distance between the protons
- Angle between the two C–H bonds
- Electronegative substituents

The coupling constants we have seen so far are all between hydrogen atoms on neighbouring carbon atoms. The coupling is through three bonds (H–C–C–H) and is designated $^3J_{HH}$. These coupling constants $^3J_{HH}$ are usually about 7 Hz in an open-chain, freely rotating system such as we have in heptanone. The C–H bonds vary little in length but the C–C bond might be a single or a double bond. In cyclohexenone it is a double bond, significantly shorter than a single bond. Couplings ($^3J_{HH}$) across double bonds are usually larger than 7 Hz (11 Hz in cyclohexenone). $^3J_{HH}$ couplings are called **vicinal** couplings because the protons concerned are on neighbouring carbon atoms.

Something else is different too: in an open-chain system we have a time average of all rotational conformations. Across a double bond there is no rotation and the angle between the two C–H bonds is fixed because they are in the same plane. In the plane of the alkene, the C–H bonds are either at 60° (*cis*) or at 180° (*trans*) to each other. Coupling constants in benzene rings are slightly less than those across *cis* alkenes because the bond is longer (bond order 1.5 rather than 2).

$^3J_{HH}$ coupling constants

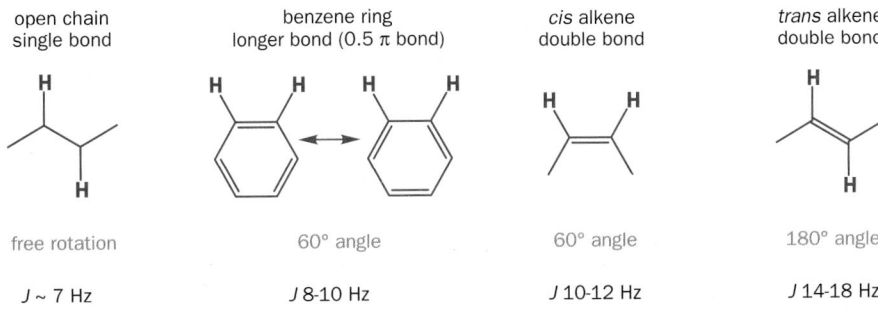

open chain single bond	benzene ring longer bond (0.5 π bond)	*cis* alkene double bond	*trans* alkene double bond
free rotation	60° angle	60° angle	180° angle
J ~ 7 Hz	J 8-10 Hz	J 10-12 Hz	J 14-18 Hz

In naphthalenes, there are unequal bond lengths around the two rings. The bond between the two rings is the shortest, and the lengths of the others are shown. Coupling across the shorter bond (8 Hz) is significantly stronger than coupling across the longer bond (6.5 Hz).

The effect of the third factor, electronegativity, is easily seen in the comparison between ordinary alkenes and enol ethers. We are going to compare two series of compounds with a *cis* or a *trans* double bond. One series has a phenyl group at one end of the alkene and the other has an OPh group. Within each box, that is for either series, the *trans* coupling is larger than the *cis*, as you would now expect. But if you compare the two series, the enol ethers have much smaller coupling constants. The *trans* coupling for the enol ethers is only just larger than the *cis* coupling for the alkenes. The electronegative oxygen atom is withdrawing electrons from the C–H bond in the enol ethers and weakening communication through the bonds.

■
Conjugation in naphthalene was discussed in Chapter 7, p. 177.

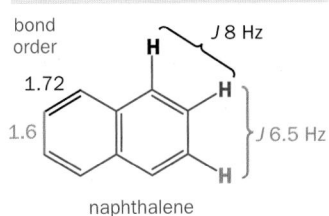

naphthalene

effect of electronegative substituents on $^3J_{HH}$ – alkenes and enol ethers

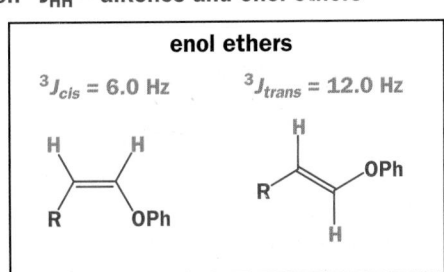

bond order

H ⌐ *J* 9 Hz
1.64
1.7
N
H ⌐ *J* 6 Hz

pyridine

meta coupling

H̶̶̶̶̶H

$0 < {}^4J_{HH} < 3$ Hz

allylic coupling

H̶̶̶̶H

Another good example is the coupling found in pyridines. Though the bond order is actually slightly less between C3 and C4, the coupling constants are about normal for an aromatic ring (compare naphthalene above), while coupling constants across C2 and C3, nearer to the electronegative nitrogen, are smaller.

When the through bond distance gets longer, coupling is not usually seen. To put it another way, four-bond coupling ${}^4J_{HH}$ is usually zero. However, it is seen in some special cases, the most important being *meta* coupling in aromatic rings and allylic coupling in alkenes. In both, the orbitals between the two hydrogen atoms can line up in a zig-zag fashion to maximize interaction. This arrangement looks rather like a letter 'W' and this sort of coupling is called **W-coupling**. Even with this advantage, values of ${}^4J_{HH}$ are usually small, about 1–3 Hz.

Meta coupling is very common when there is *ortho* coupling as well, but here is an example where there is no *ortho* coupling because none of the aromatic protons have immediate neighbours—the only coupling is *meta* coupling. There are two identical H^{A}s, which have one *meta* neighbour and appear as a 2H doublet. Proton H^X between the two MeO groups has two identical *meta* neighbours and so appears as a 1H triplet. The coupling is small (*J* ~ 2.5 Hz).

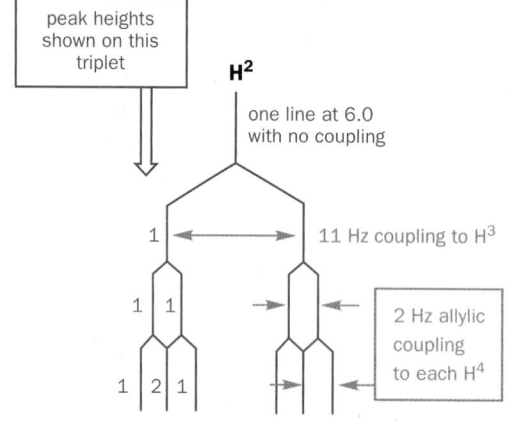

We have already seen a molecule with allylic coupling. We discussed in some detail why cyclohexenone has a double triplet for H^3. But it also has a less obvious double triplet for H^2. The triplet coupling is less obvious because *J* is small (about 2 Hz) because it is ${}^4J_{HH}$—allylic coupling to the CH$_2$ group at C4. Here is a diagram of the coupling, which you should compare with the earlier one for cyclohexenone.

O
H^2 δ 6.0
H^3 δ 7.0
H^4 H^4

peak heights shown on this triplet

H^2

one line at 6.0 with no coupling

1 ← → 1   11 Hz coupling to H^3
1 1
          2 Hz allylic coupling to each H^4
1 2 1

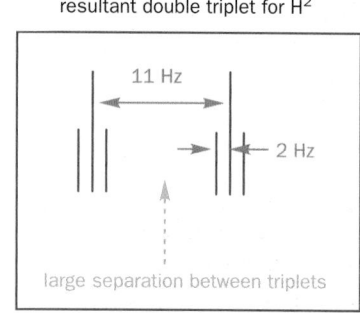

resultant double triplet for H^2

11 Hz
2 Hz
large separation between triplets

Coupling between similar protons

X
H
H
X

no coupling between these identical neighbours

one 4H singlet

We have already seen that identical protons do not couple with each other. The three protons in a methyl group may couple to some other protons, but *never* couple with each other. They are an A$_3$ system. Identical neighbours do not couple either. In the *para*-disubstituted benzenes we saw on pp. 253–254, all the protons on the aromatic rings were singlets.

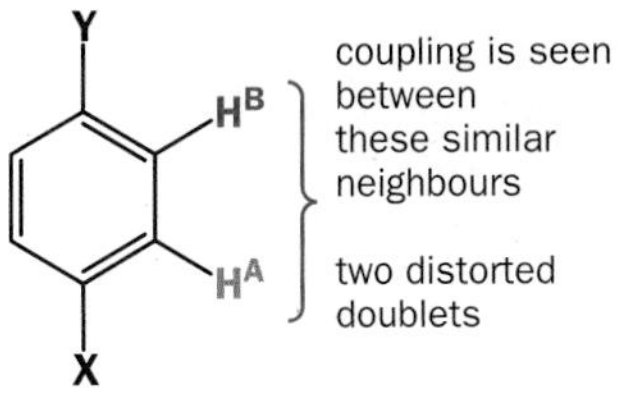

We have also seen how two different protons forming an AX system give two separate doublets. Now we need to see what happens to protons in between these two extremes. What happens to two similar neighbours? Do the two doublets of the AX system suddenly collapse to the singlet of the A_2 system? You have probably guessed that they do not. The transition is gradual. Suppose we have two different neighbours on an aromatic ring. The spectra show what we see.

coupling is seen between these similar neighbours

two distorted doublets

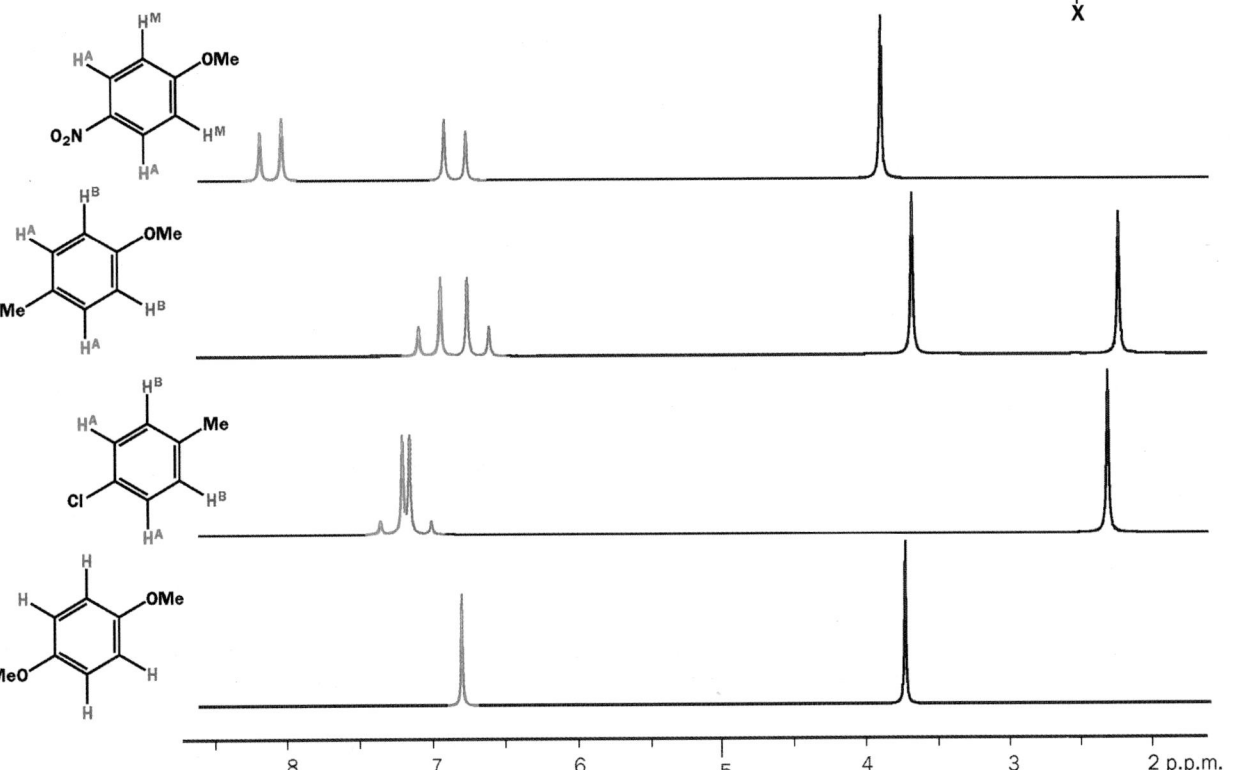

The critical factor is how the difference between the chemical shifts of the two protons ($\Delta\delta$) compares with the size of the coupling constant (J) for the machine in question. If $\Delta\delta$ is much larger than J there is no distortion: if, say, $\Delta\delta$ is 4 p.p.m. at 250 MHz (= 1000 Hz) and the coupling constant is a normal 7 Hz, then this condition is fulfilled and we have an AX spectrum of two 1:1 doublets. As $\Delta\delta$ approaches J in size, so the inner lines of the two doublets increase and the outer lines decrease until, when $\Delta\delta$ is zero, the outer lines vanish away altogether and we are left with the two superimposed inner lines—a singlet or an A_2 spectrum. You can see this progression in the diagram.

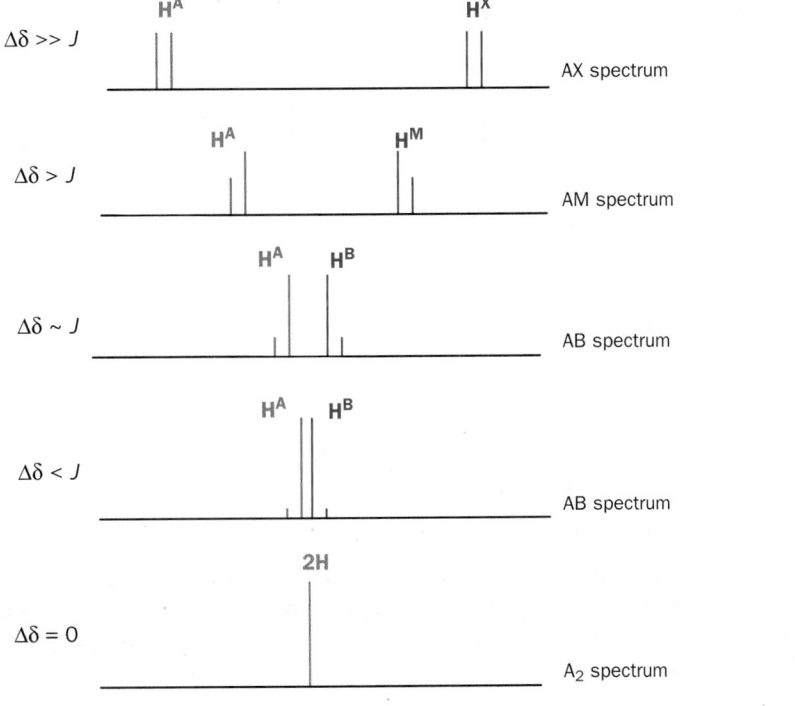

▶

You may see this situation described as an 'AB quartet'. It isn't! A quartet is an exactly equally spaced 1:3:3:1 system arising from coupling to three identical protons, and you should avoid this usage.

We call the last stages, where the distortion is great but the protons are still different, an **AB spectrum** because you cannot really talk about H^A without also talking about H^B. The two inner lines may be closer than the gap between the doublets, or the four lines may all be equally spaced. Two versions of an AB spectrum are shown in the diagram—there are many more variations.

It is a generally useful tip that a distorted doublet 'points' towards the protons with which it is coupled.

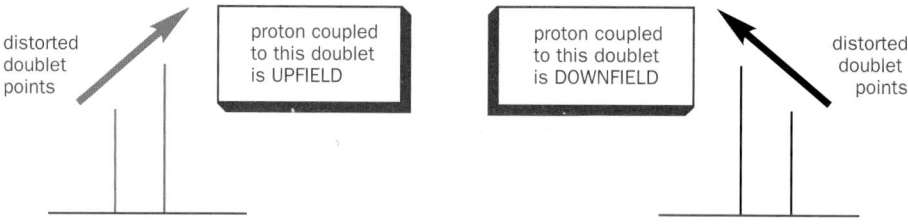

Or, to put it another way, the AB system is 'roofed' with the usual arrangement of low walls and a high middle to the roof. Look out for doublets (or any other coupled signals) of this kind.

We shall end this section with a final example illustrating *para*-disubstituted benzenes and roofing as well as an ABX system and an isopropyl group.

doublets with a roof over their heads

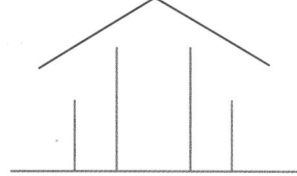

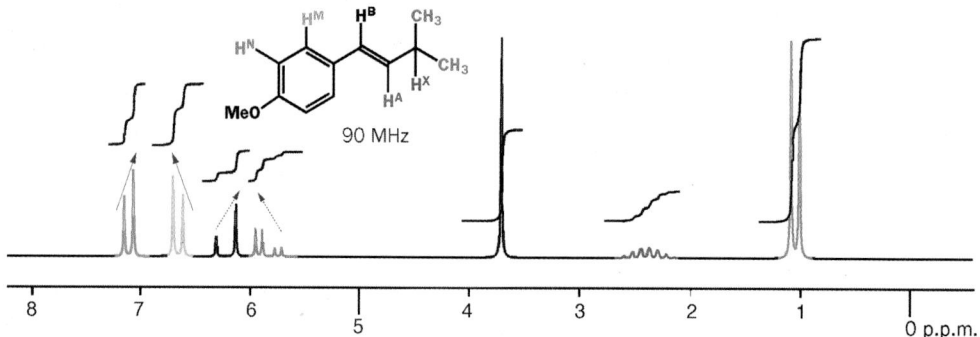

The aromatic ring protons form a pair of distorted doublets (2H each) showing that the compound is a *para*-disubstituted benzene. Then the alkene protons form the AB part of an ABX spectrum. They are coupled to each other with a large (*trans*) $J = 16$ Hz and one is also coupled to another distant proton. The large doublets are distorted (AB) but the small doublets within the right-hand half of the AB system are equal in height. The distant proton X is part of an *i*-Pr group and is coupled to H^B and the six identical methyl protons. Both Js are nearly the same so it is split by seven protons and is an octuplet. It looks like a sextuplet because the intensity ratios of the lines in an octuplet would be 1:7:21:35:35:21:7:1 (from Pascal's triangle) and it is hardly surprising that the outside lines disappear.

Coupling can occur between protons on the same carbon atom

We have seen cases where protons on the same carbon atom are different: compounds with an alkene unsubstituted at one end. If these protons are different (and they are certainly near to each other), then they should couple. They do, but in this case the coupling constant is usually very small. Here you see the example we met on p. 000.

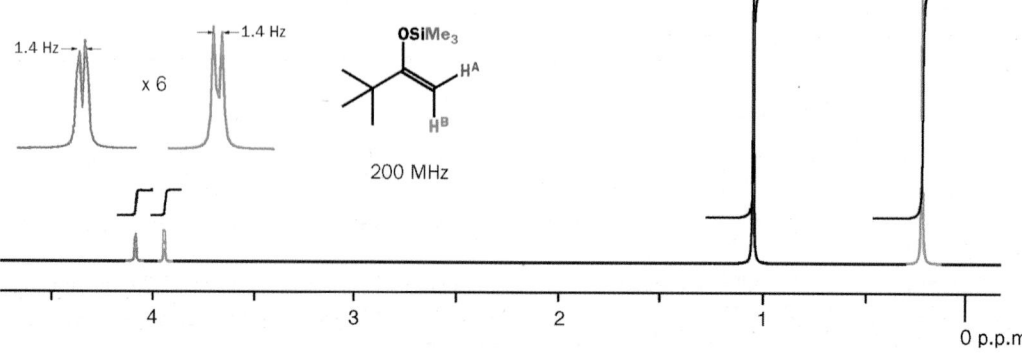

The small 1.4 Hz coupling is a $^2J_{HH}$ coupling between two protons on the same carbon that are different because there is no rotation about the double bond. $^2J_{HH}$ coupling is called **geminal** coupling.

This means that a monosubstituted alkene will have very characteristic signals for the three protons on the double bond. The three different coupling constants are very different so that this ABX system is unusually clear.

Here is an example of such a vinyl compound, ethyl acrylate (ethyl propenoate, a monomer for the formation of acrylic polymers). The spectrum looks rather complex at first, but it is easy to sort out using the coupling constants.

Proton NMR spectrum of a VINYL group

J_{AB} very small (0–2 Hz)
J_{AX} (*cis*) large (10–13 Hz)
J_{BX} (*trans*) very large (14–18 Hz)

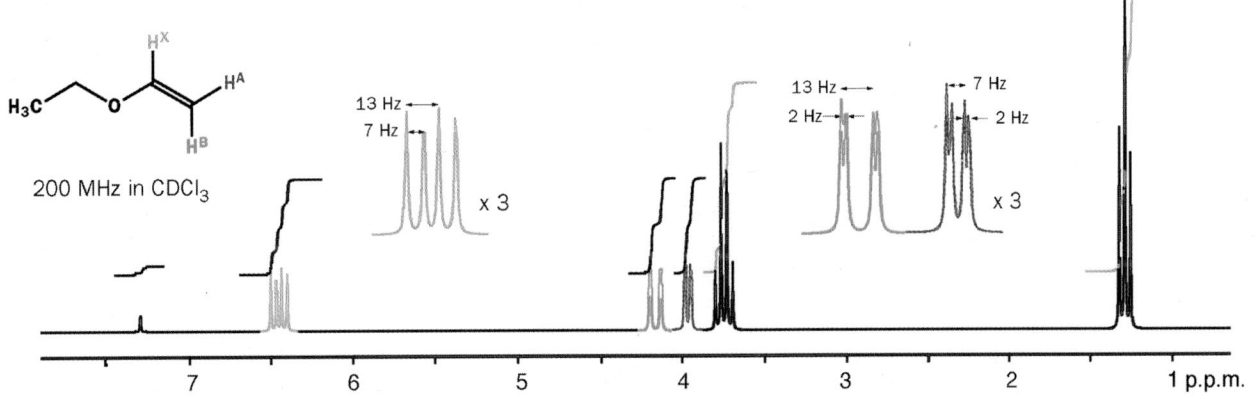

The largest J (16 Hz) is obviously between X and B (*trans* coupling), the medium J (10 Hz) is between X and A (*cis* coupling), and the small J (4 Hz) must be between A and B (geminal). This assigns all the protons: A, 5.80 p.p.m.; B, 6.40 p.p.m.; X, 6.11 p.p.m. Rather surprisingly, X comes between A and B in chemical shift. Assignments based on coupling are more reliable than those based on chemical shift alone.

An enol ether type of vinyl group is present in ethyl vinyl ether, a reagent used for the protection of alcohols. This time all the coupling constants are smaller because of the electronegativity of the oxygen atom, which is now joined directly to the double bond.

It is still a simple matter to assign the protons of the vinyl group because couplings of 13, 7, and 2 Hz must be *trans*, *cis*, and geminal, respectively. In addition, X is on a carbon atom next to oxygen and so goes downfield while A and B have extra shielding from the conjugation of the oxygen lone pairs (see p. 254).

Geminal coupling on saturated carbons can be seen only if the hydrogens of a CH_2 group are different. We have seen an example of this on the bridging CH_2 group of myrtenal (p. 248). The

myrtenal

coupling constant for the protons on the bridge, J_{AB}, is 9 Hz. Geminal coupling constants in a saturated system can be much larger (typically 10–16 Hz) than in an unsaturated one.

Typical coupling constants

Geminal $^2J_{HH}$		
saturated	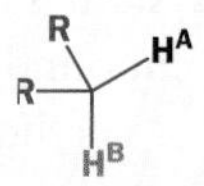	10–16 Hz
unsaturated	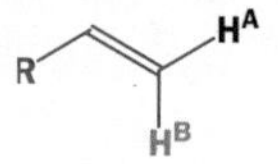	0–3 Hz
Vicinal $^3J_{HH}$		
saturated	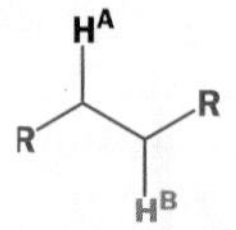	6–8 Hz
unsaturated *trans*	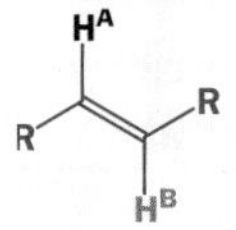	14–16 Hz
unsaturated *cis*	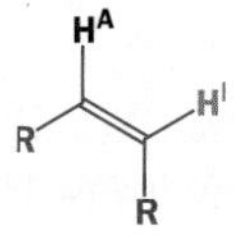	8–11 Hz
unsaturated aromatic	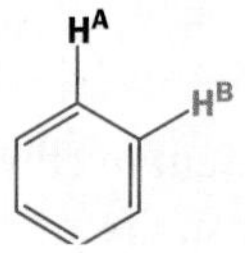	6–9 Hz
Long-range $^4J_{HH}$		
meta	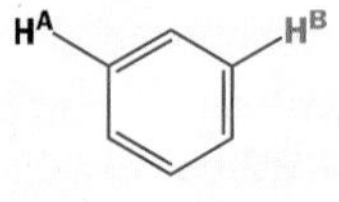	1–3 Hz
allylic	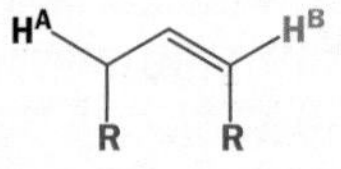	1–2 Hz

To conclude

You have now met, in Chapter 3 and this chapter, all of the most important spectroscopic techniques available for working out the structure of organic molecules. We hope you can now appreciate that proton NMR is by far the most powerful of these techniques, and we hope you will be referring back to this chapter as you read the rest of the book. We shall talk about proton NMR a lot, and specifically we will come back to it in detail in Chapter 15, where we will look at using all of the spectroscopic techniques in combination, and in Chapter 32, when we look at what NMR can tell us about the shape of molecules.

Problems

1. How many signals will there be in the ^{1}H NMR spectrum of each of these compounds? Estimate the chemical shifts of the signals.

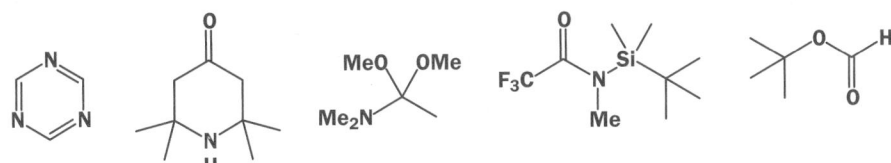

2. Comment on the chemical shifts of these three compounds and suggest whether there is a worthwhile correlation with pK_a.

Compound	δ_H, p.p.m.	pK_a
CH_3NO_2	4.33	10
$CH_2(NO_2)_2$	6.10	4
$CH(NO_2)_3$	7.52	0

3. One isomer of dimethoxybenzoic acid has the ^{1}H NMR spectrum 3.85 (6H, s), 6.63 (1H, t, J 2 Hz), 7.17 (2H, d, J 2 Hz) and one isomer of coumalic acid has the ^{1}H NMR spectrum 6.41 (1H, d, J 10 Hz), 7.82 (1H, dd, J 2, 10 Hz), 8.51 (1H, d, J 2 Hz). In each case, which isomer is it? The substituents in black can be on any carbon atoms.

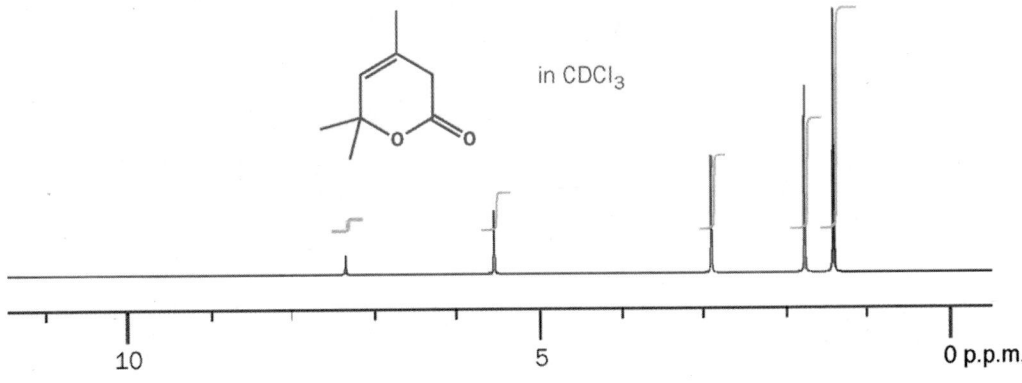

dimethoxybenzoic acid   coumalic acid

4. Assign the NMR spectra of this compound (assign means say which signal belongs to which atom) and justify your assignments.

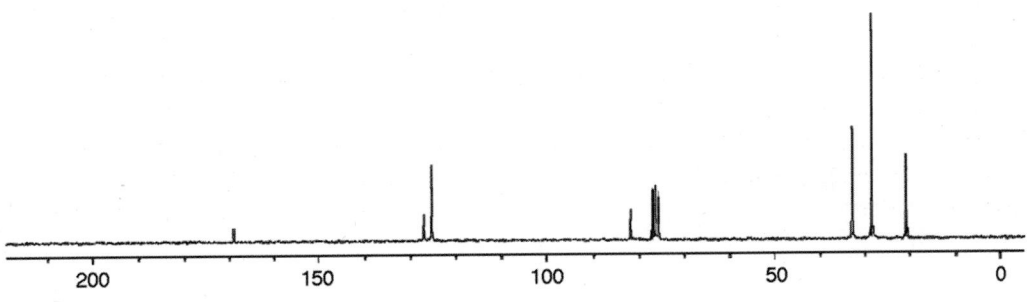

5. Assign the ^{1}H NMR spectra of these compounds and explain the multiplicity of the signals.

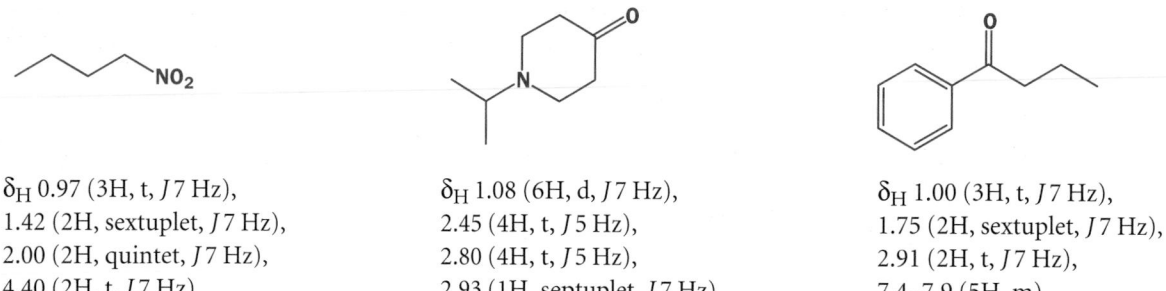

δ_H 0.97 (3H, t, J 7 Hz),
1.42 (2H, sextuplet, J 7 Hz),
2.00 (2H, quintet, J 7 Hz),
4.40 (2H, t, J 7 Hz)

δ_H 1.08 (6H, d, J 7 Hz),
2.45 (4H, t, J 5 Hz),
2.80 (4H, t, J 5 Hz),
2.93 (1H, septuplet, J 7 Hz)

δ_H 1.00 (3H, t, J 7 Hz),
1.75 (2H, sextuplet, J 7 Hz),
2.91 (2H, t, J 7 Hz),
7.4–7.9 (5H, m)

6. The reaction below was expected to give product 6A and did indeed give a product with the correct molecular formula by mass spectrometry. The ^{1}H NMR spectrum of the product was however: δ_H (p.p.m.) 1.27 (6H, s), 1.70 (4H, m), 2.88 (2H, m), 5.4–6.1 (2H, broad s, exchanges with D_2O), 7.0–7.5 (3H, m). Though the detail is missing from this spectrum, how can you already tell that this is not the compound expected?

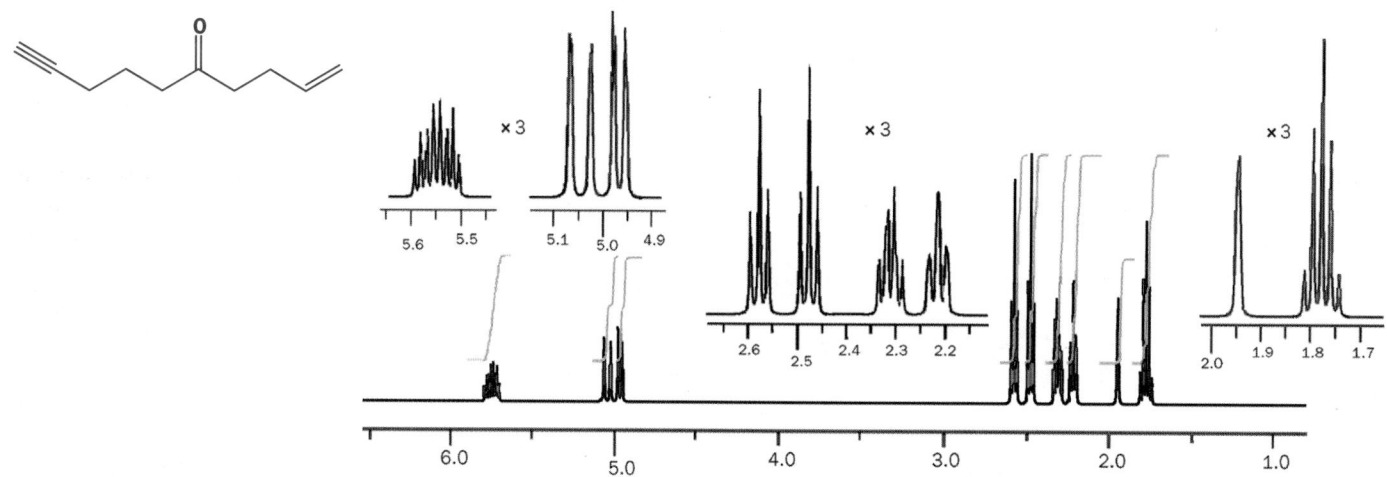

6A

7. Assign the 400 MHz ^{1}H NMR spectrum of this enynone as far as possible, justifying both chemical shifts and coupling patterns.

8. A nitration product ($C_8H_{11}N_3O_2$) of this pyridine has been isolated which has a nitro (NO_2) group somewhere on the molecule. From the 90 MHz ^{1}H NMR spectrum, deduce whether the nitro group is (a) on the ring, (b) on the NH nitrogen atom, or (c) on the aliphatic side chain and then exactly where it is. Give a full analysis of the spectrum.

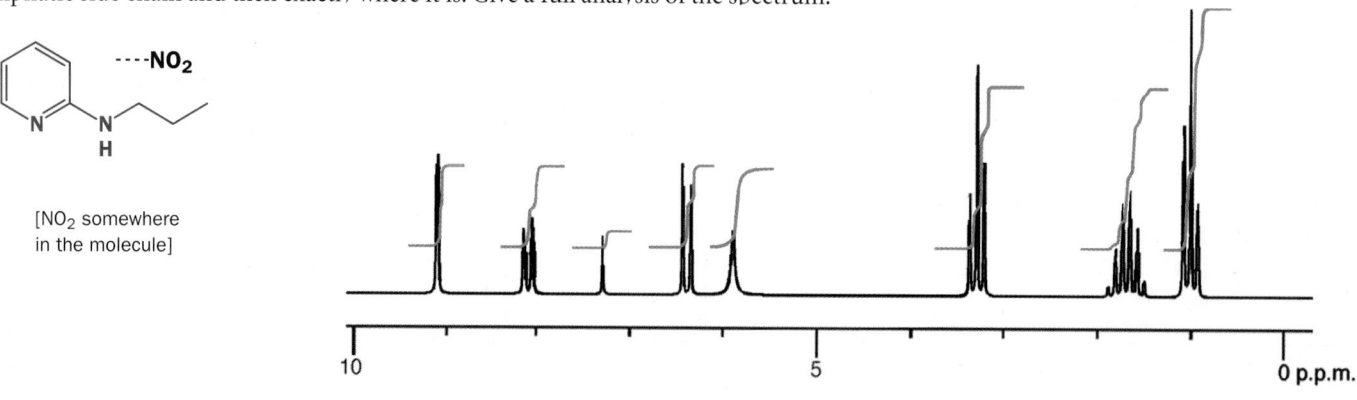

[NO$_2$ somewhere
in the molecule]

9. The natural product bullatenone was isolated in the 1950s from a New Zealand myrtle and assigned the structure 9A. Then compound 9A was synthesized and found not to be identical with natural bullatenone. Predict the expected ^{1}H NMR spectrum of 9A. Given the full spectroscopic data available nowadays, but not in the 1950s, say why 9A is definitely wrong and suggest a better structure for bullatenone.

Ph

9A
alleged bullatenone

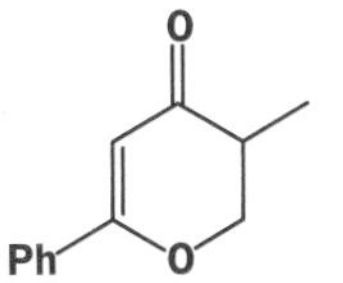

Spectra of bullatenone:
Mass spectrum: m/z 188 (10%) (high resolution confirms $C_{12}H_{12}O_2$), 105 (20%), 102 (100%), and 77 (20%)
Infrared: 1604 and 1705 cm^{-1}.
^{1}H NMR: 1.45 (6H, s), 5.82 (1H, s), 7.35 (3H, m), and 7.68 (2H, m).

10. Interpret this ^{1}H NMR spectrum.

HO
CO2Et
N
H
90 MHz
× 3
7.5
7.0
6.5
10
5
0 p.p.m.

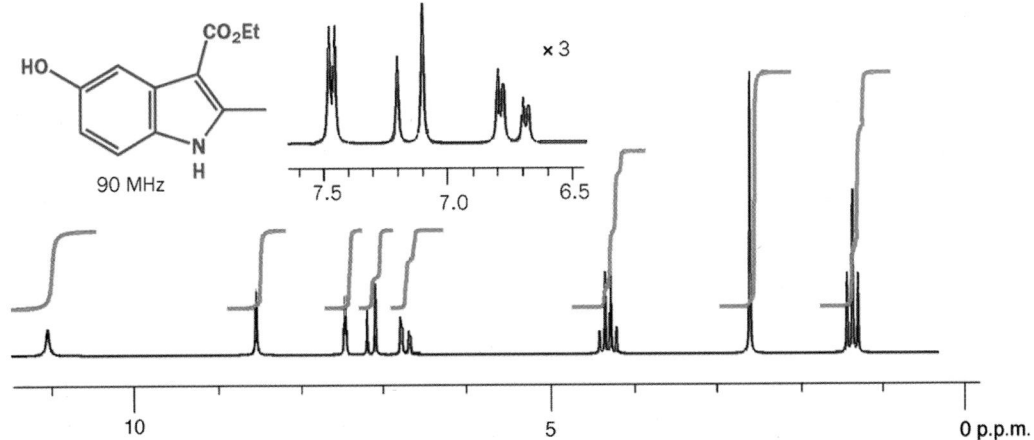

11. Suggest structures for the products of these reactions, interpreting the spectroscopic data. You are *not* expected to write mechanisms for the reactions and you should resist the temptation to work out what 'should happen' from the reactions. These are all unexpected products.

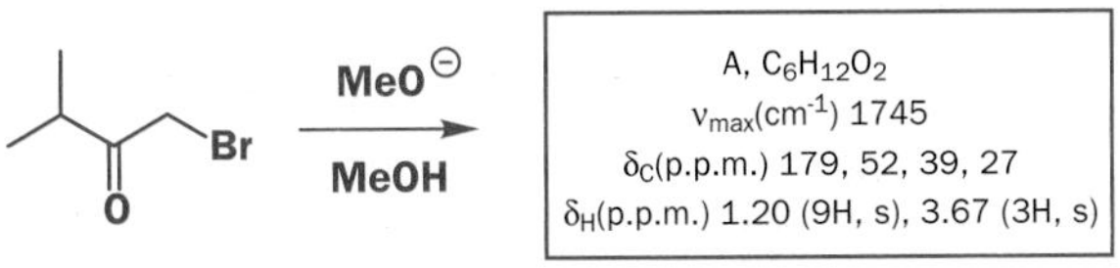

Br
MeO⊖
MeOH
A, C6H12O2
νmax(cm-1) 1745
δC(p.p.m.) 179, 52, 39, 27
δH(p.p.m.) 1.20 (9H, s), 3.67 (3H, s)

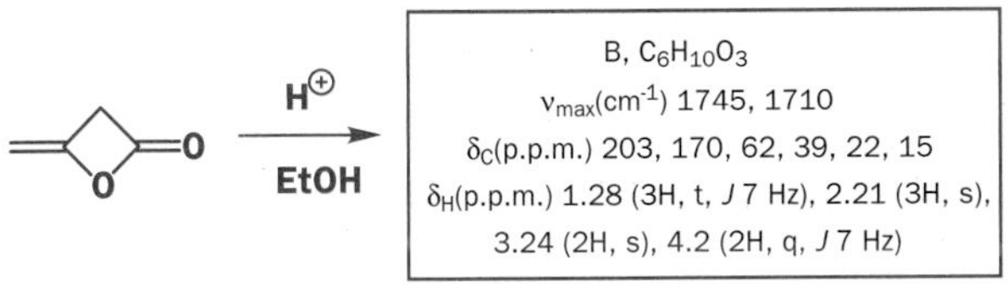

H⊕
EtOH
B, C6H10O3
νmax(cm-1) 1745, 1710
δC(p.p.m.) 203, 170, 62, 39, 22, 15
δH(p.p.m.) 1.28 (3H, t, J 7 Hz), 2.21 (3H, s), 3.24 (2H, s), 4.2 (2H, q, J 7 Hz)

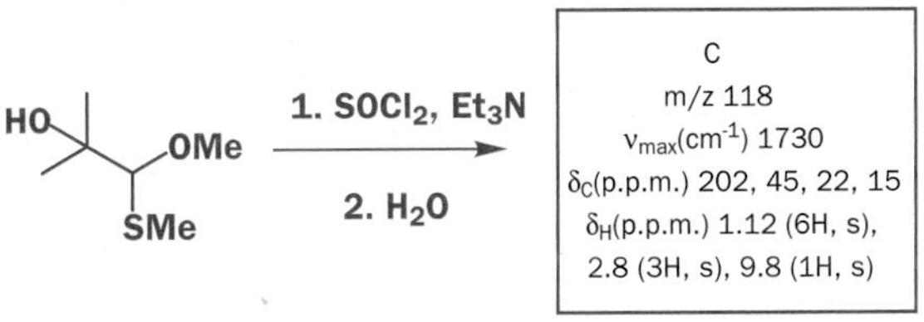

HO
OMe
SMe
1. SOCl2, Et3N
2. H2O
C
m/z 118
νmax(cm-1) 1730
δC(p.p.m.) 202, 45, 22, 15
δH(p.p.m.) 1.12 (6H, s), 2.8 (3H, s), 9.8 (1H, s)

12. Precocene is a compound that causes insect larvae to pupate and can also be found in some plants (*Ageratum* spp.) where it may act as an insecticide. It was isolated in minute amounts and has the following spectroscopic details. Propose a structure for precocene.

Spectra of precocene:
Mass spectrum: m/z (high resolution gives $C_{13}H_{16}O_3$), M–15 (100%) and M–30 (weak).
Infrared: CH and fingerprint only.
^{1}H NMR: 1.34 (6H, s), 3.80 (3H, s), 3.82 (3H, s), 5.54 (1H, d, *J* 10 Hz), 6.37 (1H, d, *J* 10 Hz), 6.42 (1H, s), and 6.58 (1H, s).

13. Suggest structures for the products of these reactions, interpreting the spectroscopic data. Though these products, unlike those in Problem 11, are reasonably logical, you will not meet the mechanisms for the reactions until Chapters 22, 29, and 23, respectively, and you are advised to solve the structures through the spectra.

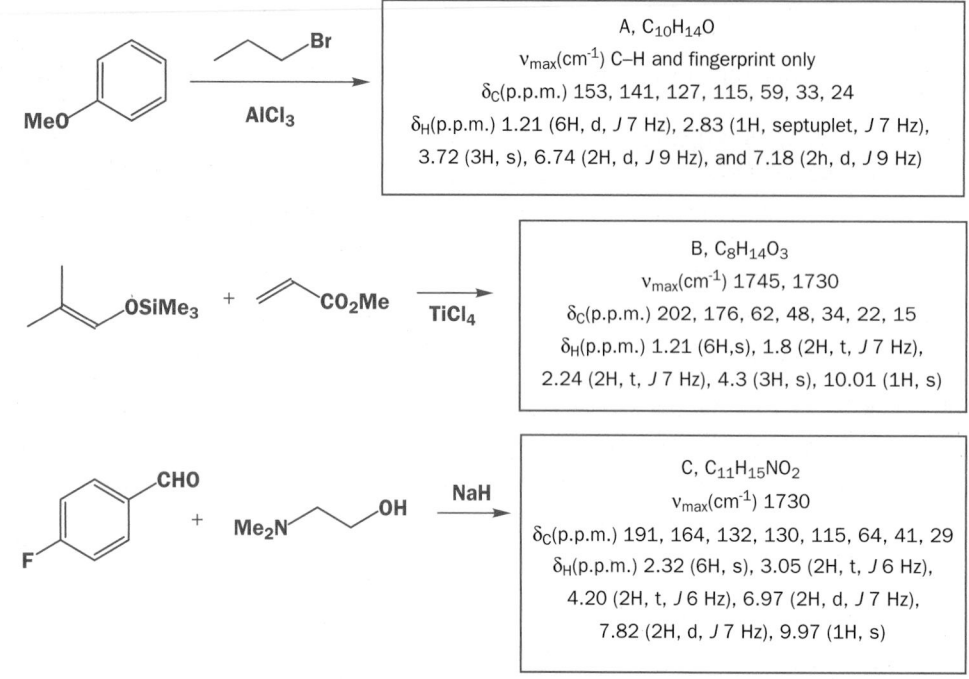

A, $C_{10}H_{14}O$
$\nu_{max}(cm^{-1})$ C–H and fingerprint only
δ_C(p.p.m.) 153, 141, 127, 115, 59, 33, 24
δ_H(p.p.m.) 1.21 (6H, d, J 7 Hz), 2.83 (1H, septuplet, J 7 Hz), 3.72 (3H, s), 6.74 (2H, d, J 9 Hz), and 7.18 (2h, d, J 9 Hz)

B, $C_8H_{14}O_3$
$\nu_{max}(cm^{-1})$ 1745, 1730
δ_C(p.p.m.) 202, 176, 62, 48, 34, 22, 15
δ_H(p.p.m.) 1.21 (6H,s), 1.8 (2H, t, J 7 Hz), 2.24 (2H, t, J 7 Hz), 4.3 (3H, s), 10.01 (1H, s)

C, $C_{11}H_{15}NO_2$
$\nu_{max}(cm^{-1})$ 1730
δ_C(p.p.m.) 191, 164, 132, 130, 115, 64, 41, 29
δ_H(p.p.m.) 2.32 (6H, s), 3.05 (2H, t, J 6 Hz), 4.20 (2H, t, J 6 Hz), 6.97 (2H, d, J 7 Hz), 7.82 (2H, d, J 7 Hz), 9.97 (1H, s)

14. The following reaction between a phosphonium salt, base, and an aldehyde gives a hydrocarbon C_6H_{12} with the 200 MHz ^{1}H NMR spectrum shown. Give a structure for the product and comment on its stereochemistry. You are not expected to discuss the chemistry!

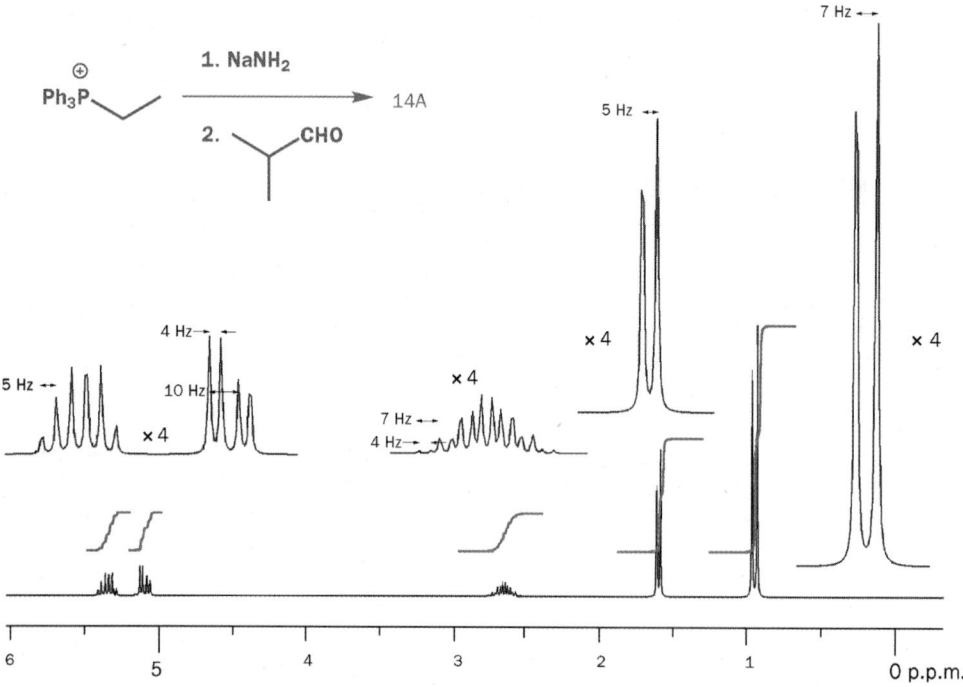

Nucleophilic substitution at the carbonyl (C=O) group

12

Connections

Building on:

- Drawing mechanisms ch5
- Nucleophilic attack on carbonyl groups ch6 & ch9
- Acidity and pK_a ch8
- Grignard and RLi addition to C=O groups ch9

Arriving at:

- Nucleophilic attack followed by loss of leaving group
- What makes a good nucleophile
- What makes a good leaving group
- There is always a tetrahedral intermediate
- How to make acid derivatives
- Reactivity of acid derivatives
- How to make ketones from acids
- How to reduce acids to alcohols

Looking forward to:

- Loss of carbonyl oxygen ch14
- Kinetics and mechanism ch13
- Reactions of enols ch21, ch26-ch29
- Synthesis in action ch25

You are already familiar with reactions of compounds containing carbonyl groups. Aldehydes and ketones react with nucleophiles at the carbon atom of their carbonyl group to give products containing hydroxyl groups. Because the carbonyl group is such a good electrophile, it reacts with a wide range of different nucleophiles: you have met reactions of aldehydes and ketones with (in Chapter 6) cyanide, water, alcohols, and (in Chapter 9) organometallic reagents (organolithiums and organomagnesiums, or Grignard reagents).

In this chapter and Chapter 14 we shall look at some more reactions of the carbonyl group—and revisit some of the ones we touched on in Chapter 6. It is a tribute to the importance of this functional group in organic chemistry that we have devoted four chapters of this book to its reactions. Just like the reactions in Chapters 6 and 9, the reactions in Chapters 12 and 14 all involve attack of a nucleophile on a carbonyl group. The difference will be that this step is followed by other mechanistic steps, which means that the overall reactions are not just *additions* but also *substitutions*.

The product of nucleophilic addition to a carbonyl group is not always a stable compound

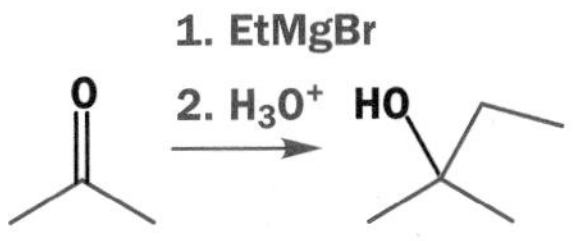

Addition of a Grignard reagent to an aldehyde or ketone gives a stable alkoxide, which can be protonated with acid to produce an alcohol (you met this reaction in Chapter 9).

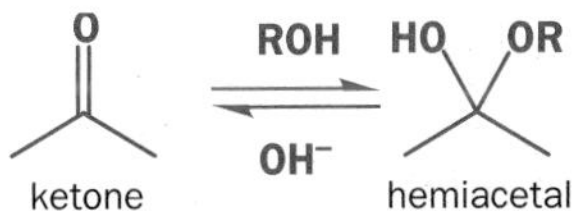

The same is not true for addition of an alcohol to a carbonyl group in the presence of base—in Chapter 6 we drew a reversible, equilibrium arrow for this transformation and said that the product, a hemiacetal, is only formed to a significant extent if it is cyclic.

The reason for this instability is that RO^- is easily expelled from the molecule. We call groups that can be expelled from molecules, usually taking with them a negative charge, **leaving groups**. We'll look at leaving groups in more detail later in this chapter and again in Chapter 17.

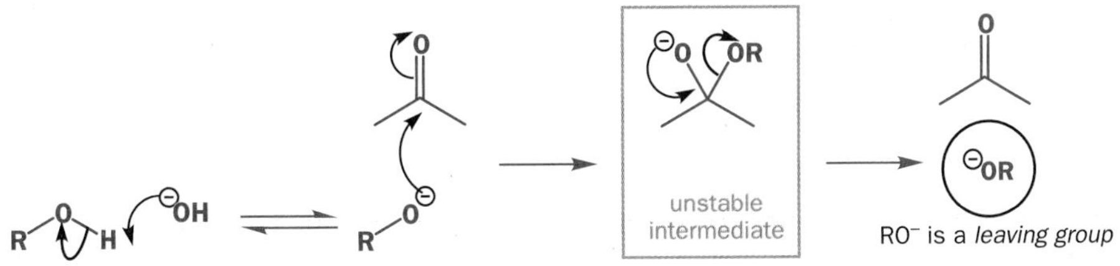

> ● **Leaving groups**
>
> Leaving groups are anions such as Cl⁻, RO⁻, and RCO_2^- that can be expelled from molecules taking their negative charge with them.

So, if the nucleophile is also a leaving group, there is a chance that it will be lost again and that the carbonyl group will reform—in other words, the reaction will be reversible. The energy released in forming the C=O bond (bond strength 720 kJ mol⁻¹) more than makes up for the loss of two C–O single bonds (about 350 kJ mol⁻¹ each), one of the reasons for the instability of the hemiacetal product in this case.

The same thing can happen if the starting carbonyl compound contains a potential leaving group. The same unstable negatively charged intermediate in the red box above is formed when a Grignard reagent is added to an ester.

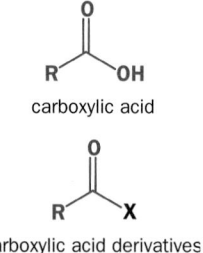

RO⁻ is a *leaving group*

unstable intermediate

ketone reacts further

Again, it collapses with loss of RO⁻ as a leaving group. This time, though, we have not gone back to starting materials: instead we have made a new compound (a ketone) by a **substitution reaction**—the OR group of the starting material has been substituted by the Me group of the product. In fact, as we shall see later, this reaction does not stop at this point because the ketone product can react with the Grignard reagent a second time.

Carboxylic acid derivatives

carboxylic acid

carboxylic acid derivatives

Most of the starting materials for, and products of, these substitutions will be carboxylic acid derivatives, with the general formula RCOX. You met the most important members of this class in Chapter 2: here they are again as a reminder.

Carboxylic acid derivatives

R–Cl acid chlorides (acyl chlorides)ᵃ	R–OR¹ esters
R–O–R' acid anhydrides	R–NH₂ amides

ᵃWe shall use these two terms interchangeably.

> ▶
>
> The reactions of alcohols with acid chlorides and with acid anhydrides are the most important ways of making esters, but not the only ways. We shall see later how carboxylic acids can be made to react directly with alcohols.

> ▶
>
> Remember the convenient organic element symbol for 'acetyl', Ac? Cyclohexyl acetate can be represented by 'OAc' but not just 'Ac'.
>
> cyclohexyl acetate can be drawn like this: OAc
>
> But NOT like this: Ac

Acid chlorides and acid anhydrides react with alcohols to make esters

Acetyl chloride will react with an alcohol in the presence of a base to give an acetate ester and we get the same product if we use acetic anhydride.

cyclohexanol

cyclohexanol

acetyl chloride

base

cyclohexyl acetate

base

acetic anhydride

In each case, a substitution (of the black part of the molecule, Cl⁻ or AcO⁻, by the orange cyclohexanol) has taken place—but how? It is important that you learn not only the *fact* that acyl chlorides and acid anhydrides react with alcohols but also the *mechanism* of the reaction. In this chapter you will meet a lot of reactions, but relatively few mechanisms—once you understand one, you should find that the rest follow on quite logically.

The first step of the reaction is, as you might expect, addition of the nucleophilic alcohol to the electrophilic carbonyl group— we'll take the acyl chloride first.

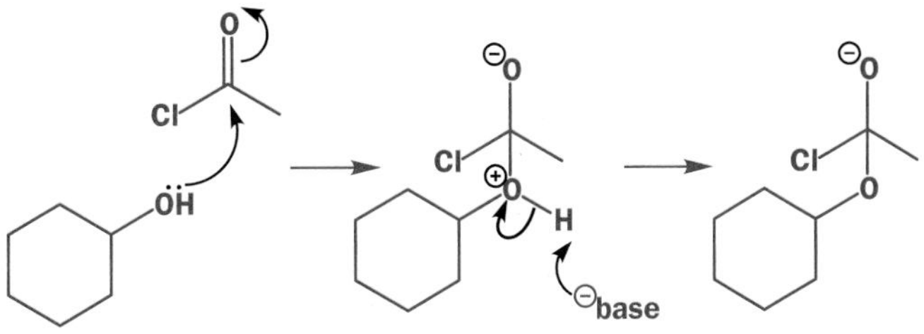

The base is important because it removes the proton from the alcohol as it attacks the carbonyl group. A base commonly used for this is pyridine. If the electrophile had been an aldehyde or a ketone, we would have got an unstable hemiacetal, which would collapse back to starting materials by eliminating the alcohol. With an acyl chloride, the alkoxide intermediate we get is also unstable. It collapses again by an elimination reaction, this time losing chloride ion, and forming the ester. Chloride is the *leaving group* here—it leaves with its negative charge.

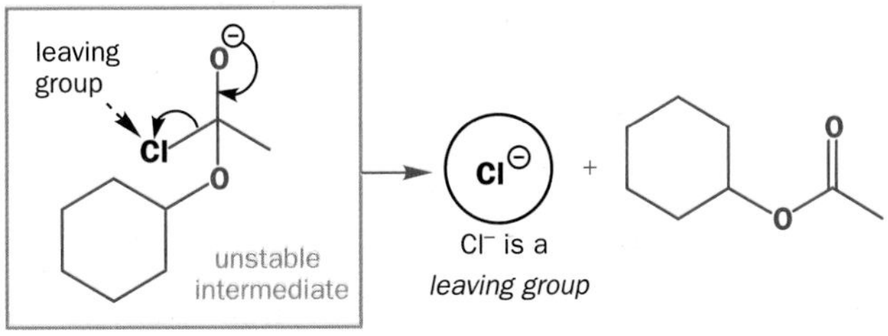

With this reaction as a model, you should be able to work out the mechanism of ester formation from acetic anhydride and cyclohexanol. Try to write it down without looking at the acyl chloride mechanism above, and certainly not at the answer below. Here it is, with pyridine as the base. Again, addition of the nucleophile gives an unstable intermediate, which undergoes an elimination reaction, this time losing a carboxylate anion, to give an ester.

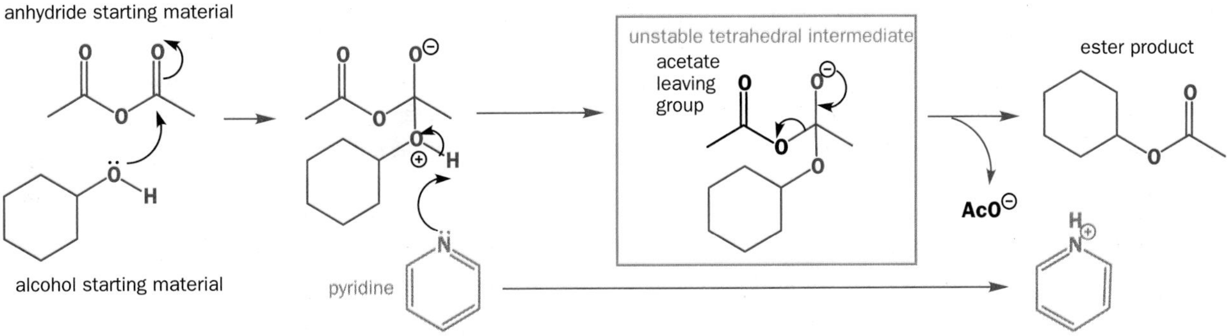

You will notice that the terms 'acid chloride' and 'acyl chloride' are used interchangeably.

We call the unstable intermediate formed in these reactions the **tetrahedral intermediate**, because the trigonal (sp^2) carbon atom of the carbonyl group has become a tetrahedral (sp^3) carbon atom.

● **Tetrahedral intermediates**

Substitutions at trigonal carbonyl groups go through a tetrahedral intermediate and then on to a trigonal product.

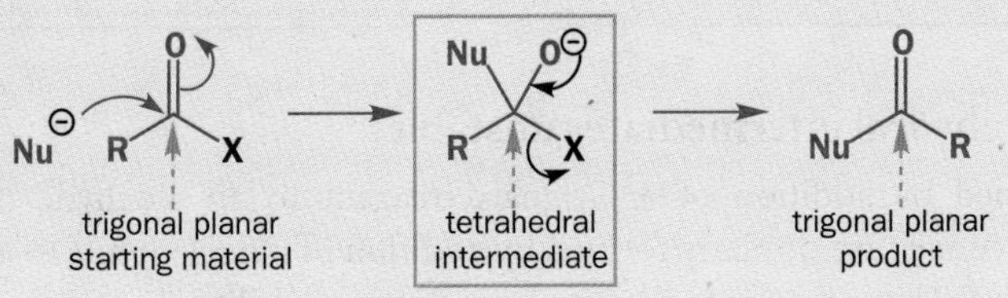

More details of this reaction

This reaction has more subtleties than first meet the eye. If you are reading this chapter for the first time, you should skip this box, as it is not essential to the general flow of what we are saying. There are three more points to notice.

1 Pyridine is consumed during both of these reactions, since it ends up protonated. One whole equivalent of pyridine is therefore necessary and, in fact, the reactions are often carried out with pyridine as solvent

2 The observant among you may also have noticed that the (weak—pyridine) base catalyst in this reaction works very slightly differently from the (strong—hydroxide) base catalyst in the hemiacetal-forming reaction on p. 279: one removes the proton after the nucleophile has added; the other removes the proton before the nucleophile has added. This is deliberate, and will be discussed further in Chapter 13

3 Pyridine is, in fact, more nucleophilic than the alcohol, and it attacks the acyl chloride rapidly, forming a highly electrophilic (because of the positive charge) intermediate. It is then this intermediate that subsequently reacts with the alcohol to give the ester. Because pyridine is acting as a nucleophile to speed up the reaction, yet is unchanged by the reaction, it is called a **nucleophilic catalyst**.

Nucleophilic catalysis in ester formation

tetrahedral intermediate — reactive trigonal intermediate — tetrahedral intermediate

How do we know that the tetrahedral intermediate exists?

We don't expect you to be satisfied with the bland statement that tetrahedral intermediates are formed in these reactions: of course, you wonder how we know that this is true. The first evidence for tetrahedral intermediates in the substitution reactions of carboxylic acid derivatives was provided by Bender in 1951. He reacted water with carboxylic acid derivatives RCOX that had been 'labelled' with an isotope of oxygen, ^{18}O.

rapid migration of the hydrogen atom between the oxygen atoms

plus $X^{\ominus}$

He then reacted these derivatives with water to make labelled carboxylic acids. However, he added insufficient water for complete consumption of the starting material. At the end of the reaction, he found that the proportion of labelled molecules in the *remaining starting material* had decreased significantly: in other words, it was no longer completely labelled with ^{18}O; some contained 'normal' ^{16}O.

> Non-radioactive isotopes are detected by mass spectrometry (Chapter 3).

This result cannot be explained by direct substitution of X by H_2O, but is consistent with the existence of an intermediate in which the unlabelled ^{16}O and labelled ^{18}O can 'change places'. This intermediate is the *tetrahedral intermediate* for this reaction.

rapid migration of protons between oxygen atoms

tetrahedral intermediate

the tetrahedral intermediate can collapse to give the carboxylic acid product

but it can also revert to unlabelled starting material

Why are the tetrahedral intermediates unstable?

The alkoxide formed by addition of a Grignard reagent to an aldehyde or ketone is stable. Tetrahedral intermediates are similarly formed by addition of a nucleophile to a carbonyl group, so why are they *unstable*? The answer is to do with **leaving group ability**.

Once the nucleophile has added to the carbonyl compound, the stability of the product (or tetrahedral intermediate) depends on how good the groups attached to the new tetrahedral carbon atom are at leaving with the negative charge. In order for the tetrahedral intermediate to collapse (and therefore be just an intermediate and not the final product) one of the groups has to be able to leave and carry off the negative charge from the alkoxide anion formed in the addition.

Here once again is the tetrahedral intermediate resulting from addition of an alcohol to an acyl chloride.

There are three choices of leaving group: Cl^-, EtO^-, and Me^-. We cannot actually make Me^- because it is so unstable, but MeLi, which is about as close to it as we can get (Chapter 9), reacts vigorously with water so Me^- must be a very bad leaving group. EtO^- is not so bad—alkoxide salts are stable, but they are still strong, reactive bases (we shall see below what pK_a has to do with this matter). But Cl^- is the best leaving group: Cl^- ions are perfectly stable and quite unreactive and happily carry off the negative charge from the oxygen atom. You probably eat several grams of Cl^- every day but you would be unwise to eat EtO^- or MeLi.

pK_{aH} is a useful guide to leaving group ability

It's useful to be able to compare leaving group ability quantitatively. This is impossible to do exactly, but a good guide is pK_{aH}. If we go back to the example of ester formation from acyl chloride plus alcohol, there's a choice of Me^-, EtO^-, and Cl^-. The leaving group with the lowest pK_{aH} is the best and so we can complete the reaction.

■ Remember that we use the term pK_{aH} to mean 'pK_a of the conjugate acid': if you need reminding about pK_a and pK_{aH}, stop now and refresh your memory by reviewing Chapter 8.

Leaving group	pK_{aH}
Me^-	48
EtO^-	16
Cl^-	−7

The same is true for the reaction of acetic anhydride with an alcohol. Possible leaving groups from this tetrahedral intermediate are the following.

Leaving group	pK_{aH}
Me^-	48
RO^-	16
$MeCO_2^-$	5

Again the group that leaves is the one with the lowest pK_{aH}.

● Leaving group ability

The lower the pK_{aH}, the better the leaving group in carbonyl substitution reactions.

Why should this be so? The ability of an anion to behave as a leaving group depends in some way on its stability—how willing it is to accept a negative charge. pK_a represents the equilibrium between an acid and its conjugate base, and is a measure of the stability of that conjugate base with respect to the acid—low pK_a means stable conjugate base, indicating a willingness to accept a negative charge. So the general trends that affect pK_a, which we discussed in Chapter 8, will also affect leaving group ability. However, you must bear in mind that pK_a is a measure of stability only with respect to the protonated form of the anion. Leaving group ability is a fundamentally different comparison between the stability of the negatively charged tetrahedral intermediate and the leaving group plus resulting carbonyl compound. But it still works as a good guide. These five values are worth learning.

increasing pK_{aH} ↑

Leaving group	pK_{aH}
R^-	50
NH_2^-	35
RO^-	16
RCO_2^-	5
Cl^-	−7

increasing leaving group ability ↓

We can use pK_a to predict what happens if we react an acyl chloride with a carboxylate salt. We expect the carboxylate salt (here, sodium formate, or sodium methanoate, HCO_2Na) to act as the nucleophile to form a tetrahedral intermediate, which could collapse in any one of three ways.

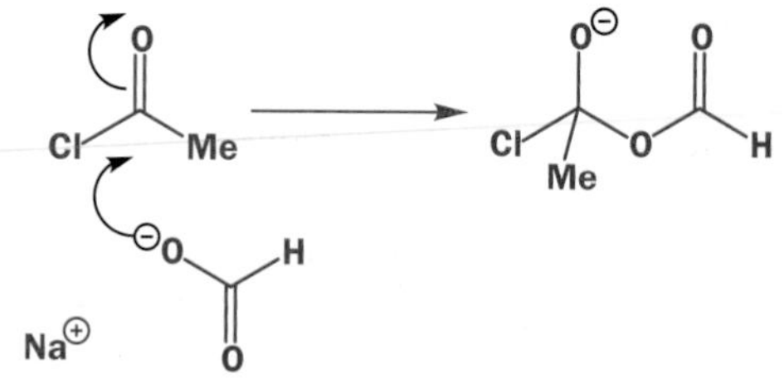

We can straight away rule out loss of Me⁻ (pK_{aH} 50), but we might guess that Cl⁻ (pK_{aH} −7) is a better leaving group than HCO_2^- (pK_a about 5), and we'd be right. Sodium formate reacts with acetyl chloride to give 'acetic formic anhydride'.

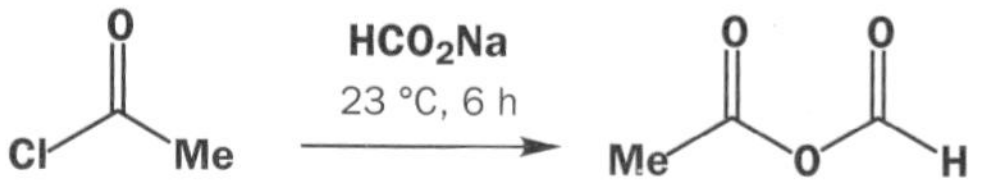

mixed anhydride
64% yield

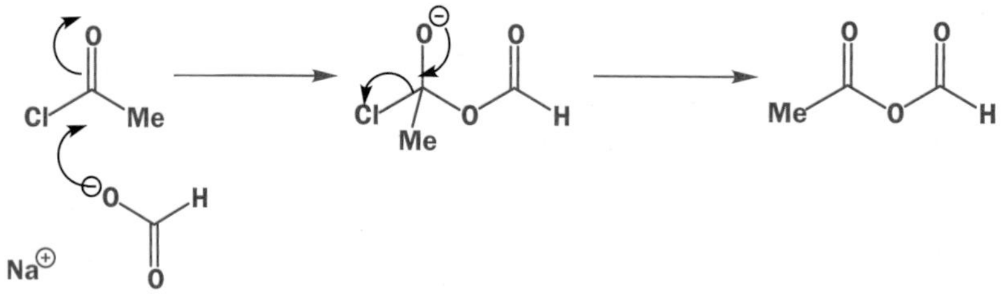

Amines react with acyl chlorides to give amides

Using the principles we've outlined above, you should be able to see how these compounds can be interconverted by substitution reactions with appropriate nucleophiles. We've seen that acid chlorides react with carboxylic acids to give acid anhydrides, and with alcohols to give esters. They'll also react with amines (such as ammonia) to give amides.

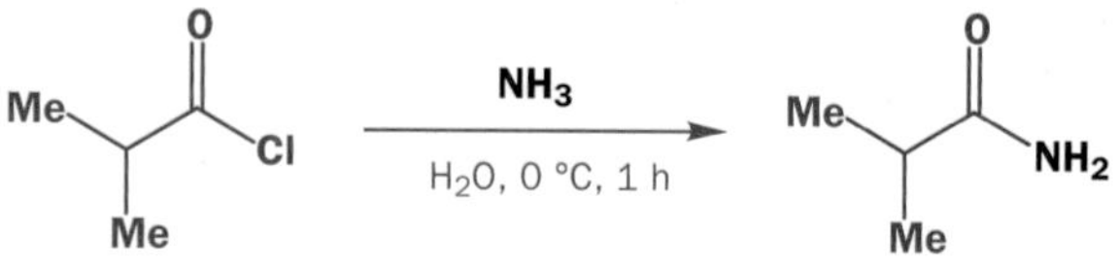

NH_3
H_2O, 0 °C, 1 h

78–83% yield

The mechanism is very similar to the mechanism of ester formation.

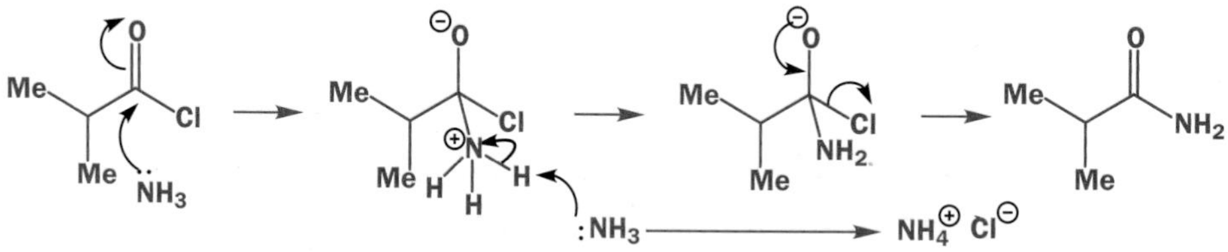

Notice the second molecule of ammonia, which removes a proton before the loss of chloride ion—the leaving group—to form the amide. Ammonium chloride is formed as a by-product in the reaction.

Here is another example, using a secondary amine, dimethylamine. Try writing down the mechanism now without looking at the one above. Again, two equivalents of dimethylamine are necessary, though the chemists who published this reaction added three for good measure.

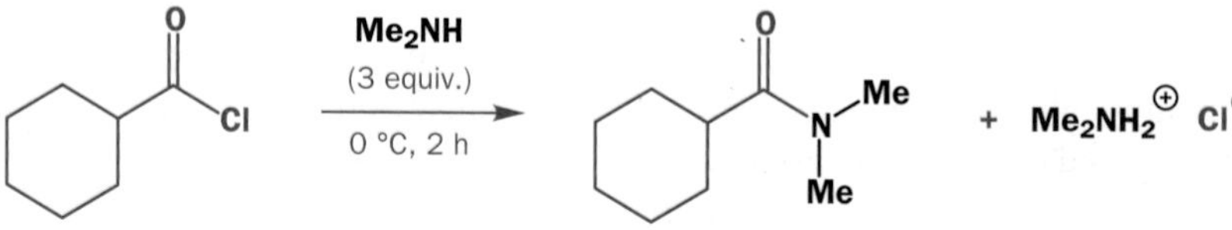

Me_2NH
(3 equiv.)
0 °C, 2 h

$+ \; Me_2NH_2^{\oplus} \; Cl^{\ominus}$

86–89% yield

Schotten–Baumann synthesis of an amide

As these mechanisms show, the formation of amides from acid chlorides and amines is accompanied by production of one equivalent of HCl, which needs to be neutralized by a second equivalent of amine. An alternative method for making amides is to carry out the reaction in the presence of another base, such as NaOH, which then does the job of neutralizing the HCl. The trouble is, OH⁻ also attacks acyl chlorides to give carboxylic acids. Schotten and Baumann, in the late nineteenth century, published a way round this problem by carrying out these reactions in *two-phase systems* of immiscible water and dichloromethane. (Carl Schotten (1853–1910) was Hofmann's assistant in Berlin and spent most of his working life in the German patent office. There is more about Hofmann in Chapter 19.) The organic amine (not necessarily ammonia) and the acyl chloride remain in the (lower) dichloromethane layer, while the base (NaOH) remains in the (upper) aqueous layer. Dichloromethane and chloroform are two common organic solvents that are *heavier* (more dense) than water. The acyl chloride reacts only with the amine, but the HCl produced can dissolve in, and be neutralized by, the aqueous solution of NaOH.

Schotten–Baumann synthesis of an amide

NaOH
H₂O, CH₂Cl₂

80% yield

upper layer:
aqueous solution
of NaOH

lower layer:
dichloromethane
solution of amine
and acid chloride

Using pK_{aH} to predict the outcome of substitution reactions of carboxylic acid derivatives

You saw that acid anhydrides react with alcohols to give esters: they will also react with amines to give amides. But would you expect esters to react with amines to give amides, or amides to react with alcohols to give esters? Both appear reasonable.

NH₃

?

MeOH

In fact only the top reaction works: amides can be formed from esters but esters cannot be formed from amides. Again, looking at pK_as can tell us why. In both cases, the tetrahedral intermediate would be the same. The possible leaving groups are shown in the table.

NH₃?

MeOH?

tetrahedral intermediate

Possible leaving groups	pK_{aH}
Ph⁻	45
NH₂⁻	35
MeO⁻	16

So RO⁻ leaves and the amide is formed. Here is an example. The base may be either the EtO⁻ produced in the previous step or another molecule of PhNH₂.

PhNH₂

135 °C 1 h

base⁻

► You will meet many more mechanisms like this, in which an unspecified base removes a proton from an intermediate. As long as you can satisfy yourself that there is a base available to perform the task, it is quite acceptable to write either of these shorthand mechanisms.

base⁻

or

Factors other than leaving group ability can be important

In fact, the tetrahedral intermediate would simply never form from an amide and an alcohol; the amide is too bad an electrophile and the alcohol not a good enough nucleophile. We've looked at leaving group ability: next we'll consider the strength of the nucleophile Y and then the strength of the electrophile RCOX.

● **Conditions for reaction**

If this reaction is to go

1. X must be a better leaving group than Y (otherwise the reverse reaction would take place)
2. Y must be a strong enough nucleophile to attack RCOX
3. RCOX must be a good enough electrophile to react with Y$^-$

pK_{aH} is a guide to nucleophilicity

We have seen how pK_a gives us a guide to leaving group ability: it is also a good guide to how strong a nucleophile will be. These two properties are the reverse of each other: good nucleophiles are bad leaving groups. A species that likes forming new bonds to hydrogen (in other words, the pK_a of its conjugate acid is high) will also like to form new bonds to carbon: it is likely to be a good nucleophile. Bases with high pK_{aH} are bad leaving groups and they are, in general, good nucleophiles towards the carbonyl group. We will come back to this concept again in Chapter 17, where you will see that it does not apply to substitution at saturated carbon atoms.

● **Guide to nucleophilicity**

In general, the higher the pK_{aH}, the better the nucleophile.

But just a moment—we've overlooked an important point. When we made acid anhydrides from acid chlorides plus carboxylate salts, we used an anionic nucleophile RCO_2^- but, when we made amides from acid chlorides plus amines, we used a neutral nucleophile NH_3, and not NH_2^-. For proper comparisons, we should include in our table ROH (pK_{aH} = −5; in other words, −5 is the pK_a of ROH_2^+) and NH_3 (pK_{aH} = 9; in other words, 9 is the pK_a of NH_4^+).

While amines react with acetic anhydride quite rapidly at room temperature (reaction complete in a few hours), alcohols react extremely slowly in the absence of a base. On the other hand, an alkoxide anion reacts with acetic anhydride extremely

Base	pK_{aH}
R$^-$	50
NH$_2^-$	35
RO$^-$	16
NH$_3$	9
RCO$_2^-$	5
ROH	−5
Cl$^-$	−7

increasing pK_{aH} → / *increasing nucleophilicity* →

rapidly—the reactions are often complete within seconds at 0 °C. We don't have to deprotonate an alcohol completely to increase its reactivity: just a catalytic quantity of a weak base can do this job by removing the alcohol's proton *as it adds* to the carbonyl group. All these observations are consistent with our table and our proposition that high pK_{aH} means good nucleophilicity.

Not all carboxylic acid derivatives are equally reactive

We can list the common carboxylic acid derivatives in a 'hierarchy' of reactivity, with the most reactive at the top and the least reactive at the bottom. Transformations are always possible moving *down*

■ You saw pyridine doing this on p. 281—it's called **general base catalysis**, and we will talk about it in more detail in Chapter 13.

the hierarchy. We've seen that this hierarchy is partly due to how good the leaving group is (the ones at the top are best), and partly due to how good the nucleophile needed to make the derivative is (the ones at the bottom are best).

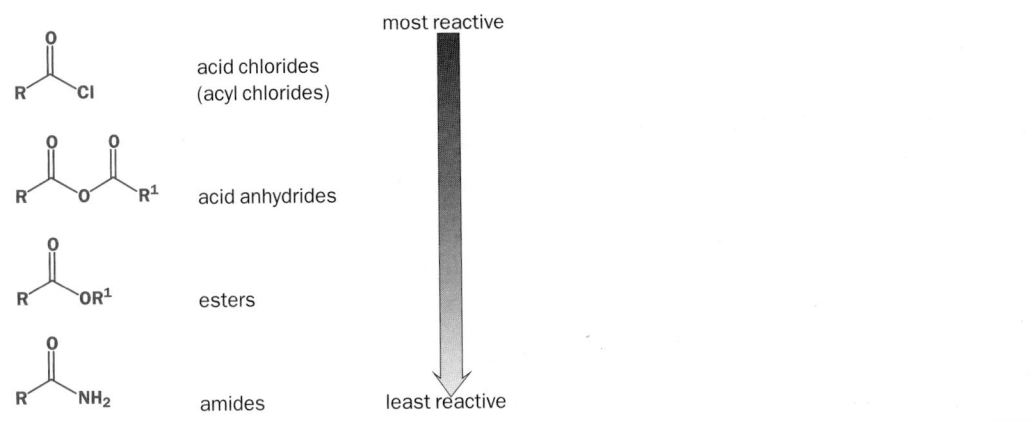

Delocalization and the electrophilicity of carbonyl compounds

All of these derivatives will react with water to form carboxylic acids, but at very different rates.

Hydrolysing an amide requires boiling in 10% NaOH or heating overnight in a sealed tube with concentrated HCl. Amides are the least reactive towards nucleophiles because they exhibit the greatest degree of delocalization. You met this concept in Chapter 7 and we shall return to it many times more. In an amide, the lone pair on the nitrogen atom can be stabilized by overlap with the π^* orbital of the carbonyl group—this overlap is best when the lone pair occupies a p orbital (in an amine, it would occupy an sp^3 orbital).

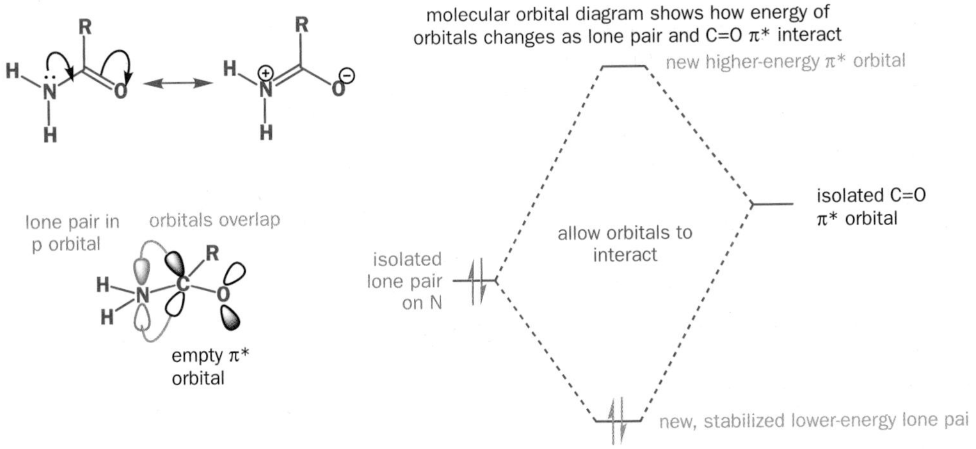

The molecular orbital diagram shows how this interaction both lowers the energy of the bonding orbital (the delocalized nitrogen lone pair), making it neither basic nor nucleophilic, and raises the energy of the π^* orbital, making it less ready to react with nucleophiles. Esters are similar but, because the oxygen lone pairs are lower in energy, the effect is less pronounced.

The greater the degree of delocalization, the weaker the C=O bond becomes. This is most clearly

■ We treat this in more detail in Chapter 15. There are two frequencies for the anhydride and the carboxylate because of symmetric and antisymmetric stretching.

evident in the stretching frequency of the carbonyl group in the IR spectra of carboxylic acid derivatives—remember that the stretching frequency depends on the force constant of the bond, itself a measure of the bond's strength (the carboxylate anion is included because it represents the limit of the series, with complete delocalization of the negative charge over the two oxygen atoms).

ν / cm^{-1}	1790–1815	1800–1850 / 1740–1790	1735–1750	1690	1610–1650 / 1300–1420

C=O   strongest ———————————————→ weakest

Amides react as electrophiles only with powerful nucleophiles such as HO$^-$. Acid chlorides, on the other hand, react with even quite weak nucleophiles: neutral ROH, for example. They are more reactive because the electron-withdrawing effect of the chlorine atom increases the electrophilicity of the carbonyl carbon atom.

Bond strengths and reactivity

You may think that a weaker C=O bond should be more reactive. This is not so because the partial positive charge on carbon is also lessened by delocalization and because the molecule as a whole is stabilized by the delocalization. Bond strength is not always a good guide to reactivity!

For example, in acetic acid the bond strengths are surprising. The strongest bond is the O–H bond and the weakest is the C–C bond. Yet very few reactions

of acetic acid involve breaking the C–C bond, and its characteristic reactivity, as an acid, involves breaking O–H, the strongest bond of them all!

The reason is that polarization of bonds and solvation of ions play an enormously important role in determining the reactivity of molecules. In Chapter 39 you will see that radicals are relatively unaffected by solvation and that their reactions follow bond strengths much more closely.

Carboxylic acids do not undergo substitution reactions under basic conditions

Substitution reactions of RCO$_2$H require a leaving group OH$^-$, with pK_{aH} = 15, so we should be able to slot RCO$_2$H into the 'hierarchy' on p. 287 just above the esters RCO$_2$R'. However, if we try to react carboxylic acids with alcohols in the presence of a base (as we would to make esters from acyl chlorides), the only thing that happens is deprotonation of the acid to give the carboxylate anion. Similarly, carboxylic acids react with amines to give not amides but ammonium carboxylate salts, because the amines themselves are basic.

■ Later in this chapter (p. 299) you will meet about the only nucleophiles that will: organolithium compounds attack lithium carboxylates.

Once the carboxylic acid is deprotonated, substitutions are prevented because (almost) no nucleophile will attack the carboxylate anion. Under neutral conditions, alcohols are just not reactive enough to add to the carboxylic acid but, with *acid* catalysis, esters can be formed from alcohols and carboxylic acids.

▶ In fact, amides *can* be made from carboxylic acids plus amines, but only if the ammonium salt is heated strongly to dehydrate it. This is not usually a good way of making amides!

Acid catalysts increase the reactivity of a carbonyl group

We saw in Chapter 6 that the lone pairs of a carbonyl group may be protonated by acid. Only strong acids are powerful enough to protonate carbonyl groups: the pK_a of protonated acetone is –7, so, for example, even 1M HCl (pH 0) would protonate only 1 in 10^7 molecules of acetone. However, even proportions as low as this are sufficient to increase the rate of substitution reactions at carbonyl groups enormously, because those carbonyl groups that are protonated become extremely powerful electrophiles.

It is for this reason that alcohols will react with carboxylic acids under acid catalysis. The acid (usually HCl, or H_2SO_4) reversibly protonates a small percentage of the carboxylic acid molecules, and the protonated carboxylic acids are extremely susceptible to attack by even a weak nucleophile such as an alcohol.

acid-catalysed ester formation: forming the tetrahedral intermediate

Acid catalysts can make bad leaving groups into good ones

This tetrahedral intermediate is unstable because the energy to be gained by re-forming a C=O bond is greater than that used in breaking two C–O bonds. As it stands, none of the leaving groups (R^-, HO^-, or RO^-) is very good. However, help is again at hand in the acid catalyst. It can protonate any of the oxygen atoms reversibly. Again, only a very small proportion of molecules are protonated at any one time but, once the oxygen atom of, say, one of the OH groups is protonated, it becomes a much better leaving group (H_2O, pK_{aH} –2, instead of HO^-, pK_{aH} 15). Loss of ROH from the tetrahedral intermediate is also possible: this leads back to starting materials—hence the equilibrium arrow in the scheme above. Loss of H_2O is more fruitful, and takes the reaction forwards to the ester product.

> Average bond strength C=O = 720 kJ mol^{-1}. Average bond strength C–O = 351 kJ mol^{-1}.

■ We shall discuss the reasons why chemists believe this to be the mechanism of this reaction later in the chapter.

acid-catalysed ester formation (continued)

tetrahedral intermediate H_2O ester product

> ● **Acid catalysts catalyse substitution reactions of carboxylic acids**
> 1 They increase the electrophilicity of the carbonyl group by protonation at *carbonyl* oxygen
> 2 They lower the pK_{aH} of the leaving group by protonation there too

Ester formation is reversible: how to control an equilibrium

Loss of water from the tetrahedral intermediate is reversible too: just as ROH will attack a protonated carboxylic acid, H_2O will attack a protonated ester. In fact, every step in the sequence from carboxylic acid to ester is an equilibrium, and the overall equilibrium constant is about 1. In order for this reaction to be useful, it is therefore necessary to ensure that the equilibrium is pushed towards the ester side by using an excess of alcohol or carboxylic acid (usually the reactions are done in a solution of the alcohol or the carboxylic acid). In this reaction, for example, using less than three equivalents of ethanol gave lower yields of ester.

RO⁓CO₂H → [3 equiv. EtOH / dry HCl gas] → RO⁓CO₂Et   68–72% yield

Lactic acid must be handled in solution in water. Can you see why, bearing in mind what we have said about the reversibility of ester formation?

Alternatively, the reaction can be done in the presence of a dehydrating agent (concentrated H_2SO_4, for example, or silica gel), or the water can be distilled out of the mixture as it forms.

lactic acid   cat. H_2SO_4 benzene (solvent) remove water by distillation   89–91% yield

AcOH   cat. H_2SO_4 silica gel (drying agent)   57% yield

▶

You have now met three ways of making esters from alcohols:

• with acyl chlorides   • with acid anhydrides   • with carboxylic acids

Try to appreciate that different methods will be appropriate at different times. If you want to make a few milligrams of a complex ester, you are much more likely to work with a reactive acyl chloride or anhydride, using pyridine as a weakly basic catalyst, than to try and distil out a minute quantity of water from a reaction mixture containing a strong acid that may destroy the starting material. On the other hand, if you are a chemist making simple esters (such as those in Chapter 2, p. 34) for the flavouring industry on a scale of many tons, you will prefer the cheaper option of carboxylic acid plus HCl in alcohol solution.

Acid-catalysed ester hydrolysis and transesterification

By starting with an ester, an excess of water, and an acid catalyst, we can persuade the reverse reaction to occur: formation of the carboxylic acid plus alcohol with consumption of water. Such a reaction is known as a **hydrolysis reaction**, because water is used to break up the ester into carboxylic acid plus alcohol (*lysis* = breaking).

excess water forces reaction forward ——— acid-catalysed ester hydrolysis ———→   ←——— acid-catalysed ester formation ——— excess ester or removal of water forces the reaction backward

ROH

The mechanisms of acid-catalysed formation and hydrolysis of esters are extremely important: you *must* learn them, and understand the reason for each step.

Acid-catalysed ester formation and hydrolysis are the exact reverse of one another: the only way we can control the reaction is by altering concentrations of reagents to drive the reaction the way we want it to go. The same principles can be used to convert to convert an ester of one alcohol into an ester of another, a process known as **transesterification**. It is possible, for example, to force this equilibrium to the right by distilling methanol (which has a lower boiling point than the other components of the reaction) out of the mixture.

+   **MeOH**

The mechanism for this transesterification simply consists of adding one alcohol (here BuOH) and eliminating the other (here MeOH), both processes being acid-catalysed. Notice how easy it is now to confirm that the reaction is *catalytic* in H^+. Notice also that protonation always occurs on the *carbonyl* oxygen atom.

irreversible because MeOH is removed from the mixture

94% yield   **MeOH**   distilled off

Polyester fibre manufacture

A transesterification reaction is used to make the polyester fibres that are used for textile production. Terylene, or Dacron, for example, is a polyester of the dicarboxylic acid terephthalic acid and the diol ethylene glycol. Polymers are discussed in more detail in Chapter 52.

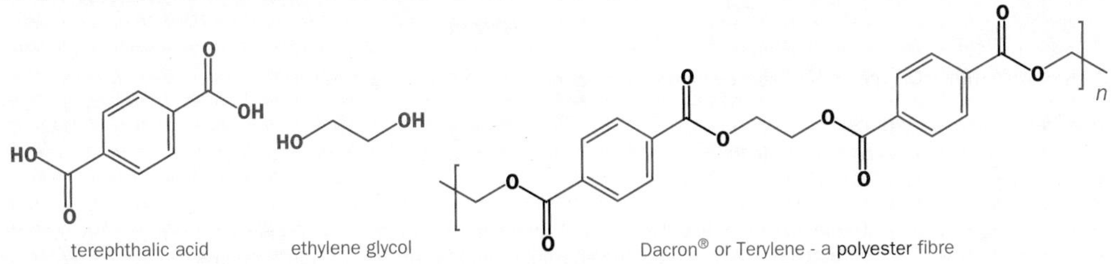

terephthalic acid   ethylene glycol   Dacron® or Terylene - a **polyester** fibre

It is made by transesterifying dimethyl terephthalate with ethylene glycol in the presence of an acid catalyst, distilling off the methanol as it forms.

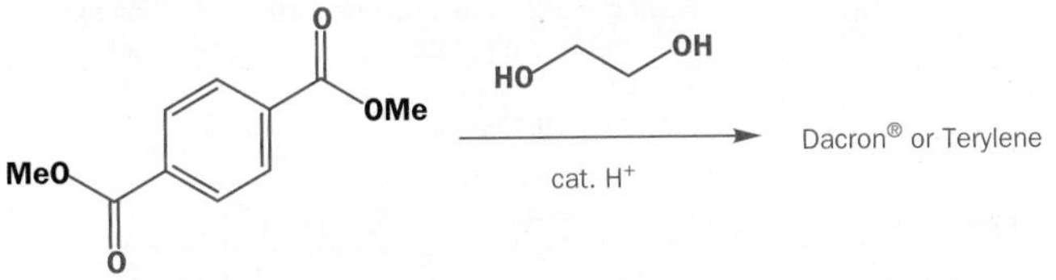

Dacron® or Terylene

cat. H+

Base-catalysed hydrolysis of esters is irreversible

You can't make esters from carboxylic acids and alcohols under basic conditions because the base deprotonates the carboxylic acid (see p. 288). However, you can reverse that reaction and hydrolyse an ester to a carboxylic acid (more accurately, a carboxylate salt) and an alcohol.

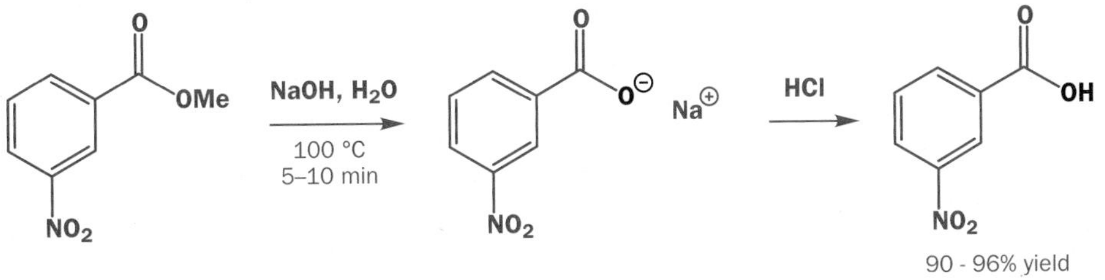

NaOH, H$_2$O

100 °C
5–10 min

HCl

90 - 96% yield

This time the ester is, of course, not protonated first as it would be in acid, but the unprotonated ester is a good enough electrophile because OH⁻, and not water, is the nucleophile. The tetrahedral intermediate can collapse either way, giving back ester, or going forward to acid plus alcohol.

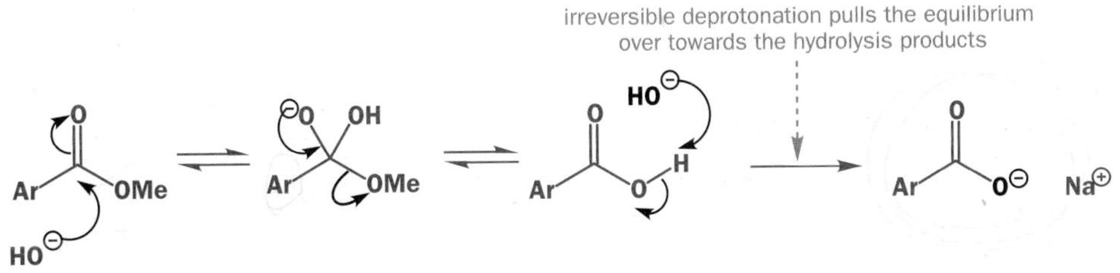

irreversible deprotonation pulls the equilibrium
over towards the hydrolysis products

Without an acid catalyst, the alcohol cannot react with the carboxylic acid; in fact, the backward reaction is doubly impossible because the basic conditions straight away deprotonate the acid to make a carboxylate salt (which, incidentally, consumes the base, making at least one equivalent of base necessary in the reaction).

How do we know this is the mechanism?

Ester hydrolysis is such an important reaction that chemists spent a lot of time and effort finding out exactly how it worked. If you want to know all the details, read a specialist textbook on physical (mechanistic) organic chemistry. Many of the experiments that tell us about the mechanism involve oxygen-18 labelling. The starting material is synthesized using as a starting material a compound enriched in the heavy oxygen isotope ^{18}O. By knowing where the heavy oxygen atoms start off, and following (by mass spectrometry—Chapter 3) where they end up, the mechanism can be established.

How do we know this is the mechanism? (continued)

1 An ^{18}O label in the 'ether' oxygen of the ester starting material ends up in the alcohol product

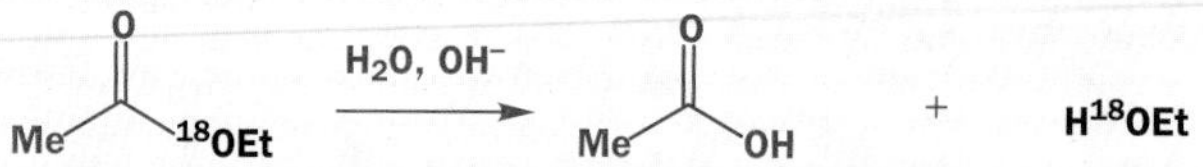

2 Hydrolysis with $^{18}OH_2$ gives ^{18}O-labelled carboxylic acid, but no ^{18}O-labelled alcohol

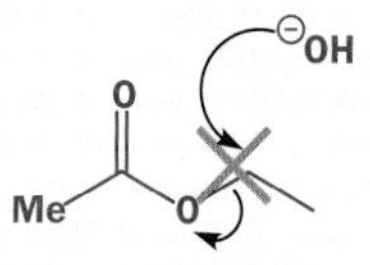

These experiments tell us that a displacement (substitution) has occurred at the carbonyl carbon atom, and rule out the alternative displacement at saturated carbon.

Having worked this out, one further labelling experiment showed that a tetrahedral intermediate must be formed: an ester labelled with ^{18}O in its carbonyl oxygen atom passes some of its ^{18}O label to the water. We discussed why this shows that a tetrahedral intermediate must be formed on p. 282.

INCORRECT

The saturated fatty acid tetradecanoic acid (also known as myristic acid) is manufactured commercially from coconut oil by base-catalysed hydrolysis. You may be surprised to learn that coconut oil contains more saturated fat than butter, lard, or beef dripping: much of it is the trimyristate ester of glycerol. Hydrolysis with aqueous sodium hydroxide, followed by reprotonation of the sodium carboxylate salt with acid, gives myristic acid. Notice how much longer it takes to hydrolyse this branched ester than it did to hydrolyse a methyl ester (p. 291).

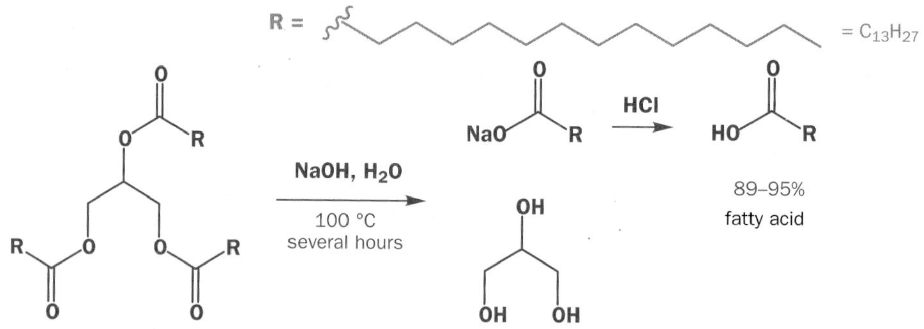

R = $C_{13}H_{27}$

NaOH, H_2O
100 °C
several hours

HCl

89–95%
fatty acid

principal component of coconut oil

glycerol

Saponification

The alkaline hydrolysis of esters to give carboxylate salts is known as **saponification**, because it is the process used to make soap. Traditionally, beef tallow (the tristearate ester of glycerol—stearic acid is octadecanoic acid, $C_{17}H_{35}CO_2H$) was hydrolysed with sodium hydroxide to give sodium stearate, $C_{17}H_{35}CO_2Na$, the principal component of soap. Finer soaps are made from palm oil and contain a higher proportion of sodium palmitate, $C_{15}H_{31}CO_2Na$. Hydrolysis with KOH gives potassium carboxylates, which are used in liquid soaps. Soaps like these owe their detergent properties to the combination of polar (carboxylate group) and nonpolar (long alkyl chain) properties.

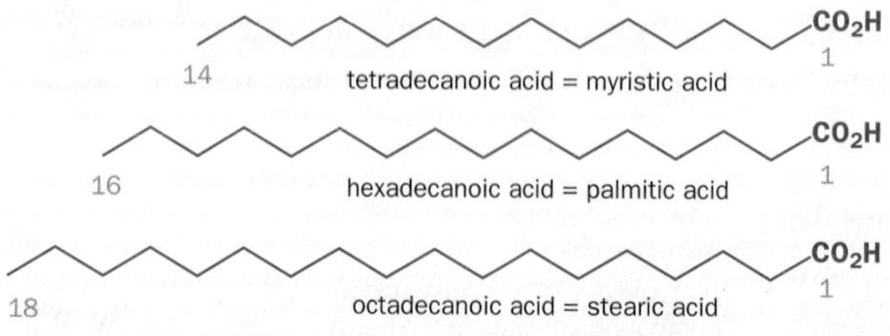

14   tetradecanoic acid = myristic acid

16   hexadecanoic acid = palmitic acid

18   octadecanoic acid = stearic acid

Amides can be hydrolysed under acidic or basic conditions too

In order to hydrolyse the least reactive of the series of carboxylic acid derivatives we have a choice: we can persuade the amine leaving group to leave by protonating it, or we can use brute force and forcibly eject it with concentrated hydroxide solution.

Amides are very unreactive as electrophiles, but they are also rather more basic than most carboxylic acid derivatives: a typical amide has a pK_{aH} of -1; most other carbonyl compounds have pK_{aH}s of around -7. You might therefore imagine that the protonation of an amide would take place on nitrogen—after all, *amine* nitrogen atoms are readily protonated. And, indeed, the reason for the basicity of amides is the nitrogen atom's delocalized lone pair, making the carbonyl group unusually electron-rich. But amides are always protonated on the oxygen atom of the carbonyl group—never the nitrogen, because protonation at nitrogen disrupts the delocalized system that makes amides so stable.

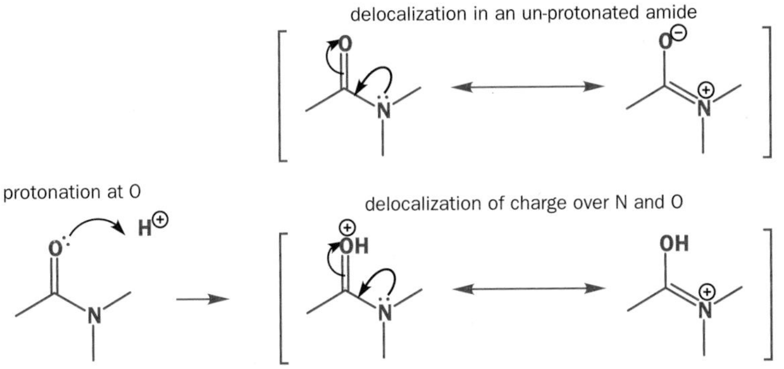

delocalization in an un-protonated amide

protonation at N (does not happen)

no delocalization possible

protonation at O

delocalization of charge over N and O

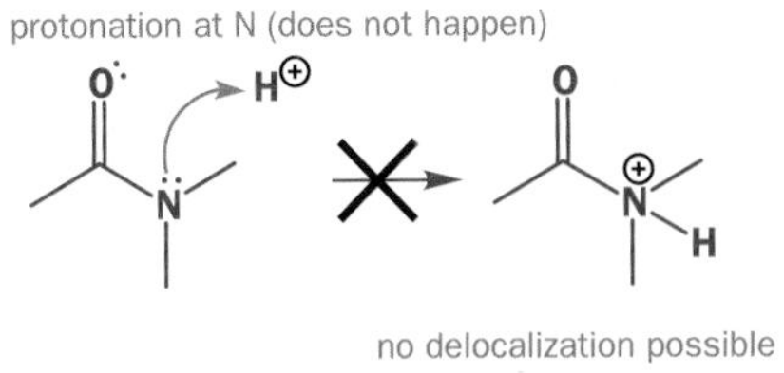

Protonation of the carbonyl group by acid makes the carbonyl group electrophilic enough for attack by water, giving a neutral tetrahedral intermediate. The amine nitrogen atom in the tetrahedral intermediate is much more basic than the oxygen atoms, so now *it* gets protonated, and the RNH_2 group becomes really quite a good leaving group. And, once it has left, it will immediately be protonated again, and therefore become completely nonnucleophilic. The conditions are very vigorous—70% sulfuric acid for 3 hours at 100 °C.

> Notice that this means that one equivalent of acid is used up in this reaction—the acid is not solely a catalyst.

amide hydrolysis in acid: 3 hours at 100 °C with 70% H_2SO_4 in water gives 70% yield of the acid

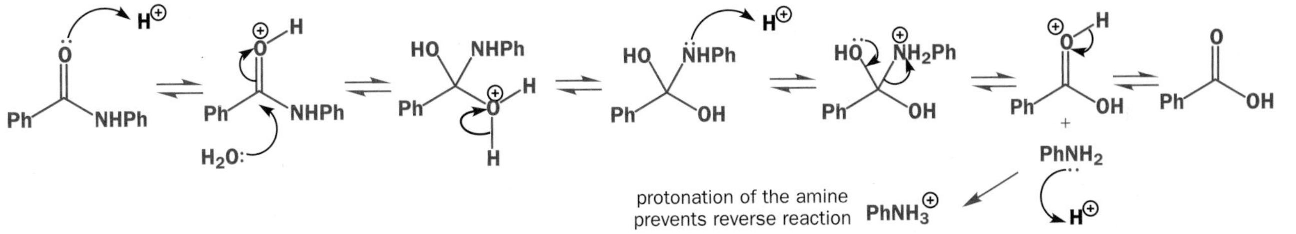

protonation of the amine prevents reverse reaction

Hydrolysis of amides in base requires similarly vigorous conditions. Hot solutions of hydroxide are sufficiently powerful nucleophiles to attack an amide carbonyl group, though even when the tetrahedral intermediate has formed, NH_2^- (pK_{aH} 35) has only a slight chance of leaving when OH^- (pK_{aH} 15) is an alternative. Nonetheless, at high temperatures, amides are slowly hydrolysed by concentrated base.

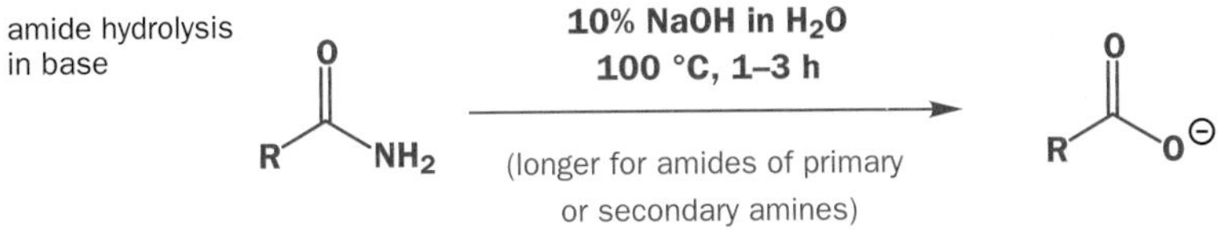

amide hydrolysis in base

10% NaOH in H$_2$O
100 °C, 1–3 h

(longer for amides of primary or secondary amines)

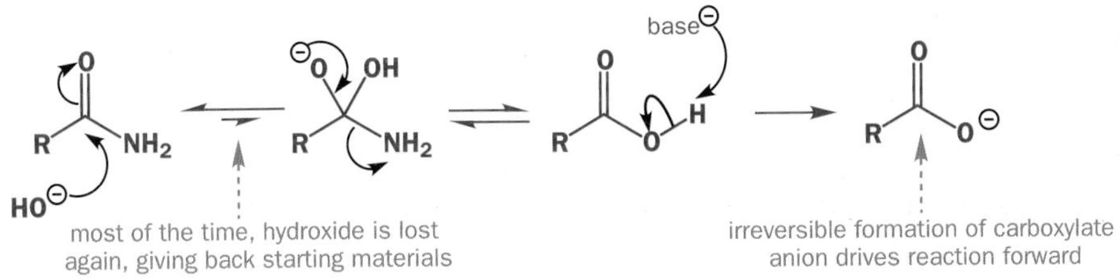

most of the time, hydroxide is lost again, giving back starting materials

irreversible formation of carboxylate anion drives reaction forward

Secondary and tertiary amides hydrolyse much more slowly under these conditions. However, with a slightly different set of reagents, even tertiary amides can be hydrolysed at room temperature.

hydrolysis of amides
using *t*-BuOK

H₂O (2 equiv.)
***t*-BuOK (6 equiv.)**

DMSO, 20 °C
then HCl (to protonate
carboxylate salt)

90%   + **Me₂NH**

85%

> You've not seen the option of O^{2-} as a leaving group before but this is what you would get if you want O^- to leave. Asking O^{2-} to be a leaving group is like asking HO^- to be an acid.

The reason is a change in mechanism. Potassium *tert*-butoxide is a strong enough base (pK_{aH} 18) to deprotonate the tetrahedral intermediate in the reaction, forming a dianion. Now that the choice is between Me_2N^- and O^{2-}, the Me_2N^- has no choice but to leave, giving the carboxylate salt directly as the product.

> The hydrolysis of some amides in aqueous NaOH probably proceeds by a similar dianion mechanism—see Chapter 13.

Me₂N⁻ has to leave - there's no alternative!

Hydrolysing nitriles: how to make the almond extract, mandelic acid

Closely related to the amides are nitriles. You can view them as primary amides that have lost one molecule of water and, indeed, they can be made by dehydrating primary amides.

They can be hydrolysed just like amides too. Addition of water to the protonated nitrile gives a primary amide, and hydrolysis of this amide gives carboxylic acid plus ammonia.

> Don't be put off by the number of steps in this mechanism—look carefully, and you will see that most of them are simple proton transfers. The only step that isn't a proton transfer is the addition of water.

H₂O, H₂SO₄

100 °C, 3 h   80%

> reminder:
> cyanohydrins from aldehydes

You met a way of making nitriles—from HCN (or NaCN + HCl) plus aldehydes—in Chapter 6: the hydroxynitrile products are known as **cyanohydrins**.

With this in mind, you should be able to suggest a way of making mandelic acid, an extract of almonds, from benzaldehyde.

This is how some chemists did it.

benzaldehyde   mandelic acid

> ■ You have just designed your first total synthesis of a natural product. We return to such things much later in this book, in Chapter 31.

synthesis of
mandelic acid
from benzaldehyde

NaCN
PhCHO ———→
H⁺

H₂O
———→
HCl

mandelic acid
50–52% yield

Acid chlorides can be made from carboxylic acids using SOCl₂ or PCl₅

We have looked at a whole series of interconversions between carboxylic acid derivatives and, after this next section, we shall summarize what you should have learned. We said that it is always easy to move down the series of acid derivatives we listed early in the chapter and, so far, that is all we have done. But some reactions of carboxylic acids also enable us to move upwards in the series. What we need is a reagent that changes the bad leaving group HO^- into a good leaving group. Strong acid does this by protonating the OH^-, allowing it to leave as H_2O. In this section we look at two more reagents, $SOCl_2$ and PCl_5, which react with the OH group of a carboxylic acid and also turn it into a good leaving group. Thionyl chloride, $SOCl_2$, reacts with carboxylic acids to make acyl chlorides.

acid chlorides are made
from carboxylic acids
with **thionyl chloride**

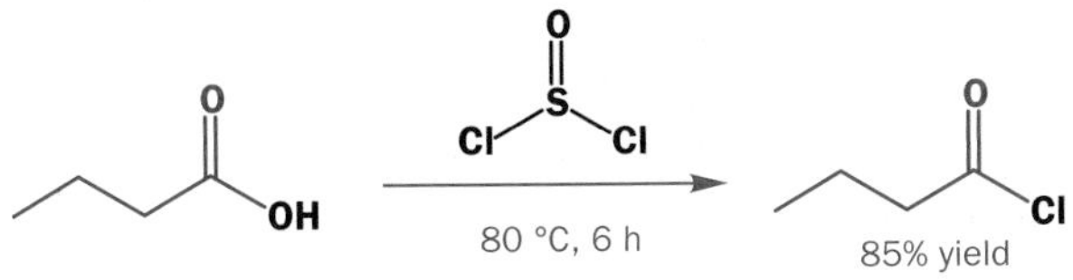

80 °C, 6 h   85% yield

This volatile liquid with a choking smell is electrophilic at the sulfur atom (as you might expect with two chlorine atoms and an oxygen atom attached) and is attacked by carboxylic acids to give an unstable, and highly electrophilic, intermediate.

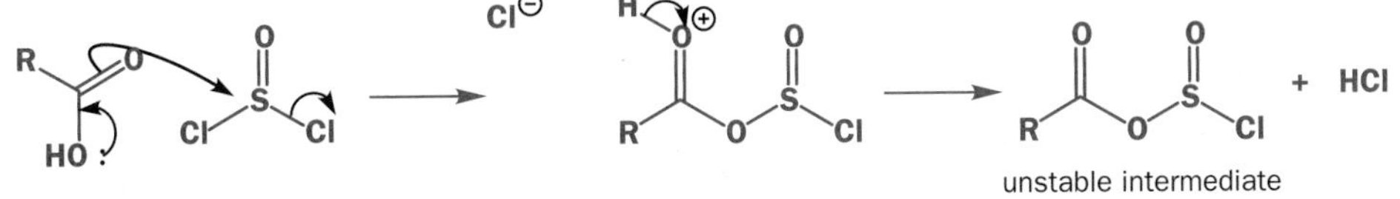

unstable intermediate

+ **HCl**

▶
You may be shocked to see the way we substituted at S=O without forming a 'tetrahedral intermediate'. Well, this trivalent sulfur atom is already tetrahedral (it still has one lone pair), and substitution can go by a direct 'S$_N$2 at sulfur' (Chapter 17).

Protonation of the unstable intermediate (by the HCl just produced) gives an electrophile powerful enough to react even with the weak nucleophile Cl^- (low pK_{aH}, poor nucleophilicity). The tetrahedral intermediate that results can collapse to the acyl chloride, sulfur dioxide, and hydrogen chloride. This step is irreversible because SO_2 and HCl are gases that are lost from the reaction mixture.

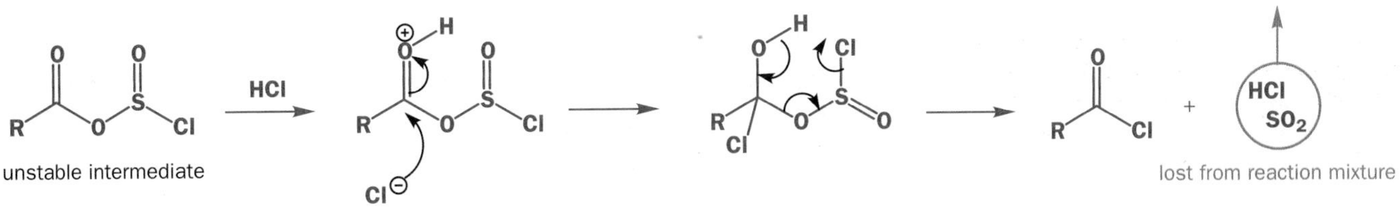

unstable intermediate   **HCl**   +   **HCl** **SO$_2$**   lost from reaction mixture

Although HCl is involved in this reaction, it cannot be used as the sole reagent for making acid chlorides. It is necessary to have a sulfur or phosphorus compound to remove the oxygen. An alternative reagent for converting RCO_2H into $RCOCl$ is phosphorus pentachloride, PCl_5. The mechanism is similar—try writing it out before looking at the scheme below.

acid chlorides are made
from carboxylic acids with
phosphorus pentachloride

PCl$_5$ 90–96% yield

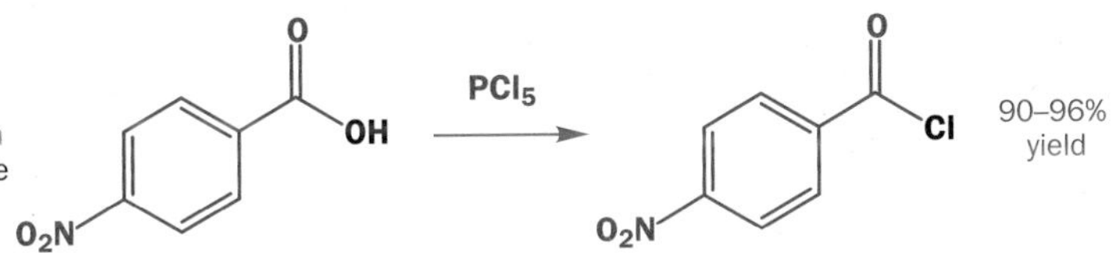

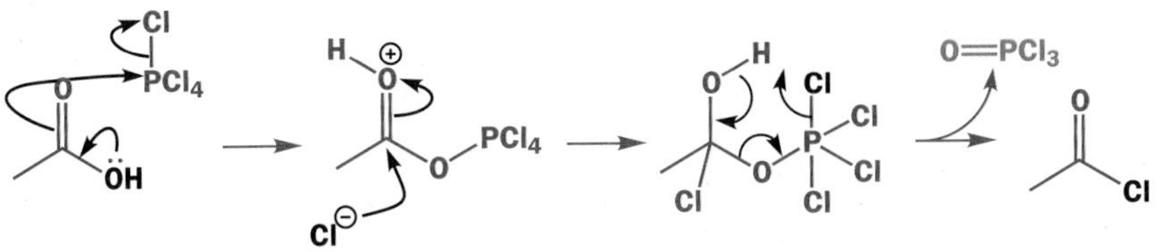

An alternative method of making acid chlorides: oxalyl chloride plus DMF

A modification of the thionyl chloride method for making acyl chlorides uses oxalyl chloride plus catalytic DMF. The oxalyl chloride reacts with the DMF in a rather remarkable way to produce a highly electrophilic cationic intermediate, plus CO and CO_2—as with the $SOCl_2$ reaction, the by-products are all gases.

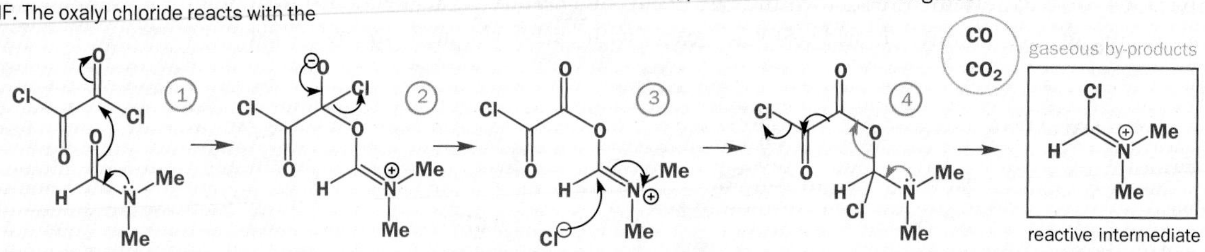

reactive intermediate

A few aspects of this mechanism need comment.

* The first two steps are simply a nucleophilic substitution of Cl at the carbonyl group, going via the now familiar tetrahedral intermediate

* Nucleophiles can attack the C=N bond (step 3) much as they might attack a C=O bond
* The black arrows in step 4 look very odd, but they are the only way we can draw the formation of carbon monoxide

The reactive intermediate is highly electrophilic and reacts rapidly with the carboxylic acid, producing another intermediate which intercepts Cl⁻ to give the acyl chloride and regenerate DMF.

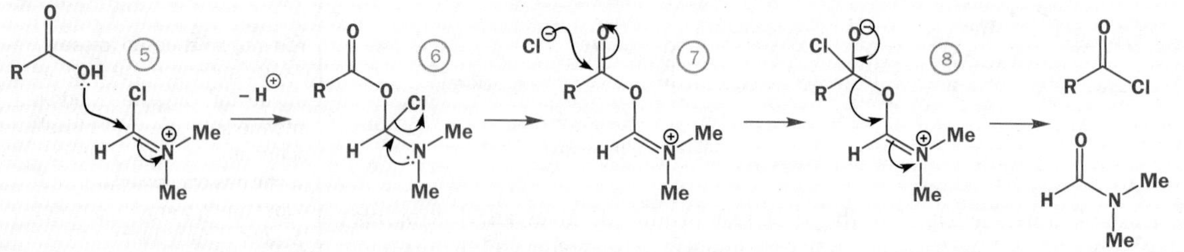

This method is usually used for producing small amounts of valuable acyl chlorides—oxalyl chloride is much more expensive than thionyl chloride. DMF will nonetheless also catalyse acyl chloride formation with thionyl chloride, though on a large scale its use may be ill advised since one of the minor by-products from these reactions is a potent carcinogen. We hope you enjoyed the eight-step mechanism.

►

Oxalyl chloride, $(COCl)_2$, is the 'double' acid chloride of oxalic acid, or ethane-1,2-dioic acid, the toxic dicarboxylic acid found in rhubarb leaves.

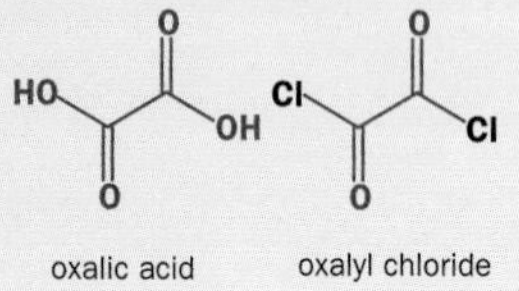

oxalic acid　　oxalyl chloride

These conversions of acids into acid chlorides complete all the methods we need to convert acids into any acid derivatives. You can convert acids directly to esters and now to acid chlorides, the most reactive of acid derivatives, and can make any other derivative from them. The chart below adds reactions to the reactivity order we met earlier.

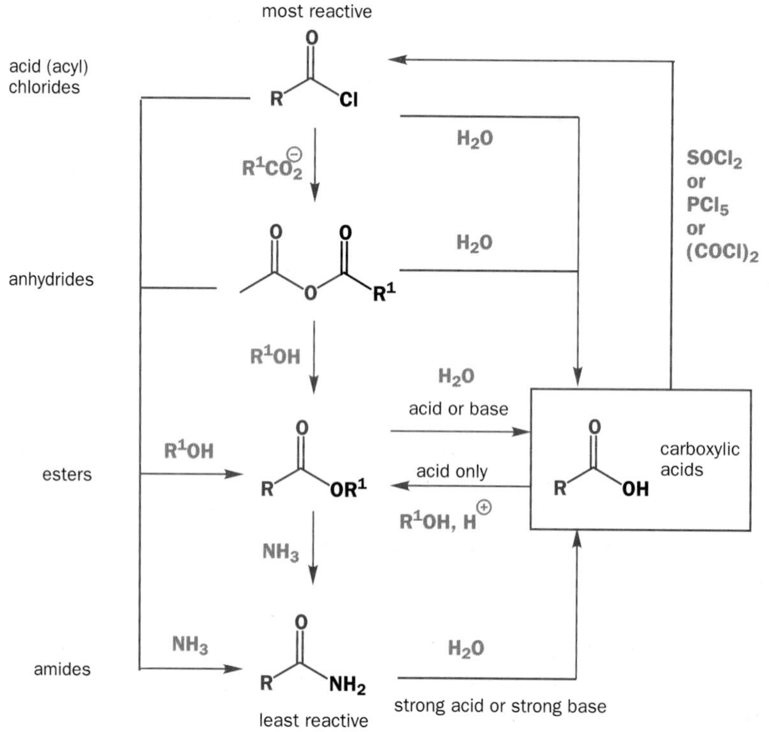

All these acid derivatives can, of course, be hydrolysed to the acid itself with water alone or with various levels of acid or base catalysis depending on the reactivity of the derivative. To climb the reactivity order therefore, the simplest method is to hydrolyse to the acid and convert the acid into the acid chloride. You are now at the top of the reactivity order and can go down to whatever level you require.

Making other compounds by substitution reactions of acid derivatives

We've talked at length about the interconversions of acid derivatives, explaining the mechanism of attack of nucleophiles such as ROH, H_2O, and NH_3 on acyl chlorides, acid anhydrides, esters, acids, and amines, with or without acid or base present. We shall now go on to talk about substitution reactions of acid derivatives that take us out of this closed company of compounds and allow us to make compounds containing functional groups at other oxidation levels such as ketones and alcohols.

■
Five 'oxidation levels'—(1) hydrocarbon; (2) alcohol; (3) aldehyde and ketone; (4) carboxylic acid; and (5) CO_2—were defined in Chapter 2.

Making ketones from esters: the problem

Substitution of the OR group of an ester by an R group would give us a ketone. You might therefore think that reaction of an ester with an organolithium or Grignard reagent would be a good way of making ketones. However, if we try the reaction, something else happens.

Two molecules of Grignard have been incorporated and we get an alcohol! If we look at the mechanism we can understand why this should be so. First, as you would expect, the nucleophilic Grignard reagent attacks the carbonyl group to give a tetrahedral intermediate. The only reasonable leaving group is RO⁻, so it leaves to give us the ketone we set out to make.

Now, the next molecule of Grignard reagent has a choice. It can either react with the ester starting material, or with the newly formed ketone. Ketones are more electrophilic than esters so the Grignard reagent prefers to react with the ketone in the manner you saw in Chapter 9. A stable alkoxide anion is formed, which gives the tertiary alcohol on acid work-up.

Making alcohols instead of ketones

In other words, the problem here lies in the fact that the ketone product is more reactive than the ester starting material. We shall meet more examples of this general problem later (in Chapter 24, for example): in the next section we shall look at ways of overcoming it. Meanwhile, why not see it as a useful reaction? This compound, for example, was needed by some chemists in the course of research into explosives.

It is a tertiary alcohol with the hydroxyl group flanked by two identical R (= butyl) groups. The chemists who wanted to make the compound knew that an ester would react twice with the same organolithium reagent, so they made it from this unsaturated ester (known as methyl methacrylate) and butyllithium.

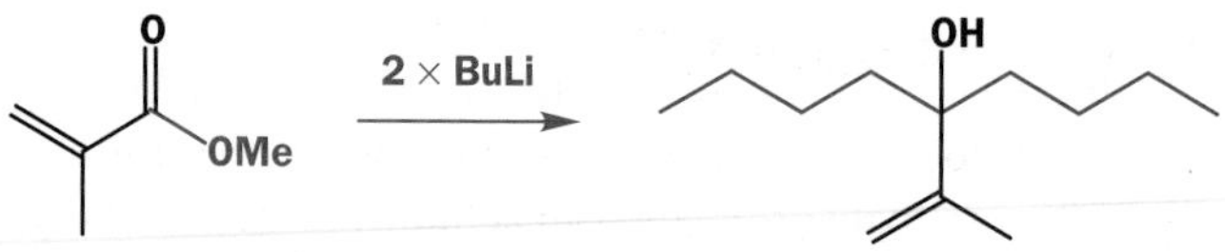

● **Tertiary alcohol synthesis**

Tertiary alcohols with two identical R^2 groups can be made from ester plus two equivalents of organolithium or Grignard reagent.

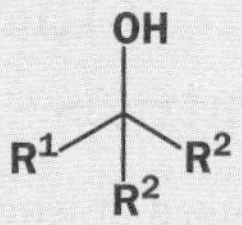

This reaction works with R=H too if we use lithium aluminium hydride. $LiAlH_4$ is a powerful reducing agent, and readily attacks the carbonyl group of an ester. Again, collapse of the tetrahedral intermediate gives a compound, this time an aldehyde, which is more reactive than the ester starting material, so a second reaction takes place and the ester is converted (reduced) into an alcohol.

reduction of esters by $LiAlH_4$

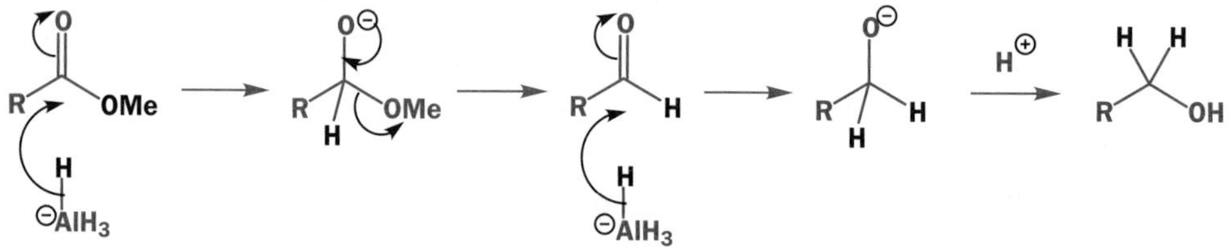

This is an extremely important reaction, and one of the best ways of making alcohols from esters. Stopping the reaction at the aldehyde stage is more difficult: we shall discuss this in Chapter 24.

A bit of shorthand

Before we go any further, we should introduce to you a little bit of chemical shorthand that makes writing many mechanisms easier.

As you now appreciate, all substitution reactions at a carbonyl group go via a tetrahedral intermediate.

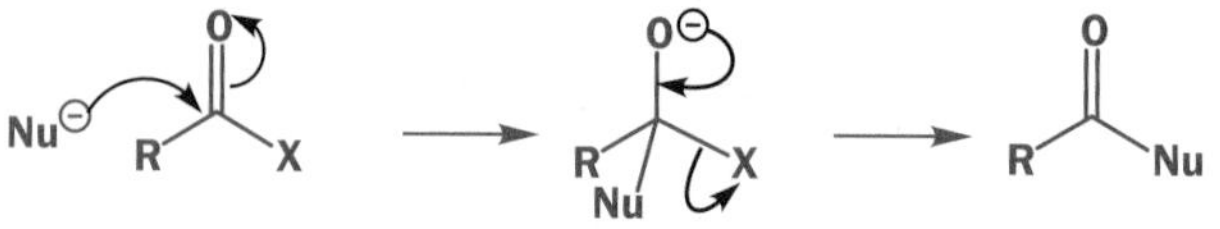

A convenient way to save writing a step is to show the formation and collapse of the tetrahedral intermediate in the same structure, by using a double-headed arrow like this.

Now, this is a useful shorthand, but it is not a substitute for understanding the true mechanism. Certainly, you must never ever write

WRONG

Here's the 'shorthand' at work in the $LiAlH_4$ reduction you have just met.

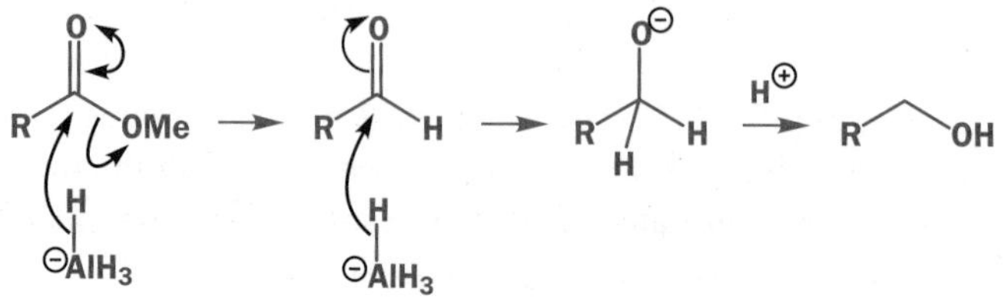

Making ketones from esters: the solution

We diagnosed the problem with our intended reaction as one of reactivity: the product ketone is more reactive than the starting ester. To get round this problem we need to do one of two things:

1 make the starting material more reactive *or*

2 make the product less reactive

Making the starting materials more reactive

A more reactive starting material would be an acyl chloride: how about reacting one of these with a Grignard reagent? This approach can work: for example, this reaction is successful.

81% yield

Often, better results are obtained by transmetallating (see Chapter 9) the Grignard reagent, or the organolithium, with copper salts. Organocopper reagents are too unreactive to add to the product ketones, but they react well with the acyl chloride. Consider this reaction, for example: the product was needed for a synthesis of the antibiotic septamycin.

97% yield

Making the products less reactive

This alternative solution is often better. With the right starting material, the tetrahedral intermediate can become stable enough not to collapse to a ketone during the reaction; it therefore remains completely unreactive towards nucleophiles. The ketone is formed only when the reaction is finally quenched with acid but the nucleophile is also destroyed by the acid and none is left for further addition.

acid quench collapses the intermediate and simultaneously destroys unreacted organolithium

choose X carefully...   ...and the tetrahedral intermediate is stable

We can illustrate this concept with a reaction of an unlikely looking electrophile, a lithium carboxylate salt. Towards the beginning of the chapter we said that carboxylic acids were bad electrophiles and that carboxylate salts were even worse. Well, that is true, but with a sufficiently powerful nucleophile (an organolithium) it is just possible to get addition to the carbonyl group of a lithium carboxylate.

tetrahedral intermediate: stable under anhydrous conditions

We could say that the affinity of lithium for oxygen means that the Li–O bond has considerable covalent character, making the CO_2Li less of a true anion. Anyway, the product of this addition is a dianion of the sort that we met during one of the mechanisms of base-catalysed amide hydrolysis. But, in this case, there is no possible leaving group, so there the dianion sits. Only at the end of the reaction, when water is added, are the oxygen atoms protonated to give a hydrated ketone, which collapses immediately (remember Chapter 6) to give the ketone that we wanted. The water quench also destroys any remaining organolithium, so the ketone is safe from further attack.

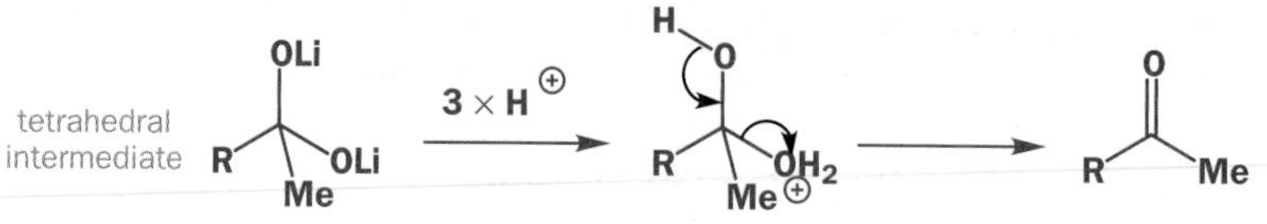

This method has been used to make some ketones that are important starting materials for making cyclic natural products known as macrolides.

> Notice that three equivalents of organolithium are needed in this reaction: one to deprotonate the acid; one to deprotonate the hydroxyl group; and one to react with the lithium carboxylate. The chemists added a further 0.5 for good measure.

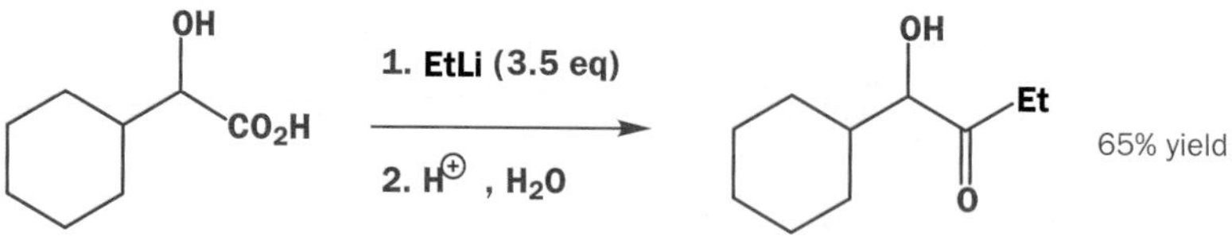

Another good set of starting materials that leads to noncollapsible tetrahedral intermediates is known as the **Weinreb amides**, after their inventor, S.M. Weinreb.

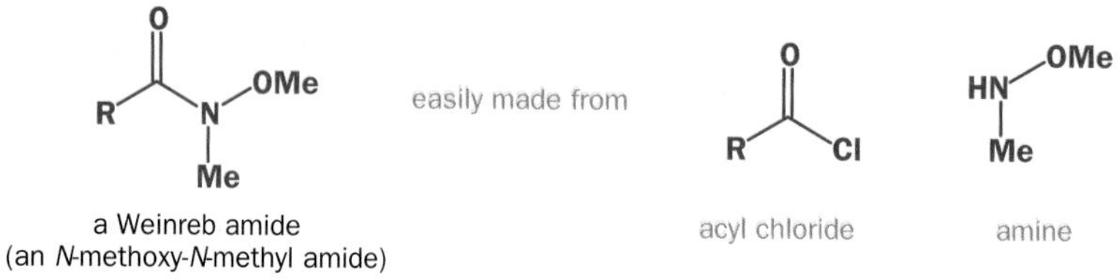

a Weinreb amide
(an *N*-methoxy-*N*-methyl amide)

easily made from

acyl chloride   amine

> **Chelation** means the coordination of more than one electron-donating atom in a molecule to a single metal atom. The word derives from *chele*, the Greek for 'claw'.

Addition of organolithium or organomagnesium reagents to *N*-methoxy-*N*-methyl amides gives a tetrahedral intermediate that is stabilized by *chelation* of the magnesium atom by the two oxygen atoms. This intermediate collapses, to give a ketone, only when acid is added at the end of the reaction.

during the reaction:

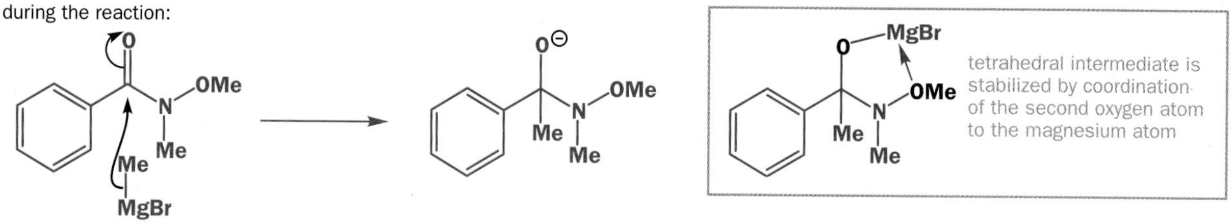

tetrahedral intermediate is stabilized by coordination of the second oxygen atom to the magnesium atom

on quenching with acid:

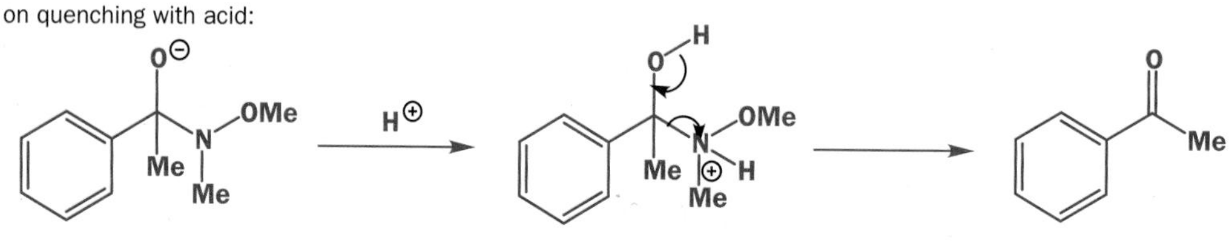

summary of reaction

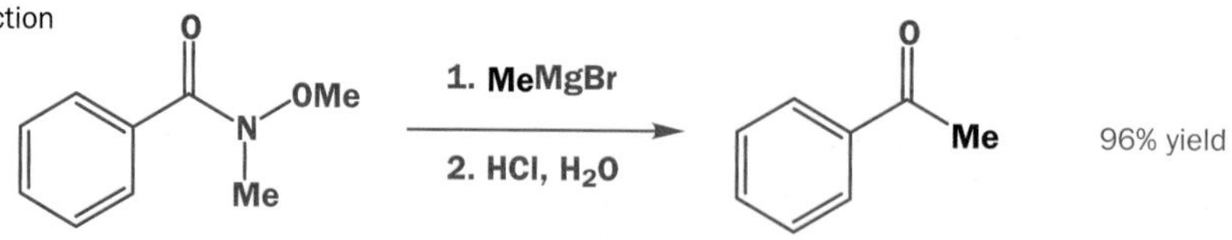

1. **MeMgBr**
2. **HCl, H₂O**

96% yield

This strategy even works for making aldehydes, if the starting material is dimethylformamide (DMF, Me₂NCHO).

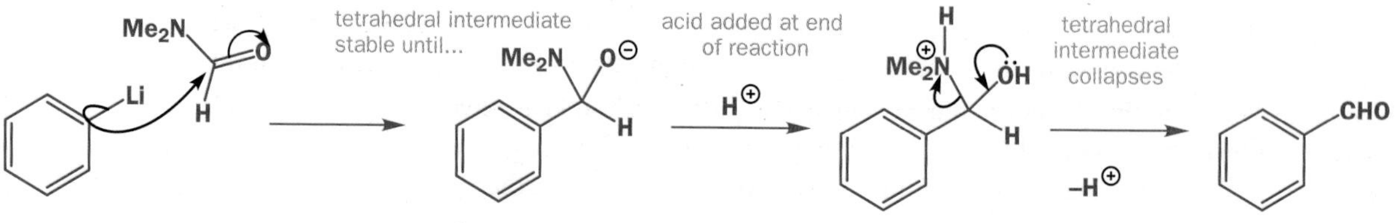

This is an extremely useful way of adding electrophilic CHO groups to organometallic nucleophiles. Here is an example. The first step is an 'ortholithiation' as described in Chapter 9.

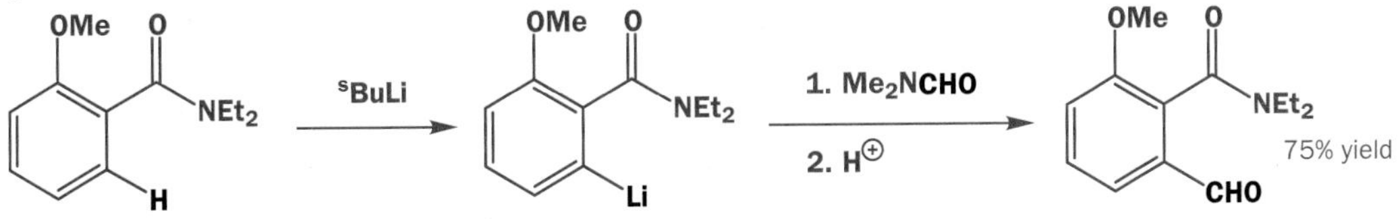

A final alternative is to use a nitrile instead of an ester.

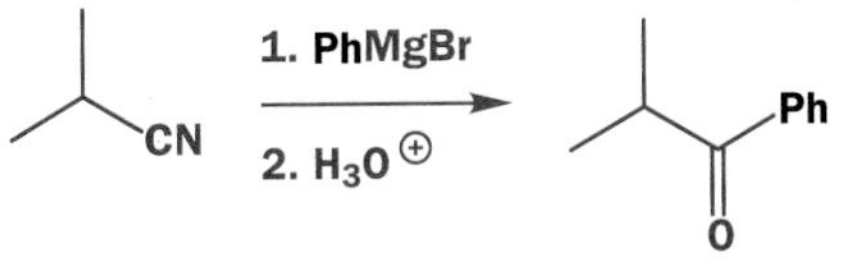

The intermediate is the anion of an imine (see Chapter 14 for more about imines), which is not electrophilic at all—in fact, it's quite nucleophilic, but there are no electrophiles for it to react with until the reaction is quenched with acid. It gets protonated, and hydrolyses (we'll discuss this in the next chapter) to the ketone.

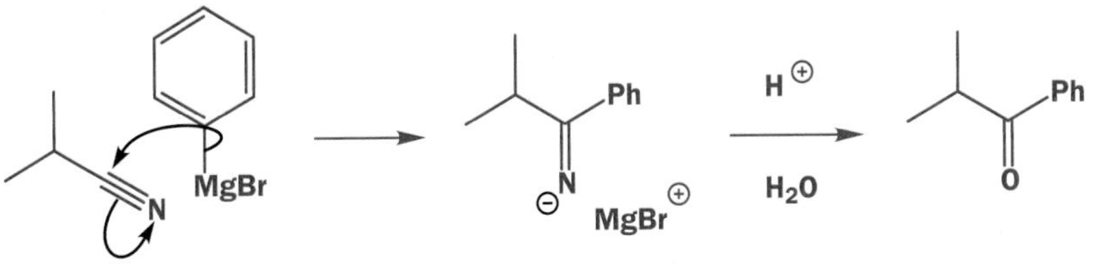

To summarize...

To finish, we should just remind you of what to think about when you consider a nucleophilic substitution at a carbonyl group.

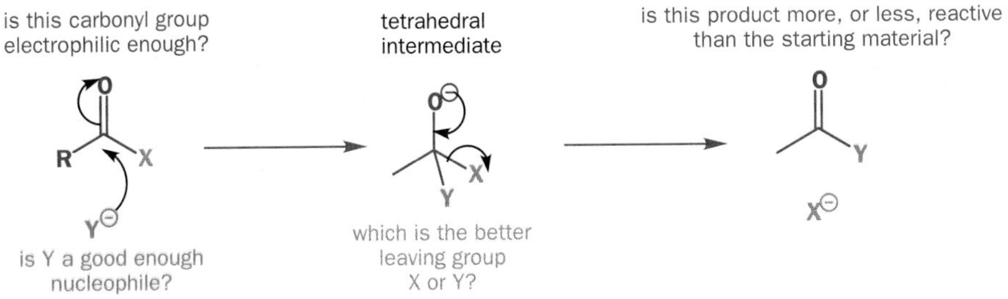

is this carbonyl group electrophilic enough?

tetrahedral intermediate

is this product more, or less, reactive than the starting material?

is Y a good enough nucleophile?

which is the better leaving group X or Y?

And to conclude...

In this chapter you have been introduced to some important reactions—you can consider them to be a series of facts if you wish, but it is better to see them as the logical outcome of a few simple mechanistic steps. Relate what you have learned to what you gathered from Chapters 6 and 9, when we first started looking at carbonyl groups. All we did in this chapter was to build some subsequent transformations on to the simplest organic reaction, addition to a carbonyl group. You should have noticed that the reactions of all acid derivatives are related, and are very easily explained by writing out proper mechanisms, taking into account the presence of acid or base. In the next two chapters we shall see more of these acid- and base-catalysed reactions of carbonyl groups. Try to view them as closely related to the ones in this chapter—the same principles apply to their mechanisms.

Problems

1. Suggest reagents to make the drug 'phenaglycodol' by the route shown.

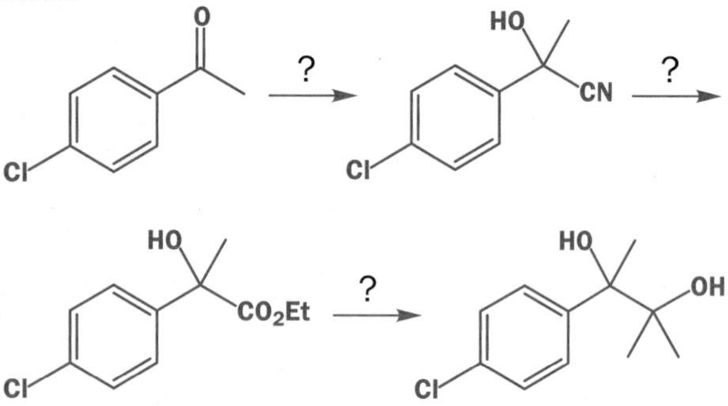

phenaglycodol

2. Direct ester formation from alcohols (R^1OH) and carboxylic acids (R^2CO_2H) works in acid solution but does not work at all in basic solution. Why not? By contrast, ester formation from alcohols (R^1OH) and carboxylic acid anhydrides, ($R^2CO)_2O$, or acid chlorides, RCOCl, is commonly carried out in the presence of amines such as pyridine or Et_3N. Why does this work?

3. Predict the success or failure of these attempted nucleophilic substitutions at the carbonyl group. You should use estimated pK_a or pK_{aH} values in your answer and, of course, draw mechanisms.

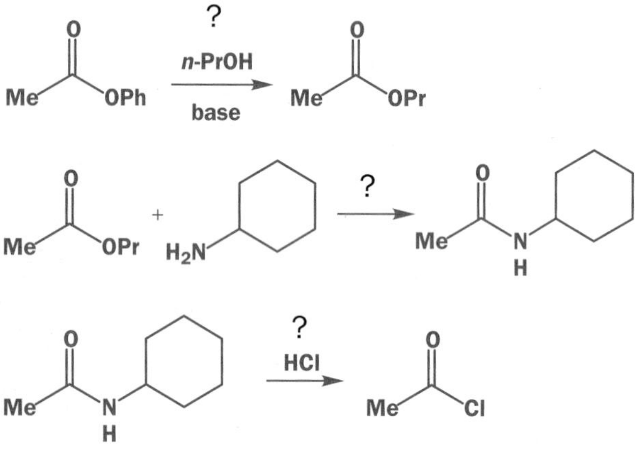

4. Suggest mechanisms for these reactions.

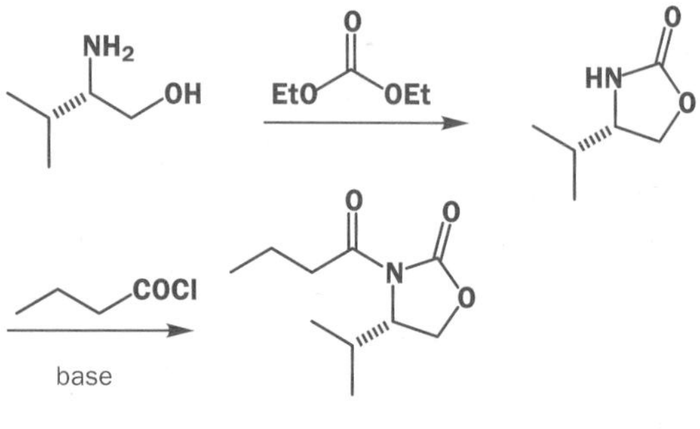

5. In making esters of the naturally occurring amino acids (general formula below) it is important to keep them as their hydrochloride salts. What would happen to these compounds if they were neutralized?

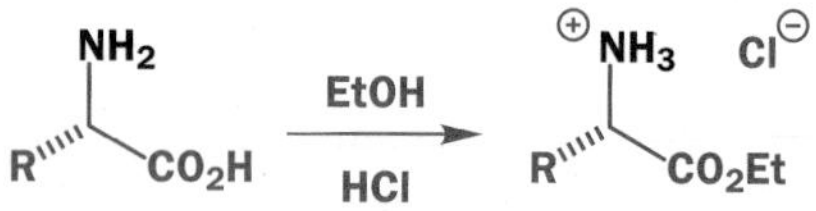

6. It is possible to make either the diester or the monoester of butanedioic acid (succinic acid) from the cyclic anhydride as shown. Why does the one method give the monoester and the other the diester?

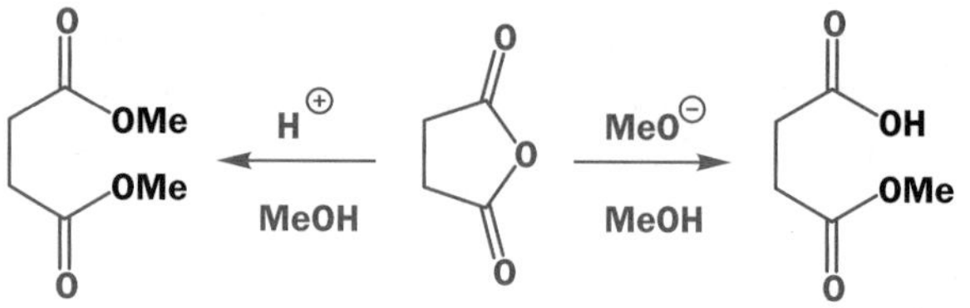

7. Suggest mechanisms for these reactions, explaining why these particular products are formed.

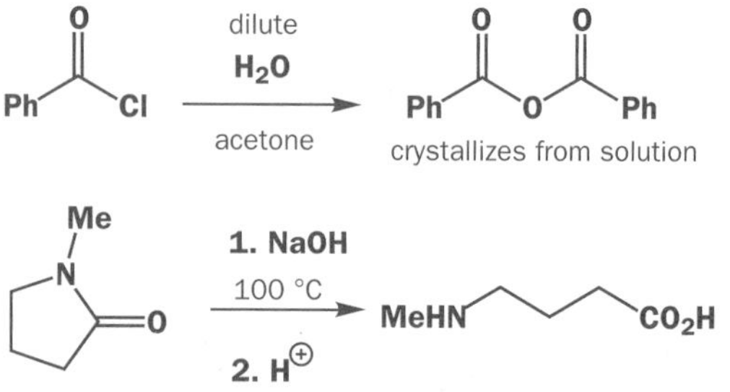

8. Here is a summary of part of the synthesis of Pfizer's heart drug Doxazosin (Cordura®). The mechanism for the first step will be a problem at the end of Chapter 17. Suggest reagent(s) for the conversion of the methyl ester into the acid chloride. In the last step, good yields of the amide are achieved if the amine is added as its hydrochloride salt in excess. Why is this necessary?

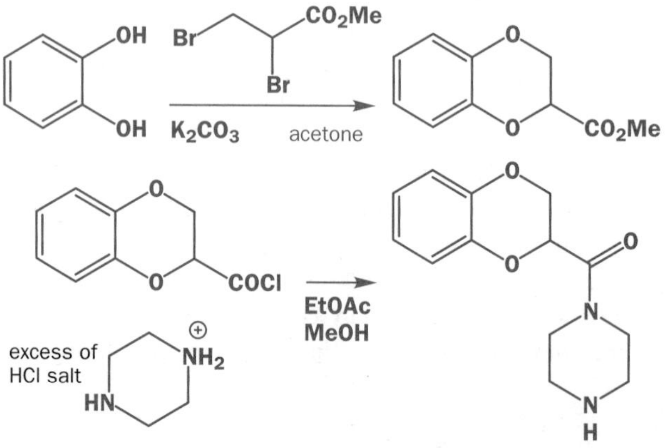

9. Esters can be made directly from nitriles by acid-catalysed reaction with the appropriate alcohol. Suggest a mechanism.

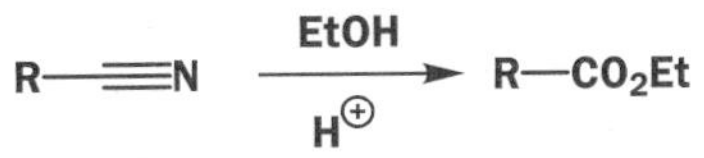

$$R-\!\!\!\equiv\!\!\!N \xrightarrow[\text{H}^{\oplus}]{\text{EtOH}} R-CO_2Et$$

10. Give mechanisms for these reactions, explaining the selectivity (or lack of it!) in each case.

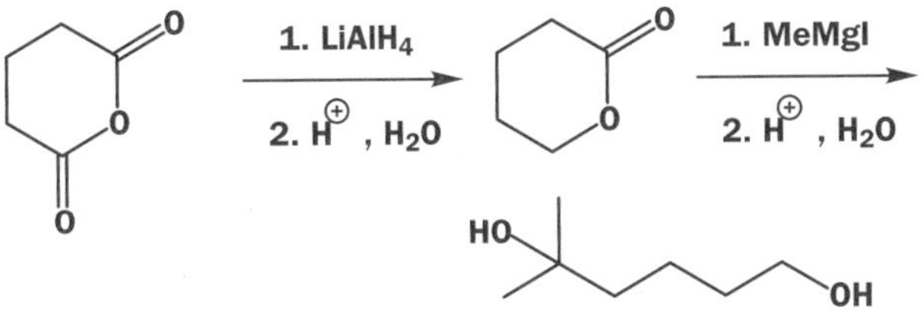

11. This reaction goes in one direction in acidic solution and in the other direction in basic solution. Draw mechanisms for the reactions and explain why the product depends on the conditions.

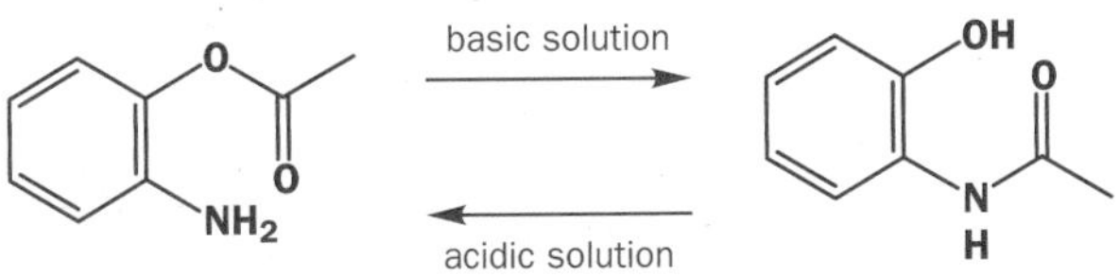

12. These reactions do not work. Explain the failures and suggest in each case an alternative method that might be successful.

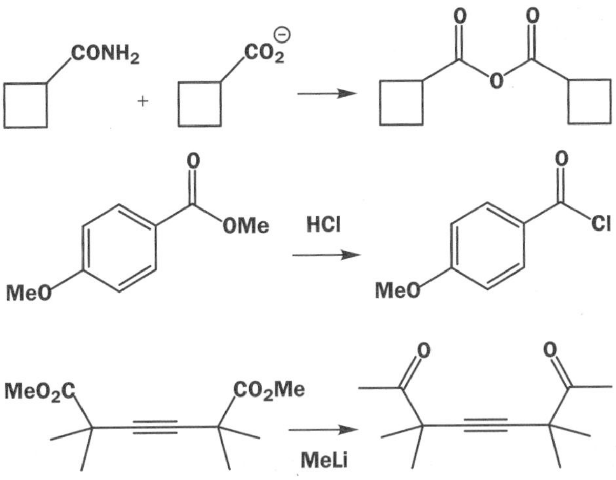

Equilibria, rates, and mechanisms: summary of mechanistic principles

13

Connections

Building on:
- Structure of molecules ch4
- Drawing mechanisms ch5
- Nucleophilic attack on carbonyl groups ch6 & ch9
- Conjugate addition ch10
- Acidity and pK_a ch8

Arriving at:
- What controls equilibria
- Enthalpy and entropy
- What controls the rates of reactions
- Intermediates and transition states
- How catalysts work
- Effects of temperature on reactions
- Why the solvent matters

Looking forward to:
- Kinetics and mechanism ch41
- Synthesis in action ch25
- How mechanisms are discovered ch41

One purpose of this chapter is to help you understand why chemists use such a vast range of different conditions when performing various organic reactions. If you go into any laboratory, you will see many reactions being heated to reflux; however, you will also see just as many being performed at −80 °C or even lower. You will see how changing the solvent in a reaction can drastically alter the time that a reaction takes or even lead to completely different products. Some reactions are over in a few minutes; others are left for hours under reflux. In some reactions the amounts of reagents are critical; in others large excesses are used. Why such a diverse range of conditions? How can conditions be chosen to favour the reaction we want? To explain all this we shall present some very basic thermodynamics but organic chemists do not want to get bogged down in algebra and energy profile diagrams will provide all the information we need.

> 'One could no longer just mix things; sophistication in physical chemistry was the base from which all chemists—including the organic—must start.' Christopher Ingold (1893–1970)

How far and how fast?

We are going to consider which way (forwards or backwards) reactions go and by how much. We are going to consider how fast reactions go and what we can do to make them go faster or slower. We shall be breaking reaction mechanisms down into steps and working out which step is the most important. But first we must consider what we really mean by the 'stability' of molecules and what determines how much of one substance you get when it is in equilibrium with another.

Stability and energy levels

So far we have been rather vague about the term **stability** just saying things like 'this compound is more stable than that compound'. What we really mean is that one compound has more or less energy than another. This comparison is most interesting when two compounds can interconvert. For example, rotation about the C–N bond of an amide is slow because conjugation (Chapter 7) gives it some double-bond character.

There is rotation, but it can be slow and can be measured by NMR spectroscopy. We can expect to find two forms of an amide of the type RNH–COR: one with the two R groups *trans* to one another, and one with them *cis*.

R groups *trans*   R groups *cis*

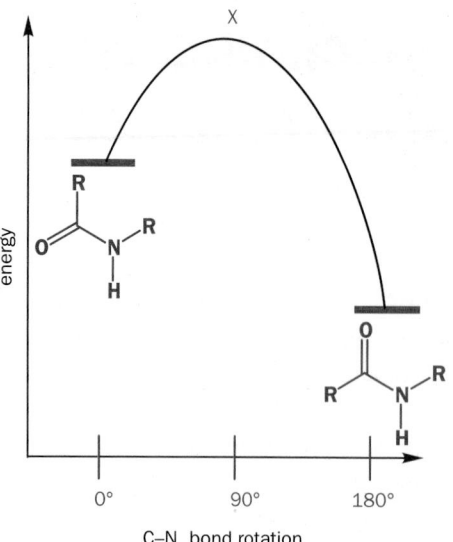

Depending on the size of R we should expect one form to be more stable than the other and we can represent this on an **energy profile diagram** showing the relationship between the two molecules in energy terms.

The two red lines show the energies of the molecules and the curved black line shows what must happen in energy terms as the two forms interconvert. Energy goes up as the C–N bond rotates and reaches a maximum at point X when rotation by 90° has removed the conjugation.

The relative energies of the two states will depend on the nature of R. The situation we have shown, with the *cis* arrangement being much less stable than the *trans*, would apply to large R groups. An extreme case would be if the substituent on nitrogen were H. Then the two arrangements would have equal energies.

The process is the same but there is now no difference between the two structures and, if equilibrium is reached, there will be an exactly 50:50 ratio of the two arrangements. The equilibrium constant is $K = 1$. In other cases, we can measure the equilibrium constant by NMR spectroscopy.

Another limit is reached if the bond is a full double bond as in simple alkenes instead of amides. Now the two states do not interconvert.

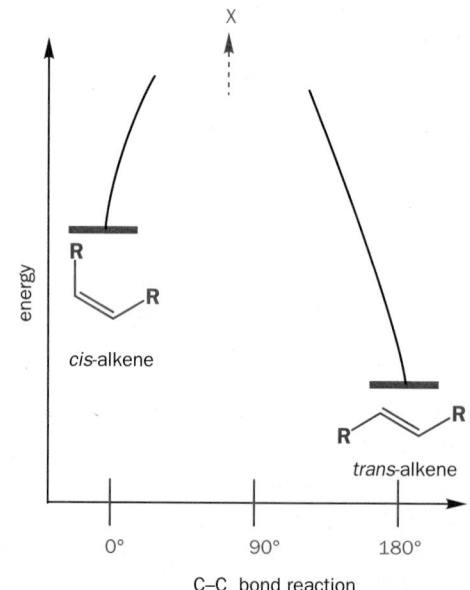

We can measure the energies of the two molecules by measuring the heat of hydrogenation of each isomer to give butane—the same product from both. The difference between the two heats of hydrogenation will be the difference in energy of *cis*- and *trans*-butene.

In more general terms, amide rotation is a simple example of an equilibrium reaction. If we replace 'rotation about the C–N bond' with 'extent of reaction' we have a picture of a typical reaction in which reagents and products are in equilibrium.

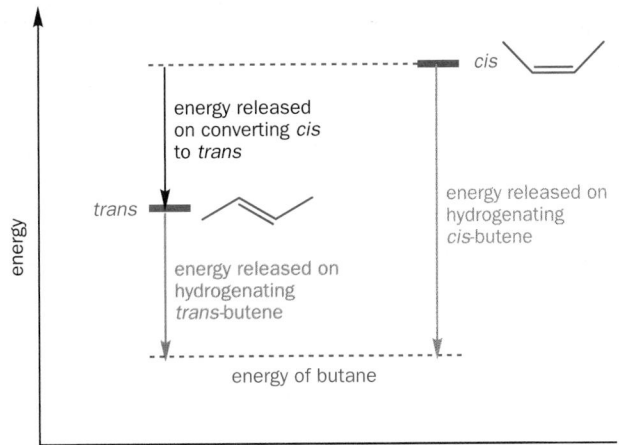

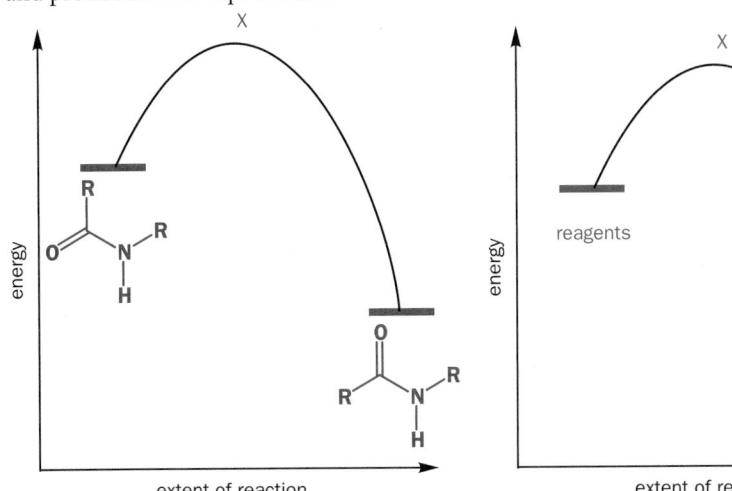

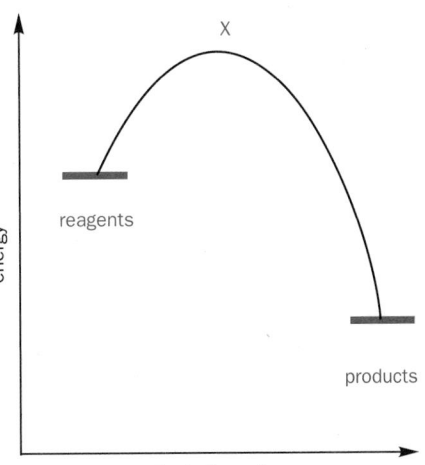

How the equilibrium constant varies with the difference in energy between reactants and products

The equilibrium constant K is related to the energy difference between starting materials and products by this equation

$$\Delta G° = -RT \ln K$$

where $\Delta G°$ (known as the **standard Gibbs energy** of the reaction) is the difference in energy between the two states (in kJ mol^{-1}), T is the temperature (in kelvin *not* °C), and R is a constant known as the **gas constant** and equal to 8.314 J K^{-1} mol^{-1}.

This equation tells us that we can work out the **equilibrium composition** (how much of each component there is at equilibrium) provided we know the difference in energy between the products and reactants. Note that this difference in energy is not the difference in energy between the starting mixture and the mixture of products but the difference in energy if one mole of reactants had been completely converted to one mole of products.

Chemical examples to show what equilibria mean

The equilibrium between isobutyr-aldehyde and its hydrate in water shows the relationship between $\Delta G°$ and K_{eq}.

isobutyraldehyde   + H$_2$O

hydrate of isobutyraldehyde

The equilibrium constant may be written to include $[H_2O]$; however, since the concentration of water effectively remains constant at 55.5 mol dm^{-3} (p. 185), it is often combined into the equilibrium constant giving

$$K_{eq} = \frac{[\text{hydrate}]_{eq}}{[\text{aldehyde}]_{eq}}$$

The concentrations of hydrate and aldehyde at equilibrium in water may be determined by measuring the UV absorption of known concentrations of aldehyde in water and comparing these with the absorptions in a solvent such as cyclohexane where no hydrate formation is possible. Such experiments reveal that the equilibrium constant for this reaction in water at 25 °C is approximately 0.5 so that there is about twice as much aldehyde as hydrate in the equilibrium mixture. The corresponding value for $\Delta G°$ is $-8.314 \times 298 \times \ln(0.5) = +1.7 \text{ kJ mol}^{-1}$. In other words, the solution of the hydrate in water is 1.7 kJ mol^{-1} higher in energy than the solution of the aldehyde in water.

We could compare this reaction to the addition of an alkyllithium reagent to the same aldehyde. You met this reaction in Chapter 9.

The difference in energy between the starting materials, the aldehyde and methyllithium, and the products is so great that at equilibrium all we have are the products. In other words, this reaction is irreversible.

The sign of $\Delta G°$ tells us whether products or reactants are favoured at equilibrium

Consider the equilibrium $A \rightleftharpoons B$. The equilibrium constant, K_{eq}, for this reaction is simply given by the expression

$$K_{eq} = \frac{[B]_{eq}}{[A]_{eq}} \quad \text{where } [A]_{eq} \text{ represents the concentration of A at equilibrium.}$$

If, at equilibrium, there is more B present than A, then K will be greater than 1. This means that the natural log of K will be positive and hence $\Delta G°$ (given by $-RT \ln K$) will be negative. Similarly, if A is favoured at equilibrium, K will be less than 1, $\ln K$ negative, and hence $\Delta G°$ will be positive. If equal amounts of A and B are present at equilibrium, K will be 1 and, since $\ln 1 = 0$, $\Delta G°$ will also be zero.

> ● $\Delta G°$ **tells us about the position of equilibrium**
> - If $\Delta G°$ for a reaction is *negative*, the *products* will be favoured at equilibrium
> - If $\Delta G°$ for a reaction is *positive*, the *reactants* will be favoured at equilibrium
> - If $\Delta G°$ for a reaction is *zero*, the equilibrium constant for the reaction will be 1

■
The sign of $\Delta G°$ for a reaction tells us whether the starting materials or products are favoured at equilibrium, but it tells us nothing about how long it will take before equilibrium is reached. The reaction could take hundreds of years! This will be dealt with later.

A small change in $\Delta G°$ makes a big difference in K

The tiny difference in energy between the hydrate and the aldehyde (1.7 kJ mol^{-1}) gave an appreciable difference in the equilibrium composition. This is because of the log term in the equation $\Delta G° = -RT \ln K$: relatively small energy differences have a very large effect on K. Table 13.1 shows the equilibrium constants, K_{eq}, that correspond to energy differences, $\Delta G°$, between 0 and 50 kJ mol^{-1}. These are relatively small energy differences—the strength of a typical C–C bond is about 350 kJ mol^{-1}—but the equilibrium constants change by enormous amounts.

In a typical chemical reaction, 'driving an equilibrium over to products' might mean getting, say, 98% of the products and only 2% of starting materials. You can see in the table that this requires an equilibrium constant of just over 50 and an energy difference of only 10 kJ mol^{-1}. This small energy difference is quite enough—after all, a yield of 98% is rather good!

Aromatic amines such as aniline (PhNH$_2$) are insoluble in water. We saw in Chapter 8 that they can be dissolved in water by lowering the pH. We are taking advantage of the equilibrium between neutral amine and its ammonium ion. So how far below the pK_{aH} of aniline do we have to go to get all of the aniline into solution?

Table 13.1 Variation of K_{eq} with $\Delta G°$

$\Delta G°$, kJ mol^{-1}	K_{eq}	% of more stable state at equilibrium
0	1.0	50
1	1.5	60
2	2.2	69
3	3.5	77
4	5.0	83
5	7.5	88
10	57	98
15	430	99.8
20	3 200	99.97
50	580 000 000	99.9999998

If the pH of a solution is adjusted to its pK_{aH}, by adding different acids there will be exactly 50% PhNH$_2$ and 50% PhNH$_3^+$. We need an equilibrium constant of about 50 to get 98% into the soluble form (PhNH$_3^+$) and we need to go only about 2 pK_a units below the pK_{aH} of aniline (4.6) to achieve this. All we need is quite a weak acid though in Chapter 8 we used HCl (pK_a −7) which certainly did the trick!

In Chapter 12 (p. 291) we looked at the hydrolysis of esters in basic solution. The decomposition of the tetrahedral intermediate could have occurred in either direction as HO$^-$ (pK_{aH} 15.7) and MeO$^-$ (pK_{aH} 16) are about the same as leaving groups. In other words K_1 and K_2 are about the same and both equilibria favour the carbonyl compound (ester or carboxylic acid).

This reaction would therefore produce a roughly 50:50 mixture of ester and carboxylic acid if this were the whole story. But it isn't because the carboxylic acid will be deprotonated in the basic solution adding a third equilibrium.

Though K_1 and K_2 are about the same, K_3 is very large (pK_a of RCO$_2$H is about 5 and pK_a of MeOH is 16 so the difference between the two K_as is about 10^{11}) and it is this equilibrium that drives the reaction over to the right. For the same reason (because K_3 is very large), it is impossible to form esters in basic solution. This situation can be summarized in an energy diagram showing that the energy differences corresponding to K_1 and K_2 ($\Delta G_1°$ and $\Delta G_2°$) are the same so that $\Delta G°$ between RCO$_2$Me + HO$^-$, on the one hand, and RCO$_2$H + MeO$^-$, on the other, is zero. Only the energy difference for K_3 provides a negative $\Delta G°$ for the whole reaction.

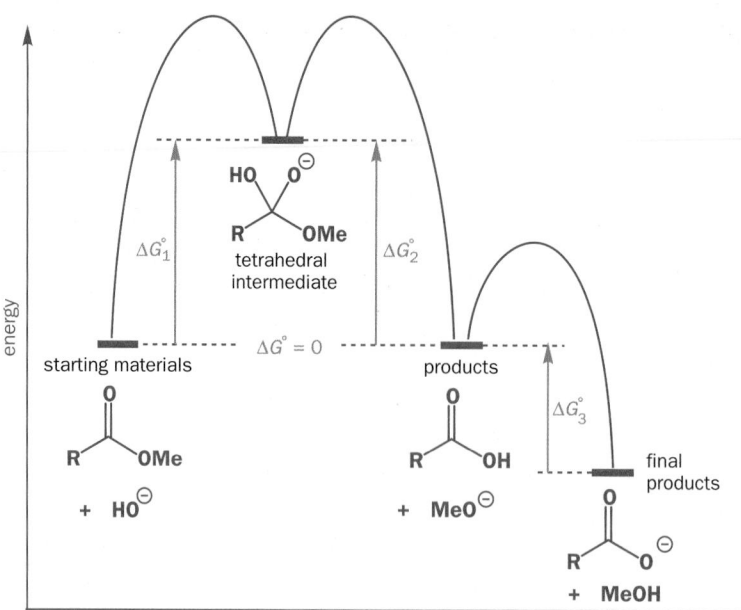

extent of reaction

How to make the equilibrium favour the product you want

The direct formation of esters

The formation and hydrolysis of esters was discussed in Chapter 12 where we established that acid and ester are in equilibrium and that the equilibrium constant is about one.

If we stew up equal amounts of carboxylic acid, alcohol, ester, and water and throw in a little acid to catalyse the reaction (we shall see exactly how this affects the reaction profile later), we find that the equilibrium mixture consists of about equal amounts of ester and carboxylic acid. The position of the equilibrium favours neither the starting materials or the products. The question now arises: how can we manipulate the conditions of the reaction if we actually want to make 100% ester?

The important point is that, at any one particular temperature, the equilibrium constant is just that—*constant*. This gives us a means of forcing the equilibrium to favour the products (or reactants) since the ratio of the two must remain constant. Therefore, if we increase the concentration of the reactants (or even that of just one of the reactants), more products must be produced to keep the equilibrium constant. One way to make esters in the laboratory is to use a large excess of the alcohol and remove water continually from the system as it is formed, for example by distilling it out. This means that in the equilibrium mixture there is a tiny quantity of water, lots of the ester, lots of the alcohol, and very little of the carboxylic acid; in other words, we have converted the carboxylic acid into the ester. We must still use an acid catalyst, but the acid must be anhydrous since we do not want any water present—commonly used acids are toluene sulfonic acid (tosic acid, TsOH), concentrated sulfuric acid (H_2SO_4), or gaseous HCl. The acid catalyst does not alter the position of the equilibrium; it simply speeds up the rate of the reaction, allowing equilibrium to be reached more quickly.

- To make the ester

  Reflux the carboxylic acid with an excess of the alcohol (or the alcohol with an excess of the carboxylic acid) with about 3–5% of a mineral acid (usually HCl or H_2SO_4) as a catalyst and distil out the water that is formed in the reaction. For example: butanol was heated under reflux with a fourfold excess of acetic acid and a catalytic amount of concentrated H_2SO_4 to give butyl acetate in a yield of 70%.

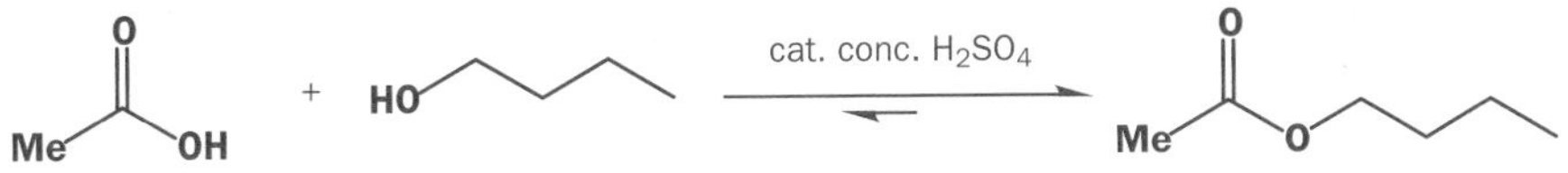

It may also help to distil out the water that is formed in the reaction: diethyl adipate (the diethyl ester of hexanedioic acid) can be made in toluene solution using a sixfold excess of ethanol, concentrated H_2SO_4 as catalyst, distilling out the water using a Dean Stark apparatus. You can tell from the yield that the equilibrium is very favourable.

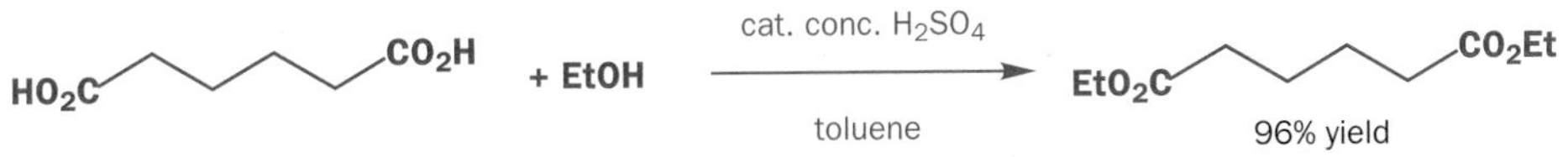

In these cases the equilibrium is made more favourable by using an excess of reagents and/or removing one of the products. The equilibrium *constant* remains the same. High temperatures and acid catalysis are used to speed up arrival at equilibrium which would otherwise take days.

- To hydrolyse the ester

Simple: reflux the ester with aqueous acid or alkali.

The equilibrium between esters and amides

If you solved Problem 12 at the end of the last chapter, you will already know of one reaction that can be driven in either direction by a selection of acidic or basic reaction conditions. The reaction is the interconversion of an ester and an amide and one would normally expect the reaction to favour the amide because of the greater stability of amides due to the more efficient conjugation of the lone pair on nitrogen.

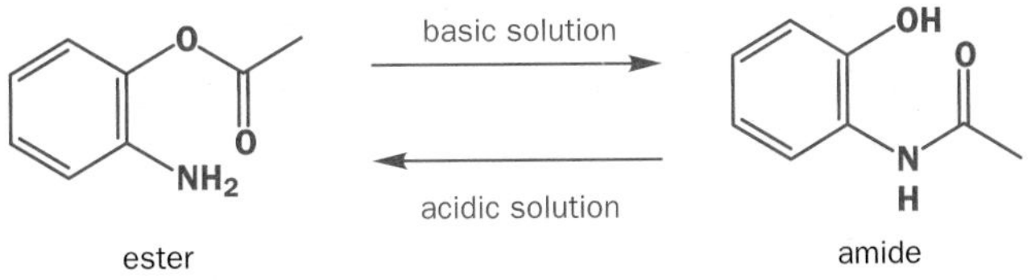

If we examine the mechanism for the reaction it is clear that ArO^- ($pK_{aH} \sim 10$) is a better leaving group than $ArNH^-$ ($pK_{aH} \sim 25$) and so the equilibria between the two compounds and the tetrahedral intermediate are like this.

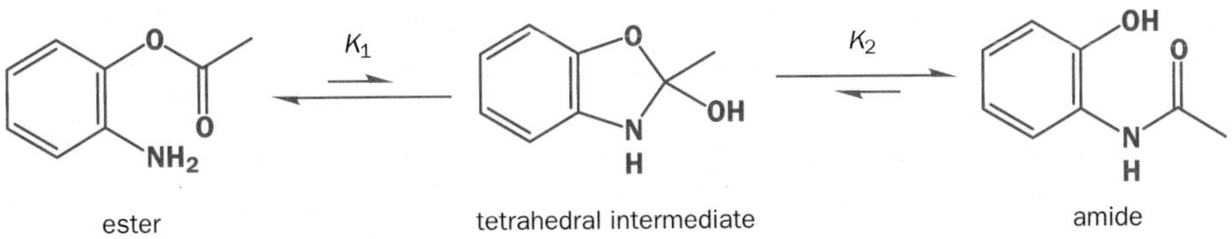

The two individual equilibria favour the carbonyl compounds over the tetrahedral intermediate but $K_1 < K_2$ so the overall equilibrium favours the amide. However, two new equilibria must be added to these if the variation of pH is considered too. In acid solution the amine will be protonated and in base the phenol will be deprotonated.

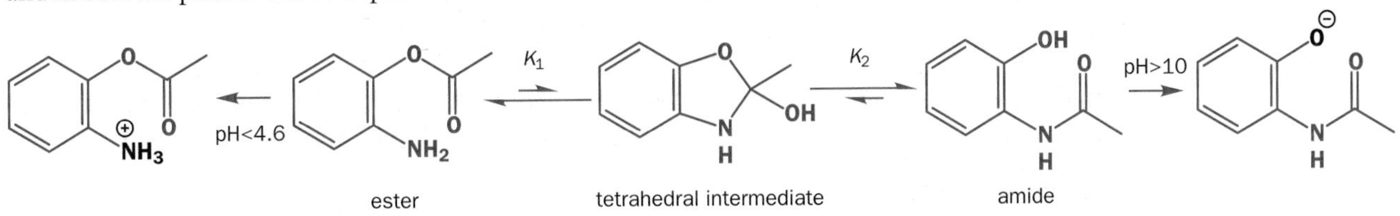

The energy profile for this equilibrium can be studied from either left or right. It is easiest to imagine the tetrahedral intermediate going to the left or to the right depending on the acidity of the solution.

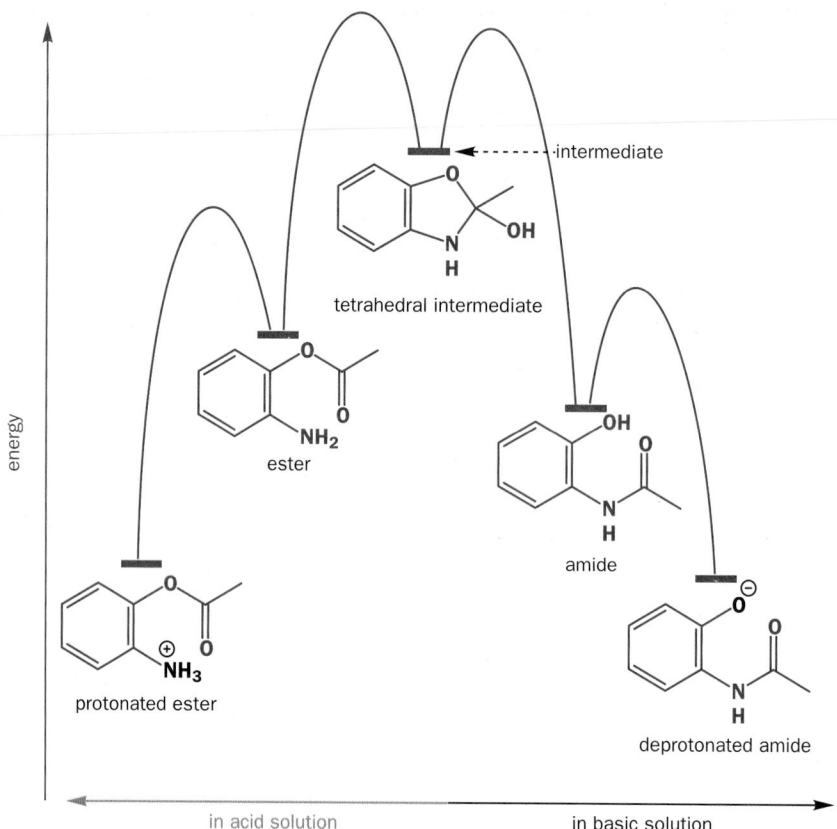

We have shown these last equilibria as reactions because they can be pushed essentially to completion by choosing a pH above 10 if we want the amide or below 4 if we want the ester. This is a relatively unusual situation but there are many other cases where reactions can be driven in either direction by choice of conditions.

Entropy is important in determining equilibrium constants

The *position* of equilibrium (that is, the equilibrium constant, which tells us in a chemical reaction whether products or reactants are favoured) is determined by the energy difference between the two possible states: in the case of the amide $RCONH_2$, there is no difference so the equilibrium constant is one; in the case of the amide RCONHR with large R groups, the arrangement with R groups *trans* is of lower energy than the state with R groups *cis*, and so the equilibrium constant is in favour of the *trans* isomer.

Even when there is a difference in energy between the two states, we still get some of the less stable state. This is because of entropy. *Why* we get the mixture of states is purely down to entropy—there is greater disorder in the mixture of states, and it is to maximize the overall entropy that the equilibrium position is reached.

Energy differences: $\Delta G°$, $\Delta H°$, and $\Delta S°$—energy, enthalpy, and entropy

Returning to that all important equation: $\Delta G° = -RT \ln K$, the sign and magnitude of the energy $\Delta G°$ are the only things that matter in deciding whether an equilibrium goes in one direction or another. If $\Delta G°$ is negative the equilibrium will favour the products (the reaction goes) and if $\Delta G°$ is large and negative the reaction goes to completion. It is enough for $\Delta G°$ to be only about -10 kJmol^{-1} to get complete reaction. The Gibbs energy, $\Delta G°$, the enthalpy of reaction, $\Delta H°$, and the entropy of reaction, $\Delta S°$, are related via the equation

$$\Delta G° = \Delta H° - T\Delta S°$$

The change in enthalpy ΔH° in a chemical reaction is the heat given out (at constant pressure). Since breaking bonds requires energy and making bonds liberates energy, the enthalpy change gives an indication of whether the products have more stable bonds than the starting materials or not. T is the temperature, in kelvin, at which the reaction is carried out. Entropy, S, is a measure of the disorder in the system. A mixture of products and reactants is more disordered than either pure products or pure reactants alone. ΔS° represents the entropy difference between the starting materials and the products.

The equation $\Delta G^\circ = \Delta H^\circ - T\Delta S^\circ$ tells us that how ΔG° varies with temperature depends mainly on the entropy change for the reaction (ΔS°). We need these terms to explain the temperature dependence of equilibrium constants and to explain why some reactions may absorb heat (endothermic) while others give out heat (exothermic).

Enthalpy versus entropy—an example

Entropy dominates equilibrium constants in the difference between inter- and intramolecular reactions. In Chapter 6 we explained that hemiacetal formation is unfavourable because the C=O double bond is more stable than two C–O single bonds. This is clearly an enthalpy factor depending simply on bond strength. That entropy also plays a part can be clearly seen in favourable intramolecular hemiacetal formation of hydroxyaldehydes. The total number of carbon atoms in the two systems is the same, the bond strengths are the same and yet the equilibria favour the reagents (MeCHO + EtOH) in the inter- and the product (the cyclic hemiacetal) in the intramolecular case.

intermolecular hemiacetal formation

intramolecular hemiacetal formation

The difference is one of entropy. In the first case two molecules would give one with an increase in order as, in general, lots of things all mixed up have more entropy than a few large things (when you drop a bottle of milk, the entropy increases dramatically). In the second case one molecule gives one molecule with little gain or loss of order. Both reactions have negative ΔS° but it is more negative in the first case.

■ There is some discussion of entropy in related reactions in Chapter 6.

The acidity of chloroacids

In Chapter 8 we saw how increasing the number of electronegative substituents on a carboxylic acid decreased the acid's pK_a, that is, increased its acidity. Acid strength is a measure of the equilibrium constant for this simple reaction.

carboxylic acid

For this equilibrium as for others, the all important equations $\Delta G^\circ = -RT\ln K$ and $\Delta G^\circ = \Delta H^\circ - T\Delta S^\circ$ apply. When the breakdown of ΔG° for acid ionization was explored, entropy proved to be more important than was expected. Take for example the series CH_3COOH, $CH_2ClCOOH$, $CHCl_2COOH$, and CCl_3COOH with pK_as 7.74, 2.86, 1.28, and 0.52, respectively. If the increase in acidity were simply due to the stabilization of the conjugate base RCO_2^- by the electronegative groups (C–Cl bonds), this would be reflected in the enthalpy difference ΔH° between the conjugate base and the acid. The enthalpy change takes into account the loss of the O–H bond on ionization of the acid and also the difference in solvations between the acid and the ions it produces (H bonds between RCO_2H and water and between RCO_2^- and water). However the data (see table below) show that the difference in equilibrium constant is determined more by entropy than by enthalpy. ΔH° changes by only 6 kJ mol^{-1} over the whole series while ΔS° changes by nearly 100 J K^{-1} mol^{-1} and the more directly comparable $T\Delta S$ changes by over 25 kJ mol^{-1}.

The entropy change depends on the difference in 'order' between the reactants and products. Going from one species (the undissociated acid) to two (the proton and conjugate base) gives an increase in entropy. This in turn makes

Acid	pK_a	ΔH°, kJ mol^{-1}	ΔS°, J K^{-1} mol^{-1}	$-T\Delta S^\circ$, kJ mol^{-1}	ΔG°, kJ mol^{-1}
CH$_3$COOH	4.76	−0.08	−91.6	27.3	27.2
CH$_2$ClCOOH	2.86	−4.6	−70.2	20.9	16.3
CHCl$_2$COOH	1.28	−0.7	−27	8.0	7.3
CCl$_3$COOH	0.52	1.2	−5.8	1.7	2.9

ΔG° more negative and so favours the dissociation. But the solvent structure also changes during the reaction. If a species is strongly solvated, it has many solvent molecules tightly associated with it; in other words, the solvent surrounding it is more ordered. As a weakly solvated neutral acid ionizes to two strongly solvated ions, the neighbouring solvent becomes more ordered and the *overall* entropy decreases.

As we expect, the pK_a decreases as more electronegative chlorines are substituted for the hydrogen atoms in acetic acid. However, the enthalpy change for the ionization remains approximately the same—the decrease in ΔG° is predominantly due to the increase in the entropy change for the reaction. With the increasing numbers of chlorine atoms, the negative charge on the conjugate base is more spread out. The less concentrated the charge, the less order is imposed on the neighbouring solvent molecules and so ΔS° becomes less negative.

> ■
> Because such trends in pK_a are often determined by the entropy change of the whole system, the order of pK_as may change in solvents where there is less solvation and be different again in the gas phase where there are no solvent effects at all. For example, whilst the pK_a of water is usually 15.74, in dimethyl sulfoxide (DMSO) it is about 29. This is because, in DMSO, the hydroxide ion is no longer as effectively solvated as it was in water and this makes the base much stronger.

Equilibrium constants vary with temperature

We have said that the equilibrium constant is a constant only so long as the temperature does not change. Exactly how the equilibrium constant varies with temperature depends on whether the reaction is exothermic or endothermic. If the reaction is **exothermic** (that is, gives out heat) then at higher temperatures the equilibrium constant will be smaller. For an **endothermic** reaction, as the temperature is increased, the equilibrium constant increases. Putting our all important equations $\Delta G^\circ = -RT \ln K$ and $\Delta G^\circ = \Delta H^\circ - T\Delta S^\circ$ together we see that $-RT \ln K = \Delta H^\circ - T\Delta S^\circ$. If we divide throughout by $-RT$ we have

$$\ln K = -\frac{\Delta H^\circ}{RT} + \frac{\Delta S^\circ}{R}$$

The equilibrium constant K can be divided into enthalpy and entropy terms but it is the enthalpy term that determines how K varies with temperature. Plotting $\ln K$ against $1/T$ would give us a straight line with slope $-\Delta H^\circ/R$ and intercept ΔS°. Since T (the temperature in Kelvin) is always positive, whether the slope is positive or negative depends on the sign of ΔH°: if it is positive then, as temperature increases, $\ln K$ (and hence K) increases. In other words, for an endothermic reaction (ΔH positive), as T increases, K ([products]/[reactants]) increases which in turn means that more products must be formed.

> ● **Thermodynamics for the organic chemist**
> - The free energy change ΔG° in a reaction is proportional to $\ln K$ (that is, $\Delta G^\circ = -RT \ln K$)
> - ΔG° and K are made up of enthalpy and entropy terms (that is, $\Delta G^\circ = \Delta H^\circ - T\Delta S^\circ$)
> - The enthalpy change ΔH° is the difference in stability (bond strength) of the reagents and products
> - The entropy change ΔS° is the difference between the disorder of the reagents and that of the products
> - The enthalpy term alone determines how K varies with temperature

Le Chatelier's principle

You may well be familiar with a rule that helps to predict how a system at equilibrium responds to a change in external conditions—**Le Chatelier's principle**. This says that if we disturb a system at equilibrium it will respond so as to minimize the effect of the disturbance. An example of a disturbance is adding more starting material to a reaction mixture at equilibrium. What happens? More product is formed to use up this extra material. This is a consequence of the equilibrium constant being, well... constant and hardly needs anybody's principle.

Another disturbance is heating. If a reaction under equilibrium is heated up, how the equilibrium changes depends on whether the reaction is exothermic or endothermic. If is exothermic (that is, gives out heat), Le Chatelier's principle would predict that, since heat is consumed in the *reverse* reaction, more of the starting materials will be formed. Again no 'principle' is needed—this change occurs because the equilibrium constant is smaller at higher temperatures in an exothermic reaction. Le Chatelier didn't know about equilibrium constants or about $-RT \ln K = \Delta H° - T\Delta S°$ so he needed a 'principle'. You know the reasons and they are more important than rules.

Some reactions are reversible on heating

Simple dimerization reactions will favour the dimer at low temperatures and the monomer at high temperatures. Two monomer molecules have more entropy than one molecule of the dimer. An example is the dimerization of cyclopentadiene. On standing, cyclopentadiene dimerizes and if monomeric material is needed the dimer must be heated and the monomer used immediately. If you lazily leave the monomer overnight and plan to do your reaction tomorrow, you will return in the morning to find dimer.

cyclopentadiene dimer   cyclopentadiene

■ This chemistry does not appear until Chapter 35 but you do not need to know the mechanism of the reaction to appreciate the idea.

This idea becomes even more pointed when we look at polymerization. Polyvinyl chloride is the familiar plastic PVC and is made by reaction of large numbers of monomeric vinyl chloride molecules. There is, of course, an enormous decrease in entropy in this reaction and any polymerization will not occur above a certain temperature. Some polymers can be depolymerized at high temperatures and this can be the basis for recycling.

■ Polymerization does not appear until Chapter 52 but you do not need to know the details to appreciate the idea.

vinyl chloride   PVC (polyvinylchloride)

► Everything decomposes at a high enough temperature eventually giving atoms. This is because the entropy for lots of particles all mixed up is much greater than that of fewer larger particles.

Making reactions go faster: the real reason reactions are heated

Although in organic laboratories you will see lots of reactions being heated, very rarely will this be to alter the equilibrium position. This is because most reactions are not carried out reversibly and so the ratio of products to reactants is not an equilibrium ratio. The main reason chemists heat up reactions is simple—it speeds them up.

How fast do reactions go?—activation energies

Using tables of thermodynamic data, it is possible to work out the energy differences for many different reactions at different temperatures. For example, for the combustion of isooctane, $\Delta G°$ (at 298 K) $= -1000$ kJ mol^{-1}.

isooctane   (l) + O_2 (l) ⇌ $8CO_2$(g) + $9H_2O$(l) $\Delta G° = -1000$ kJ mol^{-1}

■ Isooctane (2,2,4-trimethylpentane) is a major component of petrol (gasoline). Strictly speaking, if we follow the standard meaning for 'iso' (p. 29), the name isooctane should be reserved for the isomer 2-methylheptane. However, 2,2,4-trimethylpentane is by far the most important isomer of octane and so, historically, it has ended up with this name.

We have seen in Table 13.1 on p. 309 that even a difference of 50 kJ mol^{-1} gives rise to a huge equilibrium constant: -1000 kJ mol^{-1} gives an equilibrium constant of 10^{175} (at 298 K), a number too vast to contemplate (there are only about 10^{86} atoms in the observable universe). This value of $\Delta G°$ (or the corresponding value for the equilibrium constant) suggests that isooctane simply could not exist in an atmosphere of oxygen and yet we put it into the fuel tanks of our cars every day—clearly something is wrong.

Since isooctane can exist in an atmosphere of oxygen despite the fact that the equilibrium position really is completely on the side of the combustion products, the only conclusion we can draw must be that a mixture of isooctane and oxygen cannot be at equilibrium. A small burst of energy is needed to reach equilibrium: in a car engine, the spark plug provides this energy and combustion occurs. If no such burst of energy is applied, the petrol would continue to exist for a long time. The mixture of petrol and air is said to be *kinetically* stable but *thermodynamically* unstable with respect to the products of the reaction, CO_2 and H_2O. If the same small energy burst is applied to the products, they do not convert back to petrol and oxygen.

The energy required to overcome the barrier to reaction is called the **activation energy** and is usually given the symbols E_a or $\Delta G^{\ddagger}$. An energy level diagram for a reaction such as the combustion of isooctane is shown below.

> E_a and $\Delta G^{\ddagger}$ are both used for the activation energy and are almost the same. There are subtle differences that do not concern us here.

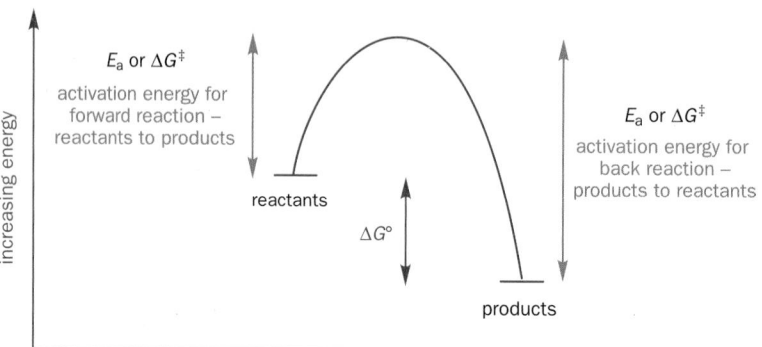

Points to notice:

- The products are lower in energy than the reactants as the equilibrium position lies in favour of the products

- The activation energy for the forward reaction is less than the activation energy for the back reaction

If a reaction cannot proceed until the reactants have sufficient energy to overcome the activation energy barrier, it is clear that, the smaller the barrier, the easier it will be for the reaction to proceed. In fact the activation energy is related to how fast the reaction proceeds by another exponential equation

$$k = Ae^{\frac{-E_a}{RT}}$$

where k is the rate constant for the reaction, R is the gas constant, T is the temperature (in kelvin), and A is a quantity known as the pre-exponential factor. This equation is called the **Arrhenius equation**. Because of the minus sign in the exponential term, the larger the activation energy, E_a, the slower the reaction but the higher the temperature, the faster the reaction.

> Svante Arrhenius (1859–1927) was one of the founders of physical chemistry. He was based at Uppsala in Sweden and won the Nobel prize in 1903 mainly for his theory of the dissociation of salts in solution.

Examples of activation energy barriers

A very simple reaction is rotation about a bond. In the compounds in the table, different amounts of energy are needed to rotate about the bonds highlighted in black. See how this activation energy barrier affects the actual rate at which the bond rotates. Approximate values for k have been calculated from the experimentally determined values for the activation energies. The **half-life**, $t_{1/2}$, is just the time needed for half of the compound to undergo the reaction.

Compound	E_a, kJ mol^{-1}	Approximate k, 298 K/s^{-1}	$t_{1/2}$ at 298 K
(ethane)	12	5×10^{10}	0.02 ns
(hexachloroethane)	45	8×10^{4}	10 μs
(N-methylformamide / acetamide)	70	3	0.2 s
(dimethyl ester alkene)	108	7×10^{-7}	11 days
(stilbene)	180	2×10^{-19}	ca. 10^{11} years[a]

[a] The age of the earth = 4.6×10^9 years.

We can see how the rate constant varies with temperature by looking at the Arrhenius equation. The pre-exponential factor, A, does not vary much with temperature, but the exponential term is a function of temperature. Once again, because of the minus sign, the greater the temperature, the greater the rate constant.

This observation is used in practice when NMR spectra give poor results because of slow rotation about bonds. Amides of many kinds, particularly carbamates, show slow rotation about the C–N bond at room temperature because of the amide delocalization. These amides have bigger bar-

NMR spectra of DMF at high and low temperature are shown on p. 165 of Chapter 7

You will see this 'Boc' group used as a protecting group for amines in Chapter 24.

riers to rotation than the 70 kJ mol^{-1} of the example in the table. The result is a poor spectrum with broad signals. In this example, the two sides of the five-membered ring are different in the two rotational isomers and give different spectra.

The solution is to run the NMR spectrum at higher temperatures. This speeds up the rotation and averages out the two structures.

A word of warning: heating is not all good for the organic chemist—not only does it speed up the reaction we want, it will also probably speed up lots of other reactions that we don't want to occur! We shall see how we can get round this, but first we shall take a closer look at what determines how fast a reaction takes place.

Rates of reaction

Suppose we have the very simple reaction of a single proton reacting with a molecule of water in the gas phase

$$H^+(g) + H_2O(g) \rightarrow H_3O^+(g)$$

We saw at the beginning of Chapter 8 that this is essentially an irreversible process, that is, $\Delta G°$ is very large and negative and therefore the equilibrium constant, K, is large and positive.

So we know that this reaction goes, but what determines how quickly it can proceed? Since the mechanism simply involves one proton colliding with one molecule of water, then clearly the rate will depend on how often the two collide. This in turn will depend on the concentrations of these species—if there are lots of protons but only a few water molecules, most collisions will be between protons. The reaction will proceed fastest when there are lots of protons and lots of water molecules.

This reaction turns two species into one, all in the gas phase. The standard entropy for the reaction must therefore be negative. In order for $\Delta G°$ to be negative, the reaction must give out heat to the surroundings. In other words, this reaction must be highly exothermic, as indeed it is.

We can express this mathematically by saying that the rate of reaction is proportional to the concentration of protons multiplied by the concentration of water molecules (the square brackets mean 'concentration of').

$$\text{rate of reaction} \propto [H^+] \times [H_2O]$$

The constant of proportionality, k, is known as the **rate constant**.

$$\text{rate of reaction} = k \times [H^+] \times [H_2O]$$

We are not very interested in reactions in the gas phase, but fortunately reactions in solution follow more or less the same laws so the reaction of a proton source like HCl and a water molecule in an inert solvent would have the rate expression: rate = $k \times$ [HCl] $\times$ [H$_2$O]. Expressing the same idea graphically requires an energy profile diagram like those we used for equilibria but concentrating rather more on $\Delta G^{\ddagger}$ than on ΔG°.

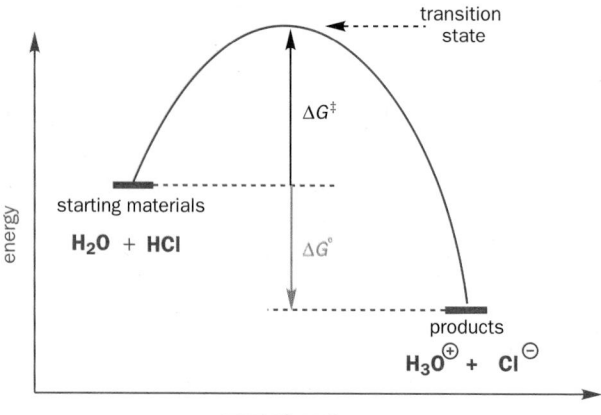

Note that the products are lower in energy than the starting materials as before. The energy barrier is now marked $\Delta G^{\ddagger}$ and the highest point on the profile is labelled *transition state*. Somewhere between the starting materials and the products there must come a point where the O–H bond is half formed. This is the least stable structure in the whole reaction scheme and would correspond to a structure about halfway between starting materials and products, something like this.

Now notice that the transition state is drawn in square brackets and marked ‡. Note the long dashed bonds not yet completely formed or not yet completely broken and the partial charges (+) and (−) meaning something about half a charge (the products have complete charges shown in circles).

● **Transition state**

A transition state is a structure that represents an energy maximum on passing from reactants to products. It is not a real molecule in that it may have partially formed or broken bonds and may have more atoms or groups around the central atom than allowed by valence bond rules. It cannot be isolated because it is an energy maximum and any change in its structure leads to a more stable arrangement. A transition state is often shown by putting it in square brackets with a double-dagger superscript.

This species is unstable—both the starting materials and the products are lower in energy. This means that it is not possible to isolate this halfway species; if the reaction proceeds just a little more forwards or backwards, the energy of the system is lowered (this is like balancing a small marble on top of a football—a small push in any direction and the marble will fall, lowering its potential energy).

Kinetics

The value of the rate constant will be different for different reactions. Consider the reaction of HCl and a water molecule discussed in the last section. Even with the same concentrations, the almost identical reaction where hydrogen is replaced by deuterium will proceed at a different rate (Chapter 19). To understand this we need to think again about what needs to happen for a reaction to occur. It is not enough for the two species to simply collide. We know that for this reaction to work the proton must come into contact with the *oxygen* atom in the water molecule, not the hydrogen atoms, that is, there is some sort of steric requirement. We have also seen that most reactions need to overcome an energy barrier. In other words, it is not enough for the two species just to collide for a reaction to proceed, they must collide in the right way and with enough force.

You can see now how the overall rate equation for our example reaction

$$\text{rate of reaction} = k \times [\text{HCl}] \times [\text{H}_2\text{O}]$$

contains all the points needed to work out how fast the reaction will proceed. The most important point concerns the concentrations of the reacting species—which are expressed directly in the rate equation. Other considerations, such as how large the species are or whether or not they collide in the right way with the right energy, are contained in the rate constant, k. Notice once again that not only is k different for different reactions (for all of the above reasons), but that it also varies with temperature. It is essential when quoting a rate constant that the temperature is also quoted. That part of chemistry that deals with reaction *rates* rather than equilibria is known as **kinetics**.

Activation barriers

In the same way that we define $\Delta G^{\ddagger}$ to be the difference in energy between the starting materials and the transition state (that is, activation energy), we can define the entropy of activation, $\Delta S^{\ddagger}$, and the enthalpy of activation, $\Delta H^{\ddagger}$, as being the entropy and enthalpy differences between the starting materials and transition sate. These quantities are directly analogous to the entropy and enthalpy of the reaction but instead refer to the difference between starting material and *transition state* rather than starting material and *products*.

In a similar manner, we could also define an equilibrium constant between the reactants and the transition state

$$K^{\ddagger} = \frac{[AB]}{[A][B]}$$

Our all-important thermodynamic equations apply equally well to these activation functions so that we may write

$$\Delta G^{\ddagger} = -RT\ln K^{\ddagger} \text{ and } \Delta G^{\ddagger} = \Delta H^{\ddagger} - T\Delta S^{\ddagger}.$$

It is possible to relate these functions with the rate constant for the reaction, k, by using a model known as **transition state theory**. We will not go into any details here, but the net result is that

$$k = \frac{k_B T}{h} K^{\ddagger}$$

where k_B and h are universal constants known as Botlzmann's constant and Planck's constant, respectively.
By substituting in the equation $K^{\ddagger} = e^{\frac{-\Delta G^{\ddagger}}{RT}}$ the rearranged form of $\Delta G^{\ddagger} = -RT\ln K^{\ddagger}$) we arrive at an equation, known as the **Eyring equation**, which relates how fast a reaction goes (k) to the activation energy ($\Delta G^{\ddagger}$)

$$k = \frac{k_B T}{h} e^{-\frac{\Delta G^{\ddagger}}{RT}}$$

This can be rearranged and the numerical values of the constants inserted to give an alternative form

$$\Delta G^{\ddagger} \text{ (in J mol}^{-1}) = 8.314 \times T \times [23.76 + \ln(T/k)]$$

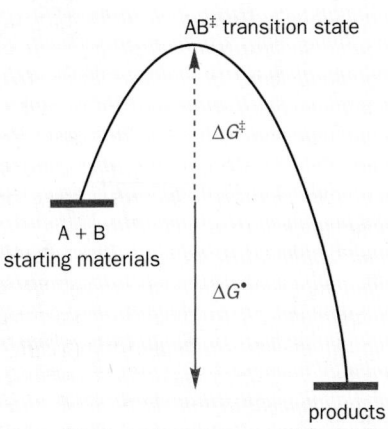

$AB^{\ddagger}$ transition state

$A + B$
starting materials

$\Delta G^{\ddagger}$

$\Delta G^{\bullet}$

products

Kinetics gives us an insight into the mechanism of a reaction

Now for some of the reactions you have seen in the last few chapters. Starting with carbonyl substitution reactions, the first example is the conversion of acid chlorides into esters. The simplest mechanism to understand is that involved when the anion of an alcohol (a metal alkoxide RO⁻) reacts with an acid chloride. The kinetics are bimolecular: rate = k[MeCOCl][RO⁻]. The mechanism is the simple addition elimination process with a tetrahedral intermediate.

The formation of the tetrahedral intermediate by the combination of the two reagents is the rate-determining step and so the highest transition state will be the one leading from the starting materials to that intermediate.

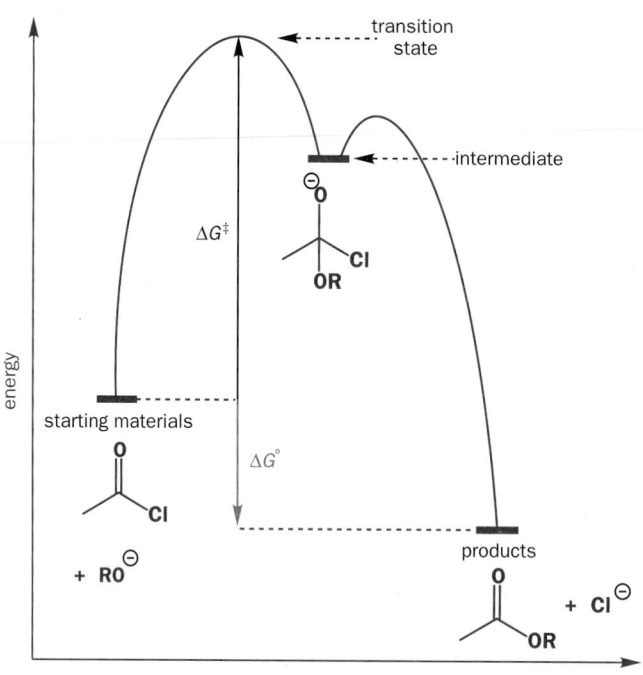

We shall return to this important mechanism in a moment after a brief mention of *first-order kinetics*. The reaction between the acid chloride and the neutral alcohol to give an ester may not have the bimolecular rate expression expected for this mechanism: rate = $k[R^1COCl][R^2OH]$.

Some such reactions have a simpler rate expression: rate = $k[R^1COCl]$ in which the alcohol does not appear at all. Evidently, no collision between the acid chloride and the alcohol is required for this reaction to go. What actually happens is that the acid chloride decomposes by itself to give a reactive cation (a cation you have already seen in mass spectrometry) with the loss of the good leaving group Cl^-.

There are three steps in this reaction scheme though the last is a trivial deprotonation. Evidently, the energy barrier is climbed in the first step, which involves the acid chloride alone. The cation is an intermediate with a real existence and reacts later with the alcohol in a step that does not affect the rate of the reaction. The easiest way to picture this detail is in an energy profile diagram (top right).
Points to notice:

- The products are again lower in energy than the starting materials
- There are three transition states in this reaction
- Only the highest-energy transition state matters in the reaction rate (here the first)
- The step leading to the highest transition state is called the rate-determining step
- The two intermediates are local minima in the reaction profile
- The highest-energy transition state is associated with the formation of the highest-energy intermediate

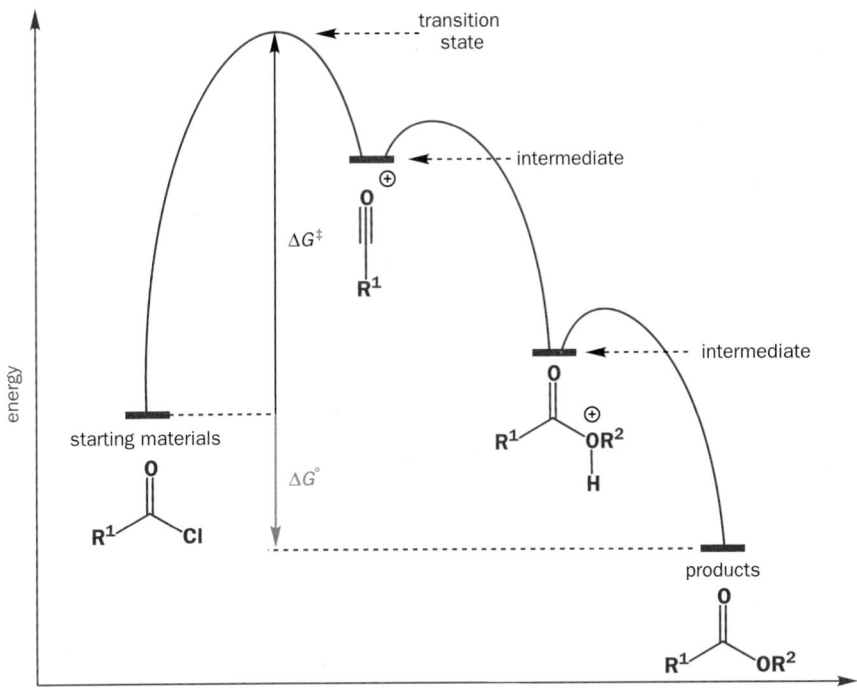

extent of reaction

● Intermediates and transition states

A transition state represents an energy maximum—any small displacement leads to a more stable product. An intermediate, on the other hand, is a molecule or ion that represents a *localized* energy minimum—an energy barrier must be overcome before the intermediate forms something more stable. As you have seen in Chapter 3, and will see again in Chapter 22, because of this energy barrier, it is even possible to isolate these reactive intermediates (RCO^+) and study their spectra.

Because the rate-determining step involves just one molecule, the rate equation shows rate = $k[R^1COCl]$, and the reaction is called a **first-order reaction** as the rate is proportional to just one concentration. A first-order reaction involves the unimolecular decomposition of something in the rate-determining step.

Second-order reactions

The unimolecular mechanism is unusual for carbonyl substitution reactions. Those in the last chapter as well as the carbonyl addition reactions in Chapter 6 all had nucleophilic addition to the carbonyl group as the rate-determining step. An example would be the formation of an ester from an anhydride instead of from an acid chloride.

The leaving group ($MeCO_2^-$) is not now good enough (pK_{aH} about 5 instead of −7 for Cl^-) to leave of its own accord so the normal *second-order* mechanism applies. The kinetics are bimolecular: rate = $k[(MeCO)_2O][ROH]$ and the rate-determining step is the formation of the tetrahedral intermediate.

All the acid derivatives (acid chlorides, anhydrides, esters, and amides) combine with a variety of nucleophiles in very similar bimolecular mechanisms.

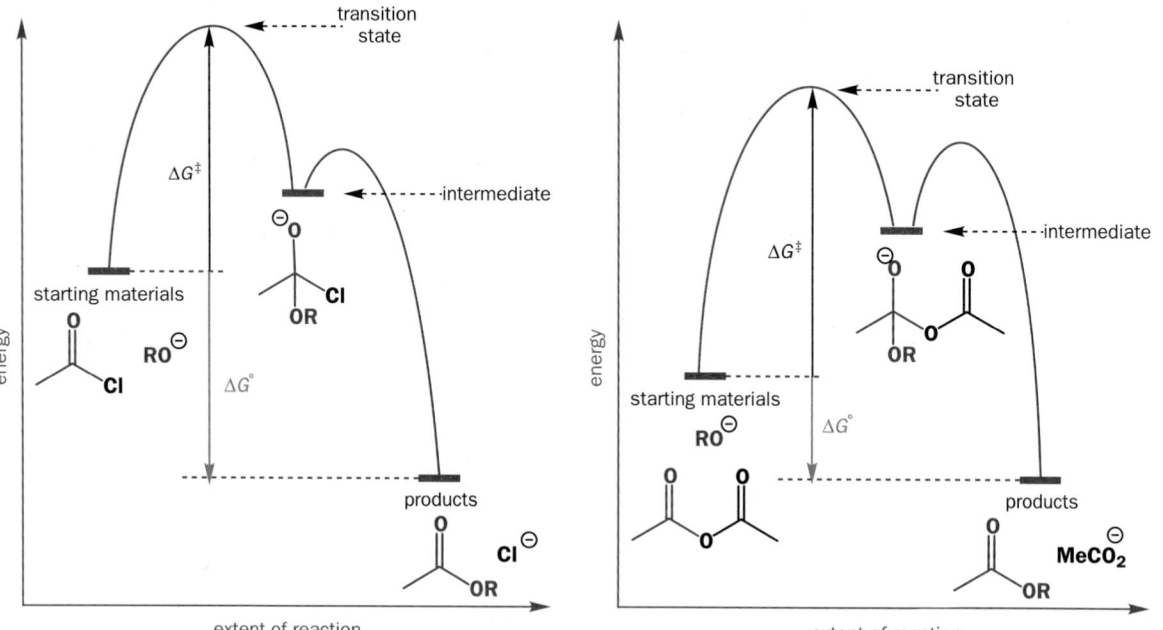

This is the simplest and the most typical bimolecular mechanism with one intermediate, and the energy profile diagrams are correspondingly easier to understand. The reactions with acid chlorides (discussed a few pages back) and anhydrides are straightforward and go in good yield.

The energy levels of the starting materials, the transition state, and the intermediate are all lower in the anhydride reaction than in the acid chloride reaction. So which goes faster? We know the answer—acid chlorides are more reactive than anhydrides towards nucleophiles. The reason is that the stability of the starting materials is determined by the interaction between the carbonyl group and the substituent attached directly to it. This is a big effect as we know from infrared spectroscopy.

The two intermediates also have different energies depending mainly on the stability of the oxyanion. This too will be affected by the substituents, Cl and OAc, but they are separated from the oxyanion by the tetrahedral carbon atom and there is no conjugation. Substituent effects on the oxyanion are smaller than they are on the starting materials so the two intermediates are similar in energy. Substituent effects on the transition state will be somewhere between the two but the transition state is nearer to the intermediate than to the starting material so substituent effects will be like those on the intermediate. The two transition states also have similar energies. The net result is that $\Delta G^{\ddagger}$ is bigger for the anhydride mainly because the energy of the starting materials is lower. This also explains why ΔG° is smaller.

The ester exchange reaction

When we move on to esters reacting with alkoxides the chart is a good deal more symmetrical. This is the reaction.

The nucleophile and the leaving group are both alkoxides, the only difference being R^1 and R^2. If R^1 and R^2 were the same, the energy profile diagram would be totally symmetrical and small differences between R^1 and R^2 are not going to affect the symmetry much.

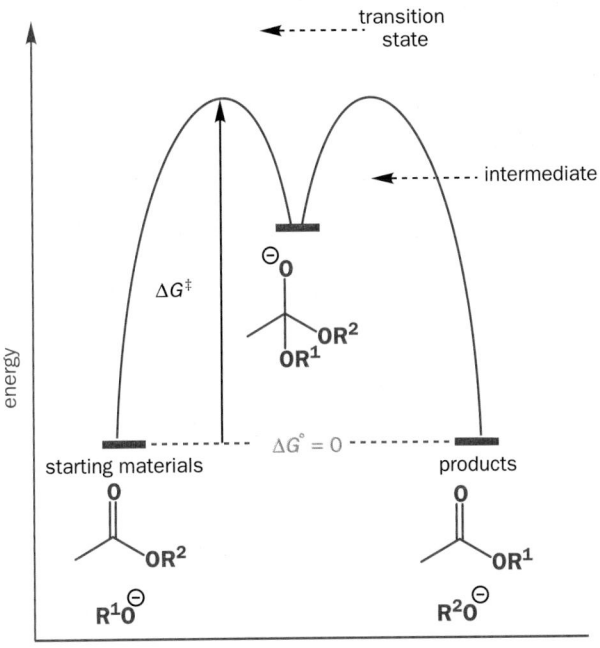

Points to notice:

- The transition states for the two steps are equal in energy
- $\Delta G^{\ddagger}$ is the same for the forward and the back reaction
- ΔG° is zero
- If $R^1 = R^2$, the intermediate has an exactly 50% chance of going forward or backward

In fact, we now have an equilibrium reaction. If R^1 and R^2 are different then the reaction is called ester exchange or transesterification and we should drive it in the direction we want by using a large excess of one of the two alcohols. If we carried out the reaction on one ester using an equivalent of the other alkoxide in that alcohol as solvent, the other ester would be formed in good yield.

Catalysis in carbonyl substitution reactions

We don't need the equivalent of alkoxide in ester exchange because alkoxide is regenerated in the second step. We need only catalytic quantities (say, 1–2% of the ester) because the role of the alkoxide is catalytic. It speeds up the reaction because it is a better nucleophile than the alcohol itself and it is regenerated in the reaction.

Making a solution more basic speeds up reactions in which alcohols act as nucleophiles because it increases the concentration of the alkoxide ion, which is more nucleophilic than the alcohol itself. The same thing happens in hydrolysis reactions. The hydrolysis of esters is fast in either acidic or basic solutions. In basic solution, hydroxide is a better nucleophile than water.

The mechanism is like that for ester exchange but hydroxide is used up in deprotonating the carboxylic acid produced so a whole equivalent of NaOH is needed. In acidic solution, protonation of the carbonyl oxygen atom makes the ester more electrophilic and attack by the weak nucleophile (water) is made faster but the acid catalyst is regenerated. In both these reactions nucleophilic attack is the rate-determining step.

So, the higher the concentration of protons, the faster the hydrolysis goes and, the higher the concentration of hydroxide ion, the faster the reaction goes. If we plot the (log of the) rate of the reaction against the pH of the solution we shall get two straight lines increasing at high and low pH and each with a slope of one. The lines intersect near neutrality when there are neither protons nor hydroxide ions. This is simple acid and base catalysis.

variation of rate of ester hydrolysis with pH

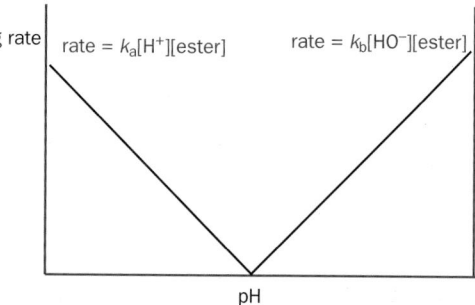

These are bimolecular reactions with bimolecular kinetics and the rate expression in each case includes the concentration of the catalyst. We can label the rate constants k_a and k_b with a suffix 'a' for acid and 'b' for base to show more clearly what we mean.

rate of ester hydrolysis in acid solution (pH < 7) = $k_a[MeCO_2R][H_3O^+]$

rate of ester hydrolysis in basic solution (pH > 7) = $k_b[MeCO_2R][HO^-]$

► You will also see rate constants labelled in other ways—this is a matter for choice. A common method is to use k_1 for unimolecular and k_2 for bimolecular rate constants.

Catalysis by weak bases

In Chapter 12 pyridine was often used as a catalyst in carbonyl substitution reactions. It can act in two ways. In making esters from acid chlorides or anhydrides pyridine can act as a nucleophile as well as a convenient solvent. It is a better nucleophile than the alcohol and this nucleophilic catalysis is discussed in Chapter 12 (p. 282). But nonnucleophilic bases also catalyse these reactions. For example, acetate ion catalyses ester formation from acetic anhydride and alcohols.

Could this be nucleophilic catalysis too? Acetate can certainly attack acetic anhydride, but the products are the same as the starting materials. This irrelevant nucleophilic behaviour of acetate ion cannot catalyse ester formation.

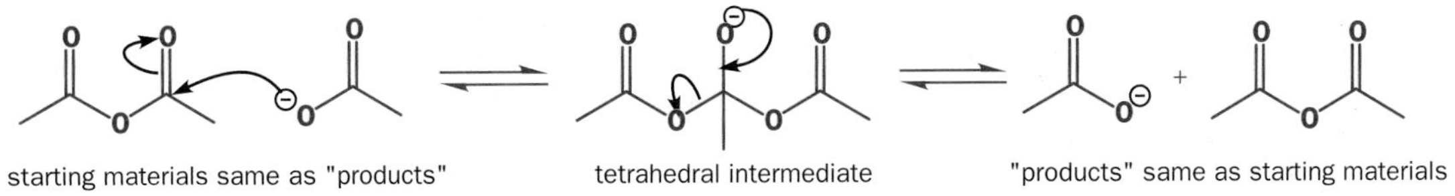

starting materials same as "products" | tetrahedral intermediate | "products" same as starting materials

Can acetate be acting as a base? With a pK_{aH} of about 5 it certainly cannot remove the proton from the alcohol (pK_{aH} about 15) before the reaction starts. What it can do is to remove the proton from the alcohol *as the reaction occurs*.

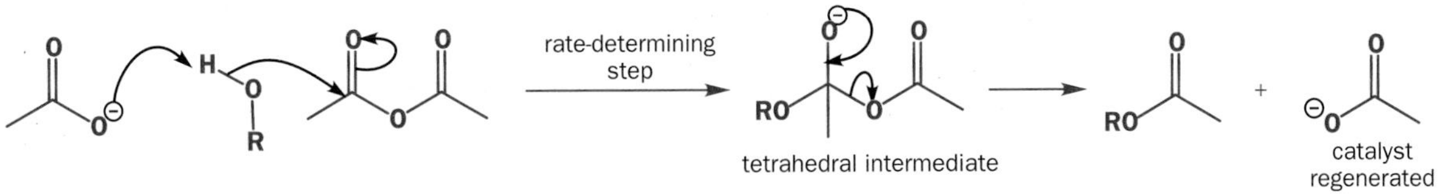

tetrahedral intermediate | catalyst regenerated

This type of catalysis, which is available to any base, not only strong bases, is called **general base catalysis** and will be discussed more in Chapters 41 and 50. It does not speed the reaction up very much but it does lower the energy of the transition state leading to the tetrahedral intermediate since that intermediate is first formed as a neutral compound instead of a dipolar species. Here is the mechanism for the uncatalysed reaction.

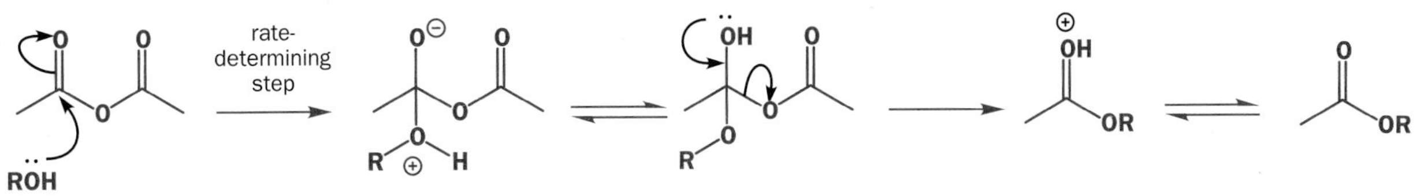

The disadvantage of general base catalysis is that the first, rate-determining, step is termolecular. It is inherently unlikely that three molecules will collide with each other simultaneously and in the next section we shall reject such an explanation for amide hydrolysis. In this case, however, if ROH is the solvent, it will always be present in any collision so a termolecular step is just about acceptable.

The hydrolysis of amides can have termolecular kinetics

When we come to reactions of amides we are at the bottom of the scale of reactivity. Because of the efficient delocalization of the nitrogen lone pair into the carbonyl group, nucleophilic attack on the carbonyl group is very difficult. In addition the leaving group (NH_2^-, pK_{aH} about 35) is very bad indeed.

■ This reaction was first discussed in Chapter 12.

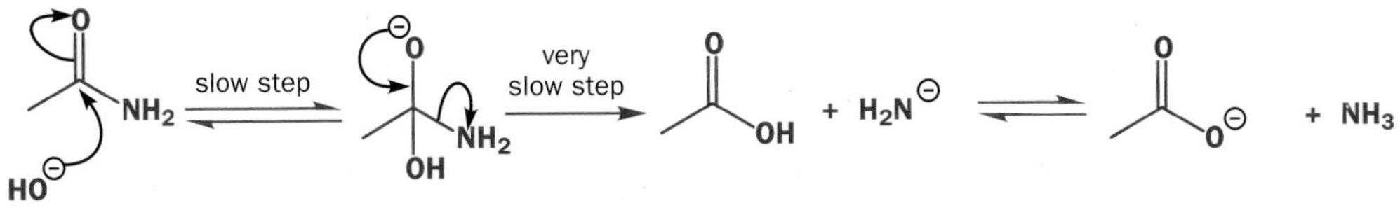

You might indeed have guessed from our previous example, the hydrolysis of esters, where the transition states for formation and breakdown of the tetrahedral intermediate had about the same energies, that in the hydrolysis of amide the second step becomes rate-determining. This offers the opportunity for further base catalysis. If a second hydroxide ion removes the proton from the tetrahedral intermediate, the loss of NH_2^- is made easier and the product is the more stable carboxylate ion.

▶ Of course the very basic leaving group (NH_2^-, pK_{aH} about 35) instantaneously reacts with water (pK_{aH} about 15) in a fast proton transfer to give NH_3 and HO^-.

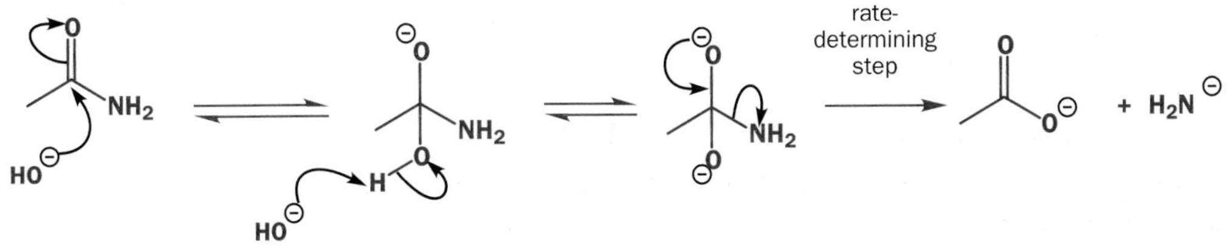

Notice that in the first mechanism the hydroxide is consumed as the product eventually emerges as an anion. In the second mechanism, one hydroxide is consumed but the second is catalytic as the NH_2^- reacts with water to give ammonia and hydroxide ion. The rate expression for the hydrolysis of amides includes a termolecular term and we shall label the rate constant k_3 to emphasize this.

$$\text{rate} = k_3[\text{MeCONH}_2][\text{HO}^-]^2$$

Where do the termolecular kinetics come from? It is, of course, extremely unlikely that three species will collide simultaneously, particularly as two of them are mutually repelling anions. The rate-determining step is actually unimolecular—the spontaneous breakdown of a dianion. But the concentration of the dianion is in the rate expression too and that depends on the reactions before the rate-determining step. With a late rate-determining step, the previous steps are in equilibrium and so we can put in some rate and equilibrium constants for each step and label the intermediates like this.

The rate of the reaction is the rate of the rate-determining step

$$\text{rate} = k[\text{dianion}]$$

We don't know the concentration of the dianion but we do know that it's in equilibrium with the monoanion so we can write

$$K_2 = \frac{[\text{dianion}]}{[\text{monoanion}][\text{HO}^-]}$$

and so $[\text{dianion}] = K_2[\text{monanion}][\text{HO}^-]$

In the same way we don't want the unknown [monoanion] in our rate expression and we can get rid of it using the first equilibrium

$$K_1 = \frac{[\text{monoanion}]}{[\text{amide}][\text{HO}^-]}$$

and so $[\text{monoanion}] = K_1[\text{amide}][\text{HO}^-]$

Substituting these values in the simple rate equation we discover that rate = $k[\text{dianion}]$ becomes

$$\text{rate} = kK_1K_2[\text{amide}][\text{HO}^-]^2$$

The termolecular kinetics result from two equilibria starting with the amide and involving two hydroxide ions followed by a unimolecular rate-determining step, and the 'termolecular rate constant' k_3 is actually a product of the two equilibrium constants and a unimolecular rate constant $k_3 = k \times K_1 \times K_2$.

We have now seen examples of unimolecular and bimolecular reactions and also how termolecular kinetics can arise from unimolecular and bimolecular reactions.

Just because a proposed mechanism gives a rate equation that fits the experimental data, it does not necessarily mean that it is the *right* mechanism; all it means is that it is consistent with the experimental facts so far but there may be other mechanisms that also fit. It is then up to the experimenter to design cunning experiments to try to rule out other possibilities.

Mechanisms are given throughout this book—eventually you will learn to predict what the mechanism for a given type of reaction is, but this is because earlier experimentalists have worked out the mechanisms by a study of kinetics and other methods (see Chapter 41 for more details on how mechanisms are elucidated). In Chapter 17 you will meet another pair of mechanisms—one first-order and one second-order—following the same pattern as these.

The *cis–trans* isomerization of alkenes

The fact that a reaction is favourable (that is, $\Delta G°$ is negative) does not mean that the reaction will go at any appreciable rate: the rate is determined by the activation energy barrier that must be overcome. Returning to the example of the *cis–trans* isomerism of butene, the energy difference between two forms is just 2 kJ mol^{-1}; the activation energy barrier is much bigger: 260 kJ mol^{-1}. The difference in energy determines the equilibrium position (2 kJ mol^{-1} corresponding to an equilibrium constant of about 2.2, or a ratio of 30:70, *cis:trans*; see table on p. 309), whilst the activation energy determines how fast the reaction occurs (260 kJ mol^{-1} means that the reaction does not happen at all at room temperature). A calculation predicts that the half-life for the reaction would be approximately 10^{25} years at room temperature, a time interval much greater than the age of the universe. At 500 °C, however, the half-life is a more reasonable 4 hours which just goes to show the power of exponentials! Unfortunately, when most alkenes are heated to these sorts of temperatures, other unwanted reactions occur.

In order to interconvert the *cis* and *trans* isomers we must use a different strategy. One method is to shine light on the molecule. If UV light is used it is of the right wavelength to be absorbed by the C=C π bond exciting one of the π electrons into the antibonding π* orbital. There is now no π bond and the molecule can rotate freely.

> The quantum symbols *hν* are conventionally used for light in a reaction and the excited state diagram is the best we can do for a molecule with one electron in the π orbital and one in the π* orbital.

trans (*E*-) alkene   alkene excited state   *cis* (*Z*-) alkene

Another approach to alkene isomerization would be to use a catalyst. Base catalysis is of no use as there are no acidic protons in the alkene. Acid catalysis can work (Chapter 19) if a carbocation is formed by protonation of the alkene.

How to catalyse the isomerization of alkenes

The rate at which a reaction occurs depends on its activation energy—quite simply, if we can decrease this, then the reaction rate will speed up. There are two ways by which the activation energy may be decreased: one way is to raise the energy of the starting materials; the other is to lower the energy of the transition state. In the *cis/trans* isomerization of alkenes, the transition state will be halfway through the twisting operation—it has p orbitals on each carbon at right angles to each other. It is the most unstable point on the reaction pathway.

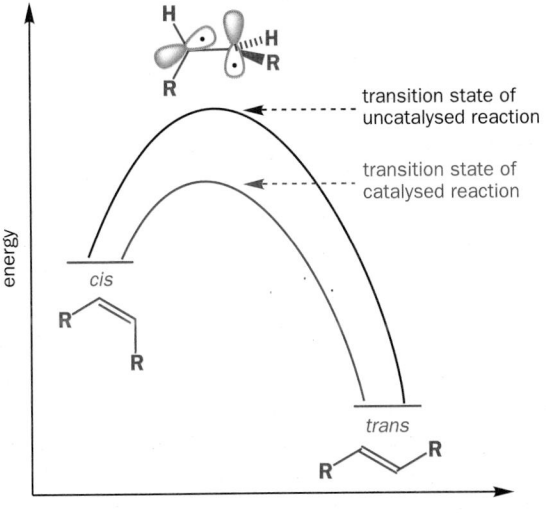

transition state of uncatalysed reaction

transition state of catalysed reaction

energy

cis

trans

extent of reaction

Lowering the energy of the transition state means stabilizing it in some way or other. For example, if there is a separation of charge in the transition state, then a more polar solvent that can solvate this will help to lower the energy of the transition state. Catalysts generally work by stabilizing the transition states or intermediates in a reaction. We shall return to this point when we have introduced kinetic and thermodynamic products.

Kinetic versus thermodynamic products

In Chapter 10 we discussed conjugate addition to unsaturated carbonyl compounds in contrast to direct addition to the carbonyl group. A classic illustration is the addition of HCN to butenone. Two products can be formed.

The 'direct' addition to the left means that cyanide ion must attack the carbonyl group directly while the 'conjugate' addition to the right means that it must attack the less electrophilic alkene. The second is a slower reaction but gives the more stable product. Both reactions have an alkoxide anion as an intermediate.

The energy profile diagram for these two reactions is quite complicated. It has the starting material in the middle, as in the mechanism above, and so extent of reaction increases both to the right for thermodynamic control and to the left for kinetic control.

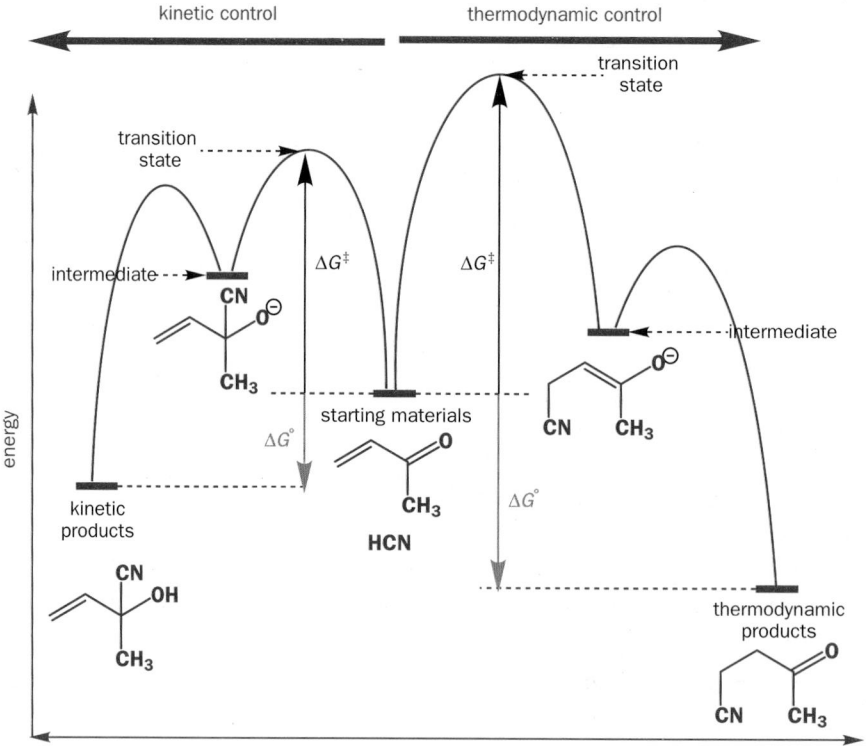

Points to notice:

- The thermodynamic product has a lower energy than the kinetic product
- The highest transition state to the right is higher than the highest to the left
- Initially the reaction will go to the left

- If there is enough energy for the kinetic product to get back to the starting materials, there will be enough energy for some thermodynamic product to be formed
- The energy needed for the thermodynamic product to get back to starting materials is very great
- The kinetic product is formed reversibly; the thermodynamic product irreversibly
- At low temperatures direct addition is favoured, but conjugate addition is favoured at high temperatures

Kinetic versus thermodynamic control in the isomerization of alkenes

Our catalyst for the isomerization of alkenes is going to be HCl absorbed on to solid alumina (aluminium oxide, Al_2O_3) and the isomerization is to occur during a reaction, the addition of HCl to an alkyne, in which the alkenes are formed as products. In this reaction the oxalyl chloride is first mixed with dried alumina. The acid chloride reacts with residual water on the surface (it is impossible to remove all water from alumina) to generate HCl, which remains on the surface.

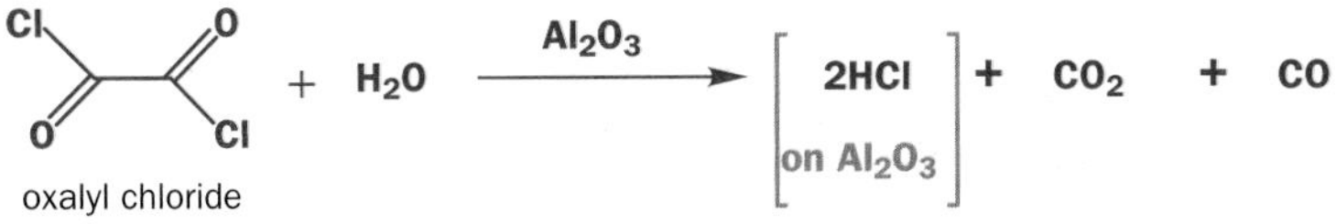

oxalyl chloride + H_2O $\xrightarrow{Al_2O_3}$ [**2HCl** on Al_2O_3] + CO_2 + **CO**

The treated alumina with HCl still attached is added to a solution of an alkyne (1-phenylpropyne) and an addition reaction occurs to produce two geometrical isomers of an alkene. One results from *cis* addition of HCl to the triple bond, and one from *trans* addition.

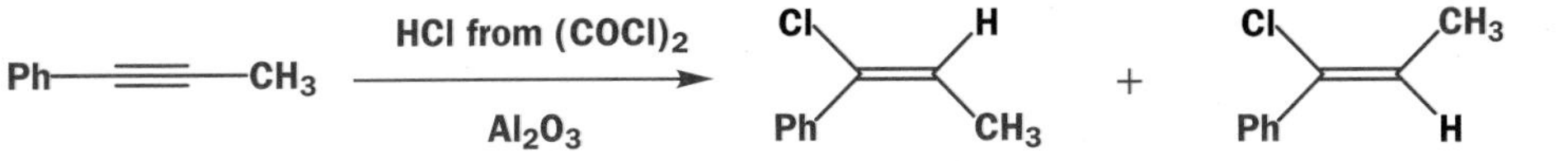

Ph——CH₃ (1-phenylpropyne) $\xrightarrow[\text{Al}_2\text{O}_3]{\text{HCl from (COCl)}_2}$

E-alkene from *cis* addition of HCl + *Z*-alkene from *trans* addition of HCl

The two alkenes are labelled E and Z. After about 2 hours the main product is the Z-alkene. However, this is not the case in the early stages of the reaction. The graph below shows how the proportions of the starting material and the two products change with time.

Points to note:
- When the alkyne concentration drops almost to zero (10 minutes), the only alkene that has been formed is the E–alkene
- As time increases, the amount of E-alkene decreases as the amount of the Z-alkene increases
- Eventually, the proportions of E- and Z-alkenes do not change

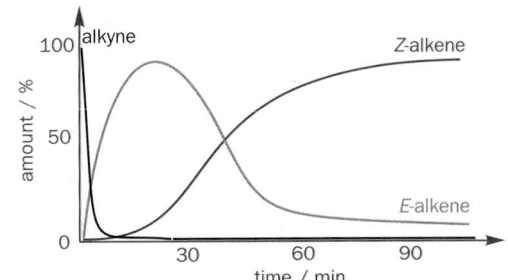

Since it is the Z-alkene that dominates at equilibrium, this must be lower in energy than the E-alkene. Since we know the ratio of the products at equilibrium, we can work out the difference in energy between the two isomers

ratio of E:Z alkenes at equilibrium = 1:35

$$K_{eq} = \frac{[Z]}{[E]} = 35$$

$$\Delta G° = -RT \ln K = -8.314 \times 298 \times \ln(35) = -8.8 \text{ kJ mol}^{-1}$$

that is, the Z-alkene is 8.8 kJ mol^{-1} lower in energy than the E-alkene.

Since the E-alkene is the quickest to form under these conditions, *cis* addition of HCl must have a smaller activation energy barrier than *trans* addition. This suggests that reaction occurs on the surface of the alumina with both the H and the Cl added to the triple bond simultaneously from the same side rather like *cis*-hydrogenation of triple bonds on a palladium catalyst (chapter 24).

> ▶
> You might normally expect an *E*-alkene to be more stable than a *Z*-alkene—it just so happens here that Cl has a higher priority than Ph and the *Z*-alkene has the two largest groups (Ph and Me) *trans*. (See p. 485 for rules of nomenclature.)

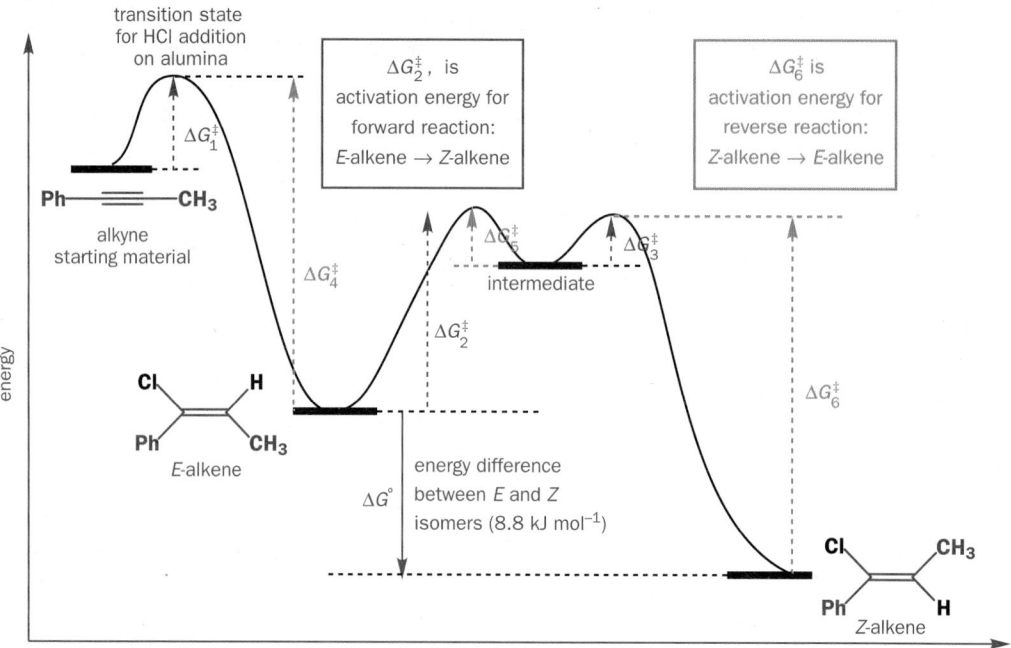

There must then be some mechanism by which the quickly formed *E*-alkene is converted into the more stable *Z*-alkene, presumably through another intermediate that is more stable than the transition state for alkene interconversion. This information is summarized on a reaction profile diagram.

Initially, the alkyne is converted into the *E*-alkene. The activation energy for this step is labelled $\Delta G_1^\ddagger$. The *E*-alkene then converts to the *Z* isomer via an intermediate. The activation energy for this step is $\Delta G_2^\ddagger$. Overall, the reaction is the addition of HCl to the alkyne to give the *Z*-alkene—we could look on the *E* isomer as just another intermediate. The only difference between the *E*-alkene and the intermediate in the isomerization reaction is the size of the activation energies; it is much easier to isolate the *E*-alkene because the activation energies to be overcome ($\Delta G_2^\ddagger$ and $\Delta G_4^\ddagger$) are both much larger than those of the intermediate ($\Delta G_5^\ddagger$ and $\Delta G_3^\ddagger$). The activation energy to be overcome to form the *E*-alkene ($\Delta G_1^\ddagger$) is less than that to be overcome to form the *Z*-alkene ($\Delta G_2^\ddagger$).

So what is this intermediate in the isomerization reaction? It is a cation from protonation of the alkene by more HCl. The cation is stabilized by delocalization into the benzene ring and can rotate as it has no double-bond character.

● **Kinetic and thermodynamic products**

The *E*-alkene is formed faster and is known as the **kinetic product**; the *Z*-alkene is more stable and is known as the **thermodynamic product**.

If we wanted to isolate the kinetic product, the *E*-alkene, we would carry out the reaction at low temperature and not leave it long enough for equilibration. If, on the other hand, we want the thermodynamic product, the *Z*-alkene, we would leave the reaction for longer at higher temperatures to make sure that the larger energy barrier yielding the most stable product can be overcome.

Low temperatures prevent unwanted reactions from occurring

So far in this chapter we have seen why chemists heat up reaction mixtures (usually because the reaction goes faster) but in the introduction we also said that, in any organic laboratory, an equal number of reactions are carried out at low temperatures. Why might a chemist want to slow a reaction down? Actually, we already hinted at the answer to this question when we said that it is possible to isolate reactive carbocations. It *is* possible to isolate these reactive intermediates but only at low temperatures. If the temperature is too high then the intermediate will have sufficient energy to overcome the energy barrier leading to the more stable products.

In our discussion of the reactions of acid chlorides, we deduced that a unimolecular reaction to give a cation must be happening. This cation cannot be detected under these conditions as it reacts too quickly with nucleophiles. If we remove reactive nucleophiles from solution, the cation is still too unstable to be isolated at room temperature. But if we go down to −120 °C we can keep the cation alive long enough to run its NMR spectrum.

Lowering the temperature lowers the energies of all of the molecules in the sample. If there are several possible reactions that might occur and if they have different activation energies, we may be able to find a temperature where the population of molecules has only enough energy to surmount the lowest of the alternative energy barriers so that only one reaction occurs. The diazotization of aromatic amines is an example. The reaction involves treating the amine with nitrous acid (HONO) made from $NaNO_2$ and HCl.

At room temperature the diazonium salt decomposes to the phenol and cannot be used but at 0–5 °C it is stable and can be reacted with other nucleophiles in useful processes discussed in Chapter 23.

Other examples you have met involve lithiated organic molecules. These are always prepared at low temperatures, often at −78 °C. The ortholithiation of aromatic amides was mentioned in Chapter 9.

> −78 °C is the convenient temperature of a bath of acetone containing pellets or slowly evaporating solid CO_2.

If the lithiation is carried out at 0 °C, each molecule of lithiated amide attacks another molecule of unlithiated amide in the substitution reaction from Chapter 12.

The situation is more critical because of the behaviour of the solvent THF. This cyclic ether is a good solvent for lithiations because it is a good ligand for lithium and it remains liquid at −78 °C. But if lithiations are attempted at higher temperatures, THF also reacts with *s*-BuLi to give surprising by-products discussed in Chapter 35.

Solvents

The nature of the solvent used in reactions often has a profound effect on how the reaction proceeds. Often we are limited in our choice of solvent by the solubilities of the reactants and products—this can also be to our advantage when trying to separate products, for example, in ether extractions. We have seen so far in this chapter that THF is a good solvent for lithiations because it coordinates to Li, that water is a good solvent for hydrolyses of carboxylic acids because it is a reagent and because it dissolves the carboxylate anion, and that alcohols are a good solvents in reactions such as transesterifications where mass action is needed to drive equilibria over towards products.

But solvents can affect reactions more drastically; for example, the reaction below gives different products depending on the choice of solvent.

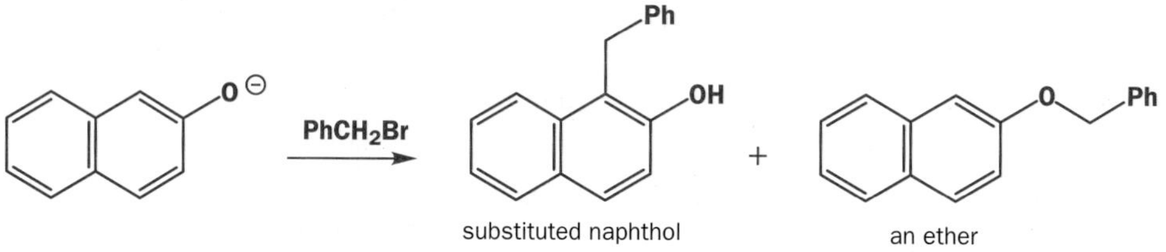

substituted naphthol   an ether

In water the product is almost all benzyl naphthol. However, in DMSO (dimethyl sulfoxide) the major product is the ether. In water the oxyanion is heavily solvated through hydrogen bonds to water molecules and the electrophile cannot push them aside to get close to O⁻ (this is an entropy effect). DMSO cannot form hydrogen bonds as it has no OH bonds and does not solvate the oxy-anion, which is free to attack the electrophile.

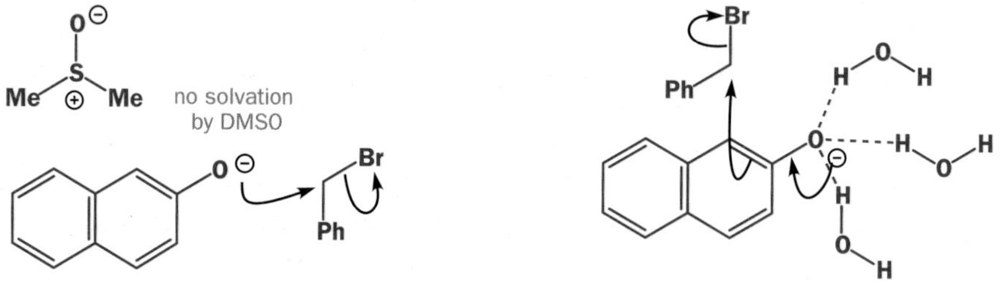

In terms of rates of reaction, where a charged intermediate is formed, a polar solvent will help to stabilize the charge by solvation. Some of this stabilization will already be present in the transition state and solvation will therefore lower the activation energy and speed up the reaction. Turning to a reaction not dealt with elsewhere in the book, an elimination of carbon dioxide, let us see how the rate constant varies with solvent.

■
We shall discuss this type of reaction—fragmentation—in Chapter 38.

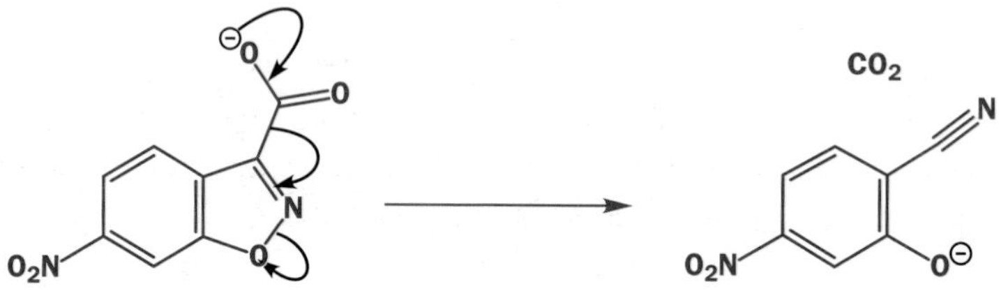

These solvents may be divided into three groups—those in which the reaction is slower than in benzene, those in which it is faster, and, of course, benzene itself. The solvents in which the reaction goes relatively slowly all have something in common—they have either O–H or N–H groups. Solvents of this kind are described as **protic solvents**, that is, they are capable of forming hydrogen bonds in solution (though none of these solvents is a good acid). Mechanistically, the important point is that these solvents solvate both cations and anions. The cations are solvated by use of the lone pairs on the oxygen or nitrogen; the anions *via* the hydrogens.

We can illustrate this with a schematic drawing of the solvation of a salt (NaBr) by water.

Rate of reaction in various solvents

Solvent	Rate[a]
H_2O	0.0015
MeOH	0.052
$HCONH_2$	0.15
C_6H_6	1
acetonitrile, CH_3CN	600
dimethyl sulfoxide, DMSO, $(CH_3)_2SO$	2 100
acetone, $(CH_3)_2CO$	5 000
dimethyl formamide, DMF, $HCON(CH_3)_2$	7 700
dimethylacetamide, $CH_3CON(CH_3)_2$	33 000
hexamethyl phosphoramide, HMPA, $[(CH_3)_2N]_3PO$	150 000

[a] Relative to reaction in benzene.

Solubilities of sodium bromide in protic solvents

Solvent	Solubility, g/100 g of solvent
H_2O	90
MeOH	16
EtOH	6

The solvents in which the reaction proceeds fastest also have something in common—they have an electronegative group (oxygen or nitrogen) but no O–H or N–H bonds. This class is known as polar **aprotic solvents**. Aprotic solvents can still solvate cations but they are unable to solvate anions.

We can now understand the observed trend in the reaction. In the aprotic solvents, the positively charged counterion is solvated and, to some extent, separated from the anion. The anion itself is not solvated and hence is not stabilized; it can therefore react very easily. In protic solvents, such as water, the anion is stabilized by solvation and so is less reactive. We could represent this information on an energy level diagram (overleaf). The main effect of the solvent is on the energy of the starting material—good solvation lowers the energy of the starting material.

The reaction in the aprotic solvent proceeds fastest because the activation energy for this reaction is smallest. This is not because the energy of the transition state is significantly different but because the energy of the starting material has been raised. You might wonder why the energy of the transition state is not stabilized to the same extent as the starting material on changing from an aprotic solvent to a protic solvent. This is because the charge is spread over a number of atoms in the transition state and so it is not solvated to the same extent as the starting material, which has its negative charge localized on the one atom. This is an important point since, if the transition state were stabilized by the same amount as the starting materials, then the reaction would proceed just as quickly in the different solvents since they would then have the same activation energy barriers.

When you meet the new reactions awaiting you in the rest of the book you should reflect that each is controlled by an energy difference. If it is an equilibrium, ΔG° must be favourable, if a kinetically controlled reaction, $\Delta G^\ddagger$ must be favourable, and either of these could be dominated by enthalpy or entropy and could be modified by temperature control or by choice of solvent.

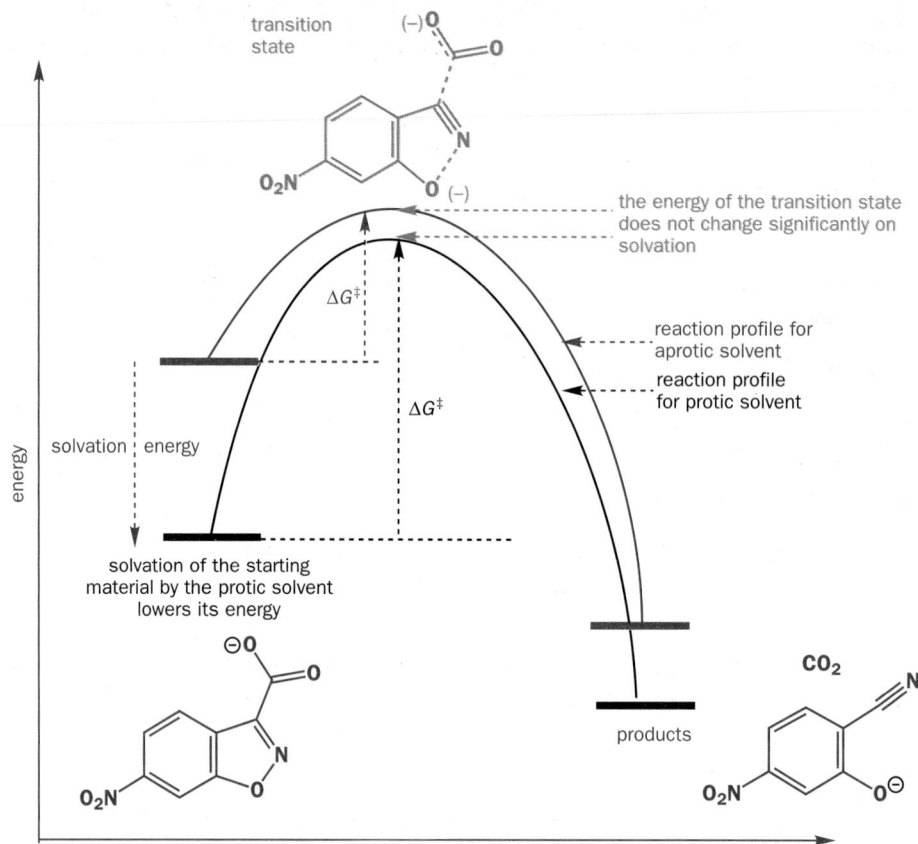

Summary of mechanisms from Chapters 6–12

We last discussed mechanisms in Chapter 5 where we introduced basic arrow-drawing. A lot has happened since then and this is a good opportunity to pull some strands together. You may like to be reminded:

1   When molecules react together, one is the *electrophile* and one the *nucleophile*

2   In most mechanisms electrons flow from an electron-rich to an electron-poor centre

3   Charge is conserved in each step of a reaction

These three considerations will help you draw the mechanism of a reaction that you have not previously met.

Types of reaction arrows

1   Simple reaction arrows showing a reaction goes from left to right or right to left

2   Equilibrium arrows showing extent and direction of equilibrium

3   Delocalization or conjugation arrows showing two different ways to draw the same molecule. The two structures ('canonical forms' or 'resonance structures') must differ only in the position of electrons

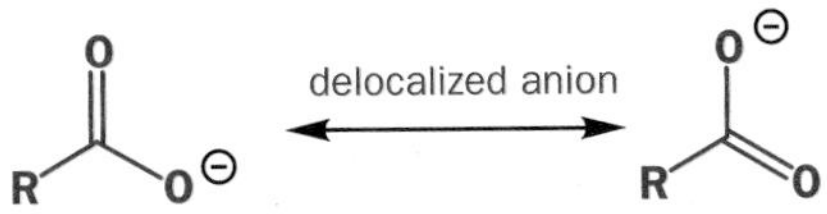

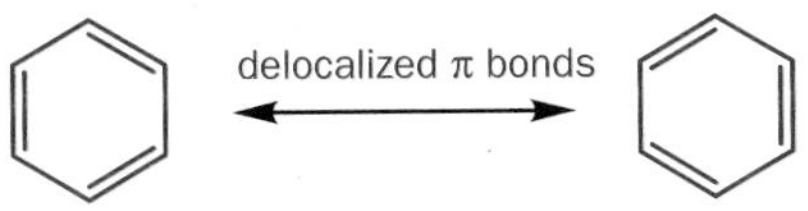

Types of curly arrows

1   The curly arrow should show clearly where the electrons come from and where they go to

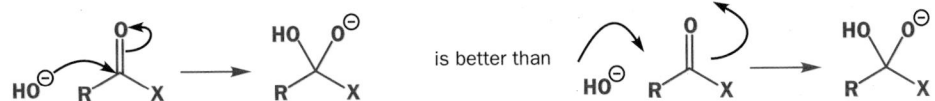

2   If electrophilic attack on a π or σ bond leads to the bond being broken, the arrows should show clearly which atom bonds to the electrophile

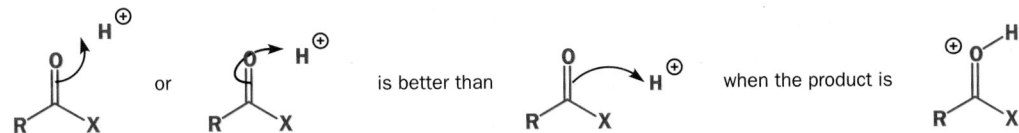

3   Reactions of the carbonyl group are dominated by the breaking of the π bond. If you use this arrow first on an unfamiliar reaction of a carbonyl compound, you will probably find a reasonable mechanism

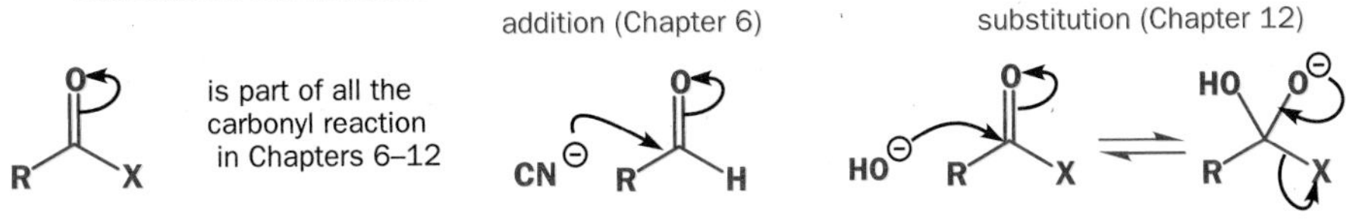

addition (Chapter 6)   substitution (Chapter 12)

is part of all the carbonyl reaction in Chapters 6–12

Short cuts in drawing mechanisms

1   The most important is the double-headed arrow on the carbonyl group used during a substitution reaction

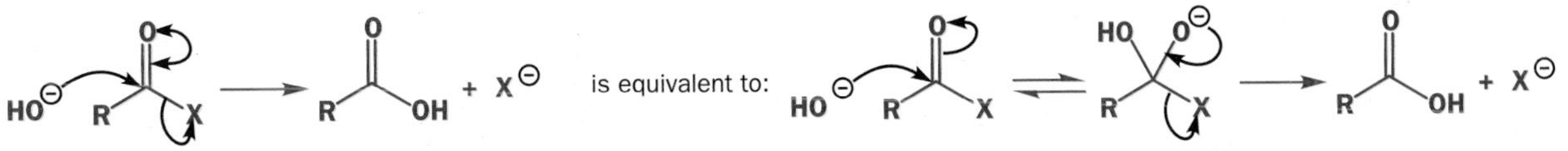

is equivalent to:

2   The symbol ±H⁺ is shorthand for the gain and loss of a proton in the same step (usually involving N, O, or S)

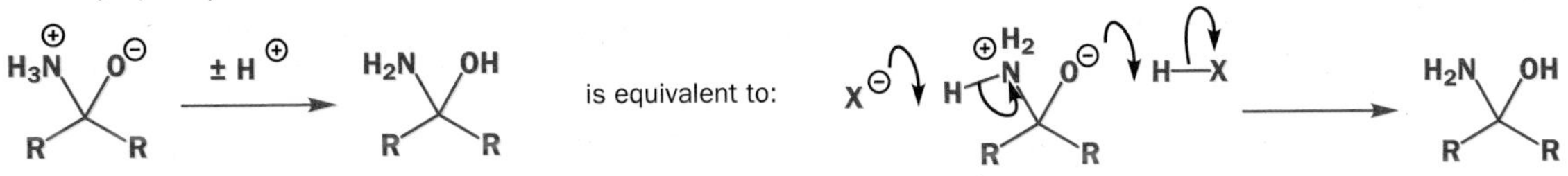

$\pm H^{\oplus}$ is equivalent to:

Problems

1. In the method for acetylating aromatic amines described in Chapter 8, p. 000 (the Lumière–Barbier method) the amine was dissolved in aqueous HCl. Using 1M amine, pK_a 4.6, and 1M HCl, pK_a −7, what will be the approximate equilibrium constant in the reaction?

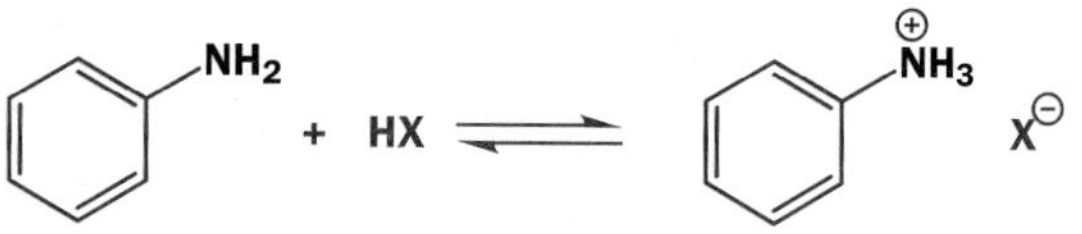

The next step in the reaction is the addition of acetic anhydride and sodium acetate. What will happen to the sodium acetate (NaOAc) in the aqueous solution of HCl? Estimate the equilibrium constant for the reaction between NaOAc and HCl. Will there be enough acid left to keep the amine in solution?

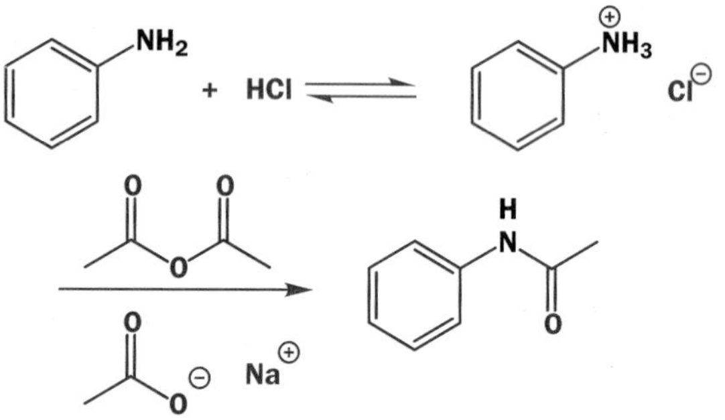

It would be simpler not to add the sodium acetate, keeping the pH low, and thus definitely keep all the amine in solution. Why is it necessary to raise the pH for the second step of the reaction (the reaction with acetic anhydride)?

2. In the comparison of stability of the last intermediates in the carbonyl substitution of acid chlorides on the one hand and anhydrides on the other to make esters we made this statement.

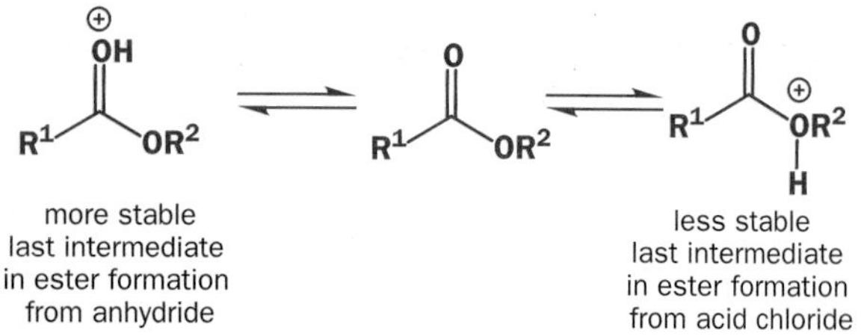

more stable last intermediate in ester formation from anhydride

less stable last intermediate in ester formation from acid chloride

Why is the one more stable than the other? If you were to react an ester with acid, which of the two would you form and why?

3. If we carry out an ester exchange reaction on a one molar solution of $MeCO_2R^2$ in an alcohol R^1OH as solvent with catalytic R^1O^- and let the mixture reach equilibrium, what will be the equilibrium composition if the solvent alcohol R^1OH is, say, 25M in itself?

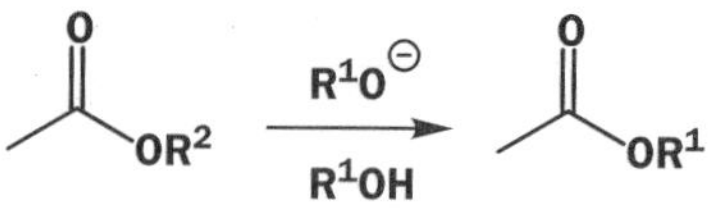

4. Write a mechanism for the reaction to give HCl on alumina. You do not need to consider the role of the alumina.

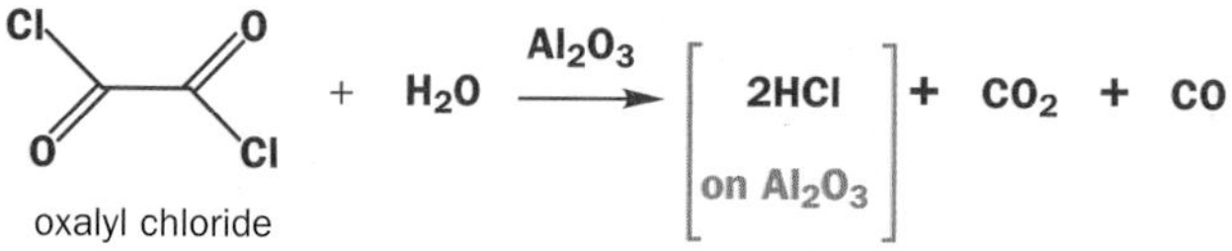

oxalyl chloride

5. Propose a mechanism for the formation of the diazonium salt referred to in the chapter. The first step is the formation of nitrous acid HONO.

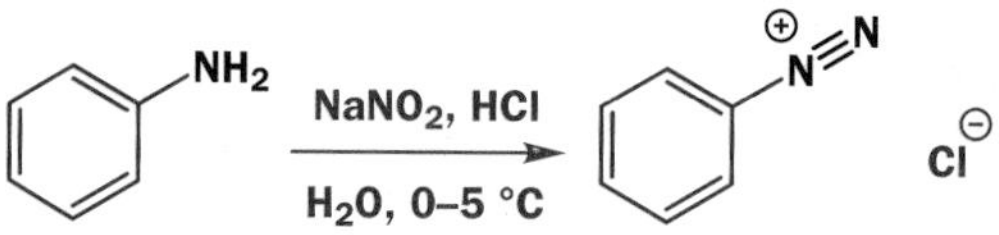

6. This reaction shows third-order kinetics as the rate expression is: rate = [ketone][HO⁻]². Suggest a mechanism that explains these observations.

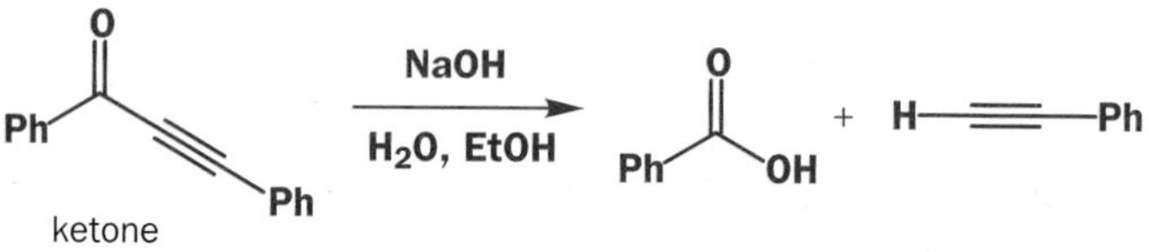

ketone

7. Draw an energy profile diagram for this reaction. You will, of course, need to draw a mechanism first. Suggest which step in the mechanism is likely to be the slow step and what kinetics would be observed.

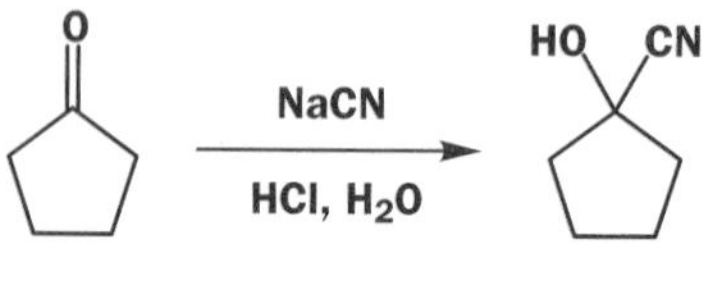

8. The equilibrium between a carbonyl compound and its hydrate usually favours the aldehyde or ketone. Draw an energy profile to express this and mark the difference in free energy between the two compounds. The hydrate of cyclopropanone is preferred to the ketone. How does the energy profile change for this compound?

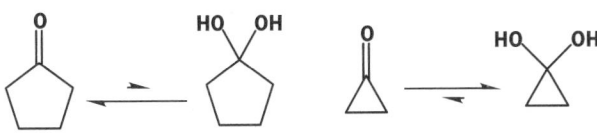

9. What would be the solvent effects on these reactions? Would they be accelerated or retarded by a change from a nonpolar to a polar solvent?

Ph₃P $\xrightarrow[\text{solvent}]{\text{Br}_2}$ Ph₃P⁺—Br + Br⁻

$\xrightarrow[\text{solvent}]{\text{heat}}$ CO₂ + NMe₃

$\xrightarrow[\text{solvent}]{\text{NH}_3}$

10. Comment on the likely effect of acid or base on these equilibria.

11. Elemental sulfur normally exists as an eight-membered ring (S₈), but it can also be found in a number of other states. How would entropy and enthalpy affect the equilibrium between sulfur in these two forms?

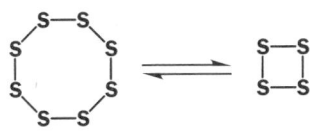

12. Draw transition states and intermediates for this reaction and fit each on an energy profile diagram. Be careful to distinguish between transition states and intermediates.

NaOH

Nucleophilic substitution at C=O with loss of carbonyl oxygen

14

Connections

Building on:	Arriving at:	Looking forward to:
• Nucleophilic attack on carbonyl groups ch6	• Replacement of carbonyl oxygen	• Protecting groups ch24
• Nucleophilic substitution at carbonyl groups ch12	• Acetal formation	• Synthesis in action ch25
	• Imine formation	• Acylation of enolates ch28
• Acidity and pK_a ch8	• Stable and unstable imines	• Synthesis of amino acids ch49
• Rate and pH ch13	• Reductive amination	• Synthesis of alkenes ch31
	• The Strecker and Wittig reactions	• Stereochemistry ch16
		• Asymmetric synthesis ch45

Introduction

Nucleophiles add to carbonyl groups to give compounds in which the trigonal carbon atom of the carbonyl group has become tetrahedral.

nucleophilic addition to a carbonyl group

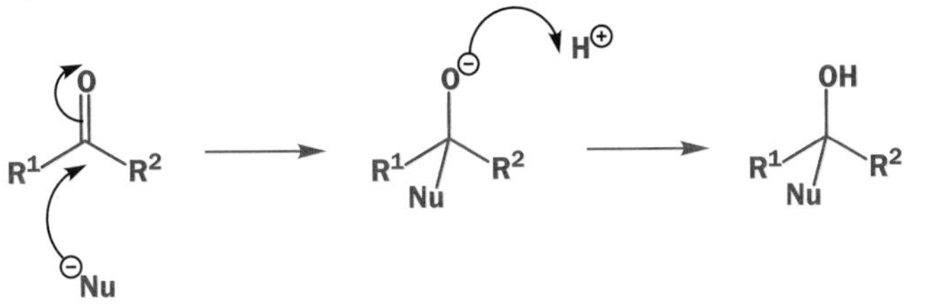

In Chapter 12 you saw that these compounds are not always stable: if the starting material contains a leaving group, the addition product is a **tetrahedral intermediate**, which collapses with loss of the leaving group to give back the carbonyl group, with overall substitution of the leaving group by the nucleophile.

nucleophilic substitution at a carbonyl group

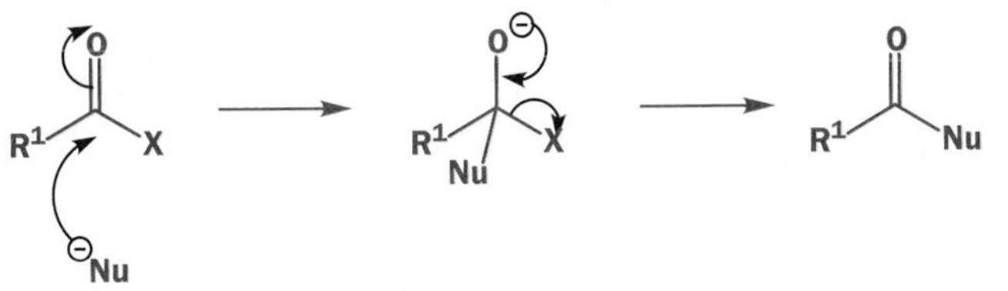

In this chapter, you will meet more substitution reactions of a different type. Instead of losing a leaving group, the carbonyl group loses its oxygen atom. Here are three examples: the carbonyl oxygen atom has been replaced by an atom of ^{18}O, a nitrogen atom, and two atoms of oxygen. Notice too the acid catalyst—we shall see shortly why it is required.

nucleophilic substitution at a carbonyl group with loss of carbonyl oxygen

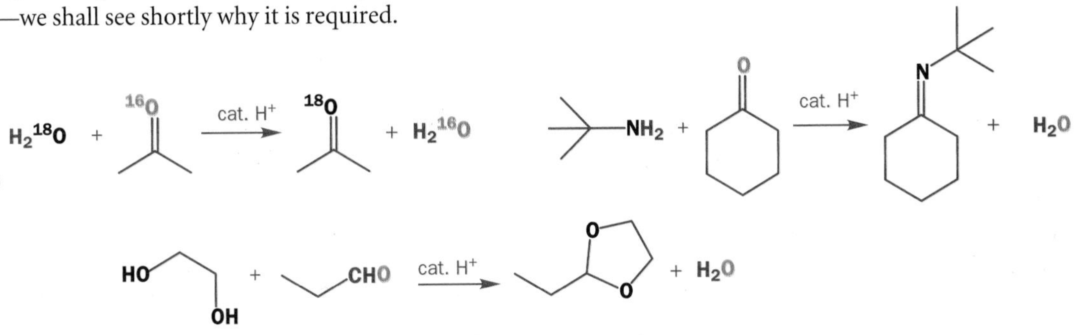

You have, in fact, already met some reactions in which the carbonyl oxygen atom can be lost, but you probably didn't notice at the time. The equilibrium between an aldehyde or ketone and its hydrate (p. 143) is one such reaction.

When the hydrate reverts to starting materials, either of its two oxygen atoms must leave: one came from the water and one from the carbonyl group, so 50% of the time the oxygen atom that belonged to the carbonyl group will be lost. Usually, this is of no consequence, but it can be useful. For example, in 1968 some chemists studying the reactions that take place inside mass spectrometers needed to label the carbonyl oxygen atom of this ketone with the isotope ^{18}O.

By stirring the 'normal' ^{16}O compound with a large excess of isotopically labelled water, $H_2{}^{18}O$, for a few hours in the presence of a drop of acid they were able to make the required labelled compound. Without the acid catalyst, the exchange is very slow. Acid catalysis speeds the reaction up by making the carbonyl group more electrophilic so that equilibrium is reached more quickly. The equilibrium is controlled by mass action—^{18}O is in large excess.

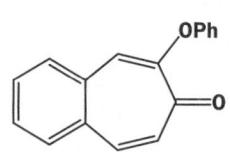

> In Chapter 13 we saw this way of making a reaction go faster by raising the energy of the starting material. We also saw that the position of an equilibrium can be altered by using a large excess of one of the reagents. This is sometimes called a **mass action effect**.

We need now to discuss hemiacetals though you may well wonder why – they retain the carbonyl oxygen and they are unstable. We need to discuss them as a preliminary to the much more important acetals. Hemiacetals are halfway to acetals.

Aldehydes can react with alcohols to form hemiacetals

When acetaldehyde is dissolved in methanol, a reaction takes place: we know this because the IR spectrum of the mixture shows that a new compound has been formed. However, isolating the product is impossible: it decomposes back to acetaldehyde and methanol.

IR:
no peak in carbonyl region 1600–1800

strong OH stretch 3000–3500

The product is in fact a hemiacetal. Like hydrates, most hemiacetals are unstable with respect to their parent aldehydes and alcohols: for example, the equilibrium constant for reaction of acetaldehyde with simple alcohols is about 0.5 as we saw in Chapter 13.

This equilibrium constant K is defined as

$$K = \frac{[\text{hemiacetal}]}{[\text{aldehyde}][\text{MeOH}]}$$

So by making [MeOH] very large (using it as the solvent, for example) we can turn most of the aldehyde into the hemiacetal. However, if we try and purify the hemiacetal by removing the methanol, more hemiacetal keeps decomposing to maintain the equilibrium constant. That is why we can never isolate such hemiacetals in a pure form.

■ These are more 'mass action' effects like the ^{18}O exchange we have just discussed.

Only a few hemiacetals are stable

Like their hydrates, the hemiacetals of most ketones (sometimes called **hemiketals**) are even less stable than those of aldehydes. On the other hand, some hemiacetals of aldehydes bearing electron-withdrawing groups, and those of cyclopropanones, are stable, just like the hydrates of the same molecules.

■ We discussed the reasons for this in Chapter 6.

a stable hemiacetal

Hemiacetals that can be formed by intramolecular cyclization of an alcohol on to an aldehyde are also often stable, especially if a five- or six-membered ring is formed. You met this in Chapter 6—many sugars (for example, glucose) are cyclic hemiacetals, and exist in solution as a mixture of open-chain and cyclic forms.

glucose: open-chain form   cyclic form

0.003%   > 99%

Why are cyclic hemiacetals stable?

Part of the reason for the stability of cyclic hemiacetals concerns *entropy*. Formation of an acyclic acetal involves a decrease in entropy ($\Delta S°$ negative) because two molecules are consumed for every one produced. This is not the case for formation of a cyclic hemiacetal. Since $\Delta G° = \Delta H° - T\Delta S°$, a reaction with a negative $\Delta S°$ tends to have a more positive $\Delta G°$; in other words, it is less favourable.

Another way to view the situation is to consider the rates of the forward and reverse processes. We can

measure the stability of a cyclic hemiacetal by the equilibrium constant K for the ring-opening reaction: a large K means lots of ring-opened product, and therefore an unstable hemiacetal, and a small K means lots of ring-closed product: a stable hemiacetal. An equilibrium constant is simply the rate constant of the forward reaction divided by the

rate constant of the reverse reaction. So, for a stable hemiacetal, we need a fast hemiacetal-forming reaction. And when the hemiacetal is cyclic that is just what we do have: the reaction is intramolecular and the nucleophilic OH group is always held close to the carbonyl group, ready to attack.

equilibrium constant

$$K = \frac{k_{\text{forward}}}{k_{\text{reverse}}}$$

k_{forward}

k_{reverse}

fast for cyclic hemiacetals

Acid or base catalysts increase the rate of equilibration of hemiacetals with their aldehyde and alcohol parents

Acyclic hemiacetals form relatively slowly from an aldehyde or ketone plus an alcohol, but their rate of formation is greatly increased either by acid or by base. As you would expect, after Chapters 12 and 13, acid catalysts work by increasing the electrophilicity of the carbonyl group.

acid-catalysed hemiacetal formation

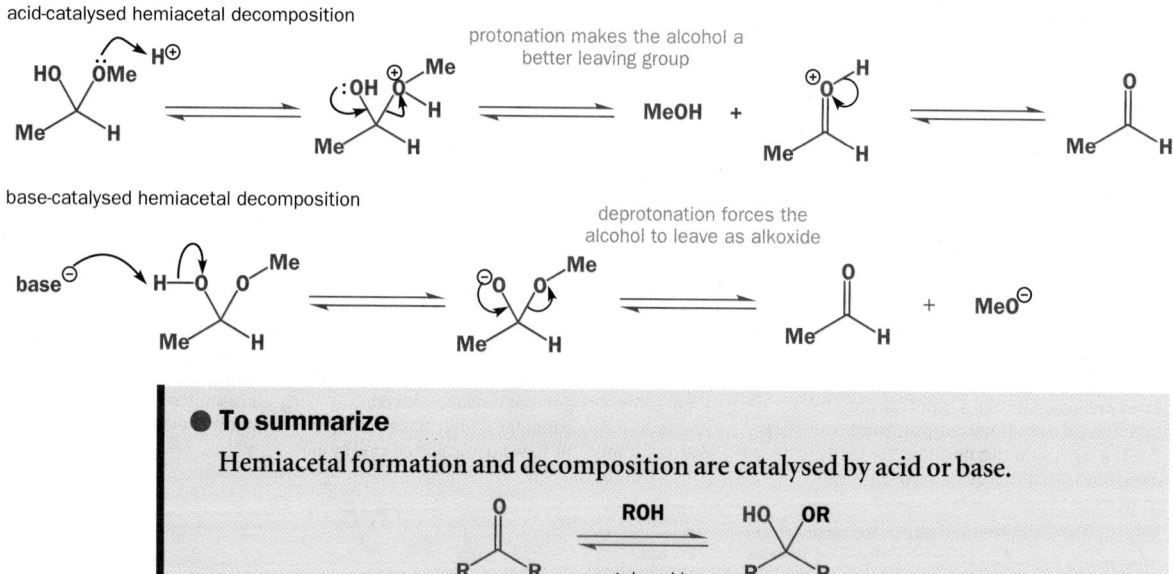

Base catalysts, on the other hand, work by increasing the nucleophilicity of the alcohol by removing the OH proton before it attacks the C=O group. In both cases the energy of the starting materials is raised: in the acid-catalysed reaction the aldehyde is destabilized by protonation and in the base-catalysed reaction the alcohol is destabilized by deprotonation.

base-catalysed hemiacetal formation

You can see why hemiacetals are unstable: they are essentially tetrahedral intermediates containing a leaving group and, just as acid or base catalyses the formation of hemiacetals, acid or base also catalyses their decomposition back to starting aldehyde or ketone and alcohol. That's why the title of this section indicated that acid or base catalysts increase the rate of equilibration of hemiacetals with their aldehyde and alcohol components—the catalysts do not change the position of that equilibrium!

acid-catalysed hemiacetal decomposition

base-catalysed hemiacetal decomposition

● **To summarize**

Hemiacetal formation and decomposition are catalysed by acid or base.

Acetals are formed from aldehydes or ketones plus alcohols in the presence of acid

We said that a solution of acetaldehyde in methanol contains a new compound: a hemiacetal. We've also said that the rate of formation of hemiacetals is increased by adding an acid (or a base) catalyst to an alcohol plus aldehyde mixture. But, if we add catalytic acid to our acetaldehyde–methanol mixture, we find not only that the rate of reaction of the acetaldehyde with the methanol increases, but also that a different product is formed. This product is an **acetal**.

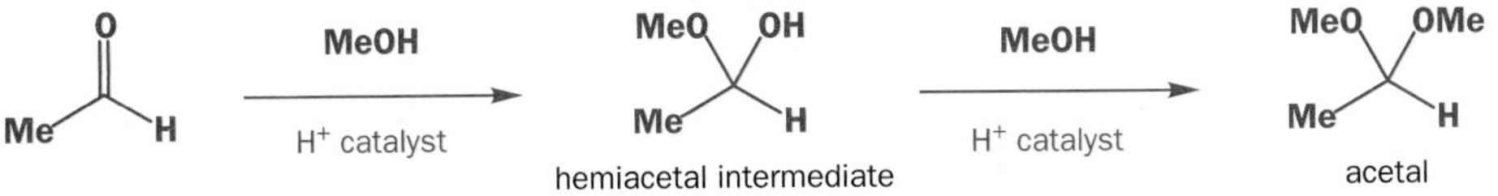

In the presence of acid (but not base!) hemiacetals can undergo an elimination reaction (different from the one that just gives back aldehyde plus alcohol), losing the oxygen atom that once belonged to the parent aldehyde's carbonyl group. The stages are:

1 Protonation of the hydroxyl group of the hemiacetal

2 Loss of water by elimination. This elimination leads to an unstable and highly reactive oxonium ion

3 Addition of methanol to the oxonium ion (breaking the π bond and not the σ bond, of course)

4 Loss of a proton to give the acetal

acid-catalysed acetal formation from hemiacetal

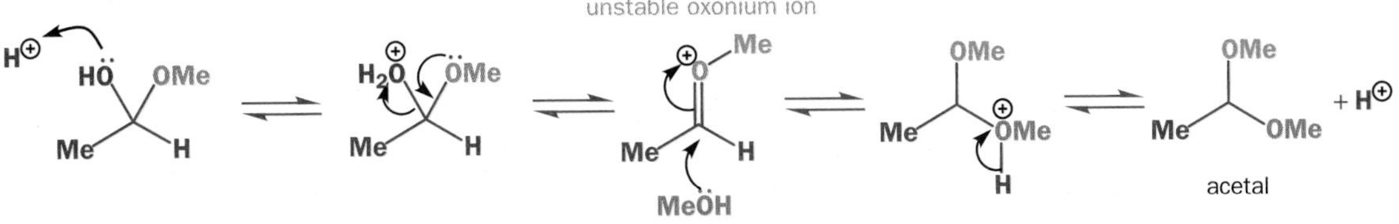

Oxonium ions

Oxonium ions have three bonds to a positively charged oxygen atom. All three bonds can be σ bonds as in H_3O^+ or Meerwein's salt, trimethyloxonium fluoroborate, a stable (though reactive) compound described in Chapter 21, or one bond can be a π bond as in the acetal intermediate. The term 'oxonium ion' describes either of these structures. They are like alkylated ethers or O-alkylated carbonyl compounds.

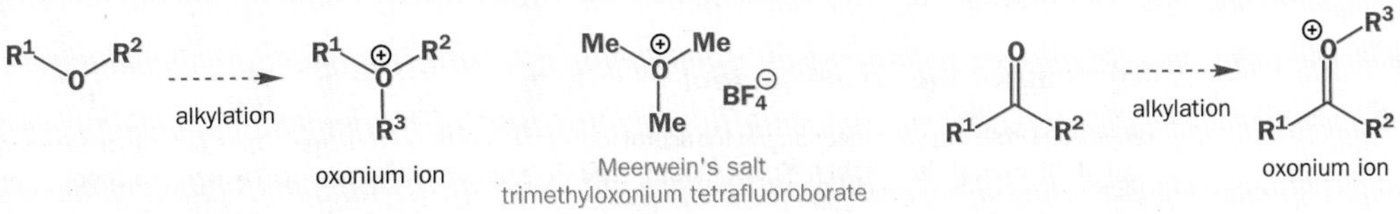

Just as protonated carbonyl groups are much more electrophilic than unprotonated ones, these oxonium ions are powerful electrophiles. They can react rapidly with a second molecule of alcohol to form new, stable compounds known as **acetals**. An oxonium ion was also an intermediate in the formation of hemiacetals in acid solution. Before reading any further, it would be worthwhile to write out the whole mechanism of acetal formation from aldehyde or ketone plus alcohol through the hemiacetal to the acetal, preferably without looking at the fragments of mechanism above, or the answer below.

● **Formation of acetals and hemiacetals**

Hemiacetal formation is catalysed by acid or base, but acetal formation is possible only with an acid catalyst because an OH group must be made into a good leaving group.

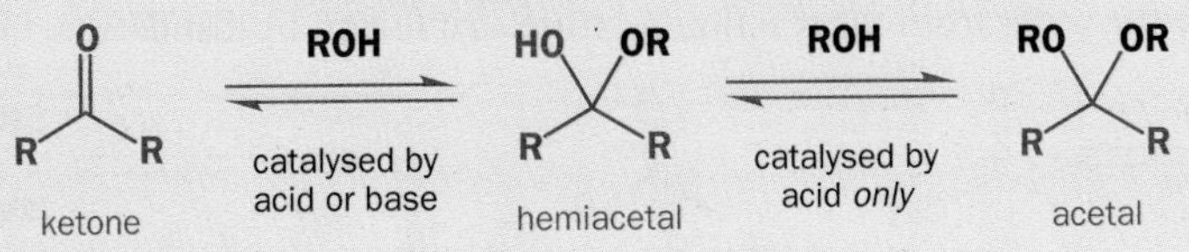

When you look at our version of this complete mechanism you should notice a remarkable degree of similarity in the two halves. The reaction starts with a protonation on carbonyl oxygen and, when you get to the temporary haven of the hemiacetal, you start again with protonation of that same oxygen. Each half goes through an oxonium ion and each oxonium ion adds the alcohol. The last step in the formation of both the acetal and the hemiacetal is the loss of a proton from the recently added alcohol.

This is about as complex a mechanism as you have seen and it will help you to recall it if you see it in two halves, each very similar to the other. First, form the hemiacetal by adding an alcohol to the C=O π bond; then lose the OH group by breaking what was the C=O σ bond to form an oxonium ion and add a second alcohol to form the acetal. From your complete mechanism you should also be able to verify that acetal formation is indeed catalytic in acid.

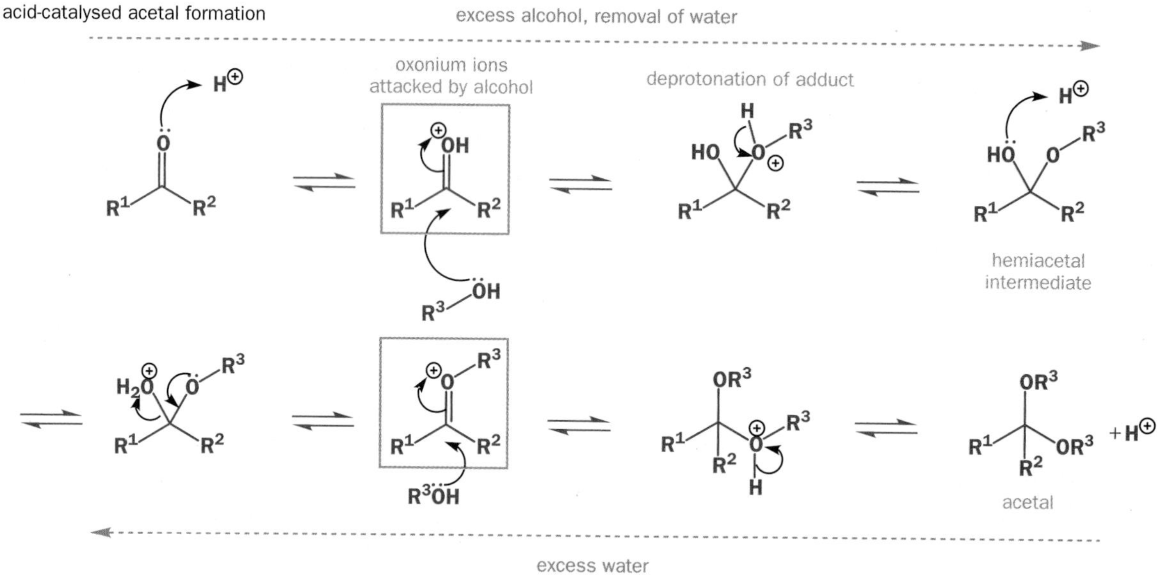

acid-catalysed acetal formation — excess alcohol, removal of water

oxonium ions attacked by alcohol

deprotonation of adduct

hemiacetal intermediate

acetal

excess water

Remember the oxonium ion!

When you wrote out your mechanism for acetal formation, we hope you didn't miss out the oxonium ion! It's easy to do so, but the mechanism most definitely does not go via a direct displacement of water by alcohol.

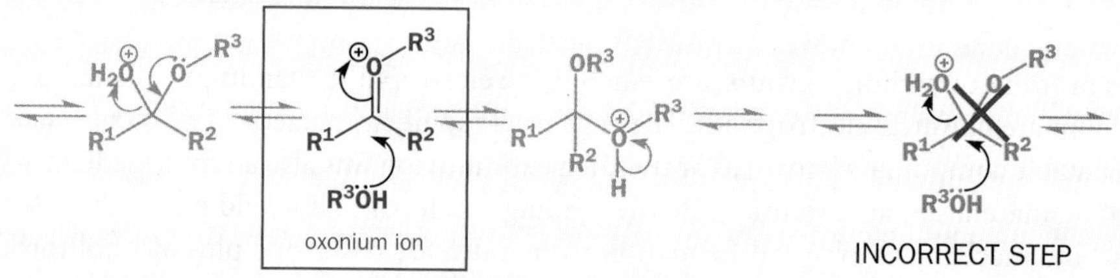

oxonium ion

INCORRECT STEP

If you wonder how we know this, consult a specialized book on organic reaction mechanisms. After you have read Chapter 17 in this book, you will be able to spot that this substitution step goes via an S_N1 and not an S_N2 mechanism.

Making acetals

Just as with the ester formation and hydrolysis reactions we discussed in Chapters 12 and 13, every step in the formation of an acetal is reversible. To make acetals, therefore, we must use an excess of alcohol or remove the water from the reaction mixture as it forms, by distillation, for example.

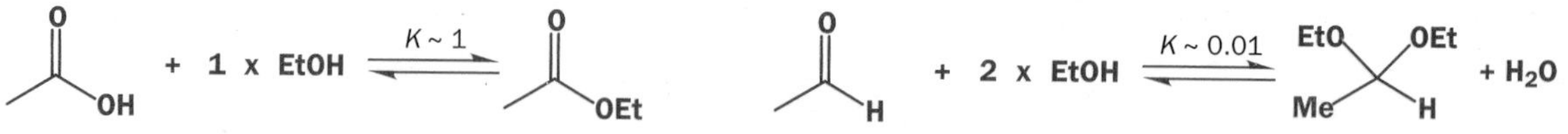

In fact, acetal formation is even more difficult than ester formation: while the equilibrium constant for acid-catalysed formation of ester from carboxylic acid plus alcohol is usually about 1, for acetal formation from aldehyde and ethanol (shown above), the equilibrium constant is $K = 0.0125$. For ketones, the value is even lower: in fact, it is often very difficult to get the acetals of ketones (sometimes called ketals) to form unless they are cyclic (we consider cyclic acetals later in the chapter). However, there are several techniques that can be used to prevent the water produced in the reaction from hydrolysing the product.

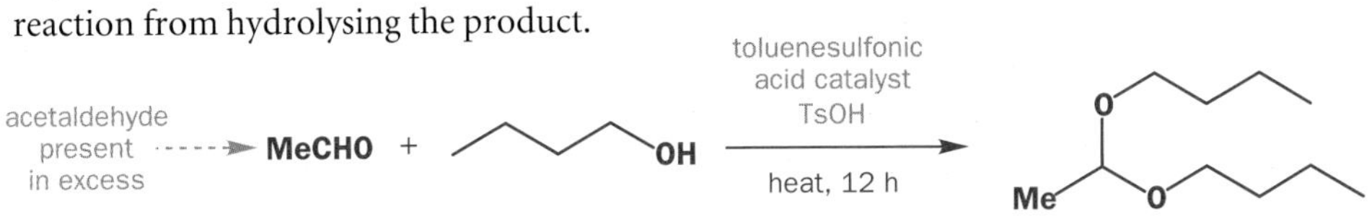

acetaldehyde present in excess - - - - ➤ **MeCHO** +

toluenesulfonic acid catalyst TsOH

heat, 12 h

50% yield of acetal

> *para*-Toluenesulfonic acid is commonly used to catalyse reactions of this sort. It is a stable solid, yet is as strong an acid as sulfuric acid. It is widely available and cheap because it is produced as a by-product in the synthesis of saccharin (for more details, see Chapter 22).

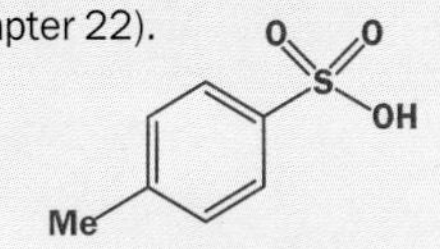

p-toluenesulfonic acid

In these two examples, with the more reactive aldehyde, it was sufficient just to have an excess of one of the reagents (acetaldehyde) to drive the reaction to completion. Dry HCl gas can work too. In the second example, with a less reactive ketone, molecular sieves (zeolite) were used to remove water from the reaction as it proceeded.

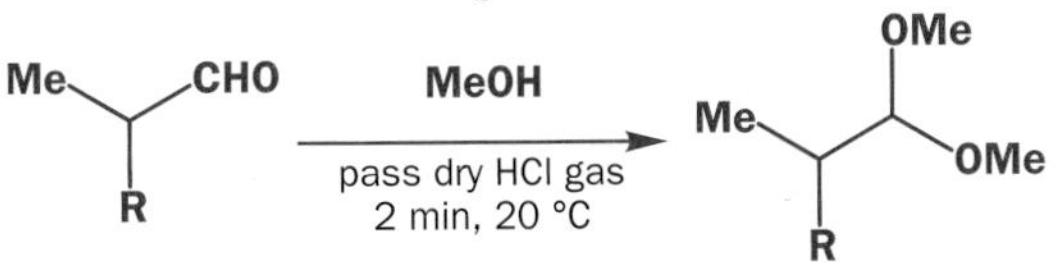

MeOH

pass dry HCl gas
2 min, 20 °C

60% yield

catalytic TsOH
molecular sieves
0 °C, 2 h

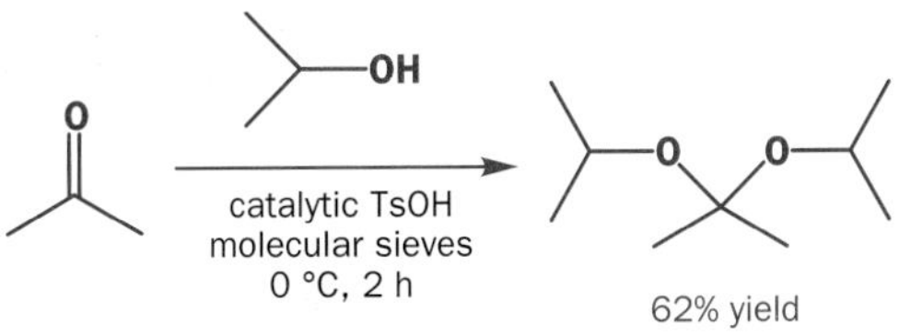

62% yield

Overcoming entropy: orthoesters

We have already mentioned that one of the factors that makes acyclic *hemi*acetals unstable is the unfavourable decrease in entropy when two molecules of starting material (aldehyde or ketone plus alcohol) become one of product. The same is true for acetal formation, when three molecules of starting material (aldehyde or ketone plus $2 \times$ alcohol) become two of product (acetal plus H_2O). We can improve matters if we tie the two alcohol molecules together in a diol and make a cyclic acetal: we discuss cyclic acetals in the next section. Alternatively, we can use an **orthoester** as a source of alcohol. Orthoesters can be viewed as the 'acetals of esters' or as the triesters of the unknown 'orthoacids'—the hydrates of carboxylic acids. They are hydrolysed by water, catalysed by acid, to ester + $2 \times$ alcohol.

orthoacids don't exist   orthoesters

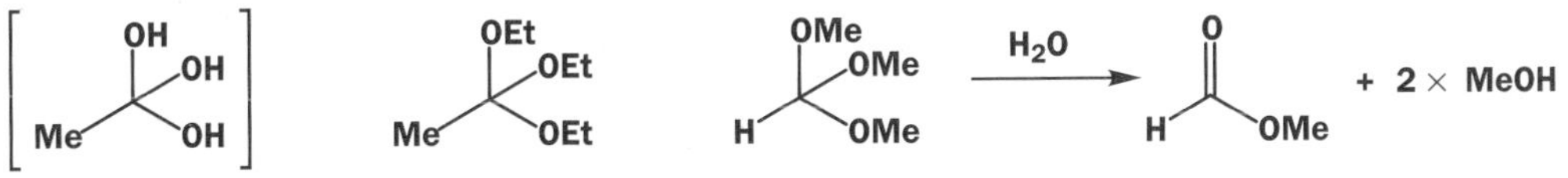

H_2O + $2 \times$ **MeOH**

orthoacetic acid   triethyl orthoacetate   trimethyl orthoformate

Here is the mechanism for the hydrolysis—you should be feeling quite familiar with this sort of thing by now.

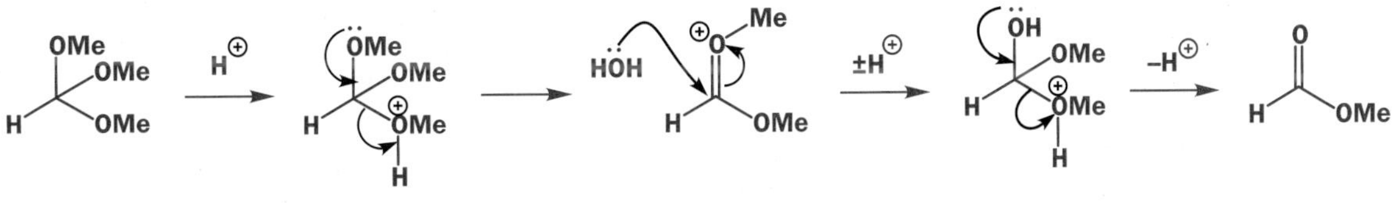

Ketones or aldehydes can undergo **acetal exchange** with orthoesters. The mechanism starts off as if the orthoester is going to hydrolyse but the alcohol released adds to the ketone and acetal formation begins. The water produced is taken out of the equilibrium by hydrolysis of the orthoester.

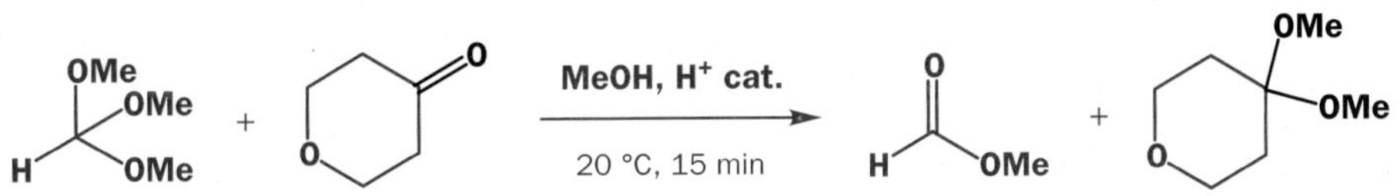

MeOH, H^+ cat.

20 °C, 15 min

trimethyl orthoformate   methyl formate

Acetals hydrolyse only in the presence of acid

Just as acetal formation requires acid catalysis, acetals can be hydrolysed only by using an acid catalyst. With aqueous acid, the hydrolysis of acyclic acetals is very easy. Our examples are the two acetals we made earlier.

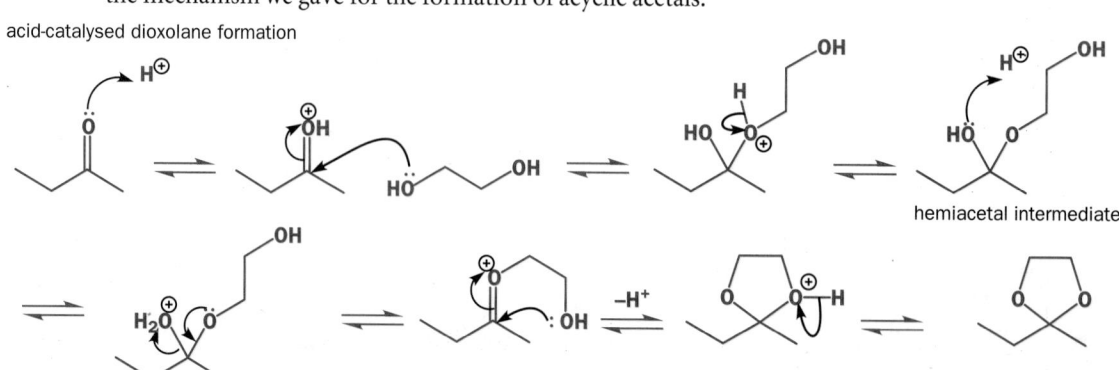

> ● **Acetal hydrolysis**
> Acetals can be hydrolysed in acid but are stable to base.

We won't go through the mechanism again—you've already seen it as the reverse of acetal formation (and you have a hint of it in the orthoester hydrolysis just discussed), but the fact that acetals are stable to base is really a very important point, which we will use on the next page and capitalize on further in Chapter 24.

Cyclic acetals are more stable towards hydrolysis than acyclic ones

Of course you want us to prove it: well—

The acetals you have met so far were formed by reaction of two molecules of alcohol with one of carbonyl compound. Cyclic acetals, formed by reaction of a single molecule of a diol, a compound containing two hydroxyl groups, are also important. When the diol is ethylene glycol (as in this example) the five-membered cyclic acetal is known as a **dioxolane**.

Before looking at the answer below, try to write a mechanism for this reaction. If you need it, use the mechanism we gave for the formation of acyclic acetals.

acid-catalysed dioxolane formation

> ▶
> We hope you didn't make the mistake of missing out the oxonium ion step!

Cyclic acetals like this are more resistant to hydrolysis than acyclic ones and easier to make—they form quite readily even from ketones. Again, we have entropic factors to thank for their stability. For the formation of a cyclic acetal, two molecules go in (ketone plus diol) and two molecules come out (acetal plus water), so the usually unfavourable $\Delta S°$ factor is no longer against us. And, as for hemiacetals (see the explanation above), equilibrium tends to lie to the acetal side because the intramolecular ring-closing reaction is fast.

Water is still generated, and needs to be got rid of: in the example above you can see that water was distilled out of the reaction mixture. This is possible with these diols because they have a boiling point above that of water (the boiling point of ethylene glycol is 197 °C). You can't distil water from a reaction mixture containing methanol or ethanol, because the alcohols distil too! One very useful piece of equipment for removing water from reaction mixtures containing only reagents that boil at higher temperatures than water is called a **Dean Stark head**.

Modifying reactivity using acetals

Why are acetals so important? Well, they're important to both nature and chemists because many carbohydrates are acetals or hemiacetals (see the box below). One important use that chemists have put them to is as *protecting groups*.

One important synthesis of the steroid class of compounds (about which more later) requires a Grignard reagent with this structure.

■ We shall discuss protecting groups in much more detail in Chapter 24.

unstable structure
– impossible to make

Yet this compound cannot exist: it would react with itself. Instead, this Grignard reagent is used, made from the same bromoketone, but with an acetal-forming step.

stable Grignard reagent

Acetals, as we stressed, are stable to base, and to basic nucleophiles such as Grignard reagents, so we no longer have a reactivity problem. Once the Grignard reagent has reacted with an electrophile, the ketone can be recovered by hydrolysing the acetal in dilute acid. The acetal is functioning here as

► Don't be confused by this statement! Acetal formation and hydrolysis are invariably carried out under thermodynamic control—what we mean here is that the equilibrium constant for acetal hydrolysis, which is a measure of rate of hydrolysis divided by rate of formation, turns out to be small because the rate of formation is large.

► **Dean Stark head**
When a mixture of toluene and water boils, the vapour produced is a constant ratio mixture of toluene vapour and water vapour known as an **azeotrope**. If this mixture is condensed, the liquid toluene and water, being immiscible, separate out into two layers with the water below. By using a Dean Stark apparatus, or Dean Stark head, the toluene layer can be returned to the reaction mixture while the water is removed. Reactions requiring removal of water by distillation are therefore often carried out in refluxing toluene or benzene under a Dean Stark head.

Acetals in nature

We showed you glucose as an example of a stable, cyclic hemiacetal. Glucose can, in fact, react with itself to form an acetal known as maltose.

glucose

maltose

cellulose

Maltose is a disaccharide (made of two sugar units) produced by the enzymatic hydrolysis of starch or cellulose, which are themselves polyacetals made up of a string of glucose units.

a **protecting group** because it protects the ketone from attack by the Grignard reagent. Protecting groups are extremely important in organic synthesis, and we will return to them in Chapter 24.

Amines react with carbonyl compounds

The ketone carbonyl group of pyruvic acid (or 2-oxopropanoic acid) has a stretching frequency of a typical ketone, 1710 cm^{-1}. When hydroxylamine is added to a solution of pyruvic acid, this stretching frequency slowly disappears. Later, a new IR absorption appears at 1400 cm^{-1}. What happens?

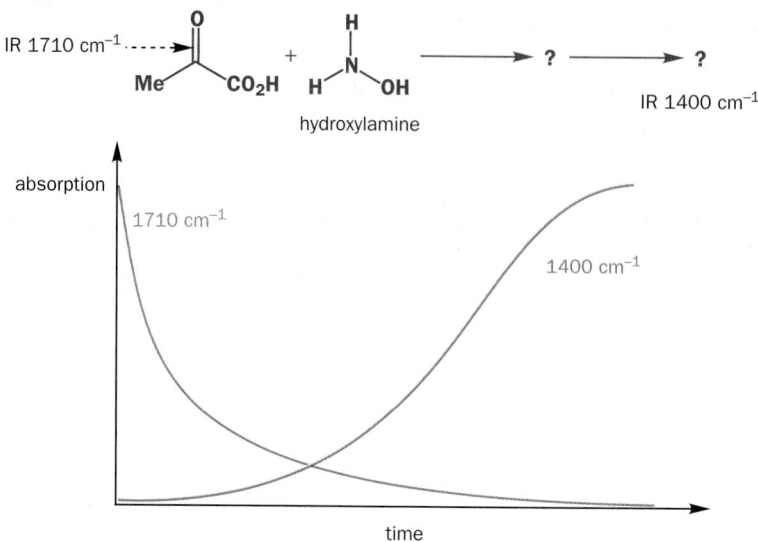

Well, you saw a diagram like this in the last chapter when we were discussing kinetic and thermo-dynamic products (p. 329) and you can probably also apply something of what you now know about the reactivity of carbonyl compounds towards nucleophiles to work out what is happening in this reaction between a carbonyl compound and an amine. The hydroxylamine first adds to the ketone to form an unstable intermediate similar to a hemiacetal.

intermediate formation

Notice that it is the more nucleophilic nitrogen atom, and not the oxygen atom, of hydroxylamine that adds to the carbonyl group. Like hemiacetals, these intermediates are unstable and can decompose by loss of water. The product is known as an **oxime** and it is this compound, with its C=N double bond, that is responsible for the IR absorption at 1400 cm^{-1}.

dehydration of the intermediate to give oxime

We know that the oxime is formed via an intermediate because the 1400 cm^{-1} absorption hardly appears until after the 1710 cm^{-1} absorption has almost completely gone. There must really be another curve to show the formation and the decay of the intermediate, the hemiacetal, just like the one in the last chapter (p. 329). The only difference is that the intermediate has no double bond to give an IR absorbance in this region of the spectrum. We come back to oximes later in the chapter.

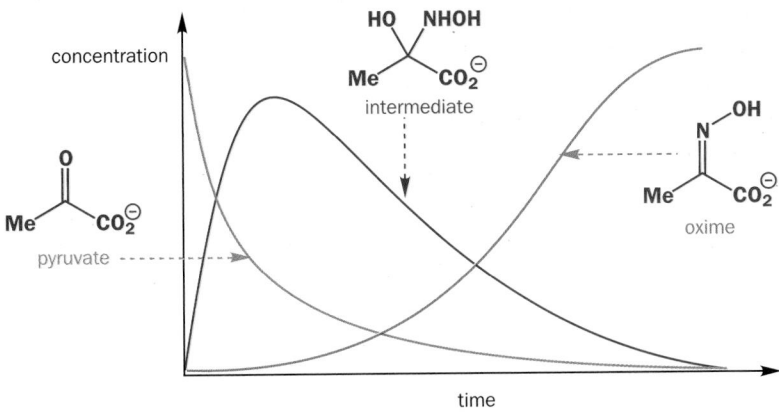

Imines are the nitrogen analogues of carbonyl compounds

In fact, the oxime formed from a ketone and hydroxylamine is just a special example of an imine.

Imines are formed when any primary amine reacts with an aldehyde or a ketone under appropriate conditions: for example, cyclohexylamine and benzaldehyde.

You shouldn't need us to tell you the mechanism of this reaction: even without looking at the mechanism we gave for the formation of the oxime it should come as no surprise to you by now. First, the amine attacks the aldehyde and the intermediate is formed. Dehydration gives the imine.

> ● **Imine formation requires acid catalysis.**

Notice that an acid catalyst is normally added for imine formation. Without an acid catalyst, the reaction is very slow, though in some cases it may still take place (oximes, for example, will form without acid catalysis, but form much faster with it). It's important to notice that acid is not needed for the addition step in the mechanism (indeed, protonation of the amine means that this step is very *slow* in strong acid), but *is* needed for the elimination of water later on in the reaction. Imine formation is in fact fastest at about pH 4–6: at lower pH, too much amine is protonated and the rate of the first step is slow; above this pH the proton concentration is too low to allow protonation of the OH leaving group in the dehydration step. Imine formation is like a biological reaction: it is fastest near neutrality.

variation of rate of imine formation with pH

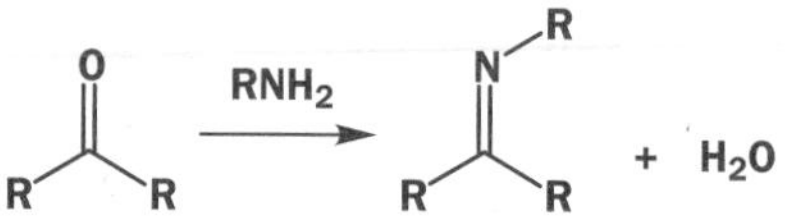

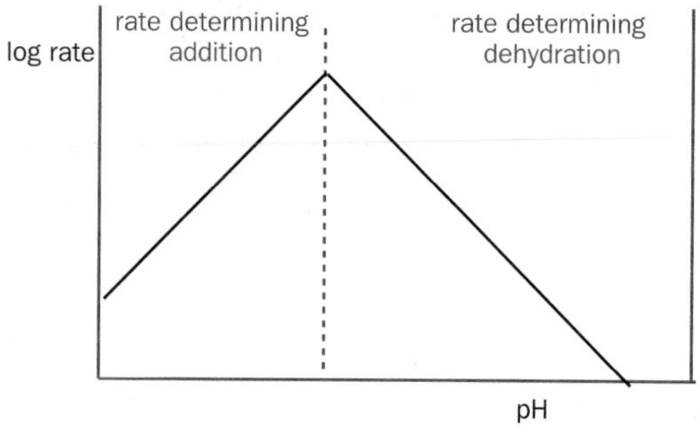

Either side of pH 5–6 the reaction goes more slowly. This is a sign of a change in rate-determining step. Where there is a choice between two rate-determining steps, the *slower* of the two determines the overall rate of the reaction. In the last chapter we saw that ester hydrolysis was a typical example of an organic reaction showing acid and base catalysis. It has a minimum rate at about neutrality showing that the mechanism must change. Where there is a choice of mechanism, the faster of the two operates. The contrast between the two is obvious from the diagrams.

variation of rate of ester hydrolysis with pH

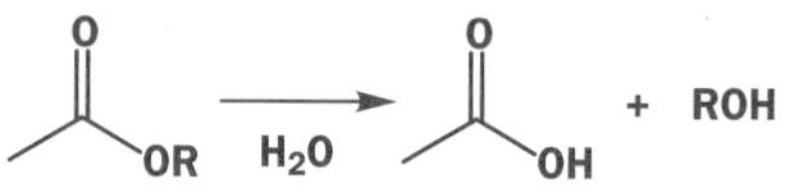

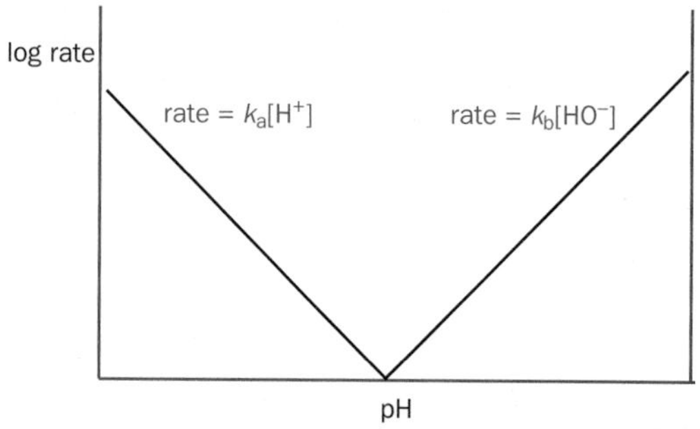

● **Multistep reaction rates**

The overall rate of a multistep reaction is decided by:

• The *faster* of two available mechanisms
• The *slower* of two rate-determining steps

Imines are usually unstable and are easily hydrolysed

Like acetals, imines are unstable with respect to their parent carbonyl compound and amine, and must be formed by a method that allows removal of water from the reaction mixture.

Because it is made from an unsymmetrical ketone, this imine can exist as a mixture of *E* and *Z* isomers, just like an alkene. When it is formed by this method, the ratio obtained is 8:1 *E:Z*. Unlike the geometrical isomers of alkenes, however, those of an imine are usually unstable and interconvert quite rapidly at room temperature. The geometrical isomers of oximes, on the other hand, are stable and can even be separated.

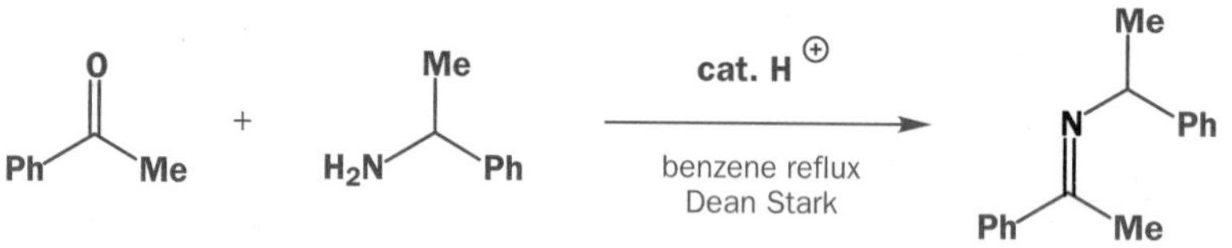

Imines are formed from aldehydes or ketones with most primary amines. In general, they are only stable enough to isolate if either the C or N of the imine double bond bears an aromatic substituent. Imines formed from ammonia are unstable, but can be detected in solution. $CH_2=NH_2$, for example, decomposes at temperatures above −80 °C, but PhCH=NH is detectable by UV spectroscopy in a mixture of benzaldehyde and ammonia in methanol.

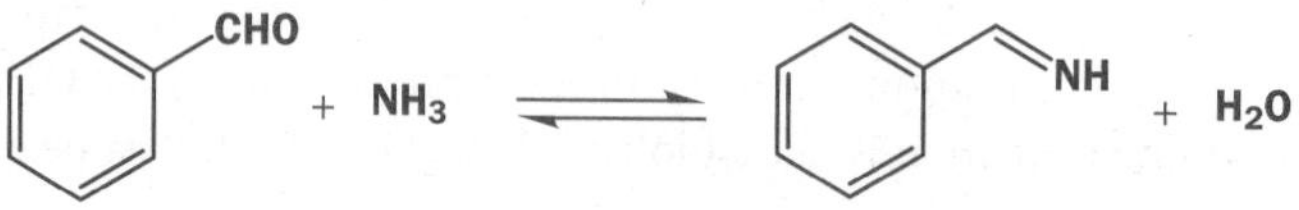

Imines are readily hydrolysed back to carbonyl compound and amine by aqueous acid—in fact, except for the particularly stable special cases we discuss below, most can be hydrolysed by water without acid or base catalysis. You have, in fact, already met an imine hydrolysis: at the end of Chapter 12 we talked about the addition of Grignard reagents to nitriles. The product is an imine that hydrolyses in acid solution to ketone plus ammonia.

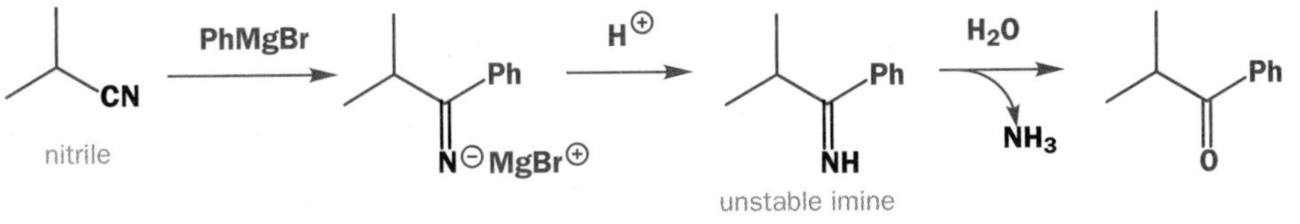

mechanism of the hydrolysis:

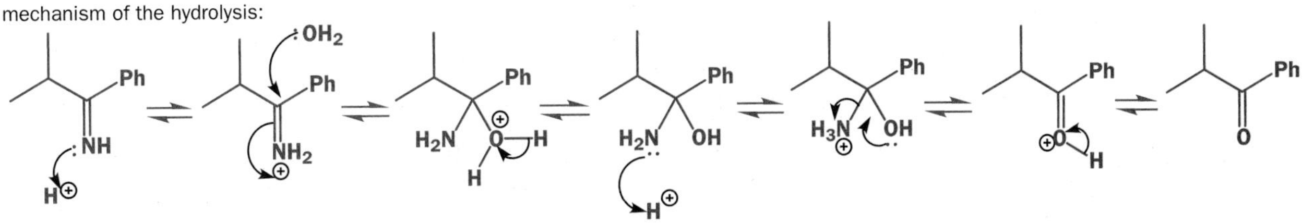

Some imines are stable

Imines in which the nitrogen atom carries an electronegative group are usually stable: examples include **oximes**, **hydrazones**, and **semicarbazones**.

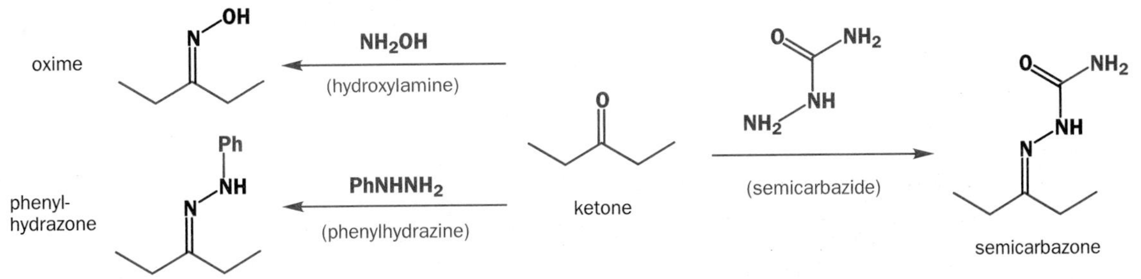

These compounds are more stable than imines because the electronegative substituent can participate in delocalization of the imine double bond. Delocalization decreases the δ+ charge on the carbon atom of the imine double bond and raises the energy of the LUMO, making it less susceptible to nucleophilic attack.

Oximes, hydrazones, and semicarbazones require acid or base catalysis to be hydrolysed.

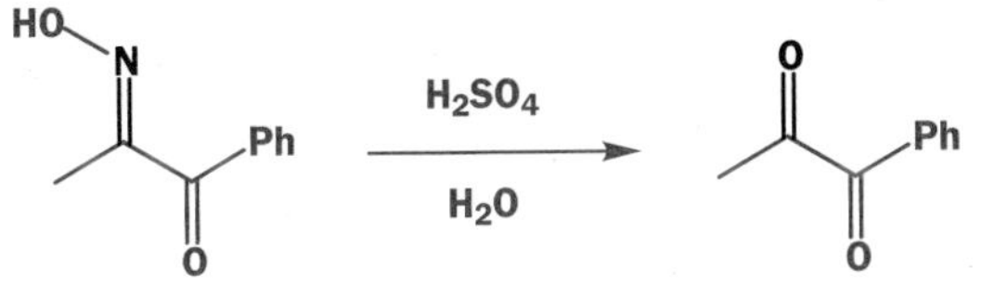

70% yield

Historical note

Because the hydrazone and semicarbazone derivatives of carbonyl compounds are often stable, crystalline solids, they used to be used to confirm the supposed identity of aldehydes and ketones. For example, the boiling points of these three isomeric five-carbon ketones are all similar, and before the days of NMR spectroscopy it would have been hard to distinguish between them.

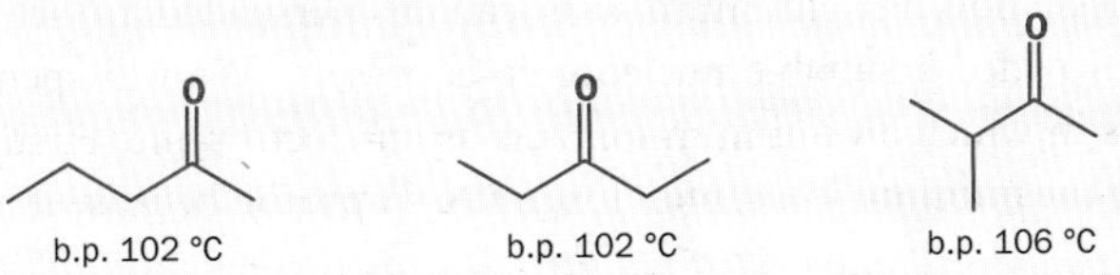

b.p. 102 °C   b.p. 102 °C   b.p. 106 °C

Historical note (continued)

Their semicarbazones and 2,4-dinitrophenylhydrazones, on the other hand, all differ in their melting points. By making these derivatives of the ketones, identification was made much easier. Of course, all of this has been totally superseded by NMR! However these crystalline derivatives are still useful in the purification of volatile aldehydes and ketones and in solving structures by X-ray crystallography.

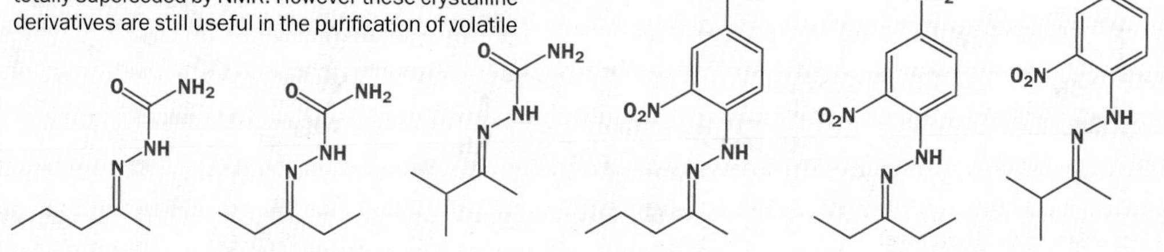

m.p. 112 °C　　m.p. 139 °C　　m.p. 157 °C　　m.p. 143 °C　　m.p. 156 °C　　m.p. 125 °C

Iminium ions and oxonium ions

Let's return to the mechanism of imine formation, and compare it for a moment with that of acetal formation. The only difference to begin with is that there is no need for acid catalysis for the addition of the amine but there is need for acid catalysis in the addition of the alcohol, a much weaker nucleophile.

acid-catalysed imine formation

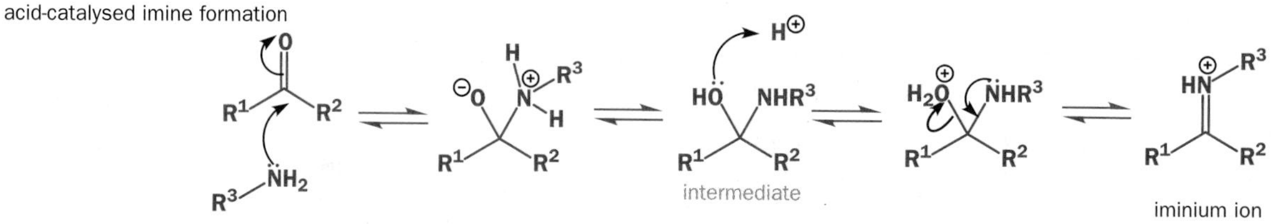

acid-catalysed acetal formation

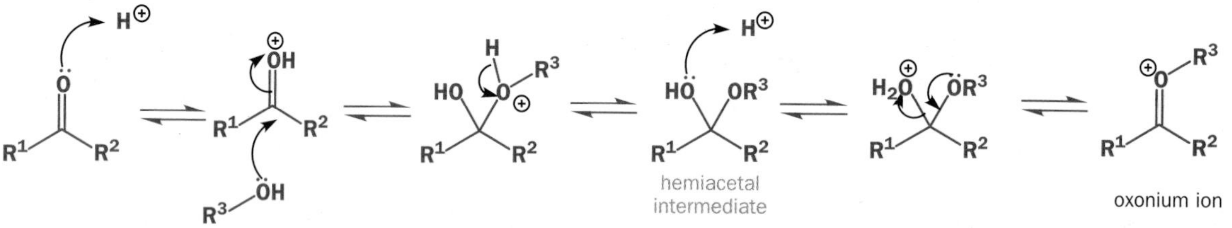

Up to this point, the two mechanisms follow a very similar path, with clear analogy between the intermediate and hemiacetal and the iminium and oxonium ion. Here, though, they diverge, because the iminium ion carries a proton, which the oxonium ion doesn't have. The iminium ion therefore acts as an acid, losing a proton to become the imine. The oxonium ion, on the other hand, acts as an electrophile, adding another molecule of alcohol to become the acetal.

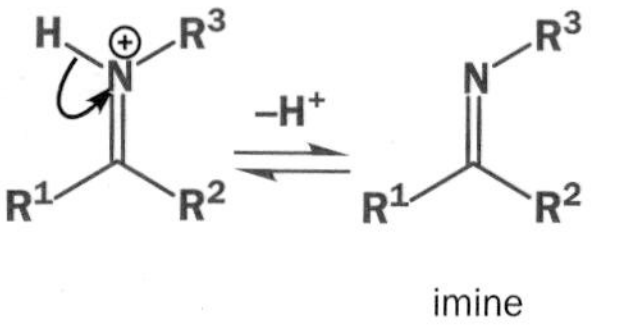

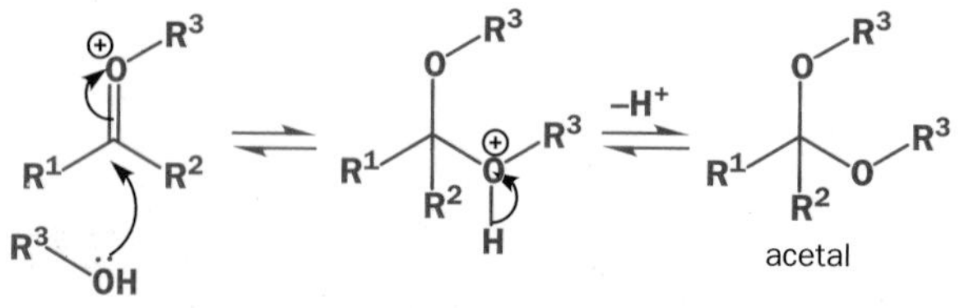

As you might guess, however, iminium ions can be persuaded to act as electrophiles, just like oxonium ions, provided a suitable nucleophile is present. We will spend the next few pages considering reactions in which an iminium ion acts as an electrophile. First, though, we will look at a reaction in which the iminium ion cannot lose an N–H proton because it has none.

Secondary amines react with carbonyl compounds to form enamines

Pyrrolidine, a secondary amine, reacts with isobutyraldehyde, under the sort of conditions you would use to make an imine, to give an **enamine**.

The name enamine combines 'ene' (C=C double bond) and 'amine'.

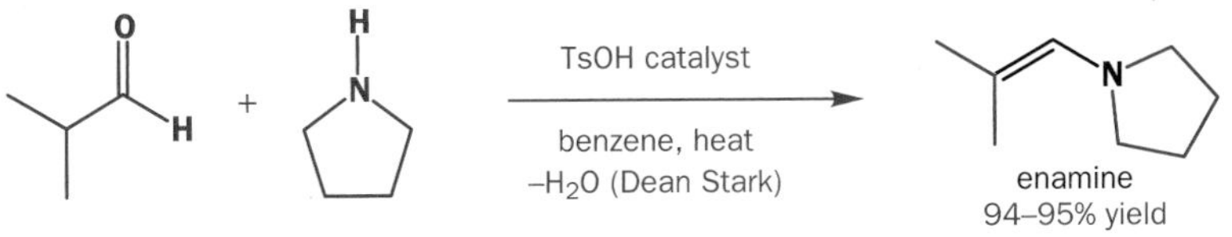

enamine
94–95% yield

The mechanism consists of the same steps as those that take place when imines form from primary amines, up to formation of the iminium ion. This iminium ion has no N–H proton to lose, so it loses one of the C–H protons next to the C=N to give the enamine. Enamines, like imines, are unstable to aqueous acid. We shall return to them in Chapter 21.

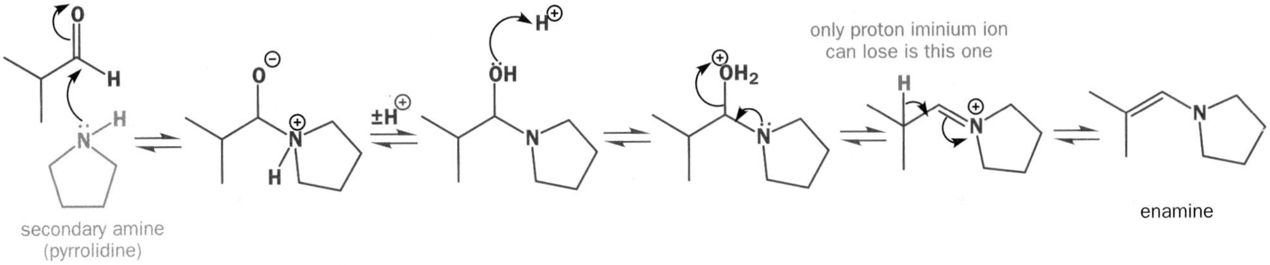

secondary amine
(pyrrolidine)

only proton iminium ion can lose is this one

enamine

● Imines and enamines

- Imines are formed from aldehydes or ketones with primary amines
- Enamines are formed from aldehydes or ketones with secondary amines
- Both require acid catalysis and removal of water

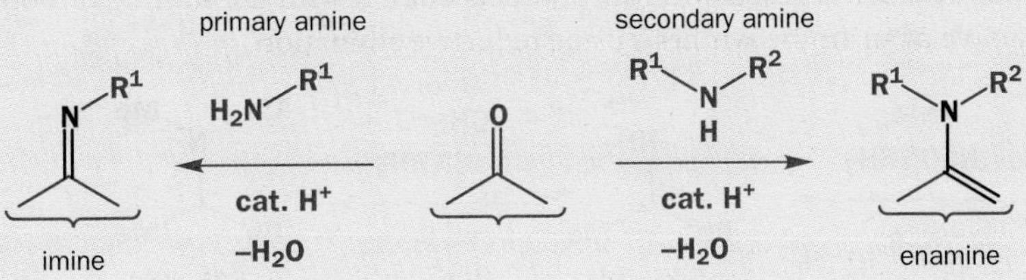

Enamines of primary amines, or even of ammonia, also exist, but only in equilibrium with an imine isomer. The interconversion between imine and enamine is the nitrogen analogue of **enolization**, which is discussed in detail in Chapter 21.

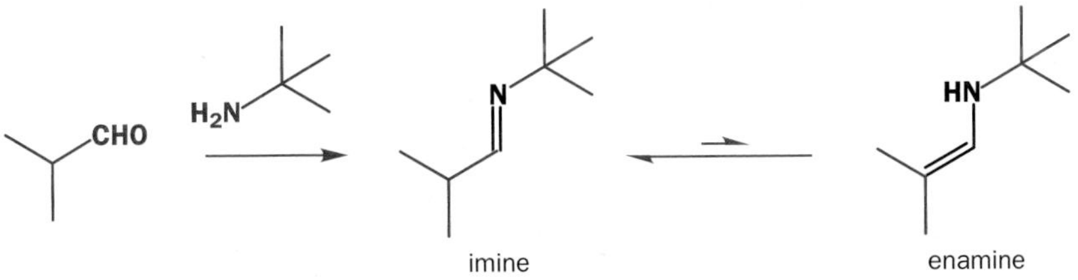

imine

enamine

Iminium ions can react as electrophilic intermediates

We made the point above that the difference in reactivity between an iminium ion and an oxonium ion is that an iminium ion can lose H$^+$ and form an imine or an enamine, while an oxonium ion reacts as an electrophile. Iminium ions can, however, react as electrophiles provided suitable nucleophiles are present. In fact, they are very good electrophiles, and are significantly more reactive than

▶ Sodium cyanoborohydride contains the cyanoborohydride anion, whose structure is

It is a 'toned down' version of sodium borohydride—the electron-withdrawing cyano group decreases the ease with which hydride is transferred.

carbonyl compounds. For example, iminium ions are reduced rapidly by the mild reducing agent sodium cyanoborohydride ($NaCNBH_3$), while carbonyl compounds are not.

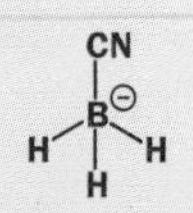

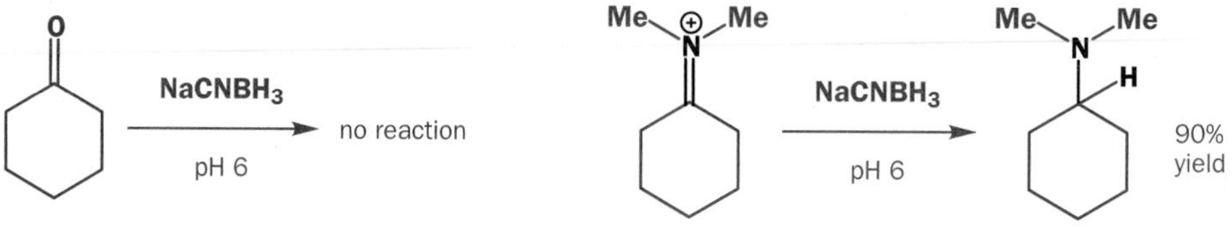

An alternative to $NaCNBH_3$ is $NaBH(OAc)_3$ (sodium triacetoxy-borohydride)—somewhat safer because strong acid can release HCN from $NaCNBH_3$.

Amines from imines: reductive amination

A useful way of making amines is by reduction of imines (or iminium ions). This overall process, from carbonyl compound to amine, is called **reductive amination.** This is, in fact, one of the few successful ways, and the best way, of making secondary amines. This should be your first choice in amine synthesis.

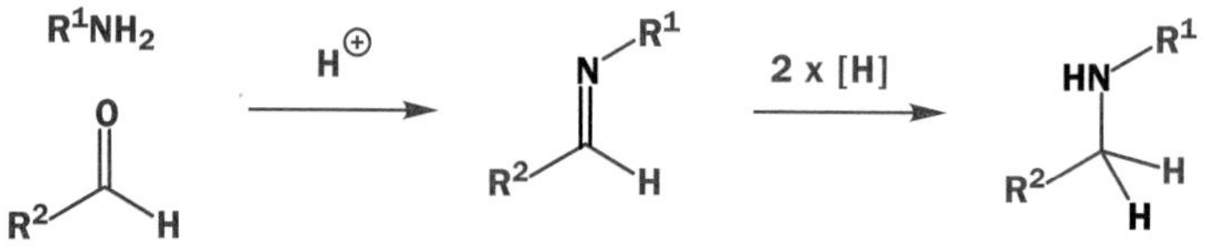

This can be done in two steps, provided the intermediate is stable, but, because the instability of many imines makes them hard to isolate, the most convenient way of doing it is to form and reduce the imine in a single reaction. The selective reduction of iminium ions (but not carbonyl compounds) by sodium cyanoborohydride makes this possible. When $NaCNBH_3$ is added to a typical imine-formation reaction it reacts with the products but not with the starting carbonyl compound. Here is an example of an amine synthesis using reductive amination.

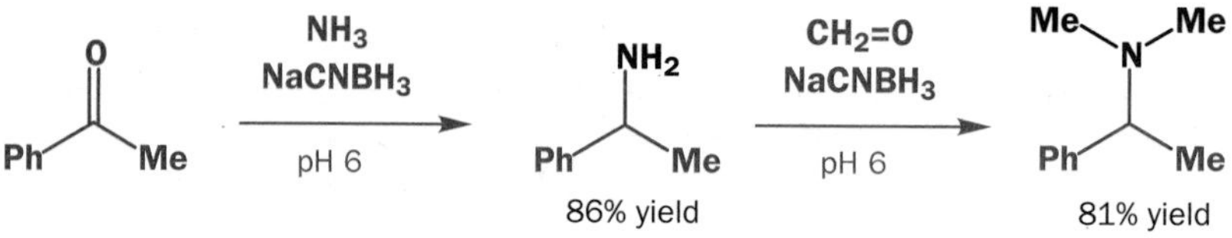

In the first step, the ketone and ammonia are in equilibrium with their imine, which, at pH 6, is partly protonated as an iminium ion. The iminium ion is rapidly reduced by the cyanoborohydride to give the amine. Reactions like this, using ammonia in a reductive amination, are often carried out with ammonium chloride or acetate as convenient sources of ammonia. At pH 6, ammonia will be mostly protonated anyway.

■ You will again meet the highly electrophilic iminium ions produced by reaction of formaldehyde with amines in Chapter 27, where we introduce you to the Mannich reaction.

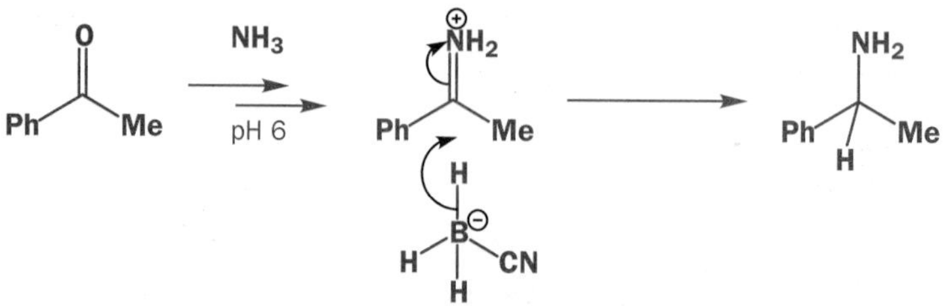

In the second step of the synthesis, amine plus formaldehyde gives an imine, present as its protonated iminium form, which gets reduced. Formaldehyde is so reactive that it reacts again with the secondary amine to give an iminium ion; again, this is reduced to the amine.

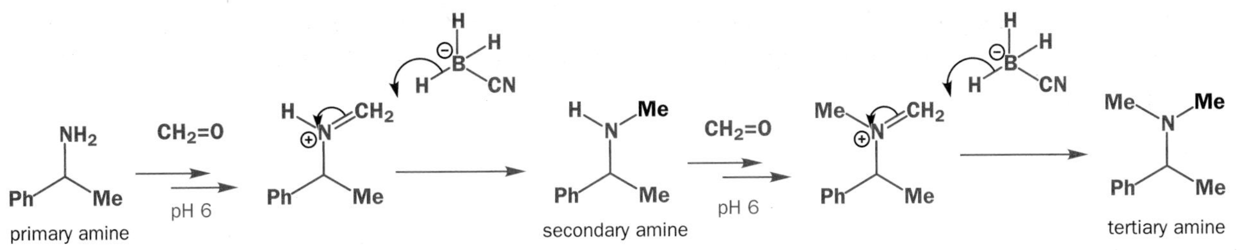

Living things make amino acids using imines

The amino acid alanine can be made in moderate yield in the laboratory by reductive amination of pyruvic acid.

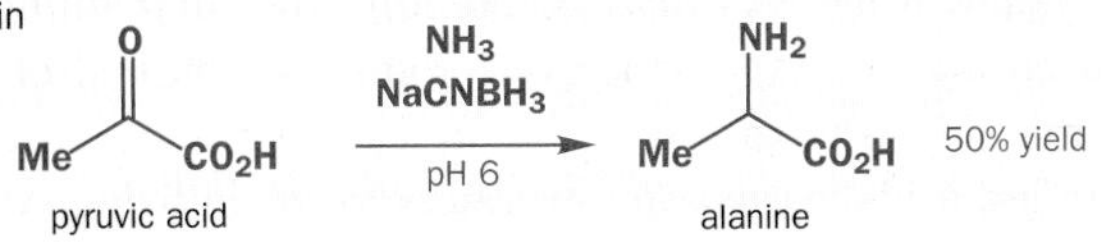

Living things use a very similar reaction to manufacture amino acids from keto acids—but do it much more efficiently. The key step is the formation of an imine between pyruvic acid and the vitamin B$_6$-derived amine pyridoxamine.

Nature's synthesis of alanine:

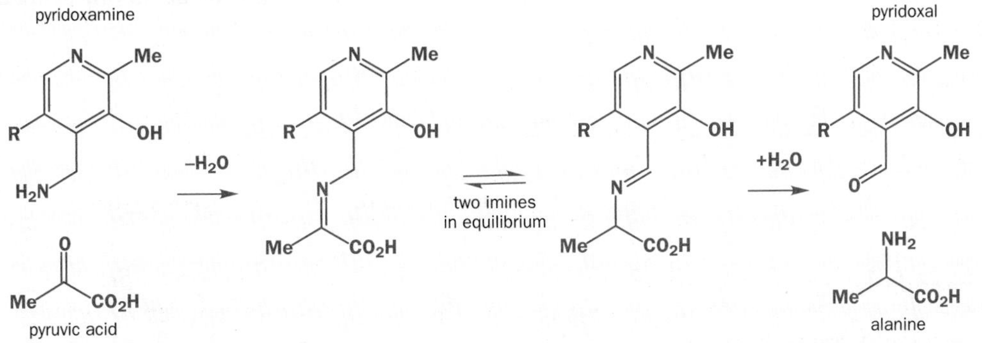

This imine (biochemists call imines **Schiff bases**) is in equilibrium with an isomeric imine, which can be hydrolysed to pyridoxal and alanine. These reactions are, of course, all controlled by enzymes, and coupled to the degradation of unwanted amino acids (the latter process

converts the pyridoxal back to pyridoxamine). Nature was doing reductive aminations a long time before sodium cyanoborohydride was invented! We will come back to this in Chapter 50.

An alternative method for reductive amination uses hydrogenation (hydrogen gas with a metal catalyst) to reduce the imine in the presence of the carbonyl compound.

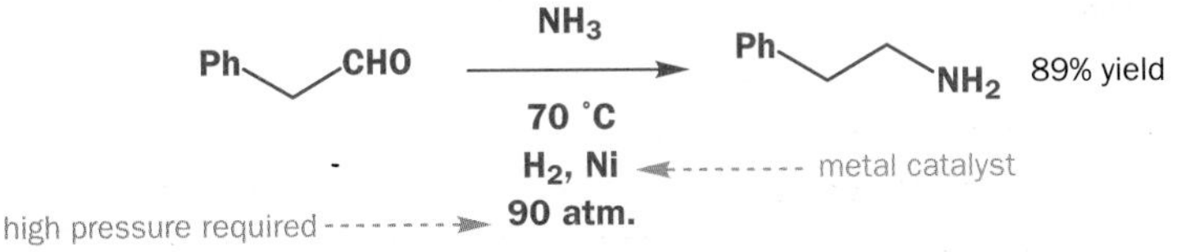

high pressure required

> Hydrogenation is a good way of reducing a number of different functional groups, but not (usually) carbonyl groups. In Chapter 24 we will look in more detail at reducing agents (and other types of reagent) that demonstrate selectivity for one functional group over another (**chemoselectivity**).

Lithium aluminium hydride reduces amides to amines

We've talked about reduction of iminium ions formed from carbonyl compounds plus amines. Iminium ions can also be formed by reducing amides with lithium aluminium hydride. A tetrahedral intermediate is formed that collapses to the iminium ion.

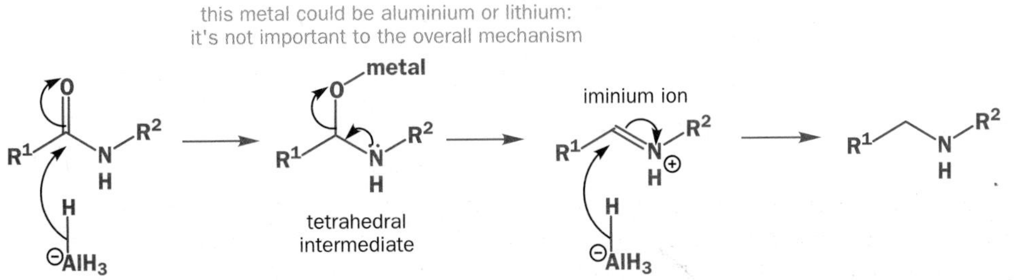

The iminium ion, is of course, more electrophilic than the starting amides (amide carbonyl groups are about the least electrophilic of any!), so it gets reduced to the secondary amine. This reaction can be used to make secondary amines, from primary amines and acyl chlorides.

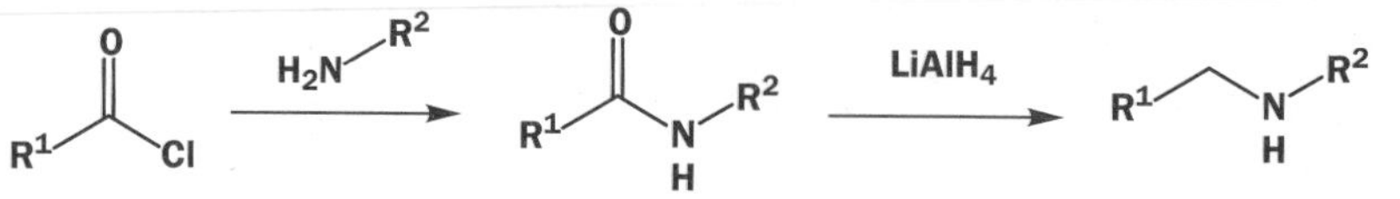

Cyanide will attack iminium ions: the Strecker synthesis of amino acids

Cyanide will react with iminium ions to form α amino nitriles. Although these compounds are relatively unimportant in their own right, a simple hydrolysis step produces α amino acids. This route to amino acids is known as the Strecker synthesis. Of course, it's not usually necessary to make the amino acids that Nature produces for us in living systems: they can be extracted from hydrolysed proteins.

This Strecker synthesis is of phenylglycine, an amino acid not found in proteins. Cyanide reacts more rapidly with the iminium ion generated in the first step than it does with the starting benzaldehyde.

▶

Make sure that you can write a mechanism for the hydrolysis of the nitrile to the carboxylic acid! (If you need reminding, it is given in Chapter 12)

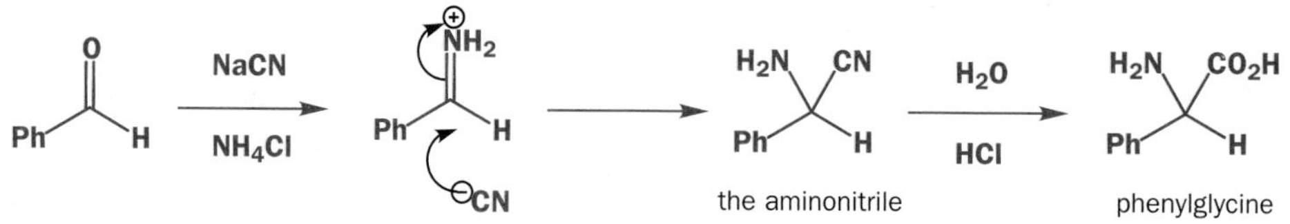

The synthesis of a spider toxin: reductive amination

This compound is the toxin used by the orb weaver spider to paralyse its prey:

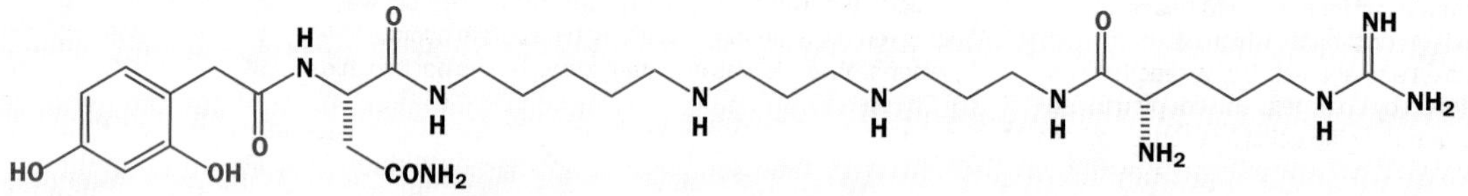

Since the spider produces only minute quantities of the compound, chemists at the University of Bath set about synthesizing it in the laboratory so that they could study its biological properties. The toxin contains several amide and amine functional groups, and the chemists decided that the best way to make it was to link two molecules together at one of the secondary amine groups using a reductive amination.

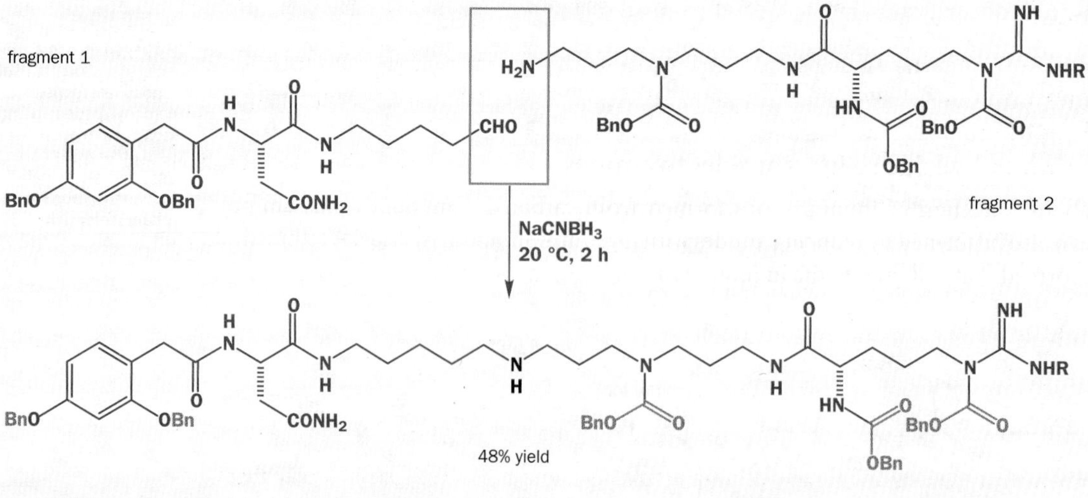

The compound made by this reaction has almost, but not exactly, the spider toxin structure. The extra groups in brown are protecting groups, and prevent unwanted side-reactions at the other amine and phenol functional groups. We will discuss protecting groups in detail in Chapters 24 and 25.

Substitution of C=O for C=C: a brief look at the Wittig reaction

Before we leave substitution reactions of carbonyl groups, there is one more reaction that we must introduce. It is an important one, and we will come back to it again later in this book, particularly in Chapter 31. It also has a rather different mechanism from most you have met in recent chapters, but we talk about it here because the overall consequence of the **Wittig reaction** is the substitution of a C=C bond for a C=O bond.

We don't normally tell you the name of a reaction before even mentioning how to do it, but here we make an exception because the reagents are rather unusual and need explaining in detail. The Wittig reaction is a reaction between a carbonyl compound (aldehyde or ketone only) and a species known as a **phosphonium ylid**. An ylid (or ylide) is a species with positive and negative charges on adjacent atoms, and a phosphonium ylid carries its positive charge on phosphorus. Phosphonium ylids are made from **phosphonium salts** by deprotonating them with a strong base.

The Wittig reaction is named after its discoverer, the Nobel Prize winner Georg Wittig (1897–1987; Nobel Prize 1979).

a phosphonium salt   methyltriphenylphosphonium bromide: an example of a phosphonium salt   phosphonium ylid

You have already met phosphonium salts in Chapter 5 where you saw the reaction of a phosphine (triphenylphosphine) with an alkyl halide (methyl iodide).

triphenylphosphine   phosphonium salt

So, here is a typical Wittig reaction: it starts with a phosphonium salt, which is treated with sodium hydride, and then with a carbonyl compound; the alkene forms in 85% yield.

a Wittig reaction

phosphonium salt   phosphonium ylid   alkene, 85% yield

Sodium hydride, Na^+H^-, is the conjugate base of H_2, and has a pK_a of about 35.

What about the mechanism? We warned you that the mechanism is rather different from all the others you have met in this chapter, but nonetheless it begins with attack on the carbonyl group by a nucleophile; the nucleophile is the carbanion part of the phosphonium ylid. This reaction generates a negatively charged oxygen that attacks the positively charged phosphorus and gives a four-membered ring called an **oxaphosphetane**.

formation of the four-membered ring

Now, this four-membered ring (like many others) is unstable, and it can collapse in a way that forms two double bonds. Here are the curly arrows: the mechanism is cyclic, and gives the alkene, which is the product of the reaction along with a **phosphine oxide**.

decomposition of
the four-membered ring

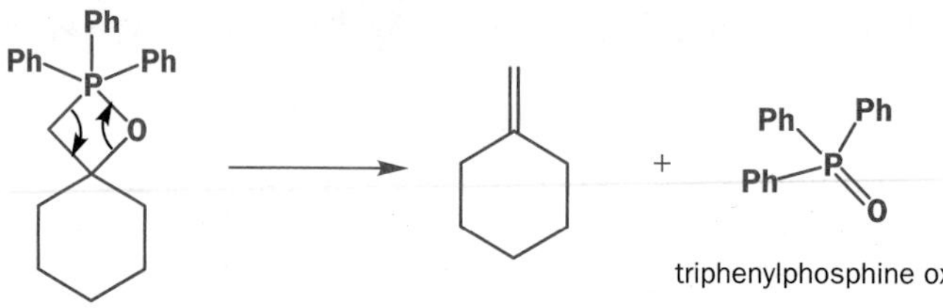

triphenylphosphine oxide

The chemistry of some elements is dominated by one particular property, and a theme running right through the chemistry of phosphorus is its exceptional affinity for oxygen. The P=O bond, with its bond energy of 575 kJ mol^{-1}, is one of the strongest double bonds in chemistry, and the Wittig reaction is irreversible and is driven forward by the formation of this P=O bond. No need here for the careful control of an equilibrium necessary when making acetals or imines. We will look at the Wittig reaction again in more detail in Chapter 31.

Summary

In this chapter, as in Chapter 12, you have met a wide variety of reactions, but we hope you have again been able to see that they are all related mechanistically. Of course, we have not been exhaustive: it would be impossible to cover every possible reaction of a carbonyl group, but having read Chapters 6, 9, 12, and 13 you should feel confident in writing a reasonable mechanism for any reaction involving nucleophilic attack on a carbonyl group. You could try thinking about this, for example.

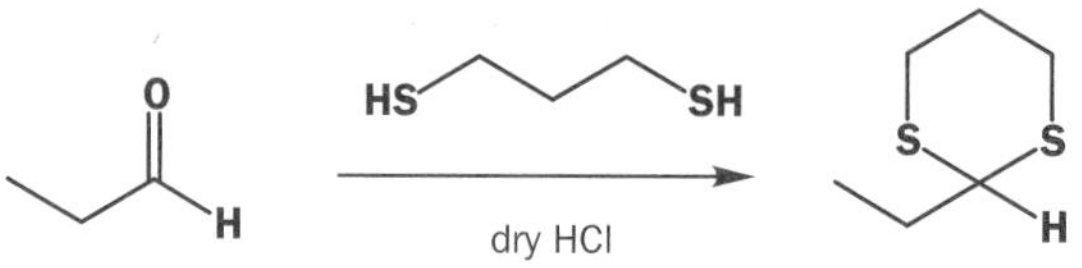

Hint. Consider sulfur's location in the periodic table.

We now take our leave of carbonyl groups until Chapter 21 when we reveal a hidden side to their character: they can be nucleophilic as well as electrophilic. Meanwhile, we shall look some more at NMR spectroscopy and what it can tell us, before applying some of the principles we've used to explain carbonyl reactions to a new type of reaction, substitution at a saturated carbon atom.

Problems

1. In the cyclization of the open-chain form of glucose to form the stable hemiacetal, it may be difficult to work out what has happened. Number the carbon atoms in the open-chain form and put the same numbers on the hemiacetal so that you can see where each carbon atom has gone. Then draw a mechanism for the reaction.

2. Draw mechanisms for these reactions, which involve the loss of carbonyl oxygen.

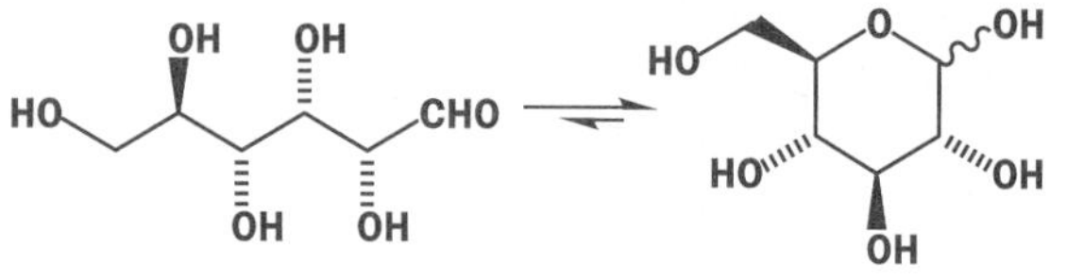

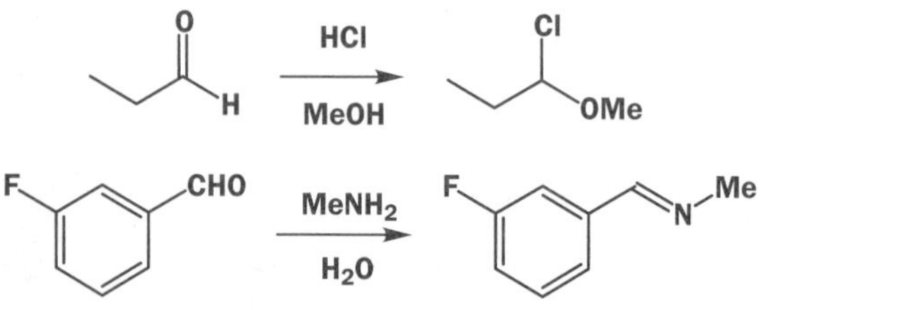

3. Each of these molecules is an acetal, that is, a compound made from an aldehyde or ketone and two alcohol groups. Which compounds were used to make these acetals?

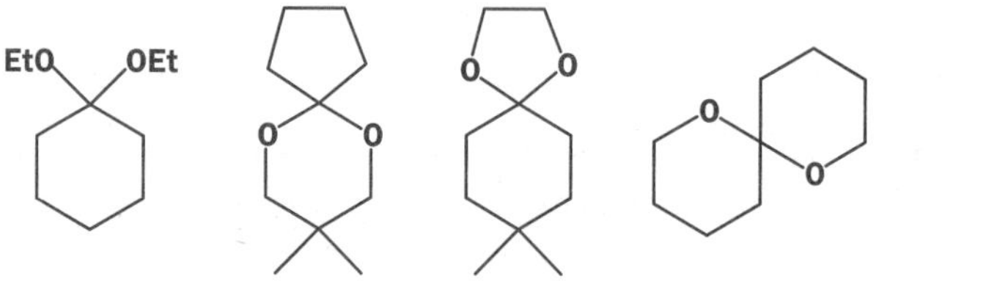

4. Each of these reactions leads to an acetal or a closely related compound and yet no alcohols are used in the first two reactions and no carbonyl group in the third. How are these acetals formed?

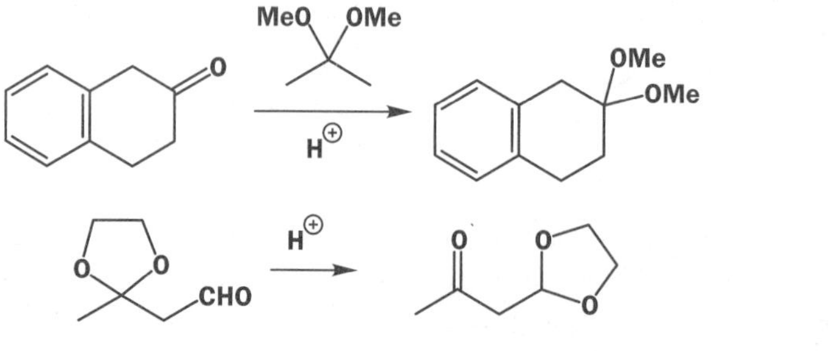

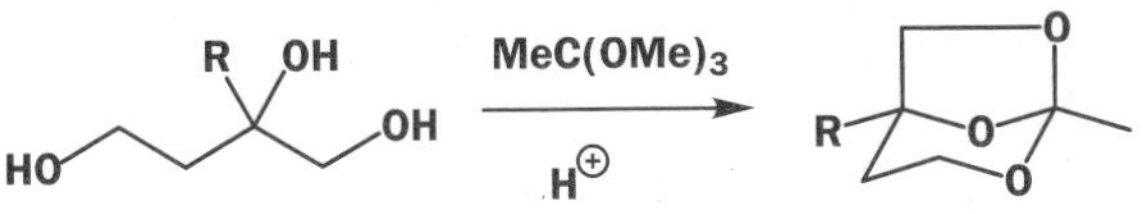

In the first and third of these two reactions, a compound, different in each case, must be distilled from the reaction mixture if the reaction is to go to completion. What are the compounds and why is this necessary? In the second case, why does the reaction go in this direction?

5. Suggest mechanisms for these two reactions of the smallest aldehyde, formaldehyde (methanal, $CH_2=O$).

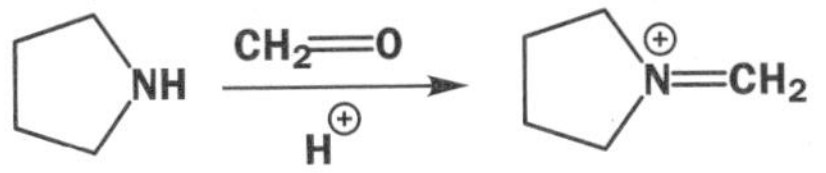

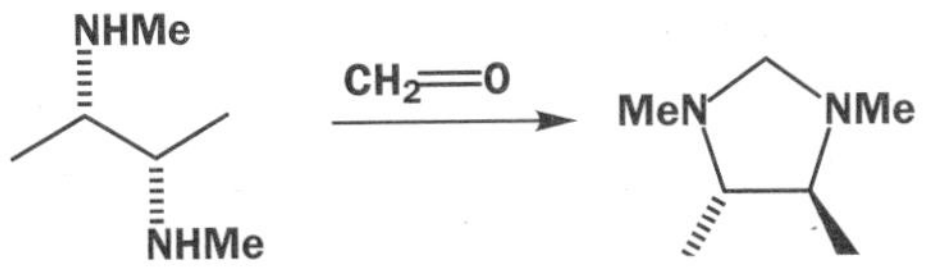

Comment on the stereochemistry of the second example.

6. Suggest mechanisms for this reaction. It first appeared in Chapter 3 where we identified the rather unexpected product from its spectra but did not attempt to draw a mechanism for the reaction.

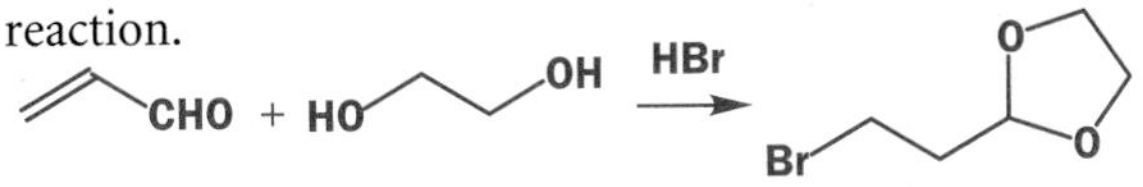

7. In Chapter 6 we described how the antileprosy drug dapsone could be made soluble by the formation of a 'bisulfite adduct'. Now that you know about the reactions described in Chapter 14, you should be able to draw a mechanism for this reaction. The adduct is described as a 'pro-drug' meaning that it can give dapsone itself in the human body. How might this happen?

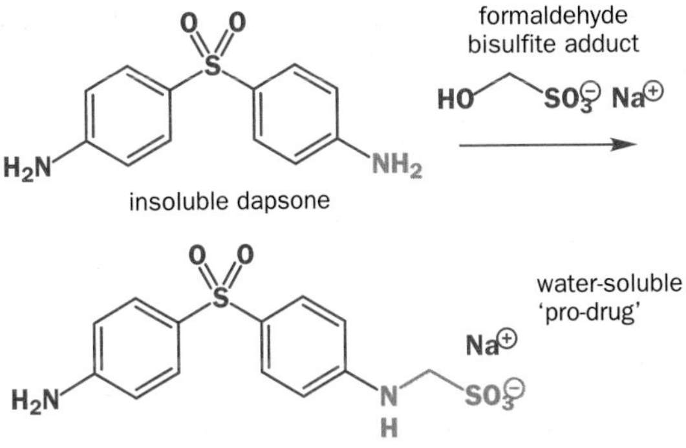

8. Suggest a detailed mechanism for the acetal exchange used in this chapter to make an acetal of a ketone from an orthoester.

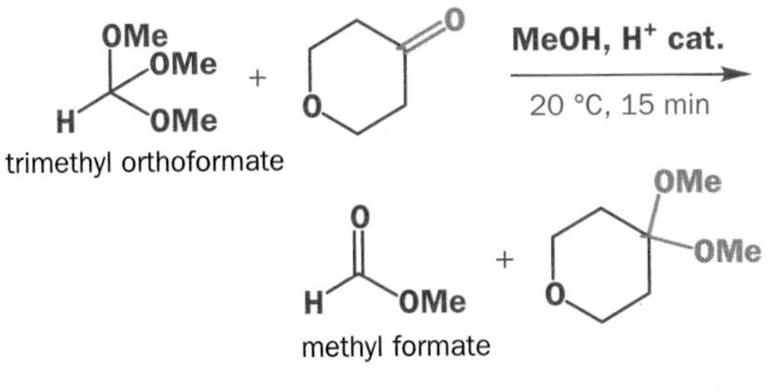

9. When we introduced cyclic acetals, we showed you this reaction.

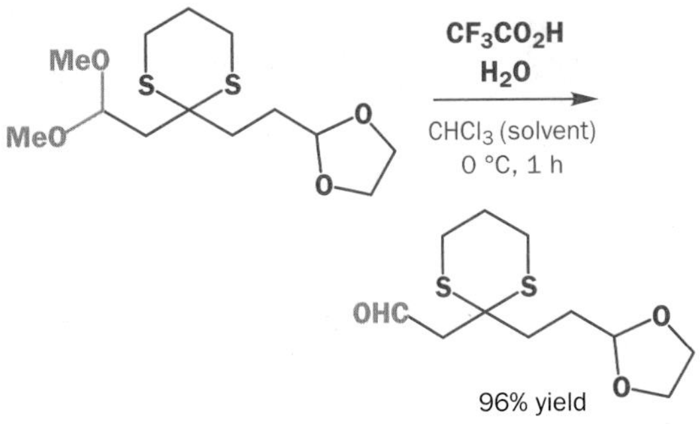

What are the two functional groups not affected by this reaction? How would you hydrolyse them?

10. What would actually happen if you tried to make the unprotected Grignard reagent shown here?

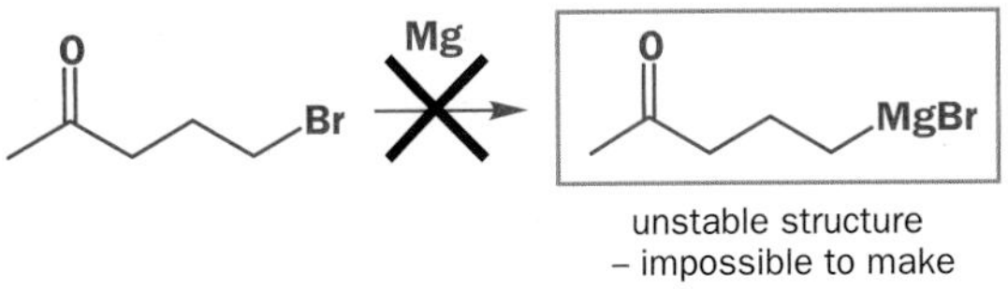

11. Find the acetals in cellulose.

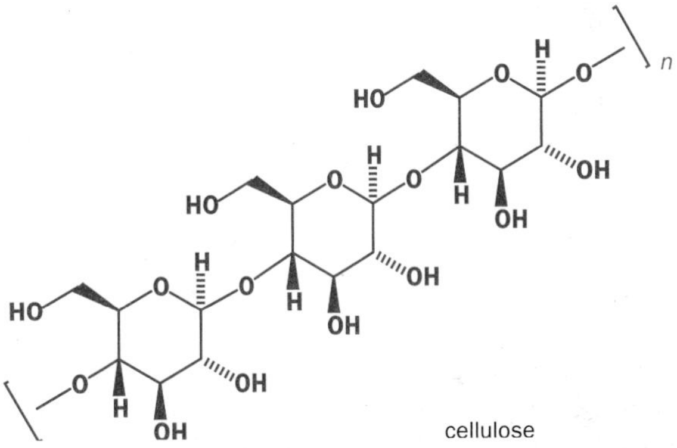

12. A stable product can be isolated from the reaction between benzaldehyde and ammonia discussed in this chapter. Suggest a mechanism for its formation.

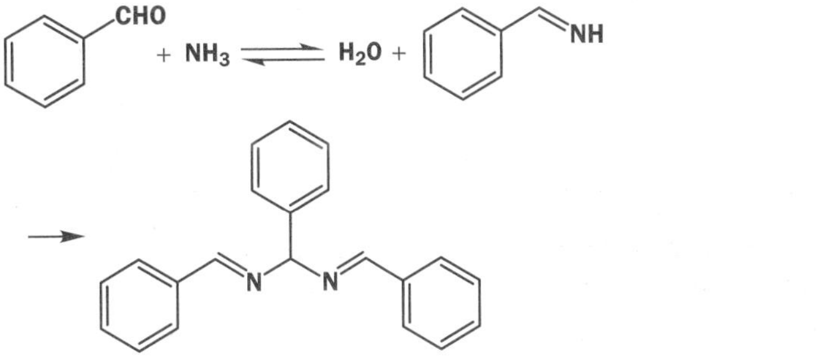

13. Suggest mechanisms for these reactions.

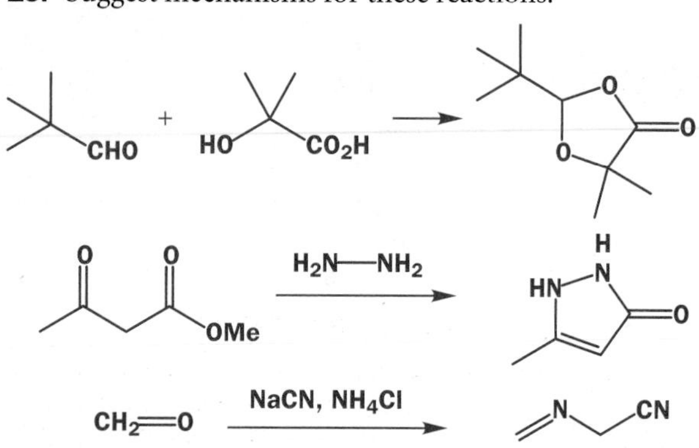

14. Finally, don't forget the problem at the end of the chapter: suggest a mechanism for this reaction.

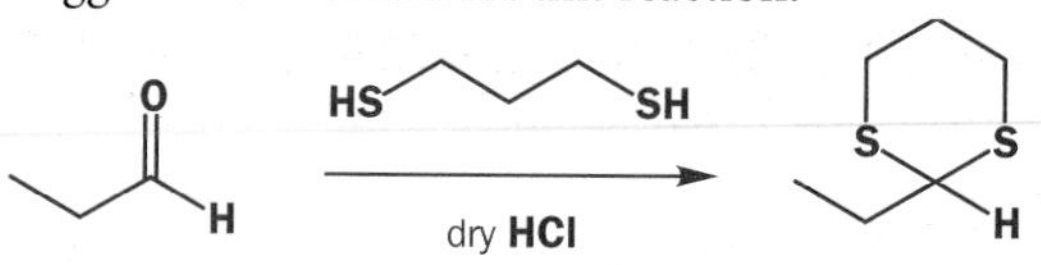

Review of spectroscopic methods

15

Connections

Building on:

- Mass spectrometry ch3
- Infrared spectroscopy ch3
- ^{13}C NMR ch3
- ^{1}H NMR ch11

Arriving at:

- How spectroscopy explains the reactions of the C=O group
- How spectroscopy tells us about the reactivity of, and reaction products from, conjugated C=C and C=O bonds
- How spectroscopy tells us about the size of rings
- How spectroscopy solves the structure of unknown compounds
- Some guidelines for solving unknown structures

Looking forward to:

- A final review of spectroscopy, including what it tells us about the stereochemistry of molecules ch32
- Spectroscopy is an essential tool and will be referred to throughout the rest of the book

This is the first of two review chapters on spectroscopic methods taken as a whole. In Chapter 32 we shall tackle the complete identification of organic compounds including the vital aspect of stereochemistry, introduced in Chapters 16 and 19. In this chapter we gather together some of the ideas introduced in previous chapters on spectroscopy and mechanism and show how they are related. We shall explain the structure of the chapter as we go along.

There are three reasons for this chapter

1. To review the methods of structure determination we met in Chapters 3 and 11, to extend them a little further, and to consider the relationships between them

2. To show how these methods may be combined to determine the structure of unknown molecules

3. To provide useful tables of data for you to use when you are yourself attempting to solve structure determination problems

The main tables of data appear at the end of the chapter so that they are easy to refer to when you are working on problems. You may also wish to look at them, along with the tables in the text, as you work through this chapter.

We shall deal with points 1 and 2 together, looking first at the interplay between the chemistry of the carbonyl group (as discussed in Chapters 12 and 14) and spectroscopy, solving some structural problems, then moving on to discuss, for example, NMR of more than one element in the same compound, doing some more problems, and so on. We hope that the lessons from each section will help in your overall understanding of structure solving. The first section deals with the assignment of carbonyl compounds to their various classes.

> ▶
>
> May we remind you that you are *not* intended to *learn* the numbers!

Does spectroscopy help with the chemistry of the carbonyl group?

As you can guess from the question, it does! Chapters 12 and 14 completed our systematic survey of carbonyl chemistry, the main chemical theme of the book so far (see also Chapters 6, 9, and 10), so this is an appropriate point to put together chemistry and spectroscopy on this most important of all functional groups.

We have divided carbonyl compounds into two main groups.

1 aldehydes (RCHO) and **ketones** ($R^1CO \cdot R^2$)
2 acids (RCO_2H) and their derivatives (in order of reactivity):
   acid chlorides (RCOCl)
   anhydrides (RCO_2COR)
   esters ($R^1CO_2R^2$)
   amides ($RCONH_2$, R^1CONMe_2, etc.)

Which spectroscopic methods most reliably distinguish these two groups? Which help us to separate aldehydes from ketones? Which allow us to distinguish the various acid derivatives? Which offer the most reliable evidence on the chemistry of the carbonyl group? These are the questions we tackle in this section.

Distinguishing aldehydes and ketones from acid derivatives

The most consistently reliable method for doing this is ^{13}C NMR.

> ● **^{13}C NMR distinguishes acid derivatives from aldehydes and ketones**
>
> The carbonyl carbons of all aldehydes and ketones resonate at about 200 p.p.m., while acid derivatives usually resonate at about 175 p.p.m.

It doesn't much matter whether the compounds are cyclic or unsaturated or have aromatic substituents; they all give carbonyl ^{13}C shifts in about the same regions. There is a selection of examples on the facing page which we now discuss. First, look at the shifts arrowed in to the carbonyl group on each structure. All the aldehydes and ketones fall between 191 and 208 p.p.m. regardless of structure, whereas all the acid derivatives (and these are very varied indeed!) fall between 164 and 180 p.p.m. These two sets do not overlap and the distinction is easily made. Assigning the spectrum of the keto-acid in the margin, for example, is easy.

The distinction can be vital in structural problems. The symmetrical alkyne diol below cyclizes in acid with Hg(II) catalysis to a compound having, by proton NMR, the structural fragments shown. The product is unsymmetrical in that the two CMe_2 groups are still present, but they are now different. In addition, the chemical shift of the CH_2 group shows that it is next to C=O but not next to oxygen. This leaves us with two possible structures. One is an ester and one a ketone. The C=O shift is 218.8 p.p.m. and so there is no doubt that the second structure is correct.

^{13}C NMR shifts of carbonyl groups

Carbonyl group	δ_C, p.p.m.
aldehydes	195–205
ketones	195–215
acids	170–185
acid chlorides	165–170
acid anhydrides	165–170
esters	165–175
amides	165–175

208.4   179.1

saturated keto-acid

a reaction with an unknown product

starting material $C_8H_{14}O_2$

Hg^{2+}, H$^\oplus$

the product is an isomer of $C_8H_{14}O_2$

^{1}H NMR shows these fragments:

[CMe$_2$, CMe$_2$, C=O, O, CH$_2$]

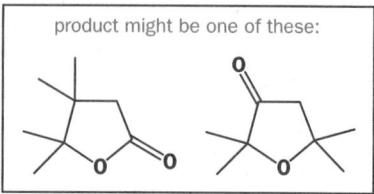

product might be one of these:

> You need not, at this stage, worry about *how* the reaction works. It is more important that you realize how spectroscopy enables us to work out *what* has happened even before we have any idea *how*. Nonetheless, it is true that the second structure here also makes more sense chemically as the carbon skeleton is the same as in the starting material.

Distinguishing aldehydes from ketones is simple by proton NMR

Now look at the first two groups, the aldehydes and ketones. The two aldehydes have smaller carbonyl shifts than the two ketones, but they are too similar for this distinction to be reliable. What distinguishes the aldehydes very clearly is the characteristic proton signal for CHO at 9–10 p.p.m. So you should identify aldehydes and ketones by C=O shifts in carbon NMR and then separate the two by proton NMR.

> ● **Aldehyde protons are characteristic**
>
> A proton at 9–10 p.p.m. indicates an aldehyde.

Identifying acid derivatives by carbon NMR is difficult

Now examine the other panels on p. 363. The four carboxylic acids are all important biologically or medicinally. Their C=O shifts are very different *from each other* as well as from those of the aldehydes or ketones.

aldehydes

aromatic aldehyde – vanillin   conjugated unsaturated aldehyde – all *trans* retinal

ketones

cyclic conjugated ketone– (–)-carvone   saturated ketone– raspberry ketone

acids

saturated: lipoic (thioctic) acid   conjugated: shikimic acid   aromatic: salicylic acid   nonconjugated: ibuprofen

acid chlorides

saturated: acetyl chloride   conjugated unsaturated

anhydrides

saturated cyclic   unsaturated conjugated cyclic: maleic anhydride   saturated cage tricyclic

esters

conjugated: methyl methacrylate   ester of aromatic acid: benzocaine

amides

simple amide: dimethyl formamide (DMF)   tetrapeptide: L-Ala-L-Ala-L-Ala-L-Ala   four C=O signals: 168.9, 171.6, 171.8, 173.8

Aldehydes and ketones

The first aldehyde is vanillin which comes from the vanilla pod and gives the characteristic vanilla flavour in, for example, ice cream. Vanilla is the seed pod of a South American orchid. 'Vanilla essence' is made with synthetic vanillin and tastes slightly different because the vanilla pod contains other flavour components in small quantities. The second aldehyde is retinal. As you look at this structure your eyes use the light reaching them to interconvert *cis* and *trans* retinal in your retina to create nervous impulses. (See also Chapter 31.)

The two ketones are all flavour compounds too. The first, (–)-carvone, is the chief component (70%) of spearmint oil. Carvone is an interesting compound: in Chapter 16 you will meet mirror-image isomers known as enantiomers, and (–)-carvone's mirror image (+)-carvone, is the chief component (35%) of dill oil. Our taste can tell the difference, though an NMR machine can't and both carvones have *identical NMR spectra*. See Chapter 16 for more detail! The second ketone is 'raspberry ketone' and is largely responsible for the flavour of

raspberries. It is entirely responsible for the flavour of some 'raspberry' foods. The signal for the aromatic carbon joined to OH is at 154.3 p.p.m. (in the 100–150 p.p.m. region because it is an unsaturated carbon atom joined to oxygen) and cannot possibly be confused with the ketone signal at 208.8 p.p.m. Both ketones have C=O shifts at about 200 p.p.m., and both lack any signals in the proton NMR of $\delta > 8$.

Acid derivatives

Lipoic acid uses its S–S bond in redox reactions (Chapter 50), while shikimic acid is an intermediate in the formation of compounds with benzene rings, such as phenylalanine, in living things (Chapter 49). Salicylic acid's ethyl ester is aspirin, which is, of course, like the last example ibuprofen, a painkiller.

The first acid chloride is a popular reagent for the synthesis of acetate esters and you have seen its reactions in Chapter 12. We used the other as an

example in Chapter 11. We have chosen three cyclic anhydrides as examples because they are all related to an important reaction (the Diels–Alder reaction), which you will meet in Chapter 35.

The first ester, methyl methacrylate is a bulk chemical. It is the monomer whose polymerization (Chapter 52) gives Perspex, the rigid transparent plastic used in windows and roofs. The second ester is an important local anaesthetic used for

minor operations.

One amide is the now-familiar DMF, but the other is a tetrapeptide and so contains one carboxylic acid group at the end (the 'C-terminus': see Chapter 52) and three amide groups. Though the four amino acids in this peptide are identical (alanine, Ala for short), the carbon NMR faithfully picks up four different C=O signals, all made different by being different distances from the end of the chain.

The first five compounds (two acid chlorides and three anhydrides) are all reactive acid derivatives, and the five esters and amides below them are all unreactive acid derivatives and yet the C=O shifts of all ten compounds fall in the same range. The C=O chemical shift is obviously *not* a good way to check on chemical reactivity.

What the carbon NMR fails to do is distinguish these types of acid derivative. There is more variation between the carboxylic acids on display than between the different classes of acid derivatives. This should be obvious if we show you some compounds containing two acid derivatives. Would you care to assign these signals?

amino acid – asparagine
177.1, 176.1

ester/acid chloride
156.1, 160.9

acid/ester – aspirin
165.6, 158.9

No, neither would we. In each case the difference between the carbonyl signals is only a few p.p.m. Though acid chlorides are extremely reactive in comparison with esters or amides, the electron deficiency at the carbon nucleus as measured by deshielding in the NMR spectrum evidently does not reflect this. Carbon NMR reliably distinguishes acid derivatives as a group from aldehydes and ketones as another group but it fails to distinguish even very reactive (for example, acid chlorides) from very unreactive (for example, amides) acid derivatives. So how do we distinguish acid derivatives?

Acid derivatives are best distinguished by infrared

A much better measure is the difference in IR stretching frequency of the C=O group. We discussed this in Chapter 12 (p. 287) where we noted a competition between conjugation by lone-pair electron donation *into* the carbonyl from OCOR, OR, or NH_2 and inductive withdrawal *from* the C=O group because of the electronegativity of the substituent. Conjugation donates electrons into the π^* orbital of the π bond and so lengthens and weakens it. The C=O bond becomes more like a single bond and its stretching frequency moves towards the single-bond region, that is, it goes *down*. The inductive effect removes electrons from the π orbital and so shortens and strengthens the π bond. It becomes more like a full double bond and moves *up* in frequency.

For a reminder of the distinction between conjugation and inductive effects, see Chapter 7, pp. 193–194.

These effects are balanced in different ways according to the substituent. Chlorine is poor at lone-pair electron donation (its lone pair is in an overly large 3p orbital and overlaps badly with the 2p orbital on carbon) but strongly electron-withdrawing so acid chlorides absorb at high frequency, almost in the triple-bond region. Anhydrides have an oxygen atom between two carbonyl groups. Inductive withdrawal is still strong but conjugation is weak because the lone pairs are pulled both ways. Esters have a well balanced combination with the inductive effect slightly stronger (oxygen donates from a compatible 2p orbital but is very electronegative and so withdraws electrons strongly as well). Finally, amides are dominated by conjugation as nitrogen is a much stronger electron donor than oxygen because it is less electronegative.

acid chlorides	anhydrides	esters	amides
inductive effect dominates	tug-of-war for lone pair: inductive effect dominates	inductive effect slightly dominates	conjugation strongly dominates
$1815\ cm^{-1}$	two peaks: ~1790, 1810 cm^{-1}	$1745\ cm^{-1}$	~1650 cm^{-1}

The two peaks for anhydrides are the symmetrical and antisymmetrical stretches for the two C=O groups; see Chapter 3, p. 71,

Conjugation with π electrons or lone pairs affects IR C=O stretches

We need to see how conjugation works when it is with a π bond rather than with a lone pair. This will make the concept more general as it will apply to aldehydes and ketones as well as to acid groups. How can we detect if an unsaturated carbonyl compound is conjugated or not? Well, compare these two unsaturated aldehydes.

pent-2-enal: conjugated   pent-4-enal: not conjugated

IR spectrum

$1620\ cm^{-1}$ strong $1690\ cm^{-1}$ strong $1640\ cm^{-1}$ weak $1730\ cm^{-1}$ strong

^{13}C NMR spectrum

152 127 192   115 137 206

^{1}H NMR spectrum

6.13 (dd)   6.92 (dt)   9.52 (d)   5.84 (ddt)   5.00 (dd)   5.04 (dd)   9.75 (t)

The key differences are the frequency of the C=O stretch (lowered by 40 cm^{-1} by conjugation) and the strength (that is, the intensity) of the C=C stretch (increased by conjugation) in the IR. In the ^{13}C NMR, C3 in the conjugated enal is moved out of the alkene region just into the carbonyl region, showing how electron-deficient this carbon atom must be. In the proton NMR there are many effects but the downfield shift of the protons on the alkene especially C3 (again!) is probably the most helpful.

Because the infrared carbonyl frequencies follow such a predictable pattern, it is possible to make a simple list of correlations using just three factors. Two are the ones we have been discussing—conjugation (frequency-lowering) and the inductive effect (frequency-raising). The third is the effect of small rings and this we next need to consider in a broader context.

We discussed the way in which conjugation affects reactivity in Chapter 10, and mentioned its spectroscopic effect there as well.

Small rings introduce strain inside the ring and higher s character outside it

Cyclic ketones can achieve the perfect 120° angle at the carbonyl group only if the ring is at least six-membered. The smaller rings are 'strained' because the orbitals have to overlap at a less than ideal angle.

The three-membered ring is, of course, flat. The others are not. Even the four-membered ring is slightly puckered, the five- and especially the six-membered rings more so. This is all discussed in Chapter 18. But you have already met the concept of ring strain in Chapter 6, where we used it to explain why cyclopropanones and cyclobutanones are readily hydrated.

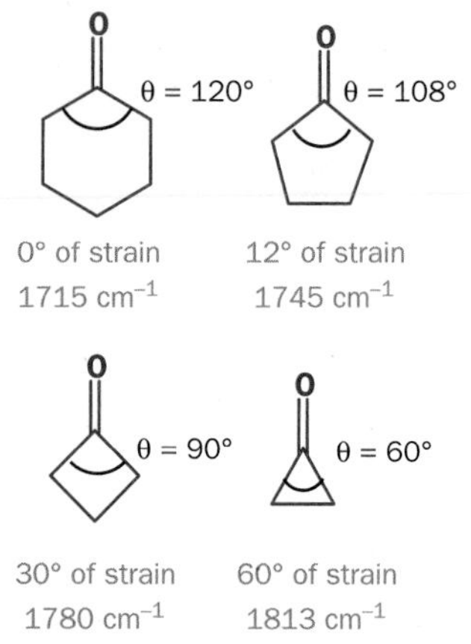

O   θ = 120°

0° of strain
1715 cm⁻¹

O   θ = 108°

12° of strain
1745 cm⁻¹

O   θ = 90°

30° of strain
1780 cm⁻¹

O   θ = 60°

60° of strain
1813 cm⁻¹

For a four-membered ring, the actual angle is 90°, so there is 120° − 90° = 30° of strain at the carbonyl group. The effects of this strain on five-, four-, and three-membered rings is shown here.

Lactam C=O stretching frequencies

A further good example is the difference between C=O stretching frequencies in cyclic amides, or **lactams**. The penicillin class of antibiotics all contain a four-membered ring amide known as a β-lactam. The carbonyl stretching frequency in these compounds is way above the 1680 cm⁻¹ of the six-membered lactam, which is what you might expect for an unstrained amide.

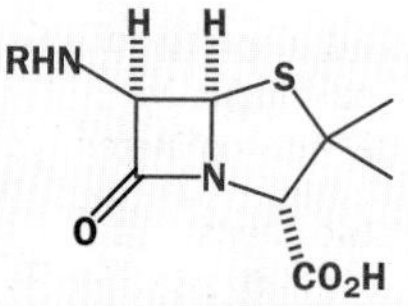

β-lactam in penicillin
1715 cm⁻¹

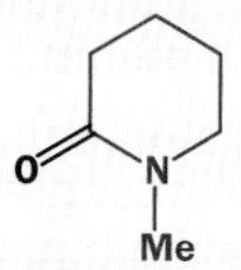

unstrained lactam
1680 cm⁻¹

But why should strain raise the frequency of a carbonyl group? It is evidently shortening and strengthening the C=O bond as it moves it towards the triple-bond region (higher frequency), not towards the single-bond region (lower frequency). In a six-membered ring, the sp² orbitals forming the σ framework around the carbonyl group can overlap perfectly with the sp³ orbitals on neighbouring carbon atoms because the orbital angle and the bond angle are the same. In a four-membered ring the orbitals do not point towards those on the neighbouring carbon atoms, but point out into space.

Ideally, we should like the orbitals to have an angle of 90° as this would make the orbital angle the same as the bond angle. In theory it *would* be possible to have a bond angle of 90° if we used pure p orbitals instead of sp² hybrid orbitals.

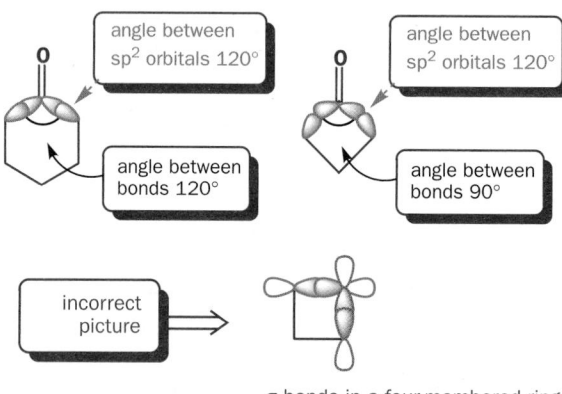

angle between sp² orbitals 120°

angle between bonds 120°

angle between sp² orbitals 120°

angle between bonds 90°

incorrect picture

σ bonds in a four-membered ring formed with pure p orbitals

If we did we should leave a pure s orbital for the σ bond to oxygen. This extreme is not possible, but a compromise is. *Some* more p character goes into the ring bonds—maybe they become s⁰·⁸p³·²—and the same amount of extra s character goes into the σ bond to oxygen. The more s character there is in the orbital, the shorter it gets as s orbitals are (much) smaller than p orbitals.

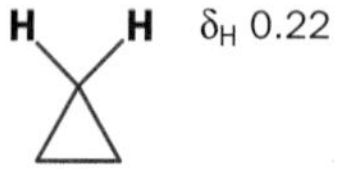

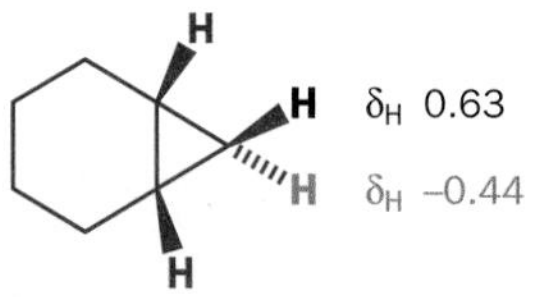

The s-character argument also explains the effects of small rings on proton NMR shifts. These hydrogens, particularly on three-membered rings, resonate at unusually high fields, between 0 and 1 p.p.m. in cyclopropanes instead of the 1.3 p.p.m. expected for CH₂ groups, and may even appear at negative δ values. High p character in the framework of small rings also means high s character in C–H bonds outside the ring and this will mean shorter bonds, greater shielding, and small δ values.

H   H   δ_H 0.22

H   δ_H 0.63
H   δ_H −0.44
H

Three-membered rings and alkynes

You have also seen the same argument used in Chapter 8 to justify the unusual acidity of C–H protons on triple bonds (such as alkynes and HCN), and alluded to in Chapter 3 to explain the stretching frequency of the same C–H bonds. Like alkynes, three-membered rings are also unusually easy to deprotonate in base.

Here is an example where deprotonation occurs at a different site in two compounds identical except for a C-C bond closing a three-membered ring. The first is an ortholithiation of the type discussed in Chapter 9.

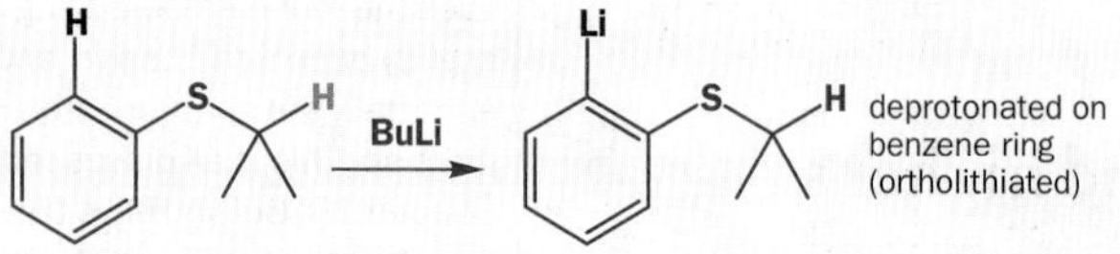

H   S   H
BuLi →

Li   S   H
deprotonated on benzene ring (ortholithiated)

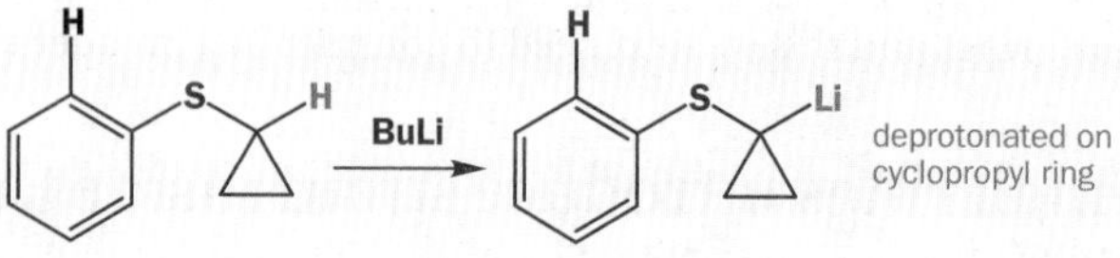

H   S   H
BuLi →

H   S   Li
deprotonated on cyclopropyl ring

NMR spectra of alkynes are related to those of small rings

Now what about the NMR spectra of alkynes? By the same argument, protons on alkynes ought to appear in the NMR at quite high field because these protons really are rather acidic (Chapter 8).

Protons on a typical alkene have δ_H about 5.5 p.p.m., while the proton on an alkyne comes right in the middle of the protons on saturated carbons at about δ_H 2–2.5 p.p.m. This is rather a large effect just for increased s character and some of it is probably due to better shielding by the triple bond, which surrounds the linear alkyne with π bonds without a nodal plane.

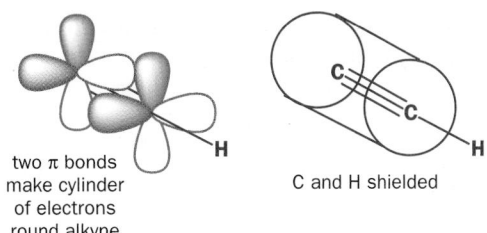

two π bonds make cylinder of electrons round alkyne

C and H shielded

This means that the carbon atoms also appear at higher field than expected, not in the alkene region but from about δ_C 60–80 p.p.m. The s-character argument is important, though, because shielding can't affect IR stretching frequencies, yet C≡C–H stretches are strong and at about 3300 cm^{-1}, just right for a strong C–H bond. The picture is consistent.

A simple example is the ether 3-methoxyprop-1-yne. Integration alone allows us to assign the spectrum, and the 1H signal at 2.42 p.p.m., the highest field signal, is clearly the alkyne proton. Notice also that it is a triplet and that the OCH$_2$ group is a doublet. This ^4J is small (about 2 Hz) and, though there is nothing like a letter 'W' in the arrangement of the bonds, coupling of this kind is often found in alkynes.

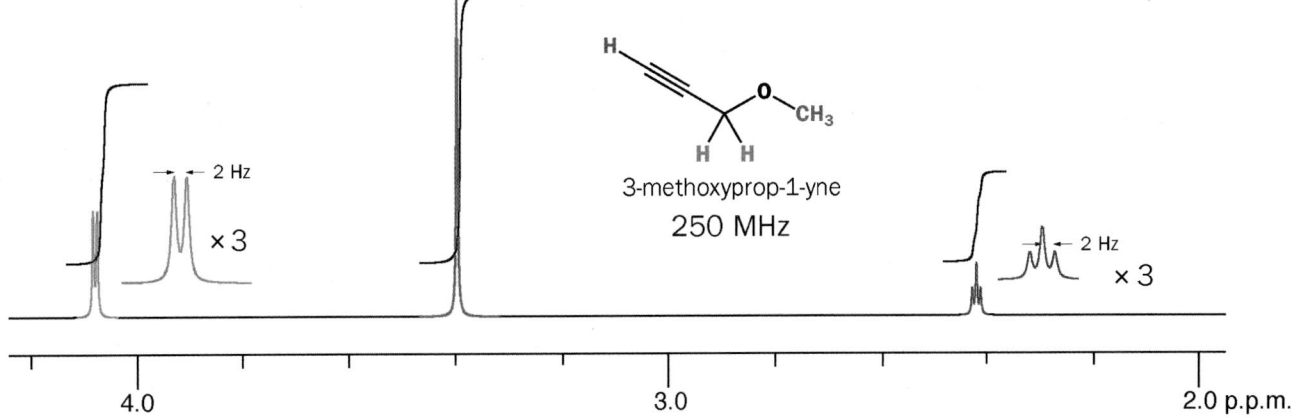

3-methoxyprop-1-yne
250 MHz

2 Hz × 3

2 Hz × 3

4.0   3.0   2.0 p.p.m.

In Chapter 11, p. 270, you saw that bonds aligned in a 'W' arrangement can give rise to $^4J_{HH}$ coupling.

A more interesting example comes from the base-catalysed addition of methanol to buta-1,3-diyne (diacetylene). The compound formed has one double and one triple bond and the ^{13}C NMR shows clearly the greater deshielding of the double bond.

buta-1,3-diyne

MeO$^{\ominus}$
MeOH

158.3
184.2
78.6
80.9
MeO
60.6
^{13}C NMR

δ_H 6.53 $^3J = 6.5$ Hz δ_H 4.52
H   H
$^5J = 1$ Hz $^4J = 2.5$ Hz
O
Me
δ_H 3.45   H δ_H 3.07
^{1}H NMR

You may have noticed that we have drawn the double bond with the *cis* (*Z*) configuration. We know that this is true because of the proton NMR, which shows a 6.5 Hz coupling between the two alkene protons (much too small for a *trans* coupling; see p. 269). There is also the longer range coupling ($^4J = 2.5$ Hz) just described and even a small very long range coupling ($^5J = 1$ Hz) between the alkyne proton and the terminal alkene proton.

Simple calculations of C=O stretching frequencies in IR spectra

The best way is to relate all our carbonyl frequencies to those for saturated ketones (1715 cm^{-1}). We can summarize what we have just learned in a table.

Notice in this simple table (for full details you should refer as usual to a specialist book) that the adjustment '30 cm^{-1}' appears quite a lot (–30 cm^{-1} for both alkene and aryl, for example), that the increment for small rings is 35 cm^{-1} each time (30 to 65 cm^{-1} and then 65 to 100 cm^{-1}), and that the extreme effects of Cl and NH$_2$ are +85 and –85 cm^{-1}, respectively. These effects are additive. If you want to estimate the C=O frequency of a proposed structure, just add or subtract all the adjustments to 1715 cm^{-1} and you will get a reasonable result.

Effects of substituents on IR carbonyl frequencies

Effect	Group	C=O stretch, cm^{-1}	Frequency change[a], cm^{-1}
inductive effect	Cl	1800	+85
	OCOR	1765, 1815	+50, +100
	OR	1745	+30
	H	1730	+15
conjugation	C=C	1685	–30
	aryl	1685	–30
	NH$_2$	1630	–85
ring strain	5-membered ring	1745	+30
	4-membered ring	1780	+65
	3-membered ring	1815	+100

[a] Difference between stretching frequency of C=O and stretching frequency of a typical saturated ketone (1715 cm^{-1}).

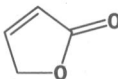

Let us try the five-membered unsaturated (and conjugated) lactone (cyclic ester) in the margin. We must add 30 cm^{-1} for the ester, subtract 30 cm^{-1} for the double bond, and add 30 cm^{-1} for the five-membered ring. Two of those cancel out leaving just 1715 + 30 = 1745 cm^{-1}. These compounds absorb at 1740–1760 cm^{-1}. Not bad!

Interactions between different nuclei can give enormous coupling constants

We have looked at coupling between hydrogen atoms and you may have wondered why we have ignored coupling between other NMR active nuclei. Why does ^{13}C not cause similar couplings? In this section we are going to consider not only couplings between the same kind of nuclei, such as two protons, called **homonuclear coupling**, but also coupling between different nuclei, such as a proton and a fluorine atom or ^{13}C and ^{31}P, called **heteronuclear coupling**.

Two nuclei are particularly important, ^{19}F and ^{31}P, since many organic compounds contain these elements and both are at essentially 100% natural abundance and have spin $I = 1/2$. We shall start with organic compounds that have just one of these nuclei and see what happens to both the ^{1}H and the ^{13}C spectra. In fact, it is easy to find a ^{19}F or a ^{31}P atom in a molecule because these elements couple to all nearby carbon and hydrogen atoms. Since they can be directly bonded to either, ^1J coupling constants such as $^1J_{CF}$ or $^1J_{PH}$ become possible, as well as the more 'normal' couplings such as $^2J_{CF}$ or $^3J_{PH}$, and these ^1J coupling constants can be enormous.

We shall start with a simple phosphorus compound, the dimethyl ester of phosphorous acid (H$_3$PO$_3$). There is an uncertainty about the structure of both the acid and its esters. They could exist as P(III) compounds with a lone pair of electrons on phosphorus, or a P(V) compounds with a P=O double bond.

phosphorous acid

P(III)
HO
‥
P—OH
HO

or

P(V)
HO
‖O
P
HO
H

dimethyl phosphite

P(III)
MeO
‥
P—OH
MeO

or

P(V)
MeO
‖O
P
MeO
H

In fact, dimethyl phosphite has a 1H doublet with the amazing coupling constant of 693 Hz: on a 250 MHz machine the two lines are over 2 p.p.m. apart and it is easy to miss that they are two halves of the same doublet. This can only be a $^1J_{PH}$ as it is so enormous and so the compound has to have a P–H bond and the P(V) structure is correct. The coupling to the methyl group is much smaller but still large for a three-bond coupling ($^3J_{PC}$ of 18 Hz).

MeO
‖O
P
MeO
H

3.80 (6H, d, $^3J_{PH}$ 9 Hz)

6.77 (1H, d, $^1J_{PH}$ 693 Hz)

Next, consider the phosphonium salt you met at the end of Chapter 14 for use in the Wittig reaction, turning aldehydes and ketones to alkenes. It has a $^2J_{PH}$ of 18 Hz. There is no doubt about this structure—it is just an illustration of coupling to phosphorus. There is coupling to phosphorus in the carbon spectrum too: the methyl group appears at δ_C 10.6 p.p.m. with a $^1J_{PC}$ of 57 Hz, somewhat smaller than typical $^1J_{PH}$. We haven't yet talked about couplings to ^{13}C: we shall now do so.

Coupling in carbon NMR spectra

We shall use coupling with fluorine to introduce this section. Fluorobenzenes are good examples because they have a number of different carbon atoms all coupled to the fluorine atom.

162.9 (d, $^1J_{FC}$ 244 Hz, *ipso*-C)
115.3 (d, $^2J_{FC}$ 21 Hz, *ortho*-C)
122.9 (d, $^3J_{FC}$ 7.5 Hz, *meta*-C)
123.9 (d, $^4J_{FC}$ 2 Hz, *para*-C)

The carbon directly joined to fluorine (the *ipso* carbon) has a very large $^1J_{CF}$ value of about 250 Hz. More distant coupling is evident too: all the carbons in the ring couple to the fluorine in PhF with steadily diminishing J values as the carbons become more distant.

Trifluoroacetic acid is an important strong organic acid (Chapter 8) and a good solvent for ^1H NMR. The carbon atom of the CF_3 group is coupled equally to all the three fluorines and so appears as a quartet with a large $^1J_{CF}$ of 283 Hz, about the same as in PhF. Even the carbonyl group is also a quartet, though the coupling constant is much smaller ($^2J_{CF}$ is 43 Hz). Notice too how far downfield the CF_3 carbon atom is!

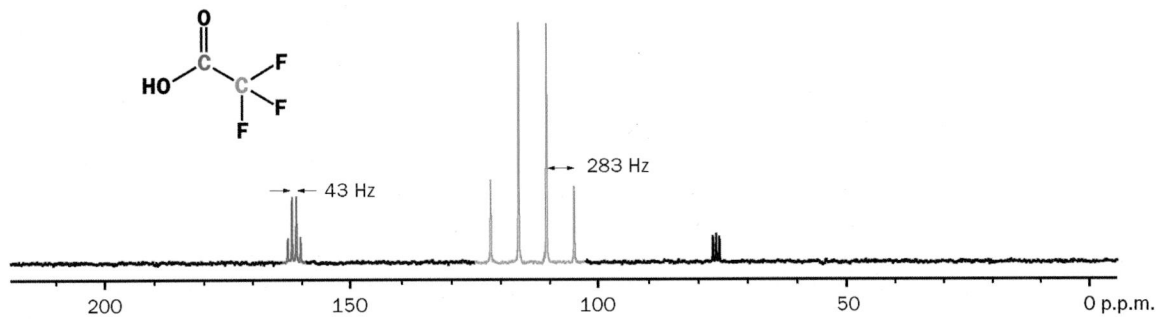

Coupling between protons and ^{13}C

In view of all this, you may ask why we don't apparently see couplings between ^{13}C and ^1H in either carbon or proton spectra. In proton spectra we don't see coupling to ^{13}C because of the low abundance (1.1%) of ^{13}C. Most protons are bonded to ^{12}C: only 1.1% of protons are bonded to ^{13}C. If you look closely at proton spectra with very flat baselines, you may see small peaks either side of strong peaks at about 0.5% peak height. These are the ^{13}C 'satellites' for those protons that are bonded to ^{13}C atoms.

As an example, look again at the 500 MHz ^1H spectrum of heptan-2-one that we saw on p. 268. When the baseline of this spectrum is vertically expanded, the ^{13}C satellites may be seen. The singlet due to the methyl protons is actually in the centre of a tiny doublet due to the 1% of protons coupling to ^{13}C. Similarly, each of the triplets in the spectrum is flanked by two tiny triplets. The two tiny triplets on either side make up a doublet of triplets with a large ^1J coupling constant to the ^{13}C (around 130 Hz) and smaller ^3J coupling to the two equivalent protons.

$Ph_3\overset{\oplus}{P}$—CH_3 $Br^{\ominus}$

methyltriphenylphosphonium bromide
aromatic protons and
δ_H 3.25 (3H, d, $^2J_{PH}$ 18 Hz)

▶
Note that these spectra with heteronuclear couplings provide the only cases where we can see *one* doublet in the proton NMR. Normally, if there is one doublet, there must be another signal with at least this complexity as all coupling appears twice (A couples to B and so B also couples to A!). If the coupling is to another element (here phosphorus) then the coupling appears once in each spectrum. The Wittig reagent has an A_3P (CH_3–P) system: proton A appears as a doublet, while the phosphorus atom appears as a quartet in the *phosphorus* spectrum at a completely different frequency.

▶
Ipso can join the list (*ortho*, *meta*, *para*) of trivial names for positions on a substituted benzene ring.

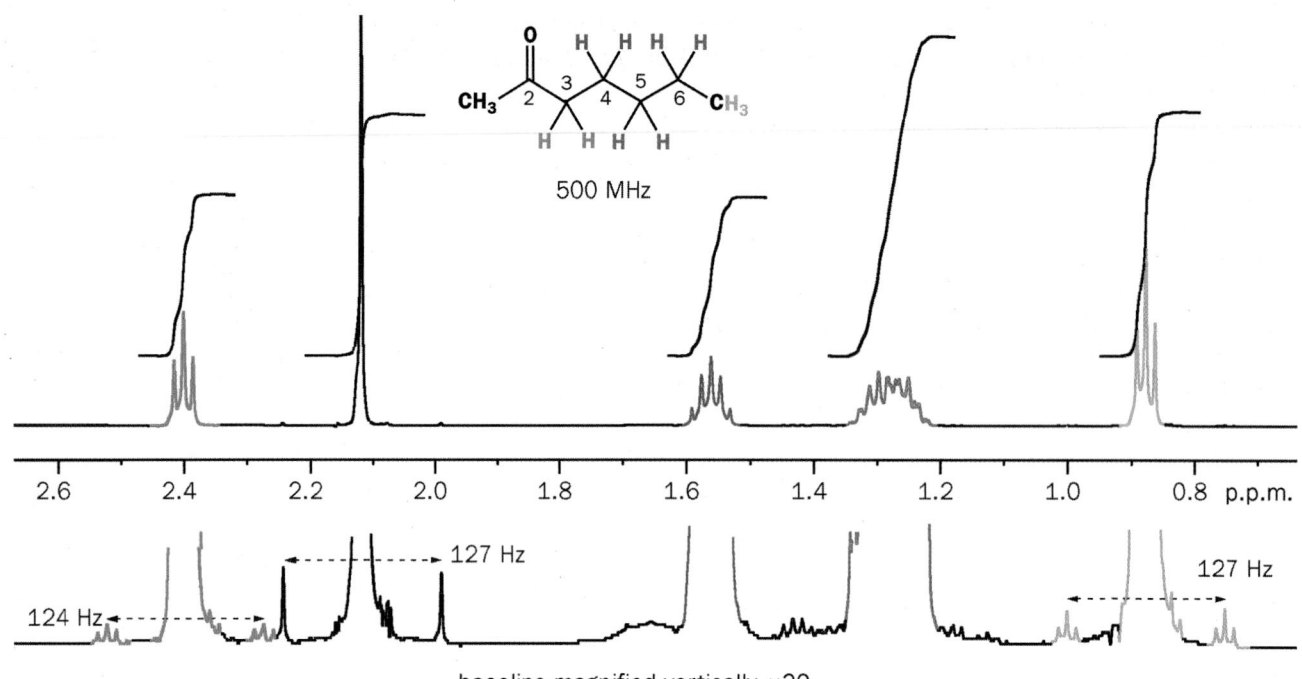

baseline magnified vertically ×30

^{13}C satellites are usually lost in the background noise of the spectrum and need concern us no further. You do, however, see coupling with ^{13}C labelled compounds where the ^{13}C abundance now approaches 100%. The same Wittig reagent we saw a moment ago shows a 3H doublet of doublets with the typically enormous $^{1}J_{CH}$ of 135 Hz when labelled with ^{13}C in the methyl group.

^{13}C-labelled phosphonium salt
δ_H 3.25 (3H, dd, $^{1}J_{CH}$ 135, $^{2}J_{PH}$ 18 Hz)

Why is there no coupling to protons in normal ^{13}C NMR spectra?

We get the singlets consistently seen in carbon spectra because of the way we record the spectra. The values of $^{1}J_{CH}$ are so large that, if we recorded ^{13}C spectra with all the coupling constants, we would

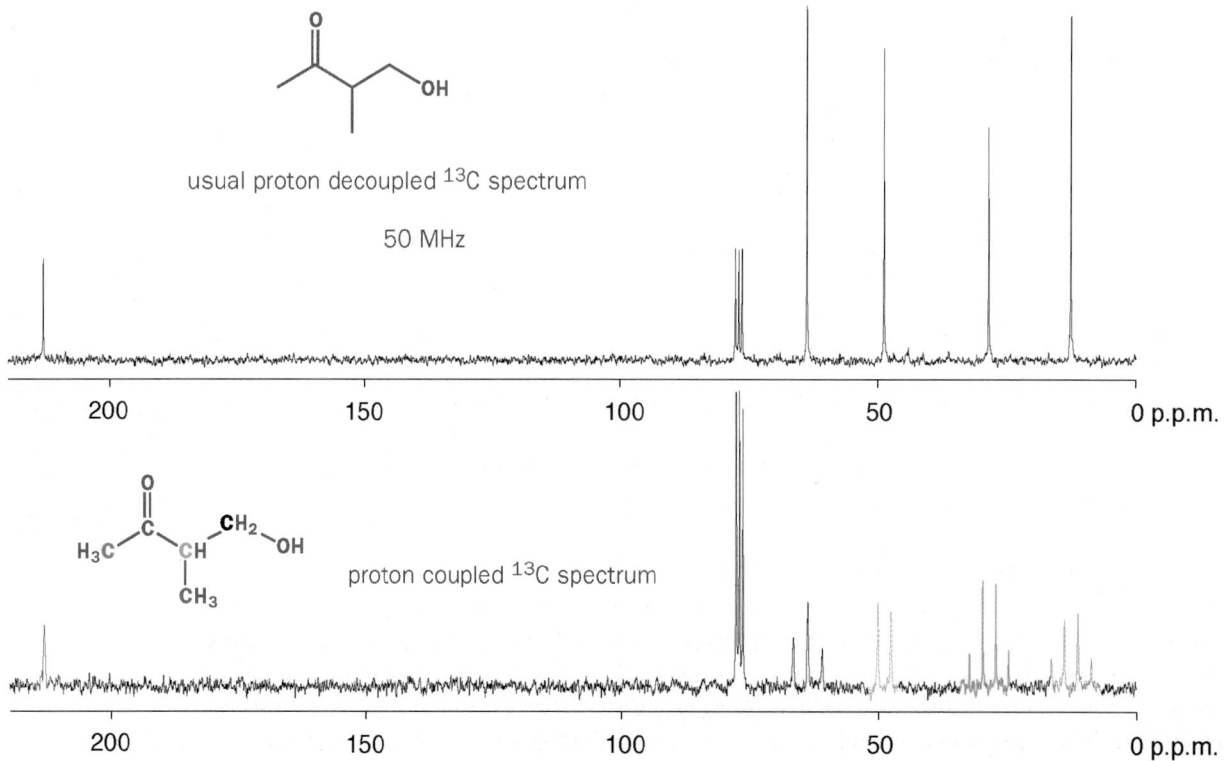

usual proton decoupled ^{13}C spectrum

50 MHz

proton coupled ^{13}C spectrum

get a mass of overlapping peaks. When run on the same spectrometer, the frequency at which ^{13}C nuclei resonate turns out to be about a quarter of that of the protons. Thus a '200 MHz machine' (remember that the magnet strength is usually described by the frequency at which the protons resonate) gives ^{13}C spectra at 50 MHz. Coupling constants ($^{1}J_{CH}$) of 100–250 Hz would cover 2–5 p.p.m. and a CH_3 group with $^{1}J_{CH}$ of about 125 Hz would give a quartet covering nearly 8 p.p.m. See the example on previous page.

Since the proton coupled ^{13}C spectrum can so easily help us to distinguish CH_3, CH_2, CH, and quaternary carbons, you might wonder why they are not used more. The above example was chosen very carefully to illustrate proton coupled spectra at their best. Unfortunately, this is not a typical example. More usually, the confusion from overlapping peaks makes this just not worthwhile. So ^{13}C NMR spectra are recorded while the whole 10 p.p.m. proton spectrum is being irradiated with a secondary radiofrequency source. The proton energy levels are equalized by this process and all coupling disappears. Hence the singlets we are used to seeing.

For the rest of this chapter, we shall not be introducing new theory or new concepts; we shall be applying what we have told you to a series of examples where spectroscopy enables chemists to identify compounds.

Identifying products spectroscopically

Conjugate or direct addition?

In Chapter 10 we were discussing the reasons for conjugate addition and direct addition to the carbonyl group. We should now consider how you find out what has happened. A famous case was the addition of hydroxylamine (NH_2–OH) to a simple enone. Nitrogen is more nucleophilic than oxygen so we expect it to add first. But will it add directly to the carbonyl group or in a conjugate fashion? Either way, an intermediate will be formed that can cyclize.

conjugate addition by the nitrogen atom of hydroxylamine

direct addition by the nitrogen atom of hydroxylamine

The two possible isomeric products were the subject of a long running controversy. Once the IR and proton NMR spectra of the product were run, doubt vanished. The IR showed no NH stretch. The NMR showed no alkene proton but did have a CH_2 group at 2.63 p.p.m. Only the second structure is possible.

We need to look now at a selection of problems of different kinds to show how the various spectroscopic methods can cooperate in structure determination.

Reactive intermediates can be detected by spectroscopy

Some intermediates proposed in reaction mechanisms look so unlikely that it is comforting if they can be isolated and their structure determined. We feel more confident in proposing an intermediate if we are sure that it can really be made. Of course, this is not necessarily evidence that the intermediate is actually formed during reactions and it certainly does *not* follow that the failure to isolate a given intermediate disproves its involvement in a reaction. We shall use **ketene** as an example.

■
Do not be concerned about the details of the mechanisms: note that we have used the '±H⁺' shorthand introduced in Chapter 13, and have abbreviated the mechanism where water is eliminated and the oxime formed—the full mechanism of imine (and oxime) formation can be found in Chapter 14, p. 349. In this chapter, we are much more concerned just with the structure of the products.

ketene

orthogonal π bonds of ketene

■ The structure of *ketene* is loosely analogous to that of *allene*, discussed in Chapter 7, p. 157.

Ketene looks pretty unlikely! It is $CH_2=C=O$ with two π bonds (C=C and C=O) to the same carbon atom. The orbitals for these π bonds must be orthogonal because the central carbon atom is sp hybridized with two linear σ bonds and two p orbitals at right angles both to the σ bonds and to each other. Can such a molecule exist? When acetone vapour is heated to very high temperatures (700–750 °C) methane is given off and ketene is supposed to be the other product. What is isolated is a ketene dimer ($C_4H_4O_2$) and even the structure of this is in doubt as two reasonable structures can be written.

heat → ketene → dimerization → cyclic ester structure for diketene / alternative cyclobuta-1,3-dione structure for diketene

The spectra fit the ester structure well, but not the more symmetrical diketone structure at all. There are *three* types of proton (cyclobuta-1,3-dione would have just *one*) with allylic coupling between one of the protons on the double bond and the CH_2 group in the ring. The carbonyl group has the shift (185 p.p.m.) of an acid derivative (not that of a ketone which would be about 200 p.p.m.) and all four carbons are different.

diketene

► **Ozonolysis** or **ozonation** is the cleavage of an alkene by ozone (O_3). The reaction and its mechanism are discussed in Chapter 35: the only point to note now is that ozone is a powerful oxidant and cleaves the alkene to make two carbonyl compounds. Again, in this chapter we are concerned only with the structure of the products and how this can be determined.

^{1}H NMR spectrum:
4.85 (1H, narrow t, J ~ 1)
4.51 (1H, s)
3.90 (2H, d, J ~ 1)

^{13}C NMR spectrum:
185.1, 147.7, 67.0, 42.4

Ozonolysis of ketene dimer gives a very unstable compound that can be observed only at low temperatures (−78 °C or below). It has two carbonyl bands in the IR and reacts with amines to give amides, so it looks like an anhydride (Chapter 12). Can it be the previously unknown cyclic anhydride of malonic acid?

The two carbonyl bands are of high frequency as would be expected for a four-membered ring—using the table on p. 368 we estimate $1715 + 50$ cm^{-1} (for the anhydride) + 65 cm^{-1} (for the four-membered ring) = 1830 cm^{-1}. Both the proton and the carbon NMR are very simple: just a 2H singlet at 4.12 p.p.m., shifted downfield by two carbonyls, a C=O group at 160 p.p.m., right for an acid derivative, and a saturated carbon shifted downfield but not as much as a CH_2O group.

► Malonic anhydride cannot be made directly from malonic acid because attempted dehydration of the acid leads to the exotic molecule carbon suboxide C_3O_2.

malonic acid
− H_2O
$O=C=C=C=O$
carbon suboxide C_3O_2

ketene dimer

O_3 / −78 °C →

anhydride of malonic acid

IR 1820, 1830 cm^{-1}
δ_H 4.12 (2H, s)
δ_C 160.3, 45.4

PhNH$_2$ →

−30 °C →

IR 2140 cm^{-1}
δ_H 2.24 (2H, s)
δ_C 193.6

All this is reasonably convincing, and is confirmed by allowing the anhydride to warm to −30 °C when it loses CO_2 (detected by the ^{13}C peak at 124.5 p.p.m.!) and gives another unstable compound with the strange IR frequency of 2140 cm^{-1}. Could this be monomeric ketene? It's certainly not either of the possible ketene dimers as we know what their spectra are like, and this is quite different: just a 2H singlet at 2.24 p.p.m. and ^{13}C peaks at 194.0 and 2.5 p.p.m. It is indeed monomeric ketene.

Squares and cubes: molecules with unusual structures

Some structures are interesting because we believe they can tell us something fundamental about the nature of bonding while others are a challenge because many people argue that they cannot be made. What do you think are the prospects of making cyclobutadiene, a conjugated four-membered ring, or the hydrocarbons tetrahedrane and cubane, which have, respectively, the shapes of the perfectly symmetrical Euclidean solids, the tetrahedron and the cube?

cyclobutadiene

With four electrons, cyclobutadiene is **anti-aromatic**—it has $4n$ instead of $4n + 2$. You saw in Chapter 7 that cyclic conjugated systems with $4n$ electrons (cyclooctatetraene, for example) avoid being conjugated by puckering into a tub shape. Cyclobutadiene cannot do this: it must be more or less planar, and so we expect it to be very unstable. Tetrahedrane has four fused three-membered rings. Though the molecule is tetrahedral in shape, each carbon atom is nowhere near a tetrahedron, with three bond angles of 60°. Cubane has six fused four-membered rings and is again highly strained.

tetrahedrane

In fact, cubane has been made, cyclobutadiene has a fleeting existence but can be isolated as an iron complex, and a few substituted versions of tetrahedrane have been made. The most convincing evidence that you have made any of these three compounds would be the extreme simplicity of the spectra. Each has only one kind of hydrogen and only one kind of carbon. They all belong to the family $(CH)_n$.

cubane

Cubane has a molecular ion in the mass spectrum at 104, correct for C_8H_8, only CH stretches in the IR at 3000 cm^{-1}, a singlet in the proton NMR at 4.0 p.p.m., and a single line in the carbon NMR at 47.3 p.p.m. A very symmetrical molecule and a stable one in spite of all those four-membered rings.

Stable compounds with a cyclobutadiene and a tetrahedrane core can be made if each hydrogen atom is replaced by a *t*-butyl group. The very large groups round the edge of the molecule repel each other and hold the inner core tightly together. Now another difficulty arises—it is rather hard to tell the compounds apart. They both have four identical carbon atoms in the core and four identical *t*-butyl groups round the edge. The starting material for a successful synthesis of both was the tricyclic ketone below identified by its strained C=O stretch and partly symmetrical NMR spectra. When this ketone was irradiated with UV light (indicated by '*hv*' in the scheme), carbon monoxide was evolved and a highly symmetrical compound $(t\text{-BuC})_4$ was formed. But which compound was it?

> ■ You can read more about the synthesis of cubane in Chapter 37, when we discuss the rearrangement reactions that were used to make it.

IR 1762 (C=O) cm^{-1}

δ_H 1.37 (18H, s), 1.27 (18H, s)

δ_C 188.7 (C=O), 60.6, 33.2, 33.1, 31.0, 30.2, 29.3

tetra-*t*-butyl tetrahedrane

tetra-*t*-butyl cyclobutadiene

The story is made more complicated (but in the end easier!) by the discovery that this compound on heating turned into another very similar compound. There are only two possible structures for $(t\text{-BuC})_4$, so clearly one compound must be the tetrahedrane and one the cyclobutadiene. The problem simplifies with this discovery because it is easier to distinguish two possibilities when you can make comparisons between two sets of spectra. Here both compounds gave a molecular ion in the mass spectrum, neither had any interesting absorptions in the IR, and the proton NMRs could belong to either compound as they simply showed four identical *t*-Bu groups. So did the carbon NMR, of course, but it showed the core too. The first product had only saturated carbon atoms, while the second had a signal at 152.7 p.p.m. for the unsaturated carbons. The tetrahedrane is formed from the tricyclic ketone on irradiation but it isomerizes to the cyclobutadiene on heating.

Identifying compounds from nature

The next molecules we need to know how to identify are those discovered from nature—natural products. These often have biological activity and many useful medicines have been discovered this way. We shall look at a few examples from different fields. The first is the sex pheromone of the

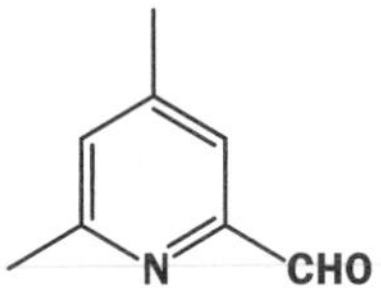

possible structure for
Lycorea sex pheromone

Trinidad butterfly *Lycorea ceres ceres*. The male butterflies start courtship by emitting a tiny quantity of a volatile compound. Identification of this type of compound is very difficult because of the minute amounts available but this compound crystallized and gave enough for a mass spectrum and an IR. The highest peak in the mass spectrum was at 135. This is an odd number so we might have one nitrogen atom and a possible composition of C_8H_9ON. The IR showed a carbonyl peak at 1680 cm^{-1}. With only this meagre information, the first proposals were for a pyridine aldehyde.

Eventually a little more compound (6 mg!) was available and a proton NMR spectrum was run. This showed at once that this structure was wrong. There was no aldehyde proton and only one methyl group. More positive information was the pair of triplets showing a –CH_2CH_2– unit between two electron-withdrawing groups (N and C=O?) and the pair of doublets for neighbouring protons on an aromatic ring, though the chemical shift and the coupling constant are both rather small for a benzene ring.

If we look at what we have got so far, we see that we have accounted for four carbon atoms in the methyl and carbonyl groups and the –CH_2CH_2– unit. This leaves only four carbon atoms for the aromatic ring. We must use nitrogen too as the only possibility is a pyrrole ring. Our fragments are now those shown below (the black dotted lines show joins to another fragment). These account for all the atoms in the molecule and suggest structures such as these.

put these fragments together to get structures such as these:

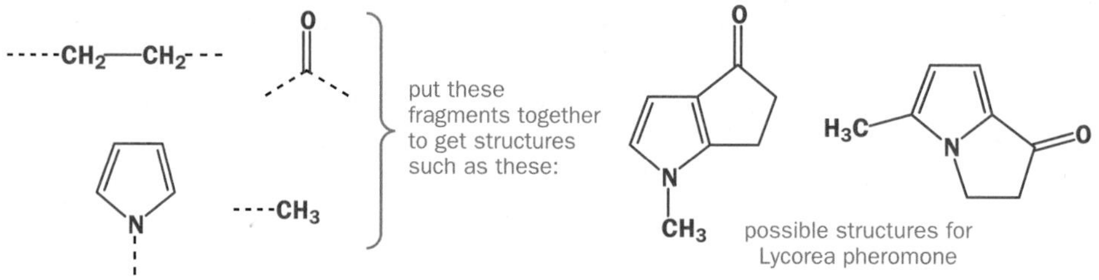

possible structures for
Lycorea pheromone

Now we need to use the known chemical shifts and coupling constants for these sorts of molecules. An N–Me group would normally have a larger chemical shift than 2.2 p.p.m. so we prefer the methyl group on a carbon atom of the pyrrole ring. Typical shifts and coupling constants around pyrroles are shown below. Chemists do not, of course, remember these numbers; we look them up in tables. Our data, with chemical shifts of 6.09 and 6.69 p.p.m. and a coupling constant of 2.5 Hz, clearly favour hydrogen atoms in the 2 and 3 positions and suggest this structure for the sex pheromone, which was confirmed by synthesis and is now accepted as correct.

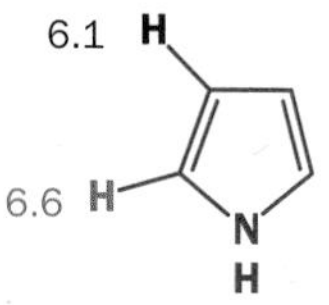

typical chemical shifts
for pyrroles

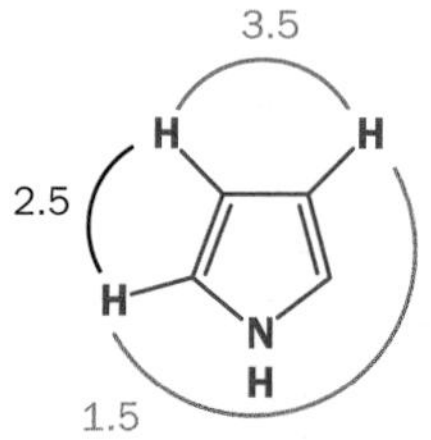

typical coupling constants
for pyrroles

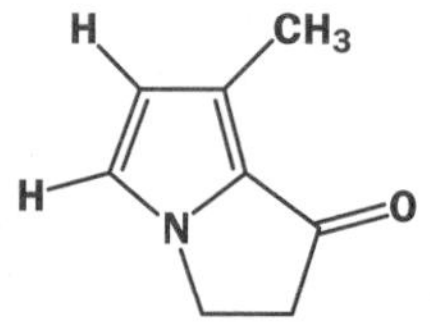

correct structure for
Lycorea sex pheromone

Tables

The final section of this chapter contains some tables of NMR data, which we hope you may want to use in solving problems. In Chapter 11 there were a few guides to chemical shift—summaries of patterns that you might reasonably be expected to remember. But we have left the main selections of hard numbers—tables that *you are not expected to remember*—until now. There are a few comments to explain the tables, but you will probably want to use this section as reference rather than bedtime reading. The first four tables give detailed values for various kinds of compounds and Table 15.5 gives a simple summary. We hope that you will find this last table particularly useful.

Effects of electronegativity

Table 15.1 shows how the electronegativity of the atom attached directly to a methyl group affects the shifts of the CH_3 protons (δ_H) and the CH_3 carbon atom (δ_C) in their NMR spectra.

Effects of functional groups

Many substituents are more complicated than just a single atom and electronegativity is only part of the story. We need to look at all the common substituents and see what shifts they cause relative to the CH skeleton of the molecule. Our zero really ought to be at about 0.9 p.p.m. for protons and at 8.4 p.p.m. for carbon, that is, where ethane (CH_3–CH_3) resonates, and not at the arbitrary zero allocated to Me_4Si. In Table 15.2 we give such a list. The reason for this is that the shifts (from Me_4Si) themselves are not additive but the shift differences (from 0.9 or 8.4 p.p.m.) are.

Table 15.1 Chemical shifts δ of methyl groups attached to different atoms

Element	Electronegativity	Compound	δ_H, p.p.m.	δ_C, p.p.m.
Li	1.0	CH_3–Li	−1.94	−14.0
Si	1.9	CH_3–$SiMe_3$	0.0	0.0
I	2.7	CH_3–I	2.15	−23.2
S	2.6	CH_3–SMe	2.13	18.1
N	3.1	CH_3–NH_2	2.41	26.9
Cl	3.2	CH_3–Cl	3.06	24.9
O	3.4	CH_3–OH	3.50	50.3
F	4.0	CH_3–F	4.27	75.2

Table 15.2 Chemical shifts δ (p.p.m.) of methyl groups bonded to functional groups

	Functional group	Compound	δ_H	δ_H − 0.9	δ_C	δ_C − 8.4
1	silane	**Me**$_4$Si	0.0	−0.9	0.0	−8.4
2	alkane	**Me–Me**	0.86	0.0	8.4	0.0
3	alkene	**Me**$_2$C=C**Me**$_2$	1.74	0.84	20.4	12.0
4	benzene	**Me**–Ph	2.32	1.32	21.4	13.0
5	alkyne	**Me**–C≡C–R^a	1.86	0.96		
6	nitrile	**Me**–CN	2.04	1.14	1.8	−6.6
7	acid	**Me**–CO_2H	2.10	1.20	20.9	11.5
8	ester	**Me**–CO_2Me	2.08	1.18	20.6	11.2
9	amide	**Me**–CONHMe	2.00	1.10	22.3	13.9
10	ketone	**Me**$_2$C=O	2.20	1.30	30.8	21.4
11	aldehyde	**Me**–CHO	2.22	1.32	30.9	21.5
12	sulfide	**Me**$_2$S	2.13	1.23	18.1	9.7
13	sulfoxide	**Me**$_2$S=O	2.71	1.81	41.0	32.6
14	sulfone	**Me**$_2$SO$_2$	3.14	2.24	44.4	36.0
15	amine	**Me**–NH_2	2.41	1.51	26.9	18.5
16	amide	MeCONH–**Me**	2.79	1.89	26.3	17.9
17	nitro	**Me**–NO_2	4.33	3.43	62.5	53.1
18	ammonium salt	**Me**$_4$–N$^+$ Cl$^-$	3.20	2.10	58.0	49.6
19	alcohol	**Me**–OH	3.50	2.60	50.3	44.3
20	ether	**Me**–OBu	3.32	2.42	58.5	50.1
21	enol ether	**Me**–OPh	3.78	2.88	55.1	46.7
22	ester	Me–CO_2**Me**	3.78	2.88	51.5	47.1
23	phosphonium salt	Ph$_3$P$^+$–**Me**	3.22	2.32	11.0	2.2

^{a}R = CH_2OH; compound is but-2-yn-1-ol.

The effects of groups based on carbon (the methyl group is joined directly to another carbon atom) appear in entries 2 to 11. All the electron-withdrawing groups based on carbonyl and cyanide have about the same effect (1.1–1.3 p.p.m. downfield shift from 0.9 p.p.m.). Groups based on nitrogen (Me–N bond) show a similar progression through amine, ammonium salt, amide, and nitro compound (entries 15–18). Finally, all the oxygen-based groups (Me–O bond) all show large shifts (entries 19–22).

Effects of substituents on CH_2 groups

It is more difficult to give a definitive list for CH_2 groups as they have two substituents. In Table 15.3 we set one substituent as phenyl (Ph) just because so many compounds of this kind are available, and give the actual shifts relative to $PhCH_2CH_3$ for protons (2.64 p.p.m.) and $PhCH_2CH_3$ for carbon (28.9 p.p.m.), again comparing the substituent with the CH skeleton.

If you compare the shifts caused on a CH_2 group by each functional group in Table 15.3 with the shifts caused on a CH_3 group by the same functional group in Table 15.2 you will see that they are broadly the same.

Table 15.3 Chemical shifts δ (p.p.m.) of CH_2 groups bonded to phenyl and functional groups

	Functional group	Compound	δ_H	δ_H – 2.64	δ_C	δ_C – 28.9
1	silane	$PhCH_2$–$SiMe_3$	?	?	27.5	–1.4
2	hydrogen	$PhCH_2$–H	2.32	–0.32	21.4	–7.5
3	alkane	$PhCH_2$–CH_3	2.64	0.00	28.9	0.0
4	benzene	$PhCH_2$–Ph	3.95	1.31	41.9	13.0
5	alkene	$PhCH_2$–CH=CH_2	3.38	0.74	41.2	12.3
6	nitrile	$PhCH_2$–CN	3.70	1.06	23.5	–5.4
7	acid	$PhCH_2$–CO_2H	3.71	1.07	41.1	12.2
8	ester	$PhCH_2$–CO_2Me	3.73	1.09	41.1	12.2
9	amide	$PhCH_2$–$CONEt_2$	3.70	1.06	?	?
10	ketone	$(PhCH_2)_2$C=O	3.70	1.06	49.1	20.2
11	thiol	$PhCH_2$–SH	3.69	1.05	28.9	0.0
12	sulfide	$(PhCH_2)_2$S	3.58	0.94	35.5	6.6
13	sulfoxide	$(PhCH_2)_2$S=O	3.88	1.24	57.2	28.3
14	sulfone	$(PhCH_2)_2SO_2$	4.11	1.47	57.9	29.0
15	amine	$PhCH_2$–NH_2	3.82	1.18	46.5	17.6
16	amide	HCONH–CH_2Ph	4.40	1.76	42.0	13.1
17	nitro[a]	$PhCH_2$–NO_2	5.20	2.56	81.0	52.1
18	ammonium salt	$PhCH_2$–NMe_3^+	4.5/4.9		55.1	26.2
19	alcohol	$PhCH_2$–OH	4.54	1.80	65.3	36.4
20	ether	$(PhCH_2)_2$O	4.52	1.78	72.1	43.2
21	enol ether	$PhCH_2$–OAr[b]	5.02	2.38	69.9	41.0
22	ester	$MeCO_2$–CH_2Ph	5.10	2.46	68.2	39.3
23	phosphonium salt	Ph_3P^+–CH_2Ph	5.39	2.75	30.6	1.7
24	chloride	$PhCH_2$–Cl	4.53	1.79	46.2	17.3
25	bromide	$PhCH_2$–Br	4.45	1.81	33.5	4.6

[a]Data from Kurz, 1978.
[b]Compound is (4-chloromethylphenyoxy)benzene.

Shifts of a CH group

We can do the same with a CH group, and in the left-hand side of Table 15.4 we take a series of iso-propyl compounds, comparing the measured shifts with those for the central proton (CHMe$_3$) or carbon (CHMe$_3$) of 2-methylpropane. We set two of the substituents as methyl groups and just vary the third. Yet again the shifts for the same substituent are broadly the same.

Table 15.4 Effects of α and β substitution on ^{1}H and ^{13}C NMR shifts on Me$_2$CHX[a]

X	Effects on C$_\alpha$ [Me$_2$CH–X], p.p.m.				Effects on C$_\beta$ [Me$_2$CH–X], p.p.m.			
	δ_H	$\delta_H - 1.68$	δ_C	$\delta_C - 25.0$	δ_H	$\delta_H - 0.9$	δ_C	$\delta_C - 8.4$
Li			10.2	−14.8			23.7	17.3
H	1.33	−0.35	15.9	−9.1	0.91	0.0	16.3	7.9
Me	1.68	0.00	25.0	0.0	0.89	0.0	24.6	16.2
CH=CH$_2$	2.28	0.60	32.0	7.0	0.99	0.09	22.0	13.6
Ph	2.90	1.22	34.1	9.1	1.24	0.34	24.0	15.6
CHO	2.42	0.74	41.0	16.0	1.12	0.22	15.5	7.1
COMe	2.58	0.90	41.7	16.7	1.11	0.21	27.4	19.0
CO$_2$H	2.58	0.90	34.0	9.0	1.20	0.30	18.8	10.4
CO$_2$Me	2.55	0.87	33.9	8.9	1.18	0.28	19.1	10.7
CONH$_2$	2.40	0.72	34.0	9.0	1.08	0.18	19.5	11.1
CN	2.71	1.03	20.0	−5.0	1.33	0.43	19.8	11.4
NH$_2$	3.11	1.43	42.8	17.8	1.08	0.18	26.2	17.8
NO$_2$	4.68	3.00	78.7	53.7	1.56	0.66	20.8	12.4
SH	3.13	1.45	30.6	5.6	1.33	0.43	27.6	19.2
SPr^i	3.00	1.32	33.5	8.5	1.27	0.37	23.7	15.3
OH	4.01	2.33	64.2	39.2	1.20	0.30	25.3	16.9
OPr^i	3.65	1.97	68.4	43.4	1.12	0.22	22.9	14.5
O$_2$CMe	5.00	3.32	67.6	42.6	1.22	0.32	21.4(8)	17.(0/4)
Cl	4.19	2.51	53.9	28.9	1.52	0.62	27.3	18.9
Br	4.29	2.61	45.4	20.4	1.71	0.81	28.5	20.1
I	4.32	2.36	31.2	6.2	1.90	1.00	21.4	13.0

[a]There is coupling between the CH and the Me$_2$ groups in the proton NMR.

Shifts in proton NMR are easier to calculate and more informative than those in carbon NMR

This final table helps to explain something we have avoided so far. Correlations of shifts caused by substituents in proton NMR really work very well. Those in ^{13}C NMR work much less well and more complicated equations are needed. More strikingly, the proton shifts often seem to fit better with our understanding of the chemistry of the compounds. There are two main reasons for this.

First, the carbon atom is much closer to the substituent than the proton. In the compounds in Table 15.2, the methyl carbon atom is directly bonded to the substituent, while the protons are separated from it by the carbon atom of the methyl group. If the functional group is based on a large electron-withdrawing atom like sulfur, the protons will experience a simple inductive electron withdrawal and have a proportional downfield shift. The carbon atom is close enough to the sulfur atom to be shielded as well by the lone-pair electrons in the large 3sp^3 orbitals. The proton shift caused by S in Me$_2$S is about the same (1.23 p.p.m.) as that caused by a set of more or less equally strong electron-withdrawing groups like CN (1.14 p.p.m.) or ester (1.18 p.p.m.). The carbon shift (9.7 p.p.m.) is less than that caused by an ester (11.2 p.p.m.) but much *more* than that caused by CN, which actually shifts the carbon upfield (–6.6 p.p.m.).

Second, the carbon shift is strongly affected not only by what is directly joined to that atom (α position), but also by what comes next (β position). The right-hand half of Table 15.4 shows what happens to methyl shifts when substituents are placed on the next carbon atom. There is very little effect on the proton spectrum: all the values are much less than the shifts caused by the same substituent on a methyl group in Table 15.2. Carbonyls give a downfield shift of about 1.2 p.p.m. when directly joined to a methyl group, but only of about 0.2 p.p.m. when one atom further away. By contrast, the shifts in the carbon spectrum are of the same order of magnitude in the two tables, and the β shift may even be greater than the α shift! The CN group shifts a directly bonded methyl group upfield (–6.6 p.p.m.) when directly bonded, but downfield (14.4 p.p.m.) when one atom further away. This is an exaggerated example, but the point is that these carbon shifts must *not* be used to suggest that the CN group is electron-donating in the α position and electron-withdrawing in the β position. The carbon shifts are erratic but the proton shifts give us useful information and are worth understanding as a guide both to structure determination and the chemistry of the compound.

When you use this table and are trying to interpret, say, a methyl group at 4.0 p.p.m. then you have no problem. Only one group is attached to a methyl group so you need a single shift value—it might be a methyl ester for example. But when you have a CH$_2$ group at 4.5 p.p.m. and you are interpreting a downfield shift of 3.2 p.p.m. you must beware. There are *two* groups attached to each CH$_2$ group and you might need a single shift of about 3 p.p.m. (say, an ester again) or two shifts of 1.5 p.p.m., and so on. The shifts are additive.

Table 15.5 Approximate additive functional group (X) shifts in ^{1}H NMR spectra

Entry	Functional group X	^{1}H NMR shift difference[a], p.p.m.
1	alkene (–C=C)	1.0
2	alkyne (–C≡C)	1.0
3	phenyl (–Ph)	1.3
4.	nitrile (–C≡N)	1.0
5	aldehyde (–CHO)	1.0
6	ketone (–COR)	1.0
7	acid (–CO$_2$H)	1.0
8	ester (–CO$_2$R)	1.0
9	amide (–CONH$_2$)	1.0
10	amine (–NH$_2$)	1.5
11	amide (–NHCOR)	2.0
12	nitro (–NO$_2$)	3.0
13	thiol (–SH)	1.0
14	sulfide (–SR)	1.0
15	sulfoxide (–SOR)	1.5
16	sulfone (–SO$_2$R)	2.0
17	alcohol (–OH)	2.0
18	ether (–OR)	2.0
19	aryl ether (–OAr)	2.5
20	ester (–O$_2$CR)	3.0
21	fluoride (–F)	3.0
22	chloride (–Cl)	2.0
23	bromide (–Br)	2.0
24	iodide (–I)	2.0

[a]To be added to 0.9 p.p.m. for MeX, 1.3 p.p.m. for CH$_2$X, or 1.7 p.p.m. for CHX.

Problems

1. A compound C_6H_5FO has a broad peak in the infrared at about 3100–3400 cm^{-1} and the following signals in its (proton decoupled) ^{13}C NMR spectrum. Suggest a structure for the compound and interpret the spectra.

δ_C (p.p.m.) 157.38 (doublet, coupling constant 229 Hz), 151.24 (singlet), 116.32 (doublet, coupling constant 7.5 Hz), 116.02 (doublet, coupling constant 23.2 Hz).

2. Suggest structures for the products of these reactions.

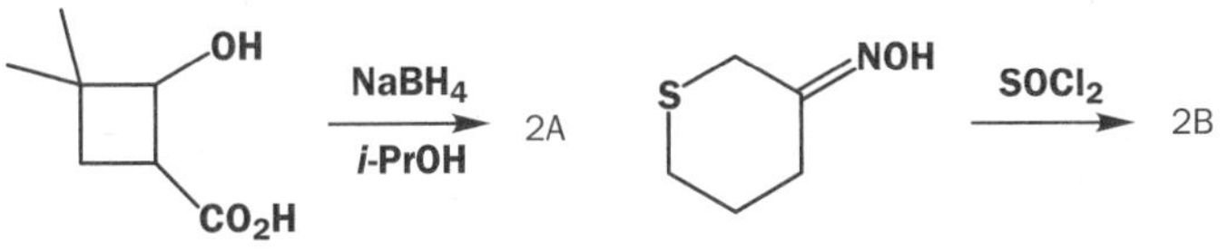

Compound 2A is $C_7H_{12}O_2$ and has IR 1725 cm^{-1}; δ_H (p.p.m.) 1.02 (6H, s), 1.66 (2H, t, J 7 Hz), 2.51 (2H, t, J 7 Hz), and 3.9 (2H, s).

Compound 2B has: m/z 149/151 (M^+ ratio 3:1); IR 2250 cm^{-1}; δ_H (p.p.m.) 2.0 (2H, q, J 7 Hz), 2.5 (2H, t, J 7 Hz), 2.9 (2H, t, J 7 Hz), and 4.6 (2H, s).

3. Two alternative structures are shown for the possible products of the following reactions. Explain in each case how you would decide which product is actually formed. Several pieces of evidence would be required and estimated values are more convincing than general statements.

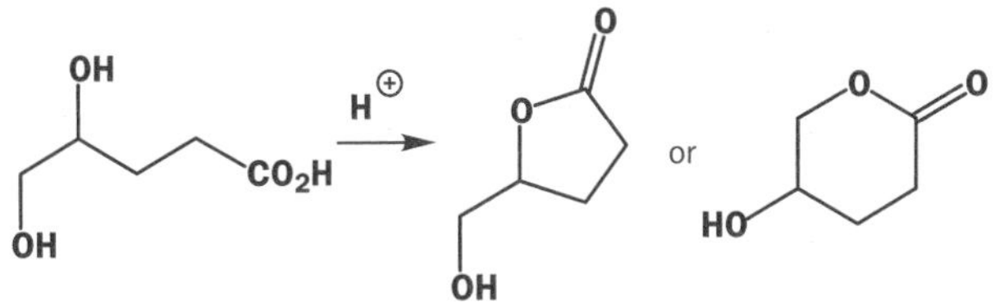

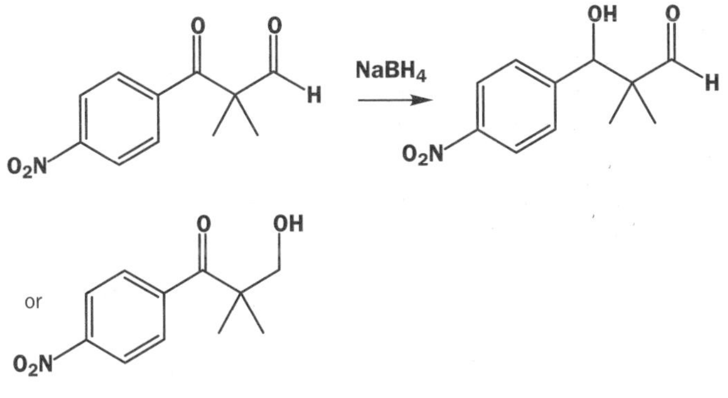

4. The following products might possibly be formed from the reaction of MeMgBr with the cyclic anhydride shown. How would you tell the difference between these compounds using IR and ^{13}C NMR spectra? With ^1H NMR available as well, how would your task be easier? Draw mechanisms for the formation of these compounds.

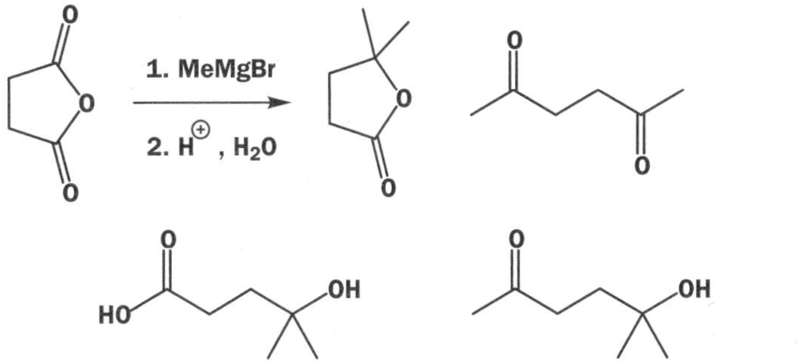

5. The NMR spectra of sodium fluoropyruvate in D_2O are given below. Are these data compatible with the structure shown? If not, suggest how the compound might exist in this solution.

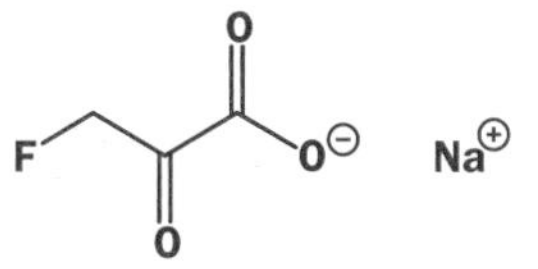

δ_H (p.p.m.) 4.43 (2H, d, J 47 Hz); δ_C 83.5 (d, J 22 Hz), 86.1 (d, J 171 Hz), and 176.1 (d, J 2 Hz).

6. An antibiotic isolated from a microorganism crystallized from water and formed (different) crystalline salts on treatment with either acid or base. The spectroscopic data were as follows.

Mass spectrum: 182 (M^+, 9%), 109 (100%), 137 (87%), and 74 (15%); δ_H (p.p.m.; in D_2O at pH < 1) 3.67 (2H, d), 4.57 (1H, t), 8.02 (2H, m), and 8.37 (1H, m); δ_C (p.p.m.; in D_2O at pH < 1) 33.5, 52.8, 130.1, 130.6, 134.9, 141.3, 155.9, and 170.2.

Suggest a structure for the antibiotic.

7. Suggest structures for the products of these two reactions.

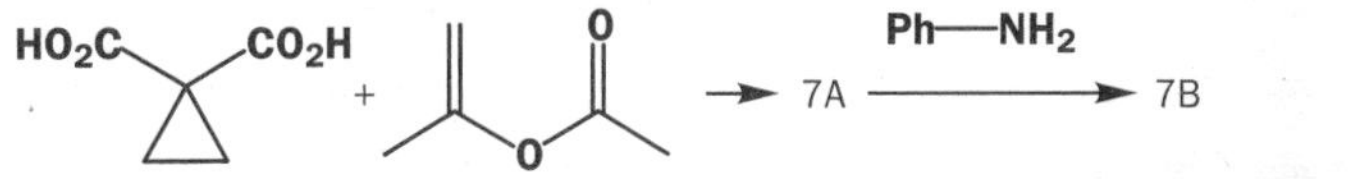

Compound 7A: m/z 170 (M^+, 1%), 84 (77%), and 66 (100%); IR 1773, 1754 cm^{-1}; δ_H(CDCl$_3$) (p.p.m.) 1.82 (6H, s) and 1.97 (4H, s); δ_C(CDCl$_3$) 22, 23, 28, 105, and 169 (the signals at 22 and 105 p.p.m. are weak).

Compound 7B: m/z 205 (M^+, 40%), 161 (50%), 160 (35%), 106 (100%), and 77 (42%); IR 1670, 1720 cm^{-1}; δ_H(CDCl$_3$) (p.p.m.) 2.55 (2H, m), 3.71 (1H, t, J 6 Hz), 3.92 (2H, m), 7.21 (2H, d, J 8 Hz), 7.35 (1H, t, J 8 Hz), and 7.62 (2H, d, J 8 Hz); δ_C(CDCl$_3$) 21, 47, 48, 121, 127, 130, 138, 170, and 172 p.p.m.

8. Treatment of the two compounds shown here with base gives an unknown compound with the spectra given here. What is its structure?

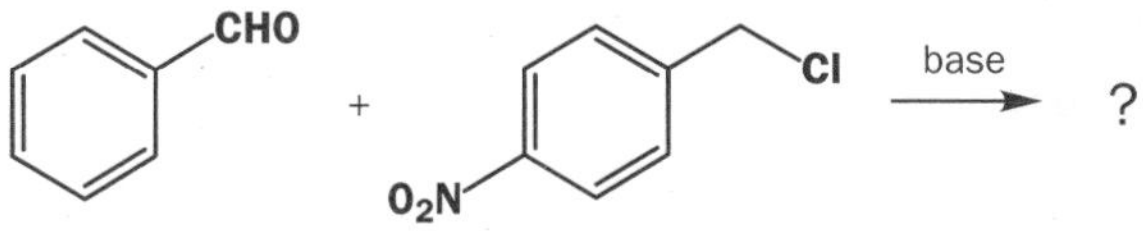

m/z 241 (M^+, 60%), 90 (100%), 89 (62%); δ_H(CDCl$_3$) (p.p.m.) 3.89 (1H, d, J 3 Hz), 4.01 (1H, d, J 3 Hz), 7.31 (5H, s), 7.54 (2H, d, J 10 Hz), and 8.29 (2H, d, J 10 Hz); δ_C(CDCl$_3$) 62, 64, 122, 125, 126, 127, 130, 136, 144, and 148 p.p.m. (the last three are weak).

9. Treatment of this epoxy-ketone gives a compound with the spectra shown below. What is its structure?

Hint. You might like to check the comments on pp. 366–7 before deciding on your answer.

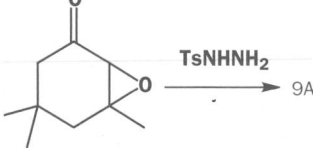

m/z 138 (M⁺, 12%), 109 (56%), 95 (100%), 81 (83%), 82 (64%), and 79 (74%); IR 3290, 2115, 1710 cm⁻¹; δ_H(CDCl₃) (p.p.m.) 1.12 (6H, s), 2.02 (1H, t, *J* 3 Hz), 2.15 (3H, s), 2.28 (2H, d, *J* 3 Hz), and 2.50 (2H, s); δ_C(CDCl₃) 26, 31, 32, 33, 52, 71, 82, and 208 p.p.m.

10. Reaction of the epoxy-alcohol below with LiBr in toluene gave a 92% yield of compound 10A. Suggest a structure for this compound.

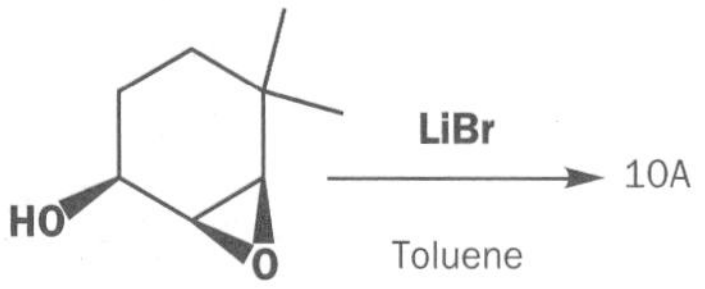

Compound 10A: *m/z* C₈H₁₂O; $\nu_{max.}$ (cm⁻¹) 1685, 1618; δ_H (p.p.m.) 1.26 (6H, s), 1.83 (2H, t, *J* 7 Hz), 2.50 (2H, dt, *J* 2.6, 7 Hz), 6.78 (1H, t, *J* 2.6 Hz), and 9.82 (1H, s); δ_C (p.p.m.) 189.2, 153.4, 152.7, 43.6, 40.8, 30.3, and 25.9.

11. Female boll weevils (a cotton pest) produce two isomeric compounds that aggregate the males for food and sex. A few mg of two isomeric active compounds, grandisol and *Z*-ochtodenol were isolated from 4.5 million insects. Suggest structures for these compounds from the spectroscopic data below. Signals marked * exchange with D₂O.

Z-Ochtodenol: *m/z* 154 (C₁₀H₁₈O), 139, 136, 121, 107, 69 (100%); ν_{max} (cm⁻¹) 3350, 1660; δ_H(p.p.m.) 0.89 (6H, s), 1.35–1.70 (4H broad m), 1.41 (1H, s*), 1.96 (2H, s), 2.06 (2H, t, *J* 6 Hz), 4.11 (2H, d, *J* 7 Hz), and 5.48 (1H, t, *J* 7 Hz).

Grandisol: *m/z* 154 (C₁₀H₁₈O), 139, 136, 121, 109, 68 (100%); ν_{max} (cm⁻¹) 3630, 3250–3550, and 1642; δ_H(p.p.m.) 1.15 (3H, s), 1.42 (1H, dddd, *J* 1.2, 6.2, 9.4, 13.4 Hz), 1.35–1.45 (1H, m), 1.55–1.67 (2H, m), 1.65 (3H, s), 1.70–1.81 (2H, m), 1.91–1.99 (1H, m), 2.52* (1H, broad t, *J* 9.0 Hz), 3.63 (1H, ddd, *J* 5.6, 9.4, 10.2 Hz), 3.66 (1H, ddd, *J* 6.2, 9.4, 10.2 Hz), 4.62 (1H, broad s), and 4.81 (1H, broad s); δ_C(p.p.m.) 19.1, 23.1, 28.3, 29.2, 36.8, 41.2, 52.4, 59.8, 109.6, and 145.1.

12. Suggest structures for the products of these reactions.

Data for compound 12A: C₁₀H₁₃OP; IR (cm⁻¹) 1610, 1235; δ_H (p.p.m.) 6.5–7.5 (5H, m), 6.42 (1H, t, *J* 17 Hz), 7.47 (1H, dd, *J* 17, 23 Hz), and 2.43 (6H, d, *J* 25 Hz).

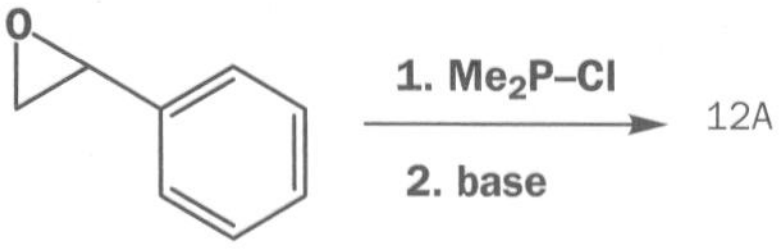

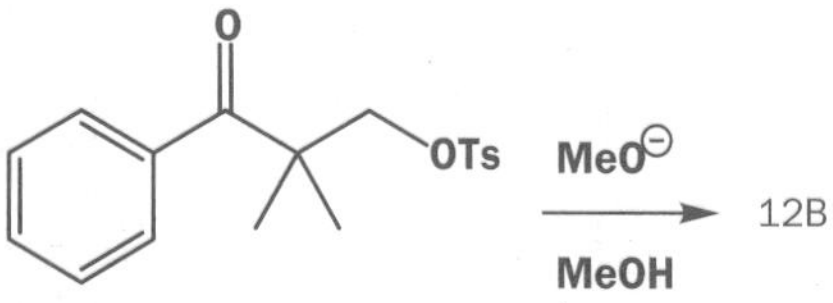

Data for compound 12B: C₁₂H₁₆O₂; IR CH and fingerprint only; δ_H (p.p.m.) 7.25 (5H, s), 4.28 (1H, d, *J* 4.8 Hz), 3.91 (1H, d, *J* 4.8 Hz), 2.96 (3H, s), 1.26 (3H, s), and 0.76 (3H, s).

13. Identify the compounds produced in these reactions. Warning! Do not attempt to deduce the structures from the starting materials but use the data! These molecules are so small that you can identify them from ¹H NMR alone.

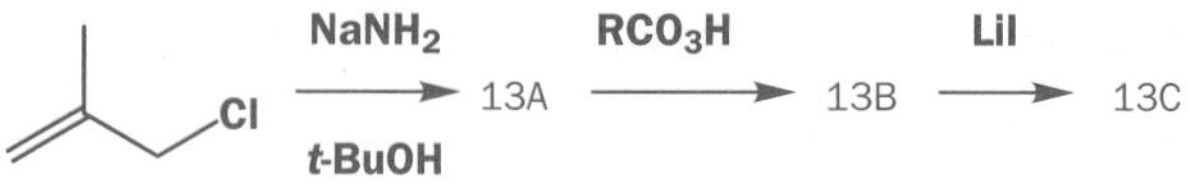

Compound 13A (C₄H₆): δ_H (p.p.m.) 5.35 (2H, s) and 1.00 (4H, s).

Compound 13B (C₄H₆O): δ_H (p.p.m.) 3.00 (2H, s), 0.90 (2H, d, *J* 3 Hz), and 0.80 (2H, d, *J* 3 Hz).

Compound 13C (C₄H₆O): δ_H (p.p.m.) 3.02 (4H, d, *J* 5 Hz) and 1.00 (2H, quintet, *J* 5 Hz)

14. The yellow crystalline antibiotic frustulosin was isolated from a fungus in 1975 and it was suggested that the structure was an equilibrium mixture of 14A and 14B. Apart from the difficulty that the NMR spectrum clearly shows one compound and not an equilibrium mixture of two compounds, what else makes you unsure of this assignment? Suggest a better structure. Signals marked * exchange with D₂O.

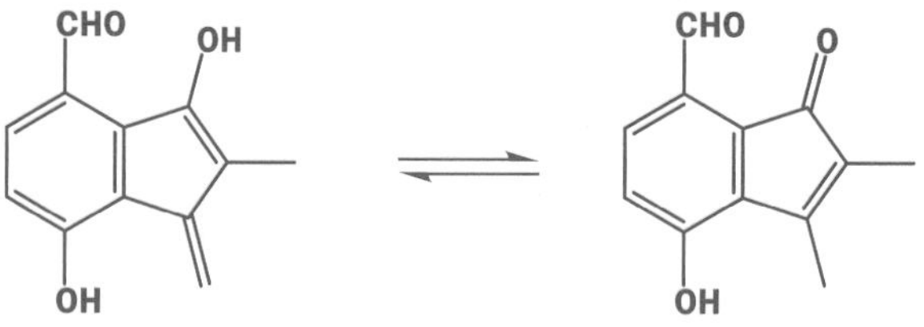

14A   14B

Frustulosin: *m/z* 202 (100%), 187 (20%), 174 (20%); ν_{max} (cm⁻¹) 3279, 1645, 1613, and 1522; δ_H (p.p.m.) 2.06 (3H, dd, *J* 1.0, 1.6 Hz), 5.44 (1H, dq, *J* 2.0, 1.6 Hz), 5.52 (1H, dq, *J* 2.0, 1.0 Hz), 4.5* (1H, broad s), 7.16 (1H, d, *J* 9.0 Hz), 6.88 (1H, dd, *J* 9.0, 0.4 Hz), 10.31 (1H, d, *J* 0.4 Hz), and 11.22* (1H, broad s); δ_C (p.p.m.) 22.8, 80.8, 100.6, 110.6, 118.4, 118.7, 112.6, 125.2, 126.1, 151.8, 154.5, and 195.6

Warning! This is difficult—after all the original authors initially got it wrong!

Hint. How might the DBEs be achieved without a second ring?

Stereochemistry

16

Connections

Building on:	**Arriving at:**	**Looking forward to:**
• Drawing organic molecules ch2	• Three-dimensional shape of molecules	• Diastereoselectivity ch34
• Organic structures ch4	• Molecules with mirror images	• Controlling alkene geometry ch31
• Nucleophilic addition to the carbonyl group ch9	• Molecules with symmetry	• Synthesis in action ch25
• Nucleophilic substitution at carbonyl groups ch12	• How to separate mirror-image molecules	• Controlling stereochemistry with cyclic compounds ch33
	• Diastereoisomers	• Asymmetric synthesis ch45
	• Shape and biological activity	• Chemistry of life ch49–51
	• How to draw stereochemistry	

Some compounds can exist as a pair of mirror-image forms

One of the very first reactions you met, back in Chapter 6, was between an aldehyde and cyanide. They give a **cyanohydrin**, a compound containing a nitrile group and a hydroxyl group.

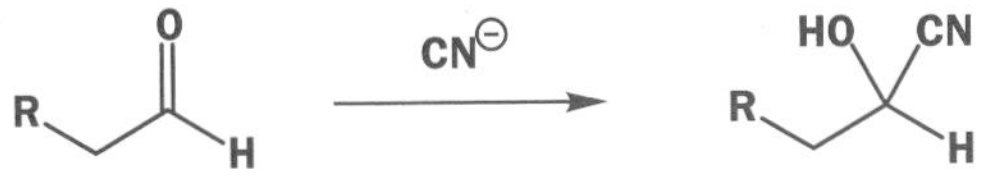

How many products are formed in this reaction? Well, the straightforward answer is one—there's only one aldehyde, only one cyanide ion, and only one reasonable way in which they can react. But this analysis is not *quite* correct. One point that we ignored when we first talked about this reaction, because it was irrelevant at that time, is that the carbonyl group of the aldehyde has two faces. The cyanide ion could attack either from the front face or the back face, giving, in each case, a distinct product.

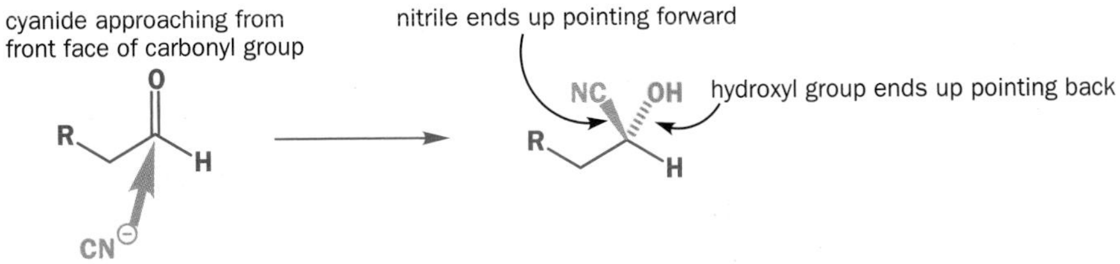

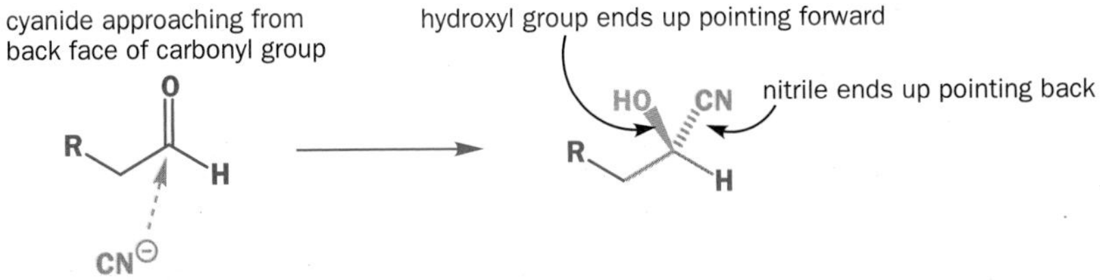

> Remember that the bold wedges represent bonds coming towards you, out of the paper, and the dashed bonds represent bonds going away from you, into the paper.

Are these two products different? If we lay them side by side and try to arrange them so that they look identical, we find that we can't—you can verify this by making models of the two structures.

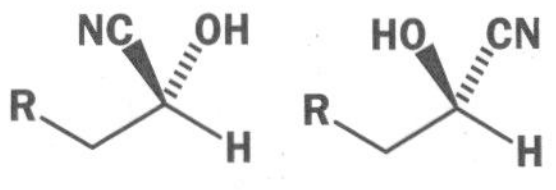

The structures are nonsuperimposable—so they are not identical. In fact, they are **mirror images** of each other: if we reflected one of the structures, A, in a mirror, we would get a structure that *is* identical with B.

> In reading this chapter, you will have to do a lot of mental manipulation of three-dimensional shapes. Because we can represent these shapes only in two dimensions, we suggest that you make models, using a molecular model kit, of the molecules we talk about. With some practice, you will be able to imagine the molecules you see on the page in three dimensions.

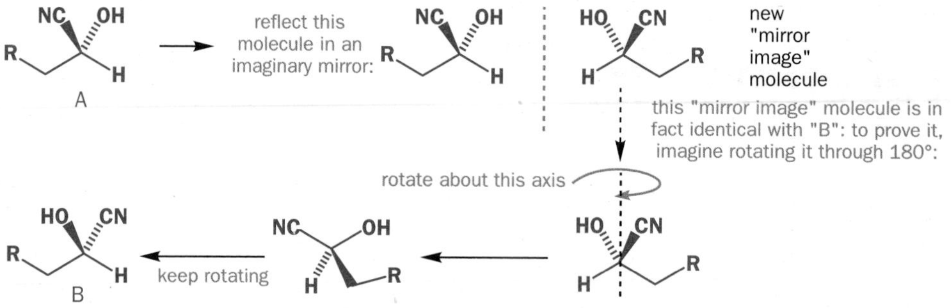

We call two structures that are not identical, but are mirror images of each other (like these two) **enantiomers**. Structures that are not superimposable on their mirror image, and can therefore exist as two enantiomers, are called **chiral**. In this reaction, the cyanide ions are just as likely to attack the 'front' face of the aldehyde as they are the 'back' face, so we get a 50:50 mixture of the two enantiomers.

● **Enantiomers and chirality**

- Enantiomers are structures that are not identical, but are *mirror images* of each other
- Structures are *chiral* if they cannot be superimposed upon their mirror image

Now consider another similar reaction, which you have also met—the addition of cyanide to acetone.

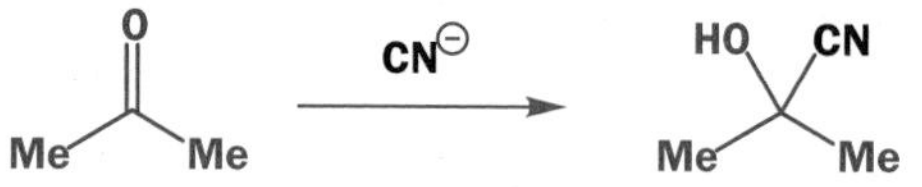

Again a cyanohydrin is formed. You might imagine that attacking the front or the back face of the acetone molecule could again give two structures, C and D.

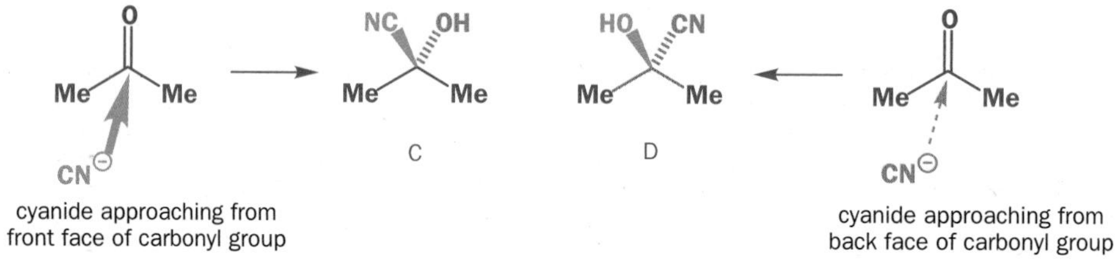

cyanide approaching from front face of carbonyl group

cyanide approaching from back face of carbonyl group

However, this time, rotating one to match the other shows that they are superimposable and therefore identical.

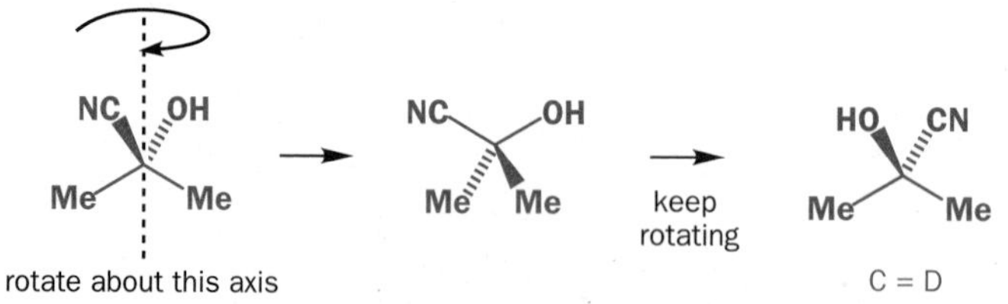

rotate about this axis

keep rotating

C = D

Make sure that you are clear about this: C and D are identical molecules, while A and B are mirror images of each other. Reflection in a mirror makes no difference to C or D; they are superimposable upon their own mirror images, and therefore cannot exist as two enantiomers. Structures that are superimposable on their mirror images are called **achiral**.

● *Achiral* structures are superimposable on their mirror images

Chiral molecules have no plane of symmetry

What is the essential difference between these two compounds that means one is superimposable on its mirror image and one is not? The answer is symmetry. Acetone cyanohydrin has a plane of symmetry running through the molecule. This plane cuts the central carbon and the OH and CN groups in half and has one methyl group on each side.

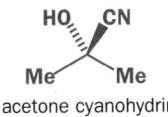

acetone cyanohydrin

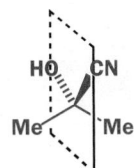

plane of symmetry
runs through
central carbon,
OH and CN

On the other hand, the aldehyde cyanohydrin has no plane of symmetry: the plane of the paper has OH on one side and CN on the other while the plane at right angles to the paper has H on one side and RCH$_2$ on the other. This compound is completely unsymmetrical and has two enantiomers.

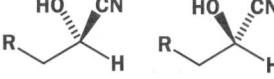

■
This statement is, in fact, slightly incomplete, but it outlines such a useful concept that for the time being we shall use it as a valuable guideline.

● Planes of symmetry and chirality

- **Any structure that has no plane of symmetry can exist as two mirror-image forms (*enantiomers*)**
- **Any structure with a plane of symmetry cannot exist as two enantiomers**

By 'structure', we don't just mean chemical structure: the same rules apply to everyday objects. Some examples from among more familiar objects in the world around us should help make these ideas clear. Look around you and find a chiral object—a car, a pair of scissors, a screw (but not the screwdriver), and anything with writing on it like this page. Look again for achiral objects with planes of symmetry—a plain mug, saucepan, chair, most man-made things without writing on them. The most significant chiral object near you is the hand you write with.

Some examples

Gloves, hands, and socks

Most gloves exist in pairs of nonidentical mirror-image forms: only a left glove fits a left hand and only a right glove fits a right hand. This property of gloves and of the hands inside them gives us the word 'chiral'—*cheir* is Greek for 'hand'. Hands and gloves are chiral; they have no plane of symmetry, and a left glove is not superimposable on its mirror image (a right glove). Feet are chiral too, as are shoes. But socks (usually!) are not. Though we all sometimes have problems finding two socks of a matching colour, once you've found them, you never have to worry about which sock goes on which foot, because socks are achiral. A pair of socks is manufactured as two identical objects, each of which has a mirror plane.

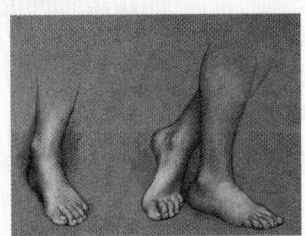

The ancient Egyptians had less care for the chirality of hands and their paintings often show people, even Pharaohs, with two left hands or two right hands—they just didn't seem to notice.

Tennis racquets and golf clubs

If you are left-handed and want to play golf, you either have to play in a right-handed manner, or get hold of a set of left-handed golf clubs. Golf clubs are clearly therefore chiral; they can exist as either of two enantiomers. You can tell this just by looking at a golf club. It has no plane of symmetry, so it must be chiral. But left-handed tennis players have no problem using the same racquets as right-handed tennis players and modern tennis players of either chirality sometimes swap the racquet from hand to hand. Look at a tennis racquet: it has a plane of symmetry, so it's achiral. It can't exist as two mirror-image forms.

> ● **To summarize**
> * A structure *with* a plane of symmetry is *achiral* and *superimposable* on its mirror image and *cannot* exist as two enantiomers
> * A structure *without* a plane of symmetry is *chiral* and *not superimposable* on its mirror image and *can* exist as two enantiomers

Stereogenic centres

Back to chemistry, and the product from the reaction of an aldehyde with cyanide. We explained above that this compound, being chiral, can exist as two enantiomers. Enantiomers are clearly isomers; they consist of the same parts joined together in a different way. In particular, enantiomers are a type of isomer called **stereoisomers**, because the isomers differ not in the connectivity of the atoms, but only in the overall shape of the molecule.

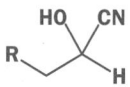

Stereoisomers and constitutional isomers

Isomers are compounds that contain the same atoms bonded together in different ways. If the connectivity of the atoms in the two isomers is different, they are **constitutional isomers**. If the connectivity of the atoms in the two isomers is the same, they are **stereoisomers**. Enantiomers are stereoisomers, and so are E and Z double bonds. We shall meet other types of stereoisomers shortly.

constitutional isomers: the way the atoms are connected up (their *connectivity*) differs

enantiomers

E/Z isomers (double bond isomers)

stereoisomers: the atoms have the same connectivity, but are arranged differently

When we don't show bold and dashed bonds to indicate the three-dimensional structure of the molecule, we mean that we are talking about both enantiomers of the molecule. Another useful way of representing this is with wiggly bonds. Wiggly bonds are in fact slightly ambiguous: chemists use them to mean, as they do here, both stereoisomers, but also to mean just one stereoisomer, but unknown stereochemistry.

We should also introduce you briefly to another pair of concepts here, which you will meet again in more detail in Chapter 17: *configuration* and *conformation*. Two stereoisomers really are different molecules: they cannot be interconverted without breaking a bond somewhere. We therefore say that they have different **configurations**. But any molecule can exist in a number of **conformations**: two conformations differ only in the temporary way the molecule happens to arrange itself, and can easily be interconverted just by rotating around bonds. Humans all have the same *configuration*: two arms joined to the shoulders. We may have different *conformations*: arms folded, arms raised, pointing, waving, etc.

● Configuration and conformation

- Changing the *configuration* of a molecule always means that bonds are broken
- A different configuration is a different molecule
- Changing the *conformation* of a molecule means rotating about bonds, but not breaking them
- Conformations of a molecule are readily interconvertible, and are all the same molecule

two configurations: going from one enantiomer to the other requires a bond to be broken

three conformations of the same enantiomer: getting from one to the other just requires rotation about a bond: all three are the same molecule

An aldehyde cyanohydrin is chiral because it does not have a plane of symmetry. In fact, it *cannot* have a plane of symmetry, because it contains a tetrahedral carbon atom carrying four different groups: OH, CN, RCH$_2$, and H. Such a carbon atom is known as a **stereogenic** or **chiral centre**. The product of cyanide and acetone is not chiral; it has a plane of symmetry, and no chiral centre because two of the groups on the central carbon atom are the same.

aldehyde cyanohydrin

four different groups

stereogenic centre or chiral centre

only three different groups

● If a molecule contains one carbon atom carrying four different groups it will not have a plane of symmetry and must therefore be chiral. A carbon atom carrying four different groups is a **stereogenic** or **chiral centre**.

We saw how the two enantiomers of the aldehyde cyanohydrin arose by attack of cyanide on the two faces of the carbonyl group of the aldehyde. We said that there was nothing to favour one face over the other, so the enantiomers must be formed in equal quantities. A mixture of equal quantities of a pair of enantiomers is called a **racemic mixture**.

> ● A **racemic mixture** is a mixture of two enantiomers in equal proportions. This principle is very important. Never forget that, if the starting materials of a reaction are achiral, and the products are chiral, they will be formed as a racemic mixture of two enantiomers.

Here are some more reactions you have come across that make chiral products from achiral starting materials. In each case, the principle must hold—equal amounts of the two enantiomers (racemic mixtures) are formed.

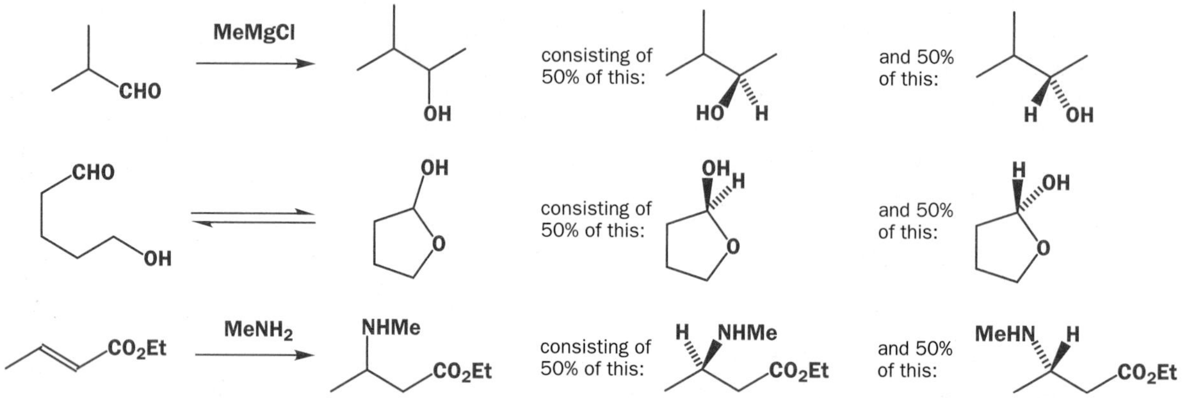

Many chiral molecules are present in nature as single enantiomers

Let's turn to some simple, but chiral, molecules—the natural amino acids. All amino acids have a carbon carrying an amino group, a carboxyl group, a hydrogen atom, and the R group, which varies from amino acid to amino acid. So unless R = H (this is the case for glycine), amino acids always contain a chiral centre and lack a plane of symmetry.

■ Molecules are chiral if they lack a plane of symmetry. You can immediately see that amino acids lack a plane of symmetry because (except glycine) they contain a chiral centre.

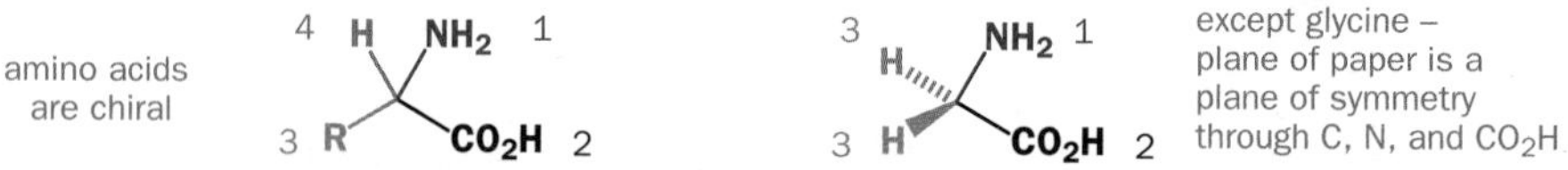

It is possible to make amino acids quite straightforwardly in the lab. The scheme below shows a synthesis of alanine, for example. It is a version of the Strecker synthesis you met in Chapter 12.

laboratory synthesis of racemic alanine from acetaldehyde

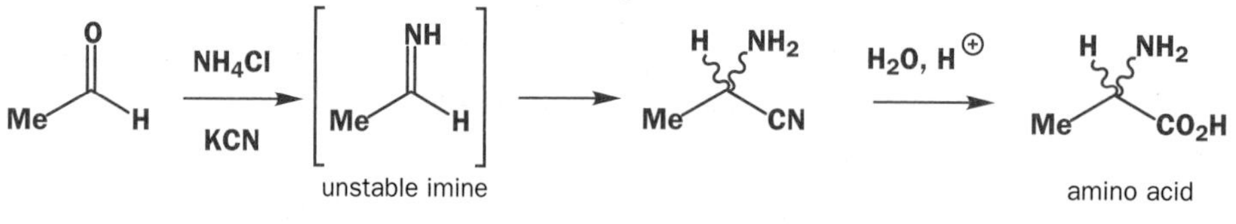

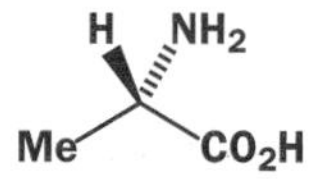

Alanine made in this way must be racemic, because the starting materials are achiral. However, if we isolate alanine from a natural source—by hydrolysing vegetable protein, for example—we find that this is not the case. Natural alanine is solely one enantiomer, the one drawn below. Samples of chiral compounds that contain only one enantiomer are called **enantiomerically pure**. We know that 'natural' alanine contains only this enantiomer from X-ray crystal structures.

alanine extracted from plants consists only of this enantiomer

Enantiomeric alanine

In fact, Nature does sometimes (but very rarely) use the other enantiomer of alanine—for example, in the construction of bacterial cell walls. Some antibiotics (such as vancomycin) owe their selectivity to the way they can recognize these 'unnatural' alanine components and destroy the cell wall that contains them.

Before we go further, we should just mention one common point of confusion. Any compound whose molecules do not have a plane of symmetry is chiral. Any sample of a chiral compound that contains molecules all of the same enantiomer is enantiomerically pure. *All* alanine is chiral (the structure has no plane of symmetry) but *lab-produced* alanine is racemic (a 50:50 mixture of enantiomers) whereas *naturally isolated* alanine is enantiomerically pure.

Most of the molecules we find in nature are chiral—a complicated molecule is much more likely not to have a plane of symmetry than to have one. Nearly all of these chiral molecules in living systems are found not as racemic mixtures, but as single enantiomers. This fact has profound implications, for example, in the chemistry of drug design, and we will come back to it later.

R and *S* can be used to describe the configuration of a chiral centre

Before going on to talk about single enantiomers of chiral molecules in more detail, we need to explain how chemists explain which enantiomer they're talking about. We can, of course, just draw a diagram, showing which groups go into the plane of the paper and which groups come out of the plane of the paper. This is best for complicated molecules. Alternatively, we can use the following set of rules to assign a letter, *R* or *S*, to describe the configuration of groups at a chiral centre in the molecule.

Here again is the enantiomer of alanine you get if you extract alanine from living things.

1   Assign a priority number to each substituent at the chiral centre. Atoms with higher atomic numbers get higher priority.

Alanine's chiral centre carries one N atom (atomic number 7), two C atoms (atomic number 6), and one H atom (atomic number 1). So, we assign priority 1 to the NH_2 group, because N has the highest atomic number. Priorities 2 and 3 will be assigned to the CO_2H and the CH_3 groups, and priority 4 to the hydrogen atom; but we need a way of deciding which of CO_2H and CH_3 takes priority over the other. If two (or more) of the atoms attached to the chiral centre are identical, then we assign priorities to these two by assessing the atoms attached to those atoms. In this case, one of the carbon atoms carries oxygen atoms (atomic number 8), and one carries only hydrogen atoms (atomic number 1). So CO_2H is higher priority that CH_3; in other words, CO_2H gets priority 2 and CH_3 priority 3.

2   Arrange the molecule so that the lowest priority substituent is pointing away from you.

In our example, naturally extracted alanine, H is priority 4, so we need to look at the molecule with the H atom pointing into the paper, like this.

3   Mentally move from substituent priority 1 to 2 to 3. If you are moving in a clockwise manner, assign the label *R* to the chiral centre; if you are moving in an anticlockwise manner, assign the label *S* to the chiral centre.

A good way of visualizing this is to imagine turning a steering wheel in the direction of the numbering. If you are turning your car to the right, you have *R*; if you are turning to the left you have *S*. For our molecule of natural alanine, if we move from NH_2 (1) to CO_2H (2) to CH_3 (3) we're going anticlockwise (turning to the left), so we call this enantiomer (*S*)-alanine.

You can try working the other way, from the configurational label to the structure. Take lactic acid as an example. Lactic acid is produced by bacterial action on milk; it's also produced in your muscles when they have to work with an insufficient supply of oxygen, such as during bursts of vigorous exercise. Lactic acid produced by fermentation is often racemic, though certain species of bacteria produce solely (*R*)-lactic acid. On the other hand, lactic acid produced by anaerobic respiration in muscles has the *S* configuration.

As a brief exercise, try drawing the three-dimensional structure of (*R*)-lactic acid. (You may find this easier if you draw both enantiomers first and then assign a label to each.)

> ▶
>
> Remember—we use the word *configuration* to describe the arrangement of bonds around an atom. Configurations cannot be changed without breaking bonds.

natural alanine

> ▶
>
> These priority rules are also used to assign *E* and *Z* to alkenes, (see p. 487) and are sometimes called the Cahn–Ingold–Prelog (CIP) rules, after their devisors.

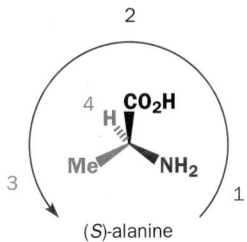

(*S*)-alanine

lactic acid

▶

Remember how, in Chapter 3, we showed you how hydrogen atoms at stereogenic centres (we didn't call them that then) could be missed out—we just assume that they take up the fourth vertex of the imagined tetrahedron at the stereogenic centre.

This also brings us to another point about drawing stereogenic centres: always try to have the carbon skeleton lying in the plane of the paper: in other words, try to draw

You should have drawn:

(*R*)-lactic acid   or   (*R*)-lactic acid

rather than, say,

(*R*)-lactic acid   (*R*)-lactic acid

Both are correct but the first will make things a lot easier when we are talking about molecules with several chiral centres!

■ The longer answer is more involved, and we go into it in more detail in Chapter 45.

Remember that, if we had made lactic acid in the lab from simple achiral starting materials, we would have got a racemic mixture of (*R*) and (*S*) lactic acid. Reactions in living systems can produce enantiomerically pure compounds because they make use of enzymes, themselves enantiomerically pure compounds of (*S*)-amino acids.

Is there a chemical difference between two enantiomers?

The short answer is *no*. Take (*S*)-alanine (in other words, alanine extracted from plants) and (*R*)-alanine (the enantiomer found in bacterial cell walls) as examples. They both have identical NMR spectra, identical IR spectra, and identical physical properties, with a single important exception. If you shine plane-polarized light through a solution of (*S*)-alanine, you will find that the light is rotated to the right. A solution of (*R*)-alanine rotates plane-polarized light to the left. Racemic alanine, on the other hand, lets the light pass unrotated.

The rotation of plane-polarized light is known as optical activity

Observation of the rotation of plane-polarized light is known as **polarimetry**; it is a straightforward way of finding out if a sample is racemic or if it contains more of one enantiomer than the other. Polarimetric measurements are carried out in a polarimeter, which has a single-wavelength (monochromatic) light source with a plane-polarizing filter, a sample holder, where a cell containing a solution of the substance under examination can be placed, and a detector with a read-out that indicates by how much the light is rotated. Rotation to the right is given a positive value, rotation to the left a negative one.

▶

Plane-polarized light can be considered as a beam of light in which all of the light waves have their direction of vibration aligned parallel. It is produced by shining light through a polarizing filter.

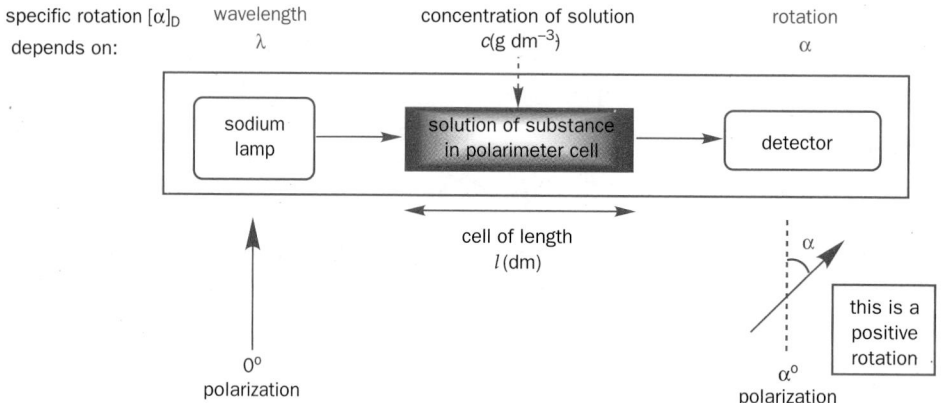

Specific rotation

The angle through which a sample of a compound (usually a solution) rotates plane-polarized light depends on a number of factors, the most important ones being the path length (how far the light has to pass through the solution), concentration, temperature, solvent, and wavelength. Typically, optical rotations are measured at 20 °C in a solvent such as ethanol or chloroform, and the light used is from a sodium lamp, with a wavelength of 589 nm.

The observed angle through which the light is rotated is given the symbol α. By dividing this value by the path length l (in dm) and the concentration c (in g cm^{-3}) we get a value, $[\alpha]$, which is specific to the compound in question. The choice of units is eccentric and arbitrary but is universal so we must live with it.

$$[\alpha] = \frac{\alpha}{cl}$$

Most $[\alpha]$ values are quoted as $[\alpha]_D$ (where the D indicates the wavelength of 589 nm, the 'D line' of a sodium lamp) or $[\alpha]_D^{20}$, the 20 indicating 20 °C. These define the remaining variables.

Here is an example. A simple acid, known as mandelic acid, can be obtained from almonds in an enantiomerically pure state.

28 mg was dissolved in 1 cm³ of ethanol and the solution placed in a 10 cm long polarimeter cell. An optical rotation α of $-4.35°$ was measured (that is, 4.35° to the left) at 20 °C with light of wavelength 589 nm.

What is the specific rotation of the acid?

First, we need to convert the concentration to grammes per cubic centimetre: 28 mg in 1 cm³ is the same as 0.028 g cm⁻³. The path length of 10 cm is 1 dm, so

$$[\alpha]_D^{20} = \frac{\alpha}{cl} = \frac{-4.35}{0.028 \times 1} = -155.4$$

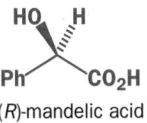

(R)-mandelic acid

▶

Note that the units of optical rotation are not degrees: by convention, $[\alpha]$ is usually quoted without units.

■

$[\alpha]_D$ values can be used as a guide to the enantiomeric purity of a sample, in other words, to how much of each enantiomer it contains. We will come back to this in Chapter 45.

Enantiomers can be described as (+) or (−)

We can use the fact that two enantiomers rotate plane-polarized light in opposite directions to assign each a label that doesn't depend on knowing its configuration. We call the enantiomer that rotates plane-polarized light to the right (gives a positive rotation) the **(+)-enantiomer** (or the *dextrorotatory* enantiomer) and the enantiomer that rotates plane-polarized light to the left (gives a negative rotation) the **(−)-enantiomer** (or the *laevorotatory* enantiomer). The direction in which light is rotated is not dependent on whether a stereogenic centre is *R* or *S*. An (*R*) compound is equally as likely to be (+) as (−)—of course, if it is (+) then its (*S*) enantiomer must be (−). The enantiomer of mandelic acid we have just discussed, for example, is *R*-(−)-mandelic acid, because its specific rotation is negative, and (*S*)-alanine happens to be *S*-(+)-alanine. The labels (+) and (−) were more useful before the days of X-ray crystallography, when chemists did not know the actual configuration of the molecules they studied, and could distinguish two enantiomers only by the signs of their specific rotations.

Enantiomers can be described as D or L

Long before the appearance of X-ray crystallography as an analytical tool, chemists had to discover the detailed structure and stereochemistry of molecules by a complex series of degradations. A molecule was gradually broken down into its constituents, and from the products that were formed the overall structure of the starting molecule was deduced. As far as stereochemistry was concerned, it was possible to measure the specific rotation of a compound, but not to determine its configuration. However, by using series of degradations it was possible to tell whether certain compounds had the same or opposite configurations.

Glyceraldehyde is one of the simplest chiral compounds in nature. Because of this, chemists took it as a standard against which the configurations of other compounds could be compared. The two enantiomers of glyceraldehyde were given the labels D (for dextro—because it was the (+)-enantiomer) and L (for laevo—because it was the (−)-enantiomer). Any enantiomerically pure compound that could be related, by a series of chemical degradations and transformations, to D-(+)-glyceraldehyde was labelled D, and any compound that could be related to L-(−)-glyceraldehyde was labelled L. The processes concerned were slow and laborious (the scheme below shows how (−)-lactic acid was shown to be D-(−)-lactic acid) and are never used today. D and L are now used only for certain well known natural molecules, where their use is established by tradition, for example, the L-amino acids or the D-sugars. These labels, D and L, are in *small capital* letters.

● Remember that the *R*/*S*, +/–, and D/L nomenclatures all arise from different observations and the fact that a molecule has, say, the *R* configuration gives no clue as to whether it will have + or – optical activity or be labelled D or L. Never try and label a molecule as D/L, or +/–, simply by working it out from the structure. Likewise, never try and predict whether a molecule will have a + or – specific rotation by looking at the structure.

The correlation between D-(–)-lactic acid and D-(+)-glyceraldehyde

Here, for example, is the way that (–)-lactic acid was shown to have the same configuration as D-(+) glyceraldehyde. We do not expect you to have come across the reactions used here.

D-(–)-lactic acid   (+)-isoserine   (–)-glyceric acid   D-(+)-glyceraldehyde

Diastereoisomers are stereoisomers that are not enantiomers

Two enantiomers are chemically identical because they are mirror images of one another. Other types of stereoisomers may be chemically (and physically) quite different. These two alkenes, for example, are geometrical isomers (or *cis–trans* isomers). Their physical chemical properties are different, as you would expect, since they are quite different in shape.

butenedioic acids

fumaric acid   maleic acid

trans-butenedioic acid (fumaric acid)   *cis*-butenedioic acid (maleic acid)
m.p. 299–300 °C   m.p. 140–142 °C

A similar type of stereoisomerism can exist in cyclic compounds. In one of these 4-*t*-butyl-cyclohexanols the two substituents are on the same side of the ring; in the other, they are on opposite sides of the ring. Again, the two compounds have chemical and physical properties that are quite different.

4-*t*-butylcyclohexanol

cis isomer   *trans* isomer
cis 4-*t*-butylcyclohexanol   *trans* 4-*t*-butylcyclohexanol
mp 82–83 °C   mp 80–81 °C
^{1}H NMR: δ_H of green proton 4.02 ^{1}H NMR: δ_H of green proton 3.50

Stereoisomers that are not mirror images of one another are called **diastereoisomers**. Both of these pairs of isomers fall into this category. Notice how the physical and chemical properties of a pair of diastereoisomers differ.

▶
The physical and chemical properties of enantiomers are identical; the physical and chemical properties of diastereoisomers differ. 'Diastereoisomer' is sometimes shortened to 'diastereomer'.

Diastereoisomers can be chiral or achiral

This pair of epoxides was produced by chemists in Pennsylvania in the course of research on drugs intended to alleviate the symptoms of asthma. Clearly, they are again diastereoisomers, and again they have different properties. Although the reaction they were using to make these compounds gave some of each diastereoisomer, the chemists working on these compounds only wanted to use the first (*trans*) epoxide. They were able to separate it from its *cis* diastereoisomer by chromatography because the diastereoisomers differ in polarity.

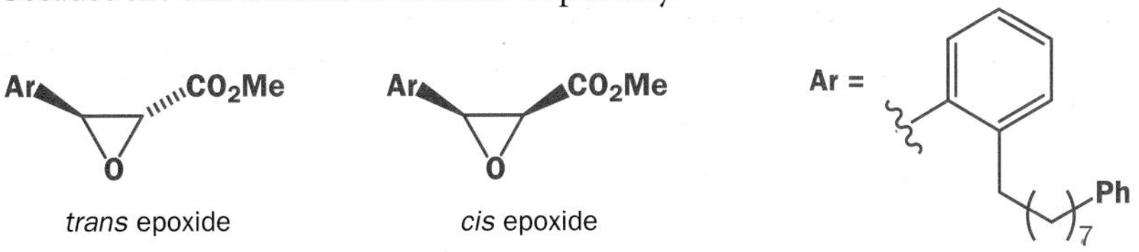

trans epoxide   cis epoxide   Ar =

This time, the diastereoisomers are a little more complex than the examples above. The first two pairs of diastereoisomers we looked at were **achiral**—they each had a plane of symmetry through the molecule.

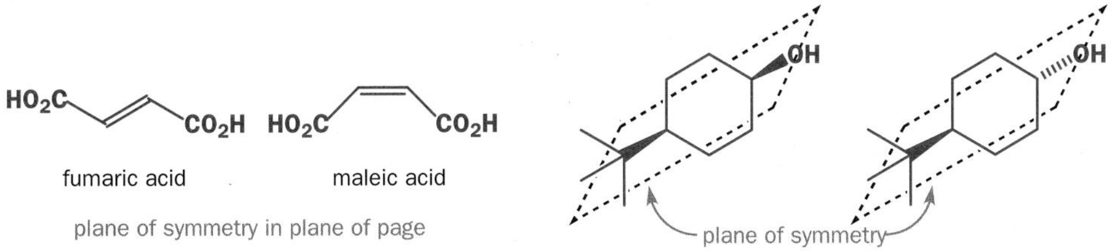

fumaric acid   maleic acid

plane of symmetry in plane of page   plane of symmetry

The last pair of diastereoisomers, on the other hand, is chiral. We know this because they do not have a plane of symmetry and we can check that by drawing the mirror image of each one: it is not superimposable on the first structure.

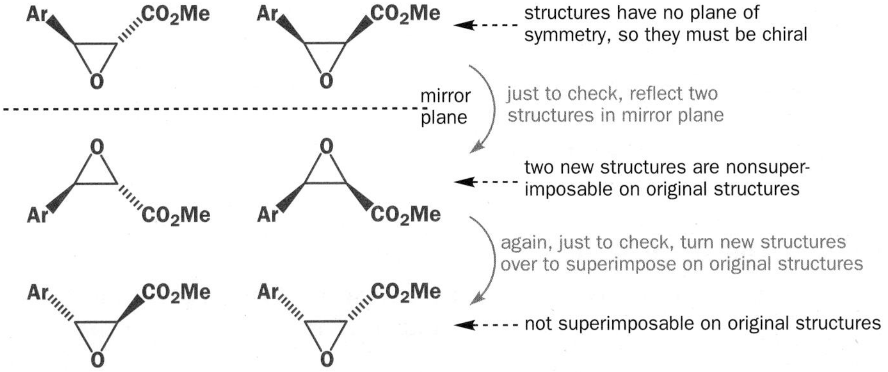

structures have no plane of symmetry, so they must be chiral

mirror plane — just to check, reflect two structures in mirror plane

two new structures are nonsuperimposable on original structures

again, just to check, turn new structures over to superimpose on original structures

not superimposable on original structures

If a compound is chiral, it can exist as two enantiomers. We've just drawn the two enantiomers of each of the diastereoisomers of our epoxide. This set of four structures contains two diastereoisomers (stereoisomers that are not mirror images). These are the two different chemical compounds, the *cis* and *trans* epoxides, that have different properties. Each can exist as two enantiomers (stereoisomers that are mirror images) indistinguishable except for rotation. We have two pairs of diastereoisomers and two pairs of enantiomers. When you are considering the stereochemistry of a compound, always distinguish the diastereoisomers first and then split these into enantiomers if they are chiral.

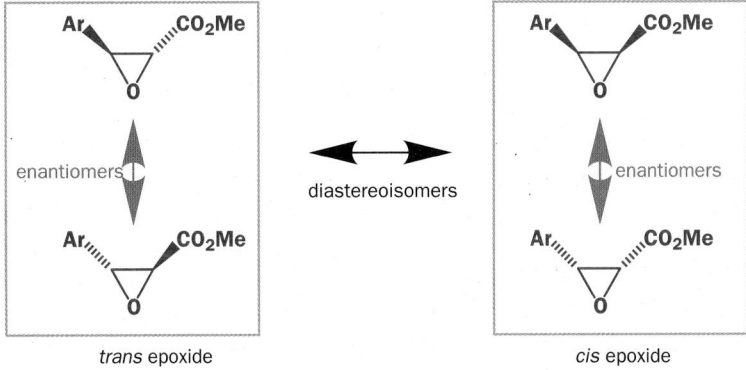

enantiomers — diastereoisomers — enantiomers

trans epoxide   cis epoxide

In fact, the chemists working on these compounds wanted only one enantiomer of the *trans* epoxide—the top left stereoisomer. They were able to separate the *trans* epoxide from the *cis* epoxide by chromatography, because they are diastereoisomers. However, because they had made both diastereoisomers in the laboratory from achiral starting materials, both diastereoisomers were racemic mixtures of the two enantiomers. Separating the top enantiomer of the *trans* epoxide from the bottom one was much harder because enantiomers have identical physical and chemical properties. To get just the enantiomer they wanted the chemists had to develop some completely different chemistry, using enantiomerically pure compounds derived from nature.

■
We shall discuss how chemists make enantiomerically pure compounds later in this chapter, and in more detail in Chapter 45.

Absolute and relative stereochemistry

When we talk about two chiral diastereoisomers, we have no choice but to draw the structure of one enantiomer of each diastereoisomer, because we need to include the stereochemical information to distinguish them, even if we're talking about a racemic mixture of the two enantiomers. To avoid confusion, it's best to write something definite under the structure, such as '±' (meaning racemic) under a structure if it means 'this diastereoisomer' but not 'this enantiomer of this diastereoisomer'.

So we should say, for example, that the chemists were able to separate these two diastereoisomers

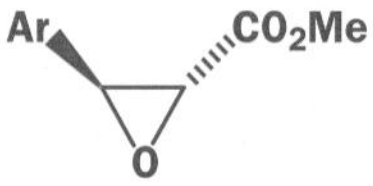

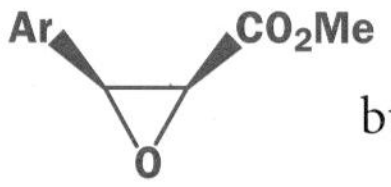

but that they wanted only this enantiomer.

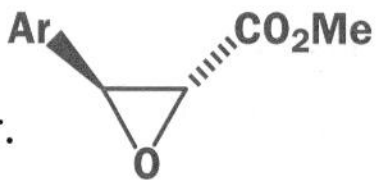

When the stereochemistry drawn on a molecule means 'this diastereoisomer', we say that we are representing **relative stereochemistry**; when it means 'this enantiomer of this diastereoisomer' we say we are representing its **absolute stereochemistry**. Relative stereochemistry tells us only how the stereogenic centres *within a molecule* relate to each other.

● Enantiomers and diastereoisomers

- **Enantiomers** are stereoisomers that are mirror images. A pair of enantiomers are mirror-image forms of the same compound and have opposite **absolute stereochemistry**

- **Diastereoisomers** are stereoisomers that are not mirror images. Two diastereoisomers are different compounds, and have different **relative stereochemistry**

Diastereoisomers may be achiral (have a plane of symmetry); for example,

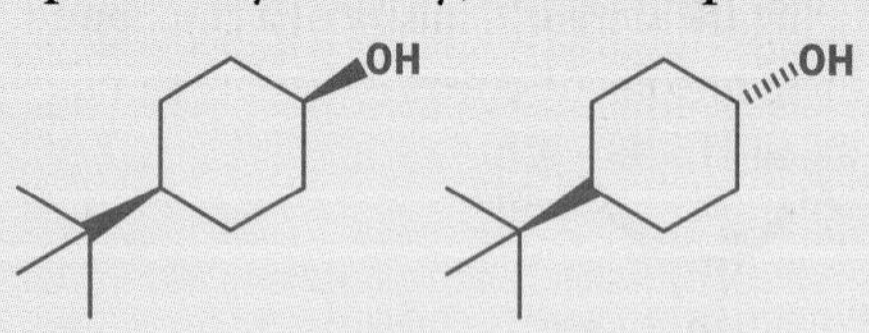

Or they may be chiral (have no plane of symmetry); for example,

Diastereoisomers can arise when structures have more than one stereogenic centre

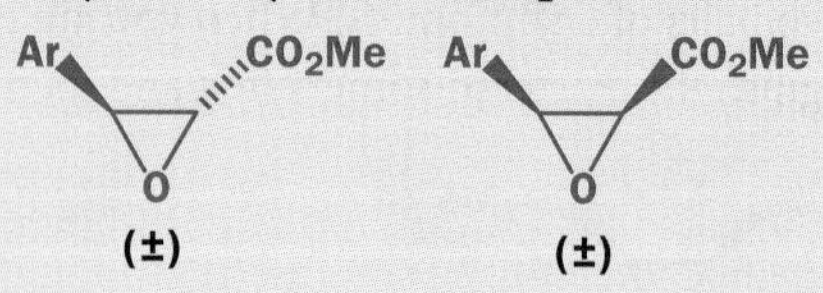

▶
You need to know, and be able to use, the rules for assigning *R* and *S*; they were explained on p. 387. If you get any of the assignments wrong, make sure you understand why.

Let's analyse our set of four stereoisomers a little more closely. You may have already noticed that these structures all contain stereogenic centres—two in each case. Go back to the diagram of the four structures at the bottom of p. 391 and, without looking at the structures below, assign an *R* or *S* label to each of these stereogenic centres.

You should have assigned *R*s and *S*s like this.

● **Converting enantiomers and diastereoisomers**

- To go from one *enantiomer* to another, *both* stereogenic centres are inverted
- To go from one *diastereoisomer* to another, only *one* of the two is inverted

All the compounds that we have talked about so far have been cyclic, because the diastereo-isomers are easy to visualize: two diastereoisomers can be identified because the substituents are either on the same side or on opposite sides of the ring (*cis* or *trans*). But acyclic compounds can exist as diastereoisomers too. Take these two, for example. Both ephedrine and pseudoephedrine are members of the amphetamine class of stimulants, which act by imitating the action of the hormone adrenaline.

Ephedrine and pseudoephedrine are stereoisomers that are clearly not mirror images of each other—only one of the two stereogenic centres in ephedrine is inverted in pseudoephedrine—so they must be diastereoisomers. Thinking in terms of stereogenic centres is useful, because, just as this compound has two stereogenic centres and can exist as two diastereoisomers, any compound with more than one stereogenic centre can exist in more than one diastereoisomeric form.

Both compounds are produced in enantiomerically pure form by plants, so, unlike the anti-asthma intermediates above, in this case we are talking about single enantiomers of single diastereoisomers.

ephedrine   pseudoephedrine

> If you are asked to explain some stereochemical point in an examination, choose a cyclic example—it makes it much easier.

Adrenaline

Adrenaline (also known as epinephrine) has a chiral structure. In nature it is a single enantiomer but it cannot have any diastereoisomers as it has only one stereogenic centre.

adrenaline

(1*R*,2*S*)-(−)-ephedrine   (1*S*,2*S*)-(+)-pseudoephedrine

Ephedrine and pseudoephedrine

Ephedrine is a component of the traditional Chinese remedy 'Ma Huang', extracted from *Ephedra* species. It is also used in nasal sprays as a decongestant. Pseudoephedrine is the active component of the decongestant Sudafed (so should that be Pseud ephed?).

> Remember that (+) and (−) refer to the sign of the specific rotation, while *R* and *S* are derived simply by looking at the structure of the compounds. There is no simple connection between the two!

The 'natural' enantiomers of the two diastereomers are (−)-ephedrine and (+)-pseudoephedrine, which does not tell you which is which, or (1*R*,2*S*)-(−)-ephedrine and (1*S*,2*S*)-(+)-pseudoephedrine, which does. From that you should be able to deduce the corresponding structures.

Here are some data on (1*R*,2*S*)-(−)-ephedrine and (1*S*,2*S*)-(+)-pseudoephedrine and their 'unnatural' enantiomers (which have to be made in the laboratory), (1*S*,2*R*)-(+)-ephedrine and (1*R*,2*R*)-(−)-pseudoephedrine.

	(1*R*,2*S*)-(−)-ephedrine	(1*S*,2*R*)-(+)-ephedrine	(1*S*,2*S*)-(+)-pseudoephedrine	(1*R*,2*R*)-(−)-pseudoephedrine
m.p.	40–40.5 °C	40–40.5 °C	117–118 °C	117–118 °C
$[\alpha]_D^{20}$	−6.3	+6.3	+52	−52

> ● **Evidently, the diastereoisomers are different compounds with different names and different properties, while the pair of enantiomers are the same compound and differ only in the direction in which they rotate polarized light.**

We can illustrate the combination of two stereogenic centres in a compound by considering what happens when you shake hands with someone. Hand-shaking is successful only if you each use the same hand! By convention, this is your right hand, but it's equally possible to shake left hands. The overall pattern of interaction between two right hands and two left hands is the same: a right-hand-shake and a left-handshake are enantiomers of one another; they differ only in being mirror images. If, however, you misguidedly try to shake your right hand with someone else's left hand you end up holding hands. Held hands consist of one left and one right hand; a pair of held hands have totally different interactions from pair of shaking hands; we can say that holding hands is a diastereoisomer of shaking hands.

We can summarize the situation when we have two hands, or two chiral centres, each one *R* or *S*.

► A sugar has the empirical formula $C_nH_{2n}O_n$, and consists of a chain of carbon atoms, one being a carbonyl group and the rest carrying OH groups. If the carbonyl group is at the end of the chain (in other words, it is an aldehyde), the sugar is an aldose. If the carbonyl group is not at the end of the chain, the sugar is a ketose. We come back to all this in detail in Chapter 49. The number of carbon atoms, *n*, can be 3–8: aldoses have *n* − 2 stereogenic centres and ketoses *n* − 3 stereogenic centres. In fact, most sugars exist as an equilibrium mixture of this open-chain structure and a cyclic hemiacetal isomer (Chapter 6).

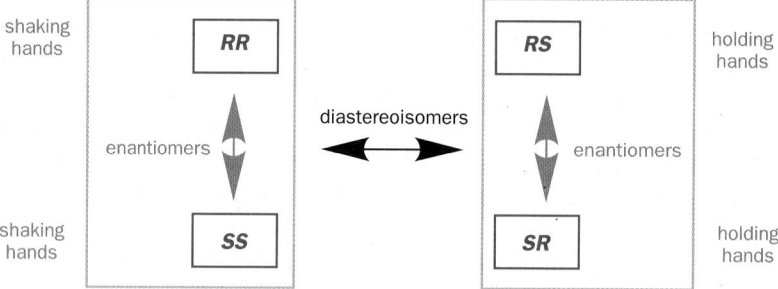

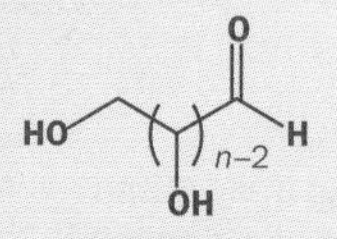

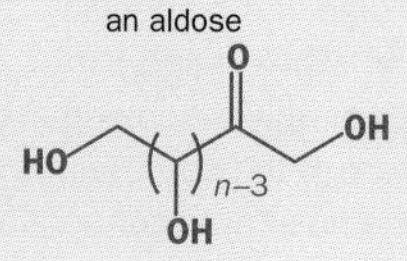

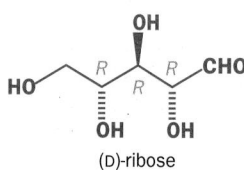

(D)-ribose

What about compounds with more than two stereogenic centres? The family of sugars provides lots of examples. Ribose is a 5-carbon sugar that contains three stereogenic centres. The enantiomer shown here is the one used in the metabolism of all living things and, by convention, is known as D-ribose. The three stereogenic centres of D-ribose have the *R* configuration.

In theory we can work out how many 'stereoisomers' there are of a compound with three stereo-genic centres simply by noting that there are 8 (=2³) ways of arranging *R*s and *S*s.

RRR	*RRS*	*RSR*	*RSS*
SSS	*SSR*	*SRS*	*SRR*

But this method blurs the all-important distinction between diastereoisomers and enantiomers. In each case, the combination in the top row and the combination directly below it are enantiomers (all three centres are inverted); the four columns are diastereoisomers. Three stereogenic centres therefore give four diastereoisomers, each a pair of two enantiomers. Going back to the example of the C₅ aldoses, each of these diastereoisomers is a different sugar. In these diagrams each diasteroisomer is in a frame but the top line shows one enantiomer (D) and the bottom line the other (L).

ribose

(D)-ribose

(L)-ribose

arabinose

(D)-arabinose

(L)-arabinose

xylose

(D)-xylose

(L)-xylose

lyxose

(D)-lyxose

(L)-lyxose

Fischer projections

The stereochemistry of sugars used to be represented by Fischer projections. The carbon backbone was laid out in a vertical line and twisted in such a way that all the substituents pointed towards the viewer.

Fischer projections are so unlike real molecules that you should never use them. However, you may see them in older books, and you should have an idea about how to interpret them. Just remember that all the branches down the side of the central trunk are effectively bold wedges (coming towards the viewer), while the central trunk lies in the plane of the paper. By mentally twisting the backbone into a realistic zig-zag shape you should end up with a reasonable representation of the sugar molecule.

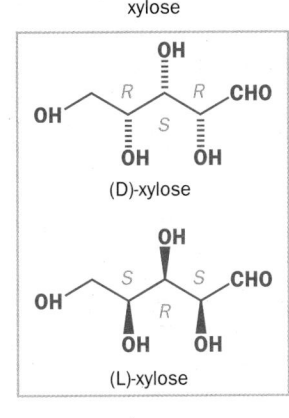

(D)-ribose

can be represented as

(D)-ribose
(Fischer projection)

equivalent to

You've probably recognized that there's a simple mathematical relationship between the number of stereogenic centres and the number of stereoisomers a structure can have. Usually, a structure with n stereogenic centres can exist as 2^n stereoisomers. These stereoisomers consist of $2^{(n-1)}$ diastereoisomers, each of which has a pair of enantiomers. This is an oversimplification to be used cautiously because it works only if all diastereoisomers are chiral. We recommend that you find out how many diastereoisomers there are in every new molecule before considering enantiomers.

Why only *usually?*—achiral compounds with more than one stereogenic centre

Sometimes, symmetry in a molecule can cause some stereoisomers to be degenerate, or 'cancel out'—there aren't as many stereoisomers as you'd expect. Take tartaric acid, for example.

This stereoisomer of tartaric acid is found in grapes, and its salt, potassium hydrogen tartrate, can precipitate out as crystals at the bottom of bottles of wine. It has two stereogenic centres, so you'd expect $2^2 = 4$ stereoisomers; two diastereoisomers, each a pair of enantiomers.

(+)-tartaric acid

OH groups *syn*

enantiomers

OH groups *anti*

diastereoisomers

?

While the pair of structures on the left are certainly enantiomers, if you look carefully at the pair of structures on the right, you'll see that they are, in fact, not enantiomers but identical structures. To prove it, just rotate the top one through 180° in the plane of the paper.

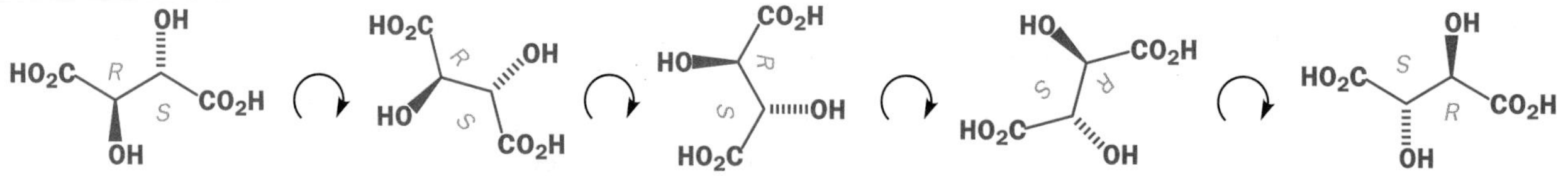

R,S-Tartaric acid and S,R-tartaric acid are not enantiomers, but they are identical because, even though they contain stereogenic centres, they are achiral. By drawing R,S-tartaric acid after a 180° rotation about the central bond, you can easily see that it has a mirror plane, and so must be achiral.

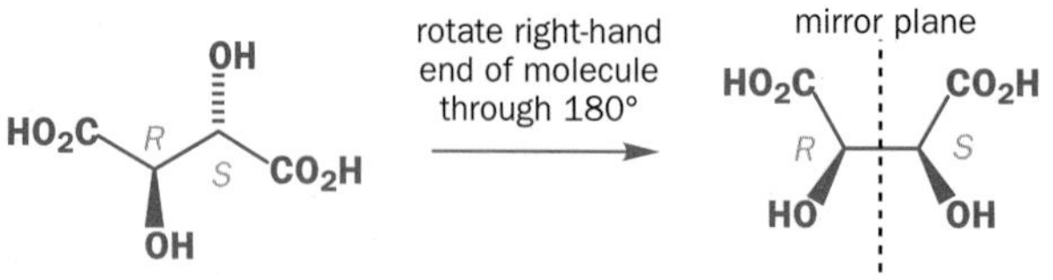

The formula stating that a compound with n stereogenic centres has 2^{n-1} diastereoisomers has worked but not the formula that states there are 2^n 'stereoisomers'. In general, it's safer not to talk about 'stereoisomers' but to talk first about diastereoisomers and then to assess each one for enantiomers. To say that a compound with two stereogenic centres has four 'stereoisomers' is rather like saying that 'four hands are getting married'. Two people are getting married, each with two hands.

> ● Compounds that contain stereogenic centres but are themselves achiral are called *meso* compounds. This means that there is a plane of symmetry with R stereochemistry on one side and S stereochemistry on the other.

Meso hand-shaking

We can extend our analogy between hand-shaking and diastereoisomers to *meso* compounds as well. Imagine a pair of identical twins shaking hands. There would be two ways for them to do it: left shakes left or right shakes right: provided you know your left from your right you could tell the two handshakes apart because they are enantiomers. But if the twins hold hands, you will not be able to distinguish left holds right from right holds left, because the twins themselves are indistinguishable—this is the *meso* hand-hold!

So tartaric acid can exist as two diastereoisomers, one with two enantiomers and the other achiral (a *meso* compound). Since the molecule has symmetry, and R is the mirror image of S, the RS diastereoisomer cannot be chiral.

	Chiral diastereoisomer		Achiral diastereoisomer
	(+)-tartaric acid	(−)-tartaric acid	*meso*-tartaric acid
$[\alpha]_D^{20}$	+12	−12	0
m.p.	168–170 °C	168–170 °C	146–148 °C

Meso diastereoisomers of inositol

Look out for *meso* diastereoisomers in compounds that have a degree of symmetry in their overall structure. Inositol, one of whose diastereomers is an important growth factor, certainly possesses some *meso* diastereoisomers.

inositol

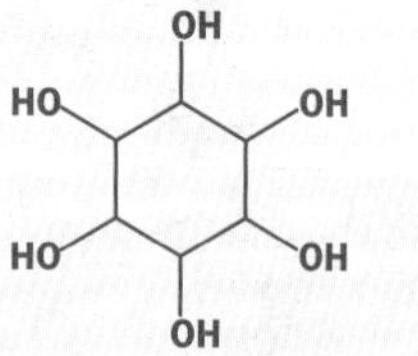

Investigating the stereochemistry of a compound

When you want to describe the stereochemistry of a compound our advice is to identify the diastereoisomers and then think about whether they are chiral or not. Here is a simple example, the linear triol 2,3,4-trihydroxypentane or pentan-2,3,4-triol.

This is what you should do.

1   Draw the compound with the carbon skeleton in the usual zig-zag fashion running across the page

2   Identify the chiral centres

3   Decide how many diastereoisomers there are by putting the substituents at those centres up or down. It often helps to give each diastereoisomer a 'tag' name. In this case there are three diastereoisomers. The three OH groups can be all on the same side or else one of the end OHs or the middle one can be on the opposite side to the rest

4   By checking on possible planes of symmetry, see which diastereoisomers are chiral. In this case only the plane down the centre can be a plane of symmetry

5   Draw the enantiomers of any chiral diastereoisomer by inverting *all* the stereogenic centres

6   Announce the conclusion

You could have said that there are four 'stereoisomers' but the following statement is much more helpful. There are three diastereoisomers, the *syn,syn*, the *syn,anti*, and the *anti,anti*. The *syn,syn* and the *anti,anti* are achiral (*meso*) compounds but the *syn,anti* is chiral and has two enantiomers.

1

2

3   all up or *syn,syn*   outside one down, others up or *anti,syn*   inside one down, others up or *anti,anti*

4   plane of symmetry
achiral (*meso*)   chiral   plane of symmetry
achiral (*meso*)

the two enantiomers of the *anti,syn* diastereoisomer

The mystery of Feist's acid

It is hard nowadays to realize how difficult structure-solving was when there were no spectra. A celebrated case was that of 'Feist's acid' discovered by Feist in 1893 from a deceptively simple reaction.

Early work without spectra led to two suggestions, both based on a three-membered ring, and this compound had some fame because unsaturated three-membered rings were rare. The favoured structure was the cyclopropene.

The argument was still going on in the 1950s when the first NMR spectrometers appeared. Though infrared appeared to support the cyclopropene structure, one of the first problems resolved by the primitive 40 MHz instruments available was that of Feist's acid, which had no methyl group signal but did have two protons on a double bond and so had to be the exomethylene isomer after all.

This structure has two chiral centres, so how will we know which diastereoisomer we have? The answer was simple: the stereochemistry has to be *trans* because Feist's acid is chiral: it can be resolved (see later in this chapter) into two enantiomers. Now, the *cis* diacid would have a plane of symmetry, and so would be achiral—it would be a *meso* compound. The *trans* acid on the other hand is chiral— it has only an axis of symmetry. If you do not see this, try superimposing it on its mirror image. You will find that you cannot.

Modern NMR spectra make the structure easy to deduce. There are only two proton signals as the

correct structure of Feist's acid

the *cis* diacid has a plane of symmetry

CO_2H protons exchange in the DMSO solvent needed. The two protons on the double bond are identical (5.60 p.p.m.) and so are the two protons on the three-membered ring which come at the expected high field (2.67 p.p.m.). There are four carbon signals: the C=O at 170 p.p.m., two alkene signals between 100 and 150 p.p.m., and the two identical carbons in the three-membered ring at 25.45 p.p.m.

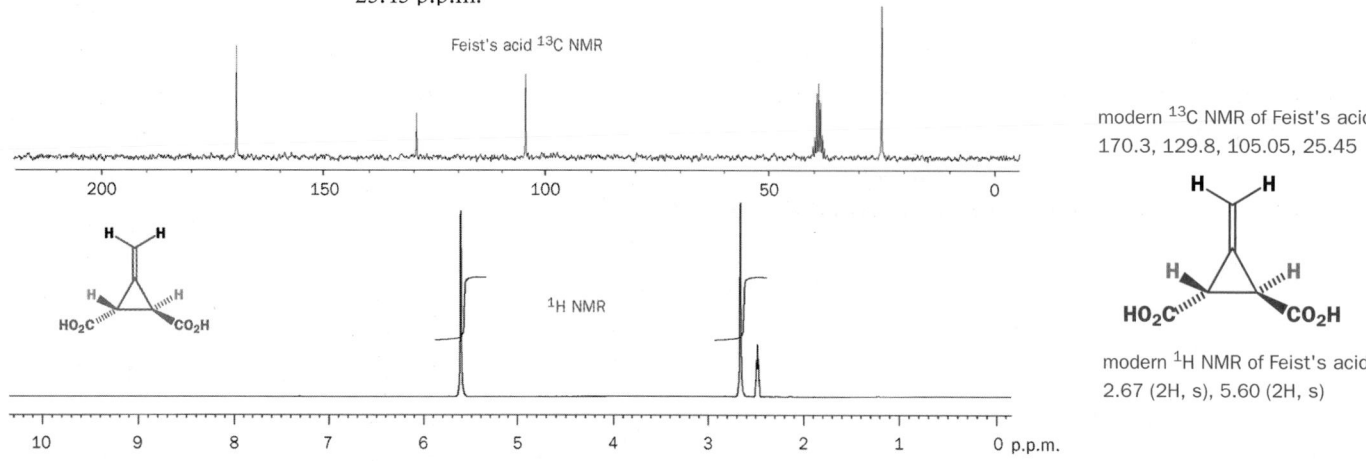

modern ^{13}C NMR of Feist's acid
170.3, 129.8, 105.05, 25.45

modern ^{1}H NMR of Feist's acid
2.67 (2H, s), 5.60 (2H, s)

Chiral compounds with no stereogenic centres

A few compounds are chiral, yet have no stereo-genic centres. We will not discuss these in detail, but try making a model of this allene, which has no stereogenic centre.

a chiral allene

These mirror images (enantiomers) are not superimposable and so the allene is chiral. Similarly, some biaryl compounds such as this important bisphosphine known as BINAP (we come back to BINAP in Chapter 45) exist as two separate enantiomers because rotation about the green bond is restricted.

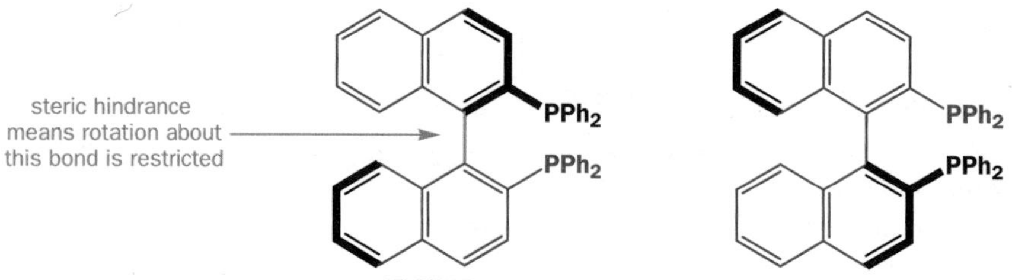

steric hindrance means rotation about this bond is restricted

(R)-BINAP

(S)-BINAP

If you were to look at this molecule straight down along the green bond, you would see that the two flat rings are at right angles to each other and so the molecule has a twist in it rather like the 90° twist in the allene.

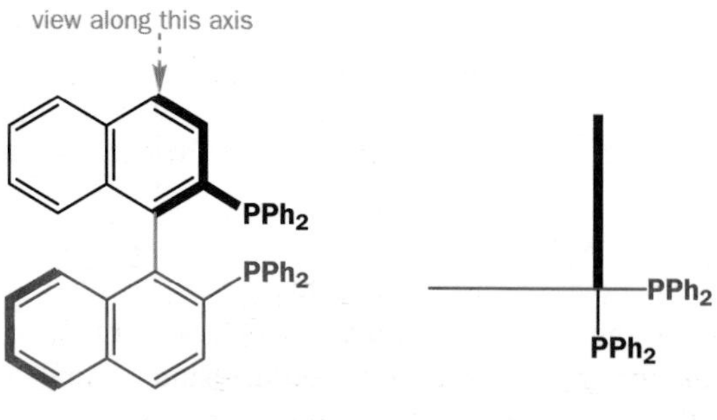

view along this axis

PPh₂

PPh₂

These two examples rely on the rigidity of π systems but this simple saturated system is also chiral. These two rings have to be orthogonal because of the tetrahedral nature of the central carbon atom. There can be no plane of symmetry here either but the central carbon is not chiral.

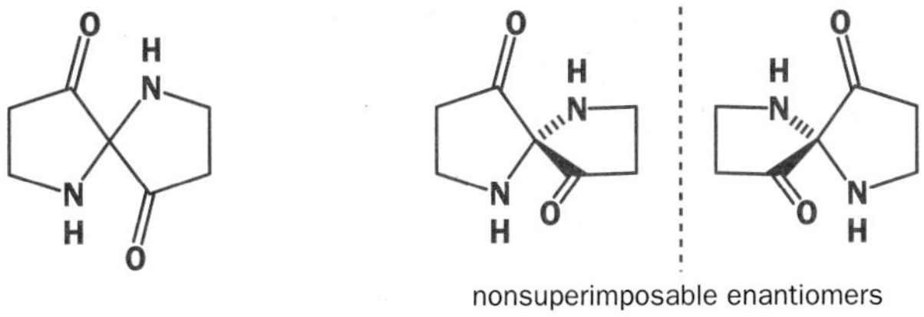

nonsuperimposable enantiomers

There are other types of chiral molecule but they all share the same feature—there is no plane of symmetry.

Separating enantiomers is called resolution

Early in this chapter, we said that most of the molecules in nature are chiral, and that Nature usually produces these molecules as single enantiomers. We've talked about the amino acids, the sugars, ephedrine, pseudoephedrine, and tartaric acid—all compounds that can be isolated from natural sources as single enantiomers. On the other hand, in the lab, if we make chiral compounds from achiral starting materials, we are doomed to get racemic mixtures. So how do chemists ever isolate compounds as single enantiomers, other than by extracting them from natural sources? We'll consider this question in much more detail in Chapter 45, but here we will look at the simplest way: using nature's enantiomerically pure compounds to help us separate the components of a racemic mixture into its two enantiomers. This process is called **resolution**.

Imagine the reaction between a chiral, but racemic alcohol and a chiral, but racemic carboxylic acid, to give an ester in an ordinary acid-catalysed esterification (Chapter 12).

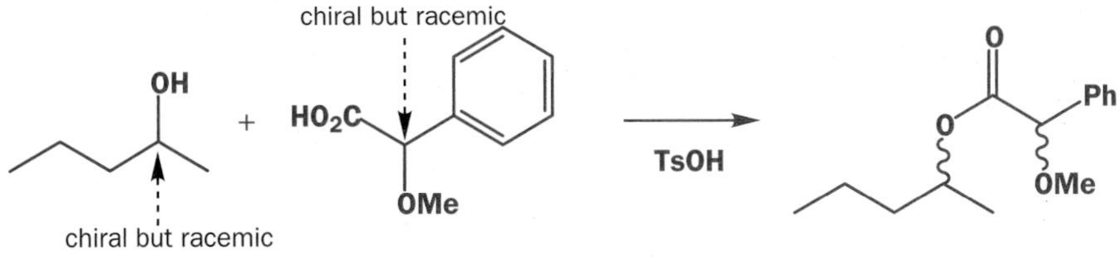

The product contains two chiral centres, so we expect to get two diastereoisomers, each a racemic mixture of two enantiomers. Diastereoisomers have different physical properties, so they should be easy to separate, for example by chromatography.

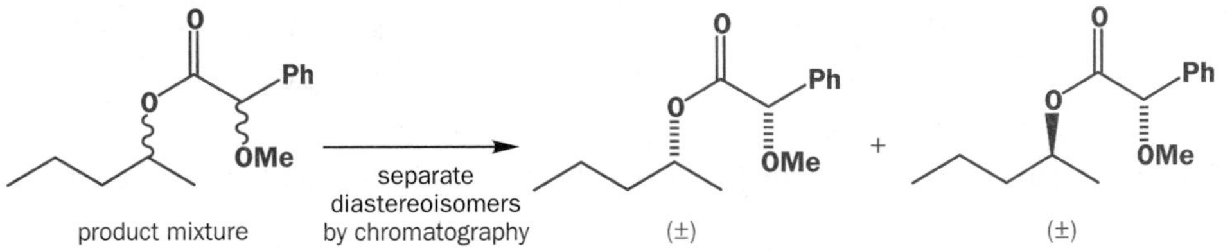

■ Remember that (±) means the compounds are racemic: we're showing only relative, not absolute, stereochemistry.

We could then reverse the esterification step, and hydrolyse either of these diastereoisomers, to regenerate racemic alcohol and racemic acid.

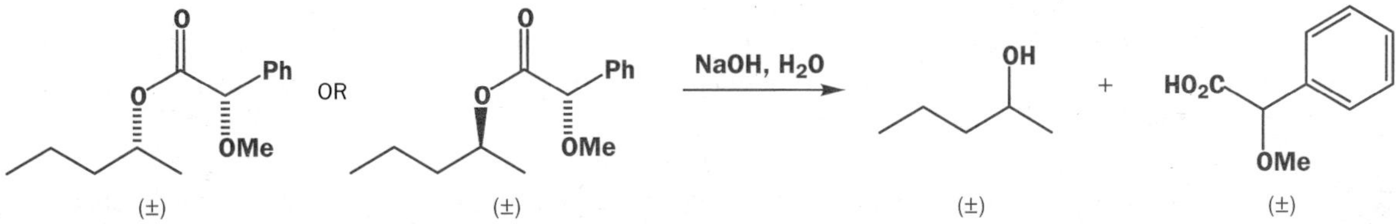

If we repeat this reaction, this time using an enantiomerically pure sample of the acid (available from (*R*)-mandelic acid, the almond extract you met on p. 294), we will again get two diastereoisomeric products, but this time each one will be enantiomerically pure.

Note that the stereochemistry shown here *is* absolute stereochemistry.

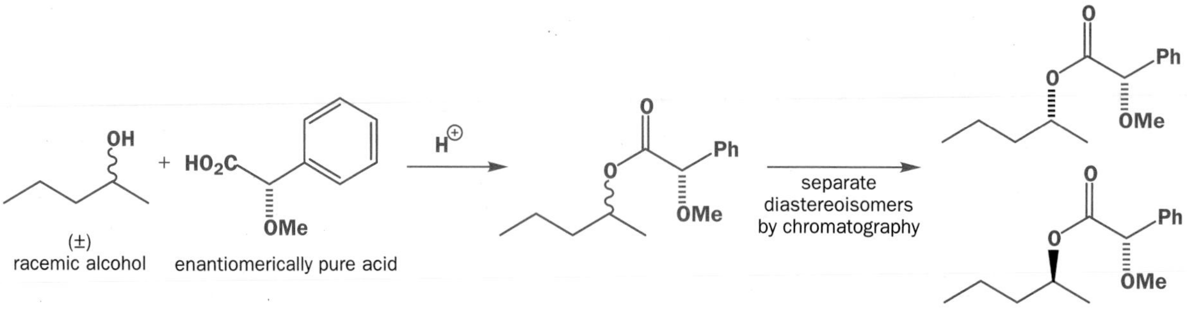

(±)
racemic alcohol   enantiomerically pure acid

separate diastereoisomers by chromatography

If we now hydrolyse each diastereoisomer separately, we have done something rather remarkable: we have managed to separate to two enantiomers of the starting alcohol.

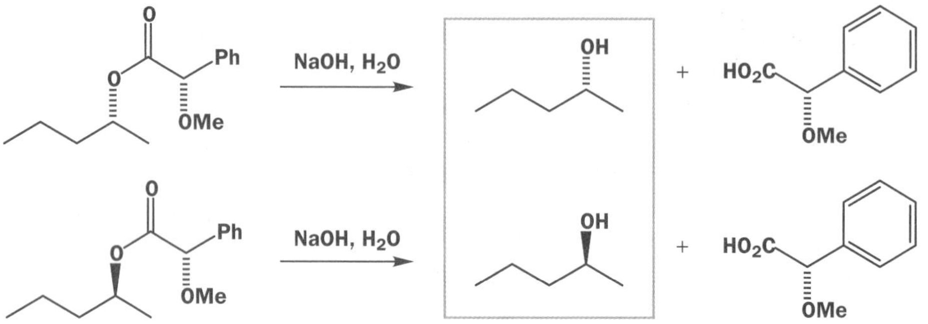

NaOH, H₂O

NaOH, H₂O

two enantiomers obtained separately: a resolution has been accomplished   acid recovered and can be recycled

A separation of two enantiomers is called a **resolution**. Resolutions can be carried out only if we make use of a component that is already enantiomerically pure: it is very useful that Nature provides us with such compounds; resolutions nearly always make use of compounds derived from nature.

Natural chirality

Why Nature uses only one enantiomer of most important biochemicals is an easier question to answer than how this asymmetry came about in the first place, or why L-amino acids and D-sugars were the favoured enantiomers, since, for example, proteins made out of racemic samples of amino acids would be complicated by the possibility of enormous numbers of diastereomers. Some have suggested that life arose on the surface of single chiral quartz crystals, which provided the asymmetric environment needed to make life's molecules enantiomerically pure. Or perhaps the asymmetry present in the spin of electrons released as gamma rays acted as a source of molecular asymmetry. Given that enantiomerically pure living systems should be simpler than racemic ones, maybe it was just chance that the L-amino acids and the D-sugars won out.

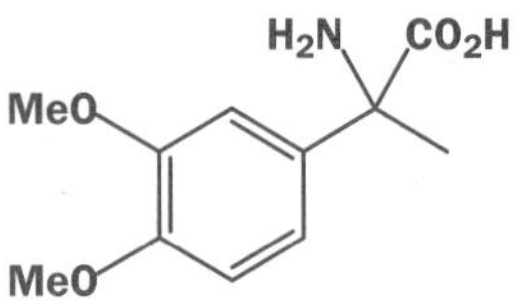

Now for a real example. Chemists studying the role of amino acids in brain function needed to obtain each of the two enantiomers of this compound.

They made a racemic sample using the Strecker synthesis of amino acids that you met in Chapter 12. The racemic amino acid was reacted with acetic anhydride to make the mixed anhydride and then with the sodium salt of naturally derived, enantiomerically pure alcohol menthol to give two diastereoisomers of the ester (see top of facing page).

One of the diastereoisomers turned out to be more crystalline (that is, to have a higher melting point) than the other and, by allowing the mixture to crystallize, the chemists were able to isolate a pure sample of this diastereoisomer. Evaporating the diastereoisomer left in solution (the 'mother liquors') gave them the less crystalline diastereoisomer.

Note that the rotations of the pure diastereoisomers were not equal and opposite. These are single enantiomers of different compounds and there is no reason for them to have the same rotation.

Next the esters were hydrolysed by boiling them in aqueous KOH. The acids obtained were enantiomers, as shown by their (nearly) opposite optical rotations and similar melting points. Finally, a more vigorous hydrolysis of the amides (boiling for 40 hours with 20% NaOH) gave them the amino acids they required for their biological studies (see bottom of facing page).

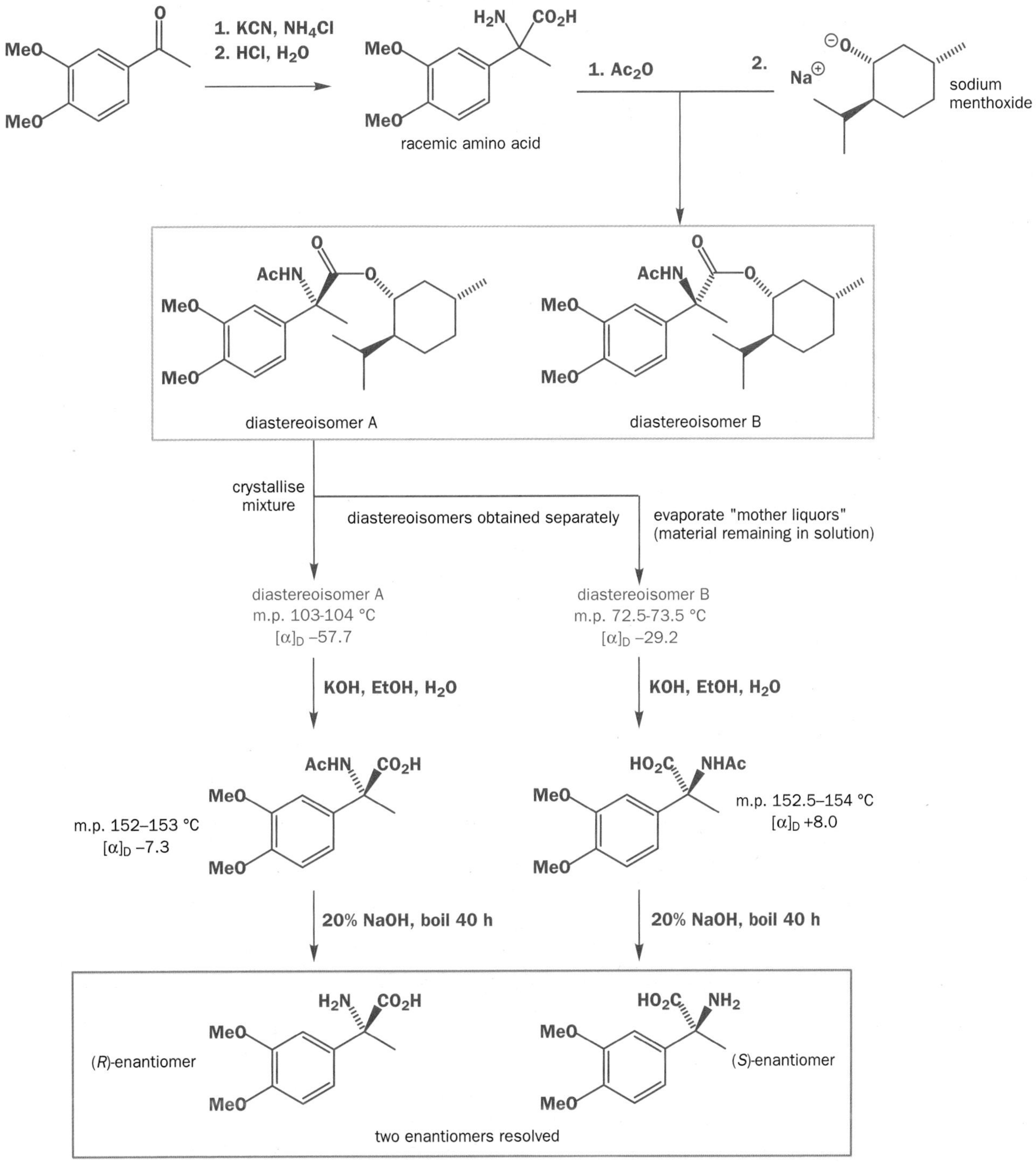

1. KCN, NH₄Cl
2. HCl, H₂O

racemic amino acid

1. Ac₂O   2.

sodium menthoxide

diastereoisomer A   diastereoisomer B

crystallise mixture

diastereoisomers obtained separately

evaporate "mother liquors" (material remaining in solution)

diastereoisomer A
m.p. 103-104 °C
$[\alpha]_D$ –57.7

diastereoisomer B
m.p. 72.5-73.5 °C
$[\alpha]_D$ –29.2

KOH, EtOH, H₂O   KOH, EtOH, H₂O

m.p. 152–153 °C
$[\alpha]_D$ –7.3

m.p. 152.5–154 °C
$[\alpha]_D$ +8.0

20% NaOH, boil 40 h   20% NaOH, boil 40 h

(R)-enantiomer   (S)-enantiomer

two enantiomers resolved

Resolutions using diastereoisomeric salts

The key point about resolution is that we must bring together two stereogenic centres in such a way that there is a degree of interaction between them: separable diastereoisomers are created from inseparable enantiomers. In the last two examples, the stereogenic centres were brought together in covalent compounds, esters. Ionic compounds will do just as well—in fact, they are often better because it is easier to recover the compound after the resolution.

An important example is the resolution of the enantiomers of naproxen. Naproxen is a member of a family of compounds known as Non-Steroidal Anti-Inflammatory Drugs (NSAIDs) which are 2-aryl propionic acids. This class also includes ibuprofen, the painkiller developed by Boots and marketed as Nurofen.

(*S*)-naproxen ibuprofen 2-arylpropionic acids

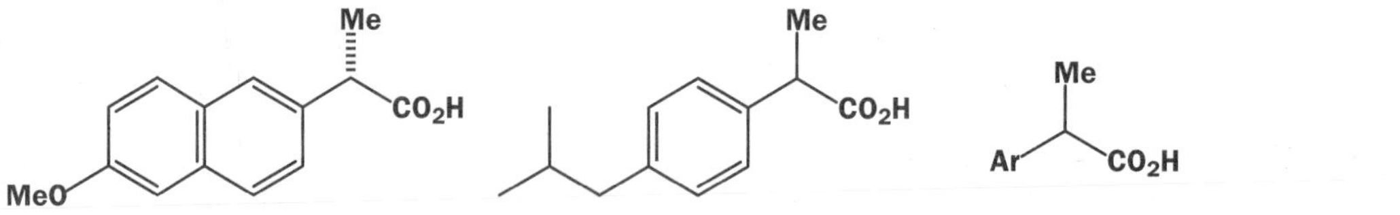

Both naproxen and ibuprofen are chiral but, while both enantiomers of ibuprofen are effective painkillers, and the drug is sold as a racemic mixture (and anyway racemizes in the body) only the (*S*) enantiomer of naproxen has anti-inflammatory activity. When the American pharmaceutical company Syntex first marketed the drug they needed a way of resolving the racemic naproxen they synthesized in the laboratory.

Since naproxen is a carboxylic acid, they chose to make the carboxylate salt of an enantiomerically pure amine, and found that the most effective was this glucose derivative. Crystals were formed, which consisted of the salt of the amine and (*S*)-naproxen, the salt of the amine with (*R*)-naproxen (the diastereoisomer of the crystalline salt) being more soluble and so remaining in solution. These crystals were filtered off and treated with base basic, releasing the amine (which can later be recovered and reused) and allowing the (*S*)-naproxen to crystallize as its sodium salt.

resolution of naproxen via an amine salt

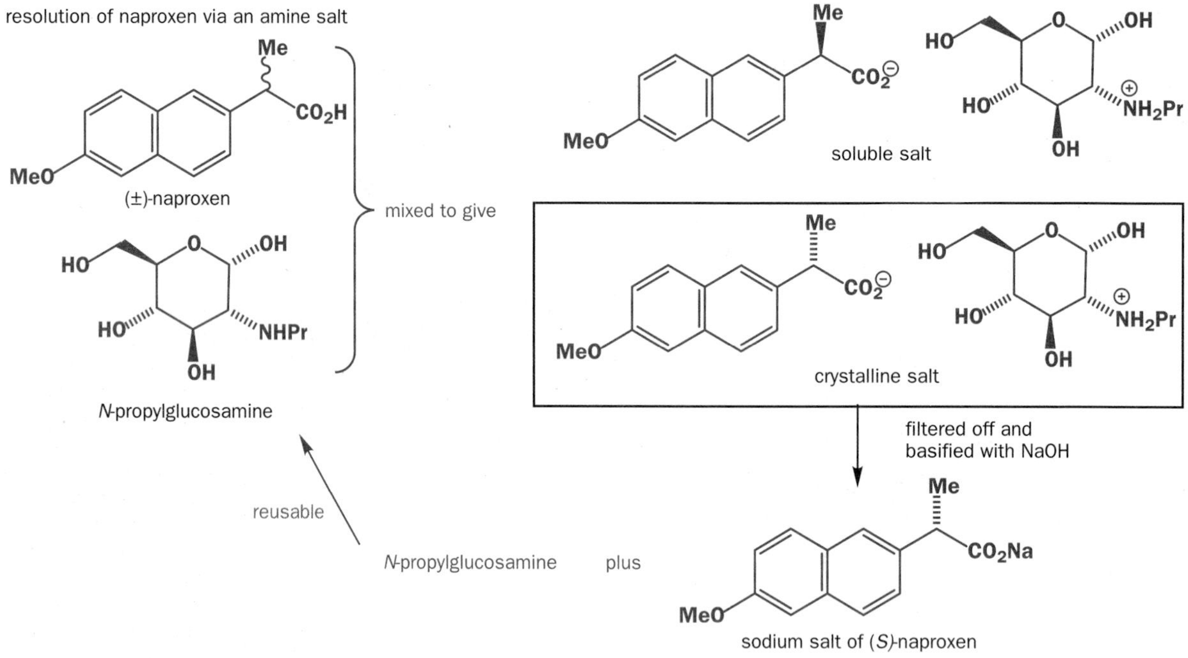

Resolutions can be carried out by chromatography on chiral materials

Interactions even weaker than ionic bonds can be used to separate enantiomers. Chromatographic separation relies on a difference in affinity between a stationary phase (often silica) and a mobile phase (the solvent travelling through the stationary phase, known as the eluent) mediated by, for example, hydrogen bonds or van der Waals interactions. If the stationary phase is made chiral by bonding it with an enantiomerically pure compound (often a derivative of an amino acid), chromatography can be used to separate enantiomers.

▶ Silica, SiO_2, is a macromolecular array of silicon and oxygen atoms. Its surface is covered with free OH groups, which can be used as an anchor for chiral derivatizing agents.

Chiral drugs

You may consider it strange that it was necessary to market naproxen as a single enantiomer, in view of what we have said about enantiomers having identical properties. The two enantiomers of naproxen do indeed have identical properties in the lab, but once they are inside a living system they, and any other chiral molecules, are differentiated by interactions with the enantiomerically pure molecules they find there. An analogy is that of a pair of gloves—the gloves weigh the same, are made of the same material, and have the same colour—in these respects they are identical. But interact them with a chiral environment, such as a hand, and they become differentiable because only one fits.

The way in which drugs interact with receptors mirrors this hand-and-glove analogy quite closely. Drug receptors, into which drug molecules fit like hands in gloves, are nearly always protein molecules, which are enantiomerically pure because they are made up of just L-amino acids. One enantiomer of a drug is likely to interact much better than the other, or perhaps in a different way altogether, so the two enantiomers of chiral drugs often have quite different pharmacological effects. In the case of naproxen, the (S)-enantiomer is 28 times as effective as the (R). Ibuprofen, on the other hand, is still marketed as a racemate because the two enantiomers have more or less the same painkilling effect.

Sometimes, the enantiomers of a drug may have completely different therapeutic properties. One example is Darvon, which is a painkiller. Its enantiomer, known as Novrad, is an anticough agent. Notice how the enantiomeric relationship between these two drugs extends beyond their chemical structures! In Chapter 45 we will talk about other cases where two enantiomers have quite different biological effects.

Darvon   Novrad

silica   chiral derivative

Chromatography on a chiral stationary phase is especially important when the compounds being resolved have no functional groups suitable for making the derivatives (usually esters or salts) needed for the more classical resolutions described above. For example, the two enantiomers of an analogue of the tranquillizer Valium were found to have quite different biological activities.

an analogue of the tranquillizer Valium

(R)-enantiomer   (S)-enantiomer   Valium

In order to study these compounds further, it was necessary to obtain them enantiomerically pure. This was done by passing a solution of the racemic compound through a column of silica bonded to an amino-acid-derived chiral stationary phase. The (R)-(–)-enantiomer showed a lower affinity for the stationary phase, and therefore was eluted from the column first, followed by the (S)-(+)-enantiomer.

> ▶
> You can think about chiral chromatography like this. Put yourself in this familiar situation: you want to help out a pensioner friend of yours who sadly lost his left leg in the war. A local shoe shop donates to you all their spare odd shoes, left and right, in his size (which happens to be the same as yours). You set about sorting the lefts from the rights, but are plunged into darkness by a power cut. What should you do? Well, you try every shoe on your right foot. If it fits you keep it; if not it's a left shoe and you throw it out.
>
> Now this is just what chromatography on a chiral stationary phase is about. The stationary phase has lots of 'right feet' (one enantiomer of an adsorbed chiral molecule) sticking out of it and, as the mixture of enantiomers of 'shoes' flows past, 'right shoes' fit, and stick but 'left shoes' do not and flow on down the column, reaching the bottom first.

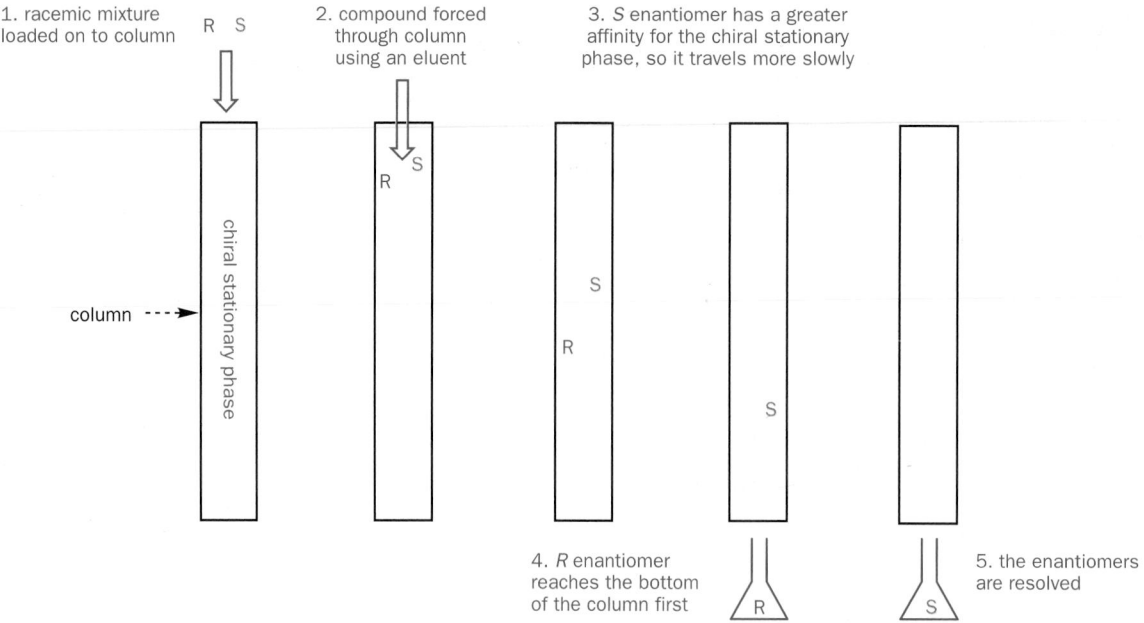

1. racemic mixture loaded on to column
2. compound forced through column using an eluent
3. *S* enantiomer has a greater affinity for the chiral stationary phase, so it travels more slowly
4. *R* enantiomer reaches the bottom of the column first
5. the enantiomers are resolved

Two enantiomers of one molecule may be the same compound, but they are clearly different, though only in a limited number of situations. They can interact with biological systems differently, for example, and can form salts or compounds with different properties when reacted with a single enantiomer of another compound. In essence, enantiomers behave identically *except* when they are placed in a chiral environment. In Chapter 45, we will see how to use this fact to make single enantiomers of chiral compounds, but next we move on to three classes of reactions in which stereochemistry plays a key role: substitutions, eliminations, and additions.

Problems

1. Assign a configuration, *R* or *S*, to each of these compounds.

2. If a solution of a compound has a rotation of +12, how could you tell if this was actually +12, or really −348, or +372?

3. Cinderella's glass slipper was undoubtedly a chiral object. But would it have rotated the plane of polarized light?

4. Are these compounds chiral? Draw diagrams to justify your answer.

5. What makes molecules chiral? Give three examples of different types of chirality. State with explanations whether the following compounds are chiral.

6. Discuss the stereochemistry of these compounds. (*Hint.* This means saying how many diastereoisomers there are, drawing clear diagrams of each, and saying whether they are chiral or not.)

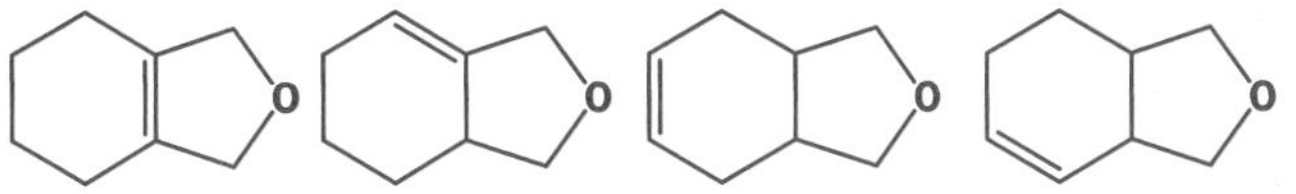

7. In each case state with explanations whether the products of these reactions are chiral and/or enantiomerically pure.

(a)

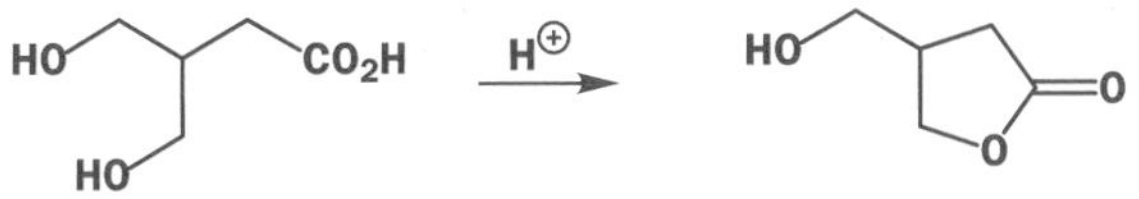

(b)

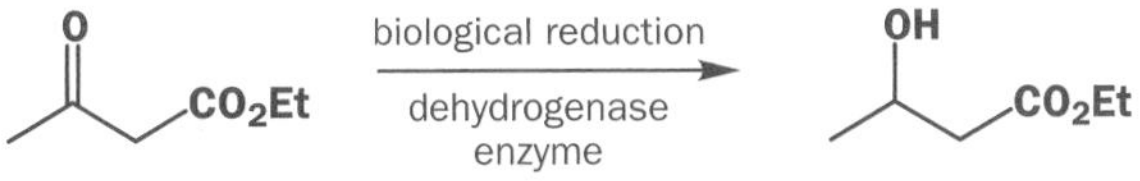

biological reduction
dehydrogenase enzyme

(c)

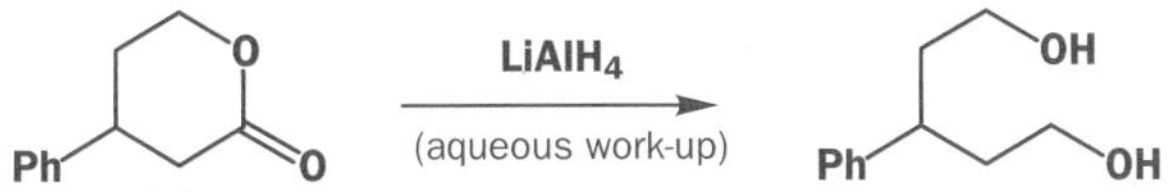

LiAlH₄
(aqueous work-up)

(d)

heat

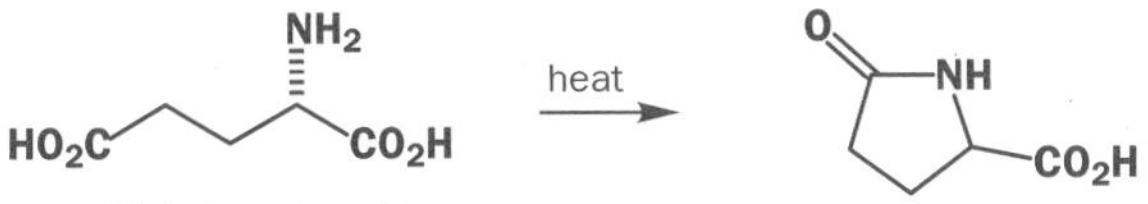

S-(+)-glutamic acid

8. Propose mechanisms for these reactions that explain the stereochemistry of the products. All compounds are enantiomercally pure.

(a)

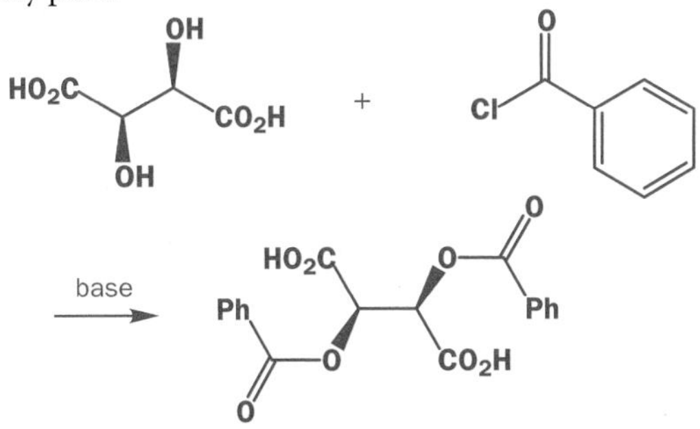

base

(b)

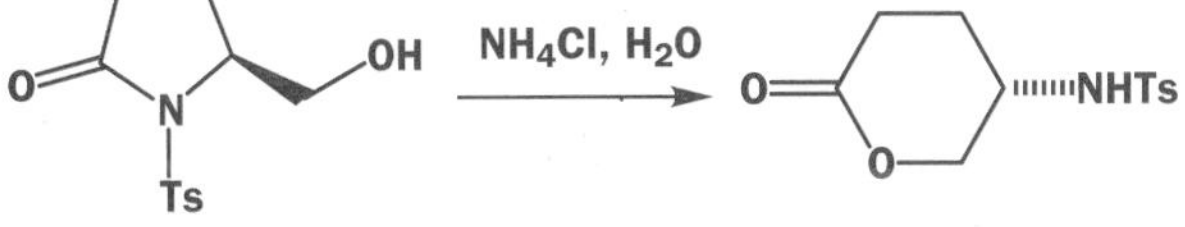

NH₄Cl, H₂O

9. Discuss the stereochemistry of these compounds. The diagrams are deliberately poor ones that are ambiguous about the stereochemistry—your answer should use good diagrams that give the stereochemistry clearly.

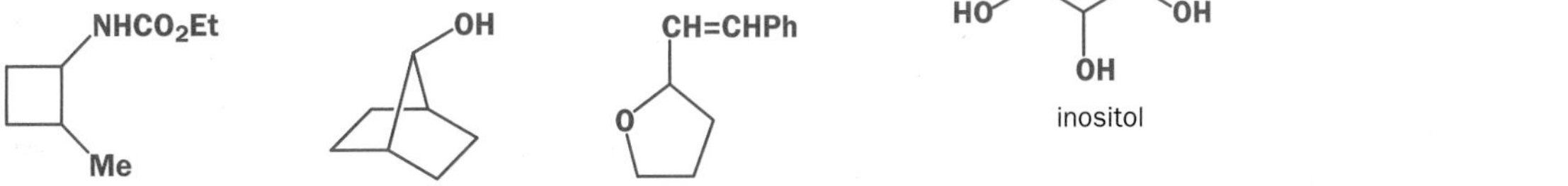

10. This compound racemizes in base. Why is that?

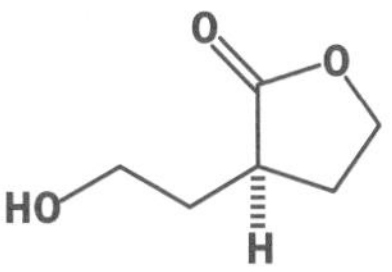

11. Draw mechanisms for these reactions. Will the products be single stereoisomers?

(a)

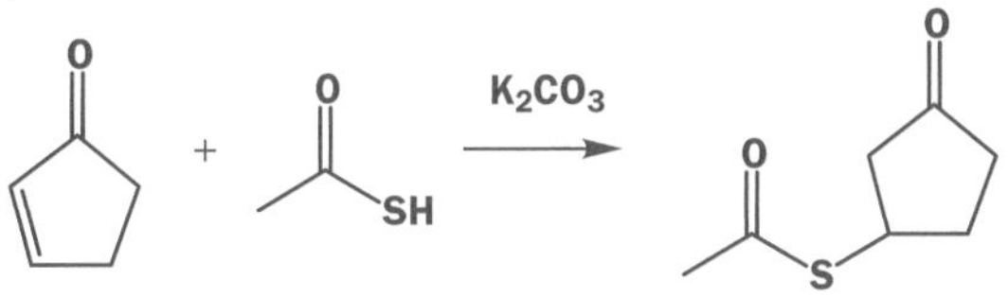

K₂CO₃

(b)

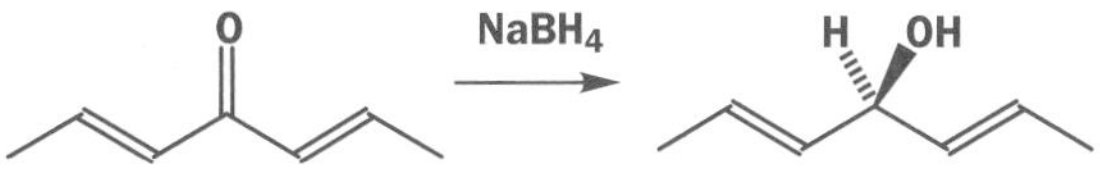

NaBH₄

(c)

NaBH₄

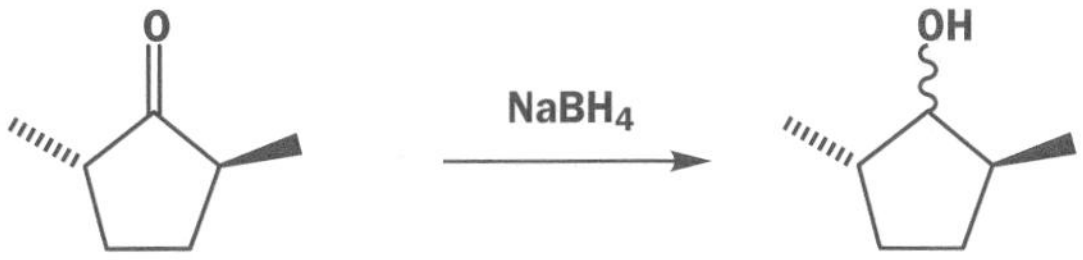

12. How many diastereoisomers of compound 1 are there? State clearly whether each diastereoisomer is chiral or not. If you had made a random mixture of stereosiomers by a chemical reaction, by what types of methods might they be separated? Which isomer(s) would be expected from the hydrogenation of compound 2?

13. Just for fun, you might like to try and work out just how many diastereoisomers inositol has and how many of them are *meso* compounds.

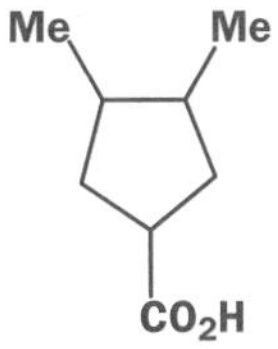

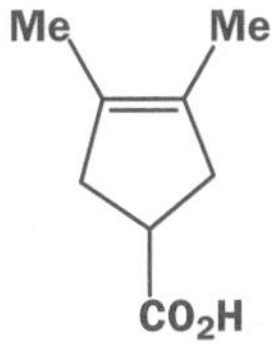

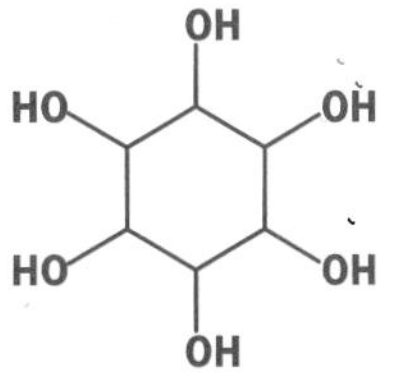

inositol

Nucleophilic substitution at saturated carbon

17

Connections

Building on:

- Attack of nucleophiles on carbonyl groups ch6, ch9, ch12, & ch14
- Attack of nucleophiles on double bonds conjugated with carbonyl groups ch10
- Substitution at carbonyl groups ch12
- Substitution of the oxygen atom of carbonyl groups ch14
- Stereochemistry ch16
- Transition states, intermediates, and rate expressions ch13

Arriving at:

- Nucleophilic attack on *saturated* carbon atoms, leading to substitution reactions
- How substitution at a saturated carbon atom differs from substitution at C=O
- Two mechanisms of nucleophilic substitution
- Intermediates and transition states in substitution reactions
- How substitution reactions affect stereochemistry
- What sort of nucleophiles can substitute, and what sort of leaving groups can be substituted
- The sorts of molecules that can be made by substitution, and what they can be made from

Looking forward to:

- Elimination reactions ch19
- Substitution reactions with aromatic compounds as nucleophiles ch22
- Substitution reactions with enolates as nucleophiles ch26
- Retrosynthetic analysis ch30

Nucleophilic substitution

Substitution is the replacement of one group by another. In Chapter 12 we discussed nucleophilic substitution at the carbonyl group, this sort of thing.

The phenyl and carbonyl groups remain in the molecule but the Cl group is replaced by the NH_2 group. We called the molecule of ammonia (NH_3) the **nucleophile** and the chloride was called the **leaving group**. In this chapter we shall be looking at similar reactions at saturated carbon atoms, this sort of thing.

During this reaction, the phenyl group remains the same and so does the CH_2 group, but the Cl group is replaced by the PhS group: it is a **substitution reaction**. The reaction happens at the CH_2 group—a *saturated* carbon atom—so the reaction is a **nucleophilic substitution at a saturated carbon atom**. This reaction and the one above may look superficially the same but they are quite different. We also changed the reagent for the substitution at a saturated carbon, because NH_3 would not give a good yield of $PhCH_2NH_2$ in the second type of reaction. The requirements for good reagents are different in substitution at the carbonyl group and at saturated carbon.

The main change is, of course, the absence of the carbonyl group. Mechanistically this is an enormous difference. The mechanism for the first reaction is:

mechanism of nucleophilic substitution at the carbonyl group

It is immediately obvious that the first step is no longer possible at a saturated carbon atom. The electrons cannot be added to a π bond as the CH_2 group is fully saturated. The nucleophile cannot add first and the leaving group go later because this would give a 5-valent carbon atom. Two new and different mechanisms become possible. Either the leaving group goes first and the nucleophile comes in later, or the two events happen at the same time. The first of these possibilities you will learn to call the S_N1 mechanism. The second mechanism, which shows that the only way the carbon atom can accept electrons is if it loses some at the same time, you will learn to call the S_N2 mechanism. You will see later that both mechanisms are possible here.

the S_N1 mechanism

the S_N2 mechanism

We shall spend some time looking at the differences between these mechanisms. But first we must establish how we know that there are two mechanisms.

If we look at a commonly used nucleophilic substitution, the replacement of OH by Br, we find that two quite different reaction conditions are used. Tertiary alcohols react rapidly with HBr to give tertiary alkyl bromides. Primary alcohols, on the other hand, react only very slowly with HBr and are usually converted to primary alkyl bromides with PBr_3.

■ The mechanism of the PBr_3 reaction will be discussed when we come to S_N2 reactions later in this chapter.

t-butanol
(2-methylpropan-2-ol)

t-butyl bromide
(2-bromo-2-methylpropane)

n-BuOH
(butan-1-ol)

n-BuBr
(1-bromobutane)

If we collect together those alcohols that react rapidly with HBr to give good yields of alkyl bromides, we find one thing in common: they can all form stable carbocations, that is, cations where the positive charge is on the carbon atom.

▶

Carbocation stability

These carbocations are *relatively* stable as far as carbocations go. But you would not be able to keep even these 'stable' carbocations in a bottle on the shelf. The concept of more and less stable carbocations is important in understanding the S_N1 reaction.

alcohols that react rapidly with HBr

tertiary alcohols

allylic alcohols

benzylic alcohols

stable carbocations

They *can* form carbocations, but *do* they? It is one thing to suggest the existence of a reactive intermediate, another to prove that it is formed. We shall spend some time showing that carbocations do really exist in solution and more time showing that they are indeed intermediates in this mechanism for substitution that you will learn to call the S_N1 mechanism.

the S$_N$1 mechanism for nucleophilic substitution at saturated carbon

stage 1: formation of the carbocation

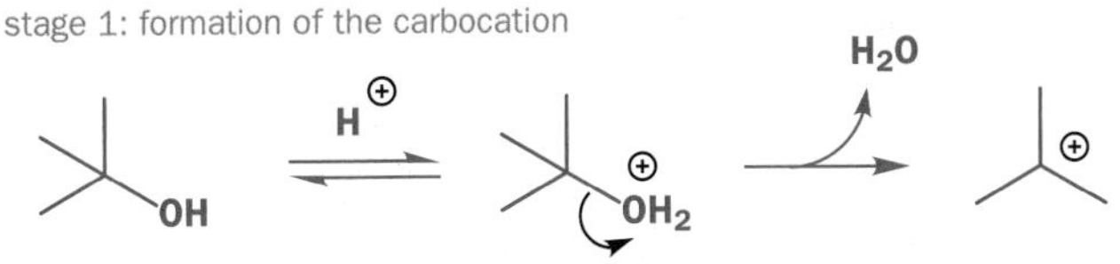

stage 2: capture of the carbocation by the nucleophile

Structure and stability of carbocations

We shall break off this mechanistic discussion to establish the nature of carbocations as ions that can be isolated and as intermediates in substitution reactions. We have seen in Chapter 3 that cations can easily be made in the gas phase by electron bombardment. We met these cations among others.

carbocations formed in the mass spectrometer

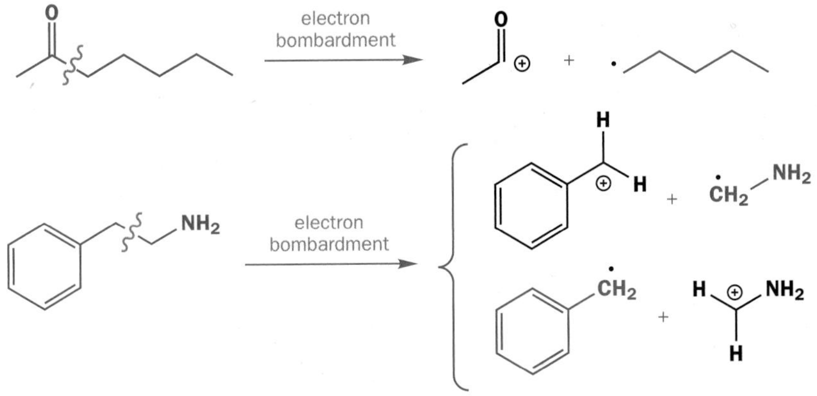

We also met the unusual cation CH_5^+. This cation shares *eight* electrons among five bonds and has a full outer shell like that of the ammonium ion NH_4^+. We call CH_5^+ a **carbonium** ion. The three ions formed in the mass spectrometer have only *three* bonds to the positively charged centre, only *six* electrons in the outer shell, and are electron-deficient. We call these ions **carbenium** ions and we may call both types **carbocations**. Table 17.1 gives a summary of the two types of carbocations.

It is the carbenium ions that interest us in this chapter because they are the intermediates in some nucleophilic substitutions. The simplest possible carbenium ion would be CH_3^+, the methyl cation, and it would be planar with an empty p orbital.

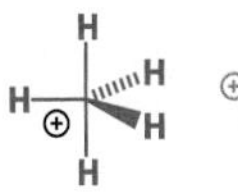

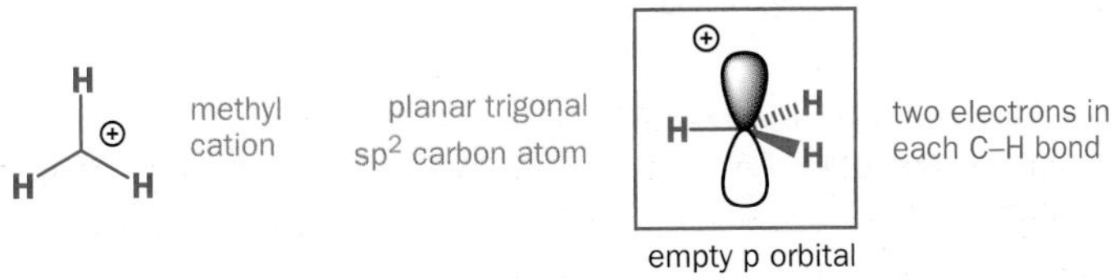

Table 17.1 Carbocations: carbenium ions and carbonium ions

Property	Carbenium ions	Carbonium ions
number of bonds to C$^+$	3	5
electrons in outer shell	6	8
empty orbital?	yes, a p orbital	no
electron-deficient?	yes	no
example		

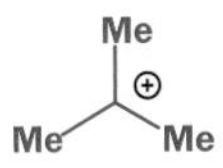

R——≡O⊕

an acylium ion

We did not meet this cation when we were discussing mass spectra, but we did meet the three ions halfway down p. 409. The methyl cation is so unstable that it is rarely formed even in the gas phase. Each of these three ions are formed because they have extra stabilization of some sort. The first is an acylium ion which is actually linear with most of the positive charge on the oxygen atom. It is more an oxonium ion than a carbocation. The third ion also has the positive charge carried by a heteroatom—this time it is nitrogen and the cation is more stable. It is much better to have a positive charge on nitrogen than on carbon. Notice that in both of the 'preferred representations' no atom is electron-deficient: all of the C, N, and O atoms have eight electrons.

this is a better representation because it shows the true linear shape of the cation

this is a better representation because it shows the true trigonal shape of the nitrogen atom

The second ion has no heteroatom but it has a benzene ring and the positive charge is delocalized around the ring, especially into the 2- and the 4- positions.

Thus, none of these three ions is a simple carbenium ion with the charge localized on an electron-deficient carbon atom. Most stable carbocations have extra stabilization of this sort. But even these relatively stable cations cannot be detected in normal solutions by NMR. This is because they are so reactive that they combine with even weak nucleophiles like water or chloride ions. Yet due to Olah's discovery of superacid (also called 'magic acid') in the 1960s we know that carbocations can exist in solution (you can read about this in the box). But are they formed as intermediates in substitution reactions?

George Olah was born in Hungary in 1927 but emigrated to the USA and did most of his work at Case Western Reserve University in Ohio. He got the Nobel prize for his work on cations in 1994. He now works at the University of Southern California.

Stable carbocations in superacid media

Olah's idea was to have a solution containing no nucleophiles. This sounds a bit tricky as any cation must have an anion to balance the charge and surely the anion will be a nucleophile? Well, nearly all anions are nucleophiles but there are some that consist of a negatively charged atom surrounded by tightly held halogen atoms. Examples include BF_4^-, PF_6^-, and SbF_6^-. The first is small and tetrahedral and the others are larger and octahedral.

nonnucleophilic anions

tetrahedral BF_4^- octahedral PF_6^- and SbF_6^-

In these anions, the fluorine atoms are very tightly held around the central atom, which carries the formal negative charge. The negative charge does not correspond to a lone pair of electrons (cf. the role of $NaBH_4$ in carbonyl reductions) and so there is nothing to act as a nucleophile. It was important too to have a nonnucleophilic solvent and low temperatures, and liquid SO_2 at –70 °C proved ideal.

With these conditions, Olah was able to make carbocations from alcohols. He treated t-butanol with SbF_5 and HF in liquid SO_2. This is the reaction.

Olah's preparation of the t-butyl cation in liquid SO_2

t-butyl cation

nonnucleophilic anion

The proton NMR of this cation showed just one signal for the three methyl groups at 4.15 p.p.m., quite far downfield for C–Me groups. The ^{13}C spectrum also showed downfield Me groups at 47.5 p.p.m., but the key evidence was the shift of the central carbon atom, which came at an amazing 320.6 p.p.m., way downfield from anything we have met before. This carbon is very deshielded—it is positively charged and electron-deficient.

Under these conditions acylium ions were also stable and their IR spectra could be run. Even crystals could be prepared so that no doubt remains that these are oxonium ions: both the bond length and the CO stretch are more triple-bond-like than carbon monoxide (see Table 17.2).

More important data were NMR spectra: both ^1H and ^{13}C NMRs could be run in liquid SO_2 at –70 °C. The proton NMR of the $MeOCH_2$ cation showed a methyl group with a large downfield shift and a CH_2 group that resembled an electron-deficient alkene rather than a saturated carbon atom. The cation is delocalized but the oxonium ion representation is better.

Table 17.2 Does the acylium ion have a triple bond?

	acylium ion Me—C≡O⊕	carbon monoxide ⊖C≡O⊕
v_{CO}, cm^{-1}	2294	2170
CO bond length, Å	1.108	1.128

oxonium ion   primary carbocation

$MeO≡CH_2$ ⟷ $MeO—CH_2$

$δ_H$ 5.6 $δ_H$ 9.9

If we mix *t*-BuOH and HBr in an NMR tube and let the reaction run inside the NMR machine, we see no signals belonging to the cation. This proves nothing. We would not expect a reactive intermediate to be present in any significant concentration. There is a simple reason for this. If the cation is unstable, it will react very quickly with any nucleophile around and there will never be any appreciable amount of cation in solution. Its rate of formation will be less, much less, than its rate of reaction. We need only annotate the mechanism you have already seen.

the S_N1 mechanism for nucleophilic substitution at saturated carbon

stage 1: formation of the carbocation

this stage is *slow*

stage 2: capture of the carbocation by the nucleophile

this stage is *fast*

It is comforting that carbocations can be prepared, even under rather artificial conditions, but we shall need other kinds of evidence to convince ourselves that they are intermediates in substitution reactions. It is time to return to the mechanistic discussion.

The S_N1 and S_N2 mechanisms for nucleophilic substitution

The evidence that convinced chemists about these two mechanisms is kinetic: it relates to the rate of the reactions. It was discovered, chiefly by Hughes and Ingold in the 1930s, that some nucleophilic substitutions are first-order, that is, the rate depends only on the concentration of the alkyl halide and *does not depend on the concentration of the nucleophile*, while in other reactions the rate depends on the concentrations of *both* the alkyl halide and the nucleophile. How can we explain this result? In the S_N2 mechanism there is just one step.

the S_N2 mechanism: reaction of *n*-BuBr with hydroxide ion

> ■ Edward David Hughes (1906–63) and Sir Christopher Ingold (1893–1970) worked at University College, London in the 1930s. They first thought of many of the mechanistic ideas that we now take for granted.

This step must therefore be the **rate-determining step**, sometimes called the slow step. The rate of the overall reaction depends only on the rate of this step. Kinetic theory tells us that the rate of a reaction is proportional to the concentrations of the reacting species such that

$$\text{rate of reaction} = k[\textit{n}\text{-BuBr}][\text{HO}^-]$$

Quantities in square brackets represent concentrations and the proportionality constant k is called the rate constant. If this mechanism is right, then the rate of the reaction will be simply and linearly proportional to both [*n*-BuBr] and to [HO$^-$]. And it is. Ingold measured the rates of reactions like these and found that they were second-order (proportional to two concentrations) and he called this mechanism Substitution, Nucleophilic, 2nd Order or S_N2 for short. The rate equation is usually given like this, with k_2 representing the second-order rate constant.

$$\text{rate} = k_2[\textit{n}\text{-BuBr}][\text{HO}^-]$$

> ■ There is more about the relationship between reaction rates and mechanisms in Chapter 13.

> ▶ Please note how this symbol is written. The S and the N are both capitals and the N is a subscript.

Usefulness and significance of the rate expression

Now what use is this equation and what does it signify? It is useful because it gives us a test for the S_N2 mechanism. It is usually carried out by varying both the concentration of the nucleophile and the concentration of the carbon electrophile in two separate series of experiments. The results of these experiments would be plotted on two graphs, one for each series. Supposing we wished to see if the reaction between NaSMe (an ionic solid—the nucleophile will be the anion MeS⁻) and MeI were indeed S_N2 as we would expect.

$$\text{MeS}^{\ominus} \quad \text{Me}{-}\text{I} \quad \longrightarrow \quad \text{MeS}{-}\text{Me} \quad + \quad \text{I}^{\ominus}$$

First, we would keep the concentration of NaSMe constant and vary that of MeI and see what happened to the rate. Then we would keep the concentration of MeI constant and vary that of MeSNa and see what happened to the rate. If the reaction is indeed S_N2 we should get a linear relationship in both cases.

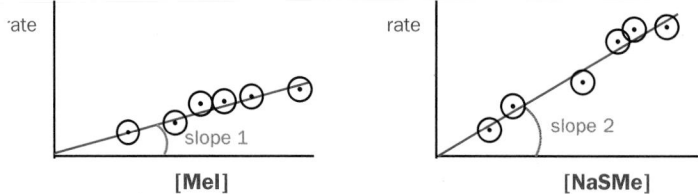

| rate ... | [MeI] | rate | [NaSMe] |

slope 1

slope 2

The first graph tells us that the rate is proportional to [MeI], that is, rate $= k_a$[MeI] and the second graph that it is proportional to [MeSNa], that is, rate $= k_b$[MeSNa]. But why are the slopes different? If you look at the rate equation for the reaction, you will see that we have incorporated a constant concentration of one of the reagents into what appears to be the rate constant for the reaction. The true rate equation is

$$\text{rate} = k_2[\text{MeSNa}][\text{MeI}]$$

If [MeSNa] is constant, the equation becomes

$$\text{rate} = k_a[\text{MeI}] \text{ where } k_a = k_2[\text{MeSNa}]$$

If [MeI] is constant, the equation becomes

$$\text{rate} = k_b[\text{MeSNa}] \text{ where } k_b = k_2[\text{MeI}]$$

If you examine the graphs you will see that the slopes are different because

$$\text{slope 1} = k_a = k_2[\text{MeSNa}], \text{ but slope 2} = k_b = k_2[\text{MeI}]$$

We can easily measure the true rate constant k_2 from these slopes because we know the constant values for [MeSNa] in the first experiment and for [MeI] in the second. The value of k_2 from both experiments should be the same! The mechanism for this reaction is indeed S_N2: the nucleophile MeS⁻ attacks as the leaving group I⁻ leaves.

$$\text{MeS}^{\ominus} \quad \text{Me}{-}\text{I} \quad \xrightarrow{S_N2} \quad \text{MeS}{-}\text{Me} \quad + \quad \text{I}^{\ominus}$$

So the usefulness of the rate equation is that it gives us a test for the S_N2 mechanism. But the equation has a meaning beyond that test.

Significance of the S_N2 rate equation

The significance of the equation is that performance of the S_N2 reaction depends both on nucleophile and on the carbon electrophile. We can make a reaction go better by changing either. If we want to displace I⁻ from MeI by an oxygen nucleophile we might consider using any of those in Table 17.3.

Table 17.3 Oxygen nucleophiles in the S_N2 reaction

Oxygen nucleophile	pK_a of conjugate acid[a]	Rate in S_N2 reaction
HO⁻	15.7 (H_2O)	fast
RCO$_2^-$	about 5 (RCO_2H)	reasonable
H_2O	–1.7 (H_3O^+)	slow
RSO$_2O^-$	0 (RSO_2OH)	slow

[a] See Chapter 8 for discussion of pK_a values.

> Each point on the slope represents a different experiment in which the rate of reaction is measured at a certain concentration of each of the reagents. All the points on the left-hand graph are measured with the concentration of NaSMe the same, but with different concentrations of MeI. On the right-hand graph, the points are measured with the concentration of MeI the same, but with different concentrations of NaSMe.

The same reasons that made hydroxide ion basic (chiefly that it is unstable as an anion and therefore reactive!) make it a good nucleophile. Basicity is just nucleophilicity towards a proton and nucleophilicity towards carbon must be related. You saw in Chapter 12 that nucleophilicity towards the carbonyl group is directly related to basicity. The same is not quite so true for nucleophilic attack on the saturated carbon atom as we shall see, but there is a relationship nonetheless. So if we want a fast reaction, we should use NaOH rather than, say, Na$_2$SO$_4$ to provide the nucleophile.

But that is not our only option. The reactivity and hence the structure of the carbon electrophile matter too. If we want reaction at a methyl group we can't change the carbon skeleton, but we can change the leaving group. Table 17.4 shows what happens if we use the various methyl halides in reaction with NaOH.

Table 17.4 Halide leaving groups in the S$_N$2 reaction

Halide X in MeX	pK_a of conjugate acid HX	Rate of reaction with NaOH
F	+3	very slow indeed
Cl	−7	moderate
Br	−9	fast
I	−10	very fast

Thus the fastest reaction will be between MeI and NaOH and will give methanol.

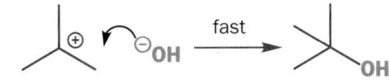

We shall discuss nucleophilicity and leaving group ability in more detail later. For the moment, the most important aspect is that the rate of an S$_N$2 reaction depends on both the nucleophile and the carbon electrophile (and hence the leaving group). Changing the nucleophile or the electrophile changes the value of k_2.

> ● **The rate of an S$_N$2 reaction depends upon:**
> - **The nucleophile**
> - **The carbon skeleton**
> - **The leaving group**

It also depends, as do all reactions, on factors like temperature and solvent.

Kinetics for the S$_N$1 reaction

We shall start with a similar reaction to the S$_N$2 reaction discussed a few pages back, but we shall replace *n*-butyl bromide with tertiary butyl bromide (*t*-BuBr).

the S$_N$1 mechanism: reaction of *t*-BuBr with hydroxide ion

stage 1: formation of the carbocation   stage 2: reaction of the carbocation

The formation of the cation is the rate-determining step. You can look at this in two ways. Either you could argue that a cation is an unstable species and so it will be formed slowly from a stable neutral organic molecule, or you could argue that the cation is a very reactive species and so all its reactions will be fast, regardless of the nucleophile. Both arguments are correct. In a reaction with an unstable intermediate, the formation of that intermediate is usually the rate-determining step.

The rate of disappearance of *t*-BuBr is simply the rate of the slow step. This is why the slow step is called the 'rate-determining' step. It is a unimolecular reaction with the simple rate equation

$$\text{rate} = k_1[t\text{-BuBr}]$$

If this is not obvious to you, think of a crowd of people trying to leave a railway station (such as a metro or underground station in a city) through the turnstiles. It doesn't matter how fast they walk away afterwards, it is only the rate of struggling through the turnstiles that determines how fast the station empties.

Once again, this rate equation is useful because we can determine whether a reaction is S$_N$1 or S$_N$2. We can plot the same graphs as we plotted before. If the reaction is S$_N$2, the graphs look like those we have just seen. But if it is S_N1, they look like this when we vary [t-BuBr] at constant [NaOH] and then vary [NaOH] at constant [t-BuBr].

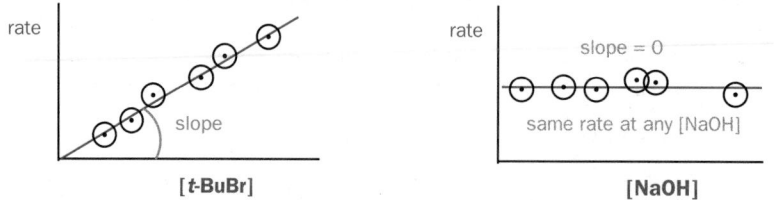

The slope of the first graph is simply the first-order rate constant because

$$\text{rate} = k_1[t\text{-BuBr}]$$

But the slope of the second graph is zero! The rate-determining step does not involve NaOH so adding more of it does not speed up the reaction. The reaction shows first-order kinetics (the rate is proportional to one concentration only) and the mechanism is called S_N1, that is, Substitution, Nucleophilic, 1st order.

This observation is very significant. It is not only the *concentration* of the nucleophile that doesn't matter—its *reactivity* doesn't matter either! We are wasting our time adding NaOH to this reaction—water will do just as well. All the oxygen nucleophiles in Table 17.3 react at the *same* rate with t-BuBr though they react at very different rates with MeI.

Nucleophiles in substitution reactions

We can see the changeover from S_N1 to S_N2 in the reactions of a single compound if we choose one that is good at both mechanisms, such as a benzyl sulfonium salt. Both mechanisms are available for this compound.

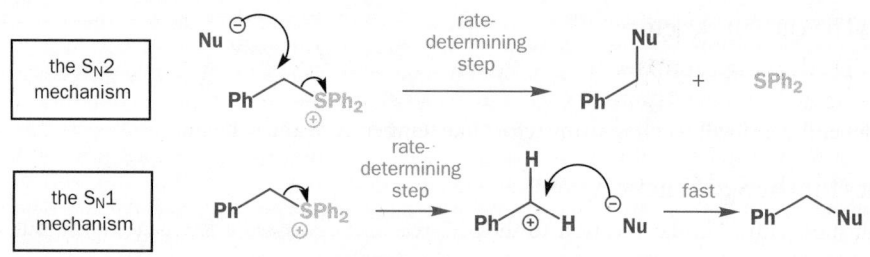

Weak nucleophiles react by the S_N1 mechanism while strong ones react by S_N2. We can tell which is which simply by looking at the rates of the reactions (see Table 17.5).

The first three nucleophiles react at the same rate within experimental error while the last two are clearly faster. The first three nucleophiles react at the same rate because they react by the S_N1 mechanism whose rate does not depend on the nucleophile. All the nucleophiles in fact react by S_N1 at the same rate (about 4.0×10^{-5} s^{-1}) but good nucleophiles also react by S_N2. The S_N2 rate for hydroxide is about 70 and for PhS$^-$ about 107. Compare these relative rates with those in Table 17.6 for reactions with MeBr where they all react at different rates by the S_N2 reaction.

Table 17.5 Rate of reaction (10^5k, s^{-1}) of nucleophiles with PhCH$_2$S$^+$Ph$_2$

Nucleophile	AcO$^-$	Cl$^-$	PhO$^-$	HO$^-$	PhS$^-$
rate	3.9	4.0	3.8	74	107

Table 17.6 Relative rate of reaction (water = 1) of nucleophiles with MeBr

Nucleophile	AcO$^-$	Cl$^-$	PhO$^-$	HO$^-$	PhS$^-$
rate	900	1100	2000	1.2×10^4	5×10^7

How can we decide which mechanism (S_N1 or S_N2) will apply to a given organic compound?

The most important factor is the structure of the carbon skeleton. A helpful generalization is that compounds that can form relatively stable cations generally do so and react by the S_N1 mechanism while the others have to react by the S_N2 mechanism.

In fact, the structural factors that make cations unstable also lead to faster S$_N$2 reactions. Cations are more stable if they are heavily substituted, that is, tertiary, but this is bad for an S$_N$2 reaction because the nucleophile would have to thread its way into the carbon atom through the alkyl groups. It is better for an S$_N$2 reaction if there are only small hydrogen atoms on the carbon atom—methyl groups react fastest by the S$_N$2 mechanism. The effects of the simplest structural variations are summarized in Table 17.7 (where R is a simple alkyl group like methyl or ethyl).

● **S$_N$1 or S$_N$2?**

Table 17.7 Simple structures and choice of S$_N$1 or S$_N$2 mechanism

structure	Me—X	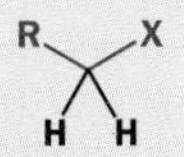 R ∨ X (H H)	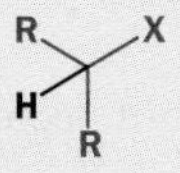 R ∨ X (H R)	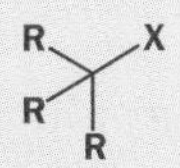 R ∨ X (R R)
type	methyl	primary	secondary	tertiary
S$_N$1 reaction?	no	no	yes	good
S$_N$2 reaction?	good	good	yes	no

The only doubtful case is the secondary alkyl derivative, which can react by either mechanism, though it is not very good at either. The first question you should ask when faced with a new nucleophilic substitution is: 'Is the carbon electrophile methyl, primary, secondary, or tertiary?' This will start you off on the right foot, which is why we introduced these important structural terms in Chapter 2.

Stability and structure of tertiary carbocations

So why are tertiary cations relatively stable whereas the methyl cation is never formed in solution? Any charged organic intermediate is inherently unstable because of the charge. A carbocation can be formed only if it has some extra stabilization. The *t*-butyl cation that we met earlier in this chapter is planar. Indeed it is a universal characteristic of carbocations that they are planar. The basic instability of the carbocation comes from its electron deficiency—it has an empty orbital. The energy of the unfilled orbital is irrelevant to the overall stability of the cation—it's only the energy of the orbitals with electrons in that matter. For any cation the most stable arrangement of electrons in orbitals results from making filled orbitals as low in energy as possible to give the most stable structure, leaving the highest-energy orbital empty. Thus, of the two structures for the *t*-butyl cation, the planar one has the lower-energy filled orbitals (sp^2) and a higher-energy empty p orbital while the tetrahedral one has higher-energy filled orbitals (sp^3) and a lower-energy empty sp^3 orbital.

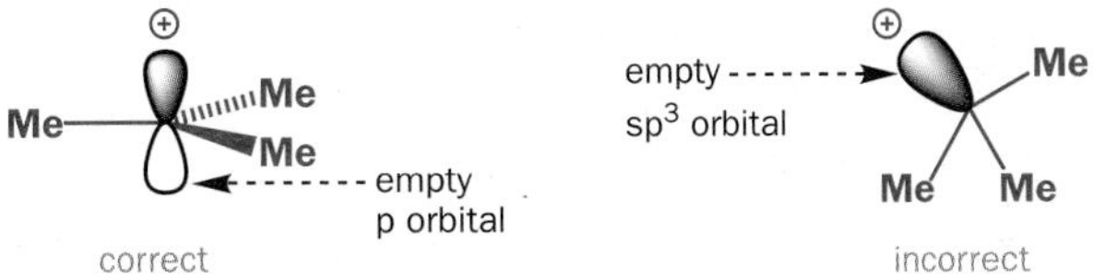

planar structure for the *t*-butyl cation

correct

less repulsion between bonding pairs of electrons

tetrahedral structure for the *t*-butyl cation

incorrect

more repulsion between bonding pairs of electrons

The diagram shows another reason why the planar structure is more stable than the tetrahedral structure for a carbocation. It is better for the filled orbitals to be:

● of the lowest possible energy (so that they contribute most to stability)

● as far from each other as possible (so that they repel each other as little as possible)

Both requirements are fulfilled in the planar structure for the carbocation.

Stabilization of tertiary carbocations by C–H or C–C bonds

Extra stabilization comes to the planar structure from weak donation of σ bond electrons into the empty p orbital of the cation. Three of these donations occur at any one time in the *t*-butyl cation. It

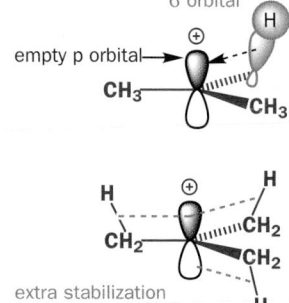

σ orbital

empty p orbital

extra stabilization
from σ donation
into empty p orbital
of planar carbocation

▶

Many textbooks say that alkyl
groups are fundamentally
electron-donating and thus
stabilize cations. This statement
does contain some truth but it is
important to understand the way
in which they really donate
electrons—weakly by σ
conjugation into empty p orbitals.

■

We discussed conjugation in allyl
cations in Chapter 7.

the allyl cation

curly arrows

the cyclohexenyl cation

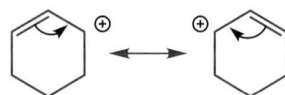

delocalized π bond

doesn't matter if the C–H bonds point up or down; one C–H bond on each methyl group must be parallel to one lobe of the empty p orbital at any one time. The top diagram shows one overlap in orbital terms and the bottom diagram three as dotted lines.

There is nothing special about the C–H bond in donating electrons into an empty orbital. A C–C bond is just as good and some bonds are much better (C–Si). But there must be a bond of some sort—a hydrogen atom by itself has no lone pairs and no σ bonds so it cannot stabilize a cation.

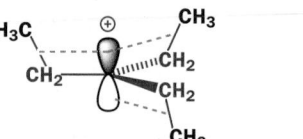

extra stabilization
from σ donation
into empty p orbital
of planar carbocation

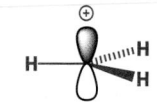

no stabilization: no electrons
to donate into empty p orbital
note: The C–H bonds are at
90° to the empty p orbital
and cannot interact with it

If a tertiary cation cannot become planar, it is not formed. A classic case is the cage halide below, which does not react with nucleophiles either by S_N1 or by S_N2. It does not react by S_N1 because the cation cannot become planar nor by S_N2 because the nucleophile cannot approach the carbon atom from the right direction (see below).

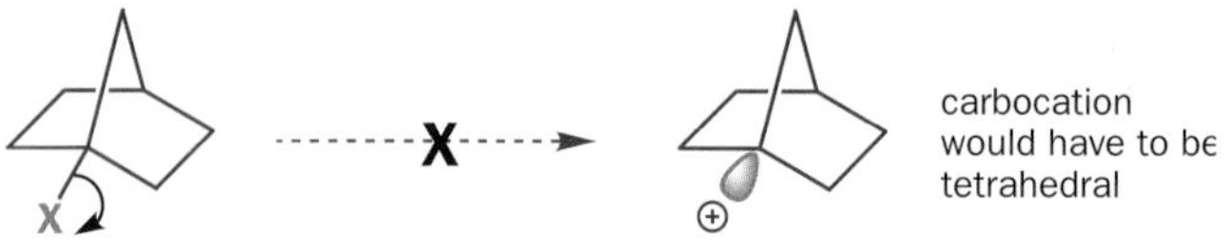

carbocation
would have to be
tetrahedral

In almost all cases, tertiary alkyl halides react rapidly with nucleophiles by the S_N1 mechanism. The nature of the nucleophile is not important: it does not affect the rate and carbocations are reactive enough to combine with even quite weak nucleophiles.

Allylic and benzylic cations

More effective stabilization is provided by genuine conjugation with π or lone-pair electrons. The allyl cation has a filled (bonding) orbital containing two electrons delocalized over all three atoms and an important empty orbital with coefficients on the end atoms only. It's this orbital that is attacked by nucleophiles and so it's the end carbon atoms that are attacked by nucleophiles. The normal curly arrow picture tells us the same thing.

molecular orbitals

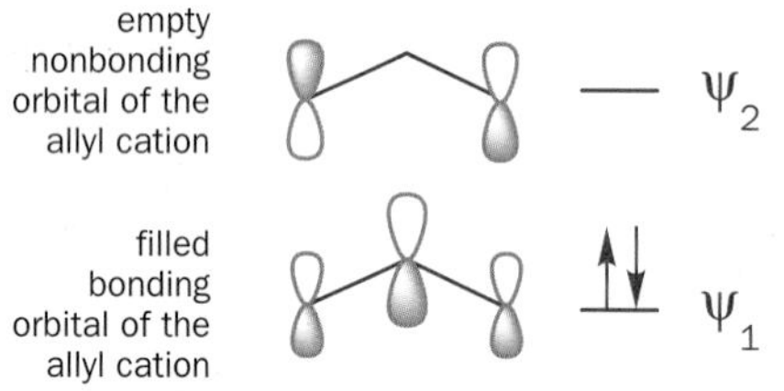

empty
nonbonding
orbital of the
allyl cation Ψ_2

filled
bonding
orbital of the
allyl cation Ψ_1

A symmetrical allyl cation can give one product only by the S_N1 reaction. We have already discussed the formation of the cyclohexenyl cation (Chapter 7) and that is a good example. The two delocalized structures are identical and the π bond is shared equally among the three atoms.

Treatment of cyclohexenol with HBr gives the corresponding allylic bromide. Only one compound is formed because attack at either end of the allylic cation gives the same product.

formation of the cyclohexenyl cation

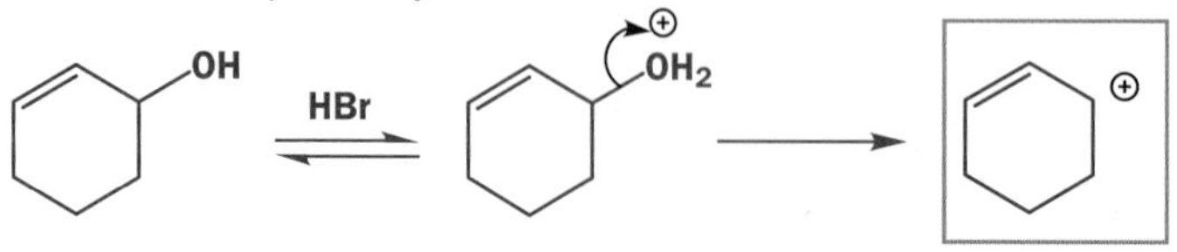

two identical reactions with bromide ion

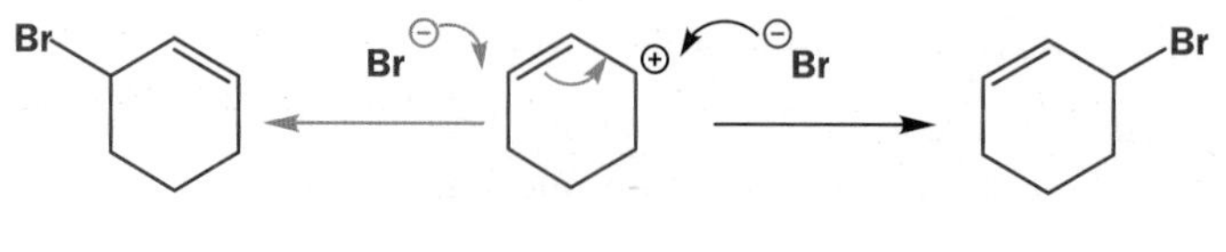

Sometimes when the allylic cation is unsymmetrical this can be a nuisance as a mixture of products may be formed. It doesn't matter which of the two butenols you treat with HBr; you get the same cation.

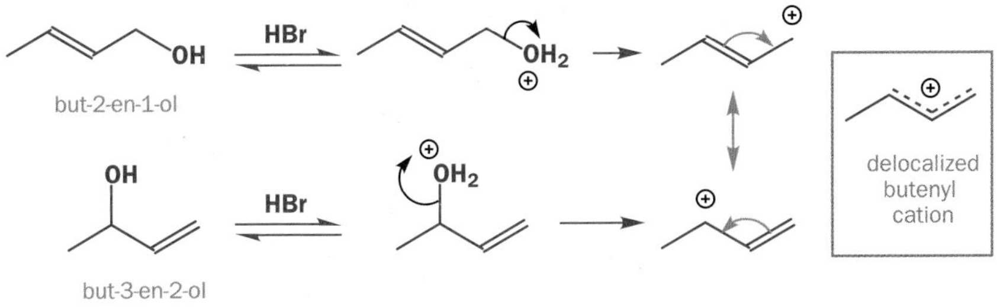

but-2-en-1-ol

but-3-en-2-ol

delocalized butenyl cation

When this cation reacts with Br$^-$, about 80% goes to one end and 20% to the other, giving a mixture of butenyl bromides. Notice that we have chosen one localized structure for our mechanisms. The choice is meaningless since the other structure would have done as well. It's just rather too difficult to draw mechanisms on the delocalized structure.

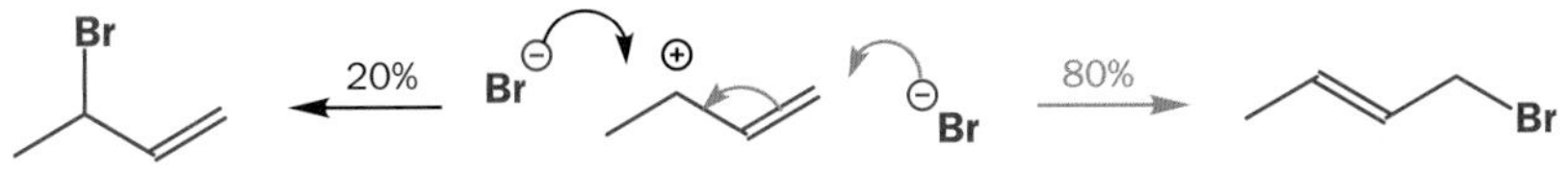

20%   80%

Sometimes this ambiguity is useful. The tertiary allylic alcohol 2-methylbut-3-en-2-ol is easy to prepare and reacts well by the S$_N$1 mechanism because it is both tertiary and allylic. The allylic carbocation intermediate is very unsymmetrical and reacts only at the less substituted end to give 'prenyl bromide'.

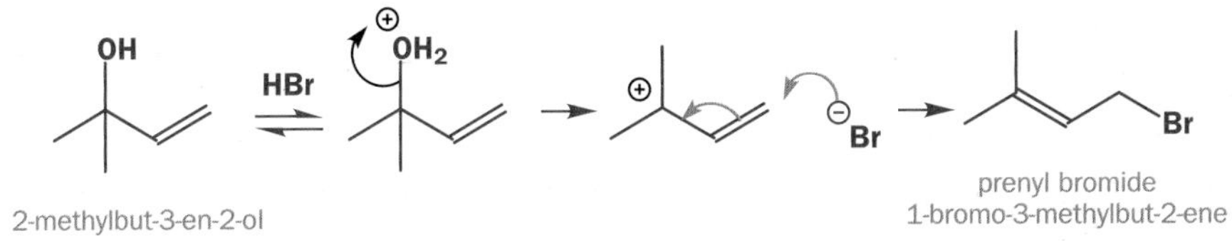

2-methylbut-3-en-2-ol

prenyl bromide
1-bromo-3-methylbut-2-ene

> The **regioselectivity** (where the nucleophile attacks) is determined by steric hindrance: attack is faster at the less hindered end of the allylic system.

■ Prenyl bromide is a building block for making the class of natural products known as terpenes and discussed in Chapter 49. We come back to reactions of allylic compounds in Chapter 23.

The benzyl cation is about as stable as the allyl cation but lacks its ambiguity of reaction. Though the positive charge is delocalized around the benzene ring, the benzyl cation almost always reacts on the side chain.

formation and reaction of the benzyl cation

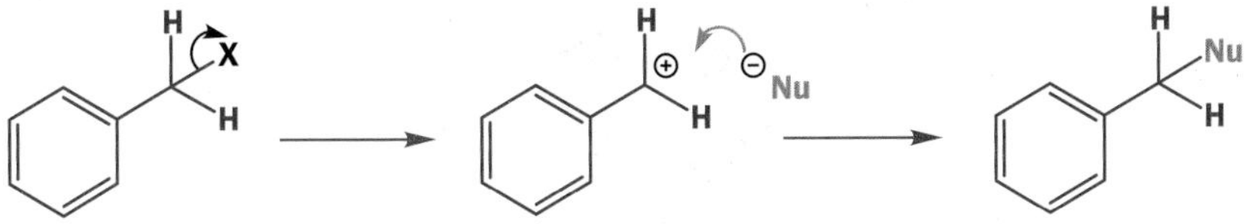

If you draw the arrows for the delocalization, you will see that the positive charge is spread right round the ring, to three positions in particular.

delocalization in the benzyl cation

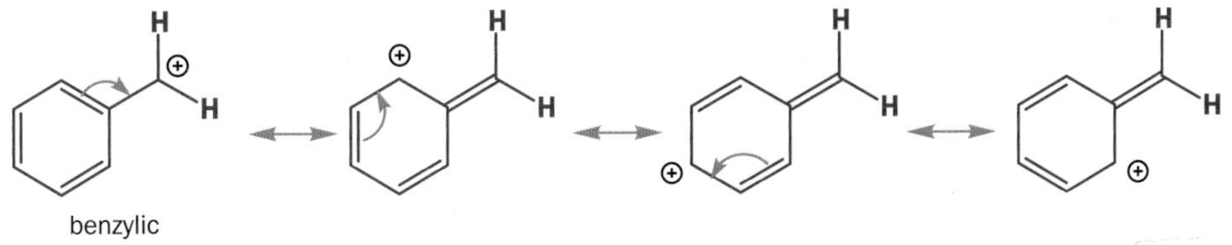

benzylic

■ This sort of delocalization will be given special importance in Chapter 22

An exceptionally stable cation is formed when three benzene rings can help to stabilize the same positive charge. The result is the triphenylmethyl cation or, for short, the trityl cation. The symbol Tr (another of these 'organic elements') refers to the group Ph$_3$C. Trityl chloride is used to form an ether with a primary alcohol group by an S$_N$1 reaction. Here is the reaction.

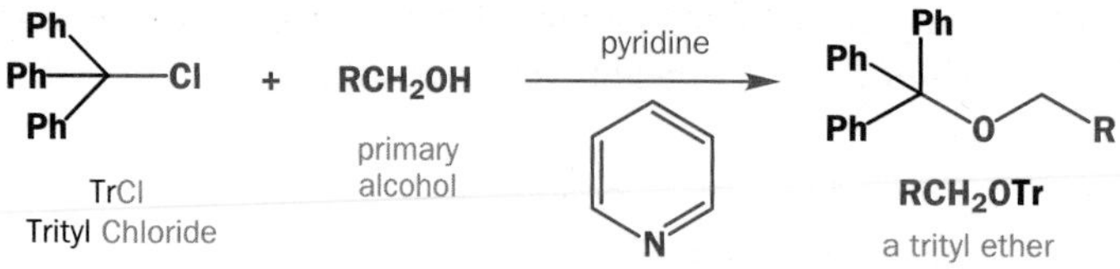

You will notice that pyridine is used as solvent for the reaction. Pyridine (a weak base, pK_{aH} 5.5; see Chapter 8) is not strong enough to remove the proton from the primary alcohol (pK_a about 15), and there would be no point in using a base strong enough to make RCH$_2$O$^-$ as the neutral alcohol is as good in an S$_N$1 reaction. Instead the TrCl ionizes first to trityl cation, which now captures the primary alcohol and finally pyridine is able to remove the proton from the oxonium ion. Pyridine does not catalyse the reaction; it just stops it becoming too acidic by removing the HCl formed. Pyridine is also a convenient polar organic solvent for ionic reactions.

S$_N$1 formation of trityl ethers:

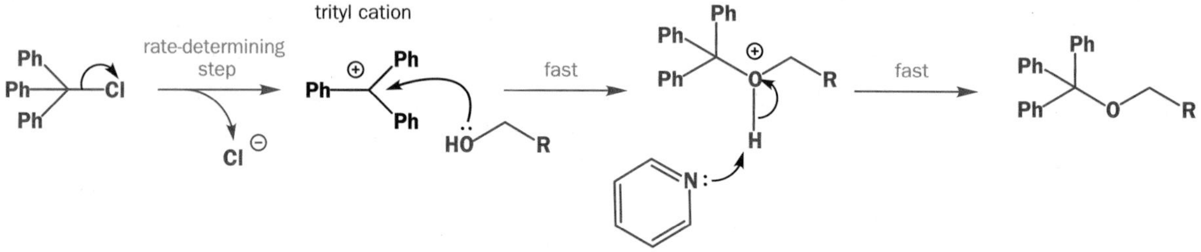

Rate data for substituted allylic chlorides compared with benzylic chlorides and simple alkyl chlorides on solvolysis in 50% aqueous ethanol give us some idea of the magnitude of stabilization (Table 17.8). These rates are mostly S$_N$1, but there will be some S$_N$2 creeping in with the primary compounds. Note the wide range of rates.

> ■
> A **solvolysis reaction** is a reaction in which the solvent is also the nucleophile.

Table 17.8 Rates of solvolysis of alkyl chlorides in 50% aqueous ethanol at 44.6 °C

Compound	Relative rate	Comments
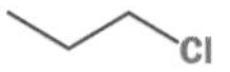	0.07	primary chloride: probably all S$_N$2
	0.12	secondary chloride: can do S$_N$1 but not very well
	2 100	tertiary chloride: very good at S$_N$1
	1.0	primary but allylic: S$_N$1 all right
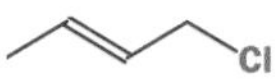	91	allylic cation is secondary at one end
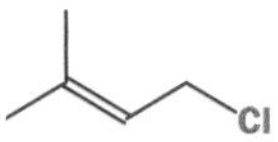	130 000	allylic cation is tertiary at one end: compare with 2100 for simple tertiary
	7 700	primary but allylic and benzylic

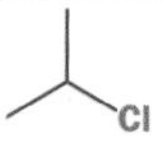

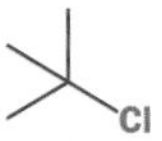

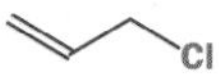

One type of carbocation remains to be discussed, the type with an electron-donating group on the same atom as the leaving group. A classic case is MeOCH$_2$Cl, which loses chloride ion in polar solvents and which can be converted in good yield (89%) to a stable cation using Olah's methods described on p. 410. Even though it is primary (so you might expect S$_N$2), substitution reactions of this chloroether, 'methoxymethyl chloride' (or 'MOM chloride') follow the S_N1 mechanism and go via this cation.

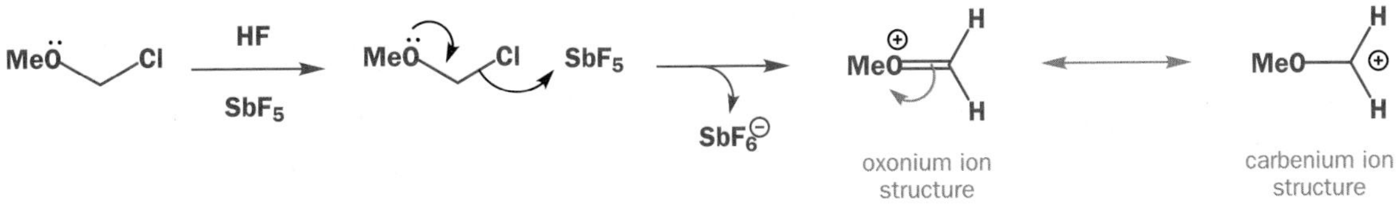

The methoxymethyl cation

This cation can be drawn either as an oxonium ion or as a primary carbenium ion. The oxonium ion structure is the more realistic. Primary carbenium ions are not known in solution, let alone as isolable intermediates, and the proton NMR spectrum of the cation compared with that of the isopropyl cation (this is the best comparison we can make) shows that the protons on the CH_2 group resonate at 9.9 p.p.m. instead of at the 13.0 p.p.m. of the true carbenium ion.

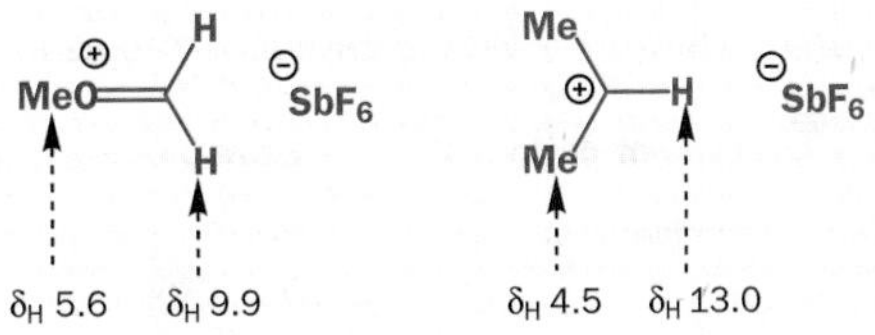

The first step in the hydrolysis of acetals is similar. One alkoxy group is replaced by water to give a hemiacetal.

hydrolysis of acetals – the first step

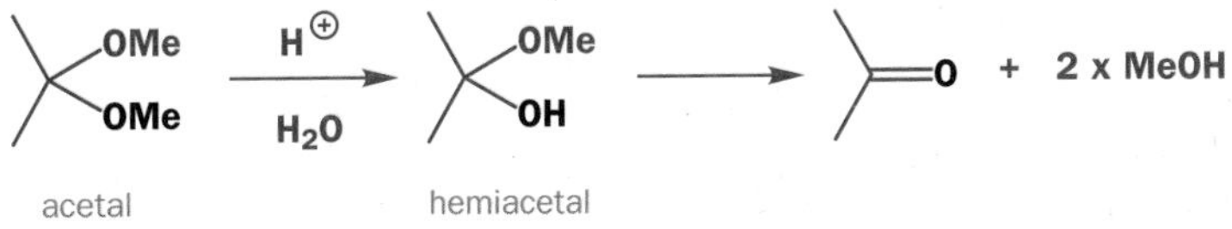

We considered the mechanism for this reaction in Chapter 14 but did not then concern ourselves with a label for the first step. It has, in fact, an S_N1 style of rate-determining step: the decomposition of the protonated acetal to give an oxonium ion. If you compare this step with the decomposition of the chloroether we have just described you will see that they are very similar.

hydrolysis of acetals – S_N1 mechanism for the first step

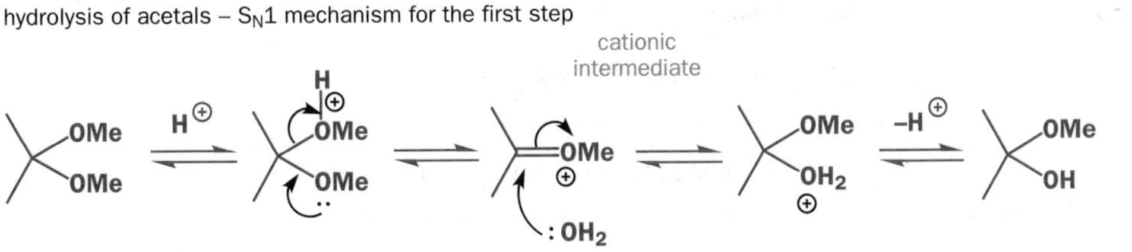

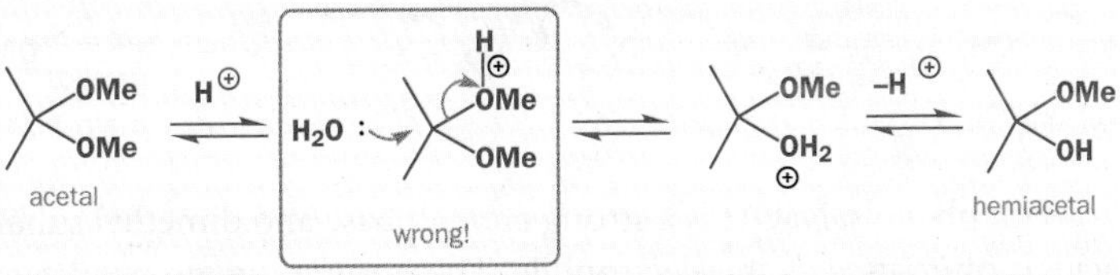

A common mistake

Students of organic chemistry often make a mistake with this mechanism and draw the displacement of the first molecule of methanol by water as an S_N2 reaction.

hydrolysis of acetals – *incorrect* S_N2 mechanism for the first step

When we discuss the S_N2 reaction shortly you will see that an S_N2 mechanism is unlikely at such a crowded carbon atom. However, the main reason why the S_N2 mechanism is wrong is that the S_N1 mechanism is so very efficient with a neighbouring MeO group. The S_N2 mechanism doesn't get a chance.

This mechanism for the S_N1 replacement of one electronegative group at a carbon atom by a nucleophile where there is another electronegative group at the same carbon atom is very general. You should look for it whenever there are two atoms such as O, N, S, Cl, or Br joined to the same carbon atom. The better leaving groups (such as the halogens) need no acid catalyst but the less good ones (N, O, S) usually need acid. Here is a summary diagram and a specific example.

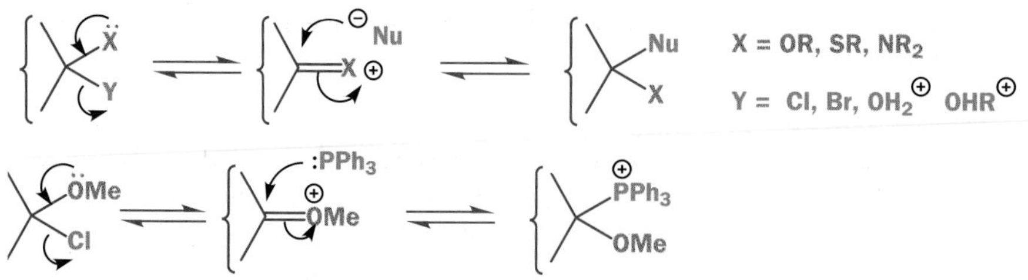

$X = OR, SR, NR_2$

$Y = Cl, Br, OH_2^{\oplus}\ OHR^{\oplus}$

We now have in Table 17.9 a complete list of the sorts of structures that normally react by the S_N1 mechanism rather than by the S_N2 mechanism.

Table 17.9 Stable carbocations as intermediates in S_N1 reactions

Type of cation	Example 1	Example 2
simple alkyl	tertiary (good) *t*-butyl cation 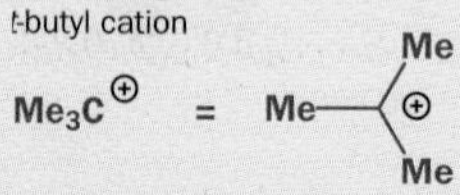	secondary (not so good) *i*-propyl cation 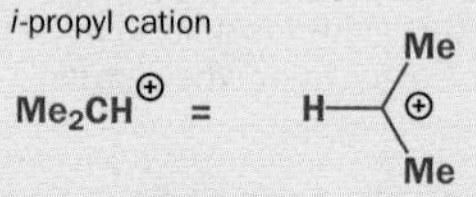
conjugated	allylic 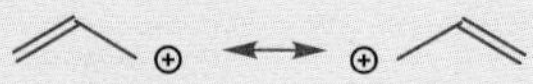	benzylic 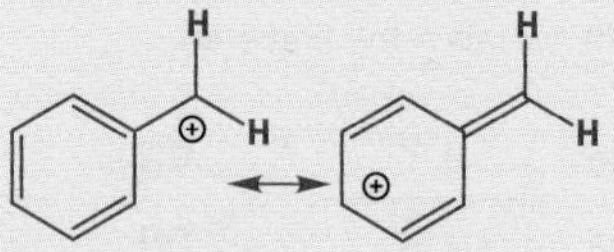
heteroatom-stabilized	oxygen-stabilized (oxonium ions) 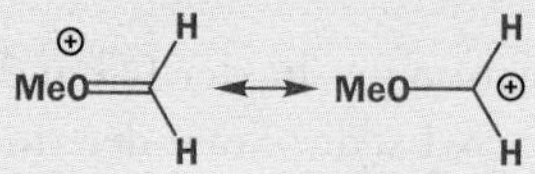	nitrogen-stabilized 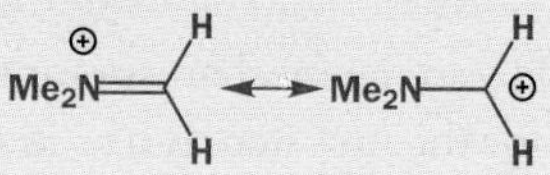

The S_N2 reaction

Small structures that favour the S_N2 reaction

Among simple alkyl groups, methyl and primary alkyl groups always react by the S_N2 mechanism and never by S_N1. This is partly because the cations are unstable and partly because the nucleophile can push its way in easily past the hydrogen atoms.

Thus, a common way to make ethers is to treat an alkoxide anion with an alkyl halide. If the alkyl halide is a methyl compound, we can be sure that this will be by the S_N2 mechanism. A strong base, here NaH, will be needed to form the alkoxide ion (Chapter 6) and methyl iodide is a suitable electrophile.

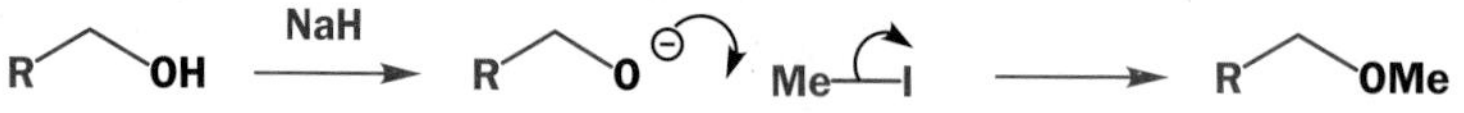

With phenols, NaOH is a strong enough base and dimethyl sulfate, the dimethyl ester of sulfuric acid, is often used as the electrophile. These variations do not affect the mechanism. As long as we have a good nucleophile (here reactive RO^-), a methyl electrophile, and a good leaving group (here an iodide or a sulfate anion), the S_N2 mechanism will work well.

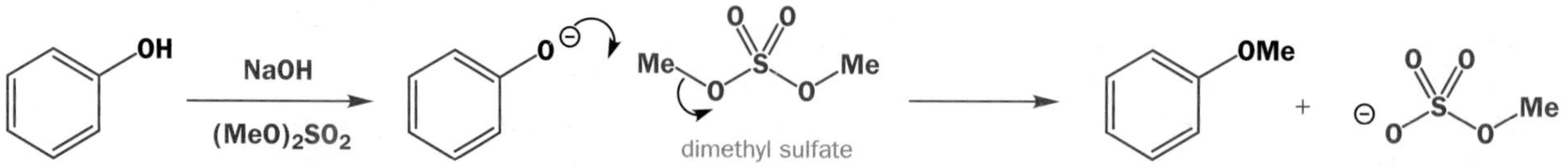

Notice that we said *simple* alkyl groups: of course, primary allylic, benzylic, and RO or R_2N substituted primary derivatives may react by S_N1!

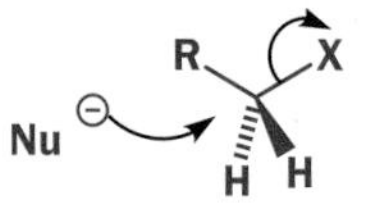

uncluttered approach for nucleophile in S_N2 reactions of methyl compounds (R=H) and primary alkyl compounds (R=alkyl)

The nature of the nucleophile and the leaving group and the structure of the compound under attack all affect the S_N2 mechanism because its rate expression is

rate = k_2[nucleophile][MeX]

This expression shows that the rate of an S_N2 reaction is proportional both to the concentration of the nucleophile and to the concentration of the alkyl halide (MeX). The alkyl halide combines the carbon skeleton and the leaving group in the same molecule. We must consider all three factors (nucleophile, carbon skeleton, and leaving group) in an S_N2 reaction. So it was worth removing the proton from the alcohol or the phenol in these ether syntheses because we get a better nucleophile that way. We established on p. 414 that this was not worth doing in an S_N1 reaction because the nucleophile is not involved in the rate-determining step.

The transition state for an S_N2 reaction

Another way to put this would be to say that the nucleophile, the methyl group, and the leaving group are all present in the transition state for the reaction as explained in Chapter 13. This is the point about halfway through the slow step where the combined reagents reach their highest energy.

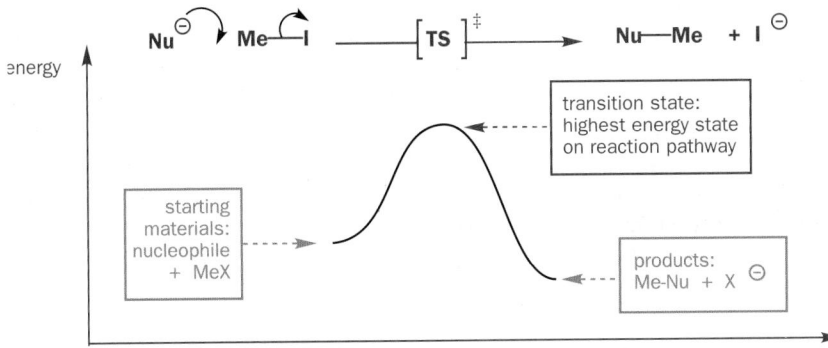

energy diagram for an S_N2 reaction

A transition state is not an intermediate. It can never be isolated because any change in its structure leads to a lower-energy state. In an S_N2 reaction any molecule at the transition state cannot stay there—it must roll down the slope towards products or back to starting materials. So what does it look like and why are we interested in it? The transition state in an S_N2 reaction is about halfway between the starting materials and the products. The bond to the nucleophile is partly formed and the bond to the leaving group is partly broken. It looks like this.

starting materials   transition state   products

The dashed bonds indicate partial bonds (the C—-Nu bond is partly formed and the C—-X bond partly broken) and the charges in brackets indicate substantial partial charges (about half a minus charge each in this case as they must add up to one!). Transition states are often shown in square brackets and marked with the symbol ‡. Another way to look at this situation is to consider the orbitals. The nucleophile must have lone-pair electrons, which will interact with the σ* orbital of the C–X bond.

filled orbital   empty σ* orbital   new σ bond   p orbital   old σ bond   new σ bond
of nucleophile   of C–X bond   being formed   on C atom   being broken

In the transition state there is a p orbital at the carbon atom in the middle that shares one pair of electrons between the old and the new bonds. Both these pictures suggest that the transition state for an S_N2 reaction has a more or less planar carbon atom at the centre with the nucleophile and the leaving group arranged at 180° to each other.

Stereochemistry and substitution

If this is true, it has a very important consequence. The nucleophile attacks the carbon atom on the opposite side from the leaving group and the carbon atom turns inside out as the reaction goes along, just like an umbrella in a high wind. If the carbon atom under attack is a stereogenic centre (Chapter 16), the result will be inversion of configuration. This is easily proved by a simple sequence of reactions. We start by looking at the stereochemistry of an S_N1 reaction.

(+)-(S)-sec-butanol   secondary   racemic
                                     butyl cation   (±)-sec-butanol

Starting with the optically active secondary alcohol *sec*-butanol (or butan-2-ol, but we want to emphasize that it is *secondary*), the secondary cation can be made by the usual method and has a characteristic ^{13}C NMR shift. Quenching this cation with water regenerates the alcohol but without any optical activity. Water has attacked the two faces of the planar cation with exactly equal probability as we described in Chapter 16. The product is an exactly 50:50 mixture of (S)-butanol and (R)-butanol. It is *racemic*.

■
TsCl and its synthesis is discussed
later in this chapter (p. 433) and in
Chapter 22.

If, however, we first make the *para*-toluene sulfonate ('tosylate') by nucleophilic attack of the OH group on the sulfonyl chloride TsCl in pyridine solution, the sulfonate will be formed with retention as no bonds have been formed or broken at the chiral carbon atom. This is a substitution reaction too, but at sulfur rather than at carbon.

(+)-(S)-sec-butanol   TsCl   (+)-(S)-sec-butyl
                        para-toluene   *para*-toluene sulfonate
                        sulfonyl chloride   [(+)-(S)-sec-butyl tosylate]

Now we can carry out an S_N2 reaction on the sulfonate with a carboxylate anion. A *tetra*-alkyl ammonium salt is often used in the polar solvent DMF to get a clean reaction. This is the key step and we don't want any doubt about the outcome.

S_N2 reaction   DMF   optically active
                                 dimethylformamide   *sec*-butyl acetate

The product is optically active and we can measure its rotation. But this tells us nothing. Unless we know the true rotation for pure *sec*-butyl acetate, we don't yet know whether it is optically pure nor even whether it really is inverted. But we can easily find out. All we have to do is to hydrolyse the ester and get the original alcohol back again. We know the true rotation of the alcohol—it was our starting material—and we know the mechanism of ester hydrolysis (Chapter 12)—nucleophilic attack occurs at the carbonyl carbon and retention must be the stereochemical outcome as no reaction occurs at the stereogenic centre.

an optically active
sec-butyl acetate

tetrahedral
intermediate

(−)-*sec*-butanol

Now we really know where we are. This new sample of *sec*-butanol has the same rotation as the original sample, *but with the opposite sign*. It is (−)-(*R*)-*sec*–butanol. It is optically pure and inverted. Somewhere in this sequence there has been an inversion, and we know it wasn't in the formation of the tosylate or the hydrolysis of the acetate as no bonds are formed or broken at the stereogenic centre in these steps. It must have been in the S$_N$2 reaction itself.

● **This is a general conclusion.**
● The S$_N$2 reaction goes with inversion of configuration at the carbon atom under attack but the S$_N$1 reaction generally goes with racemization

Substitution reactions at other elements

S$_N$2 reactions can occur at elements other than carbon. Common examples in organic chemistry are silicon, phosphorus, sulfur, and the halogens. The formation of the tosylate above by attack of the alcohol on TsCl is an example of an S$_N$2 reaction at sulfur. Later in this chapter you will see that alcohols attack phosphorus very easily and that we use the reaction between ROH and PBr$_3$ to make alkyl bromides. Alcohols also react rapidly with Si–Cl compounds such as Me$_3$SiCl to give silyl ethers by an S$_N$2 reaction at silicon. You have already seen several examples of silyl ether formation (p. 240, for example), though up to this point we have not discussed the mechanism. Here it is: B: represents a base such as triethylamine.

For an example of an S$_N$2 reaction at chlorine we can choose a reaction we will need later in the book. Triphenyl phosphine reacts with CCl$_4$ to give a phosphonium salt by what looks like an S$_N$2 reaction at carbon.

In fact there is no room around the carbon atom of CCl$_4$ for any nucleophile, let alone such a large one as PPh$_3$ and the reaction occurs by two separate S$_N$2 steps: one at chlorine and one at phosphorus.

Structural variation and the S_N2 mechanism

We have already established that methyl and primary alkyl compounds react well by the S_N2 mechanism, while secondary alkyl compounds *can* do so. There are other important structural features that also encourage the S_N2 mechanism. Two, allyl and benzyl compounds, also encourage the S_N1 mechanism.

Here you see a typical S_N2 reaction of allyl bromide. We have drawn the transition state for this reaction. This is not because we want to encourage you to do this for all S_N2 reactions but so that we can explain the role of the allyl system. Allyl compounds react rapidly by the S_N2 mechanism because the double bond can stabilize the transition state by conjugation.

The benzyl group acts in much the same way using the π system of the benzene ring for conjugation with the p orbital in the transition state.

Since the p orbital in question has electrons in it—it shares a pair of electrons with the nucleophile and the leaving group—more effective conjugation is possible with an electron-deficient π bond. The most important example is the carbonyl group: carbon electrophiles like those in the margin give the fastest S_N2 reactions.

With α-bromo carbonyl compounds, substitution leads to two electrophilic groups on neighbouring carbon atoms. Each has a low-energy empty orbital, π^* from C=O and σ^* from C–Br (this is what makes them electrophilic), and these can combine to form a molecular LUMO ($\pi^* + \sigma^*$) lower in energy than either. Nucleophilic attack will occur easily where this new orbital has its largest coefficients, shown in orange on the diagram.

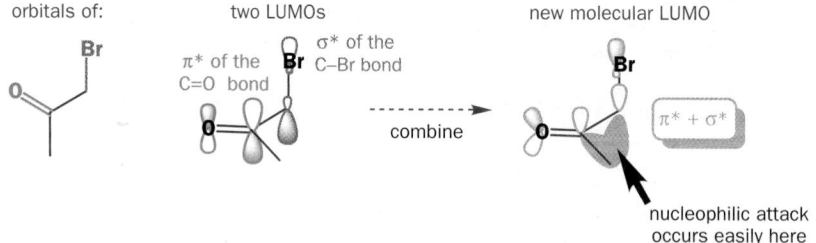

This orange area is on one side of the carbonyl group and in the usual place at the back of the C–Br bond. Each group has become more electrophilic because of the presence of the other—the C=O group makes the C–Br bond more reactive and the Br makes the C=O group more reactive. Another way to put this is that the carbonyl group stabilizes the transition state by overlap of its π^* orbital with the full p orbital of the carbon atom under attack. The nucleophile may well attack the carbonyl group but this will be reversible whereas displacement of bromide is irreversible.

transition state for nucleophilic attack on an α-bromo-ketone

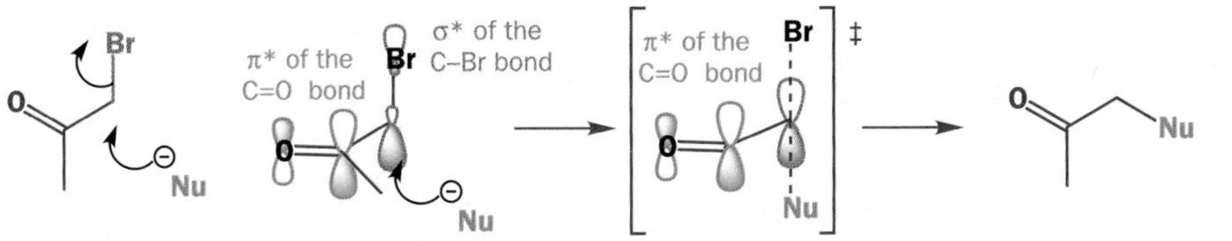

There are many examples of this type of reaction. Reactions with amines go well and the amino-ketone products are widely used in the synthesis of drugs.

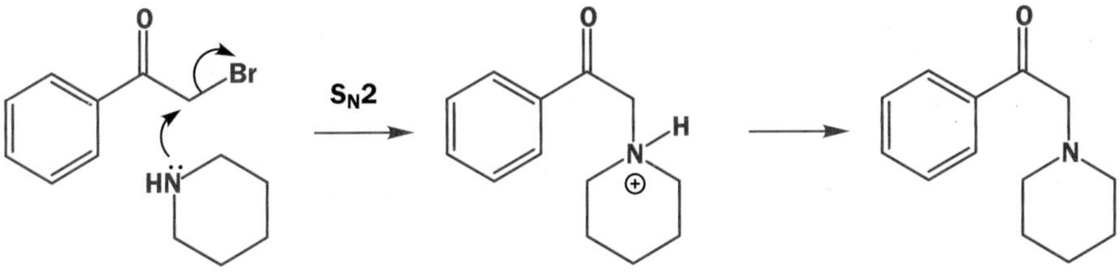

an amino-ketone

Variation of rate with structure

Some actual data may help at this point. The rates of reaction of the following alkyl chlorides with KI in acetone at 50 °C broadly confirm the patterns we have just analysed. These are relative rates with respect to n-BuCl as a 'typical primary halide'. You should not take too much notice of precise figures but rather observe the trends and notice that the vartiations are quite large—the full range from 0.02 to 100 000 is eight powers of ten.

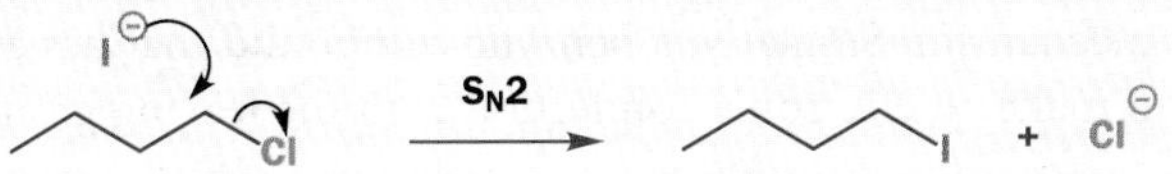

Table 17.10 Relative rates of S$_N$2 reactions of alkyl chlorides with the iodide ion

Alkyl chloride	Relative rate	Comments
Me—Cl	200	least hindered alkyl chloride
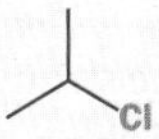 (isopropyl chloride)	0.02	secondary alkyl chloride; slow because of steric hindrance
 (allyl chloride)	79	allyl chloride accelerated by π conjugation in transition state
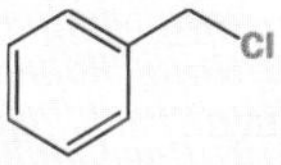 (benzyl chloride)	200	benzyl chloride slightly more reactive than allyl: benzene ring better at π conjugation than isolated double bond
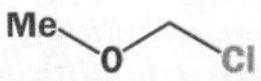 Me-O-Cl	920	conjugation with oxygen lone pair accelerates reaction
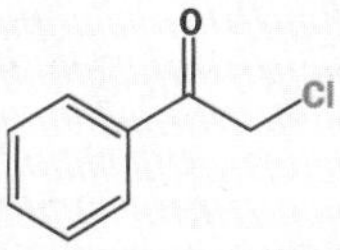 (phenacyl chloride)	100 000	conjugation with carbonyl group much more effective than with simple alkene or benzene ring. These α-carbonyl halides are the most reactive of all

Summary of structural variations and nucleophilic substitution

We are now in a position to summarize those effects we have been discussing over the last few pages on both mechanisms. It is simplest to list the structural types and rate each reaction qualitatively.

● **Table 17.11** Structural variations for the S_N1 and S_N2 reactions

Type of electrophilic carbon atom	S_N1 reaction	S_N2 reaction
methyl (CH_3–X)	no	very good
primary alkyl (RCH_2–X)	no	good
secondary alkyl (R_2CH–X)	yes	yes
tertiary alkyl (R_3C–X)	very good	no
allylic (CH_2=CH–CH_2–X)	yes	good
benzylic ($ArCH_2$–X)	yes	good
α-carbonyl ($RCO\cdot CH_2$–X)	no	excellent
α-alkoxy ($RO\cdot CH_2$–X)	excellent	good
α-amino ($R_2N\cdot CH_2$–X)	excellent	good

You must not regard this list as fixed and inflexible. The last five types will also be either primary, secondary, or tertiary. If they are primary, as shown, they will favour S_N2 more, but if they are tertiary they will all react by the S_N1 mechanism except the tertiary α-carbonyl ($RCO\cdot CR_2$–X) compounds, which will still react by the S_N2 mechanism, if rather slowly. If they are secondary they might react by either mechanism. Similarly, a benzylic compound that has a well placed electron-donating group able to make an electronic connection with the leaving group will favour the S_N1 mechanism.

a benzylic chloride that favours the S_N1 mechanism

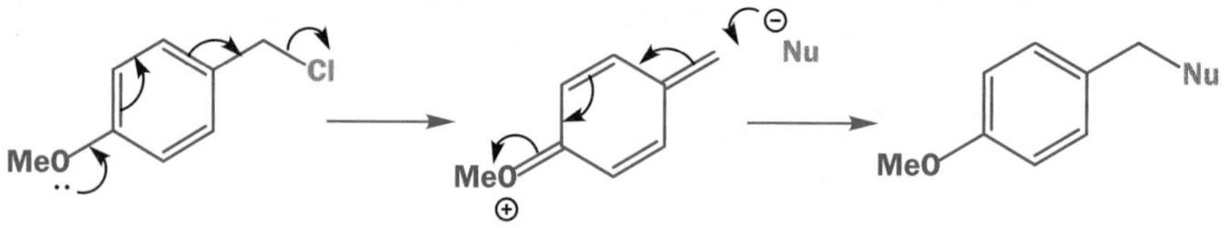

On the other hand, a 4-nitrobenzyl chloride is likely to react by the S_N2 mechanism as the strongly electron-withdrawing nitro group would destabilize the carbocation intermediate of the S_N1 mechanism.

a benzylic chloride that disfavours the S_N1 mechanism

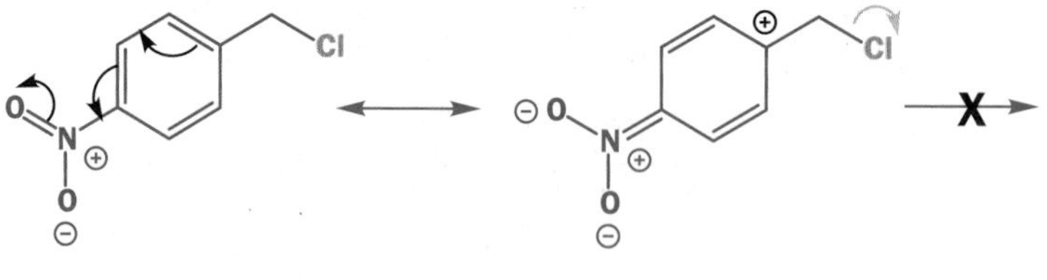

no S_N1 found

electron-withdrawing nitro group would destabilize cation intermediate

the same benzylic chloride that favours the S_N2 mechanism

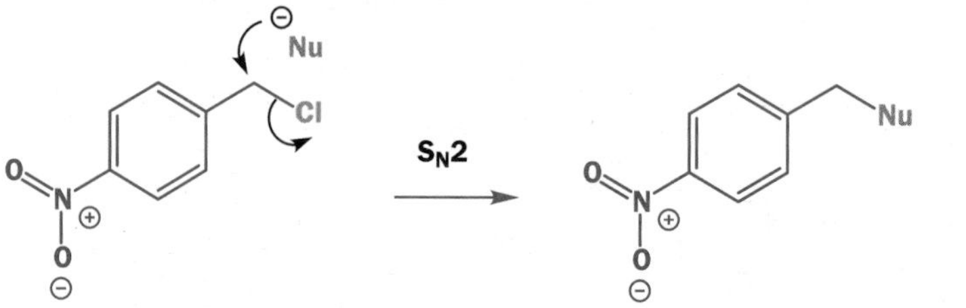

transition state stabilized by electron-withdrawing nitro group

> ► Rate measurements for these two compounds are very revealing. We can force them to react by S_N1 by using methanol as the solvent (p. 414). If we set the rate of substitution of the benzyl compound with methanol at 25 °C at 1.0, then the 4-MeO benzyl compound reacts about 2500 times faster and the 4-NO_2 benzyl compound about 3000 times more slowly.

Steric hindrance in nucleophilic substitution

We have already considered the inversion of stereochemistry necessary in an S$_N$2 mechanism, but there is another steric effect, the rather cruder steric hindrance. In the approach to the S$_N$2 transition state, the carbon atom under attack gathers in another ligand and becomes (briefly) five-coordinate. The angles between the substituents decrease from tetrahedral to about 90°.

steric hindrance in the S$_N$2 reaction

tetrahedral –
all angles 109°

trigonal bipyramid –
three angles of 120°
six angles of 90°

rate-determining step

tetrahedral –
all angles 109°

In the starting material there are four angles of about 109°. In the transition state (enclosed in square brackets and marked ‡ as usual) there are three angles of 120° and six angles of 90°, a significant increase in crowding. The larger the substituents R, the more serious this is. We can easily see the effects of steric hindrance if we compare these three structural types:

- methyl: CH$_3$–X: very fast S$_N$2 reaction
- primary alkyl: RCH$_2$–X: fast S$_N$2 reaction
- secondary alkyl: R$_2$CH–X: slow S$_N$2 reaction

The opposite is true of the S$_N$1 reaction. The slow step is simply the loss of the leaving group. The starting material is again tetrahedral (four angles of about 109°) and in the intermediate cation there are just three angles of 120°—fewer and less serious interactions. The transition state will be on the way towards the cation, rather closer to it than to the starting material.

steric acceleration in the S$_N$1 reaction

planar trigonal –
three angles 120°

tetrahedral –
all angles 109°

rate-determining step

120°

Even in the transition state, the angles are increasing towards 120° and all interactions with the leaving group are diminishing as it moves away. There is steric *acceleration* in the S$_N$1 reaction rather than steric *hindrance*. This, as well as the stability of *t*-alkyl cations, is why *t*-alkyl compounds react by the S$_N$1 mechanism.

Rates of S$_N$1 and S$_N$2 reactions

Here is a simple illustration of these effects. The green curve in Figure 17.1 (next page) shows the rates (k_1) of an S$_N$1 reaction: the conversion of alkyl bromides to alkyl formate esters in formic acid at 100 °C. Formic acid is very polar and, though a weak nucleophile, is adequate for an S$_N$1 reaction.

The red curve in Figure 17.1 shows the rates of displacement of Br⁻ by radioactive ^{82}Br⁻ in acetone at 25 °C by the S$_N$2 mechanism, the rates (k_2) being multiplied by 10^5 to bring both curves on to the same

graph. The actual values of the rate constants are not important. Table 17.12 gives the relative rates compared with that of the secondary halide, *i*-PrBr, set at 1.0 in each case.

> You will often read that *t*-alkyl compounds do not react by the S$_N$2 mechanism because the steric hindrance would be too great. This is a reasonable assumption given that secondary alkyl compounds are already reacting quite slowly. The truth is that *t*-alkyl compounds react so fast by the S$_N$1 mechanism that the S$_N$2 mechanism wouldn't get a chance *even if it went as fast as it goes with methyl compounds*. The nucleophile would have to be about 100 molar in concentration to compensate for the difference in rates and this is impossible! Even pure water is only 55 molar (Chapter 8). You see only the faster of the two possible mechanisms.
>
> - If there are two *steps* in a single mechanism, the *slower* of the two determines the rate of the overall reaction
> - If there are two different *mechanisms* available under the reaction conditions, only the *faster* of the two actually occurs.

Rates of S$_N$1 and S$_N$2 reactions (contd)

Both curves are plotted on a log scale, the $\log_{10}$ of the actual rate being used on the y-axis. The x-axis has no real significance; it just shows the four points corresponding to the four basic structures: MeBr, MeCH$_2$Br, Me$_2$CHBr, and Me$_3$CBr. The values plotted are given in Table 17.12

Table 17.12 Rates of S$_N$1 and S$_N$2 reactions of simple alkyl bromides

alkyl bromide	CH$_3$Br	CH$_3$CH$_2$Br	(CH$_3$)$_2$CHBr	(CH$_3$)$_3$CBr
type	methyl	primary	secondary	tertiary
k_1, s^{-1}	0.6	1.0	26	10^8
$10^5 k_2$ (lm^{-1}s^{-1})	13 000	170	6	0.0003
relative k_1	2×10^{-2}	4×10^{-2}	1	4×10^6
relative k_2	6×10^3	30	1	5×10^{-5}

The reactions were chosen to give as much S$_N$1 reaction as possible in one case and as much S$_N$2 reaction as possible in the other case. Formic acid is a very polar solvent but a poor nucleophile; this gives the maximum opportunity for a cation to form. Bromide ion is a good nucleophile and acetone is polar enough to dissolve the reagents but not so polar that ionization is encouraged. Of course, you will understand that we cannot prevent the molecules doing the 'wrong' reaction! The values for the 'S$_N$1' reaction of MeBr and MeCH$_2$Br are actually the low rates of S$_N$2 displacement of the bromide ion by the weak nucleophile HCO$_2$H, while the 'S$_N$2' rate for t-BuBr may be the very small rate of ionization of t-BuBr in acetone.

Figure 17.1: S$_N$1 and S$_N$2 rates for simple alkyl bromides

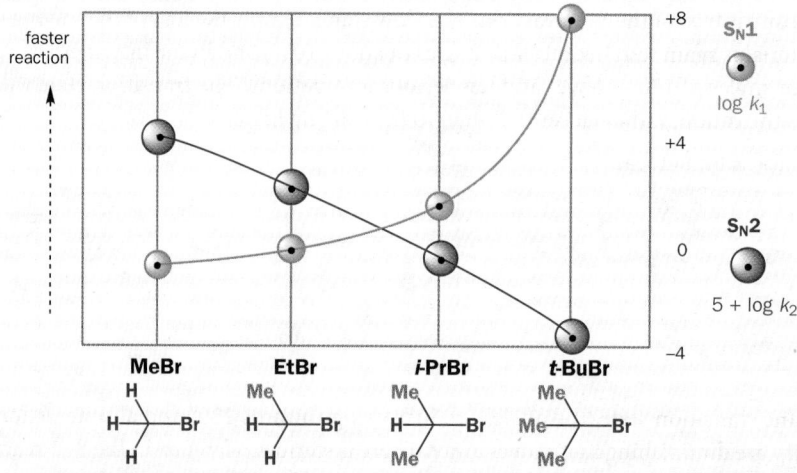

The actual values of the rate constants are not important. The graph in Figure 17.1 has been plotted to put the rates of the S$_N$2 and S$_N$1 reactions of the secondary alkyl bromide at about the same level to give a graphical illustration of the *relative* speed of the S$_N$2 reaction with MeBr and the *relative* speed of the S$_N$1 reaction of t-BuBr.

►
Solvating polar compounds or transition states

Three things are important:
- Polarity—simply measured by dipole moment. The + end of the dipole stabilizes full or partial anions and the – end of the dipole stabilizes full or partial cations
- Electron donation to cationic centres by lone-pair electrons
- Hydrogen bonding to stabilize full or partial anions

Solvent effects

In the box above, you can see acetone used as a solvent for an S$_N$2 reaction and formic acid (HCO$_2$H) as solvent for the S$_N$1 reaction. These are typical choices: a less polar solvent for the S$_N$2 reaction (just polar enough to dissolve the ionic reagents) and a polar protic solvent for the S$_N$1 reaction. The S$_N$1 reaction fairly obviously needs a polar solvent as the rate-determining step usually involves the formation of ions and the rate of this process will be increased by a polar solvent. More precisely, the transition state is more polar than the starting materials and so is stabilized by the polar solvent. Hence solvents like water or carboxylic acids (RCO$_2$H) are ideal.

It is less obvious why a less polar solvent is better for the S$_N$2 reaction. The most common S$_N$2 reactions use an anion as the nucleophile and the transition state is less polar than the localized anion as the charge is spread between two atoms.

preparation of alkyl iodides by the S$_N$2 reaction

$$RCH_2Br + NaI \xrightarrow{\text{solvent} \quad \text{acetone}} RCH_2I + NaBr\downarrow$$

charge localized on one atom (I)

charge spread over two atoms (I and Br)

A polar solvent solvates the anionic nucleophile and slows the reaction down. A nonpolar solvent destabilizes the starting materials more than it destabilizes the transition state and speeds up the reaction. There is another reason for using acetone for this particular reaction. NaI is very soluble in acetone but NaBr is rather insoluble. The NaBr product precipitates out of solution which helps to drive the reaction over to the right.

If an S$_N$2 reaction has neutral starting materials and an ionic product, then a polar solvent is better. A good choice is DMF, a polar aprotic solvent often used for the synthesis of phosphonium salts by the S$_N$2 reaction.

$$Ph_3P + MeI \xrightarrow[\substack{\text{polar} \\ \text{aprotic} \\ \text{solvent}}]{\text{DMF}} Ph_3P{\overset{\oplus}{-}}Me \quad I^{\ominus}$$

a phosphine

a phosphonium salt

Ph$_3$P : ⤳ Me—I

nonpolar starting materials

dipolar transition state

Polar aprotic solvents

Water, alcohols, and carboxylic acids are polar protic solvents able to form hydrogen bonds (**hydroxylic solvents**). They solvate both cations and anions well. A nucleophilic reagent such as bromide ion must be accompanied by a cation, say, the sodium ion, and hydroxylic solvents dissolve salts such as NaBr by hydrogen bonding to the anion and electron donation to the cation. This is solvation by a polar protic solvent. These solvents do not 'ionize' the salt, which already exists in the solid state as ions; they separate and solvate the ions already present.

Polar aprotic solvents, on the other hand, have dipole moments and are still able to solvate cations by electron donation from an oxygen atom, but they lack the ability to form hydrogen bonds because any hydrogen atoms they may have are on carbon. Examples include DMF and DMSO (dimethyl sulfoxide).

solvation of salts by hydrophilic solvents

cation solvated by electron-donation from oxygen atom

anion solvated by electron-acceptance through hydrogen bonding

there are more than the one solvent molecule shown for each ion

solvation of salts by polar aprotic solvents

cation solvated by electron-donation from oxygen atom

anion not solvated - no hydrogen bonding is possible
the anion is "naked" and hence more nucleophilic

We have considered the important effects of the basic carbon skeleton on the S$_N$1 and S$_N$2 reactions and we shall now consider the remaining two possible structural variations: the nucleophile and the leaving group. We shall tackle the leaving group first because it plays an important role in both S$_N$1 and S$_N$2 reactions.

The leaving group

We have mostly seen halides and water from protonated alcohols as leaving groups in both S$_N$1 and S$_N$2 reactions. Now we need to establish the principles that make for good and bad leaving groups. We might be considering an S$_N$1 reaction.

Or we might be considering an S_N2 reaction—both have a leaving group, which we are representing as 'X' in these mechanisms. In both cases the C–X bond is breaking in the slow step.

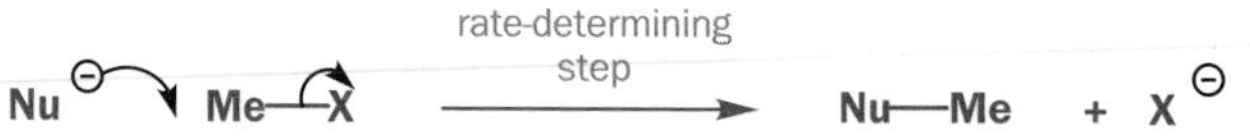

Starting with the halides, two main factors are at work: the strength of the C–halide bond and the stability of the halide ion. The strengths of the C–X bonds have been measured and are listed in Table 17.13. How shall we measure anion stability? One way, which you met in Chapter 8, was to use the pK_a values of the acids HX. We established in Chapter 8 that bond strength can be used to explain pK_a values so these two factors are not independent.

Table 17.13 Halide leaving groups in the S_N1 and S_N2 reactions

Halide (X)	Strength of C–X bond, kJ mol^{-1}	pK_a of HX
fluorine	118	+3
chlorine	81	–7
bromine	67	–9
iodine	54	–10

It is clearly easiest to break a C–I bond and most difficult to break a C–F bond. Iodide sounds like the best leaving group. We get the same message from the pK_a values: HI is the strongest acid, so it must ionize easily to H^+ and I^-. This result is quite correct—iodide is an excellent leaving group and fluoride a very bad one with the other halogens in between.

Nucleophilic substitutions on alcohols

Now what about leaving groups joined to the carbon atom by a C–O bond? There are many of these but the most important are OH itself, the carboxylic esters, and the sulfonate esters. First we must make one thing clear. In spite of what you may suppose, alcohols do *not* react with nucleophiles. Why not? Hydroxide ion is very basic, very reactive, and a bad leaving group. If the nucleophile were strong enough to produce hydroxide ion, it would be more than strong enough to remove the proton from the alcohol.

S_N2 displacement of hydroxide ion is *not* a known reaction

if the nucleophile reacts, it attacks the *proton* instead

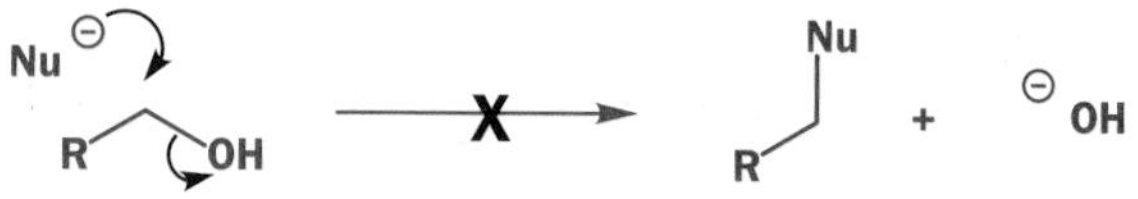

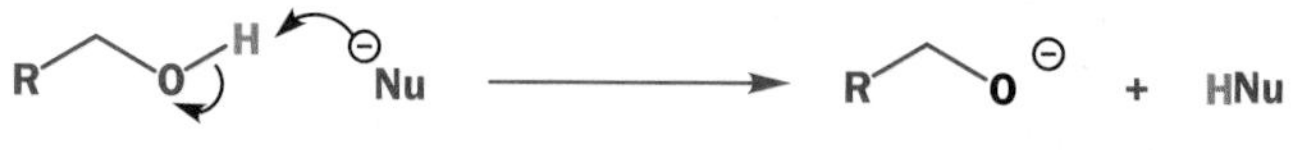

But we want to use alcohols in nucleophilic substitution reactions because they are easily made. The simplest answer is to protonate the OH group with strong acid. This will work only if the nucleophile is compatible with strong acid, but many are. The preparation of *t*-BuCl from *t*-BuOH simply by shaking it with concentrated HCl is a good example. This is obviously an S_N1 reaction with the *t*-butyl cation as intermediate.

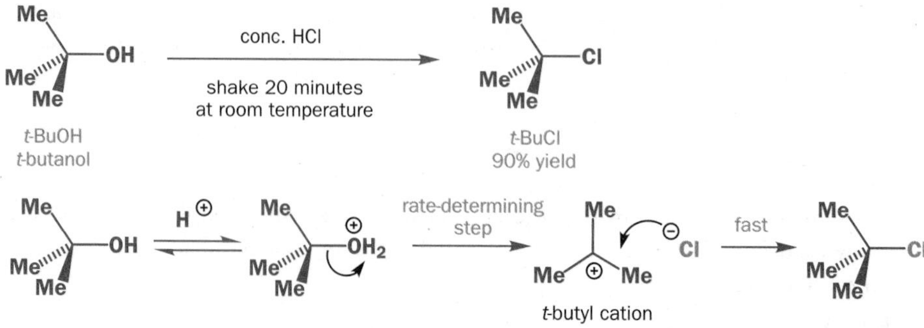

Similar methods can be used to make secondary alkyl bromides with HBr alone and primary alkyl bromides using a mixture of HBr and H_2SO_4. The second is certainly an S_N2 reaction and we show just one stage in a two-step process that is very efficient.

substituting a secondary alcohol in acid

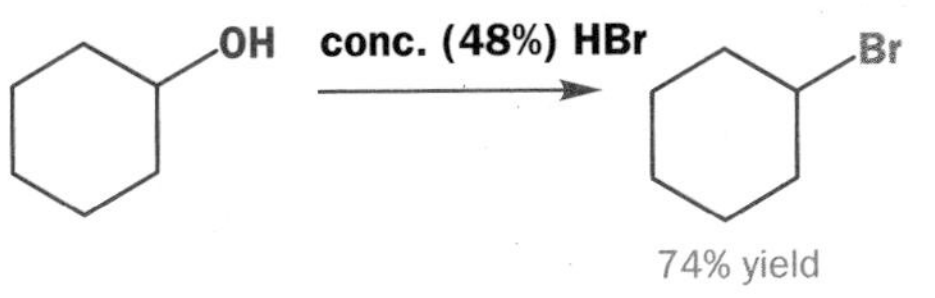

substituting a primary alcohol in acid

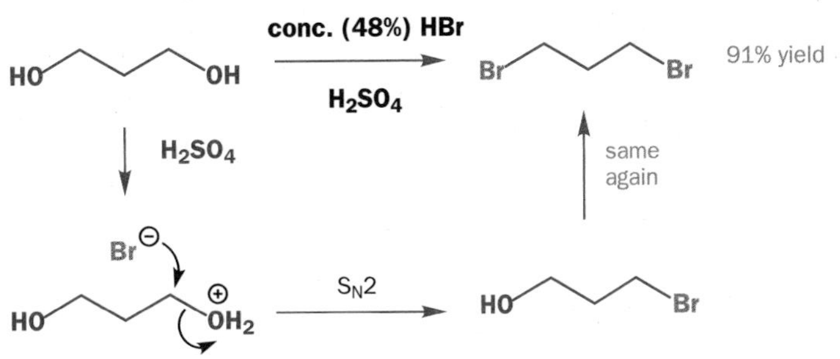

Another way is to convert the OH group into a better leaving group by combination with an element that forms very strong bonds to oxygen. The most popular choices are phosphorus and sulfur. Making primary alkyl bromides with PBr_3 usually works well.

The phosphorus reagent is first attacked by the OH group (an S_N2 reaction at phosphorus) and the displacement of an oxyanion bonded to phosphorus is now a good reaction because of the anion stabilization by phosphorus.

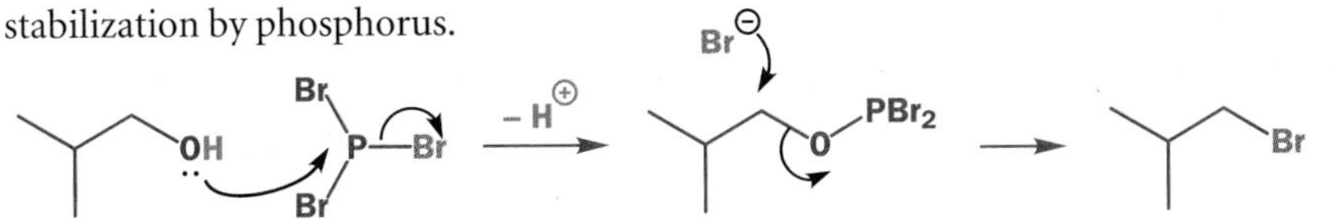

The Mitsunobu reaction is a modern S_N2 reaction using phosphorus chemistry

So far we have seen methods of displacing the OH group by first converting it to something else—a better leaving group like Br, for example. There is one recent invention that allows us to put an alcohol straight into a reaction mixture and get an S_N2 product in one operation. This is the **Mitsunobu reaction**. The alcohol becomes the electrophile, the nucleophile can be whatever you choose, and there are two other reagents.

a Mitsunobu reaction

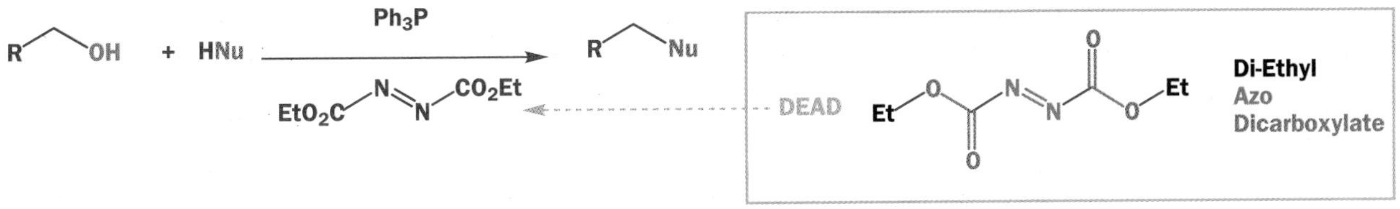

One of these reagents, Ph_3P, triphenylphosphine, is a simple phosphine, rather like an amine but with P instead of N. The other deserves more comment. Its full name is diethyl azodicarboxylate, or DEAD.

Oyo Mitsunobu was born in 1934 in Japan and works at the Aoyama Gakuin University in Tokyo. He is one of the few modern chemists to have a famous reaction named after him. Please note the spelling of his name: MitsUnObU.

Azo compounds

The 'azo' in the name of DEAD refers to two nitrogen atoms joined together by a double bond and compounds such as azobenzene are well known. Many dyestuffs have an azo group in them—Bismarck Brown (mentioned in Chapter 1) is used to dye kippers.

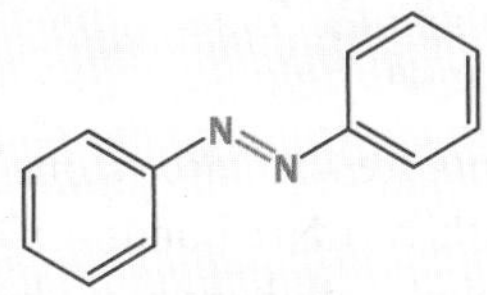

azobenzene

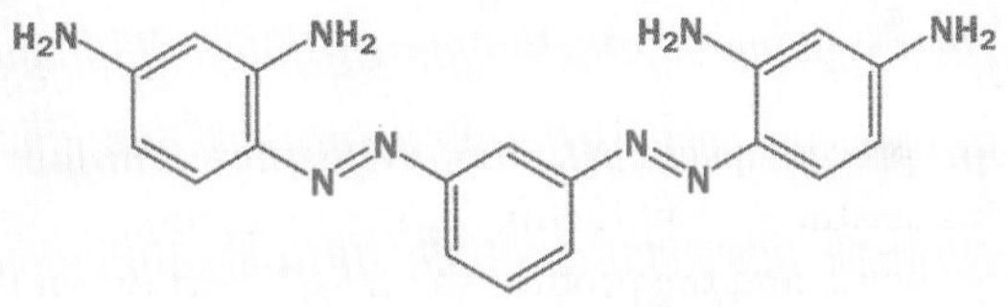

Bismarck Brown Y: an azo dye

So how does the Mitsunobu reaction work? The first step involves neither the alcohol nor the nucleophile. The phosphine adds to the weak N=N π bond to give an anion stabilized by one of the ester groups.

stage 1 of the Mitsunobu reaction

stabilization of the nitrogen anion by the ester group

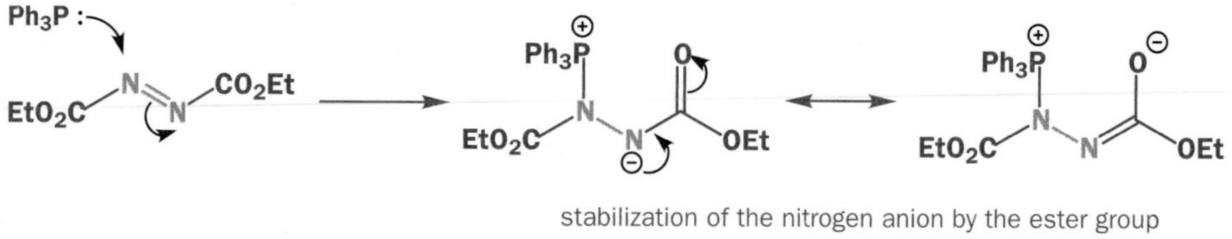

The anion produced by this first stage is basic enough to remove a proton from the alcohol. This is always what will happen if a strong nucleophile is combined with an alcohol and previously this was a fatal disadvantage when we wanted an S_N2 reaction. But wait and see.

stage 2 of the Mitsunobu reaction

alkoxide ion

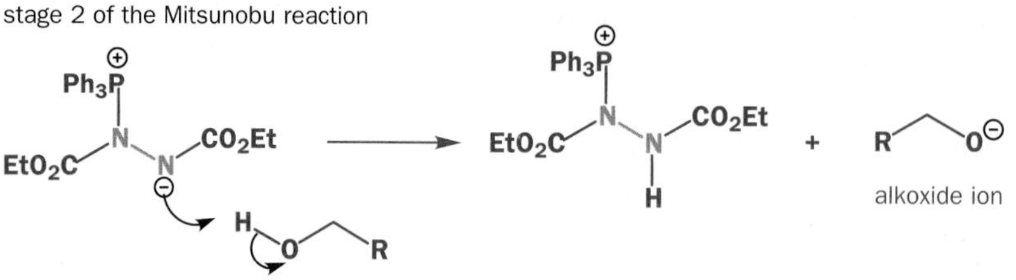

Oxygen and phosphorus have a strong affinity as we saw in the conversion of alcohols to bromides with PBr_3 and in the Wittig reaction (Chapter 14, p. 359) and so the new alkoxide ion immediately attacks the positively charged phosphorus atom displacing a second nitrogen anion stabilized in the same way as the first. This is an S_N2 reaction at phosphorus.

stage 3 of the Mitsunobu reaction

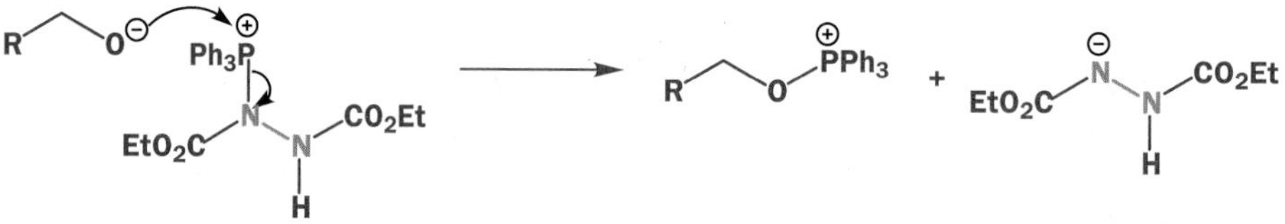

The second basic nitrogen anion removes a proton from the nucleophile, which has been patiently waiting in disguised form as HNu while all this is going on. The true nucleophile is now revealed as an anion.

stage 4 of the Mitsunobu reaction

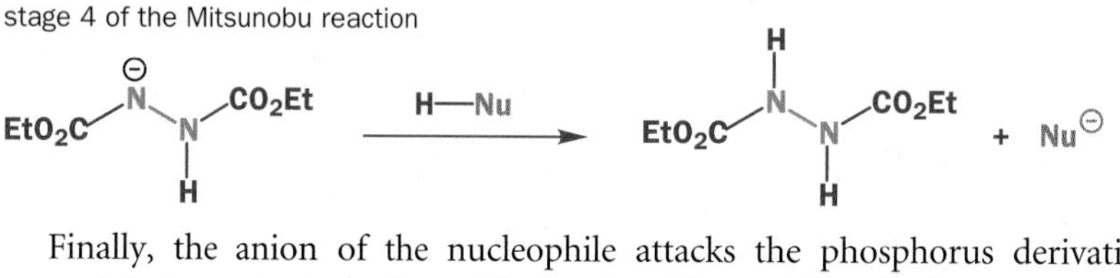

Finally, the anion of the nucleophile attacks the phosphorus derivative of the alcohol in a normal S_N2 reaction at carbon with the phosphine oxide as the leaving group. We have arrived at the products.

stage 5 of the Mitsunobu reaction

S_N2 product   phosphine oxide

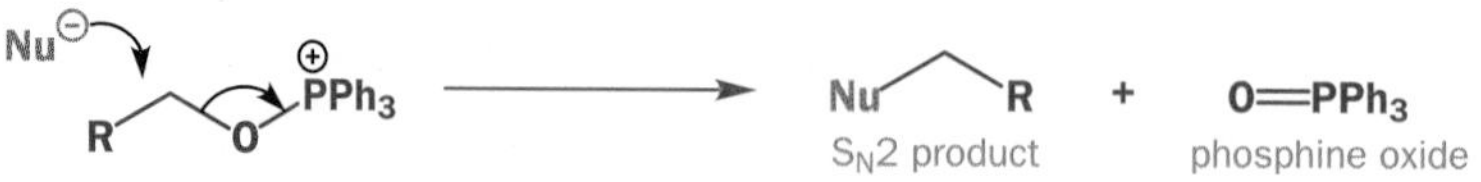

The whole process takes place in one operation. The four reagents are all added to one flask and the products are the phosphine oxide, the reduced azo diester with two NH bonds replacing the N=N double bond, and the product of an S_N2 reaction on the alcohol. Another way to look at this reaction is that a molecule of water must formally be lost: OH must be removed from the alcohol and H from the nucleophile. These atoms end up in very stable molecules—the P=O and N–H bonds are very stable while the N=N bond was weak. This compensates for the sacrifice of the strong C–O bond in the alcohol.

the Mitsunobu reaction – summary

problem: strong bond to be broken
problem: acidic proton

the reaction we want to happen

solution: strong bond formed

Ph₃P ⟶ Ph₃P=O

solution: strong bonds formed

solution: weak bond sacrificed

If this is all correct, then the vital S_N2 step should lead to inversion as it always does in S_N2 reactions. This turns out to be one of the great strengths of the Mitsunobu reaction—it is a reliable way to replace OH by a nucleophile with inversion of configuration. The most dramatic example is probably the formation of esters from secondary alcohols with inversion. Normal ester formation leads to retention as the C–O bond of the alcohol is not broken.

ester formation from a secondary alcohol with retention

- **The Mitsunobu reaction is used to replace OH by another group with inversion of configuration.**

In the Mitsunobu reaction, the C–O bond of the alcohol is broken because the alcohol becomes the electrophile and the acid derivative must be a nucleophile so an acid is better than an acid chloride. The ester is formed with inversion. Note the fate of the oxygen atoms.

ester formation from a secondary alcohol with inversion by the Mitsunobu reaction

The Mitsunobu reaction is by no means the only way to turn OH groups into leaving groups and a method based on sulfur chemistry is as important.

Tosylate, TsO⁻, is an important leaving group made from alcohols

The most important of all these leaving groups are those based on sulfonate esters. The intermediates in the PBr₃ reaction are unstable, but it is usually easy to make stable, usually crystalline toluene-*para*-sulfonates from primary and secondary alcohols. We met these derivatives on p. 422. These isolable but reactive compounds are so popular that they have been given a trivial name ('tosylates') and the functional group has been allocated an 'organic element' symbol Ts. This is what it means.

Sulfonic acids are strong acids (pK_a from Chapter 8) and so any sulfonate is a good leaving group. Another closely related leaving group, methane sulfonate or MsO⁻ is discussed in Chapter 19 under elimination reactions.

TsCl
toluene-*para*-sulfonyl chloride

RCH₂OTs
alkyl toluene-*para*-sulfonate

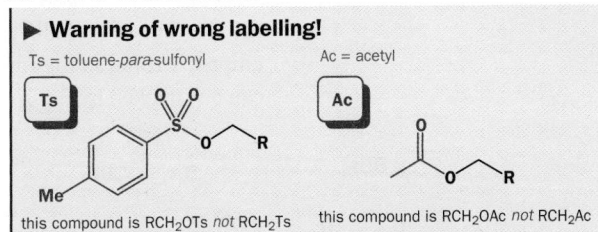

▶ **Warning of wrong labelling!**

Ts = toluene-*para*-sulfonyl   Ac = acetyl

this compound is RCH₂OTs *not* RCH₂Ts   this compound is RCH₂OAc *not* RCH₂Ac

The leaving groups are toluene-*para*-sulfon**ate**, TsO⁻, and acet**ate**. AcO⁻, but the substituents are toluene-*para*-sulfon**yl**, Ts–, and acet**yl**, Ac–.

You have already seen the tosyl group used in the inversion sequence on p. 422, where it was displaced by as weak a nucleophile as acetate. This should alert you to the fact that TsO^- can be displaced by almost anything. We choose some examples in which new carbon–carbon bonds are formed. This will be an important topic later in the book when we meet enolate anions (Chapter 21) but our two examples here use sp anions derived from nitriles and acetylenes.

Cyanide ion is a good small nucleophile and displaces tosylate from primary carbon atoms and adds one carbon atom to the chain. As the cyanide (nitrile) group can be converted directly to a carboxylic acid or ester (Chapter 14) this sequence is a useful chain extension.

Elias J. Corey (1928–), PhD from MIT, works at Harvard University. He invented the disconnection approach to the design of organic synthesis. His group has invented many of the most important modern methods of synthesis, and have made an enormous number of complex compounds. He won the Nobel prize in 1990.

Corey's synthesis of leukotrienes, human metabolites that control many important natural defence reactions like inflammation, involves the lithium derivative of an alkyne prepared by deprotonation with the very strong base butyllithium. The tosyl derivative of a primary alcohol reacts with this lithium derivative and a perfectly normal S_N2 reaction follows. The alkyne provides the carbanion (Chapter 8) for the displacement of the tosylate.

Leukotrienes will be discussed in detail in Chapter 51. They are C_{20} chain compounds, normally with three double bonds as their name suggests.

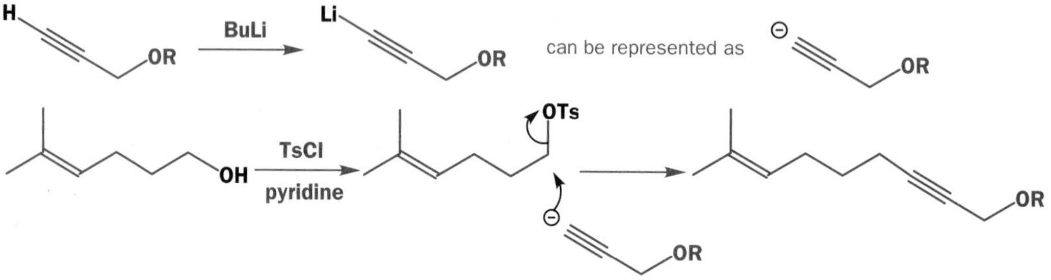

Ethers as electrophiles

Ethers are stable molecules, which do not react with nucleophiles: they must be stable because THF and Et_2O are used as solvents. But we can make them react by using an acid with a nucleophilic counterion (HBr or HI, for example) and then nucleophilic attack will occur preferentially at the more susceptible carbon atom. Aryl alkyl ethers cleave only on the alkyl side. We shall explain in Chapter 23 why nucleophilic attack does not occur on a benzene ring.

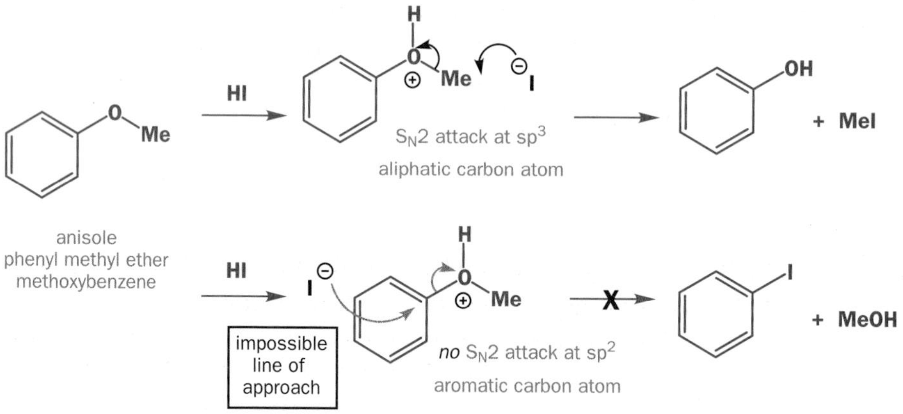

So far we have used only protic acids to help oxygen atoms to leave. Lewis acids work well too, and the cleavage of aryl alkyl ethers with BBr_3 is a good example. Trivalent boron compounds have an empty p orbital so they are very electrophilic and prefer to attack oxygen. The resulting oxonium ion can be attacked by Br^- in an S_N2 reaction.

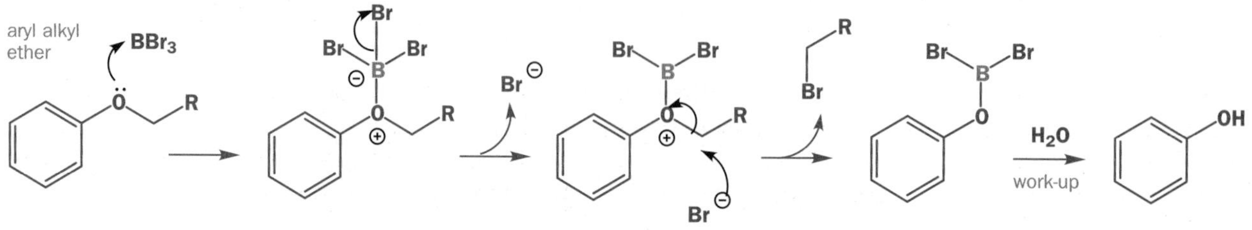

Epoxides

One type of ether reacts in nucleophilic substitution without acids or Lewis acids. The leaving group is genuinely an alkoxide anion RO⁻. Obviously, some extra special feature must be present in these ethers making them unstable and this feature is ring strain. They are the three-membered cyclic ethers called **epoxides** (or oxiranes). You will see how to make these compounds in Chapter 20. The ring strain comes from the angle between the bonds in the three-membered ring which has to be 60° instead of the ideal tetrahedral angle of 109°. You could subtract these numbers and say that there is '49° of strain' at each carbon atom, making about 150° of strain in the molecule. This is a lot. The idea of strain is that the molecule wants to break open and restore the ideal tetrahedral angle at all atoms. This can be done by one nucleophilic attack.

We first discussed the idea of ring strain in Chapter 6, pp. 144–145. The true origin of strain is the poor overlap between the orbitals forming the σ bonds inside the three-membered ring. This is discussed in Chapter 15 where another piece of evidence for ring strain is the peculiar chemical shifts in the proton NMR spectra of epoxides and other three-membered rings.

S_N2 attack on epoxides relieves ring strain

60° bond angle inside the ring   3 × 60° bond angles   all bond angles normal

Epoxides react cleanly with amines to give amino-alcohols. We have not so far featured amines as nucleophiles because their reactions with alkyl halides are often bedevilled by overreaction (see the next section), but with epoxides they give good results.

It is easy to see that inversion occurs in these S_N2 reactions if we put the epoxide on the side of another ring. With a five-membered ring only *cis*-fusion of the epoxide is possible and nucleophilic attack with inversion gives the *trans* product. As the epoxide is *up*, attack has to come from underneath. Notice that the new C–N bond is *down* and that the H atom at the site of attack was *down* in the epoxide but is *up* in the product. Inversion has occurred.

The product of this reaction is used in the manufacture of the antidepressant drug eclanamine by the Upjohn Company. Because the starting material must be a single diastereoisomer (the *cis* or *syn* isomer) and inversion has occurred at one carbon atom, the product must be the *trans* or *anti* diastereoisomer. The starting material cannot be a single enantiomer as it is not chiral (it has a plane of symmetry). Though the product *is* chiral, it cannot be optically active as no optically active reagents have gone into the reaction (Chapter 15). The biological activity in the drug requires this diastereoisomer.

Esters

Nucleophilic attack on esters in acidic or basic solution normally occurs at the carbonyl group (Chapter 12). We are going to concentrate here on what happens to the hydrolysis of simple esters in acid solution as the alkyl group varies in size.

The slow step is the addition of water, which increases the crowding at the central carbon atom. As the alkyl group R is made larger, the reaction gets slower and slower. Then a dramatic thing happens. If the alkyl group R is made *tertiary*, the reaction suddenly becomes very fast indeed—faster than when R was methyl under the same conditions. Clearly, the mechanism has changed. It is no

normal ester hydrolysis in acid solution

tetrahedral intermediate

▶

t-Butyl esters

If you have several ester groups in a molecule and want to remove one without disturbing the others, then a t-butyl ester is the answer as it can be 'hydrolysed' in acid solution under very mild conditions. t-Butyl esters are used in protecting groups because they are so easily hydrolysed and this aspect of their chemistry is discussed in Chapter 24.

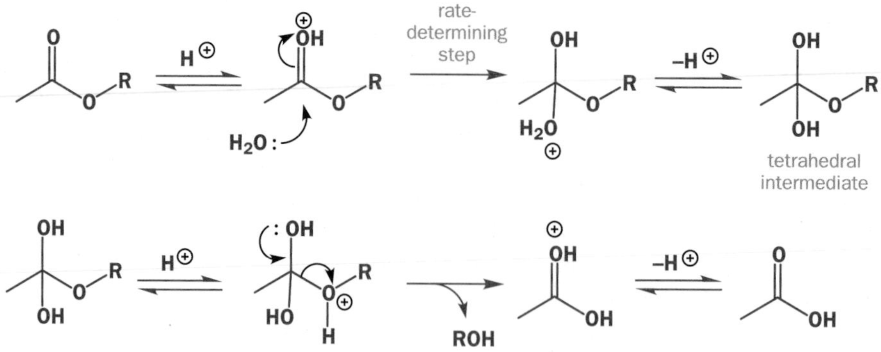

longer the normal ester hydrolysis but has become an S_N1 reaction at the alkyl group. It is still a substitution reaction but at the saturated carbon atom rather than at the carbonyl group. The first step is the same, but the protonated ester is a good leaving group and so the intermediate decomposes to the t-alkyl cation without needing water at all.

the S_N1 mechanism for t-alkyl ester hydrolysis in acid solution

same intermediate in normal hydrolysis

t-butyl cation

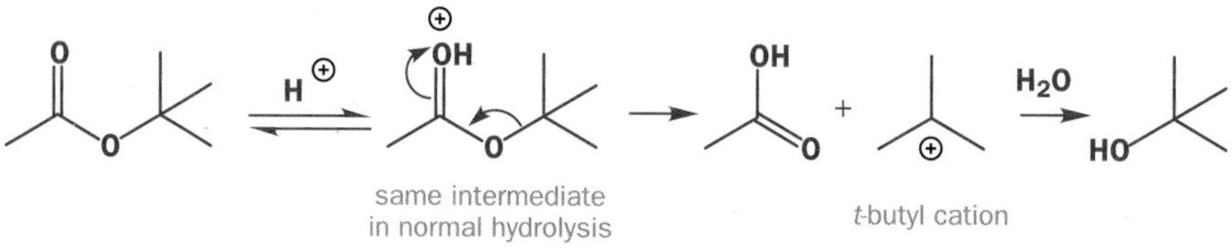

Nucleophiles

We have established that the nucleophile is not important in the *rate* of an S_N1 reaction. We need now to discuss two ways in which it is important. Both concern the nature of the product. A better nucleophile will not accelerate the S_N1 reaction but it may determine which product is formed. In the reactions of tertiary alcohols with concentrated HCl or HBr there is always more water than halide ion present and yet the t-alkyl halide is formed in good yield.

reaction of tertiary alcohols with hydrogen halides

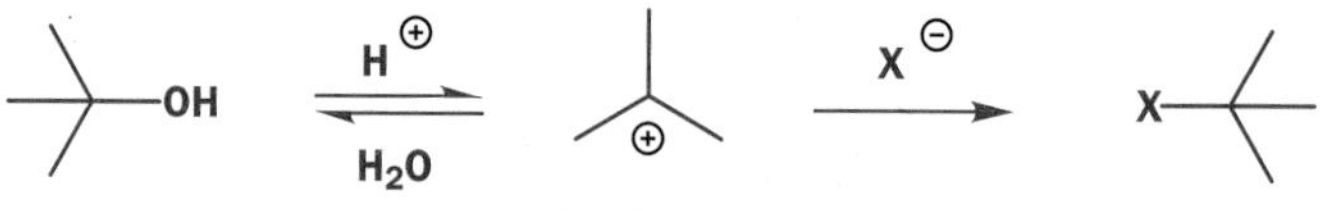

t-butyl cation

This is partly because the halide ion is a better nucleophile than water for a carbocation as both are charged and partly because, if water does act as a nucleophile, it merely regenerates the starting material, which may react again.

A more interesting result of the unimportance of the nucleophile in the rate is that very poor nucleophiles indeed may react in the absence of anything better. In Chapter 8 we established that nitriles are only weakly basic because the lone pair of electrons on the nitrogen atom is in a low-energy sp orbital. They are not good nucleophiles either.

nitriles

lone pair in sp orbital

nitriles are only weakly basic

no reaction

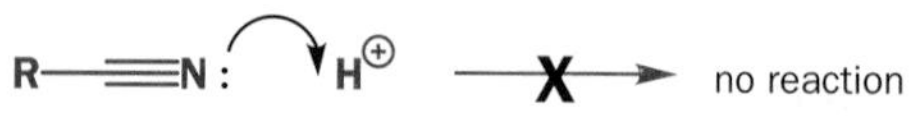

If we dissolve t-butanol in a nitrile as solvent and add strong acid, a reaction does take place. The acid does not protonate the nitrile, but does protonate the alcohol to produce the t-butyl cation in the usual way. This cation is reactive enough to combine with even such a weak nucleophile as the nitrile.

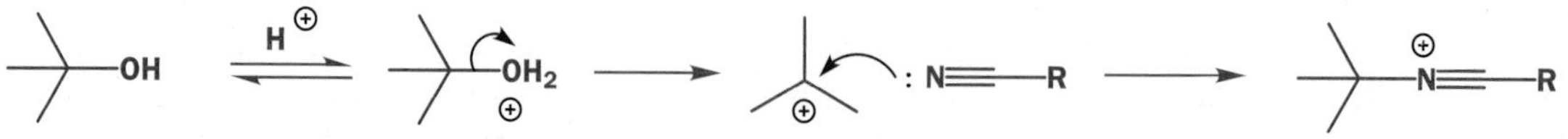

t-butyl cation

The resulting cation is captured by the water molecule released in the first step and an exchange of protons leads to an amide.

The overall process is called the **Ritter reaction** and is one of the few reliable ways to make a C–N bond to a tertiary centre.

Nucleophiles in the S$_N$2 reaction

Nitrogen nucleophiles

Reactions between ammonia and alkyl halides rarely lead to single products. The problem is that the primary amine product is at least as nucleophilic as the starting material and is formed in the reaction mixture so that it in turn reacts with the alkyl halide.

alkylation of ammonia

primary amine formed
in reaction mixture

alkylation of the primary amine

secondary amine formed
in reaction mixture

Even this is not all! If the alkylation were to continue, the secondary and the tertiary amines would be produced all together in the reaction mixture. The reaction comes to an end only when the *tetra-alkylammonium* salt R$_4$N$^+$ is formed. This salt could be the product if a large excess of alkyl halide RI is used, but other more controlled methods are needed for the synthesis of primary, secondary, and tertiary amines.

alkylation of the secondary amine

tertiary amine formed
in reaction mixture

alkylation of the tertiary amine

quaternary ammonium salt
no proton can be removed
end of the line!

One solution for primary amines is to replace ammonia with azide ion N$_3^-$. This is a linear triatomic species, nucleophilic at both ends—a little rod of electrons able to insert itself into almost any electrophilic site. It is available as the water-soluble sodium salt NaN$_3$.

Azide reacts only once with alkyl halides because the product, an alkyl azide, is no longer nucleophilic.

nucleophilic azide ion N$_3^-$ neutral alkyl azide RN$_3$

structure of azide ion N$_3^-$

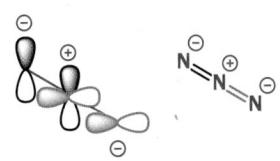

You should compare the structure of azide with those of ketene (p. 372) and allene (p. 157).

▶ **A warning about azides**

Azides can be converted by heat—or even sometimes just by a sharp blow—suddenly into nitrogen gas. In other words they are potentially explosive, particularly inorganic (that is, ionic) azides and small covalent organic azides.

The alkyl azide produced can be reduced to the primary amine by a number of methods such as catalytic hydrogenation (Chapter 24) or $LiAlH_4$ (Chapter 12). This method has a similar philosophy to the reductive amination discussed in Chapter 14.

$$RX + NaN_3 \longrightarrow RN_3 \xrightarrow{LiAlH_4} RNH_2$$

Azide reacts cleanly with epoxides too: here is an example with some stereochemistry in an open-chain epoxide.

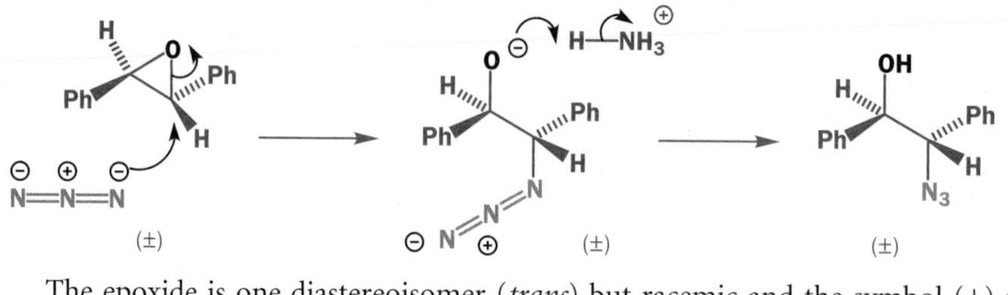

The epoxide is one diastereoisomer (*trans*) but racemic and the symbol (±) under each structure reminds you of this (Chapter 15). Azide attacks at either end of the three-membered ring (the two ends are the same) to give the hydroxy-azide. The reaction is carried out in a mixture of water and an organic solvent with ammonium chloride as buffer to provide a proton for the intermediate.

Next, triphenylphosphine in water was used for reduction to the primary amine. This process might remind you of the Mitsunobu reaction earlier in this chapter.

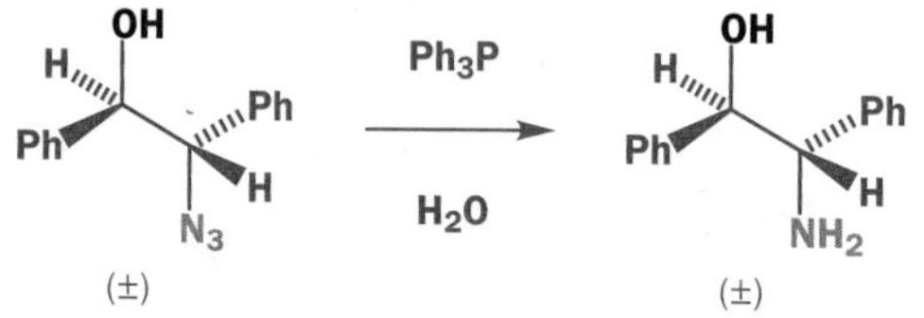

The probable mechanism follows. Notice the similarity with the Wittig reaction (p. 357). What is certainly true is that a molecule of nitrogen is lost and a molecule of water is 'dismembered' and shared between the reagents. The phosphorus atom gets the oxygen and the nitrogen atom gets the two hydrogens. These (P=O and N–H rather than N–O and P–H) are the stronger bonds.

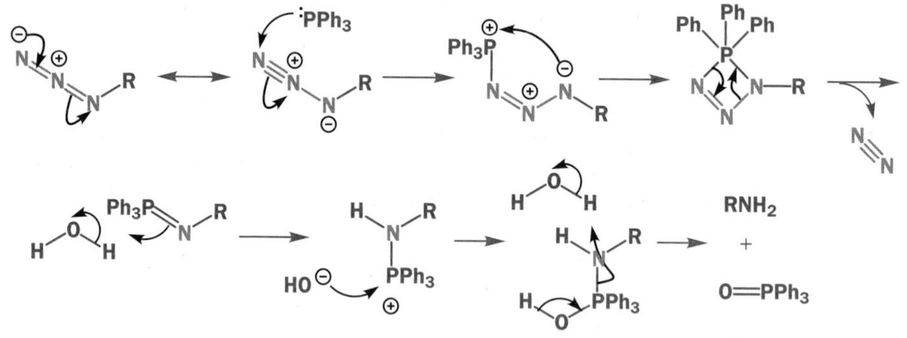

Sulfur nucleophiles are better than oxygen nucleophiles in S_N2 reactions

Thiolate anions make excellent nucleophiles in S_N2 reactions on alkyl halides. It is enough to combine the thiol, sodium hydroxide, and the alkyl halide to get a good yield of the sulfide.

$$PhSH + NaOH + n\text{-}BuBr \longrightarrow PhSBu + NaBr$$

There is no competition between hydroxide and thiol because thiols are more acidic than water (pK_a of RSH is typically 9–10, pK_a of PhSH is 6.4, pK_a of H_2O is 15.7; Chapter 8) and there is a rapid proton transfer from sulfur to oxygen.

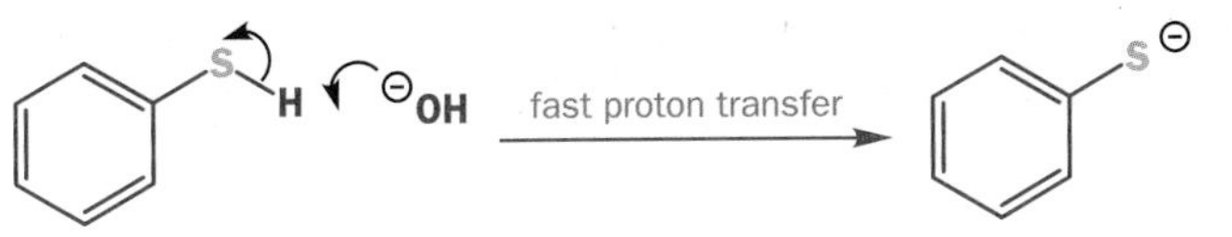

The thiolate anion produced then acts as a nucleophile in the S$_N$2 reaction.

the S$_N$2 reaction with a thiolate anion as nucleophile

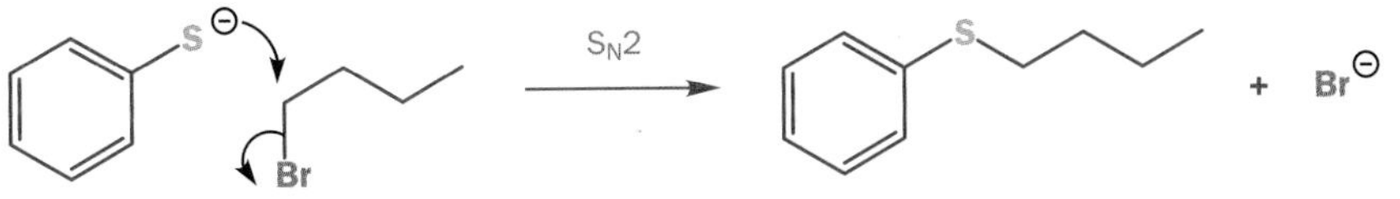

But how do you make a thiol in the first place? The obvious way to make aliphatic thiols would be by an S$_N$2 reaction using NaSH on the alkyl halide.

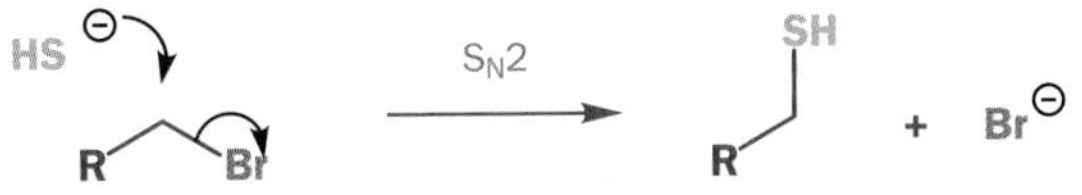

This works well but, unfortunately, the product easily exchanges a proton and the reaction normally produces the symmetrical sulfide—this should remind you of what happened with amines!

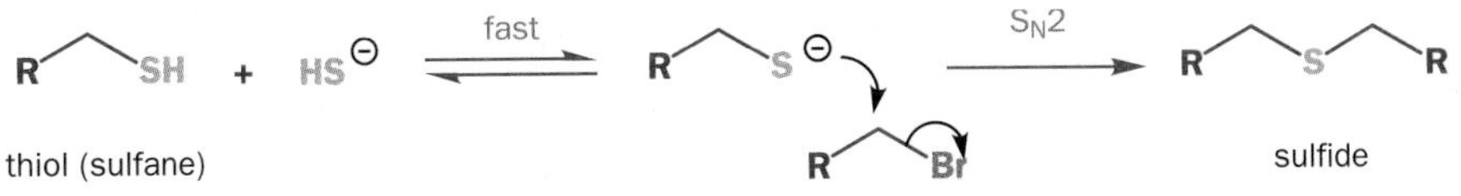

The solution is to use the anion of thiolacetic acid, usually the potassium salt. This reacts cleanly through the more nucleophilic sulfur atom and the resulting ester can be hydrolysed in base to liberate the thiol.

the S$_N$2 reaction with a thiolacetate anion as nucleophile

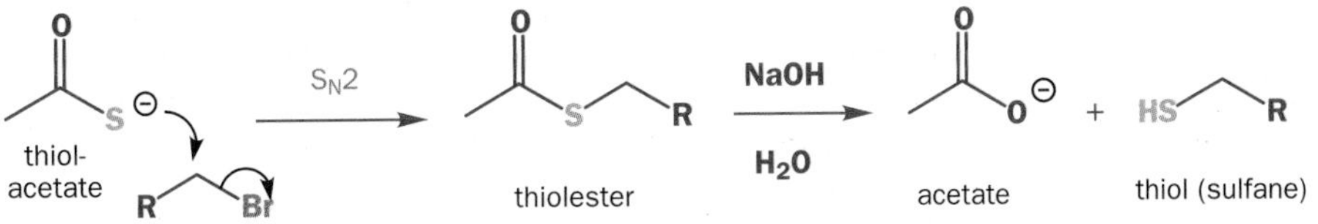

Effectiveness of different nucleophiles in the S$_N$2 reaction

Just to remind you of what we said before: basicity is nucleophilicity towards protons and nucleophilicity towards the carbonyl group parallels basicity almost exactly.

During this chapter you have had various hints that nucleophilicity towards saturated carbon is not so straightforward. Now we must look at this question seriously and try to give you helpful guidelines.

1   If the atom that is forming the new bond to carbon is the same over a range of nucleophiles—it might be oxygen, for example, and the nucleophiles might be HO⁻, PhO⁻, AcO⁻, and TsO⁻—then nucleophilicity does parallel basicity. The anions of the weakest acids are the best nucleophiles. The order for the nucleophiles we have just mentioned will be: HO⁻ > PhO⁻ > AcO⁻ > TsO⁻. The actual values for the rates of attack of the various nucleophiles on MeBr in EtOH relative to the rate of reaction with water (=1) are given in Table 17.14

This was discussed in Chapter 12.

Table 17.14 Relative rates (water = 1) of reaction with MeBr in EtOH

Nucleophile X	pK_a of HX	Relative rate
HO⁻	15.7	1.2×10^4
PhO⁻	10.0	2.0×10^3
AcO⁻	4.8	9×10^2
H$_2$O	−1.7	1.0
ClO$_4^-$	−10	0

2   If the atoms that are forming the new bond to carbon are *not* the same over the range of

nucleophiles we are considering, then another factor is important. In the very last examples we have been discussing we have emphasized that RS⁻ is an excellent nucleophile for saturated carbon. Let us put that another way. RS⁻ is a better nucleophile for saturated carbon than is RO⁻, even though RO⁻ is more basic than RS⁻ (Table 17.15).

Table 17.15 Relative rates (water = 1) of reaction with MeBr in EtOH

Nucleophile X	pK_a of HX	Relative rate
PhS⁻	6.4	5.0×10^7
PhO⁻	10.0	2.0×10^3

You might have noticed that the thiolacetate ion could have reacted with an alkyl halide through sulfur or through oxygen:

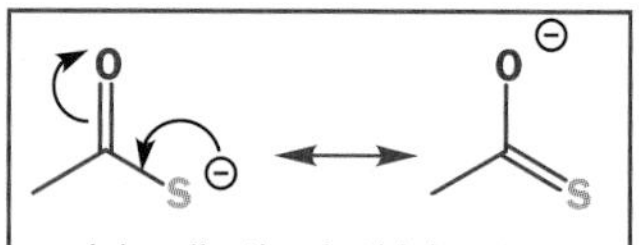

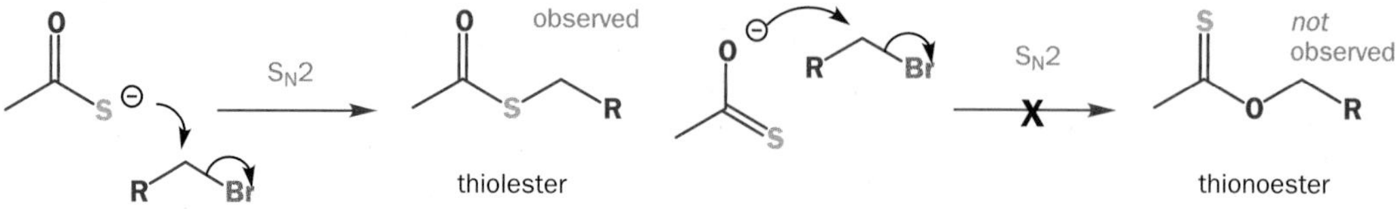

It is clear then that sulfur is a better nucleophile than is oxygen for saturated carbon. Why should this be? There are two main factors controlling bimolecular reactions: electrostatic attraction (simple attraction of opposite charges) and productive interactions between the HOMO of the nucleophile and the LUMO of the electrophile.

Reactions of nucleophiles with protons and with carbonyl groups are heavily influenced by electrostatic attraction (as well as by HOMO–LUMO interactions). The proton is, of course, positively charged. The carbonyl group too has a substantial positive charge on the carbon atom, which comes from the uneven distribution of electrons in the C=O π bond (Chapter 4).

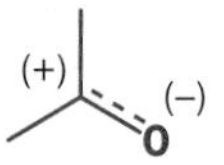

There is, of course, also some polarity in the bond between a saturated carbon atom and a leaving group, say, a bromine atom, but this is a much smaller effect leading only to very small charge separation represented as δ+. In alkyl iodides, one of the best electrophiles in S$_N$2 reactions, there is in fact almost no dipole at all—the electronegativity of C is 2.55 and that of I is 2.66. Electrostatic attraction is unimportant in S$_N$2 reactions.

So what does matter? Only HOMO–LUMO interactions matter. In nucleophilic attack on the carbonyl group, the nucleophile added in to the low-energy π* orbital. In attack on a saturated carbon atom, the nucleophile must donate its electrons to the σ* orbital of the C–X bond as we discussed in Chapter 10.

We had a similar discussion in Chapter 10 when we were considering nucleophiles attacking conjugated C=C–C=O systems. Attack at C=O in these systems tends to be electrostatically controlled, while nucleophilic attack at C=C is under orbital (HOMO–LUMO) control.

considerable polarization in the C=O group

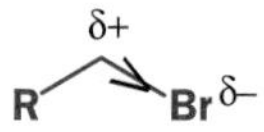

very little polarization in the C–Br bond

typical arrangement of molecular energy levels

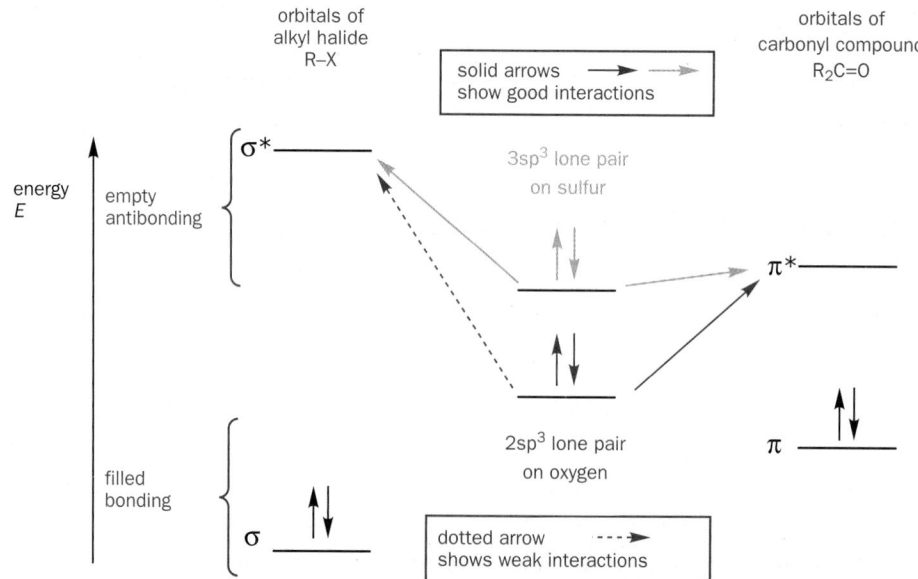

The higher-energy ($3sp^3$) lone-pair electrons on sulfur overlap better with the high-energy σ^* orbital of the C–X bond than do the lower-energy ($2sp^3$) lone-pair electrons on oxygen because the higher energy of the sulfur electrons brings them closer in energy to the C–X σ^* orbital. Notice that both elements overlap well with the lower-energy π^* orbital. The conclusion is that nucleophiles from lower down the periodic table are more effective in S_N2 reactions than those from the top few rows. Typically, nucleophilic power towards saturated carbon goes like this.

$I^- > Br^- > Cl^- > F^-$

$RSe^- > RS^- > RO^-$

$R_3P: > R_3N:$

Nucleophiles in substitution reactions

Some rates (relative to that of water = 1) of various nucleophiles towards methyl bromide in ethanol are shown in Table 17.16.

Table 17.16 Relative rates (water = 1) of reaction of nucleophiles with MeBr in EtOH

nucleophile	F^-	H_2O	Et_3N	Br^-	PhO^-	EtO^-	I^-	PhS^-
relative rate	0.0	1.0	1400	5000	2.0×10^3	6×10^4	1.2×10^5	5.0×10^7

You have met a similar sequence before in Chapter 10, and it would be useful to review the terms we used then. Nucleophiles like R_3P: and RS^-, the ones that react well with saturated carbon, are referred to as **soft** nucleophiles and those that are more basic and react well with carbonyl groups referred to as **hard** nucleophiles. These are useful and evocative terms because the soft nucleophiles are rather large and flabby with diffuse high-energy electrons while the hard nucleophiles are small with closely held electrons and high charge density. When we say 'hard' (nucleophile or electrophile) we refer to species whose reactions are dominated by electrostatic attraction and when we say 'soft' (nucleophile or electrophile) we refer to species whose reactions are dominated by HOMO–LUMO interactions.

> Just to remind you: reactions dominated by electrostatic attraction also need to pass electrons from HOMO to LUMO, but reactions that are dominated by HOMO–LUMO interactions need have *no* contribution from electrostatic attraction.

● **It is worth summarizing the characteristics of the two types of nucleophile.**

Hard nucleophiles X	Soft nucleophiles Y
small	large
charged	neutral
basic (HX weak acid)	not basic (HY strong acid)
low-energy HOMO	high-energy HOMO
like to attack C=O	like to attack saturated carbon
such as RO^-, NH_2^-, MeLi	such as RS^-, I^-, R_3P

Nucleophiles and leaving groups compared

In nucleophilic attack on the carbonyl group, a good nucleophile is a bad leaving group and vice versa because the intermediate chooses to expel the best leaving group. If that is the nucleophile, it just goes straight back out again.

Chloride ion will always be the best leaving group from the intermediate, however it is formed, and the attempt to make an acid chloride from an ester with NaCl is doomed. Chloride is a good leaving group from C=O and a bad nucleophile towards C=O while EtO⁻ is a bad leaving group from C=O and a good nucleophile towards C=O.

The S_N2 reaction is different because it does not have an intermediate. Therefore anything that lowers the energy of the transition state will speed up both the forward and the back reactions. We need to consider two results of this: the rate of the reaction and which way it will go.

Iodide ion is one of the best nucleophiles towards saturated carbon because it is at the bottom of its group in the periodic table and its lone-pair electrons are very high in energy. This is in spite of the very low basicity of iodide (Table 17.17). It reacts rapidly with a variety of alkyl derivatives and alkyl iodides can be made by displacement of chloride or tosylate by iodide.

Table 17.17 Relative rates (water = 1) of reaction with MeBr in EtOH

Nucleophile X	pK_a of HX	Relative rate
I⁻	−10	1.2×10^5
Br⁻	−9	5.0×10^3
Cl⁻	−7	1.1×10^3
F⁻	+3	0

> The first of these reactions is assisted by precipitation of NaCl from acetone, which drives the reaction along.

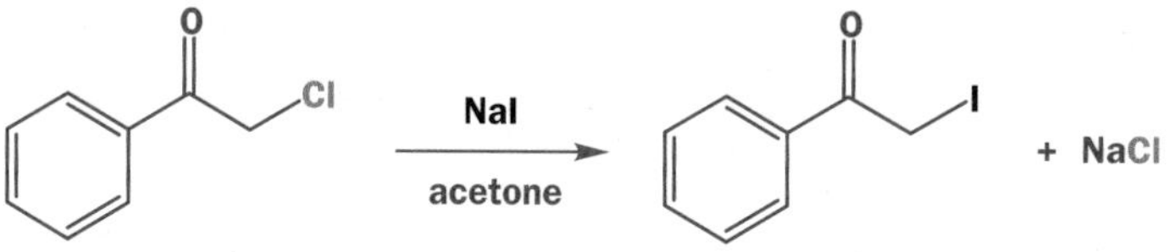

$$\text{ROTs} + \text{NaI} \longrightarrow \text{RI} + \text{NaOTs}$$

But why are these alkyl iodides made? They are needed for reactions with other nucleophiles in which iodide is again displaced. As well as being one of the best nucleophiles for saturated carbon, iodide ion is one of the best leaving groups from saturated carbon (see p. 430). Yields are often higher if the alkyl iodide is prepared than if the eventual nucleophile is reacted directly with the alkyl tosylate or chloride.

An example is the synthesis of the phosphonium salt used by Corey in a synthesis of terpenes (Chapter 51). An unsaturated primary alcohol was first made into its tosylate, the tosylate was converted into the iodide, and the iodide into the phosphonium salt.

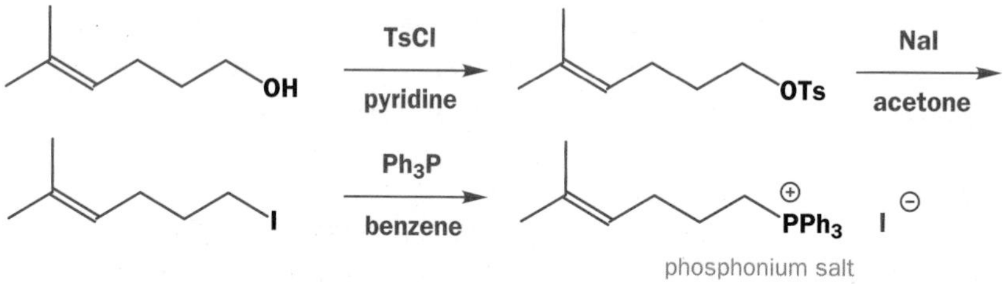

phosphonium salt

However, iodine is expensive and a way round that problem is to use a catalytic amount of iodide. The next phosphonium salt is formed slowly from benzyl bromide but the addition of a small amount of LiI speeds up the reaction considerably.

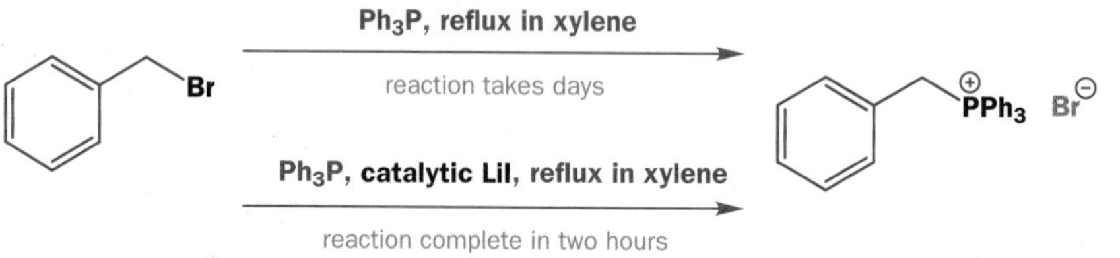

Xylenes

The solvent 'xylene' needs some explanation. Xylene is the trivial name for dimethyl benzene and there are three isomers. Mixed xylenes are isolated cheaply from oil and often used as a relatively high boiling solvent (b.p. about 140 °C) for reactions at high temperature. In this case, the starting materials are soluble in xylene but the product is a salt and conveniently precipitates out during the reaction.

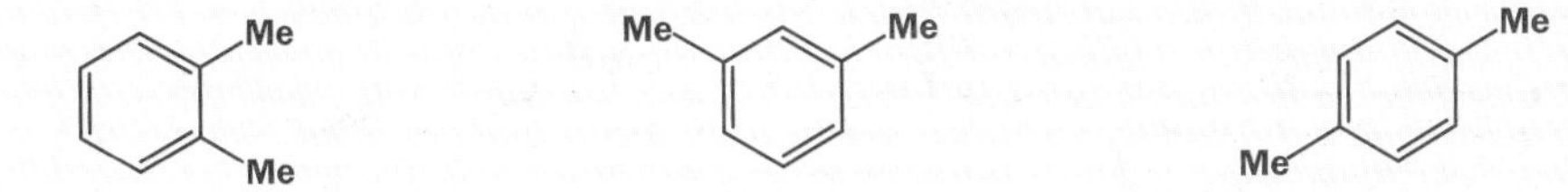

ortho-xylene
1,2-dimethyl benzene

meta-xylene
1,3-dimethyl benzene

para-xylene
1,4-dimethyl benzene

The iodide reacts as a better nucleophile than Ph_3P and then as a better leaving group than Br^-. Each iodide ion goes round and round many times as a **nucleophilic catalyst**.

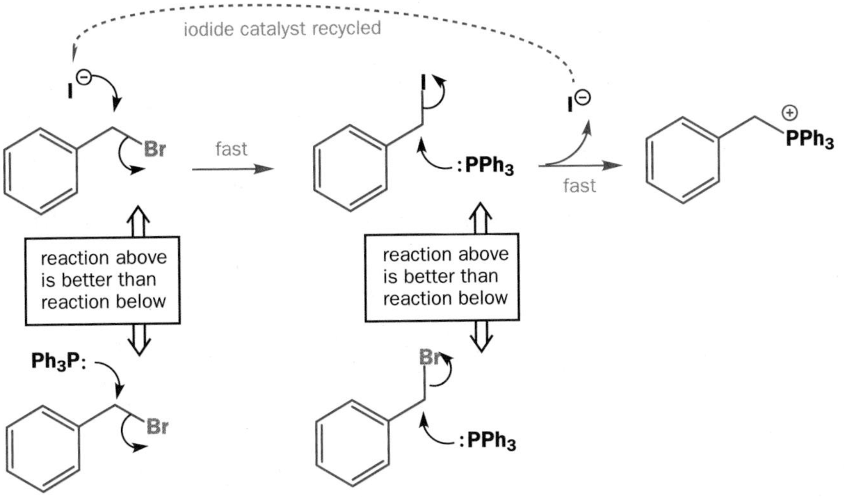

Looking forward: elimination and rearrangement reactions

Simple nucleophilic substitutions at saturated carbon atoms are fundamental reactions found wherever organic chemistry is practised. They are used in industry on an enormous scale to make 'heavy chemicals' and in pharmaceutical laboratories to make important drugs. They are worth studying for their importance and relevance.

There is another side to this simple picture. These were among the first reactions whose mechanisms were thoroughly investigated by Ingold in the 1930s and since then they have probably been studied more than any other reactions. All our understanding of organic mechanisms begins with S_N1 and S_N2 reactions and you need to understand these basic mechanisms properly. Some of the more sophisticated investigations into nucleophilic substitutions have clouded the main issues by looking at minute details and we shall not discuss these.

We shall, however, be returning to this sort of chemistry in several further chapters. The carbocations you met in this chapter are reactive species. One of the most convincing pieces of evidence for their formation is that they undergo reactions other than simple addition to nucleophiles. The carbon skeleton of the cation may rearrange.

a rearrangement reaction

secondary cation   tertiary cation

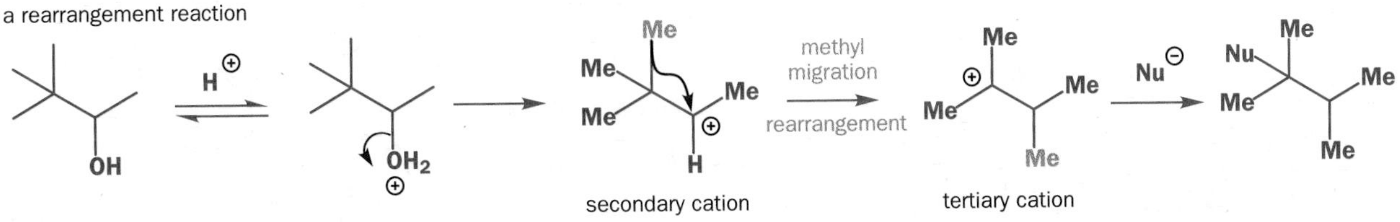

You will meet rearrangements in several chapters later in the book especially Chapter 37. Another common fate of cations, and something that may also happen instead of an intended S_N1 or S_N2 reaction, is an elimination reaction where an alkene is formed by the nucleophile acting as a base to remove HX instead of adding to the molecule.

an elimination reaction (E1)

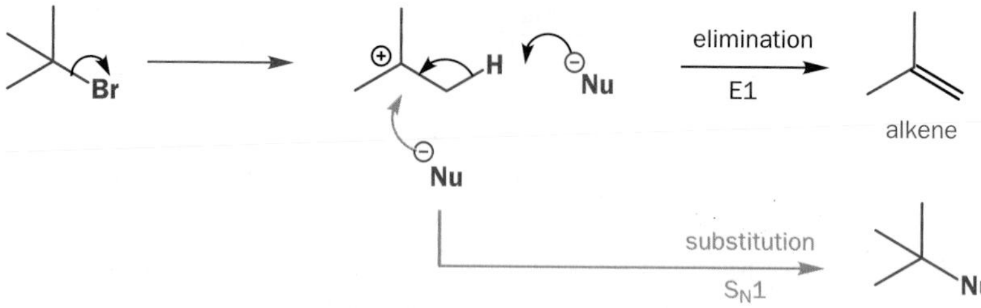

You will meet elimination reactions in the next chapter but one (19) after some further exploration of stereochemistry.

Problems

1. Suggest mechanisms for the following reactions, commenting on your choice of S_N1 or S_N2.

(a)

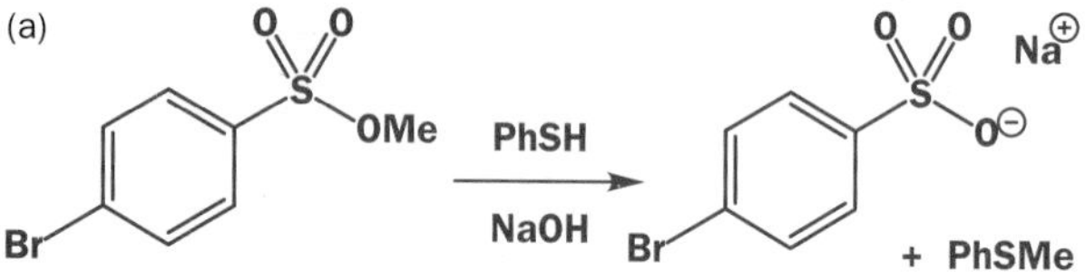

(b)

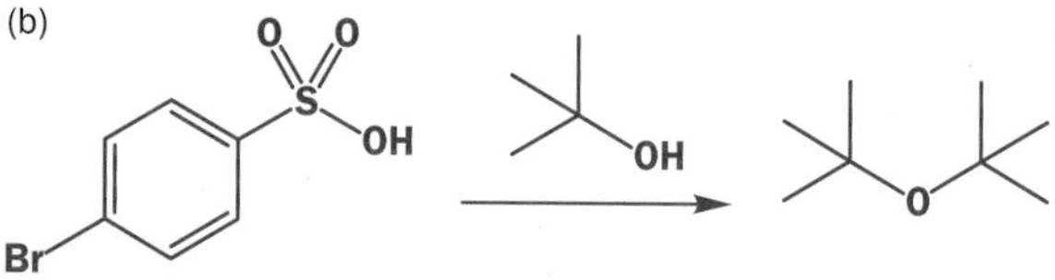

2. Draw mechanisms for the following reactions. Why were acidic conditions chosen for the first reaction and basic conditions for the second?

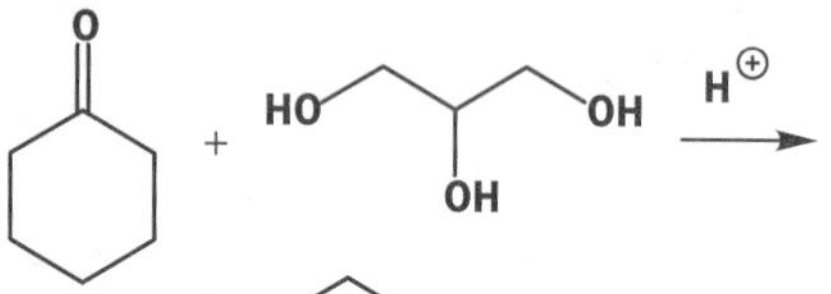

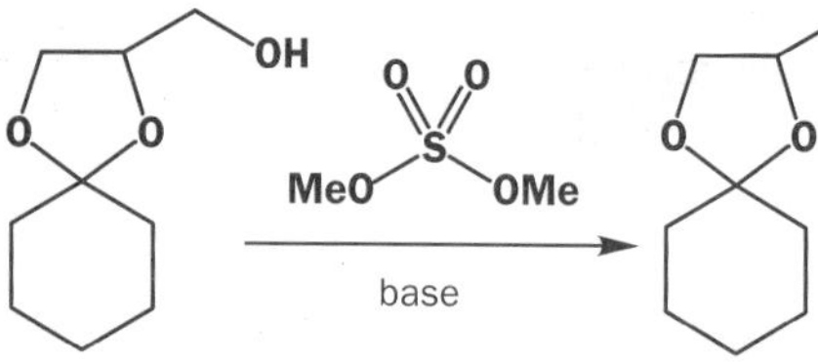

3. Draw mechanisms for these reactions, explaining why these particular products are formed.

(a)

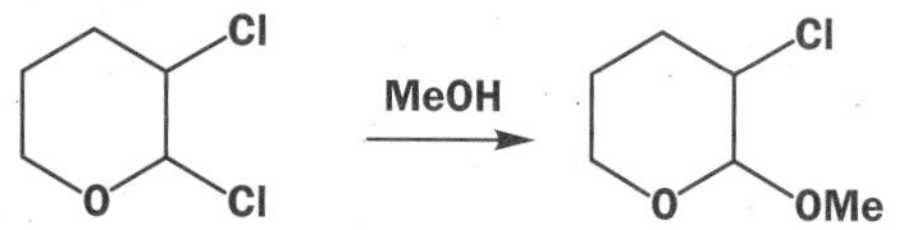

(b)

4. The chemistry shown here is the first step in the manufacture of Pfizer's doxazosin (Cardura), a drug for hypertension. Draw mechanisms for the reactions involved and comment on the bases used.

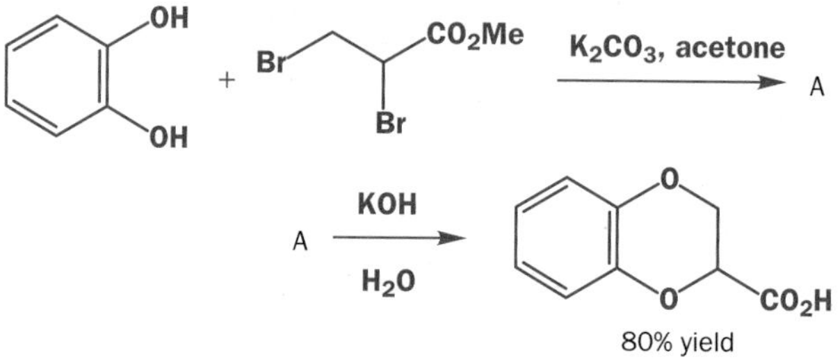

5. Suggest mechanisms for these reactions, commenting on the choice of reagents and solvents. How would you convert the final product into diethyl hexanedioate [diethyl adipate, $EtO_2C(CH_2)_4CO_2Et$]?

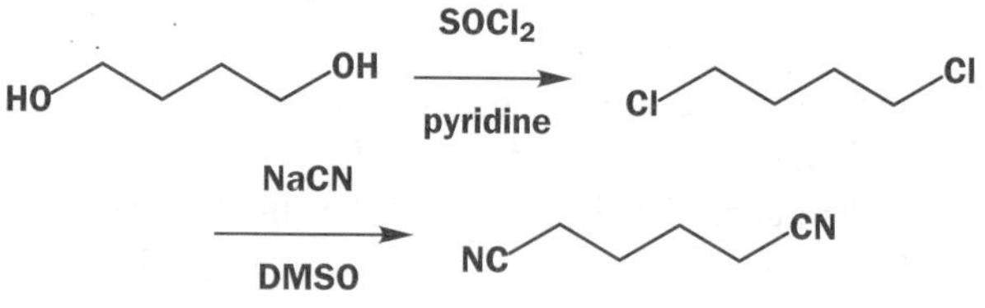

6. Draw mechanisms for these reactions and describe the stereochemistry of the product.

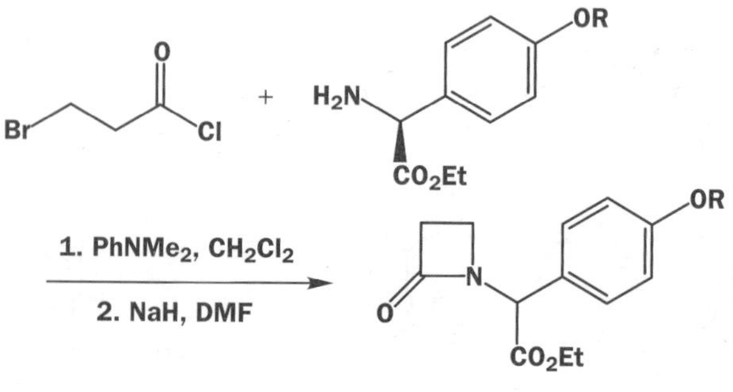

7. Suggest a mechanism for this reaction. You will find it helpful first of all to draw good diagrams of reagents and products.

t-BuNMe$_2$ + (MeCO)$_2$O → Me$_2$NCOMe + *t*-BuO$_2$CMe

8. Predict the stereochemistry of these products. Are they single diastereoisomers, enantiomerically pure, or racemic, or something else?

(a)

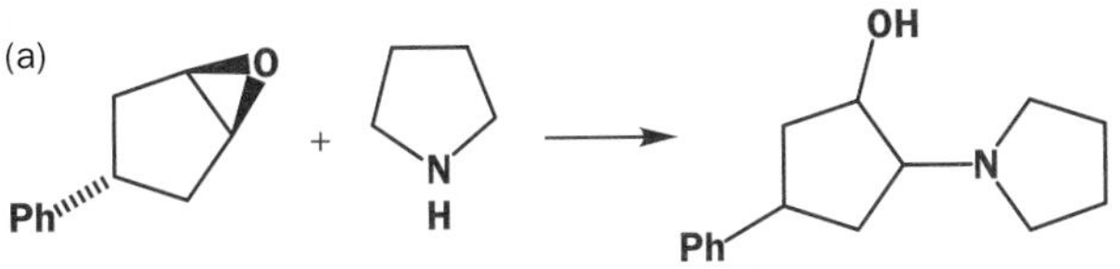

(b)

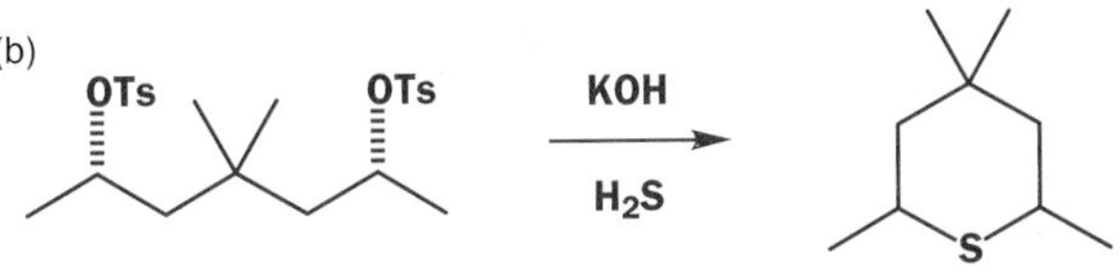

9. What are the mechanisms of these reactions, and what is the role of the ZnCl$_2$ in the first step and the NaI in the second?

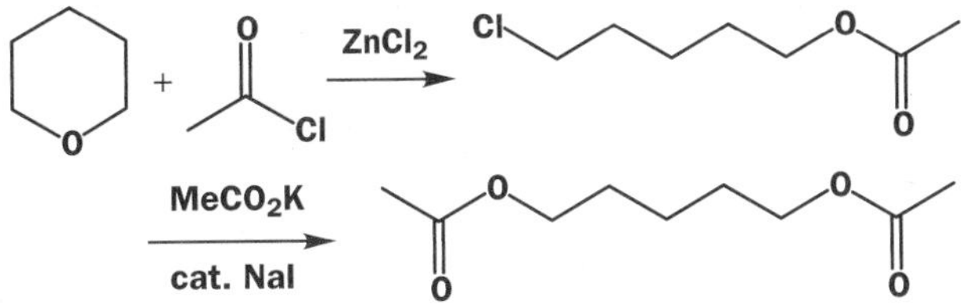

10. Describe the stereochemistry of the products of these reactions.

(±)

enantiomerically pure

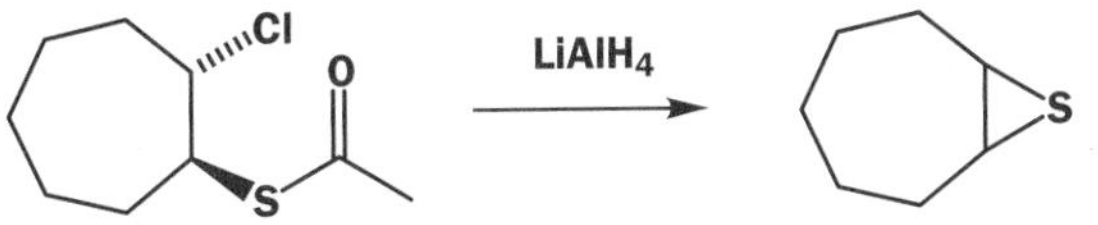

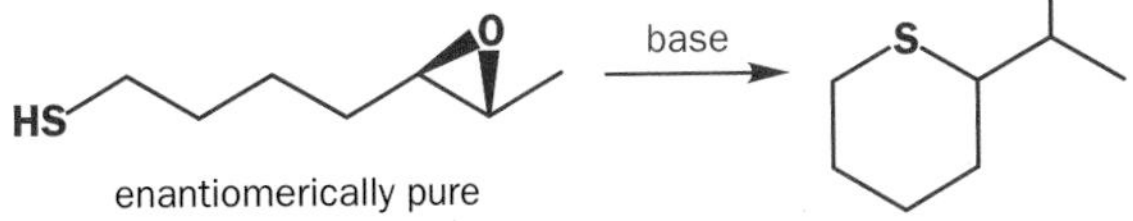

11. Identify the intermediates in these syntheses and give mechanisms for the reactions.

(a)

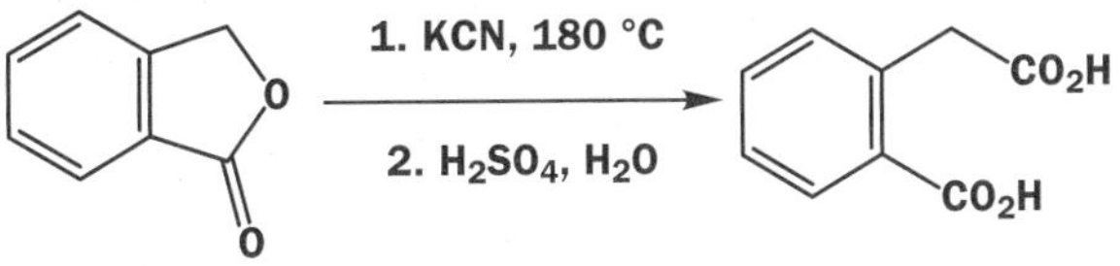

(b)

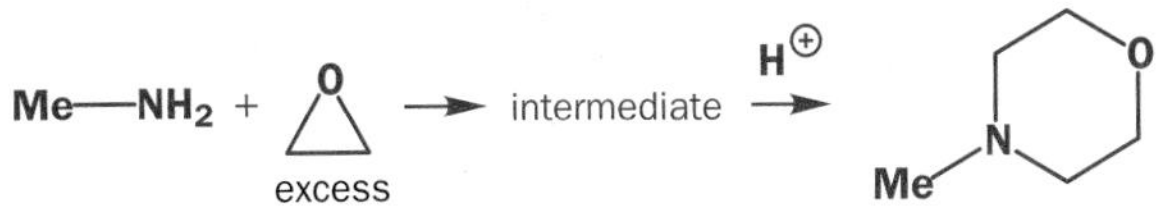

12. State with reasons whether these reactions will be either S$_N$1 or S$_N$2.

(a)

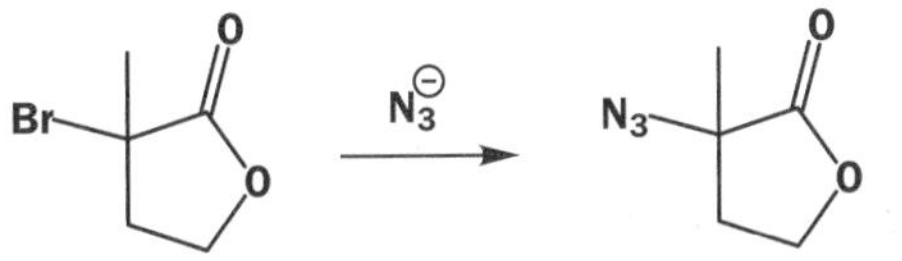

(b)

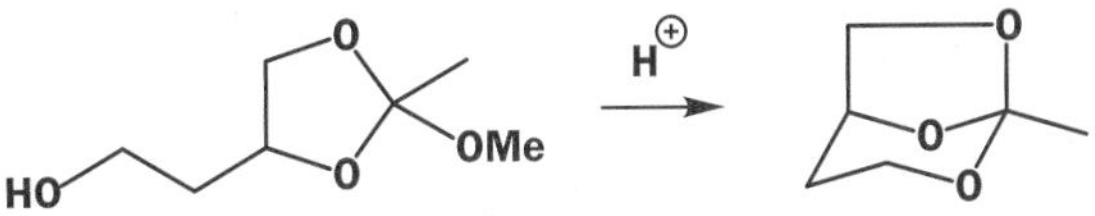

(c)

(±) (±)

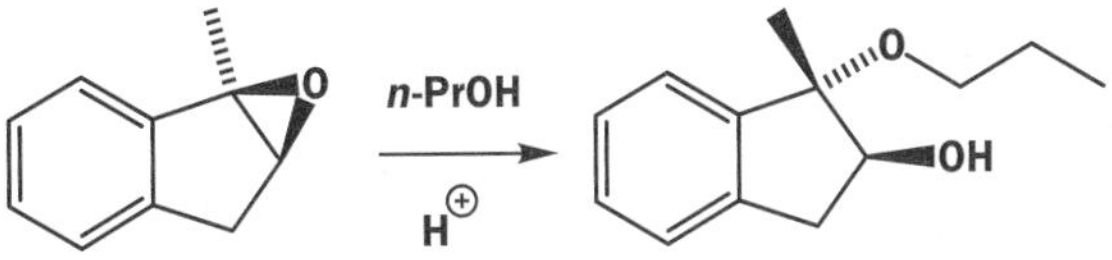

(d)

(±) (±)

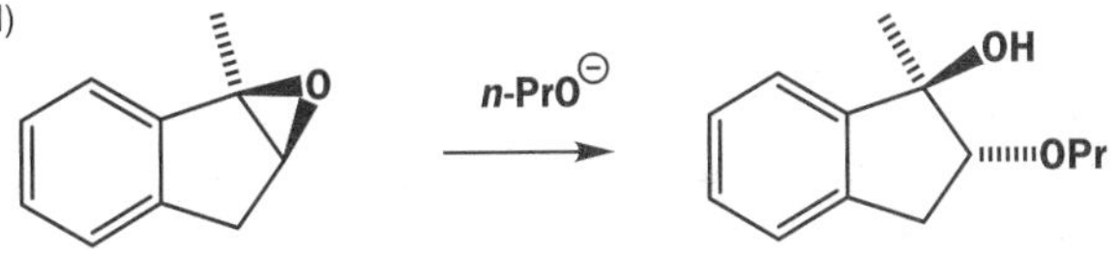

Conformational analysis

18

Connections

Building on:

- How to determine a molecule's structure ch3, ch11, & ch15
- How some molecules can exist as stereoisomers ch16

Arriving at:

- If I could see a molecule, what would its three-dimensional shape (conformation) be?
- What effect does a molecule's shape have on its reactions?
- How single bonds are free to rotate, but spend most of their time in just two or three well-defined arrangements
- How rings of atoms are usually not planar, but 'puckered'
- How 'puckered' six-membered rings have the most well-defined arrangements of atoms
- How to draw six-membered rings accurately
- How to use the known arrangements of the atoms in a six-membered ring to predict and explain their reactions

Looking forward to:

- How conformation, and the alignment of atoms, can affect elimination reactions ch19
- How NMR spectroscopy backs up what we have said in this chapter ch32
- How the conformation of molecules dictates how they react—e.g. from which direction they will be attacked by reagents ch33 & ch45
- How the alignment of bonds can allow groups in molecules to move around (rearrangement reactions) or allow C–C bonds to break (fragmentation reactions) ch37 & ch38
- How the alignment of orbitals controls reactivity (stereoelectronics) ch42
- The accurate drawing of rings as transition states is necessary ch35 & ch36

Bond rotation allows chains of atoms to adopt a number of conformations

Several chapters of this book have considered how to find out the structure of molecules. We have seen X-ray crystallography pictures, which reveal exactly where the atoms are in crystals; we have looked at IR spectroscopy, which gives us information about the bonds in the molecule, and at NMR spectroscopy, which gives us information about the atoms themselves. Up to now, we have mainly been interested in determining which atoms are bonded to which other atoms and also the shapes of small localized groups of atoms. For example, a methyl group has three hydrogen atoms bonded to one carbon atom and the atoms around this carbon are located at the corners of a tetrahedron; a ketone consists of a carbon atom bonded to two other carbon atoms and doubly bonded to an oxygen atom with all these atoms in the same plane.

But, on a slightly larger scale, shape is not usually so well defined. Rotation is possible about single bonds and this rotation means that, while the localized arrangement of atoms stays the same (every saturated carbon atom is still always tetrahedral), the molecule as a whole can adopt a number of different shapes. Shown on the next page are several snapshot views of one molecule—it happens to be a pheromone used by pea moths to attract a mate. Although the structures look dissimilar, they differ from one another only by rotation about one or more single bonds. Whilst the overall shapes differ, the localized structure is still the same: tetrahedral sp^3 carbons; trigonal planar sp^2 carbons. Notice another point too, which we will pick up on later: the arrangement about the double bond always remains the same because double bonds can't rotate.

At room temperature in solution, all the single bonds in the molecule are constantly rotating—the chances that two molecules would have exactly the same shape at any one time are quite small.

Yet, even though no two molecules have exactly the same shape at any one time, they are still all the same chemical compound—they have all the same atoms attached in the same way. We call the different shapes of molecules of the same compound different **conformations**.

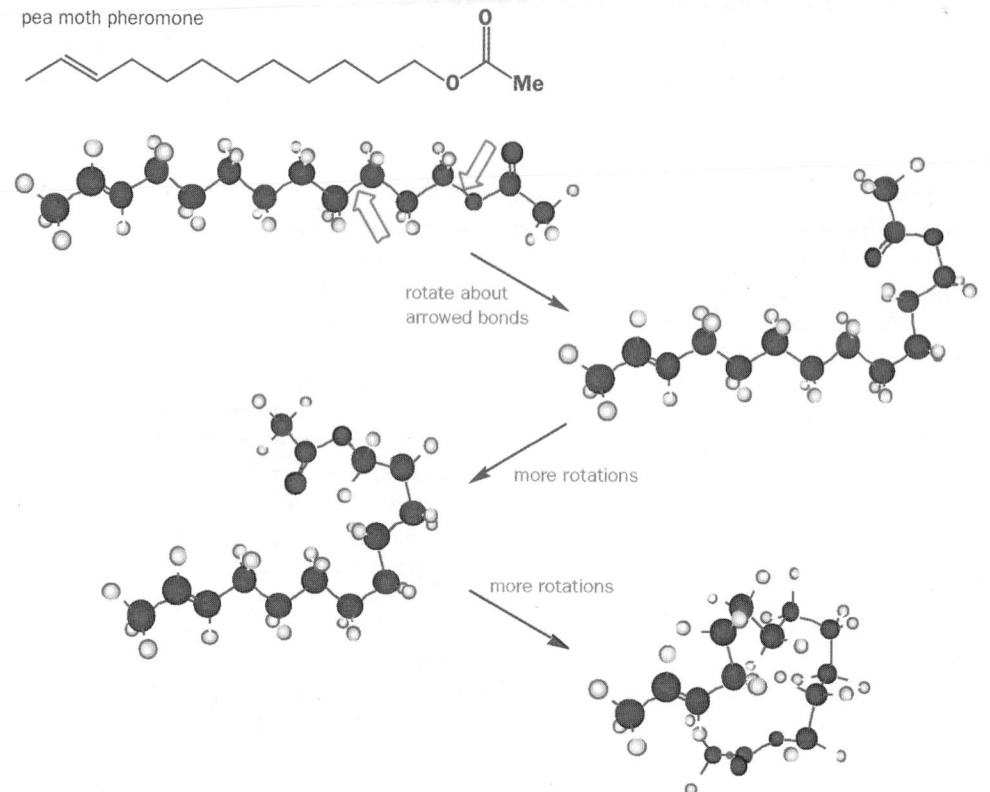

pea moth pheromone

rotate about arrowed bonds

more rotations

more rotations

► **Make models**

If you find this hard to see, get a set of molecular models and build the first one of each pair. You should be able to rotate it straightforwardly into the second without breaking your model. Our advice throughout this chapter, certainly with things that you find difficult to understand from the two-dimensional drawings to which we are limited, is to *make models*.

Conformation and configuration

To get from one conformation to another, we can rotate about as many single bonds as we like. The one thing we can't do though is to break any bonds. This is why we can't rotate about a double bond—to do so we would need to break the π bond. Below are some pairs of structures that can be interconverted by rotating about single bonds: they are all different conformations of the same molecule.

three compounds, each shown in two conformations

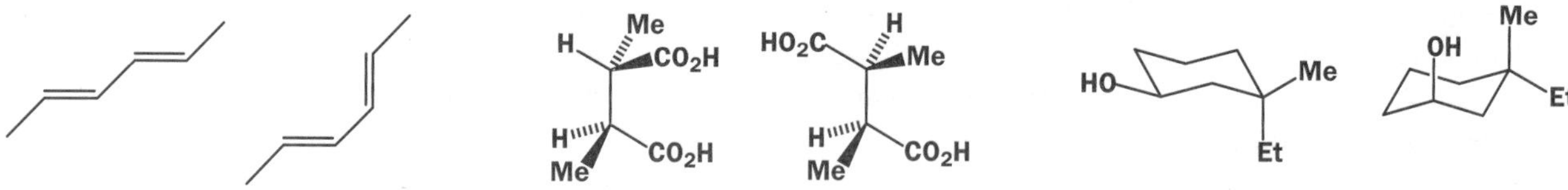

The next block of molecules is something quite different: these pairs can only be interconverted by breaking a bond. This means that they have different **configurations**—configurations can be interconverted only by breaking bonds. Compounds with different configurations are called **stereoisomers** and we dealt with them in Chapter 16.

three pairs of stereoisomers: each member of a pair has a different configuration

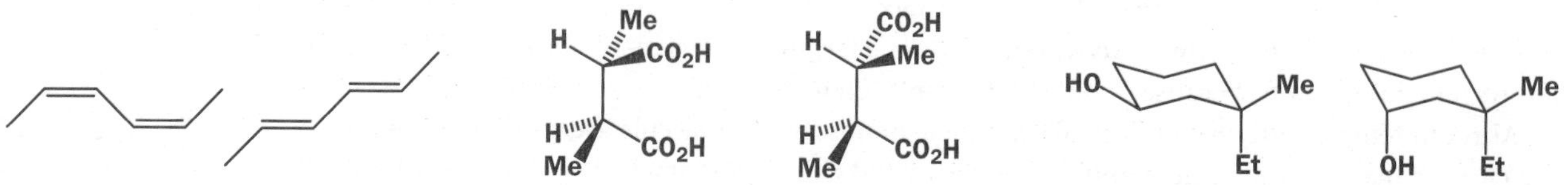

● **Rotation or bond breaking?**

- Structures that can be interconverted simply by rotation about single bonds are **conformations** of the same molecule
- Structures that can be interconverted only by breaking one or more bonds have different **configurations**, and are stereoisomers

Conformation and configuration

Different conformations of a person – some being more stable than others . . .

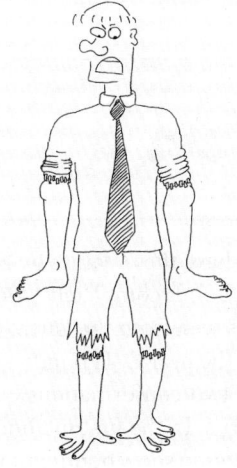

A different configuration!

Barriers to rotation

We saw in Chapter 7 that rotation about the C–N bond in an amide is relatively slow at room temperature—the NMR spectrum of DMF clearly shows two methyl signals (p. 165). In Chapter 13 you learned that the rate of a chemical process is associated with an energy barrier (this holds both for reactions and simple bond rotations): the lower the rate, the higher the barrier. The energy barrier to the rotation about the C–N bond in an amide is usually about 80 kJ mol^{-1}, translating into a rate of about 0.1 s^{-1} at 20 °C. Rotation about single bonds is much faster than this at room temperature, but there is nonetheless a barrier to rotation in ethane, for example, of about 12 kJ mol^{-1}.

▶ **Barriers to rotation about different types of bond**

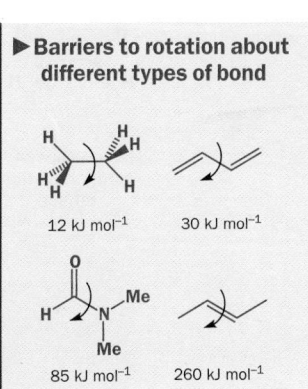

12 kJ mol^{-1} 30 kJ mol^{-1}

85 kJ mol^{-1} 260 kJ mol^{-1}

Rates and barriers

It can be useful to remember some simple guidelines to the way in which energy barriers relate to rates of rotation. For example:

- A barrier of 73 kJ mol^{-1} allows one rotation every second at 25 °C (that is, the rate is 1 s^{-1})
- Every 6 kJ mol^{-1} changes the rate at 25 °C by about a factor of 10
- To see signals in an NMR spectrum for two different conformations, they must interconvert no faster than

(very roughly) 1000 s^{-1}—a barrier of about 55 kJ mol^{-1} at 25 °C. This is why NMR shows two methyl signals for DMF, but only one set of signals for butadiene. See p. 461 for more on this

- For conformations to interconvert slowly enough for them to exist as different compounds, the barrier must be over 100 kJ mol^{-1}. The barrier to rotation about a C=C double bond is 260 kJ mol^{-1}—which is why we can separate *E* and *Z* isomers

Conformations of ethane

Why should there be an energy barrier in the rotation about a single bond? In order to answer this question, we should start with the simplest C–C bond possible—the one in ethane. Ethane has two extreme conformations called the **staggered** and **eclipsed conformations**. Three different views of these are shown below.

the two extreme conformations of ethane, staggered and eclipsed, each shown from three different viewpoints

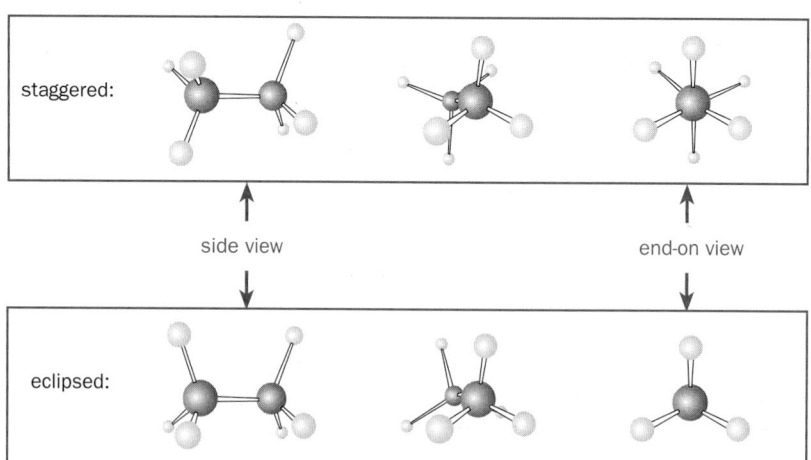

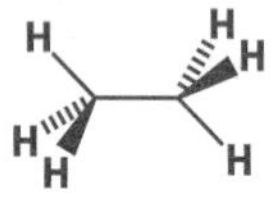

You can see why the conformations have these names by looking at the end-on views in the diagram. In the eclipsed case the near C–H bonds completely block the view of the far bonds, just as in a solar eclipse the moon blocks the sun as seen from the Earth. In the staggered conformation, the far C–H bonds appear in the gaps between the near C–H bonds—the bonds are staggered.

Chemists often want to draw these two conformations quickly and two different methods are commonly used, each with its own merits. In the first method, we simply draw the side view of the molecule and use wedged and hashed lines to show bonds not in the plane of the paper (as you saw in Chapter 16). Particular attention must be paid to which of the bonds are in the plane and which go into and out of the plane.

In the second method we draw the end-on view, looking along the C–C bond. This view is known as a **Newman projection**, and Newman projections are subject to a few conventions:

- The carbon atom nearer the viewer is at the junction of the front three bonds
- The carbon further away (which can't in fact be seen in the end-on view) is represented by a large circle. This makes the perspective inaccurate—but this doesn't matter
- Bonds attached to this further carbon join the *edge* of the circle and do not meet in the centre
- Eclipsed bonds are drawn slightly displaced for clarity—as though the bond were rotated by a tiny fraction

Newman projections for the staggered and eclipsed conformations of ethane are shown below.

the staggered conformation of ethane

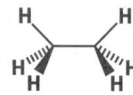

the eclipsed conformation of ethane

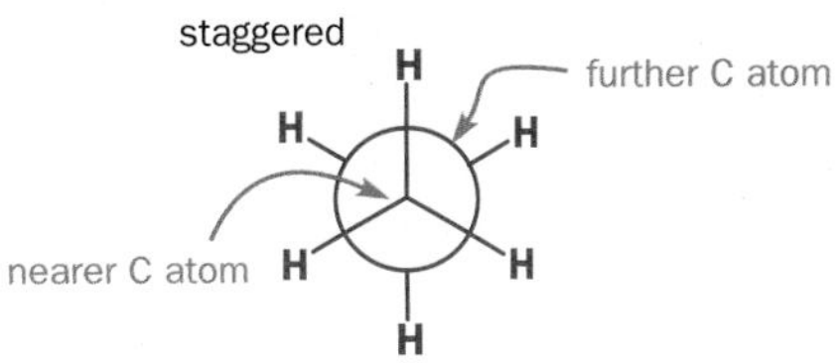

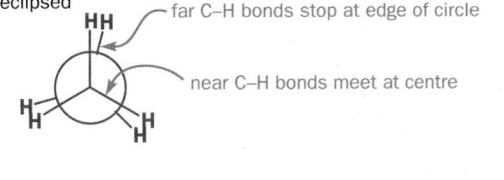

staggered — further C atom — nearer C atom

eclipsed — far C–H bonds stop at edge of circle — near C–H bonds meet at centre

The staggered and eclipsed conformations of ethane are not identical in energy: the staggered conformation is lower in energy than the eclipsed by 12 kJ mol^{-1}, the value of the rotational barrier. Of course, there are other possible conformations too with energies in between these extremes, and we can plot a graph to show the change in energy of the system as the C–C bond rotates. We define the **dihedral angle**, θ (sometimes called the torsion angle), to be the angle between a C–H bond at the nearer carbon and a C–H bond at the far carbon. In the staggered conformation, $\theta = 60°$ whilst in the eclipsed conformation, $\theta = 0°$.

The energy level diagram shows the staggered conformation as a potential energy minimum whilst the eclipsed conformation represents an energy maximum. This means that the eclipsed conformation is not a stable conformation since any slight rotation will lead to a conformation lower in energy. The molecule will actually spend the vast majority of its time in a staggered or nearly staggered conformation and only briefly pass through the eclipsed conformation *en route* to another staggered conformation. It might help to compare the situation here with that of a marble in an egg-box. The marble will sit at the bottom of one of the wells. Rock the egg-box about gently, and the marble will stay in the well but it will roll around a bit, perhaps making its way a centimetre or so up the side. Shake the egg-box more vigorously and eventually the marble will go all the way over the side and down into a new well. One thing is certain: it won't sit on top of the ridge, and the amount of time it will spend there is insignificant.

But *why* is the eclipsed conformation higher in energy than the staggered conformation? At first glance it might seem reasonable to suggest that there is some steric interaction between the hydrogen atoms in the eclipsed conformation that is reduced in the staggered conformation. However, this is not the case, as is shown by these space-filling models. The hydrogen atoms are just too small to get in each other's way. It has been estimated that steric factors make up less than 10% of the rotational barrier in ethane.

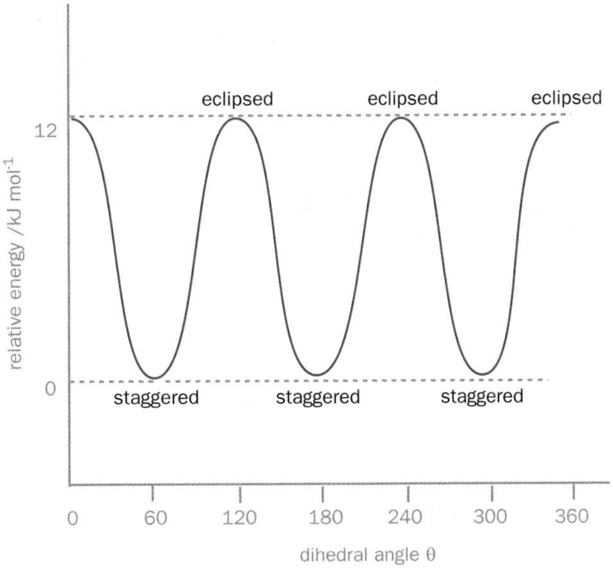

▶
Picturing dihedral angles is sometimes hard—one way to do it is to imagine the two C–H bonds drawn on to two facing pages of a book. The dihedral angle is then the angle between the pages, measured perpendicular to the spine. See page 825.

relative energy / kJ mol^{-1}

eclipsed   eclipsed   eclipsed

12

0

staggered   staggered   staggered

0   60   120   180   240   300   360

dihedral angle θ

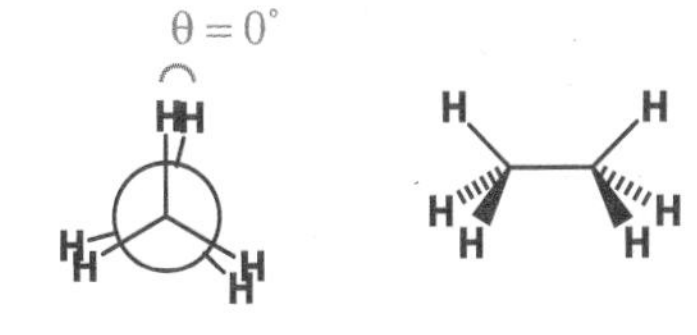

$\theta = 0°$

in the eclipsed conformation, $\theta = 0, 120,$ or $240°$

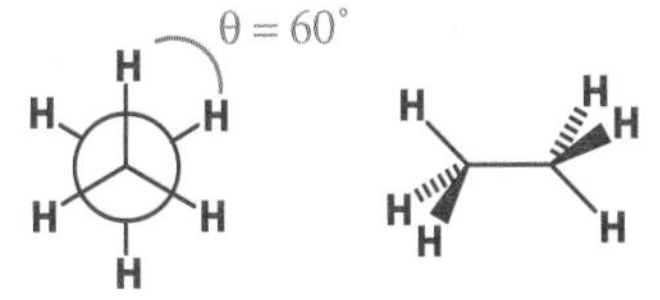

$\theta = 60°$

in the staggered conformation, $\theta = 60, 180,$ or $300°$

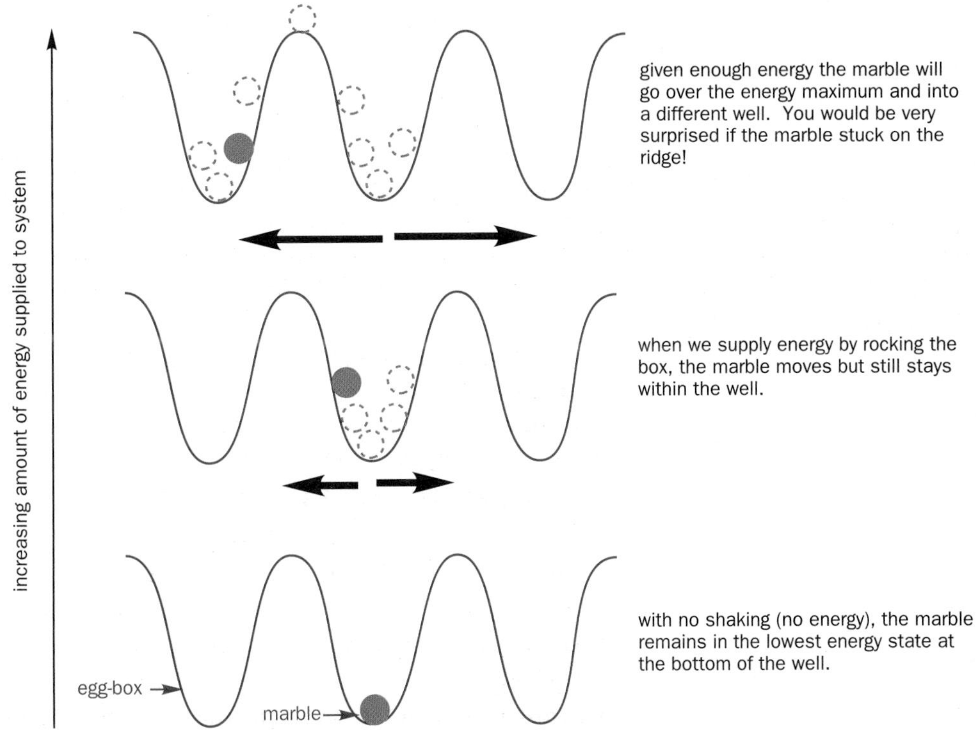

increasing amount of energy supplied to system

given enough energy the marble will go over the energy maximum and into a different well. You would be very surprised if the marble stuck on the ridge!

when we supply energy by rocking the box, the marble moves but still stays within the well.

with no shaking (no energy), the marble remains in the lowest energy state at the bottom of the well.

egg-box →

marble →

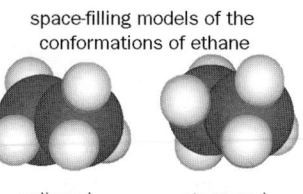

space-filling models of the
conformations of ethane

eclipsed   staggered

■

Space-filling models represent atoms
as spheres with dimensions
determined by the van der Waals radius
of the atom.

There are two more important reasons why the staggered conformation of ethane is lower in energy than the eclipsed conformation. The first is that the electrons in the bonds repel each other and this repulsion is at a maximum in the eclipsed conformation. The second is that there may be some stabilizing interaction between the C–H σ bonding orbital on one carbon and the C–H σ*antibonding orbital on the other carbon, which is greatest when the two orbitals are exactly parallel: this only happens in the staggered conformation.

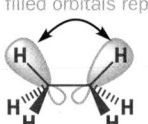

eclipsed:

filled orbitals repel

staggered:

stabilizing interaction between
filled C–H σ bond...

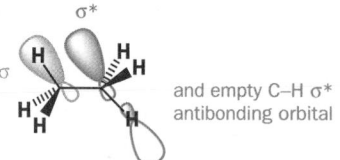

σ*

σ

and empty C–H σ*
antibonding orbital

Of course, the real picture is probably a mixture of all three effects, each contributing more or less depending on the compound under consideration.

Hexachloroethane

Compare ethane with hexachloroethane, C_6Cl_6. The chlorine atoms are much larger than hydrogen atoms (van der Waals radius: H, about 130 pm; Cl, about 180 pm) and now they do physically get in the way of each other. This is reflected in the increase in the rotational barrier from 12 kJ mol^{-1} in C_2H_6 to 45 kJ mol^{-1} in C_2Cl_6 (although other factors also contribute).

space-filling models for the eclipsed and staggered conformations
of hexachloroethane drawn to the same scale as the ethane
models

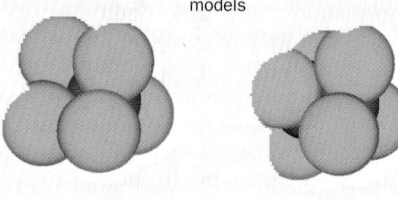

eclipsed   staggered

Conformations of propane

Propane is the next simplest hydrocarbon. Before we consider what conformations are possible for propane we should first look at its geometry. The C–C–C bond angle is not 109.5° (the tetrahedral angle—see Chapters 2 and 4) as might expect but 112.4°. Consequently, the H–C–H bond angle on the central carbon is smaller than the ideal angle of 109.5°, only 106.1°. Once more, this does not necessarily mean that the two methyl groups on the central carbon clash in some way, but instead that two C–C bonds repel each other more than two C–H bonds do.

As in the case of ethane, two extreme conformations of propane are possible—in one the C–H and C–C bonds are staggered; in the other they are eclipsed.

112.4°

106.1°

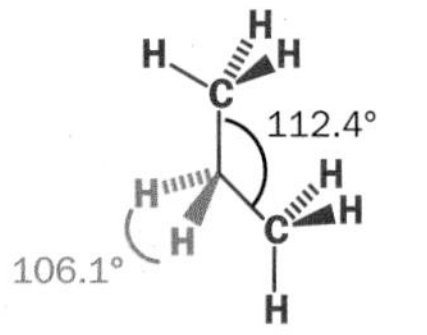

there is greater repulsion between two
C–C bonds than between two C–H bonds

▶

Notice that when we draw the
eclipsed conformation we have to
offset the front and back bonds
slightly to see the substituents
clearly. In reality, one is right
behind the other.

the staggered conformation of propane   the eclipsed conformation of propane

The rotational barrier is now slightly higher than for ethane: 14 kJ mol^{-1} as compared to 12 kJ mol^{-1}. This again reflects the greater repulsion of electrons in the coplanar bonds in the eclipsed conformation rather than any steric interactions. The energy graph for bond rotation in propane would look exactly the same as that for ethane except that the barrier is now 14 kJ mol^{-1}.

Conformations of butane

With butane things start to get slightly more complicated. Now we have effectively replaced two hydrogen atoms in ethane by larger methyl groups. These *are* large enough to get in the way of each other, that is, steric factors become a significant contribution to the rotational energy barriers. However, the main complication is that, as we rotate about the central C–C bond, not all the staggered conformations are the same, and neither are all the eclipsed conformations. The six conformations that butane can adopt as the central C–C bond is rotated in 60° intervals are shown below.

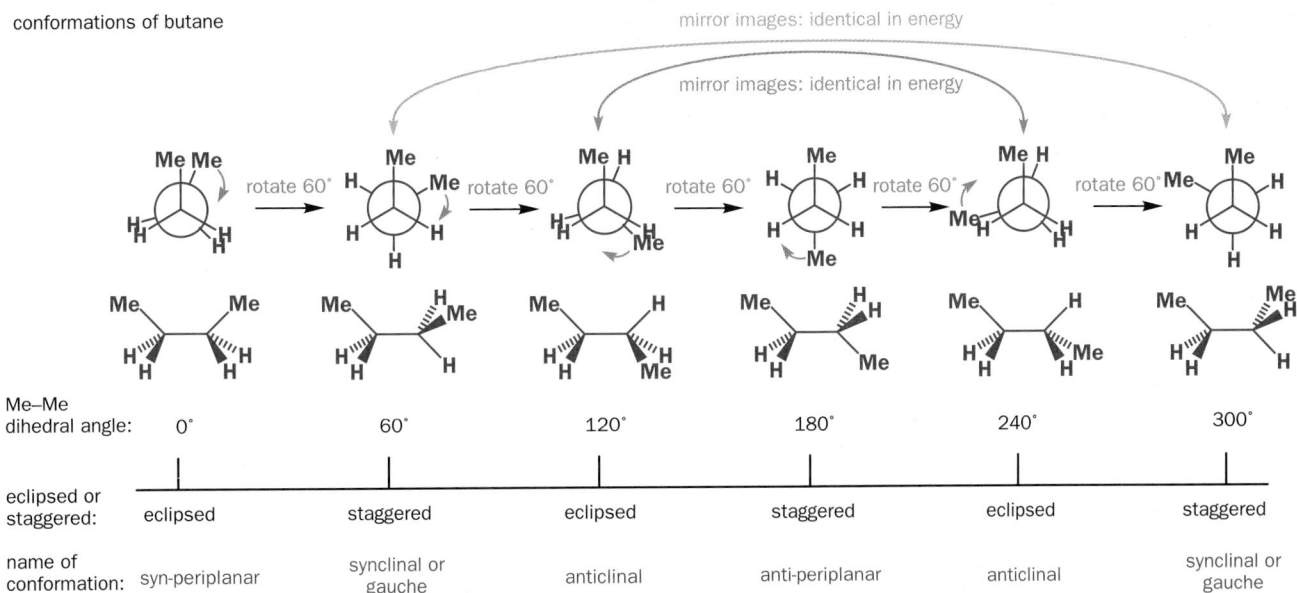

conformations of butane

Me–Me dihedral angle:	0°	60°	120°	180°	240°	300°
eclipsed or staggered:	eclipsed	staggered	eclipsed	staggered	eclipsed	staggered
name of conformation:	syn-periplanar	synclinal or gauche	anticlinal	anti-periplanar	anticlinal	synclinal or gauche

Look closely at these different conformations. The conformations with dihedral angles 60° and 300° are actually mirror images of each other, as are the conformations with angles 120° and 240°. This means that we really only have four different maxima or minima in energy as we rotate about the central C–C bond: two types of eclipsed conformations, which will represent maxima in the energy-rotation graph, and two types of staggered conformations, which will represent minima. These four different conformations have names, shown in the bottom row of the diagram. In the **syn-periplanar** and **anti-periplanar** conformations the two C–Me bonds lie in the same plane; in the **synclinal** (or gauche) and **anticlinal** conformations they slope towards (*syn*) or away from (*anti*) one another.

Before we draw the energy-rotation graph, let's just stop and think what it might look like. Each of the eclipsed conformations will be energy maxima but the syn-periplanar conformation ($\theta = 0°$) will be higher in energy than the two anticlinal conformations ($\theta = 120°$ and $240°$): in the syn-periplanar-conformation two methyl groups are eclipsing each other whereas in the anticlinal conformations each methyl group is eclipsing only a hydrogen atom. The staggered conformations will be energy minima but the two methyl groups are furthest from each other in the anti-periplanar conformation so this will be a slightly lower minimum than the two synclinal (gauche) conformations.

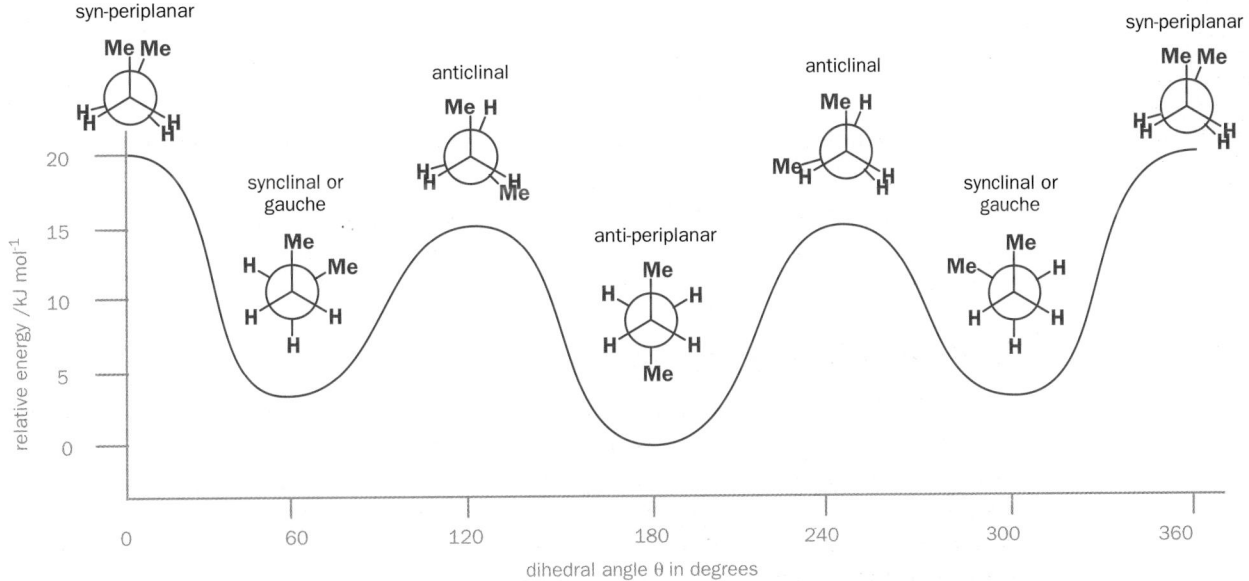

The rotation is very rapid indeed: the barrier of 20 kJ mol^{-1} corresponds to a rate at room temperature of 2×10^9 s^{-1}. This is far too fast for the different conformers to be detected by NMR (see p. 461): the NMR spectrum of butane shows only one set of signals representing an average of the two conformations.

You now have a more thorough explanation of the zig-zag arrangement of carbon chains, first introduced in Chapter 2 when we showed you how to draw molecules realistically. This is the shape you get if you allow all the C–C bonds to take up the anti-periplanar conformation, and will be the most stable conformation for any linear alkane.

We have used ring strain a number of times to explain the reactivity and spectra of cyclic molecules.

Number of atoms in ring	Internal angle in planar ring	109.5°— internal angle[a]
3	60°	49.5°
4	90°	19.5°
5	108°	1.5°
6	120°	−10.5°
7	128.5°	−19°
8	135°	−25.5°

[a] A measure of strain per carbon atom.

As in ethane, the eclipsed conformations are not stable since any rotation leads to a more stable conformation. The staggered conformations are stable since they each lie in a potential energy well. The anti-periplanar conformation, with the two methyl groups opposite each other, is the most stable of all. We can therefore think of a butane molecule as rapidly interconverting between synclinal and anti-periplanar conformations, passing quickly through the eclipsed conformations on the way. The eclipsed conformations are energy maxima, and therefore represent the transition states for interconversion between conformers.

If we managed to slow down the rapid interconversions in butane (by cooling to very low temperature, for example), we would be able to isolate the three stable conformations—the anti-periplanar and the two synclinal conformations. These different stable conformations of butane are some sort of isomers. They are called *confor*mational iso*mers* or **conformers** for short.

● **Conformations and conformers**
Butane can exist in an infinite number of *conformations* (we have chosen to show only the six most significant) but has only three *conformers* (potential energy minima)—the two synclinal (gauche) conformations and the anti-periplanar conformation.

You will see why such detailed conformational analysis of acyclic compounds is so important in Chapter 19 on eliminations where the products of the reactions can be explained only by considering the conformations of the reactants and the transition states. But first we want to use these ideas to explain another branch of organic chemistry—the conformation of ring structures.

Ring strain

Up to now, we haven't given an entirely accurate impression of rings. We have been drawing them all as if they were planar—though this is actually not the case. In this section you will learn how to draw rings more accurately and understand the properties of the different conformations adopted.

If we assume that in fully saturated carbocyclic rings each carbon is sp^3 hybridized, then each bond angle would ideally be 109.5°. However, in a planar ring, the carbon atoms don't have the luxury of choosing their bond angles: internal angle depends only on the number of atoms in the ring. If this angle differs from the ideal 109.5°, there will be some sort of strain in the molecule. This is best seen in the picture below where the atoms are forced planar. The more strained the molecules are, the more the bonds curve—in a strain-free molecule, the bonds are straight.

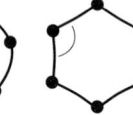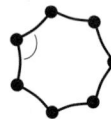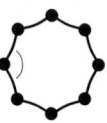

all internal angles 109.5°

Notice how in the smaller rings the bonds curve outwards, whilst in the larger rings the bonds curve inwards. The table gives values for the internal angles for regular planar polygons and an indication of the strain per carbon atom due to the deviation of this angle from the ideal tetrahedral angle of 109.5°.

This data is best presented as a graph and the ring strains per carbon atom in planar rings for ring sizes up to seventeen are shown on the next page. Whether the bonds are strained inwards or outwards is not important so only the magnitude of the strain is shown.

From these figures (represented in the graph on p. 455), note:

- The ring strain is largest for three-membered rings but rapidly decreases through a four-membered ring and reaches a minimum for a five-membered ring
- A planar five-membered ring is predicted to have the minimum level of ring strain
- The ring strain keeps on increasing (although less rapidly) as the rings get larger after the minimum at 5

But what we really need is a measure of the strain in actual compounds, not just a theoretical prediction in planar rings, so that we can compare this with the theoretical angle strain. A good measure of the strain in real rings is obtained using heats of combustion. Look at the following heats of combustion for some straight-chain alkanes. What is striking is that the difference between any two in the series is very nearly constant at around -660 kJ mol^{-1}.

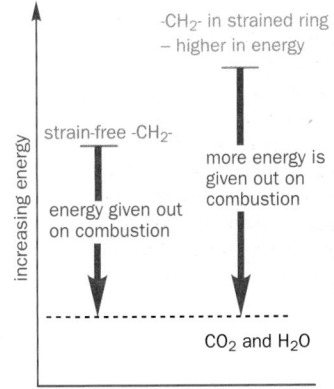

A similar measurement was used in Chapter 7 to demonstrate the stabilization of benzene due to its aromaticity.

Heats of combustion for some straight-chain alkanes

Straight-chain alkane	$CH_3(CH_2)_nCH_3$: $n =$	$-\Delta H_{combustion}$, kJ mol^{-1}	Difference, kJ mol^{-1}
ethane	0	1560	
propane	1	2220	660
butane	2	2877	657
pentane	3	3536	659
hexane	4	4194	658
heptane	5	4853	659
octane	6	5511	658
nonane	7	6171	660
decane	8	6829	658
undecane	9	7487	658
dodecane	10	8148	661

If we assume (as is reasonable) that there is no strain in the straight-chain alkanes, then each extra methylene group, $-CH_2-$, contributes on average an extra 658.7 kJ mol^{-1} to the heat of combustion for the alkane. A cycloalkane $(CH_2)_n$ is simply a number of methylene groups joined together. If the cycloalkane is strain-free, then its heat of combustion should be $n \times 658.7$ kJ mol^{-1}. If, however, there is some strain in the ring that makes the ring less stable (that is, raises its energy) then more energy is given out on combustion.

Now, let's put all this together in a graph showing, for each ring size: (a) angle strain per CH_2 group; and (b) heat of combustion per CH_2 group.

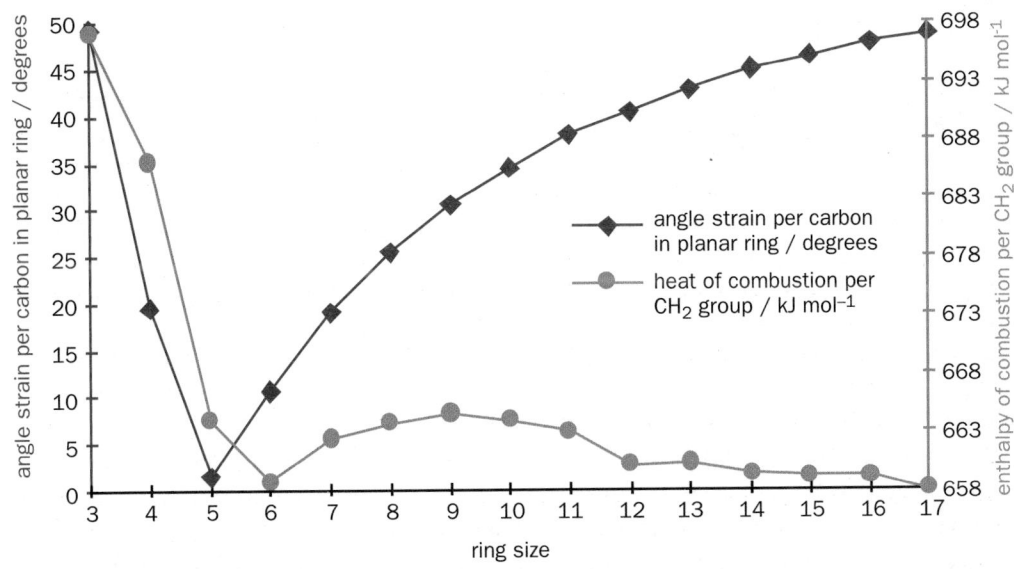

Points to notice in the green-coloured graph:

- The greatest strain by far is in the three-membered ring, cyclopropane ($n = 3$)

► Chemists class rings as small, normal, medium, and large depending on their size.
• small, $n = 3$ or 4
• normal, $n = 5$, 6, or 7
• medium, $n = 8$–about 14
• large, $n >$ about 14
This is because these different classes all have different properties and synthetic routes to making them. The groupings are evident in the graph.

• The strain decreases rapidly with ring size but reaches a minimum for cyclohexane *not* cyclopentane as you might have predicted from the angle calculations
• The strain then increases but not nearly as quickly as the angle calculation suggested: it reaches a maximum at around $n = 9$ and then decreases once more
• The strain does not go on increasing as ring size increases but instead remains roughly constant after about $n = 14$
• Cyclohexane ($n = 6$) and the larger cycloalkanes ($n \geq 14$) all have heats of combustion per $-CH_2-$ group of around 658 kJ mol^{-1}, the same value as that of a $-CH_2-$ group in a straight-chain alkane, that is, *they are essentially strain-free*

Why are there discrepancies between the two graphs? Specifically:

• Why are six-membered rings and large rings virtually strain-free?
• Why is there still some strain in five-membered rings even though the bond angles in a planar structure are almost 109.5°?

The answer to the first point, as you may already have guessed, is that the assumption that the rings are planar is simply not correct. It is easy to see how large rings can fold up into many different conformations as easily as acyclic compounds do. It is less clear to predict what happens in six-membered rings.

Six-membered rings

If you were to join six tetrahedral carbon atoms together, you would probably find that you ended up with a shape like this.

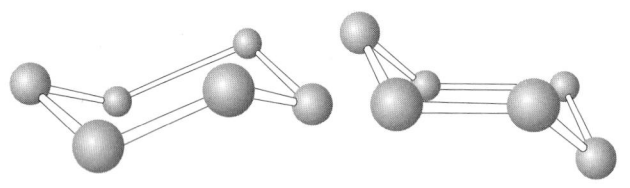

the carbon skeleton for cyclohexane   cyclohexane as a "chair"

► By far the easiest way to get to grips with these different shapes is by building models. We strongly recommend you do this!

All the carbon atoms are certainly not in the same plane, and there is no strain because all the bond angles are 109.5°. If you squash the model against the desk, forcing the atoms to lie in the same plane, it springs back into this shape as soon as you let go. If you view the model from one side (the second picture above) you will notice that four carbon atoms lie in the same plane with the fifth above the plane and the sixth below it (though it's important to realize that all six are identical—you can check this by rotating your model). The slightly overly imaginative name for this conformation—the **chair conformation**—derives from this view.

There is another conformation of cyclohexane that you might have made that looks like this. This conformation is know as the **boat conformation**. In this conformation there are still four carbon atoms in one plane, but the other two are both above this plane. Now all the carbon atoms are not the same—the four in the plane are different from the ones above. However, this is not a stable conformation of cyclohexane, even though there is no bond angle strain (all the angles are 109.5°). In order to understand why not, we must go back a few steps and answer our other question: why is cyclopentane strained even though a planar conformation has virtually no angle strain?

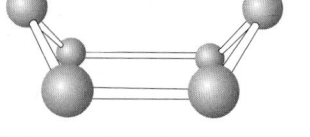

the boat conformation of cyclohexane

Smaller rings (three, four, and five members)

The three carbon atoms in cyclopropane must lie in a plane since it is always possible to draw a plane through any three points. All the C–C bond lengths are the same which means that the three carbon atoms are at the corners of an equilateral triangle. From the large heat of combustion per methylene group (p. 455) we know that there is considerable strain in this molecule. Most of this is due to the bond angles deviating so greatly from the ideal tetrahedral value of 109.5°. Most but not all. If we view along one of the C–C bonds we can see a further cause of strain—all the C–H bonds are eclipsed.

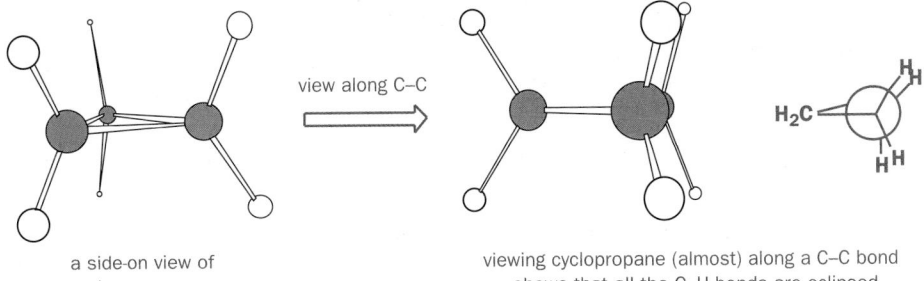

a side-on view of
cyclopropane

viewing cyclopropane (almost) along a C–C bond
shows that all the C–H bonds are eclipsed

The eclipsed conformation of ethane is an energy maximum and any rotation leads to a more stable conformation. In cyclopropane it is not possible to rotate any of the C–C bonds and so all the C–H bonds are forced to eclipse their neighbours.

In fact, in any planar conformation all the C–H bonds will be eclipsed with their neighbours. In cyclobutane, the ring distorts from a planar conformation in order to reduce the eclipsing interactions, even though this reduces the bond angles further and so increases the bond angle strain. Cyclobutane adopts a puckered or 'wing-shaped' conformation.

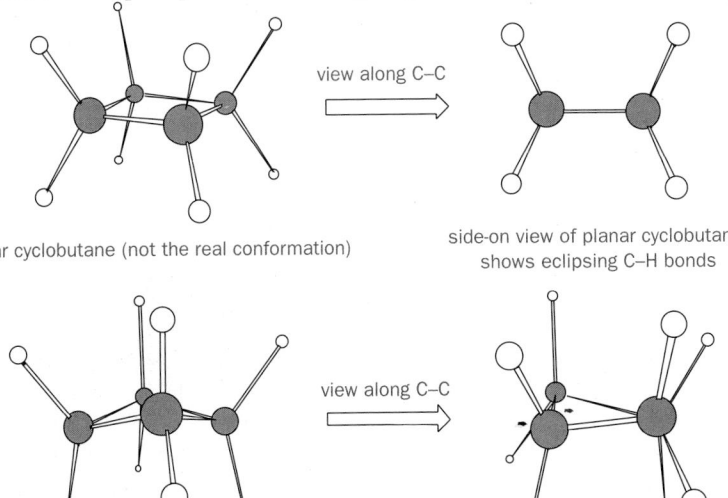

planar cyclobutane (not the real conformation)

side-on view of planar cyclobutane
shows eclipsing C–H bonds

the puckered 'wing' conformation of cyclobutane

C–H bonds no longer fully eclipsed

This explains why cyclopentane is not entirely strain-free even though in a planar conformation the C–C–C bond angles are close to 109.5°. The heat of combustion data give us an indication of the total strain in the molecule, not just the contribution of angle strain. There is strain in planar cyclopentane caused by the eclipsing of adjacent C–H bonds. As in cyclobutane, the ring distorts to reduce the eclipsing interactions but this increases the angle strain. Whatever happens, there is always going to be some strain in the system. The minimum energy conformation adopted is a balance of the two opposing effects. Cyclopentane adopts a shape approximating to an 'open envelope', with four atoms in a plane and one above or below it. The atoms in the ring rapidly take turns not to be in the plane, and cyclopentanes have much less well-defined conformational properties than cyclohexanes, to which we shall now return.

A closer look at cyclohexane

The heats of combustion data show that cyclohexane is virtually strain-free. This must include strain from eclipsing interaction as well as angle strain. A model of the chair conformation of cyclohexane including all the hydrogen atoms looks like this.

We shall consider the conformations, and reactions, of cyclopentanes in Chapter 33.

"open envelope" conformation of cyclopentane

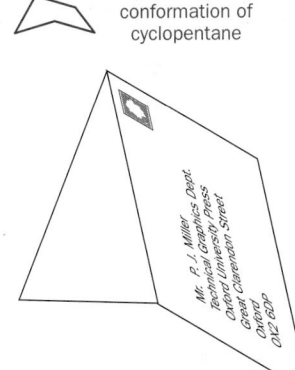

A side-on view of the chair conformation of cyclohexane   A view of cyclohexane looking along two of the C–C bonds.   A Newman projection of the same view

The view along two of the C–C bonds clearly shows that there are no eclipsing C–H bonds in the chair conformation of cyclohexane—in fact, all the bonds are fully staggered, giving the lowest energy possible. This is why cyclohexane is strain-free.

Contrast this with the boat conformation. Now all the C–H bonds are eclipsed, and there is a particularly bad interaction between the 'flagstaff' C–H bonds.

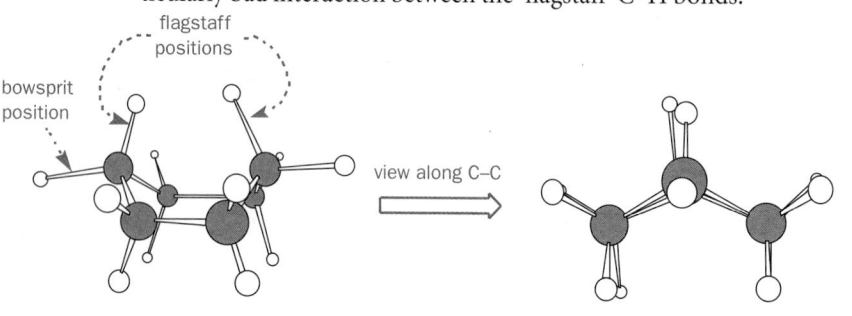

a side-on view of the boat conformation of cyclohexane   a view of the boat conformation looking along two of the C–C bonds   Newman projection of the same view

This explains why the boat conformation is much less important than the chair conformation. Even though both are free from angle strain, the eclipsing interactions in the boat conformation make it approximately 25 kJ mol^{-1} higher in energy than the chair conformation. In fact, as we shall see later, the boat conformation represents an energy maximum in cyclohexane whilst the chair conformation is an energy minimum. Earlier we saw how the eclipsing interactions in planar cyclobutane and cyclopentane could be reduced by distortion of the ring. The same is true for the boat conformation of cyclohexane. The eclipsing interactions can be relieved slightly if the two 'side' C–C bonds twist relative to each other.

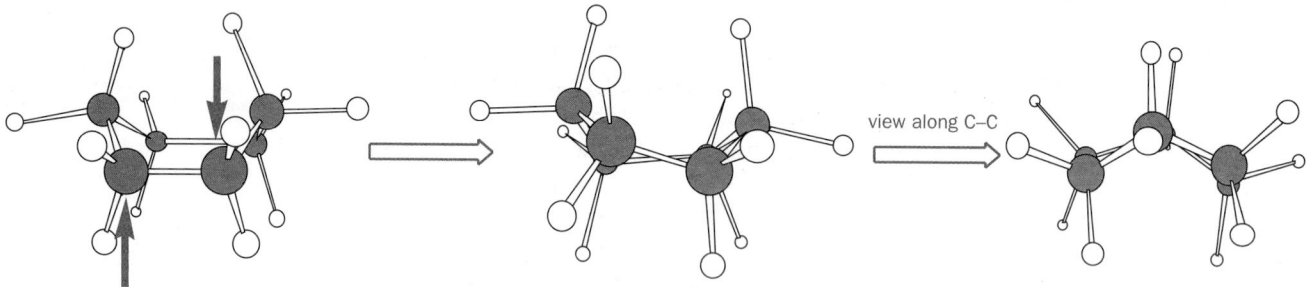

pushing these two carbon atoms in the direction shown....   ...gives a slightly different conformation in which the eclipsing interactions have been reduced: the "twist-boat" conformation   an end-on view of the twist-boat conformation shows how the eclipsing interactions have been reduced

▶

A **local energy minimum** is the bottom of the potential energy well, but not necessarily the deepest possible well, which is the **global energy minimum**. Small changes in conformation will increase the energy, although a large change may be able to decrease the energy further. As an example, the synclinal (gauche) conformation of butane is a local energy minimum; the anti-periplanar conformation is the global energy minimum.

This twisting gives rise to a slightly different conformation of cyclohexane called the **twist-boat conformation**, which, although not as low in energy as the chair form, is lower in energy (by 4 kJ mol^{-1})

than the boat form and is a local energy minimum as we shall see later. Cyclohexane has two stable conformers, the chair and the twist boat. The chair form is approximately 21 kJ mol^{-1} lower in energy than the twist-boat form.

Drawing cyclohexane

Take another look at the chair conformation on p. 458. All six carbon atoms are identical, but there are two types of protons—one type stick either vertically up or down and are called **axial** hydrogen atoms; the other sort stick out sideways and are called **equatorial** hydrogen atoms.

As you go round the ring, notice that each of the CH_2 groups has one hydrogen sticking up and one sticking down. However, all the 'up' ones alternate between axial and equatorial, as do all the 'down' ones.

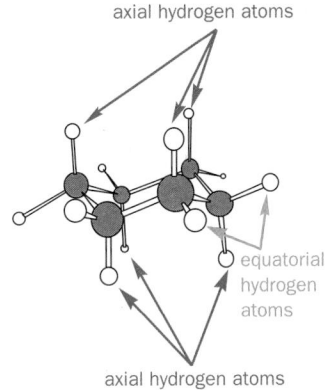

Compare the equator and axis of the earth: equatorial bonds are around the equator of the molecule. Note the spelling.

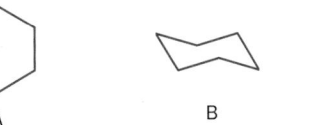

these hydrogen atoms are all 'up' relative to their partners on the same C carbon

these hydrogens are all 'down' relative to their partners on the same C atom

Before going any further, it's important that you learn how to draw cyclohexane properly. Without cluttering the structure with Cs and Hs, a chemist would draw cyclohexane as one of these three structures.

A   B   C

Up to now, we have simply used the hexagon A to represent cyclohexane. We shall see that, whilst this is not strictly accurate, it is nonetheless still useful. The more correct structures B and C (which are actually just different views of the same molecule) take some practice to draw properly. A recommended way of drawing cyclohexane is shown in the box.

Guidelines for drawing cyclohexane

The carbon skeleton

Trying to draw the chair conformation of cyclohexane in one continuous line can lead to some dreadful diagrams. The easiest way to draw a chair conformation is by starting off with one end.

Next draw in two parallel lines of equal length.

At this stage, the top of the new line should be level with the top of the original pair.

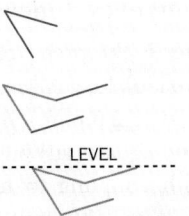

LEVEL

Finally, the last two lines should be added. These lines should be parallel to the first pair of lines as shown

these lines should be parallel

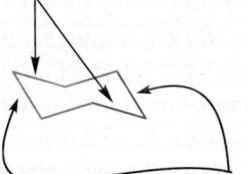

these lines should be parallel

and the lowest points should also be level.

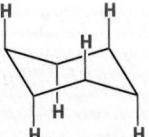

LEVEL

Adding the hydrogen atoms

This is often the trickiest part. Just remember that you are trying to make each of the carbon atoms look tetrahedral. (Note that we don't normally use wedged and hashed bonds; otherwise things get really messy.)

The axial bonds are relatively easy to draw in. They should all be vertically aligned and alternate up and down all round the ring.

Guidelines for drawing cyclohexane (contd)

The equatorial bonds require a little more care to draw.
The thing to remember is that each equatorial bond must
be parallel to two C–C bonds.

notice the 'W' shape here. . .

put in all 6 equatorial
C–H bonds

in each diagram, all the red bonds are parallel

. . . and the 'M' shape here

The complete diagram with all the hydrogen atoms should
look like this.

Common mistakes

If you follow all the guidelines above, you will soon be
drawing good conformational diagrams. However, a few

common mistakes have been included to show you what
not to do!

how *not* to draw cyclohexanes...

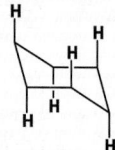

the chair has been drawn with the
middle bonds horizontal, so the upper
points of the chair are not level. This
means the axial hydrogens can no
longer be drawn vertical

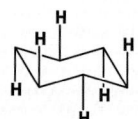

the axial hydrogens have been
drawn alternating up and down on
the wrong carbons. This structure
is impossible because none of
the carbons can be tetrahedral

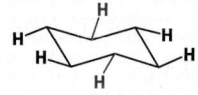

the red hydrogens have been
drawn at the wrong angles – look
for the parallel lines and the
'W' and 'M'

> There is only one type of equatorial
> conformer, and one type of axial
> conformer. Convince yourself that
> these drawings are exactly the
> same conformation just viewed
> from different vantage points.

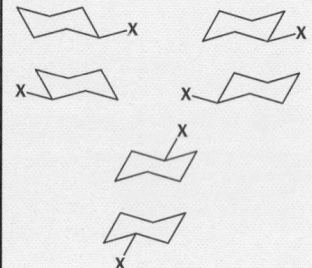

substituent equatorial

substituent axial

> Make a model of cyclohexane and
> try the ring inversion for yourself.

The ring inversion (flipping) of cyclohexane

Given that this chair conformer is the preferred conformation for cyclohexane, what would you
expect its ^{13}C NMR spectrum to look like? All six carbon atoms are the same so there should only be
one signal (and indeed there is, at 25.2 p.p.m.). But what about the ^{1}H NMR spectrum? The two different sorts of protons (axial and equatorial) ought to resonate at different frequencies, so two signals should be seen (each with coupling to neighbouring protons). In fact, there is only *one*
resonance in the proton spectrum, at 1.40 p.p.m.

In a monosubstituted cyclohexane, there should be two isomers detectable—one with the substituent axial, the other with the substituent equatorial. But again at room temperature only one set
of signals is seen.

This changes when the NMR spectrum is run at low temperature. Now two isomers are visible,
and this gives us a clue as to what is happening: the two isomers are conformers that interconvert—
rapidly at room temperature, but more slowly when the temperature is lowered. Recall that NMR
does not distinguish between the three different stable conformers of butane (two synclinal and one
anti-periplanar) because they are all rapidly interconverting so fast that only an average is seen. The
same happens with cyclohexane—just by rotating bonds (that is, without breaking any!) cyclohexane can **ring invert** or 'flip'. After ring inversion has taken place, all the bonds that were axial are now
equatorial and vice versa.

ring inversion of a monosubstituted cyclohexane
notice that the hydrogen atom shown changes from axial to equatorial

The whole inversion process can be broken down into the conformations shown below. The green arrows show the direction in which the individual carbon atoms should move in order to get to the next conformation.

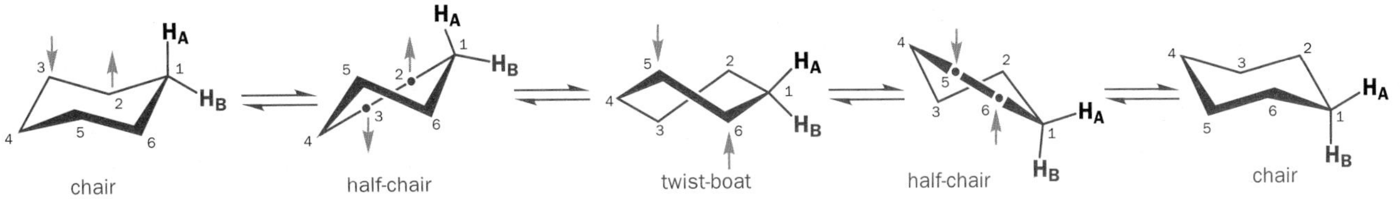

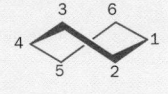

chair ⇌ half-chair ⇌ twist-boat ⇌ half-chair ⇌ chair

The energy profile for this ring inversion shows that the half-chair conformation is the energy maximum on going from a chair to a twist boat. The true boat conformation is the energy maximum on interchanging between two mirror-image twist-boat conformers, the second of which is converted to the other chair conformation through another half-chair.

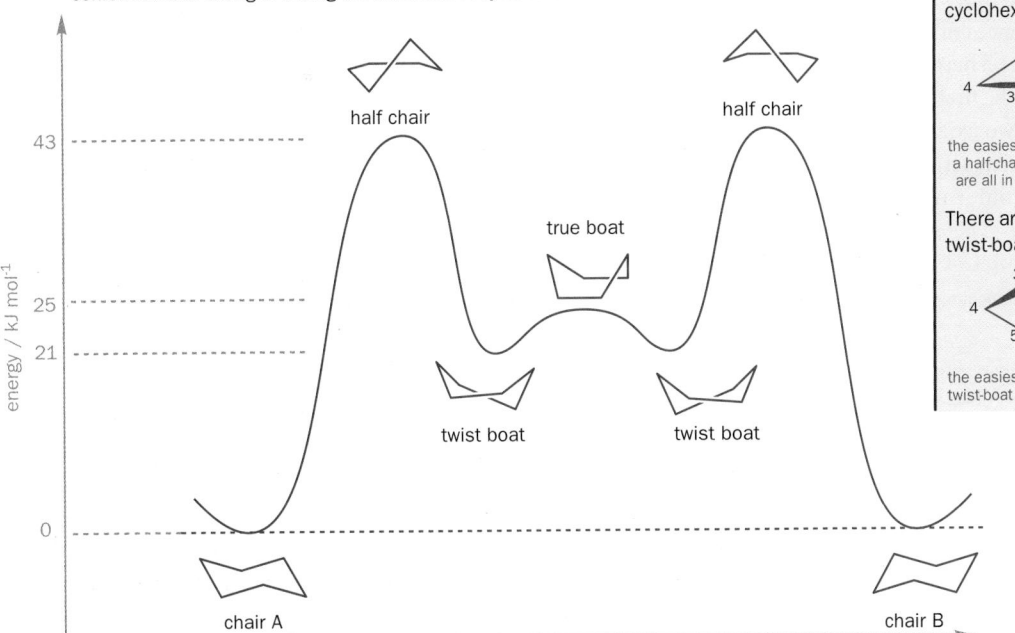

conformational changes during the inversion of cyclohexane

energy / kJ mol⁻¹

43

25

21

0

half chair   true boat   half chair

twist boat   twist boat

chair A   chair B

reaction coordinate

> In the **half-chair** conformation of cyclohexane, four *adjacent* carbon atoms are in one plane with the fifth above this plane and the sixth below it. You will this conformation again later—it represents the energy minimum for cyclohexene, for example. 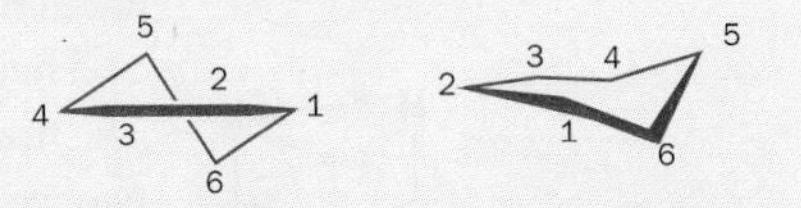
>
> the easiest way of drawing a half-chair. Carbons 1–4 are all in the same plane
>
> an alternative perspective of a half-chair conformation
>
> There are also a number of ways of drawing a twist-boat conformer.
>
> the easiest way to draw a twist-boat conformation. . .
>
> . . . although it's easier to see why it's called a twist boat from this viewpoint

> This would be a good point to remind you again of Chapter 13. This energy profile shows the conversion of one chair to another via two twist-boat *intermediates* (local energy minima). In between the energy minima are energy maxima, which are the *transition states* for the process. The progress of the ring-flipping 'reaction' is shown along an arbitrary 'reaction coordinate'.

It's clear from the diagram that the barrier to ring inversion of cyclohexane is 43 kJ mol⁻¹, or a rate at 25 °C of about 2×10^5 s⁻¹. Ring inversion also interconverts the axial and equatorial protons, so these are also exchanging at a rate of 2×10^5 s⁻¹ at 25 °C—too fast for them to be detected individually by NMR, which is why they appear as an averaged signal.

Rates and spectroscopy

NMR spectrometers behave like cameras with a shutter speed of *about* 1/1000 s. Anything happening faster than that, and we get a blurred picture; things happening more slowly give a sharp picture. In fact, a more exact number for the 'shutter speed' of an NMR machine (not a real shutter speed—just figuratively speaking!) is given by the equation

$$k = \pi \, \Delta\nu \, / \, \sqrt{2} = 2.22 \times \Delta\nu$$

where k is the fastest exchange rate that still gives individual signals and $\Delta\nu$ is the separation of those signals in the NMR spectrum measured in hertz. For example, on a 200 MHz spectrometer, two signals separated by 0.5 p.p.m. are 100 Hz apart, so any process

exchanging with a rate slower than 222 s⁻¹ will still allow the NMR machine to show two separate signals; if they exchange with a rate faster than 222 s⁻¹ only an averaged signal will be seen.

The equation above holds for any spectroscopic method, provided we think in terms of differences between signals or peaks measured in hertz. So, for example, a difference between two IR absorptions of 100 cm⁻¹ can be represented as a wavelength of 0.01 cm (1×10^{-4} m) or a frequency of 3×10^{12} s⁻¹. IR can detect changes happening a lot faster than NMR can—its 'shutter speed' is of the order of one-trillionth of a second.

Substituted cyclohexanes

In a monosubstituted cyclohexane, there can exist two different chair conformers: one with the substituent axial, the other with it equatorial. The two chair conformers will be in rapid equilibrium (by the process we have just described) but they will not have the same energy. In almost all cases, *the conformer with the substituent axial is higher in energy*, which means there will be less of this form present at equilibrium.

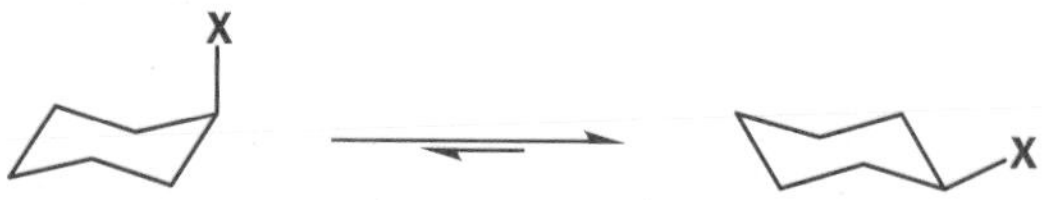

this conformation is
lower in energy

For example, in methylcylcohexane ($X = CH_3$), the conformer with the methyl group axial is 7.3 kJ mol^{-1} higher in energy than the conformer with the methyl group equatorial. This energy difference corresponds to a 20:1 ratio of equatorial:axial conformers at 25 °C.

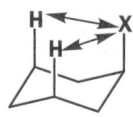

There are two reasons why the axial conformer is higher in energy than the equatorial conformer. The first is that the axial conformer is destabilized by the repulsion between the axial group X and the two axial hydrogen atoms on the same side of the ring. This interaction is known as the **1,3-diaxial interaction**. As the group X gets larger, this interaction becomes more severe and there is less of the conformer with the group axial.

The second reason is that in the equatorial conformer the C–X bond is anti-periplanar to two C–C bonds, while, for the axial conformer, the C–X bond is synclinal (gauche) to two C–C bonds.

equatorially substituted cyclohexane: axially substituted cyclohexane:

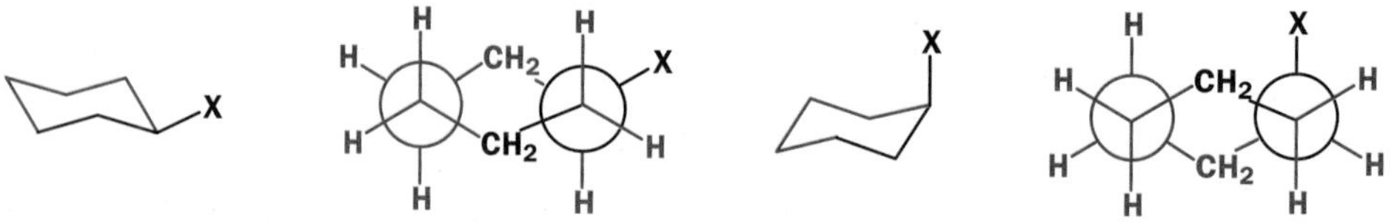

the black bonds are anti-periplanar   the black bonds are synclinal (gauche)
(only one pair shown for clarity)   (only one pair shown for clarity)

The table shows the preference of a number of substituted cyclohexanes for the equatorially substituted conformer over the axially substituted conformer at 25 °C.

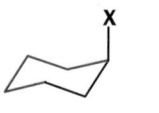

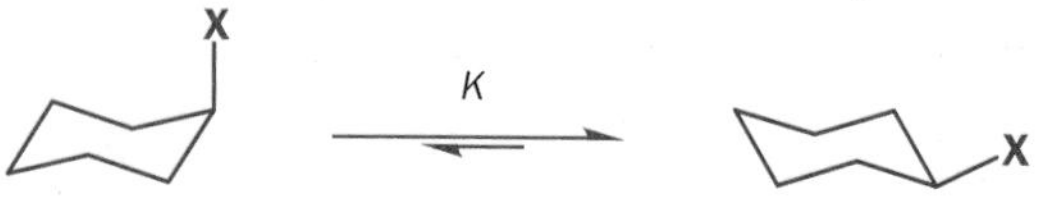

$$K = \frac{\text{concentration of equatorial conformer}}{\text{concentration of axial conformer}}$$

X	Equilibrium constant, K	Energy difference between axial and equatorial conformers, kJ mol^{-1}	% with substituent equatorial
H	1	0	50
Me	19	7.3	95
Et	20	7.5	95
i-Pr	42	9.3	98
t-Bu	>3000	>20	>99.9
OMe	2.7	2.5	73
Ph	110	11.7	99

Note the following points.

• The three columns in the table are three different ways of expressing the same information. However, just looking at the percentages column, it is not immediately obvious to see how much more of the equatorial conformer there is—after all, the percentages of equatorial conformer for methyl, ethyl, isopropyl, *t*-butyl, and phenyl-cyclohexanes are all 95% or more. Looking at the equilibrium constants gives a much clearer picture.

- The amount of equatorial conformer present does increase in the order Me < Et < *i*-Pr < *t*-Bu, but perhaps not quite as expected. The ethyl group *must* be physically larger than a methyl group but there is hardly any difference in the equilibrium constants. The increase in the proportion of equatorial conformer on going from Et to *i*-Pr is only a factor of two but for *t*-butylcyclohexane, it is estimated that there is about 3000 times more of the equatorial conformer than the axial conformer

- The same anomaly occurs with the methoxy group—there is a much greater proportion of the conformer with a methoxy group axial than with a methyl group axial. This is despite the fact that the methoxy group is physically larger than a methyl group

The equilibrium constant does not depend on the actual size of the substituent, but rather its interaction with the neighbouring axial hydrogens. In the case of the methoxy group, the oxygen acts as link and removes the methyl group away from the ring, lessening the interaction. The groups Me, Et, *i*-Pr, and *t*-Bu all need to point some atom towards the other axial hydrogens, and for Me, Et, and *i*-Pr this can be H. Only for *t*-Bu must a methyl group be pointing straight at the axial hydrogens, so *t*-Bu has a much larger preference for the equatorial position than the other alkyl groups. In fact, the interactions between an axial *t*-butyl group and the axial hydrogen atoms are so severe that the group virtually always stays in the equatorial position. As we shall see later, this can be very useful.

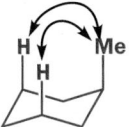

in the axial conformer of methylcyclohexane, there is a direct interaction between the methyl group and the axial hydrogen atoms

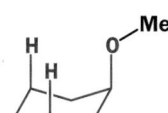

in methoxycyclohexane, the methyl group is removed somewhat from the ring

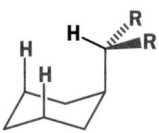

when a methyl, ethyl or *i*-propyl group is axial, only a hydrogen atom need lie directly over the ring

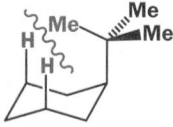

the steric requirements for putting a *t*-butyl group axial are enormous since now there is a severe interaction between a methyl group and the axial protons

What happens with more than one substituent on the ring?

When there are two or more substituents on the ring, stereoisomerism is possible. For example, there are two isomers of 1,4-cyclohexanediol—in one (the *cis* isomer) both the substituents are either above or below the cyclohexane ring; in the other (the *trans* isomer) one hydroxyl group is above the ring whilst the second is below. For a *cis*-1,4-disubstituted cyclohexane with both the substituents the same, ring inversion leads to a second identical conformation, while for the *trans* configuration there is one conformation with both groups axial and one with both groups equatorial.

cis-1,4-cyclohexanediol

both the OH groups occupy positions on the upper side of the cyclohexane ring

ring inversion

both the Hs occupy positions on the lower side of the ring

in *trans*-1,4-cyclohexanediol, one OH group is above the plane of the ring in either conformation. .

ring inversion

. . . whilst the other is below, in either conformation

trans-1,4-cyclohexanediol

conformer with both OHs axial

the more stable conformer with both OHs equatorial

> Ring inversion interconverts all of the axial and equatorial substituents, but it does not change which face of the ring a substituent is on. If an equatorial substituent starts off above the ring (that is, 'up' relative to its partner on the same C atom) it will end up above the ring, but now axial. Axial and equatorial are *conformational* terms; which side of the ring a substituent is on depends on the compound's *configuration*.

►

The *cis* and *trans* compounds are different diastereoisomers. Consequently, they have different chemical and physical properties and cannot interconvert simply by rotating bonds.

m.p. 113–114 °C   m.p. 143–144 °C

This contrasts with the two *conformers* of *trans*-1,4-dimethoxycyclohexane (diaxial or diequatorial), which rapidly interconvert at room temperature without breaking any bonds.

►

It is not always easy to decide if an equatorial substituent is 'up' or 'down'. The key is to compare it with its axial partner on the same C atom—axial substituents very clearly point 'up' or 'down'. If the axial partner is 'up', the equatorial substituent must be 'down' and vice versa.

The chair-structure diagrams contain much more information than the simple 'hexagon' diagrams that we have used up to now. The former show both configuration and conformation—they show which stereoisomer (*cis* or *trans*) we are talking about and also (for the *trans* compound) the conformation adopted (diaxial or the more stable diequatorial). In contrast, the simpler hexagon diagrams carry no information about the conformation—only information about which isomer we are dealing with. This can be useful, because it enables us to talk about one configuration of a compound without specifying the conformation. When you are solving a problem requiring conformational diagrams to predict the configuration of a product, always start and finish with a configurational (hexagon) drawing.

The chair conformer of *cis*-1,4-disubstituted cyclohexane has one substituent equatorial, the other axial. This will not necessarily be the case for other substitution patterns; for example, the chair conformer of a *cis*-1,3-disubstituted cyclohexane has either both substituents axial or both equatorial. Remember, the '*cis*' and '*trans*' prefixes merely indicate that both groups are on the same 'side' of the cyclohexane ring. Whether the substituents are both axial/equatorial or one axial and the other equatorial depends on the substitution pattern. Each time you meet a molecule, draw the conformation or make a model to find out which bonds are axial and equatorial.

cis-1,3-disubstituted cyclohexane   in both conformers, both substituents are 'up'

trans-1,3-disubstituted cyclohexane   in both conformers one substituent is 'up', the other 'down'

What if the two substituents on the ring are different? For the *cis* 1,3-disubstituted example above, there is no problem, because the favoured conformation will still be the one that places these two different substituents equatorial. But when one substituent is axial and the other equatorial (as they happen to be in the *trans* diastereoisomer above) the preferred conformation will depend on what those substituents are. In general, the favoured conformation will place the maximum number of substituents equatorial. If both conformations have the same number of equatorial substituents, the one with the larger substituent equatorial will win out, and the smaller group will be forced to be axial. Various possibilities are included in the examples below.

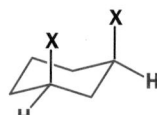
two substituents equatorial none axial   favoured   no substituents equatorial two axial

one substituents equatorial one axial (smaller OH)   favoured   one substituents equatorial one axial (large Br)

isomenthol

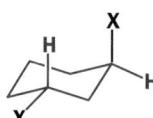
two substituents equatorial one axial   favoured   one substituent equatorial two axial

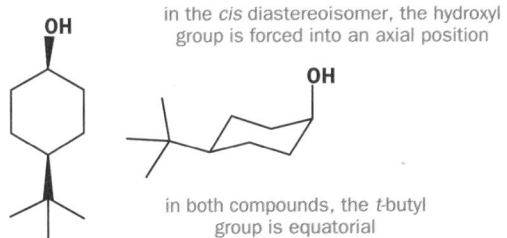

two substituents equatorial
two axial

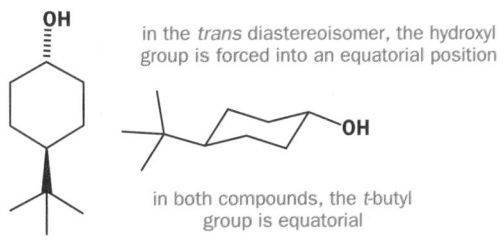

favoured

two substituents equatorial
two axial (including large phenyl)

This is only a guideline, and in many cases it is not easy to be sure. Instead of concerning ourselves with these uncertainties, we shall move on to some differentially substituted cyclohexanes for which it is absolutely certain which conformer is preferred.

Locking groups—*t*-butyl groups, decalins, and steroids

t-Butyl groups

We have already seen how a *t*-butyl group always prefers an equatorial position in a ring. This makes it very easy to decide which conformation the two different compounds below will adopt.

cis-4-*t*-butylcyclohexanol

in the *cis* diastereoisomer, the hydroxyl group is forced into an axial position

in both compounds, the *t*-butyl group is equatorial

trans-4-*t*-butylcyclohexanol

in the *trans* diastereoisomer, the hydroxyl group is forced into an equatorial position

in both compounds, the *t*-butyl group is equatorial

Cis-1,4-di-*t*-butylcyclohexane

An axial *t*-butyl group really is very unfavourable. In *cis*-1,4-di-*t*-butylcyclohexane, one *t*-butyl group would be forced axial if the compound existed in a chair conformation. To avoid this, the compound prefers to pucker into a twist boat so that the two large groups can both be in equatorial positions (or 'pseudoequatorial', since this is not a chair).

the twist-boat conformer (with both *t*-butyl groups in pseudoequatorial positions) is lower in energy than the chair conformer.

cis-1,4-di-*t*-butylcyclohexane

Decalins

It is also possible to lock the conformation of a cyclohexane ring by joining another ring to it. **Decalin** is two cyclohexane rings fused at a common C–C bond. Two diastereoisomers are possible, depending on whether the hydrogen atoms at the ring junction are *cis* or *trans*. For *cis*-decalin, the second ring has to join the first so that it is axial at one point of attachment and equatorial at the other; for *trans*-decalin, the second ring can be joined to the first in the equatorial position at both attachment points.

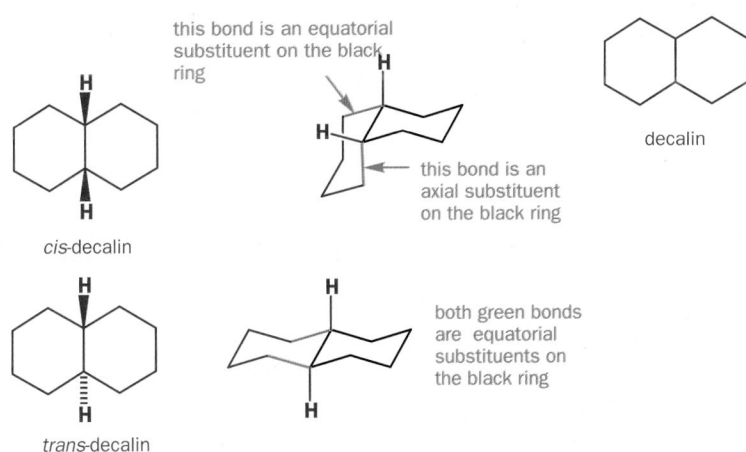

this bond is an equatorial substituent on the black ring

this bond is an axial substituent on the black ring

cis-decalin

both green bonds are equatorial substituents on the black ring

trans-decalin

decalin

When a cyclohexane ring inverts, the substituents that were equatorial become axial and vice versa. This is fine for *cis*-decalin, which has an axial–equatorial junction, but it means that ring inversion is not possible for *trans*-decalin. For *trans*-decalin to invert, the junction would have to become axial–axial, and it's not possible to link the axial positions to form a six-membered ring. *Cis*-decalin, on the other hand, ring inverts just as fast as cyclohexane.

ring inversion of *cis*-decalin

green H starts axial on black ring
yellow H starts equatorial on black ring

after ring inversion,
green H is equatorial on black ring
yellow H is axial on black ring

no ring inversion in *trans*-decalin

impossible to join two axial positions into six-membered ring

If you find it hard to visualize the ring inversion of *cis*-decalin, you are not alone! The best way to think about it is to ignore the second ring till the very end: just concentrate on what happens to one ring (black in this diagram), the hydrogens at the ring junction, and the (orange) bonds next to these hydrogens that form the 'stumps' of the second ring. Flip the black ring, and the 'stumps', and the hydrogens swap from axial to equatorial and vice versa. Draw the result, but don't fill in the second ring yet or it will usually just come out looking like a flat hexagon (as in diagram A). Instead, rotate the complete (black) ring 60° about a vertical axis so that both of the orange 'stumps' can form part of a chair, which can now be filled in (diagram B). To make a chair (and not a hexagon) they must be pointing in a convergent direction, as the orange bonds are in B but not A.

the steroid skeleton

Steroids

Steroids are an important class of compounds occurring in all animals and plants and have many important functions from regulating growth (anabolic steroids) and sex drive (all sex hormones are steroids) to acting as a self-defence mechanism in plants, frogs, and even sea cucumbers. A steroid is defined by its structure: all steroids contain a basic carbon framework consisting of four fused rings—three cyclohexane rings and one cyclopentane ring—labelled and joined together as shown in the margin.

Just as in the decalin system, each ring junction could be *cis* or *trans*, but it turns out that all steroids have all *trans*-junctions except where rings A and B join which is sometimes *cis*. Examples are cholestanol (all *trans*) and coprostanol (A and B fused *cis*).

equatorial hydroxyl   cholestanol

axial hydroxyl   coprostanol

It was a desire to explain the reactions of steroids that led Sir Derek Barton (1918–98) to discover, in the 1940s and 1950s, the principles of conformational analysis described in this chapter. It was for this work that he shared the Nobel prize in 1969. We will come back to steroids in more detail in Chapter 51.

Because steroids (even those with a *cis* A–B ring junction) are essentially substituted *trans*-decalins they can't ring flip. This means, for example, that the hydroxyl group in cholestanol is held equatorial on ring A while the hydroxyl group in coprostanol is held axial on ring A. The steroid skeleton really is remarkably stable—samples of sediment 1.5×10^9 years old have been found to contain steroids still with the same ring-junction stereochemistry.

Axially and equatorially substituted rings react differently

We shall be using ring structures throughout the rest of the book, and you will learn how the conformation affects chemistry extensively. Here we shall give a few examples in which the outcome of a reaction may depend on whether a functional group is axial or equatorial. In many of the examples, the functional group will be held in its axial or equatorial position by 'locking' the ring using a *t*-butyl group or a fused ring system such as *trans*-decalin.

Nucleophilic substitution

In the last chapter we looked at two mechanisms for nucleophilic substitution: S_N1 and S_N2. We saw that the S_N2 reaction involved an inversion at the carbon centre. Recall that the incoming nucleophile had to attack the σ^* orbital of the C–X bond. This meant that it had to approach the leaving group directly from behind, leading to inversion of configuration.

inversion during nucleophilic substitution at saturated carbon

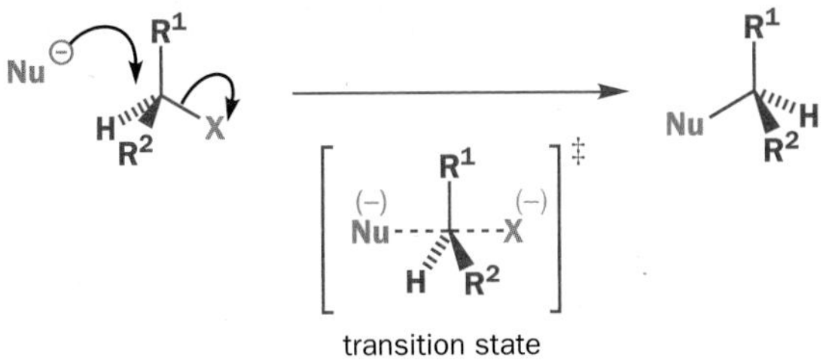

transition state

What do you think would happen if a cyclohexane derivative underwent an S_N2 reaction? If the conformation of the molecule is fixed by a locking group, the inversion mechanism of the S_N2 reaction, means that, if the leaving group is axial, then the incoming nucleophile will end up equatorial and vice versa.

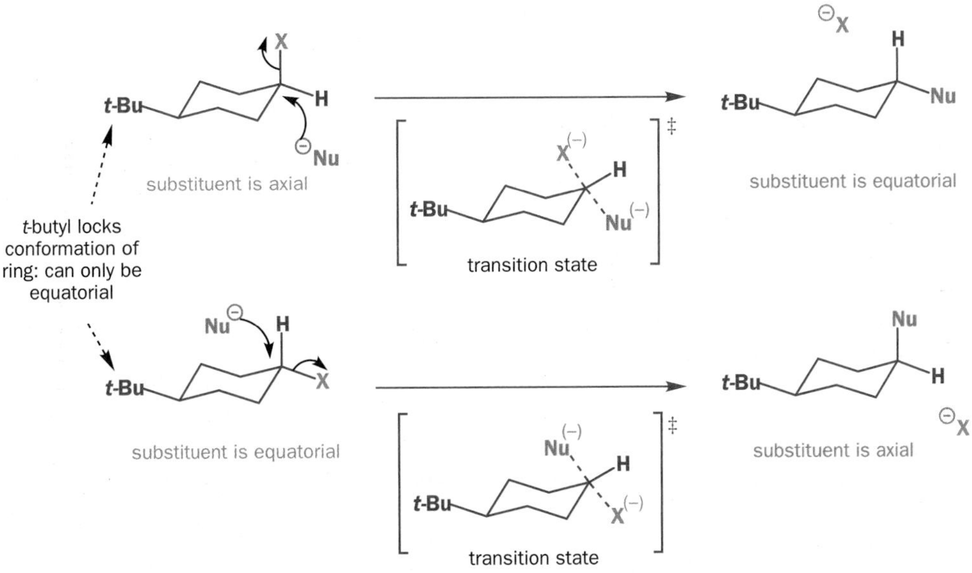

t-butyl locks conformation of ring: can only be equatorial

substituent is axial

transition state

substituent is equatorial

substituent is equatorial

transition state

substituent is axial

Substitution reactions are not very common for substituted cyclohexane. The substituted carbon in a cyclohexane ring is a secondary centre—in the last chapter, we saw that secondary centres do not react well via either S_N1 or S_N2 mechanisms (p. 426). To encourage an S_N2 mechanism, we need a good attacking nucleophile and a good leaving group. One such example is shown—the substitution of a tosylate by PhS⁻.

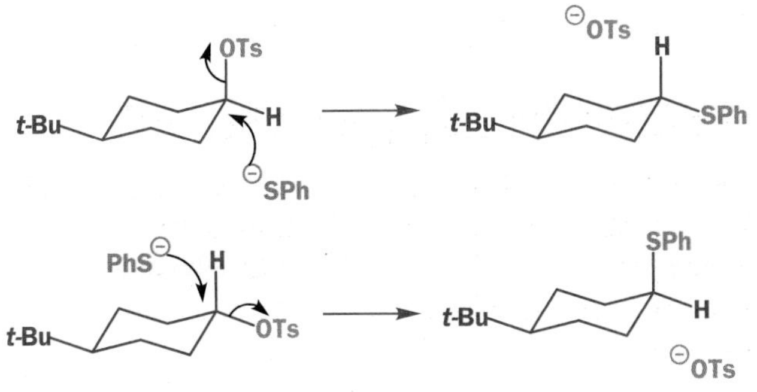

axial leaving group is substituted 31 times faster than equatorial leaving group

It is found that the substitution of an axial substituent proceeds faster than the substitution of an equatorial substituent. There are several contributing factors making up this rate difference, but probably the most important is the direction of approach of the nucleophile. The nucleophile must attack the σ* of the leaving group, that is, directly behind the C–X bond. In the case of an equatorially substituted compound, this line of attack is hindered by the (green) axial hydrogens—it passes directly through the region of space they occupy. For an axial leaving group, the direction of attack is parallel with the (orange) axial hydrogens anti-periplanar to the leaving group, and approach is much less hindered.

equatorial leaving group

axial leaving group

approach hindered by green axial Hs

less hindered approach

We must assume that this holds even for simple unsubstituted cyclohexanes, and that substitution reactions of cyclohexyl bromide, for example, occur mainly on the minor, axial conformer. This slows down the reaction because, before it can react, the prevalent equatorial conformer must first flip axial.

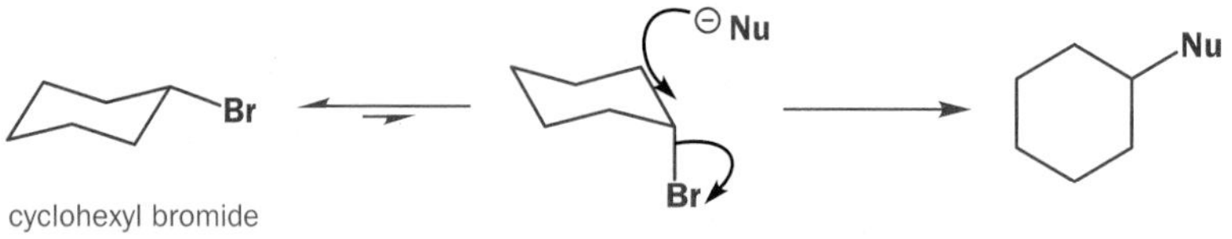

cyclohexyl bromide

substitution faster from axial conformer

Epoxides

In the last chapter you met epoxides as electrophiles reacting with nucleophiles such as amines and azide, and we shall look at this sort of reaction again in a few pages time. Epoxides can be formed from compounds containing an adjacent hydroxyl group and a leaving group by treatment with base. The reaction is essentially the reverse of their ring-opening reaction with nucleophiles.

epoxide ring-opening reaction

epoxide ring-closing reaction

intramolecular S_N2 nucleophilic attack by RO^- on alkyl chloride

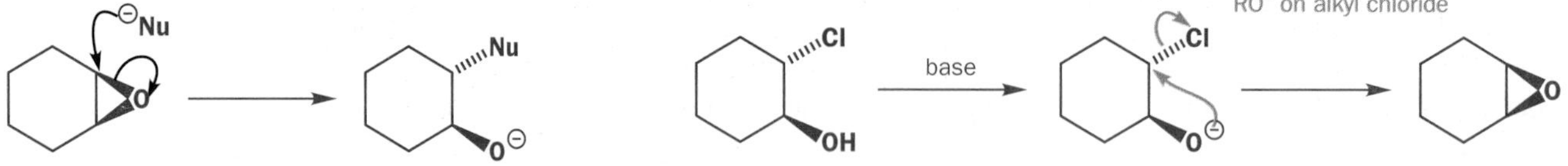

base

■

In Chapter 37 you will meet the alternative **rearrangement reactions** that occur if you try and force *cis*-substituted compounds like these to react.

As for intermolecular substitutions, the incoming nucleophile must still attack into the σ* orbital of the leaving group. In the formation of an epoxide, such an attack can take place only if both groups are axially substituted. As a consequence, only a *trans* 2-chloro cyclohexanol can form an epoxide, and then only when in the less energetically favourable conformation with both groups axial. Of course, as the diaxial conformer reacts, rapid ring inversion of the major equatorial isomer ensures that it is replaced.

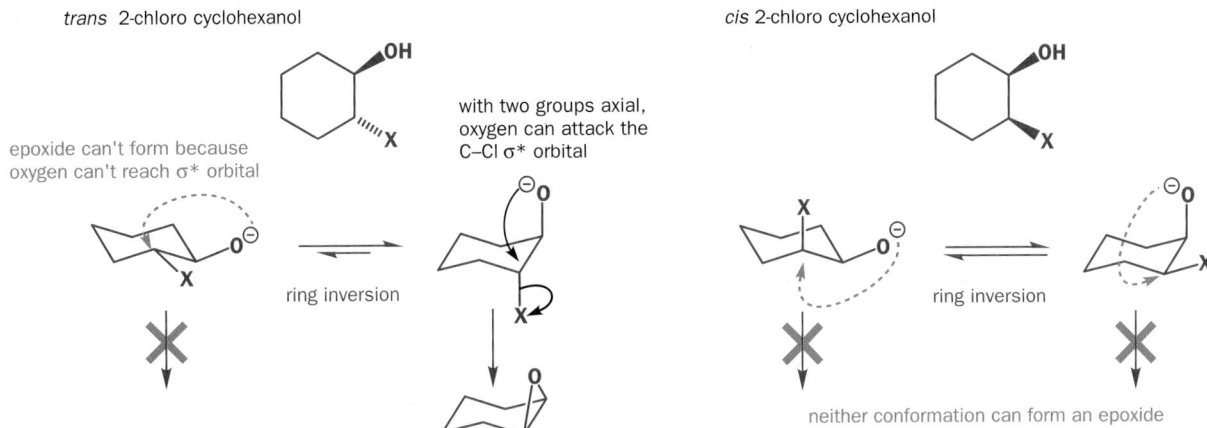

trans 2-chloro cyclohexanol

cis 2-chloro cyclohexanol

epoxide can't form because oxygen can't reach σ* orbital

with two groups axial, oxygen can attack the C–Cl σ* orbital

ring inversion

ring inversion

neither conformation can form an epoxide

It is impossible for the CO bonds of the product epoxide ring to adopt perfectly axial and equatorial positions. If you make a model of cyclohexene oxide you will see that the ring is a slightly deformed chair—it is more of a half-chair conformation in which four of the carbon atoms are in the same plane (you met this on p. 461).

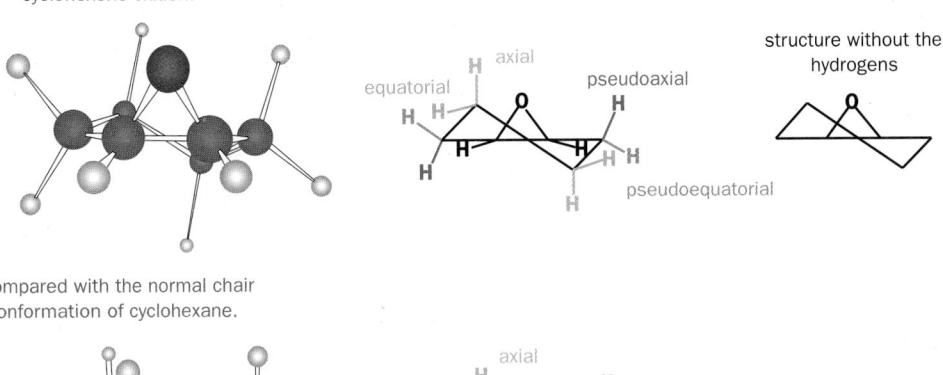

the half-chair conformation of cyclohexene oxide...

axial

equatorial

pseudoaxial

structure without the hydrogens

pseudoequatorial

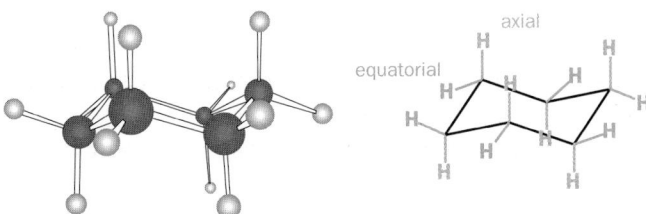

...compared with the normal chair conformation of cyclohexane.

axial

equatorial

The usual way of drawing cyclohexene oxide is shown: notice that the distortion due to the three-membered ring changes the orientation of the axial and equatorial hydrogens next to the ring—they are **pseudoaxial** and **pseudoequatorial**. The hydrogens on the back of the ring (this part of the ring remains about the same as in the chair conformation) can be still considered as 'normal' axial and equatorial hydrogens.

We said that the epoxide-forming reaction is essentially the reverse of the epoxide-opening reaction. If we took a snapshot of the transition state for either reaction, we would not be able to tell whether it was the RO⁻ that was attacking the C–X σ* to form the epoxide or the X⁻ attacking the C–O σ* of the epoxide to form a ring-opened alcohol. In other words, the transition state is the same for both reactions.

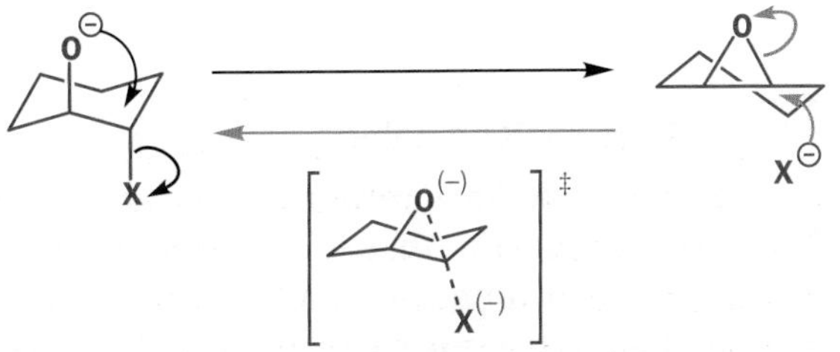

this transition state is the same for both formation and ring opening of the epoxide

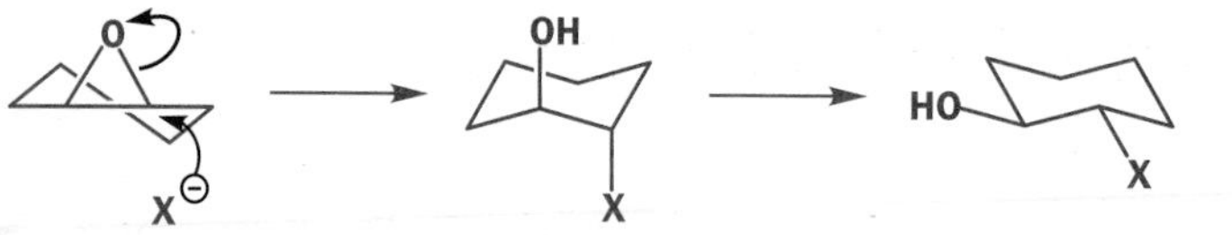
OH
HO
X
X
X

initial product of ring opening is diaxial

diaxial can flip to diequatorial

Since ring closure is only possible when the starting material is diaxially substituted, this has to mean that ring opening is similarly only possible if the *product* is diaxial. This is a general principle: *ring opening of cyclohexene oxides always leads directly to diaxial products.* The diaxially substituted product may then subsequently flip to the diequatorial one.

How do we know this to be true? If the ring bears a *t*-butyl substituent, ring flipping is impossible, and the diaxial product has to stay diaxial. An example is nucleophilic attack of halide on the two epoxides shown below.

▶ Ph_3P/X_2 is a way of making reactive, unsolvated X^- in nonpolar solvent, favouring S_N2.

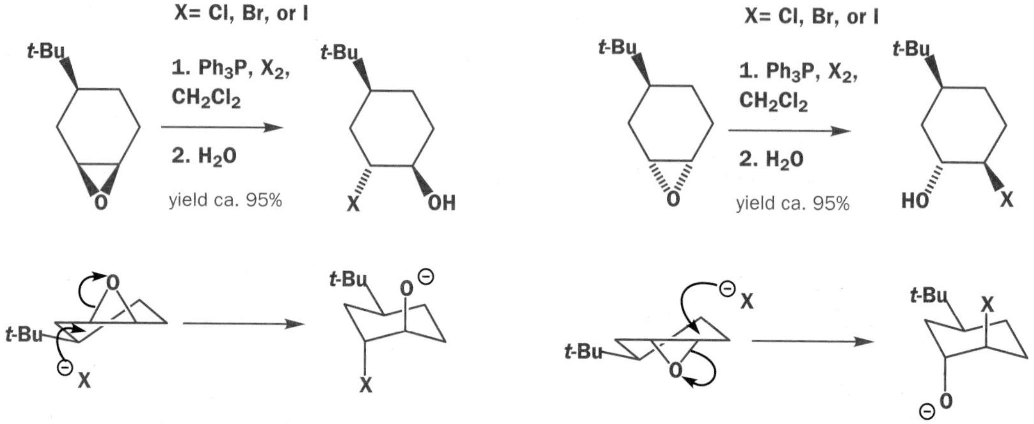
X= Cl, Br, or I
t-Bu
1. Ph3P, X2, CH2Cl2
2. H2O
yield ca. 95%
t-Bu
X
OH
O
X= Cl, Br, or I
t-Bu
1. Ph3P, X2, CH2Cl2
2. H2O
yield ca. 95%
t-Bu
HO
X
O
t-Bu
X
t-Bu
O
X
X
t-Bu
t-Bu
X
O

Points to note:

- The *t*-butyl group locks the conformation of the epoxide. Whereas cyclohexene oxide can flip (see above), enabling the nucleophile to attack either of the epoxide carbon atoms, here the ring is conformationally rigid
- The nucleophile must attack from the opposite side of the epoxide into the C–O σ*. This means that the nucleophile and hydroxyl group end up *trans* in the product
- In each case the epoxide opens only at the end that gives the diaxially substituted chair. Ring opening at the other end would still give a diaxially substituted product, but it is a diaxially substituted high-energy twist-boat conformation. The twist boat can, in fact, flip to give an all-equatorial product, but this is a kinetically controlled process, and it is the barrier to reaction that matters, not the stability of the final product

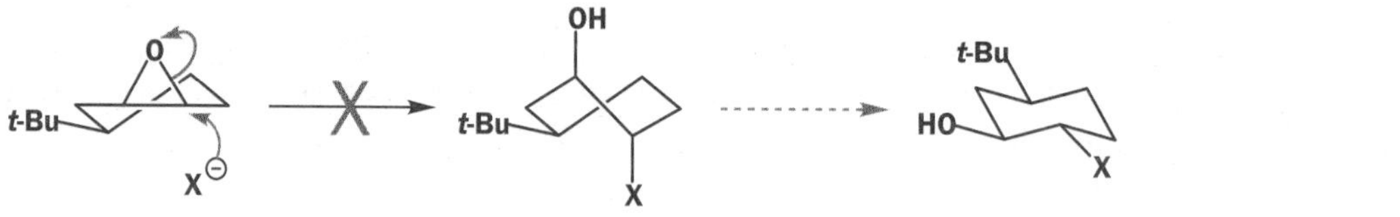
O
t-Bu
X
OH
t-Bu
X
t-Bu
HO
X

ring opening the wrong way would give a twist boat, which is too high in energy to form

even though it could ring flip to the stable all-equatorial chair if it got the chance

● **Axial attack on half-chairs**

Epoxide openings are not alone in always giving diaxial products. We can give the general guideline that, for any reaction on a six-membered ring that is not already in the chair conformation, axial attack is preferred. You will see in later chapters that this is true for cyclohexenes, which also have the half-chair conformation described in the next section. Cyclohexanones, on the other hand, already have a chair conformation, and so can be attacked axially or equatorially.

Rings containing sp^2 hybridized carbon atoms: cyclohexanone and cyclohexene

Every ring you've seen in this chapter has been fully saturated. You've seen the distortion to a half-chair resulting from fusion of a six-membered ring with an epoxide—what happens if some of the tetrahedral carbons are replaced with trigonal (sp^2) hybridized ones? Well, for one sp^2 carbon atom the simple answer is nothing—the conformation is not significantly altered by the presence of just one sp^2 centre in a ring. The conformations of methylenecyclohexane and cyclohexanone—along with a model of cyclohexanone—are shown below.

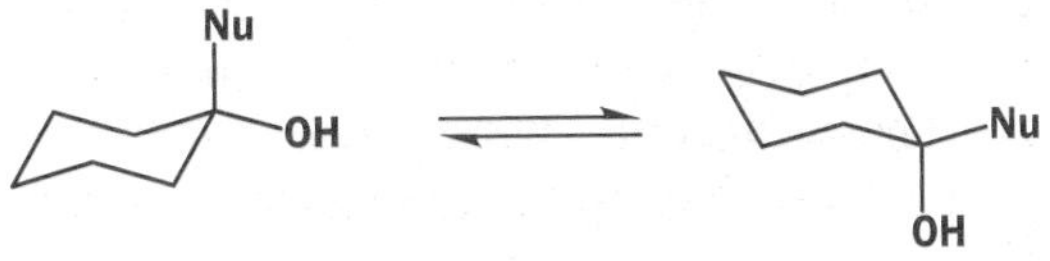

▶ **Drawing cyclohexanones**
Make sure you point the ketone in the right direction! It should bisect the angle there would be between the axial and equatorial substituents, if the carbon atom were tetrahedral. It's always best to put the carbonyl group at one of the 'end' carbons of the ring: it's much harder to get it right if you join it to one of the middle ones.

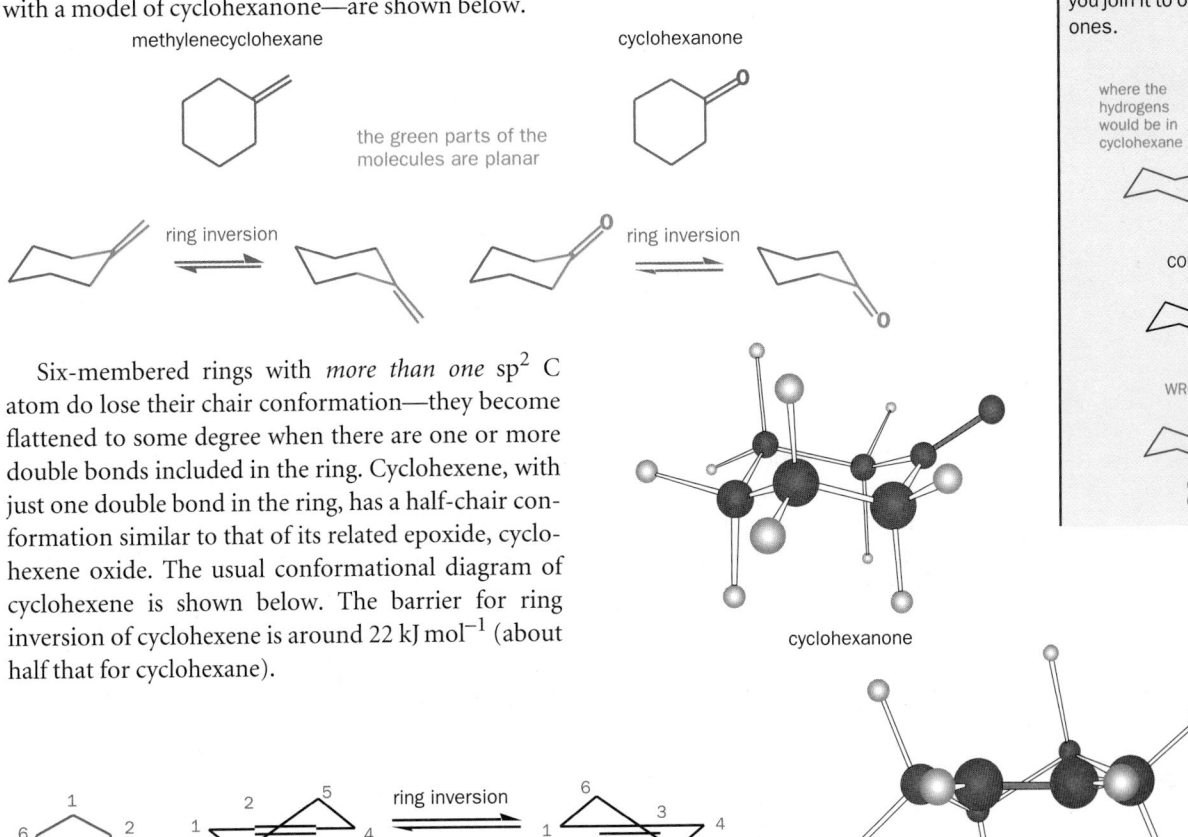

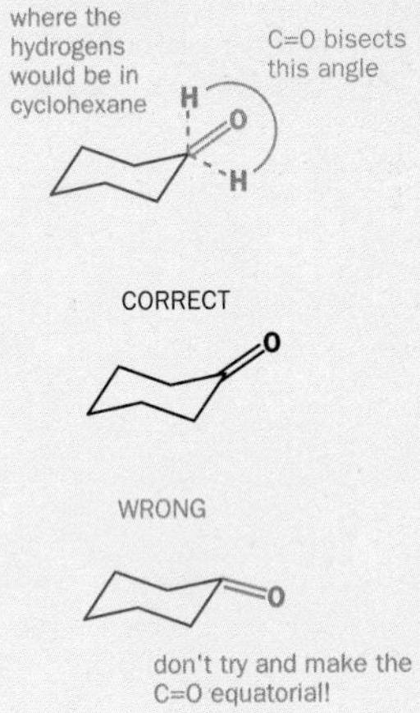

Six-membered rings with *more than one* sp^2 C atom do lose their chair conformation—they become flattened to some degree when there are one or more double bonds included in the ring. Cyclohexene, with just one double bond in the ring, has a half-chair conformation similar to that of its related epoxide, cyclohexene oxide. The usual conformational diagram of cyclohexene is shown below. The barrier for ring inversion of cyclohexene is around 22 kJ mol^{-1} (about half that for cyclohexane).

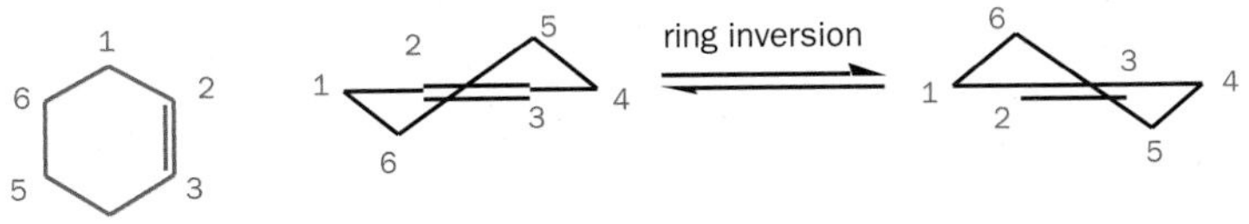

carbon atoms 1, 2, 3, and 4 are all in the same plane

We will look more closely at the reactions of cyclohexene along with other alkenes in later chapters. For now, we return to the chemistry of cyclohexanones. Before you had read this chapter you might simply have drawn the mechanism for nucleophilic attack on cyclohexanone as shown.

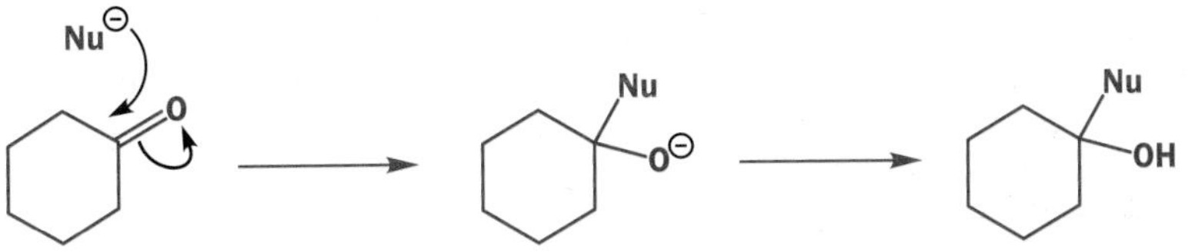

The product contains the two functional groups Nu and OH, which you now know can be arranged in two conformations: one in which the alcohol is axial and one in which it is equatorial. But we can't predict which conformation is more favourable without knowing what the group Nu is: if Nu is smaller than

OH (H, say) then the conformation with the hydroxyl group equatorial will be lower in energy; if Nu is large then the most stable conformation will have the alcohol group axial and Nu equatorial.

Now think of a nucleophile attacking 4-*t*-butylcyclohexanone. Since the *t*-butyl group locks the ring, whether Nu is axial or equatorial will depend only on which face of the C=O group it attacked. Attack on the same face as the *t*-butyl group leaves the nucleophile axial and the hydroxyl group equatorial; attack on the opposite face leaves the nucleophile equatorial and the hydroxyl group axial. The nucleophile is said to attack either in an axial or equatorial manner, depending on where it ends up. It's easier to see this in a diagram.

■
Remember the guideline in the summary box on p. 470: unlike cyclohexene oxides, cyclohexanones are already chairs, so they can be attacked from the axial *or* the equatorial direction.

axial attack of the nucleophile

equatorial attack of the nucleophile

Now for the observation—we'll try and explain it later. In general, large nucleophiles attack equatorially and small nucleophiles attack axially. For example, reduction of 4-*t*-butylcyclohexanone with lithium aluminium hydride in Et_2O gives 90% of the *trans* alcohol: 90% of the hydride has added axially. AlH_4^- is quite small as nucleophiles go: to make more of the *cis* alcohol we need a larger nucleophile—lithium tri-*sec*-butylborohydride, for example, sold under the name of L-selectride®. This is so large that it only attacks equatorially, yielding typically 95% of the *cis* alcohol.

large nucleophile: 96% equatorial attack   small nucleophile: 90% axial attack

Carbon-centred nucleophiles follow the same trend—the table shows that, as size increases from the slender ethynyl anion through primary and secondary organometallics to *t*-BuMgBr, the axial selectivity drops off correspondingly.

Now the difficult part—why? This is a question that is very difficult to answer because the answer really is not known for certain. It's certainly true that the direction of approach for axial attack is more hindered than for equatorial attack, and this is certainly the reason large nuclophiles prefer to attack equatorially.

■
The diagrams on p. 468 make this clear.

▶
Ph is flat and can slip through.

	% of product resulting from	
Nucleophile	**Axial attack**	**Equatorial attack**
HC≡CNa	88	12
MeLi	35	65
PhLi	42	58
MeMgBr	41	59
EtMgBr	29	71
i-PrMgBr	18	82
t-BuMgBr	0	100

But if this is the case, why do small ones actually *prefer* to attack axially? There must be another factor that favours axial attack for those nucleophiles small enough to avoid the bad interactions with the other axial hydrogens. At the transition state, the forming $-O^-$ oxygen substituent is moving in either an axial or an equatorial direction. Just as the axial substituent is less favourable than an equatorial one, so is the transition state leading there, and the route leading to the equatorial hydroxyl group is favoured.

Multiple rings

Cyclohexane sometimes adopts a twist-boat conformation, but never a true boat structure, which represents an energy maximum. But boat structures are important in some bicyclic compounds where the compound simply doesn't have any choice in the conformation it adopts. The simplest compound locked into a boat structure is norbornane. The CH_2 bridge *has* to be diaxial (otherwise it can't reach), which means that the cyclohexane ring part of the structure has no choice but to be a boat.

norbornane

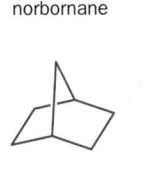

without the hydrogens

with hydrogens included

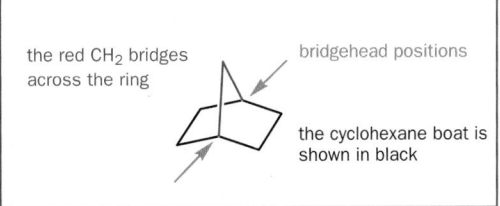

the red CH_2 bridges across the ring

bridgehead positions

the cyclohexane boat is shown in black

Nor-

The *nor-* prefix has a number of meanings in 'trivial' organic nomenclature. Here it tells us that this structure is like that of the parent compound but less one or more alkyl groups—that is, *no R* groups. This isn't the derivation of the word though—historically it comes from the German

Nitrogen ohne Radikal ('nitrogen without R-groups')—it was used first for amines such noradrenaline (also known as norephinephrine) and norephedrine. You met ephedrine in Chapter 16.

camphor
(bornane skeleton in red)

norbornane

adrenaline

noradrenaline

Look closely at the structure of norbornane with its full quota of hydrogen atoms, and you will see that all of the hydrogen atoms on the six-membered ring (except those on the bridgehead carbons) eclipse hydrogens on neighbouring carbon atoms. There is some evidence that the next member in this series of bicyclic alkanes, [2.2.2]-bicyclooctane, flexes slightly to avoid the eclipsing interactions.

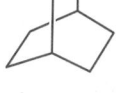

[2.2.2]bicyclooctane

[2.2.2]-bicyclooctane

It is worth briefly explaining this systematic name. *Octane* is obvious—it's C_8. And *bicyclo* is the minimum two rings required to define the structure. [2.2.2] means that each linking chain from one bridgehead to the other is two carbon atoms long. This system of nomenclature allows norbornane to be given the systematic (and less memorable) name [2.2.1]-bicycloheptane. In

Chapter 8 you met the bases DBU (1,8-diazabicyclo[5.4.0]undecene-7) and DBN (1,5-diazabicyclo[3.4.0]nonene-5) named in the same way—and you will meet them again in the very next chapter, as they are particularly good bases for the promotion of elimination reactions.

To conclude...

You may wonder why we have spent most of this chapter looking at six-membered rings, ignoring other ring sizes almost totally. Apart from the fact that six is the most widespread ring size in organic chemistry, the reactions of six-membered rings are also the easiest to explain and to understand. The conformational principles we have outlined for six-membered rings (relief of ring strain, staggered favoured over eclipsed, equatorial favoured over axial, direction of attack) hold, in modified form, for other ring sizes as well. These other rings are less well-behaved than six-membered rings because they lack the well-defined strain-free conformations that cyclohexane is blessed with. We shall now leave sterochemistry in rings for some time, but we come back to these more difficult rings—and how to tame them—in a whole chapter on controlling stereochemistry with cyclic compounds, Chapter 33.

Problems

1. Identify the chair or boat six-membered rings in the following structures and say why that particular shape is adopted.

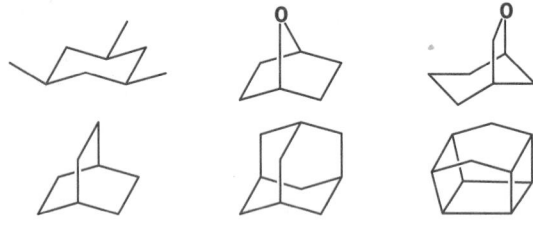

2. Draw clear conformational drawings for these molecules, labelling each substituent as axial or equatorial.

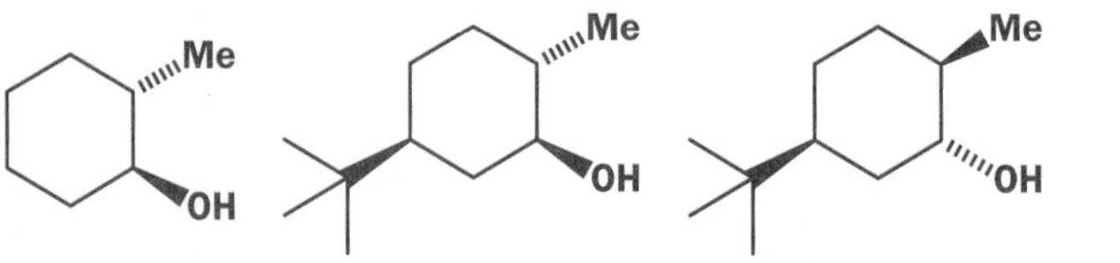

3. Would the substituents in these molecules be axial or equatorial or a mixture of the two?

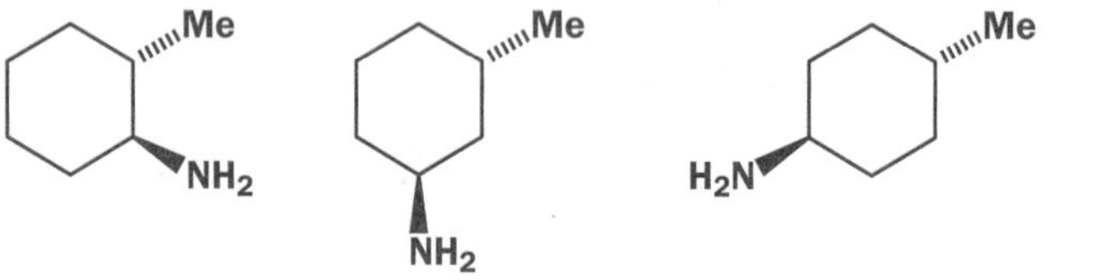

4. Why is it difficult for cyclohexyl bromide to undergo an E2 reaction? When it is treated with base, it does undergo an E2 reaction to give cyclohexene. What conformational changes must occur during this reaction?

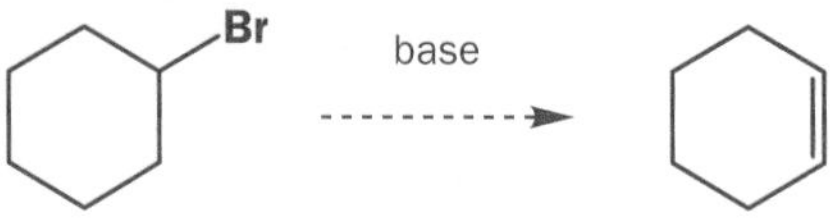

5. Treatment of this diketoalcohol with base causes an elimination reaction. What is the mechanism, and which conformation must the molecule adopt for the elimination to occur?

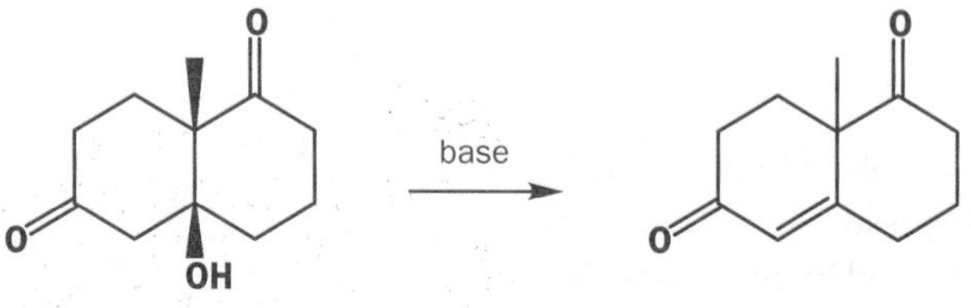

6. Which of these two compounds would form an epoxide on treatment with base?

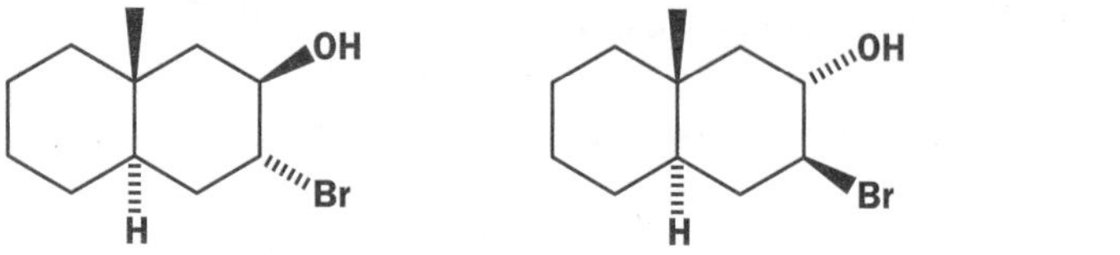

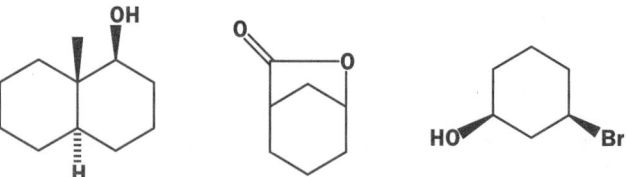

7. Draw conformational diagrams for these compounds. State in each case why the substituents have the positions you state. To what extent could you confirm your predictions experimentally?

8. It is more difficult to form an acetal of compound 8A than of 8B. Why is this?

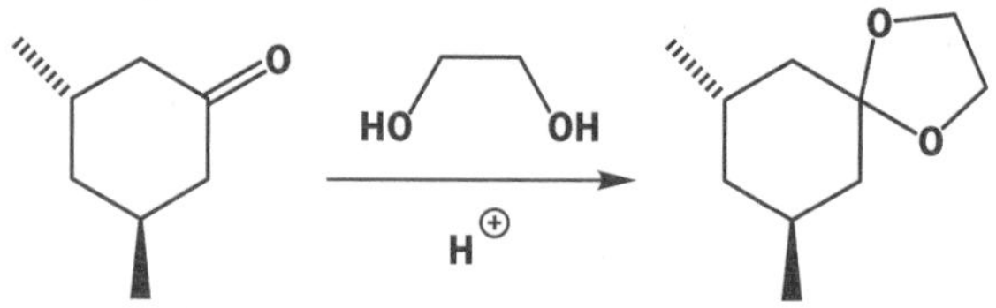

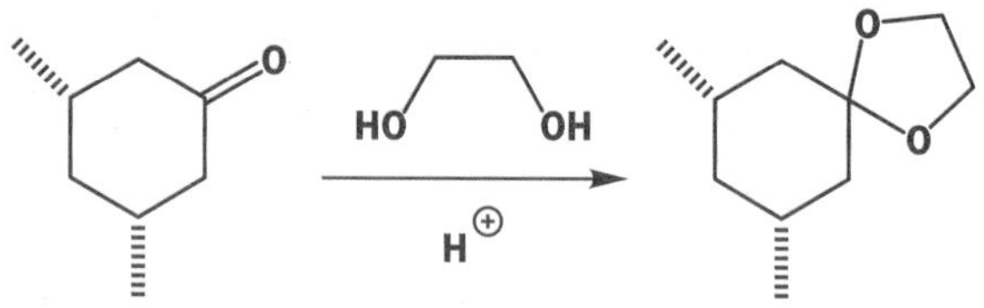

9. Predict which products would be formed on opening these epoxides with nucleophiles, say, cyanide ion.

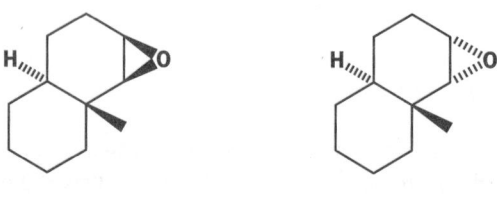

10. These two sugar analogues are part of the structure of two compounds used to treat poultry diseases. Which conformations would they prefer?

11. Hydrolysis of the tricyclic bromide shown here in water gives an alcohol. What is the conformation of the bromide and what will be the stereochemistry of the alcohol?

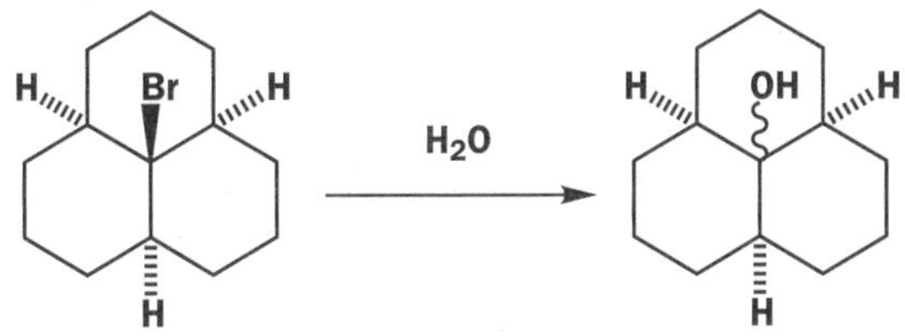

12. Treatment of the triol 12A with benzaldehyde in acid solution produces one diastereoisomer of the acetal 12B and none of the alternative acetal. Why is this acetal preferred? (*Hint.* What controls acetal formation?) What is the stereochemistry of the undefined centre in 12B?

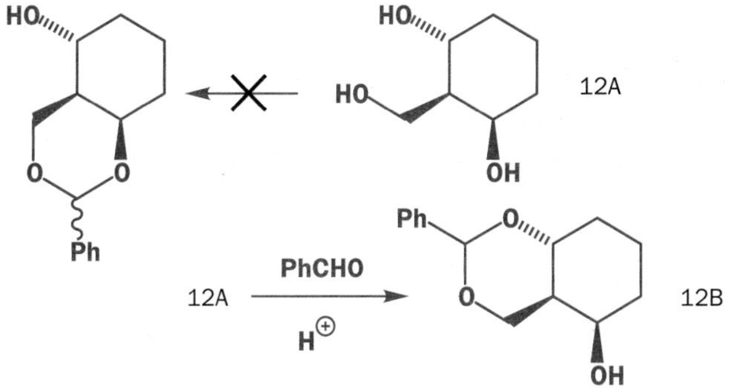

Elimination reactions

19

Connections

Building on:

- **Mechanisms of nucleophilic substitution at saturated carbon ch17**
- **Conformation ch18**

Arriving at:

- **Elimination reactions**
- **What factors favour elimination over substitution**
- **The three important mechanisms of elimination reactions**
- **The importance of conformation in elimination reactions**
- **How to use eliminations to make alkenes (and alkynes)**

Looking forward to:

- **Electrophilic additions to alkenes (the reverse of the reactions in this chapter) ch20**
- **How to control double-bond geometry ch31**

Substitution and elimination

Substitution reactions of *t*-butyl halides, you will recall from Chapter 17, invariably follow the S_N1 mechanism. In other words, the rate-determining step of their substitution reactions is unimolecular—it involves only the alkyl halide. And this means that, no matter what the nucleophile is, the reaction goes at the same rate. You can't speed this S_N1 reaction up, for example, by using hydroxide instead of water, or even by increasing the concentration of hydroxide. 'You'd be wasting your time,' we said (p. 414).

■ Remember the turnstiles at the railway station (p. 413).

nucleophilic substitution reactions of *t*-BuBr

rate = k[*t*-BuBr]

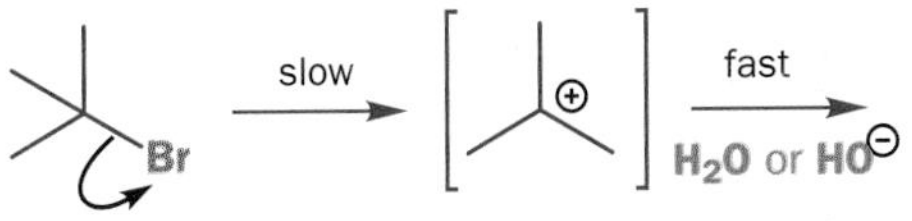

reaction goes at the same rate whatever the nucleophile

t-butyl bromide

t-butanol

You'd also be wasting your alkyl halide. This is what actually happens if you try the substitution reaction with a *concentrated* solution of sodium hydroxide.

reaction of *t*-BuBr with concentrated solution of NaOH

rate = k[*t*-BuBr][HO^-]

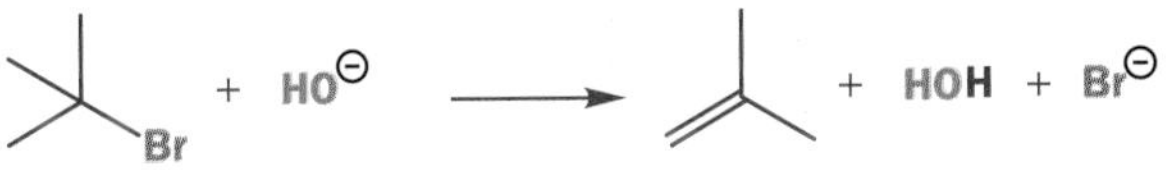

elimination reaction forms alkene

t-butyl bromide

isobutene (2-methylpropene)

The reaction stops being a substitution and an alkene is formed instead. Overall, HBr has been lost from the alkyl halide, and the reaction is called an **elimination**.

In this chapter we will talk about the mechanisms of elimination reactions—as in the case of substitutions, there is more than one mechanism for eliminations. We will compare eliminations with substitutions—either reaction can happen from almost identical starting materials, and you will learn how to predict which is the more likely. Much of the mechanistic discussion relates very closely to Chapter 17, and we suggest that you should make sure you understand all of the points in that chapter before tackling this one. This chapter will also tell you about uses for elimination reactions. Apart from a brief look at the Wittig reaction in Chapter 14, this is the first time you have met a way of making alkenes.

▶

The correlation is best for attack at C=O. In Chapter 17, you met examples of nucleophiles that are good at substitution at saturated carbon (such as I^-, Br^-, PhS^-) but that are not strong bases.

Elimination happens when the nucleophile attacks hydrogen instead of carbon

The elimination reaction of *t*-butyl bromide happens because the nucleophile is *basic.* You will recall from Chapter 12 that there is *some* correlation between basicity and nucleophilicity: strong bases are usually good nucleophiles. But being a good nucleophile doesn't get hydroxide anywhere in the substitution reaction, because it doesn't appear in the first-order rate equation. But being a good base does get it somewhere in the elimination reaction, because hydroxide is involved in the rate-determining step of the elimination, and so it appears in the rate equation. This is the mechanism.

E2 elimination

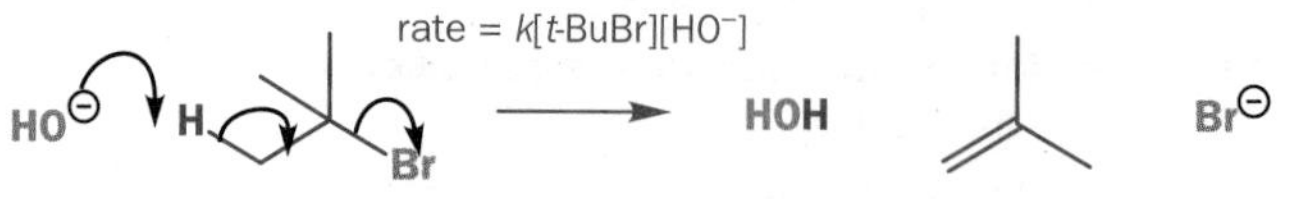

2 molecules involved in rate-determining step

The hydroxide is behaving as a base because it is attacking the hydrogen atom, instead of the carbon atom it would attack in a substitution reaction. The hydrogen atom is not acidic, but proton removal can occur because bromide is a good leaving group. As the hydroxide attacks, the bromide is forced to leave, taking with it the negative charge. Two molecules—*t*-butyl bromide and hydroxide—are involved in the rate-determining step of the reaction. This means that the concentrations of both appear in the rate equation, which is therefore second-order

$$\text{rate} = k_2[t\text{-BuBr}][HO^-]$$

■

Note. No subscripts or superscripts, just plain E2.

and this mechanism for elimination is termed **E2**, for *elimination, bimolecular.*

Now let's look at another sort of elimination. We can approach it again by thinking about an S_N1 substitution reaction. It is another one you met early in Chapter 17, and it is the reverse of the one at the beginning of this chapter.

nucleophilic substitution of *t*-BuOH with HBr

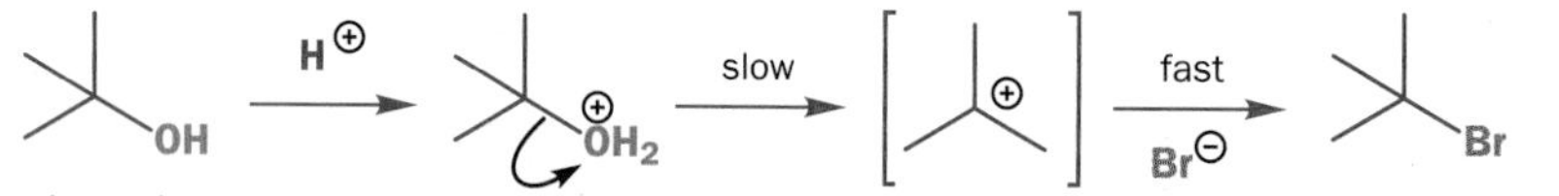

t-butanol *t*-butyl bromide

Bromide, the nucleophile, is not involved in the rate-determining step, so we know that the rate of the reaction will be independent of the concentration of Br^-. But what happens if we use an acid whose counterion is such a weak nucleophile that it doesn't even attack the carbon of the carbocation? Here is an example—*t*-butanol in sulfuric acid doesn't undergo substitution, but undergoes elimination instead.

E1 elimination of *t*-BuOH in H_2SO_4

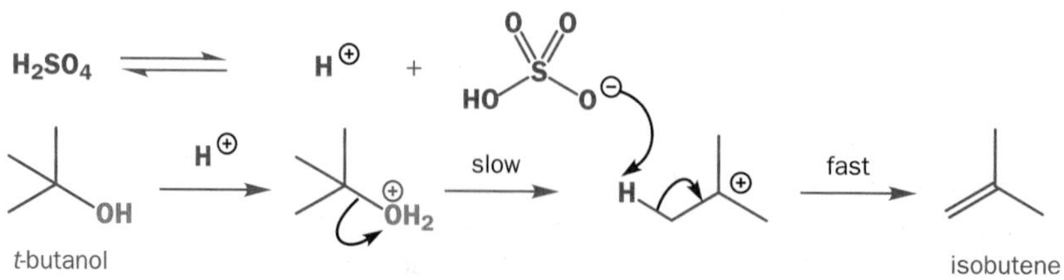

t-butanol isobutene (2-methylpropene)

Now, the HSO_4^- is not involved in the rate-determining step—HSO_4^- is not at all basic and only behaves as a base (that is, it removes a proton) because it is even more feeble as a nucleophile. The rate equation will not involve the concentration of HSO_4^-, and the rate-determining step is the same as that in the S_N1 reaction—unimolecular loss of water from the protonated *t*-BuOH. This elimination mechanism is therefore called **E1**.

We will shortly come back to these two mechanisms for elimination, plus a third, but first we need to answer the question: when does a nucleophile start behaving as a base?

Elimination in carbonyl chemistry

We have left detailed discussion of the formation of alkenes till this chapter, but we used the term elimination in Chapters 12 and 14 to describe the loss of a leaving group from a tetrahedral intermediate. For example, the final steps of the acid-catalysed ester hydrolysis shown below involve E1 elimination of ROH to leave a double bond: C=O rather than C=C.

E1 elimination of ROH during ester hydrolysis

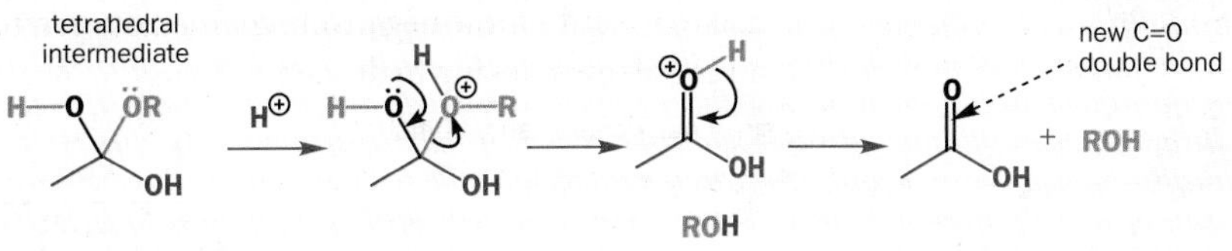

In Chapter 14, you even saw an E1 elimination giving an alkene. That alkene was an enamine—here is the reaction.

E1 elimination of H_2O during enamine formation

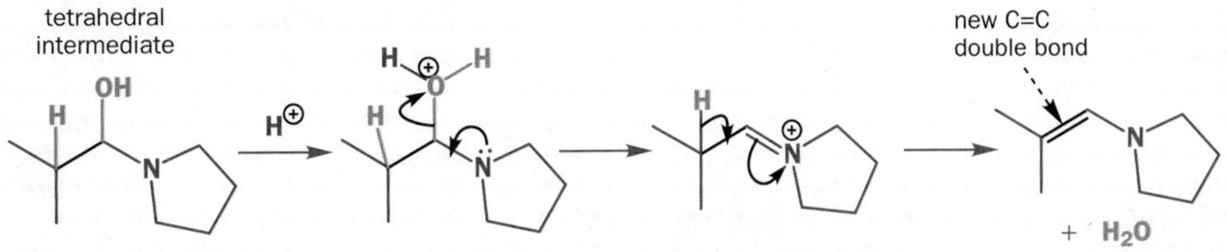

How the nucleophile affects elimination versus substitution

Basicity

You have just seen molecules bearing leaving groups being attacked at two distinct electrophilic sites: the carbon to which the leaving group is attached, and the hydrogen atoms on the carbon adjacent to the leaving group. Attack at carbon leads to substitution; attack at hydrogen leads to elimination. Since strong bases attack protons, it is generally true that, the more basic the nucleophile, the more likely that elimination is going to replace substitution as the main reaction of an alkyl halide.

Here is an example of this idea at work.

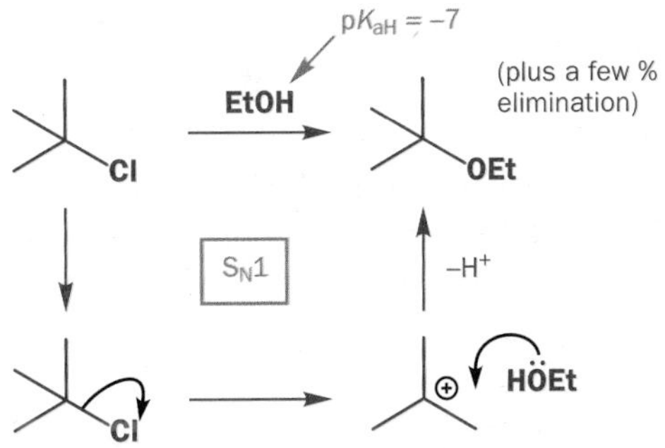

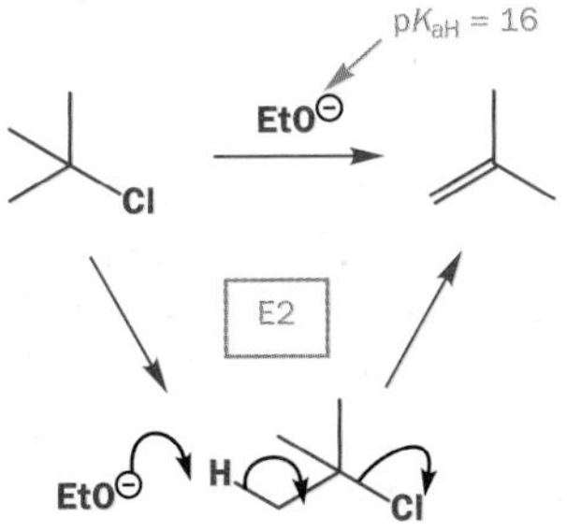

Elimination, substitution, and hardness

We can also rationalize selectivity for elimination versus substitution, or attack of H versus attack on C in terms of hard and soft electrophiles (pp. 237–238). In an S_N2 substitution, the carbon centre is a soft electrophile—it is essentially uncharged, and with leaving groups such as halide the C–X σ* is a relatively low-energy LUMO. Substitution is therefore favoured by nucleophiles whose HOMOs are best able to interact with this LUMO—in other words soft nucleophiles. In contrast, the C–H σ* is higher in energy because the atoms are less electronegative. This, coupled with the hydrogen's small size, makes the C–H bond a hard electrophilic site, and as a result hard nucleophiles favour elimination.

Size

For a nucleophile, attacking a carbon atom means squeezing past its substituents—and even for unhindered primary alkyl halides there is still one alkyl group attached. This is one of the reasons that S_N2 is so slow on hindered alkyl halides—the nucleophile has difficulty getting to the reactive centre. Getting at a more exposed hydrogen atom in an elimination reaction is much easier, and this means that, as soon as we start using hard, basic nucleophiles that are also bulky, elimination becomes preferred over substitution, even for primary alkyl halides. One of the best bases for promoting elimination and avoiding substitution is potassium *t*-butoxide. The large alkyl substituent makes it hard for the negatively charged oxygen to attack carbon in a substitution reaction, but it has no problem attacking hydrogen.

small nucleophile: substitution

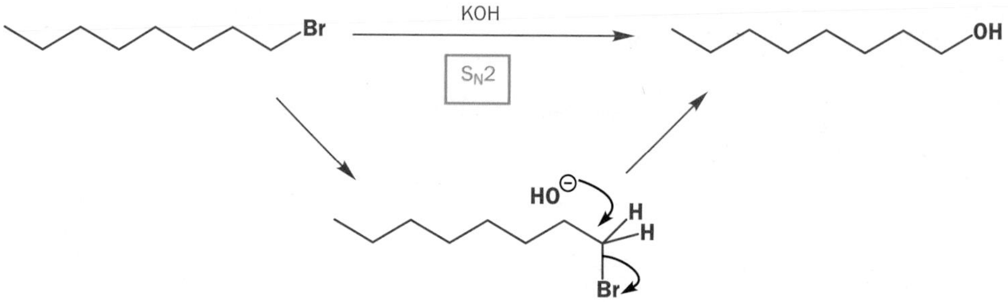

large nucleophile: elimination

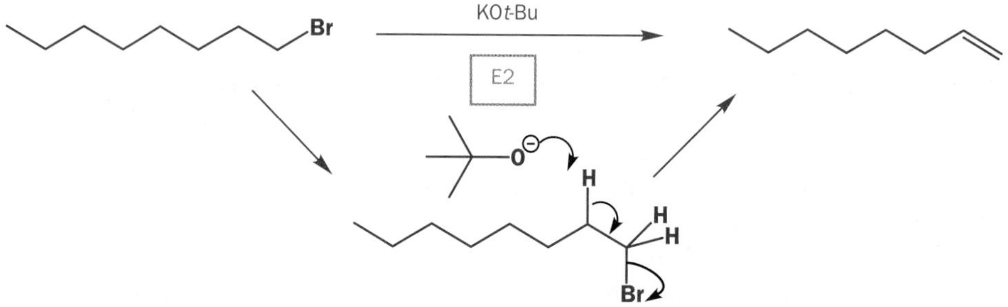

Temperature

Temperature has an important role to play in deciding whether a reaction is an elimination or a substitution. In an elimination, two molecules become three. In a substitution, two molecules form two new molecules. The two reactions differ therefore in the change in entropy during the reaction: ΔS is greater for elimination than for substitution. In Chapter 13, we discussed the equation

$$\Delta G = \Delta H - T\Delta S$$

This equation says that a reaction in which ΔS is positive is more exothermic at higher temperature. Eliminations should therefore be favoured at high temperature, and this is indeed the case: most eliminations you will see are conducted at room temperature or above.

■
This explanation is simplified, because what matters is the rate of the reaction, not the stability of the products. A detailed discussion is beyond the scope of the book, but the general argument still holds.

● **To summarize these three effects:**

- Nucleophiles that are strong bases favour elimination over substitution
- Nucleophiles (or bases) that are bulky favour elimination over substitution
- High temperatures favour elimination over substitution

E1 and E2 mechanisms

Now that you have seen a few examples of elimination reactions, it is time to return to our discussion of the two mechanisms for elimination. To summarize what we have said so far:

- E1 describes an elimination reaction (E) in which the rate-determining step is unimolecular (1) and does not involve the base. The leaving group leaves in this step, and the proton is removed in a separate second step

general mechanism for E1 elimination

rate = k[alkyl halide]

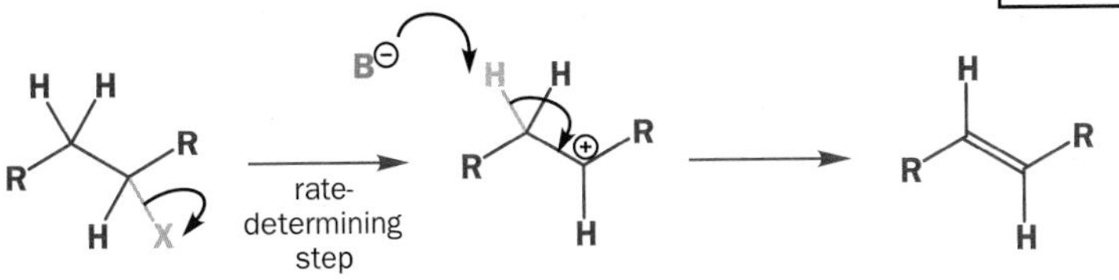

rate-
determining
step

The loss of the leaving group and removal of the proton are **concerted**.

- E2 describes an elimination (E) that has a bimolecular (2) rate-determining step that must involve the base. Loss of the leaving group is simultaneous with removal of the proton by the base

general mechanism for E2 elimination

rate = k[B⁻][alkyl halide]

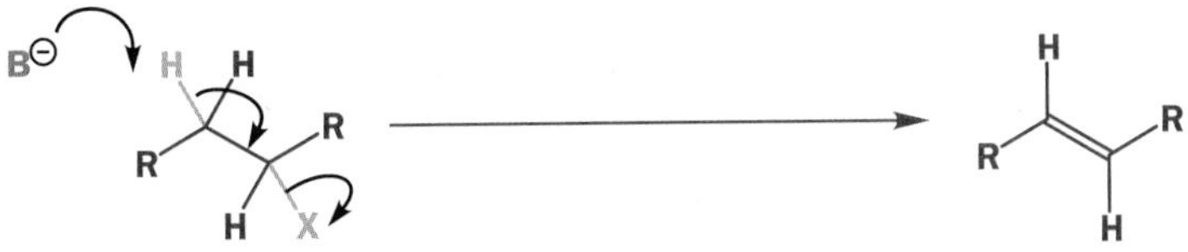

There are a number of factors that affect whether an elimination goes by an E1 or E2 mechanism. One is immediately obvious from the rate equations: only the E2 is affected by the concentration of base, so at high base concentration E2 is favoured. The rate of an E1 reaction is not even affected by what base is present—so E1 is just as likely with weak as with strong bases, while E2 goes faster with strong bases than weak ones: strong bases at whatever concentration will favour E2 over E1. If you see a strong base being used for an elimination, it is certainly an E2 reaction. Take the first elimination in this chapter as an example.

reaction of *t*-butyl bromide with concentrated hydroxide

HOH

Br⁻

With less hindered alkyl halides hydroxide would not be a good choice as a base for an elimination because it is rather small and still very good at S_N2 substitutions (and even with tertiary alkyl halides, substitution outpaces elimination at low concentrations of hydroxide). So what are good alternatives?

We have already mentioned the bulky *t*-butoxide—ideal for promoting E2 as it's both bulky and a strong base (pK_{aH} = 18). Here it is at work converting a dibromide to a diene with two successive E2 eliminations. Since dibromides can be made from alkenes (you will see how in the next chapter), this is a useful two-step conversion of an alkene to a diene.

synthesis of a diene by a double E2 elimination

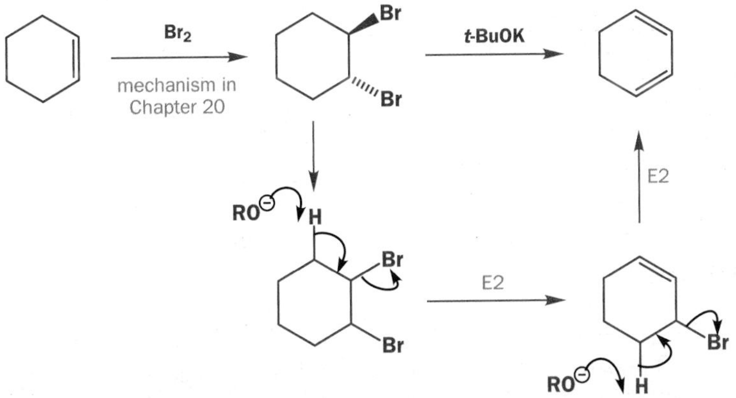

The product of the next reaction is a 'ketene acetal'—you met ketene, $CH_2=C=O$, in Chapter 15. Unlike most acetals, this one can't be formed directly from ketene (ketene is too unstable), so instead, the acetal is made by the usual method from bromoacetaldehyde, and then HBr is eliminated using *t*-BuOK.

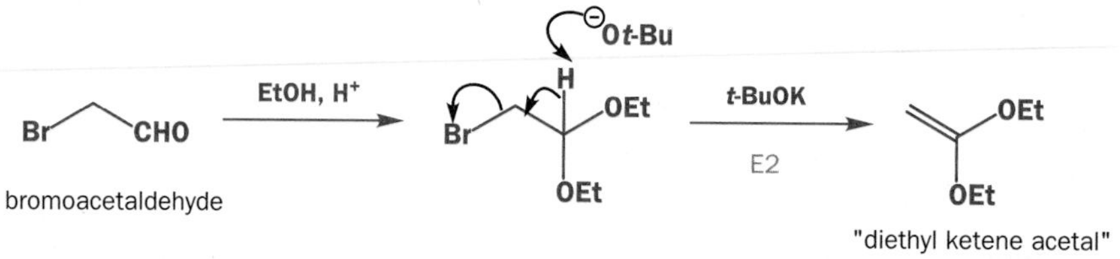

Among the most commonly used bases for converting alkyl halides to alkenes are two that you met in Chapter 8 and that received a mention at the end of Chapter 18: DBU and DBN. These two bases are amidines—delocalization of one nitrogen's lone pair on to the other, and the resulting stabilization of the protonated amidinium ion, makes them particularly basic, with pK_{aH}s of about 12.5. There is not much chance of getting those voluminous fused rings into tight corners—so they pick off the easy-to-reach protons rather than attacking carbon atoms in substitution reactions.

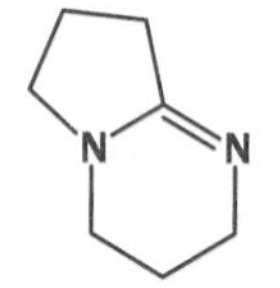

DBN
1,5-diazabicyclo-
[3.4.0]nonene-5

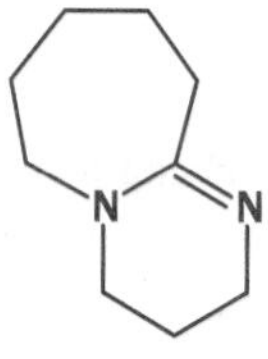

DBU
1,8-diazabicyclo-
[5.4.0]undecene-7

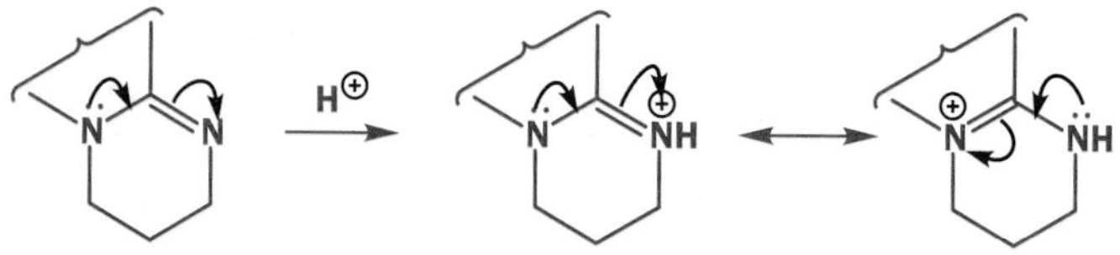

delocalization in the
amidine system

delocalization stabilizes the
protonated amidinium ion

DBU or DBN will generally eliminate HX from alkyl halides to give alkenes. In these two examples, the products were intermediates in the synthesis of natural products.

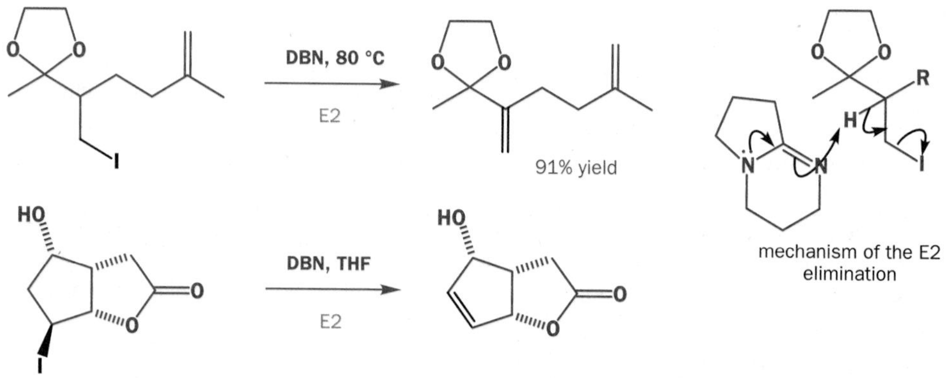

91% yield

mechanism of the E2
elimination

Substrate structure may allow E1

The first elimination of the chapter (*t*-BuBr plus hydroxide) illustrates something very important: the starting material is a tertiary alkyl halide (and would therefore *substitute* only by S$_N$1) it can *eliminate* by either E2 (with strong bases) or E1 (with weak bases). The steric factors that disfavour S$_N$2 at hindered centres don't exist for eliminations. Nonetheless, E1 can occur *only* with substrates that can ionize to give relatively stable carbocations—tertiary, allylic or benzylic alkyl halides, for example. Secondary alkyl halides may eliminate by E1, while primary alkyl halides only ever eliminate by E2 because the primary carbocation required for E1 would be too unstable. The chart on the facing page summarizes the types of substrate that can undergo E1—but remember that any of these substrates, under the appropriate conditions (in the presence of strong bases, for example), may also undergo E2. For completeness, we have also included in this chart three alkyl halides that cannot eliminate by either mechanism simply because they do not have any hydrogens to lose from carbon atoms adjacent to the leaving group.

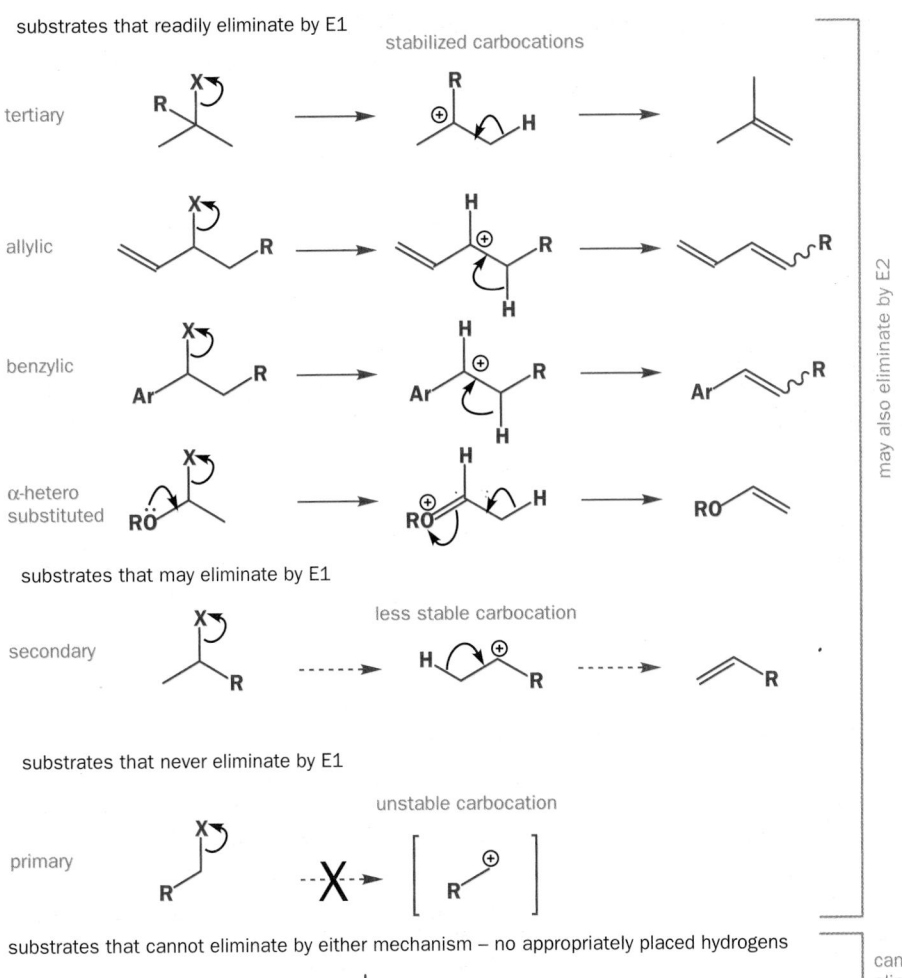

substrates that readily eliminate by E1

stabilized carbocations

tertiary

allylic

benzylic

α-hetero
substituted

substrates that may eliminate by E1

less stable carbocation

secondary

substrates that never eliminate by E1

unstable carbocation

primary

substrates that cannot eliminate by either mechanism – no appropriately placed hydrogens

Me—X   Ph⁀X   ⁀⁀X

may also eliminate by E2

cannot
eliminate
by E2

Polar solvents also favour E1 reactions because they stabilize the intermediate carbocation. E1 eliminations from alcohols in aqueous or alcohol solution are particularly common, and very useful. An acid catalyst is used to promote loss of water, and in dilute H_2SO_4 or HCl the absence of good nucleophiles ensures that substitution does not compete. Under these conditions, the secondary alcohol cyclohexanol gives cyclohexene.

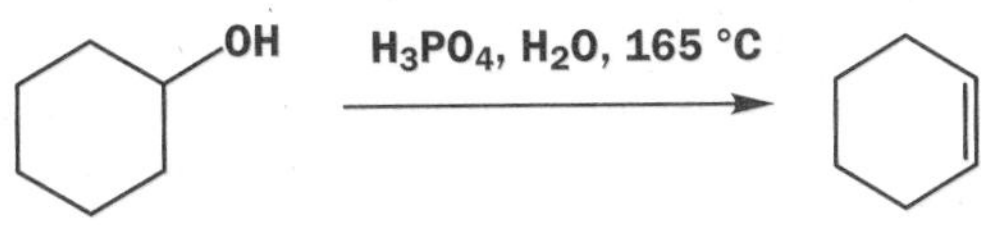

But the best E1 eliminations of all are with tertiary alcohols. The alcohols can be made using the methods of Chapter 9: nucleophilic attack by an organometallic on a carbonyl compound. Nucleophilic addition, followed by E1 elimination, is the best way of making this substituted cyclohexene, for example. Note that the proton required in the first step is recovered in the last—the reaction requires only catalytic amounts of acid.

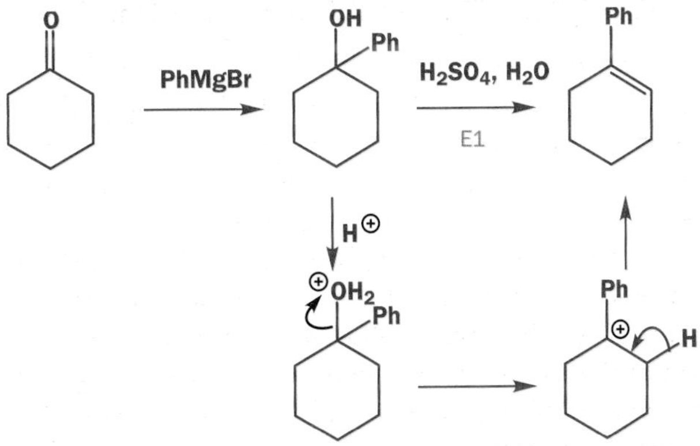

$H_3PO_4, H_2O, 165 °C$

PhMgBr

H_2SO_4, H_2O

E1

$H^⊕$

> In E1 mechanisms, once the leaving group has departed almost anything will serve as a base to remove a proton from the intermediate carbocation. Weakly basic solvent molecules (water or alcohols), for example, are quite sufficient, and you will often see the proton just 'falling off' in reaction mechanisms. We showed the loss of a proton like this in the last example, and in the chart on this page. The superacid solutions we described in Chapter 17 were designed with this in mind—the counterions BF_4^- and SbF_6^- are not only nonnucleophilic but also nonbasic.
>
> solvent:
>
> is equivalent to

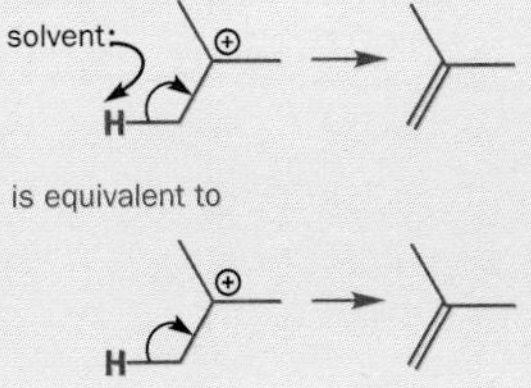

Cedrol is important in the perfumery industry—it has a cedar wood fragrance. Corey's synthesis includes this step—the acid (toluenesulfonic acid) catalyses both the E1 elimination and the hydrolysis of the acetal.

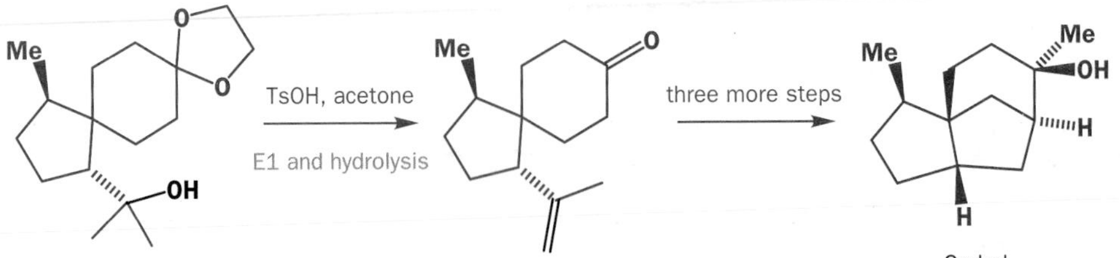

Cedrol

At the end of the last chapter you met some bicyclic structures. These sometimes pose problems for elimination reactions. For example, this compound will not undergo elimination by either an E1 or an E2 mechanism. We shall see shortly what the problem with E2 is, but for E1 the hurdle to be overcome is the formation of a planar carbocation. The bicyclic structure prevents the bridgehead carbon becoming planar so, although the cation would be tertiary, it is very high in energy and does not form. You could say that the nonplanar structure forces the cation to be an empty sp³ orbital instead of an empty p orbital, and we saw in Chapter 4 that it is always best to leave the orbitals with the highest possible energy empty.

'Bridgehead' was defined on p. 473.

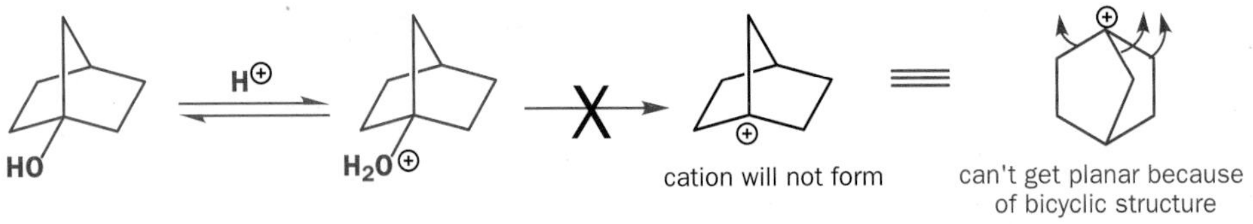

cation will not form — can't get planar because of bicyclic structure

Bredt's rule

The impossibility of planar bridgehead carbons means that double bonds can never be formed to bridgehead carbons in bicyclic systems. This principle is known as 'Bredt's rule', but, as with all rules, it is much more important to know the reason than to know the name, and Bredt's rule is simply a consequence of the strain induced by a planar bridgehead carbon.

The role of the leaving group

We haven't yet been very adventurous with our choice of leaving groups for eliminations: all you have seen so far are E2 from alkyl halides and E1 from protonated alcohols. This is deliberate: the vast majority of the two classes of eliminations use one of these two types of starting materials. Since the leaving group is involved in the rate-determining step of both E1 and E2, in general, any good leaving group will lead to a fast elimination. You may, for example, see amines acting as leaving groups in eliminations of quaternary ammonium salts.

eliminations from quaternary ammonium salts

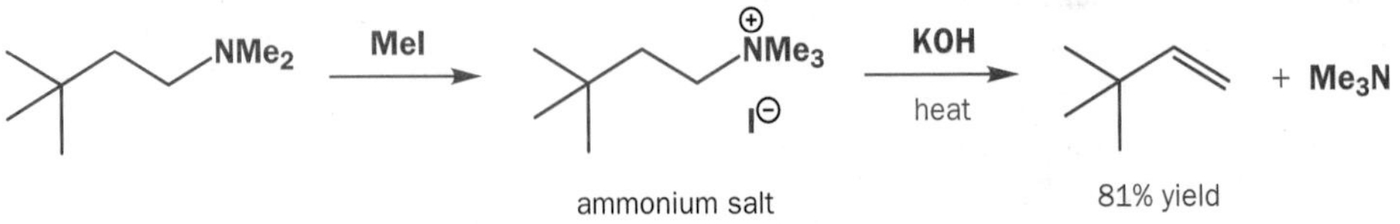

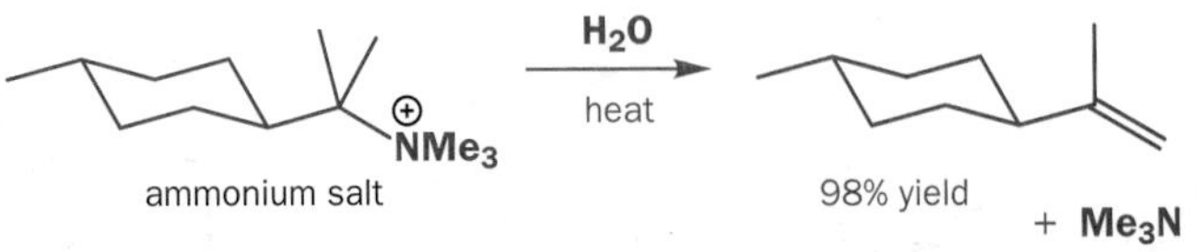

Both E1 and E2 are possible, and from what you have read so far you should be able to spot that there is one of each here: in the first example, a stabilized cation cannot be formed (so E1 is impossible), but a strong base is used, allowing E2. In the second, a stabilized tertiary cation could be formed (so *either* E1 or E2 might occur), but no strong base is present, so the mechanism must be E1.

E2 elimination

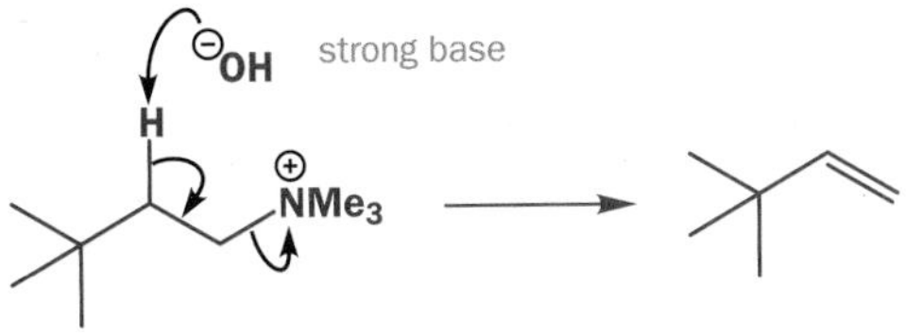

E1 elimination

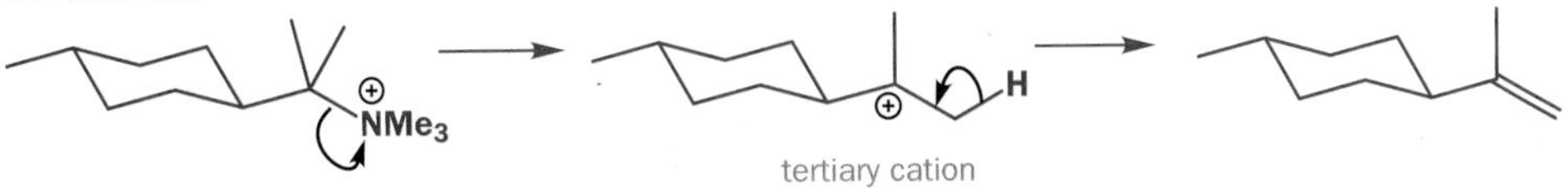

tertiary cation

You have just seen that hydroxyl groups can be turned into good leaving groups in acid, but this is only useful for substrates that can react by E1 elimination. The hydroxyl group is *never* a leaving group in E2 eliminations, since they have to be done in base.

> ● OH⁻ is never a leaving group in an E2 reaction.

For primary and secondary alcohols, the hydroxyl is best made into a leaving group for elimination reactions by sulfonylation with toluene-*para*-sulfonyl chloride (tosyl chloride, TsCl) or methanesulfonyl (mesyl chloride, MeSO$_2$Cl or MsCl).

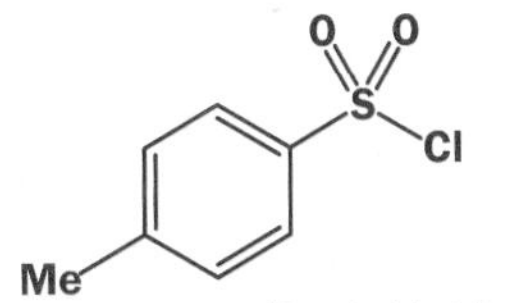

toluene-*para*-sulfonyl chloride
(tosyl chloride, TsCl)

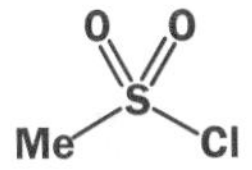

methanesulfonyl chloride
(mesyl chloride, MsCl)

> ►
> There is a new 'organic element' here: –Ms = –SO$_2$Me.

Toluenesulfonate esters (tosylates) can be made from alcohols (with TsCl, pyridine). You have already met tosylates in Chapter 17 because they are good electrophiles for substitution reactions with *nonbasic* nucleophiles. With strong bases such as *t*-BuOK, NaOEt, DBU, or DBN they undergo very efficient elimination reactions. Here are two examples.

E2 eliminations of tosylates

Methanesulfonyl chloride may be a new reagent to you. In the presence of a base (usually triethylamine, Et$_3$N) it reacts with alcohols to give methanesulfonate esters, but the mechanism differs from the mechanism with TsCl. The first step is an elimination of HCl from the sulfonyl chloride (this can't happen with TsCl, because there are no available protons) to give a **sulfene**. The sulfene is highly electrophilic at sulfur, and will react with any alcohol (including tertiary alcohols, which react very slowly with TsCl). Here are the two mechanisms compared.

formation of toluenesulfonates (tosylates): reagents ROH + TsCl + pyridine

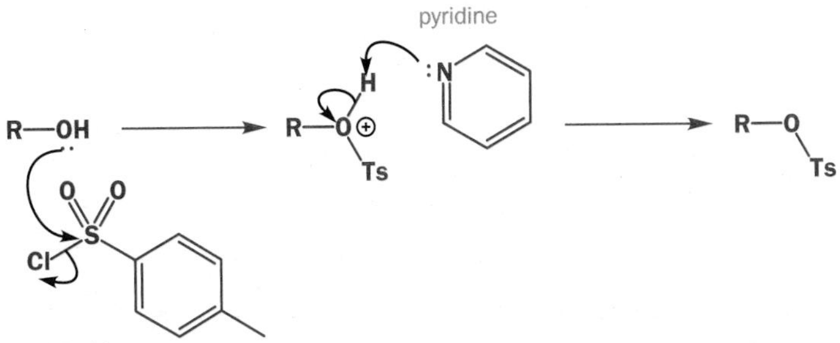

tosyl chloride

formation of methanesulfonates (mesylates): reagents ROH + MsCl + triethylamine

Methanesulfonyl esters (or **mesylates**) can be eliminated using DBU or DBN, but a good way of using MsCl to convert alcohols to alkenes is to do the mesylation and elimination steps in one go, using the same base (Et₃N) for both. Here are two examples making biologically important molecules. In the first, the mesylate is isolated and then eliminated with DBU to give a synthetic analogue of uracil, one of the nucleotide bases present in RNA. In the second, the mesylate is formed and eliminated in the same step using Et₃N, to give a precursor to a sugar analogue.

> ■ More about RNA bases and sugars in Chapter 49.

The second example here involves (overall) the elimination of a tertiary alcohol—so why couldn't an acid-catalysed E1 reaction have been used? The problem here, nicely solved by the use of the mesylate, is that the molecule contains an acid-sensitive acetal functional group. An acid-catalysed reaction would also have risked eliminating methanol from the other tertiary centre.

How to distinguish E1 from E2: kinetic isotope effects

We have told you what sorts of starting materials and conditions favour E1 or E2 reactions, but we haven't told you how we know this. E1 and E2 differ in the order of their rate equations with respect to the base, so one way of finding out if a reaction is E1 or E2 is to plot a graph of the variation of rate with base concentration. But this can be difficult with E1 reactions because the base (which need be only very weak) is usually the solvent. More detailed evidence for the differences between reaction mechanisms comes from studying the rates of elimination in substrates that differ only in that one or more of the protons have been replaced by deuterium atoms. These differences are known as **kinetic isotope effects**.

Up to now you have probably (and rightly) been told that isotopes of an element (that is, atoms that differ only in the number of neutrons their nuclei contain) are chemically identical. It may come as a surprise to find that this is not quite true: isotopes do differ chemically, but

> ■ The calculations that give this result are beyond the scope of this book, but you can find them in textbooks on physical organic chemistry.

this difference is only significant for hydrogen—no other element has one isotope twice as massive as another! Kinetic isotope effects are the changes in rate observed when a (^{1}H) hydrogen atom is replaced by a (^{2}H) deuterium atom in the same reaction. For any reaction, the kinetic isotope effect is defined as

$$KIE = \frac{k_H \leftarrow \text{rate with substrate containing } ^1H}{k_D \leftarrow \text{rate with substrate containing } ^2H}$$

Changing H for D can affect the rate of the reaction only if that H (or D) is involved in the rate-determining step. The theoretical maximum is about 7 for reactions at room temperature in which a bond to H or D is being broken. For example, the rates of these two eliminations can be compared, and k_H/k_D turns out to be 7.1 at 25 °C.

The kinetic isotope effect tells us that the C–H (or C–D) bond is being broken during the rate-determining step, and so the reaction must be an E2 elimination. It's evidence like this that allows us to piece together the mechanisms of organic reactions.

How do kinetic isotope effects come about? Even in its lowest energy state a covalent bond never stops vibrating. If it did it would violate a fundamental physical principle, Heisenberg's uncertainty principle, which states that position and momentum cannot be known exactly at the same time: a nonvibrating pair of atoms have precisely zero momentum and precisely fixed locations. The minimum vibrational energy a bond can have is called the zero point energy (E_0) – given by the expression $E_0 = \frac{1}{2}h\nu$.

In order to break a covalent bond, a certain amount of energy is required to separate the nuclei from their starting position. This energy has to raise the vibration state of the bond from the zero point energy to the point where it breaks. Because the zero point energy of a C–H bond is higher than that for a C–D bond, the C–H bond has a head start in energy terms. The energy required to break a C–H bond is less than that required to break a C–D bond, so reactions breaking C–H bonds go faster than those breaking C–D bonds, provided bond breaking is occurring in the rate-determining step. This is only the case in E2 reactions, not E1 reactions, so the general rule is that, if changing C–H for C–D changes the rate of the elimination, the reaction must be E2 and not E1.

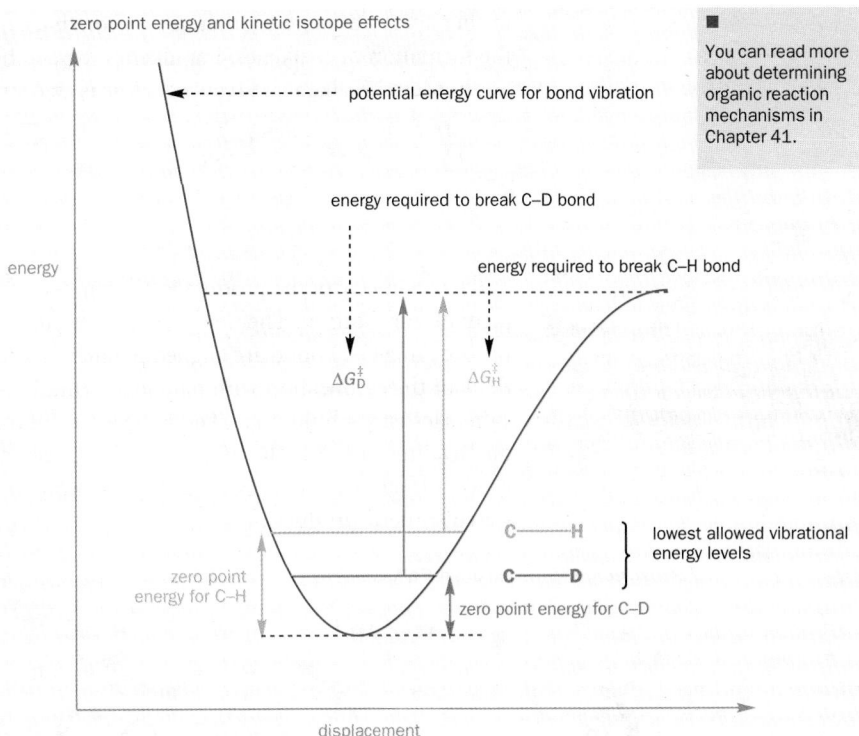

zero point energy and kinetic isotope effects

■ You can read more about determining organic reaction mechanisms in Chapter 41.

E1 reactions can be stereoselective

For some eliminations only one product is possible. For others, there may be a choice of two (or more) alkene products that differ either in the location or stereochemistry of the double bond. We shall now move on to discuss the factors that control the stereochemistry (geometry) and regiochemistry (that is, where the double bond is) of the alkenes, starting with E1 reactions.

only one alkene possible

two regioisomeric alkenes possible

trisubstituted alkene   disubstituted alkene   regioisomers

two stereoisomeric alkenes possible

E-alkene   Z-alkene   stereoisomers (geometrical isomers)

▶ **E and Z alkenes**

The E/Z nomenclature was introduced in Chapter 7, and now that you have read Chapter 16 we can be more precise with our definition. For disubstituted alkenes, E corresponds to *trans* and Z corresponds to *cis*. To assign E or Z to tri- or tetrasubstituted alkenes, the groups at either end of the alkene are given an order of priority according to the same rules as those outlined for R and S in Chapter 16. If the two higher priority groups are *cis*, the alkene is Z; if they are *trans* the alkene is E. Of course, molecules don't know these rules, and sometimes (as in the second example here) the E alkene is less stable than the Z.

For steric reasons, *E*-alkenes (and transition states leading to *E*-alkenes) are usually lower in energy than *Z*-alkenes (and the transition states leading to them) because the substituents can get farther apart from one another. A reaction that can choose which it forms is therefore likely to favour the formation of *E*-alkenes. For alkenes formed by E1 elimination, this is exactly what happens: the less hindered *E*-alkene is favoured. Here is an example.

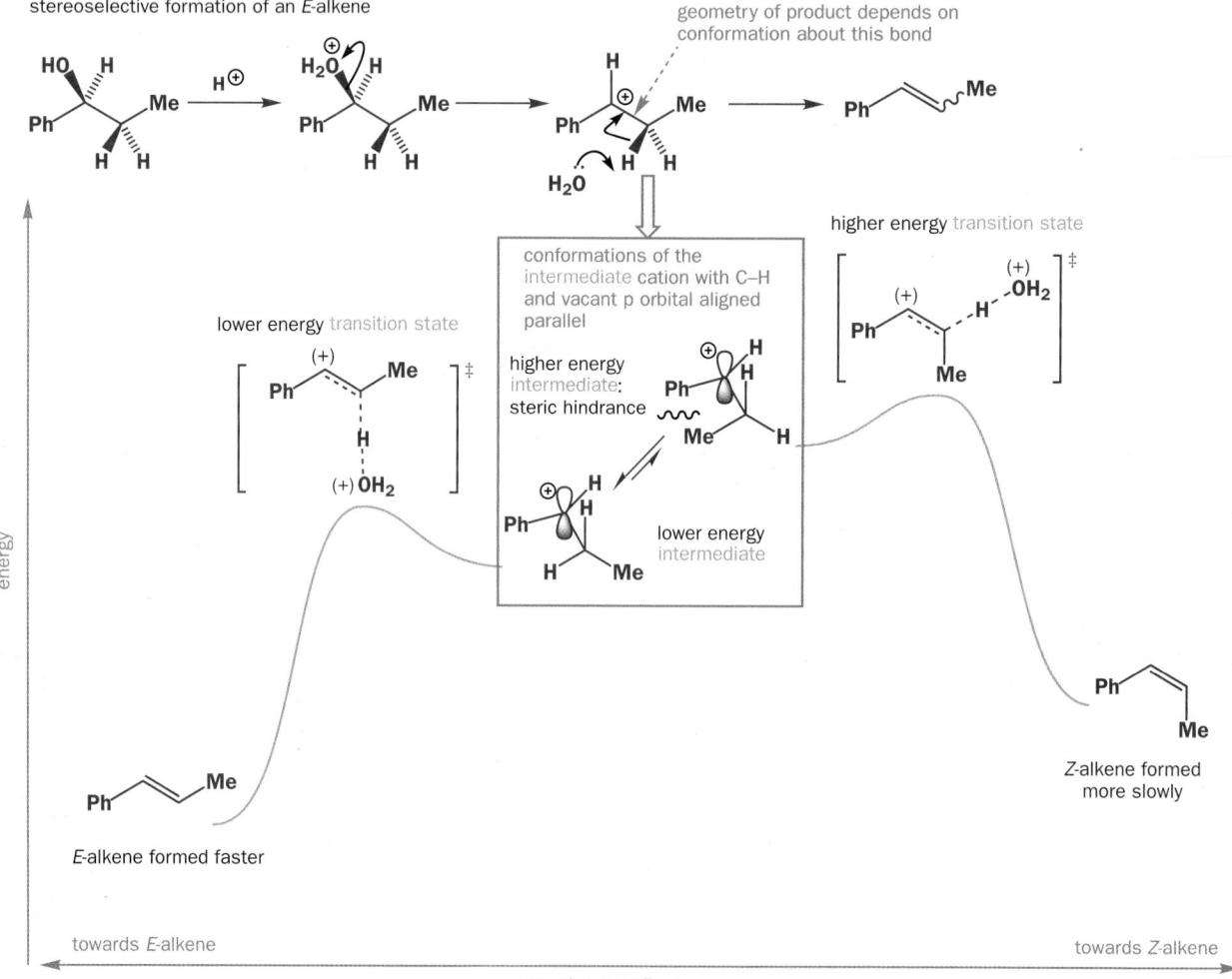

95% *E*-alkene   5% *Z*-alkene

The geometry of the product is determined at the moment that the proton is lost from the intermediate carbocation. The new π bond can only form if the vacant p orbital of the carbocation and the breaking C–H bond are aligned parallel. In the example shown there are two possible conformations of the carbocation with parallel orientations, but one is more stable than the other because it suffers less steric hindrance. The same is true of the transition states on the route to the alkenes—the one leading to the *E*-alkene is lower in energy and more *E*-alkene than *Z*-alkene is formed. The process is steroselective, because the reaction chooses to form predominantly one of two possible stereoisomeric products.

■ In Chapter 41, we shall discuss why the transition states for decomposition of high-energy intermediates like carbocations are very similar in structure to the carbocations themselves.

Tamoxifen is an important drug in the fight against breast cancer, one of the most common forms of cancer. It works by blocking the action of the female sex hormone oestrogen. The tetra-substituted double bond can be introduced by an E1 elimination: there is no ambiguity about where the double bond goes, though the two stereoisomers form in about equal amounts.

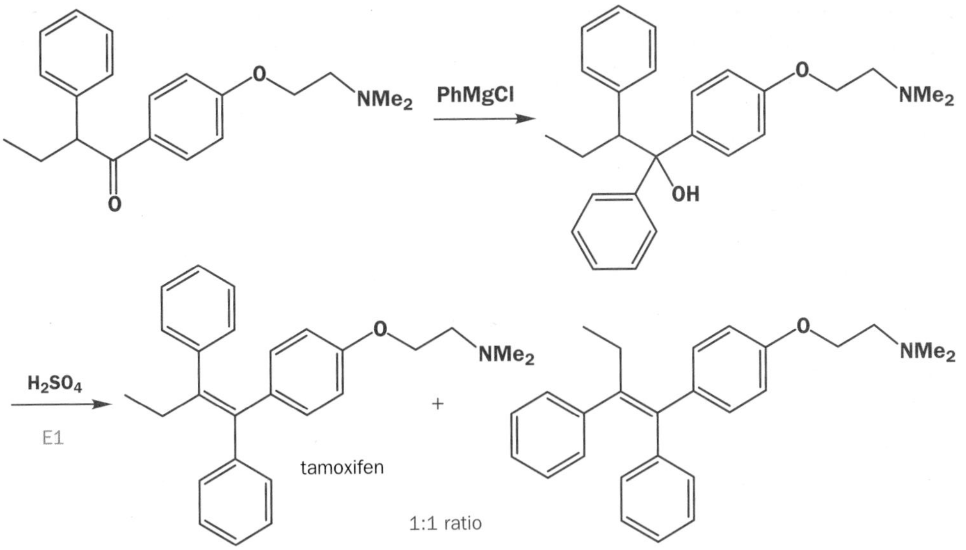

tamoxifen

+

1:1 ratio

E1 reactions can be regioselective

We can use the same ideas when we think about E1 eliminations that can give more than one regio-isomeric alkene. Here is an example.

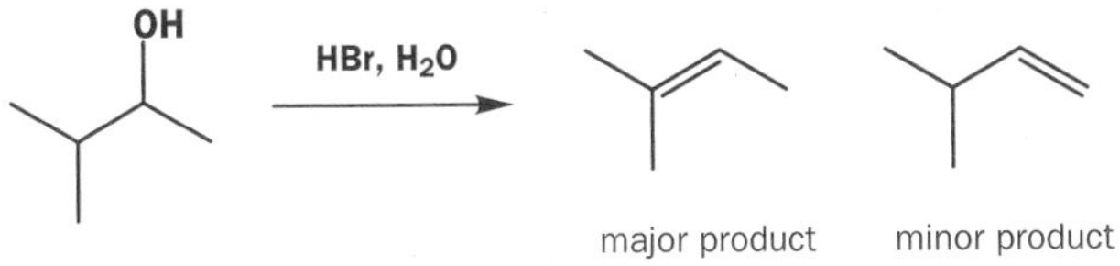

major product   minor product

The major product is the alkene that has the more substituents, because this alkene is the more stable of the two possible products.

> ● **More substituted alkenes are more stable.**

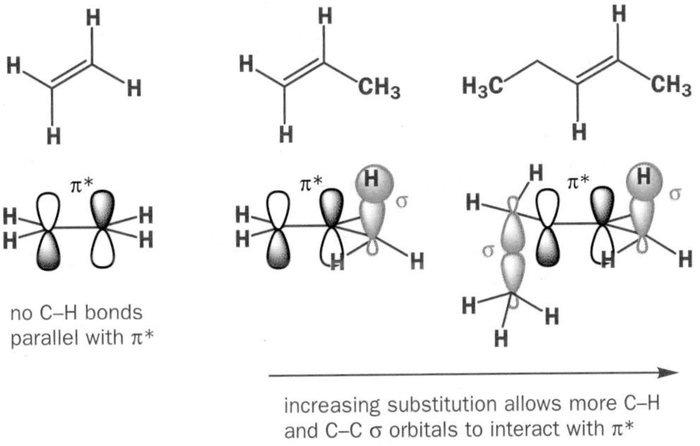

This is quite a general principle. But why should it be true? The reason for this is related to the reason why more substituted carbocations are more stable. In Chapter 17 we said that the carbocation is stabilized when its empty p orbital can interact with the filled orbitals of parallel C–H and C–C bonds. The same is true of the π system of the double bond—it is stabilized when the empty π^* antibonding orbital can interact with the filled orbitals

no C–H bonds parallel with π^*

increasing substitution allows more C–H and C–C σ orbitals to interact with π^*

of parallel C–H and C–C bonds. The more C–C or C–H bonds there are, the more stable the alkene.

The more substituted alkene is more stable, but this does not necessarily explain why it is the one that forms faster. To do that, we should look at the transition states leading to the two alkenes. Both form from the same carbocation, but which one we get depends on which proton is lost. Removal of the proton on the right (brown arrow) leads to a transition state in which there is a monosubstituted double bond partly formed. Removal of the proton on the left (orange arrow) leads to a partial double bond that is trisubstituted. This is more stable—the transition state is lower in energy, and the more substituted alkene forms faster.

regioselective formation of the more substituted alkene

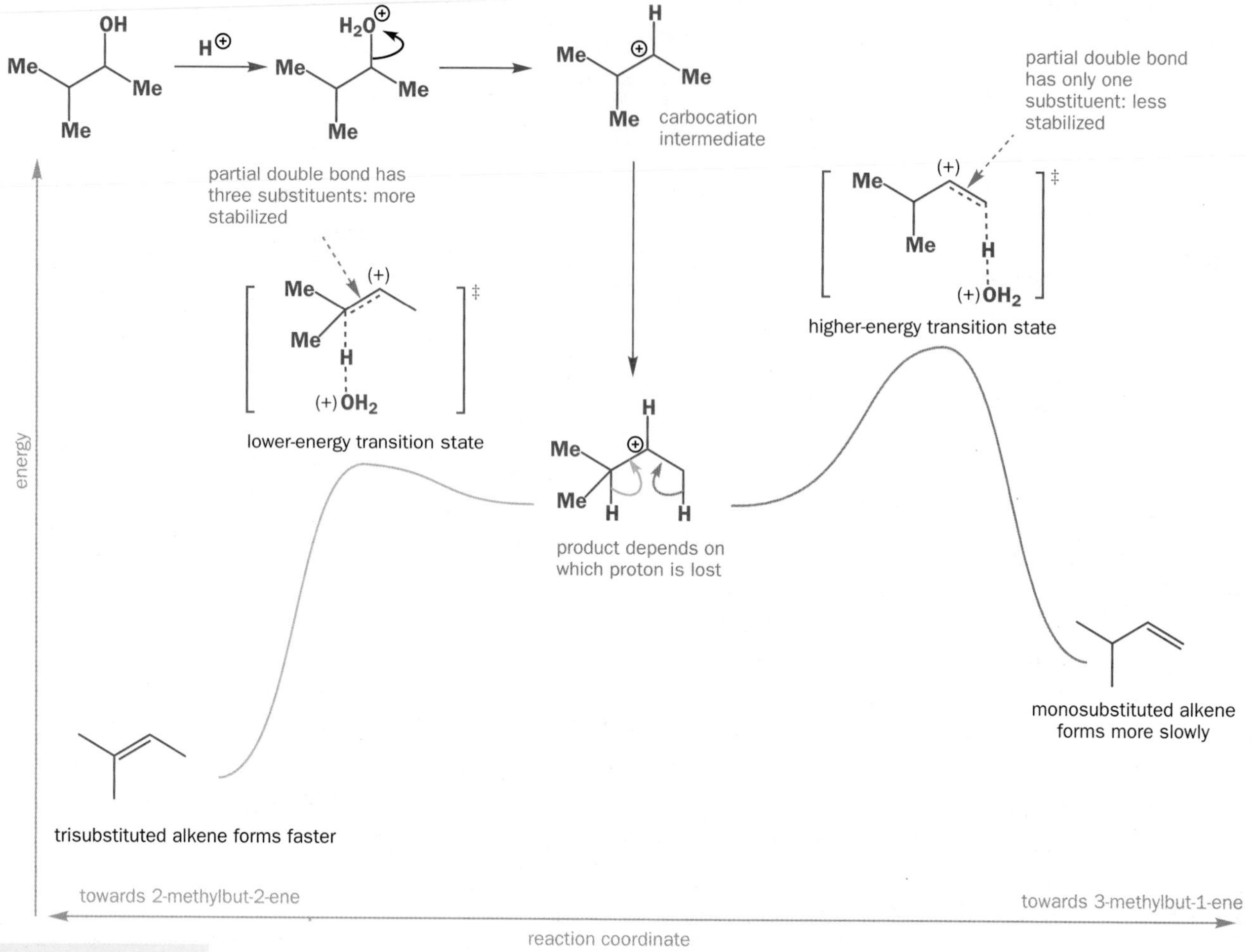

[Energy diagram showing the reaction coordinate with carbocation intermediate. Left branch "towards 2-methylbut-2-ene", right branch "towards 3-methylbut-1-ene". Left transition state labeled "partial double bond has three substituents: more stabilized" and "lower-energy transition state". Right transition state labeled "partial double bond has only one substituent: less stabilized" and "higher-energy transition state". Center labeled "carbocation intermediate" and "product depends on which proton is lost". Bottom left "trisubstituted alkene forms faster", bottom right "monosubstituted alkene forms more slowly". Vertical axis: energy; horizontal axis: reaction coordinate.]

Although E1 reactions show some stereo- and regioselectivity, the level of selectivity in E2 reactions can be much higher because of the more stringent demands on the transition state for E2 elimination. We will come back to the most useful ways of controlling the geometry of double bonds in Chapter 31.

E2 eliminations have anti-periplanar transition states

In an E2 elimination, the new π bond is formed by overlap of the C–H σ bond with the C–X σ* anti-bonding orbital. The two orbitals have to lie in the same plane for best overlap, and now there are two conformations that allow this. One has H and X syn-periplanar, the other anti-periplanar. The anti-periplanar conformation is more stable because it is staggered (the syn-periplanar conformation is eclipsed) but, more importantly, only in the anti-periplanar conformation are the bonds (and therefore the orbitals) truly parallel.

two conformations with H and X coplanar

syn-periplanar (eclipsed)

bonds fully parallel

anti-periplanar (staggered)

E2 eliminations therefore take place from the anti-periplanar conformation. We shall see shortly how we know this to be the case, but first we consider an E2 elimination that gives mainly one of two possible stereoisomers. 2-Bromobutane has two conformations with H and Br anti-periplanar, but the one that is less hindered leads to more of the product, and the *E*-alkene predominates.

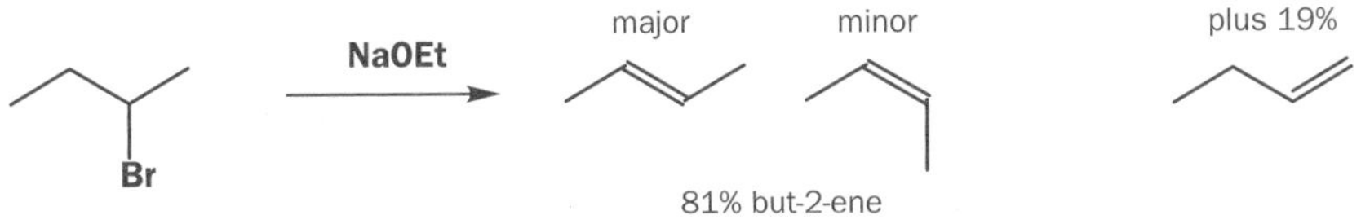

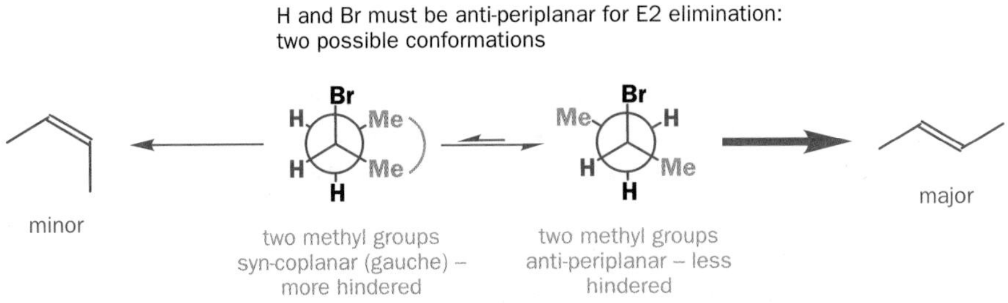

H and Br must be anti-periplanar for E2 elimination:
two possible conformations

There is a choice of protons to be eliminated—the stereochemistry of the product results from which proton is anti-periplanar to the leaving group when the reaction takes place, and the reaction is stereoselective as a result.

E2 eliminations can be stereospecific

In the next example, there is only one proton that can take part in the elimination. Now there is no choice of anti-periplanar transition states. Whether the product is *E* or *Z*, the E2 reaction has only one course to follow. And the outcome depends on which diastereoisomer of the starting material is used. When the first diastereoisomer is drawn with the proton and bromine anti-periplanar, as required, and in the plane of the page, the two phenyl groups have to lie one in front and one behind the plane of the paper. As the hydroxide attacks the C–H bond and eliminates Br⁻, this arrangement is preserved and the two phenyl groups end up *trans* (the alkene is *E*). This is perhaps easier to see in the Newman projection of the same conformation.

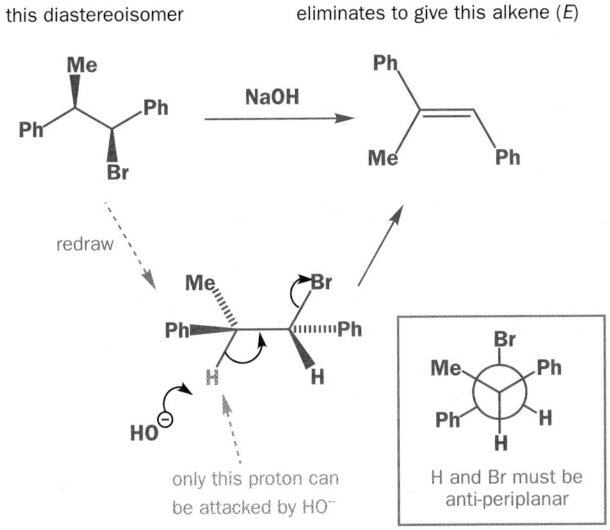

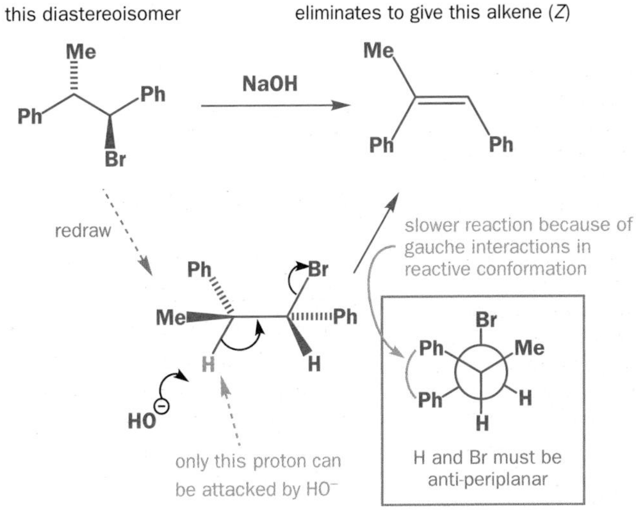

The second diastereoisomer forms the *Z*-alkene for the same reasons: the two phenyl groups are now on the same side of the H–C–C–Br plane in the reactive anti-periplanar conformation (again, this is clear in the Newman projection) and so they end up *cis* in the product. Each diastereoisomer gives a different alkene geometry, and they do so at different rates. The first reaction is about ten times as fast as the second because, although this anti-periplanar conformation is the only reactive one, it is not necessarily the most stable. The Newman projection for the second reaction shows clearly that the two phenyl groups have to lie synclinal (gauche) to one another: the steric interaction between these large groups will mean that, at any time, a relatively small proportion of molecules will adopt the right conformation for elimination, slowing the process down.

Reactions in which the stereochemistry of the product is determined by the stereochemistry of the starting material are called **stereospecific**.

● **Stereoselective or stereospecific?**

- Stereoselective reactions give one predominant product because the reaction pathway has a choice. Either the pathway of lower activation energy is preferred (kinetic control) or the more stable product (thermodynamic control)

- Stereospecific reactions lead to the production of a single isomer as a direct result of the mechanism of the reaction and the stereochemistry of the starting material. There is no choice. The reaction gives a different diastereoisomer of the product from each stereoisomer of the starting material

E2 eliminations from cyclohexanes

The stereospecificity of the reactions you have just met is very good evidence that E2 reactions proceed through an anti-periplanar transition state. We know with which diastereoisomer we started, and we know which alkene we get, so there is no question over the course of the reaction.

More evidence comes from the reactions of substituted cyclohexanes. You saw in Chapter 18 that substituents on cyclohexanes can be parallel with one another only if they are both axial. An equatorial C–X bond is anti-periplanar only to C–C bonds and cannot take part in an elimination. For unsubstituted cyclohexyl halides treated with base, this is not a problem because, although the axial conformer is less stable, there is still a significant amount present (see the table on p. 462), and elimination can take place from this conformer.

equatorial X is anti-periplanar only to C–C bonds and cannot be eliminated by an E2 mechanism

axial X is anti-periplanar to C–H bonds, so E2 elimination is possible

● For E2 elimination in cyclohexanes, both C–H and C–X must be axial.

These two diastereoisomeric cyclohexyl chlorides derived from menthol react very differently under the same conditions with sodium ethoxide as base. Both eliminate HCl but diastereoisomer A reacts rapidly to give a mixture of products, while diastereoisomer B (which differs only in the configuration of the carbon atom bearing chlorine) gives a single alkene product but very much more slowly. We can safely exclude E1 as a mechanism because the same cation would be formed from both diastereoisomers, and this would mean the ratio of products (though not necssarily the rate) would be the same for both.

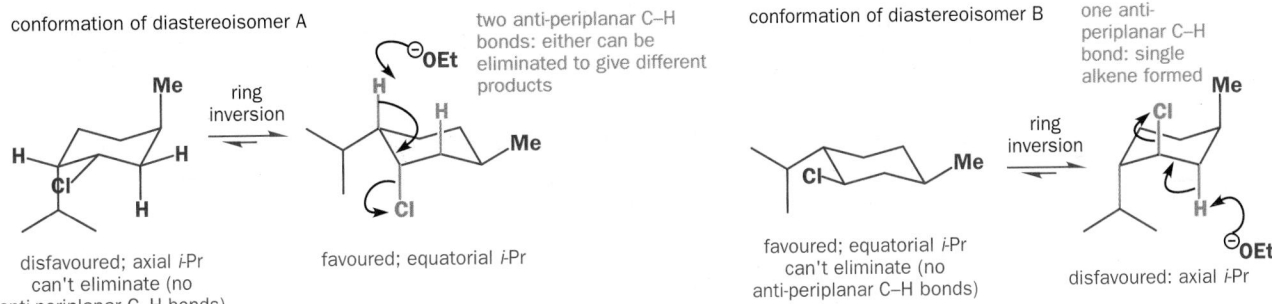

The key to explaining reactions like this is to draw the conformation of the molecules. Both will adopt a chair conformation, and generally the chair having the largest substituent equatorial (or the largest *number* of substituents equatorial) is the more stable. In these examples the isopropyl group is most influential—it is branched and will have very severe 1,3-diaxial interactions if it occupies an axial position. In both diastereoisomers, an equatorial *i*-Pr also means an equatorial Me: the only difference is the orientation of the chlorine. For diastereoisomer A, the chlorine is forced axial in the major conformer: there is no choice, because the relative configuration is fixed in the starting material. It's less stable than equatorial Cl, but is ideal for E2 elimination and there are two protons that are anti-periplanar available for removal by the base. The two alkenes are formed as a result of each of the possible protons with a 3:1 preference for the more substituted alkene (see below).

For diastereoisomer B, the chlorine is equatorial in the lowest-energy conformation. Once again there is no choice. But equatorial leaving groups cannot be eliminated by E2: in this conformation there is no anti-periplanar proton. This accounts for the difference in rate between the two diastereoisomers. A has the chlorine axial virtually all the time ready for E2, while B has an axial leaving group only in the minute proportion of the molecules that happen not to be in the lowest-energy conformation, but that have all three substituents axial. The all-axial conformer is much higher in energy, but only in this conformer can Cl⁻ be eliminated. The concentration of reactive molecules is low, so the rate is also low. There is only one proton anti-periplanar and so elimination gives a single alkene.

E2 elimination from vinyl halides: how to make alkynes

An anti-periplanar arrangement of C–Br and C–H is attainable with a vinylic bromide too, provided the Br and H are *trans* to one another. E2 elimination from the *Z* isomer of a vinyl bromide gives an alkyne rather faster than elimination from the *E* isomer, because in the *E* isomer the C–H and C–Br bonds are syn-periplanar.

The base used here is LDA (lithium diisopropylamide) made by deprotonating *i*-Pr₂NH with BuLi. LDA is very basic (pK_a about 35) but too hindered to be nucleophilic—ideal for promoting E2 elimination.

Vinyl bromides can themselves be made by elimination reactions of 1,2-dibromoalkanes. Watch what happens when 1,2-dibromopropane is treated with three equivalents of R_2NLi: first, elimination to the vinyl halide; then, elimination of the vinyl halide to the alkyne. The terminal alkyne is amply acidic enough to be deprotonated by R_2NLi, and this is the role of the third equivalent. Overall, the reaction makes a lithiated alkyne (ready for further reactions) from a fully saturated starting material. This may well be the first reaction you have met that makes an alkyne from a starting material that doesn't already contain a triple bond.

making an alkyne from 1,2-dibromopropane

The regioselectivity of E2 eliminations

Here are two deceptively similar elimination reactions. The leaving group changes and the reaction conditions are very different but the overall process is elimination of HX to produce one of two alkenes.

In the first example acid-catalysed elimination of water from a tertiary alcohol produces a trisubstitued alkene. Elimination of HCl from the corresponding tertiary alkyl chloride promoted by a very hindered alkoxide base (more hindered than *t*-BuOK because all the ethyl groups have to point away from one another) gives exclusively the less stable disubstituted alkene.

The reason for the two different regioselectivities is a change in mechanism. As we have already discussed, acid-catalysed elimination of water from tertiary alcohols is usually E1, and you already know the reason why the more substituted alkene forms faster in E1 reactions (p. 490). It should come to you as no surprise now that the second elimination, with a strong, hindered base, is an E2 reaction. But why does E2 give the less substituted product? This time, there is no problem getting C–H bonds antiperiplanar to the leaving group: in the conformation with the Cl axial there are two equivalent ring hydrogens available for elimination, and removal of either of these would lead to the trisubstituted alkene. Additionally, any of the three equivalent methyl hydrogens are in a position to undergo E2 elimination to form the disubstituted alkene whether the Cl is axial or equatorial—and yet it is these and only these that are removed by the hindered base. The diagram summarizes two of the possibilities.

two ring hydrogens anti-periplanar to Cl

ring hydrogens more hindered: no reaction

methyl hydrogen anti-periplanar to Cl

methyl hydrogens more accessible: bulky base prefers to form less substituted alkene

The base attacks the methyl hydrogens because they are less hindered—they are attached to a primary carbon atom, well away from the other axial hydrogens. E2 eliminations with hindered bases typically give the less substituted double bond, because the fastest E2 reaction involves deprotonation at the least substituted site. The hydrogens attached to a less substituted carbon atom are also more acidic. Think of the conjugate bases: a *t*-butyl anion is more basic (because the anion is destabilized by the three alkyl groups) than a methyl anion, so the corresponding alkane must be less acidic. Steric factors are evident in the following E2 reactions, where changing the base from ethoxide to *t*-butoxide alters the major product from the more to the less substituted alkene.

	69%	31%
NaOEt	69%	31%
t-BuOK	28%	73%

● **Elimination regioselectivity**

- E1 reactions give the more substituted alkene
- E2 reactions may give the more substituted alkene, but become more regioselective for the less substituted alkene with more hindered bases

Hofmann and Saytsev

Traditionally, these two opposite preferences—for the more or the less substituted alkenes—have been called 'Saytsev's rule' and 'Hofmann's rule', respectively. You will see these names used (along with a number of alternative spellings—acceptable for Saytsev, whose name is transliterated from Russian, but not for Hofmann: this Hofmann had one f and two n's), but there is little point remembering which is which (or how to spell them)—it is far more important to understand the reasons that favour formation of each of the two alkenes.

Anion-stabilizing groups allow another mechanism—E1cB

To finish this chapter, we consider a reaction that at first sight seems to go against what we have told you so far. It's an elimination catalysed by a strong base (KOH), so it looks like E2. But the leaving group is hydroxide, which we categorically stated cannot be a leaving group in E2 eliminations.

The key to what is going on is the carbonyl group. In Chapter 8 you met the idea that negative charges are stabilized by conjugation with carbonyl groups, and the table on p. 193 demonstrated how acidic a proton adjacent to a carbonyl group is. The proton that is removed in this elimination reaction is adjacent to the carbonyl group, and is therefore also rather acidic (pK_a about 20). This means that the base can remove it without the leaving group departing at the same time—the anion that results is stable enough to exist because it can be delocalized on to the carbonyl group.

best representation of anion adjacent to
C=O delocalized on to oxygen

green proton acidified (p*K*a ca. 20)
by adjacent carbonyl group

> ▶
> This delocalized anion is called an **enolate**, and we will discuss enolates in more detail in Chapter 21 and beyond.

Although the anion is stabilized by the carbonyl group, it still prefers to lose a leaving group and become an alkene. This is the next step.

This step is also the rate-determining step of the elimination—the elimination is unimolecular, and so is some kind of E1 reaction. But the leaving group is not lost from the starting molecule, but from the *conjugate base* of the starting molecule, so this sort of elimination, which starts with a deprotonation, is called **E1cB** (cB for conjugate Base). Here is the full mechanism, generalized for other carbonyl compounds.

■ *Note.* E1cB has no super- or subscripts, a lower-case c, and an upper-case B.

the E1cB mechanism

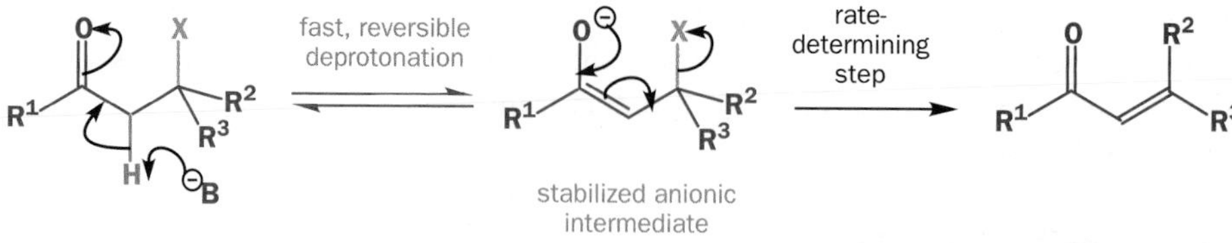

It's important to note that, while HO^- is never a leaving group in E2 reactions, it can be a leaving group in E1cB reactions. The anion it is lost from is already an alkoxide—the oxyanion does not need to be created. The establishment of conjugation also assists loss of HO^-. As the scheme above implies, other leaving groups are possible too. Here are two examples with methanesulfonate leaving groups.

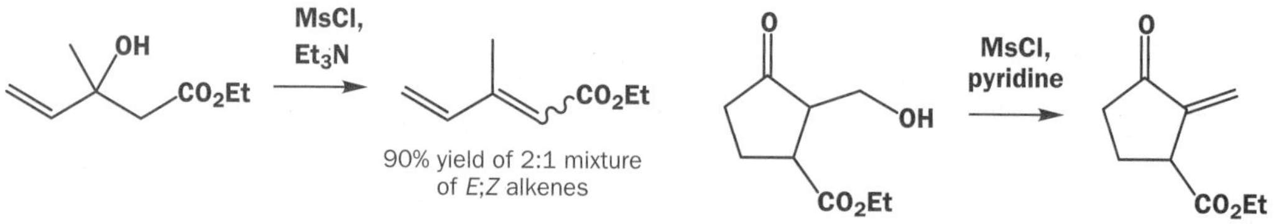

The first looks E1 (stabilized cation); the second E2—but in fact both are E1cB reactions. The most reliable way to spot a likely E1cB elimination is to see whether the product is a conjugated carbonyl group. If it is, the mechanism is probably E1cB.

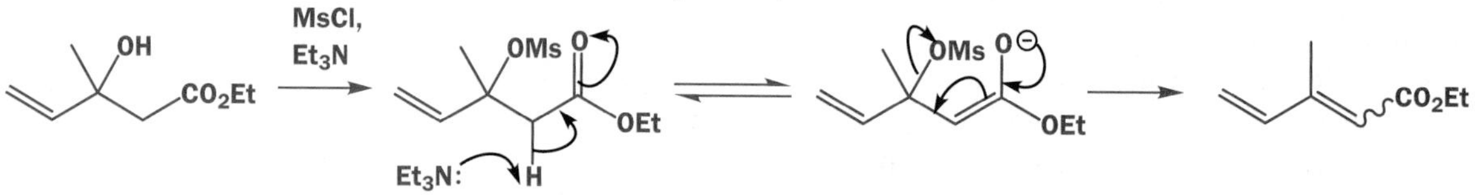

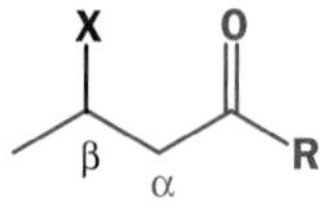
a β-halocarbonyl compound

β-Halocarbonyl compounds can be rather unstable: the combination of a good leaving group and an acidic proton means that E1cB elimination is extremely easy. This mixture of diastereoisomers is first of all lactonized in acid (Chapter 12), and then undergoes E1cB elimination with triethylamine to give a product known as **butenolide**. Butenolides are widespead structures in naturally occurring compounds.

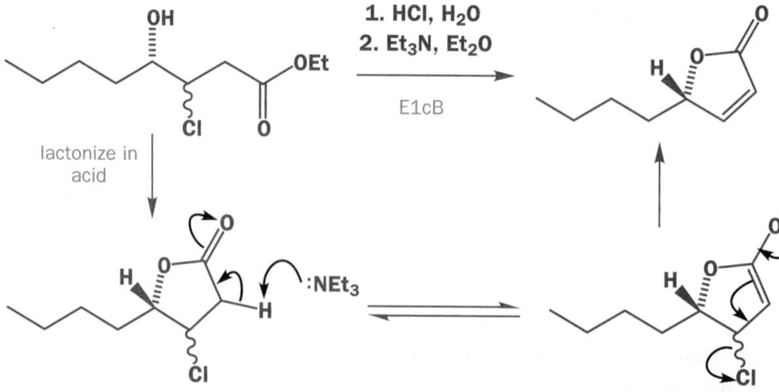

You will have noticed that we have shown the deprotonation step in the last few mechanisms as an equilibrium. Both equilibria lie rather over to the left-hand side, because neither triethylamine (pK_{aH} about 10) nor hydroxide (pK_{aH} = 15.7) is basic enough to remove completely a proton next to a carbonyl group ($pK_a \geq 20$). But, because the loss of the leaving group is essentially irreversible, only a small amount of deprotonated carbonyl compound is necessary to keep the reaction going. The important point about substrates that undergo E1cB is that there is some form of anion-stabilizing group next to the proton to be removed—it doesn't have to stabilize the anion very well but, as long as it makes the proton more acidic, an E1cB mechanism has a chance. Here is an important example with two phenyl rings helping to stabilize the anion, and a carbamate anion (R_2N—CO_2^-) as the leaving group.

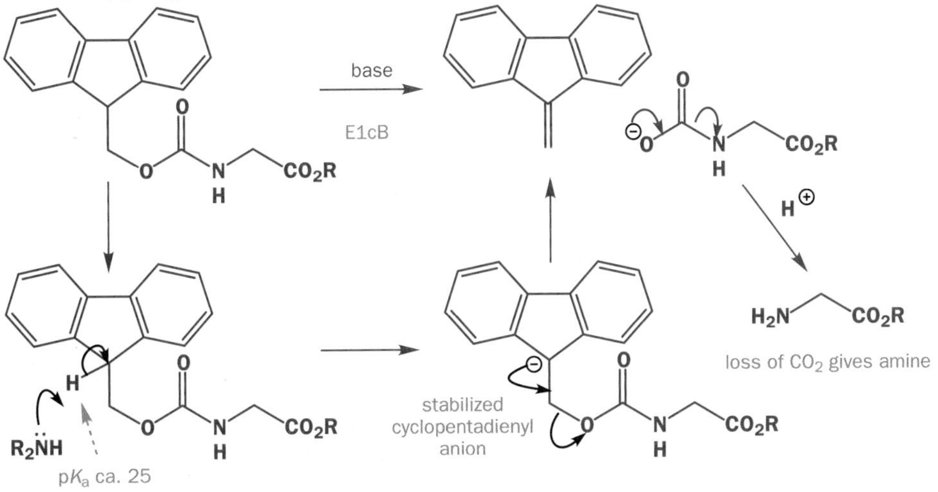

The proton to be removed has a pK_a of about 25 because its conjugate base is an aromatic cyclopentadienyl anion (we discussed this in Chapter 8). The E1cB elimination takes place with a secondary or tertiary amine as the base. Spontaneous loss of CO_2 from the eliminated product gives an amine, and you will meet this class of compounds again shortly in Chapter 25 where we discuss the Fmoc protecting group.

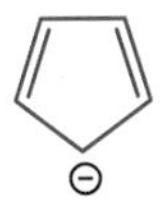

6 π electrons
aromatic
cyclopentadienyl
anion

The E1cB rate equation

The rate-determining elimination step in an E1cB reaction is unimolecular, so you might imagine it would have a first-order rate equation. But, in fact, the rate is also dependent on the concentration of base. This is because the unimolecular elimination involves a species—the anion—whose concentration is itself determined by the concentration of base by the equilibrium we have just been discussing. Using the following general E1cB reaction, the concentration of the anion can be expressed as shown.

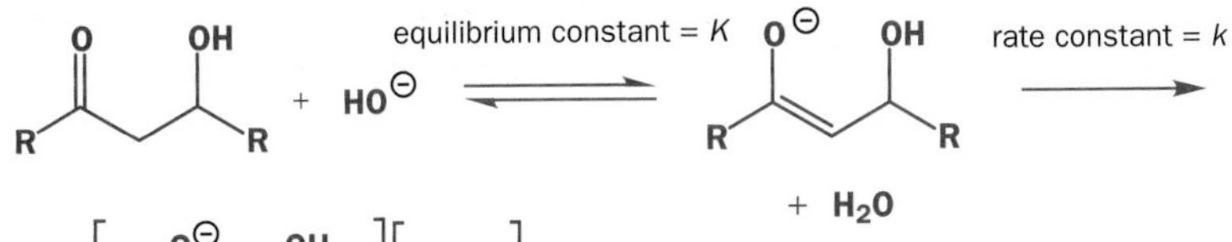

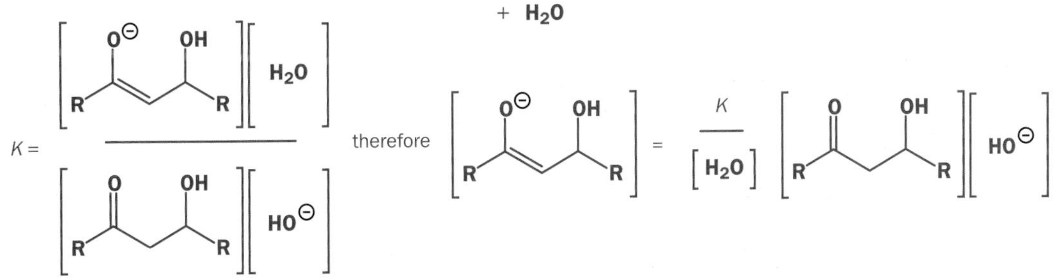

The rate is proportional to the concentration of the anion, and we now have an expression for that concentration. We can simplify it further because the concentration of water is constant.

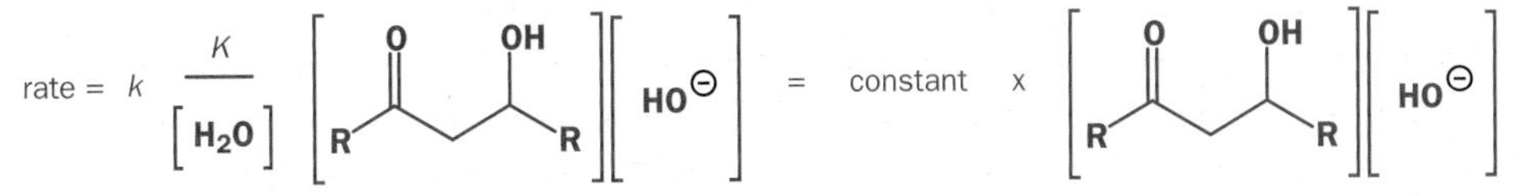

Just because the base (hydroxide) appears in this rate equation doesn't mean to say it is involved in the rate-determining step. Increasing the concentration of base makes the reaction go faster by increasing the amount of anion available to eliminate.

■

You met this idea in chapter 13, pp. 325–326.

● **For reactions with several steps in which the rate-determining step is not the first, the concentrations of species involved in those earlier steps will appear in the rate equation, even though they take no part in the rate-determining step itself.**

E1cB eliminations in context

It is worthwhile comparing the E1cB reaction with some others with which you are familiar: for a start, you may have noticed that it is the reverse of the conjugate addition reactions we introduced in Chapter 10. In Chapter 10, conjugated carbonyl compounds were the starting materials; now they are the products—but both reactions go through a stabilized anion intermediate. E1cB reactions are so general that they are by far the most common way of making the enone starting materials for conjugate additions.

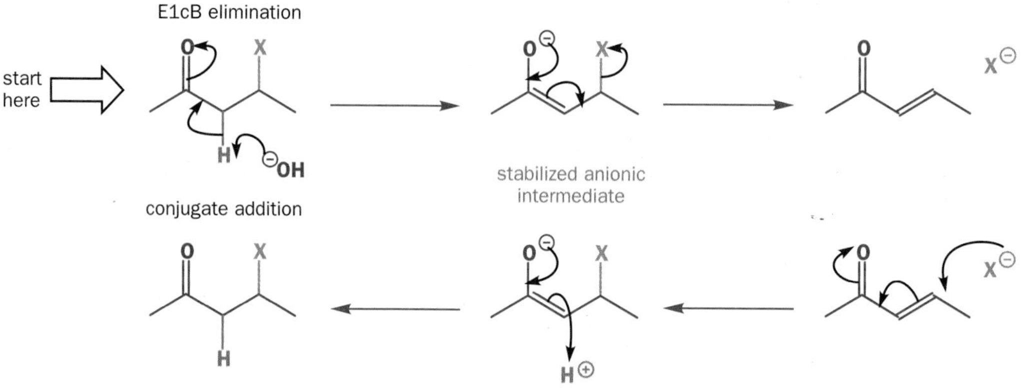

E1cB elimination

start here

stabilized anionic intermediate

conjugate addition

A deceptive S$_N$2 substitution

In some rare cases, you may see E1cB elimination and conjugate addition taking place in a single reaction. Look at this 'substitution' reaction, for example. Apparently, the ammonium salt has been substituted by the cyanide in what looks to be an S$_N$2 reaction.

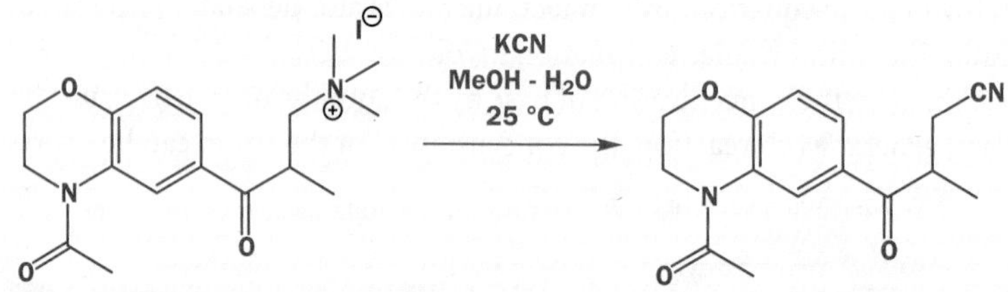

KCN
MeOH - H$_2$O
25 °C

A little consideration will tell you that it can't be S$_N$2 though, because, if it were, it would go like this.

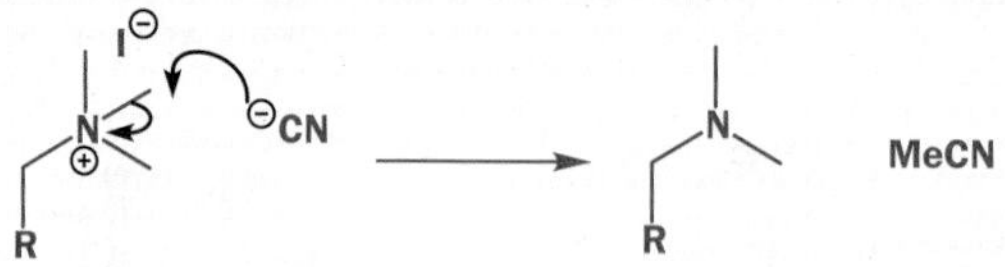

MeCN

Instead, the mechanism is first an E1cB elimination, followed by conjugate addition.

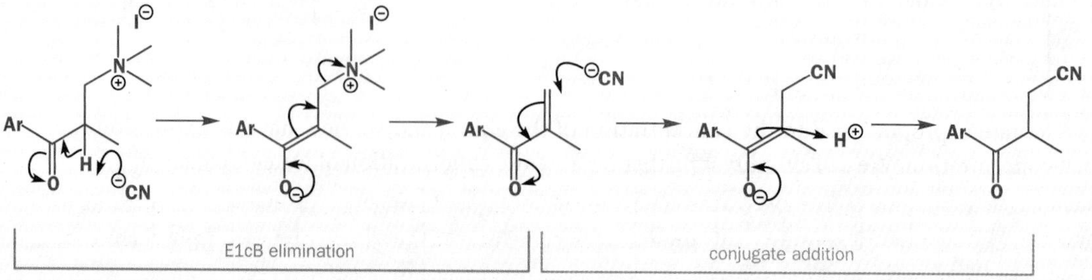

E1cB elimination   conjugate addition

We can also compare it with the other elimination reactions you have met by thinking of the relative timing of proton removal and leaving group departure. E1 is at one end of the scale: the leaving group goes first, and proton removal follows in a second step. In E2 reactions, the two events happen at the same time: the proton is removed as the leaving group leaves. In E1cB the proton removal moves in front of leaving group departure.

E1cB elimination
deprotonation first
leaving group second

E2 elimination
deprotonation and loss
of leaving group
simultaneous

E1 elimination
leaving group first
deprotonation second

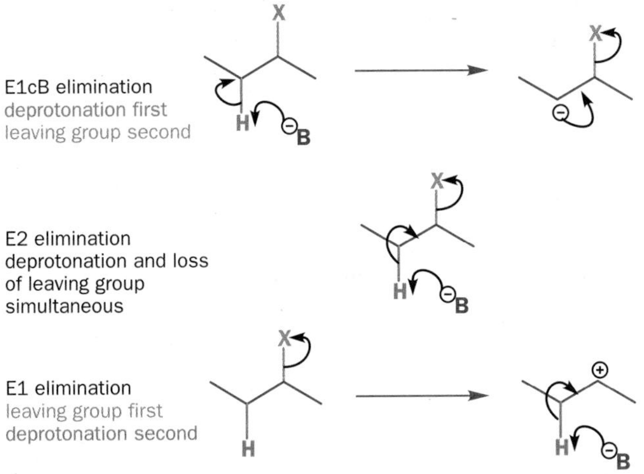

We talked about regio- and stereoselectivity in connection with E1 and E2 reactions. With E1cB, the regioselectivity is straightforward: the location of the double bond is defined by the position of: (a) the acidic proton and (b) the leaving group.

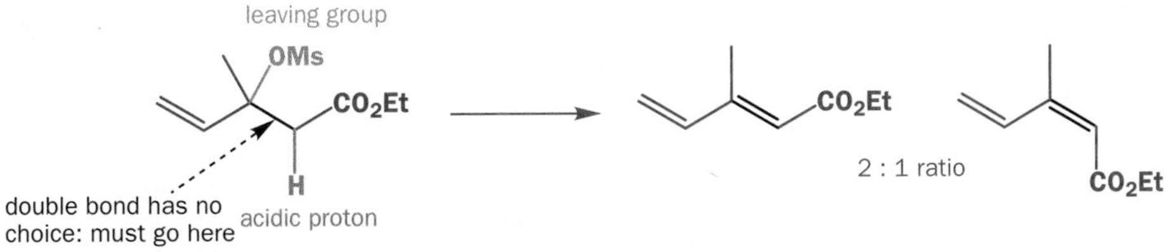

leaving group
OMs

double bond has no choice: must go here   acidic proton

2 : 1 ratio

E1cB reactions may be stereoselective—this one, for example, gives mainly the *E*-alkene product (2:1 with *Z*). The intermediate anion is planar, so the sterochemistry of the starting materials is irrelevant, the less sterically hindered (usually *E*) product is preferred. This double E1cB elimination, for example, gives only the *E,E*-product.

To finish this chapter we need to tell you about two E1cB eliminations that you may meet in unexpected places. We have saved them till now because they are unusual in that the leaving group is actually part of the anion-stabilizing group itself. First of all, try spotting the E1cB elimination in this step from the first total synthesis of penicillin V in 1957.

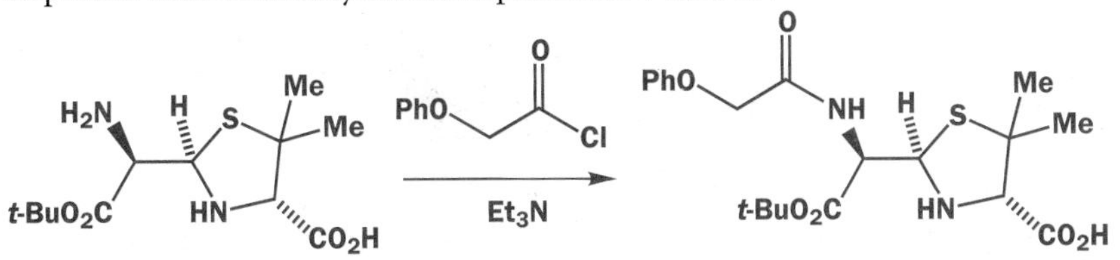

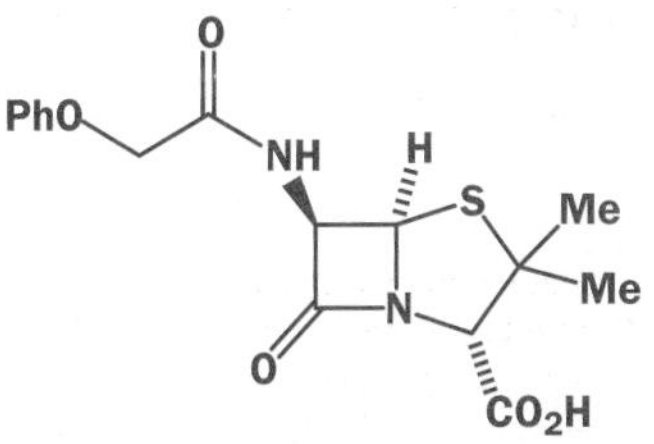

penicillin V

The reaction is deceptively simple—formation of an amide in the presence of base—and you would expect the mechanism to follow what we told you in Chapter 12. But the acyl chloride is, in fact, set up for an E1cB elimination—and you should expect this whenever you see an acyl chloride *with acidic protons next to the carbonyl group* used in the presence of triethylamine.

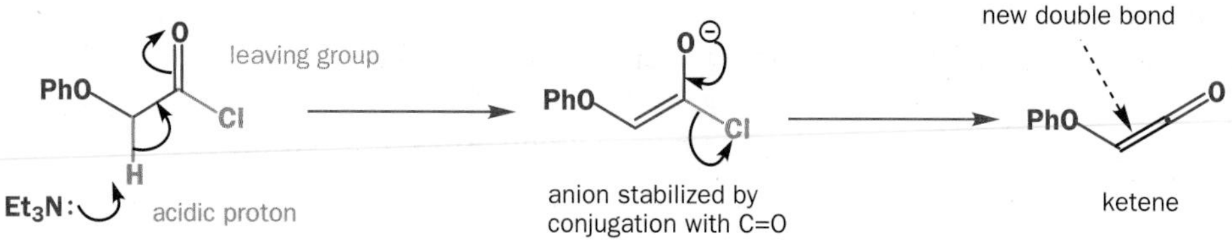

The product of the elimination is a substituted ketene—a highly reactive species whose parent ($CH_2=C=O$) we talked about in Chapter 15. It is the ketene that reacts with the amine to form the amide.

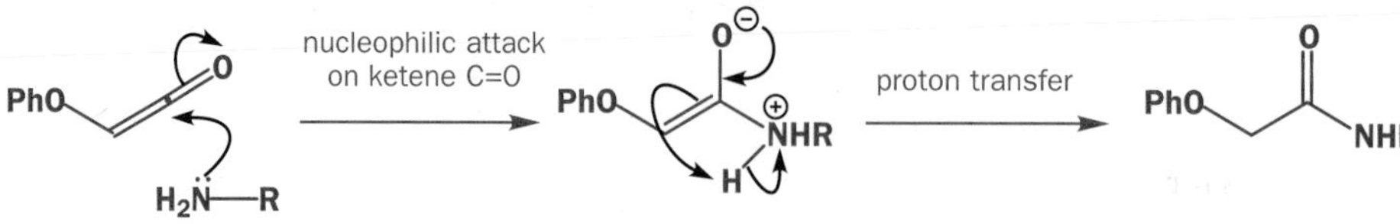

The second 'concealed' E1cB elimination is in the elimination of HCl from MsCl, which we showed you on p. 486 of this chapter. You can now see the similarity with the acyl chloride mechanism above.

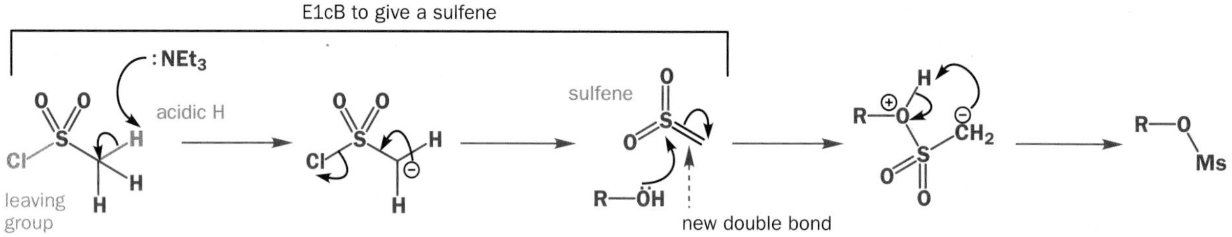

E1cB to give a sulfene

To conclude...

The table summarizes the general pattern of reactivity expected from various structural classes of alkyl halides (or tosylates, mesylates) in reactions with a representative range of nucleophiles (which may behave as bases).

	Poor nucleophile (e.g. H_2O, ROH)[a]	Weakly basic nucleophile (e.g. I⁻, RS⁻)	Strongly basic, unhindered nucleophile (e.g. RO⁻)	Strongly basic, hindered nucleophile (e.g. DBU, DBN, t-BuO⁻)
methyl 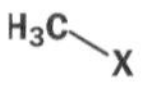 H_3C—X	no reaction	S_N2	S_N2	S_N2
primary (unhindered) 	no reaction	S_N2	S_N2	E2
primary (hindered) 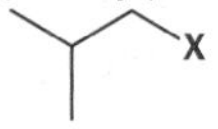	no reaction	S_N2	E2	E2
secondary 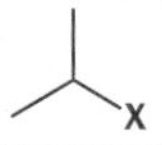	S_N1, E1 (slow)	S_N2	E2	E2
tertiary 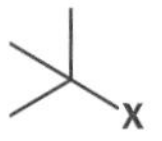	E1 or S_N1	S_N1, E1	E2	E2
β to anion-stabilizing group 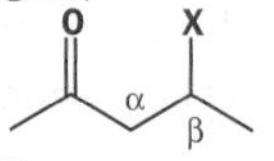	E1cB	E1cB	E1cB	E1cB

[a] Acid conditions.

Some points about the table:

- Methyl halides cannot eliminate as there are no appropriately placed protons
- Increasing branching favours elimination over substitution and strongly basic hindered nucleophiles always eliminate unless there is no option
- Good nucleophiles undergo substitution by S_N2 unless the substrate is tertiary and then the intermediate cation can eliminate by E1 as well as substitute by S_N1
- High temperatures favour elimination by gearing up the importance of entropy in the free energy of reaction ($\Delta G = \Delta H - T\Delta S$). This is a good way of ensuring E1 in ambiguous cases

Problems

1. Draw mechanisms for these reactions, which were listed in the chapter.

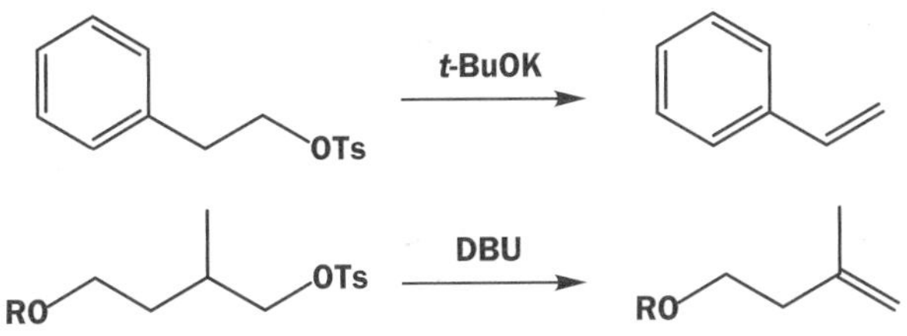

2. Give a mechanism for the elimination reaction in the formation of tamoxifen from p. 489 and comment on the fact that it gives a mixture of *cis* and *trans* alkenes.

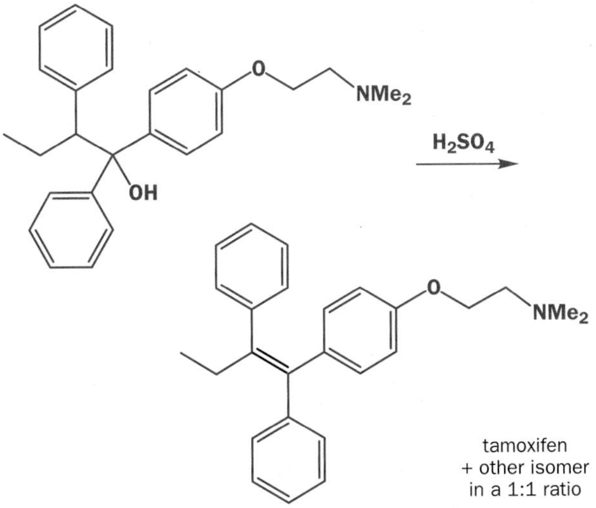

tamoxifen
+ other isomer
in a 1:1 ratio

3. Account for the contrasting results in these two reactions.

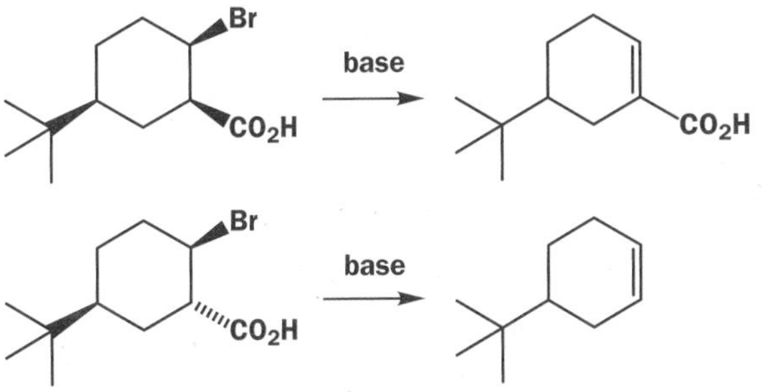

4. Draw mechanisms for these two elimination reactions of epoxides.

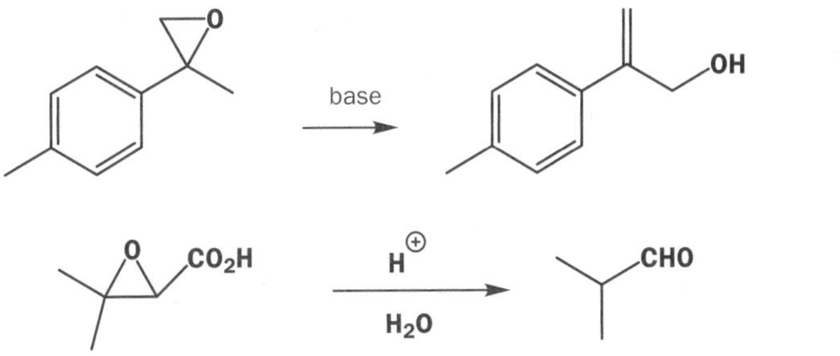

5. Only one of the diastereomeric bromides shown here eliminates to give alkene A. Why? Neither bromide gives alkene B. Why not?

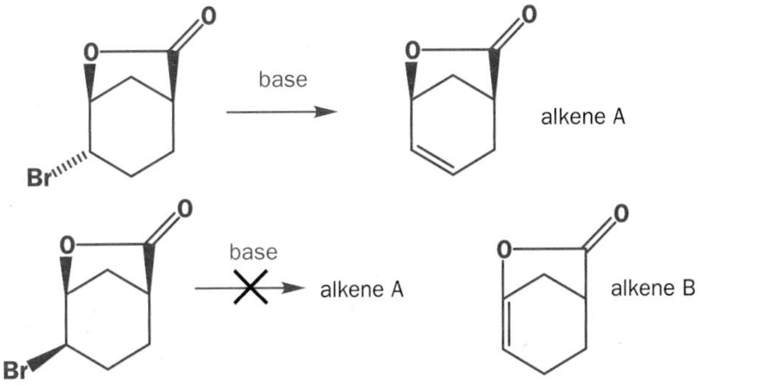

6. Suggest mechanisms for these reactions, paying particular attention to the elimination steps.

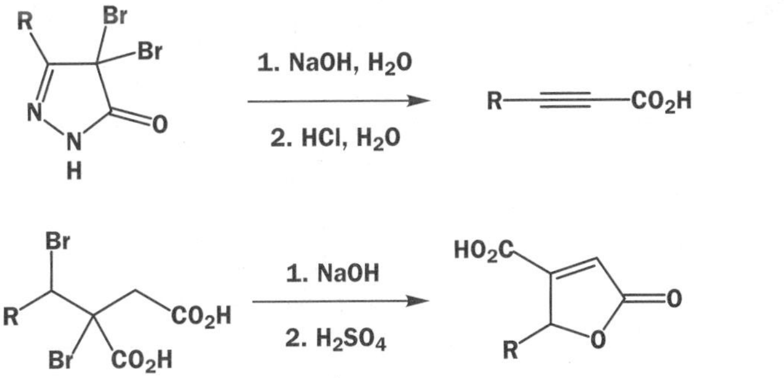

7. Suggest mechanisms for these two eliminations. Why does the first one give a mixture and the second a single product?

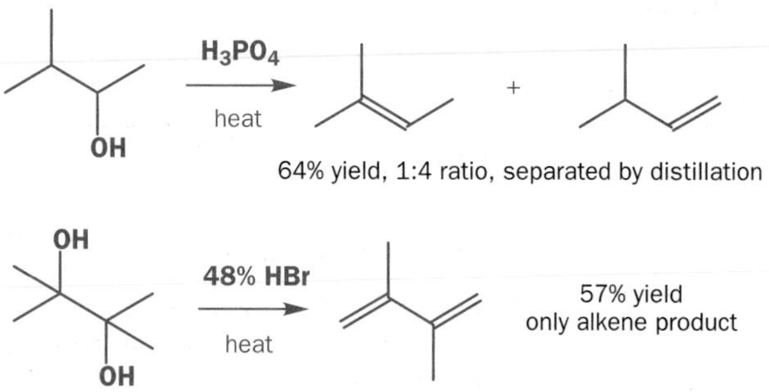
OH
H_3PO_4
heat
+
64% yield, 1:4 ratio, separated by distillation
OH
OH
48% HBr
heat
57% yield
only alkene product

8. Explain the position of the double bond in the products of these reactions. The starting materials are enantiomerically pure. Are the products also enantiomerically pure?

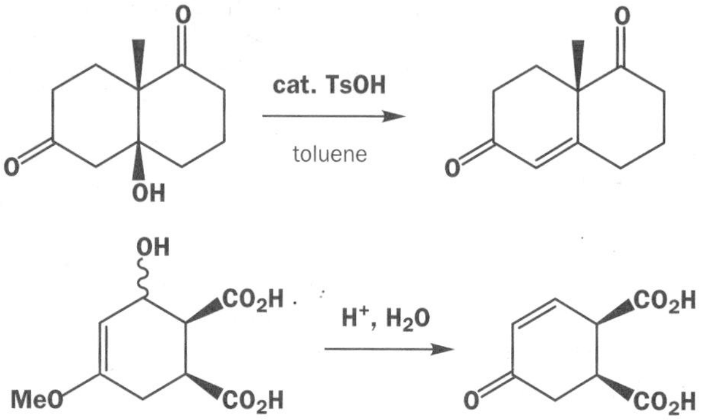
O
O
OH
cat. TsOH
toluene
O
O
OH
CO_2H
MeO
CO_2H
H^+, H_2O
CO_2H
O
CO_2H

9. Explain the stereochemistry of the double bonds in the products of these reactions.

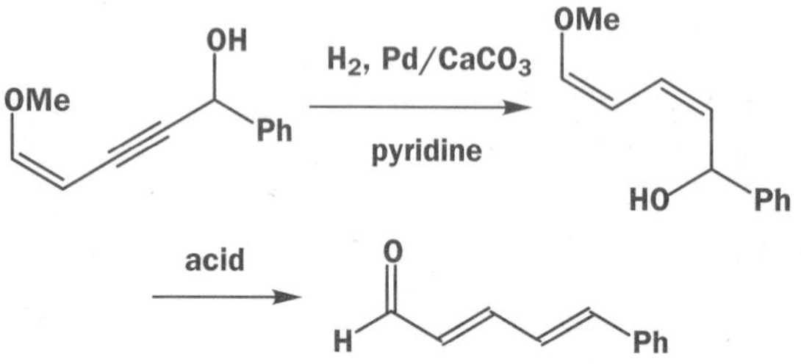
OMe
OH
Ph
H_2, Pd/$CaCO_3$
pyridine
OMe
HO
Ph
acid
O
H
Ph

10. Why is elimination preferred to hemiacetal formation in the acid-catalysed cyclization of this ketone?

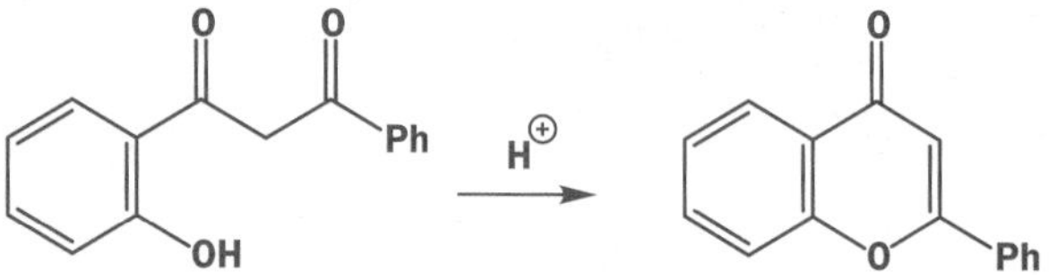
O
O
Ph
OH
$H^{\oplus}$
O
O
Ph

11. Comment on the position taken up by the alkene in these elimination reactions.

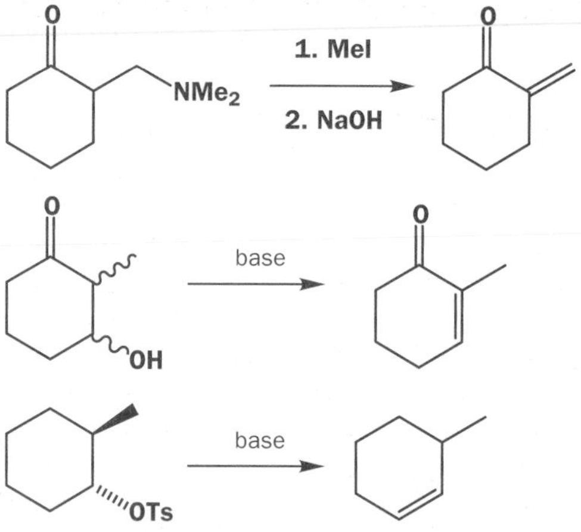
O
NMe_2
1. MeI
2. NaOH
O
O
OH
base
O
OTs
base

12. Give mechanisms for these reactions drawn from Chapter 19 and comment on the stereochemistry.

(a)

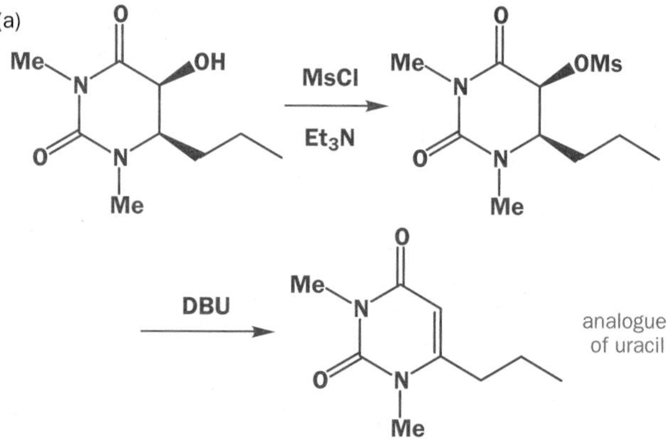
O
Me
N
OH
O
N
Me
MsCl
Et_3N
O
Me
N
OMs
O
N
Me
DBU
O
Me
N
O
N
Me
analogue of uracil

(b)

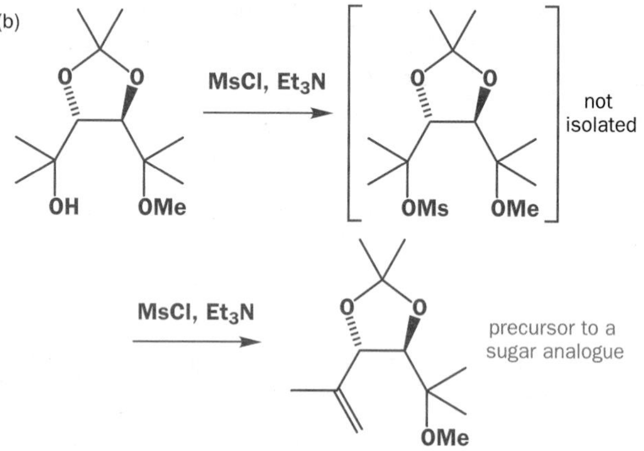
O
O
OH
OMe
MsCl, Et_3N
O
O
OMs
OMe
not isolated
MsCl, Et_3N
O
O
OMe
precursor to a sugar analogue

Electrophilic addition to alkenes

20

Connections

Building on:

- Elimination reactions that form alkenes ch19
- Stability of carbocations, and their reactions during the S_N1 reaction ch17
- *Nucleophilic* addition to conjugated alkenes ch10

Arriving at:

- Reactions of simple, unconjugated alkenes with *electrophiles*
- Converting C=C double bonds to other functional groups by electrophilic addition
- How to predict which end of an unsymmetrical alkene reacts with the electrophile
- Stereoselective and stereospecific reactions of alkenes
- How to make alkyl halides, epoxides, alcohols, and ethers through electrophilic addition

Looking forward to:

- Electrophilic addition to alkenes carrying oxygen substituents (enols and enolates) ch21
- Electrophilic addition to aromatic rings ch22
- Reactions of alkenes by pericyclic reactions ch35
- Reactions of alkenes with boranes ch47

Alkenes react with bromine

Bromine (Br_2) is brown, and one of the classic tests for alkenes is that they turn a brown aqueous solution of bromine colourless. Alkenes decolourize bromine water: alkenes react with bromine. The product of the reaction is a dibromoalkane, and the reaction below shows what happens with the simplest alkene, ethylene (ethene).

In order to understand this reaction, and the other similar ones you will meet in this chapter, you need to think back to Chapter 5, where we started talking about reactivity in terms of nucleophiles and electrophiles. As soon as you see a new reaction, you should immediately think to yourself, 'Which reagent is the nucleophile; which reagent is the electrophile?' Evidently, neither the alkene nor bromine is charged, but Br_2 has a low-energy empty orbital (the Br–Br σ^*), and is therefore an electrophile. The Br–Br bond is exceptionally weak, and bromine reacts with nucleophiles like this.

Chloramines

Have you ever wondered why conventional wisdom (and manufacturers' labels) warns against mixing different types of cleaning agent? The danger arises from nucleophilic attack on another halogen, chlorine. Some cleaning solutions contain chlorine (bleach, to kill moulds and bacteria, usually for the bathroom) while others contain ammonia (to dissolve fatty deposits, usually for the kitchen). Ammonia is nucleophilic, chlorine electrophilic, and the products of their reaction are the highly toxic and explosive chloramines NH_2Cl, $NHCl_2$, and NCl_3.

The alkene must be the nucleophile, and its HOMO is the C=C π bond. This is a very important point, because in the first reactions of alkenes you met, in Chapter 10, the conjugated alkene was an *electrophile*. We told you about conjugated alkenes first, because their chemistry is very similar to the chemistry of the carbonyl group. But normal, simple, unconjugated alkenes are electron-rich—they have no nearby carbonyl group to accept electrons—and they typically act as *nucleophiles* and attack *electrophiles*.

> ● **Simple, unconjugated alkenes are nucleophilic and react with electrophiles.**

When it reacts with Br_2, the alkene's filled π orbital (the HOMO) will interact with the bromine's empty σ^* orbital to give a product. But what will that product be? Look at the orbitals involved.

alkene = nucleophile

HOMO = filled π orbital

Br_2 = electrophile

LUMO = empty σ^* orbital

The highest electron density in the π orbital is right in the middle, between the two carbon atoms, so this is where we expect the bromine to attack. The only way the π HOMO can interact in a bonding manner with the σ^* LUMO is if the Br_2 approaches end-on—and this is how the product forms. The symmetrical three-membered ring product is called a **bromonium ion**.

electrophilic attack by Br_2 on ethylene

bonding interaction

HOMO = filled π orbital

LUMO = empty σ^* orbital

bromonium ion

In Chapter 17, you saw many examples of S_N2 reactions at carbon, and some at Si, P, and S. This reaction is also a nucleophilic substitution (from the point of view of the bromine) at Br. Just replace the alkene with another nucleophile, and Br–Br with Me–Br, and you are on familiar ground again.

bonding interaction

HOMO = filled sp^3 orbital

LUMO = empty σ^* orbital

How shall we draw curly arrows for the formation of the bromonium ion? We have a choice. The simplest is just to show the middle of the π bond attacking Br–Br, mirroring what we know happens with the orbitals.

But there is a problem with this representation: because only one pair of electrons is moving, we can't form two new C–Br bonds. We should really then represent the C–Br bonds as partial bonds. Yet the bromonium ion is a real intermediate with two proper C–Br bonds (read the box below for evidence of this). So an alternative way of drawing the arrows is to involve a lone pair on bromine.

We think the first way represents more accurately the key orbital interaction involved, and we shall use that one, but the second is acceptable too.

Another way of thinking about bromonium ions

You can think of the bromonium ion as a carbocation that has been stabilized by interaction with a nearby bromine atom. You have seen a similar effect with oxygen—this 'oxonium ion' was an intermediate, for example, in the S_N1 substitution of MOM chloride on p. 419 of Chapter 17.

oxygen's lone pair stabilizes "primary" cation

MOM chloride (methoxymethyl chloride)

"oxonium ion" – better representation of cation

The bromine is one atom further away but, with bromine being lower in the periodic table and having more diffuse lone pairs, it can have a similar stabilizing effect, despite the angle strain in a three-membered ring.

unstable primary cation

"bromonium ion" – correct representation of cation

this is not a reaction – just a demonstration of how bromine's lone pair stabilizes the cation

The two types of stabilization are not equivalent: the cation and the bromonium ion are different molecules with different shapes, while the two representations of the oxonium ion are just that—they aren't different molecules. This stabilization of an adjacent cationic centre by a heteroatom with at least one lone pair to form a three-membered ring intermediate is not restricted to bromine or the other halogens but is also an important aspect of the chemistry of compounds containing oxygen, sulfur, or selenium, as you will see in Chapter 46.

Of course, the *final* product of the reaction isn't the bromonium ion. The second step of the reaction follows on at once: the bromonium ion is an electrophile, and it reacts with the bromide ion lost from the bromine in the addition step. We can now draw the correct mechanism for the whole reaction, which is termed **electrophilic addition** to the double bond, because bromine is an electrophile. Overall, the molecule of bromine *adds across* the double bond of the alkene.

■
Compare the second step with the way nucleophiles attack epoxides, Chapter 17, p. 435.

electrophilic addition of bromine to ethylene

bromonium ion

Attack of Br⁻ on a bromonium ion is a normal S_N2 substitution—the key orbitals involved are the HOMO of the bromide and the $\sigma*$ of one of the two carbon–bromine bonds in the strained three-membered ring. As with all S_N2 reactions, the nucleophile maintains maximal overlap with the $\sigma*$ by approaching in line with the leaving group but from the opposite side, resulting in inversion at the carbon that is attacked. The stereochemical outcome of more complicated reactions (discussed below) is important evidence for this overall reaction mechanism.

orbitals involved in the opening of the bromonium ion

LUMO = empty $\sigma*$ orbital

HOMO = filled n orbital

Why doesn't the bromine simply attack the positive charge and re-form the bromine molecule? Well, in fact, it does and the first step is reversible.

How do we know bromonium ions exist?

Very hindered alkenes form bromonium ions that are resistant to nucleophilic attack. In one very hindered case, the bromine ion just can't get at the bromonium ion to attack it, and the bromonium ion is sufficiently stable to be characterized by X-ray crystallography. In another case, the use of superacid systems (Chapter 17) has allowed direct NMR observation of the bromonium ion intermediate to the bromination of propene.

crystalline solid

can be observed by NMR

Oxidation of alkenes to form epoxides

The electrophilic addition of bromine to alkenes is an oxidation. The starting alkene is at the alcohol oxidation level, but the product has two carbons at the alcohol oxidation level—the elimination reactions of dibromides to give alkynes that you met in the last chapter (p. 493) should convince you of this. There are a number of other oxidants containing electrophilic oxygen atoms that react with nucleophilic alkenes to produce epoxides (oxiranes). You can view epoxides as the oxygen analogues of bromonium ions, but unlike bromonium ions they are quite stable.

The simplest epoxide, ethylene oxide (or oxirane itself), can be produced on the tonne scale by the direct oxidation of ethene by oxygen at high temperature over a silver catalyst. These conditions are hardly suitable for general lab use, and the most commonly used epoxidizing agents are peroxy-carboxylic acids. **Peroxy-acids** (or **peracids**) have an extra oxygen atom between the carbonyl group and their acidic hydrogen—they are half-esters of hydrogen peroxide (H_2O_2). They are rather less acidic than carboxylic acids because their conjugate base is no longer stabilized by delocalization into the carbonyl group reagent. But they are electrophilic at oxygen, because attack there by a nucleophile displaces carboxylate, a good leaving group. The LUMO of a peroxy-carboxylic acid is the $\sigma*$ orbital of the weak O–O bond.

■
You have met epoxides being formed by intramolecular substitution reactions, but the oxidation of alkenes is a much more important way of making them. Their alternative name *derives* from a systematic way of naming rings: 'ox' for the O atom, 'ir' for the three-membered ring, and 'ane' for full saturation. You may meet oxetane (remember the oxaphosphetane in the Wittig reaction, Chapter 14, p. 357) and, while THF is never called oxolane, dioxolane is another name for five-membered cyclic acetals.

oxirane oxetane dioxolane

peroxy-carboxylic acid

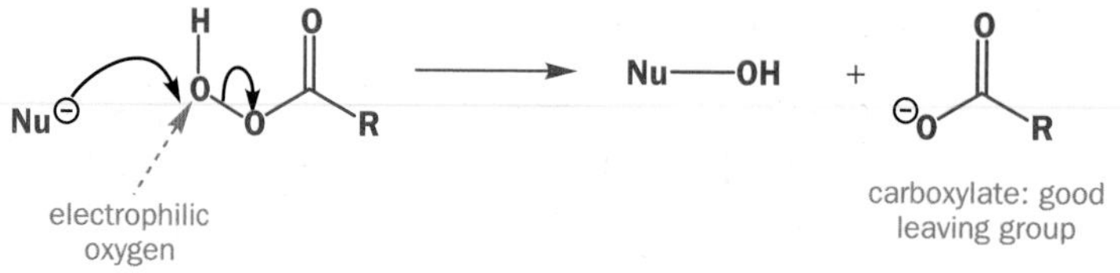

electrophilic oxygen

carboxylate: good leaving group

Making peroxy-acids

Peroxy-acids are prepared from the corresponding acid anhydride and high-strength hydrogen peroxide. In general, the stronger the parent acid, the more powerful the oxidant (because the carboxylate is a better leaving group): one of the most powerfully oxidizing peroxy-acids is peroxy-trifluoroacetic acid. Hydrogen peroxide, at very high concentrations (> 80%), is explosive and difficult to transport.

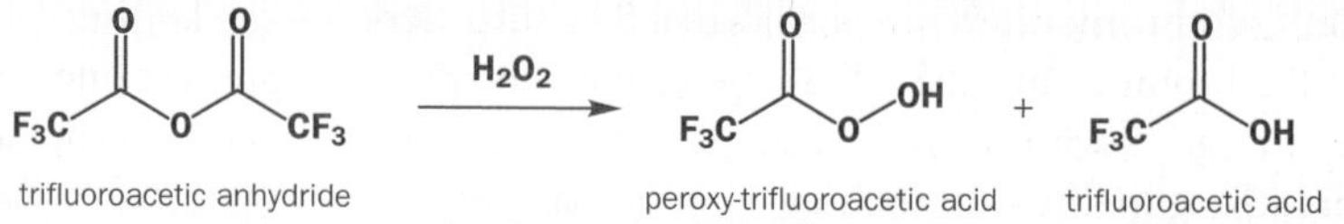

trifluoroacetic anhydride

peroxy-trifluoroacetic acid

trifluoroacetic acid

The most commonly used peroxy-acid is known as *m*-CPBA, or *meta*-ChloroPeroxyBenzoic Acid. *m*-CPBA is a safely crystalline solid. Here it is, reacting with cyclohexene, to give the epoxide in 95% yield.

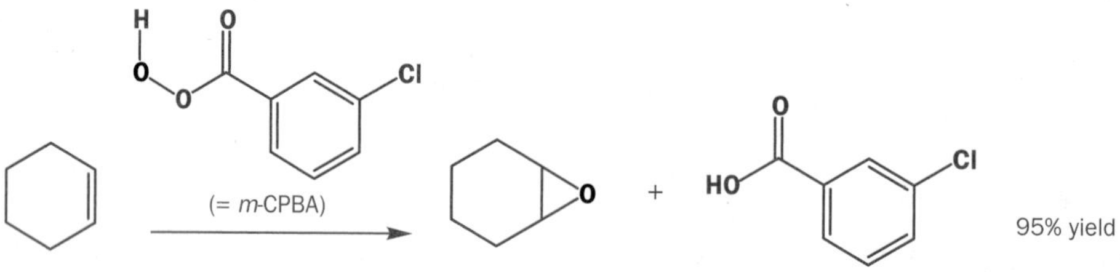

(= *m*-CPBA)

95% yield

As you will expect, the alkene attacks the peroxy-acid from the centre of the HOMO, its π orbital. First, here is the orbital involved.

electrophilic attack by a peroxy-acid on an alkene

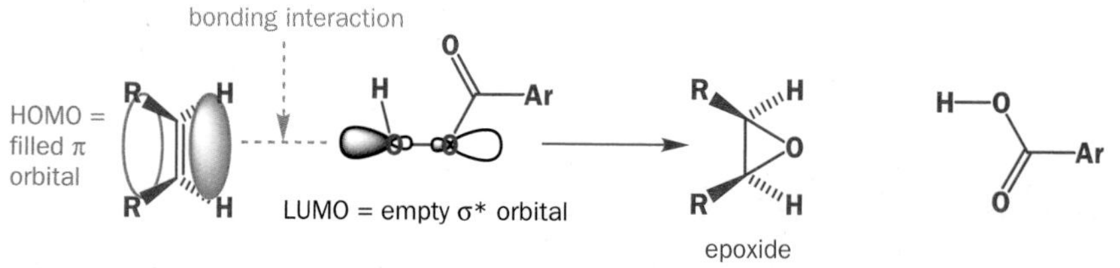

bonding interaction

HOMO = filled π orbital

LUMO = empty σ* orbital

epoxide

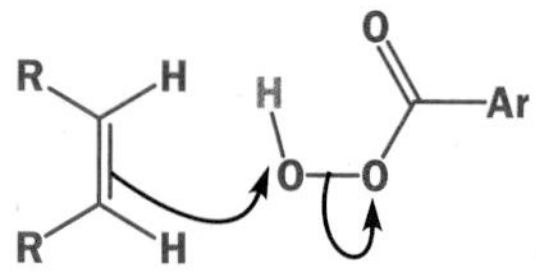

And now the curly arrow mechanism. The essence of the mechanism is electrophilic attack by the weak, polarized O–O bond on the π orbital of the alkene, which we can represent most simply as shown in the margin. But, in the real reaction, a proton (shown in brown in this mechanism) has transferred from the epoxide oxygen to the carboxylic acid by-product. You can represent this all in one step if you draw the arrows carefully. Start with the nucleophilic π bond: send the electrons on to oxygen, breaking O–O and forming a new carbonyl bond. Use those electrons to pick up the proton, and use the old O–H bond's electrons to make the second new C–O bond. Dont' be put off by the spaghetti effect—each arrow is quite logical when you think the mechanism through. The transition state for the reaction makes the bond-forming and -breaking processes clearer.

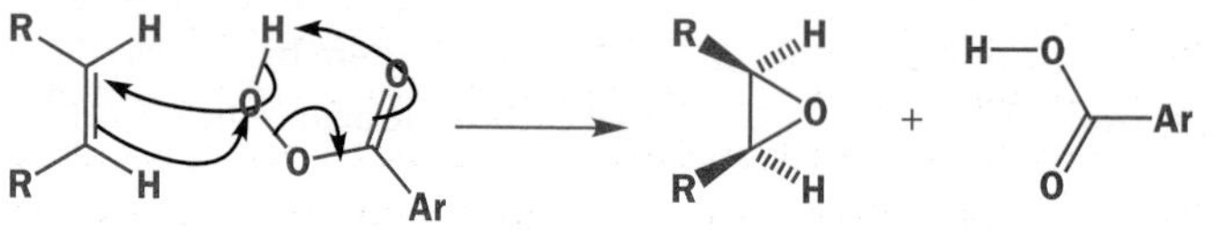

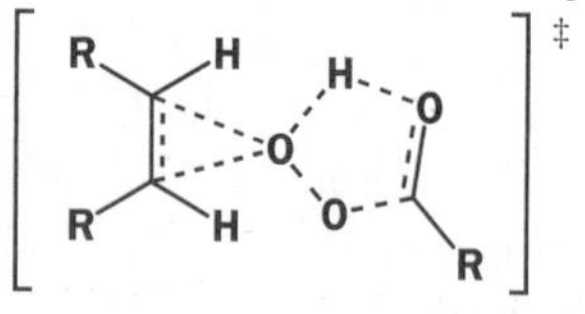

transition state for epoxidation

Epoxidation is stereospecific

Because both new C–O bonds are formed on the same face of the alkene's π bond, the geometry of the alkene is reflected in the stereochemistry of the epoxide. The reaction is therefore stereospecific. Here are two examples demonstrating this: *cis*-alkene gives *cis*-epoxide and *trans*-alkene gives *trans*-epoxide.

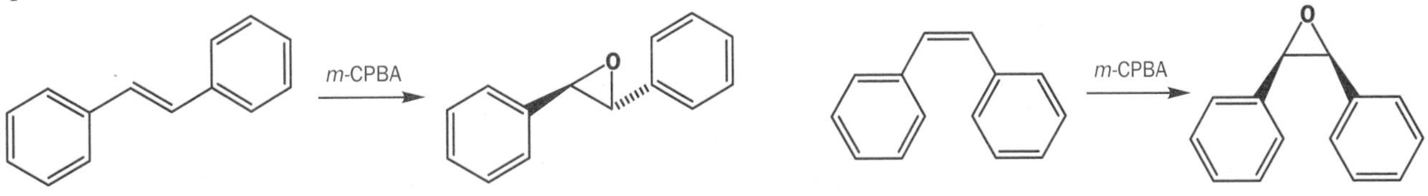

trans-stilbene   *trans*-stilbene oxide   *cis*-stilbene   *cis*

More substituted alkenes epoxidize faster

Peracids give epoxides from alkenes with any substitution pattern (except ones conjugated with electron-withdrawing groups, for which a different reagent is required: see Chapter 23) but the chart alongside shows how the rate varies according to the number of substituents on the double bond.

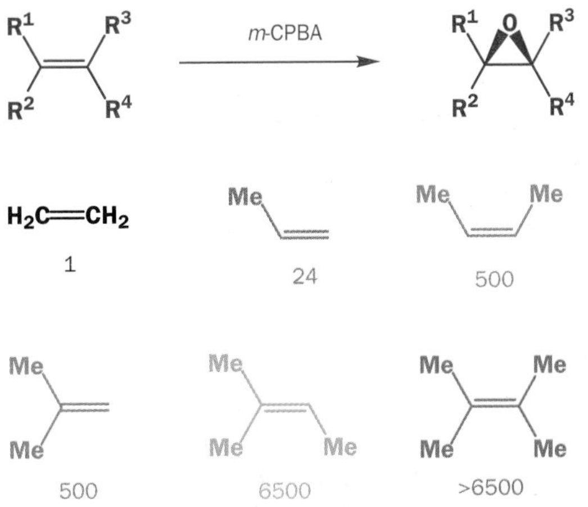

Not only are more substituted double bonds more stable (as you saw in Chapter 19), but they are more nucleophilic. We showed you in Chapter 17 that alkyl groups are electron-donating because they stabilize carbocations. This same electron-donating effect raises the energy of the HOMO of a double bond, and makes it more nucleophilic. You can think of it this way: every C–C or C–H bond that can allow its σ orbital to interact with the π orbital of the alkene will raise the HOMO of the alkene slightly, as shown by the energy level diagram. The more substituents the alkene has, the more the energy is raised.

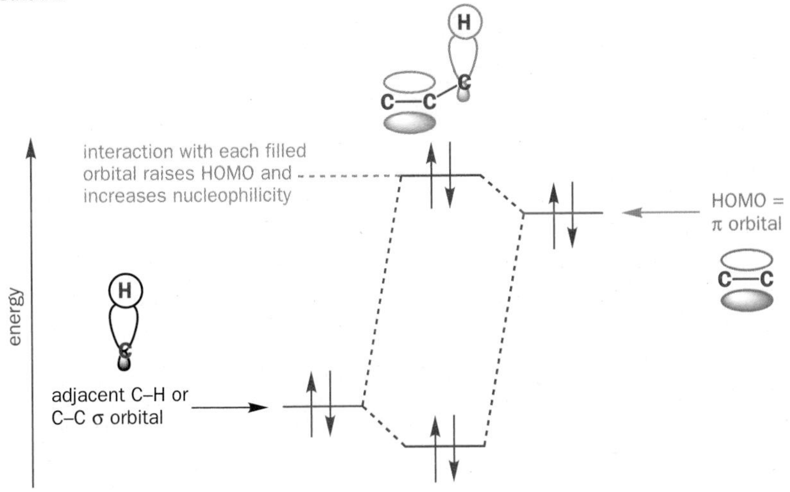

relative rates of reaction of alkenes with *m*-CPBA

The differences in reactivity between alkenes of different substitution patterns can be exploited to produce the epoxide only of the more reactive alkene of a pair, provided the supply of oxidant is limited. In the first example below, a tetrasubstituted alkene reacts in preference to a *cis* disubstituted one. Even when two alkenes are equally substituted, the effect of epoxidizing one of them is to reduce the nucleophilicity of the second (the new oxygen atom is electron-withdrawing, and dienes are in

▶

You may notice additives in peroxy-acid epoxidations, such as sodium carbonate/sodium acetate here. These are buffers, added to prevent the reaction mixture becoming too acidic— remember, the carboxylic acid is a by-product of the epoxidation. Some epoxides are unstable in acid, as we shall see shortly.

▶

Spiro compounds have two rings joined at a single atom. Compare **fused** rings (joined at two adjacent atoms) and **bridged** rings (joined two nonadjacent atoms).

general more nucleophilic than alkenes: see below). The monoepoxide of cyclopentadiene is a useful intermediate and can be prepared by direct epoxidation of the diene under buffered conditions.

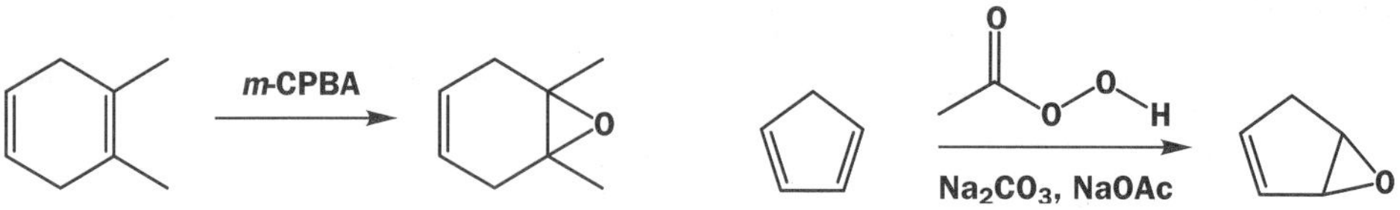

p-Nitroperoxybenzoic acid is dangerously explosive, but it is sufficiently reactive to produce this remarkable and highly strained *spiro* epoxide (oxaspiropentane), which was made in order to study its reactions with nucleophiles.

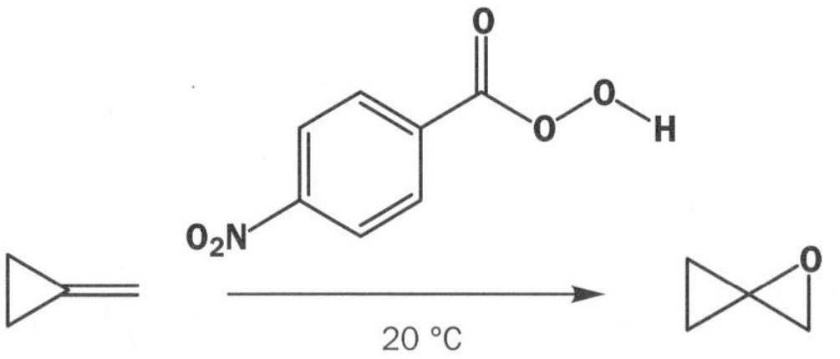

Dimethyldioxirane and carcinogenic epoxides

Certain fungi, especially the mould *Aspergillus sp.* (which grows on damp grain), produce a group of the most carcinogenic substances known to man, the aflatoxins. One of the toxins (which are, of course, entirely natural) is metabolized in the human body to the epoxide shown below. Some American chemists decided to synthesize this epoxide to investigate its reaction with DNA, hoping to discover exactly how it causes cancer. The epoxide is far too reactive to be made using a

peroxy-acid (because of the acid by-product), and instead these chemists used a relatively new reagent called dimethyldioxirane. Dimethyldioxirane is made by oxidizing acetone with $KHSO_5$, but is too reactive to be stored for more than a short period in solution. After it has transferred an oxygen atom in the epoxidation step, only innocuous acetone is left, as shown by the mechanism below.

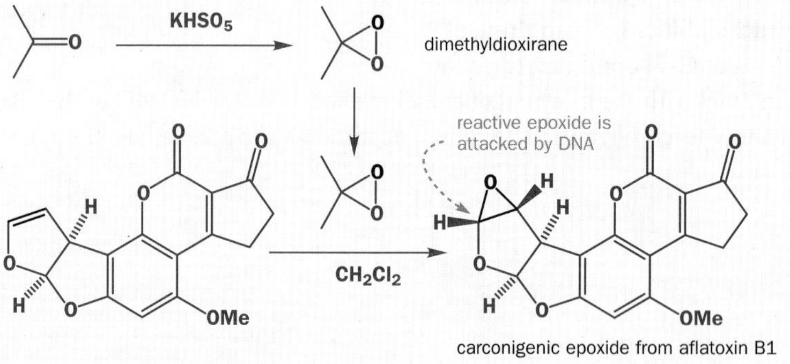

carconigenic epoxide from aflatoxin B1

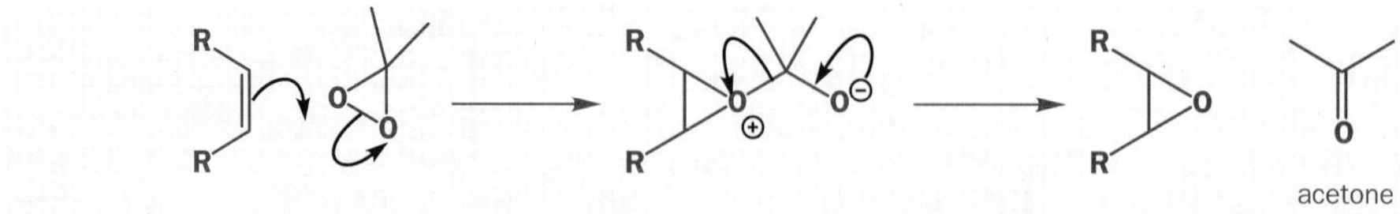

acetone

The liver is home to a wide variety of enzymes that carry out oxidation—the aim is to make unwanted water-insoluble molecules more polar and therefore soluble by peppering them with hydroxyl groups. Unfortunately, some of the intermediates in the oxidation processes are highly

reactive epoxides that damage DNA. This is the means by which benzene and other aromatic hydrocarbons cause cancer, for example. Note that it is very hard to epoxidize benzene by chemical (rather than biological) methods.

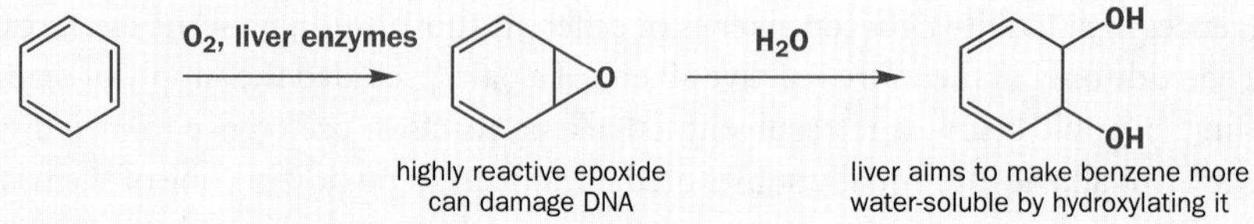

highly reactive epoxide can damage DNA

liver aims to make benzene more water-soluble by hydroxylating it

Electrophilic addition to unsymmetrical alkenes is regioselective

In epoxidation reactions, and in electrophilic additions of bromine, each end of the alkene is joined to the same sort of atom (Br or O). But in the addition reactions of other electrophiles, H–Br for example, there is a choice: which carbon gets the H and which gets the Br? You will need to be able to predict, and to explain, reactions of unsymmetrical alkenes with HBr, but we should start by looking at the reaction with a symmetrical alkene—cyclohexene. This is what happens. When H–Br reacts as an electrophile, it is attacked at H, losing Br^-. Unlike a bromine atom, a hydrogen atom can't form a three-membered ring cation—it has no lone pairs to use. So electrophilic addition of a proton (which is what this is) to an alkene gives a product best represented as a carbocation. This carbocation rapidly reacts with the bromide ion just formed. Overall, H–Br adds across the alkene. This is a useful way of making simple alkyl bromides.

electrophilic addition of HBr to cyclohexene

cyclohexene | cyclohexyl bromide

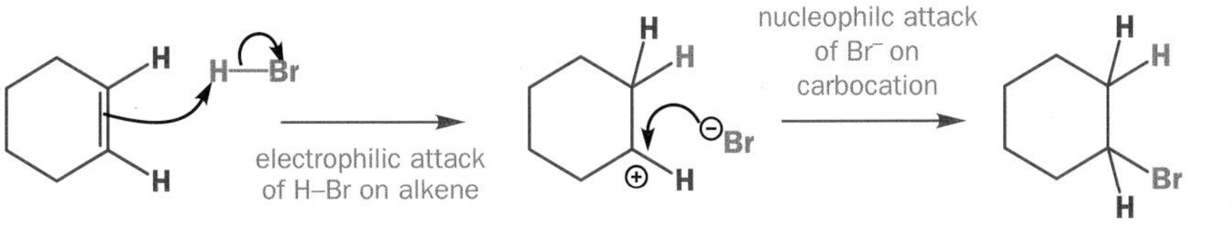

Here are two more syntheses of alkyl bromides, but this time we need to ask our question about which end of the alkene is attacked, because the alkenes are unsymmetrical (they have different substituents at each end). First, the results.

styrene | 1-bromo-1-phenylethane | isobutene | *t*-butyl bromide

In each case, the bromine atom ends up on the more substituted carbon, and the mechanism explains why. There are the two possible outcomes for protonation of styrene by HBr, but you should immediately be able to spot which is preferred, even if you don't know the outcome of the reaction. Protonation at one end gives a stabilized, benzylic cation, while protonation at the other would give a highly unstable primary cation, and therefore does not take place. The benzylic cation gives the benzylic alkyl bromide.

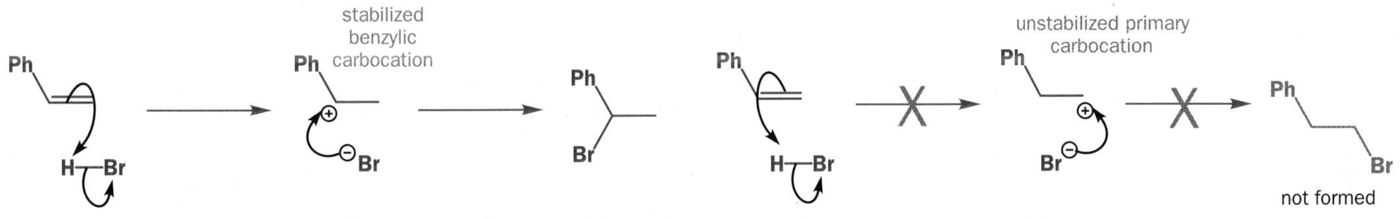

You get the same result with isobutene: the more stable, tertiary cation leads to the product; the alternative primary cation is not formed.

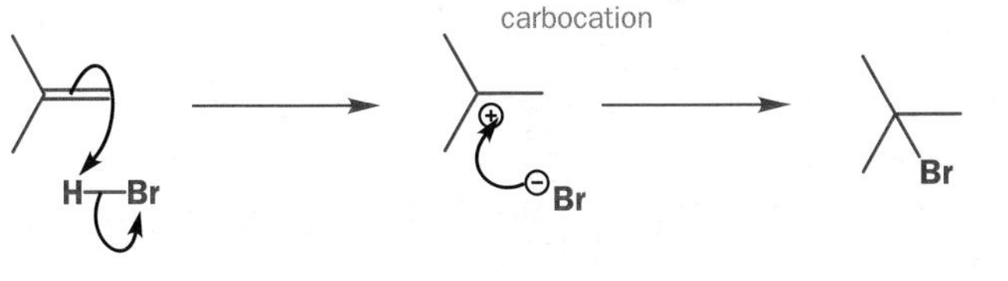

Markovnikov's rule

There is a traditional mnemonic called 'Markovnikov's rule' for electrophilic additions of H–X to alkenes, which can be stated as 'The hydrogen ends up attached to the carbon of the double bond that had more hydrogens to start with.' We don't suggest you learn this rule, though you may hear it referred to. As with all 'rules' it is much more important to understand the reason behind it. For example, *you* can now predict the product of the reaction below. Markovnikov couldn't.

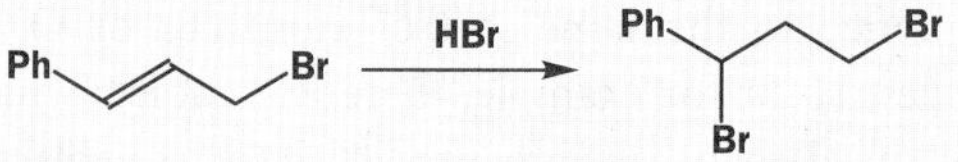

The protonation of alkenes to give carbocations is quite general. The carbocations may trap a nucleophile, as you have just seen, or they may simply lose a proton to give back an alkene. This is just the same as saying the protonation is reversible, but it needn't be the same proton that is lost. A more stable alkene may be formed by losing a different proton, which means that acid can catalyse the isomerization of alkenes—both between *Z* and *E* geometrical isomers and between regioisomers.

isomerization of an alkene in acid

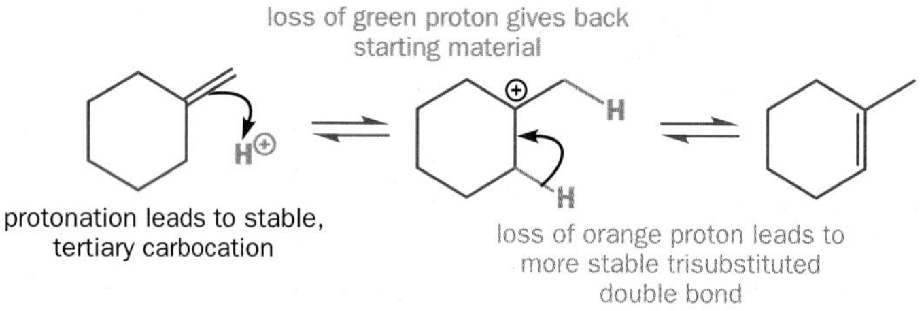

loss of green proton gives back starting material

protonation leads to stable, tertiary carbocation

loss of orange proton leads to more stable trisubstituted double bond

E1 and isomerization

The isomerization of alkenes in acid is probably a good part of the reason why E1 eliminations in acid generally give *E*-alkenes. In Chapter 19, we explained how *kinetic* control could lead to *E*-alkenes: interconversion of *E*- and *Z*-alkenes under the conditions of the reaction allows the *thermodynamic* product to prevail.

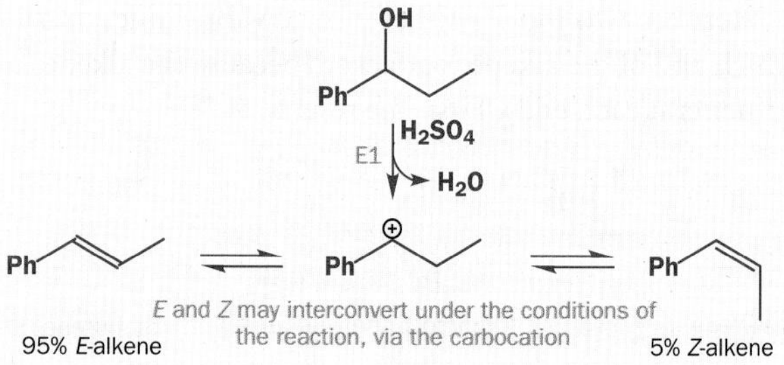

E and *Z* may interconvert under the conditions of the reaction, via the carbocation

95% *E*-alkene   5% *Z*-alkene

Other nucleophiles may also intercept the cation: for example, alkenes can be treated with HCl to form alkyl chlorides, with HI to form alkyl iodides, and with H_2S to form thiols.

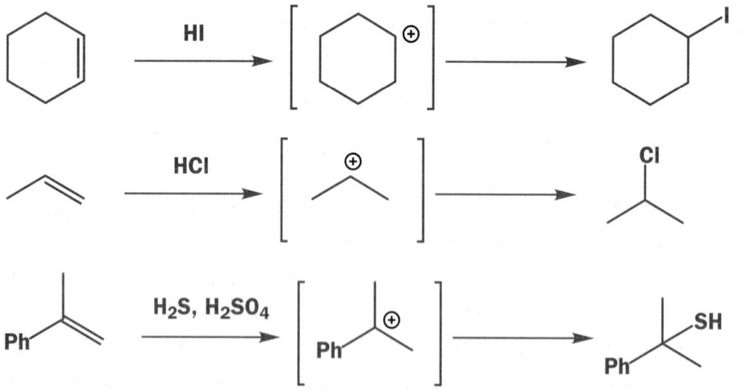

Electrophilic addition to dienes

Earlier in the chapter you saw the epoxidation of a diene to give a monoepoxide: only one of the double bonds reacted. This is quite a usual observation: dienes are more nucleophilic than isolated alkenes. This is easy to explain by looking at the relative energy of the HOMO of an alkene and a diene—this discussion is on p. 168 of Chapter 7. Dienes are therefore very susceptible to protonation by acid to give a cation. This is what happens when 2-methylbuta-1,3-diene (isoprene) is treated with acid. Protonation gives a stable delocalized allylic cation.

isoprene

Why protonate this double bond and not the other one? The cation you get by protonating the other double bond is also allylic, but it cannot benefit from the additional stabilization from the methyl group because the positive charge is not delocalized on to the carbon carrying the methyl.

positive charge not delocalized on to this carbon, so Me cannot contribute to stability of cation

isoprene   prenyl bromide

If the acid is HBr, then nucleophilic attack by Br⁻ on the cation follows. The cation is attacked at the less hindered end to give the important compound prenyl bromide. This is very much the sort of reaction you met in Chapter 17—it is the second half of an S_N1 substitution reaction on an allylic compound.

Overall, the atoms H and Br are added to the ends of the diene system. The same appears to be the case when dienes are brominated with Br_2.

Changing the conditions slightly gives a different outcome. If the reaction is done at lower temperatures, the bromine just adds across one of the double bonds to give a 1,2-dibromide.

This compound turns out to be the kinetic product of the bromination reaction. The 1,4-dibromide is formed only when the reaction is heated, and is the thermodynamic product. The mechanism is electrophilic attack on the diene to give a bromonium ion, which bromide opens to give the dibromide. We have shown the bromide attacking the more substituted end of the bromide—though we can't know this for sure (attack at either end gives the same product), you are about to see (in the next section) evidence that this is the usual course of reactions of unsymmetrical bromonium ions.

This 1,2-dibromide can still react further, because it can undergo nucleophilic substitution. Bromide is a good nucleophile and a good leaving group and, with an allylic system like this, S_N1 can take place in which both the nucleophile and the electrophile are bromine. The intermediate is a cation, but here the carbocation is disguised as the bromonium ion because bromine's lone pair can help stabilize the positive charge. Bromide can attack where it left, returning to starting material, but it can also attack the far end of the allylic system, giving the 1,4-dibromide. The steps are all reversible at higher temperatures, so the fact that the 1,4-dibromide is formed under these conditions must mean it is more stable than the 1,2-dibromide. It is not hard to see why: it has a more substituted double bond and the two large bromine atoms are further apart.

cation intermediate

Unsymmetrical bromonium ions open regioselectively

We ignored the issue of symmetry in the alkene when we discussed the bromination of alkenes, because even unsymmetrical alkenes give the same 1,2-dibromides whichever way the bromide attacks the bromonium ion.

bromination of isobutene

same product whichever end
is attacked

But when a bromination is done in a nucleophilic solvent—water or methanol, for example—solvent molecules compete with the bromide to open the bromonium ion. As you know, alcohols are much worse nucleophiles than bromide but, because the concentration of solvent is so high (remember—the concentration of water in water is 55M), the solvent gets there first most of the time. This is what happens when isobutene is treated with bromine in methanol. An ether is formed by attack of methanol only at the *more substituted* end of the bromonium ion.

methanol attacks at the more
substituted end of the
bromonium ion

Methanol is attacking the bromonium ion where it is most hindered, so there must be some effect at work more powerful than steric hindrance. One way of looking at this is to reconsider our assumption that bromonium ion opening is an S_N2 process. Here, it hardly looks S_N2. We have a tertiary centre, so naturally you expect S_N1, via the cation below. But we have already said that cations like this can be stabilized by formation of the three-membered bromonium ion and, if we let this happen, we have to attack the bromonium ion which gets us back to where we started: an S_N2 mechanism!

two limiting mechanisms for substitution on bromonium ion

The answer to the conundrum is that substitution reactions don't always go by pure S_N1 or pure S_N2 mechanisms: sometimes the mechanism is somewhere in between. Perhaps the leaving group starts to leave, creating a partial positive charge on carbon which is intercepted by the nucleophile. This provides a good explanation of what is going on here. The bromine begins to leave, and a partial positive charge builds up at carbon. The departure of bromine can get to a more advanced state at the tertiary end than at the primary end, because the substituents stabilize the build-up of positive charge. The bromonium ion can be more accurately represented as shown in the margin, with one C–Br bond longer than the other, and more polarized than the other.

build-up of
partial positive
charge

longer, weaker bond

The nucleophile now has a choice: does it attack the more accessible, primary end of the bromonium ion, or does it attack the more charged end with the weaker C–Br bond? Here, the latter is clearly the faster reaction. The transition state has considerable positive charge on carbon, and is known as a **loose S_N2 transition state**.

loose S_N2 transition state

The products of bromination in water are called **bromohydrins**. They can be treated with base, which deprotonates the alcohol. A rapid intramolecular S_N2 reaction follows: bromide is expelled as a leaving group and an epoxide is formed. This can be a useful alternative synthesis of epoxides avoiding peroxy-acids.

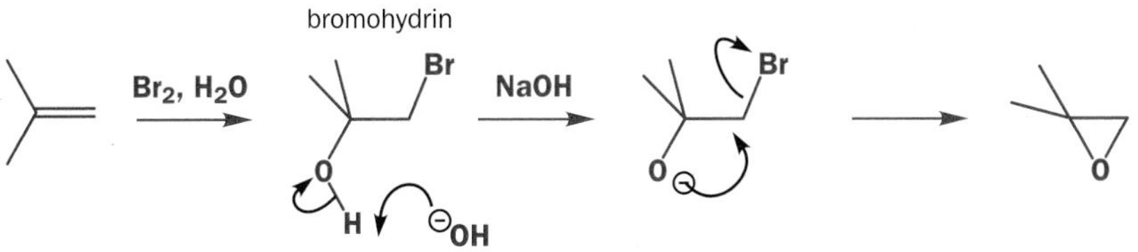

Rates of bromination of alkenes

The pattern you saw for epoxidation with peroxy-acids (more substituted alkenes react faster) is followed by bromination reactions too. The bromonium ion is a reactive intermediate, so the rate-determining step of the brominations is the bromination reaction itself. The chart shows the effect on the rate of reaction with bromine in methanol of increasing the number of alkyl substituents from none (ethylene) to four. Each additional alkene substituent produces an enormous increase in rate. The degree of branching (Me versus n-Bu versus t-Bu) within the substituents has a much smaller, negative effect (probably of steric origin) as does the geometry (E versus Z) and substitution pattern (1,1-disubstituted versus 1,2-disubstituted) of the alkene.

relative rates of reaction of alkenes with bromine in methanol solvent.

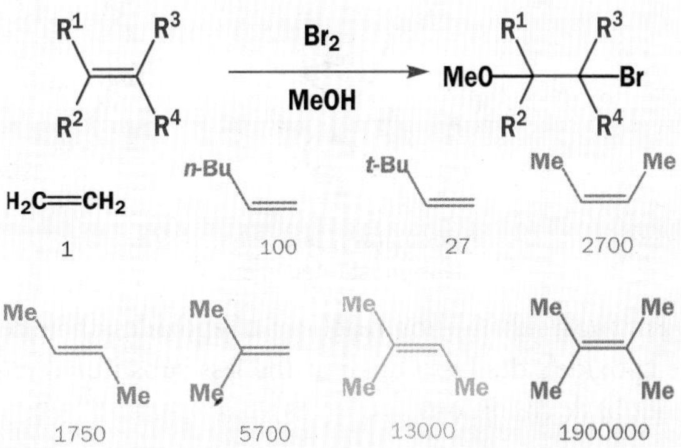

The regioselectivity of epoxide opening can depend on the conditions

Although epoxides, like bromonium ions, contain strained three-membered rings, they require either acid catalysis or a powerful nucleophile to react well. Compare these two reactions of a 1,1,2-trisubstituted epoxide. They are nucleophilic substitutions related to those we introduced in Chapter 17 (p. 435) but in that chapter we carefully avoided discussing epoxides of the unsymmetrical variety. In this example, the regiochemistry reverses with the reaction conditions. Why?

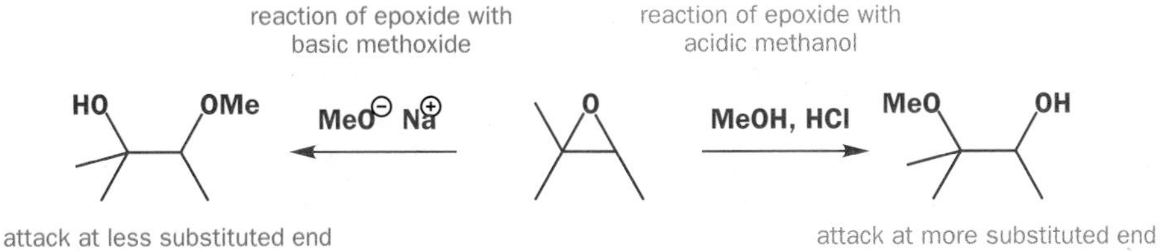

reaction of epoxide with basic methoxide

reaction of epoxide with acidic methanol

attack at less substituted end

attack at more substituted end

We'll start with the acid-catalysed reaction, because it is more similar to the examples we have just been discussing—opening happens at the more substituted end. Protonation by acid produces a positively charged intermediate that bears some resemblance to the corresponding bromonium ion. The two alkyl groups make possible a build-up of charge on the carbon at the tertiary end of the protonated epoxide, and methanol attacks here, just as it does in the bromonium ion.

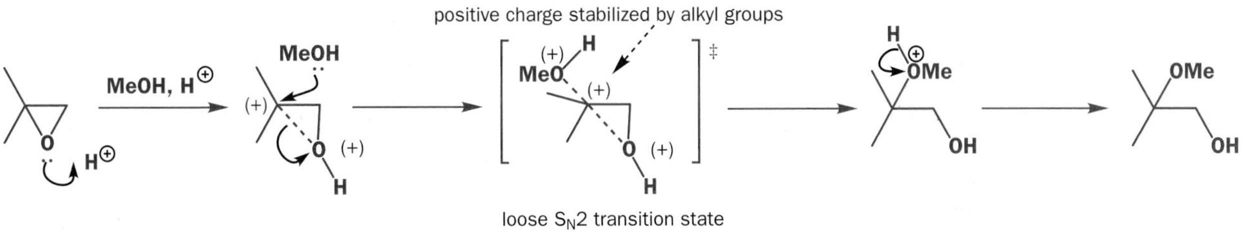

positive charge stabilized by alkyl groups

loose S_N2 transition state

■
Remember, S$_N$1 can be fast only with good leaving groups.

In base, there can be no protonation of the epoxide, and no build-up of positive charge. Without protonation, the epoxide oxygen is a poor leaving group, and leaves only if pushed by a strong nucleophile: the reaction becomes pure S$_N$2. Steric hindrance becomes the controlling factor, and methoxide attacks only the primary end of the epoxide.

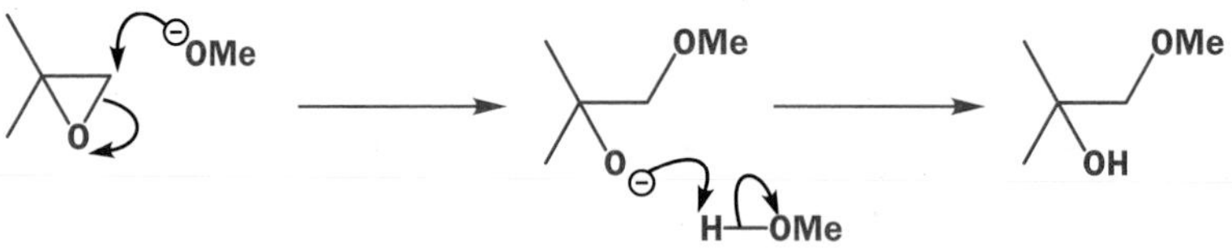

This example makes the matter look deceptively clear-cut. But with epoxides, regioselectivity is not as simple as this because, even with acid catalysts, S$_N$2 substitution at a primary centre is very fast. For example, Br⁻ in acid attacks this epoxide mainly at the less substituted end, and only 24% of the product is produced by the 'cation-stabilized' pathway. It is very difficult to override the preference of epoxides unsubstituted at one end to react at that end.

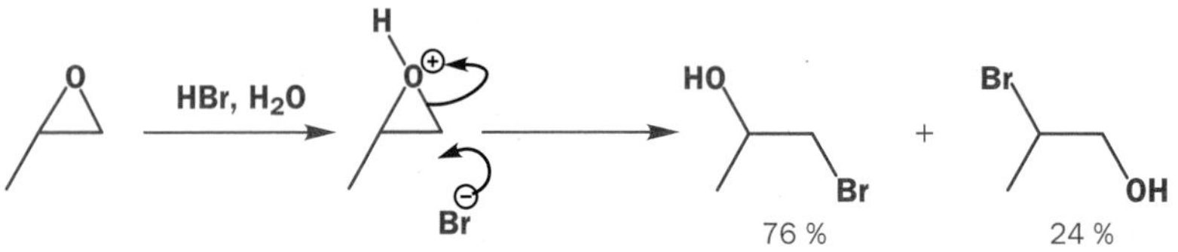

major pathway is S$_N$2 on
less substituted end

For most substitution reactions of epoxides, then, regioselectivity is much higher if you give in to the epoxide's desire to open at the less substituted end, and enhance it with a strong nucleophile under basic conditions.

Electrophilic additions to alkenes can be stereoselective

▶
The reaction is stereospecific because it's the stereochemistry of the epoxide that determines the outcome of the reaction. The S$_N$2 reaction has no choice but to go with inversion. We discussed the terms stereospecific and stereoselective on p. 492.

Although they really belong in Chapter 17 with other nucleophilic substitution reactions, we included the last few examples of epoxide-opening reactions here because they have many things in common with the reactions of bromonium ions. Now we are going to make the analogy work the other way when we look at the stereochemistry of the reactions of bromonium ions, and hence at the stereoselectivity of electrophilic additions to alkenes. We shall first remind you of an epoxide reaction from Chapter 17, where you saw this.

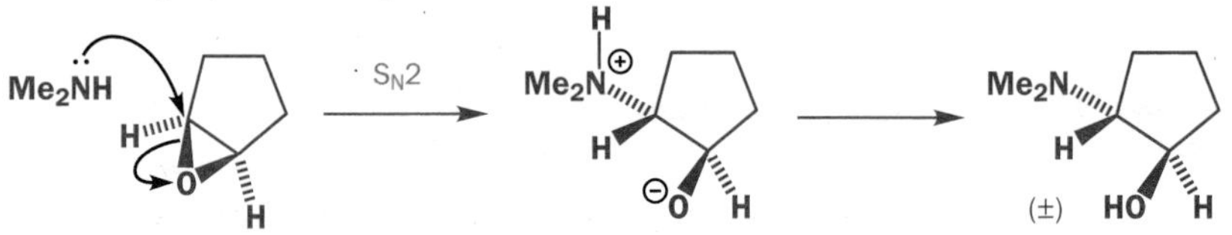

▶
Notice the all important (±) symbol below the products in the diagram. They are single diastereoisomers, but they are *necessarily* formed as racemic mixtures, as we discussed in Chapter 16. You can look at it this way: the Me$_2$NH will attack the two identical ends of the epoxide with precisely equal probability. Both give the same *anti* diastereoisomer, but each gives an opposite enantiomer. The two enantiomers will be formed in precisely equal amounts.

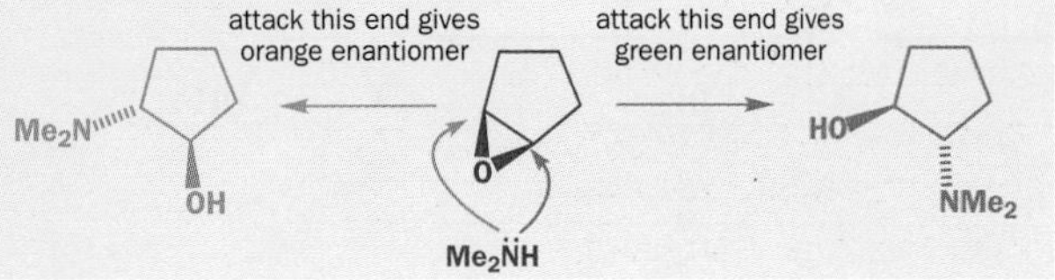

The epoxide ring opening is stereospecific: it is an S$_N$2 reaction, and it goes with inversion. The epoxide starts on the top face of the ring, and the amino group therefore ends up on the bottom face. In other words, the two groups end up *anti* or *trans* across the ring. You now know how to make this epoxide—you would use cyclopentene and *m*-CPBA, and in two steps you could 'add' an OH group and a Me$_2$N group *anti* across the double bond.

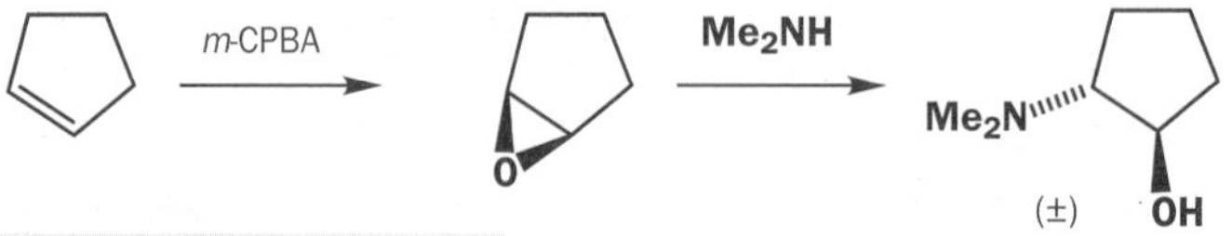

Now we can move on to look at the stereochemistry of electrophilic addition to alkenes.

attack this end gives
orange enantiomer

attack this end gives
green enantiomer

Electrophilic addition to alkenes can produce stereoisomers

When cyclohexene is treated with bromine in carbon tetrachloride, the racemic *anti*-1,2-dibromocyclohexane is obtained exclusively.

exclusively this
diastereoisomer
formed

none of this
diastereoisomer
formed

> We don't need to write (±) next to the isomer that isn't formed, because it is an achiral structure—it has a plane of symmetry and is a *meso* compound. See p. 396.

The result is no surprise if we think first of the formation of the bromonium ion that is opened with inversion in an S_N2 reaction. Here is the mechanism drawn 'flat', which is all we need to explain the stereochemistry of the product. The fact that this reaction (like other similar ones) gives a single diastereoisomer is one of the best pieces of evidence that electrophilic additions of Br_2 to alkenes proceed through a bromonium ion.

intermediate bromonium ion

But these compounds are six-membered rings, so we will get a more accurate picture of what is going on if we draw them in their correct conformation. Cyclohexene is a flattened chair, as you saw in Chapter 18, and the bromonium ion can be drawn as a flattened chair too, like an epoxide (p. 469). Bromonium opening mirrors epoxide opening closely and, for the same reason, it will open only to give the diaxial product. In the absence of a locking group, the diaxial 1,2-dibromocyclohexane rapidly flips to the diequatorial conformation. This, of course, has no effect on the relative configuration, which will always be *anti*.

> ■ This part of the discussion is a revision of the material in Chapter 18. When dealing with six-membered rings, you should always aim to draw their conformation, though in this case you can explain the result adequately without conformational diagrams.

diaxial product formed

rapid ring inversion to
diequatorial conformation

Bromination of alkenes is stereospecific, because the geometry of the starting alkene determines which product diastereoisomer is obtained. We couldn't demonstrate this with cyclohexene, because only a *Z* double bond is possible in a six-membered ring. But bromination or chlorination of *Z*- and *E*-2-butene in acetic acid produces a single diastereoisomer in each case, and they are different from each other. *Anti* addition occurs in both cases—more evidence that a bromonium ion is the intermediate. In the scheme below, the product of each reaction is shown in three different ways. Firstly, the two new C–Br bonds are shown in the plane of the paper to highlight the inversion of configuration during the bromonium opening step. Secondly, this diagram has been rotated to place the carbon chain in the plane of the paper and highlight the fact that these are indeed two different diastereoisomeric products. In this conformation you can clearly see that there has been an *anti*-addition across the *E* double bond. Thirdly, the middle bond has been rotated 180° to give an (unrealistically) eclipsed conformation. We show this conformation for two reasons: it makes it clear that the addition across the *Z*-butene is stereospecific and *anti* too, and it also makes it quite clear that the product of the *E*-butene bromination is achiral: you can see the plane of symmetry in this conformation, and this is why we haven't placed (±) signs next to the products from the *E*-alkene. Note that in all three different views of each product the same stereoisomer is represented. There is no change of configuration, only changes of conformation to help you understand what is going on. If you cannot follow any of the 'redrawing' steps, make a model. With practice, you will soon learn to manipulate mental models in your head, and to see what happens to substituents when bonds are rotated. Most importantly, don't let all of this more subtle stereochemical discussion cloud the simple message: addition of Br_2 to alkenes is stereospecific and *trans*.

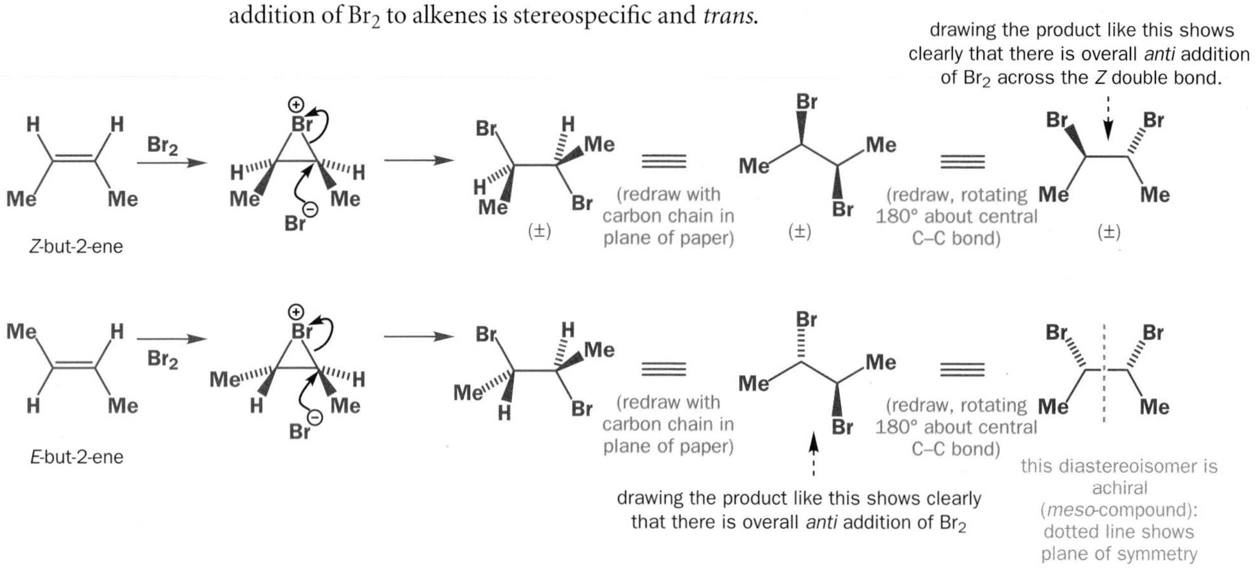

drawing the product like this shows clearly that there is overall *anti* addition of Br_2 across the *Z* double bond.

drawing the product like this shows clearly that there is overall *anti* addition of Br_2

this diastereoisomer is achiral (*meso*-compound): dotted line shows plane of symmetry

Bromonium ions as intermediates in stereoselective synthesis

You will not be surprised to learn that the other nucleophiles (water and alcohols) you saw intercepting bromonium ions earlier in the chapter also do so stereospecifically. The following reaction can be done on a large scale, and produces a single diastereoisomer of the product (racemic, of course) because water opens the bromonium ion with inversion.

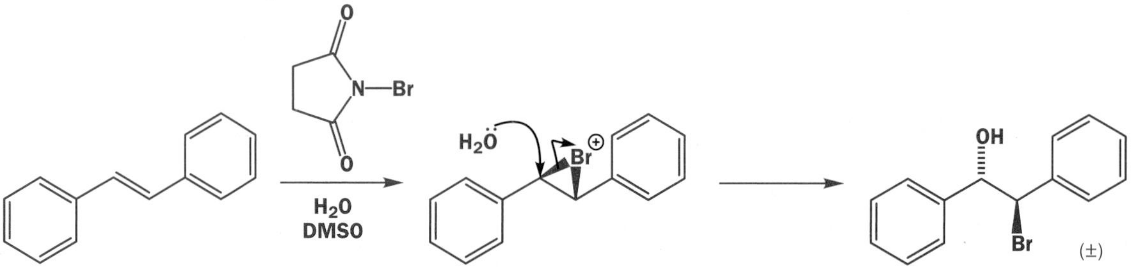

The reagent used to form the bromonium ion here is not bromine, and may be new to you. It is called *N*-bromosuccinimide, or NBS for short. Unlike the noxious brown liquid bromine, NBS is an easily handled crystalline solid, and is perfect for electrophilic addition of bromine to alkenes when the bromonium ion is not intended to be opened by Br^-. It works by providing a very small concentration of Br_2 in solution: a small amount of HBr is enough to get the reaction going, and thereafter every addition reaction produces another molecule of HBr which liberates more Br_2 from NBS. In a sense, NBS is a source of 'Br^+'.

▶

NBS is known to act as a source of Br_2 because the results of reactions of NBS and of Br_2 in low concentration are identical.

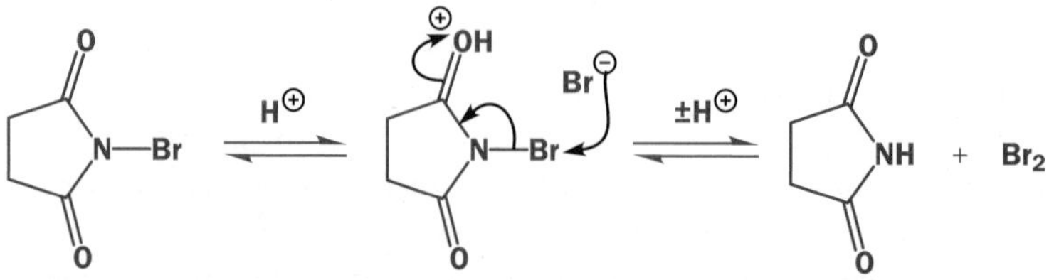

N-bromosuccinimide (NBS)

With NBS, the concentration of Br⁻ is always low, so alcohols compete with Br⁻ to open the epoxide even if they are not the solvent. In the next example, the alcohol is 'propargyl alcohol', prop-2-yn-1-ol. It gives the expected *anti*-disubstituted product with cyclohexene and NBS.

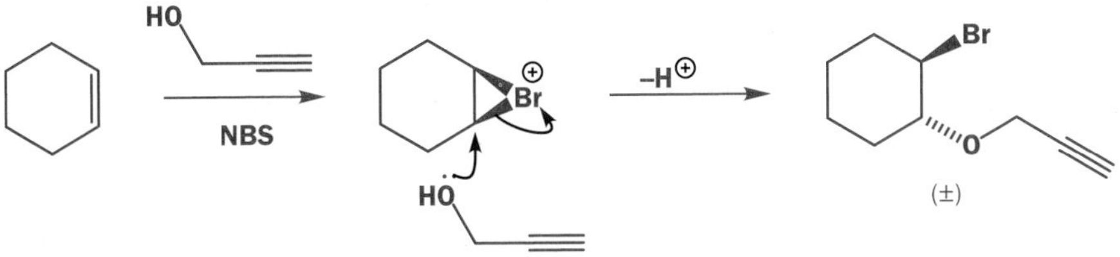

When 1-methylcyclohexene is used as the starting material, there is additionally a question of regioselectivity. The alcohol attacks the more hindered end of the bromonium ion—the end where there can be greatest stabilization of the partial positive charge in the 'loose S$_N$2' transition state. This reaction really does illustrate the way in which a mechanism can lie in between S$_N$1 and S$_N$2. We see a configurational inversion, indicative of an S$_N$2 reaction, happening at a tertiary centre where you would usually expect S$_N$1.

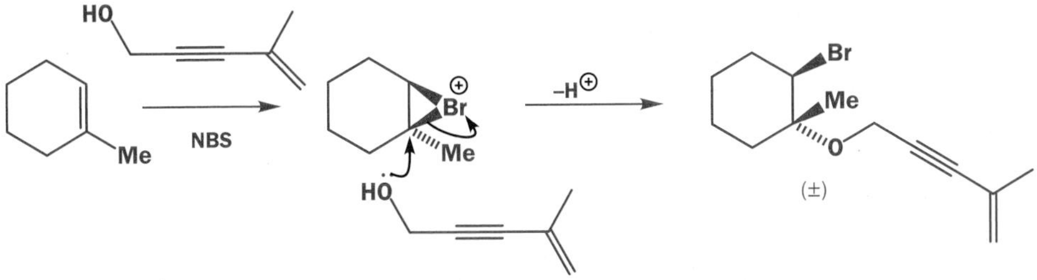

Iodolactonization and bromolactonization make new rings

To finish our discussion of bromonium ions, you need to know about one more important class of reactions, those in which the nucleophile is located within the same molecule as the bromonium ion. Here is an example: the nucleophile is a carboxylate, and the product is a lactone (a cyclic ester). This type of reaction—the cyclization of an unsaturated acid—is known as a bromolactonization. Intermolecular attack on the bromonium ion by bromide ion does not compete with the intramolecular cyclization step.

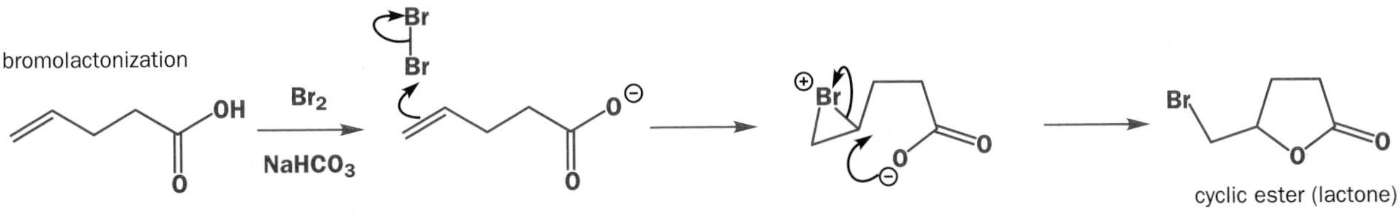

cyclic ester (lactone)

Every example of electrophilic addition of a halogen to an alkene that we have shown you so far has been with bromine. This is quite representative: bromine is the most widely used halogen for electrophilic addition, since its reactivity is second only to iodine, yet the products are more stable. However, in these lactonization reactions, iodine is the more commonly used reagent, and the products of iodolactonizations are important intermediates (you will meet them again in Chapter 33). In the next example, the iodolactonization product is treated with sodium methoxide, which appears (a) to hydrolyse the lactone, and (b) to substitute the iodide for OMe. In fact, there is a little more to this than meets the eye.

> It should be mentioned at this point that five-membered ring formation is the norm in iodolactonizations—you will need to wait until Chapter 42 to hear the full details why—but here this preference is reinforced by the preference for opening at the more substituted end of an iodonium or bromonium ion.

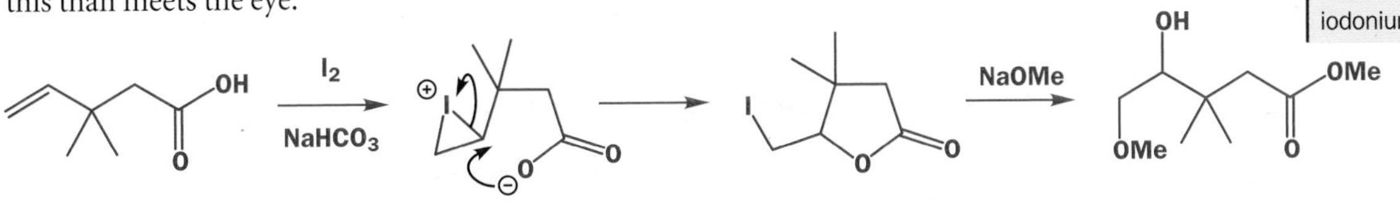

The first step is now familiar to you: electrophilic attack of iodine to form an iodonium ion, which cyclizes to the iodolactone; the key step of the mechanism is shown above.

Methoxide must attack the carbonyl group, liberating an alkoxide that immediately cyclizes, with the iodide as a leaving group, to form an epoxide. Finally, methoxide attacks the epoxide at the less hindered end. Contrast the regioselectivities for attack on the iodonium ion with attack on the epoxide.

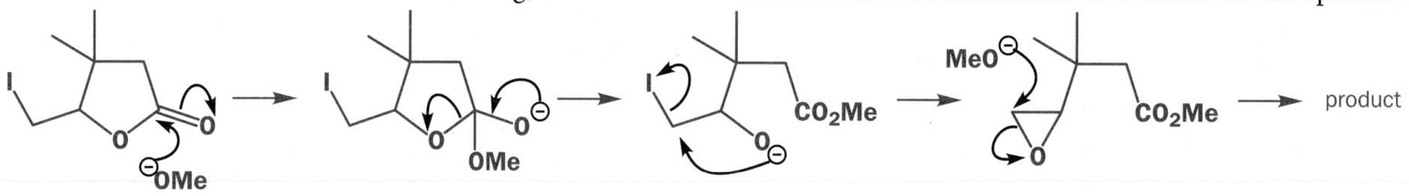

How to add water across a double bond

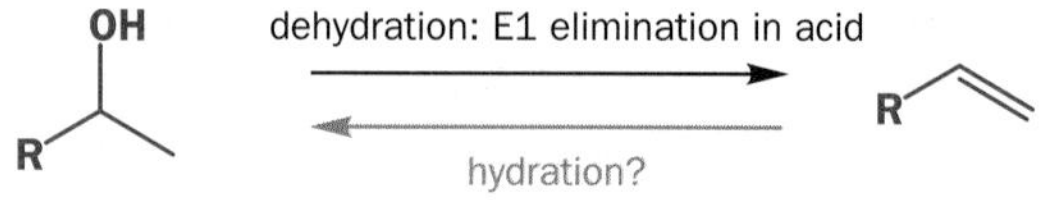

In the last chapter, you saw alkenes being made from alcohols by E1 elimination—dehydration—under acid catalysis. The question we are going to answer in this section is: how can you make this elimination run backwards—in other words, how can you hydrate a double bond?

It is possible on occasion simply to use aqueous acid to do this. The reaction works only if protonation of the alkene can give a stable, tertiary cation. The cation is then trapped by the aqueous solvent.

acid-catalysed hydration of an alkene

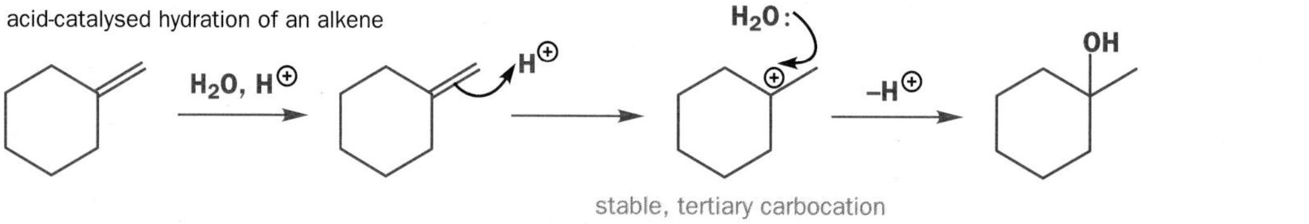

In general, though, it is very difficult to predict whether aqueous acid will hydrate the alkene or dehydrate the alcohol. The method we are about to introduce is much more reliable. The key is to use a transition metal to help you out. Alkenes are soft nucleophiles (p. 237) and interact well with soft electrophiles such as transition metal cations. Here, for example, is the complex formed between an alkene and mercury(II) cation. Don't be too concerned about the weird bond growing from the middle of the alkene: this is a shorthand way of expressing the rather complex bonding interaction between the alkene and mercury. An alternative, and more useful, representation is the three-membered ring on the right.

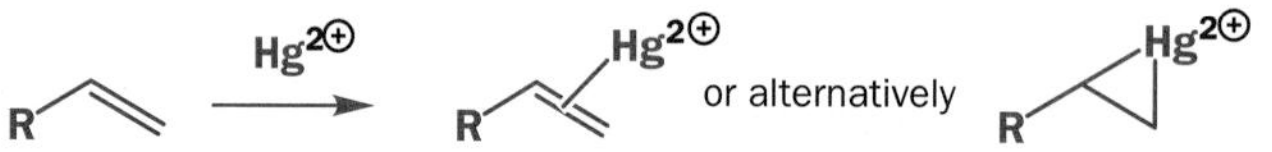

■
There is more detail on organometallic chemistry in Chapter 48.

The complex should remind you of a bromonium ion, and rightly so, because its reactions are really rather similar. Even relatively feeble nucleophiles such as water and alcohols, when used as the solvent, open the 'mercurinium' ion and give alcohols and ethers. In the next scheme, the mercury(II) is supplied as mercury(II) acetate, $Hg(OAc)_2$, which we shall represent with two covalent Hg–O bonds (simply because it helps with the arrows and with electron-accounting to do so). Unsurprisingly, water attacks at the more substituted end of the mercurinium ion.

oxymercuration

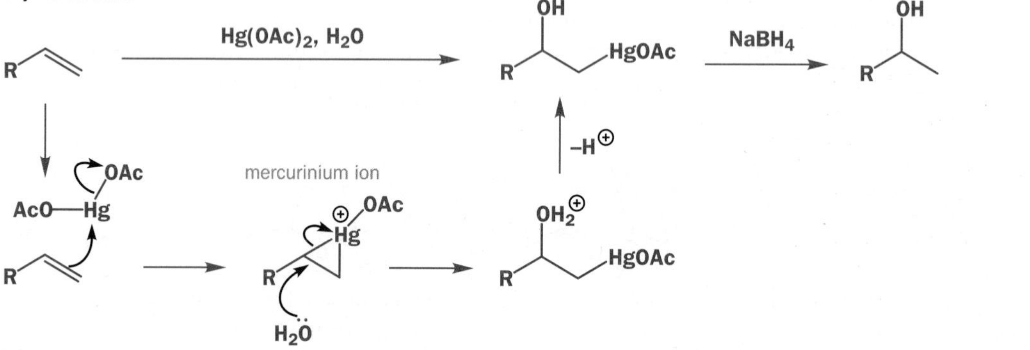

We've added OH and Hg(II) across the alkene, and the reaction is termed an **oxymercuration**. But a problem remains: how to get rid of the metal. The C–Hg bond is very weak and the simplest way to replace Hg with H is to cleave with a reducing agent. NaBH$_4$ works fine. Here is an example of oxymercuration–demercuration at work—the intermediate organomercury is not isolated.

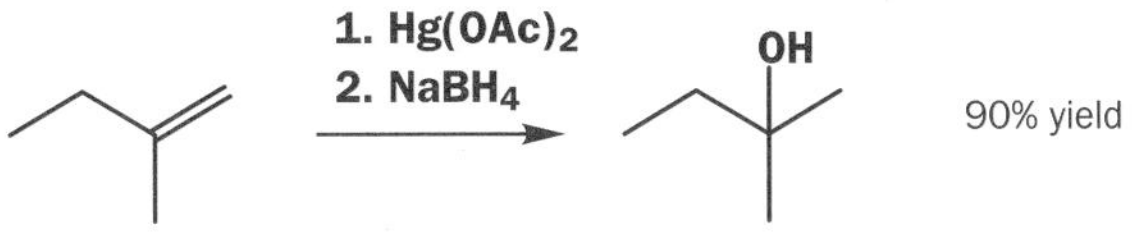

1. Hg(OAc)$_2$
2. NaBH$_4$

90% yield

This reaction is discussed in more detail in Chapter 39.

Hydration of alkynes

Oxymercuration works particularly well with alkynes. Here are the conditions, and the product, following the analogy of alkene hydration, should be the compound shown at the right-hand end of the scheme below.

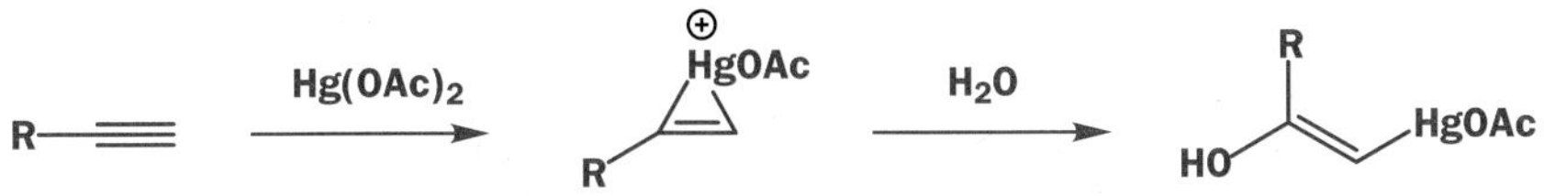

But the product isolated from an alkyne oxymercuration is in fact a ketone. You can see why if you just allow a proton on this initial product to shift from oxygen to carbon—first protonate at C then deprotonate at O. C=O bonds are stronger than C=C bonds, and this simple reaction is very fast.

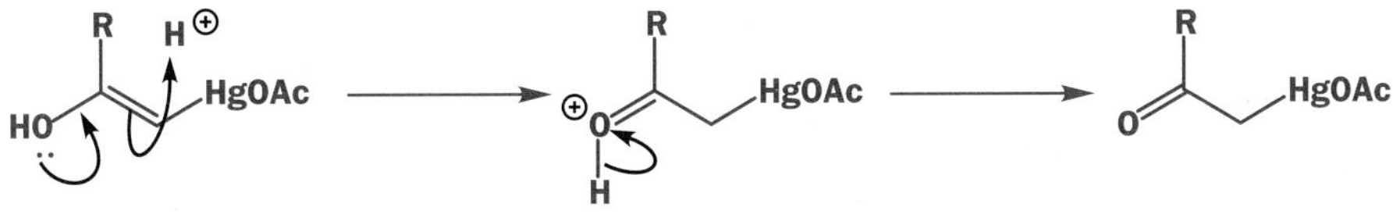

We now have a ketone, but we also still have the mercury. That is no problem when there is a carbonyl group adjacent, because any weak nucleophile can remove mercury in the presence of acid as shown below. Finally, another proton transfer (from O to C again) gives the real product of the reaction: a ketone.

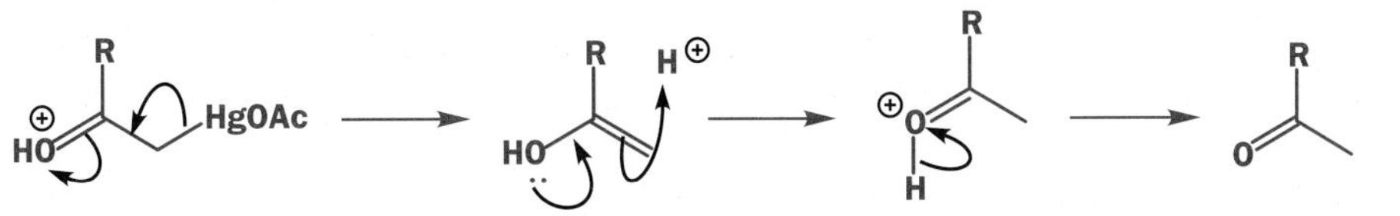

This is truly a very useful way of making methyl ketones, because terminal alkynes can be made using the methods of Chapter 9 (addition of metallated alkynes to electrophiles).

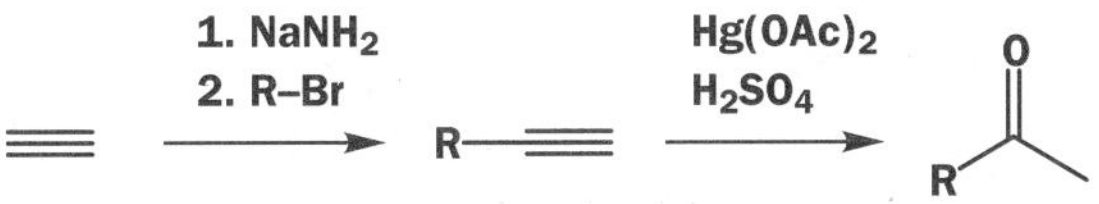

1. NaNH$_2$
2. R–Br

Hg(OAc)$_2$
H$_2$SO$_4$

Anticancer compounds

The anthracyclinone class of anticancer compounds (which includes daunomycin and adriamycin) can be made using a mercury(II)-promoted alkyne hydration. You saw the synthesis of alkynes in this class on Chapter 9 where we discussed additions of metallated alkynes to ketones. Here is the final step in a synthesis of the anticancer compound deoxydaunomycinone: the alkyne is hydrated using Hg^{2+} in dilute sulfuric acid; the sulfuric acid also catalyses the hydrolysis of the phenolic acetate to give the final product.

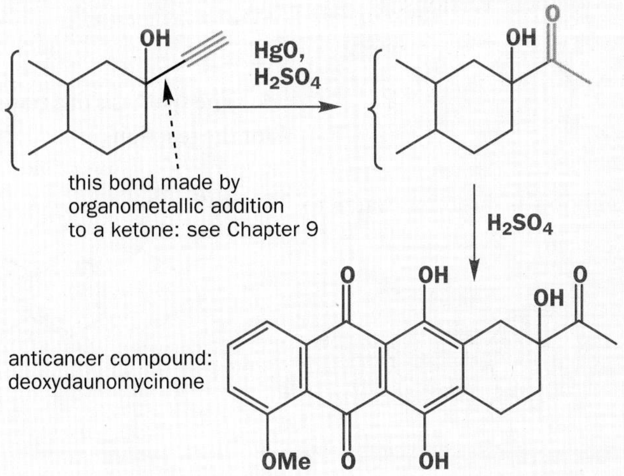

this bond made by organometallic addition to a ketone: see Chapter 9

HgO,
H$_2$SO$_4$

H$_2$SO$_4$

anticancer compound: deoxydaunomycinone

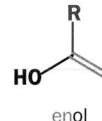
enol

Those alkenes carrying hydroxyl groups are called **enols** (ene + ol), and they are among the most important intermediates in chemistry. They happen to be involved in this reaction, and this was a good way to introduce you to them but, as you will see in the next chapter and beyond, enols (and their deprotonated sisters, enolates) have far reaching significance in chemistry.

To conclude...

Electrophilic addition to double bonds gives three-membered ring intermediates with Br_2, with Hg^{2+}, and with peroxy-acids (in which case the three-membered rings are stable and are called epoxides). All three classes of three-membered rings react with nucleophiles to give 1,2-difunctionalized products with control over (1) regioselectivity and (2) stereoselectivity. Protonation of a double bond gives a cation, which also traps nucleophiles, and this reaction can be used to make alkyl halides. Some of the sorts of compounds you can make by the methods of this chapter are shown below.

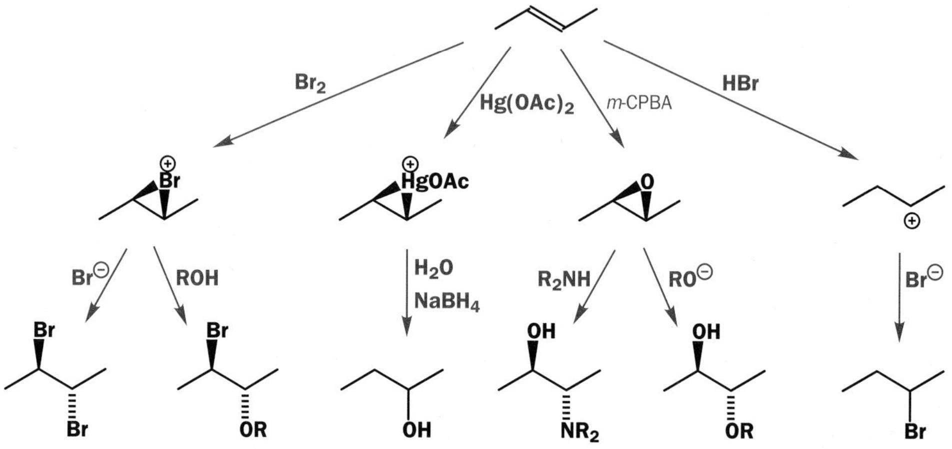

Problems

1. Predict the orientation in HCl addition to these alkenes.

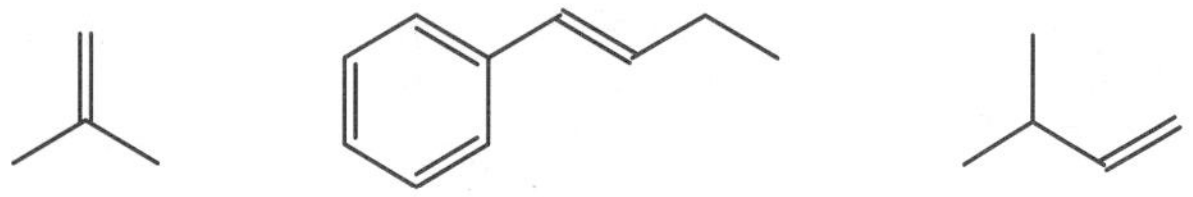

2. Suggest mechanisms and products for these reactions.

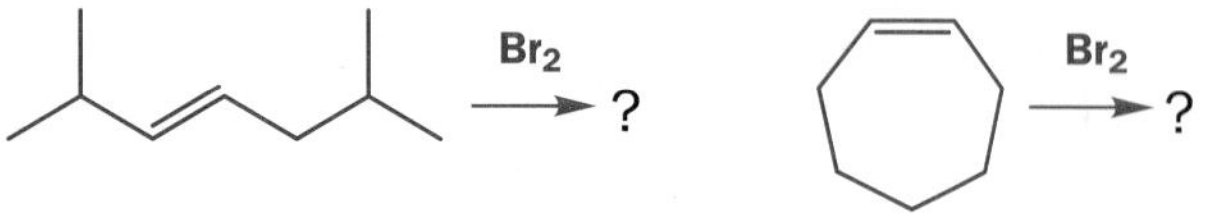

3. What will be the products of addition of bromine water to these alkenes?

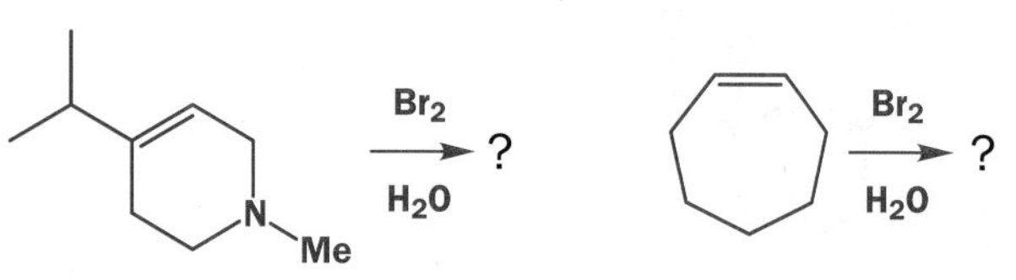

4. By working at low temperature with one equivalent of a buffered solution of a peroxy-acid, it is possible to prepare the monoepoxide of cyclopentadiene. Why are the precautions necessary and why does the epoxidation not occur again?

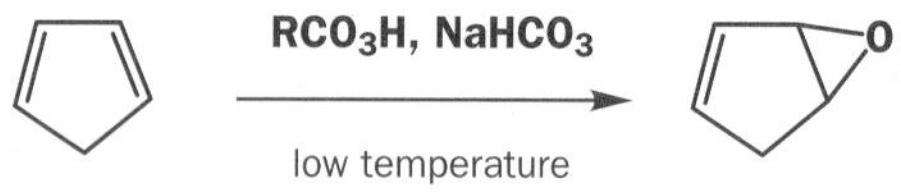

5. The synthesis of a tranquillizer uses this step. Give mechanisms for the reactions.

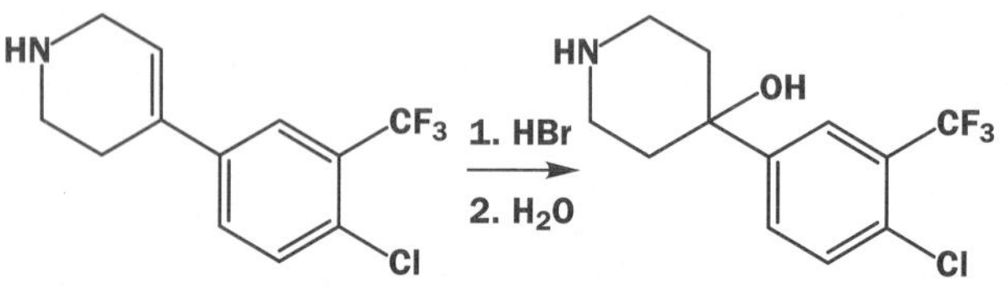

6. Explain this result.

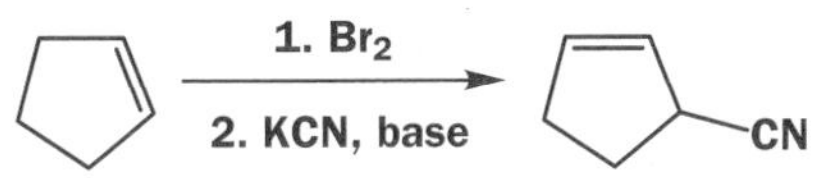

1. Br$_2$
2. KCN, base

7. Bromination of this alkene in water gives a single product in good yield. What is the structure and stereochemistry of this product?

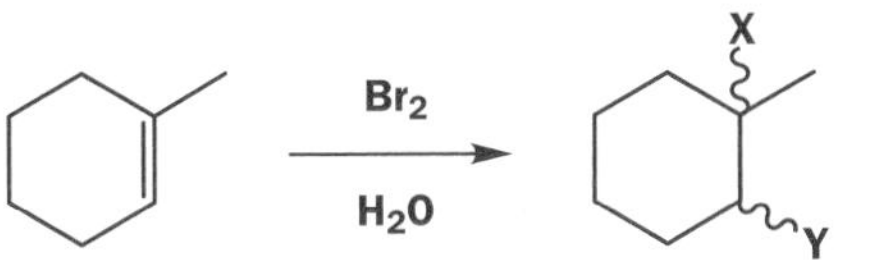

Br$_2$

H$_2$O

X, Y = Br or OH

8. Suggest mechanisms for these reactions.

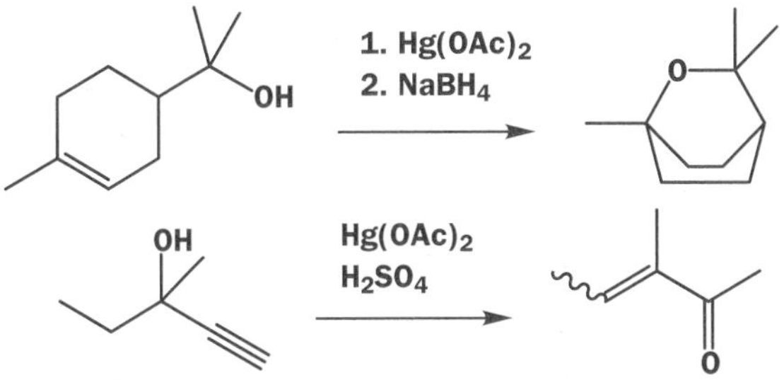

1. Hg(OAc)$_2$
2. NaBH$_4$

Hg(OAc)$_2$
H$_2$SO$_4$

9. Comment on the formation of a single diastereoisomer in this reaction.

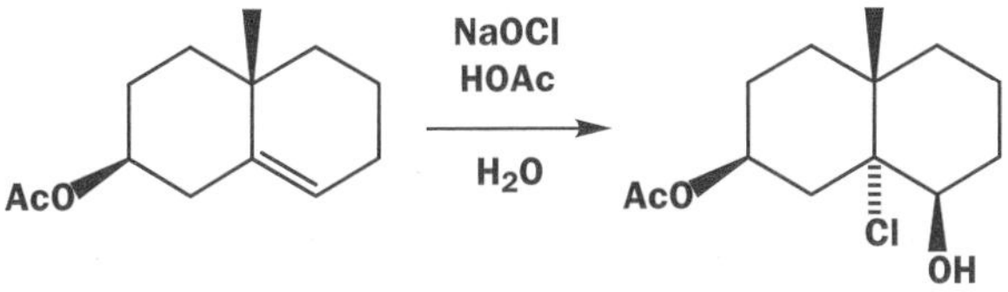

NaOCl
HOAc

H$_2$O

10. Chlorination of this triarylethylene leads to a chloro-alkene rather than a dichloroalkane. Suggest a mechanism and an explanation.

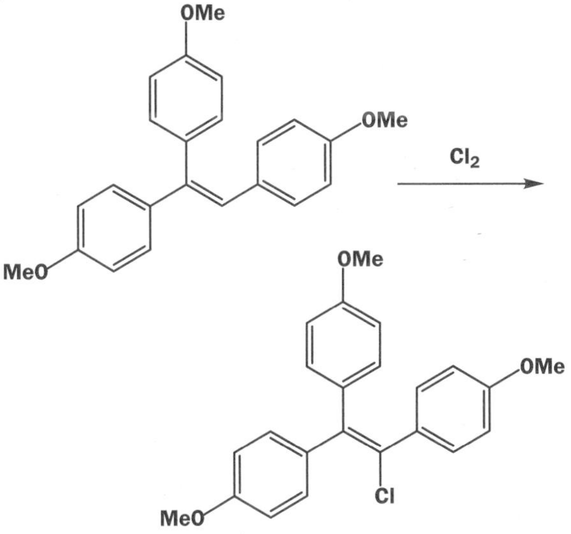

Cl$_2$

11. Revision problem. Give mechanisms for each step in this synthesis and explain any regio- and stereochemistry.

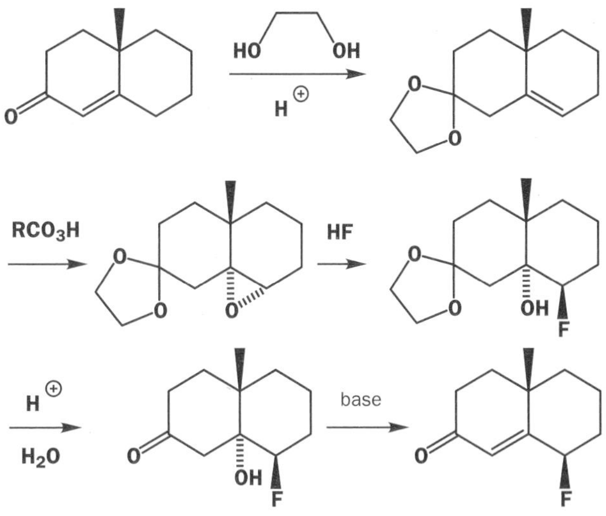

HO   OH

H$^{\oplus}$

RCO$_3$H

HF

H$^{\oplus}$

H$_2$O

base

12. Suggest a mechanism for the following reaction. What is the stereochemistry and conformation of the product?

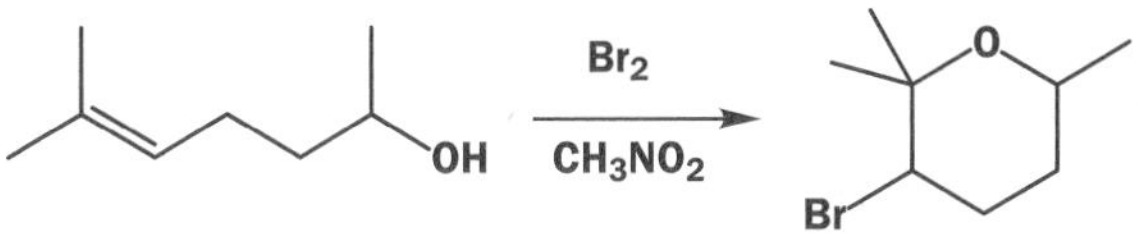

Br$_2$

CH$_3$NO$_2$

12A

12A has these signals in its NMR spectrum: δ_H 3.9(IH, ddq, J12,4,7) and δ_H 4.3 (IH, dd, J11,3).

13. Give a mechanism for this reaction and show clearly the stereochemistry of the product.

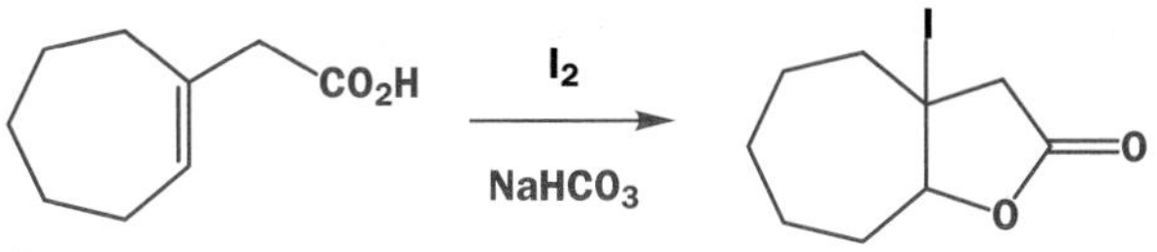

I$_2$

NaHCO$_3$

Formation and reactions of enols and enolates

21

Connections

Building on:	Arriving at:	Looking forward to:
• Carbonyl chemistry ch6, ch9–ch10, ch12, & ch14 • Electrophilic additions to alkenes ch20	• How carbonyl compounds exist in equilibrium with isomers called enols • How acid or base promotes the formation of enols and their conjugate bases, enolates • How enols and enolates have inherent nucleophilic reactivity • How this reactivity can be exploited to allow the introduction of functional groups next to carbonyl groups • How silyl enol ethers and lithium enolates can be used as stable enolate equivalents	• Aromatic compounds as nucleophiles ch22 • The use of enolates in the construction of C–C bonds ch26–ch29 • The central position of enolate chemistry in the chemist's methods of making molecules ch30

We make no apologies for the number of pages we have devoted to carbonyl chemistry. The first reactions you met, in Chapter 6, involved carbonyl compounds. Then in Chapters 9, 10, 12, and 14 we considered different aspects of nucleophilic attack on electrophilic carbonyl compounds. But carbonyl compounds have two opposed sides to their characters. They can be nucleophilic as well: *electrophilic* attack on aldehydes, ketones, and acid derivatives is a useful reaction too. How can the same class of compound be subject both to nucleophilic and to electrophilic attack? The resolution of this paradox is the subject of this chapter where we shall see that most carbonyl compounds exist in two forms—one electrophilic and one nucleophilic. The electrophilic form is the carbonyl compound itself and the nucleophilic form is called the enol.

Would you accept a mixture of compounds as a pure substance?

You can buy dimedone (5,5-dimethylcyclohexane-1,3-dione) from chemical suppliers. If, as is wise when you buy any compound, you run an NMR spectrum of the compound to check on its purity, you might be inclined to send the compound back. In CDCl$_3$ solution it is clearly a mixture of two compounds. Overleaf you can see ^{1}H and ^{13}C NMR spectra of the mixture with the peaks of the dione in red.

The majority of the sample is indeed 5,5-dimethylcyclohexane-1,3-dione. What is the rest? The other component has a similar spectrum and is clearly a similar compound: it has the 6H singlet for the CMe$_2$ group and the two CH$_2$ groups at the side of the ring; it also has five signals in its ^{13}C NMR spectrum. But it has a broad signal at δ_H 8.15, which looks like an OH group, and a sharp signal at δ_H 5.5 in the double-bond region. It also has *two different* sp^2 carbon atoms. All this fits the *enol* structure.

"dimedone"
5,5-dimethylcyclohexane-1,3-dione

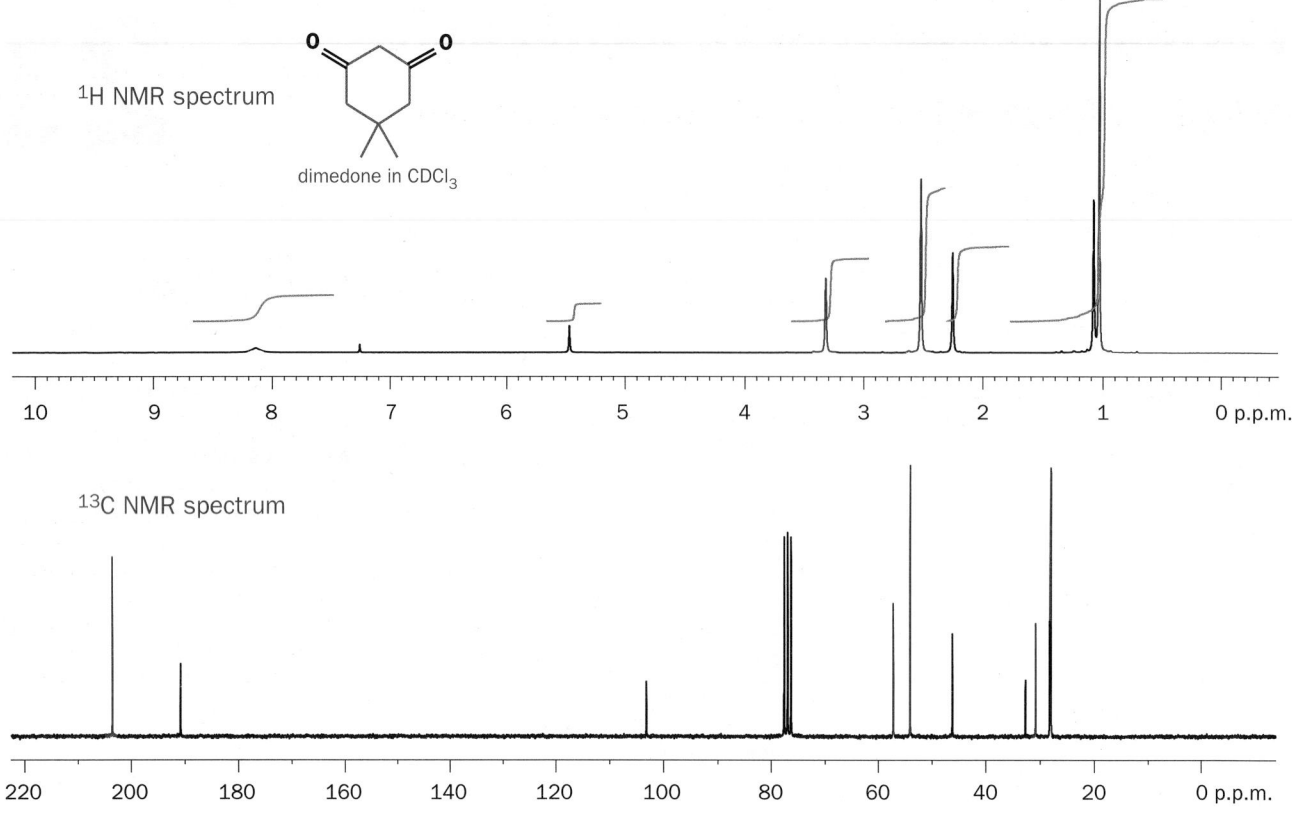

¹H NMR spectrum

dimedone in CDCl₃

¹³C NMR spectrum

keto form of dimedone

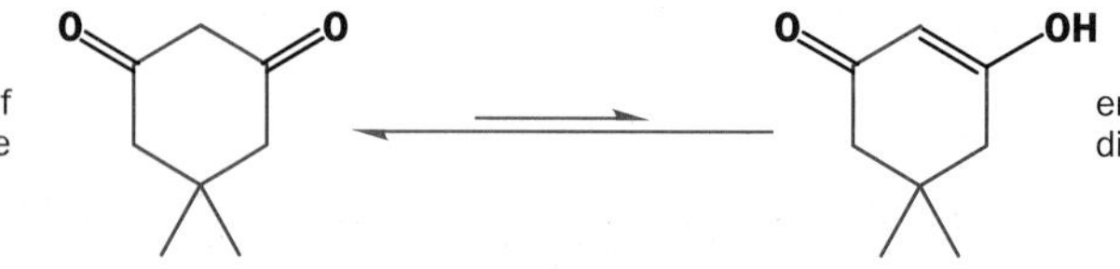

enol form of dimedone

These forms are in equilibrium and cannot be separated at room temperature. The equilibrium is nothing to do with the two methyl groups at C5. And yet the 2,2-dimethyl compound is a perfectly normal diketone with all the expected peaks in the NMR. You will see later that it is only the relative position (1,3) of the carbonyl groups and the presence of at least one hydrogen at C2 that matter.

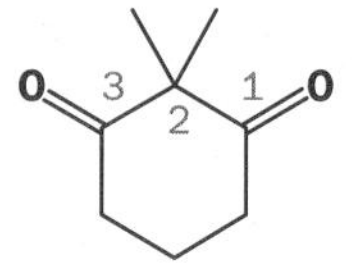

2,2-dimethylcyclohexane-1,3-dione

Tautomerism: formation of enols by proton transfer

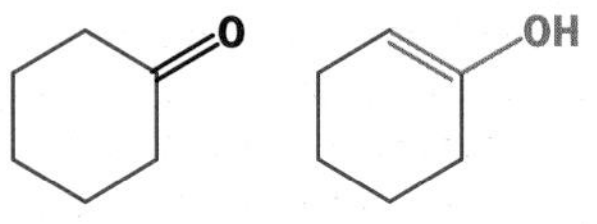

keto form of cyclohexanone

enol form of cyclohexanone

An enol is exactly what the name implies: an ene-ol. It has a C=C double bond and an OH group joined directly to it. Simple carbonyl compounds have enols too—in the margin is the enol of cyclohexanone (just dimedone without the extras).

In the case of dimedone, the enol must be formed by a transfer of a proton from the central CH₂ group of the keto form to one of the OH groups.

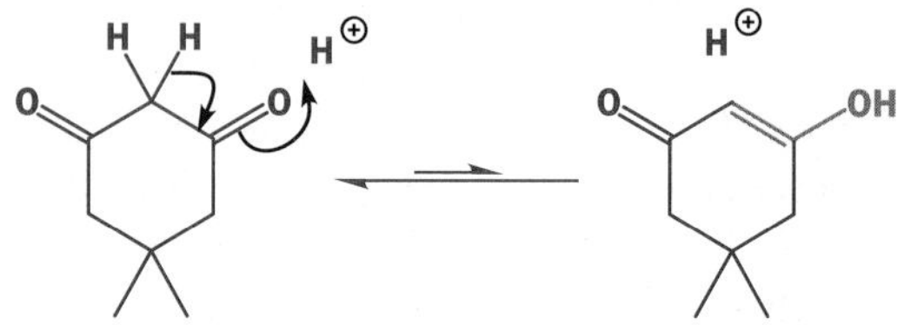

Notice that there is no change in pH—a proton is lost from carbon and gained on oxygen. The reaction is known as **enolization** as it is the conversion of a carbonyl compound into its enol. It is a strange reaction in which little happens. The product is almost the same as the starting material since the only change is the transfer of one proton and the shift of the double bond. Reactions like this are given the name **tautomerism**.

Why don't simple aldehydes and ketones exist as enols?

When we were looking at spectra of carbonyl compounds in Chapter 15 we saw no signs of enols in IR or NMR spectra. Dimedone is exceptional—although any carbonyl compound with protons adjacent to the carbonyl group can enolize, simpler carbonyl compounds like cyclohexanone or acetone have only a trace of enol present under ordinary conditions. The equilibrium lies well over towards the keto form (the equilibrium constant K for acetone is about 10^{-6}).

keto form of acetone   enol form of acetone

This is because the combination of a C=C double bond and an O–H single bond is (slightly) less stable than the combination of a C=O double bond and a C–H single bond. The balance between the bond energies is quite fine. On the one hand, the O–H bond in the enol is a stronger bond than the C–H bond in the ketone but, on the other hand, the C=O bond of the ketone is much more stable than the C=C bond of the enol. Here are some average values for these bonds.

Typical amounts of enols in solution are about one part in 10^5 for normal ketones. So why do we think they are important? *Because enolization is just a proton transfer, it is occurring all the time even though we cannot detect the minute proportion of the enol.* Let us look at the evidence for this statement.

> Any reaction that simply involves the intramolecular transfer of a proton is called a tautomerism. Here are two other examples.
>
>
> tautomerism of a carboxylic acid. The mixture must be exactly 50:50 as the two forms (tautomers) are identical
>
>
> tautomerism of an imidazole. The mixture need not be exactly 50:50 as the two forms are not identical
>
> This sort of chemistry was discussed in Chapter 8 where the acidity and the basicity of atoms were the prime considerations. In the first case the two tautomers are the same and so the equilibrium constant must be exactly 1 or, if you prefer, the mixture must be exactly 50:50. In the second case the equilibrium will lie on one side or the other depending on the nature of R.

Typical bond strengths (kJ mol^{-1}) in keto and enol forms

	Bond to H	π bond	Sum
keto form	(C–H) 440	(C=O) 720	1160
enol form	(O–H) 500	(C=C) 620	1120

Evidence for equilibration of carbonyl compounds with enols

If you run the NMR spectrum of a simple carbonyl compound (for example, 1-phenyl-propan-1-one, 'propiophenone') in D_2O, the signal for protons next to the carbonyl group very slowly disappears. If the compound is isolated from the solution afterwards, the mass spectrum shows that those hydrogen atoms have been replaced by deuterium atoms: there is a peak at $(M + 1)^+$ or $(M + 2)^+$ instead of at M^+. To start with, the same keto–enol equilibrium is set up.

1-phenylpropan-1-one

keto form of   enol form of
1-phenylpropan-1-one   1-phenylpropan-1-one

But, when the enol form reverts to the keto form, it picks up a deuteron instead of a proton because the solution consists almost entirely of D_2O and contains only a tiny amount of DOH (and no H_2O at all).

enolization   return to keto form

> Notice that the double bond in this enol could be either *E* or *Z*. It is drawn as *Z* here, but in reality is probably a mixture of both—though this is irrelevant to the reaction. We shall not be concerned with the geometry of enols in this chapter, but there are some reactions that you will meet in later chapters where it is important, and you need to appreciate that the issue exists.

The process can now be repeated with the other hydrogen atom on the same carbon atom.

There are, of course, eight other hydrogens in the molecule but they are not affected. In the NMR spectrum we see the slow disappearance of the 2H signal for the protons on C2 next to the carbonyl group.

Enolization is catalysed by acids and bases

Enolization is, in fact, quite a slow process in neutral solution, even in D_2O, and we would catalyse it with acid or base if we really wanted it to happen. In the acid-catalysed reaction, the molecule is first protonated on oxygen and then loses the C–H proton in a second step. We shall use a different example here to show that aldehydes form enols too.

acid-catalysed enolization of an aldehyde

"keto" form of aldehyde   enol form of aldehyde

This is a better mechanism for enolization than those we have been drawing because it shows that something (here a water molecule) must actually be removing the proton from carbon. Though this reaction will occur faster than the uncatalysed enolization, the equilibrium is not changed and we still cannot detect the enol spectroscopically.

In the base-catalysed reaction the C–H proton is removed first by the base, say, a hydroxide ion, and the proton added to the oxygen atom in a second step.

base-catalysed enolization of an aldehyde

"keto" form of aldehyde   enol form of aldehyde

This is a good mechanism too because it shows that something must remove the proton from carbon and something (here a water molecule—we can't, of course, have protons in basic solution) must put the proton on the oxygen atom. The concentration of free protons in water is vanishingly small (Chapter 8).

Notice that both of these reactions are genuinely catalytic. You get the proton back again at the end of the acid-catalysed mechanism.

enol form of aldehyde

Something else will happen to the proton NMR spectrum. The signal for the CH_3 group was a triplet in the original ketone, but when those two Hs are replaced by Ds, it becomes a singlet. In the carbon spectrum, coupling to deuterium appears: remember the shape of the $CDCl_3$ peak (Chapter 3)?

In Chapter 19 (p. 483) we discussed the equivalence of mechanisms showing protons just 'falling off' with those in which basic solvent molecules are involved to remove a proton. In this chapter, and in the rest of the book, you will see both variants in use according to the context. They mean exactly the same thing.

And you get the hydroxide ion back again at the end of the base-catalysed mechanism.

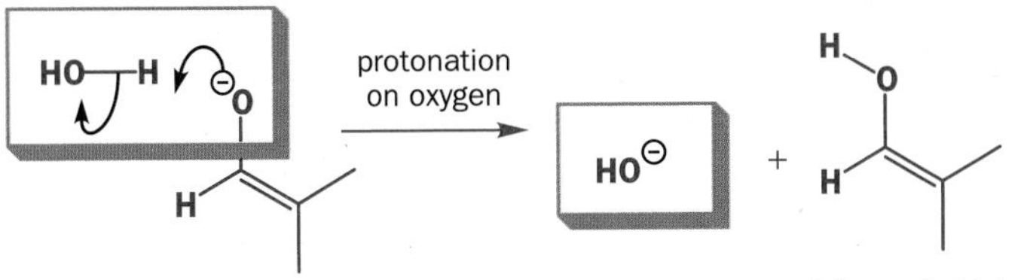

protonation on oxygen

enol form of aldehyde

The intermediate in the base-catalysed reaction is the enolate ion

There are more insights to be gained from the base-catalysed reaction. The intermediate anion is called the **enolate ion**. It is the conjugate base of the enol and can be formed either directly from the carbonyl compound by the loss of a C–H proton or from the enol by loss of the O–H proton.

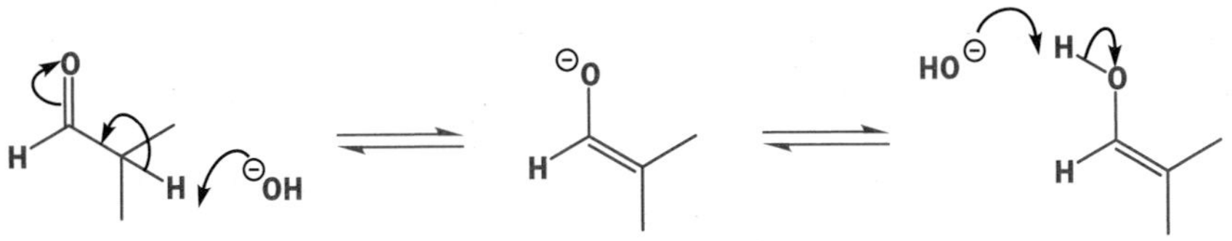

"keto" form of aldehyde   the enolate ion   enol form of aldehyde

The enol form is more acidic than the keto form

The enol is less stable than the aldehyde and both lose a proton to give the same enolate ion. It follows that the enol is the more acidic. Make sure you understand this. Think of it this way: the keto/enol equilibrium constant is small.

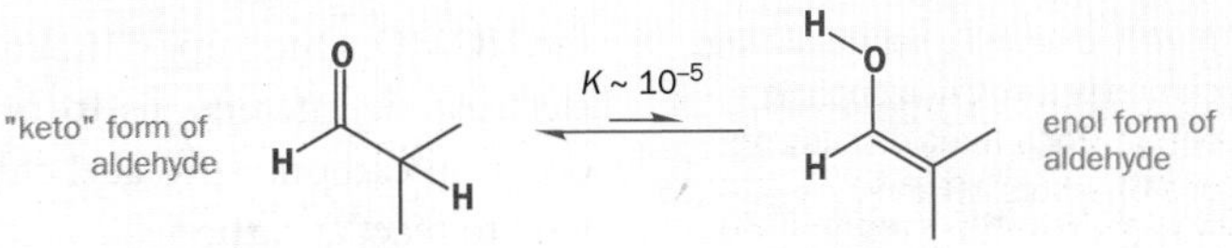

"keto" form of aldehyde $K \sim 10^{-5}$ enol form of aldehyde

The acidity equilibrium constants for each form (with the enolate ion) are both small, but they are not the same.

If the keto form is more stable than the enol form, then K_a(keto) must be smaller than K_a(enol): the enol form gives *more* of the enolate ion. The acidity of each form is measured by pK_a which is just $-\log_{10} K_a$ so if K_a(keto) $< K_a$(enol) then pK_a(keto) $>$ pK_a(enol) and the keto form is less acidic.

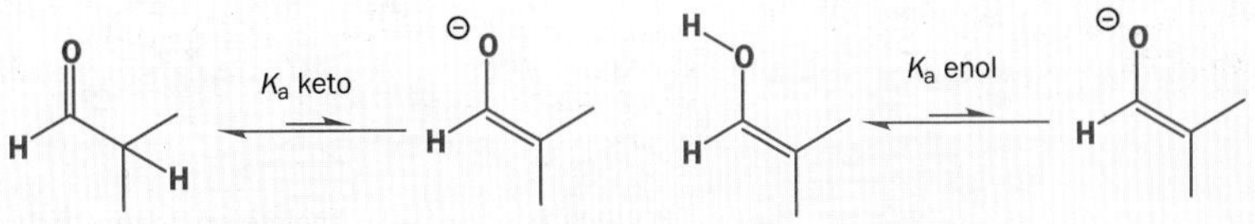

K_a keto K_a enol

The enolate ion is an alkoxide ion as we have drawn it, but it is more stable than the corresponding saturated structure because it is conjugated.

The enolate ion is one of those three-atom four-electron systems related to the allyl anion that we met in Chapter 7. The negative charge is mainly on oxygen, the most electronegative atom. We can show this with curly arrows using the simplest enolate possible (from MeCHO).

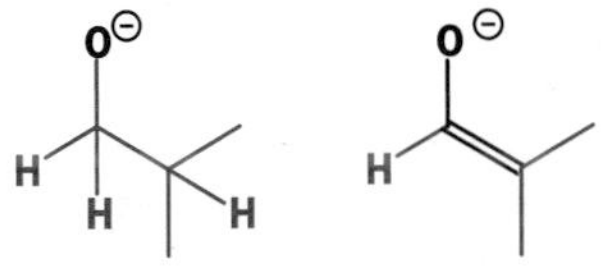

simple alkoxide anion: unconjugated and less stable

enolate anion: conjugated and more stable

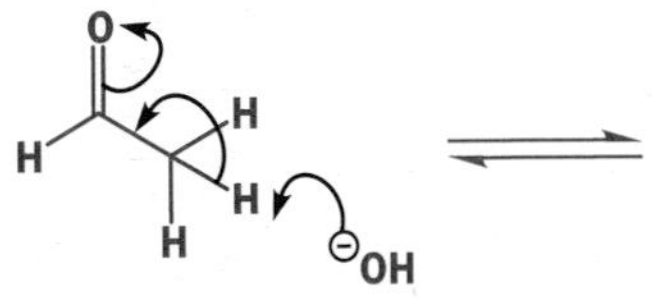

formation of the enolate ion of acetaldehyde (ethanal)

delocalization of the enolate anion

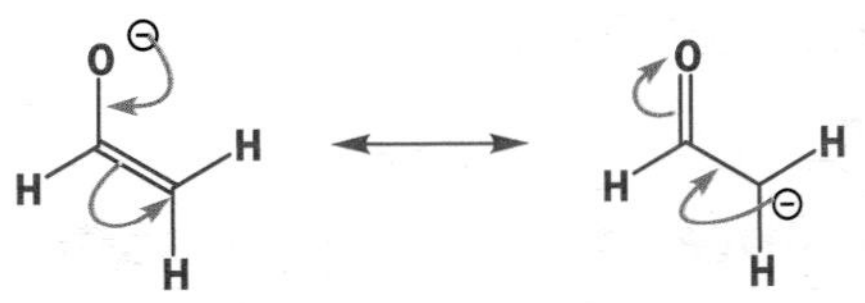

oxyanion   carbanion

▶

It is important that you appreciate one key difference between the enolate and enol forms: the enolate is a delocalized system, with negative charge carried on both C and O—we use a double-headed conjugation arrow to connect these two representations. But for the proton to move from C to O in the enol form requires σ bonds to break and form, and this is a real equilibrium, which must be represented by equilibrium arrows.

Remember that the oxyanion and carbanion structures are just two different ways to represent the same thing. We shall usually prefer the oxyanion structure as it is more realistic. You can say the same thing in orbitals.

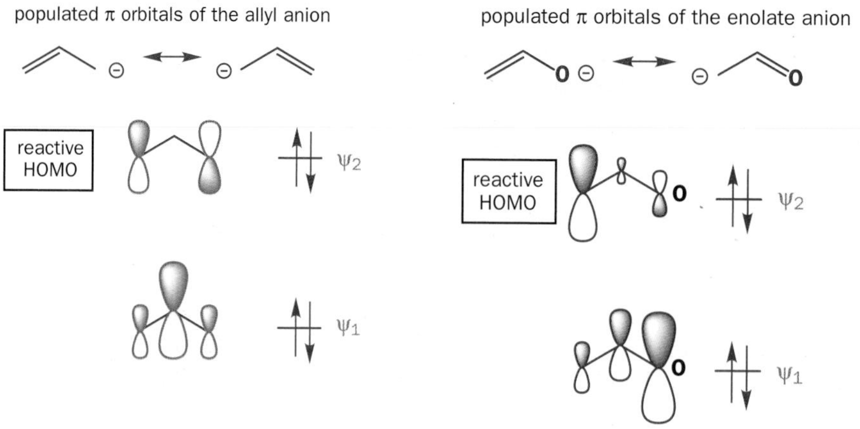

populated π orbitals of the allyl anion

populated π orbitals of the enolate anion

■
Refer to Chapter 7 if you fail to see where these orbitals come from.

On the left you see the populated orbitals of the allyl anion and on the right the corresponding orbitals of the enolate ion. The allyl anion is, of course, symmetrical. Two changes happen when we replace one carbon by an oxygen atom. Because oxygen is more electronegative, both orbitals go down in energy. The orbitals are also distorted. The lower-energy atomic orbital of the more electronegative oxygen contributes more to the lower-energy orbital (ψ_1) and correspondingly less to ψ_2. The charge distribution comes from both populated orbitals so the negative charge is spread over all three atoms, but is mostly on the ends. The important reactive orbital is the HOMO (ψ_2) which has the larger orbital on the terminal carbon atom.

In the enolate, the oxygen atom has more of the negative charge, but the carbon atom has more of the HOMO. One important consequence is that we can expect reactions dominated by charges and electrostatic interactions to occur on oxygen and reactions dominated by orbital interactions to occur on carbon. Thus acyl chlorides tend to react at oxygen to give enol esters, while alkyl halides tend to react at carbon.

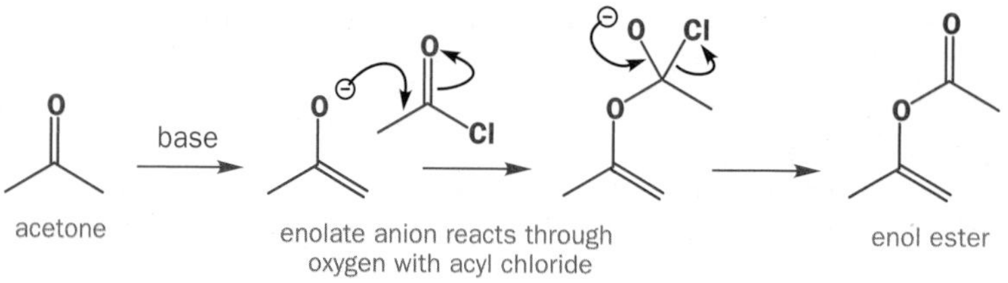

▶
In other words, the oxygen is a *hard* nucleophilic centre and the carbon is a *soft* nucleophilic centre. For related discussions, see Chapters 10 and 17.

acetone

enolate anion reacts through oxygen with acyl chloride

enol ester

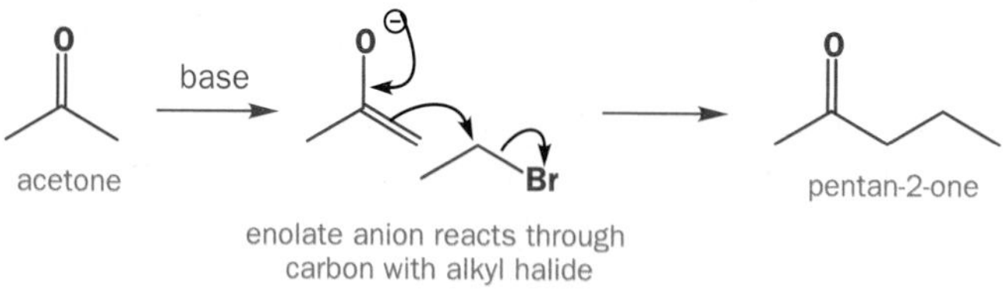

▶
Notice that in drawing this mechanism it is *not* necessary to locate the negative charge on the carbon atom. You should always draw enolate mechanisms using the better oxyanion structure.

acetone

enolate anion reacts through carbon with alkyl halide

pentan-2-one

We shall be looking at these reactions in Chapter 26. For the rest of this chapter we are going to look at some simpler consequences of enolization and some reactions of enolates with heteroatom nucleophiles.

Summary of types of enol and enolate

In this section, the hydrogen atom lost in the enolization is shown in green. First let us summarize the various kinds of enol and enolate we can have from carbonyl compounds. We have seen such compounds from aldehydes and ketones already, but here are some variants. Cyclic ketones form enols and enolates just like open-chain compounds.

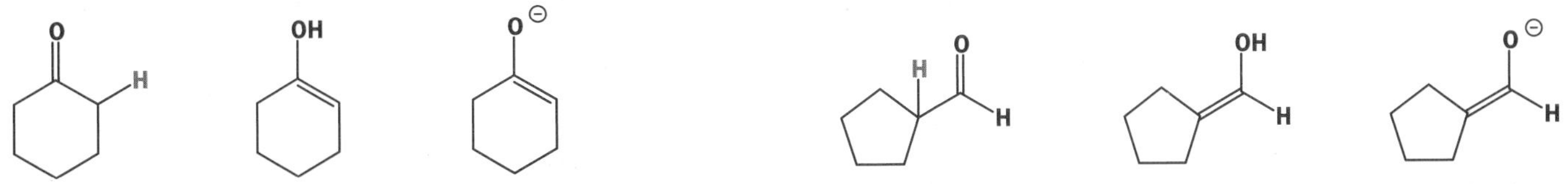

cyclohexanone enol enolate ion cyclopentane aldehyde enol enolate ion

You can have a cyclic aldehyde only if the carbonyl group is outside the ring and cyclic aldehydes too form enols and enolates.

All the acid derivatives can form enols of some kind. Those of esters are particularly important and either enols or enolates are easily made. It is obviously necessary to avoid water in the presence of acid or base, as esters hydrolyse under these conditions. One solution is to use the alkoxide belonging to the ester (MeO⁻ with a methyl ester, EtO⁻ with an ethyl ester, and so on) to make enolate ions.

> Note that the aldehyde proton itself (CHO) is *never* enolized. Try to draw the curly arrows and you will see that they don't work.

Then, if the alkoxide does act as a nucleophile, no harm can be done as the ester is simply regenerated.

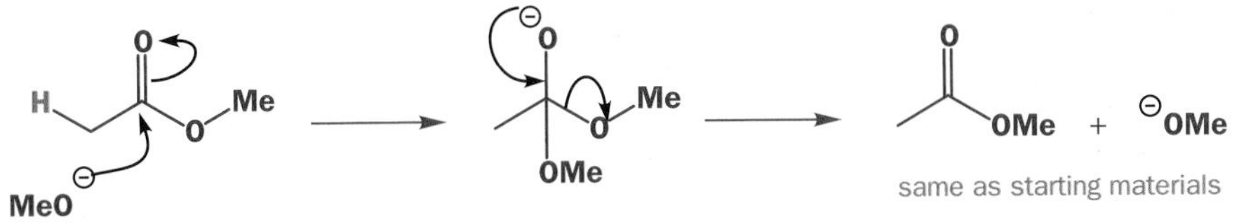

same as starting materials

The carbonyl group is accepting electrons both in the enolization step and in the nucleophilic attack. The same compounds that are the most electrophilic are also the most easily enolizable. This makes acyl chlorides very enolizable. To avoid nucleophilic attack, we cannot use chloride ion as base since chloride is not basic, so we must use a nonnucleophilic base such as a tertiary amine.

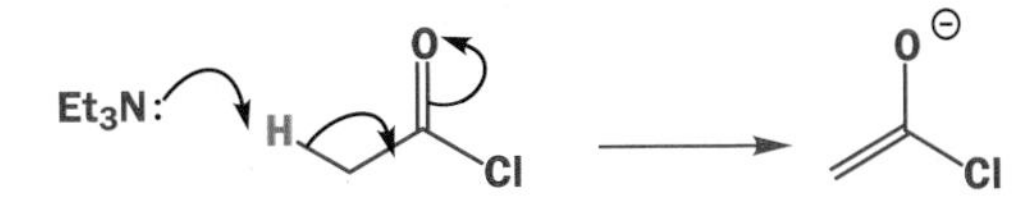

The resulting enolate is not stable as it can eliminate chloride ion, a good leaving group, to form a ketene. This works particularly well in making dichloroketene from dichloroacetyl chloride as the proton to be removed is very acidic.

> ■ This is an E1cB elimination, and you saw this sort of chemistry in Chapter 19.

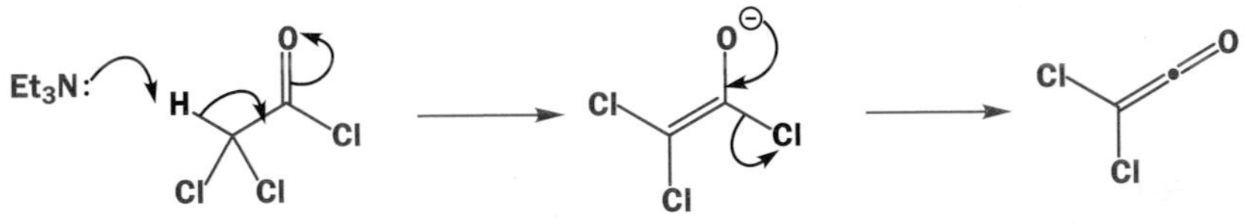

unstable enolate dichloroketene

Carboxylic acids do not form enolate anions easily as the base first removes the acidic OH proton. The same thing protects acids from attack by nucleophiles.

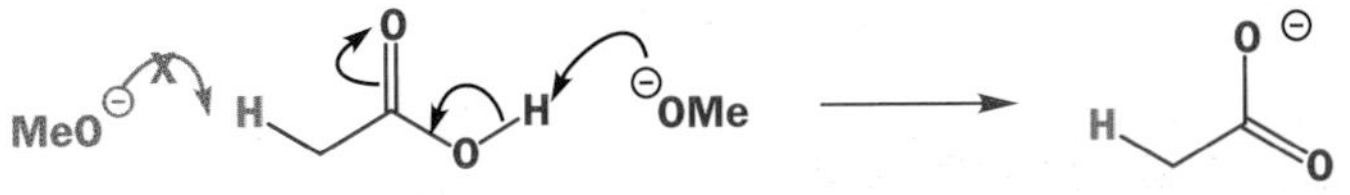

stable carboxylate anion

In acid solution, there are no such problems and 'ene-diols' are formed. The original OH group of the carboxylic acid and the new OH group of the enol are equivalent.

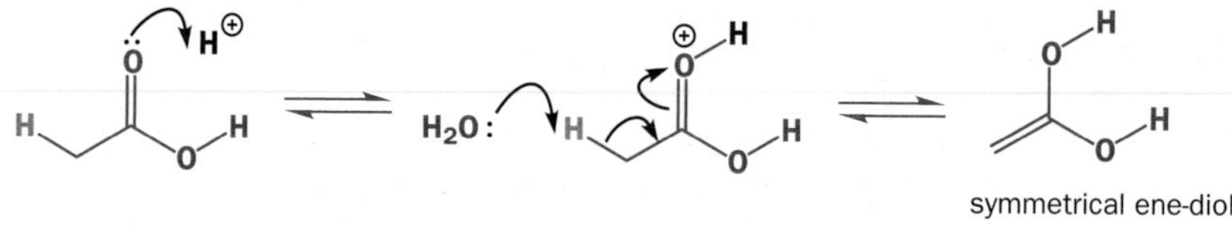

symmetrical ene-diol

Amides also have rather acidic protons, though not, of course, as acidic as those of carboxylic acids. Attempted enolate ion formation in base removes an N–H proton rather than a C–H proton. Amides are also the least reactive and the least enolizable of all acid derivatives, and their enols and enolates are rarely used in reactions.

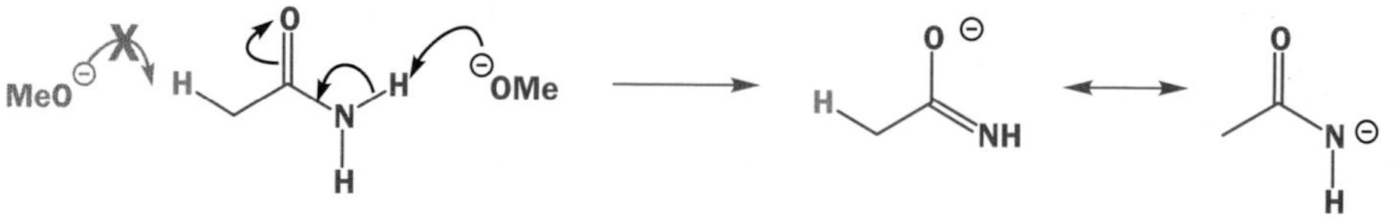

It is not even necessary to have a carbonyl group to observe very similar reactions. Imines and enamines are related by the same kind of tautomeric equilibria.

■ You should make sure you can write mechanisms for these reactions: we discussed them in Chapter 14.

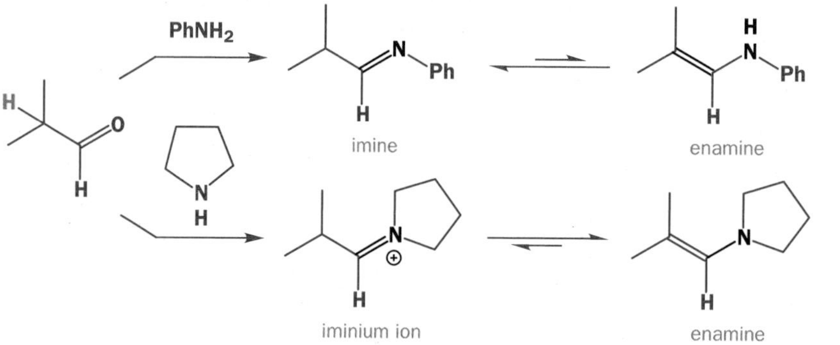

imine

enamine

iminium ion

enamine

With a primary amine (here PhNH$_2$) a reasonably stable imine is formed, but with a secondary amine (here a simple cyclic amine) the imine itself cannot be formed and the iminium salt is less stable than the enamine.

Just as enamines are the nitrogen analogues of enols, **aza-enolates** are the nitrogen analogues of enolates. They are made by deprotonating enamines with strong base. You will see both enamines and aza-enolates in action in Chapters 26 and 27.

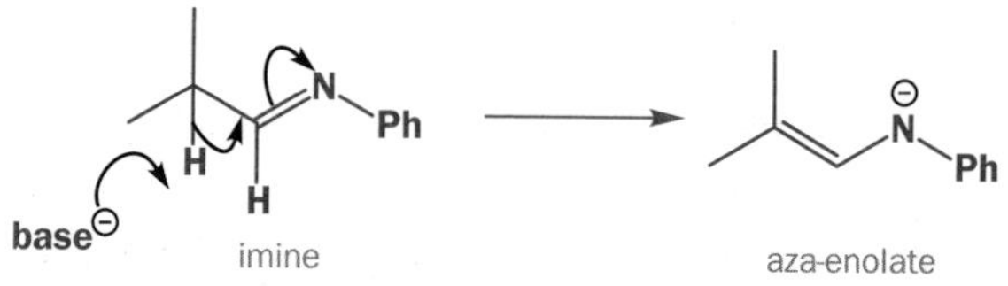

base   imine

aza-enolate

■ Deprotonation of nitroalkanes is discussed in detail in Chapter 8.

Nitroalkanes form enolate-like anions in quite weak base. As in base-catalysed enolization, a proton is removed from a carbon atom and a stable oxyanion is formed.

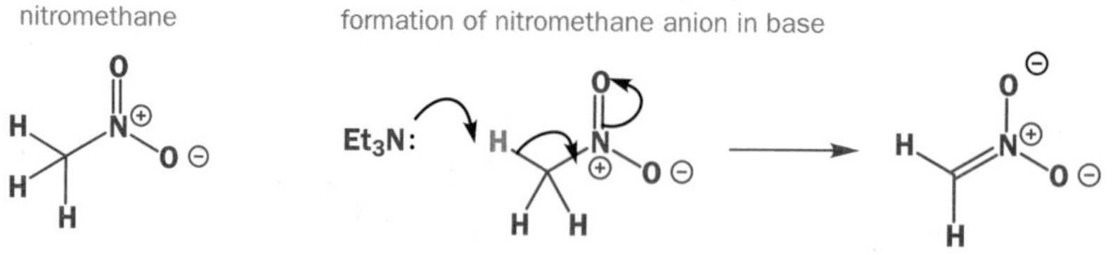

nitromethane   formation of nitromethane anion in base

Nitriles (cyanides) also form anions but require stronger base as the negative charge is delocalized on to a single nitrogen atom rather than on to two oxygens. The negative charge is mostly on a nitrogen atom and the anion is a three-atom four-electron system like ketene or allene.

benzyl cyanide
(phenylacetonitrile)

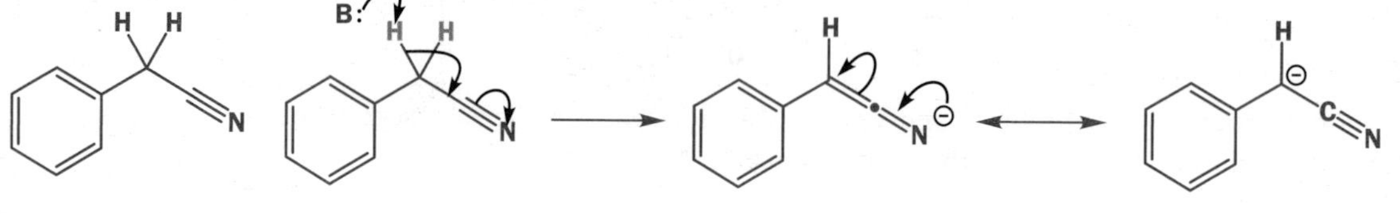

● **Requirement for enolization**

In summary, any organic compound with an electron-withdrawing functional group, with at least one π bond joined to a saturated carbon atom having at least one hydrogen atom, may form an enol in neutral or acid solution. Many also form enolates in basic solution (exceptions are carboxylic acids and primary and secondary amides).

The enols will probably not be detectable in solution (only about one part in 10^4–10^6 is enol for most compounds). Some compounds by contrast form stable enols.

Stable enols

Kinetically stable enols

We have established that enols are, in general, less stable than the keto form of the molecule. We might hope to see stable enols if we changed that situation by adding some feature to the molecule that stabilized the enol thermodynamically. Or we might try to create an enol that would revert only slowly to the keto form—in other words, it would be *kinetically* stable. We shall look at this type first.

We have established that the formation of enols is catalysed by acids and bases. The reverse of this reaction—the formation of ketone from enol—must therefore also be catalysed by the same acids and bases. If you prepare simple enols in the strict absence of acid or base they have a reasonable life-time. A famous example is the preparation of the simplest enol, vinyl alcohol, by heating ethane-1,2-diol (glycol—antifreeze) to very high temperatures (900 °C) at low pressure. Water is lost and the enol of acetaldehyde is formed. It survives long enough for its proton NMR spectrum to be run, but gives acetaldehyde slowly.

vinyl alcohol (ethenol)
enol form of acetaldehyde

acetaldehyde

NMR spectrum of vinyl alcohol

The spectrum fits the enol perfectly. The alkene proton next to OH is deshielded and the two alkene protons on the other carbon atom shielded as we should expect from the feeding of electrons into the double bond by the OH group.

The coupling constants across the double bond are as expected too. The *trans* coupling is large (14.0 Hz) and the *cis* coupling smaller (6.5 Hz). The geminal coupling is very small as is usually the case for a CH_2 group on a double bond.

Other enols can be made that are stable because it is very difficult for the carbon atom to be protonated. This example is very crowded by two substituted benzene rings.

electron-rich
shielded

smaller δ_H value

electron-deficient
(aldehyde-like)   larger δ_H value
deshielded

both faces of the
enol are hindered
by the methyl groups

The enol would have to be protonated at the C2 to form the aldehyde but this is not possible because the two benzene rings are twisted out of the plane of the double bond by the interference of the *ortho* methyl groups. The view down the double bond shows that both faces are blocked by one of the *ortho* methyl groups and an acid cannot approach close enough to deliver its proton.

Enols of 1,3-dicarbonyl compounds: thermodynamically stable enols

We started this chapter by looking at a molecule that contained about 33% enol in solution—dimedone. In fact, this is just one example of the class of 1,3-dicarbonyl compounds (also called β-dicarbonyls) all of which contain substantial amounts of enol and may even be completely enolized in polar solvents.

keto form
of dimedone
(67% in CDCl$_3$)

enol form
of dimedone
(33% in CDCl$_3$)

We need now to examine why these enols are so stable. The main reason is that this unique (1,3) arrangement of the two functional groups leads to enols that are conjugated rather like a carboxylic acid.

delocalization in the enol form of dimedone

delocalization in a carboxylic acid

Did you notice when we were looking at the NMR spectrum of dimedone (p. 522) that the two CH$_2$ groups in the ring seemed to be the same, though they are different (a and b) and the delocalization we have just looked at does not make them the same? This must mean that the enol is in *rapid* equilibrium with another identical enol. This is *not* delocalization—a proton is moving—so it is **tautomerism**.

equilibration of the enol form of dimedone

equilibration of a carboxylic acid

Once again, this is very like the situation in a carboxylic acid. Thus the two enols equilibrate fast with each other in CDCl$_3$ solution but equilibrate slowly enough with the keto form for the two spectra to be recorded at the same time. If equilibration with the keto form were fast, we should see a time-averaged spectrum of the two. In CD$_3$OD solution the ^{1}H and ^{13}C NMR spectra show that only the enol form exists, presumably stabilized by hydrogen bonding.

Other 1,3-dicarbonyl compounds also exist largely in the enol form. In some examples there is an additional stabilizing factor, intramolecular hydrogen bonding. Acetylacetone (propane-2,4-dione) has a symmetrical enol stabilized by conjugation. The enol form is also stabilized by a very favourable intramolecular hydrogen bond in a six-membered ring.

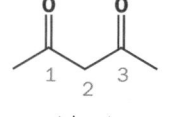

acetylacetone

■ This hydrogen bond was not possible in dimedone.

enol form of acetylacetone stabilized by an intramolecular hydrogen bond

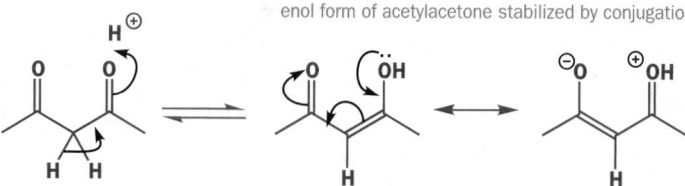

enol form of acetylacetone stabilized by conjugation

This allows interconversion of the two identical enol structures by proton transfer, that is, by tautomerism.

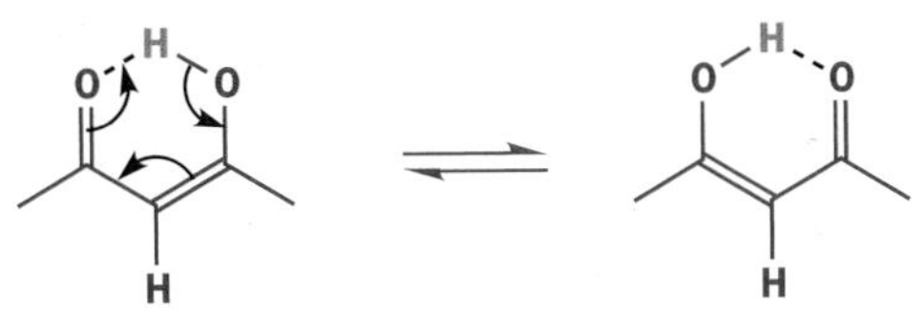

The 1,3-dicarbonyl compound need not be symmetrical and if it is not two different enol forms will interconvert by proton transfer. Here is a cyclic keto-aldehyde as an example. It exists as the rapidly equilibrating enol. The proportions of the three species can be measured by NMR: there is 0% keto-aldehyde, 76% of the first enol, and 24% of the second.

tautomerism in the enol form of acetylacetone

> Again, note carefully the difference between this **tautomerism** in which a proton is moved around the molecule and the structures are linked by *equilibrium* arrows and the **delocalization** (conjugation) where only electrons are 'moved' (no actual movement occurs, of course) and the two structures are linked by one *double-headed* arrow as they are just two ways of drawing the same thing.

1,3-dicarbonyl (keto-aldehyde)   *two different stable enols rapidly interconverting by tautomerism*

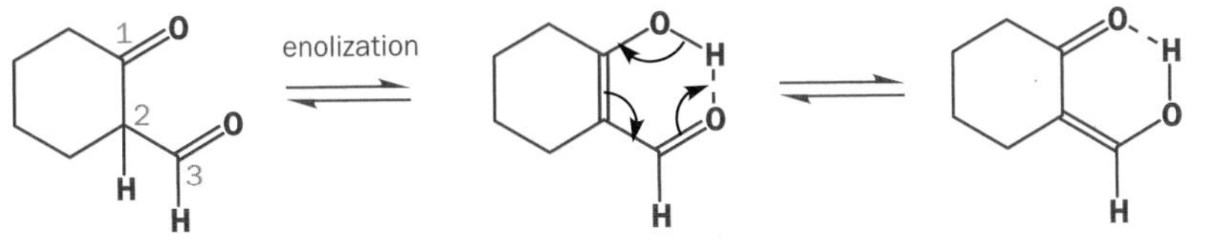

Enols occur in nature too. Vitamin C has a five-membered ring containing two carbonyl groups but normally exists as a very conjugated ene-diol.

one unstable keto form   *stable ene-diol form of vitamin C*   *another unstable keto form*

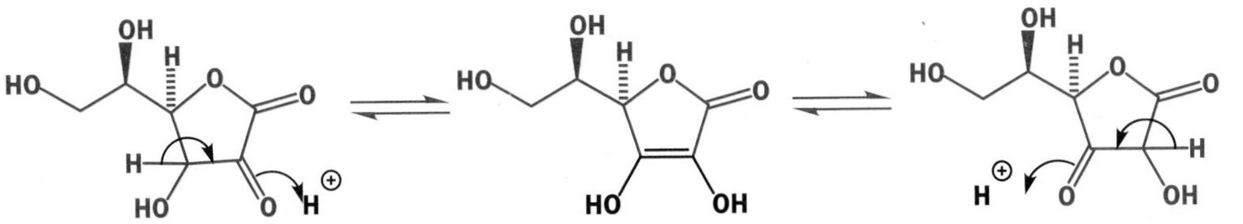

The enol is stable; it is delocalized. We can show the delocalization and explain why vitamin C is called ascorbic *acid* at the same time. The black enol proton is acidic because the anion is delocalized over the 1,3-dicarbonyl system.

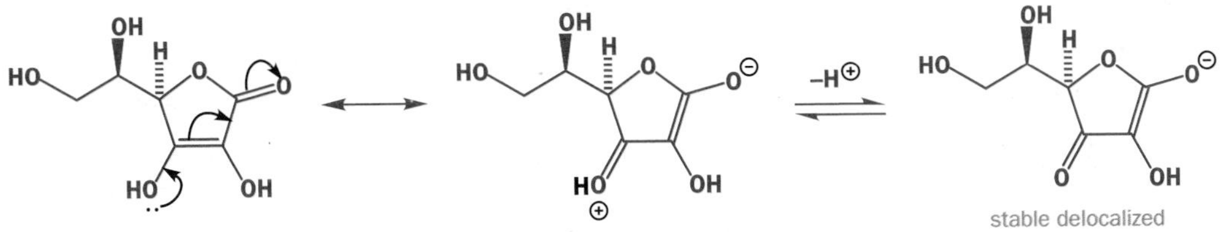

stable delocalized

The ultimate in stable enols has to be the Ph-enol, the aromatic alcohols or phenols, which prefer the substantial advantage of aromaticity to the slight advantage of a C=O over a C=C double bond. They exist entirely in the phenol form.

"ketonization" of phenol (not observed)

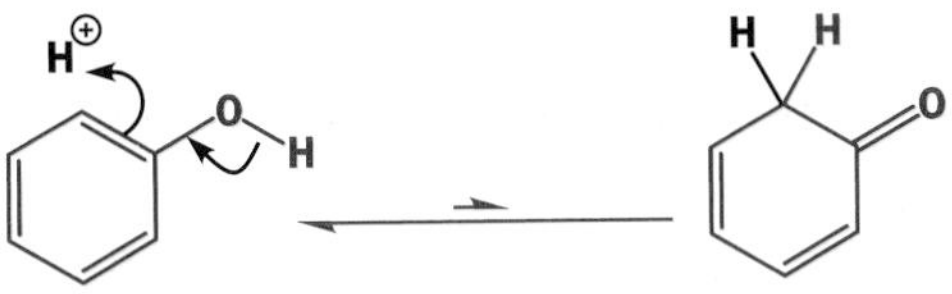

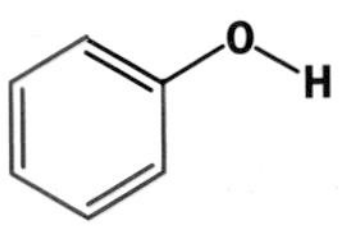

phenol

Even so, you will see in Chapter 22 that intermediates with this 'keto' structure are formed in reactions on the benzene ring of phenols. Like ascorbic acid, phenol is also quite acidic (pK_a 10) and used to be called carbolic acid.

Stable enols

Pfizer's antiinflammatory drug 'Feldene' (used to treat arthritis) is a stable enol based on a 1,3-dicarbonyl compound. It also has amide and sulfonamide groups in its structure but you should be able to pick out the enol part.

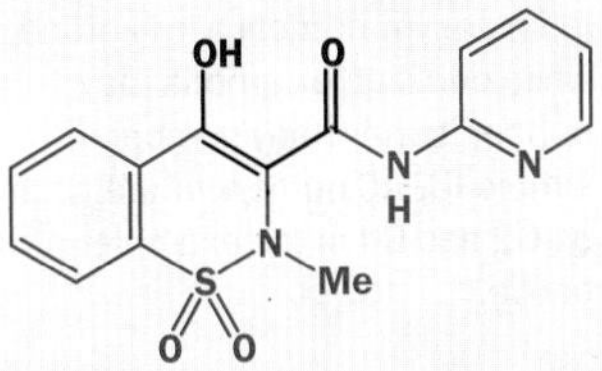

Pfizer's piroxicam or Feldene once-a-day treatment for arthritis

When speaking generally of
carbonyl compounds, the Greek
letters α, β, γ, and so on are used
to designate the positions along
the chain from the carbonyl group.
Of course, if the compound is an
aldehyde or an acid derivative, we
can use the normal numbers
instead as the carbonyl group will
normally be C1, but this will not
usually be the case for ketones. It
is useful to have a general
method of describing the
positions where enolization may
take place and so they are called
the α positions.

labelling carbon atoms
in carbonyl compounds

acid derivative
(ester)

ketone

An enolizable position is always
α, even if there are two of them as
in an unsymmetrical alkyl ketone,
while enolizable carbons may
happen to have any number in
simple IUPAC nomenclature such
as C2 and C4 in the example
above.

A planar molecule *must* have a
plane of symmetry.

a natural (S) amino acid

Note the use of α: the amino group is
on the α position with respect to the
carboxylic acid.

Consequences of enolization

Unsaturated carbonyl compounds prefer to be conjugated

It is difficult to keep a β,γ-unsaturated carbonyl compound because the double bond tends to move into conjugation with the carbonyl group in the presence of traces of acid or base. The intermediate is, of course, an enol in acid solution but an enolate ion in base.

Protonation at the α position takes the molecule back to the unconjugated ketone, but protonation in the γ position gives the more stable conjugated isomer. All the reactions are equilibria so the conjugated isomer gradually predominates.

Racemization

Any stereogenic centre next to a carbonyl group is precarious because enolization will destroy it. It would be foolish to try and make optically active β-keto esters whose only stereogenic centre was between the two carbonyl groups.

Though the keto-ester is chiral, the enol is flat and cannot be chiral. The two forms are in rapid equilibrium so all optical activity would quickly be lost.

Compounds with one carbonyl group next to the stereogenic centre can be made but care still needs to be taken. The α amino acids, the component parts of proteins, are like this. They are perfectly stable and do not racemize in aqueous acid or base. In base they exist as carboxylate anions that do not enolize, as explained above. Enolization in acid is prevented by the NH_3^+ group, which inhibits the second protonation necessary for enol formation.

Amino acids can be converted into their *N*-acetyl derivatives with acetic anhydride. These *N*-acetyl amides can be racemized on recrystallization from hot acetic acid, no doubt by enolization. The amino group is no longer basic, is not protonated in acid, and so protonation on the carbonyl group and hence enolization is now possible.

You may think it a crazy idea to *want* to racemize an amino acid. Supposing, however, that you are preparing pure (*S*)-amino acid by resolution. Half your material ends up as the wrong (*R*)-enantiomer and you don't want just to throw it away. If you racemize it you can put it back into the next resolution and convert half of it into the (*S*)-acid. Then you can racemize what remains and so on.

Some compounds may be racemized inside the human body. Bacterial cell walls are built partly from 'unnatural' (*R*)-amino-acids and we can't digest these. Instead, we use enzymes designed to racemize them. These also work by enolization, though it is the imine–enamine type from p. 528.

There is an important group of analgesic (pain-killing) drugs such as ibuprofen based on the aryl-propionic acid structure. Ibuprofen is given to arthritis sufferers as 'Brufen' and can be bought over the counter in chemists' shops as the headache remedy 'Nurofen'. Only one enantiomer actually cures pain but the compound is administered as the racemate. The body does the rest, racemizing the compound by enolizing it.

We discussed **resolution**, the separation of enantiomers by the formation of diastereoisomers with an optically active resolving agent, in Chapter 16.

There are details of this reaction in Chapter 50.

the biologically inactive enantiomer of ibuprofen — flat achiral enol — the biologically active enantiomer of ibuprofen — ibuprofen

Reaction with enols or enolates as intermediates

We have already seen that exchange of hydrogen for deuterium, movement of double bonds into conjugation, and racemization can occur with enols or enolates as intermediates. These are chemical reactions of a sort, but it is time to look at some reactions that make significant changes to the carbonyl compound.

Halogenation

Carbonyl compounds can be halogenated in the α position by halogens (such as bromine, Br_2) in acidic or basic solutions. We shall look at the acid-catalysed reaction first because it is simpler. Ketones can usually be cleanly brominated in acetic acid as solvent.

acetone — bromoacetone

The first step is acid-catalysed enolization and the electrophilic bromine molecule then attacks the nucleophilic carbon of the enol. The arrows show why this particular carbon is the one attacked.

enol

Notice that the acid catalyst is regenerated at the end of the reaction. The reaction need not be carried out in an acidic solvent, or even with a protic acid at all. Lewis acids make excellent catalysts for the bromination of ketones. This example with an unsymmetrical ketone gives 100% yield of the bromoketone with catalytic $AlCl_3$ in ether as solvent.

Br_2, 0.75 mol% $AlCl_3$, Et_2O — 100% yield

Bromination occurs nowhere else in the molecule—not on the benzene ring (which, as you will see in the next chapter, it easily might under these conditions), nor on any other atom of the aliphatic side chain. This is because only one position can form an enol and the enol is more reactive towards bromine than the aromatic ring.

We have introduced a slight but unimportant variation into this mechanism. In the previous mechanism, we used the lone pairs on oxygen to assist attack on Br_2 and then lost the acid catalyst in a separate step. Here, we have written $AlCl_3$ leaving as the Br_2 is attacked. The difference is not significant, and you will see mechanisms written in both ways. The second way saves a step, of course.

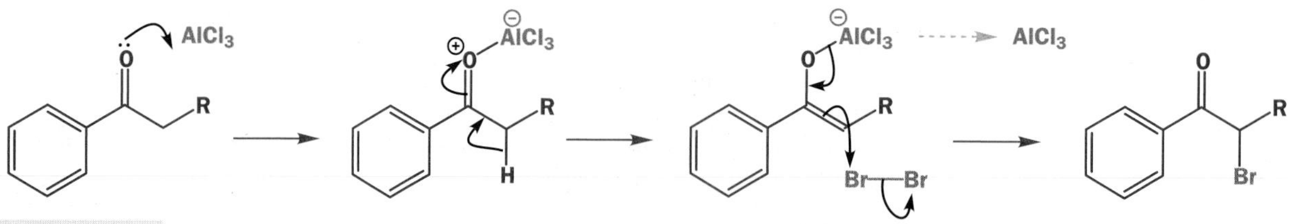

These mechanisms should remind you of the mechanism of alkene bromination (p. 504)—except that here the attack on the bromine is assisted by an electron pair on oxygen. The product, instead of being a bromonium ion (which would undergo further reactions), loses a proton (or the Lewis acid) to give a ketone.

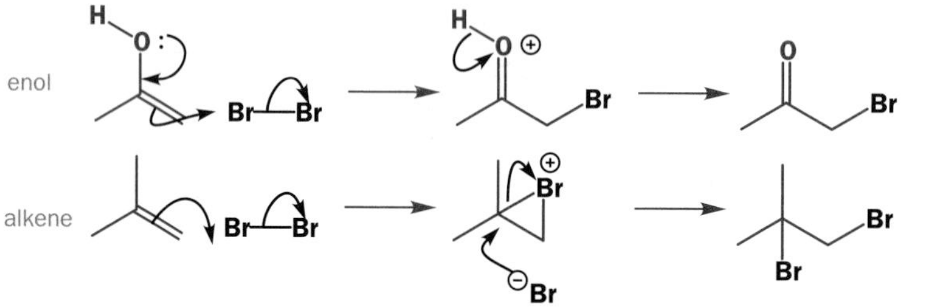

Enols are more nucleophilic than simple alkenes—the HOMO is raised by the interaction with the oxygen's lone pairs and looks not unlike the HOMO of the enolate anion we discussed on p. 528.

Bromination of acid derivatives is usually carried out not on the acid itself but by converting it to an acyl bromide or chloride, which is not isolated but gives the α-bromoacyl halide via the enol. This used to be done in one step with red phosphorus and bromine, but a two-step process is usually preferred now, and the bromoester is usually made directly without isolating any of the intermediates. We can summarize the overall process like this.

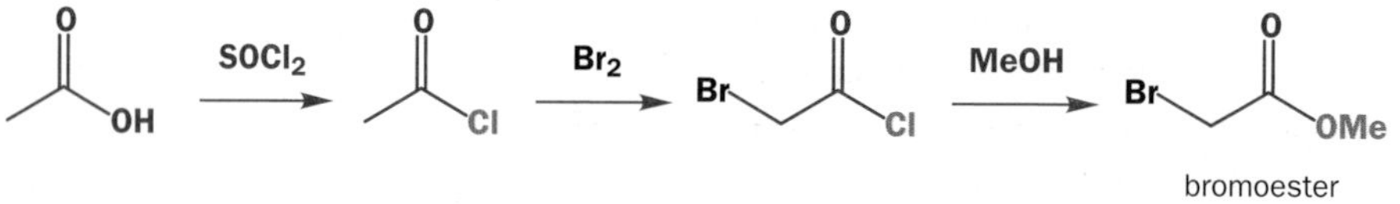

The formation of the acyl chloride with $SOCl_2$ and the conversion of the α-bromoacyl chloride into the bromoester with MeOH are simple nucleophilic substitutions at the carbonyl group, just like the synthesis of esters from acyl chlorides in Chapter 12. The intermediate stage, the bromination of the very easily enolized acyl chloride, is a typical enol bromination.

Hard and soft nucleophiles in substitution reactions are discussed in Chapter 17.

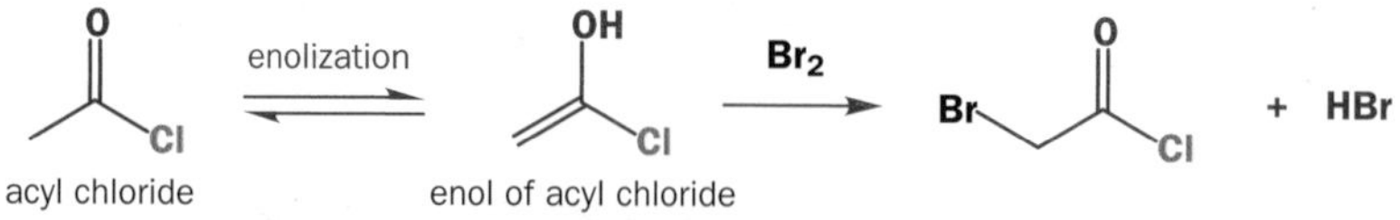

In the reaction of the bromoacyl chloride with methanol, attack occurs at the carbonyl group with an alcohol because oxygen nucleophiles are 'hard' nucleophiles (controlled by charge interactions). If we want to displace the α-bromo group we can use any 'soft' (orbital-dominated) nucleophile. Triphenylphosphine Ph_3P is particularly important—the product is a phosphonium salt, employed in Wittig reactions and discussed in Chapter 31.

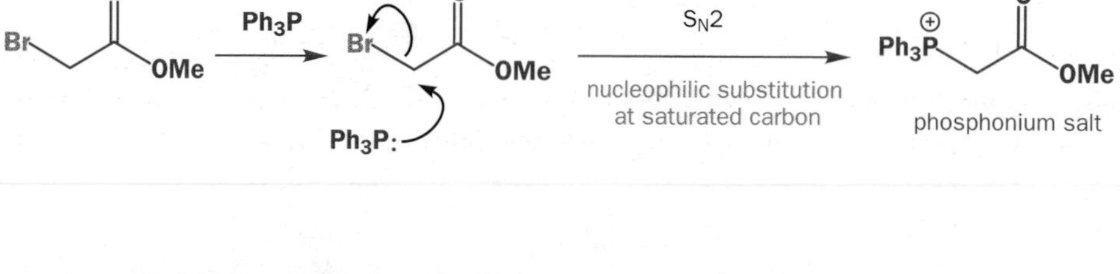

Base-catalysed halogenation

This is different and more complicated because it usually won't stop at the introduction of one halogen atom. If we go back to the bromination of acetone, the first step will now be a base-catalysed enolization to give the enolate ion instead of the enol. The enolate ion can attack a bromine molecule in a very similar way to the attack of the enol on bromine. The enolate will, of course, be even more reactive than the enol was (the enolate carries a negative charge).

The problem is that the reaction does not stop at this point. The first step was the removal of a proton and the protons between the carbonyl group and the bromine atom in the product are *more* acidic than those in the original acetone because of the electron-withdrawing bromine atom. Bromoacetone forms an enolate faster than acetone does.

Dibromoacetone is formed. Now we have one remaining proton in between the carbonyl group and two bromine atoms. It is even more acidic and so forms a new enolate ion even more quickly. The first product we can see in any amount is tribromoacetone.

But even this is not the end of the story. To see why, we need to backtrack a bit. You may already have asked yourself, 'Why doesn't the hydroxide ion, being a nucleophile, attack the carbonyl group?' This is a general question you might ask about all base-catalysed enolizations. The answer is that it does. The reaction is shown in the margin. A tetrahedral intermediate forms.

What can happen now? This tetrahedral intermediate will revert to a carbonyl compound by expelling the best leaving group—and in Chapter 12 we saw that this is usually the group with the lowest pK_{aH}. But Me$^-$ can never act as a leaving group ($pK_{aH} > 48$). Indeed the only possible leaving group is the hydroxide ion ($pK_{aH} = 15.7$), so it just drops out again.

This state of affairs continues until we reach the tribromoketone. In Chapter 8, you saw that the pK_a of CHBr$_3$ is only 9: the CBr$_3^-$ group is a better leaving group than hydroxide since the carbanion is stabilized by three bromine atoms. So now a real reaction occurs.

These initial products exchange a proton to reveal the true products of the reaction—the anion of a carboxylic acid and tribromomethane (CHBr$_3$).

The same thing happens with iodine, and we can summarize the whole process with iodine using a general structure for a carbonyl compound bearing a methyl group. It must be a methyl group because three halogens are necessary to make the carbanion into a leaving group.

Notice that the hydroxide ion is *not* regenerated in this reaction—bromide ion is not basic and does not react with water to regenerate hydroxide ion (Chapter 8). So we need to add a whole equivalent of hydroxide.

This reaction is often called the 'iodoform' reaction. Iodoform was an old name for tri-iodomethane, just as chloroform is still used for trichloromethane. It is one of the rare cases where nucleophilic substitution at a carbonyl group results in the cleavage of a C–C single bond.

● **Acid mediated halogenation is best**

Halogenation of carbonyl compounds should be carried out in acid solution. Attempts in basic solution lead to multiple substitutions and C–C bond cleavage.

Why does acid-catalysed halogenation work better?

The reason why halogenation in base continues until all the hydrogens have been replaced is clear: each successive halide makes the remaining proton(s) more acidic and the next enolization easier. But why does acid-catalysed halogenation stop after the introduction of one halogen? It would be more accurate to say that it *can be made to stop* after one halogen is introduced if only one equivalent of halogen is used. Acid-catalysed halogenation *will* continue if there is more halogen available.

However, the second halogen goes on the other side of the carbonyl group, if it can. It is evidently the case that the second halogenation is slower than the first. This is firstly because most of the intermediates are positively charged and hence destabilized by the presence of a halogen. The bromoketone is less basic than acetone so less of the reactive protonated form is present. This slows down any further electrophilic attack.

The second step is the rate-determining step, and the presence of a bromine atom at the α position slows this step down still further: if a proton can be lost from a different α position—one without a Br atom—it will be lost. The transition state for proton removal illustrates why bromine slows this step down. The part of the structure close to the bromine atom is positively charged.

We can add a useful piece of evidence to this weak-sounding explanation. The halogenation of an unsymmetrical dialkyl ketone gives different results in acid and in base. In base halogenation occurs preferentially on a methyl group, that is, on the less highly substituted side. In acid solution by contrast, halogenation occurs on the more substituted side of the carbonyl

group. Alkyl groups have the opposite effect to bromine atoms—they stabilize positive charges. So the reactions of an enol, with a positively charged transition state, are faster at more highly substituted positions. Enolates react through negatively charged transition states, and are faster at less highly substituted carbon atoms.

Nitrosation of enols

Now for a reaction with nitrogen as an electrophile that illustrates enol reactivity and reminds us that tautomerism applies to functional groups other than the carbonyl. Let us suppose you have a carbonyl compound and wish to introduce another carbonyl group next to the first. One way you might go about it is this.

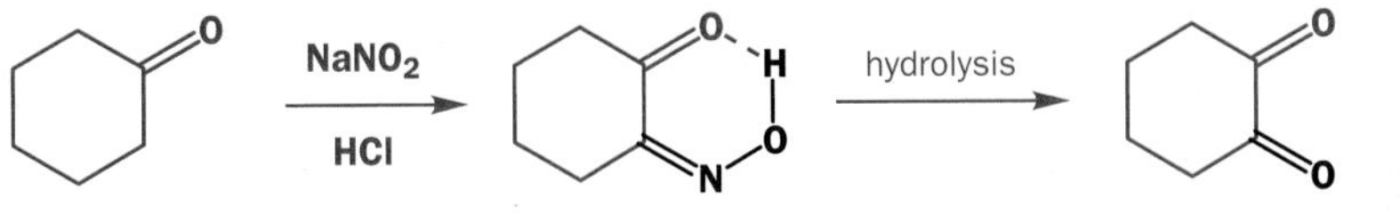

α-diketone or 1,2-diketone

The first step involves the formation of the weak acid nitrous acid (HNO_2 or, more helpfully, HONO) from the sodium salt and the strong acid HCl. Nitrous acid is itself protonated and then loss of water creates the reactive electrophile NO^+.

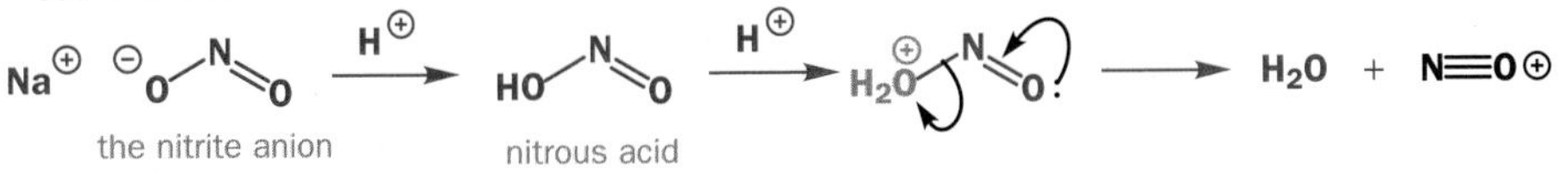

the nitrite anion   nitrous acid

This diatomic cation, isoelectronic with carbon monoxide, is electrophilic at nitrogen and attacks the enol of the ketone to form an unstable nitroso compound.

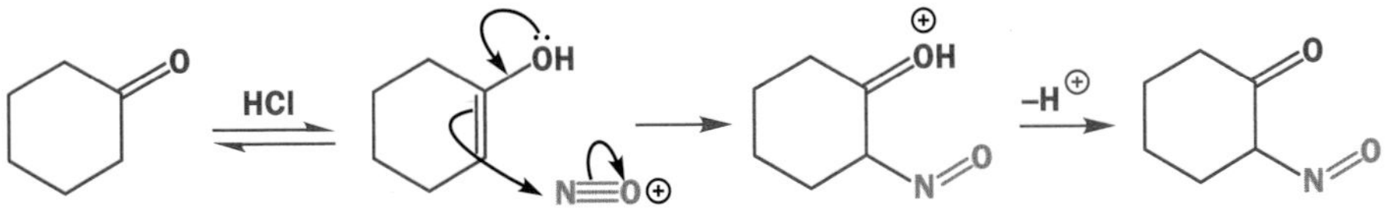

nitroso-ketone

■ The nitroso functional group, –N=O, may be new to you.

The nitroso compound is unstable because it can tautomerize with the transfer of a proton from carbon to the oxygen of the nitroso group. This process is exactly like enolization but uses an N=O instead of a C=O group. It gives a more familiar functional group from Chapter 14, the oxime, as the stable 'enol'. The second structure shows how the oxime's O–H can form an intramolecular hydrogen bond with the ketone carbonyl group. Hydrolysis of the oxime reveals the second ketone.

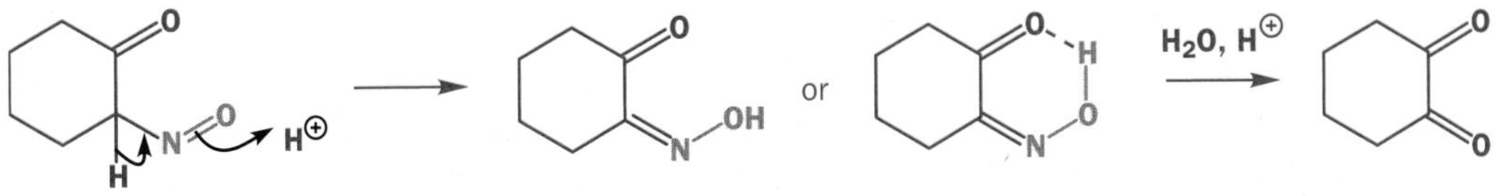

oximino-ketone   hydrogen-bonded oxime

■ Imine (and therefore oxime) hydrolysis was discussed in Chapter 14.

If the ketone is unsymmetrical, this reaction will occur on the more substituted side, for the same reason that acid-catalysed enol bromination gives the more substituted α-bromocarbonyl compound (see the box on p. 536).

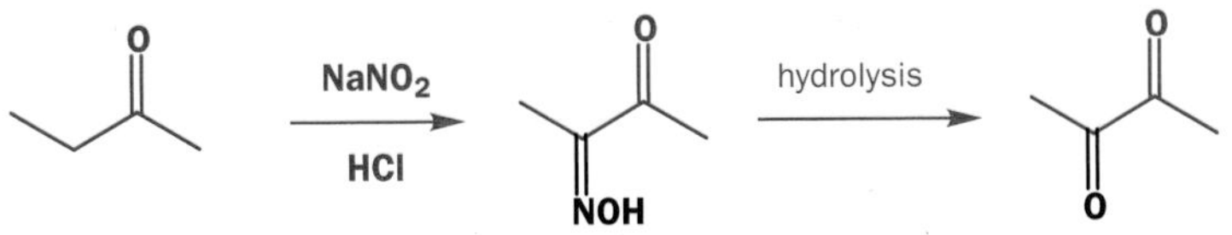

Before we move on to any more reactions, we want you to take away this message from the reactions of enols and enolates with Br_2 and with NO^+.

● **Enols and enolates generally react with electrophiles at *carbon*.**

The nitroso group

The difference between the nitro and nitroso groups is one of oxidation state and conjugation. The nitroso group contains trigonal trivalent nitrogen with a lone pair in the plane and is not delocalized. The much more stable nitro group has trigonal N$^+$ with no lone pair and is delocalized. Both can form 'enols' but the equilibria are biased in different directions.

nitroso-alkane   oxime   nitro-alkane   "enol" of nitro-alkane

no delocalization possible   complete delocalization in nitro-alkane

Stable enolate equivalents

Even with fairly strong bases such as hydroxides or alkoxides, most carbonyl compounds are converted to their enolates only to a very small extent. A typical pK_a for the protons next to a carbonyl group is 20–25, while the pK_a of methoxide is around 16, so we can only hope for about 1 part enolate in 10^4 parts carbonyl compound. With a much stronger base, this all changes, and the enolate is formed quantitatively from the carbonyl compound. This is a very important result which we shall capitalize on in Chapters 26 and 27. The base usually used is LDA (Lithium Di-isopropyl Amide), and it works like this.

LDA = Lithium Di-isopropyl Amide

■ You have already met LDA in Chapter 19 promoting elimination reactions (p. 493), but no other use of this base compares in importance with what we are telling you now. By far the most important use of LDA is for making lithium enolates.

▶ Never try to use BuLi to deprotonate a carbonyl compound! BuLi almost invariably adds to carbonyl groups as a *nucleophile*.

lithium enolate

silyl enol ether

■ The TriMethyl Silyl group is sometimes abbreviated to TMS. As this is not much shorter than Me$_3$Si, we shall use the real thing instead of the abbreviation.

LDA is bulky, so it does not take part in nucleophilic attack at the carbonyl group, and it is basic—the pK_a of diisopropylamine is about 35—plenty basic enough to deprotonate next to any carbonyl group. The lithium enolate is stable at low temperature (–78 °C) but reactive enough to be useful. Lithium enolates are the most commonly used stable enolate equivalents in chemistry.

Second only to lithium enolates in usefulness are silyl enol ethers. Silicon is less electropositive than lithium, and silyl enol ethers are more stable, but less reactive, than lithium enolates. They are made by treating an enolate with a silicon electrophile. Silicon electrophiles invariably react with enolates at the oxygen atom firstly because they are hard (see p. 237) and secondly because of the very strong Si–O single bond. The most common silicon electrophile is trimethylsilyl chloride (Me$_3$SiCl), an intermediate made industrially in bulk and used to make the NMR standard tetramethyl silane (Me$_4$Si).

Silicon–oxygen bonds are so strong that silicon reacts with carbonyl compounds on oxygen even without a strong base to form the enolate: the reaction probably goes through the small amount of enol present in neutral solution, and just needs a weak base (Et$_3$N) to remove the proton from the product. An alternative view is that the silicon reacts with oxygen first, and the base just converts to oxonium ion to the silyl enol ether. Both mechanisms are given below—either might be correct. This is one of the two best ways to make a stable enol derivative from virtually any enolizable carbonyl compound.

another possible mechanism:

Silyl enol ethers can also be made from lithium enolates just by treating them with trimethylsilyl chloride.

lithium enolate   silyl enol ether

> ■ Although we didn't discuss the details at that stage, you first met silyl enol ethers in Chapter 10, where you saw that adding Me₃SiCl is a good way of ensuring high yields in the addition of cuprates to unsaturated carbonyl compounds.

Occasionally, it can be useful to run this reaction in reverse, generating the lithium enolate from the silyl enol ether. This can be done with methyllithium, which takes part in nucleophilic substitution at silicon to generate the lithium enolate plus tetramethylsilane. The reason why you might want to carry out this seemingly rather pointless transformation will become clear in Chapters 26 and 27.

silyl enol ether   lithium enolate

We shall be returning to silyl enol ethers and lithium enolates later in the book, but for the moment you should view them simply as enol derivatives that are stable enough to be isolated. This is important because it means that we do not have to content ourselves simply with small, equilibrium concentrations of enol or enolate for our reactions: we can actually prepare enolate derivatives like these in quantitative yield, and use them in a separate step.

Enol and enolate reactions at oxygen: preparation of enol ethers

You have just seen that silyl enol ethers are easy to make. But, if enolate ions have most of their negative charge on the oxygen atom, it ought to be possible to make ordinary enol ethers from them. It is—but only under strange conditions. Normally, enols and enolate ions prefer to react with alkyl electrophiles at carbon, as we shall see in Chapter 26. If enolate ions are prepared with potassium bases in dipolar aprotic solvents (such as dimethyl sulfoxide, DMSO) that cannot solvate the oxygen anion, and are reacted with dimethyl sulfate or trimethyloxonium ion—powerful methylating agents that react best with charged atoms—some at least of the enol ether is formed. The Me₃O⁺ ion is found in the stable (though reactive) compound trimethyloxonium tetrafluoroborate, or 'Meerwein's salt', Me₃O⁺BF₄⁻. This compound and dimethylsulfate, Me₂SO₄, are hard electrophiles with highly polarized C–O bonds and therefore react at hard O rather than soft C.

free enolate
not bound by potassium
not solvated by DMSO   enol ether   C-alkylated by-product

The yield in this reaction is about 60–70% of enol ether, the rest being mainly C-alkylated product. A more reliable method is the acid-catalysed decomposition of an acetal in the strict absence of water. Here is an example.

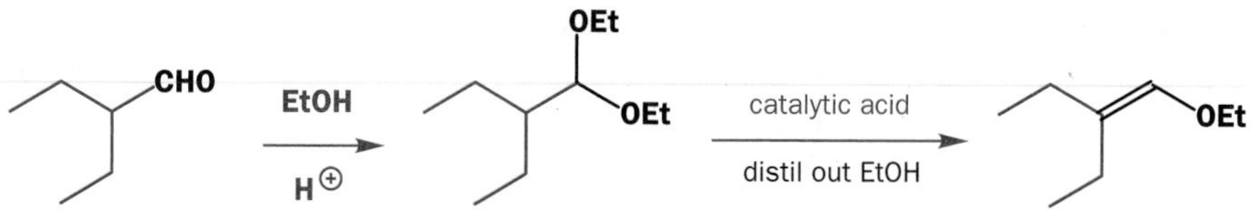

aldehyde · acetal · enol ether

The reaction starts as though the acetal were being hydrolysed, but there is no water to continue the hydrolysis, so a proton is lost instead. In other words, with no suitable nucleophile for S$_N$1 substitution, E1 elimination takes place.

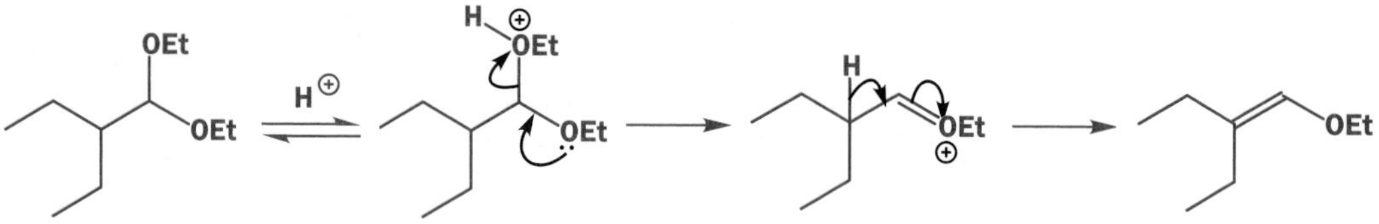

These enol ethers are rather unstable, particularly towards acid-catalysed hydrolysis (next section) and are not as useful as the silyl enol ethers. We shall next look at the enol-like reactions of both groups of enol ethers.

Reactions of enol ethers

Hydrolysis of enol ethers

Enols have an OH group and are alcohols of a sort. Normal alcohols form stable ethers that are difficult to convert back to the alcohol. Powerful reagents such as HI or BBr$_3$ are required and these reactions were discussed in Chapter 17. The reaction with HI is an S$_N$2 attack on the methyl group of the protonated ether and that is why a good nucleophile for saturated carbon, such as iodide or bromide, is needed for the reaction.

conversion of normal ether to alcohol with HI

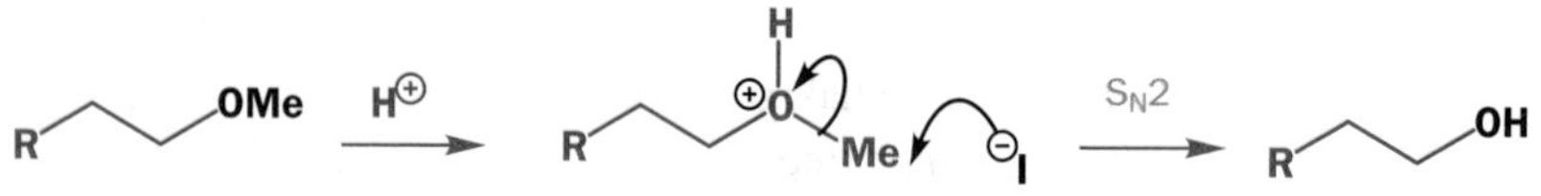

Enol ethers, by contrast, are relatively unstable compounds that are hydrolysed back to the carbonyl compound simply with aqueous acid.

Hydrolysis of enol ether with aqueous acid

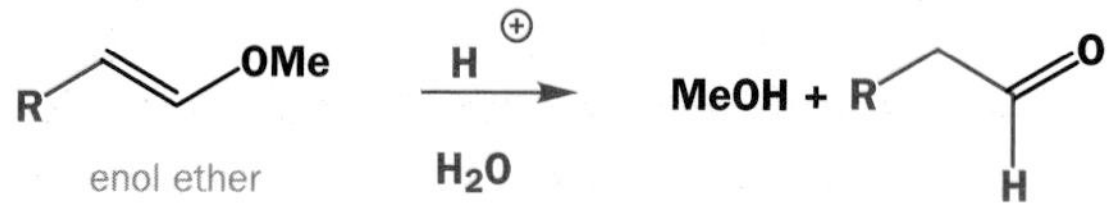

enol ether

Why the big difference? The reason is that the enol ether can be protonated at carbon using the delocalization of the oxygen lone pair in the enol derivative to produce a reactive oxonium ion.

enol ether · oxonium ion

This oxonium ion could be attacked on the methyl group in the same way that the ordinary ether was attacked.

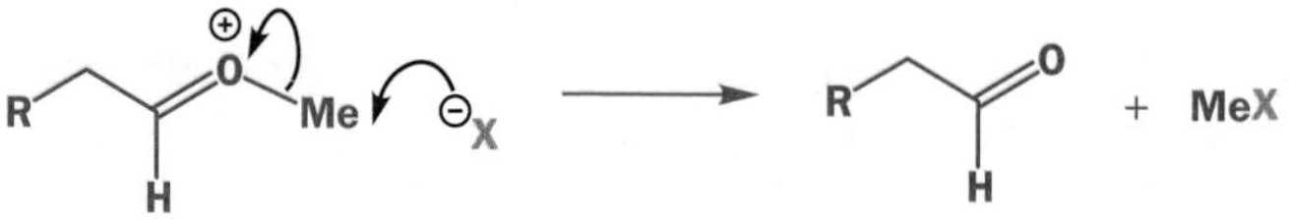

Acetal hydrolysis is discussed in Chapter 14.

We wouldn't really expect this reaction to happen much faster than the same reaction on an ordinary ether. So there must be another better and faster mechanism. It is attack on the π bond instead of on the σ bond.

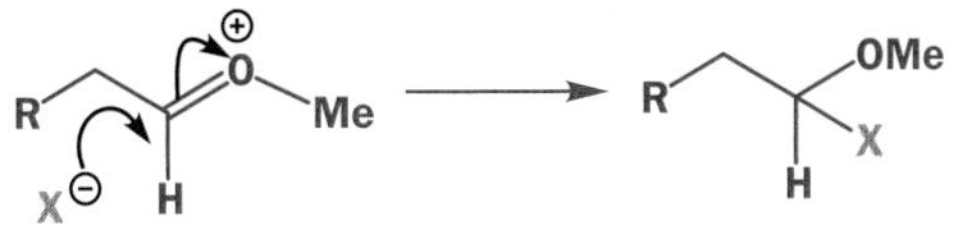

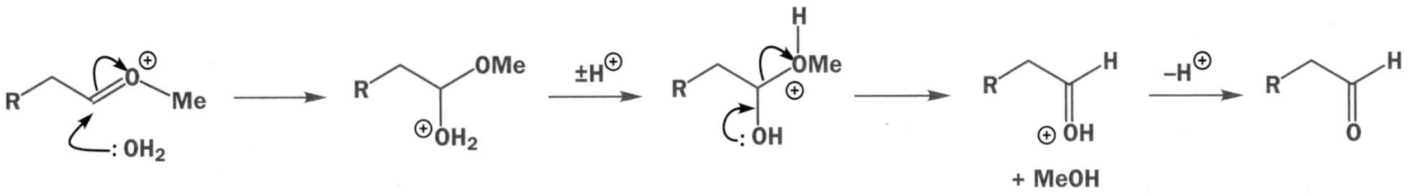

▶ Attacks on π bonds are inherently faster than attacks on σ bonds as the more weakly held π electrons are more polarized by the difference in electronegativity between C and O.

In aqueous acid the nucleophile X^- is just water and we find ourselves in the middle of the mechanism of hydrolysis of acetals (Chapter 14). The oxonium ion is a common intermediate to both mechanisms.

A similar reaction occurs when enol ethers react with alcohols in acid solution and in the absence of water, but now we are starting in the middle of the acetal hydrolysis mechanism and going the other way, in the direction of the acetal. A useful example is the formation of THP (= TetraHydroPyranyl) derivatives of alcohols from the enol ether dihydropyran. You will see THP derivatives of alcohols being used as 'protecting groups' in Chapter 24.

DHP
DiHydroPyran

the TetraHydroPyranyl derivative of the alcohol

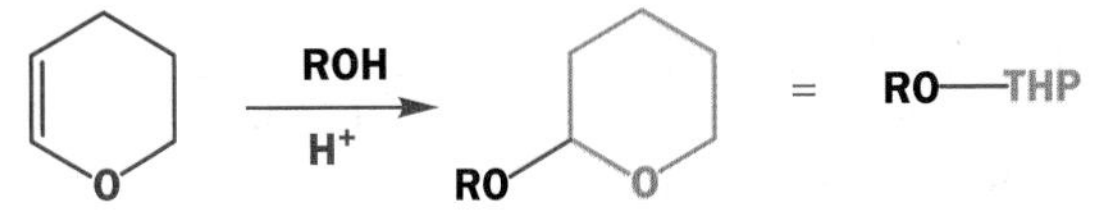

Pyrans

The naming of these compounds is a bit odd. Pyran refers to the six-membered oxygen-containing heterocyclic ring system with two double bonds. It is not aromatic though compounds like pyrones are. The compound with only one double bond is therefore dihydropyran, and the saturated ring system is tetrahydropyran.

(2*H*)-pyran (4*H*)-pyran 2-pyrone DHP DiHydroPyran THP TetraHydroPyran

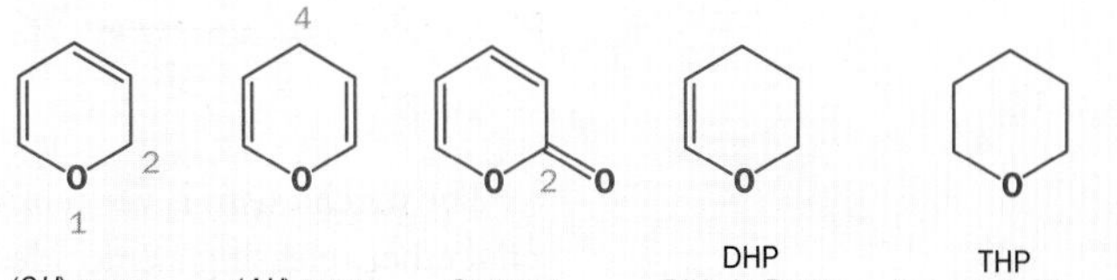

Silyl enol ethers hydrolyse by a slightly different mechanism, though the first step is the same—protonation at carbon using the lone pair on oxygen. We have already seen how easy it is to attack silicon with nucleophiles, especially those with oxygen or a halogen as the nucleophilic atom. This tips the balance towards attack by water at silicon for the next step.

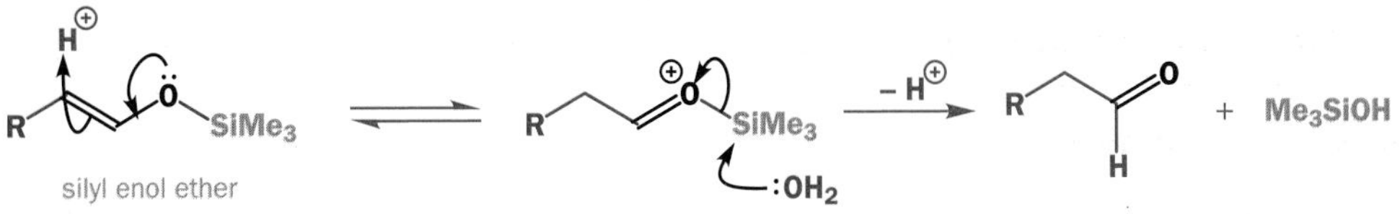

The aldehyde is formed immediately. What happens to the other product illustrates again just how easy nucleophilic substitution at silicon can be. Two of these compounds combine together to give a disilyl ether, called a disiloxane.

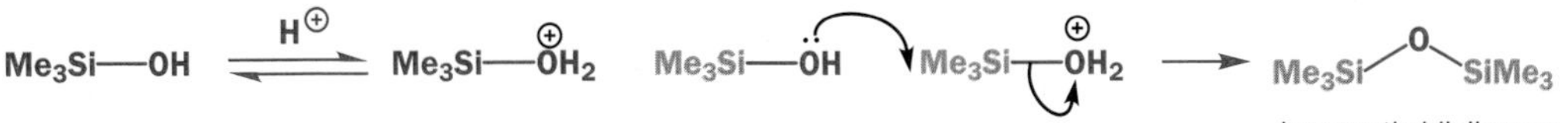

Reactions of enol ethers with halogen and sulfur electrophiles

In comparison with other ethers, enol ethers of all kinds are rather unstable. As alkenes they are also more reactive than normal alkenes because of the lone pair of electrons on the oxygen atom. They react with electrophiles like bromine or chlorine on the α carbon atom, behaving like enol derivatives and not like alkenes.

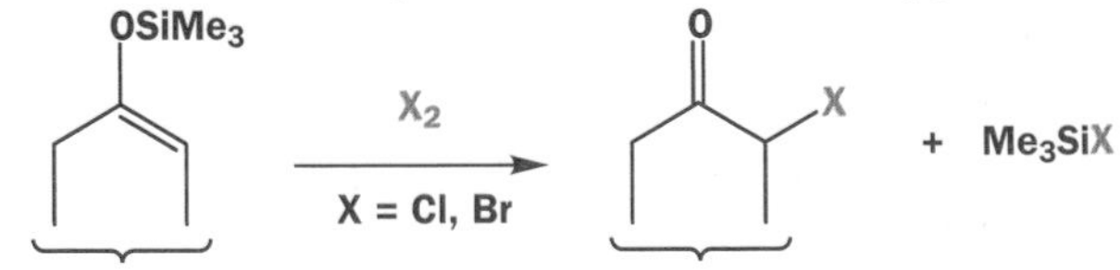

Electrophilic attack occurs at the α carbon atom and the halide ion released in this step then attacks the silicon atom to release the product and a molecule of Me₃SiX, which will be hydrolysed during the work-up.

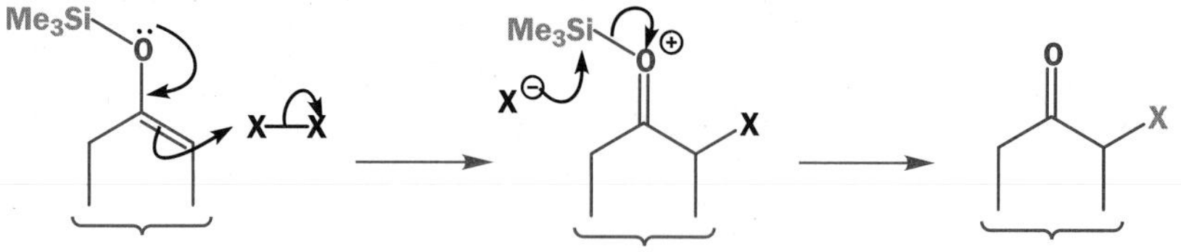

This procedure avoids the difficulties we outlined earlier in the direct halogenation of aldehydes and ketones. It allows the preparation of haloketones on the less substituted side of the carbonyl group, for instance.

LDA removes the least hindered proton

A similar method with the good soft electrophiles RSCl allows sulfenylation next to the carbonyl group.

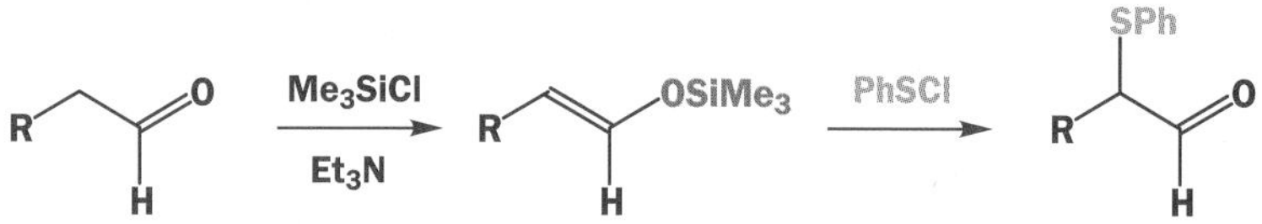

The mechanism is very similar: the electrophilic sulfur atom attacks the α carbon atom of the silyl enol ether releasing a chloride ion that removes the Me₃Si group from the intermediate.

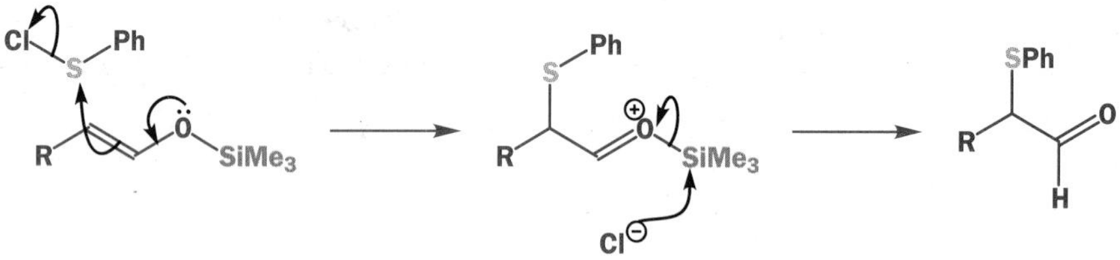

To conclude...

You have now seen how enols and enolates react with electrophiles based on hydrogen (deuterium), carbon, halogens, silicon, sulfur, and nitrogen. What remains to be seen is how new carbon–carbon bonds can be formed with alkyl halides and carbonyl compounds in their normal electrophilic mode. These reactions are the subject of Chapters 26–29. We must first look at the ways aromatic compounds react with electrophiles. You will see similarities with the behaviour of enols.

Problems

1. Draw all the possible enol forms of these carbonyl compounds and comment on the stability of the various enols.

2. The proportions of enol in a neat sample of the two ketones below are shown. Why are they so different?

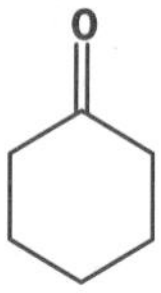

4 × 10⁻⁴% enol 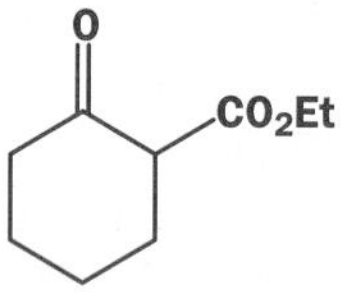   62% enol

3. Draw mechanisms for these reactions using just enolization and its reverse.

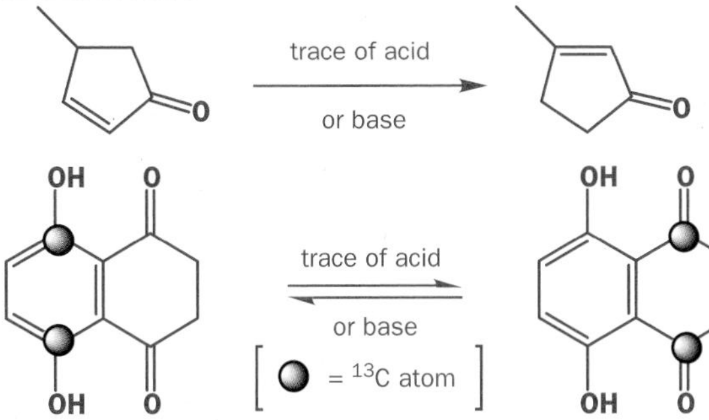

$$[\bigcirc = {}^{13}C \text{ atom}]$$

4. The NMR spectrum of this dimethyl ether is complicated—the two MeO groups are different as are all the hydrogen atoms on the rings. However, the diphenol has a very simple NMR—there are only two types of protons (marked a and b) on the rings. Explain.

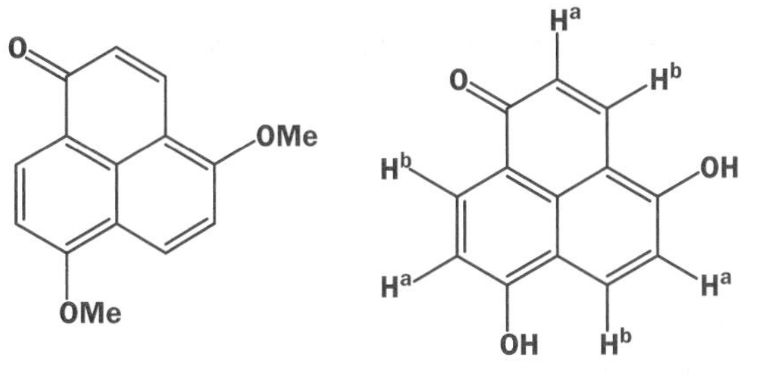

dimethyl ether   diphenol

5. Suggest mechanisms for these reactions.

(i)

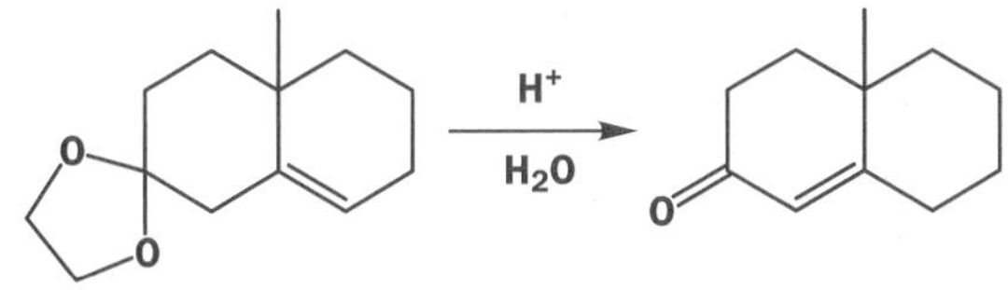

(ii)

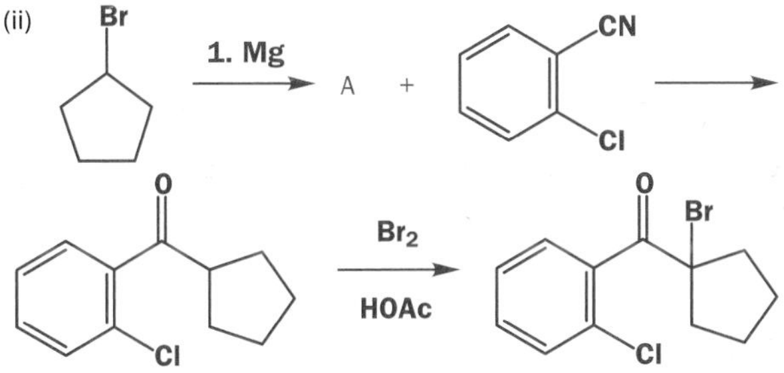

6. Treatment of this ketone with basic D_2O leads to rapid replacement of two hydrogen atoms by deuterium. Then, more slowly, all the other nonaromatic hydrogens *except* the one marked 'H' are replaced. How is this possible?

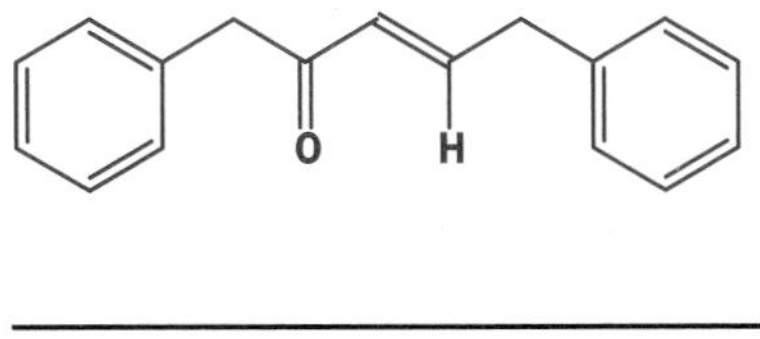

7. A red alga growing in sea water produces an array of bromine-containing compounds including $CHBr_3$, CBr_4, and $Br_2C=CHCO_2H$. The brominating agent is believed to be derived by the oxidation of bromide ion (Br^-) and can be represented as Br-OX. Suggest mechanistic details for the proposed biosynthesis of $CHBr_3$ in the alga.

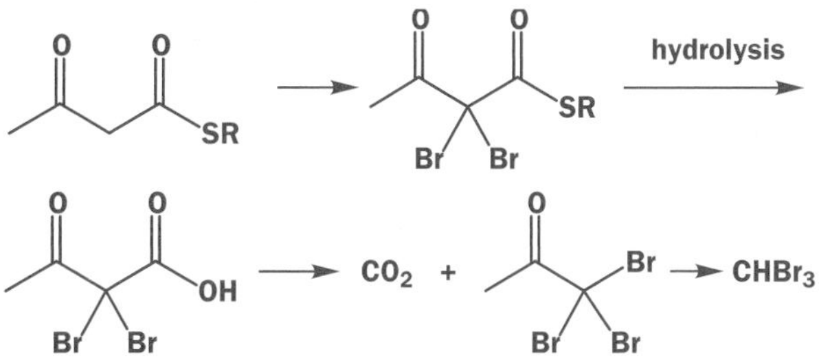

8. Suggest mechanisms for these reactions and explanations as to why these products are formed.

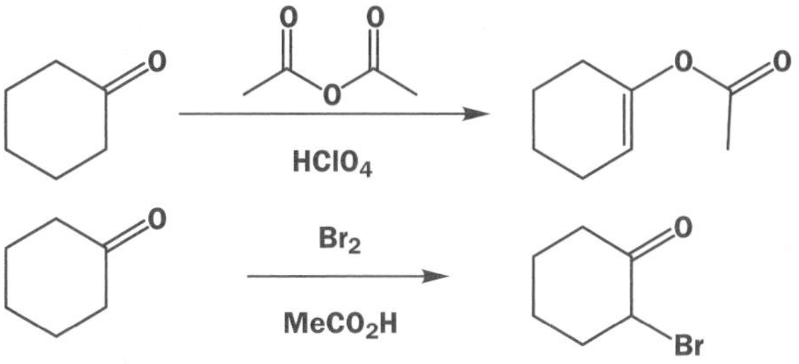

9. 1,3-Dicarbonyl compounds such as A are usually mostly enolized. Why is this? Draw the enols available to compounds B–E and explain why B is 100% enol but C, D, and E are 100% ketone.

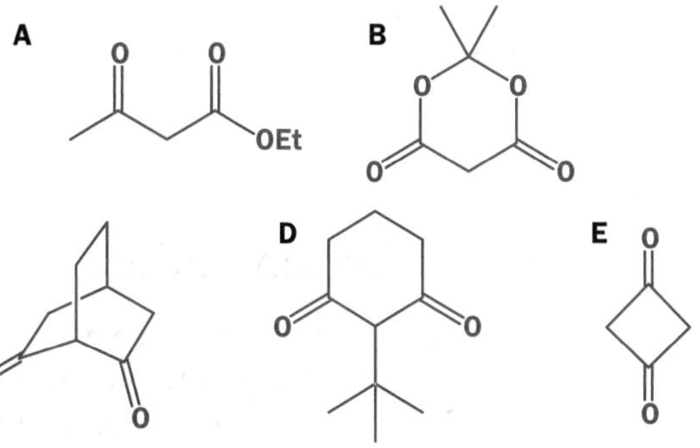

10. Bromination of ketones can be carried out with molecular bromine in a carboxylic acid solution. Give a mechanism for the reaction.

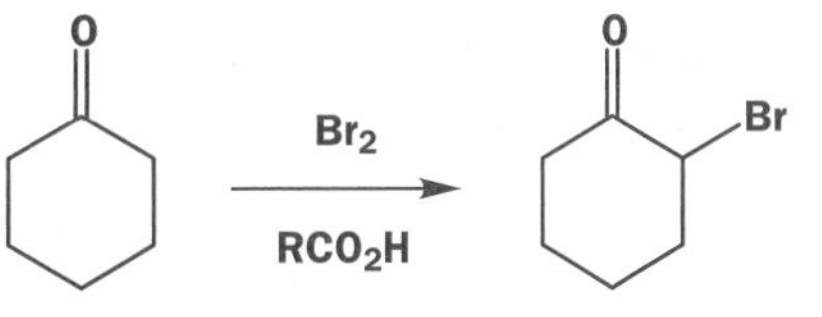

The rate of the reaction is *not* proportional to the concentration of bromine $[Br_2]$. Suggest an explanation. Why is the bromination of ketones carried out in acidic and not in basic solution?

Electrophilic aromatic substitution

22

Connections

Building on:

- **Structure of molecules ch4**
- **Conjugation ch7**
- **Mechanisms and catalysis ch13**
- **Electrophilic addition to alkenes ch20**
- **Enols and enolates ch21**

Arriving at:

- **Phenols as aromatic enols**
- **Benzene and alkenes compared**
- **Electrophilic attack on benzene**
- **Activation and deactivation**
- **Position of substitution**
- **Competition and cooperation**
- **Problems with some reactions**

Looking forward to:

- **Nucleophilic aromatic substitution ch23**
- **Oxidation and reduction ch24**
- **Synthesis in action ch25**
- **Retrosynthetic analysis ch30**
- **Aromatic heterocycles ch43–ch44**
- **Rearrangements ch37**

Introduction: enols and phenols

You have seen how, in acid or base, the protons in the α positions of ketones could be replaced by deuterium atoms (Chapter 21). In each case the reaction goes via the enol tautomer of the ketone (or enolate). For example, in base:

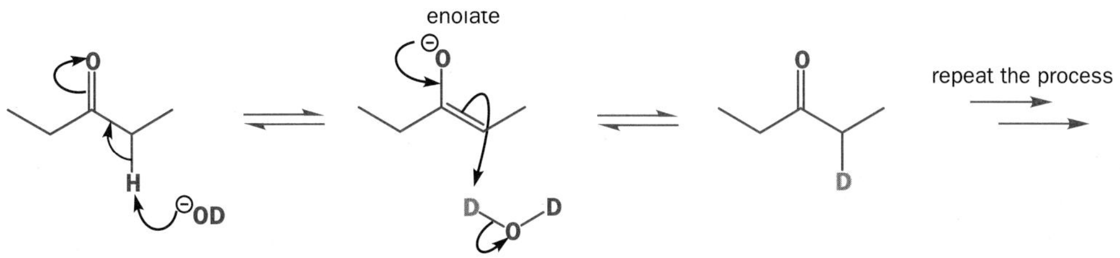

If you did Problem 6 in Chapter 21 (if you didn't, now would be a good time!), you also saw how conjugated ketones could be deuterated at positions other than the α positions; for example, in acid we now need to concentrate on the conjugated enol formed by loss of a more remote proton.

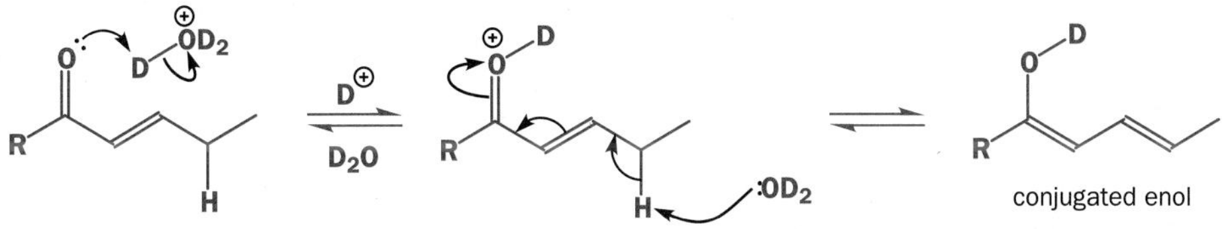

The conjugated enol could react with D_3O^+ at the normal α position, but it could also react at the more remote γ position, the position from which we have just removed a proton.

If you don't see how the third deuterium atom gets into the molecule, see the solution to Problem 21.6 in the Solutions Manual.

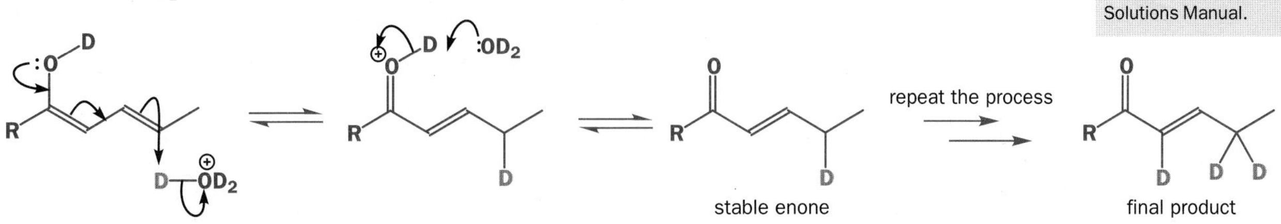

Here, as usual, the keto/enol equilibrium lies well over in favour of the ketone, but in this chapter we shall be discussing one very stable enol—phenol, PhOH. The proton NMR spectrum for phenol is shown below. Also shown below is the proton NMR after shaking phenol with acidic D_2O. Can you assign the spectra?

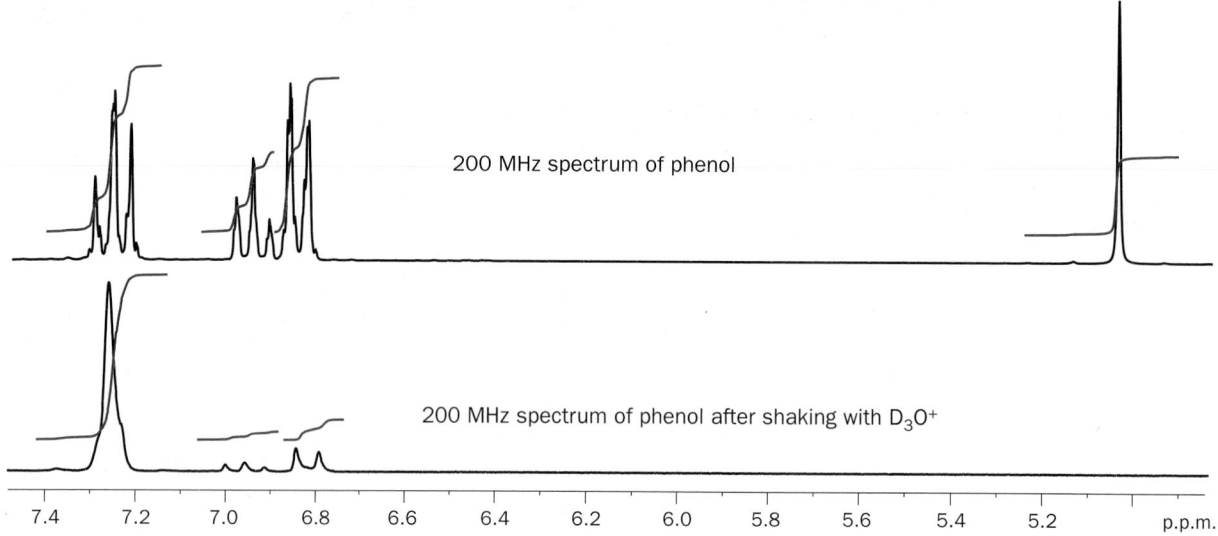

200 MHz spectrum of phenol

200 MHz spectrum of phenol after shaking with D_3O^+

Phenol is deuterated in exactly the same way as any other conjugated enol except that the final product remains as the very stable enol form of phenol rather than the keto form. The enol form of phenol is so stable because of the aromaticity of the benzene ring. The first step is addition of D_3O^+ to the 'enol'.

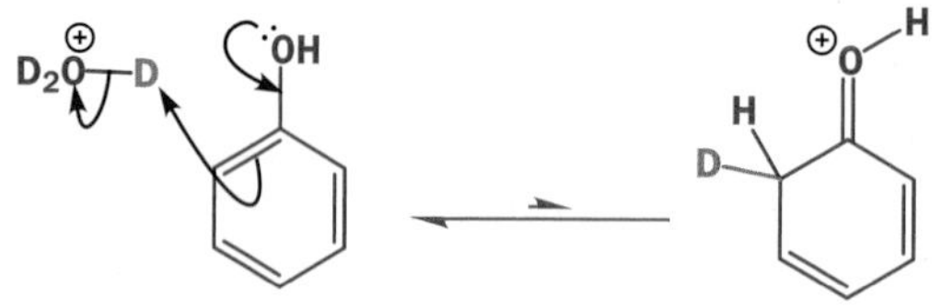

Now the intermediate cation could lose the proton from oxygen to leave a ketone or it could lose the proton from carbon to leave the phenol, or it could lose the deuteron and go back to the starting material.

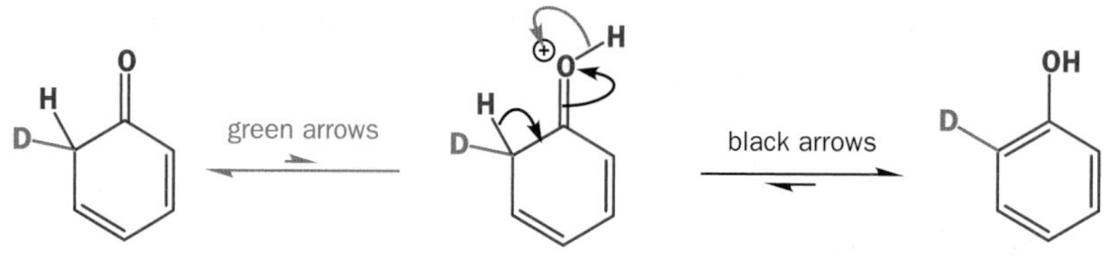

less stable 'keto' form   green arrows   black arrows   very stable 'enol' form of phenol

The end product on treating phenol with D_3O^+ simply has certain H atoms replaced by deuterium atoms. We say 'certain H atoms' but exactly which ones? The most acidic proton is the phenolic proton, the OH (pK_a about 10); this will rapidly be exchanged. The other protons that will be replaced will be the ones in the 2, 4, and 6 positions (that is, the *ortho* and *para* positions). Below is a reminder of the names we give to the positions around a benzene ring relative to any substituent.

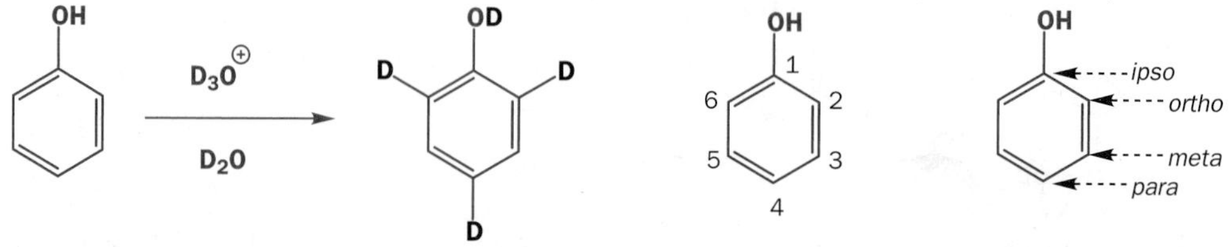

OH → $D_3O^{\oplus}$ / D_2O → (deuterated phenol)

positions: OH, 1, 2, 3, 4, 5, 6

ipso, *ortho*, *meta*, *para*

Phenol initially behaves like a conjugated enol in its reactions with electrophiles but, instead of giving a ketone product, the enol is formed because the very stable aromatic ring is regained. This chapter is about the reactions of phenols and other aromatic compounds with electrophiles. You will see phenols reacting like enols, except that the final product is also an enol, and you will also see simple benzenes reacting like alkenes, except that the result is substitution rather than addition. We shall start with a discussion of the structure of benzene and of aromaticity.

■ The details of the orbitals of benzene appeared in Chapter 7 (p. 174).

Benzene and its reaction with electrophiles

Benzene is a planar symmetrical hexagon with six trigonal (sp^2) carbon atoms, each having one hydrogen atom in the plane of the ring. All the bond lengths are 1.39 Å (compare C–C 1.47 Å and C=C 1.33 Å). All the ^{13}C shifts are the same (δ_C 128.5 p.p.m.).

The special stability of benzene (aromaticity) comes from the six π electrons in three molecular orbitals made up by the overlap of the six atomic p orbitals on the carbon atoms. The energy levels of these orbitals are arranged so that there is exceptional stability in the molecule (a notional 140 kJ mol^{-1} over a molecule with three conjugated double bonds), and the shift of the six identical hydrogen atoms in the NMR spectrum (δ_H 7.2 p.p.m.) is evidence of a ring current in the delocalized π system.

■ This section revises material from Chapters 4, 8, and 11 where more details can be found.

δ_H 7.2

δ_C 128.5

Drawing benzene rings

Benzene is symmetrical and the circle in the middle best represents this. However, it is impossible to draw mechanisms on that representation so we shall usually use the **Kekulé** form with three double bonds. This does

not mean that we think the double bonds are localized but just that we need to draw curly arrows. It makes no difference which Kekulé structure you draw—the mechanism can be equally well drawn on either.

this circle structure best represents the six delocalized π electrons

these Kekulé structures are best for drawing curly arrows. They are equivalent

When we move away from benzene itself to discuss molecules such as phenol, the bond lengths are no longer exactly the same. However, it is still all right to use either representation, depending on the purpose of the drawing. With some aromatic compounds, such as naphthalene, it

does matter as there is some bond alternation. Either representation is still all right, but only one Kekulé representation shows that the central bond is the strongest and shortest in the molecule and that the C1–C2 bond is shorter that the C2–C3 bond.

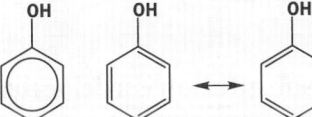

three acceptable drawings of phenol. The Kekulé drawings are equivalent

two acceptable drawings of naphthalene. Only one Kekulé drawing is satisfactory

▶ Not everyone agrees that the two circles are all right for naphthalene. If each circle represents six electrons, this representation is wrong as there are only ten electrons altogether. If you don't interpret the circles quite so strictly they're all right.

Electrophilic attack on benzene and on cyclohexene

Simple alkenes, including cyclohexene, react rapidly with electrophiles such as bromine or peroxy-acids (Chapter 20). Bromine gives a product of *trans* addition, peracids give epoxides by *cis* addition. Under the same conditions benzene does not react with either reagent.

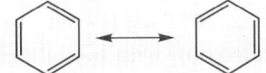

Benzene can be persuaded to react with bromine if a Lewis acid catalyst such as $AlCl_3$ is added. The product contains bromine but is not from either *cis* or *trans* addition.

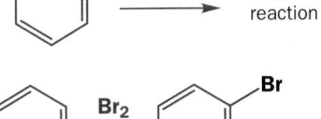

The bromine atom has replaced an atom of hydrogen and so this is a substitution reaction. The reagent is electrophilic bromine and the molecule is aromatic so the reaction is **electrophilic aromatic substitution** and that is the subject of this chapter. We can compare the bromination of cyclohexene and of benzene directly.

electrophilic addition
to an alkene

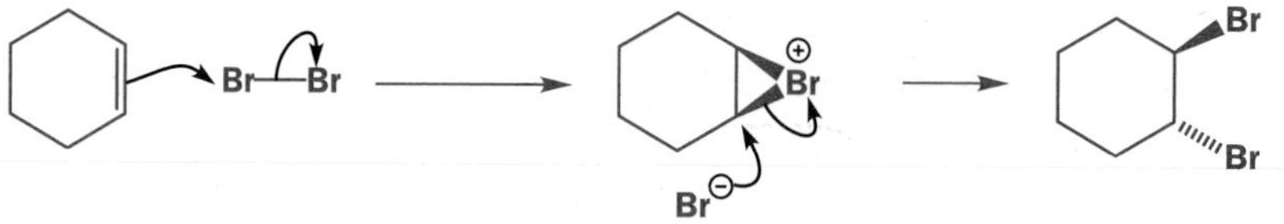

electrophilic substitution
on benzene

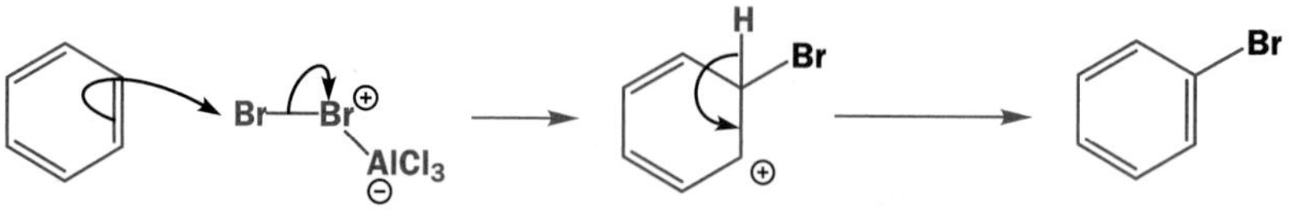

The intermediate in both reactions is a cation but the first (from cyclohexene) adds an anion while the second (from benzene) loses a proton so that the aromatic system can be restored. Notice also that neutral bromine reacts with the alkene but the cationic AlCl$_3$ complex is needed for benzene. Another way to produce a more electrophilic source of bromine is to use a pyridine catalyst. Pyridine attacks the bromine molecule producing a cationic bromine compound.

formation of reactive
brominating agent

pyridine catalyst
is recycled

electrophilic substitution
on benzene

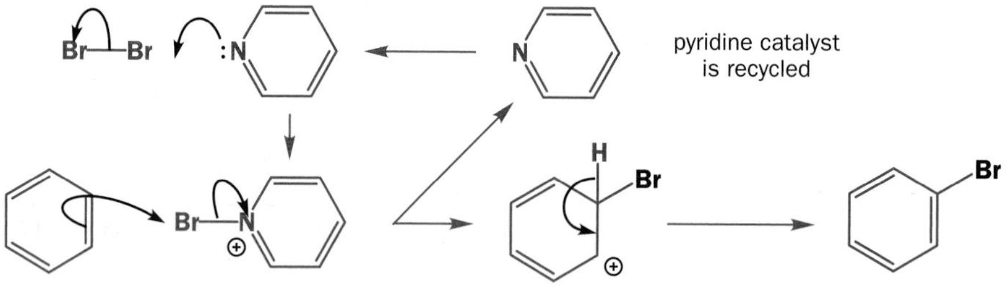

Bromine itself is a very reactive electrophile. It is indeed a dangerous compound and should be handled only with special precautions. Even so it does not react with benzene. It is very difficult to get benzene to react with anything.

⬤ **Benzene is very unreactive**

- It combines only with very reactive (usually cationic) electrophiles
- It gives substitution and not addition products

The intermediate in electrophilic aromatic substitution is a delocalized cation

These two brominations are examples of the mechanism of electrophilic aromatic substitution, which, in many different guises, will return again and again during this chapter. In its most general form the mechanism has two stages: attack by an electrophile to give an intermediate cation and loss of a proton from the cation to restore the aromaticity.

mechanism for electrophilic
aromatic substitution

addition of
electrophile

loss of
proton

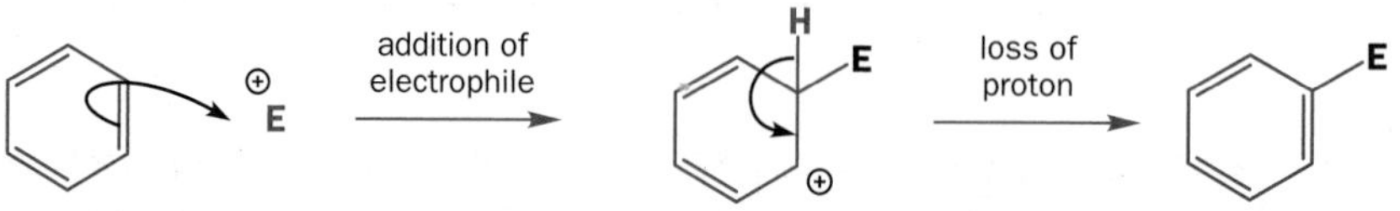

The cationic intermediate is, of course, less stable than the starting materials or the product but as a cation it is reasonably stable because of delocalization around the six-membered ring. The charge can be delocalized to the two *ortho* positions and to the *para* position or can be drawn as a delocalized structure with dotted bonds and about one-third of a plus charge (+) at three atoms.

the intermediate in electrophilic aromatic substitution

In strong acid, the electrophile would be a proton and the reaction would be the exchange of the protons in the benzene ring in the style of the proton exchange on phenol with which we started this chapter. In D_3O^+, this would ultimately lead to C_6D_6 which is a useful solvent in NMR. As with the bromination reaction, the first step in the mechanism is the formation of a cationic intermediate.

It is actually possible to observe this cationic intermediate. The trick is to pick a nonnucleophilic and nonbasic counterion X^-, such as antimony hexafluoride SbF_6^-. In this octahedral anion, the central antimony atom is surrounded by the fluorine atoms and the negative charge is spread over all seven atoms. The protonation is carried out using FSO_3H and SbF_5 at $-120\,°C$.

■ This trick was used to show the existence of simple carbocations as intermediates in the S_N1 mechanism in Chapter 17.

Under these conditions it is possible to record the ^1H and ^{13}C NMR spectra of the cation. The shifts show that the positive charge is spread over the ring but is greatest (or the electron density is least) in the *ortho* and *para* positions. Using the data for the ^1H and ^{13}C NMR shifts (δ_H and δ_C, respectively), a charge distribution can be calculated.

Compound	Position	δ_H, p.p.m.	δ_C, p.p.m.	Calculated charge distribution
	1	5.6	52.2	
	2, 6	9.7	186.6	0.26 ... 0.26
	3, 5	8.6	136.9	0.09 ... 0.09
	4	9.3	178.1	0.30
benzene		7.33	129.7	

Curly arrows also predict the same electron distribution for all these intermediates, whether the electrophile is a proton or any of the other reagents we will meet in this chapter. The cation can be represented as three different delocalized structures that show clearly the electron-deficient atoms, or by a structure with partial bonds that shows the delocalization but is of no use for drawing mechanisms.

It is not surprising that the formation of the cationic intermediate is the rate-determining step, as aromaticity is temporarily lost in this step. The mechanism of the fast proton loss from the intermediate is shown in three ways just to prove that it doesn't matter which of the delocalized structures you choose. A useful piece of advice is that, when you draw the intermediate in any electrophilic aromatic substitution, you should always draw in the hydrogen atom at the point of substitution, just as we have been doing.

Nitration of benzene

■ In Chapter 2 the structure of the nitro group was discussed. It has a delocalized structure with two charges as nitrogen cannot have five bonds as it would then have to have ten electrons.

Perhaps the most important of all the reactions in this chapter is nitration, the introduction of a nitro (NO_2) group, into an aromatic system, as it provides a general entry into aromatic nitrogen compounds. This reaction is not available for aliphatic nitrogen compounds, which are usually made with nucleophilic nitrogen reagents. Aromatic nitration requires very powerful reagents, the most typical being a mixture of concentrated nitric and sulfuric acids.

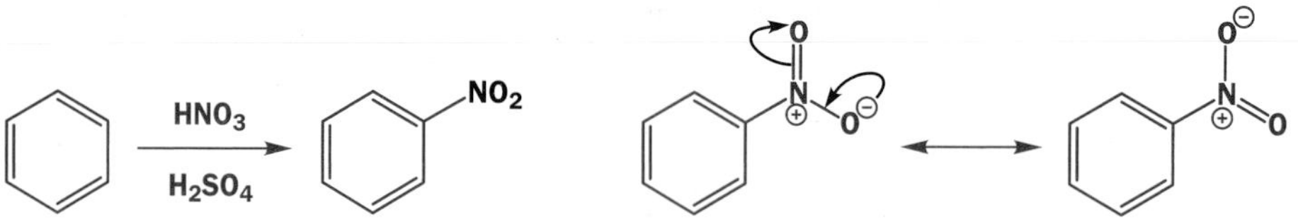
HNO3
H2SO4
NO2

The first steps are the formation of a very powerful electrophile, none other than NO_2^+, by the interaction of the two strong acids. Sulfuric acid is the stronger and it protonates the nitric acid on the OH group so that a molecule of water can leave.

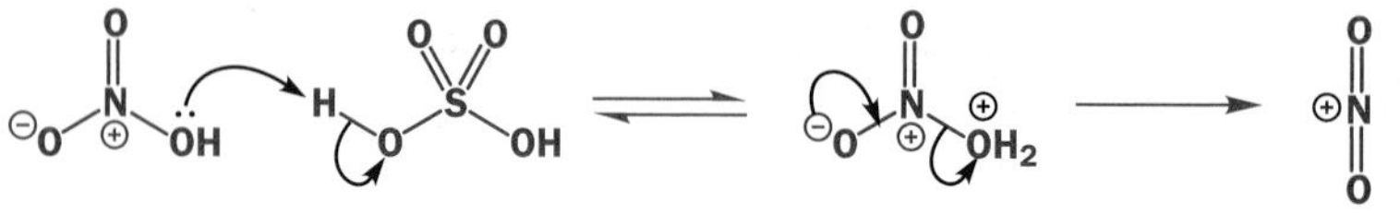

nitric acid   sulfuric acid   nitronium ion

Notice that the nitronium ion (NO_2^+) is linear with an sp hybridized nitrogen at the centre. It is isoelectronic with CO_2. It is also very reactive and combines with benzene in the way we have just described. Benzene attacks the positively charged nitrogen atom but one of the N=O bonds must be broken at the same time to avoid five-valent nitrogen.

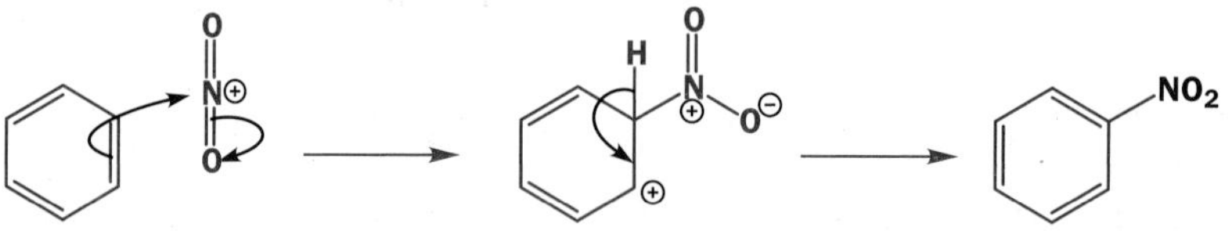

● **Nitration converts aromatic compounds (ArH) into nitrobenzenes ($ArNO_2$) using NO_2^+ from $HNO_3 + H_2SO_4$.**

Sulfonation of benzene

Benzene reacts slowly with sulfuric acid alone to give benzenesulfonic acid. The reaction starts with the protonation of one molecule of sulfuric acid by another and the loss of a molecule of water. This is very similar to the first steps in nitration.

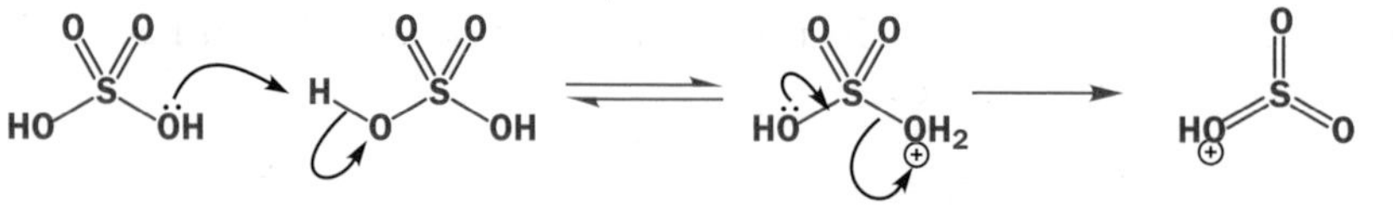

The cation produced is very reactive and combines with benzene by the same mechanisms we have seen for bromination and nitration: slow addition to the aromatic π system followed by rapid loss of a proton to regenerate the aromaticity. The product contains the sulfonic acid functional group $-SO_2OH$.

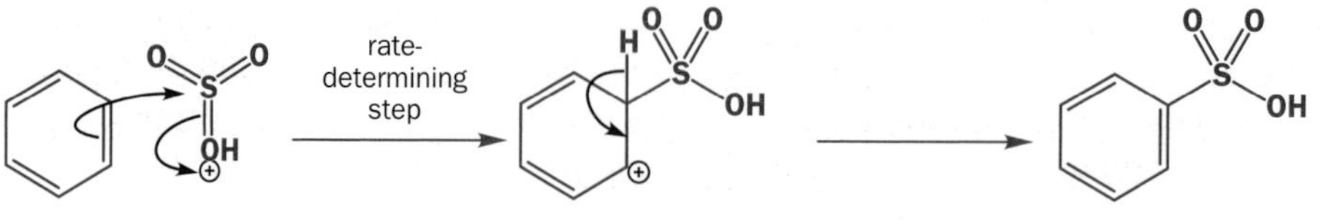
rate-
determining
step

benzenesulfonic acid

The cationic intermediate can also be formed by the protonation of sulfur trioxide, SO_3, and another way to do sulfonations is to use concentrated sulfuric acid with SO_3 added. These solutions have the industrial name **oleum**. It is possible that the sulfonating agent in all these reactions is not protonated SO_3 but SO_3 itself.

Sulfonic acids are strong acids, about as strong as sulfuric acid itself. They are stronger than HCl, for example, and can be isolated from the reaction mixture as their crystalline sodium salts if an excess of NaCl is added. Not many compounds react with NaCl.

benzenesulfonic acid

crystalline sodium
benzene sulfonate

> ● Sulfonation with H_2SO_4 or SO_3 in H_2SO_4 converts aromatic compounds (ArH)
> into aromatic sulfonic acids ($ArSO_2OH$). The electrophile is SO_3 or SO_3H^+.

Alkyl and acyl substituents can be added to a benzene ring by the Friedel–Crafts reaction

So far we have added heteroatoms only—bromine, nitrogen, or sulfur. Adding carbon electrophiles requires reactive carbon electrophiles and that means carbocations. In Chapter 17 you learned that any nucleophile, however weak, will react with a carbocation in the S_N1 reaction and even benzene rings will do this. The classic S_N1 electrophile is the *t*-butyl cation generated from *t*-butanol with acid.

This is, in fact, an unusual way to carry out such reactions. The **Friedel–Crafts alkylation**, as this is known, usually involves treating benzene with a *t*-alkyl chloride and the Lewis acid $AlCl_3$. Rather in the manner of the reaction with bromine, $AlCl_3$ removes the chlorine atom from *t*-BuCl and releases the *t*-Bu cation for the alkylation reaction.

■ Charles Friedel (1832–99), a French chemist, and James Crafts (1839–1917), an American mining engineer, both studied with Wurtz and then worked together in Paris where in 1877 they discovered the Friedel–Crafts reaction.

We have not usually bothered with the base that removes the proton from the intermediate. Here it is chloride ion as the by-product is HCl, so you can see that even a very weak base will do. Anything, such as water, chloride, or other counterions of strong acids, will do this job well enough and you need not in general be concerned with the exact agent.

A more important variation is the Friedel–Crafts acylation with acid chlorides and AlCl₃. As you saw in Chapter 13, acid chlorides can give the rather stable acylium ions even in hydrolytic reactions and they do so readily with Lewis acid catalysis. Attack on a benzene ring then gives an aromatic ketone. The benzene ring has been acylated.

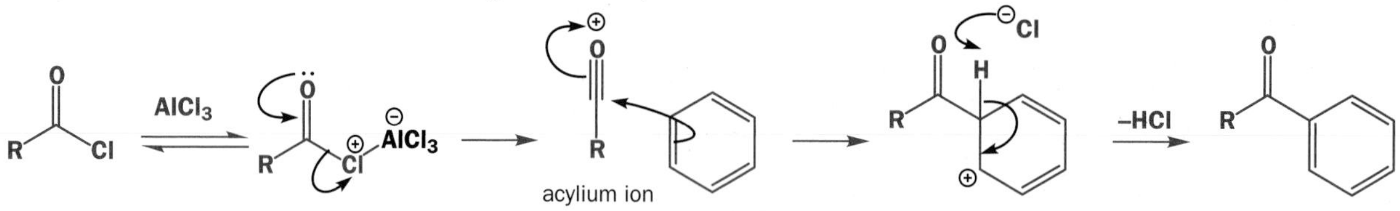

acylium ion

The acylation is better than the alkylation because it does not require any particular structural feature in the acyl chloride—R can be almost anything. In the alkylation step it is essential that the alkyl group can form a cation; otherwise the reaction does not work very well. In addition, for reasons we are about to explore, the acylation stops cleanly after one reaction whereas the alkylation often gives mixtures of products.

● Friedel–Crafts reactions

Friedel–Crafts alkylation with *t*-alkyl chlorides and Lewis acids (usually AlCl₃) gives *t*-alkyl benzenes. The more reliable Friedel–Crafts acylation with acid chlorides and Lewis acids (usually AlCl₃) gives aryl ketones.

Summary of electrophilic substitution on benzene

This completes our preliminary survey of the most important reactions in aromatic electrophilic substitution. We shall switch our attention to the benzene ring itself now and see what effects various types of substituent have on these reactions. During this discussion we will return to each of the main reactions and discuss them in more detail. Meanwhile, we leave the introduction with an energy profile diagram in the style of Chapter 13 for a typical substitution.

mechanism for electrophilic aromatic substitution

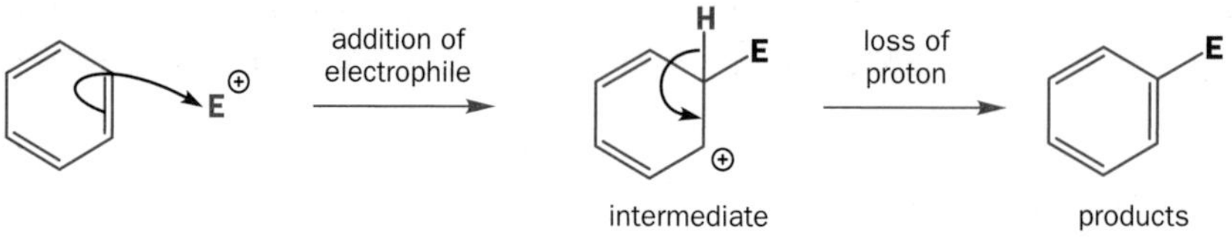

▶

This argument is based on the **Hammond postulate,** which suggests that structures close in energy that transform directly into each other are also similar in structure.

Since the first step involves the temporary disruption of the aromatic π system, and is therefore rate-determining, it must have the higher-energy transition state. The intermediate is unstable and has a much higher energy that either the starting material or the products, close to that of the transition states. The two transition states will be similar in structure to the intermediate and we shall use the intermediate as a model for the important first transition state.

The reaction is so slow and the transition state so high because the only HOMO available is a pair of very low-energy bonding electrons in the benzene ring and because the uniquely stable aromatic π system is already disrupted in the transition state.

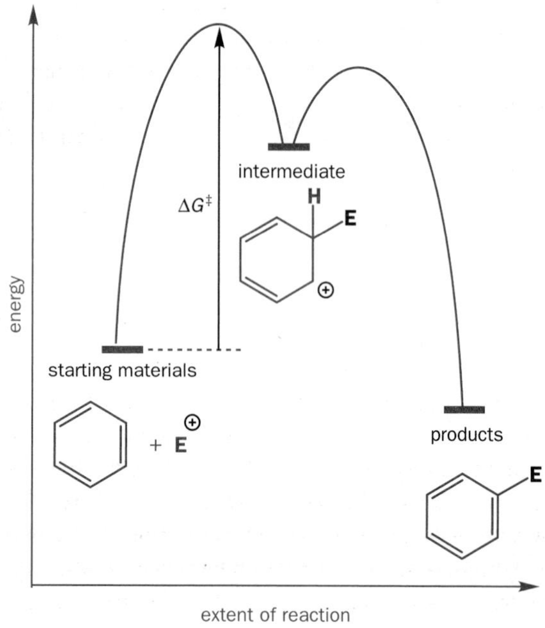

● **Summary of the main electrophilic substitutions on benzene**

Reaction	Reagents	Electrophile	Products
bromination	Br_2 and Lewis acid, e.g. $AlCl_3$, $FeBr_3$, Fe powder	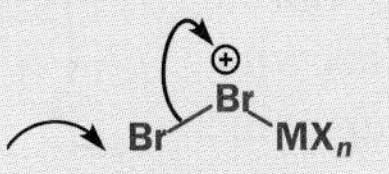	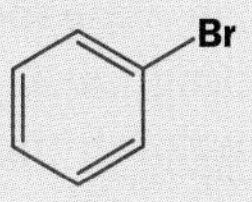
nitration	$HNO_3 + H_2SO_4$	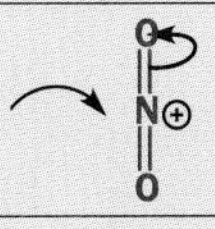	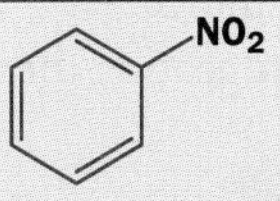
sulfonation	concentrated H_2SO_4 or $H_2SO_4 + SO_3$ (oleum)	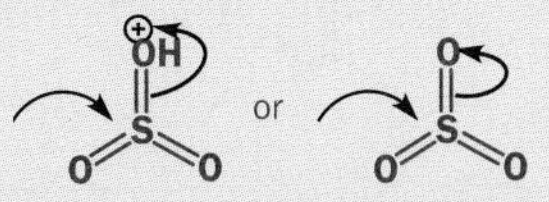	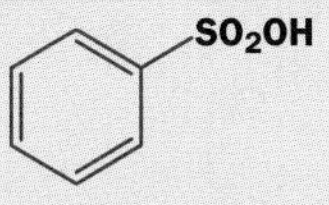
Friedel–Crafts alkylation	RX + Lewis acid usually $AlCl_3$	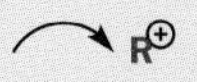	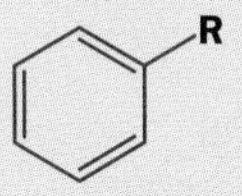
Friedel–Crafts acylation	RCOCl + Lewis acid usually $AlCl_3$	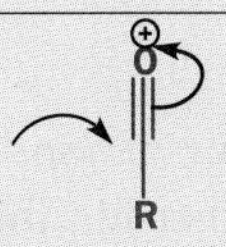	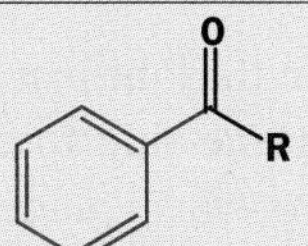

Electrophilic substitution on phenols

We started this chapter by comparing phenols with enols (Ph-enol is the phenyl enol) and now we return to them and look at electrophilic substitution in full detail. You will find that the reaction is much easier than it was with benzene itself because phenols are like enols and the same reactions (bromination, nitration, sulfonations, and Friedel–Crafts reactions) occur more easily. There is a new question too: the positions round the phenol ring are no longer equivalent—where does substitution take place?

Phenols react rapidly with bromine

Benzene does not react with bromine except with Lewis acid catalysis. Phenols react in a very different manner: no Lewis acid is needed, the reaction occurs very rapidly, and the product contains three atoms of bromine in specific positions. All that needs to be done is to add bromine dropwise to a solution of phenol in ethanol. Initially, the yellow colour of the bromine disappears but, if, when the colour just remains, water is added, a white precipitate of 2,4,6-tribromophenol is formed.

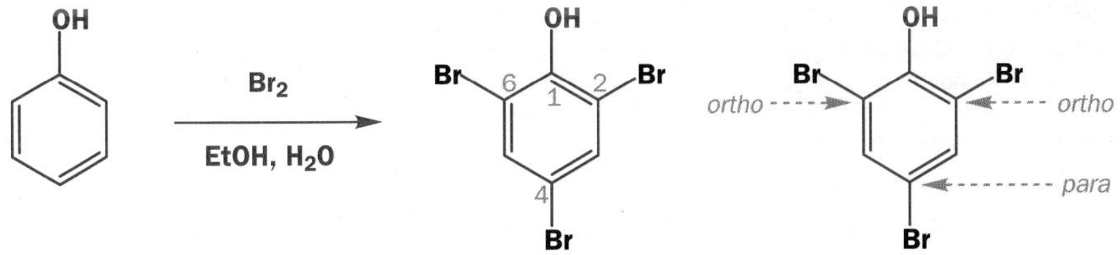

The product shows that bromination has occurred at the *para* position and at both *ortho* positions. What a contrast to benzene! Phenol reacts three times as rapidly without catalysis at room temperature. Benzene reacts once and needs a Lewis acid to make the reaction go at all. The difference is, of course, the enol nature of phenol. The highest-energy electrons in phenol are no longer those in the benzene ring but the lone pairs on oxygen. These nonbonding electrons contribute to a much higher-energy HOMO than the very low-energy bonding electrons in the aromatic ring. We should let our mechanism show this. Starting in the *para* position:

► This is not strictly *catalysis* as a stoichiometric amount of Lewis acid is needed and cannot be recovered.

Notice that we start the chain of arrows with the lone pair electrons on the OH group and push them through the ring so that they emerge at the *para* position to attack the bromine molecule. The benzene ring is acting as a conductor allowing electrons to flow from the OH group to the bromine molecule.

para (4-) bromophenol

Now repeating the reaction but this time at one of the two equivalent *ortho* positions:

para (4-) bromophenol   2,4-dibromophenol   2,4,6-tribromophenol

Again the lone pair electrons on the OH group are the HOMO and these electrons are fed through the benzene ring to emerge at the *ortho* position. A third bromination in the remaining *ortho* position—you could draw the mechanisms for this as practice—gives the final product 2,4,6-tribromophenol.

The OH group is said to be **ortho, para-directing** towards electrophiles. No substitution occurs in either *meta* position. We can understand this by looking at the curly arrow mechanisms or by looking at the molecular orbitals. In Chapter 21 (p. 526) we looked at the π system of an enolate and saw how the electron density is located mainly on the end atoms (the oxygen and the carbon). In phenol it is the *ortho* and *para* positions that are electron-rich (and, of course, the oxygen itself). We could show this using curly arrows.

The curly arrows actually give an indication of the electron distribution in the HOMO of the molecule. The reason is that the HOMO has large coefficients at *every other* atom, just as the allyl anion had large coefficients at its ends but not in the middle (Chapter 7).

Benzyl anion HOMO – a model for phenol

A better analogy for phenol is the benzyl anion. The benzyl anion is simpler because we do not have the added complication of the differences in electronegativities between the oxygen and carbon atoms. According to simple calculations, the highest occupied molecular orbital (HOMO) for the benzyl anion is a nonbonding molecular orbital (MO) with the distribution like this.

the benzyl anion   the charge distribution in the HOMO for the benzyl anion

In this MO there are no bonding interactions between adjacent atoms so the HOMO for the benzyl anion is actually a nonbonding MO. Most of the electron density is on the benzylic carbon atom not in the ring, but there is also significant electron density on the ring carbon atoms in the *ortho* and *para* positions. The distribution for phenol will be different because it is not an anion and the oxygen atom is more electronegative than carbon but the overall distribution will be as predicted by the curly arrows—most on the oxygen and on the *ortho* and *para* carbon atoms.

NMR can give us some confirmation of the electron distribution

The ^{1}H NMR shifts of phenol give us an indication of the electron distribution in the π system. The more electron density that surrounds a nucleus, the more shielded it is and so the smaller the shift (see Chapter 11). All the shifts for the ring protons in phenol are less than those for benzene (7.26 p.p.m.), which means that overall there is greater electron density in the ring. There is little difference between the *ortho* and the *para* positions: both are electron-rich.

The shifts are smallest in the *ortho* and *para* positions so these are where there is greatest electron density and hence these are the sites for electrophilic attack. The shifts in the *meta* positions are not significantly different from those in benzene. If you want to put just one bromine atom into a phenol, you must work at low temperature (< 5 °C) and use just one equivalent of bromine. The best solvent is the rather dangerously inflammable carbon disulfide (CS_2), the sulfur analogue of CO_2. Under these conditions, *para* bromophenol is formed in good yield as the main product, which is why we started the bromination of phenol in the *para* position. The minor product is *ortho* bromophenol.

proton NMR shifts in phenol (benzene, 7.26 p.p.m.)

4-bromophenol
85% yield

● Electrophilic attack on phenols

- OH groups on benzene rings are *ortho*, *para*-directing and activating
- You will get the right product if you start your arrows at a lone pair on the OH group

Benzene is less reactive than phenol towards electrophiles

To brominate phenol, all we had to do was to mix bromine and phenol—if we do this with benzene itself, nothing happens. We therefore say that, relative to benzene, the OH group in phenol *activates* the ring towards electrophilic attack. The OH group is activating and *ortho*, *para*-directing. Benzene *will* undergo electrophilic aromatic substitution as we have seen in a variety of reactions with catalysis by strong protic acids or Lewis acids such as $AlCl_3$. It is the donation of electrons on the oxygen into the aromatic ring that makes phenol so much more reactive than benzene towards electrophiles. Other groups that can donate electrons also activate and direct *ortho*, *para*. Anisole (methoxybenzene) is the 'enol ether' equivalent of phenol. It reacts faster than benzene with electrophiles.

The multiple chlorination of another activated compound, phenoxyacetic acid, leads to a useful product. This compound is made industrially by an S_N2 reaction (Chapter 17) on chloroacetic acid (made by chlorination of acetic acid, Chapter 21) with phenol in alkaline solution. Reaction occurs at the oxygen atom rather than on the ring.

phenoxyacetic acid

The herbicide '2,4-D' is 2,4-dichlorophenoxy acetic acid and is made, again industrially, by chlorination of the acid with two equivalents of chlorine. The first probably goes into the *para* position and the second into one of the equivalent *ortho* positions.

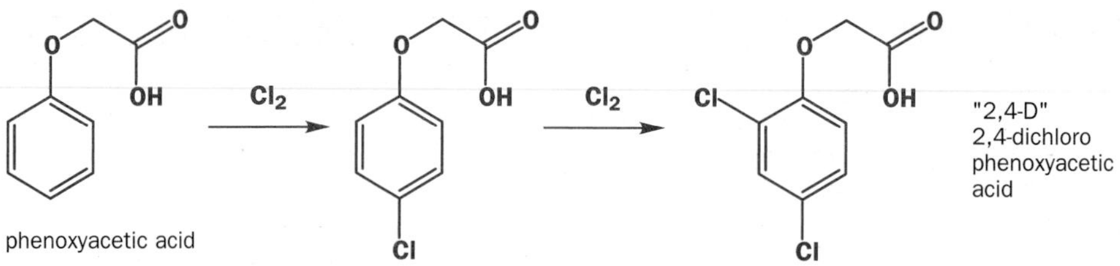

phenoxyacetic acid

"2,4-D"
2,4-dichloro
phenoxyacetic
acid

Salicylic acid is 2-hydroxybenzoic acid and is named after the willow trees (genus *Salix*) from which it was first isolated.

The phenoxide ion is even more reactive towards electrophilic attack than phenol. It will even react with such weak electrophiles as carbon dioxide. This reaction, known as the Kolbe–Schmitt process, is used industrially to prepare salicylic acid, a precursor in making aspirin.

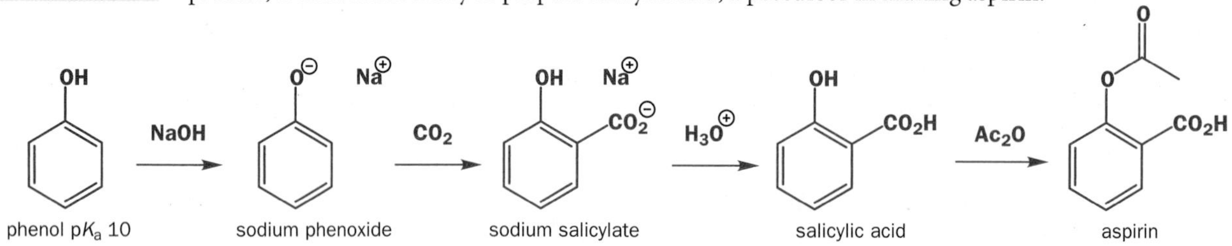

phenol pK$_a$ 10   sodium phenoxide   sodium salicylate   salicylic acid   aspirin

The O$^-$ substituent is *ortho, para*-directing but the electrophilic substitution step with CO_2 gives mostly the *ortho* product so there must be some coordination between the sodium ion and two oxygen atoms, one from the phenoxide and one from CO_2. The electrophile is effectively delivered to the *ortho* position.

We shall return to reactions of phenols and phenyl ethers when we consider directing effects in electrophilic aromatic substitution in other reactions and in Friedel–Crafts reactions in particular.

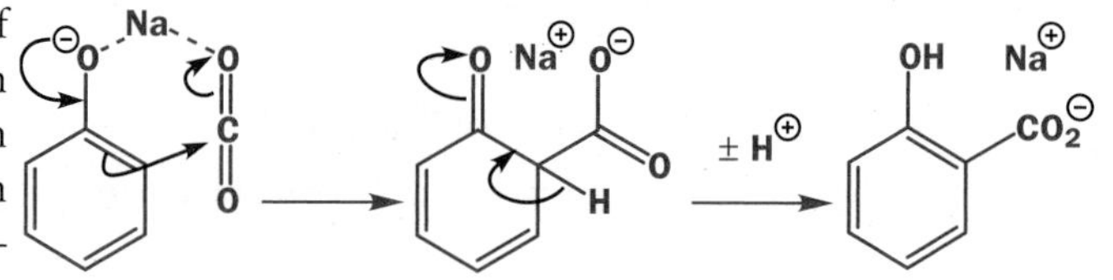

A nitrogen lone pair activates even more strongly

Aniline (phenylamine) is even more reactive towards electrophiles than phenols, phenyl ethers, or phenoxide ions. Because nitrogen is less electronegative than oxygen, the lone pair is higher in energy and so more available to interact with the π system than is the lone pair on oxygen (look back to p. 287 where we compare the reactivity of amides and esters). Reaction with bromine is very vigorous and rapidly gives 2,4,6-tribromoaniline. The mechanism is very similar to the bromination of phenol so we show only one *ortho* substitution.

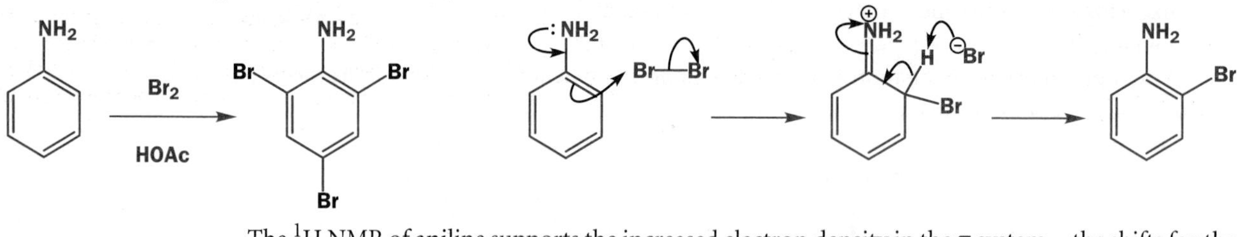

The ^{1}H NMR of aniline supports the increased electron density in the π system—the shifts for the aromatic protons are even smaller than those for phenol showing greater electron density in the *ortho* and *para* positions.

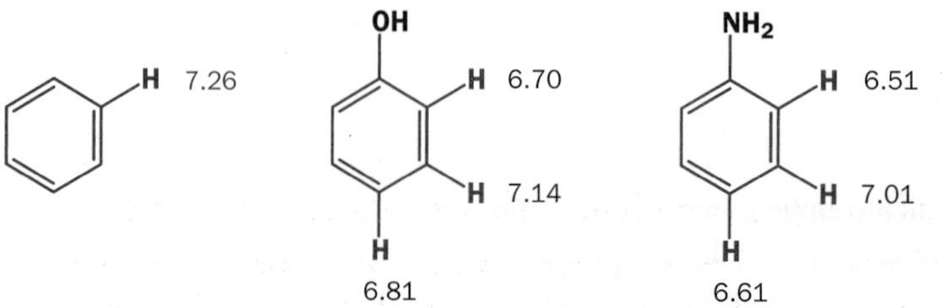

Just how good nitrogen is in donating electrons into the π system is shown by comparing the relative rates for the bromination of benzene, methoxybenzene (anisole), and N,N-dimethylaniline.

Compound	Rate of bromination relative to benzene
Benzene	1
Methoxybenzene (anisole)	10^9
N,N-dimethylaniline	10^{14}

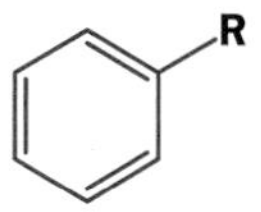

R = H; benzene
R = OMe; anisole
R = NMe$_2$; N,N-dimethylaniline

Making amines less reactive

The high reactivity of aniline can actually be a problem. Suppose we wanted to put just one bromine atom on to the ring. With phenol, this is possible (p. 557)—if bromine is added slowly to a solution of phenol in carbon disulfide and the temperature is kept below 5 °C, the main product is *para*-bromophenol. Not so if aniline is used—the main product is the triply substituted product.

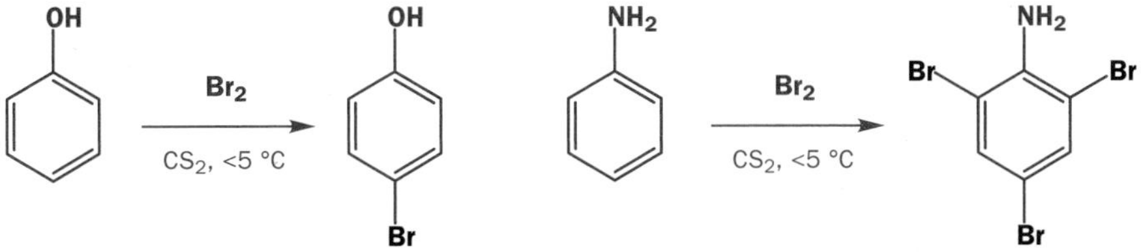

How then could we prevent oversubstitution from occurring? What we need is a way to make aniline less reactive by preventing the nitrogen lone pair from interacting so strongly with the π system of the ring. Fortunately, it is very simple to do this. In Chapter 8 (p. 201) we saw how the nitrogen atom in an amide is much less basic than a normal amine because it is conjugated with the carbonyl group. This is the strategy that we will use here—simply acylate the amine to form an amide. The amide nitrogen can still donate electrons into the ring, but much less efficiently than the amine and so the electrophilic aromatic substitution is more controlled. After the reaction, the amide can be hydrolysed back to the amine.

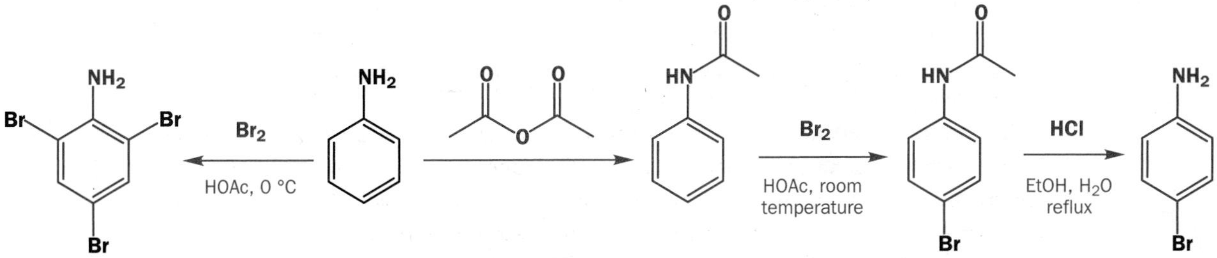

The lone pair electrons on the nitrogen atom of the amide are conjugated with the carbonyl group as usual but they are also delocalized into the benzene ring, though more weakly than in the amine. Reaction still occurs in the *ortho* and *para* positions (mainly *para*) but it occurs once only.

▶

Compounds formed by the acylation of ammonia are familiar to you as amides and those formed by the acylation of anilines are sometimes called *anilides*. If they are acetyl derivatives they are called *acetanilides*. We shall not use these names but you may see them in some books.

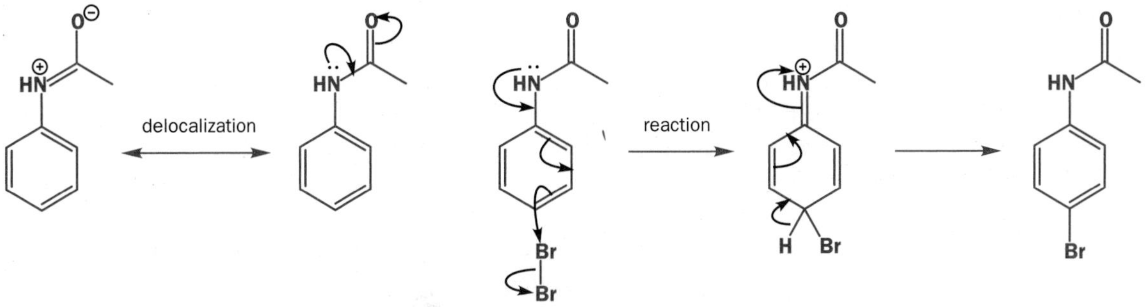

Selectivity between *ortho* and *para* positions is determined by steric hindrance

Phenols and anilines react in the *ortho* and/or *para* positions for electronic reasons. These are the most important effects in deciding where an electrophilic substitution will occur on a benzene ring.

When it comes to choosing between *ortho* and *para* positions we need to consider steric effects as well. You will have noticed that we have seen one *ortho* selective reaction—the formation of salicylic acid from phenol—and several *para* selective reactions such as the bromination of an amide just discussed.

If the reactions occurred merely statistically, we should expect twice as much *ortho* as *para* product because there are two *ortho* positions. However, we should also expect more steric hindrance in *ortho* substitution since the new substituent must sit closely beside the one already there. With large substituents, such as the amide, steric hindrance will be significant and it is not surprising that we get more *para* product.

A closer look at the transition state

We haven't given the whole picture as to why groups with a lone pair that can conjugate into the ring make the ring so much more reactive towards electrophilic attack. What we have said so far is that the starting material is more reactive because of the increased electron density in the ring. This is true, but what we should really be concerned with is the activation energy for the reaction. The energy profile for an electrophilic substitution reaction with 'E$^+$' on a phenyl ether looks rather like the one we showed earlier for benzene.

We need to understand how the activation energy, $\Delta G^{\ddagger}$, changes when R is an electron-donating substituent and so we really need to know the relative energy of the transition state. We do not know the energy of the transition state, or even

> A diagram such as this shows a molecule with two substituents but either the relationship between them is not known or the compound is a mixture.

> This argument is based on the Hammond postulate, which suggests that structures close in energy that transform directly into each other are also similar in structure (Chapter 41).

exactly what it looks like (Chapter 13), but we can assume that the transition state looks more like the intermediate than like the starting material because it is close in energy to the unstable intermediate. It will help to look at the different intermediates that could be formed by attacking in the *ortho*, *meta*, and *para* positions and try to work out which of these, and hence which transition states, might be higher in energy.

For an electrophile attacking a benzene ring containing an electron-donating group (here OR), the following intermediates are possible, depending on whether the electrophile attacks *ortho*, *meta*, or *para* to the group already present. The intermediate in *para* substitution is not drawn since it has the same stabilization as the *ortho* intermediate.

stabilization of the intermediate in *ortho* (or *para*) substitution   stabilization of the intermediate in *meta* substitution

Each intermediate is stabilized by delocalization of the positive charge to three carbon atoms in the ring. If the electrophile attacks *ortho* (or *para*) to the electron-donating group, OR, the positive charge is further delocalized directly on to OR, but the intermediate in *meta* substitution does not enjoy this extra stabilization. We can assume that the extra stabilization in the intermediate in *ortho* (or *para*) substitution means that the transition state is similarly lower in energy than that in *meta* substitution. Not only is there more electron density in the *ortho* and *para* positions in the starting material (and hence a good interaction between these sites and the electrophile) but also the transition states resulting from *ortho* and *para* attack are lower in energy than the transition state for *meta* attack. These points both mean that $\Delta G^{\ddagger}$ is smaller for *ortho*/*para* attack and that the reaction is faster than *meta* attack.

Alkyl benzenes react at the *ortho* and *para* positions: σ donor substituents

The rate constant for the bromination of toluene (methylbenzene) is about 4000 times that for benzene (this may sound like a lot, but the rate constant for *N*,*N*-dimethylaniline is 10^{14} times greater). The methyl group also directs electrophiles mostly into the *ortho* and *para* positions. These two observations together suggest that alkyl groups may also increase the electron density in the π system of the benzene ring, specifically in the *ortho* and *para* positions, rather like a weakened version of an OR group.

about 60% *ortho*   about 35% *para*   about 5% *meta*

There is a small inductive effect between any sp^2 and sp^3 carbon atoms (Chapter 8) but, if this were the only effect, then the carbon to which the alkyl group is attached (the *ipso* carbon) should have the greatest electron density, followed by the *ortho* carbons, then the *meta* carbons, and finally the carbon atom furthest from the substituent, in the *para* position.

The ^{1}H NMR spectrum for toluene suggests that there is slightly more electron density in the *para* position than in the *meta* positions. All the shifts are smaller than those of benzene but not by much and the shielding is much less than it is in phenols or anilines. The methyl group donates electrons weakly by conjugation. In phenol, a lone pair on oxygen is conjugated into the π system. In toluene there is no lone pair but one of the C–H σ bonds can interact with the π system in a similar way. This interaction,

one of the C–H σ bonds can overlap with the π system of the ring

know as σ **conjugation**, is not as good as the full conjugation of the oxygen lone pair, but it is certainly better than no interaction at all.

Just as the conjugation of the oxygen lone pair increases the electron density at the *ortho* and *para* positions, so too does σ conjugation, but more weakly. However, it does not provide another pair of electrons to act as the HOMO. Toluene uses π electrons, which are slightly higher in energy than those of benzene. It is best to regard alkyl benzenes as rather reactive benzenes. We have to draw the mechanism using the π electrons as the nucleophile.

The positive charge in the intermediate is delocalized over three carbons as usual and we can study the intermediate by protonation in superacid as we did with benzene. The result is more revealing because protonation actually occurs in the *para* position.

The *ortho* (to the Me group) carbon has a shift (139.5 p.p.m.) only 10 p.p.m. greater than that of benzene (129.7 p.p.m.) but the *ipso* and *meta* carbons have the very large shifts that we associate with cations. The charge is mainly delocalized to these carbons but the greatest charge is at the *ipso* carbon. Electrophilic attack occurs on alkyl benzenes so that the positive charge can be delocalized to the carbon bearing the alkyl group. This carbon is tertiary and so cations there are stable (Chapter 17) and they can enjoy the σ conjugation from the alkyl group. This condition is fulfilled if toluene is attacked at the *ortho* or *para* positions as you have seen but not if it is attacked at the *meta* position.

Now the charge is delocalized to the three carbon atoms that do not include the *ipso* carbon and no σ conjugation from the alkyl group is possible. The situation is no worse than that of benzene, but toluene reacts some 10^3 faster than benzene at the *ortho* and *para* positions. The stability of the transition states for electrophilic attack on toluene can again be modelled on these intermediates, so they follow the same pattern. The transition states for *ortho* and *para* attack have some positive charge at the *ipso* carbon but that for *meta* substitution does not.

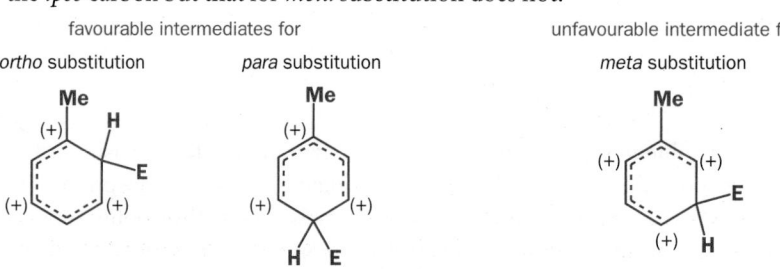

favourable intermediates for

ortho substitution   *para* substitution

unfavourable intermediate for

meta substitution

The sulfonation of toluene

Direct sulfonation of toluene with concentrated sulfuric acid gives a mixture of *ortho* and *para* sulfonic acids from which about 40% of toluene *para* sulfonic acid can be isolated as the sodium salt. The free acid is important as a convenient solid acid, useful when a strong acid is needed to catalyse a reaction. Being much more easily handled than oily and corrosive sulfuric acid or syrupy phosphoric acid, it is useful for acetal formation (Chapter 14) and eliminations by the E1 mechanism on alcohols (Chapter 19). It is usually called tosic acid, TsOH, or PTSA (*para* toluene sulfonic acid).

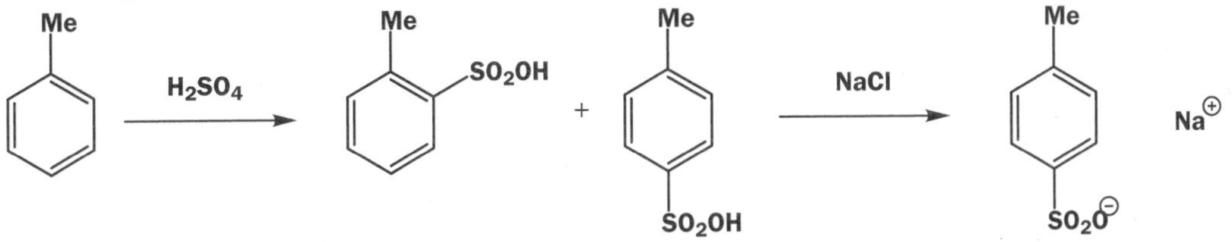

about 40% *para* isolated as sodium salt

We shall use SO_3 as the electrophile in this case and draw the intermediate with the charge at the *ipso* carbon to show the stabilization from the methyl group. We shall see later that these steps are reversible.

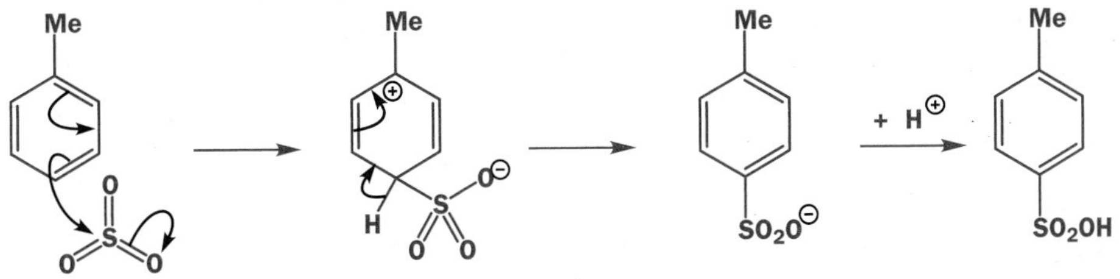

The toluene-*para*-sulfonate group (OTs) is important as a leaving group if you want to carry out an S_N2 reaction on an alcohol (Chapter 17) and the acid chloride (tosyl chloride, TsCl) can be made from the acid in the usual way with PCl_5. It can also be made directly from toluene by sulfonation with chlorosulfonic acid $ClSO_2OH$. This reaction favours the *ortho* sulfonyl chloride which is isolated by distillation.

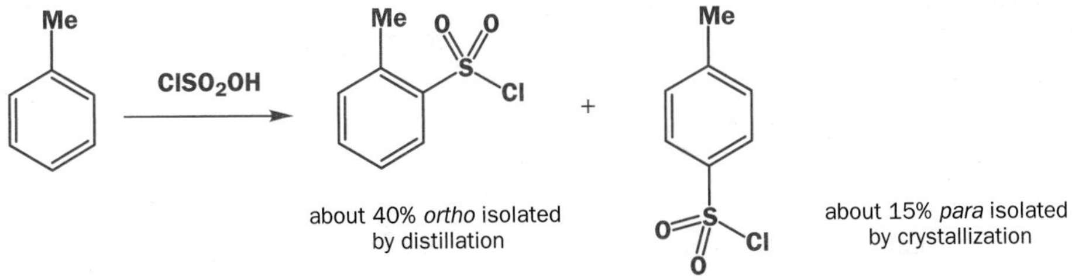

about 40% *ortho* isolated by distillation

about 15% *para* isolated by crystallization

No Lewis acid is needed because chlorosulfonic acid is a very strong acid indeed and protonates itself to give the electrophile. This explains why OH is the leaving group rather than Cl and why chlorosulfonation rather than sulfonation is the result.

In drawing the mechanism, we again put the positive charge on the *ipso* atom. No treatment with NaCl is needed in this reaction as the major product (the *ortho* acid chloride) is isolated by distillation.

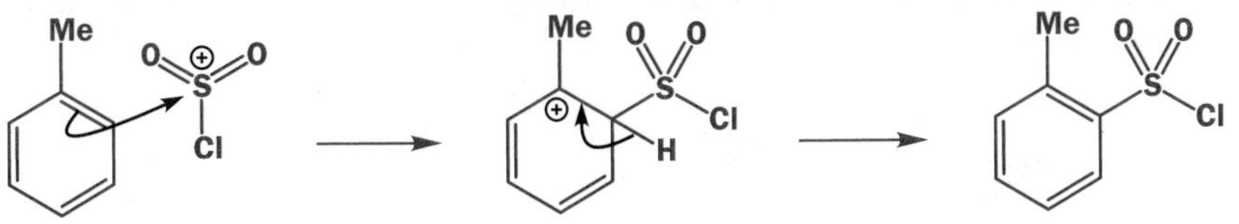

The preference for *para* product in the sulfonation and *ortho* product in the chlorosulfonation is the first hint that sulfonation is reversible and this point is discussed later. It is fortunate that the *ortho* acid chloride is the major product in the chlorosulfonation because it is needed in the synthesis of saccharin, the first and still one of the best of the non-fattening sweeteners.

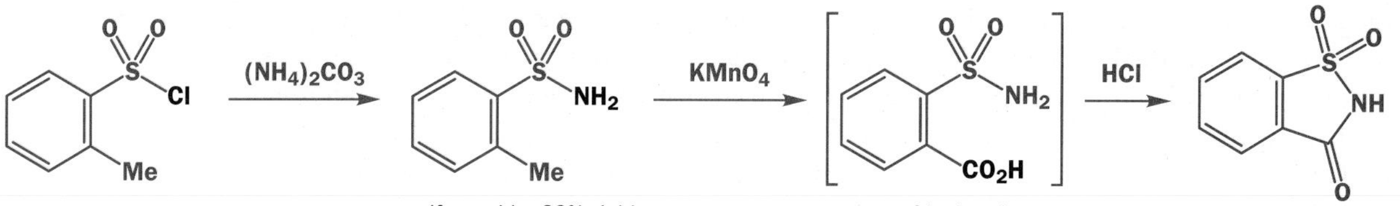

sulfonamide, 89% yield   not isolated   saccharin, 58% yield

These are all reactions that you know, with the exception of the oxidation with $KMnO_4$ (Chapter 25) to carboxylic acids but the formation of sulfonamides is like that of ordinary amides. This synthesis is discussed in Chapter 25.

Electron-withdrawing substituents give *meta* products

A few substituents (Z) exert an electronic effect on the benzene ring simply by polarization of the Ar–Z σ bond because of the electronegativity of Z. The most important is the CF_3 group, but ammonium (R_3N^+) and phosphonium (R_3P^+) fall into the same category. The $Ar-N^+$ and $Ar-P^+$ bonds are obviously polarized towards the positively charged heteroatom and the Ar–C bond in $Ar-CF_3$ is polarized towards the CF_3 group because of the three very electronegative fluorine atoms polarizing the C–F bonds so much that the Ar–C bond is polarized too.

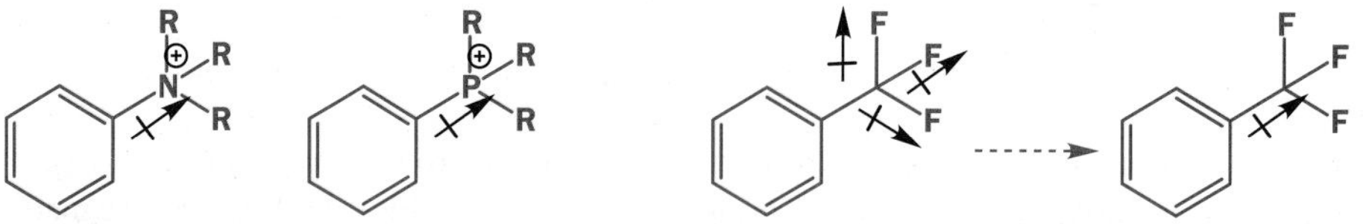

These groups direct electrophiles to the *meta* position and reduce reactivity. Nitration of trifluoromethyl benzene gives a nearly quantitative yield of *meta* nitro compound so there cannot be any significant *ortho* or *para* by-products. This reaction is important because reduction of the product (Chapter 24) gives the amine, also in very good yield.

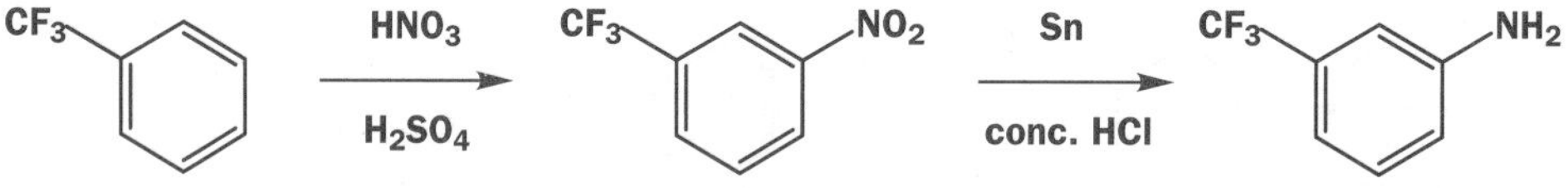

96% yield   95% yield

In drawing the mechanism we need to produce the intermediate in which the cation is not delocalized to the carbon atom bearing the electron-withdrawing group. In other words, the situation with electron-withdrawing CF_3 is the opposite to that with electron-donating CH_3. The CF_3 group is deactivating and *meta*-directing.

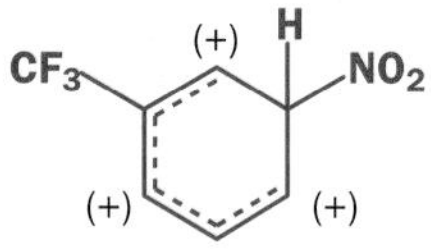

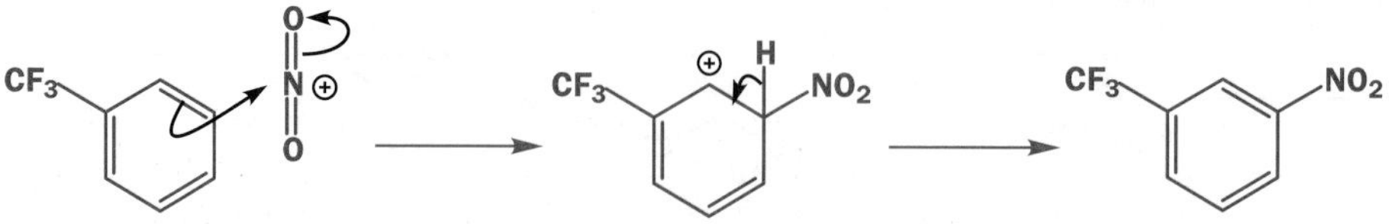

In the nitration of the phenyltrimethylammonium ion, 90% of the product is *meta*-substituted (with 10% *para*) and kinetic studies show that the nitration proceeds approximately 10^7 times more slowly than the nitration of benzene.

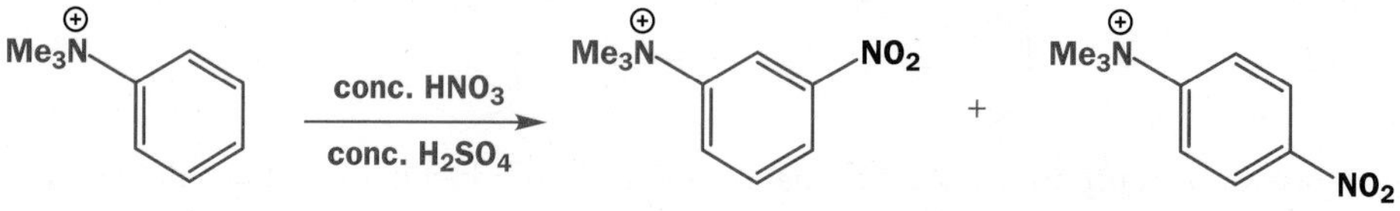

90% *meta* product   10% *para* product

Some substituents withdraw electrons by conjugation

Aromatic nitration is important, because it is a convenient way of adding an amino group to the ring and because it stops cleanly after one nitro group has been added. Further nitration is possible but stronger conditions must be used—fuming nitric acid instead of normal concentrated nitric acid and the mixture refluxed at around 100 °C. The second nitro group is introduced *meta* to the first, that is, the nitro group is deactivating and *meta*-directing.

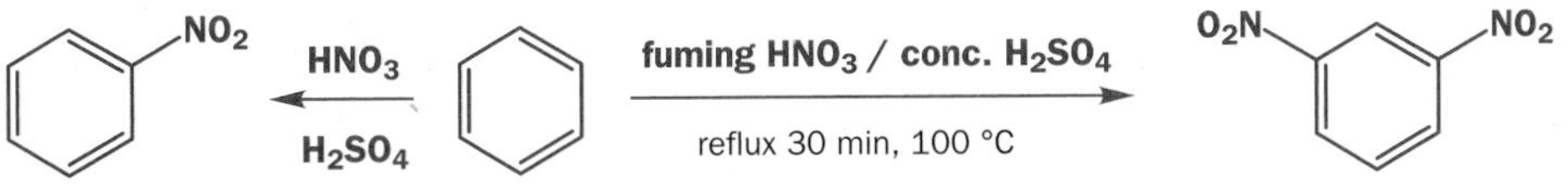

75% yield *meta*-dinitrobenzene

The nitro group is conjugated with the π system of the benzene ring and is strongly electron-*withdrawing*—and it withdraws electrons specifically from the *ortho* and *para* positions. We can use curly arrows to show this.

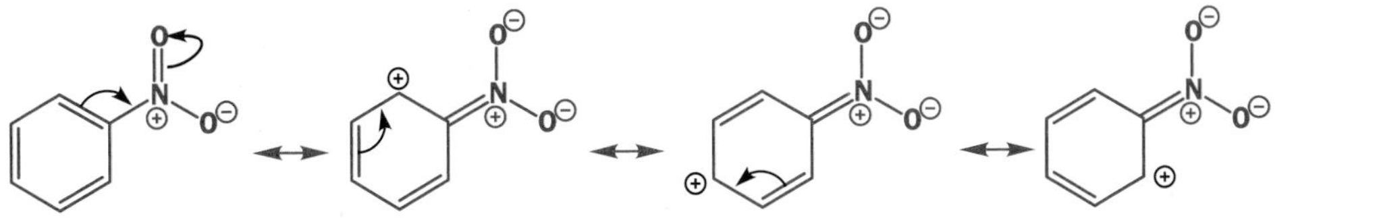

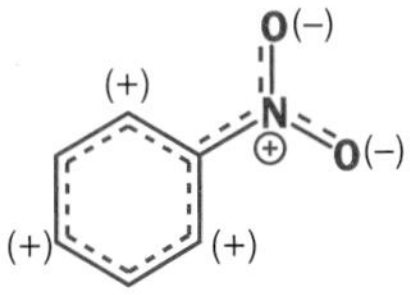

The nitro group withdraws electron density from the π system of the ring thereby making the ring less reactive towards something wanting electrons, an electrophile. Hence the nitro group is deactivating towards electrophilic attack. Since more electron density is removed from the *ortho* and *para* positions, the least electron-deficient position is the *meta* position. Hence the nitro group is *meta*-directing. In the nitration of benzene, it is much harder to nitrate a second time and, when we insist, the second nitro group goes in *meta* to the first.

Other reactions go the same way so that bromination of nitrobenzene gives *meta*-bromonitrobenzene in good yield. The combination of bromine and iron powder provides the necessary Lewis acid ($FeBr_3$) while the high temperature needed for this unfavourable reaction is easily achieved as the boiling point of nitrobenzene is over 200 °C.

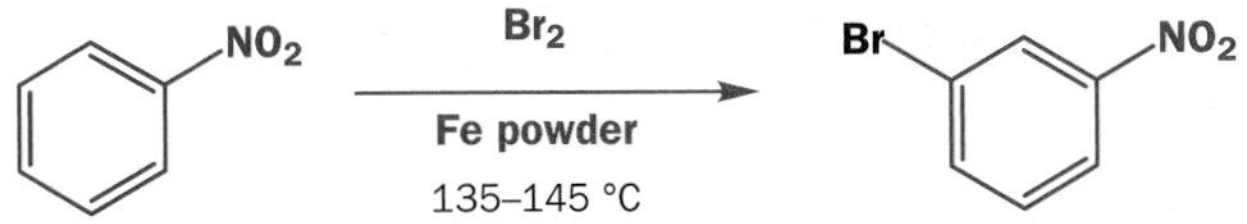

74% yield
meta-bromonitrobenzene

In drawing the mechanism it is best to draw the intermediate and to emphasize that the positive charge must not be delocalized to the carbon atom bearing the nitro group.

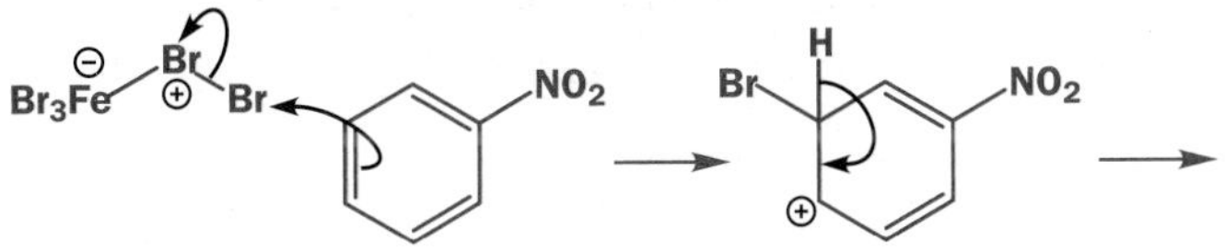

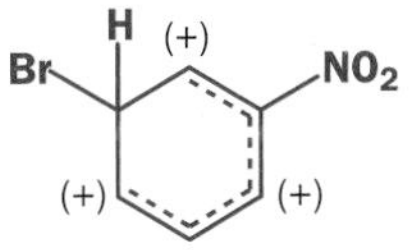

Nitro is just one of a number of groups that are also deactivating towards electrophiles and *meta*-directing because of electron withdrawal by conjugation. These include carbonyl groups (aldehydes, ketones, esters, etc.), cyanides, and sulfonates and their ^{1}H NMR shifts confirm that they remove electrons from the *ortho* and *para* positions.

nitrobenzene   benzaldehyde   methyl benzoate   methyl benzenesulfonate   benzonitrile

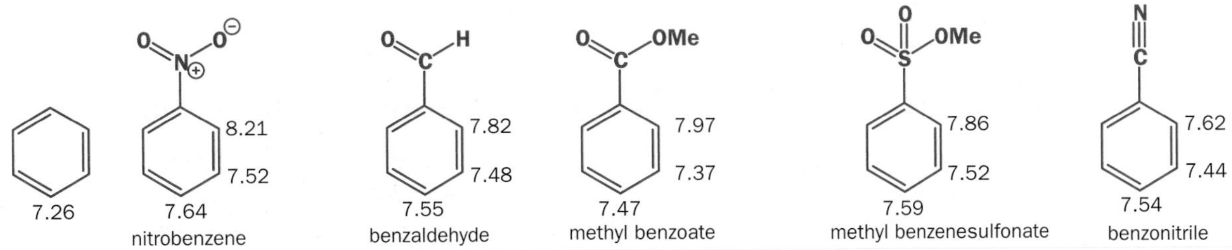

Points to note:

- Each of the compounds contains the unit Ph–X=Y, where Y is an electronegative element, usually oxygen
- In each compound, all the protons resonate further downfield relative to benzene (that is, they have larger chemical shifts)
- The protons are less shielded than those of benzene because the electron density at carbon is less
- The protons in the *meta* position have the smallest shift and so the greatest electron density

Nitro is the most electron-withdrawing of these groups and some of the other compounds are nearly as reactive (in the *meta* position, of course) as benzene itself. It is easy to nitrate methyl benzoate and the *m*-nitro ester can then be hydrolysed to *m*-nitrobenzoic acid very easily.

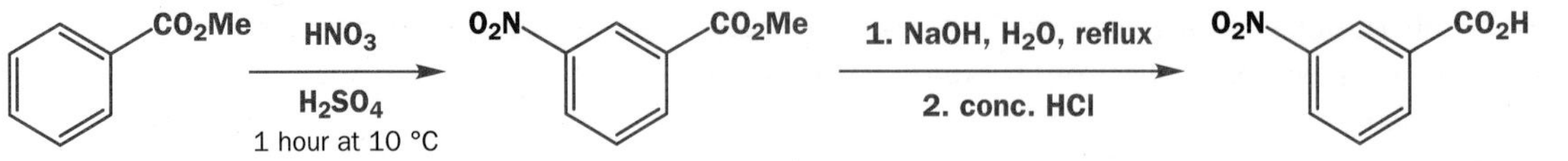

84% yield methyl *m*-nitrobenzoate

96% yield *m*-nitrobenzoic acid

An interesting example of a reaction with a ketone is the sulfonation of anthraquinone. Many dyestuffs contain this unit and the sulfonate group makes them soluble in water. Oleum at 160 °C must be used for the sulfonation, which goes in one of the four equivalent positions on the two benzene rings, *meta* to one carbonyl group but *para* to the other.

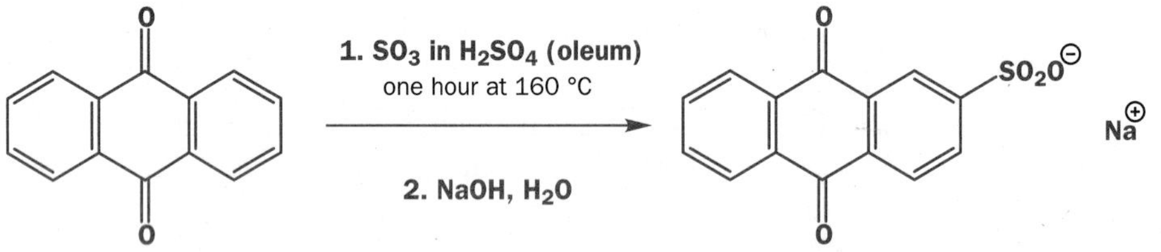

anthraquinone

60% yield sodium anthraquinone-2-sulfonate

The yield is not wonderful and the main by-product is unchanged anthraquinone showing how unreactive this compound is even under these forcing conditions. In Chapter 7 we saw how dyes are highly conjugated molecules, often containing aromatic rings. Here are two common water-soluble dyes containing sulfonate groups.

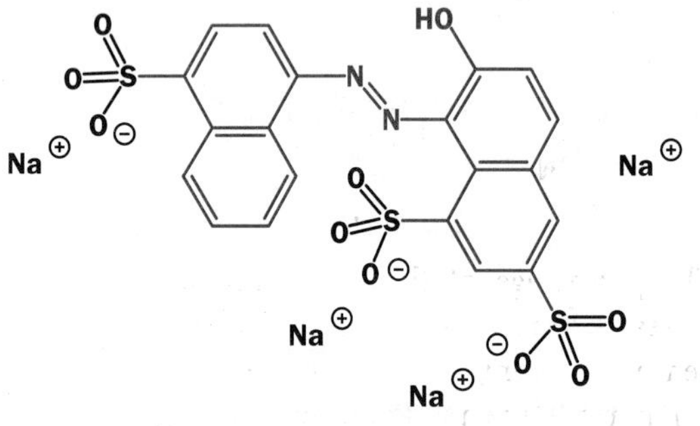

Brilliant scarlet 4R; E124

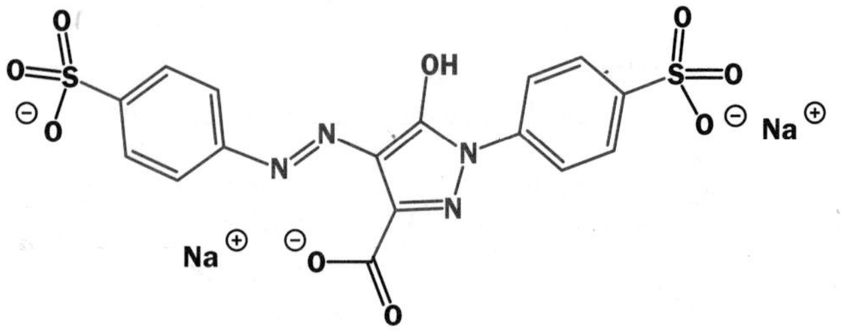

Tartrazine (yellow); E102

These dyes also contain the diazo group (–N=N–) and we shall return to that soon. One group of substituents remains and they are slightly odd. They are *ortho*, *para*-directing but they are also deactivating. These are the halogens.

Halogens (F, Cl, Br, and I) both withdraw and donate electrons

The halogens deactivate the ring towards electrophilic attack but direct *ortho* and *para*. The only way this makes sense is if there are two opposing effects—electron donation by conjugation and electron withdrawal by induction. The halogen has three lone pairs, one of which may conjugate with the ring just like in phenol or aniline. However, there are two mismatching aspects to this conjugation: lone pair orbital size and electronegativity.

When Cl, Br, or I is the substituent, there is a size mismatch, and therefore a poor overlap, between the 2p orbitals from the carbon atoms and the p orbitals from the halogen (3p for chlorine, 4p for bromine, and 5p for iodine). This size mismatch is clearly illustrated by comparing the reactivities of aniline and chlorobenzene: chlorine and nitrogen have approximately the same electronegativity, but aniline is much more reactive than chlorobenzene because of the better overlap between the carbon and nitrogen 2p orbitals.

Fluorine 2p orbitals are the right size to overlap well with the carbon 2p orbitals, but the orbitals of fluorine are much lower in energy than the orbitals of carbon since fluorine is so electronegative. Also, the more electronegative a substituent, the better it is at withdrawing electrons by induction. When we looked at aniline and phenol, we didn't mention any electron withdrawal by induction, even though both oxygen and nitrogen are very electronegative. The conjugative electron donation was clearly more important since both compounds are much more reactive towards electrophiles than benzene. However, we did point out that aniline is more reactive than phenol because nitrogen is less electronegative than oxygen and so better able to donate electrons into the π system.

With this in mind, how would you expect fluorobenzene to react? Most electron density is removed first from the *ortho* positions by induction, then from the *meta* positions, and then from the *para* position. Any conjugation of the lone pairs on fluorine with the π system would increase the electron density in the *ortho* and *para* positions. Both effects favour the *para* position and this is where most substitution occurs. But is the ring more or less reactive than benzene? This is hard to say and the honest answer is that sometimes fluorobenzene is more reactive in the *para* position than benzene (for example, in proton exchange and in acetylation—see later) and sometimes it is less reactive than benzene (for example, in nitration). In all cases, fluorobenzene is significantly more reactive than the other halobenzenes. We appreciate that this is a rather surprising conclusion, but the evidence supports it.

Data for the rate and the products of nitration of halobenzenes show these opposing effects clearly.

Compound	Products formed (%)			Nitration rate
	ortho	*meta*	*para*	(relative to benzene)
PhF	13	0.6	86	0.18
PhCl	35	0.9	64	0.064
PhBr	43	0.9	56	0.060
PhI	45	1.3	54	0.12

- The percentage of the *ortho* product increases from fluorobenzene to iodobenzene. We might have expected the amount to decrease as the size of the halide increases because of increased steric hindrance at the *ortho* position but this is clearly not the case. The series can be explained by the greater inductive effect of the more electronegative atoms (F, Cl) withdrawing electron density mostly from the *ortho* positions

- The relative rates follow a U-shaped sequence; fluorobenzene nitrates most quickly (but not as fast as benzene), followed by iodo-, then chloro-, and then bromo-benzenes. This is a result of two opposing effects: electron donation by conjugation and electron withdrawal by inductive effect

In practical terms, it is usually possible to get high yields of *para* products from these reactions. Both nitration and sulfonation of bromobenzene give enough material to make the synthesis worthwhile. Though mixtures of products are always bad in a synthesis, electrophilic aromatic substitution is usually simple to carry out on a large enough scale to make separation of the major product a workable method.

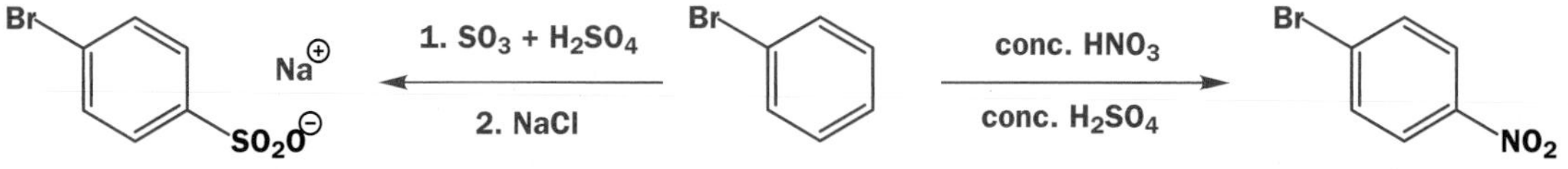

A 68% yield of sodium *p*-bromobenzenesulfonate can be achieved by recrystallization of the sodium salt from water and a 70% yield of *p*-bromonitrobenzene by separation from the *ortho* isomer by recrystallization from EtOH.

● Summary of directing and activating effects

Now we can summarize the stage we have reached in terms of *activation* and *direction*.

Electronic effect	Example	Activation	Direction
donation by conjugation	–NR$_2$, –OR	very activating	*ortho*, *para* only
donation by inductive effect	alkyl	activating	mostly *ortho*, *para* but some *meta*
donation by conjugation *and* withdrawal by inductive effect	F, Cl, Br, and I	deactivating	*ortho* and (mostly) *para*
withdrawal by inductive effect	–CF$_3$, –NR$_3^+$	deactivating	*meta* only
withdrawal by conjugation	–NO$_2$, –CN, –COR, –SO$_3$R	very deactivating	*meta* only

Why do some reactions stop cleanly at monosubstitution?

Reactions such as nitration, sulfonation, and Friedel–Crafts acylation add a very deactivating substituent. They stop cleanly after a single substitution unless there is also a strongly activating substituent. Even then it may be possible to stop after a single substitution. Nitration of phenol is difficult to control because the OH group is very activating and because concentrated nitric acid oxidizes phenol. The solution is to use dilute nitric acid. The concentration of NO$_2^+$ will be small but that does not matter with such a reactive benzene ring.

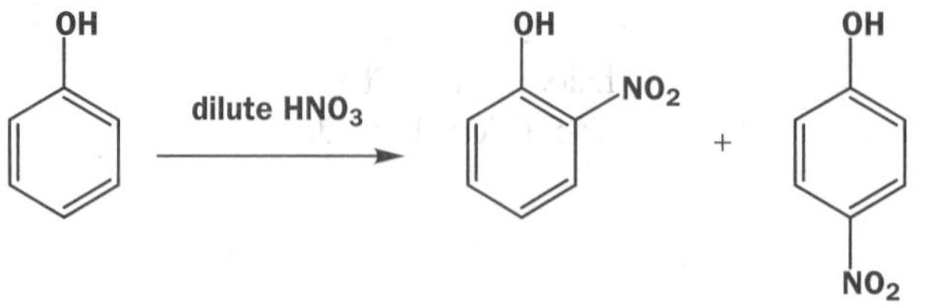

ortho, 36%   para, 25% 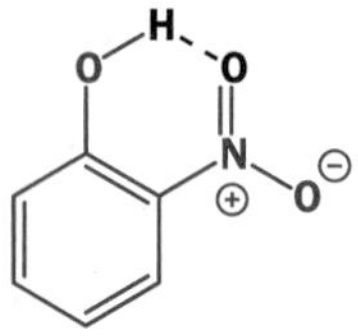   strong intramolecular H bond

The product is a mixture of *ortho*- and *para*-nitrophenol from which the *ortho* compound can be separated by steam distillation. A strong intramolecular hydrogen bond reduces the availability of the OH group for intermolecular hydrogen bonds so the *ortho* compound has a lower boiling point. The remaining *para*-nitrophenol is used in the manufacture of the painkiller paracetamol.

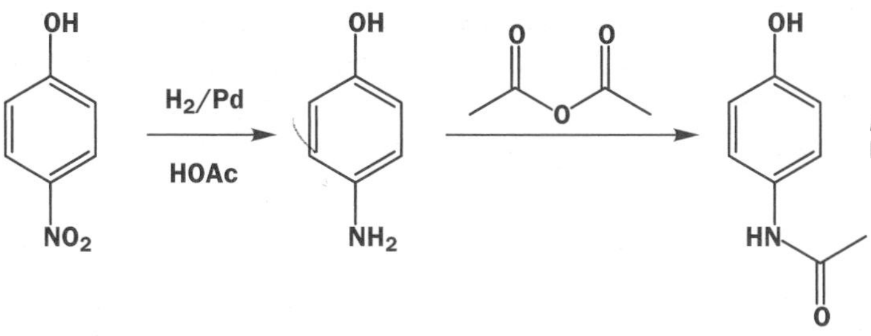

para acetylamino phenol
paracetamol

Weakly electron-withdrawing substituents like the halogens can be added once, but multiple substitution is common with strongly activating substituents like OH and NH$_2$. When electron-donating substituents are added, multiple substitution is always a threat. As it happens, this threat is not serious as there are no good reagents for adding strongly activating substituents such as 'HO$^+$' or 'H$_2$N$^+$' to aromatic systems. Now you see why adding nitrogen as the deactivating nitro group is such an advantage. The only reactions of this kind where multiple substitution is a genuine problem are likely to be Friedel–Crafts alkylation reactions. Preparation of diphenylmethane from benzene and benzyl chloride is a fine reaction but the product has two benzene rings, each more reactive than benzene itself. A 50% yield is the best we can do and that requires a large excess of benzene to ensure that it competes successfully with the product for the reagent.

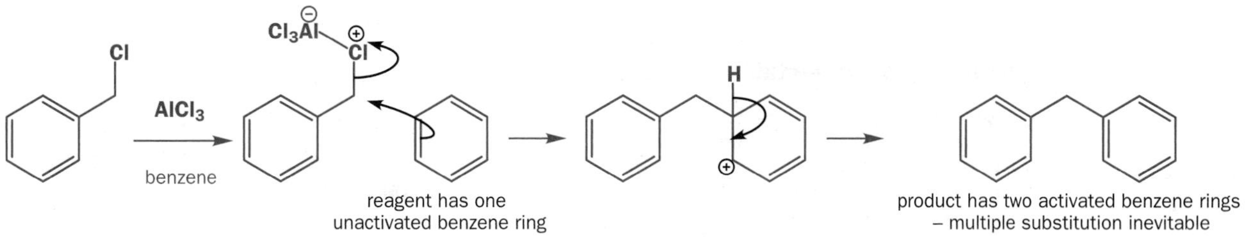

reagent has one
unactivated benzene ring

product has two activated benzene rings
– multiple substitution inevitable

We have drawn the substitution at the benzylic centre as an S$_N$2 reaction as it would normally be with a primary alkyl halide, though it could be S$_N$1 in this case as the benzylic cation is stable. Friedel–Crafts alkylation works well with relatively stable cations especially tertiary cations. The cation can be generated in a number of ways such as the protonation of an alkene, the acid-catalysed decomposition of a tertiary alcohol, or the Lewis-acid-catalysed decomposition of a *t*-alkyl chloride.

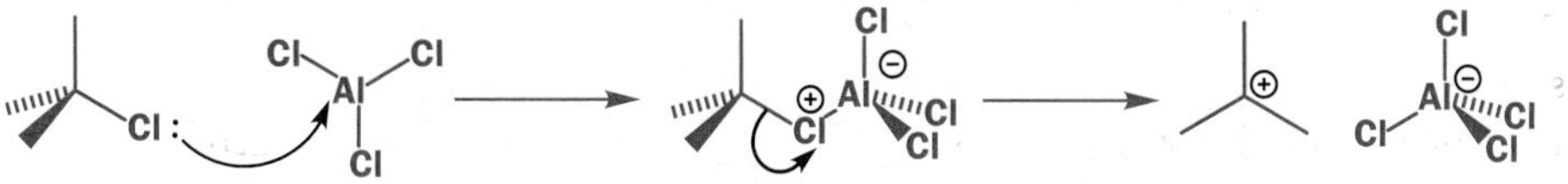

Two or more substituents may cooperate or compete

We can, in a qualitative way, combine the directing effects of two or more substituents. In some cases the substituents both direct to the same positions, as in the syntheses of bromoxynil and ioxynil, contact herbicides especially used in spring cereals to control weeds resistant to other weedkillers. They are both synthesized from *p*-hydroxybenzaldehyde by halogenation. The aldehyde directs *meta* and the OH group directs *ortho* so they both direct to the same position. The aldehyde is deactivating but the OH is activating.

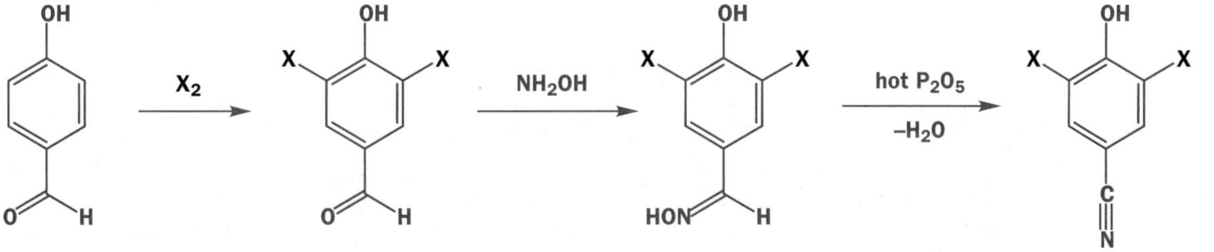

bromoxynil (X=Br)
ioxynil (X=I)

The reaction with NH_2OH is the formation of an oxime from the aldehyde and hydroxylamine and was dealt with in Chapter 14. The reaction with P_2O_5 is a dehydration—phosphorus is used to remove water from the oxime.

In other cases substituents compete by directing to different positions. For example, in the synthesis of the food preservative BHT (p. 30) from 4-methylphenol (*p*-cresol) by a Friedel–Crafts alkylation, the methyl and OH groups each direct *ortho* to themselves. The –OH group is much more powerfully directing than the methyl group because it provides an extra pair of electrons, so it 'wins' and directs the electrophile (a *t*-butyl cation) *ortho* to itself. The *t*-butyl cation can be made from the alkene or *t*-butanol with protic acid or from *t*-butyl chloride with $AlCl_3$.

BHT
Butylated Hydroxy Toluene

BHT—a case of mistaken identity?

When BHT and other similar phenols were first prepared in the 1940s, chemists were not sure of their structures. The chemical formulae could be determined by elemental analysis, but NMR, which would have instantly revealed the structure, had not yet been discovered. The problem arose because the compound exhibited none of the normal reactions or 'tests' for phenols; for example, it was not soluble in alkali. The chemists thought the second *t*-butyl group had added to oxygen to make an ether. BHT does not behave like other phenols because the –OH group is hindered by the two large *t*-butyl groups.

BHT   supposed BHT

▶

If you are in a bar and someone picks a fight with you, it is no help that an inoffensive little man in the corner would prefer not to pick a fight. Aggressive $-NR_2$ and $-OR$ groups are not much affected by inoffensive $-Br$ or carbonyl groups in another corner of the molecule.

Even a watered-down activating group like the amide –NHCOMe, which provides an extra pair of electrons, will 'win' over a deactivating group or an activating alkyl group. Bromination of this amide goes *ortho* to the –NHCOMe group but *meta* to the methyl group.

When looking at any compound where competition is an issue it is sensible to consider electronic effects first and then steric effects. For electronic effects, in general, any activating effects are more important than deactivating ones. For example, the aldehyde below has three groups—two methoxy groups that direct *ortho* and *para* and an aldehyde that directs *meta*.

main nitration product

3,4-dimethoxybenzaldehyde

the green methoxy group directs here

the green methoxy group directs here

the carbonyl deactivates the ring and directs here

the red methoxy group directs here

Despite the fact that the aldehyde group withdraws electron density from positions 2 and 6, C6 is still the position for nitration. The activating methoxy groups dominate electronically and the choice is really between C2, C5, and C6. Now consider steric factors—the –OMe groups block the positions *ortho* to them more than the carbonyl does because reaction at C2 or C5 would lead to three adjacent substituents which is why substitution occurs at position 6.

Review of important reactions including selectivity

We shall now return to the main reactions and consider important examples including selectivity.

Sulfonation

The exact nature of the electrophile in sulfonation reactions seems to vary with the amount of water present. Certainly for oleum (fuming sulfuric acid, concentrated sulfuric acid with added sulfur trioxide) and solutions of sulfur trioxide in organic solvents, the electrophile is sulfur trioxide itself, SO_3. With more water around, $H_3SO_4^+$ and even $H_2S_2O_7$ have been suggested. One important difference between sulfonation and other examples of electrophilic substitution is that sulfonation is reversible. This can be useful because large sulfonic acid groups can act as blocking groups and be removed later. Mixing bromine and phenol at low temperatures produces mainly *p*-bromophenol. At higher temperatures, the tribromo product is formed. The *ortho*-substituted product can be made with the aid of sulfonation.

In stage 1 the phenol is sulfonated twice—the first sulfonic acid group (which adds *para* to the OH group) deactivates the ring, making the introduction of the second group (which goes *ortho* to the OH and *meta* to the first sulfonic acid) harder and that of the third group harder still, which is why we can isolate the disulfonated phenol. In the second stage, the bromination, the OH directs to the *ortho* and *para* positions, but only one *ortho* position is vacant, so the bromine attacks there. Sodium hydroxide is needed to deprotonate the sulfonic acid groups to make them less deactivating. The sulfonation reaction is reversible, and in the third stage it is possible to drive the reaction over by distilling out relatively volatile 2-bromophenol.

Direct sulfonation of aromatic amines is even possible. This is very surprising because in sulfuric acid essentially all the amine will be protonated. The protonated amine would react in the *meta* position just like Ph—NMe_3^+ but in these reactions the *para*-sulfonic acid is formed.

■ The reversibility of sulfonation with sulfuric acid may account for the higher yield of *para* product in the sulfonation of toluene with H_2SO_4 as compared with $ClSO_2OH$ (p. 563).

50–60% yield sulfanilic acid

There are two possible explanations for this. Either the very tiny amount of unprotonated amine reacts very rapidly with SO_3 in the *para* position or the reaction is reversible and the *para*-sulfonic acid is formed because it is stabilized by delocalization and least hindered. The product is important because the amides derived from it (sulfanilamides) were the first antibiotics, the 'sulfa' drugs.

sulfanilamide

sulfapyridine, a sulfa drug

Aromatic nitration and diazo-coupling

We have already described how nitration leads eventually to aromatic amines by reduction of the nitro group. In the next chapter you will meet the further development of these amines into diazonium salts as reagents for nucleophilic aromatic substitution by the S_N1 mechanism with loss of nitrogen. In this chapter we need to address their potential for electrophilic aromatic substitution without the loss of nitrogen as this leads to the important azo dyes. Treatment of the amine with nitrous acid (HON=O) at around 0 °C gives the diazonium salt.

■ The mechanism of formation of NO⁺ is discussed in Chapter 21.

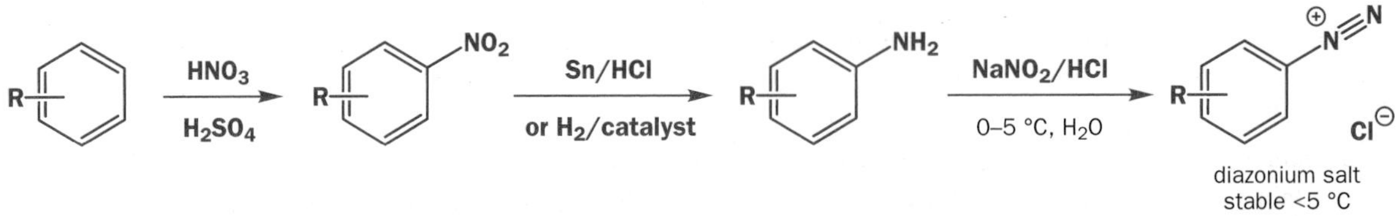

diazonium salt
stable <5 °C

These diazonium salts are good electrophiles for activated aromatic rings, such as amines and phenols, and this is how azo dyes are prepared. Diazotization of the salt of sulfanilic acid, which we have just made by sulfonation of aniline, gives an inner salt that combines with *N,N*-dimethylaniline to form the water-soluble dye, methyl orange.

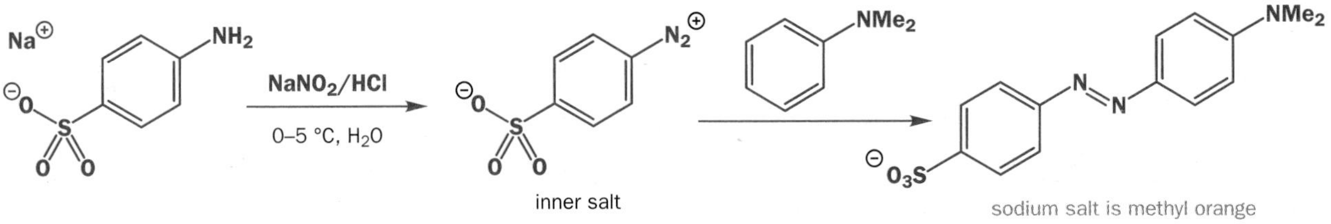

inner salt

sodium salt is methyl orange

The electrophilic substitution is straightforward, occurring in the *para* position on the activated hindered dialkylamine. Notice that nucleophilic attack must occur on the end nitrogen atom of the diazonium salt to avoid forming pentavalent nitrogen.

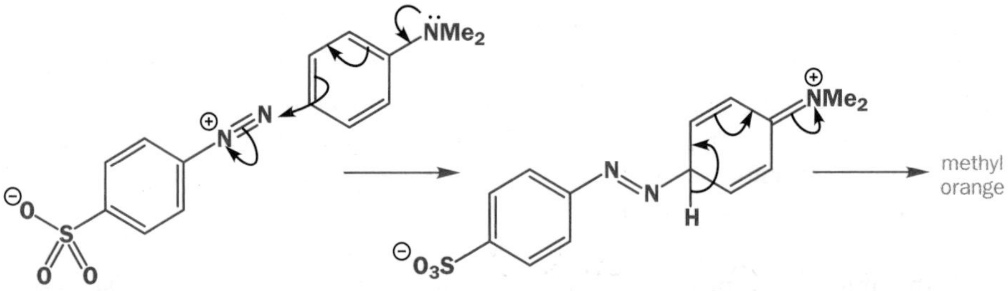

methyl orange

Oxygen and nitrogen can also complex to the catalyst

In Friedel–Crafts alkylation using alkenes and alcohols with strong acids, OH and NH_2 groups activate towards electrophilic attack and direct to the *ortho* or *para* positions. However, in Friedel–Crafts alkylations using *t*-alkyl chlorides and $AlCl_3$, reaction does not proceed much faster than the alkylation of unsubstituted benzene, that is, the –OH group seems to have very little effect on the reaction. This is because oxygen can also complex with the Lewis acid. The Friedel–Crafts alkylation of amines is even worse and normally does not proceed at all—nitrogen forms an even stronger complex with the Lewis acid than oxygen does. This complex then withdraws electrons from the ring, rather than donating electrons as the neutral nitrogen did.

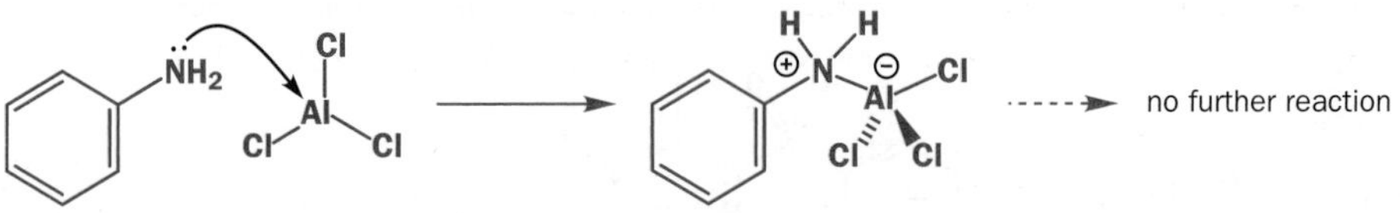

no further reaction

Friedel–Crafts alkylations are especially useful for forming polycyclic compounds. These are usually intramolecular reactions in which the electrophile and the aromatic system are all part of the same compound. Fairly elaborate examples are discussed in Chapter 51. A simple example reveals the basic plan: an intramolecular Friedel–Crafts alkylation that will be faster than any other, inevitably intermolecular, side reaction.

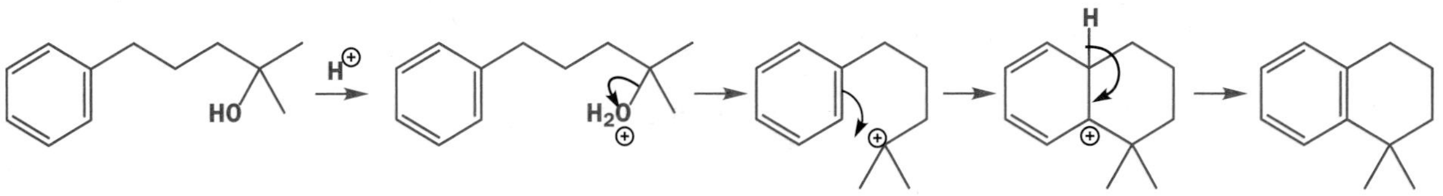

Friedel–Crafts alkylation cannot be used with primary alkyl halides

Even if you successfully prevent multiple substitution from occurring, there is a second and more serious problem—the alkyl cations often rearrange to yield more stable cations. We shall look into such rearrangements more closely in Chapter 37 but for the moment we shall just consider Friedel–Crafts alkylation with primary halides.

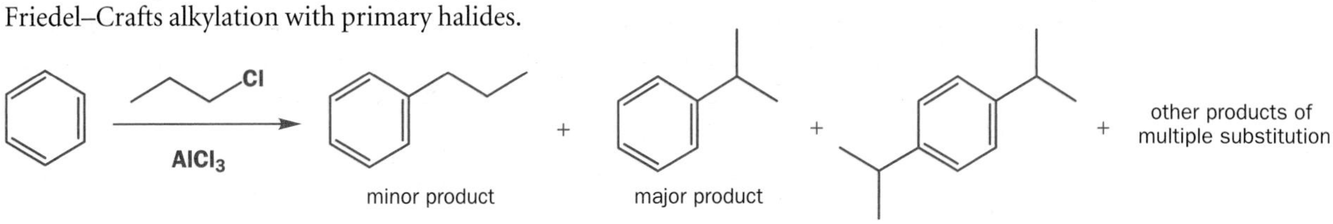

minor product   major product   other products of multiple substitution

The major product is isopropyl benzene—approximately twice as much as *n*-propyl benzene. The rearrangement in this mechanism occurs because primary cations do not exist in solution (Chapter 17) so that the alkyl halide–AlCl$_3$ complex must either react directly or rearrange to the more stable secondary carbocation.

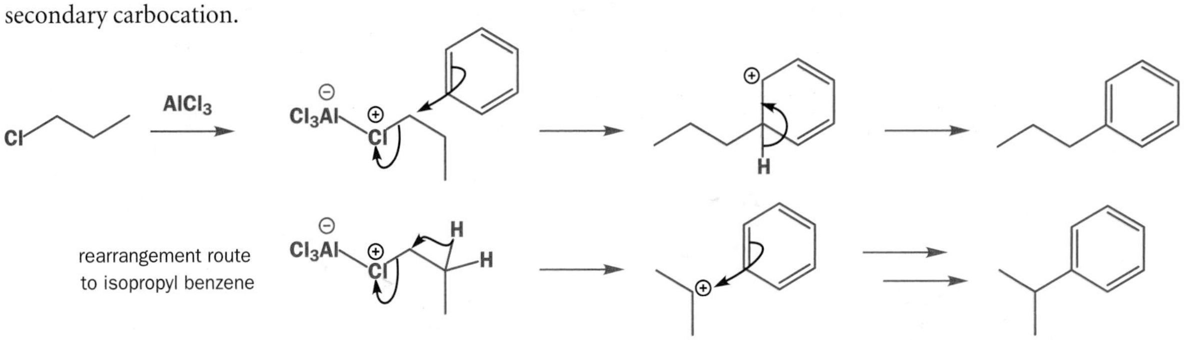

rearrangement route to isopropyl benzene

Friedel–Crafts *acylation* is much more reliable

Of more use than Friedel–Crafts alkylation is **Friedel–Crafts acylation**, the introduction of an acyl group (RCO–) on to the ring. Instead of using an alkyl chloride, an acyl chloride (acid chloride) or an acid anhydride is used together with the Lewis acid to produce the reactive acylium ion. We have seen an acid chloride in action (p. 553); here is an anhydride.

anhydride   acylium ion

▶

In Friedel–Crafts alkylations using an alkyl chloride, the Lewis acid is used in catalytic quantities. In an acylation, however, the Lewis acid can also complex to any oxygen atoms present, to the carbonyl in the product, for example. As a result, in acylation reactions, more Lewis acid is required—just over one equivalent per carbonyl group.

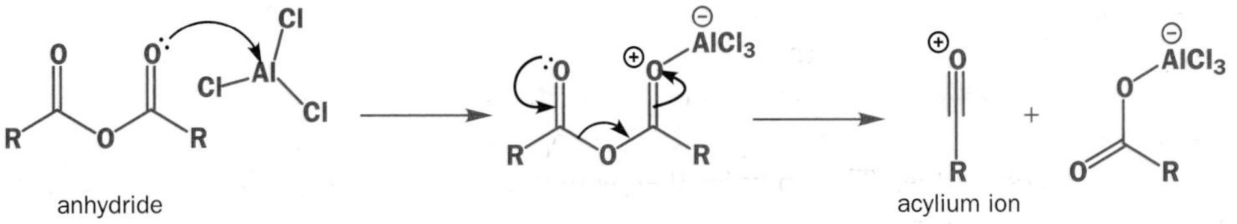

The acylium ion is then attacked by the aromatic system in the usual way. Multiple substitution is rarely a problem because the deactivated conjugated ketone is much less reactive than benzene.

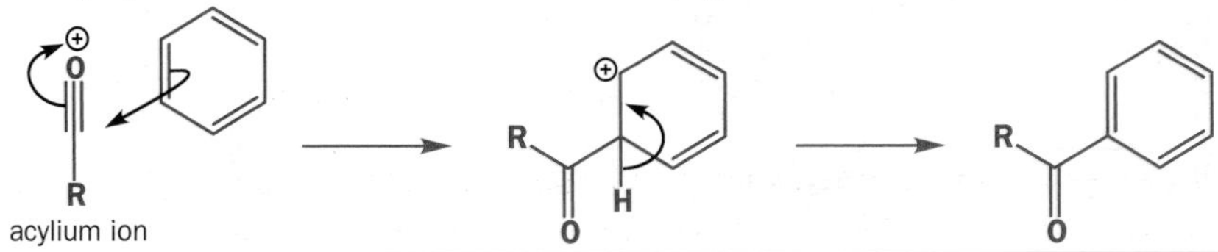

acylium ion

Cyclic anhydrides can be used to make keto-acids. Either carbonyl group is used for the acylation and the other becomes an AlCl₃ complex until work-up. Thus 3-benzoylpropanoic acid can be prepared from benzene and succinic anhydride.

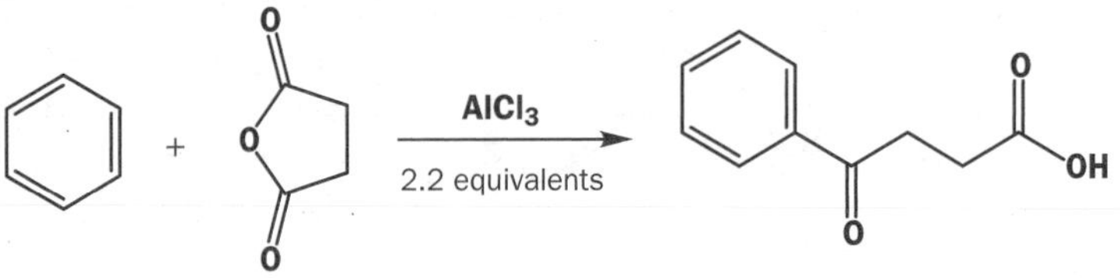

succinic anhydride   3-benzoylpropanoic acid

The advantages of acylation over alkylation

Two problems in Friedel–Crafts alkylation do not arise with acylation.

- The acyl group in the product withdraws electrons from the π system making multiple substitutions harder. Indeed, if the ring is too deactivated to start off with, Friedel–Crafts acylation may not be possible at all—nitrobenzene is inert to Friedel–Crafts acylation and is often used as a solvent for these reactions

- Rearrangements are also no longer a problem because the electrophile, the acylium cation, is already relatively stable

Because the acylation reaction is so much more reliable than Friedel–Crafts alkylation, a common method to alkylate is actually to acylate first and then reduce the carbonyl to a methylene group (–CH₂–). For example, the 3-benzoylpropanoic acid just made can be reduced to 4-phenylbutanoic acid using acid and zinc amalgam. This sort of reaction is discussed in Chapter 24. We could go one step further with the 4-phenylbutanoic acid and do an intramolecular Friedel–Crafts acylation. Intramolecular reactions are easy to do and, when starting from carboxylic acids, polyphosphoric acid (represented in the diagram as H_3PO_4) is commonly used to make the OH group into a better leaving group.

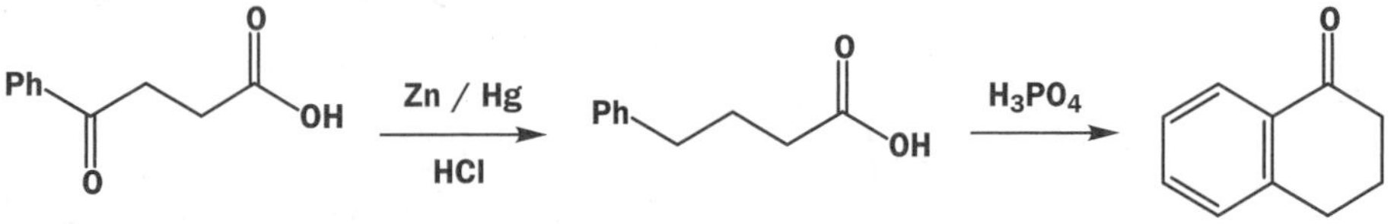

3-benzoylpropanoic acid   4-phenylbutanoic acid

One-carbon electrophiles are difficult to use

When R–C≡O⁺ is used as the electrophile a ketone is produced. If an aldehyde were wanted, H–C≡O⁺ would have to be used but it cannot be made from HCOCl because that is unstable. Instead, it can be generated by passing carbon monoxide and hydrogen chloride through a mixture of the aromatic hydrocarbon, a Lewis acid, and a co-catalyst, usually copper (I) chloride. Copper(I) chloride is known to form a complex with carbon monoxide and this probably speeds up the protonation step.

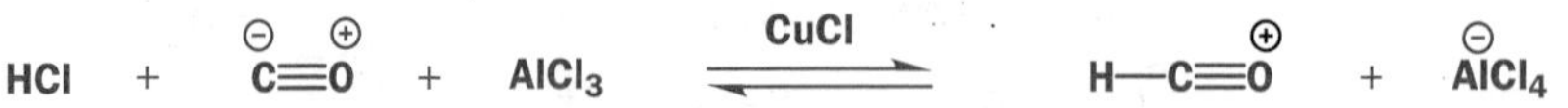

carbon monoxide

Ludwig Gatterman (1860–1920) worked at Freiburg and had a taste for danger. He made and studied the dangerously explosive NCl₃ and noticed the strange taste that gaseous HCN gave to a cigar.

This reaction, known as the **Gatterman–Koch reaction**, does not work with phenolic or amino aromatic species due to complex formation with the Lewis acid. It does work well with aromatic hydrocarbons and is used industrially to prepare benzaldehyde and, as here, *p*-tolualdehyde.

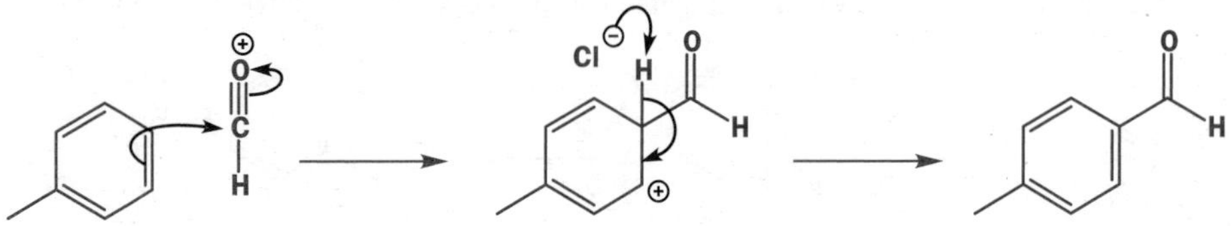

about 50% yield

For more reactive aromatic systems such as phenols (but still not amines) a variation of this reaction, called the **Gatterman reaction**, can be useful in preparing aldehydes. Instead of using protonated carbon monoxide, protonated hydrogen cyanide is used (the two are isoelectronic). The reaction goes via an imine intermediate, $ArCH=NH$, which under the conditions of the reaction is hydrolysed to the aldehyde (see p. 351). When such reactive aromatic species as phenols are involved, the Lewis acid need not be so strong and zinc chloride is often used. With less reactive systems, $AlCl_3$ is needed. The zinc chloride can be conveniently generated from zinc cyanide, $Zn(CN)_2$, and HCl. This has the added advantage of also generating the necessary HCN *in situ* as well.

In a variation of the Gatterman reaction an alkyl cyanide RCN is used in place of HCN as a useful way of preparing ketones from reactive aromatic species that do not react well under Friedel–Crafts conditions. The electrophile involved is effectively $R-C\equiv NH^+$, although, perhaps, the imino chloride, $R(C=NH)Cl$, the analogue of an acyl chloride, RCOCl, is also involved. As in the Gatterman reaction, the imine is an intermediate.

These reactions work even when there are three hydroxyls on the benzene ring.

about 60% yield

We have already seen how salicylic acid can be made by reaction of the sodium salt of phenol (PhONa) with CO_2. More important than these reactions is **chloromethylation**, a way of adding a single carbon atom at the alcohol oxidation level. A combination of formaldehyde ($CH_2=O$) and HCl provides the one-carbon electrophile.

Chloromethylation is an efficient process but it has a serious drawback. Small amounts of the very carcinogenic (cancer-causing) bis(chloromethyl)ether are formed in the reaction mixture so that the process has fallen out of favour.

warning: carcinogenic

bis(chloromethyl)ether

One-carbon electrophiles: summary of methods

Reaction	Substrate	Reagents	Electrophile	Intermediate	Product
Gatterman–Koch	hydrocarbons	CO, HCl, $AlCl_3$, CuCl	$H-C\equiv O^+$		ArCHO
Gatterman	phenols	$Zn(CN)_2$, HCl	$H-C\equiv NH^+$	ArCH=NH	ArCHO
Hoesch	phenols	RCN, HCl, Zn(II)	$R-C\equiv NH^+$	ArRC=NH	ArCOR
chloromethylation	any	$CH_2=O$, HCl	$H_2C=OH^+$	$ArCH_2OH$	$ArCH_2Cl$
Kolbé–Schmidt	phenoxides	NaOH, CO_2	CO_2	$ArCO_2Na$	$ArCO_2H$
Reimer–Tiemann	phenols	$CHCl_3$, NaOH	CCl_2	$ArCHCl_2$	ArCHO

■ The Reimer–Tiemann reaction has dichlorocarbene (CCl_2) as an intermediate and is discussed in Chapter 40.

Electrophilic substitution is the usual route to substituted aromatic compounds

A group of potent anti-leukaemia compounds (the maytansinoids) has an aromatic ring as part of a complex large-ring structure. The synthesis of these molecules could be imagined as starting from a simple aromatic ring with four different substituents in the right positions.

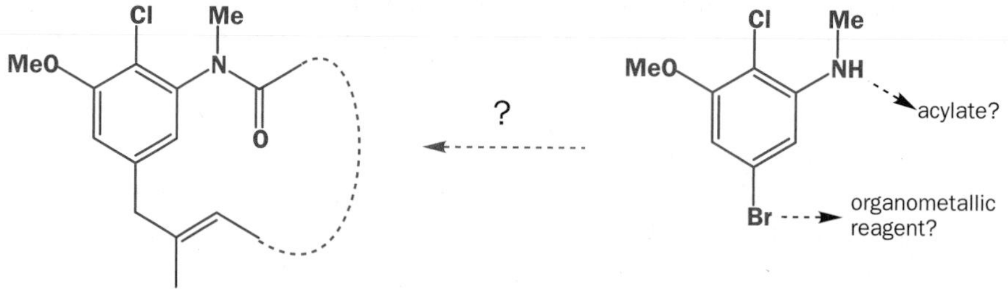

One complete synthesis is shown as the conclusion of this chapter. It is here to demonstrate that manipulation of simple aromatic rings is very much part of modern organic chemistry and because almost all the reactions are ones you have seen so far in the book.

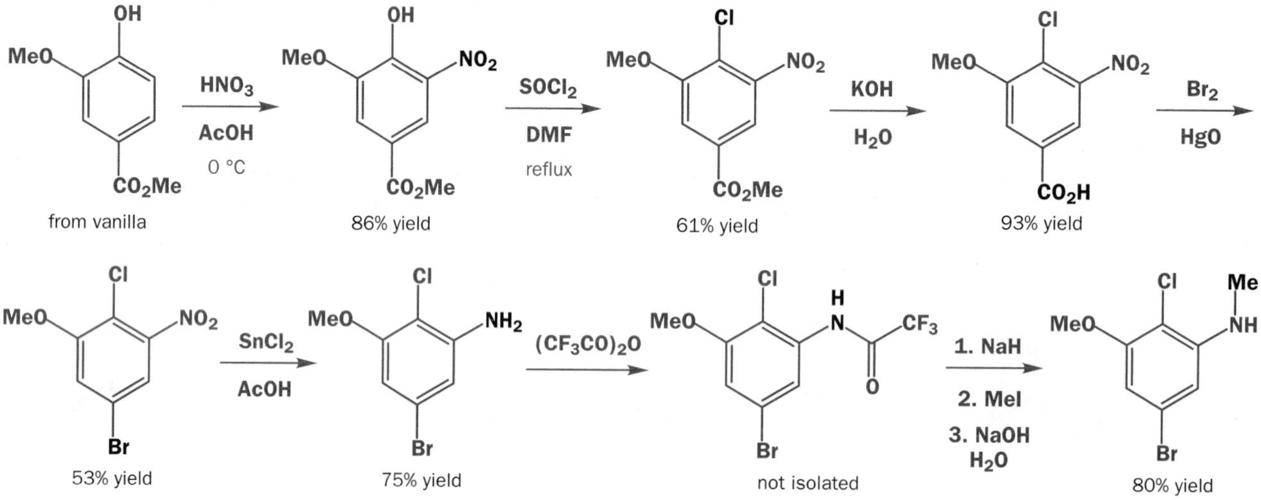

Points to notice:

1   The starting material was chosen because it was cheap. It has the right number of substituents in the right places but only one (MeO–) is still there at the end

2   Nitration is used to put in the nitrogen atom as NO$_2$, later reduced to the required amino group. The nitro group goes in *ortho* to the OH group and *meta* to the CO$_2$Me group as you might have predicted

3   Step 3, the hydrolysis of the ester, and step 6, amide formation, are familiar reactions

4   Step 2, the replacement of OH by Cl, will be discussed in Chapter 23 as it is a *nucleophilic* aromatic substitution

5   Step 4 is an unusual type of electrophilic aromatic substitution. The leaving group is CO$_2$ rather than the usual proton and occurs at the only place it can (though it is *meta* to NO$_2$ and *para* to Cl)

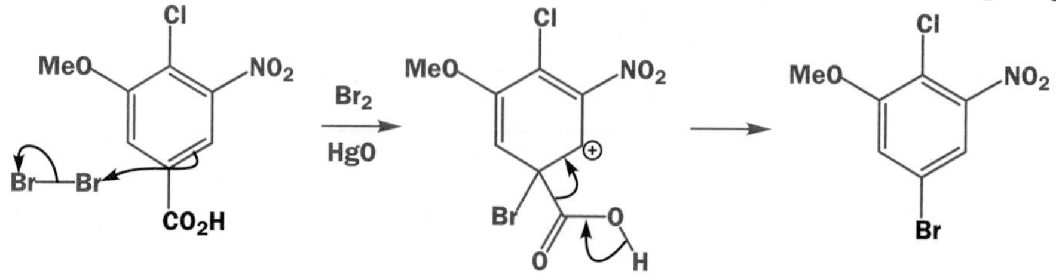

6 The last step is a way to achieve monomethylation of an amino group. Problem 13 gives you a chance to try your hand at a mechanism

Problems

1. All you have to do is to spot the aromatic rings in these compounds. It may not be as easy as you think and you should state some reasons for your choice!

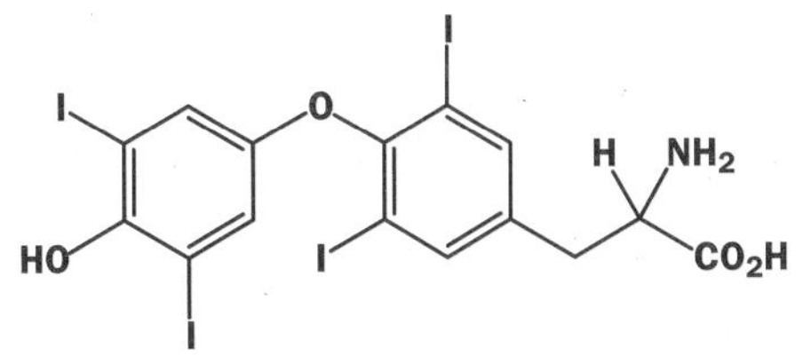

thyroxine
(human hormone)
iodine carrier
in thyroid gland

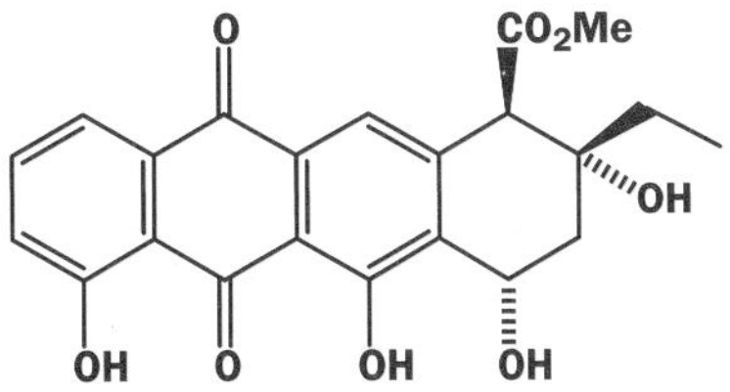

aklavinone
tetracycline antibiotic
[why tetracycline?]

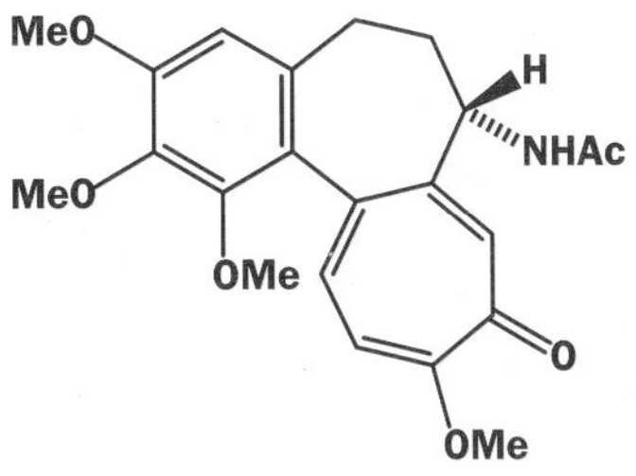

colchicine
compound from
autumn crocus
used to treat gout

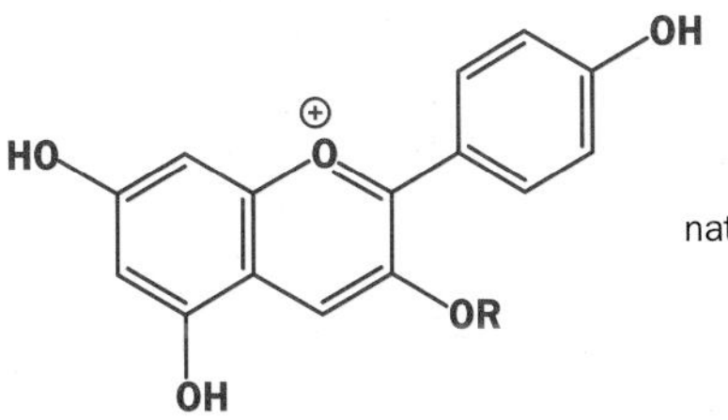

callistephin
natural red flower pigment

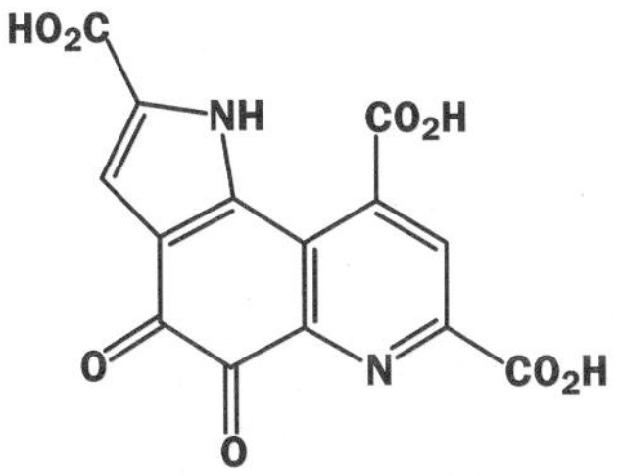

methoxatin
coenzyme from bacteria
living on methane

2. Just to remind you—write out a detailed mechanism for these steps.

$$HNO_3 + H_2SO_4 \longrightarrow {}^{\oplus}NO_2$$

In a standard nitration reaction with, say, HNO_3 and H_2SO_4, each of these compounds forms a single mono-nitration product. What is its structure? Justify your answer with a mechanism.

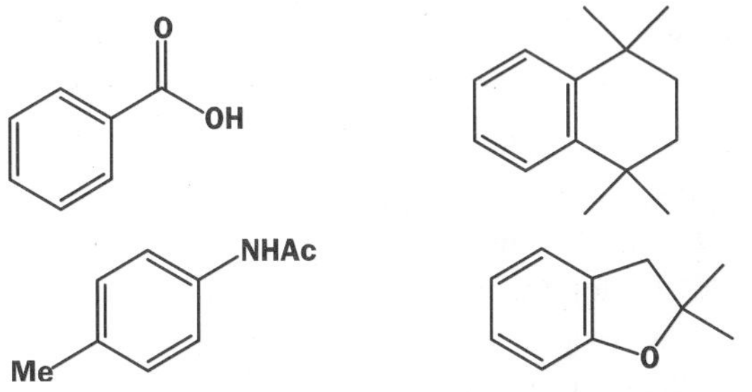

3. Write mechanisms for these reactions, justifying the position of substitution.

(i)

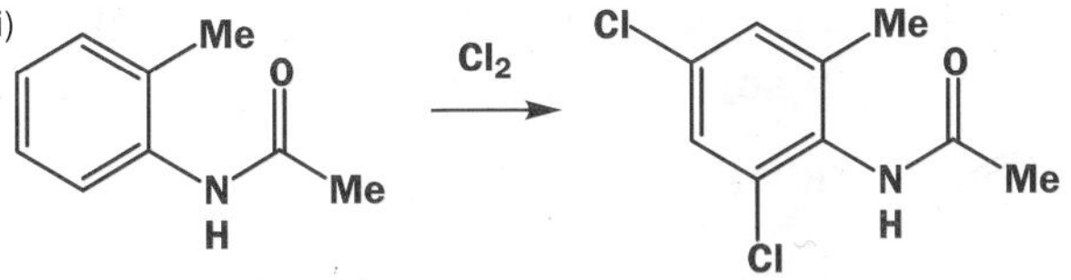

(ii)

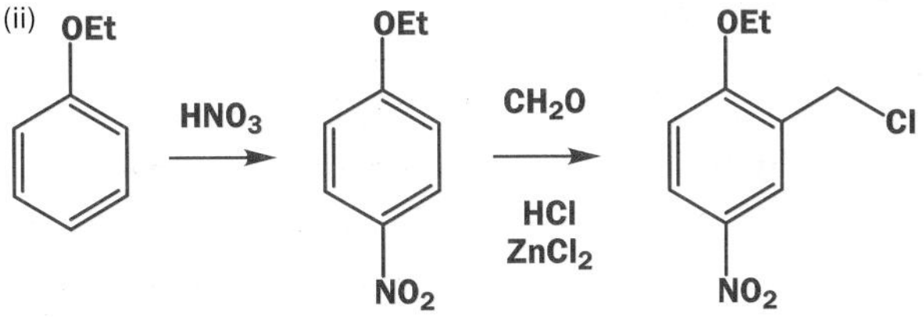

4. How reactive are the different sites in toluene? Nitration of toluene produces the three possible products in the ratios shown. What would be the ratio of products if all sites were equally reactive? What is the actual relative reactivity of the three sites? (You could express this as $x{:}y{:}1$ or as $a{:}b{:}c$ where $a + b + c = 100$.) Comment on the ratio you deduce.

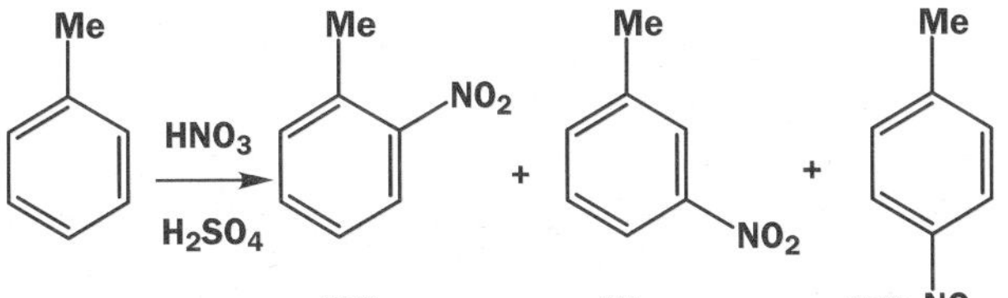

toluene   59%   4%   37%

578 **22 · Electrophilic aromatic substitution**

5. Revision problem. The local anaesthetic proparacaine is made by this sequence of reactions. Deduce a structure for each product. Draw a mechanism for each step and explain why it gives that particular product.

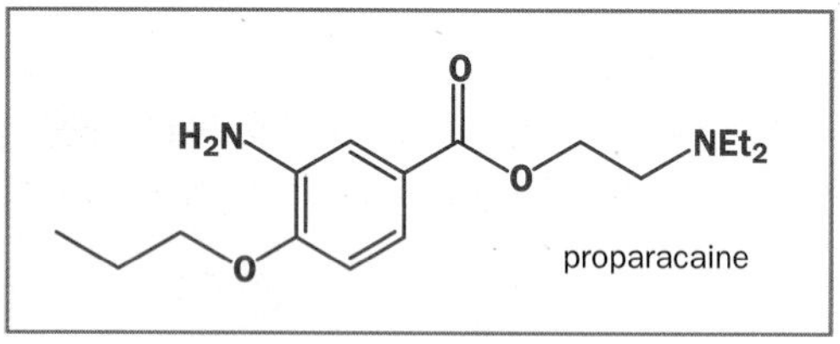

6. In the chapter, we established that electron-withdrawing groups direct *meta*. Among such reactions is the nitration of trifluoromethyl benzene. Draw out the detailed mechanism for this reaction and also for a reaction that does not happen—the nitration of the same compound in the *para* position. Draw all the delocalized structures of the intermediates and convince yourself that the intermediate for *para* substitution is destabilized by the CF₃ group while that for *meta* substitution is not.

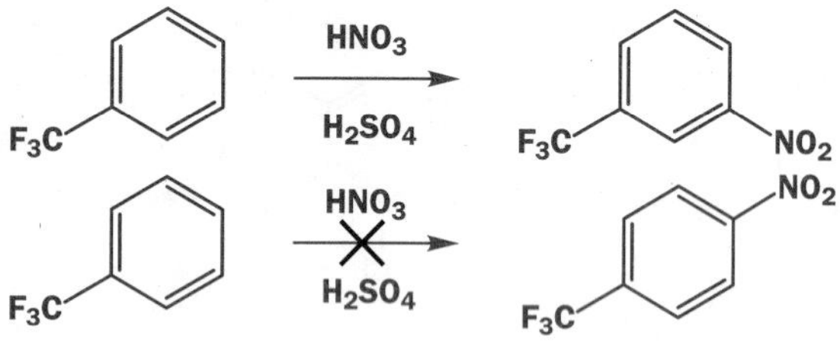

7. Draw mechanisms for the following reactions and explain the position(s) of substitution.

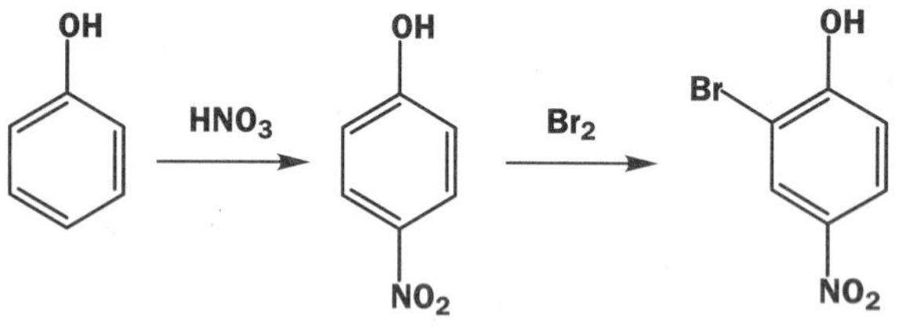

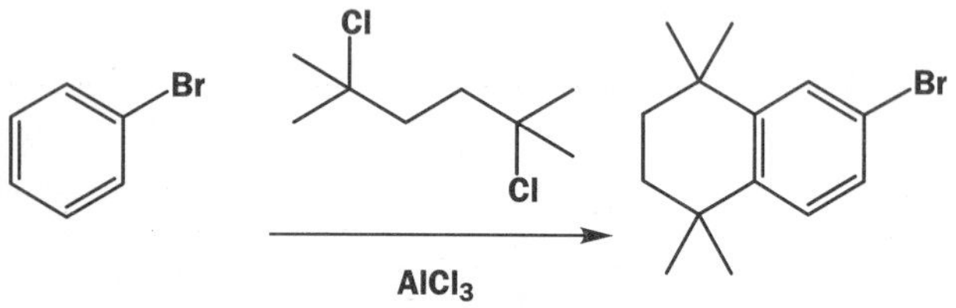

8. Nitration of these compounds gives products with the proton NMR spectra shown. Deduce the structures of the products from the NMR and explain the position of substitution.

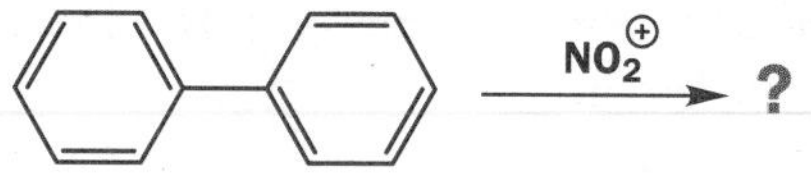

δH
7.77 (4H, d, J 10)
8.26 (4H, d, J 10)

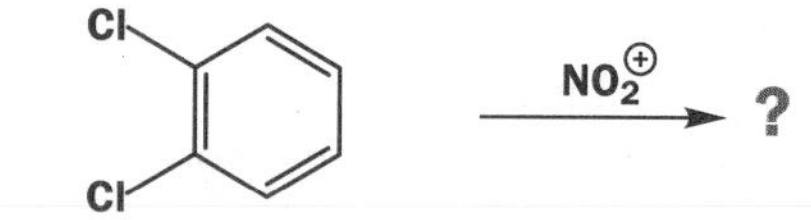

δH
7.6 (1H, d, J 10)
8.1 (1H, dd, J 10, 2)
8.3 (1H, d, J 2)

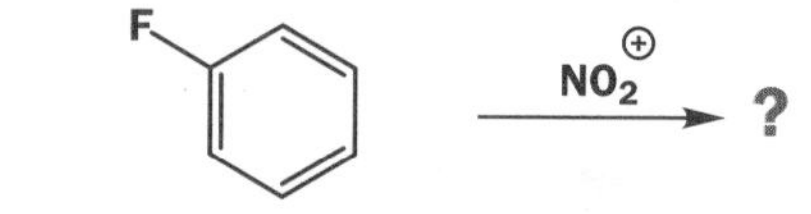

δH
7.15 (2H, dd, J 7,8)
8.19 (2H, dd, J 6,8)

9. Attempted Friedel–Crafts acylation of benzene with *t*-BuCOCl gives some of the expected ketone, as a minor product, and also some *t*-butyl benzene, but the major product is the disubstituted compound C. Explain how these compounds are formed and suggest the order in which the two substituents are added to form compound C.

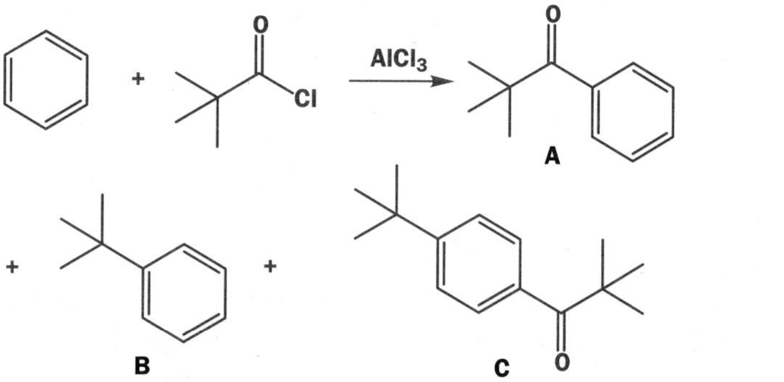

10. Draw mechanisms for the following reactions.

(a) (b) 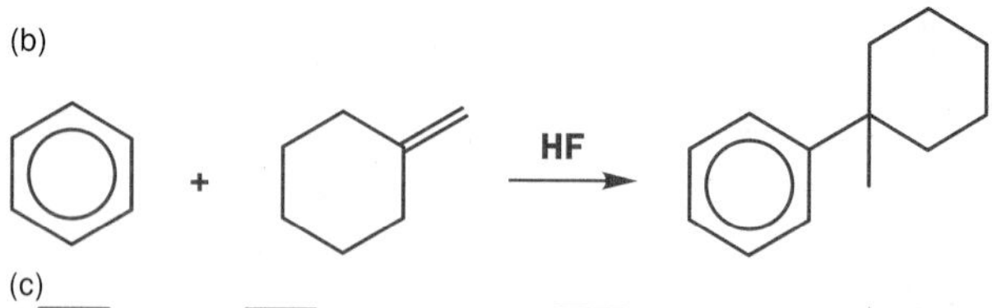 (c) 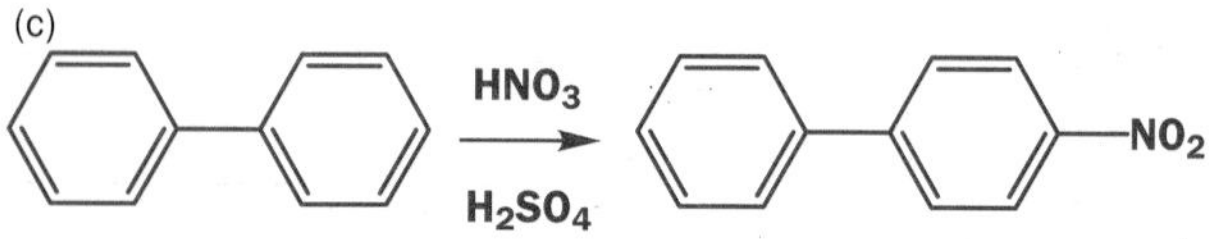 (d)

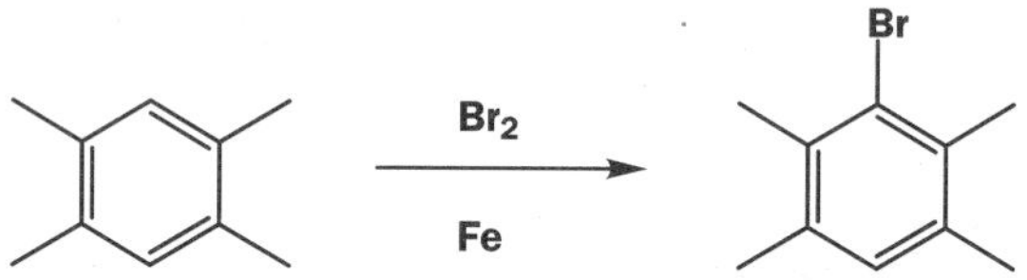

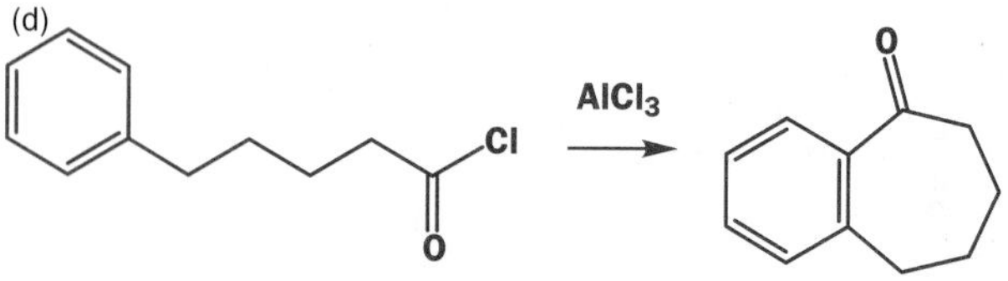

11. Nitration of this aromatic heterocycle with the usual mixture of HNO_3 and H_2SO_4 gives a product whose NMR spectrum is given. Though you have not yet met heterocycles you should be able to deduce the structure of the product and explain why it is formed.

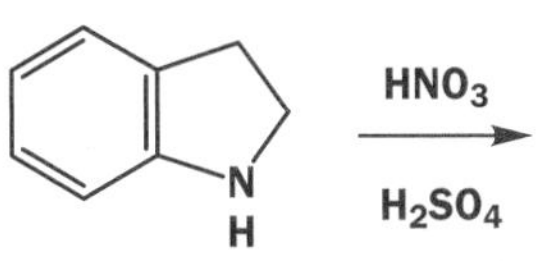

$C_8H_8N_2O_2$
δ_H 3.04 (2H, t, J 7Hz)
3.68 (2H, t, J 7Hz)
6.45 (1H, d, J 8Hz)
7.28 (1H, broad s)
7.81 (1H, d, J 1 Hz)
7.90 (1H, dd, J 8, 1 Hz)

12. Explain the position of substitution in the following reactions and predict the structure of the final product. Why is a Lewis acid necessary for the second bromination but not for the first?

(a)

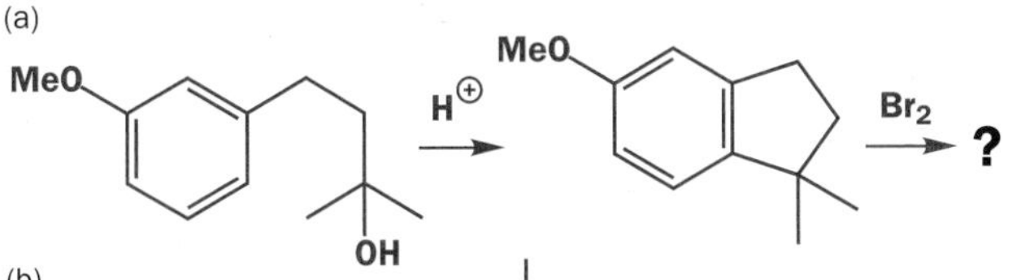

(b)

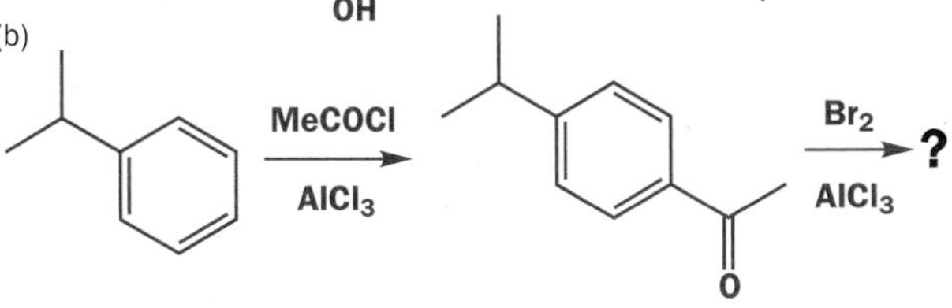

13. Suggest mechanisms for the methylation step at the end of the synthesis that concludes the chapter. Why is it necessary to go to these lengths rather than just react with MeI?

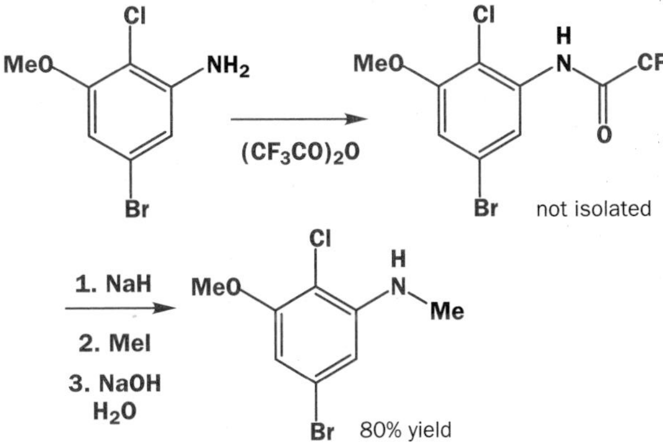

14. So what happens if we force phenol to react again with bromine? Will reaction then occur in the *meta* positions? It is possible to brominate 2,4,6-tribromophenol if we use bromine in acetic acid. Account for the formation of the product.

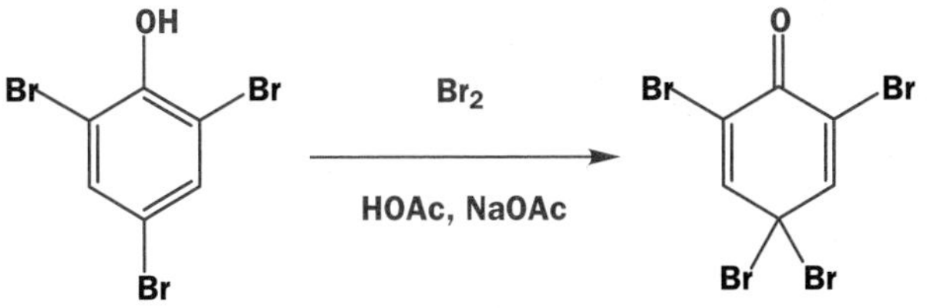

2,4,6-tribromophenol

This product can be used for bromination as in the mono-bromination of this amine. Suggest a mechanism and explain the selectivity.

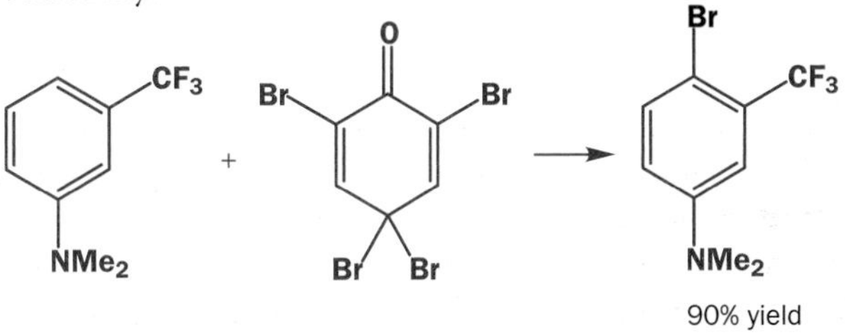

90% yield

Electrophilic alkenes

23

Connections

Building on:

- Conjugate addition ch10
- Electrophilic additions to alkenes ch20
- Electrophilic substitution on aromatic rings ch22

Arriving at:

- Conjugate addition: conjugation of alkenes with electron-withdrawing groups other than C=O (CN and NO_2), which makes them electrophilic and allows nucleophilic attack
- Conjugate substitution: electrophilic alkenes bearing leaving groups can promote substitution reactions at C=C related to those at C=O
- Nucleophilic aromatic substitution: electron-poor aromatic rings that allow substitution reactions with nucleophiles rather than the usual electrophiles
- Special leaving groups and nucleophiles that allow nucleophilic aromatic substitution on electron-rich rings
- Allylic systems: how double bonds adjacent to leaving groups share the electrophilic character of the carbon atom carrying the leaving group, and may allow more than one product to form

Looking forward to:

- Conjugate addition of enolate-type nucleophiles ch29
- Reactions of heterocyclic aromatic compounds ch43

This chapter is also the last chapter in the second cycle of chapters within this book, with which we complete our survey of the important elementary types of organic reactions. We follow it with two review chapters, before looking in more detail at enolate chemistry and how to make molecules.

Introduction—electrophilic alkenes

Alkenes are nucleophilic. Almost regardless of their substituents, they react with electrophiles like bromine to form adducts in which the π bond of the alkene has been replaced by two σ bonds.

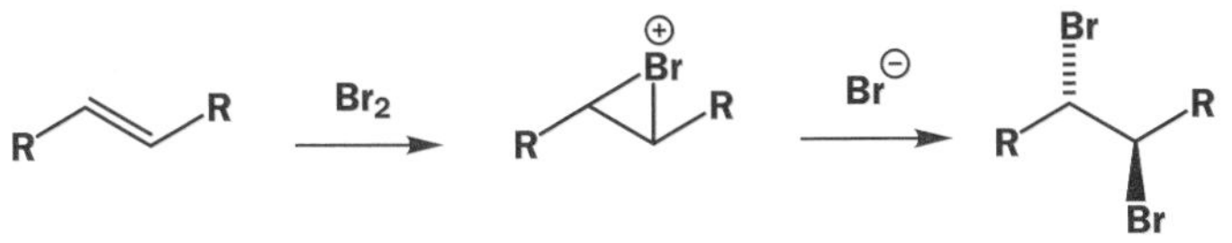

Reactions like this were discussed in Chapter 20.

Even when the alkene is conjugated with an electron-withdrawing group, bromine addition still occurs, though less readily. As we said, alkenes are nucleophilic.

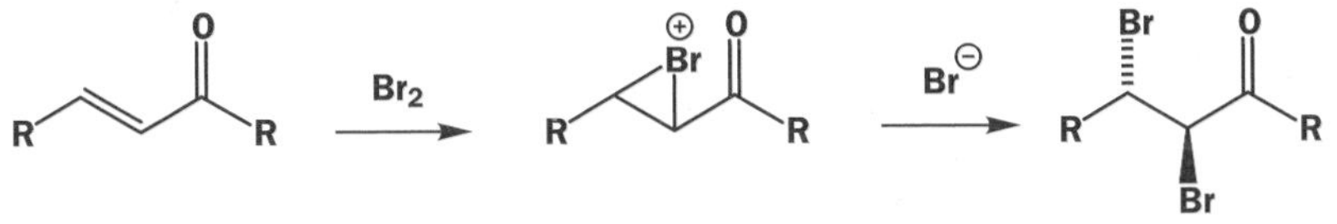

■ We saw examples of this reaction in Chapter 10.

But this last type of alkene is also electrophilic. The carbonyl group dominates the alkene in the interaction between the two groups and nucleophiles add so that the enolate is an intermediate and the negative charge resulting from conjugate addition is stabilized by conjugation. This intermediate is protonated on carbon to give the conjugate addition product—the result of a nucleophilic addition of HX to the alkene. The final product has an unchanged carbonyl group but without that carbonyl group no nucleophilic addition could have occurred.

We are going to extend this idea now and show that other groups besides the carbonyl group can promote nucleophilic addition to alkenes and then extend the idea further into the reactions of allylic and aromatic compounds. First of all we are going to look at other conjugating electron-withdrawing groups.

Nucleophilic conjugate addition to alkenes

Unsaturated nitriles

The essential requirement for these reactions is a conjugating substituent that is about as anion-stabilizing as a carbonyl group. One we have seen before is cyanide and we shall look first at conjugated nitriles. The simplest is acrylonitrile. This compound adds amines readily.

■ Like many simple acrylic derivatives, this nitrile is readily available as it is manufactured on a large scale for polymer synthesis. Superglue is a polymerized acrylonitrile. There is more about this in Chapter 52.

The amine first attacks the alkene in a typical conjugate addition to make a stable anion. Notice that the nucleophile must attack the far end of the alkene to do this—attack next to the electron-withdrawing group would not work.

The anion can have its charge drawn on the nitrogen atom but it is really delocalized over the two neighbouring carbon atoms and is very like an enolate. Do not be put off by the odd appearance of the 'enolate'. The dot between the two double bonds is a reminder that there is a linear sp carbon atom at this point.

Protonation at carbon restores the cyanide and gives the product—an amino-nitrile. The whole process adds a 2-cyano-ethyl group to the amine and is known industrially as **cyanoethylation**.

▶ You will see a few mechanisms in this chapter where we have written an intramolecular deprotonation. This saves writing two steps—protonation of the enolate and deprotonation of N (here)—but quite possibly this is not the actual mechanism by which the proton transfer takes place. Any proton will do, as will any base—do not take the arrows here too literally.

With a primary amine, the reaction need not stop at that stage as the product is still nucleophilic and a second addition can occur to replace the second hydrogen atom on nitrogen.

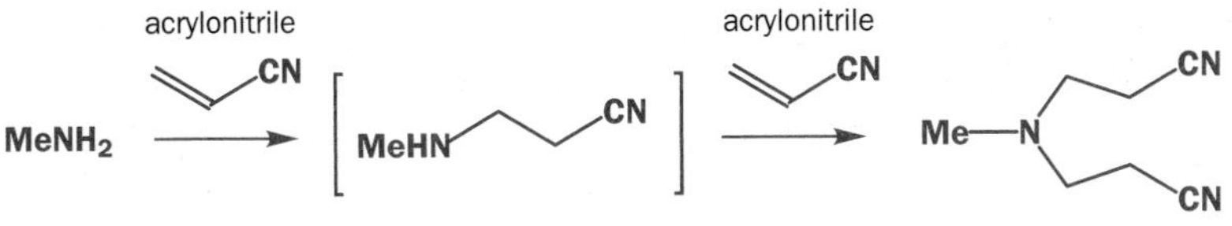

Other elements add too. Phenyl phosphine can undergo a double addition just as in the last example, but alcohols can add only once.

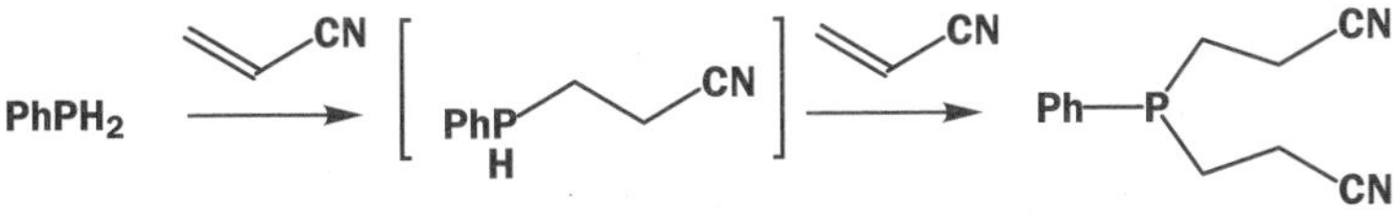

If there is a competition between a second row (for example, N or O) and a third row (for example, S or P) element, the third row element normally wins. The lone pair electrons are of higher energy ($3sp^3$) in the third-row element than in the second-row element ($2sp^3$).

The cyanide group is a typical group for promoting conjugate addition. It is possible for nucleophiles to attack directly at the CN group but it is not very electrophilic so that these reactions tend to be thermodynamically controlled and attack is preferred in the conjugate position.

Unsaturated nitro compounds

The nitro group (NO_2) is extremely electron-withdrawing—about twice as electron-withdrawing as a carbonyl group. This should theoretically make it prefer direct attack rather than conjugate attack but in practice direct attack at NO_2 is almost unknown. The products from direct attack are very unstable compounds and revert to starting materials easily. You may rely on conjugate addition to nitro-alkenes.

■
We summarized factors favouring direct versus conjugate attack in Chapter 10, p. 240.

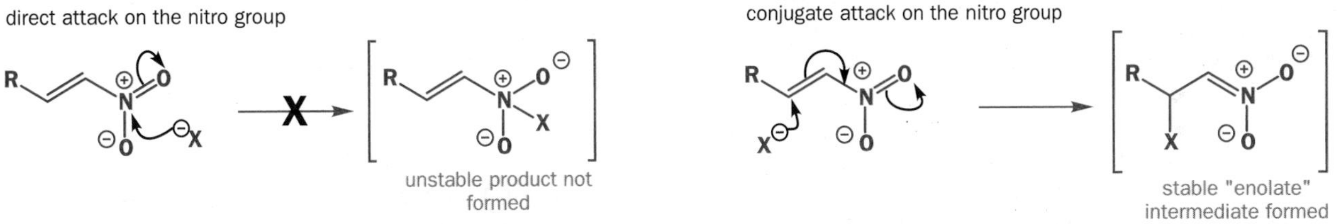

The intermediate is rather like an enolate anion, with a negatively charged oxygen atom conjugated to a (N=C) double bond. It reacts like an enolate, picking up a proton on carbon to re-form the nitro group and give a stable product—the result of conjugate addition of HX. Here is the full mechanism with borohydride acting as the nucleophile, reducing the nitroalkene to a nitro-alkane.

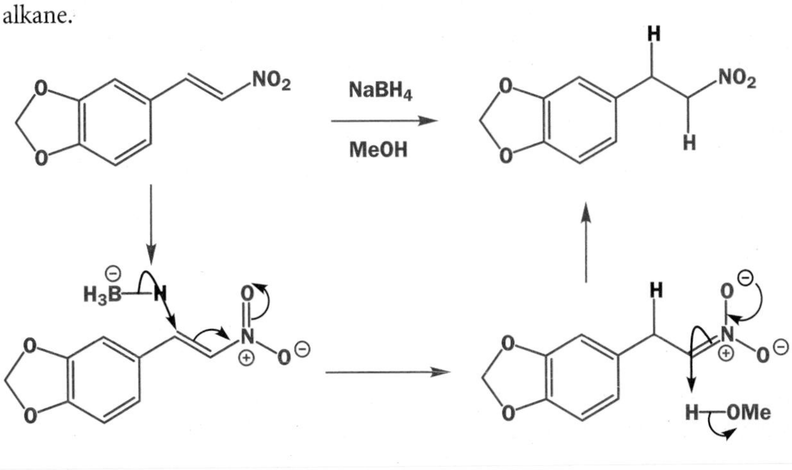

Michael acceptors are dangerous

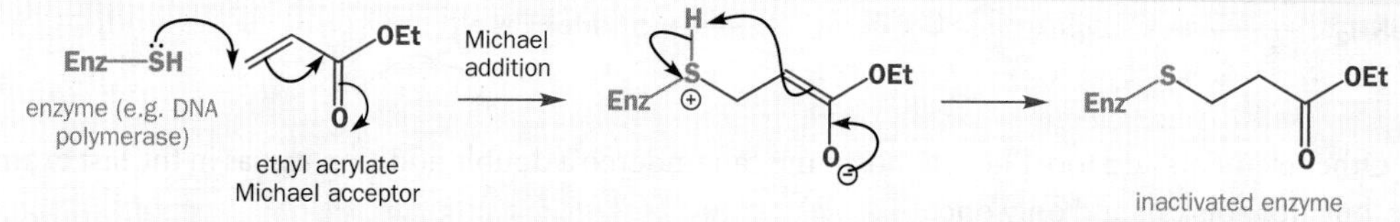

Any compound capable of conjugate addition (a **Michael acceptor**—conjugate additions are also known as **Michael additions**) is potentially dangerous to living things. Even simple compounds like ethylacrylate are labelled 'cancer suspect agent'. They attack enzymes, particularly the vital DNA polymerase involved in cell division by conjugate addition to thiol and amino groups in the enzyme.

Any compound that is good at conjugate addition is probably toxic and carcinogenic (cancer-causing). In Chapter 10, we mentioned some anticancer drugs that work by this same mechanism, but do it more selectively in rapidly proliferating cancer cells. Most Michael acceptors are less benign, and damage the DNA replication process unselectively. Fortunately, we are offered some degree of protection by an important compound present in most tissues. The compound is glutathione, a tripeptide—a compound made from three amino acids. We shall discuss such compounds in more detail later in the book (Chapter 49) but notice for the moment that this compound can be divided into three at the two amide bonds.

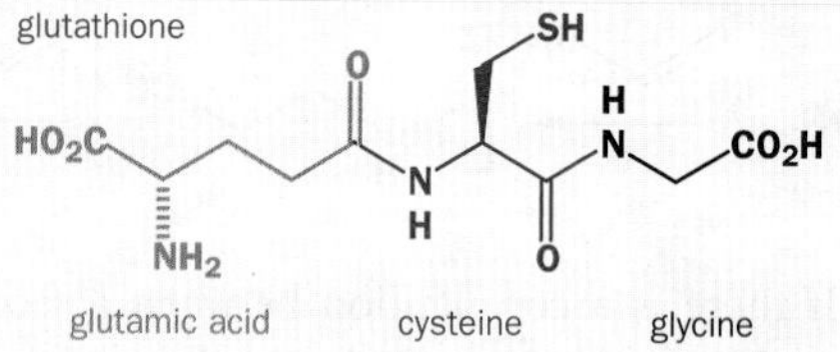

The business end of glutathione is the thiol (SH) group, which scavenges carcinogenic compounds by conjugate addition. If we use an 'exomethylene lactone'—a highly reactive Michael acceptor—as an example and represent glutathione as RCH_2SH, you can see the sort of thing that happens.

If the normally abundant glutathione is removed by such processes as oxidation (Chapter 46) and cannot any longer scavenge toxins, then the organism is in danger. This is one reason why vitamin C is so beneficial—it removes stray oxidizing agents and protects the supply of glutathione. Keep eating the fruit and vegetables!

detoxification of carcinogens by glutathione

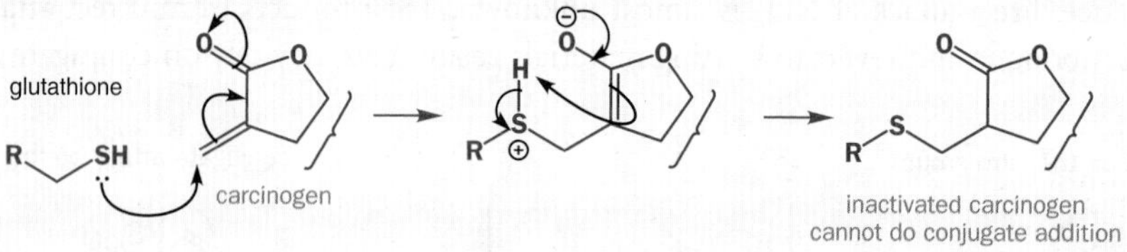

Other nucleophiles in conjugate addition

Since we introduced conjugate addition in Chapter 10, a number of new reactions have been covered and a number of new nucleophiles introduced. Some of these can lead to conjugate addition. One important new reaction is electrophilic aromatic substitution, which we met in the last chapter. Michael acceptors can combine with Lewis acids to provide electrophiles for reactions with benzene derivatives.

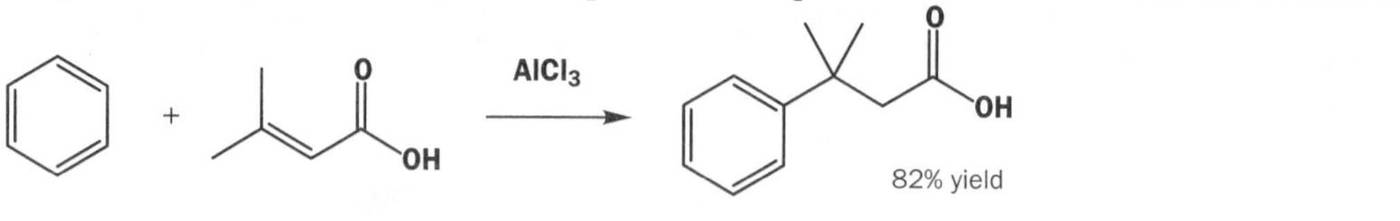

82% yield

The Lewis acid ($AlCl_3$) must combine with the carboxylic acid to create a reactive electrophile that is attacked by a benzene molecule. The first step is just like the reactions of benzene we discussed in the last chapter.

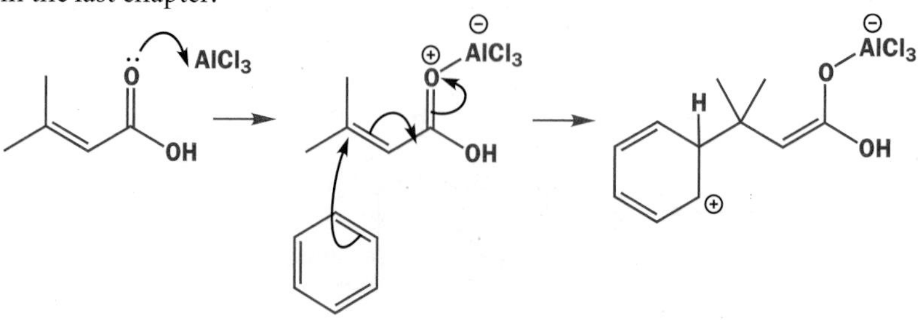

The next step must be the restoration of the aromaticity of the ring by the removal of the proton at the site of attack. This gives the aluminium enolate of the carboxylic acid. There is a proton now available to convert the aluminium enolate to the acid and this is the final product. This is a useful reaction because it has added a benzene ring to a quaternary carbon atom—conjugate addition has overcome steric hindrance.

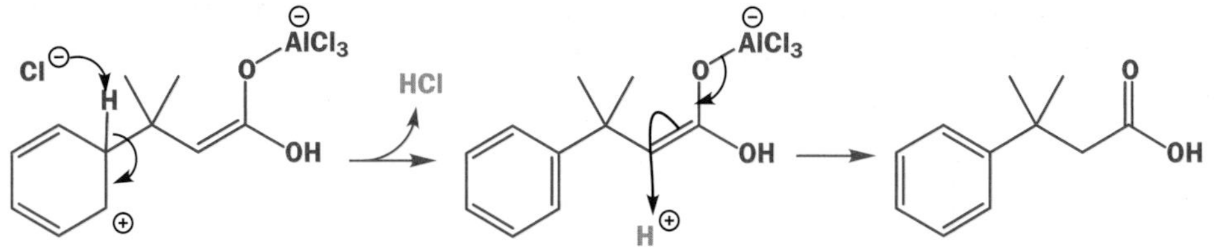

Another less common class of nucleophile that does conjugate addition is nitriles. We used unsaturated nitriles a moment ago as Michael acceptors, and nitriles are usually electrophiles rather than nucleophiles. We did see in Chapter 17 that nitriles will act as nucleophiles in the S_N1 reaction (the Ritter reaction). The next reaction is related to the Ritter reaction.

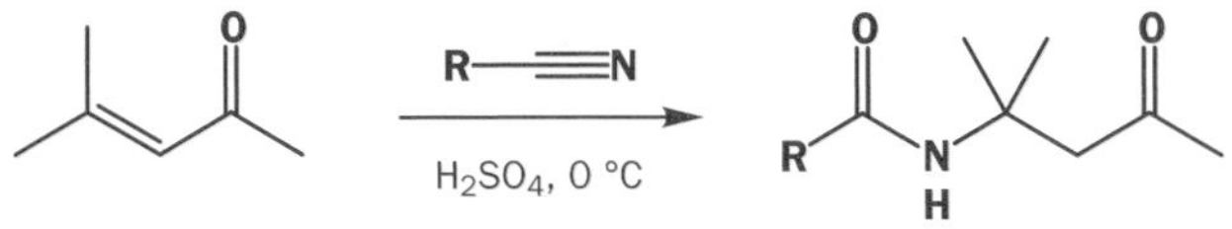

Protonation of the carbonyl group gives a very electrophilic cation that is reactive enough to persuade the nitrile to do conjugate addition.

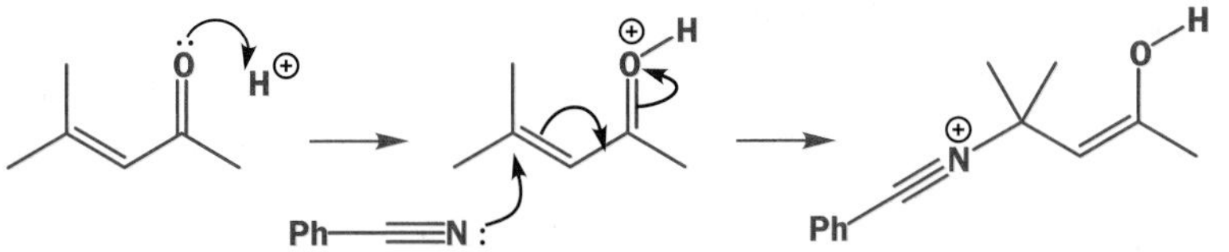

Tautomerization of the enol to a ketone, addition of water, and another tautomerization to an amide complete the mechanism. Notice here that a nitrogen has been added to a tertiary centre—this is not an easy result to accomplish and it is worth noting that conjugate addition is a good way to make bonds to crowded centres.

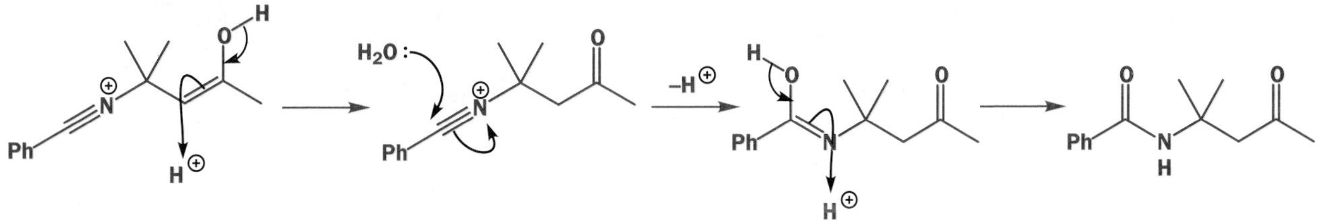

Conjugate substitution reactions

Just as direct addition to C=O (Chapter 6) becomes substitution at C=O (Chapter 12) when there is a leaving group at the carbonyl carbon, so conjugate *addition* becomes conjugate *substitution* if there is a leaving group, such as Cl, at the β carbon atom. Here is an example: substitution has replaced Cl with OMe, just as it would have done in a reaction with an acyl chloride.

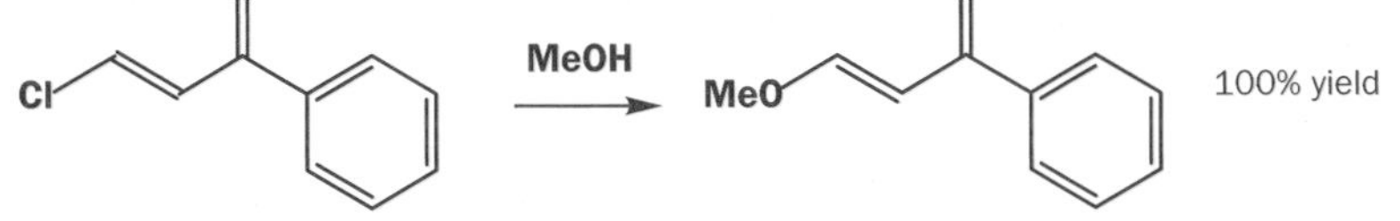

This apparently simple substitution does *not* involve a direct displacement of the leaving group in a single step! As you will see again shortly, S_N2 reactions do not occur at sp^2 hybridized carbon.

formation of the enol intermediate

The mechanism starts in exactly the same way as for conjugate addition, giving an enol intermediate.

Now the leaving group can be expelled by the enol: the double bond moves back into its original position in this step, which is exactly the same as the final step of an E1cB reaction (Chapter 19). The 'new' double bond usually has the *E* configuration as the molecule can choose which of the two possible perpendicular conformations to eliminate.

expulsion of the leaving group from the enol intermediate

Halogens are excellent leaving groups and are often used in conjugate substitution reactions. In the next example, two consecutive conjugate substitution reactions give a diamine.

98% yield

At first sight, the product looks rather unstable—sensitive to water, or traces of acid perhaps. But, in fact, it is remarkably resistant to reaction with both. The reason is conjugation: this isn't really an amine (or a diamine) at all, because the lone pairs of the nitrogen atoms are delocalized into the carbonyl group, very much as they are in an amide. This makes them less basic, and makes the carbonyl group less electrophilic.

delocalization of the nitrogen's lone pair

delocalization in an amide

Compounds like this are for this reason known as **vinylogous amides**—the C=C bond between the N and C=O allows conjugation still to take place but at a greater distance. This is the essence of vinylogous behaviour.

Just as the cyanide (CN) and nitro (NO_2) groups can be used to bring about conjugate addition, so also they can initiate conjugate substitution. Examples of these reactions play vital roles in the synthesis of two of the most important drugs known—the anti-ulcer drugs Tagamet (SmithKline Beecham) and Zantac (GlaxoWellcome).

Preventing ulcers (1): Tagamet

One cause of ulcers is excess acid secretion by the stomach, and one method of prevention is to stop this by blocking the acid-releasing action of histamine. You can see here the resemblance between histamine and Tagamet (generic name cimetidine).

histamine   cimetidine—SmithKline Beecham's Tagamet   guanidine

▶ **Vinylogous behaviour**

The conjugated double bond serves as an electronic linker between the carbonyl group and the halogen or other heteroatom, which makes the chemical and spectroscopic behaviour of the composite functional group similar to that of the simple relative. You could think of the β-chloro enone at the beginning of this section as a vinylogous acyl chloride that reacts with methanol to give a vinylogous ester.

is the vinylogous version of, and reacts similarly to,

As well as the histamine-like portion of the molecule, Tagamet has a sulfur atom and then, at the end of a short chain of carbon atoms, a complicated functional group based on guanidine. It is easy to add the sulfur atom and the short carbon chain to the heterocyclic building block (see Chapter 43 for more about this) so that the only problem is how to build on the guanidine at the end of the molecule.

■ Guanidine is an organic base, as strong as NaOH, and it was discussed in Chapter 8.

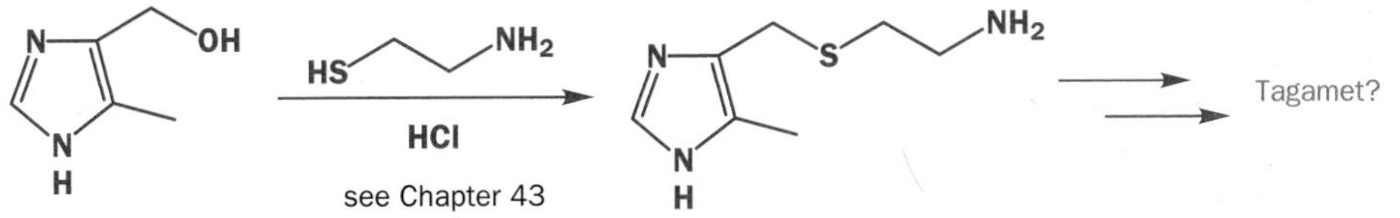

Now enter the star of the show! This simple cyanoimine, with two SMe groups as built-in leaving groups, is readily available and reacts with amines to give guanidines in two stages.

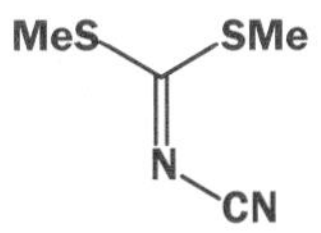

Each of the reactions is a conjugate substitution. It will be clearer if we draw the reaction with a generalized primary amine RNH$_2$ first.

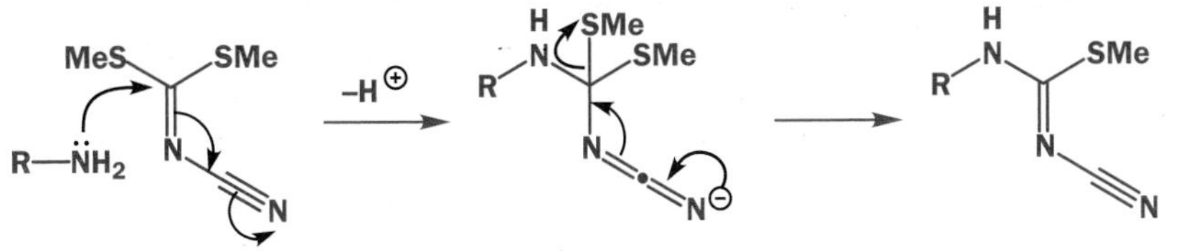

The first step is conjugate addition, exactly as we saw with acrylonitrile at the beginning of this chapter. The second step shows the return of the negative charge and the expulsion of the best leaving group. Thiols are acidic compounds, and MeS$^-$ is a better leaving group than RNH$^-$.

The reaction stops cleanly at this point and more vigorous conditions are required to displace the second MeS$^-$ group. This is because the first product is less reactive than the starting material. Why is this? The introduced amino group is electron-donating and a strong conjugation is established between it and the cyano group.

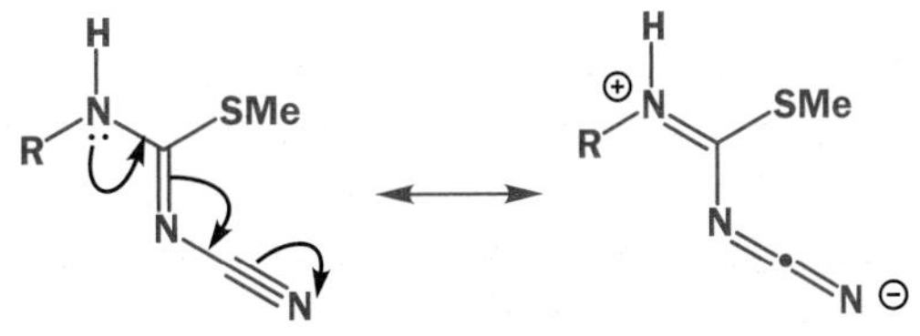

Now a second and different amine can be introduced and the second MeS$^-$ group displaced. In the Tagamet synthesis; the second amine is MeNH$_2$, and the synthesis is complete.

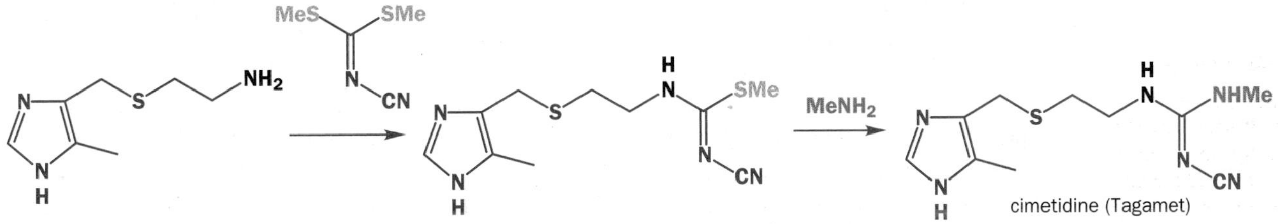

cimetidine (Tagamet)

Preventing ulcers (2): the best selling drug of all time—GlaxoWellcome's Zantac

This anti-ulcer drug has some obvious similarities to Tagamet, and some differences too. Here are the two structures side-by-side.

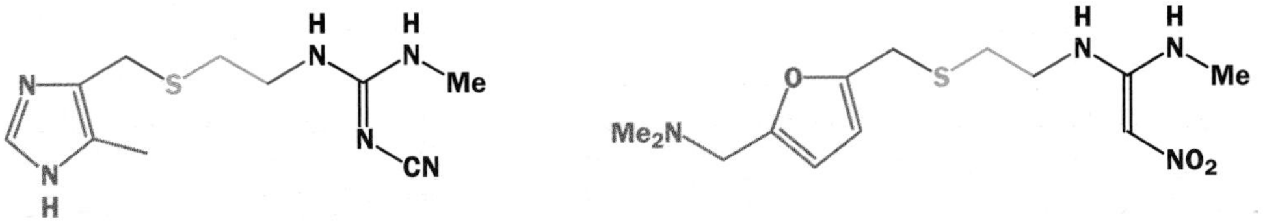

cimetidine—SmithKline Beecham's Tagamet   ranitidine—GlaxoWellcome's Zantac

The heterocyclic ring is still there but it is very different. The sulfur and its surrounding CH$_2$ groups are the same and the guanidine seems to be still there. But it isn't. Look closely at this 'guanidine' and you will see that there are only *two* nitrogen atoms around the central carbon atom instead of the *three* in a guanidine. This is an **amidine**. The nitrile has also been replaced by a nitro group. The synthesis is, however, remarkably similar to that of the real guanidine in Tagamet. Two conjugate substitutions use MeS⁻ as leaving groups and amine as nucleophiles. Here is the first, with mechanism.

The first step is conjugate addition, just like the conjugate additions to nitroalkenes at the beginning of this chapter, and the second step brings the negative charge back and expels the best leaving group. Again the reaction can be made to stop at this stage because this product is stabilized by conjugation between the green amino group and the nitro group. A second substitution puts together the two halves of the drug.

ranitidine (Zantac)

Nucleophilic epoxidation

The conjugate substitutions we have just been discussing rely on a starting material containing a leaving group. In this section we are going to look at what happens if the leaving group is not attached to the unsaturated carbonyl compound, but instead is attached to the nucleophile. We shall look at this class of compounds—nucleophiles with leaving groups attached—in more detail in Chapter 40, but for the moment the most important will be hydroperoxide, the anion of hydrogen peroxide.

hydrogen peroxide

hydroperoxide anion

Hydroperoxide is a good nucleophile because of the **alpha effect**: interaction of the two lone pairs on adjacent oxygen atoms raises the HOMO of the anion and makes it a better and softer nucleophile than hydroxide.

Hydroperoxide is also less basic than hydroxide because of the inductive electron-withdrawing effect of the second oxygen atom. Basicity and nucleophilicity usually go hand in hand—not here though. This means that the hydroperoxide anion can be formed by treating hydrogen peroxide with aqueous sodium hydroxide.

The same effect explains why hydroxylamine and hydrazine are more nucleophilic than ammonia: p. 351.

The pKₐ of hydrogen peroxide is 11.6.

new, higher energy HOMO of hydroperoxide anion

This is what happens when this mixture is added to an enone. First, there is the conjugate addition.

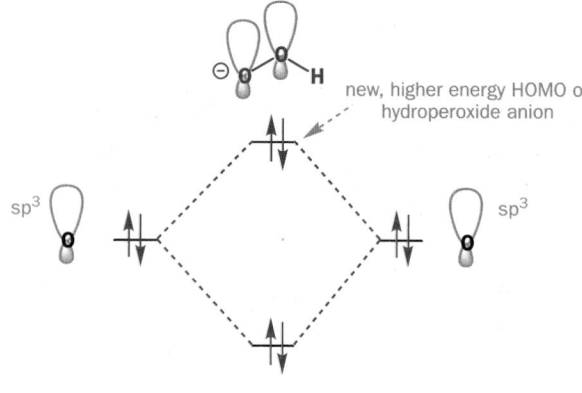

The product is not stable, because hydroxide can be lost from the oxygen atom that was the nucleophile. Hydroxide is fine as a leaving group here—after all, hydroxide is lost from enolates in E1cB eliminations, and here the bond breaking is a weak O–O bond. The product is an epoxide.

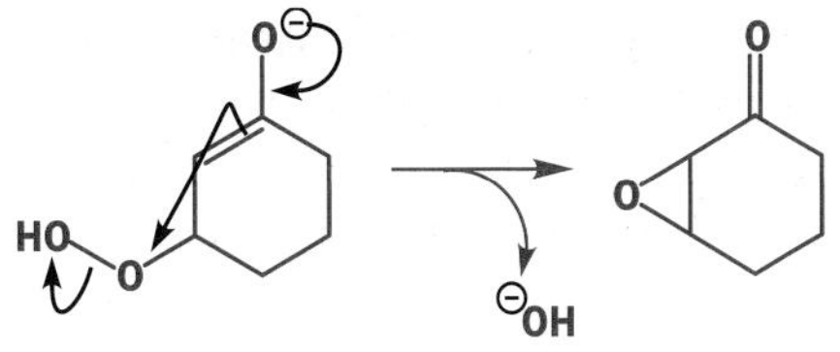

The electrophilic epoxidizing agents such as *m*-CPBA, which you met in Chapter 20, are less good with electron-deficient alkenes: we need a nucleophilic epoxidizing agent instead. There is another significant difference between hydrogen peroxide and *m*-CPBA, highlighted by the pair of reactions below.

cis-alkene   *trans*-epoxide   *cis*-alkene   *cis*-epoxide

m-CPBA epoxidation is stereospecific because the reaction happens in one step. But nucleophilic epoxidation is a two-step reaction: there is free rotation about the bond marked in the anionic intermediate, and the more stable, *trans*-epoxide results, whatever the geometry of the starting alkene.

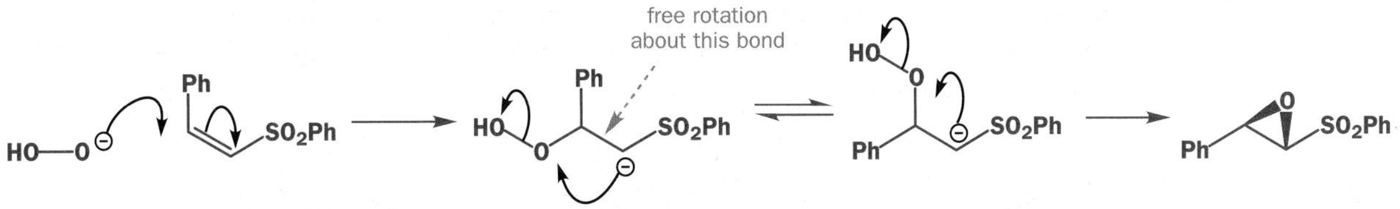

free rotation about this bond

In general, conjugate substitution is not nearly as important as the next topic in this chapter—nucleophilic aromatic substitution. Before we describe in detail those reactions that do occur, we need to explain why the most obvious reactions do not occur.

Nucleophilic aromatic substitution

The simplest and most obvious nucleophilic substitutions on an aromatic ring, such as the displacement of bromide from bromobenzene with hydroxide ion, do *not* occur.

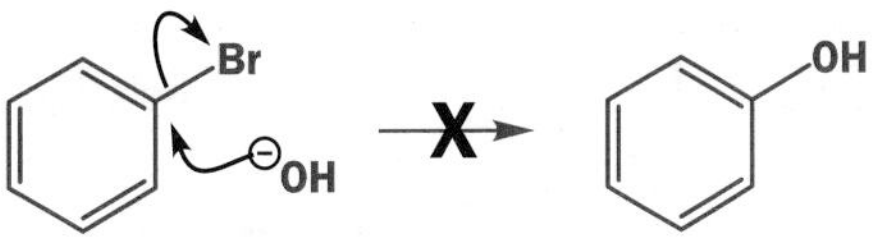

reaction doesn't happen

Please note—this mechanism is *wrong*! No such reactions are known. You might well ask, 'Why not?' The reaction looks all right and, if the ring were saturated, it *would* be all right.

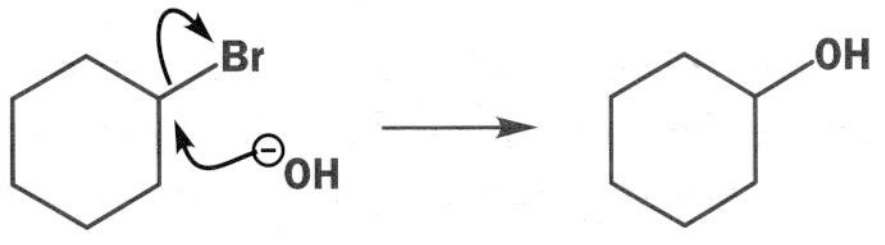

This is an S_N2 reaction, and we know (Chapter 17) that attack must occur in line with the C–Br bond

reaction does happen

from the back, where the largest lobe of the $\sigma*$ orbitals lies. That is perfectly all right for the aliphatic ring because the carbon atom is tetrahedral and the C–Br bond is not in the plane of the ring. Substitution of an equatorial bromine goes like this.

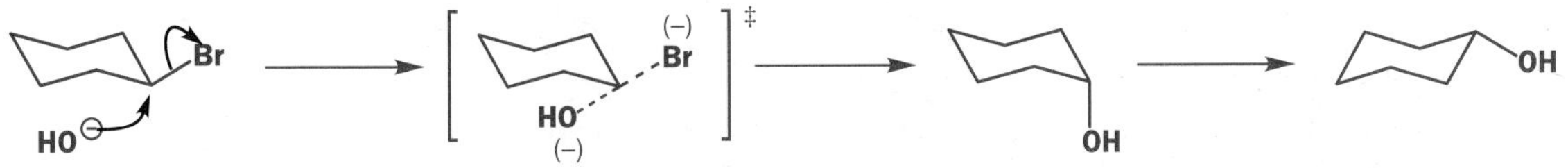

line of attack is not in plane of ring

But in the aromatic compound, the C–Br bond is in the plane of the ring as the carbon atom is trigonal. To attack from the back, the nucleophile would have to appear inside the benzene ring and invert the carbon atom in an absurd way. This reaction is not possible!

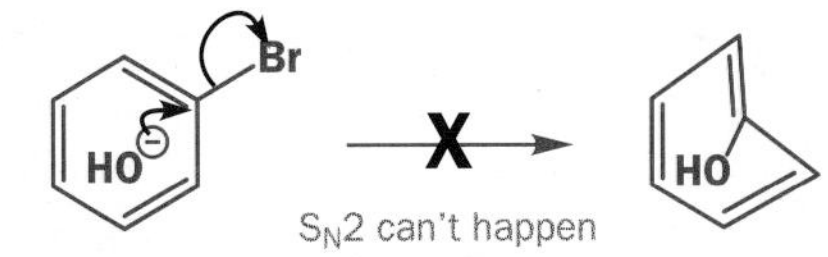

S_N2 can't happen

This is another example of the general rule.

● S_N2 at sp^2 C does *not* occur.

If S_N2 is impossible, what about S_N1? This is possible but very unfavourable. It would involve the unaided loss of the leaving group and the formation of an aryl cation. All the cations we saw as intermediates in the S_N1 reaction (Chapter 17) were planar with an empty p orbital. This cation is planar but the p orbital is full—it is part of the aromatic ring—and the empty orbital is an sp^2 orbital outside the ring.

▶
In fact, the mechanism *can* occur, but only with the best leaving group—a molecule of gaseous nitrogen—as we shall see later.

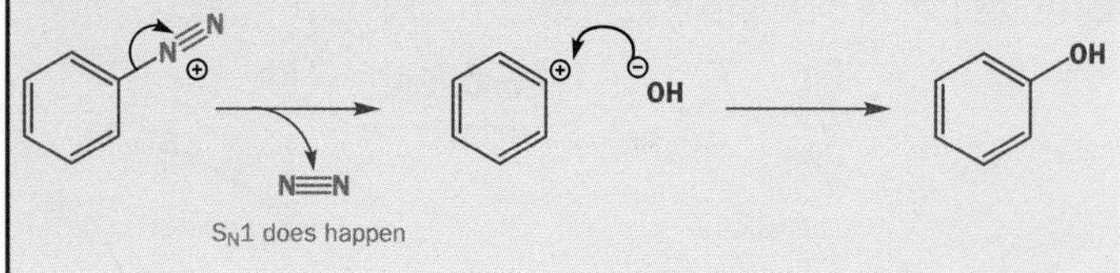

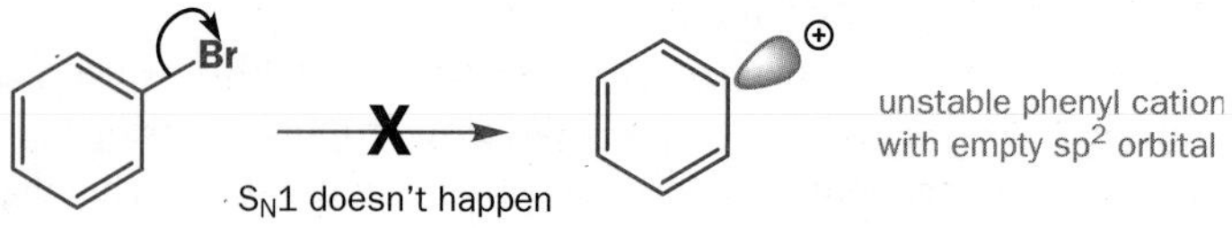

S_N1 does happen

S_N1 doesn't happen

unstable phenyl cation with empty sp^2 orbital

The most important mechanism for aromatic nucleophilic substitution follows directly from conjugate substitution and we shall introduce it that way. It is called the 'addition–elimination mechanism'.

The addition–elimination mechanism

Imagine a cyclic β-fluoro-enone reacting with a secondary amine in a conjugate substitution reaction. The normal addition to form the enolate followed by return of the negative charge to expel the fluoride ion gives the product.

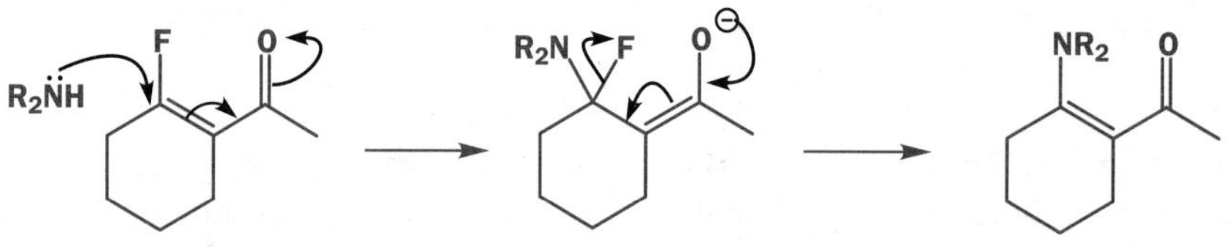

Now imagine just the same reaction with two extra double bonds in the ring. These play no part in our mechanism; they just make what was an aliphatic ring into an aromatic one. Conjugate substitution has become nucleophilic aromatic substitution.

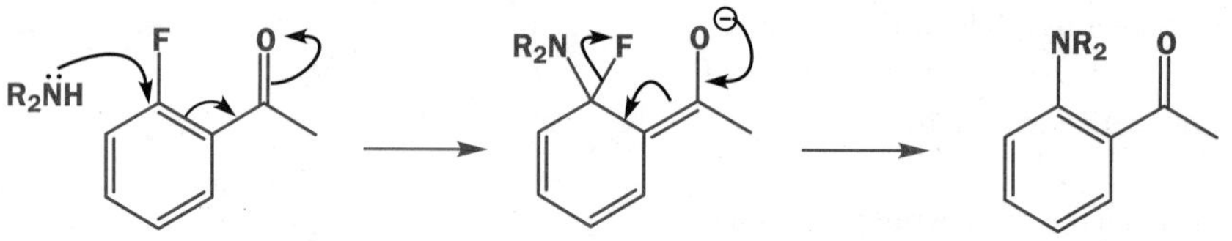

The mechanism involves *addition* of the nucleophile followed by *elimination* of the leaving group—the **addition–elimination mechanism**. It is not necessary to have a carbonyl group—any electron-withdrawing group will do—the only requirement is that the electrons must be able to get out of the ring into this anion-stabilizing group. Here is an example with a *para*-nitro group.

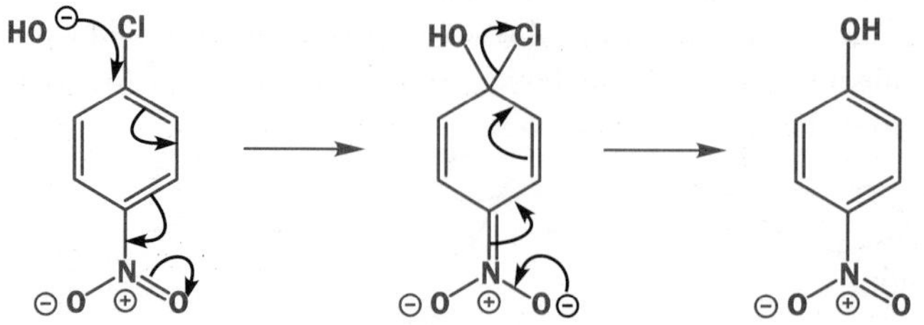

Everything is different about this example—the nucleophile (HO⁻), the leaving group (Cl⁻), the anion-stabilizing group (NO₂), and its position (*para*)—but the reaction still works. The nucleophile is a good one, the negative charge can be pushed through on to the oxygen atom(s) of the nitro group, and chloride is a better leaving group than OH.

A typical nucleophilic aromatic substitution has:

• an oxygen, nitrogen, or cyanide nucleophile

- a halide for a leaving group
- a carbonyl, nitro, or cyanide group *ortho* and/or *para* to the leaving group

Since the *nitro* group is usually introduced by electrophilic aromatic substitution (Chapter 22) and halides direct *ortho/para* in nitration reactions, a common sequence is nitration followed by nucleophilic substitution.

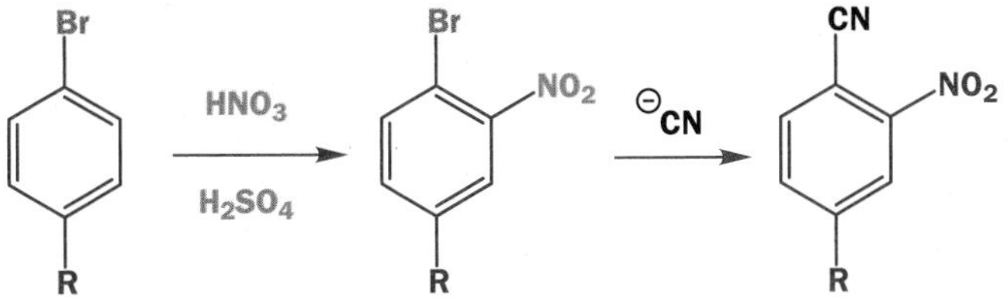

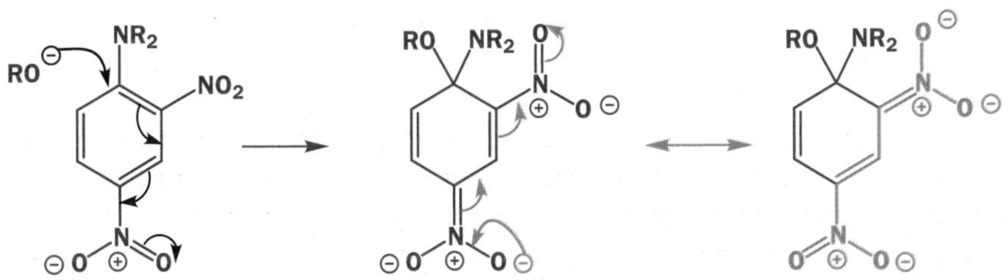

If you try and do the same reaction with a *meta* anion-stabilizing group, it doesn't work. You can't draw the arrows to push the electrons through on to the oxygen atom. Try it yourself.

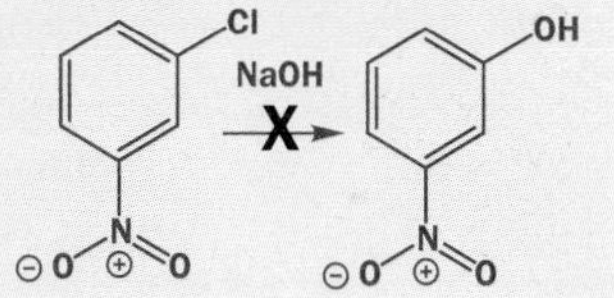

This sequence is useful because the nitro group could not be added directly to give the final product as nitration would go in the wrong position. The cyanide is *meta*-directing, while the alkyl group (R) is *ortho, para*-directing.

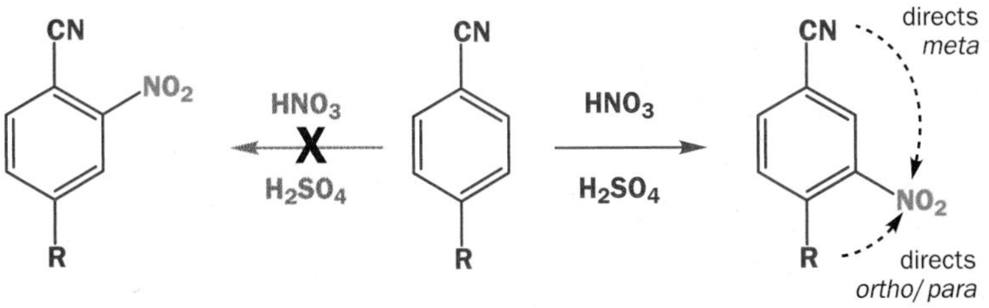

Two activating electron-withdrawing groups are better than one and dinitration of chlorobenzene makes a very electrophilic aryl halide. Reaction with hydrazine gives a useful reagent.

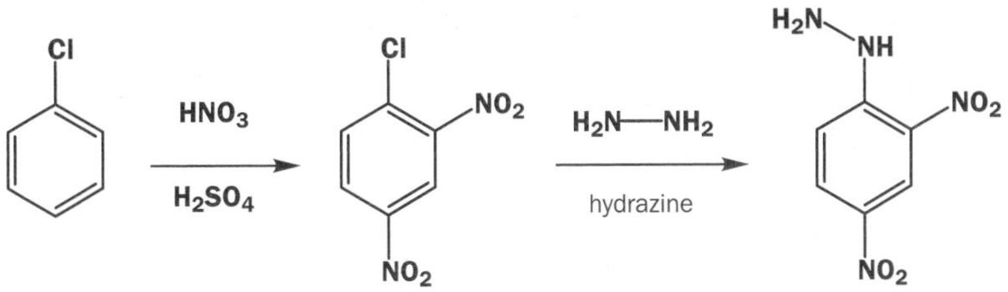

It also makes a very toxic one! The reason is the same as with Michael acceptors—this compound is carcinogenic.

This compound forms coloured crystalline imines (hydrazones) with most carbonyl compounds—before the days of spectroscopy these were used to characterize aldehydes and ketones (see p. 351).

The intermediate in the addition–elimination mechanism

What evidence is there for intermediates like the ones we have been using in this section? When reactions like this last example are carried out, a purple colour often appears in the reaction mixture and then fades away. In some cases the colour is persistent and thought to be due to the intermediate. Here is an example with RO⁻ attacking a nitrated aniline.

delocalization of the negative charge in the intermediate

This intermediate is persistent because neither potential leaving group (NR_2 or OR) is very good. If the nucleophile is part of the same molecule, the intermediate becomes a stable cyclic compound and can be isolated. It is more stable because neither leaving group can get away from the molecule as it is tethered by the rest of the ring. Notice that there are *three* active nitro groups in this molecule all stabilizing the negative charge.

What is the nature of this intermediate? We can best answer that by comparing the ^{13}C NMR spectra of three species: benzene itself; the simplest version of our carbanion intermediate (that is, with no substituents); and the simplest version of the cationic intermediate in electrophilic aromatic substitution. Direct protonation of benzene gives this last compound.

The intermediate in nucleophilic substitution cannot be made by adding H^- to benzene as no reaction occurs. Olah, the carbocation pioneer (p. 408), managed to make it by treating dihydrobenzene (cyclohexadiene) with a strong base. Deprotonation creates the anion.

Here are the details of the NMR spectra side-by-side with those of benzene. We shall use a summary structure for each ion showing delocalized charges around the five trigonal atoms in the ring. You may judge whether the NMR spectra justify these structures.

These results are very striking. The shifts of the *meta* carbons in both ions are very slightly different from those of benzene itself (about 130 p.p.m.). But the *ortho* and *para* carbons in the cation have gone downfield to much larger shifts while the *ortho* and *para* carbons in the anion have gone upfield to much smaller shifts.

The differences are very great—about 100 p.p.m. between the cation and the anion! It is very clear from these spectra that the ionic charge is delocalized almost exclusively to the *ortho* and *para* carbons in both cases. The alternative structures in the margin show this delocalization.

This means that stabilizing groups, such as nitro or carbonyl in the case of the anion, must be on the *ortho* or *para* carbons to have any effect. A good illustration of this is the selective displacement of

► *A reminder.* A larger shift means less electronic shielding and a smaller shift more electronic shielding.

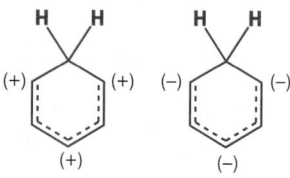

one chlorine atom out of these two. It is the *ortho* chlorine group that is lost and the *meta* one that is retained.

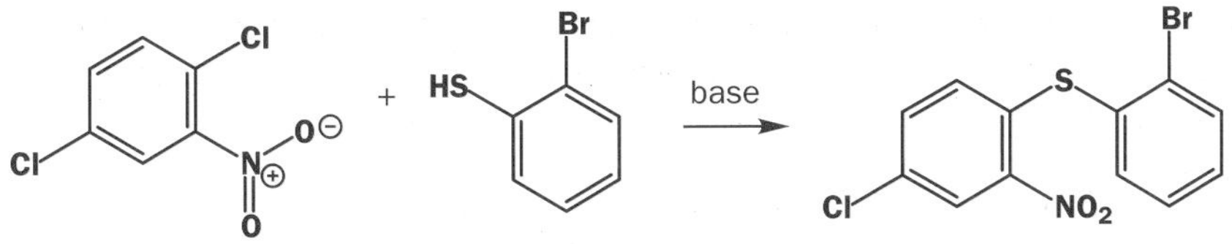

The mechanism works well if we attack the chlorine position *ortho* to the nitro group with the anion of the thiol nucleophile as the negative charge can then be pushed into the nitro group. Satisfy yourself that you cannot do this if you attack the other chlorine position. This is a very practical re-action and is used in the manufacture of a tranquillizing drug.

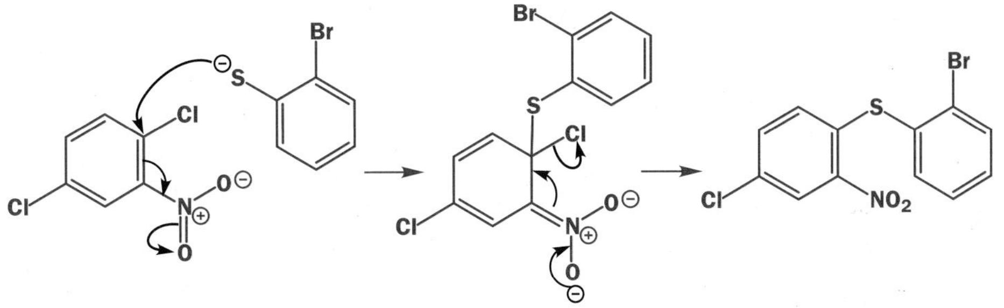

The leaving group and the mechanism

In the first nucleophilic aromatic substitution that we showed you, we used fluoride ion as a leaving group. Fluoride works very well in these reactions, and even such a simple compound as 2-nitro-fluorobenzene reacts efficiently with a variety of nucleophiles, as in these examples.

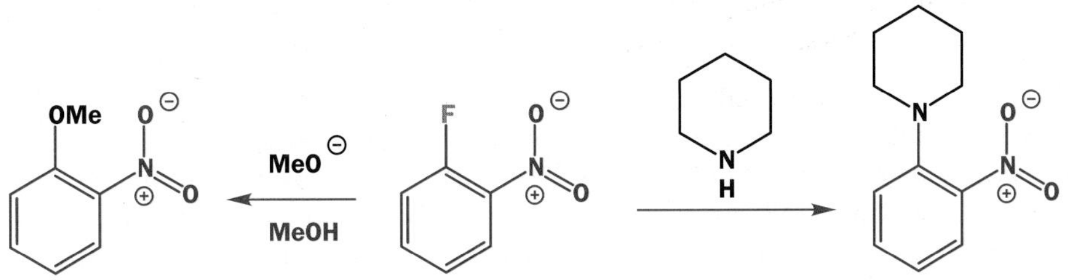

The same reactions happen with the other 2-nitro-halobenzenes but less efficiently. The fluoro-compound reacts about 10^2–10^3 times faster than the chloro- or bromo-compounds and the iodo-compound is even slower.

reactivity of 2-halo-1-nitrobenzenes in nucleophilic aromatic substitution

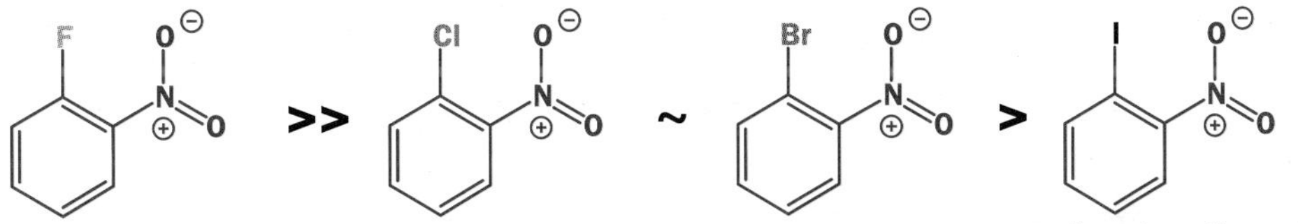

fastest reactions   slowest reactions

This ought to surprise you. When we were looking at other nucleophilic substitutions such as those at the carbonyl group or saturated carbon, we never used fluoride as a leaving group! The C–F bond is very strong—the strongest of all the single bonds to carbon—and it is difficult to break.

these reactions are not used:

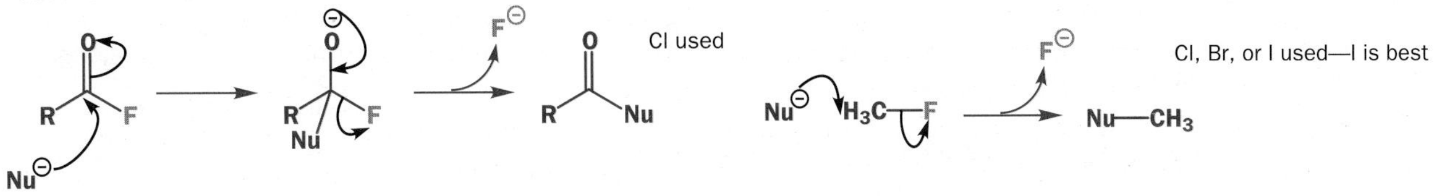

►

Azide is a good nucleophile because of its shape—it is like a needle—and because it is equally nucleophilic at either end. We discussed azides in Chapter 17, p. 000.

So why is fluoride often preferred in nucleophilic aromatic substitution and why does it react faster than the other halogens when the reverse is true with other reactions? You will notice that we have *not* said that fluoride is a better leaving group in nucleophilic aromatic substitution. It isn't! The explanation depends on a better understanding of the mechanism of the reaction. We shall use azide ion as our nucleophile because this has been well studied, and because it is one of the best.

The mechanism is exactly the same as that we have been discussing all along—a two-stage addition–elimination sequence. In a two-step mechanism, one step is slower and rate-determining; the other is unimportant to the rate. You may guess that, in the mechanism for nucleophilic aromatic substitution, it is the first step that is slower because it disturbs the aromaticity. The second step restores the aromaticity and is faster. The effect of fluoride, or any other leaving group, can only come from its effect on the first step. How good a leaving group it might be does not matter: the rate of the second step—the step where fluoride leaves—has no effect on the overall rate of the reaction.

fluoride accelerates this step because it is very electronegative   the intermediate

►

Note carefully that this is an *inductive* effect: there are no arrows to be drawn to show how fluorine withdraws electrons—it does it just by polarizing C–F bonds towards itself. Contrast the electron-withdrawing effect of the nitro group, which works (mainly) by conjugation.

Fluoride does, in fact, slow down the second step (relative to Cl⁻, say), but it accelerates the first step simply by its enormous inductive effect. It is the most electronegative element of all and it stabilizes the anionic intermediate, assisting the acceptance of electrons by the benzene ring.

A dramatic illustration of the effect of fluorine is the reactions of benzene rings with more than one fluorine substituent. These undergo nucleophilic substitution without any extra conjugation from electron-withdrawing groups. All the fluorine atoms that are not reacting help to stabilize the negative charge in the intermediate.

delocalized negative charge
stabilized by two fluorines on ring

Intellectual health warning!

Some textbooks tell you that nucleophilic aromatic substitution doesn't happen with ordinary aryl halides because of conjugation between the lone pairs of the halide and the aromatic system.

This is supposed to stop the reaction by making the C–Br bond stronger. This is nonsense. The reaction doesn't happen on simple aryl halides because there is no available mechanism. It is easy to show that the false textbook reason is wrong. The conjugation in this nitro compound is much better than in bromobenzene, so it should be even less reactive.

In fact, as you now know, this compound is much *more* reactive towards nucleophiles. The false textbook reason would also suggest that fluoride would work really badly because this same conjugation is stronger with fluoride than with the other halogens as its p orbitals are the right size ($2sp^2$) to conjugate with carbon p orbitals. Again, you already know the opposite to be true.

The strength of the bond to the leaving group does not affect the efficiency of nucleophilic aromatic substitution because that bond is not broken in the rate-determining step. Understand the mechanism and it all becomes clear.

The activating anion-stabilizing substituent

We have used nitro groups very extensively so far and that is only right and proper as they are the best at stabilizing the anionic intermediate. Others that work include carbonyl, cyanide, and sulfur-based groups such as sulfoxides and sulfones. Here is a direct comparison for the displacement of bromide ion by the secondary amine piperidine. First the reaction with a carbonyl group.

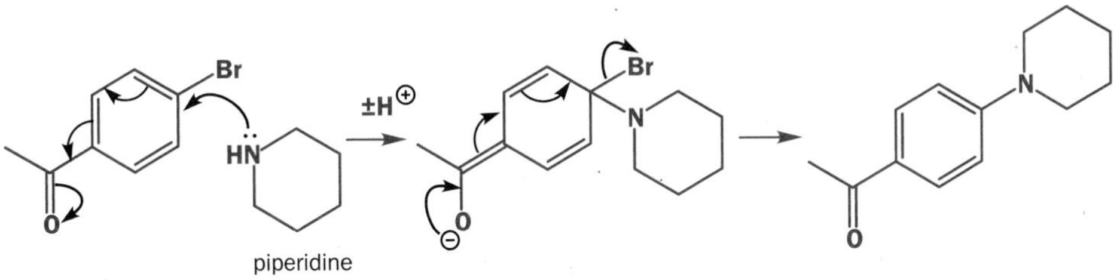

piperidine

Now we are going to give the rates for the same reaction but with different activating groups. The mechanism is the same in each case; the only difference is the electron-withdrawing power of the activating group. You recall that this is vital for the rate-determining first step and for stabilization of the intermediate. The symbol Z represents the anion-stabilizing group and the margin shows what Z might be. The numbers are the relative rates compared with Z = nitro.

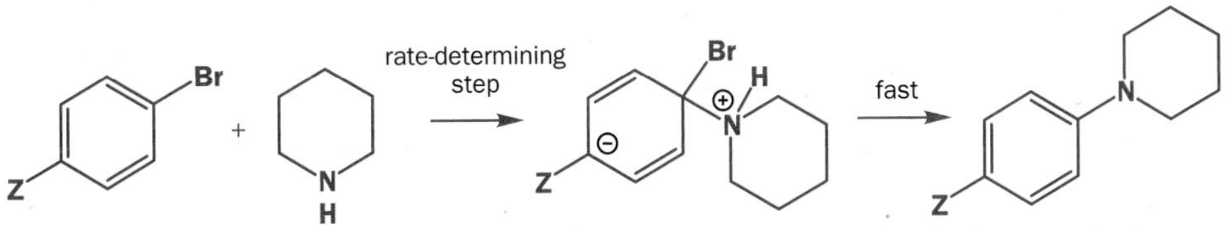

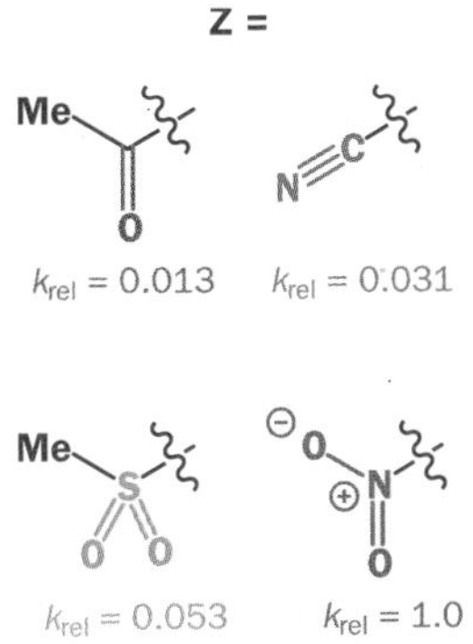

$Z =$

$k_{rel} = 0.013$ $k_{rel} = 0.031$

$k_{rel} = 0.053$ $k_{rel} = 1.0$

All the compounds react more slowly than the nitro-compound. We have already mentioned (Chapters 8 and 22) the great electron-withdrawing power of the nitro group—here is a new measure of that power. The sulfone reacts twenty times slower, the nitrile thirty times slower, and the ketone a hundred times slower.

The nitro is the best activating group, but the others will all perform well especially when combined with a fluoride rather than a bromide as the leaving group. Here are two reactions that work well in a preparative sense with other anion-stabilizing groups. Note that the trifluoromethyl group works by using only its powerful inductive effect.

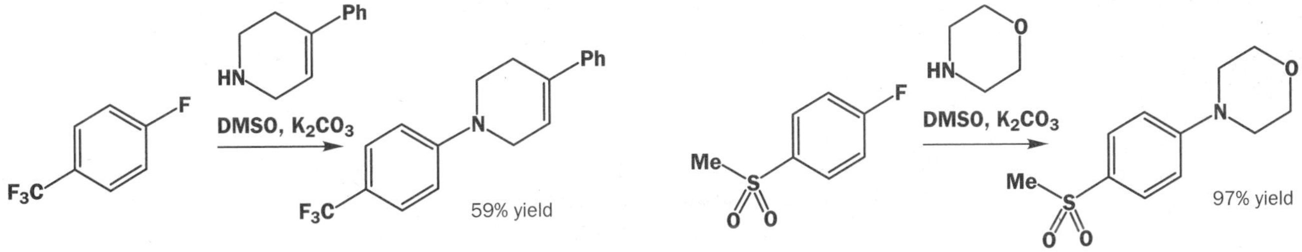

59% yield

97% yield

● **To summarize**

Any anion-stabilizing (electron-withdrawing) group *ortho* or *para* to a potential leaving group can be used to make nucleophilic aromatic substitution possible.

Some medicinal chemistry—preparation of an antibiotic

We want to convince you that this chemistry is useful and also that it works in more complicated molecules so we are going to describe in part the preparation of a new antibiotic, ofloxacin. The sequence starts with an aromatic compound having four fluorine atoms. Three are replaced specifically by different nucleophiles and the last is present in the antibiotic itself. As a reminder of the first section of this chapter, the preparation also involves a conjugate substitution. The structure of ofloxacin (below) highlights the remaining fluorine atom and (in black) the four bonds made by reactions discussed in this chapter. Underneath is the starting material with its four fluorine atoms.

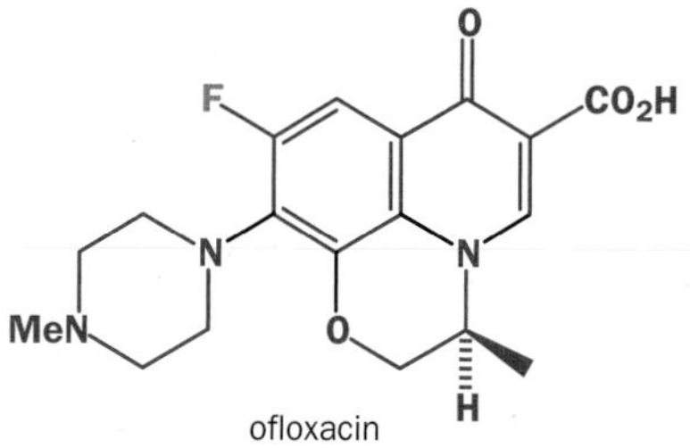

ofloxacin

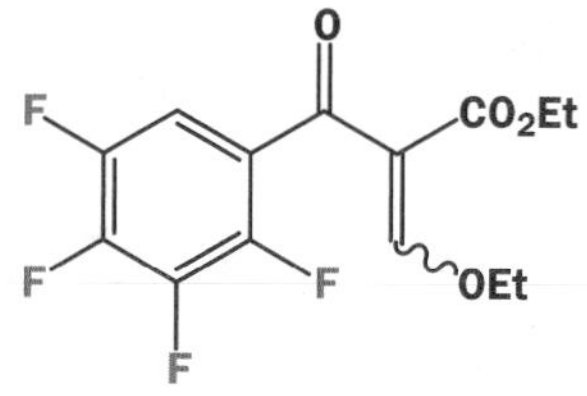

starting material for
ofloxacin synthesis

The preparation of the starting material involves reactions that we will meet later in the book and is described in Chapter 28. The next reaction is the conjugate substitution. An amino alcohol is used as the nucleophile and it does a conjugate addition to the double bond. Notice that it is the more nucleophilic amino group that adds to the alkene, not the hydroxyl group. And, when the negative charge comes back to complete the conjugate substitution, the better leaving group is alkoxide rather than a very unstable amine anion.

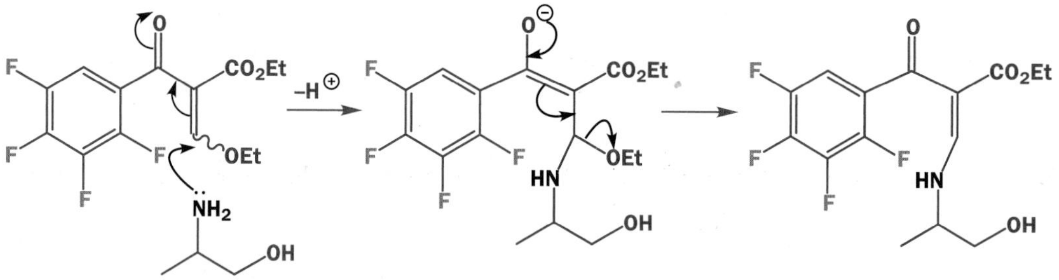

The next step is the first nucleophilic aromatic substitution. The amino group attacks in the position *ortho* to the carbonyl group so that an enolate intermediate can be formed. When the charge returns, the first fluoride is expelled.

> We had to draw the arrows going the long way round the ring in the addition step because we happened to draw the double bonds in those positions in the starting material: this is the sort of thing you will find happens when you write mechanisms—there is no significance in it and it doesn't matter which way round the arrows go.

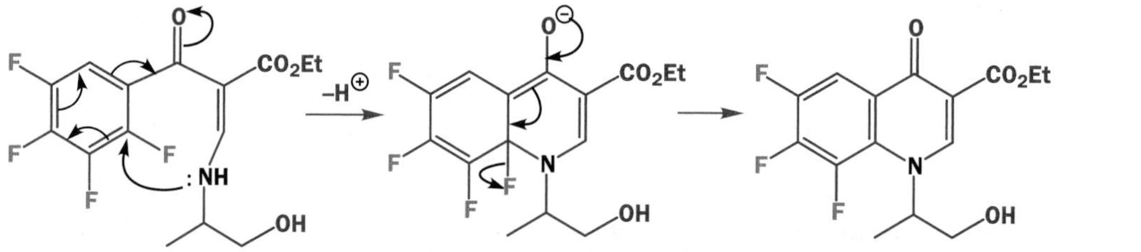

Treatment with base (NaH can be used) now converts the OH group into an alkoxide and it does the next aromatic nucleophilic substitution. In this reaction we are attacking the position *meta* to the ketone so we cannot put the negative charge on the oxygen atom. The remaining three fluorines must stabilize it by the inductive effect we described earlier.

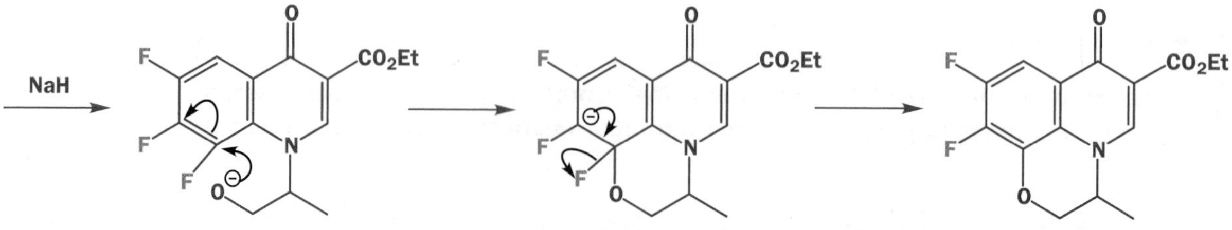

negative charge stabilized by
inductive effect only

When this charge returns to restore the benzene ring, the second fluoride is expelled and only two are left. One of these is now displaced by, for the first time, an external nucleophile—an amine. It is easy to predict which one because of the need to stabilize the charge in the intermediate.

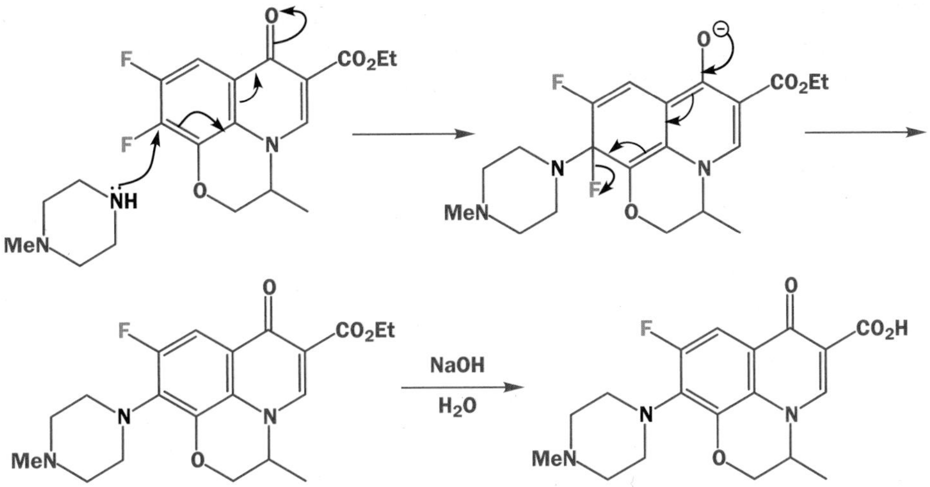

This displaces the third fluorine and all that is left is to hydrolyse the ester to the free acid with aqueous base (Chapter 12). Every single reaction in this quite complicated sequence is one that you have met earlier in the book, and it forms a fitting climax to this section on the addition–elimination mechanism for aromatic nucleophilic substitution. We now need to mention two other less important possibilities.

The S_N1 mechanism for nucleophilic aromatic substitution— diazonium compounds

When primary amines are treated with nitrous acid (HONO), or more usually with a nitrite salt or an alkyl nitrite in acid solution, an unstable **diazonium salt** is formed. You met diazonium salts in Chapter 22 undergoing coupling reactions to give azo compounds, but they can do other things as well. First, a reminder of the mechanism of formation of these diazonium salts. The very first stage is the formation of the reactive species NO$^+$.

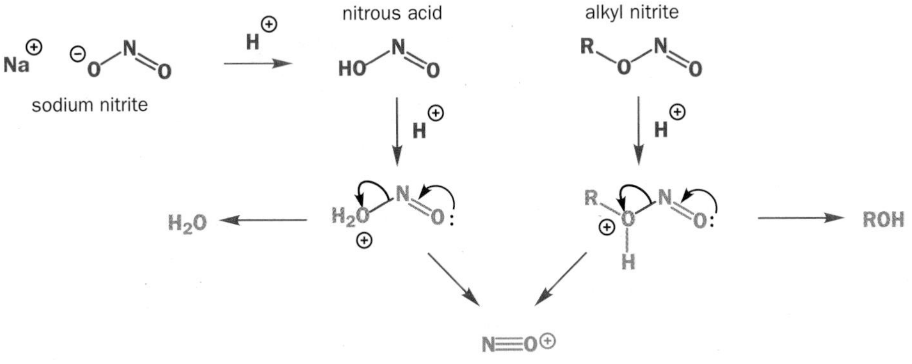

The NO$^+$ cation then attacks the lone pair of the amine and dehydration follows. The mechanism is quite simple—it just involves a lot of proton transfers! There is, of course, an anion associated with the nitrogen cation, and this will be the conjugate base (Cl$^-$ usually) of the acid used to form NO$^+$.

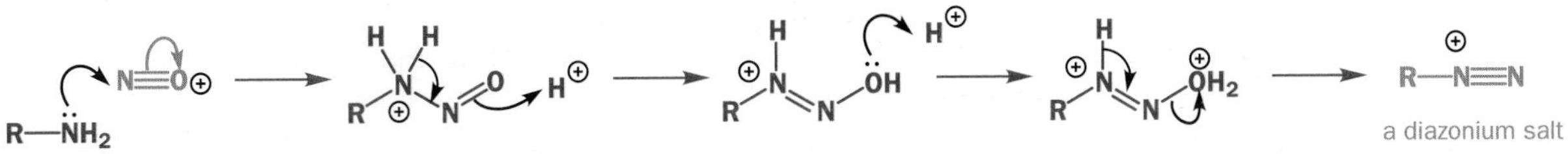

If R is an alkyl group, this diazonium salt is very unstable and immediately loses nitrogen gas to give a planar carbocation, which normally reacts with a nucleophile in an S_N1 process (Chapter 17) or loses a proton in an E1 process (Chapter 19). It may, for example, react with water to give an alcohol.

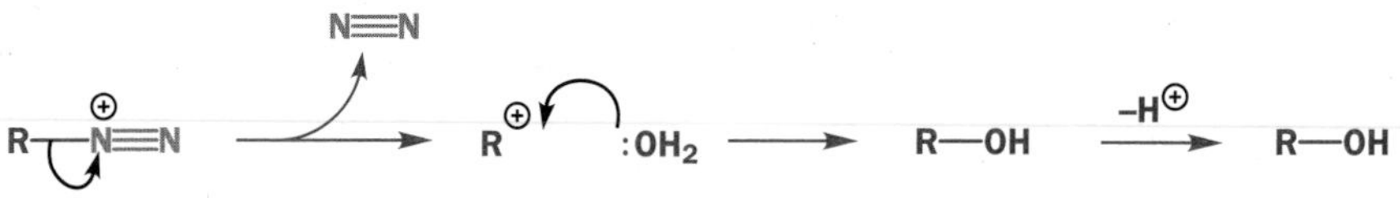

If R is an aryl group, the carbocation is much less stable (for the reasons we discussed earlier—chiefly that the empty orbital is an sp^2 rather than a p orbital) and that makes the loss of nitrogen slower. If the diazotization is done at lowish temperatures (just above 0 °C, classically at 5 °C), the diazonium salt is stable and can be reacted with various nucleophiles.

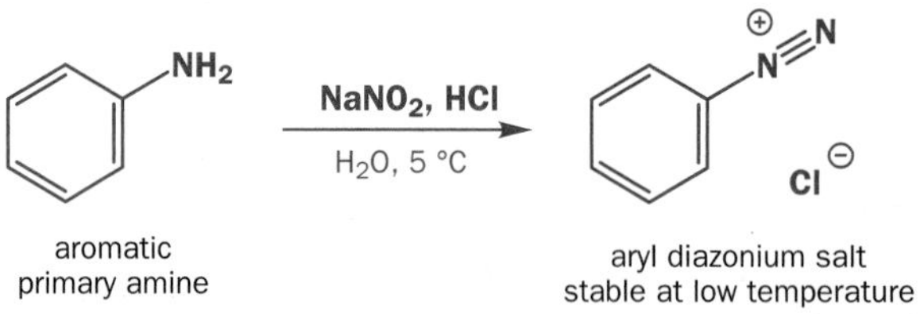

aromatic
primary amine

aryl diazonium salt
stable at low temperature

If the aqueous solution is heated, water again acts as the nucleophile and a phenol is formed from the amine. The aryl cation is an intermediate and this is an S_N1 reaction at an aromatic ring.

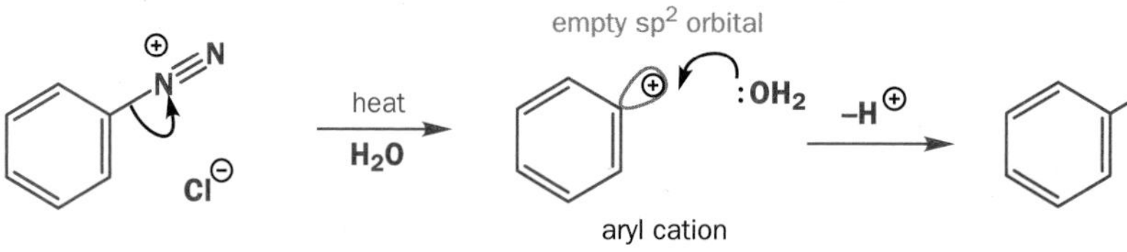

aryl cation

The point of this reaction is that it is rather difficult to add an oxygen atom to a benzene ring by the normal electrophilic substitution as there is no good reagent for 'OH$^+$'. A nitrogen atom can be added easily by nitration, and reduction and diazotization provide a way of replacing the nitro group by a hydroxyl group.

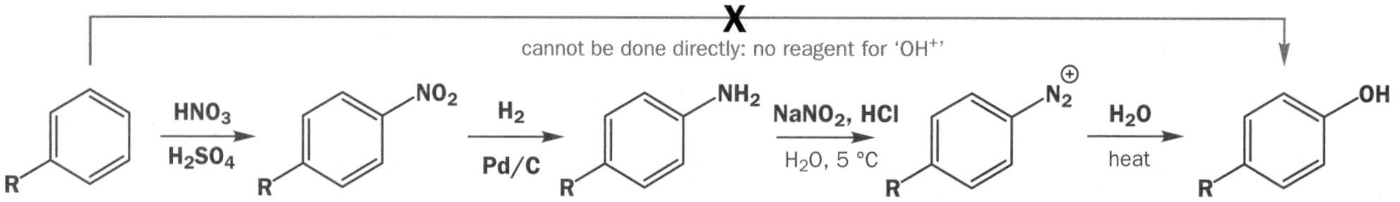

This is a practical sequence and is used in manufacturing medicines. An example is the drug thymoxamine (Moxysylyte), which has a simple structure with ester and ether groups joined to a benzene ring through their oxygen atoms.

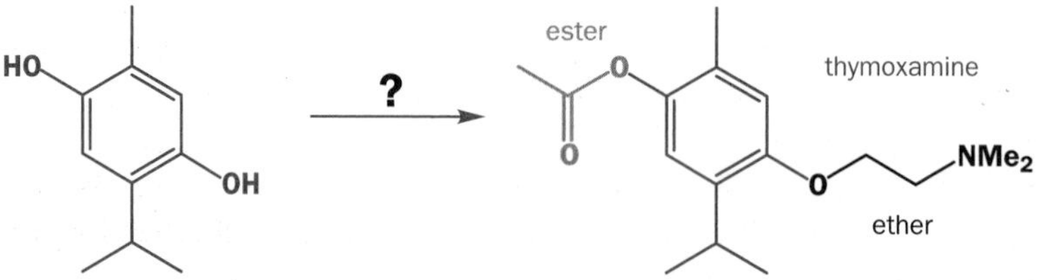

It seems obvious to make this compound by alkylation and acylation of a dihydroxybenzene. But how are we to make sure that the right phenol is acylated and the right phenol alkylated? French pharmaceutical chemists had an ingenious answer. Start with a compound having only one OH group, alkylate that, and only then introduce the second using the diazonium salt method. They used a simple phenol and introduced nitrogen as a nitroso (NO) rather than a nitro (NO$_2$) group. This means using the same reagent, HONO, as we used for the diazotization. These were the first two steps.

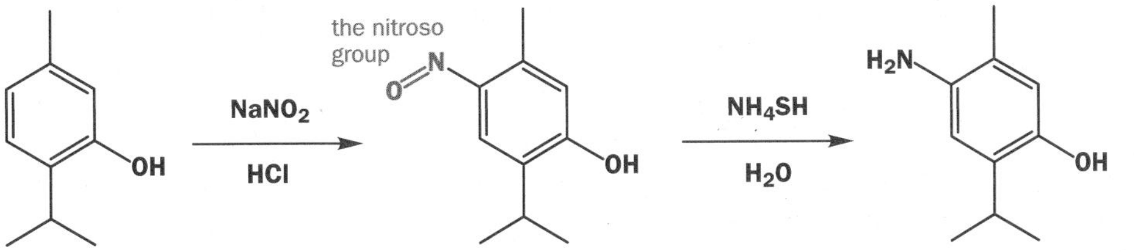

The reduction of NO is easier than that of NO$_2$, and HS$^-$ is enough to do the job. The amine can now be converted to an amide to lessen its nucleophilicity so that alkylation of the phenol occurs cleanly.

Finally, the amide must be hydrolysed, the amino converted into an OH group by diazotization and hydrolysis, and the new phenol acetylated.

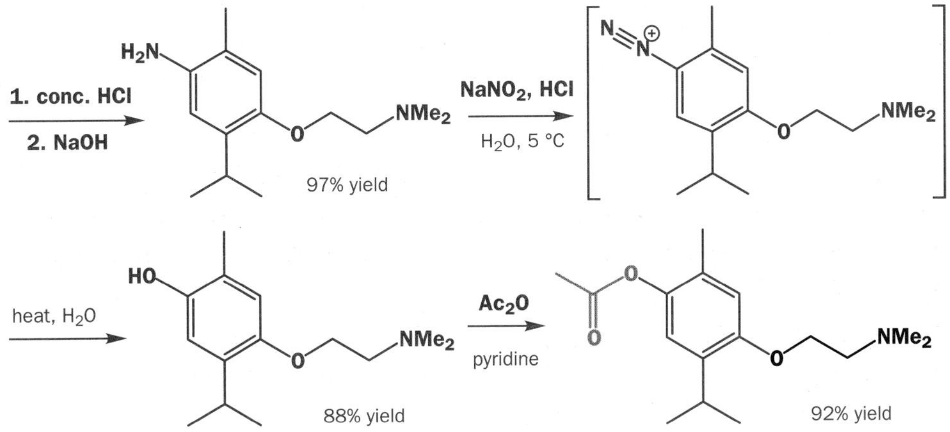

This is yet another synthesis in which almost every step is a reaction that you have already met in this book! There are three nucleophilic substitutions at the carbonyl group, one S$_N$2 reaction, one electrophilic and one nucleophilic aromatic substitution (the latter being an S$_N$1 reaction), and a reduction. The chemistry you already know is enough for a patented manufacture of a useful drug.

Other nucleophiles

Because aryl diazonium salts are reasonably stable, other nucleophiles may be introduced to capture the aryl cation when the diazonium salt is heated. Among these, iodide ion is important as it allows the preparation of aryl iodides in good yield. These compounds are not so easy to make by electrophilic substitution (Chapter 22) as aryl chlorides or bromides because iodine is not reactive enough to attack benzene rings. Aryl iodides are useful in the more modern palladium chemistry of the Heck reaction, which you will meet in Chapter 48.

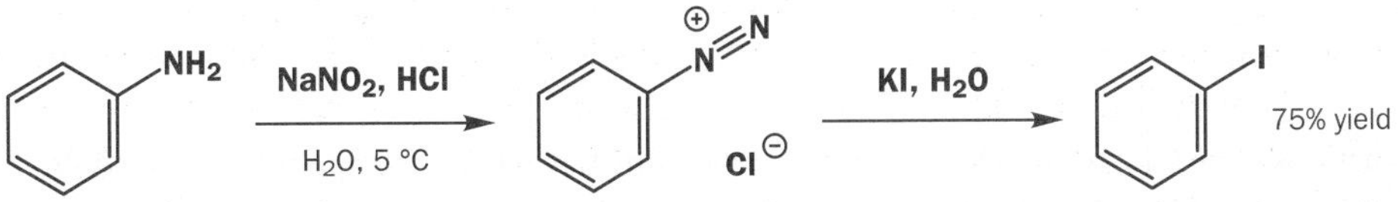

Other nucleophiles, such as chloride, bromide, and cyanide, are best added with copper(I) salts. These reactions are almost certainly radical in character (Chapter 39). Since aromatic amines

The nitrosation uses the same intermediate (NO$^+$) used in the diazotization and is really very like the nitrosation of the enol we described on p. 000. There it tautomerized to an oxime—it can't do that here. Make sure you can draw a mechanism for this reaction and explain why the NO group goes in that position.

Reduction of –NO$_2$ groups is discussed in Chapter 24, p. 626.

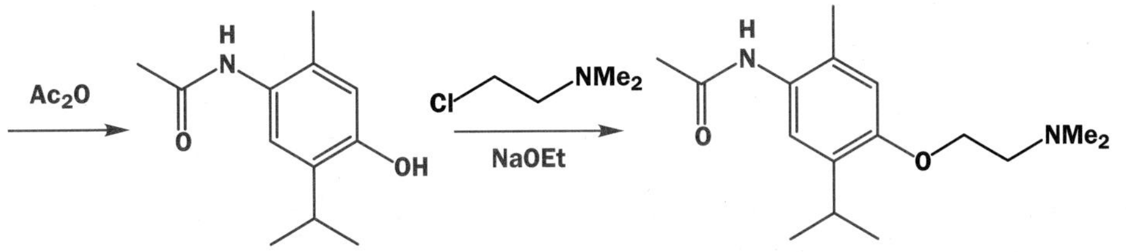

This is 'protection' of the amine as an amide: protecting groups are discussed in the next chapter. The amino group is more nucleophilic than the phenol so it would be alkylated if we did not protect it. Protection is selective for the same reason—the amino group attacks the anhydride preferentially. Now the amide is less nucleophilic than the phenol so alkylation occurs at oxygen. You should draw mechanisms for all these steps and make sure you understand why they happen in the way that they do.

are usually made by reduction of nitro-compounds, a common sequence of reactions goes like this.

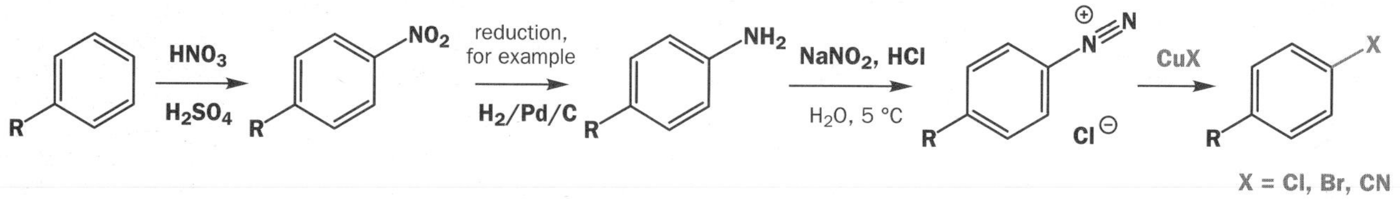

X = Cl, Br, CN

A reaction that may seem rather pointless is the reduction of diazonium salts, that is, the replacement of N_2^+ by H. A good reagent is H_3PO_2.

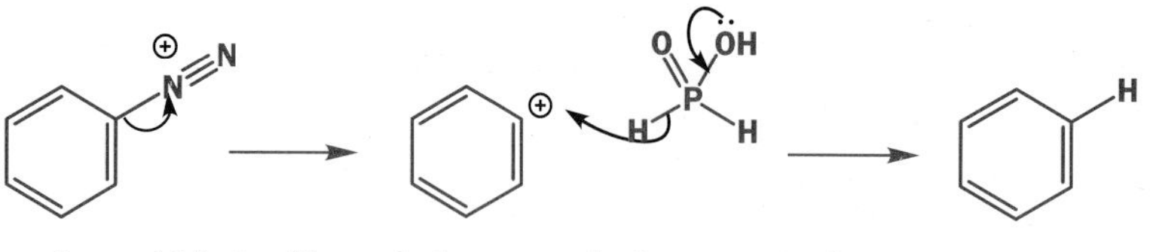

It would indeed be pointless to make benzene in this way, but this reaction allows the introduction of an amino group for the purpose of directing an electrophilic substitution and then its removal once its job is done. Here is a famous example.

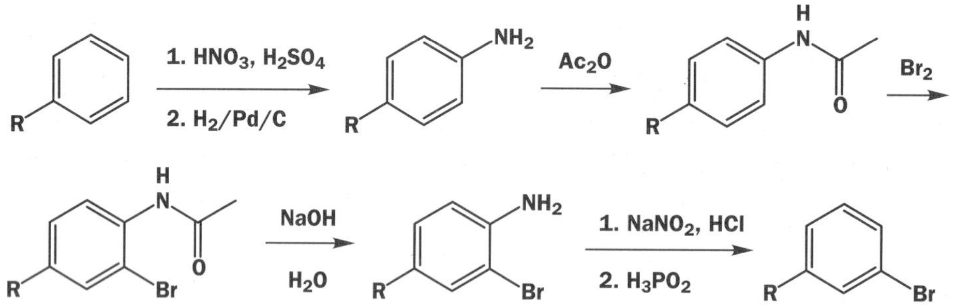

This chemistry is very long-winded, and now rather old-fashioned. Difficult-to-make substitution patterns are more usually set up using variants of the directed metallation (ortholithiation) chemistry we introduced in Chapter 9.

Nitration puts in a substituent *para* to the alkyl group, which, after reduction, becomes a powerful *ortho* director so that the bromine is directed *meta* to the original alkyl group (Chapter 22). Removal of the amino group by reduction allows the preparation of *meta* bromo alkyl benzenes that cannot be made directly.

The benzyne mechanism

There is one last mechanism for aromatic nucleophilic substitution and you may well feel that this is the weirdest mechanism you have ever seen with the most unlikely intermediate ever! For our part, we hope to convince you that this mechanism is not only possible but useful.

At the start of the section on 'Nucleophilic aromatic substitution' we said that 'the displacement of bromide from bromobenzene with hydroxide ion does not occur'. That statement is not quite correct. Substitution by hydroxide on bromobenzene can occur but only under the most vigorous conditions—such as when bromobenzene and NaOH are melted together (fused) at very high temperature. A similar reaction with the very powerful reagent $NaNH_2$ (which supplies NH_2^- ion) also happens, at rather lower temperature.

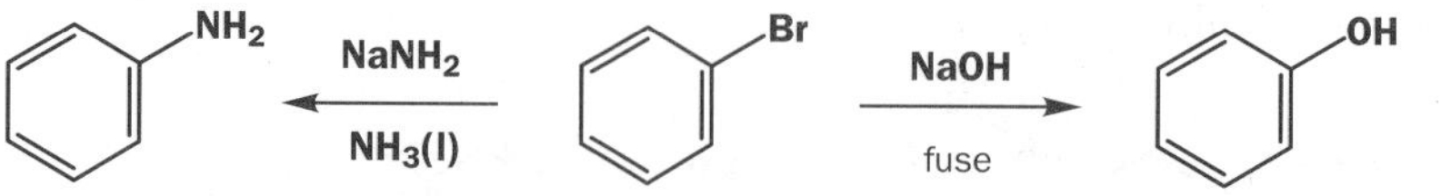

These reactions were known for a long time before anyone saw what was happening. They do not happen by an S_N2 mechanism, as we explained at the start of the section, and they can't happen by the addition–elimination mechanism because there is nowhere to put the negative charge in the intermediate. The first clue to the true mechanism is that all the nucleophiles that react in this way are very basic, and it was suggested that they start the reaction off by removing a proton *ortho* to the leaving group.

The carbanion is in an sp² orbital in the plane of the ring. Indeed, this intermediate is very similar to the aryl cation intermediate in the S$_N$1 mechanism from diazonium salts. That had no electrons in the sp² orbital; the carbanion has two.

full sp² orbital

Why should this proton be removed rather than any other? The bromine atom is electronegative and the C–Br bond is in the plane of the sp² orbital and removes electrons from it. The stabilization is nonetheless weak and only strong bases will do this reaction.

The next step is the loss of bromide ion in an elimination reaction. This is the step that is difficult to believe as the intermediate we are proposing looks impossible. The orbitals are bad for the elimination too—it is a syn- rather than an anti-periplanar elimination. But it happens.

benzyne

The intermediate is called benzyne as it is an alkyne with a triple bond in a benzene ring. But what does this triple bond mean? It certainly isn't a normal alkyne as these are linear. In fact one π bond is normal—it is just part of the aromatic system. One π bond—the new one—is abnormal and is formed by overlap of two sp² orbitals outside the ring. This external π bond is very weak and benzyne is a very unstable intermediate. Indeed, when the structure was proposed few chemists believed it and some pretty solid evidence was needed before they did. We shall come to that shortly, but let us first finish the mechanism. Unlike normal alkynes, benzyne is electrophilic as the weak third bond can be attacked by nucleophiles.

Notice the symmetry in this mechanism. Benzyne is formed from an *ortho* carbanion and it gives an *ortho* carbanion when it reacts with nucleophiles. The whole mechanism from bromobenzene to aniline involves an elimination to give benzyne followed by an addition of the nucleophile to the triple bond of benzyne. In many ways, this mechanism is the reverse of the normal addition–elimination mechanism for nucleophilic aromatic substitution and it is sometimes called the **elimination–addition mechanism**.

the elimination step

ortho-carbanion   benzyne

the addition step

benzyne   *ortho*-carbanion

Any nucleophile basic enough to remove the *ortho* proton can carry out this reaction. Known examples include oxyanions, amide anions (R$_2$N⁻), and carbanions. The rather basic alkoxide *t*-butoxide will do the reaction on bromobenzene if the potassium salt is used in the dipolar aprotic solvent DMSO to maximize reactivity.

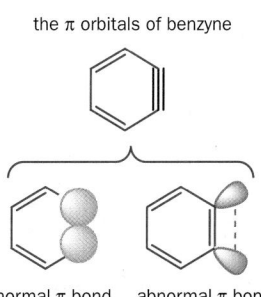

the π orbitals of benzyne

normal π bond   abnormal π bond
two p orbitals   two sp² orbitals
inside the ring outside the ring

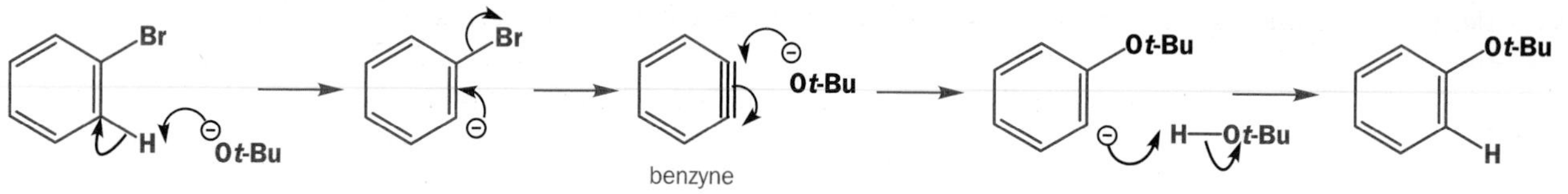

Evidence for benzyne as an intermediate

As you would expect, the formation of benzyne is the slow step in the reaction so there is no hope of isolating benzyne from the reaction mixture or even of detecting it spectroscopically. However, it can be made by other reactions where there are no nucleophiles to capture it. The most important is a diazotization reaction.

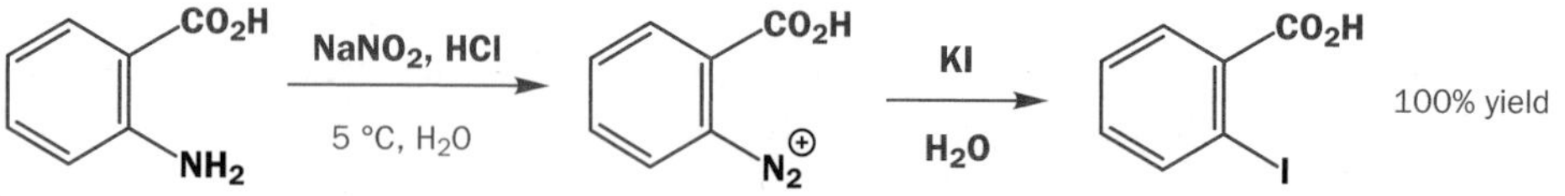

This diazotization is particularly efficient as you can see by the quantitative yield of the *ortho*-iodo-acid on capture of the diazonium salt with iodide ion. However, if the diazonium salt is neutralized with NaOH, it gives a zwitterion with the negative charge on the carboxylate balancing the positive charge on the diazonium group. This diazotization is usually done with an alkyl nitrite in an organic solvent (here, dimethoxyethane, DME) to avoid the chance that nucleophiles such as chloride or water might capture the product. When the zwitterion is heated it decomposes in an entropically favourable reaction to give carbon dioxide, nitrogen, and benzyne.

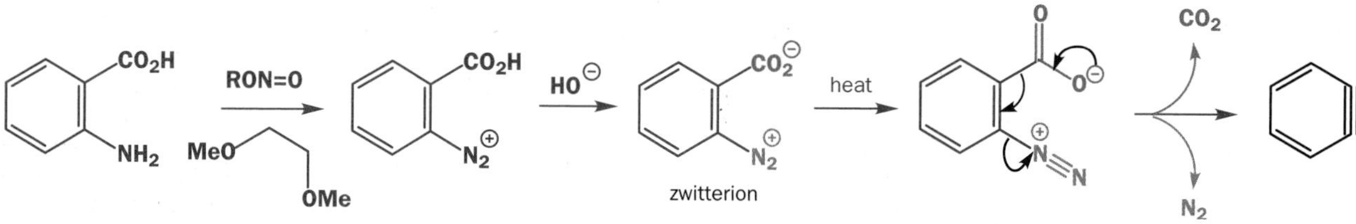

You can't isolate the benzyne because it reacts with itself to give a benzyne dimer having a four-membered ring between two benzene rings. If the zwitterion is injected into a mass spectrometer, there is a peak at 152 for the dimer but also a strong peak at 76, which is benzyne itself. The lifetime of a particle in the mass spectrometer is about 20 ns (nanosecond = 10^{-9} second) so benzyne can exist for at least that long in the gas phase.

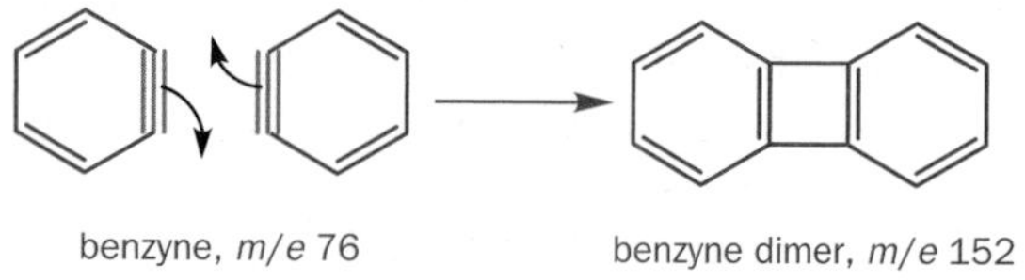

benzyne, *m/e* 76   benzyne dimer, *m/e* 152

Benzyne produced from the zwitterion can also be captured by dienes in a Diels–Alder reaction (see Chapter 35). But this merely shows that benzyne can exist for a short time. It does not at all prove that benzyne is an intermediate in aromatic substitution reactions. Fortunately, there is very convincing evidence for this as well.

There is one very special feature of the benzyne mechanism. The triple bond could be attacked by nucleophiles at either end. This is of no consequence when we are dealing with bromobenzene as the products would be the same, but we can make the ends of the triple bond different and then we see something interesting. *ortho*-Chloro aryl ethers are easy to prepare by chlorination of the ether (Chapter 22). When these compounds are treated with NaNH$_2$ in liquid ammonia, a single amine is formed in good yield.

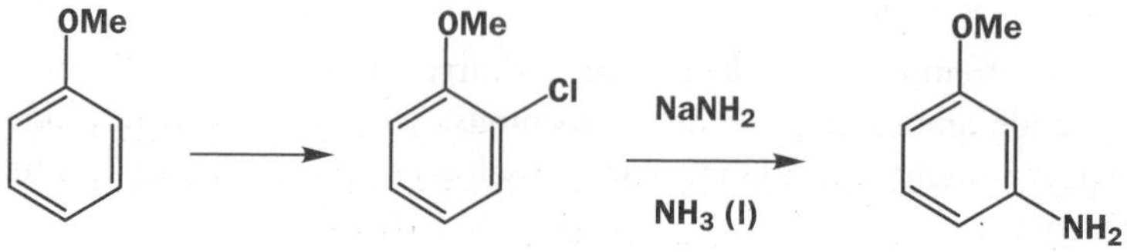

There is no mistake in this scheme. The amine is really at the *meta* position even though the chlorine was at the *ortho* position. It would be very difficult to explain this by any other mechanism but very easy to explain using a benzyne mechanism. Using the same two steps that we have used before, we can write this.

the elimination step

the addition step

That shows *how* the *meta* product might be formed, but *why* should it be formed? Attack could also occur at the *ortho* position, so why is there no *ortho* product? There are two reasons: electronic and steric. Electronically, the anion next to the electronegative oxygen atom is preferred, because oxygen is inductively electron-withdrawing. The same factor facilitates deprotonation next to Cl in the formation of the benzyne. Sterically, it is better for the amide anion to attack away from the OMe group rather than come in alongside it. Nucleophilic attack on a benzyne has to occur in the plane of the benzene ring because that is where the orbitals are. This reaction is therefore very sensitive to steric hindrance as the nucleophile must attack in the plane of the substituent as well.

> Oxygen is an electron-withdrawing group here because the anion is formed in the plane of the ring and has nothing to do with the benzene's π orbitals. Of course, as far as the π orbitals are concerned, oxygen is electron-*donating* because of its lone pairs.

> Steric hindrance is not nearly as important in electrophilic substitution or in nucleophilic substitution by the addition–elimination mechanism. In both of these reactions, the reagent is attacking the p orbital *at right angles* to the ring and is some distance from an *ortho* substituent.

attack close to OMe

no stabilization from OMe group

stabilization from OMe group

This is a useful way to make amino ethers with a *meta* relationship as both groups are *ortho, para*-directing and so the *meta* compounds cannot be made by electrophilic substitution. The alternative is the long-winded approach using a diazonium salt that was described in the previous section.

para-Disubstituted halides can again give only one benzyne and most of them give mixtures of products. A simple alkyl substituent is too far away from the triple bond to have much steric effect.

only one benzyne possible

about 50:50

If the substituent is an electron-repelling anion, then the *meta* product is formed exclusively because this puts the product anion as far as possible from the anion already there. This again is a useful result as it creates a *meta* relationship between two *ortho, para*-directing groups.

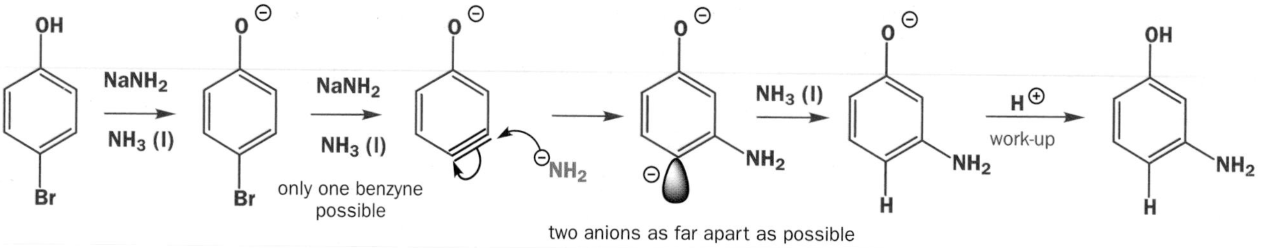

only one benzyne possible

two anions as far apart as possible

One case where selectivity of attack is no problem is in reactions with intramolecular nucleophiles. These cyclizations simply give the only possible product—the result of cyclization to the nearer end of the triple bond. One important example is the making of a four-membered ring. Only one benzyne can be formed.

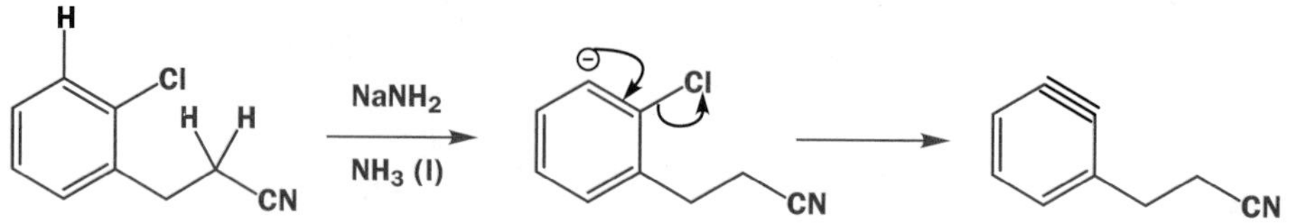

There are acidic protons next to the cyanide, and the amide ion is strong enough to form an 'enolate' by the removal of one of those. The enolate cyclizes on to the benzyne to give a four-membered ring. As it happens, the nucleophile adds to the position originally occupied by the chlorine, but that is not necessary.

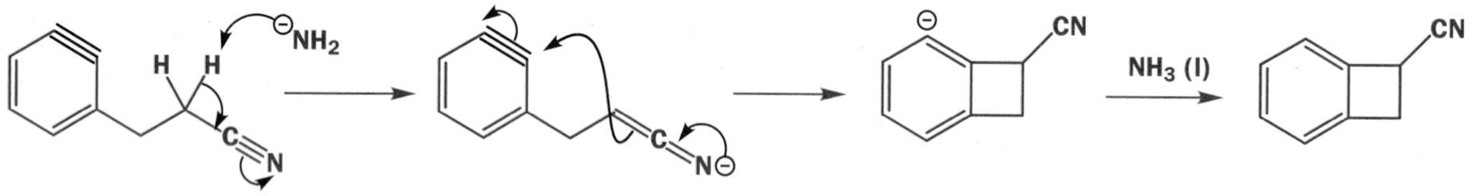

Nucleophilic attack on allylic compounds

We shall finish this chapter with some alkenes that are electrophilic, not because they are conjugated with another π system, but because they have a leaving group adjacent to them. We shall start with some substitution reactions with which you are familiar from Chapter 17. There we said that allyl bromide is about 100 times more reactive towards simple S_N2 reactions than is propyl bromide or other saturated alkyl halides.

The double bond stabilizes the S_N2 transition state by conjugation with the p orbital at the carbon atom under attack. This full p orbital (shown in yellow in the diagram below) forms a partial bond with the nucleophile and with the leaving group in the transition state. Any stabilization of the transition state will, of course, accelerate the reaction by lowering the energy barrier.

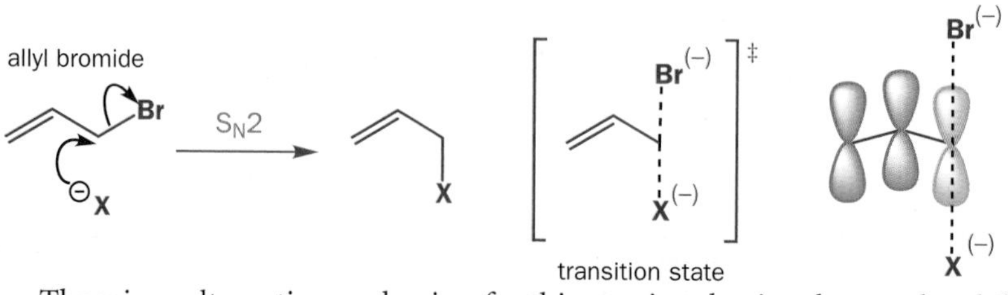

allyl bromide

transition state

There is an alternative mechanism for this reaction that involves nucleophilic attack on the alkene instead of on the saturated carbon atom. This mechanism leads to the same product and is often called the S_N2' (pronounced 'S-N-two-prime') mechanism.

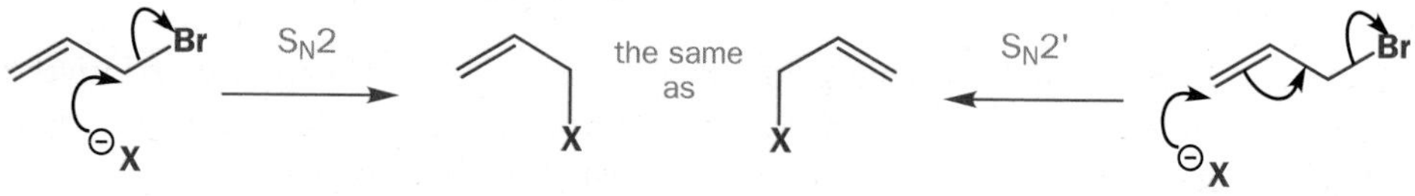

the same as

We can explain both mechanisms in a unified way if we look at the frontier orbitals involved. The nucleophile must attack an empty orbital (the LUMO) which we might expect to be simply σ* (C–Br) for the S_N2 reaction.

But this ignores the alkene. The interaction between π* (C=C) and the adjacent σ* (C–Br) will as usual produce two new orbitals, one higher and one lower in energy. The lower-energy orbital, π* + σ*, will now be the LUMO. To construct this orbital we must put all the atomic orbitals parallel and make the contact between π* + σ* a bonding contact.

LUMO constructed from π* + σ*

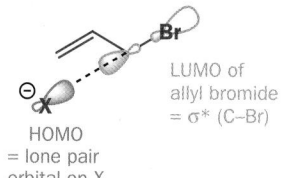

orbitals treated as for 'simple' S_N2

LUMO of allyl bromide = σ* (C–Br)

HOMO = lone pair orbital on X

π* (C=C) σ* (C–Br) → add π* + σ* → molecular LUMO → S_N2' nucleophilic attack can occur at the points marked with dotted black arrows S_N2' S_N2

LUMOs of localized bonds

▶
So far we have used the word 'allyl' to describe these compounds. Strictly, that word applies only to specific compounds CH_2=CH–CH_2X with no substituents other than hydrogen. Allyl is often used loosely to describe any compound with a functional group on the carbon atom *next to* the alkene. We shall use 'allylic' for that and 'allyl' only for the unsubstituted version.

If the allylic halide is unsymmetrically substituted, we can tell which process occurs and the normal result is that nucleophilic attack occurs at the less hindered end of the allylic system whether that means S_N2 or S_N2'. This important allylic bromide, known as 'prenyl bromide', normally reacts entirely via the S_N2 reaction.

prenyl bromide reacts like this   and not like this

The two ends of the allylic system are contrasted sterically: direct (S_N2) attack is at a primary carbon while allylic (S_N2') attack is at a tertiary carbon atom so that steric hindrance favours the S_N2 reaction. In addition, the number of substituents on the alkene product means that the S_N2 product is nearly always preferred—S_N2 gives a trisubstituted alkene while the S_N2' product has a less stable monosubstituted alkene.

An important example is the reaction of prenyl bromide with phenols. This is simply carried out with K_2CO_3 in acetone as phenols are acidic enough ($pK_a \sim 10$) to be substantially deprotonated by carbonate. The product is essentially entirely from the S_N2 route, and is used in the Claisen rearrangement (Chapter 36).

If we make the two ends of the allyl system more similar, say one end primary and one end secondary, things are more equal. We could consider the two isomeric butenyl chlorides.

1-chlorobut-2-ene S_N2

3-chlorobut-1-ene S_N2'

1-chlorobut-2-ene S_N2'

3-chlorobut-1-ene S_N2

All routes look reasonable, though we might again prefer attack at the primary centre kinetically and the disubstituted alkene thermodynamically and this is the usual outcome. The reactions in the left-hand box are preferred to those in the right-hand box. But there is no special preference for the S_N2 over the S_N2' mechanism or vice versa—the individual case decides. If we react the secondary butenyl chloride with an amine we get the S_N2' mechanism entirely.

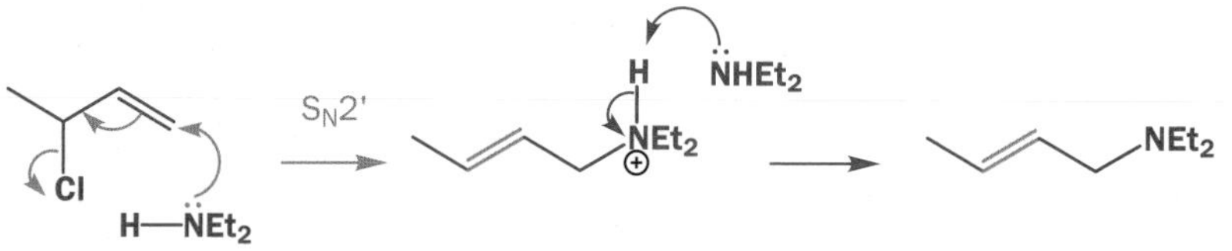

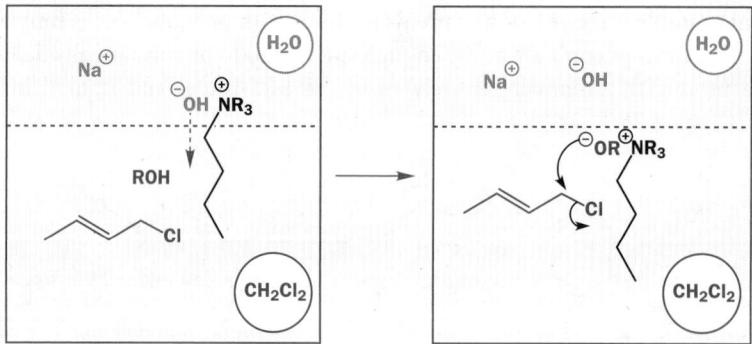

If the primary chloride is used, only the S_N2 reaction normally occurs so that once again we get nucleophilic attack at the primary centre and the more stable product with the more highly substituted alkene. Here is a slightly more advanced example.

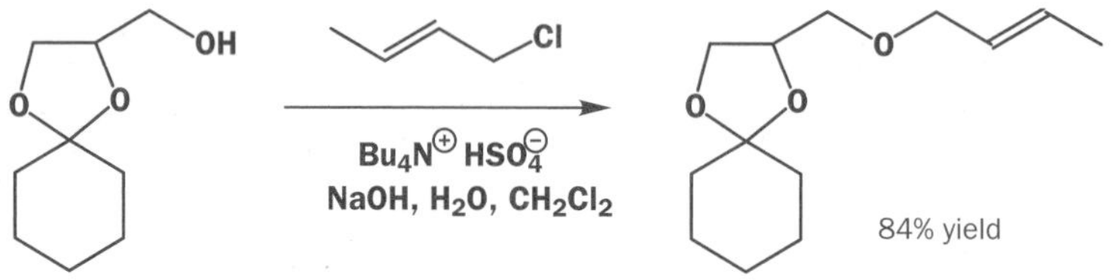

84% yield

Phase transfer catalysis

The last example is interesting because the starting material contains an acetal as well as a primary alcohol group. Acetals are very easily destroyed by acid so the conditions must be kept strictly alkaline. Sodium hydroxide does this but it is insoluble in organic solvents. The method shown here uses a two-phase system of water and dichloromethane (CH_2Cl_2). The organic molecules are in the CH_2Cl_2 layer and the NaOH is in the water layer. The tetraalkyl ammonium salt has a polar group (N^+) and hydrocarbon side chains (butyl groups). These chains mean that, although it is charged, $Bu_4N^+HO^-$ ion pairs are soluble in the organic layer. The ammonium salt allows a low concentration of hydroxide ions to pass into the CH_2Cl_2 layer where they act as a base catalyst for the reaction. Here are the layers shown schematically.

This method is called **phase transfer catalysis** because the tetraalkyl ammonium salt acts as a phase transfer agent, allowing ions to pass into the organic phase. The ether product is, of course, soluble in the organic phase and the work-up is very simple—separation of the phases removes unchanged NaOH and the inorganic by-product, NaCl.

Notice that these reactions take place with allylic *chlorides*. We should not expect an alkyl chloride to be particularly good at S_N2 reactions as chloride ion is only a moderate leaving group and we should normally prefer alkyl bromides or iodides. *Allylic* chlorides are more reactive because of the alkene. Even if the reaction occurs by a simple S_N2 mechanism without rearrangement, the alkene is still making the molecule more electrophilic.

You might ask a very good question at this point. How do we know that these reactions really take place by S_N2 and S_N2' mechanisms and not by an S_N1 mechanism via the stable allyl cation? Well in the case of prenyl bromide, we don't! In fact, we suspect that the cation probably *is* an intermediate, because prenyl bromide and its allylic isomer are in rapid equilibrium in solution at room temperature.

The equilibrium is entirely in favour of prenyl bromide because of its more highly substituted double bond. Reactions on the tertiary allylic isomer are very likely to take place by the S_N1 mechanism: the cation is stable because it is tertiary and allylic and the equilibration tells us it is already there. Even if the reactions were bimolecular, no S_N2' mechanism would be necessary for the tertiary bromide because it can equilibrate to the primary isomer more rapidly than the S_N2 or S_N2' reaction takes place.

Even the secondary system we also considered is in rapid equilibrium when the leaving group is bromide. This time both allylic isomers are present, and the primary allylic isomer (known as 'crotyl bromide') is an *E/Z* mixture. The bromides can be made from either alcohol with HBr, and the same ratio of products results, indicating a common intermediate in the two mechanisms. You saw at the beginning of Chapter 17 that this reaction (Chapter 16) is restricted to alcohols that can react by S_N1.

Displacement of the bromide by cyanide ion, using the copper(I) salt as the nucleophile, gives a mixture of nitriles in which the more stable primary nitrile predominates even more. These can be separated by a clever device. Hydrolysis in concentrated HCl is successful with the predominant primary nitrile but the more hindered secondary nitrile does not hydrolyse. Separation of compounds having two different functional groups is easy: in this case the acid can be extracted into aqueous base, leaving the neutral nitrile in the organic layer.

Once again, we do not know for sure whether this displacement by cyanide goes by the S_N1, S_N2, or S_N2' mechanism, as the reagents equilibrate under the reaction conditions. However, the chlorides do *not* equilibrate and so, if we want a clear cut result on a single well-defined starting material, the chlorides are the compounds to use.

Regiospecific preparation of allylic chlorides

Allylic alcohols are good starting materials for making allylic compounds with control over where the double bond and the leaving group will be. Allylic alcohols are easily made by addition of Grignard reagents or organolithium compounds to enals or enones (Chapter 9) or by reduction of enals or enones (Chapter 24). More to the point, they do not equilibrate except in strongly acidic solution, so we know which allylic isomer we have.

allylic chlorides do not equilibrate

▶

By analogy with *stereospecific*, we can define **regiospecific** to mean a reaction where the regiochemistry (that is, the location of the functional groups) of the product is determined by the regiochemistry of the starting material.

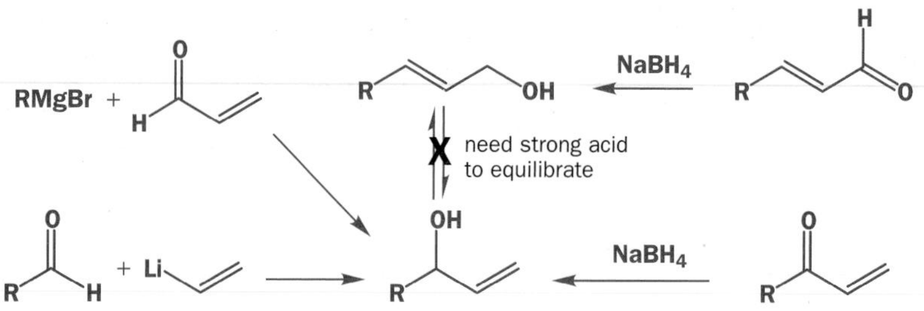

Conversion of the alcohols into the chlorides is easier with the primary than with the secondary alcohols. We need to convert OH into a leaving group and provide a source of chloride ion to act as a nucleophile. One way to do this is with methanesulfonyl chloride ($MeSO_2Cl$) and LiCl.

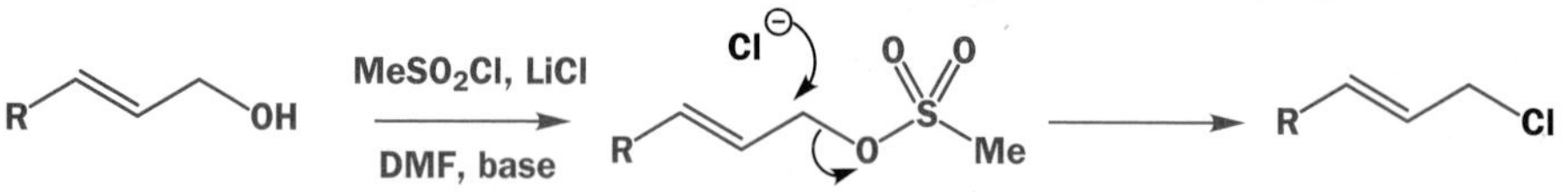

This result hardly looks worth reporting and, anyway, how do we know that equilibration or S_N1 reactions aren't happening? Well, here the mechanism must be S_N2 because the corresponding Z-allylic alcohol preserves its alkene configuration. If there were equilibration of any sort, the Z-alkene would give the E-alkene because E and Z allylic cations are not geometrically stable.

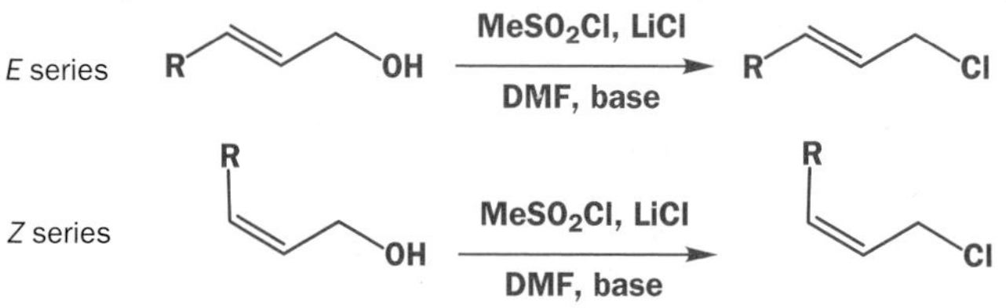

E series

Z series

Sadly, this method fails to preserve the integrity of the secondary allylic alcohol, which gives a mixture of allylic chlorides.

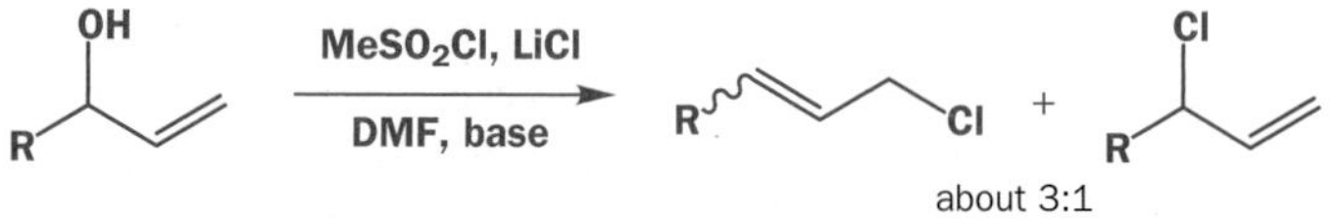

about 3:1

The Mitsunobu reaction was discussed in Chapter 17, p. 429. Mitsunobu chemistry involves using a phosphorus atom to remove the OH group, after the style of PBr_3 as a reagent to make alkyl bromides from alcohols.

Reliable clean S_N2 reactions with secondary allylic alcohols can be achieved only with Mitsunobu chemistry. Here is a well-behaved example with a Z-alkene. The reagents have changed since your last encounter with a Mitsunobu-type reaction: instead of DEAD and a carboxylic acid we have hexachloroacetone.

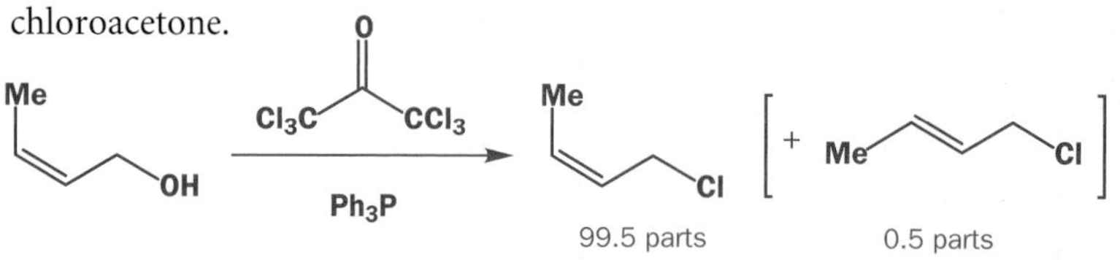

99.5 parts   0.5 parts

The first thing that happens is that the lone pair on phosphorus attacks one of the chlorine atoms in the chloroketone. The leaving group in this S_N2 reaction at chlorine is an enolate, which is a basic species and can remove the proton from the OH group in the allylic alcohol.

Phosphorus doing a substitution at a C–Cl bond the wrong way round! But P is soft, so it cares little about the polarization of the bond, only about the energy of the C–Cl σ^*. The energy is the same whichever end of the bond is attacked. You may see similar reactions of PPh_3 with CBr_4 or CCl_4: all produce stabilized carbanions.

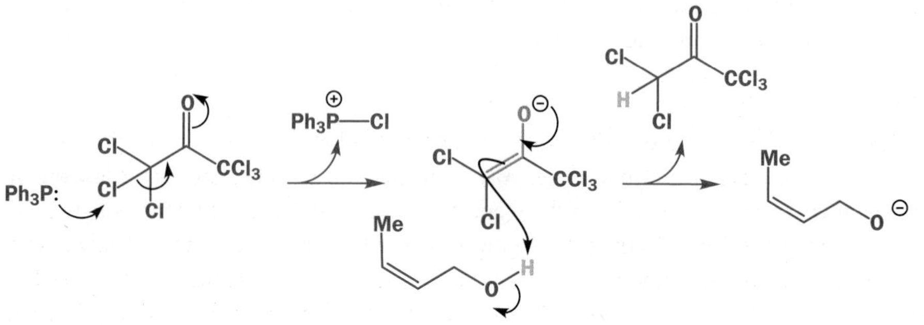

Now the alkoxide anion can attack the positively charged phosphorus atom. This is a good reaction in two ways. First, there is the obvious neutralization of charge and, second, the P–O bond is very strong. This reaction, which we have drawn as an S_N2 reaction at phosphorus, really goes through a pentacovalent intermediate shown to the right, but you will usually see it drawn in a concerted fashion.

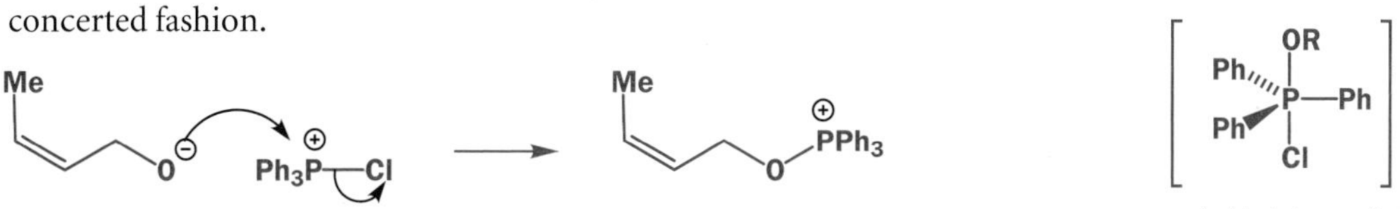

probable intermediate

The next step is a true S_N2 reaction at carbon as the very good leaving group is displaced. The already strong P–O single bond becomes an even stronger P=O double bond to compensate for the loss of the strong C–O single bond.

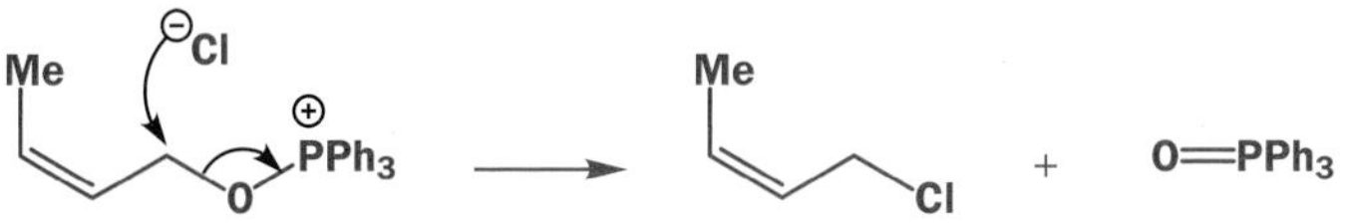

There is obviously no S_N1 component in this displacement (otherwise the Z-alkene would have partly isomerized to the E-alkene) and very little S_N2' as only 0.5% of the rearrangement product is formed. These displacements of $Ph_3P=O$ are often the 'tightest' of S_N2 reactions. Now for the really impressive result. Even if the alcohol is secondary, and the rearranged product would be thermodynamically more stable, very little of it is formed and almost all the reaction is clean S_N2.

94% yield + 6% yield

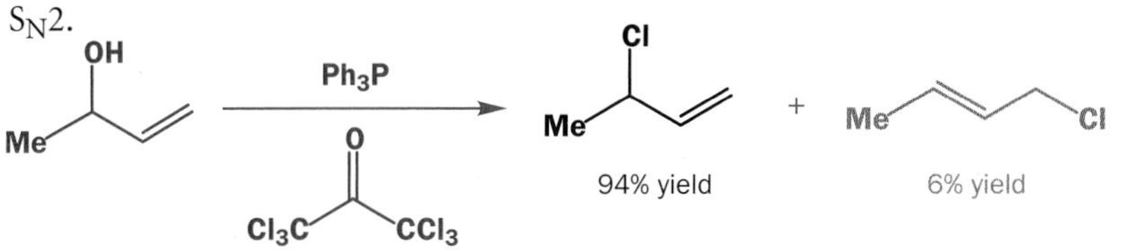

There is a bit more rearrangement than there was with the other isomer but that is only to be expected. The very high proportion of direct S_N2 product shows that there is a real preference for the S_N2 over the S_N2' reaction in this displacement.

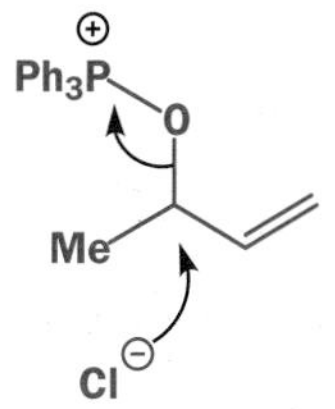

More evidence for S_N2 on the phosphonium intermediate

It is possible to show that the stereochemistry of the double bond is not affected during this reaction and that it goes with clean inversion by using an optically active alcohol with a labelled hydrogen (deuterium) on the alkene.

Note the inversion at the stereogenic centre (see discussion of this as a criterion of the S_N2 reaction in Chapter 17) but retention in the geometry of the alkene. This is clear evidence for an S_N2 reaction at the secondary centre.

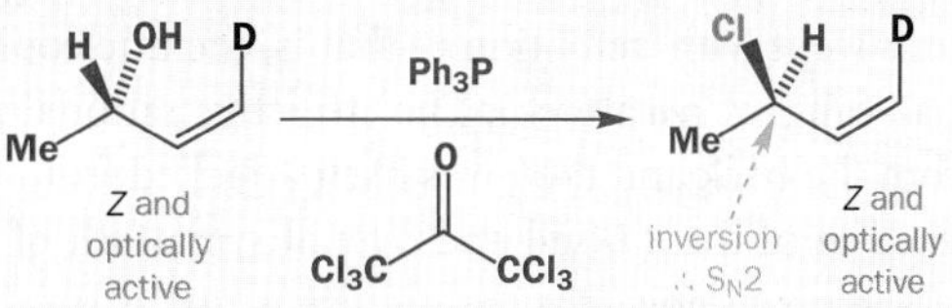

Now that we know how to make allylic chlorides of known structure—whether primary or secondary—we need to discover how to replace the chlorine with a nucleophile with predictable regioselectivity. We have said little so far about carbon nucleophiles (except cyanide ion) so we shall concentrate on simple carbon nucleophiles in the S_N2' reaction of allylic chlorides.

The S_N2' reaction of carbon nucleophiles on allylic chlorides

Ordinary carbon nucleophiles such as cyanide or Grignard reagents or organolithium compounds fit the patterns we have described already. They usually give the more stable product by S_N2 or S_N2' reactions depending on the starting material. If we use copper compounds, there is a tendency—no more than that—to favour the S_N2' reaction. You will recall that copper(I) was the metal we used to ensure conjugate addition to enones (Chapter 10) and its use in S_N2' reactions is obviously related.

▶

Probably all 'S_N2 reactions' at Si, P, and S go through addition intermediates because these elements can sustain five full bonds. The substitution mechanism is then: (1) addition to give an anionic species; (2) elimination of the best leaving group. Do you see an analogy with some reactions from earlier in this chapter?

■

We looked at the converse—'loose' S_N2 transition states with considerable S_N1 character—in the reactions of bromonium ions and protonated epoxides in Chapter 19.

S_N2 preferred to S_N2'

Simple alkyl copper reagents (RCu, known as Gilman reagents) generally favour the S_N2' reaction but we can do much better by using RCu complexed with BF_3.

major product   minor product

generally >98:2

The nature of metal–alkene complexes is discussed in Chapter 48.

The copper must complex to the alkene and then transfer the alkyl group to the S_N2' position as it gathers in the chloride. This might well be the mechanism, though it is often difficult to draw precise mechanisms for organometallic reactions.

The secondary allylic isomer also gives almost entirely the rearranged product. This is perhaps less surprising, as the major product is the more stable isomer, but it means that either product can be formed in high yield simply by choosing the right (or should we say *wrong*, since there is complete allylic rearrangement during the reaction) isomer. The reaction is regio*specific*.

major product   minor product

generally >96:4

The most remarkable result of all is that prenyl chloride gives rearranged products in good yield. This is about the only way in which these compounds suffer attack at the tertiary centre by S_N2' reaction when there is the alternative of an S_N2 reaction at a primary centre.

major product   minor product

95:5

Stereochemistry of the S_N2' reaction

There is some controversy over this issue. There is, of course, none over the S_N2 reaction on these allylic compounds—inversion occurs as in all S_N2 reactions. It used to be supposed than S_N2' reactions went with 'retention'—that is, the nucleophile attacked the same face of the allylic system (we shall call this *syn* attack). The attractive rationalization was that the π bond attacked the C–Br bond from the back and then was itself attacked from the back by the nucleophile. This results in an *anti* reaction of the π bond and overall *syn* attack of the nucleophile with respect to the leaving group.

We now know that the picture is not as simple as this. *syn* S_N2' reactions are preferred but *anti* S_N2' reactions are also possible and the result found depends on the molecule under observation. Here is a convincing example of S_N2' reactions going with *syn* stereochemistry. The molecule is a planar cyclobutene, which makes the stereochemistry easy to see.

retention in S_N2' reactions?

X^- does 'S_N2'' on π bond with inversion

π bond does 'S_N2' with inversion

S_N2

The deuterium labels are there so that we can see that the S_N2' reaction is indeed taking place. This reaction is entirely *syn* even though the methoxide nucleophile must attack alongside the other chlorine atom. The reaction does not stop there since a second methoxide displaces the other chloride—also in a *syn* fashion. Here too there must be considerable resistance to *syn* attack as the second methoxide anion must approach alongside the first.

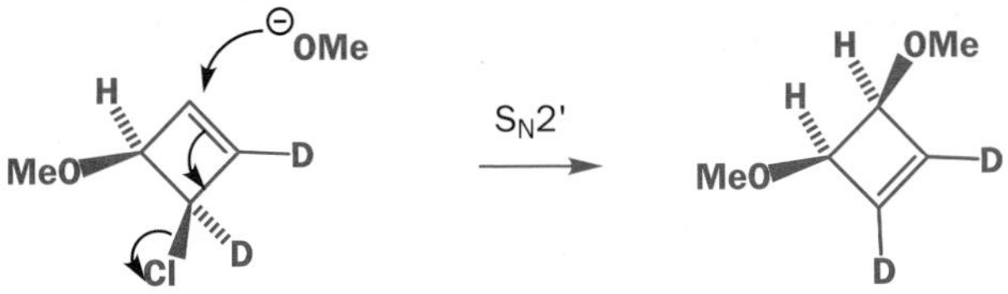

In other cases, especially in open-chain compounds, the stereochemical outcome is not so clear cut and mixtures are often formed. The best generalization is that the S_N2' reaction prefers *syn* stereochemistry but that *anti* stereochemistry is also possible. In the absence of other evidence, you should first suggest a *syn* course for the reaction—but do not be surprised if your suggestion turns out to be wrong.

To conclude...

This chapter is about electrophilic alkenes. We started by saying that alkenes are really nucleophilic and not electrophilic but in this chapter (and in Chapter 10) we have managed to find a remarkable collection of electrophilic alkenes from various types of chemistry. Here is a summary chart.

Page no.	Type of alkene	Examples	Reaction
ch. 10	unsaturated carbonyl compounds	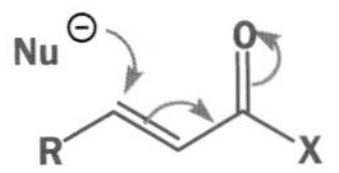	conjugate addition
583	unsaturated nitriles and nitroalkenes	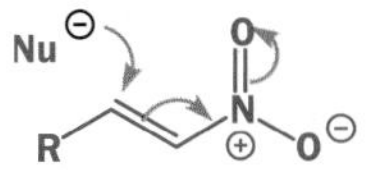	conjugate addition
585	enones, etc. with β-leaving group	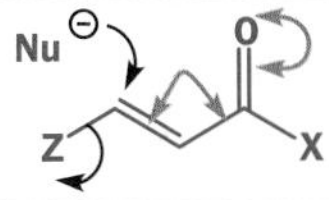	conjugate substitution
586–588	guanidines, amidines, and nitroalkenes with β-leaving group	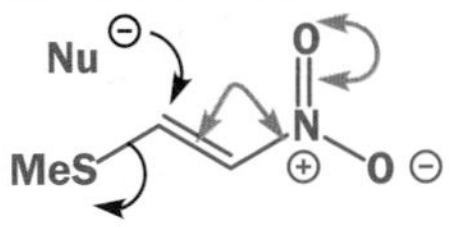	conjugate substitution
590–597	benzene rings with electron-withdrawing substituents and leaving groups	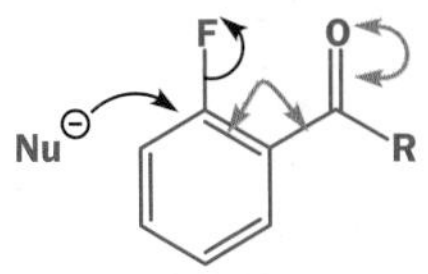	nucleophilic aromatic substitution: addition–elimination mechanism
597–600	aryl cations	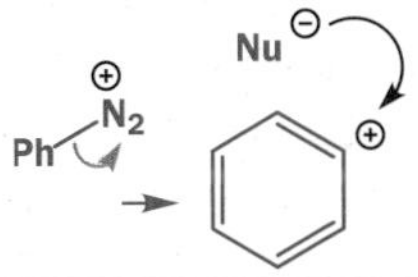	nucleophilic aromatic substitution: S_N1 mechanism
600–604	benzyne	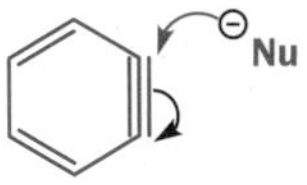	nucleophilic aromatic substitution: elimination–addition mechanism
604–11	allylic halides and esters of allylic alcohols	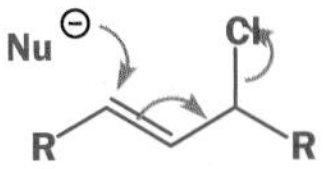	nucleophilic substitution (S_N2 and S_N2')

Still to come:

ch. 29	enolates and enolate equivalents as nucleophiles	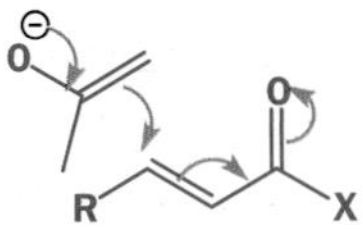	conjugate addition

Problems

1. What is the structure of the product of this reaction and how is it formed?

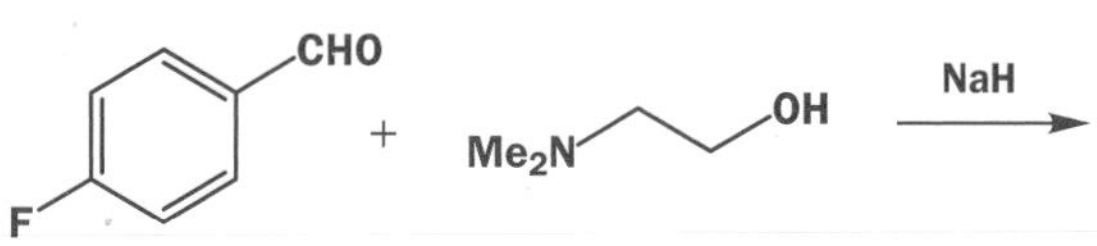

C$_{11}$H$_{15}$NO$_2$
ν_{max}(cm^{-1}) 1730
δ_C(p.p.m.) 191, 164, 132, 130, 115, 64, 41, 29
δ_H(p.p.m.) 2.32 (6H, s), 3.05 (2H, t, J 6 Hz),
4.20 (2H, t, J 6 Hz), 6.97 (2H, d, J 7 Hz),
7.82 (2H, d, J 7 Hz), 9.97 (1H, s)

2. Draw a detailed mechanism for this reaction. Note that no base is added to the mixture. Why is base unnecessary?

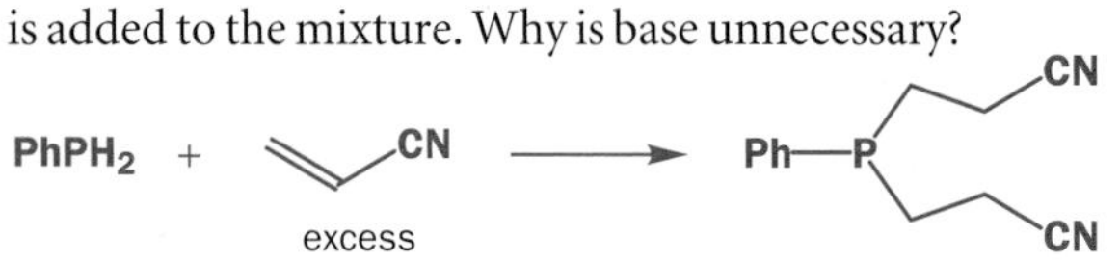

3. Which of the two routes suggested here would actually lead to the product? What might happen in the other sequence?

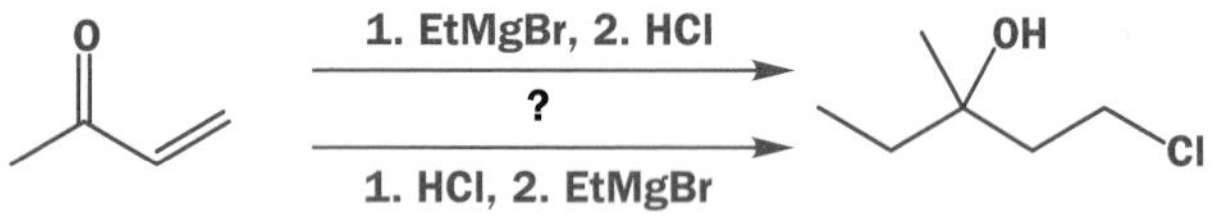

4. Suggest reasons for the different outcome of each of these reactions. Your answer must, of course, include a mechanism for each reaction.

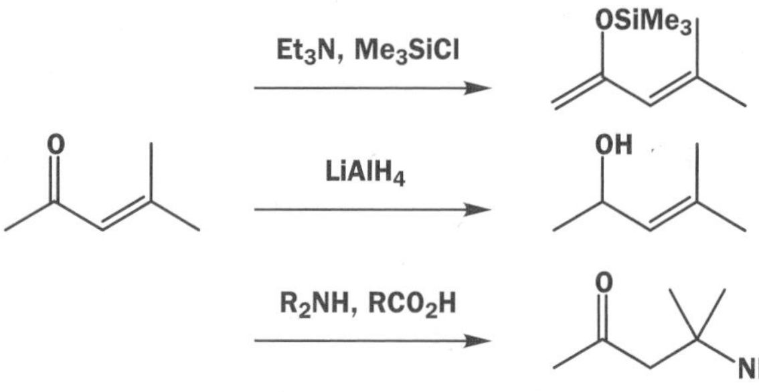

5. Suggest mechanisms for these reactions. You should explain why one of the cyanides is lost but not the other.

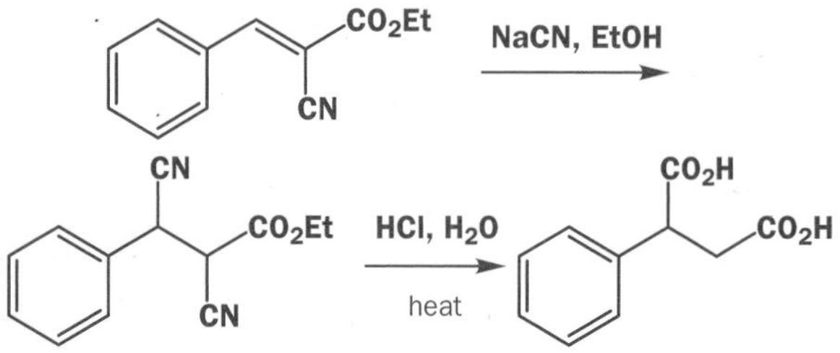

6. Suggest a mechanism for this reaction.

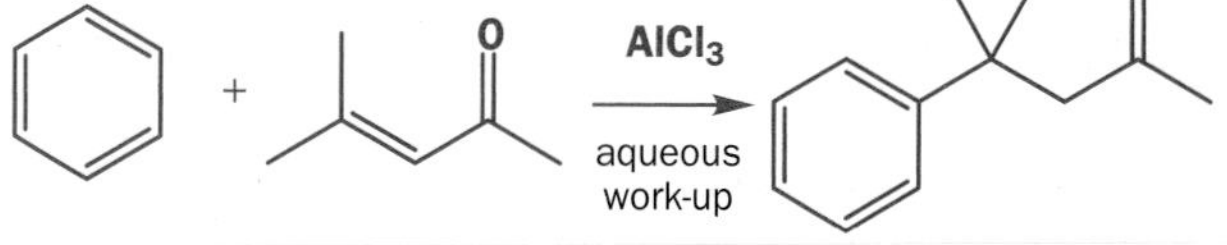

7. Suggest a mechanism for this reaction explaining the selectivity.

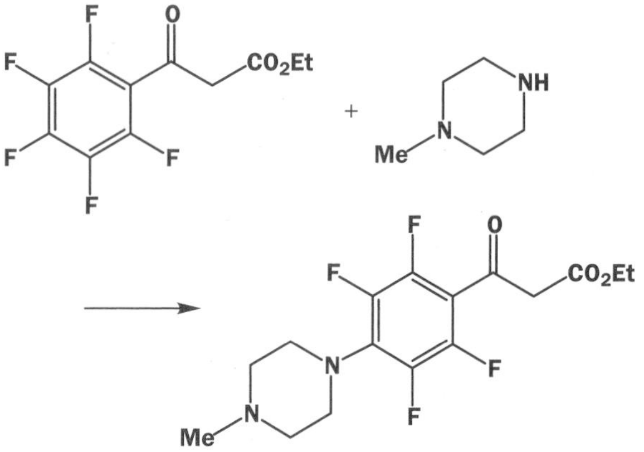

8. Suggest mechanisms for all of the steps in this synthesis of 2,4-dinitrophenylhydrazine given in the chapter.

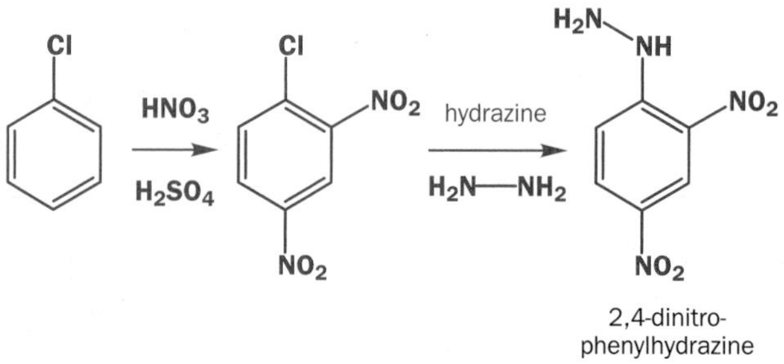

2,4-dinitro-phenylhydrazine

9. Pyridine is a six-electron aromatic system like benzene. You have not yet been taught anything systematic about pyridine but see if you can work out why 2- and 4-chloropyridines react with nucleophiles but 3-chloropyridine does not.

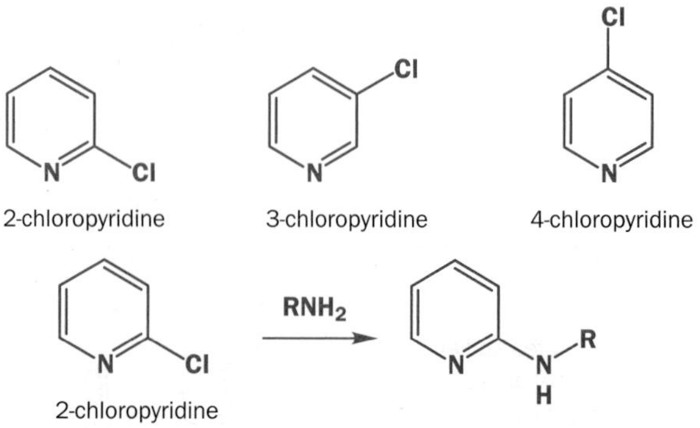

2-chloropyridine   3-chloropyridine   4-chloropyridine

2-chloropyridine

10. Draw detailed mechanisms for the last two steps in the ranitidine synthesis that involve conjugate substitution. Why is it possible to replace one MeS group at a time?

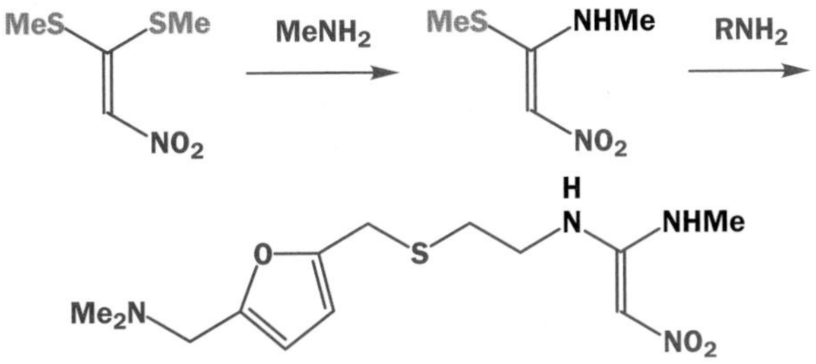

ranitidine - GlaxoWellcome's Zantac

11. How would you convert this aromatic compound into the two derivatives shown?

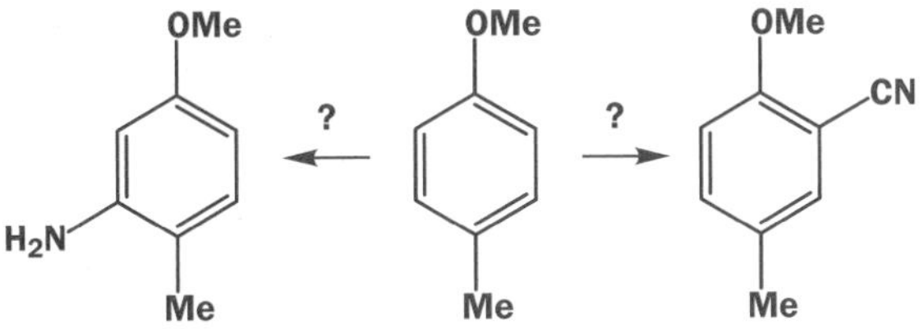

12. Comment on the selectivity shown in these reactions.

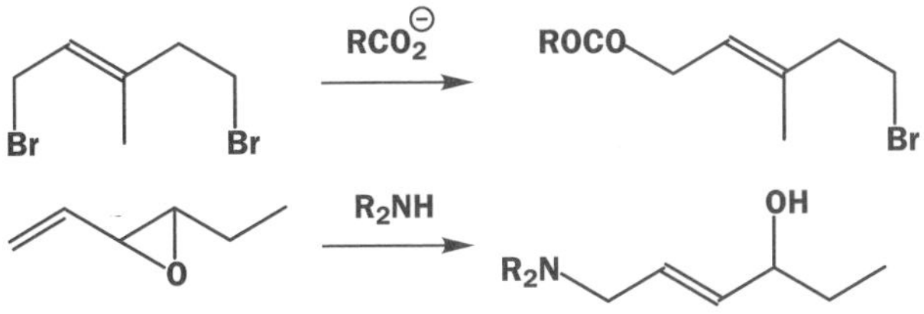

13. Suggest what products might be formed from the unsaturated lactone and the various reagents given and comment on your choice.

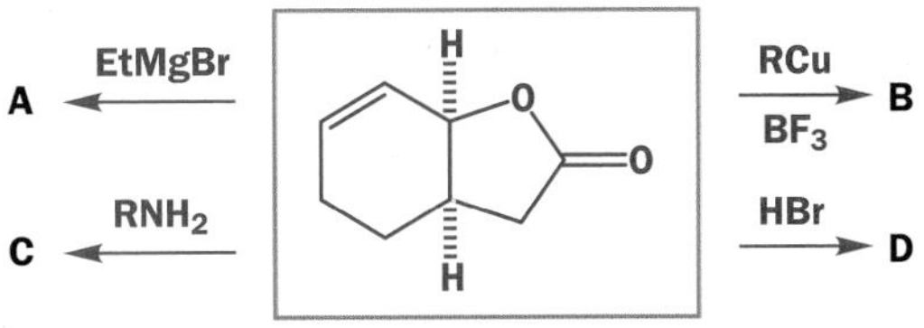

14. Suggest mechanisms for these reactions, pointing out what guided you to choose these pathways.

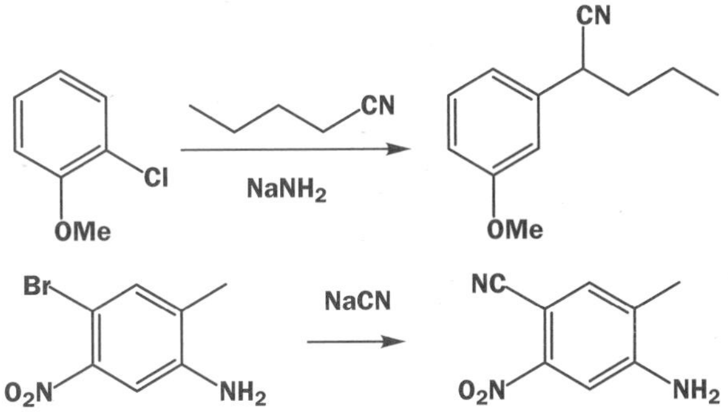

Chemoselectivity: selective reactions and protection

24

Connections

Building on:
- Carbonyl addition and substitution ch6, ch12, & ch14
- Conjugate addition ch10
- Mechanisms and catalysis ch13
- Electrophilic addition to alkene ch20
- Nucleophilic aromatic substitution ch23

Arriving at:
- Regio-, stereo-, and chemoselectivity
- Reagents for reduction of alkenes and carbonyl compounds
- Removal of functional groups
- Reduction of benzene rings
- Protection of aldehydes, ketones, alcohols, and amines
- Reagents for oxidation of alcohols

Looking forward to:
- Synthesis in action ch25
- Enolates especially aldol chemistry ch26–ch29
- Retrosynthetic analysis ch30
- Cycloadditions ch35
- Rearrangements ch37
- Sulfur chemistry ch46

Selectivity

Most organic molecules contain more than one functional group, and most functional groups can react in more than one way, so organic chemists often have to predict *which* functional group will react, *where* it will react, and *how* it will react. These questions are what we call **selectivity**.

Selectivity comes in three sorts: chemoselectivity, regioselectivity, and stereoselectivity. **Chemoselectivity** is *which* group reacts; **regioselectivity** is *where* it reacts. **Stereoselectivity** is *how* the group reacts with regard to the stereochemistry of the product.

> ● **There are three main types of selectivity**
>
> - Chemoselectivity: *which* functional group will react
> - Regioselectivity: *where* it will react
> - Stereoselectivity: *how* it will react (stereochemistry of the products)

We talked a lot about regioselectivity two chapters ago, when you learned how to predict and explain which product(s) you get from electrophilic aromatic substitution reactions. The functional group is the aromatic ring: *where* it reacts is the reaction's regioselectivity. Going back further, one of the first examples of regioselectivity you came across was nucleophilic addition to an unsaturated ketone. Addition can take place in a 1,2- or a 1,4-fashion—the question of which happens (*where* the unsaturated ketone reacts) is a question of regioselectivity, which we discussed in Chapters 10 and 23. We shall leave all discussion of stereoselectivity until Chapters 31–34.

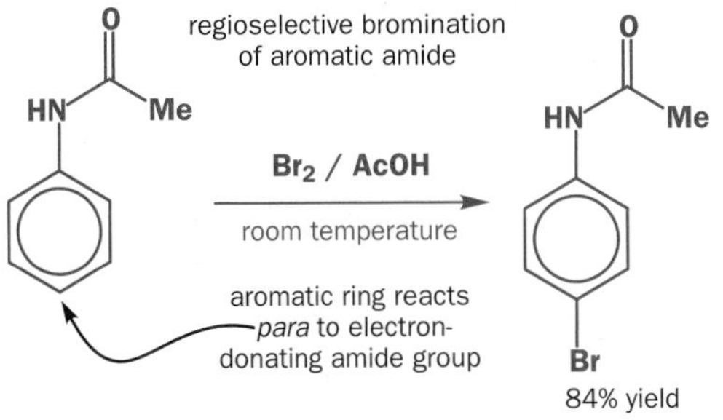

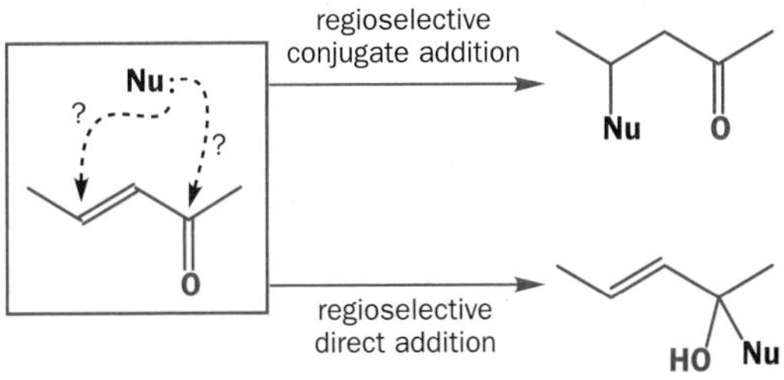

regioselective bromination of aromatic amide

Br_2 / AcOH

room temperature

aromatic ring reacts *para* to electron-donating amide group

84% yield

regioselective conjugate addition

regioselective direct addition

This chapter is about chemoselectivity—in a compound with more than one functional group, which group reacts? Let's start with a straightforward example—the synthesis of paracetamol briefly described in Chapter 22. 4-Aminophenol could react with acetic anhydride on both nitrogen and oxygen to give a compound containing an amide and an ester functional group. This is what happens on heating with excess Ac_2O in toluene.

But with just one equivalent of acetic anhydride in the presence of a base (pyridine) only the NH_2 group is acylated, and paracetamol is the product. This is chemoselectivity, and it is to be expected that the NH_2 group is more nucleophilic than the OH group. It is even possible to hydrolyse the doubly acetylated product to paracetamol with aqueous sodium hydroxide. The ester is more reactive than the amide and hydrolyses much more easily (Chapter 12).

We know that ketones are more reactive towards Grignard reagents and organolithiums than esters because you can't isolate a ketone from the reaction of an ester with a Grignard reagent or an organolithium (in Chapter 12 we devoted some time to what you *can* react with an organometallic compound to get a ketone—p. 299). So it should come as no surprise that, when some chemists at Pfizer were developing anticonvulsants related to the tranquillizer oblivon by adding lithium acetylide to ketones, they were successful in making a tertiary alcohol by chemoselective reaction of a ketone in the presence of an ester.

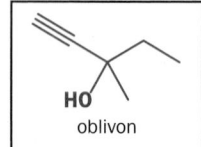
oblivon

ketone is more electrophilic than ester

These reactions work because, although each starting material contains two carbonyl groups, one is more electrophilic and therefore more reactive towards nucleophiles (OH⁻ in the first case; lithium acetylide in the second) than the other. We can order carbonyl compounds into a sequence in which it will *usually* be possible to react those on the left with nucleophiles in the presence of those on the right.

reactivity towards nucleophiles

aldehyde		ketone		ester		amide		carboxylate
R H	>	R R	>	R OR	>	R NR_2	>	R $O^{\ominus}$

We've already discussed this sequence of reactivity in relation to acid derivatives in Chapters 12 and 14—make sure you understand the reason for the ordering of ester > amide > carboxylate. Here we're adding on aldehyde (the most reactive, for steric reasons—it is the least hindered) and ketone (more reactive than esters because the carbonyl group is not stabilized by conjugation with a lone pair).

Reducing agents

Chemists at Glaxo exploited this reactivity sequence in their synthesis of the anti-asthma drug, salmefamol (sister of the best seller salbutamol, which will be discussed in Chapter 25). Three reducing agents are used in the sequence: sodium borohydride ($NaBH_4$); lithium aluminium hydride ($LiAlH_4$); and hydrogen gas over a palladium catalyst.

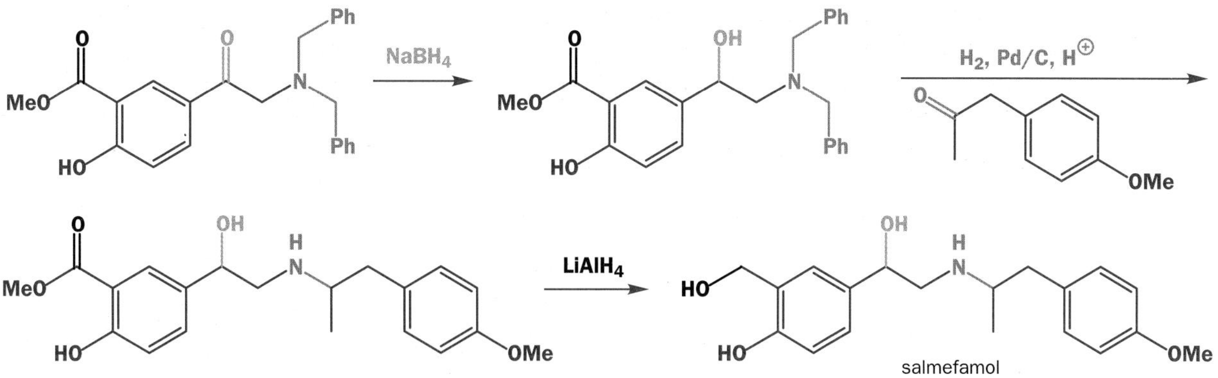

We shall use this synthesis as a basis for discussion on chemoselectivity in reductions. In the first step, sodium borohydride leaves the black carbonyl group of the ester untouched while it reduces the ketone (in yellow); in the last step, lithium aluminium hydride reduces the ester (in black). These chemoselectivities are typical of these two most commonly used reducing agents: borohydride can usually be relied upon to reduce an aldehyde or a ketone in the presence of an ester, while lithium aluminium hydride will reduce almost any carbonyl group.

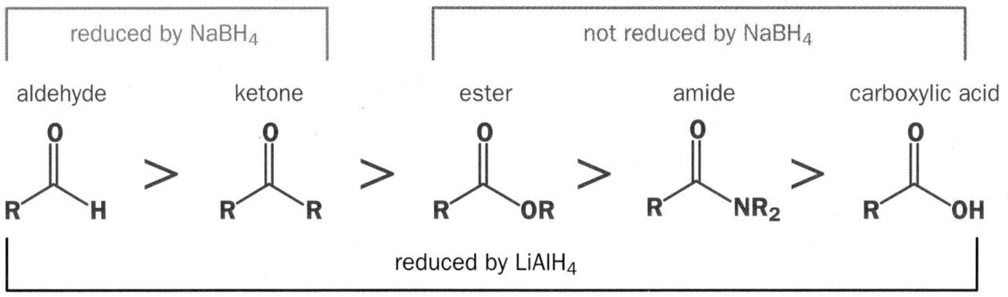

Each reduction gives an alcohol, apart from the reduction of an amide with $LiAlH_4$, which gives an amine, which we shall explain next. We shall return to the salmefamol synthesis later to explain the reductions with hydrogen gas catalysed by palladium.

> In general, it's best to use the mildest conditions possible for any particular reaction—the potential for unwanted side-reactions is lessened. What is more, $NaBH_4$ is a lot easier to handle than $LiAlH_4$—for example, it simply dissolves in water while $LiAlH_4$ catches fire if it gets wet. $NaBH_4$ is usually used to reduce aldehydes and ketones, even though $LiAlH_4$ also works.

Reduction of carbonyl groups

We should now look in detail at reductions of carbonyl compounds, and in doing so we shall introduce a few more specialized reducing agents. Then we will come back to the other type of reduction in the salmefamol synthesis—catalytic hydrogenation.

How to reduce aldehydes and ketones to alcohols

We don't need to spend much time on this—sodium borohydride does it very well, and is a lot easier to handle than lithium aluminium hydride. It is also more selective: it will reduce this nitroketone, for example, where $LiAlH_4$ would reduce the nitro group as well.

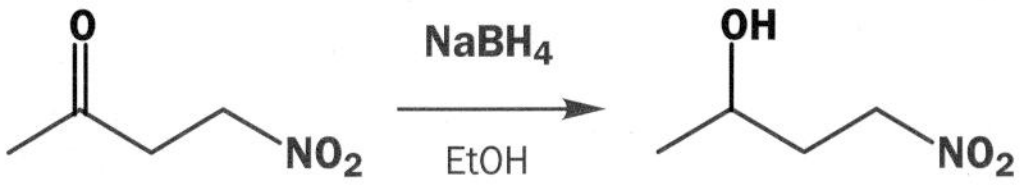

You met borohydride in Chapter 6, where we discussed the mechanism of its reactions. Sodium borohydride will reduce only in protic solvents (usually ethanol, methanol, or water) or in the presence of electrophilic metal cations such as Li^+ or Mg^{2+} ($LiBH_4$ can be used in THF, for example). The precise mechanism, surprisingly, is still unclear, but follows a course something like this with the dotted lines representing some association, perhaps coordination or bond formation.

The essence of the reaction is the transfer of a hydrogen atom with two electrons (called **hydride transfer** though no hydride ion is involved). In addition, the developing negative charge on oxygen gets help from the alcohol or the sodium ion or both and a molecule of alcohol adds to the boron during or immediately after the reduction. The by-product, an alkoxyborohydride anion, is itself a reducing agent, and can go on to reduce three more molecules of carbonyl compound, transferring step-by-step all of its hydrogen atoms.

How to reduce esters to alcohols

LiAlH$_4$ is often the best reagent, and gives alcohols by the mechanism we discussed in Chapter 12. As a milder alternative (LiAlH$_4$ has caused countless fires through careless handling), lithium borohydride in alcoholic solution will reduce esters—in fact, it has useful selectivity for esters over acids or amides that LiAlH$_4$ does not have. Sodium borohydride reduces most esters only rather slowly.

> Why not try writing the mechanism out now to make sure you understand it, before checking back to p. 298. In a moment, we will show you a slightly more sophisticated version, in which we account for the fate of the Li and Al species.

How to reduce amides to amines

Again, LiAlH$_4$ is a good reagent for this transformation. The mechanism follows very much the same course as the reduction of esters, but there is a key difference at the steps boxed in yellow and in green.

> The ester mechanism has rather more detail than the simplified one we presented to you in Chapter 12.

How to reduce carboxylic acids to alcohols

The best reagent for this is borane, BH$_3$. Borane is, in fact, a gas with the structure B$_2$H$_6$, but it can be 'tamed' as a liquid by complexing it with ether (Et$_2$O), THF, or dimethyl sulfide (DMS, Me$_2$S).

> These complexes are Lewis salts: BH$_3$ is a Lewis acid that accepts a lone pair of electrons from the basic ether or sulfide.

Although borane appears superficially similar to borohydride, it is not an ion and that makes all the difference to its reactivity. Whereas borohydride reacts best with the most electrophilic carbonyl groups, borane's reactivity is dominated by its desire to accept an electron pair into its empty p orbital. In the context of carbonyl group reductions, this means that it reduces electron-rich carbonyl groups fastest. The carbonyl groups of acyl chlorides and esters are relatively electron-poor (Cl and OR are very electronegative); borane will not touch acyl chlorides and reduces esters only slowly. But it will reduce amides.

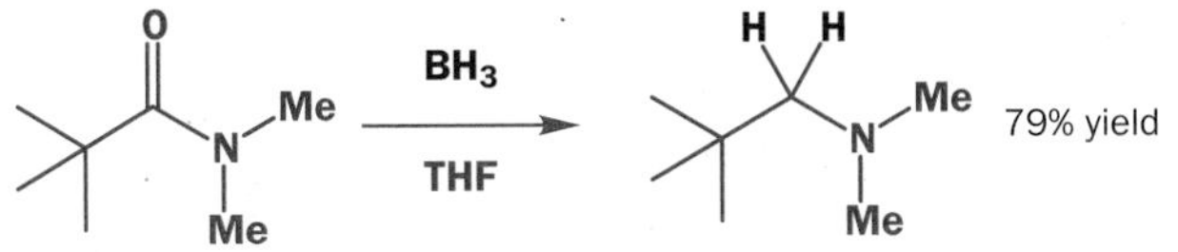

The Lewis basic carbonyl group forms a complex with the empty p orbital of the Lewis acidic borane. Hydride transfer is then possible from anionic boron to electrophilic carbon. The resulting tetrahedral intermediate collapses to an iminium ion that is reduced again by the borane.

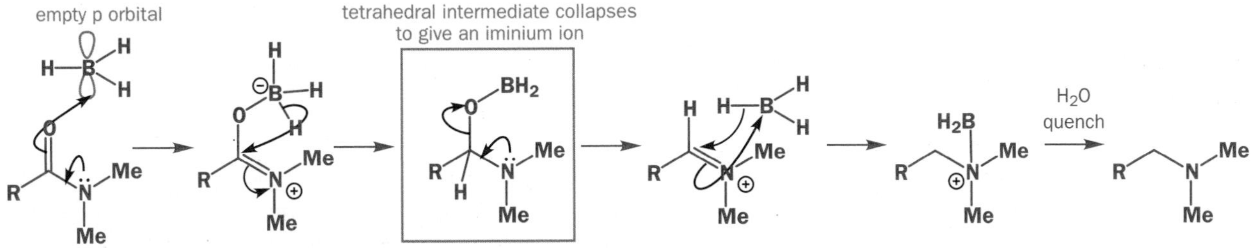

Borane also makes a good alternative to LiAlH$_4$ for reducing amides as the two reagents have slightly different chemoselectivity—in this example borane reduces an amide in the presence of an ester.

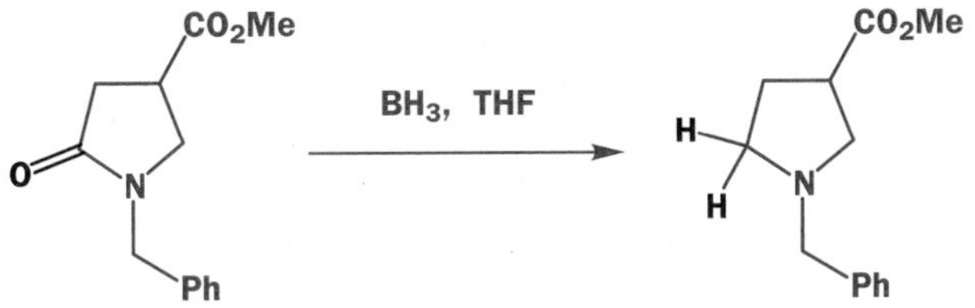

Borane is an excellent reagent for reducing carboxylic acids. It reacts with them first of all by forming triacylborates, with evolution of hydrogen gas. Esters are usually less electrophilic than ketones because of conjugation between the carbonyl group and the lone pair of the sp^3 hybridized oxygen atom—but, in these boron esters, the oxygen next to the boron has to share its lone pair between the carbonyl group and the boron's empty p orbital, so they are considerably more reactive than normal esters, or the lithium carboxylates formed from carboxylic acids and LiAlH$_4$.

oxygen donates lone pair electrons into boron's empty p orbital

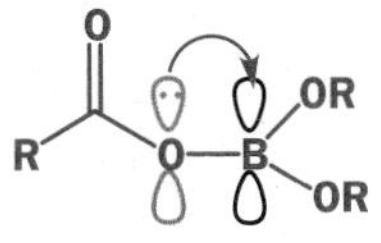

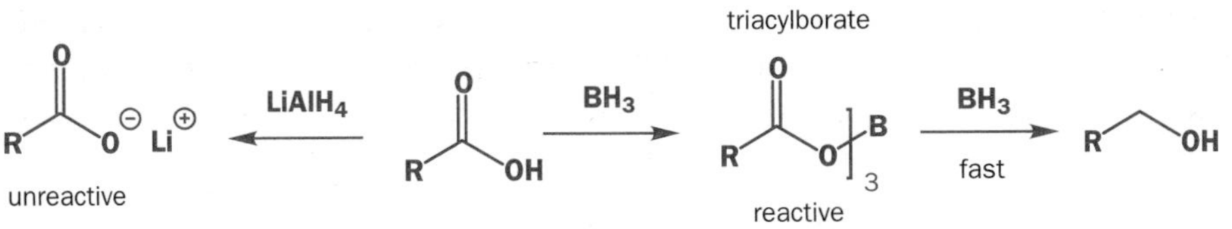

Borane is a highly chemoselective reagent for the reduction of carboxylic acids in the presence of other reducible functional groups such as esters, and even ketones.

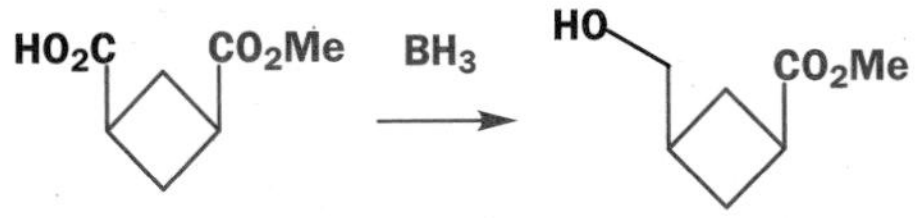

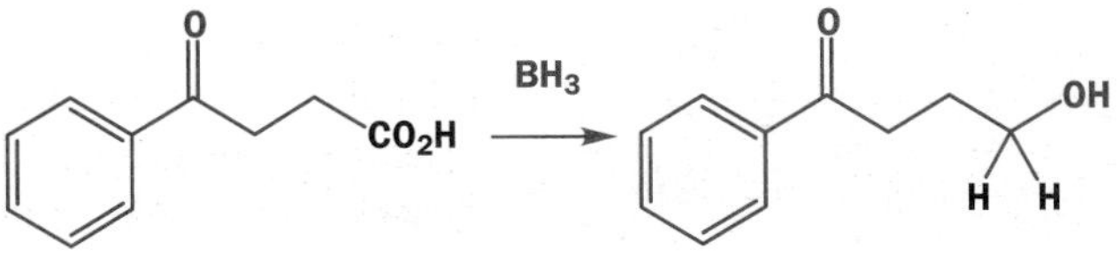

■
This type of asymmetric synthesis is discussed in Chapter 45.

Borane and lithium borohydride are a most useful pair of reducing agents, with opposite selectivities. Japanese chemists used an enzyme to make a single enantiomer of the acid below, and were able to reduce either the ester or the carboxylic acid by choosing lithium borohydride or borane as their reagent. Check for yourself that the lactones (cyclic esters) in black frames are enantiomers.

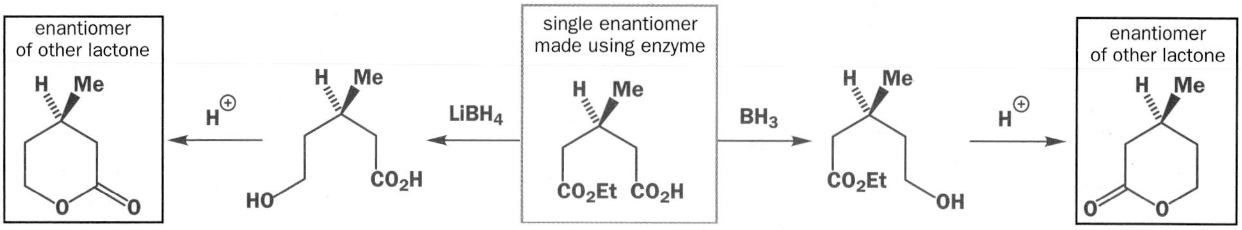

How to reduce esters and amides to aldehydes

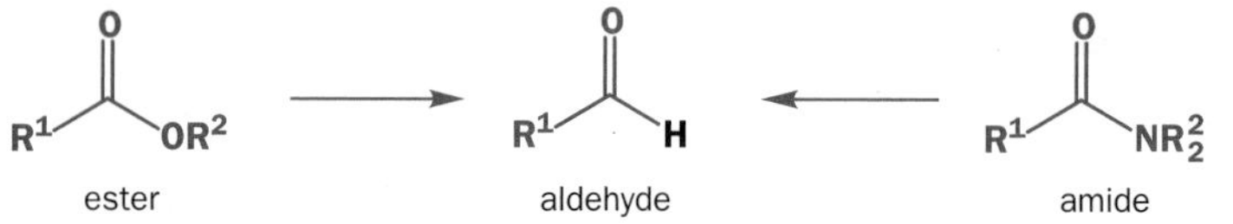

The step boxed in yellow in the ester reduction scheme on p. 618 gave an aldehyde. The aldehyde is more readily reduced than the ester, so the reduction doesn't stop there, but carries on to the alcohol oxidation level. How, then, can you reduce an ester to an aldehyde? This is a real problem in synthetic chemistry—the ester below, for example, is easy to make by methods you will meet in Chapter 27. But an important synthesis of the antibiotic monensin requires the aldehyde.

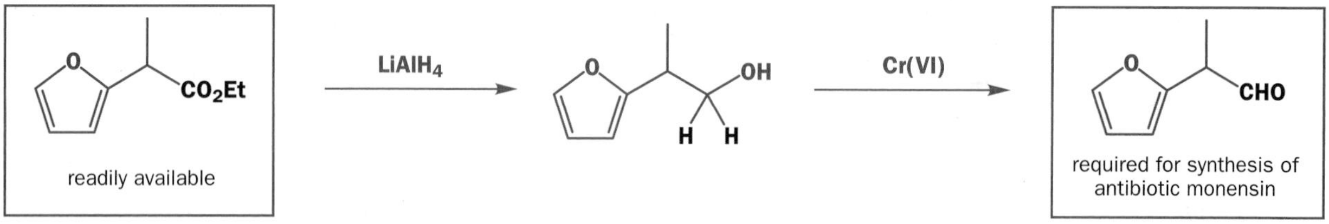

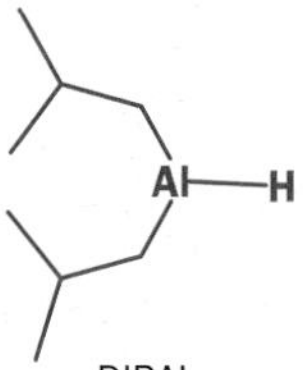

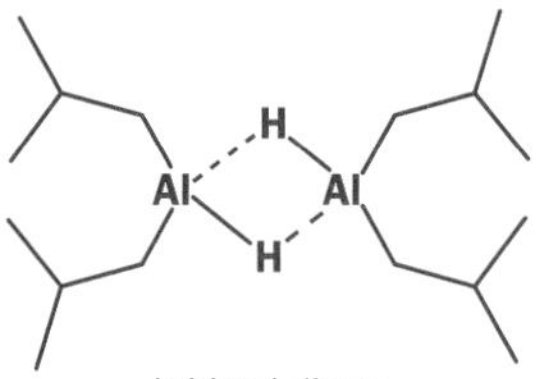

In this case, the chemists decided simply to put up with the fact that $LiAlH_4$ gives the alcohol, and re-oxidize the alcohol back to the aldehyde using chromium(VI) (see later for details of this step). There is, however, a reagent that will sometimes do the job in a single step, though you must bear in mind that this is not at all a general reaction. The reagent is known as DIBAL (or DIBAH or DIBALH—diisobutyl aluminium hydride, $i\text{-}Bu_2AlH$).

DIBAL is in some ways like borane—it exists as a bridged dimer, and it becomes a reducing agent only after it has formed a Lewis acid–base complex, so it too reduces electron-rich carbonyl groups most rapidly. DIBAL will reduce esters even at −70 °C, and at this temperature the tetrahedral intermediate may be stable. Only in the aqueous work-up does it collapse to the aldehyde when excess DIBAL has been destroyed so that no further reduction is possible.

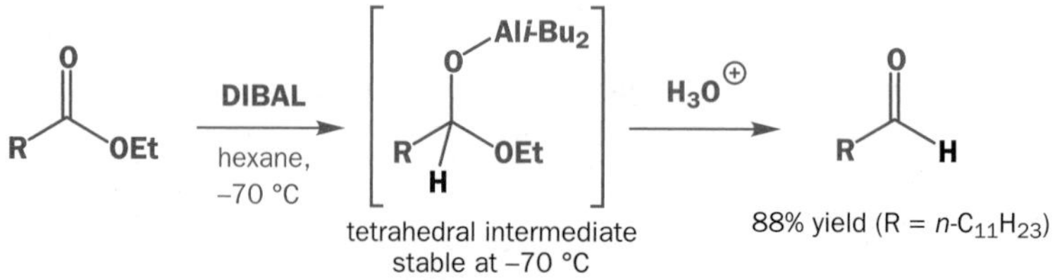

A stable tetrahedral intermediate is more likely in the reduction of lactones, and DIBAL is most reliable in the reduction of lactones to lactols (cyclic hemiacetals), as in E.J. Corey's synthesis of the prostaglandins. The key step, the hydride transfer from Al, is shown in the green frame.

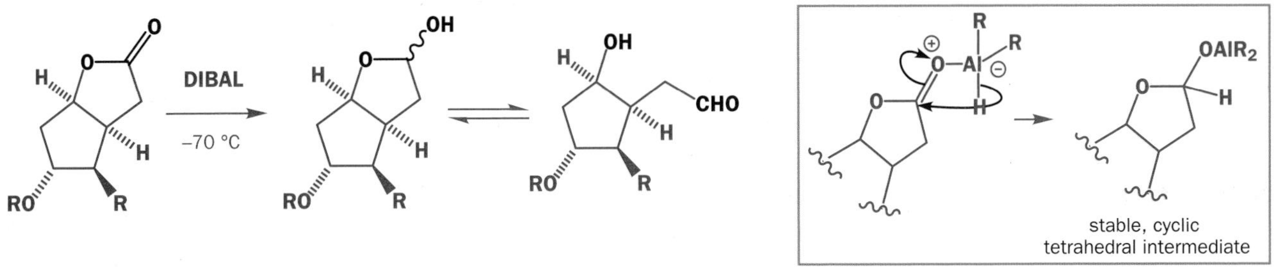

In the amide reduction scheme on p. 618, the step framed in green gives an iminium ion. Stopping the reaction here would therefore provide a way of making aldehydes from amides. Because these tetrahedral intermediates are rather more stable than those from ester reduction, this can often be achieved simply by carrying out the amide reduction, and quenching, at 0 °C (–70 °C is usually needed to stop esters overreducing to alcohols).

tetrahedral intermediate
stable at 0 °C

80% yield

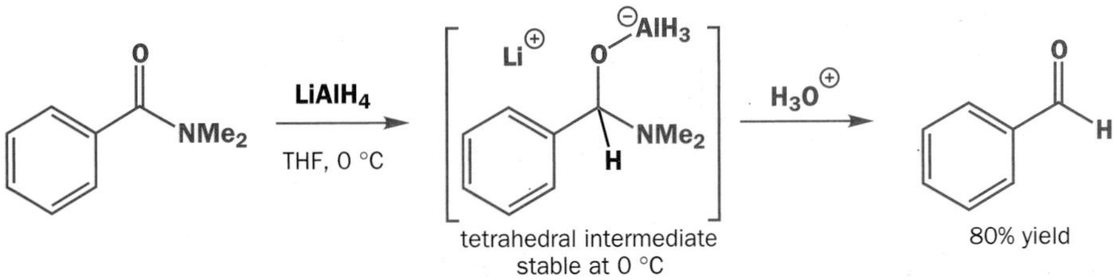

DIBAL is also good for reducing nitriles to aldehydes. Indeed, this reaction and the reduction of lactones to lactols are the best things that DIBAL does.

1. DIBAL, –70 °C

2. H_2O, $H^{\oplus}$

96% yield

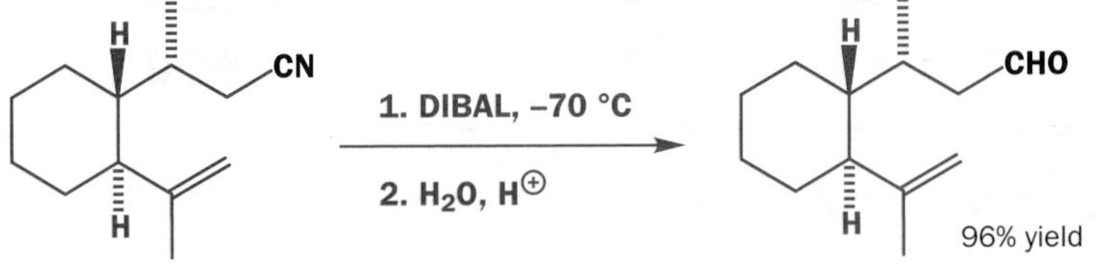

Now, let's go back to the salmefamol synthesis we started with on p. 617. The other reducing agent used in the sequence is hydrogen gas over a palladium catalyst. Catalytic hydrogenation has two functions here: firstly, it removes the two benzyl groups from the nitrogen, revealing a primary amine (this reaction is discussed later in this chapter), and, secondly, it reduces the imine that forms between this amine and the ketone added in this second step—an instance of **reductive amination**. We shall consider the second first, because it is another example of chemoselectivity in the reduction of a carbonyl-like group. You met reductive amination in Chapter 14, but as a reminder, here is the process again.

> *Reminder.* Cyclic hemiacetals are more stable than acyclic ones. Note how the product stays as a lactol—an acyclic hemiacetal would revert to alcohol plus aldehyde.

> Carboxylic acids can be reduced to aldehydes via their acyl chlorides using the *Rosenmund reaction*—see below.

> 'Pd/C' means palladium metal dispersed on a charcoal support—usually 5–10% by mass Pd and 90–95% C. It is made by suspending charcoal powder in a $PdCl_2$ solution, and then reducing the $PdCl_2$ to Pd metal, usually with H_2 gas, but sometimes with formaldehyde, HCHO (which becomes oxidized to formic acid, HCO_2H). The palladium metal precipitates on to the charcoal, which can be filtered off and dried. The fine Pd particles present maximum surface area to the reaction they catalyse and, while Pd is an expensive metal, it is recyclable since the Pd/C is insoluble and can be recovered by filtration.

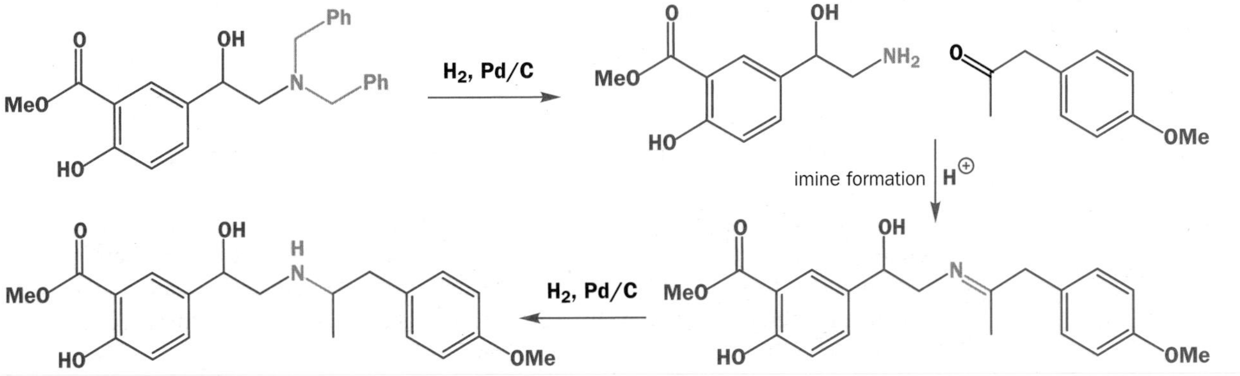

Catalytic hydrogenation reduces the imine (as the protonated iminium ion) but not the ketone from which it is formed. This chemoselectivity (reduction of iminium ions but not ketones) is also displayed by sodium cyanoborohydride and we can add NaCNBH$_3$ to complete our table of reactivity, if we insert imines at the left-hand end.

● **Carbonyl reductions using hydride reducing agents**

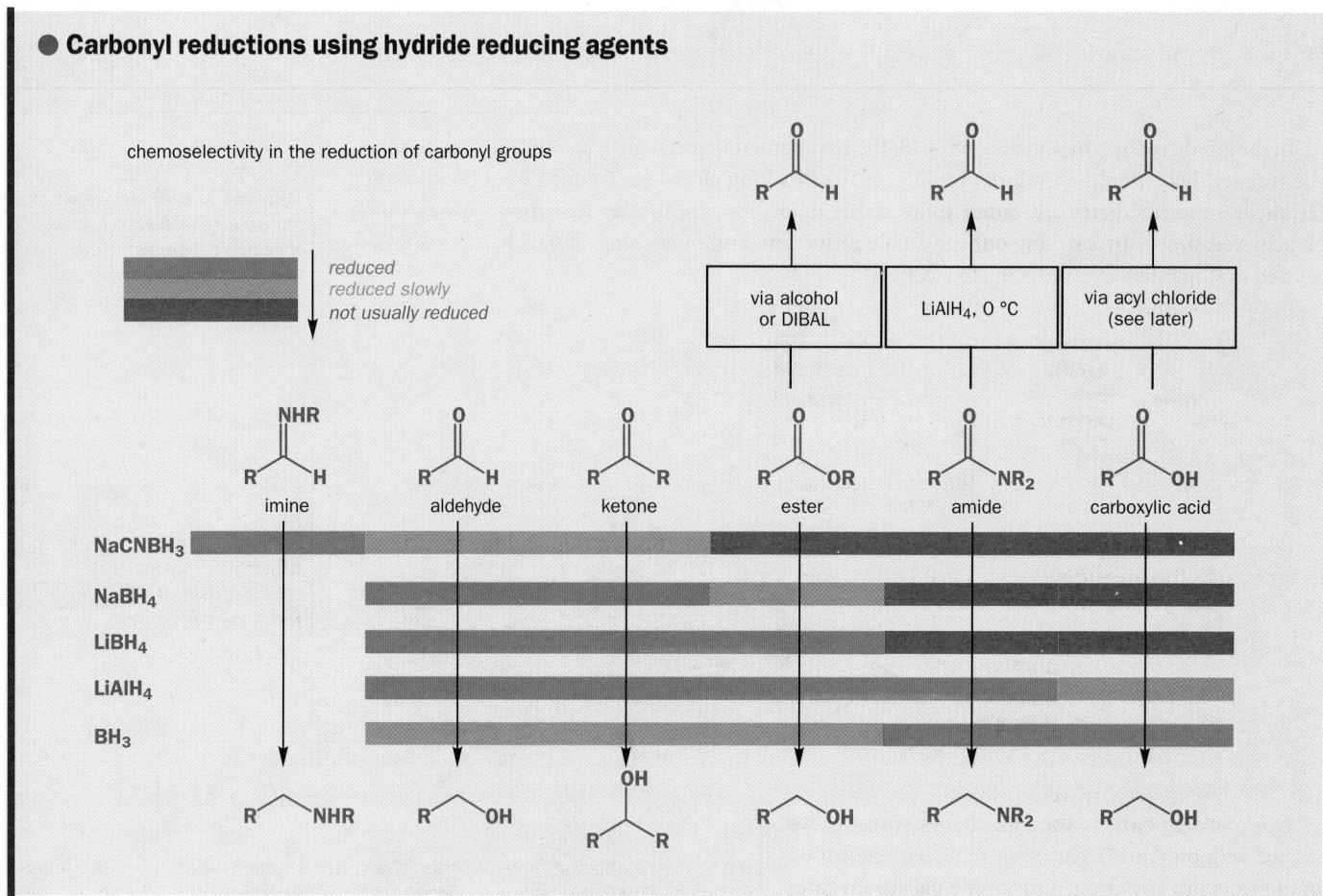

Now, what about the removal of the *N*-benzyl groups? This reaction is a **hydrogenolysis**—a cleavage of a C–X single bond by addition of hydrogen—and is just one of the many reactions hydrogen will do over metal catalysts. The 'mechanism' probably goes something like this.

We put 'mechanism' in inverted commas because this isn't really a proper chemical mechanism, more a scheme with a suggested sequence of events. The key points are that the benzyl amine co-ordinates to the metal catalyst via the electron-rich aromatic ring. The C–N bond is now in close proximity to the palladium-bound hydrogen atoms, and is reduced.

Because of the need for initial coordination with the catalyst, only benzylic or allylic C–X bonds can be reduced, but the X can be oxygen as well as nitrogen. We will come back to benzyl groups, and their hydrogenolysis, as a means for temporary protection of amines and alcohols later in the chapter. For the moment, though, we should take a broader look at catalytic hydrogenation as our second (after hydride reduction) important class of reductions.

Catalytic hydrogenation

You need to know about three sorts of hydrogenation reactions: the hydrogenation of a triple bond to a *Z*-alkene using 'Lindlar's catalyst', a poisoned form of palladium on barium sulfate; the hydrogenation of alkenes (including the imine above); and the hydrogenolysis of benzyl ethers and amines. We shall discuss each of these. The mechanism of hydrogenations is quite different from that of reductions by nucleophilic reducing agents like borohydride and, for this reason, catalytic hydrogenations have a totally different chemoselectivity. For example, it is quite possible to hydrogenate double bonds in the presence of aldehydes.

Even aromatic rings can be reduced by hydrogenation: in these examples the carbonyl groups survive while phenyl is reduced to cyclohexyl.

The catalyst in each of these three reductions is a different metal. Palladium and platinum are the most commonly used metal catalysts for hydrogenation, but hydrogenation can also work with nickel, rhodium, or ruthenium. The choice of catalyst depends on the compound to be reduced.

Substrate	Usual choice of metal
benzyl amine or ether	Pd
alkene	Pd, Pt, or Ni
aromatic ring	Pt or Rh, or Ni under high pressure

▶

Some hydrogenations require high pressures of hydrogen gas to get them to go at a reasonable rate. They are usually done in a sealed apparatus known as a **Parr hydrogenator**.

Catalytic hydrogenation is often chosen as a method for reduction because of its chemoselectivity for C=C double bonds and benzylic C–X bonds over C=O groups. The most important hydrogenation involving a carbonyl compound is not actually a reduction of the C=O double bond. Hydrogenation of acyl chlorides gives aldehydes in a reaction known as the **Rosenmund reaction**—really a hydrogenolysis of a C–Cl bond.

This is a good way of reducing compounds at the carboxylic acid oxidation level to aldehydes, which is why we included it in the table of carbonyl reductions on p. 622. The tertiary amine is needed both to neutralize the HCl produced in the reaction and to moderate the activity of the catalyst (and prevent overreduction). You will notice too that the catalyst support is different: Pd/BaSO₄ rather than Pd/C. BaSO₄ (and CaCO₃) are commonly used as supports with more easily reduced substrates because they allow the products to escape from the catalyst more rapidly and prevent overreduction. Acyl chlorides are among the easiest of all compounds to hydrogenate—look at this example.

the tertiary amine

quinoline

acyl chloride reduced   BaSO₄ is the support   aldehyde not reduced

H₂, Pd, BaSO₄

quinoline

aromatic rings not reduced

74–81% yield

Although aromatic rings can be hydrogenated, as you saw above, neither they nor the aldehyde product are reduced under these conditions and, as with hydride reductions of carbonyl compounds, we can draw up a sequence of reactivity towards hydrogenation. The precise ordering varies with the catalyst, especially with regard to the interpolation of the (less important, because other methods are usually better) carbonyl reductions (in yellow). Some catalysts are particularly selective towards certain classes of compound—for example, Pt, Rh, and Ru will selectively hydrogenate aromatic rings in the presence of benzylic C–O bonds, while with Pd catalysts the benzylic C–O bonds are hydrogenolysed faster.

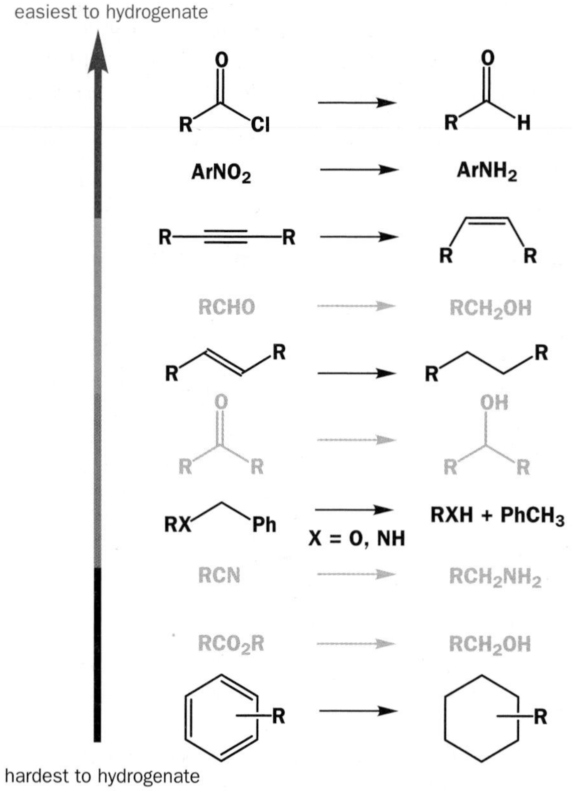

easiest to hydrogenate

hardest to hydrogenate

Like hydrogenolysis, the mechanism of the hydrogenation of C=C double bonds starts with coordination of the double bond to the catalyst surface.

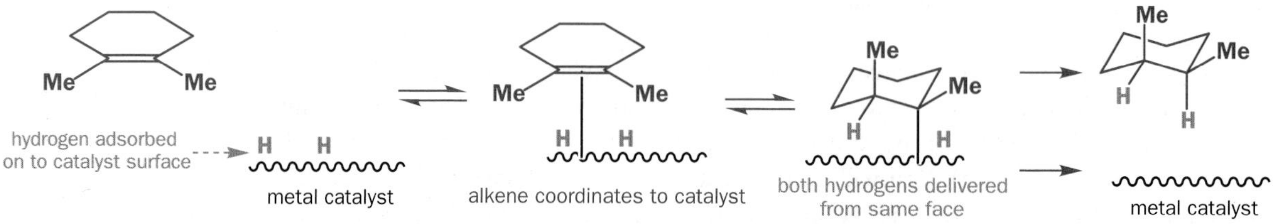

Two hydrogen atoms are transferred to the alkene, and they are often both added to the same face of the alkene. In Chapter 20 you met other reactions of alkenes: some, like bromination, were *anti*-selective, but others like epoxidation were *syn*-selective like hydrogenation.

> This cannot be relied upon though! The same reaction with Pd as catalyst gives mainly the *trans* isomer, because of the reversibility of the hydrogenation process. This intermediate can easily escape from the catalyst as an isomeric alkene, which can be re-hydrogenated from the other face. Isomerizations of this sort sometimes accompany hydrogenations.

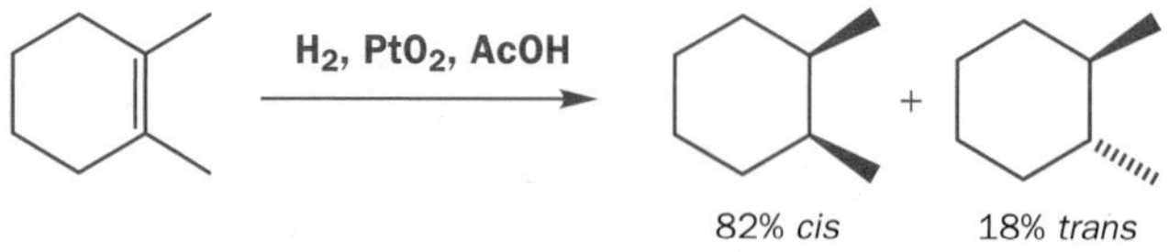

H_2, PtO_2, AcOH

82% *cis*   +   18% *trans*

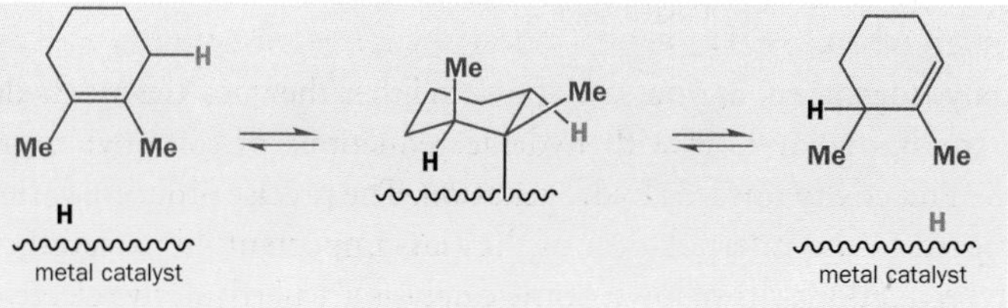

Hydrogenated vegetable oil

Plants such as soya, rapeseed, cottonseed, and sunflower are useful sources of edible vegetable oils, but these oils are unsuitable as 'butter substitutes' because of their low melting points. Their low melting points relative to animal fats are largely due to *cis* double bonds that disrupt the packing of the alkyl chains in the solid state. Treating the crude vegetable oil with hydrogen over a metal catalyst removes some of these double bonds, increases the proportion of saturated fat in the oil, and raises its melting point, making it suitable for making margarine.

Not all the double bonds are hydrogenated, of course: margarine manufacturers are desperate to tell us that their products are still 'high in unsaturated fatty acids'. Many also advertise that they are 'low in *trans* unsaturated fatty acids', because of a suggested link between incidence of coronary heart disease and *trans* unsaturated fatty acid intake.

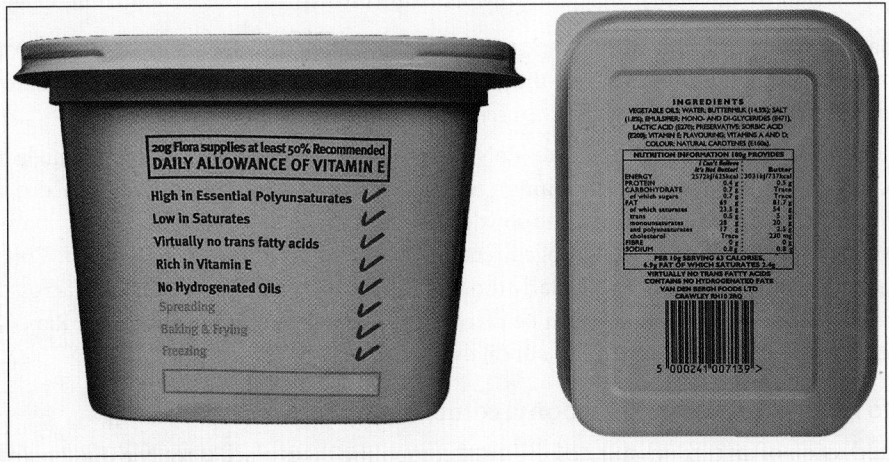

Where have the *trans* double bonds come from? Well, partial hydrogenation can lead to significant double-bond isomerization, not just to regioisomers (as in the example in the marginal box above) but to geometrical isomers too.

In Chapter 31 we shall come back to double-bond geometry and how to control it. There is more on fats in Chapter 49.

cis-unsaturated fat (ester of oleic acid)

↓ H₂, catalyst

saturated fat (ester of stearic acid)

trans-unsaturated fat (ester of elaidic acid)

A note on some catalysts

Catalytic hydrogenations take place only on the surface of the particles of a metal catalyst. The metal must therefore be very finely divided and is often mixed with a **support**—this is what Pd/C or Pd/BaSO₄ means—palladium particles deposited on a support of powdered charcoal or barium sulfate. Palladium on charcoal is probably the most commonly used catalyst, but three others deserve special mention.

1 You will meet **Lindlar's catalyst** in Chapter 31 but we will mention it now because of its special chemoselectivity. Unlike the other hydrogenations we have described, the Lindlar catalyst will hydrogenate alkynes to alkenes, rather than alkenes to alkanes. This requires rather subtle chemoselectivity: alkenes are usually hydrogenated at least as easily as alkynes, so we need to be sure the reaction stops once the alkene has been formed. The Lindlar catalyst is a palladium catalyst (Pd/CaCO₃) deliberately poisoned with lead. The lead lessens the activity of the catalyst and makes further reduction of the alkene product slow: most palladium catalysts would reduce alkynes all the way to alkanes. Best selectivities are obtained if quinoline is added to the reaction, just as in the Rosenmund reaction, and, in fact, alkyne to alkene reductions work with Pd/BaSO₄ + quinoline too. Even so, Lindlar reactions often have to be monitored carefully to make sure that overreduction is not taking place

Lindlar's catalyst = **Pd, CaCO₃, Pb(OAc)₂**

product of the reaction

reduction of alkene slow with Lindlar's catalyst

2 **Adams's catalyst** is formally PtO_2, and you have already seen this at work in one or two examples. The actual catalyst is, however, not the oxide of platinum, but the platinum metal that forms by reduction of PtO_2 to Pt *during the hydrogenation*

3 **Raney nickel** (often abbreviated to RaNi) is a finely divided form of nickel made from a nickel–aluminium alloy. The aluminium is dissolved away using concentrated aqueous sodium hydroxide, leaving the nickel as a fine powder. The process liberates H_2 (check this for yourself— on paper!), and some of this hydrogen remains adsorbed on to the nickel catalyst. This means that some hydrogenations, particularly those of C–S bonds, which you will come across later in this chapter and in Chapter 46, can be carried out just by using freshly prepared Raney nickel, with no added H_2 (RaNi as reagent, not catalyst)

How to reduce unsaturated carbonyl compounds

Where reduction of an α,β-unsaturated carbonyl compound takes place is really a question of regio-selectivity, not chemoselectivity, but it's useful to discuss the problem here having just introduced you to these hydrogenation methods. When we first covered conjugate addition in Chapter 10, we pointed out that hydride reducing agents are not good choices for the selective reduction of the C=O bond of unsaturated carbonyl compounds because they tend to add to the double bond as well, giving first the saturated carbonyl compound, which is then reduced to the alcohol. The way to get regioselective addition directly to the carbonyl group is to add a hard, Lewis-acidic metal salt, such as $CeCl_3$.

It should not surprise you that regioselective reduction of the C=C double bond alone is best done using catalytic hydrogenation as the C=C bond is weaker than the C=O bond. The flavouring compound known as 'raspberry ketone' is made by this method.

raspberry ketone

Nitro group reduction

Near the top of the list of reactivity towards hydrogenation lies the NO_2 group and in Chapter 22 we saw how the sequence of nitration of aromatic rings followed by reduction was a useful route to aromatic amines. The reduction can be carried out by Sn/HCl but catalytic hydrogenation is much simpler. The reaction is usually done in ethanol with a Pd or Pt catalyst and it may be necessary to add a weak acid to prevent the amine produced from poisoning the catalyst.

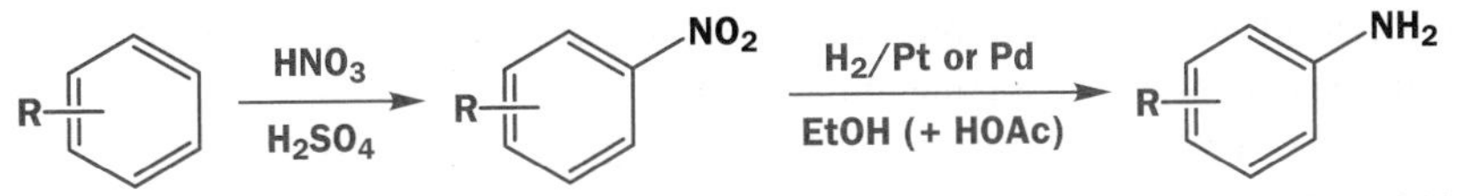

usually ~100% yield

The real gain over the Sn/HCl method is in the work-up. Instead of separating and disposing of voluminous toxic tin residues, a simple filtration to remove the catalyst, evaporation, and crystallization or distillation gives the amine.

Getting rid of functional groups

Functional groups can be useful for putting a molecule together, but their presence may not be required in the final product. We need ways of getting rid of them. Hydrogenation of alkenes is one way that you have seen, and alcohols can be got rid of either by elimination and then hydrogenation or by tosylation and substitution using borohydride to provide a nucleophilic hydrogen atom.

> Lithium triethylborohydride is used here, but other powerful hydride reducing agents would do as well.

Removal of carbonyl groups is harder, though there are several possible methods. C–O bonds are strong, but C–S bonds are much weaker, and are often easily reduced with Raney nickel (we come back to this in Chapter 46). We can get rid of aldehyde and ketone carbonyl groups by making them into **thioacetals**, sulfur analogues of acetals, formed in a reaction analogous to acetal formation (ch. 14) but using a dithiol with a Lewis acid catalyst. Freshly prepared Raney nickel carries enough H$_2$ (p. 626) to reduce the thioacetal without added hydrogen.

> This is sometimes known as the **Mozingo reaction**.

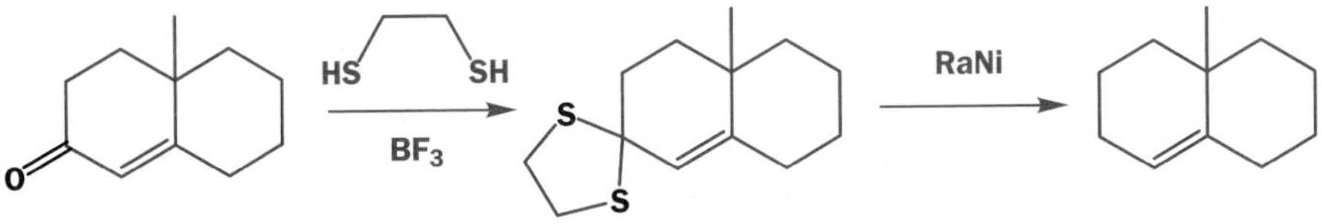

A slightly more vigorous method, known as the **Wolff–Kishner reduction**, is driven by the elimination of nitrogen gas from a hydrazone. Hot concentrated sodium hydroxide solution deprotonates the hydrazone, which can then eliminate an alkyl anion—a reaction you would usually be wary of writing, but which is made possible by the thermodynamic stability of N$_2$.

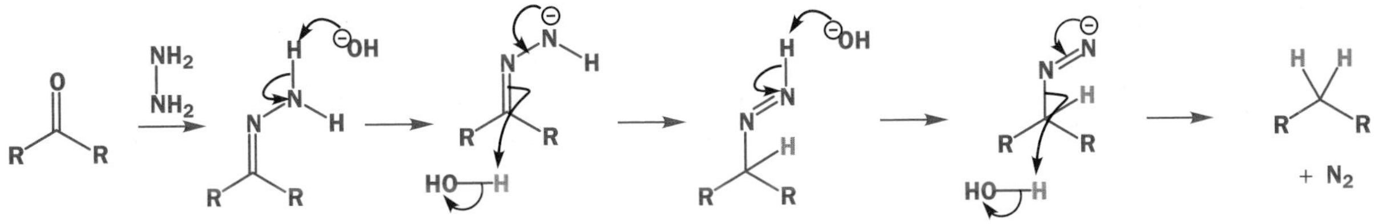

The third method is the simplest to do, but has the most complicated mechanism. The **Clemmensen reduction** is also rather violent, and really reasonable only for compounds with just the one functional group. It uses zinc metal dissolv*ing* in hydrochloric acid. As the metal dissolves, it gives up two electrons—in the absence of something else to do, these electrons would reduce the H$^+$ in the acid to H$_2$, and give ZnCl$_2$ and H$_2$. But in the presence of a carbonyl compound, the electrons go to reduce the C=O bond.

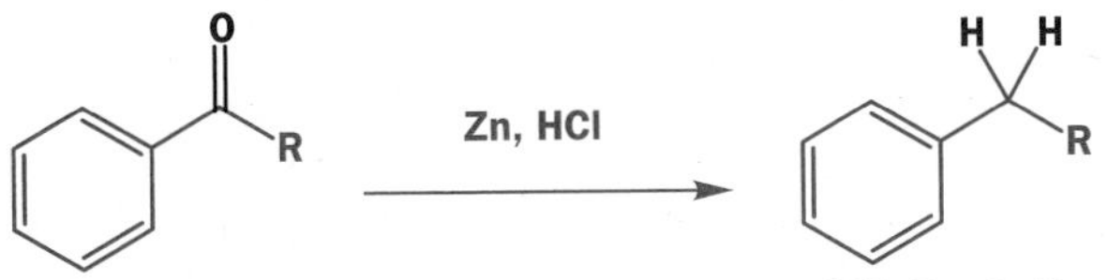

88%, R = C$_{17}$H$_{35}$

The mechanism has a good deal in common with a whole class of reductions, of which the Clemmensen is a member, known as **dissolving metal reductions**. We shall now look at these as our third (after metal hydrides and catalytic hydrogenation) important class of reducing agents.

Dissolving metal reductions

Group 1 metals, such as sodium or lithium, readily give up their single outer-shell electron as they dissolve in solvents such as liquid ammonia or ethanol. Electrons are the simplest reducing agents, and they will reduce carbonyl compounds, alkynes, or aromatic rings—in fact any functional group with a low-energy π^* orbital into which the electron can go.

We shall start by looking at the dissolving metal reduction of aromatic rings, known as the **Birch reduction**. Here is the reaction of benzene with lithium in liquid ammonia. At first sight, this reaction looks quite improbable, with an aromatic ring ending up as an unconjugated diene! The mechanism explains why we get this regiochemistry, and also why the reaction stops there—in other words why the dissolving lithium reduces an aromatic ring more readily than an alkene.

The first thing to note is that when lithium or sodium dissolve in ammonia they give an intense blue solution. Blue is the colour of solvated electrons: these group 1 metals ionize to give Li^+ or Na^+ and $e^-(NH_3)_n$—the gaps between the ammonia molecules are just the right size for an electron. With time, the blue colour fades, as the electrons reduce the ammonia to NH_2^- and hydrogen gas. Sodium amide, $NaNH_2$, the base you met early in this book, is made by dissolving Na in liquid NH_3 *and then waiting till the solution is no longer blue.*

Birch reductions use those blue solutions, with their solvated electrons, as reducing agents. The reduction of NH_3 to NH_2^- and H_2 is quite slow, and a better electron acceptor will get reduced in preference. In the example above, the electrons go into benzene's lowest lying antibonding orbital (its LUMO). The species we get can be represented in several ways, all of them radical anions (molecules with one excess, unpaired electron).

The radical anion is very basic, and it picks up a proton from the ethanol that is in the reaction mixture. The molecule is now no longer anionic, but it is still a radical. It can pick up another electron, which pairs with the radical to give an anion, which is quenched again by the proton source (ethanol).

The regiochemistry of the reaction is determined at the final protonation step—the anion itself is of course delocalized and could react at either end to give a conjugated diene, which would be more stable. Why then does it choose to pick up a proton in the middle and give a less stable isomer? Well, the full explanation is beyond the scope of this book, but suffice it to say that kinetically controlled reactions of pentadienyl anions with electrophiles typically take place at this central carbon.

Further questions of regioselectivity arise when there are substituents around the aromatic ring. Here are two examples. The second product was used by Evans in his synthesis of the alkaloid lucidu-line. These examples serve to illustrate the general principle that electron-withdrawing groups pro-mote *ipso, para* reduction while electron-donating groups promote *ortho, meta* reduction.

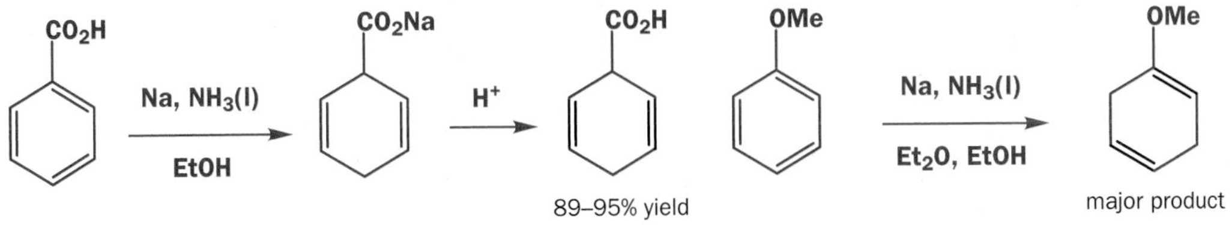

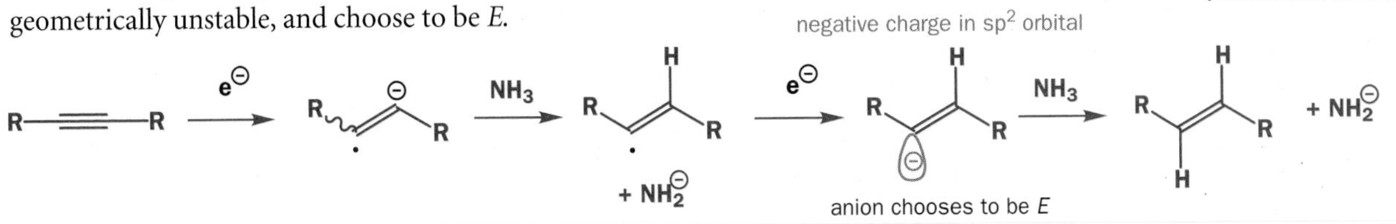

You can read more in Ian Fleming (1976). *Frontier orbitals and organic reaction mechanisms.* Wiley, Chichester.

Alkaloids appear in Chapter 51.

89–95% yield   major product

The explanation must lie in the distribution of electron density in the intermediate radical anions. Electron-withdrawing groups stabilize electron density at the *ipso* and *para* positions, and protona-tion occurs *para,* while electron-donating groups stabilize *ortho* and *meta* electron density.

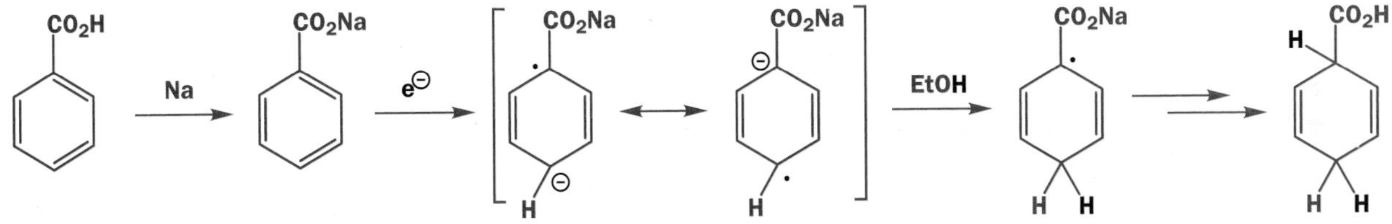

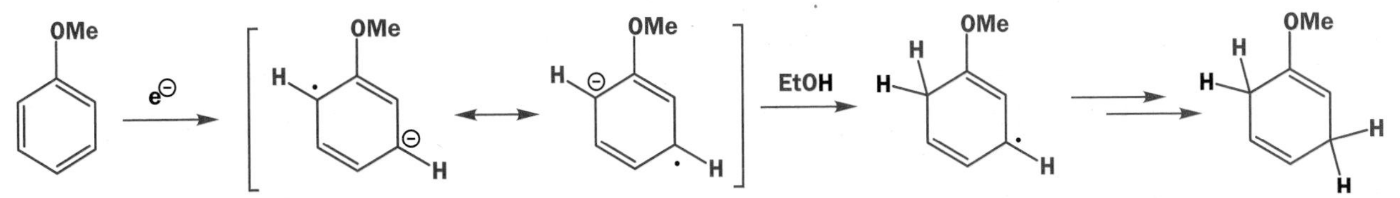

If you want the conjugated dienes as products, it is quite a simple matter to isomerize them using an acid catalyst. In fact, a small amount (about 20%) of the con-jugated product is produced anyway in the reaction of anisole above.

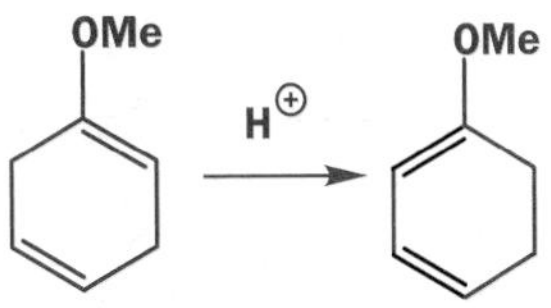

With anilines, it is impossible to stop the isomeriza-tion taking place during the reaction, and Birch reduc-tion always gives conjugated enamines.

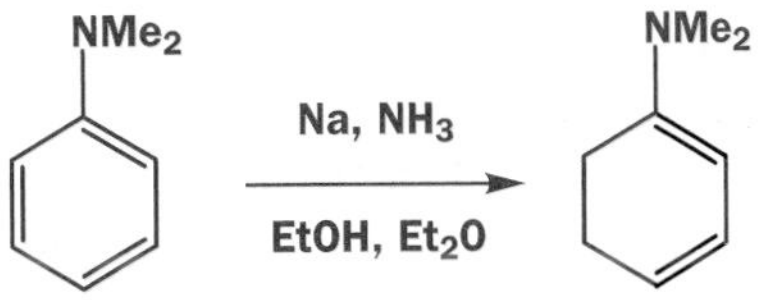

Birch reduction works for alkynes too, and is a good way of reducing them, to *trans* double bonds (the best way to reduce them to *cis*-alkenes is via H_2 and the Lindlar catalyst).

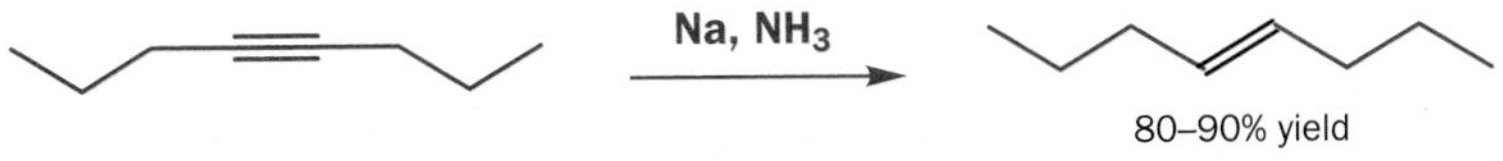

Make sure you can write a mechanism for this isomerization. *Hint.* Start as though you were protonating an enol ether on carbon. You saw this sort of thing in Chapter 21.

Birch-style reduction of α,β-unsaturated carbonyl compounds is described in Chapter 26.

► We come back to dissolving metals in Chapter 39, where we will also introduce another type of reduction—a good way of reducing C–halogen bonds to C–H.

80–90% yield

The mechanism follows the same course as the reduction of aromatic rings, but the vinyl anion is basic enough to deprotonate ammonia, so no added proton source is required. Vinyl anions are geometrically unstable, and choose to be *E.*

negative charge in sp² orbital

anion chooses to be *E*

One functional group may be more reactive than another for *kinetic* or for *thermodynamic* reasons

We hope that our survey of the important methods for reduction has shown you that, by choosing the right reagent, you can often react the functional group you want. The chemoselectivity you obtain is kinetic chemoselectivity—reaction at one functional group is simply faster than at another. Now look at the acylation of an amino alcohol (which is, in fact, a synthesis of the painkiller isobucaine) using benzoyl chloride under *acid* conditions. The hydroxyl group is acylated to form an ester. Yet under *basic* conditions, the selectivity is quite different, and an amide is formed.

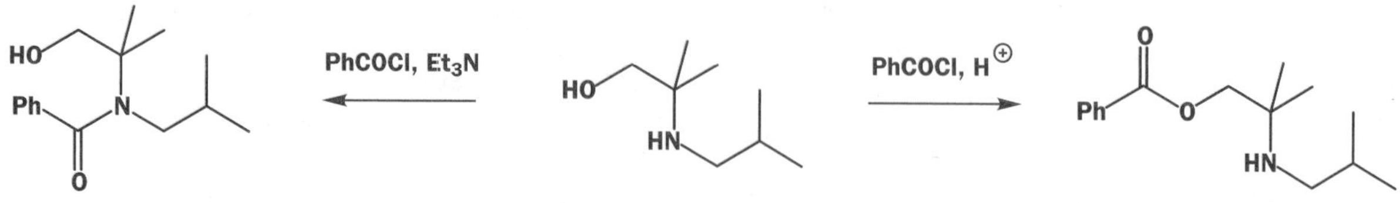

hydroxy amide   amino alcohol   amino ester

A clue to why the selectivity reverses is shown below—it is, in fact, possible to interconvert the ester and the amide simply by treating either with acid or with base.

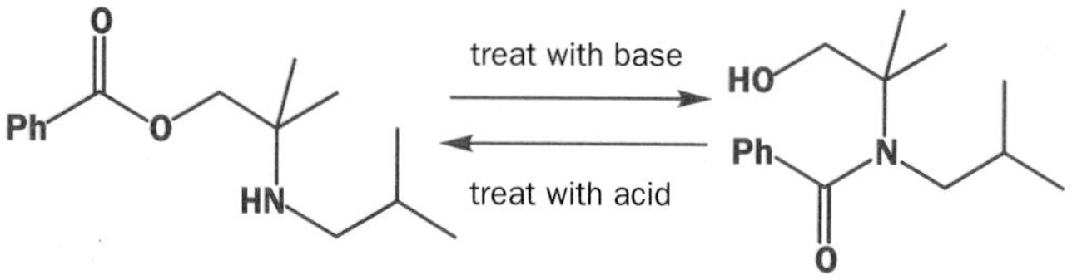

■

We first met examples of kinetic and thermodynamic control in Chapter 13.

The selectivity in these reactions is *thermodynamic* chemoselectivity. Under conditions in which the ester and amide can equilibrate, the product obtained is the more stable of the two, not necessarily the one that is formed faster. In base the more stable amide predominates, while in acid the amine is protonated, which prevents it from acting as a nucleophile and removes it from the equilibrium, giving the ester.

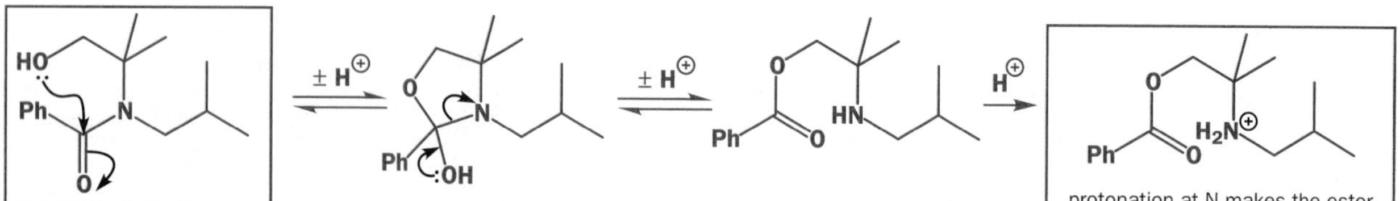

the amide is the thermo-dynamic product in base   reversible acyl migration between N and O   protonation at N makes the ester the thermodynamic product in acid

How to react the less reactive group (I)

The relative reactivity of the alcohol and amine in the example just given could be overturned by conducting a reaction under thermodynamic control. In kinetically controlled reactions, the idea that you can conduct chemoselective reactions on the more reactive of a pair of functional groups—carbonyl-based ones, for example—is straightforward. But what if you want to react the less reactive of the pair? There are two commonly used solutions. The first is illustrated by a compound needed by chemists at Cambridge to study an epoxidation reaction. They were able to make the following diol, but wanted to acetylate only the more hindered secondary hydroxyl group.

more hindered, less reactive secondary hydroxyl group   selective acetylation of secondary hydroxyl group required

more reactive primary hydroxyl group

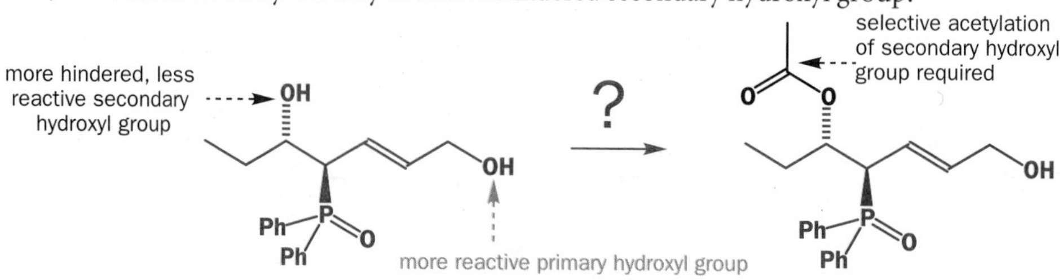

Treatment with one equivalent of an acyl chloride agent is no good because the primary hydroxyl group is more reactive; instead, the chemists acetylated both hydroxyl groups, and then treated the bis-acetate with mildly basic methanol (K_2CO_3, MeOH, 20 °C), which reacted only at the less hindered acetoxy group and gave the desired compound in 65% yield.

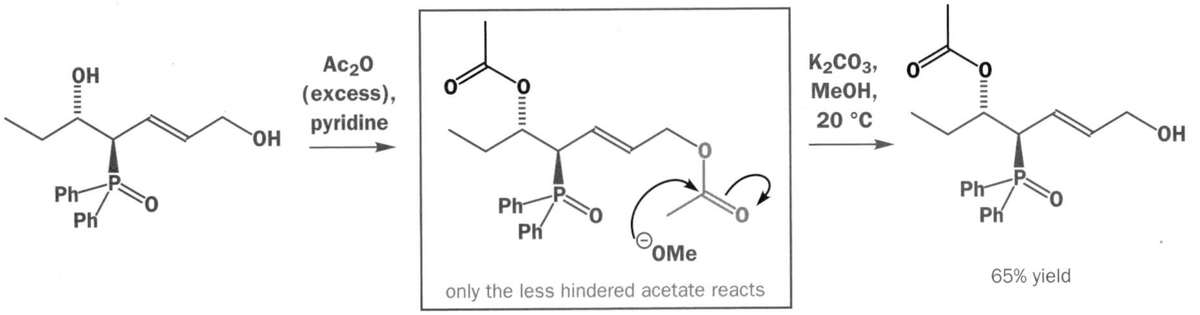

only the less hindered acetate reacts

65% yield

In other words, start by letting both groups react, and then go backwards but reverse the reaction at only one of the groups. Steric hindrance meant that the less favourable reaction (in other words, reaction at the less reactive group) was less readily reversed.

Chemoselectivity in the reactions of dianions

The converse of this idea is central to a useful bit of chemoselectivity that can be obtained in the reactions of dianions. 1-Propynol can be deprotonated twice by strong bases—first, at the hydroxyl group to make an alkoxide anion (the pK_a of the OH group is about 16) and, secondly, at the alkyne (pK_a of the order of 25) to make a 'dianion'. When this dianion reacts with electrophiles it always reacts at the alkynyl anion and not at the alkoxide.

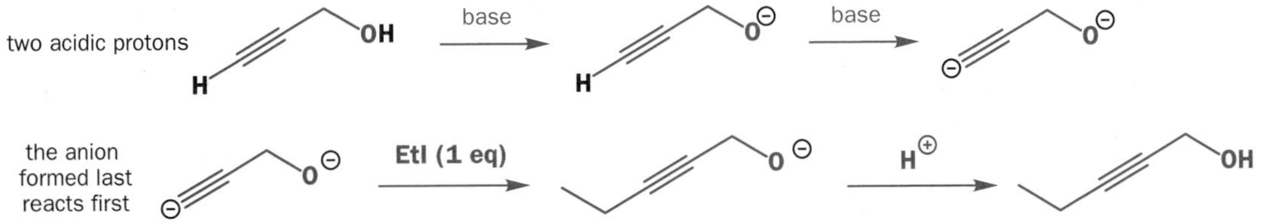

two acidic protons

the anion formed last reacts first

This reaction is important in a synthesis of the perfumery compound *cis*-jasmone. The alkyne is the precursor to *cis*-jasmone's alkene side chain.

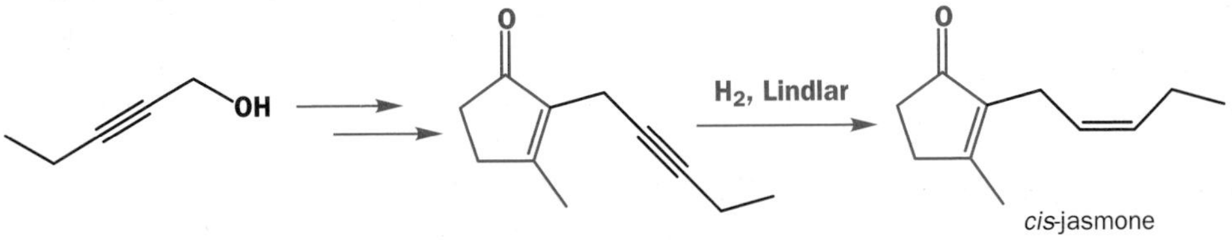

cis-jasmone

● **Reactivity of dianions**

The anion that is formed *last* reacts *first*.

Vollhardt used this sort of chemoselectivity in his 1977 synthesis of the female sex hormone oestrone. He needed an alkyl iodide, which could be made by reacting an anion of a bis-alkyne with ethylene oxide.

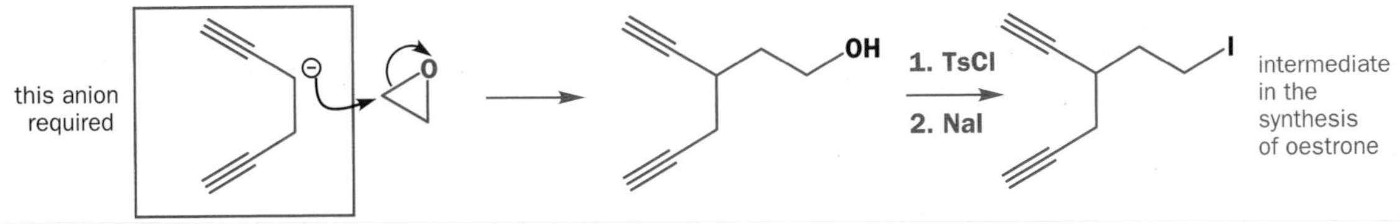

this anion required

intermediate in the synthesis of oestrone

Although anions can often be formed straightforwardly next to alkynes, there are two other more acidic protons (green) in the molecule that would be removed by base before the yellow proton. However, treatment with *three* equivalents of butyl lithium removes all three, and the trianion reacts with ethylene oxide at the last-formed anionic centre to give the required compound.

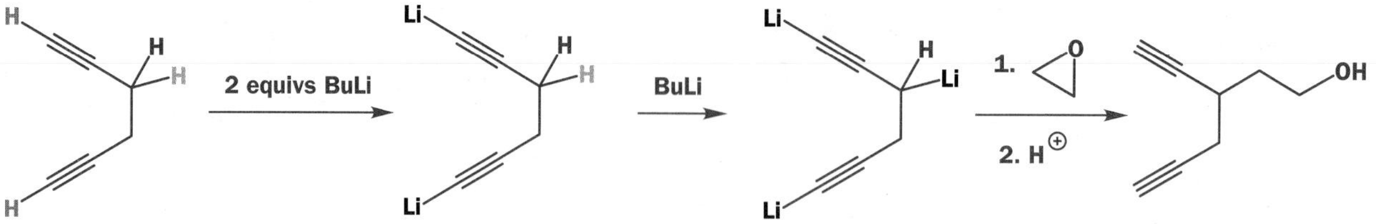

How to react the less reactive group (II): protecting groups

The usual way of reacting a less reactive group in the presence of a more reactive one is to use a protecting group. This tertiary alcohol, for example, could be made from a keto-ester if we could get phenylmagnesium bromide to react with the ester rather than with the ketone.

As you would expect, simply adding phenylmagnesium bromide to ethyl acetoacetate leads mainly to addition to the more electrophilic ketone.

52% yield

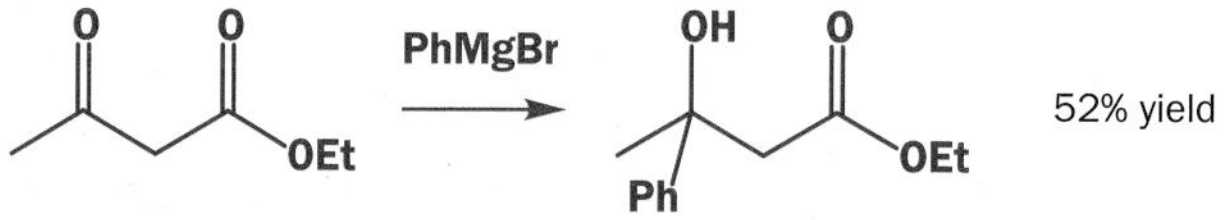

One way of making the alcohol we want is to protect the ketone as an acetal. An **acetal-protecting group** (shown in black) is used.

> Five-membered cyclic acetals like these are known as dioxolanes. You met them first in Chapter 14 when we were discussing acetal formation and hydrolysis.

> This table of protecting groups will grow, line by line, as we move through this chapter and the next.

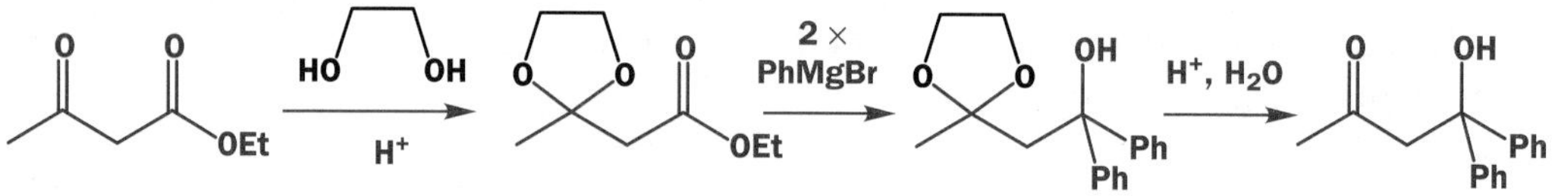

The first step puts the protecting group on to the (more electrophilic) ketone carbonyl, making it no longer reactive towards nucleophilic addition. The Grignard then adds to the ester, and finally a 'deprotection' step, acid-catalysed hydrolysis of the acetal, gives us back the ketone. An acetal is an ideal choice here—acetals are stable to base (the conditions of the reaction we want to do), but are readily cleaved in acid.

Protecting group	Structure	Protects	From	Protection	Deprotection
acetal (dioxolane)	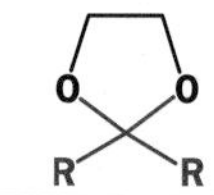	ketones, aldehydes	nucleophiles, bases	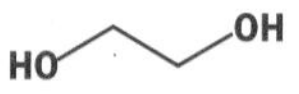 HO—⌒—OH	water, H⁺ cat.

By protecting sensitive functional groups like ketones it becomes possible to make reagents that would otherwise be unstable. In a synthesis of the natural product porantherine, a compound based on this structure was needed.

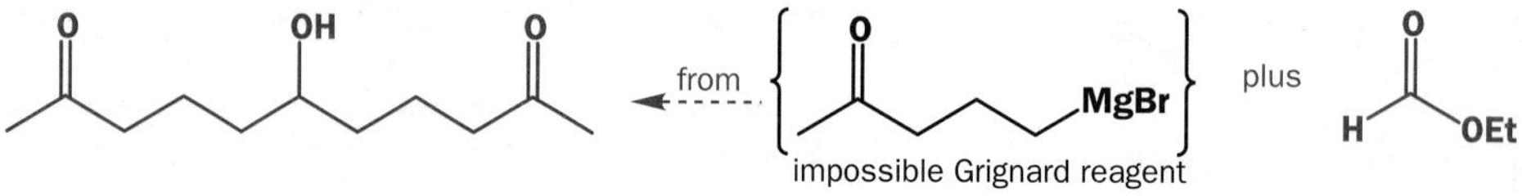

One way to make it is to add a Grignard reagent twice to ethyl formate. But, of course, a ketone-containing Grignard is an impossibility as it would self-destruct, so an acetal-protected compound was used.

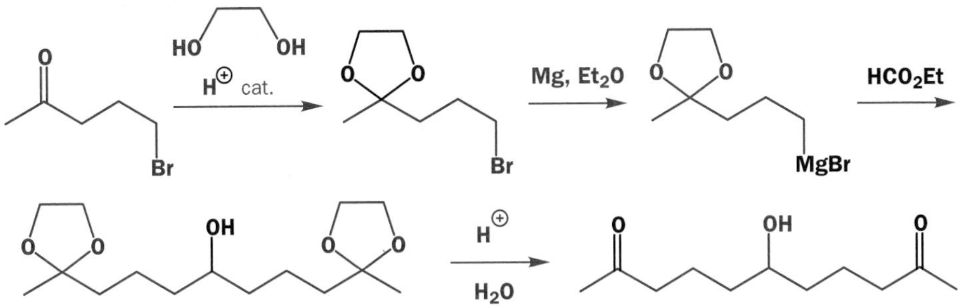

Strongly nucleophilic reagents like Grignard reagents and organolithiums are also strong bases, and may need protecting from acidic protons as well as from electrophilic carbonyl groups. Among the most troublesome are the protons of hydroxyl groups. When some American chemists wanted to make the antiviral agent Brefeldin A, they needed a simple alkynol.

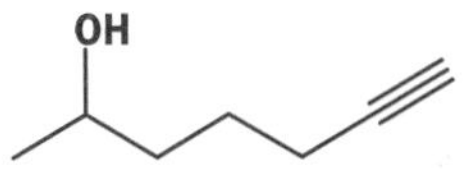

A synthesis could start with the same bromoketone as the one above: reduction gives an alcohol, but alkylation of an alkynyl anion with this compound is not possible, because the anion will just deprotonate the hydroxyl group.

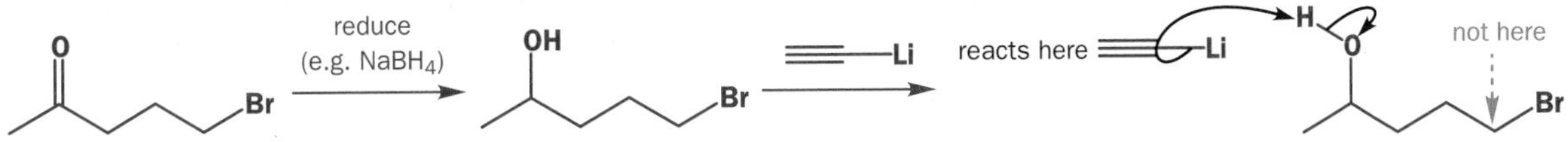

deprotonation of hydroxyl group by strongly basic reagent

The answer is to protect the hydroxyl group, and the group chosen here was a **silyl ether**. Such ethers are made by reacting the alcohol with a trialkylsilyl chloride (here *t*-butyl dimethyl silyl chloride, or TBDMSCl) in the presence of a weak base, usually imidazole, which also acts as a nucleophilic catalyst (Chapter 12).

imidazole = (a weak base)

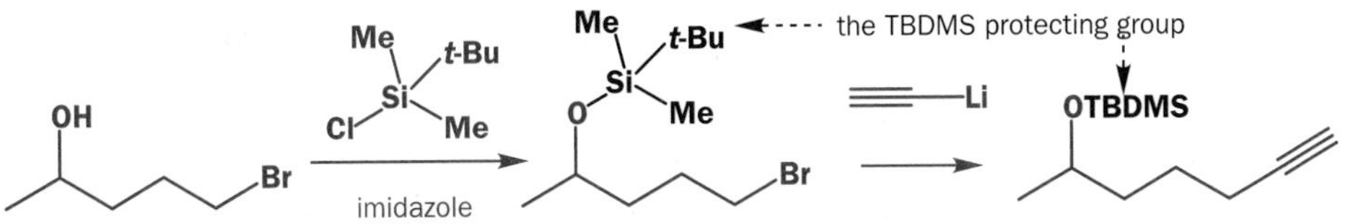

Silicon has a strong affinity for electronegative elements, particularly O, F, and Cl, so trialkyl-silyl ethers are attacked by hydroxide ion, water, or fluoride ion but are more stable to carbon or nitrogen bases or nucleophiles. They are usually removed with aqueous acid or fluoride salts, particularly $Bu_4N^+F^-$ which is soluble in organic solvents. In fact, TBDMS is one member of a whole family of trialkylsilyl protecting groups and their relative stability to nucleophiles of various kinds is determined by the three alkyl groups carried by silicon. The most labile, trimethylsilyl (TMS), is removed simply on treatment with methanol, while the most stable require hydrofluoric acid.

Although not important to our discussion here, these substitution reactions are not the simple S_N2 reactions (Chapter 17) they might appear to be. The nucleophile adds to silicon first to form a five-valent anion which decomposes with the loss of the alcohol (Chapter 21).

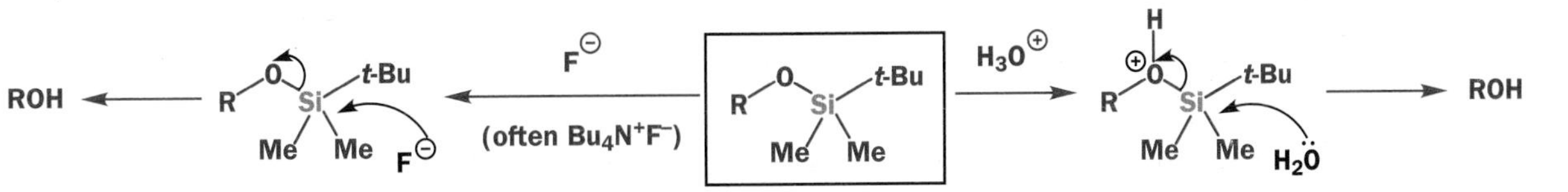

Protecting group	Structure	Protects	From	Protection	Deprotection
trialkylsilyl (R_3Si-, e.g. TBDMS)	RO—$SiMe_3$ RO—$SiMe_2Bu^t$	alcohols (OH in general)	nucleophiles, C or N bases	R_3SiCl, base	H^+, H_2O, or F^-

Why can't we just use a simple alkyl ether (methyl, say) to protect a hydroxyl group? There is no problem making the ether, and it will survive most reactions—but there *is* a problem getting an ether off again. This is always a consideration in protecting group chemistry—you want a group that is stable to the conditions of whatever reaction you are going to do (in these examples, strong bases and nucleophiles), but can then be removed under mild conditions that do not result in total decomposition of a sensitive molecule. What we need then, is an ether that has an 'Achilles' heel'—a feature that makes it susceptible to attack by some specific reagent or under specific conditions. One such group is the tetrahydropyranyl (THP) group. Although it is stable under basic conditions, as an ether would be, it is an acetal—the presence of the second oxygen atom is its 'Achilles' heel' and makes the THP protecting group susceptible to hydrolysis under acidic conditions. You could see the lone pair on the second oxygen atom as a 'safety catch' that is released only in the presence of acid.

Making the THP acetal has to be done in a slightly unusual way because the usual carbonyl compound plus two alcohols is inappropriate. Alcohols are protected by reacting them with an enol ether, dihydropyran, under acid catalysis. Notice the oxonium intermediate (formed by a familiar mechanism from Chapter 14)—just as in a normal acetal-forming reaction. In this example the THP group is at work preventing a hydroxyl group from interfering in the reduction of an ester.

> Some chemistry of enol ethers is in Chapter 21.

> ▶
> A little further inspection will show you that the THP group here is not just stopping the OH interfering with the LiAlH₄ reduction, but is also crucial to the preservation of the chirality of this compound. The wedged bond shows you that the starting material is a single enantiomer: without a protecting group on one of the hydroxyls, they would be identical and the compound would no longer be chiral. More detailed inspection shows that the THP group also complicates the situation by introducing an extra chiral centre, and hence the potential for two diastereoisomers, which we will ignore.

Protecting group	Structure	Protects	From	Protection	Deprotection
tetrahydropyranyl (THP)		alcohols (OH in general)	strong bases	dihydropyran and acid	H⁺, H₂O

The THP-protected compound above is an intermediate in a synthesis of the insecticide milbemycin as a single enantiomer. It needs to be converted to this alkyne—and now the *other* hydroxyl group will need protecting.

This time, though, TBDMS will not do, because the protecting group needs to withstand the acidic conditions needed to remove the THP protecting group! What is more, the protecting group needs to be able to survive acid conditions in later steps of the synthesis of the insecticide. The answer is to use a third type of hydroxyl-protecting group, a benzyl ether. Benzyl (Bn) protecting groups are put on using strong base (usually sodium hydride) plus benzyl bromide, and are stable to both acid and base.

> Note the abbreviation for a benzyl ether, ROCH$_2$Ph, is RO**Bn**. Contrast this with benz*oyl* esters, ROCOPh, which may be abbreviated RO**Bz**.

the benzyl (Bn) protecting group

The benzyl ether's Achilles' heel is the aromatic ring and, after reading the first half of this chapter, you should be able to suggest conditions that will take it off again: hydrogenation (hydrogenolysis) over a palladium catalyst.

> It must be a *palladium* catalyst—platinum would catalyse hydrogenation of the aromatic ring.

benzyl ether deprotection: catalytic hydrogenation

Benzyl ethers can sometimes be removed by acid, if the acid has a *nucleophilic* conjugate base. HBr, for example, will remove a benzyl ether because Br$^-$ is a good enough nucleophile to displace ROH, though only at the reactive, benzylic centre.

benzyl ether deprotection: acid with nucleophilic counterion

HBr in acetic acid (just the solvent) is used to remove the benzyl ether protecting groups in this example, which forms part of a synthesis of the alkaloid galanthamine.

> Alkaloids appear in Chapter 51.

galanthamine

Protecting group	Structure	Protects	From	Protection	Deprotection
benzyl ether (OBn)	RO—CH$_2$Ph / ROBn	alcohols (OH in general)	almost everything	NaH, BnBr	H$_2$, Pd/C, or HBr
methyl ether (ArOMe)	MeO—Ar—R	phenols (ArOH)	bases	NaH, MeI, or (MeO)$_2$SO$_2$	BBr$_3$, HBr, HI, Me$_3$SiI

We said earlier that simple methyl ethers are inappropriate as protecting groups for OH because they are too hard to take off again. That is usually true, but not if the OH is phenolic—ArOH is an

Alternatives to HBr include BBr₃, usually the favoured reagent, HI, and Me₃SiI. You met the reaction of phenyl ethers with BBr₃ in Chapter 17.

even better leaving group than ROH, so HBr will take off methyl groups from aryl methyl ethers too. You will see an example in Chapter 25.

deprotection of aryl methyl ethers

ArOMe $\xrightarrow{\text{HBr}}$ **ArOH** + **MeBr**

Protecting groups may be useful, but they are also wasteful—both of time, because there are two extra steps to do (putting the group on and taking it off), and of material, because these steps may not go in 100% yield. Here's one way to avoid using them. During the development of the best-selling anti-asthma drug salbutamol, the triol boxed in green was needed. With large quantities of salbutamol already available, it seemed most straightforward to make the triol by adding phenylmagnesium bromide to an ester available from salbutamol. Unfortunately, the ester also contains three acidic protons, making it look as though the hydroxyl and amine groups all need protecting. But, in fact, it was possible to do the reaction just by adding a large excess of Grignard reagent: enough to remove the acidic protons *and* to add to the ester.

This strategy is easy to try, and, providing the Grignard reagent isn't valuable (you can buy PhMgBr in bottles), is much more economical than putting on protecting groups and taking them off again. But it doesn't always work—there is no way of telling whether it will until you try the reaction in the lab. In this closely related reaction, for example, the same chemists found that they needed to protect both the phenolic hydroxyl group (but not the other, normal alcohol OH!) as a benzyl ether and the amine NH as a benzyl amine. Both protecting groups come off in one hydrogenation step.

This is the last appearance of the table of protecting groups in this chapter but it is extended in Chapter 25.

Benzyl groups are one way of protecting secondary amines against strong bases that might deprotonate them. But it is the nucleophilicity of amines that usually poses problems of chemoselectivity, rather than the acidity of their NH groups, and we come back to ways of protecting them from electrophiles when we deal with the synthesis of peptides in Chapter 25.

Bergamotene

An acidic proton posed a potential problem during E.J. Corey's synthesis of bergamotene (a component of the fragrance of Earl Grey tea). You met the Wittig reaction in Chapter 14, and phosphonium ylids are another type of basic, nucleophilic reagent that –OH groups often need protecting against. But, in this synthesis, a successful Wittig reaction was carried out even in the presence of a carboxylic acid, again by using an excess of the phosphonium ylid. We talk about carboxylic acid protection in the next chapter. In fact the carboxylate anion is itself a kind of protecting group as it discourages the rather basic Wittig reagent from removing a proton to form an enolate.

Protecting group	Structure	Protects	From	Protection	Deprotection
acetal (dioxolane)	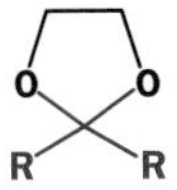 (structure)	ketones, aldehydes	nucleophiles, bases	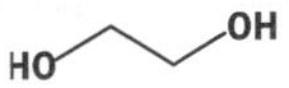 (structure) HO—OH	water, H$^+$ cat.
trialkylsilyl (R$_3$Si-, e.g. TBDMS)		alcohols (OH in general)	nucleophiles, C or N bases	R$_3$SiCl, base	H$^+$, H$_2$O, or F$^-$
tetrahydropyranyl (THP)	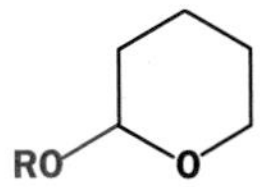 (structure) RO—O	alcohols (OH in general)	strong bases	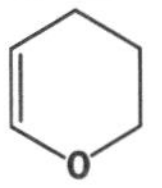 (structure) dihydro-pyran and acid	H$^+$, H$_2$O
benzyl ether (OBn)	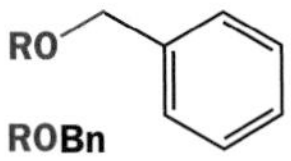 (structure) RO—⬡ **ROBn**	alcohols (OH in general)	almost everything	NaH, BnBr	H$_2$, Pd/C, or HBr
methyl ether (ArOMe)	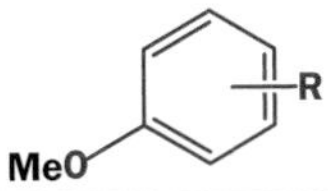 (structure) MeO—⬡—R	phenols (ArOH)	bases	NaH, MeI, or (MeO)$_2$SO$_2$	BBr$_3$, HBr, HI, Me$_3$SiI
benzyl amine (NBn)	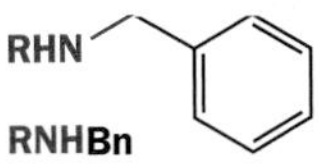 (structure) RHN—⬡ **RNHBn**	amines	strong bases	BnBr, K$_2$CO$_3$	H$_2$, Pd

We have dealt with protecting groups for C=O, OH, and NH that resist nucleophiles, acids, and base. Sometimes functional groups need protecting against oxidation, and we finish our introduction to protecting groups with an example. During a synthesis of the bacterial product rapamycin, an epoxy alcohol needed converting to a ketone through a sequence that involves selective oxidation of only one of two hydroxyl groups. The group to be oxidized is there in the starting material, so it can be protected straight away. The protecting group (Bn) needs to be acid-stable, because the next step is to open the epoxide with methanol, revealing the second hydroxyl group. This then needs protecting—TBDMS was chosen, so as to be stable to hydrogenolysis, which deprotects the hydroxyl that we want to oxidize. Finally, oxidation gives the ketone.

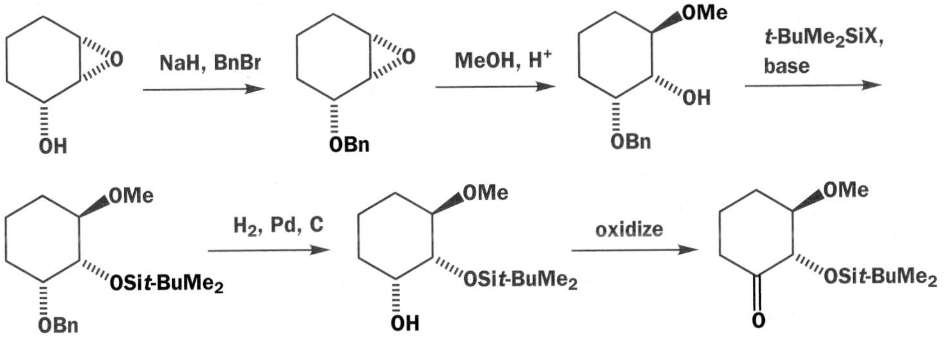

In this chapter we have talked about most of the steps in this sequence, except the epoxide-opening reaction (for which read Chapters 17 and 18) and the oxidation step. Which reagent would a chemist choose to oxidize the alcohol to the ketone, and why? We shall now move on to look at oxidizing agents in detail.

Oxidizing agents

We dealt in detail earlier in the chapter with reducing agents and their characteristic chemoselectivities. Oxidizing agents are equally important, and in the chapter on electrophilic addition to alkenes we told you about peracids as oxidizing agents for C=C double bonds—they give epoxides. But

▶ In Chapter 37 you will find out that peracids also react with ketones, but that need not concern us here.

peracids do not react with alcohols: they are chemoselective oxidants of C=C double bonds only. Later in the book, you will meet more oxidizing agents, such as osmium tetroxide (OsO_4) and ozone (O_3)—these are also chemoselective for double bonds, because they react with the C=C π bond, and we shall leave them until Chapter 35. In this section we will be concerned only with oxidizing agents that oxidize alcohols and carbonyl compounds.

The most commonly used methods for oxidizing alcohols are based around metals in high oxidation states, often chromium(VI) or manganese(VII), and you will see that mechanistically they are quite similar—they both rely on the formation of a bond between the hydroxyl group and the metal. Another class of oxidations, those that use halogens, sulfur, or nitrogen in high oxidation states, we will deal with relatively briefly.

● Oxidizing agents

Chemoselective for C=C double bonds[a]	Chemoselective for alcohols or carbonyl compounds
peracids, RCO_3H (Chapter 20)	Cr(VI) compounds
osmium tetroxide, OsO_4 (Chapter 35)	Mn(VII) compounds
ozone, O_3 (Chapter 35)	some high oxidation state Hal, N, or S compounds

[a]not dealt with in this chapter.

How to oxidize secondary alcohols to ketones

We start with this, because overoxidation is difficult. Provided the alcohol is not acid-sensitive, a good method is sodium dichromate in dilute sulfuric acid. This is usually added to a solution of the alcohol in acetone, and is known as the **Jones oxidation**.

The mechanism starts with the formation of $HCrO_4^-$ ions, that is, Cr(VI), from dichromate ion in solution. In acid, these form chromate esters with alcohols. The esters (boxed in black) decompose by elimination of the Cr(IV) $HCrO_3^-$, which subsequently reacts with a Cr(VI) species to yield $2 \times$ Cr(V). These Cr(V) species can oxidize alcohols in the same way, and are thereby reduced to Cr(III) (the final metal-containing by-product). Cr(VI) is orange and Cr(III) is green, so the progress of the reaction is easy to follow by colour change.

oxidation of alcohols with Cr(VII)

chromate ester

Chromic acid is best avoided if acid-sensitive alcohols are to be oxidized, and an alternative reagent for these is PCC (pyridinium chlorochromate), which can be used in dichloromethane.

pyridinium chlorochromate, PCC

How to oxidize primary alcohols to aldehydes

Aqueous methods like the Jones oxidation are no good for this, since the aldehyde that forms is further oxidized to acid via its hydrate. The oxidizing agent treats the hydrate as an alcohol, and oxidizes it to the acid.

overoxidation of aldehydes

The key thing is to avoid water—so PCC in dichloromethane works quite well. The related reagent PDC (pyridinium dichromate) is particularly suitable for oxidation to aldehydes.

Some very mild oxidizing agents are being more and more widely used for the synthesis of very sensitive aldehydes. One of these is known as TPAP (tetra-*n*-propylammonium perruthenate, pronounced 'tee-pap').

TPAP can be used catalytically, avoiding the large amounts of toxic heavy metal by-products generated by most chromium oxidations. The stoichiometric oxidant in this reaction is 'NMO' (*N*-methylmorpholine-*N*-oxide), which is reduced to the amine, reoxidizing the ruthenium back to Ru(VI).

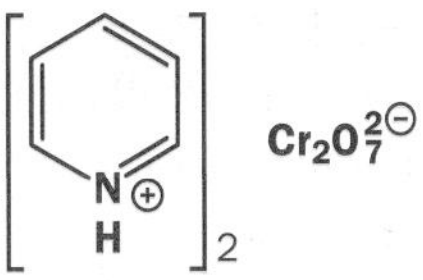

pyridinium dichromate, PDC

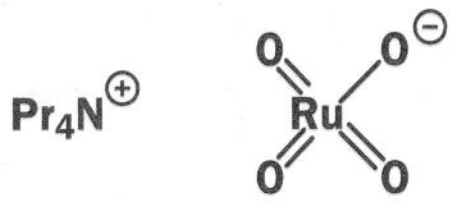

TPAP
tetrapropylammonium perruthenate

> Notice that the two protecting groups, both acid-sensitive, survive these conditions very well.

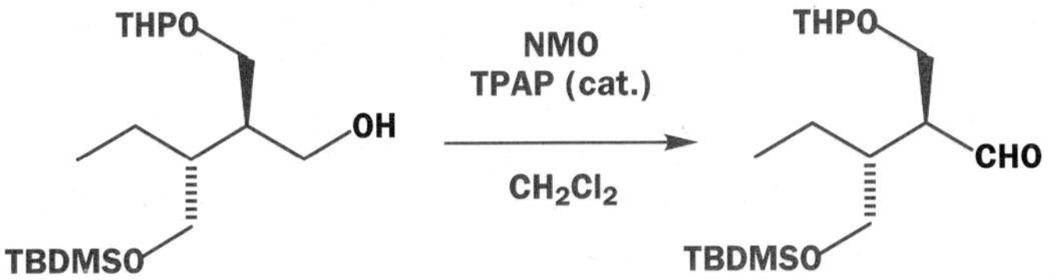

97% yield

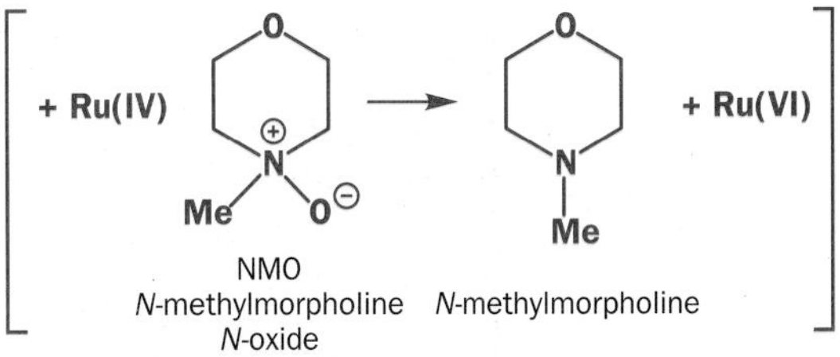

NMO
N-methylmorpholine *N*-oxide   *N*-methylmorpholine

Another important modern reagent (discovered in 1983) is known as the **Dess–Martin periodinane**, and is an iodine compound that can be made from 2-iodobenzoic acid, itself available from anthranilic acid via the diazonium salt route, as described in the last chapter.

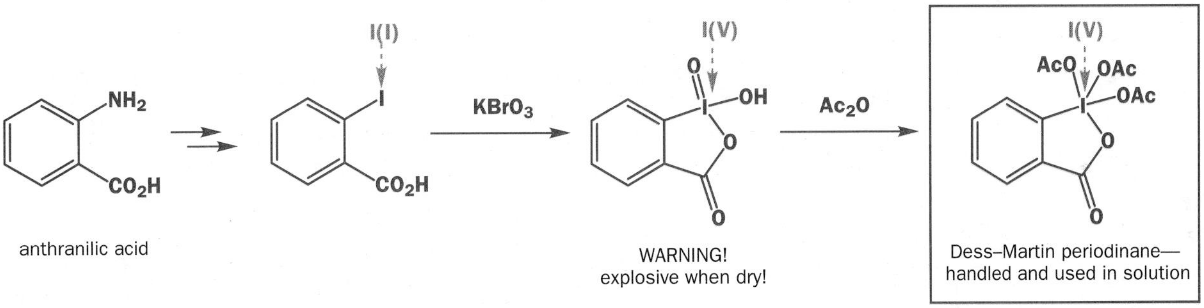

anthranilic acid

WARNING!
explosive when dry!

Dess–Martin periodinane—
handled and used in solution

It will oxidize even very sensitive alcohols to carbonyl compounds—few others, for example, would give a *cis*-α,β-unsaturated aldehyde from a *cis*-allylic alcohol without isomerizing it to *trans*, or producing other by-products.

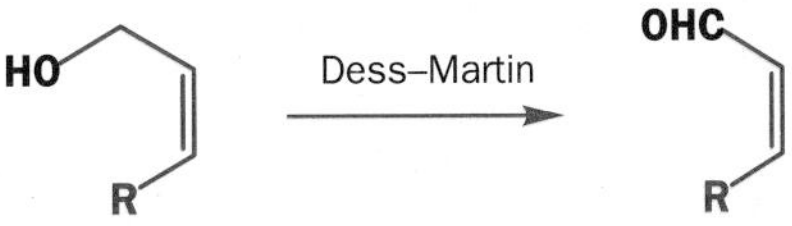

We shall leave detailed discussion of one more method till much later, in Chapter 46, since the mechanism involves some sulfur chemistry you will meet there. But we introduce it here because of its synthetic importance. Known as the **Swern oxidation**, it uses a sulfoxide [S(IV)] as the oxidizing agent. The sulfoxide is reduced to a sulfide, while the alcohol is oxidized to an aldehyde.

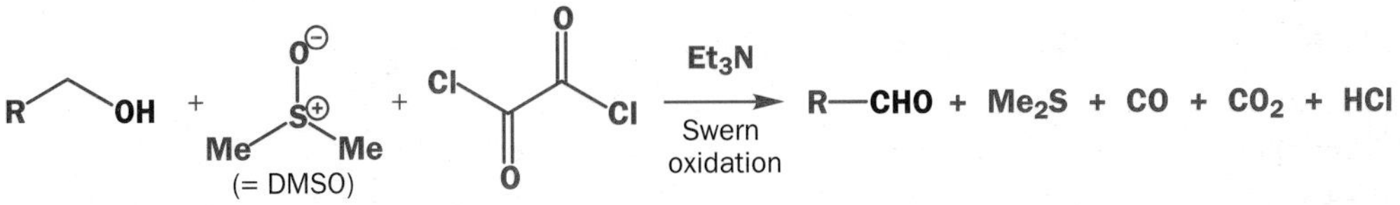

(= DMSO)   Swern oxidation

$R—CHO + Me_2S + CO + CO_2 + HCl$

How to oxidize primary alcohols or aldehydes to carboxylic acids

This is the 'overoxidation' we were trying to avoid in oxidizing alcohols to aldehydes, and is best done with an aqueous solution of Cr(VI) or Mn(VII). Acidic or basic aqueous potassium permanganate is often a good choice. From alcohols in acidic solution the mechanism follows very much the lines of the chromic acid mechanism; from aldehydes, the mechanism is very similar.

oxidation of aldehydes with Mn(VII)

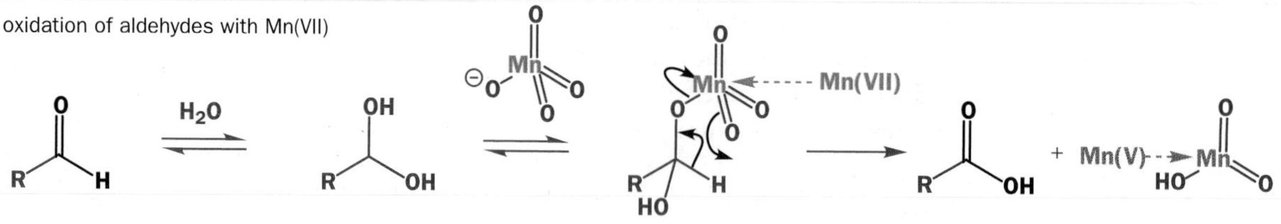

To conclude...

In the next chapter we will look at the ways in which the ideas and principles we have talked about in this chapter, and the reactions you have met in the 23 preceding ones, can be used in a practical way to make useful and interesting molecules. We will look at the synthesis of some of the molecules found in nature, such as hormones, plant-derived products with medicinal properties, and insect pheromones, as well as others that Nature has not made but that for one reason or another man has chosen to make.

Problems

1. How would you convert this bromoaldehyde chemoselectively into the two products shown?

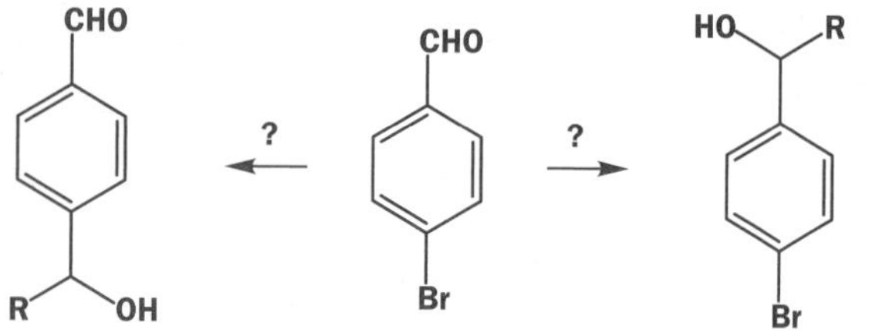

2. Explain the chemoselectivity of these reactions. What is the role of the Me₃SiCN?

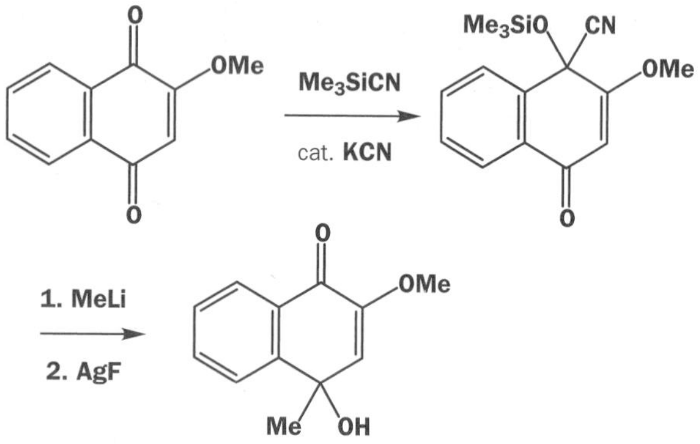

3. How would you convert this lactone selectively either into the hydroxy-acid or into the unfunctionalized acid?

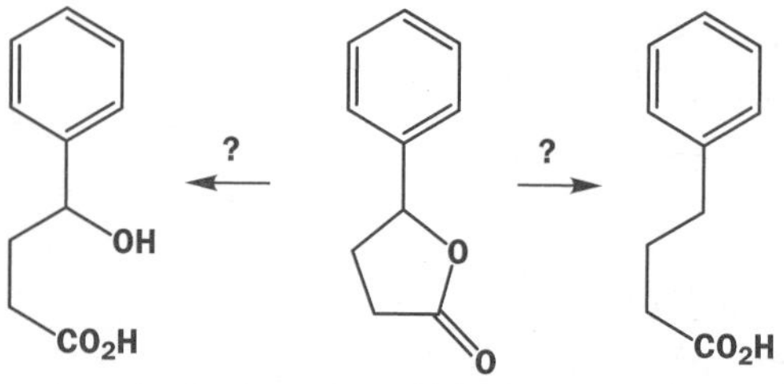

4. Predict the products of Birch reduction of these aromatic compounds.

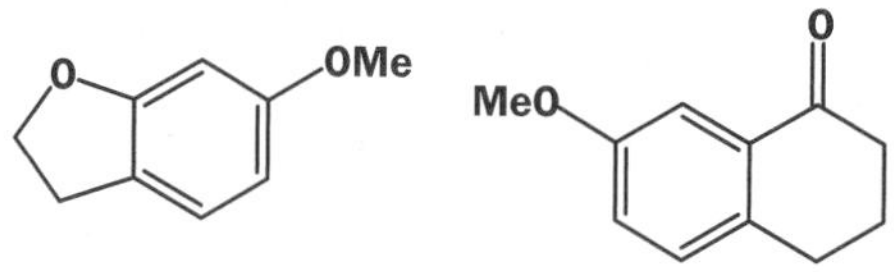

5. How would you carry out these reactions? In some cases more than one step may be needed.

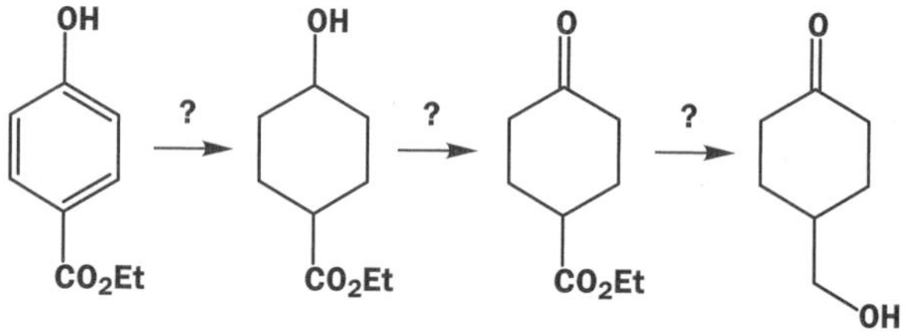

6. How would you convert this nitro compound into the two products shown? Explain the order of events with special regard to reduction steps.

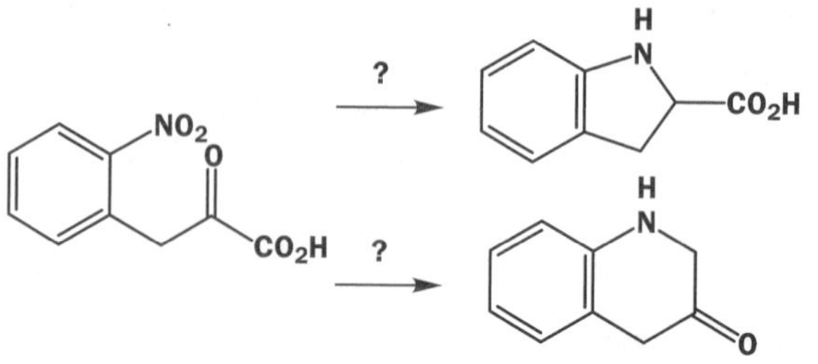

7. What kinds of selectivity are operating in these reactions and how do they work?

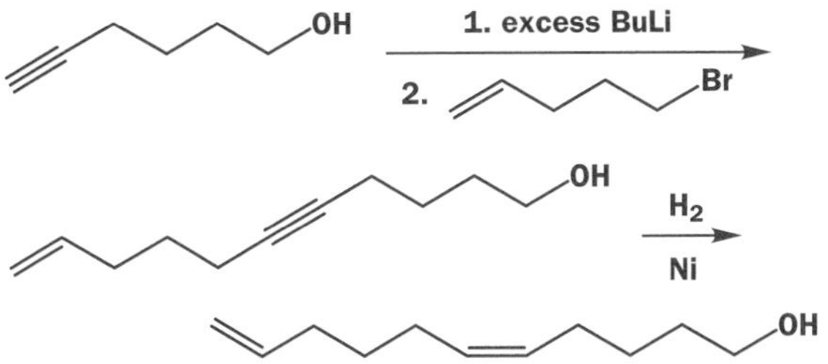

8. These two Wittig reactions (Chapter 14) give very different results. The first gives a single alkene in high yield (which?). The second gives a mixture from which one alkene can be separated with difficulty and in low yield. Why are they so different?

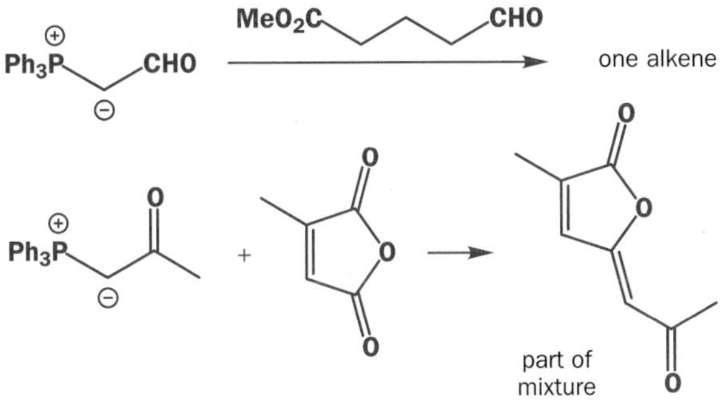

9. Why is this particular amine formed by reductive amination?

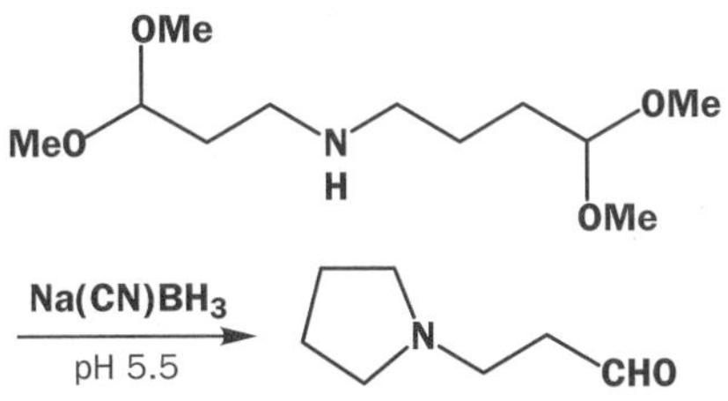

10. Account for the chemoselectivity of the first reaction and the stereoselectivity of the second. A conformational drawing of the intermediate is essential.

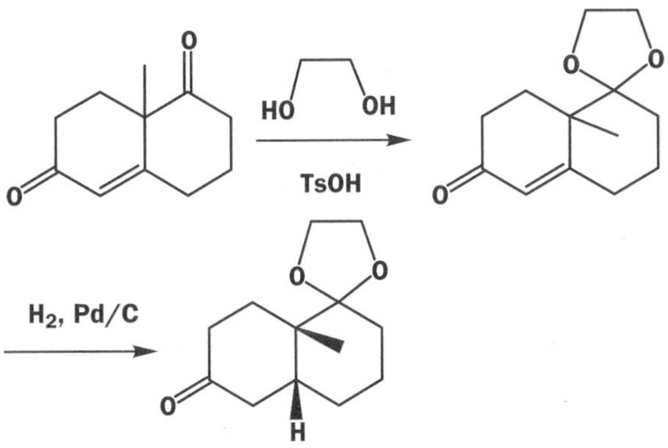

11. Account for the chemoselectivity of these reductions.

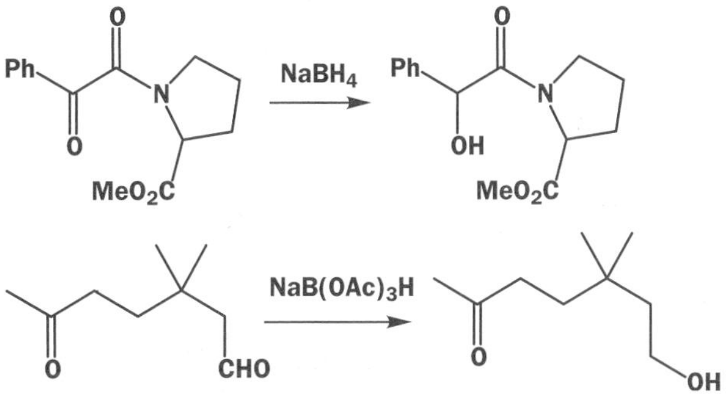

12. How would you carry out the following conversions? More than one step may be needed and you should comment on any chemoselective steps.

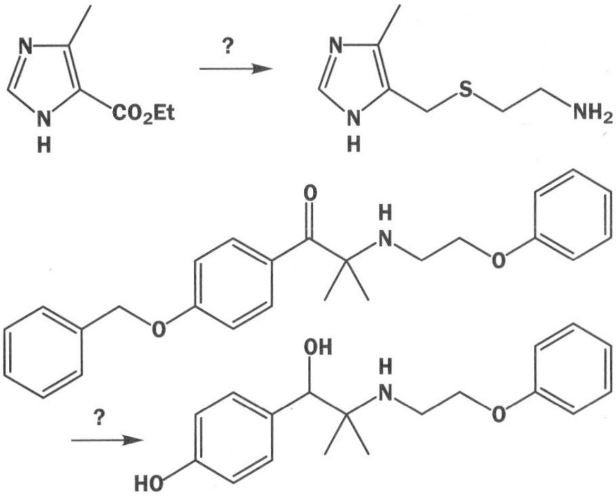

Synthesis in action

25

Connections

Building on:
- Carbonyl addition and substitution ch6, ch12, & ch14
- Mechanisms and catalysis ch13
- S_N1 and S_N2 mechanisms ch17
- Electrophilic aromatic substitution ch22
- Chemoselectivity ch24
- Protecting groups ch24
- Oxidation and reduction ch24

Arriving at:
- Introduction to synthesis
- More chemoselectivity
- Combining reactions from all previous chapters in practical applications
- Further protection of amines and carboxylic acids
- When to avoid protecting groups
- Synthesis of peptide hormones
- Solid phase chemistry

Looking forward to:
- Chemistry of enolates ch26–ch29
- Retrosynthetic analysis ch30
- Diastereoselectivity ch33–ch34
- Synthesis of aromatic heterocycles ch43
- Asymmetric synthesis ch45
- The chemistry of life ch49
- Natural products ch51
- Organic synthesis ch53

Introduction

In the last chapter, you saw examples of groups of sequential reactions used together to construct more complex organic molecules. We call these sequences **syntheses**, and our aim in this chapter is to show you how the reactions you have met in the first 24 chapters of this book can be used to make molecules.

Why make molecules?

Making molecules, the job of the synthetic chemist, developed from a rather random process in the nineteenth century into a well-ordered and well-understood science during the course of the twentieth century. Syntheses can even be planned (and, in some specialized cases, executed) by computers. But why do it?

Historically, the first reason was to prove structures. If you make a compound by a series of known reactions, and understand what happened at each step, you can compare the compound of known structure that you have made with, say, a compound extracted from a plant whose structure you do not know. As methods like NMR arrived on the scene, this became less and less necessary—structures could be deduced spectroscopically. Instead chemists started making molecules in order to do things—to combat diseases, for example, or to develop new fragrances or materials. Many drugs are the product of 'fine tuning' of a naturally occurring compound to alter its properties and, in the course of the development of a drug, an enormous variety of compounds are made by chemists. Some drugs are themselves natural products, but are available in quantities too small to be widely used—so chemists are called upon to make them in gram, kilo, and eventually tonne quantities. Other chemists make molecules in order to find out about the molecules themselves, perhaps because the molecules have particular theoretical interest or because they shed light on the mechanism of a chemical (or biochemical) reaction. Finally, chemists make molecules simply because they are not there (yet) but are a challenge to make. Many of the great advances in the science of synthesis have occurred during the synthesis of natural products, and a frequent test of a new synthetic method is—can it be used to make a natural product?

In this chapter we will look in detail at a few syntheses of important molecules. We hope you will appreciate that the chemistry you encountered in the first 24 chapters is being used all the time in chemical and pharmaceutical labs, in hospitals, and in industrial plants across the world to make valuable, sometimes life-saving, compounds. We start with two simple compounds made from one starting material: toluene.

Benzocaine

Benzocaine is a local anaesthetic with a range of applications (see box). It is manufactured from toluene in a few steps using some quite simple chemistry.

Uses of benzocaine

Benzocaine has been used as a component of appetite suppressants; astringents; analgesics; burn and sunburn remedies; cough tablets, drops, and lozenges; haemorrhoidal creams, suppositories, and enemas; oral

and gingival products for teething, toothaches, canker sores, and denture irritation; and in oral antibacterial agents; treatments for athlete's foot, corns, calluses, and warts; and sore throat sprays and lozenges.

First, one of the classical reactions of aromatic chemistry: the nitration of toluene. The methyl group directs the nitration to the *para* position, so we get the right substitution pattern for benzocaine. But we also get the wrong oxidation levels: first, the nitro group needs reducing to NH_2: this can be done with catalytic hydrogenation (Chapters 22 and 24).

> Formation of the acid chloride followed by reaction with the alcohol would not be suitable here, for chemical reasons as well as economical ones. Why not?

Benzocaine needs an ester (CO_2Et) in place of our methyl group: an oxidation is needed, and the reagent used is $KMnO_4$. This rather odd-looking oxidation is worth remembering: $KMnO_4$ oxidizes aromatic methyl groups (in other words, methyl groups attached directly to benzene rings) to carboxylic acids. And, finally, the esterification: heating with an excess of ethanol in acid gives benzocaine.

> Notice that a synthesis is drawn out as a scheme showing starting materials, reagents, and products connected by reaction arrows. Intermediates that are formed but not isolated are usually shown [in square brackets]. Reagents shown on one arrow are all present at the same time. The most important are usually on top of the arrow—those underneath may be catalyst or solvent but this is not a universal convention. Conditions and yields may be added if important or interesting. We do not usually show mechanisms in the scheme but, if an explanation of an unusual reaction or selectivity is needed, then a mechanism might be included.

Saccharin

Saccharin is, of course, the famous artificial sweetener. It was discovered at Johns Hopkins University in 1879 in the days before disposable gloves. Ira Remsen (1846–1927) asked a research fellow Constantin Fahlberg (1850–1910) to oxidize a sulfonamide he had made. Fahlberg did so and found that evening that the food he was eating tasted remarkably sweet. Saccharin is a cyclic imide with a nitrogen atom acylated on one side by a sulfonic acid and on the other by a carboxylic acid.

■ Some of the reactions used in the synthesis were discussed in Chapter 22.

■ The details of this reaction are analysed in Chapter 22.

The first step in the synthesis of saccharin is an electrophilic substitution reaction, like the first step of the benzocaine synthesis, but this time we want the *ortho*-substituted product. Chlorosulfonic acid gives a mixture of *ortho* and *para* products—it is impossible to find conditions that completely avoid forming the *para*-toluenesulfonyl chloride. However, you may recognize an old friend here—the by-product is, of course, TsCl. You may have wondered why we always use TsCl and not $PhSO_2Cl$ to make OH into a leaving group: now you know.

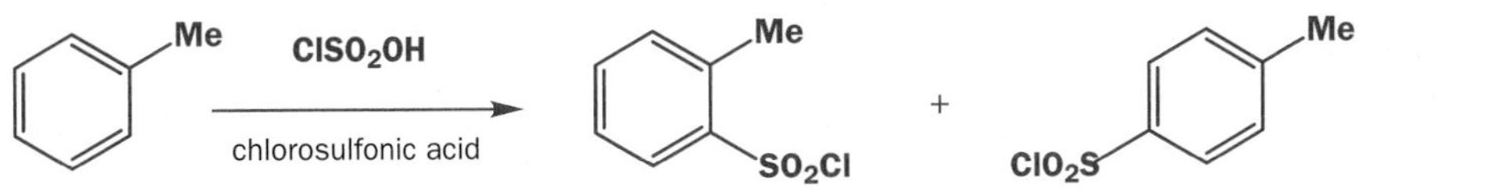

ortho-toluenesulfonyl chloride   para-toluenesulfonyl chloride = TsCl

The sulfonyl chlorides react with ammonia to give sulfonamides. Notice that this compound's aromatic methyl group is at the wrong oxidation level, so we again use $KMnO_4$ to make the acid before dehydrating to give saccharin.

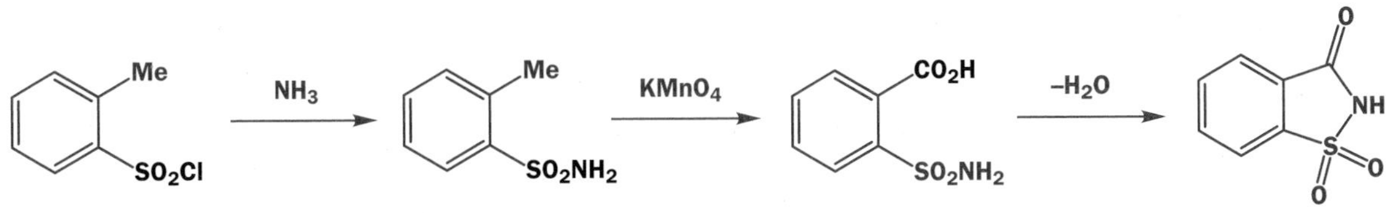

saccharin

Salbutamol

Anti-asthma drugs work by dilating the air passages of the lungs, releasing the constriction that characterizes the disease. Salbutamol does this by imitating the action of the hormone adrenaline (epinephrine). Adrenaline has other effects—it increases heart rate for example—but the medicinal chemists at Glaxo working on asthma found that adding on the extra carbon atom avoided dangerous side-effects on the heart. The *t*-butyl group increases the stability of the drug, so its effects last longer.

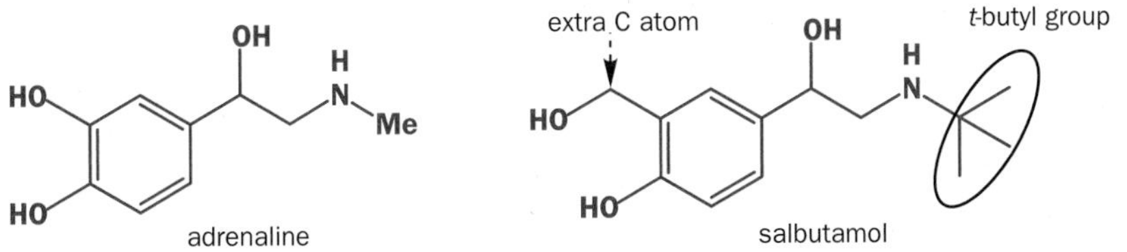

adrenaline   salbutamol

Salbutamol is made from aspirin, itself simply the acetate ester of the natural product salicylic acid, by a series of substitution reactions. The first is a Friedel–Crafts acylation (an electrophilic substitution) in which aspirin itself is the acylating agent: it is an isomerization in which the acetyl group gets transferred from O to C. Acylation occurs *para* to the electron-donating alkoxy substituent, and gives this ketone.

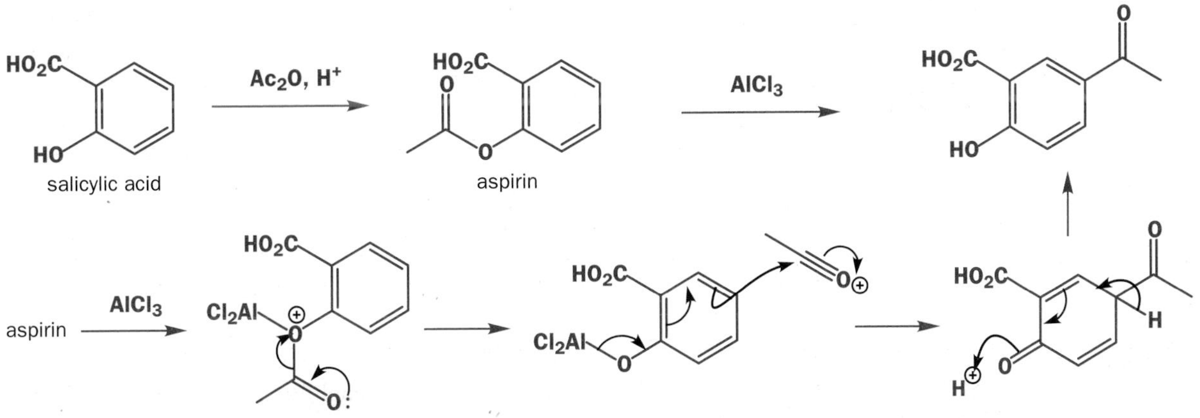

salicylic acid   aspirin

Because this is an unusual Friedel–Crafts acylation, we think it worthwhile to draw a mechanism in the description of a synthesis. This is just such a situation as we described above. Another electrophilic substitution occurs when this ketone reacts with bromine via its enol (Chapter 21).

▶
It is given another name, the **Fries rearrangement**.

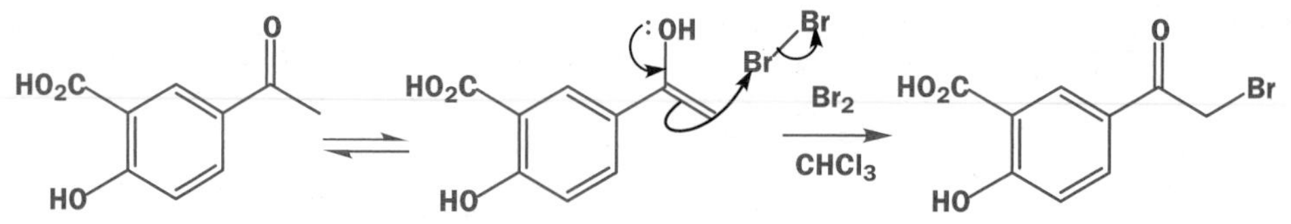

Next, a nucleophilic substitution reaction at saturated carbon. α-Halo ketones are excellent electrophiles and react rapidly with nucleophiles, such as this secondary amine, by the S_N2 mechanism (Chapter 17). All that remains is to reduce the ketone and the acid to alcohols and remove the benzyl protecting group (both discussed in Chapter 24).

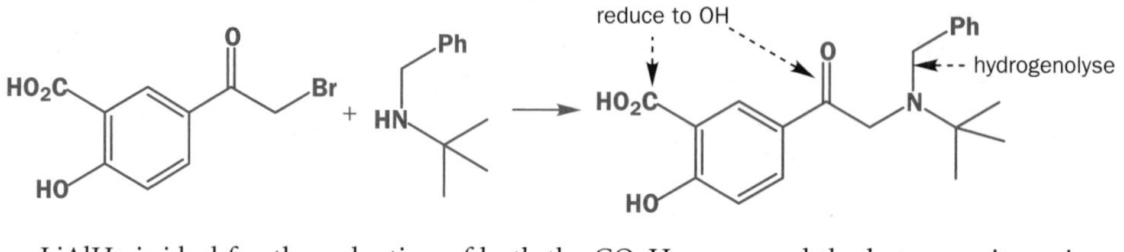

$LiAlH_4$ is ideal for the reduction of both the CO_2H group and the ketone as it carries out both reductions in a single step. The other reduction is a hydrogenolysis of the benzyl group, for which we need catalytic hydrogenation.

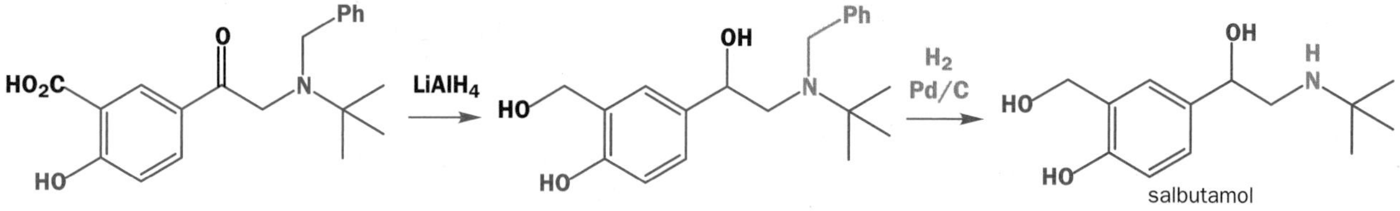

Thyroxine

Salbutamol works by imitating the action of a hormone: thyroxine *is* a hormone—it is part of the body's control over its metabolic rate. Lack of thyroxine (or rather, of the iodine needed to make it) causes hyperthyroidism, or goitre. Our next synthesis is one that has been used on an industrial scale for the manufacture of synthetic thyroxine (identical with, but less macabre than, naturally extracted thyroxine).

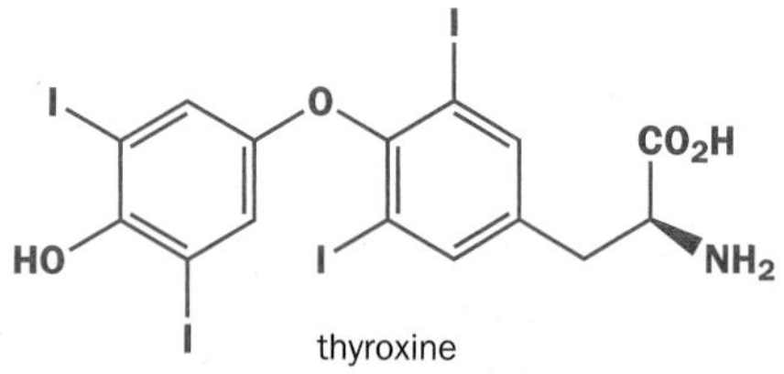

Thyroxine has two aromatic rings, and you should be prepared to draw upon what you learned about aromatic chemistry in Chapters 22 and 23. It is also an amino acid and, in order to make the synthesis as cheap as possible, the chemists at Glaxo who developed the method used the amino acid tyrosine as a starting material. Nitration of tyrosine puts two nitro groups *ortho* to the OH group in an electrophilic aromatic substitution (make sure that you understand why!).

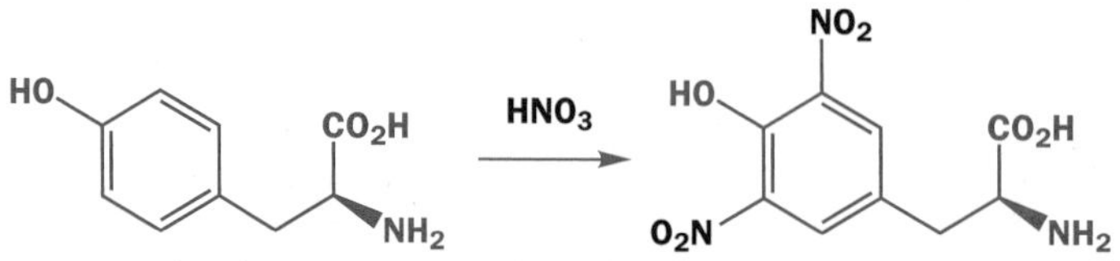

These make the aromatic ring electron-poor, ready for a nucleophilic aromatic substitution that will introduce the other aromatic ring of the target. We need to displace OH, but OH⁻ is a bad leaving group, so we must first make it into a tosylate. The trouble is—there is a free amine in the starting material and we do not want that to react with the TsCl. The answer—as you should be able to predict after Chapter 24—is to protect the amine.

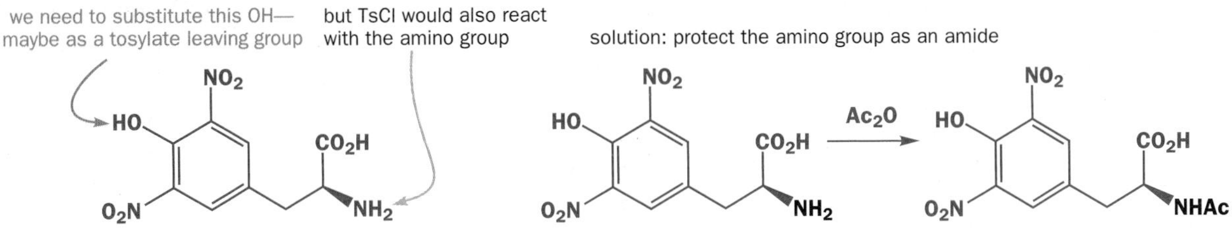

we need to substitute this OH— maybe as a tosylate leaving group · but TsCl would also react with the amino group · solution: protect the amino group as an amide

We haven't yet discussed amine protection (that will come later in this chapter) but, since it is the amino group's nucleophilicity that is the problem, it makes sense to react it with an acylating agent: an amide is much less nucleophilic than an amine because the nitrogen's lone pair is involved in conjugation with the carbonyl group. The same method was used to reduce the nucleophilicity of aromatic amines in bromination (Chapter 22). The carboxylic acid also needs protecting, and it is made into an ethyl ester.

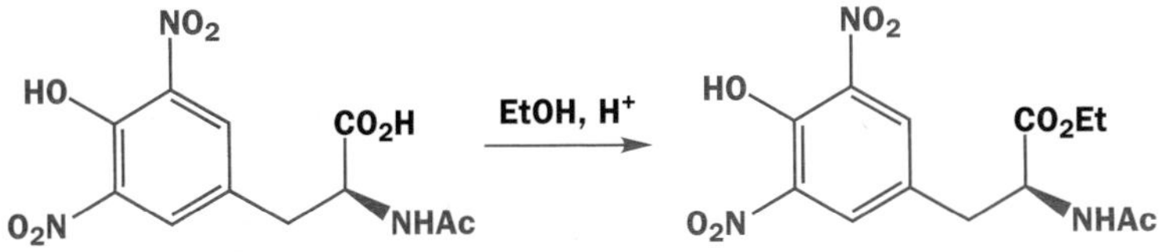

Now the tosylation—under the usual conditions—followed by the nucleophilic aromatic substitution (Chapter 23). The leaving group is *ortho* to two electron-withdrawing groups, and so the substitution pattern is right for nucleophilic aromatic substitution. The nucleophile is 4-methoxyphenol, deprotonated by pyridine.

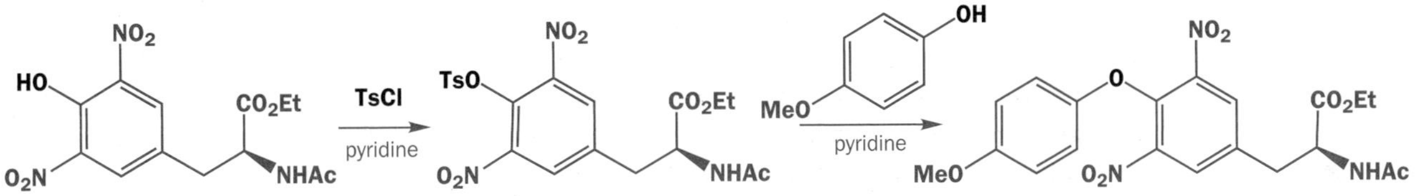

The nitro groups need replacing by iodine atoms, and you should not be surprised that they were reduced to amino groups by hydrogenation over palladium and then diazotized. Sodium iodide substitutes I^- for N_2^+.

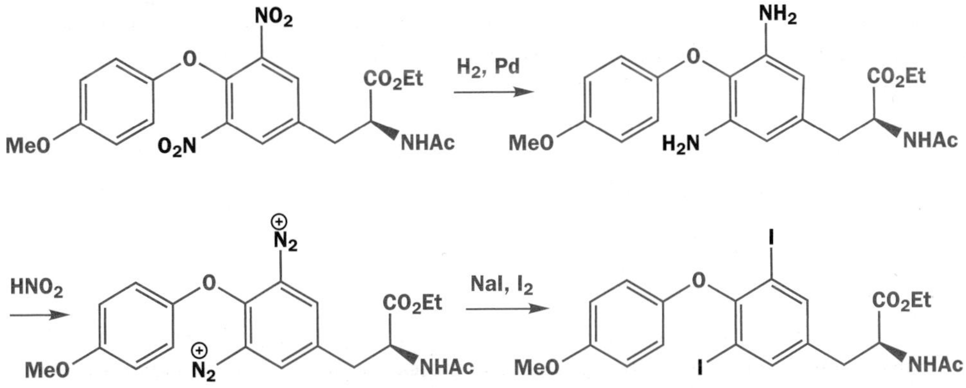

The methyl ether is really a protected version of the phenolic OH we need in thyroxine, and its deprotection uses a method that you met in Chapter 24. Most ethers are very hard to cleave—phenyl

ethers are a bit easier, because phenols are reasonable leaving groups. HI in AcOH protonates the oxygen atom, and now attack of the good nucleophile I⁻ on the electrophilic Me centre can kick out the phenol.

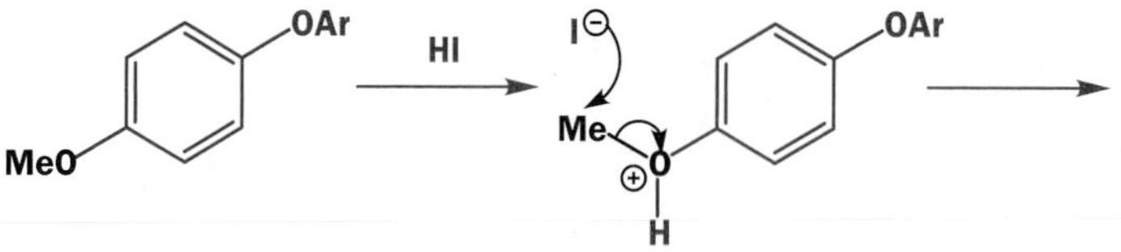

Remarkably, the same conditions hydrolyse the amide- and ester-protecting groups too—very useful for an industrial process where every step means another reaction vessel.

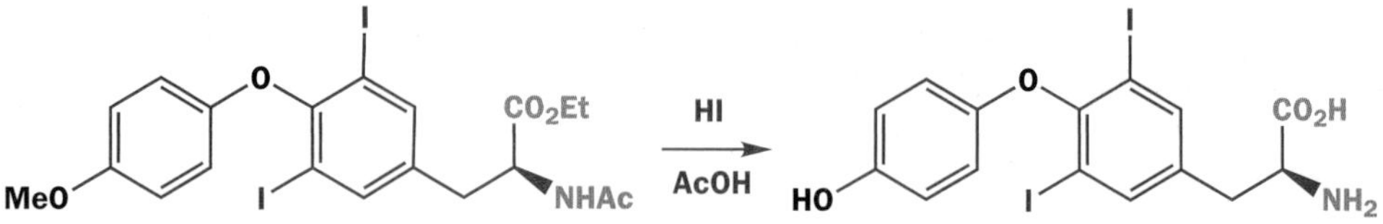

Finally, electrophilic substitution on the left-hand ring in the manner of the first nitration step puts in the third and fourth iodine substituents of thyroxine. Notice that the free O⁻ (the phenol is ionized with Et₂NH) group is more activating (electron-donating) than the ether oxygen atom.

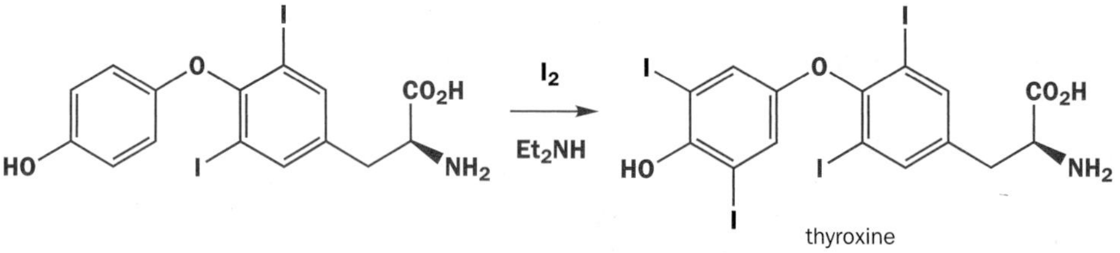

thyroxine

This synthesis shows how important electrophilic substitution in aromatic compounds is in industrial processes. It involves four separate such reactions as well as three nucleophilic aromatic substitutions. The chemistry of Chapters 22 and 23 is well represented here.

Muscalure: the sex pheromone of the house-fly

Many insects attract a mate by releasing a volatile organic compound known as a pheromone. Pheromones are highly specific to species, and provide a cunning means of controlling pests: place a pad of cotton wool soaked in male pheromone inside a trap, and in drop all the female pests—no next generation. If insect control is to rely on a supply of the pheromone, that supply has to be synthetic—it takes enormous numbers of squashed insects to provide even a few milligrams of most pheromones.

We will start by looking at two syntheses of the very simple pheromone of a very common insect—the house-fly. The pheromone, known as **muscalure**, is a Z-alkene.

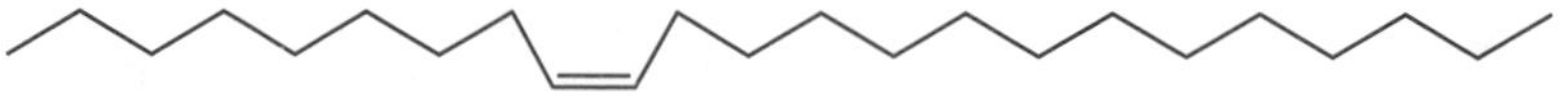

muscalure, the pheromone of the house-fly

One approach, used by some American chemists in the early 1970s, was very simple. These chemists noted the similarity between the structures of muscalure and the fatty acid known as erucic acid, which is abundant in rapeseed oil, and decided to make muscalure from erucic acid. They first reacted the acid with two equivalents of methyllithium—the first equivalent deprotonates the acid to make a lithium carboxylate salt, while the second reacts with the lithium carboxylate to make a ketone.

> ▶
> The name muscalure comes from the generic name of the house-fly, *Musca*.

> ■
> You met this reaction in Chapter 12 as one of the few ways of adding a nucleophile to a carboxylic acid derivative to give a ketone.

erucic acid, a fatty acid
extracted from rapeseed oil

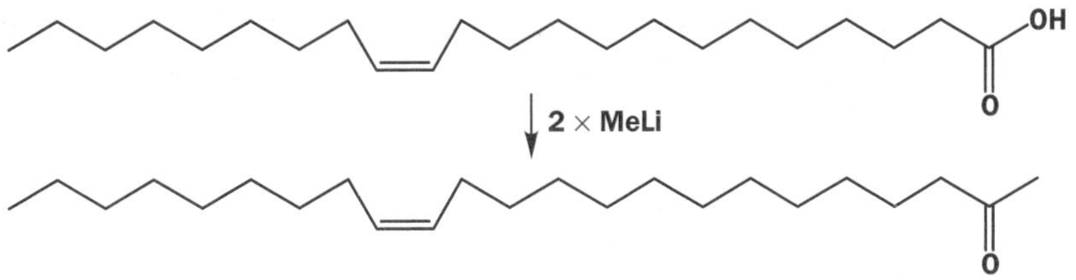

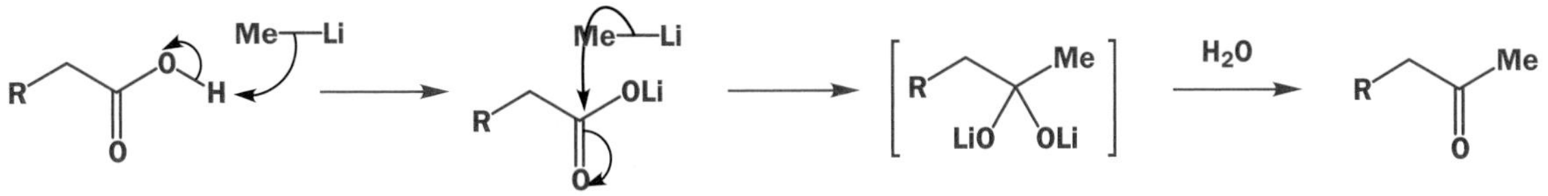

2 × MeLi

H₂O

The next step is to remove the ketone functional group. You met a few ways of doing this in the last chapter; here the method chosen was to make a hydrazone and heat in the presence of base. Muscalure is the product.

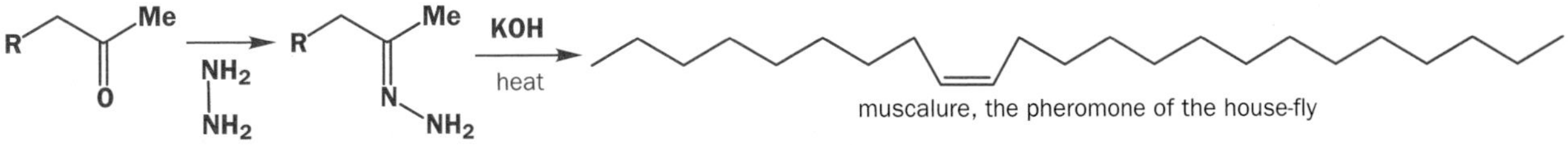

KOH
heat

muscalure, the pheromone of the house-fly

In 1977, some Russian chemists made the same compound by a different route. They chose to introduce the *Z* double bond by hydrogenation of an alkyne over Lindlar's catalyst. To make the alkyne they needed, they took 1-decyne, treated it with $LiNH_2$ to remove the acidic terminal proton, and reacted the anion with an *n*-alkyl bromide.

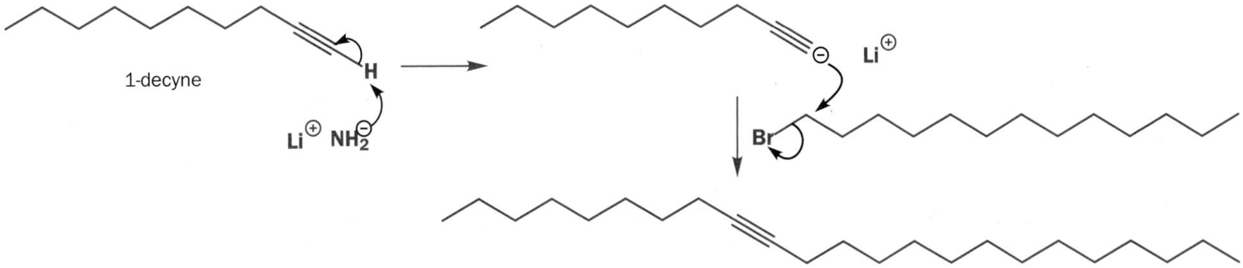

1-decyne

$Li^{\oplus}\ NH_2^{\ominus}$

$Li^{\oplus}$

Br

By stirring the alkyne with Lindlar's catalyst under an atmosphere of hydrogen they were able to make muscalure.

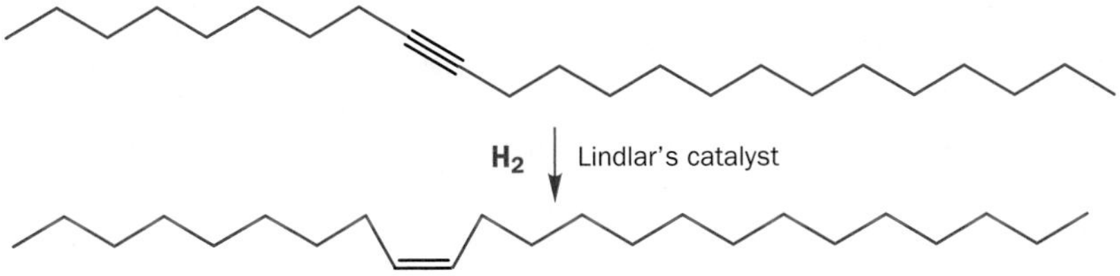

■
In Chapter 24 you saw that Lindlar's catalyst is a weak catalyst that only allows hydrogenation of alkynes, and gives *Z* double bonds.

H_2 | Lindlar's catalyst

Grandisol—the sex pheromone of the male cotton boll weevil

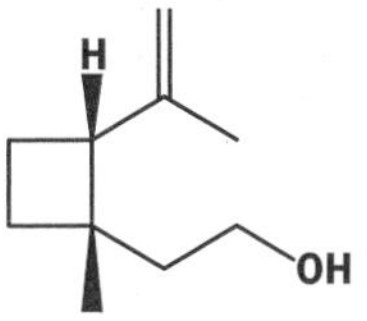

House-flies are irritating and a minor health hazard, but the cotton boll weevil is an enormously destructive pest of the American cotton crop and is responsible for vast economic losses. The weevil has a pheromone called grandisol. The structure and synthesis of grandisol are rather more complicated than the syntheses of muscalure, but *all the reactions are ones you have met in the first 24 chapters of the book.*

You saw in Chapter 21 how carbonyls form enolates when they are treated with base. On p. 528, you met nitriles doing something very similar, and the first step of the grandisol synthesis is the reaction of the 'enolate' of a nitrile with an electrophile—an alkyl bromide.

grandisol

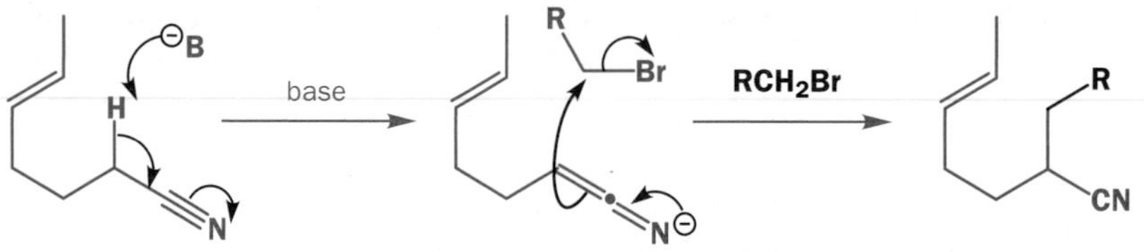

The electrophile required carries a hydroxyl group. But this is no good because the acidic OH proton will react with the basic enolate—we need a protecting group, and the one chosen here was THP. So, here is the first step presented in a way that you will find useful. Because the base is added first and the alkyl bromide afterwards when all the base has reacted, the synthesis is written as : 1. base; 2. RCH_2Br.

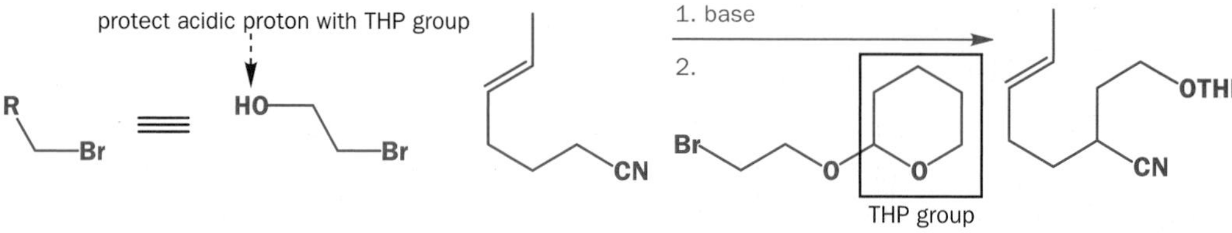

Next, the double bond is made into an epoxide with *m*-CPBA. Epoxides react with nucleophiles, and this is the way that the four-membered ring of grandisol was formed: the nitrile still has a proton next to it, and a strong base will remove this proton as before to give an 'enolate'. The enolate reacts with the epoxide to give a four-membered ring.

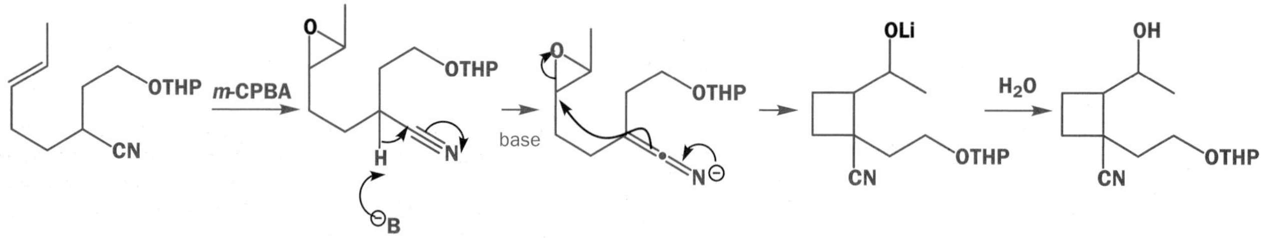

You can now clearly see the similarity with our 'target molecule', grandisol, but there are several more steps to carry out yet. The nitrile needs to be got rid of completely—we showed you a few ways of getting rid of functional groups in the last chapter, and the one used here was the Wolff–Kishner reduction of an aldehyde. The aldehyde comes from reduction of the nitrile with DIBAL.

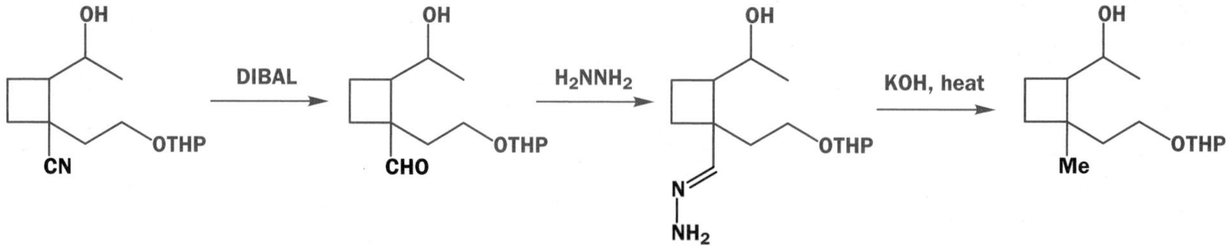

Now we need to put in the C=C double bond, using a Wittig reaction. The Wittig reaction turns aldehydes or ketones into alkenes, turning the C=O bond into a C=C bond (Chapter 14). Of the many methods for oxidizing secondary alcohols to ketones (Chapter 24), these chemists chose CrO_3. Finally, the protecting group needs to come off. THP-protected alcohols are acetals, and THP groups are removed in aqueous acid: the product is grandisol.

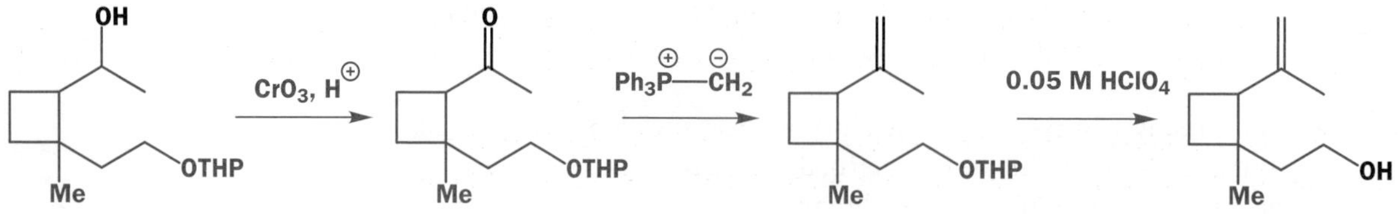

We hope that you can see from this example that all the steps in this important synthesis use chemistry that you have already met. The art in synthesis is to put the steps together in the right order, and we aren't asking you to think about that (that can wait until Chapter 30)—at this stage see these syntheses as a way of revising what you have already learned.

▶

Neither at this stage are we asking you to think about the stereochemistry of the molecules we are making. Grandisol is, in fact, a single diastereoisomer and a single enantiomer, as shown on p. 29, and this synthesis makes a racemic mixture of a single diastereoisomer. Why the product is racemic should be evident to you after reading Chapter 16; why it gives a single diastereoisomer you might like to think about once you have read beyond Chapter 33.

Peptide synthesis: carbonyl chemistry in action

In this part of the chapter we will talk in detail about the synthesis of a single class of biologically important molecules, peptides. In doing so, we will introduce you to protecting groups for two more important functional groups: amines and carboxylic acids. The ability to control the reactivity of these groups is vital to the controlled synthesis of peptides. This field has grown vastly since the introduction of the Z or Cbz protecting group (which you will meet shortly) in 1932, and today machines can be programmed to synthesize peptides automatically.

■

The names, structures, and abbreviations for the natural amino acids are in Chapter 49.

Let's start by thinking how you might react two amino acids together, to make a dipeptide—leucine and glycine, for example. If we want the NH_2 group of glycine to react with the CO_2H group of leucine we will first have to activate the carboxylic acid towards nucleophilic substitution—by making the acyl chloride, say, or a particularly reactive ester.

▶

In fact, the use of acyl chlorides in peptide synthesis is rare because of the potential for unwanted side-reactions, including racemization and dimerization to form by-products known as diketopiperazines. We will represent our activated acid as RCOX, where X is some other good leaving group such as a phenol.

The main problem, though, is that there is another free CO_2H, which could react with the COX group to form an anhydride, and two different free amines, either of which might react, giving both LeuLeu (which we don't want) and LeuGly (which we do).

For this reason, we need to protect both the NH_2 group of leucine and the CO_2H group of glycine. What sort of protecting groups do they need to be? We will need to be able to take them off again once they have done their job, so there is no point using, say, an amide to protect the amine since we would have great difficulty hydrolysing the amide in the presence of the amide bond we are trying to form. Ideally, not only do we want the protecting groups to be removable under mild conditions that will not destroy the rest of the molecule, but we want two groups (one for each of NH_2 and CO_2H) which we can take off under *different* conditions. We then have the opportunity to modify either end of the dipeptide at will.

A good choice for a pair of conditions might be acid and base—we might protect the NH_2 group with a protecting group we can remove only in acid, and the CO_2H group with protection we can remove only in base.

> ▶ ***C*-terminus** means the end of the peptide that carries the terminal CO_2H group. The other end, carrying the NH_2 group, is the ***N*-terminus**. By convention, we write the *N*-terminus on the left and the *C*-terminus on the right. The structure of oxytocin is represented here in terms of the three-letter codes for amino acids—there is a full list in Chapter 49.

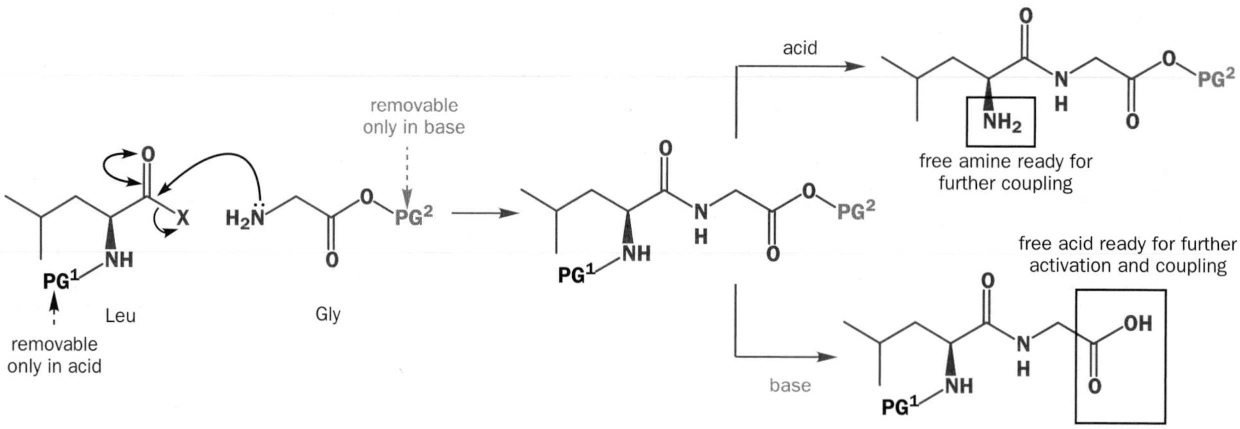

The Cbz protecting group—oxytocin

We introduced the dipeptide LeuGly as an example because it forms the *C*-terminus of the peptide hormone oxytocin.

H_2N–Cys-Tyr-Ile-Gln-Asn-Cys-Pro-Leu-Gly–$CONH_2$

oxytocin

Synthetic hormones

Oxytocin is a hormone involved in controlling the onset of labour in women and the subsequent release of milk. It was the first peptide hormone to be synthesized, in 1953, and the synthetic 'version' of the hormone, Syntocinon (identical, of course, with the 'natural' version isolated from human placentas, though probably purer), is regularly used in modern obstetrics to induce labour in women whose babies are overdue. Several other peptide hormones have been synthesized in the lab, and we will come back to one of them, gastrin, shortly.

> ■ du Vigneaud won the Nobel Prize for Chemistry in 1955 for his work on the synthesis of peptides.

> ■ Problem 12.5 depends on understanding this.

The first step in the synthesis of oxytocin is indeed the coupling of glycine (through its amino group) with leucine. This is how it was done by du Vigneaud and Bodanszky. First, the carboxylic acid of the glycine was protected as an ethyl ester. Making an ester is the obvious way to stop CO_2H groups interfering as acids or as nucleophiles. However, simple methyl and ethyl esters may pose problems—they can still react with such nucleophiles as amines. Ethyl esters of amino acids are therefore stable only if the NH_2 group is protected. The glycine ethyl ester had to be stored as its hydrochloride salt: in effect, the $-NH_2$ group is 'protected' as $-NH_3^+$.

> ▶ This method is preferable in this case to the usual way of making esters—from acyl chloride plus alcohol—since steric hindrance makes that a very slow reaction with *t*-butanol.

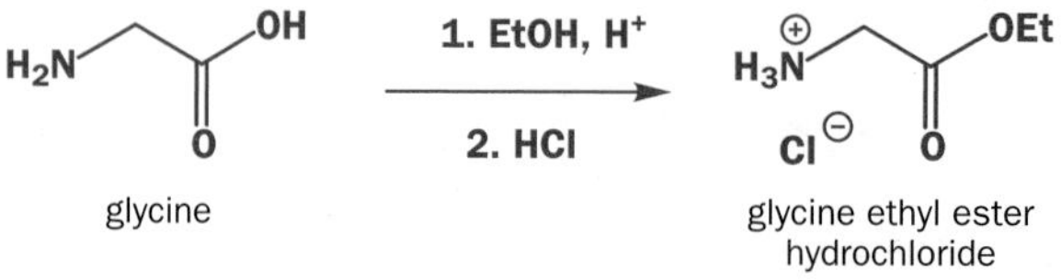

If du Vigneaud and Bodanszky had wanted a carboxylic-acid-protecting group that was more stable towards attack by nucleophiles, they could have made a *t*-butyl ester with isobutene in sulfuric acid.

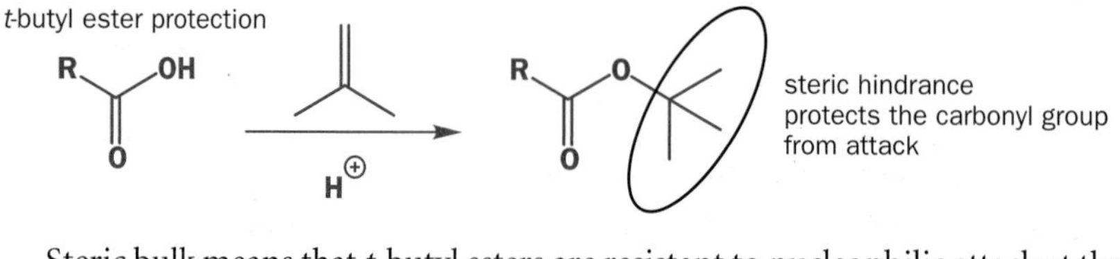

Steric bulk means that *t*-butyl esters are resistant to nucleophilic attack at the carbonyl group, and that includes hydrolysis under basic conditions (nucleophilic attack by HO^-). But they do hydrolyse relatively easily in acid, because the mechanism of hydrolysis of *t*-butyl esters in acid is quite different. It does not involve nucleophilic attack at the carbonyl group and is a favourable S_N1 reaction at the *t*-butyl group (Chapter 17).

hydrolysis of *t*-butyl esters in acid: *t*-Bu–O bond breaks in S_N1 reaction
(compare usual ester hydrolysis)

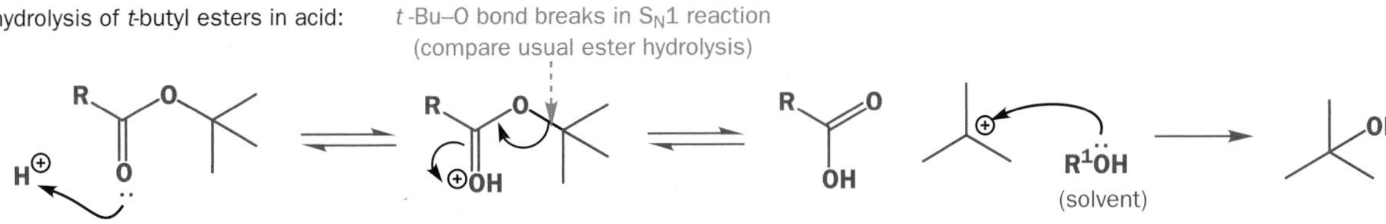

This is a good point to continue our growing table of protecting groups started in Chapter 24. We need only one new entry for *t*-butyl esters.

Protecting group	Structure	Protects	From	Protection	Deprotection
t-butyl ester (CO$_2$Bu-*t*)	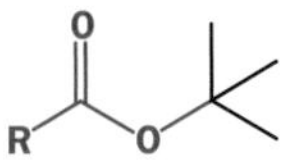	carboxylic acid (RCO$_2$H)	bases, nucleophiles	isobutene, H$^+$	H$_3$O$^+$

In the event, the chemists needed a group that they could later react with ammonia to make the amide that is present in oxytocin. They also wanted a group that was stable to mild acid—so they chose the ethyl ester. As for the leucine residue, it had to have its NH$_2$ group protected using a base-stable protecting group, because base would be needed to release the NH$_2$ group of the glycine hydrochloride salt. The group that was used is one of the most important nitrogen-protecting groups and is known, rather uninformatively, as the **Z group** (also known as **Cbz**, or carboxybenzyl). Cbz (Z) groups are put on by treating with benzyl chloroformate (BnOCOCl) and weak base.

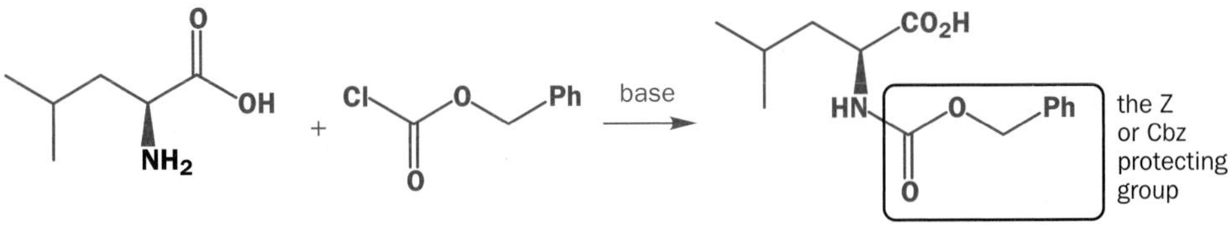

leucine (Leu)   (= BnOCOCl)   Z-Leu or Cbz-Leu

Cbz-protected amines behave like amides—they are no longer nucleophilic, because the nitrogen's lone pair is tied up in conjugation with the carbonyl group. They are resistant to both aqueous acid and aqueous base, but they have, to use the analogy we developed in the last chapter, an Achilles' heel or safety catch—the benzyl ester. The same conditions that removed benzyl ethers in Chapter 24 will remove Cbz: HBr or hydrogenolysis.

> You can tell the difference between Z as a protecting group and *Z* as a label for the stereochemistry of an alkene because the latter is in *italics*. It's less confusing to use Cbz for the protecting group and *Z* for the alkene.

cleavage of Cbz (Z) in HBr/AcOH

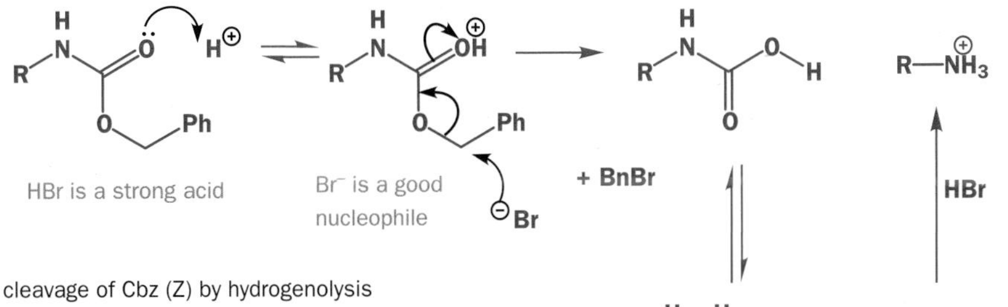

HBr is a strong acid   Br$^-$ is a good nucleophile   + BnBr   HBr

cleavage of Cbz (Z) by hydrogenolysis

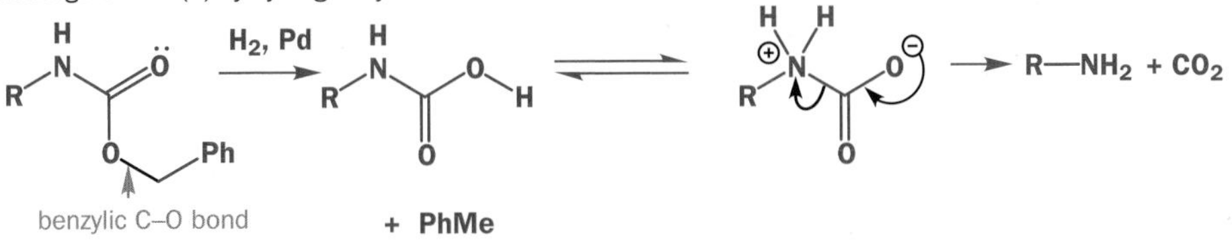

benzylic C–O bond   + PhMe

Protecting group	Structure	Protects	From	Protection	Deprotection
Cbz (Z) (OCOBn)	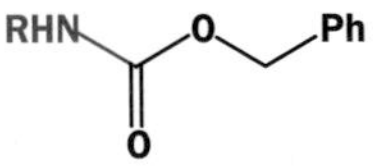	amines	electrophiles	BnOCOCl, base	HBr, AcOH or H$_2$, Pd

The Cbz-protected leucine next had to be activated so that it would react with the glycine. The acyl chloride won't do as it is unstable, and a common alternative in peptide chemistry is to make a *p*-nitrophenyl or 2,4,6-trichlorophenyl ester. Phenoxide, especially when substituted with electron-withdrawing substituents, is a good leaving group, and Cbz-leucine *p*-nitrophenyl ester reacts with the glycine hydrochloride ethyl ester in the presence of a weak base (triethylamine, to release the glycine's NH$_2$ group). Notice the chemoselectivity in this step—the glycine's NH$_2$ group has three carbonyl groups to choose from, but reacts only with the most electrophilic—the one bearing the best leaving group.

The dipeptide is now coupled—but is still protected. Deprotection (HBr/AcOH) gave the HCl salt of LeuGly ethyl ester for further reaction. The rest of the peptide was built up in much the same way—each amino acid being introduced as the Cbz-protected *p*-nitrophenyl ester before being deprotected ready for the next coupling, until all nine of oxytocin's amino acids had been introduced.

The *t*-Boc protecting group—gastrin and aspartame

H$_2$N–Trp-Met-Asp-Phe–CONH$_2$
gastrin *C*-terminal tetrapeptide

Gastrin is a hormone released from the stomach that controls the progress of digestion. Early work on the hormone showed that only the four *C*-terminal amino acids of the peptide were necessary for its physiological activity.

The synthesis starts with the coupling of two more amino acids: aspartic acid and phenylalanine. As you would expect, the carboxylic acid group of phenylalanine is protected, this time as a methyl ester, and the NH$_2$ group of aspartic acid is protected as a Cbz-derivative. Since aspartic acid has two carboxylic acid groups, one of these also has to be protected. Here is the method—first the Cbz-group is put on; then both acids are protected as *benzyl* esters. Then just one of the benzyl esters is hydrolysed. It may seem surprising to you that this chemoselective hydrolysis is possible, and you could not have predicted that it would work, without trying it out in the lab.

■
Again—note the chemoselectivity! Phenylalanine's NH$_2$ group attacks only one of the four carbonyl groups in this molecule.

The protected acid is activated as its 2,4,6-trichlorophenyl ester, ready for coupling with the phenylalanine methyl ester in base. Now you see why the benzyl ester was chosen to protect Asp's side-chain carboxylic acid group—hydrogenolysis can be used to cleave both the Cbz-group and the benzyl ester at the same time.

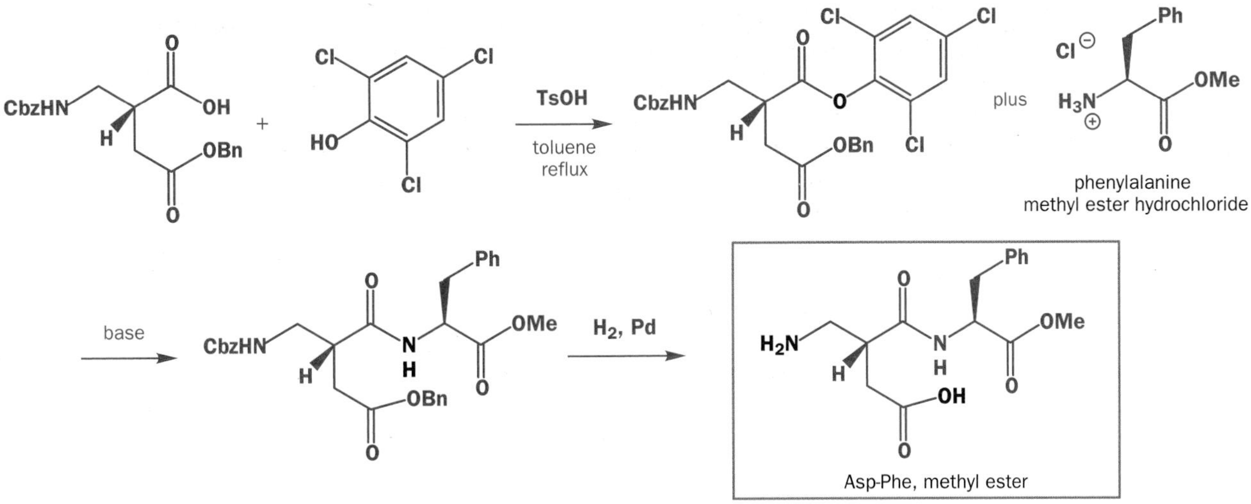

At this point in one synthesis of the tetrapeptide in the laboratories of Searle, the American pharmaceutical company, a remarkable discovery occurred. The AspPhe methyl ester was accidentally found to taste sweet: extremely sweet—about 200 times as sweet as sucrose. AspPhe is now known as aspartame, marketed under the brand name Nutrasweet.

The next amino acid in the peptide is methionine, and it will of course need N-protecting and C-activating. The N-protecting group used this time was different—still a carbamate, not Cbz or Z but t-Boc (or just Boc or BOC)—standing for t-butyloxycarbonyl and pronounced 'bock' or 'tee-bock'.

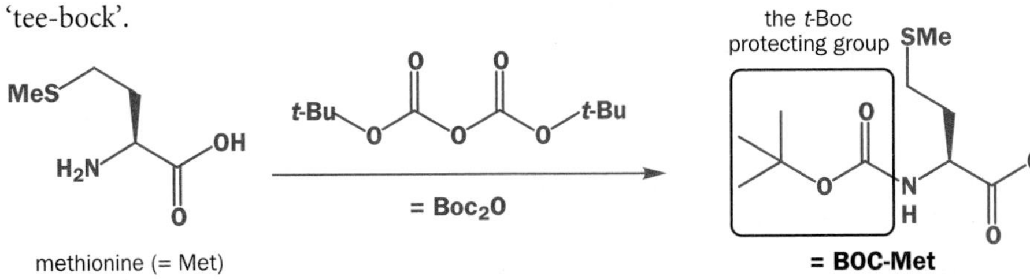

Like Cbz, the t-Boc group is a carbamate protecting group. But, unlike Cbz, it can be removed simply with dilute aqueous acid. Just 3M HCl will hydrolyse it, again by protonation, loss of t-butyl cation, and decarboxylation.

> The mechanism for this hydrolysis is comparable to the acid-catalysed cleavage of Cbz groups, but remember that here the t-Bu group leaves in an S_N1 step. Cbz-groups are cleaved by using a good nucleophile, Br$^-$, because an S_N2 step is involved.

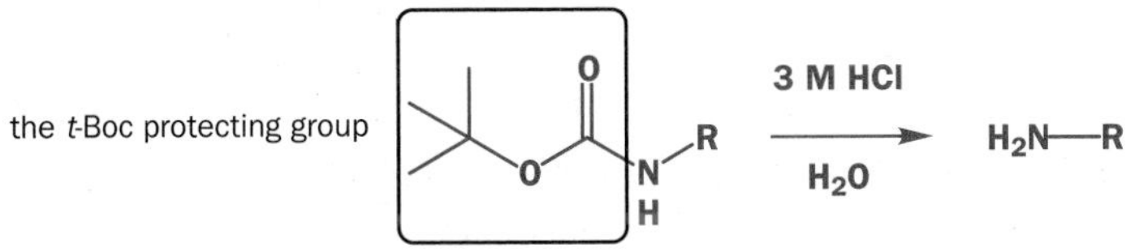

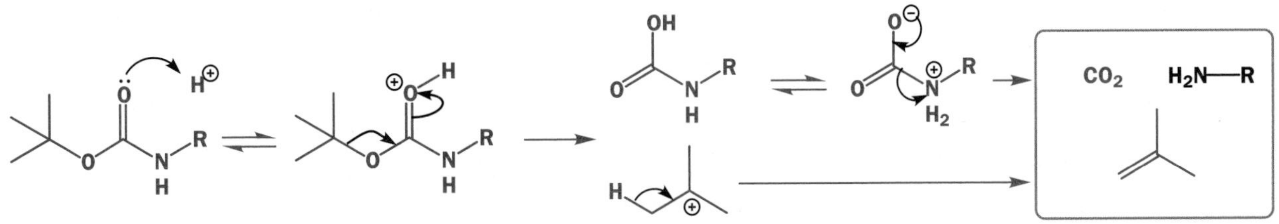

Base, on the other hand, cannot touch the t-Boc group—the carbonyl group is too hindered to be attacked even by OH$^-$, and t-Boc is strongly resistant to basic hydrolysis: again, another example of an amide with an Achilles' heel. The obvious way to make carbamates from amines is to react them with a carbamoyl chloride—this is how Z-groups are usually put on. Unfortunately, t-BuOCOCl is unstable, and we have to use some other electrophilic derivative—usually the anhydride Boc$_2$O as here, for example.

Protecting group	Structure	Protects	From	Protection	Deprotection
t-Boc (OCOBu-*t*)	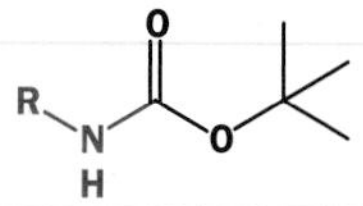	amines	electrophiles	(*t*-BuOCO)$_2$O, base	H$^+$, H$_2$O

Meanwhile, back at the tetrapeptide synthesis, methionine (Met) has been BOC-protected, and is ready for activation—as a 2,4,6-trichlorophenyl ester (Cp) this time and coupling with the deprotected Asp-Phe-OMe. Aqueous acid takes off the BOC group without hydrolysing peptide or ester bonds, and a repeat of this cycle with BOC-tryptophan trichlorophenyl ester (BOC-Trp-OCp) finally gives the tetrapeptide.

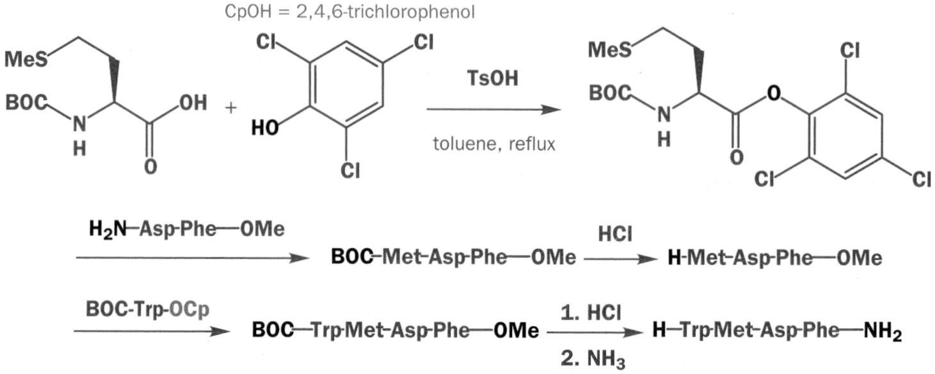

CpOH = 2,4,6-trichlorophenol

The Fmoc protecting group—solid-phase synthesis

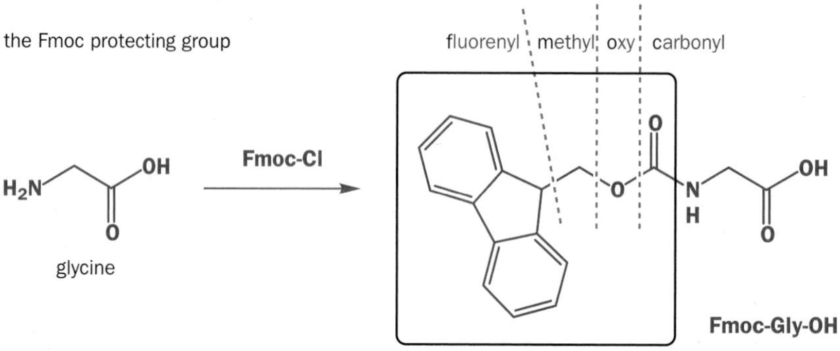

You have already met our next and last amine-protecting group in Chapter 8.

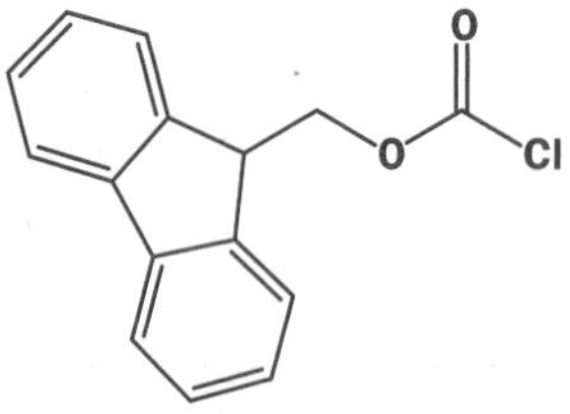

It is called Fmoc (pronounced 'eff-mock'), for fluorenylmethyloxycarbonyl, and has a susceptibility inverse to that of *t*-Boc. It cannot be lost by substitution in the manner of Cbz or *t*-Boc because neither S$_N$1 nor S$_N$2 mechanisms can operate at the ringed carbon atom: it is both primary *and* hindered.

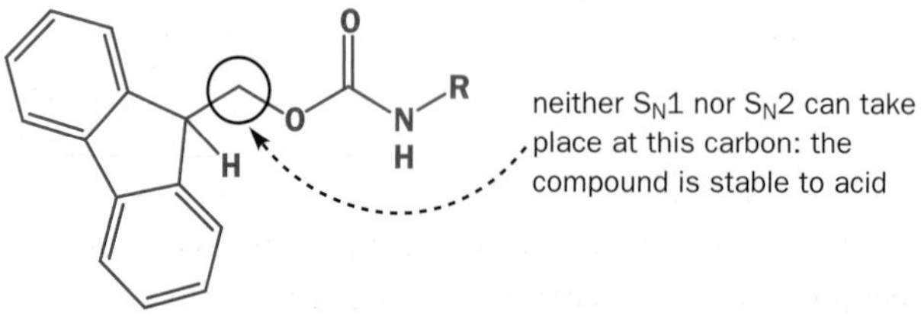

neither S$_N$1 nor S$_N$2 can take place at this carbon: the compound is stable to acid

So, where is the safety catch? The important point about Fmoc is that is has a rather acidic proton (pK_a about 25), shown in black. The proton is the Achilles' heel: treatment of Fmoc-protected amines with base eliminates a fulvene to reveal the NH$_2$ group.

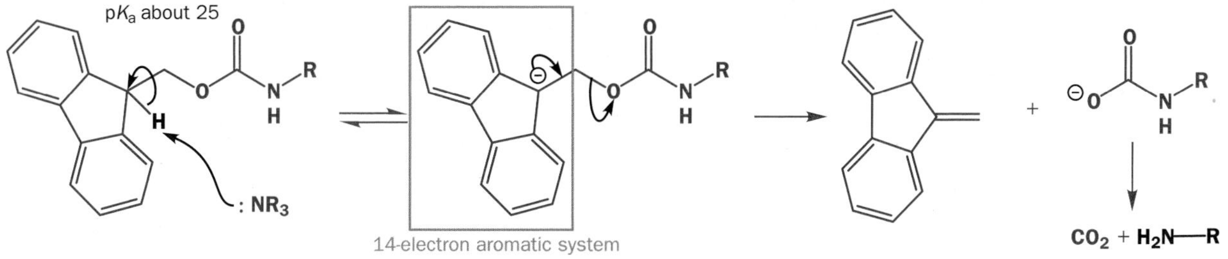

The table of protecting groups, built up slowly over this chapter and the last, is now complete.

Protecting group	Structure	Protects	From	Protection	Deprotection
acetal (dioxolane)	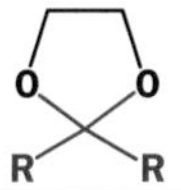	ketones, aldehydes	nucleophiles, bases	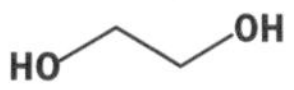 HO–CH2CH2–OH	water, H$^+$ cat.
trialkylsilyl (R$_3$Si-, e.g. TBDMS)	RO—SiMe$_3$ RO—SiMe$_2$Bu^t	alcohols (OH in general)	nucleophiles, C or N bases	R$_3$SiCl, base	H$^+$, H$_2$O, or F$^-$
tetrahydropyranyl (THP)	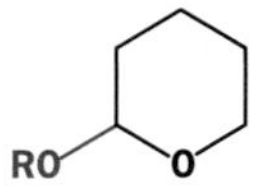	alcohols (OH in general)	strong bases	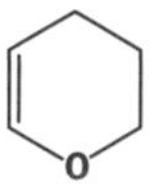 dihydropyran and acid	H$^+$, H$_2$O
benzyl ether (OBn)	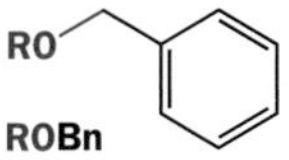 ROBn	alcohols (OH in general)	almost everything	NaH, BnBr	H$_2$, Pd/C, or HBr
methyl ether (ArOMe)	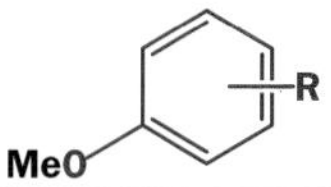	phenols (ArOH)	bases	NaH, MeI, or (MeO)$_2$SO$_2$	BBr$_3$, HBr, HI, Me$_3$SiI
benzyl amine (NBn)	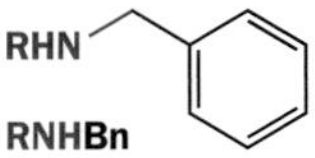 RNHBn	amines	strong bases	BnBr, K$_2$CO$_3$	H$_2$, Pd
Cbz (Z) (OCOBn)	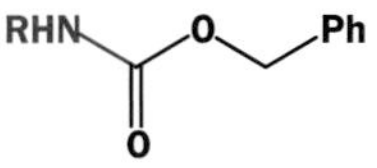 RHN–CO–O–Ph	amines	electrophiles	BnOCOCl, base	HBr, AcOH, or H$_2$, Pd
t-Boc (OCOBu-t)	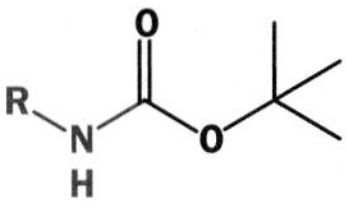	amines	electrophiles	(t-BuOCO)$_2$O, base	H$^+$, H$_2$O
Fmoc fluoroenyloxycarbonyl	see text	amines	electrophiles,	Fmoc-Cl	base, e.g. amine
t-butyl ester (CO$_2$Bu-t)	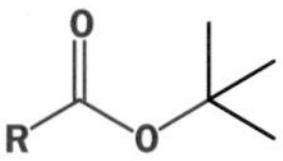	carboxylic acid (RCO$_2$H)	bases, nucleophiles	isobutene, H$^+$	H$_3$O$^+$

The synthesis of peptides on a solid support, usually beads of either polystyrene (the **Merrifield approach**) or polyamide (the **Sheppard approach**) resins has become extremely important, because it allows peptides to be synthesized by machines, and a key feature of the Sheppard approach is the use of Fmoc-protected amino acid residues. The idea is that the *C*-terminus amino acid is tethered to the resin by means of a carbamate linker that is stable to mild acid or base. The peptide chain is then built up using the sorts of methods we have been discussing and, when complete, is released by cleaving the linker with strong acid.

The side chains of the amino acids in this approach are also protected with acid-labile groups (*t*-butyl esters and BOC, for example), so that they too are revealed only in the final deprotection step.

Acid cannot therefore be used for protection for the *N*-terminus of the chain as it grows, so the solution is to use Fmoc. Each amino acid is introduced as its Fmoc-protected pentafluorophenyl ester (yet another electrophilically activated electron-poor phenyl ester), and then the Fmoc group is cleaved with piperidine ready for the next residue to be added. The green blob in the diagram represents a polystyrene or polyamide bead, each of which carries many linkers and many growing peptide chains.

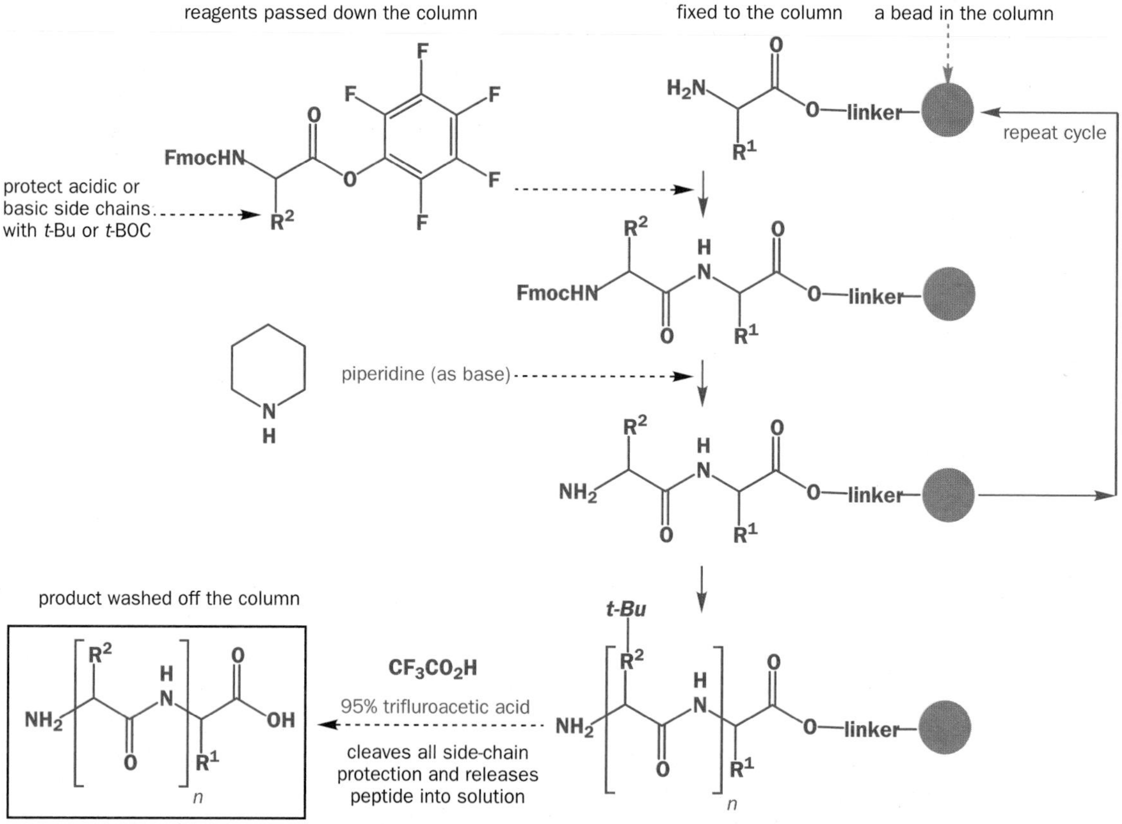

Once the first amino acid is fixed to the column, reagents are added simply by passing solutions down the column. Any excess or by-products are washed off. Finally, the product is released by passing a solution of CF_3CO_2H down the column. The simplicity and reliability of this type of simple iterative process, with two steps per cycle, has made automated peptide synthesis common laboratory practice.

The synthesis of dofetilide, a drug to combat erratic heartbeat

The chapter ends with a complete synthesis of an important new drug. Cardiac arrhythmia (erratic and inefficient heart action) is a major problem in the modern world causing poor lifestyle (exhaustion) and death by blood clots. A new drug dofetilide (Tikosyn®) is being introduced by Pfizer to treat this problem. It works by blocking the passage of potassium ions out of heart muscle and so delays the onset of an irregular beat until the next normal beat takes over.

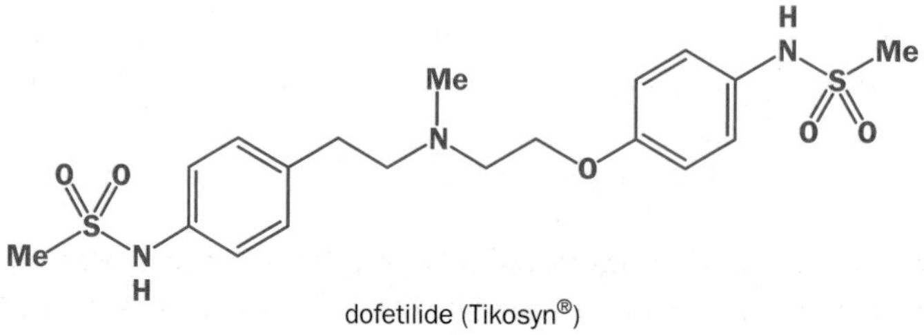

dofetilide (Tikosyn®)

We are going to do a little more than simply give the reactions that eventually made up the synthesis of dofetilide. We are going to put ourselves in the place of the chemists who invented the synthesis and try to see what led them to the reactions they chose. First, we should inspect the structure of the molecule. There are two sulfonamides, one at each end. We have seen how to make sulfonamides earlier in this chapter when saccharin was being discussed. The usual way is to react the amine with a sulfonyl chloride. In this case we shall need to react methane sulfonyl chloride ($MeSO_2Cl$ or MsCl) with the aromatic amines. This is a well-known reaction and should work well here. The other functional groups—tertiary amine and alkyl aryl ether—should not interfere so no protection is needed.

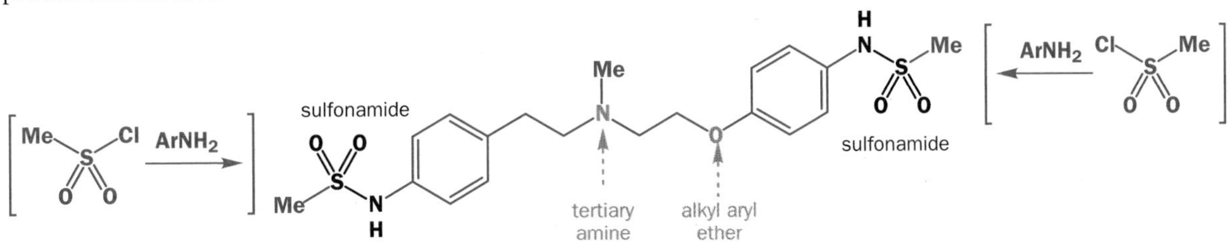

So we need to make the required aromatic diamine. We might guess from what we did earlier in this chapter as well as from Chapter 22 that this is likely to be achieved by nitration and reduction so we should check that double nitration of our proposed starting material will occur in the right positions (regioselectivity). Remember that at this stage we are just making proposals—we can only predict whether the reactions will actually occur or not.

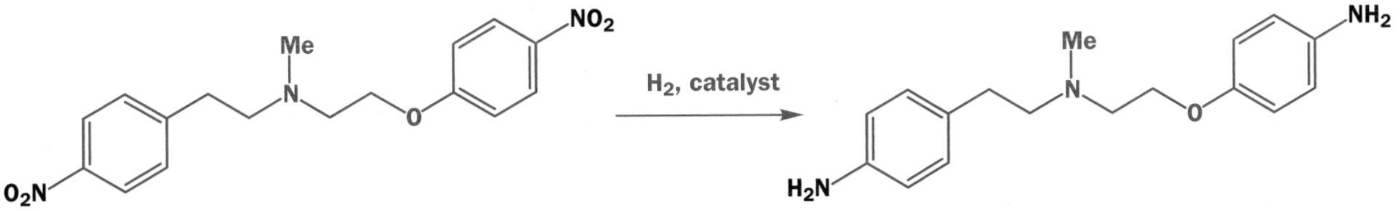

The substituents on both rings are activating and *ortho, para*-directing so there is reasonable hope that *para* selectivity can be achieved in both cases. However the left-hand ring is only weakly activated by an alkyl group whereas the ring on the right is strongly activated by the oxygen atom. It might be difficult to get the left-hand ring to react even once before the right-hand ring reacts three times (Chapter 22). A good solution would be to build the dinitro compound with each separate ring already previously nitrated once only. So we need to think how to link the rings together. The most obvious approach is to combine some organic electrophiles (alkyl halides, carbonyl compounds) with nitrogen and oxygen nucleophiles. We might, for example, join up the right-hand ring by nucleophilic aromatic substitution using the convenient *para* nitro group.

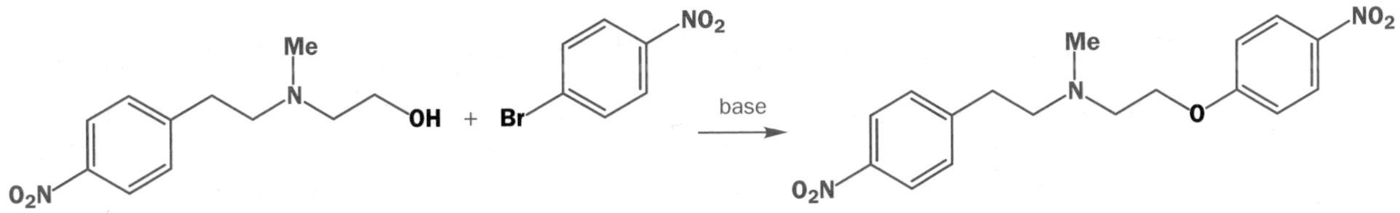

Then we could make the amino alcohol by adding an amine to an epoxide—that N–CH$_2$–CH$_2$–OH group looks as though it comes from ethylene oxide and an amine.

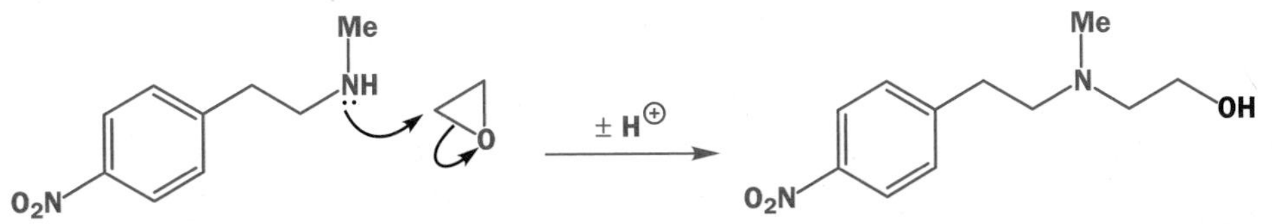

In its turn the amine could come from a reductive amination (Chapter 24) or by an alkylation, using in both cases MeNH$_2$ as the nucleophile and an aldehyde or an alkyl halide as the electrophile.

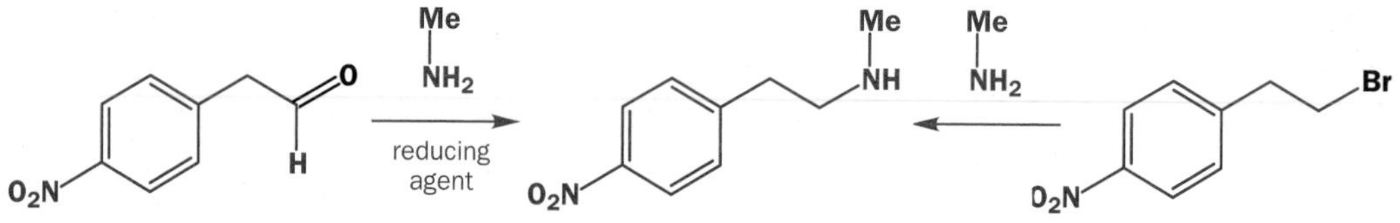

Now we have a selection of possible starting materials and we should consider which might be available commercially as that will make the job so much easier. In fact, two of the nitro compounds we want can be made so easily by direct nitration that they are available commercially.

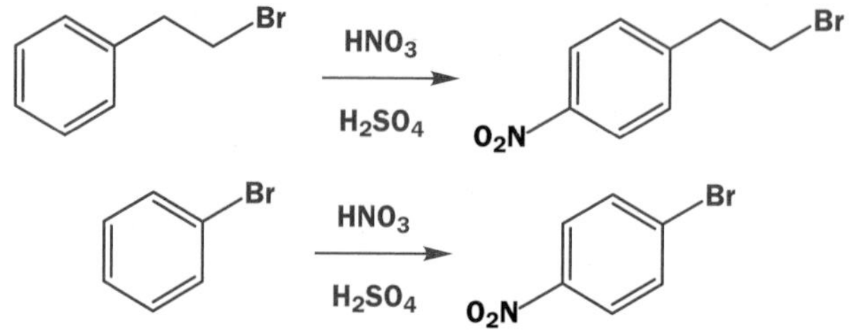

Only the aldehyde is not a commercial product and we might guess that the oxidizing power of nitric acid might convert the aldehyde into an acid. An even cheaper compound is *para*-nitro phenol, which can be made very easily from phenol and dilute nitric acid (Chapter 22).

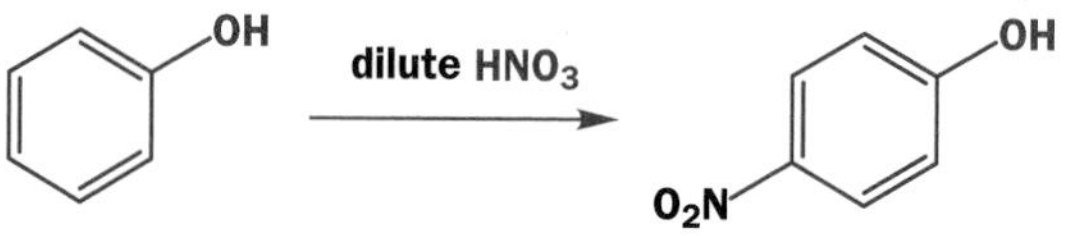

From the many possible approaches, Pfizer chose the one summarized below. The question of cheapness and availability of starting materials matters more in large-scale manufacture than in laboratory work but we can only guess at some of the choices. The last stages are as we suggested but the first stages are not. The ether is made from *para*-nitro phenol and 2-chloroethanol. This is an unusual electrophile and this reaction forms the subject of a problem at the end of the chapter. The resulting alcohol is converted into the chloride with SOCl$_2$, a standard method from Chapter 17. The yields are excellent and this too is an important consideration in manufacture.

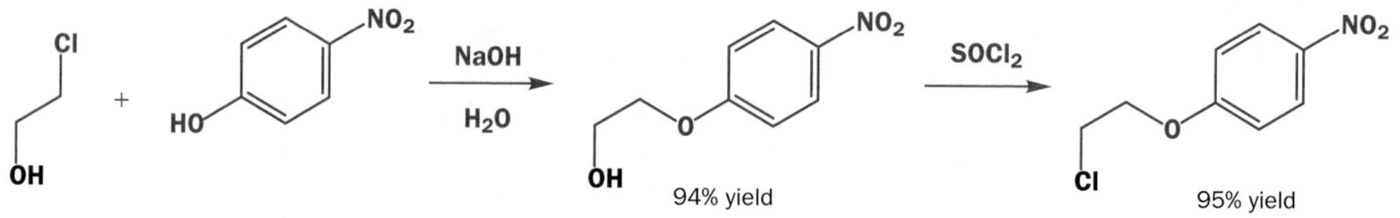

The amine is made by a simple alkylation reaction on methylamine. This choice is made chiefly because of the cheapness of the alkyl bromide and the good yield. There could be a real problem here with further alkylation of the product but they probably use a large excess of methylamine to prevent that.

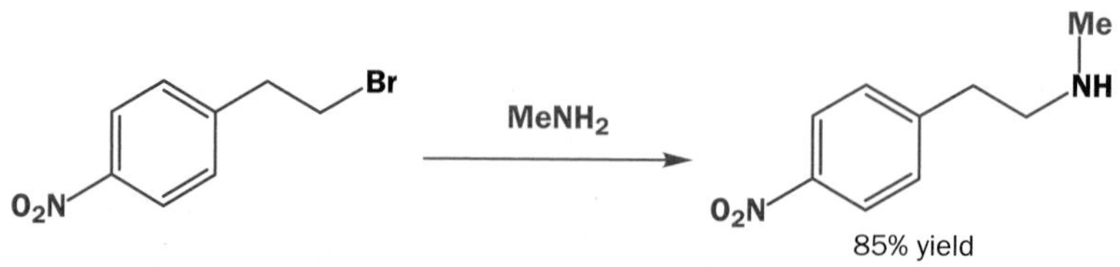

Now that the parts have been assembled, they can be joined together and these last steps follow the plan we outlined earlier. The worst step in the synthesis is the joining together of the two halves and even that gives a respectable 64% yield. This approach to synthesis—analysing the problem first and then proposing solutions—will be the subject of Chapter 30.

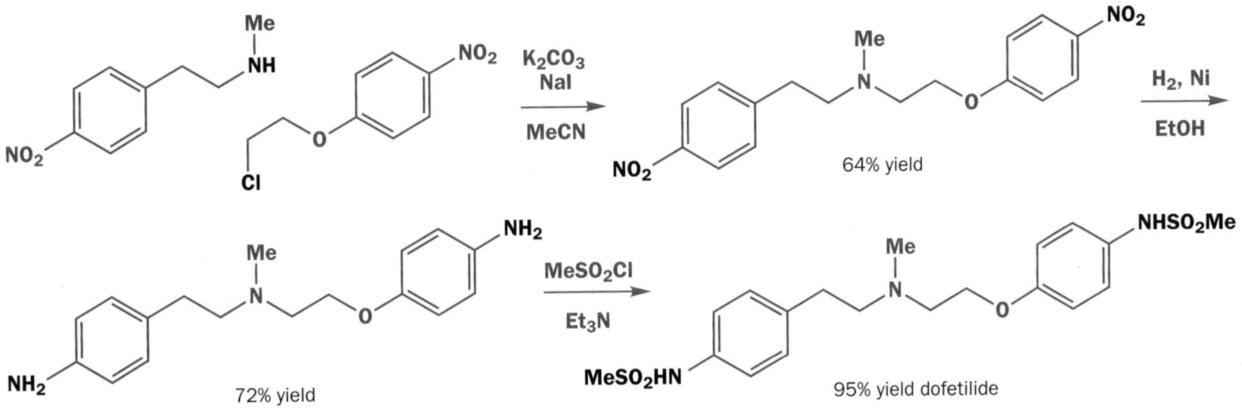

This is a commercial synthesis of an important new compound and uses only chemistry that you have met in the first 24 chapters of the book. Though new syntheses are completed and new methods invented daily, the basic organic reaction types remain the foundation on which these inventions are constructed.

Looking forward

So far, most of the reactions presented in the book that are useful in synthesis have made C–O, C–N, or C–halogen bonds and only a few (Wittig, Friedel–Crafts, and reactions of cyanides and alkynes) make C–C bonds. This limitation has severely restricted the syntheses that we can discuss in this chapter. This is by design as we wanted to establish the idea of synthesis before coming to more complicated chemistry. The next four chapters introduce the main C–C bond-forming reactions in the chemistry of enols and enolates. You met these valuable intermediates in Chapter 21 but now you are about to see how they can be alkylated and acylated and how they add directly to aldehydes and ketones and how they do conjugate addition to unsaturated carbonyl compounds. Then in Chapter 30 we return to a more general discussion of synthesis and develop a new approach in the style of the last synthesis in this chapter.

Problems

1. Suggest two different syntheses for these ethers and say which you prefer (and why!).

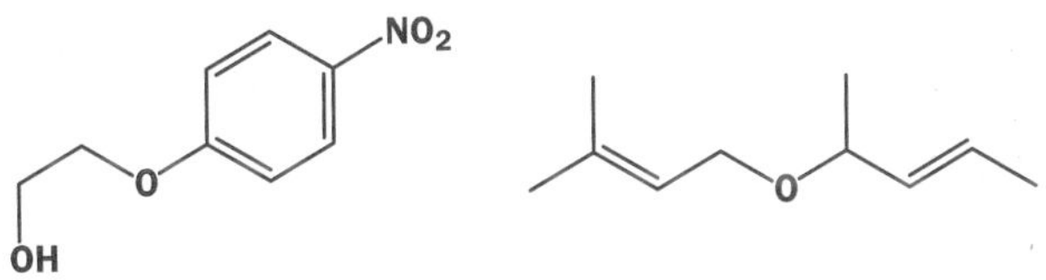

2. Suggest syntheses for these esters. The starting materials might also need to be made.

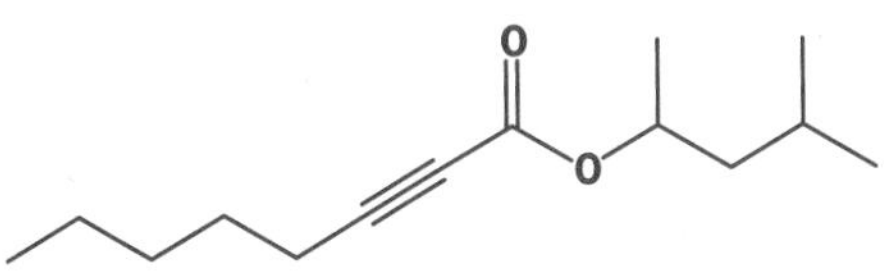

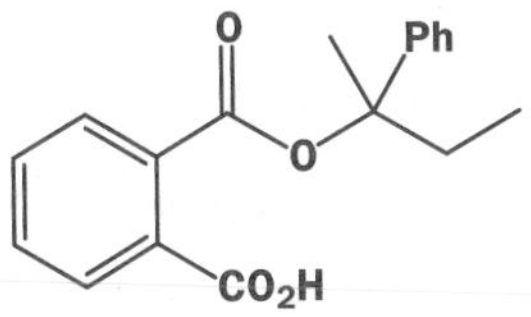

3. Suggest a synthesis for this local anaesthetic.

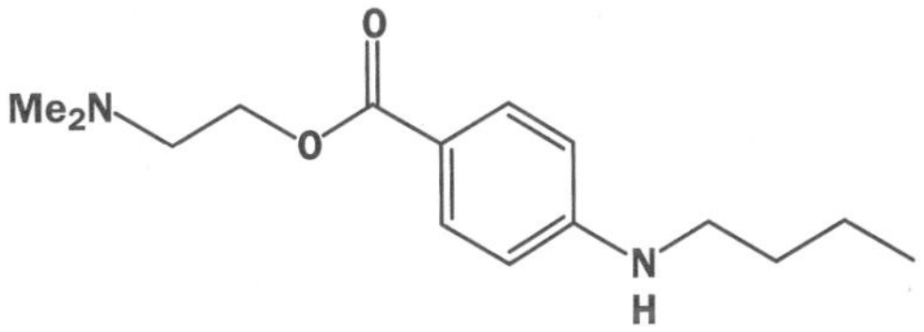

4. Suggest syntheses for this simple compound. What selectivity problems must be overcome?

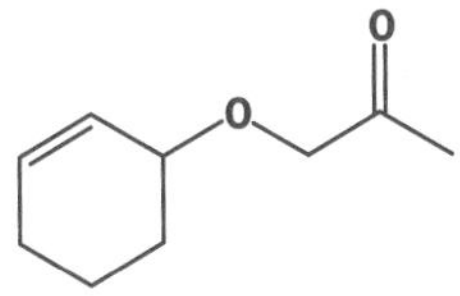

5. Suggest a synthesis for this compound. Justify your choice of methods and reagents.

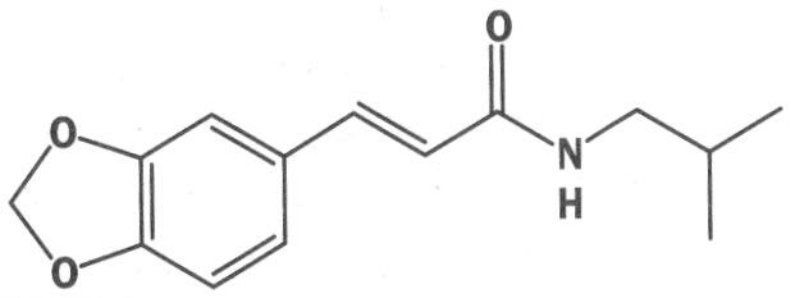

6. Suggest how these amines might be synthesized.

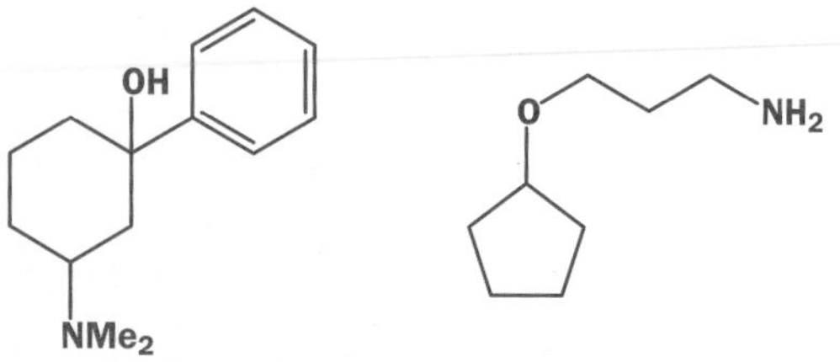

7. This hexa-alcohol can be deprotected, one OH at a time, by the sequence of reagents shown below. Explain how each reagent works, stating, of course, which protecting group it removes! Would any other order of events be successful?

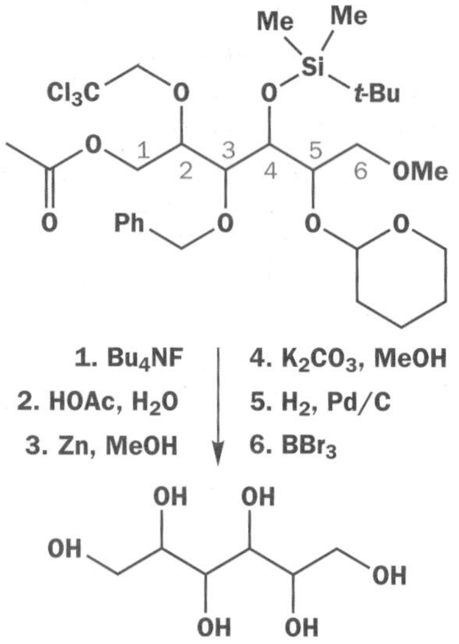

1. Bu₄NF | 4. K₂CO₃, MeOH
2. HOAc, H₂O | 5. H₂, Pd/C
3. Zn, MeOH | 6. BBr₃

8. Suggest a synthesis of the starting material and give mechanisms for the reactions. Why does the last step go under such unusual conditions?

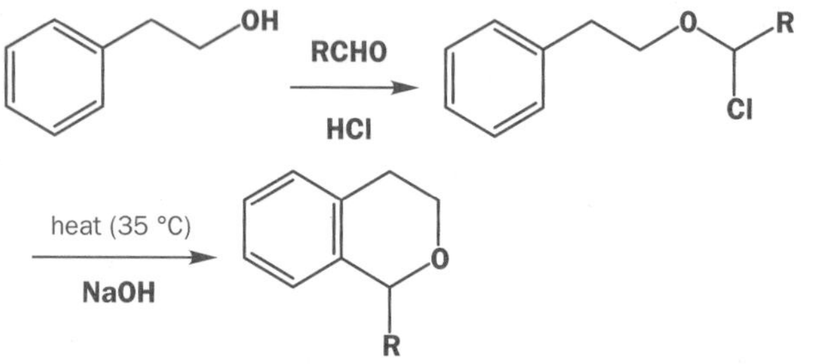

9. Esters are normally made from alcohols and activated acids. This one is made by a completely different method. Why?

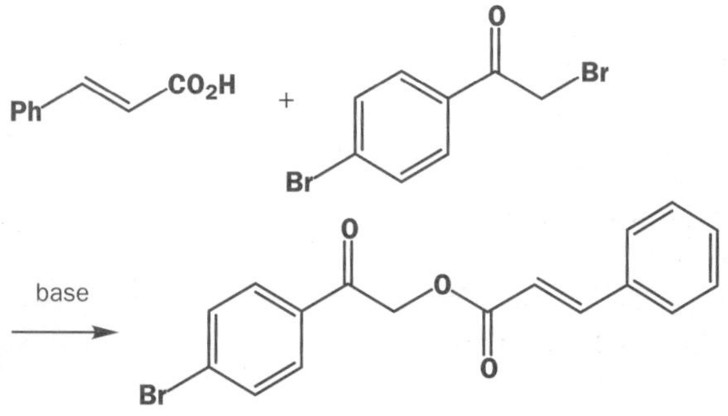

10. Suggest a synthesis of this non-protein peptide, emphasizing the choice of protecting groups.

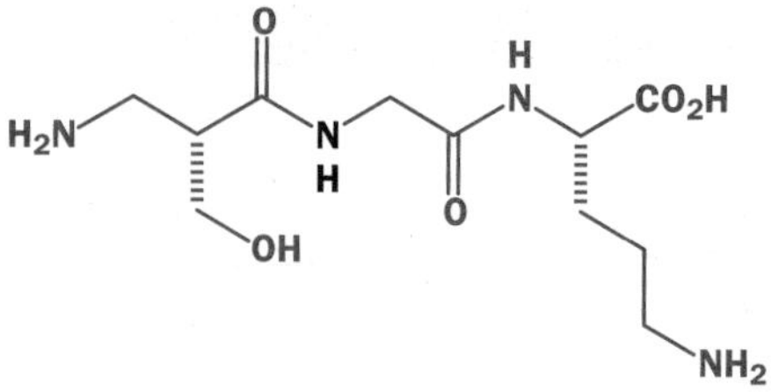

11. The β-iodoethoxycarbonyl group has been suggested as a protecting group for amines. It is removed with zinc in methanol. How would you add this protecting group to an amine and how does the deprotection occur? What other functional groups might survive the deprotection?

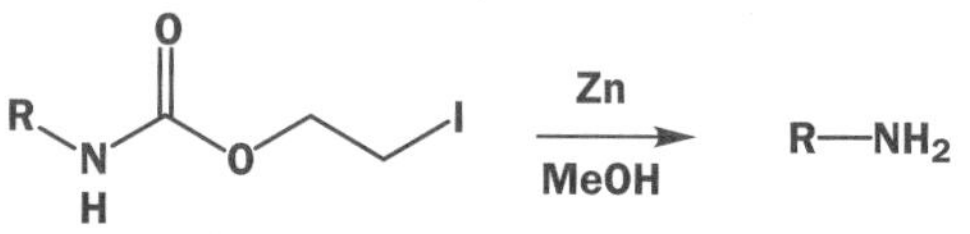

12. Revision of Chapters 10 and 16. Give mechanisms for this synthesis and suggest why this route was followed.

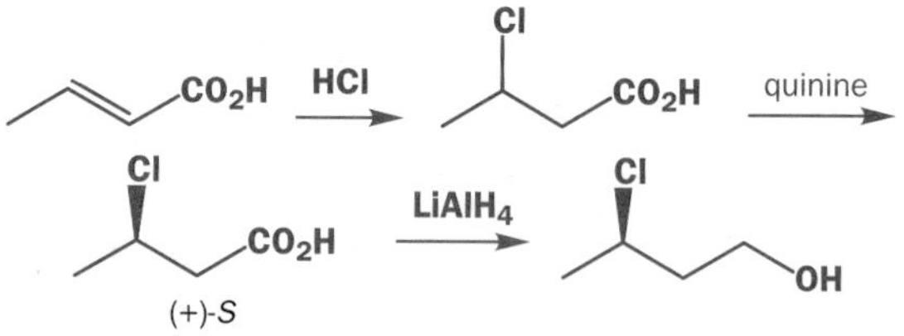

13. Revision of Chapters 9 and 19. Draw the structures of the intermediates in this synthesis of a diene and comment on the selectivity of the last step.

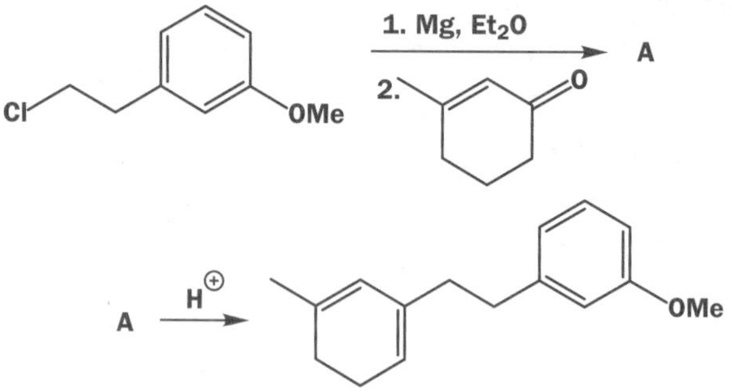

14. Suggest ways to make these compounds.

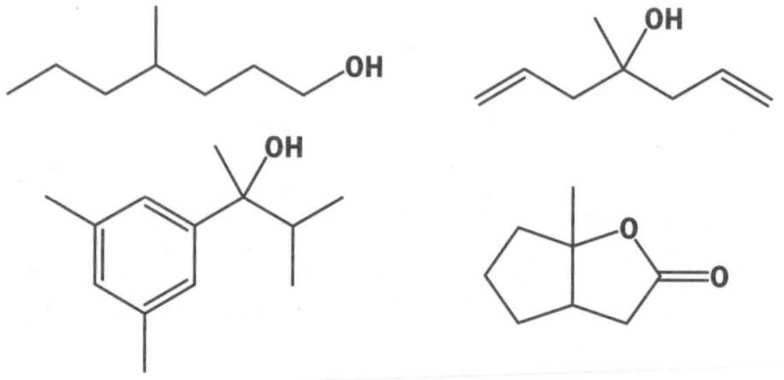

Connections

Building on:

- Enols and enolates ch21
- Electrophilic addition to alkenes ch20
- Nucleophilic substitution reactions ch17

Arriving at:

- How to make new C–C bonds using carbonyl compounds as nucleophiles
- How to prevent carbonyl compounds reacting with themselves

Looking forward to:

- Forming C–C bonds by reacting nucleophilic enolates with electrophilic carbonyl compounds ch27
- Forming C–C bonds by reacting nucleophilic enolates with electrophilic carboxylic acid derivatives ch28
- Forming C–C bonds by reacting nucleophilic enolates with electrophilic alkenes ch29
- Retrosynthetic analysis ch30

Chapters 26–29 continue the theme of synthesis that started with Chapter 24 and will end with Chapter 30. This group of four chapters introduces the main C–C bond-forming reactions of enols and enolates. We develop the chemistry of Chapter 21 with a discussion of enols and enolates attacking to alkylating agents (Chapter 26), aldehydes and ketones (Chapter 27), acylating agents (Chapter 28), and electrophilic alkenes (Chapter 29).

Carbonyl groups show diverse reactivity

In earlier chapters we discussed the two types of reactivity displayed by the carbonyl group. We first described reactions that involve **nucleophilic attack** on the carbon of the carbonyl, and in Chapter 9 we showed you that these are among the best ways of making new C–C bonds. In this chapter we shall again be making new C–C bonds, but using **electrophilic attack** on carbonyl compounds: in other words, the carbonyl compound will be reacting as the nucleophile in the reaction. We introduced the nucleophilic forms of carbonyl compounds—enols, and enolates—in Chapter 21. There you saw them reacting with heteroatomic electrophiles, but they will also react well with carbon electrophiles provided the reaction is thoughtfully devised. Much of this chapter will concern that phrase, 'thoughtfully devised'.

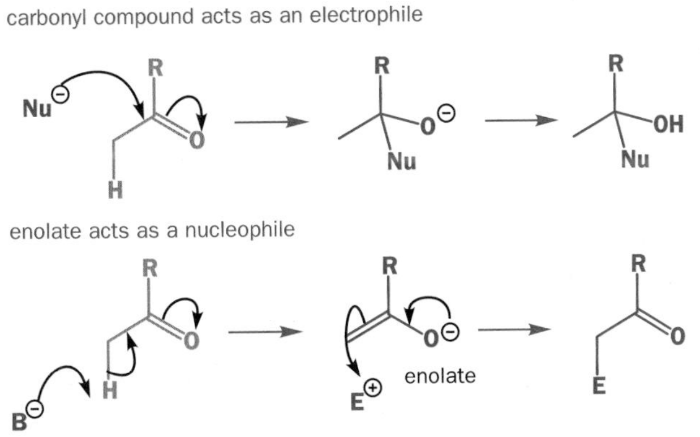

carbonyl compound acts as an electrophile

enolate acts as a nucleophile

Thought is needed to ensure that the carbonyl compound exhibits the right sort of reactivity. In particular, the carbonyl compound must not act as an electrophile when it is intended to be a nucleophile. If it does, it may react with itself to give some sort of dimer—or even a polymer—rather than neatly attacking the desired electrophile. This chapter is devoted to ways of avoiding this: in Chapter 27 we shall talk about how to promote and control the dimerization, known as the **aldol reaction**.

aim to avoid an unwanted dimerization: the aldol reaction

Fortunately, over the last three decades lots of thought has *already* gone into the problem of controlling the reactions of enolates with carbon electrophiles. This means that there are many excellent solutions to the problem: our task in this chapter is to help you understand which to use, and when to use them, in order to design useful reactions.

Some important considerations that affect all alkylations

These reactions consist of two steps. The first is the formation of a stabilized anion—usually (but not always) an enolate—by deprotonation with base. The second is a substitution reaction: attack of the nucleophilic anion on an electrophilic alkyl halide. All the factors controlling S_N1 and S_N2 reactions, which we discussed at length in Chapter 17, are applicable here.

In each case, we shall take one of two approaches to the choice of base.

- A strong base can be chosen to deprotonate the starting material completely. There is complete conversion of the starting material to the anion before addition of the electrophile, which is added in a subsequent step

- Alternatively, a weaker base may be used *in the presence of the electrophile*. The weaker base will not deprotonate the starting material completely: only a small amount of anion will be formed, but that small amount will react with the electrophile. More anion is formed as alkylation uses it up

The second approach is easier practically (just mix the starting material, base, and electrophile), but works only if the base and the electrophile are compatible and don't react together. With the first approach, which is practically more demanding, the electrophile and base never meet each other, so their compatibility is not a concern. We shall start with some compounds that avoid the problem of competing aldol reactions completely, because they are not electrophilic enough to react with their own nucleophilic derivatives.

Nitriles and nitroalkanes can be alkylated

Problems that arise from the electrophilicity of the carbonyl group can be avoided by replacing C=O by functional groups that are much less electrophilic but are still able to stabilize an adjacent anion. We shall consider two examples, both of which you met in Chapter 21.

Alkylation of nitriles

Firstly, the nitrile group, which mirrors the carbonyl group in general reactivity but is much less easily attacked by nucleophiles (N is less electronegative than O).

You met nitrile hydrolysis and addition reactions, for example, in Chapter 12.

The anion formed by deprotonating a nitrile using strong base will not react with other molecules of nitrile but will react very efficiently with alkyl halides. The slim, linear structure of the anions makes them good nucleophiles for S$_N$2 reactions.

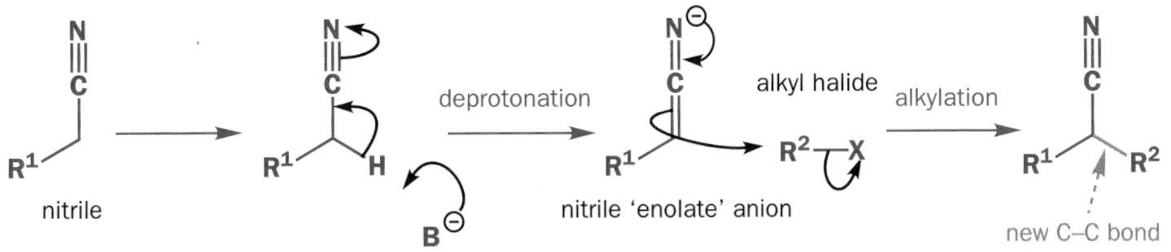

The nitrile does not have to be deprotonated completely for alkylation: with sodium hydroxide only a small amount of anion is formed. In the example below, such an anion reacts with propyl bromide to give 2-phenylpentanenitrile.

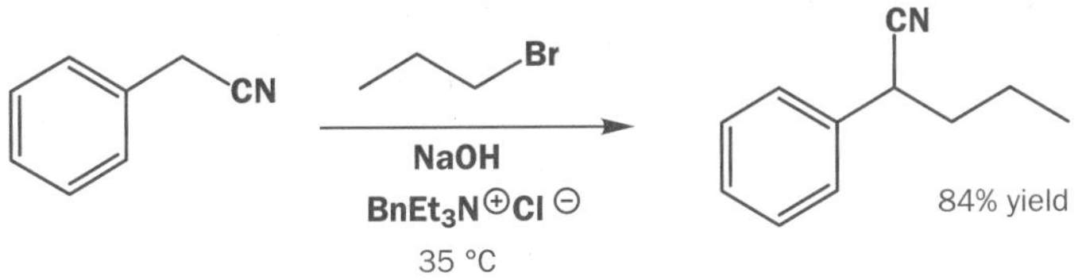

This reaction is carried out in a two-phase mixture (water + an immiscible organic solvent) to prevent the hydroxide and propyl bromide merely reacting together in an S$_N$2 reaction to give propanol. The hydroxide stays in the aqueous layer, and the other reagents stay in the organic layer. A tetraalkylammonium chloride (benzyltriethylammonium chloride BnEt$_3$N$^+$Cl$^-$) is needed as a **phase transfer catalyst** to allow sufficient hydroxide to enter the organic layer to deprotonate the nitrile.

Nitrile-stabilized anions are so nucleophilic that they will react with alkyl halides rather well even when a crowded quaternary centre (a carbon bearing no H atoms) is being formed. In this example the strong base, sodium hydride, was used to deprotonate the branched nitrile completely and benzyl chloride was the electrophile. The greater reactivity of benzylic electrophiles compensates for the poorer leaving group. In DMF, the anion is particularly reactive because it is not solvated (DMF solvates only the Na$^+$ cation).

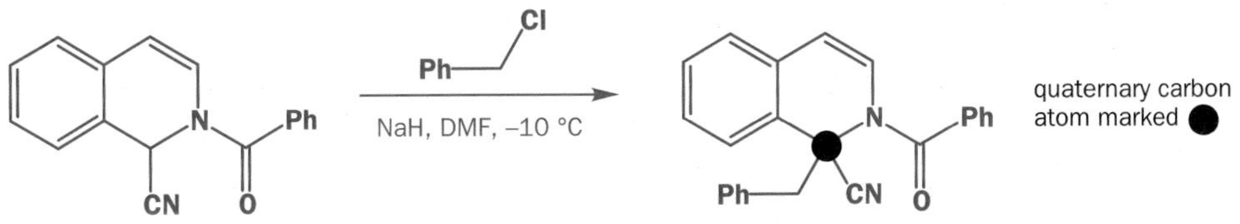

The compatibility of sodium hydride with electrophiles means that, by adding two equivalents of base, alkylation can be encouraged to occur more than once. This dimethylated acid was required in the synthesis of a potential drug, and it was made in two steps from a nitrile. Double alkylation with two equivalents of NaH in the presence of excess methyl iodide gave the methylated nitrile which was hydrolysed to the acid. The monoalkylated product is not isolated—it goes on directly to be deprotonated and react with a second molecule of MeI.

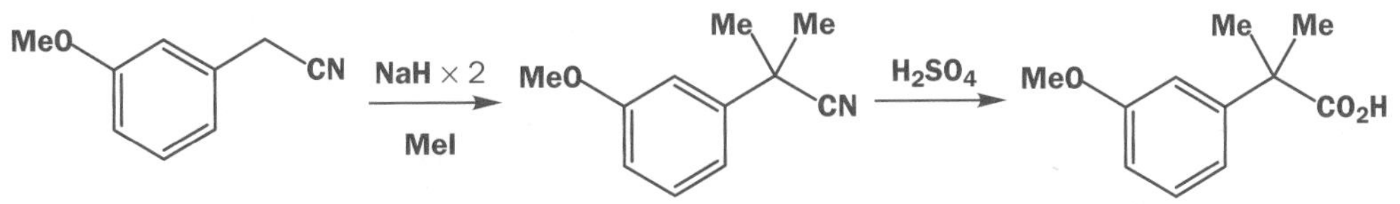

You met phase transfer catalysis in Chapter 23, p. 606.

Remember our discussion about the lack of nucleophilicity of hydride (H$^-$) in Chapter 6? Here is hydride acting as a base even in the presence of the electrophile: there was no need to do this reaction in two steps because the base and electrophile cannot react together.

first alkylation

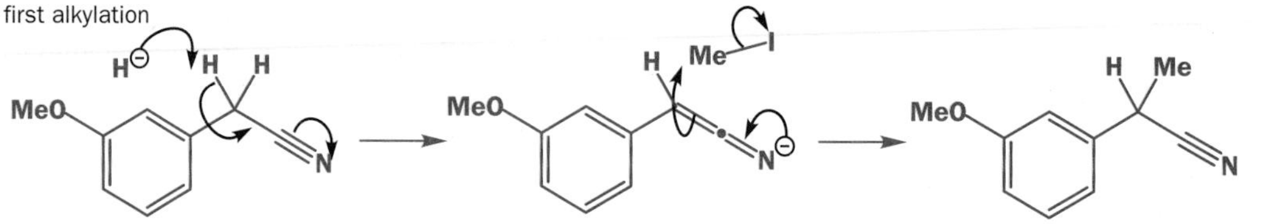

second alkylation

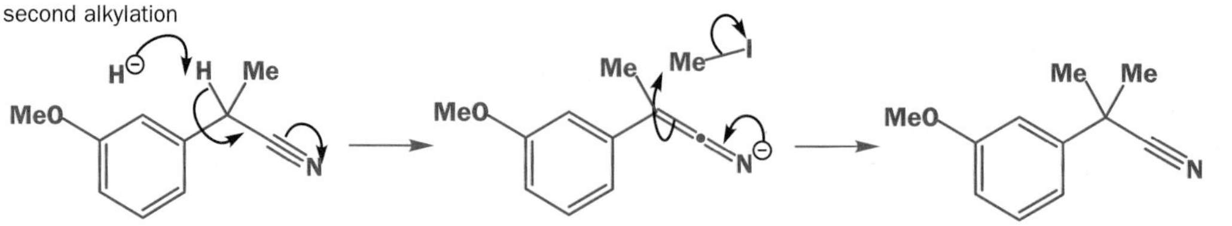

> Multiple alkylation is not always desirable, and one of the side-reactions in alkylations that are intended to go only once is the formation of doubly, or in special cases triply, alkylated products. These arise when the first alkylation product still has acidic protons and can be deprotonated to form another anion. This may in turn react further. Clearly, this is more likely to be a problem if the base is present in excess and can usually be restricted by using only one equivalent of the electrophile.

With two nitrile groups, the delocalized anion is so stable that even a weak, neutral amine (triethylamine) is sufficiently basic to deprotonate the starting material. Here double alkylation again takes place: note that the electrophile is good at S_N2, and the solvent is dipolar and aprotic (DMSO and DMF have similar properties). The doubly alkylated quaternary product was formed in 100% yield.

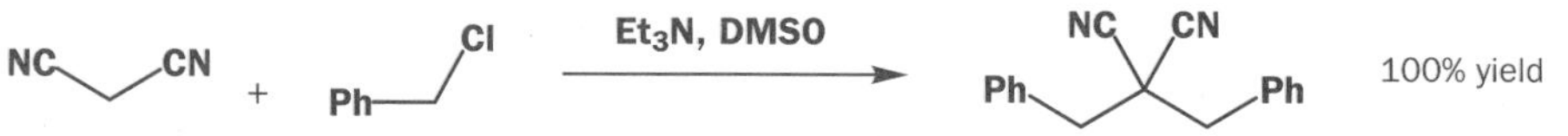

If the electrophile and the nitrile are in the same molecule and the spacing between them is appropriate, then intramolecular alkylation will lead to cyclization to form rings that can have anything from three to six members. The preparation of a cyclopropane is shown using sodium hydroxide as the base and chloride as a leaving group. With an intramolecular alkylation, the base and the electrophile have to be present together, but the cyclization is so fast that competing S_N2 with HO^- is not a problem.

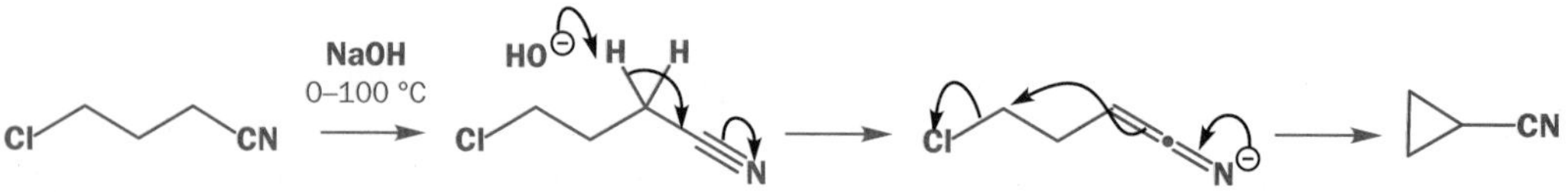

Alkylation of nitroalkanes

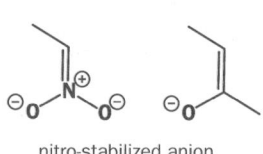

nitro-stabilized anion
—compare enolate

■ Nitro-stabilized anions also undergo additions to aldehydes, ketones, and electrophilic alkenes: these reactions appear in Chapters 27 and 29.

The powerful electron-withdrawing nature of the nitro group means that deprotonation is possible even with very mild bases (the pK_a of $MeNO_2$ is 10). The anions react with carbon electrophiles and a wide variety of nitro-containing products can be produced. The anions are not, of course, enolates, but replacing the nitrogen with a carbon should help you to recognize the close similarity of these alkylations with the enolate alkylations described later.

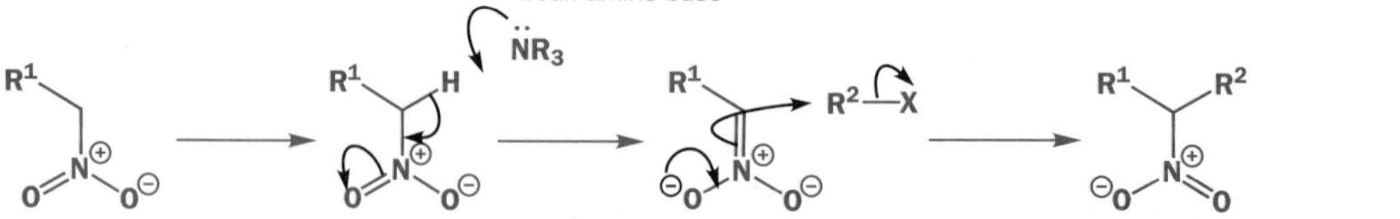

weak amine base

nitroalkane   nitro-stabilized anion

Surprisingly few simple nitroalkanes are commercially available but more complex examples can be prepared readily by alkylation of the anions derived from nitromethane, nitroethane, and 2-nitropropane. Deprotonation of nitroalkanes with butyllithium followed by the addition of alkyl halides gives the alkylated nitroalkanes in good yield. Some examples of this general method are shown below. These reactions really do have to be done in two steps: BuLi is not compatible with alkyl halides!

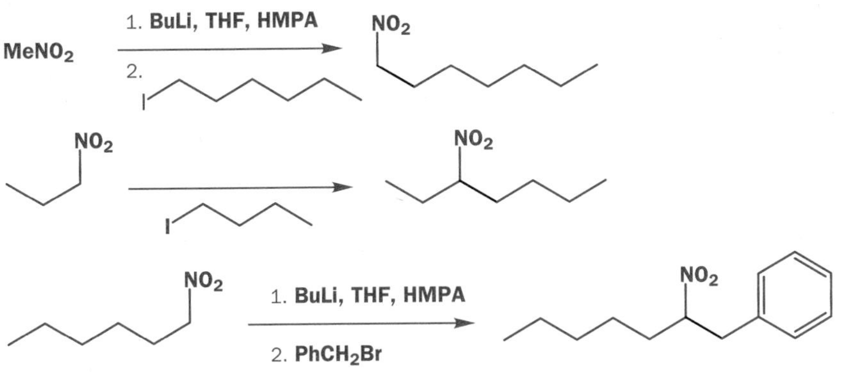

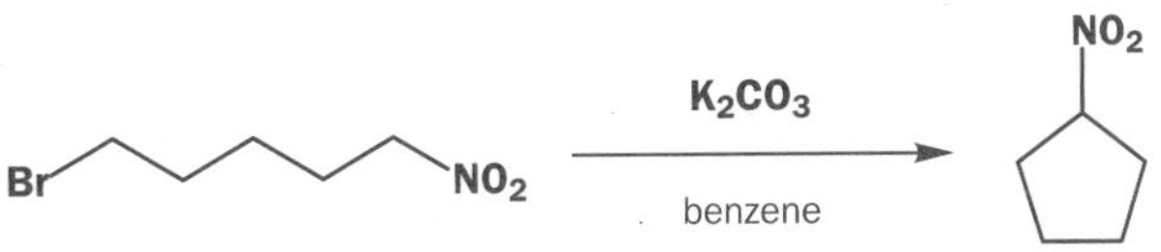

► **Hexamethylphosphoramide (HMPA)**

HMPA has the structure shown below, and the basic oxygen atom coordinates to lithium extremely powerfully. The cation is solvated, leaving the anion unsolvated and more reactive. HMPA is known to cause cancer, and should not be confused with its less common cousin HMPT (hexamethyl-phosphorous triamide).

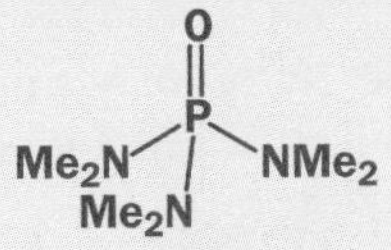

Nitroalkanes can be alkylated in a single step with hydroxide as a base: phase transfer conditions keep the HO⁻ and the electrophile apart, preventing alcohol formation. This compound forms despite its quaternary carbon atom.

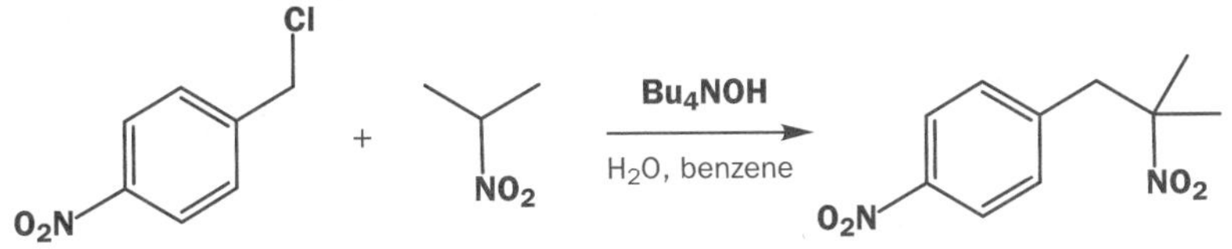

Cyclic nitroalkanes can be prepared by intramolecular alkylation provided that the ring size is appropriate (3–7 members). Now there really is no alternative: the base and electrophile must cohabit in the reaction mixture, so a weaker base such as potassium carbonate must be used—amines are no good here because they undergo substitution reactions with the halide.

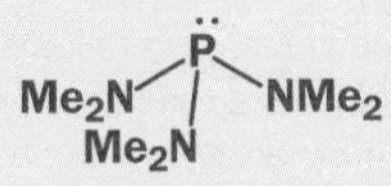

Choice of electrophile for alkylation

Enolate alkylations are S_N2 reactions (polar solvents, good charged nucleophile) so the electrophile needs to be S_N2-reactive if the alkylation is to succeed: primary and benzylic alkyl halides are among the best alkylating agents. More branched halides tend to prefer to undergo unwanted E2 elimination (Chapter 19), because the anions themselves are rather basic. As a result, tertiary halides are useless for enolate alkylation. We shall see a way round this problem later in the chapter.

methyl	allyl	benzyl	primary alkyls	secondary alkyls	tertiary alkyls
alkylate very well			alkylate well	alkylate slowly	do *not* alkylate

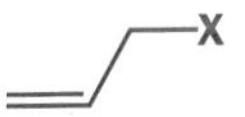

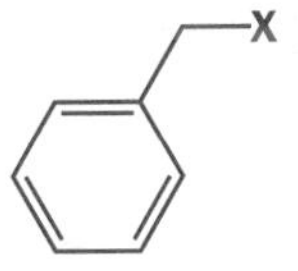

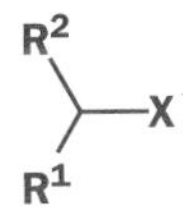

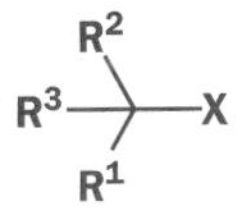

Lithium enolates of carbonyl compounds

The problem of self-condensation of carbonyl compounds (that is, enolate reacting with unenolized carbonyl) under basic conditions does not exist if there is absolutely no unenolized carbonyl compound present. One way to achieve this is to use a base sufficiently strong (pK_a at least 3 or 4 units higher than pK_a of the carbonyl compound) to ensure that all of the starting carbonyl is converted into the corresponding enolate. This will work only if the resulting enolate is sufficiently stable to survive until the alkylation is complete. As you saw in Chapter 21, lithium enolates are stable, and are among the best enolate equivalents for use in alkylation reactions.

■
LDA is described on p. 538.

a reminder: how to make LDA

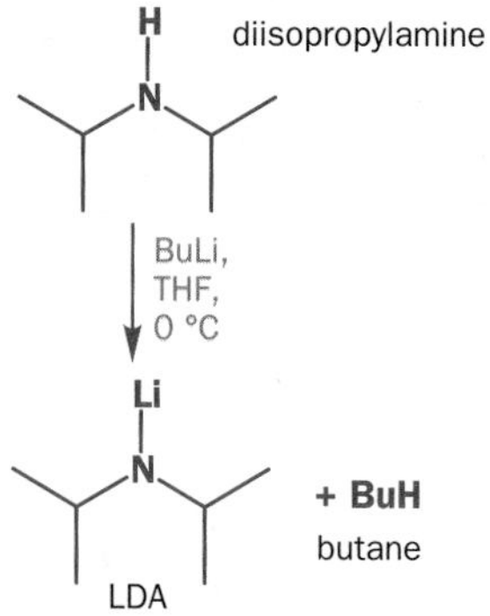

diisopropylamine

BuLi,
THF,
0 °C

+ **BuH**
butane

LDA

▶

Enolates are a type of alkene, and there are two possible geometries of the enolate of an ester. The importance of enolate geometry is discussed in Chapter 34 and will not concern us here. More important is the question of regioselectivity when unsymmetrical ketones are deprotonated. We shall discuss this aspect later in the chapter.

The best base for making lithium enolates is usually LDA, made from diisopropylamine (i-Pr$_2$NH) and BuLi. LDA will deprotonate virtually all ketones and esters that have an acidic proton to form the corresponding lithium enolates rapidly, completely, and irreversibly even at the low temperatures (about –78°C) required for some of these reactive species to survive.

Deprotonation occurs through a cyclic mechanism illustrated below for ketones and esters. The basic nitrogen anion removes the proton as the lithium is delivered to the forming oxyanion.

deprotonation of an ester

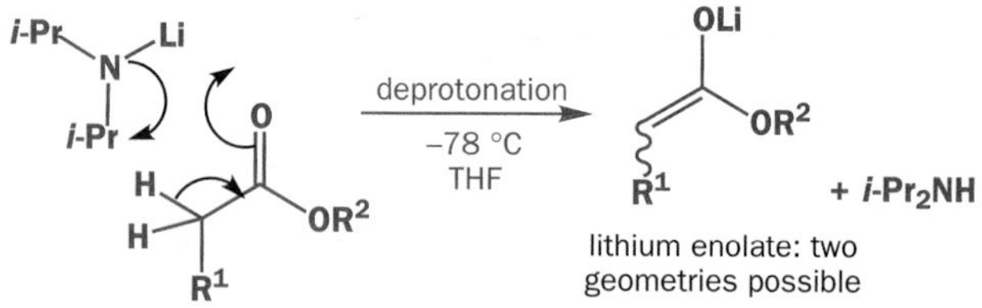

deprotonation

–78 °C
THF

lithium enolate: two geometries possible

+ i-Pr$_2$NH

deprotonation of a ketone

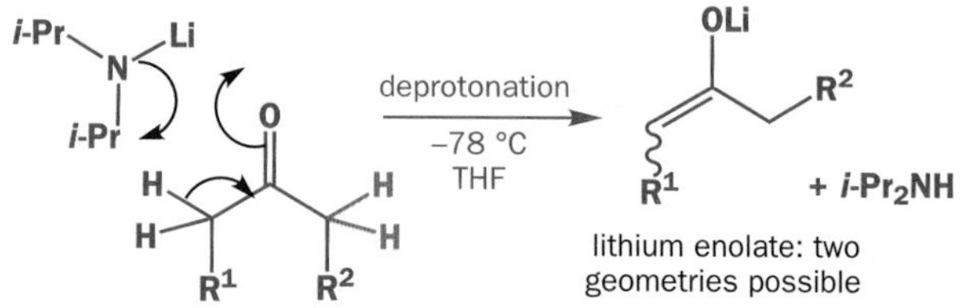

deprotonation

–78 °C
THF

lithium enolate: two geometries possible

+ i-Pr$_2$NH

if R^1 ≠ R^2, removal of the green protons gives a different enolate

Variations on a theme

LDA came into general use in the 1970s, and you may meet more modern variants derived from butyllithium and isopropylcyclohexylamine (lithium isopropylcyclo-hexylamide, LICA) or 2,2,6,6-tetramethylpiperidine (lithium tetramethylpiperidide, LTMP) or hexamethyldisilazane (lithium hexamethyldisilazide, LHMDS), which are even more hindered and are even less nucleophilic as a result.

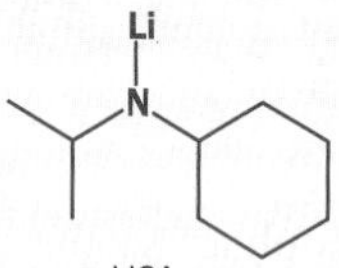

LICA 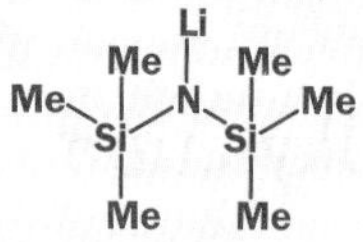   LHMDS 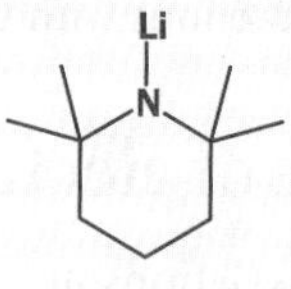   LTMP or LiTMP

Alkylations of lithium enolates

The reaction of these lithium enolates with alkyl halides is one of the most important C–C bond-forming reactions in chemistry. Alkylation of lithium enolates works with both acyclic and cyclic ketones as well as with acyclic and cyclic esters (lactones). The general mechanism is shown below.

alkylation of an ester enolate   alkylation of a ketone enolate

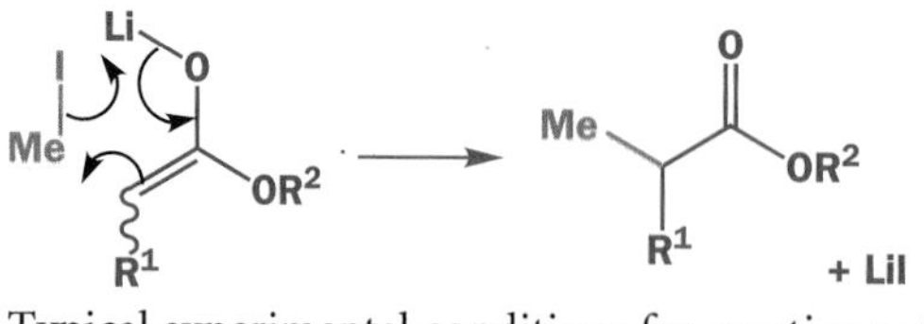

+ LiI   + LiI

Typical experimental conditions for reactions of kinetic enolates involve formation of the enolate at very low temperature (–78°C) in THF. Remember, the strong base LDA is used to avoid self-condensation of the carbonyl compound but, while the enolate is forming, there is always a chance that self-condensation will occur. The lower the temperature, the slower the self-condensation reaction, and the fewer by-products there are. Once enolate formation is complete, the electrophile is added (still at –78°C: the lithium enolates may not be stable at higher temperatures). The reaction mixture is then usually allowed to warm up to room temperature to speed up the rate of the S$_N$2 alkylation.

Alkylation of ketones

Precisely this sequence was used to methylate the ketone below with LDA acting as base followed by methyl iodide as electrophile.

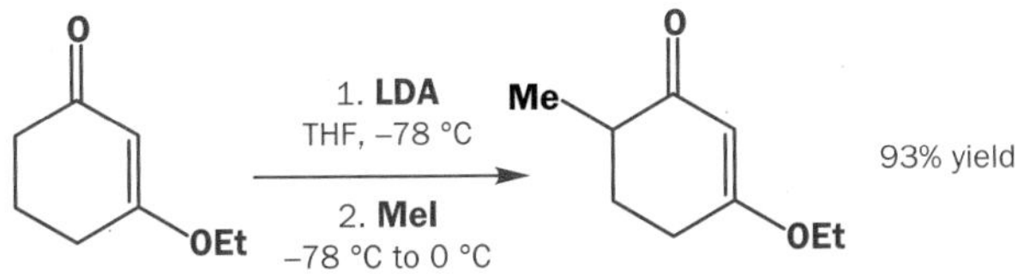

93% yield

In Chapter 17 you saw epoxides acting as electrophiles in S_N2 reactions. They can be used to alkylate enolates providing epoxide opening is assisted by coordination to a Lewis acidic metal ion: in this case the lanthanide yttrium(III). The new C–C bond in the product is coloured black. Note that the ketone starting material is unsymmetrical, but has protons only to one side of the carbonyl group, so there is no question over which enolate will form. The base is one of the LDA variants we showed you on p. 668—LHMDS.

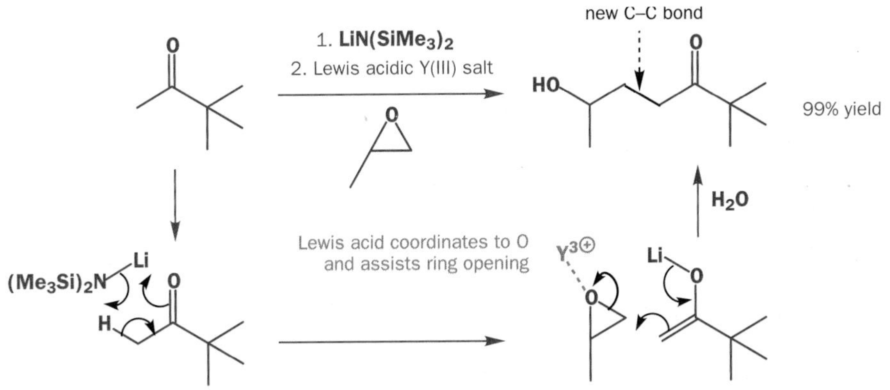

Sodium and potassium also give reactive enolates

Their stability at low temperature means that lithium enolates are usually preferred, but sodium and potassium enolates can also be formed by abstraction of a proton by strong bases. The increased separation of the metal cation from the enolate anion with the larger alkali metals leads to more reactive but less stable enolates. Typical very strong Na and K bases include the hydrides (NaH, KH) or amide anions derived from ammonia (NaNH$_2$, KNH$_2$) or

hexamethyldisilazane (NaHMDS, KHMDS). The instability of the enolates means that they are usually made and reacted in a single step, so the base and electrophile need to be compatible. Here are two examples of cyclohexanone alkylation: the high reactivity of the potassium enolate is demonstrated by the efficient tetramethylation with excess potassium hydride and methyl iodide.

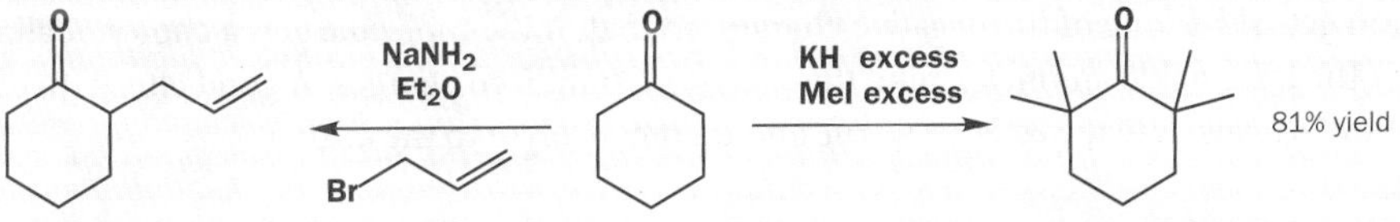

81% yield

Alkylation of esters

In Chapter 28 you will meet the reaction of an ester with its own enolate: the **Claisen condensation**. This reaction can be an irritating side-reaction in the chemistry of lithium ester enolates when alkylation is desired, and again it can be avoided only if the ester is converted entirely to its enolate under conditions where the Claisen condensation is slow. A good way of stopping this happening is to add the ester *to the solution of LDA* (and not the LDA to the ester) so that there is never excess ester for the enolate to react with.

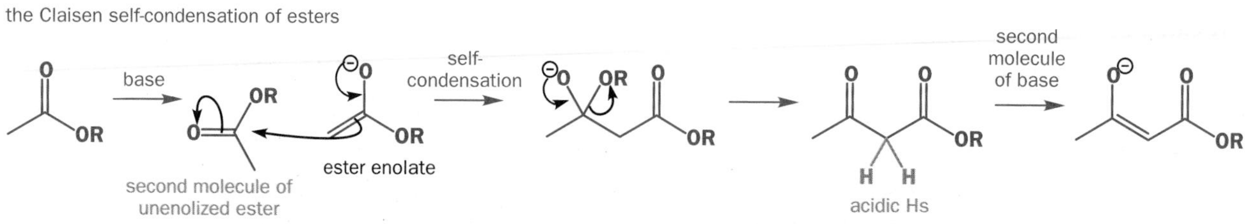

the Claisen self-condensation of esters

Another successful tactic is to make the group R as large as possible to discourage attack at the carbonyl group. Tertiary butyl esters are particularly useful in this regard, because they are readily made, *t*-butyl is extremely bulky, and yet they can can still be hydrolysed in aqueous acid under mild conditions by the method discussed on pp. 652–3. In this example, deprotonation of *t*-butyl acetate with LICA (lithium isopropylcyclohexylamide) gives a lithium enolate that reacts with butyl iodide as the reaction mixture is warmed to room temperature.

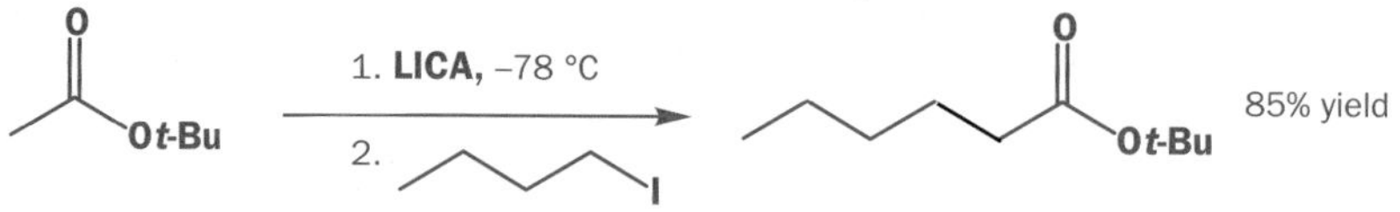

85% yield

Alkylation of carboxylic acids

The lithium enolates of carboxylic acids can be formed if two equivalents of base are used. Carboxylic acids are very acidic so it is not necessary to use a strong base to remove the first proton but, since the second deprotonation requires a strong base such as LDA, it is often convenient to use two equivalents of LDA to form the dianion. With carboxylic acids, even BuLi can be used on occasion because the intermediate lithium carboxylate is much less electrophilic than an aldehyde or a ketone.

> ■ Why doesn't BuLi add to the carboxylate as you saw in Chapter 12 to form the ketone? Presumably in this case the aromatic ring helps acidify the benzylic protons to tip the balance towards deprotonation. Even with carboxylic acids, LDA would be the first base you would try.

> ▶ You saw this sort of reactivity with dianions in Chapter 24: the last anion to form will be the most reactive.

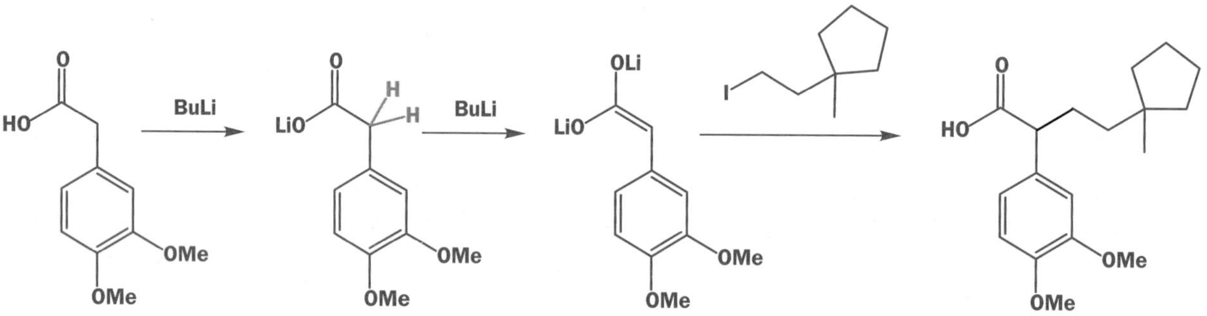

The next alkylation of an acid enolate is of a carbamate-protected amino acid, glycine. As you saw in Chapter 25, carbamates are stable to basic reaction conditions. Three acidic protons are removed by LDA, but alkylation takes place only at carbon—the site of the last proton to be removed. Alkylation gets rid of one of the negative charges, so that, if the molecule gets a choice, it alkylates to get rid of the least stable anion, keeping the two more stabilized charges. A good alternative to using the dianion is to alkylate the ester or nitrile and then hydrolyse to the acid.

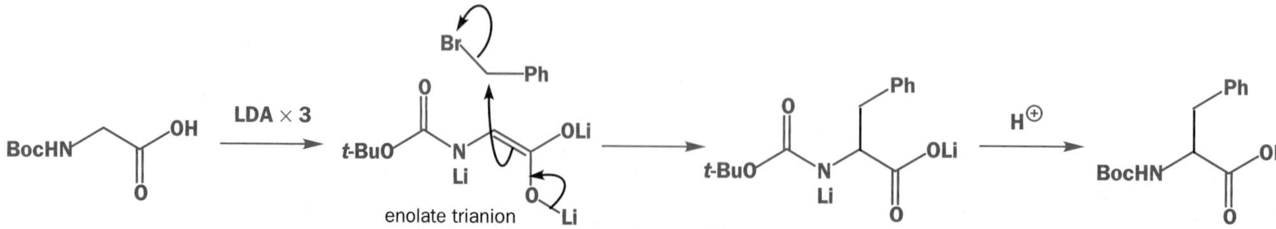

enolate trianion

● **Alkylation of ketones, esters, and carboxylic acids is best carried out using the lithium enolates.**

Why do enolates alkylate on carbon?

Enolates have two nucleophilic sites: the carbon and the oxygen atoms: on p. 526 we showed that:

- Carbon has the greater coefficient in the HOMO, and is the softer nucleophilic site
- Oxygen carries the greater total charge and is the harder nucleophilic site

In Chapter 21 you saw that hard electrophiles prefer to react at oxygen—that is why it is possible to make silyl enol ethers, for example. Some carbon electrophiles with very good leaving groups also tend to react on carbon, but soft electrophiles such as alkyl halides react at carbon, and you will see only this type of electrophile in this chapter.

In general:

- Hard electrophiles, particularly alkyl sulfates and sulfonates (mesylates, tosylates), tend to react at oxygen
- Soft electrophiles, particularly alkyl halides (I > Br > Cl), react at carbon
- Polar aprotic solvents (HMPA, DMF) promote O-alkylation by separating the enolate anions from each other and the counterion (making the bond more polar and increasing the charge at O) while ethereal solvents (THF, DME) promote C-alkylation
- Larger alkali metals (Cs > K > Na > Li) give more separated ion pairs (more polar bonds) which are harder and react more at oxygen

hard electrophiles react at O

$X = OMs, OSO_2OMe, {}^{+}OMe_2$

soft electrophiles react at C

$X = I, Br, Cl$

Alkylation of aldehydes

Aldehydes are so electrophilic that, even with LDA at –78°C, the rate at which the deprotonation takes place is not fast enough to outpace reactions between the forming lithium enolate and still-to-be-deprotonated aldehyde remaining in the mixture. Direct addition of the base to the carbonyl group of electrophilic aldehydes can also pose a problem.

reactions which compete with aldehyde enolate formation

deprotonation
–78 °C
THF

lithium enolate

aldol self-condensation

addition
–78 °C
THF

● Avoid using lithium enolates of aldehydes.

Using specific enol equivalents to alkylate aldehydes and ketones

These side-reactions mean that aldehyde enolates are not generally useful reactive intermediates. Instead, there are a number of aldehyde enol and enolate equivalents in which the aldehyde is present only in masked form during the enolization and alkylation step. The three most important of these **specific enol equivalents** are:

- enamines
- silyl enol ethers
- aza-enolates derived from imines

You met all of these briefly in Chapter 21, and we shall discuss how to use them to alkylate aldehydes shortly. All three types of specific enol equivalent are useful not just with aldehydes, but with ketones as well, and we shall introduce each class with examples for both types of carbonyl compound.

Enamines are alkylated by reactive electrophiles

Enamines are formed when aldehydes or ketones react with secondary amines. The mechanism is given in Chapter 14. The mechanism below shows how they react with alkylating agents to form new carbon–carbon bonds: the enamine here is the one derived from cyclohexanone and pyrrolidine. The product is at first not a carbonyl compound: it's an iminium ion or an enamine (depending on whether an appropriate proton can be lost). But a mild acidic hydrolysis converts the iminium ion or enamine into the corresponding alkylated carbonyl compound.

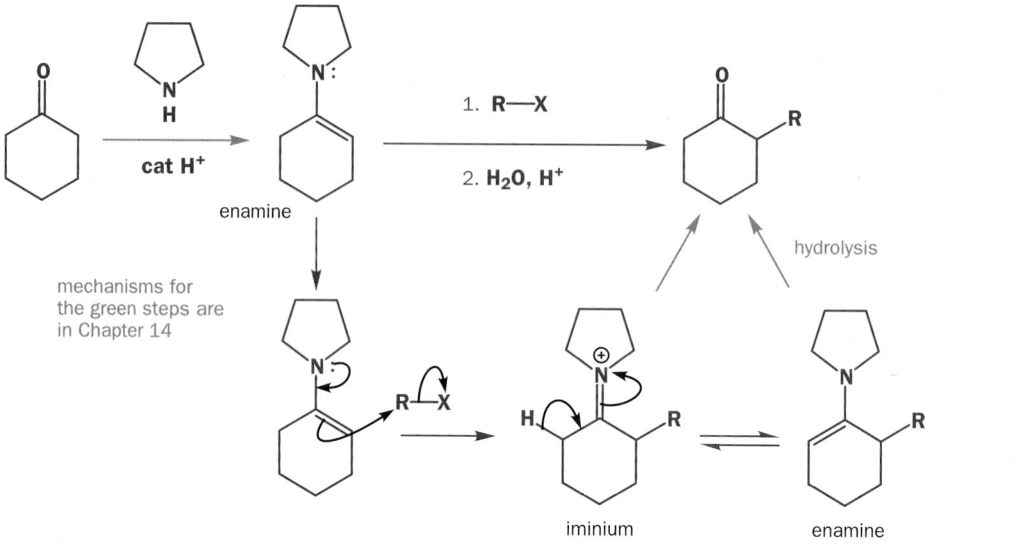

The overall process, from carbonyl compound to carbonyl compound, amounts to an enolate alkylation, but no strong base or enolates are involved so there is no danger of self-condensation. The example below shows two specific examples of cyclohexanone alkylation using enamines. Note the relatively high temperatures and long reaction times: enamines are among the most reactive of neutral nucleophiles, but they are still a lot less nucleophilic than enolates.

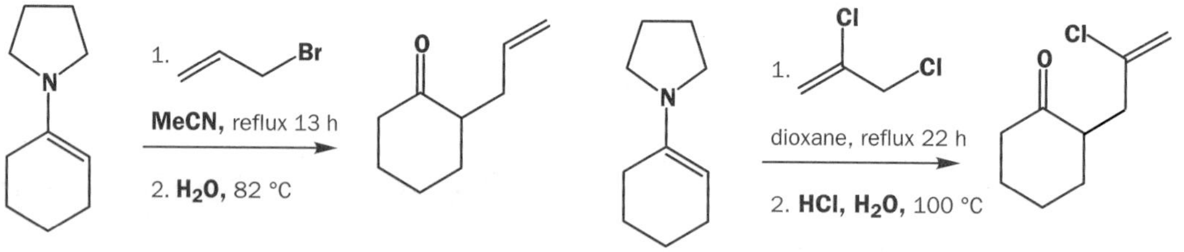

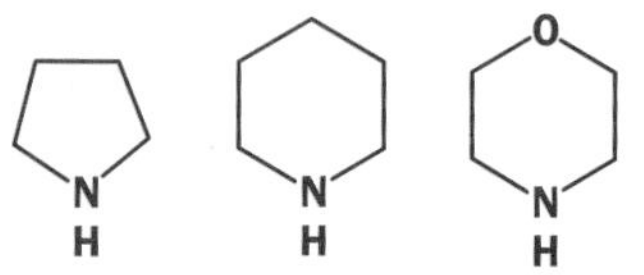

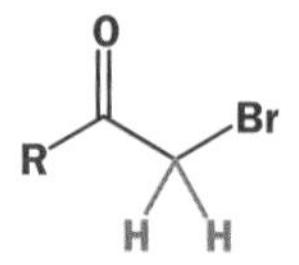

The choice of the secondary amine for formation of the enamine is not completely arbitrary even though it does not end up in the final alkylated product. Simple dialkyl amines can be used but cyclic amines such as pyrrolidine, piperidine, and morpholine are popular choices as the ring structure makes both the starting amine and the enamine more nucleophilic (the alkyl groups are 'tied back' and can't get in the way). The higher boiling points of these amines allow the enamine to be formed by heating.

α-Bromo carbonyl compounds are excellent electrophiles for S_N2 reactions because of the rate-enhancing effect of the carbonyl group (Chapter 17). The protons between the halogen and the carbonyl are significantly more acidic than those adjacent to just a carbonyl group and there is a serious risk of an enolate nucleophile acting as a base. Enamines are only very weakly basic, but react well as a nucleophile with a-bromo carbonyl compounds, and so are a good choice.

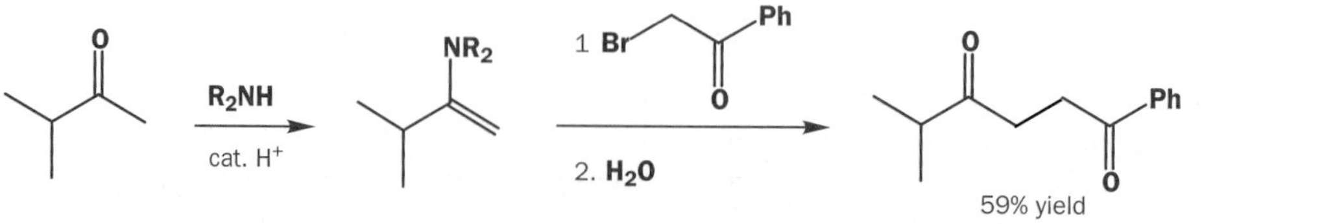

The original ketone here is unsymmetrical, so two enamines are possible. However, the formation of solely the *less* substituted enamine is typical. The outcome may be explained as the result of thermodynamic control: enamine formation is reversible so the less hindered enamine predominates.

For the more substituted enamine, steric hindrance forces the enamine to lose planarity, and destabilizes it. The less substituted enamine, on the other hand, is rather more stable. Note how the preference for the less substituted enamine is opposite to the preference for a *more* substituted enol.

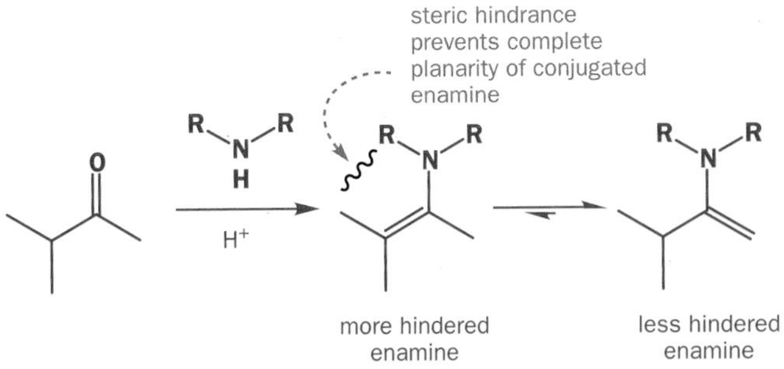

There is, however, a major problem with enamines: reaction at nitrogen. Less reactive alkylating agents—simple alkyl halides such as methyl iodide, for example—react to a significant degree at N rather than at C. The product is a quaternary ammonium salt, which hydrolyses back to the starting material and leads to low yields.

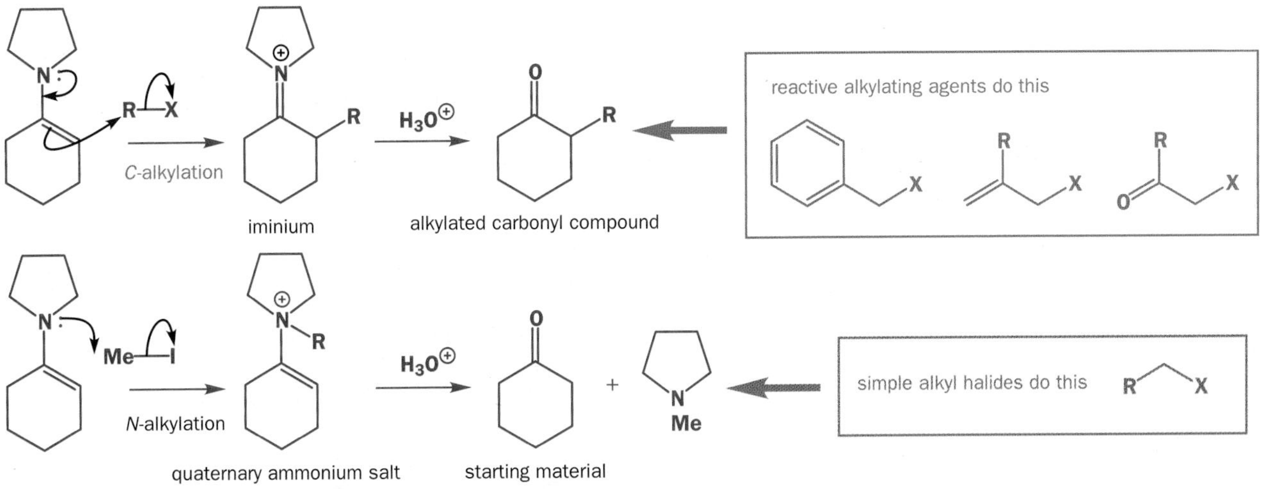

- **Enamines can be used only with reactive alkylating agents.**

- allylic halides
- benzyl halides
- α-halo carbonyl compounds

That said, enamines are a good solution to the aldehyde enolate problem. Aldehydes form enamines very easily (one of the advantages of the electrophilic aldehyde) and these are immune to attack by nucleophiles—including most importantly the enamines themselves. Below are two examples of aldehyde alkylation using the enamine method.

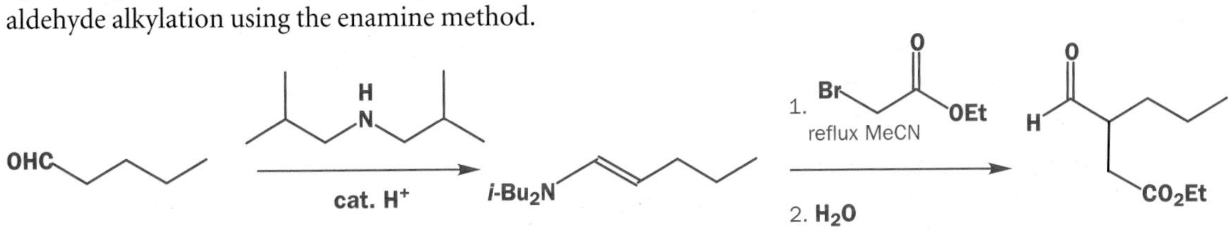

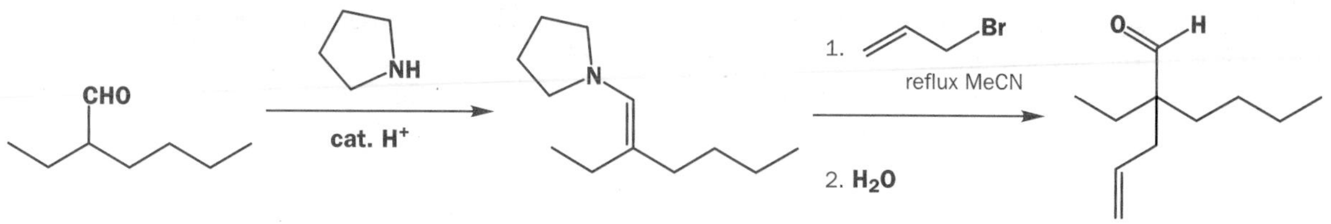

Both again use highly S_N2-reactive electrophiles, and this is the main drawback of enamines. In the next section we consider a complementary class of enol equivalents that react only with highly S_N1-reactive electrophiles.

Silyl enol ethers are alkylated by S_N1-reactive electrophiles in the presence of Lewis acid

You saw the quantitative formation of carbocations by this method in Chapter 17.

Enamines are among the most powerful neutral nucleophiles and react spontaneously with alkyl halides. Silyl enol ethers are less reactive and so require a more potent electrophile to initiate reaction. Carbocations will do, and they can be generated *in situ* by abstraction of a halide or other leaving group from a saturated carbon centre by a Lewis acid.

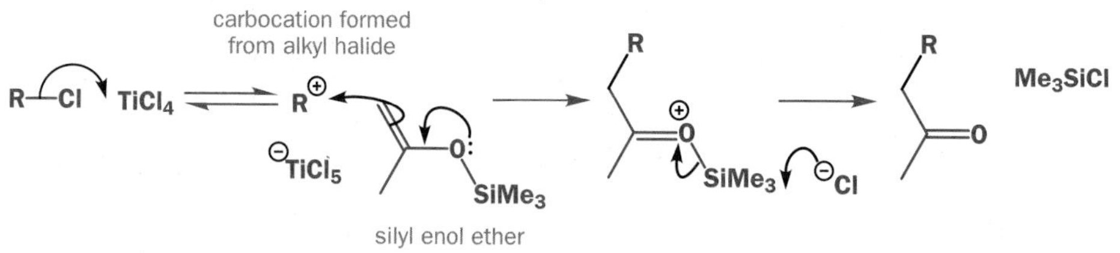

The best alkylating agents for silyl enol ethers are tertiary alkyl halides: they form stable carbocations in the presence of Lewis acids such as $TiCl_4$ or $SnCl_4$. Most fortunately, this is just the type of compounds that is unsuitable for reaction with lithium enolates or enamines, as elimination results rather than alkylation: a nice piece of complementary selectivity.

Below is an example: the alkylation of cyclopentanone with 2-chloro-2-methylbutane. The ketone was converted to the trimethylsilyl enol ether with triethylamine and trimethylsilylchloride: we discussed this step on p. 538 (Chapter 21). Titanium tetrachloride in dry dichloromethane promotes the alkylation step.

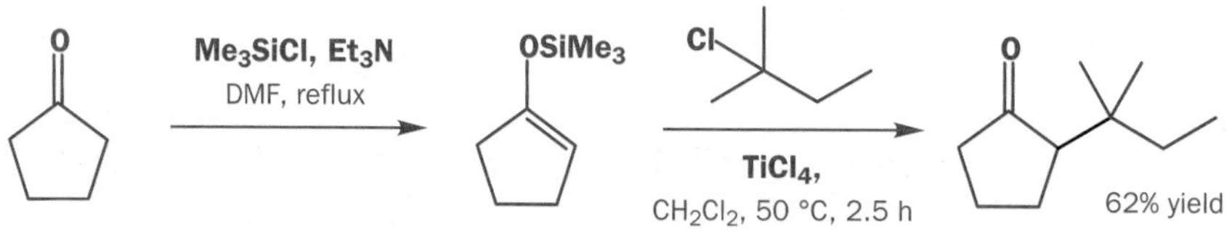

Aza-enolates react with S_N2-reactive electrophiles

Enamines are the nitrogen analogues of enols and provide one solution to the aldehyde enolate problem when the electrophile is reactive. Imines are the corresponding nitrogen analogues of aldehydes and ketones: a little lateral thinking should therefore lead you to expect some useful reactivity from the nitrogen equivalents of enolates, known as aza-enolates. Aza-enolates are formed when imines are treated with LDA or other strong bases.

In basic or neutral solution, imines are less electrophilic than aldehydes: they react with organolithiums, but not with many weaker nucleophiles (they are more electrophilic in acid when they are protonated). So, as the aza-enolate forms, there is no danger at all of self-condensation.

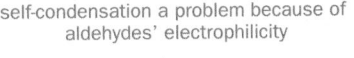

The overall sequence involves formation of the imine from the aldehyde that is to be alkylated—usually with a bulky primary amine such as t-butyl- or cyclohexylamine to discourage even further nucleophilic attack at the imine carbon. The imine is not usually isolated, but is deprotonated directly with LDA or a Grignard reagent (these do not add to imines, but they will deprotonate them to give magnesium aza-enolates).

> ▶
> *Note.* Aza-enolates are formed from imines, which can be made only from *primary* amines. Enamines are made from aldydes or ketones with *secondary* amines.

The resulting aza-enolate reacts like a ketone enolate with S_N2-reactive alkylating agents—here, benzyl chloride—to form the new carbon–carbon bond and to re-form the imine. The alkylated imine is usually hydrolysed by the mild acidic work-up to give the alkylated aldehyde.

In the next example, a lithium base (lithium diethylamide) is used to form the aza-enolate. The ease of imine cleavage in acid is demonstrated by the selective hydrolysis to the aldehyde without any effect on the acetal introduced by the alkylation step. The product is a mono-protected dialdehyde—difficult to prepare by other methods.

● Aldehyde alkylation

Aza-enolates are the best general solution for alkylating aldehydes with most electrophiles. With very S_N2-reactive alkylating agents, enamines can be used, and with very S_N1-reactive alkylating agents, silyl enol ethers must be used

Aza-enolate alkylation is so successful that it has been extended from aldehydes, where it is essential, to ketones where it can be a useful option. Cyclohexanones are among the most electrophilic simple ketones and can suffer from undesirable side-reactions. The imine from cyclohexanone and cyclohexylamine can be deprotonated with LDA to give a lithium aza-enolate. In this example, iodomethylstannane was the alkylating agent, giving the tin-containing ketone after hydrolysis.

● Specific enol equivalents for aldehydes and ketones

To summarize:

- **Lithium enolates can be used with S_N2-reactive electrophiles, but cannot be made from aldehydes**
- **Aza-enolates of aldehydes or ketones can be used with the same S_N2-reactive electrophiles, but *can* be made from aldehydes**
- **Enamines of aldehydes or ketones can be used with allylic, benzylic, or α-halocarbonyl compounds**
- **Silyl enol ethers of aldehydes or ketones can be used with S_N1-reactive (tertiary, allylic or benzylic) alkyl halides**

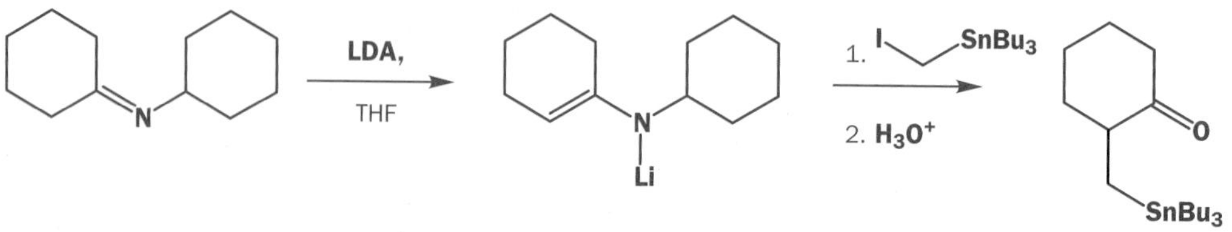

Alkylation of β-dicarbonyl compounds

▶ Typical electron-withdrawing groups include COR, CO_2R, CN, $CONR_2$, SO_2R, P=O(OR)$_2$.

The presence of two, or even three, electron-withdrawing groups on a single carbon atom makes the remaining proton(s) appreciably acidic (pK_a 10–15), which means that even mild bases can lead to complete enolate formation. With bases of the strength of alkoxides or weaker, only the multiply stabilized anions form: protons adjacent to just one carbonyl group generally have a pK_a > 20. The most important enolates of this type are those of 1,3-dicarbonyl (or β-dicarbonyl) compounds.

alkylation of a 1,3-dicarbonyl compound (or β-dicarbonyl compound)

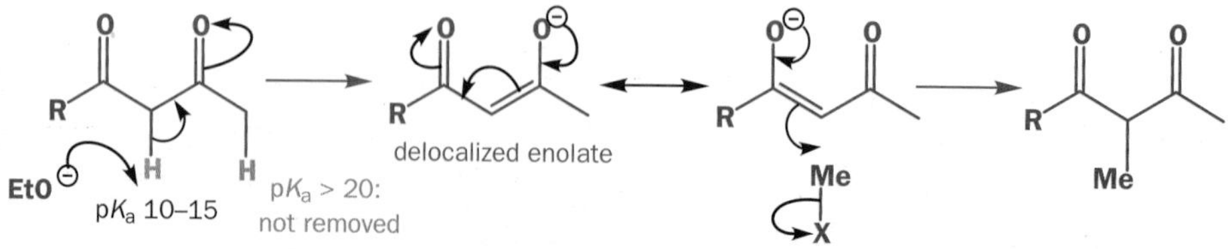

The resulting anions are alkylated very efficiently. This diketone is enolized even by potassium carbonate, and reacts with methyl iodide in good yield. Carbonate is such a bad nucleophile that the base and the electrophile can be added in a single step.

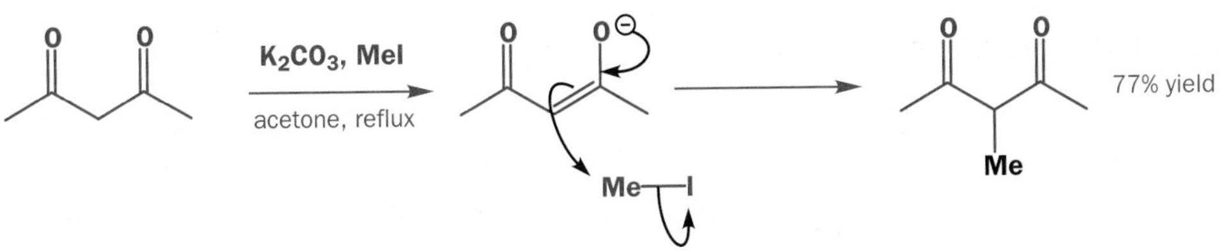

Among the β-dicarbonyls, two compounds stand out in importance—diethyl (or dimethyl) malonate and ethyl acetoacetate. You should make sure you remember their structures and trivial names.

You met these compounds, and their stable enols, in Chapter 21.

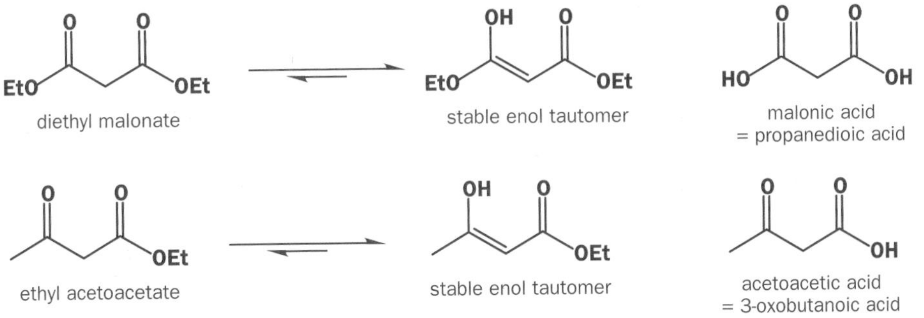

diethyl malonate — stable enol tautomer — malonic acid = propanedioic acid

ethyl acetoacetate — stable enol tautomer — acetoacetic acid = 3-oxobutanoic acid

With these two esters, the choice of base is important: nucleophilic addition can occur at the ester carbonyl, which could lead to transesterification (with alkoxides), hydrolysis (with hydroxide), or amide formation (with amide anions). The best choice is usually an alkoxide identical with the alkoxide component of the ester (that is, ethoxide for diethyl malonate; methoxide for dimethyl malonate). Alkoxides (pK_a 16) are basic enough to deprotonate between two carbonyl groups but, should substitution occur at C=O, there is no overall reaction.

In this example the electrophile is the allylic cyclopentenyl chloride, and the base is ethoxide in ethanol—most conveniently made by adding one equivalent of sodium metal to dry ethanol.

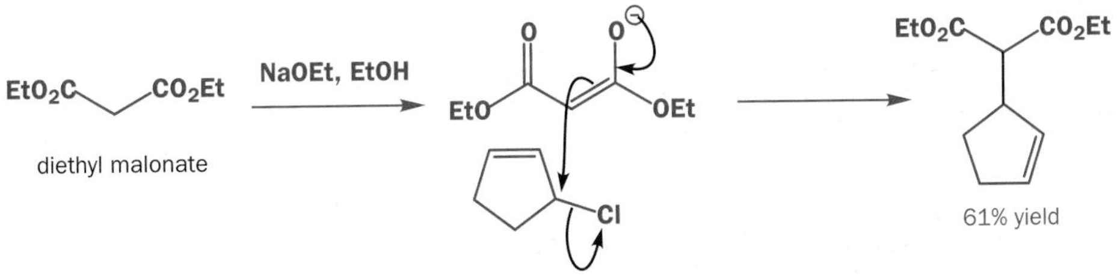

diethyl malonate

NaOEt, EtOH

61% yield

The same base is used in the alkylation of ethyl acetoacetate with butyl bromide.

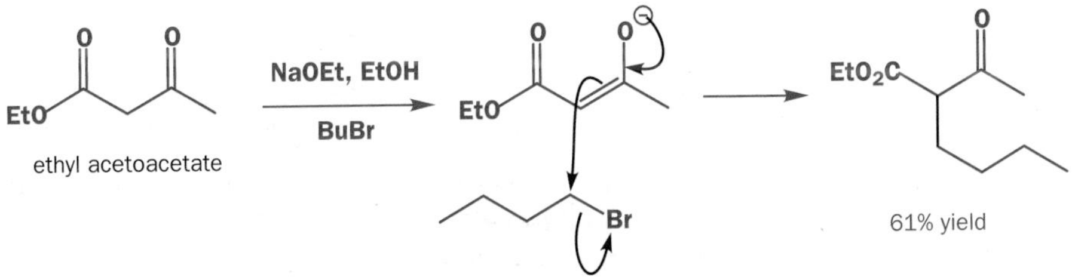

ethyl acetoacetate

NaOEt, EtOH

BuBr

61% yield

Various electron-withdrawing groups can be used in almost any combination with good results. In this example an ester and a nitrile cooperate to stabilize an anion. Nitriles are not quite as anion-stabilizing as carbonyl groups so this enolate requires a stronger base (sodium hydride) in an aprotic solvent (DMF) for success. The primary alkyl tosylate serves as the electrophile.

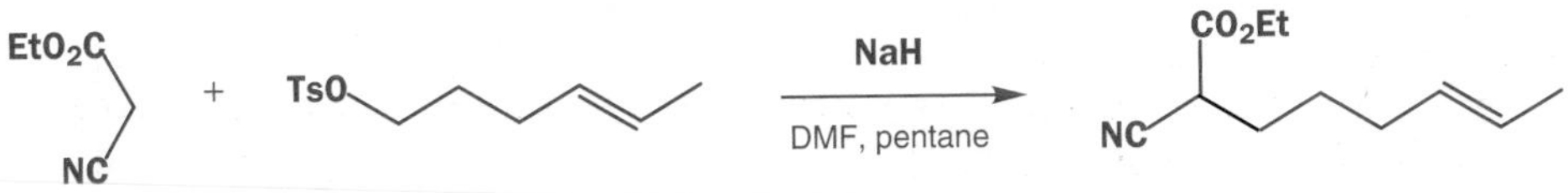

NaH

DMF, pentane

These doubly stabilized anions are alkylated so well that it is common to carry out an alkylation between two carbonyl groups, only to remove one of them at a later stage. This is made possible by the fact that carboxylic acids with a β-carbonyl group **decarboxylate** (lose carbon dioxide) on heating. The mechanism below shows how. After alkylation of the dicarbonyl compound the unwanted ester is first hydrolysed in base. Acidification and heating lead to decarboxylation via a six-membered cyclic transition state in which the acid proton is transferred to the carbonyl group as the key bond breaks, liberating a molecule of carbon dioxide. The initial product is the enol form of a carbonyl compound that rapidly tautomerizes to the more stable keto form—now with only one carbonyl group. Using this technique, β-keto-esters give ketones while malonate esters give simple carboxylic acids (both ester groups hydrolyse but only one can be lost by decarboxylation). Decarboxylation can occur only with a second carbonyl group appropriately placed β to the acid, because the decarboxylated product must be formed as an enol.

decarboxylation of acetoacetate derivatives to give ketones

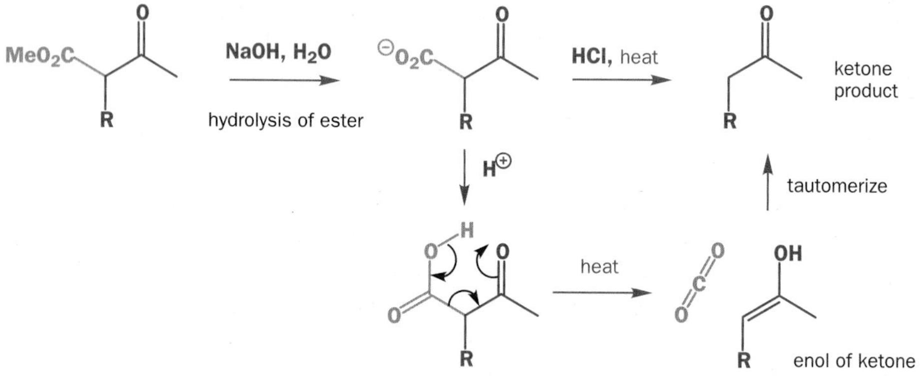

decarboxylation of malonate derivatives to give carboxylic acids

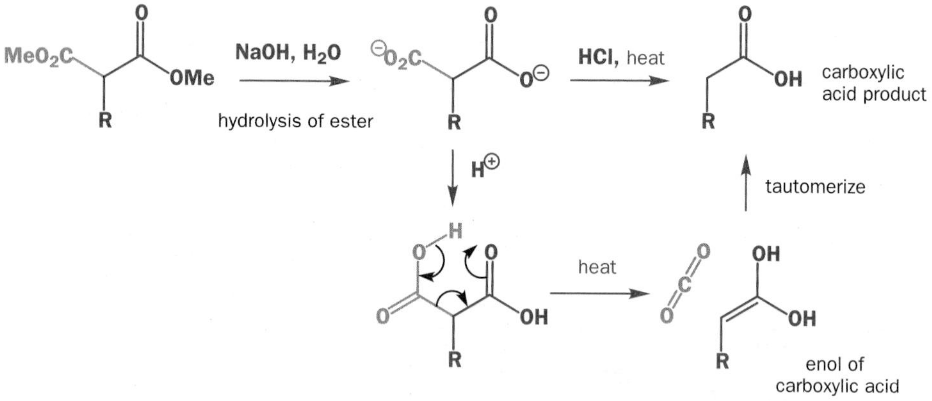

The alkylation of ethyl acetoacetate with butyl bromide on p. 677 was done with the expressed intention of decarboxylating the product to give hexan-2-one. Here are the conditions for this decarboxylation: the heating step drives off the CO_2 by increasing the gearing on the entropy term ($T\Delta S^{\ddagger}$) of the activation energy (two molecules are made from one).

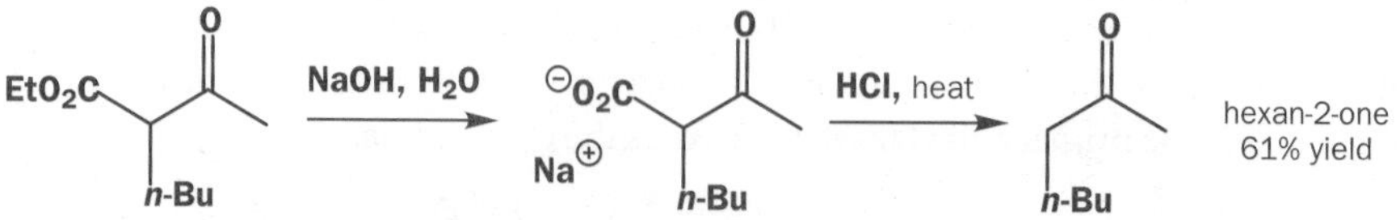

Esters are much easier to work with than carboxylic acids, and a useful alternative procedure removes one ester group without having to hydrolyse the other. The malonate ester is heated in a

polar aprotic solvent—usually DMSO—in the presence of sodium chloride and a little water. No acid or base is required and, apart from the high temperature, the conditions are fairly mild. The scheme below shows a dimethyl malonate alkylation (note that NaOMe is used with the dimethyl ester) and removal of the methyl ester.

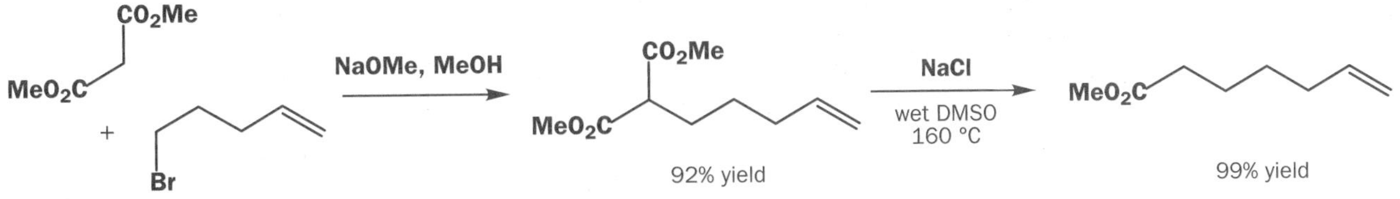

The mechanism is a rather unusual type of ester cleavage reaction. You met, in Chapter 17 and again in Chapter 25, the cleavage of *t*-butyl esters in acid solution via an S_N1 mechanism. In the reaction we are now considering, the same bond breaks (O–alkyl)—but not, of course, via an S_N1 mechanism because the alkyl group is Me. Instead the reaction is an S_N2 substitution of carboxylate by Cl⁻.

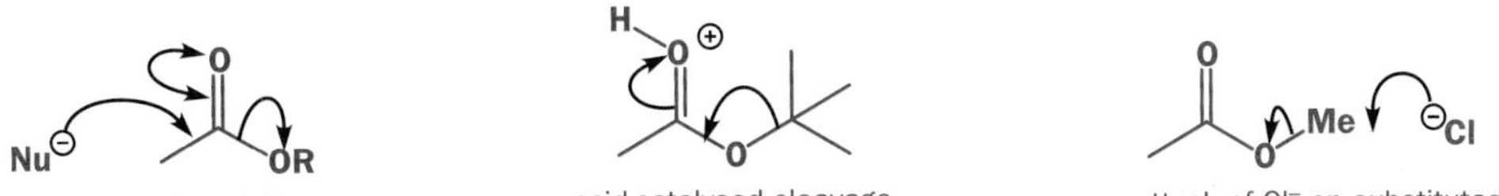

normal nucleophilic attack on C=O of ester

acid-catalysed cleavage of *t*-butyl esters: S_N1

attack of Cl⁻ on substituted dimethylmalonates: S_N2

Chloride is a poor nucleophile, but it is more reactive in DMSO by which it cannot be solvated. And, as soon as the carboxylate is substituted, the high temperature encourages (entropy again) irreversible decarboxylation, and the other by-product, MeCl, is also lost as a gas. The 'decarboxylation' (in fact, removal of a CO_2Me group, not CO_2) is known as the **Krapcho decarboxylation**. Because of the S_N2 step, it works best with *methyl* malonate esters.

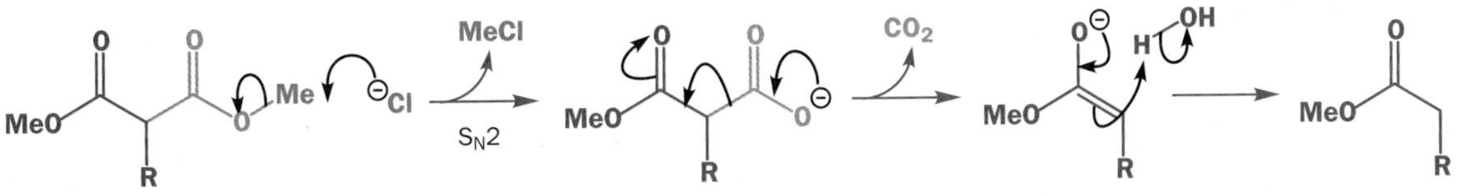

We have only looked at single alkylations of dicarbonyl compounds, but there are two acidic protons between the carbonyl groups and a second alkylation is usually possible. Excess of base and alkyl halide gives two alkylations in one step. More usefully, it is possible to introduce two different alkyl groups by using just one equivalent of base and alkyl halide in the first step.

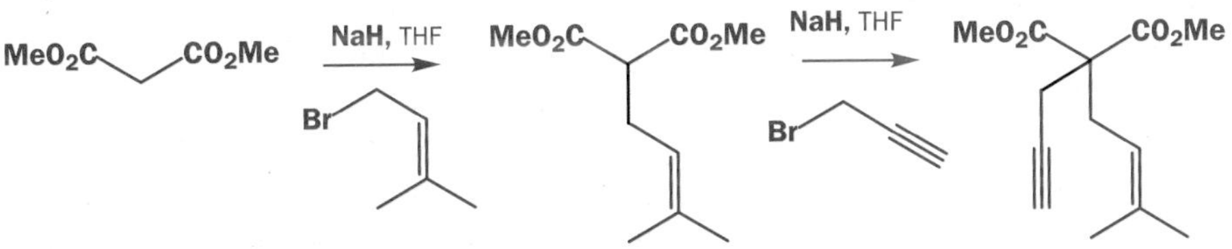

With a dihaloalkane, rings can be formed by two sequential alkylation reactions: this is an important way of making cycloalkanecarboxylic acids. Even the usually more difficult (see Chapter 42) four-membered rings can be made in this way.

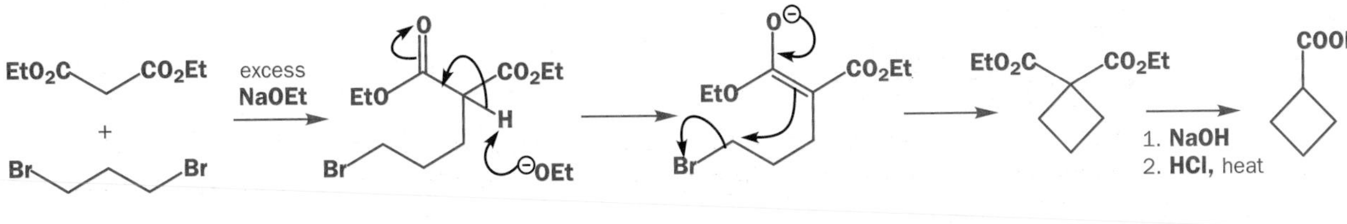

Ketone alkylation poses a problem in regioselectivity

Ketones are unique because they can have enolizable protons on both sides of the carbonyl group. Unless the ketone is symmetrical, or unless one side of the ketone happens to have no enolizable protons, two regioisomers of the enolate are possible and alkylation can occur on either side to give regioisomeric products. We need to be able to control which enolate is formed if ketone alkylations are to be useful.

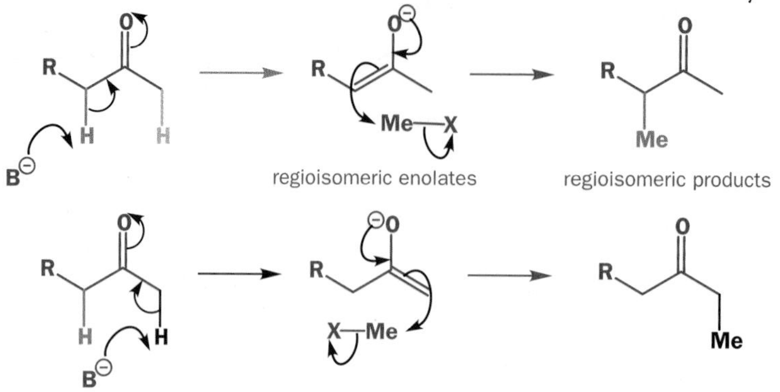

regioisomeric enolates   regioisomeric products

Thermodynamically controlled enolate formation

Selective enolate formation is straightforward if the protons on one side of the ketone are significantly more acidic than those on the other. This is what you have just seen with ethyl acetoacetate: it is a ketone, but with weak bases ($pK_aH < 18$) it only ever enolizes on the side where the protons are acidified by the second electron-withdrawing group. If two new substituents are introduced, in the manner you have just seen, they will always both be joined to the same carbon atom. This is an example of thermodynamic control: only the more stable of the two possible enolates is formed.

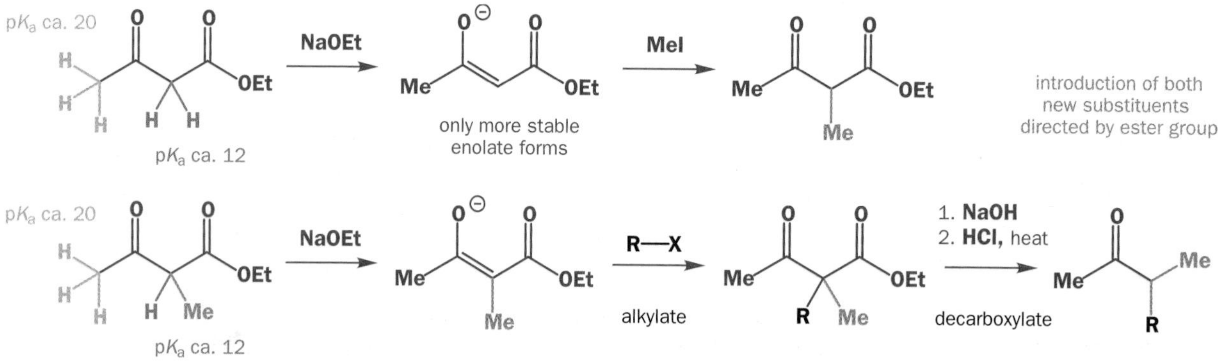

only more stable
enolate forms

introduction of both
new substituents
directed by ester group

This principle can be extended to ketones whose enolates have less dramatic differences in stability. We said in Chapter 21 that, since enols and enolates are alkenes, the more substituents they carry the more stable they are. So, in principle, even additional alkyl groups can control enolate formation under thermodynamic control. Formation of the more stable enolate requires a mechanism for equilibration between the two enolates, and this must be proton transfer. If a proton source is available—and this can even be just excess ketone—an equilibrium mixture of the two enolates will form. The composition of this equilibrium mixture depends very much on the ketone but, with 2-phenylcyclohexanone, conjugation ensures that only one enolate forms. The base is potassium hydride: it's strong, but small, and can be used under conditions that permit enolate equilibration.

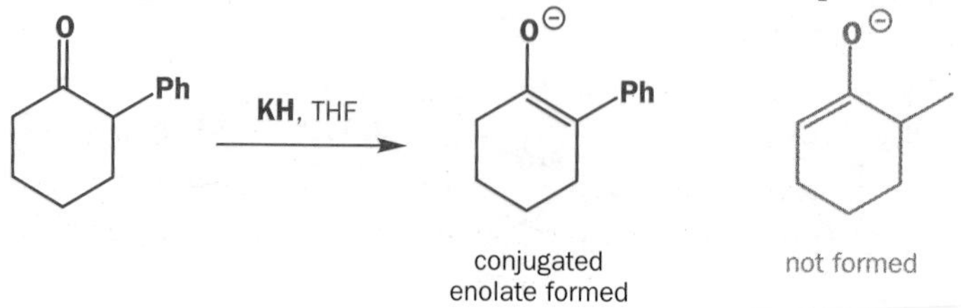

conjugated
enolate formed   not formed

The more substituted lithium enolates can also be formed from the more substituted silyl enol ethers by substitution at silicon—a reaction you met in Chapter 21. The value of this reaction now becomes clear, because the usual way of making silyl enol ethers (Me₃SiCl, Et₃N) typically produces, from unsymmetrical ketones, the more substituted of the two possible ethers.

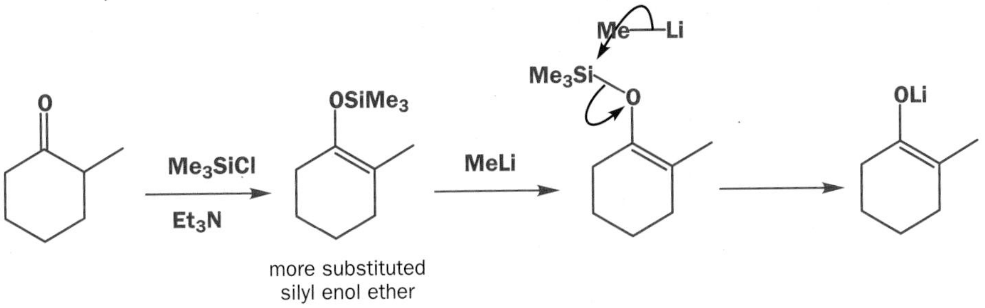

more substituted
silyl enol ether

One possible explanation for the thermodynamic regioselectivity in the enol ether-forming step is related to our rationalization of the regioselectivity of bromination of ketones in acid on p. 536. Triethylamine (pK_{aH} 10) is too weak a base to deprotonate the starting carbonyl compound (pK_a ca. 20), and the first stage of the reaction is probably an oxygen–silicon interaction. Loss of a proton now takes place through a cationic transition state, and this is stabilized rather more if the proton being lost is next to the methyl group: methyl groups stabilize partial cations just as they stabilize cations.

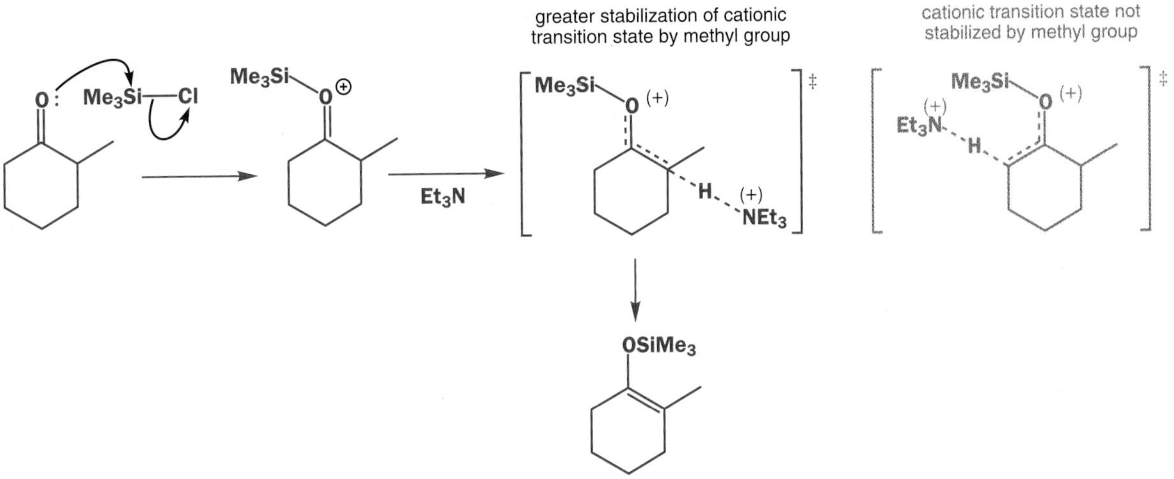

greater stabilization of cationic
transition state by methyl group

cationic transition state not
stabilized by methyl group

An alternative view is that reaction takes place through the enol: the Si–O bond is so strong that even neutral enols react with Me₃SiCl, on oxygen, of course. The predominant enol is the more substituted, leading to the more substituted silyl enol ether.

minor enol: less substituted

major enol: more substituted

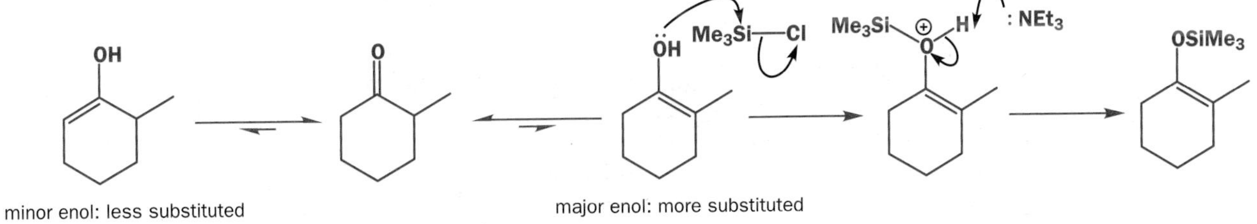

Kinetically controlled enolate formation

LDA is too hindered to attack C=O, so it attacks C–H instead. And, if there is a choice of C–H bonds, it will attack the least hindered possible. It will also prefer to attack more acidic C–H bonds, and C–H bonds on less substituted carbons are indeed more acidic. Furthermore, statistics helps, since a less substituted C atom has more protons to be removed (three versus two in this example) so, even if the rates were the same, the less substituted enolate would predominate.

► Think of base strengths: MeLi is a weaker base than t-BuLi, so the conjugate acid must be a stronger acid.

▶

There must never be more ketone in the mixture than base, or exchange of protons between ketone and enolate will lead to equilibration. Kinetic enolate formations with LDA must be done by adding the ketone *to* the LDA so that there is excess LDA present throughout the reaction.

These factors multiply to ensure that the enolate that forms will be the one with the fewer substituents—provided we now prevent equilibration of the enolate to the more stable, more substituted one. This means keeping the temperature low, typically –78 °C, keeping the reaction time short, and using an excess of strong base to deprotonate irreversibly and ensure that there is no remaining ketone to act as a proton source. The enolate that we then get is the one that formed faster—the kinetic enolate—and not necessarily the one that is more stable.

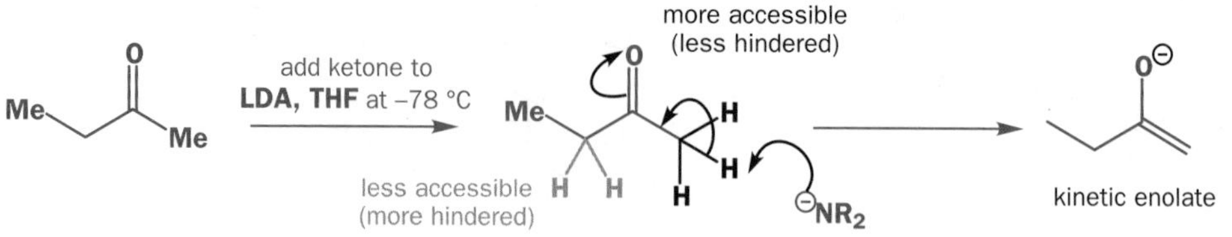

In general, this effect is sufficient to allow selective kinetic deprotonation of methyl ketones, that is, where the distinction is between Me and alkyl. In this example, unusually, MeLi is used as a base: LDA was probably tried but perhaps gave poorer selectivity. The first choice for getting kinetic enolate formation should always be LDA.

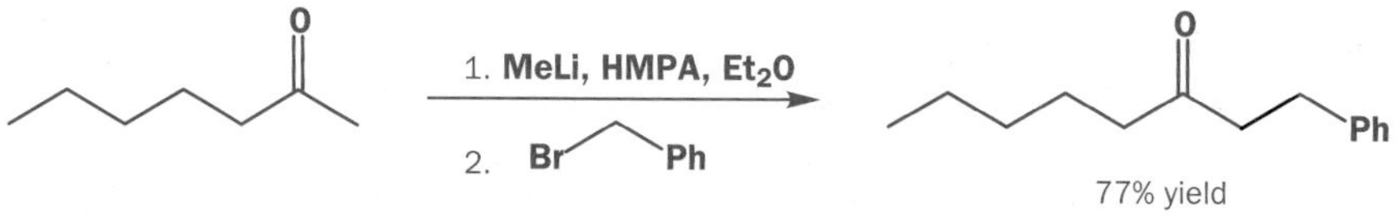

The same method works very well for 2-substituted cyclohexanones: the less substituted enolate forms. Even with 2-phenylcyclohexanone, which, as you have just seen, has a strong thermodynamic preference for the conjugated enolate, only the less substituted enolate forms.

2-Methylcyclohexanone can be regioselectively alkylated using LDA and benzyl bromide by this method.

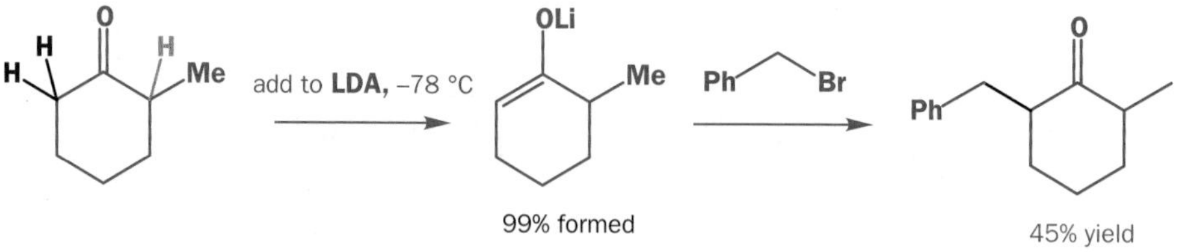

● **Regioselective formation of enolates from ketones**

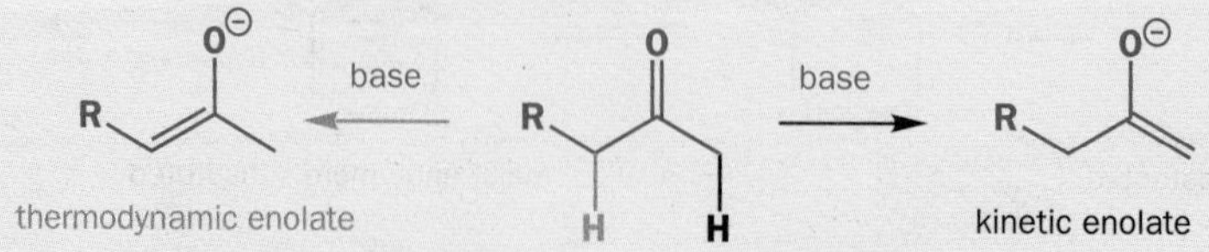

Thermodynamic enolates are:
- more substituted
- more stable
- favoured by excess ketone, high temperature, long reaction time

Kinetic enolates are:
- less substituted
- less stable
- favoured by strong, hindered base, low temperature, short reaction time

Dianions allow unusual regioselectivity in alkylations of methyl acetoacetate

In Chapter 24, we introduced the idea that the last-formed anion in a dianion or trianion is the most reactive. Methyl acetoacetate is usually alkylated on the central carbon atom because that is the site of the most stable enolate. But methyl acetoacetate dianion—formed by removing a second proton from the usual enolate with a very strong base (usually butyllithium)—reacts first on the less stable anion: the terminal methyl group. Protonation of the more stable enolate then leads to the product. Butyllithium can be used as a base because the anionic enolate intermediate is not electrophilic.

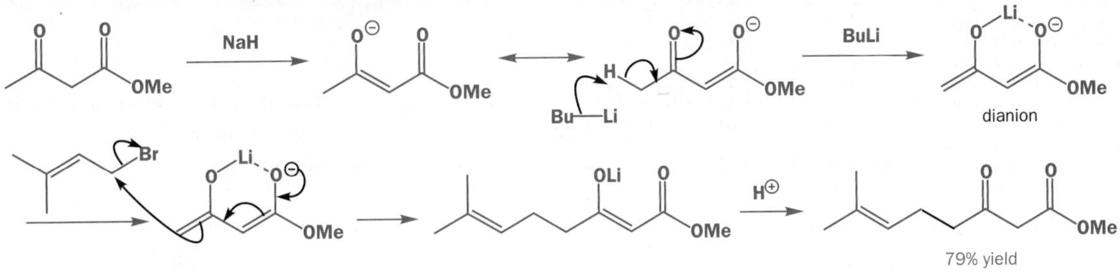

Enones provide a solution to regioselectivity problems

Enolates can be made regiospecifically from, for example, silyl enol ethers or enol acetates just by treating them with an alkyllithium. These are both substitution reactions in which RLi displaces the enolate: one is $S_N2(Si)$ and the other is attack at C=O. Provided there is no proton source, the enolate products have the same regiochemistry as their stable precursors, and single enolate regioisomers are formed. But there is a problem: forming enol ethers or enol esters will usually itself require a regioselective enolization! There are two situations in which this method is nonetheless useful: when the *more* substituted lithium enolate (which is hard to make selectively otherwise) is required, and when a silyl enol ether can be formed by a method not involving deprotonation. These methods are what we shall now consider.

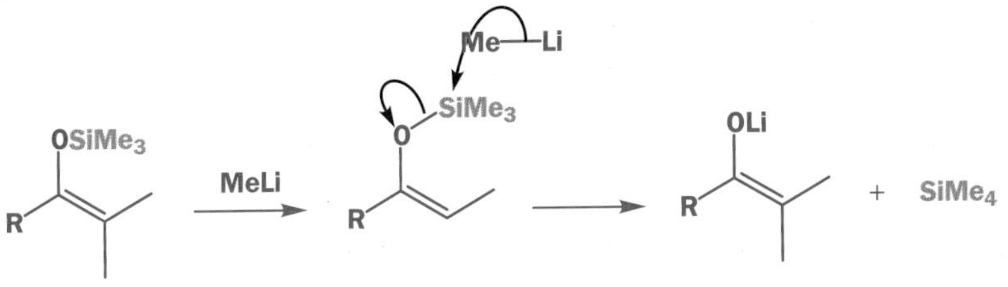

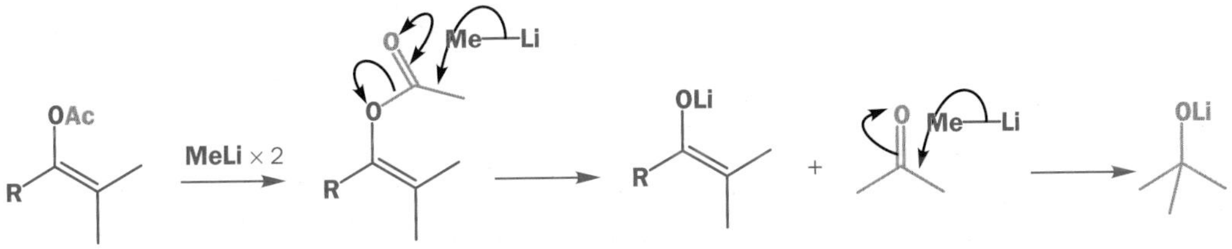

Dissolving metal reduction of enones gives enolates regiospecifically

In Chapter 24 you met the **Birch reduction**: the use of dissolving metals (K, Na, or Li in liquid ammonia, for example) to reduce aromatic rings and alkynes. The dissolving metal reduction of enones by lithium metal in liquid ammonia is similar to these reactions—the C=C bond of the enone is reduced, with the C=O bond remaining untouched. An alcohol is required as a proton source and, in total, two electrons and two protons are added in a stepwise manner giving net addition of a molecule of hydrogen to the double bond.

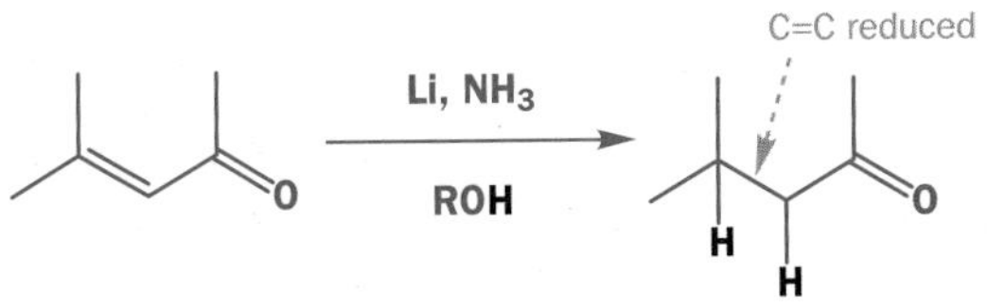

The mechanism follows that described on p. 628: transfer of an electron forms a radical anion that is protonated by the alcohol to form a radical. A second electron transfer forms an anion that can undergo tautomerization to an enolate.

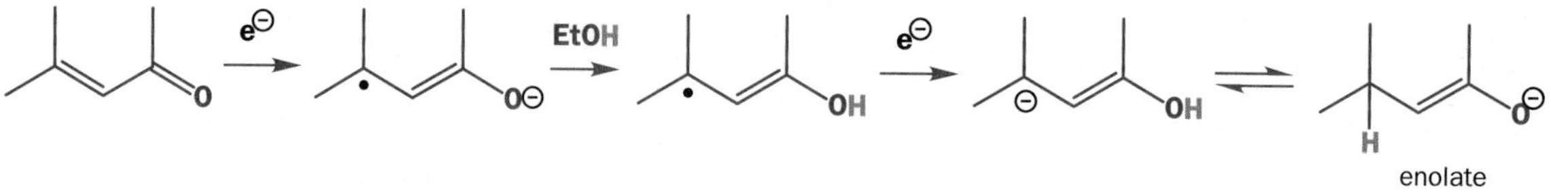

The enolate is stable to further reduction, and protonation during the work-up will give a ketone. But reaction with an alkyl halide is more fruitful: because the enolate forms only where the double bond of the enone was, regioselective alkylation becomes possible.

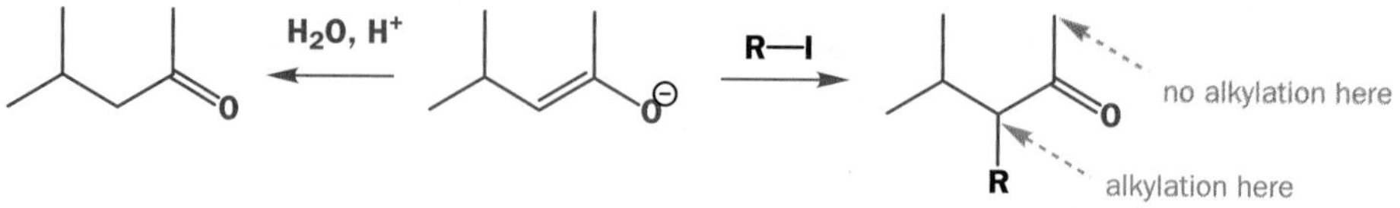

You saw above that an equilibrium mixture of the enolates of 2-methylcyclohexanone contains only about a 4:1 ratio of regioisomers. By reducing an enone to an enolate, only 2% of the unwanted regioisomer is formed.

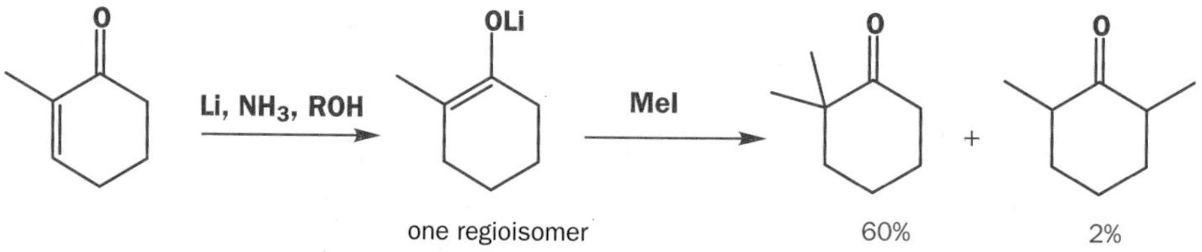

The transfer of electrons is not susceptible to steric hindrance so substituted alkenes pose no problem. In the next example, the enolate reacts with allyl bromide to give a single stereoisomer of the product (the allyl bromide attacks from the face opposite the methyl group). Naturally, only one regioisomer is formed as well, and it would be a tall order to expect formation of this single enolate regioisomer by any form of deprotonation method.

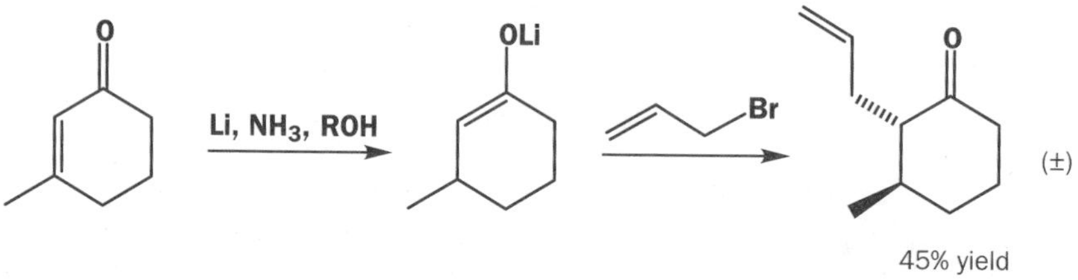

Conjugate addition to enones gives enolates regiospecifically

Although we did not talk in detail about them at that time, you will recall from Chapter 10 that conjugate addition to enones generates first an enolate, which is usually protonated in the work-up. But, again, more fruitful things can be done with the enolate under the right conditions.

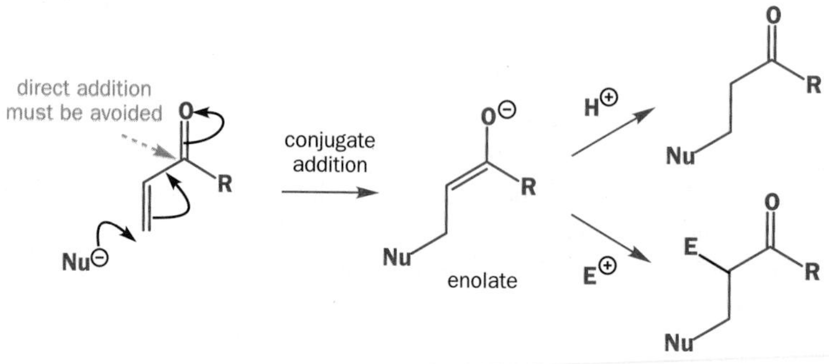

The simplest products are formed when Nu = H, but this poses a problem of regioselectivity in the nucleophilic attack step: a nucleophilic hydride equivalent that selectively undergoes conjugate addition to the enone is required. This is usually achieved with extremely bulky hydride reagents such as lithium or potassium tri(sec-butyl)borohydride (often known by the trade names of L- or K-Selectride, respectively). In this example, K-Selectride reduces the enone to an enolate that is alkylated by methyl iodide to give a single regioisomer. The reaction also illustrates the difference in reactivity between conjugated and isolated double bonds.

bulky reducing agents

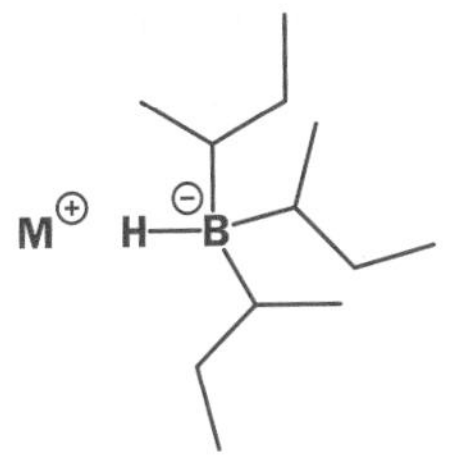

M = Li: lithium tri-*sec*-butyl-borohydride (L-Selectride)

M = K: potassium tri-*sec*-butyl-borohydride (K-Selectride)

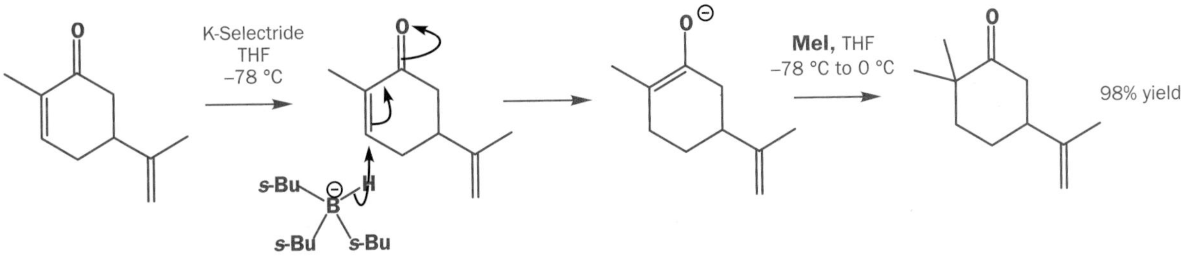

With organocopper reagents, conjugate addition introduces a new alkyl group and, if the resulting enolates are themselves alkylated, two new C–C bonds can be formed in a single step (a **tandem reaction**: one C–C bond-formation rides behind another). In Chapter 10 we explained that the best organocuprate additions are those carried out in the presence of Me₃SiCl: the product of these reactions is a silyl enol ether, formed regioselectively (the 'enol' double bond is always on the side where the enone used to be).

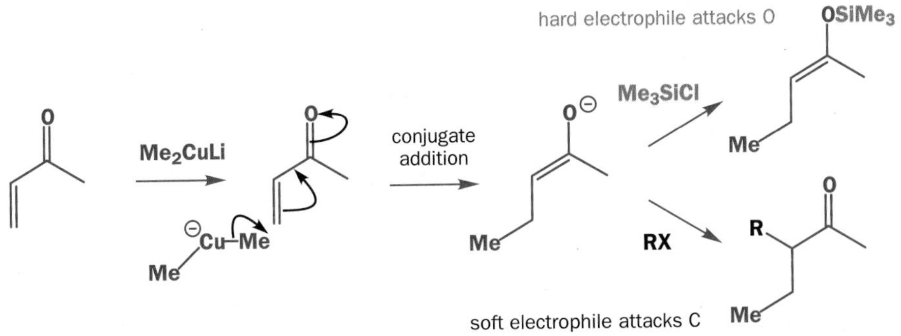

The silyl enol ethers are too unreactive for direct alkylation by an alkyl halide, but by converting them to lithium enolates all the usual alkylation chemistry becomes possible. This type of reaction forms the key step in a synthesis of the natural product α-chamigrene. Conjugate addition of Me₂CuLi gives an enolate that is trapped with trimethylsilyl chloride. Methyllithium converts the resulting silyl enol ether into a lithium enolate (by S_N2 at Si). The natural product has a *spiro* six-membered ring attached at the site of the enolate, and this was made by alkylating with a dibromide (you saw this done on p. 679). The first substitution is at the more reactive allylic bromide. A second enolization is needed to make the ring, but this can be done under equilibrating conditions because the required six-membered ring forms much faster than the unwanted eight-membered ring that would arise by attack on the other side of the ketone.

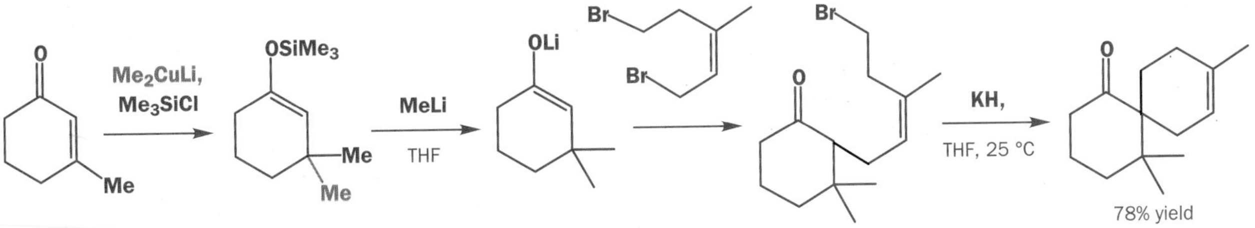

Among the most important of these tandem conjugate addition–alkylation reactions are those of cyclopentenones. With cyclopentenone itself, the *trans* diastereoisomer usually results because the alkylating agent approaches from the less hindered face of the enolate.

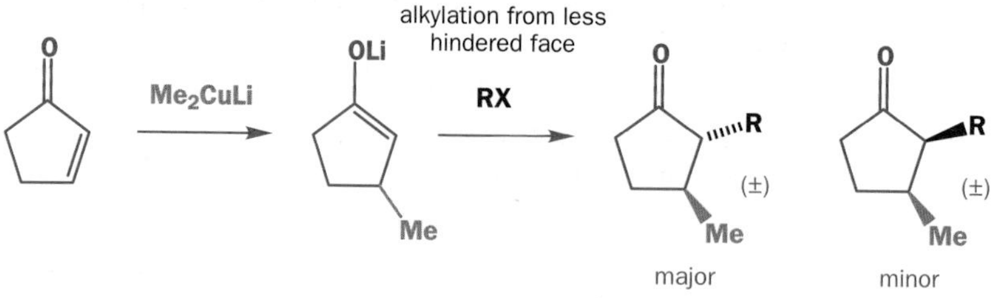

This is the sort of selectivity evident in the next example, which looks more complicated but is really just addition of an arylcopper reagent followed by alkylation (*trans* to the bulky Ar group) with an iodoester.

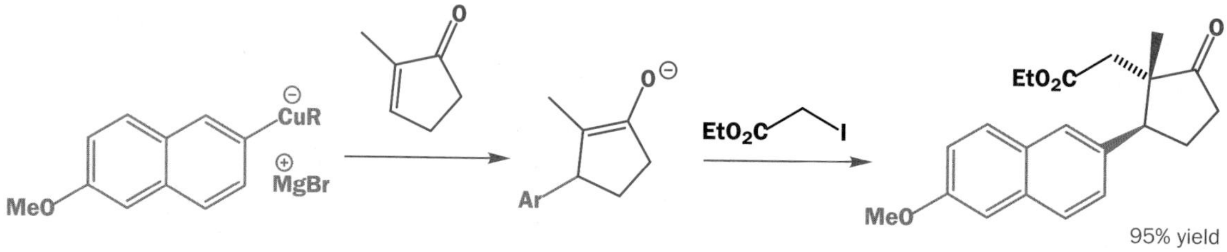

95% yield

Ryoji Noyori works at the University of Nagoya in Japan. He has introduced many methods for making molecules, the most important of which allow the formation of single enantiomers using chiral catalysts. You will meet some more of his chemistry in Chapter 45.

One of the most dramatic illustrations of the power of conjugate addition followed by alkylation is the short synthesis of the important biological molecule prostaglandin E$_2$ by Ryoji Noyori in Japan. The organocopper reagent and the alkylating agent contain all the functionality required for both side chains of the target in protected form. The required *trans* stereochemistry is assembled in the key step, which gives a 78% yield of a product requiring only removal of the silyl ether and ester protecting groups. The organometallic nucleophile was prepared from a vinyl iodide by halogen–metal exchange (Chapter 9). In the presence of copper iodide this vinyllithium adds to the cyclopentenone in a conjugate sense to give an intermediate enolate. Because in this case the starting enone already has a stereogenic centre, this step is also stereoselective: attack on the less hindered face (opposite the silyl ether) gives the *trans* product. The resulting enolate was alkylated with the allylic iodide containing the terminal ester: once again the *trans* product was formed. It is particularly vital that enolate equilibration is avoided in this reaction to prevent the inevitable E1cB elimination of the silyloxy group that would occur from the other enolate.

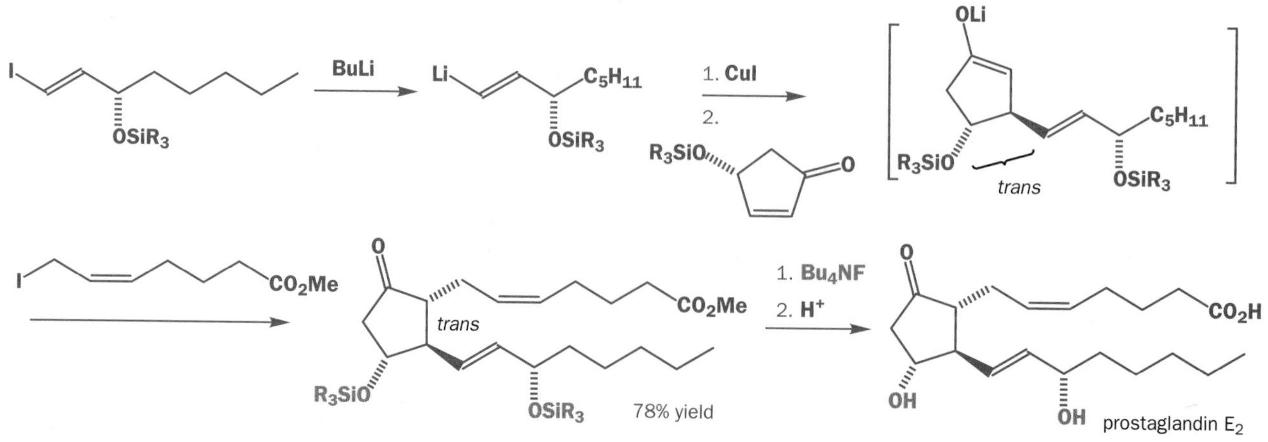

To conclude...

We have considered the reactions of enolates and their equivalents with alkyl halides. In the next chapter we move on to consider the reactions of the same types of enolate equivalents with a different class of electrophiles: carbonyl compounds themselves.

Summary of methods for alkylating enolates

Specific enol equivalent	Notes
To alkylate esters	
• LDA → lithium enolate	
• use diethyl- or dimethylmalonate and decarboxylate	gives acid (NaOH, HCl) or ester (NaCl, DMSO)
To alkylate aldehydes	
• use enamine	with reactive alkylating agents
• use silyl enol ether	with S_N1-reactive alkylating agents
• use aza-enolate	with S_N2-reactive alkylating agents
To alkylate symmetrical ketones	
• LDA → lithium enolate	
• use acetoacetate and decarboxylate	equivalent to alkylating acetone
• use enamine	with reactive alkylating agents
• use silyl enol ether	with S_N1-reactive alkylating agents
• use aza-enolate	with S_N2-reactive alkylating agents
To alkylate unsymmetrical ketones on more substituted side	
• Me_3SiCl, Et_3N → silyl enol ether	with S_N1-reactive alkylating agents
• Me_3SiCl, Et_3N → silyl enol ether → lithium enolate with MeLi	with S_N2-reactive alkylating agents
• alkylate acetoacetate twice and decarboxylate	two successive alkylations of ethyl acetoacetate
• addition or reduction of enone to give specific lithium enolate or silyl enol ether	
To alkylate unsymmetrical ketones on less substituted side	
• LDA → kinetic lithium enolate	with S_N2-reactive electrophiles
• LDA then Me_3SiCl → silyl enol ether	with S_N1-reactive electrophiles
• use dianion of alkylated acetoacetate and decarboxylate	two successive alkylations of ethyl acetoacetate
• use enamine	with reactive electrophiles

Problems

1. Suggest how the following compounds might be made by the alkylation of an enol or enolate.

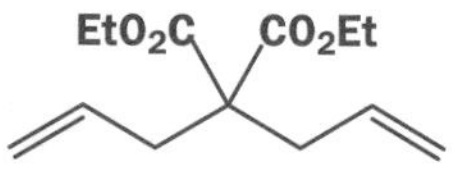

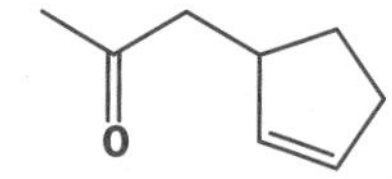

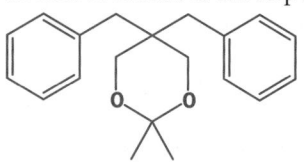

2. And how might these compounds be made using alkylation of an enol or enolate as one step in the synthesis?

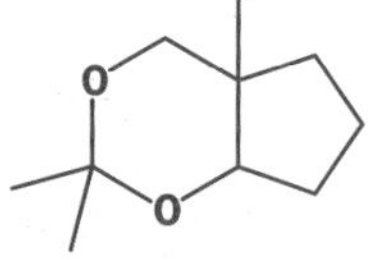

3. And, further, how might these amines by synthesized using alkylation reactions of the enolate style as part of the synthesis?

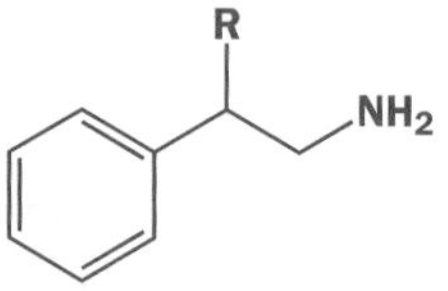

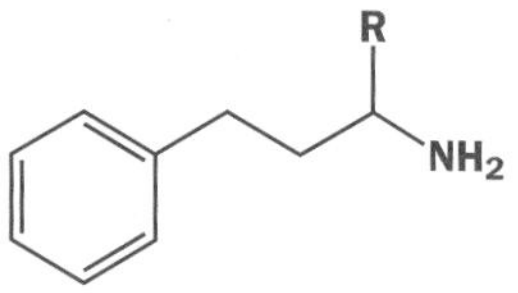

4. This attempted enolate alkylation does not give the required product. What goes wrong? What products would be expected from the reaction?

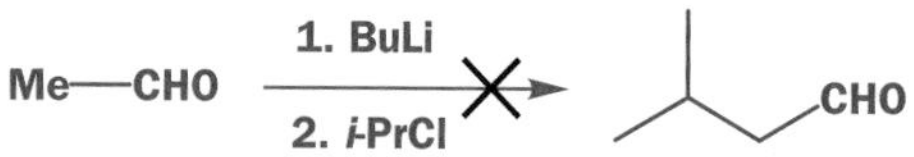

Me—CHO $\xrightarrow[\text{2. } i\text{-PrCl}]{\text{1. BuLi}}$ ✗ $\qquad$ CHO

5. Draw mechanisms for the formation of this enamine, its reaction with the alkyl halide shown, and the hydrolysis of the product.

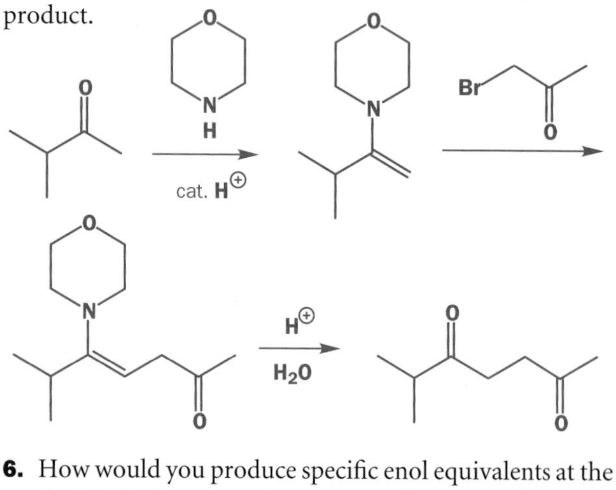

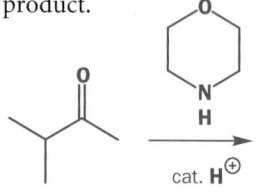

6. How would you produce specific enol equivalents at the points marked with the arrows (not necessarily starting from the simple carbonyl compound shown)?

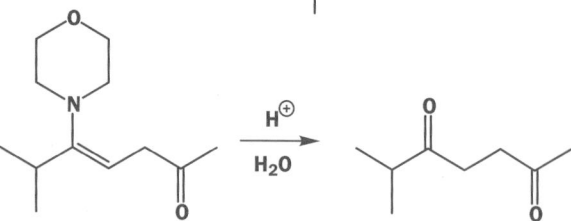

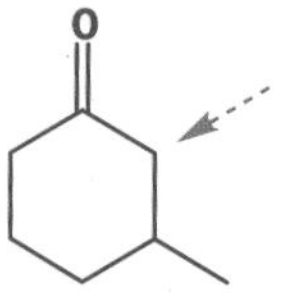

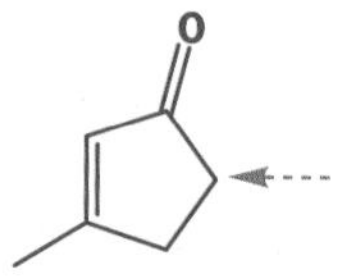

7. How would the reagents you have suggested in Problem 6 react with: (a) Br_2; (b) a primary alkyl halide RCH_2Br?

8. Draw a mechanism for the formation of the imine from cyclohexylamine and the following aldehyde.

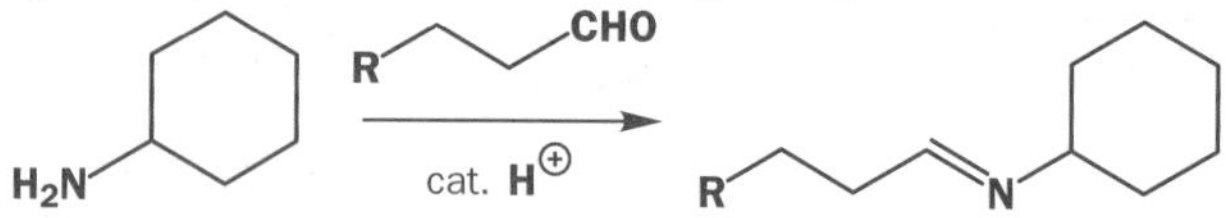

9. How would the imine from Problem 8 react with LDA followed by *n*-BuBr? Draw mechanisms for each step: reaction with LDA, reaction of the product with *n*-BuBr, and the work-up.

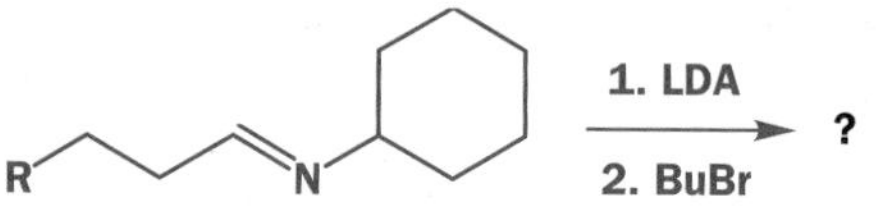

$$\xrightarrow[\text{2. BuBr}]{\text{1. LDA}} \ ?$$

10. What would happen if this short cut for the reaction in Problems 8 and 9 were tried?

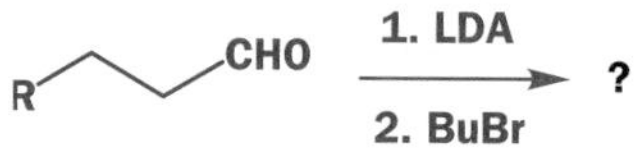

$$\text{R} \diagup \text{CHO} \xrightarrow[\text{2. BuBr}]{\text{1. LDA}} \ ?$$

11. Suggest mechanisms for these reactions.

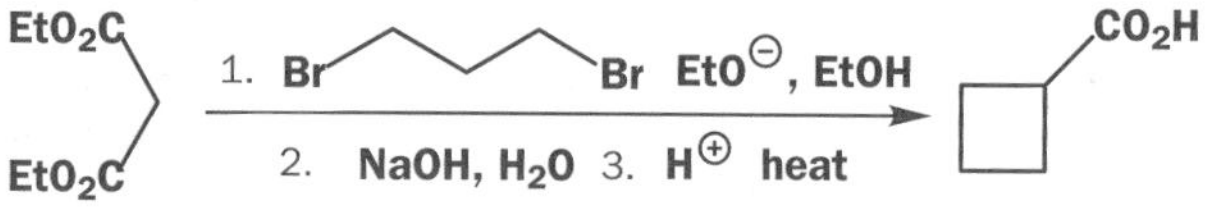

12. How does this method of making cyclopropyl ketones work? Give mechanisms for all the reactions.

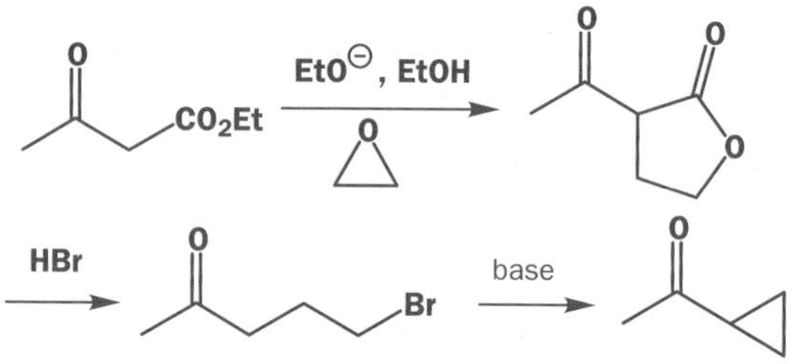

13. Give the structures of the intermediates in the following reaction sequence and mechanisms for the reactions. Comment on the formation of this particular product.

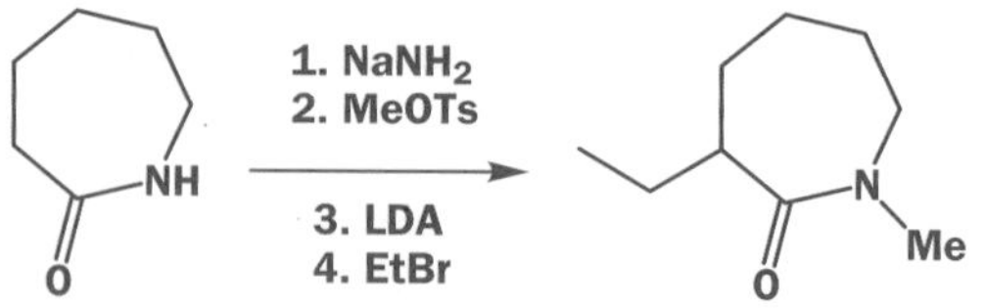

14. Suggest how the following products might be made using enol or enolate alkylation as at least one step. Explain your choice of specific enol equivalents.

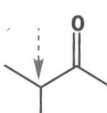

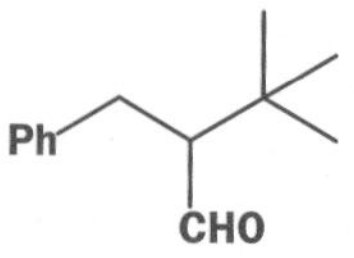

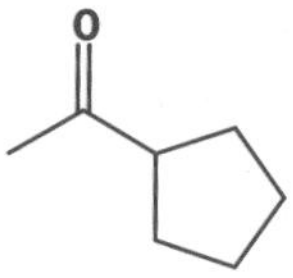

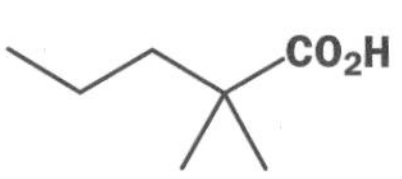

Reactions of enolates with aldehydes and ketones: the aldol reaction

27

Connections

Building on:

- Carbonyl compounds reacting with cyanide, borohydride, and bisulfite nucleophiles ch6
- Carbonyl compounds reacting with organometallic nucleophiles ch9
- Carbonyl compounds taking part in nucleophilic substitution reactions ch12 & ch14
- How enols and enolates react with heteroatomic electrophiles such as Br_2 and NO^+ ch21
- How enolates and their equivalents react with alkylating agents ch26

Arriving at:

- Reactions with carbonyl compounds as both nucleophile and electrophile
- How to make hydroxy-carbonyl compounds or enones by the aldol reaction
- How to be sure that you get the product you want from an aldol reaction
- The different methods available for doing aldol reactions with enolates of aldehydes, ketones, and esters
- How to use formaldehyde as an electrophile
- How to predict the outcome of intramolecular aldol reactions

Looking forward to:

- Enolates taking part in a substitution at C=O ch28
- Enolates undergoing conjugate addition ch29
- Synthesis of aromatic heterocycles ch44
- Asymmetric synthesis ch45
- Biological organic chemistry ch49–ch51

Introduction: the aldol reaction

The simplest enolizable aldehyde is acetaldehyde (ethanal, CH_3CHO). What happens if we add a small amount of base, say NaOH, to this aldehyde? Some of it will form the enolate ion.

acetaldehyde

enolate ion

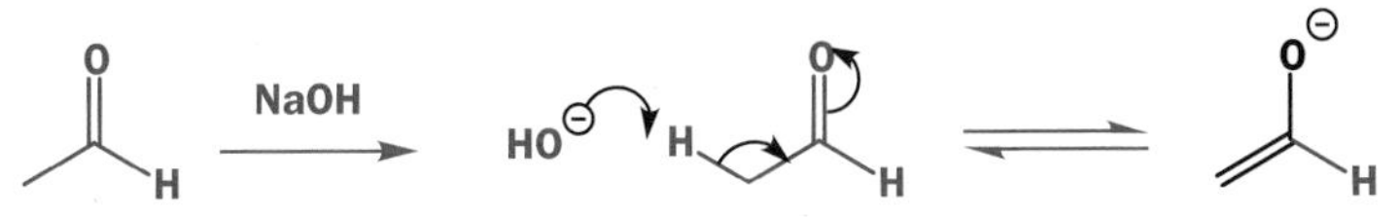

Only a small amount of the nucleophilic enolate ion is formed: hydroxide is not basic enough to enolize an aldehyde completely. Each molecule of enolate is surrounded by molecules of the aldehyde that are not enolized and so still have the electrophilic carbonyl group intact. Each enolate ion will attack one of these aldehydes to form an alkoxide ion, which will be protonated by the water molecule formed in the first step.

enolate ion

unenolized aldehyde

'aldol'
3-hydroxybutanal

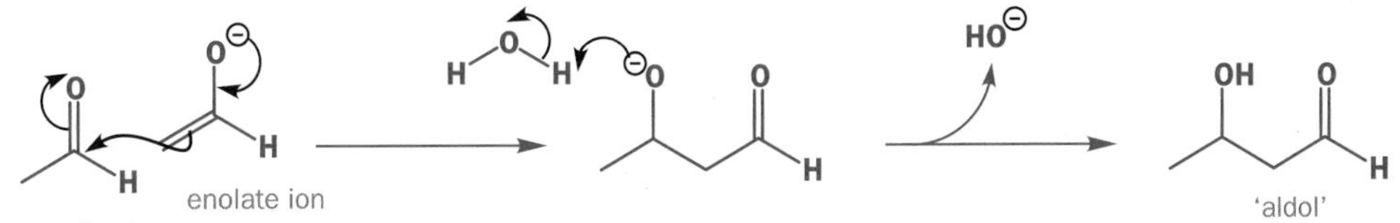

The product is an aldehyde with a hydroxy (ol) group whose trivial name is **aldol**. The name aldol is given to the whole class of reactions between enolates (or enols) and carbonyl compounds even if in most cases the product is not a hydroxy-aldehyde at all. Notice that the base catalyst (hydroxide ion) is regenerated in the last step, so it is truly a catalyst.

This reaction is so important because of the carbon–carbon bond formed when the nucleophilic enolate attacks the electrophilic aldehyde. This bond is shown as a black bond in this version of the key step.

electrophilic   nucleophilic   newly formed
aldehyde   enolate   carbon–carbon bond

The rate equation for the aldol reaction

Not only is this step the most important; it is usually the rate-determining step. The rate expression for the aldol reaction at *low concentrations* of hydroxide is found experimentally to be

rate = $k_2[CH_3CHO] \times [HO^-]$

rate-determining step at
low hydroxide concentration

enolate ion

showing that the formation of the enolate ion is rate-determining. Though this is a proton transfer, which we normally expect to be fast, the proton is being removed from a carbon atom. Proton transfers to and from carbon atoms can be slow.

At higher hydroxide ion concentration, the rate expression becomes termolecular (k_3 expresses this) with the aldehyde concentration being squared.

rate = $k_3[CH_3CHO]^2 \times [HO^-]$

rate-determining step at
normal hydroxide concentration

The mechanism does not, of course, involve three molecules colliding together. The rate-determining step has changed, and is now the second step.

But this does not obviously give a termolecular rate expression. The rate expression for this step is

rate = $k_2[CH_3CHO] \times$ [enolate ion]

We cannot easily measure the concentration of the enolate, but we can work it out because we know that the enolate and the aldehyde are in equilibrium.

aldehyde   enolate ion

So we can express the enolate concentration using K_1 as the equilibrium constant and omitting the water concentration. We can write

$$K_1 = \frac{[\text{enolate ion}]}{[\text{MeCHO}][\text{HO}^-]}$$

Or, rearranging this to get the enolate ion concentration,

[enolate ion] = $K_1[CH_3CHO] \times [HO^-]$

And, substituting this in the rate expression,

rate = $k_2[CH_3CHO] \times$ [enolate ion]

 = $k_2[CH_3CHO] \times K_1[CH_3CHO] \times [HO^-] = k_2K_1[CH_3CHO]^2 \times [HO^-]$

This is what is observed, if we can remind you:

rate = $k_3[CH_3CHO]^2 \times [HO^-]$

It just turns out that the 'termolecular rate constant' k_3 is actually the product of an equilibrium constant K_1 and a genuine bimolecular rate constant k_2 such that $k_3 = K_1 \times k_2$. You saw a similar thing in the rate expressions for amide hydrolysis (Chapter 13) and E1cB elimination (Chapter 19, p. 497)

The reaction occurs with ketones as well. Acetone is a good example for us to use at the start of this chapter because it gives an important product and, as it is a symmetrical ketone, there can be no argument over which way it enolizes.

the enolization step

acetone   enolate ion
                                     of acetone

the carbon–carbon bond-forming step

second molecule
of acetone

'aldol' product from acetone
4-hydroxy-4-methylpentan-2-one

Each step is the same as the aldol sequence with acetaldehyde, and the product is again a hydroxy-carbonyl compound, but this time a hydroxy-ketone.

The acetaldehyde reaction works well when one drop of dilute sodium hydroxide is added to acetaldehyde. The acetone reaction is best done with insoluble barium hydroxide, Ba(OH)$_2$. Both approaches keep the base concentration low. Without this precaution, the aldol products are not the compounds isolated from the reaction. With more base, further reactions occur, because the aldol products dehydrate rather easily under the reaction conditions to give stable conjugated unsaturated carbonyl compounds.

These are elimination reactions, and you met them in Chapter 19. You cannot normally eliminate water from an alcohol in basic solution and it is the carbonyl group that allows it to happen here. A second enolization reaction starts things off, and these are E1cB reactions.

In the examples that follow in the rest of the chapter you will see that base-catalysed aldol reactions sometimes give the aldol and sometimes the elimination product. The choice is partly based on conditions—the more vigorous conditions (stronger base, higher temperatures, longer reaction time) tend to give the elimination product—and partly on the structure of the reagents: some combinations are easy to stop at the aldol stage, while some almost always give the elimination reaction as well. You do not, of course, need to learn the results: if you ever need to do an aldol reaction you can consult the massive review in the 1968 volume of *Organic Reactions* to find the best conditions for getting the result you want.

The elimination is even easier in acid solution and acid-catalysed aldol reactions commonly give unsaturated products instead of aldols. In this simple example with a symmetrical cyclic ketone, the enone is formed in good yield in acid or base. We shall use the acid-catalysed reaction to illustrate the mechanism. First the ketone is enolized under acid catalysis as you saw in Chapter 21.

acid-catalysed enolization step

Then the aldol reaction takes place. Enols are less nucleophilic than enolates, and the reaction occurs because the electrophilic carbonyl component is protonated: the addition is acid-catalysed. An acid-catalysed aldol reaction takes place.

acid-catalysed aldol addition step

the 'aldol' product

With the acetone reaction a further trick is required to ensure that the aldol product does not meet the base. The apparatus is arranged so that, on heating, the volatile acetone is condensed into a vessel containing the insoluble base. The less volatile aldol product is kept away from it.

See p. 495 for a discussion of the E1cB mechanism.

The aldol is a tertiary alcohol and would be likely to eliminate by an E1 mechanism in acid even without the carbonyl group. But the carbonyl ensures that only the stable conjugated enone is formed. Notice that the dehydration too is genuinely acid-catalysed as the acid reappears in the very last step.

the acid-catalysed dehydration step (E1 elimination)

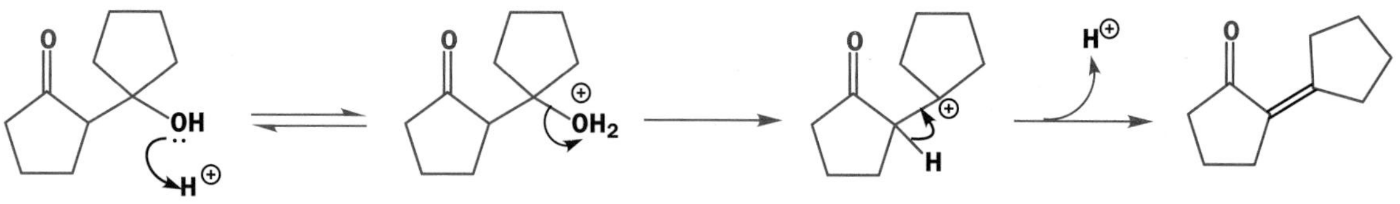

> ▶ **Condensation reactions**
>
> condensation of cyclopentanone 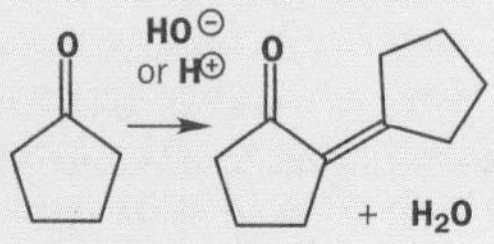
>
>
> The term condensation is often used of reactions like this. Condensations are reactions where two molecules combine with the loss of another small molecule—usually water. In this case, two ketones combine with the loss of water. This reaction is called an **aldol condensation** and chemists may say 'two molecules of cyclopentanone condense together to give a conjugated enone'. You will also find the term 'condensation' used for all aldol reactions whether they occur with dehydration or not. The distinction is no longer important.

None of these intermediates is detected or isolated in practice—simple treatment of the ketone with acid gives the enone in good yield. A base-catalysed reaction gives the same product via the aldol–E1cB elimination mechanism.

> - Base-catalysed aldol reactions may give the aldol product, or may give the dehydrated enone or enal by an E1cB mechanism
> - Acid-catalysed aldol reactions may give the aldol product, but usually give the dehydrated enone or enal by an E1 mechanism
>

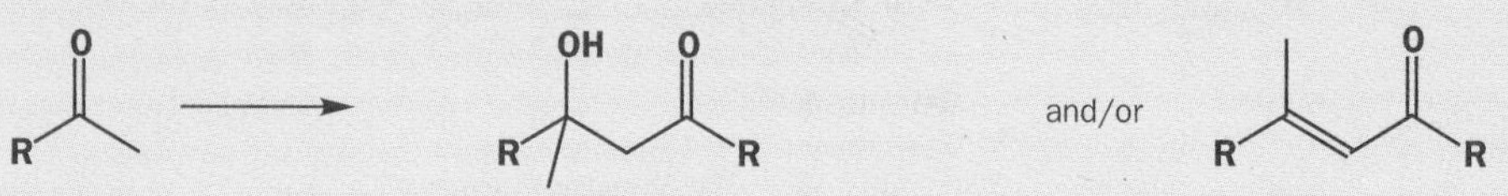

Aldol reactions of unsymmetrical ketones

If the ketone is blocked on one side so that it cannot enolize—in other words it has no α protons on that side—only one aldol reaction is possible. Ketones of this type might bear a tertiary alkyl or an aryl substituent. *t*-Butyl methyl ketone (3,3-dimethylbutan-2-one), for example, gives aldol reactions with various bases in 60–70% yield. Enolization cannot occur towards the *t*-butyl group and must occur towards the methyl group instead.

ketones which can enolize only one way:

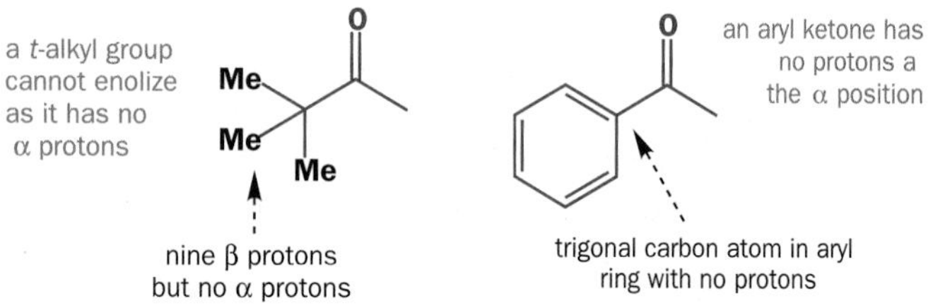

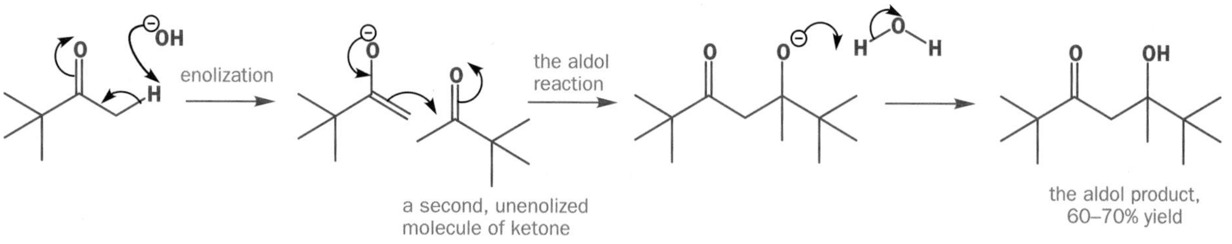

A specially interesting case of the blocked carbonyl compound is the lactone or cyclic ester. Open-chain esters do not give aldol reactions: they prefer a different reaction that is the subject of the next chapter. But lactones are in some ways quite like ketones and give unsaturated carbonyl products under basic catalysis. Enolization is unambiguous because the ester oxygen atom blocks enolization on one side.

enolate formation from a lactone (cyclic ester)

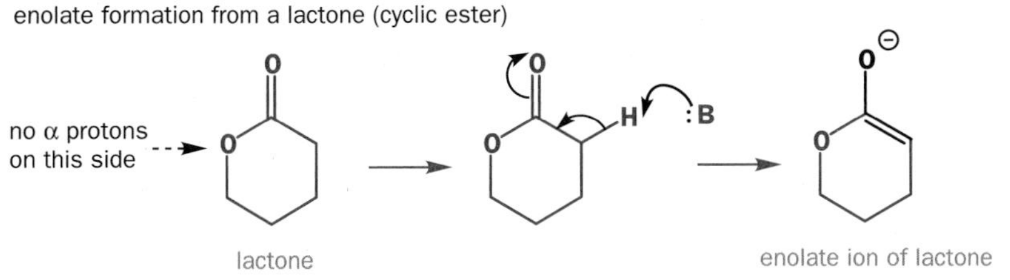

B in this scheme means 'base'.

no α protons on this side

lactone

enolate ion of lactone

The enolate then attacks the carbonyl group of an unenolized lactone just as we have seen with aldehydes and ketones.

aldol reaction of a lactone (cyclic ester)

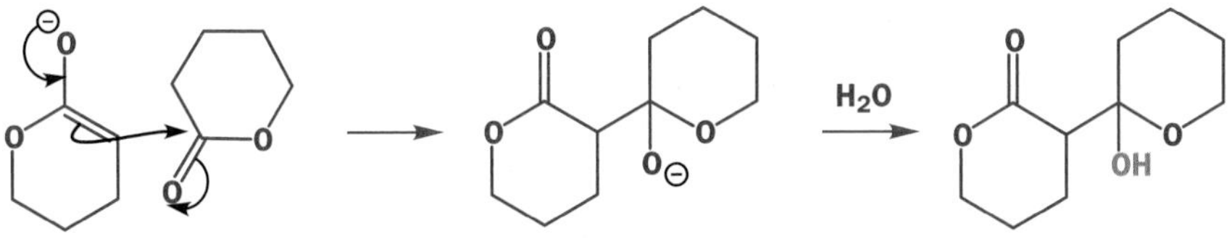

The last step is the familiar dehydration. As this reaction is being carried out in base we had better use the E1cB mechanism via the enolate of the aldol product.

the dehydration step

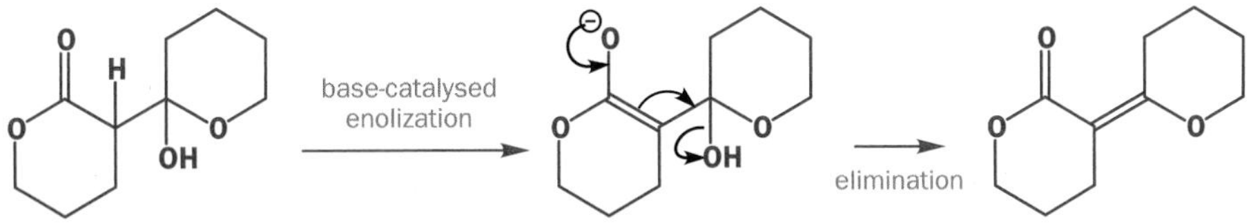

base-catalysed enolization

elimination

You might have been surprised that the intermediate in the aldol step of this reaction did not decompose. This intermediate could be described as a tetrahedral intermediate in a nucleophilic substitution at a carbonyl group (Chapter 12). Why then does it not break down in the usual way?

possible breakdown of a tetrahedral intermediate in a lactone aldol reaction

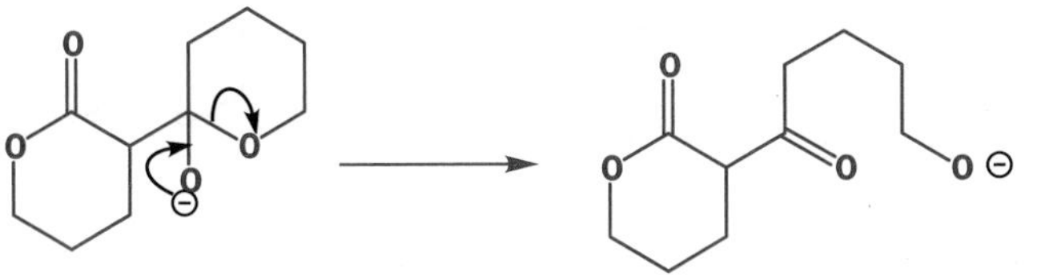

The best leaving group is the alkoxide and the product is quite reasonable. But what is it to do now? The only reasonable next step is for it to close back up again. Because the lactone is a *cyclic* ester, the leaving group cannot really leave—it must stay attached to the molecule. This reaction is reversible, but dehydration is effectively irreversible because it gives a stable conjugated product. This is the true situation.

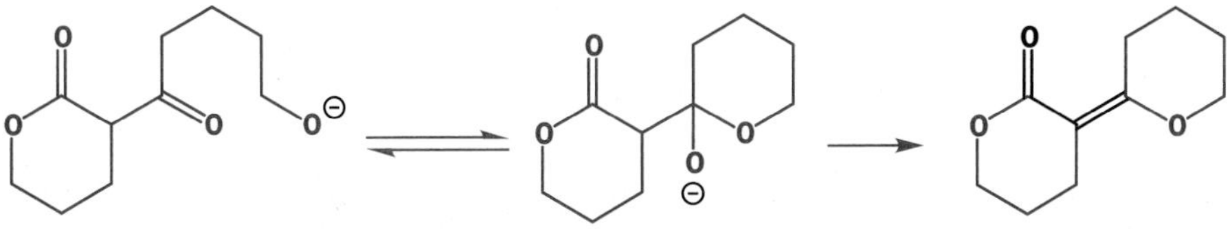

The equilibrium on the left does not affect the eventual product; it simply withdraws some of the material out of the productive reaction. We call this sort of equilibrium a **parasitic equilibrium** as it has no real life of its own—it just sucks the blood of the reaction.

Normal, acyclic esters are different: their alkoxide leaving groups *can* leave, and the result is a different sort of reaction, which you will meet in the next chapter.

Cross-condensations

So far we have considered only 'self-condensations'—dimerization reactions of a single carbonyl compound. These form only a tiny fraction of known aldol reactions. Those that occur between two different carbonyl compounds, one acting as a nucleophile in its enol or enolate form, and the other as an electrophile, are called **cross-condensations**. They are more interesting than self-condensations, but working out what happens needs more thought.

We shall start with an example that works well. The ketone PhCOMe reacts with 4-nitrobenzaldehyde in aqueous ethanol under NaOH catalysis to give a quantitative yield of an enone.

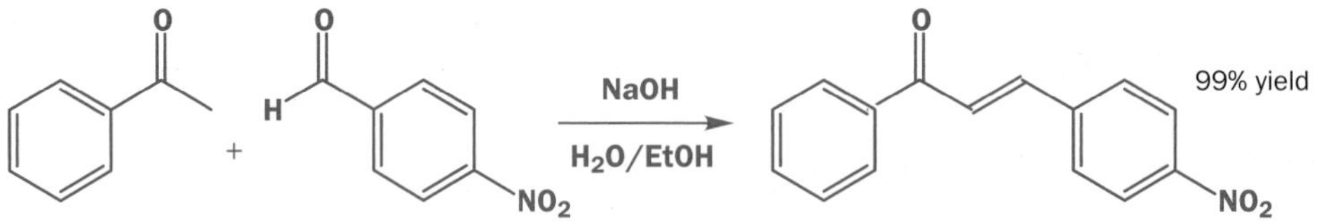

The first step must be the formation of an enolate anion using NaOH as a base. Though both carbonyl compounds are unsymmetrical, there is only one site for enolization as there is only one set of α protons, on the methyl group of the ketone. The aldehyde has no α protons at all.

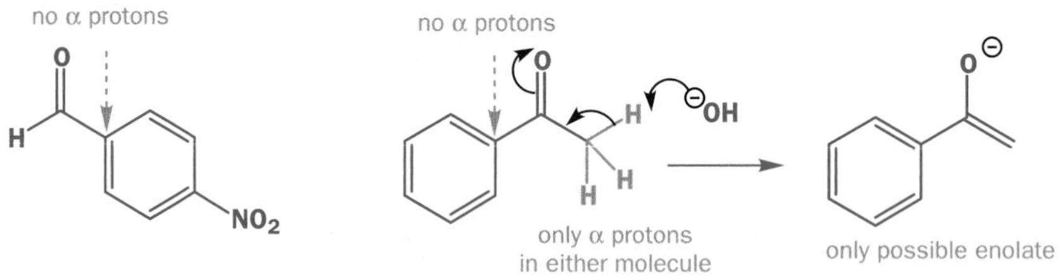

To get the observed product, the enolate obviously attacks the aldehyde to give an aldol, which then dehydrates by the E1cB mechanism.

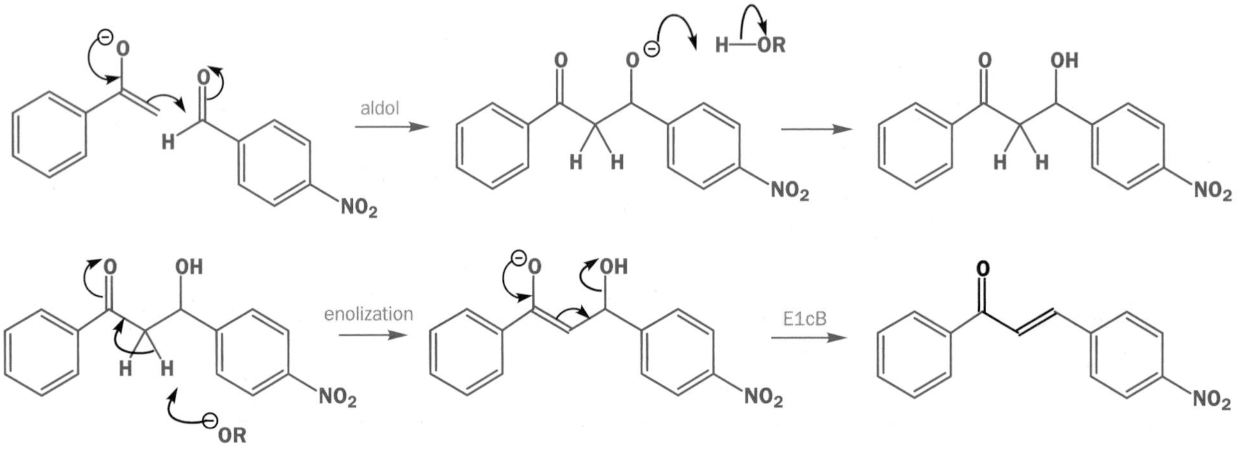

Now, in this step there was a choice. The enolate could have attacked another molecule of unenolized ketone. It didn't, because ketones are less reactive than aldehydes (Chapter 6). In this case the aldehyde has an electron-withdrawing nitro substituent too, making it even more reactive. The enolate selects the better electrophile, that is, the aldehyde.

In other cases the balance may shift towards self-condensation. You might think that a crossed aldol reaction between acetaldehyde and benzophenone (diphenylketone Ph$_2$C=O) should work well.

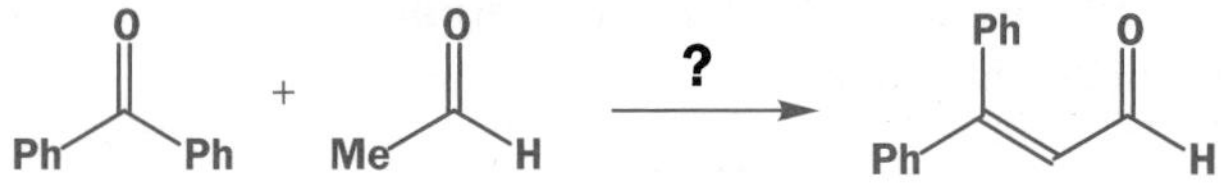

After all, only the aldehyde can enolize and the enolate could attack the ketone.

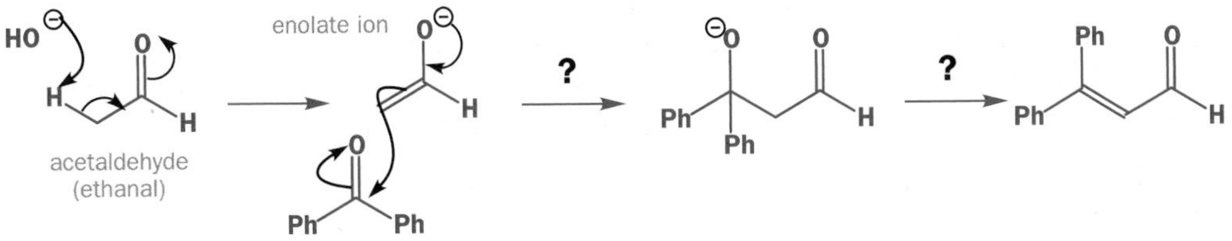

But it won't work. The ketone is very hindered and very conjugated. It is less electrophilic than a normal ketone and normal ketones are less reactive than aldehydes. Given a choice between attacking this ketone and attacking another (but unenolized) molecule of acetaldehyde, the enolate will choose the aldehyde every time. The reaction at the start of the chapter occurs and the ketone is just a spectator.

● Successful crossed aldol reactions

For this kind of crossed aldol reaction to work well we must have two conditions.
- One partner only must be capable of enolization
- The other partner must be incapable of enolization and be *more electrophilic than the enolizable partner.*

Everyone remembers the first of these conditions, but it is easy to forget the second.

Here follows a list of carbonyl substituents that prevent enolization. They are arranged roughly in order of reactivity with the most reactive towards nucleophilic attack by an enolate at the top. You do, of course, need two substituents to block enolization so typical compounds also appear in the list.

Carbonyl substituents that block enolization

	Substituent	Typical compounds
most reactive[a]	H	[b] 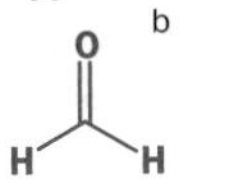
	CF$_3$, CCl$_3$	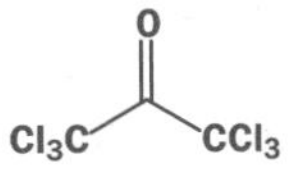 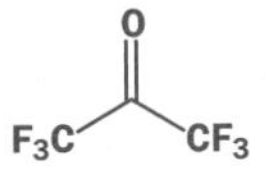
	t-alkyl	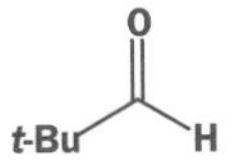
	alkenyl	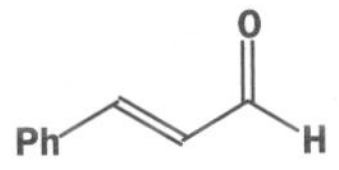
	aryl	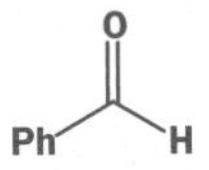
least reactive[a]	NR$_2$	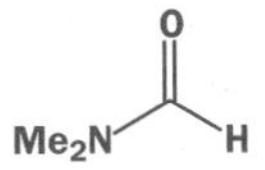
	OR	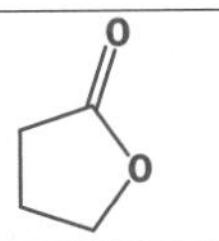

[a] Reactivity towards nucleophilic attack by an enolate.
[b] This compound needs special methods, discussed in the section on the Mannich reaction, p. 712.

nitromethane

'enolization'   'enolate' ion

Compounds that can enolize but that are not electrophilic

We can complement this type of selectivity with the opposite type. Are there any compounds that can enolize but that cannot function as electrophiles? No carbonyl compound can fill this role, but in Chapter 21 we met some 'enolizable' compounds that lacked carbonyl groups altogether. Most notable among these were the nitroalkanes. Deprotonation of nitroalkanes is not enolization nor is the product an enolate ion, but the whole thing is so similar to enolization that it makes sense to consider them together. The anions, sometimes called **nitronates**, react well with aldehydes and ketones.

anion of nitromethane

This particular example, using cyclohexanone as the electrophile and nitromethane itself as the source of the 'enolate', works quite well with NaOH as the base in methanol solution to give the 'aldol' in reasonable yield. Once again this reaction involves choice. Either compound could enolize, and, indeed, cyclohexanone reacts well with itself under essentially the same conditions.

70% yield

Although cyclohexanone forms an enolate in the absence of nitromethane, when both ketone and nitroalkane are present the base prefers to remove a proton from nitromethane. This is simply a question of pK_a values. The pK_a of a typical ketone is about 20 but that of nitromethane is 10. It is not even necessary to use as strong a base as NaOH ($pK_{aH} = 15.7$) to deprotonate nitromethane: an amine will do (pK_{aH} about 10) and secondary amines are often used.

The elimination step also occurs easily with nitro compounds and is difficult to prevent in reactions with aromatic aldehydes. Now you can see how the useful nitroalkene Michael acceptors in Chapter 23 were made.

85% yield

Nitroalkenes as termite defence compounds

Termites are social insects, and every species has its own 'soldier' termites that defend the nest. Soldier termites of the species *Prorhinotermes simplex* have huge heads from which they spray a toxic nitroalkene on their enemies.

defensive nitroalkene from termite soldiers

R = *n*-dodecyl

Though this compound kills other insects and even other species of termites, it has no effect on the workers of the same species. To find out why this was so, Prestwich made some radioactive compound using the aldol reaction. First, the right aldehyde was made using an S_N2 reaction with radioactive (^{14}C) cyanide ion on a tosylate followed by DIBAL reduction (Chapter 24) of the nitrile. The position of the ^{14}C atom in each compound is shown in black.

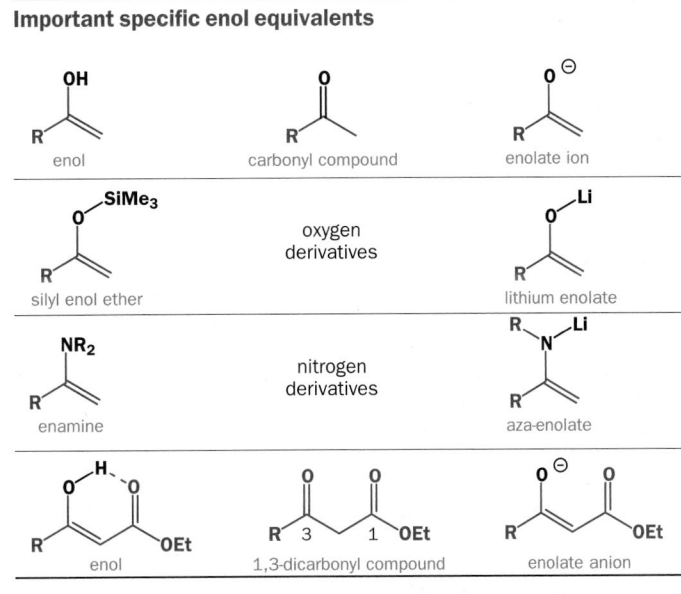

Then the aldol reaction was carried out with nitromethane and sodium methoxide to give the nitro aldol. Elimination using acetic anhydride in pyridine gave the defence compound (*E*-1-nitropentadec-1-ene) in 37% yield over the four steps.

It was found that, if the worker termites were sprayed with the labelled compound, they were able to make it harmless by using an enzyme to reduce the nitroalkene to a nitroalkane. The labelled nitroalkane could only be re-isolated from workers of the same species: other insects do not have the enzyme.

If an aldol reaction can be done with

- only one enolizable component
- only one set of enolizable protons
- a carbonyl electrophile more reactive than the compound being enolized

then you are lucky and the crossed aldol method will work. But most aldol reactions aren't like this: they are cross-condensations of aldehydes and ketones of various reactivities with several different enolizable protons. Crossed aldols on most pairs of carbonyl compounds lead to hopeless mixtures of products. In all cases that fail to meet these three criteria, a specific enol equivalent will be required: one component must be turned quantitatively into an enol equivalent, which will be reacted in a separate step with an electrophile. That is what the next section is about—and you will find that some of the methods have a lot in common with those we used for alkylating enolates in Chapter 26.

Controlling aldol reactions with specific enol equivalents

In Chapter 26 we saw that the alkylation of enolates was most simply controlled by preparing a specific enol equivalent from the carbonyl compound. The same approach is the most powerful of all the ways to control the aldol reaction. The table is a reminder of some of the most useful of these specific enol equivalents.

Specific enol equivalents are intermediates that still have the reactivity of enols or enolates but are stable enough to be prepared in good yield from the carbonyl compound. That was all we needed to know in Chapter 26. Now we know that a further threat is the reaction of the partly formed enol derivative with its unenolized parent and we should add that 'no aldol reaction should occur during the preparation of the specific enol equivalent'.

> ● Specific enol equivalents are intermediates that still have the reactivity of enols or enolates but are stable enough to be prepared in good yield from the carbonyl compound *without any aldol reaction.*

Sensible choice of an appropriate specific enol equivalent will allow almost any aldol reaction to be performed successfully. The first two compounds in our list, the silyl enol ethers and the lithium enolates, have a specially wide application and we should look first at the way these work. As the table suggests, silyl enol ethers are more like enols: they are nonbasic and not very reactive. Lithium enolates are more like enolate anions: they are basic and reactive. Each is appropriate in different circumstances.

Lithium enolates in aldol reactions

Lithium enolates are usually made at low temperature in THF with a hindered lithium amide base (often LDA) and are stable under those conditions because of the strong O–Li bond. The formation of the enolate begins with Li–O bond formation before the removal of the proton from the α position by the basic nitrogen atom.

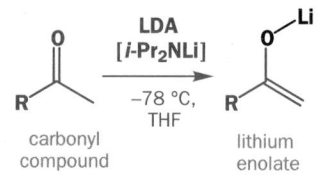

carbonyl compound → lithium enolate

(LDA [*i*-Pr₂NLi], −78 °C, THF)

■ The formation of lithium enolates was discussed in Chapter 26.

▶ Aldehydes are an exception. You can make lithium enolates from some aldehydes such as *i*-PrCHO, but generally self-condensation is too fast, so unwanted aldol self-condensation products are produced during the formation of the lithium enolate. To make specific enolates of aldehydes we need to use another type of derivative: see later.

▶ There are four coordination sites on the lithium atom—those we do not show are occupied by THF molecules. Before the aldol reaction can take place, one of the THFs must be displaced by the electrophilic carbonyl partner.

the tetrahedral structure of a lithium enolate in THF

This reaction happens very quickly—so quickly that the partly formed enolate does not have a chance to react with unenolized carbonyl compound before proton removal is complete.

Now, if a second carbonyl compound is added, it too complexes with the same lithium atom. This allows the aldol reaction to take place by a cyclic mechanism in the coordination sphere of the lithium atom.

aldol reaction with a lithium enolate

1. electrophilic carbonyl compound forms a complex with the lithium atom
2. cyclic mechanism gives the aldol product
3. aqueous work-up

lithium enolate

The aldol step itself is now a very favourable intramolecular reaction with a six-membered cyclic transition state. The product is initially the lithium alkoxide of the aldol, which gives the aldol on work-up.

This reaction works well even if the electrophilic partner is an enolizable aldehyde. In this example, an unsymmetrical ketone (blocked on one side by an aromatic ring) as the enol partner reacts in excellent yield with a very enolizable aldehyde. This is the first complete aldol reaction we have shown you using a specific enol equivalent: notice the important point that it is done in two steps—first, form the specific enol equivalent (here, the lithium enolate); *then* add the electrophile. Contrast the crossed aldols earlier in the chapter, where enolizable component, base, and electrophile were all mixed together in one step.

first form the enolate   then add the electrophile   94% yield of aldol

The next example is particularly impressive. The enol partner is a symmetrical ketone that is very hindered—there is only one α hydrogen on either side. The electrophilic partner is a conjugated enal that is not enolizable but that might accept the nucleophile in a conjugate manner. In spite of these potential problems, the reaction goes in excellent yield.

> Because of the six-membered ring mechanism for the addition, lithium enolates don't usually do conjugate additions. For enol equivalents that do, see Chapter 29.

first form the enolate   then add the electrophile   82% yield of aldol

You may wonder why we did not mention the stereochemistry of the first of these two products. Two new stereogenic centres are formed and the product is a mixture of diastereoisomers. In fact, both of these products were wanted for oxidation to the 1,3-diketone so the stereochemistry is irrelevant. This sequence shows that the aldol reaction can be used to make diketones too.

two diastereoisomers   both give the same diketone

> The symbol [O] denotes oxidation by one of the very general but ill-defined oxidizing agents from the laboratory of the famous Welsh chemist Owen Bracketts. Here the Swern reagents were the best (see Chapter 24).

Silyl enol ethers in aldol reactions

The silyl enol ether can be prepared from its parent carbonyl compound by forming a small equilibrium concentration of enolate ion with weak base such as a tertiary amine and trapping the enolate with the very efficient oxygen electrophile Me$_3$SiCl. The silyl enol ether is stable enough to be isolated but is usually used immediately without storing.

You should look upon silyl enol ethers as rather reactive alkenes that combine with things like protons or bromine (Chapter 21) but do not react with aldehydes and ketones without catalysis: they are much less reactive than lithium enolates. As with alkylation (p. 674), a Lewis acid catalyst is needed to get the aldol reaction to work, and a Ti(IV) compound such as TiCl$_4$ is the most popular.

carbonyl compound   silyl enol ether

silyl ether of aldol   aldol

The immediate product is actually the silyl ether of the aldol but this is hydrolysed during work-up and the aldol is formed in good yield. The Lewis acid presumably bonds to the carbonyl oxygen atom of the electrophile.

Now the aldol reaction can occur: the positive charge on the titanium-complexed carbonyl oxygen atom makes the aldehyde reactive enough to be attacked even by the not very nucleophilic silyl enol ether. Chloride ion removes the silyl group and the titanium alkoxide captures it again. This last step should not surprise you as any alkoxide (MeOLi for example) will react with Me$_3$SiCl to form a silyl ether.

Lewis acid binds to the carbonyl oxygen atom

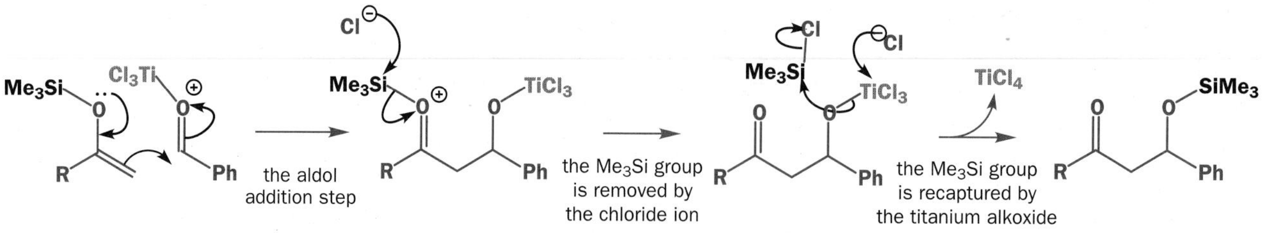

This mechanism looks complicated, and it is. It is, in fact, not clear that the details of what we have written here are right: the titanium may well coordinate to *both* oxygens through the reaction, and some of the steps that we have represented separately probably happen simultaneously. However, all reasonable mechanisms will agree on two important points, which you must understand:

- Lewis acid is needed to get silyl enol ethers to react
- The key step is an aldol attack of the silyl enol ether with the Lewis-acid complexed electrophile

The use of silyl enol ethers can be illustrated in a synthesis of manicone, a conjugated enone that ants use to leave a trail to a food source. It can be made by an aldol reaction between the pentan-3-one (as the enol component) and 2-methylbutanal (as the electrophile). Both partners are enolizable so we shall need to form a specific enol equivalent from the ketone. The silyl enol ether works well.

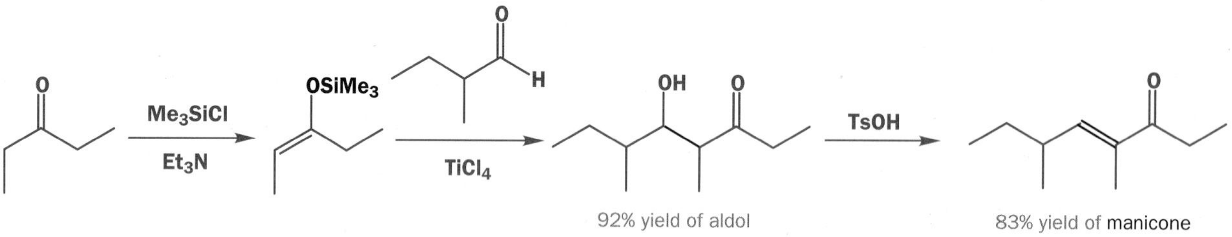

92% yield of aldol

83% yield of manicone

The silyl enol ether is not isolated but reacted immediately with the aldehyde to give an excellent yield of the aldol. Dehydration in acid solution with toluene sulfonic acid (TsOH) gives the enone. You can see by the high yield in the aldol reaction that there is no significant self-condensation of either partner in the aldol reaction.

Conjugated Wittig reagents as specific enol equivalents

When the Wittig reaction was introduced (Chapter 14) we saw it simply as an alkene synthesis. Now if we look at one group of Wittig reagents, those derived from α-halo-carbonyl compounds, we can see that they behave as specific enol equivalents in making unsaturated carbonyl compounds.

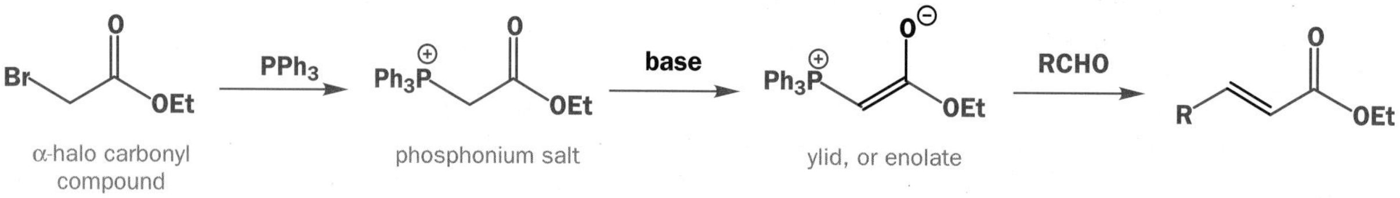

α-halo carbonyl compound

phosphonium salt

ylid, or enolate

You notice that we have drawn the intermediate ylid as an enolate just to emphasize that it is an enolate derivative: it can also be represented either as the ylid or as a C=P 'phosphorane' structure. If we look at the details of this sort of Wittig reaction, we shall see that ylid formation is like enolate anion formation (indeed it *is* enolate anion formation). Only a weak base is needed as the enolate is stabilized by the Ph_3P^+ group as well.

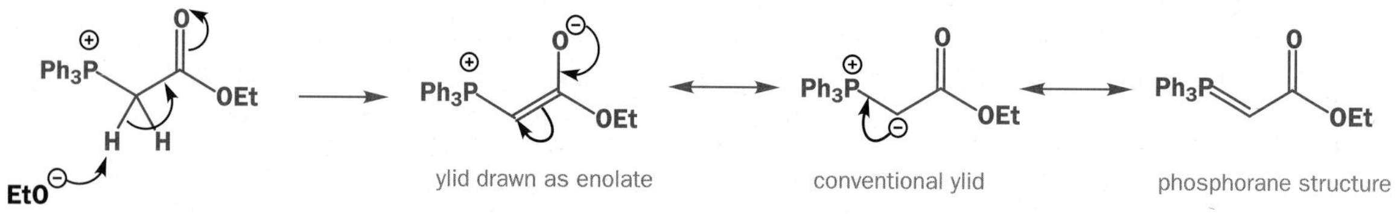

ylid drawn as enolate

conventional ylid

phosphorane structure

The first step of the Wittig reaction proper is just like an aldol reaction as it consists of an enolate attacking an electrophilic carbonyl compound. But, instead of forming an 'aldol' product, this adduct goes on to form an unsaturated carbonyl compound directly.

The final stages follow the mechanism of the Wittig reaction you met in Chapter 14: you see them as a special case of dehydration made favourable by the formation of a phosphine oxide as well as an unsaturated carbonyl compound.

The conjugated ylides derived from aldehydes, ketones, and esters are all sufficiently stable to be commercially available as the ylids—one of the few examples of specific enol equivalents that you can actually buy. The ylid corresponding to the enolate of acetaldehyde is a solid, m.p. 185–188 °C that reacts well with other aldehydes, even if they are enolizable.

solid, m.p. 185–188 °C
commercially available

enolizable aldehyde

We haven't yet considered in detail the geometry of the double bonds arising from aldol condensations. Those that are E1 or E1cB eliminations give mainly the more stable *E*-alkene products for the reasons described in Chapter 19. These Wittig variants are usually highly *E*-selective: we shall consider why in Chapter 31, where we deal with the question of how to control double bond geometry.

The Wittig equivalent of an aldol reaction with a ketone enolate can be illustrated by the synthesis of a compound in juniper berries, junionone, with a four-membered ring.

junionone

No base was needed in either of the last two examples: the stable ylid itself was used as a reagent. The stability of the enolate ylid means that the Wittig reagent must act as the enol partner and the other compound as the electrophile.

The stability of the phosphonium-stabilized enolates also means that, although they react well with aldehydes, their reactions with ketones are often poor, and it is better in these cases to use phosphonate-stabilized enolates. Being anionic, rather than neutral, these are more reactive. If an ester enolate equivalent is being used, the best base is the alkoxide ion belonging to the ester; with a ketone enolate equivalent, use sodium hydride or an alkoxide.

trimethyl
phosphonoacetate

phosphonate-stabilized enolate

α,β-unsaturated ester

The 'brace' device here is commonly used rather like 'R'—it means that the rest of the molecule is unimportant to the reaction in question and could be anything.

dimethyl
phosphonoacetone

phosphonate-stabilized enolate

α,β-unsaturated ketone

These last reagents, where the anion is stabilized both by the adjacent carbonyl group (as an enolate) and by the adjacent P=O group, are just one of many examples of enolate anions stabilized by two electron-withdrawing groups. The most important members of this class, enolates of 1,3-dicarbonyl compounds, are the subject of the next section.

Specific enol equivalents from 1,3-dicarbonyl compounds

Though these are the oldest of the specific enol equivalents, they are still widely used because they need no special conditions—no low temperatures or strictly anhydrous solvents. The two most important are derived from malonic acid and ethyl acetoacetate.

These compounds are largely enolized under normal conditions. So, you might ask, why don't they immediately react with themselves by the aldol reaction? There are two aspects to the answer. First, the enols are very stable (see Chapter 21 for a full discussion) and, secondly, the carbonyl groups in the unenolized fraction of the sample are poorly electrophilic ester and ketone groups. The second carbonyl group of the enol is not electrophilic because of conjugation.

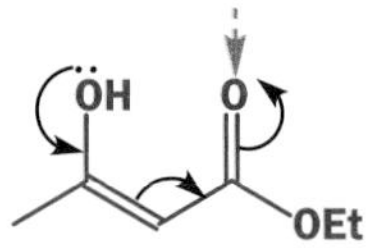

When a normal carbonyl compound is treated with catalytic acid or base, we have a small proportion of reactive enol or enolate in the presence of large amounts of unenolized electrophile. Aldol reaction (self-condensation) occurs. With 1,3-dicarbonyl compounds we have a small proportion of not particularly reactive unenolized compound in the presence of large amounts of stable (and hence unreactive) enol. No aldol occurs.

If we want a **crossed aldol reaction**, we simply add a second, electrophilic carbonyl compound such as an aldehyde, along with a weak acid or base. Often a mixture of a secondary amine and a carboxylic acid is used.

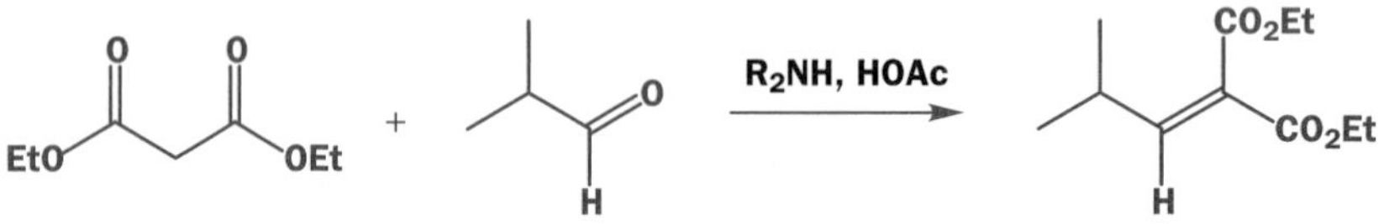

Reaction no doubt occurs via the enolate ion generated by the amine while the carboxylic acid buffers the solution, neutralizing the product, and preventing enolization of the aldehyde. The amine (pK_{aH} about 10) is a strong enough base to form a significant concentration of enolate from the 1,3-dicarbonyl compound (pK_a about 13) but not strong enough to form the enolate from the aldehyde (pK_a about 20). The formation of the enolate can be drawn from either tautomer of the malonate.

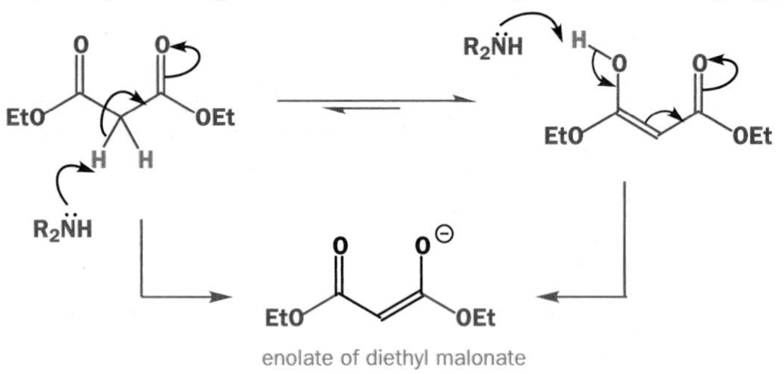

Now the enolate ion can attack the aldehyde in the usual way, and the buffer action of the acid produces the aldol in the reaction mixture.

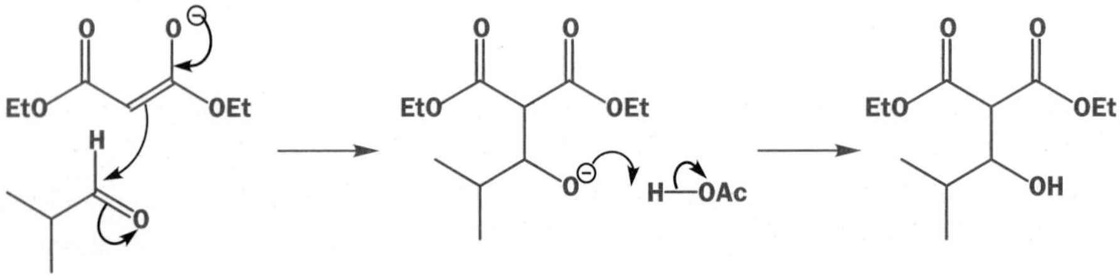

ethyl acetoacetate
(ethyl 3-oxobutanoate)

diethyl malonate
(diethyl propanedioate)

electrons are fed into this carbonyl group making it less electrophilic

92% yield

enolate of diethyl malonate

> **Tautomers** are isomers related to one another by tautomerism: see Chapter 21, p. 522.

There is still one proton between the two carbonyl groups so enolate anion formation is again easy and dehydration follows to give the unsaturated product.

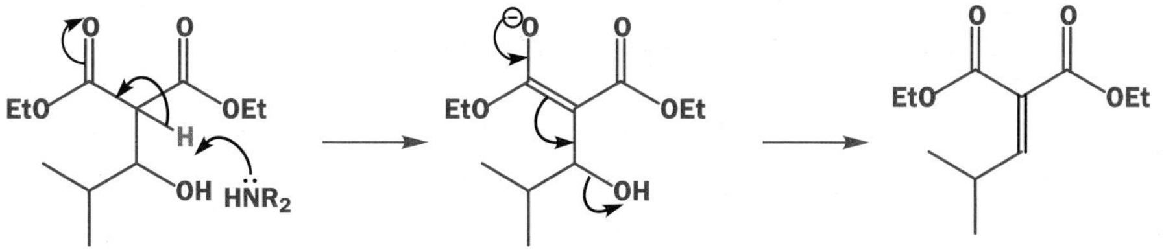

You may not want a product with both ester groups present, and we discussed in Chapter 26 how one of two 1,3-related ester groups may be removed by hydrolysis and decarboxylation. There is a simpler route with the aldol reaction. If, instead of the malonate diester, malonic *acid* is used, the decarboxylation occurs spontaneously during the reaction. The catalysts this time are usually a more basic mixture of piperidine and pyridine.

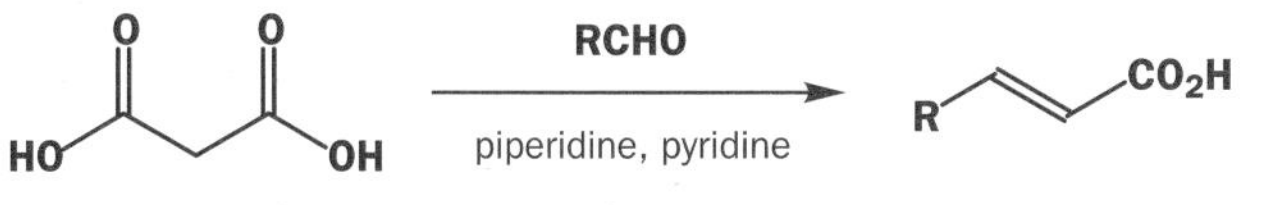

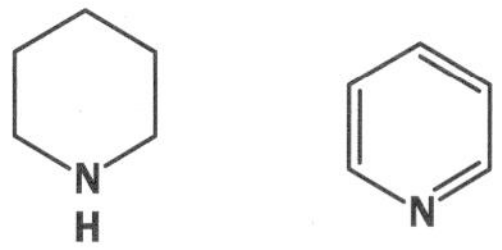

malonic acid

piperidine
pK_{aH} 11

pyridine
pK_{aH} 5.5

The reaction under these conditions is sometimes called the **Knoevenagel reaction** after its nineteenth century inventor, and presumably uses the enolate anion of the monocarboxylate of the malonic acid. Though this enolate is a dianion, its extensive delocalization and the intramolecular hydrogen bond make it really quite stable.

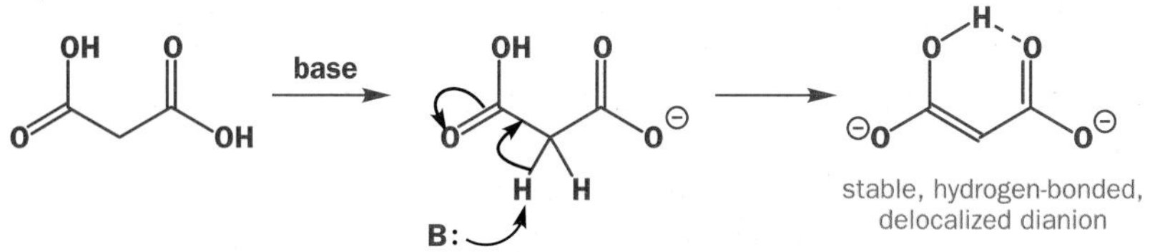

stable, hydrogen-bonded,
delocalized dianion

Next comes the aldol step. The dianion attacks the aldehyde, and after proton exchange the aldol is formed (still as the monocarboxylate in this basic solution).

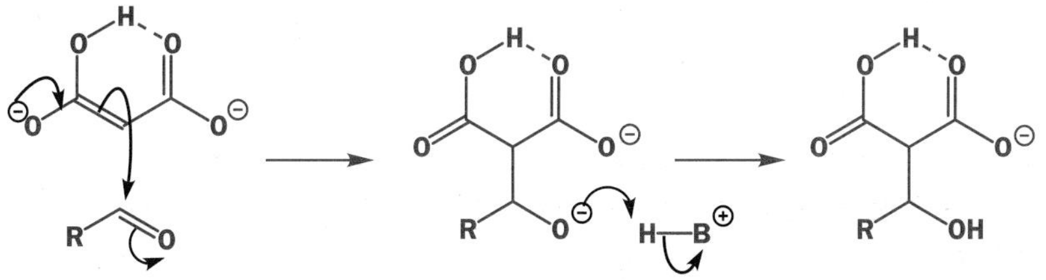

Finally comes the decarboxylation step, which can occur though a cyclic mechanism (compare the decarboxylation mechanisms in Chapter 26). The decarboxylation could give either *E* or *Z* double bond depending on which acid group is lost as CO_2, but the transition state leading to the more stable *E* product must be lower in energy since the product has *E* geometry.

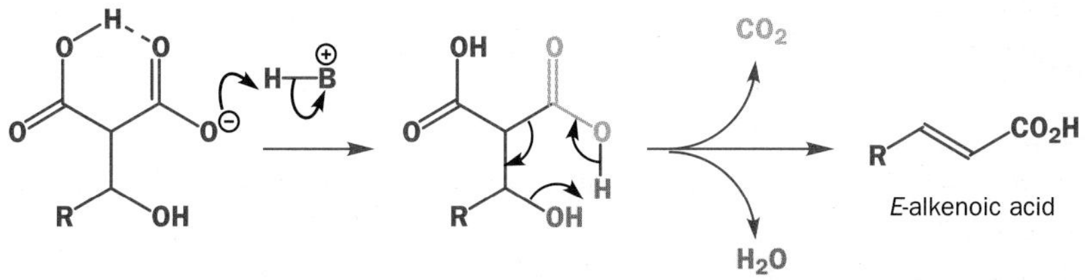

E-alkenoic acid

● We have now completed our survey of the most important types of aldol reaction and of the varieties of specific enol equivalents available. We shall now move on to look at carbonyl compounds type by type, and consider the best options for making specific enol equivalents of each.

Specific enol equivalents for carboxylic acid derivatives

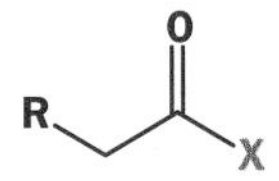

electrophilic reactivity for X =
Cl > OCOR > OR > NR$_2$

We established in Chapter 12 a hierarchy for the electrophilic reactivity of acid derivatives that should by now be very familiar to you—acyl chlorides at the top to amides at the bottom. But what about the reactivity of these same derivatives towards enolization at the α position, that is, the CH$_2$ group between R and the carbonyl group in the various structures? You might by now be able to work this out. The principle is based on the mechanisms for the two processes.

mechanism of nucleophilic attack

mechanism of enolate formation

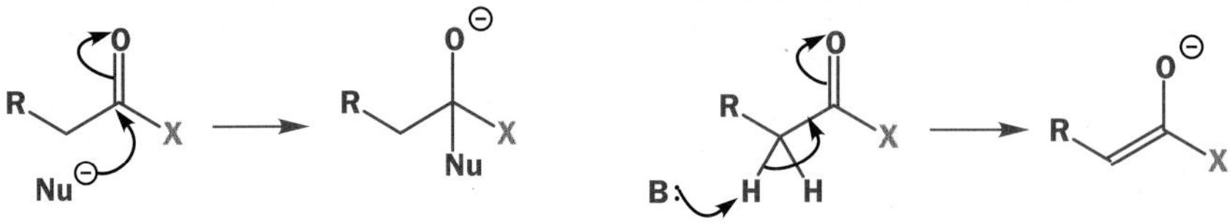

See how similar these two mechanisms are. In particular, they are the same at the carbonyl group itself. Electrons move into the C=O π* orbital: the C=O bond becomes a C–O single bond as a negative charge develops on the oxygen atom. It should come as no surprise that *the order of reactivity for enolization is the same as the order of reactivity towards nucleophilic attack.*

Enolate formation and electrophilic reactivity of acid derivatives

Electrophilic reactivity	Derivative	Structure	Reactivity towards enolate formation
very high	acid chloride	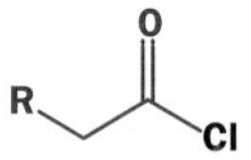	very high
high	anhydride	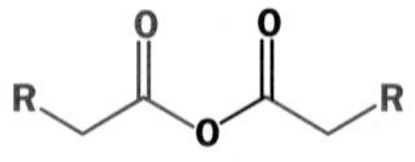	high
low	ester	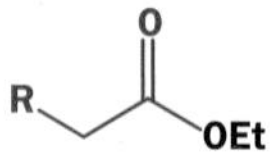	low
very low	amide	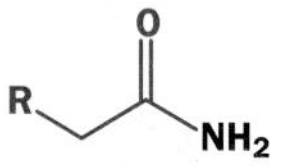	very low

In Chapter 21 we established that enolates can be formed from acid chlorides, but that they decompose to ketenes. Enolates can be formed from amides with difficulty, but with primary or secondary amides one of the NH protons is likely to be removed instead.

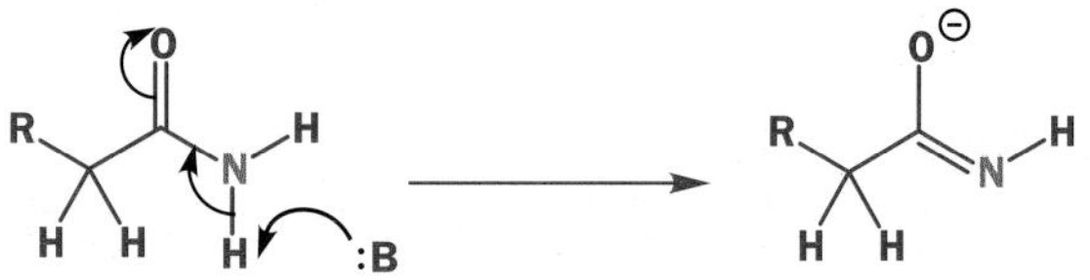

For the remainder of this section we shall look at how to make specific enol equivalents of the remaining carboxylic acid derivatives.

Enols and enolates from acid anhydrides

Enols or enolates from anhydrides are not used very often in aldol reactions other than in one important application, usually known as the **Perkin reaction**. An acid anhydride, such as acetic anhydride, is combined with a non-enolizable aldehyde and a weak base, usually the salt of the acid. This base is used so that nucleophilic attack on the anhydride does no harm, simply regenerating the anhydride.

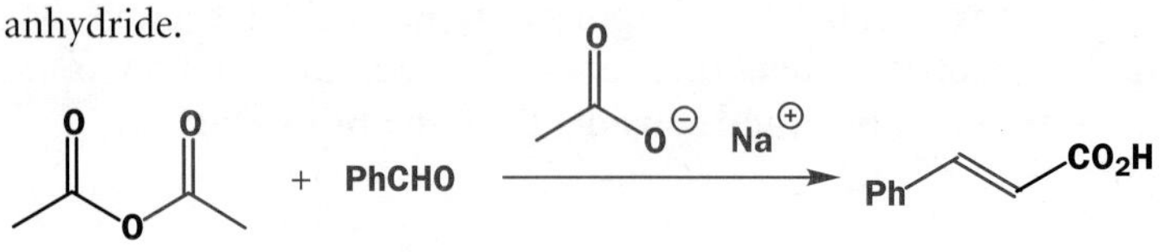

+ PhCHO → Ph⁀CO$_2$H

The low equilibrium concentration of the enolate attacks the aldehyde.

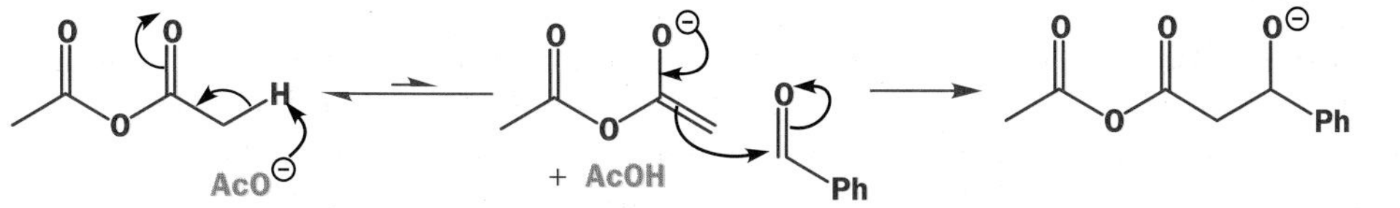

Thus far the reaction is a normal aldol reaction, but now something quite different happens. Six atoms along the molecule from the alkoxide ion is the carbonyl group of an anhydride. An intramolecular acylation is inevitable, given that anhydrides acylate alcohols even if the two groups are in different molecules.

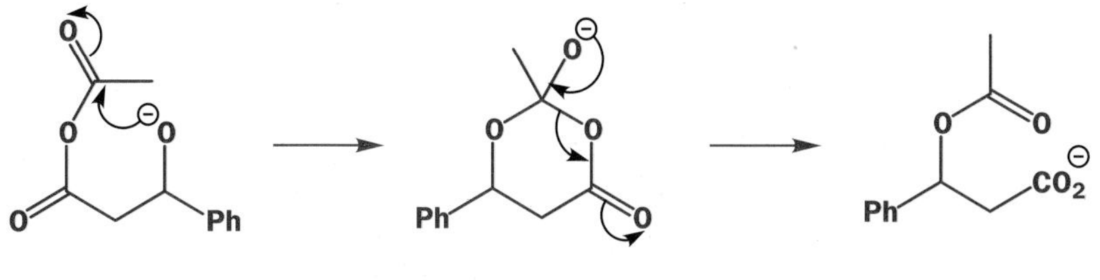

carboxylate is the best leaving group
from this tetrahedral intermediate

Next, acetic acid is lost. Just as acetate is a better leaving group than hydroxide, this step is much more favourable than the usual dehydration at the end of an aldol condensation. Elimination of acetic acid may occur either from the carboxylic acid itself or from the mixed anhydride formed from one more molecule of the acetic anhydride. Whichever route is followed, the unsaturated acid is formed in a single step with the anhydride assisting both the aldol and the dehydration steps.

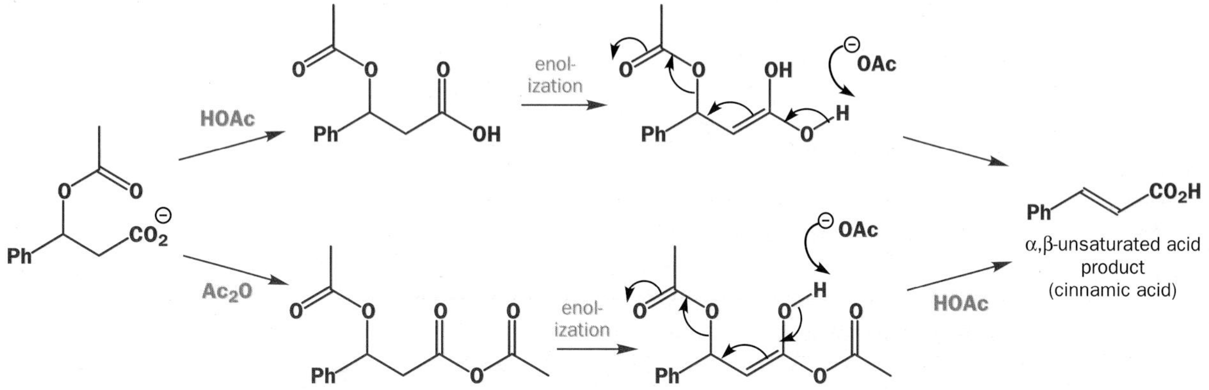

α,β-unsaturated acid
product
(cinnamic acid)

Enols and enolates from esters

Among the enolates of carboxylic acid derivatives, esters are the most widely used. Ester enolates cannot be used in crossed aldols with aldehydes because the aldehyde is both more enolizable and more electrophilic than the ester. It will just condense with itself and ignore the ester. The same is true for ketones. A specific enol equivalent for the ester will therefore be needed for a successful ester aldol reaction.

Fortunately, because this is a classic problem, many solutions are available. You can use the lithium enolate, or the silyl enol ether, usually made best via the lithium enolate.

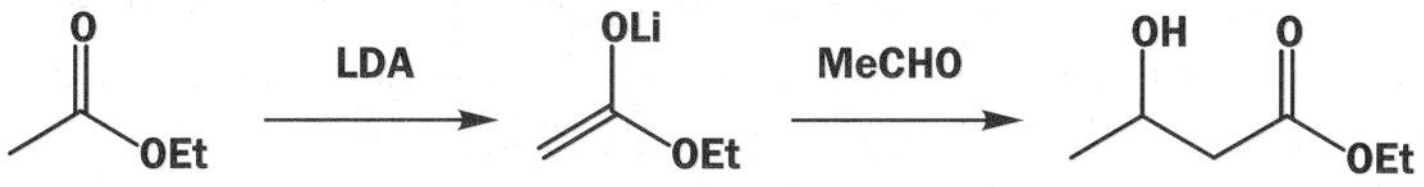

■

We have already discussed the special examples of malonate and phosphonoacetate esters. Now we need to consider ester enolates more generally.

▶

Forgive the reminder that a Lewis acid is necessary with silyl enol ethers.

lithium enolate of ester

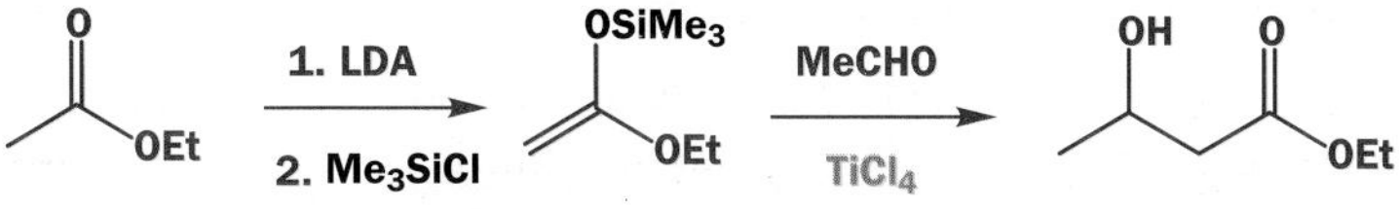

silyl enol ether of ester

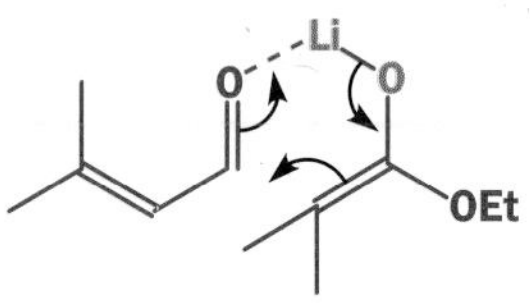

cyclic mechanism for ester aldol reaction

A good example is the first step in a synthesis of the natural product himalchene by Oppolzer and Snowden. Even though the ester and the aldehyde are both crowded with substituents, the aldol reaction works well with the lithium enolate of the ester. The cyclic mechanism ensures that the enolate adds directly to the carbonyl group of the aldehyde and not in a conjugate (Michael) fashion.

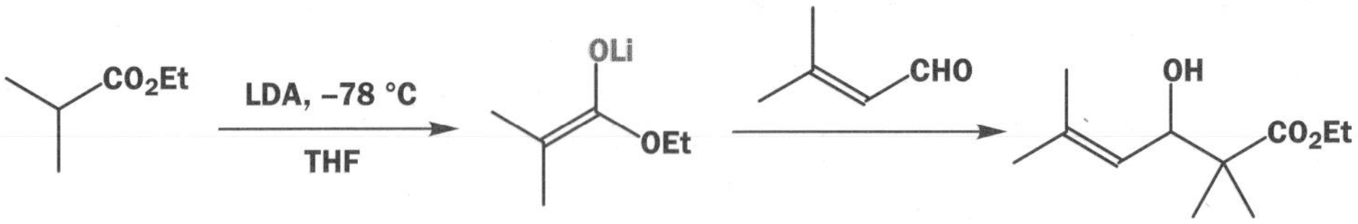

CO_2Et →(LDA, −78 °C / THF)→   OLi ... OEt (lithium enolate)   →(CHO)→   OH ... CO_2Et **72% yield**

Zinc enolates, made from the bromoesters, are a good alternative to lithium enolates of esters. The mechanism for zinc enolate formation should remind you of the formation of a Grignard reagent.

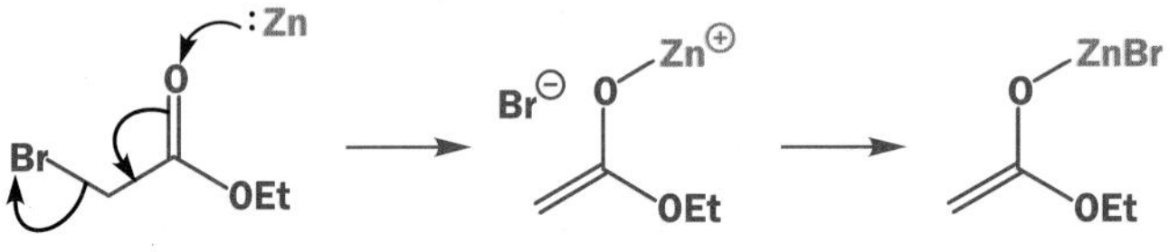

Br ... OEt + :Zn → $Br^{\ominus}$ ... $Zn^{\oplus}$ ... OEt → ZnBr ... OEt (zinc enolate)

There is no danger of self-condensation with zinc enolates as they do not react with esters. But they do react cleanly with aldehydes and ketones to give aldols on work-up. You will appreciate that the use of zinc enolates is therefore special to esters: you cannot make a zinc enolate from a 2-bromoaldehyde or an α-bromoketone as then you *would* get self-condensation.

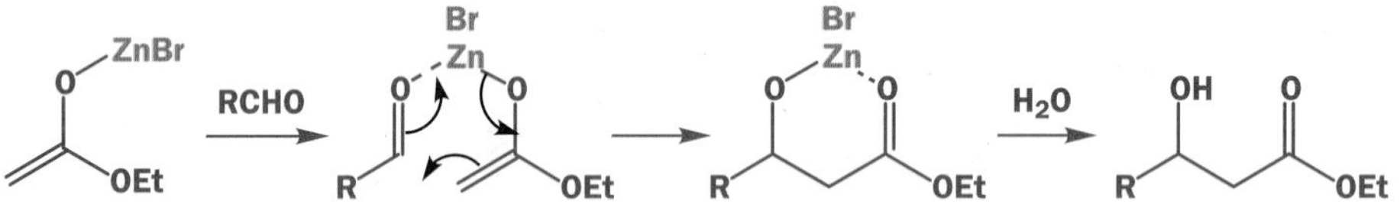

ZnBr ... OEt   →(RCHO)→   R ... (Br, Zn) ... OEt   →   R ... (Br, Zn) ... OEt   →(H_2O)→   OH ... R ... OEt

> Zinc, like magnesium, is a two-electron donor and likes to be oxidized from Zn(0) to Zn(II). This enolate is often called the **Reformatsky reagent** after its inventor, which is fine, and often drawn as a C–Zn compound, which is not fine because it isn't one.
>
> BrZn ... OEt
> bad structure

> The dehydration product from this aldol product is best made directly by one of the Wittig variants we discussed earlier. The same bromoester is of course the starting material for the ylid synthesis.

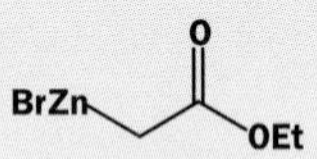

● **Ester enolate equivalents**

For aldol reactions with an ester enolate equivalent, use

● lithium enolates *or*

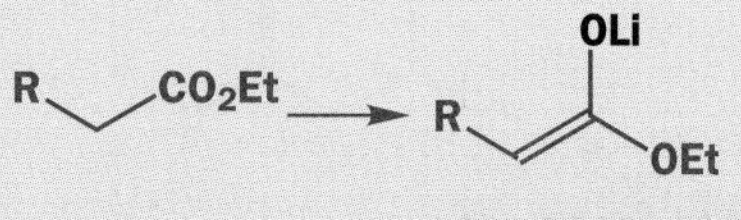

R ... CO_2Et → R ... OLi ... OEt
lithium enolate

● silyl enol ethers *or*

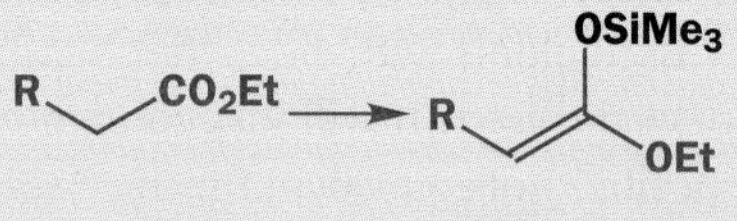

R ... CO_2Et → R ... $OSiMe_3$ ... OEt
silyl enol ether

● zinc enolates

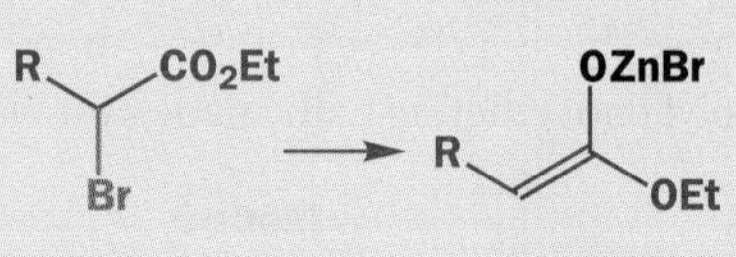

R ... CO_2Et (Br) → R ... OZnBr ... OEt
zinc enolate

Enols and enolates from free carboxylic acids

You might think that the presence of the acidic proton in a carboxylic acid would present an insuperable barrier to the formation and use of any enol derivatives. In fact, this is not a problem with either the lithium enolates or the silyl enol ethers. Addition of BuLi or LDA to a carboxylic acid immediately results in the removal of the acidic proton and the formation of the lithium salt of the carboxylic acid. If BuLi is used, the next step is addition of BuLi to the carbonyl group and the eventual formation of a ketone (see Chapter 12, p. 299). But, if LDA is used, it is possible to form the lithium enolate of the lithium derivative of the carboxylic acid.

reaction with BuLi

lithium carboxylate

butyl ketone

reaction with LDA

lithium carboxylate

lithium enolate

silyl derivative

The enolate derivative is rather strange as it has two OLi groups on the same double bond, but it can be cleanly converted to the corresponding silyl enol ether. Both lithium enolates and silyl enol ethers from acids can be used in aldol reactions.

Ketene acetals

Because these compounds have two identical OR groups joined to the same end of the same double bond, you will see them called 'ketene acetals' or, here, 'silyl ketene acetals'. This is a reasonable description as you can imagine the carbonyl group of a ketene forming an acetal in the same way as an aldehyde. In fact, they cannot be made this way.

aldehyde   acetal   ketene   'ketene acetal'
             (imaginary reaction)

Specific enol equivalents for aldehydes

Aldehydes enolize very readily but also self-condense rather easily. Lithium enolates can't be made cleanly, because the self-condensation reaction happens even at −78 °C and is as fast as the enolization by LDA. Silyl enol ethers are a much better choice. They clearly must not be made via the lithium enolate, and amine bases are usually used. As each molecule of enolate is produced in the equilibrium, it is efficiently trapped by the silylating agent.

weak base used with aldehyde   low concentration of enolate   efficient trapping by oxygen-loving silicon electrophile

isobutyraldehyde 3-phenylpropanal

acid or base

mixture of self-condensation and cross-coupling products from both aldehydes

These silyl enol ethers are probably the best way of carrying out crossed aldol reactions with an aldehyde as the enol partner. An example is the reaction of the enol of the not very enolizable isobutyraldehyde with the very enolizable 3-phenylpropanal. Mixing the two aldehydes and adding base would of course lead to an orgy of self-condensation and cross-couplings.

Preliminary formation of the silyl enol ether from either aldehyde, *in the absence of the other*, would be trouble-free as Me₃SiCl captures the enolate faster than self-condensation occurs. Here we need the silyl enol ether from isobutyraldehyde. The other aldehyde is now added along with the necessary Lewis acid, here TiCl$_4$. The mechanism described on p. 699 gives the aldol after work-up in an excellent 95% yield. No more than 5% of other reactions can have occurred.

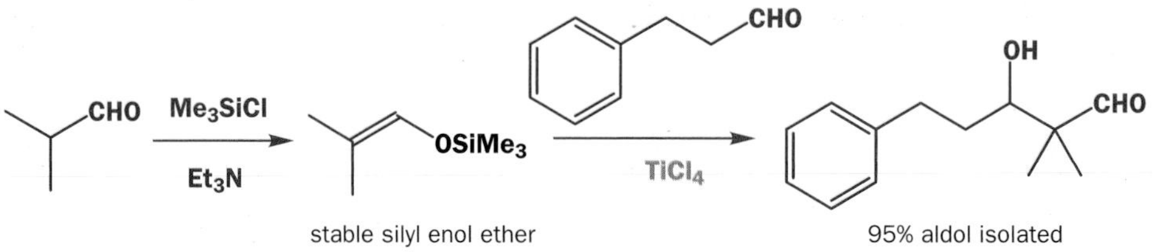

stable silyl enol ether

95% aldol isolated

Other useful specific enol equivalents of aldehydes and ketones are enamines and aza-enolates, which you saw in use in alkylation reactions in Chapter 26. Aza-enolates—the lithium enolates of imines—derived from aldehydes are useful too in aldol reactions.

Cyclohexylamine gives a reasonably stable imine even with acetaldehyde and this can be isolated and lithiated with LDA to give the aza-enolate. The mechanism is similar to the formation of lithium enolates and the lithium atom binds the nitrogen atom of the aza-enolate, just as it binds the oxygen atom of an enolate.

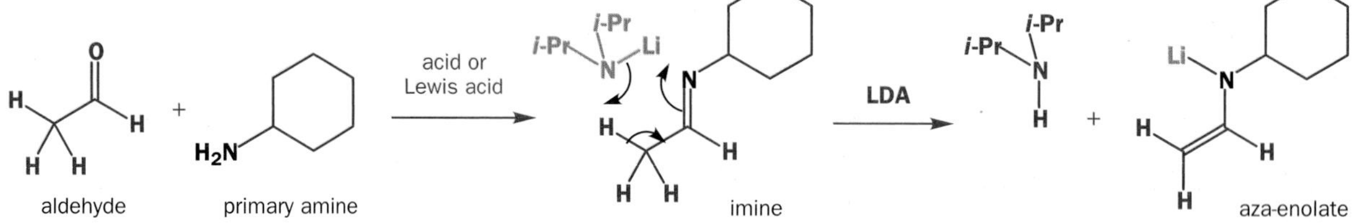

aldehyde   primary amine   imine   aza-enolate

▶

Imines are susceptible to hydrolysis and they are best not stored but used at once. To understand fully these reactions you should ensure you are familiar with the mechanisms of imine formation and hydrolysis from Chapter 14.

The aza-enolate reacts cleanly with other aldehydes or ketones to give aldol products. Even the most challenging of cross-couplings—attack on another similar enolizable aldehyde—occurs in good yield.

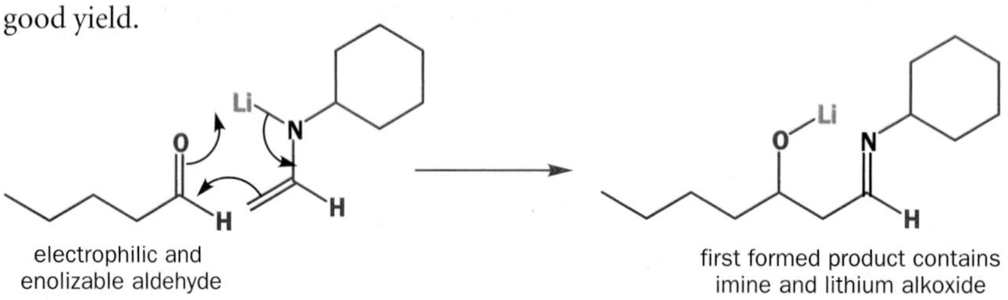

electrophilic and enolizable aldehyde

first formed product contains imine and lithium alkoxide

The initial product is a new imine, which is easily hydrolysed during acidic aqueous work-up. The alkoxide is protonated, the imine hydrolysed, and finally the aldol is dehydrated to give the enal—65% overall yield in this case.

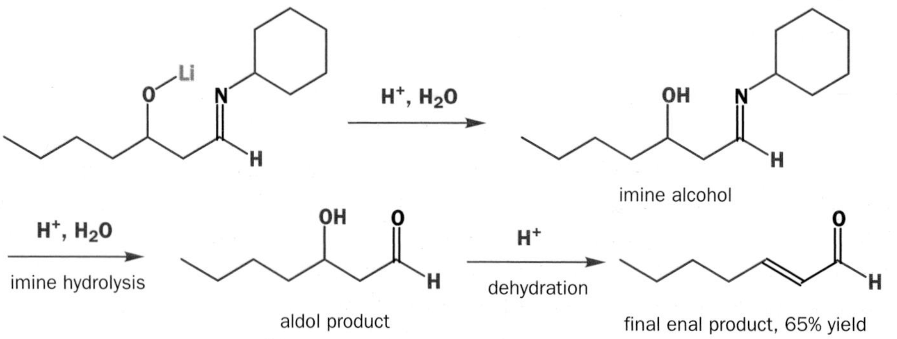

imine alcohol

aldol product

final enal product, 65% yield

The key to the success of the aza-enolates is that the imine is first formed from the aldehyde with the primary amine, a relatively weak base, and under these conditions imine formation is faster than self-condensation. Only after the imine is formed is LDA added when self-condensation cannot occur simply because no aldehyde is left.

Enamines are not generally used in aldol condensations, partly because they are not reactive enough, but mainly because they are too much in equilibrium with the carbonyl compound itself and exchange would lead to self-condensation and the wrong cross-couplings. You will see in the next chapter that enamines come into their own when we want to acylate enols with the much more reactive acid chlorides.

● **For crossed aldol reactions with an aldehyde as the enol partner, use**

● **silyl enol ethers** *or*

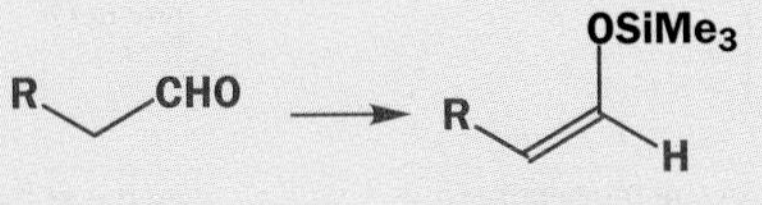

silyl enol ether

● **aza-enolates**

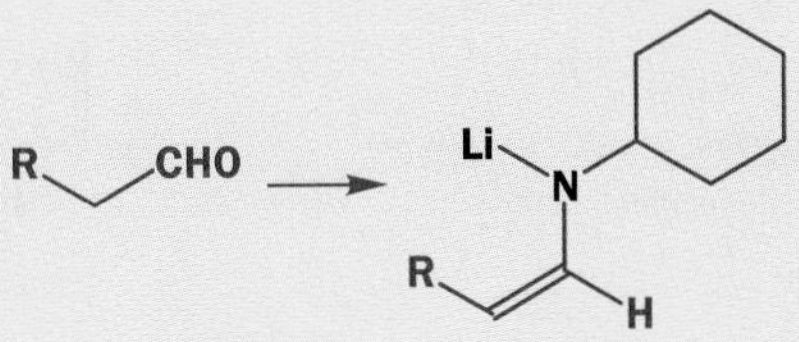

aza-enolate

Specific enol equivalents for ketones

The enolization of ketones, unless they are symmetrical, poses a special problem. Not only do we need to prevent them self-condensing (though this is less of a problem than with aldehydes), but we also need to control which side of the carbonyl group the ketone enolizes. In this section we shall introduce aldol reactions with unsymmetrical ketones where one of two possible enols or enolates must be made.

Making the less substituted enolate equivalent: kinetic enolates

Treatment of methyl ketones with LDA usually gives only the lithium enolate on the methyl side. This is the enolate that forms the fastest, and is therefore known as the **kinetic enolate**. It is formed faster *because*:

- the protons on the methyl group are more acidic
- there are three of them as against two on the other side, and
- there is steric hindrance to attack by LDA on the other side of the carbonyl group

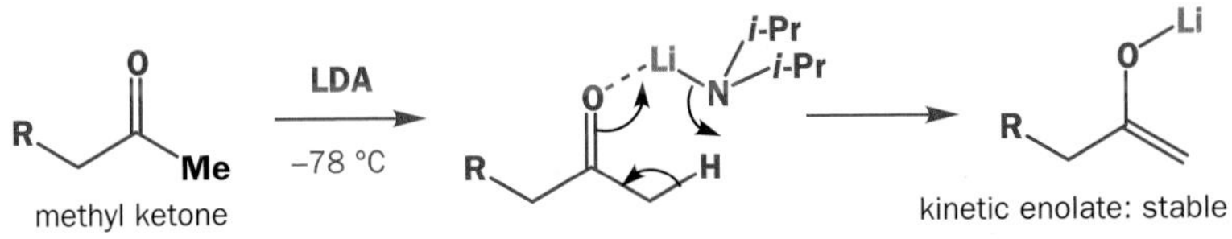

methyl ketone   kinetic enolate: stable at −78 °C

A simple example from the first report of this reaction by Gilbert Stork and his group in 1974 is the condensation of pentan-2-one with butanal to give the aldol and then the enone oct-4-en-3-one by acid-catalysed dehydration. The yields may seem disappointing, but this was the first time anyone had carried out a crossed aldol reaction like this with an unsymmetrical ketone and an enolizable aldehyde and got just one aldol product in any reasonable yield at all.

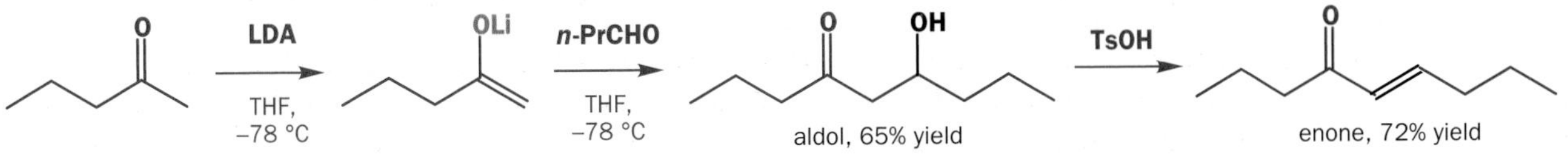

aldol, 65% yield   enone, 72% yield

■ Kinetic and thermodynamic enolates were introduced in Chapter 26, p. 682.

■ Gilbert Stork was born in Brussels and became an assistant professor of chemistry at Harvard in 1948. Since 1953, Stork has been at Columbia University in New York. Since the 1950s, he has pioneered new synthetic methods, among them many involving enolates and enamines.

An uncontrolled ketone aldol

A more typical result from the days before specific enol equivalents had been invented is this attempted crossed condensation between butanone and butanal with catalytic base. Two products were isolated in low yield.

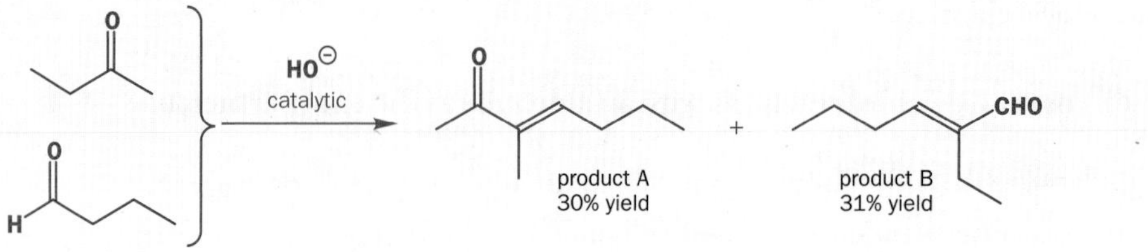

product A
30% yield

product B
31% yield

Product A is from the enolate of the more substituted side of the ketone reacting with the aldehyde, and product B is just the self-condensation product from the aldehyde.

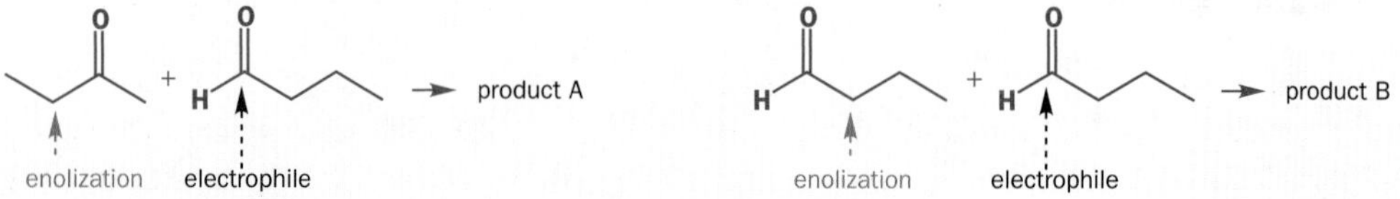

enolization   electrophile → product A

enolization   electrophile → product B

These kinetic lithium enolates are stable in THF at –78 °C for a short time but can be preserved at room temperature in the form of their silyl ethers.

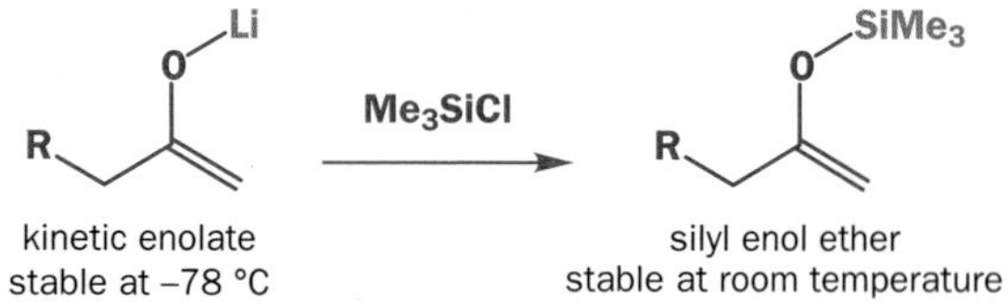

kinetic enolate
stable at –78 °C

silyl enol ether
stable at room temperature

Aldol reactions can be carried out with either the lithium enolate or the silyl enol ether. As an example we shall use the synthesis of a component of the flavour of ginger. The hotness of ginger comes from 'gingerol'—the 'pungent principle' of ginger. Gingerol is a 3-hydroxyketone, so we might consider using an aldol reaction to make it. We shall need the enol (or enolate) on the methyl side of an unsymmetrical ketone to react with a simple aldehyde (pentanal) as the electrophilic partner in the aldol reaction. Pentanal is an enolizable aldehyde, so we must stop it enolizing. The diagram summarizes the proposed aldol reaction.

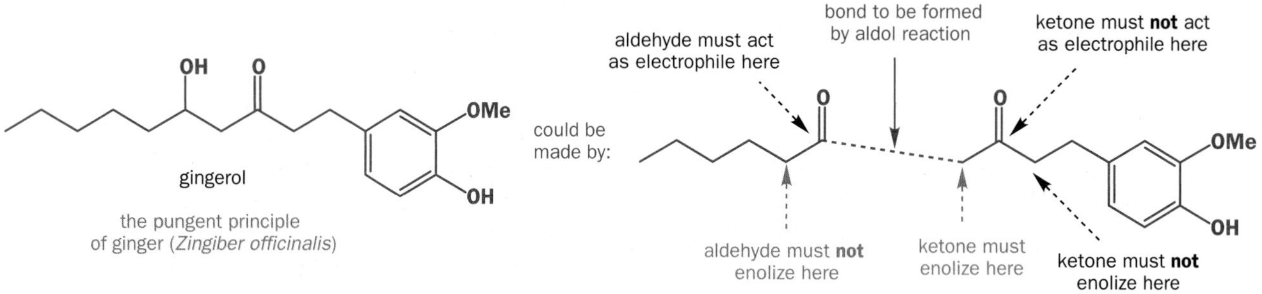

gingerol
the pungent principle
of ginger (*Zingiber officinalis*)

could be
made by:

bond to be formed
by aldol reaction

aldehyde must act
as electrophile here

ketone must **not** act
as electrophile here

aldehyde must **not**
enolize here

ketone must
enolize here

ketone must **not**
enolize here

We might consider using the lithium enolate or the silyl enol ether. As we need the kinetic enolate (the enolate formed on the less substituted side of the ketone), we shall be using the lithium enolate to make the silyl enol ether, so it would make sense to try that first.

There is another problem too. The ketone has a free OH group on the far side of the ring that will interfere with the reaction. We must protect that first as an ordinary silyl ether (not a silyl *enol* ether).

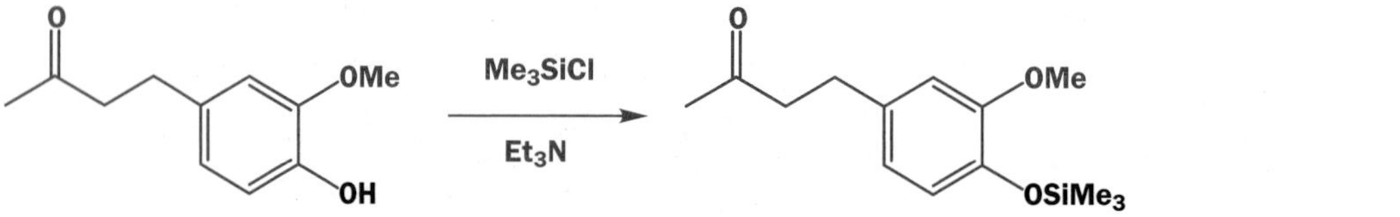

Now we can make the kinetic lithium enolate with a hindered lithium amide base. In fact, the one chosen here was even more hindered than LDA as it has two Me₃Si groups on the nitrogen atom.

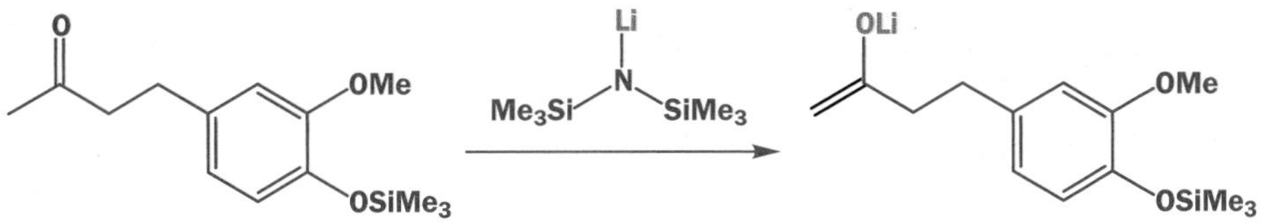

kinetic lithium enolate

Lithium hexamethyldisilazide

Lithium hexamethyldisilazide (LiHMDS) is a little more hindered than LDA and a little less basic. It is made by deprotonating hexamethyldisilazane with BuLi.

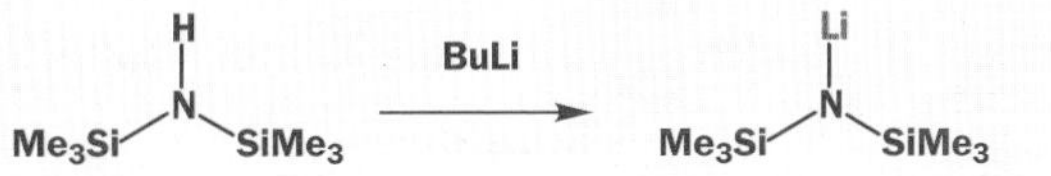

hexamethyldisilazane

lithium hexamethyldisilazide (LiHMDS)

An aldol reaction with this lithium enolate on pentanal was successful and the protecting group (the silyl ether) conveniently fell off during work-up to give gingerol itself. However, the yield was only 57%. When the silyl enol ether was used with $TiCl_4$ as the Lewis acid catalyst, the yield jumped to 92%. This is one of the many successful uses of this style of aldol reaction by Mukaiyama, the inventor of the method.

■ Teruaki Mukaiyama, of the Science University of Tokyo (and formerly of the Tokyo Institute of Technology and the University of Tokyo) is one of the foremost Japanese chemists, whose work has had a significant impact on the development of the aldol reaction and on other areas of organic synthesis.

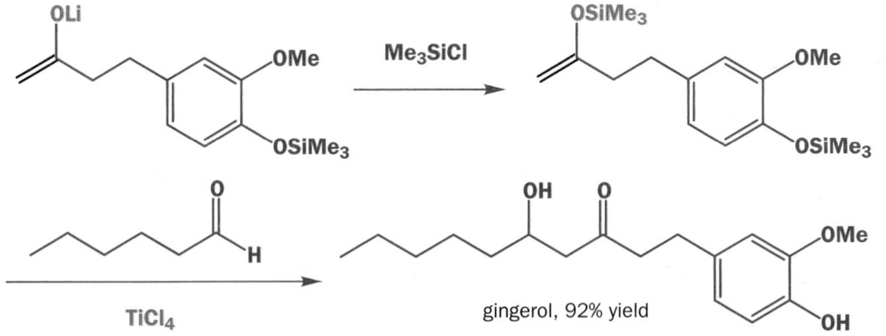

gingerol, 92% yield

Making the more substituted enolate equivalent: thermodynamic enolates

Being an alkene, an enol or enolate is more stable if it has more substituents. So the way to make the more substituted enolate equivalent is to make it under conditions where the two enolates can interconvert: equilibration will give the more stable. You have seen in Chapter 26 (p. 681) how the silyl enol ether on the more substituted side of a ketone can be made by treating the ketone with Me_3SiCl and a weak base, but these thermodynamic silyl enol ethers have been little used in aldol reactions. One successful example is the thermodynamic silyl enol ether of 1-phenylpropan-2-one: enolization on the conjugated side is overwhelmingly favoured thermodynamically. The aldol reaction with a 2-keto-aldehyde goes exclusively for the more reactive aldehyde group.

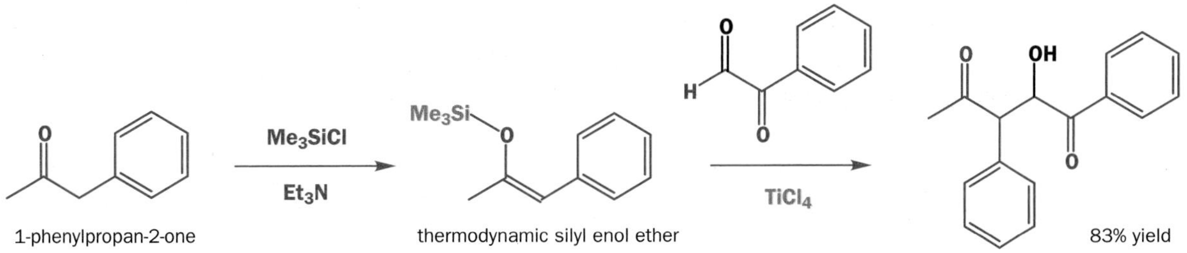

1-phenylpropan-2-one — thermodynamic silyl enol ether — 83% yield

● **Summary of the last four sections on specific enol equivalents.**

Useful enolates for the aldol reaction

Enolate type	Aldehyde	Ketone	Ester	Acid
lithium enolate	×	✓	✓	✓
silyl enol ether	✓	✓	✓	✓
enamine	✓	✓	×	×
aza-enolate	✓	✓	×	×
zinc enolate	×	×	✓	×

This concludes our general survey of the aldol reaction. Two special topics remain, both important, one dealing with an awkward and difficult reagent and one with a collection of aldol reactions that are particularly easy to do.

The Mannich reaction

At first sight formaldehyde (methanal, $CH_2=O$) seems the ideal electrophilic partner in a mixed aldol reaction. It cannot enolize. (Usually we are concerned with α hydrogen atoms in an aldehyde. Formaldehyde does not even have α carbon atoms.) And it is a super aldehyde. Aldehydes are more electrophilic than ketones because a hydrogen atom replaces one of the alkyl groups. Formaldehyde has two hydrogen atoms.

The trouble is that it is too reactive. It tends to react more than once and to give extra unwanted reactions as well. You might think that condensation between acetaldehyde and formaldehyde in base would be quite simple. The acetaldehyde alone can form an enolate, and this enolate will attack the more electrophilic carbonyl group, which is formaldehyde, like this.

crossed aldol reaction between acetaldehyde and formaldehyde

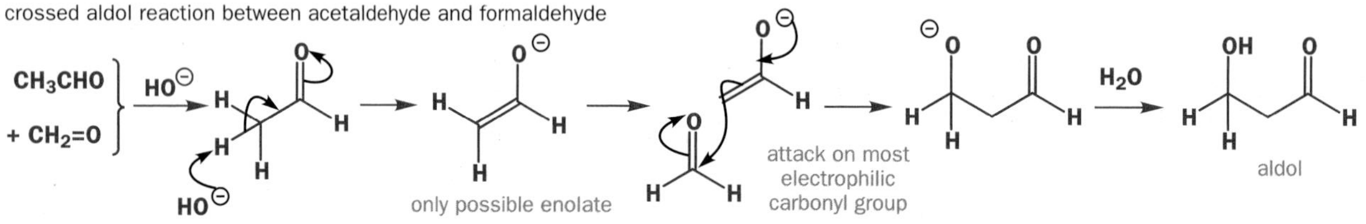

This aldol is formed all right but it is not the final product of the reaction because, with an electrophile as powerful as formaldehyde, a second and a third aldol follow swiftly on the heels of the first. Here is the mechanism of the second aldol.

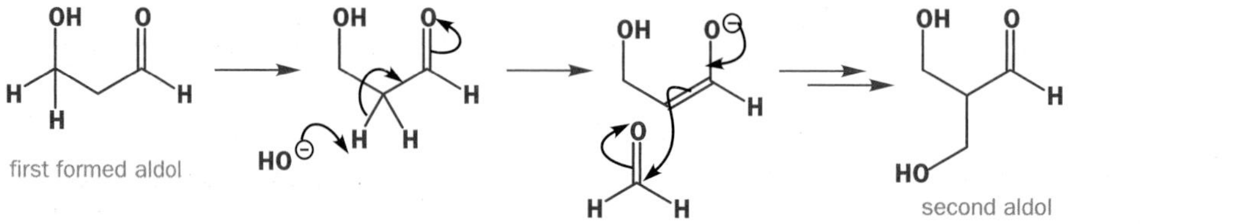

In each reaction the only possible enolate attacks another molecule of formaldehyde. By now you have got the idea so we simply draw the next enolate and the structure of the third aldol.

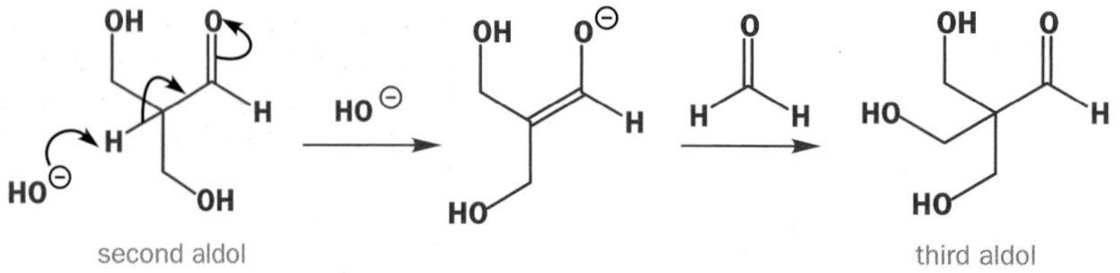

Even this is not all. A fourth molecule of formaldehyde reacts with hydroxide ion and then reduces the third aldol. This reduction is known as the **Cannizzaro reaction**, and is described in the box. The final product is the highly symmetrical 'pentaerythritol', $C(CH_2OH)_4$, with four CH_2OH groups joined in a tetrahedral array about the same carbon atom.

reduction by the Cannizzaro reaction

dianion formed by attack of
hydroxide on formaldehyde:
see box for details

+

formate ion

pentaerythritol

> ■ Pentaerythritol is a useful industrial product in, for example, the cross-linking of polymers: see Chapter 52.

The overall reaction uses four molecules of formaldehyde and can give a high yield (typically 80% with NaOH but as much as 90% with MgO) of the product.

The Cannizzaro reaction

As you know, aldehydes are generally at least partly hydrated in water. Hydration is catalysed by base, and we can represent the hydration step in base like this. The hydration product is an anion but, if the base is sufficiently strong (or concentrated) and as long as the aldehyde cannot be enolized, at least some will be present as a dianion.

hydration in base
(hydroxide as
nucleophile)

deprotonation of
hydrate (hydroxide
as base)

aldehyde   hydrate anion   hydrate dianion

The dianion is very unstable, and one way in which it can become much more stable is by behaving like a tetrahedral intermediate. Which is the best leaving group? Out of a choice of O^{2-}, R^-, and H^-, it's H^- that (if reluctantly) has to go. Hydride is, of course, too unstable to be released into solution but, if there is a suitable electrophile at hand (another molecule of aldehyde, for example), it is transferred to the electrophilic centre in a mechanism that bears some resemblance to a borohydride reduction.

hydride is least bad
leaving group

carboxylate anion

alcohol

compare...

The dianion becomes a much more stable carboxylate monoanion, and a second molecule of aldehyde has been reduced to an alcohol. This is the Cannizzaro reaction: in this case it takes the form of a disproportionation of two molecules of aldehyde to one of carboxylate and one of alcohol.

HO^- as nucleophile

HO^- as base

reduces third aldol product
from acetaldehyde +
formaldehyde to give
pentaerythritol

In the pentaerythritol case, the dianion reducing agent is formed from formaldehyde: first hydroxide attacks it as a nucleophile, then as a base. The dianion transfers 'hydride' to a different aldehyde, the third aldol product, to make pentaerythritol. The Cannizzaro reaction waits till this point because only after the third aldol does the aldehyde lose its ability to enolize, and the reaction works **only with unenolizable aldehydes**.

If you want a more controlled reaction with addition of formaldehyde to an aldehyde or ketone without the reduction step, you can sometimes succeed with a weaker base such as potassium carbonate. Typically in these reactions *all* the enolizable hydrogen atoms (green) are replaced by molecules of formaldehyde (black).

> ▶
> Formaldehyde is not available as a pure monomer because it forms trimers and tetramers in the pure state (Chapter 52). The aqueous solution 'formalin' used to preserve biological specimens is available—it is 37% formaldehyde and mostly consists of the hydrate $CH_2(OH)_2$; see Chapter 6. A pure dry polymer 'paraformaldehyde' is also available and was mentioned in Chapter 9. Neither of these is particularly useful in aldol reactions. The aqueous solution is used in the Mannich reaction that we describe shortly. It is possible to make the short-lived monomer and capture it with a lithium enolate, but this is not trivial experimentally.

$CH_2=O$

K_2CO_3

90% yield

But a more general solution is to use the **Mannich reaction**. A typical example is shown here: the reaction involves an enolizable aldehyde or ketone (here we use cyclohexanone), a secondary amine (here dimethylamine), formaldehyde as its aqueous solution, and catalytic HCl. The product is an amino-ketone from the addition of one molecule each of formaldehyde and the amine to the ketone.

the Mannich reaction

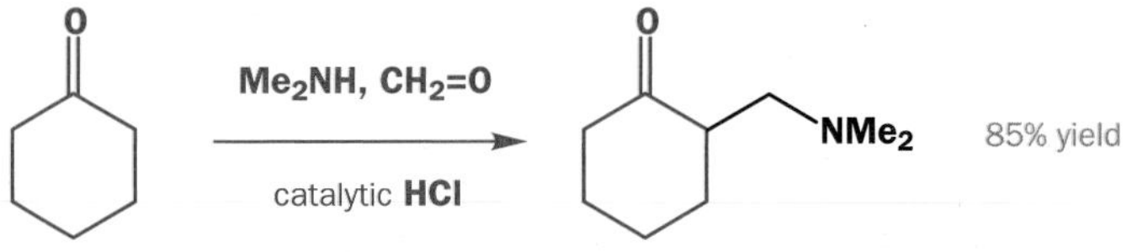

Me₂NH, CH₂=O

catalytic **HCl**

85% yield

The mechanism involves the preliminary formation of an imine salt from the amine and formaldehyde. The amine is nucleophilic and attacks the more electrophilic of the two carbonyl compounds available. That is, of course, formaldehyde. No acid is needed for this addition step, but acid-catalysed dehydration of the addition product gives the imine salt. In the normal Mannich reaction, this is just an intermediate but it is quite stable and the corresponding iodide is sold as 'Eschenmoser's salt' for use in Mannich reactions.

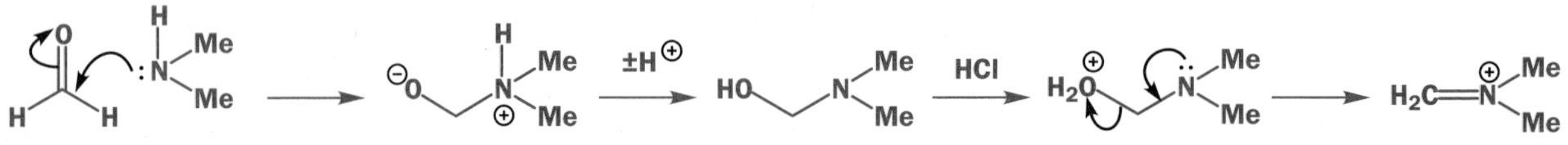

nucleophilic attack on more
electrophilic C=O group

imine salt

The electrophilic salt can now add to the enol (we are in acid solution) of the ketone to give the product of the reaction, an amine sometimes called a **Mannich base**.

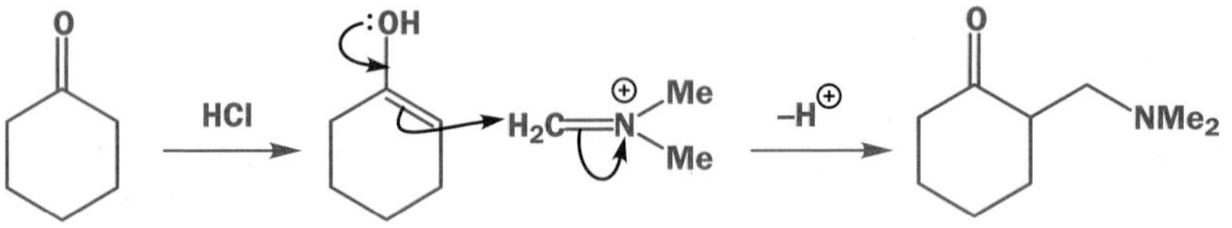

By using this reaction, you can add one molecule of formaldehyde—one only—to carbonyl compounds. You might, of course, reasonably object that the product is not actually an aldol product at all—indeed, if you wanted the aldol product, the Mannich reaction would be of little use to you. It nevertheless remains a very important reaction. First of all, it is a simple way to make amino-ketones and many drug molecules belong to this class. Secondly, the Mannich products can be converted to enones. We will discuss this reaction next.

The most reliable method for making the enone is to alkylate the Mannich base with MeI and then treat the ammonium salt with base. Enolate ion formation leads to an E1cB reaction rather like the dehydration of aldols, but with a better leaving group.

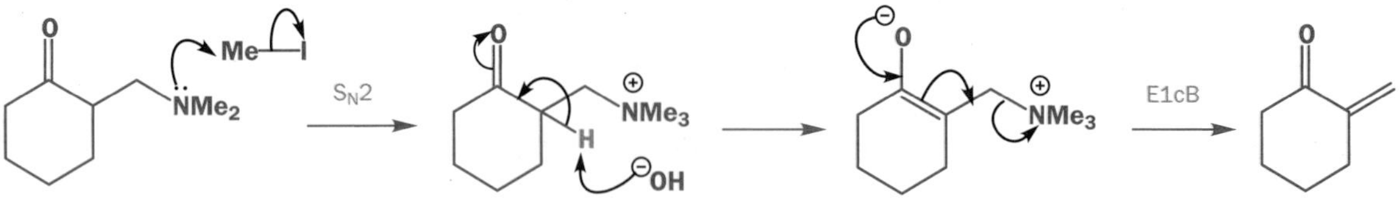

1. alkylate amine to give ammonium salt   2. treat with base: E1cB elimination gives enone

Enones like this, with two hydrogen atoms at the end of the double bond, are called **exo-methylene compounds**; they are very reactive, and cannot easily be made or stored. They certainly cannot be made by aldol reactions with formaldehyde alone as we have seen. The solution is to make the Mannich base, store that, and then to alkylate and eliminate only when the enone is needed. We shall see how useful this is in the Michael reaction in Chapter 29.

If the enone is wanted, the secondary amine does not end up in the molecule so the more convenient (less volatile and less smelly) cyclic amines, pyrrolidine and piperidine, are often used. Enones with monosubstituted double bonds can be made in this way.

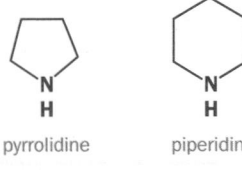

pyrrolidine   piperidine

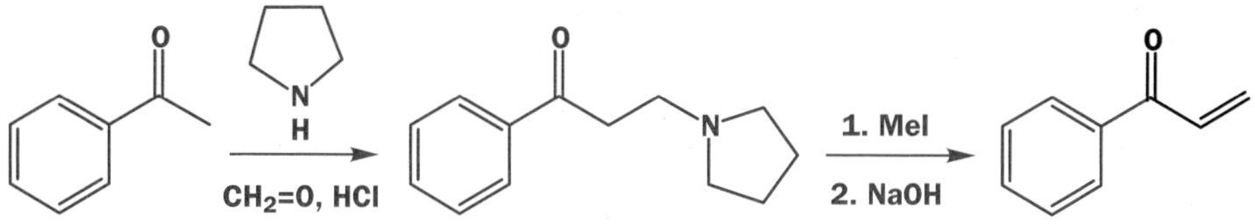

Intramolecular aldol reactions

Now for something easy. When an aldol reaction can form a five- or six-membered ring, you need no longer worry about specific enols or anything like that. Equilibrium methods with weak acids or bases are quite enough to give the cyclic product by an intramolecular aldol reaction because intramolecular reactions are faster than intermolecular ones. We shall illustrate intramolecular reactions by looking at the cyclization of a series of diketones of increasing complexity starting with one that can form four equivalent enols: cyclodeca-1,6-dione.

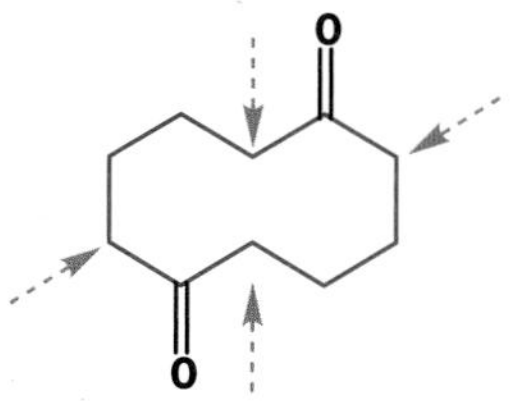

It doesn't matter where enolization occurs, because the same enol is formed. And once the enol is formed, there is only one thing it can reasonably do: attack the other ketone to form a stable five-membered ring. It also gives a reasonably stable seven-membered ring, but that is by the way. In weak acid or base, only a small proportion of carbonyl groups will be enolized, so the chance of two being in the same molecule is very low. No intermolecular condensation is found and the yield of the bicyclic enone from the intramolecular reaction is almost 100% (96% with Na_2CO_3).

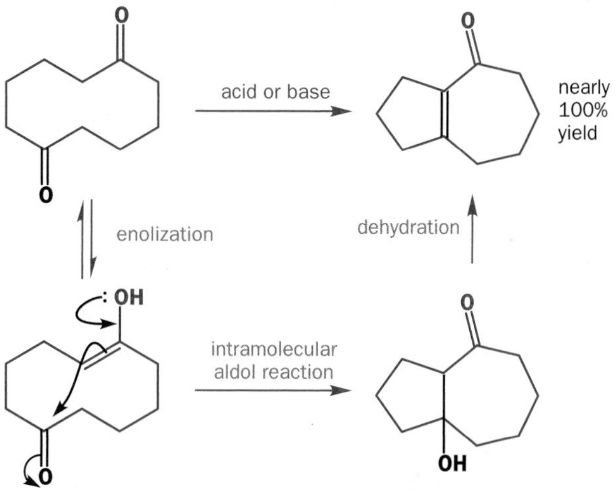

cyclodeca-1,6-dione: four identical positions for enolization (- - - ▶)

■ Ring size and stability were discussed in Chapter 18.

This may look like a long stretch for the enol to reach across the ten-membered ring to reach the other ketone, but the conformational drawing in the margin shows just how close they can be. You should compare this conformation with that of a decalin (Chapter 18).

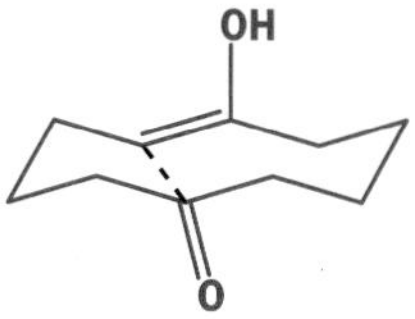

The key point to remember with intramolecular aldols is this.

> ● Intramolecular reactions giving five- or six-membered rings are preferred to those giving strained three- or four-membered rings on the one hand or medium rings (eight- to thirteen-membered) on the other.

Acid-catalysed cyclization of the symmetrical diketone nona-2,8-dione could give two enols.

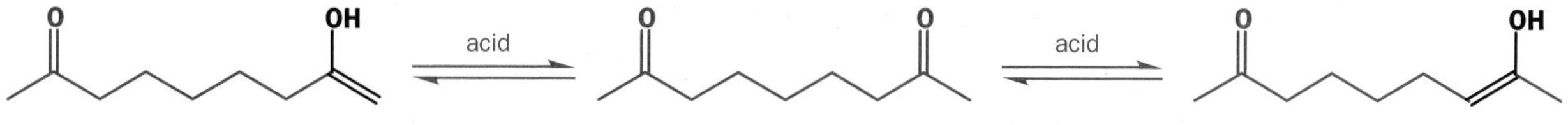

nona-2,8-dione

One enol can cyclize through an eight-membered cyclic transition state and the other through a six-membered ring. In each case the product would first be formed as an aldol but would dehydrate to the cyclic enone having the same ring size as the transition state. In practice, only the less strained six-membered ring is formed and the enone can be isolated in 85% yield.

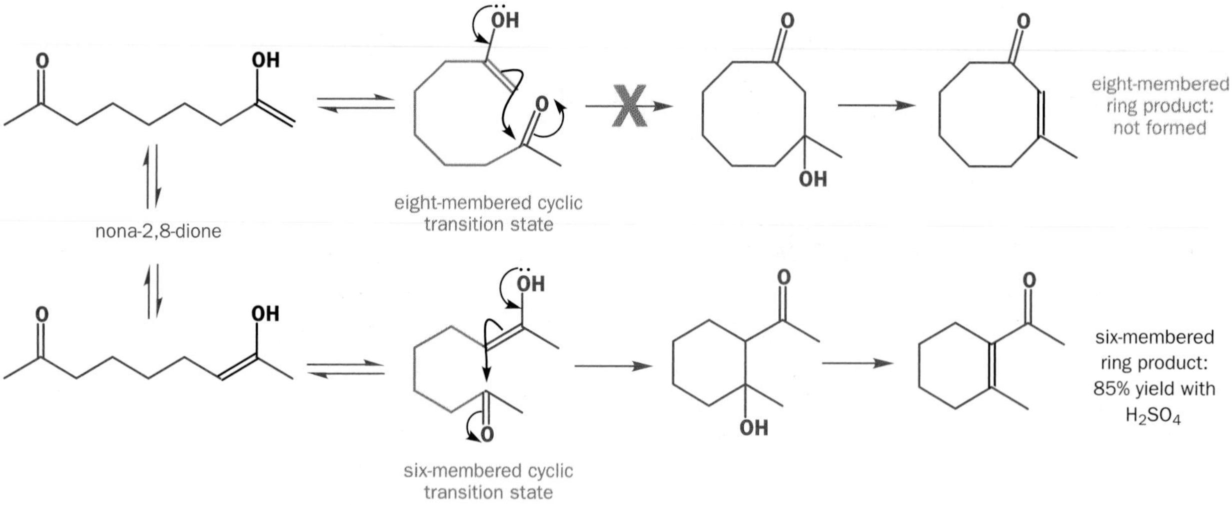

nona-2,8-dione

eight-membered cyclic
transition state

eight-membered
ring product:
not formed

six-membered cyclic
transition state

six-membered
ring product:
85% yield with
H_2SO_4

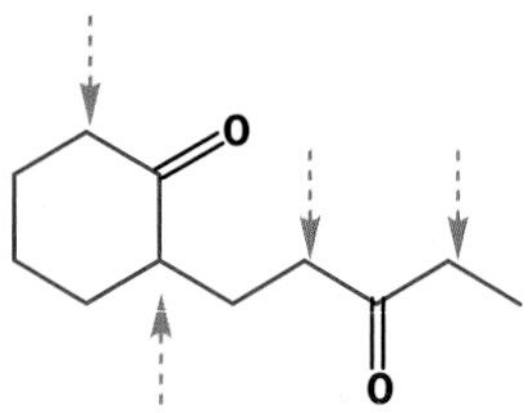

four different positions where
enolization is possible

Most diketones lack symmetry, and will potentially have four different sites for enolization. Consider what might happen when this diketone is treated with KOH. There are four different places where an enolate anion might be formed as there are four different α carbon atoms. There are also two different electrophilic carbonyl groups so that there are many possibilities for inter- and intramolecular condensation. Yet only one product is formed, in 90% yield.

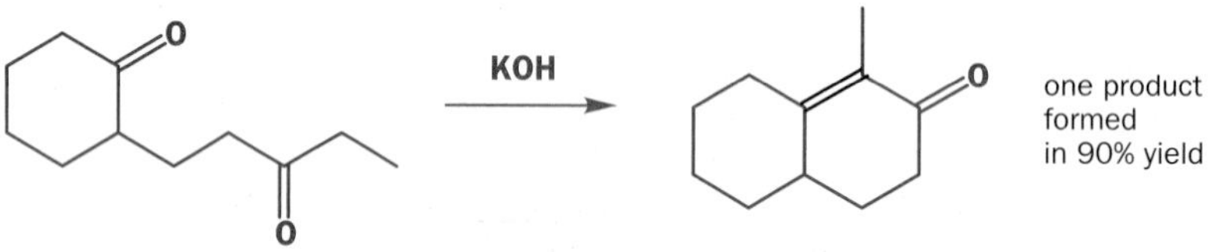

KOH

one product
formed
in 90% yield

We can deduce the mechanism of the reaction simply from the structure of the product by working backwards. The double bond is formed from an aldol whose structure we can predict and hence we can see which enolate anion was formed and which ketone acted as the electrophilic partner.

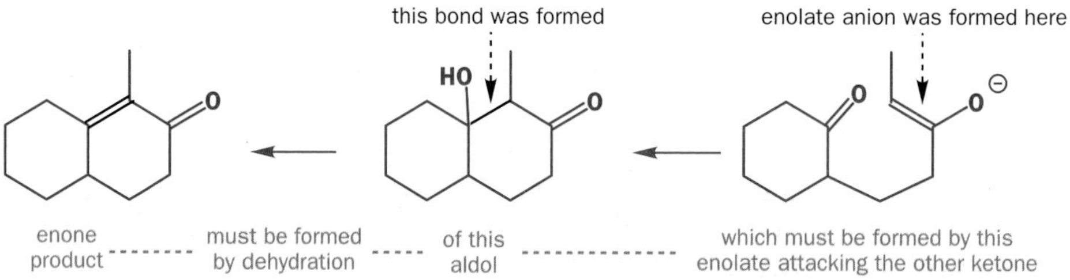

this bond was formed

enolate anion was formed here

enone
product

must be formed
by dehydration

of this
aldol

which must be formed by this
enolate attacking the other ketone

Must we argue that this one enolate is more easily formed than the other three? No, of course not. There is little difference between all four enolates and almost no difference between the three enolates from CH_2 groups. We *can* argue that this is the only aldol reaction that leads to a stable conjugated enone in a stable six-membered ring. This must be the mechanism; protonation and dehydration follow as usual.

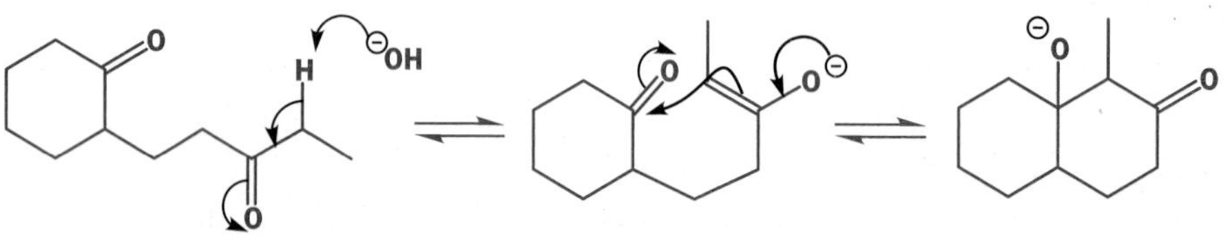

Now try one of the alternatives in which the same ketone forms an enolate on the other side.

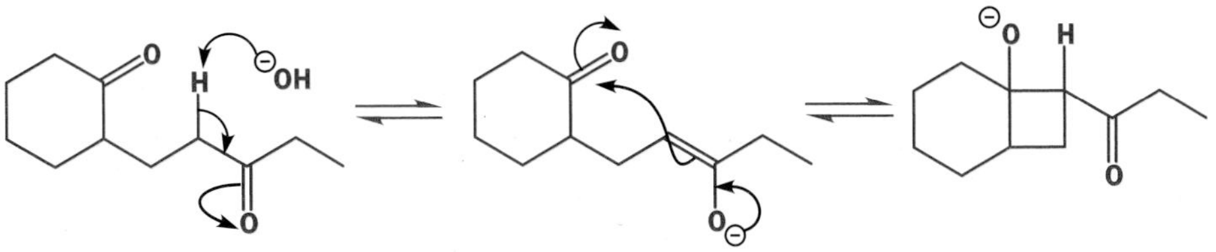

This reaction gives an unstable four-membered ring that would revert to the enolate. Providing the reaction is done under equilibrating conditions, the whole process would go into reverse back to the original diketone and the observed (six-membered ring) cyclization would eventually predominate. There is one alternative cyclization to give a six-membered ring and this does not occur for an interesting reason. Here is the reaction.

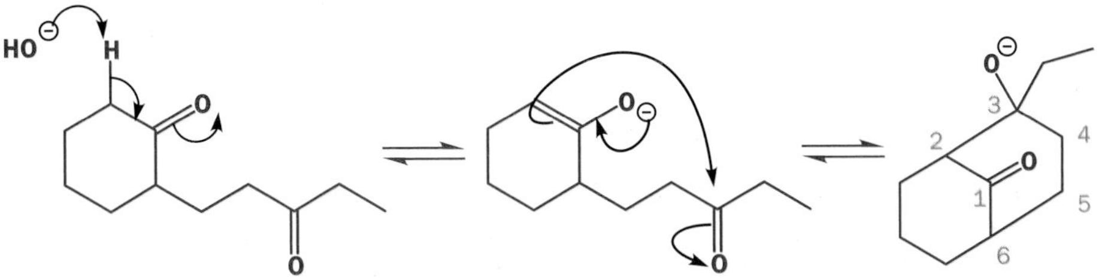

The new ring is a six-membered ring and we have numbered it to convince you. It is, of course, a rather strained bridged compound, but the key point is that dehydration is impossible. No enolate can form at the bridgehead, because bridgehead carbons cannot be planar (see Chapter 19) and the enone product cannot exist for the same reason: the carbons marked (•) in the brown structure would all have to lie in the same plane. The aldol has a perfectly acceptable conformation but that elimination is impossible. The aldol product remains in equilibrium with the alternative aldol products, but only one elimination is possible—and that is irreversible, so eventually all the material ends up as the one enone.

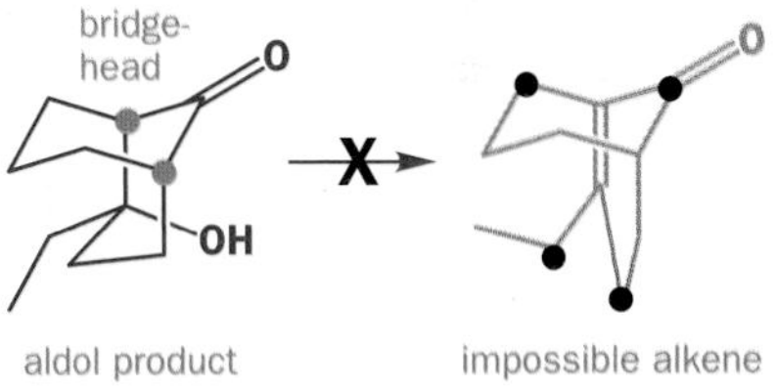

bridge-
head

aldol product

impossible alkene

Even without the constraint of avoiding a bridgehead alkene, some completely unsymmetrical diketones give single products in high yield. Here are two related examples with similar structures.

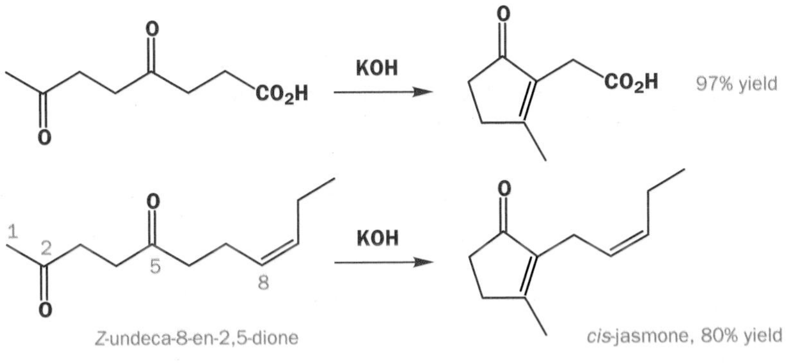

97% yield

Z-undeca-8-en-2,5-dione

cis-jasmone, 80% yield

The first of these is impressive for the high yield and the lack of interference by the carboxylic acid group. The second is important because the product is the perfumery compound *cis*-jasmone found naturally in jasmine flowers, and is formed in good yield with no change in the position or geometry of the *Z* double bond.

In these reactions there is some selectivity between two possible five-membered rings, both of which can easily dehydrate to give an enone. These are the alternatives, using a general structure where R might be CH_2CO_2H in the first or the unsaturated chain in the second example.

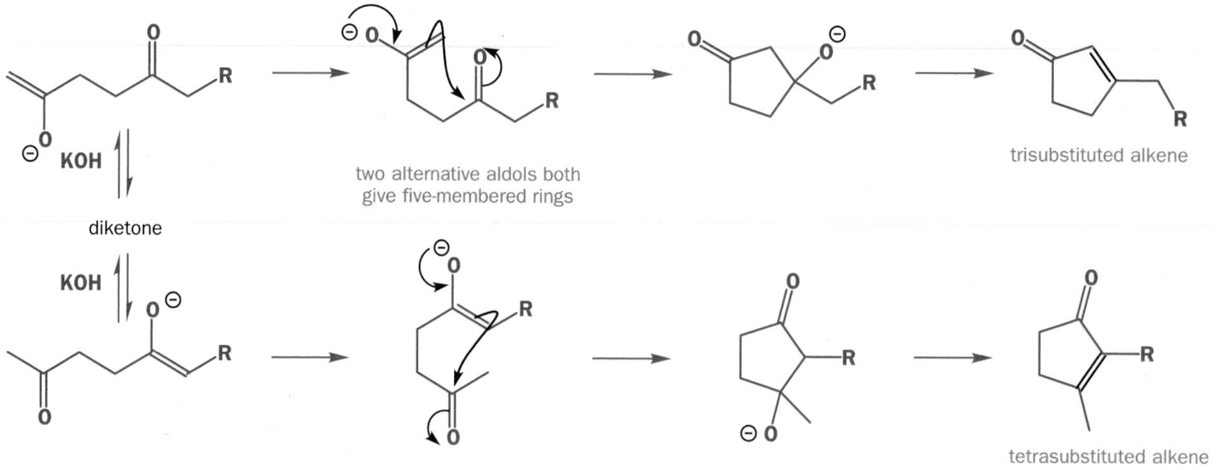

two alternative aldols both
give five-membered rings

trisubstituted alkene

diketone

tetrasubstituted alkene

So far it is very difficult to see much difference between the two routes. Indeed, we might have argued that the upper route is better because enolization is faster at a methyl group. But this is wrong because the reaction is not under kinetic but rather under thermodynamic control. The two products differ by the number of substituents on the double bond, and the more substituents there are on a double bond, the more stable it is. This factor is discussed in Chapter 19. It is the only difference between these two products and it controls the reaction very effectively.

To conclude: a summary of equilibrium and directed aldol methods

As we leave this chapter, it is important to make sure that you understand the two different approaches to controlled aldol reactions that we have been considering. The two methods ensure in their different ways that only one carbonyl group gives only one enol or enolate as the nucleophilic partner in the aldol reaction while only one carbonyl compound acts as the electrophilic partner.

● **Equilibrium control**

In the equilibrium method, the carbonyl compound(s) must be treated with weak, usually aqueous or alcoholic, acid or base and allowed to equilibrate with all possible enols or enolates. Either only one product is possible (due to symmetry or blocking of α positions) or some thermodynamic factor (such as the formation of a stable conjugated enone) ensures that the reaction goes down one preferred route.

In the equilibrium method, 'weak' acid or base means too weak to ensure complete conversion to enol or enolate. The method works only if enol and carbonyl compound are in equilibrium. Typical examples are shown in the table.

Types of aldol reaction under thermodynamic control

Type of reaction	Typical conditions	Example
self-condensation of aldehydes	2% NaOH aqueous ethanol	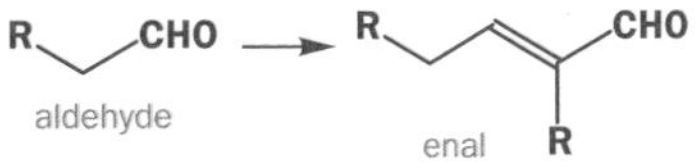 $R\text{-CH}_2\text{-CHO}$ (aldehyde) → enal
self-condensation of ketones	HCl, Al(OR)$_3$, NaOH, or KOH	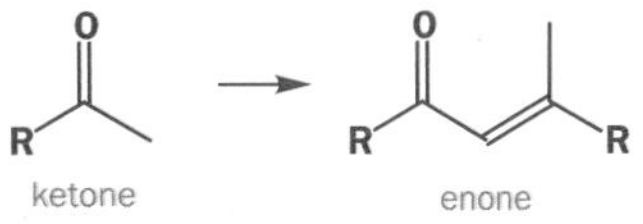 ketone → enone
cross-condensations of an enolizable ketone and a non-enolizable aldehyde	NaOH, KOH, Na$_2$CO$_3$, HCl, or H$_2$SO$_4$	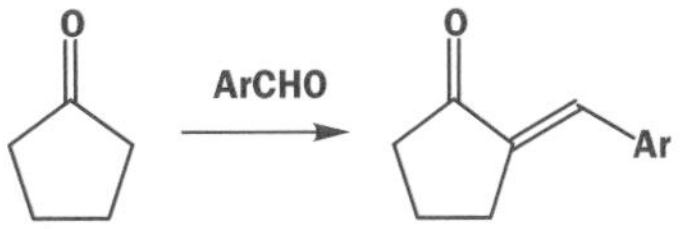 cyclopentanone $\xrightarrow{\text{ArCHO}}$ benzylidene cyclopentanone (=CH–Ar)
cross-condensations of aryl methyl ketones and non-enolizable aldehydes	dilute HCl or NaOH	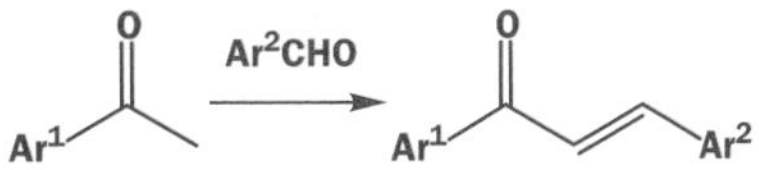 $Ar^1\text{COCH}_3 \xrightarrow{Ar^2\text{CHO}} Ar^1\text{CO–CH=CH–}Ar^2$
cyclization reactions	2% NaOH aqueous ethanol, or HCl, or H$_2$SO$_4$	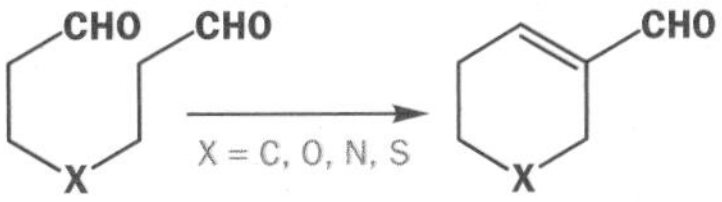 dialdehyde (X = C, O, N, S) → cyclic enal

Similar conditions are used for condensations where 1,3-dicarbonyl compounds provide the enol partner. The differences are that now the weak acid or base is strong enough to convert the 1,3-dicarbonyl compound essentially completely into enol or enolate, and that enolate (enolization between the two carbonyl groups) is highly favoured over all others. In a way these are intermediate between the two kinds of control, though they really belong to the directed aldol category.

Aldol reactions with highly enolizable compounds

1,3-Dicarbonyl compound	Conditions	Example
malonic acid CH$_2$(CO$_2$H)$_2$	piperidine, DMSO	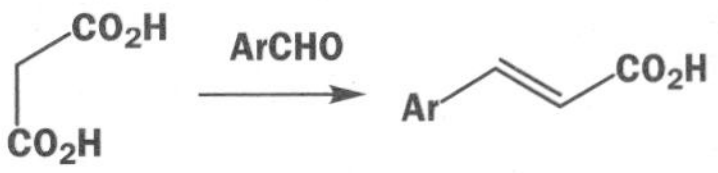 $\text{CH}_2(\text{CO}_2\text{H})_2 \xrightarrow{Ar\text{CHO}} Ar\text{–CH=CH–CO}_2\text{H}$
malonic esters CH$_2$(CO$_2$Et)$_2$	NH$_4^+$AcO$^-$	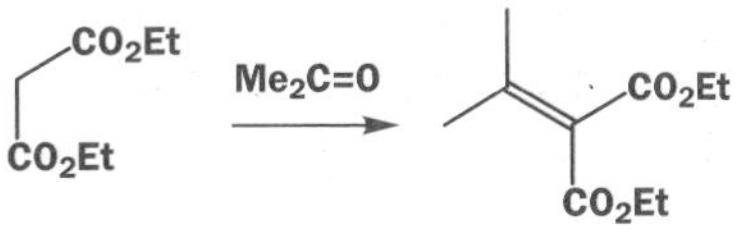 $\text{CH}_2(\text{CO}_2\text{Et})_2 \xrightarrow{\text{Me}_2\text{C=O}} \text{Me}_2\text{C=C(CO}_2\text{Et})_2$
acetoacetates CH$_3$CO·CH$_2$CO$_2$Et	piperidine, EtOH, room temperature	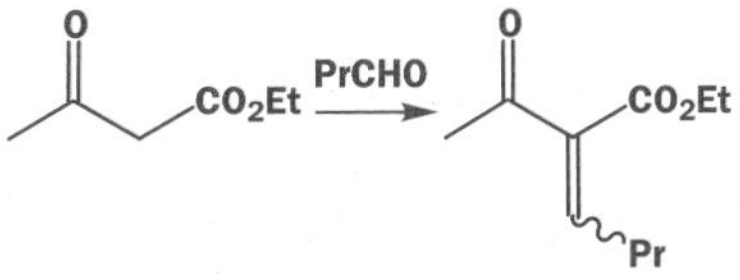 $\text{CH}_3\text{COCH}_2\text{CO}_2\text{Et} \xrightarrow{Pr\text{CHO}} \text{CH}_3\text{CO–C(=CH–Pr)–CO}_2\text{Et}$
nitro compounds[a] RCH$_2$NO$_2$	NaOH, H$_2$O	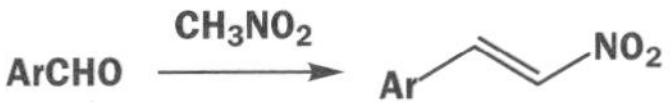 $\xrightarrow[Ar\text{CHO}]{\text{CH}_3\text{NO}_2} Ar\text{–CH=CH–NO}_2$
Wittig reagents[a] 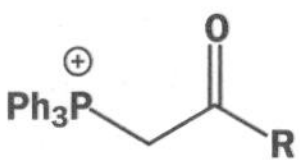 Ph$_3$P$^\oplus$–CH$_2$–CO–R	NaOMe, MeOH	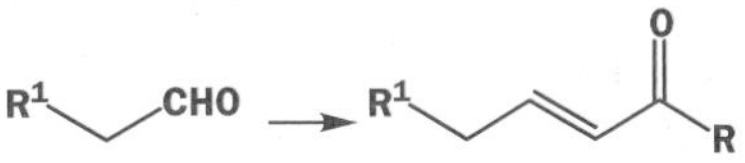 $R^1\text{–CH}_2\text{–CHO} \rightarrow R^1\text{–CH=CH–CH=CH–CO–R}$

[a] These are not, of course, 1,3-dicarbonyl compounds but they have pK_as of about 10–12 and do form enolates with weak bases.

● Directed aldol reactions

In the directed aldol reaction, one component is first converted into a specific enol equivalent and *only then* combined with the electrophilic partner.

These are the most versatile methods and can be used to make essentially any aldol or any conjugated unsaturated carbonyl compound. The disadvantages are that an extra step is inevitably introduced (the making of the specific enol equivalent), that strong bases or powerful Lewis acids must be used, and that strictly anhydrous conditions in organic solvents are usually required.

The specific enol equivalents are used only when necessary. Check first whether you might be able to get away with an equilibrium method before planning a directed aldol reaction. Directed aldol reactions are among the greatest achievements of modern organic chemistry, but simpler methods still have their place.

The table gives some details of the conditions used for directed aldol reactions. You should refer to the table on p. 712 to see which specific enol equivalents are appropriate to which types of carbonyl compounds.

Specific enol equivalent	Conditions	Example
lithium enolate	1. LDA, THF, −78 °C, 2. aldehyde, 3. NH₄Cl, H₂O	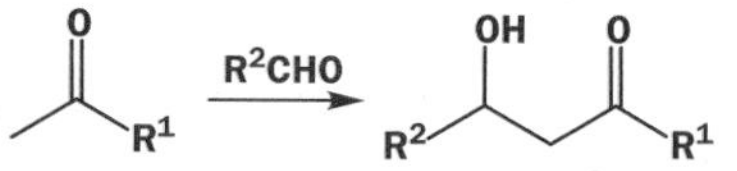
silyl enol ether	TiCl₄, CH₂Cl₂, −78 °C, 1 hour, under argon	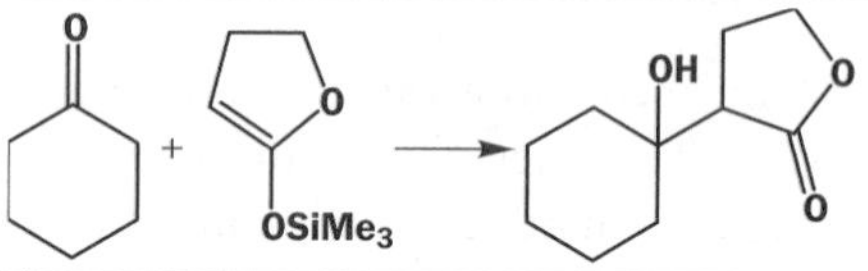
enamine	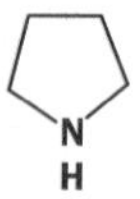 (pyrrolidine) heat	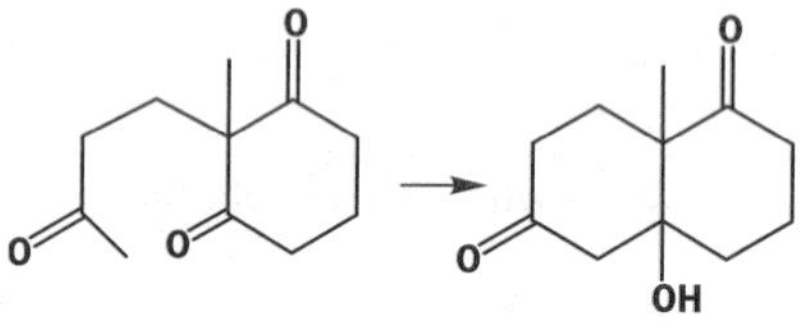
aza-enolate	1. RNH₂, 2. LDA, 3. ketone, 4. dilute H₂SO₄	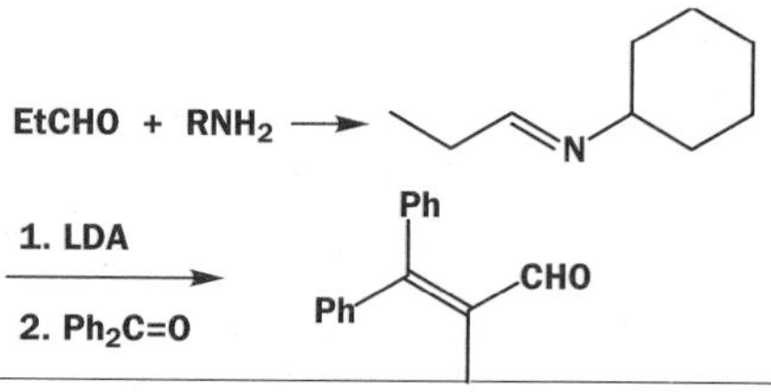
zinc enolate (Reformatsky)	1. Zn, 2. aldehyde or ketone	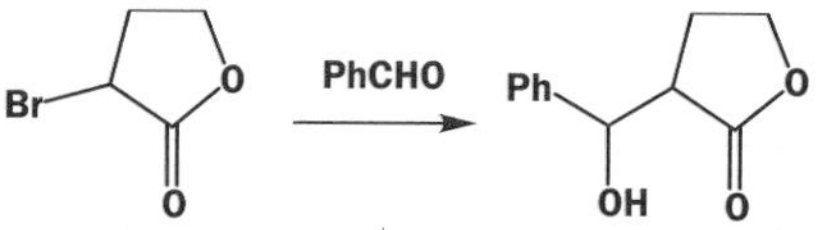

We have spent some considerable time and effort in understanding the aldol reaction simply because it is one of the most important reactions in organic chemistry. In the next chapter you will see how these ideas can be extended with almost no addition of principles to the acylation of enolates—the reaction of enols, enolates, and specific enol equivalents with acid chlorides and esters. We hope that you will see that the ideas introduced in this chapter find immediate application in the next.

Problems

1. Propose mechanisms for the 'aldol' and dehydration steps in the termite defence compound synthesis presented in the chapter.

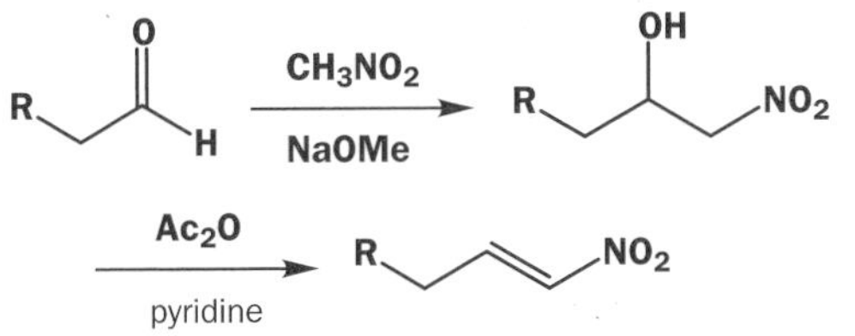

2. The aldehyde and ketone below are self-condensed with aqueous NaOH so that an unsaturated carbonyl compound is the product. Give a structure for each product and explain why you think this product is formed.

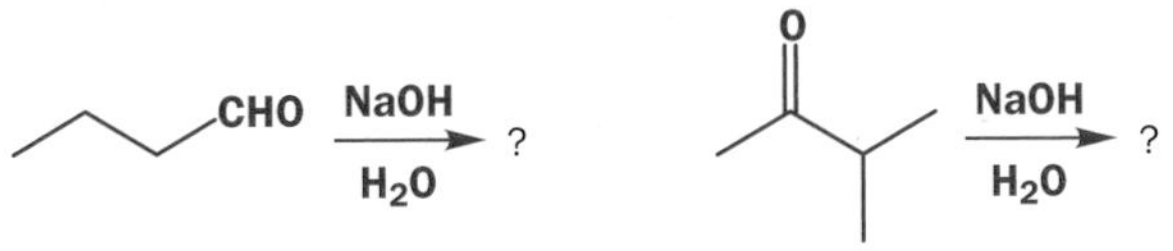

3. How would you synthesize the following compounds?

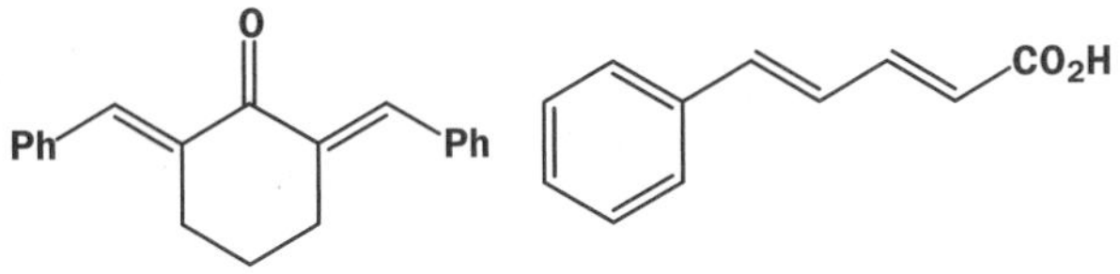

4. How would you use a silyl enol ether to make this aldol product? Why is it necessary to use this particular intermediate? What would the products be if the two carbonyl compounds were simply mixed and treated with base?

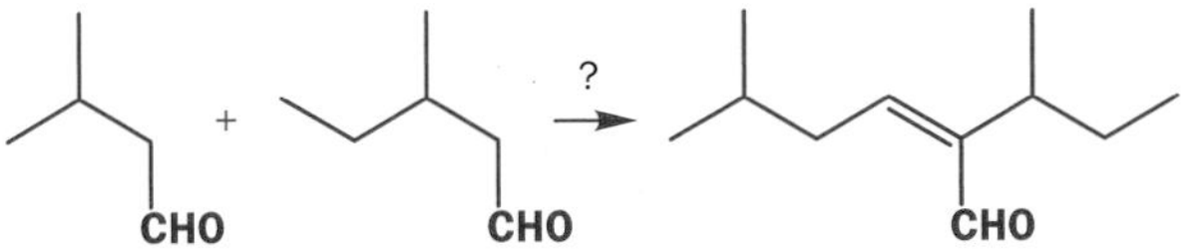

5. In what way does this reaction resemble an aldol reaction? How could the same product be made without using phosphorus chemistry? Comment on the choice of base.

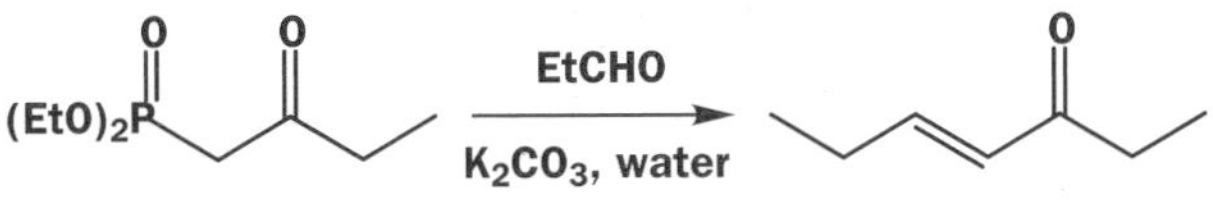

6. Suggest a mechanism for this attempted aldol reaction. How could the aldol product actually be made?

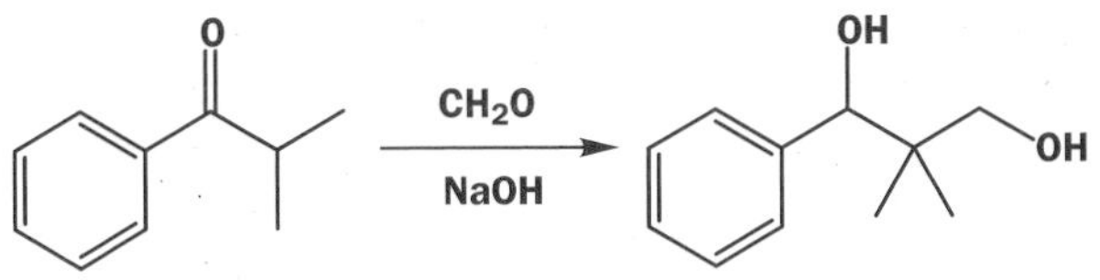

7. What are the structures of the intermediates and the mechanisms of the reactions leading to this simple cyclohexenone?

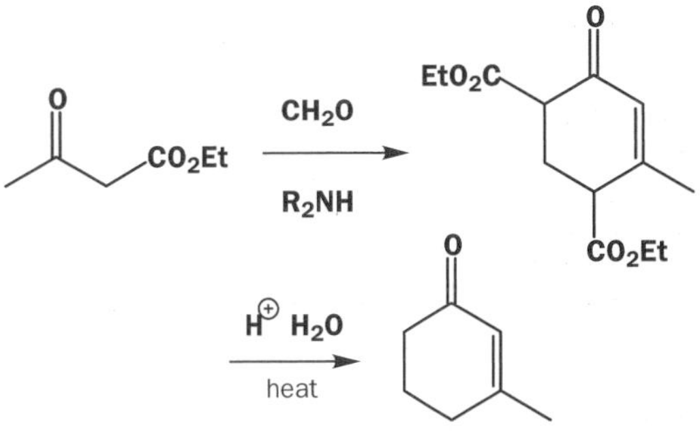

8. How would you convert the product of that last reaction into these two products?

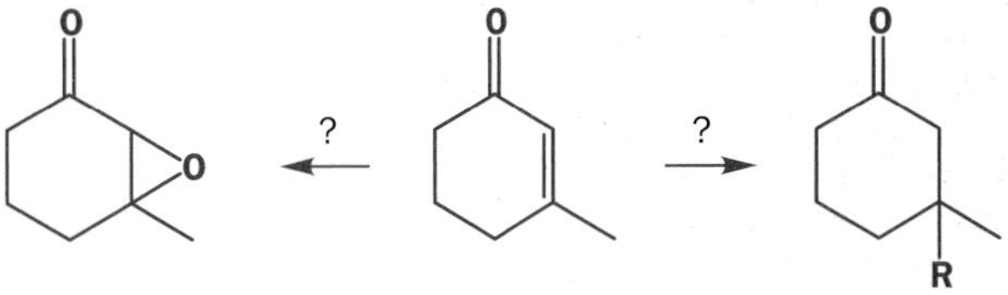

9. Comment on the selectivity shown in these two cyclizations.

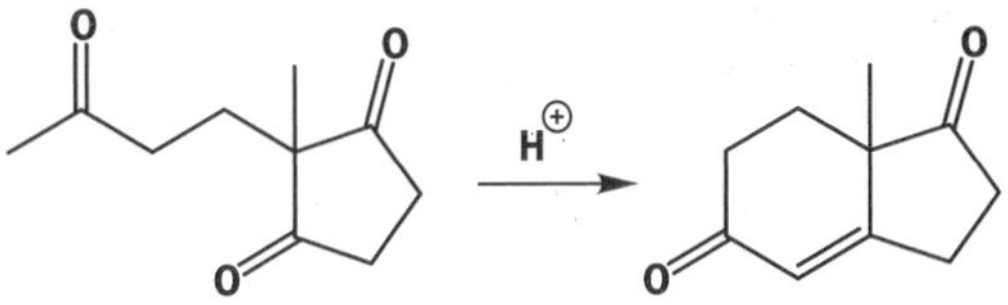

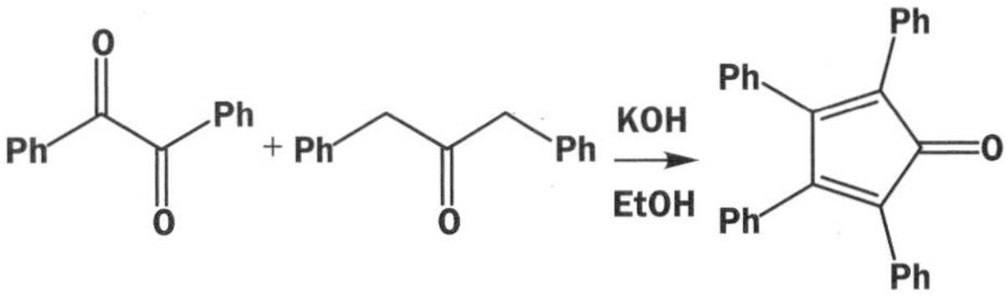

10. Using the Mannich reaction as a guide, suggest a mechanism for this reaction.

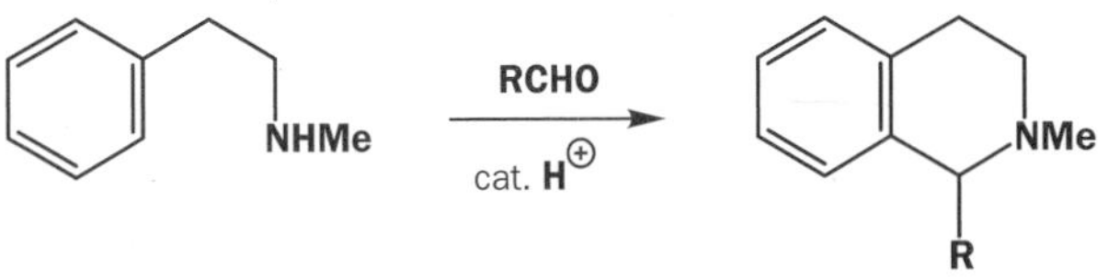

11. Suggest mechanisms for this reaction. One of the by-products is carbon dioxide.

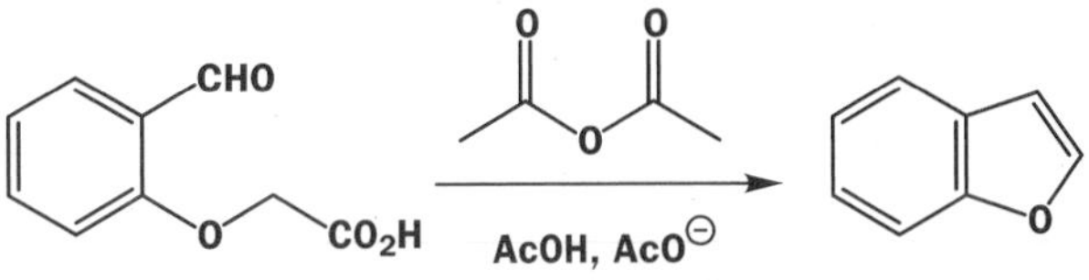

12. Treatment of this keto-aldehyde with KOH gives a compound $C_7H_{10}O$ with the spectroscopic data shown. What is its structure and how is it formed? You should, of course, assign the NMR spectrum and give a mechanism for the reaction.

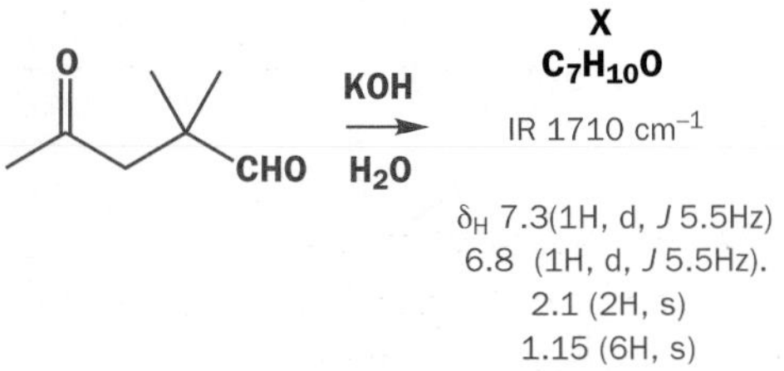
O
CHO
KOH
H2O
X
C7H10O
IR 1710 cm−1
δH 7.3(1H, d, J 5.5Hz)
6.8 (1H, d, J 5.5Hz).
2.1 (2H, s)
1.15 (6H, s)

13. Predict which enone product would be formed in this intramolecular aldol reaction.

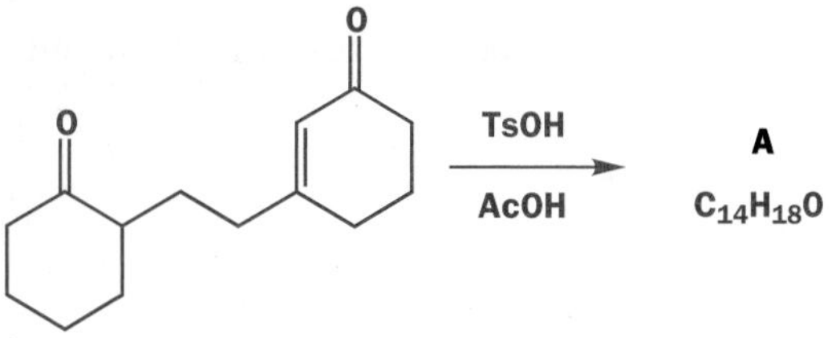
O
O
TsOH
AcOH
A
C14H18O

14. The unstable liquid diketone 'biacetyl' deposits crystals of a dimer slowly on standing or more quickly with traces of base. On longer standing the solution deposits crystals of a trimer. Suggest mechanisms for the formation of the dimer and the trimer. Why are they more stable than the monomer?

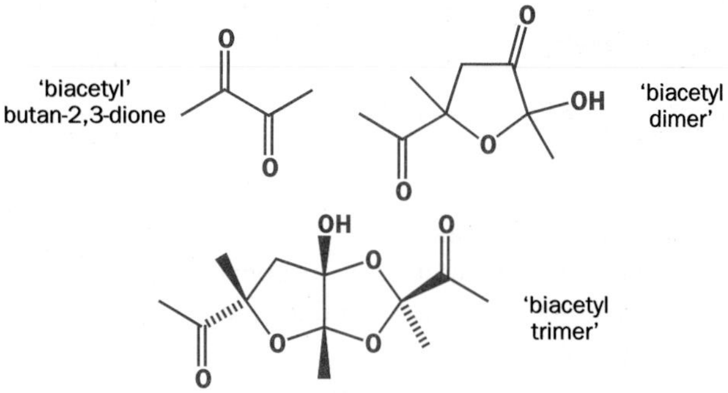
'biacetyl'
butan-2,3-dione
O
O
O
O
OH
O
'biacetyl
dimer'
OH
O
O
O
O
O
'biacetyl
trimer'

Acylation at carbon

28

Connections

Building on:

- **Enols and enolates ch21**
- **Alkylation of enolates ch26**
- **The aldol reaction ch27**
- **Nucleophilic substitution at C=O ch12**

Arriving at:

- **How esters react with enolates: the Claisen condensation**
- **How to acylate the enolates of esters and ketones**
- **How to get *C*-acylation and avoid *O*-acylation**
- **How to make cyclic ketones by intramolecular acylation**
- **Enamines in acylation reactions**
- **Modelling acylation on nature**

Looking forward to:

- **Michael additions of enolates ch29**
- **Retrosynthetic analysis ch30**
- **Biological chemistry ch49–ch51**

Introduction: the Claisen ester condensation compared to the aldol reaction

We began the last chapter with the treatment of acetaldehyde with base. This led initially to the formation of an enolate anion and then to the aldol reaction. We are going to start this chapter with the treatment of ethyl acetate with base. To start with, there is hardly any difference. We shall use ethoxide as base rather than hydroxide as hydroxide would hydrolyse the ester, but otherwise the first steps are very similar. Here they are, one above the other.

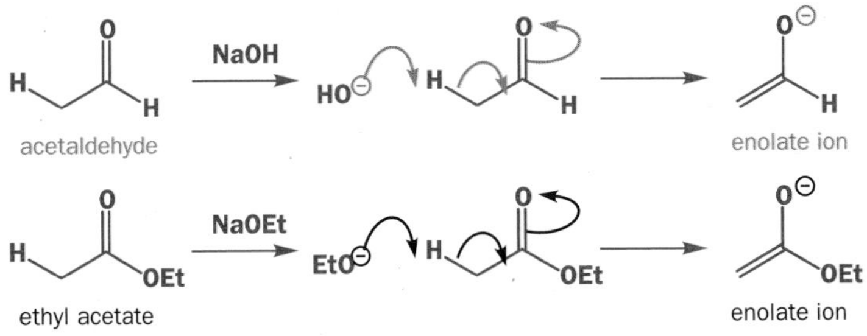

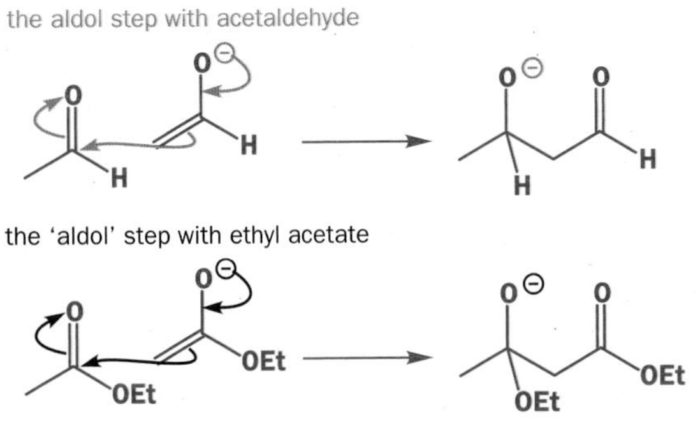

The next step in both cases is nucleophilic attack by the enolate ion on unenolized carbonyl compound. The concentration of enolate is low and each enolate ion is surrounded by unenolized aldehyde or ester molecules, so this reaction is to be expected. Here is that step, again shown for both aldehyde and ester.

Only now does something different happen. The aldehyde dimer simply captures a proton from the solvent to give an aldol product. The 'aldol' from the ester (not, in fact, an aldol at all) has a leaving group, EtO^-, instead of a hydrogen atom and is actually the tetrahedral intermediate in a nucleophilic substitution at the carbonyl group. Compare the two different steps again.

completion of the aldol with acetaldehyde

3-hydroxybutanal ('aldol')

the Claisen condensation with ethyl acetate

ethyl 3-oxobutanoate
(ethyl acetoacetate)

Even though the last step is different, the two products are quite similar. Both are dimers of the original two-carbon chain and both have carbonyl groups at the end of the chain and oxygen substituents at position three. The two reactions obviously belong to the same family but are usually given different names. The ester reaction is sometimes known as the **Claisen ester condensation** and sometimes as the **Claisen–Schmidt reaction**. More important than remembering the name is being familiar with the reaction and its mechanism. Here is a summary.

This is another of those reactions where the base is not strong enough to transform the ester entirely into the enolate. Only a small equilibrium concentration is produced, which reacts with the ester electrophile. The by-product from the reaction is ethoxide ion and so it looks at first sight as though we get our catalyst back again—the aldol, if you remember, is catalytic in base. But not the Claisen reaction. The second step of the reaction is also really an equilibrium, and the reaction works only because the product can be irreversibly deprotonated by the ethoxide by-product, consuming ethoxide in the process. You recall that the aldol reaction often works best when there is an extra driving force to push it across—dehydration to an enone, for example. Similarly, the ester dimerization works best when the product reacts with the ethoxide ion to give a stable enolate ion.

ethyl acetate

ethyl 3-oxobutanoate
(ethyl acetoacetate)

reactive enolate

irreversible
deprotonation

stable enolate

The point is that the base used, ethoxide ion EtO⁻, is too weak (EtOH has a pK_a of about 16) to remove the proton completely from ethyl acetate (pK_a about 25), but is strong enough to remove a proton from the acetoacetate product (pK_a about 10). Under the conditions of the reaction, a small amount of the enolate of ethyl acetate is produced—just enough to let the reaction happen—but the product is completely converted into its enolate. The neutral product, ethyl acetoacetate itself, is formed on acidic work-up.

the complete Claisen ester condensation

stable enolate

ethyl acetoacetate

The final product has been formed by the acylation at carbon of the enolate of an ester. This general process—acylation at carbon—is the subject of this chapter. It so happened in this case that the acylating agent was another molecule of the same ester, but the general process we shall consider is the acylation of enolates at carbon. We shall use a variety of enols, enolates, and specific enol equivalents and a variety of acylating agents, but the basic idea is this.

enolate acylation at C

overall, this acyl group has been added to the C atom of the carbonyl compound

X = a leaving group

new C–C bond

Problems with acylation at carbon

The main problem with the acylation of enolates is that reaction tends to occur at oxygen rather than at carbon.

enolate acylation at O

new C–O bond

overall, the acyl group ha been added to the O atom the carbonyl compound

You have seen reaction at oxygen before. Enolates react on oxygen with silicon electrophiles and we found the products, silyl enol ethers, useful in further reactions. Enol esters also have their uses—as precursors of lithium enolates, for example. You saw one being used like this on p. 683.

The product of acylation on oxygen is an **enol ester**. The tendency to attack through oxygen is most marked with reactive enolates and reactive acylating agents. The combination of a lithium enolate and an acid chloride, for example, is pretty certain to give an enol ester.

making an enol ester

enol acetate

If we want acylation at carbon we must use *either*

- less reactive specific enol equivalents, such as enamines or silyl enol ethers, with reactive acylating agents such as acid chlorides *or*

- reactive enols, such as the enolate anions themselves, with less reactive acylating agents such as esters

We introduced this chapter with an example of the second type of reaction, and we shall continue with a more detailed consideration of the Claisen ester condensation and related reactions.

Reaction at oxygen

In Chapter 27, we mentioned no trouble with reaction at oxygen in the aldol reaction. This may now seem surprising, in view of what we have said about esters, as the electrophiles were aldehydes and ketones—not so very different from esters. We can resolve this by looking at what would happen if an aldehyde did attack an enolate on the oxygen atom.

only possible leaving group

$H^{\ominus}$ is not a leaving group

The only plausible leaving group from the intermediate is the enolate anion itself: the reaction just reverses. It may well be that aldol reactions *do* involve attack through oxygen. But no products can be formed from this reversible pathway: only when the electrophile has a leaving group is reaction at oxygen productive.

Acylation of enolates by esters

The Claisen ester condensation and other self-condensations

▶

We spent some time in Chapters 26 and 27 considering the reactions of ethyl acetoacetate: now you see how it is made.

ethyl acetoacetate

The self-condensation of ethyl acetate, with which we opened this chapter, is the most famous example of the Claisen ester condensation and it works in good yield under convenient conditions. The product (ethyl acetoacetate) is commercially available—and cheap too—so you are unlikely to want to do this particular example.

A more generally useful reaction is the self-condensation of simple substituted acetates RCH_2CO_2Et. These work well under the same conditions (EtO⁻ in EtOH). The enolate anion is formed first in low concentration and in equilibrium with the ester. It then carries out a nucleophilic attack on the more abundant unenolized ester molecules.

These steps are all unfavourable equilibria and, on their own, would give very little product. However, as we mentioned before, the reaction works because the equilibrium is driven over by the essentially irreversible formation of a stable, delocalized enolate from the product.

stable delocalized enolate

final work-up with HCl

3-oxo-ester or β-keto-ester

■

We shall discuss the significance of the 1,3-relationship in Chapter 30.

Finally, the reaction is worked up in acid and the β keto-ester product is formed. Notice that all products of Claisen ester condensations have a 1,3-dicarbonyl relationship. These compounds are useful in the preparation of specific enol equivalents and you have seen them in action in Chapters 21, 26, and 27.

How do we know that deprotonation drives the reaction?

If the original ester has two substituents on the α carbon atom (C2 of the ester), the formation of the stable enolate of the product is no longer possible as there are no hydrogen atoms left to remove.

unfavourable equilibrium

can't be deprotonated: no Hs left

As you might expect, all the equilibria are now unfavourable, and this reaction does not go well under the normal equilibrating conditions (EtO⁻ in EtOH). It can be made to go in reasonable yield if a stronger base is used. Traditionally, triphenylmethyl sodium is chosen. This is made from Ph_3CCl and sodium metal and is a very conjugated carbanion.

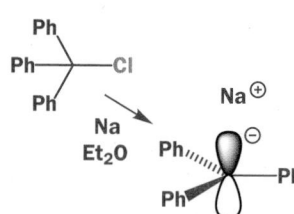

Triphenylmethyl carbanion is a strong enough base to convert an ester entirely into its enolate. Reaction of the enolate with a second molecule of ester then gives the keto-ester in good yield.

74% yield

Intramolecular acylation: the Dieckmann reaction

Intramolecular acylations often go very well indeed when a five- or a six-membered ring is being formed. A classic case is the cyclization of the diethyl ester of adipic acid (diethyl hexanedioate), a component in nylon manufacture.

diethyl hexanedioate

It doesn't matter which ester group forms the enolate anion as they are the same. The cyclization has to give a five-membered ring.

As in the intermolecular version, the product under the reaction conditions is the stable enolate but work-up in acid forms the keto-ester as final product.

We can simultaneously prove that the enolate really is formed under the reaction conditions and demonstrate the usefulness of the process by trapping the enolate with an alkyl halide before work-up.

This sequence was used to prepare the important flavouring compound 'Corylone' which has, it is claimed, a 'sweet and powerful spicy–coffee–caramel odour'. You may imagine how popular it is with food-additive chemists and this sequence provides a short process for its manufacture.

The intramolecular version of the Claisen ester condensation is sometimes known as the **Dieckmann reaction**. It provides an excellent route to heterocyclic ketones (cyclic ketones with heteroatoms in the ring: very important in drug manufacture). The starting diester can be made by two Michael additions to conjugated esters (see Chapter 10).

Treatment with base under the usual equilibrating conditions allows an efficient intramolecular condensation by the usual mechanism. Both ester groups are again identical and, since you should by now be accustomed to this mechanism, we just show the key step.

The β keto-esters can be easily hydrolysed and decarboxylated by the methods of Chapter 26 to give the symmetrical cyclic ketone. The carboxylate anion is reasonably stable, but the free acid cannot usually be isolated as it loses carbon dioxide easily and gives the enol of the final product.

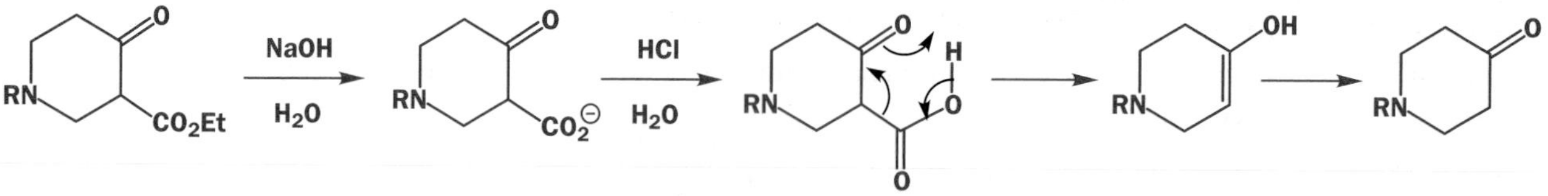

Crossed ester condensations

Much the same type of arguments applies here as applied in the crossed aldol reaction (Chapter 27). We must be quite sure that we know which compound is going to act as the enol partner and which as the acylation partner.

Reactive esters that cannot enolize

There are several useful esters of this kind, of which these four are the most important. They cannot act as the enol partner, and the first three are more electrophilic than most esters, so they should acylate an ester enolate faster than the ester being enolized can.

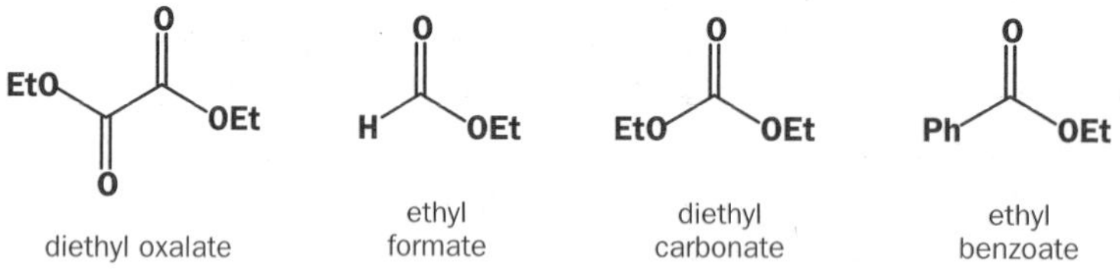

diethyl oxalate   ethyl formate   diethyl carbonate   ethyl benzoate

These four are arranged in order of reactivity towards nuclophiles, the most electrophilic first and the least electrophilic last. Oxalates are very reactive because each carbonyl group makes the other more electrophilic. The molecular LUMO is the *sum* of the two π* orbitals and is lower in energy than either.

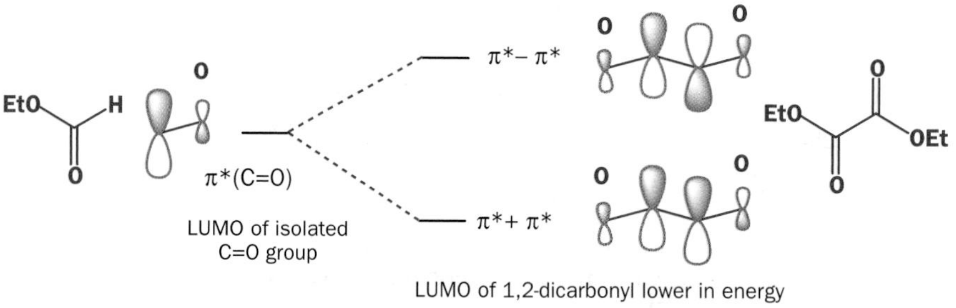

LUMO of isolated C=O group

LUMO of 1,2-dicarbonyl lower in energy

Formate esters look a bit like aldehydes but their ester character dominates. The hydrogen atom just makes them very electrophilic as they lack the σ conjugation (and steric hindrance) of simple esters.

Carbonates are particularly useful as they introduce a CO_2R group on to an enolate. It is not immediately obvious why they are more electrophilic than simple esters. Normal esters are (slightly) less electrophilic than ketones bcause the deactivating lone pair donation by the oxygen atom is more important than the inductive effect of the electronegative oxygen atom.

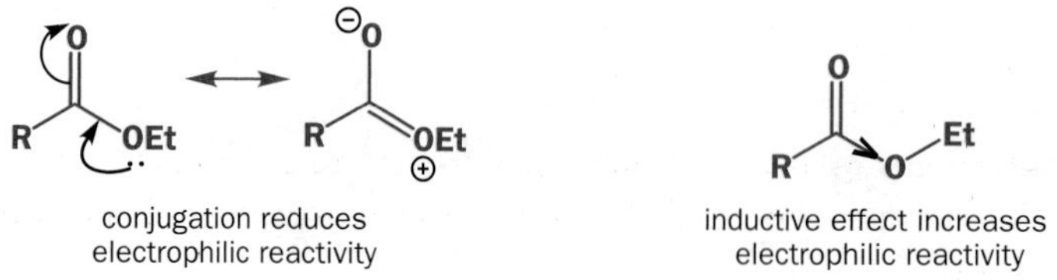

conjugation reduces electrophilic reactivity

inductive effect increases electrophilic reactivity

▶

σ conjugation from adjacent C–H bonds raises the LUMO of most esters, just as an adjacent nitrogen lone pair raises the LUMO of an amide, but to a lesser extent.

The result is a small difference between two large effects. In carbonate esters there are two oxygen atoms on the same carbonyl group. Both can exert their full inductive effect but the lone pairs are trying to overlap with the same π^* orbital. The balance is changed—the summed inductive effects win out—and carbonates are more electrophilic than ordinary esters.

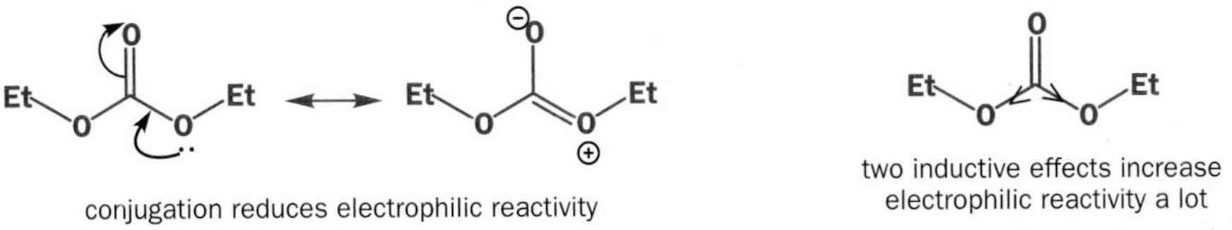

conjugation reduces electrophilic reactivity

two inductive effects increase electrophilic reactivity a lot

Finally, esters of aromatic acids cannot enolize but are less reactive than ordinary esters because of conjugation from the aromatic ring. These compounds may still be useful as we shall see.

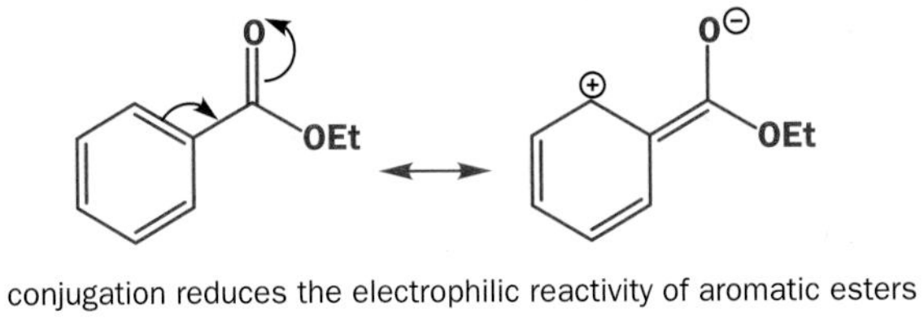

conjugation reduces the electrophilic reactivity of aromatic esters

Crossed Claisen ester condensations between two different esters

We shall now give a few examples of crossed Claisen ester condensations between ordinary esters and the compounds we have just discussed. First, a reaction between a simple linear ester and diethyl oxalate performed under equilibrating conditions with ethoxide as the base.

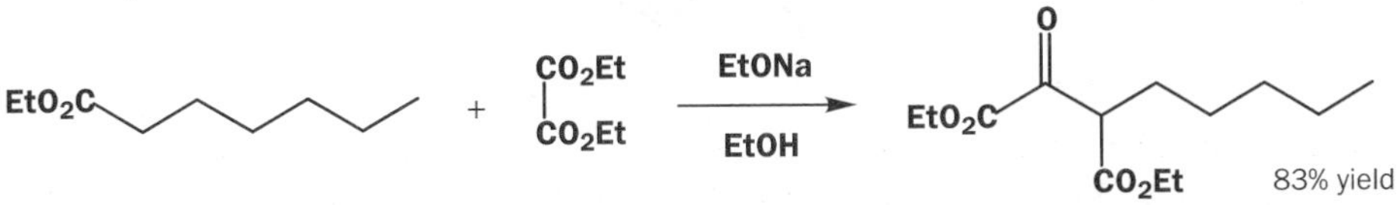

83% yield

Only the simple ester can give an enolate, and the low concentration of this enolate reacts preferentially with the more electrophilic diethyl oxalate in a typical acylation at carbon.

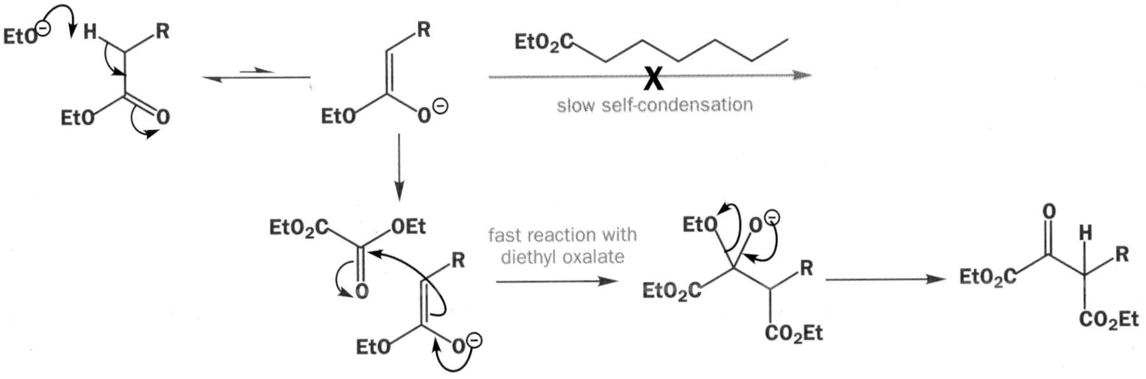

slow self-condensation

fast reaction with diethyl oxalate

The product has an acidic hydrogen atom so it is immediately converted into a stable enolate, which is protonated on work-up in aqueous acid to give the tricarbonyl compound back again.

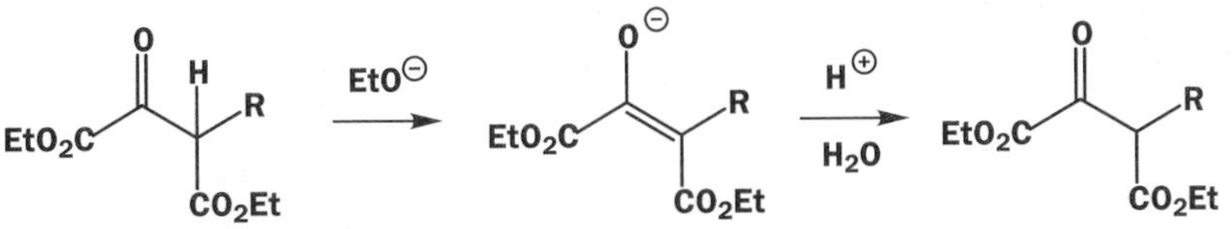

This compound was made because it was needed in a synthesis of multicolanic acid, a metabolite of a penicillium mould. It is easy to see which atoms of the natural product were provided by the compound we have just made in a single easy step.

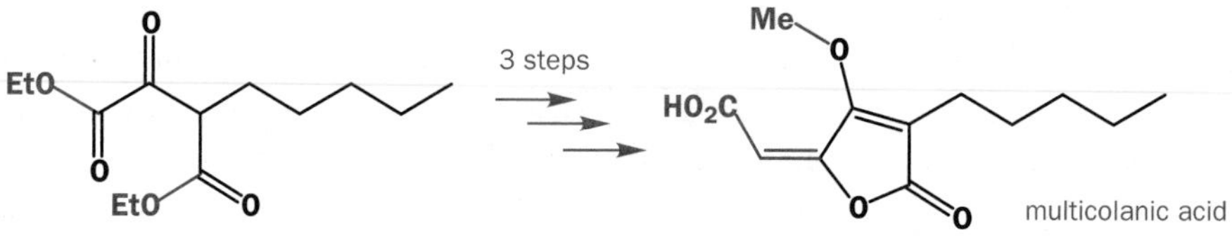

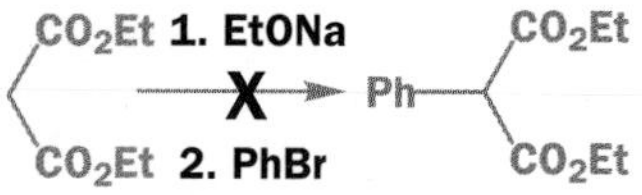

multicolanic acid

Another important example leads to the preparation of diethyl phenylmalonate. This compound cannot be made by 'alkylation' of diethyl malonate as aryl halides do not undergo nucleophilic substitution (Chapter 23).

A crossed Claisen ester condensation between very enolizable ethyl phenylacetate and unenolizable but electrophilic diethyl carbonate works very well indeed under equilibrating conditions.

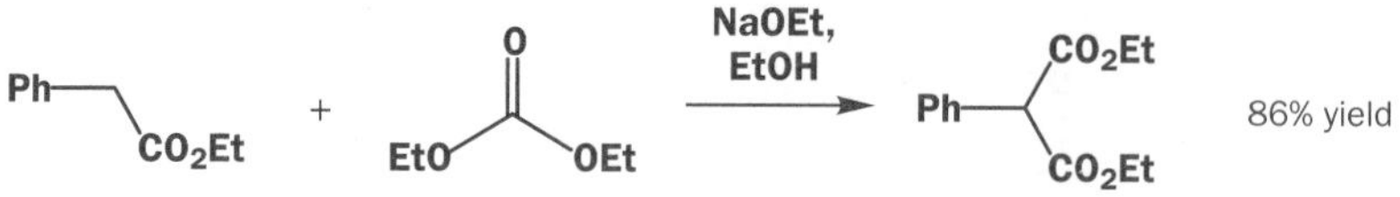

86% yield

Claisen condensations between ketones and esters

Claisen condensations always involve esters as the electrophilic partner, but enolates of other carbonyl compounds—ketones, for example—may work equally well as the enol partner. In a reaction with a carbonate, only the ketone can enolize and the reactive carbonate ester is more electrophilic than another molecule of the ketone. A good example is this reaction of cyclooctanone. It does not matter which side of the carbonyl group enolizes—they are both the same.

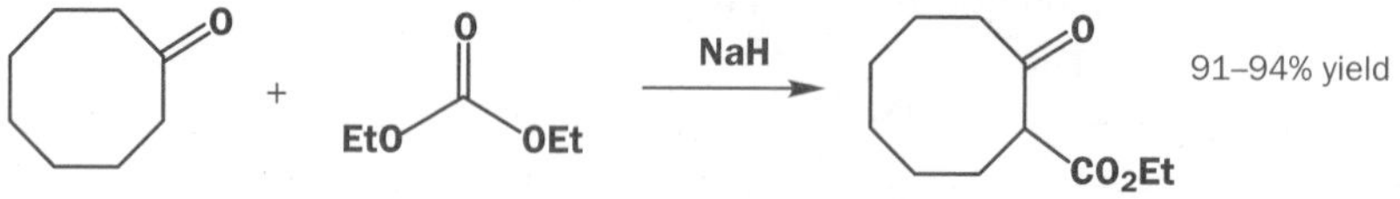

91–94% yield

The alternative route to this cyclic dicarbonyl—Dieckmann condensation—would be a bad choice in this case. Dieckmann condensation works well for five- and six-membered rings, reasonably well for seven-membered rings, but not very well at all for eight-membered rings. The yield is almost exactly half what the ketone–carbonate reaction gives.

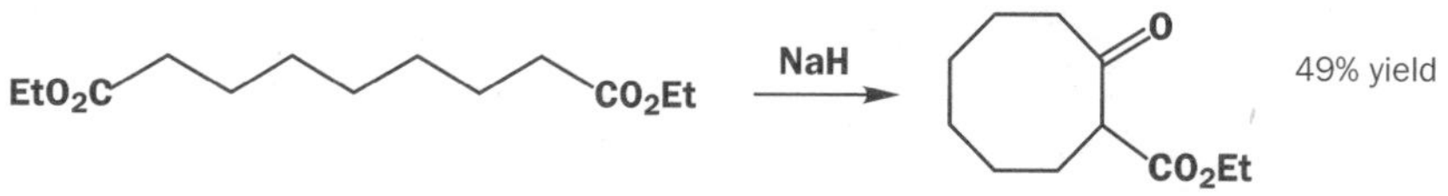

49% yield

Unsymmetrical ketones often give a single product, even without the use of a specific enol equivalent, as reaction usually occurs on the less substituted side. This is another consequence of the final enolization being the irreversible step. In this example, both possible products may form, but only one of them can enolize. Under the equilibrating conditions of the reaction, only the enolate is stable, and all the material ends up as the isomer shown.

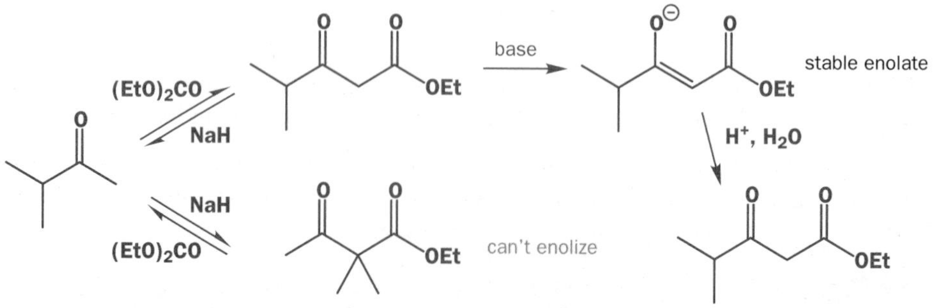

Unsymmetrical ketones work well even when one side is a methyl group and the other a primary alkyl chain. This example gives an impressive yield and shows that, as expected, a remote alkene does not affect the reaction.

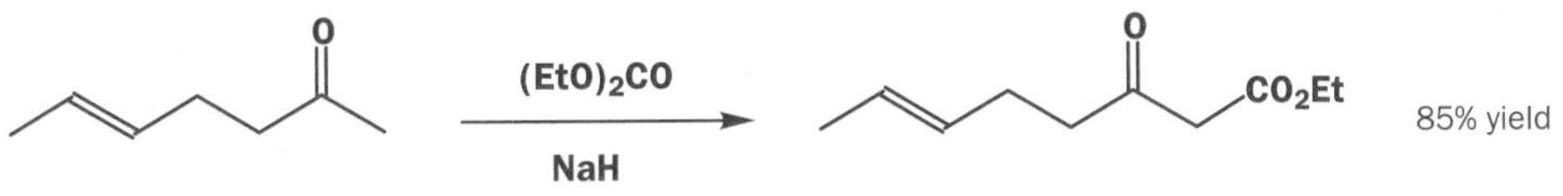

Even when both enolates can form, the less substituted dicarbonyl enolate is preferred because it constrains fewer groups to lie in the hindered plane of the tetrasubstituted enolate double bond.

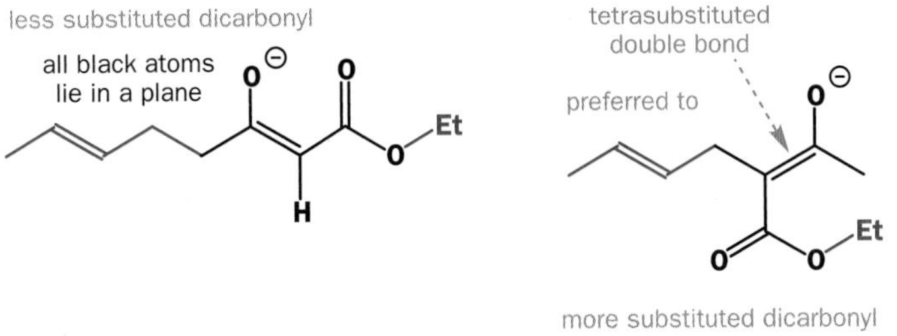

Diethyl oxalate also gives well-controlled condensations with ketones and we shall take the synthesis of a new drug as an example. One way to try and prevent heart disease is to reduce the amount of 'bad' lipoproteins in the blood. The drug Acifran does this, and a key step in its synthesis is the base-catalysed reaction between diethyl oxalate and a methyl ketone.

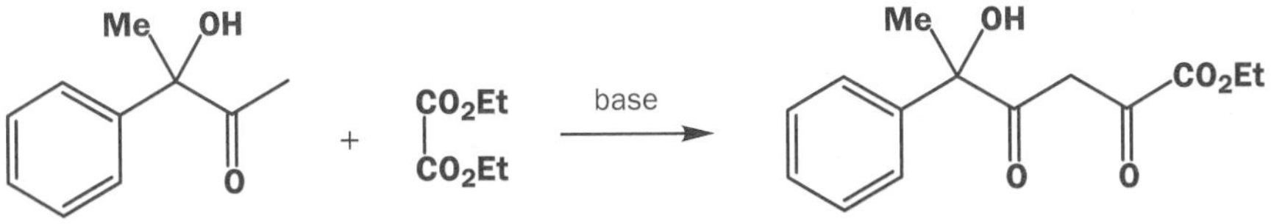

Notice that the hydroxyl group on the ketone does not interfere with the reaction. No doubt the first molecule of base removes the OH proton and the second molecule forms the enolate (the only possible enolate in either molecule). Fast condensation with highly electrophilic diethyl oxalate follows. The drug itself results from simple acid treatment of this product.

■
Try to write a mechanism for the cyclization step.

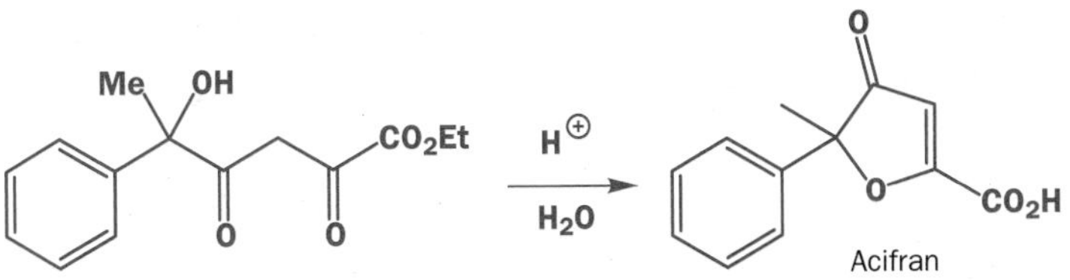

The other two unenolizable esters we mentioned on p. 728 undergo cross-condensations with ketones. Unlike formaldehyde, formate esters are well behaved—no special method is necessary to correspond with the Mannich reaction in aldol chemistry. Here is what happens with cyclohexanone.

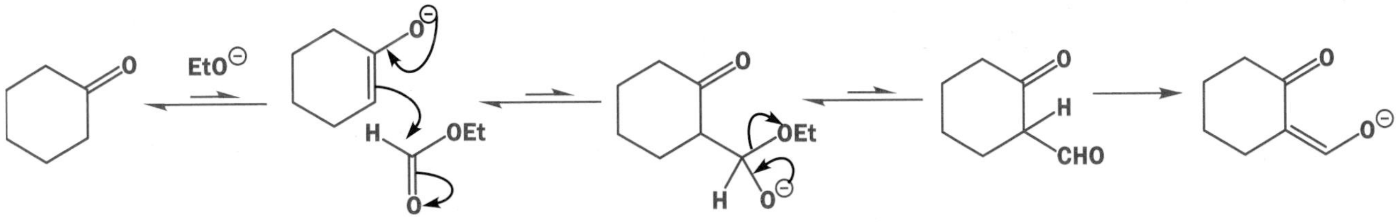

The product aldehyde is not at risk from nucleophilic attack, as it appears to be, because it immediately enolizes in base. The product is formed as a stable enol with an intramolecular hydrogen bond.

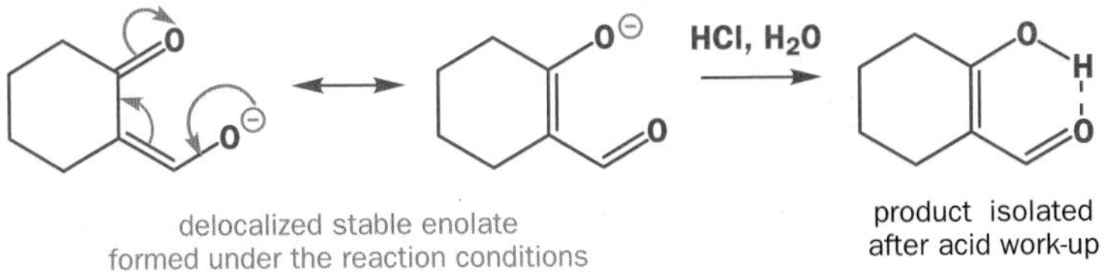

delocalized stable enolate
formed under the reaction conditions

product isolated
after acid work-up

Esters of aromatic acids are used rather less frequently in this manner because they are considerably less reactive than carbonates or formates. This simple example works quite well—admittedly the ketone is very enolizable.

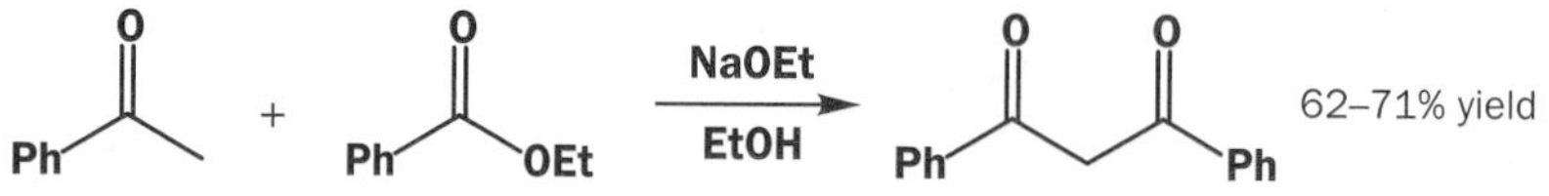

62–71% yield

A more important example is the synthesis of the rat poison 'Pival'. An enolizable ketone that is blocked on one side by a tertiary butyl group reacts with diethyl phthalate to give a five-membered cyclic diketone in one reaction by two Claisen ester condensations.

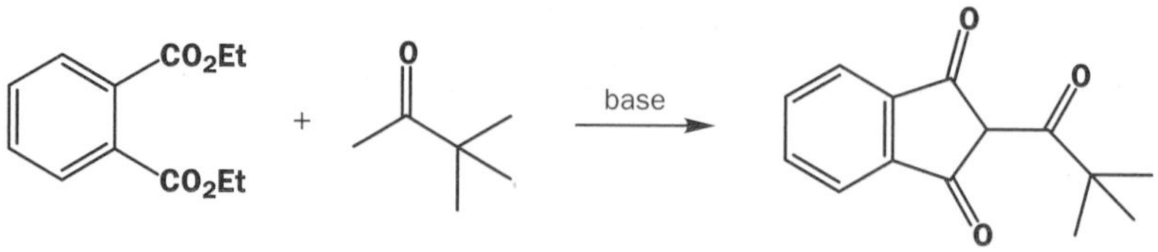

Only one enolate can be formed and this attacks either of the two aromatic ester groups to give a 1,3-diketone by a crossed Claisen condensation.

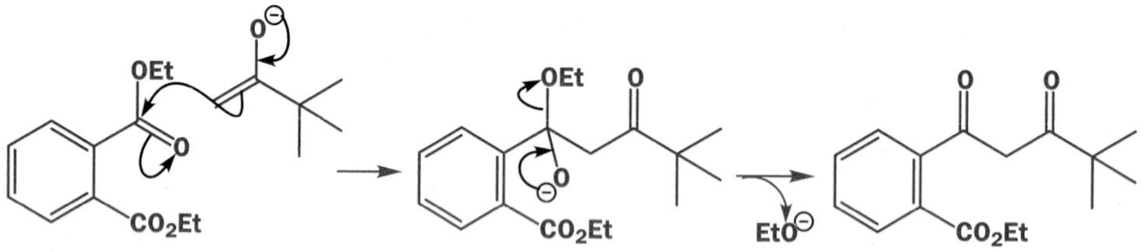

The ethoxide ion released in this first reaction will, as usual, form a stable enolate from the 1,3-diketone but this now cyclizes in a second Claisen condensation on to the second ester group.

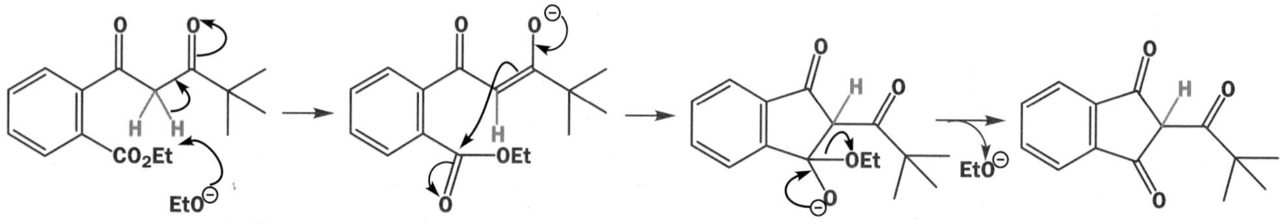

The product has an exceptionally acidic hydrogen atom, shown in green, on a carbon atom between three carbonyl groups. Under the reaction conditions this will of course be lost to form an enolate, and after protonation Pival itself exists as a mixture of enol forms.

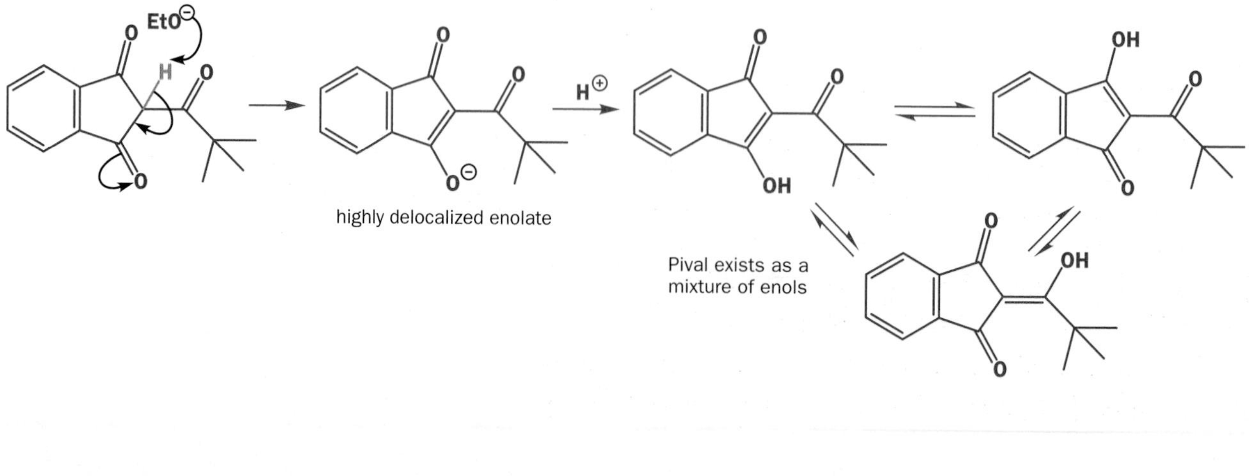

highly delocalized enolate

Pival exists as a
mixture of enols

Summary of preparation of keto-esters by the Claisen reaction

It is worth pausing at this moment to summarize which keto-esters can be made easily by the two methods we have discussed, namely

- Claisen ester condensation
- acylation of ketones with enolates

Ethyl acetoacetate (ethyl 3-oxobutyrate) can of course be made by the self-condensation of ethyl acetate.

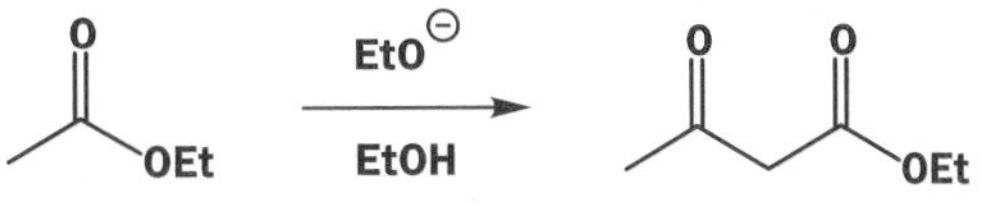

This ester is cheap to buy but homologues, available by the self-condensation of other esters, are usually made in the laboratory. Which esterifying group is used (OEt, OMe, etc.) is not important so long as the same alkoxide is used as the base.

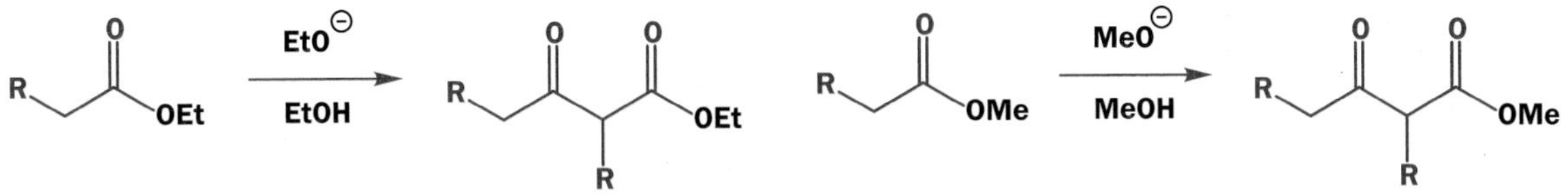

Compounds with only one of the 'R' substituents in this structure are also easy to make. If the 'R' substituent is at C2, it is best introduced by alkylation of the unsubstituted ester.

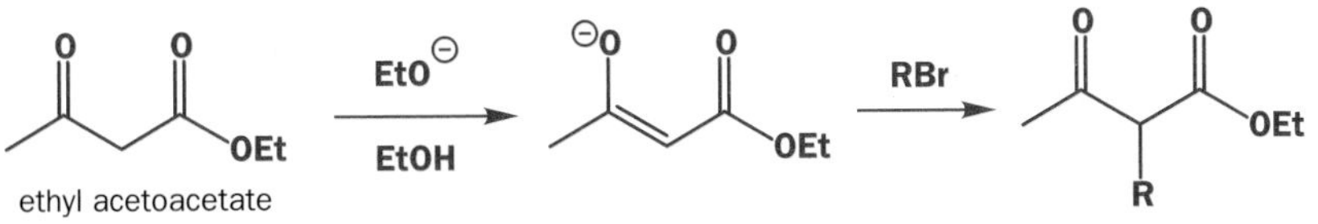

Attempts to make this compound by the Claisen ester condensation would require one of the approaches in the diagram below. The dashed curly arrows suggest the general direction of the condensation required and the coloured bonds are those that would be formed *if* the reaction worked.

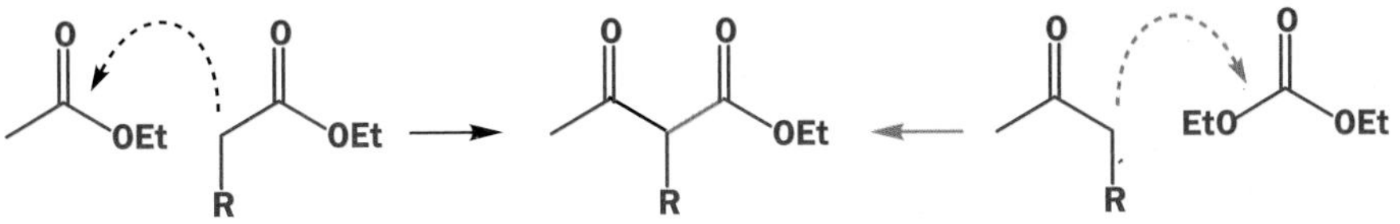

Unfortunately neither reaction will work! The black route requires a controlled condensation between two different enolizable esters—a recipe for a mixture of products. The simple alkylation route above removes the need for control. The green route requires a condensation between an unsymmetrical ketone and diethyl carbonate. This condensation will work all right, but not to give this product. As you saw on p. 730, Claisen condensations prefer to give the less substituted dicarbonyl compound, and condensation would occur at the methyl group of the ketone on the right to give the other unsymmetrical keto-ester.

> We shall see later in this chapter how such reactions *can* be controlled.

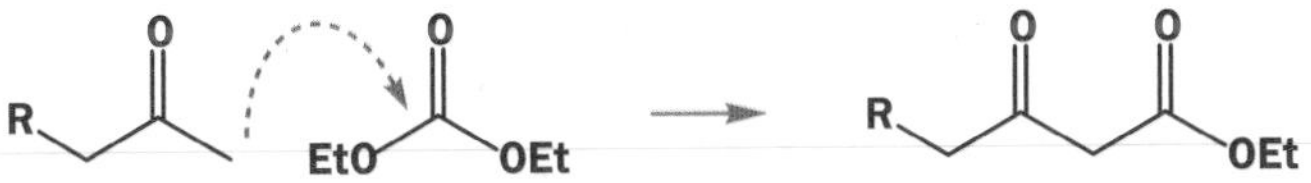

● Making β keto-esters: a check-list

A combination of self-condensation, condensation with diethyl carbonate, and alkylation of keto-esters prepared by one of these means will allow us to make most β keto-esters that we are likely to want. Look out for all the usual problems of enolate chemistry.

- Will the right carbonyl compound enolize?
- If it is a ketone, will it enolize in the right way?
- Will the enolate react with the right acylation partner?

If any of these poses problems, try using an alkylation step.

Intramolecular crossed Claisen ester condensations

As usual with intramolecular condensations, we do not have to worry so much about controlling where enolization occurs providing that one product is more stable than the others—for example, it might have a five- or a six-membered ring (rather than a four- or eight-membered one)—and we carry out the reaction under equilibrating conditions. A couple of examples should show what we mean.

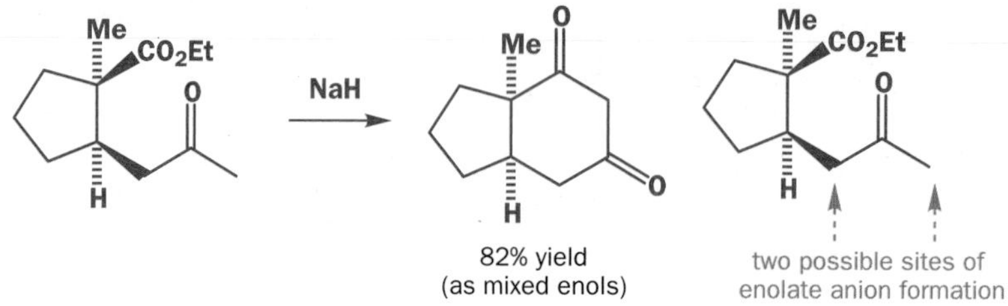

82% yield
(as mixed enols)

two possible sites of
enolate anion formation

Though there are two sites for enolate anion formation, one would give a four-membered ring and can be ignored. Only enolization of the methyl group leads to a stable six-membered ring.

91% yield
(as enol)

This time the two possible sites for enolate anion formation would both lead to stable five-membered rings, but one product cannot form a stable enolate anion under the reaction conditions so the other is preferred.

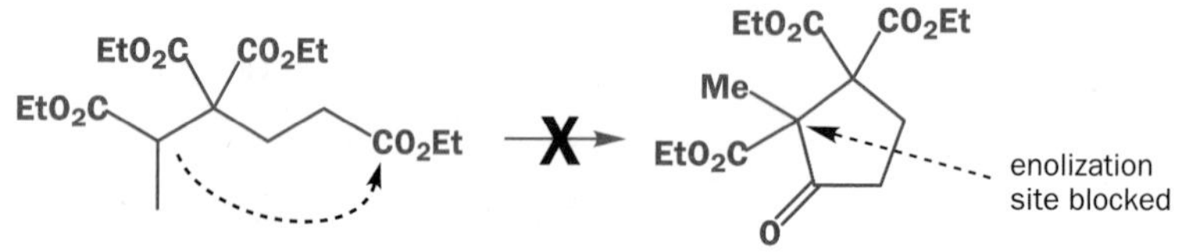

enolization
site blocked

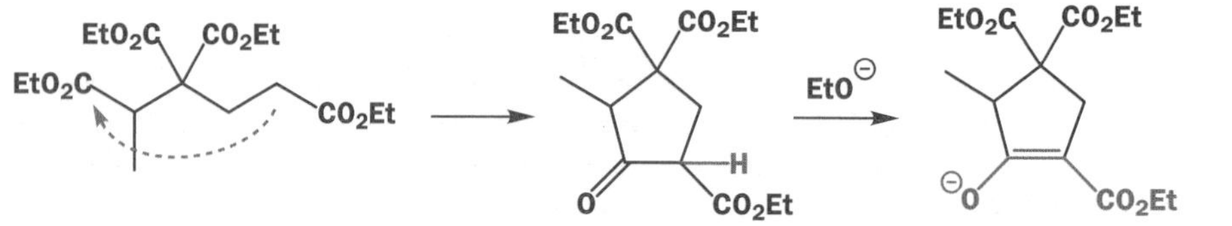

In the next example, there are three possible sites for enolate anion formation, but only one product is formed and in good yield too.

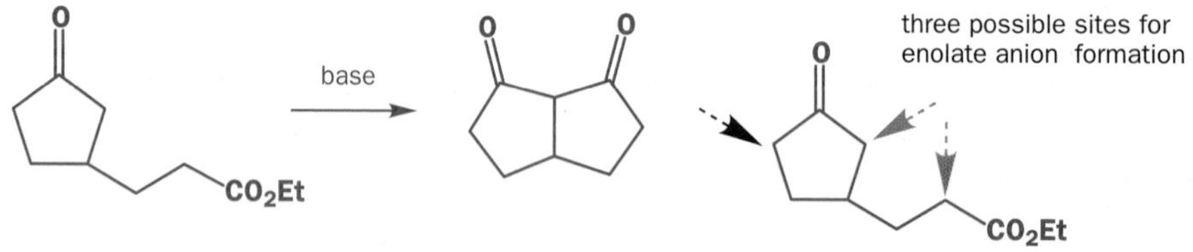

three possible sites for
enolate anion formation

If we consider all three possible enolate anions, the choice is more easily made. First, the reaction that *does* happen. An enolate anion is formed from the ketone at the green site and acylation at carbon follows.

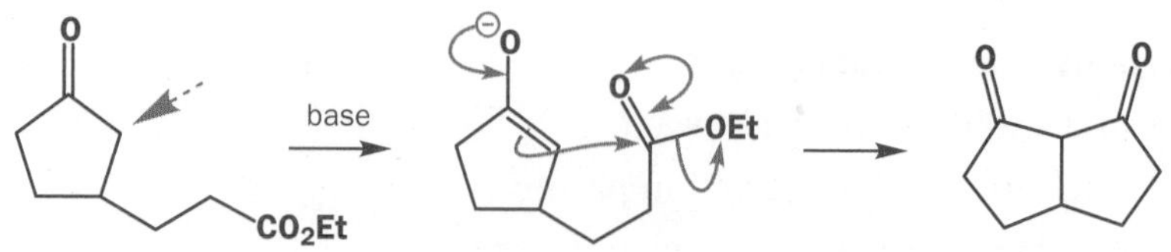

We could form the enolate anion on the other side of the ketone and attack the ester in the same way using the black arrows. The product is an attractive bicyclic diketone, but it is not formed.

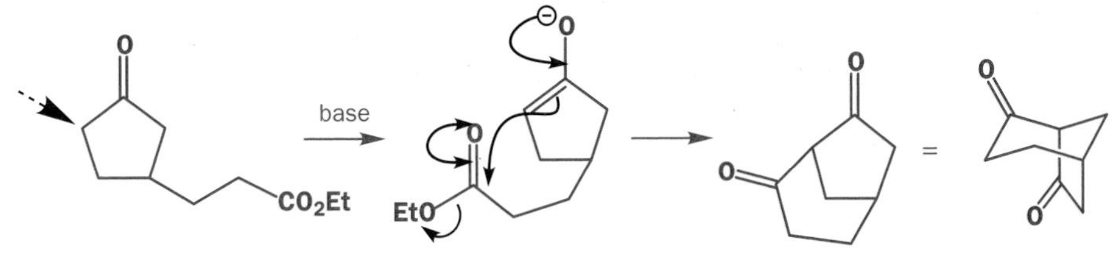

The third cyclization mode (brown arrows) would be to form an enolate from the ester and attack the ketone. This would be an aldol rather than a Claisen reaction.

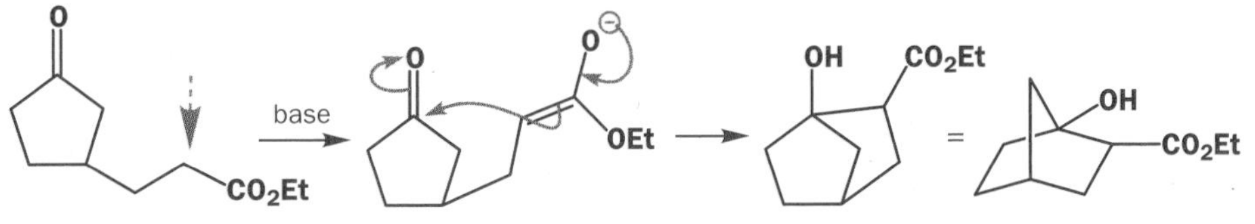

This is another bicyclic compound but again it is not formed. The choice is made by considering what can happen to the three products under the reaction conditions. The aldol product cannot dehydrate nor can the black Claisen product form a stable enolate because both would have an impossible double bond at a bridgehead position.

We discussed bicyclic compounds and bridgehead double bonds in Chapter 19.

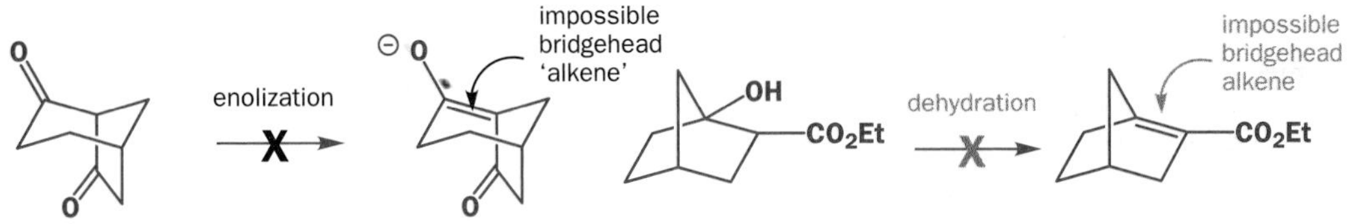

The product from the green Claisen reaction is, on the other hand, a fused rather than a bridged bicyclic structure. It can easily form a stable enolate anion.

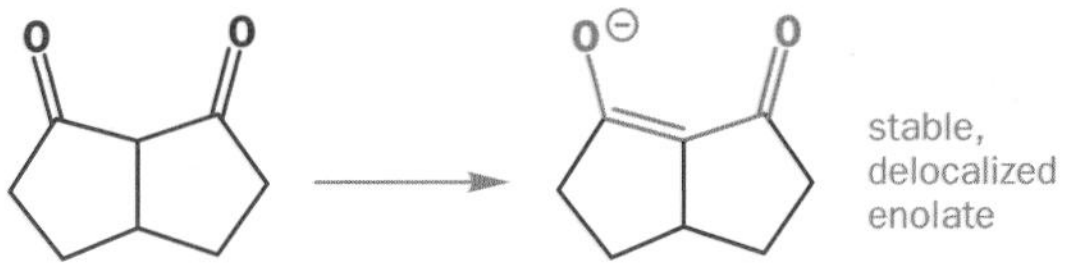

Remind yourself (p. 508) of the difference between fused compounds (one bond in common), *spiro* compounds (one atom in common), and bridged compounds (rings joined at two non-adjacent atoms). Each of these three examples has two five-membered rings

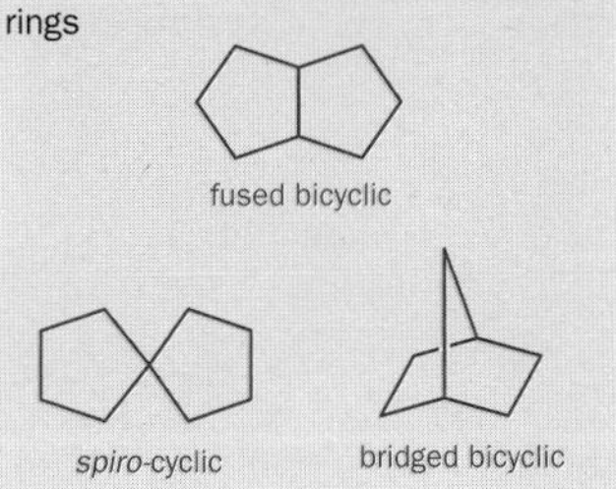

fused bicyclic

spiro-cyclic   bridged bicyclic

Symmetry in intramolecular crossed Claisen condensations

If cyclization is to be followed by decarboxylation, a cunning plan can be set in motion. Addition of an amine by an S_N2 reaction to an α halo-ester followed by conjugate addition to an unsaturated ester gives a substrate for Claisen ester cyclization.

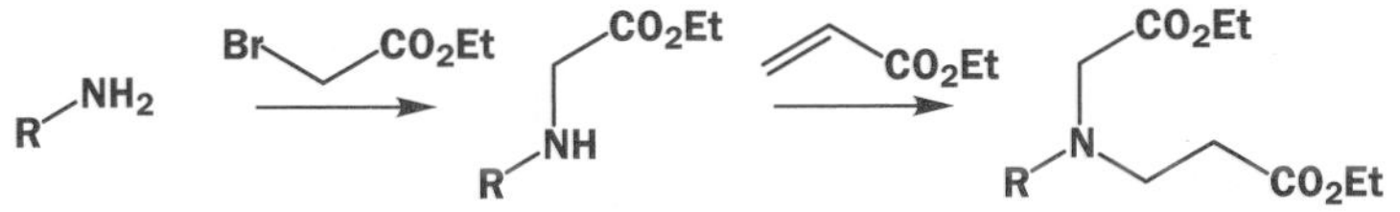

This diester is unsymmetrical so cyclization is likely to lead to two different keto-esters. Either can form a stable enolate so both are indeed formed. This sounds like very bad news since it gives a mixture of products.

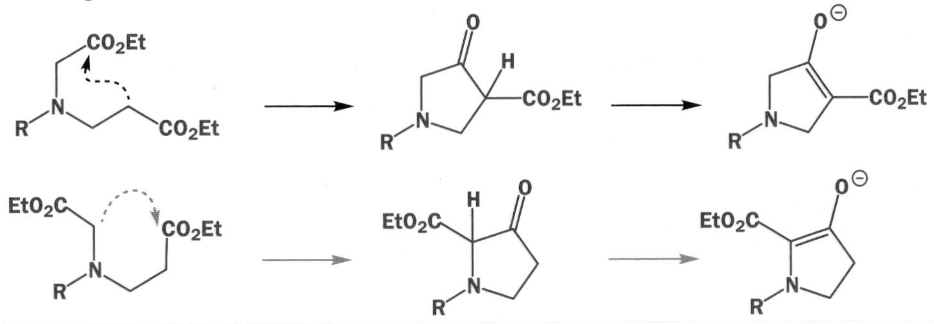

The cunning plan is that the relative positions of the ketone and the nitrogen atom in the five-membered ring are the same in both products. All that differs is the position of the CO_2Et group. When the two different products are hydrolysed and decarboxylated they give the same amino-ketone!

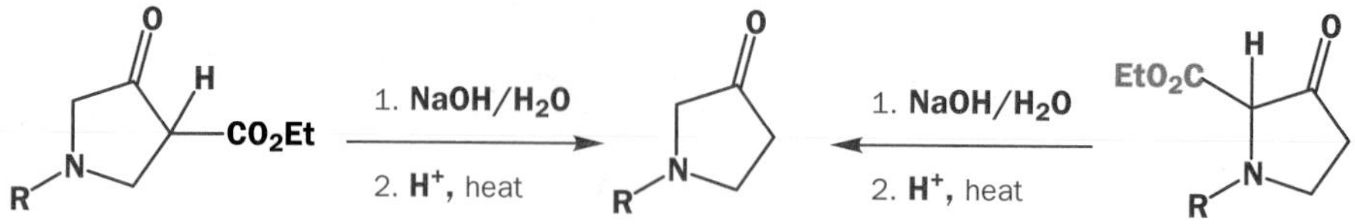

Just occasionally it is possible to carry out cross-condensations between two different enolizable molecules under equilibrating conditions. A notable example is the base-catalysed reaction between methyl ketones and lactones. With sodium hydride—a strong base that can convert either starting material entirely into its enolate anion—good yields of products from the attack of the enolate of the ketone on the electrophilic lactone can be obtained.

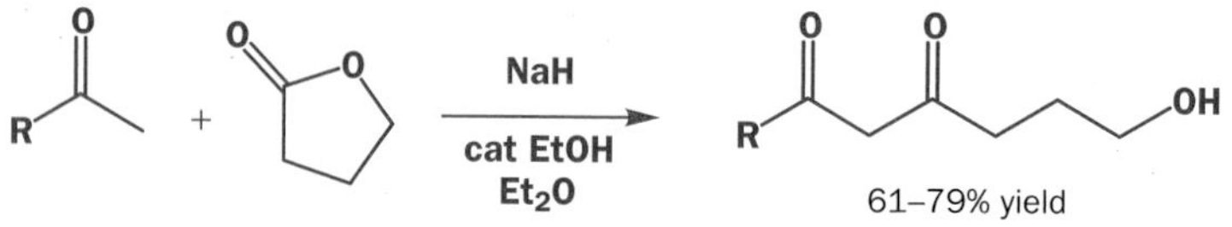

61–79% yield

Kinetic enolate formation must occur at the methyl group of the ketone followed by acylation with the lactone. Lactones are rather more electrophilic than noncyclic esters, but the control in this sequence is still remarkable. Notice how a stable enolate is formed by proton transfer within the first-formed product.

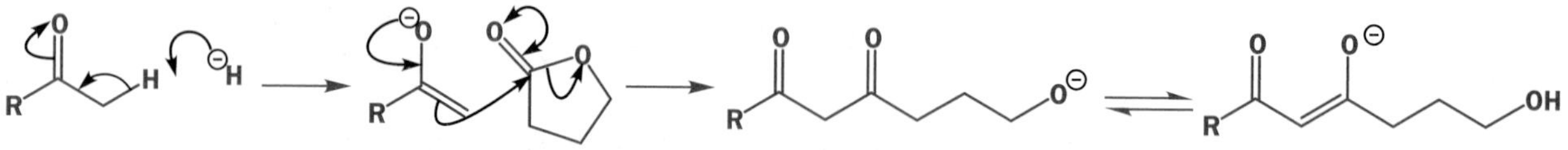

All these reactions have depended for their selectivity on the spontaneous behaviour of the molecules. It is time now to look at some reactions that cannot be controlled in that way—reactions where we must impose our will on the molecules by using specific enol equivalents.

Directed *C*-acylation of enols and enolates

The danger we have to face is that acylation is inclined to occur on oxygen rather than on carbon. In the extreme case, naked enolates (those with completely non-coordinating cations) acylate cleanly on oxygen with anhydrides or acid chlorides.

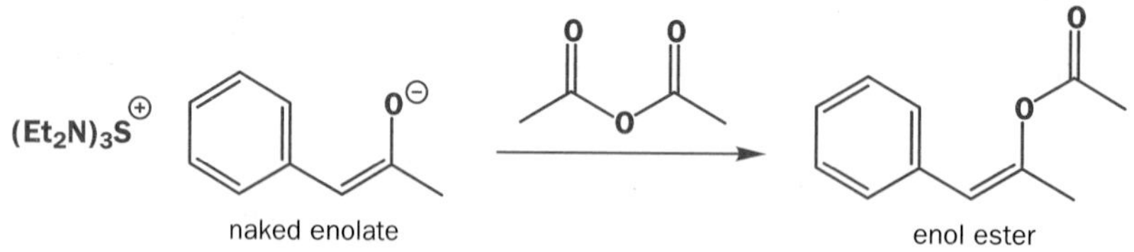

naked enolate　　　　　　　　　　　　　　　　　　enol ester

Alkali metal enolates (Li, Na, or K) tend to acylate on oxygen with acid chlorides too and it is often necessary to use magnesium enolates, particularly those of 1,3-dicarbonyl compounds, if reliable *C*-acylation is wanted. The magnesium atom bonds strongly to both oxygens, lessening their effective negative charge.

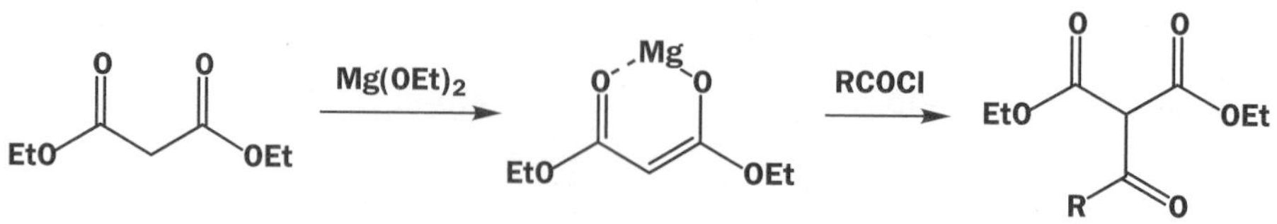

Hydrolysis and decarboxylation in the usual way lead to keto-esters or keto-acids. Of the more common metals used to form enolates, lithium is the most likely to give good *C*-acylation as it, like magnesium, forms a strong O–Li bond. It is possible to acylate simple lithium enolates with enolizable acid chlorides.

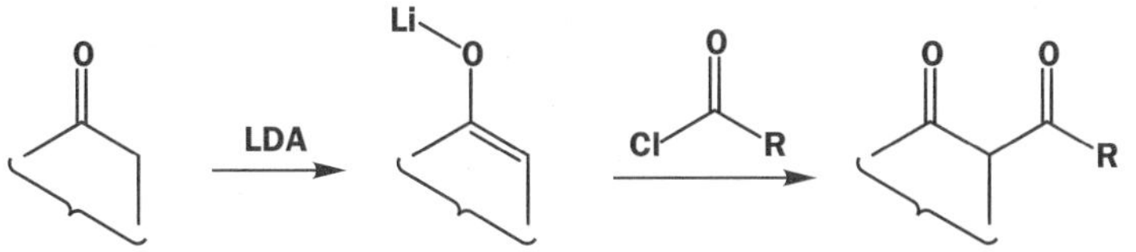

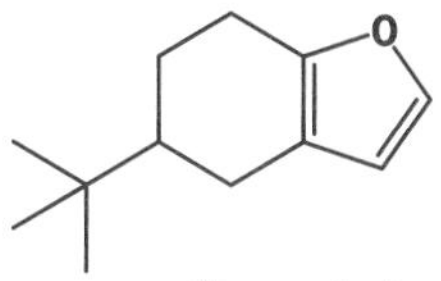

We shall describe two examples of this reaction being used as part of the synthesis of natural products. The first is pallescensin A, a metabolite of a sponge.

It is quite a simple compound and some chemists in Milan conceived that it might be made from the chloro-diketone shown below by alkylation of the enolate and subsequent reduction and dehydration of the remaining ketone.

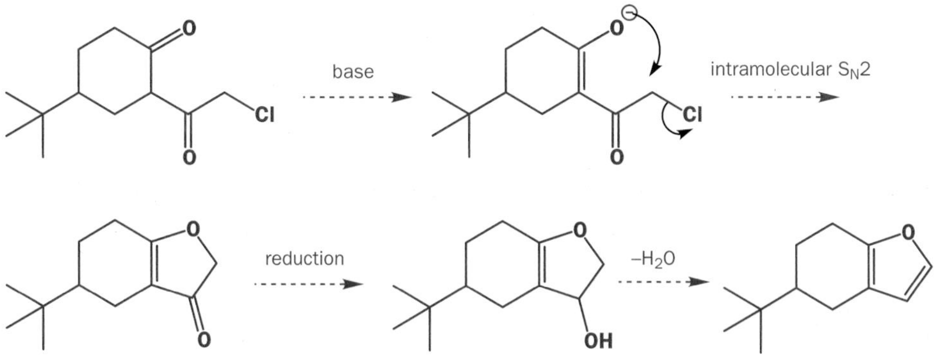

pallescensin A

To test this idea, the chloro-diketone must be made and the route chosen was to react the lithium enolate of 4-*t*-butyl cyclohexanone with the correct acid chloride.

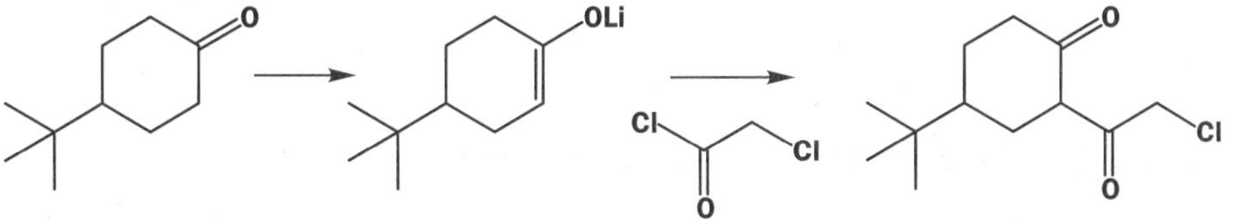

This reaction worked well, as did the rest of the synthesis of pallescensin A which was first made by this route. The key step, the acylation of the lithium enolate, is interesting because it could have alkylated instead. The acid chloride is more electrophilic than the alkyl chloride in this reaction, though alkylation does occur in the next step. Notice how the lithium atom holds the molecules together during the reaction.

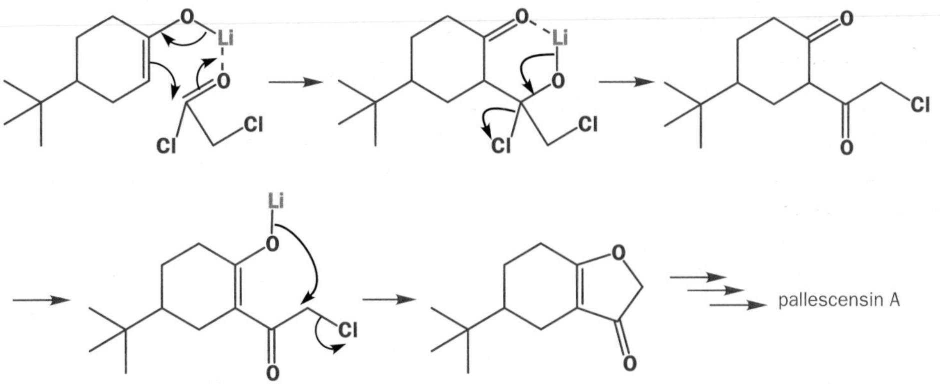

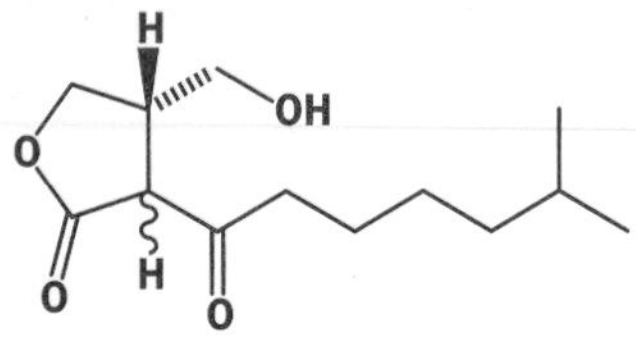

A-factor

Our second example is from the chemistry of microorganisms. The antibiotic streptomycin is produced rather erratically by the microorganism *Streptomyces griseus*. It has now been discovered that another compound, called 'A-factor', stimulates the microorganism into streptomycin production. Synthetic A-factor can be used to switch on antibiotic synthesis in the microorganism.

A-factor is an optically active compound, but notice that one stereogenic centre is not specified in the structure (H with a wavy line). This is because it is a 1,3-dicarbonyl compound and is therefore in equilibrium with its stable enol which has a trigonal centre at that point. The obvious way to complete the synthesis is to acylate an enolate of the lactone with an acid chloride.

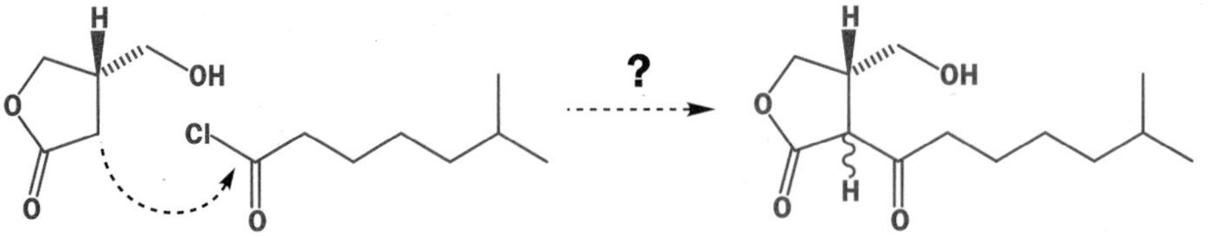

It will not be possible to have a free OH group on the lactone during this step as the acid chloride would, of course, react there too. In practice, protection as a silyl ether (Chapter 24) was enough and the lithium enolate was then used for the acylation reaction. Aqueous ethanol work-up removed the silyl protection.

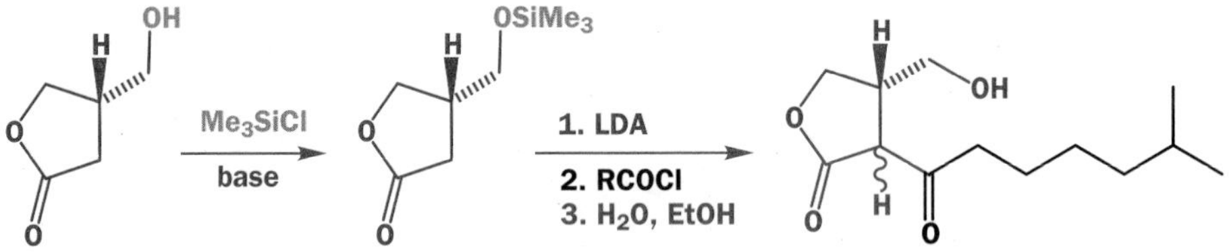

The preparation of the starting material is worth a closer look because it too involved a cross-condensation between two esters. Here it is in full. You have met all of these reactions in earlier chapters of this book.

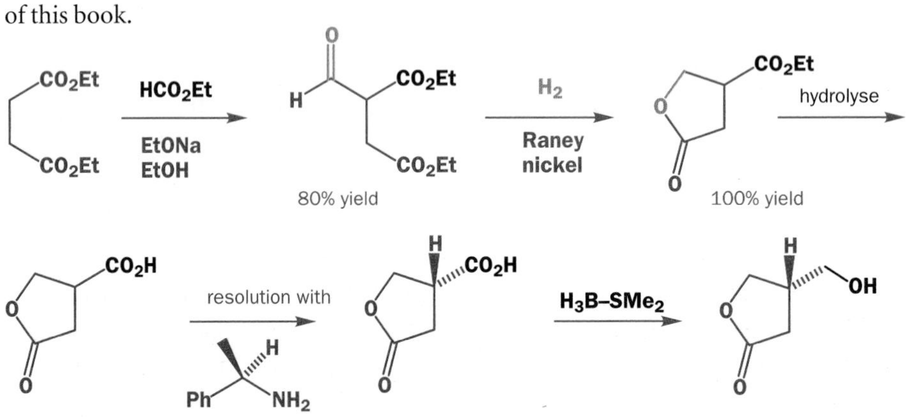

Even the dilithio derivatives of carboxylic acids, made by treating a carboxylic acid with two molecules of LDA, can give good reactions with acid chlorides. In these reactions it is not necessary to have a proton remaining between the two carbonyl groups of the product as the reaction is between a strong nucleophile and a strong electrophile and is under kinetic control.

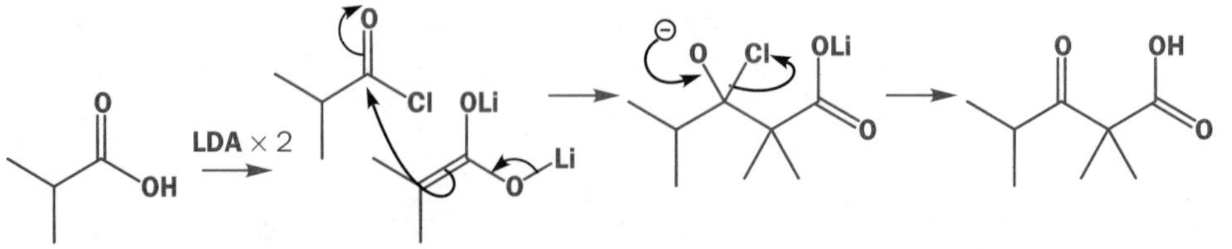

It is rather more common to use enamines or silyl enol ethers in acylations with acid chlorides. These are more general methods—enamines work well for aldehydes and ketones while silyl enol ethers work for all classes of carbonyl compounds. It is possible to combine two enolizable molecules quite specifically by these methods, and we shall consider them next.

The acylation of enamines

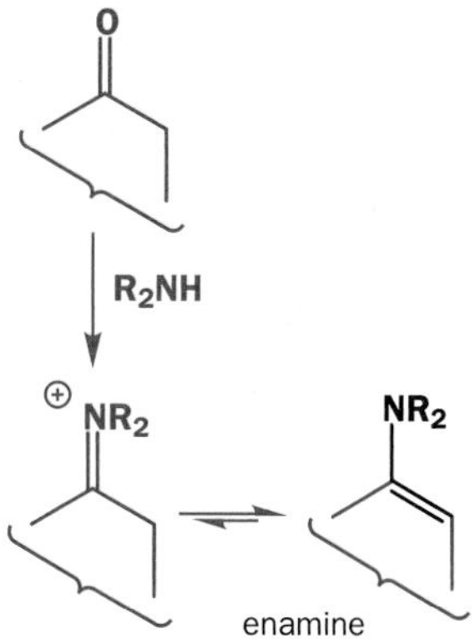

Enamines are made from secondary amines and aldehydes or ketones via the iminium salt: you met them in Chapter 14 and have seen them in action in Chapters 21, 26, and 27.

In Chapter 26 we saw that reliable *C*-alkylation occurs with reactive allyl halides and α halo-carbonyl compounds, but that unwanted *N*-alkylation often competes with simple alkyl halides.

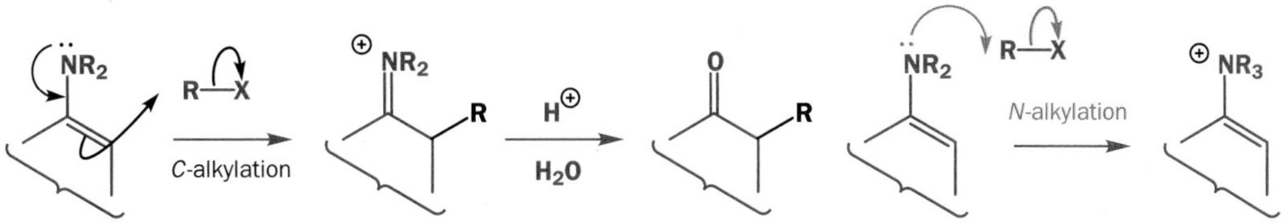

Acylation with acid chlorides could follow the same two pathways, but with one big difference. The products of *N*-acylation are unstable salts and *N*-acylation is reversible. Acylation on carbon, on the other hand, is irreversible. For this reason enamines end up acylated reliably on carbon.

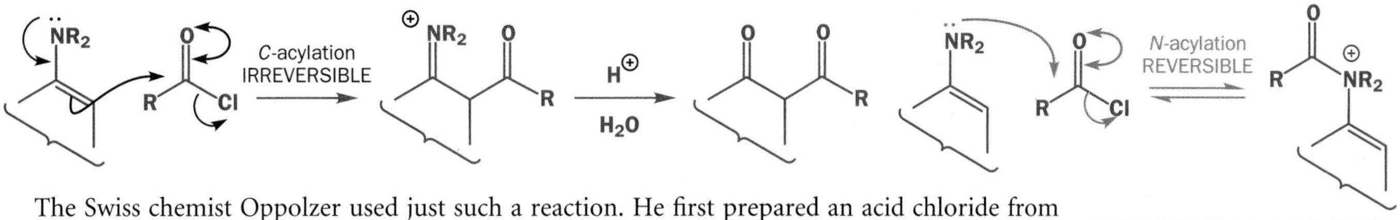

The Swiss chemist Oppolzer used just such a reaction. He first prepared an acid chloride from cyclopentadiene, and the enamine from cyclopentanone and the secondary amine morpholine.

> Morpholine is frequently used in the preparation of enamines—see p. 672.

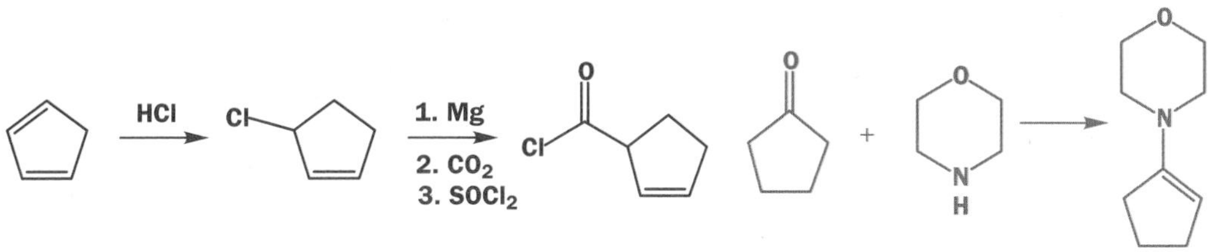

Combining the enamine with the acid chloride led to a clean acylation at carbon in 82% yield and eventually to a successful synthesis of the natural product longifolene.

82% yield

> ■ We shall revisit this synthesis in Chapter 35 when we discuss [2 + 2] cycloadditions.

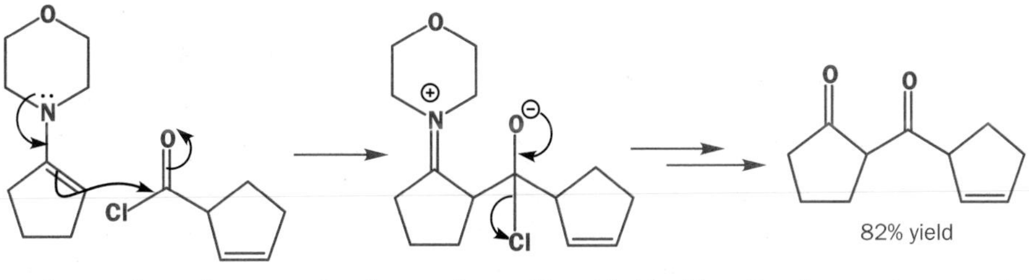

Aza-enolates also react cleanly at carbon with acid chlorides. Good examples come from dimethyl-hydrazones of ketones. When the ketone is unsymmetrical, the aza-enolate forms on the less substituted side, even when the distinction is between primary and secondary carbons. The best of our previous regioselective acylations have distinguished only methyl from more highly substituted carbon atoms.

> Hydrazones, as we explained on p. 351 of Chapter 14, are much less electrophilic than ketones. Even BuLi can be used as a base: it does not attack the C=N bond.

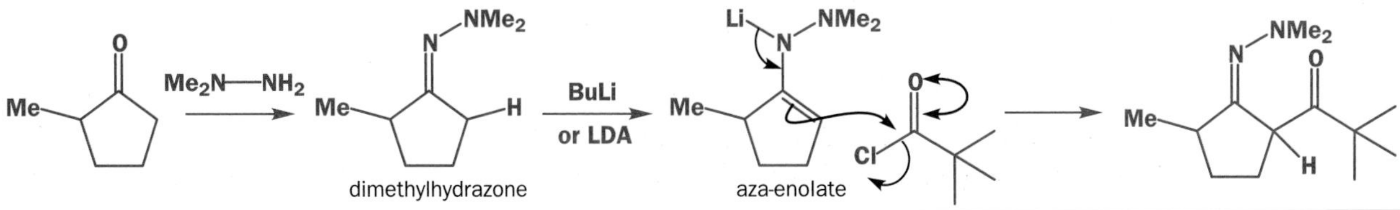

dimethylhydrazone   aza-enolate

You will not be surprised to find that the immediate product tautomerizes to an acyl-enamine further stabilized by an internal hydrogen bond. Mild acidic work-up releases the diketone product. The overall procedure may sound complicated—Me$_2$NNH$_2$ then base then acyl chloride then acidic methanol—but it is performed in a single flask and the products, the 1,3-diketones, are formed in excellent yield—in this case 83% overall.

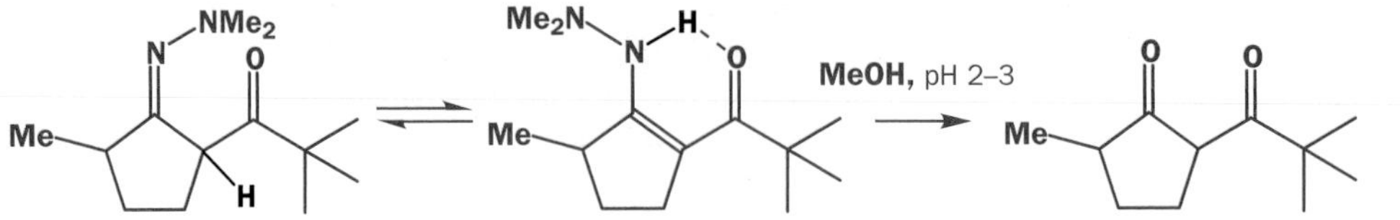

83% yield from starting ketone

Acylation of enols under acidic conditions

Under strongly acidic anhydrous conditions, carboxylic acids dehydrate to give the acylium ions, which you met as intermediates in the Friedel–Crafts reaction (Chapter 22).

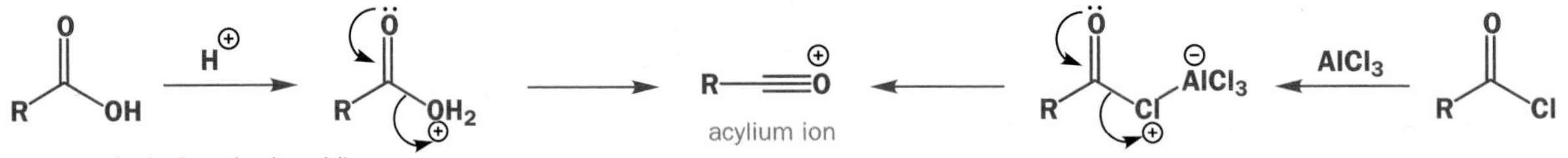

(polyphosphoric acid)   acylium ion

With another enolizable carbonyl group in the molecule, cyclization may occur to give a new 1,3-dicarbonyl compound. Popular conditions for this reaction are

PPA, 100 °C

HOAc

75% yield

polyphosphoric acid (PPA—partly dehydrated and polymerized H$_3$PO$_4$) in acetic acid as solvent.

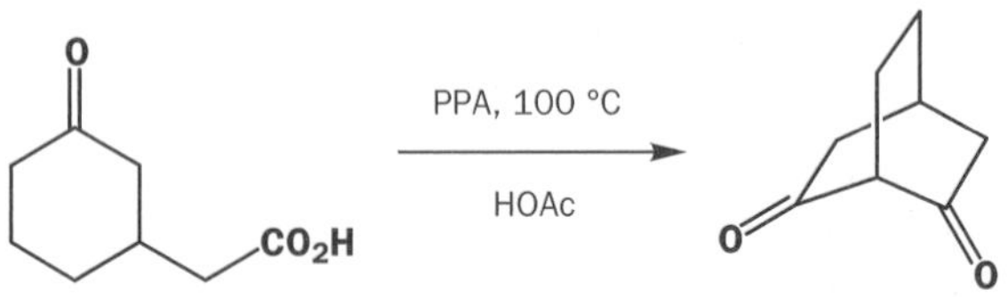

The first step is the formation of the acylium ion, which cyclizes on to one of the two possible enols of the ketone.

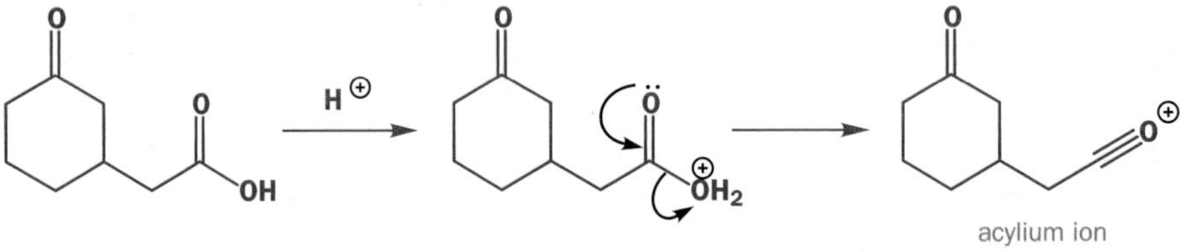

acylium ion

Though the cyclization looks awkward—the product is a bridged bicyclic diketone—the alternative would give a strained four-membered ring and does not occur.

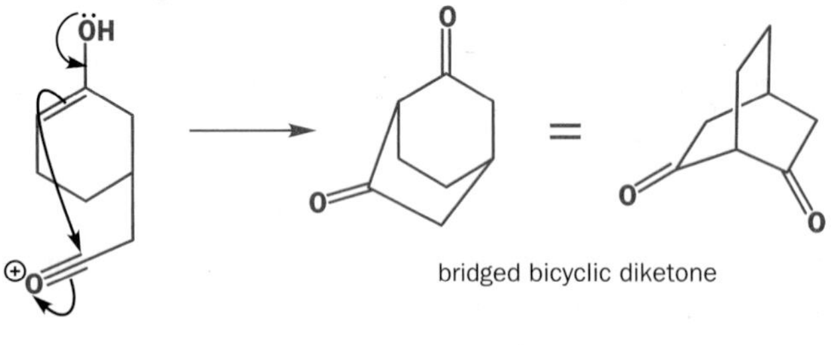

bridged bicyclic diketone

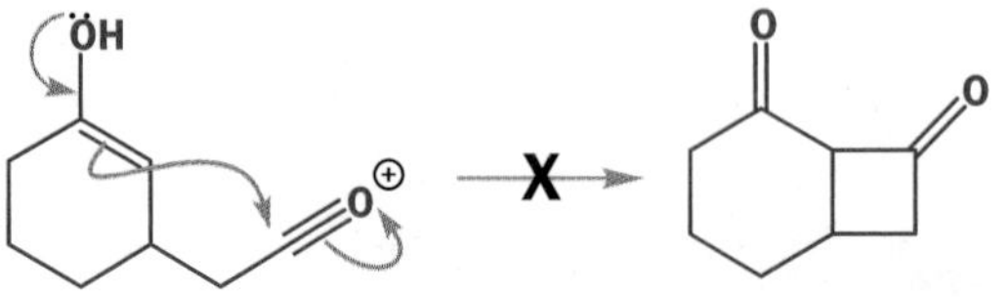

four-membered ring
—not formed

This cyclization is particularly impressive as the corresponding base-catalysed reaction on the keto-ester does not occur because a stable enolate cannot be formed—it would have an impossible bridgehead double bond.

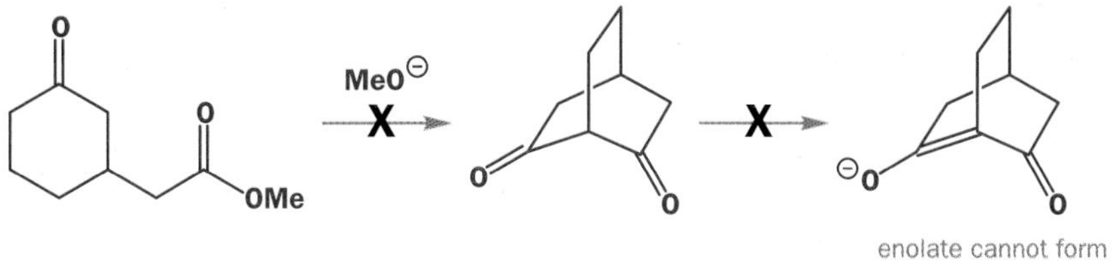

enolate cannot form

Evidently it is not necessary to form a stable conjugated enol in this acid-catalysed cyclization of keto-acids, and the reaction can even be used to make 1,3-diketones with no hydrogen atoms between the two carbonyl groups.

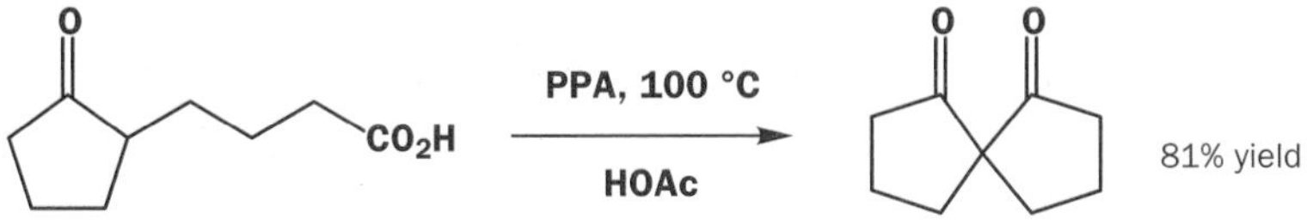

81% yield

This time the *spiro*-bicyclic diketone (one C atom common to both rings) is preferred to the alternative bridged bicyclic compound because both rings are five-membered.

Lewis acid-catalysed acylation of enols

Acylations of ketone enols with anhydrides are catalysed by Lewis acids such as BF$_3$. This process will remind you of Friedel–Crafts acylation but a better analogy is perhaps the aldol reaction where metals such as lithium hold the reagents together so that reaction can occur around a six-membered ring.

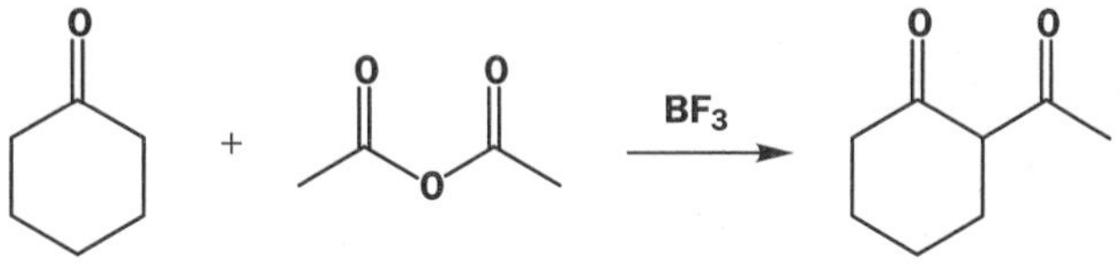

The mechanism obviously involves attack by the enol (or 'boron enolate') of the ketone on the anhydride, catalysed by the Lewis acid. Probably BF$_3$ or BF$_2$ groups (fluoride can come and go from boron easily) hold the reagent together at all times, much like lithium in the aldol reaction (p. 698).

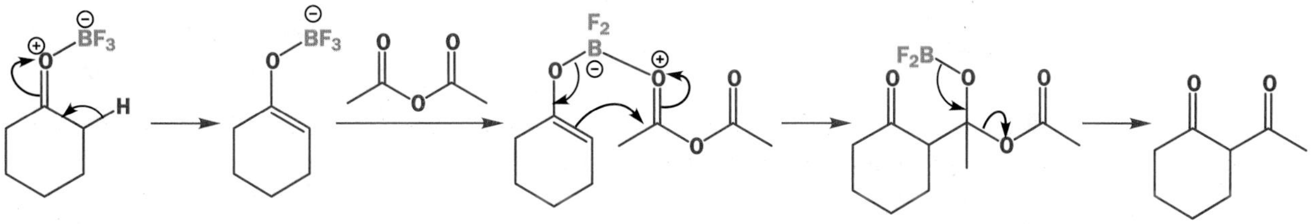

Under the conditions of the reaction, the product forms a stable boron enolate, which needs to be decomposed to the diketone with refluxing aqueous sodium acetate.

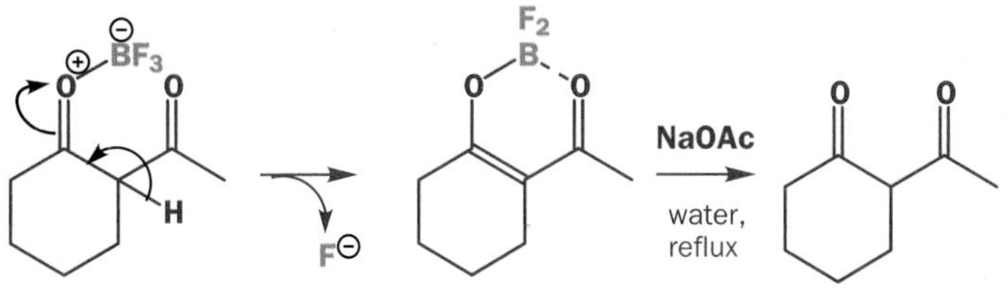

Preparation of a modern antibiotic—ofloxacin

As resistance to familiar antibiotics such as penicillin grows, it becomes even more important to discover not only new antibiotics but new classes of antibiotics having totally new structures. Ofloxacin is a member of one such class—the quinolone antibiotics.

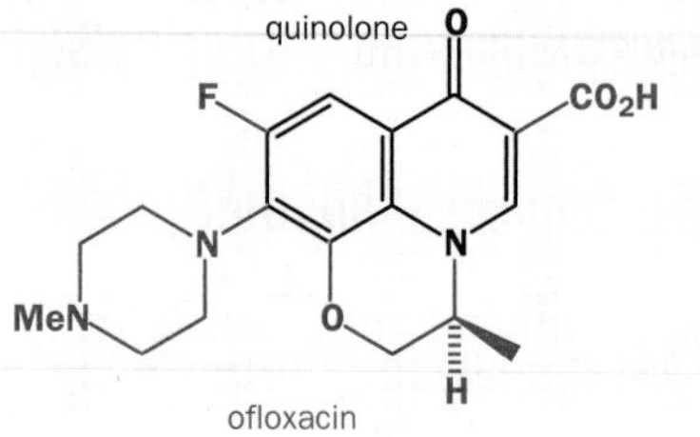

ofloxacin

Members of this class usually have an amine and a fluorine atom on the benzene ring as well as other embellishments as in ofloxacin, a recent example. The preparation of ofloxacin starts with two enolate acylations.

The first is the acylation of the magnesium derivative of diethyl malonate. The magnesium atom prevents O-acylation with acid chlorides, and decarboxylation (p. 678) removes the redundant ester group.

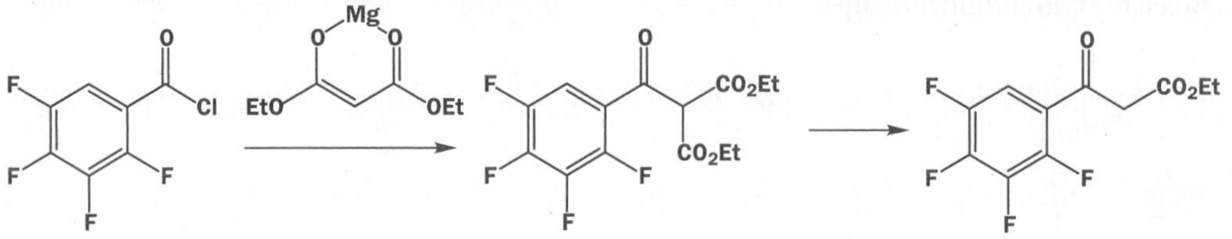

Later steps in the synthesis of ofloxacin were discussed in Chapter 23.

Acylation at nucleophilic carbon (other than enols and enolates)

We should not leave the subject of acylation at carbon without considering a problem that affects all such reactions to some degree. It can be understood most easily if we imagine some functional group Z that is able to stabilize a carbanion, and the acylation of that carbanion with an acid chloride—something like this.

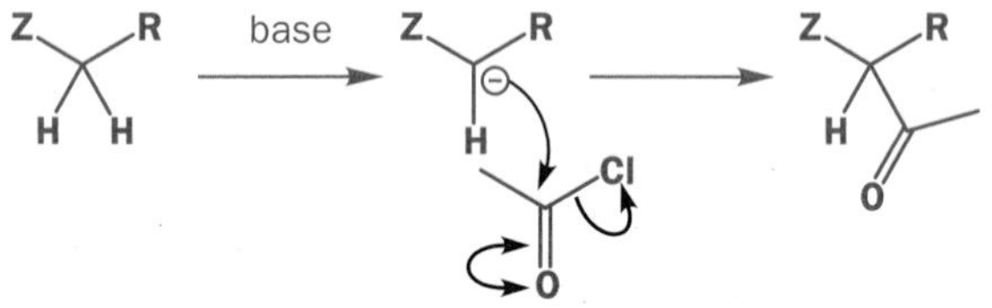

All looks well until we consider what might happen to the product under the reaction conditions. It too can form an anion, and a very stable one at that, because, not only is it stabilized by Z, but it is also an enolate.

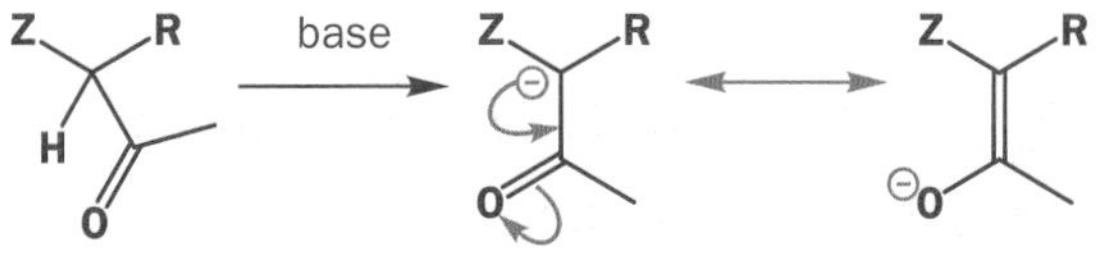

stabilized enolate anion

Since this anion is *more stable* (less basic) than the original anion, if there is an equilibrium between the two carbanions in the reaction mixture, the original carbanion will be sufficiently basic to act as the base that removes the proton from the product.

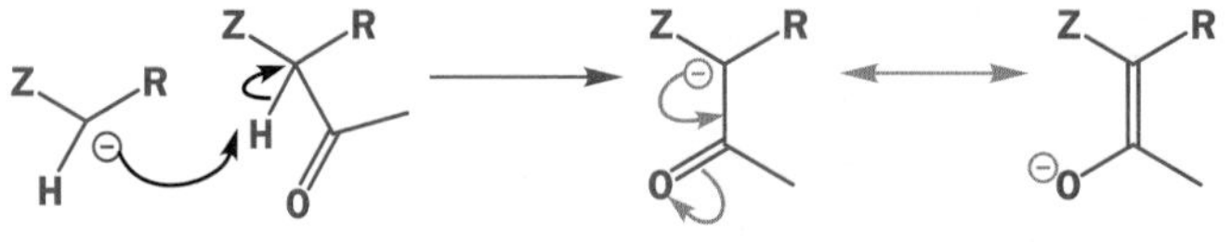

stabilized enolate anion

So, instead of being acylated, the starting anion is protonated. This side-reaction could reduce the maximum possible yield in the acylation reaction to 50%: half the starting material forms the product by acylation, while the other half simply *deprotonates* the product. How is this to be avoided?

In most of this chapter, we used enolates as our nucleophiles and worked under equilibrating conditions with alkoxide bases. There was alkoxide base present throughout the reaction, so the enolate didn't get used up deprotonating the product or, if it did, it could be re-deprotonated by the alkoxide. But the problem does arise in reactions such as the acylation of simple phosphorus ylids. Here two equivalents of ylid must be used to give a good yield of product. This does not matter in this case, because the ylid is cheap and disposable.

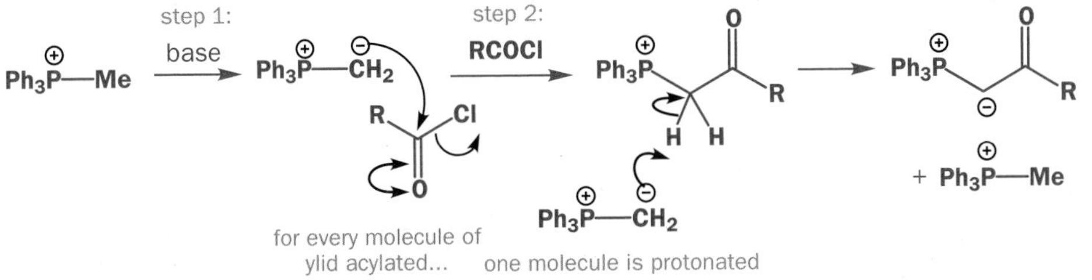

for every molecule of
ylid acylated...   one molecule is protonated

This is a good way of making stabilized ylides for the Wittig reaction: see p. 817.

If the reagent is too precious to waste, another device is to use two molecules of base for every one of the compound. In this way there is always a molecule of strong base waiting to remove a proton from the product. The acylation of sulfones with esters is a good example.

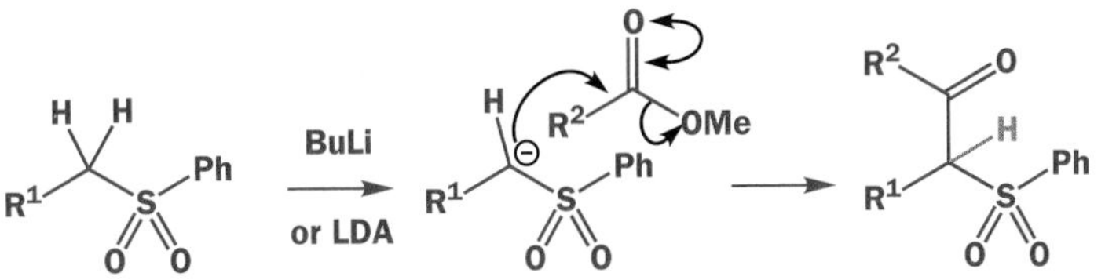

The product has a more acidic hydrogen atom (green) than any in the starting material so it could protonate the original anion. Using two equivalents of base avoids this.

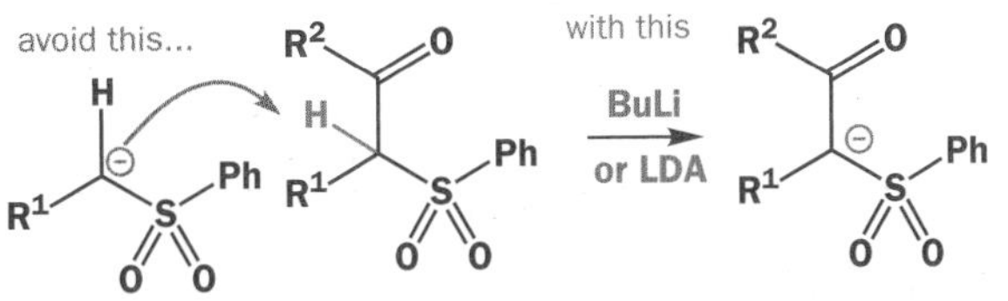

avoid this...   with this

By this device good yields of keto-sulfone can be formed even when both partners are aliphatic compounds with acidic protons.

79% yield

How Nature makes fatty acids

Fatty acids are big news, whether saturated or unsaturated. Too much saturated fatty acid seems to be bad for us, clogging arteries, while some unsaturated fatty acids seem to protect us against that fatal condition. There are hundreds of fatty acids in living things but most have one special characteristic—they have an even number of carbon atoms. Here are two of the most frequently found fatty acids.

palmitic acid (saturated fatty acid)

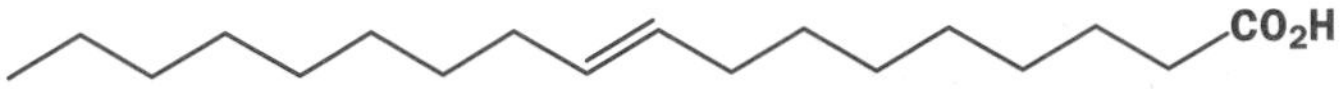

elaidic acid 'monounsaturate'

They have an even number of carbon atoms because they are made in living things by Claisen ester style condensations of acetic acid derivatives. In fact, at some stage in the biosynthesis of palmitic acid, there was a carbonyl group at each atom marked with a green blob here.

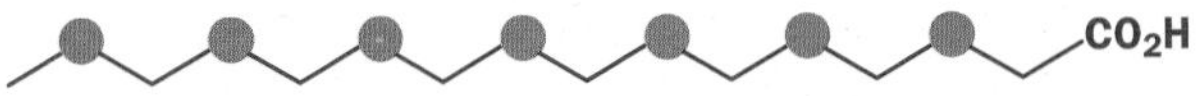

Nature takes the trouble to remove all these carbonyl groups. So why were they put there in the first place? It is because these long chains are much easier to assemble by the Claisen ester conjugation than by alternatives such as alkylation. Other natural products in this group show more obvious traces of carbonyl groups. Orsellinic acid, for example, is clearly formed directly by an aldol-style cyclization of this tetracarbonyl precursor.

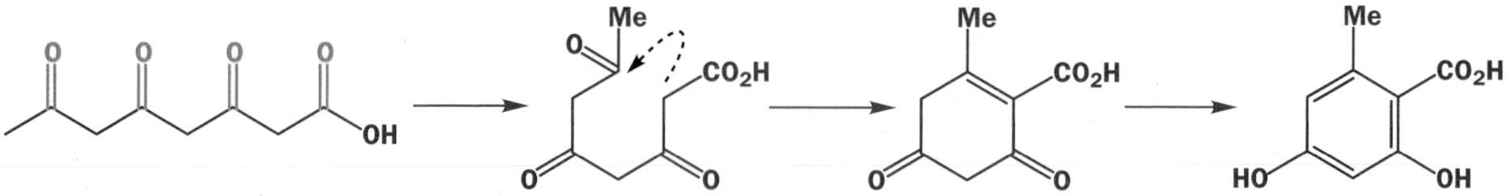

The straight-chain triketo-acid wraps itself round and cyclizes by a simple aldol reaction. Enolization of the two remaining ketones gives a benzene ring. So how does Nature assemble these chains in the first place?

The reactions use thiol esters rather than ordinary esters. The esterifying group is a thiol called coenzyme A, and we shall just represent this molecule as R (you can find its full structure on p. 1386). The first reaction is between a malonate half-thioester and an acetate thioester of coenzyme A. Look at the mechanism and you will see how similar it is to the Claisen ester condensation.

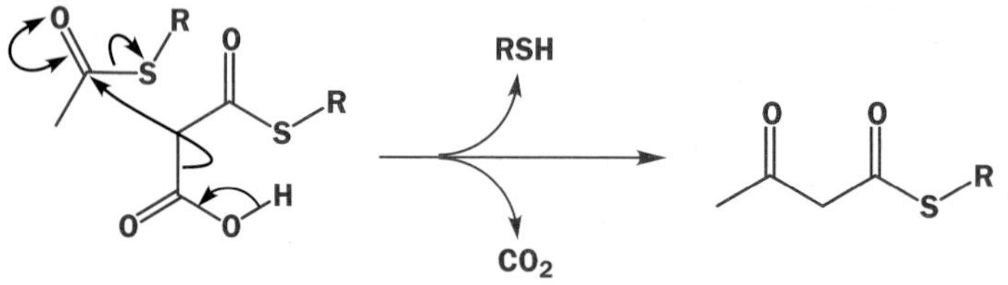

The main difference is that no discrete enol or enolate is actually formed. Instead CO_2 is lost from the malonate as the acylation occurs. This is an improvement from Nature's point of view—it is much easier to lose a proton from a carboxylic acid than from a CH_2 group. This reaction joins two C_2 units together and the whole process can be repeated as many times as necessary.

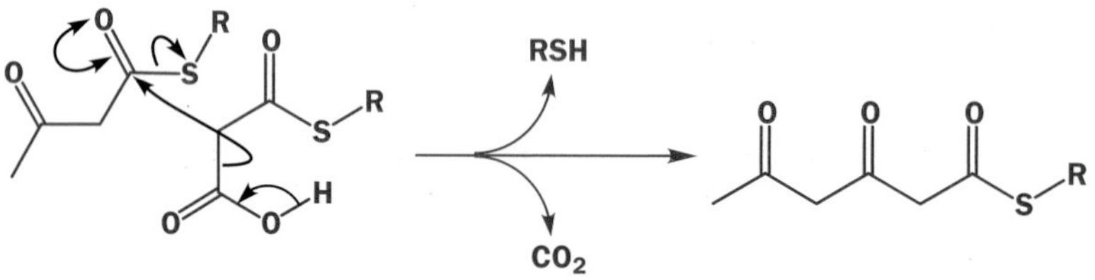

Because of the ketone group on every other carbon atom in the growing chain, these compounds are known collectively as **polyketides**. To make a saturated fatty acid, the ketone needs to be selectively reduced to an alcohol, water needs to be eliminated, and the conjugated double bond reduced. All these steps have simple chemical analogies.

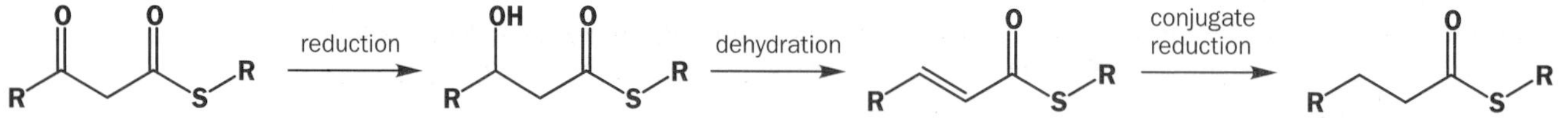

Polyketides of enormous variety are known with all these groups present in the chain at the various stages of reduction. But all are made by Nature's version of the Claisen ester condensation.

Learning from Nature

So what is so special about thiol esters? The main difference from ordinary esters is that the lone pairs on the sulfur atom are in 3p orbitals instead of 2p orbitals. These orbitals are too large to overlap efficiently with the 2p orbital on the carbon atom of the carbonyl group, so thiol esters have less conjugation than ordinary esters.

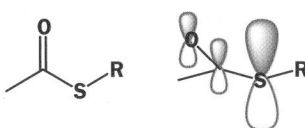
thiol ester

This difference affects each stage of the Claisen ester condensation in the same way. Thiol esters are more easily converted to enolate anions, they are more easily attacked by nucleophiles, and RS⁻ is a better leaving group than RO⁻. In each case the reaction is better (faster or equilibrium further towards product).

the Claisen thiol ester condensation

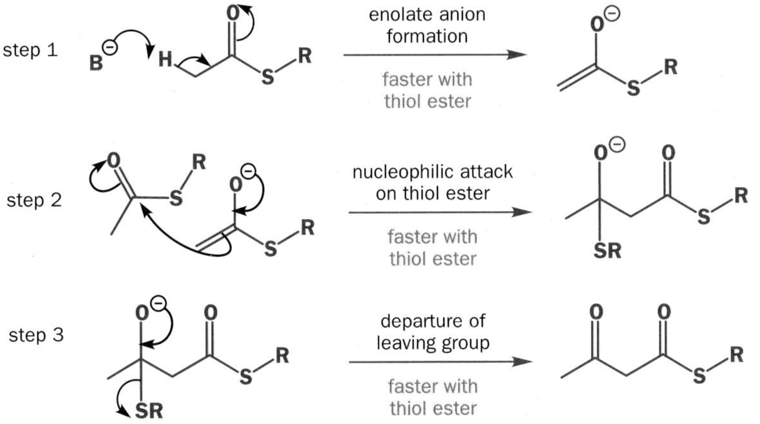

step 1 — enolate anion formation — faster with thiol ester

step 2 — nucleophilic attack on thiol ester — faster with thiol ester

step 3 — departure of leaving group — faster with thiol ester

We can learn from Nature by using thiol esters in simple Claisen condensations. Cyclization of this COSEt diester rather than the CO₂Et diester needs milder conditions (2 hours at room temperature in dimethoxyethane) and gives better yields.

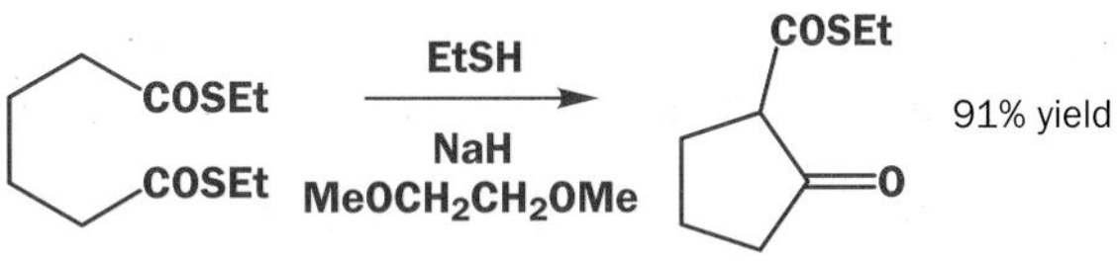

COSEt, COSEt → (EtSH, NaH, MeOCH₂CH₂OMe) → 91% yield

The thiol ester group can be removed, if necessary, by using Raney nickel, a good reducing agent for C–S bonds (see Chapter 46). Decarboxylation follows.

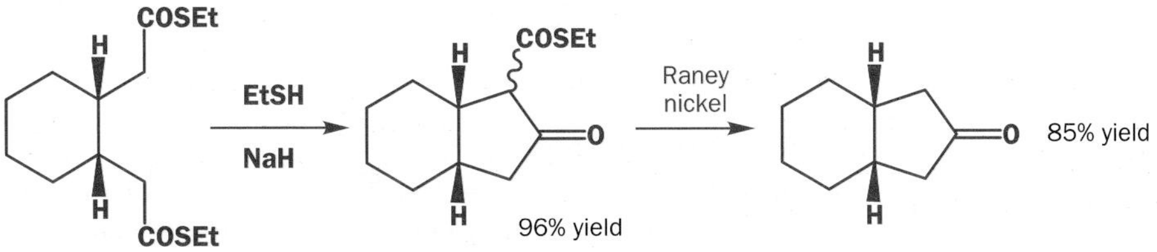

EtSH, NaH → 96% yield → Raney nickel → 85% yield

If we copy Nature rather more exactly, the Claisen ester condensation can be carried out under neutral conditions. This requires rather different reagents. The enol component is the magnesium salt of a malonate mono-thiol-ester, while the electrophilic component is an **imidazolide**—an amide derived from the heterocycle imidazole. Imidazole has a pK_a of about 7. Imidazolides are therefore very reactive amides, of about the same electrophilic reactivity as thiol esters. They are prepared from carboxylic acids with 'carbonyl diimidazole' (CDI).

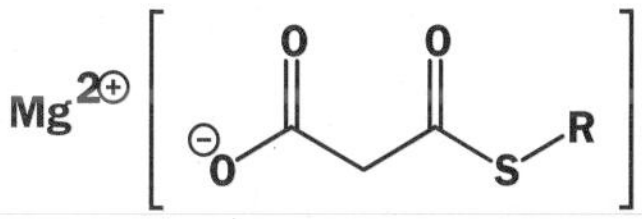

malonate mono-thiol ester magnesium salt

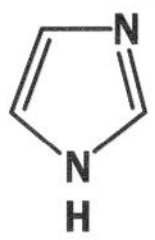

imidazole

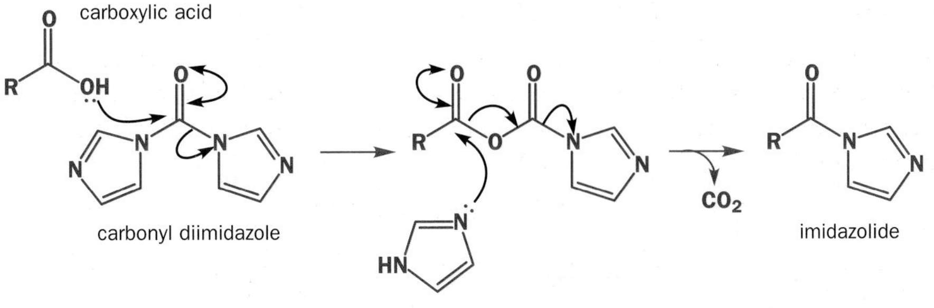

carboxylic acid

carbonyl diimidazole → → CO_2 → imidazolide

Reactions like these are said to be **bio-mimetic** because they draw their inspiration from Nature even if the imitation is not exact. We shall be discussing other important reactions carried out in Nature at various points of the book and collecting these ideas together in Chapters 49–51.

Combining the two reagents at neutral pH gives clean specific acylation at carbon. This is very like the biological reaction as CO_2 is lost *during* acylation.

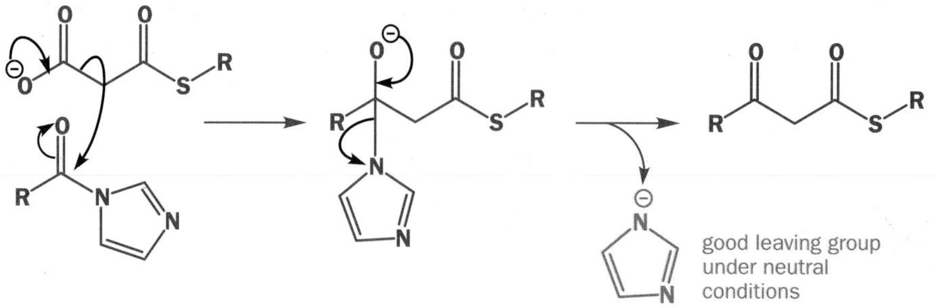

good leaving group under neutral conditions

To conclude...

You have now met enols and enolates doing nearly all of the things that other nucleophiles do:

- taking part in nucleophilic substitution reactions at saturated C (Chapter 26)
- adding to C=O groups (the aldol reaction, Chapter 27)
- substituting at C=O groups (Chapter 28)

There is one more aspect of enolate chemistry left to discuss:

- conjugate addition

It follows in the next chapter.

Problems

1. Attempted acylation at carbon often fails. What would be the true products of these attempted acylations, and how would you actually make the target molecules?

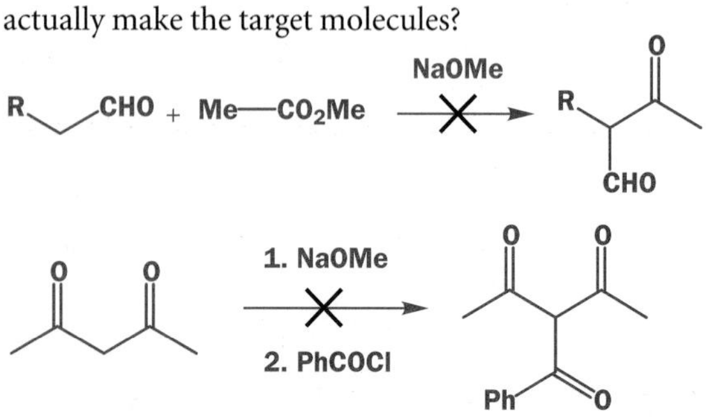

2. The synthesis of six-membered heterocyclic ketones by intramolecular Claisen condensation was described in the chapter and we pointed out that it doesn't matter which way round the cyclization happens as the product is the same. For example:

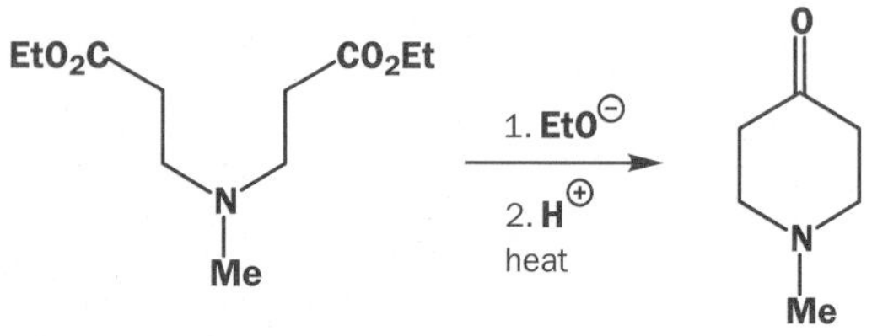

Strangely enough, five-membered heterocyclic ketones can be made by a similar process. The starting material is not symmetrical and two possible cyclized products can be formed. Draw

structures for these two products and explain why it is unimportant which is formed.

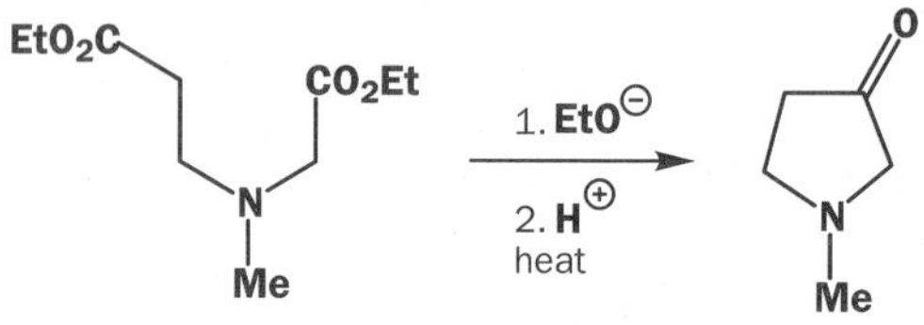

3. The synthesis of corylone was outlined in the chapter but no mechanistic details were given. Suggest mechanisms for the first two steps. The last step is a very unusual type of reaction and you have not met anything quite like it before. However, organic chemists should be able to draw mechanisms for new reactions and you might like to try your hand at this one. There are several steps.

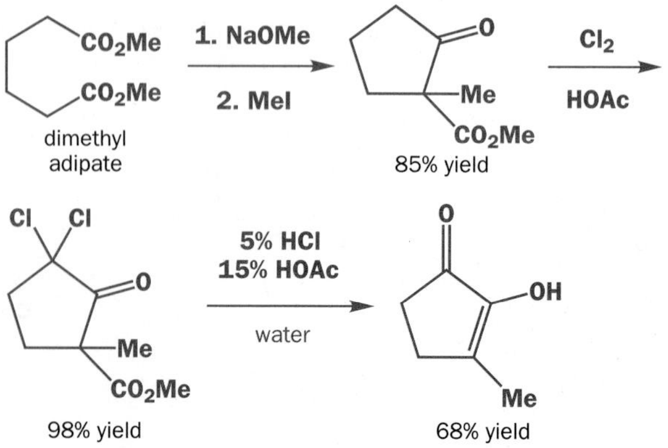

4. Acylation of the phenolic ketone gives a compound A, which is converted into an isomeric compound B in base. Cyclization of B in acid gives the product shown. Suggest mechanisms for the reactions and structures for A and B.

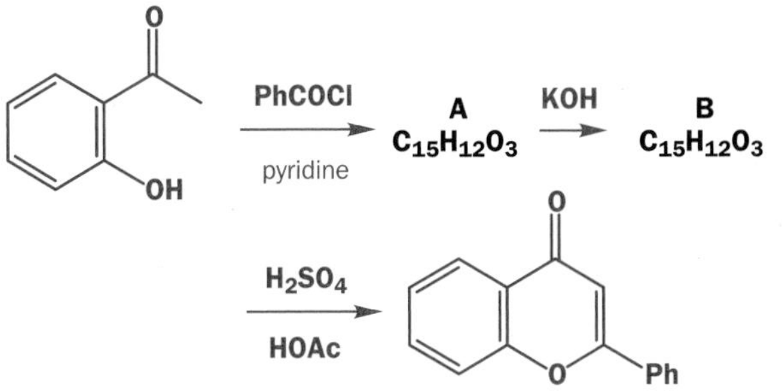

5. How could these compounds be made using the acylation of an enol or enolate as a key step?

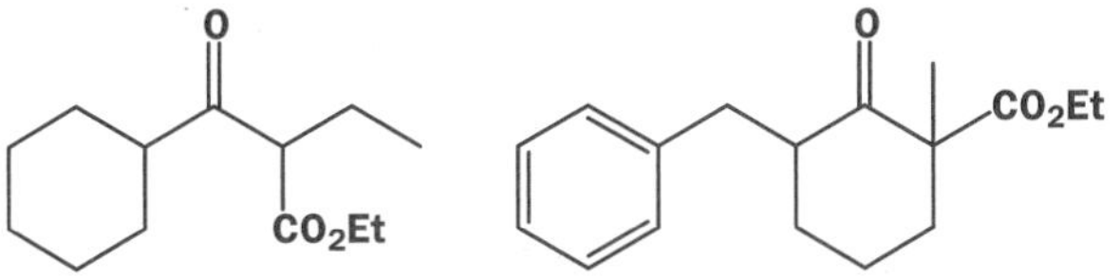

6. In a synthesis of cubane, a key step was the intramolecular acylation of this symmetrical diester. Explain why a strong base (the anion of DMSO, MeSO.CH$_2^-$, was actually used) is necessary for this cyclization.

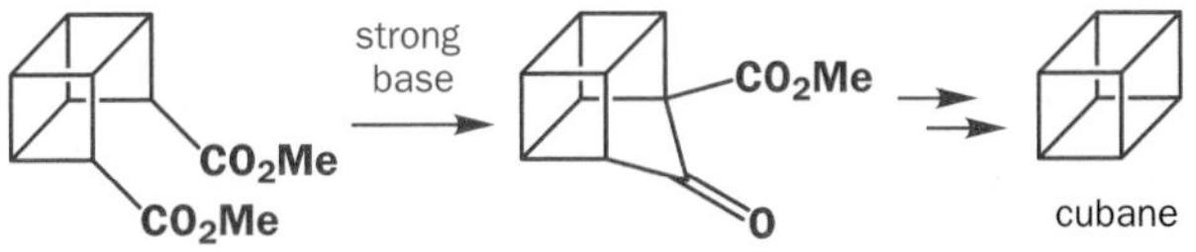

The starting material had both of the ester groups on the outside of the molecule so that cyclization is impossible. What preliminary step must first occur for it to become possible?

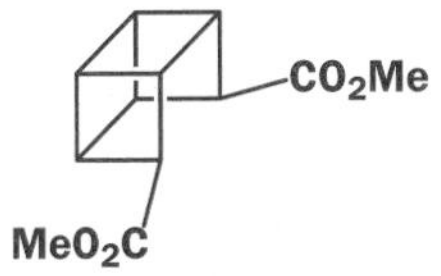

7. Suggest mechanisms for this sequence leading to a bicyclic compound with four- and seven-membered rings *cis*-fused to each other.

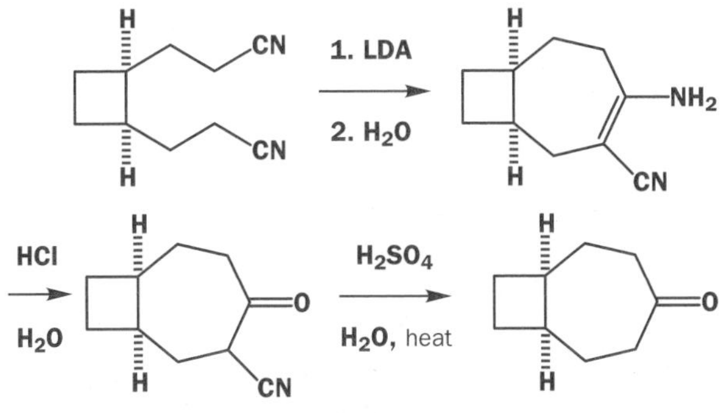

8. Give mechanisms for the steps used in this synthesis of the natural product bullatenone. Comment on the reagents used for the acylation step, on the existence of the first intermediate as 100% enol, on the mechanism of the cyclization, and on how the decarboxylation is possible.

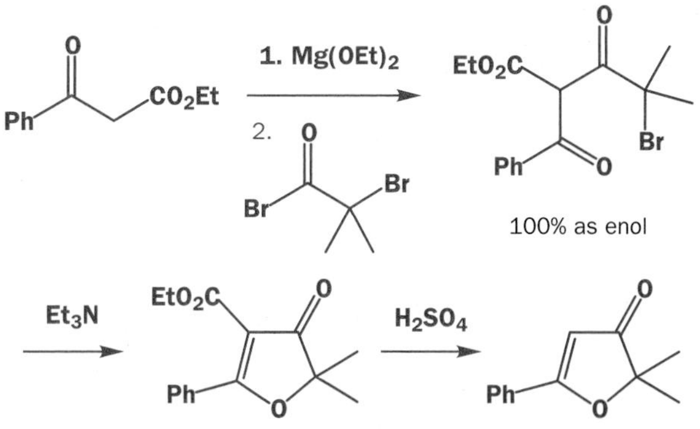

9. Suggest how the following reactions might be made to work. You will probably have to select a specific enol equivalent.

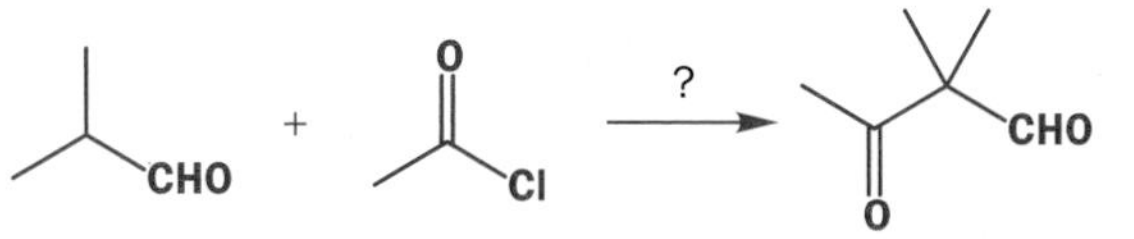

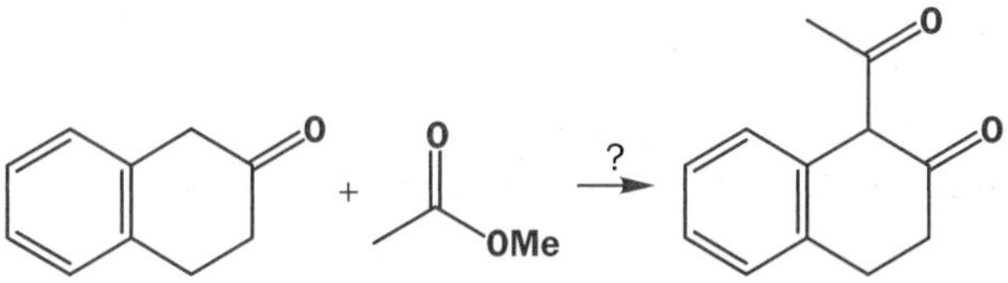

10. Suggest mechanisms for these reactions, explaining why these particular products are formed.

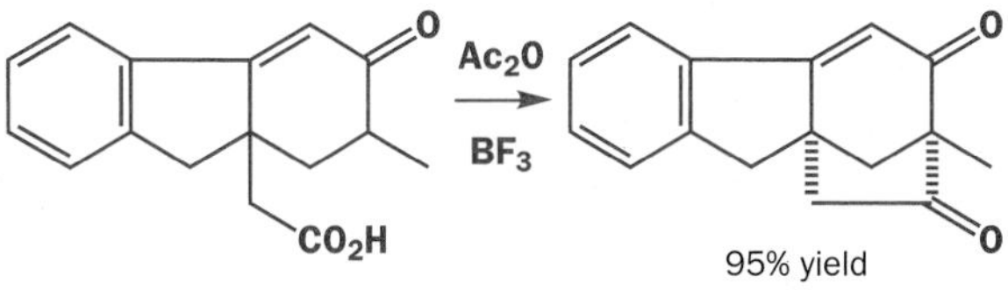

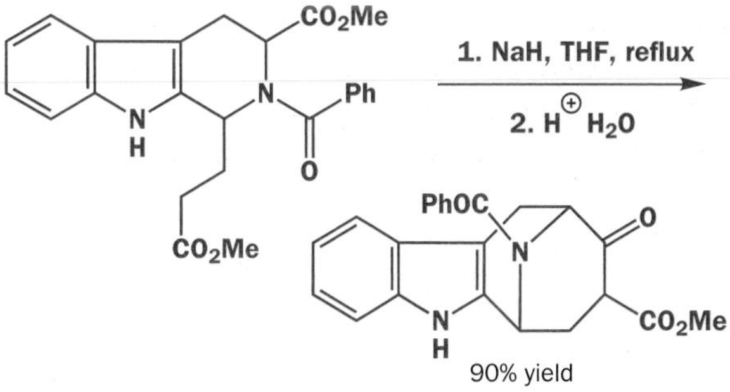

11. Sodium enolates generally react with acid chlorides to give enol esters. Give a mechanism for this reaction and explain the selectivity.

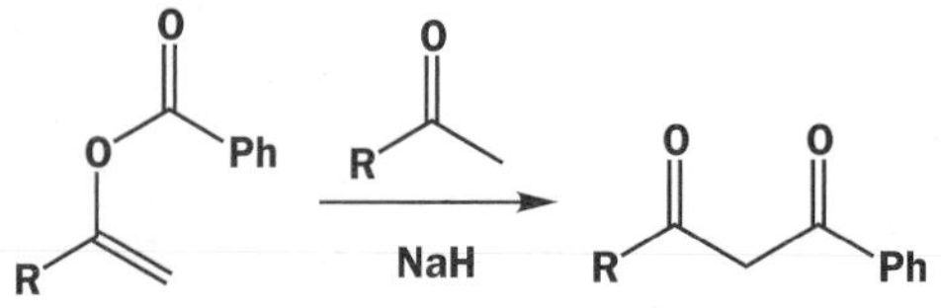

If the enol ester is treated with an excess of the sodium enolate, *C*-acylation occurs. Give a mechanism for this reaction. Why does the *C*-acylated product predominate?

12. This is a *C*-acylation route to a simple ketone. Why was NaH chosen as the base? Why did *O*-acylation not occur? Why were *t*-butyl esters used? What would probably have happened if the more obvious Friedel–Crafts (Chapter 22) route were tried instead?

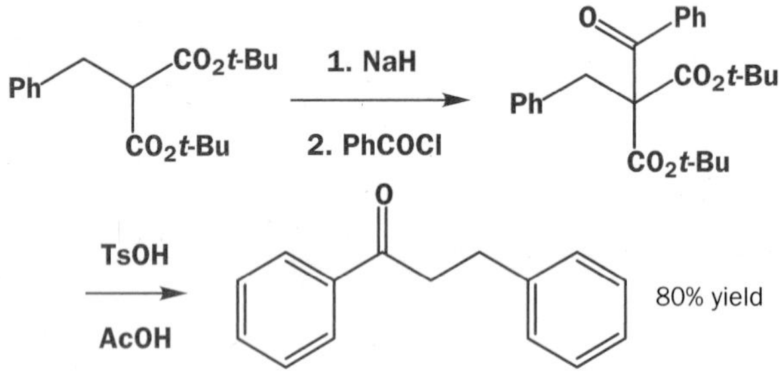

13. Base-catalysed reaction between these two esters allows the isolation of one product in 82% yield. Predict its structure.

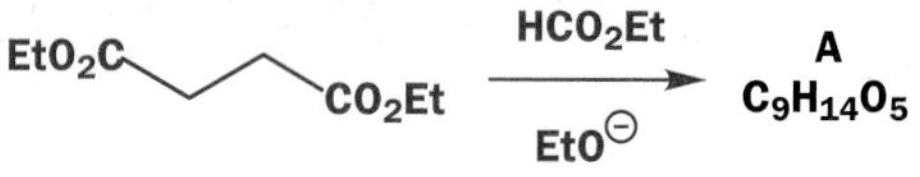

The NMR spectrum of the product shows that two species are present. Both show two 3H triplets at about $\delta_H = 1$ and two 2H quartets at about $\delta_H = 3$. One has a very low field proton and an ABX system at 2.1–2.9 with J_{AB} 16 Hz, J_{AX} 8 Hz, and J_{BX} 4 Hz. The other has a 2H singlet at 2.28 and two protons at 5.44 and 8.86 coupled with J 13 Hz. One of these protons exchanges with D_2O. Any attempt to separate the mixture (for example, by distillation or chromatography) gives the same mixture. Both compounds, or the mixture, on treatment with ethanol in acid solution give the same product. What are these compounds?

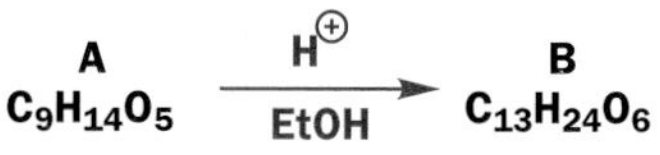

Compound B has IR 1740 cm^{-1}, δ_H (p.p.m.) 1.15–1.25 (four t, each 3H), 3.45 (2H, q), 3.62 (2H, q), 4.1 (two 2, each 2H), 2.52 (2H, ABX system, J_{AB} 16 Hz), 3.04 (1H, X of ABX split into a further doublet by J 5 Hz), and 4.6 (1H, d, J 5 Hz). The couplings between A and X and between B and X are not quoted in the paper. Nevertheless, you should be able to work out a structure for compound B.

Conjugate addition of enolates

29

Connections

Building on:
- Carbonyl chemistry ch6, ch12, & ch14
- Conjugate addition ch10
- Enols and enolates ch21
- Nucleophilic attack on electrophilic alkenes ch23
- Synthesis in action ch25
- Chemistry of enol(ate)s ch26–ch29

Arriving at:
- Convergent plans for synthesis
- Thermodynamic control
- Selection of reagents for enol(ate) conjugate addition
- Tandem reactions and Robinson annelation
- Substitution may be elimination–conjugate addition in disguise
- Nitriles and nitro compounds

Looking forward to:
- Synthesis and retrosynthesis ch30l
- Diastereoselectivity ch33–ch34l
- Saturated and unsaturated heterocycles ch42 & ch44
- Main group chemistry ch46–ch47
- Asymmetric synthesis ch45
- Natural products ch51

Introduction: conjugate addition of enolates is a powerful synthetic transformation

The product of a conjugate addition of an enolate or enol equivalent to an α,β-unsaturated carbonyl compound will necessarily be a dicarbonyl compound or an equivalent derivative. As the carbonyl group occupies such a central position in synthesis it will come as no surprise that these intermediates, with two carbonyl groups, are very widely used.

> Conjugate addition is also called Michael addition and is described in Chapters 10 and 23.

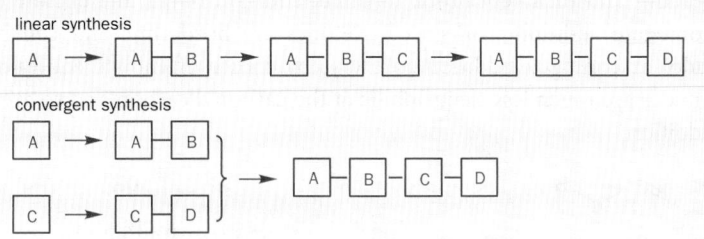

The other important feature of this conjugate addition reaction is that the two carbonyl groups in the product are reasonably far apart while the newly formed bond is in the middle of the molecule. This means that Michael addition can be a *convergent* route to the product—a feature that usually maximizes synthetic efficiency.

Linear vs. convergent syntheses

A **convergent synthesis** joins large fragments that have been assembled beforehand rather than adding together many small fragments in a linear fashion. The overall yield will generally be higher.

linear synthesis

A → A—B → A—B—C → A—B—C—D

convergent synthesis

A → A—B
C → C—D
} → A—B—C—D

Conjugate addition of enolates is the result of thermodynamic control

Enolate nucleophiles have exactly the same opportunity to attack the carbonyl group directly as do the simple nucleophiles discussed in Chapter 10 and the same factors govern the eventual outcome

We discussed the reason for this in Chapters 10 and 23. The main reason that the conjugate addition product is more stable is that it has a C=O group while the direct addition product has a C=C group

of the reaction. Thermodynamic control leads to conjugate addition but kinetic control leads to direct addition. The key to successful conjugate addition is to ensure that direct addition to the carbonyl (an aldol reaction, Chapter 28) is reversible. This enables the conjugate addition to compete and, as its product is more stable, it eventually becomes the sole product. This is thermodynamic control at its best!

A **retro-aldol reaction** is just an aldol reaction in the reverse direction. You will meet other 'retro' reactions later in the book, such as the important retro-Diels–Alder reaction in Chapter 35.

The aldol product is more sterically hindered than the conjugate addition product so increased branching on the nucleophile tends to accelerate the retro-aldol process, which releases steric strain and favours equilibration to the thermodynamic product. Perhaps more important is the stability of the enolate: the more stable the starting enolate, the easier it is to reverse both reactions and this favours the more stable conjugate addition product. One of the most important ways of stabilizing an enolate—using another electron-withdrawing group such as CO_2Et—achieves both of these enhancements at the same time as branching inevitably accompanies the extra anion stabilization.

There is also a frontier orbital effect that assists conjugate addition over the aldol reaction. You will recall that the carbonyl carbon is a relatively hard centre, whereas the β carbon of an enone is soft. As the nucleophilic enolate becomes more stabilized with extra electron-withdrawing groups, it becomes increasingly soft and hence more likely to attack the β carbon.

The unsaturated component plays an important role

These factors are discussed in Chapters 10 and 23.

The nature of the carbonyl group in the α,β-unsaturated electrophile is also important as the more electrophilic carbonyl groups give more direct addition and the less electrophilic carbonyl groups (esters, amides) give more conjugate addition. Aldehydes are unhindered and very reactive and thus very prone to direct addition but, if the enolate equivalent is carefully chosen, conjugate addition works well. Ketones are borderline and can be pushed towards either the aldol or conjugate addition pathways by choice of enolate equivalent as we shall see. Esters and amides are much less electrophilic at the carbonyl carbon and so are good substrates for conjugate addition.

decreasing reactivity of carbonyl with nucleophiles

increasing tendency to conjugate addition

● **Conjugate addition is *thermodynamically* controlled; direct addition is *kinetically* controlled**

Stable enolates promote conjugate addition by:
- making the aldol reaction more reversible
- making the enolate anion softer

Less reactive Michael acceptors promote conjugate addition by:
- making the aldol reaction more reversible
- making the carbonyl group less electrophilic

Esters are excellent anion-stabilizing groups on enolate or Michael acceptors

β-Diesters (malonates and substituted derivatives) combine three useful features in conjugate addition reactions: they form stable enolate anions that undergo clean conjugate addition; if required, one of the ester groups can be removed by hydrolysis and decarboxylation; and, finally, the remaining acid or ester is ideal for conversion into other functional groups.

■ Hydrolysis and decarboxylation and the choice of base were discussed in Chapter 26.

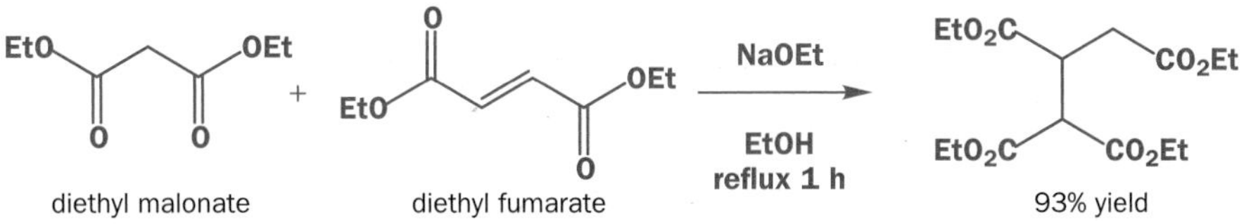

diethyl malonate   diethyl fumarate   93% yield

Diethyl malonate adds to diethyl fumarate in a conjugate addition reaction promoted by sodium ethoxide in dry ethanol to give a tetraester. Diethyl fumarate is an excellent Michael acceptor because two ester groups withdraw electrons from the alkene. The mechanism involves deprotonation of the malonate, conjugate addition, and reprotonation of the product enolate by ethanol solvent. In this reaction two ester groups stabilize the enolate and two more promote conjugate addition.

stabilized enolate   tetraester product

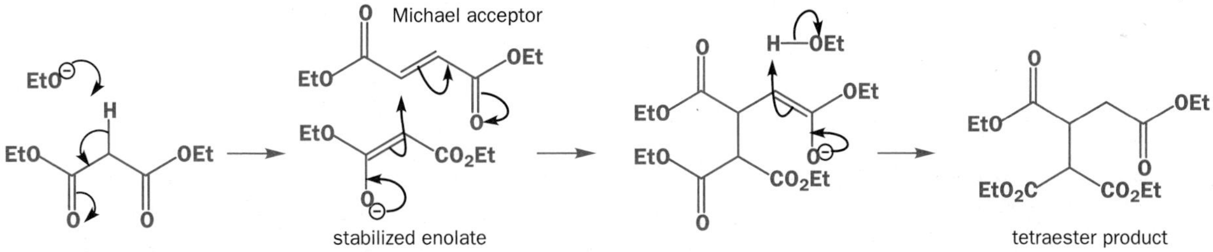

The value of malonate esters is illustrated in this synthesis of a substituted cyclic anhydride by conjugate addition to ethyl crotonate, hydrolysis, and decarboxylation, followed by dehydration with acetic anhydride. This route is very general and could be used to make a range of anhydrides with different substituents simply by choosing an appropriate unsaturated ester.

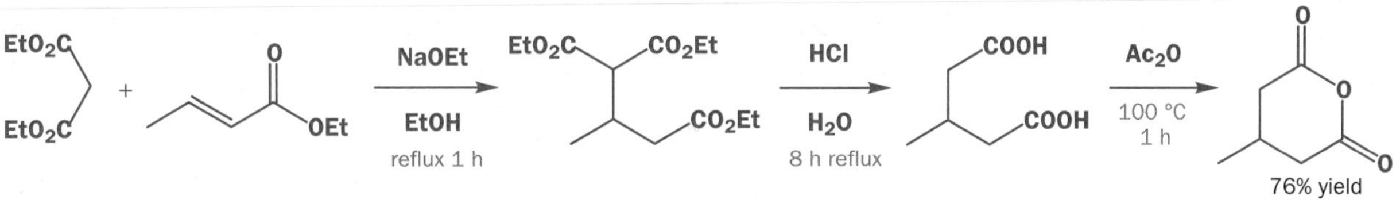

76% yield

The mechanism of the conjugate addition is the same as that in the previous example and the mechanism for ester hydrolysis was covered in Chapter 12. The key step in the dehydration reaction is the formation and cyclization of the mixed anhydride formed from the diacid and acetic anhydride. Both steps have the same mechanism, attack of an acid on an anhydride, but the second step is intramolecular. Like most cyclizations the reaction is entropically favoured as two molecules react to give three—the cyclic anhydride and two molecules of acetic acid.

> ● **Use of electron-withdrawing groups to favour conjugate addition**
>
> Conjugate addition of enolates is promoted by electron-withdrawing groups (for example, CO_2Et), especially by:
> - two electron-withdrawing groups stabilizing the enolate
> - two electron-withdrawing groups conjugated with the alkene
>
> It is not necessary to have both features in the same reaction.

Alkali metal (Li, Na, K) enolates can undergo kinetic conjugate addition

It is not essential to have two anion-stabilizing groups for successful conjugate addition and it is even possible with simple alkali metal (Li, Na, and K) enolates. Lithium enolates are not ideal nucleophiles for thermodynamically controlled conjugate addition. Better results are often observed with sodium or potassium enolates, which are more dissociated and thus more likely to revert. Lithium binds strongly to oxygen and so tends to prevent reversible aldol addition, which leads to loss of conjugate addition product. Potassium t-butoxide is the ideal base for this example as it is hindered and so will not attack the ester but is basic enough to deprotonate the ketone to a certain extent.

Two enolates are possible but, under the equilibrating conditions, the more stable and more reactive enolate is the important intermediate leading to the more interesting product with a quaternary carbon atom.

If the conditions are right, good yields are sometimes observed from *kinetically* controlled conjugate addition even with lithium enolates. This unlikely outcome is favoured by hindered nucleophiles and conjugated or hindered carbonyls. In these cases the lack of reversibility is not an issue as the aldol product is never formed. In this example the enolate of the t-butyl ketone is the hindered nucleophile and the conjugated ketone is rather unreactive.

However, most successful conjugate additions use stable enol or enolate equivalents and we shall continue to discuss them in the next section.

Conjugate addition can be catalytic in base

As the penultimate product in a conjugate addition is an enolate anion, if the pK_a of the nucleophile is appropriate, only a catalytic quantity of base is required to initiate the reaction. The enolate anion of the product is protonated by a molecule of starting material to give the neutral final product and another enolate anion of starting material. The reversible reaction sequence, including the unwanted aldol equilibrium, can be forced over towards the conjugate addition product. The balance of pK_as is likely to be right for nucleophiles with two electron-withdrawing groups when adding to a double bond conjugated to a single carbonyl group.

pK_{aH} 20–25 pK_a 10–15 pK_a 20–25 pK_{aH} 10–15

This proton exchange sets up a catalytic cycle. The cycle is started by an external base removing a proton from the most acidic species present in the reaction mixture at the start which is the nucleophile. This is an important condition for success of the catalytic method and the reason that all the reactants can be mixed together at the start of the reaction with no adverse effects. There is no need to form the nucleophilic enolate quantitatively; more is formed as the reaction proceeds. The advantages of this way of running a conjugate addition are that strongly basic conditions are avoided so that mild bases such as tertiary amines (for example, Et_3N) or fluorides (for example, Bu_4NF) can be employed successfully.

Hydrogen fluoride is a weak acid in aqueous solution, $pK_a = 3.45$, due to the strength of the H–F bond. This bond strength also accounts for the basicity of the fluoride ion.

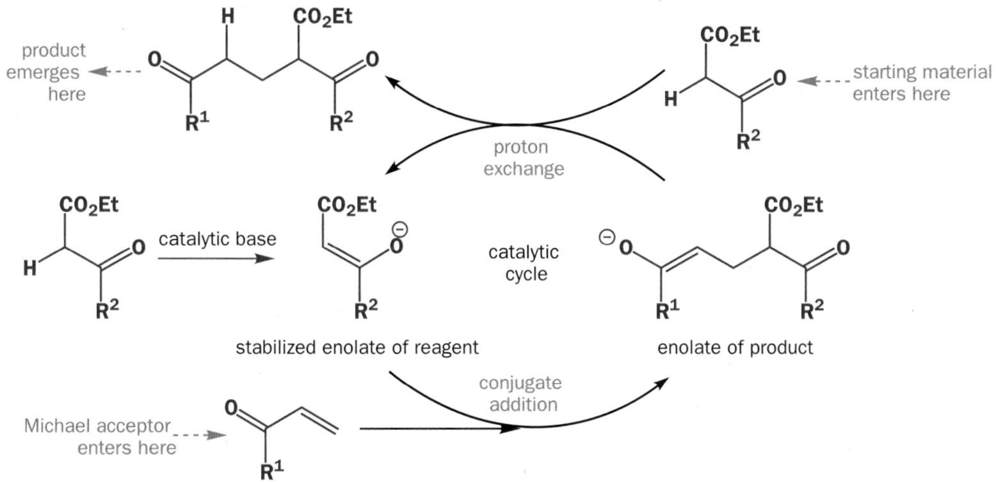

Diagrams of catalytic cycles are not always easy to understand. The main cycle rotates anticlockwise round the centre of the diagram with the starting materials entering top right and bottom left with the product emerging top left. The first molecules of enolate enter middle left. It would be helpful if you were to follow the formation of one molecule of product on the diagram and see how it sets off the next cycle. It is very important that you do not allow catalytic cycles to replace mechanisms in your understanding of chemical reactions.

The catalytic approach to conjugate addition is illustrated by the addition of a β-diketone to an aromatic enone catalysed by potassium hydroxide and benzyltriethylammonium chloride, which is a phase transfer catalyst. Once again, the catalytic cycle is initiated by deprotonation of the most acidic component in the reaction mixture, acetyl acetone, which is followed by a cycle of conjugate addition and proton exchange leading inexorably to the product.

The origins of the benefits of phase transfer catalysis (PTC) were presented in Chapters 23 and 26.

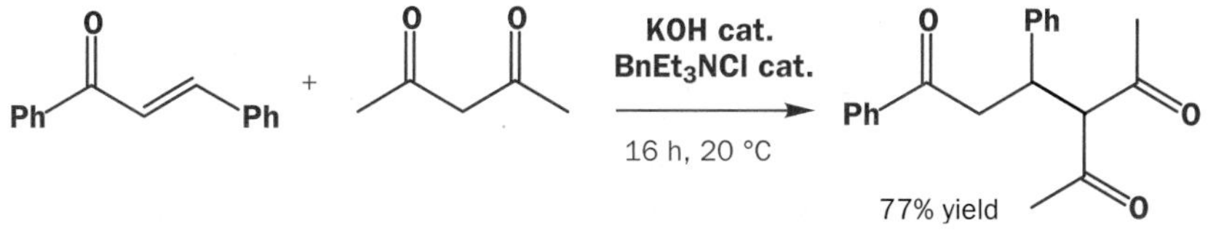

Enols are more likely than enolates to undergo direct conjugate addition

Base catalysis is not required for conjugate addition. If the nucleophile is sufficiently enolized under the reaction conditions then the enol form is perfectly able to attack the unsaturated carbonyl compound. Enols are neutral and thus soft nucleophiles favouring conjugate attack, and β-dicarbonyl compounds are enolized to a significant extent (Chapter 21). Under acidic conditions there can be absolutely no base present but conjugate addition proceeds very efficiently. In this way methyl vinyl ketone (butenone) reacts with the cyclic β-diketone promoted by acetic acid to form a quaternary centre. The yield is excellent and the triketone product is an important intermediate in steroid synthesis as you will see later in this chapter.

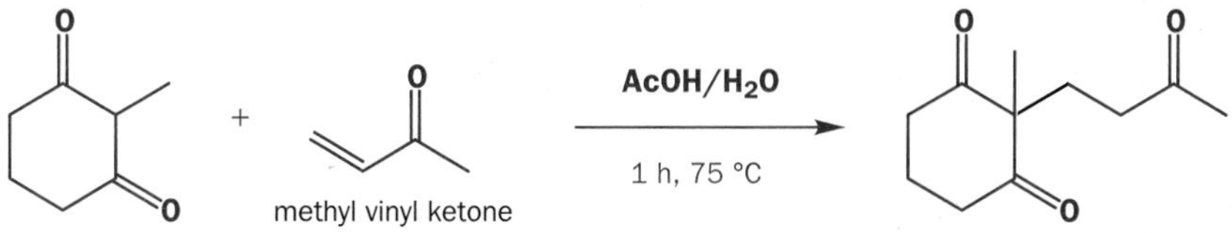

The mechanism involves acid-catalysed conversion of the keto form of the cyclic β-diketone into the enol form, which is able to attack the protonated enone. The mechanistic detail is precisely analogous to the attack of an enolate shown above; the only difference is that both reactants are protonated. The product is the enol form of the triketone, which rapidly tautomerizes to the more stable keto form.

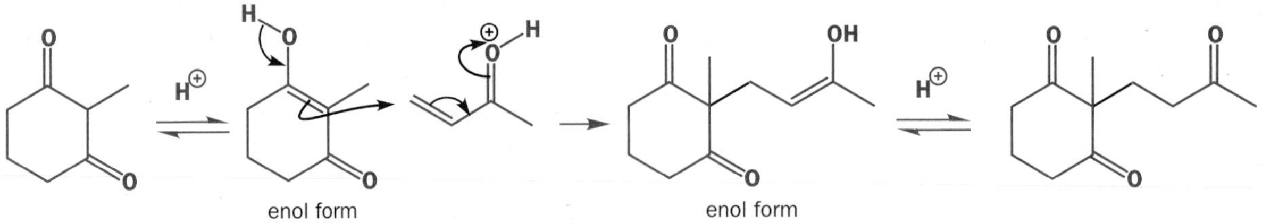

enol form   enol form

The thermodynamic control of conjugate addition allows even enals that are very electrophilic at the carbonyl carbon to participate successfully. Any aldol reaction, which must surely occur, is reversible and 1,4-addition eventually wins out. Acrolein combines with this five-membered di-ketone under very mild conditions to give a quantitative yield of product. The mechanism is analogous to that shown above.

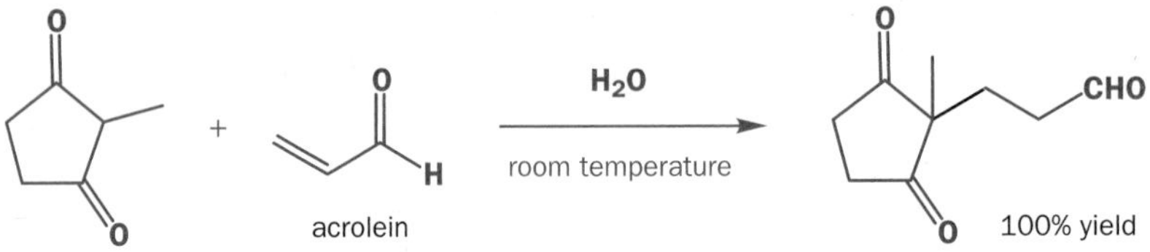

acrolein   100% yield

Enamines are convenient stable enol equivalents for conjugate addition

If you want to do a conjugate addition of a carbonyl compound without having a second anion-stabilizing group, you need some stable and relatively unreactive enol equivalent. In Chapters 27 and 28 you saw how enamines are useful in alkylation reactions. These neutral species are also perfect for conjugate addition as they are soft nucleophiles but are more reactive than enols and can be prepared quantitatively in advance. The reactivity of enamines is such that heating the reactants together, sometimes neat, is all that is required. Protic or Lewis acid catalysis can also be used to catalyse the reaction at lower temperature.

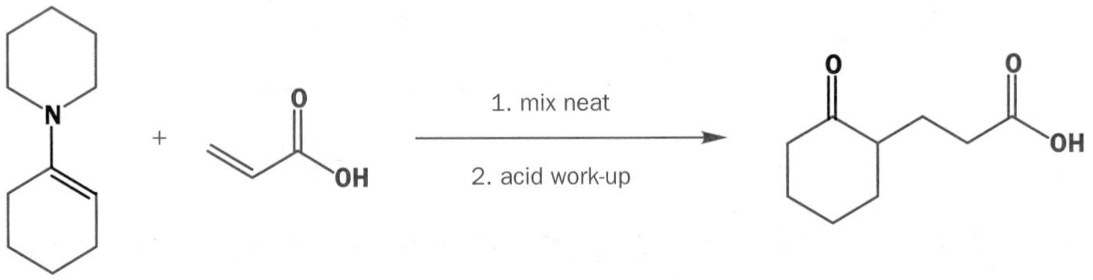

The mechanism is rather like enol addition. The differences are that the enamine is more nucleo-philic because of the nitrogen atom and that the product is an enamine, which can be converted into the corresponding carbonyl by mild acidic hydrolysis. This is usually performed during the work-up and so does not really constitute an extra step. The amine is washed out as the hydrochloride salt so isolation is straightforward. After conjugate addition the resulting enolate-iminium ion undergoes proton transfer rapidly to produce the more stable carbonyl-enamine tautomer. This is shown as an intramolecular process but it could just as easily be drawn with an external base and source of protons. The resulting enamine is then stable until aqueous acid is added at the end of the reaction. Hydrolysis occurs via the iminium ion to reveal the second carbonyl group and release the secondary amine.

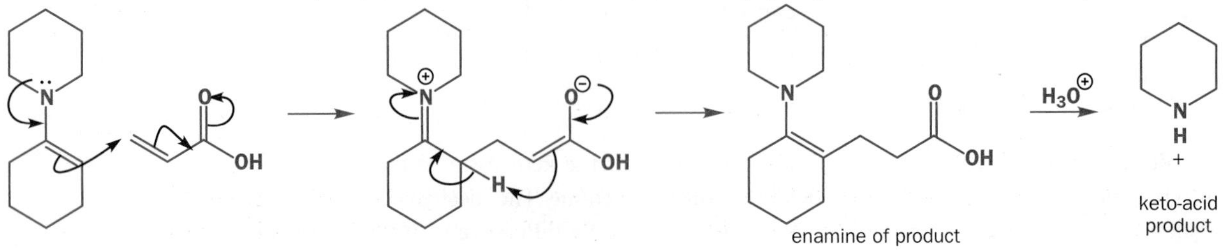

enamine of product   keto-acid product

A range of secondary amines can be used to form the enamines but those formed from piperidine, pyrrolidine, and morpholine combine reduced steric demands at the reactive double bond with good availability of the nitrogen lone pair. The electronic nature of the other substituents on the key double bond can vary without affecting the success of the conjugate addition. In these two examples enamines from cyclohexanone formed with pyrrolidine and morpholine add in good yield to an α,β-unsaturated carbonyl compound with an extra electron-withdrawing methylthio or phenylsulfonyl group.

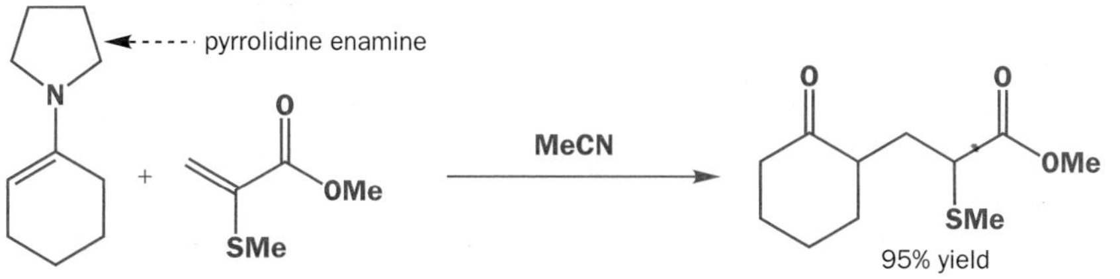

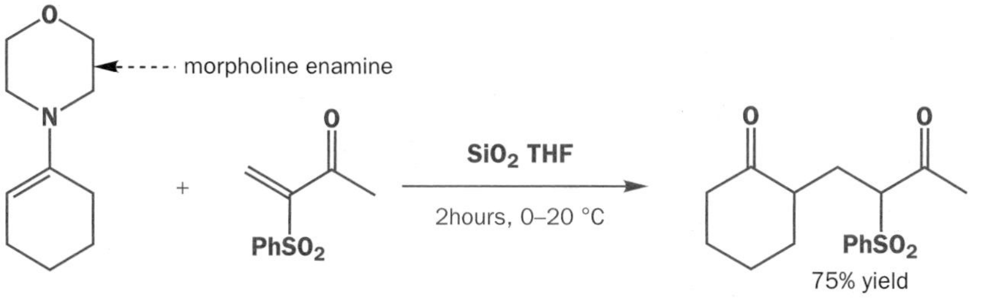

Conjugate addition of silyl enol ethers leads to the silyl enol ether of the product

The best alternatives to enamines for conjugate addition of aldehyde, ketone, and acid derivative enols are silyl enol ethers. Their formation and some uses were discussed in Chapters 21 and 26–28, but these stable neutral nucleophiles also react very well with Michael acceptors either spontaneously or with Lewis acid catalysis at low temperature.

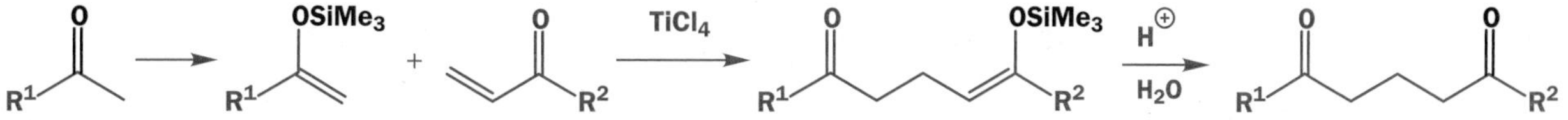

If the 1,5-dicarbonyl compound is required, then an aqueous work-up with either acid or base cleaves the silicon–oxygen bond in the product but the value of silyl enol ethers is that they can undergo synthetically useful reactions other than just hydrolysis. Addition of the silyl enol ether derived from acetophenone (PhCOMe) to a disubstituted enone promoted by titanium tetrachloride is very rapid and gives the diketone product in good yield even though a quaternary carbon atom is created in the conjugate addition. This is a typical example of this very powerful class of conjugate addition reactions.

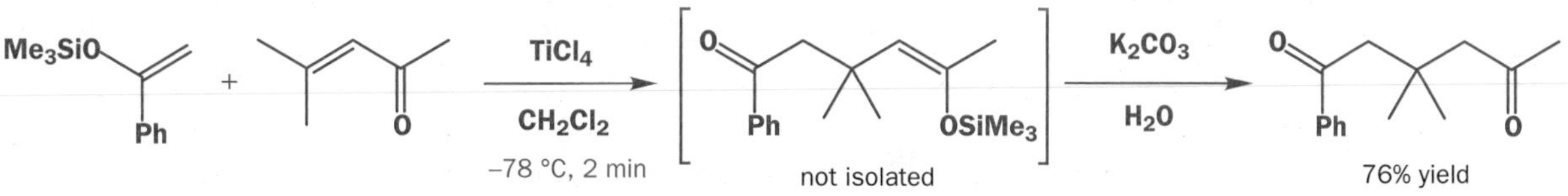

Silyl ketene acetals are even more nucleophilic than ordinary silyl enol ethers and react spontaneously with acyl chlorides. The intermediate enol ether of the acid chloride was not isolated but converted directly into a methyl ester with methanol.

The synthesis and reactivity of silyl ketene acetals are described in Chapters 21, 26, and 27.

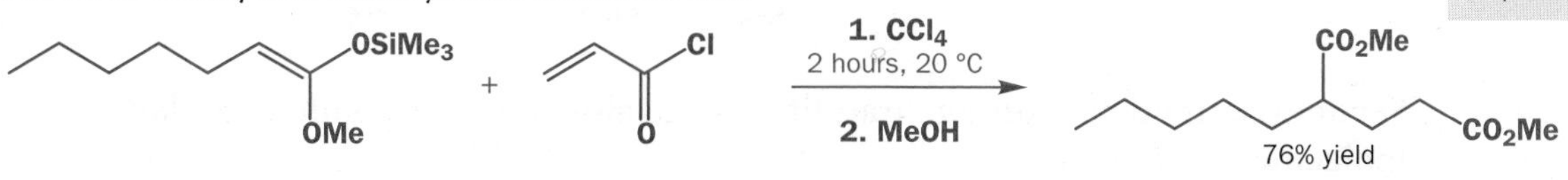

The mechanism, in the absence of a catalyst, can be written as a cyclic process involving direct transfer of silicon from the nucleophile to the electrophile but it might actually be stepwise. The soft nature of the silyl enol ether is demonstrated by the choice of soft double bond over hard carbonyl carbon as the electrophilic partner even though the carbonyl compound is an acid chloride.

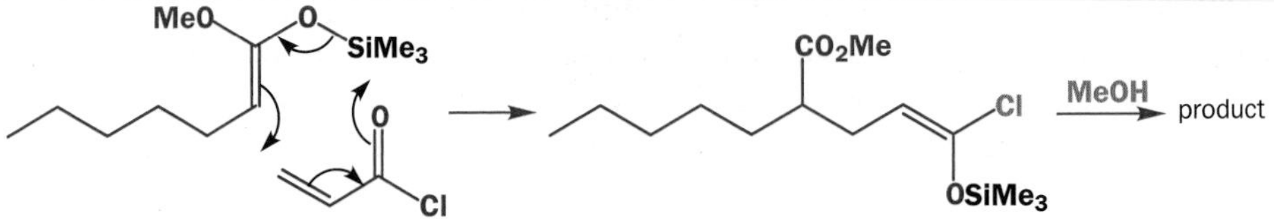

Lewis acid catalysis (TiCl₄) is normally required for silyl enol ether reactions

Conventional Lewis acid catalysis using a mixture of titanium tetrachloride and titanium isopropoxide is used to promote the addition of the silyl ketene acetal to methyl vinyl ketone. The key step in the mechanism is the conjugate addition of the silyl ketene acetal to the enone to form the bond shown in black in the product. The catalysis allows the reaction to proceed at much lower temperature, −78 °C. Do not be confused by the second SiMe₃ group. This is not an *O*-SiMe₃ group but a *C*-SiMe₃ group and plays no active part in the reaction.

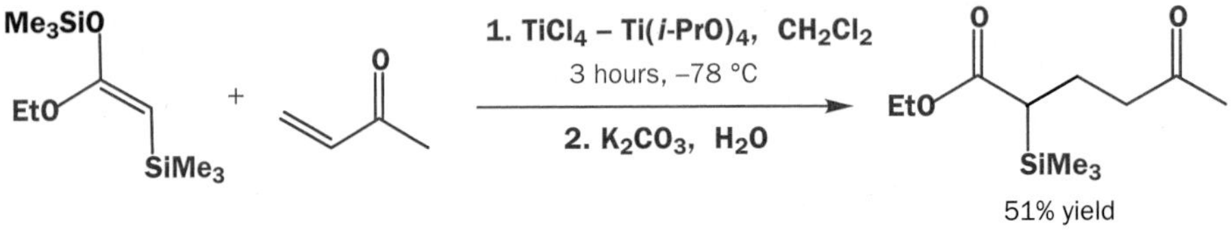

The electrophile coordinates to the Lewis acid first producing an activated enone that is attacked by the silylated nucleophile. It is difficult to determine at what stage the trimethylsilyl group moves from its original position and whether it is transferred intramolecularly to the product. In many cases the anion liberated from the Lewis acid (Cl⁻, RO⁻, Br⁻) is a good nucleophile for silicon so it is reasonable to assume that there is a free trimethylsilyl species (Me₃SiX) that captures the titanium enolate (Chapter 28).

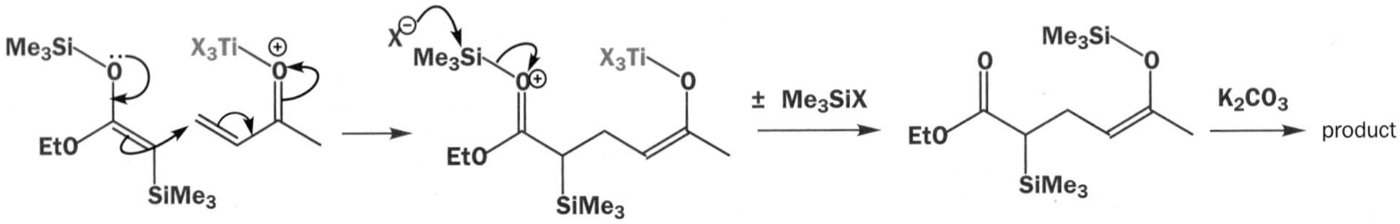

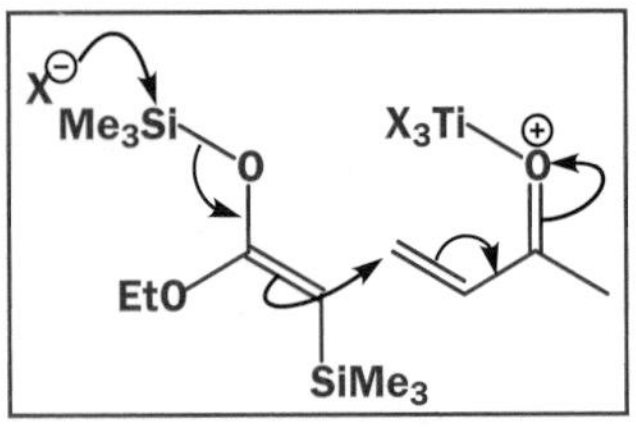

The mechanism can be drawn in a more concise form as shown in the frame. This gives the essence of the reaction but the details of the transfer of the TiX₃ and SiMe₃ groups are not shown and are in any case uncertain. The *C*-SiMe₃ group survived the mild basic treatment that cleaved the silyl enol ether formed by initial conjugate addition.

It is even possible to use a silyl enol ether to create a new C–C bond that joins two new quaternary centres. In this example the silyl ketene acetal does conjugate addition on an unsaturated ketone catalysed by the usual Lewis acid (TiCl₄) for such reactions.

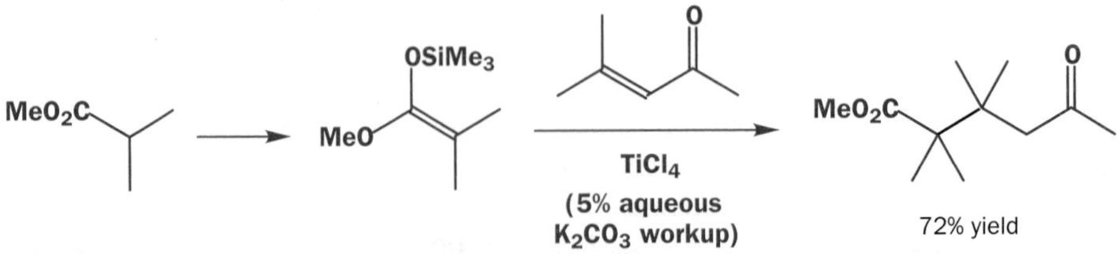

Sequential (tandem) conjugate additions and aldol reactions build complex molecules in a few steps

The silyl enol ether that is the initial product from conjugate addition of a silyl enol ether or silyl ketene acetal need not be hydrolysed but can also be used in aldol reactions. This example uses trityl perchlorate (trityl = Ph₃C), which is a convenient source of the trityl cation, as catalyst rather than a metal-based Lewis acid. The very stable Ph₃C⁺ cation carries a full positive charge and presumably functions in the same way as a Lewis acid. The combination of a silyl ketene acetal, cyclohexenone, and benzaldehyde gives a highly chemoselective and stereoselective conjugate addition–aldol sequence.

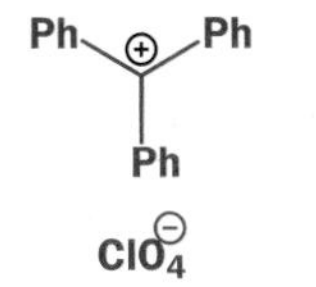

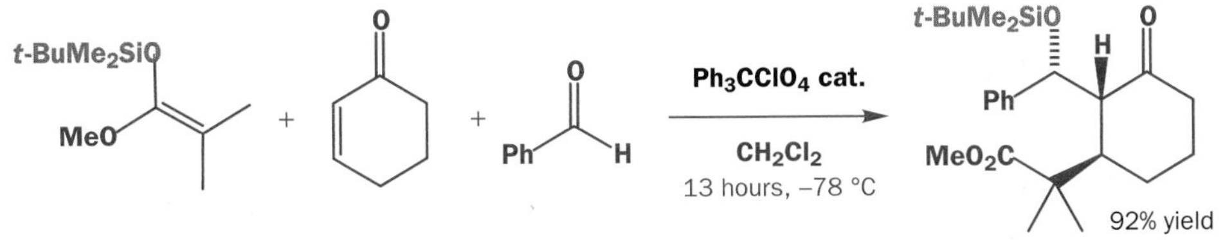

First, chemoselective (Chapter 24) conjugate addition of the silyl ketene acetal on the enone is preferred to direct aldol reaction with the aldehyde. Then an aldol reaction of the intermediate silyl enol ether on the benzaldehyde follows. The stereoselectivity results, firstly, from attack of benzaldehyde on the less hindered face of the intermediate silyl enol ether, which sets the two side chains *trans* on the cyclohexanone, and, secondly, from the intrinsic diastereoselectivity of the aldol reaction (this is treated in some detail in Chapter 34). This is a summary mechanism.

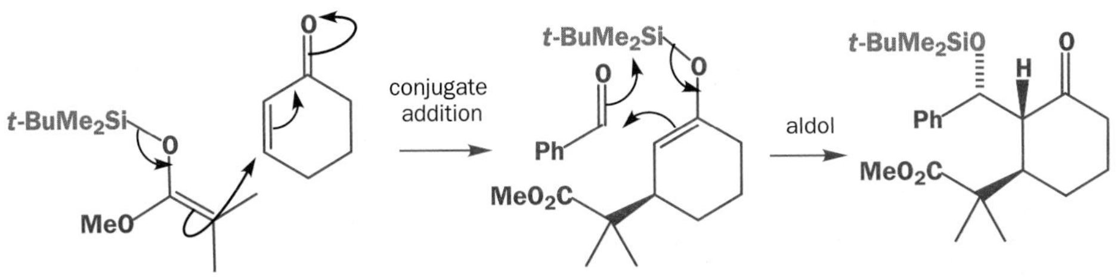

A variety of electrophilic alkenes will accept enol(ate) nucleophiles

The simplest and best Michael acceptors are those α,β-unsaturated carbonyl compounds with exposed unsaturated β carbon atoms, such as *exo*-methylene ketones and lactones and vinyl ketones, and we shall see in the next section that these need to have their high reactivity moderated in most applications.

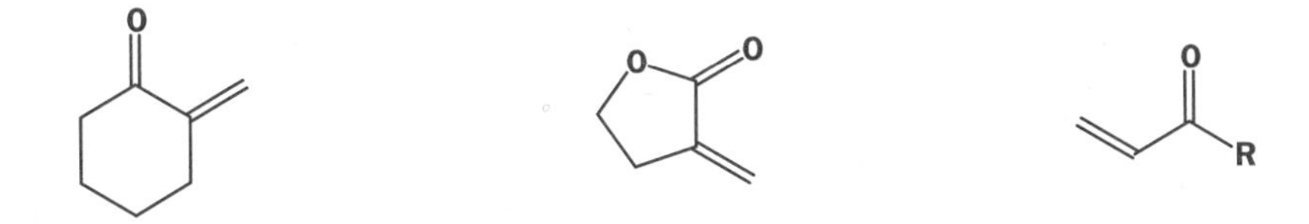

exo-methylene ketones   *exo*-methylene lactones   vinyl ketones

These Michael acceptors react with most enol equivalents to give good yields of conjugate addition products. Before discussing them we shall first briefly discuss other good Michael acceptors that are not so important but have their uses. Esters are good Michael acceptors because they are not very electrophilic. Unsaturated amides are even less electrophilic and will even give conjugate addition products with lithium enolates.

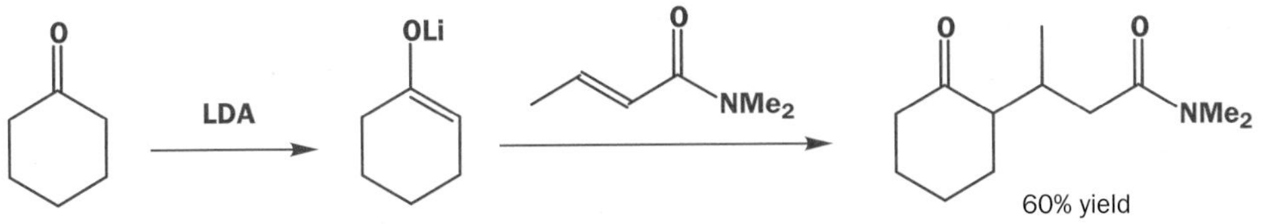

The fact that this is an *N,N*-dimethyl amide should remind you of the use of this kind of saturated amide in carbonyl substitution reactions with RLi in Chapter 12.

If all else fails, the trick to persuade a stubborn enolate to do conjugate rather than direct substitution is to add an extra anion-stabilizing substituent in the α position. Here is a selection of reagents that do this. In each case the extra group (CO₂Et, SPh, SOPh, SO₂Ph, SiMe₃, and Br) can be removed after the conjugate addition is complete.

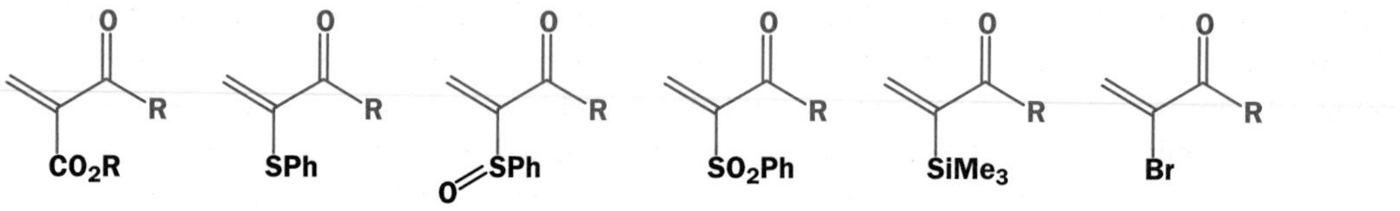

However, most α,β-unsaturated *ketones* can be made to do conjugate addition by suitable choice of enol(ate) equivalent and conditions. Now we need to look at the best Michael acceptors, their reactions, and how to make them.

The Mannich reaction provides stable equivalents of *exo*-methylene ketones

The key substrates for conjugate addition are the α,β-unsaturated carbonyl compounds. When the double bond is inside a chain or ring these compounds are available via a wide variety of routes including the aldol reaction and are generally stable intermediates that can be stored for use at will. When the double bond is *exo* to the ring or chain (*exo*-methylene compounds), the unhindered nature of the double bond makes them especially susceptible to attack by nucleophiles (and radicals). This reactivity is needed for conjugate additions but the compounds are unstable and polymerize or decompose rather easily.

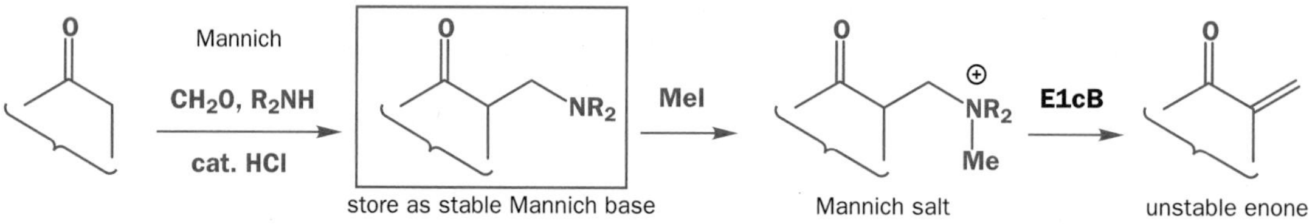

The preferred synthetic route to these important intermediates is the Mannich reaction (Chapter 27). The compound is stored as the stable Mannich base and the unstable enone released by elimination of a tertiary amine with mild base. The same conditions are right for this elimination and for conjugate addition. Thus the *exo*-methylene compounds can be formed in the flask for immediate reaction with the enol(ate) nucleophile. The overall reaction from β-amino carbonyl to 1,5-dicarbonyl appears to be a substitution but the actual mechanism involves elimination and conjugate addition.

> ■ The mechanism for the elimination is given in Chapter 27 and the mechanism for conjugate addition in Chapter 19 and earlier in this chapter.

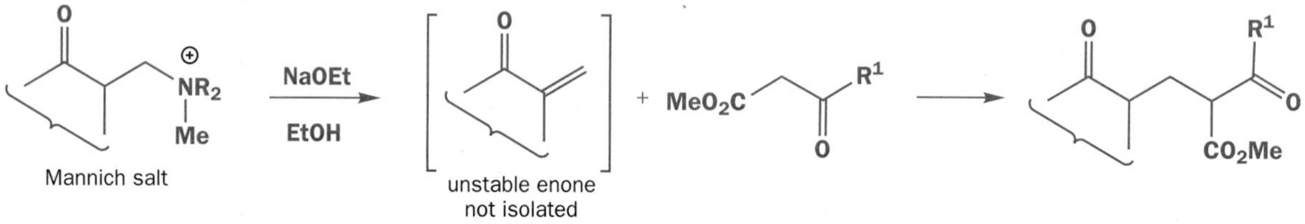

Using the Mannich reaction in conjugate addition

Either the tertiary amine or the quaternary ammonium salt can be stored as a stable equivalent of the *exo*-methylene compound. In our first example, the Mannich base with dimethylamine is first methylated with methyl iodide and then added to the conjugate addition reaction. Elimination of trimethylamine, which escapes from the refluxing ethanol as a gas, reveals the *exo*-methylene ketone in which the methylene group is *exo* to a chain. Fast conjugate addition of the stabilized enolate of diethyl malonate produces the product.

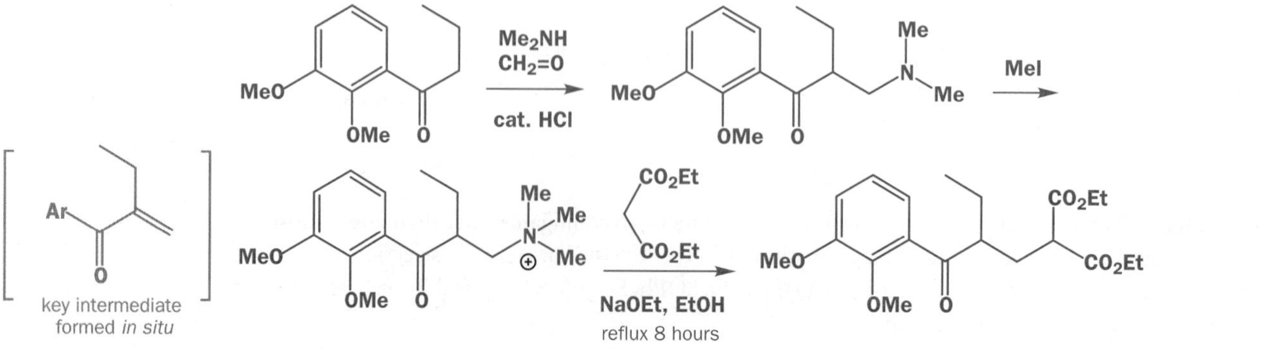

Cyclic ketones with *exo* cyclic methylenes can be prepared in just the same way and used *in situ*. Morpholine is often used as a convenient secondary amine for the Mannich reaction and the resulting amino-ketones can be methylated and undergo elimination–addition reactions with stabilized enolates such as that derived from ethyl acetoacetate. This starting material was prepared from natural menthone and the mixture of diastereoisomers produced is unimportant because the product is to be used in a Robinson annelation (see below).

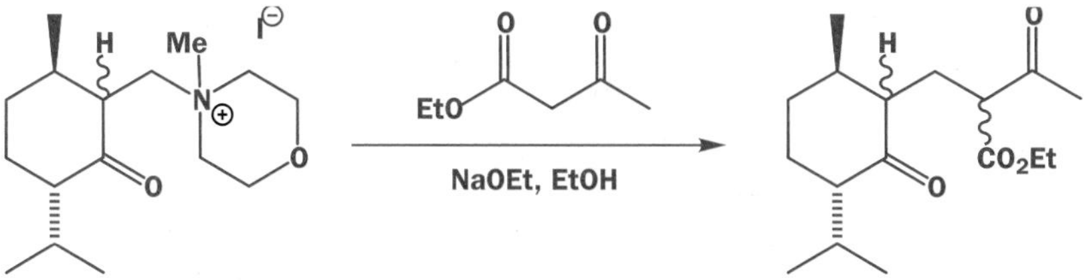

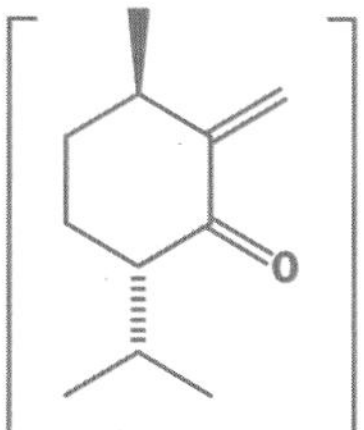

key intermediate
formed *in situ*

α,β-Unsaturated nitriles are ideal for conjugate addition

The nitrile group is not as reactive towards direct attack by nucleophiles as its carbonyl cousins but is equally able to stabilize an adjacent negative charge in the style of enolates. Alkenes conjugated with nitriles are thus activated towards nucleophilic attack without the complications of competing direct addition to the activating group.

The selective activation achieved by a nitrile group was also exploited in enolate alkylation (Chapter 27).

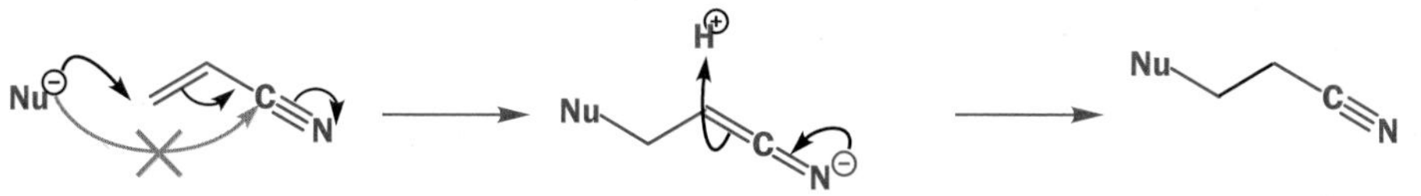

The regioselectivity of enolate formation is governed by the usual factors so that methyl benzyl ketone forms the more stable enolate with sodium metal. This undergoes smooth and rapid conjugate addition to acrylonitrile, which is unsubstituted at the β position and so very reactive.

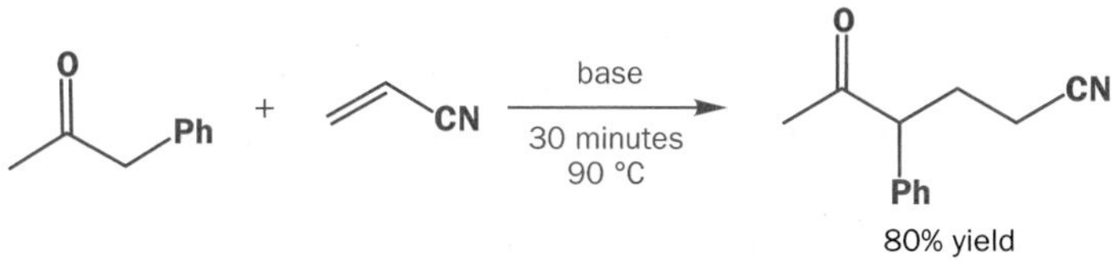

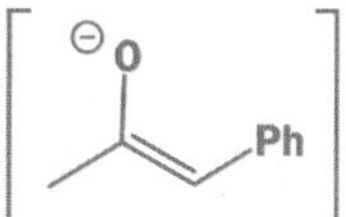

80% yield

intermediate
thermodynamic
enolate

The cyanide group can also act as an anion-stabilizing group in the nucleophile. In combination with an ester group, the enolizable proton is acidified to such an extent that potassium hydroxide can be used as base.

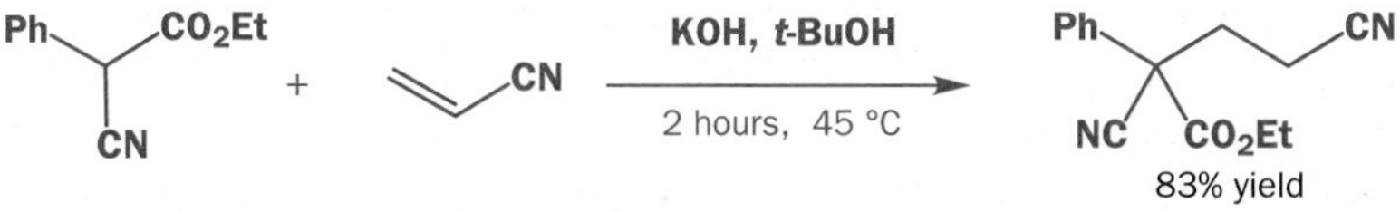

83% yield

Acrylonitrile CH$_2$=CHCN is one of the best Michael acceptors for enol(ate)s. The reaction is known as **cyanoethylation** as it adds a –CH$_2$CH$_2$CN group to the enol(ate).

The simplest amino acid, glycine, would be an ideal starting material for the synthesis of more complicated amino acids but it does not easily form enols or enolates. The methyl ester of the benzaldehyde imine has two electron-withdrawing groups to help stabilization of the enolate and conjugate addition of acrylonitrile is now possible. The base used was solid potassium carbonate with a quaternary ammonium chloride as phase transfer catalyst. Simple hydrolysis of the alkylated product leads to the extended amino acid.

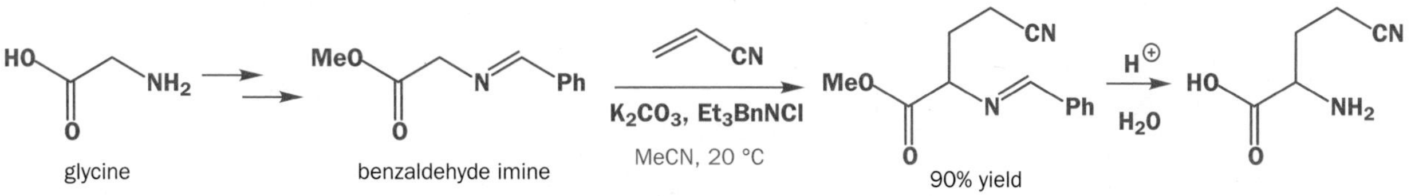

glycine   benzaldehyde imine   90% yield

Nitro is more powerful than carbonyl in directing conjugate addition

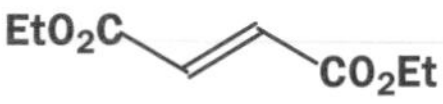

We have seen how two ester groups in fumarate diesters encourage conjugate addition, but what if there are two *different* groups at the ends of the Michael acceptor? Then you must make a judgement as to which is more electron-withdrawing. One case is clear-cut. The nitro group is worth two carbonyl groups (p. 193) so that conjugate addition occurs β to the nitro group in this case.

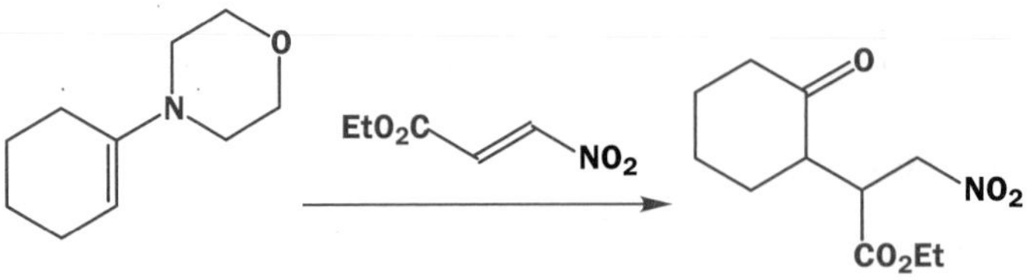

Conjugate addition followed by cyclization makes six-membered rings

The product of Michael addition of an enolate to an α,β-unsaturated carbonyl compound will normally be a 1,5-dicarbonyl compound. The two reactive carbonyl groups separated from one another by three carbon atoms present the opportunity for ring formation by intramolecular aldol condensation. If one of the carbonyls acts as an electrophile while the other forms a nucleophilic enolate, this cyclization gives a six-membered ring.

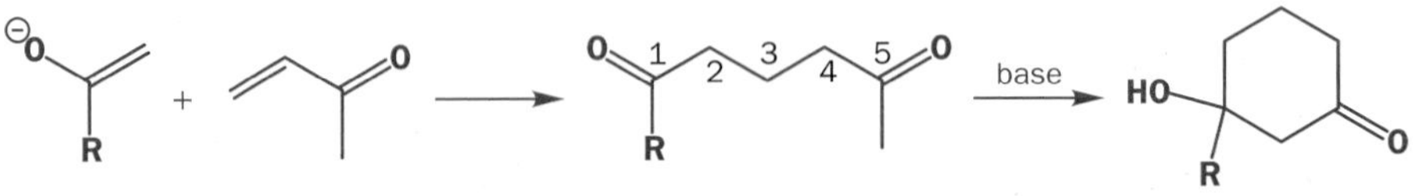

Drawing out the curly arrows for the formation is not easy as the chain has to fold back on itself which is hard to represent in two dimensions. However, remembering that the actual structure of a six-membered ring is a chair is extremely helpful. By using the structure of the product as a template for the transition state and reactive conformation of the starting material a clear representation is achieved.

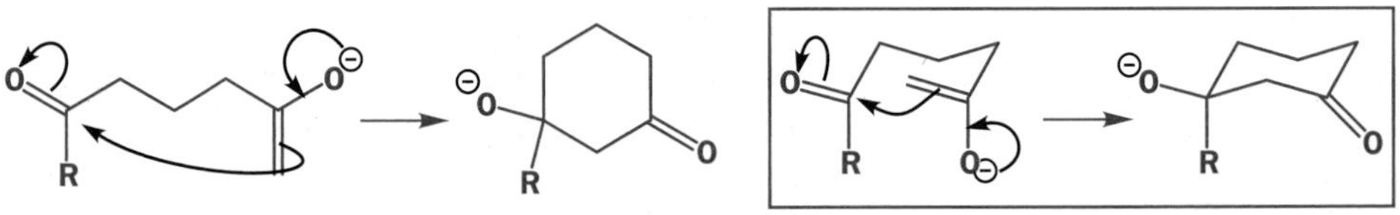

mechanism drawn on molecule in shape of product

The precise nature of the carbonyl groups determines what happens next. If R is a leaving group (OR, Cl, etc.), the tetrahedral intermediate collapses to form a ketone and the product is a 1,3-diketone. The synthesis of dimedone (later in this chapter) is an example of this process where an alkoxy group is the leaving group. Alternatively, if R is an alkyl or aryl group, loss of R is not an option and the cyclization is an intramolecular aldol reaction. Dehydration produces an α,β-unsaturated ketone, which is a stable final product.

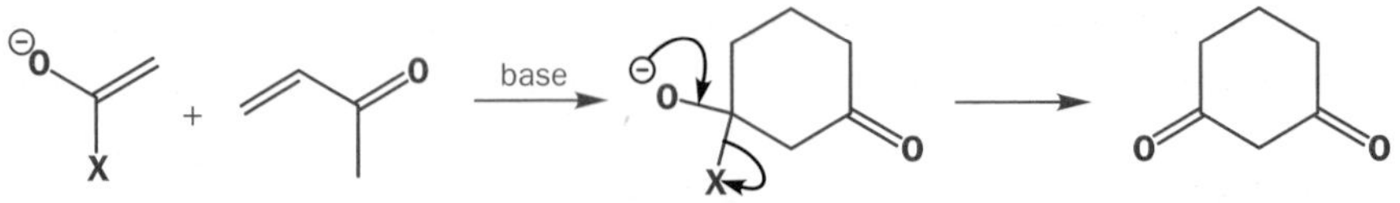

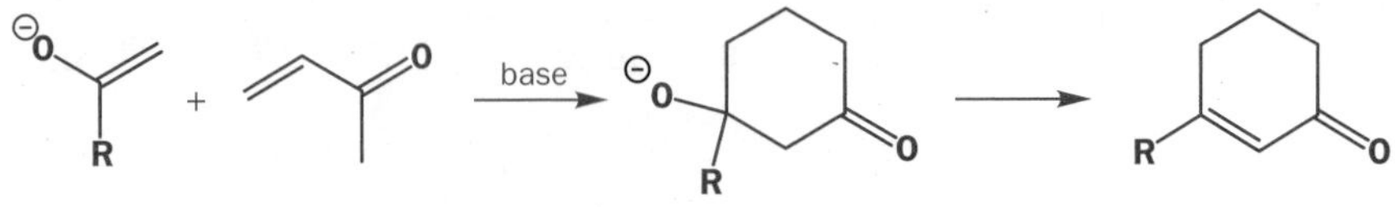

The Robinson annelation is the result of conjugate addition followed by aldol cyclization

Conditions for aldol reactions are very similar to those required for conjugate addition so that it is not unusual for conjugate addition and cyclization to occur sequentially without isolation of any intermediates. When we described one Michael addition a few pages back, we were not telling you the whole truth. The product isolated from this reaction was actually the enone from cyclization.

This sequential process of Michael–aldol reaction leading to a new six-membered ring is known as the **Robinson annelation**. It was, in fact, Robinson who invented the idea of using a Mannich product in conjugate additions because he wanted to develop this important reaction. There are now thousands of examples used to make all kinds of compounds, especially steroids (Chapter 49).

The essential requirement for a Robinson annelation is a Michael addition of an enolate to an enone that has a second enolizable group on the other side of the ketone. The classic enone is butenone (methyl vinyl ketone) and the classic Robinson annelation is the synthesis of rings A and B of the steroid nucleus.

> ▶
> **Annelation** describes the formation of a ring. You may also see the term spelt 'annulation'.

> ■
> Sir Robert Robinson (1886–1975) carried out many famous syntheses at Liverpool and Oxford and has two reactions, this annelation and the tropinone synthesis (Chapter 51), named after him. He won the Nobel prize in 1947. He was brilliantly inventive and the first person to work out mechanistically how to do syntheses.

the Robinson annelation   product of conjugate addition   product of intramolecular aldol

the steroid ring system

The Robinson annelation mechanism has three familiar stages

The mechanism combines two important reactions and we shall take it step by step. The first stage is the formation of the stable enolate, here of the 1,3-diketone, and the conjugate addition to the enone. The enolate of the product is in equilibrium with the triketone.

the Robinson annelation: mechanism—stage 1: the conjugate addition

The second stage is the formation of a new enolate on the other side of the ketone from the first. Note that the original enolate, the intermediate in the conjugate addition, can cyclize to give only an unstable four-membered ring so this cyclization would be reversible. The next intermediate, the aldol product, is often isolated from Robinson annelations.

the Robinson annelation: mechanism—stage 2: the intramolecular aldol reaction

The final stage is dehydration of the aldol and an E1cB reaction that involves the carbonyl group as in a standard aldol reaction (Chapter 27). Another enolate must form in the same position as the last.

the Robinson annelation: mechanism—stage 3: the E1cB dehydration

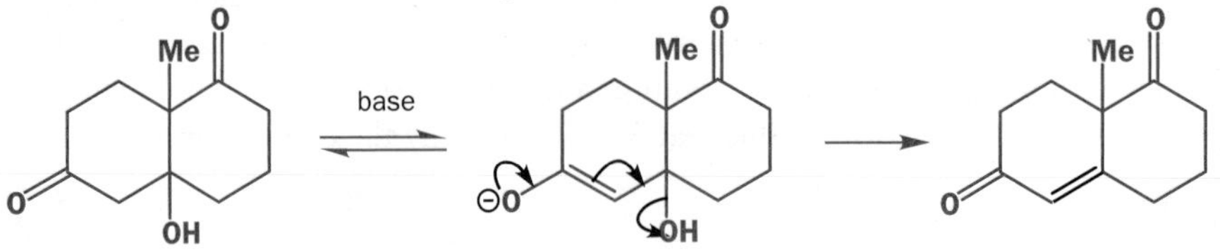

Enantioselective Robinson annelation

This particular product of the Robinson annelation is an important intermediate for the synthesis of natural products. The natural products exist as a single enantiomer so to be useful this material must also be a single enantiomer. A remarkably efficient preparation employs (S)-proline as the catalyst for the asymmetric

aldol reaction is the final stage. The product was isolated enantiomerically enriched. Presumably, the proline forms an iminium ion that is the electrophile in the aldol reaction and the source of asymmetry. You will meet further examples of asymmetric synthesis in Chapter 45.

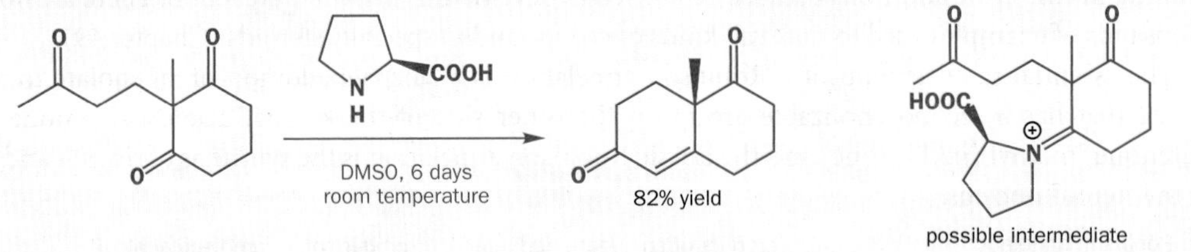

Each step in the Robinson annelation is controlled by the various devices you have already met. In the conjugate addition step, the α,β-unsaturated carbonyl compound is usually butenone or another ketone and they are suitable Michael acceptors. There is much more variation in the enol equivalent. Compounds with 1,3-dicarbonyl groups are popular so ester groups can be added to ketones and removed afterwards by hydrolysis and decarboxylation. Keto-esters react well in the Robinson annelation. The ester group stabilizes the enolate but is not very electrophilic. In this example MeOK is the base for the conjugate addition and a weaker base is used for the aldol.

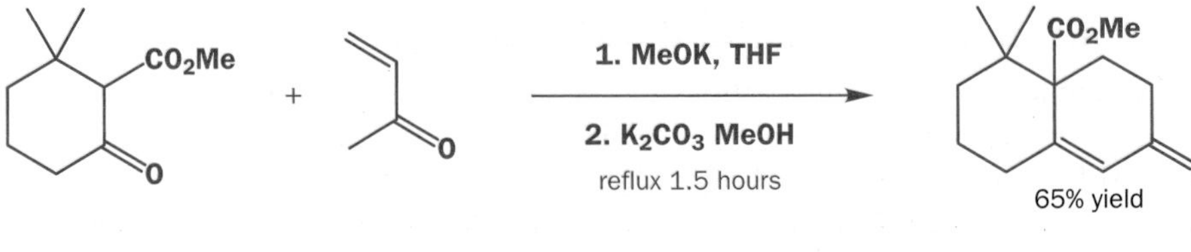

In fact, even very weak bases are enough for most 1,3-dicarbonyl compounds and piperidine and acetic acid combine to form a mild buffered system that facilitates both conjugate addition and aldol reactions via enol intermediates. The trifluoromethyl ketone is extremely electrophilic so the aldol reaction proceeds very smoothly.

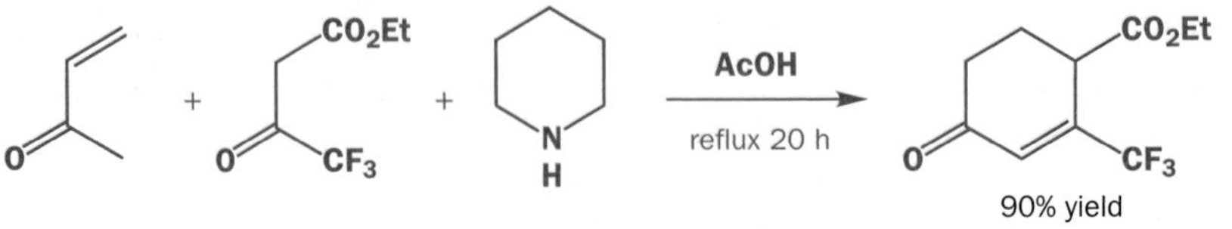

Enamines are good enol equivalents for Robinson annelation

If the enol component is an aldehyde, none of these methods will do and enamines or silyl enol ethers are the best choice. Enamines are excellent nucleophilic components and the iminium ion that is formed in the conjugate addition can provide the electrophilic component in a cyclization reaction. Acid-catalysed hydrolysis of the β amino-ketone liberates the amine that was used to form the enamine at the start revealing the cyclohexenone product. In this example a quaternary centre is formed in the new ring.

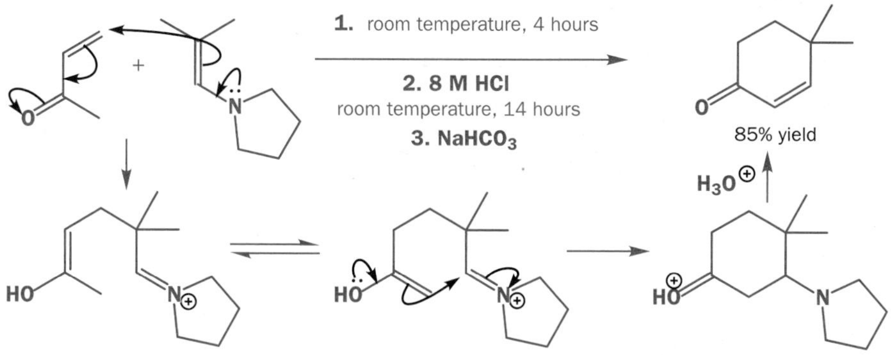

85% yield

The addition of an anion-stabilizing group to the α,β-unsaturated component at the α carbon promotes conjugate addition and allows a wider range of enolate nucleophiles to be used. In particular, enolates that are prone to equilibration to regioisomers can be used because conjugate addition becomes essentially irreversible. Trimethylsilyl has proved very effective because it stabilizes the enolate intermediate in the conjugate addition and is easily removed during the later stages of the reaction. Conjugate addition of Me₂CuLi to the cyclohexenone in our next example produces a new carbon–carbon bond and a regiodefined enolate. The presence of a proton source would allow equilibration of the enolate to the less hindered position but the trimethylsilyl enone was used to trap the enolate without equilibration, creating the two adjacent stereocentres in the Robinson annelation.

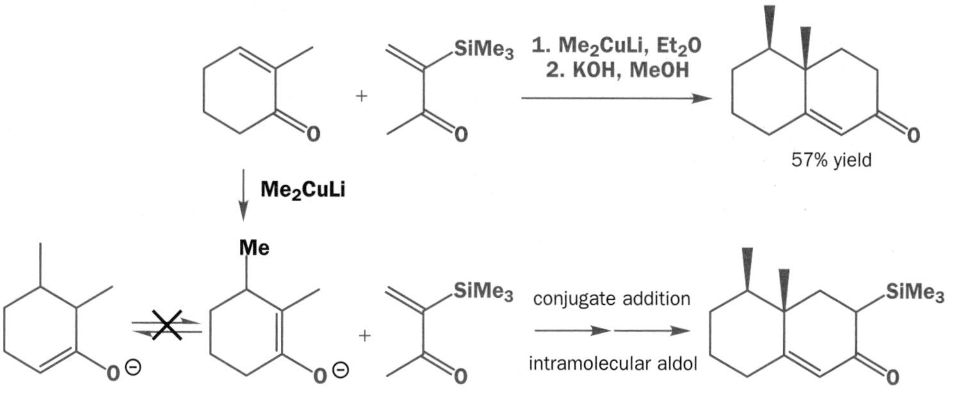

57% yield

A more common method of ensuring that the conjugate addition step is free from side-reactions is to use the method Robinson himself invented—replace the enone by the Mannich base or Mannich salt as we have discussed already in this chapter. This ensures that the enone need have only a very short lifetime in the reaction mixture.

The aldol cyclization step and the dehydration are sometimes separated from the conjugate addition and from each other and sometimes not. It depends to some extent on the conditions. Very mild conditions in this example allowed each step to be performed separately and in good yield but notice the exceptionally mild conditions for the conjugate addition (just mix in water!) which are possible only because of the two carbonyl groups in the enol component.

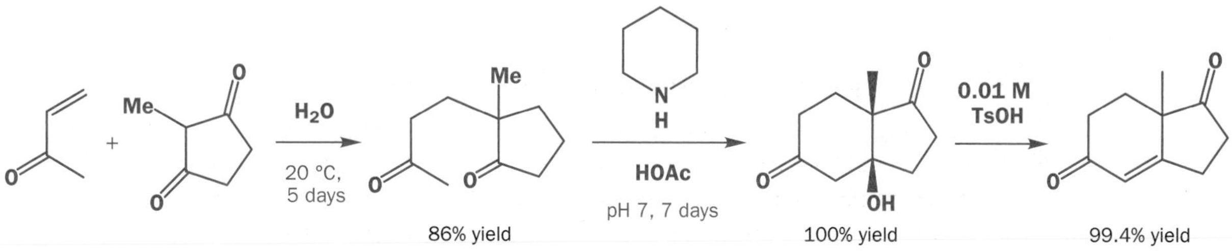

86% yield   100% yield   99.4% yield

We have devoted a lot of space to the Robinson annelation because it is so important. For a multi-stage reaction, it is easy to understand because each step is a well-known step in its own right. It is because the second step is an *intra*molecular aldol condensation that it occurs so easily.

Conjugate addition followed by Claisen ester cyclization gives cyclic diketones

The first enol you saw at the start of Chapter 21 was the stable enol of 'dimedone', 5,5,-dimethyl-cyclohexa-1,3-dione. This six-membered ring is made by a close analogue of the Robinson annelation. The only difference is in the cyclization step, which is a Claisen ester condensation rather than an aldol reaction.

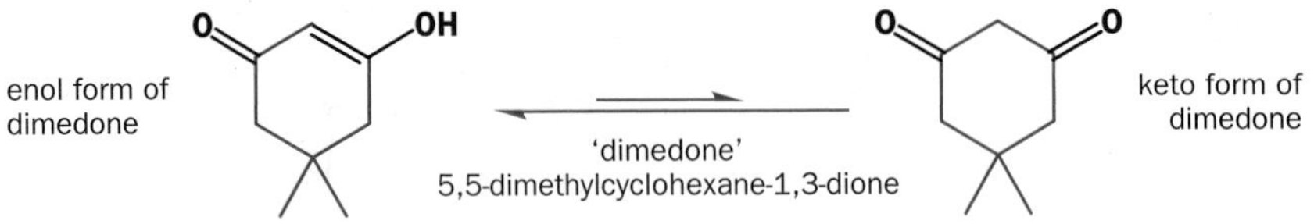

Dimedone has a trivial name because its preparation is so easy that it was discovered early in the history of organic chemistry. The first step is a conjugate addition of diethyl malonate to the unsaturated ketone 'mesityl oxide' (4-methylpent-3-en-2-one; given a trivial name for the same reason). Ethoxide ion is the base for the usual reason that nucleophilic substitution at the ester group simply regenerates starting material.

> Exceptionally we have drawn the enolate of diethyl malonate as a carbanion. This is not generally recommended but you will see it and in this case, with the negative charge delocalized over the two ester groups, it is relatively harmless.

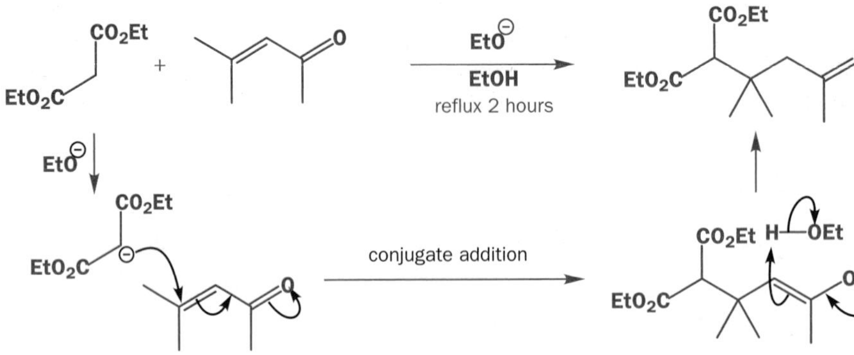

Under the reaction conditions the product will exist as a stable enolate but cyclization of this enolate would lead to a four-membered ring so it is reversible. The alternative enolate on the methyl group at the other end of the chain leads to a six-membered ring so this is what happens.

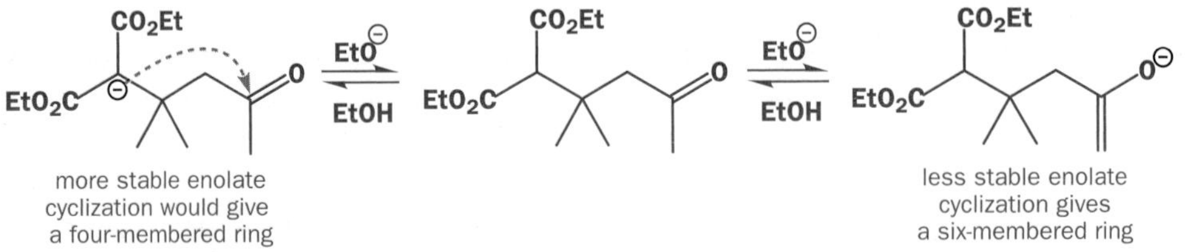

So far the mechanism is almost the same as that of the Robinson annelation but the cyclization is now the attack of a ketone enolate on an ester group (it doesn't matter which one as they are equivalent) and so it is an intramolecular Claisen ester condensation (Chapter 28). The intermediate must be redrawn to allow cyclization.

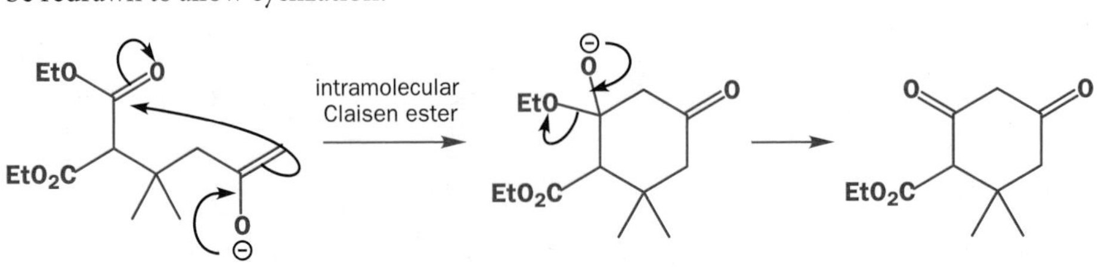

This intermediate will exist as a stable enolate under the reaction conditions. Now aqueous KOH is added to the reaction mixture, which is refluxed to hydrolyse the remaining ester. On acidification with HCl decarboxylation occurs and dimedone is released.

■ Decarboxylation of the free acid is faster than that of the anion (Chapter 26).

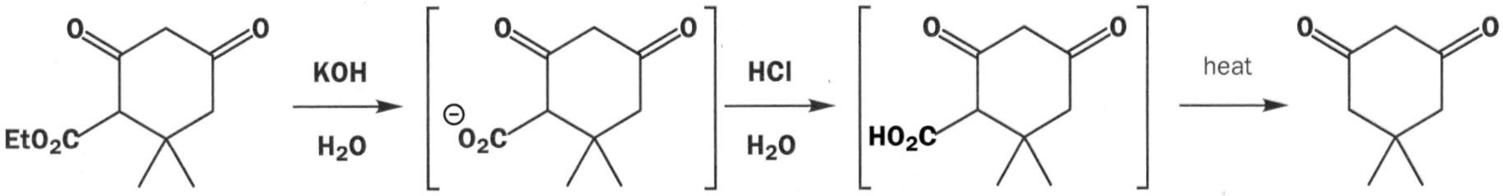

The whole operation is conducted in one flask, just as for the Robinson annelation, and dimedone is isolated as the crystalline enol in 84% yield. This reaction has not enjoyed such wide application as the Robinson annelation but it has been used to make an aromatic compound that is a starting material for the synthesis of maytensine, which we discussed at the end of Chapter 22.

■ This aromatic compound is not the same as the one suggested at the end of Chapter 22: it resembles the natural product rather more and is an alternative starting material.

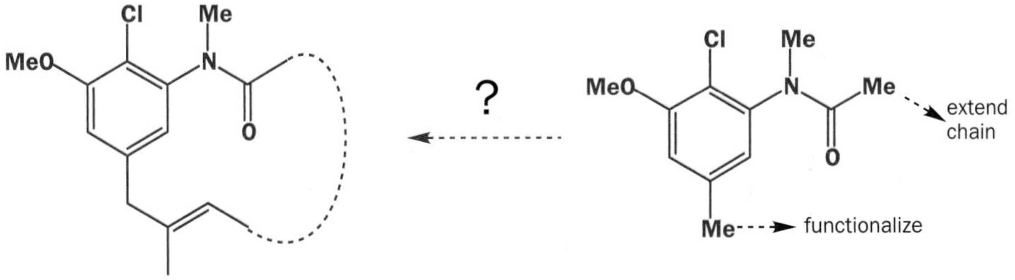

The clue to the synthesis of this compound using a dimedone-style condensation is the 1,3,5-relationship between OMe, N, and Me around the ring. If we carry out the conjugate addition on an enone with only one methyl group at the end of the double bond, this is what we will get.

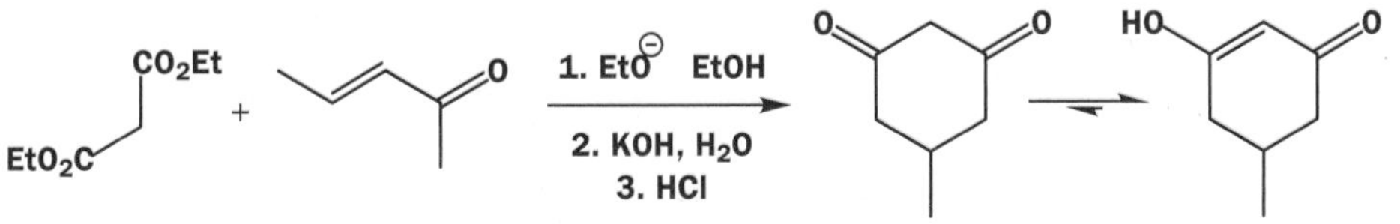

Particularly in the enol form, this is beginning to look something like what is needed. The next step is to add $MeNH_2$. Even in aqueous solution ($MeNH_2$ is available as a 40% aqueous solution) the enamine forms very easily because it is conjugated, like the enol but more so. This is again a crystalline compound and formed in 70% yield.

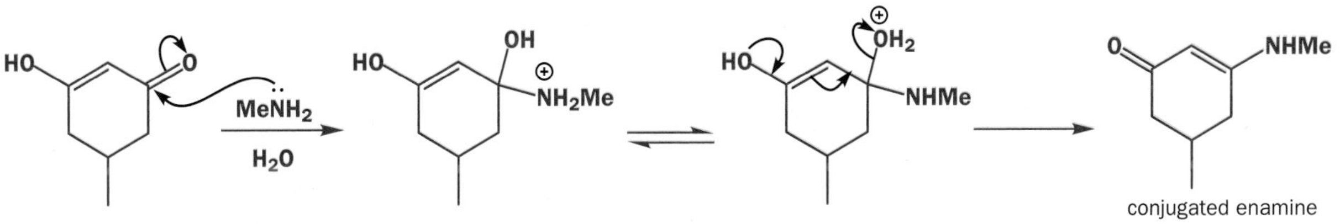

conjugated enamine

The chlorine atom can now be introduced by direct chlorination of the enamine with *N*-chlorosuccinimide. This electrophilic chlorine source reacts via the mechanism that enols follow when they react with halogens (Chapter 21).

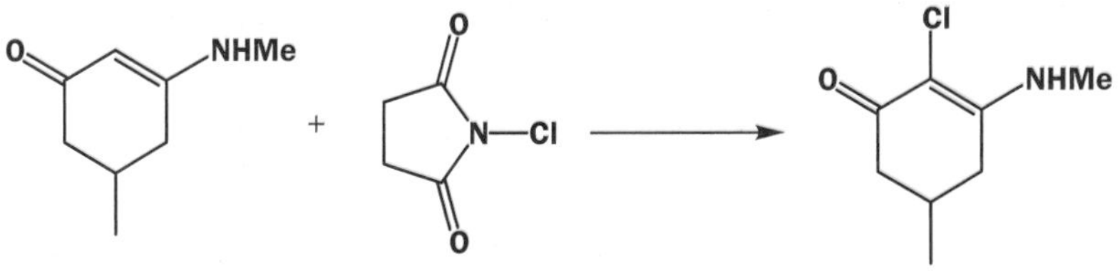

conjugated enamine   *N*-chlorosuccinimide (NCS)   72% yield

Now it is time to aromatize the ring. If you imagine that the ketone in its enol form would already be two double bonds in the ring, bromination and elimination of HBr would give the third. This can be done with bromine followed by acetic anhydride, which gives the benzene ring and acetylates the amine at one go.

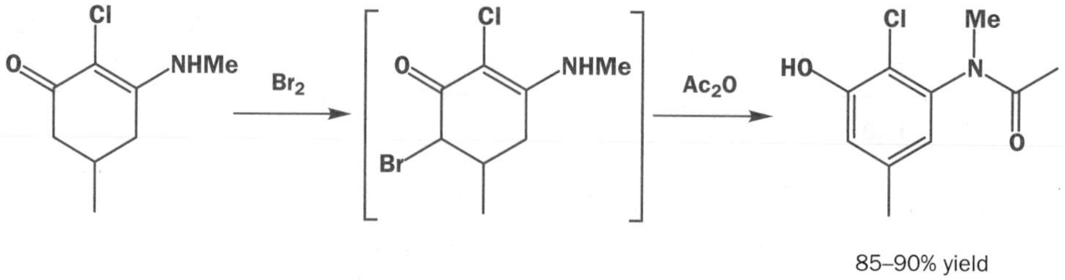

85–90% yield

Nitroalkanes are superb at conjugate addition

In this chapter so far we have concentrated on anions stabilized by carbonyl groups for use in conjugate addition. Anions that are well stabilized, such as those from β-dicarbonyl compounds, are the usual nucleophiles for this important class of reaction. The key to their success is the pK_a of the acidic proton, which allows initial enolate anion formation, helps to reverse the unwanted alternative aldol pathway, and facilitates proton transfer in the catalytic version of the reaction. The nitro group is so powerfully electron-withdrawing that just one is equivalent to two carbonyls in pK_a terms (Chapter 26). Thus if β-dicarbonyls are good for conjugate addition and our analysis of the reasons for this is correct, you might expect nitroalkanes to undergo conjugate addition in just the same way. The good news is that they do, very well. The first stage is a base-catalysed conjugate addition.

▶

Always draw out the nitro group in full when using it in mechanisms.

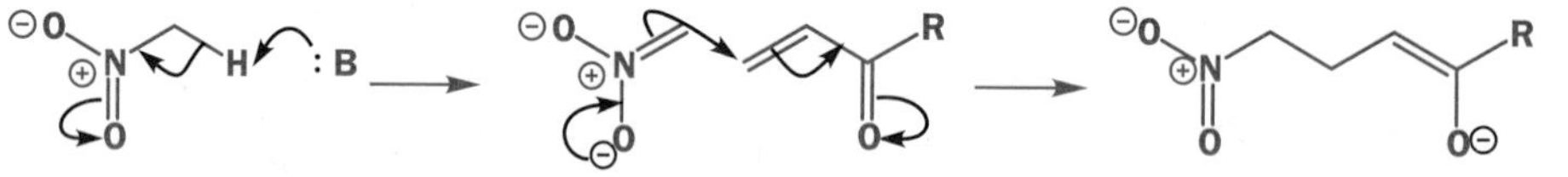

The enolate ion intermediate is now much more basic than the anion of the nitro compound so it removes a proton from the nitro compound and provides another molecule of anion for the second round of the reaction.

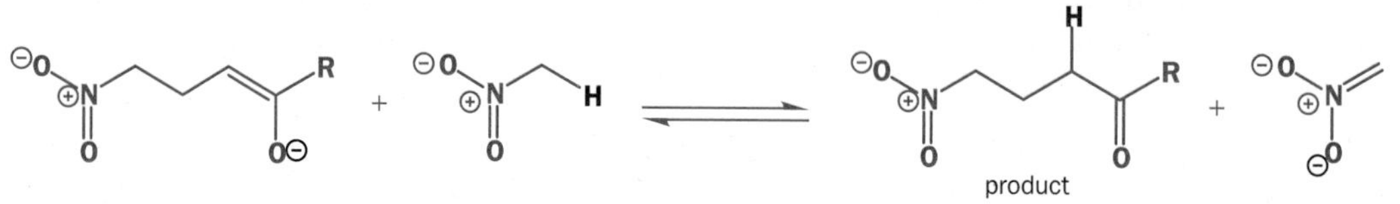

The acidifying effect of the nitro group is so profound that very mild bases can be used to catalyse the reaction. This enables selective removal of the proton next to the nitro group and helps to avoid side-reactions involving aldol condensations of the carbonyl component. Common examples include amines, quaternary ammonium hydroxides, and fluorides. Even basic alumina is sufficient to catalyse virtually quantitative addition of this benzylic nitroalkane to cyclohexenone at room temperature!

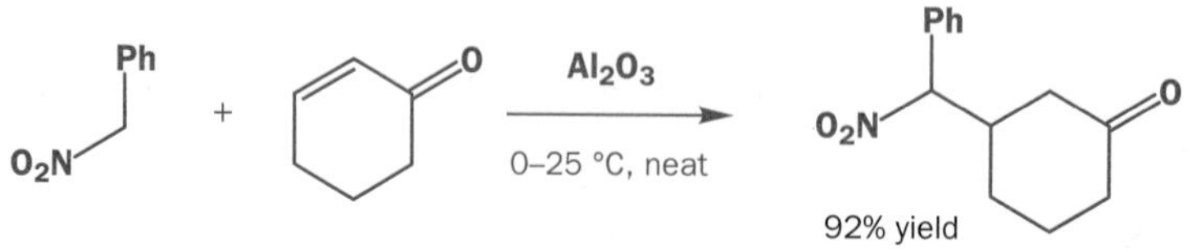

92% yield

Anions of nitro compounds form quaternary centres with ease in additions to α,β-unsaturated mono- and diesters. The difference between acidity of the protons next to a nitro group and those next to the esters in the products combined with the very mild basic conditions ensure that no unwanted Claisen condensations occur.

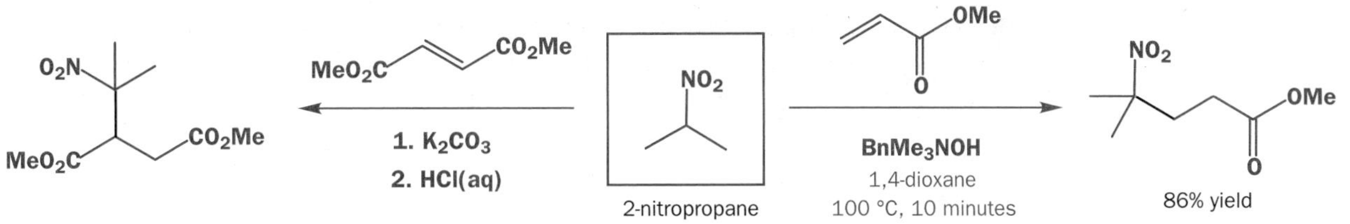

Nitromethane readily undergoes multiple conjugate additions under more forcing conditions with excess ester.

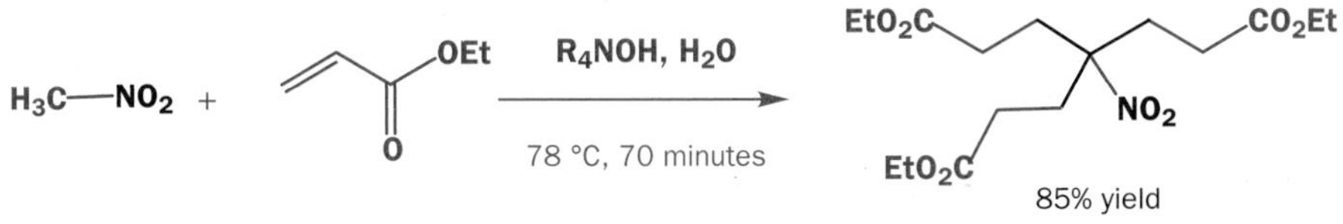

Nitroalkane conjugate addition can be combined with other reactions

The effectiveness of nitro compound conjugate addition makes it ideal for use in combination with other reactions in making several bonds in one pot. The last example showed triple conjugate addition. The next example combines conjugate addition and intramolecular conjugate addition to make a six-membered ring. The base used for both steps is Cs_2CO_3. Caesium, the most electropositive of readily available metals, forms ionic compounds only so that the carbonate ion can exert its full basicity. Deprotonation of the conjugate addition product next to the nitro group produces a second anion, which does an intramolecular S_N2 displacement of iodide to form a six-membered ring.

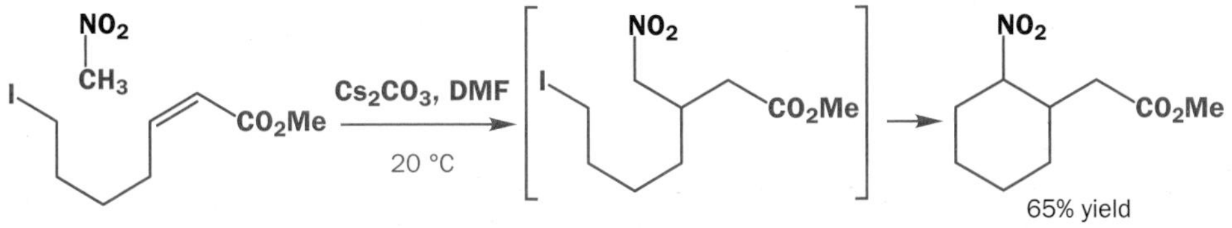

The nitro group can be converted into other useful functional groups following conjugate addition. Reduction gives primary amines while hydrolysis reveals ketones. The hydrolysis is known as the **Nef reaction** and used to be achieved by formation of the nitro-stabilized anion with a base such as sodium hydroxide followed by hydrolysis with sulfuric acid. These conditions are rather unforgiving for many substrates (and products) so milder methods have been developed. One of these involves ozonolysis of the nitro 'enolate' at low temperature rather than treatment with acid.

Base-catalysed conjugate addition of nitropropane to methyl vinyl ketone occurred smoothly to give the nitroketone. Formation of the salt with sodium methoxide was followed by oxidative cleavage of the C=N linkage with ozone. The product was a 1,4-diketone which was isolated without further aldol reaction by this route.

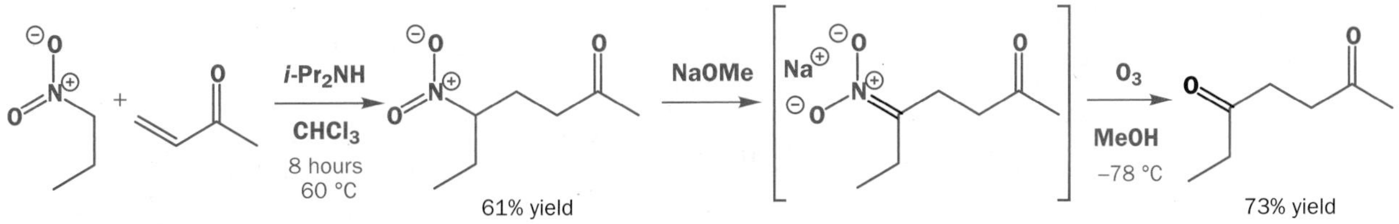

■ Ozonolysis is described in Chapter 35.

This is a good general method for the synthesis of 1,4-diketones, which can be otherwise difficult to make, and additional substituents are easily accommodated on the enone—a characteristic of conjugate addition.

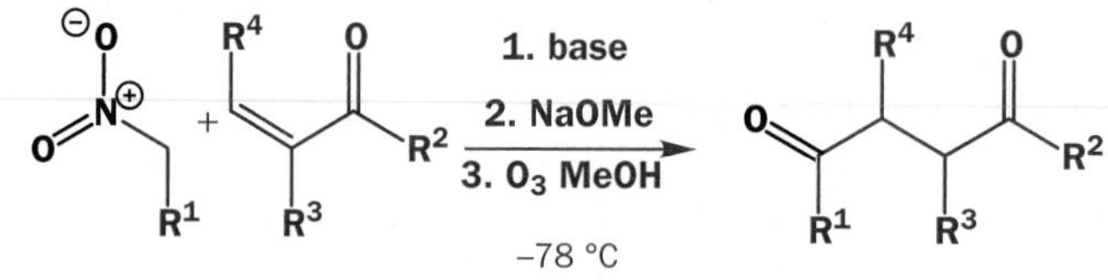

The synthesis of a drug that acts on brain chemistry

We end this chapter with a simple commercial synthesis of a drug molecule described as a dopaminergic antagonist. It uses four reactions that you have met: conjugate addition of an enolate to acrylonitrile; reduction of CN to a primary amine; alkylation; and reduction of the amide. There is another reaction involved—cyclization to an amide—but this occurs spontaneously. These reactions may be simple but they are important.

synthesis of a drug molecule

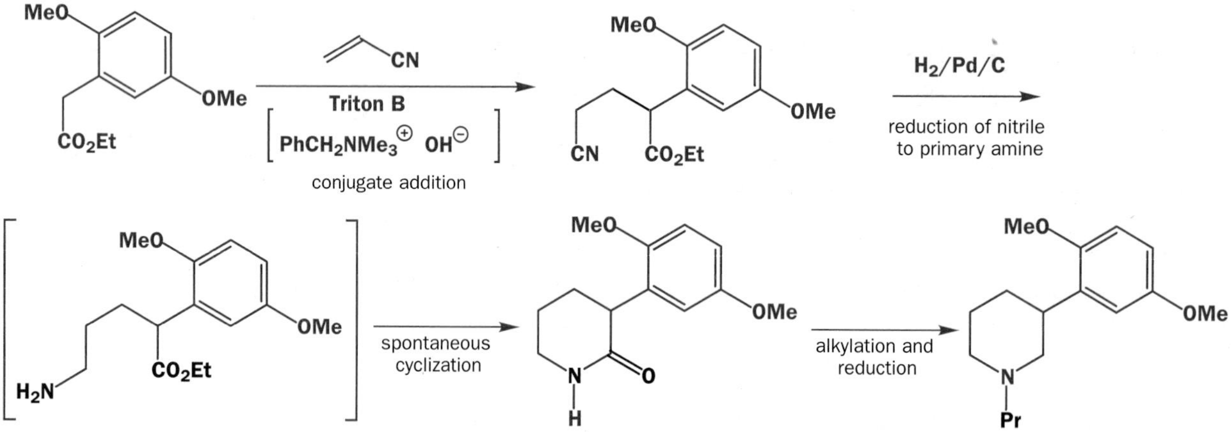

This was the last chapter in our sequence (Chapters 26–28) devoted to the chemistry of enols and enolates and, in particular, to their use in making new C–C bonds. In the next chapter we shall be using these reactions when we introduce you to synthetic planning. We shall be answering questions such as, 'how was the synthesis of this drug planned?'

Problems

1. Write full mechanisms for these reactions mentioned earlier in the chapter.

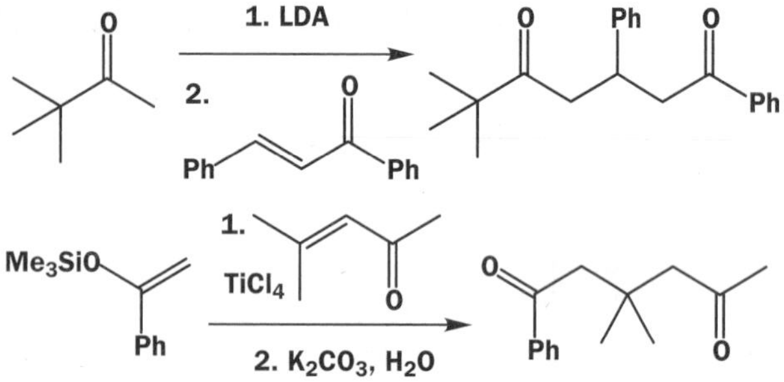

2. Suggest syntheses for these compounds.

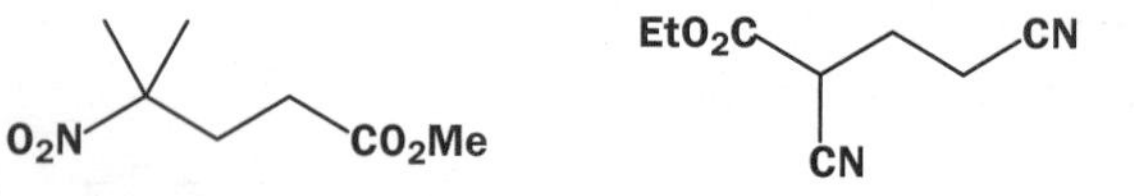

3. Suggest two different approaches to these compounds by conjugate addition of an enol(ate). Which do you prefer?

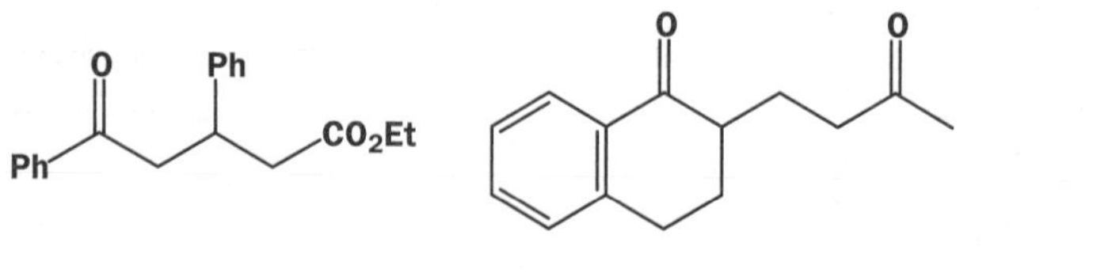

4. How could you use the Robinson annelation to make these compounds?

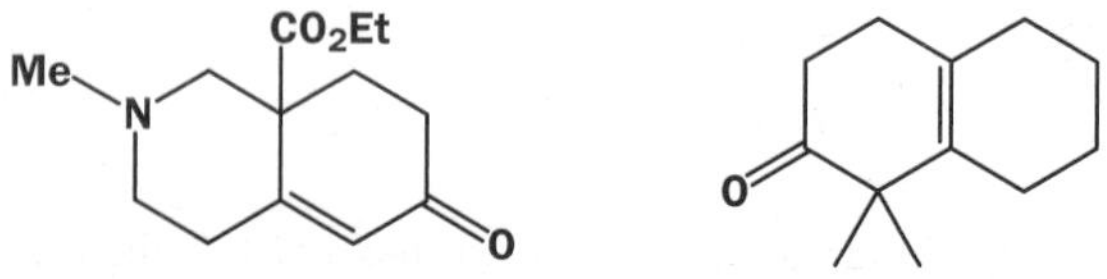

5. Predict the product that would be formed in these conjugate additions.

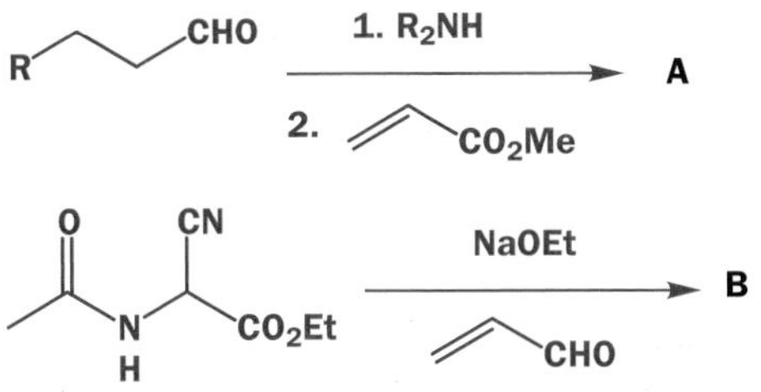

6. Suggest mechanisms for this reaction, commenting on any selectivity.

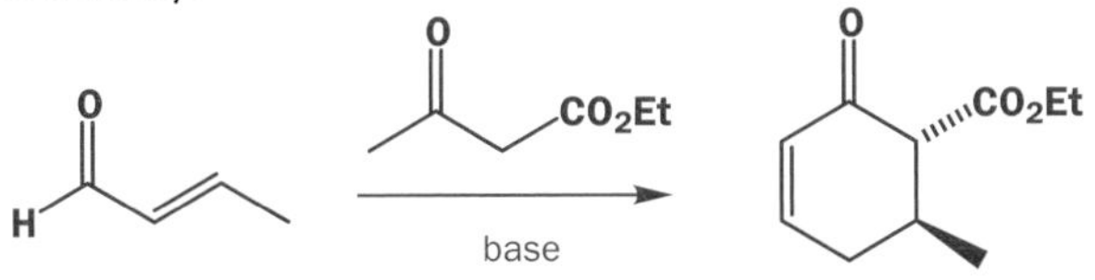

7. This example of the use of the Mannich reaction was given in the chapter. Draw detailed mechanisms for the two key steps shown here.

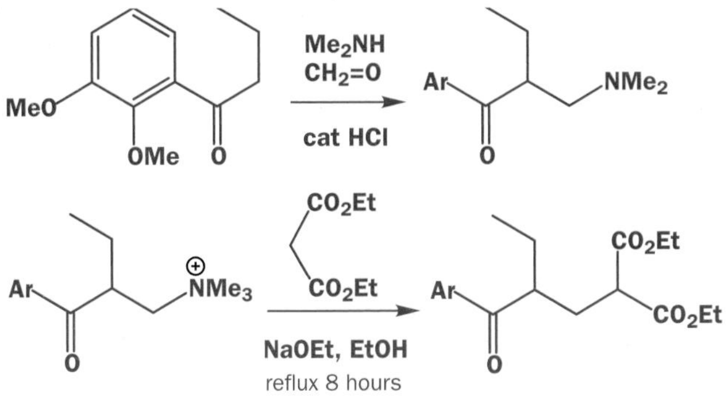

8. This symmetrical bicyclic ketone can easily be synthesized in two steps from simple precursors. What is the structure of the intermediate and what are the mechanisms of the reactions?

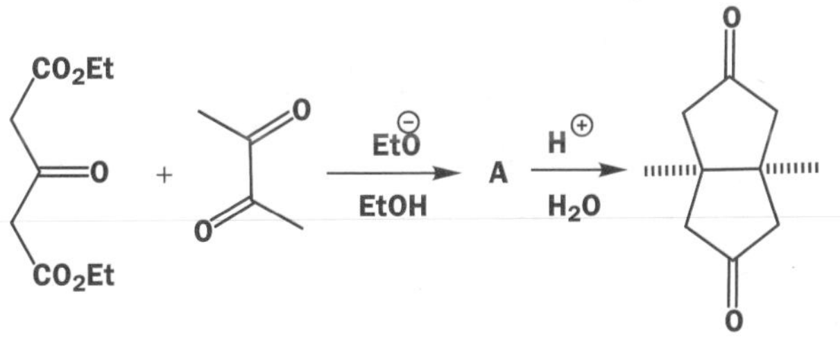

9. Suggest ways to make these compounds using conjugate addition of enol(ate)s.

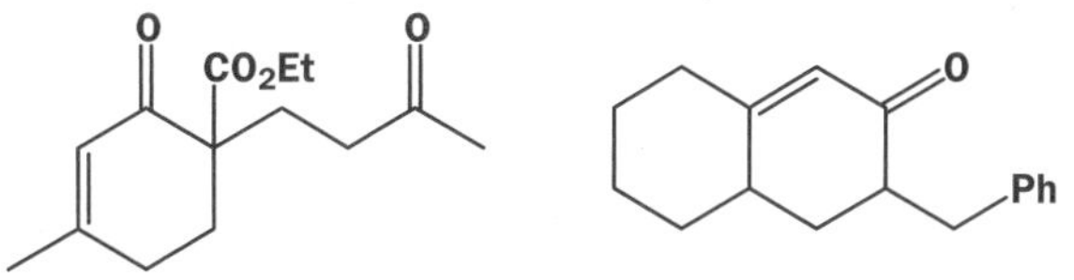

10. Identify the product of this reaction and propose a mechanism for its formation.

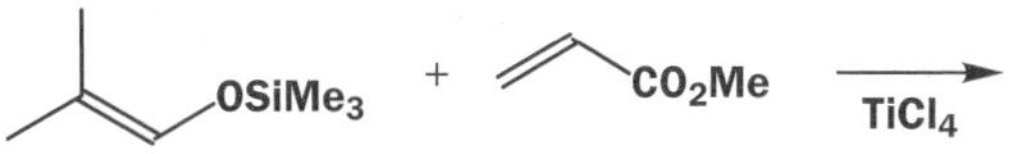

B, C$_8$H$_{14}$O$_3$
v_{max}(cm^{-1}) 1745, 1730
δ_C(p.p.m.) 202, 176, 62, 48, 34, 22, 15
δ_H(p.p.m.) 1.21 (6H,s), 1.8 (2H, t, J 7 Hz),
2.24 (2H, t, J 7 Hz), 4.3 (3H, s), 10.01 (1H, s)

11. Suggest a synthesis for the starting material for this reaction, a mechanism for the reaction, and an explanation for the selectivity.

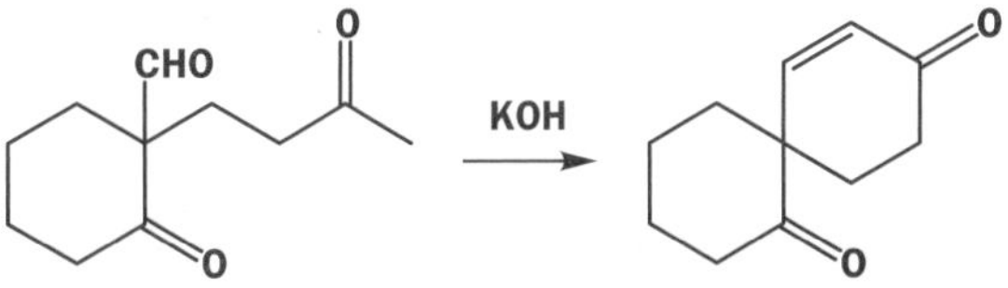

12. Suggest a mechanism for this reaction.

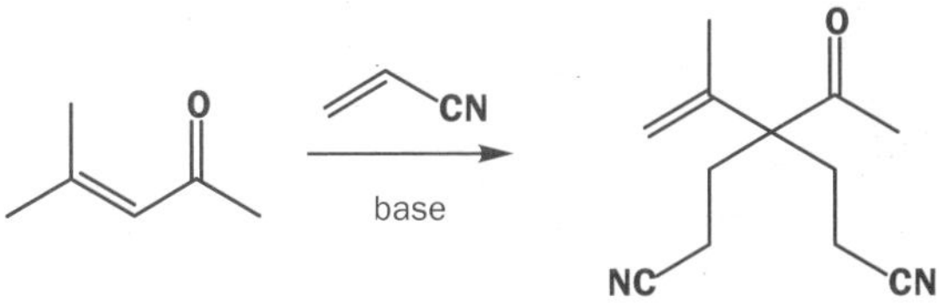

13. Suggest a mechanism for this reaction. How would you convert the product into the antibiotic anticapsin?

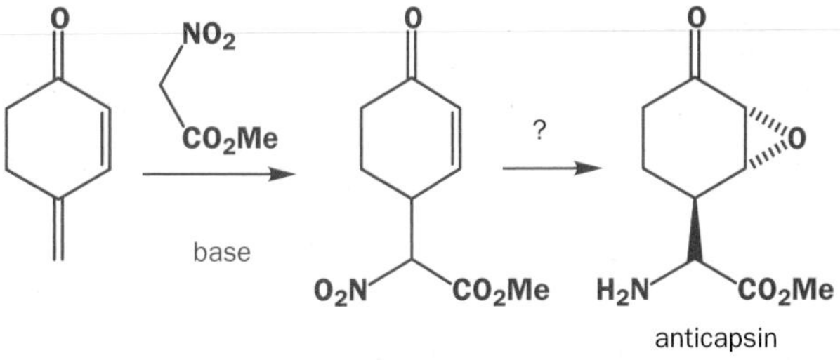

anticapsin

Retrosynthetic analysis

30

Connections

Building on:
- Carbonyl chemistry ch6, ch12, & ch14
- Conjugate addition ch10
- S_N1 and S_N2 reactions ch17
- Electrophilic aromatic substitution ch22

Arriving at:
- Synthesis and retrosynthesis
- Thinking backwards
- How to make amines and ethers
- What are synthons?
- Choosing which C–C bonds to make
- Two-group disconnections are best
- Logical planning in enolate chemistry

Looking forward to:
- Diastereoselectivity ch33–ch34
- Pericyclic reactions ch35–ch36
- Synthesis of aromatic heterocycles ch44
- Asymmetric synthesis ch45
- Natural products ch51

Creative chemistry

Chemistry is above all a creative science. Nearly all that you have learned so far in this book has had one underlying aim: to teach you how to make molecules. This is after all what most chemists do, for whatever reason. Small amounts of many drugs can be isolated from plants or marine animals; much greater quantities are made by chemists in laboratories. A limited range of dyes can be extracted from plants; many more vivid and permanent ones are made by chemists in the laboratory. Synthetic polymers, created by chemists, have replaced more expensive and less durable alternatives like rubber. Despite the bad press it has received, the use of PVC as insulating material for electric wires has prevented numerous fires and saved many lives. Eating is cheap and people live longer because pesticides allow agriculture to supply copious quantities of food to the shelves of our shops, markets, and supermarkets. Most of the improvements in the quality of life over the last 50 to 100 years can be traced to new molecules created by chemists.

ICI-D7114

But, faced with the challenge of making a new compound, how do chemists go about deciding how to make it? This molecule is known as ICI-D7114, and was identified as a possible anti-obesity drug. To test its efficacy, several hundred grams of it had to be made, and overleaf is how it was done.

The chemists who made this molecule could have chosen any route—any starting materials and any sequence of reactions. All that mattered was the final product—what we will call the **target molecule.** Synthetic planning starts with the product, which is fixed and unchangeable, and works backwards towards the starting materials. This process is called **retrosynthesis**, and the art of planning the synthesis of a target molecule is called **retrosynthetic analysis**. The aim of this chapter is to introduce you to the principles of retrosynthetic analysis: once you have read and understood it you will be well on the way to designing your own organic syntheses.

■ Of course, in a general text like this we are limited in the amount of detail we can cover—if you want to know more then read a specialized text.

> ▶ You now know four types of reaction arrow: the simple reaction arrow → meaning 'reacts to give', the delocalization arrow ↔ meaning 'two different ways to draw the same delocalized structure', the equilibrium arrow ⇌ meaning 'these two structures are interconverting', and now the retrosynthesis arrow ⇒ meaning 'could be made from'.

> ▶ This chapter will rely heavily on the reactions you have met earlier in the book, and should therefore provide you with the opportunity to revise them and check you understand how they work. If you come across a reaction you aren't familiar with, look it up before carrying on to the next one.

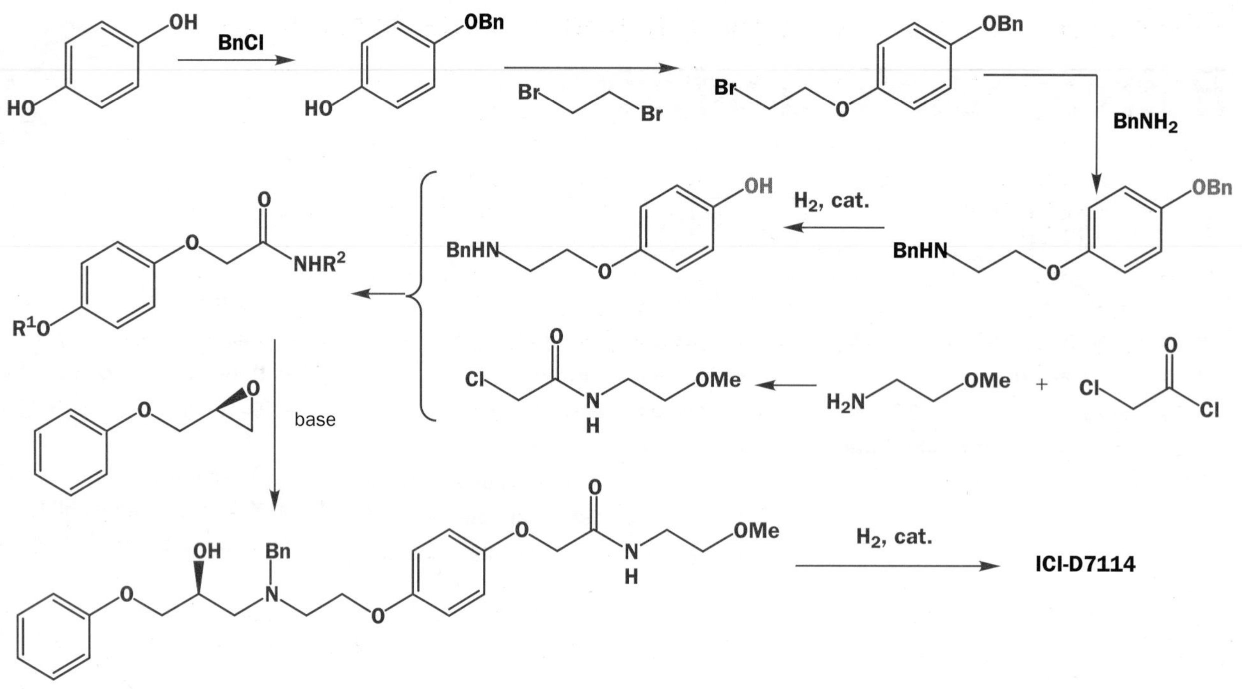

Retrosynthetic analysis: synthesis backwards

Most of the chemistry you have learned so far has concentrated on *reactions* (questions like 'what do you need to add to X to get Y?') or on *products* (questions like 'what will happen if X and Y react together?'). Now we're looking at **starting materials** (questions like 'what X and Y do you need to react together to make Z?'). We're looking at reactions in reverse, and we have a special symbol for a reverse reaction called a **retrosynthetic arrow** (the 'implies' arrow from logic).

A scheme with a retrosynthetic arrow **Z** $\Longrightarrow$ **X + Y** means 'Z could be made from X plus Y'.

a retrosynthetic arrow

This compound is used as an insect repellent. As it's an ester, we know that it can be made from alcohol plus acyl chloride, and we can represent this using a retrosynthetic arrow.

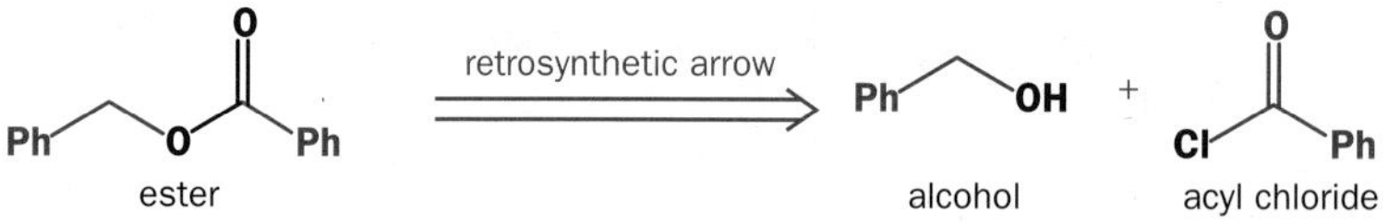

The aromatic amide amelfolide is a cardiac antiarrhythmic agent. Because we see that it is an amide, we know that it can be made quite simply from *p*-nitrobenzoyl chloride and 2,6-dimethylaniline—again, we can represent this using a retrosynthetic arrow. Mentally breaking a molecule into its component parts like this is known as **disconnection**, and it's helpful to indicate the site of the disconnection with a wiggly line as we have here.

product or target molecule amelfolide

proposed starting materials

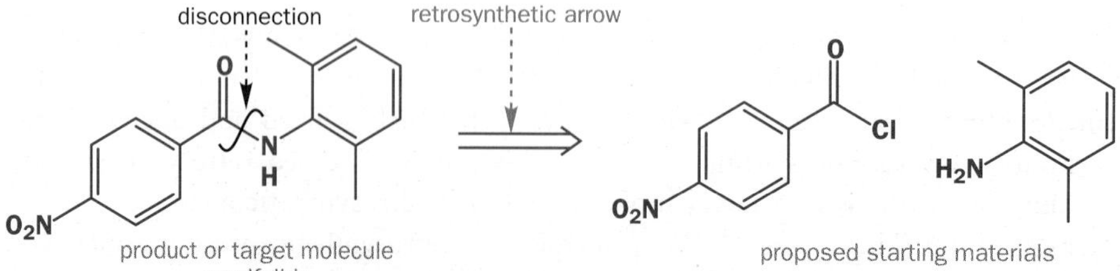

Disconnections must correspond to known, reliable reactions

The chemists who first made amelfolide chose to make it from an amine and an acyl chloride because they knew that this reaction, the standard way of making an amide, had a very good chance of success. They chose to disconnect the C–N bond because this disconnection corresponds to a reliable reaction in a way that no other possible disconnection of this molecule does.

Now that you've seen the principle of retrosynthetic analysis at work, you should be able to suggest a reasonable disconnection of this compound, which is known as daminozide.

You probably spotted immediately that daminozide is again an amide, so the best disconnection is the C–N bond, which could take us back to acyl chloride and dimethylhydrazine. This time we've written 'C–N amide' above the retrosynthetic arrow as a reminder of why we've made the disconnection and we advise you to follow this practice.

Daminozide is an agrochemical used to stunt the growth of chrysanthemums and dwarf fruit trees artificially.

Now, in fact, there is a problem with this acyl chloride—it would be unstable as it can cyclize to an anhydride. But this poses no problem for the synthesis of daminozide—we could just use the anhydride instead, since the reaction should be just as reliable. A better retrosynthesis therefore gives the anhydride and indeed this is how daminozide is made.

You will find that you learn much more and much faster if you try to do the retrosynthetic analyses in this chapter as you read it, before looking at the suggested solutions. Use a piece of paper to cover up the rest of the page as you read, and write some ideas down on another piece of paper. Don't just say 'oh I can do that' and move on—you'll miss out on the chance of teaching yourself a lot of chemistry. Don't waste the opportunity! Next time you read this chapter you'll have your memory as an aid—and retrosynthetic analysis isn't about remembering; it's about deducing. Another important thing about retrosynthetic analysis is that there is rarely one single 'right' answer, so even if your suggestions don't match up with ours, don't be discouraged. Aim to learn from the points where your attempts differ from our suggestions.

Synthons are idealized reagents

In the synthesis of daminozide an anhydride is used out of necessity rather than out of choice, but it often turns out that there are several alternative reagents all corresponding to the same disconnection. Paracetamol, for example, is an amide that can be disconnected either to amine + acyl chloride or to amine + anhydride.

Which reagent is best can often only be determined by experimentation—commercially, paracetamol is made from *para*-aminophenol and acetic anhydride largely because the by-product, acetic acid, is easier to handle than HCl. In a retrosynthetic analysis, we don't really want to be bothered by this sort of decision, which is best made later, so it's useful to have a single way of representing the key attributes of alternative reagents. We can depict both anhydride and acyl chloride in this scheme as an 'idealized reagent'—an electrophilic acetyl group MeCO$^+$.

We call such idealized reagents **synthons**. Synthons are fragments of molecules with an associated polarity (represented by a '+' or '−') which stand for the reagents we are going to use in the forward synthesis. They are not themselves reagents, though they may occasionally turn out to be intermediates along the reaction pathway. By disconnecting bonds to synthons rather than to actual reagents we can indicate the polarity of the bond-forming reaction we are going to use without having to specify details of the reagents.

synthon

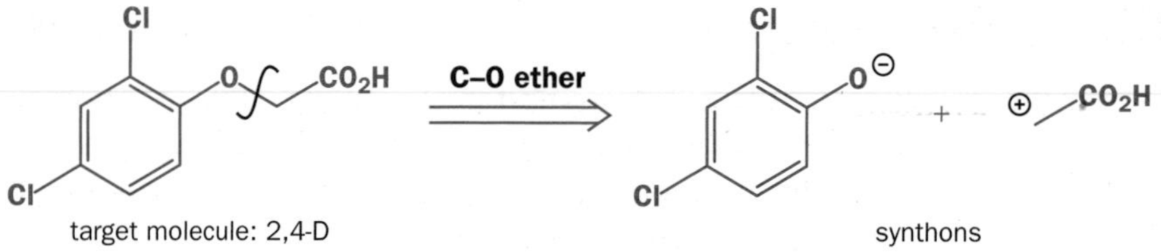

target molecule: 2,4-D   synthons

We can apply these ideas to the synthesis of the herbicide 2,4-D (2,4-dichlorophenoxyacetic acid). The most reasonable disconnection of an ether is the C–O bond because we know that ethers can be made from <u>alkyl halides</u> by substitution with an <u>alkoxide anion</u>. We don't at this stage need to decide exactly which alkyl halide or alkoxide to use, so we just write the synthons.

Once the retrosynthetic analysis is done, we can go back and use our knowledge of chemistry to think of reagents corresponding to these synthons. Here, for example, we should certainly choose the anion of the phenol as the nucleophile and some functionalized acetic acid molecule with a leaving group in the α position.

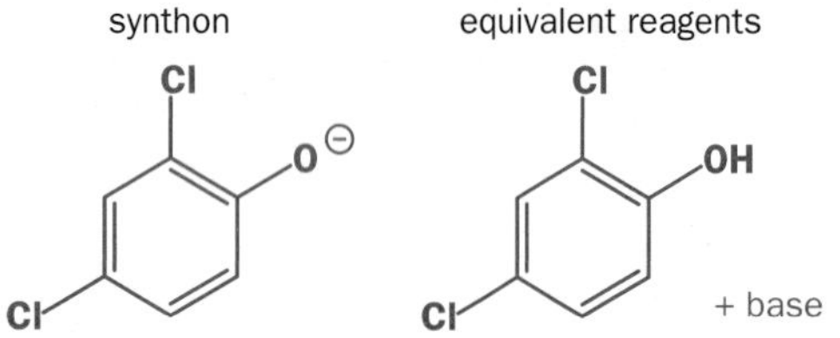

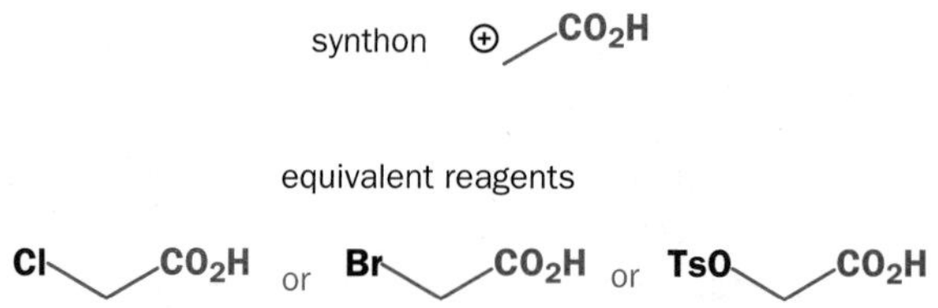

We can then write out a suggested synthesis in full from start to finish. It isn't reasonable to try to predict exact conditions for a reaction: to do that you would need to conduct a thorough search of the chemical literature and do some experiments. However, all of the syntheses in this chapter are real examples and we shall often give full details of conditions to help you become familiar with them.

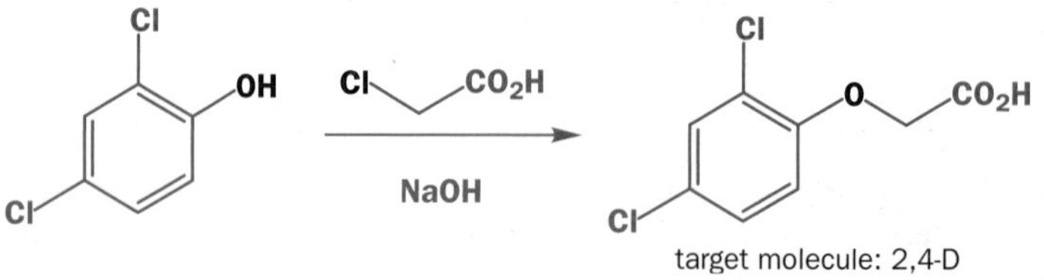

target molecule: 2,4-D

● Some definitions of terms used in synthesis

- **target molecule (or TM)** — the molecule to be synthesized
- **retrosynthetic analysis or retrosynthesis** — the process of mentally breaking down a molecule into starting materials
- **retrosynthetic arrow** — an open-ended arrow, $\Rightarrow$, used to indicate the reverse of a synthetic reaction
- **disconnection** — an imaginary bond cleavage, corresponding to the reverse of a real reaction
- **synthon** — idealized fragments resulting from a disconnection. *Synthons* need to be replaced by *reagents* in a suggested synthesis
- **reagent** — a real chemical compound used as the equivalent of a synthon

Choosing a disconnection

The hardest task in designing a retrosynthetic analysis is spotting where to make the disconnections. We shall offer some guidelines to help you, but the best way to learn is through experience and practice. The overall aim of retrosynthetic analysis is to get back to starting materials that are available from chemical suppliers, and to do this as efficiently as possible.

● Guideline 1

Disconnections must correspond to known, reliable reactions

We have already mentioned that disconnections must correspond to known reliable reactions and it's the most important thing to bear in mind when working out a retrosynthesis. When we disconnected the ether 2,4-D we chose to disconnect next to the oxygen atom because we know about the synthesis of ethers. We chose *not* to disconnect on the aryl side of the oxygen atom because we know of no reliable reaction corresponding to nucleophilic attack of an alcohol on an unactivated aromatic ring.

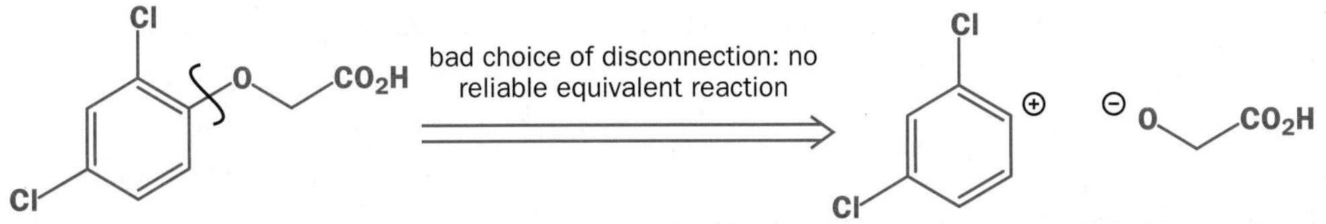

We talked about cases where nucleophilic aromatic substitution *is* possible in Chapter 23.

● Guideline 2

For compounds consisting of two parts joined by a heteroatom, disconnect next to the heteroatom

In all the retrosynthetic analyses you've seen so far there is a heteroatom (N or O) joining the rest of the molecule together, and in each case we made the disconnection next to that N or O. This guideline works for esters, amides, ethers, amines, acetals, sulfides, and so on, because these compounds are often made by a substitution reaction.

Chlorbenside is used to kill ticks and mites. Using Guideline 2 we can suggest a disconnection next to the sulfur atom; using Guideline 1 we know that we must disconnect on the alkyl and not on the aryl side.

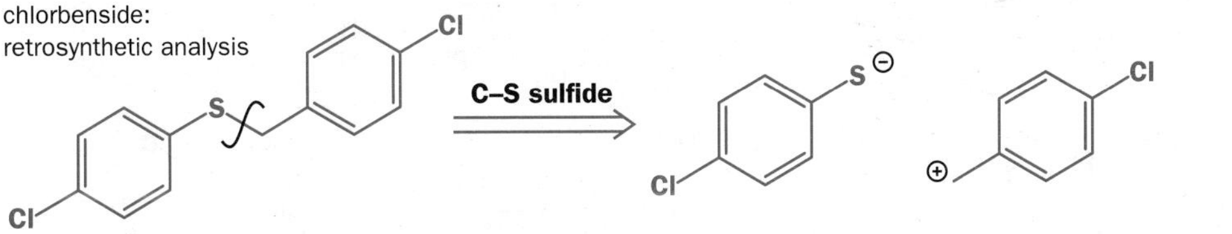

We can now suggest reagents corresponding to the synthons, and propose a synthetic scheme.

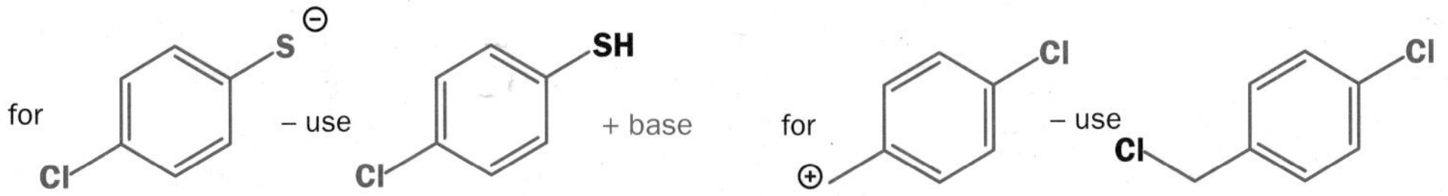

▶

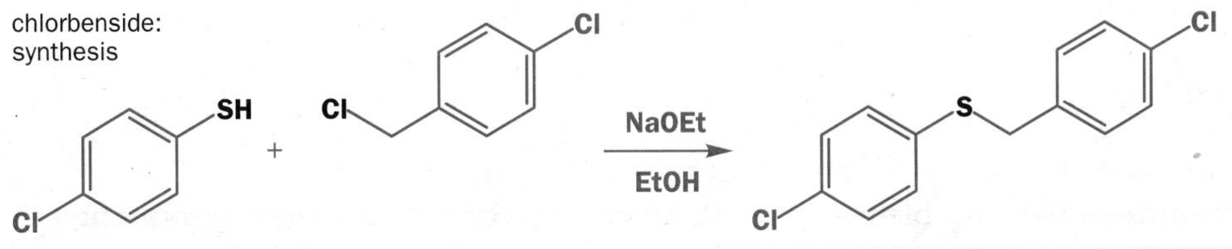

You shouldn't have expected to predict that sodium ethoxide would be the base used for this reaction, but you should have been aware that a base is needed, and have had some idea of the base strength required to deprotonate a thiol.

The next example is the ethyl ester of, and precursor to, cetaben, a drug that can be used to lower blood lipid levels. It is an amine, so we disconnect next to the nitrogen atom.

You don't always need to write out the synthons first—here the reagents are simple so we just write those instead.

cetaben ethyl ester:
retrosynthetic analysis
(R = n-C$_{15}$H$_{31}$)

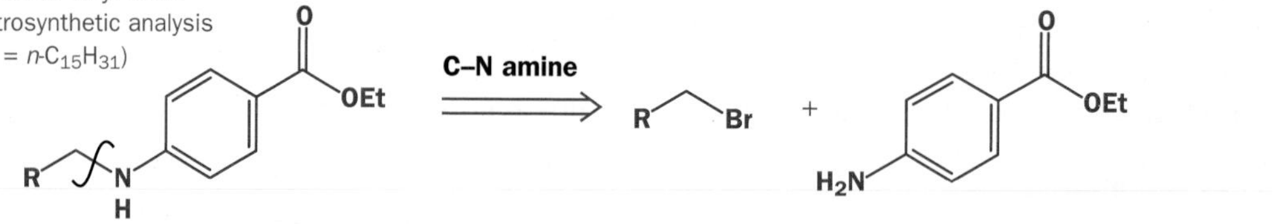

C–N amine

The alkyl bromide is available but we shall need to make the aromatic amino-ester and the best disconnection for an ester is the C–O bond between the carbonyl group and the esterifying group.

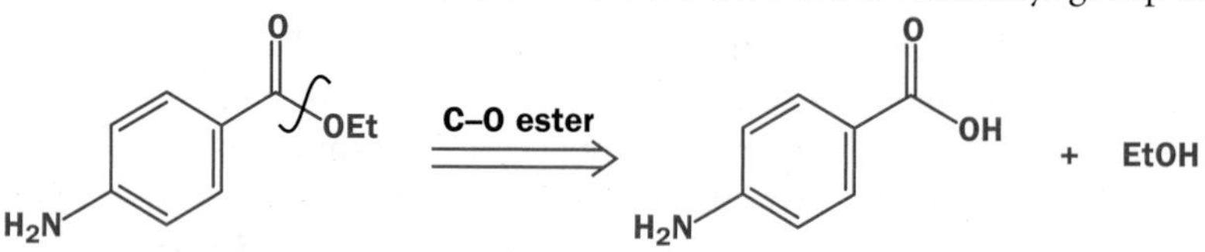

C–O ester

+ EtOH

We have now designed a two-step synthesis of our target molecule, and this is how it was carried out.

cetaben ethyl ester:
synthesis
(R = n-C$_{15}$H$_{31}$)

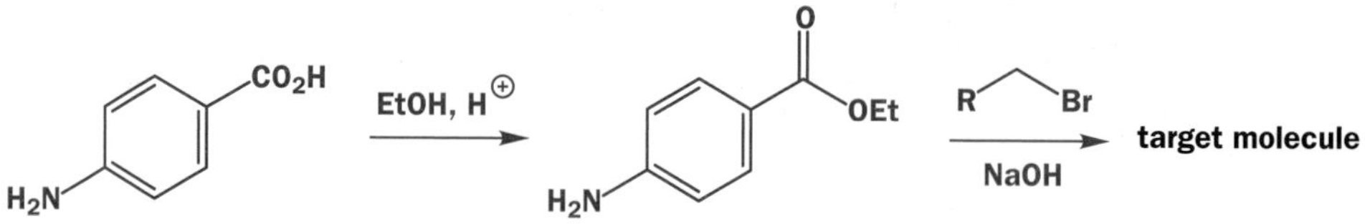

EtOH, H$^{\oplus}$

R   Br
—————→
NaOH   **target molecule**

Multiple step syntheses: avoid chemoselectivity problems

This compound was an intermediate in the synthesis of the potential anti-obesity drug ICI-D7114 you met at the beginning of the chapter. You can spot that, with two ethers and an amine functional group, it requires several disconnections to take it back to simple compounds. The question is which do we do first? One way to solve the problem is to write down all the possibilities and see which looks best. Here there are four reasonable disconnections: one at each of the ether groups (*a* and *b*) or on either side of the amine (*c* and *d*).

ICI-D7114 intermediate: retrosynthetic analysis

possible disconnections

We talked about this type of thing in Chapter 24.

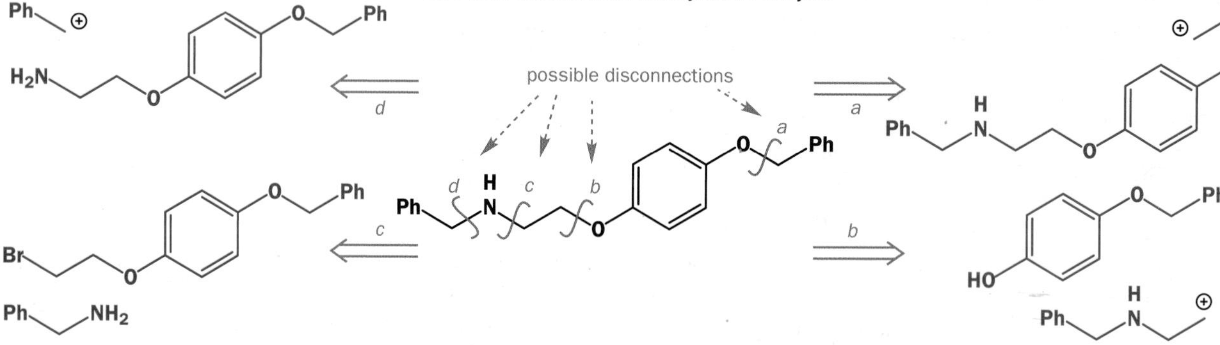

Both (*a*) and (*b*) pose problems of chemoselectivity as it would be hard to alkylate the phenol in the presence of the basic nitrogen atom. Between (*c*) and (*d*), (*c*) appears to be the better choice because the next disconnection after (*d*) will have to be an alkylation of O in the presence of an NH$_2$ group. To avoid chemoselectivity problems like this, we want to try and *introduce reactive groups late in the synthesis*. In terms of retrosynthetic analysis, then, we can formulate another guideline.

● **Guideline 3**

Consider alternative disconnections and choose routes that avoid chemoselectivity problems—often this means disconnecting reactive groups first

This guideline helps us in the next retrosynthetic step for the ICI-D7114 intermediate. Disconnection (c) gave us a compound with two ethers that might be disconnected further by disconnection (e) or (f).

ICI-D7114 intermediate: retrosynthetic analysis

target molecule

C–O ether

C–O ether

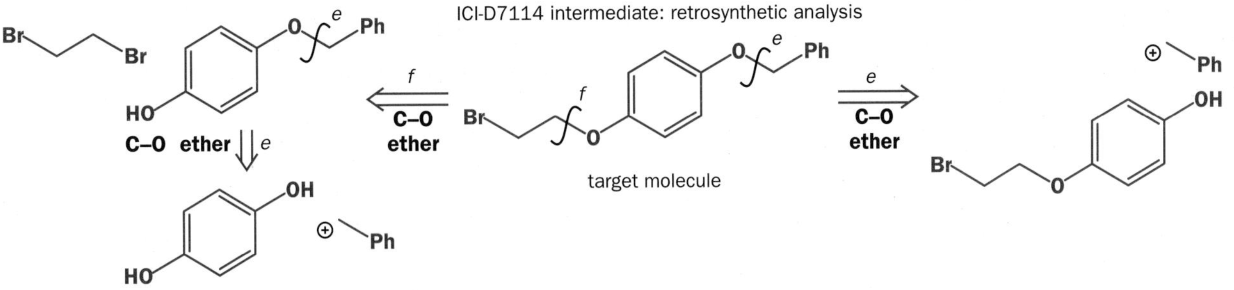

Disconnection (e) requires alkylation of a compound that is itself an alkylating agent. Disconnection (f) is much more satisfactory, and leads to a compound that is easily disconnected to 4-hydroxyphenol (*para*-cresol) and 1,2-dibromoethane. Using Guideline 3, we can say that it's best to disconnect the bromoethyl group (f) before the benzyl group because the bromoethyl group is more reactive and more likely to cause problems of chemoselectivity.

ICI-D7114 intermediate: synthesis

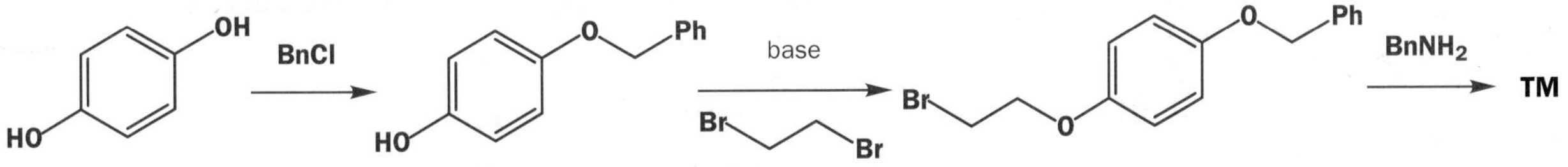

Functional group interconversion

The antihypertensive drug ofornine contains an amide and an amine functional group, and we need to decide which to disconnect first. If we disconnect the secondary amine first (b), we will have chemoselectivity problems constructing the amide in the presence of the resulting NH_2 group.

ofornine: retrosynthetic analysis

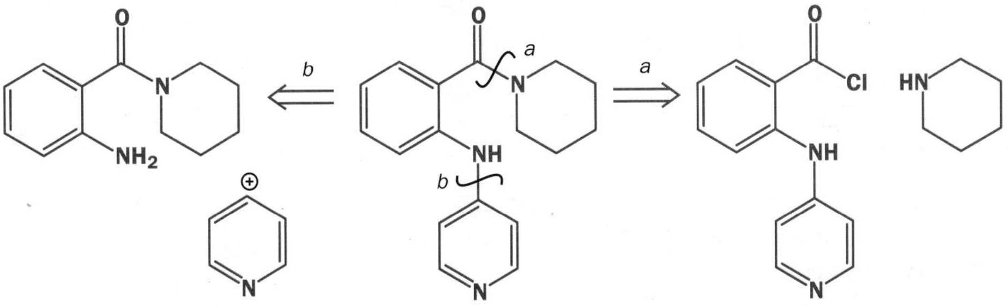

Yet disconnection (a), on the face of it, seems to pose an even greater problem because we now have to construct an amine in the presence of an acyl chloride! However, we shall want to make the acyl chloride from the carboxylic acid, which can then easily be disconnected to 2-aminobenzoic acid (anthranilic acid) and 4-chloropyridine.

We discussed nucleophilic substitutions on electron-poor aromatic rings like this in Chapter 23 and there is more detail on chloropyridines in Chapter 43.

ofornine:
retrosynthetic analysis

FGI   C–N amine

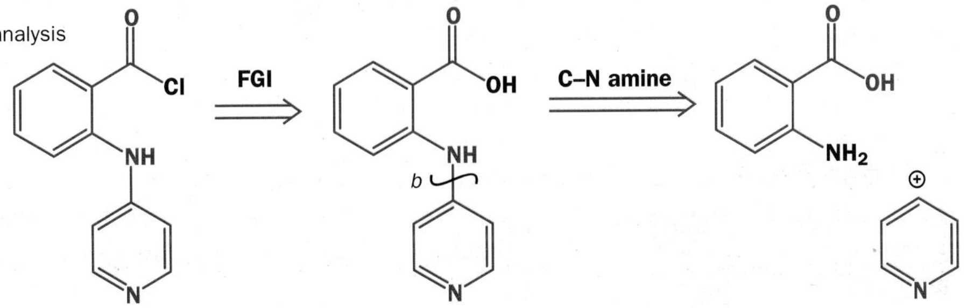

The retrosynthetic transformation of an acyl chloride to a carboxylic acid is not really a disconnection because nothing is being disconnected. We call it instead a **functional group interconversion**, or **FGI**, as written above the retrosynthetic arrow. Functional group interconversions often aid disconnections because the sort of reactive functional groups (acyl chlorides, alkyl halides) we want in starting materials are not desirable in compounds to be disconnected because they pose chemoselectivity problems. They are also useful if the target molecule contains functional groups that are not easily disconnected.

ofornine: synthesis

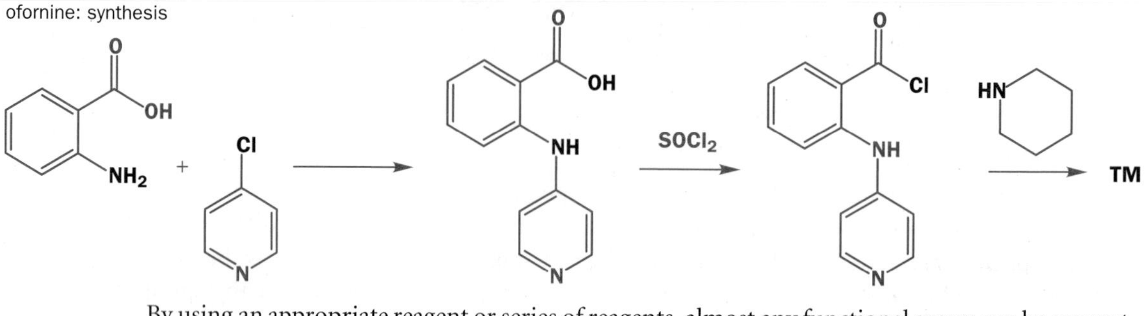

By using an appropriate reagent or series of reagents, almost any functional group can be converted into any other. You should already have a fair grasp of reasonable functional group interconversions. They mostly fall into the categories of oxidations, reductions, and substitutions (Chapters 12, 14, 17, and 24).

Amine synthesis using functional group interconversions

The synthesis of amines poses a special problem because only in certain cases is the obvious disconnection successful.

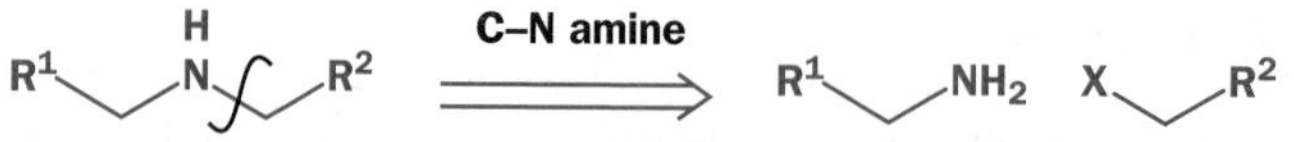

We discussed this in Chapters 14 and 24.

The problem is that the product is usually more reactive than the starting material and there is a danger that multiple alkylation will take place.

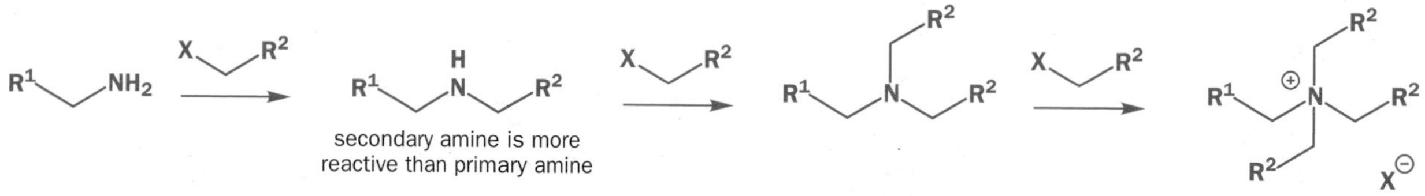

secondary amine is more reactive than primary amine

The few successful examples you have seen so far in this chapter have been exceptions, either for steric or electronic reasons, and from now on we advise you to avoid disconnecting an amine in this way. Sometimes further alkylation is made unfavourable by the increased steric hindrance that would result: this is probably the case for the cetaben ethyl ester we made by this reaction.

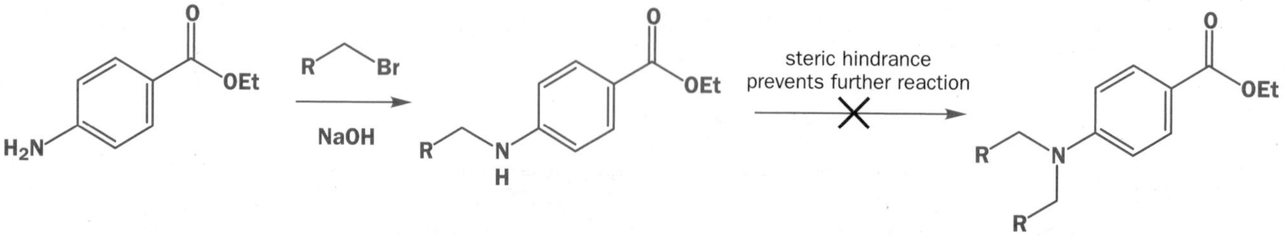

steric hindrance prevents further reaction

If the alkylating agent contains an inductive electron-withdrawing group, the product may be less reactive than the starting material—benzylamine was only alkylated once by the alkyl bromide in the synthesis of ICI-D7114 on p. 772 because of the electron-withdrawing effect of the aryloxy group.

What are the alternatives? There are two main ones, and both involve functional group inter-conversion, with the reactive amine being converted to a less reactive derivative before disconnection. The first solution is to convert the amine to an amide and then disconnect that. The reduction of amide to amine is quite reliable, so the FGI is a reasonable one.

amines: retrosynthetic analysis 1

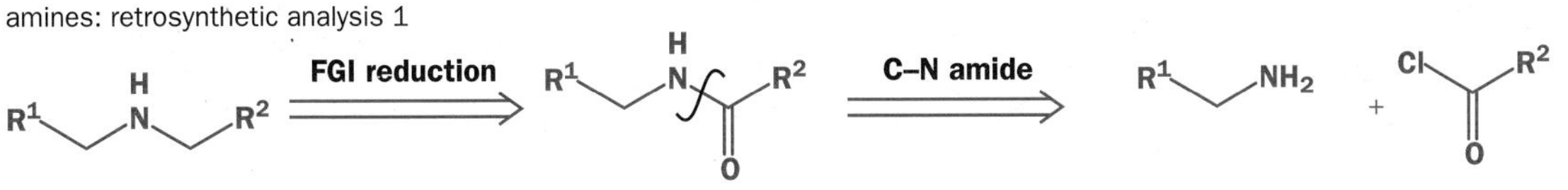

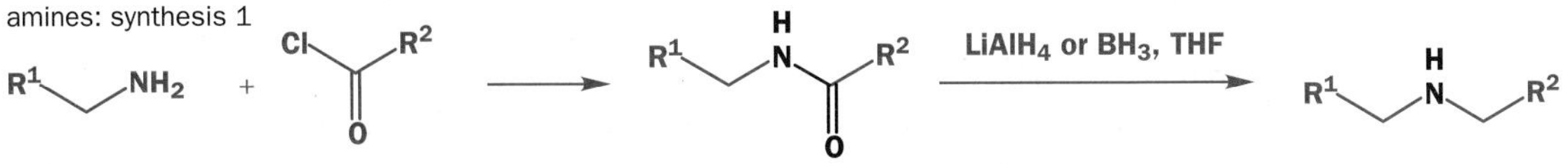

> Notice that we write 'FGI reduction' above the arrow because we are talking about the *forward* reaction we are going to do at this step.

amines: synthesis 1

This approach was used in a synthesis of this amine, though in this case catalytic hydrogenation was used to reduce the amide.

retrosynthetic analysis:
R = C$_5$H$_{11}$

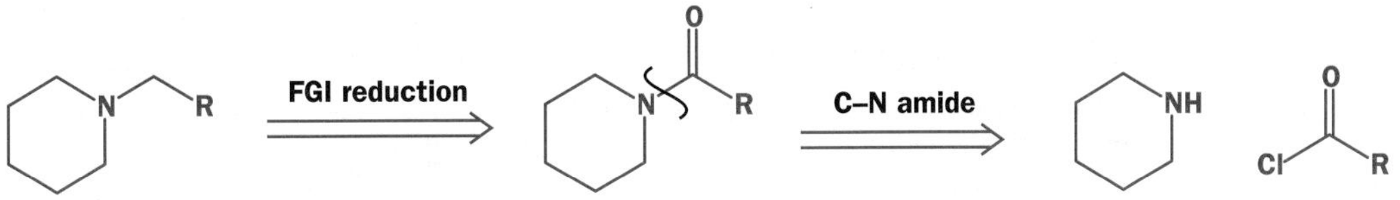

synthesis:

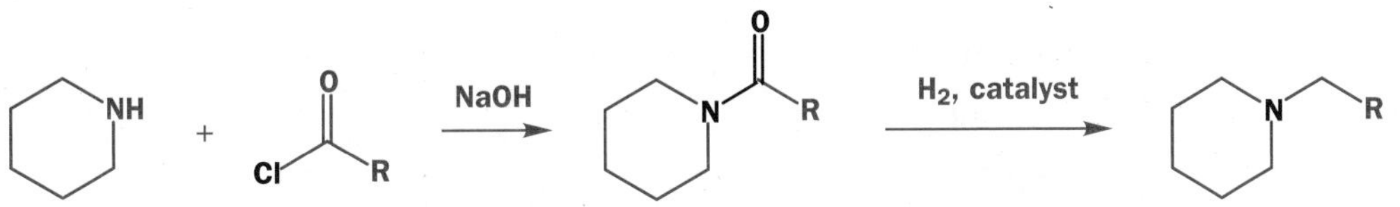

The second alternative is to convert to an imine, which can be disconnected to amine plus carbonyl compound. This approach is known as **reductive amination**, and we discussed it in detail in Chapter 14.

amines: retrosynthetic analysis 2

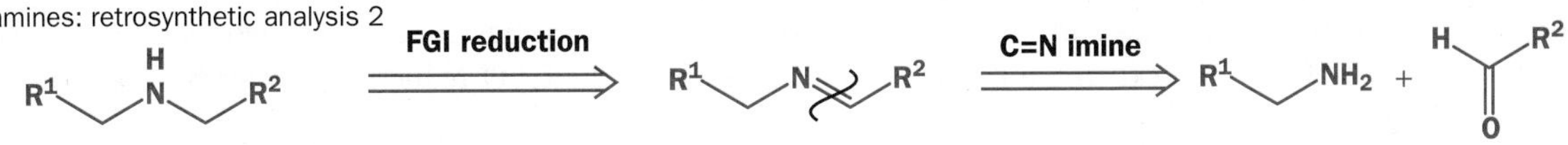

amines: synthesis 2 (reductive amination)

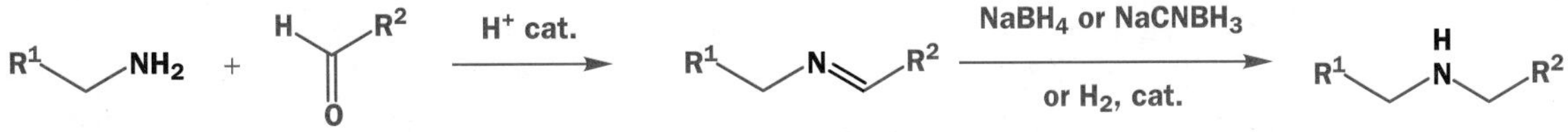

Ocfentanil is an opioid painkiller that lacks the addictive properties of morphine. Disconnection of the amide gives a secondary amine that we can convert to an imine for disconnection to a ketone plus 2-fluoro aniline.

ocfentanil:
retrosynthetic analysis

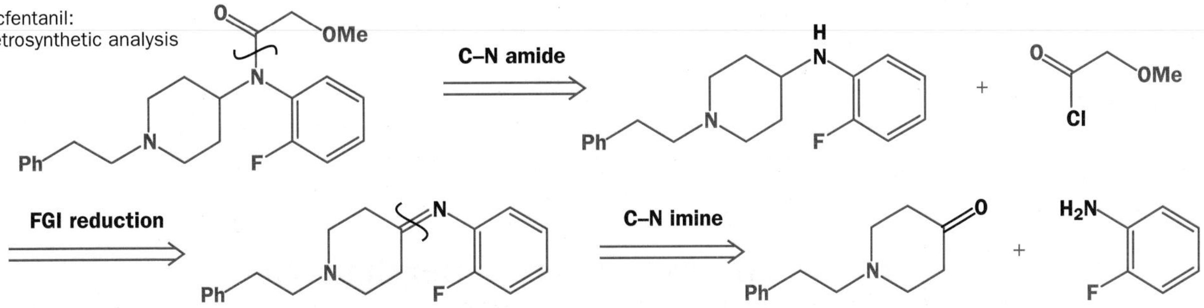

The synthesis is straightforward: a reductive amination followed by acylation of the only remaining NH group. The tertiary amine in the left-hand ring interferes with neither of these reactions.

ocfentanil: synthesis

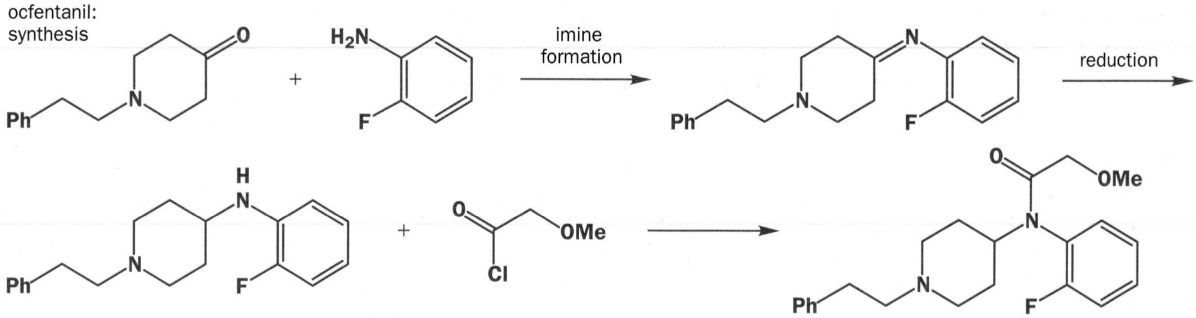

There are several conceivable routes to the neuroactive drug fenfluramine—one analysis, which uses both the amide and the imine FGI methods, is shown below and this was the route used to make the drug. Notice that the oxime was used instead of the imine. *N*-unsubstituted imines are very unstable, and the much more stable, indeed isolable oxime serves the same purpose. Oximes are generally reduced with LiAlH$_4$.

fenfluramine: retrosynthetic analysis

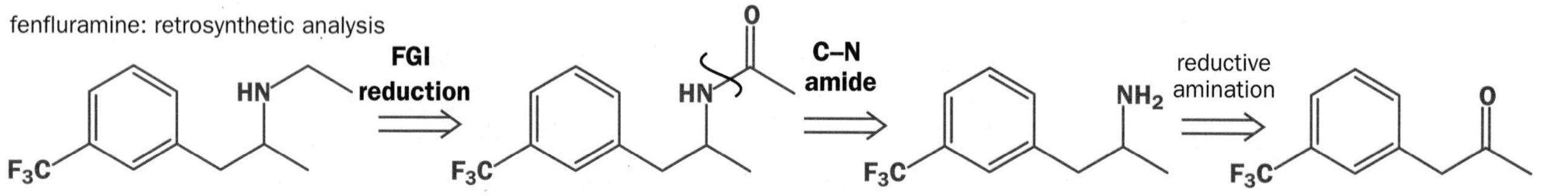

fenfluramine: synthesis

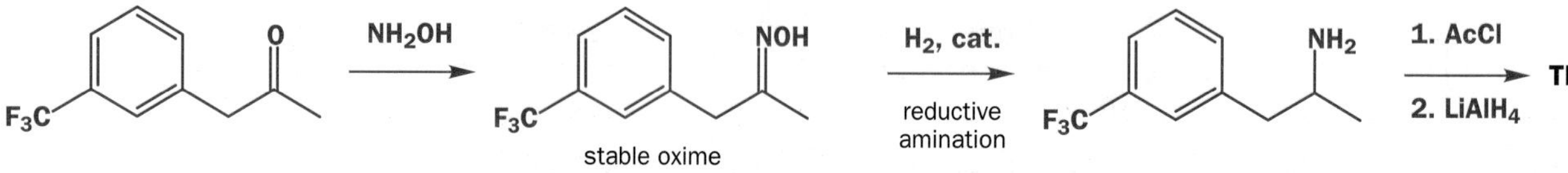

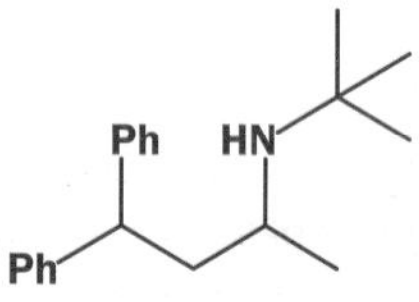

You should now be able to suggest a plausible analysis of the secondary amine terodilin. This is the structure; write down a retrosynthetic analysis and suggested synthesis before looking at the actual synthesis below.

You should find yourself quite restricted in choice: the amide route clearly works only if there is a CH$_2$ group next to the nitrogen (this comes from the C=O reduction), so we have to use an imine.

terodilin

terodilin: retrosynthetic analysis

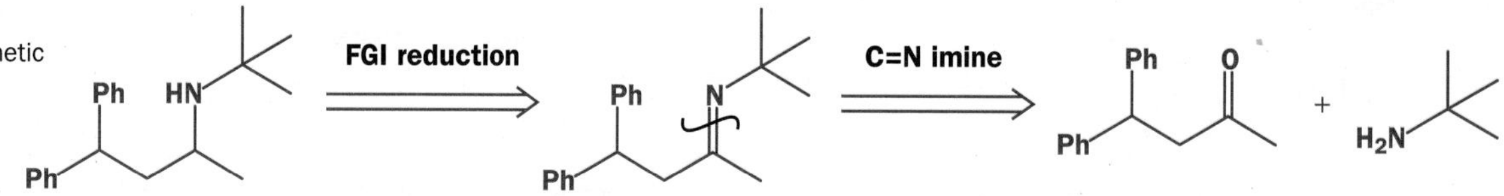

terodilin: synthesis

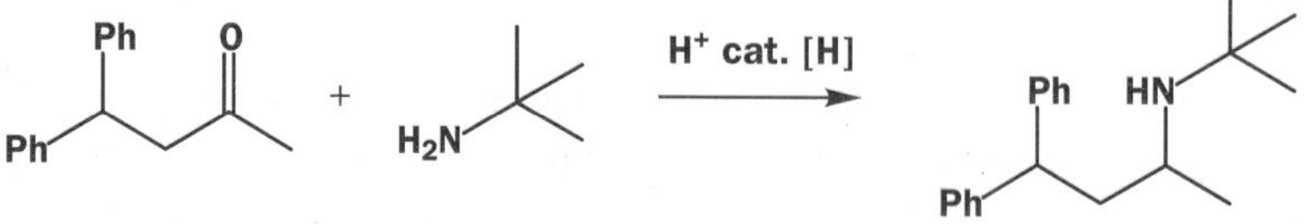

See Chapter 24 for more on this.

In the synthesis of terodilin, it was not necessary to isolate the imine—reduction of imines is faster than reduction of ketones, so formation of the imine in the presence of a mild reducing agent (usually NaCNBH$_3$ or catalytic hydrogenation) can give the amine directly.

Two-group disconnections are better than one

This compound was needed for some research into the mechanisms of rearrangements. We can disconnect on either side of the ether oxygen atom, but (*b*) is much better because (*a*) does not correspond to a reliable reaction: it might be hard to control selective alkylation of the primary hydroxyl group in the presence of the secondary one.

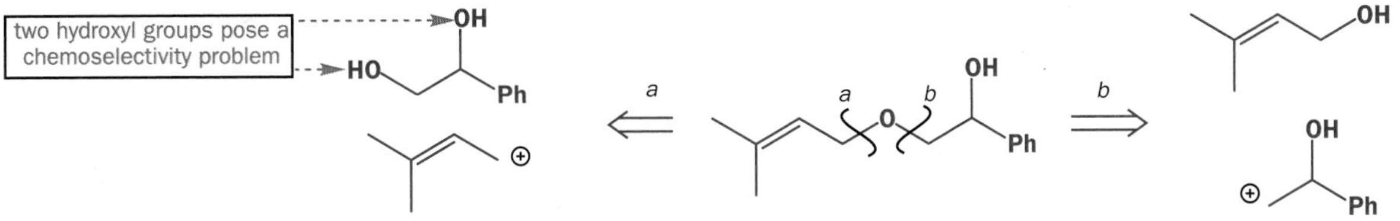

two hydroxyl groups pose a chemoselectivity problem

You might think that the best reagent to use as the equivalent of the synthon:

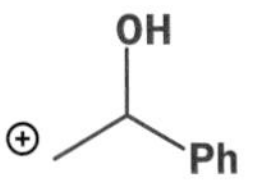

would be

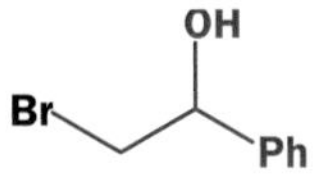

Be more ingenious! A much better solution is to use an epoxide

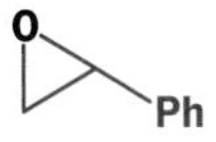

Nucleophile attack on the less hindered terminal carbon atom of the epoxide gives us the type of compound we want, and this was how the target molecule was made.

synthesis:

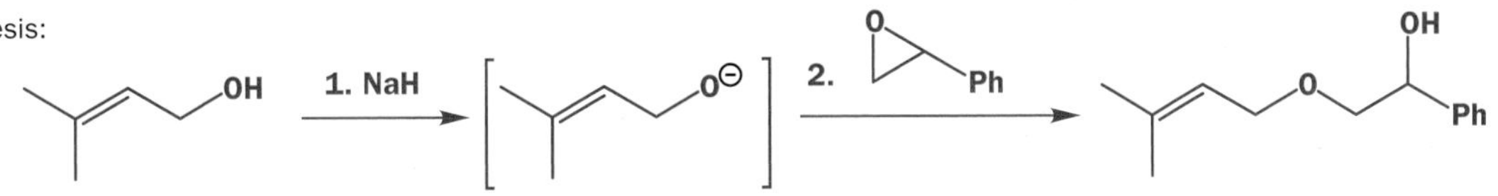

In using the epoxide we have gone one step beyond all the disconnections we have talked about so far, because we have *used one functional group to help disconnect another*—in other words, we noticed the alcohol adjacent to the ether we wanted to disconnect, and managed to involve them both in the disconnection. Such disconnections are known as **two-group disconnections**, and you should always be on the look-out for opportunities of using them because they are an efficient way of getting back to simple starting materials. We call this epoxide disconnection a 1,2-disconnection because the two functional groups in the two-group disconnection are in a 1,2-relationship.

Drug molecules often have 1,2-related functional groups: 2-amino alcohols form one important class. Phenyramidol, for example, is a muscle relaxant. A simple two-group disconnection takes it straight back to 2-amino pyridine and styrene oxide.

phenyramidol: retrosynthetic analysis

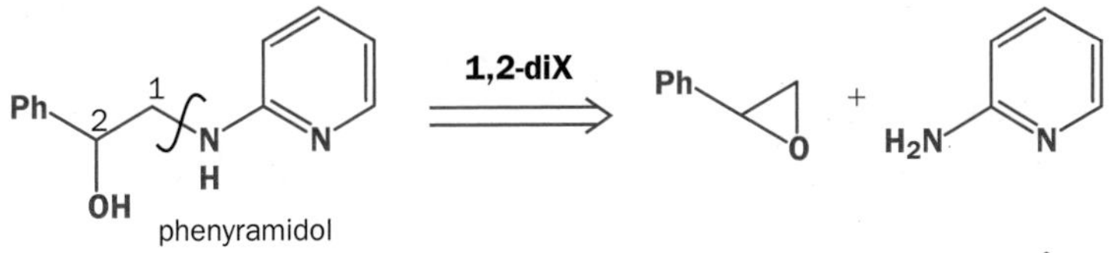

phenyramidol: synthesis

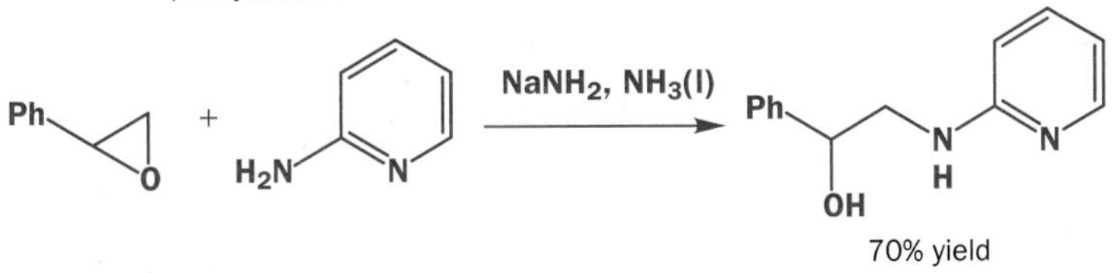

70% yield

Notice that we have written '1,2-diX' above the arrow to show that it's a two-group ('diX') disconnection—we've also numbered the carbon atoms in the starting material to show the 1,2-relationship. It may seem trivial in such a simple example, but it's a useful part of the process of writing retrosynthetic analyses because it helps you to spot opportunities for making two-group disconnections.

> ▶
> The observant among you may now be questioning why this synthesis is successful—after all, we have made a secondary amine by alkylating a primary one with an epoxide—exactly the sort of thing we advised against on p. 778. Alkylations with epoxides usually stop after the first step because the inductively electron-withdrawing hydroxyl group in the product makes it *less* nucleophilic than the starting material. In the synthesis of ICI-D7114 on p. 772, it's this same effect that prevents the amine being multiply alkylated.

Propranolol is one of the top heart drugs

The Zeneca drug propranolol is a **beta-blocker** that reduces blood pressure and is one of the top drugs worldwide. It has two 1,2-relationships in its structure but it is best to disconnect the more reactive amine group first.

propranolol: retrosynthetic analysis

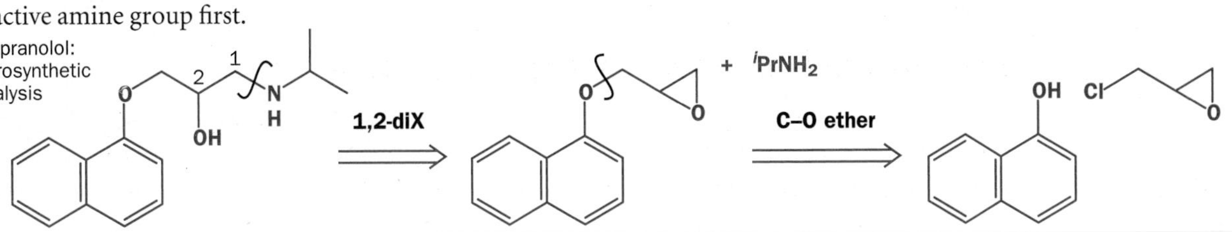

The second disconnection can't make use of an epoxide, but a simple ether disconnection takes us back to 1-naphthol and epichlorohydrin, a common starting material for this type of compound.

propranolol:
synthesis

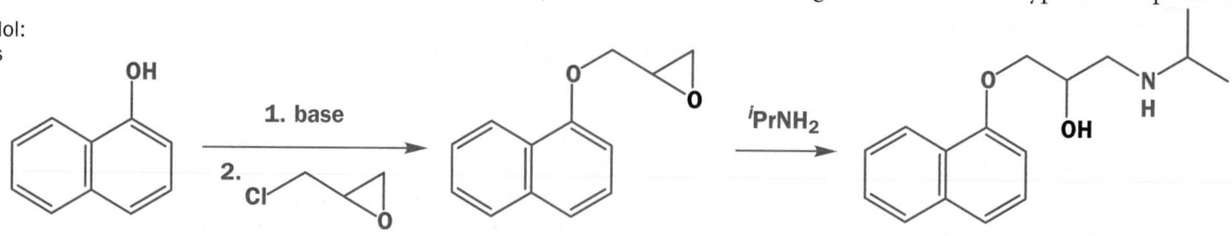

> Epichlorohydrin is a useful starting material for 1,2,3-substituted compounds. The epoxide is more electrophilic than the C–Cl bond, and the mechanism of the first step of the synthesis is surprising. 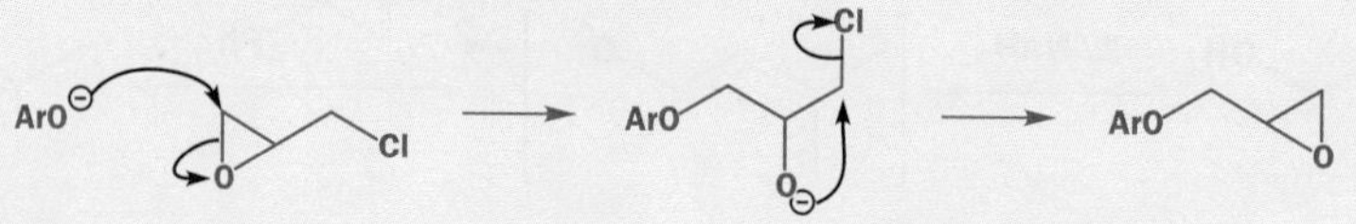
>
>
> How would you verify this experimentally? Think about what would happen if the epichlorohydrin were enantiomerically pure.

Moxnidazole can be made with epichlorohydrin

Moxnidazole is an antiparasitic drug, and our next target molecule is an important intermediate in its synthesis. The obvious first disconnection is of the carbamate group, revealing two 1,2 relationships. A 1,2-diX disconnection gives an epoxide that can be made by alkylation of morpholine with epichlorohydrin.

moxnidazole intermediate: retrosynthesis

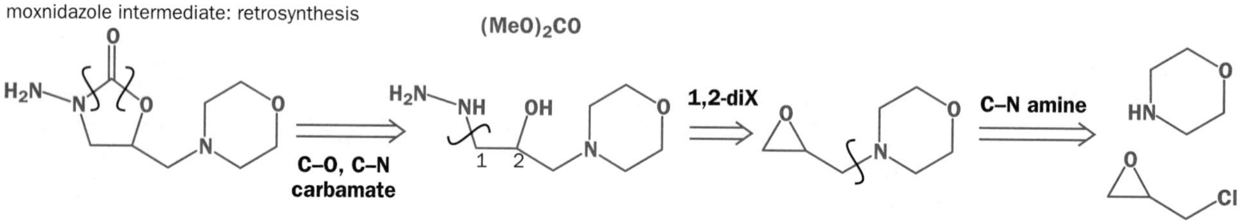

moxnidazole intermediate: synthesis

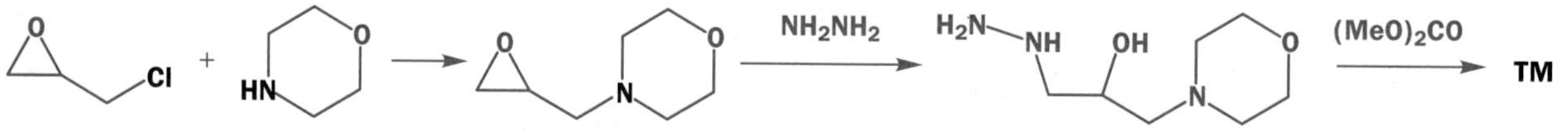

At the carbonyl oxidation level another synthon is needed for 1,2-diX disconnections

Just as epoxides are useful reagents for this synthon: α halocarbonyl compounds are useful reagents for the carbonyl equivalent:

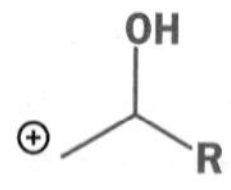

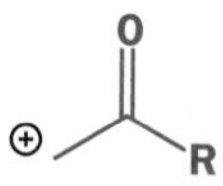

We can consider disconnection to this synthon to be a two-group disconnection because the α halocarbonyl equivalents are easily made by halogenation of a ketone, ester, or carboxylic acid (see Chapter 21) and the carbonyl group adjacent to the halide makes them extremely reactive electrophiles (Chapter 17).

Nafimidone is an anticonvulsant drug with an obvious two-group disconnection of this type.

nafimidone: retrosynthetic analysis

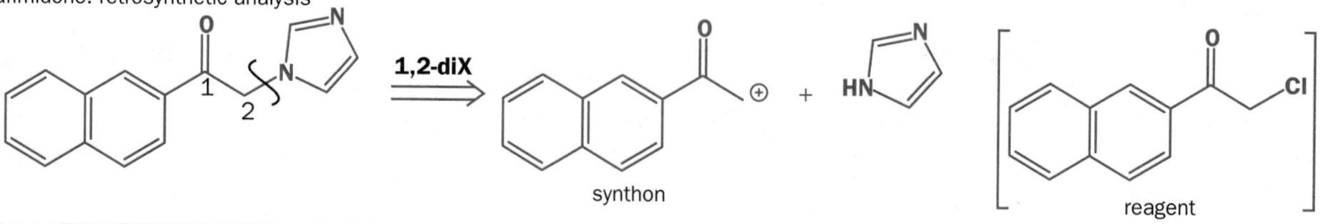

The α chloroketone is simply made by chlorination, and substitution is rapid and efficient even with the weakly basic (Chapter 8) heterocyclic amine.

nafimidone: synthesis

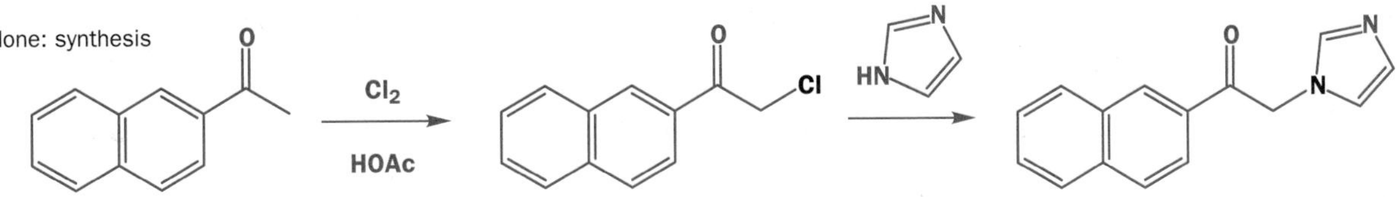

The aldehyde below was needed by ICI when they were developing a thromboxane antagonist. Two-group disconnection gives a 2-halo-aldehyde that can be made from isobutyraldehyde.

ICI aldehyde: retrosynthetic analysis

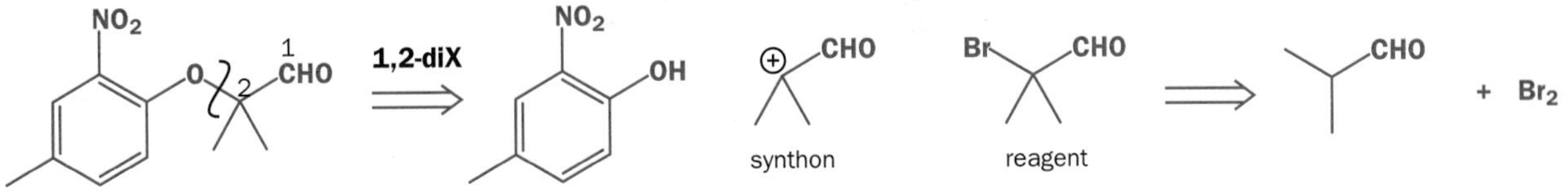

The synthesis requires a normal bromination of a carbonyl compound in acid solution but the next step is a most unusual S_N2 reaction at a *tertiary* centre. This happens because of the activation by the aldehyde group (Chapter 17) and is further evidence that the functional groups must be thought of as working together in this type of synthesis.

ICI aldehyde: synthesis

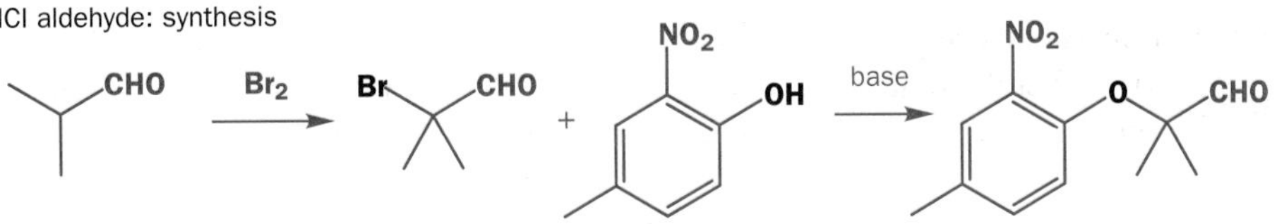

1,3-Disconnections

In Chapter 10 you saw how α,β-unsaturated carbonyl compounds undergo conjugate additions—reactions like this.

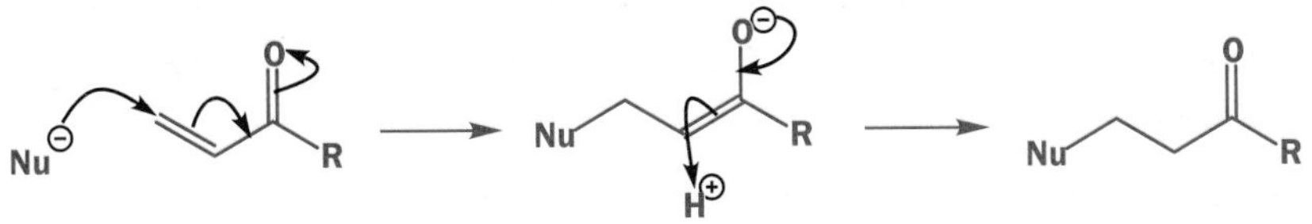

Two-group 1,3-disconnections are therefore possible because they correspond to this forward reaction. These **Michael acceptors** have an electrophilic site two atoms away from the carbonyl group, and are therefore the reagents corresponding to this synthon.

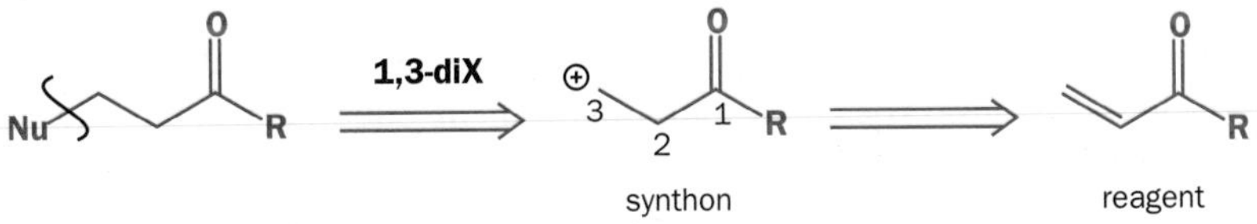

This type of reaction is available only when the alkene is conjugated to an electron-withdrawing group—usually carbonyl (Chapter 10) but it can be nitro, cyanide, etc. (Chapter 23). This disconnection is available only at this oxidation level unlike the last. We can do a two-group 1,3-disconnection on this sulfide, for example.

retrosynthetic analysis

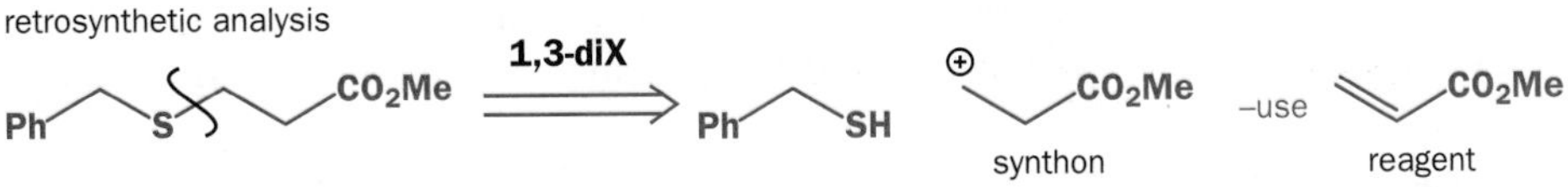

synthesis

Remember that not all nucleophiles will successfully undergo Michael additions—you must bear this in mind when making a 1,3-disconnection of this type. Most reliable are those based on nitrogen, sulfur, and oxygen (Chapter 10).

Our second example is an amine structurally similar to the 'deadly nightshade' drug, atropine, which has the ability to calm involuntary muscle movements. There is a 1,3-relationship between the amine and ketone functional groups, and 1,3-disconnection takes us back to piperidine and an unsaturated ketone.

■ We shall discuss ways of disconnecting this starting material, and other α,β-unsaturated carbonyl compounds, later in the chapter.

▶ Don't be tempted to try using β haloketones as equivalents for this synthon! They are hard to make and highly unstable and they undergo rapid E1cB elimination (see Chapter 19).

atropine mimic:
retrosynthetic analysis

atropine mimic:
synthesis

To summarize...

Before we leave C–X disconnections and go on to look at C–C disconnections we should just review some important points. We suggested three guidelines for choosing disconnections and now that you have met the principle of two-group disconnections, we can add a fourth:

● **Guidelines for good disconnections**

1. **Disconnections must correspond to known, reliable reactions**

2. **For compounds consisting of two parts joined by a heteroatom, disconnect next to the heteroatom**

3. **Consider alternative disconnections and choose routes that avoid chemoselectivity problems—often this means disconnecting reactive groups first**

4. **Use two-group disconnections wherever possible**

Two-group disconnections reduce the complexity of a target molecule more efficiently than one-group disconnections, and you should always be on the look-out for them. You will meet more two-group disconnections in the next section, which deals with how to disconnect C–C bonds.

C–C disconnections

The disconnections we have made so far have all been of C–O, C–N, or C–S bonds, but, of course, the most important reactions in organic synthesis are those that form C–C bonds. We can analyse C–C disconnections in much the same way as we've analysed C–X disconnections. Consider, for example, how you might make this simple compound, which is an intermediate in the synthesis of a carnation perfume.

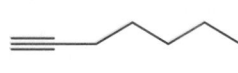

The only functional group is the triple bond, and we shall want to use the chemistry of alkynes to show us where to disconnect. You know that alkylation of alkynes is a reliable reaction, so a sensible disconnection is next to the triple bond.

carnation perfume intermediate: retrosynthetic analysis

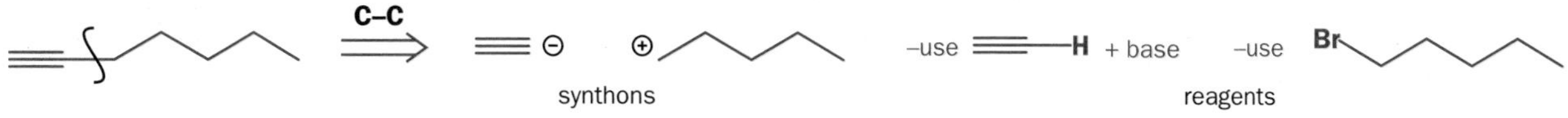

carnation perfume intermediate: synthesis

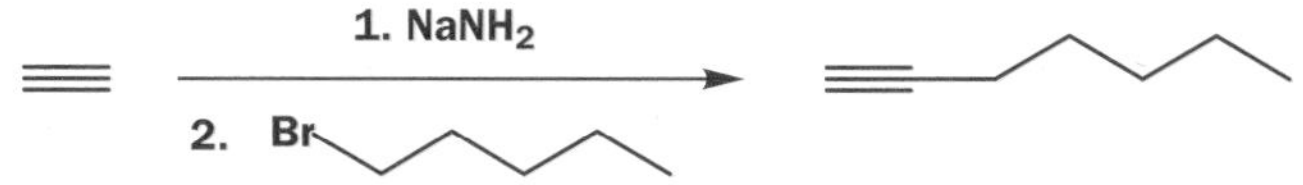

Alkynes are particularly valuable as synthetic intermediates because they can be reduced either to *cis* or to *trans* double bonds.

cis (Z)-alkene

trans (E)-alkene

> You met these reductions in Chapter 24, and we will talk about them again in the context of double bond synthesis in Chapter 31.

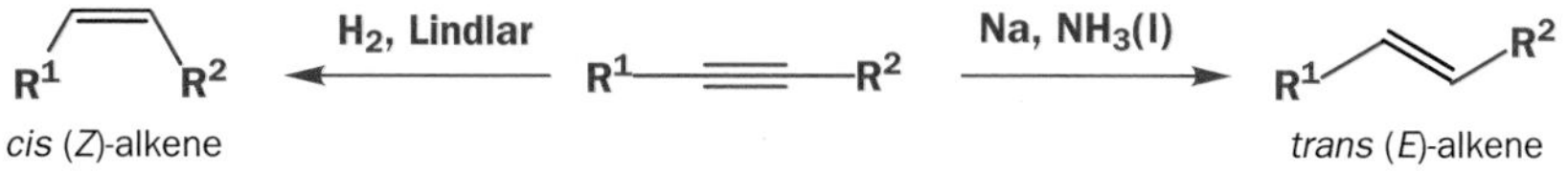

It's often a good idea to start retrosynthetic analysis of target molecules containing isolated double bonds by considering FGI to the alkyne because C–C disconnections can then become quite easy.

This *cis*-alkene is a component of violet oil, and is an intermediate in the synthesis of a violet oil component. FGI to the alkyne reveals two further disconnections that make use of alkyne alkylations. The reagent we need for the first of these is, of course, the epoxide as there is a 1,2-relationship between the OH group and the alkyne.

> There are, of course, many other ways of disconnecting double bonds: you are about to meet an important disconnection of double bonds conjugated with carbonyl groups. Chapter 31 is devoted to the alternative methods available for making double bonds and controlling their stereochemistry.

violet oil component: retrosynthetic analysis

cis (Z)-alkene

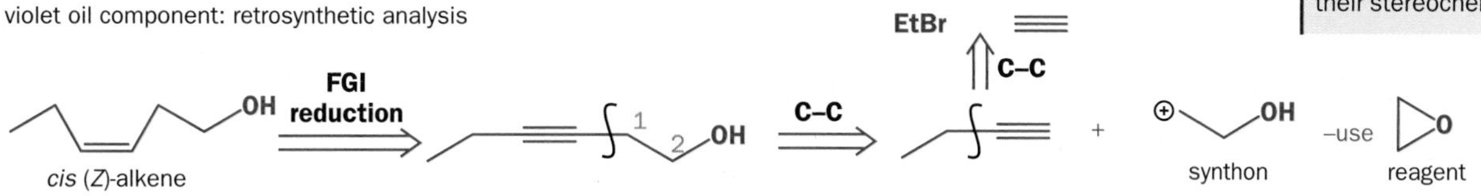

violet oil component: synthesis

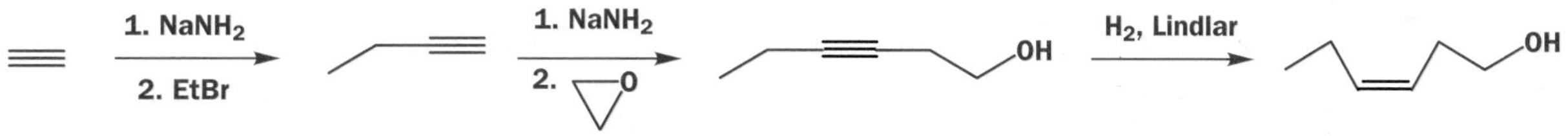

The next example is the pheromone of the pea-moth, and can be used to trap the insects (see the introduction to Chapter 24). After disconnecting the ester, FGI on the *trans* double bond gives an alkyne.

pea-moth pheromone: retrosynthetic analysis

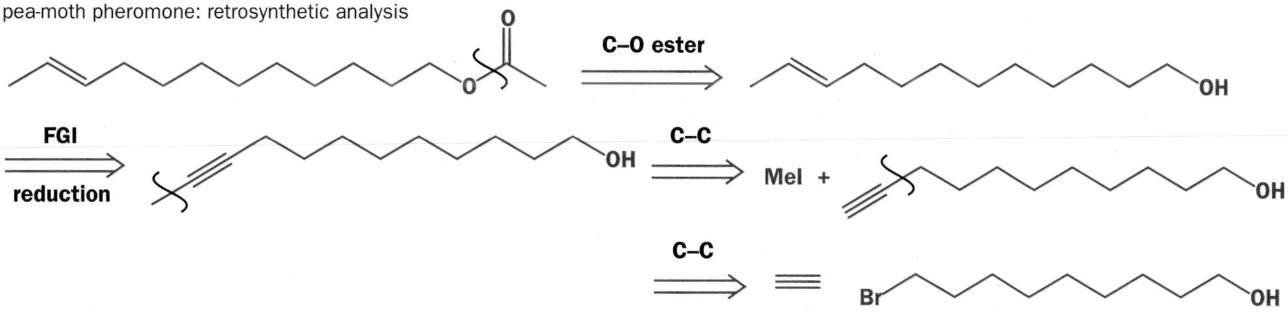

Disconnection on either side of the alkyne leads us back to a bromo-alcohol alkylating agent. In the synthesis of the pheromone, it turned out to be best if the hydroxyl group was protected as its THP ether. You should be able to think of other alkylation-type reactions that you have met that proceed reliably and therefore provide a good basis for a disconnection—the alkylation of enolates of esters or ketones, for example (Chapter 26).

> Protecting groups were discussed in detail in Chapter 24.

pea-moth pheromone: synthesis

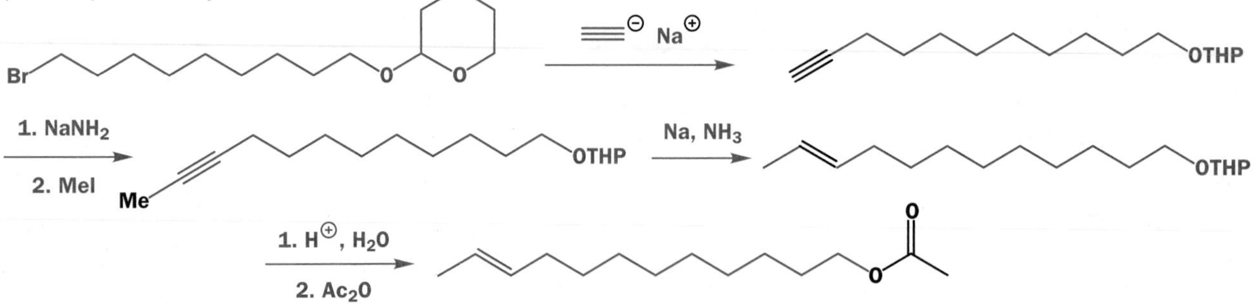

This next ester was needed for a synthesis of the sedative rogletimide (see later for the full synthesis). The ethyl group is disconnected because it can be readily introduced by alkylation of the ester enolate.

rogletimide intermediate: retrosynthetic analysis   rogletimide intermediate: synthesis

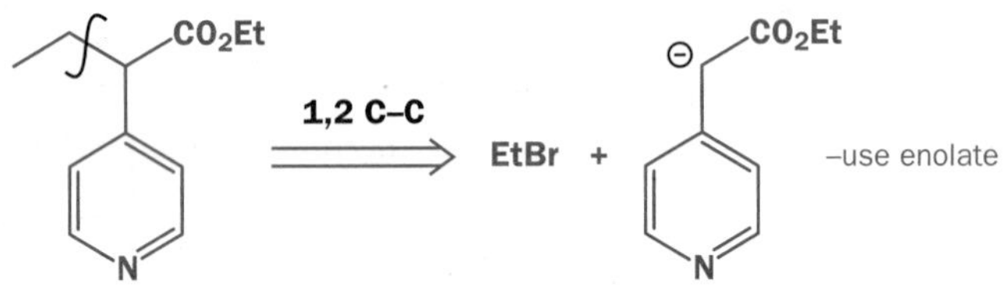

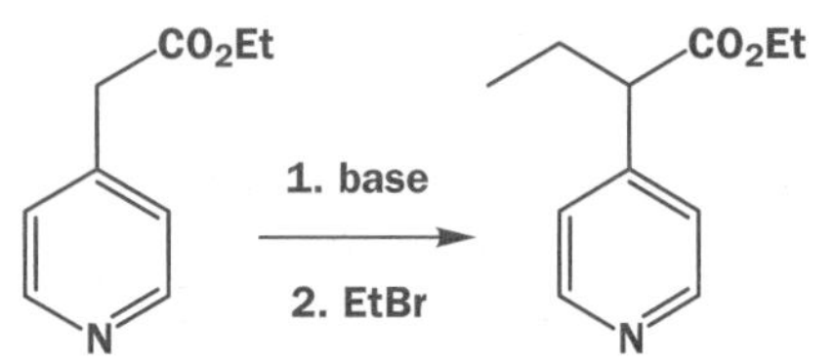

We have labelled the disconnection '1,2 C–C' because the new C–C bond is forming two atoms away from the carbonyl group. To spot disconnections of this sort, you need to look for alkyl groups in this 2-position.

Arildone is a drug that prevents polio and herpes simplex viruses from 'unwrapping' their DNA, and renders them harmless. It has just the structural characteristic you should be looking for: a branch next to a carbonyl group.

arildone: retrosynthetic analysis

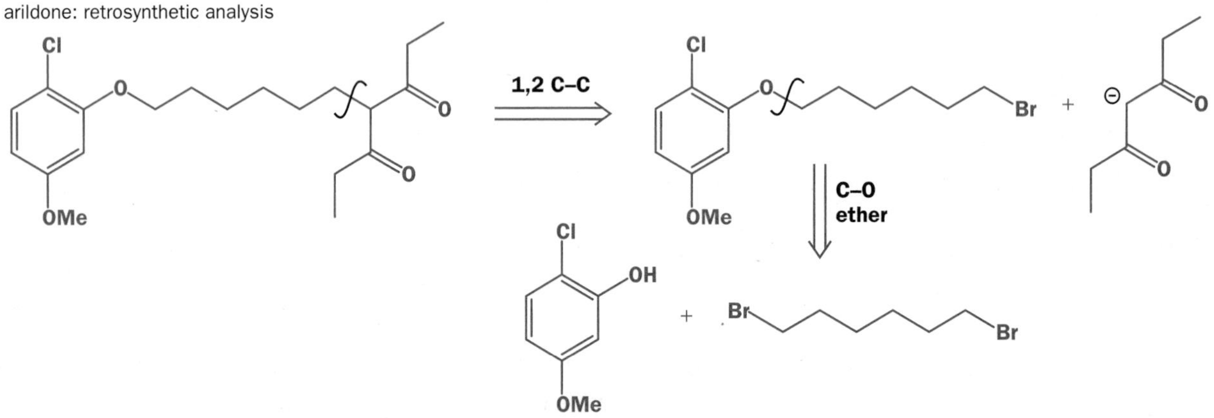

Look back to Chapter 26 if you don't understand why.

With two carbonyl groups, the alkylation should be particularly straightforward since we can use a base like methoxide. The ether disconnection is then immediately obvious. In the synthesis of arildone the alkyl iodide was used for the alkylation.

arildone: synthesis

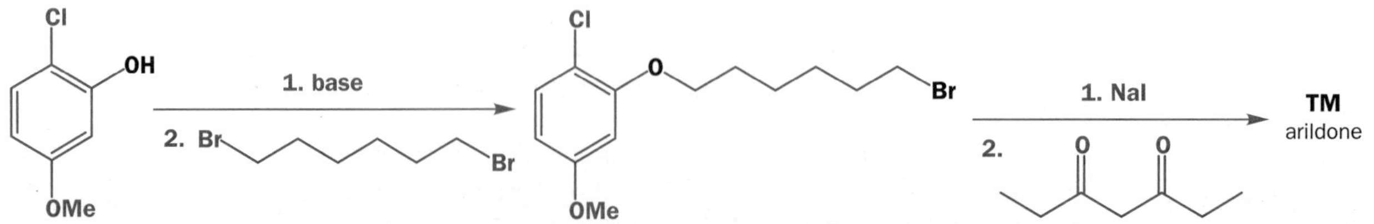

We introduced the chemistry of malonate esters in Chapters 21 and 26 as a useful way of controlling the enolization of carbonyl compounds. Alkylation followed by decarboxylation means that we can treat acetoacetate and malonate esters as equivalent for these synthons.

This unsaturated ketone is an important industrial precursor to β-carotene, vitamin A, and other similar molecules. Disconnection using the carbonyl group gives a synthon for which a good reagent will be acetoacetate.

▶

Having read Chapter 27, you should be able to suggest why the enolate of acetone itself would not be a good choice in this reaction.

carotene precursor: retrosynthetic analysis

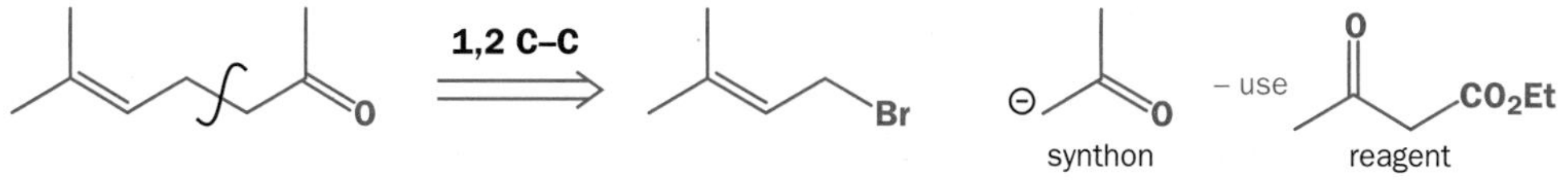

carotene precursor: synthesis

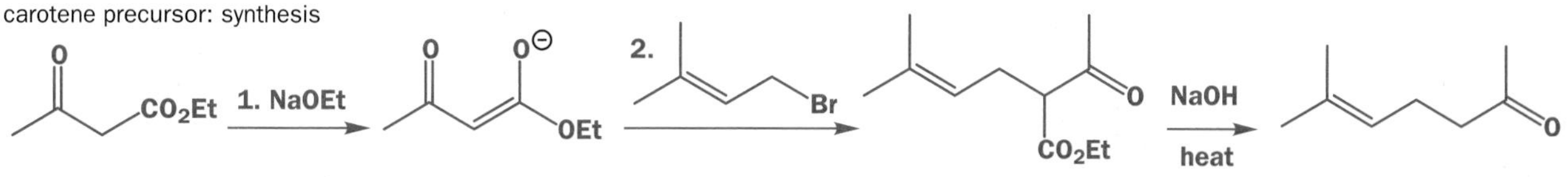

This organophosphorus compound, belfosil, is a Ca^{2+} channel blocker. You haven't met many phosphorus compounds yet, but you should be able to reason that a good disconnection will be the C–P bond by analogy with the sulfides you met earlier in the chapter. We could use bromide as a leaving group, but alkyl bromides are inconvenient to disconnect further, so we go back to the more versatile diol—in the forward synthesis we shall need a way of making the OH groups into good leaving groups. There is still no obvious disconnection of the diol, but FGI to the ester oxidation level reveals a malonate derivative.

belfosil: retrosynthetic analysis

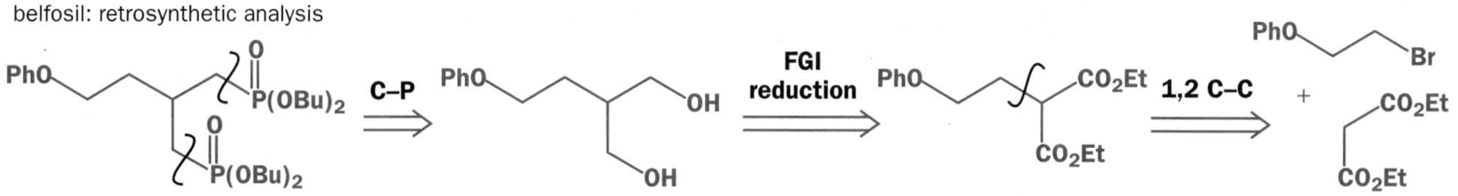

In the synthesis, the diol was converted to the bis-tosylate (see Chapter 17 if you've forgotten about tosylates and mesylates) and reacted with a phosphorus nucleophile.

belfosil: synthesis

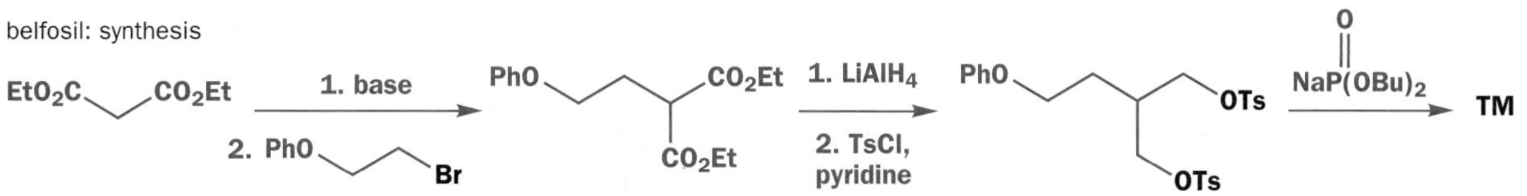

Notice how we disconnected the phosphorus-based functional groups straight back to alcohols in the retrosynthetic analysis, and not, say, to alkyl halides. Oxygen-based functional groups (alcohols, aldehydes, ketones, esters, and acids) have one important property in common—versatility. They are easily converted into each other by oxidation and reduction, and into other groups by substitution. What is more, many of the C–C disconnections you will meet correspond to reactions of oxygen-based groups, and particularly carbonyl groups. Faced with an unusual functional group in a target molecule the best thing to do is convert it to an oxygen-based group at the same oxidation level—it usually makes subsequent C–C disconnections simpler. So we add a new guideline.

● **Guideline 5**

Convert to oxygen-based functional groups to facilitate C–C disconnections

Looking for 1,2 C–C disconnections

In each of the cases you have met so far, we have used a functional group present in the molecule to help us to disconnect the C–C bond using a 1,2 C–C disconnection. You can look for 1,2 C–C disconnections in alkynes, carbonyl compounds, and alkylated aromatic rings. And, if the target isn't a carbonyl compound, consider what would be possible if functional groups such as hydroxyl groups were converted to carbonyl groups (just as we did with belfosdil).

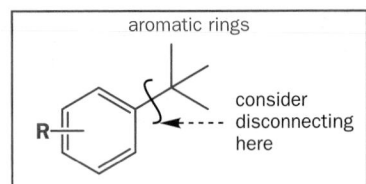

All of these disconnections relied on the reaction of a carbon electrophile with a nucleophilic functional group. The alternative, reaction of a carbon nucleophile (such as a Grignard reagent) with an electrophilic functional group, allows us to do C–C disconnections on alcohols. For example, this compound, which has a fragrance reminiscent of lilac, is a useful perfume for use in soap because (unlike many other perfumes that are aldehydes or ketones) it is stable to alkali.

We look to the one functional group, the hydroxyl, to tell us where to disconnect, and disconnection next to the OH group gives two synthons for which sensible reagents are a Grignard reagent and acetone. The perfume is made from benzyl chloride and acetone in this way. Notice that we label these disconnections 1,1 C–C because the bond being disconnected is attached to the same carbon atom as the hydroxyl functional group.

This similar alcohol has a 'peony-like fruity odour' and could be disconnected in three ways.

Disconnection (*c*) leads back to a ketone, which is cheaply made starting from acetone and benzaldehyde, and this was the route that was chosen for the synthesis.

> ■ The synthesis of this starting material involves an aldol reaction between acetone and benzaldehyde of the sort discussed in Chapter 27 followed by hydrogenation of the double bond.

Available starting materials

Although any of the three routes to the fruity peony perfume would give an acceptable synthesis, the key factor in choosing route (c) was the ease of synthesis of the starting materials from available compounds. But how can you know which materials will be available? So far in this chapter we have avoided this question, and often our retrosynthetic analyses have been incomplete because the suggested starting materials must themselves be synthesized in the laboratory. From now on, though,

we will take every analysis back to available starting materials to help you get a feel for what is, and is not, available.

The only way to be absolutely sure what you can buy is to look up a compound in a supplier's catalogue, and this is what a chemist would do when assessing possible alternative synthetic routes. A good rule of thumb is that **compounds with up to about six carbon atoms and with *one* functional**

group (alcohol, aldehyde, ketone, acid, amine, double bond, or alkyl halide) **are usually available.** This is less true for heavily branched compounds, but most straight-chain compounds with these functional groups are available up to eight or more carbon atoms. Cyclic compounds with one functional group from five- to eight-membered are also available. Of course, many other compounds are available too, including some difunctional compounds. Here are a few of them.

You will soon start to appreciate what is available as you see which compounds we use as starting materials. Supplier's catalogues are available free for the asking and make quite useful textbooks. You could consider getting one. In addition, on-line and CD catalogues are available in most chemistry departments and can be searched by structure.

acetoacetates   malonates   acrylates (R = H); methacrylates (R = Me)

R = H, Me, Et   R = H, Me, Et

Some starting materials become available because other chemists have made them

Stop

violet leaf alcohol precursor

two steps

violet leaf alcohol

Our next target is an allylic alcohol that produces the perfumery compound 'violet leaf alcohol' by a rearrangement step . Two disconnections are possible, but one of them, (a), leads back to a Grignard reagent that can be made by FGI on the violet oil component whose synthesis we described on p. 785.

violet leaf alcohol precursor: retrosynthetic analysis

a   1,1 C–C   b   1,1 C–C

FGI

violet oil component: synthesis known

The synthesis was best carried out using the alkylmagnesium iodide and the iodide was made from the alcohol via the chloride.

violet leaf alcohol precursor: synthesis

1. SOCl₂
2. NaI

1. Mg, Et₂O
2. OHC

Linalool is another perfumery compound. Disconnection of the vinyl group leads to the ketone you met on p. 787, best made by alkylation of acetoacetate, an acetone enolate equivalent.

linalool: retrosynthetic analysis

1,1 C–C   1,2 C–C

–use acetoacetate

On an industrial scale it was best to introduce the vinyl anion synthon as acetylene and then hydrogenate the alkyne. The unsaturated ketone was chosen as the starting material because its synthesis was already known.

linalool: synthesis

H₂

Lindlar

SOCl₂ – thionyl chloride

Double disconnections can be a short cut

Tertiary alcohols with two identical groups next to the hydroxyl group are often made by attack of two equivalents of a Grignard reagent on an ester. The synthesis of the antihistamine compound fenpiprane provides an example: the tertiary alcohol is a precursor to the drug and can be disconnected to ester + Grignard reagent because of the two Ph groups. The ester required has a 1,3 functional group relationship, and can be disconnected to amine plus Michael acceptor.

fenpiprane precursor: retrosynthetic analysis

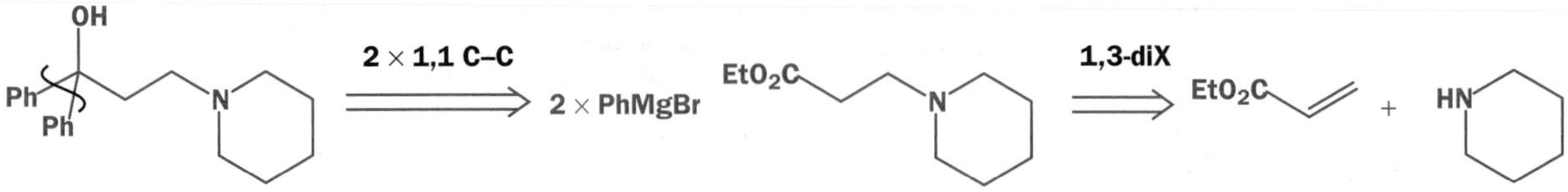

fenpiprane precursor: synthesis

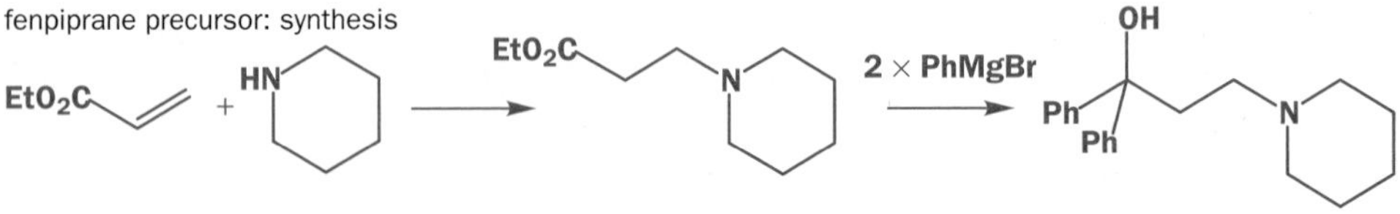

The fact that Grignard reagents add twice to esters means that disconnection of a *ketone* in this way is often not reliable. We talked about a few ways of doing this type of reaction in Chapter 12.

An alternative is to first convert to the alcohol oxidation level, then disconnect. This was the method chosen for this starting material for the synthesis of chlorphedianol.

chlorphedianol starting material: retrosynthetic analysis

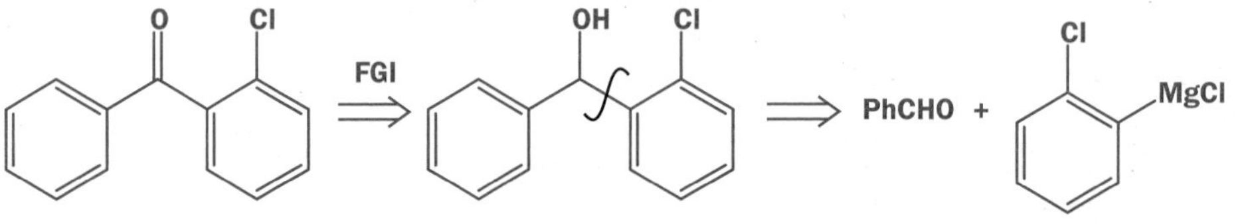

chlorphedianol starting material: synthesis

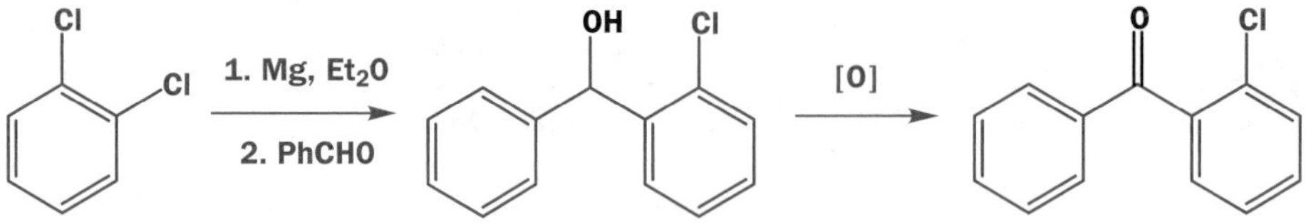

Summary: 1,1 disconnections using Grignard reagents

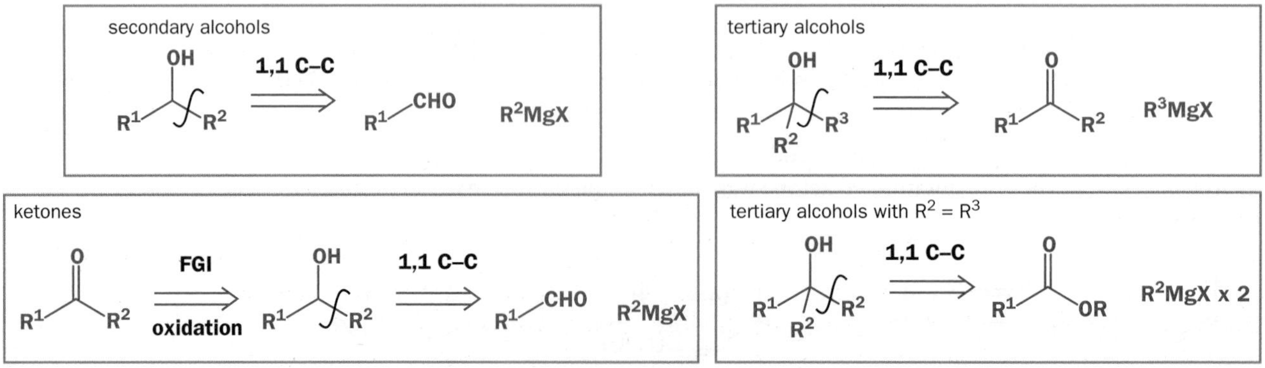

PhMgBr — Phenyl Magnesium Bromide

Donor and acceptor synthons

You've now met a variety of synthons and it's useful to be able to classify them as donor or acceptor synthons. We call a negatively polarized synthon a **donor synthon** and give it the symbol 'd'. Positively polarized synthons are called **acceptor synthons** and are given the symbol 'a'.

We can classify the synthons further according to where the functional group is in relation to the reactive site. The first synthon in the diagram below, which corresponds to an aldehyde, we call an a^1 **synthon**, because it is an acceptor that carries a functional group on the same carbon as its reactive centre. The second is a d^2 **synthon** because it is a donor whose reacting site is in the 2-position relative to the carbonyl group. Earlier you met two other types of synthon, corresponding to epoxide and Michael acceptor, and we can now classify these as a^2 and a^3 **synthons**.

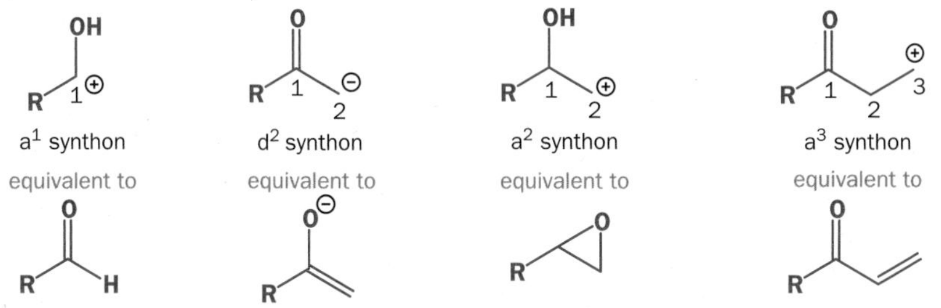

This terminology is useful because it reduces synthons to the bare essentials: what polarity they are and where the polarity is sited. The actual functional group they carry is, as you now appreciate, less important because FGI will usually allow us to turn one FG into another.

> ● **Synthons are classified as a (acceptor) or d (donor)**
>
> - A number shows the position of the acceptor or donor site relative to a functional group
> - An a^1 synthon is a carbonyl compound and a d^2 synthon an enolate

Two-group C–C disconnections

1,3-Difunctionalized compounds

It's not only Grignard reagents that will react with aldehydes or ketones to make alcohols: enolates will too—we spent Chapters 27 and 28 discussing this reaction, the aldol reaction, its variants, and ways to control it.

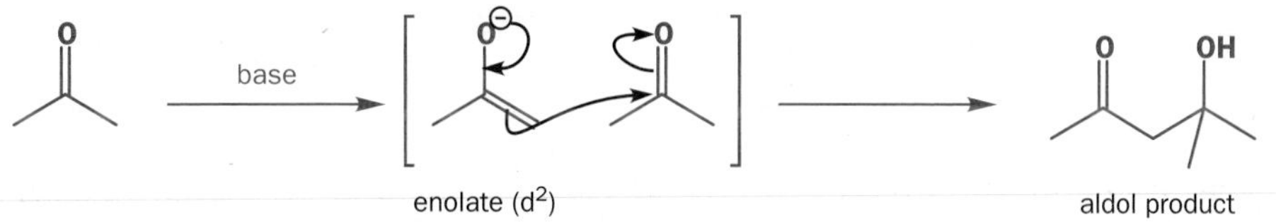

The aldol reaction is extremely important in organic synthesis because it makes compounds with two functional groups in a 1,3-relationship. Whenever you spot this 1,3-relationship in a target molecule—think aldol! In disconnection terms we can represent it like this.

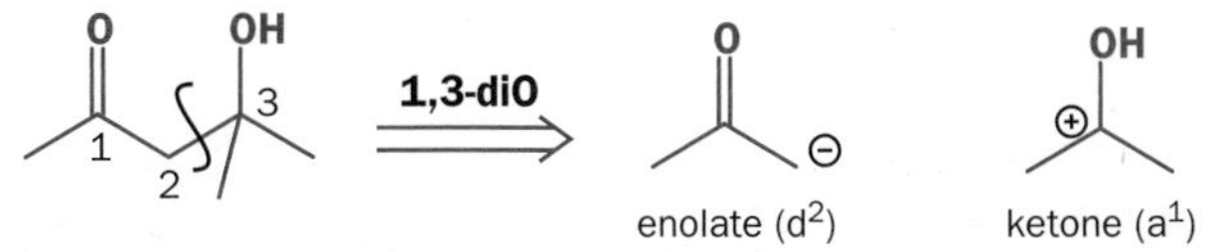

We call this disconnection a **two-group C–C disconnection**, because we are using the OH and the C=O groups together to guide our disconnection. The disconnection gives us a d^2 synthon for which we shall use an enolate equivalent, and an a^1 synthon, for which we shall use an aldehyde or a ketone.

Chapter 27 has many examples and perhaps gingerol is the best. As soon as you see the 1,3-relationship, the disconnection should be obvious.

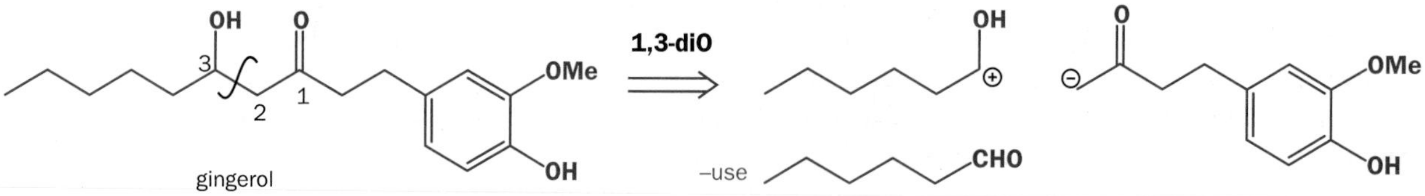

The β-hydroxy carbonyl products of aldol reactions are often very easily dehydrated to give α,β-unsaturated carbonyl compounds and, if you spot an α,β-unsaturated carbonyl group in the molecule, you should aim to make it by an aldol reaction. You will first need to do an FGI to the β-hydroxy carbonyl compound, then disconnect as before.

> The elimination is easy because it goes by an E1cB mechanism—see Chapter 18.

oxanamide intermediate: retrosynthetic analysis

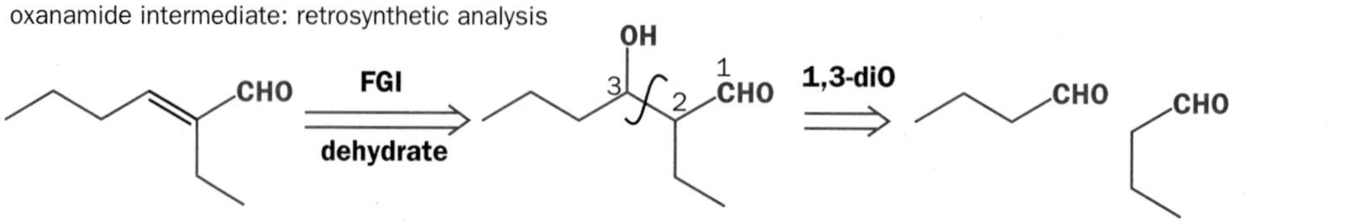

This aldehyde is an intermediate in the synthesis of the tranquillizer oxanamide. Because both components of the aldol reaction are the same, no special precautions need to be taken to prevent side-reactions occurring. In the synthesis, the dehydration happened spontaneously.

oxanamide intermediate: synthesis

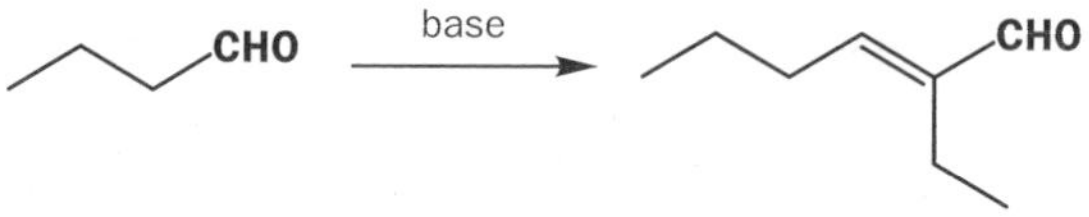

Because this disconnection of unsaturated carbonyl compounds is so common, it's often written using a shorthand expression.

oxanamide intermediate: retrosynthetic analysis

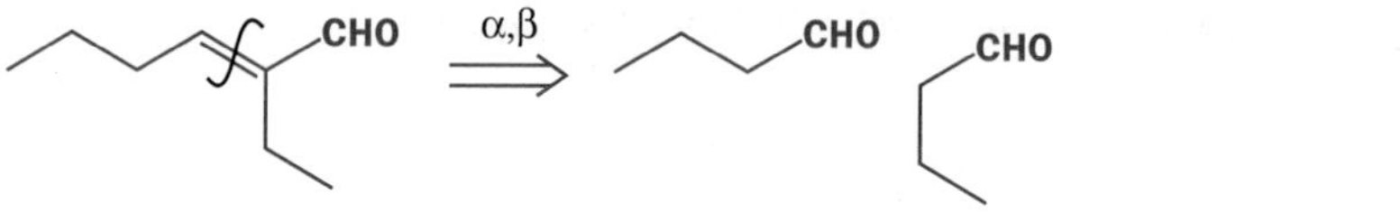

The next compound was needed for an early synthesis of carotene. Again, it's an α,β-unsaturated ketone so we can disconnect using the same 'α,β' disconnection.

carotene intermediate: retrosynthetic analysis

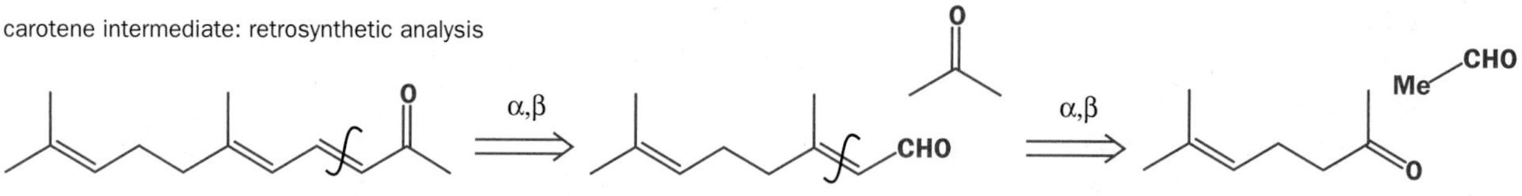

The aldehyde generated by this first disconnection is also α,β-unsaturated, so we can do another α,β disconnection, back to a ketone whose synthesis we have already discussed (p. 787).

An aldol reaction using the enolate of acetaldehyde and requiring it to react with a ketone is doomed to failure: acetaldehyde itself is far too good an electrophile. In the forward synthesis, therefore, this first step was carried out at the ester oxidation level (using a Reformatsky reaction), and the ester was subsequently converted to the aldehyde by a reduction of the kind discussed in Chapter 24.

carotene intermediate: synthesis

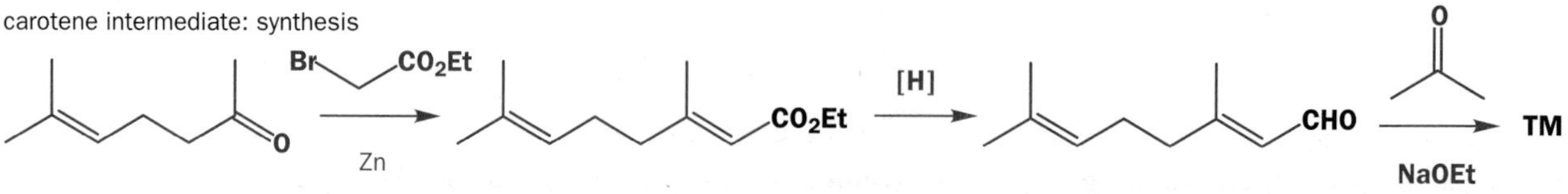

There was no problem with selectivity in the second aldol reaction because the aldehyde is not enolizable. The Reformatsky reaction in this sequence illustrates the fact that, of course, aldol-type reactions happen at the ester oxidation level as well, and you should equally look to disconnect β-hydroxy or α,β-unsaturated esters, acids, or nitriles in this way. Just remember to look for 1,3-relationships, convert the functional groups to oxygen-based ones, and disconnect them to d^2 plus a^1 synthons.

■
If you don't understand what we are saying here, you must go back and read Chapter 27 on selectivity in the aldol reaction.

The next compound was needed by ICI when chemists there were developing a thromboxane antagonist to inhibit blood clot formation. You can immediately spot the 1,3-relationship between the ester and the hydroxyl group, so 1,3-diO disconnection is called for.

thromboxane antagonist intermediate: retrosynthetic analysis

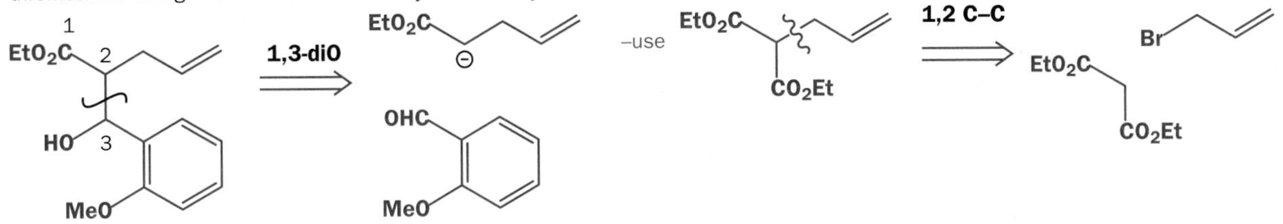

A good equivalent for the 'ester enolate' d^2 synthon is a β-dicarbonyl compound, because it can easily be disconnected to diethyl malonate and an alkylating agent.

thromboxane antagonist intermediate: synthesis

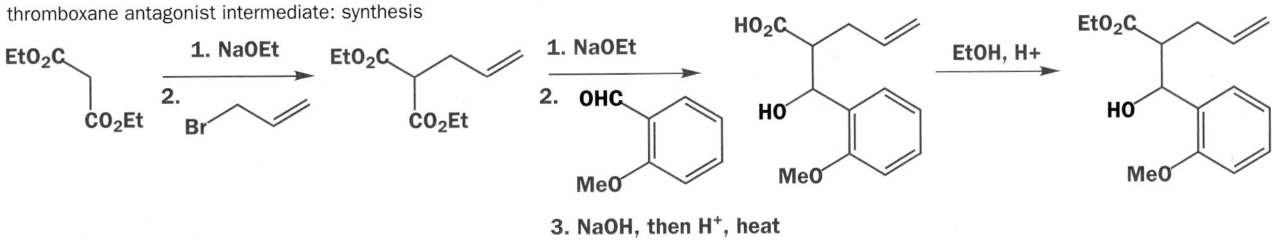

3. NaOH, then H$^+$, heat

This unsaturated amide is known as cinflumide and is a muscle relaxant. Disconnection of the amide gives an acid chloride that we can make by FGI from the acid. You should then spot the α,β-unsaturated carbonyl disconnection, a masked 1,3-diO disconnection, back to *m*-fluoro-benzaldehyde.

cinflumide: retrosynthetic analysis

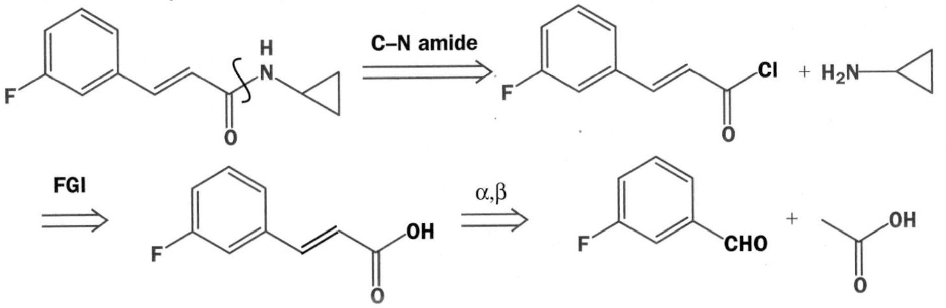

Again, the forward reaction was best done using malonate chemistry but the variant with malonic acid was used. The cyclopropyl amine unit (here as an amide) is present in many biologically active compounds and the free amine is available.

cinflumide: synthesis

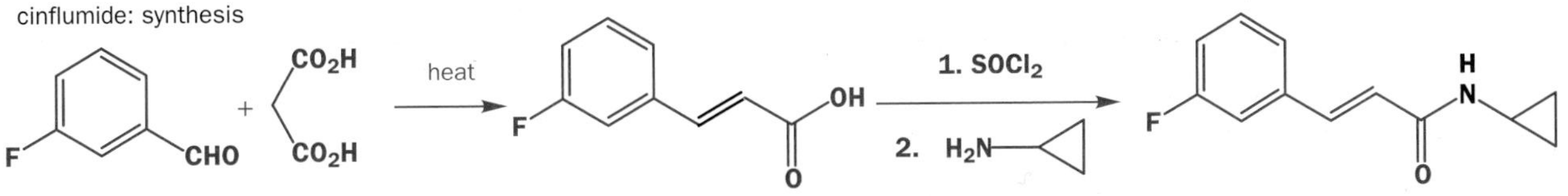

Functional group relationships may be concealed by protection

The analgesic doxpicomine is a more difficult problem than those you have seen so far. At first sight it has no useful disconnections especially as there are no carbonyl groups. However, removal of the acetal reveals a 1,3-diol that could be formed by reduction of a much more promising diester.

doxpicomine: retrosynthetic analysis I

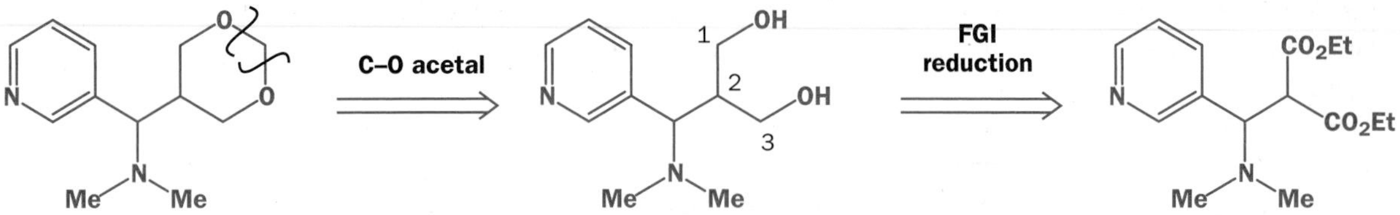

The diester has a 1,3-diCO relationship and could be disconnected but we have in mind using malonate so we would rather disconnect the alternative 3-amino carbonyl compound (the Me$_2$N group has a 1,3-relationship with both ester groups) by a 1,3-diX disconnection giving an unsaturated ester. This α,β-unsaturated ester disconnects nicely to a heterocyclic aldehyde and diethyl malonate.

doxpicomine: retrosynthetic analysis II

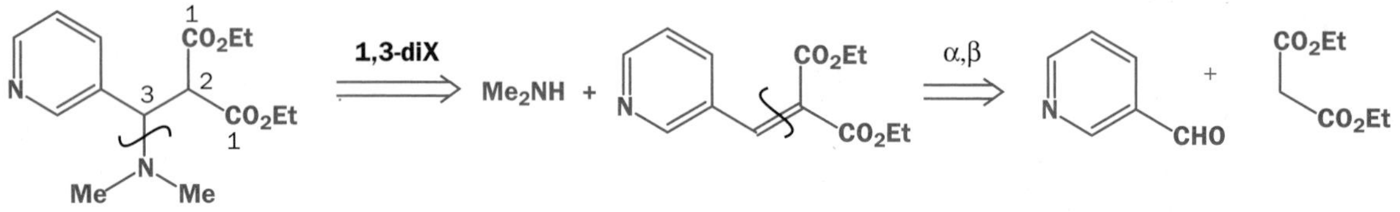

► It is interesting to note that acetals, usually employed for protection, can be useful in their own right as in this drug.

The synthesis is shorter than the retrosynthetic analysis and involves only three steps. Good retrosynthetic analysis, using two-group disconnections, should lead to short syntheses.

doxpicomine: synthesis

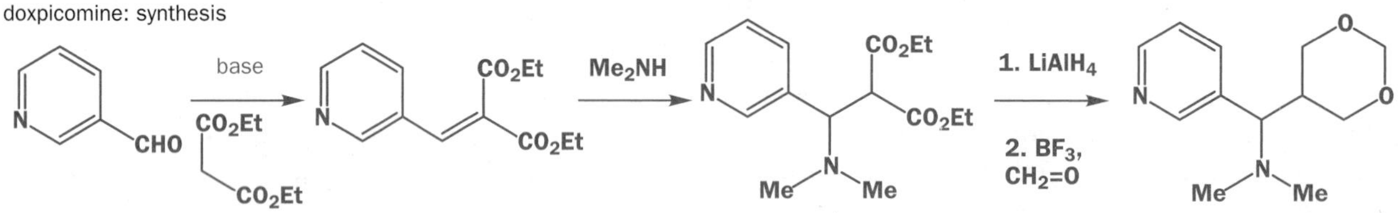

Aldol-style disconnections with N and O in a 1,3-relationship: I

Another important class of compounds that undergo aldol-type additions to aldehydes and ketones is nitriles. Because nitriles can be reduced to amines, this reaction provides another useful route to 3-amino-alcohols.

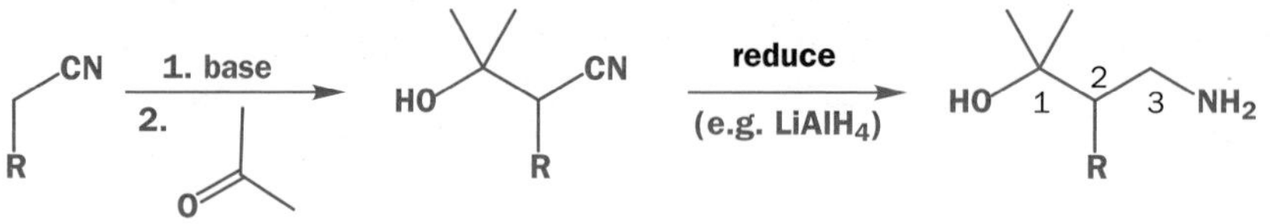

This reaction, coupled with the reduction of cyanohydrins (Chapter 6), means that compounds with either a 1,3- or a 1,2-relationship between N and O can be made from cyanides.

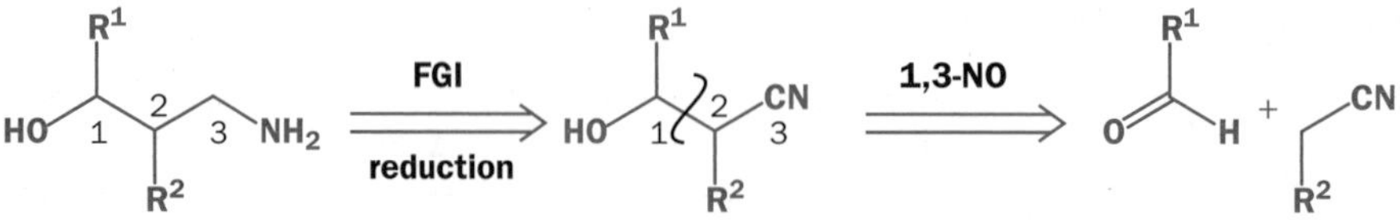

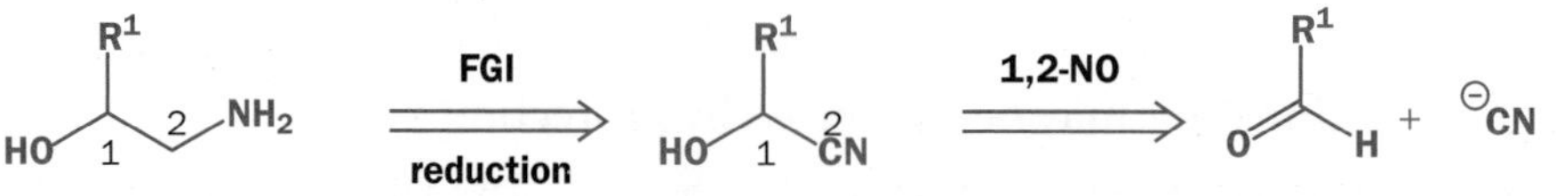

Venlafaxine is an antidepressant and, like many neuroactive agents, it is an amino-alcohol. In this case, the two functional groups are 1,3-related, so we aim to use a 1,3-diO disconnection. Usually, you would convert the amine to an alcohol to simplify the disconnection, but by spotting the opportunity for using a nitrile you can avoid the need for this extra step. A preliminary removal of the two *N*-Me groups is necessary.

venlafaxine: retrosynthetic analysis

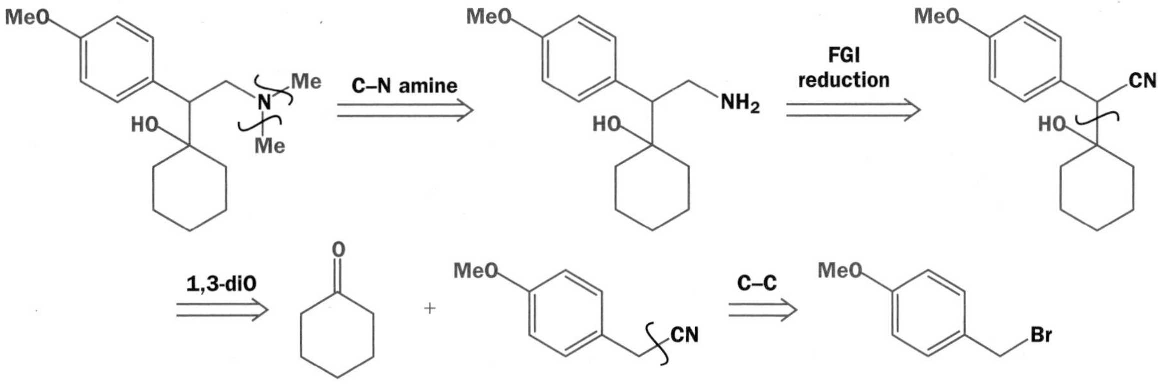

In the forward synthesis, it turned out that the nitrile reduction was best done using hydrogen and a metal (Rh) catalyst. The final methylation of the primary amine had to be done via the imine and iminium ion (see Chapter 24) to prevent further unwanted alkylations. The reagent was an excess of formaldehyde (methanal $CH_2=O$) in the presence of formic acid (HCO_2H).

venlafaxine: synthesis

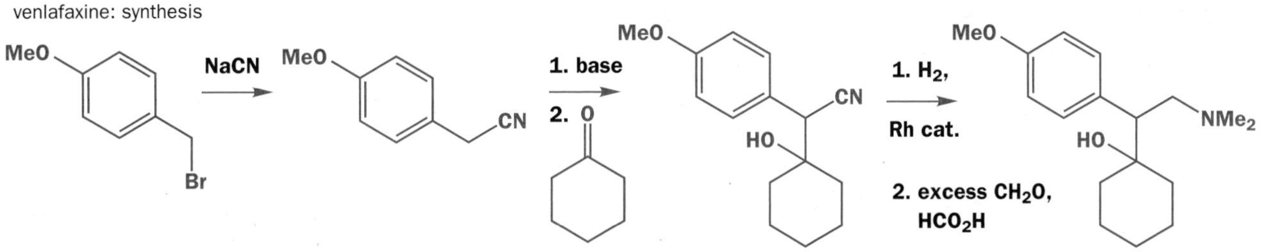

Aldol-style disconnections with N and O in a 1,3-relationship: II—the Mannich reaction

Another important reaction for making amines with a 1,3-relationship to a carbonyl group is the Mannich reaction. You met this in Chapter 27 as a way of doing otherwise unreliable aldol additions to formaldehyde. Because the amine is introduced directly and not by reduction of a nitrile, it can have two alkyl groups from the start. Compare this scheme with the one above using a nitrile group as the source of the amine.

the Mannich disconnection

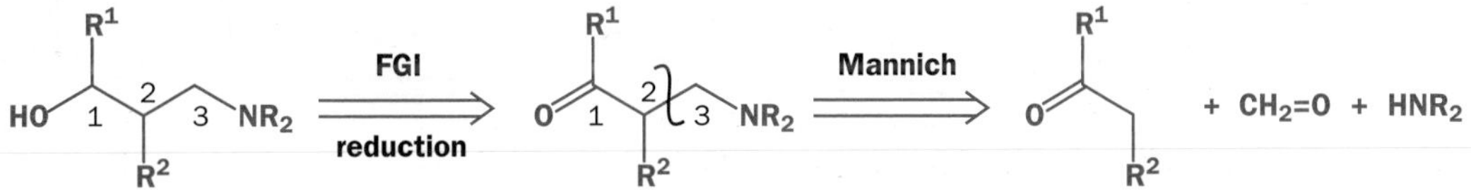

Our example is clobutinol—an antitussive (cough medicine). A preliminary 1,1 C–C disconnection of the tertiary alcohol is necessary to provide a 3-amino ketone that we can make by a Mannich reaction.

clobutinol: retrosynthetic analysis

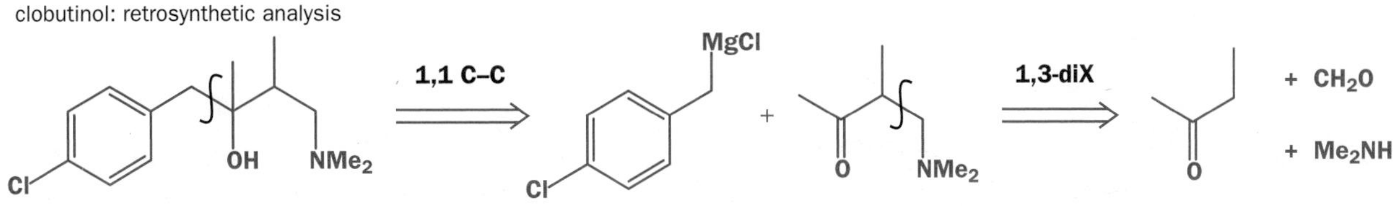

clobutinol: synthesis

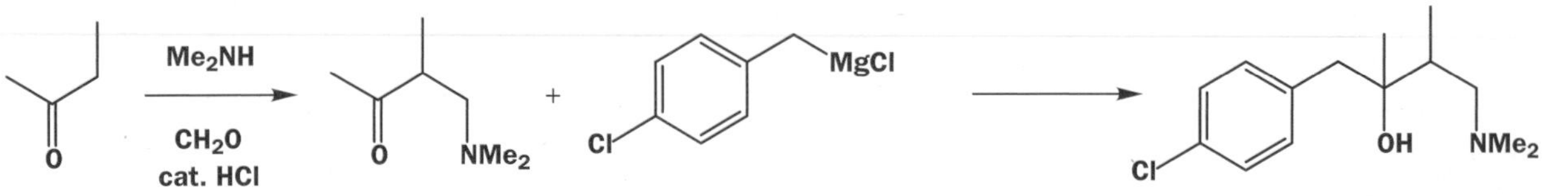

You can immediately spot the 1,3 relationship in this analogue of the antidepressant, nisoxetine, but, unfortunately, it can't be disconnected straight back to an amino-alcohol because that would require nucleophilic substitution on an electron-rich aromatic ring. We have to disconnect the ether on the other side, giving an alkyl chloride.

nisoxetine analogue: retrosynthetic analysis

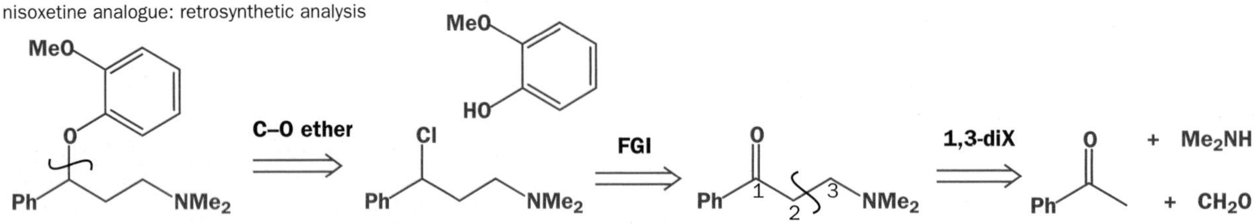

Using guideline 5 (p. 787) we want to convert the halide to an oxygen-based group, and a sensible solution is to choose the ketone. 1,3-Disconnection of this compound corresponds to a Mannich reaction. This is another case where FGI of the amine to an alcohol is not desirable, because the Mannich reaction will produce the amine directly.

nisoxetine analogue: synthesis

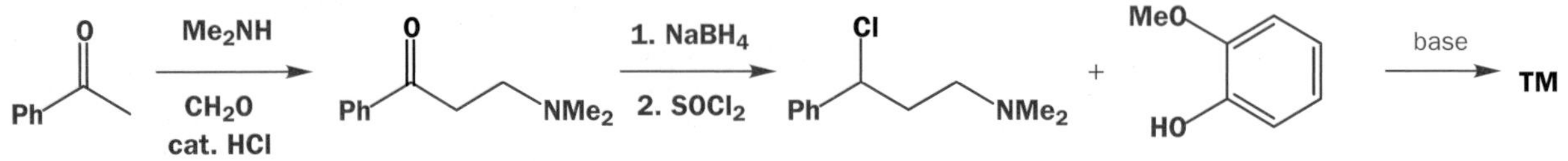

The Claisen ester disconnection: a 1,3-diO relationship needing two carbonyl groups

1,3-Diketones can be disconnected in a similar way: this time the disconnection corresponds to a Claisen condensation, but it's still 1,3-diO, and again you need to look out for the 1,3 relationship. The synthons are still d^2 plus a^1 but the a^1 synthon is used at the ester oxidation level. This diketone is the starting material for the synthesis of the antidepressant tazadolene. With 1,3-diketones, there's always a choice where to disconnect, and you should be guided by which disconnection (1) corresponds to the most reliable reaction and (2) gives the simplest starting materials. In this case, it's much better to disconnect back to cyclohexanone.

tazadolene starting material: retrosynthetic analysis

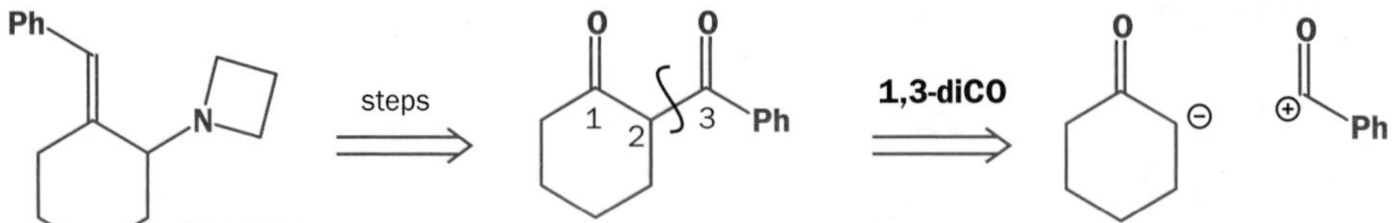

The synthesis is interesting because, after the acylation of the enamine, the amino group is introduced by a clever reductive amination with benzylamine ($PhCH_2NH_2$) that forms the C–N bond, reduces the ketone, and hydrogenolyses the N–benzyl bond (Chapter 24). Dehydration and double alkylation then give tazadolene.

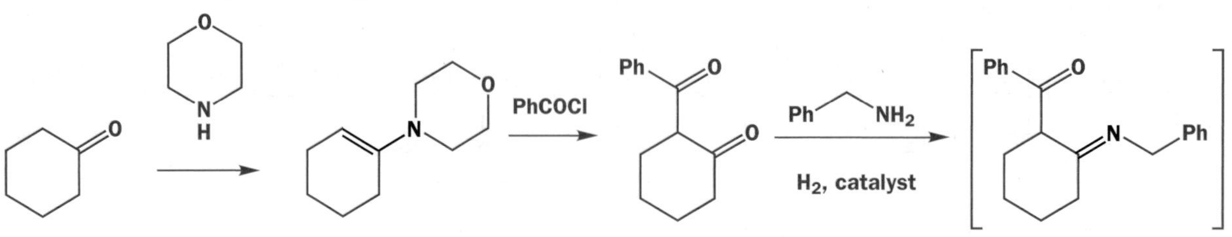

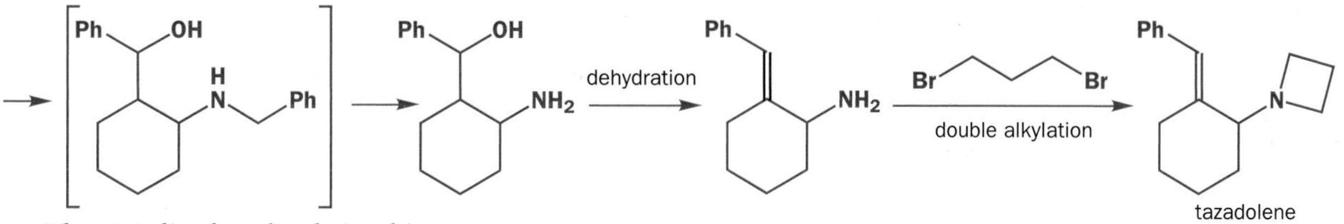

dehydration

double alkylation

tazadolene

The 1,3-dicarbonyl relationship may not be revealed in the target molecule and C–heteroatom disconnections or FGIs may be needed before the 1,3-diO C–C disconnection. Bropirimine is a bromine-containing antiviral and anticancer drug. The bromine atom can be put in last of all by electrophilic bromination.

bropirimine: retrosynthetic analysis

C–Br

electrophilic substitution

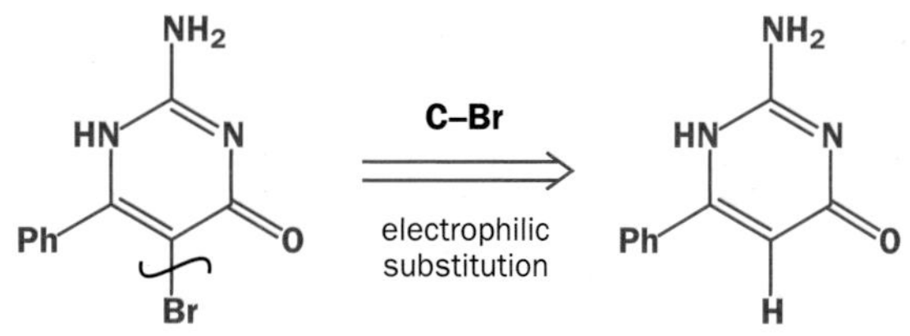

Disconnection of two C–N bonds removes a molecule of guanidine and reveals a 1,3-dicarbonyl relationship with a straightforward disconnection.

bropirimine: retrosynthetic analysis

C–N × 2

enol

1,3-diO

–use acyl chloride   –use malonate

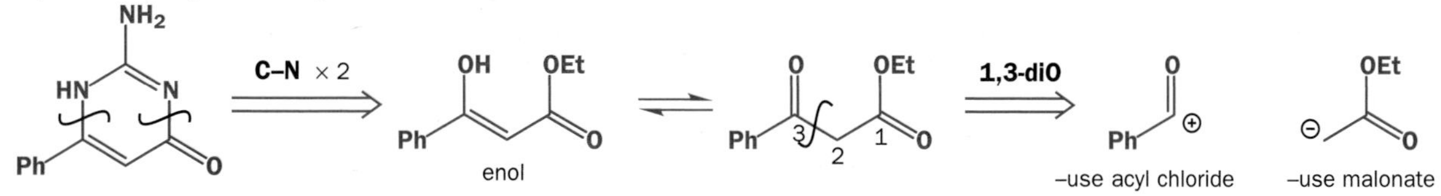

In the event, the 1,3-dicarbonyl was made using malonate chemistry with an unusual twist: the lithium derivative gave *C*-acylation in good yield. Simply refluxing the product with guanidine formed the heterocycle and bromination gave bropirimine.

■ Guanidine is the strong delocalized organic base mentioned in Chapter 8.

Summary: 1,3-diO disconnections

3-hydroxy carbonyls and α,β-unsaturated carbonyls: use the aldol reaction

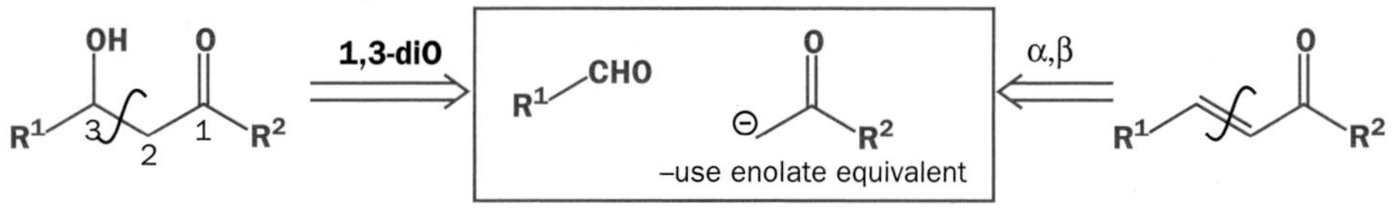

1,3-diO   –use enolate equivalent   **α,β**

3-amino ketones and alcohols: use Mannich or nitrile aldol

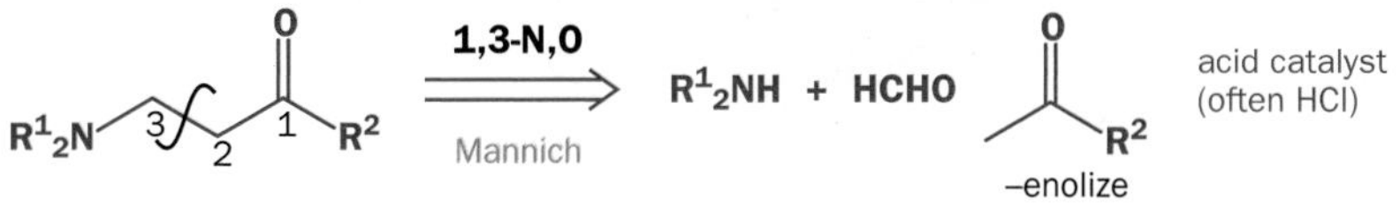

1,3-N,O   Mannich R^1_2NH + HCHO   acid catalyst (often HCl)   –enolize

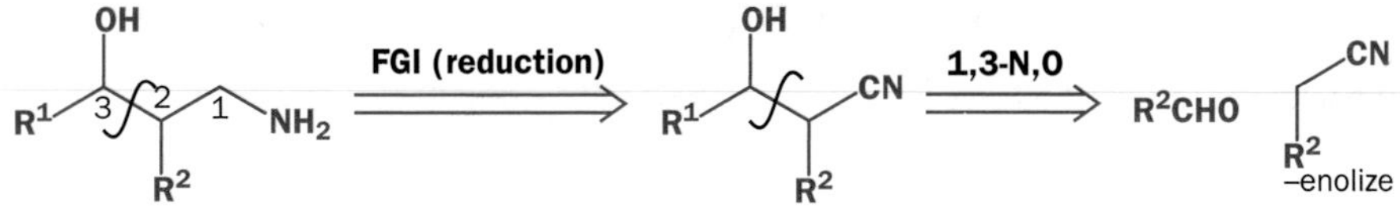

FGI (reduction)   **1,3-N,O** R^2CHO –enolize

1,3-diketones: use the Claisen condensation

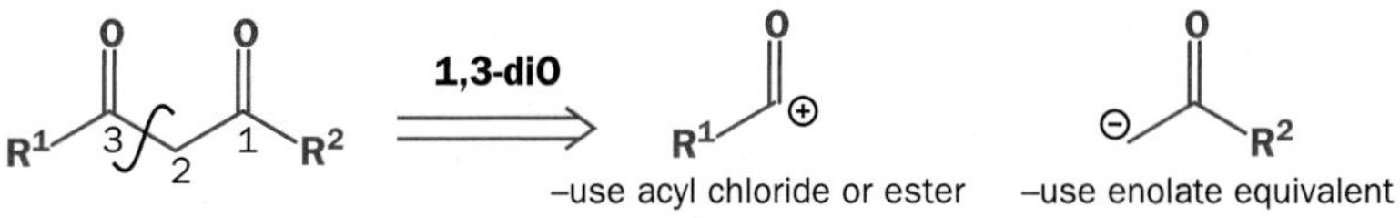

1,3-diO   –use acyl chloride or ester   –use enolate equivalent

✳ 1,5-Related functional groups

This compound has a 1,5 rather than a 1,3 relationship between two carbonyl groups. Disconnection to give an enolate as one reagent therefore requires an a^3 rather than an a^1 synthon: in other words a **Michael acceptor.**

1,5-dicarbonyl compounds:
retrosynthetic analysis

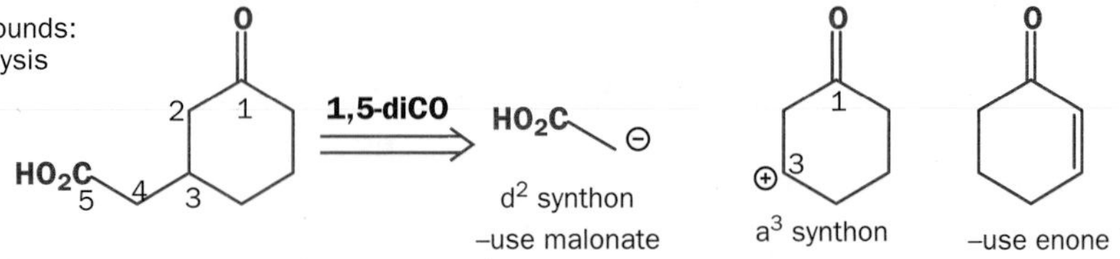

The synthesis will be successful only if (1) the right reagent enolizes and (2) the nucleophile undergoes conjugate (and not direct 1,2-) addition to the unsaturated carbonyl compound (Chapter 29). Malonate derivatives enolize easily *and* do Michael additions and are therefore a good choice for this type of reaction.

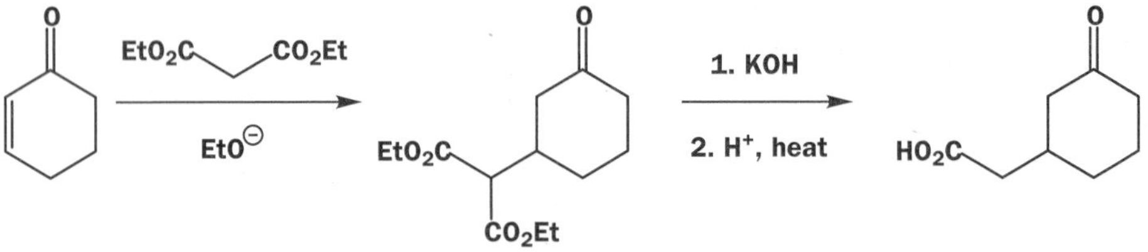

Michael addition of enolates to α,β-unsaturated compounds is a good way of making 1,5-difunctionalized compounds, and you should look for these 1,5-relationships in target molecules with a view to making them in this way. Our example is rogletimide, a sedative that can be disconnected to a 1,5-diester. Further 1,5-diCO disconnection gives a compound we made earlier by ethylation of the ester enolate.

rogletimide: retrosynthetic analysis

> ■
> There are many examples of conjugate addition of enolates in Chapter 29.

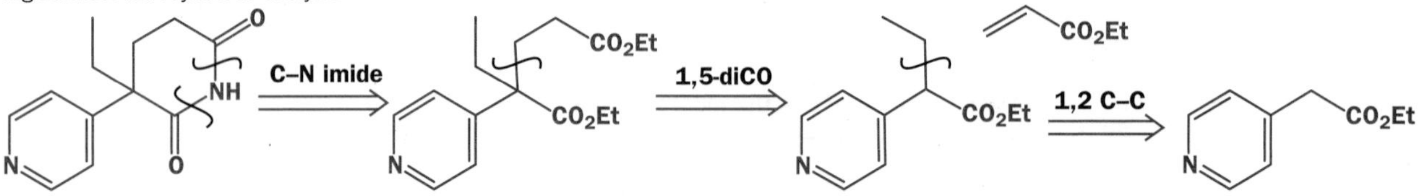

The synthesis was most efficient with an unsaturated amide as Michael acceptor.

rogletimide: synthesis

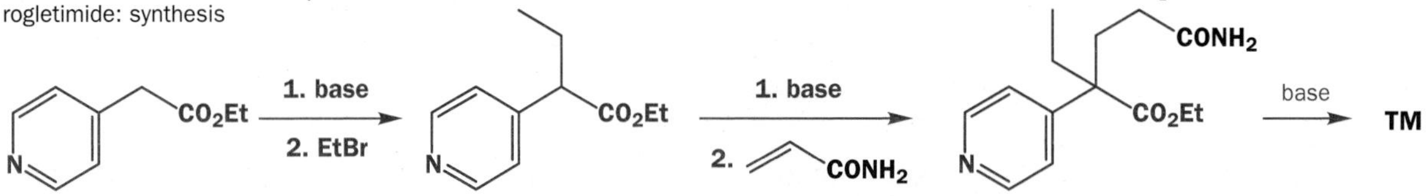

'Natural reactivity' and 'umpolung'

Cast your mind back over the synthons we have used in these two-group C–C disconnections.

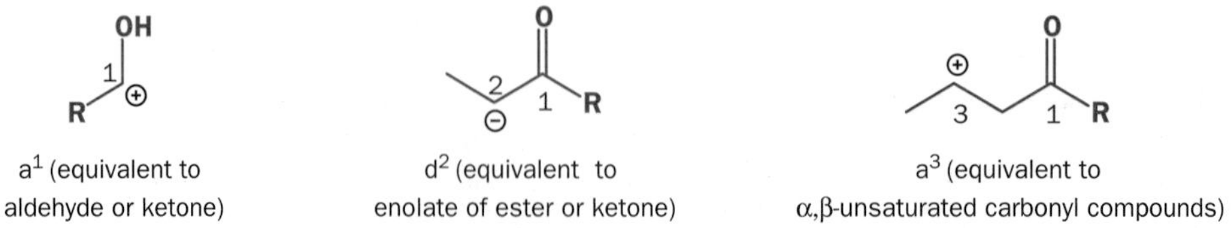

Notice that the acceptor synthons have odd numbers; the donor synthon has an even number: donor and acceptor properties alternate along the chain as we move away from a carbonyl group. This 'natural reactivity' of carbonyl compounds explains why we find it easy to discuss ways of making 1,3- and 1,5-difunctionalized compounds, because they arise from $a^1 + d^2$ and from $a^3 + d^2$. Reagents corresponding to synthons like d^1 or a^2 are rarer, and therefore compounds with 1,2- or 1,4- related functional groups require special consideration retrosynthetically.

You have in fact met one example of each of the 'unnatural' synthons with a^2 and d^1 reactivity. Such synthons are given the German name *Umpolung,* meaning 'inverse polarity' because their natural reactivity is reversed, and **umpolung reagents** are the key to the synthesis of 1,2- and 1,4-difunctionalized compounds.

two umpolung reagents

$^\ominus$CN
1

d^1 synthon
(cyanide anion)

OH
R $\underset{1}{}\overset{\oplus}{}$
2

a^2 synthon
(equivalent to epoxide)

We shall finish this chapter by looking at disconnections of 1,2- and 1,4-difunctionalized compounds because these require us to use reagents with umpolung equivalent to d^1, d^3, a^2, and a^4 synthons. There are very many reagents for these synthons—if you are interested to learn more, consult a specialized book.

1,2-Difunctional compounds

You met ways of making 1,2-difunctionalized compounds when we first talked about two-group disconnections, and we used an epoxide as an a^2 synthon. Epoxides are, of course, also 1,2-functionalized, and in fact this is often the key to making 1,2-functionalized compounds: use something with the 1,2 relationship already in place. You saw lots of examples of this type of strategy earlier in this chapter. Perhaps the simplest approach is electrophilic addition to alkenes. If the alkene is made by a Wittig reaction, the disconnection is (eventually) between the two functionalized carbon atoms in the target molecule. This example shows dihydroxylation as the electrophilic addition but there is also epoxidation, bromination, and bromination in water to give Br and OH as the functional groups.

A normal C–C disconnection is also a possibility, but disconnection to the 'natural' a^1 synthon and the umpolung d^1 is necessary. One very useful umpolung reagent is cyanide, and you can see it in action in this synthesis of the tranquillizer phenaglycodol. The tertiary alcohol with two R groups the same should prompt you to think of doing a double Grignard addition to an ester. FGI then reveals the nitrile functional group necessary for a 1,2-diX disconnection to cyanide plus ketone.

phenaglycodol: retrosynthetic analysis

The starting material is obviously available by a Friedel–Crafts acylation of chlorobenzene and the rest of the synthesis follows. Note that the nitrile can be converted directly into the ester with acidic ethanol and that an excess of Grignard reagent is needed because the free OH group destroys some of it.

phenaglycodol: synthesis

1,4-Difunctional compounds

There are more possibilities here and we shall finish this chapter with a brief analysis of them to show you how much of this subject lies beyond what we can do in this book. If we start with a 1,4-dicarbonyl compound we might consider first disconnection of the central bond.

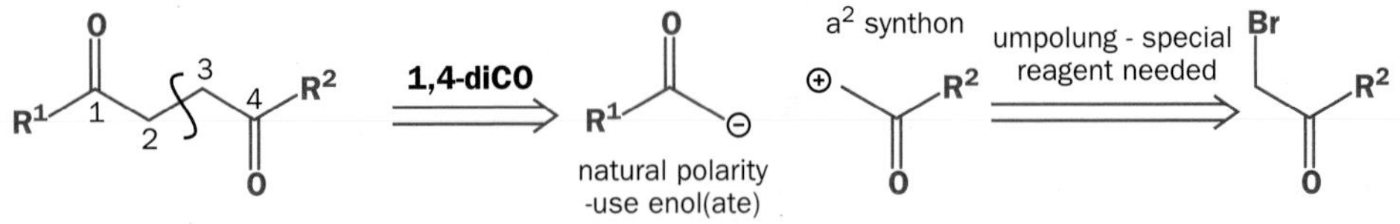

We can use an enolate for one reagent but the other will have to have umpolung. This is not a very serious kind of umpolung as an α-bromo carbonyl compound will do the job nicely if we select our enol(ate) equivalent carefully. In Chapter 26 we suggested enamines for this job. The synthesis becomes:

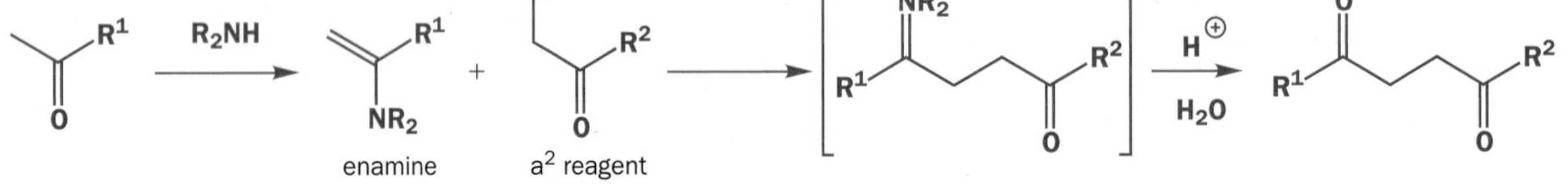

If we attempt the disconnection of one of the other bonds, two possibilities are available because the two fragments are different. We can use either a $d^1 + a^3$ strategy or an $a^1 + d^3$ strategy. In each case we have one natural synthon and one with umpolung.

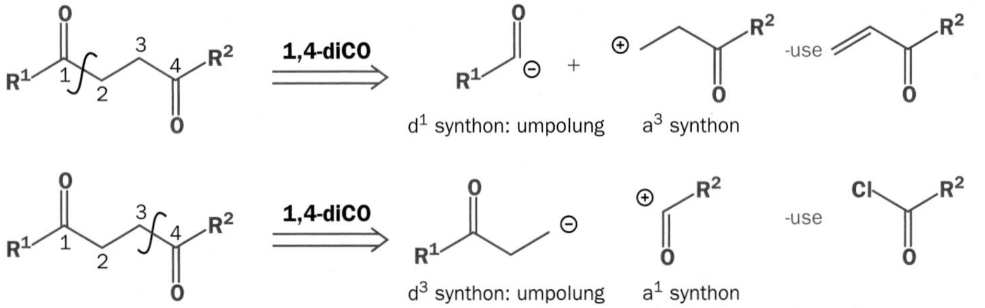

These strategies are more difficult to realize with the reagents you have met so far but conjugate addition of a cyanide to an unsaturated carbonyl compound would be an example of the $d^1 + a^3$ strategy. We have included these to try to convince you that there is no escape from umpolung in the synthesis of a 1,4-dicarbonyl compound. If you were making this keto-ester you would have to understand two of the three strategies.

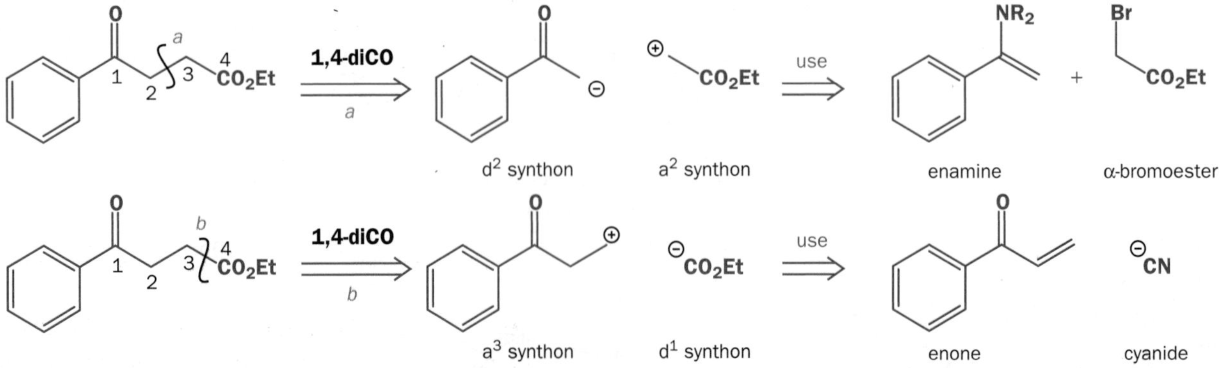

There is one way to avoid umpolung and that is to make the disconnection outside the 1,4 relationship. As it happens, we have already seen this strategy in action (p. 799). It involves a Friedel–Crafts acylation of benzene (Chapter 22) with a cyclic anhydride and leads directly to this

■
Another approach using the nitro group and the Nef reaction appears at the end of Chapter 29.

product by quite a short route. This strategy is available only if there happens to be a starting materi-al available to suit any particular case.

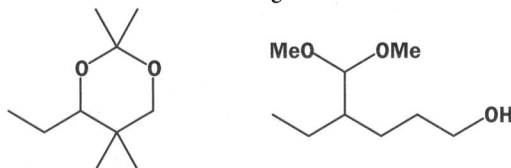

This chapter is meant to give you just the basic ideas of retrosynthetic analysis. They are impor-tant because they reinforce the concept that the combination of electrophile and nucleophile is the basis for the understanding of organic reactions. Synthesis and reactions are two sides of the same coin. From now on we shall use the methods introduced in this chapter when we think that they will help you to develop your understanding.

Problems

1. Suggest ways to make these two compounds. Show your dis-connections and don't forget to number the relationships.

2. Propose syntheses of these two compounds, explaining your choice of reagents and how the necessary selectivity is achieved.

3. The reactions to be discussed in this problem were planned to give syntheses of these three target molecules.

In the event, each reaction gave a different product shown below. What went wrong? Suggest syntheses that would give the target molecules.

4. The natural product nuciferal was synthesized by the route summarized here.

nuciferal

(**a**) Suggest a synthesis of the starting material A.

(**b**) Suggest reagents for each step.

(**c**) Draw out the retrosynthetic analysis giving the disconnec-tions that you consider the planners had in mind and label them suitably.

d) What synthon does the starting material A represent?

5. A synthesis of the enantiomerically pure ant pheromone is required. One suitable starting material might be the enantiomer-ically pure alkyl bromide shown. Suggest a synthesis of the pheromone based on this or another starting material.

(S)-(–)-alkyl bromide   (S)-(–)-ant pheromone

6. Show how the relationship between the alkene and the car-boxylic acid influences your suggestions for a synthesis of these unsaturated acids.

7. How would you make these compounds?

8. Show how the relationship between the two functional groups influences your suggestions for a synthesis of these diketones.

9. Suggest syntheses for these compounds. (*Hint.* Look out for a 1,4-dicarbonyl intermediate.)

10. Suggest a synthesis of this diketo-ester from simple starting materials.

11. Explain what is happening in this reaction. Draw a scheme of retrosynthetic analysis corresponding to the synthesis. How would you make the starting materials?

12. These diketones with different aryl groups at the ends were needed for a photochemical experiment. The compounds could be prepared by successive Friedel–Crafts acylations with a diacid dichloride but the yields were poor. Why is this a bad method? Suggest a better synthesis.

13. This is a synthesis for the ladybird defence compound coccinelline.

Suggest reagents for the reactions marked '?' (several steps may be needed) and give mechanisms for those that are not.

14. Suggest syntheses for these compounds.

Controlling the geometry of double bonds

31

Connections

Building on:
- Carbonyl chemistry ch6, ch12, & ch14
- Kinetic and thermodynamic control lch13l
- Wittig reaction ch14
- Conjugate addition ch10
- Stereochemistry ch16
- Elimination reactions ch19
- Reduction ch24
- Chemistry of enol(ate)s ch26–ch29

Arriving at:
- What makes *E*- and *Z*-alkenes different?
- Why *E/Z* control matters
- Eliminations are not stereoselective
- Cyclic alkenes are *cis*
- Equilibration of alkenes gives *trans*
- Effects of light and how we see
- Julia olefination and the Wittig reaction at work
- Reliable reduction of alkynes

Looking forward to:
- Diastereoselectivity ch33–ch34
- Pericyclic reactions ch35–ch36
- Fragmentations ch38
- Radicals and carbenes ch39–ch40
- Main group chemistry ch46–ch47
- Asymmetric synthesis ch45
- Polymerization ch52
- Organic synthesis ch53

The properties of alkenes depend on their geometry

You have met alkenes participating in reactions in a number of chapters, but our discussion of how to *make* alkenes has so far been quite limited. Chapter 19 was about elimination reactions, and there you met E1 and E2 reactions.

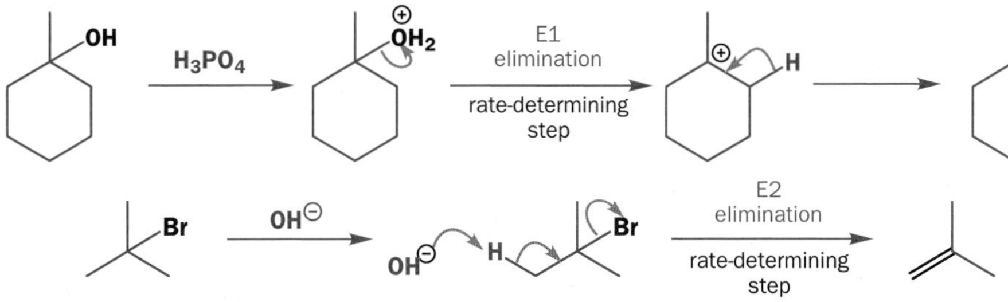

In Chapter 14, you met an important reaction known as the **Wittig reaction**, which also forms alkenes.

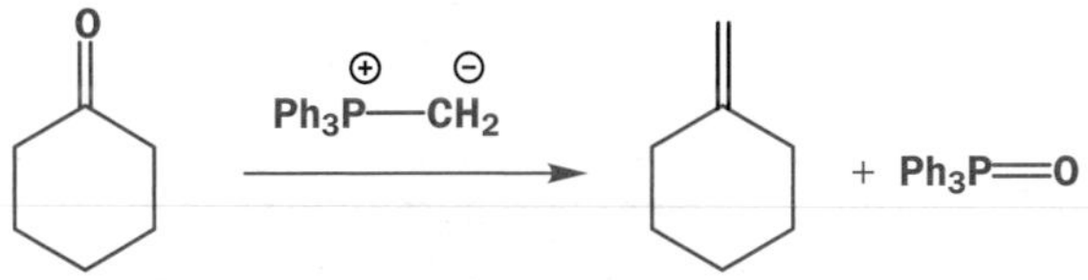

Different physical properties: maleate and fumarate

These two compounds, (*Z*)- and (*E*)-dimethyl but-2-enedioate, are commonly known as dimethyl maleate and dimethyl fumarate. They provide a telling example of how different the physical properties of geometrical isomers

can be. Dimethyl maleate is a liquid with a boiling point of 202 °C (it melts at –19 °C), while dimethyl fumarate is a crystalline compound with a melting point of 103–104 °C.

dimethyl maleate
m.p. –19 °C

dimethyl fumarate
m.p. 103–104 °C

In this chapter we shall talk about reactions similar to the ones on the previous page and we shall be interested in *how to control the geometry of double bonds*. Geometrical isomers of alkenes are different compounds with different physical, chemical, and biological properties. They are often hard to separate by chromatography or distillation, so it is important that chemists have methods for making them as single isomers.

Why is double bond control important?

The activity of the fungicide diniconazole is dependent on the geometry of its double bond: the *E*-isomer disrupts fungal metabolism, while the *Z*-isomer is biologically inactive.

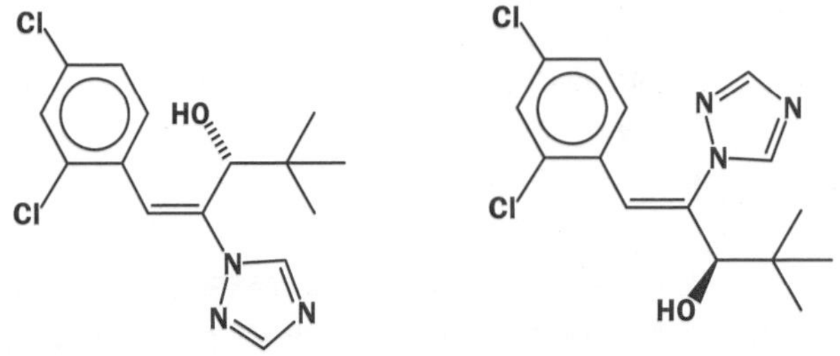

diniconazole: *E*-isomer has fungicidal activity   *Z*-isomer is inactive

If insect pests can be prevented from maturing they fail to reproduce and can thus be brought under control. Juvenile insects control their development by means of a 'juvenile hormone', one of which is the monoepoxide of a triene.

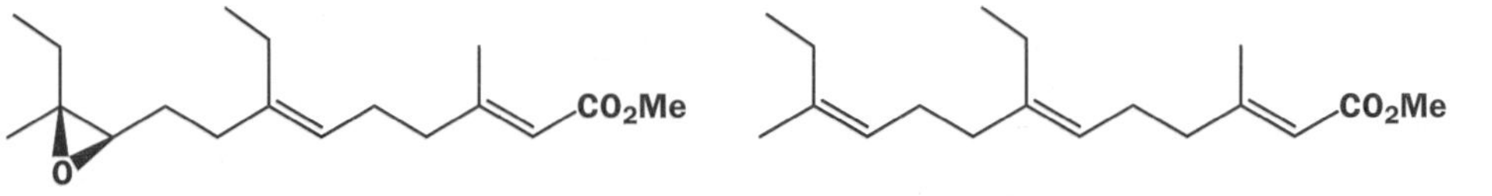

cecropia juvenile hormone: activity = 1000   the *Z, E, E*-triene; activity = 100

Synthetic analogues of this compound, such as the trienes, are also effective at arresting insect development, *providing that the double bond geometry is controlled*. The *Z,E,E* geometrical isomer of the triene is over twice as active as the *E,E,E*-isomer, and over 50 times as active as the *E,Z,Z*- or *Z,E,Z*-isomers.

activity of juvenile hormone analogues (natural hormone = 1000)

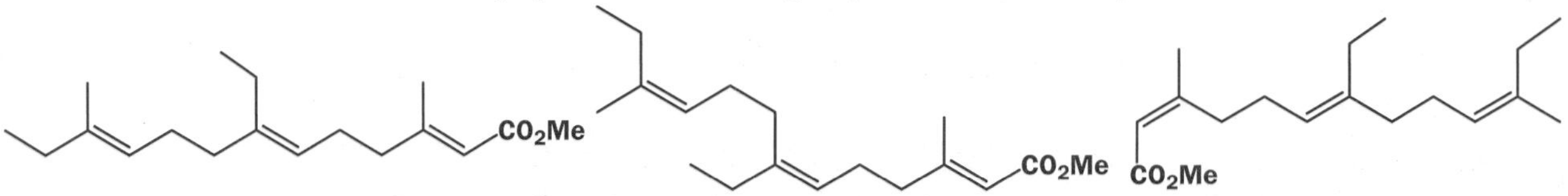

E,E,E-triene: 40   *Z,Z,E*-triene: <2   *Z,E,Z*-triene: <2

These are, of course, just two out of very many examples of compounds where the *E*- and *Z*-isomers have sufficiently different properties that it's no good having one when you need the other.

Chemical reactions on *E*- and *Z*-isomers usually give the same type of product, though often with different stereochemistry. The two geometrical isomers may also react at very different rates. For example, the reaction of these conjugated *E*- and *Z*-enones with alkaline hydrogen peroxide gives in each case an epoxide, but with different stereochemistry and at very different rates.

▶

We shall see later how to make these isomers.

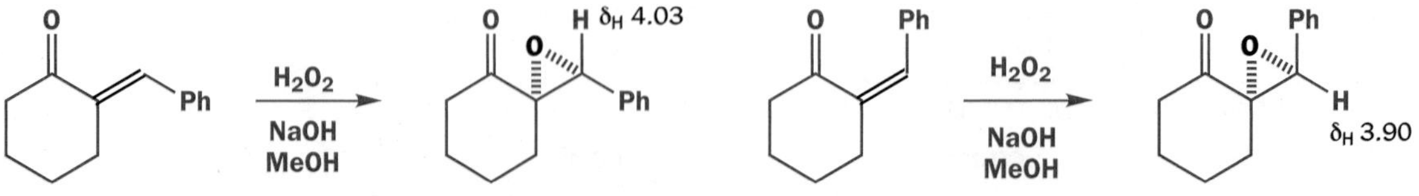

E-enone   *Z*-enone

Epoxidation of the *E*-enone is complete in 2 hours and the epoxide can be isolated in 78% yield. The reaction on the *Z*-enone is very slow—only 50% is converted to the epoxide under the same conditions in 1 week. The mechanism involves conjugate addition and ring closure with cleavage of the weak O–O bond (Chapter 23). The closure of the three-membered ring is fast enough to preserve the stereochemistry of the intermediate enolate.

Elimination reactions are often unselective

You saw in Chapter 19 that elimination reactions can be used to make alkenes from alcohols using acid or from alkyl halides using base. The acid-catalysed dehydration of tertiary butanol works well because the double bond has no choice about where to form. But the same reaction on *s*-butanol is quite unselective—as you would expect, the more substituted alkene is formed (almost solely, as it happens) but even then it's a mixture of geometrical isomers.

In Chapter 19 we explained why more substituted double bonds are formed preferentially (p. 489) and why *E*-alkenes are more stable than *Z*-alkenes (p. 487).

How, then, can we use elimination reactions to give single geometrical isomers? You have, in fact, already met one such reaction, on p. 491, and in this chapter we shall cover other reactions that do just this. These reactions fall into four main classes, and we shall look at each in turn before summarizing the most important methods at the end of the chapter.

Some people call geometrical isomers diastereoisomers, which they are in a sense: they are stereoisomers that are not mirror images. However, we shall avoid this usage in the chapter since for most chemists the word *diastereoisomer* carries implications of three-dimensional stereochemistry.

> ● **Ways of making single geometrical isomers of double bonds**
>
> **1.** Only one geometrical isomer is possible (for example, a *cis* double bond in a six-membered ring)
>
> **2.** The geometrical isomers are in equilibrium and the more stable (usually *E*) is formed
>
> **3.** The reaction is stereoselective and the *E*-alkene is formed as the main product by kinetic control
>
> **4.** The reaction is stereospecific and the alkene geometry depends on the stereochemistry of the starting materials and the mechanism of the reaction

In three- to seven-membered rings, only *cis*-alkenes are possible

In Chapter 28 you met the Robinson annelation as a method of making cyclohexenones. The product of the elimination step contains a double bond, but there is no question about its geometry because in a six-membered ring only a *cis* double bond can exist—a *trans* one would be far too strained.

These reactions fall into class (1) of the list above.

The same is true for three-, four-, five-, and seven-membered rings, though *trans*-cycloheptene has been observed fleetingly. An eight-membered ring, on the other hand, is just about large enough to accommodate a *trans* double bond, and *trans*-cyclooctene is a stable compound, though still less stable than *cis*-cyclooctene.

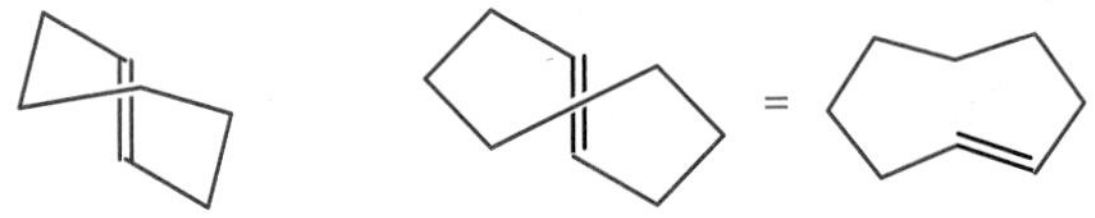

trans-cycloheptene very unstable

trans-cyclooctene: stable liquid

▶ Beware! The terms *cis* and *trans* do not always translate directly into *Z* and *E*. Consider the preparation of an enamine from cyclohexanone, which forms a double bond that you'd probably call *cis* (it's in a ring). But applying the rigorous rules laid down for *E/Z* nomenclature (p. 487), it is *E*. As for the useful terms *syn* and *anti* (Chapter 16), there are no rigid rules for deciding whether a double bond is *cis* or *trans*.

double bond *E* but *cis*?

■ Ozone is a reagent for the oxidative cleavage of C=C double bonds. The products have carbonyl groups at the ends of the old alkene. The mechanism is described in Chapter 35.

You may think that this method is rather too trivial to be called a method for controlling the geometry of double bonds, as it's only of any use for making cyclic alkenes. Well, chemists are more ingenious than that! Corey needed this *cis*-alkene as an intermediate in his synthesis of the juvenile hormone we talked about above (it forms the left-hand end of the structure as shown there).

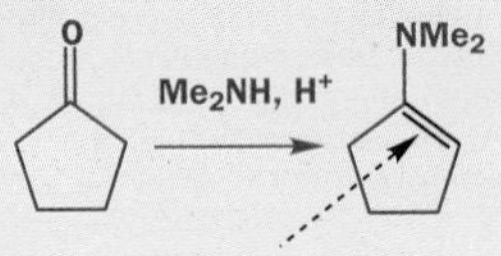

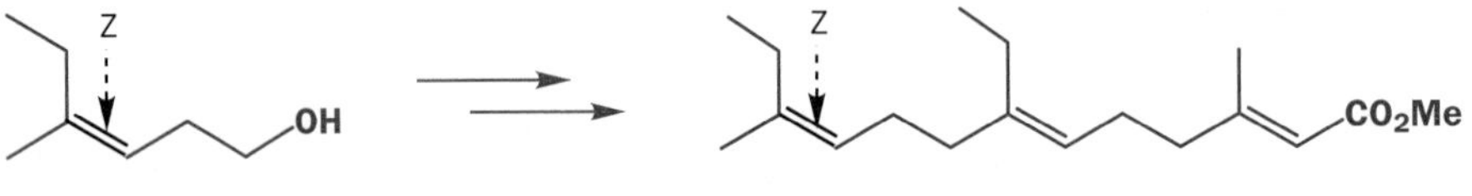

He realized that the *Z* double bond would be easy to make if he were to start with a cyclic molecule (in which only *cis* double bonds are possible) which could be ring-opened to the compound he needed. This is how he did it.

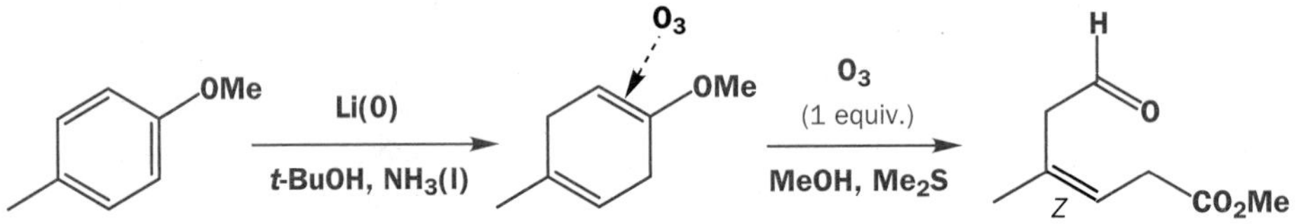

Birch reduction (Chapter 24) of a simple aromatic ether generated two *cis* double bonds (notice that one of these is actually *E*!). The more reactive (because it is more electron-rich) of these reacts first with ozone to give an aldehyde-ester in which the *Z* geometry is preserved. $NaBH_4$ reduces the aldehyde group to a hydroxyl group, which needs to be got rid of: a good way to do this is to tosylate and reduce with $LiAlH_4$, which substitutes H for OTs. The $LiAlH_4$ also does the job of reducing the ester to an alcohol, giving the compound that Corey needed.

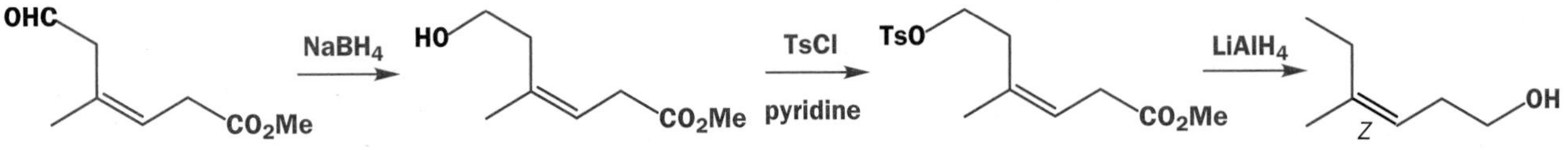

It is not necessary to have an all-carbon ring to preserve the *cis* geometry of a double bond. Lactones (cyclic esters) and cyclic anhydrides are useful too. A double bond in a five- or six-membered compound must have a *cis* configuration and compounds like these are readily made. Dehydration of this hydroxylactone can give only a *cis* double bond and ring-opening with a nucleophile (alcohol, hydroxide, amine) gives an open-chain compound also with a *cis* double bond. The next section starts with an anhydride example.

■ These reactions fall into class (2) of the list on p. 805.

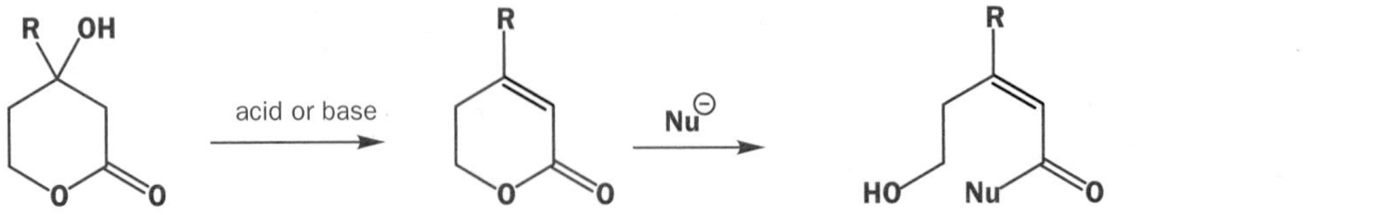

Equilibration of alkenes to the thermodynamically more stable isomer

Acyclic *E*-alkenes are usually more stable than acyclic *Z*-alkenes because they are less sterically hindered. Yet *Z*-alkenes do not spontaneously convert to *E*-alkenes because the π bond prevents free rotation: the energy required to break the π bond is about 260 kJ mol^{-1} (rotation about a σ bond requires about 10 kJ mol^{-1}). You may therefore find the following result surprising. Dimethyl maleate is easily made by refluxing maleic anhydride in methanol with an acid catalyst.

maleic anhydride

MeOH, H$_2$SO$_4$

dimethyl maleate

This reaction is, of course, another simple example of the type we have just been discussing: the Z-alkene arises from the cyclic starting material.

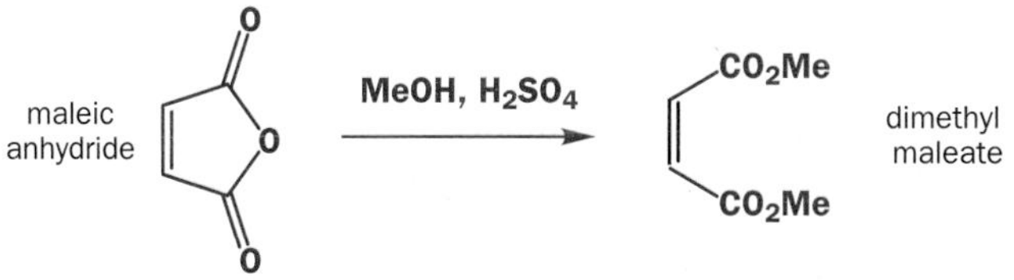

If the product is isolated straight away, a liquid boiling at 199–202 °C is obtained. This is dimethyl maleate. However, if the product is left to stand, crystals of *dimethyl fumarate* (the *E*-isomer of dimethyl maleate) form. How has the geometry been inverted so easily?

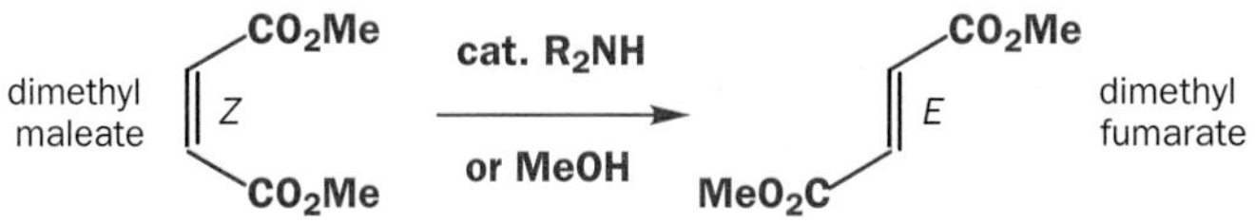

dimethyl maleate

cat. R$_2$NH

or MeOH

dimethyl fumarate

A clue is that the process is accelerated enormously by a trace of amine. Michael addition of this amine, or of methanol, or any other nucleophile, provides a chemical mechanism by which the π bond can be broken. There is free rotation in the intermediate, and re-elimination of the nucleophile can give either *E*- or *Z*-alkene. The greater stability and crystallinity of the *E*-alkene means that it dominates the equilibrium. Michael addition therefore provides a mechanism for the equilibration of *Z*-alkenes to *E*-alkenes.

For this reason, it can be very difficult to make *Z*-alkenes conjugated to reactive electrophilic groups such as aldehydes.

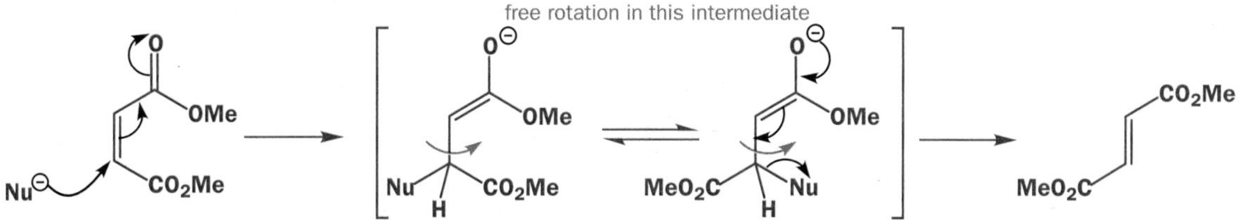

free rotation in this intermediate

Similar mechanisms account for the double bond geometry obtained in aldol reactions followed by dehydration to give α,β-unsaturated carbonyl compounds. Any *Z*-alkene that is formed is equilibrated to *E* by reversible Michael addition during the reaction.

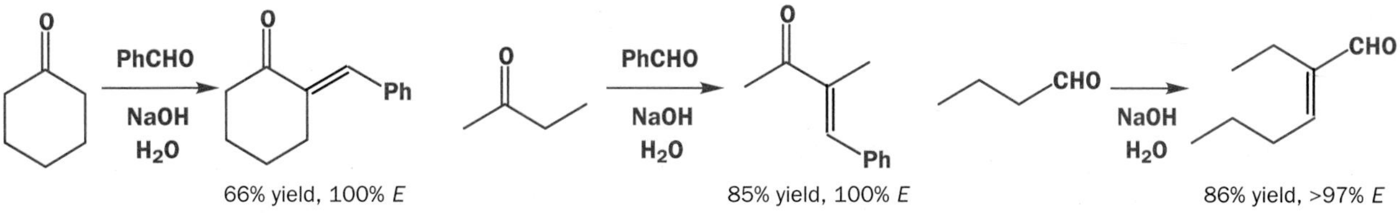

PhCHO / NaOH / H$_2$O

66% yield, 100% *E*

PhCHO / NaOH / H$_2$O

85% yield, 100% *E*

NaOH / H$_2$O

86% yield, >97% *E*

The double aldol product from acetone and benzaldehyde, known as dibenzylidene acetone (dba), is a constituent of some sun-protection materials and is used in organometallic chemistry as a metal ligand. It is easily made geometrically pure by a simple aldol reaction—again, reversible Michael addition equilibrates any *Z* product to *E*.

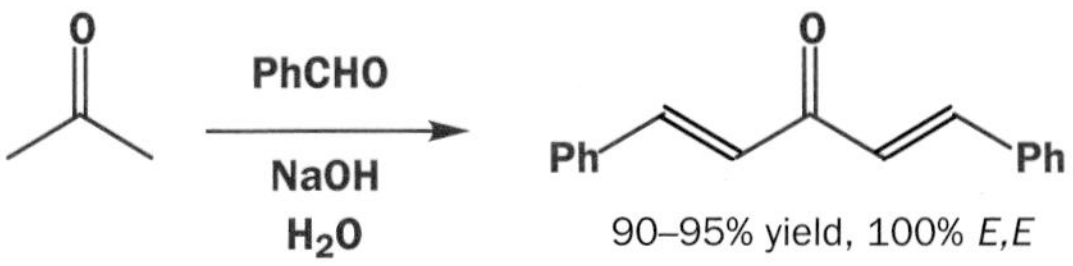

PhCHO / NaOH / H$_2$O

90–95% yield, 100% *E,E*

Equilibration of alkenes not conjugated with carbonyl groups requires different reagents

Iodine will add reversibly not only to Michael acceptors but also to most other alkenes. It can therefore be a useful reagent for equilibrating double bond geometrical isomers.

Some Japanese chemists needed the *E,E*-diene below for a synthesis of a neurotoxic compound that they had isolated from poison dart frogs. Unfortunately, their synthesis (which used a Wittig reaction—Chapter 14 and later in this chapter) gave only 4:1 *E* selectivity at one of the double bonds. To produce pure *E,E*-diene, they equilibrated the *E,Z*-diene to *E,E* by treating with iodine and irradiating with a sun-lamp.

The chemistry of vision

The human eye uses a *cis*-alkene, 11-*cis*-retinal, to detect light, and a *cis–trans* isomerism reaction is at the heart of the chemical mechanism by which we see. The light-sensitive pigment in the cells of the retina is an imine, formed by reaction of 11-*cis*- retinal with a lysine residue of a protein, opsin. Absorption of light by the opsin–retinal compound, known as rhodopsin, promotes one of the electrons in the conjugated polyene system to an antibonding orbital. Free rotation in this excited state allows the *cis* double bond to isomerize to *trans*, and the conformational changes in the protein molecule that result trigger a cascade of reactions that ultimately leads to a nerve signal being sent to the brain.

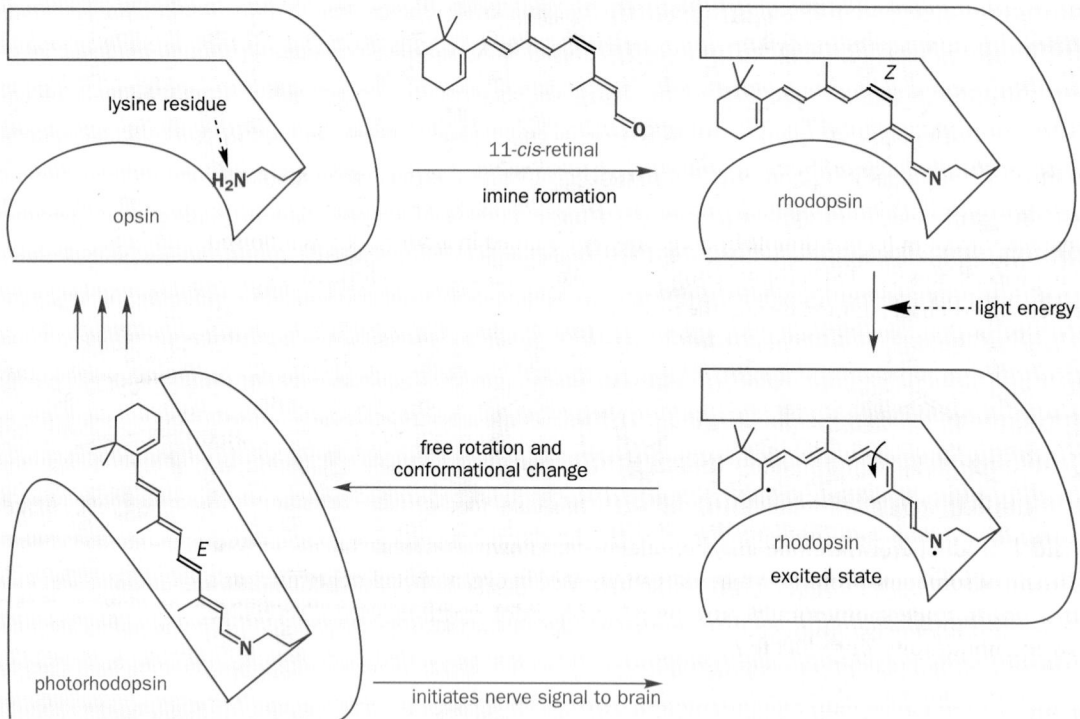

Using light to make *Z*-alkenes from *E*-alkenes

Light allows the equilibration of the two isomers of an alkene, by promoting a π electron into the π* orbital, but does not necessarily favour either isomer. One difference between *cis*- and *trans*-alkenes is that the *trans*-alkenes often absorb light better than the *cis*-alkenes do. They absorb light of a higher wavelength and they absorb more of it. This is particularly true of alkenes conjugated with carbonyl groups. Steric hindrance often forces the *cis*-alkene to twist about the σ bond joining the alkene to the carbonyl group and conjugation is then less efficient. A good example is the enone we saw a few pages back. Aldol condensation of cyclohexanone and benzaldehyde gives pure E-alkene. Irradiation with longer-wavelength UV light equilibrates this to the Z-alkene in excellent yield.

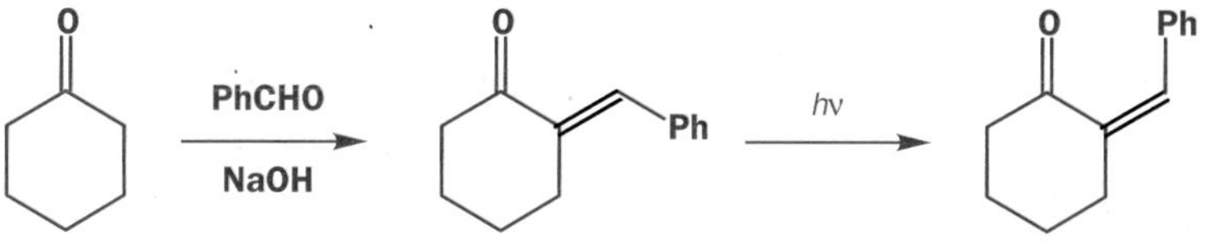

E-enone → Z-enone; 85% yield

It is not possible for the benzene ring and the enone system to be planar in the Z-enone and so they twist and conjugation is not as good as in the E-enone. Longer-wavelength light is absorbed only by the E-enone, which is continually equilibrated back to the excited state. Eventually, all the E-enone is converted to the Z-enone, which is not as efficiently excited by the light.

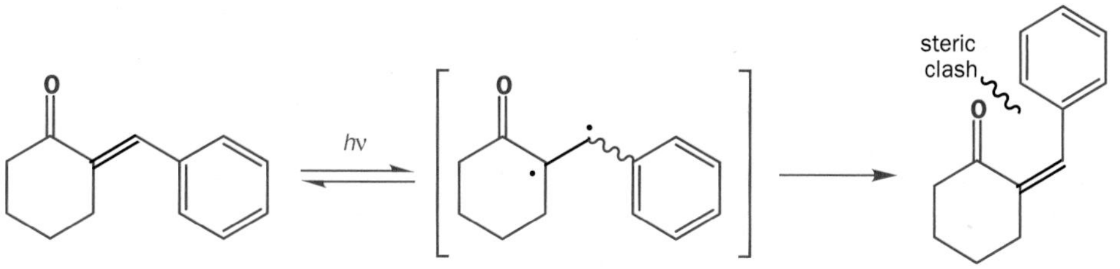

E-enone — π–π* excited state — Z-enone

This twisting and loss of conjugation is also the cause of the very slow epoxidation of the Z-enone discussed above. Conjugate addition is obviously best when there is good conjugation between the alkene and the carbonyl group. The rate-determining step in the epoxidation is conjugate addition.

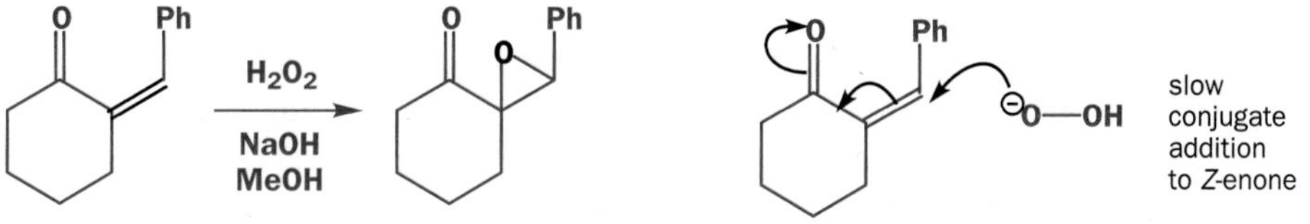

Z-enone

slow conjugate addition to Z-enone

Predominantly E-alkenes can be formed by stereoselective elimination reactions

In Chapter 19 you saw that E1 elimination reactions usually give mainly E-alkenes (there's an example earlier in this chapter) because the transition state leading to an E double bond is lower in energy than that leading to a Z double bond. In other words, E1 reactions are **stereoselective**, and their stereoselectivity is **kinetically controlled**. E2 reactions are similar if there is a choice of protons that can be removed. Treatment of 2-pentyl bromide with base gives about three times as much E-alkene as Z-alkene because the transition state leading to the E-alkene, which resembles conformation (i) below, is lower in energy than the transition state leading to the Z-alkene, resembling conformation (ii). Again, this is kinetic control.

2-bromopentane → E-alkene 51%, Z-alkene 18%, terminal alkene 31%

H and Br must be anti-periplanar for elimination to occur

conformation (i) — conformation (ii) — this conformation disfavoured by steric hindrance

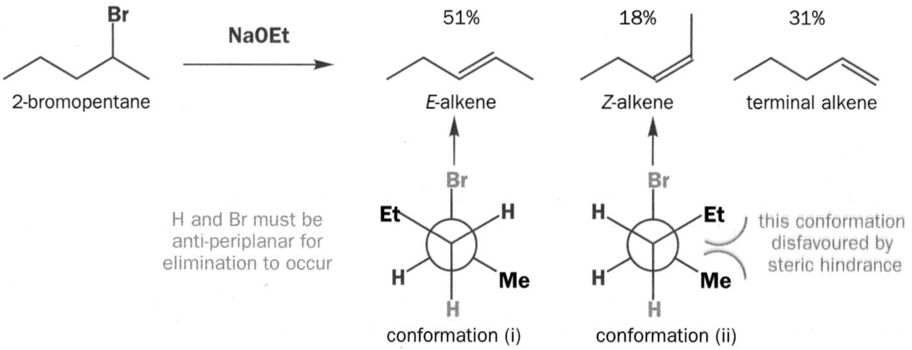

However, in neither this E2 reaction nor the E1 reaction on p. 487 is the stereoselectivity very good, and in this reaction the regioselectivity is bad too. The root of the problem is that one of the groups lost is always H (either as HBr or H_2O in these cases), and in most organic molecules there are lots of Hs to choose from!

Both stereo- and regioselectivity are better in E1cB reactions, such as the opening of this unsaturated lactone in base. The double bond inside the ring remains Z but the new one, formed as the ring opens, prefers the E geometry. The transition state for the elimination step already has a product-like shape and prefers this for simple steric reasons.

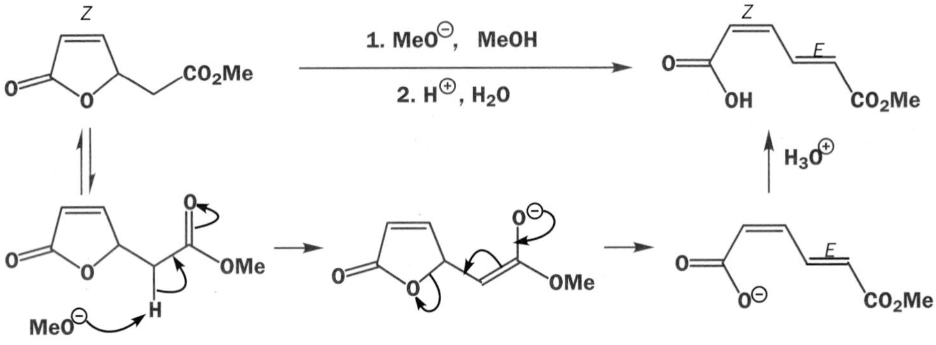

The Julia olefination is regiospecific and connective

This reaction is an elimination—the phenylsulfonyl ($PhSO_2$) and benzoate ($PhCO_2$) groups in the starting material are lost to form the double bond—but it is completely regioselective. Only the alkene shown is formed, with the double bond joining the two carbons that carried the $PhSO_2$ and $PhCO_2$ groups. This elimination is promoted by a reducing agent, usually sodium amalgam (a solution of sodium metal in mercury) and works for a variety of compounds providing they have a phenylsulfonyl group adjacent to a leaving group. It is called the **Julia olefination** after Marc Julia (1922–) who did his PhD at Imperial College, London, with Sir Derek Barton and now works at the École Normale in Paris and is best known for his work on sulfones.

> **Olefin** is an alternative name for alkene and **olefination** simply means alkene synthesis, usually by the formation of both σ and π bonds.

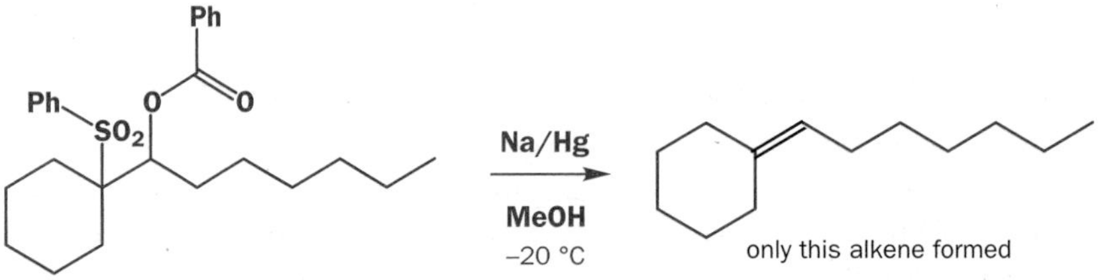

The most common leaving groups are carboxylates such as acetate or benzoate, and the starting materials are very easily made. As you will see in Chapter 46, sulfones are easily deprotonated next to the sulfur atom by strong bases like butyllithium or Grignard reagents, and the sulfur-stabilized anion will add to aldehydes. A simple esterification step, which can be done in the same reaction vessel as the addition, introduces the acetate or benzoate group. This is how the starting material for the elimination above was made.

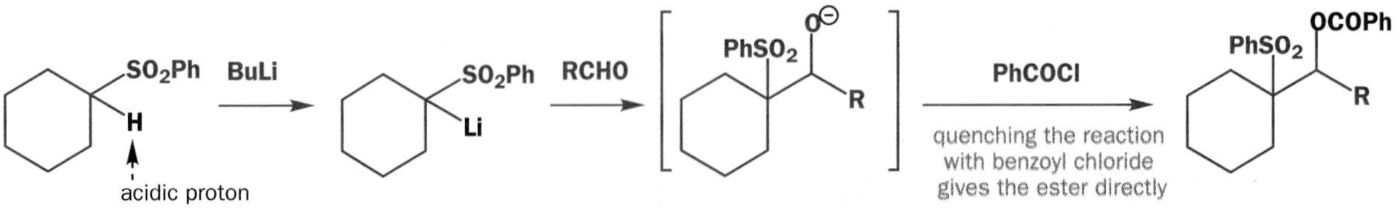

The short sequence of steps (starting with sulfone plus aldehyde and leading through to alkene) is known as the Julia olefination. It is our first example of a **connective double bond synthesis**—in other words, the double bond is formed by joining two separate molecules together (the aldehyde and sulfone). You will be reminded of the most important connective double-bond forming reaction, the Wittig reaction, later in the chapter.

The Julia olefination is stereoselective

Here are the results of a few simple Julia olefinations.

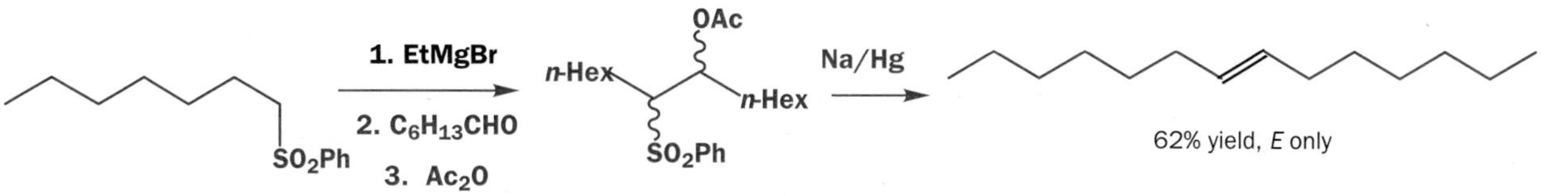

62% yield, E only

Notice that deprotonations can be with BuLi or EtMgBr and that the acylation step works with acetic anhydride or with benzoyl chloride. As you can see, they are all highly stereoselective for the E-isomer, and the Julia olefination is one of the most important ways of making E double bonds connectively.

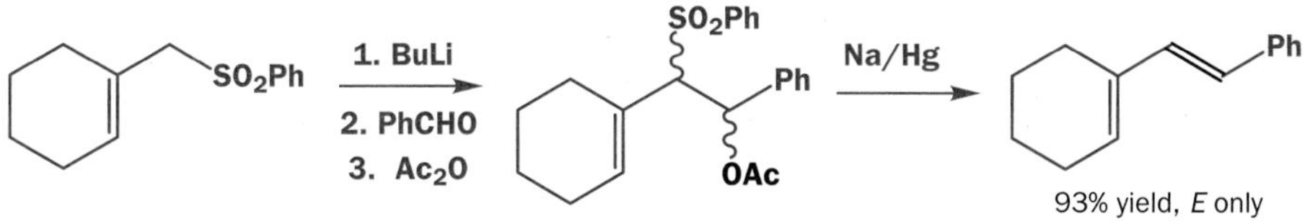

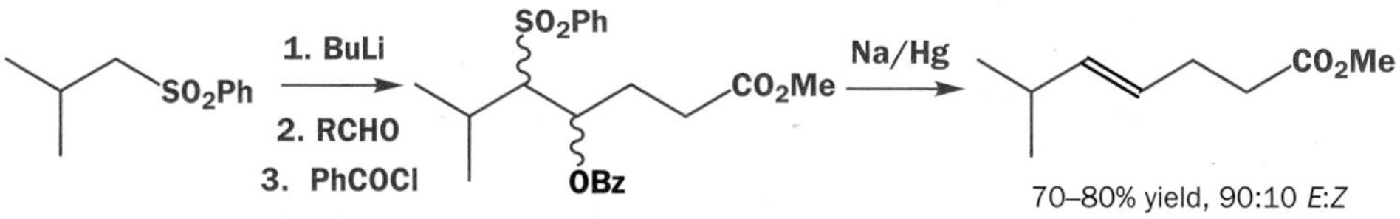

93% yield, E only

70–80% yield, 90:10 E:Z

Further example—preparation of sphingosine

In 1987, American chemists were studying the synthesis of some biological molecules using enzymes. One of the compounds they were interested in was sphingosine, an amino-alcohol that forms the backbone of sphingolipids (fat-like molecules found in cell membranes). They wanted to compare the enzyme-produced material with an authentic sample, which they made by using a Julia olefination to introduce the E double bond.

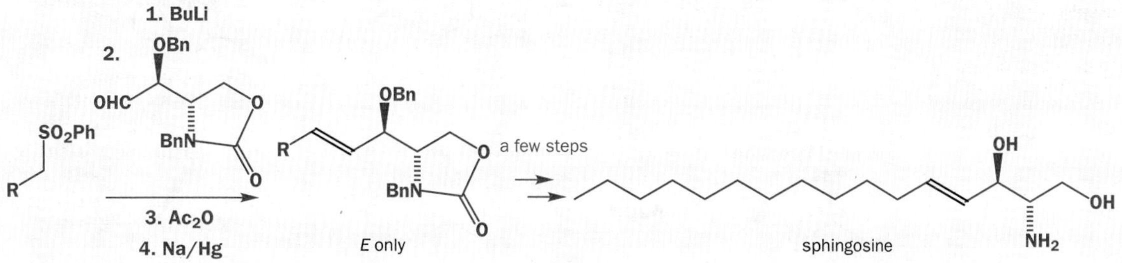

E only

sphingosine

The Julia olefination is stereoselective and not stereospecific

The reason for the E selectivity lies in the mechanism of the elimination. The first step is believed to be two successive electron transfers from the reducing agent (sodium metal) to the sulfone. Firstly, a radical anion is formed, with one extra unpaired electron, and then a dianion, with two extra electrons and therefore a double negative charge. The dianion fragments to a transient carbanion that expels acetate or benzoate to give the double bond.

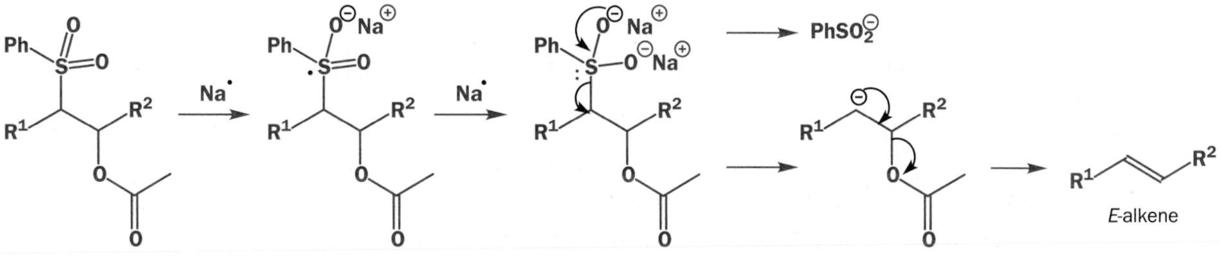

E-alkene

▶

A single-step E2 elimination
would have to go via an anti-
periplanar transition state and
would be stereospecific. You will
be able compare this
stereoselective Julia olefination
with the stereospecific Peterson
elimination shortly.

We know that there must be an anion intermediate because the elimination is *not stereospecific*—in other words, it doesn't matter which diastereoisomer of the starting material you use (all of the examples in this section have been mixtures of diastereoisomers) you always get the *E*-alkene product. The intermediate anion must have a long enough lifetime to choose its conformation for elimination.

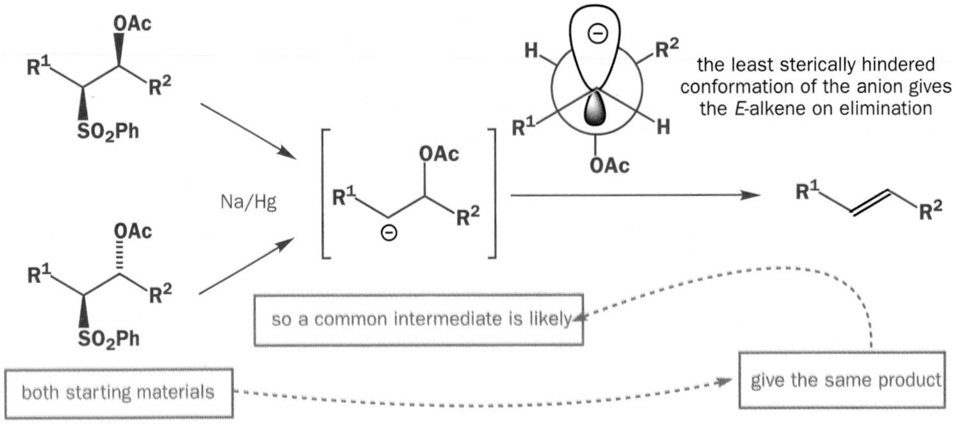

■

These reactions fall into class (4) of the
list on p. 805.

Stereospecific eliminations can give pure single isomers of alkenes

You met a stereospecific elimination in Chapter 19. The requirement for the H and the Br to be anti-periplanar in the E2 transition state meant that the two diastereoisomers of this alkyl bromide eliminated to alkenes with different double bond geometries (p. 491).

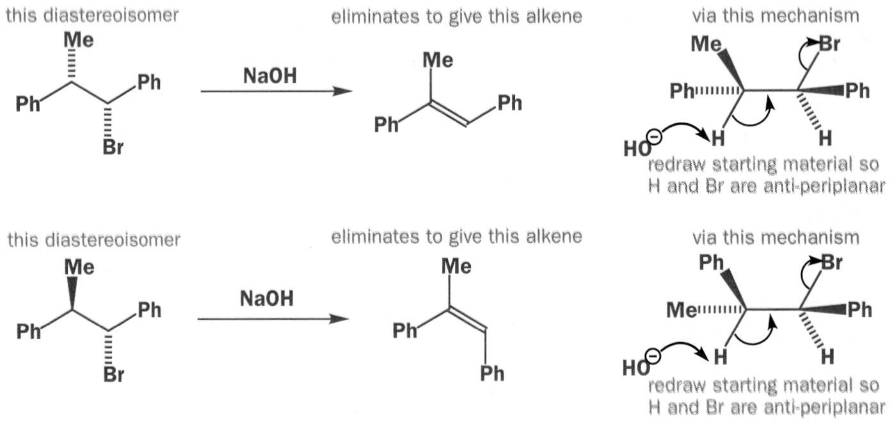

However, reactions like this are of limited use—their success relies on the base's lack of choice of protons to attack: provide an alternative H and we are back with the situation in the reaction on p. 810. Logic dictates, therefore, that only trisubstituted double bonds can be made stereospecifically in this way, because the reaction must not have a choice of hydrogen atoms to participate in the elimination. The answer is, of course, to move away from eliminations involving H, as we did with the Julia olefination. We shall look at this type of reaction for much of the rest of this chapter.

The Peterson reaction is a stereospecific elimination

There are many reactions in organic chemistry in which an Me_3Si group acts like a proton—Chapter 47 will detail some more reactions of silicon-containing compounds. Just as acidic protons are removed by bases, silicon is readily removed by hard nucleophiles, particularly F^- or RO^-, and this can promote an elimination. An example is shown here.

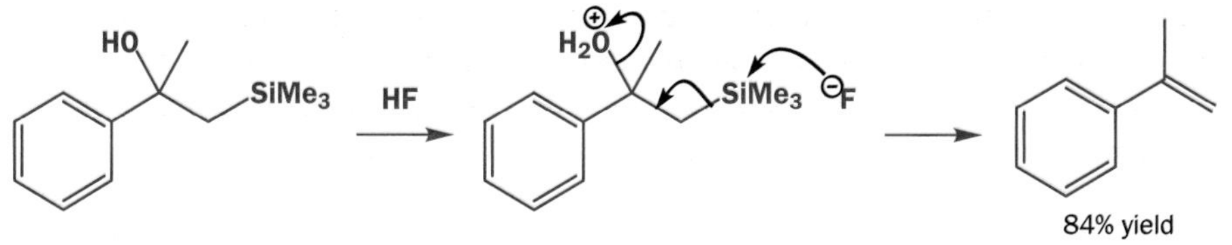

84% yield

The reaction is known as the **Peterson reaction**. It is rather like those we discussed right at the beginning of this chapter—eliminations of alcohols under acidic conditions to give alkenes. But, unlike those reactions, it is fully regioselective (like the Julia olefination), and so is particularly useful for making double bonds where other elimination methods might give the wrong regioisomer or mixtures of regioisomers. In this next example only one product is formed, in high yield, and it has an exocyclic double bond. Just think what would have happened without the silicon atom (ignore the one attached to the oxygen—that's just a protecting group). This compound is, in fact, an intermediate in a synthetic route to the important anticancer compound Taxol.

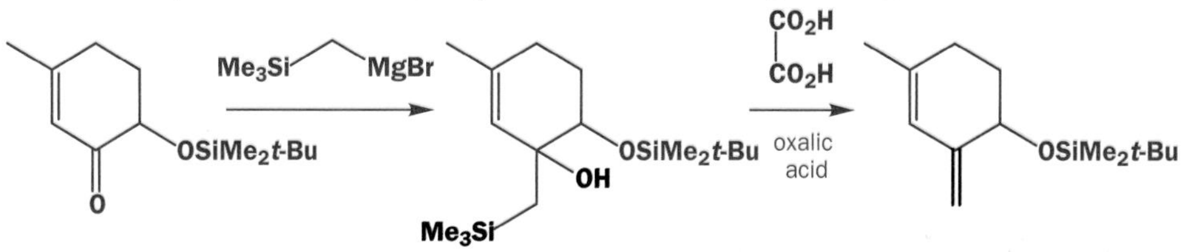

You've probably spotted that this is another connective alkene synthesis. The Peterson reaction is particularly useful for making terminal or exocyclic double bonds connectively because the starting material (the magnesium derivative shown above) is easily made from available Me_3SiCH_2Br. The reaction is also stereospecific, because it is an E2 elimination proceeding via an anti-periplanar transition state. In principle, it can therefore be used to make single geometrical isomers of alkenes, the geometry depending on the relative stereochemistry of the starting material. However, this use of the Peterson reaction is limited by difficulties in making diastereoisomerically pure starting materials.

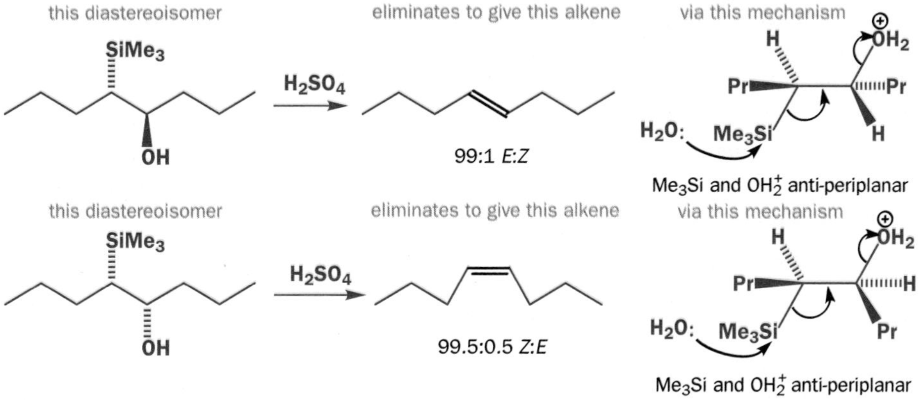

There is another, complementary version of the Peterson reaction that uses base to promote the elimination. The starting materials are the same as for the acid-promoted Peterson reaction. When base (such as sodium hydride or potassium hydride) is added, the hydroxyl group is deprotonated, and the oxyanion attacks the silicon atom *intramolecularly*. Elimination takes place this time via a *syn-periplanar* transition state—it has to because the oxygen and the silicon are now bonded together, and it is the strength of this bond that drives the elimination forward.

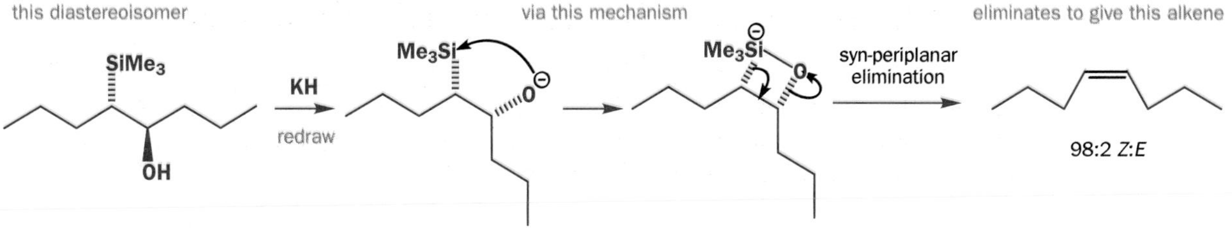

In Chapter 19 you saw that anti-periplanar transition states are usually preferred for elimination reactions because this alignment provides the best opportunity for good overlap between the orbitals involved. *Syn*-periplanar transition states can, however, also lead to elimination—and this particular case should remind you of the Wittig reaction (Chapter 14) with a four-membered cyclic intermediate.

The two versions of the Peterson reaction give opposite geometrical isomers from the same diastereoisomer of the starting material, so from any single diastereomer of hydroxy silane we can make either geometrical isomer of alkene product by choosing whether to use acid or base. The problem is still making those single diastereoisomers!

Perhaps the most important way of making alkenes—the Wittig reaction

The Wittig reaction is another member of the class we have been talking about—it's an elimination that does not involve loss of H. You met it in Chapter 14, where we gave a brief outline of its mechanism.

Conceptually, the Wittig reaction is like the base-promoted Peterson reaction: it is a *syn* elimination, driven by the strength of an oxygen–heteroatom bond, but in this case the heteroatom is phosphorus. But, unlike the other eliminations described above, the elimination step of the Wittig reaction occurs only from an intermediate and not from isolated starting materials. This intermediate is made *in situ* in the reaction and decomposes spontaneously: the Wittig reaction is therefore another connective alkene-forming reaction but, unlike either the Julia or Peterson reactions, it goes in one step, and for this reason is much more widely used.

We must start at the beginning. Phosphorus atoms, especially those that are positively charged or that carry electronegative substituents, can increase the acidity of protons adjacent to them on the carbon skeleton. Phosphonium salts (made in a manner analogous to the formation of ammonium salts from amines, in other words, by reaction of an alkyl halide with a phosphine) can therefore be deprotonated by a moderately strong base to give a species known as an **ylid**, carrying (formally) a positive and a negative charge on adjacent atoms. Ylids can alternatively be represented as doubly bonded species, called **phosphoranes**.

▶

Chemists are still unsure about the exact mechanism of this reaction, and what we have described here is certainly a very simplified picture of what actually happens.

Ylids can be isolated, but are usually used in reactions immediately they are formed. They are nucleophilic species that will attack the carbonyl groups of aldehydes or ketones, generating the four-membered ring oxaphosphetane intermediates. Oxaphosphetanes are unstable: they undergo elimination to give an alkene (65% yield for this particular example) with a phosphine oxide as a by-product. The phosphorus–oxygen double bond is extremely strong and it is this that drives the whole reaction forward.

Stereoselectivity in the Wittig reaction depends on the ylid

The Wittig reactions below were all used in the synthesis of natural products. You will notice that some reactions are *Z* selective and some are *E* selective. Look closer, and you see that the stereoselectivity is dependent on the *nature of the substituent* on the carbon atom of the ylid.

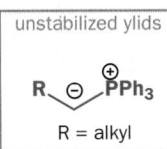

We can divide ylids into two types: those with conjugating or anion-stabilizing substituents adjacent to the negative charge (such as carbonyl groups) and those without. We call the first sort **stabilized ylids**, because the negative charge is stabilized not only by the phosphorus atom but by the adjacent functional group—we can draw an alternative enolate-type structure to represent this extra stabilization. The rest we call **unstabilized ylids**.

> ● **The stereochemistry of the Wittig reaction**
>
> The general rule is:
> - with *stabilized* ylids the Wittig reaction is *E* selective
> - with *unstabilized* ylids the Wittig reaction is *Z* selective

The *Z* selective Wittig reaction

The *Z* selectivity observed with simple alkyl R groups is nicely complementary to the *E* selectivity observed in the Julia olefination. This complementarity was exploited by some chemists who wanted to make isomers of capsaicin (the compound that gives chilli peppers their 'hotness') after suggestions that capsaicin might be carcinogenic.

The key intermediates in the synthesis of the *E*- and the *Z*-isomers of capsaicin were the *E* and *Z* unsaturated esters shown below. By using a Wittig reaction with an unstabilized ylid it was possible to make the *Z*-isomer selectively, whilst the Julia olefination gave the *E*-isomer.

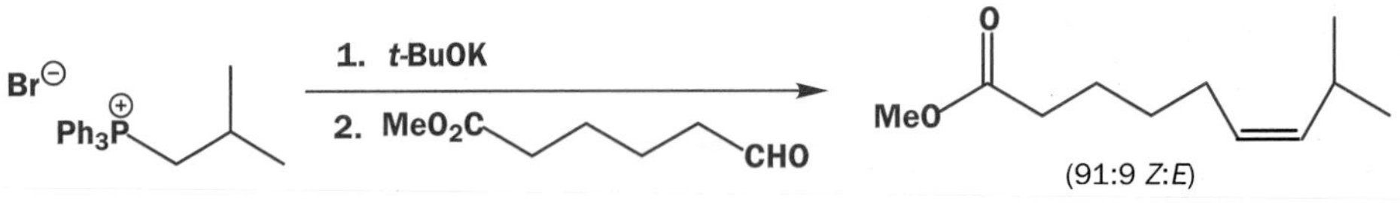

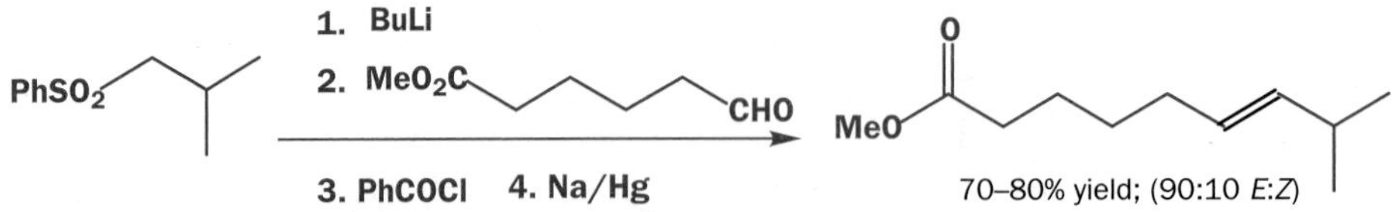

(91:9 *Z:E*)

70–80% yield; (90:10 *E:Z*)

How can the *Z* selectivity in Wittig reactions of unstabilized ylids be explained? We have a more complex situation in this reaction than we had for the other eliminations we considered, because we have two separate processes to consider: formation of the oxaphosphetane and decomposition of the oxaphosphetane to the alkene. The elimination step is the easier one to explain—it is stereospecific, with the oxygen and phosphorus departing in a syn-periplanar transition state (as in the base-catalysed Peterson reaction). Addition of the ylid to the aldehyde can, in principle, produce two diastereomers of the intermediate oxaphosphetane. Provided that this step is irreversible, then the stereospecificity of the elimination step means that the ratio of the final alkene geometrical isomers will reflect the stereoselectivity of this addition step. This is almost certainly the case when R is not conjugating or anion-stabilizing; the *syn* diastereoisomer of the oxaphosphetane is formed preferentially, and the predominantly *Z*-alkene that results reflects this. The *Z* selective Wittig reaction therefore consists of a kinetically controlled stereoselective first step followed by a stereospecific elimination from this intermediate.

alkene geometry is determined by the stereoselectivity of
the oxaphosphetane-forming step, which gives this
diastereoisomer of oxaphosphetane as the kinetic product

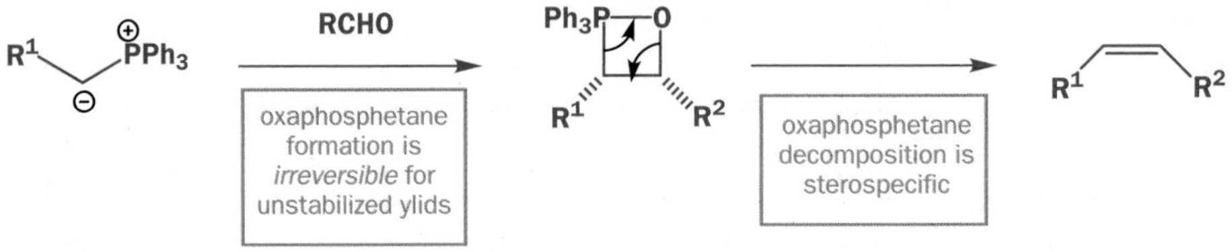

Why is formation of the *syn* oxaphosphetane favoured?

This question is the subject of much debate, because the mechanism by which the oxaphosphetane is formed is not entirely understood. One possible explanation relies on rules of orbital symmetry, which you will meet in Chapters 35 and 36—we need not explain them in detail here but suffice it to say that there is good reason to believe that, if the ylid and carbonyl compound react together to give the oxaphosphetane in one step, they will do so by approaching one another at right angles. Keeping the large substituents apart produces a transition state like that shown below, which (correctly) predicts that the oxaphosphetane will have *syn* stereochemistry.

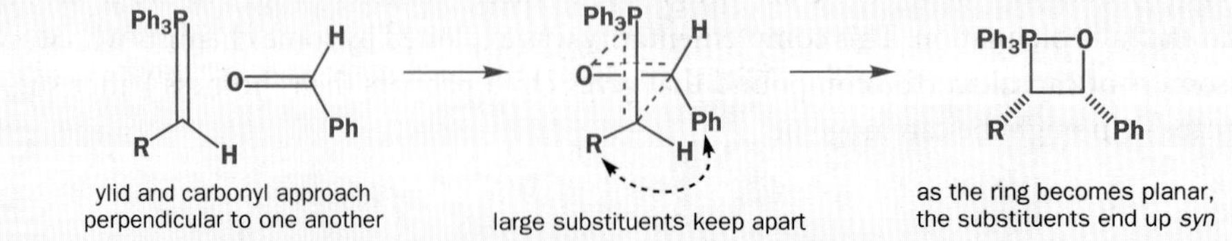

ylid and carbonyl approach
perpendicular to one another

large substituents keep apart

as the ring becomes planar,
the substituents end up *syn*

The *E* selective Wittig reaction

Stabilized ylids, that is ylids whose anion is stabilized by further conjugation, usually within a carbonyl group, give *E*-alkenes on reaction with aldehydes. These ylids are also enolates and were discussed in Chapter 27.

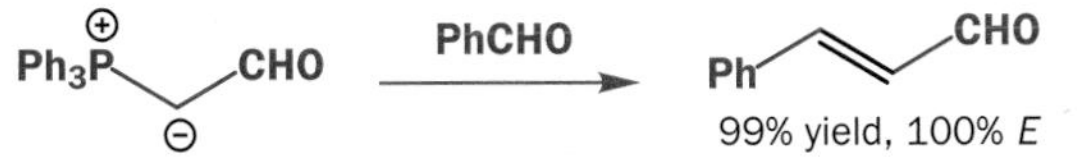

99% yield, 100% *E*

These stabilized ylids really are stable—this one, for example, can be recrystallized from water. This stability means though that they are not very reactive, and often it is better not to use the phosphonium salt but a phosphonate instead.

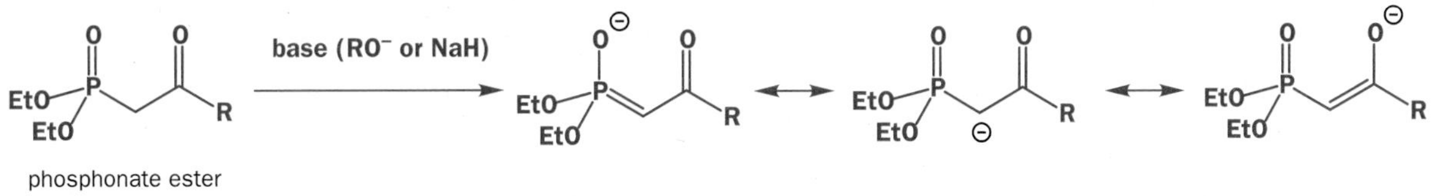

phosphonate ester

Phosphonate esters can be deprotonated with sodium hydride or alkoxide anions to give enolate-type anions that react well with aldehydes or ketones to give *E*-alkenes. Alkene-forming reactions with phosphonates are called **Horner–Wadsworth–Emmons** (or Horner–Emmons, Wadsworth–Emmons, or even Horner–Wittig) **reactions**. This example is a reaction that was used by some Japanese chemists in the synthesis of polyzonimine, a natural insect repellent produced by millipedes.

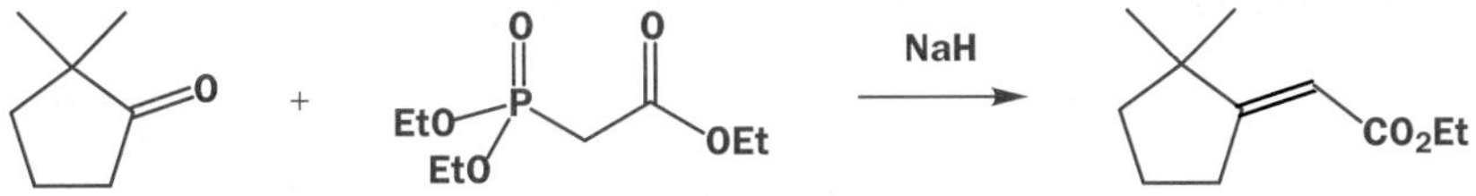

So why the change to *E* stereoselectivity when the ylid is stabilized? Again, chemists disagree about the details but a likely explanation is that the extra stability given to the ylid starting materials makes the reaction leading to the oxaphosphetane reversible. Stereoselectivity in this step is therefore no longer kinetically controlled but is **thermodynamically controlled**: reversal to starting materials provides a mechanism by which the oxaphosphetane diastereoisomers can interconvert. Providing the rate of interconversion is faster than the rate of elimination to alkene, the stereospecific step will no longer reflect the initial kinetic ratio of oxaphosphetane diastereoisomers. It is not unreasonable to suppose that the thermodynamically more stable of the oxaphosphetanes is the *trans*-diastereoisomer, with the two bulky groups on opposite sides of the ring, and that elimination of this gives *E*-alkene. What is more, the rate of elimination to give an *E*-alkene ought to be significantly faster than the rate of elimination to give a *Z*-alkene, simply by virtue of steric crowding in their respective transition states. The *anti* diastereoisomer is therefore 'siphoned off' to give *E*-alkene more rapidly than the *syn* diastereoisomer gives *Z*-alkene. Meanwhile equilibration of the two oxaphosphetane diastereomers via starting material replenishes the supply of *anti* diastereoisomer, and virtually only *E*-alkene is produced.

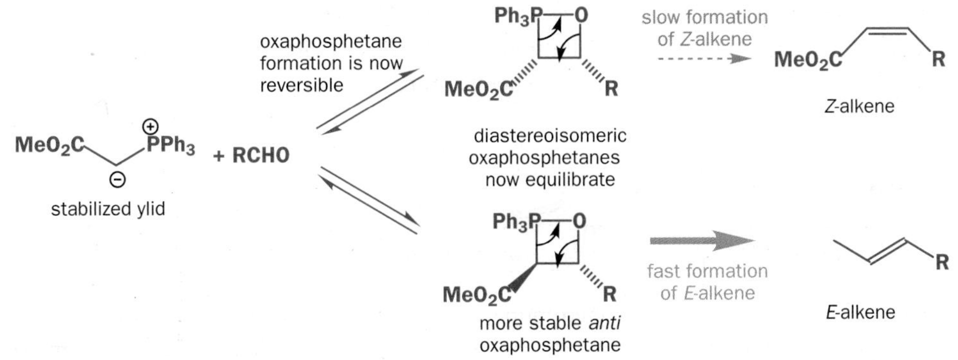

An *E,Z*-diene by two successive Wittig reactions

The female silkworm moth attracts mates by producing a pheromone known as bombykol. Bombykol is an *E,Z*-diene, and in this synthesis (dating from 1977) two successive Wittig reactions exploit the stereoselectivity obtained with stabilized and unstabilized ylids, respectively, to control the stereochemistry of the product.

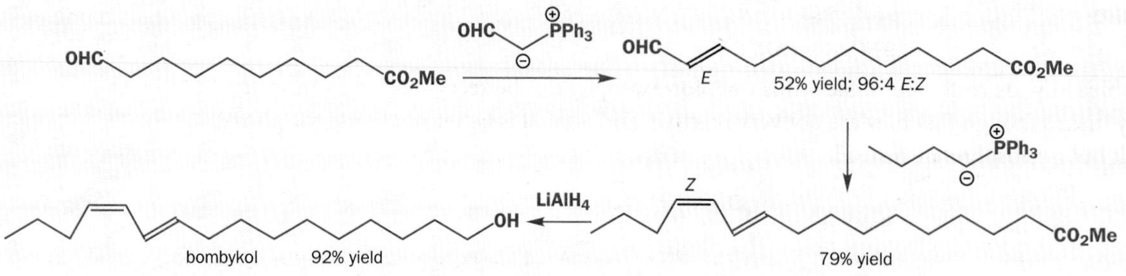

E- and *Z*-alkenes can be made by stereoselective addition to alkynes

In this last section of the chapter we shall leave elimination reactions to look at addition reactions. Alkynes react with some reducing agents stereoselectively to give either the *Z* double bond or the *E* double bond. Some of these reactions were described briefly in Chapter 24.

Z selective reduction of alkynes uses Lindlar's catalyst

This pure *Z*-alkene was needed for studies on the mechanism of a rearrangement reaction. In Chapter 24 you met catalytic hydrogenation as a means of reducing alkenes to alkanes, and we introduced Lindlar's catalyst (palladium and lead acetate on a support of calcium carbonate) as a means of controlling chemoselectivity so that *alkynes* could be reduced to *alkenes*. What we did not empha-

> The reason that catalytic hydrogenation often results in *syn addition* of hydrogen to *alkenes* was discussed in Chapter 24.

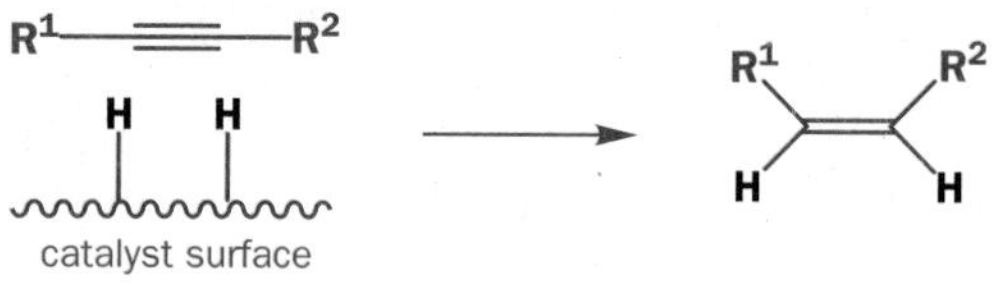

size then was that the two hydrogen atoms add to the alkyne in a *syn* fashion and the alkene produced is a *Z*-alkene. The stereoselectivity arises because two hydrogen atoms, bound to the catalyst, are delivered simultaneously to the alkyne.

You can compare this method of forming *Z*-alkenes directly with the Wittig reaction in these two syntheses of another insect pheromone, that of the Japanese beetle.

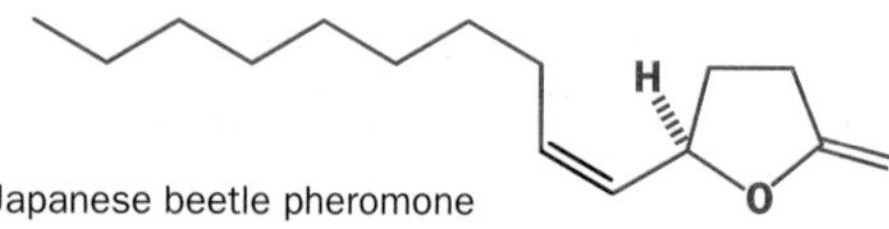

In this case, the Wittig reaction is not entirely *Z*-selective, and it generates some *E*-isomer. Lindlar-catalysed reduction, on the other hand, generates pure *Z*-alkene.

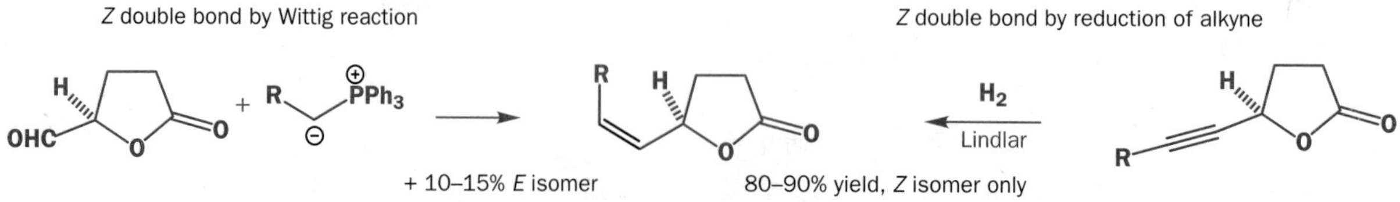

For a biologically active sample of this pheromone, it is better that the stereochemistry is the same as that of the natural compound—the *E* double bond isomer is more or less inactive. Even more important is the configuration at the chiral centre in the pheromone—the wrong enantiomer is not only inactive, but it also inhibits the male beetles' response to the natural stereoisomer. In Chapter 45 we shall talk about ways of making single enantiomers selectively.

E selective reduction of alkynes uses sodium in liquid ammonia

The best way of ensuring *anti* addition of hydrogen across any triple bond is to treat the alkyne with sodium in liquid ammonia.

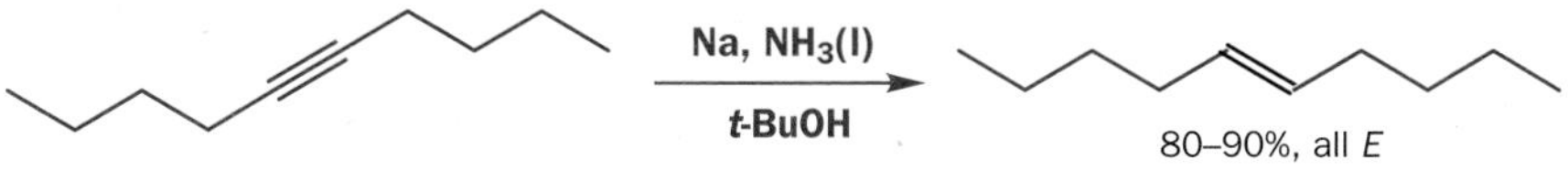

The sodium donates an electron to the LUMO of the triple bond (one of the two orthogonal π^* orbitals). The resulting radical anion can pick up a proton from the ammonia solution to give a vinyl radical. A second electron, supplied again by the sodium, gives an anion that adopts the more stable *trans* geometry. A final proton quench by a second molecule of ammonia or by an added proton source (*t*-butanol is often used, as in the Birch reduction) forms the *E*-alkene.

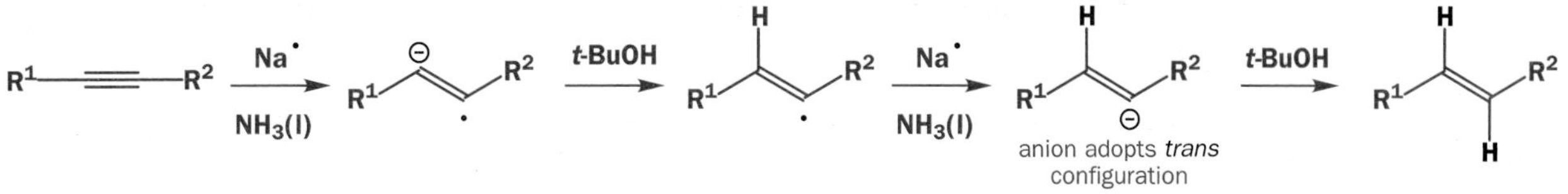

An alternative, and more widely used, method is to reduce alkynes with LiAlH$_4$. This reaction works only if there is a hydroxy or an ether functional group near to the alkyne, because it relies on delivery of the reducing agent to the triple bond through complexation to this oxygen atom.

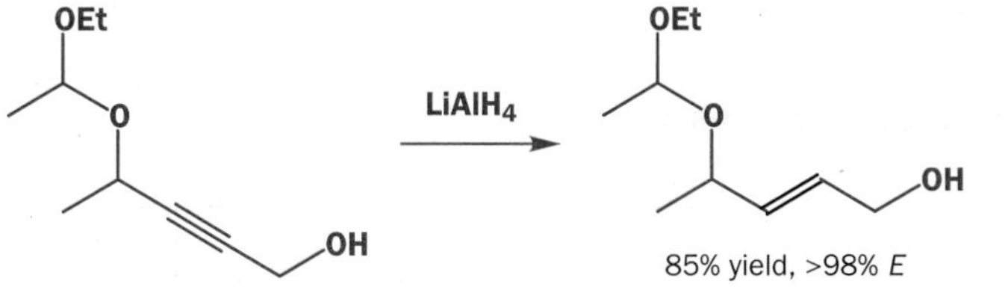

Making alkenes by addition to alkynes offers two distinct advantages. Firstly, although the reaction is not connective in the sense that the Wittig and Julia reactions are, the starting materials can often be made straightforwardly by alkylation of alkynyl anions. Secondly, the same alkyne can be used to make either *E*- or *Z*-alkene—an advantage shared with the Peterson reaction but here the starting material is much easier to make. In some early work on sphingosine (a constituent of cell membranes), some Swiss chemists needed to make both *E*- and *Z*-isomers of the naturally occurring compound. This was an easy task once they had made the alkyne.

■
You saw a Julia reaction used to make sphingosine on p. 811.

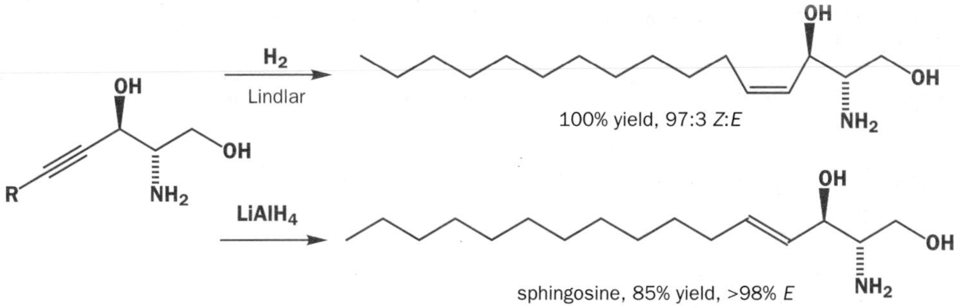

Addition of nucleophiles to alkynes

This rarer, and rather surprising, approach to *Z*-alkenes sometimes gives excellent results particularly in the addition of nucleophiles to butadiyne. The base-catalysed addition of methanol gives an excellent yield of Z-1-methoxybut-1-en-3-yne. This reaction is so easy to do that the product is available commercially.

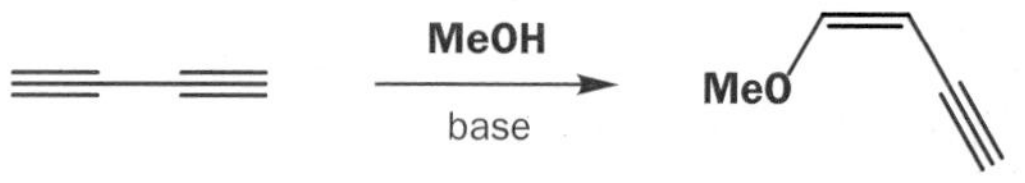

Notice that methanol adds once only: you would not expect nucleophiles to add to a simple alkyne and it is the conjugation that makes addition possible. Methoxide ion adds to one of the alkynes to give a conjugated anion.

The anion is linear with the negative charge delocalized along the conjugated system and the charge is therefore in a p orbital in the plane of the molecule. The other p orbital is involved in π bonding as well but at right angles to the plane of the molecule. When the anion reacts with a molecule of methanol, protonation occurs on the lobe of the p orbital away from the MeO group and the Z-alkene is formed. This product is mentioned in other chapters of the book: now you know why it is available.

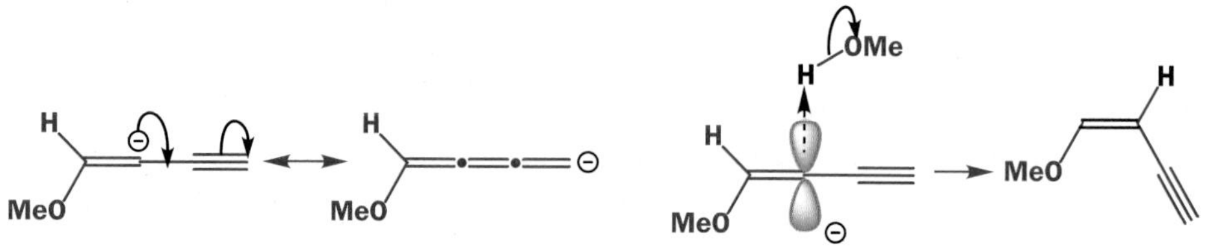

● **Summary of methods for making alkenes stereospecifically**

Here is a summary of the most important methods for making double bonds stereoselectively.

To make *cis* (*Z*)-alkenes

- Wittig reaction of *unstabilized* ylid
- Constrain the alkene in a ring

- *syn* addition of hydrogen across an alkyne
- Peterson elimination

To make *trans* (*E*)-alkenes

- Wittig reaction of *stabilized* ylid
- Equilibration to the more stable isomer
- Julia olefination
- Simple unselective elimination reactions
- *trans* selective reduction of alkyne
- Peterson elimination

In this chapter we have dealt for the first time with the problem of producing compounds as single stereoisomers—the stereoisomers concerned were geometrical isomers of alkenes. The next two chapters will look in more detail at making stereoisomers, but we shall move out of two dimensions into three and consider reactions that have diastereoselectivity. The two subjects are closely related, since often single diastereoisomers are made by addition reactions of single geometrical isomers of double bonds and, as you saw with the Peterson and Wittig reactions, single diastereoisomers can lead stereospecifically to single geometrical isomers.

Problems

1. Deduce the structure of the product of this reaction from the spectra and explain the stereochemistry. Compound A has δ_H 0.95 (6H, d, J 7 Hz), 1.60 (3H, d, J 5 Hz), 2.65 (1H, double septuplet, J 4 and 7 Hz), 5.10 (1H, dd, J 10 and 4 Hz), and 5.35 (1H, dq, J 10 and 5 Hz).

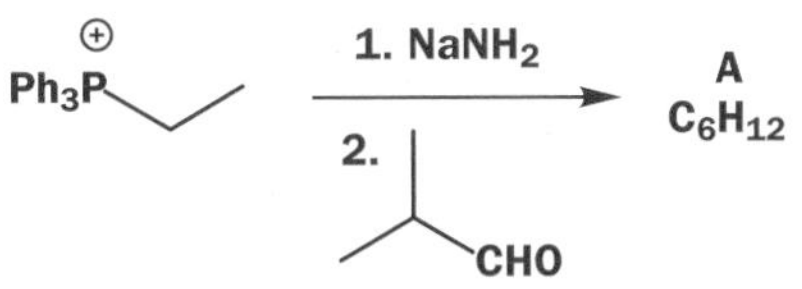

2. A single diastereoisomer of an insect pheromone was prepared in the following way. Which isomer is formed and why? Outline a synthesis of one other isomer.

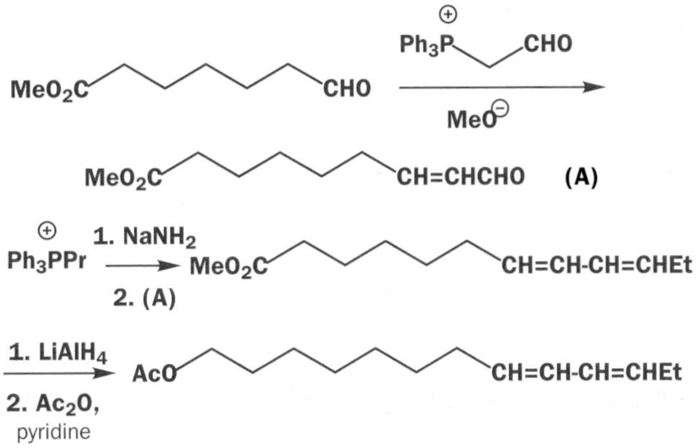

3. How would you prepare samples of both geometrical isomers of this compound?

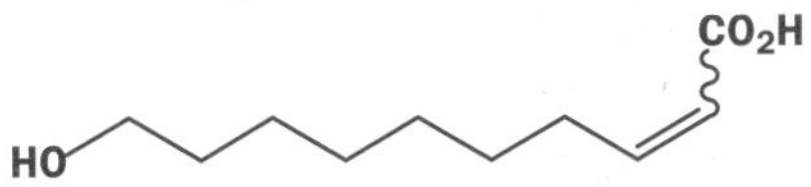

4. Decomposition of this diazocompound in methanol gives an unstable alkene A ($C_8H_{14}O$) whose NMR spectrum contains these signals: δ_H (p.p.m.) 5.80 (1H, ddd, J 17.9, 9.2, and 4.3 Hz), 5.50 (1H, dd, J 17.9 and 7.9 Hz), 4.20 (1H, m), 3.50 (3H, s), and 1.3–2.7 (8H, m). What is its structure and geometry? You are not expected to work out a mechanism for the reaction.

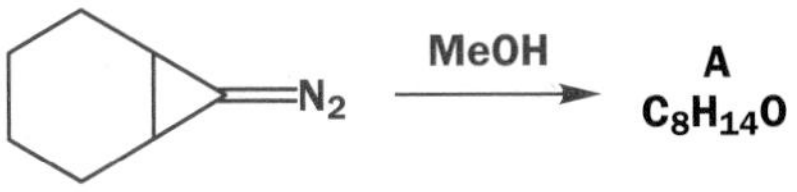

5. Why do these reactions give different alkene geometries?

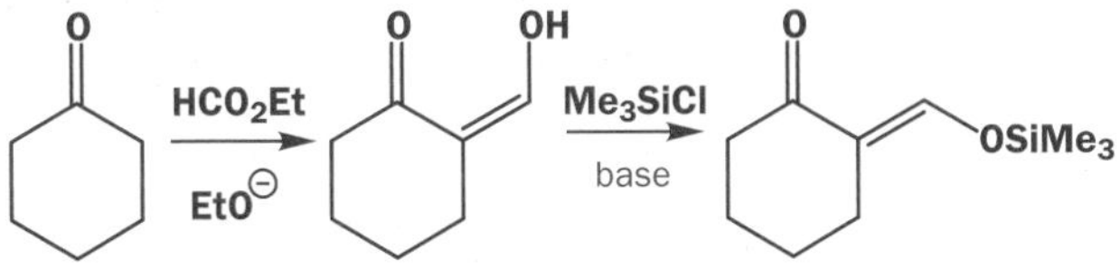

6. Here is a synthesis of a prostaglandin analogue. Suggest reagents for the steps marked '?', give mechanisms for those not so marked, and explain any control of alkene geometry.

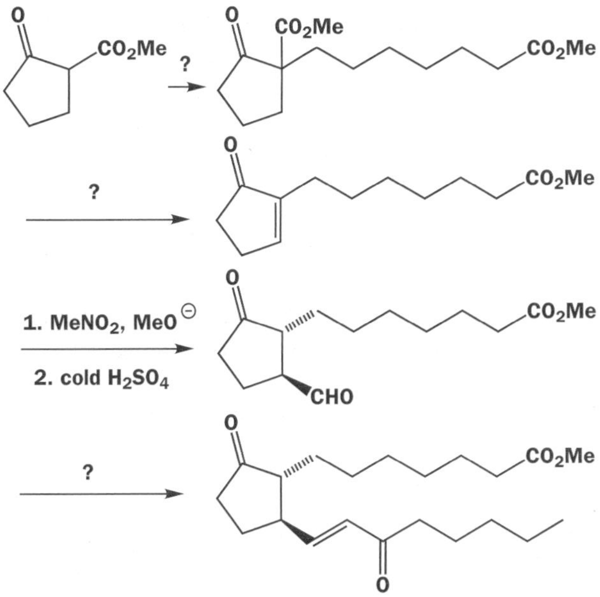

7. Isoeugenol, the flavouring principle of cloves, occurs in the plant in both the E (solid) and Z (liquid) forms. How would you prepare a pure sample of each and how would you purify each from any of the other isomer?

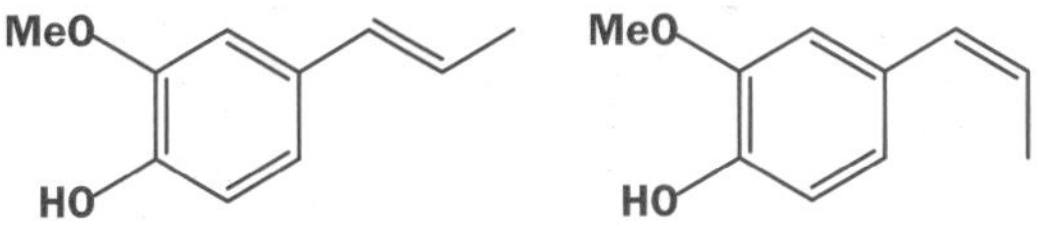

8. Thermal decomposition of this lactone gives mainly the Z-alkene shown with minor amounts of the E-alkene and an unsaturated acid. Suggest a mechanism for the reaction that explains these results.

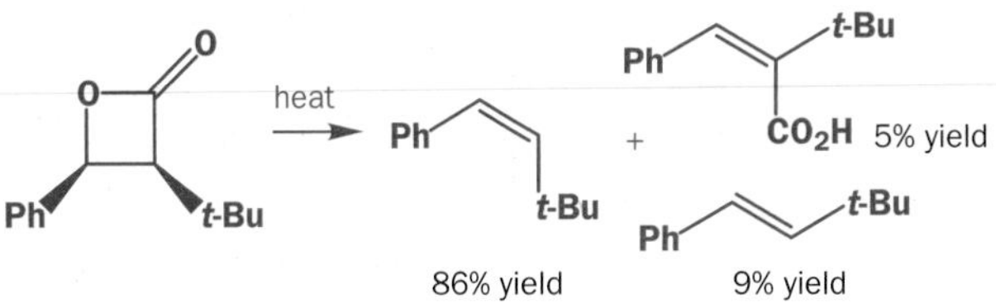

9. What controls the double bond geometry in these examples? In the second example, one alkene is not defined by the drawing.

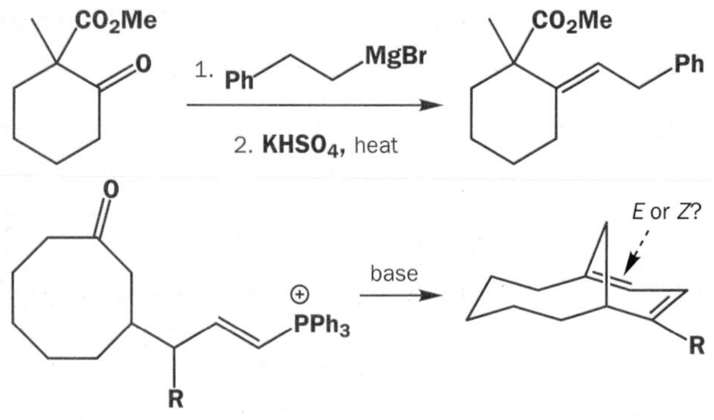

10. Treatment of this epoxide with base gives the same *E*-alkene regardless of the stereochemistry of the epoxide. Comment.

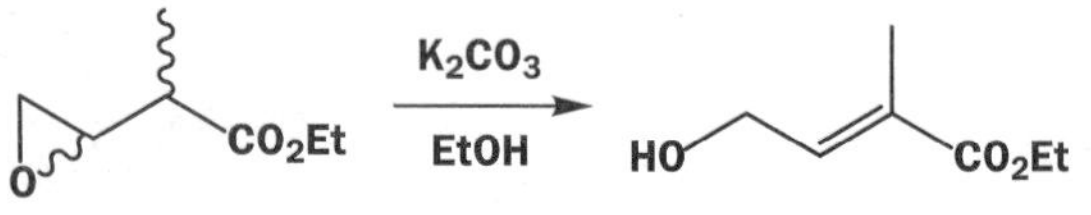

11. Which alkene would be formed in each of the following reactions? Explain your answer mechanistically.

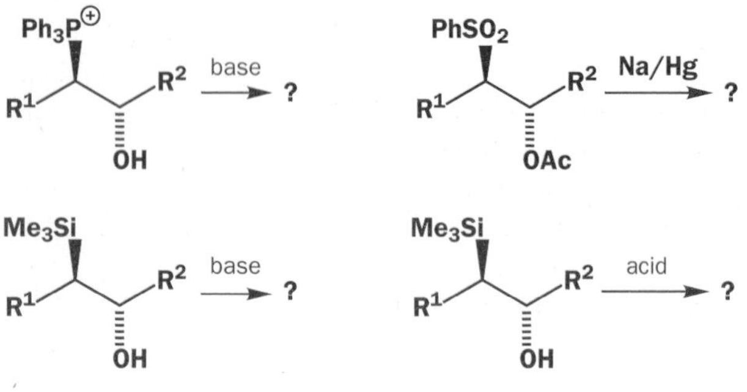

12. Comment on the difference between these two reactions.

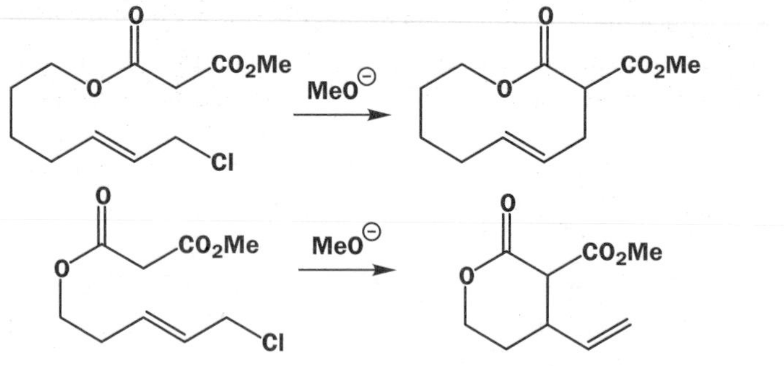

13. The elimination of alcohols 13A to give cinnamic acids *trans*-13B in aqueous sulfuric acid has been studied. If optically active 13A is used and the reaction stopped at 10% completion, the starting material is found to be completely racemized. What can you deduce about the mechanism of the elimination step?

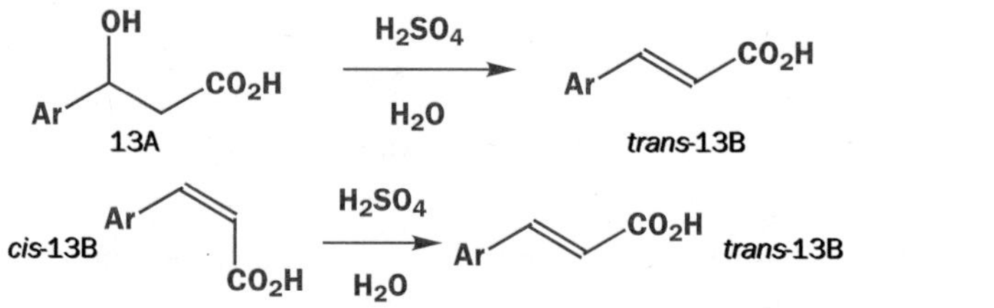

The *cis* cinnamic acids *cis*-13B also isomerize to the *trans* acids under the same conditions but more slowly. What is the mechanism of this reaction?

14. Give mechanisms for these stereospecific reactions on single geometrical isomers of alkenes.

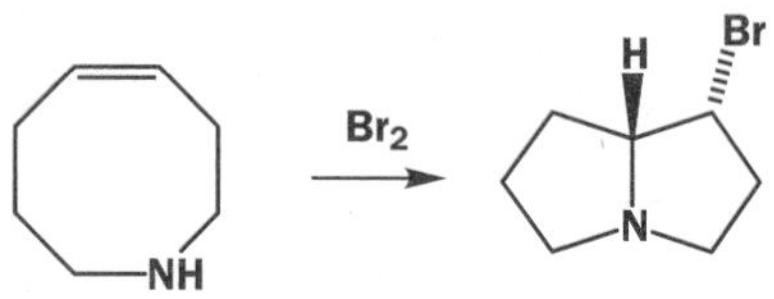

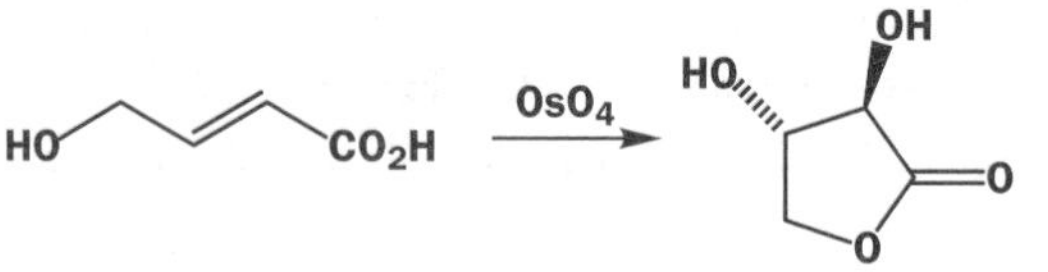

Determination of stereochemistry by spectroscopic methods

32

Connections

Building on:

- Determining organic structures ch3
- Proton NMR spectroscopy ch11
- Review of spectroscopic methods ch15
- Stereochemistry ch16
- Conformation ch18
- Controlling double bond geometry ch31

Arriving at:

- How coupling varies with the angle between bonds
- How ring size affects coupling
- How electronegative atoms reduce coupling
- How π systems increase geminal coupling
- How protons attached to the same carbon can be different, and can couple to one another
- What homotopic, enantiotopic, and diastereotopic mean
- The nuclear Overhauser effect: what it is and how to exploit it

Looking forward to:

- Controlling stereochemistry with rings ch33
- Diastereoselectivity ch34
- Saturated heterocycles ch42
- Asymmetric synthesis ch45
- Organic synthesis ch 53

Introduction

From time to time throughout the book we have spread before your eyes some wonderful structures. Some have been very large and complicated (such as palytoxin, p. 19) and some small but difficult to believe (such as tetra-*t*-butyl tetrahedrane, p. 373). They all have one thing in common. Their structures were determined by spectroscopic methods and everyone believes them to be true. Among the most important organic molecules today is Taxol, an anticancer compound from yew trees. Though it is a 'modern' compound, in that chemists became interested in it only in the 1990s, its structure was actually determined in 1971.

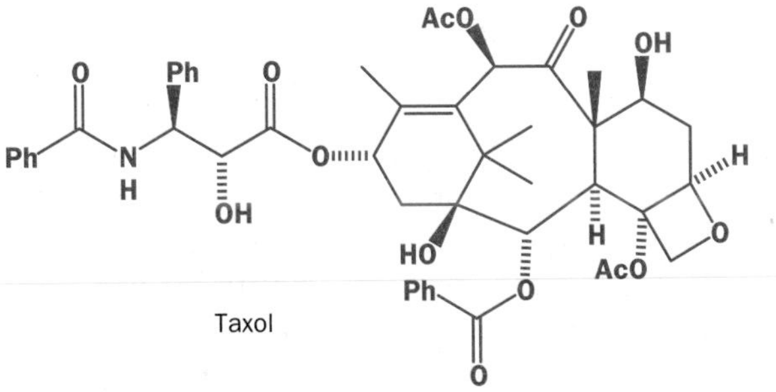

Taxol

No one argued with this structure because it was determined by reliable spectroscopic methods—NMR plus an X-ray crystal structure of a derivative. This was not always the case. Go back another 25 years to 1946 and chemists argued about structures all the time. An undergraduate and an NMR spectrometer can solve in a few minutes structural problems that challenged teams of chemists for years half a century ago. In this chapter we will combine the knowledge presented systematically in Chapters 3, 11, and 15, add your more recently acquired knowledge of stereochemistry (Chapters 16, 18, and 31), and show you how structures are actually determined in all their stereochemical detail using all the evidence available.

In general, we will not look at structures as complex as Taxol. But it is worth a glance at this stage to see what was needed. The basic carbon skeleton contains one eight- and two six-membered rings. These can be deduced from proton and carbon NMR. There is a four-membered heterocyclic ring—a feature that caused a lot of argument over the structure of penicillin. The four-membered cyclic ether in Taxol is easily deduced from proton NMR as we will see soon. There are ten functional groups (at least—it depends on how you count) including six carbonyl groups. These are easily seen in the carbon NMR and IR spectra. Finally, there is the stereochemistry. There are eleven stereogenic centres, which were deduced mostly from the proton NMR and the X-ray crystal structure of a closely related compound (Taxol itself is not crystalline).

New structures are being determined all the time. A recent issue of one important journal (*Tetrahedron Letters* No. 14 of 1996) has a paper on Taxol but also reports the discovery and structure determination of the two new natural products in the margin. Both compounds were discovered in ocean sponges, one from Indonesia and one from a fungus living in a sponge common in the Pacific and Indian oceans. Both structures were determined largely by NMR and in neither case was an X-ray structure necessary. You should feel a bit more in tune with the chemists who deduced these structures as they look much simpler than Taxol or even than penicillin. We hope you will feel by the end of this chapter that you can tackle structural problems of this order of complexity with some confidence. You will need practice, and in this area above all it is vital that you try plenty of problems. Use the examples in the text as worked problems: try to solve as much as you can before reading the answer—you can do this only the first time you read because next time you will have your memory as a prompt.

The stereochemistry at two of the stereogenic centres of chlorocarolide was unknown when this structure was published—stereochemistry is one of the hardest aspects of structure to determine. Nonetheless, NMR is second only to X-ray in what it tells us of stereochemistry, and we shall look at what coupling constants (*J* values) reveal about configuration, conformation, and reactivity. The first aspect we consider is the determination of conformation in six-membered rings.

two recently discovered simple natural products

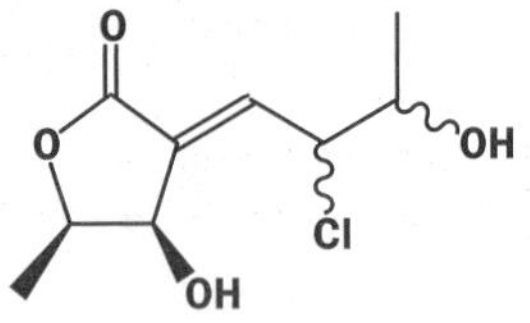

chlorocarolide A

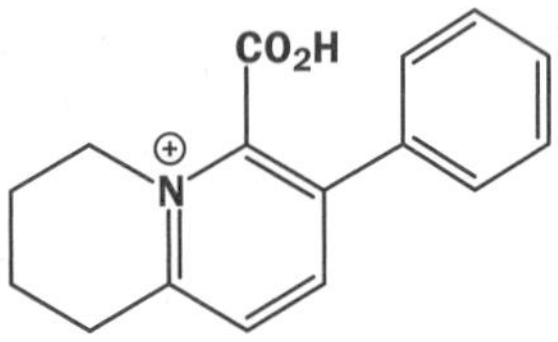

ciathryimine A

■ It would be wise to review Chapter 18 now if what we said there is not fresh in your mind.

^3J values vary with H–C–C–H dihedral angle

● Remember

Parallel orbitals interact best.

In the last chapter, we looked at some stereospecific eliminations to give double bonds, and you know that E2 elimination reactions occur best when there is an anti-periplanar arrangement between the proton and the leaving group.

best arrangement for E2 elimination

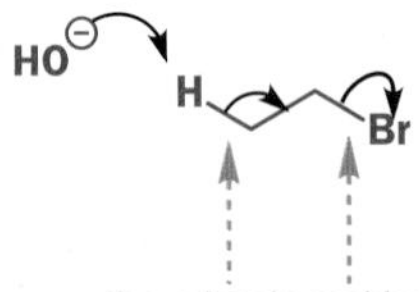

the orbitals making up these bonds are parallel

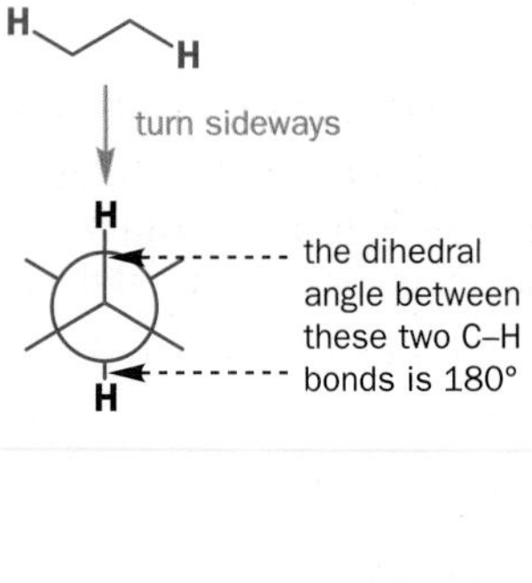

In the NMR spectrum, coupling between protons arises from through-bond and not through-space interactions: *trans* coupling in alkenes is *bigger* than *cis* coupling (see Chapter 11, p. 269). So the same arrangement that leads to the best reaction ought also to lead to the largest coupling constant. In other words, if we replace 'Br' in the diagram with a second hydrogen atom but keep the orbital alignment the same, we ought to get the biggest possible coupling constant for a saturated system.

largest ^3J from parallel orbitals

$^3J_{HH}$~ 10 Hz

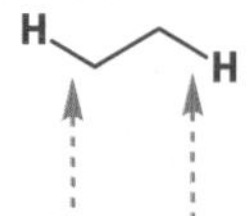

the orbitals making up these bonds are parallel

turn sideways

the dihedral angle between these two C–H bonds is 180°

The usual description of this situation is in terms of the dihedral angle between the H–C–C–H bonds. The dihedral angle is obvious in the Newman projection as it is the angle between the two C–H bonds projected on a plane orthogonal to the C–C bond. In a Newman projection this plane is the plane of the paper, and here the angle is 180°.

When the dihedral angle is zero, the two C–H bonds are again in the same plane but not perfectly parallel. The coupling constant is again large, but not so large as in the previous case. In fact, the two arrangements are very like *cis* and *trans* double bonds, but the C atoms are tetrahedral not trigonal.

You may guess that, when the dihedral angle is 90°, the coupling constant is zero. What happens in between these extremes was deduced by Karplus in the 1960s and the relationship is usually known as the **Karplus equation**. It is easiest to understand from a graph of *J* against dihedral angle.

Examine this graph carefully and note the basic features as you will need them as we go through the chapter. These features are:

- Coupling is largest at 180° when the orbitals of the two C–H bonds are perfectly parallel
- Coupling is nearly as large at 0° when the orbitals are in the same plane but not parallel
- Coupling is zero when the dihedral angle is 90°—orthogonal orbitals do not interact
- The curve is flattened around 0°, 90°, and 180°—*J* varies little in these regions from compound to compound

the Karplus relationship: *J* vs. dihedral angle

- The curve slopes steeply at about 60° and 120°—*J* varies a lot in this region with small changes of angle and from compound to compound
- Numerical values of *J* vary with substitution, ring size, etc., but the Karplus relationship still works—it gives good *relative* values

These ideas come to life in the determination of conformation in six-membered rings. *Trans* diaxial hydrogen atoms are aligned with a dihedral angle of 180° and give the largest *J* values.

The other two situations, where one or both hydrogen atoms are equatorial, both have angles of about 60°, though axial/equatorial couplings are usually slightly larger than equatorial/equatorial ones.

Now for some illustrations. The simple cyclohexyl ester has just one substituent, which we expect to be equatorial (Chapter 18). The black hydrogen therefore has four neighbours—two axial Hs and two equatorial Hs. We expect to see a triplet from each and that the axial/axial coupling constant will be large. In fact, there is a 1H signal at δ 4.91, it is a tt (triplet of triplets) with *J* = 8.8 and 3.8 Hz. Only an axial H can have couplings as big as 8.8 Hz, so now we *know* that the ester is equatorial.

By contrast, the next ester, which also has only one substituent, has a 1H signal at δ 6.0 p.p.m. which is a simple triplet with *J* = 3.2 Hz. With no large couplings this cannot be an axial proton and the *substituent* must now be axial. It so happens that the small equatorial/axial and equatorial/equatorial couplings to the green hydrogens are the same. This is not so surprising as the dihedral angles are both 60°.

None of the dihedral angles in a six-membered ring are 90°, but in some bicyclic systems they are. Norbornane-type structures (bicyclo[2.2.0]heptanes), for example, typically have couplings of 0 Hz between the protons shown in black and green because the H–C–C–H dihedral angle is 90°.

The determination of *conformation* by NMR may more importantly allow us to

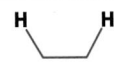 two C-H bonds are eclipsed in the plane of the paper – dihedral angle = 0°

▶ As a reminder, the dihedral angle is most easily visualized by imagining the C–C bond lying along the spine of a partially-opened book. If the C–H bonds are written one on one page and the other on the other, then the dihedral angle is the angle between the pages of the book.

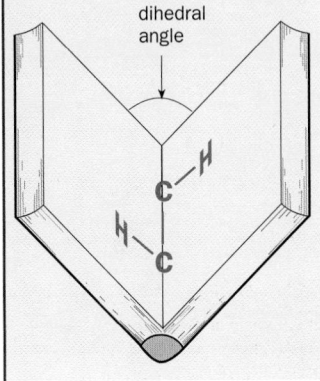

diaxial Hs dihedral angle 180° $^3J \sim$ 10–12 Hz

diequatorial Hs dihedral angle 60° $^3J \sim$ 2–3 Hz

axial/equatorial Hs dihedral angle 60° $^3J \sim$ 3–5 Hz

■ We discuss in Chapter 42 why this substituent prefers to be axial.

determine *configuration* at the same time. This often occurs when there are two or more substituents on the ring. Here is a simple example: you saw in Chapter 18 that the reduction of 4-*t*-butylcyclohexanone can be controlled by choice of reagent to give either a *cis* or a *trans* alcohol. It is easy to tell them apart as the *t*-butyl group will always be equatorial.

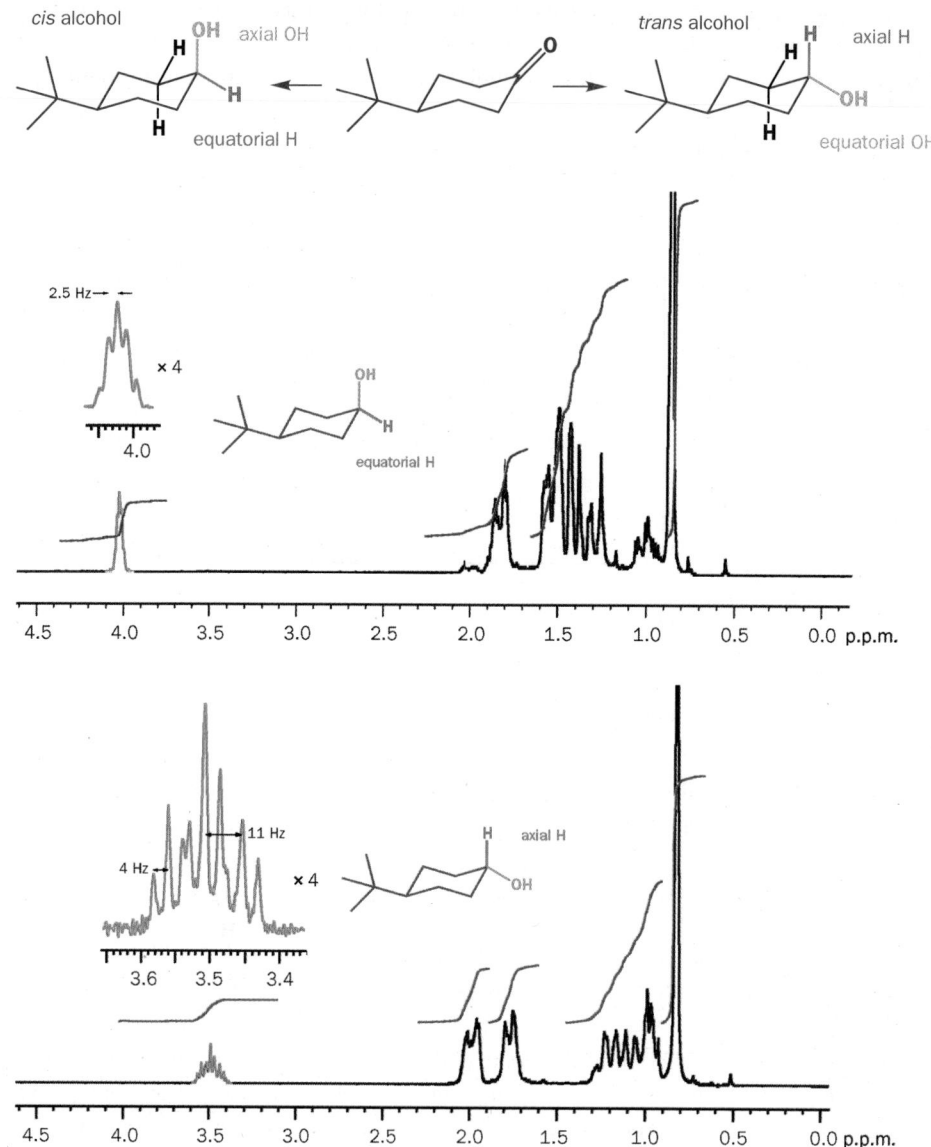

►

You can draw a general conclusion from this observation: an NMR signal is roughly as wide as the sum of all its couplings. In any given compound, an axial proton will have a much *wider* signal than an equatorial proton.

The NMR spectrum of the green H is quite different in the two cases. Each has two identical axial neighbours and two identical equatorial neighbours (two are shown in black—there are two more at the front). Each green H appears as a triplet of triplets. In the *cis* alcohol both couplings are small (2.72 and 3.00 Hz) but in the *trans* alcohol the axial/axial coupling is much larger (11.1 Hz) than the axial/equatorial (4.3 Hz) coupling.

Hydrogenation of the double bond in this unsaturated acetal gives the saturated compound as a single isomer. But which one? Are the two substituents, Me and OEt, *cis* or *trans*?

The appearance of the two black hydrogens in the NMR spectrum reveals the answer and also shows what conformation the molecule adopts. There is a 1H signal at 3.95 p.p.m. (which is therefore next to oxygen) and it is a double quartet. It must be the hydrogen next to the methyl group because of the quartet coupling. The quartet coupling constant has the 'normal' *J* value of 6.5 Hz. The doublet coupling is 9 Hz and this is too large to be anything other than an axial/axial coupling. This hydrogen is axial.

There is another 1H signal at 4.40 p.p.m. (next to *two* oxygens) which is a double doublet with *J* = 9 and 2 Hz. This must also be an axial proton as it shows an axial/axial (9 Hz) and an axial/equatorial coupling. We now know the conformation of the molecule.

Both black hydrogens are axial so both substituents are equatorial. That also means in this case that they are *cis*. But note that this is because they are both on the same, upper side of the ring, not because they are both equatorial! The hydrogen at the front has two neighbours—an axial (brown) H, *J* = 9, and an equatorial (green) H, *J* = 2 Hz. All this fits the Karplus relationship as expected. You may have spotted that the H at the back appears to be missing a small coupling to its equatorial neighbour. No doubt it does couple, but that small coupling is not noticed in the eight lines of the double quartet. Small couplings can easily be overlooked.

When this compound is allowed to stand in slightly acidic ethanol it turns into an isomer. This is the *trans* compound and its NMR spectrum is again very helpful. The proton next to the methyl group is more or less the same but the proton in between the two oxygen atoms is quite different. It is at 5.29 p.p.m. and is an unresolved signal of width about 5 Hz. In other words it has no large couplings and must be an equatorial proton. The conformation of the *trans* compound is shown in the margin.

Now for a surprising product, whose structure and stereochemistry can be determined by NMR. Normally, reaction of a symmetrical ketone such as acetone with an aromatic aldehyde and base gives a double aldol condensation product in good yield.

But in one particular case, the reaction between pentan-2-one and 4-chlorobenzaldehyde, a different product is formed. The mass spectrum shows that two aldehydes have reacted with one ketone as usual, but that only one molecule of water has been lost. Some of what we know about this compound is shown in the scheme.

The ^{13}C NMR spectrum shows that there is one ketone carbonyl group, as expected, but no alkene carbons. There is only one set of ^{13}C signals for the 4-Cl-phenyl ring and only two other carbons. This must mean that the molecule is symmetrical.

The three molecules must be joined up somewhere in the region marked. But how can we lose only one molecule of water and keep the symmetry?

The proton NMR spectrum gives the answer. Both methyl groups are still there, and they are identical, so we have two identical MeCH fragments. These CH protons (black) are *double* quartets so they have another neigh-

δ_H 3.95, 1H, dq, δ_H 4.40, 1H, dd,
J 9 and 6.5 Hz   *J* 9 and 2 Hz

this H has only small couplings

■ You met this reaction in Chapter 28.

A = C$_{19}$H$_{18}$Cl$_2$O$_2$

m.p. 189–190 °C

IR: 1705, 1600 cm^{-1}

δ_H 7.73 (8H, s)
4.49 (2H, d, *J* 10.4)
2.98 (2H, dq, *J* 10.4, 6.6)
1.08 (6H, d, *J* 6.6)

A must be symmetrical

joined up in this region

δ 2.98, dq
δ 4.49, d

bour, the only remaining aliphatic proton (actually again two identical protons, in green) at δ_H 4.49 p.p.m. These protons must be next to both oxygen and the aromatic ring to have such a large shift. But there is only one spare oxygen atom so the protons at 4.49 p.p.m. must be next to the same oxygen atom—the structure is shown on the previous page.

All that remains is the stereochemistry. There are four stereogenic centres but because of the symmetry only two structures are possible. Both methyl groups must be on the same side and both aryl rings must be on the same side.

The coupling constant between the hydrogen atoms is 10.4 Hz and so they must both be axial.

Me and Ar *trans* Me and Ar *cis*

This means that the molecule has this structure and it is the *trans* compound: all the substituents are equatorial so it is the most stable structure possible.

Only fully saturated six-membered rings are really chairs or boats. Even with one double bond in the ring, the ring is partly flattened: here we will look at an even flatter example. A unique antibiotic has been discovered in China and called 'chuangxinmycin' (meaning 'a new kind of mycin' where mycin = antibiotic). It is unique because it is a sulfur-containing indole: few natural products and no other antibiotics have this sort of structure.

The structure itself was easy to elucidate, but the stereochemistry of the two black hydrogens was not so obvious. The coupling constant (^3J) was 3.5 Hz. During attempts to synthesize the compound, Kozikowski hydrogenated the alkene ester below to give an undoubted *cis* product.

chuangxinmycin

chuangxinmycin minor by-product: *trans* stereoisomer

The ^3J coupling between the black hydrogens in this compound was 4.1 Hz, much the same as in the antibiotic and, when the ester group was hydrolysed in aqueous base, the main product was identical to natural chuangxinmycin. However, there was a minor product, which was the *trans* isomer. It had ^3J = 6.0 Hz. Note how much smaller this value is than the axial/axial couplings of 10 Hz or more in saturated six-membered rings. The flattening of the ring reduces the dihedral angle, reducing the size of J.

Stereochemistry of fused rings

Where rings are fused together (that is, have a common bond) determination of conformation may allow the determination of ring junction stereochemistry as well. Both isomers of this bicyclic ether were formed as a mixture and then separated.

We looked at the conformations of cyclohexenes and cyclohexene oxides in Chapter 18, and we will look again at the stereochemistry of reactions of six-membered rings containing double bonds in Chapter 33.

► Hydrogenation is *cis*-selective: see Chapter 24.

One proton at the ring junctions appears clearly in the NMR spectrum as it is next to two oxygen atoms (shown in black on the conformational diagrams alongside). In one compound it is a doublet, $J = 7.1$ Hz, and in the other a doublet $J = 1.3$ Hz. Which is which?

which is which?

black H is either d, $J = 7.1$ Hz, or d, $J = 1.3$ Hz

trans ring fusion

cis ring fusion

The coupling is to the green proton in each case and the dihedral angles are 180° for the *trans* compound but only 60° for the *cis* one, so the smaller coupling belongs to the *cis* compound. We shall discuss below why the *absolute* values are so low: this example illustrates how much easier stereochemical determination is if you have both stereoisomers to compare.

In the next example, unlike the last one, it eventually proved possible to make both compounds in high yield. But first the story: reaction of an amino-ketone with benzaldehyde in base gave a mixture of diastereoisomers of the product.

In unravelling the mechanism of the reaction, chemists protected the nitrogen atom with Boc (Chapter 25) before the reaction with benzaldehyde and found that a new product was formed that was clearly an *E*-alkene as its NMR spectrum contained δ_H 6.73 (1H, d, J 16). This is too large a coupling constant even for axial/axial protons and can be only *trans* coupling across a double bond. They quickly deduced that a simple aldol reaction had happened.

Couplings around double bonds were discussed in Chapter 11, p. 269.

When the Boc protecting group was removed, the cyclization reaction occurred under very mild conditions but now a *single* diastereoisomer of the product was formed.

This isomer had one proton that could be clearly seen at δ_H 4.27 p.p.m.—well away from all the rest. This is the proton marked in black between nitrogen and the phenyl group. It was a double doublet with $J = 6$ and 4 Hz. Neither of these is large enough to be an axial/axial coupling but 6 Hz is within the range for axial/equatorial and 4 Hz for equatorial/equatorial coupling. The compound must have the conformation shown in the margin.

Treatment of this product with stronger base (NaOH) isomerized it to a compound in which the same proton, now at δ_H 3.27 p.p.m., was again a double doublet but with $J = 10$ and 5 Hz. It is now an axial proton so the new conformation is this.

NaOH

Notice that we have confidently assigned the configuration of these compounds without ever being able to 'see' the yellow proton at the ring junction. Since nitrogen can invert rapidly, we know that this decalin-like structure will adopt the more stable *trans* arrangement at the ring junction.

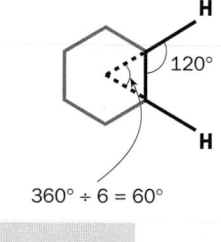

360° ÷ 6 = 60°

■
These would be the angles
if the structures were
regular, planar polyhedrals

360° ÷ 4 = 90°

The dihedral angle is not the only angle worth measuring

We should also consider how the two C–H bonds are spread out in space. The dihedral angle is what we see when we look down the spine of the book in our earlier analogy (p. 825)—now we want to look at the pages in the normal way, at right angles to the spine, as if we were going to read the book. We can show what we mean by fixing the dihedral angle at 0° (the C–H bonds are in the same plane) and looking at the variation of J with the ring size of cyclic alkenes.

120°	90°	72°	60°	51°
$J = 0.5–1.5$	$J = 2.5–4.0$	$J = 5.0–7.0$	$J = 8.5–11.0$	$J = 9.0–12.5$

The wider apart the hydrogens are spread, the smaller the coupling constant. Remember, the dihedral angle stays the same (0°)—we are just varying the angle in the plane. A dramatic illustration of this comes with the product of dehydrogenation of the natural product guaiol with elemental sulfur. From the brown, smelly reaction mixture, guaiazulene, a deep blue oil, can be distilled.

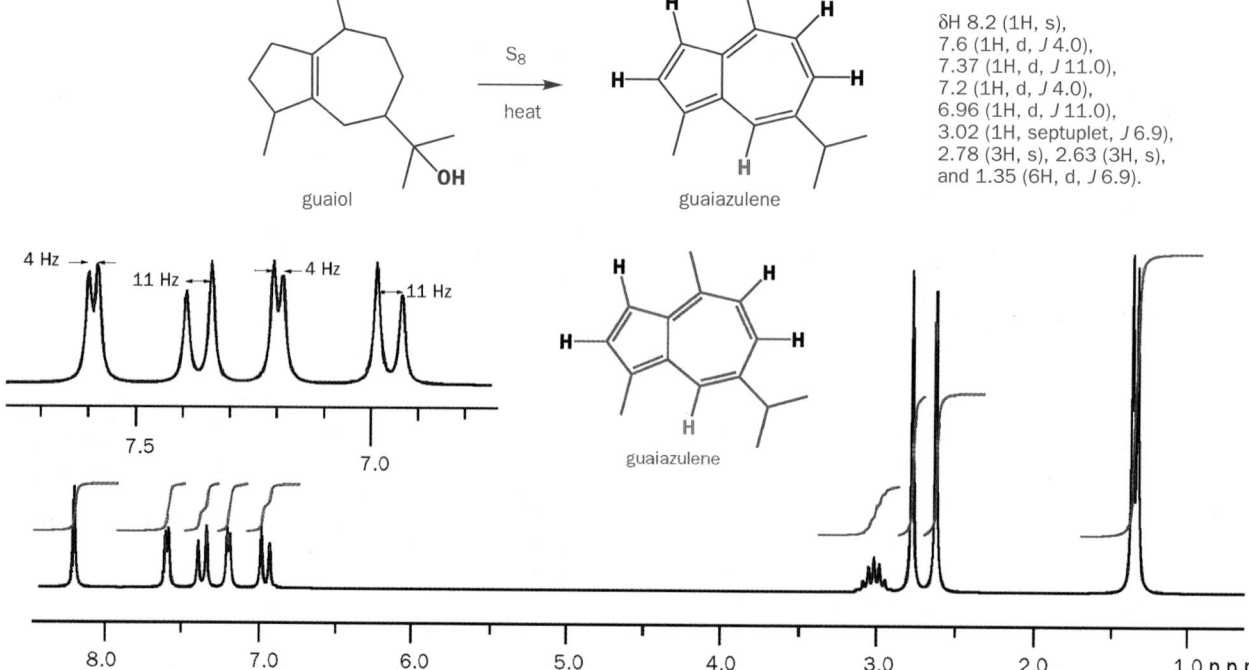

δH 8.2 (1H, s),
7.6 (1H, d, J 4.0),
7.37 (1H, d, J 11.0),
7.2 (1H, d, J 4.0),
6.96 (1H, d, J 11.0),
3.02 (1H, septuplet, J 6.9),
2.78 (3H, s), 2.63 (3H, s),
and 1.35 (6H, d, J 6.9).

guaiol   guaiazulene

guaiazulene

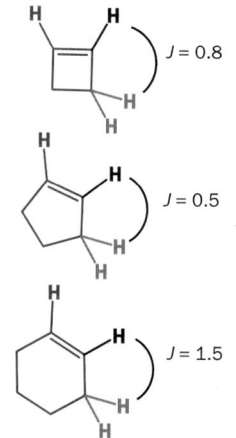

$J = 0.8$

$J = 0.5$

$J = 1.5$

Some assignments are clear. The 6H doublet and the 1H septuplet are the isopropyl group and the two 3H singlets belong to the two methyl groups—we can't really say which belongs to which. The 1H singlet must be the green hydrogen as it has no neighbours and that leaves us with two coupled pairs of protons. One pair has $J = 4$ Hz and the other $J = 11$ Hz. We expect to find larger coupling where the H–C–C–H angle is smaller, so we can say that the 4 Hz coupling is between the pair on the five-membered ring and the 11 Hz coupling is between the pair on the seven-membered ring.

When protons on a double bond in a ring have neighbours on saturated carbon, the coupling constants are all small and for the same reason—the angles in the plane of the ring are approaching 90° even though the dihedral angles are 45–60° in these examples. A bizarre result of this is that the ^3J coupling between the red and black hydrogens is often about the same as the allylic (^4J) coupling between the red and the green hydrogens. An example follows in a moment.

Vicinal (³J) coupling constants in other ring sizes

The 'spreading out' effect also affects vicinal (³J) couplings in simple saturated rings. No other ring size has so well defined a conformation as that of the six-membered ring. We can still note useful trends as we move from 6 to 5 to 4 to 3. Briefly, in five-membered rings, *cis* and *trans* couplings are about the same. In four- and three-membered rings, *cis* couplings are larger than *trans*. But in all cases the absolute values of *J* go down as the ring gets smaller and the C–H bonds are 'spread out' more. Indeed, you can say that *all* coupling constants are smaller in small rings, as we shall see. We need to examine these cases a bit more.

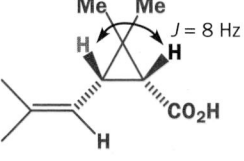

cis chrysanthemic acid

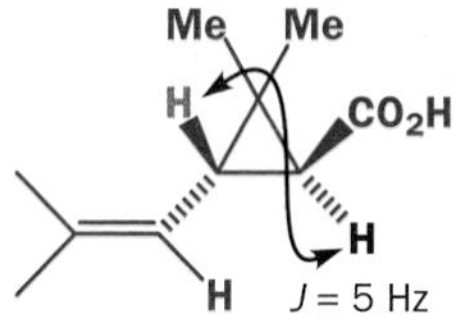

trans chrysanthemic acid

Three-membered rings

Three-membered rings are flat with all bonds eclipsed so the dihedral angle is 0° for *cis* Hs and 109° for *trans* Hs. Looking at the Karplus curve, we expect the *cis* coupling to be larger, and it is. A good example is chrysanthemic acid, which is part of the pyrethrin group of insecticides found in the pyrethrum plant. Both *cis* and *trans* chrysanthemic acids are important.

In both isomers the coupling between the green proton on the ring and its red neighbour on the double bond is 8 Hz. In the *cis* compound, the green proton is a triplet so the *cis* coupling in the ring is also 8 Hz. In the *trans* compound it is a double doublet with the second coupling, *trans* across the ring to the black H, of 5 Hz.

The most important three-membered rings are the epoxides. You saw in Chapter 11 (p. 269) that electronegative atoms reduce coupling constants by withdrawing electron density from the bonds that transmit the coupling 'information'. This means that epoxide couplings are very small—much smaller than those of their closely related alkenes, for example. Compare the four coupling constants in the diagram: for the epoxide, all couplings are small, but *cis* coupling is larger than *trans* coupling. In alkenes, *trans* coupling is larger (Chapter 11, p. 269). The table summarizes the coupling constants for alkenes, epoxides, and cyclopropanes.

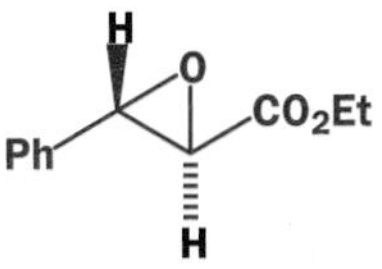

J 2.0 Hz

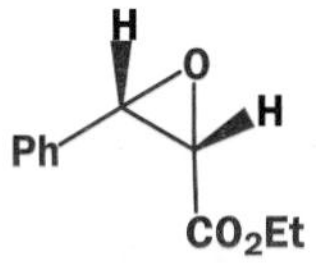

J 5.0 Hz

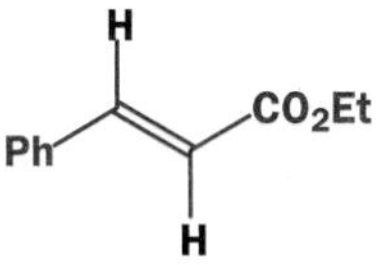

J 16 Hz

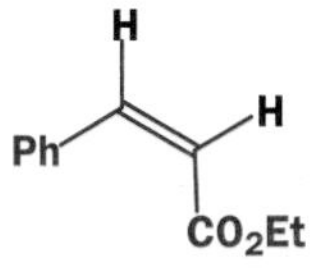

J 11 Hz

Coupling constants J, Hz

Stereochemistry	Alkene	Cyclopropane	Epoxide
cis	10–12	8	5
trans	14–18	5	2

▶
The epoxides have much smaller coupling constants because: (1) the C–C bond is longer; (2) there is an electronegative element; and (3) the 'spreading out' effect of the small ring comes into play.

Cerulenin

The natural product cerulenin is an antibiotic containing a *cis* epoxide. The coupling constant between the black hydrogens is 5.5 Hz.

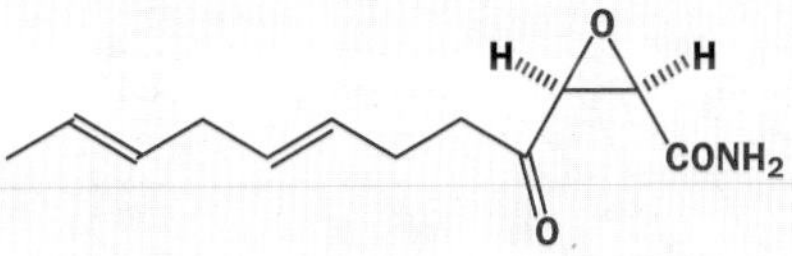

cerulenin

The compound has been made from an unsaturated lactone by epoxidation and ring opening. Follow what happens to the coupling constant between the black hydrogens as this sequence develops.

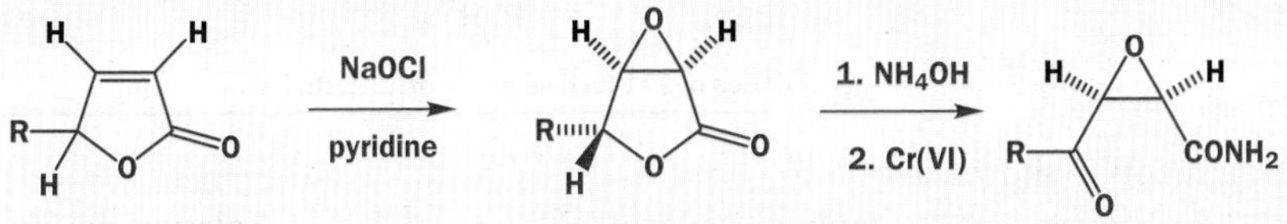

$^3J = 6$ Hz $\qquad$ NaOCl / pyridine $\qquad$ $^3J = 2.5$ Hz $\qquad$ 1. NH₄OH / 2. Cr(VI) $\qquad$ $^3J = 5.5$ Hz

The *cis* coupling in the alkene is small because it is in a five-membered ring. It gets smaller in the bicyclic epoxide because the black Hs are now in both five- and three- membered rings and both are next to oxygen, but it gets larger in cerulenin itself because the five-membered ring has been opened.

thienamycin

Four-membered rings

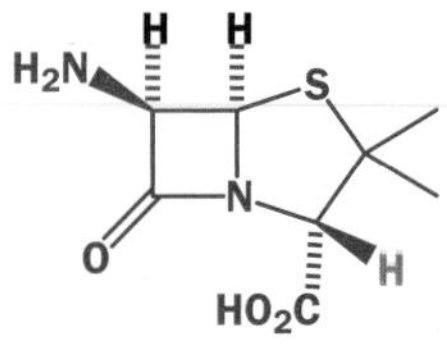

A similar situation exists with four-membered rings—the *cis* coupling is larger than the *trans* but they are generally both smaller than those in larger rings. A good example is the amino acid in the margin, the skeleton of the penicillins. The NMR spectrum contains three 1H signals in the middle regions. There is a singlet at δ_H 4.15 p.p.m. that clearly belongs to the isolated green proton and two doublets at δ_H 4.55 and 5.40 p.p.m. that must belong to the black protons. The coupling constant between them is 5 Hz and they are *cis*-related.

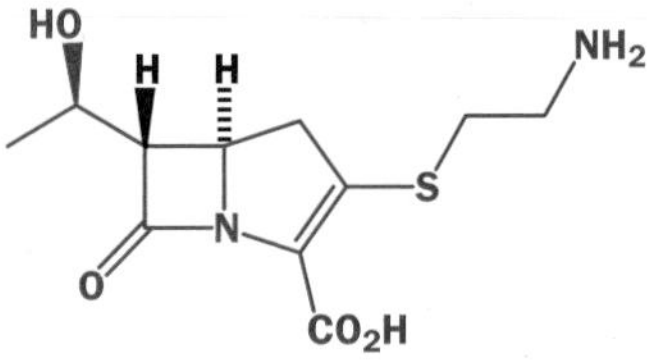

There are now large numbers of β-lactam antibiotics known and one family has the opposite (*trans*) stereochemistry around the four-membered ring. The typical member is thienamycin. We will analyse the spectrum in a moment, but first look at the differences—apart from stereochemistry—between this structure and the last. The sulfur atom is now outside the five-membered ring, the acid group is on a double bond in the same ring, and the amino group has gone from the β-lactam to be replaced by a hydroxyalkyl side chain.

Turning to the spectrum and the key question of stereochemistry, this is what the Merck discoverers said in their original article: '^{1}H NMR spectra of thienamycin (and derivatives)...show small vicinal coupling constants $J \leq 3$ Hz for the two β-lactam hydrogens. Past experience with penicillins...shows the *cis* relationship of the β-lactam hydrogens to be always associated with the larger coupling.' As we have just seen penicillins have $J \sim 5$ Hz for these hydrogens.

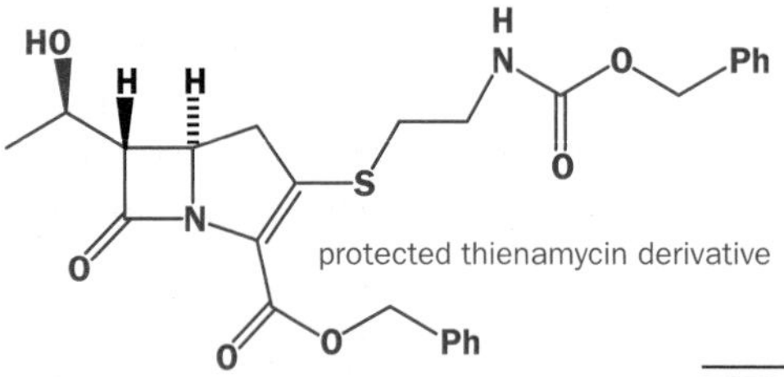

protected thienamycin derivative

The NMR spectrum of a thienamycin derivative with protecting groups on the amine and carboxylic acids is shown below. Try your hand at interpreting it before you read the explanation below. Your aim is to find the coupling constant across the four-membered ring.

The simple answer is 2.5 Hz. The signals at 3.15 and 4.19 p.p.m. are the protons on the β-lactam ring and the 9 Hz extra coupling is to the CH_2 in the five-membered ring. If you went into this spectrum in detail you may have been worried about the 12.5 and especially the 18 Hz couplings. These are ^2J (geminal) couplings and we will discuss them in the next section.

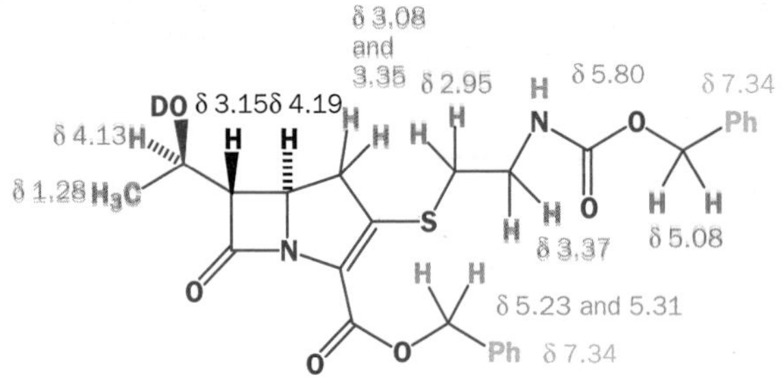

NMR spectrum of thienamycin derivative in CD$_3$OD

Shift (δ_H), p.p.m.	Integration	Multiplicity	Coupling constants (J), Hz
1.28	3H	d	6.5
2.95	2H	m	not resolved
3.08	1H	dd	9, 18
3.15	1H	dd	2.5, 7
3.35	1H	dd	9, 18
3.37	2H	m	not resolved
4.13	1H	dq	7, 6.5
4.19	1H	dt	2.5, 9
5.08	2H	s	—
5.23 and 5.31	2H	AB system[a]	12.5
5.80	1H	broad	—
7.34	10 H	m	not resolved

[a] See p. 271 for discussion of AB systems.

assignment of protons

The full assignment is shown above.

We should emphasize that a coupling constant of 5 or 2.5 Hz in isolation would not allow us to assign stereochemistry across the four-membered ring but, when we have both, we can say with confidence that the larger coupling is between *cis* Hs and the smaller coupling between *trans* Hs.

Five-membered rings

You can visualize the conformation of a five-membered ring simply as a chair cyclohexane with one of the atoms deleted. But this picture is simplistic because the five-membered ring flexes (rather than flips) and any of the carbon atoms can be the one out of the plane. All the hydrogen atoms are changing positions rapidly and the NMR spectrum 'sees' a time-averaged result. Commonly, both *cis* and *trans* couplings are about 8–9 Hz in this ring size.

The best illustration of the similarity of *cis* and *trans* couplings in five-membered rings is a structure that was incorrectly deduced for that very reason. Canadensolide is an antifungal compound found in a *Penicillium* mould. The gross structure was quite easy to deduce from the mass spectrum, which gave the formula $C_{11}H_{14}O_4$ by exact mass determination; the infrared, which showed (at 1780 and 1667 cm^{-1}) a conjugated 5-ring lactone; and some aspects of the proton NMR. The proposed structure is shown alongside.

The stereochemistry of the ring junction Hs (shown in black and green) is not in question. They are certain to be *cis* as it is virtually impossible for two five-membered rings to be fused *trans*. The stereochemistry in question involves the third stereogenic centre on the left-hand ring. The coupling constant between the black and green Hs is 6.8 Hz, while that between the green and brown Hs is 4.5. Is this different enough for them to be *trans*? The original investigators decided that it was.

The mistake emerged when some Japanese chemists made this compound by an unambiguous route. The NMR spectrum was quite like that of canadensolide, but not the same. In particular, the coupling between the green and brown Hs was 1.5 Hz—quite different! So they also made the other possible diastereoisomer and found that it was identical to natural canadensolide. The details are in the margin.

An example of vicinal coupling in structural analysis: aflatoxins

We can bring together a lot of these points in the structure of one compound, the dreaded aflatoxin. Aflatoxin B_1 is an example.

The four red protons on saturated carbons in the five-membered ring in the margin appear as two triplets: δ_H 2.61 (2H, t, *J* 5 Hz) and δ_H 3.42 (2H, t, *J* 5 Hz). The *cis* and *trans* couplings are the same. The yellow proton on the left, on the junction between the two five-membered cyclic ethers, is a doublet δ_H 6.89 (1H, d, *J* 7 Hz). This is, of course, the *cis* coupling to the black hydrogen. The black hydrogen has this coupling too, but it appears as a doublet of triplets with a triplet coupling of 2.5 Hz: δ_H 4.81 (1H, dt, *J* 7, 2.5, 2.5 Hz). These small couplings can only be to the two green hydrogens: the ^3J and ^4J couplings are indeed the same.

Finally there is another strange coincidence—each green hydrogen appears as a triplet with 2.5 Hz couplings. Evidently, the *cis* coupling across the double bond is also 2.5 Hz. We expect *cis* coupling in a cyclopentene to be small (it was 4 Hz in the azulene on p. 830), but not that small—it must be the electronegative oxygen atom that is reducing the value still further.

proposed structure for canadensolide

■ You can convince yourself of this by making a model.

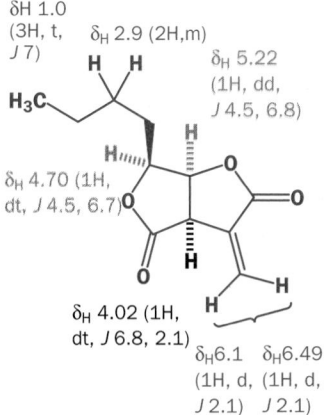

δ_H 1.0 (3H, t, *J* 7)
δ_H 2.9 (2H, m)
δ_H 5.22 (1H, dd, *J* 4.5, 6.8)
δ_H 4.70 (1H, dt, *J* 4.5, 6.7)
δ_H 4.02 (1H, dt, *J* 6.8, 2.1)
δ_H 6.1 (1H, d, *J* 2.1)
δ_H 6.49 (1H, d, *J* 2.1)

▶ **Aflatoxins**

Aflatoxins were mentioned in Chapter 20: they occur in moulds, including those that grow on some foods, and cause liver cancer. These slow-acting poisons are among the most toxic compounds known.

aflatoxin B_1

Coupling in furans

The size of coupling constants in five-membered rings containing oxygen is illustrated clearly in furfuraldehyde (furan-2-aldehyde): note how small the couplings are.

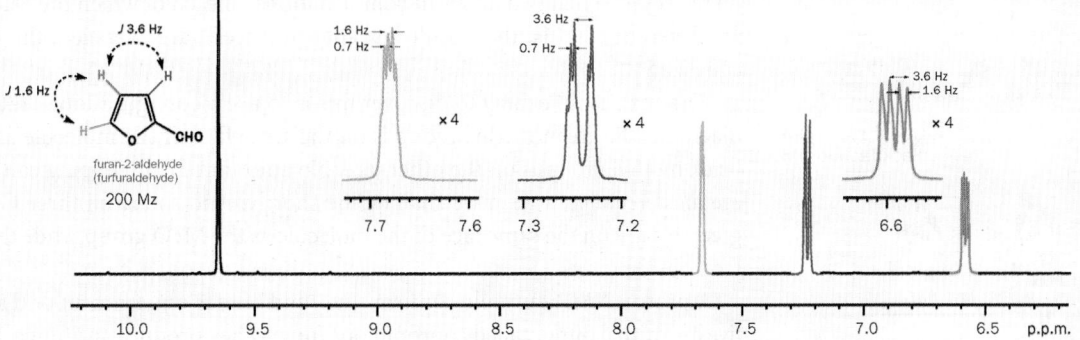

J 3.6 Hz
J 1.6 Hz
furan-2-aldehyde (furfuraldehyde)
200 Mz

1.6 Hz
0.7 Hz
×4

3.6 Hz
0.7 Hz
×4

3.6 Hz
1.6 Hz
×4

Geminal (^2J) coupling

For coupling to be seen, the two hydrogen atoms in question must have different chemical shifts. For ^2J couplings the two hydrogen atoms are on the same carbon atom, so in order to discuss geminal coupling we must first consider what leads the two hydrogens of a CH_2 group to have different shifts.

To introduce the topic, an example. It may seem to you that any six-membered ring might show different chemical shifts for axial and equatorial groups. But this doesn't happen. Consider the result of this Robinson annelation reaction.

The two methyl groups at C4 give rise to a single signal in the ^{13}C NMR at 27.46 p.p.m. Even though one of them is (pseudo)axial and one (pseudo)equatorial, the molecule exists in solution as a rapidly equilibrating mixture of two conformations. The axial green methyl in the left-hand conformer becomes equatorial in the right-hand conformer, and vice versa for the black methyl group. This exchange is rapid on the NMR time-scale and the equilibrium position is 50:50. Time averaging equalizes the chemical shifts of the two methyl groups, and the same is true for the CH_2 groups around the back of the ring.

However, the enone is not the only product of this reaction. A methanol adduct is also formed by Michael addition of methanol to the conjugated enone.

This product has two methyl signals at 26.1 and 34.7 p.p.m. If we examine the molecule by conformational analysis as we did for the first product we see a similar situation.

Similar but not the same. This time, the two conformations are not identical. One has the OMe group equatorial and the other has it axial. Even the two methyl groups do not entirely change places in the two conformations. True, the green methyl is axial on the left and equatorial on the right, but it has a gauche (dihedral angle 60°) relationship with the OMe group in *both* conformations. The black Me group is gauche to OMe on the left but anti-periplanar to the OMe group on the right. When two different conformations, in each of which the black and green methyl groups are different (that is, they don't just change places), are averaged, the two methyl groups are not equalized.

Perhaps a simpler way to discover this is to use a configurational, rather than a conformational, diagram. The green methyl group is on the same face of the molecule as the MeO group, while the black methyl group is on the other face. No amount of ring flipping can make them the same. They are *diastereotopic*, a term we shall define shortly. And so are all three CH_2 groups in the ring. The green Hs are on the same face of the molecule as the MeO group while the black Hs are on the other face.

A proton NMR example confirms this, and here is one from an odd source. There are fungi that live on animal dung, called coprophilous fungi. They produce antifungal compounds, presumably to

green: *syn* to OMe
black: *anti* to OMe

fight off competition! Anyway, in 1995 two new antifungal compounds were discovered in a fungus living on lemming dung. They were named coniochaetones A and B and their structures were deduced with the usual array of mass and NMR spectra. The proton spectra, run on a 600 MHz machine, are shown below, and they reveal considerable detail.

Coniochaetone A		Coniochaetone B	
δ_H, p.p.m.	Coupling	δ_H, p.p.m.	Coupling
2.41 (3H)	s	2.38 (3H)	s
		5.43 (1H)	ddd, J 1.4, 3.3, 7.6 Hz
2.70 (2H)	m	2.49 (1H)	m
		2.03 (1H)	m
3.07 (2H)	m	3.10 (1H)	dddd, J 1.4, 5.1, 9.4, 18 Hz
		2.81 (1H)	ddd, J 5.1, 9.3, 18 Hz
6.77 (1H)	broad s	6.70 (1H)	broad s
6.69 (1H)	broad s	6.62 (1H)	broad s
12.21 (1H)[a]	s	12.25 (1H)[a]	s

[a] Exchanges with D$_2$O.

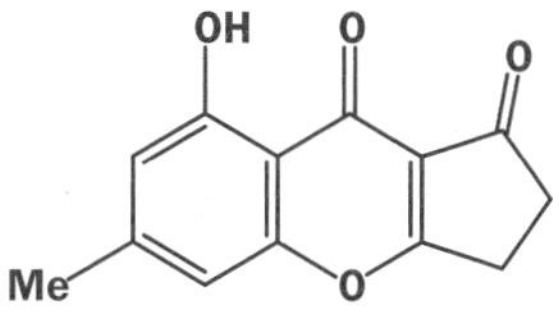

coniochaetone A

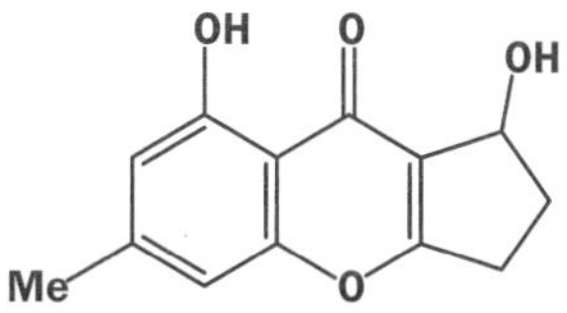

coniochaetone B

Some of the spectrum is essentially the same for the two compounds, but other parts are quite different. Coniochaetone A has a very simple spectrum, very easily assigned.

Coniochaetone B is rather more interesting. The spectrum is much more complicated, even though it has only one more C–H than coniochaetone A. The reason is that addition of that H atom creates a stereogenic centre and makes the top and bottom faces of the molecule different. Both CH$_2$ groups become diastereotopic.

The green Hs are coupled to each other (J = 18 Hz) and to each of the black Hs with a different coupling constant. One of the green hydrogens also shows a long-range (^4J = 1.4 Hz) W-coupling to the red H. The black Hs are too complex to analyse, even at 600 MHz, but the different couplings to the red hydrogen are shown by the signal at 5.43 p.p.m.

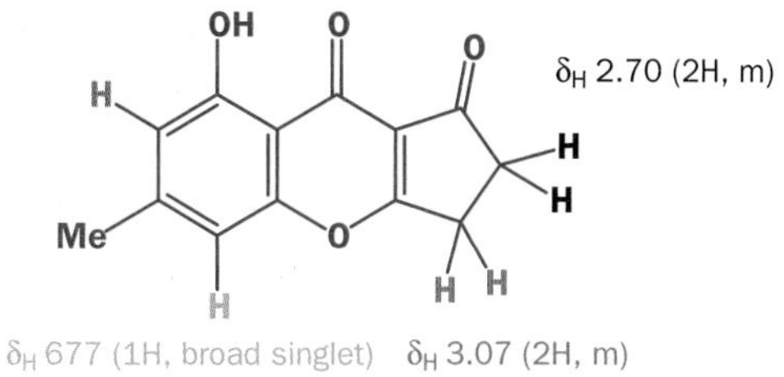

δ_H 2.70 (2H, m)
δ_H 677 (1H, broad singlet) δ_H 3.07 (2H, m)
coniochaetone A

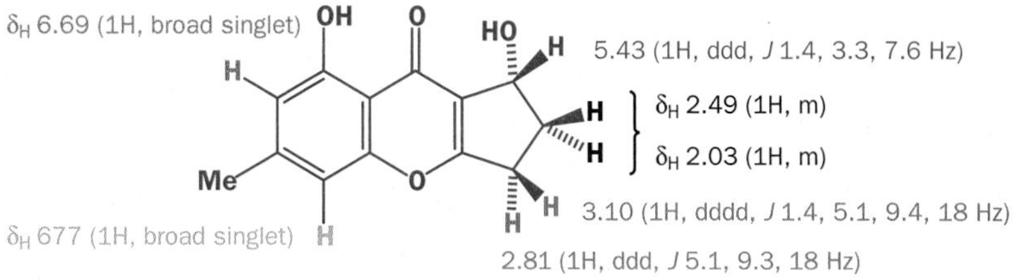

δ_H 6.69 (1H, broad singlet)
5.43 (1H, ddd, J 1.4, 3.3, 7.6 Hz)
δ_H 2.49 (1H, m)
δ_H 2.03 (1H, m)
3.10 (1H, dddd, J 1.4, 5.1, 9.4, 18 Hz)
2.81 (1H, ddd, J 5.1, 9.3, 18 Hz)
δ_H 677 (1H, broad singlet)
coniochaetone B

Diastereotopic CH$_2$ groups

The green protons in the last example couple to one another, so they must be different. Until this chapter, you may have thought it self-evident that two protons attached to the same carbon would be identical, but you have now seen several examples where they are not. It is now time to explain more rigorously the appearance of CH$_2$ groups in NMR spectra, and you will see that there are *three* possibilities. To do this, we shall have to discuss some aspects of symmetry that build on what you learned in Chapter 16.

First, an example in which the two hydrogens are indeed the same. We may draw one hydrogen coming towards us and one going away, but the two Hs are the same. This is easy to demonstrate. If we colour one H black and one green, and then rotate the molecule through 180°, the black H appears in the place of the green H and vice versa. The rotated molecule hasn't changed because the *other* two substituents (OMe here) are also the same.

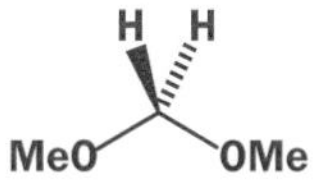

If we had given out uncoloured models of this molecule with this book, and asked each reader to paint one H green and one H black, we would have no way at all of giving instructions about which to paint what colour. But it wouldn't matter because, even without these instructions, every reader would produce an identical model, whichever way they painted their Hs.

The correct description for this pair of hydrogen atoms is **homotopic**. They are the same (*homo*) topologically and cannot be distinguished by chemical reagents, enzymes, NMR machines, or human beings. The molecule is achiral—it has no asymmetry at all.

> ● **Homotopic groups**
>
> Homotopic groups cannot be distinguished by any means whatsoever: they are chemically entirely identical.

What happens when the other two substituents are different? At first sight the situation does not seem to have changed. Surely the two hydrogens are still the same as one another?

In fact, they aren't—not quite. If we had given out uncoloured models of this molecule and just said 'paint one H green and one H black', we would not have got just one type of model.

We would have got about 50% looking like this: and 50% looking like this:

But this time, we *could* give instructions about which H we wanted which colour. To get the first of these two, we just need to say 'Take the MeO group in your *left* hand and the Ph group in your *right*, kink the carbon chain upwards. The hydrogen coming towards you is to be painted black.' All the models produced by readers would then be identical—*as long as the readers knew their left from their right*. This is a very important point: the green and black hydrogens in this molecule (unlike the first one) can be described only in phrases incorporating the words 'left' or 'right', and are distinguishable only by a system that knows its left from its right.

Human beings are such a system: so are enzymes, and the asymmetric reagents you will meet in Chapter 45. But NMR machines are not. NMR machines cannot distinguish right and left—the NMR spectra of two enantiomers are identical, for example. It is not a matter of enantiomers in the molecule in question—it has a plane of symmetry and is achiral. Nonetheless, the relationship between these two hydrogens is rather like the relationship between enantiomers (the two possible ways of colouring the Hs are enantiomers—mirror images) and so they are called **enantiotopic**. Enantiotopic protons appear identical in the NMR spectrum.

> ● **Enantiotopic groups**
>
> Enantiotopic groups can be distinguished by systems that can tell right from left, but are still magnetically equivalent and appear identical in the NMR spectrum.

The third situation usually arises when the molecule has a stereogenic centre. As an example we can take the Michael product from the beginning of this section.

It is now very easy to distinguish the two hydrogens on each ring carbon atom and, if we want to give instructions on how to paint a model of this molecule, we can just say 'Make all the Hs on the same side of the ring as OMe green, and the ones on the opposite side to OMe black.' We do not need to use the words 'right' or 'left' in the instructions, and it is not necessary to

▶ To understand this discussion, it is very important that you appreciate points such as this which we covered in Chapter 16. You may need to refresh your memory of the stereochemical points there before you read further.

know your right from your left to tell the two types of Hs apart. Ordinary chemical reagents and NMR machines can do it. These Hs are different in the way that diastereoisomers are different and they are **diastereotopic**. We expect them to have different chemical shifts in the proton NMR spectrum.

The same is true of the methyl groups: they too are diastereotopic and we expect them to have different shifts.

> ● **Diastereotopic groups**
>
> Diastereotopic groups are chemically different: they can be distinguished even by systems that cannot tell right from left, and they appear at different chemical shifts in the NMR spectrum.

How to tell if protons are homotopic, enantiotopic, or diastereotopic

What we have said so far explains to you why homotopic and enantiotopic groups appear identical in the NMR spectrum, but diastereotopic protons may not. Now we will give a quick guide to determining what sort of pair you are dealing with in a given molecule.

The key is to turn your molecules into two molecules. Replace one of the Hs (we'll assume we're looking at Hs, but the argument works for other groups too—Me groups, for example, as in the last example above) with an imaginary group 'G'. Write down the structure you get, with stereochemistry shown. Next, write down the structure you get by replacing the other H with the group G. Now the more difficult bit: identify the stereochemical relationship between the two molecules you have drawn.

- If they are identical molecules, the Hs are homotopic
- If they are enantiomers, the Hs are enantiotopic
- If they are diasatereoisomers, the Hs are diastereotopic

This is really just a simpler way of doing what we did with black and green above, but it is easy to do for any molecule. Take the first of our examples, and replace each H in turn by G.

These two molecules are identical, because just turning one over gives the other: the protons are homotopic. Now for the next example.

The two molecules are not identical: to make one into the other you need to reflect in the plane of the paper, so they are enantiomers, and the Hs are *enantiotopic*. There is another term we must introduce you to in relation to this molecule, which will become useful in the next chapter, and that is 'prochiral'. The molecule we started with here was not chiral—it had a plane of symmetry. But by changing just one of the Hs to a different group we have made it chiral. Molecules that are achiral but can become chiral through one simple change are called **prochiral**.

Now we will choose one of the three pairs of Hs in the cyclohexanone example. The starting molecule is, of course, now chiral, and the two molecules we get when we replace each H by G are now diastereoisomers: one has G and OMe *anti*, the other *syn*, and the pairs of hydrogens are *diastereotopic*.

Finally, one last look at symmetry in the three molecules. We will consider two planes as potential planes of symmetry—the plane that bisects the H–C–H angle of the two Hs we are interested in (this is the plane of the paper as we have drawn all three molecules), and a plane at right angles to that plane, passing through the carbon atom and both hydrogen atoms. This second plane is marked on the diagrams in yellow.

This molecule, the most symmetrical of the three, is achiral. The central carbon atom is completely nonstereogenic. Both planes are planes of symmetry and the hydrogens are homotopic. They are chemically and magnetically equivalent.

> ▶
>
> NMR machines *can* tell the difference, but it does not follow that they *will*. There are many examples of protons that are different but have the same chemical shift (toluene, PhMe, shows a singlet in the NMR for all its aromatic protons even though they are of three different kinds). Sometimes diastereotopic protons have the same chemical shift, sometimes slightly different chemical shifts, and sometimes very different chemical shifts.

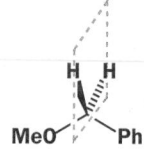

This slightly less symmetrical molecule is not chiral but *prochiral*. The carbon atom is a prochiral (or prostereogenic) centre. The plane of the paper is still a plane of symmetry, but the yellow plane containing the two H atoms is not and the hydrogen atoms are enantiotopic. They are magnetically equivalent and can be distinguished only by humans, enzymes, and other asymmetric reagents.

This least symmetrical molecule is chiral as it has a chiral (stereogenic) centre. The carbon atom we are discussing is not a stereogenic centre but is again a prochiral centre. Neither plane is a plane of symmetry and the hydrogen atoms are diastereotopic. They are chemically and magnetically different and can be distinguished by NMR or by chemical reagents.

Look back at the structures we have just been discussing and you should see that both the enone used to produce this molecule and coniochaetone A have a plane of symmetry bisecting their CH_2 groups while coniochaetone B does not. This gives another easy way of telling if a pair of groups will appear different in the NMR spectrum. If the plane passing through the carbon atom and bisecting the H–C–H bond angle (the plane of the paper in these diagrams) is a plane of symmetry, then the two Hs (which are reflected in that plane) are magnetically equivalent. (If they also lie in a plane of symmetry, they are homotopic; if they don't, they are enantiotopic.)

The shape of the NMR signal

A prochiral CH_2 group with diastereotopic Hs isolated from any other Hs will give rise to two signals, one for each H, and they will couple to each other so that the complete signal is a pair of doublets. You would expect geminal coupling constants to be larger than vicinal ones simply because the Hs are closer—we are talking about ^2J instead of ^3J couplings. A typical vicinal (^3J) coupling constant for a freely rotating open-chain system without nearby electronegative atoms would be 7 Hz. A typical geminal (^2J) coupling constant is just twice this, 14 Hz.

The chemical shift differences ($\Delta\delta$) between Hs on the same carbon atom tend to be small —usually less than 1 p.p.m.—and the coupling constants J tend to be large so the signals usually have $\Delta\delta \sim J$ and are distorted into an AB pattern. The signal may have any of the forms indicated here, depending on the relative sizes of $\Delta\delta$ (the chemical shift difference between the peaks) and J.

> The shape of NMR signals where J and the chemical shift difference are of the same order of magnitude were discussed in Chapter 11. The arguments apply to any coupled protons of similar chemical shift— there we used disubstituted aromatic rings as the example—but are particularly relevant here.

> ▶
> It is not usually easy to decide which proton gives rise to which signal. This is not important in assigning the structure, but may be important in assigning stereochemistry. We shall discuss how to assign the protons shortly in relation to the conformation of six-membered rings, and then again later using the nuclear Overhauser effect.

some different shapes of an AB system in the proton NMR spectrum from a diastereotopic CH_2 group

$\Delta\delta > J$ $\Delta\delta \sim J$ $\Delta\delta < J$ $\Delta\delta = 0$

δ_H δ_H δ_H δ_H δ_H δ_H

The coupling constant is always the difference in Hz between the two lines of the same colour in these diagrams, but the chemical shifts are not so easily measured. The chemical shift of each proton is at the weighted mean of the two lines—the more distorted the signal, the nearer the chemical shift to that of the larger inner line.

Examples of AB systems from diastereotopic CH_2 groups

It is time to look at some examples. The insect pheromone frontalin can be drawn like this.

There is nothing wrong with this drawing except that it fails to explain why the black and green hydrogens are different and give a pair of doublets at δ_H 3.42 and 2.93 p.p.m., each 1H, J 7 Hz (an AB

frontalin

system) in the proton NMR. These protons must be diastereotopic. A conformational diagram should help.

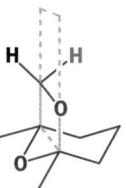

The vital H atoms are on a diaxial bridge across the six-membered ring. Under the black H is an oxygen atom, while under the green H is a three-carbon link. If there were a plane of symmetry between these two Hs, it would have to be the plane marked by the dashed yellow lines in the second diagram. This is *not* a plane of symmetry and the two Hs *are* diastereotopic. They have no neighbours, so they give a simple AB system. The coupling constant here is small for ^2J—only 7 Hz—but that should not surprise you since we have a five-membered ring and a nearby oxygen atom.

The same principles apply to open-chain compounds, such as amino acids. All of the amino acids in proteins except glycine are chiral. Glycine has a prochiral CH_2 group that gives a singlet in the NMR spectrum as the Hs are enantiotopic. Similarly, the *N*-benzyl derivative of glycine has a second prochiral CH_2 group (NCH_2Ph) that gives another singlet in the NMR spectrum as these Hs too are enantiotopic.

glycine

N-benzyl glycine

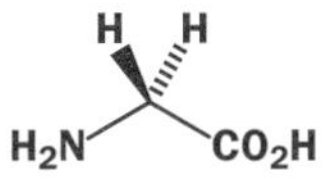

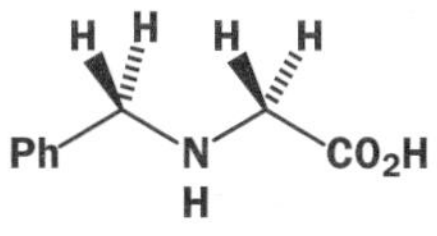

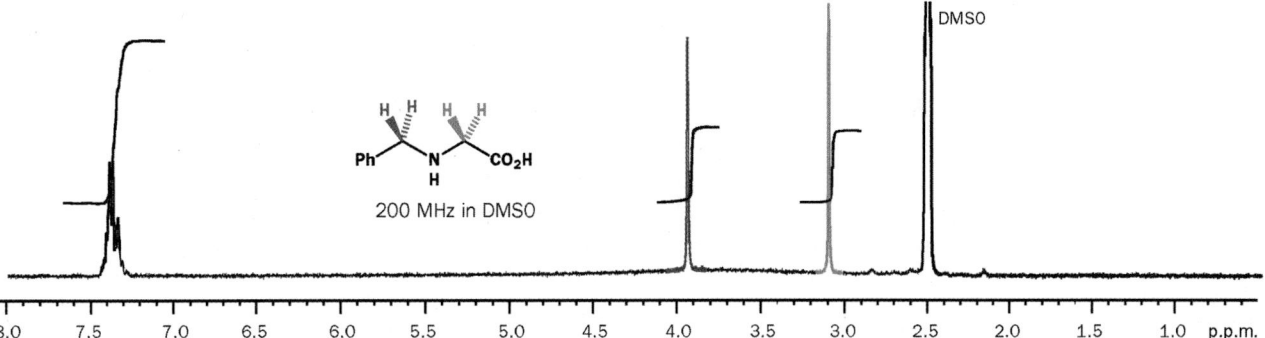

The plane of the paper is a plane of symmetry for both these CH_2 groups in the way they are drawn here. The *N*-benzyl derivatives of the other amino acids are quite different. Each shows an AB signal for the NCH_2Ph group because these molecules have stereogenic centres and there are no planes of symmetry. The Hs of the NCH_2Ph group are diastereotopic.

alanine

N-benzyl alanine

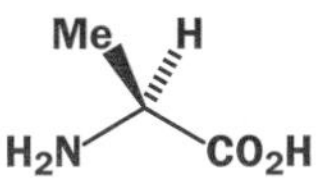

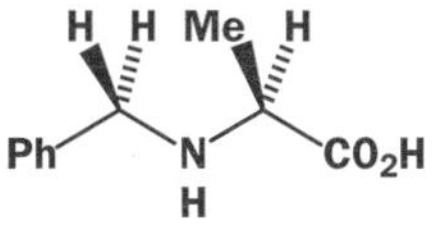

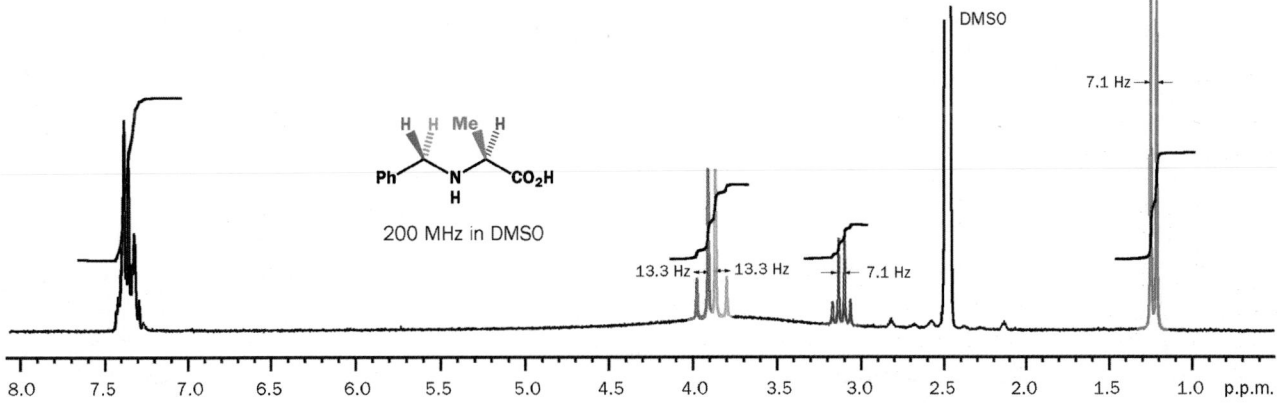

In the way in which the molecule is drawn, the brown H is on the same side as the Me group and the yellow H on the other. It does not matter that there is free rotation in this molecule—there is no conformation you can draw in which the important plane, passing between the diastereotopic Hs through their carbon atom, is a plane of symmetry.

■
See p. 271 for the use of A, B, X, etc. to describe protons.

The ABX system

It is more common to find diastereotopic CH_2 groups with neighbours, and the most common situation is that in which there is one neighbour, giving an **ABX system**. We will outline diagrammatically what we expect . Let's start with the AB system for the diastereotopic CH_2 group and the singlet for the neighbour, which we call 'X' because it's at a quite different chemical shift.

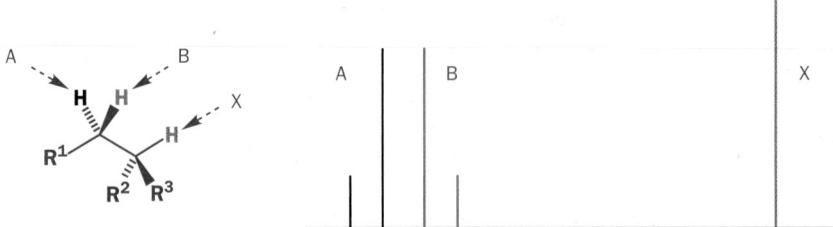

Now we must add the coupling between A and X and between B and X. Since A and B are different, there is no reason why J_{AX} and J_{BX} should be the same. One is normally larger than the other, and both are normally smaller than J_{AB}, since J_{AX} and J_{BX} are vicinal ^3J couplings while J_{AB} is a geminal ^2J coupling. We shall arbitrarily put $J_{AX} > J_{BX}$ in this example.

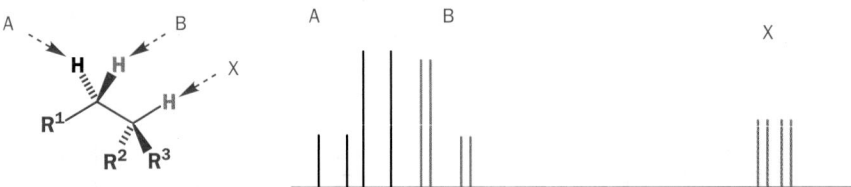

You can read J_{AX} and J_{BX} from the AB part of the signal quite easily by measuring the distance between each pair of lines, in Hz. If you want to read them from the X part, remember that it is made up like this.

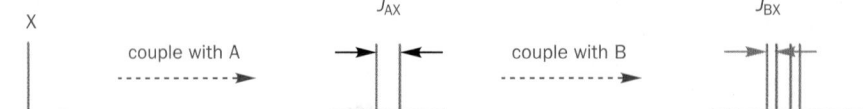

In the signal for X, the larger coupling, J_{AX}, is the spacing between lines 1 and 3 or between lines 2 and 4 while the smaller coupling, J_{BX}, is the spacing between lines 1 and 2 or 3 and 4. Naturally, J_{AX} and J_{BX} are the same whether you measure them in the AB signal or in the X signal.

When aspartic acid is dissolved in D_2O with NaOD present, all OH and NH_2 protons are exchanged for deuterium atoms and do not show up in the spectrum—the molecule exists as its dianion.

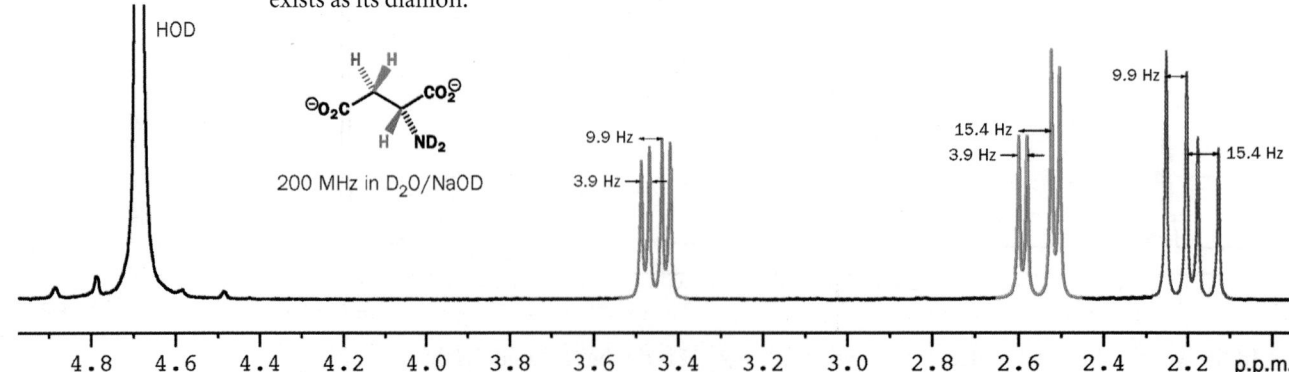

The spectrum consists of a beautiful ABX system with the brown proton as a double doublet at δ_H 3.45 p.p.m. and the black and green protons as an AB pair between 2 and 3 p.p.m. The coupling between red and green is typical: 15 Hz.

More complex examples

We have stressed all along that diastereotopic CH_2 groups may be separated in the proton NMR but need not be. It may just happen that the chemical shift difference is zero giving an A_2 system. It is not possible to predict which diastereotopic CH_2 groups will be revealed in the NMR spectrum as AB systems and which as A_2. Both may even appear in the same molecule. As an example, consider the compound shown below. The brown hydrogen has a very complicated signal, coupling to four other hydrogens. The spectrum for these four hydrogens is also complicated but may be simplified by irradiating the brown hydrogen to remove any coupling to it. Then we can clearly see that one CH_2 group shows itself as diastereotopic while the other does not. From the chemical shifts we may guess that the CH_2Cl group is the A_2X system at 3.7 p.p.m. and that it is the one in the ring that gives the ABX system.

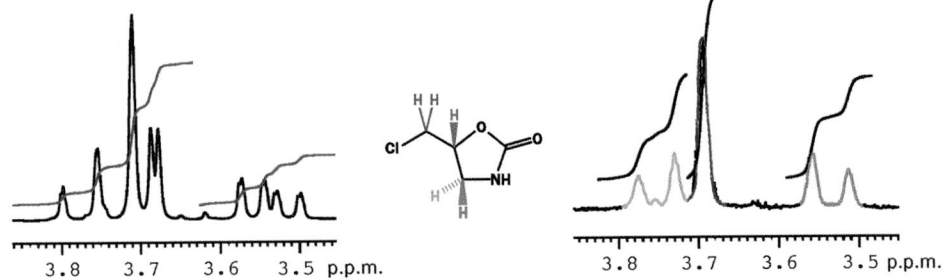

expanded region of 200 MHz NMR spectrum   same region with the coupling to the brown proton removed

As a general guide, CH_2 groups close to a stereogenic centre are more likely to be revealed as diastereotopic than those further away. Those in part of a structure with a fixed conformation are more likely to be revealed as diastereotopic than those in a flexible, freely rotating part of the molecule.

In this molecule, all three marked CH_2 groups are diastereotopic, but it is more likely that the ones next to the stereogenic centre, whether in the ring or in the open chain, will show up as AB systems in the NMR. The remote CH_2 group at the end of the chain is more likely to be A_2 in the NMR, but one cannot be sure. You must be able to recognize diastereotopic CH_2 groups and to interpret AB and ABX systems in the NMR. You must also not be surprised when a diastereotopic CH_2 group appears in the NMR spectrum as an A_2 or A_2X system.

stereogenic centre

■ For another example, look back at thienamycin on p. 832. Compare the two OCH_2Ph groups: both have a diastereotopic CH_2 pair, but one appears as a singlet and one as an AB system.

Geminal coupling in six-membered rings

While we were discussing coupling in rings earlier in the chapter we avoided the question of geminal coupling by never considering the CH_2 groups in the ring. In practice there will often be diastereotopic CH_2 groups in six-membered rings. As an example, we will look at a problem in structure determination of a rather complex molecule. It is pederin, the toxic principle of the blister beetle *Paederus fuscipes*. After some incorrect early suggestions, the actual structure of the compound was eventually deduced.

We are not going to discuss the full structure elucidation, but will concentrate on the stereochemistry of the right-hand ring. You can see that there is a CH_2 group in this ring and it has, of course, diastereotopic Hs. At first the OH group was placed at the wrong position on the ring, but a careful analysis of the NMR spectrum put this right and also gave the stereochemistry. The five (green) protons on the ring gave these signals (left-hand part of the molecule omitted for clarity).

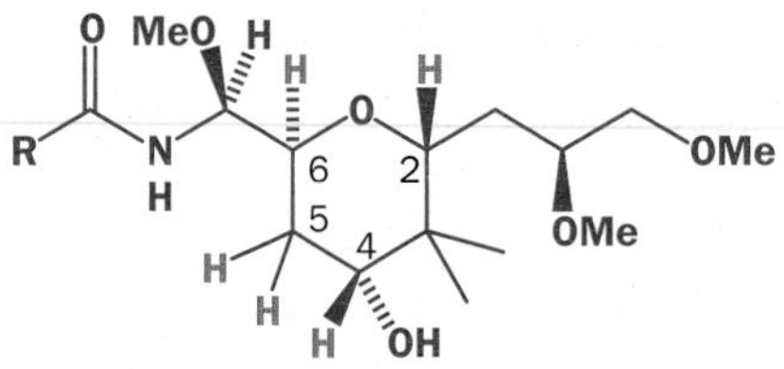

δ_H 1.85 (1H, ddd, J 5, 10, 12)

2.10 (1H, ddd, J 3, 4, 12)

3.75 (1H, dd, J 4, 10)

3.85 (1H, ddd, J 3, 5, 8)

4.00 (1H dd, J 3, 7)

Three of the protons have shifts δ_H 3–4 p.p.m. and are obviously on carbons attached to oxygen atoms. The other two, δ_H about 2 p.p.m., must be the diastereotopic pair at C5. The coupling of 12 Hz, which appears in both signals, must be the geminal coupling and the other couplings are found in the signals at δ_H 3.75 and 3.85 p.p.m. The signal at δ_H 3.75 p.p.m. has no other couplings and must be from C4 so that leaves δ_H 3.85 p.p.m. for the hydrogen atom at C6 which is also coupled to the hydrogen in the side chain. The 10 Hz coupling is axial/axial but the others are all much smaller so we can draw the conformation immediately.

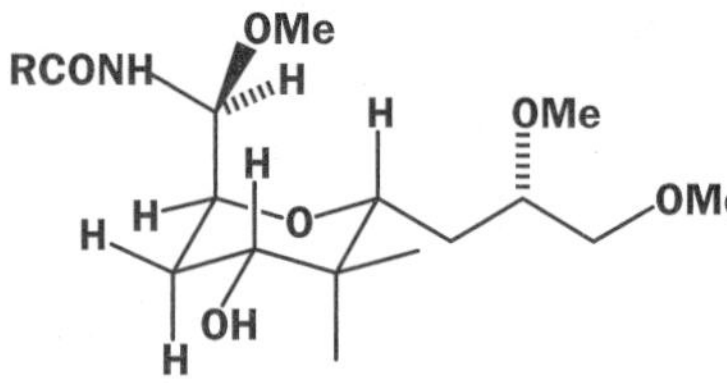

There is just the one axial/axial coupling and so the left-hand side chain must occupy an axial position. This is perhaps a bit surprising—it's large and branched—but the molecule has no choice but to place one of the two side chains axial.

A surprising reaction product

Chapter 26 revealed that sodium chloride can be a surprisingly powerful reagent. It removes ester groups from malonate derivatives, like this.

However, using this reaction to decarboxylate the malonate shown here did not merely remove the CO_2Me group. Instead, a compound was formed with a much more complicated NMR spectrum than that of the expected product (which was known as it could be made another way). The NMR data for both compounds are detailed below.

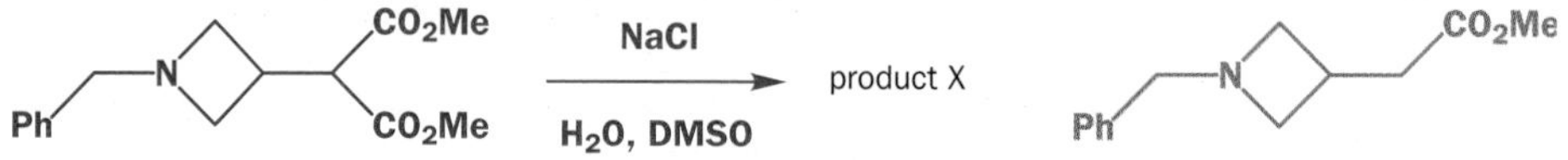

product X		product X			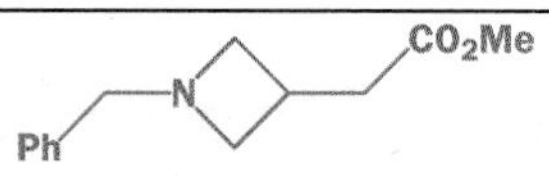
C_{14}H_{15}NO_3		C_{14}H_{15}NO_3		C_{13}H_{17}NO_2	
δ_H	7.35–7.25 (3H, m)	δ_C	169.1	δ_H	7.2–7.4 (5H, m, Ph)
	7.2 (2H, d, J 7)		169.0		3.65 (3H, s, OMe)
	4.45 (1H, d, J 14)		136.2		3.45 (2H, t, J 7)
	4.3 (1H, d, J 14)		128.6		2.95–2.85 (2H, m)
	3.8 (3H, s, OMe)		128.1		2.85–2.75 (1H, m)
	3.45 (1H, dd, J 7, 10)		127.6		2.6 (2H, t, J 7)
	3.1 (1H, d, J 10)		52.4		
	2.35–2.25 (1H, m)		46.45		
	1.9 (1H, dd, J 5, 10)		46.4		
	1.1 (1H, t, J 5)		31.5		
			22.8		
			20.7		

The unknown product has lost MeOH but retained both carbonyl groups (δ_C 169.1, 169.0 p.p.m. typical for acid derivatives). In the ^{1}H NMR, the phenyl ring and one OMe group are still there. The other striking thing about the ^{1}H NMR is the presence of so many couplings. It looks as if all the hydrogens are magnetically distinct. Indeed we can see one diastereotopic CH$_2$ at 4.45 and 4.3 p.p.m. with ^2J = 14 Hz. This is the 'normal' value and would fit well for the NCH$_2$Ph group. But note the chemical shift! For δ_H to be so large the nitrogen atom must be part of an amide, which would also explain the two acid derivative C=O groups. So we have the partial structure on the right.

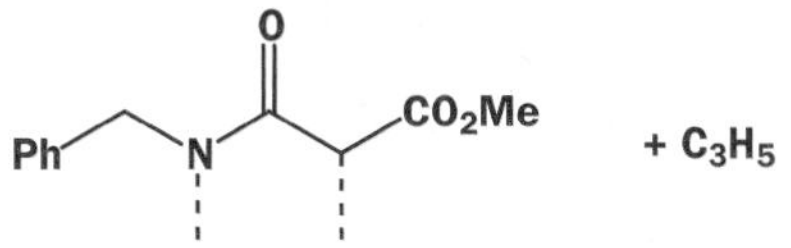

All that is left is C$_3$H$_5$ and this must be fitted in where the dotted lines go. One reasonable interpretation from the NMR would be two diastereotopic CH$_2$ groups, one with ^2J = 10 and one with ^2J = 5 Hz, linked by a CH group.

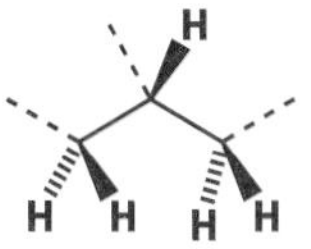

If this is the case, what has brought the values of ^2J down from 14 to 10 and even 5 Hz? Electronegative elements can't be the culprits as the only one is nitrogen, but small rings could. If, in fact, we simply join these two fragments together in rather a surprising way (the dotted lines show how), we get the correct structure.

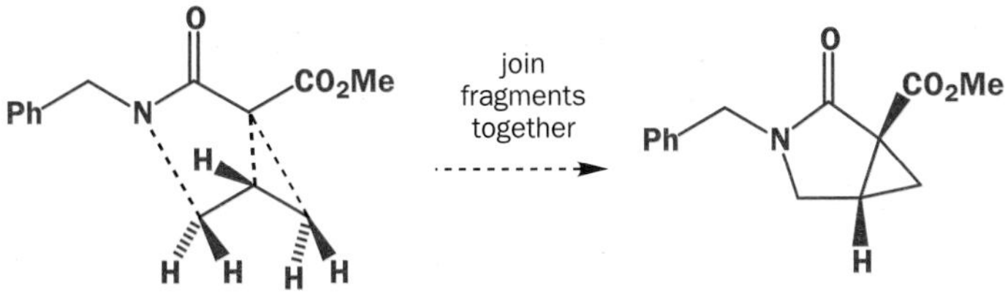

In this case, the geminal couplings do not help to assign the stereochemistry—the three- and five-membered rings can only be fused *cis* (just try making a model of the *trans* compound!)—but they do help in assigning the structure.

We should at this point just recap what we have done here—we made no attempt to work out the structure by thinking about what the mechanism of the reaction might be. We used, purely and simply, NMR to work out fragments of the structure which we then put together in a logical way. Considering reasonable mechanisms can be a help in structure determination—but it can also be a hindrance. If the product is unexpected, it follows that the mechanism is unexpected too.

For an example with a four-membered ring, we go back to β-lactams. A serious problem with β-lactam antibiotics is that bacteria develop resistance by evolving enzymes called β-lactamases, which break open the four-membered ring. In 1984, a team from Beechams reported the exciting discovery of some very simple inhibitors of these enzymes all based on the core structure named clavulanic acid. This too was a β-lactam but a much simpler one than the penicillins we saw earlier.

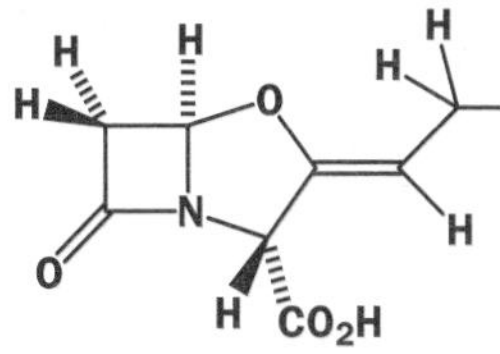

The structure elucidation used all the usual spectroscopic techniques as well as X-ray crystallography, but it is the ^{1}H NMR that is particularly interesting to us here. Here it is, with the assignments shown.

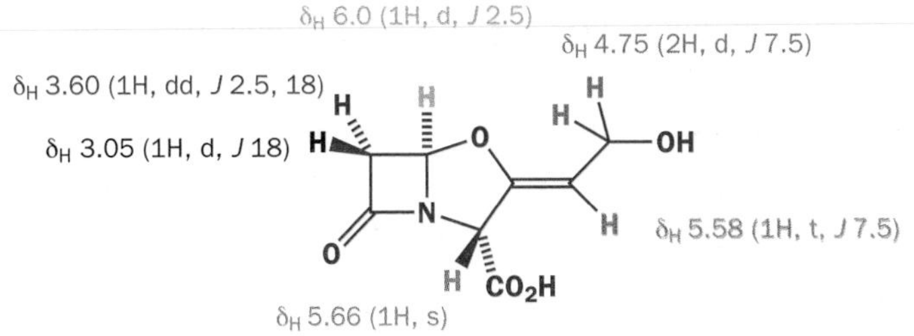

clavulanic acid

δ_H 6.0 (1H, d, J 2.5)

δ_H 4.75 (2H, d, J 7.5)

δ_H 3.60 (1H, dd, J 2.5, 18)

δ_H 3.05 (1H, d, J 18)

δ_H 5.58 (1H, t, J 7.5)

δ_H 5.66 (1H, s)

Notice the very large geminal coupling between the red and the black hydrogens (more of this later) and the fact that the green hydrogens, though actually diastereotopic, resonate at the same chemical shift. The *cis* coupling across the four-membered ring is larger (2.5 Hz) than the *trans* coupling (0 Hz) as expected.

The π contribution to geminal coupling

We began this chapter with a diagram of Taxol. This molecule is rather too complex for us to analyse in detail, but the geminal couplings of an important closely related compound are worth noting. Here are the details.

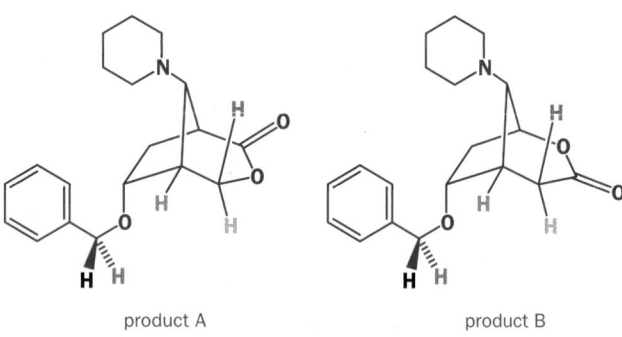

The coupling between the black Hs is 20 Hz while that between the green Hs is 6 Hz. This is a rather extreme example as the green Hs are in a four-membered ring and next to an oxygen atom, so they are expected to show a small J value, while the black Hs are in a six-membered ring and not next to an electronegative element. Nevertheless, 20 Hz is a very large coupling constant. The reason is the adjacent π bond. If a CH_2 group is next to an alkene, aromatic ring, C=O group, CN group, or any other π-bonded functional group, it will have a larger geminal coupling constant. This effect is quite clear in both Taxol and clavulanic acid.

The oxidation of the bicyclic amino-ketone shown in the margin demonstrates how useful this effect can be. This is the Baeyer–Villiger rearrangement, which you will meet in Chapter 37. The mechanism is not important here: all you need to know is that it inserts an oxygen atom on one side or the other of the ketone C=O group. The question is—which side?

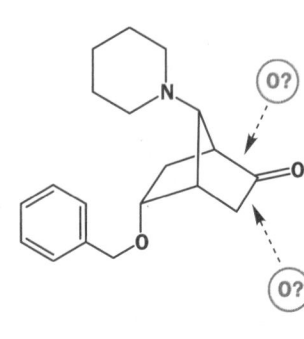

In fact, both lactones were isolated and the problem then became—which was which? In both NMR spectra there were AB systems at 4.6–4.7 for diastereotopic CH_2 groups isolated from the rest of the molecule, with $^2J = 11.8$ Hz. These are clearly the black and green hydrogens on the benzyl groups. The coupling constant is reduced by the oxygen atom and increased by the phenyl's π contribution, so it ends up about average.

product A   product B

Both lactones also had clear ABX systems in the NMR corresponding to the yellow, brown, and orange protons. In one compound $^2J = 10.8$ Hz and in the other $^2J = 18.7$ Hz. The smaller value has been reduced by neighbouring oxygen and this must be compound A. The larger value has been increased by the π contribution from the carbonyl group and this must be compound B.

● **The size of ^2J and ^3J coupling constants**

We have now covered all of the important influences on the size of coupling constants. They are:

- dihedral angle: ^3J greatest at 180° and 0°; about 0 Hz at 90°
- ring size, which leads to 'spreading out' of bonds and lower ^2J and lower ^3J in small rings
- electronegative atoms, which decrease ^2J and ^3J coupling constants between protons
- π systems, which increase ^2J coupling constants between protons

---

The nuclear Overhauser effect

■
We looked at the stereoselectivity of electrophilic additions to double bonds in Chapter 20.

Many occasions arise when even coupling constants do not help us in our quest for stereochemical information. Consider this simple sequence. Bromination of the alkene gives as expected *trans* addition and a single diastereoisomer of the dibromide.

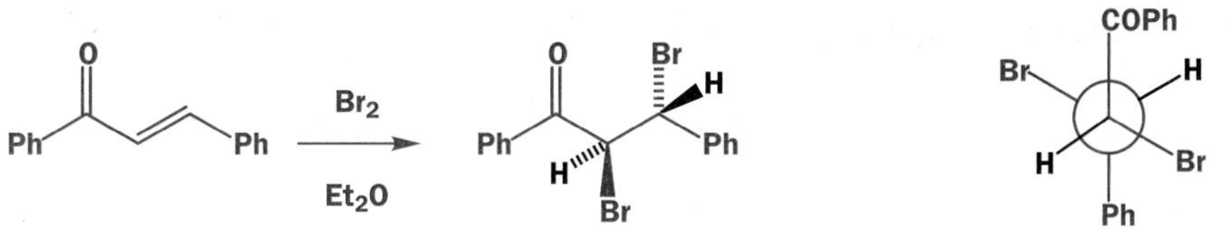

Newman projection of product

The vicinal (^3J) coupling constant between the two black Hs is 11 Hz. This is rather large and can be explained by a predominant conformation shown in the Newman projection, with the two large groups (PhCO and Ph) as far from each other as possible, the two medium groups (Br) as distant as possible, and the two black Hs in the places which are left. The dihedral angle between the black Hs is then 180° (they are anti-periplanar) and a large J is reasonable.

But now see what happens when we react the dibromide with piperidine. A single diastereoisomer of an amine is formed, and there is good evidence that it has the opposite configuration from the dibromide; in other words, replacement of Br by N has occurred with inversion.

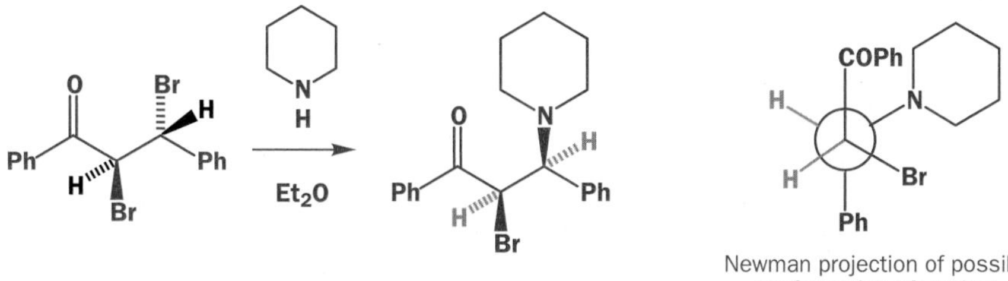

Newman projection of possible conformation of product

We might expect that the conformation would now be different and that, since inversion has occurred, the two green Hs would now be gauche instead of anti-periplanar. With a dihedral angle of 60° the coupling constant would be much less. But it isn't. The coupling constant between the green Hs is exactly the same (11 Hz) as the coupling constant between the black Hs in the starting material. Why? The new substituent (piperidine) is very big, much bigger than Br and probably bigger in three dimensions than a flat Ph group. The conformation must change (all we are doing is rotating the back carbon atom by 120°) so that the two green Hs also have a dihedral angle of 180°.

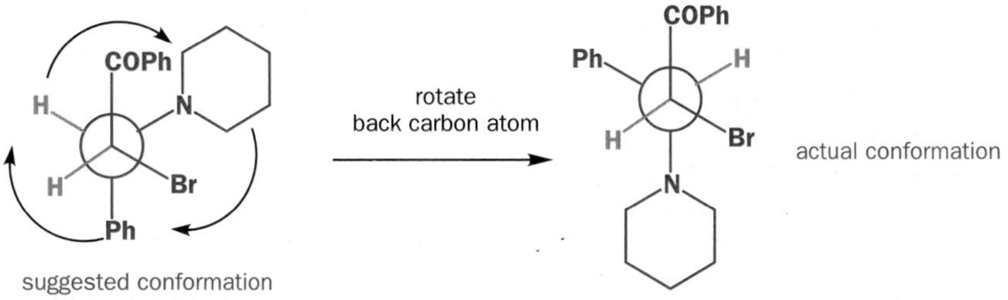

suggested conformation   rotate back carbon atom   actual conformation

A more serious situation arises when we treat this product with base. An unusual elimination product is formed, in which the amine group has moved next to the ketone. The reaction is interesting for this point alone, and one of the problems at the end of the chapter asks you to suggest a mechanism. But there is added interest, because the product is also formed as a single geometrical isomer, *E* or *Z*. But which one? There is a hydrogen atom at one end of the alkene but not at the other so we can't use ^3J coupling constants to find out as there aren't any.

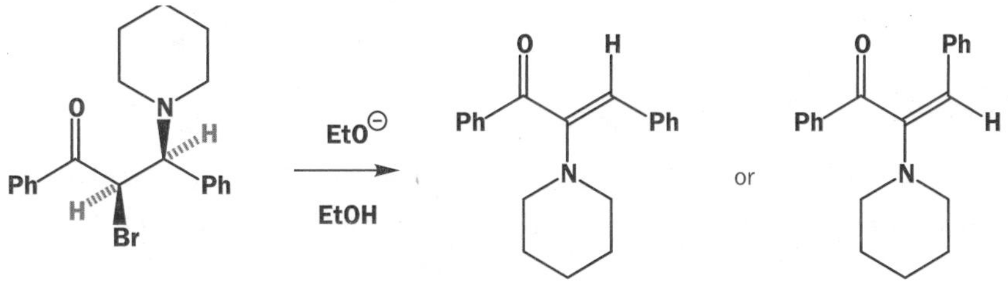

► **Why you can't integrate
 ^{13}C NMR spectra**

Relaxation is the real reason why
you can't integrate ^{13}C signals.
Relaxation of ^{13}C is slow, but is
fastest with lots of nearby
protons. This is the reason that
you will often find that –CH_3
groups show strong signals in the
^{13}C NMR, while quaternary
carbons, with no attached
protons, show weak ones:
quaternary carbons relax only
slowly, so we don't detect such
an intense peak. Allowing plenty
of time for all ^{13}C atoms to relax
between pulses gives more
proportionally sized peaks, but at
the expense of a very long NMR
acquisition time.

What we need is a method that allows us to tell which groups are close to one another in space
(though not necessarily through bonds) even when there are no coupling constants to help out. Very
fortunately, an effect in NMR known as the **nuclear Overhauser effect** allows us to do this.

The details of the origin of the nuclear Overhauser effect are beyond the scope of this book, but we
can give you a general idea of what the effect is. As you learned from Chapter 11, when a proton NMR
spectrum is acquired, a pulse of radiofrequency electromagnetic radiation jolts the spins of the protons
in the molecule into a higher energy state. The signal we observe is generated by those spins dropping
back to their original states. In Chapter 11 it sufficed to assume that the drop back down was spontaneous, just like a rock falling off a cliff. In fact it isn't—something needs to 'help' the protons to drop
back again—a process called **relaxation**. And that 'something' is other nearby magnetically active
nuclei—usually more protons. Notice *nearby*—nearby in space not through bonds. With protons, relaxation is fast, and the number of nearby protons does not affect the appearance of the NMR spectrum.

We find that, although peak intensity is independent of the number of nearby protons, by using
methods whose description is beyond the scope of this book, it is possible to make the intensity
respond, to a small extent, to those protons that are nearby. The idea is that as certain protons (or
groups of identical protons) are irradiated selectively (in other words, they are jolted into their high-
energy state and held there by a pulse of radiation at exactly the right frequency—not the broad pulse
needed in a normal NMR experiment). Under the conditions of the experiment, this causes protons
that *were* relying on the irradiated protons to relax them to appear as a slightly more intense (up to a
few per cent) peak in the NMR spectrum. This effect is known as the nuclear Overhauser effect, and the
increase in intensity of the peak the nuclear Overhauser enhancement. Both are shortened to 'NOE'.

All you need to be aware of at this stage is that irradiating protons in an NOE experiment gives
rise to enhancements at other protons that are nearby in space—no coupling is required, and NOE is
not a through-bond phenomenon. The effect also drops off very rapidly: the degree of enhancement
is proportional to $1/r^6$ (where r is the distance between the protons) so moving two protons twice as
far apart decreases the enhancement one can give to the other by a factor of 64. NOE spectra are usually presented as differences: the enhanced spectrum minus the unenhanced, so that those protons
that change in intensity can be spotted immediately.

Applying NOE to the problem in hand solves the structure. If the protons next to the nitrogen
atom in the piperidine ring are irradiated, the signal for the alkene proton increases in intensity, so
these two groups of protons must be near in space. The compound is the *E*-alkene.

Data from NOE experiments nicely supplement information from coupling constants in the
determination of three-dimensional stereochemistry too. Reduction of this bicyclic ketone with a
bulky hydride reducing agent gives one diastereoisomer of the alcohol, but which? Irradiation of the
proton next to the OH group leads to an NOE to the green proton.

This suggests that the two protons are on the same side of the molecule and that reduction has
occurred by hydride delivery to the face of the ketone opposite the two methyl groups on the three-
membered ring.

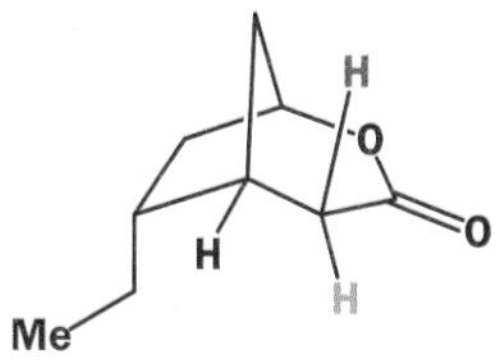

For a more complex example we can return to a lactone (shown in the margin) obtained by oxidation of a bicyclic ketone similar to the one we mentioned earlier (p. 844). When this compound was made, two questions arose. What was the stereochemistry of the ethyl group, and which signal in the NMR spectrum belonged to which hydrogen atom? In particular, was it possible to distinguish the signals of the diastereotopic brown and yellow Hs? Three experiments were carried out, summarized in the diagrams below. First the CH_2 and then the CH_3 protons of the ethyl group were irradiated and the other protons were observed. Finally, the green proton was irradiated.

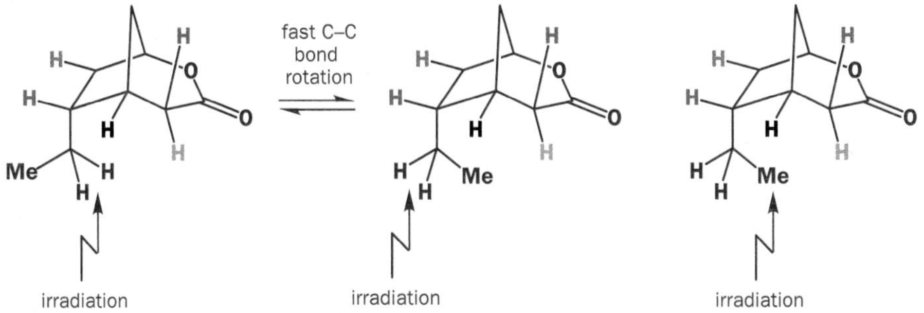

In the first experiment, enhancement of the signals of the black, yellow, and green Hs was observed. The ethyl group can rotate rapidly on the NMR time-scale so all the enhancements can be explained by the first two conformations. An NOE effect to the yellow but not to the brown H is particularly significant. Irradiation of the methyl group led to enhancement of the yellow proton but not the brown. Clearly, the ethyl group is in the position shown.

Irradiation of the green proton, whose stereochemistry is now clear, enhanced the orange proton and allowed its chemical shift to be determined. Previously, it had been lost in the many CHs in the rings.

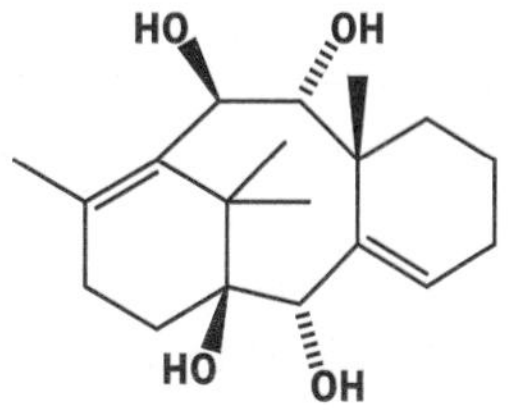

We shall finish this chapter by returning to Taxol once more. The tricyclic compound drawn here was made in 1996 as an intermediate for Taxol synthesis. The stereochemistry and the conformation of the molecule were deduced by a series of NOE experiments.

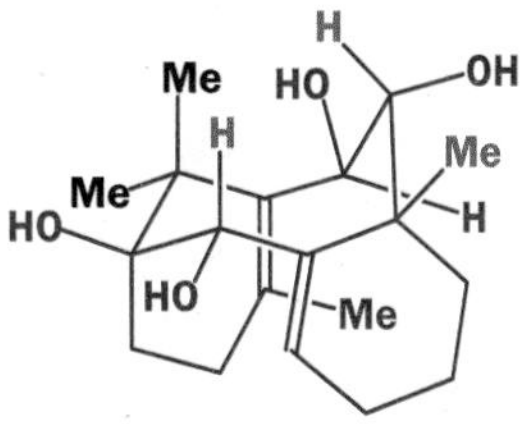

Four NOE experiments were carried out, summarized two at a time in the diagrams on the right. Irradiation of the methyl groups established that the black pair were on the same carbon atom and hence allowed assignment of the spectrum. Then irradiation of the remaining methyl group on saturated carbon established the proximity of the green hydrogens and gave the stereochemistry at three centres.

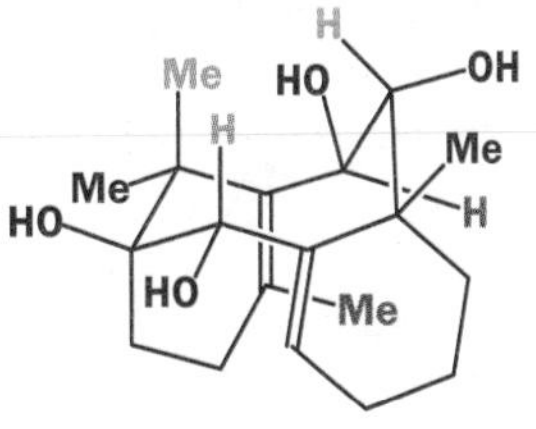

Next irradiation of the brown methyl group on a double bond showed it was close to the brown hydrogen and gave the stereochemistry at that centre. Finally, irradiation at one of the two methyl groups of the CMe_2 group (yellow) showed that it was close to the two green hydrogens and hence all these three groups were clustered in the centre of the molecule. It's important here to draw a conformational diagram as they do not look very close in the flat diagram shown.

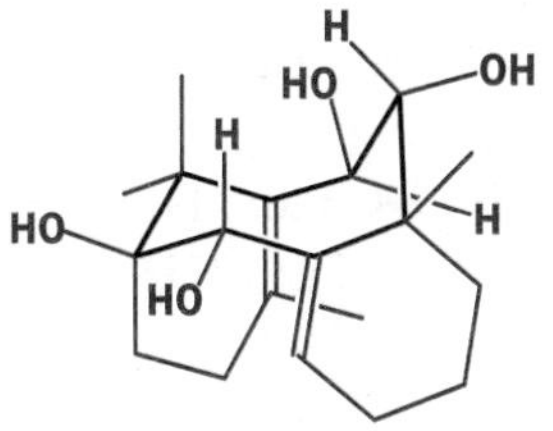

These experiments fixed not only the stereochemistry at all the stereogenic centres but also allowed the conformation of the central eight-membered ring to be deduced. This ring is outlined in black on the diagram in the margin and has two chair-like sections. It is no trivial matter to work out such conformations without X-ray data and the NOE result tells us about the more important conformation *in solution*, rather than in the crystal. The alliance between coupling constants and NOE gives us a powerful method for structural determination.

To conclude...

As you leave this chapter, you should carry the message that, while X-ray crystallography is the 'final appeal' with regard to determining configuration, NMR can be a very powerful tool too. Analysis of coupling constants and nuclear Overhauser effects allows:

- determination of configuration, even in noncrystalline compounds
- determination of conformation in solution

As you embark on the next two chapters, which describe how to make molecules stereoselectively, bear in mind that many of the stereochemical outcomes were deduced using the techniques we have described in this chapter.

Problems

Note. All NMR shifts are in p.p.m. and coupling constants are quoted in hertz (Hz). The usual abbreviations are used: d = doublet; t = triplet; and q = quartet.

1. A revision problem to start you off easily. A Pacific sponge contains 2.8% dry weight of a sweet-smelling oil with the following spectroscopic details. What is its structure and stereochemistry?

Mass spectrum gives formula: $C_9H_{15}O$

IR 1680, 1635 cm^{-1}

δ_H 0.90 (6H, d, J 7), 1.00 (3H, t, J 7), 1.77 (1H, m), 2.09 (2H, t, J 7), 2.49 (2H, q, J 7), 5.99 (1H, d, J 16), and 6.71 (1H, dt, J 16, 7)

δ_C 8.15 (q), 22.5 (two qs), 28.3 (d), 33.1 (t), 42.0 (t), 131.8 (d), 144.9 (d), and 191.6 (s)

2. Reaction between this aldehyde and ketone in base gives a compound A with the ^{1}H NMR spectrum: δ 1.10 (9H, s), 1.17 (9H, s), 6.4 (1H, d, J 15) and 7.0 (1H, d, J 15). What is its structure? (Don't forget stereochemistry!) When this compound reacts with HBr it gives compound B with this NMR spectrum: δ 1.08 (9H, s), 1.13 (9H, s), 2.71 (1H, dd, J 1.9, 17.7), 3.25 (dd, J 10.0, 17.7), and 4.38 (1H, dd, J 1.9, 10.0). Suggest a structure, assign the spectrum, and give a mechanism for the formation of B.

3. One of the sugar components in the antibiotic kijanimycin has the gross structure and NMR spectrum shown below. What is its stereochemistry? All couplings in Hz; signals marked * exchange with D_2O.

δ_H 1.33 (3H, d, J 6), 1.61* (1H, broad s), 1.87 (1H, ddd, J 14, 3, 3.5), 2.21 (1H, ddd, J 14, 3, 1.5), 2.87 (1H, dd, J 10, 3), 3.40 (3H, s), 3.47 (3H, s), 3.99 (1H, dq, J 10, 6), 4.24 (1H, ddd, J 3, 3, 3.5), and 4.79 (1H, dd, J 3.5, 1.5)

4. Two diastereoisomers of this cyclic keto-lactam have been prepared. The NMR spectra have many overlapping signals but the proton marked in green can clearly be seen. In isomer A it is δ_H 4.12 (1H, q, J 3.5), and isomer B has δ_H 3.30 (1H, dt, J 4, 11, 11). Which isomer has which stereochemistry?

5. How would you determine the stereochemistry of these two compounds?

6. The structure and stereochemistry of the anti-fungal antibiotic ambruticin was in part deduced from the NMR spectrum of this simple cyclopropane. Interpret the NMR spectrum and show how it gives definite evidence on the stereochemistry.

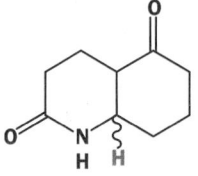

δ_H 1.21 (3H, d, J 7 Hz), 1.29 (3H, t, J 9), 1.60 (1H, t, J 6), 1.77 (1H, ddq, J 6, 13, 7), 2.16 (1H, dt, J 6, 13), 4.18 (2H, q, J 9), 6.05 (1H, d, J 20), and 6.62 (1H, dd, J 13, 20).

7. In Chapter 20 we set a problem asking you what the stereochemistry of a product was. Now we can give you the NMR spectrum of the product and ask: how do we *know* the stereochemistry of the product? You need only the partial NMR spectrum: δ_H 3.9 (1H, ddq, J 12, 4, 7) and 4.3 (1H, dd, J 11, 3).

8. The structure of a Wittig product intended as a prostaglandin model was established by the usual methods—except for the geometry of the double bond. Irradiation of a signal at 3.54 (2H, t, J 7.5) led to an enhancement of another signal at δ_H 5.72 (1H, t, J 7.1) but not to a signal at δ_H 3.93 (2H, d, J 7.1). What is the stereochemistry of the alkene? How is the product formed?

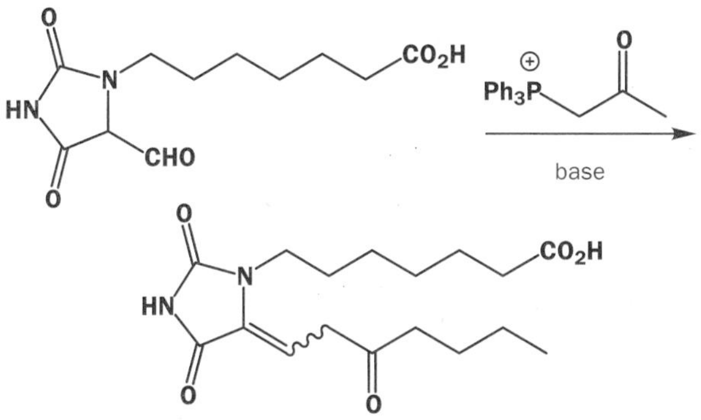

9. How would you determine the stereochemistry of this cyclopropane? The NMR spectra of the three protons on the ring are given: δ_H 1.64 (1H, dd, J 6, 8), 2.07 (1H, dd, J 6, 10), and 2.89 (1H, dd, J 10, 8).

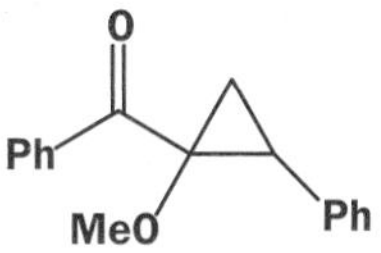

10. A chemical reaction produces two diastereoisomers of the product. Isomer A has δ_H 3.08 (1H, dt, J 4, 9, 9) and 4.32 (1H, d, J 9) while isomer B has δ_H 4.27 (1H, d, J 4). The other protons overlap. Isomer B is converted into isomer A on treatment with base. What is the stereochemistry of A and B?

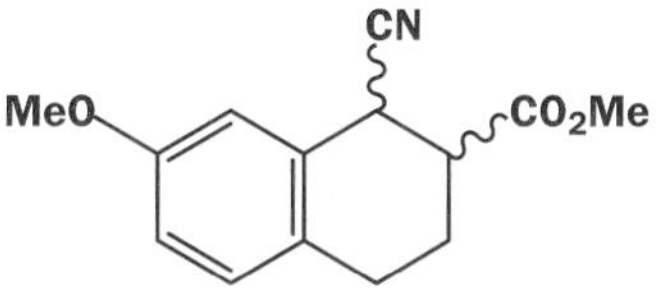

11. Muscarine, the poisonous principle of the death cap mushroom, has the following structure and proton NMR spectrum. Assign the spectrum. Can you see definite evidence for the stereochemistry? All couplings in Hz; signals marked * exchange with D_2O.

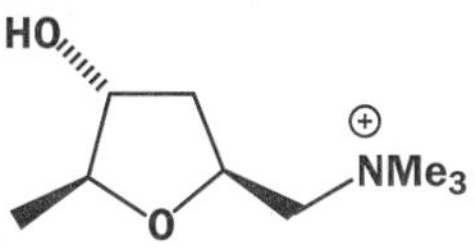

δ_H 1.16 (3H, d, J 6.5), 1.86 (1H, ddd, J 12.5, 9.5, 5.5), 2.02 (1H, ddd, J 12.5, 2.0, 6.0), 3.36 (9H, s), 3.54 (1H, dd, J 13, 9.0), 3.74 (1H, dd, J 13, 1.0), 3.92 (1H, dq, J 2.5, 6.5), 4.03 (1H, m), 4.30* (1H, d, J 3.5), and 4.68 (1H, m).

12. An antifeedant compound that deters insects from eating food crops has the gross structure shown below. Some of the NMR signals that can clearly be made out are also given. Since NMR coupling constants are clearly useless in assigning the stereochemistry, how would you set about it?

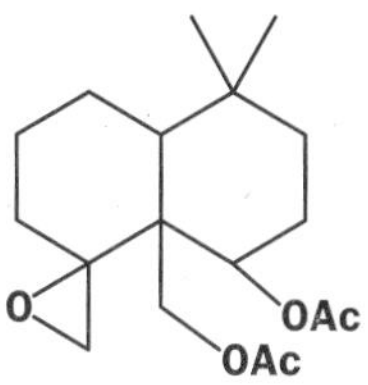

δ_H 2.22 (1H, d, J 4), 2.99 (1H, dd, J 4, 2.4), 4.36 (1H, d, J 12.3), 4.70 (1H, dd, J 4.7, 11.7), 4.88 (1H, d, J 12.3)

13. The seeds of the Costa Rican plant *Ateleia herbert smithii* are avoided by all seed eaters (except a weevil that adapts them for its defence) because they contain two toxic amino acids (IR spectra like other amino acids). Neither compound is chiral. What is the structure of these compounds? They can easily be separated because one (A) is soluble in aqueous base but the other (B) is not.

A is $C_6H_9NO_4$ (mass spectrum) and has δ_C 34.0 (d), 40.0 (t), 56.2 (s), 184.8 (s), and 186.0 (s). Its proton NMR has three exchanging protons on nitrogen and one on oxygen and two complex signals at δ_H 2.68 (4H, A_2B_2 part of A_2B_2X system) and 3.37 (X part of A_2B_2X system) with J_{AB} 9.5, J_{AX} 9.1, and J_{BX} small.

B is $C_6H_9NO_2$ (mass spectrum) and has δ_C 38.0 (d), 41.3 (t), 50.4 (t), 75.2 (s), and 173.0 (s). Its proton NMR spectrum contains two exchanging protons on nitrogen and δ_H 1.17 (2H, ddd, J 2.3, 6.2, 9.5), 2.31 (2H, broad m), 2.90 (1H, broad t, J 3.2), and 3.40 (2H, broad s).

Because the coupling pattern did not show up clearly as many of the coupling constants are small, decoupling experiments were used. Irradiation at δ_H 3.4 simplifies the δ_H 2.3 signal to (2H, ddd, J 5.8, 3.2, 2.3), sharpens each line of the ddd at 1.17, and sharpens the triplet at 2.9.

Irradiation at 2.9 sharpens the signals at 1.17 and 2.9 and makes the signal at 2.31 into a broad doublet, J about 6. Irradiation at 2.31 sharpens the signal at 3.4 slightly and reduces the signals at 2.9 and 1.17 to broad singlets. Irradiation at 1.17 sharpens the signal at 3.4 slightly so that it is a broad doublet, J about 1.0, sharpens the signal at 2.9 to a triplet, and sharpens up the signal at 2.31 but irradiation here had the least effect.

This is quite a difficult problem but the compounds are so small (C_6 only), have no methyl groups, and have some symmetry so you should try drawing structures at an early stage.

Stereoselective reactions of cyclic compounds

33

Connections

Building on:	**Arriving at:**	**Looking forward to:**
• Stereochemistry ch16	• Stereoselectivity in cyclic systems is easy to understand	• Diastereoselectivity ch34
• Conformational analysis ch18	• Flattened four- and five-membered rings are attacked *anti* to large substituents	• Asymmetric synthesis ch45
• Determination of stereochemistry by spectroscopy ch32	• Flattened six-membered rings are attacked from an axial direction	• Organic synthesis ch53
	• Bicyclic structures are attacked on the outside face	• Pericyclic reactions ch35–ch36
	• Tethering together nucleophile and electrophile forces one stereochemical outcome	
	• Hydrogen-bonding can reverse the normal stereochemical outcome of a reaction	

Introduction

This chapter is about rings and stereochemistry. Stereochemistry is easier to understand in cyclic compounds and that alone might make a separate chapter worthwhile. But there is something much more fundamental behind this chapter. Stereochemistry is better behaved in cyclic compounds. Suppose you were to reduce this ketone to one of the corresponding alcohols.

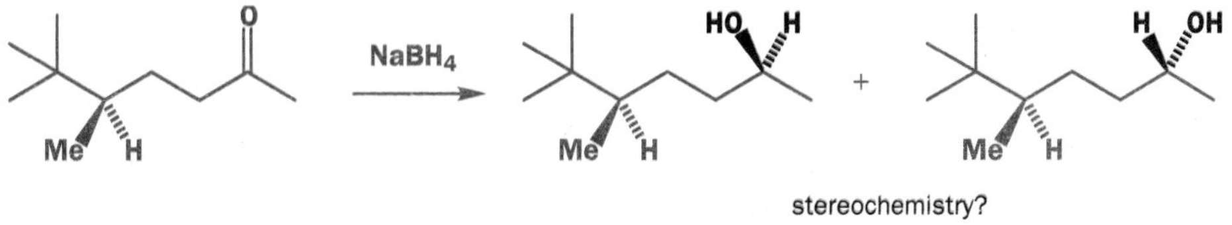

stereochemistry?

There would be very little chance of any control of stereochemistry at the new stereogenic centre (shown in black). A more or less 50:50 mixture of the two diastereoisomers would be expected. However, if we join up the molecule into a ring, things are suddenly quite different. (This is not, of course, a chemical reaction—just a thought process!)

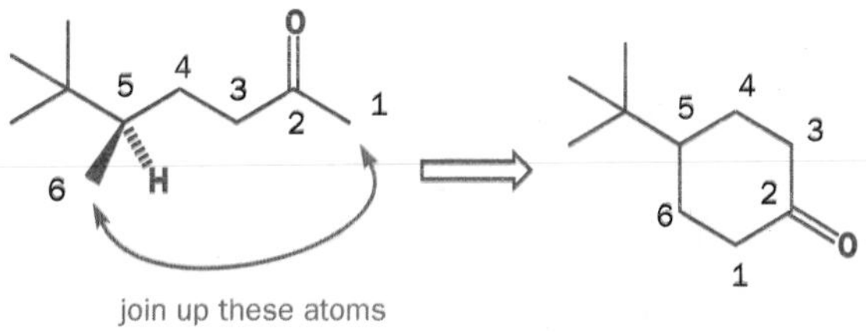

join up these atoms

The cyclic ketone has a fixed conformation controlled by the determination of the *t*-butyl group to be equatorial. Reduction can be controlled to give almost exclusively either the axial or the equatorial alcohol as we explained in Chapter 18. Large reagents prefer to approach equatorially while small reagents like to put the new OH group into an equatorial position. These are stereo-*selective* reactions, and, because the two different outcomes are diastereoisomers, we can call them **diastereoselective**.

■
If your memory of Chapter 18's discussion of axial and equatorial attack on cyclohexanones is dim, you should refresh it now. We shall use several examples that build on what we said there (p. 470).

The key to the difference is in the conformations. The cyclic ketone has one conformation and the two approaches to the faces of the ketone are very different. The open-chain compound has an indefinite number of conformations as rotation about all the C–C bonds is possible. In any one conformation, attack on one face of the ketone or the other may happen to be preferred, but on average there will be very little difference. There is all the difference in the world between cyclic and open-chain compounds when it comes to stereoselective reactions. This is why we have made this topic into two chapters: this one (33) dealing with rings, the next (34) with what happens without rings.

In this chapter we shall look at reactions happening to cyclic compounds, reactions that close rings (cyclizations), and reactions with cyclic intermediates and with cyclic transition states. We shall investigate what happens to stereochemistry when two (or even more) rings are joined together at a bond or at an atom. We shall see how stereochemical effects change as the ring size increases from three atoms to eight or more. You will find that you have met some of the reactions before in this book. This chapter collects them together and explains the principles of stereochemical control in cyclic systems as well as introducing some new reactions.

Reactions on small rings

Four-membered rings can be flat

> We introduced to you the usual conformations of small rings in Chapter 18. We will briefly revisit that material in this section, and show you how it affects the stereoselectivity of the reactions of rings.

The smallest ring that we can conveniently work on is four-membered. Saturated four-membered rings have a slightly bent conformation but four-membered lactones are flat. The enolates of these lactones can be made in the usual way with LDA at −78 °C and are stable at that temperature.

The formation of the lithium enolate is straightforward but it might be expected to be unstable because of a simple elimination reaction. It is not possible to make open-chain lithium enolates with β oxygen substituents like this because they do undergo elimination.

> This is a **stereoelectronic effect**, due to the spatial arrangement of orbitals, and we will discuss more of these in Chapters 38 and 42.

But, in the four-membered ring, the p orbitals of the enolate and the C–O single bond are orthogonal (see drawing in margin) so that no interaction between them, and no elimination, can occur. The enolate can be combined with electrophiles in the usual way (Chapters 26 and 27).

bond and orbitals are orthogonal

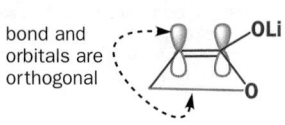

If the β-lactone has a substituent already then there may be a choice as to which face of the enolate is attacked by an electrophile. Simple alkylation with a variety of alkyl halides gives essentially only one diastereoisomer of the product.

>98:2 R *trans* to *i*-Pr

► The stereoselectivity we are discussing in this chapter is diastereoselectivity: we are not concerned with enantiomers, and all of our discussions are equally valid whether the starting materials are racemic or enantiomerically pure. The product here, as in many other examples in the chapter, is racemic, so we could write (±) underneath the structure.

The enolate, as we have seen, is planar, the phenyl group is in the plane (so it doesn't matter which of the two possible diastereo-isomers of the starting material is used), and the isopropyl group is the only thing out of the plane. The electrophile simply adds to the face of the enolate not blocked by the iso-propyl group. This is a very simple case of a diastereoselective reaction.

Reduction of substituted four-membered ring ketones is usually reasonably stereoselec-tive. If the substituent is in the 3-position and small reagents like NaBH₄ are used, the *cis* isomer is favoured.

This result sounds very like the results already noted for six-membered rings and the expla-nation is similar. Saturated four-membered rings—even the ketones—are slightly puckered to reduce eclipsing interactions between hydrogen atoms on adjacent carbon atoms, and 'axial' attack by the small nucleophile gives the more stable *cis* product having both substituents 'equa-torial'.

'equatorial attack'

'axial attack'

Five-membered rings are flexible

We discussed the conformation of some five-membered rings in Chapter 32: a saturated five-mem-bered ring has a conformation variously called a 'half-chair' or an 'envelope'. It does look a bit like an opened envelope with one atom at the point of the flap, or it looks like most of (five-sixths rather than half?) a chair cyclohexane.

cyclohexane   cyclopentane

At any one moment, one of the carbon atoms is at the point of the envelope but rapid ring flip-ping equilibrates all these conformers so that all five atoms are, on average, the same. Substituted cyclopentanes can have substituents in pseudoaxial or pseudoequatorial positions or on the point position, like this.

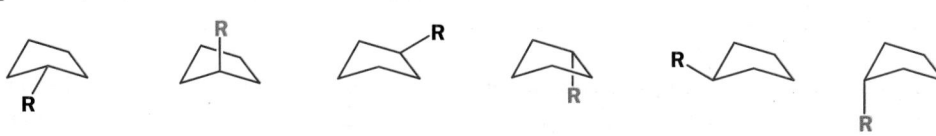

pseudoequatorial and pseudoaxial
substituents on cyclopentanes

You may recall (Chapter 32) that *cis* and *trans* couplings in proton NMR spectra of five-membered rings are often the same.

The result is a very flexible system that often behaves in stereoselective reactions as if the two positions on any carbon atom are the same.

As you can see, reduction of 2-substituted cyclopentanones may not be very stereoselective. The substituent probably occupies a pseudoequatorial position and the two faces of the ketone are very similar.

What selectivity there is (about 3:1) favours pseudoaxial attack in the conformation drawn as is reasonable for a small nucleophile. The use of a much more bulky reducing agent such as LiBH(s-Bu)$_3$ dramatically reverses and increases the stereoselectivity. Essentially only the *cis* compound is formed because the bulky reagent attacks the side of the carbonyl opposite to the methyl group.

If the conformation of the five-membered ring is fixed in a bridged (caged) structure, the stereoselectivity dramatically increases, even with LiAlH$_4$, as you will see later in the chapter (p. 862)

When there are two or three trigonal carbons in the ring, the ring is flatter, and reactions such as enolate alkylation and conjugate addition give excellent stereoselectivity even with a simple cyclopentane ring. Unsaturated five-membered lactones ('butenolides') give a very clear illustration of stereochemically controlled conjugate addition. There is only one possible stereogenic centre and the ring is almost planar so we expect nucleophilic attack to occur from the less hindered face. Cuprates are good nucleophiles for this reaction and here Me$_2$CuLi adds to the unsaturated lactone.

The starting material was a single enantiomer and hence so is the product—an insect pheromone.

This would be a good point at which to remind you of what we stressed in Chapter 16. If all the starting materials are achiral or racemic, the products must be too. That has been the case in many of the reactions so far in this chapter: we haven't put in (±) under every compound but we could have done. But here we do have a single enantiomer of starting material, so we get a single enantiomer of product. Diastereoselectivity is the same whether the starting material is enantiomerically pure or racemic.

single enantiomer of starting material

66% yield
(3S,4R)-(+)-eldanolide

It is not even necessary to have a stereogenic centre in an unsaturated ring if we want to create stereochemistry. A tandem conjugate addition and alkylation creates two new stereogenic centres in one operation. The conjugate addition of a lithium cuprate makes a lithium enolate, which will react in turn with an alkyl halide. The product is usually *trans*.

The conjugate addition forms a lithium enolate regiospecifically, and that was why you met this sequence in Chapter 26. We showed you a dramatic use of the stereoselectivity there as well, in a synthesis of a prostaglandin (p. 686).

The key step is the alkylation of the enolate intermediate. Enolates in five-membered rings are almost flat and the incoming alkyl halide prefers the less hindered face away from the recently added group R. The example below shows that, if both new groups have double bonds in their chains, it is easier to add a vinyl group as the nucleophile and an allyl group as an electrophile.

Our main example of enolate reactions in five-membered rings is one of some general importance. It illustrates how stereochemical information can be transmitted across a ring even though the original source of that information may be lost during the reaction. That may sound mysterious, but all will become clear. The first reaction is to make a five-membered cyclic acetal from an optically active hydroxy-acid. Our example shows (*S*)-(+)-mandelic acid reacting with *t*-BuCHO.

(*S*)-(+)-mandelic acid   24:1 *cis:trans*

Acetal formation involves nucleophilic attack of the OH group on the aldehyde so there is no change at the stereogenic centre. The stereochemistry of the new (acetal) centre may surprise you—why should the *cis*-isomer be so favoured? This is a conformational effect as both substituents can occupy pseudoequatorial positions.

Now, if we make the lithium enolate with LDA, the original stereogenic centre is destroyed as that carbon becomes trigonal. The only stereogenic centre left is the newly introduced one at the acetal position.

Check that you can write the mechanisms for acetal formation (Chapter 14). Acetal formation is under thermodynamic control (Chapter 13), so the product produced is the more stable.

both substituents
pseudoequatorial

The ring is now flattened by the alkene and reaction of the enolate with an electrophile is again a simple matter of addition to the face of the enolate opposite to the *t*-butyl group.

84% yield
>97:3 diastereoselectivity

If the acetal is now hydrolysed, the new stereogenic centre is revealed as an alkylated version of the starting material. It may appear that the alkylation has happened stereospecifically with retention, but what has really happened is that the new stereogenic centre in the acetal intermediate has relayed the stereochemical information through the reaction.

Five-membered rings also allow us to explore electrophilic attack on alkenes. A simple 4-substituted cyclopentene has two different faces—one on the same side as the substituent and one on the opposite side. Epoxidation with a peroxy-acid occurs preferentially on the less hindered face.

▶

Note that this reaction is diastereoselective—but neither starting material nor products are chiral. Diastereoselectivity need have nothing to do with chirality!

■

The mechanism of RCO₃H epoxidation was discussed on p. 506.

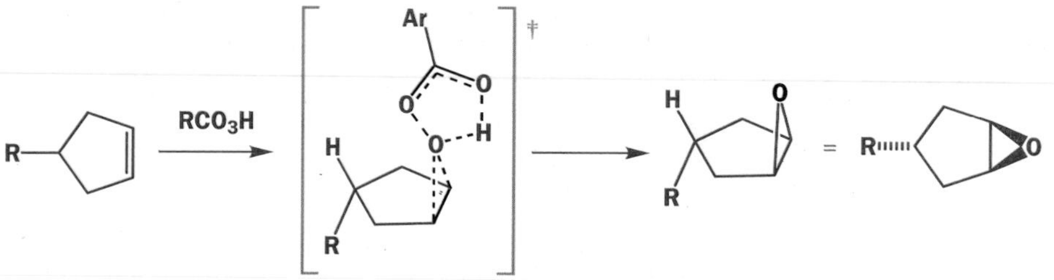

In the transition state (marked ‡) the peroxyacid prefers to be well away from R, even if R is only a methyl group. The selectivity is 76:24 with methyl. The opposite stereoselectivity can be achieved by bromination in water. The bromonium ion intermediate is formed stereoselectively on the less hindered side and the water is forced to attack stereospecifically in an S_N2 reaction from the more hindered side.

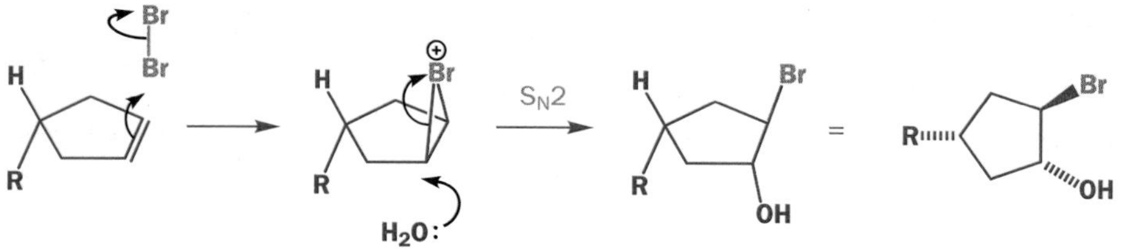

Treatment of the product with base (NaOH) gives an epoxide by another S_N2 reaction in which oxygen displaces bromide. This is again stereospecific and gives the epoxide on the same side as the group R.

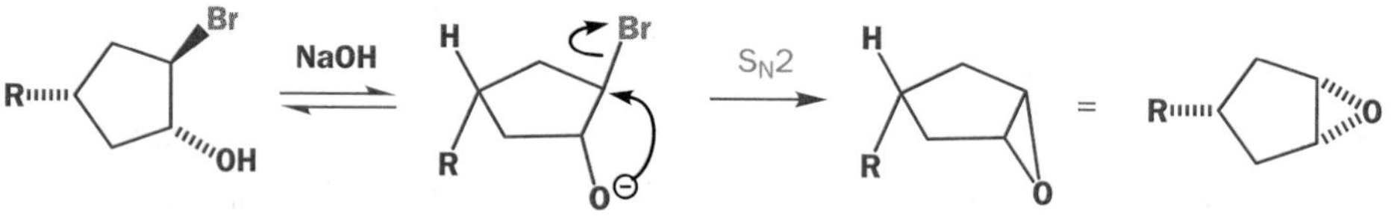

Two substituents on the *same side* of a five-membered ring combine to dictate approach from the other side by any reagent, and the two epoxides can be formed each with essentially 100% selectivity.

■

Remember—NBS acts as a source of electrophilic bromine: see p. 516 of Chapter 20.

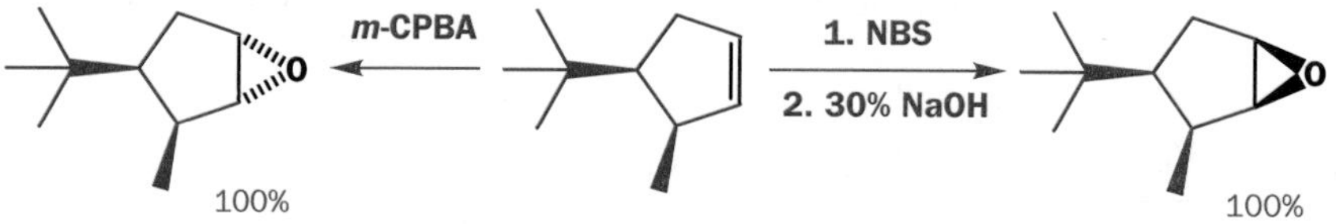

Stereochemical control in six-membered rings

From five-membered rings we move on naturally to six-membered rings. As well as the opportunity for more stereogenic centres around the larger ring, we have the additional prospect of conformational control—something special to six-membered rings because of their well-defined conformational properties. We shall start with simple reactions occurring on the opposite face to existing substituents and move on to conformational control, particularly to one theme—axial addition.

First, something about thermodynamic control. Because of the strong preference for substituents to adopt the equatorial position, diastereoisomers may equilibrate by processes such an enolization. For example, this fine perfumery material is made worthless by enolization.

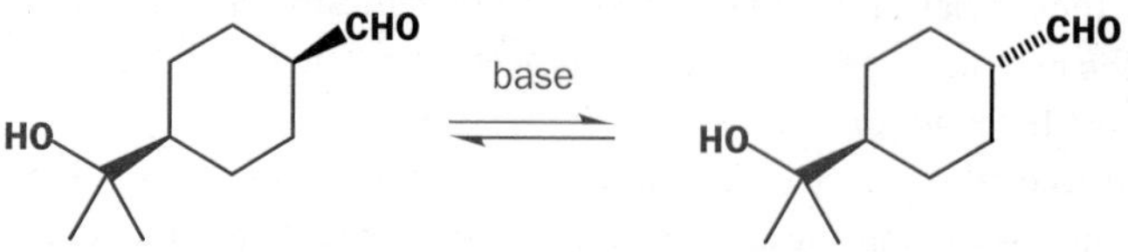

intense flowery perfume   odourless

The situation is bad because the worthless compound is preferred in the equilibrium mixture (92:8). This is because the two substituents are both equatorial in the *trans*-isomer.

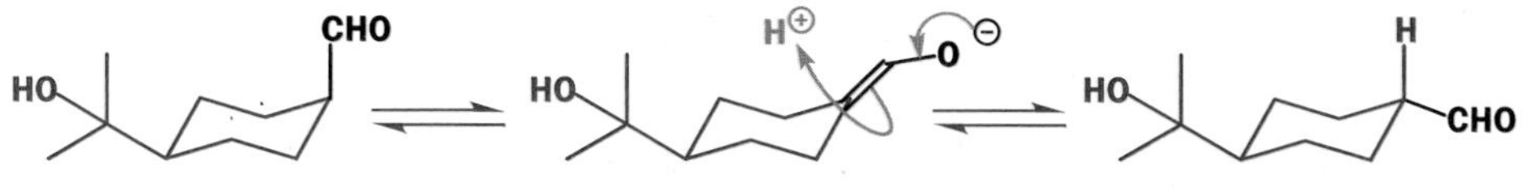

enolate

Although a disadvantage here, in other cases equilibration to the more stable all-equatorial conformation can be a useful source of stereochemical control. You will very shortly see an example of this.

Stereoselectivity in reactions of six-membered rings

We discussed the reduction of cyclohexanones in Chapter 18 and established that reducing agents prefer the equatorial approach while small reagents may prefer to put the OH group in the more stable equatorial position. If the nucleophile is not H but something larger than OH then we can expect equatorial attack to dominate both because of ease of approach and because of product stability.

A simple example is the addition of PhLi to the heterocyclic ketone below which has one methyl group next to the carbonyl group. This methyl group occupies an equatorial position and the incoming phenyl group also prefers the equatorial approach so that good stereoselectivity is observed.

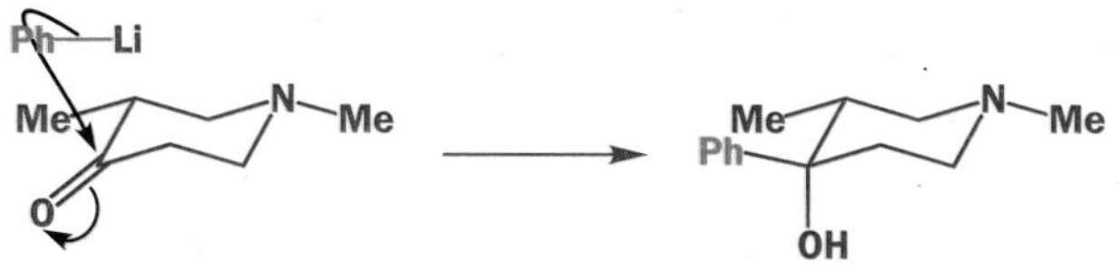

This product was used in the preparation of the analgesic drug alphaprodine. We shall represent the reaction now in configurational terms. It is important for you to recognize and be able to draw both configurational (as below) and conformational (as above) diagrams.

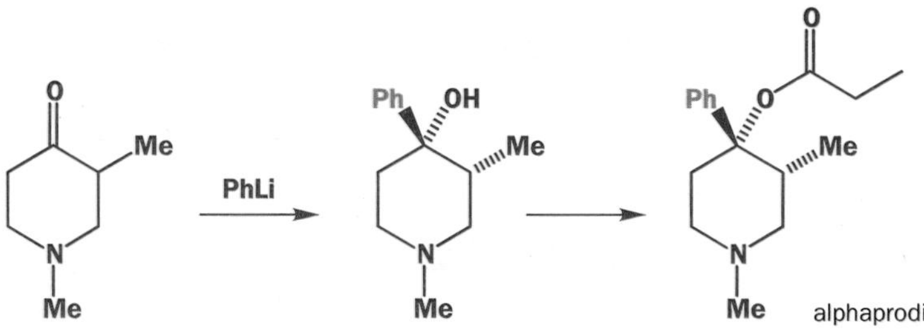

When the stereogenic centre is further away from the site of attack, the stereoselectivity may not be so good. Zeneca have announced the manufacture of a drug by the addition of a lithiated thiophene to another heterocyclic ketone, which initially gave a mixture of diastereoisomers.

mixture of diastereoisomers

Such a mixture is no good for manufacture of a pure drug, but the compound can be equilibrated in dilute acid by repeated S$_N$1 formation of a tertiary benzylic cation and recapture by water so that the required product (which is more stable as it has both Me and the thiophene equatorial) dominates by 92:8 and can be purified by crystallization. The unwanted isomer can be recycled in the next batch.

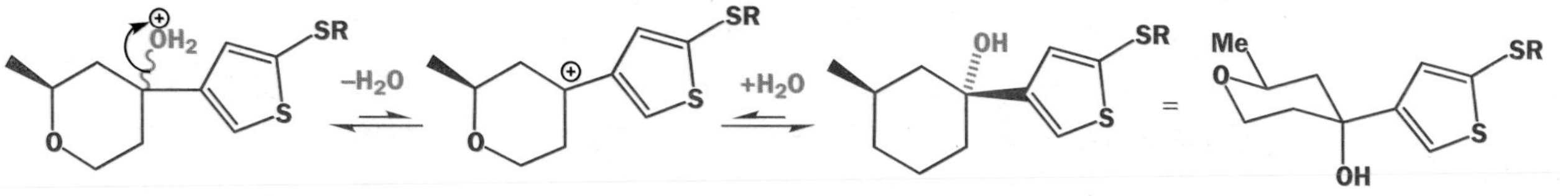

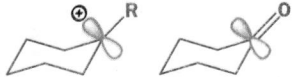

only one trigonal (sp²) atom in the ring:
chair conformation

■

We introduced this idea in Chapter 18, and we shall now develop it further.

In these reactions the molecule has a free choice whether to place a substituent in an axial or equatorial position and this is the only consideration because the starting materials in the reactions—ketones or carbocations—have six-membered rings that are already in the chair conformation even though they have one trigonal (sp²) atom in the ring.

Axial attack is preferred with unsaturated six-membered rings

When the starting material for a reaction has two or more trigonal (sp²) atoms in the ring, it is no longer in the chair conformation. In these cases, the stereochemistry of the reaction is likely to be driven by the need for the transition state and product to have a chair rather than a boat conformation. This can override the preference for substituents to go into equatorial positions. This is the basis for axial attack on enolates, cyclohexenes, and enones.

● **The number of trigonal carbon atoms in the ring is important**

- Six-membered rings with one trigonal (sp²) carbon atom can undergo axial or equatorial attack
- Six-membered rings with two or more trigonal carbon atoms undergo *axial attack* in order to form chairs rather than boats. The final product may end up with axial or equatorial substitution, but this is not a consideration in the reaction itself

Alkylations of enolates, enamines, and silyl enol ethers of cyclohexanone usually show substantial preference for axial attack. The enamine of 4-*t*-butylcyclohexanone, which has a fixed conformation because of the *t*-butyl group, gives 90% axial alkylation and only 10% equatorial alkylation with *n*-PrI.

It is a simple matter to show that the preferred product has the new propyl group in the axial position because both the starting ketone and the product have chair conformations with the *t*-butyl equatorial.

To get at the explanation we need to look at the conformation of the enamine intermediate. At this point we shall generalize a bit more and write a structure that represents any enol derivative where X may be OH, O⁻, OSiMe₃, NR₂, and so on. The conformation has a double bond in the ring, and is a partially flattened chair, as described in Chapter 18.

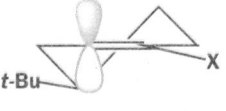

The *t*-butyl group is in an equatorial position at the back of the ring. The electrophile must approach the enol derivative from more or less directly above or below because only then can it attack one of the lobes of the p orbital at the enol position shown in yellow. The top of the molecule looks to be more open to attack so we shall try that approach first.

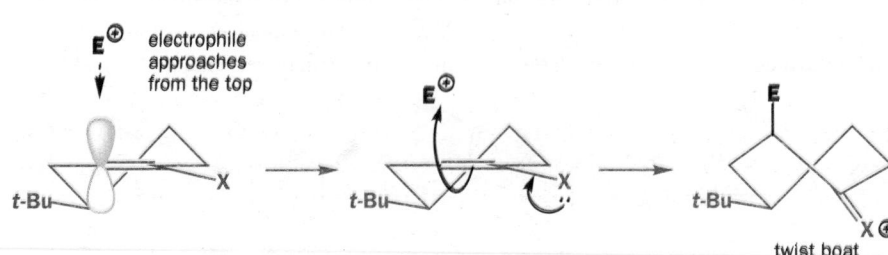

As the electrophile bonds to the trigonal carbon atom, that atom must become tetrahedral and it does so by forming a vertical bond upwards. The result is shown in the diagram—the ring turns into a twist-boat conformation. Now, of course, after the reaction is over, the ring can flip into a chair conformation and the new substituent will then be equatorial, but that information is not present in the transition state for the reaction. We could say that, at the time of reaction, the molecule doesn't 'know' it can later be better off and get the substituent equatorial: all it sees is the formation of an unstable twist boat with a high-energy transition state leading to it.

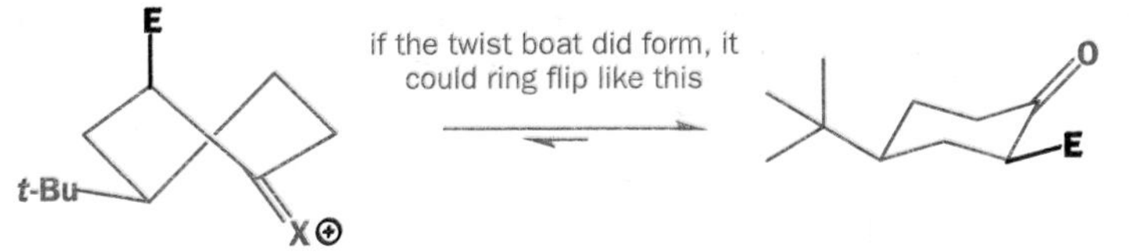

Attack from the apparently more hindered bottom face makes the trigonal carbon atom turn tetrahedral in the opposite sense by forming a vertical bond to the electrophile downwards. The ring goes directly to a chair form with the electrophile in the axial position.

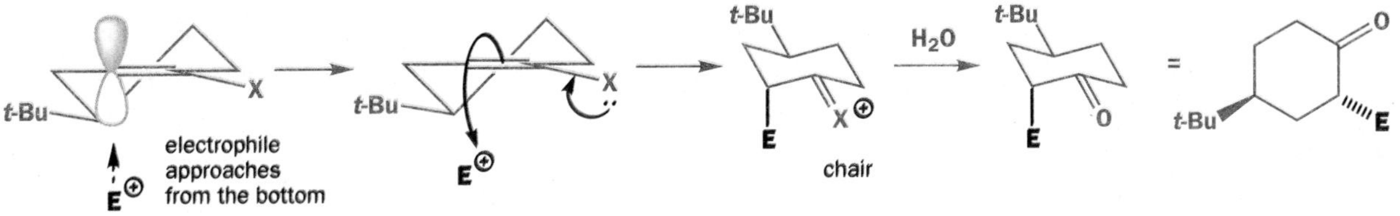

When the carbonyl group is restored by hydrolysis (if necessary—X may be O already) the ring need not flip: it's already a chair with the *t*-butyl equatorial, and the new substituent is axial on the chair. This is the observed product of the reaction.

It's important that you understand what is going on here. The reagent *has* to attack from an axial direction to interact with the p orbital. If it attacks from above, the new substituent is axial on an unstable twist boat. If it attacks from below, the new substituent is axial on a chair—granted, this is not as good as equatorial on a chair, but that's not an option—it has to be axial on something, and a chair is better than a twist boat. So this is the product that forms. It's just hard luck for the substituent that it can't know that *if it did* weather it out on the twist boat it could later get equatorial—it plumps for life on the easy chair and so has to be content with ending up axial.

Here is an example with an unsaturated carbonyl compound as an electrophile: the reaction is Michael addition. The ketone here is slightly different—it has the *t*-butyl group in the 3- rather than the 4-position and the reacting centre becomes quaternary during the Michael reaction. But the result is still axial attack.

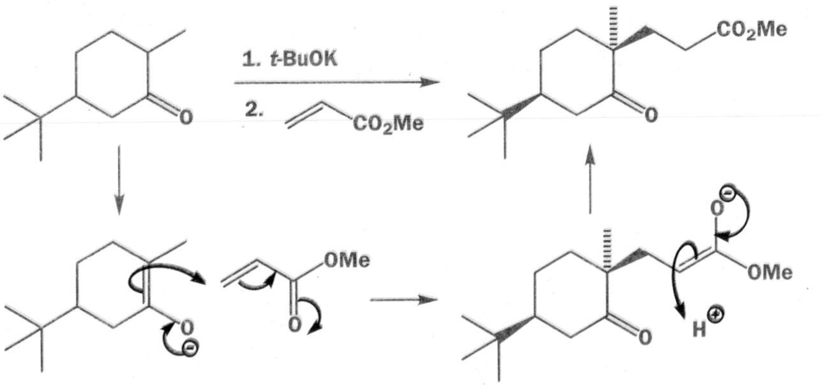

This result is more impressive because the large electrophile ends up on the *same* side of the ring as the *t*-butyl group, so the stereoselectivity cannot be based on any simple idea of reaction on the less hindered side of the ring. It is genuine axial attack, as the conformational diagram of the product confirms.

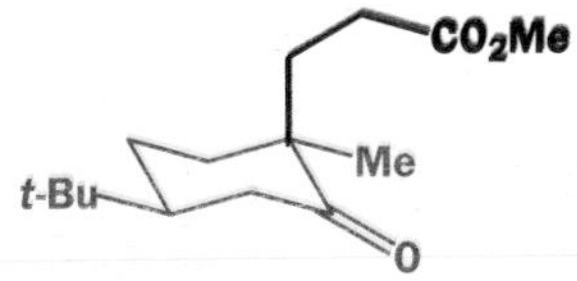

Cyclohexenones are even flatter than cyclohexenes, but it is convenient to draw them in a similar conformation. Conjugate addition to this substituted cyclohexenone gives the *trans* product.

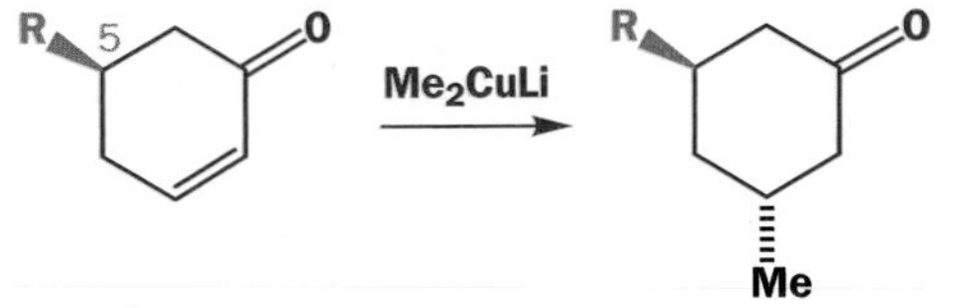

▶

Beware: you also get the right answer for the wrong reason by saying that the nucleophile approaches from the less hindered side.

This is also axial addition to form a chair directly (rather than a twist boat) with the nucleophile approaching from the bottom. We must draw the ring as a flattened chair.

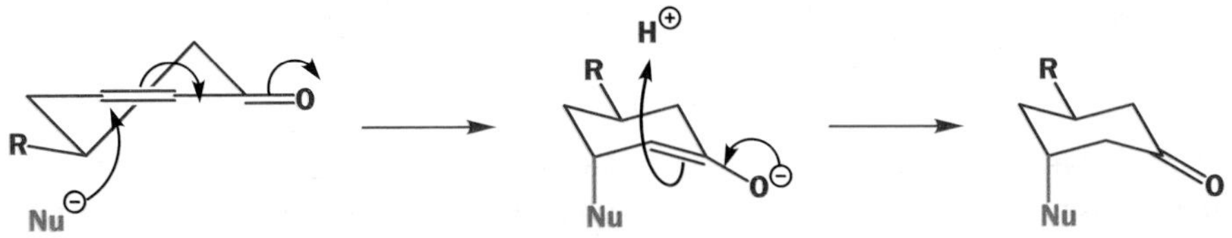

The 5-alkyl cyclohexenone that we have chosen as our example gives the best results. The mechanism suggests that the enolate intermediate is protonated on the top face (axial addition again) though we cannot tell this. But, if we carry out a tandem reaction with the enolate trapped by a different electrophile, the product is again that of axial attack.

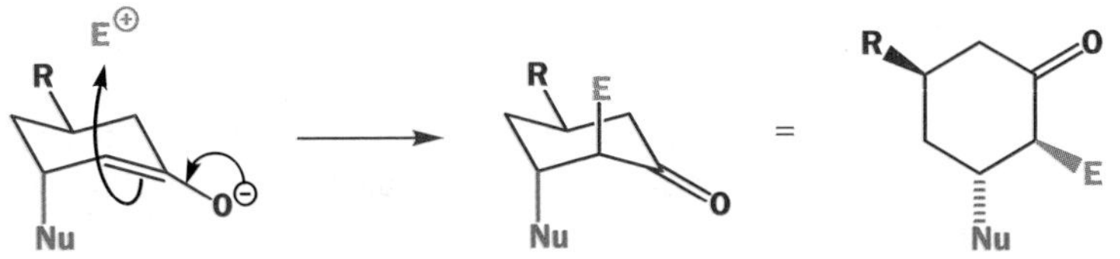

■

You will see 8-phenylmenthol used as a 'chiral auxiliary' in Chapter 45.

We shall end this section on conformational control in six-membered rings with the preparation of a useful chiral molecule 8-phenylmenthol from the natural product (*R*)-(+)-pulegone. The first step is a conjugate addition to an exocyclic alkene. A new stereogenic centre is formed by protonation of the enolate intermediate but with virtually no stereoselectivity.

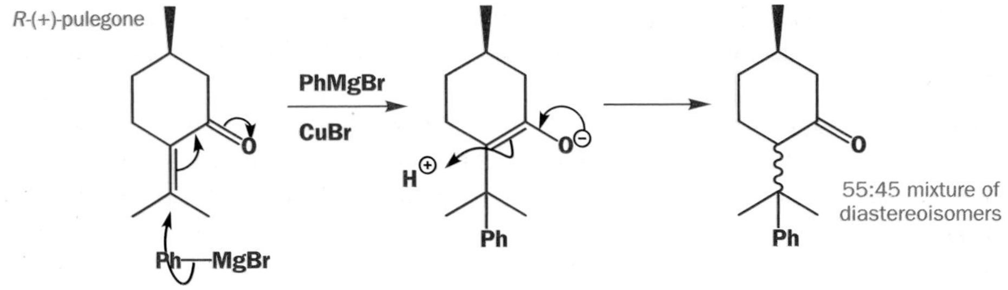

Now thermodynamic control can be brought into play. The position next to the ketone can be epimerized via the enolate to give the more stable isomer with both substituents equatorial. This improves the ratio of diastereoisomers from 55:45 to 87:13.

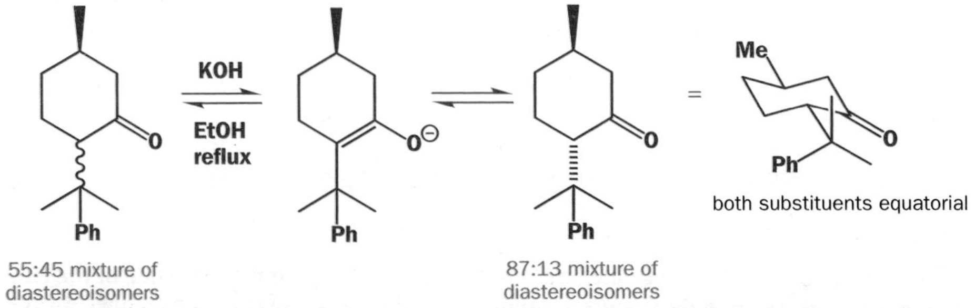

Now the ketone can be reduced with a small reagent—Na in *i*-PrOH works well—to put the hydroxyl group equatorial. This means that all the product has OH *trans* to the large group next to the ketone, though it is still an 87:13 mixture of diastereoisomers with respect to the relative configuration at the centre bearing Me.

Na in *i*-PrOH is a single-electron Birch-type reduction (Chapter 25). You can't get much smaller than an electron!

87:13 mixture of diastereoisomers

Na, *i*-PrOH
reflux in toluene

trans   87%   trans   13%

These alcohols can be separated (they are, of course, diastereoisomers and not enantiomers) and the major, all-equatorial one is the useful one (see Chapter 45). This is an impressive example of conformational control by thermodynamic and by kinetic means using only a distant methyl group in a six-membered ring.

Conformational control in the formation of six-membered rings

In Chapter 32 we solved a structural problem from the aldol reaction of pentan-3-one and 4-chlorobenzaldehyde in basic solution. The product turned out to be a six-membered cyclic keto-ether.

Once you know the gross structure of the product, the stereochemistry should be no surprise. This is a typical thermodynamically controlled formation of a six-membered ring with all the substituents equatorial.

Any reaction that is reversible and that forms a six-membered ring can be expected to put as many substituents as possible in the thermodynamically favourable equatorial position. This principle can be used in structure determination too. Suppose you have one diastereoisomer of a 1,3-diol and you want to find out which stereoisomer it is.

Having read Chapter 32 you might think of using the NMR coupling constants of the two black protons. But that will do no good because the molecule has no fixed conformation. Free rotation about all the σ bonds means that the Karplus equation cannot be used as a time-averaged J value of about 6–7 Hz will probably be observed for both protons regardless of stereochemistry. But suppose we make an acetal from the 1,3-diol with benzaldehyde.

This may not seem to help much. But acetal formation is under thermodynamic control, so the most stable possible conformation will result with the large phenyl group equatorial and the two R groups either both equatorial or one equatorial and one axial, depending on which diastereoisomer you started with.

this diastereoisomer   gives an acetal in this conformation   this diastereoisomer   gives an acetal in this conforma

Now the molecule has a fixed conformation and the coupling constants of the black Hs to the neighbouring CH_2 group can be determined—an axial H will show one large J value, an equatorial H only small J values.

This section has been strong on thermodynamic control but weak on the more common kinetic control. This will be remedied in Chapter 35 where you will meet the most important cyclization reaction of all—the Diels–Alder reaction. It is under kinetic control and there is a great deal of stereochemistry associated with it.

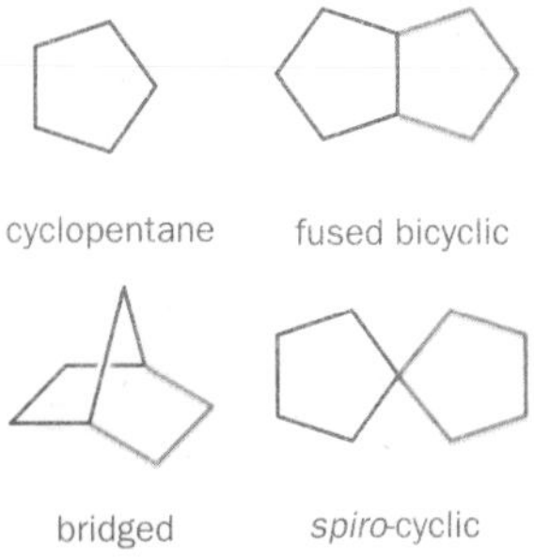

Stereochemistry of bicyclic compounds

There are broadly three kinds of bicyclic compounds, some of which you have met before (Chapter 18, for example). If we imagine adding a new five-membered ring to one already there, we could do this in a bridged, fused, or *spiro* fashion. Bridged bicyclic compounds are just what the name implies—a bridge of atom(s) is thrown across from one side of the ring to the other. Fused bicyclic compounds have one *bond* common to both rings, while *spiro* compounds have one *atom* common to both rings.

You will notice that these three types of bicyclic compounds with five-membered rings have different numbers of atoms added to a 'parent' five-membered ring. The bridged compound has two extra atoms, the fused compound three, and the *spiro* compound four. These are marked in green with the original five-membered ring in red. We shall consider stereoselectivity in each of these types of bicyclic ring systems, starting with bridged structures.

A selection of important bridged bicyclic compounds is shown below, with the various ring sizes indicated in black.

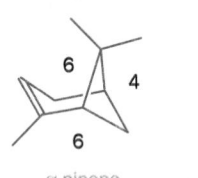
α-pinene

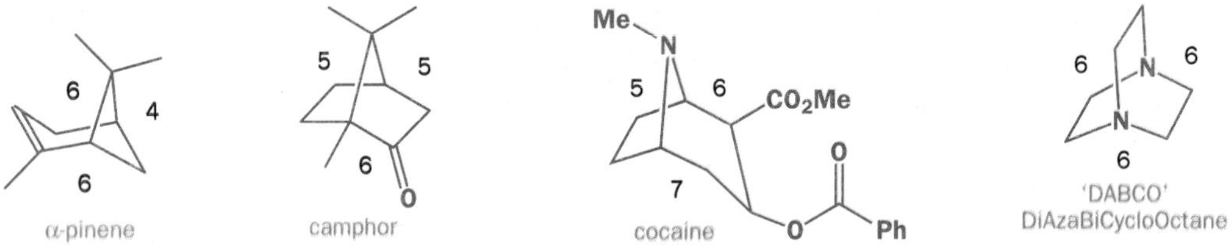

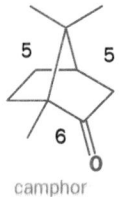
camphor

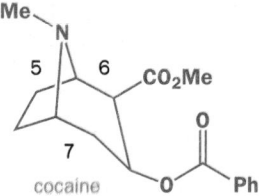
cocaine

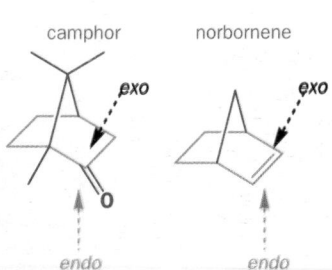

Bridged structures (sometimes called cage structures) are generally very rigid—the only exception among these examples is the bottom right-hand portion of cocaine. This rigidity is reflected in the stereochemistry of their reactions.

Attack on this unsubstituted bridged ketone—norbornanone—occurs predominantly from the side of the one-atom bridge rather than the two-atom bridge.

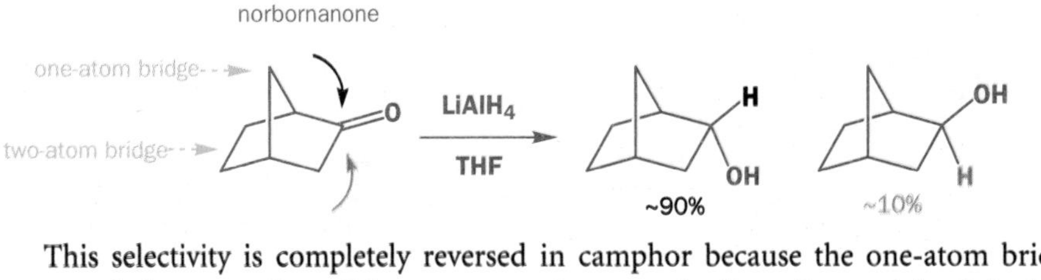

This selectivity is completely reversed in camphor because the one-atom bridge then carries two methyl groups. One of these must project over the line of approach of the hydride reducing agent.

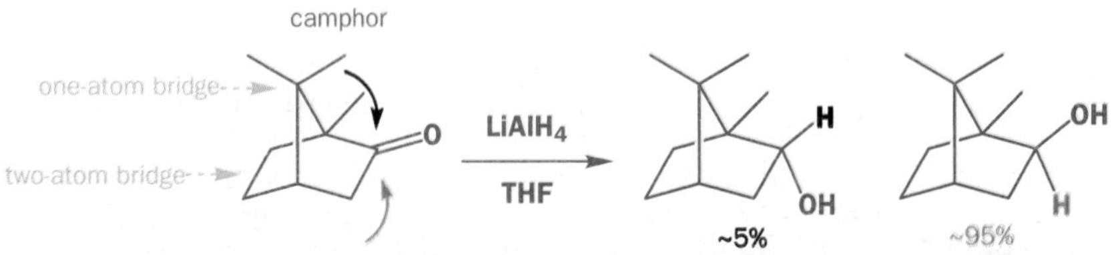

The two methyl groups on the bridge of the camphor molecule are key features in stereoselective reactions—take them away and the result often changes dramatically. This bicyclic system, with and without methyl groups, has been so widely used to establish stereochemical principles that the two faces of, say, the ketone group in camphor, or the alkene in norbornene, have been given the names *endo* and *exo*. These refer to inside (*endo*) and outside (*exo*) the boat-shaped six-membered ring highlighted in orange.

Like LiAlH$_4$ reduction, addition of a Grignard reagent to camphor occurs almost entirely from the *endo* face, but almost entirely from the *exo* face with norbornanone.

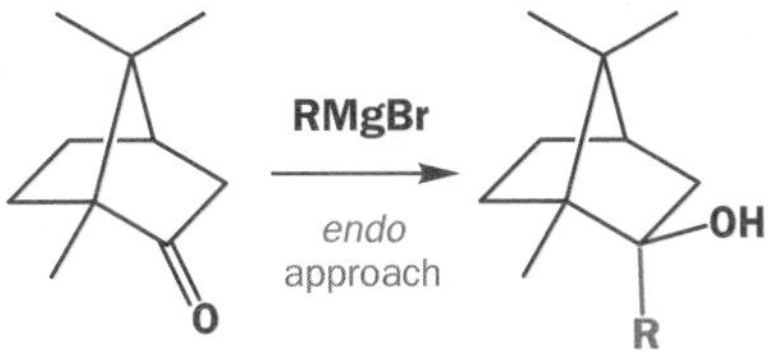

camphor

norbornanone

In a similar style, epoxidation of the two alkenes is totally stereoselective, occurring *exo* in norbornene and *endo* when methyl groups are present on the bridge. These stereoselectivities would be remarkable in a simple monocyclic compound, but in a rigid bridged bicyclic structure they are almost to be expected.

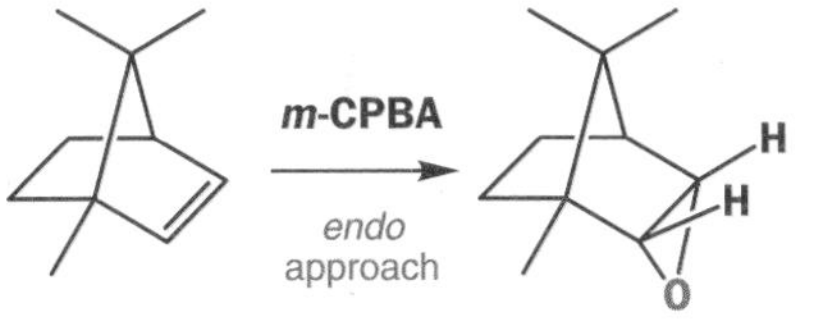

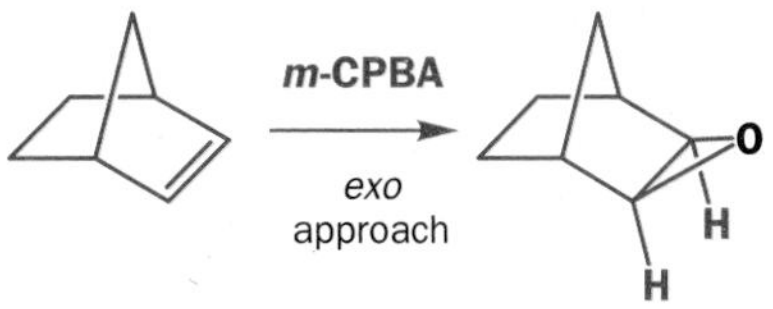

Reactions that break open bridged molecules preserve stereochemistry

Some powerful oxidizing agents are able to cleave C–C bonds, as you will see in Chapter 35. Oxidation of camphor in this way produces a diacid known as camphoric acid. The usual reagent is nitric acid (HNO$_3$) and oxidation goes via camphor's enol.

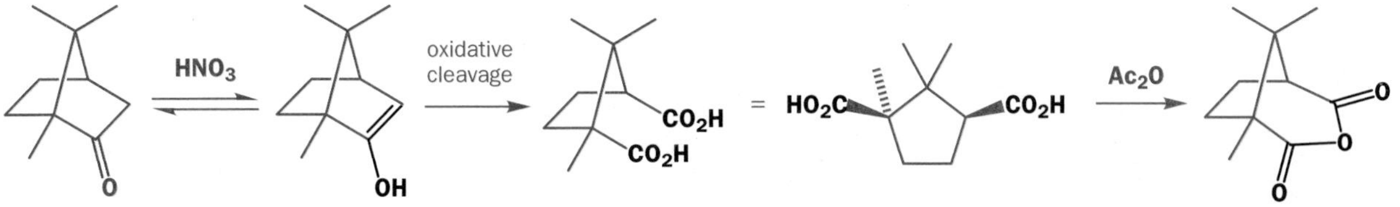

camphor   enol   camphoric acid   camphoric anhydride

Because the bridge holds the molecule in a fixed conformation, the cleaved diacid has to have a specific stereochemistry. There is no change at the stereogenic centres, so the reaction must give retention of configuration. We can confidently write the structure of camphoric acid with *cis*-CO$_2$H groups, but any doubt is dispelled by the ability of camphoric acid to form a bridged bicyclic anhydride.

> ▶
> Anhydride formation with acetic anhydride goes via attack of one acid group on Ac$_2$O to form a mixed anhydride, followed by displacement of AcOH by the other acid group.

Fused bicyclic compounds

trans-Fused rings

The ring junction of a fused 5/6-membered ring system can have *cis* or *trans* stereochemistry, and so can any pair of larger rings. For smaller rings, *trans* 5/5- and 4/6-ring junctions can be made, with difficulty, but with smaller rings *trans* ring junctions are essentially impossible.

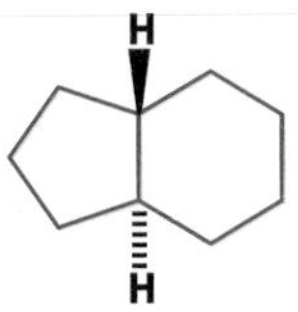

The *trans*-fused 6/6 systems—***trans*-decalins**—have been very widely studied because they appear in steroids (Chapter 51). Their conformation is discussed in Chapter 18 and conformational control simply extends what we saw with simple six-membered rings.

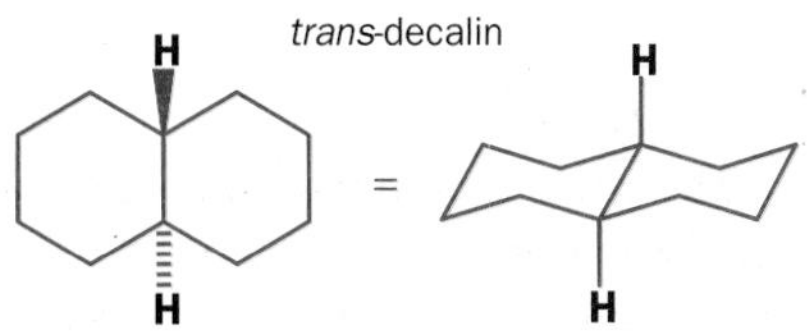

A 6/6 fused system will prefer a *trans* ring junction as *trans*-decalins (Chapter 18) have all-chair structures with every bond staggered from every other bond, as you can see from the diagram alongside. We can show

trans-decalin

6,6 can be *trans* or *cis*, but prefers *trans*

5,6 can be *trans* or *cis*

this by giving a 6/6 system the choice: reducing this enone with lithium metal gives a lithium enolate (Chapter 26). Protonation of this anion with the solvent (liquid ammonia) gives a *trans* ring junction.

In this scheme, and the next, the methyl group attached at the yellow p orbital has been omitted for clarity.

The lithium enolate remains and can be alkylated with an alkyl halide in the usual way. When there are hydrogen atoms at both ring junction positions, axial alkylation occurs just as you should now expect, and a new ketone with three stereogenic centres is formed with >95% stereoselectivity.

>95% this diastereoisomer

However, if there is anything else—even a methyl group—at the ring junction, so that axial approach would give a bad 1,3-diaxial interaction in the transition state, the stereoselectivity switches to >95% equatorial alkylation. This unexpected reversal of normal stereoselectivity is a result of the extra rigidity of the *trans*-decalin system.

>95% this diastereoisomer

In most reactions of *trans*-decalins, the conformational principles of simple six-membered rings can be used, but you may expect tighter control from the greater rigidity. If you wish to design a molecule where you are quite certain of the conformation, a *trans*-decalin is a better bet than even a *t*-butyl cyclohexane as *trans*-decalins cannot flip.

cis-Fused rings

bicyclo[1.1.0]butane

You met catalytic hydrogenation in Chapter 24.

For a reminder of what **stereoselective** and **stereospecific** mean, see p. 492.

Almost any *cis*-fused junction from 3/3 upwards can be made. Bicyclo[1.1.0]butane exists, though it is not very stable. *cis*-Fused 4/5, 4/6, and 5/5 systems are common and are much more stable than their *trans*-isomers.

cis-fused 5/4, 6/4, and 5/5 bicyclic rings

Any method of making such bicyclic compounds will automatically form this stereochemistry. An important method of stereochemical control that we have not used so far in this chapter is catalytic hydrogenation of alkenes, which adds a molecule of hydrogen stereospecifically *cis*. If the reaction also makes a fused ring system, it may show stereoselectivity too. Here is an example with 5/5 fused rings.

The two new hydrogen atoms (shown in black) must, of course, add *cis* to one another: this is a consequence of the stereospecificity of the reaction. What is interesting is that they have also added *cis* to the green hydrogen atom that was already there. This approach does give the more stable *cis* ring junction but the stereochemistry really arises because the other ring hinders approach to the other face of the alkene. Think of it this way: the alkene has two different faces. On one side there is the green hydrogen atom, and on the other the black parts of the second ring. To get hydrogenated, the alkene must lie more or less flat on the catalyst surface and that is easier on the top face as drawn.

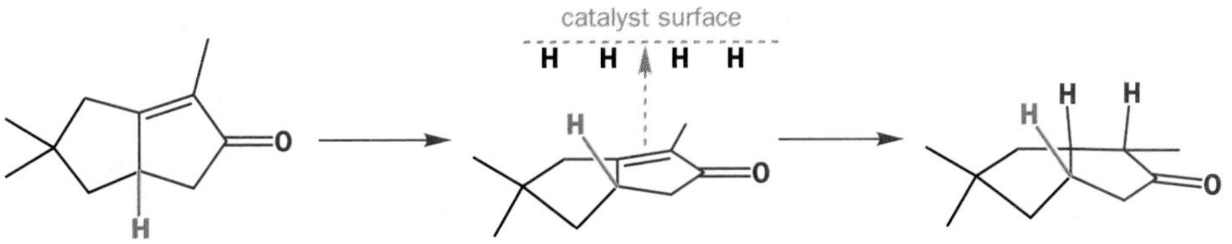

If one of the ring junctions is a nitrogen atom, we might think that there is no question of stereochemistry because pyramidal nitrogen inverts rapidly. So it does, but if it is constrained in a small ring, it usually chooses one pyramidal conformation and sticks to it. The next case is rather like the last.

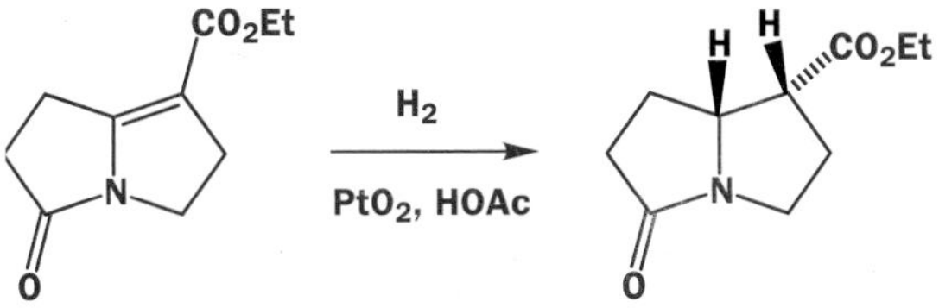

Here again the two black hydrogens have added stereospecifically *cis*, but there is no stereogenic centre in the starting material to control stereoselectivity. So what is there to discuss? If the product is treated with a tertiary amine base (actually DBN is used), it equilibrates to the other diastereoisomer via the ester enolate.

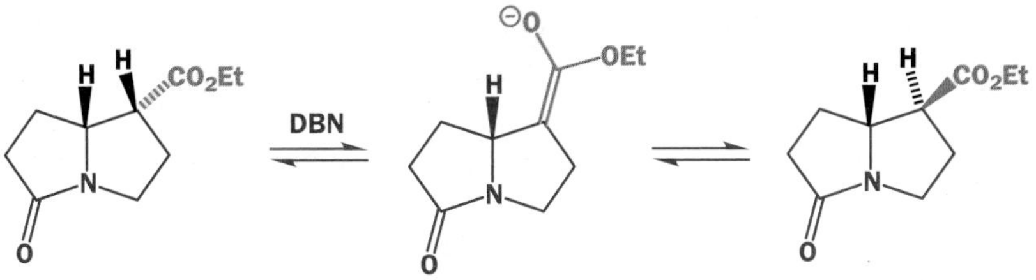

It is easy to see *how* the equilibration happens as the enolate can be protonated at the front or the back, but *why* should it prefer the second structure? This is thermodynamic control and results from the 'disguised' *cis* ring junction. Because it is more stable to have two five-membered rings *cis*-fused, the nitrogen atom is slightly (only slightly, because it is part of an amide) pyramidalized in that direction.

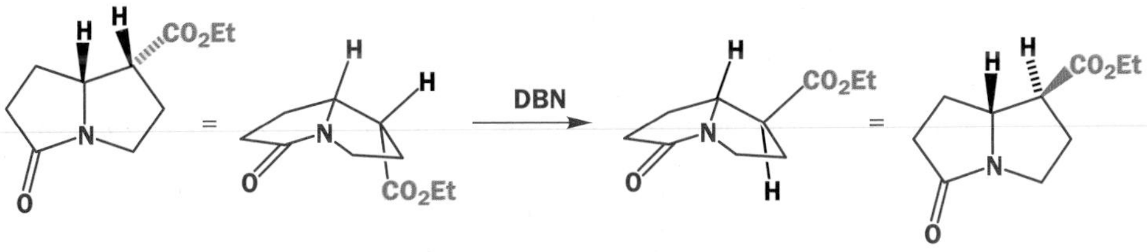

The molecule folds along the C–N bond common to both rings so that it looks rather like that half-opened book that you put face downwards on the table while you answered the phone. The ester group much prefers to be in free space outside the folded rings and not cramped inside them.

This is the key to *cis*-fused bicyclic rings—everything happens on the outside (on the cover of the book). Nucleophiles add to carbonyl groups from the outside, enolates react with alkyl halides or Michael acceptors on the outside, and alkenes react with peroxyacids on the outside. Notice that this means the same side as the substituents at the ring junction. The rings are folded away from these substituents that are on the outside.

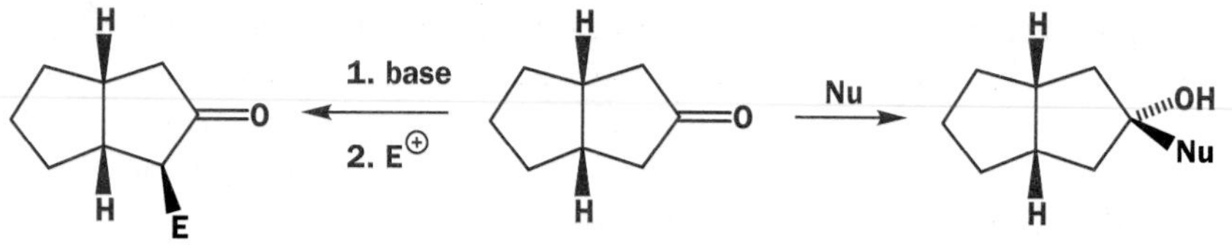

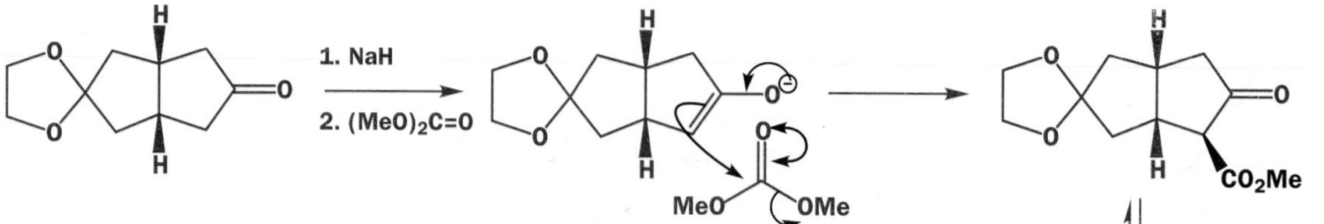

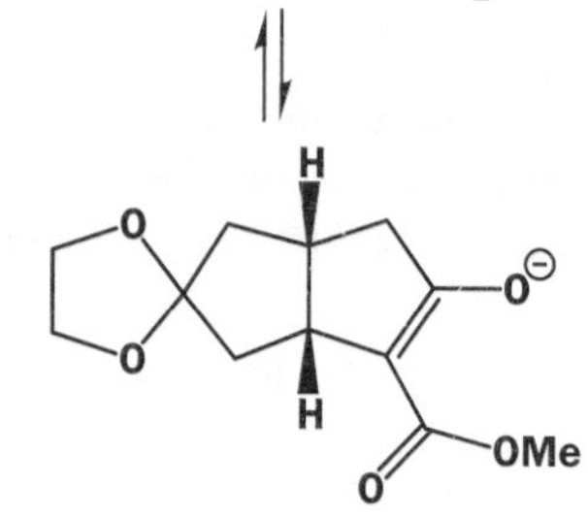

A real example comes in the acylation (Chapter 28) of the enolate from the keto-acetal above and alongside. The molecule is folded downwards and the enolate is essentially planar. Addition presumably occurs entirely from the outside, though the final stereochemistry of the product is controlled thermodynamically because of reversible enolization of the product: whatever the explanation, the black ester group prefers the outside.

Reduction of the ketone product also occurs exclusively from the outside and this has the ironic effect of pushing the new OH group into the inside position. Attack from the inside is very hindered in this molecule because one of the acetal oxygen atoms is right on the flight path. You will see more in a moment on how to force groups into the inside.

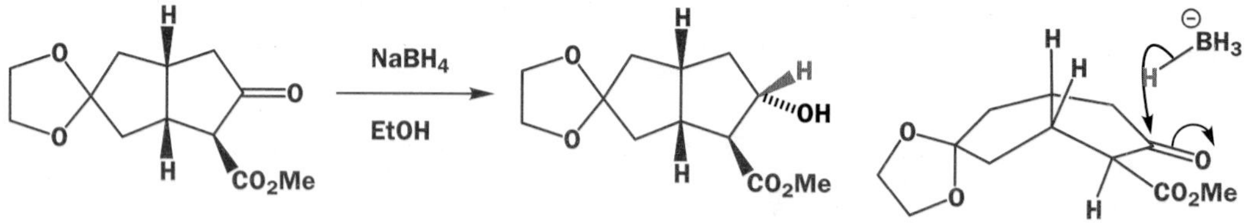

A simple example of epoxidation occurs on a cyclobutane fused to a five-membered ring. This is a very rigid system and attack occurs exclusively from the outside to give a single epoxide in good yield.

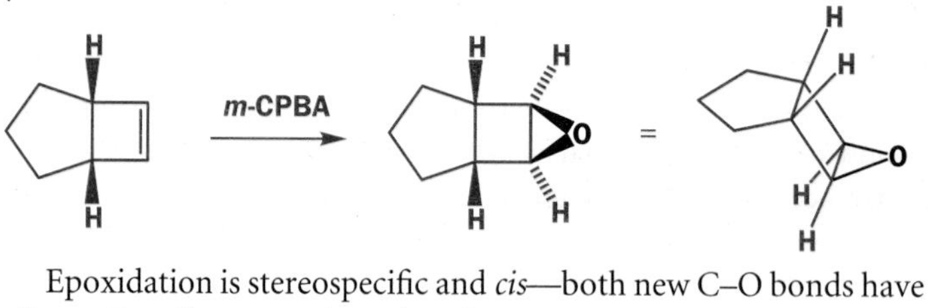

Epoxidation is stereospecific and *cis*—both new C–O bonds have to be on the same face of the old alkene. But Chapter 20 introduced you to several electrophilic additions to alkenes that were stereospecific and *trans*, many of them proceeding through a bromonium ion. If stereospecific *trans* addition occurs on a *cis*-fused bicyclic alkene, the electrophile will first add to the outside of the fold, and the nucleophile will then be forced to add from the inside. A telling example occurs when the 4/5 fused unsaturated ketone below is treated with *N*-bromoacetamide in water.

N-bromoacetamide is another source of electrophilic Br.

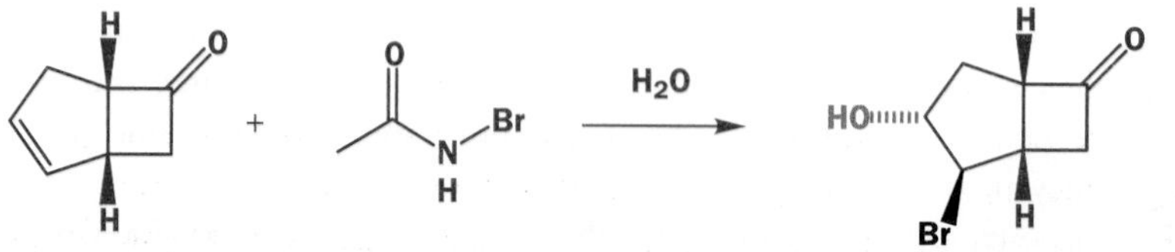

The bromonium ion is formed on the outside of the rigid structure and the water is then forced to add from the inside to get *trans* addition. As well as exhibiting stereospecificity (*trans* addition) and stereoselectivity (bromonium forms on outside), this reaction also exhibits regioselectivity in the attack of water on the bromonium ion. Water must come from inside, but it attacks the less hindered end of the bromonium ion, keeping as far from the 'spine of the half-open book' as possible.

After protection of the OH group, treatment with base closes a three-membered ring to give a remarkably strained molecule. The ketone forms an enolate and the enolate attacks the alkyl bromide intramolecularly to close the third ring. This enolate is in just the right position to attack the C–Br bond from the back, precisely because of the folding of the molecule.

Inside/outside selectivity may allow the distinction between two otherwise similar functional groups. The *cis*-fused bicyclic diester below may look at first rather symmetrical but ester hydrolysis leaves one of the two esters alone while the other is converted to an acid.

Only the outside ester—on the same side as the ring junction Hs—is hydrolysed. In the mechanism for ester hydrolysis, the rate-determining step is the attack by the hydroxide ion so the functional group *increases* in size in the vital step. This will be much easier for the free outside CO_2Et group than for the one inside the half-open book.

The end result is that the larger of the two groups is on the inside! There are other ways to do this too. If we alkylate the enolate of a bicyclic lactone, the alkyl group (black) goes on the outside as expected. But what will happen if we repeat the alkylation with a different alkyl group? The new enolate will be flat and the stereochemistry at the enolate carbon will be lost. When the new alkyl halide comes in, it will approach from the outside (green) and push the alkyl group already there into the inside.

alkylation on the outside face

R^1 forced on to the inside face

Should you wish to reverse the positions of the two groups, you simply add them in the reverse order. Whichever group is added first finishes on the inside; the other finishes on the outside.

Before we move on to *cis*-decalins, here is a sequence of reactions that starts with a symmetrical eight-membered ring with no stereogenic centres and ends with two fused five-membered rings with five stereogenic centres, all controlled by stereospecific reactions, some with stereoselective aspects controlled by *cis*-fused rings.

> This molecule now has three-, four-, and five-membered rings fused together in a tricyclic cage structure. This is nowhere near the limit for cage molecules. You saw tetra-*t*-butyl tetrahedrane in Chapter 15, and you will see in Chapter 37 how even molecules such as cubane can be made.
>
>
>
> cubane

The first step is a reaction you haven't yet met—it comes in Chapter 47. All you need to know now is that the reagent, a boron-containing compound called 9-borabicyclononane (9-BBN), hydrates one of the double bonds in the reverse fashion to what you would expect with acid or Hg^{2+} (Chapter 20) and stereospecifically (H and OH go in *cis*). The resulting alcohol is mesylated (p. 486) in the usual way. This puts in H and OMs stereospecifically *cis* to each other.

- This is a hydroboration reaction, and the oxygen comes from the peroxide used to work up the reaction. You can read more about 9-BBN, and see its structure, on p. 1283 of Chapter 47.

hydroboration: OH and H add *cis*

Now comes the first really interesting step. The other alkene does an intramolecular S_N2 reaction to displace the mesylate with inversion and form two fused five-membered rings. The ring junction is *cis*, of course.

▶ If you find it hard to see that the reaction has gone with inversion, look at the kink in the ring highlighted in green. The old OMs comes downwards from that kink and towards you, out of the page. The new bond forms upward from the kink, more or less 180° from where the old CO bond was, and the remaining three bonds accommodate this by 'inverting' like an inside-out umbrella.

The resulting tertiary cation is not isolated but quenched in the reaction mixture with water. One new stereogenic centre is set up in the cyclization and another in the reaction with water. In the cyclization the molecule prefers to fold in such a way that the new ring junction is *cis*.

Addition of water to the cation occurs from the outside—but, in fact, this is unimportant as that stereogenic centre is about to be lost anyway. Treatment with TsCl causes an E2 *anti* elimination. The only proton *anti* to the OTs group is away from the ring junction, so this is where the new double bond goes.

- The elimination is regioselective because of its stereospecificity. You saw a similar example on p. 493 of Chapter 19.

Finally, a second hydroboration with 9-BBN occurs regiospecifically and on the outside of the folded molecule. This reaction adds the last two centres making five in all.

- Another feature of hydroboration that is relevant here and discussed in Chapter 47 is the fact that replacement of B by O in the second step goes with retention. The mechanism will come on p. 1284.

cis-Decalins: *cis*-fused six-membered rings

First a brief reminder of the conformation of *cis*-decalins (see Chapter 18). Unlike *trans*-decalins, which are rigid, they can flip rapidly between two all-chair conformations. During the flip, all substitutents change their conformation. The substituent R is axial on ring B in the first conformation but equatorial in the second. The ring junction Hs are always axial on one ring and equatorial on the other. The green hydrogen is equatorial on ring A and axial on ring B in the first conformation and vice versa in the second. Of course, they are *cis* in both. Because R gets equatorial, the second conformation is preferred in this case.

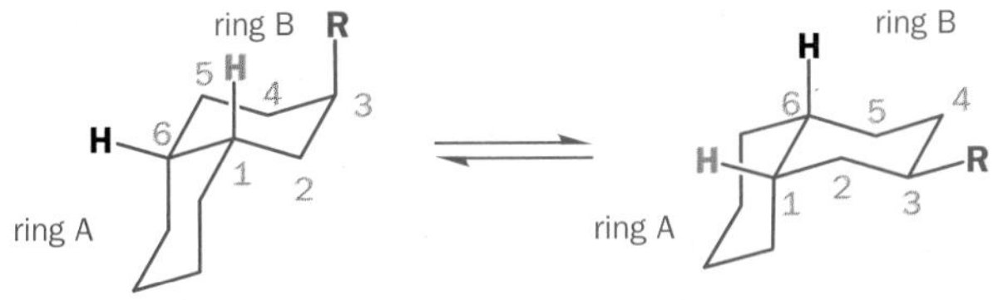

A standard reaction that gives substituted decalins is the Robinson annelation (Chapter 29). A Robinson annelation product available in quantity is the keto-enone known sometimes as the **Wieland–Miescher ketone** and used widely in steroid synthesis. The nonconjugated keto group can be protected or reduced without touching the more stable conjugated enone.

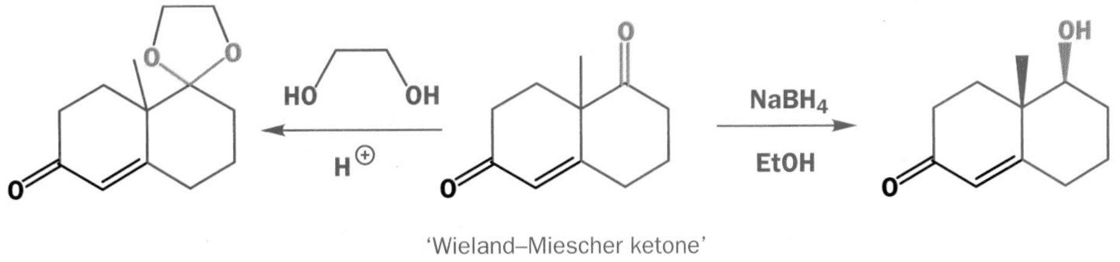

'Wieland–Miescher ketone'

The synthesis of this ketone can be found in Chapter 29, p. 761.

If either of these products is reduced with hydrogen and a Pd catalyst (the alcohol is first made into a tosylate), the *cis*-decalin is formed. We saw a few pages back that the same kind of enones can be reduced with lithium metal in liquid ammonia and that then the more stable *trans*-decalin results.

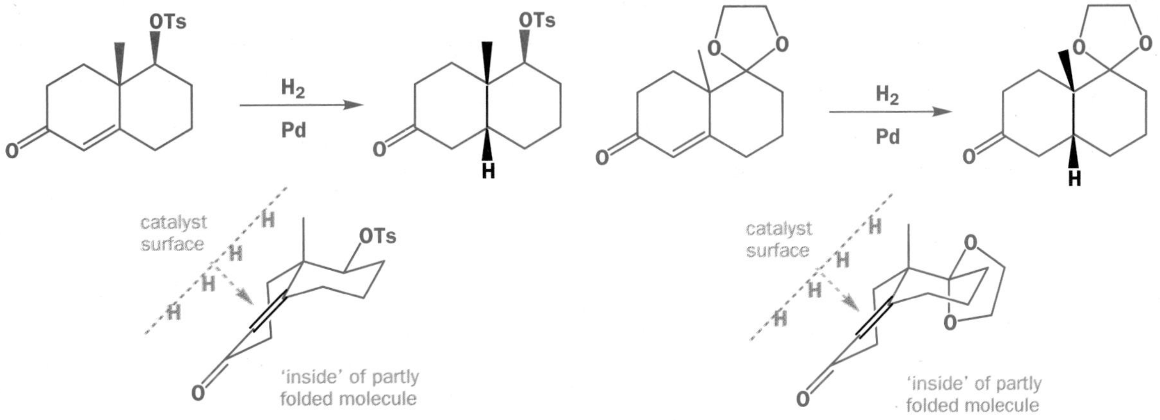

The *cis*-decalin is formed because the enone, though flattened, is already folded to some extent. A conformational drawing of either molecule shows that the top surface is better able to bind to the flat surface of the catalyst. Each of these products shows interesting stereoselective reactions. The ketal can be converted into an alkene by Grignard addition and E1 elimination and then epoxidized. Everything happens from the outside as expected with the result that the methyl group is forced inside at the epoxidation stage.

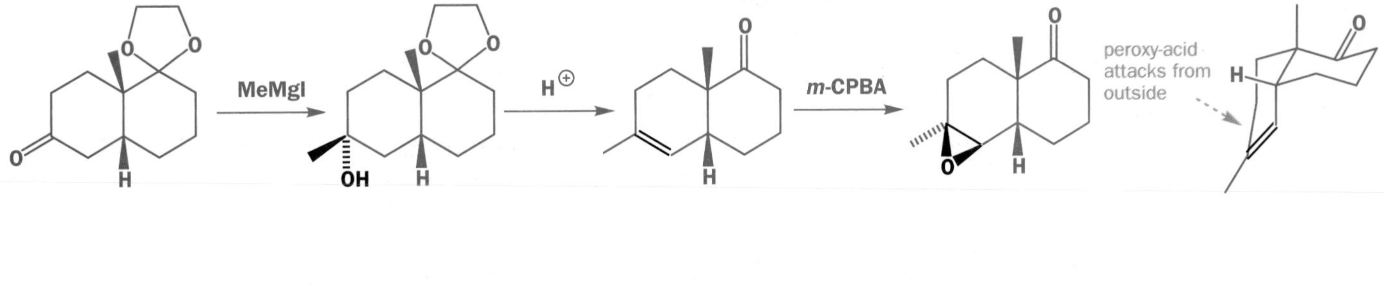

Treatment of the other product, the keto-tosylate, with base leads to an intramolecular enolate alkylation—a cyclization on the inside of the folded molecule that actually closes a four-membered ring. The reaction is easily seen in conformational terms and the product cannot readily be drawn in conventional diagrams.

A similar reaction happens on the epoxide to produce a beautiful cage structure. This time it is a five-membered ring that is formed, but the principle is the same—the molecule closes across the fold rather easily. The new stereogenic centres can only be formed the way they are.

> ▶
> Notice how the right-hand ring in the starting material has to go into a boat conformation for cyclization to be possible. This is unfavourable but still better than any intermolecular reaction.

> ● **A summary of stereoselective reactions that occur on the *cis*-fused rings**
>
> **1.** Reactions on the outside
> • Nucleophilic additions to carbonyl groups in the ring
> • Reactions of enolates of the same ketones with electrophiles: alkyl halides, aldols, Michael additions
> • *cis*-Additions to cyclic alkenes: hydrogenation, hydroboration, epoxidation
> **2.** Reactions on the outside and the inside
> • *trans*-Additions to cyclic alkenes: bromination, epoxide openings
> **3.** Reactions on the inside
> • Bond formation across the ring(s)

Spirocyclic compounds

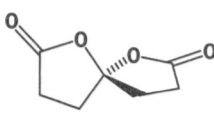

These rings meet at an atom alone. This means that the two rings are orthogonal about the tetrahedral atom that is common to both. Even symmetrical-looking versions are unexpectedly chiral. The compound in the margin, for example, is not superimposable on its mirror image, and its chirality is rather similar to that of an allene.

These sorts of compounds may look rather difficult to come by, but some simple ones are simply made. Cyclization of this keto-acid with polyphosphoric acid leads to a spirocyclic diketone.

The *spiro* compound is formed because the more substituted enol is preferred in acid solution. In a different case, with an enamine, a bridged product is preferred.

After the first alkylation, the enamine prefers to re-form on the less substituted side so that the second alkylation occurs on the other side of the ketone from the first. The spirocyclic compound is further disfavoured as it would have a four-membered ring in this case.

A less substituted enamine is usually preferred to a more substituted one: see p. 672.

It is much more difficult to pass stereochemical information from one ring to the other in spirocyclic compounds because each ring is orthogonal to the other. Nonetheless, some reactions are surprisingly stereoselective—one such is the reduction of the spirocyclic diketone that we made a moment ago. Treatment with LiAlH$_4$ gives one diastereoisomer of the spirocyclic diol.

1,6-*spiro*[4.4]nonadiene

The diol was resolved and used to make the very simple *spiro*-diene as a single enantiomer. It is chiral even though it has no chiral centre because it does not have a plane of symmetry.

In Chapter 16 we explained that planes of symmetry, not chiral centres, are the things to look for when deciding whether or a not a compound is chiral.

Reactions with cyclic intermediates or cyclic transition states

Rings are so good at controlling stereochemistry (as you have seen) that it's well worth introducing them where they are not really necessary in the final product, simply in order to enjoy those high levels of stereochemical control. In the rest of this chapter we shall consider the use of temporary rings in stereochemical control: these might be cyclic intermediates in a synthetic pathway, or cyclic reaction intermediates, or even merely cyclic transition states. All aid good stereocontrol. We shall concentrate on examples where the ring reverses the normal stereoselectivity so that some different result is possible.

Tethered functional groups can reach only one side of the molecule

The proverbial donkey starved to death in the field with two heaps of hay because it could not decide which one to go for first. If the donkey had been tethered to a stake near one heap it would have been able to reach that heap alone and it could have feasted happily.

This principle is often applied to molecules. If a nucleophile is joined to the carbonyl group it is to attack by a short chain of covalent bonds, it may be able to reach only one side of the carbonyl group. An example from a familiar reaction concerns the Robinson annelation. The first step, Michael addition, creates a stereogenic centre but no relative stereochemistry. It is in the second step—the aldol cyclization—that the stereochemistry of the ring junction is decided.

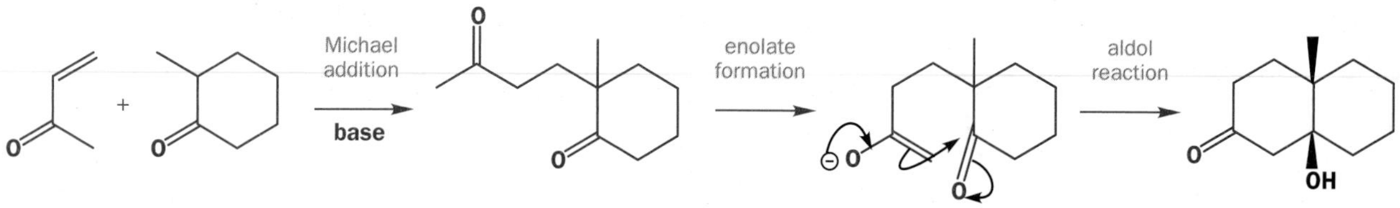

The enolate is tethered to the atom next to the ketone in the other ring. It can attack easily from the side to which it is attached through a stable chair-like transition state. Attacking the other face of the ketone (to give a *trans*-decalin) is much more difficult, even though it would give the thermodynamically more stable product.

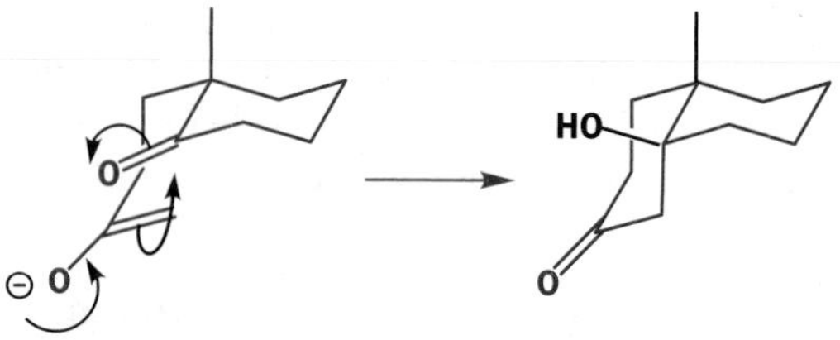

In fact, this is not such a good example because the aldol product is normally dehydrated and the second stereogenic centre is lost. More important examples are those in which a ring is formed but can later be cleaved, and among the best of this type of reaction are iodolactonizations, which you first met in Chapter 20. To remind you, iodolactonization involves treating a nonconjugated unsaturated acid with iodine in aqueous $NaHCO_3$. The product is an iodolactone.

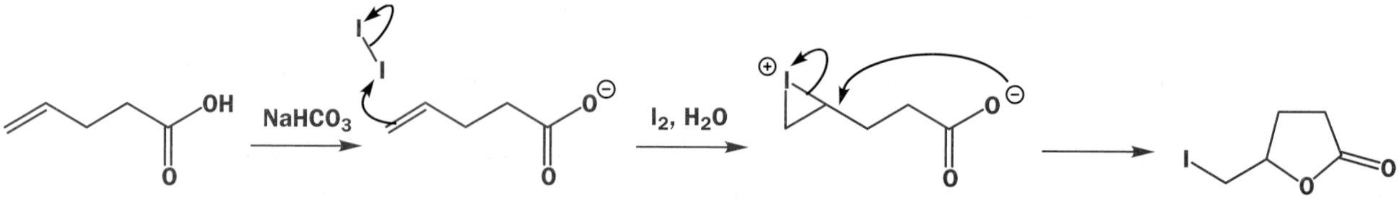

The cyclization reaction is a typical two-stage electrophilic addition to an alkene (Chapter 20) with attack by the nucleophile at the more substituted end of the intermediate halonium ion. The iodonium ring opening is a stereospecific S_N2 and, in the simplest cases where stereochemistry can be observed, the stereochemistry of the alkene will be reproduced in the product.

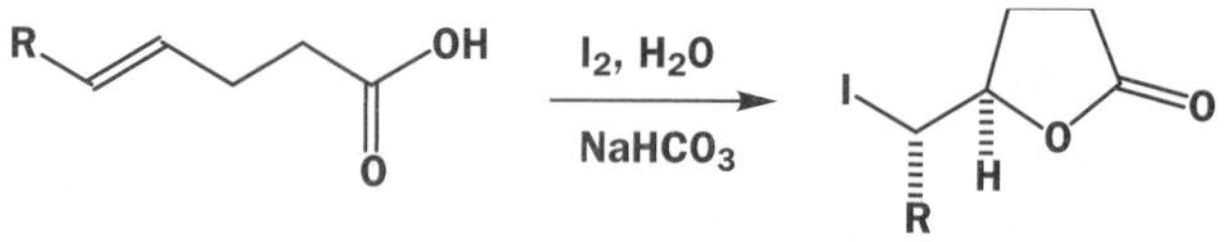

The starting acid contains an *E*-alkene that gives a *trans* iodonium ion. Inversion occurs in the attack of the carboxylate anion on the iodonium ion and we have shown this by bringing the nucleophile in at 180° to the leaving group with both bonds in the plane of the paper. A single diastereoisomer of the iodolactone results from this stereospecific reaction.

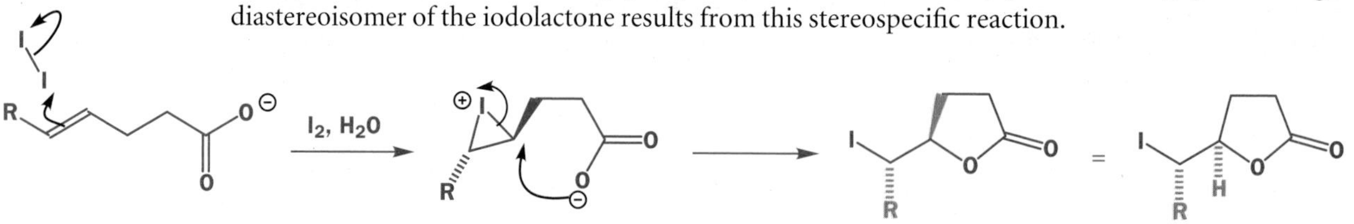

■
Chapter 35 describes how to make the unsaturated six-membered starting material.

▶

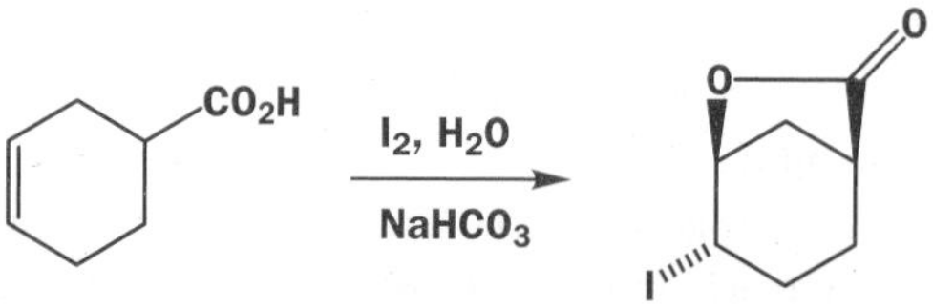

Try making a model of, or even drawing, a two-atom bridge axial–equatorial or equatorial–equatorial between these two carbons and you will find that it's impossible.

The following cyclic example illustrates the stereoselective aspect of iodolactonization.

The relationship between the two stereogenic centres on the old alkene is not an issue—that aspect of the reaction is stereospecific. A more interesting question is the relationship with the third centre. One way to look at this question would be to say that the structure shown is the only possible one. The lactone bridge has to be diaxial (and hence *cis*) if it is to exist and the O and I atoms have to be *trans*. End of story.

But it is still interesting to see how the product arises as it gives us insight into other less clear-cut reactions. The –CO₂H group is too far away for us to argue seriously that the two faces of the alkene are sufficiently different for the iodine to attack one only. A more reasonable explanation is that iodine attacks both faces reversibly but that only the iodonium ion with the I and CO_2H groups *trans* to each other can cyclize. This turns out to be a general rule—iodolactonizations are reversible and under thermodynamic control.

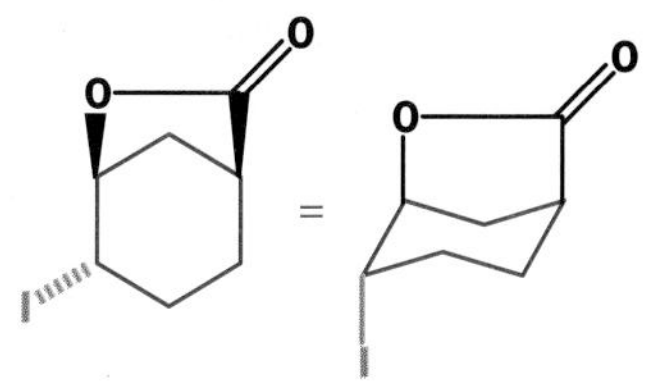

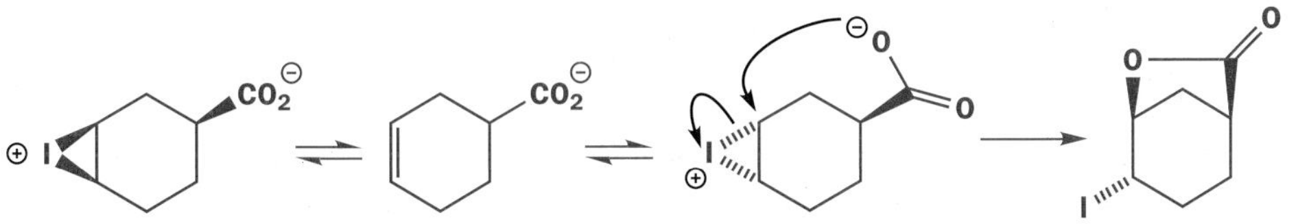

bridge must be diaxial

I must be *trans* to O

One of the simplest open-chain examples is 2-methylbut-3-enoic acid, which cyclizes in >95% yield to a single iodolactone with three stereogenic centres. Two come from stereospecific *trans* addition to the *E*-alkene but the third reveals that iodine attacked the face of the alkene opposite the green methyl group in the conformation that can cyclize.

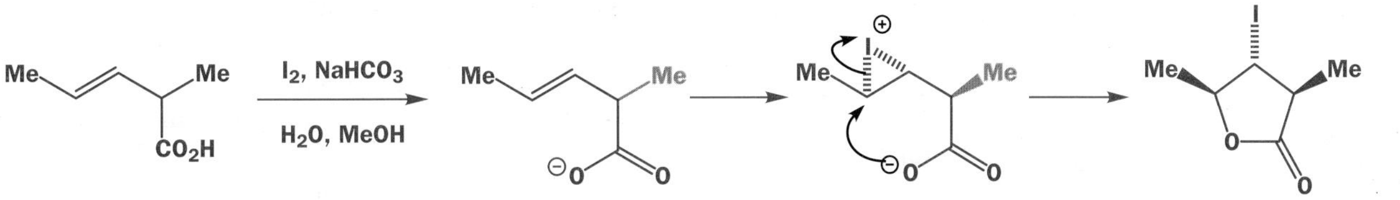

We have said little in this chapter about the stereospecific transformation of one ring into another but we now have an opportunity to remedy that defect. Iodolactonization of a terminal alkene with a stereogenic centre next to it is as stereoselective as (if not more than) the example we have just seen. The two side chains on the ring end up *trans* to one another as we should expect. This is a purely stereoselective process as the alkene has no geometry.

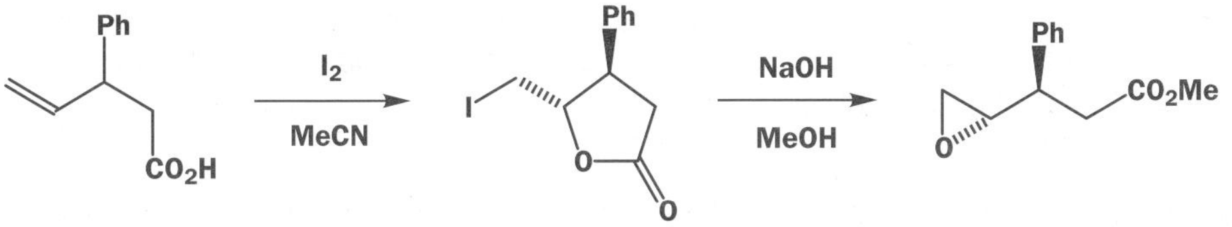

97% yield, 10:1 *trans:cis*

Reaction of the iodolactone product with alkaline methanol transforms it stereospecifically into the methyl ester of an epoxy acid. There is no change in stereochemistry here: methoxide opens the lactone and the oxyanion released carries out an internal S_N2 reaction on the primary alkyl iodide.

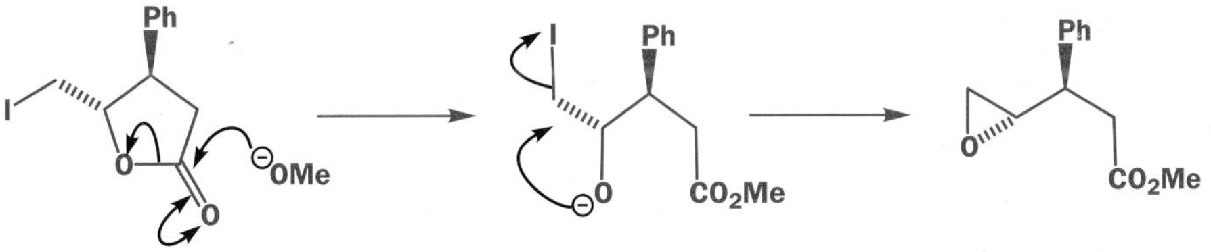

The more obvious way to make this epoxide would be by epoxidation of the ester of the original unsaturated acid. However, the stereoselectivity in that reaction is nowhere near as good as in the iodolactonization. We shall return to this subject when we discuss reactions in acyclic systems in the next chapter.

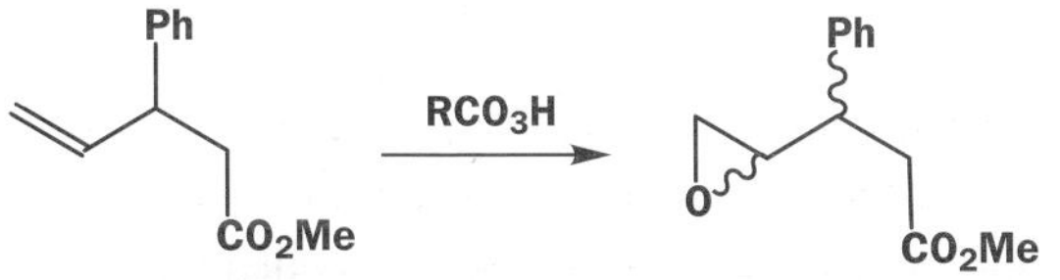

■
There is a brief introduction to steroids in Chapter 18, p. 466. Chapter 49 contains much more detail.

A general problem in the synthesis of steroid compounds is the construction of a diketone with 5/6 *trans*-fused rings and a quaternary carbon atom at the ring junction. **Tethering** can solve this problem, and we will present two strategies—one using a lactone derived from an iodolactonization reaction, and one using a sulfur atom.

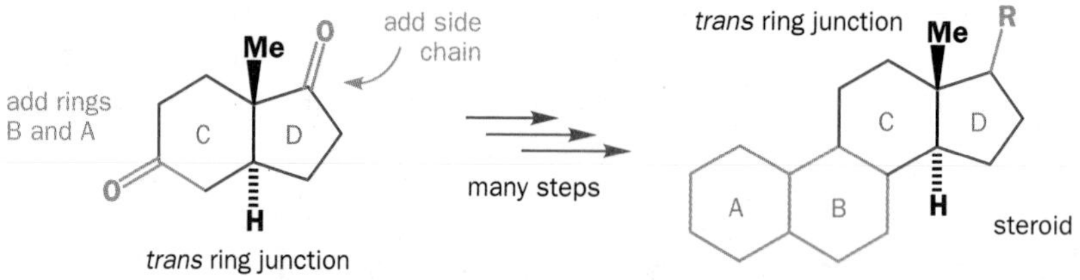

A lactone makes a good temporary tether because it can be hydrolysed or reduced to break the ring at the C–O bond and reveal new stereogenic centres on the old structure. In this sequence a lactone, formed by iodolactonization, controls all the subsequent stereochemistry of the molecule in two ways: it fixes the conformation rigidly in one chair form—hence forcing the iodide to be axial—and it blocks one face of the ring. The iodolactonization is very similar to one you saw on p. 872. Next, an alkene is introduced by E2 reaction on the iodide. This stereospecific reaction requires an anti-periplanar H atom so it has to take the only available neighbouring axial hydrogen atom—furthermore, reaction the other way would produce a bridgehead alkene.

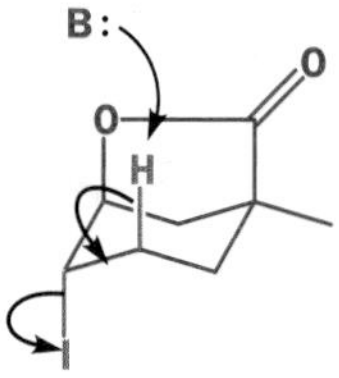

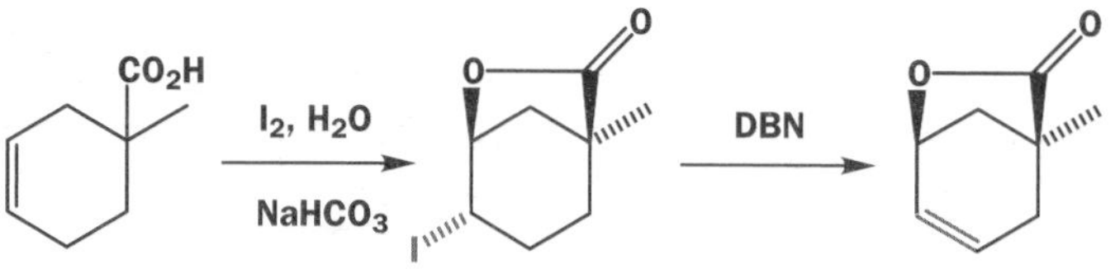

The resulting alkene has its top face blocked by the bridge so a *cis* addition reaction, such as epoxidation, will occur entirely from the bottom face.

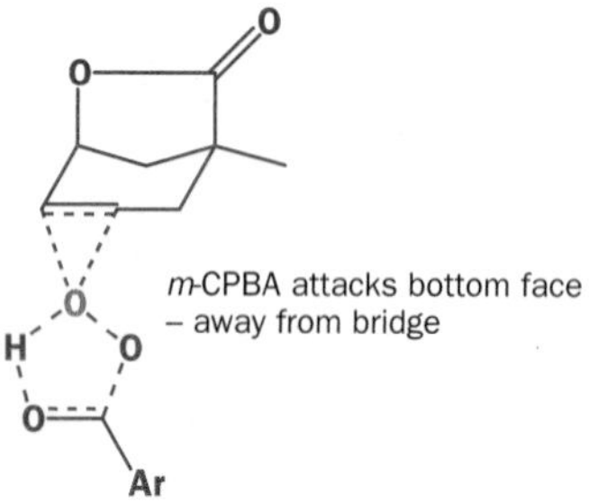

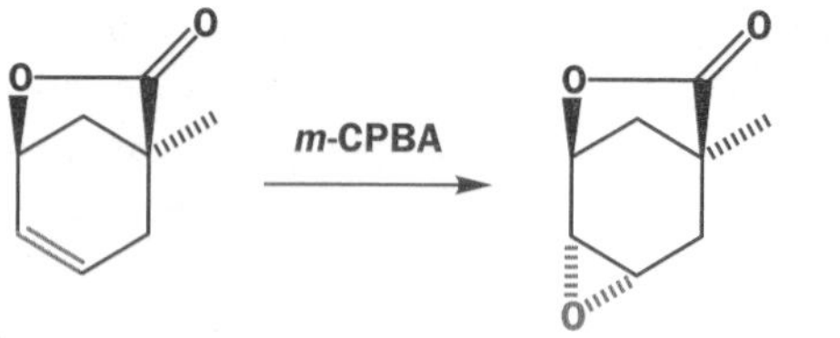

Now the epoxide is opened with HBr to give the only possible *trans* diaxial product (Chapter 18). The role of the bridge in fixing the conformation of the ring is more important in this stereospecific reaction because the bromide ion is forced to attack from the top face. The alcohol is protected as a silyl ether.

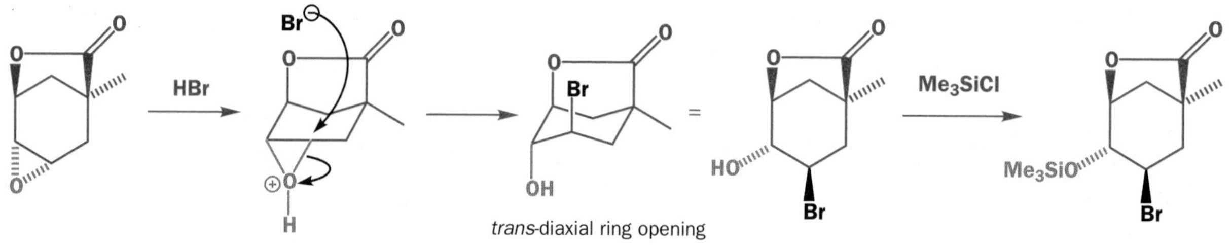

trans-diaxial ring opening

Do you see how the functional groups are being pushed round the ring? This process is extended further by a second elimination also with DBN, which this time really does have to seek out the only neighbouring axial hydrogen: there's no bridgehead to take the decision for it. Acid removes the silyl protecting group.

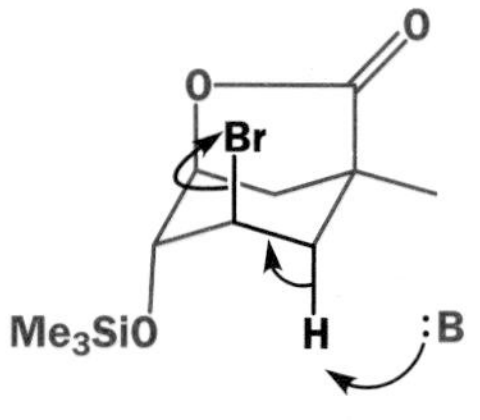

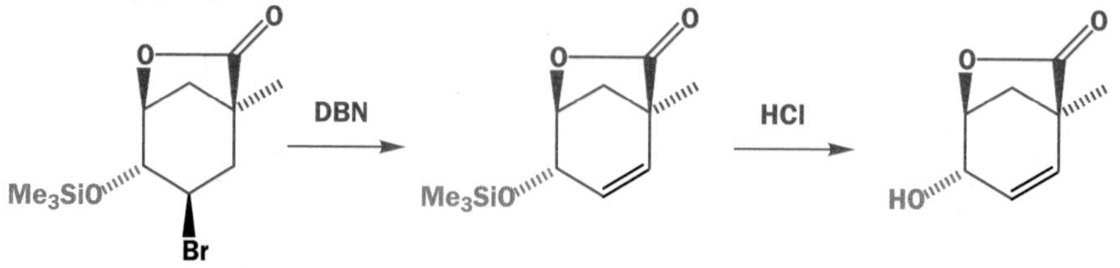

The next important reaction is a Michael addition so the alcohol must first be oxidized to a ketone. As it is an allylic alcohol, it can be oxidized by manganese dioxide. The ring is further flattened as three atoms are now trigonal. But-3-enyl Grignard reagent is next added with Cu(I)

catalysis to make sure that conjugate addition occurs. Conjugate addition normally gives the axial product as we saw earlier and fortunately this is not the direction blocked by the bridge.

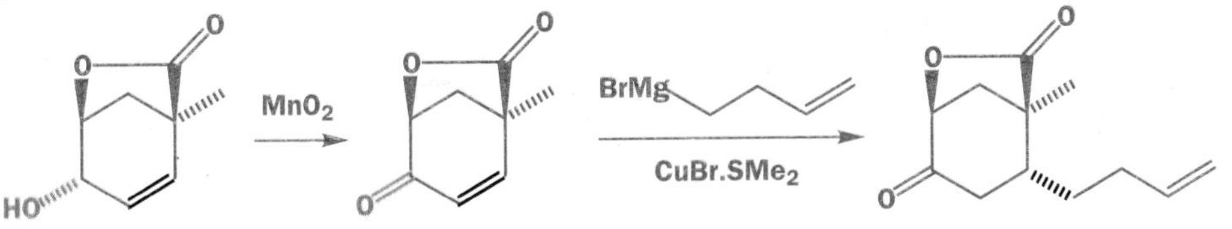

▶

Manganese dioxide is a reagent that oxidizes only allylic or benzylic hydroxyl groups to ketones.

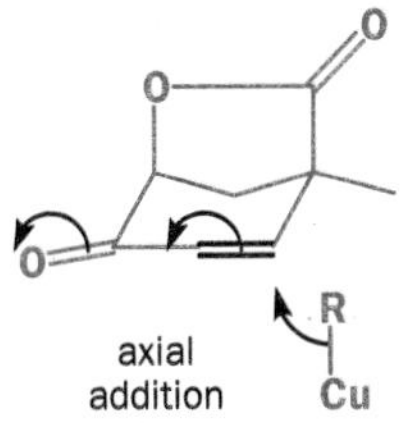

The bridge has now done its work and is removed by zinc metal reduction. This reaction removes leaving groups on the atoms next to carbonyl groups. In this case it is the axial carboxylate that is driven out by the zinc. The released carboxyl group is esterified.

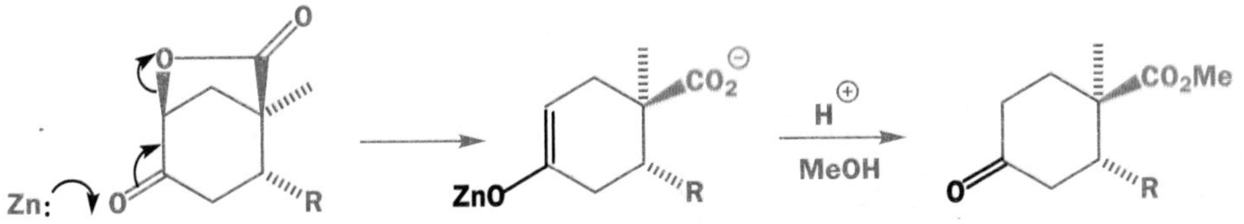

▶

This may look like a new reaction, but think back to the Reformatsky reaction (Chapter 27). Both form zinc enolates from carbonyl compounds with adjacent leaving groups.

The last stages are shown below. The ketone is protected, and the alkene oxidized to a carbonyl group, cleaving off one of the C atoms (you will meet this reaction—ozonolysis—in Chapter 35). The diester can be cyclized by a Claisen ester condensation. The stereogenic centres in the ring are not affected by any of these reactions so a *trans* ring junction must result from this reaction.

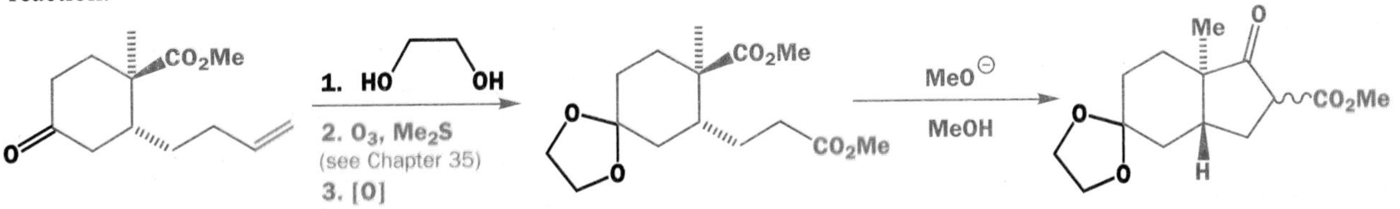

Finally, after ester hydrolysis, HCl decarboxylates the product and removes the protecting group. As we saw earlier, it is not easy to get a *trans*-fused 5/6 system. In this sequence the molecule is effectively tricked into making the *trans* ring junction by the work done with the blocking lactone bridge.

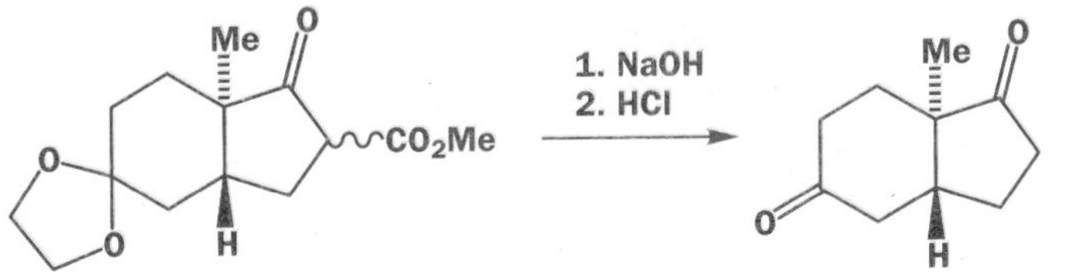

Sulfur as a tether

An even more versatile tether is a sulfur atom, which can be removed completely with Raney nickel (which reduces C–S to C–H). The sulfur atom makes the tether easy to assemble too. Here is the essence of the idea.

■

You met Raney nickel in Chapter 24 and you will see more of it in Chapter 46.

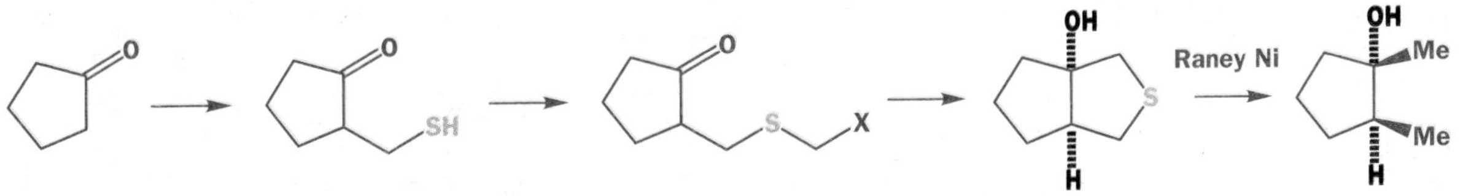

In this second synthesis of the problematic steroid *trans* ring junction, the idea is to make the five-membered ring by a Claisen ester condensation and to direct the stereochemistry by tethering the *cis* groups with a sulfur atom. We can represent this easily in disconnection terms (Chapter 31). The *cis*-carbons to be joined through sulfur are shown in black.

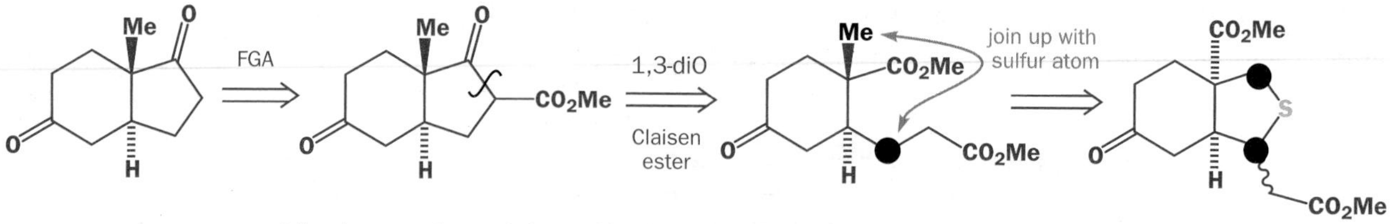

The preparation of the sulfur heterocycle uses reactions you have met before—first a five-membered ring ketone is formed, which is reduced, lactonized, and eliminated.

> You should be able to write mechanisms for all of the reactions in this sequence except the Diels–Alder reaction (Chapter 35). You are asked to do so in one of the problems at the end of the chapter.

> The interpretation of these reactions will be one of the problems we set in Chapter 35.

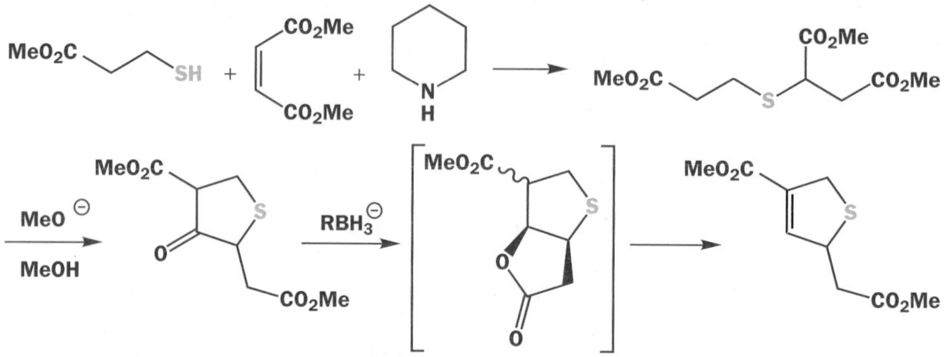

The next steps involve the Diels–Alder reaction, which you will meet in Chapter 35, so we will have no detailed discussion here, just giving the reactions, and pointing out that the product necessarily has a *cis* 6/5 ring junction.

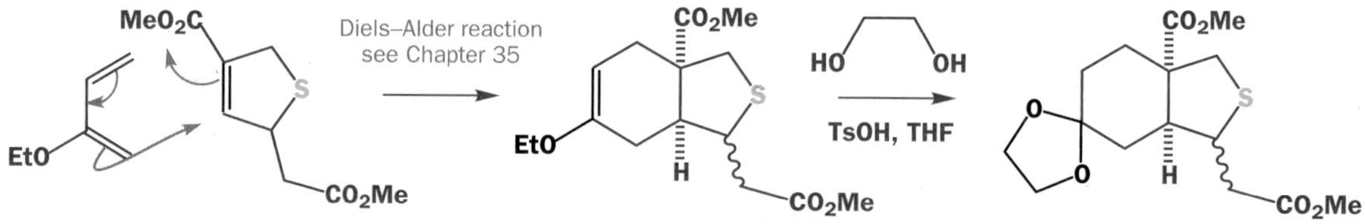

Now the ring has done its work, the two necessary stereogenic centres are fixed, and the sulfur atom can be removed with Raney nickel. The third, undefined, stereogenic centre becomes a CH_2 group in this operation, so the lack of stereocontrol at this centre during the Diels–Alder reaction is of no consequence.

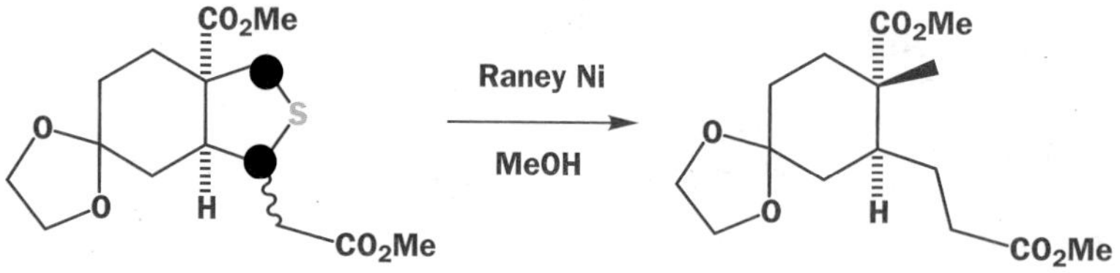

The Claisen ester condensation involves the only possible enolate attacking the only possible electrophilic carbonyl group. The stereochemistry of the ring junction cannot be changed by the reaction, and the two ester groups that started *trans* must end up *trans* in the product.

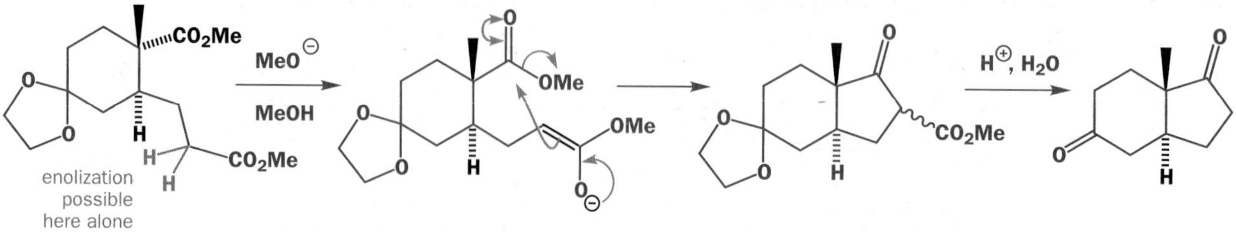

Cyclic transition states can reverse normal stereoselectivity

We have considered what happens when there is a ring present in the starting material, or where we encourage formation of a ring in an intermediate as a means of controlling stereochemistry. In this final section of this chapter we shall consider some examples where stereoselectivity arises because of a ring formed only transiently during a reaction in a cyclic transition state.

We'll start with some epoxidation reactions. Of course these *form* rings, and you have seen, in Chapter 20, epoxidations of alkenes such as cyclohexene. We said in Chapter 20 that epoxidation was stereospecific because both new C–O bonds form to the same face of the alkene.

If we block one face of the ring with a substituent—even quite a small one, such as an acetate group—epoxidation becomes stereoselective for the face *anti* to the substituent already there.

allylic acetate   *anti* epoxide

peroxy-acid approaches less hindered face of ring

With one exception—when the substituent is a hydroxyl group. When an allylic alcohol is epoxidized, the peroxy-acid attacks the face of the alkene *syn* to the hydroxyl group, even when that face is more crowded. For cyclohexenol the ratio of *syn* epoxide to *anti* epoxide is 24:1 with *m*-CPBA and it rises to 50:1 with CF_3CO_3H.

allylic alcohol   *cis* epoxide

hydrogen bond

green hydrogen bond favours attack on same face as OH

The reason is shown in the transition state: the OH group can hydrogen bond, through the H of the alcohol, to the peroxy-acid, stabilizing the transition state when the epoxidation is occurring *syn*. This hydrogen bond means that peroxy-acid epoxidations of alkenes with adjacent hydroxyl groups are much faster than epoxidations of simple alkenes, even when no stereochemistry is involved.

Peroxy-acids work for epoxidizing allylic alcohols *syn* to the OH group, but another reagent is better when the OH group is further from the alkene. 4-Hydroxycyclopentene, for example, can be converted into either diastereomer of the epoxide. If the alcohol is protected with a large group such as TBDMS (*t*-butyl-

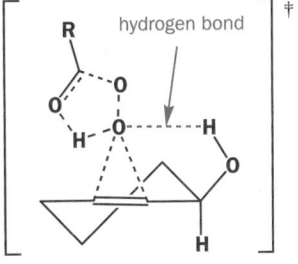

83:17 *anti:syn*

dimethylsilyl) it becomes a simple blocking group and the epoxide is formed on the opposite face of the alkene. The selectivity is reasonable (83:17) given that the blocking group is quite distant.

If the OH group is not blocked at all but left free, and the epoxidation reagent is the vanadium complex VO(acac)$_2$ combined with *t*-BuOOH, the *syn* epoxide is formed instead. The vanadyl group chelates reagent and alcohol and delivers the reactive oxygen atom to the same face of the alkene.

VO(acac)₂

Vanadyl (acac) ₂ is a square pyramidal complex of two molecules of the enolate of 'acac' (acetyl acetone, pentan-2,4-dione) and the vanadyl (V=O) dication. It can easily accept another ligand to form an octahedral complex so there is

plenty of room for the alcohol to add and for the *t*-BuOOH to displace one of the 'acac' ligands to give some complex with the essential ingredients for the reaction as shown above.

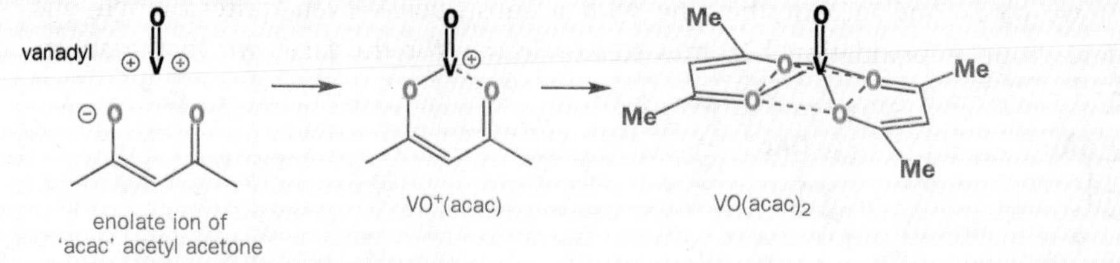

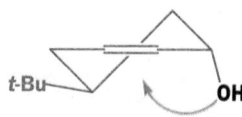

The delivery of an oxygen atom through a cyclic transition state by vanadyl complexes is also particularly effective with allylic alcohols. Here is a simple example—the green arrow shows merely the directing effect and is not a mechanism. Delivery of oxygen from OH through a VO complex is particularly effective when the OH group is pseudoaxial and the *t*-Bu group ensures this.

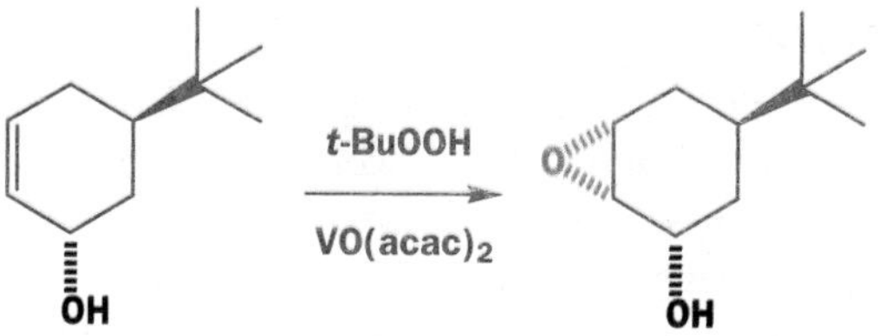

In both epoxidation examples, the stereoselectivity is due to the cyclic nature of the transition state: the fact that there is a hydrogen bond or O–metal bond 'delivering' the reagent to one face of the alkene. This is a very important concept, and we revisit it in the next chapter: cyclic transition states are the key to getting good stereoselectivity in reactions of acyclic compounds.

Before we move on, we leave you with one final example. Stereoselectivity in the epoxidation of lactone-bridged alkenes related to those we saw earlier (p. 874) can be completely reversed if the lactone is hydrolysed, revealing a hydroxyl group. In this bicyclic example, the hydroxyl group delivers the peroxy-acid from the bottom face of the alkene. First, the lactone bridge is used to introduce the alkene as before.

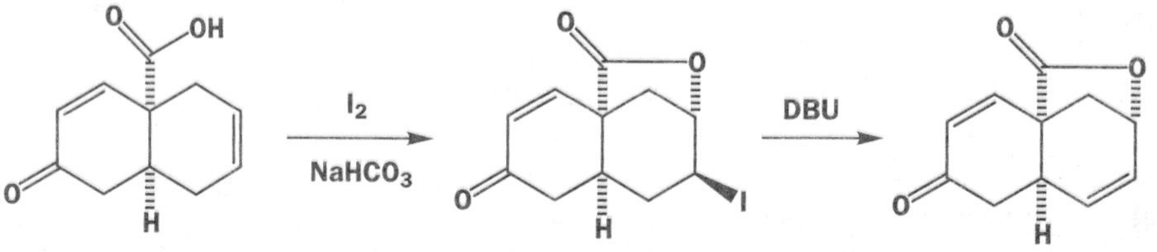

Now the critical steps—the lactone bridge is hydrolysed, the epoxide added from the bottom face by a peroxy-acid hydrogen bonded to the OH group, and the lactone bridge reinstated.

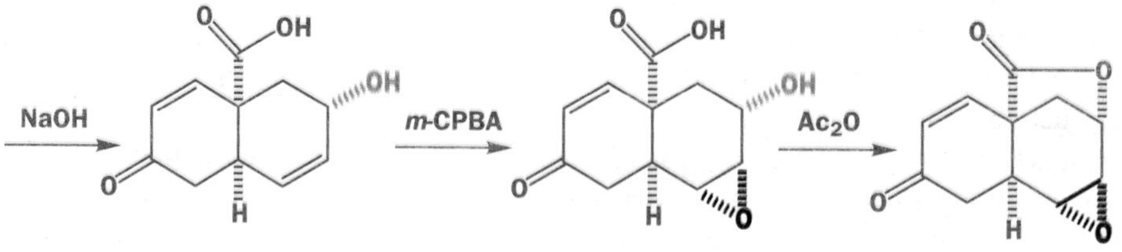

The second ring in these compounds is actually a tether, and it enables two more functional groups to be introduced in a *cis* fashion by oxidation of the remaining alkene.

> You have met several methods for cleaving C=C bonds in this chapter, including this one and also ozone. These reactions will be discussed in Chapter 35.

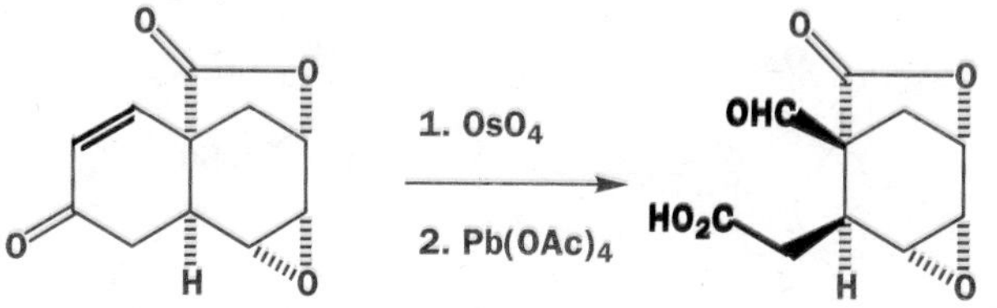

To conclude ...

Diastereoselectivity in rings generally follows a few simple principles:

- Flattened three-, four-, or five-membered rings, especially ones with two or more trigonal carbons in the ring, are generally attacked from the less hindered face
- Flattened six-membered rings with two or more trigonal carbons in the ring (that is, which are not already a chair—so six-membered rings with one trigonal C atom don't count here) react in such a way that the product becomes an axially substituted chair
- Bicyclic compounds react on the outside face
- Reaction on the more hindered face can be encouraged by: (1) tethered nucleophiles, or (2) cyclic transition states

Diastereoselectivity in compounds without rings is different: it is less well controlled, because there are many more conformations available to the molecule. But even in acyclic compounds, rings can still be important, and some of the best diastereoselectivities arise when there is a ring formed temporarily in the transition state of the reaction. With or without cyclic transition states, in some cases we have good prospects of predicting which diastereoisomer will be the major reaction product, or explaining the diastereoselectivity if we already know this. That is the subject of the next chapter.

Problems

1. Comment on the control over stereochemistry achieved in this sequence.

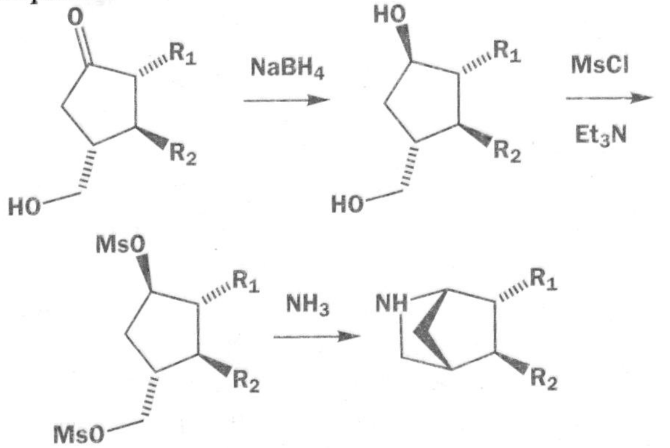

2. Explain the stereochemistry of this sequence of reactions, noting the second step in particular.

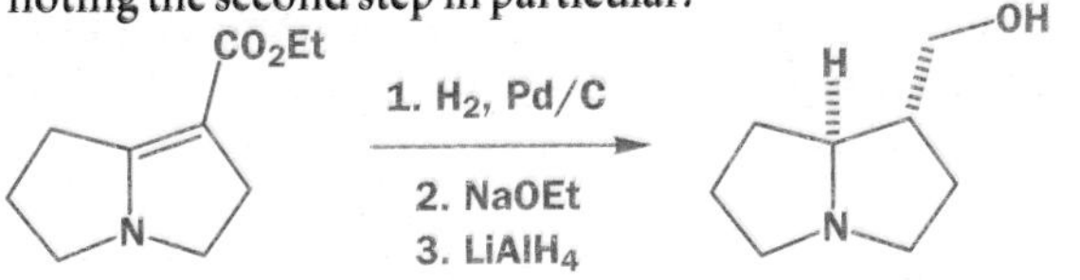

3. Explain how the stereo- and regiochemistry of these compounds are controlled. Why is the epoxidation only moderately stereoselective, and why does the amine attack where it does?

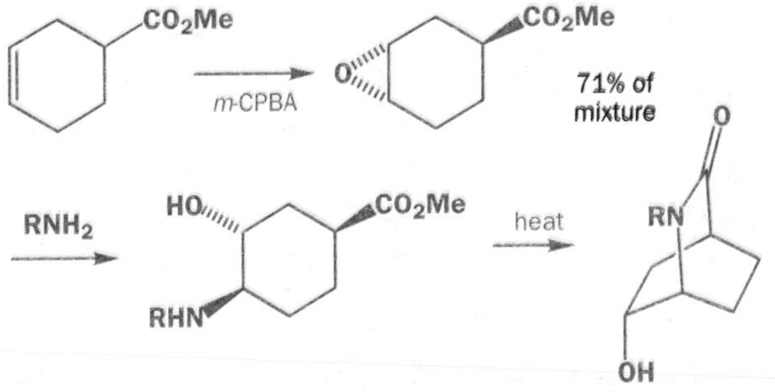

4. What controls the stereochemistry of this product? You are advised to draw a mechanism first and then consider the stereochemistry.

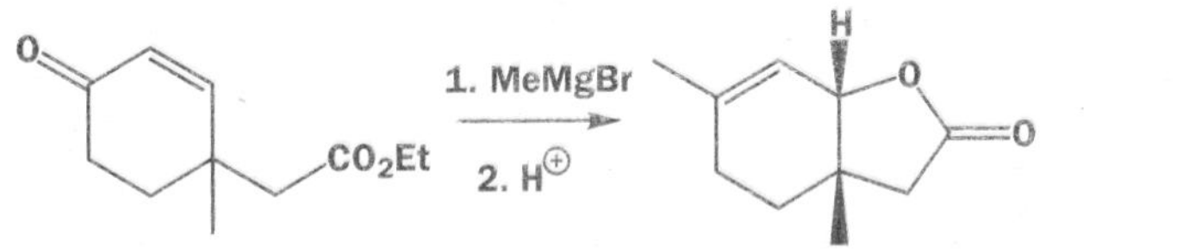

5. Why is one of these esters more reactive than the other?

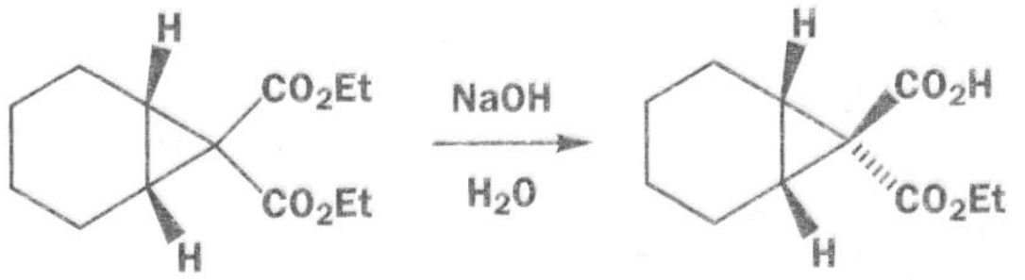

6. Explain the stereoselectivity in these reactions.

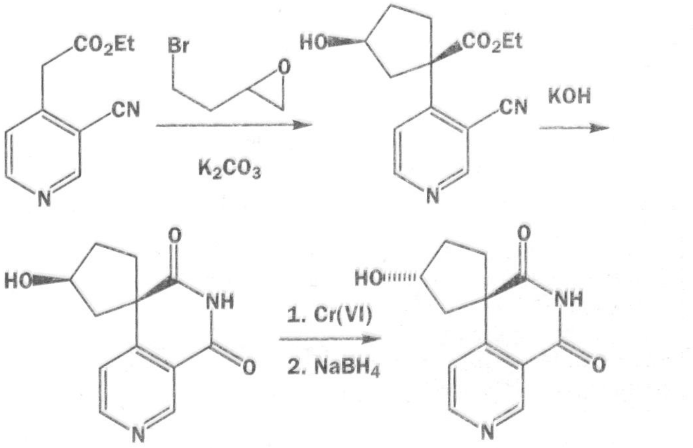

7. A problem from the chapter. Draw a mechanism for this reaction and explain why it goes so much better than the elimination on a β-lactone.

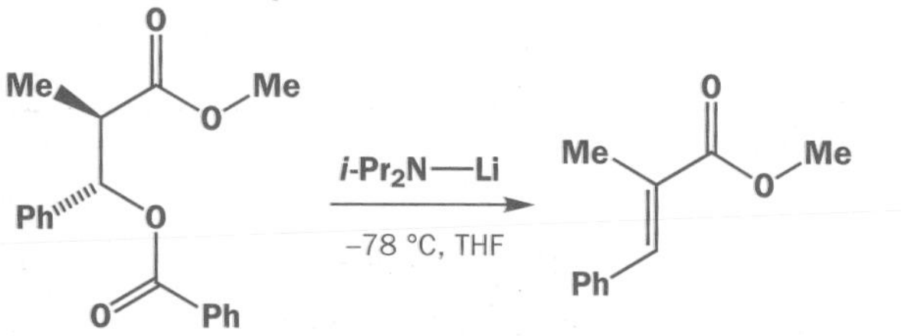

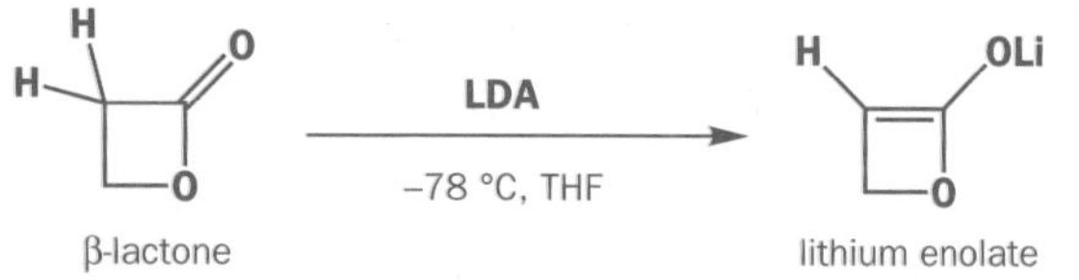

8. Another problem from the chapter. The synthesis of the starting material for this reaction is a good example of how cyclic compounds can be used in a simple way to control stereochemistry. Draw mechanisms for each reaction and explain the stereochemistry.

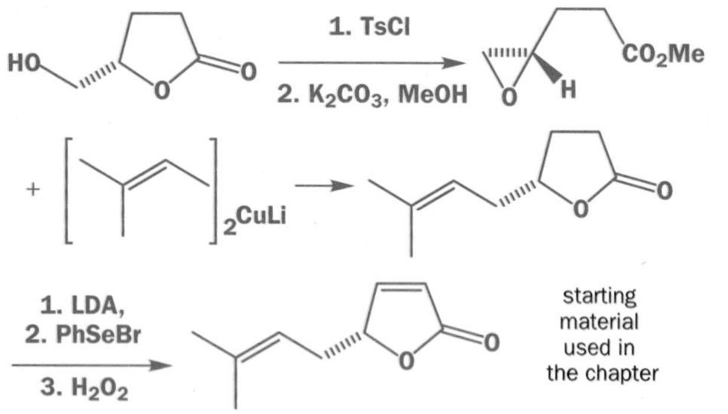

9. A revision problem. Suggest mechanisms for the reactions used to make this starting material used in the chapter.

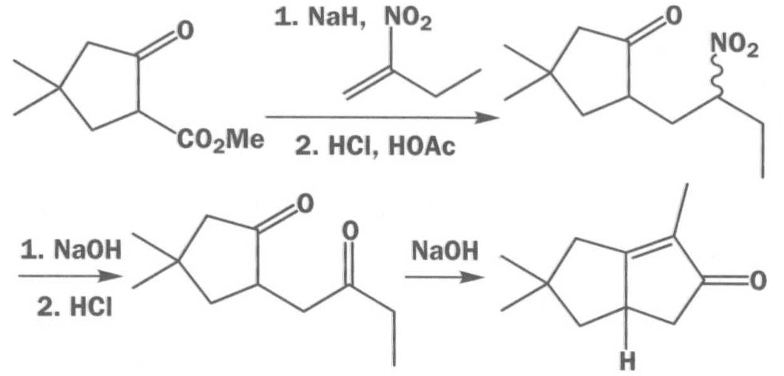

10. And another problem from the chapter. Here also draw a mechanism for the formation of the starting material. You have never seen the cyclopropane reagent, but think how it might react . . .

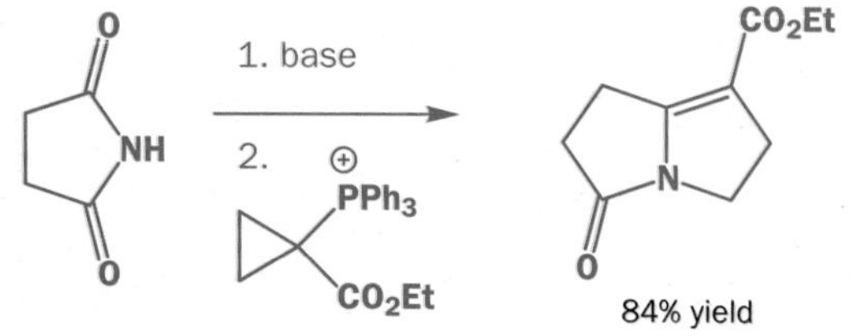

84% yield

Stuck? The first step opens the three-membered ring and the second step is a well-known alkene-forming reaction . . .

11. In the chapter we introduced the selective reduction of the Wieland–Miescher ketone. The problem is: can you suggest a reason for this stereoselectivity?

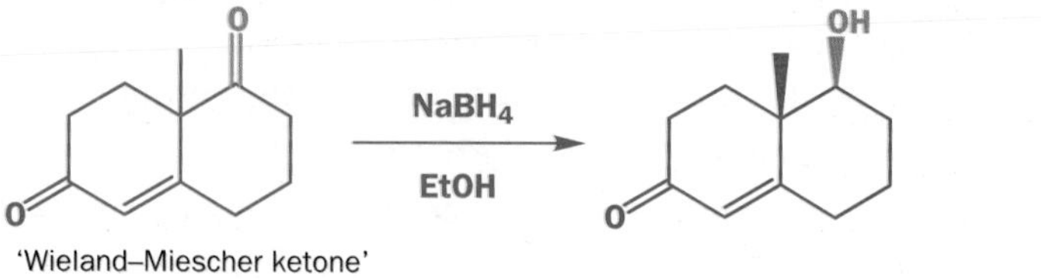

'Wieland–Miescher ketone'

12. Suggest mechanisms for these reactions and explain the stereochemistry.

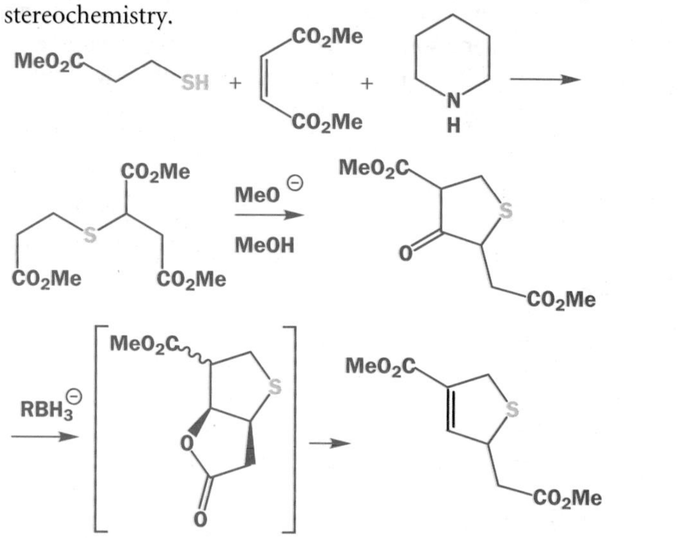

13. Hydrolysis of a bis-silylated ene-diol gives a hydroxy-ketone A whose stereochemistry is supposed to be as shown. Reduction of A gives a diol B. The ^{13}C NMR spectrum of B has five signals: one in the 100–150 p.p.m. range, one in the 50–100 p.p.m. range, and three below 50 p.p.m. The proton NMR of the three marked hydrogens in A is given below with some irradiation data. Does this information give you confidence in the stereochemistry assigned to A? You may wish to consider the likely stereochemical result of the reduction of A.

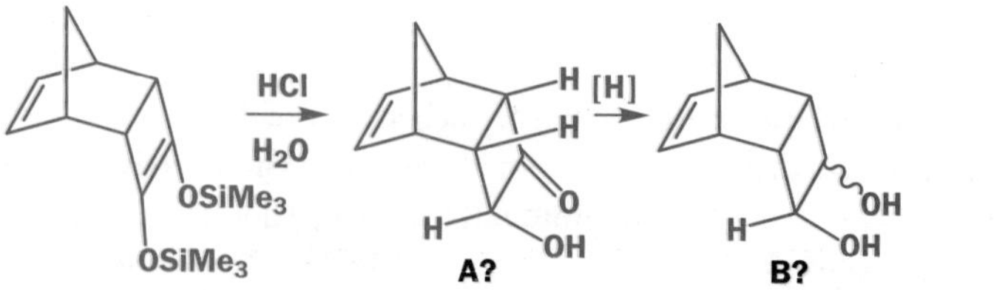

A has δ_H 4.46 (1H, dd, J 9.0, 3.8 Hz), 3.25 (1H, ddd, J 9.0, 7.5, 4.5 Hz), and 3.48 (1H, ddd, J 7.5, 5.5, 3.8 Hz). Irradiation at 3.48 p.p.m. collapses the signal at 4.46 to (d, J 9.0 Hz) and the signal at 3.25 to (dd, J 9.0, 4.5 Hz); irradiation at 4.46 collapses the signal at 3.48 to (dd, J 7.5, 5.5) and the signal at 3.25 to (dd, J 7.5, 4.5).

Diastereoselectivity

34

Connections

Building on:

- **Stereochemistry** ch16
- **Conformation** ch18
- **Controlling double bond stereochemistry** ch31
- **Determining stereochemistry by NMR** ch32
- **Controlling stereochemistry in cyclic compounds** ch33

Arriving at:

- **How to make single diastereoisomers from single geometrical isomers**
- **How to predict and explain the reactions of chiral carbonyl compounds**
- **How chelation to metal ions can change stereoselectivity**
- **How to predict and explain the reactions of chiral alkenes**
- **Stereoselectivity in the aldol reaction**
- **How to make *syn* aldol products**
- **How to make *anti* aldol products**

Looking forward to:

- **Saturated heterocycles** ch42
- **Asymmetric synthesis** ch45
- **Organic synthesis** ch53

Looking back

You have had three chapters in a row about stereochemistry: this is the fourth, and it is time for us to bring together some ideas from earlier in the book. We aim firstly to help you grasp some important general concepts, and secondly to introduce some principles in connection with stereoselective reactions in acyclic systems. But, first, some revision.

We introduced the stereochemistry of structures in Chapter 16. We told you about two types of stereoisomers.

> **● Enantiomers and diastereoisomers**
>
> - **Enantiomers**—stereoisomers that are mirror images of one another
> - **Diastereoisomers**—stereoisomers that are not mirror images of one another

In this chapter we shall talk about how to make compounds as single diastereoisomers. Making single enantiomers is treated in Chapter 45. Chapter 33 was also about making single diastereoisomers, and we hope that, having read that chapter, you are used to thinking stereochemically.

In this chapter we shall talk about two different ways of making single diastereoisomers.

> **● Reactions that make single diastereoisomers**
>
> - **Stereospecific reactions**—reactions where the mechanism means that the stereochemistry of the starting material determines the stereochemistry of the product and there is no choice involved
> - **Stereoselective reactions**—reactions where one stereoisomer of product is formed predominantly because the reaction has a choice of pathways, and one pathway is more favourable than the other

These terms were introduced in Chapter 19 in connection with elimination reactions, and many of the reactions we mention will be familiar from earlier chapters (particularly Chapters 17–20 and 26–27).

Making single diastereoisomers using stereospecific reactions of alkenes

The essence of the definition we have just reminded you of is much easier to grasp with some familiar examples. Here are two.

- S$_N$2 reactions are stereospecific: they proceed with inversion so that the absolute stereochemistry of the starting material determines the absolute stereochemistry of the product

This is discussed in Chapter 17, p. 422.

- E2 reactions are stereospecific: they proceed through an anti-periplanar transition state, with the relative stereochemistry of the starting material determining the geometry of the product

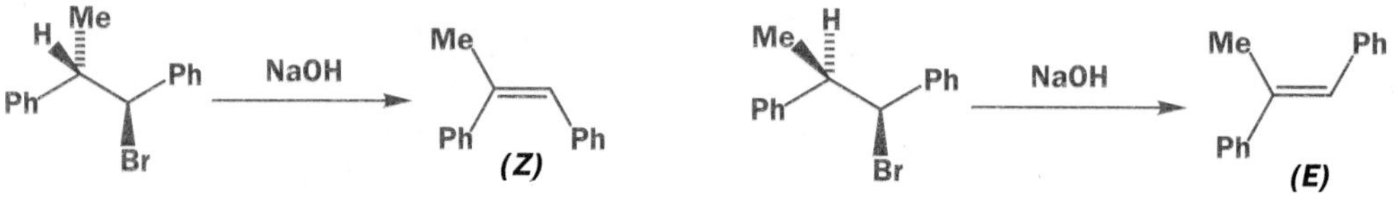

Both of these examples are very interesting because they show how, once we have some stereochemistry in a molecule, we can change the functional groups but keep the stereochemistry—this is the essence of a stereospecific reaction. In the second example, we change the bromide to a double bond, but we keep the stereochemistry (or 'stereochemical information') because the geometry of the double bond tells us which bromide we started with.

This is a good place to begin if we want to make single diastereoisomers, because we can reverse this type of reaction: instead of making a single geometry of alkene from a single diastereoisomer, we make a single diastereoisomer from a single geometry of double bond. Here is an example of this—again, one you have already met (Chapter 19). Electrophilic addition of bromine to alkenes is stereospecific and leads to *anti* addition across a double bond. So if we want the *anti* dibromide we choose to start with the *trans* double bond; if we want the *syn* dibromide we start with the *cis* double bond. The geometry of the starting material determines the relative stereochemistry of the product.

Chapter 31 described the methods available for controlling the geometry of double bonds.

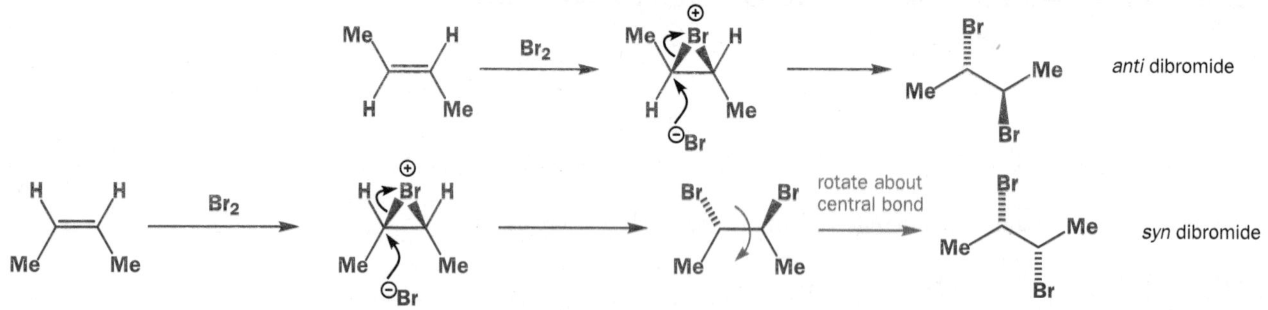

Iodolactonization has a similar mechanism; notice how in these two examples the geometry of the double bond in the starting material defines the relative stereochemistry highlighted in black in the product.

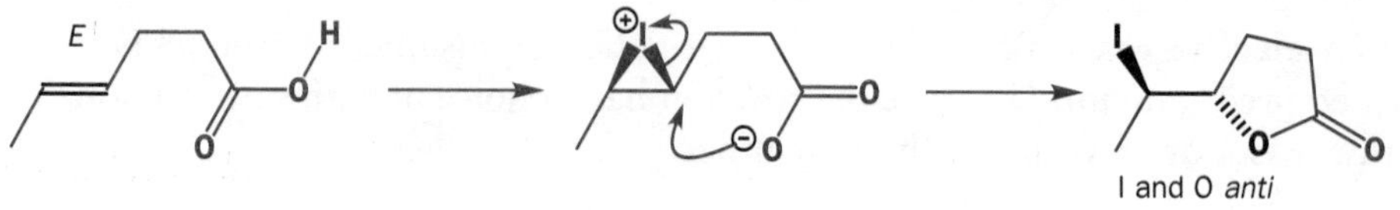

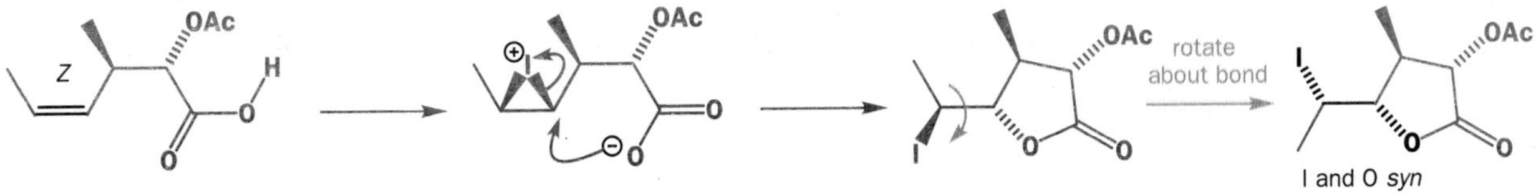

I and O *syn*

For a stereospecific alkene transformation, choose the right geometry of the starting material to get the right diastereoisomer of the product. Don't try to follow any 'rules' over this—just work through the mechanism.

Now for some examples with epoxides. Epoxides are very important because they can be formed stereospecifically from alkenes: *cis*-alkenes give *cis* (or *syn*) -epoxides and *trans*-alkenes give *trans* (or *anti*) -epoxides.

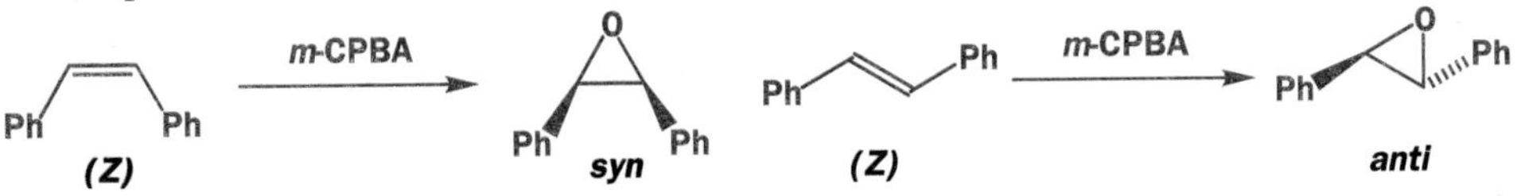

Epoxides also react stereospecifically because the ring-opening reaction is an S_N2 reaction. A single diastereoisomer of epoxide gives a single diastereoisomer of product.

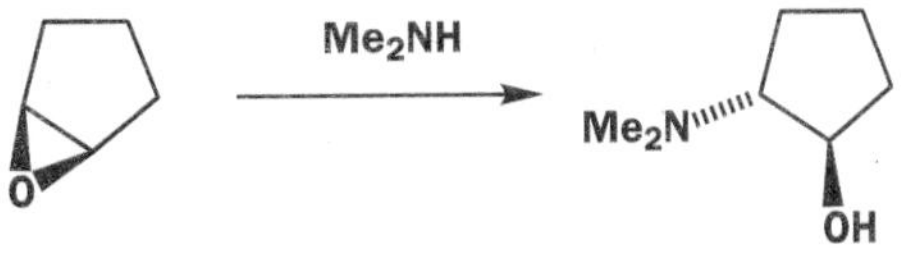

We have mentioned **leukotrienes** before: they are important molecules that regulate cell and tissue biology. Leukotriene C_4 (LTC_4) is a single diastereoisomer with an *anti* 1,2 S,O functional group relationship. In nature, this single diastereoisomer is made by an epoxide opening: since the opening is S_N2 the epoxide must start off *anti* and, indeed, the epoxide precursor is another leukotriene, LTA_4.

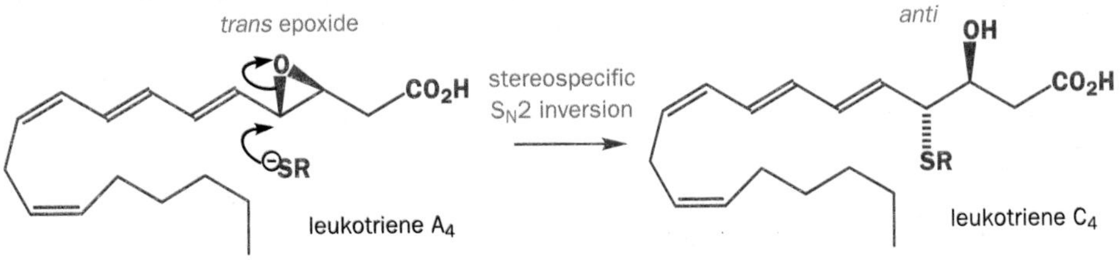

When Corey was making these compounds in the early 1980s he needed to be sure that the relative stereochemistry of LTC_4 would be correctly controlled, and to do this he had to make a *trans* epoxide. Disconnecting LTA_4 as shown led back to a simpler epoxide.

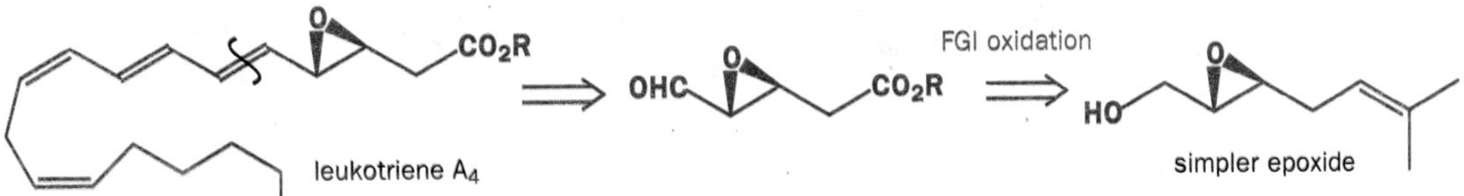

The *trans* allylic alcohol needed to make this compound was made using one of the methods we introduced in Chapter 31: reduction of an alkynyl alcohol with $LiAlH_4$. Here is the full synthesis: alkylation of an ester enolate with prenyl bromide gives a new ester, which itself is turned into an alkylating agent by reduction and tosylation. The alkyne is introduced as its lithium derivative with the alcohol protected as a THP acetal. Hydrolysis of the acetal with aqueous acid gives the hydroxyalkyne needed for reduction to the *E* double bond, which is then epoxidized.

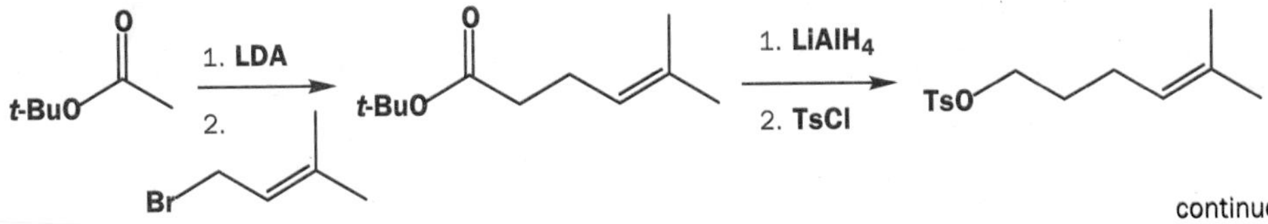

continued overleaf

There are two more stereogenic centres in the second example here and, although they do not affect the relative stereochemistry shown in black, they do affect how those two new stereogenic centres relate to the two that are already present in the starting material. We discuss how later in the chapter.

Chapter 20, p. 507.

Chapter 17, p. 435.

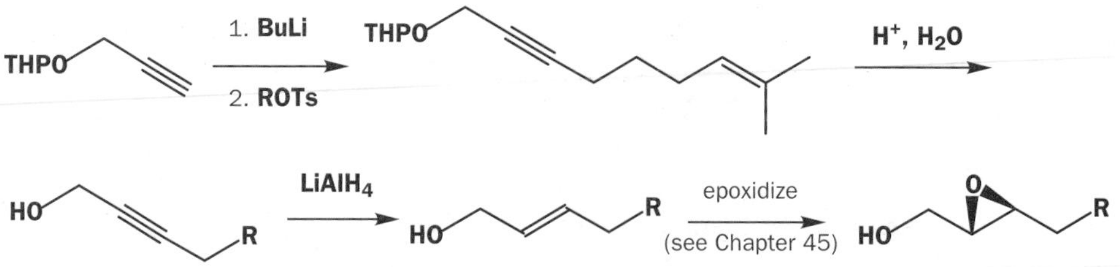

The epoxide was, in fact, made as a single enantiomer using the Sharpless epoxidation, which we will describe in Chapter 45.

Stereoselective reactions

For most of the rest of the chapter we shall discuss stereoselective reactions. You have already met several examples and we start with a summary of the most important methods.

Chapter 19, p. 487.

- E1 reactions are stereoselective: they form predominantly the more stable alkene

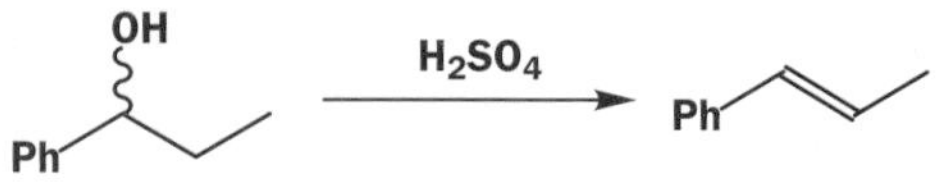

Chapter 18, p. 472.

- Nucleophilic attack on six-membered ring ketones is stereoselective: small nucleophiles attack axially and large ones equatorially

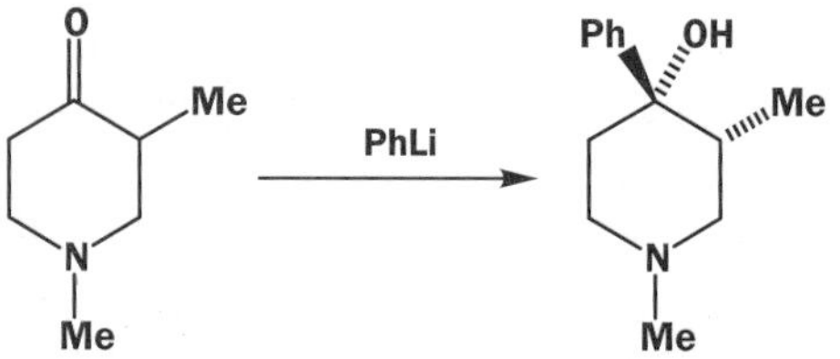

Chapter 33, p. 852.

- Alkylation of cyclic enolates is stereoselective, with reaction taking place on the less hindered face (four- or five-membered rings) or via axial attack (six-membered rings)

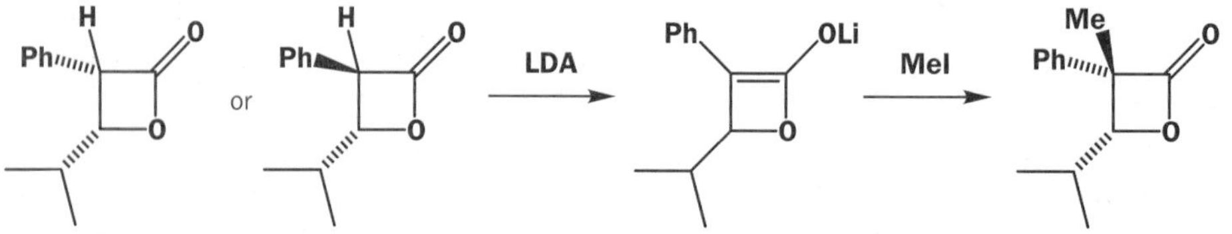

For a *stereoselective* reaction we can specify two different stereoisomers of the starting material and get the same product (first and third examples). In a *stereospecific* reaction, different starting material stereochemistry means different product stereochemistry.

- Epoxidation of cyclic alkenes is stereoselective, with reaction taking place on the less hindered face, or directed by hydrogen bonding to a hydroxyl group

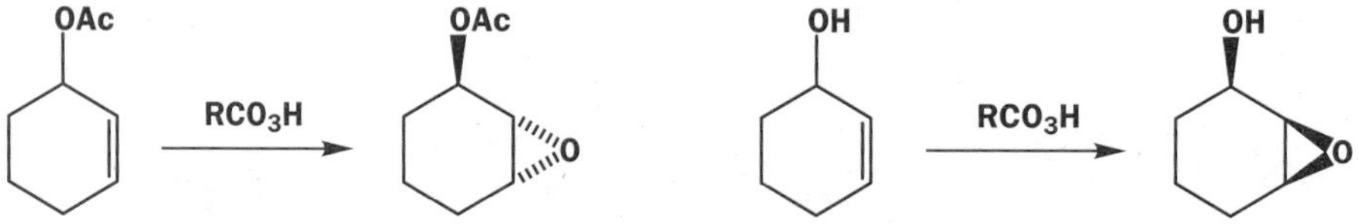

Chapter 33, p. 877.

A common misapprehension is that stereospecific means merely very stereoselective. It doesn't—the two terms describe quite different properties of the stereochemistry of a reaction.

Prochirality

Take another look at the reactions in the chapter so far—in particular those that give single diastereoisomers (rather than single enantiomers or geometrical isomers)—in other words, those that are **diastereoselective**. They all involve the creation of a new, tetrahedral stereogenic centre at a carbon that was planar and trigonal. This leads us to our first new definition. Trigonal carbons that aren't stereogenic (or chiral) centres but can be made into them are called **prochiral**.

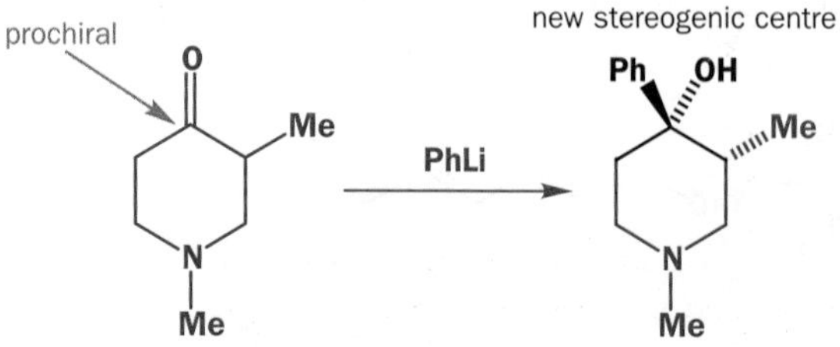

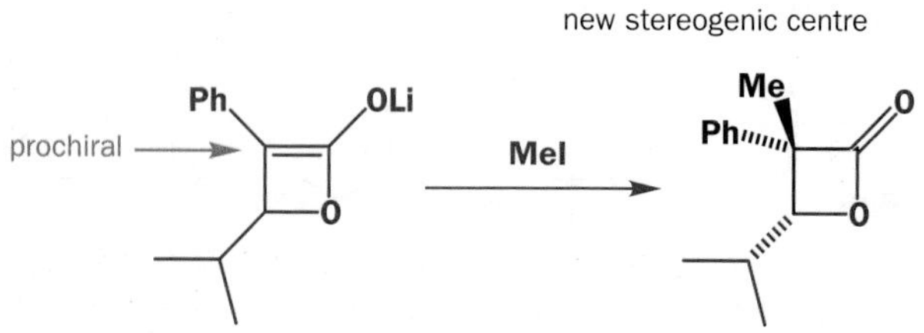

At the very start of Chapter 17, we introduced stereochemistry by thinking about the reactions of two sorts of carbonyl compounds. They are shown again here: the first has a prochiral carbonyl group. The second, on the other hand, is not prochiral because no stereogenic centre is created when the compound reacts.

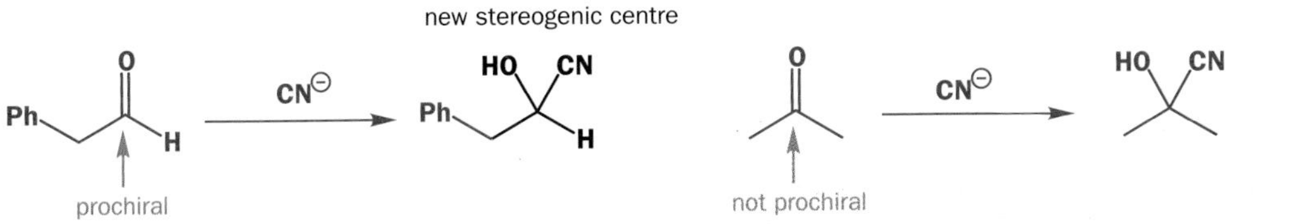

Tetrahedral carbon atoms can be prochiral too—if they carry two identical groups (and so are not a chiral centre) but replacement of one of them leads to a new chiral centre, then the carbon is prochiral.

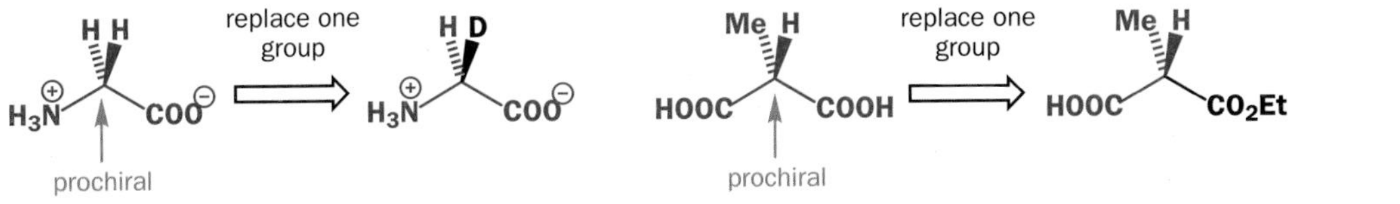

Glycine is the only α amino acid without a chiral centre, but replacing one of the two protons on the central carbon with, say, deuterium creates one: the CH_2 carbon is prochiral. Similarly, converting malonate derivate into its monoester makes a chiral centre where there was none: the central C is prochiral.

Now, does this ring any bells? It should remind you very much of the definitions in Chapter 32 of *enantiotopic* and *diastereotopic* in connection with NMR spectra. Replacing one of two enantiotopic groups with another group leads to one of two enantiomers; replacing one of two diastereotopic groups with another group leads to one of two diastereoisomers. Diastereotopic groups are chemically different; enantiotopic groups are chemically identical.

Exactly the same things are true for the faces of a prochiral carbonyl group or double bond. If reaction on one of two faces of the prochiral group generates one of two enantiomers, the faces are enantiotopic; if the reaction generates one of two diastereoisomers, the faces are diastereotopic. We will now apply this thinking to the first few reactions in this chapter: they are shown again below. The first two examples have prochiral C=C or C=O bonds with diastereotopic faces: choosing which face of the double bond or carbonyl group to react on amounts to choosing which diastereoisomer to form. In the third example, the faces of the prochiral carbonyl group are enantiotopic: choosing which face to attack amounts to choosing which enantiomer to form. In the fourth example, the two faces of C=O are **homotopic**: an identical product is formed whichever face is attacked.

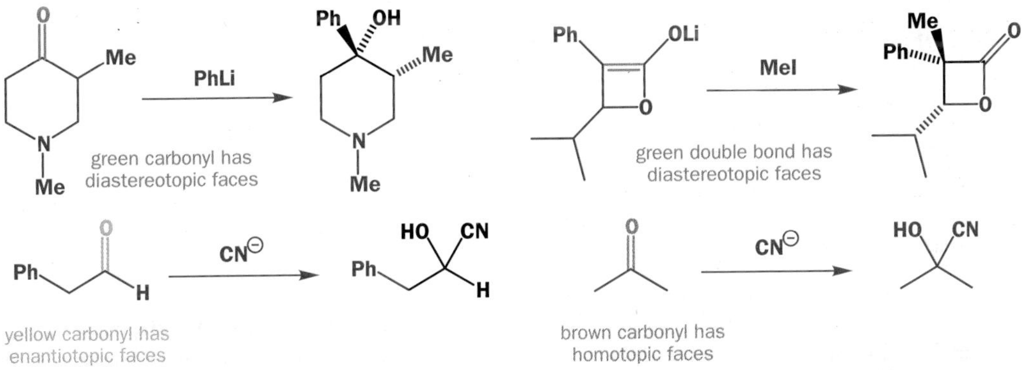

■ Enantiotopic and diastereotopic protons and groups are discussed in Chapter 32, p. 837.

Knowing this throws some new light on the last chapter. Almost without exception, every stereoselective reaction there involved a double bond (usually C=C; sometimes C=O) with diastereotopic faces. The diastereotopic faces were distinguished by steric hindrance, or by a nearby hydrogen-bonding group, and so were able to react differently with an incoming reagent.

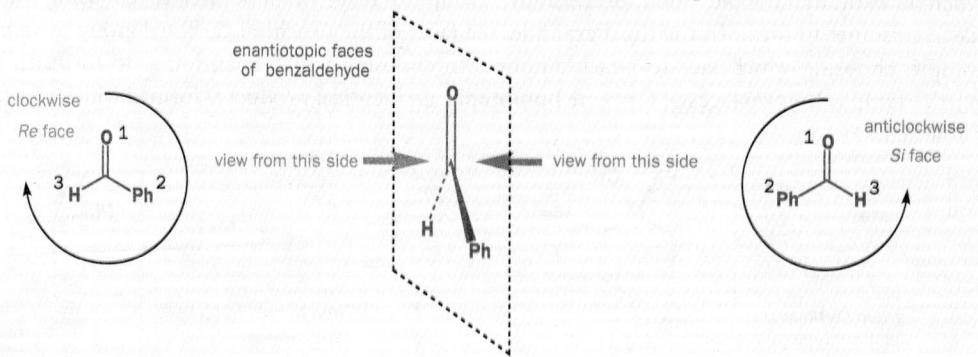

Using an *R/S*-type system to name prochiral faces and groups

Just as stereogenic centres can be described as *R* or *S*, it is possible to assign labels to the enantiotopic groups at prochiral tetrahedral carbon atoms or the enantiotopic faces of prochiral trigonal carbon atoms. The basis of the system is the usual *R,S* system for stereogenic centres, but *pro-R* and *pro-S* are used for groups and *Re* and *Si* for faces.

Pro-R and *pro-S* can be assigned to a pair of enantiotopic groups simply by using the usual rules to assign *R* or *S* to the centre created *if* the group in question is artificially elevated to higher priority than its enantiotopic twin. We'll use G to replace H as we did in Chapter 32: just assume that G has priority immediately higher than H. The method is illustrated for glycine.

Faces of a prochiral trigonal carbon atom are assigned *Re* and *Si* by viewing the carbon from that side and counting down the groups in priority 1–3. Counting round to the right (clockwise) means the face is *Re*; counting round to the left (anticlockwise) means it's *Si*. Remember our advice from Chapter 16: think of turning a steering wheel in the direction of the numbers: does the car go to the right or the left?

Like *R* and *S*, these stereochemical terms are merely labels: they are of no consequence chemically.

Just like diastereotopic signals in an NMR spectrum, diastereotopic faces are always different in principle, but sometimes not so in practice. The very first reaction of Chapter 33 is a case in point: this C=O group has two diastereotopic faces, which, due to free rotation about single bonds, average out to about the same reactivity, so we cannot expect any reasonable level of diastereoselectivity.

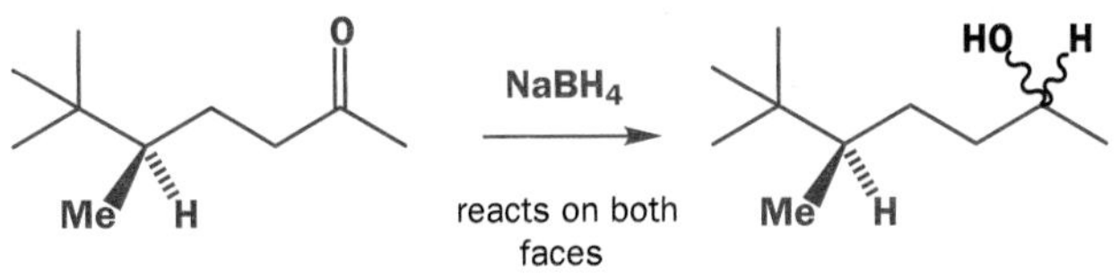

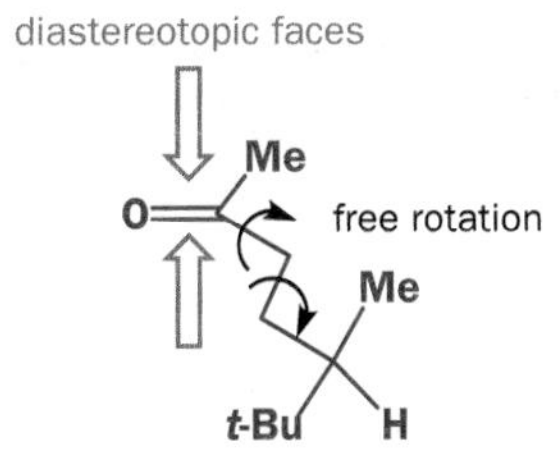

We put Chapter 33 first because in rings conformation is well defined, and this 'averaging' effect is held at bay. We are about to let it out again, but we will show you how it can be tamed to surprisingly good effect.

Additions to carbonyl groups can be diastereoselective even without rings

What happens if we bring the stereogenic centre closer to the carbonyl group than it was in the last example? You might expect it to have a greater influence over the carbonyl group's reactions. And it does. Here is an example.

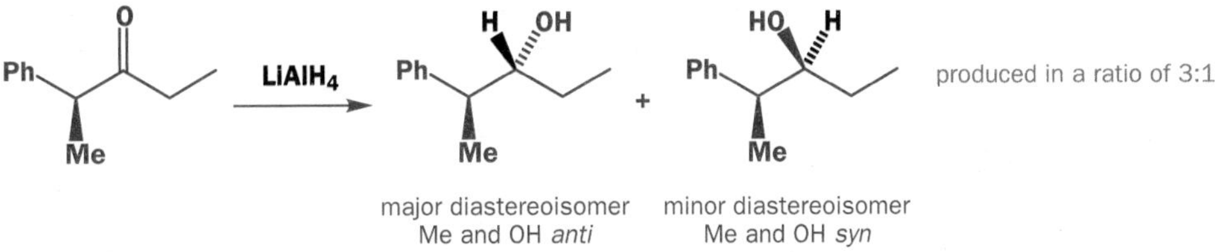

produced in a ratio of 3:1

major diastereoisomer
Me and OH *anti*

minor diastereoisomer
Me and OH *syn*

There is three times as much of one of the two diastereoisomeric products as there is of the other, and the major (*anti*) diastereoisomer is the one in which the nucleophile has added to the front face of the carbonyl group as drawn here. We can make these same two diastereoisomers by addition of an organometallic to an aldehyde. For example, this Grignard reagent gives three times as much of the *syn* diastereoisomer as the *anti* diastereoisomer. The major product has changed, but the product still arises from attack on the front face of the carbonyl as shown.

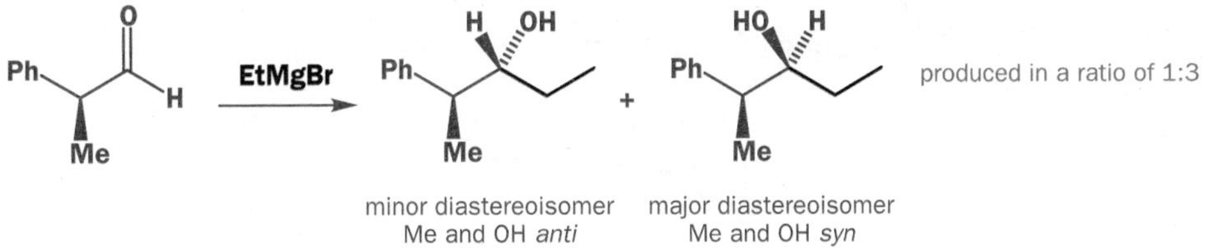

produced in a ratio of 1:3

minor diastereoisomer
Me and OH *anti*

major diastereoisomer
Me and OH *syn*

Drawing diastereoisomers of acyclic molecules

If you find it hard to see that these are still the same two diastereoisomers, try mentally rotating the right-hand half of the molecule about the bond shown below. The next three structures all show the same diastereoisomer (the major product from the last reaction), but in three different conformations (we are just rotating about a bond to get from one to another).

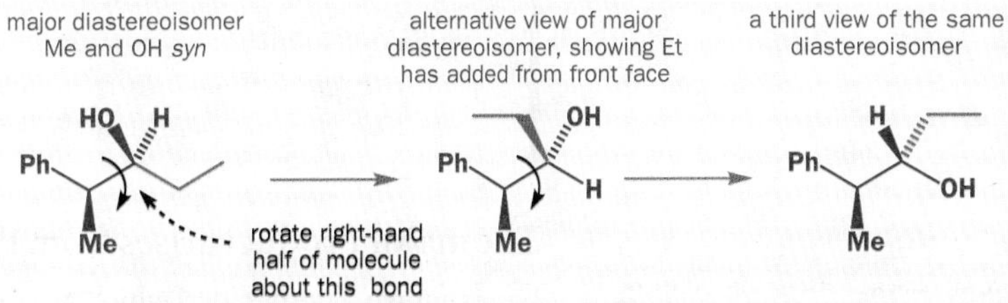

major diastereoisomer
Me and OH *syn*

rotate right-hand half of molecule about this bond

alternative view of major diastereoisomer, showing Et has added from front face

a third view of the same diastereoisomer

Which is the best? A good guideline, which we suggested in Chapter 16, is to place the longest carbon chain zig-zagging across the page in the plane of the paper, and allow all the smaller substituents to extend above or below that chain. The first structure here is drawn like that. But this is only a guideline, and the second structure here is a bit more informative regarding the reaction because, when it is drawn like this, you can clearly see from which direction the ethyl group has attacked the carbonyl. Our advice would be that you first of all

draw the product of any reaction in more or less the same conformation as the starting material to ensure you make no mistakes, and then rotate about a single bond to place the longest chain in the plane of the paper.

If you still have problems manipulating structures mentally—for example, if you find it hard to work out whether the substituents that aren't in the plane should be in front of or behind the page—build some models.

diastereotopic faces

free rotation

> In Chapter 32 we showed that homotopic and enantiotopic protons are identical by NMR. Similarly, homotopic faces or groups are always chemically identical. Enantiotopic faces are also chemically identical, provided that all the reagents in the reaction in question are achiral or racemic. In Chapter 45, we will consider what happens to enantiotopic faces when enantiomerically pure reagents are used.

> We have termed the major diastereoisomer *anti* because the two substituents (Me and OH) are on opposite sides of the chain as drawn. There is no formal definition of *anti* and *syn*: they can only really be used in conjunction with a structural drawing.

no eclipsing interactions

largest group, Ph, is furthest from O and H

Newman projection of one possible conformation

These two reactions are not nearly as diastereoselective as most of the reactions of cyclic compounds you met in the last chapter. But we do now need to explain why they are diastereoselective at all, given the free rotation possible in an acyclic molecule. The key, as much with acyclic as with cyclic molecules, is **conformation**.

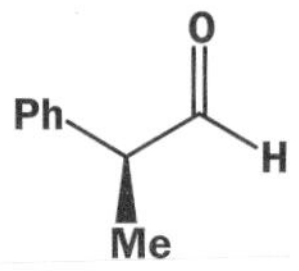

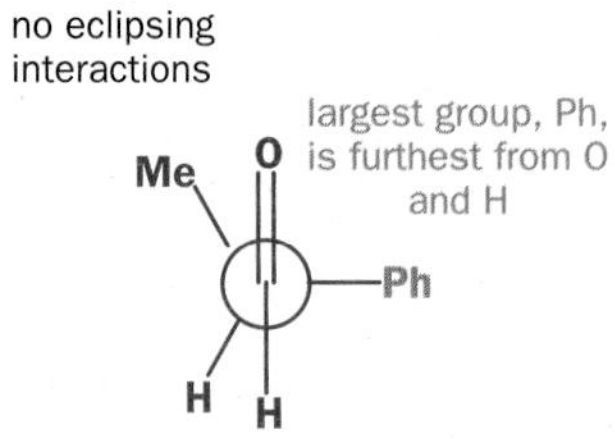

The conformation of a chiral aldehyde

What will be the conformation of the aldehyde in the margin? Using the principles we outlined in Chapter 17, we can expect it to be staggered, with no eclipsing interactions, and also with large substituents as far apart from one another as possible. A Newman projection of one of the possible conformers might look like the one shown in the margin. There are no eclipsing interactions, and the large phenyl group is held satisfactorily far away from the O and the H atoms of the aldehyde.

By rotating about the central bond of the aldehyde (the one represented by a circle in the Newman projection) we can suggest a series of possible conformations. Provided we move in 60° steps, none of them will have any eclipsing interactions. The full set of six conformers is shown here. Look at them for a moment, and notice how they differ.

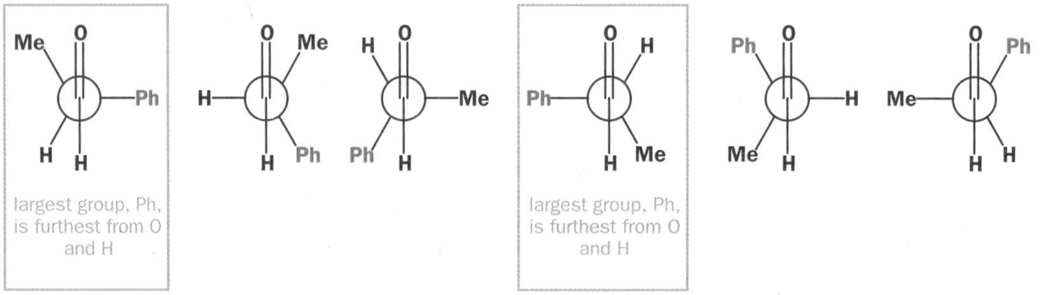

largest group, Ph, is furthest from O and H

largest group, Ph, is furthest from O and H

Only two of them, boxed in yellow, place the large Ph group perpendicular to the carbonyl group. These yellow boxed conformations are therefore the lowest-energy conformers and, for the purpose of the discussion that follows, they are the only ones whose reactions we need to consider.

● Lowest energy conformations of a carbonyl compound

The most important conformations of a carbonyl compound with a stereogenic centre adjacent to the carbonyl group are those that place the largest group perpendicular to the carbonyl group.

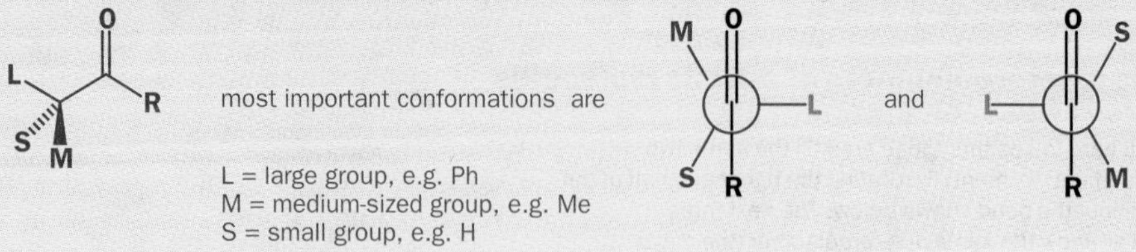

most important conformations are

L = large group, e.g. Ph
M = medium-sized group, e.g. Me
S = small group, e.g. H

The major product arises from the most reactive conformer

Now that we have decided which are the important conformations, how do we know which gives the product? We need to decide which is the *most reactive*. All we need to do is to remember that any nucleophile attacking the carbonyl group will do so from the Bürgi–Dunitz angle—about 107° from the C=O bond. The attack can be from either side of C=O, and the following diagrams show the possible trajectories superimposed on the two conformations we have selected, which are in equilibrium with one another.

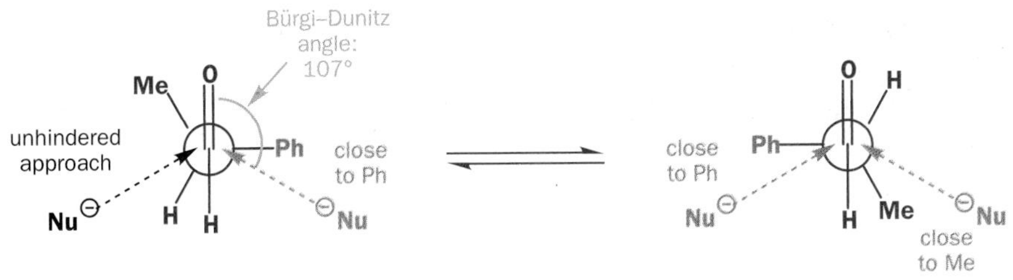

Bürgi–Dunitz angle: 107°

unhindered approach

close to Ph

close to Ph

close to Me

the black flight path is the best

the three brown flight paths are hindered by Ph or Me

Not all four possible 'flight paths' for the nucleophile are equally favourable. For the three shown in brown, the nucleophile passes within 30° or so of another substituent. But, for the one shown in black, there is no substituent nearby except H to hinder attack: the conformation on the left is the most reactive one, and it reacts to give the diastereoisomer shown below.

rotate to view from this direction

unhindered approach

redraw

Remember our guideline: draw the product in a conformation similar to that of the starting material; then redraw to put the longest chain in the plane of the paper. Here, this just means drawing the view from the top of the Newman projection–there is no need to rotate any bonds in this case.

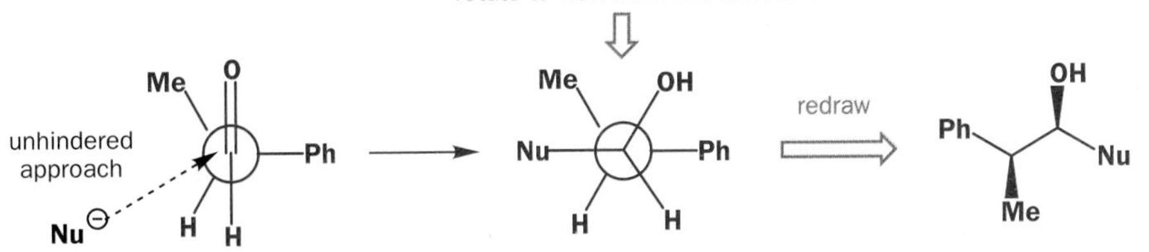

With Nu = Et we have the right product and, more importantly, we can be pretty sure it is for the right reason: this model of the way a nucleophile attacks a carbonyl compound, called the **Felkin-Anh model**, is supported by theoretical calculations and numerous experimental results. Notice that we don't have to decide which is the lower energy of the two conformations: this is not necessary because the attack in black will occur even if the conformer on the left is the minor one in the mixture.

This is an example of the **Curtin–Hammett principle**, which says that it is the relative energies of the transition states that control selectivity, not the relative energies of the starting materials. It's really more of a reminder not to make a mistake than a principle.

Cram's rule

You may hear 'Cram's rule' used to explain the outcome of reactions involving attack on chiral carbonyl compounds. Cram was the first to realize that these reactions could be predicted, but we now know why these compounds react in a predictable way. We will not describe Cram's rule because, although it often does predict the right product, in this case it does so for the wrong reason. Explanations and clear logical thinking are more important than rules, and you must be able to account for and predict the reactions of chiral aldehydes and ketones using the Felkin–Anh model.

The same reasoning accounts for the diastereoselectivity of the reduction on p. 887: first we need to draw the two important conformers of the ketone; the ones that have the large group (Ph) perpendicular to the C=O group.

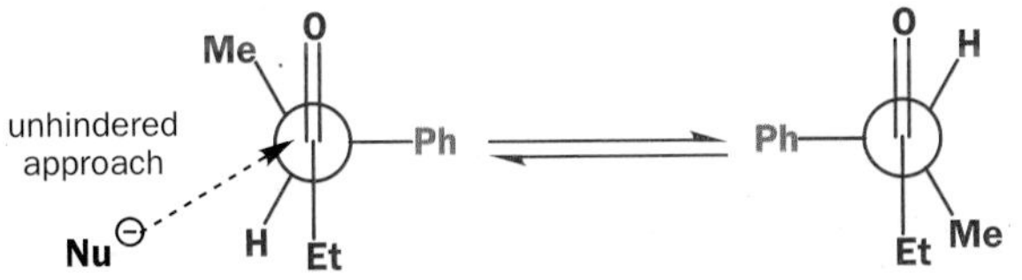

unhindered approach

Now choose the angle of attack that is the least hindered, and draw a Newman proejction of the product. Finally, redraw the Newman projection as a normal structure, preferably with the longest chain in the plane of the paper.

unhindered approach

redraw

redraw, rotating about central bond to put longest chain in plane of paper

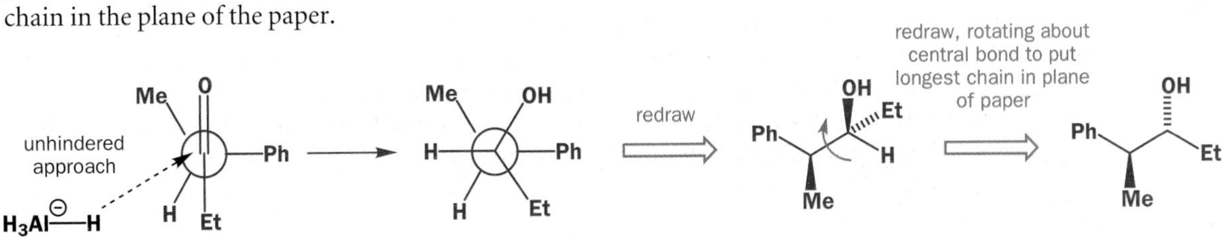

The effect of electronegative atoms

One of the most powerful anticancer agents known is dolastatin, isolated from the sea-hare *Dolabella*. Dolastatin contains an unusual amino acid, with three stereogenic centres, and chemists in Germany managed to exploit Felkin–Anh control very effectively to make it from the much more widespread amino acid isoleucine. This is the sequence of reactions.

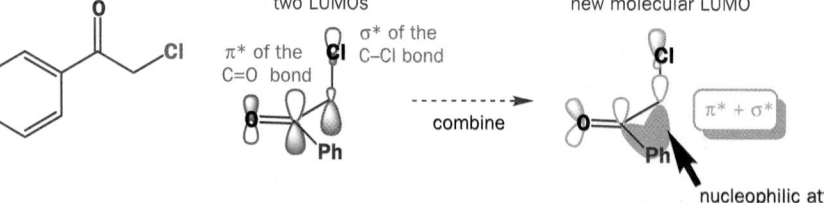

> Try for yourself putting alkyl perpendicular to C=O: you will get the wrong diastereoisomer.

The key step is the aldol reaction of the enolate of methyl acetate with the protected amino aldehyde. To rationalize the stereoselectivity, we first need to draw the two most important conformations of this aldehyde with the large group perpendicular to C=O. The trouble is—which do we choose as 'large': the –NBn$_2$ group or the branched alkyl group? Since we know which diastereoisomer is produced we can work backwards to find that it must be the NBn$_2$ group that sits perpendicular to C=O in the reactive transition state, and not alkyl.

> When you see a selectivity given as 'greater than' something, it means that the other diastereoisomer was undetectable, but here 96:4 was the limit of detection by the method used—possibly NMR.

unhindered attack alongside H

Now look at the diastereoselectivity of the reaction: it is much greater than the 3:1 we saw before—more like 20:1. This really does suggest that there is a further factor at work here, and that further factor is the electronegative N atom.

Carbonyl groups increase the reactivity of adjacent leaving groups towards nucleophilic substitution by several orders of magnitude. This was an effect that we noted in Chapter 17, where we showed that the ketone below reacts by the S$_N$2 mechanism 5000 times as fast as methyl chloride itself.

> ■
> This is discussed on p. 424 of Chapter 17.

We explained this effect by saying that the π^* of the C=O and the σ^* of C–Cl overlap to form a new, lower-energy (and therefore more reactive) LUMO. What we did not note then, because it was not relevant, is that this overlap can only occur when the C–Cl bond is perpendicular to the C=O bond, because only then are the π^* and σ^* orbitals aligned correctly.

The same thing happens even with electronegative atoms X that are not leaving groups in the S$_N$2 reaction (for example, X = OR, NR$_2$, SR, etc.). The π^* and σ^* orbitals add together to form a new, lower-energy molecular orbital, more susceptible to nucleophilic attack. But, if X is not a leaving group, attack on this orbital will result not in nucleophilic substitution but in addition to the carbonyl group. Again, this effect will operate only when the C–X and C=O bonds are perpendicular so that the orbitals align correctly.

X is an electronegative group but not a leaving group (OR, NR₂, SR, etc.)

two LUMOs

σ* of the C–X bond

π* of the C=O bond

combine

new molecular LUMO

π* + σ*

addition to C=O

nucleophilic attack occurs easily here

Nu

in energy terms:

σ* of the C–X bond

π* of the C=O bond

π* + σ*

new molecular LUMO – lower energy; more reactive

What does this mean for stereoselectivity? Conformations of the chiral carbonyl compound that place an electronegative atom perpendicular to the C=O bond will be more reactive—size doesn't matter. So, in the dolastatin amino acid example, the conformations with NBn₂ perpendicular to C=O are the only conformations we need to consider.

● Using the Felkin–Anh model

To predict or explain the stereoselectivity of reactions of a carbonyl group with an adjacent stereogenic centre, use the Felkin–Anh model.

- Draw Newman projections of the conformations of the starting material that place a large group or an electronegative group perpendicular to C=O
- Allow the nucleophile to attack along the least hindered trajectory, taking into account the Bürgi–Dunitz angle
- Draw a Newman projection of the product that arises from attack in this way
- Carefully flatten the Newman projection on to the page to produce a normal structure, preferably with the longest chain of C atoms in the plane of the page. Check that you have done this last step correctly: it is very easy to make mistakes here. Use a model if necessary, or do the 'flattening out' in two stages—first view the Newman projection from above or below and draw that; then rotate some of the molecule about a bond if necessary to get the long chain into the plane of the page.

As an illustration of two sorts of diastereoselectivity, our next example is a natural product called penaresidin A. It was isolated from a Japanese sponge in 1991, and has the structure shown below

penaresidin A

or something like this, because at the time of writing the relative stereochemistry between the two remotely related groups of chiral centres is still not known for sure. What is sure is the stereochemistry around the ring: NMR (the methods of Chapter 32) gives that. What Mori and his co-workers set out to do was to make, using unambiguous stereoselective methods, all the possible diastereoisomers of penaresidin A to discover which was the same as the natural product. It was fairly straightforward to get to the target molecule from the structure below and overleaf, so that's the compound whose synthesis we need to consider. If we imagine getting the E-alkene by stereoselective reduction of the alkyne, disconnection to an alkynyl anion equivalent reveals an aldehyde with a chiral centre next to the carbonyl group.

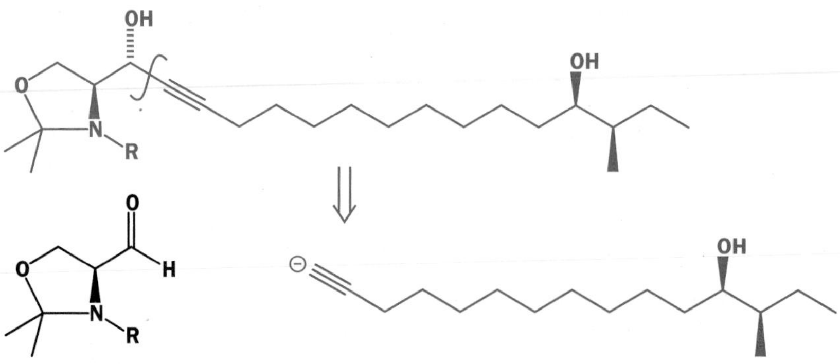

chiral aldehyde

How will this aldehyde (which can be made from the amino acid serine) react with nucleophiles such as lithiated alkynes? Consider a Felkin–Anh transition state: again, we know that the nitrogen, being electronegative, will lie perpendicular to the carbonyl group in the most reactive conformation, so we need only consider these two. The least hindered direction of attack is shown, and that indeed gives the required product.

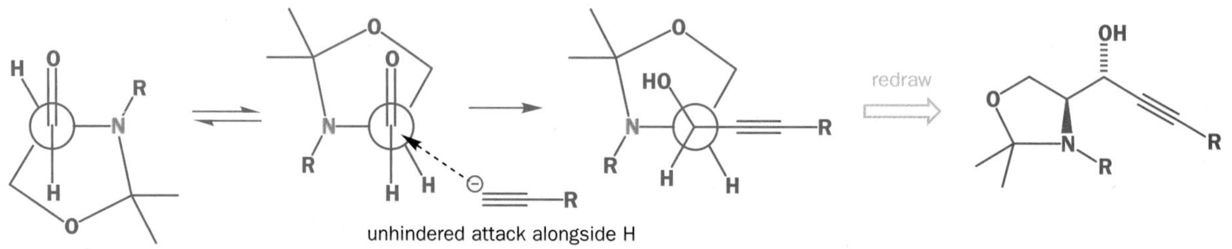

unhindered attack alongside H

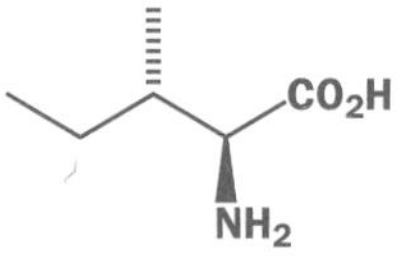

isoleucine

The other two chiral centres need to be controlled separately. The *trans* relative configuration could be obtained from another amino acid, which itself has two stereogenic centres—isoleucine. The *cis* was harder. The chemists decided to make it by starting with the diol shown, which could come from ring opening of an epoxide with an aluminium reagent. Since the ring opening goes with inversion, the epoxide need to be *cis*, so the ultimate starting material was chosen to be a *cis* allylic alcohol. It turned out that the *cis* stereochemistry was right.

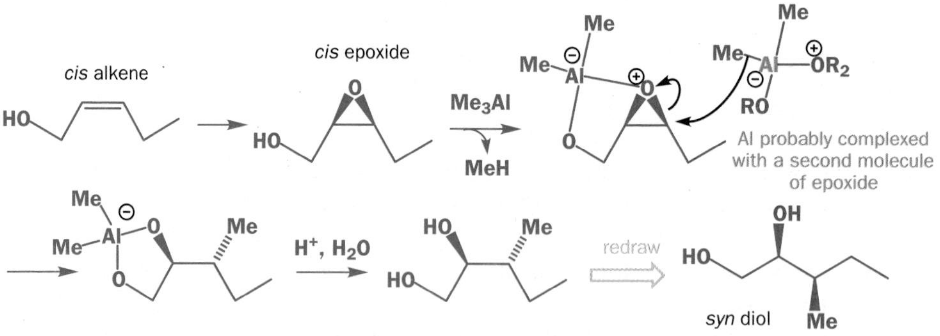

cis alkene

cis epoxide

Al probably complexed with a second molecule of epoxide

syn diol

Chelation can reverse stereoselectivity

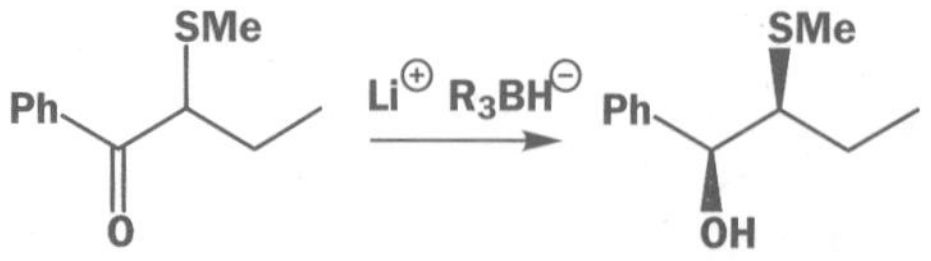

You should now be in a position to explain the outcome of this reaction without much difficulty. Sulfur is the electronegative atom, so the conformations we need to consider are the two following. Unhindered attack on the second gives the diastereoisomer shown.

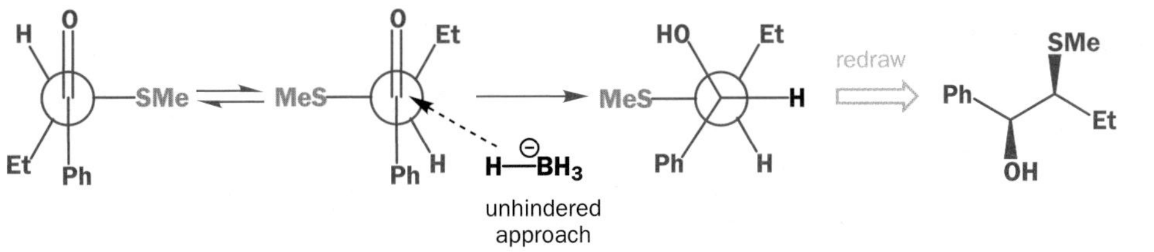

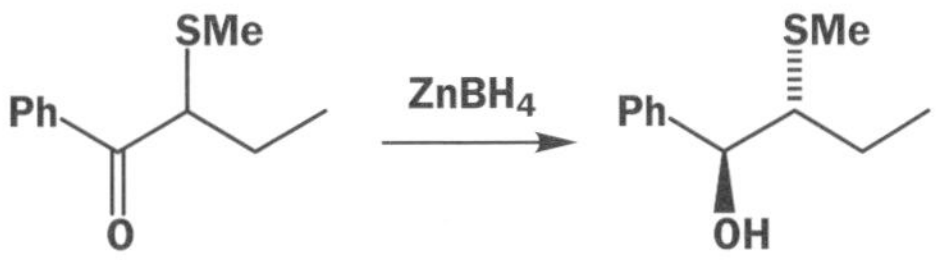

But, from what we have told you so far, the next reaction would present a problem: changing the metal from sodium to zinc has reversed the stereoselectivity. Using the simple Felkin–Anh model now does not work: it gives the wrong answer.

The reason is that zinc can chelate sulfur and the carbonyl group. **Chelation** is the coordination of two heteroatoms carrying lone pairs to the same metal atom, and here it changes the conformation of the starting material. No longer does the most reactive or most populated conformation place the electronegative S atom perpendicular to C=O; instead it prefers S to lie as close to the carbonyl oxygen as possible so that Zn can bridge between S and O, like this.

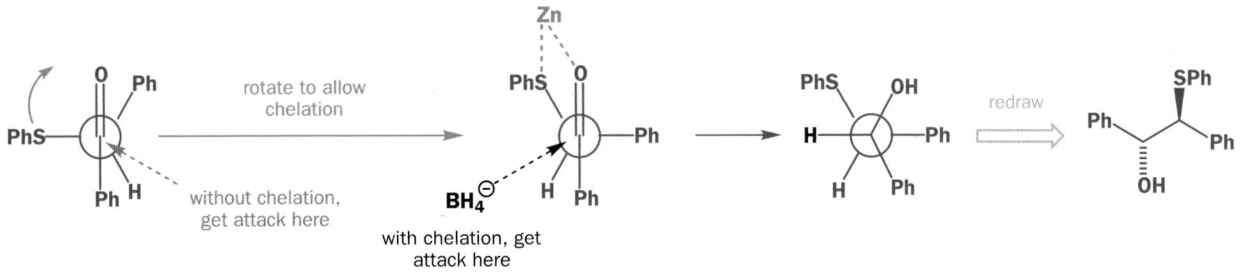

When chelation is possible, this is the conformation to consider—the one with the carbonyl O and the other chelating atom almost eclipsing one another. It is the most populated because it is stabilized by the chelation, and it is also the most reactive, because the Lewis-acidic metal atom increases the reactivity of the carbonyl group. Attack is still along the less hindered pathway, but this now leads to the other face of the carbonyl group, and the stereochemical outcome is reversed.

Two things are needed for chelation to occur:

- a heteroatom with lone pairs available for coordination to a metal
- a metal ion that prefers to coordinate to more than one heteroatom at once. These are mainly more highly charged ions as shown in the table

Here is another example of a reversal in selectivity that can be explained using a nonchelated Felkin–Anh model with Na^+ and a chelated model with Mg^{2+}.

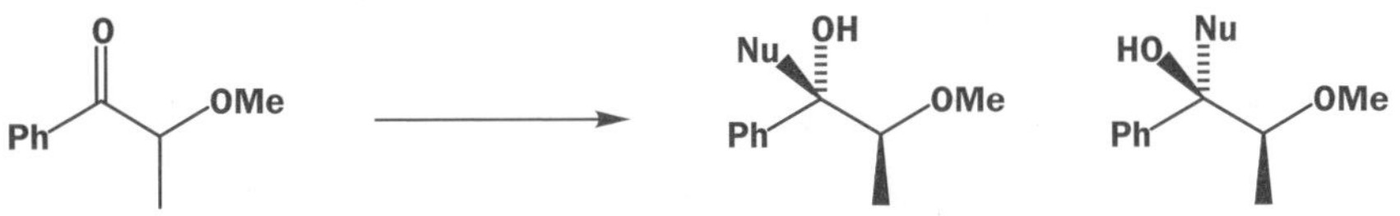

Metals commonly involved in chelation	Metals not usually involved in chelation
Li^+ sometimes	Li^+ often
Mg^{2+}	Na^+
Zn^{2+}	K^+
Cu^{2+}	
Ti^{4+}	
Ce^{3+}	
Mn^{2+}	

NaBH₄ (Nu = H)	73%	27%
Me₂Mg (Nu = Me)	1%	99%

Not only does chelation control reverse the stereoselectivity, but it gives a much higher *degree* of stereoselectivity. Stereoselectivities in chelation-controlled additions to C=O groups are typically >95:5. But this fits in nicely with the ideas we presented at the end of the last chapter: stereoselectivity is likely to be high if a cyclic transition state is involved. Chelation involves just such a transition state, so it should be no surprise that it lets us achieve much higher levels of control than the acyclic Felkin–Anh model does.

Chelation, rate, and stereoselectivity

The correlation of rate of addition with diastereoselectivity was demonstrated in a series of experiments that involved reacting Me$_2$Mg with protected α-hydroxy-ketones. As the protecting group was changed from a methyl ether to a trimethylsilyl ether and then through a series of increasingly bulky silyl ethers, both the rate of the reaction and the diastereoselectivity decreased. With small protecting groups, the reaction takes place through the chelated transition state—the selectivity shows this—and the rate is faster because of the activating effect of the Lewis-acidic magnesium ion. But with larger protecting groups, chelation of Mg^{2+} between the two oxygen atoms is frustrated: the rate drops off, and the selectivity becomes more what would be expected from the Felkin–Anh model.

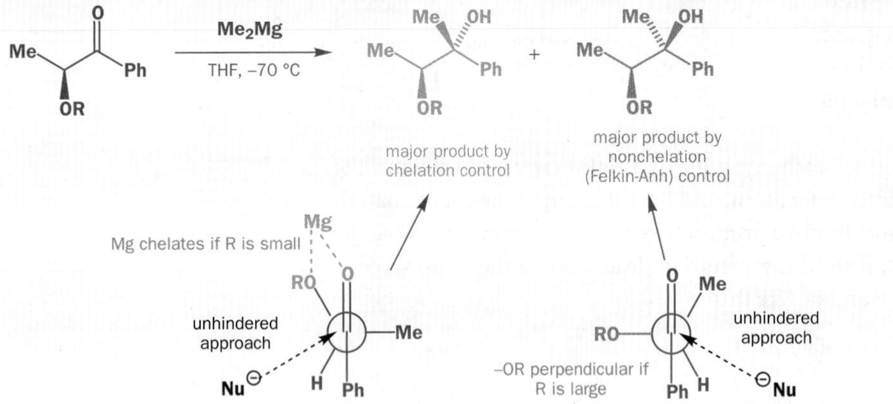

major product by chelation control

major product by nonchelation (Felkin-Anh) control

Mg chelates if R is small

unhindered approach

–OR perpendicular if R is large

unhindered approach

R	Ratio	Relative rate
Me	>99:1	1000
SiMe$_3$	99:1	100
SiEt$_3$	96:4	8
SiMe$_2$t-Bu	88:12	2.5
SiPh$_2$t-Bu	63:37	0.82
Si(i-Pr)$_3$	42:58	0.45

● **Chelation**

- may change the direction of diastereoselectivity
- leads to high levels of diastereoselectivity
- increases the rate of the addition reaction

Chelation is possible through six- as well as five-membered rings, and the reduction of the ketone below is a nice example of the reversal of diastereoselectivity observed when chelating Ce^{3+} ions are added to a normal sodium borohydride reduction. The products were important for making single geometrical isomers of alkenes in a modification of the Wittig reaction (Chapter 31). Notice too how the rate must change: with Ce^{3+} the reaction can be done at −78°C.

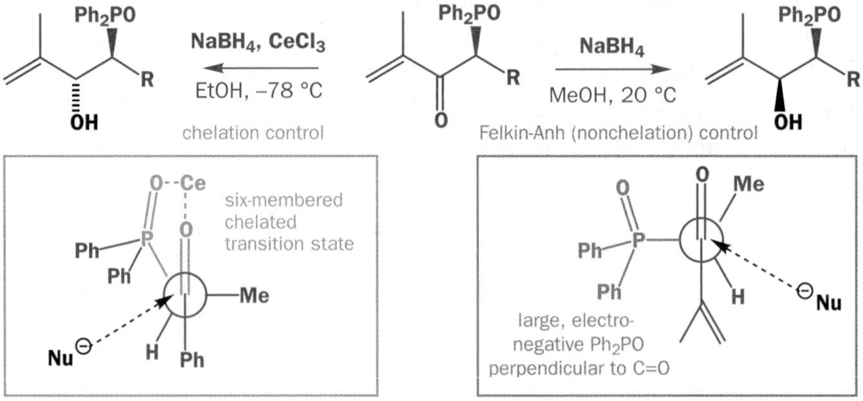

NaBH$_4$, CeCl$_3$
EtOH, −78 °C
chelation control

NaBH$_4$
MeOH, 20 °C
Felkin-Anh (nonchelation) control

six-membered chelated transition state

large, electro-negative Ph$_2$PO perpendicular to C=O

Attack on α chiral carbonyl compounds: summary

The flow chart summarizes what you should consider when you need to predict or explain the stereochemical outcome of nucleophilic attack on a chiral carbonyl compound.

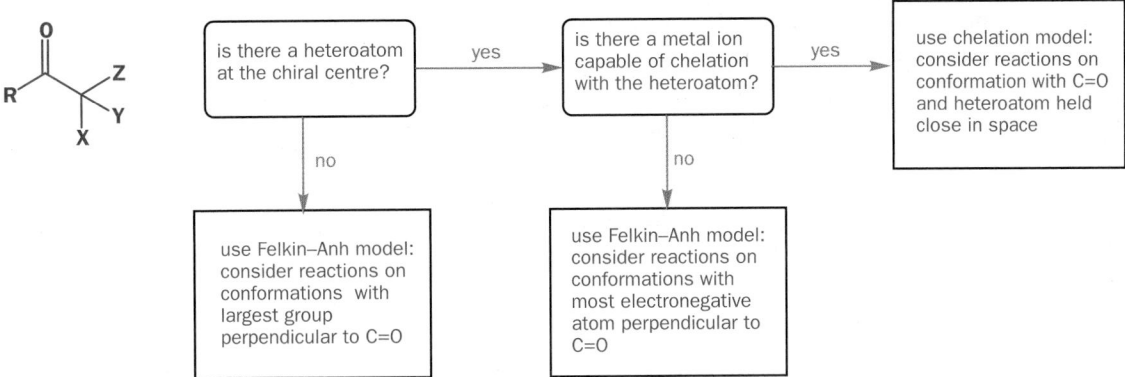

Stereoselective reactions of acyclic alkenes

Earlier the chapter we discussed how to make single diastereoisomers by stereospecific additions to double bonds of fixed geometry. But if the alkene also contains a chiral centre there will be a stereoselective aspect to its reactions too: its faces will be diastereotopic, and there will be two possible outcomes even if the reaction is fully stereospecific. Here is an example where the reaction is an epoxidation.

epoxidation is stereospecific: both epoxides retain the *cis* geometry of the starting alkene

epoxidation is stereoselective: >95% of the product is this diastereoisomer

this diastereoisomer arises from attack on the other diastereotopic face

this diastereoisomer arises from attack on one diastereotopic face

The Houk model

In order to explain reactions of chiral alkenes like this, we need to assess which conformations are important, and consider how they will react, just as we have done for chiral carbonyl compounds. Much of the work on alkene conformations was done by K.N. Houk using theoretical computer models, and we will summarize the most important conclusions of these studies. The theoretical studies looked at two model alkenes, shown in the margin.

The calculations found that the low-energy conformations in each case were those in which a substituent eclipses the double bond. For the simple model alkene 1, the lowest-energy conformation is the one that has the proton in the plane of the alkene. Another low-energy conformation—only 3.1 kJ mol^{-1} higher—has one of the methyl groups eclipsing the double bond, so that when we start looking at reactions of this type of alkene, we shall have to consider both conformations.

K.N. Houk works at the University of California in Los Angeles. He has provided explanations for a number of stereochemical results by using powerful computational methods.

model alkene **1**   model alkene **2**

this alkene has two low-energy conformations

lowest energy: H eclipses plane of double bond

slightly higher energy: Me eclipses plane of double bond

> This effect—the control of conformation by a *cis* substituent—is known as **allylic strain** or A^{1,3} strain. The groups involved are on carbons 1 and 3 of an allylic system.

For the model alkene **2**, with a *cis* substituent, the conformation is more predictable and the only low-energy conformer is the one with the hydrogen eclipsing the double bond. There is no room for a methyl group to eclipse the double bond because if it did it would get too close to the *cis* substituent at the other end of the double bond.

this alkene has only one low-energy conformation

only important conformer: H eclipses double bond

high-energy conformation due to Me–Me interaction

The message from the calculations is this:

• The lowest-energy conformation of a chiral alkene will have H eclipsing the double bond

• If there is a *cis* substituent on the alkene, this will be the only important conformation; if there is no *cis* substituent, other conformations may be important too

Now we can apply the theoretical model to some real examples.

Stereoselective epoxidation

We started this section with a diastereoselective epoxidation of an alkene. The alkene was this one, and it has a substituent *cis* to the stereogenic centre. We can therefore expect it to have one important conformation, with H eclipsing the double bond. When a reagent—*m*-CPBA here—attacks this conformation, it will approach the less hindered face, and the outcome is shown.

cis-substituted alkene

this face hindered by large SiMe₂Ph group

only important conformer has H eclipsing double bond

m-CPBA attacks the less hindered face

redraw

> Again—draw the product in the same conformation as the starting material, then flatten into the plane of the page.

Without the *cis* substituent, selectivity is much lower.

m-CPBA

61:39 ratio of diastereoisomers

m-CPBA still attacks the less hindered face of the alkene, but with no *cis* substituent there are two low-energy conformations: one with H eclipsing the double bond, and one with Me eclipsing. Each gives a different stereochemical result, explaining the low stereoselectivity of the reaction.

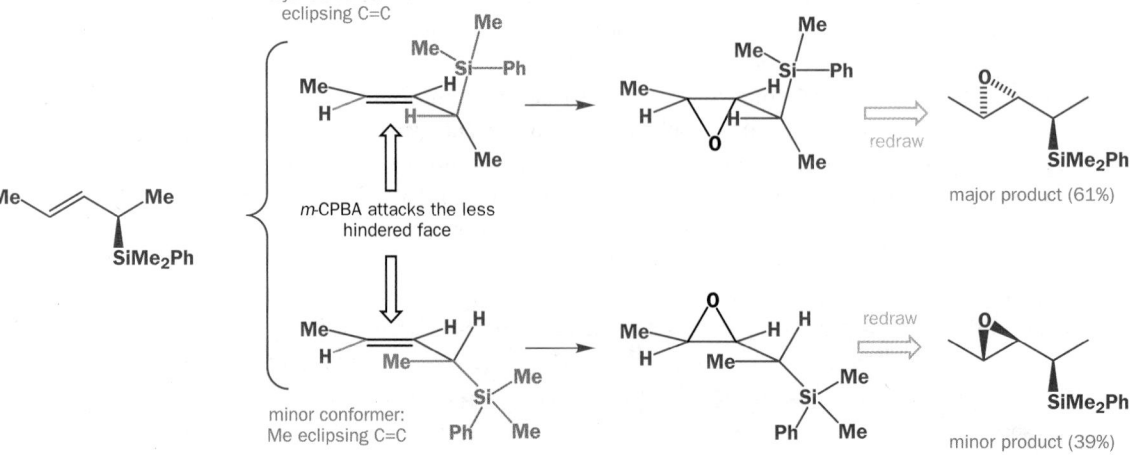

major conformer: H eclipsing C=C

m-CPBA attacks the less hindered face

redraw

major product (61%)

minor conformer: Me eclipsing C=C

redraw

minor product (39%)

You saw at the end of the last chapter that the reactions of *m*-CPBA can be directed by hydroxyl groups, and the same thing happens in the reactions of acyclic alkenes. This allylic alcohol epoxidizes to give a 95:5 ratio of diastereoisomers.

95:5 ratio of diastereoisomers

Drawing the reactive conformation explains the result. The thing that counts is the *cis* methyl group: the fact that there is a *trans* one too is irrelevant as it is just too far away from the stereogenic centre to have an effect on the conformation.

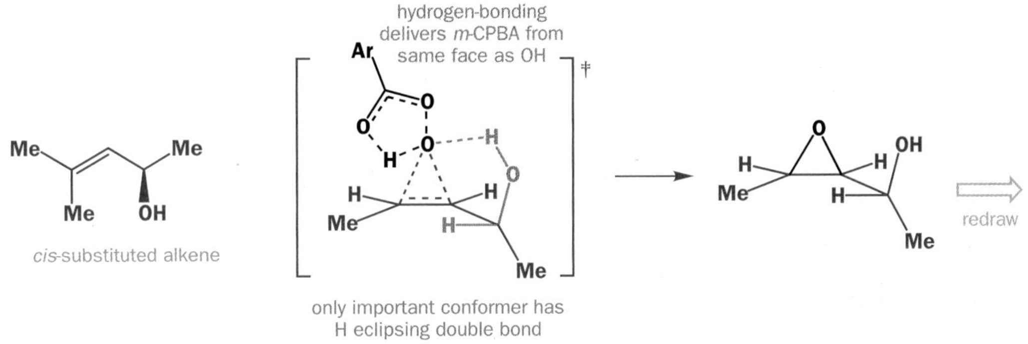

To explain the stereoselectivity of reactions of chiral alkenes:

- Draw the conformation with H eclipsing the double bond
- Allow the reagent to attack the less hindered of the two faces or, if co-ordination is possible, to be delivered to the face *syn* to the coordinating group
- Draw the product in the same conformation as the starting material
- Redraw the product as a normal structure with the longest chain in the plane of the paper

Stereoselective enolate alkylation

Chiral enolates can be made from compounds with a stereogenic centre β to a carbonyl group. Once the carbonyl is deprotonated to form the enolate, the stereogenic centre is next to the double bond and in a position to control the stereoselectivity of its reactions. The scheme below shows stereoselectivity in the reactions of some chiral enolates with methyl iodide.

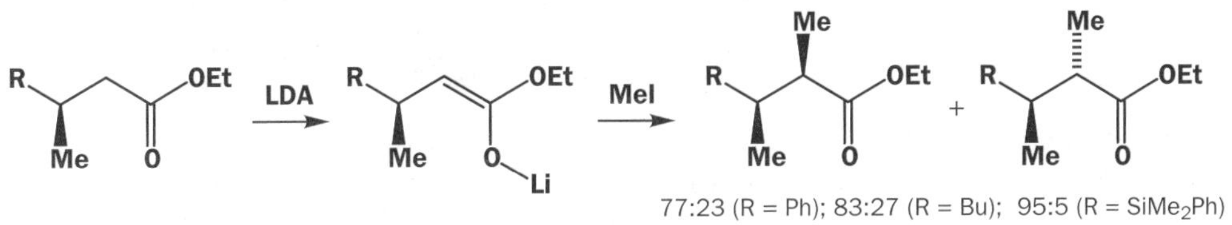

77:23 (R = Ph); 83:27 (R = Bu); 95:5 (R = SiMe₂Ph)

The enolate is a *cis*-substituted alkene, because either O⁻ or OEt must be *cis* to the stereogenic centre, so that to explain the stereoselectivity, we need consider only the conformation with H eclipsing the double bond. Notice how the diastereoselectivity increases at the group R gets bigger, because there is then more contrast between the size of Me and R. In each case, the electrophile adds to the less hindered face, opposite R.

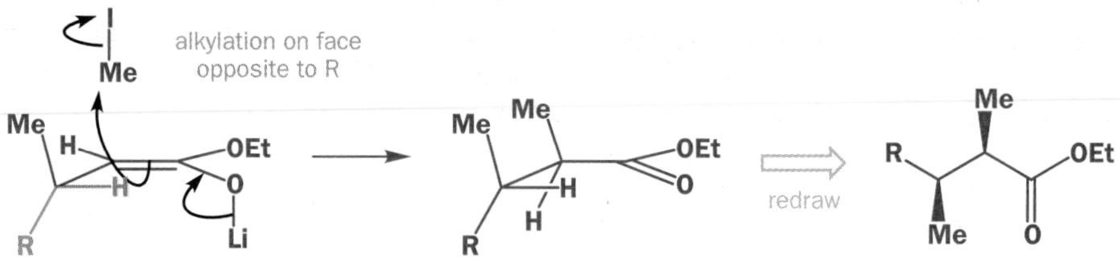

alkylation on face opposite to R

redraw

The other diastereoisomer can be made just by having the methyl group in place first and then protonating the enolate. The selectivities are lower (because a proton is small), but this does illustrate the way in which reversing the order of introduction of two groups can reverse the stereochemical outcome of the reaction.

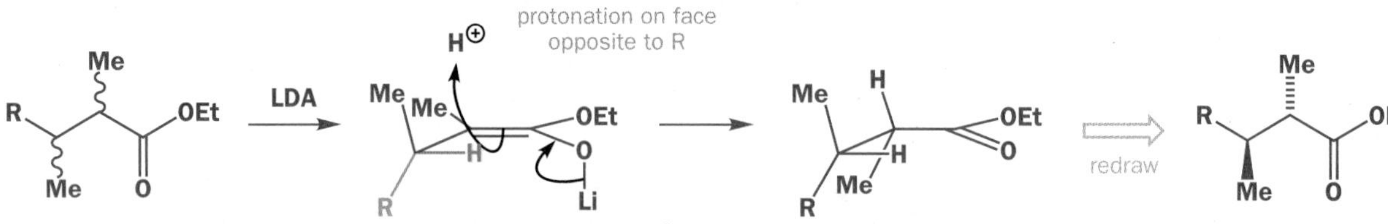

protonation on face opposite to R

redraw

Aldol reactions can be stereoselective

In Chapter 27 you met the **aldol reaction**: reaction of an enolate with an aldehyde or a ketone. Many of the examples you saw approximated to this general pattern.

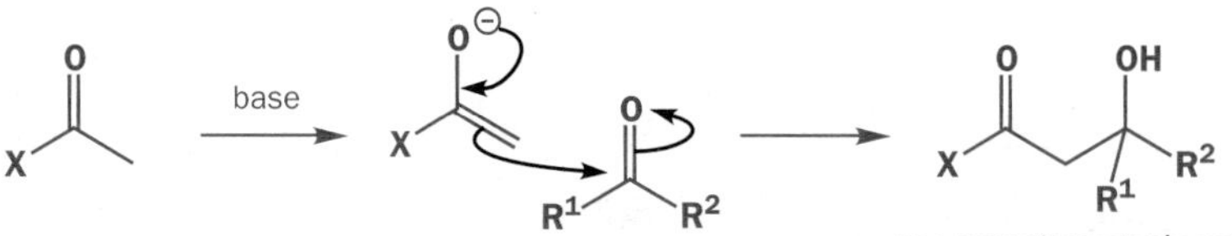

one new stereogenic centre: no diastereoselectivity involved

Only one new stereogenic centre is created, so there is no question of diastereoselectivity. But with substituted enolates, two new stereogenic centres are created, and we need to be able to predict which diastereoisomer will be formed. Here is an example from p. 699. We did not consider stereochemistry at that stage, but we can now reveal that the *syn* diastereoisomer is the major product of the reaction.

LDA, −78 °C, THF

enolate bears a Me substituent

syn aldol (major product)

anti aldol (minor product)

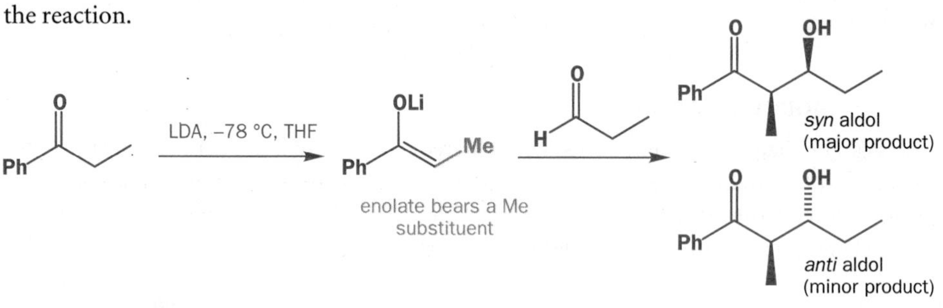

two new stereogenic centres: two diastereoisomers possible

The important point about substituted enolates is that they can exist as two geometrical isomers, *cis* or *trans*. Which enolate is formed is an important factor controlling the diastereoselectivity because it turns out that, in many examples of the aldol reaction, *cis*-enolates give *syn* aldols preferentially and *trans*-enolates give *anti* aldols preferentially.

● **Diastereoselectivity in aldol reactions**

Generally (but certainly not always!) in aldol reactions:

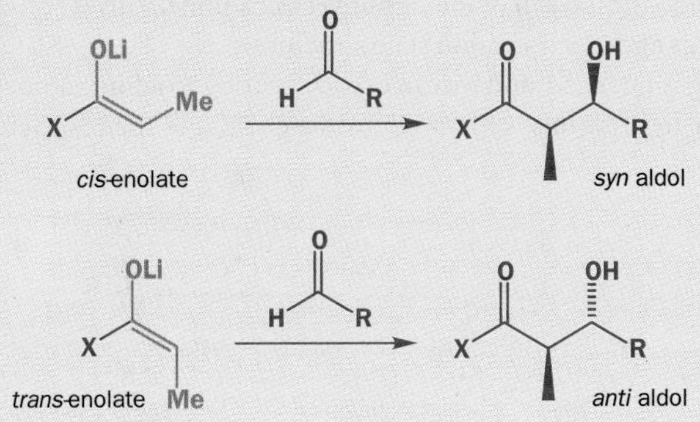

cis-enolate → *syn* aldol

trans-enolate → *anti* aldol

Let's start by showing some examples and demonstrating how we know this to be the case. Some enolates can only exist as *trans*-enolates because they are derived from cyclic ketones. This enolate, for example, reacts with aldehydes to give only the *anti* aldol product.

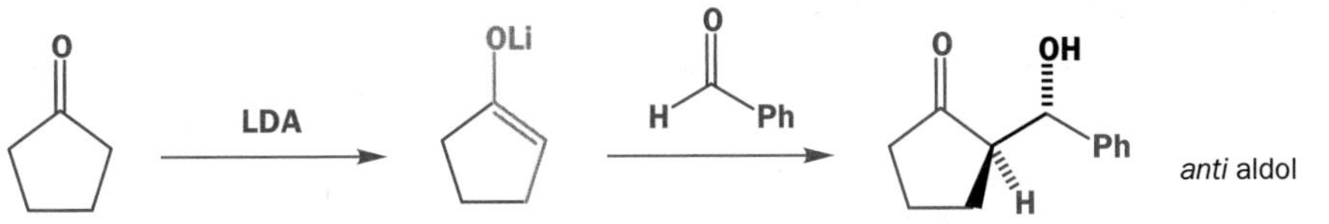

only *trans*-enolate can form

anti aldol

If we choose the group 'X', next to the carbonyl group, to be large, then we can be sure of getting just the *cis*-enolate. So, for example, the lithium enolate of this *t*-butyl ketone forms just as one geometrical isomer, and reacts with aldols to give only the *syn* aldol product.

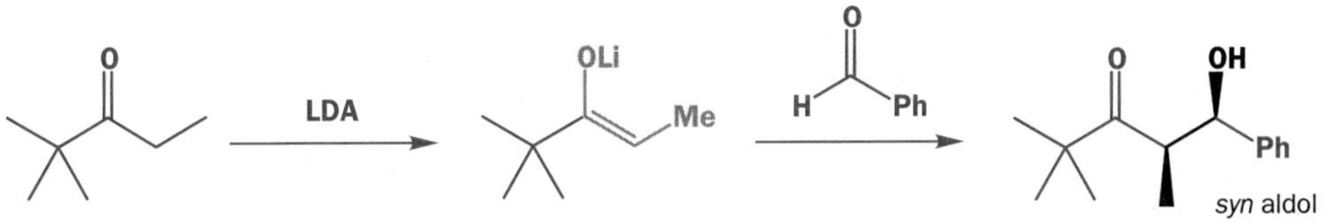

cis-enolate avoids Me and *t*-Bu coming into contact

syn aldol

cis and *trans*, *E* and *Z*, *syn* and *anti*

Before going further, there are two points we must clarify. The first is a problem of nomenclature, and concerns the enolates of esters. Here are two closely related ester enolate equivalents, drawn with the same double bond geometry. Is it *E* or *Z*?

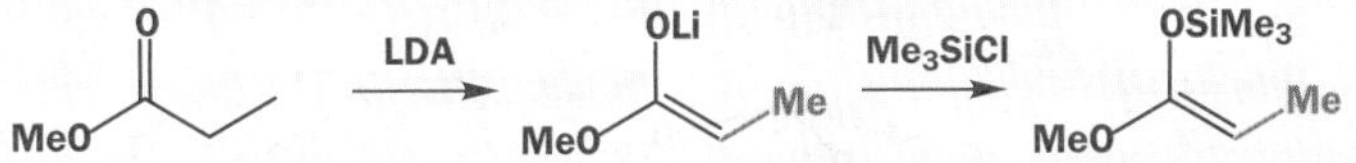

The answer is both! For the Li enolate, the usual rule makes OLi of lower priority than OMe, so it's *E*, while the silyl enol ether (or 'silyl ketene acetal') has OSi of higher priority than OMe, so it's *Z*. This is merely a nomenclature problem, but it would be irritating to have to reverse all our arguments for lithium enolates simply because lithium is of lower atomic number than carbon. So, for the sake of consistency, it is much better to avoid the use of *E* and *Z* with enolates and instead use *cis* and *trans*, which then always refer to the relationship between the substituent and the anionic oxygen (bearing the metal).

The other point concerns *syn* and *anti*. We said earlier that there is no precise definition of these terms: they are a useful way of distinguishing two diastereoisomers provided the structure of at least one of them is presented in diagrammatic form. For aldol products the convention is that *syn* or *anti* refers to the enolate substituent (the green Me in the last example) and the new hydroxyl group, provided the main chain is in the plane of the paper, the way we have encouraged you to draw molecules.

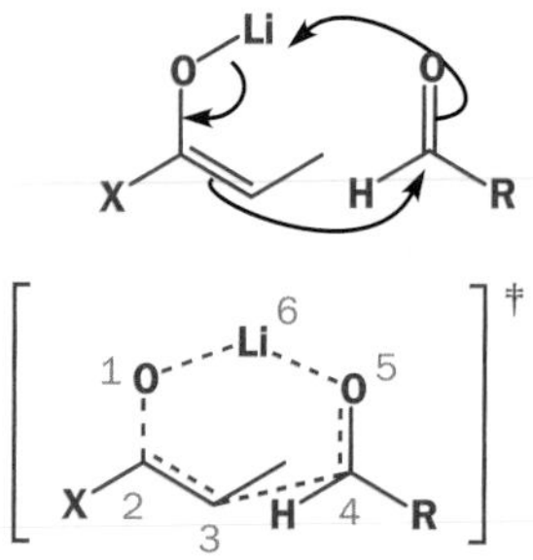

> The six-membered ring transition state for the aldol reaction was proposed by Zimmerman and Traxler and is sometimes called the **Zimmerman–Traxler transition state**.

The aldol reaction has a chair-like transition state

These are the experimental facts: how can we explain them? Aldol reactions are another class of stereoselective process with a cyclic transition state. During the reaction, the lithium is transferred from the enolate oxygen to the oxygen of the carbonyl electrophile. This is represented in the margin both in curly arrow terms and as a transition state structure.

A six-membered ring is involved, and we can expect this ring to adopt more or less a chair conformation. The easiest way to draw this is first to draw the chair, and then convert atoms to O or Li as necessary. Here it is.

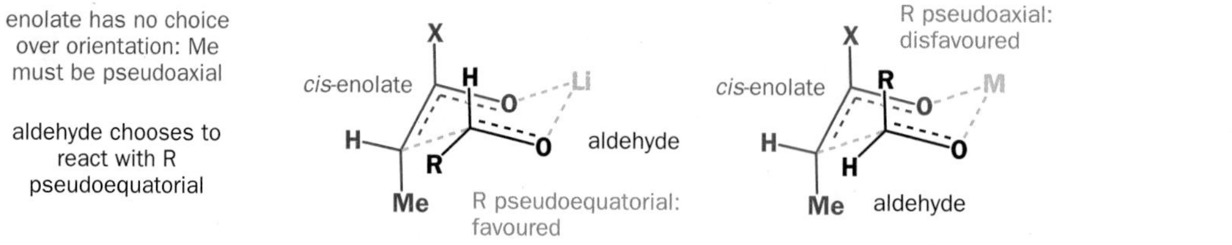

In drawing this chair, we have one choice: do we allow the aldehyde to place R equatorial or axial? Both are possible but, as you should now expect, there are fewer steric interactions if R is equatorial. Note that the enolate doesn't have the luxury of choice. If it is to have three atoms in the six-membered ring, as it must, it can do nothing but place the methyl group pseudoaxial.

The aldol formed from the favoured transition state structure, with R pseudoequatorial, is shown below—first in the conformation of the transition state, and then flattened out on to the page, and it is *syn*.

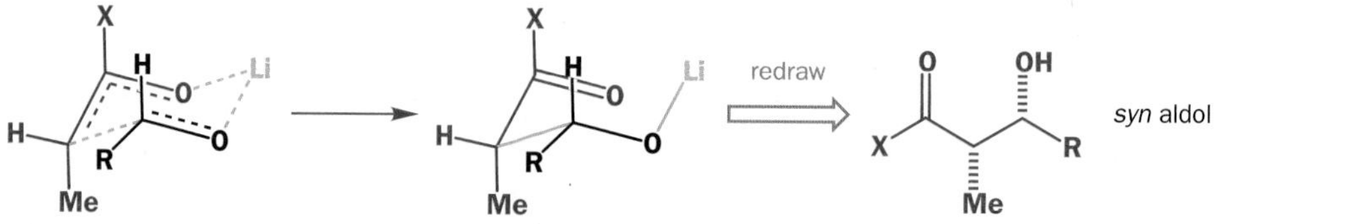

We can do the same for a *trans*-enolate. The enolate has no choice but to put its methyl substituent pseudoequatorial, but the aldehyde can choose either pseudoequatorial or pseudoaxial. Again, pseudoequatorial is better

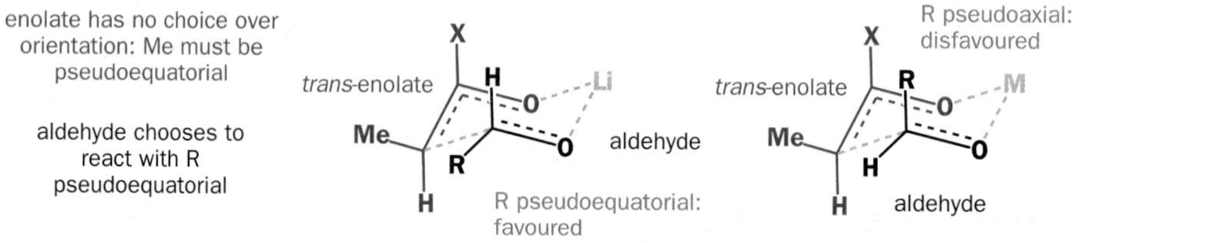

and the reaction gives the product shown—the *anti* aldol.

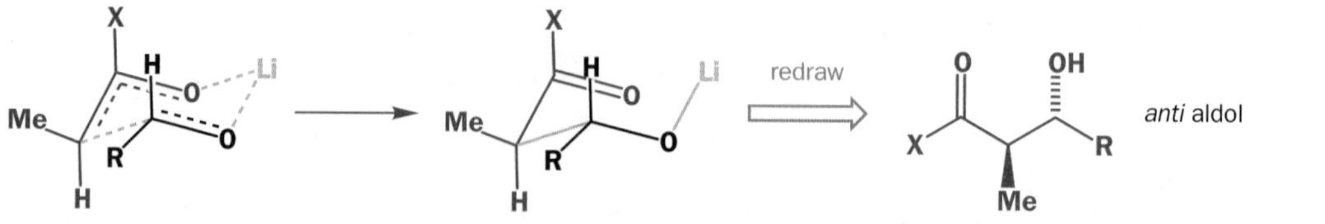

Stereoselective enolization is needed for stereoselective aldols

The cyclic transition state explains how enolate geometry controls the stereochemical outcome of the aldol reaction. But what controls the geometry of the enolate? For lithium enolates of ketones the most important factor is the size of the group that is not enolized. Large groups force the enolate to adopt the *cis* geometry; small groups allow the *trans*-enolate to form. Because we can't separate the lithium enolates, we just have to accept that the reactions of ketones with small R will be less diastereoselective.

With *boron* enolates, we don't have to rely on the structure of the substrate—we choose the groups on boron—and we can get either *cis* or *trans* depending on which groups these are. Boron enolates are made by treating the ketone with an amine

R = t-Bu	98%	2%
R = Et	30%	70%

base (often Et$_3$N or *i*-PrNEt$_2$) and R$_2$B–X, where X$^-$ is a good leaving group such as chloride or triflate (CF$_3$SO$_2^-$). With bulky groups on boron, such as two cyclohexyl groups, a *trans*-enolate forms from most ketones. The boron enolate reacts reliably with aldehydes to give *anti* aldol products through the same six-membered transition state that you saw for lithium enolates.

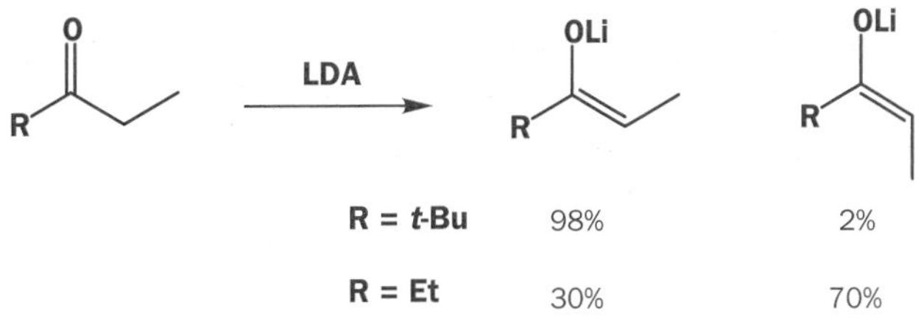

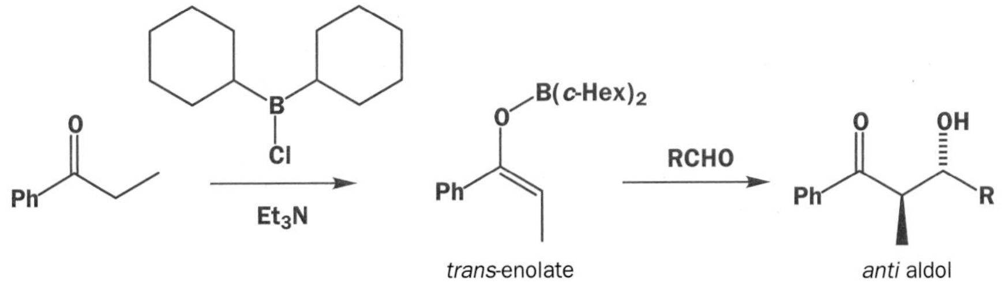

With smaller B substituents, the *cis*-enolate forms selectively. Here, the boron is part of a bicyclic structure known as 9-BBN (9-borabicyclononane—you will meet this in Chapter 47). The bicyclic part may look large but, as far as the rest of the molecule is concerned, it's 'tied back' behind the boron, and the methyl group can easily lie *cis* to oxygen. The *cis*-enolate then gives *syn* aldol products. Di-*n*-butylboron triflate (Bu$_2$BOTf) also gives *cis*-enolates.

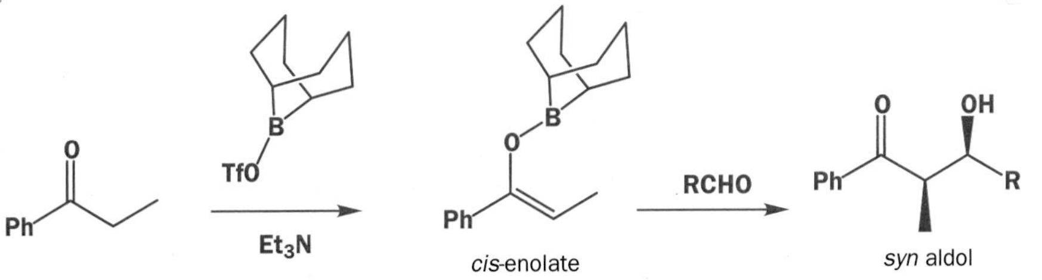

Stereoselective ester aldols

We have talked mainly about aldol reactions of ketones (as the enolate component). Esters usually form the *trans* lithium enolates quite stereoselectively. You might therefore imagine that their aldol reactions would be stereoselective for the *anti* product. Unfortunately, this is not the case, and even pure *trans*-enolate gives about a 1:1 mixture of *syn* and *anti* aldols.

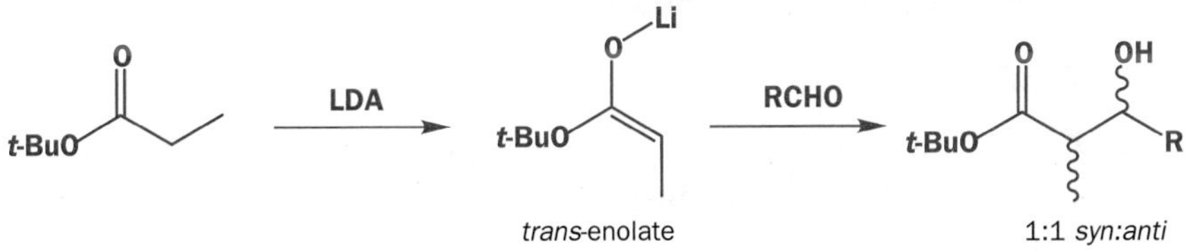

There is one important exception, and that is a class of esters of hindered phenols. The *trans*-enolates of these compounds react selectively with aldehydes to give the *anti* aldol products.

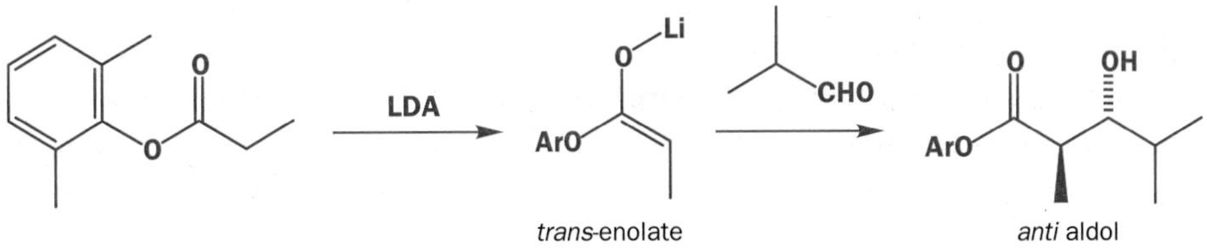

> In fact, geometrically defined boron enolates give the aldol products with greater stereospecificity than do lithium enolates, possibly because the B–O bonds are shorter than Li–O bonds, so the six-membered ring is 'tighter'.

hindered phenols:

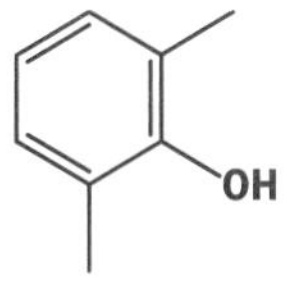

2,6-dimethylphenol

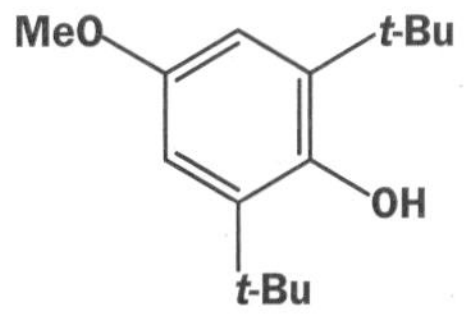

'butylated hydroxyanisole', BHA

An ingenious way of getting a *syn* ester aldol product is to do the more reliable ketone *syn* aldol with a bulky group (to ensure the *cis*-enolate is formed) and then to oxidize off the bulky group. Here's what we mean. The starting material is very like the *t*-butyl ketone that you saw enolize stereoselectively above: only the *cis*-enolate can form. The enolate reacts highly *syn* selectively with the aldehyde, via the six-membered transition state.

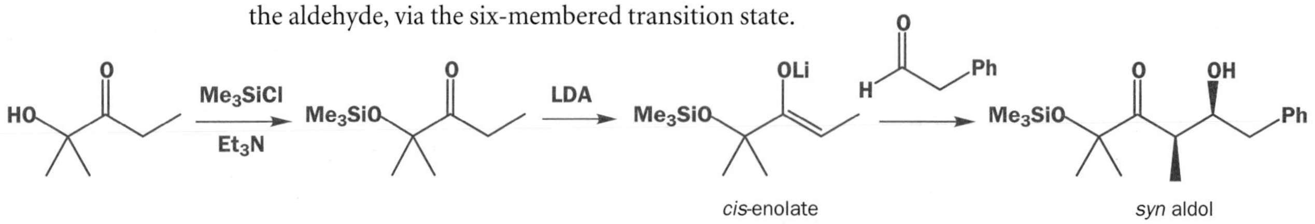

At this point, the bulky group is no longer needed. The oxygen is deprotected in acid and, in the same step, periodate ions oxidatively cleave the C–C bond between the two oxygen substituents. The product is the acid parent of a *syn* ester aldol product.

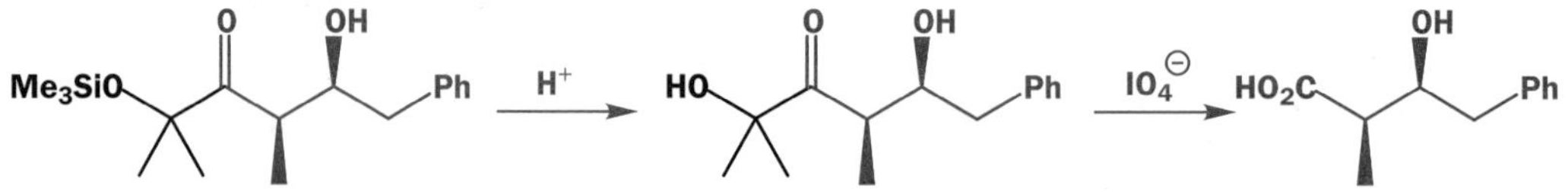

We shall show you the mechanism of the cleavage, because it leads us nicely into the next chapter. The first step is rather like the first step of many oxidations—formation of an inorganic ester (here a periodate). The periodate can form a cyclic ester by attack on the carbonyl group. Next, we can push the arrows round the ring to reduce the iodine from I(VII) to I(V), cleave the double bond, and generate acetone and the acid.

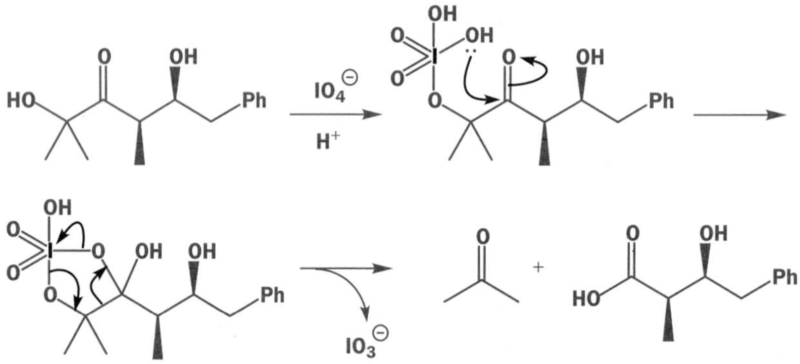

You will see many more cyclic mechanisms in the next two chapters, including some more C–C cleavage reactions.

● Summary: How to make *syn* and *anti* aldols

To make *syn* aldols of ketones:

- with a ketone RCOEt with bulky R, use lithium enolate
- use boron enolate with 9-BBN-OTf or Bu₂BOTf

To make *syn* aldols of esters:

- use a bulky 2-alkoxyketone and cleave to an acid

To make *anti* aldols of ketones:

- with a cyclic ketone, use lithium enolate
- use boron enolate with dicyclohexylboron chloride

To make *anti* aldols of esters:

- use the ester of a hindered phenol

Problems

1. How would you make each diastereoisomer of this product from the same alkene?

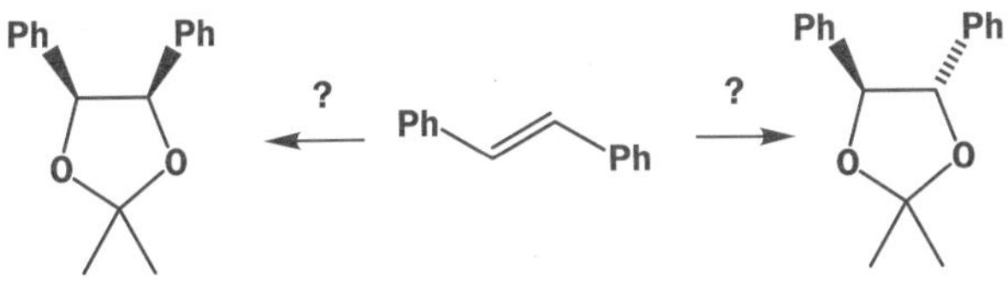

2. Explain the stereoselectivity shown in this sequence of reactions.

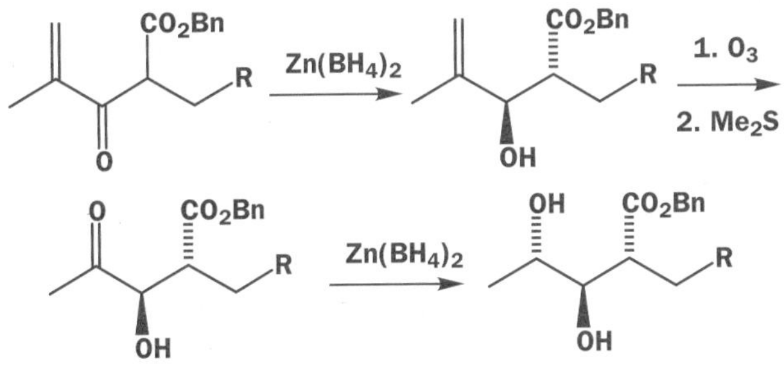

3. How is the relative stereochemistry of this product controlled? Why was this method chosen?

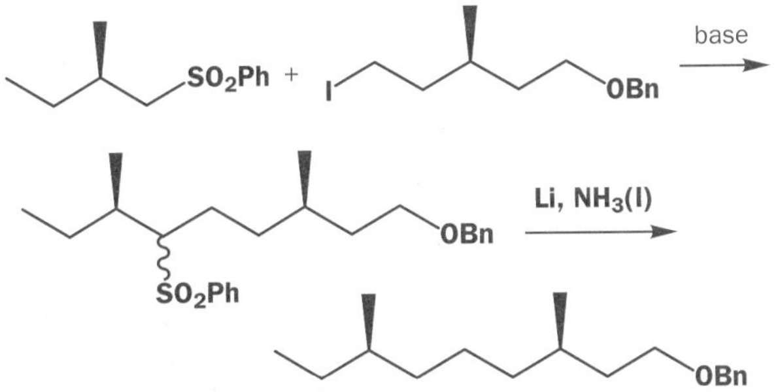

4. Explain the stereochemical control in this reaction, drawing all the intermediates.

5. When this hydroxy-ester is treated with a twofold excess of LDA and then alkylated, one diastereoisomer of the product predominates. Why?

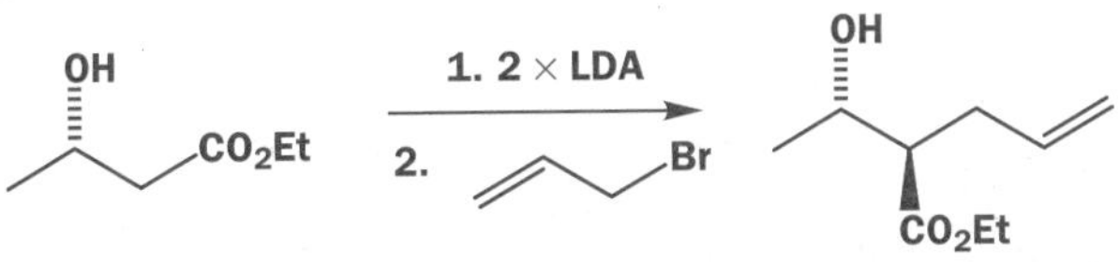

6. Explain how the stereochemistry of this epoxide is controlled.

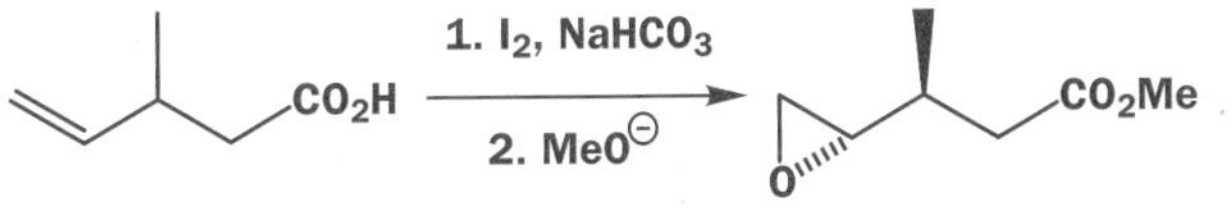

7. Explain how these two reactions give different diastereoisomers of the product.

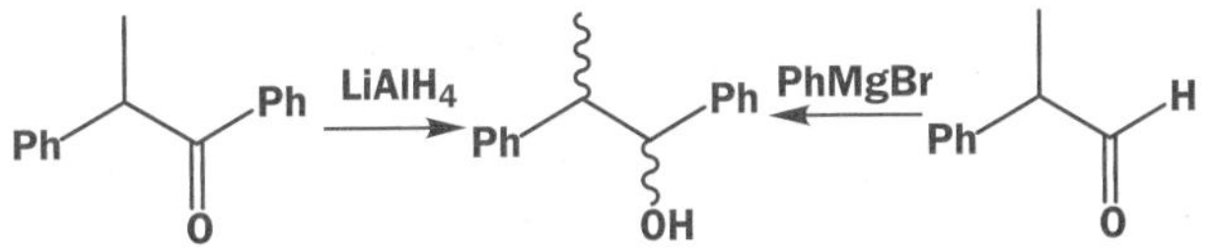

8. Explain the stereoselectivity in this reaction. What isomer of an epoxide would be produced on treatment of the product with base?

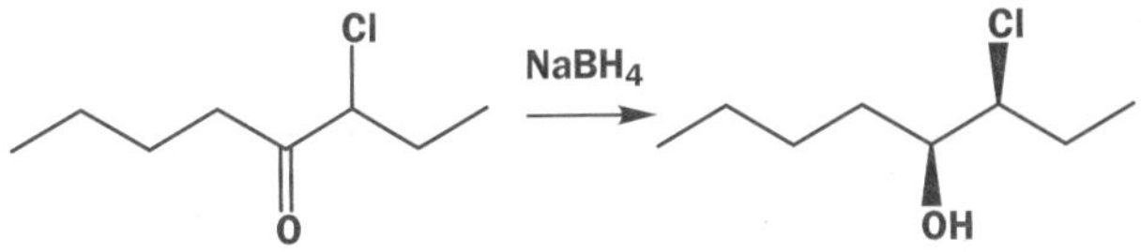

9. How could this cyclic compound be used to produce the open-chain compound with correct relative stereochemistry?

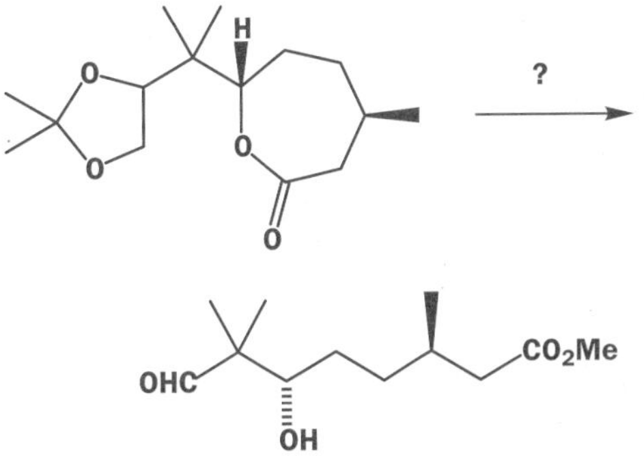

10. How would you transform this alkene stereoselectively into either of the diastereoisomers of the amino-alcohol?

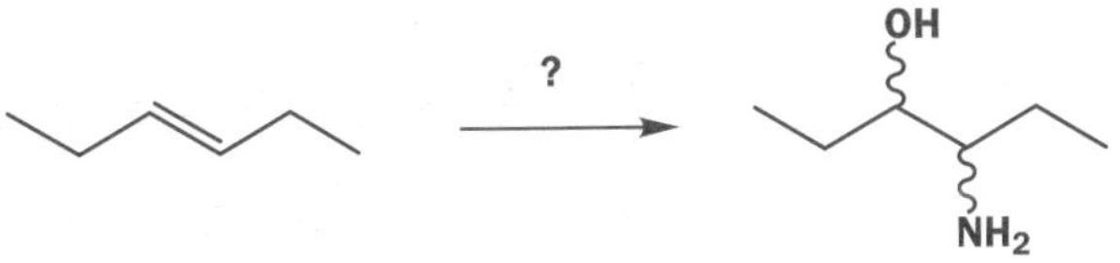

11. Explain the formation of essentially one stereoisomer in this reaction.

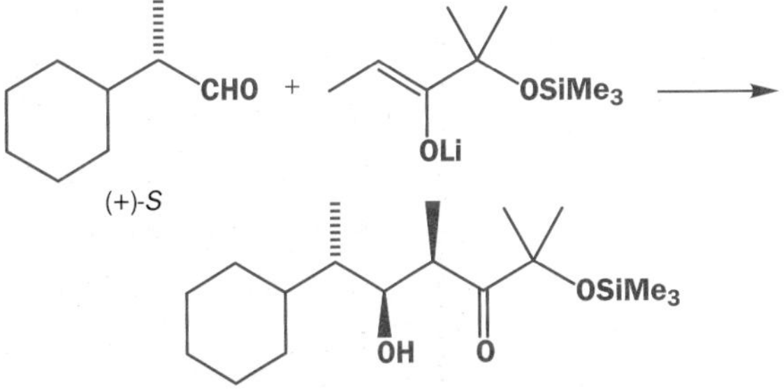

12. How would you attempt to transform this allylic alcohol into both diastereoisomers of the epoxide stereoselectively? You are not expected to estimate the degree of success.

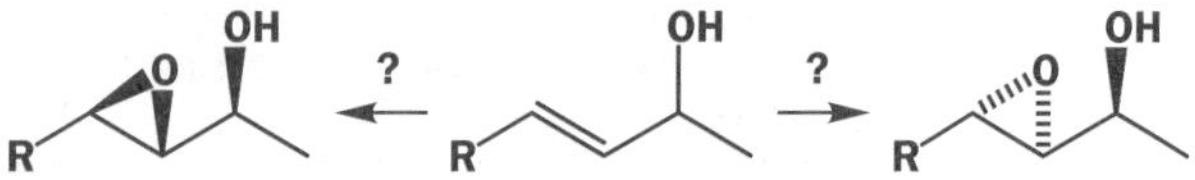

13. Revision. Here is an outline of the AstraZeneca synthesis of a thromboxane analogue. Explain the reactions, giving mechanisms for each step, and explain how the stereochemistry is controlled. In what way could this be considered an example of the control of open-chain stereochemistry when all of the molecules are cyclic?

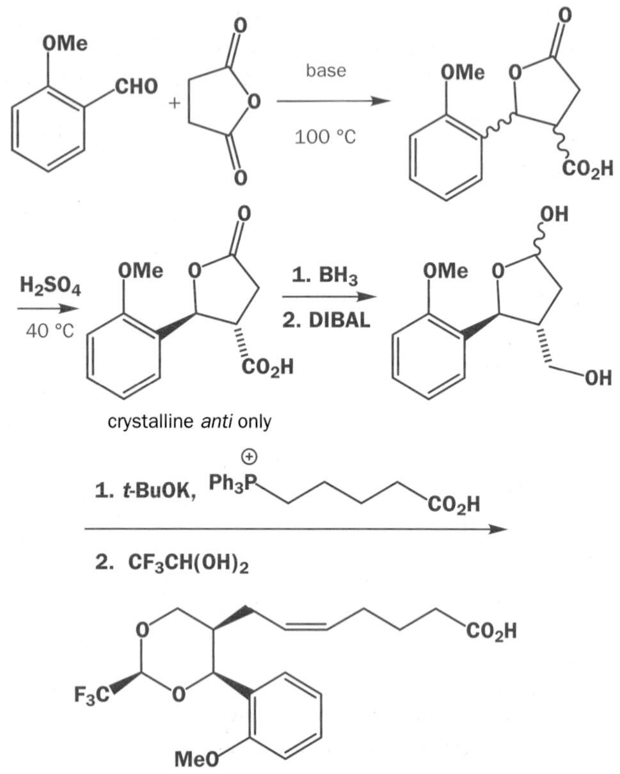

Pericyclic reactions 1: cycloadditions $\quad$ 35

Connections

Building on:

- Structure of molecules ch4
- Reaction mechanisms ch5
- Conjugation and delocalization ch7

Arriving at:

- In cycloadditions electrons move in a ring
- In cycloadditions more than one bond is formed simultaneously
- There are no intermediates in cycloadditions
- Cycloadditions are a type of pericyclic reaction
- The rules that govern cycloadditions: how to predict what will and will not work
- Photochemical reactions: reactions that need light
- Making six-membered rings by the Diels–Alder reaction
- Making four-membered rings by [2 + 2] cycloaddition
- Making five-membered rings by 1,3-dipolar cycloaddition
- Using cycloaddition to functionalize double bonds stereospecifically
- Using ozone to break C=C double bonds

Looking forward to:

- Electrocyclic reactions and sigmatropic rearrangements ch36
- Radical reactions ch39
- Aromatic heterocycles ch43–ch44
- Asymmetric synthesis ch45
- Organic synthesis ch53

A new sort of reaction

Most organic reactions are ionic. Electrons move from an electron-rich atom towards an electron-poor atom: anions or cations are intermediates. Formation of a cyclic ester (a lactone) is an example.

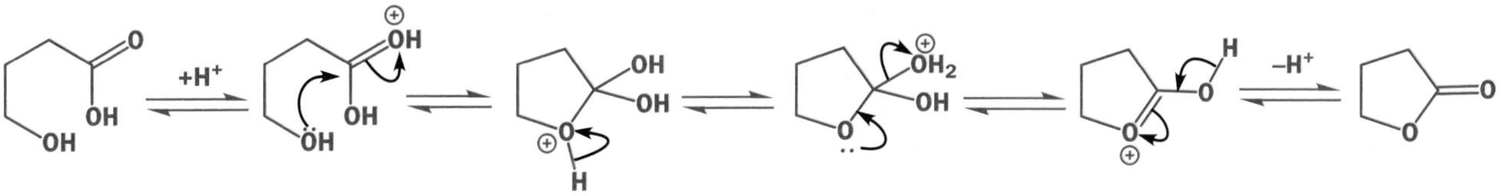

The reaction involves five steps and four intermediates. The reaction is acid-catalysed and each intermediate is a cation. Electrons flow in one direction in each step—towards the positive charge. This is an ionic reaction.

This chapter is about a totally different reaction type. Electrons move round a circle and there are no positive or negative charges on any intermediates—indeed, there are no intermediates at all. This type of reaction is called **pericyclic**. The most famous example is the **Diels–Alder reaction**.

> In Chapter 39 you will meet a third category—radical reactions—in which one electron instead of two is on the move.

> Otto Diels (1876–1954) and his research student Kurt Alder (1902–58) worked at the University of Kiel and discovered this reaction in 1928. They won the Nobel Prize in 1950. Diels also discovered the existence of carbon suboxide, C_2O_3 (see p. 372).

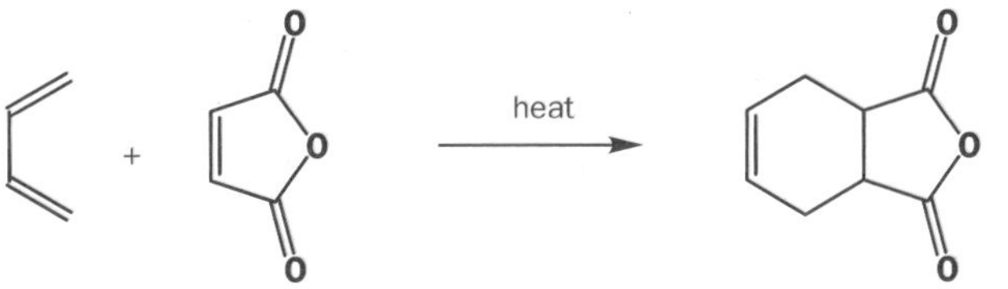

This reaction goes in a single step simply on heating. We can draw the mechanism with the electrons going round a six-membered ring.

Each arrow leads directly to the next, and the last arrow connects to the first. We have drawn the electrons rotating clockwise, but it would make no difference at all if we drew the electrons rotating anticlockwise.

Both mechanisms are equally correct. The electrons do not really rotate at all. In reality two π bonds disappear and two σ bonds take their place by the electrons moving smoothly out of the π orbitals into the σ orbitals. Such a reaction is called a **cycloaddition**. We must spend some time working out how this could happen.

First, just consider the orbitals that overlap to form the new bonds. Providing the reagents approach in the right way, nothing could be simpler.

The black p orbitals are perfectly aligned tol make a new σ bond as are the two green orbitals, while the two brown orbitals are exactly right for the new π bond at the back of the ring. As this is a one-step reaction there are no intermediates but there is one transition state looking something like this.

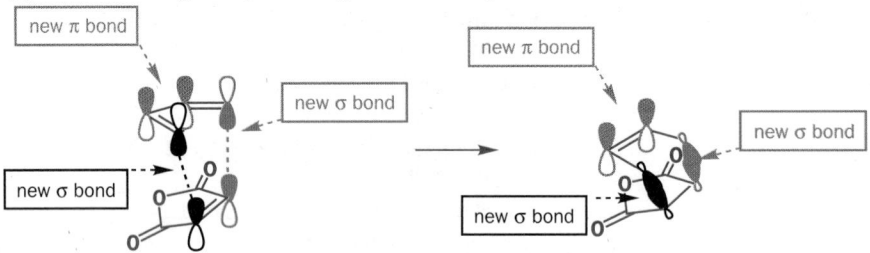

transition state has six delocalized π electrons

One reason that the Diels–Alder reaction goes so well is that the transition state has six delocalized π electrons and thus is aromatic in character, having some of the special stabilization of benzene. You could look at it as a benzene ring having all its π bonds but missing two σ bonds. This simple picture is fine as far as it goes, but it is incomplete. We shall return to a more detailed orbital analysis when we have described the reaction in more detail.

> Cycloadditions are the first of three classes of pericyclic reactions, and the whole of this chapter will be devoted to cycloadditions. The other two—sigmatropic and electrocyclic reactions—are discussed in Chapter 36.

Captan

One important industrial application of the Diels–Alder reaction we have been discussing is in the synthesis of the agriculturla fungicide Captan.

Captan

General description of the Diels–Alder reaction

Diels–Alder reactions occur between a **conjugated diene** and an alkene, usually called the **dienophile**. Here are some examples: first an open-chain diene with a simple unsaturated aldehyde as the dienophile.

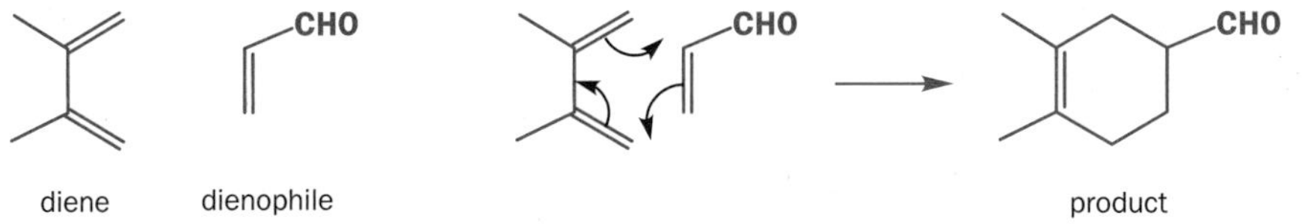

The mechanism is the same and a new six-membered ring is formed having one double bond. Now a reaction between a cyclic diene and a nitroalkene.

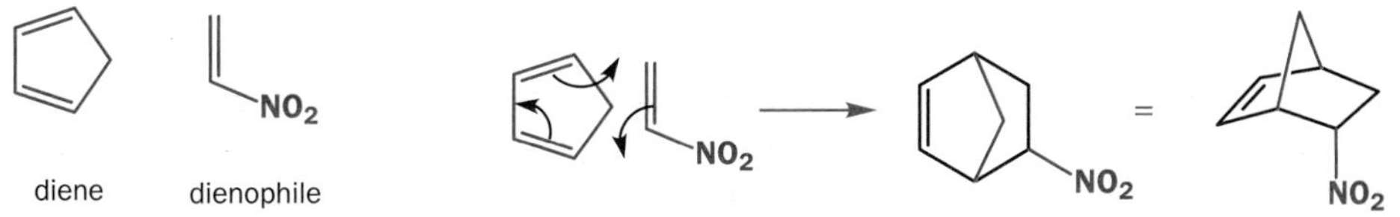

The mechanism leads clearly to the first drawing of the product but this is a cage structure and the second drawing is better. The new six-membered ring is outlined in black in both diagrams. Now a more elaborate example to show that quite complex molecules can be quickly assembled with this wonderful reaction.

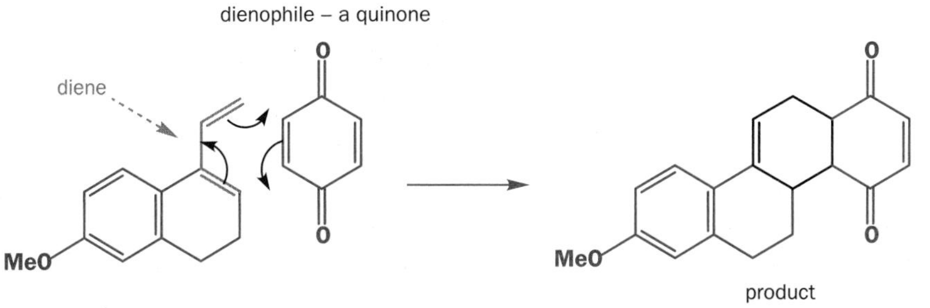

The diene

The diene component in the Diels–Alder reaction can be open-chain or cyclic and it can have many different kinds of substituents. There is only one limitation: it must be able to take up the conformation shown in the mechanism. Butadiene normally prefers the **s-*trans*** conformation with the two double bonds as far away from each other as possible for steric reasons. The barrier to rotation about the central σ bond is small (about 30 kJ mol^{-1} at room temperature: see Chapter 18) and rotation to the less favourable but reactive **s-*cis*** conformation is rapid.

> The 's' in the terms 's-*cis*' and 's-*trans*' refers to a σ bond and indicates that these are conformations about a single bond and not configurations about a double bond.

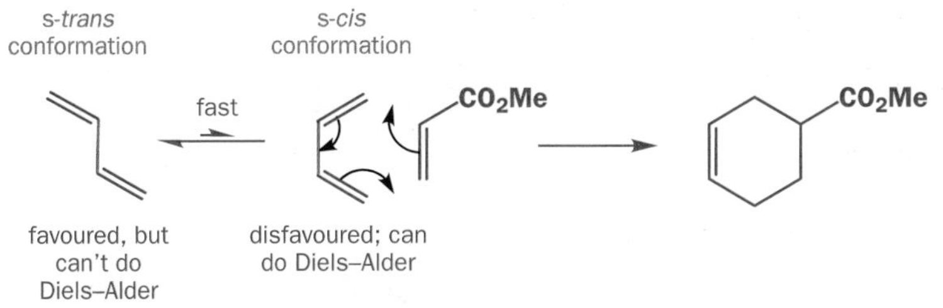

Cyclic dienes that are permanently in the s-*cis* conformation are exceptionally good at Diels–Alder reactions—cyclopentadiene is a classic example—but cyclic dienes that are permanently in the s-*trans* conformation and cannot adopt the s-*cis* conformation will not do the Diels–Alder reaction at all. The two ends of these dienes cannot get close enough to react with

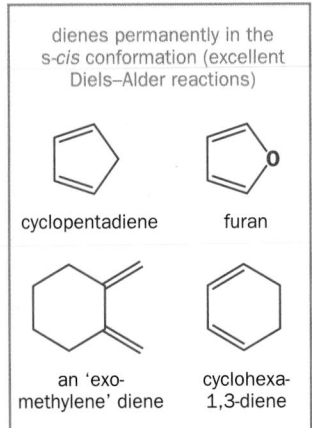

dienes permanently in the s-*cis* conformation (excellent Diels–Alder reactions)

cyclopentadiene   furan

an 'exo-methylene' diene   cyclohexa-1,3-diene

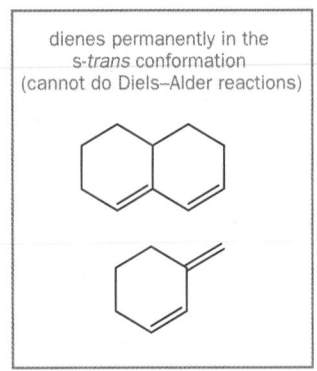

dienes permanently in the s-*trans* conformation (cannot do Diels–Alder reactions)

an alkene and, in any case, the product would have an impossible *trans* double bond in the new six-membered ring. (In the Diels–Alder reaction, the old σ bond in the centre of the diene becomes a π bond in the product and the conformation of that σ bond becomes the configuration of the new π bond in the product.)

> ● **The diene**
>
> The diene must have the s-*cis* conformation.

The dienophile

The dienophiles you have seen in action so far all have one thing in common. They have an electron-withdrawing group conjugated to the alkene. This is a common though not exclusive feature of Diels–Alder dienophiles. There must be some extra conjugation—at least a phenyl group or a chlorine atom—or the cycloaddition does not occur. You will often see the reaction between butadiene and a simple alkene (even ethylene) given in books as the basic Diels–Alder reaction. This occurs in only poor yield. Attempts to combine even such a reactive diene as cyclopentadiene with a simple alkene lead instead to the dimerization of the diene. One molecule acts as the diene and the other as the dienophile to give the cage structure shown.

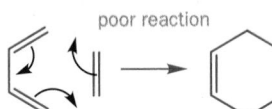

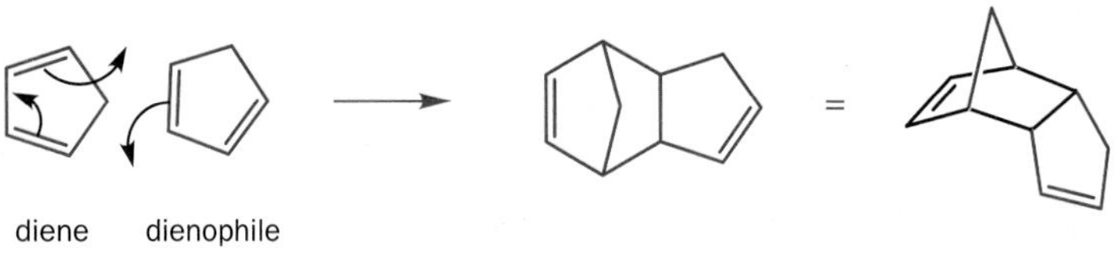

poor reaction

diene   dienophile

Cyclopentadiene

Cyclopentadiene is formed in considerable amounts during the refining of petroleum. It exists as its dimer at room temperature but can be dissociated into the monomer on heating—the effect of the increased importance of entropy at higher temperatures (Chapter 13). It can be chlorinated to give hexachlorocyclopentadiene, and the Diels–Alder product of this diene with maleic anhydride is a flame retardant.

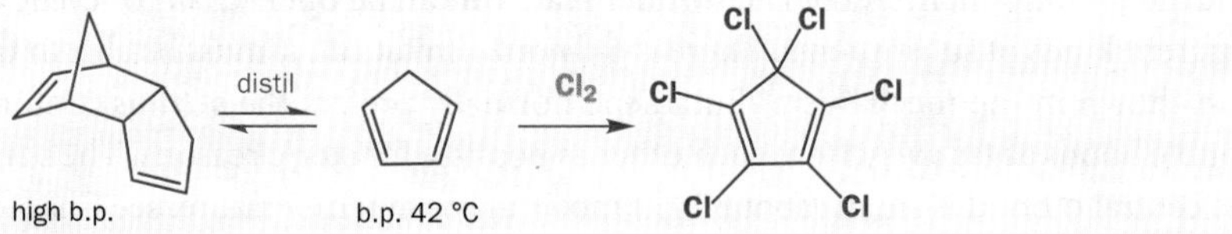

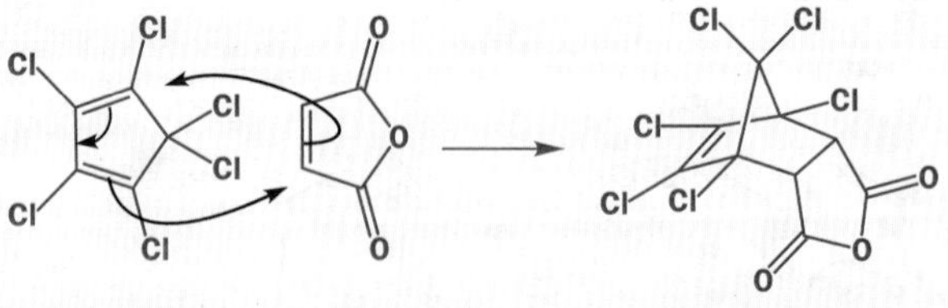

high b.p.   distil   b.p. 42 °C   Cl₂

flame retardant

some dienophiles for the Diels–Alder reaction

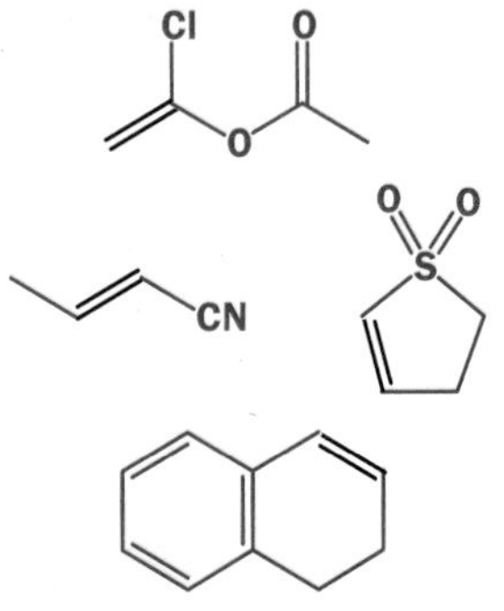

Simple alkenes that do undergo the Diels–Alder reaction include conjugated carbonyl compounds, nitro compounds, nitriles, sulfones, aryl alkenes, vinyl ethers and esters, haloalkenes, and dienes. In addition to those you have seen so far, a few examples are shown in the margin. In the last example it is the isolated double bond in the right-hand ring that accepts the diene. Conjugation with the left-hand ring activates this alkene. But what exactly do we mean by 'activate' in this sense? We shall return to that question in a minute.

Dieldrin and Aldrin

In the 1950s two very effective pesticides were launched and their names were 'Dieldrin' and 'Aldrin'. As you may guess they were made by the Diels–Alder reaction. Aldrin is derived from two consecutive Diels–Alder reactions. In the first, cyclopentadiene reacts with acetylene to give a simple symmetrical cage molecule 'norbornadiene'

(bicyclo[2.2.1]heptadiene). Norbornadiene is not conjugated and cannot take part in a Diels–Alder reaction as a *diene*. However, it is quite strained because of the cage and it reacts as a *dienophile* with perchlorocyclopentadiene to give Aldrin.

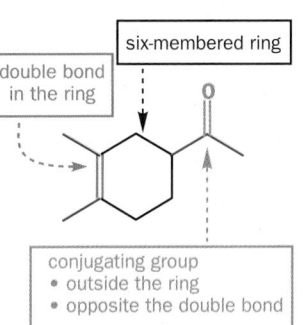

norbornadiene

Aldrin

This is quite a complex product but we hope you can see how it is made up by looking at the two new bonds marked in black. Dieldrin is the epoxide of Aldrin. The use of these compounds, like that of many organochlorine compounds,

was eventually banned when it was found that chlorine residues were accumulating in the fat of animals high up in the food chain such as birds of prey and humans.

The product

Recognizing a Diels–Alder product is straightforward. Look for the six-membered ring, the double bond inside the ring, and the conjugating group outside the ring and on the opposite side of the ring from the alkene. These three features mean that the compound is a possible Diels–Alder product.

The simplest way to find the starting materials is to carry out a disconnection that is closer to a real reaction than most. Just draw the reverse Diels–Alder reaction. To do this, draw three arrows going round the cyclohexene ring starting the first arrow in the middle of the double bond. It doesn't, of course, matter which way round you go.

recognizing a Diels–Alder product:

six-membered ring

double bond in the ring

conjugating group
• outside the ring
• opposite the double bond

the disconnection is the imaginary reverse Diels–Alder reaction

start first arrow in the middle of the double bond

Diels–Alder

+

The reaction couldn't be simpler—just heat the components together without solvent or catalyst. Temperatures of around 100–150 °C are often needed and this may mean using a sealed tube if the reagents are volatile, as here.

+

100 °C

sealed tube
no solvent

Stereochemistry

The Diels–Alder reaction is stereospecific. If there is stereochemistry in the dienophile, then it is faithfully reproduced in the product. Thus *cis* and *trans* dienophiles give different diastereoisomers of the product. Esters of maleic and fumaric acids provide a simple example.

CO_2Me

CO_2Me

CO_2Me

CO_2Me

CO_2Me

CO_2Me

dimethyl maleate

In both cases the ester groups simply stay where they are. They are *cis* in the dienophile in the first reaction and remain *cis* in the product. They are *trans* in the dienophile in the second reaction and remain *trans* in the product. The second example may look less convincing—may we remind you that the diene actually comes down on top of the dienophile like this.

One of the CO_2Me groups is tucked under the diene in the transitions state and then, when the product molecule is flattened out in the last drawing, that CO_2Me group appears underneath the ring. The orange hydrogen atom remains *cis* to the other CO_2Me group.

The search by the Parke–Davis company for drugs to treat strokes provided an interesting application of dienophile stereochemistry. The kinds of compound they wanted were tricyclic amines. They don't look like Diels–Alder products at all. But if we insert a double bond in the right place in the six-membered ring, Diels–Alder (D–A) disconnection becomes possible.

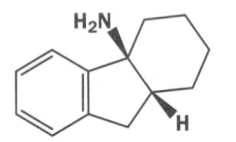

potential drugs to treat stroke

Butadiene is a good diene, but the enamine required is not a good dienophile. An electron-with-drawing group such as a carbonyl or nitro group is preferable: either would do the job. In the event a carboxylic acid that could be converted into the amine by a rearrangement with Ph_2PON_3 (see Chapter 40) was used.

The stereochemistry at the ring junction must be *cis* because the cyclic dienophile can have only a *cis* double bond. Hydrogenation removes the double bond in the product and shows just how useful the Diels–Alder reaction is for making saturated rings, particularly when there is some stereochemistry to be controlled.

> ►
> You can add the Diels–Alder reaction to your mental list of reactions to consider for making a single diastereoisomer from a single geometrical isomer of an alkene: see Chapter 34.

Stereochemistry of the diene

This is slightly more complicated as the diene can be *cis*, *cis*, or *cis*, *trans* (there are two of these if the diene is unsymmetrical) or *trans*, *trans*. We shall look at each case with the same dienophile, an acetylenedicarboxylate, as there is then no stereochemistry in the triple bond! Starting with *cis*, *cis*-dienes is easy if we make the diene cyclic.

The diene has two sets of substituents—inside and outside. The inside one is the bridging CH_2 group and it has to end up on one side of the molecule (above in the last diagram) while the two green hydrogens are outside and remain so. In the final diagram they are below the new six-membered ring.

With a *trans, trans*-diene we simply exchange the two sets of substituents, in this example putting Ph where H was and putting H where the bridging CH_2 group was. This is the reaction.

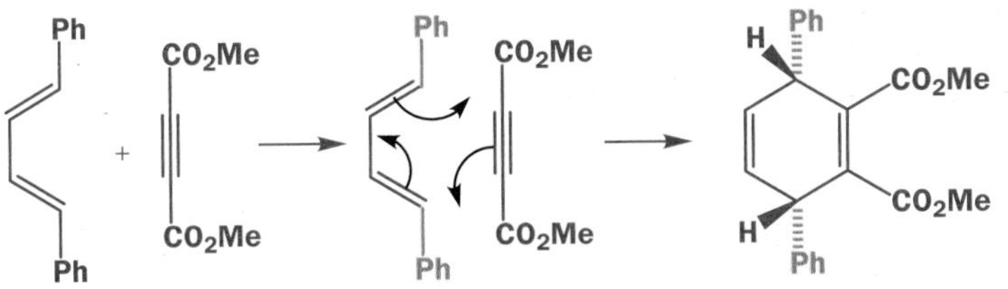

The green Ph groups end up where the hydrogens were in the first example—beneath the new six-membered ring—and the hydrogens end up above. It may seem puzzling at first that a *trans, trans*-diene gives a product with the two phenyls *cis*. Another way to look at these two reactions is to consider their symmetry. Both have a plane of symmetry throughout and the products must have this symmetry too because the reaction is concerted and no significant movement of substituents can occur. The black dotted line shows the plane of symmetry, which is at right angles to the paper.

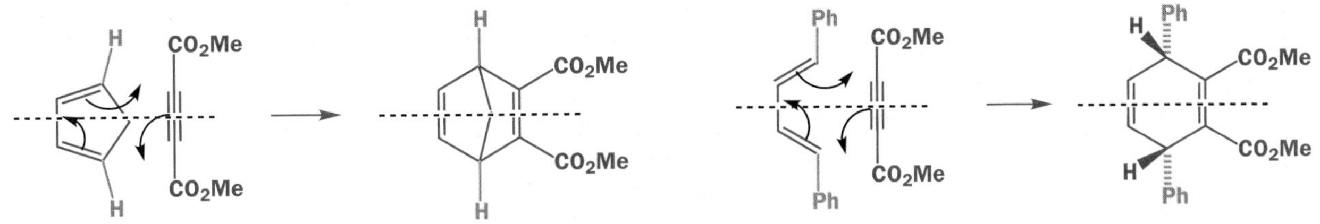

The remaining case—the *cis, trans*-diene—is rarer than the first two, but is met sometimes. This is the unsymmetrical case and the two substituents clearly end up on opposite sides of the new six-membered ring.

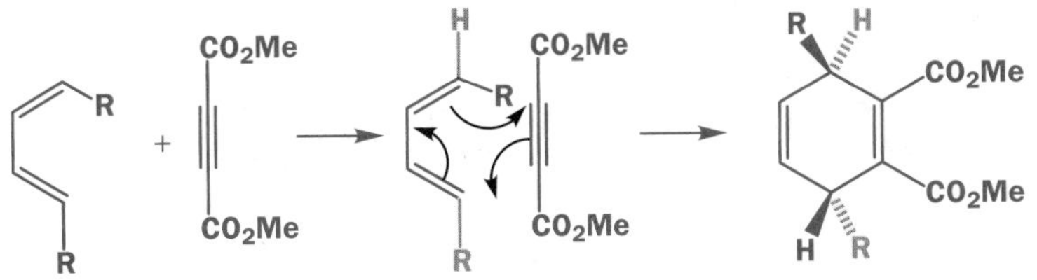

The red R group may seem to get in the way of the reaction but, of course, the dienophile is not approaching in the plane of the diene but from underneath. It is difficult to find a convincing example of this stereochemistry as there are so few known, partly because of the difficulty of making *E,Z*-dienes. One good approach uses two reactions you met in Chapter 31 for the control of double bond geometry. The *cis* double bond is put in first by the addition of methanol to butadiyne and the *trans* double bond then comes from $LiAlH_4$ reduction of the intermediate acetylenic alcohol.

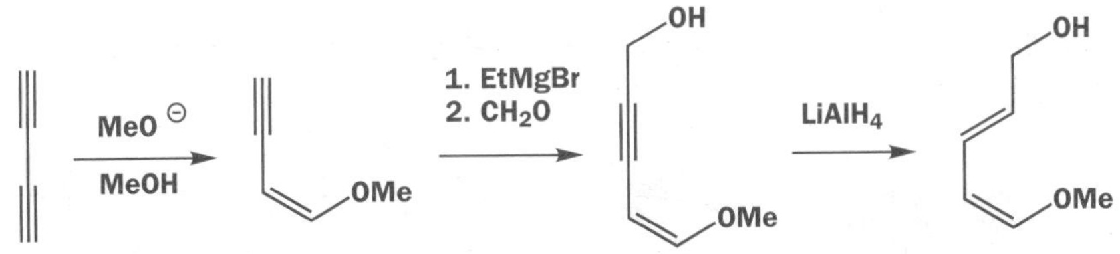

■ The mechanism for these reactions is given on pp. 819 and 820.

The acetate of this alcohol is used in a Diels–Alder reaction with the interesting dienophile DEAD (diethyl azodicarboxylate—in orange).

■

DEAD is a key component of the Mitsunobu reaction: see p. 431.

89% yield

The product is formed in excellent yield and has the *trans* stereochemistry that was predicted. Do not be misled into thinking that DEAD is being shown with stereochemistry—it has none—and in the product the amide nitrogen atoms are planar and there is no stereochemistry there.

Now to the most interesting cases of all, when both the diene and the dienophile have stereochemistry.

The *endo* rule for the Diels–Alder reaction

It is probably easier to see this when both the diene and the dienophile are cyclic. All the double bonds are *cis* and the stereochemistry is clearer. In the most famous Diels–Alder reaction of all time, that between cyclopentadiene and maleic anhydride, there are two possible products that obey all the rules we have so far described.

the '*endo*' adduct (formed)   the '*exo*' adduct (not formed)

The two green hydrogen atoms must be *cis* in the product but there are two possible products in which these Hs are *cis*. They are called *exo* and *endo*.

The product is, in fact, the *endo* compound. This is impressive not only because only one diastereoisomer is formed but also because it is the less stable one. How do we know this? Well, if the Diels–Alder reaction is reversible and therefore under thermodynamic control, the *exo* product is formed instead. The best known example results from the replacement of cyclopentadiene with furan in reaction with the same dienophile.

►

These names arise from the relationship in space between the carbonyl groups on the dienophile and the newly formed double bond in the middle of the old diene. If these are on the same side they are called *endo* (inside) and if they are on opposite sides they are called *exo* (outside).

the '*endo*' adduct
(less stable)

furan

the '*exo*' adduct
(more stable)

Why is the *exo* product the more stable? Look again at these two structures. On the left-hand side of the molecules, there are two bridges across the ends of the new bonds (highlighted in black): a one-C-atom bridge and a two-C-atom bridge. There is less steric hindrance if the smaller (that is, the one-atom) bridge eclipses the anhydride ring.

The *endo* product is less stable than the *exo* product and yet it is preferred in irreversible Diels–Alder reactions—it must be the kinetic product of the reaction. It is preferred because there is a bonding interaction between the carbonyl groups of the dienophile and the developing π bond at the back of the diene. (The black bonds are the new σ bonds between the two reagents.)

bonding interaction
in transition state
between C=O groups
and back of diene

new double bond and C=O
groups end up on same side
of molecule: *endo*

'*endo*' adduct

The same result is found with noncyclic dienes and dienophiles—normally one diastereoisomer is preferred and it is the one with the carbonyl groups of the dienophile closest to the developing π bond at the back of the diene. Here is an example.

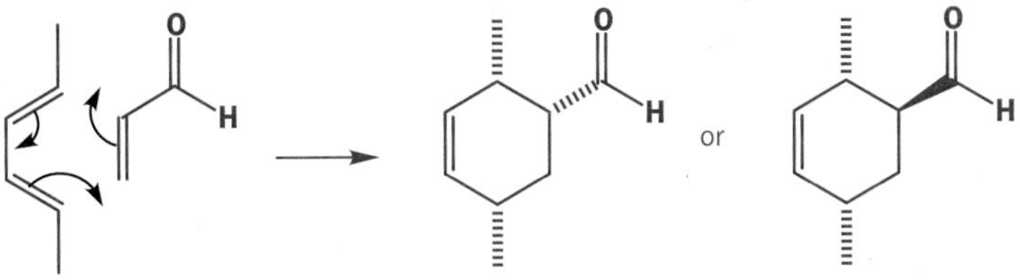

From our previous discussion we expect the two methyl groups to be *cis* to each other and the only question remaining is the stereochemistry of the aldehyde group—up or down? The aldehyde will be *endo*—but which compound is that? The easiest way to find the answer is to draw the reagents coming together in three dimensions. Here is one way to do this.

1. Draw the mechanism of the reaction and diagrams of the product to show what you are trying to decide. Put in the known stereochemistry if you wish

■
This we have just done.

2. Draw both molecules in the plane of the paper with the diene on top and the carbonyl group of the dienophile tucked under the diene so it can be close to the developing π-bond

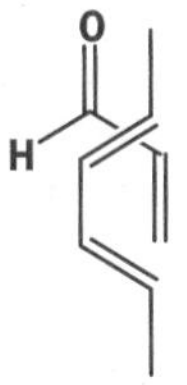

3. Now draw in all the hydrogen atoms on the carbon atoms that are going to become stereogenic centres, that is, those shown in green here

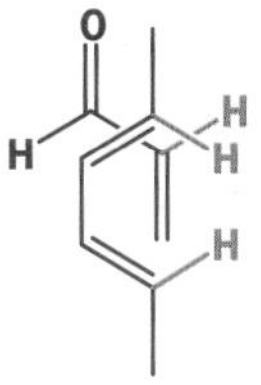

4. Draw a diagram of the product. All the substituents to the right in the previous diagram are on one side of the new molecule. That is, all the green hydrogen atoms are *cis* to each other

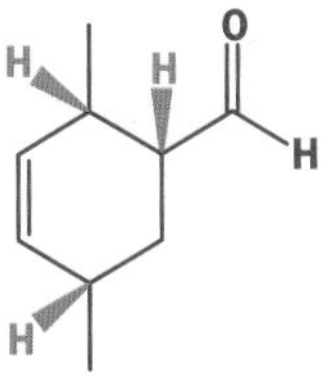

5. Draw a final diagram of the product with the stereochemistry of the other substituents shown too in the usual way. This is the *endo* product of the Diels–Alder reaction

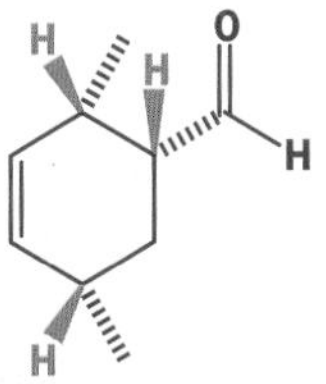

If you prefer, you may draw a three-dimensional representation of the reagents coming together, rather like the ones we have been drawing earlier in the chapter. You may indeed prefer to invent a method of your own—it does not matter which method you choose providing that you can quickly decide on the structure of the endo adduct in any given Diels–Alder reaction.

Time for some explanations

We have accumulated rather a lot of unexplained results.

- Why does the Diels–Alder reaction work so well?
- Why must we have a conjugating group on the dienophile?
- Why is the stereochemistry of each component retained so faithfully?
- Why is the *endo* product preferred kinetically?

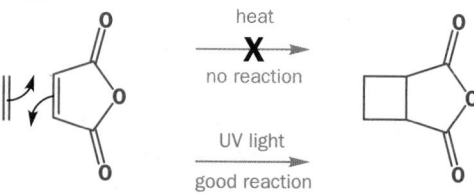

There is more. The simpler picture we met earlier in this chapter also fails to explain why the Diels–Alder reaction occurs simply on heating while attempted additions of simple alkenes (rather than dienes) to maleic anhydride fail on heating but succeed under irradiation with UV light.

We shall now explain all this in one section using frontier molecular orbitals. Of all the kinds of organic reactions, pericyclic ones are the most tightly controlled by orbitals, and the development of the ideas we are about to expound is one of the greatest triumphs of modern theoretical chemistry. It is a beautiful and satisfying set of ideas based on very simple principles.

The frontier orbital description of cycloadditions

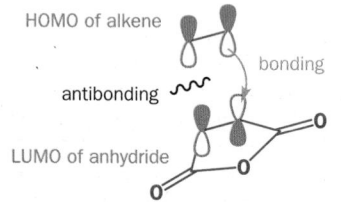

■
You may need to remind yourself about the orbitals of conjugated π systems by re-reading Chapter 7.

When an ionic cyclization reaction occurs, such as the lactonization at the head of this chapter, one important new bond is formed. It is enough to combine one full orbital with one empty orbital to make the new bond. But in a cycloaddition two new bonds are formed at the same time. We have to arrange for two filled p orbitals and two empty p orbitals to be available at the right place and with the right symmetry. See what happens if we draw the orbitals for the reaction above. We could try the HOMO (π) of the alkene and the LUMO (π^*) of the double bond in the anhydride.

This combination is bonding at one end, but antibonding at the other so that no cylcoaddition reaction occurs. It obviously doesn't help to use the other HOMO/LUMO pair as they will have the same mismatched symmetry.

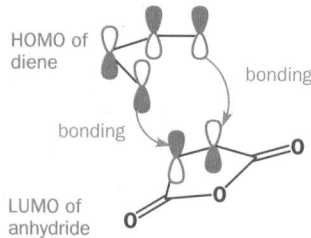

Now see what happens when we replace the alkene with a diene. We shall again use the LUMO of the electron-poor anhydride.

Now the symmetry is right because there is a node in the middle of the HOMO of the diene (the HOMO is Ψ_2 of the diene) just as there is in the LUMO of the dienophile. If we had tried the opposite arrangement, the LUMO of the diene and the HOMO of the dienophile, the symmetry would again be right.

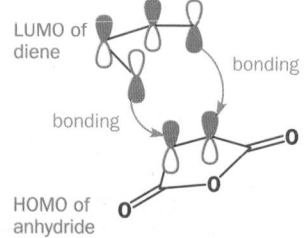

Now the LUMO of the diene has two nodes and gives the same symmetry as the HOMO of the dienophile, which has no nodes. So either combination is excellent. In fact most Diels–Alder reactions use electron-deficient dienophiles and electron-rich dienes so we prefer the first arrangement. The electron-deficient dienophile has a low-energy LUMO and the electron-rich diene has a high-energy HOMO so that this combination gives a better overlap in the transition state. The energy levels will be like this.

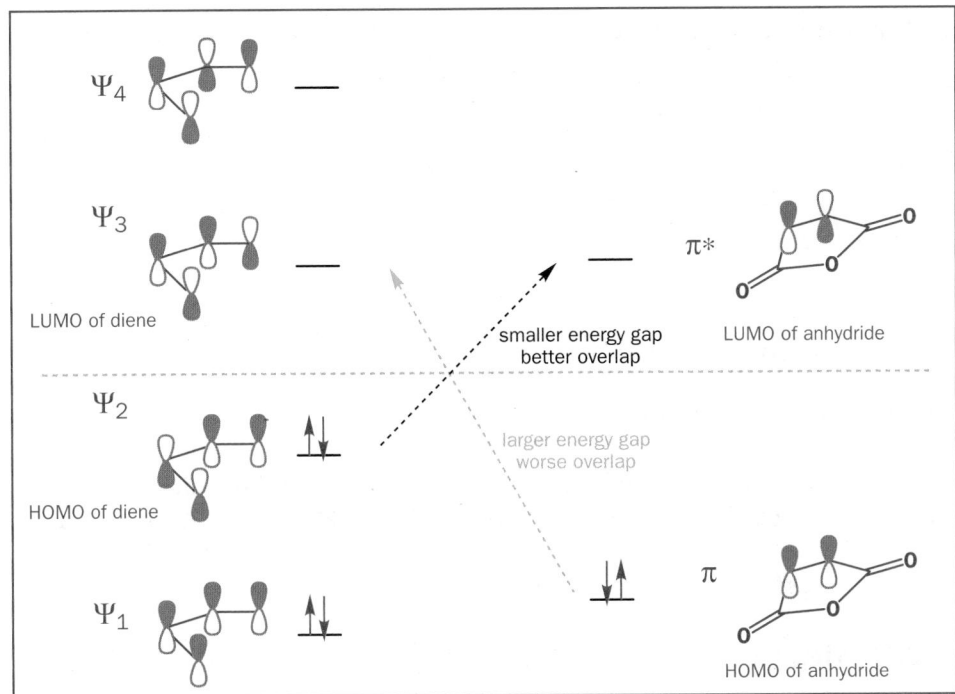

This is why we usually use dienophiles with conjugating groups for good Diels–Alder reactions. Dienes react rapidly with electrophiles because their HOMOs are relatively high in energy, but simple alkenes have relatively high-energy LUMOs and do not react well with nucleophiles. The most effective modification we can make is to lower the alkene LUMO energy by conjugating the double bond with an electron-withdrawing group such as carbonyl or nitro. These are the most common type of Diels–Alder reactions—between electron-rich dienes and electron-deficient dienophiles.

Dimerizations of dienes by cycloaddition reactions

Because dienes have relatively high-energy HOMOs and low-energy LUMOs they should be able to take part in cycloadditions with themselves. And they do. What they cannot do is form an eight-membered ring in one step (though this is possible photochemically or with transition metal catalysis as we shall see later).

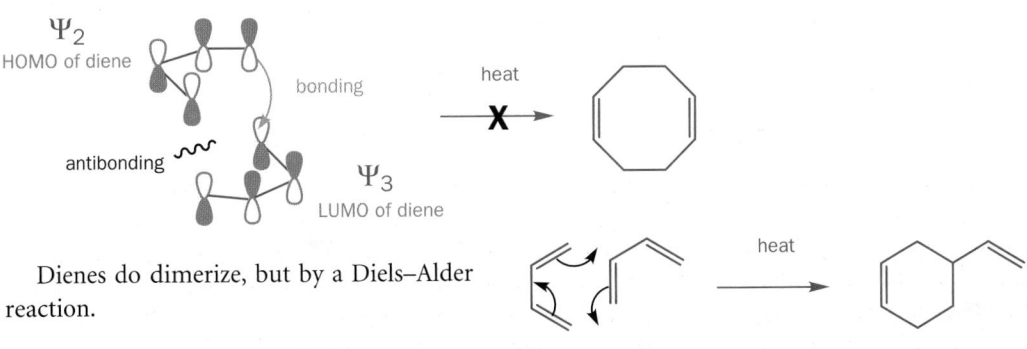

You should have expected this failure because the ends of the required orbitals must again have the wrong symmetry, just as they had when we tried the alkene dimerization.

Dienes do dimerize, but by a Diels–Alder reaction.

> A rarer type is the **reverse electron demand Diels–Alder reaction** in which the dienophile has electron-donating groups and the diene has a conjugated electron-withdrawing group. These reactions use the HOMO of the dienophile and the LUMO of the diene. This combination still has the right orbital symmetry.

One molecule of the diene acts as a dienophile. Now the symmetry is correct again.

Ψ_2
HOMO of diene

bonding

bonding

heat

Ψ_3
LUMO of diene

● **Count the number of π electrons**

- The cycloadditions that *do* occur thermally, for example, the Diels–Alder reaction, have $(4n + 2\pi)$ electrons in their 'aromatic' transition states
- The cycloadditions that do *not* occur thermally, for example the dimerization of alkenes and of dienes, have $4n\pi$ electrons in their 'anti-aromatic' transition states

---

The Diels–Alder reaction in more detail

The orbital explanation for the *endo* rule in Diels–Alder reactions

We are going to use a diene as dienophile to explain the formation of *endo* products. The diene serves as a good model for the very wide variety of dienophiles because the one thing they all have in common is a conjugating group and a second alkene is the simplest of these. To make matters even easier we shall look at the dimerization of a cyclic diene; we might almost say *the* cyclic diene—cyclopentadiene. We introduced this reaction above where we simply stated that there was a favourable electronic interaction between the conjugating group on the dienophile and the back of the diene in the *endo* product though we did not explain it at the time.

endo
relationship
between two alkenes

If we now draw the frontier orbitals in the two components as they come together for the reaction, we can see first of all that the symmetry is correct for bond formation.

Now we shall look at that same diagram again but replace with orange dashed lines the orbitals that are overlapping to form the new σ bonds so that we can see what is happening at the back of the diene.

The symmetry of the orbitals is correct for a bonding interaction at the back of the diene too. This interaction does not lead to the formation of any new bonds but it leaves its imprint in the stereochemistry of the product. The *endo* product is favoured because of this favourable interaction across the space between the orbitals even though no bonds are formed.

bonding   bonding
bonding

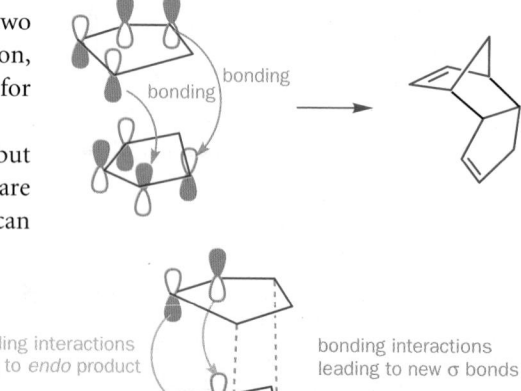

bonding interactions
leading to *endo* product

bonding interactions
leading to new σ bonds

Entropy and the *endo* rule

Another way to look at this result comes from recognizing the special entropy problem involved in cycloaddition reactions. A very precise orientation of the two molecules is required for two bonds to be formed at once. These reactions have large negative entropies of activation (Chapter 41)—order must be created at the transition state as the two components align with one another. The through-space attractive HOMO/LUMO interaction between the two molecules can lead to an initial association that can be compared to a squishy sandwich with too much mayonnaise. The cyclopentadiene rings are the slices of bread and the electrons are the filling that holds them together but still allows them to rotate until the right atoms come together for bonding.

Rotation about a vertical axis through the centre of the sandwich eventually brings the right atoms together for bond formation. At that moment the backs of the rings are still stuck together by the 'mayonnaise' and the *endo* product results.

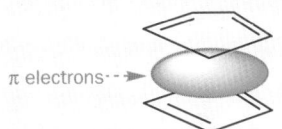
π electrons

The solvent in the Diels–Alder reaction

We discussed some effects of varying the solvent in Chapter 13, and we shall now introduce a remarkable and useful special solvent effect in the Diels–Alder reaction. The reaction does not *need* a solvent and often the two reagents are just mixed together and heated. Solvents can be used but, because there are no ionic intermediates, it seems obvious that *which* solvent is unimportant—any solvent that simply dissolves both reagents will do. This is, in general, true and hydrocarbon solvents are often the best.

However, in the 1980s an extraordinary discovery was made. Water, a most unlikely solvent for most organic reactions, has a large accelerating effect on the Diels–Alder reaction. Even some water added to an organic solvent accelerates the reaction. And that is not all. The *endo* selectivity of these reactions is often superior to those in no solvent or in a hydrocarbon solvent. Here is a simple example.

endo product   *exo* product

Solvent	Relative rate	*endo:exo* ratio
hydrocarbon (isooctane)	1	80:20
water	700	96:4

The suggestion is that the reagents, which are not soluble in water, are clumped together in oily drops by the water and forced into close proximity. Water is not exactly a solvent—it is almost an anti-solvent!

Water-soluble dienes are also used in Diels–Alder reactions in water and they too work very well. Sodium salts of carboxylic acids and protonated amines both behave well under these conditions. Presumably, the soluble tail is in the water but the diene itself is inside the oily drops with the dienophile. In this example an aminodiene reacts with a quinone dienophile.

water-soluble dienes

soluble in basic solution

soluble in acidic solution

H₂O, room temperature

98% yield

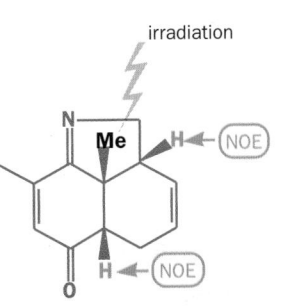
irradiation

A single regio- and stereoisomer was formed in essentially quantitative yield and the stereochemistry was easily proved by NMR using NOE (Chapter 32). Irradiation at the black methyl group in the middle of the molecule gave strong NOEs to the two green hydrogen atoms, which must therefore be on the same side of the molecule as the methyl group.

Intramolecular Diels–Alder reactions

When the diene and the dienophile are already part of the same molecule it is not so important for them to be held together by bonding interactions across space and the *exo* product is often preferred.

■ We discuss regioselectivity in Diels–Alder reactions later in the chapter (p. 919).

Indeed, it seems that intramolecular Diels–Alder reactions are governed more by normal steric considerations than by the *endo* rule.

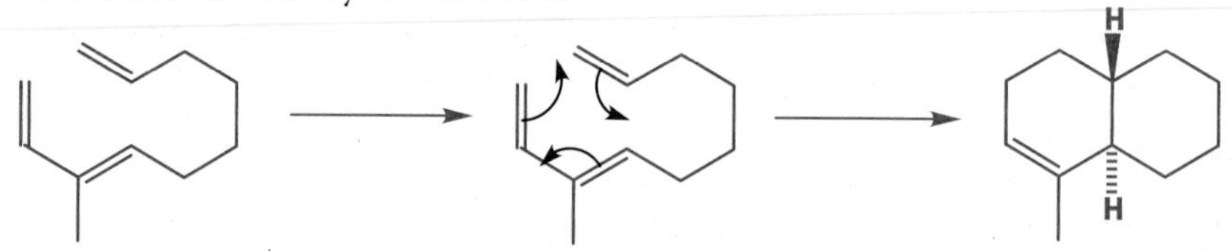

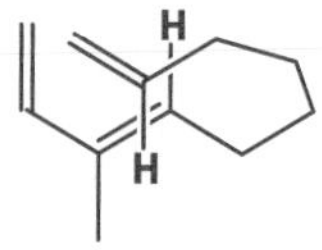

This reaction happens only because it is intramolecular. There is no conjugating group attached to the dienophile and so there are no orbitals to overlap with the back of the diene. The molecule simply folds up in the sterically most favourable way (as shown in the margin, with the linking chain adopting a chair-like conformation) and this leads to the *trans* ring junction.

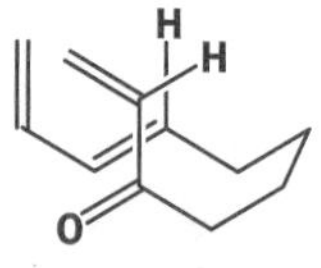

In the next example there is a carbonyl group conjugated with the dienophile. Now the less stable *cis* ring junction is formed because the molecule can fold so that the carbonyl group can enjoy a bonding overlap with the back of the diene. This time the linking chain has to adopt a boat-like conformation.

▶

If you think about the way a Diels–Alder reaction goes, the forming ring must *always* adopt a boat-like conformation. This is clear if you make a model.

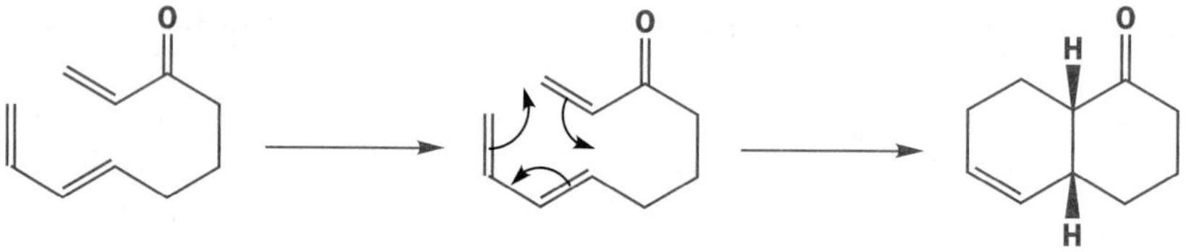

If, on the other hand, we give the dienophile a conjugating group at the other end of the double bond, stereoselectivity is lost.

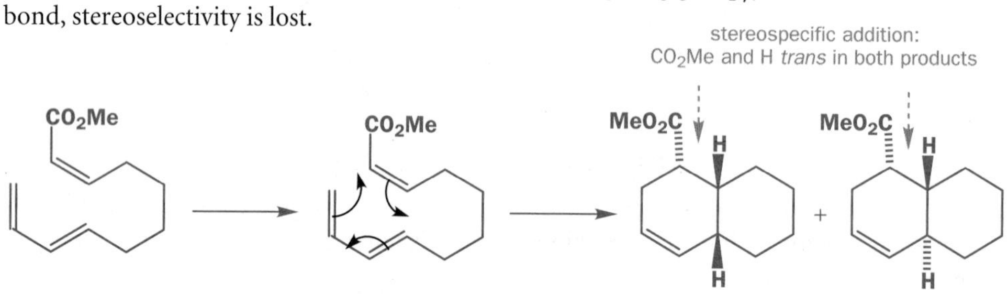

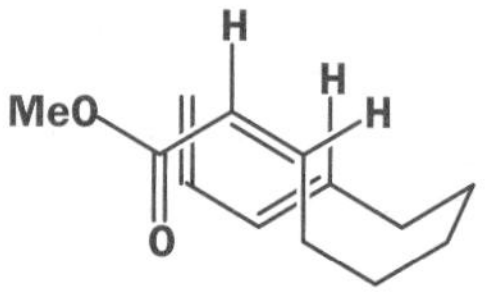

stereospecific addition:
CO$_2$Me and H *trans* in both products

The *cis*-alkene dienophile gives stereospecific addition—in each product the CO$_2$Me is *cis* to the alkyl chain (and therefore *trans* to the H atom). But we get about a 50:50 mixture of *endo* and *exo* products. This does not seem to be because there is anything wrong with the transition state for *endo* addition, which leads in this case to *cis*-fused rings.

Similarly, with the *trans*-alkene, two products are formed and both retain the *trans* geometry of the dienophile. But once again a nearly 50:50 mixture of *endo* and *exo* products is formed.

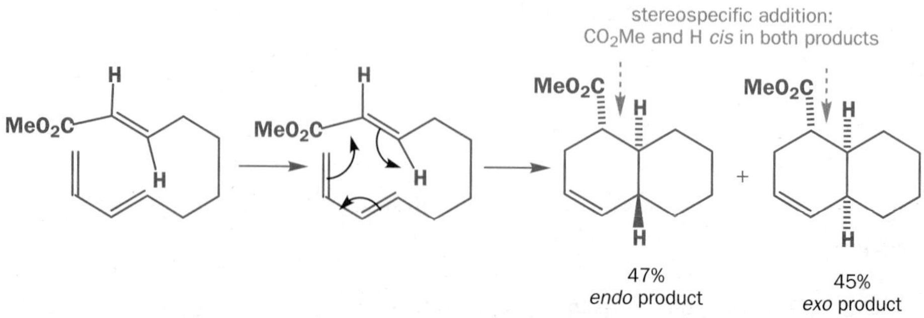

endo folding for
the *cis*-alkene

stereospecific addition:
CO$_2$Me and H *cis* in both products

47%
endo product

45%
exo product

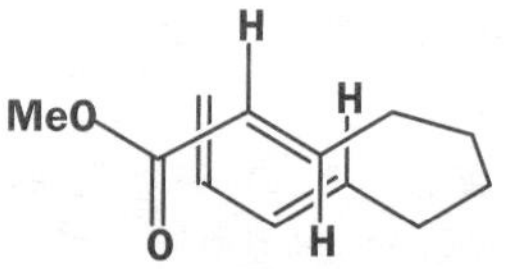

Folding the molecule so that the *endo* product would be formed does not again seem to present any problem. Presumably, either the carbonyl group of the ester is too far away from the diene to be effective or else it is simply that the advantage of the *endo* arrangement is not worth having in intramolecular Diels–Alder reactions.

endo folding for
the *trans*-alkene

● **Intramolecular Diels–Alder**

Intramolecular Diels–Alder reactions may give the *endo* product or they may not!
Be prepared for either *exo* or *endo* products or a mixture.

Regioselectivity in Diels–Alder reactions

The compounds that we are now calling dienophiles were the stars of Chapters 10, 23, and 29 where
we called them **Michael acceptors** as they were the electrophilic partners in conjugate addition reactions. Nucleophiles always add to the β carbon atoms of these alkenes because the product is then a
stable enolate. Ordinary alkenes do not react with nucleophiles.

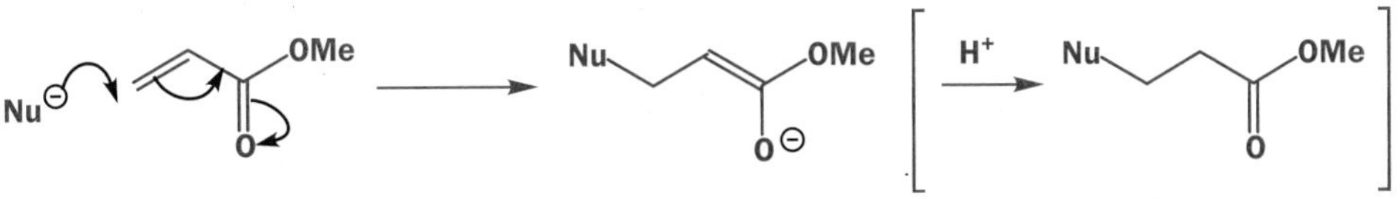

In frontier orbital terms this is because conjugation with a carbonyl group lowers the energy of
the LUMO (the π* orbital of the alkene) and at the same time distorts it so that the coefficient on the
β carbon atom is larger than that on the α carbon atom. Nucleophiles approach the conjugated
alkene along the axis of the large p orbital of the β carbon atom.

■ This is discussed in Chapter 10.

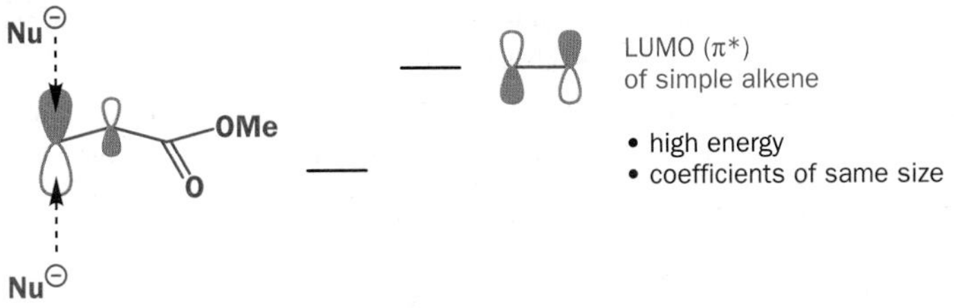

LUMO of an unsaturated
carbonyl compound

• lower energy
• unequal coefficients

LUMO (π*)
of simple alkene

• high energy
• coefficients of same size

These same features can ensure regioselective Diels–Alder reactions. The same orbital of the
dienophile is used and, if the HOMO of the diene is also unsymmetrical, the regioselectivity of the
reaction will be controlled by the two largest coefficients bonding together.

So what about distortion of the HOMO in the diene? If a diene reacts with an electrophile, the
largest coefficient in the HOMO will direct the reaction. Consider the attack of Hbr on a diene. We
should expect attack at the ends of the diene because that gives the most stable possible cation—an
allyl cation as an intermediate.

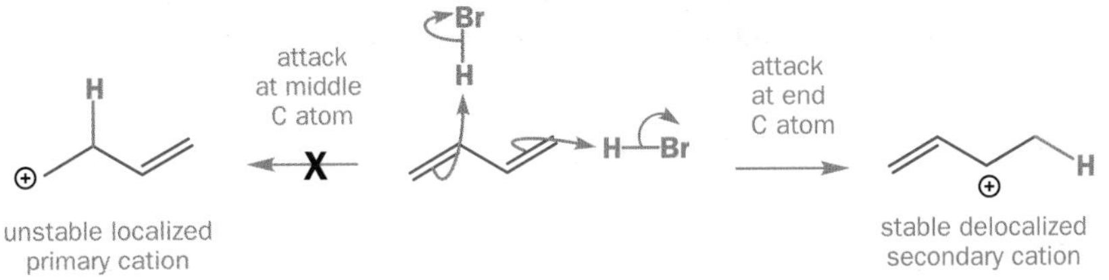

unstable localized
primary cation

attack
at middle
C atom

attack
at end
C atom

stable delocalized
secondary cation

In orbital terms attack occurs at the ends of the diene because the coefficients in the HOMO are
larger there. We need simply to look at the HOMO (Ψ_2) of butadiene to see this.

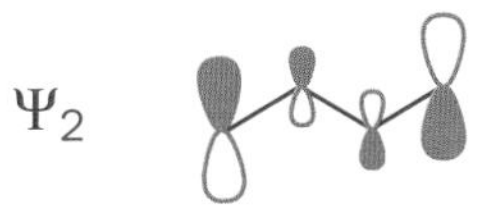

So it is not surprising that the dienes react in the Diels–Alder reaction through their end carbons.
But supposing the two ends are different—which reacts now? We can again turn to the reaction with
HBr as a guide. Addition of HBr to an unsymmetrical diene will give the more stable of the two pos-
sible allyl cations as the intermediate.

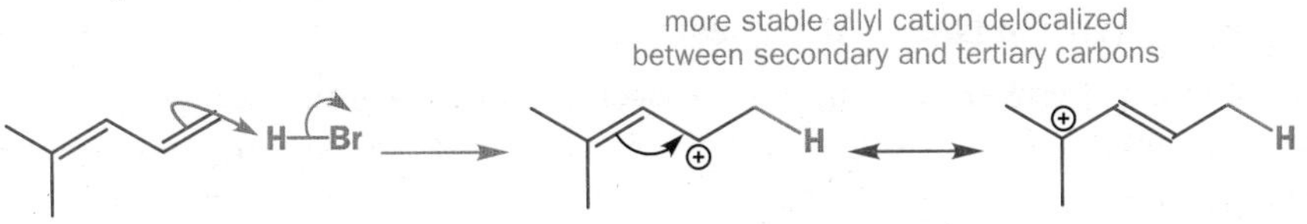

HOMO of butadiene

Ψ_2

more stable allyl cation delocalized
between secondary and tertiary carbons

less stable allyl cation delocalized
between secondary and primary carbons

HOMO of 1,1-dimethylbutadiene

Ψ_2

▶

It is not 'cheating' to use the regioselectivity of chemical reactions to tell us about the coefficients in orbitals. Chemistry is about using experimental evidence to find out about the theoretical background and not about theory telling us what *ought* to happen. In fact, theoretical chemists have calculated the HOMO energies and coefficients of unsymmetrical dienes and they have reached the same conclusions.

In orbital terms, this clearly means that the HOMO of the diene is distorted so that the end that reacts has the larger coefficient.

When the unsymmetrical diene and the unsymmetrical dienophile combine in a Diels–Alder reaction, the reaction itself becomes unsymmetrical. It remains concerted but, in the transition state, bond formation between the largest coefficients in each partner is more advanced and this determines the regioselectivity of the reaction.

little bond formation
in transition state

CO_2Me

‡

CO_2Me

bond formation almost
complete in transition state

The simplest way to decide which product will be formed is to draw an 'ionic' stepwise mechanism for the reaction to establish which end of the diene will react with which end of the dienophile. Of course this stepwise mechanism is not completely correct but it does lead to the correct orientation of the reagents and you can draw the right mechanism afterwards. As an example we shall look at a diene with a substituent in the middle. This is the reaction.

OMe

+

CN

Diels–Alder ⟶ **?**

First decide where the diene will act as a nucleophile and where the diene will act as an electrophile.

reaction of the diene with
an electrophile

:ÖMe

$E^{\oplus}$

reaction of the dienophile
with a nucleophile

$Nu^{\ominus}$

N

■

The two circles represent the largest coefficients of the HOMO and the LUMO.

Now draw the reagents in the correct orientation for these two ends to combine and draw a concerted Diels–Alder reaction.

CN

MeO

Diels–Alder ⟶

CN

MeO

This is an important example because an enol ether functional group is present in the product and this can be hydrolysed to a ketone in aqueous acid (see Chapter 21).

enol
ether

CN

MeO

$H^{\oplus}$
H_2O ⟶

CN

O

Summary of regioselectivity in Diels–Alder reactions

The important substitution patterns are: a diene with an electron-donating group (X) at one end or in the middle and a dienophile with an electron-withdrawing group (Z) at one end. These are the products formed.

X = electron-donating group such as

alkyl, aryl
RO, Me$_3$SiO
R$_2$N

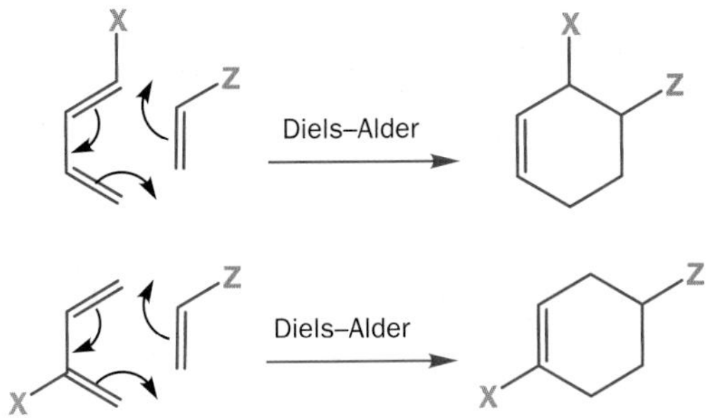

Diels–Alder

Z = electron-withdrawing (or conjugating) group such as

CHO, COR, CO$_2$H, CO$_2$R
CN, NO$_2$
halogen
alkenyl, aryl

Diels–Alder

● **A useful mnemonic**

If you prefer a rule to remember, try this one.
● The Diels–Alder reaction is a cycloaddition with an aromatic transition state that is *ortho* and *para* directing

You can see that this mnemonic works if you look at the two products above: the first has the two substituents X and Z on neighbouring carbon atoms, just like *ortho* substituents on a benzene ring, while the second has X and Z on opposite sides of the ring, just like *para* substituents.

Lewis acid catalysis in Diels–Alder reactions

Where the reagents are unsymmetrical, a Lewis acid that can bind to the electron-withdrawing group of the dienophile often catalyses the reaction by lowering the LUMO of the dienophile still further. It has another important advantage: it increases the difference between the coefficients in the LUMO (a Lewis-acid complexed carbonyl group is a more powerful electron-withdrawing group) and may increase regioselectivity.

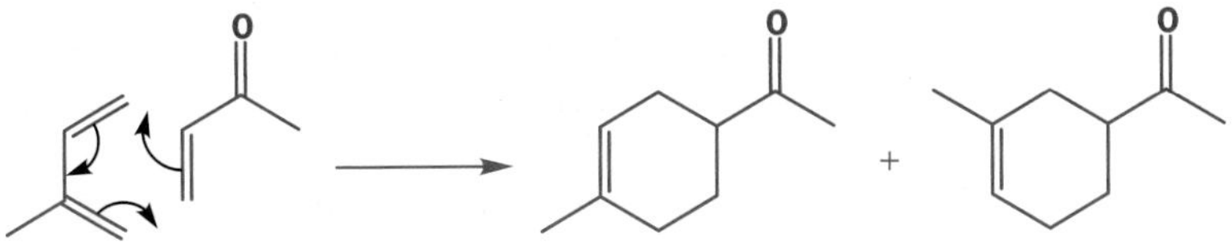

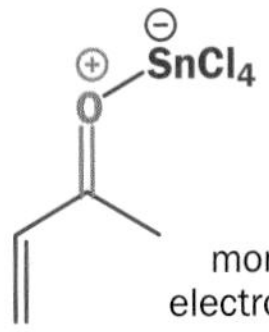

more powerfully electron-withdrawing

| heat in toluene at 120 °C in a sealed tube | 71: | :29 |
| with SnCl$_4$·5H$_2$O at 0 °C | 93: | :7 |

This Diels–Alder reaction is useful because it produces a substitution pattern ('*para*') common in natural terpenes (Chapter 51). But the regioselectivity introduced by one methyl group on the diene is not very great—this reaction gives a 71:29 mixture when the two compounds are heated together at 120 °C in a sealed tube. In the presence of the Lewis acid (SnCl$_4$) the reaction can be carried out at lower temperatures (below 25 °C) without a sealed tube and the regioselectivity improves to 93:7.

Regioselectivity in intramolecular Diels–Alder reactions

Just as the stereoselectivity may be compromised in intramolecular reactions, so may the regioselectivity. It may be simply impossible for the reagents to get together in the 'right' orientation. The examples below have a very short chain—just three carbon atoms—joining diene to dienophile and so the same regioselectivity is found regardless of the position of the conjugating carbonyl group.

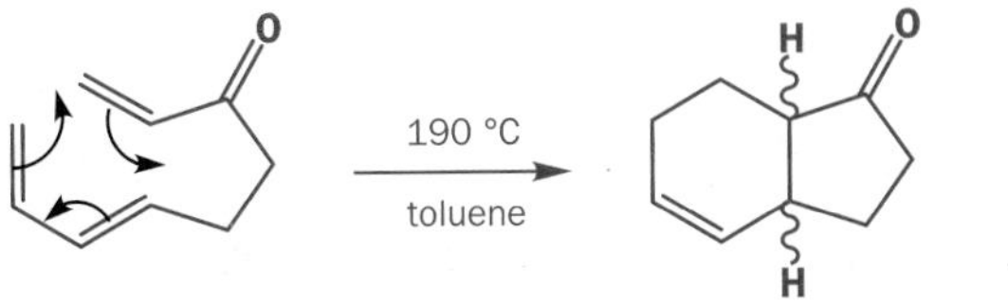

190 °C
toluene

100% yield
70:30 *cis:trans* mixture

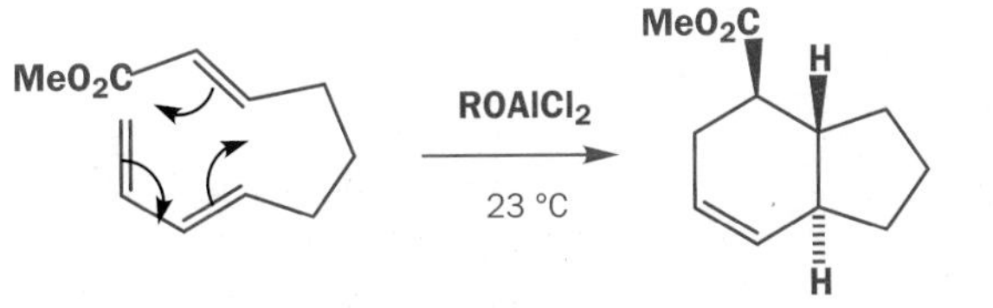

ROAlCl$_2$
23 °C

only product; 72% yield

The first example has the 'right' orientation ('*ortho*') but the second has the 'wrong' orientation ('*meta*'). In real life there is no prospect of any other orientation and, as the reaction is intramolecular, it goes anyway. Notice the lower temperature required for the Lewis acid (ROAlCl$_2$)-catalysed reaction.

The Woodward–Hoffmann description of the Diels–Alder reaction

Kenichi Fukui and Roald Hoffmann won the Nobel prize in 1981 (Woodward died in 1979 and so couldn't share this prize: he had already won the Nobel prize in 1965 for his work on synthesis) for the application of orbital symmetry to pericyclic reactions. Theirs is an alternative description to the frontier orbital method we have used and you need to know a little about it. They considered a more fundamental correlation between the symmetry of all the orbitals in the starting materials and all the orbitals in the products. This is rather too complex for our consideration here, and we shall concentrate only on a summary of the conclusions—the **Woodward–Hoffmann rules**. The most important of these states:

> ● **Woodward–Hoffmann rules**
>
> In a thermal pericyclic reaction the total number of $(4q + 2)_s$ and $(4r)_a$ components must be odd.

This needs some explanation. A **component** is a bond or orbital taking part in a pericyclic reaction as a single unit. A double bond is a $\pi2$ component. The number 2 is the most important part of this designation and simply refers to the number of electrons. The prefix π tells us the type of electrons. A component may have any number of electrons (a diene is a $\pi4$ component) but may not have mixtures of π and σ electrons. Now look back at the rule. Those mysterious designations $(4q + 2)$ and $(4r)$ simply refer to the number of electrons in the component where q and r are integers. An alkene is a $\pi2$ component and so it is of the $(4q + 2)$ kind while a diene is a $\pi4$ component and so is of the $(4r)$ kind.

Now what about the suffixes 's' and 'a'? The suffix 's' stands for suprafacial and 'a' for antarafacial. A **suprafacial** component forms new bonds on the same face at both ends while an **antarafacial** component forms new bonds on opposite faces at both ends. See how this works for the Diels–Alder reaction. Here is the routine.

1. Draw the mechanism for the reaction (we shall choose a general one)

2. Choose the components. All the bonds taking part in the mechanism must be included and no others

3. Make a three-dimensional drawing of the way the components come together for the reaction, putting in orbitals at the ends of the components (only!)

4. Join up the components where new bonds are to be formed. Coloured dotted lines are often used.

▶
You have already seen the significance of $4n$ and $4n + 2$ numbers in aromaticity.

▶
Please note—these orbitals are just p orbitals, and do *not* make up HOMOs or LUMOs or any particular molecular orbital. Do *not* attempt to mix frontier orbital and Woodward–Hoffmann descriptions of pericyclic reactions.

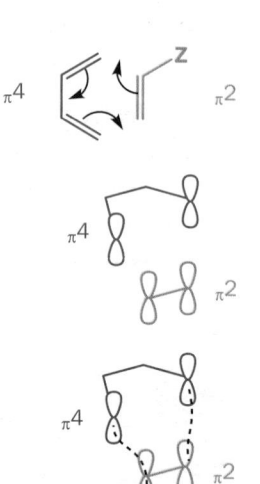

5. Lable each component s or a depending on whether new bonds are formed on the same or on opposite sides.

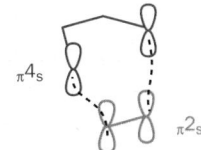

$\pi 4_s$

$\pi 2_s$

6. Count the number of $(4q + 2)_s$ and $(4r)_a$ components. If the total count is odd, the reaction is allowed

■ There is *one* $(4q + 2)_s$ component (the alkene) and *no* $(4r)_a$ components. Total = 1 so it is an allowed reaction

▶ Components of the other symmetry, that is $(4q + 2)_a$ and $(4r)_s$ components, do not count. You can have as many of these as you want!

You may well feel that there is very little to be gained from the Woodward–Hoffmann treatment of the Diels–Alder reaction. It does not explain the *endo* selectivity nor the regioselectivity. However, the Woodward–Hoffmann treatment of other pericyclic reactions (particularly electrocyclic reactions, in the next chapter) is helpful. You need to know about this treatment because the Diels–Alder reaction is often described an an **all-suprafacial [4 + 2] cycloaddition**. Now you know what that means.

Trapping reactive intermediates by Diels–Alder reactions

In Chapter 23 we met the remarkable intermediate benzyne and mentioned that convincing evidence for its existence was the trapping by a Diels–Alder reaction. An ideal method for generating benzyne for this purpose is the diazotization of anthranilic acid (2-aminobenzoic acid).

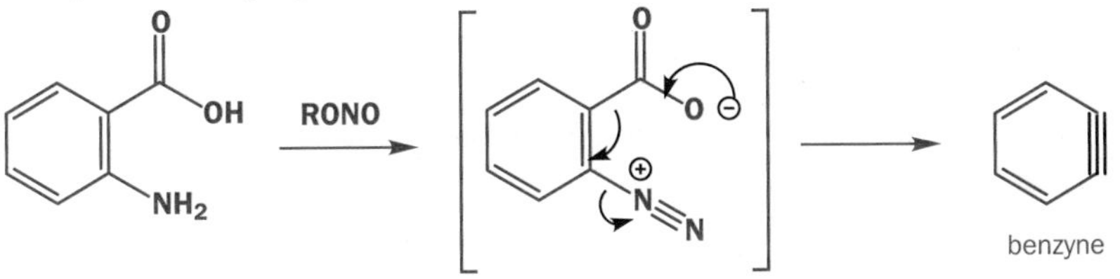

benzyne

Benzyne may not look like a good dienophile but it is an unstable electrophilic molecule so it must have a low-energy LUMO (π^* of the triple bond). If benzyne is generated in the presence of a diene, efficient Diels–Alder reactions take place. Anthracene gives a specially interesting product with a symmetrical cage structure.

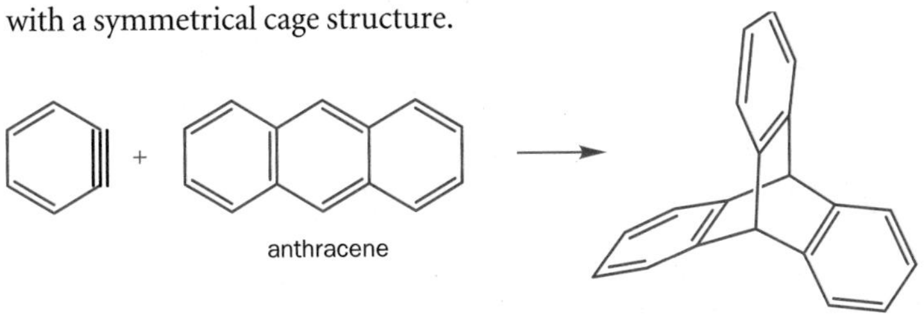

anthracene

It is difficult to draw this mechanism convincingly. The two flat molecules approach each other in orthogonal planes, so that the oribitals of the localized π bond of benzyne bond with the p orbitals on the central ring of anthracene.

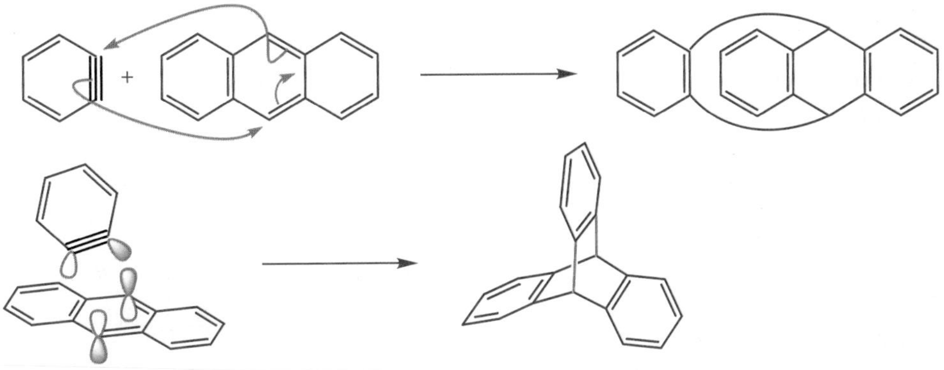

Another intermediate for which Diels–Alder trapping provided convincing evidence is the oxy-allyl cation. This compound can be made from α,α'-dibromoketones on treatment with zinc metal. The first step is the formation of a zinc enolate (compare the Reformatsky reaction), which can be drawn in terms of the attack of zinc on oxygen or bromine. Now the other bromine can leave as an anion. It could not do so before because it was next to an electron-withdrawing carbonyl group. Now it is next to an electron-rich enolate so the cation is stabilized by conjugation.

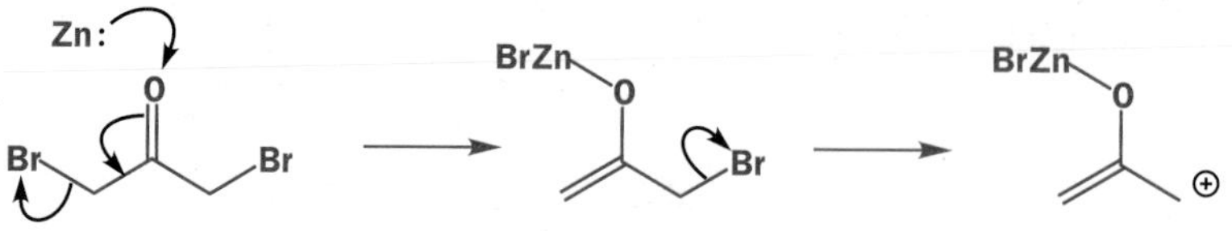

The allyl cation has three atoms but only two electrons so it can take part in cycloadditions with dienes—the total number of electrons is the required six. This is one of the few reactions that works only to produce a seven-membered ring.

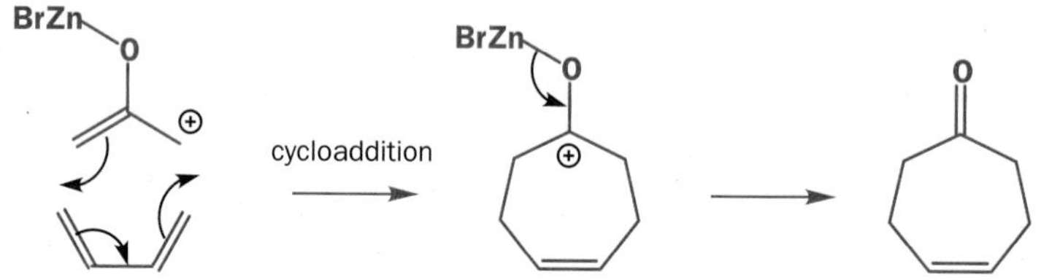

Other thermal cycloadditions

Six is not the only $(4n + 2)$ number and there are a few cycloadditions involving ten electrons. These are mostly diene + triene, that is, $_\pi4_s + _\pi6_s$ cycloadditions. Here are a couple of examples.

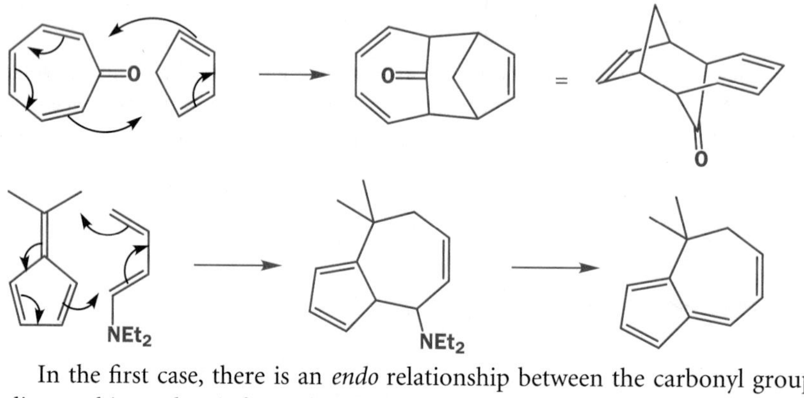

In the first case, there is an *endo* relationship between the carbonyl group and the back of the diene—this product is formed in 100% yield. In the second case Et_2NH is lost from the first product under the reaction conditions to give the hydrocarbon shown. This type of reaction is more of an oddity: by far the most important type of cycloaddition is the Diels–Alder reaction.

The Alder 'ene' reaction

The Diels–Alder reaction was originally called the 'diene reaction' so, when half of the famous team (K. Alder) discovered an analogous reaction that requires only one alkene, it was called the **Alder ene reaction** and the name has stuck. Compare here the Diels–Alder and the Alder ene reactions.

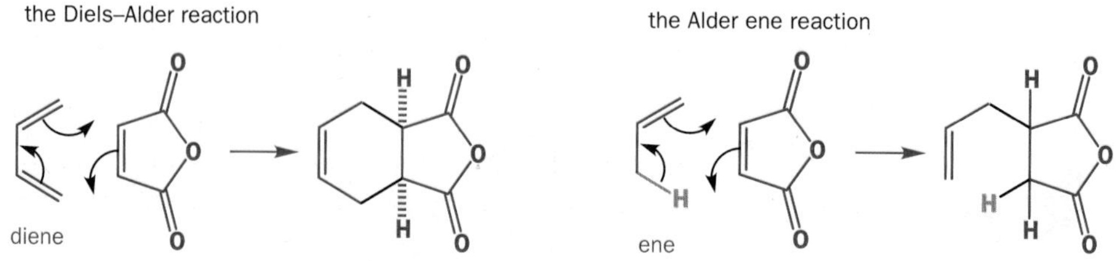

the Diels–Alder reaction

the Alder ene reaction

The simplest way to look at the ene reaction is to picture it as a Diels–Alder reaction in which one of the double bonds in the diene has been replaced by a C–H bond (green). The reaction does not form a new ring, the product has only one new C–C bond (shown in black on the product), and a hydrogen atom is transferred across space. Otherwise, the two reactions are remarkably similar.

The ene reaction is rather different in orbital terms. For the Woodward–Hoffmann description of the reaction we must use the two electrons of the C–H bond to replace the two electrons of the double bond in the Diels–Alder reaction, but we must make sure that all the orbitals are parallel, as shown.

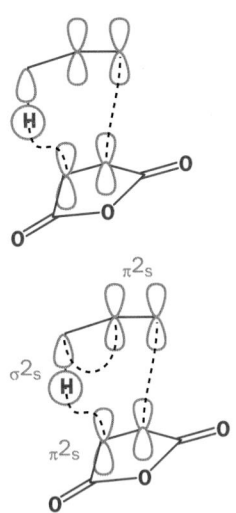

The C–H bond is parallel with the p orbitals of the ene so that the orbitals that overlap to form the new π bond are already parallel. The two molecules approach one another in parallel planes so that the orbitals that overlap to form the new σ bonds are already pointing towards each other. Because the electrons are of two types, π and σ, we must divide the ene into two components, one $_\pi 2$ and one $_\sigma 2$. We can then have an all-suprafacial reaction with three components.

All three components are of the $(4q + 2)_s$ type so all count and the total is three—an odd number—so the reaction is allowed. We have skipped the step-by-step approach we used for the Diels–Alder reaction because the two are so similar, but you should convince yourself that you can apply it here.

In frontier orbital terms we shall want again to use the LUMO of the anhydride so we need to construct the HOMO of the ene component. This must be the HOMO of the π bond and σ bond (C–H) combined. These two bonds can combine in a bonding way (σ + π) or in an antibonding fashion (σ − π). The second is higher in energy than the first and since there are a total of four electrons (two in the σ bond and two in the π bond), it is the molecular HOMO. The HOMO of the ene is bonding at both ends with the LUMO of the anhydride and the reaction is favourable.

▶

We discuss in more detail in Chapter 36 how to assign s or a with σ bonds. Here the σ bond reacts suprafacially because the 1s orbital of H has no nodes.

construction of HOMO of ene

HOMO of ene

σ   π   σ − π

combine these orbitals in two different ways   σ + π

bonding

bonding

LUMO of anhydride

Now for some real examples. Most ene reactions with simple alkenes are with maleic anhydride. Other dienophiles—or **enophiles** as we should call them in this context—do not work very well. However, with one particular alkene, the natural terpene β-pinene from pine trees, reaction does occur with enophiles such as acrylates.

ene reaction with β-pinene

shape of β-pinene

The major interaction between these two molecules is between the nucleophilic end of the exocyclic alkene and the electrophilic end of the acrylate. These atoms have the largest coefficients in the HOMO and LUMO, respectively, and, in the transition state, bond formation between these two will be more advanced than anywhere else. For most ordinary alkenes and enophiles, Lewis acid catalysis to make the enophile more electrophilic, or an intramolecular reaction (or both!), is necessary for an efficient ene reaction.

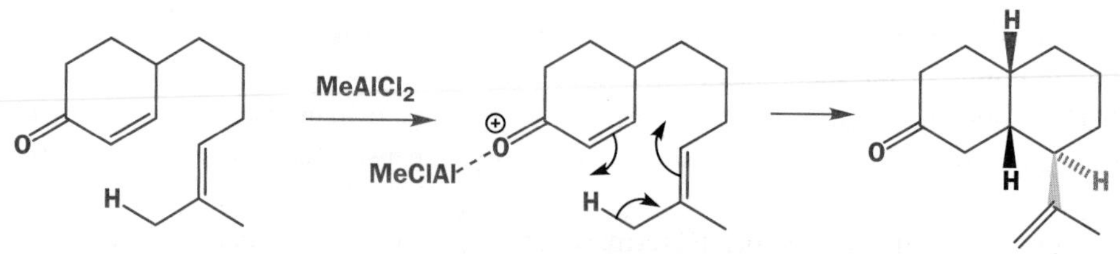

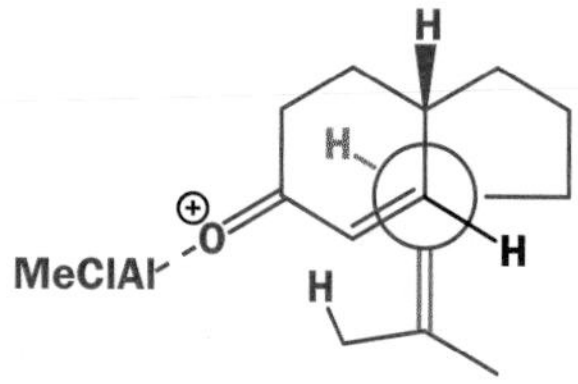

The ene is delivered to the bottom face of the enone, as its tether (Chapter 33) is too short for it to reach the top face, and a *cis* ring junction is formed. The stereochemistry of the third centre is most easily seen by a Newman projection of the reaction. In the diagram in the margin we are looking straight down the new C–C bond and the colour coding should help you to see how the stereochemistry follows.

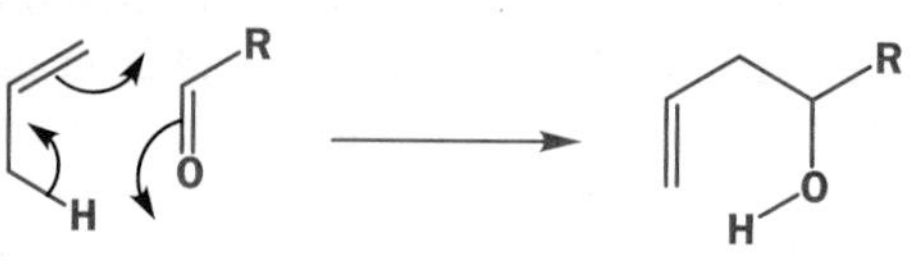

Since the twin roles of the enophile are to be attacked at one end by a C=C double bond and at the other by a proton, a carbonyl group is actually a very good enophile. These reactions are usually called **carbonyl ene reactions**.

The important interaction is between the HOMO of the ene system and the LUMO of the carbonyl group—and a Lewis–acid catalyst can lower the energy of the LUMO still further. If there is a choice, the more electrophilic carbonyl group (the one with the lower LUMO) reacts.

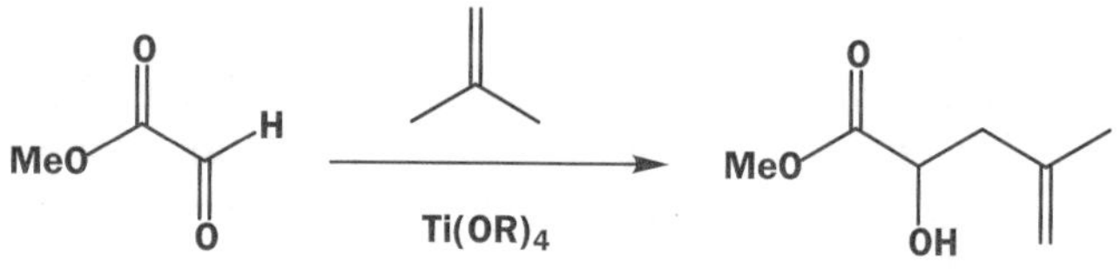

It is not obvious that an ene reaction has occurred because of the symmetry of the ene. The double bond in the product is not, in fact, in the same place as it was in the starting material.

the carbonyl ene reaction

One carbonyl ene reaction is of commercial importance as it is part of a process for the production of menthol used to give a peppermint smell and taste to many products. This is an intramolecular ene reaction on another terpene derivative.

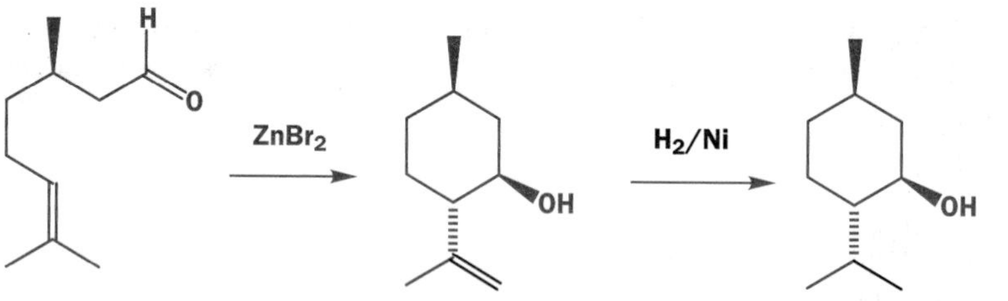

(R)-citronellal → isopulegol → (−)-menthol

It is not obvious what has happened in the first step, but the movement of the alkene and the closure of the ring with the formation of one (not two) new C–C bonds should give you the clue that this is a Lewis-acid-catalysed carbonyl ene reaction.

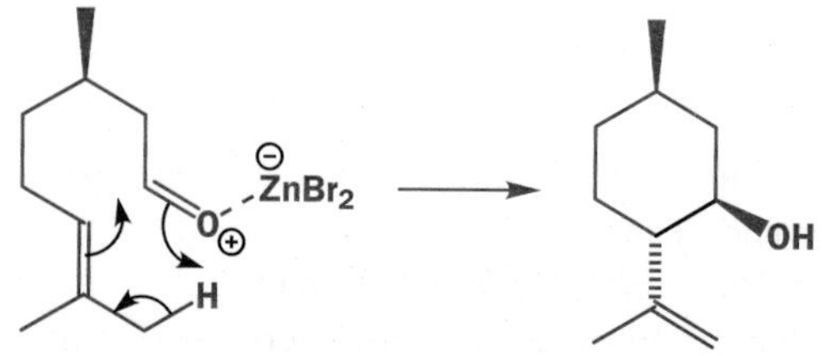

The stereochemistry comes from an all-chair arrangement in the conformation of the transition

state. The methyl group will adopt an equatorial position in this conformation, fixing the way the other bonds are formed. Again, colour coding should make it clearer what has happened.

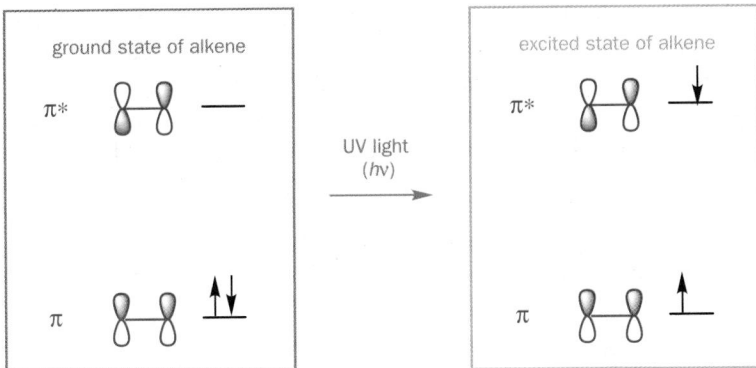

Menthol manufacture

It may seem odd to you to have a chemical process to produce menthol, which would be available naturally from mint plants. This process is now responsible for about half the world's menthol production so it must make some sort of sense! The truth is that menthol *cultivation* is wasteful in good land that could produce food crops such as rice while the starting material for menthol *manufacture* is the same β-pinene we have just met. This is available in large quantities from pine trees grown on poor land for paper and furniture. The early stages of the process are discussed in Chapter 45.

Photochemical [2 + 2] cycloadditions

We shall now leave six-electron cycloadditions such as the Diels–Alder and ene reactions and move on to some four-electron cycloadditions. Clearly, four is not a $(4n + 2)$ number, but when we told you in the box on p. 916 that only cycloadditions with $(4n + 2)$ electrons are allowed we used the term 'thermally'. Cycloadditions with $4n$ electrons *are* allowed if the reaction is not thermal (that is, driven by heat energy) but **photochemical** (that is, driven by light energy). All the cycloadditions that are not allowed thermally are allowed photochemically. The problem of the incompatible symmetry in trying to add two alkenes together is avoided by converting one of them into the excited state photochemically. First, one electron is excited by the light energy from the π to the π* orbital.

Now, combining the excited state of one alkene with the ground state of another solves the symmetry problem. Mixing the two π orbitals leads to two molecular orbitals and two electrons go down in energy while only one goes up. Mixing the two π* orbitals is as good—one electron goes down in energy and none goes up. The result is that three electrons go down in energy and only one goes up. Bonding can occur.

Alkenes can be dimerized photochemically in this way, but reaction between two different alkenes is more interesting. If one alkene is bonded to a conjugating group, it alone will absorb UV light and

In Chapter 7 we discussed why conjugated systems absorb UV light more readily than unconjugated ones do.

be excited while the other will remain in the ground state. It is difficult to draw a mechanism for these reactions as we have no simple way to represent the excited alkene. Some people draw it as a diradical (since each electron is in a different orbital); others prefer to write a concerted reaction on an excited alkene marked with an asterisk.

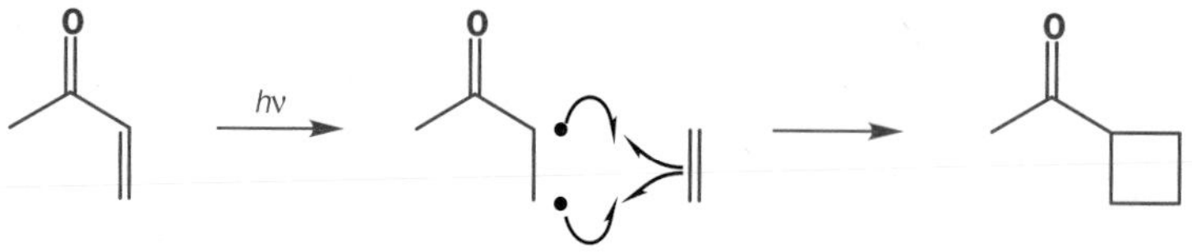

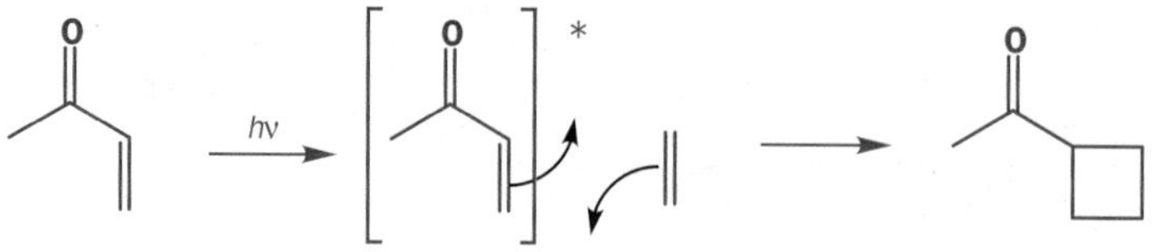

The reaction is stereospecific within each component but there is no *endo* rule—there is a conjugating group but no 'back of the diene'. The least hindered transition state usually results.

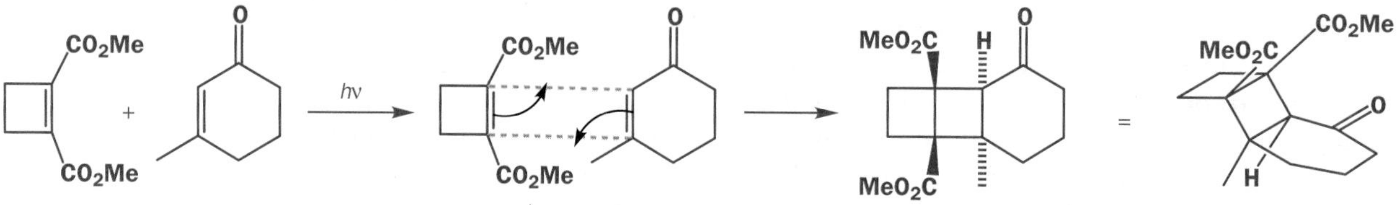

The dotted lines on the central diagram simply show the bonds being formed. The two old rings keep out of each other's way during the reaction and the conformation of the product looks reasonably unhindered.

You may be wondering why the reaction works at all, given the strain in a four-membered ring: why doesn't the product just go back to the two starting materials? This reverse reaction is governed by the Woodward–Hoffmann rules, just like the forward one, and to go back again the four-membered ring products would have to absorb light. But since they have now lost their π bonds they have no low-lying empty orbitals into which light can promote electrons (see Chapter 7). The reverse photochemical reaction is simply not possible because there is no mechanism for the compounds to absorb light.

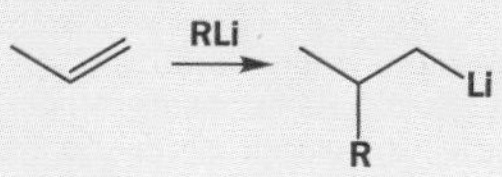

Regioselectivity in photochemical [2 + 2] cycloadditions

The observed regioselectivity is of this kind.

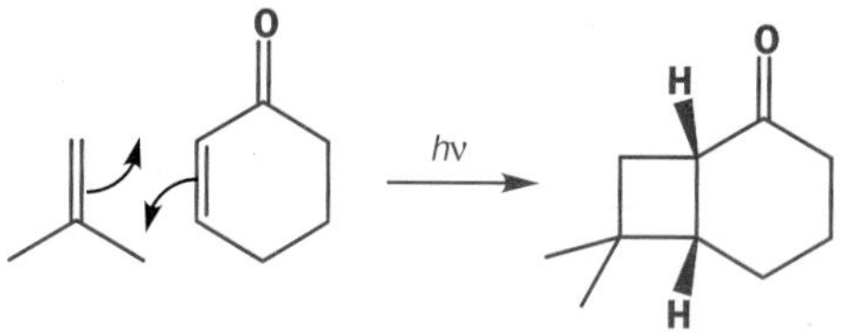

If we had combined the HOMO of the alkene with the LUMO of the enone, as we should in a thermal reaction, we would expect the opposite orientation so as to use the larger coefficients of the frontier orbitals and to maximize charge stabilization in the transition state.

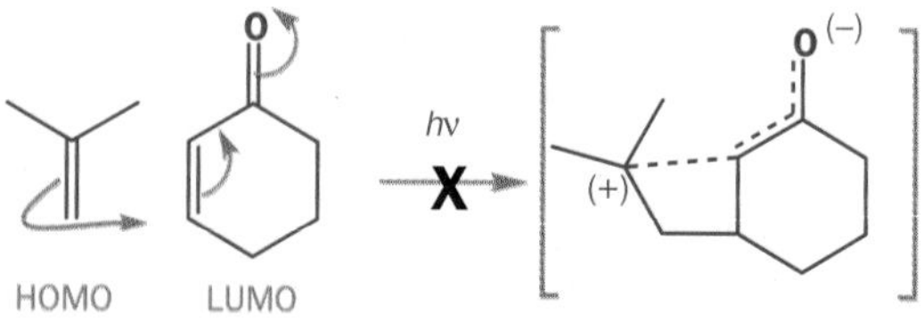

But we are not doing a thermal reaction. If you look back at the orbital diagram above, you will see that it is the HOMO/HOMO and LUMO/LUMO interactions that now matter in the reactions of the excited state. The sizes of the coefficients in the LUMO of the alkene are the other way round to those in the HOMO. There is one electron in this pair of orbitals—in the LUMO of the enone in fact, as the enone has been excited by the light—so overlap between the two LUMOs (shown in the frame)

> ►
> It may not be immediately obvious why the sizes of the coefficients swap round, but you can think of it as we did before by considering an ionic reaction of the alkene. If we want to know about the alkene's LUMO you have to consider what *would* happen *if* you could add nucleophiles to it. Of course, with an electron-rich alkene this is a very rare reaction because the LUMO is high in energy. But some organometallic additions to unactivated alkenes are known, and they attack the more substituted end in order to locate a C–metal bond at the less substituted carbon. The LUMO has a greater coefficient at the more substituted carbon.
>

is bonding and leads to the observed product. The easiest way to work it out quickly is to draw the product you do *not* expect from a normal HOMO/LUMO or curly arrow controlled reaction.

Thermal [2 + 2] cycloadditions

Despite what we have told you, there are some thermal [2 + 2] cycloadditions giving four-membered rings. These feature a simple alkene reacting with an electrophilic alkene of a peculiar type. It must have two double bonds to the *same* carbon atom. The most important examples are ketenes and isocyanates. The structures have two π bonds at right angles.

Here are typical reactions of dimethyl ketene to give a cyclobutanone and chlorosulfonyl isocyanate to give a β-lactam.

To understand why these reactions work, we need to consider a new and potentially fruitful way for two alkenes to approach each other. Thermal cycloadditions between two alkenes do not work because the HOMO/LUMO combination is antibonding at one end.

If one alkene turns at 90° to the other, there is a way in which the HOMO of one might bond at both ends to the LUMO of the other. First we turn the HOMO of one alkene so that we are looking down on the p orbitals.

Now we add the LUMO of the other alkene on top of this HOMO and at 90° to it so that there is the possibility of bonding overlap at both ends.

This arrangement looks quite promising until we notice that there is antibonding at the other two corners! Overall there is no net bonding.

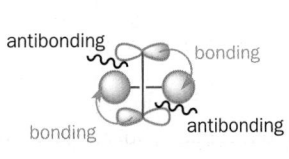

We can tilt the balance in favour of bonding by adding a p orbital to one end of the LUMO and at a right angle to it so that both orbitals of the HOMO can bond to this extra p orbital. There are now four bonding interactions but only two antibonding. The balance is in favour of a reaction. This is also quite difficult to draw!

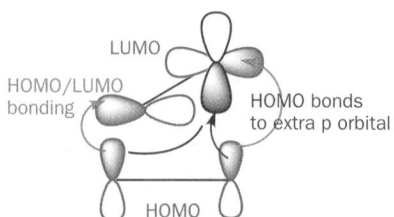

If you find this drawing difficult to understand, try a three-dimensional representation.

Ketenes have a central sp carbon atom with an extra π bond (the C=O) at right angles to the first alkene—perfect for thermal [2 + 2] cycloadditions. They are also electrophilic and so have suitable low-energy LUMOs.

Ketene [2 + 2] cycloadditions

Ketene itself is usually made by high-temperature pyrolysis of acetone but some ketenes are easily made in solution. The very acidic proton on dichloroacetyl chloride can be removed even with a tertiary amine and loss of chloride ion then gives dichloroketene in an ElcB elimination reaction.

If the elimination is carried out in the presence of cyclopentadiene a very efficient regio- and stereospecific [2 + 2] cycloaddition occurs.

The most nucleophilic atom on the diene adds to the most electrophilic atom on the ketene and the *cis* geometry at the ring junction comes from the *cis* double bond of cyclopentadiene. It is impressive that even this excellent diene undergoes no Diels–Alder reaction with ketene as dienophile. The [2 + 2] cycloaddition must be much faster.

Using the products

Dichloroketene is convenient to use, but the two chlorine atoms are not usually needed in the product. Fortunately, these can be removed by zinc metal in acetic acid solution. Zinc forms a zinc enolate, which is converted into the ketone by the acid. Repetition removes both chlorine atoms. You saw the reductive formation of a zinc enolate earlier in the chapter (p. 924) and in the Reformatsky reaction (Chapter 26, p. 706).

But what do we do if we *want* the product of a ketene [4 + 2] cycloaddition? We must use a compound that is not a ketene but that can be transformed into a ketone afterwards—a **masked ketene** or a **ketene equivalent**. The two most important types are nitroalkenes and compounds such as the 'cyanohydrin ester' in the second example.

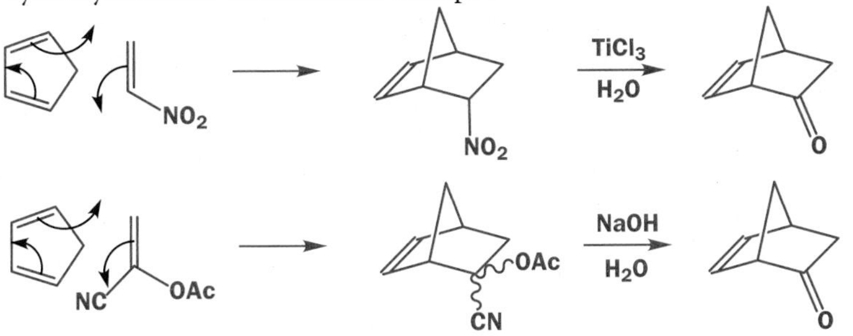

The conversion of nitro compounds to ketones by TiCl₃ is an alternative to the Nef reaction that you met in Chapter 29 (p. 767), and you should be able to write a mechanism for the last reaction in the scheme yourself.

Finding the starting materials for a cyclobutane synthesis

The disconnection of a four-membered ring is very simple—you just split in half and draw the two alkenes. There may be two ways to do this.

Both sets of starting materials look all right—the regiochemistry is correct for the first and doesn't matter for the second. However, we prefer the second because we can control the stereochemistry by using *cis*-butene as the alkene and we can make the reaction work better by using dichloroketene instead of ketene itself, reducing out the chlorine atoms with zinc.

Synthesis of β-lactams by [2 + 2] cycloadditions

Now the disconnections are really different—one requires addition of a ketene to an imine and the other the addition of an isocyanate to an alkene. Isocyanates are like ketenes, but have a nitrogen atom instead of the end carbon atom. Otherwise the orbitals are the same.

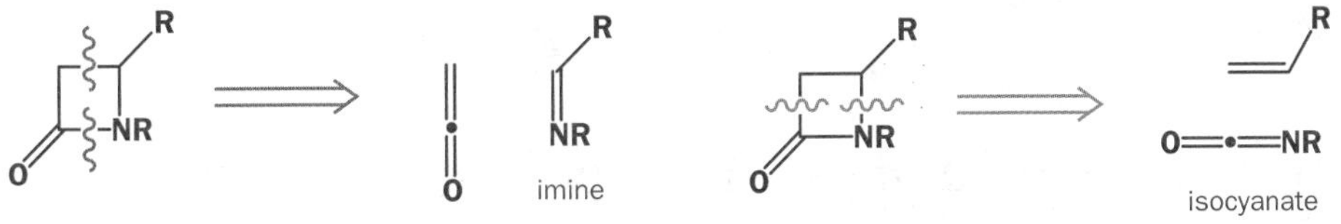

And the good news is that both work, providing we have the right substituents on nitrogen. The dichloroacetyl chloride trick works well with imines and, as you ought to expect, the more nucleophilic nitrogen atom attacks the carbonyl group of the ketene so that the regioselectivity is right to make β-lactams.

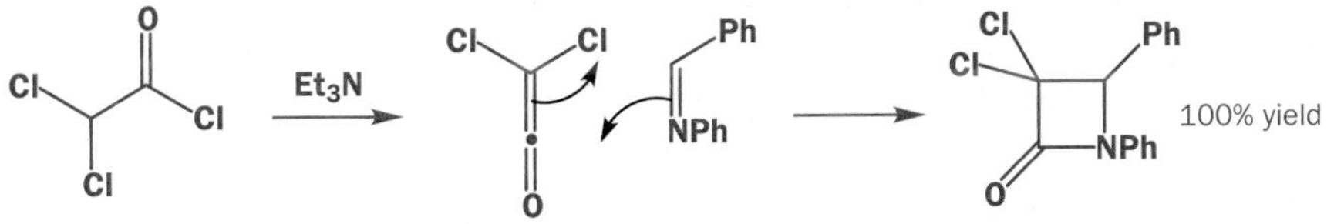

If both components have one substituent, these will end up *trans* on the four-membered ring just to keep out of each other's way. This example has more functionality and the product could be used to make β-lactams with antibiotic activity, such as analogues of the β-lactamase inhibitor, clavulanic acid (Chapter 32).

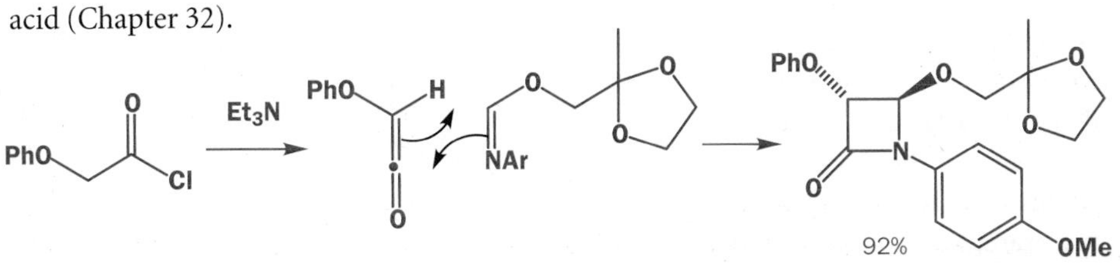

an isocyanate

chlorosulfonyl isocyanate

You will notice that in both of these examples there is an aryl substituent on the nitrogen atom of the imine. This is simply because imines are rather unstable and cannot normally be prepared with a hydrogen atom on the nitrogen. *N*-Aryl imines are quite stable (Chapter 14, p. 349).

When we wish to make β-lactams by the alternative addition of an isocyanate to an alkene, a substituent on nitrogen is again required, but for quite a different reason. Because alkenes are only moderately nucleophilic, we need a strongly electron-withdrawing group on the isocyanate that can be removed after the cycloaddition, and the most popular by far is the chlorosulfonyl group. The main reason for its popularity is the commercial availability of chlorosulfonyl isocyanate. It reacts even with simple alkenes.

The alkene's HOMO interacts with the isocyanate's LUMO, and the most electrophilic atom is the carbonyl carbon so this is where the terminal carbon atom of the alkene attacks. The chlorosulfonyl group can be removed simply by hydrolysis under mild conditions via the sulfonic acid.

With a more electron-rich alkene—an enol ether, for example, or the following example with its sulfur analogue, a vinyl sulfide—the reaction ceases to be a concerted process and occurs stepwise. We know this must be the case in the next example because, even though the starting material is an *E/Z* mixture, the product has only *trans* stereochemistry: it is stereoselective rather than stereospecific, indicating the presence of an intermediate in which free rotation can take place.

> The lack of stereospecificity in some nonconcerted reactions is discussed in Chapter 40 in relation to carbenes.

free rotation means lack of stereospecificity

Making five-membered rings—1,3-dipolar cycloadditions

We have seen how to make four-membered rings by [2 + 2] cycloadditions and, of course, how to make six-membered rings by [4 + 2] cycloadditions. Now what about five-membered rings? It sounds at first impossible to make an odd-numbered ring in this way. However, all we need is a three-atom, four-electron 'diene' and we can do a Diels–Alder reaction. Impossible? Not at all—the molecules are called 1,3-dipoles and are good reagents for cycloadditions. Here is an example.

The molecule containing N and O atoms labelled 'four-electron component' is the 1,3-dipole. It has a nucleophilic end (O⁻) and an electrophilic end—the end of the double bond next to the central N⁺. These are 1,3-related so it is indeed a 1,3-dipole. This functional group is known as a **nitrone**. You could also think of it as the *N*-oxide of an imine.

> The charges make it look like a 1,2-dipole, but nucleophilic attack on N⁺ is impossible.

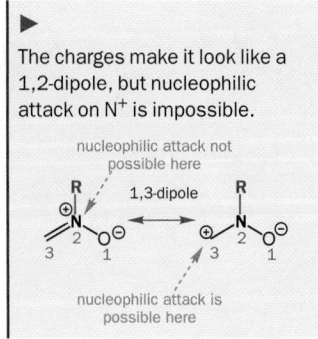

nucleophilic attack not possible here

1,3-dipole

nucleophilic attack is possible here

four-electron component

two-electron component

The nitrone gets its four electrons in this way: there are two π electrons in the N=C double bond and the other two come from one of the lone pairs on the oxygen atom. The two-electron component is a simple alkene in this example. In a Diels–Alder

four-electron component

two-electron component

reaction it would be called the dienophile. Here it is called the **dipolarophile**. Simple alkenes (which are bad dienophiles) are good dipolarophiles and so are electron-deficient alkenes.

The difference between dienes and 1,3-dipoles is that dienes are nucleophilic and prefer to use their HOMOs in cycloadditions with electron-deficient dienophiles while 1,3-dipoles, as their name implies, are both electrophilic and nucleophilic. They can use either their HOMOs or their LUMOs depending on whether the dipolarophile is electron-deficient or electron-rich.

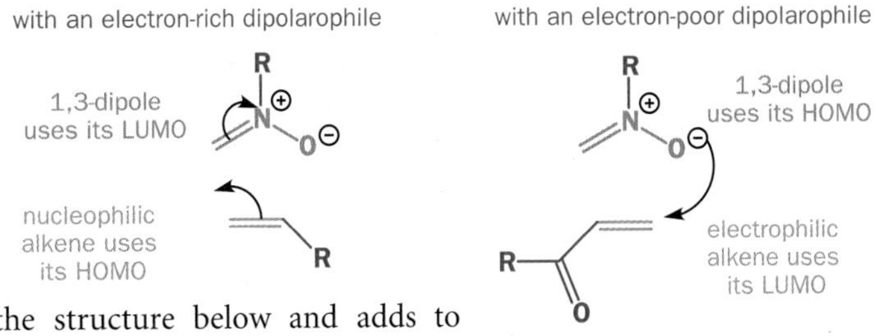

One important nitrone is a cyclic compound that has the structure below and adds to dipolarophiles (essentially any alkene!) to give two five-membered rings fused together. The stereochemistry comes from the best approach with the least steric hindrance, as shown. There is no *endo* rule in these cycloadditions as there is no conjugating group to interact across space at the back of the dipole or dipolarophile. The product shown here is the more stable *exo* product.

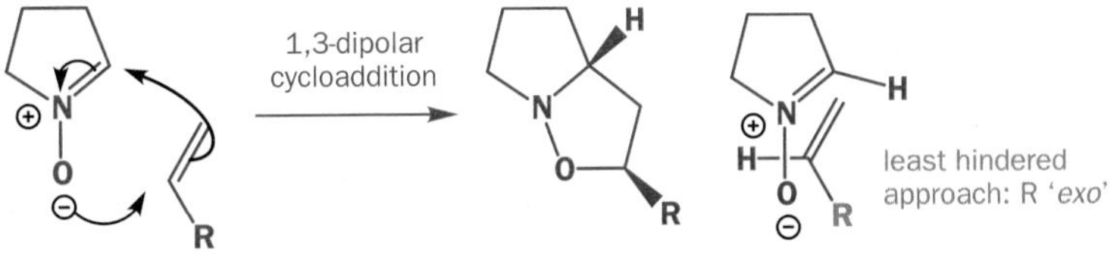

If the alkene is already joined on to the nitrone by a covalent bond so that the dipolar cycloaddition is an intramolecular reaction, one particular outcome may be dictated by the impossibility of the alternatives. Here is a simple case where an allyl group is joined to the same ring as in the previous example. The product has a beautifully symmetrical cage structure and the mechanism shows the only way in which the molecule can fold up to allow a 1,3-dipolar cycloaddition to occur.

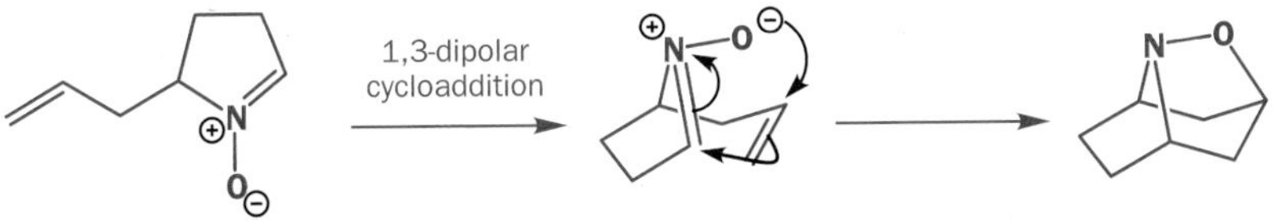

Making nitrones

There are two important routes to nitrones: both start from hydroxylamines. Open-chain nitrones are usually made simply by imine formation between a hydroxylamine and an aldehyde.

The cyclic nitrones are made from simple tertiary amines by oxidation and then cyclic elimination to give a hydroxylamine. This is oxidized again with Hg(II) to give the nitrone.

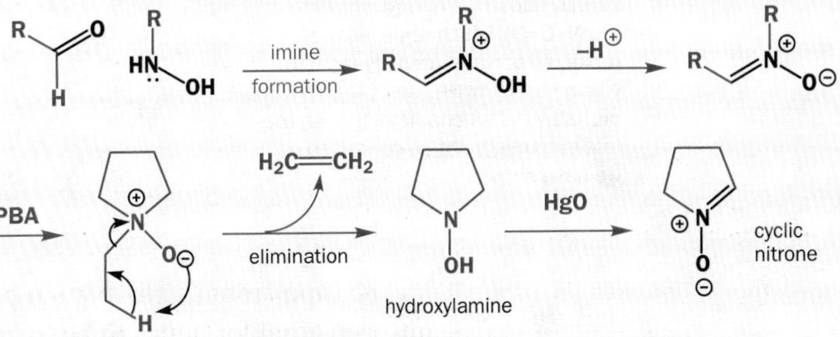

The importance of the Diels–Alder reaction is that it makes six-membered rings with control over stereochemistry. The importance of 1,3-dipolar cycloadditions is not so much in the heterocyclic products but in what can be done with them. Almost always, the first formed heterocyclic ring is broken down in some way by carefully controlled reactions. The nitrone adducts we have just seen contain a weak N–O single bond that can be selectively cleaved by reduction. Reagents such as LiAlH$_4$ or zinc metal in various solvents (acetic acid is popular) or hydrogenation over catalysts such as nickel reduce the N–O bond to give NH and OH functionality without changing the structure or stereochemistry of the rest of the molecule. From the examples above, we get these products.

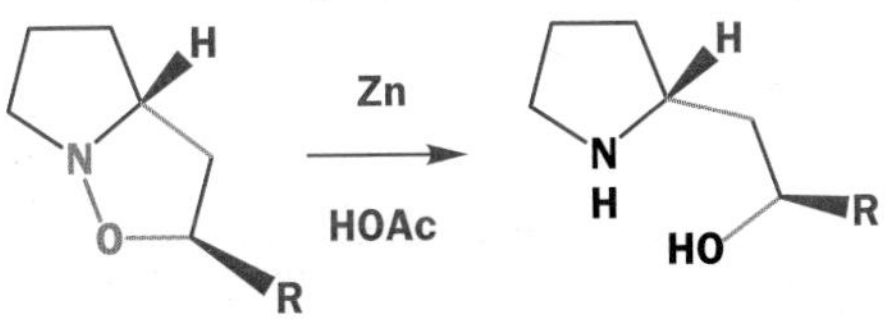

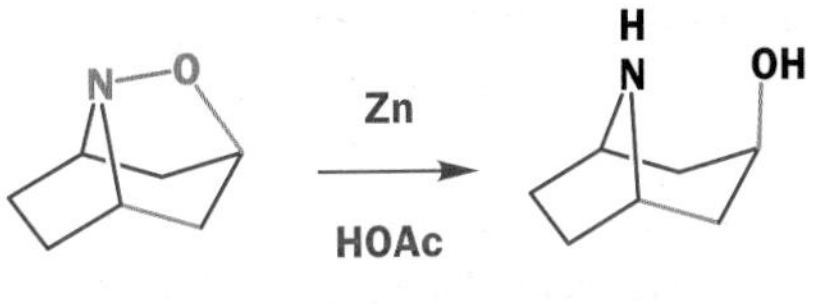

In each cycloaddition, one permanent C–C and one C–O bond (shown in orange) were made. These were retained while the N–O bond present in the original dipole was discarded. The final product is an amino-alcohol with a 1,3-relationship between the OH and NH groups.

Linear 1,3-dipoles

In the Diels–Alder reaction, the dienes had to have an s-*cis* conformation about the central single bond so that they were already in the shape of the product. Many useful 1,3-dipoles are actually linear and their 1,3-dipolar cycloadditions look very awkward. We shall start with the nitrile oxides, which have a triple bond where the nitrone had a double bond.

R——N
a nitrile

R——N⊕—O⊖
a nitrile oxide

a 1,3-dipolar cycloaddition with a nitrile oxide

Making nitrile oxides

There are two important routes to these compounds, both of which feature interesting chemistry. Oximes, easily made from aldehydes with hydroxylamine (NH_2-OH), are rather enol-like and can be chlorinated on carbon.

Treatment of the chloro-oxime with base (Et_3N is strong enough) leads directly to the nitrile oxide with the loss of HCl. This is an elimination of a curious kind as we cannot draw a connected chain of arrows for it. We must use two steps—removal of the OH proton and then loss of chloride. It is a γ elimination rather than the more common β elimination.

oxime

The other method starts from nitroalkanes and is a dehydration. Inspect the two molecules and you will see that the nitro compound contains H_2O more than the nitrile oxide. But how to remove the molecule of water? The reagent usually chosen is phenyl isocyanate (Ph-N=C=O), which removes the molecule of water atom by atom to give aniline ($PhNH_2$) and CO_2. This is probably the mechanism, though the last step might not be concerted as we have shown.

$-H_2O$?

LUMO of nitrile oxide

HOMO of alkene

The dipolarophile (here a simple alkene) has to approach uncomfortably close to the central nitrogen atom for bonds to be formed. Presumably, the nitrile oxide distorts out of linearity in the transition state. As you should expect, this is a reaction between the HOMO of the alkene and the LUMO of the nitrile oxide so that the leading interaction that determines the structure of the product is the one in the margin.

If there is stereochemistry in the alkene, it is faithfully reproduced in the heterocyclic adduct as we should expect for a concerted cycloaddition.

Z-alkene   *cis* substituents

cycloaddition of nitrile oxide and alkyne

an isoxazole

There is much more about making heterocycles in Chapter 44.

Both partners in nitrile oxide cycloadditions can have triple bonds—the product is then a stable aromatic heterocycle called an **isoxazole**.

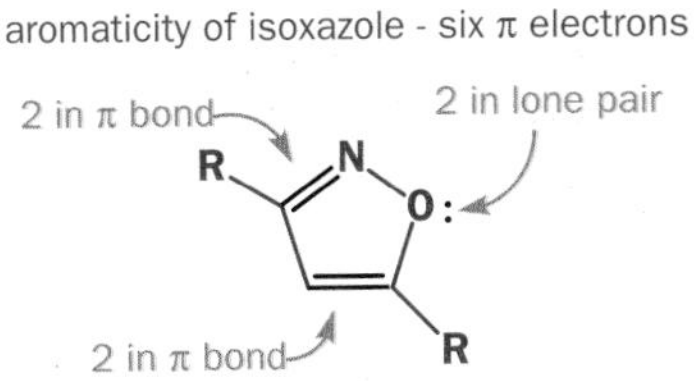

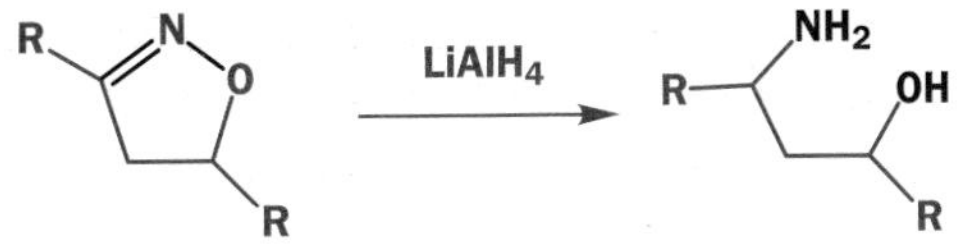

aromaticity of isoxazole - six π electrons

2 in π bond

2 in lone pair

R

N

O:

2 in π bond

R

R

N

O

R

LiAlH₄

R

NH₂

OH

R

Though isoxazoles have some importance, the main interest in nitrile oxide cyclo[
additions lies again in the products that are formed by reduction of the N–O bond and by the C=N
double bond. This produces amino-alcohols with a 1,3-relationship between the two functional
groups.

The N–O bond is the weaker of the two and it is possible to reduce that and leave the C=N bond
alone. This leaves an imine that usually hydrolyses during work-up.

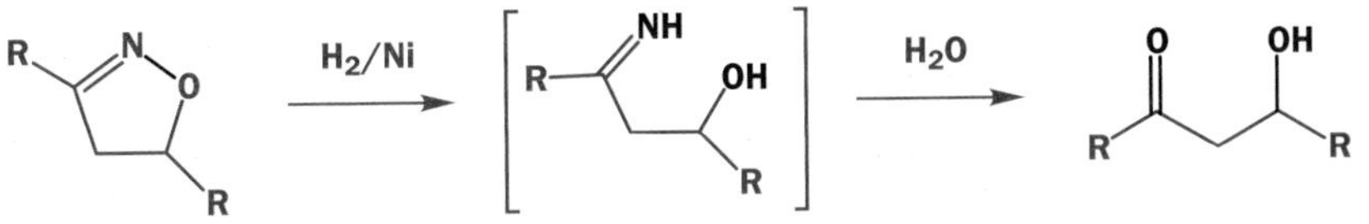

R

N

O

R

H₂/Ni

⎡
R

NH

OH

R
⎤

H₂O

O

R

OH

R

Any stereochemistry in the adduct is preserved right through this reduction and hydrolysis
sequence: you might like to compare the products with the products of the stereoselective aldol
reactions you saw in Chapter 34.

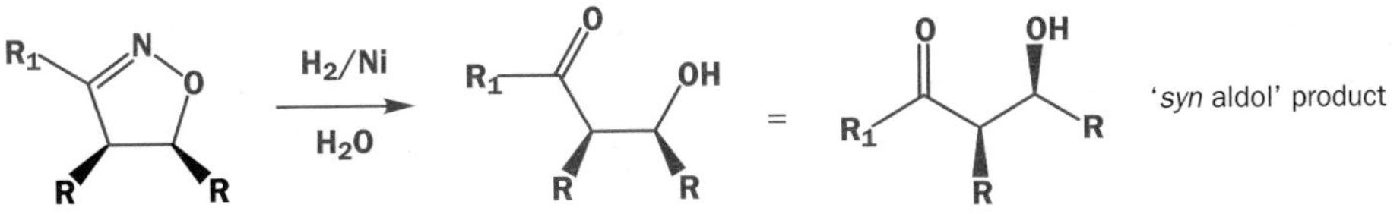

R₁

N

O

R

R

H₂/Ni

H₂O

R₁

O

R

OH

R

=

R₁

O

OH

R

R

'syn aldol' product

cis substituents

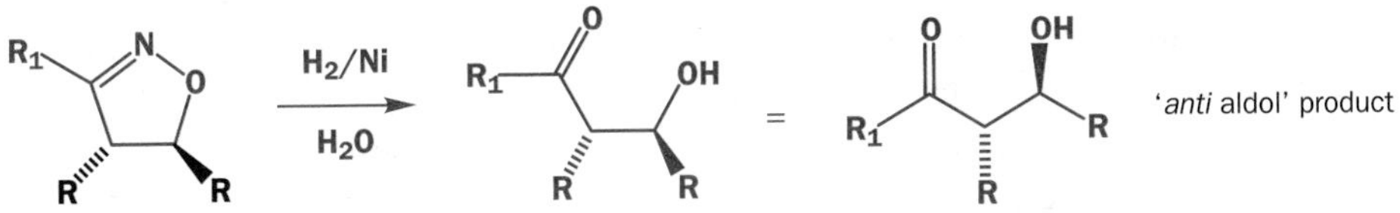

R₁

N

O

R

R

H₂/Ni

H₂O

R₁

O

R

OH

R

=

R₁

O

OH

R

R

'anti aldol' product

trans substituents

Biotin

Biotin is an enzyme cofactor that activates and
transports CO_2 for use as a electrophile in
biochemical reactions.

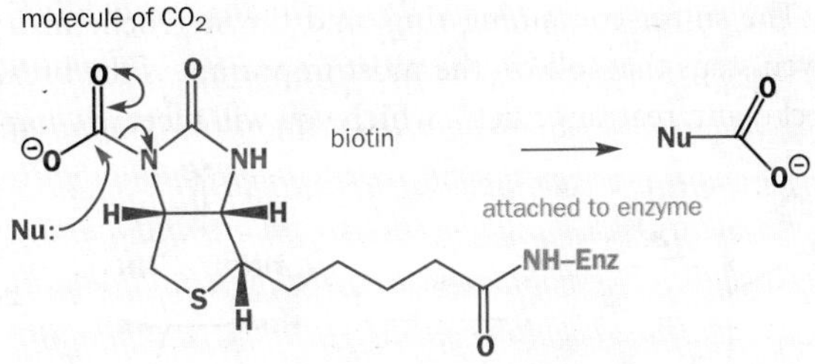

molecule of CO_2

⊖O

O

O

N

NH

Nu:

H

H

S

H

biotin

→

Nu

O

O⊖

attached to enzyme

NH–Enz

O

We shall end this section with a beautiful illustration of an intramolecular 1,3-dipolar cycloaddition of a nitrile oxide that was used in the synthesis of the vitamin biotin. Starting at the beginning
of the synthesis will allow you to revise some reactions from earlier chapters. The starting material is a
simple cyclic allylic bromide that undergoes an efficient S_N2 reaction with a sulfur nucleophile. In
fact, we don't know (or care!) whether this is an S_N2 or S_N2' reaction as the product of both reactions
is the same. This sort of chemistry was discussed in Chapter 23 if you need to check up on it. Notice
that it is the sulfur atom that does the attack—it is the soft end of the nucleophile and better at S_N2
reactions. The next step is the hydrolysis of the ester group to reveal the thiolate anion.

This step is strictly an ester exchange rather than a hydrolysis and is discussed in Chapter 12. Next the nucelophilic thiolate anion does a conjugate addition (Chapters 10 and 23) on to a nitroalkene.

Now comes the exciting moment. The nitroalkene gives the nitrile oxide directly on dehydration with PhN=C=O and the cycloaddition occurs spontaneously in the only way it can, given the intramolecular nature of the reaction.

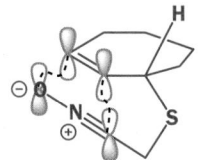

We have drawn the reaction with the nitrile oxide coming up from the underside of the seven-membered ring, pushing all the hydrogen atoms at the ring junctions upwards and making all the rings join up in a *cis* fashion.

Next the cycloadduct is reduced completely with LiAlH$_4$ so that both the N–O and C=N bonds are cleaved. This step is very stereoselective so the C=N reduction probably precedes the N–O cleavage and the hydride has to attack from the outside (top) face of the molecule. These considerations are explored more thoroughly in Chapter 33.

The sulfur-containing ring, and the stereochemistry, of biotin are already defined and, in the seven steps that follow, the most important is the breaking open of the seven-membered ring by a Beckmann rearrangement, which you will meet in Chapter 37.

Two very important synthetic reactions: cycloaddition of alkenes with osmium tetroxide and with ozone

We shall end this chapter with two very important reactions, both of which we have alluded to earlier in the book. These reactions are very important not just because of their mechanisms, which you must be aware of, but even more because of their usefulness in synthetic chemistry, and in that regard they are second only to the Diels–Alder reaction when considering all the reactions in this chapter. They are both oxidations—one involves osmium tetroxide (OsO_4) and one involves ozone (O_3) and they both involve cycloaddition.

OsO_4 adds two hydroxyl groups *syn* to a double bond

We emphasized the fact that cycloadditions, being concerted, are stereospecific with regard to the geometry of the double bond. One very important example of this is the stereospecific reaction of an alkene with OsO_4. First, we give you the result of the reaction—the overall outcome is that two hydroxyl groups are added *syn* to the double bond.

They add *syn* whether the double bond is *E* or *Z*, and, by redrawing the second example in a different conformation, you can see how defining the geometry of the starting material defines which diastereoisomer of the product is obtained.

Now for the mechanism. We must admit before we start that this is a reaction about which there is still some controversy, and we give you the simplest reasonable view of the mechanism. Future results may show this mechanism to be wrong, but it will certainly do to explain any result you might meet. The first step is a cycloaddition between the osmium tetroxide and the alkene. You can treat the OsO_4 like a dipole—it isn't drawn as one because osmium has plenty of orbitals to accommodate four double bonds.

The product of the stereospecific cycloaddition is an 'osmate ester'. This isn't the required product, and the reaction is usually done in the presence of water (the usual solvent is a *t*-BuOH–water mixture), which hydrolyses the osmate ester to the diol. Because both oxygen atoms were added in one concerted step during the cycloaddition, their relative stereochemistry must remain *syn*.

The osmium starts as Os(VIII) and ends up as Os(VI)—the reaction is, of course, an oxidation, and it's one that is very specific to C=C double bonds (as we mentioned in Chapter 24). As written, it would involve a whole equivalent of the expensive, toxic, and heavy metal osmium, but it can be made catalytic by introducing a reagent to oxidize Os(VI) back to Os(VIII). The usual reagent is *N*-methylmorpholine-*N*-oxide (NMO) or Fe(III), and typical conditions for an osmylation, or dihydroxylation, reaction are shown in the scheme alongside.

In behaviour that is typical of a 1,3-dipolar cycloaddition reaction, OsO_4 reacts almost as well with electron-poor as with electron-rich alkenes. OsO_4 simply chooses to attack the alkene HOMO or its LUMO depending on which gives the best interaction. This is quite different from the electrophilic addition of *m*-CPBA or Br$_2$ to alkenes.

syn and anti addition of hydroxyl groups

It is important that you note the link between the OsO$_4$ reaction and the stereospecific transformations that we highlighted at the beginning of Chapter 34. In particular, you now know ways to add two hydroxyl groups both *syn* and *anti* across a double bond: the *syn* addition uses OsO$_4$ and the anti addition uses epoxidation followed by ring opening with HO⁻.

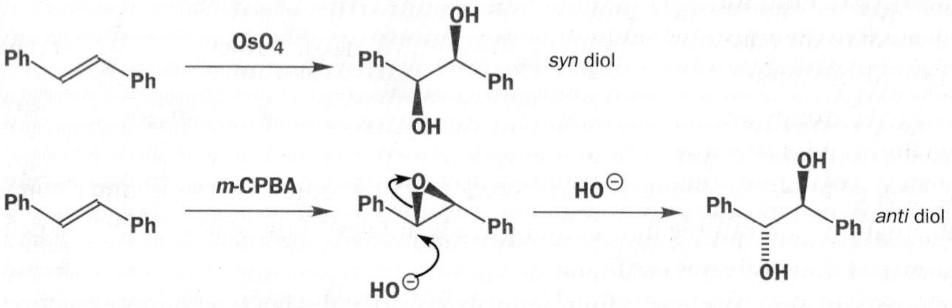

A cycloaddition that destroys bonds—ozonolysis

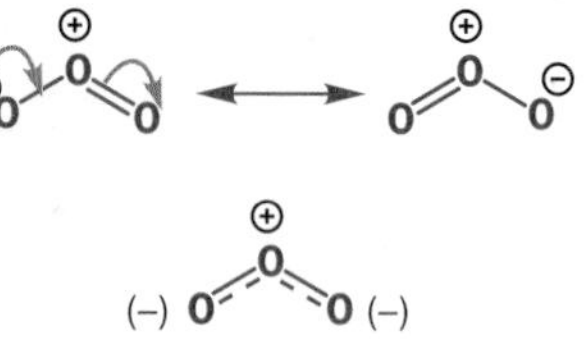

Our last type of cycloaddition is most unusual. It starts as a 1,3-dipolar cycloaddition but eventually becomes a method of cleaving π bonds in an oxidative fashion so that they end up as two carbonyl groups. The reagent is ozone, O$_3$.

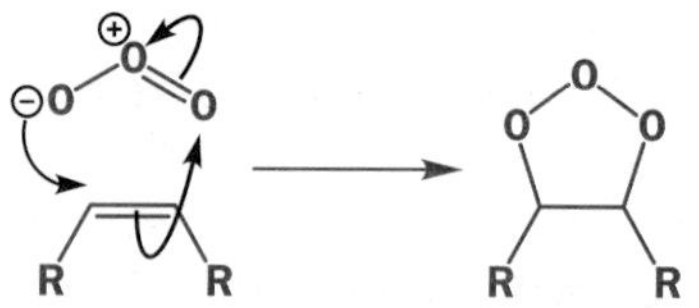

structure of ozone

Ozone is a symmetrical bent molecule with a central positively charged oxygen atom and two terminal oxygen atoms that share a negative charge. It is a 1,3-dipole and does typical 1,3-dipolar cycloadditions with alkenes.

The product is a very unstable compound. The O–O single bond (bond energy 140 kJ mol^{-1}) is a very weak bond—much weaker than the N–O bond (180 kJ mol^{-1}) we have been describing as weak in previous examples—and this heterocycle has two of them. It immediately decomposes—by a *reverse* 1,3-dipolar cycloaddition.

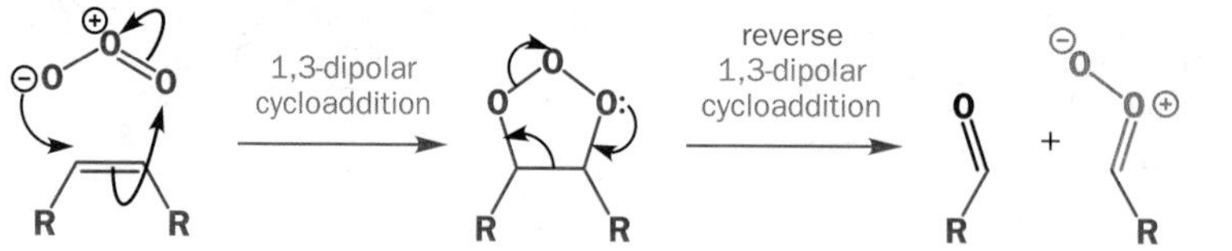

The products are a simple aldehyde on the left and a new, rather unstable looking molecule—a 1,3-dipole known as a carbonyl oxide—on the right. At least it no longer has any true O–O single bonds (the one that looks like a single bond is part of a delocalized system like the one in ozone). Being a 1,3-dipole, it now adds to the aldehyde in a third cycloaddition step. It might just add back the way it came, but it much prefers to add in the other way round with the nucleophilic oxyanion attacking the carbon atom of the carbonyl group like this.

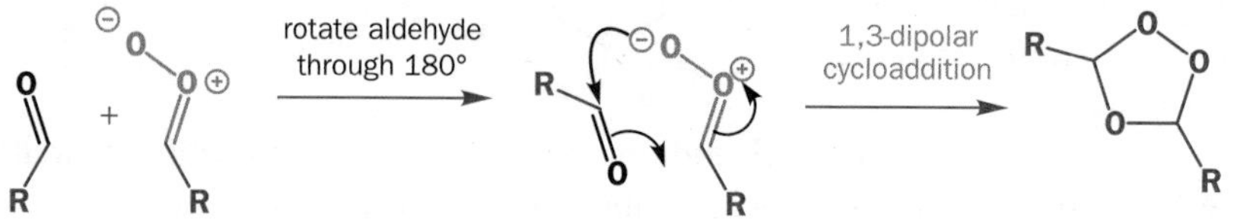

This compound—known as an **ozonide**—is the first stable product of the reaction with ozone. It is the culmination of two 1,3-dipolar cycloadditions and one reverse 1,3-dipolar cycloaddition. It is still not that stable and is quite explosive, so for the reaction to be of any use it needs decomposing. The way this is usually done is with dimethylsulfide, which attacks the ozonide to give DMSO and two molecules of aldehyde.

Ph₃P is also used.

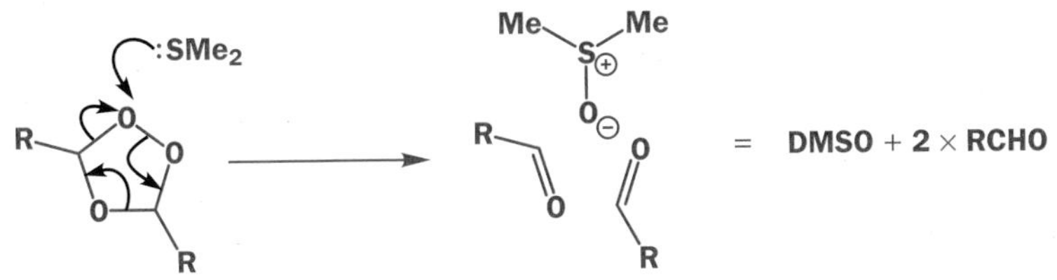

The ozonide will also react with oxidizing agents such as H_2O_2 to give carboxylic acids, or with more powerful reducing agents such as $NaBH_4$ to give alcohols. Here are the overall transformations—each cleaves a double bond—it is called an **ozonolysis**.

ozonolysis of alkenes to...

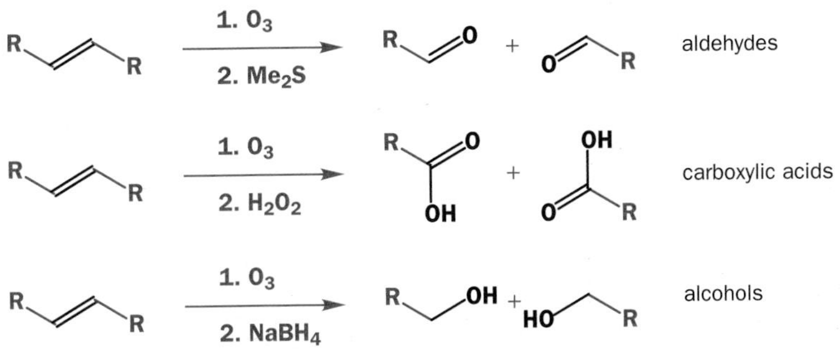

Ozonolysis of cyclohexenes is particularly useful as it gives 1,6-dicarbonyl compounds that are otherwise difficult to make. In the simplest case we get hexane 1,6-dioic acid (adipic acid) a monomer for nylon manufacture.

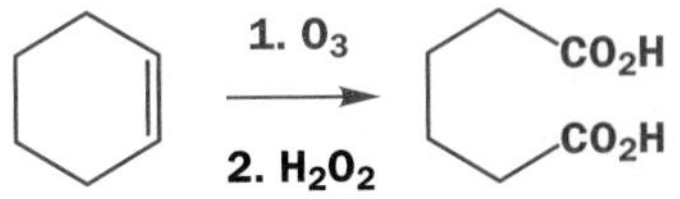

More interesting cases arise when the products of Birch reduction (Chapter 24) are treated with ozone. Here it is the electron-rich enol ether bond that is cleaved, showing that ozone is an electrophilic partner in 1,3-dipolar cycloadditions. If the ozonide is reduced, a hydroxy ester is formed whose trisubstituted bond's Z geometry was fixed by the ring it was part of (see Chapter 31).

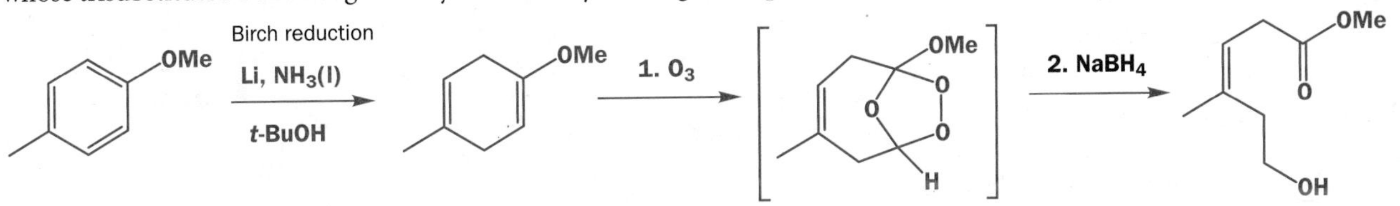

An alternative method of cleaving C=C bonds is to use OsO_4 in conjunction with $NaIO_4$. The diol product forms a periodate ester, which decomposes to give two molecules of aldehyde. These are themselves oxidized by the periodate to carboxylic acids.

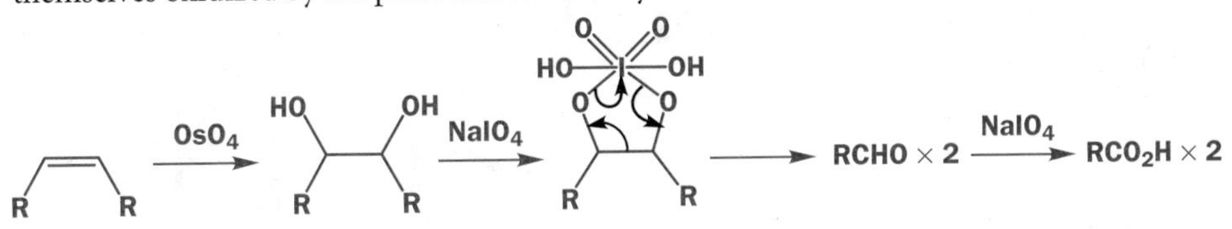

You saw periodate being used to cleave C–C bonds in this way at the end of Chapter 34, p. 902.

Summary of cycloaddition reactions

- A cycloaddition is a one-step ring-forming reaction between two conjugated π systems in which two new σ bonds are formed joining the two reagents at each end. The mechanism has one step with no intermediates, and all the arrows start on π bonds and go round in a ring.

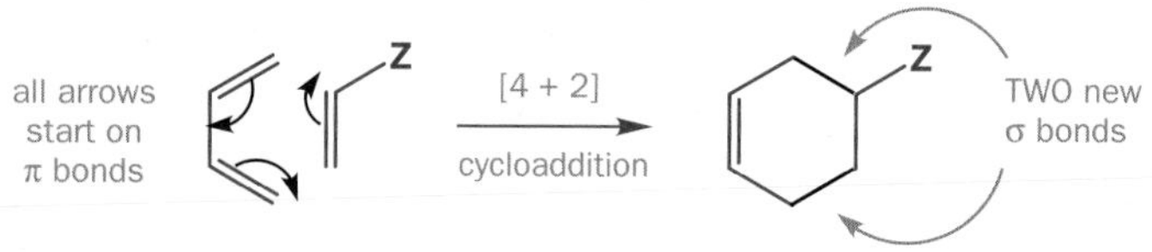

- The cycloadditions are supra-facial—they occur on one face only of each π system—and for a thermally allowed reaction there should be $4n + 2$ electrons in the mechanism, but $4n$ in a photochemical cycloaddition. These rules are dictated by orbital symmetry.

- Cycloaddition equilibria generally lie over on the right-hand side in a thermal reaction because C–C σ bonds are stronger than C–C π bonds. In a photochemical cycloaddition, the product loses its π bonds and therefore its means of absorbing energy. It is the kinetic product of the reaction even if it has a strained four-membered ring.

- The stereochemistry of each component is faithfully reproduced in the product—the reactions are stereospecific—and the relationship between their stereochemistries may be governed by orbital overlap to give an *endo* product.

Problems

1. Give mechanisms for these reactions, explaining the stereo-chemistry.

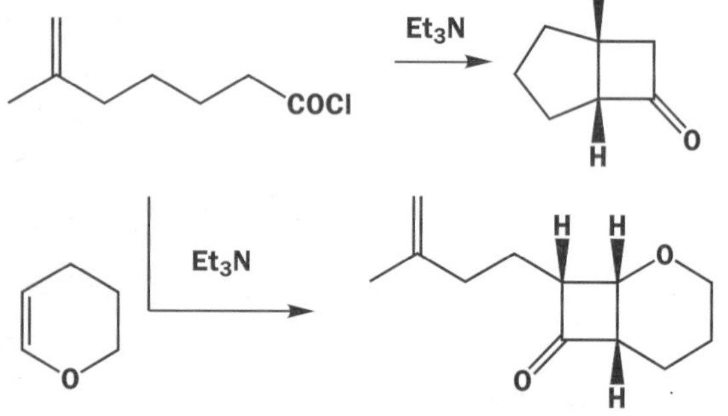

2. Predict the structure of the product of this Diels–Alder reaction.

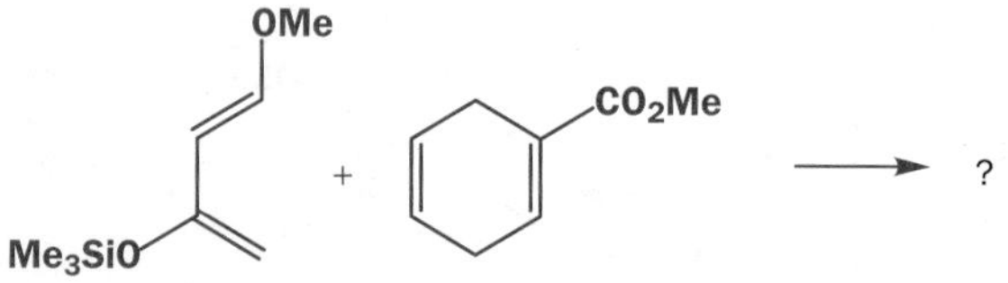

3. Comment on the difference in rate between these two reactions. It is estimated that the second goes about 10^6 times faster than the first.

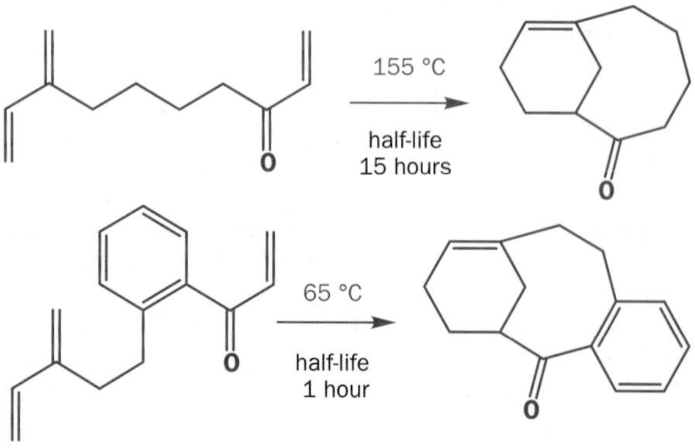

4. Justify the stereoselectivity in this intramolecular Diels–Alder reaction.

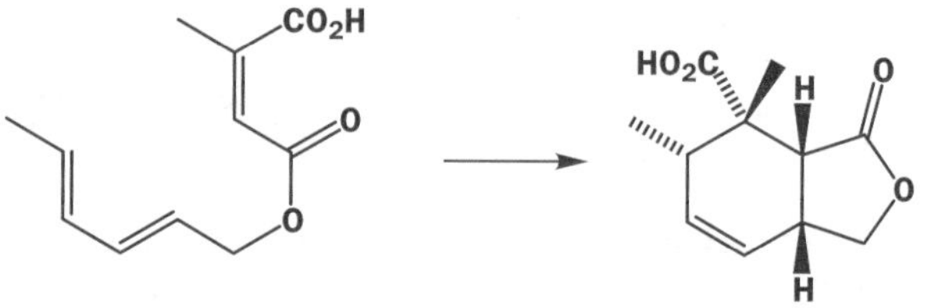

5. Explain the formation of single adducts in these reactions.

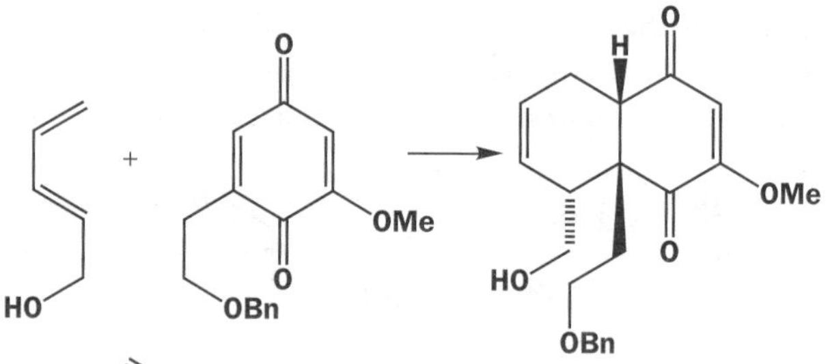

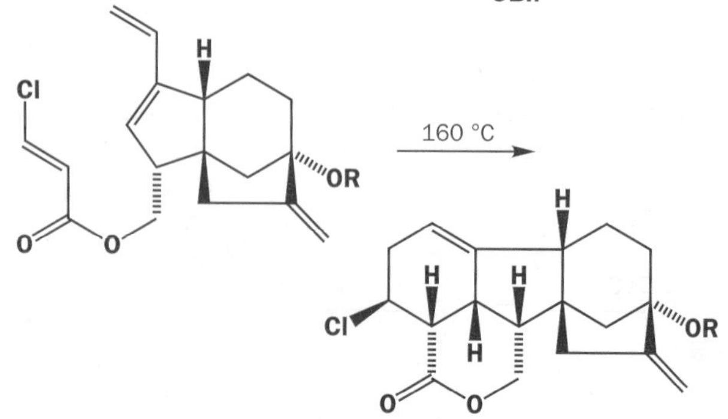

6. Revision elements. Suggest two syntheses of this spirocyclic ketone from the starting materials shown. Neither starting material is available.

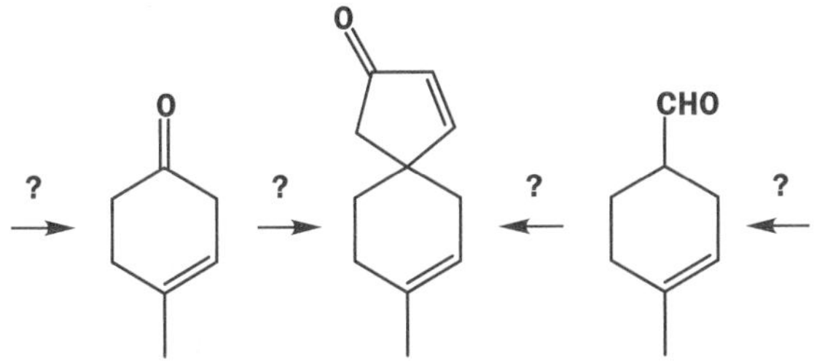

7. This reaction appeared in Chapter 33. Account for the selectivity.

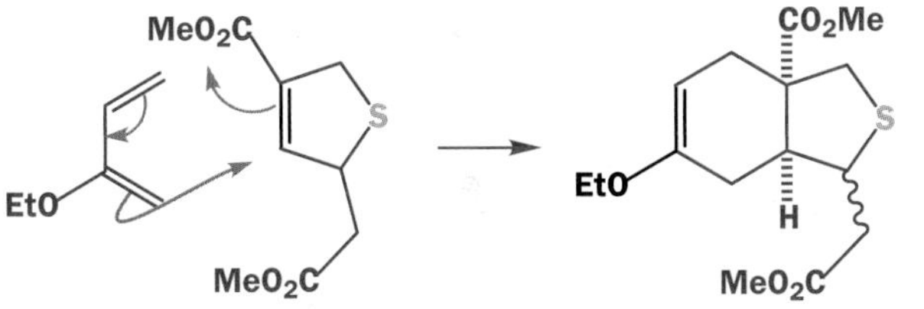

8. Draw mechanisms for these reactions and explain the stereo-chemistry.

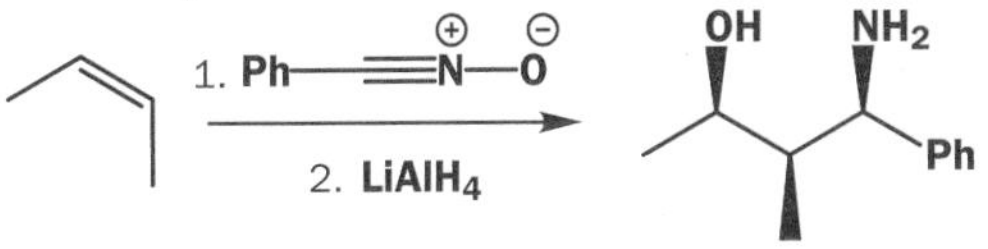

9. Revision. One of the nitrones used as an example in the chapter was prepared by this route. Explain what is happening and give details of the reactions.

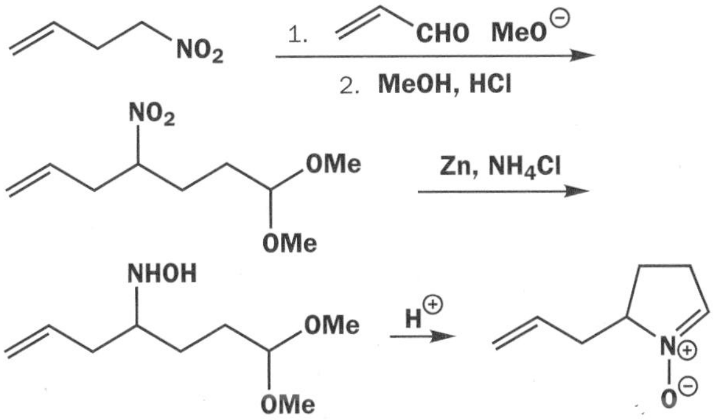

10. Explain why this Diels–Alder reaction gives total regio-selectivity and stereospecificity but no stereoselectivity. What is the mechanism of the second step? What alternative route might you have considered if you wanted to make this final product and why would you reject it?

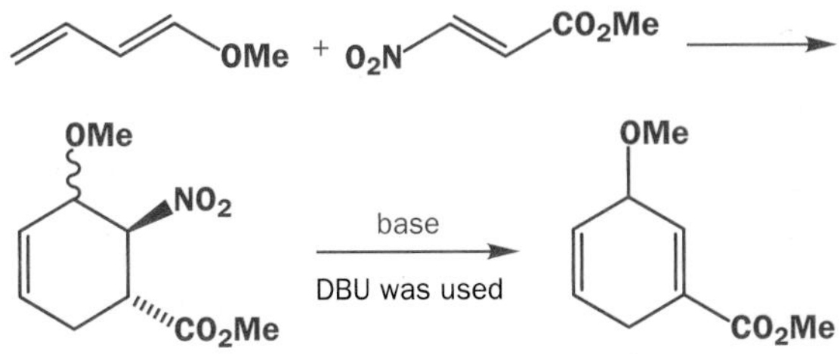

11. Give mechanisms for these reactions and explain the regio- and stereochemical control (or the lack of it!).

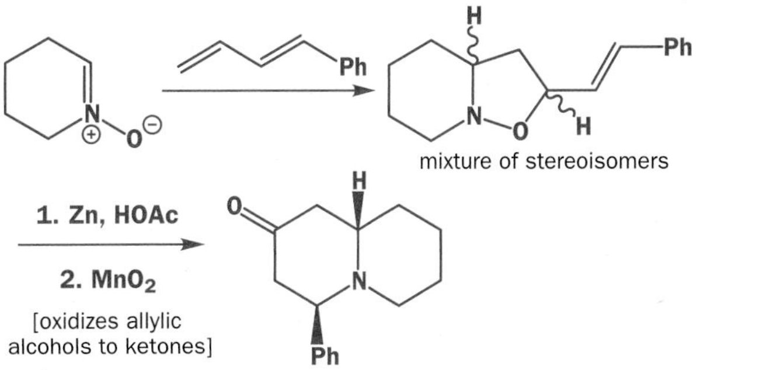

1. Zn, HOAc

2. MnO₂

[oxidizes allylic alcohols to ketones]

12. Suggest a mechanism for this reaction and explain the stereo- and regiochemistry. How would you prepare the unsaturated ketone starting material?

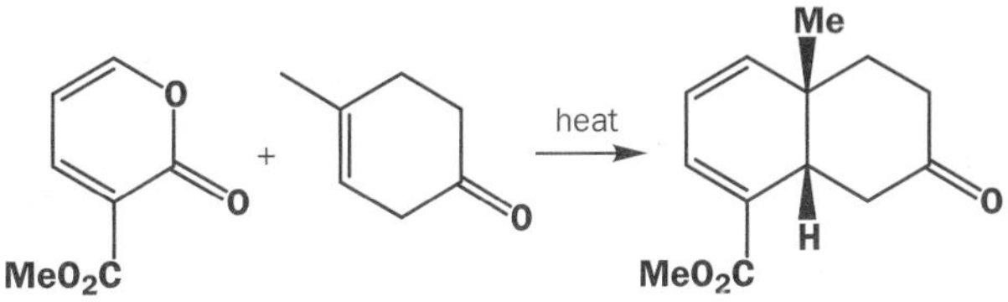

13. Photochemical cycloaddition of these two compounds is claimed to give the single diastereoisomer shown. The chemists who did this work claim that the stereochemistry of the adduct is simply provided by its conversion into a lactone on reduction. Comment on the validity of this deduction and explain the stereochemistry of the cycloaddition.

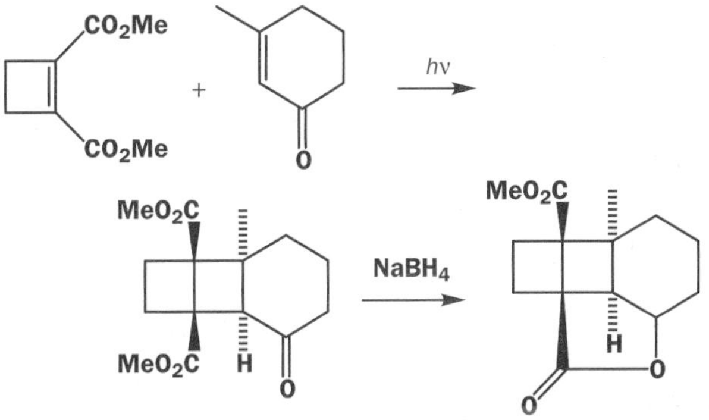

14. Thioketones, with a C=S bond, are not usually stable as we shall see in Chapter 46. However, this thioketone is quite stable and undergoes reaction with maleic anhydride to give the product shown. Comment on the stability of the starting material, the mechanism of the reaction, and the stereochemistry of the product.

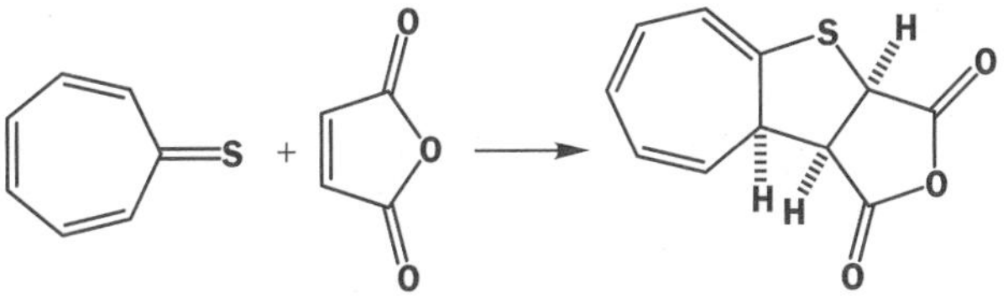

15. This unsaturated alcohol is perfectly stable until it is oxidized with Cr(VI): it then immediately cyclizes to the product shown. Explain.

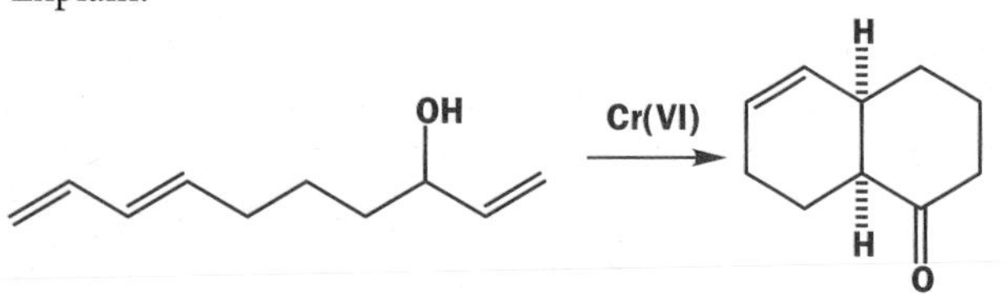

16. Suggest mechanisms for these reactions and comment on the stereochemistry of the first product.

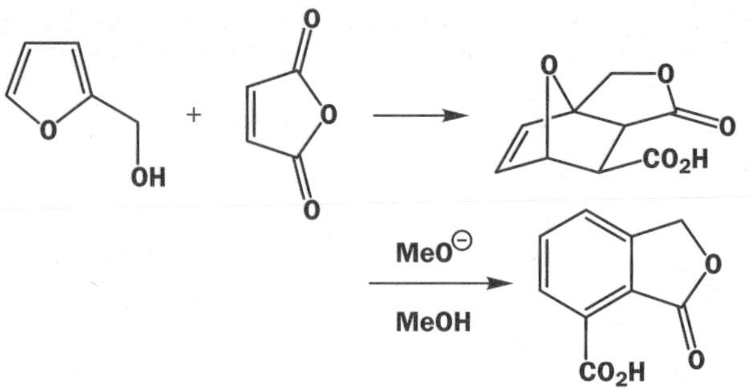

Pericyclic reactions 2: Sigmatropic and electrocyclic reactions

36

Connections

Building on:

- Cycloadditions and the principles of pericyclic reactions (essential reading!) ch35
- Acetal formation ch14
- Conformational analysis ch18
- Elimination reactions ch19
- Controlling alkene geometry ch31

Arriving at:

- The second and third types of pericyclic reaction
- Stereochemistry from chair-like transition states
- Making γ,δ-unsaturated carbonyl compounds
- What determines whether these pericyclic reactions go 'forwards' or 'backwards'
- Special chemistry of N, S, and P
- Why substituted cyclopentadienes are unstable
- What 'con'- and 'dis'-rotatory mean
- Reactions that open small rings and close larger rings

Looking forward to:

- Rearrangements ch37
- Synthesis of aromatic heterocycles ch44
- Main group chemistry ch46–ch47
- Asymmetric synthesis ch45
- Natural products ch51

Cycloadditions, the subject of the last chapter, are just one of the three main classes of pericyclic rearrangement. In this chapter, we consider the other two classes—sigmatropic rearrangements and electrocyclic reactions. We will analyse them in a way that is similar to our dealings with cycloadditions.

Sigmatropic rearrangements

The Claisen rearrangement was the first to be discovered

The original sigmatropic rearrangement occurred when an aryl allyl ether was heated without solvent and an *ortho*-allyl phenol resulted. This is the **Claisen rearrangement**.

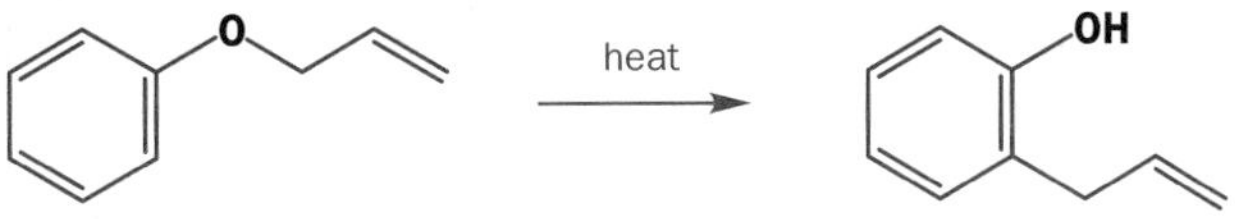

The first step in this reaction is a pericyclic reaction of a type that we will learn to call a [3,3]-**sigmatropic rearrangement**.

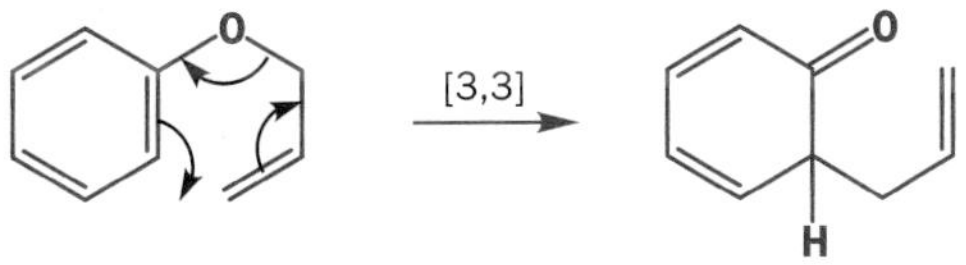

This is a one-step mechanism without ionic intermediates or any charges, just like a cycloaddition. The arrows go round in a ring. The difference between this and a cycloaddition is that one of the arrows starts on a σ bond instead of on a π bond. The second step in the reaction is a simple ionic proton transfer to regenerate aromaticity.

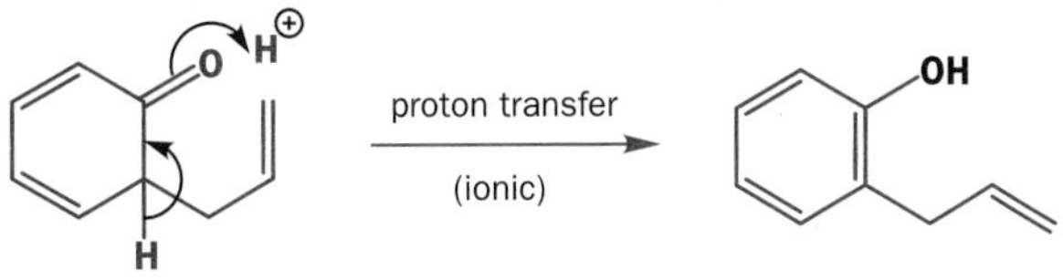

How do we know that this is the mechanism? If the allyl ether is unsymmetrical, it turns 'inside-out' during Claisen rearrangement, as required by the mechanism. Check for yourself that this is right.

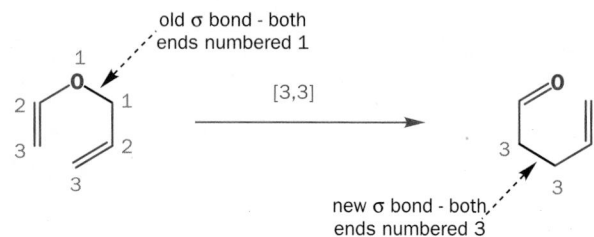

The aliphatic Claisen rearrangement also occurs

It was later found that the same sort of reaction occurs without the aromatic ring. This is called either the **aliphatic Claisen rearrangement** or the **Claisen–Cope rearrangement**. Here is the simplest possible example.

These reactions are called **sigmatropic** because a σ bond appears to move from one place to another during the reaction. The important bonds are coloured black here.

This particular reaction is called a **[3,3]-sigmatropic rearrangement** because the new σ bond has a 3,3 relationship to the old σ bond. You can see this if you number the ends of the old σ bond '1' and '1' and count round to the ends of the new σ bond in the product. You will find that the ends of the new σ bond both have the number '3'.

old σ bond - both ends numbered 1

[3,3]

new σ bond - both ends numbered 3

These [3,3]-sigmatropic rearrangements happen through a chair-like transition state, which allows us both to get the orbitals right and to predict the stereochemistry (if any) of the new double bond. The orbitals look something like this.

old bond broken here   old bond breaking here

[3,3]

new bond formed here   new bond forming here

Note that these do not represent any specific frontier orbitals, they simply show that, in this conformation, the new σ bond is formed from two p orbitals that point directly at each other and that the two new π bonds are formed from orbitals that are already parallel.

Alkene stereochemistry in the Claisen rearrangement comes from a chair-like transition state

Stereochemistry may arise if there is a substituent on the saturated carbon atom next to the oxygen atom. If there is, the resulting double bond strongly favours the *trans* (E) geometry. This is because the substituent prefers an equatorial position on the chair transition state.

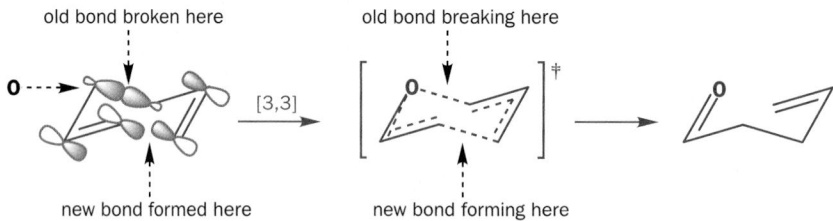

The substituent R prefers an equatorial position as the molecule reacts and R retains this position in the product. The new alkene bond is shown in black and the substituents in green. Notice that the *trans* geometry of the alkene in the product is already there in the conformation chosen by the starting material and in the transition state.

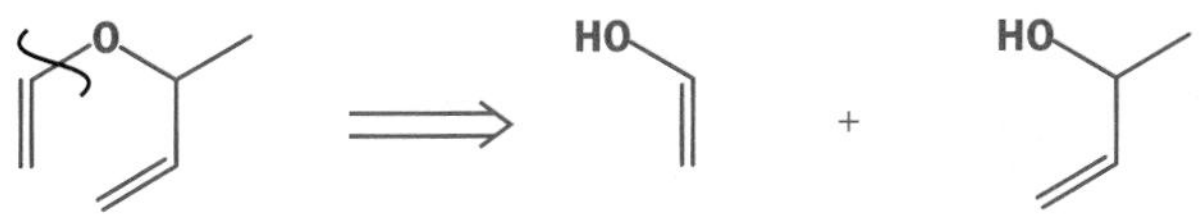

"vinyl alcohol"
= enol of MeCHO

substituted
allyl alcohol

The starting material for these aliphatic Claisen rearrangements consists of ethers with one allyl and one vinyl group. We need now to consider how such useful molecules might be made. There is no problem about the allyl half—allylic alcohols are stable easily made compounds. But what about the vinyl half? 'Vinyl alcohols' are just the enols of aldehydes (MeCHO). The solution is to use an acetal of the aldehyde in an acid-catalysed exchange process with the allylic alcohol.

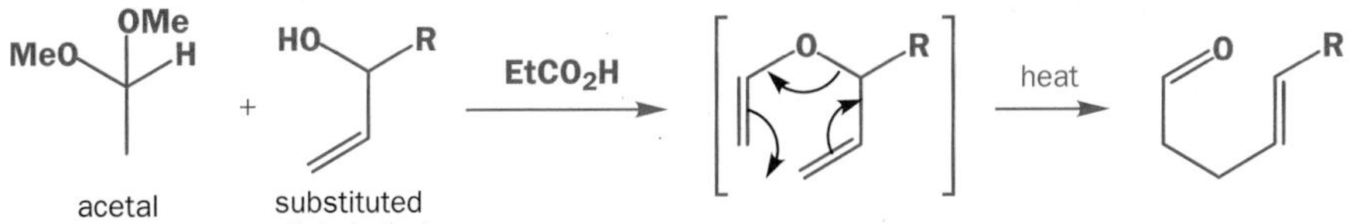

acetal
of MeCHO

substituted
allyl alcohol

It is not necessary to isolate the allyl vinyl ether as long as some of it is formed and rearranges into the final product. The acid catalyst usually used, propanoic acid, has a conveniently high boiling point so that the whole mixture can be equilibrated at high temperature. The first step is an acetal exchange in which the allylic alcohol displaces methanol.

the acetal exchange step

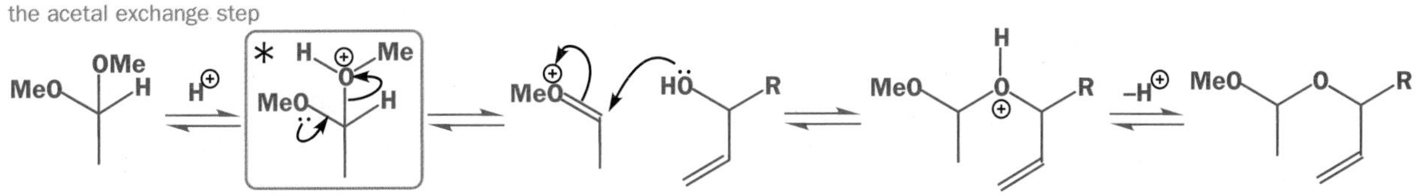

The methanol is distilled off as it is the most volatile of the components in this mixture. A second molecule of methanol is now lost in an acid-catalysed elimination reaction to give the vinyl group.

the E1 elimination step

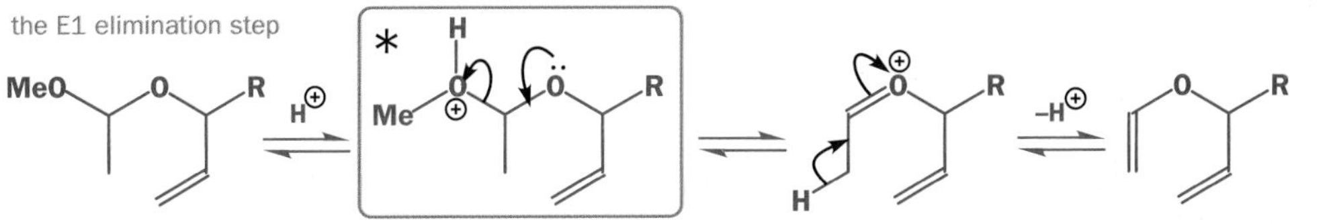

> Note that the first molecule of methanol was displaced in an S_N1 reaction and the second lost in an E**1** reaction. The chemistry of acetals is dominated by the loss of protonated OR or OH groups as in the steps marked *. Never be tempted to use S_N2 mechanisms with acetals.

The Claisen rearrangement is a general synthesis of γ,δ-unsaturated carbonyl compounds

Finally, the [3,3]-sigmatropic rearrangement can be carried out by heat as part of the same step or as a separate step depending on the compounds. This is a very flexible reaction sequence and can be used for aldehydes (as shown above), ketones, esters, or amides. In each case acetal-like compounds are used—acetals themselves for aldehydes and ketones; orthoesters and orthoamides for the other two (though the orthoamides are often called 'amide acetals').

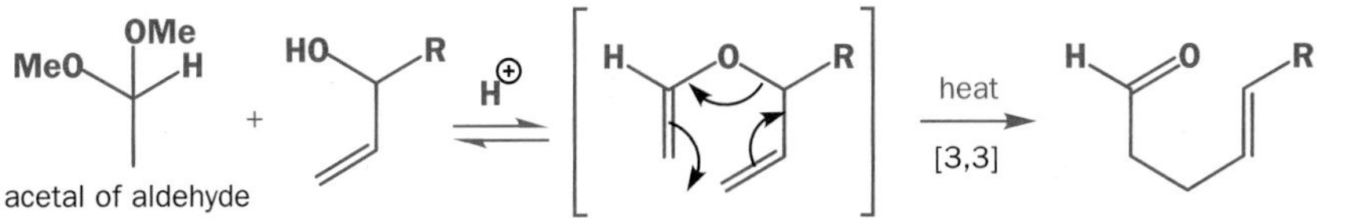

acetal of aldehyde

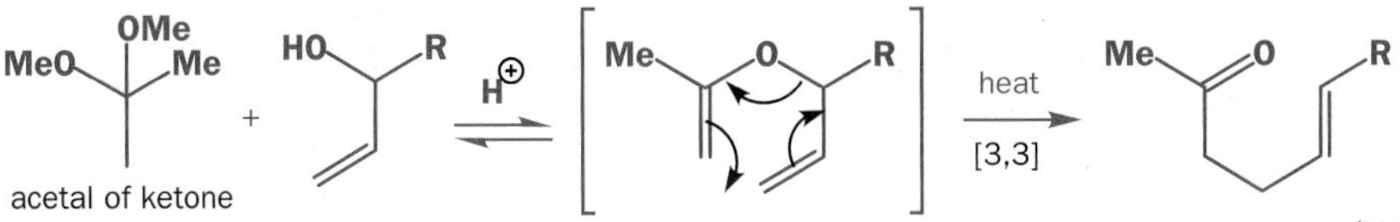

acetal of ketone

(continued overleaf)

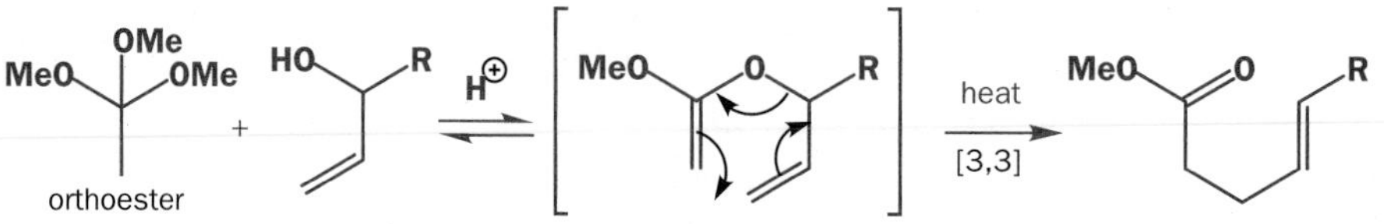

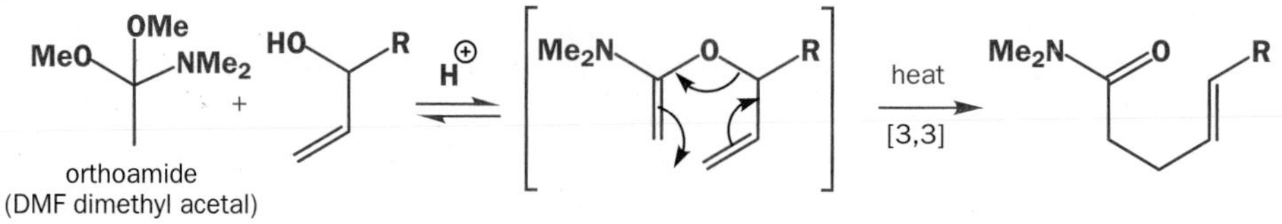

The common feature in the products of these Claisen rearrangements is a γ,δ-unsaturated carbonyl group. If this is what you need in a synthesis, make it by a Claisen rearrangement.

Orbital descriptions of [3,3]-sigmatropic rearrangements

It is possible to give a frontier orbital description of a [3,3]-sigmatropic rearrangement but this is not a very satisfactory treatment because two reagents are not recognizing each other across space as they were in cycloadditions. There are *three* components in these reactions—two nonconjugated π bonds that do have to overlap across space and a σ bond in the chain joining the two π bonds.

The Woodward–Hoffmann rules give a more satisfying description and we shall follow the routine outlined for cycloadditions. Note that for stage 3, we can use the three-dimensional diagram we have already made.

> In a thermal pericyclic reaction the total number of $(4q + 2)_s$ and $(4r)_a$ components must be odd.

1 Draw the mechanism for the reaction (we shall stay with a familiar one)

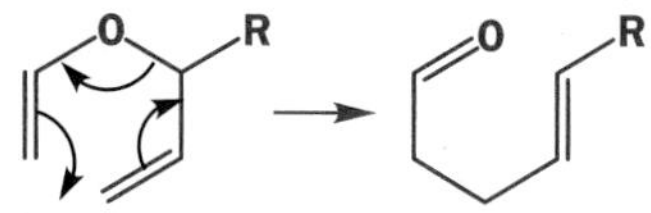

2 Choose the components. All the bonds taking part in the mechanism must be included and no others

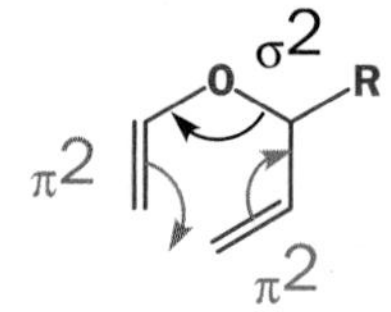

> Note that we have dropped the shading in the orbital from the previous diagrams earlier in the chapter.

3 Make a three-dimensional drawing of the way the components come together for the reaction, putting in orbitals at the ends of the components (only!)

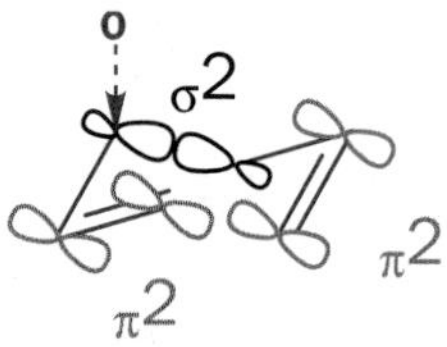

4 Join up the components where new bonds are to be formed. Make sure you join orbitals that are going to form new bonds

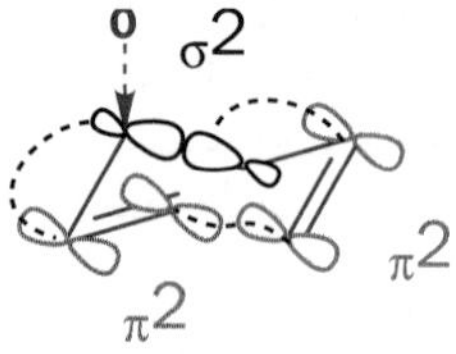

5 Label each component s or a depending whether new bonds are formed on the same or on opposite sides. See below for the σ bond symmetry

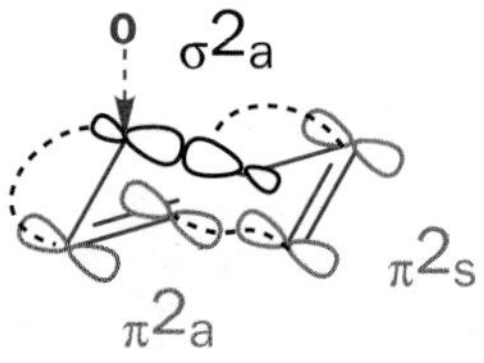

6 Add up the number of $(4q + 2)_s$ and $(4r)_a$ components. If the sum is *odd*, the reaction is allowed

■ There is *one* ($4q + 2)_s$ component (one alkene) and *no* $(4r)_a$ components. Total = 1 so this is an allowed reaction. The $_\pi 2_a$ and $_\sigma 2_a$ components have irrelevant symmetry and are not counted (see Chapter 35 for a full explanation).

$_\pi 2_s$ 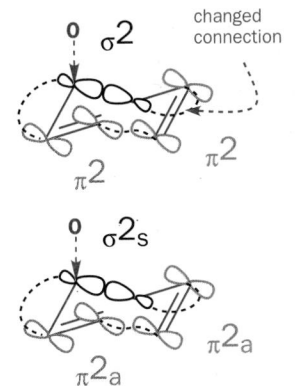 $_\pi 2_a$

■ If you are interested in the frontier orbital approach to [3,3]-sigmatropic reactions, you could read about it in Ian Fleming (1976). *Frontier orbitals and organic chemical reactions*. Wiley, Chichester. We shall use that approach when we come to [1,5]-sigmatropic rearrangements.

One new aspect of orbital symmetry has appeared in this diagram—how did we deduce a or s symmetry in the way the σ bond reacted? For π bonds it is simple—if both bonds are formed on the same side of the old π bond, it has reacted suprafacially; if on opposite sides, antarafacially.

With a σ bond the symmetry is not so obvious. We want to know if it does the *same* thing at each end (s) or a *different* thing (a). But what is the 'thing' it does? It reacts using the large lobe of the sp³ orbital (retention) or the small lobe (inversion). If it reacts with retention at both ends or inversion at both ends, it reacts *supra*facially, while if it reacts with retention at one end and inversion at the other, it reacts *antara*facially. There are four possibilities.

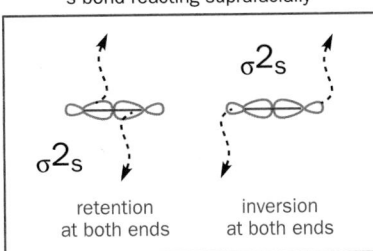

s bond reacting suprafacially

$_\sigma 2_s$ — retention at both ends

$_\sigma 2_s$ — inversion at both ends

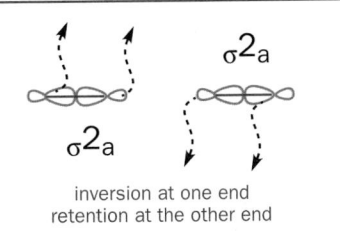

s bond reacting antarafacially

$_\sigma 2_a$

$_\sigma 2_a$ — inversion at one end retention at the other end

In the routine above, we chose to use our σ bond so that we got inversion at one end and retention at the other. That was why we identified it as an antarafacial component. If we had chosen another style we should have got different descriptions of the components, but the reaction would still have been allowed—for example, changing just one connecting line.

This changes the symmetry of the σ bond so that it becomes a $_\sigma 2_s$ component but it also changes the symmetry of one of the π bonds so that it becomes a $_\pi 2_a$ component. The net result is still only one component of the Woodward–Hoffmann symmetry, the sum is still one, and the reaction still allowed.

changed connection

$_\sigma 2$

$_\pi 2$ $_\pi 2$

$_\sigma 2_s$

$_\pi 2_a$ $_\pi 2_a$

number of $(4q + 2)_s$ components: 1
number of $(4r)_a$ components: 0
sum = 1
so reaction is allowed thermally

The direction of [3,3]-sigmatropic rearrangements

Orbital symmetry tells us that [3,3]-sigmatropic rearrangements are allowed but says nothing about which way they will go. They are allowed in either direction. So why does the Claisen–Cope rearrangement always go in this direction?

Think back to our discussion on enols and you may recall that the combination of a carbonyl group and a C–C σ bond made the keto form more stable than the enol form with its combination of a C=C π bond and a C–O σ bond. The same is true here. It is the formation of the carbonyl group that drives the reaction to the right.

The **Cope rearrangement** is a [3,3]-sigmatropic rearrangement with only carbon atoms in the ring. In its simplest version it is not a reaction at all.

heat
[3,3]

Directing the Cope rearrangement by the formation of a carbonyl group

The starting material and the product are the same. We can drive this reaction too by the formation of a carbonyl group if we put an OH substituent in the right place.

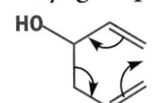

heat
[3,3]

The product of the sigmatropic step is the enol of the final product. It turns out that the reaction is accelerated if the starting alcohol is treated with base (KH is the best) to make the alkoxide. The product is then the potassium enolate, which is more stable than the simple potassium alkoxide starting material. As the reaction proceeds, conjugation is growing between O^- and the new π bond.

Bredt's rule forbidding bridgehead alkenes and the reasons are discussed in Chapter 19.

Some remarkable compounds can be made by this method. One of the strangest—a 'bridgehead' alkene—was made by a potassium-alkoxide-accelerated Cope rearrangement in which a four-membered ring was expanded into an eight-membered ring containing a *trans* double bond.

A combination of an oxygen atom in the ring and another one outside the ring is very powerful at promoting [3,3]-sigmatropic rearrangements and easy to arrange by making the lithium enolate of an ester of an allylic alcohol.

Sometimes it is better to convert the lithium enolate into the silyl enol ether before heating to accomplish the [3,3]-sigmatropic rearrangement. In any case, both products give the unsaturated carboxylic acid on work-up.

This reaction is known as the **Ireland–Claisen rearrangement** as it was a variation of the Claisen rearrangement invented by R.E. Ireland in the 1970s and widely used since. If the substituents are suitably arranged, it shows the same *E* selectivity as the simple Claisen rearrangement and for the same reason.

In some cases simple Cope rearrangements without any oxygen atoms at all can be directed by an unstable starting material or a stable product. The instability might be strain and the stability might simply be more substituents on the double bonds. In this case the driving force is the breaking of a weak σ bond in a three-membered ring. This reaction goes in 100% yield at only just above room temperature, so it is very favourable.

In this second example, the trisubstituted double bonds inside the five-membered rings of the product are more stable than the exomethylene groups in the starting material.

An industrial synthesis of citral

'Citral' is a key intermediate in the synthesis of vitamin A, and in Chapter 31 you had a go at designing a synthesis of it. BASF manufacture citral by a remarkable process that involves two successive [3,3]-sigmatropic rearrangements, a Claisen followed by a Cope.

The allyl vinyl ether needed for the Claisen rearrangement is an enol ether of an unsaturated aldehyde with an unsaturated alcohol. The two starting materials are themselves derived from a common precursor, making this a most efficient process! Heating the enol ether promotes [3,3]-sigmatropic rearrangement propelled by the formation of a carbonyl group.

But the product of this rearrangement is now set up for a second [3,3]-sigmatropic rearrangement, this time made favourable by a shift into conjugation and the formation of two trisubstituted double bonds from two terminal ones. Overall, the prenyl group walks from one end of the molecule to the other, inverting twice as it goes.

Sex for seaweeds censored by a [3,3]-sigmatropic rearrangement

In order to reproduce, the female gametes of marine brown algae must attract mobile male gametes. This they do by releasing a pheromone, long thought to be the cycloheptadiene ectocarpene. In 1995 results were published that suggested that, in fact, the pheromone was a cyclopropane, and that ectocarpene was ineffective as a pheromone

How had the confusion arisen? Well, the remarkable thing is that the cyclopropyl pheromone inactivates itself, with a half-life of several minutes at ambient temperature, by [3,3]-sigmatropic rearrangement to the cycloheptadiene, driven by release of strain from the three-membered ring. This not only confused the earlier pheromone chemists, but it also provides a marvellously precise way for the algae to signal their presence and readiness for reproduction without saturating the sea water with meaningless pheromone.

Applications of [3,3]-sigmatropic rearrangements using other elements

There is no need to restrict our discussion to carbon and oxygen atoms. We shall finish this section with two useful reactions that use other elements. The most famous synthesis of indoles is a nineteenth century reaction discovered by Emil Fischer—the **Fischer indole synthesis**—and it would be a remarkable discovery even today. Reaction of phenylhydrazine with a ketone in slightly acidic solution gives an imine (Chapter 14) called a phenylhydrazone.

phenylhydrazine

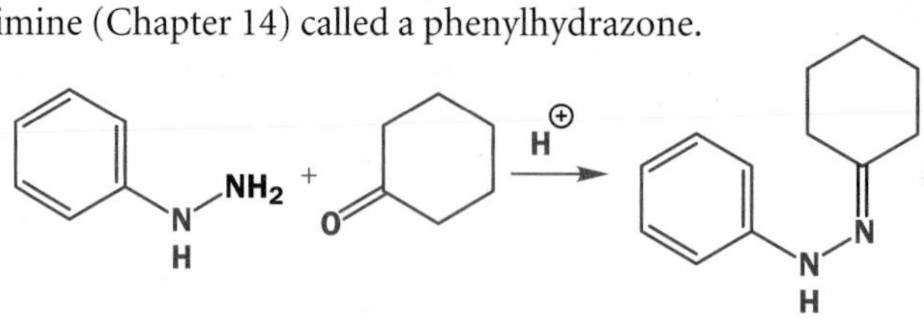

a phenylhydrazone

If the ketone is enolizable, this imine is in equilibrium with the corresponding enamine. The important bonds are given in black in the diagram.

a phenylhydrazone

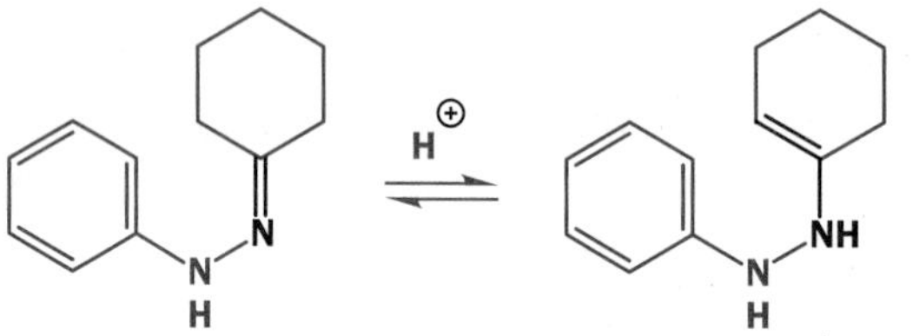

an enamine

The enamine is ideally set up for a [3,3]-sigmatropic rearrangement in which the σ bond to be broken is the weak N–N σ bond and one of the π bonds is in the benzene ring.

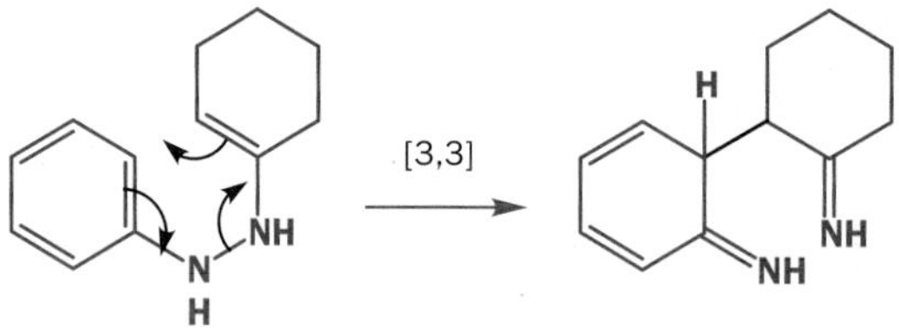

The product is a highly unstable double imine. Aromaticity is immediately restored and a series of proton shifts and C–N bond formation and cleavage give the aromatic indole. In the last diagram the ten-π-electron indole is outlined in black.

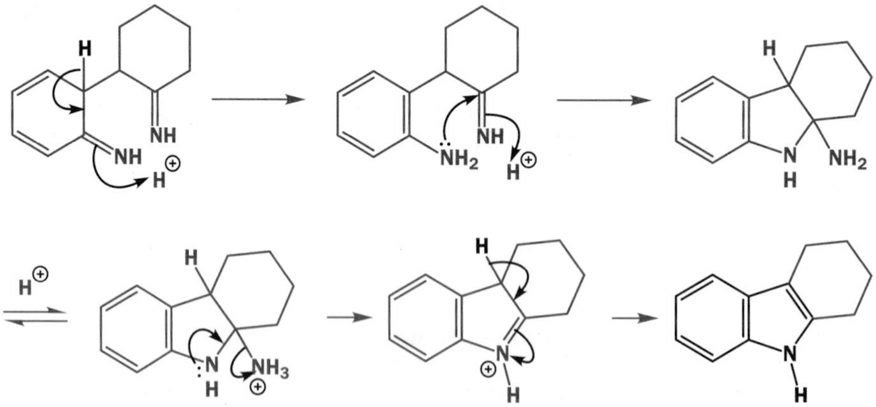

A detailed discussion of this reaction as a synthesis of indoles appears in Chapter 44.

Indoles are of some importance in biology and medicine and the Fischer indole synthesis is widely used. Sometimes the complete reaction occurs, as in this example, under the slightly acidic conditions needed to make the phenylhydrazone. More commonly, the phenylhydrazone is isolated and converted into the indole with a Lewis acid such as $ZnCl_2$.

That was a [3,3]-sigmatropic reaction involving two nitrogens. There follows one with two oxygens and a chromium atom. When tertiary allylic alcohols are oxidized with CrO_3 in acid solution, no direct oxidation can take place, but a kind of conjugate oxidation occurs.

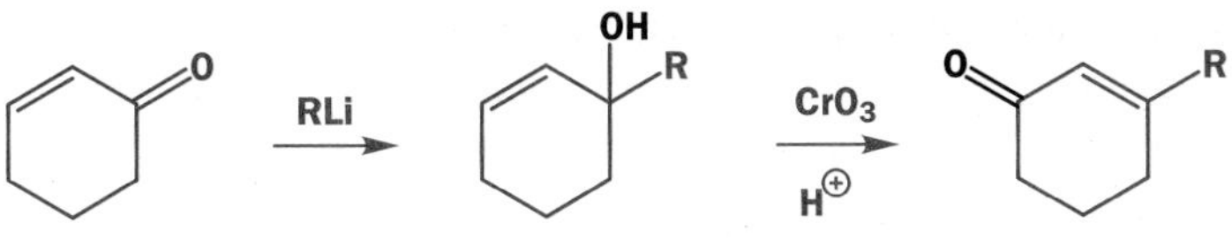

The first step in Cr(VI) oxidations can take place to give a chromate ester (Chapter 24) but this intermediate has no proton to lose so it transfers the chromate to the other end of the allylic system where there is a proton. The chromate transfer can be drawn as a [3,3]-sigmatropic rearrangement.

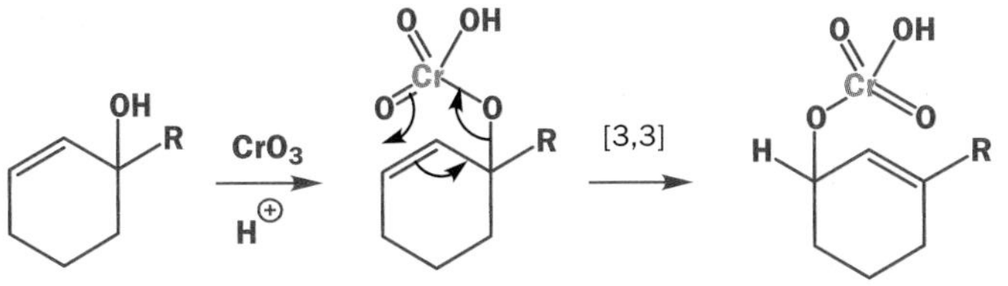

The final step is the normal oxidation (Chapter 24) in which chromium drops down from orange Cr(VI) to Cr(IV) and eventually by disproportionation to green Cr(III).

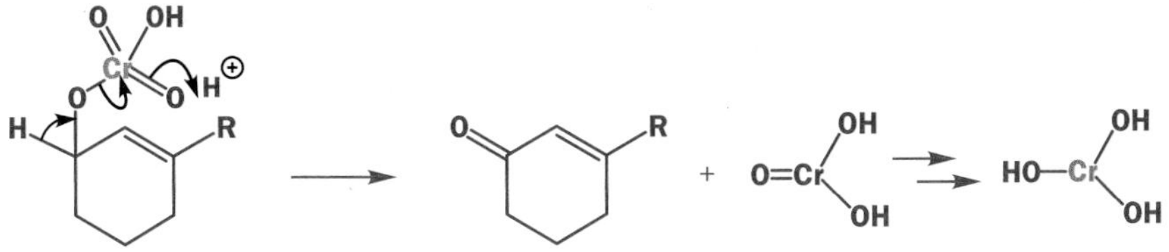

[2,3]-Sigmatropic rearrangements

All [3,3]-sigmatropic rearrangements have six-membered cyclic transition states. It is no accident that the size of the ring is given by the sum of the two numbers in the square brackets as this is universally the case for sigmatropic rearrangements. We are now going to look at [2,3]-sigmatropic rearrangements so we will be needing five-membered cyclic transition states. There is a problem here. You cannot draw three arrows going round a five-membered ring without stopping or starting on an atom. One way to do this is to use a carbanion.

The starting material is a benzyl allyl ether and undergoes [2,3]-sigmatropic rearrangement to make a new C–C σ bond at the expense of a C–O σ bond—a bad bargain this as the C–O bond is stronger. The balance is tilted by the greater stability of the oxyanion in the product than of the carbanion in the starting material. The new bond has a 2,3 relationship to the old and the transition state is a five-membered ring.

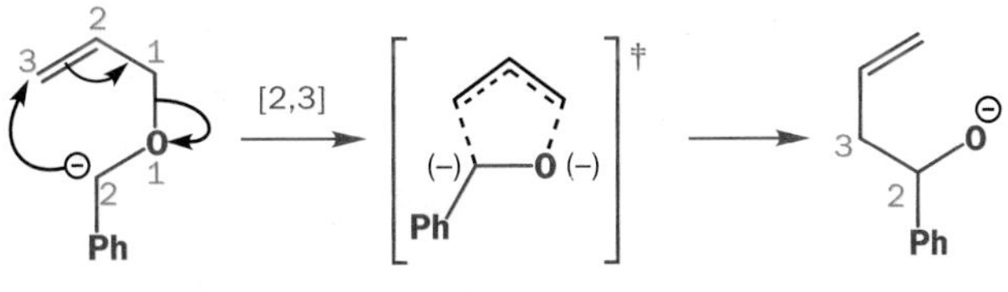

The transition state can be quite chair-like so that the new π bond will be *trans* if it has a choice. There will be a choice if the ether has been made from a substituted allyl alcohol.

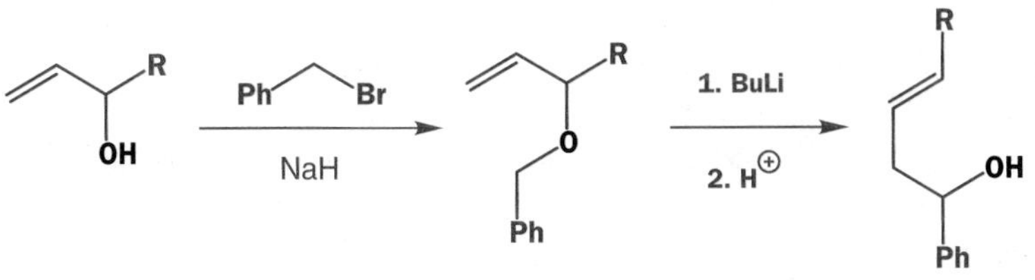

We cannot draw a complete chair as we would need a six-membered ring for that (see discussion of [3,3]-sigmatropic rearrangements above), but the part that is to become the new π bond can be in a chair-like part of the five-membered ring. The substituent R prefers an equatorial position and the resulting *trans* arrangement of the groups is outlined in black.

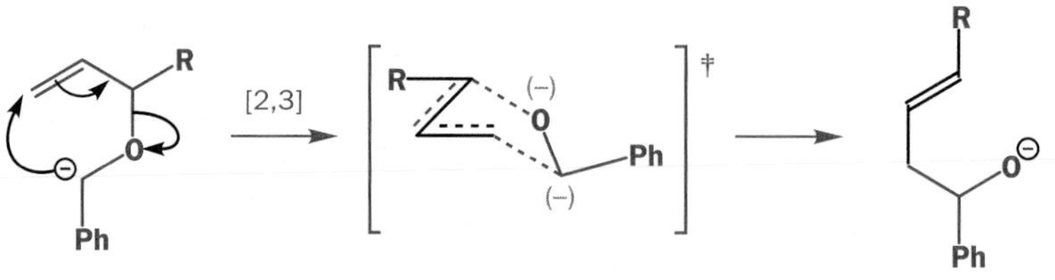

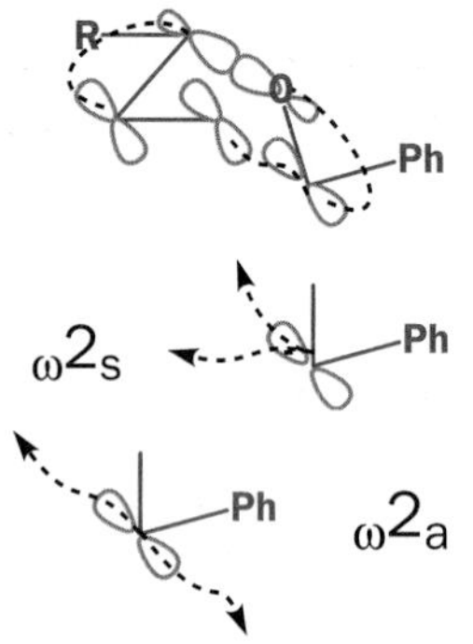

We can use the same conformational diagram to show how the orbitals overlap as the new bond is formed.

When we come to use the Woodward–Hoffmann rules on these [2,3]-sigmatropic rearrangements, we find something new. We have a π bond and a σ bond and a carbanion. How are we to represent a carbanion (or a carbocation) that is just a p orbital on an atom? The new symbol we use for a simple p orbital is ω. A carbanion is an $_{\omega}2$ component and a carbocation is an $_{\omega}0$ component as it has zero electrons. If the two new bonds are formed to the same lobe of the p orbital of the carbanion, we have an $_{\omega}2_s$ component but, if they are formed to different lobes, we have an $_{\omega}2_a$ component.

Without going through the whole routine again, the [2,3]-sigmatropic rearrangement we have been discussing can be described as an $_{\omega}2_a + _{\sigma}2_s + _{\pi}2_a$ reaction. There is one $(4q + 2)_s$ and no $(4r)_a$ components so the reaction is thermally allowed.

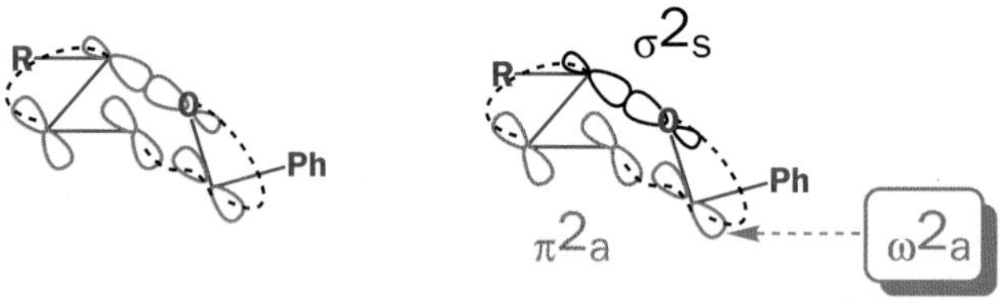

Sulfur is good at [2,3]-sigmatropic rearrangements

There are many [2,3]-sigmatropic rearrangements involving a variety of heteroatoms as well as carbon. We shall describe just one more because it involves no ions at all. The key is an element that is prepared to change its oxidation state by two so that we can start and finish an arrow on that element. The element is sulfur, which can form stable compounds at three oxidation states: S(II), S(IV), or S(VI).

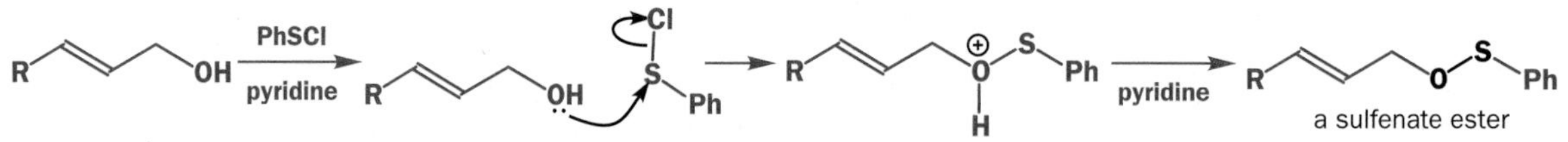

Reaction of an allylic alcohol with PhSCl gives an unstable sulfenate ester that rearranges on heating to an allylic sulfoxide by a [2,3]-sigmatropic rearrangement involving both O and S.

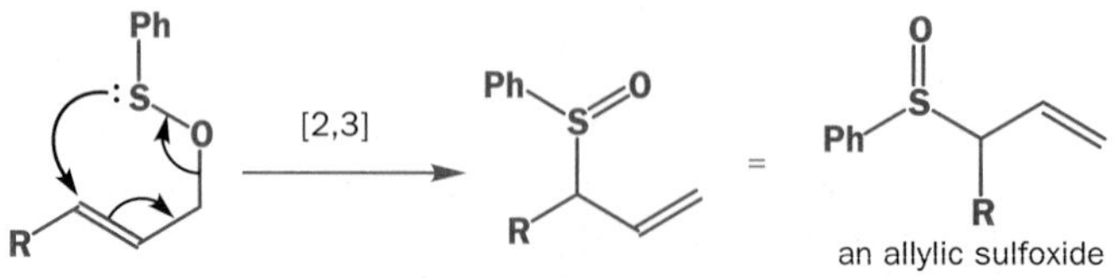

Notice that arrows both start and stop on the sulfur atom, which changes from S(II) to S(IV) during the reaction. The new functional group with an S=O bond is called a **sulfoxide**. This is a good preparation of allylic sulfoxides. The product forms an anion stabilized by sulfur, which can be alkylated.

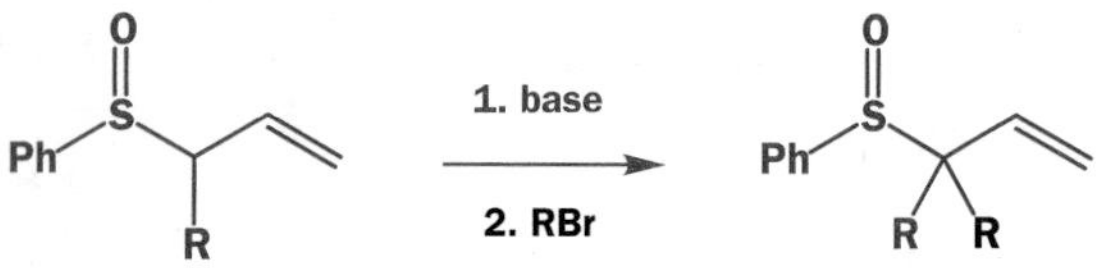

We have said that all these sigmatropic rearrangements are reversible but now we can prove it. If this product is heated in methanol with a nucleophile such as $(MeO)_3P$, which has a liking for sulfur, the [2,3]-sigmatropic rearrangement runs backwards and a sulfenate ester is again formed.

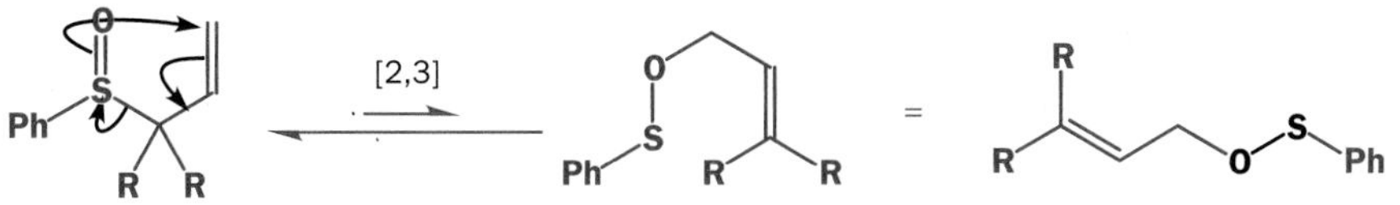

This is an unfavourable reaction, because the equilibrium lies over on the sulfoxide side. But the nucleophile traps the sulfenate ester and the methanol ensures that the alkoxide ion formed is immediately protonated so that we get another allylic alcohol.

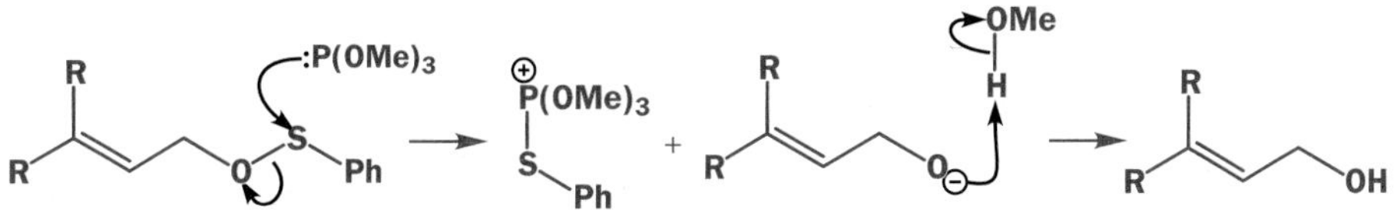

> The other products are actually PhSMe and $(MeO)_3P=O$. You might like to work out a mechanism for these stages of the reaction.
>

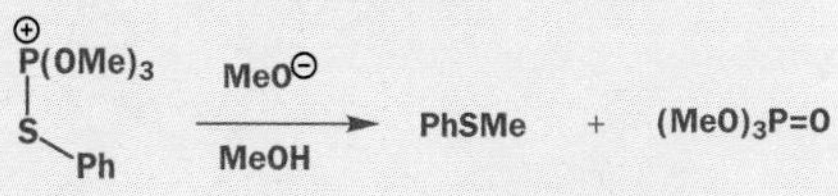

So what is the point of going round in circles like this? The net result is the alkylation of an allylic alcohol in a position where alkylation would not normally be considered possible.

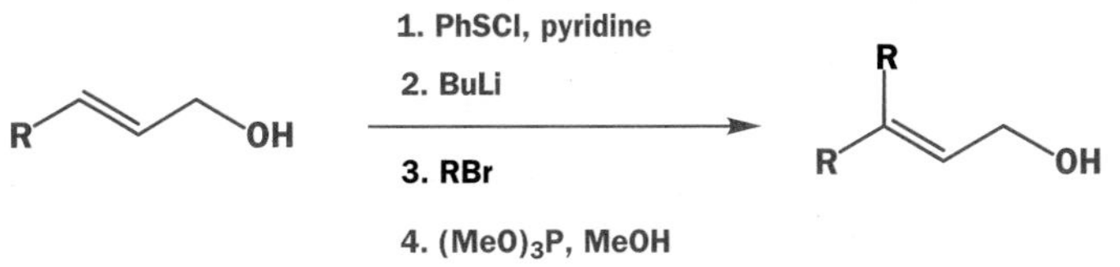

1. PhSCl, pyridine
2. BuLi
3. RBr
4. $(MeO)_3P$, MeOH

[1,5]-Sigmatropic hydrogen shifts

When one of the numbers in square brackets is '1', the old and new σ bonds are to the same atom, so we are dealing with the migration of a group around a conjugated system. In the case of a [1,5] shift the transition state is a six-membered ring (remember—just add together the numbers in square brackets). Here is an important example.

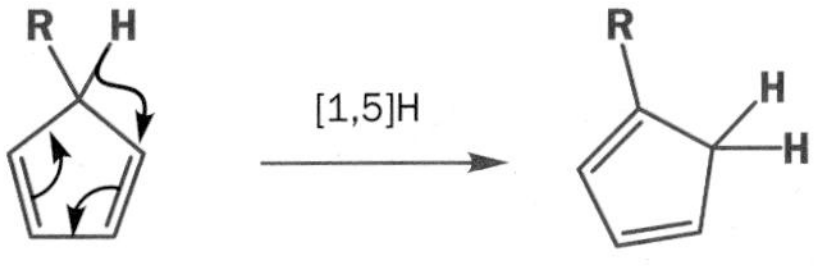

Let us first check that this is indeed a [1,5]-sigmatropic rearrangement by numbering the position of the new σ bond with respect to the old. Note that we must go the long way round the five-membered ring because that is the way the mechanism goes.

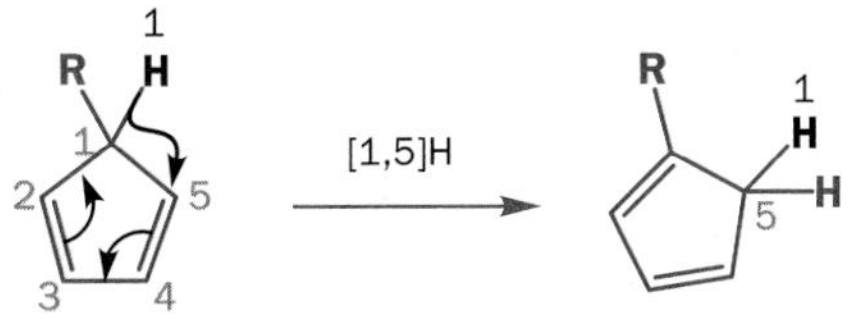

It *is* a [1,5]-sigmatropic rearrangement. The figure '1' in the square brackets shows that the same atom is at one end of the new σ bond as was at one end of the old σ bond. One atom has moved in a 1,5 manner and these are often called [1,5]-sigmatropic *shifts*. This is often abbreviated to **[1,5]H shift** to show which atom is moving. This particular example is important because sadly it prohibits a most attractive idea. The cyclopentadiene anion is very stable (Chapter 8) and can easily be alkylated. The sequence of alkylation and Diels–Alder reaction looks very good.

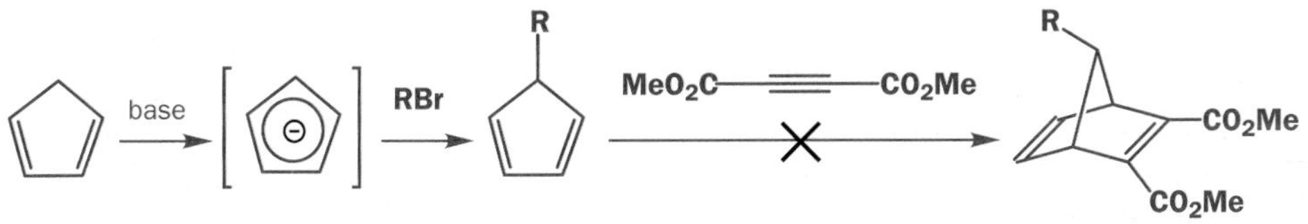

Sadly this sequence is, in fact, no good at all. A mixture of three Diels–Alder adducts is usually obtained resulting from addition to the three cyclopentadienes present in solution as the result of rapid [1,5]H shifts. The one drawn above is a minor product because there is more of the other two dienes, which have an extra substituent on the double bonds.

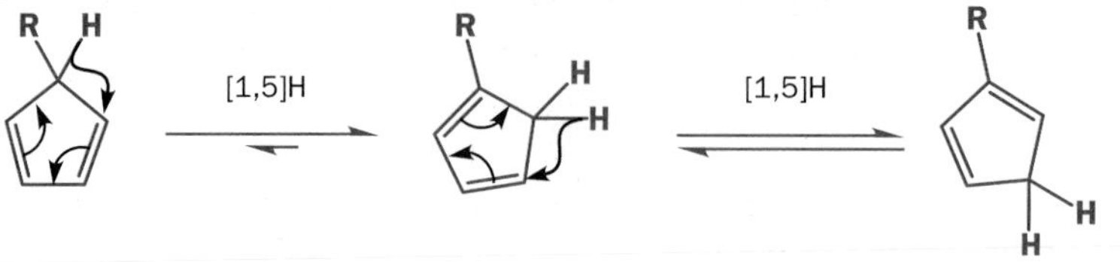

An excellent example comes from the intramolecular Diels–Alder reactions explored by Dreiding in 1983. One particular substituted cyclopentadiene was made by a fragmentation reaction (see Chapter 38). It might have been expected to give a simple Diels–Alder adduct.

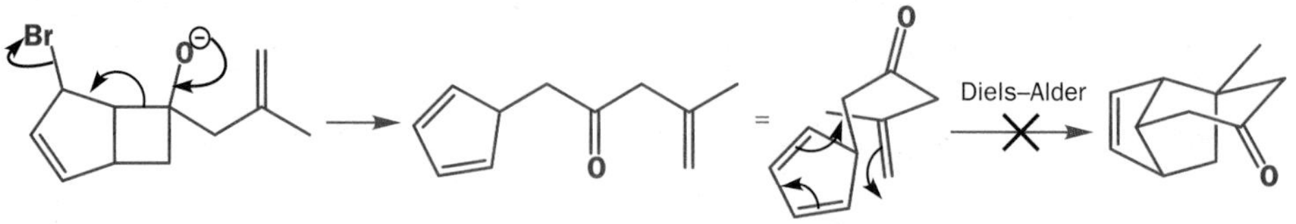

There is nothing wrong with this reaction; indeed, the product looks beautifully stable, but it is not formed because the [1,5]H shift is too quick and gives a more stable cyclopentadiene with more substituents on a double bond. *Then* it does the Diels–Alder reaction.

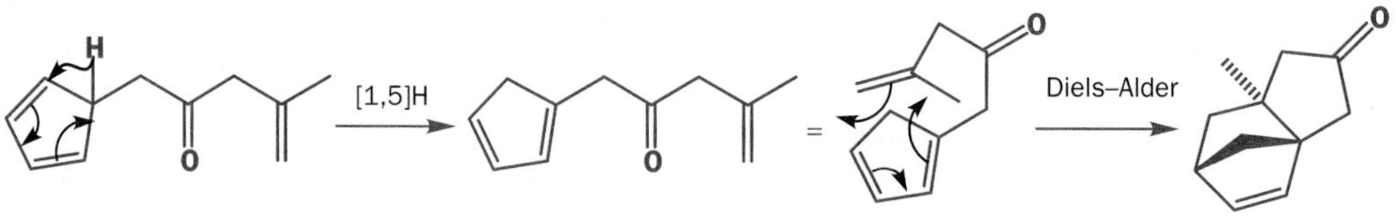

Notice that in these compounds the ketone is not conjugated to any of the alkenes and so does not influence the reaction. If we increase the reactivity of the dienophile by putting an ester group in conjugation with it, most of the compound does the Diels–Alder reaction *before* it does the [1,5]H shift.

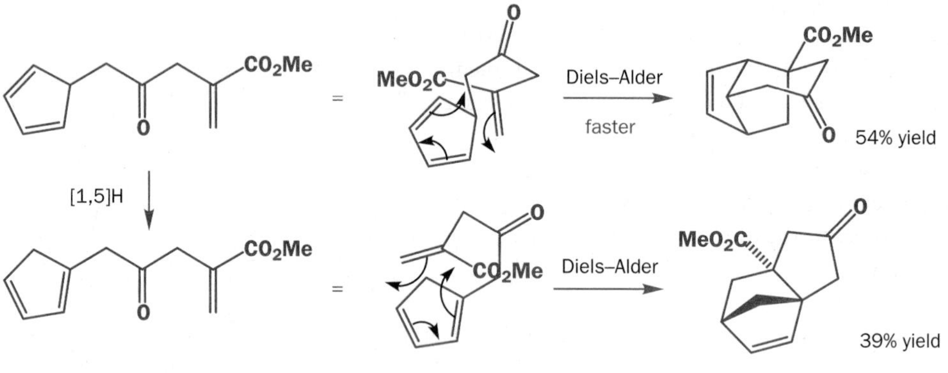

Orbital description for the [1,5]H sigmatropic shift

▶

You should satisfy yourself that the other frontier orbital combination—HOMO of the diene and LUMO of the C–H bond— works equally well.

It is equally satisfactory to use frontier orbitals or the Woodward–Hoffmann rules for these reactions. We can take the diene as one component (HOMO or LUMO or $\pi 4$) and the C–H bond as the other (LUMO or HOMO or $\sigma 2$). Let us start by using the LUMO of the diene and the HOMO of the C–H bond.

If the circle around the H atom surprised you, perhaps it will also remind you that hydrogen has only a 1s orbital which is spherical. You can probably see already that all the orbitals are correctly lined up for the reaction.

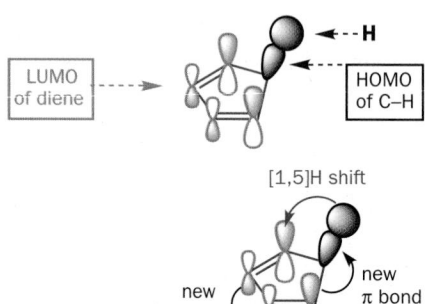

The hydrogen atom slides across the top face of the planar cyclopentadiene ring. We call this a **suprafacial migration**. This name has got nothing to do with the components in the Woodward–Hoffmann rules—it just means that the migrating group leaves from one face of the π system and rejoins that same face (the top face in this example). **Antarafacial migration** would mean leaving the top face and rejoining the bottom face—a clear impossibility here.

If you use the Woodward–Hoffmann rules, you need to note that the hydrogen atom must react with retention. The 1s orbital is spherically symmetrical and has no node, so wherever you draw the dotted line from that orbital it always means retention. Choosing the components is easy—the diene is a $_\pi 4$ and the C–H bond a $_\sigma 2$ component.

The easiest way to join them up is to link the hydrogen atom's 1s orbital to the top lobe of the p orbital at the back of the diene and the black sp^3 orbital to the top lobe at the front of the diene. This gives us $_\pi 4_s$ and $_\sigma 2_s$ components and there is one $(4q + 2)_s$ and no $(4r)_a$ components so the sum is odd and the reaction is allowed. Both approaches give us the same picture—a suprafacial migration of the hydrogen atom with (inevitably) retention at the migrating group.

These [1,5]-sigmatropic shifts are not restricted to cyclopentadienes. In Chapter 35 we bemoaned the lack of Diels–Alder reactions using *E,Z*-dienes. One reason for this dearth is that such dienes undergo [1,5]H shifts rather easily and mixtures of products result.

The complete rules for sigmatropic hydrogen shifts are simple. In thermal reactions, [1,5]H shifts occur suprafacially but [1,3]H and [1,7]H shifts must be antarafacial. It is just as well that antarafacial [1,3]H shifts are impossible (though allowed) as otherwise double bonds would wander about organic molecules like this.

Antarafacial [1,3]H shifts are impossible because a rigid three-carbon chain is too short to allow the H atom to transfer from the top to the bottom—the H atom just can't reach. When we come to [1,7]H shifts, the situation is different. Now the much longer chain is just flexible enough to allow the transfer.

The hydrogen atom leaves the top side of the triene and adds back in on the bottom side. Antarafacial migration is allowed and possible.

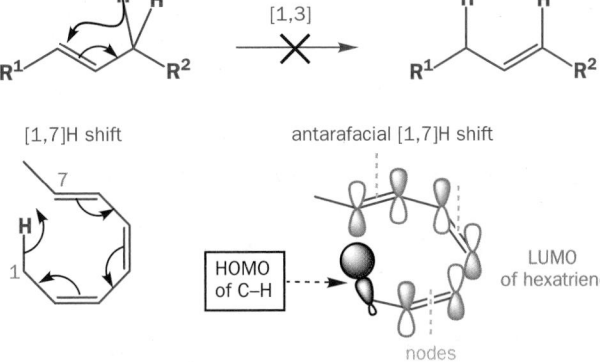

Summary of thermal sigmatropic hydrogen shifts

	[1,3]H shift	[1,5]H shift	[1,7]H shift
stereochemistry	antarafacial	suprafacial	antarafacial
feasibility	impossible	easy	possible

Photochemical [1,*n*]H sigmatropic shifts follow the opposite rules

As you should by now expect (p. 927), all this is reversed in photochemical reactions. Here is an example of a [1,7]H shift that cannot occur antarafacially because the molecule is a rigid ring, but that can and does occur photochemically.

A [1,7]H shift occurs in the final stages of the human body's synthesis of vitamin D from cholesterol. Here is the last step of the biosynthesis.

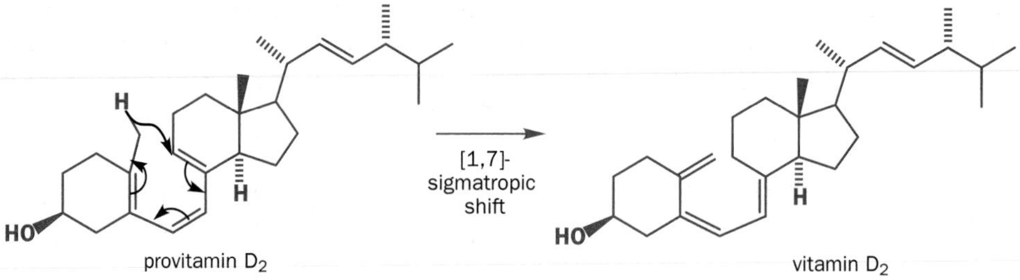

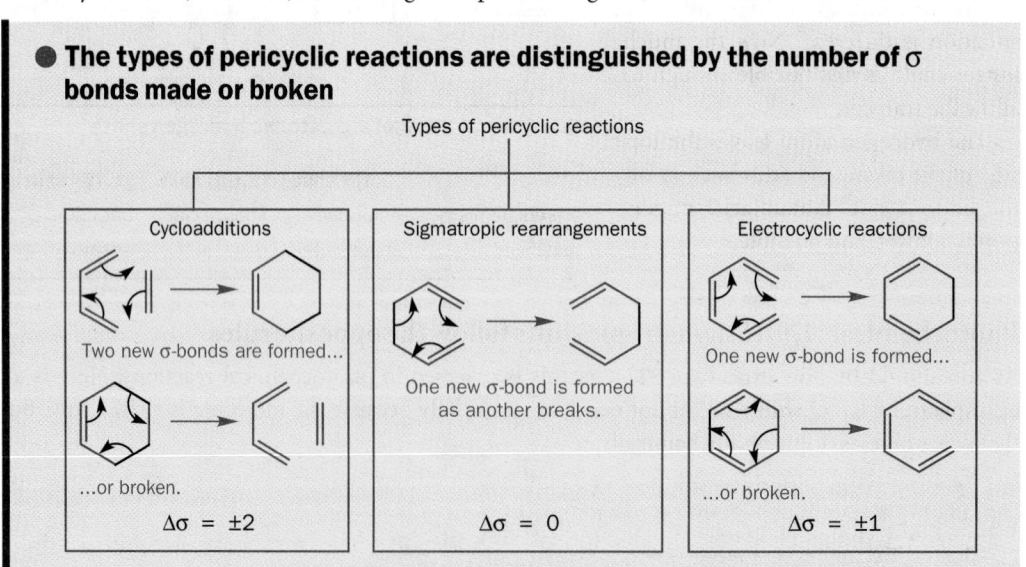

provitamin D₂ → [1,7]-sigmatropic shift → vitamin D₂

This step happens spontaneously, without the need for light, so the shift must be antarafacial. The reason the body *does* need light to make vitamin D is the previous step, which only occurs when light shines on the skin.

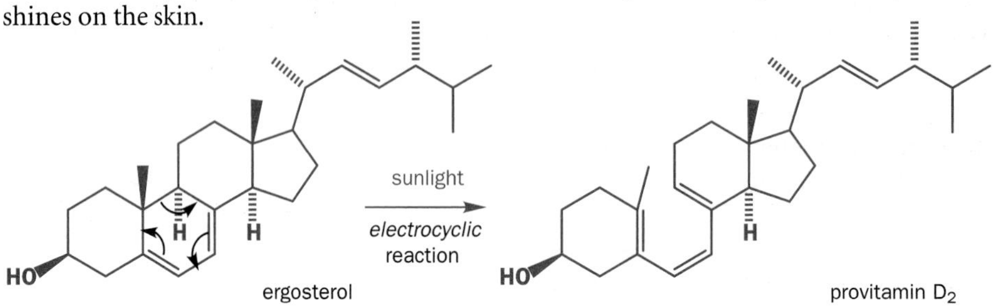

ergosterol → sunlight *electrocyclic reaction* → provitamin D₂

This ring opening is clearly pericyclic—the electrons go round in a ring, and the curly arrows could be drawn either way—but it is neither a cycloaddition (only one π system is involved) nor a sigmatropic rearrangement (a σ bond is broken rather than moved). It is, in fact, a member of the third and last kind of pericyclic reaction, an *electrocyclic reaction.*

Electrocyclic reactions

In an **electrocyclic reaction** a ring is always broken or formed. Rings may, of course, be formed by cycloadditions as well, but the difference with electrocyclic reactions is that just one new σ bond is formed (or broken) across the ends of a single conjugated π system. In a cycloaddition, two new σ bonds are always formed (or broken), and in a sigmatropic rearrangement one σ bond forms while one breaks.

● **The types of pericyclic reactions are distinguished by the number of σ bonds made or broken**

Types of pericyclic reactions

Cycloadditions	Sigmatropic rearrangements	Electrocyclic reactions
Two new σ-bonds are formed...	One new σ-bond is formed as another breaks.	One new σ-bond is formed...
...or broken.		...or broken.
Δσ = ±2	Δσ = 0	Δσ = ±1

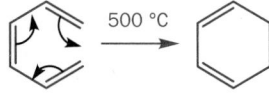

500 °C

One of the simplest electrocyclic reactions occurs when hexatriene is heated to 500 °C.

It is a pericyclic reaction because the electrons go round in a ring (you could equally draw the arrows going the other way); it's electrocyclic because a new σ bond is formed across the ends of a π system. The reaction goes because the σ bond that is formed is stronger than the π bond that is lost. The opposite is true for the electrocyclic reaction shown in the margin—ring strain in the four-membered ring means that the reverse (ring-opening) reaction is preferred to ring closure.

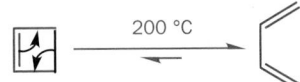

Rules for electrocyclic reactions

Whether they go in the direction of ring opening or ring closure, electrocyclic reactions are subject to the same rules as all other pericyclic reactions—you saw the same principle at work in Chapter 35 where we applied the Woodward–Hoffmann rules both to cycloadditions and to reverse cycloadditions. With most of the pericyclic reactions you have seen so far, we have given you the choice of using either HOMO–LUMO reasoning or the Woodward–Hoffmann rules. With electrocyclic reactions, you really have to use the Woodward–Hoffmann rules because (at least for the ring closures) there is only one molecular orbital involved.

▶

In one famous case, the release of ring strain is almost exactly counterbalanced by the formation of a σ bond at the expense of a π bond. Cycloheptatriene exists in equilibrium with a bicyclic isomer known as norcaradiene. Usually cycloheptatriene is the major component of the equilibrium, but the norcaradiene structure is favoured if R is an electron-withdrawing group.

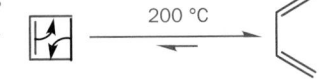

cycloheptatriene (R = H)   norcaradiene (R = H)

● **Electrocyclic reactions**

- An **electrocyclic reaction** is the formation of a new σ bond across the ends of a conjugated polyene or the reverse

  It is important that you do not confuse electrocyclic reactions with pericyclic reactions. Pericyclic is the name for the family of reactions involving no charged intermediates in which the electrons go round the outside of the ring. *Electrocyclic* reactions, *cycloadditions*, and *sigmatropic* rearrangements are the three main classes of *pericyclic* reactions.

Let's start with the hexatriene ring closure, first looking at the orbitals, and then following the same procedure that we taught you for cycloadditions and sigmatropic rearrangements to see what the Woodward–Hoffmann rules have to say about the reaction. As a preliminary, we should just note that hexatriene is, of course, a 6 π electron ($_\pi 6$) conjugated system and, on forming cyclohexadiene, the end two orbitals have to form a σ bond.

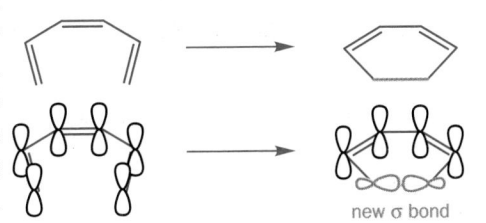

new σ bond

▶

Reminder. In a thermal pericyclic reaction the total number of $(4q + 2)_s$ and $(4r)_a$ components must be odd.

So, now for the Woodward–Hoffmann treatment.

1 Draw the mechanism for the reaction

2 Choose the components. All the bonds taking part in the mechanism must be included and no others

$\pi 6$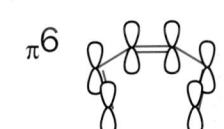

3 Make a three-dimensional drawing of the way the components come together for the reaction, putting in orbitals at the ends of the components (only!)

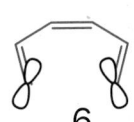
$\pi 6$

4 Join up the components where new bonds are to be formed. Make sure you join orbitals that are going to form new bonds

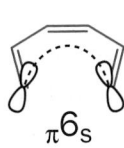
$\pi 6_s$

5   Label each component s or a depending on whether new
bonds are formed on the same or on opposite sides

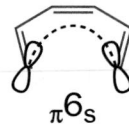

6   Add up the number of $(4q + 2)_s$ and $(4r)_a$ components. If the
sum is *odd*, the reaction is allowed

> ■
> There is **one** $(4q + 2)_s$ component and
> **no** $(4r)_a$ components. Total = 1 so this
> is an allowed reaction

Notice that we called the reaction 's' because the top halves of the two π orbitals were
joining together. We can give the same treatment to the cyclobutene ring-opening reaction—
the Woodward–Hoffmann rules tell us nothing about which way the reaction will go, only if
the reaction is allowed, and it is invariably easier with electrocyclic reactions to consider the ring-
closing reaction even if the ring opening is favoured thermodynamically. This is the process we need
to consider.

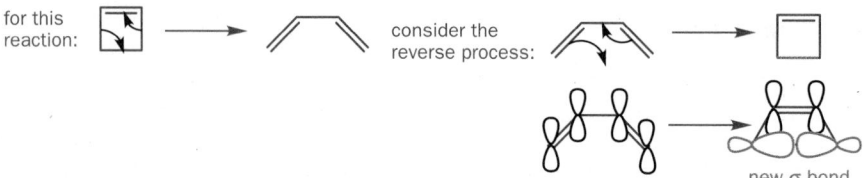

new σ bond

And the Woodward–Hoffmann treatment again.

1   Draw the mechanism for the reaction

2   Choose the components. All the bonds taking part in the
mechanism must be included and no others

$\pi 4$

3   Make a three-dimensional drawing of the way the
components come together for the reaction, putting in
orbitals at the ends of the components (only!)

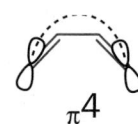

$\pi 4$

4   Join up the components where new bonds are to be formed.
Make sure you join orbitals that are going to form new bonds

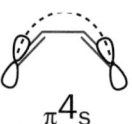

$\pi 4$

5   Label each component s or a depending whether new bonds
are formed on the same or on opposite sides

$\pi 4_s$

6   Add up the number of $(4q + 2)_s$ and $(4r)_a$ components. If the
sum is *odd*, the reaction is allowed.

> ■
> There are **no** $(4q + 2)_s$ components and
> **no** $(4r)_a$ components. Total = 0 so this
> is a *disallowed* reaction.

Oh dear! We know that the reaction works, so something must be wrong. It certainly isn't
Woodward and Hoffmann's Nobel-prize-winning rules—it's our way of drawing the orbital overlap
that is at fault. We were fine till stage 3 (we had no choice till then)—but look at what happens if we
make the orbitals overlap in a different way.

1   As before

2   As before

3 Make a three-dimensional drawing of the way in which the components come together for the reaction, putting in orbitals at the ends of the components (only!)

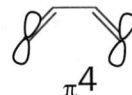

4 Join up the components where new bonds are to be formed. Make sure you join orbitals that are going to form new bonds

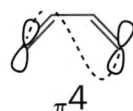

5 Label each component s or a depending on whether new bonds are formed on the same or on opposite sides

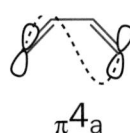

6 Add up the number of $(4q + 2)_s$ and $(4r)_a$ components. If the sum is *odd*, the reaction is allowed.

> ■ There are **no** $(4q + 2)_s$ components and **one** $(4r)_a$ component. Total = 1 so this is an allowed reaction.

Now it works! In fact, extension of this reasoning to other electrocyclic reactions tells you that they are *all* allowed—provided you choose to make the conjugated system react with itself *suprafacially* for $(4n + 2)$ π systems and *antarafacially* for $(4n)$ π systems. This may not seem particularly informative, since how you draw the dotted line has no effect on the reaction product in these cases. But it can make a difference. Here is the electrocyclic ring closure of an octatriene, showing the product from (a) suprafacial reaction and (b) antarafacial reaction.

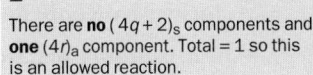

one methyl groups rotates upwards and one downwards to allow orbitals to overlap

methyl groups both rotate upwards to allow orbitals to overlap

$\pi 6_s$ ALLOWED $\pi 6_a$ DISALLOWED

The meanings of con- and disrotation

Whether the reaction is supra- or antarafacial ought to be reflected in the relative stereochemistry of the cyclized products—and indeed it is. This reaction gives solely the diastereoisomer on the left, with the methyl groups *syn*—clear proof that the reaction is suprafacial. This is a difficult result to explain without the enlightenment provided by the Woodward–Hoffmann rules!

This electrocyclic cyclobutene ring opening also gives the product as a single stereoisomer.

syn-1,2-dimethylcyclobutene → heat → *E, Z*-hexa-2,4-diene

Again, if we draw the reverse reaction, we can see that the reaction required has to be antarafacial for the stereochemistry to be right.

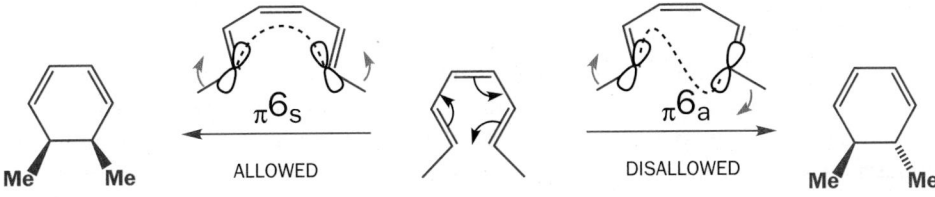

both methyl groups rotate upwards to allow orbitals to overlap

only this geometrical isomer... $\pi 4_a$ ...can give this diastereoisomer

We have drawn little green arrows on the two diagrams to show how the methyl groups move as the new σ bonds form. For the allowed suprafacial reaction of the 6π electron system they rotate in

> ► The green arrows in this and subsequent diagrams are merely mechanical devices to show the way in which the substituents move. They are nothing to do with real, mechanistic curly arrows.

opposite directions so the reaction is called **disrotatory** (yes, they both go up, but one has to rotate clockwise and one anticlockwise) while for the allowed antarafacial reaction of the 4π electron system they rotate in the same direction so the reaction is called **conrotatory** (both clockwise as drawn, but they might equally well have both gone anticlockwise). We can sum up the course of all electrocyclic reactions quite simply using these words.

> ### ● Rules for electrocyclic reactions
>
> * All electrocyclic reactions are allowed
> * Thermal electrocyclic reactions involving $(4n + 2)$ π electrons are *disrotatory*
> * Thermal electrocyclic reactions involving $(4n)$ π electrons are *conrotatory*
> * In *conrotatory* reactions the two groups rotate in the *same* way: *both* clockwise or *both* anticlockwise
> * In *disrotatory* reactions, *one* group rotates *clockwise* and *one anticlockwise*

This rotation is the reason why you must carefully distinguish electrocyclic reactions from all other pericyclic reactions. In cycloadditions and sigmatropic rearrangements there are small rotations as bond angles adjust from 109° to 120° and vice versa, but in electrocyclic reactions, rotations of nearly 90° are required as a planar polyene becomes a ring, or vice versa. These rules follow directly from application of the Woodward–Hoffmann rules—you can check this for yourself.

Electrocyclic reactions occur in nature

A beautiful example of electrocyclic reactions at work is provided by the chemistry of the endiandric acids. This family of natural products, of which endiandric acid D is one of the simplest, is remarkable in being racemic—most chiral natural products are enantiomerically pure (or at least enantiomerically enriched) because they are made by enantiomerically pure enzymes (we discuss all this in Chapter 45). So it seemed that the endiandric acids were formed by non-enzymatic cyclization reactions, and in the early 1980s their Australian discoverer, Black, proposed that their biosynthesis might involve a series of electrocyclic reactions, starting from an acyclic polyene precursor.

What made his proposal so convincing was that the stereochemistry of the endiandric acid D is just what you would expect from the requirements of the Woodward–Hoffmann rules. The first step from the precursor is an 8π electrocyclic reaction, and would therefore be conrotatory.

This sets up a new 6π system, which can undergo an electrocyclic reaction in disrotatory fashion. Because there are already chiral centres in the molecule, there are, in fact, two possible diastereoisomeric products from this reaction, both arising from disrotatory cyclization. One is endiandric acid D; one is endiandric acid E.

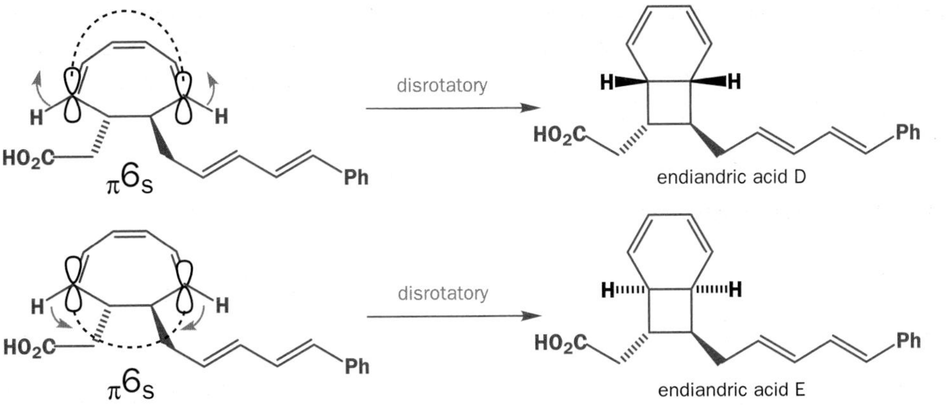

endiandric acid D

endiandric acid E

Of course, this was only a theory—until in 1982 K.C. Nicolaou's group synthesized the proposed endiandric acid precursor polyene—and in one step made both endiandric acids D and E, plus endiandric acid A, which arises from a further pericyclic reaction, an intramolecular Diels–Alder cycloaddition of the acyclic diene on to the cyclohexadiene as dienophile.

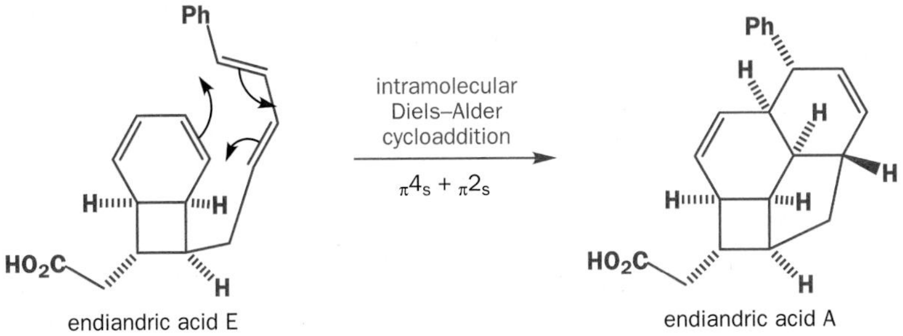

Endiandric acid A has four rings and eight stereogenic centres and yet is formed as a single diastereoisomer in one step from an acyclic polyene! And it's all controlled by pericyclic reactions.

Photochemical electrocyclic reactions

After your experience with cycloadditions and sigmatropic rearrangements, you will not be surprised to learn that, in photochemical electrocyclic reactions, the rules regarding conrotatory and disrotatory cyclizations are reversed.

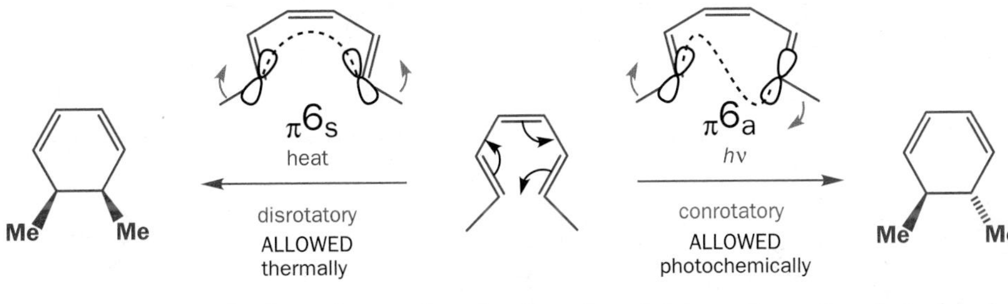

We can now go back to the reaction that introduced this section—the photochemical electrocyclic ring opening of ergosterol to give provitamin D_2. By looking at the starting material and product we can deduce whether the reaction is conrotatory or disrotatory.

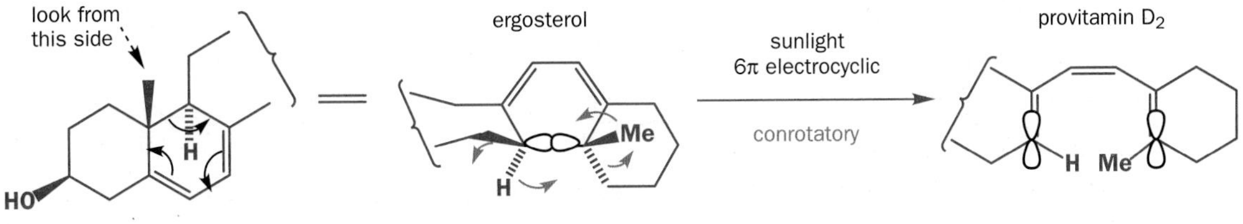

K.C. Nicolaou (1946–) was born in Cyprus, did his PhD in London, and has worked mainly at University of Pennsylvania and the Scripps Institute in California. His group was the first to synthesize some of the most complex natural products ever made by people.

It's clearly conrotatory, and a little more thought will tell you why it has to be—a disrotatory thermal 6π cyclization would put an impossible *trans* double bond into one of the two six-membered rings. Vitamin D deficiency is endemic in those parts of the world where sunlight is scarce for many months of the year—and all because of orbital symmetry.

Cations and anions

What we have just been telling you should convince you that the two reactions below are electrocyclic reactions, not least because the stereochemistry reverses on going from thermal to photochemical reaction.

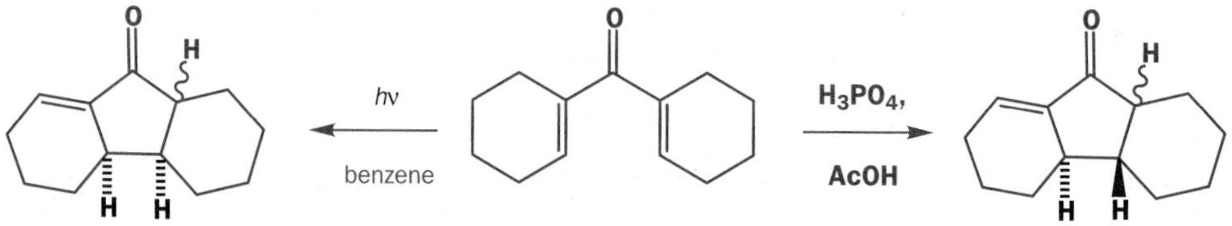

They are examples of what is known, after its Russian discoverer, as the **Nazarov cyclization**. In its simplest form, the Nazarov cyclization is the ring closure of a doubly α,β-unsaturated ketone to give a cyclopentenone.

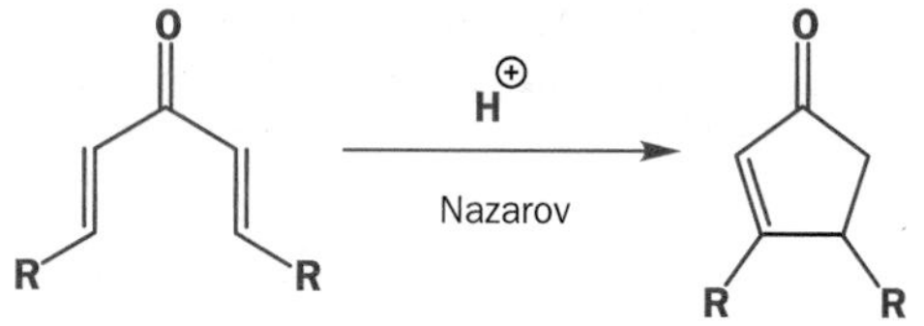

Nazarov cyclizations require acid, and protonation of the ketone sets up the conjugated π system required for an electrocyclic reaction.

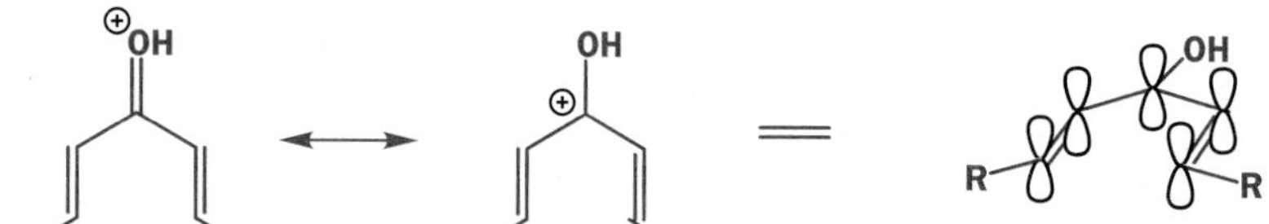

array of five p orbitals containing 4π electrons

One of the five π orbitals involved is empty—so the cyclization is a 4π electrocyclic reaction, and the orbitals forming the new σ bond must interact antarafacially. Loss of a proton and tautomerism gives the cyclopentenone.

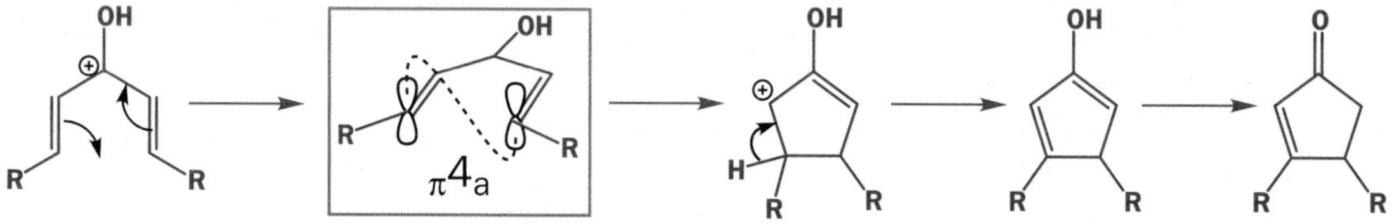

The real example above confirms that the reaction is thermally conrotatory and photochemically disrotatory.

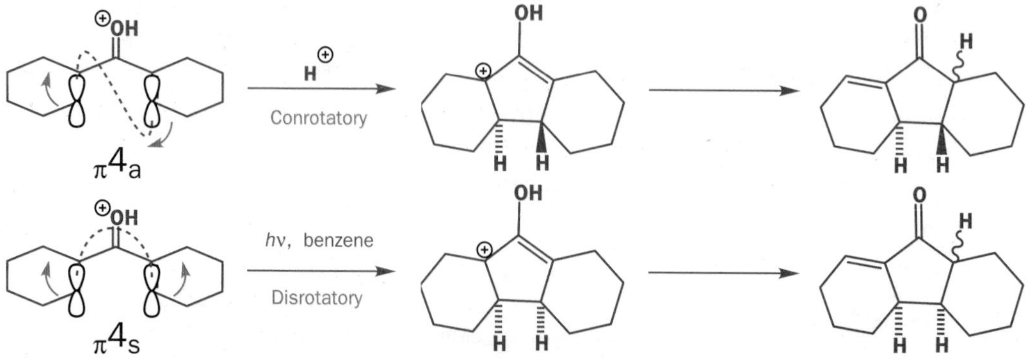

Dienyl cations and dienyl anions both undergo electrocyclic ring closure—a nice example occurs when cyclooctadiene is deprotonated with butyllithium.

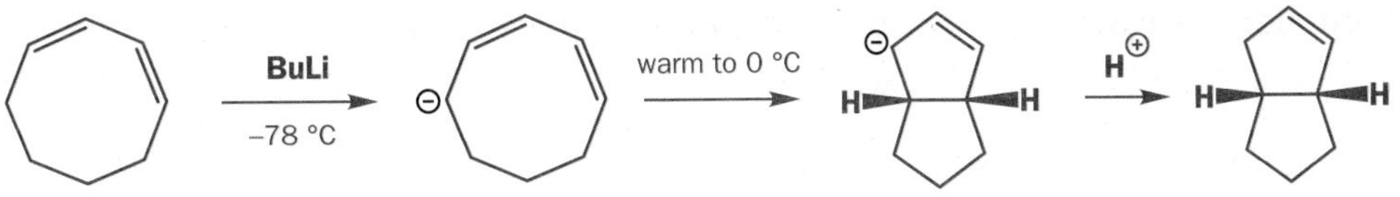

There are still five p orbitals involved in the cyclization, but now there are six π electrons, so the reaction is disrotatory.

In this case, it is the conrotatory *photochemical* cyclization that is prevented by strain (it was tried—cyclooctadienyl anion is stable for at least a week at −78 °C in broad daylight) as the product would be a 5,5 *trans*-fused system. The same strain prevents thermal electrocyclic ring closure of cyclooctadienyl cations.

● **All electrocyclic reactions are allowed**

It would be a good point here to remind you that, although all electrocyclic reactions are allowed both thermally and photochemically providing the rotation is right, the steric requirements for con- or disrotatory cyclization or ring opening may make one or both modes impossible.

Small rings are opened by electrocyclic reactions

Ring strain is important in preventing a reaction that would otherwise change your view of a lot of the chemistry you know. Allyl cations are conjugated systems containing 2π electrons, so if you knew no other chemistry than what is in this chapter you might expect them to cyclize via disrotatory electrocyclic ring closure.

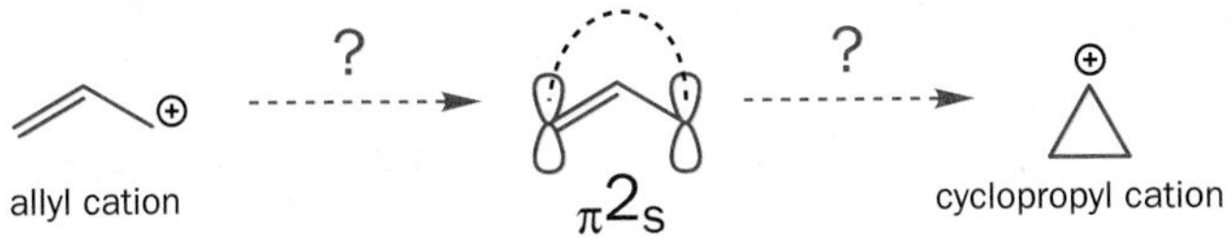

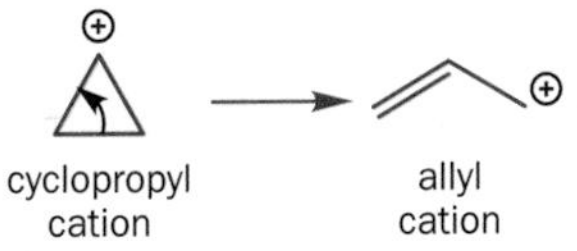

The product would be a cyclopropyl cation. Now, in fact, it is the cyclopropyl cations that undergo this reaction (very readily indeed—cyclopropyl cations are virtually unobservable) because ring strain encourages them to undergo electrocyclic ring opening to give allyl cations.

The instability of cyclopropyl cations means that, even as they start to form as intermediates, they spring open to give allyl cation-derived products. Try nucleophilic substitution on a cyclopropane ring and this happens.

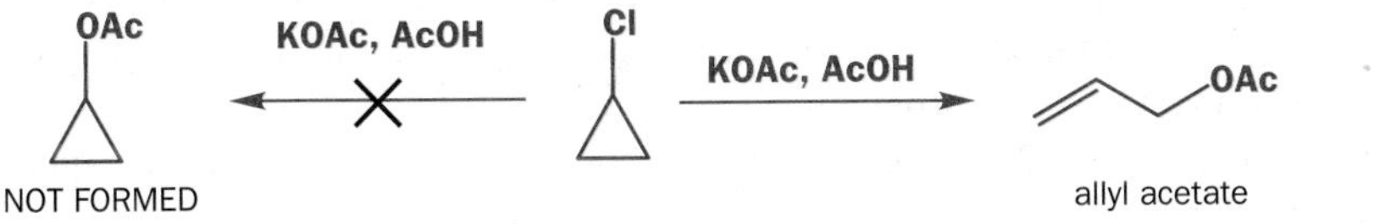

Although the initial product of the ring opening is a cation, and therefore a hard-to-observe reactive intermediate, some nice experiments in 'superacid' media (Chapters 17 and 22) have proven that cyclopropyl cation ring openings are indeed disrotatory.

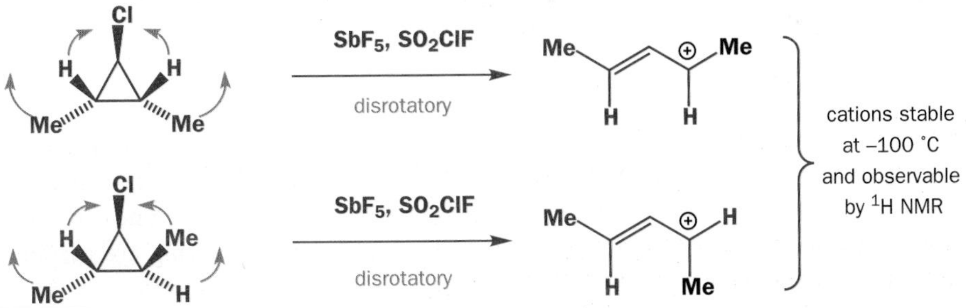

The stereochemistry of aziridine opening is predictable

One last type of three-membered ring whose electrocyclic ring opening does tell us about the stereochemistry of the process is the aziridine. Many aziridines are stable compounds, but those bearing electron-withdrawing groups are un-stable with respect to electrocyclic ring opening.

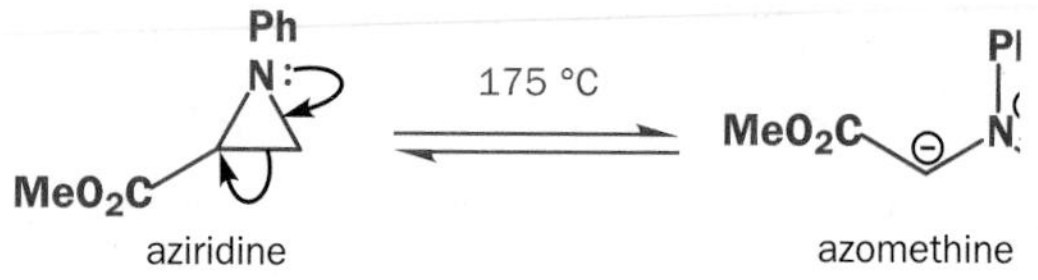

The products are azomethine ylids, and can be trapped by [3+2] cycloaddition reactions with dipolarophiles (look back at Chapter 35).

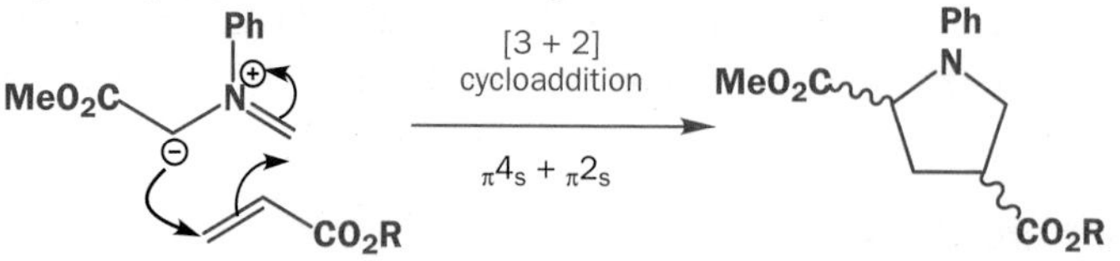

Because the cycloaddition is stereospecific (suprafacial on both components), the stereochemistry of the products can tell us the stereochemistry of the intermediate ylid, and confirms that the ring opening is conrotatory (the ylid is a 4π electron system).

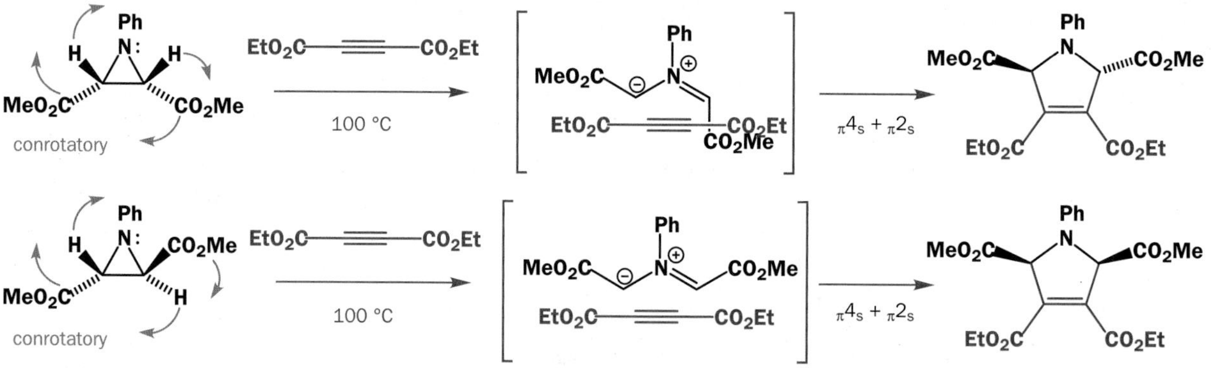

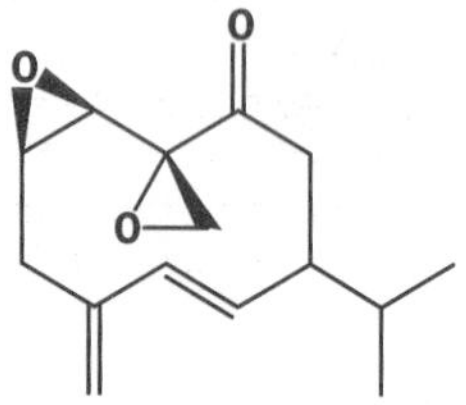

The synthesis of a cockroach pheromone required pericyclic reactions

We finish this pair of chapters about pericyclic reactions with a synthesis whose simplicity is out-classed only by its elegance. Periplanone B is a remarkable bis-epoxide that functions as the sex pheromone of the American cockroach. Insect sex pheromones often have economic importance because they can form the key to remarkable effective traps for insect pests.

In 1984, Schreiber published a synthesis of the pheromone in which the majority of steps involve pericyclic reactions. Make sure you understand each one as it appears—re-read the appropriate part of Chapter 35 or this chapter if you have any problems.

The first step is a photochemical [2+2] cycloaddition. You could not have predicted the regio-chemistry, but it is typical of the cycloaddition of allenes with unsaturated ketones.

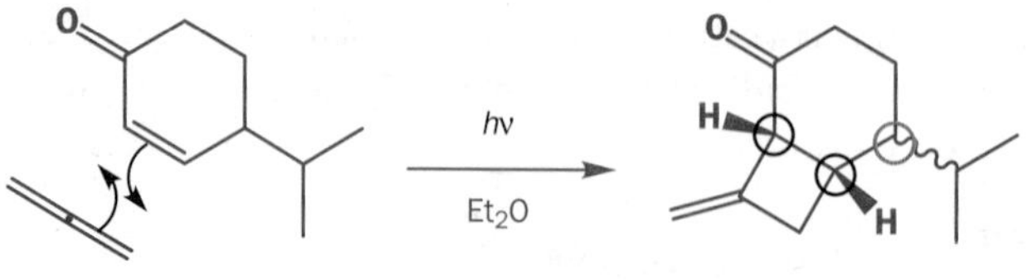

periplanone B

Stuart Schreiber (1956–) did his PhD at Harvard University where he is now a professor. One of the modern style of organic chemists who is equally at home with synthesis and biology.

[2 + 2] cycloaddition   72% yield

The product is a mixture of diastereoisomers because of the chiral centre already in the molecule (ringed in green), but it is, of course, fully stereospecific for the two new black chiral centres in the four-membered ring. The next step adds vinylmagnesium bromide to the ketone—again a mixture of diastereoisomers results. Now all the carbons in the 12-membered ring are present, and they are sorted out by the two steps that follow. The first is a Cope rearrangement: a [3,3]-

sigmatropic rearrangement, accelerated as we have described (p. 948) by the presence of an alkoxide substituent.

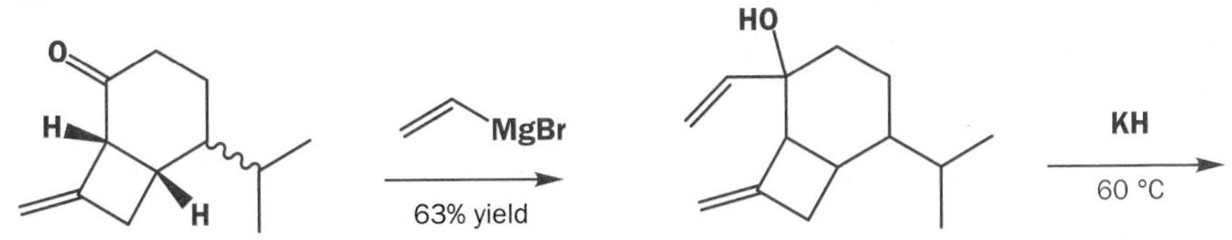

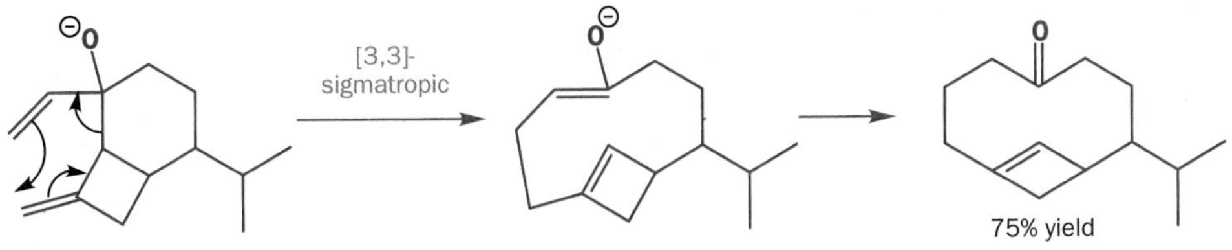

The six-membered ring has expanded to a ten-membered ring. Now for a second ring-expansion step—heating the compound to 175 °C makes it undergo electrocyclic ring opening of the four-membered ring, giving the 12-membered ring we want. Or rather not quite—the new double bond in the ring is formed as a mixture of *cis* and *trans* isomers, but irradiation isomerizes the less stable *cis* to the more stable *trans* double bond.

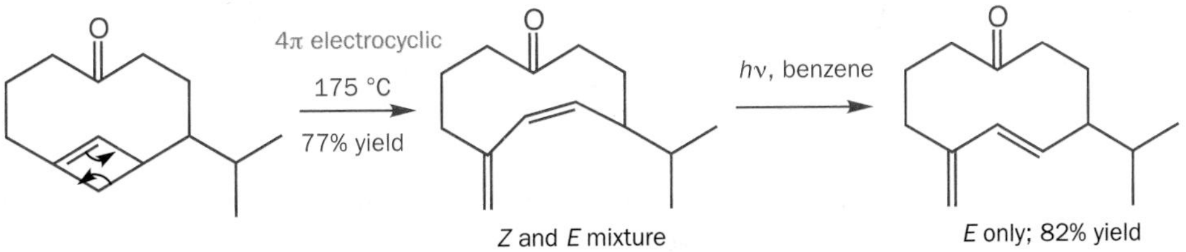

The remaining steps in the synthesis use chemistry not yet introduced in this book but involve the insertion of another (*Z*) alkene and two epoxides. Pericyclic reactions are particularly valuable in the synthesis and manipulation of rings.

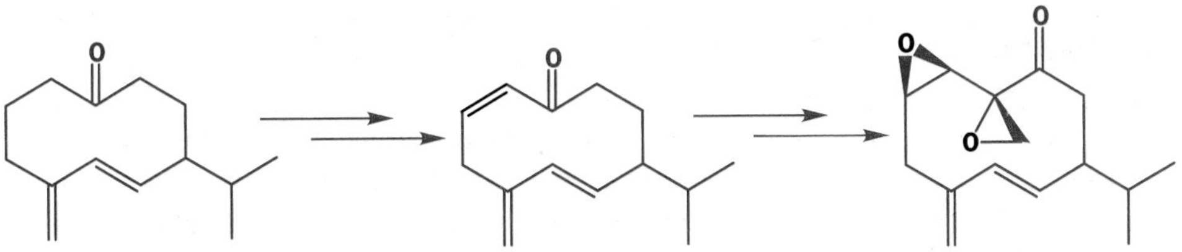

periplanone B

We must now take our leave of this trio of pericyclic reactions and move on to two reaction classes that have appeared frequently in these two chapters, but that involve mechanisms other than pericyclic ones and deserve chapters of their own: *rearrangements* and *fragmentations*.

▶
There are two things to note here—firstly, the geometry of the double bond is nothing to do with whether the reaction is conrotatory or disrotatory. As you know, this 4π electron electrocyclic ring opening must be conrotatory, but as there is no substituent on the other end of the diene product we can't tell. Secondly, notice that, in this 12-membered ring, a *trans* double bond is not only possible, but probably preferred. We introduced irradiation as a means of interconverting double bond isomers in Chapter 31.

Problems

1. Give mechanisms for these steps, commenting on the regio-selectivity of the pericyclic step and the different regioselectivity of the two metals.

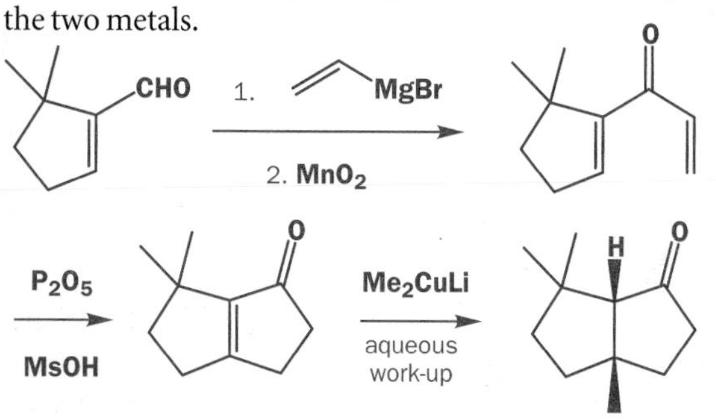

2. Predict the product of this reaction.

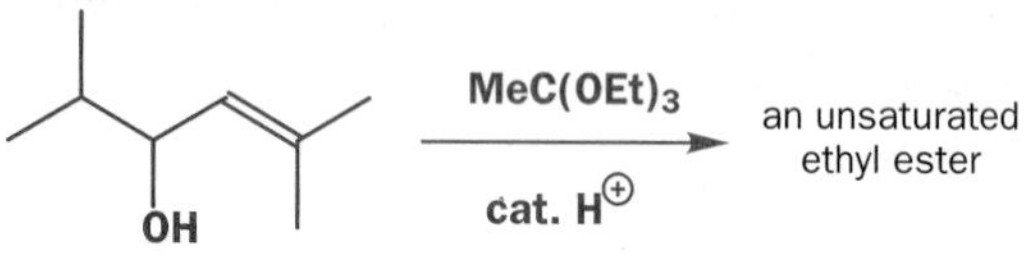

3. Give mechanisms for this alternative synthesis of two fused five-membered rings.

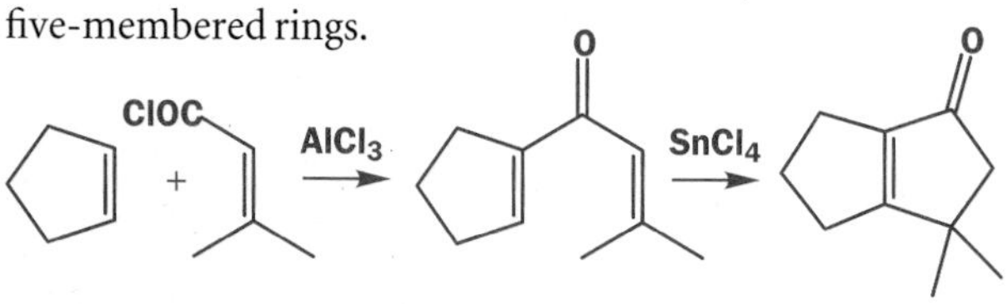

4. Explain what is going on here.

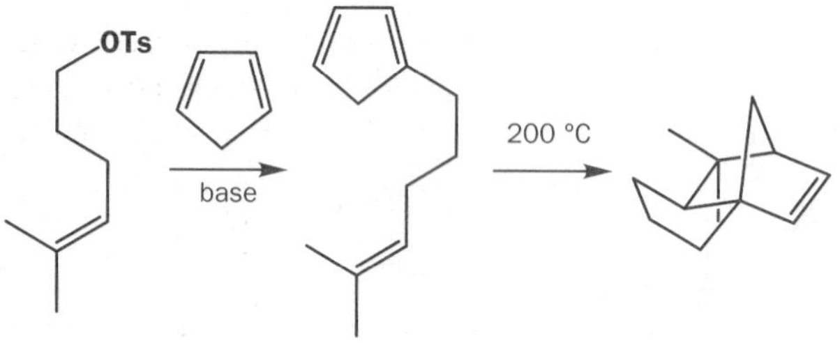

5. In Chapter 33, Problem 13, we used a tricyclic hydroxy-ketone whose stereochemistry had been wrongly assigned. Now we are going to show you how it was used and you are going to interpret the results. This is the correct result.

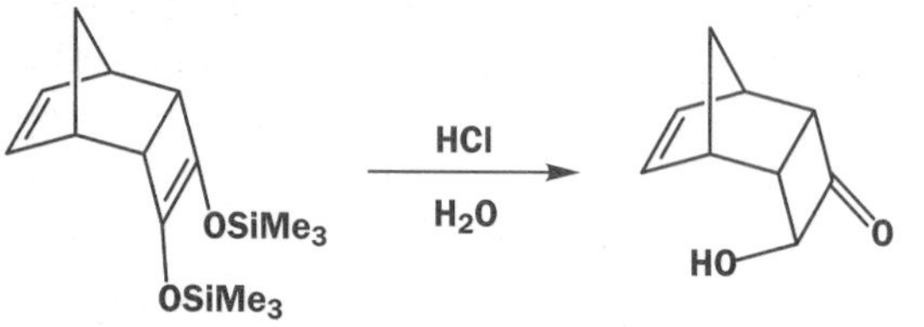

The hydroxy-ketone was first converted into a compound with PhS and OAc substituents. Explain the stereochemistry of this process.

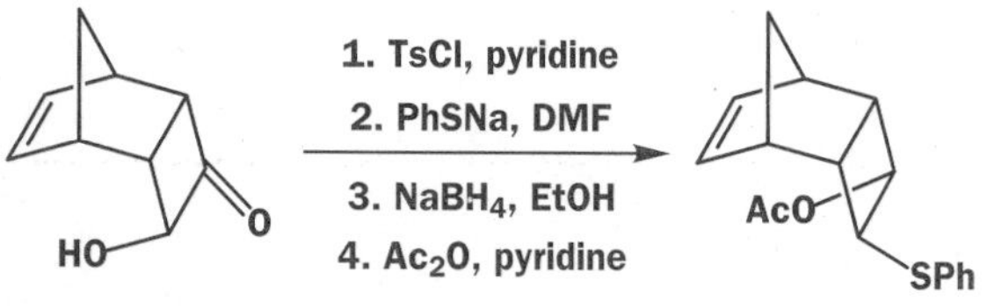

Pyrolysis of this compound at 460 °C gave a diene whose NMR spectrum included δ_H (p.p.m.) 6.06 (1H, dd, J 10.3, 12.1 Hz), 6.23 (1H, dd, J 10.3, 14.7 Hz), 6.31 (1H, d, J 14.7 Hz), and 7.32 (1H, d, J 12.1 Hz). Does this agree with the structure given? How is this diene formed and why does it have that stereochemistry?

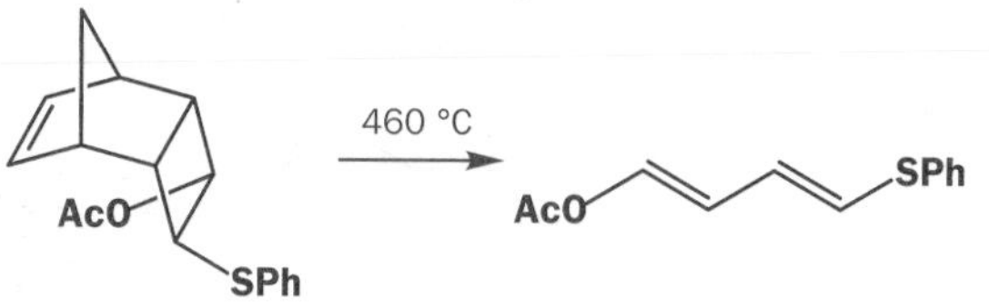

6. Careless attempts to carry out a Claisen rearrangement on this allyl ether often give the compound shown instead of the expected product. What is the expected product? How is the unwanted product formed? Addition of a small amount of a weak base, such as PhNMe$_2$, helps to prevent the unwanted reaction. How?

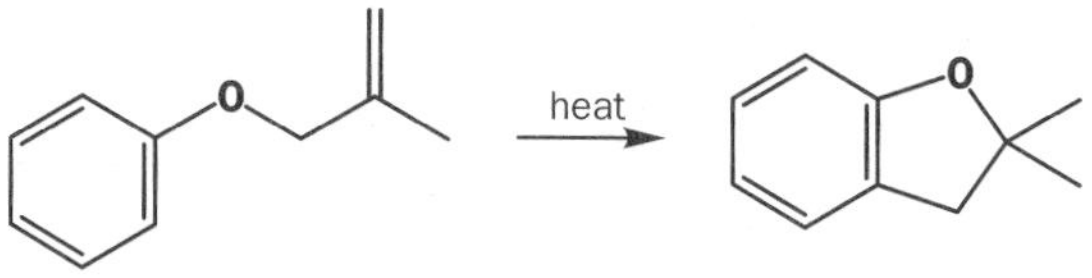

7. Treatment of this imine with base followed by an acidic work-up gives a cyclic product with two phenyl groups *cis* to one another. Why is this?

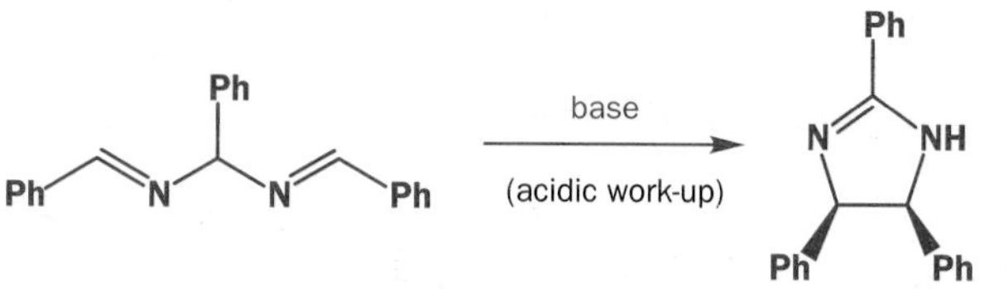

8. This question concerns the structure and chemistry of an unsaturated nine-membered ring. Comment upon its structure. Explain its different behaviour under thermal or photochemical conditions.

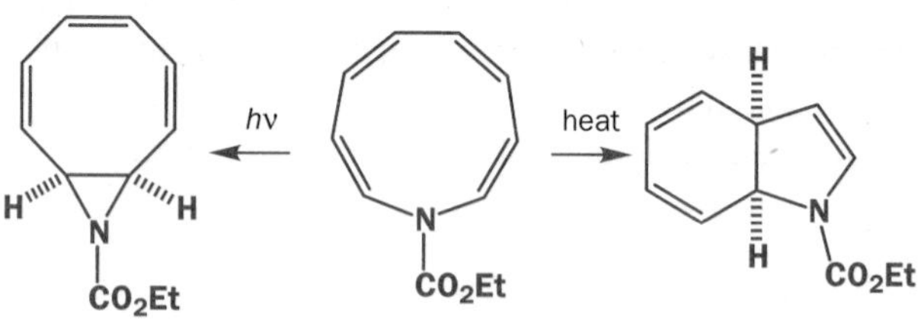

9. Propose a mechanism for this reaction that accounts for the stereochemistry of the product.

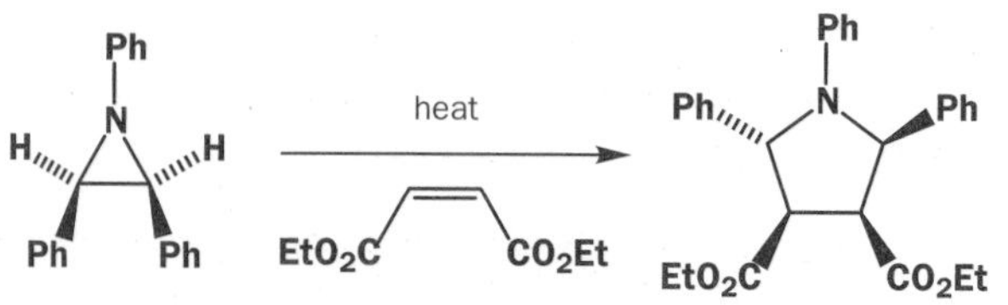

10. Treatment of cyclohexa-1,3-dione with this acetylenic amine gives a stable enamine in good yield. Refluxing this enamine in nitrobenzene gives a pyridine after a remarkable series of reactions. Fill in the details: give mechanisms for the reactions, structures for any intermediates, and suitable explanations for each pericyclic step. A mechanism is not required for the last step (nitrobenzene acts as an oxidant).

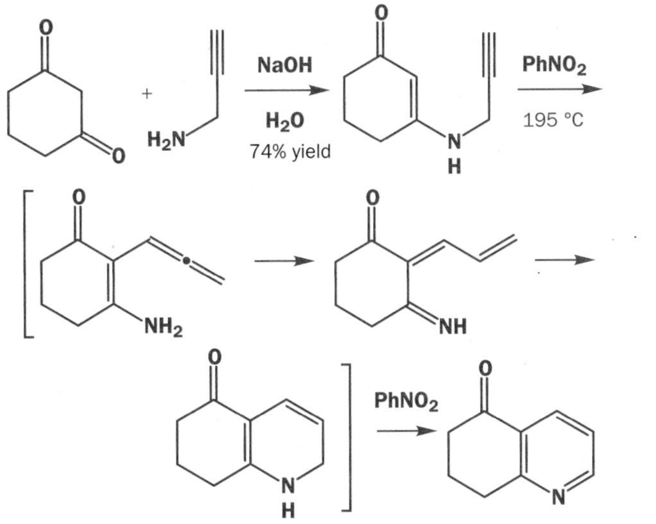

11. Problem 11 in Chapter 32 was concerned with two diastereoisomers of this compound that were formed in 'a chemical reaction'.

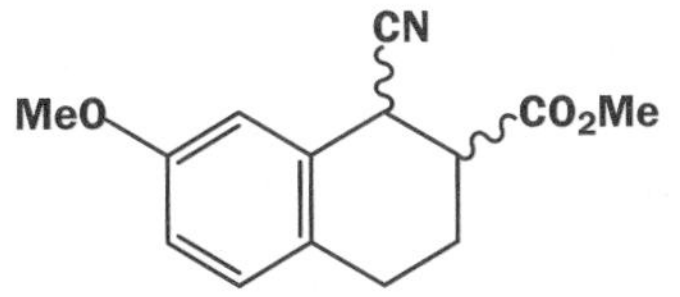

We can now let you into the secret of that 'chemical reaction'. A benzocyclobutene was heated with methyl acrylate to give a 1:1 mixture of the two isomers. What is the mechanism of the reaction and why is only one regioisomer but a mixture of stereoisomers formed? Isomer B is converted into isomer A on treatment with base. What is the stereochemistry of A and B?

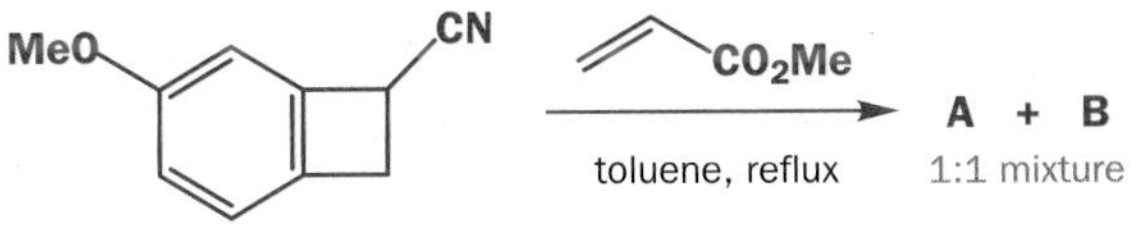

12. Treatment of this amine with base at low temperature gives an unstable anion that isomerizes to another anion above −35 °C. Aqueous work-up gives a bicyclic amine. What are the two anions? Explain the stereochemistry of the product. Revision of NMR. In the NMR spectrum of the product the two green hydrogens appear as an ABX system with J_{AB} 15.4 Hz. Comment.

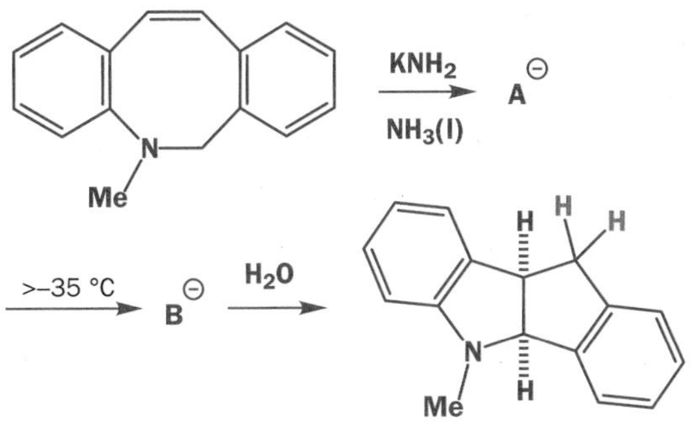

13. How would you make the starting material for these reactions? Treatment of the anhydride with butanol gives an ester that gives two inseparable compounds on heating. On treatment with an amine, an easily separable mixture of an acidic and a neutral compound is formed. What are the components of the first mixture and how are they formed?

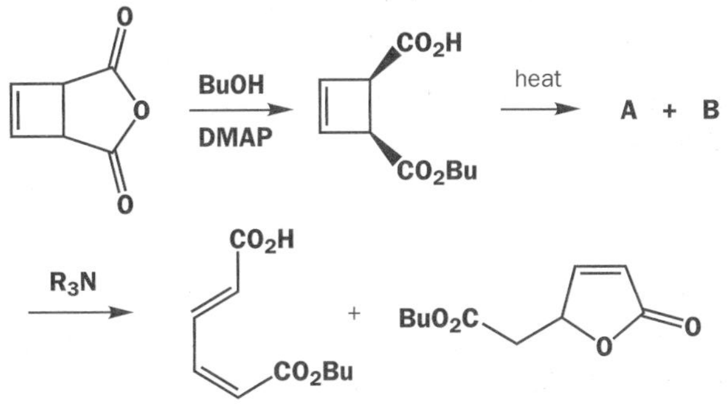

14. Treatment of this keto-aldehyde (which exists largely as an enol) with the oxidizing agent DDQ (a quinone—see p. 1192) gives an unstable compound that converts into the product shown. Explain the reactions and comment on the stereochemistry.

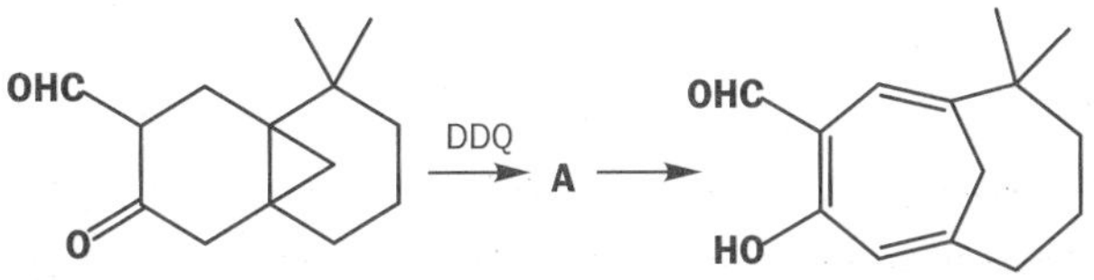

15. Explain the following observations. Heating this phenol brings it into rapid equilibrium with a bicyclic compound that does not spontaneously give the final aromatic product unless treated with acid.

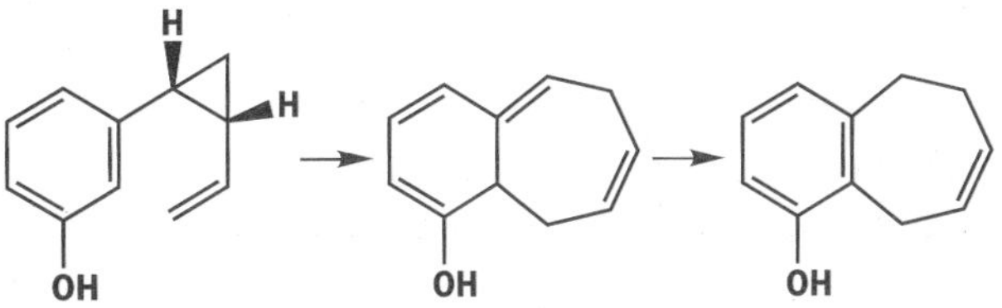

Rearrangements

37

Connections

Neighbouring groups can accelerate substitution reactions

Compare the rates of the following substitution reactions. Each of these reactions is a substitution of the leaving group (OTs or Cl) by solvent, known as a **solvolysis**.

■ A solvolysis was defined in Chapter 17 as 'a reaction in which the solvent is also the nucleophile'.

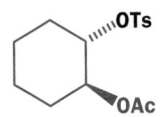

reacts with water
600 times faster than

reacts with CF$_3$CO$_2$H
3000 times faster than

reacts with acetic acid
10^{11} times faster than

reacts with acetic acid
670 times faster than

Nearby groups can evidently increase the rate of substitution reactions significantly. Now, you may be thinking back to Chapter 17 and saying 'yes, yes, we know that'—when we were discussing the mechanisms of substitution reactions we pointed out that a cation-stabilizing group at the reaction centre makes S$_N$1 reactions very fast: for example—

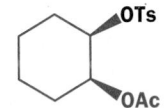

reacts with nucleophiles 10^6 times as fast as

reacts with nucleophiles 10^5 times as fast as

In the four examples above, though, it is not at the reaction centre itself that the functional groups change but at the carbon *next* to the reaction centre, and we call these groups **neighbouring groups**.

Neighbouring group participation is occasionally called **anchimeric assistance** (Greek *anchi* = neighbouring; *mer* = part).

The mechanism by which they speed up the reactions is known as **neighbouring group participation.** Compare the reaction of this ether and this sulfide with an alcohol.

S_N1 reaction of ethoxymethyl chloride

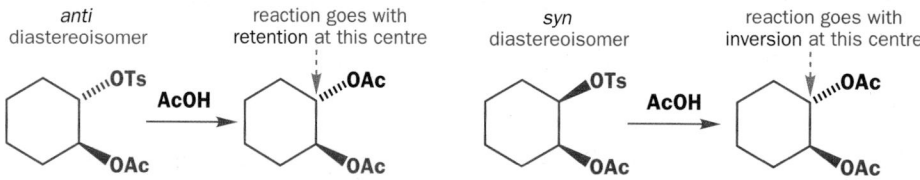

π-bonded cationic intermediate

neighbouring group participation of a sulfide

three-membered ring intermediate

In both cases, ionization of the starting material is assisted by the lone pair of an electron-rich functional group. The ether in the first example assists by forming a π bond, the sulfide assists by forming a three-membered ring, and a common feature of all mechanisms involving neighbouring group participation is the formation of a cyclic intermediate.

Stereochemistry can indicate neighbouring group participation

How do we know that neighbouring group participation is taking place? Well, the first bit of evidence is the *increase in rate*. The neighbouring groups will become involved only if they can increase the rate of the substitution reaction—otherwise the mechanism will just follow the ordinary S_N2 pathway. But more important information comes from reactions where stereochemistry is involved, and one of these is the last of the four examples above. Here it is again in more detail. Not only does the first of these reactions go faster than the second—its stereochemical course is different too.

| *anti* diastereoisomer | reaction goes with retention at this centre | *syn* diastereoisomer | reaction goes with inversion at this centre |

Although one starting material has *syn* and the other *anti* stereochemistry, the products have the same (*anti*) stereochemistry: one substitution goes with retention and one goes with inversion. Again, neighbouring group participation is the reason. To explain this, we should first draw the six-membered rings in their real conformation. For the *anti* compound, both substituents can be equatorial.

However, not much can happen in this conformation—but, if we allow the ring to flip, you can see immediately that the acetate substituent is ideally placed to participate in the departure of the tosylate group.

If you are unsure what we are talking about, go back and read Chapter 18 now!

both substituents equatorial

AcO / OTs both substituents equatorial

ring flip → both substituents axial → symmetrical intermediate

While the mechanism of this first step of the substitution reaction is S_N2 in appearance—a nucleophile (the acetate group) arrives just as a leaving group (the tosylate group) departs—it is also, of course, only unimolecular.

What results is an entirely symmetrical intermediate—the positive charge on one of the oxygens is, of course, delocalized over both of them. The intramolecular S_N2 reaction takes place with inversion, as required by the orbitals, so now the junction of the two rings is *cis*.

The next step is attack of acetic acid on the intermediate. This is another S_N2 reaction, which also proceeds with inversion and gives back a *trans* product.

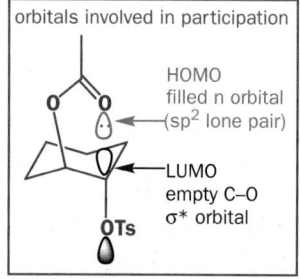

orbitals involved in participation

HOMO filled n orbital (sp² lone pair)

LUMO empty C–O σ* orbital

OTs

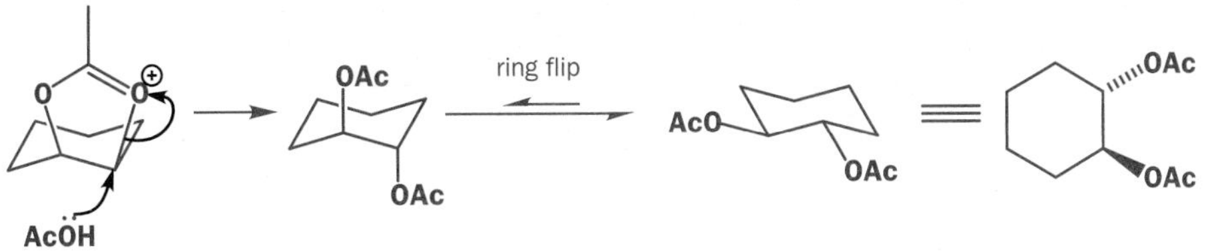

Overall, we have *retention* of stereochemistry. As you know, S_N2 reactions go with inversion, and S_N1 reactions with loss of stereochemical information—so this result is possible only if we have two sequential S_N2 reactions taking place—in other words neighbouring group participation.

Why, then, does the other diastereoisomer react with inversion of stereochemistry? Well, try drawing the mechanism for intramolecular displacement of the tosyl group. Whether you put the tosylate or the acetate group equatorial doesn't matter; there is no way in which the acetate oxygen's lone pair can reach the σ^* orbital of the tosylate C–O bond.

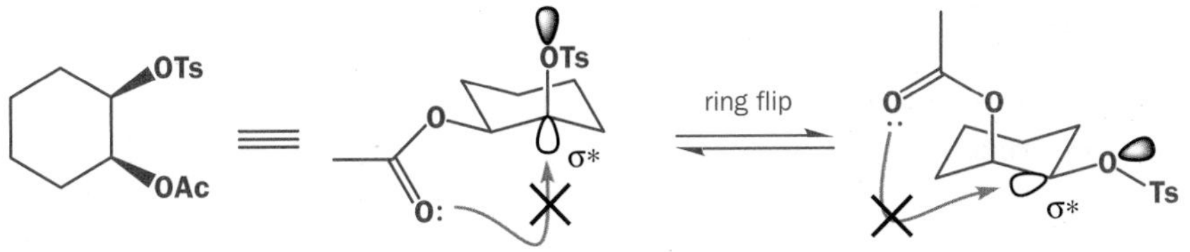

Neighbouring group participation is impossible, and substitution goes simply by intermolecular displacement of OTs by AcOH. Just one S_N2 step means overall inversion of configuration, and no participation means a slower reaction.

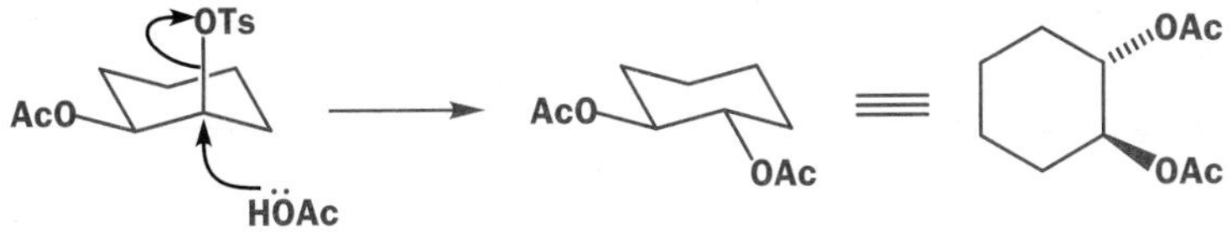

Retention of configuration is an indication of neighbouring group participation

Enantiomerically pure (*S*)-2-bromopropanoic acid reacts with concentrated sodium hydroxide to give (*R*)-lactic acid. The reaction goes with inversion and is a typical S_N2 reaction—and a good one too, since the reaction centre is adjacent to a carbonyl group (see Chapter 17).

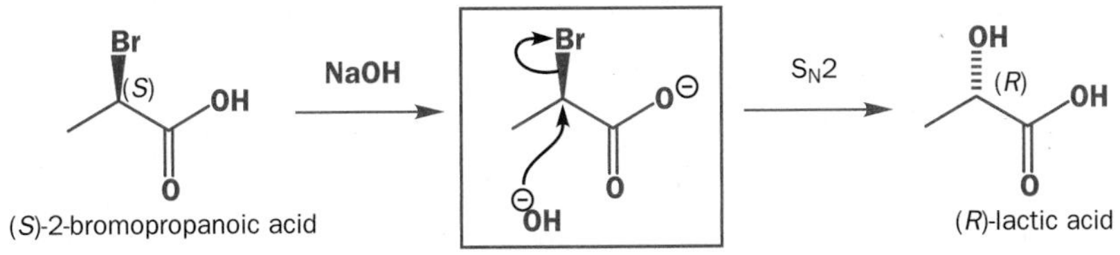

If, on the other hand, the reaction is run using Ag_2O and a low concentration of sodium hydroxide, (*S*)-lactic acid is obtained—there is overall *retention* of stereochemistry.

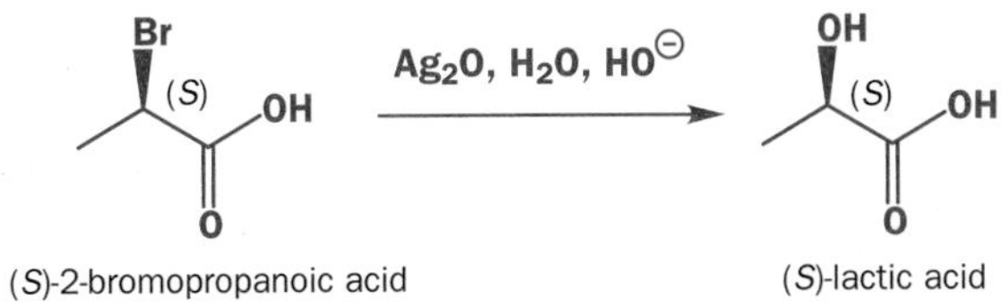

Nucleophilic substitution reactions that go with retention of stereochemistry are rather rare and mostly go through two successive inversions with neighbouring group participation, like the example you saw in the last section. This time the neighbouring group is carboxylate: the silver oxide is important because it encourages the ionization of the starting material by acting as a halogen-selective Lewis acid.

A three-membered ring intermediate forms, which then gets opened by hydroxide in a second S_N2 step.

> Lactones (that is, cyclic esters) don't usually react with hydroxide by this mechanism, and you might expect this intermediate (which is a cyclic ester) to hydrolyse by attack of hydroxide at the C=O group. You might like to think about why this doesn't happen in this case.

● **Retention suggests participation**

If you see a substitution reaction at a stereogenic saturated carbon atom that goes with retention of stereochemistry, look for neighbouring group participation!

Why does the carboxylate group participate only at low HO^- concentration and in the presence of Ag^+? You can think of the situation in these two reactions in terms of the factors that favour S_N1 and S_N2 reactions. In the first, we have conditions suited to an S_N2 reaction: a very good nucleophile (HO^-) and a good leaving group (Br^-). Improve the leaving group by adding Ag^+ (Ag^+ assists Br^-'s departure much as H^+ assists the departure of OH^- by allowing it to leave as H_2O), and worsen the nucleophile (H_2O instead of HO^-, of which there is now only a low concentration), and we have the sorts of conditions that *would* favour an S_N1 reaction. The trouble is, without neighbouring group participation, the cation here would be rather unstable—right next to a carbonyl group. The carboxylate saves the day by participating in the departure of the Br^- and forming the lactone. The key thing to remember is that a reaction always goes by the mechanism with the fastest rate.

● Neighbouring groups participate only if they speed up the reaction.

What sorts of groups can participate?

You've already met the most important ones—sulfides, esters, carboxylates. Ethers and amines (you will see some of these shortly) can also assist substitution reactions through neighbouring group participation. The important thing that they have in common is an electron-rich heteroatom with a lone pair that can be used to form the cyclic intermediate. Sulfides are rather better than ethers—this sulfide reacts with water much faster than *n*-PrCl but the ether reacts with acetic acid four times more *slowly* than *n*-PrOSO₂Ar.

sulfide participation **PhS**⏤⏤Cl   reacts with H_2O 600 times faster than ⏤⏤Cl

ether participation? **MeO**⏤⏤OSO₂Ar   reacts with AcOH 4 times slower than ⏤⏤OSO₂Ar

The OMe group slows the reaction down just because it is electronegative more than it accelerates it by participation. A more distant OMe group can participate: this 4-MeO alkyl sulfonate reacts with alcohols 4000 times faster than the *n*-Bu sulfonate.

ether participation ⏤⏤⏤OSO₂Ar   reacts with ROH   ⏤⏤⏤OSO₂Ar
                    OMe   4000 times faster than

Again neighbouring group participation is involved, but this time through a five- rather than a three-membered ring. Participation is most commonly through three- and five-membered rings, less often six-membered ones, and very rarely four- or more than seven-membered ones.

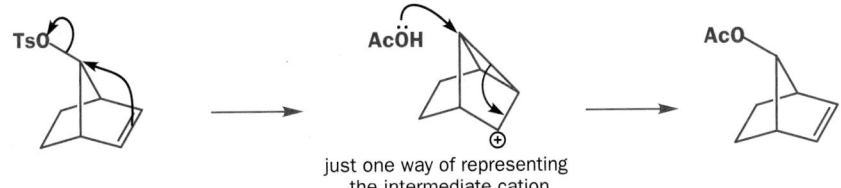

5-membered ring intermediate

▶

Why these ring sizes? Well, the underlying reasons are the same as those we discussed in Chapter 13 when we talked about the kinetics (rates) of formation and thermodynamics (stability) of different ring sizes: three- and five-membered rings form particularly rapidly in any reaction. See also Chapter 42.

Mustard gas

Participation of sulfides through three-membered rings was used to gruesome effect in the development of mustard gas during the Second World War. Mustard gas itself owes its toxicity to the neighbouring group participation of sulfur, which accelerates its alkylation reactions.

mustard gas

Not all participating groups have lone pairs

Another of the four examples we started with shows that even the π electrons of a C=C double bond can participate. Retention of stereochemistry in the product (the starting tosylate and product acetate are both *anti* to the double bond) and the extremely fast reaction (10^{11} times that of the saturated analogue) are tell-tale signs of neighbouring group participation.

just one way of representing
the intermediate cation

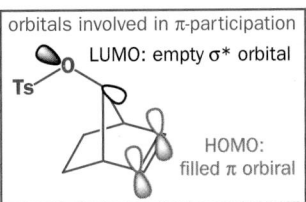

orbitals involved in π-participation
LUMO: empty σ* orbital

HOMO:
filled π orbiral

What is the structure of the intermediate?

During the 1950s and 1960s, this sort of question provoked a prolonged and acrimonious debate, which we have no intention of stirring up, and all we will do is point out that the intermediate in this reaction is not fully represented by the structure we have here: it is symmetrical and could be represented by two structures with three-membered rings or by a delocalized structure in which two electrons are shared between three atoms. The difference need not concern us.

Aryl participation is more common than simple alkene participation

Finally, an example with a neighbouring phenyl group. Participation is hinted at by the retention of relative stereochemistry.

Again, π electrons are involved, but the reaction is now electrophilic aromatic substitution (Chapter 22) rather like an intramolecular Friedel–Crafts alkylation with a delocalized intermediate often termed a **phenonium ion**.

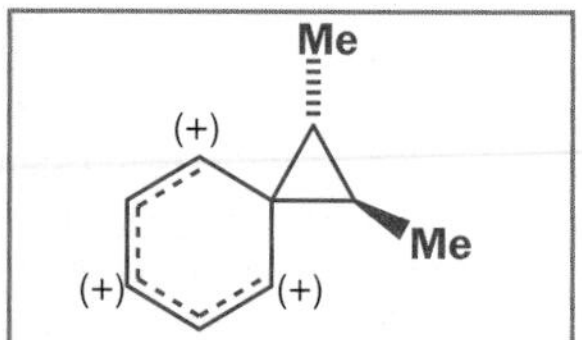

the delocalized phenonium ion

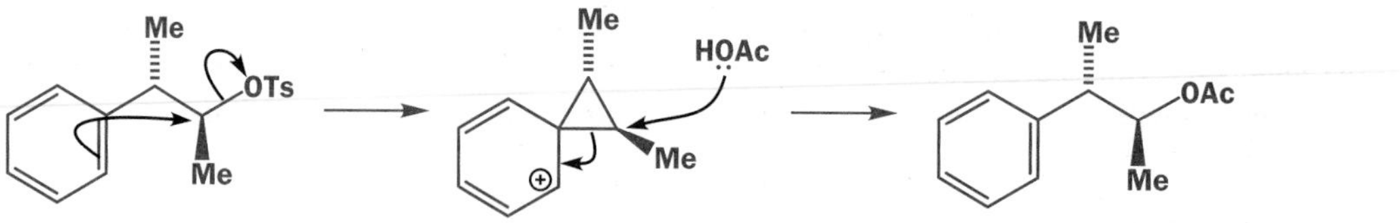

More stereochemical consequences of neighbouring group participation

The phenonium ion is symmetrical. The acetic acid can attack either atom in the three-membered ring to give the same product.

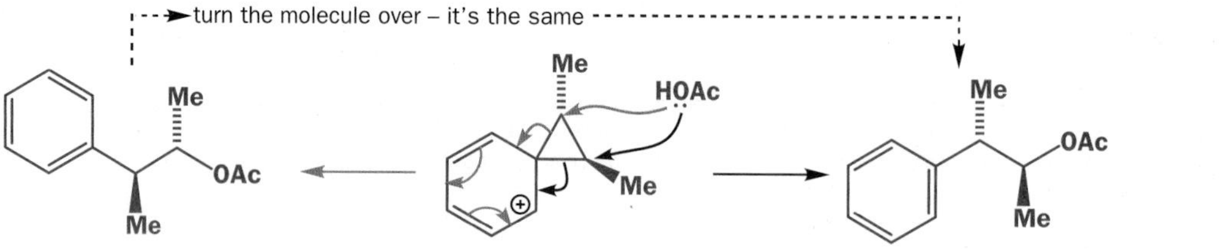

turn the molecule over – it's the same

The phenonium ion is nonetheless still chiral, since it has an axis (and not a plane or centre) of symmetry, so if we use an enantiomerically pure starting material we get an enantiomerically pure product.

start with this enantiomer of tosylate . . . we get this phenonium ion . . . and therefore this enantiomer of product
whichever end the acid attacks

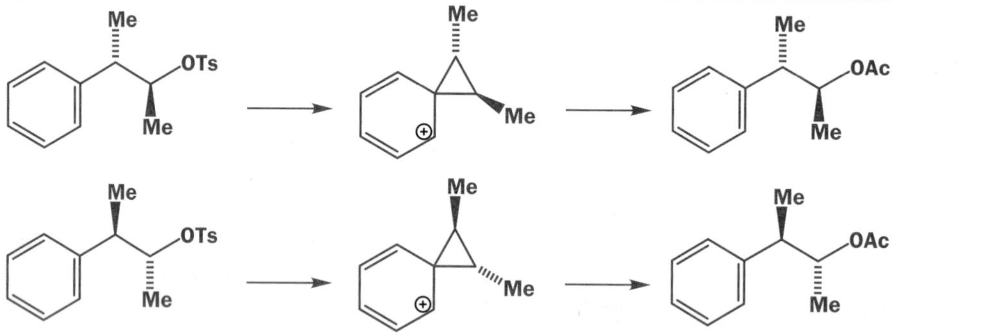

Not so with the other diastereoisomer of this compound! Now, the phenonium ion is symmetrical with a plane of symmetry—it is therefore achiral, and the same whichever enantiomer we start from. Attack on each end of the phenonium ion gives a different enantiomer, so whichever enantiomer of starting material we use we get the same racemic mixture of products. You can compare this reaction with the loss of stereochemical information that occurs during an S_N1 reaction of enantiomerically pure compounds. Both reactions pass through an achiral intermediate.

start with either enantiomer . . . we get the same achiral phenonium ion . . . and therefore racemic product

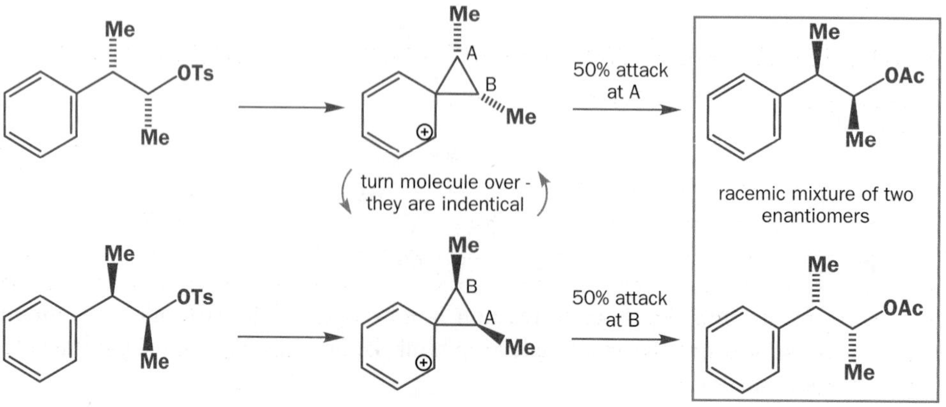

50% attack at A

turn molecule over - they are indentical

50% attack at B

racemic mixture of two enantiomers

There is a subtlety here that you should not overlook and that makes this study, which was carried out by Cram in 1949, exceedingly elegant. Both of these reactions are stereospecific: the *relative* stereochemistry of the products depends on the *relative* stereochemistry of the starting materials. Yet, while the absolute stereochemistry of the starting materials is retained in one case (we get a single enantiomer of a single diastereoisomer), it is lost in the other (we get a racemic mixture of both enantiomers of a single diastereoisomer). These are important distinctions, and if you are in any doubt about them, re-read Chapters 16 and 34. Donald Cram (1919–) of UCLA was awarded the Nobel prize in 1987 jointly with Jean-Marie Lehn (1939–) of Strasbourg and Paris and Charles Pedersen (a Norwegian born in Korea in 1904) of DuPont for 'their development and use of molecules with structure-specific interactions of high selectivity'.

Direct cation trapping is not observed

You may be wondering why acetic acid does not intercept the phenonium ion directly at one of the positively charged carbon atoms.

The problem is that the product would not then be aromatic and would contain a strained three-membered ring. The same sort of intermediates occur in electrophilic aromatic substitution (Chapter 22) and addition to the cation does not occur there either. The reaction that does occur here is a fragmentation: a C–C bond is broken. In the next chapter we will look at fragmentations in more detail.

The same loss of absolute stereochemical information (but retention of relative stereochemistry) occurs in another reaction that you met at the start of this chapter. We then emphasized two features: the acceleration in rate and the retention of stereochemistry.

The intermediate oxonium ion is delocalized and achiral. If a single enantiomer of the starting material is used, racemic product is formed through this achiral intermediate. Attack at one carbon atom gives one enantiomer; attack at the other gives the mirror image.

In this case the neighbouring group can be caught in the act—when the rearrangement is carried out in ethanol, the intermediate is trapped by attack at the central carbon atom. It is as though someone switched the light on while the acetate's fingers were in the biscuit tin (the cookie jar).

The product is an orthoester and is achiral too. This chemistry should remind you of the formation of acetals as described in Chapter 14.

Rearrangements occur when a participating group ends up bonded to a different atom

Because the intermediates in these examples are symmetrical, 50% of the time one substituent ends up moving from one carbon atom to another during the reaction. This is clearer in the following example: the starting material is prepared such that the carbon atom carrying the phenyl group is an unusual isotope—carbon-14. This doesn't affect the chemistry, but means that the two carbon atoms are easily distinguishable. Reacting the compound with trifluoroacetic acid scrambles the label between the two positions: the intermediate is symmetrical and, in the 50% of reactions with the nucleophile that take place at the labelled carbon atom, the phenyl ends up migrating to the unlabelled carbon atom in a rearrangement reaction.

> Labelling an atom with an unusual isotope is a standard way to probe the details of a reaction. Radioactive ^{3}H (tritium) or ^{14}C used to be used but, with the advent of high-field NMR, non-radioactive ^{2}H (deuterium) and ^{13}C have become more popular. These methods are treated more thoroughly in Chapter 41.

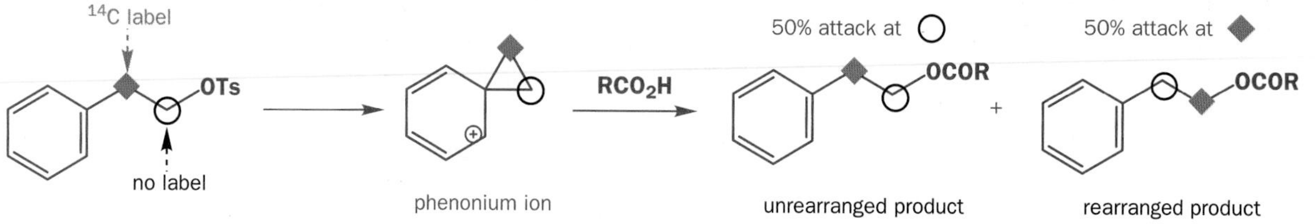

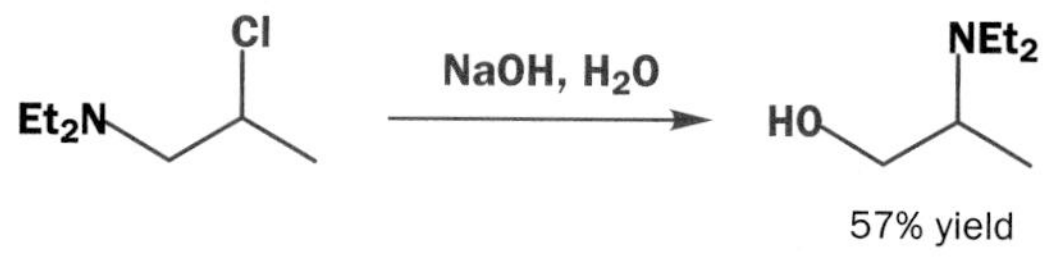

Now, consider this substitution reaction in which OH replaces Cl but with a change in the molecular structure. The substitution goes with complete rearrangement—the amine ends up attached to a different carbon atom.

We can easily see why if we look at the mechanism. The reaction starts off looking like a neighbouring group participation of the sort you are now familiar with (the carbon atoms are numbered for identification).

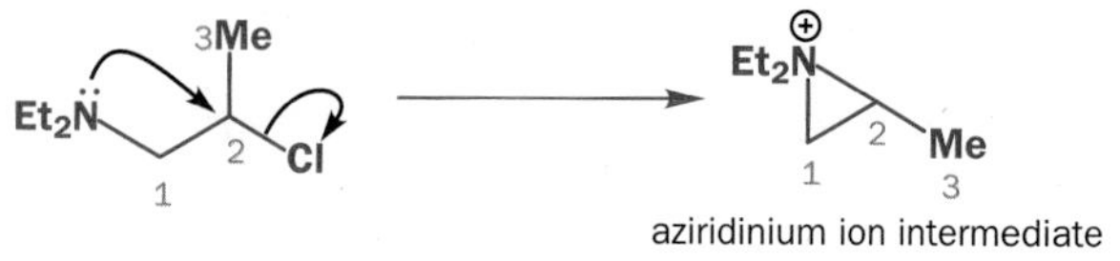

The intermediate is an aziridinium ion (**aziridines** are three-membered rings containing nitrogen—the nitrogen analogues of epoxides). The hydroxide ion chooses to attack only the less hindered terminal carbon 1, and a rearrangement results—the amine has migrated from carbon 1 to carbon 2.

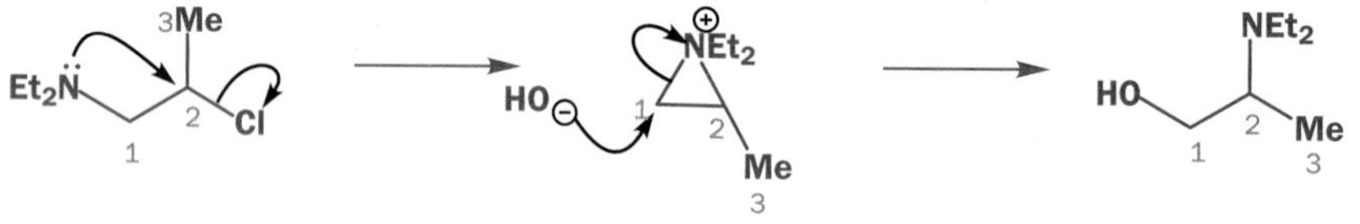

We should just pause here for a moment to consider why this rearrangement works. We start with a secondary alkyl chloride that contains a very bad leaving group (Et_2N) and a good one (Cl^-)—but the good one is hard for HO^- to displace because it is at a secondary centre (remember—secondary alkyl halides are slow to react by S_N1 or S_N2). But the NEt_2 can participate to make an aziridinium intermediate—now there is a good leaving group ($RNEt_2$ without the negative charge) at the primary as well as the secondary carbon, so HO^- does a fast S_N2 reaction at the primary carbon.

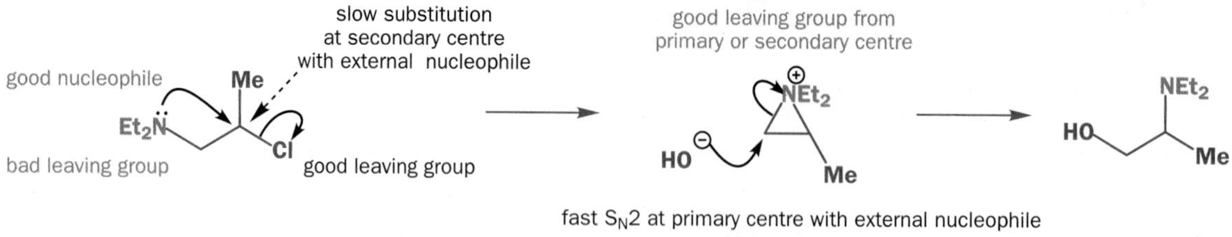

Another way to look at this reaction is to see that the good internal nucleophile Et_2N will compete successfully for the electrophile with the external nucleophile HO^-. Intramolecular reactions are usually faster than bimolecular reactions.

● Intramolecular reactions, including participation, that give three-, five-, or six-membered rings are usually faster than intermolecular reactions.

The Payne rearrangement

The reaction of an epoxy alcohol in base does not always give the expected product.

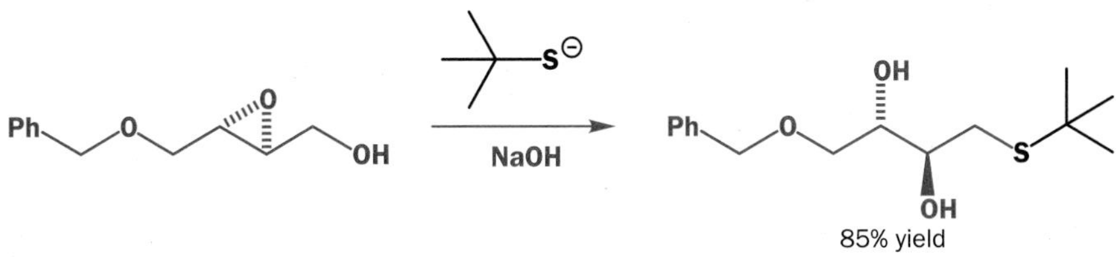

85% yield

The thiolate nucleophile has not opened the epoxide directly, but instead *appears* to have displaced HO⁻—a very bad leaving group. Almost no nucleophile will displace OH⁻, so we need an alternative explanation. This comes in the form of another rearrangement, this time involving oxygen, but otherwise rather similar to the ones you have just met. Again, our epoxide, though reactive as an electrophile, suffers from being secondary at both electrophilic centres. *t*-BuS⁻ is a bulky nucleophile, so direct attack on the epoxide is slow. Instead, under the basic conditions of the reaction, the neighbouring alkoxide group attacks intramolecularly to make a new, rearranged epoxy alcohol. This rearrangement is called the **Payne rearrangement**.

the Payne rearrangement

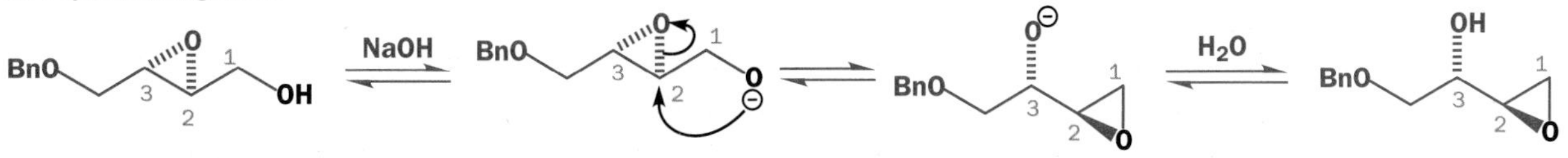

Now we do have a reactive, primary electrophilic site, which undergoes an S_N2 reaction with the *t*-BuS⁻ under the conditions of the rearrangement. Notice how the black OH, which started on the carbon labelled 1, has ended up on carbon 2.

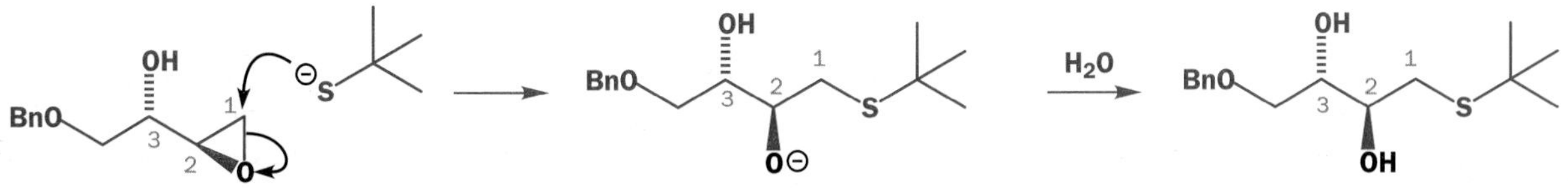

The direction of rearrangement can depend on the nucleophile

Compare these reactions: you saw the first on p. 976 but the second is new.

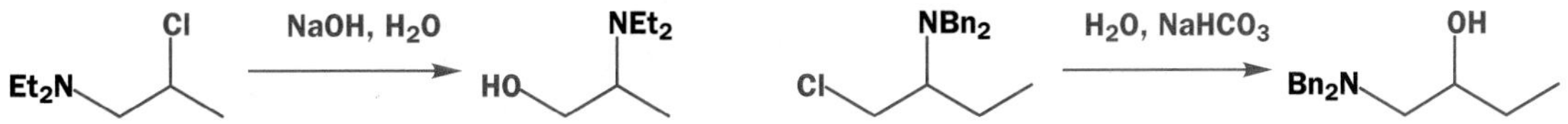

In the first reaction, the amine migrates from the primary to the secondary position; in the other from secondary to primary. Both go through very similar aziridinium intermediates, so the difference must be due to the regioselectivity with which this aziridinium opens in each case.

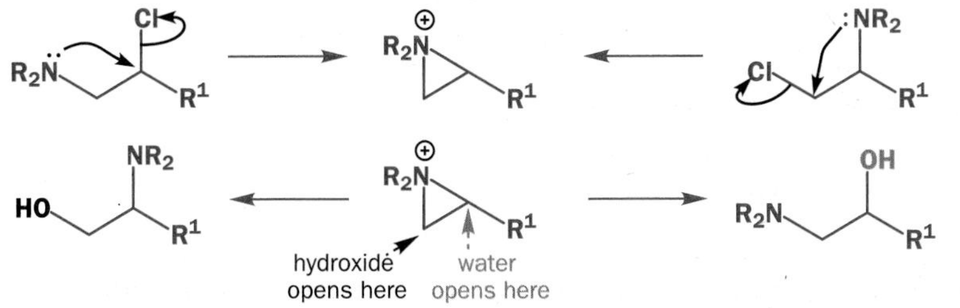

hydroxide opens here   water opens here

> When a group migrates from a primary to a secondary carbon, we say the rearrangement has a primary **migration origin** and a secondary **migration terminus**. The migrating group moves from the migration origin to the migration terminus.

The only important difference is the nucleophile used in the reaction. Hydroxide opens the aziridinium at the less hindered end; water opens the aziridinium ion at the more hindered (more substituted) end. Why?

We can think of the aziridinium ion as a compound containing two alternative leaving groups—one from a primary centre and one from a secondary one. Primary centres can take part in fast S_N2 reactions, but cannot undergo S_N1. Secondary centres can undergo either S_N1 or S_N2 reactions, but, in general, do neither very well. Now, the rate of an S_N2 reaction depends on the nucleophile, so a good nucleophile (like HO^-) can do fast S_N2 reactions, while a bad one (like H_2O) cannot. The fastest reaction HO^- can do then is S_N2 at the primary centre (remember: you see only the reaction that goes by the fastest mechanism). Water, on the other hand, takes part only reluctantly in substitution reactions—but this does not matter if they are S_N1 reactions because their rates are independent of nucleophile. H_2O waits until the leaving group has left of its own accord, to give a cation, which rapidly grabs *any* nucleophile—water will do just as well as HO^-. This can happen *only* at the secondary centre because the primary cation is too unstable to form.

All the rearrangements you have met so far occurred during substitution reactions. All happened because reaction *with* rearrangement is faster than reaction *without* rearrangement—in other words, rearrangement occurs because of a kinetic preference for the rearrangement pathway. You could see these reactions as 'special case' examples of neighbouring group participation—in both participation and rearrangement, the neighbouring group speeds up the reaction, but in rearrangement reactions the neighbouring group gets rather more than it bargained for, and ends up elsewhere in the molecule. Both proceed through a cyclic transition state or intermediate, and it is simply the way in which that transition state or intermediate collapses that determines whether rearrangement occurs.

Rearrangement can involve migration of alkyl groups

You have seen reactions in which the lone pairs of N, O, and S atoms participate, and reactions in which the π orbitals of alkenes and aromatic groups participate, and participation can lead to rearrangement for any of these groups. Alkyl groups too may rearrange. This example is a nucleophilic substitution under conditions (Ag^+, H_2O) designed to encourage S_N1 reactions (excellent leaving group, poor nucleophile). First of all, this is what does not happen (and indeed without Ag^+ *nothing happens at all*).

▶
The *t*-butylmethyl group is also called 'neopentyl'.

t-butylmethyl iodide
2,2-dimethyliodopropane
- neopentyl iodide

the neopentyl group

Compounds like this, with a *t*-butyl group next to the electrophilic centre, are notoriously slow to undergo substitution reactions. They can't do S_N2, they are too hindered; they can't do S_N1, the cation you would get is primary.

In fact, a rearrangement occurs. One of the methyl groups moves ('migrates') from carbon 2 to carbon 1, the new OH group taking its place at carbon 2.

How has this happened? Well, firstly, our principle (p. 972) tells us that it has happened because S_N1 and S_N2 are both so slow that this new rearrangement mechanism is faster than either. Adding Ag^+ makes I^- desperate to leave, but unassisted this would mean the formation of a primary carbocation. The molecule does the only thing it can to stop this happening, and uses the electrons in an adjacent C–C bond to assist the departure of I⁻.

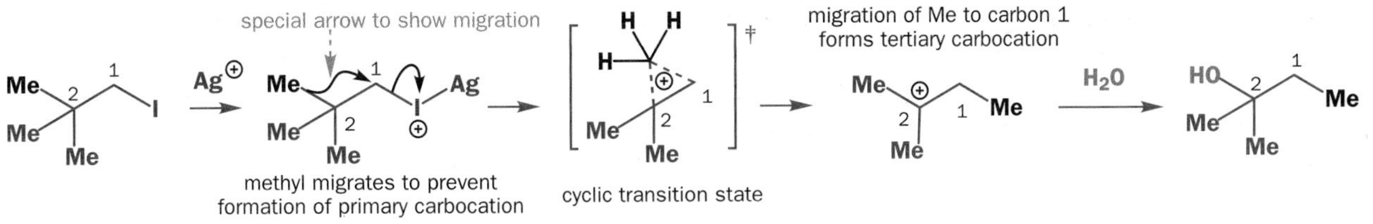

Having participated, the methyl group continues to migrate to carbon 1 because by doing so it allows the formation of a stable tertiary carbocation, which then captures water in a step reminiscent of the second half of an S_N1 reaction.

In the migration step we used a slightly unusually curved curly arrow to represent the movement of a group (Me) along a bond taking its bonding electrons with it. We shall use this type of arrow when a group migrates from one atom to another during a rearrangement.

Often, you will see this rearrangement represented in a different way. Both are correct, but we feel that the first is more intuitively descriptive.

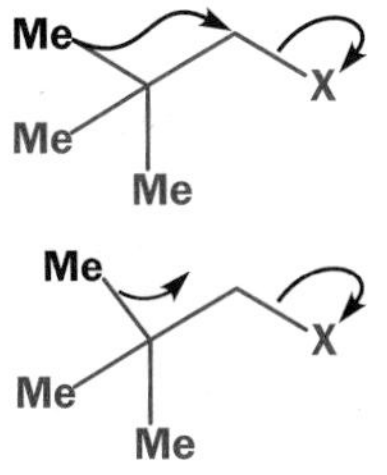

> Some of the cyclic species you have seen so far (aziridinium ions, epoxides) are intermediates; this cyclic species is probably only a transition state.

Carbocations readily rearrange

In Chapter 17 we showed you that it is possible to run the NMR spectra of carbocations by using a polar but nonnucleophilic solvent such as liquid SO_2 or SOClF. Treating an alkyl halide RX with the powerful Lewis acid SbF_5 under these conditions gives a solution of carbocation: the carbocation reacts neither with solvent nor the SbF_5X^- counterion because neither is nucleophilic. We know, for example, that the chemical shifts in both the ^{13}C and ^1H NMR spectra of the *t*-butyl cation are very large, particularly the ^{13}C shift at the positively charged centre.

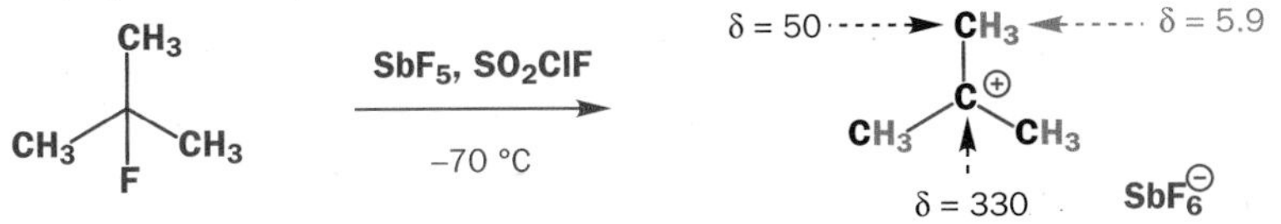

NMR can be used to follow the course of rearrangement reactions involving carbocations too. We can illustrate this with an experiment that tries to make the neopentyl cation by the substitution reaction you have just seen. This time the starting material and solvent are slightly different, but the outcome is nonetheless most revealing. Dissolving neopentyl tosylate in fluorosulfonic acid (a strong, nonnucleophilic acid) at −77 °C gives a 77% yield of a cation whose spectrum is shown below. Assigning the peaks is not hard once you know that the same spectrum is obtained when 2,2-dimethyl-2-butanol is dissolved in fluorosulfonic acid with SbF_5 added.

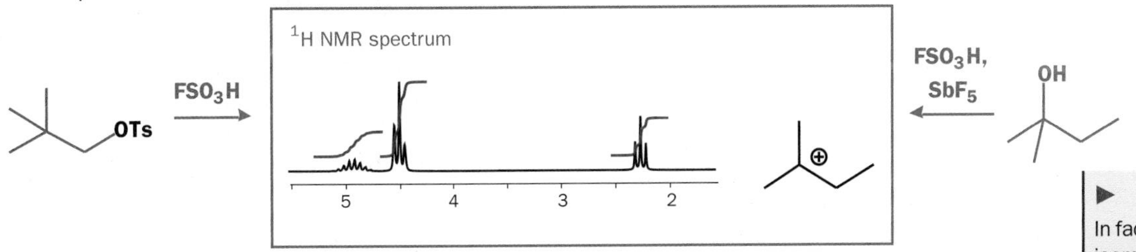

Clearly, both spectra are of the tertiary 2-methylbutyl cation and the neopentyl cation never saw the light of day. The reaction is the same rearrangement that you saw in the substitution reaction of neopentyl iodide, but here the rate of rearrangement can be measured and it is extremely fast. Neopentyl tosylate reacts to form a cation under these conditions about 10^4 times as fast as ethyl tosylate, even though both tosylates are primary. This massive rate difference shows that if migration of an alkyl group can allow rearrangement to a more stable carbocation, it will happen, and happen rapidly.

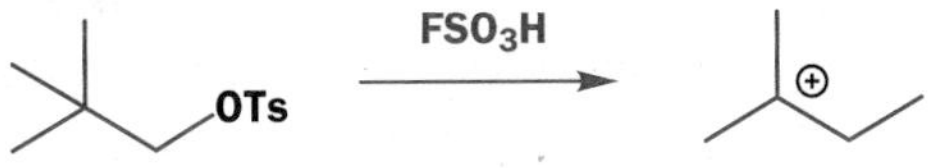

> In fact, all *seven* possible isomers of pentyl alcohol ($C_5H_{11}OH$) give this same spectrum under these conditions at temperatures greater than −30 °C.

The distinction here is quite subtle and need not detain us long. We know that a secondary cation is formed in this case because we can see it by NMR; it subsequently rearranges to a tertiary cation. As we can never see primary cations, we don't know that they are ever formed, and the most reasonable explanation for rearrangements of the type you saw on p. 978 is that migration of the alkyl group begins *before* the leaving group is fully gone. This has been proven in a few cases, but we will from now on not distinguish between the two alternatives.

Primary cations can never be observed by NMR—they are too unstable. But secondary cations can, provided the temperature is kept low enough. *sec*-Butyl chloride in SO_2ClF at −78 °C gives a stable, observable cation. But, as the cation is warmed up, it rearranges to the *t*-butyl cation. Now this rearrangement truly is a carbocation rearrangement: the starting material is an observable carbocation, and so is the product, and we should just look at the mechanism in a little more detail.

With rearrangements like this it is best to number the C atoms so you can see clearly what moves where. If we do this, we see that the methyl group we have labelled 4 and the H on C3 have changed places. (Note that C3 starts off as a CH_2 group and ends up as CH_3.)

● Top tip for rearrangements

Number the carbon atoms in starting material and product before you try to work out the mechanism.

You will see why Me has to migrate first if you try drawing the mechanism out with H migrating first instead.

Using the sort of arrows we introduced on p. 979, we can draw a mechanism for this in which first the Me migrates, and then the hydride. We say **hydride** migration rather than *hydrogen* (or *proton*) because the H atom migrates *with* its pair of electrons.

As these rearrangements are a new type of reaction, we should just spend a moment looking at the molecular orbitals that are involved. For the first step, migration of the methyl group, the LUMO must clearly be the empty p orbital of the cation, and the HOMO is the C–C σ bond, which is about to break.

The methyl group migrates smoothly from one orbital to another—there are bonding interactions all the way. The next step, migration of H, is just the same—except that the HOMO is now a C–H σ bond. The methyl migration is unfavourable as it transforms a secondary cation into an unstable primary cation but the hydride migration puts that right as it gives a stable tertiary cation. The whole reaction is under thermodynamic control.

Wagner–Meerwein rearrangements

Carbocation rearrangements involving migration of H or alkyl groups don't just happen in NMR machines. They happen during normal reactions too. For example, acid-catalysed dehydration of the natural product camphenilol gives the alkene santene (a key component of the fragrance of sandalwood oil) in a reaction involving migration of a methyl group.

camphenilol

santene

The mechanism shows why the rearrangement happens: the first-formed cation cannot eliminate H^+ in an E1 reaction because loss of the only available proton would give a very strained alkene (make a model and see!).

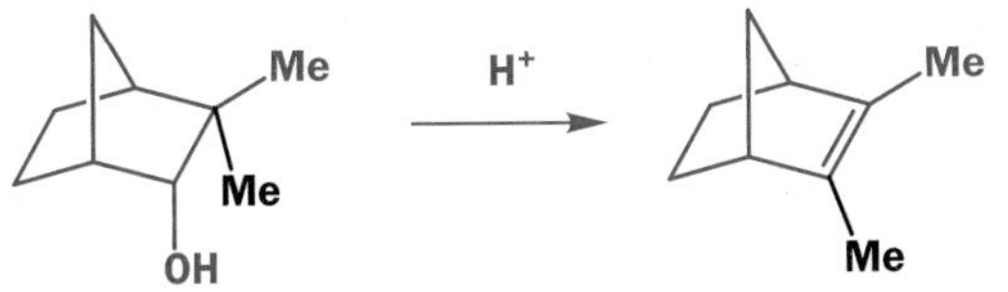

Bredt's rule is discussed in Chapter 19 and essentially forbids bridgehead alkenes.

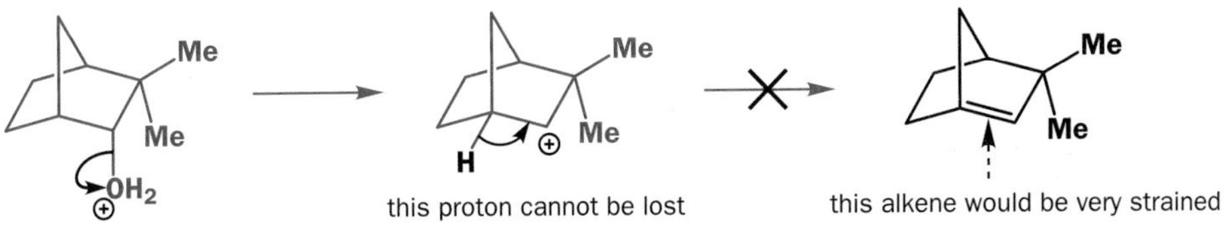

However, migration of a methyl group both stabilizes the cation—it becomes tertiary instead of secondary—and allows E1 elimination of H^+ to take place to give a stable alkene.

secondary carbocation

tertiary carbocation

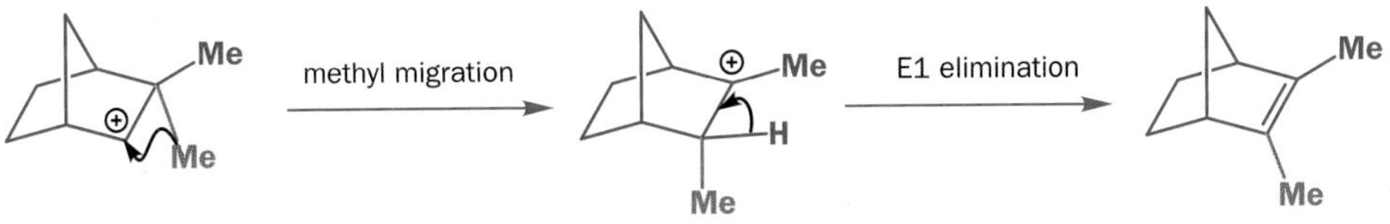

The migration of an alkyl group to a cationic centre is known as a **Wagner–Meerwein rearrangement** or **Wagner–Meerwein shift**, and this migration is, of course, a synthetic manifestation of the rearrangement we have just been looking at in NMR spectra. Wagner–Meerwein shifts have been studied extensively in the class of natural products to which both of these natural products belong—terpenes—and we will come back to them in Chapter 51 (natural products). For the moment, though, we will just illustrate this type of reaction with one more example—another acid-catalysed dehydration, of isoborneol to give camphene.

isoborneol

camphene

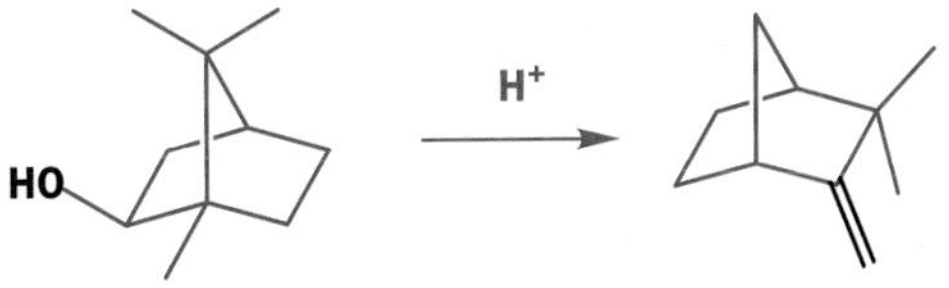

This one *seems* much more complicated—but, in fact, only one alkyl migration is involved. To see what has happened, remember the 'top tip'—number the carbons. You can number the starting material any way you choose—we've started with the *gem*-dimethyl group because it will be easy to spot in the product. The numbers just follow round the ring, with C8 being the methyl group attached to C5.

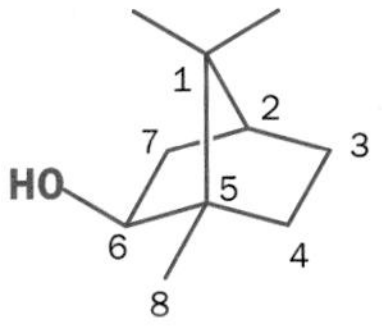

Now for the hard bit—we need to work out which carbon in the starting material becomes which carbon in the product. The best thing is just have a go—mistakes will soon become obvious, and you can always try again.

- Use the substituents to help you—some will have changed, but most will be the same or similar—for example, C1 is still easy to spot as the carbon carrying the *gem*-dimethyl group

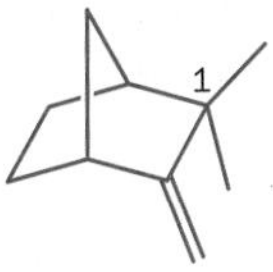

- Use connectivity to help you—again, a C–C bond or two may have broken or formed, but most of the C–C bonds in the starting material will be there in the product. C1 and C2 will probably still be next door to one another—C2 was a bridgehead carbon in the starting material, and there is a bridgehead C attached to C1 in the product; assume that's C2

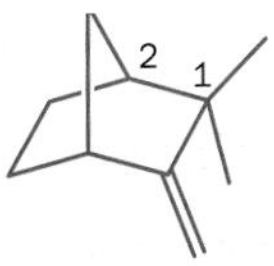

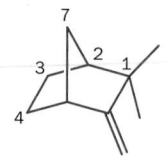

- C3 and C4 were unsubstituted carbons in the starting material, and are identifiable in the product too. The other easily spotted atom is C7—an unsubstituted C attached to C2

- C5, C6, and C8 are harder. We can assume that C8 is the =CH$_2$ carbon—it was a methyl group but perhaps has become involved in an elimination. C5 was attached to C1, C4, C6, and C8: one of the remaining carbons is attached to C1 and C8, so that seems more likely to be C5, which leaves C6 as the bridgehead, attached as before to C7 and C5

Now we have the whole picture and we can assess what has happened in the reaction—which old bonds have been broken and which new bonds have been formed.

Numbering the atoms this way identifies the likely point of rearrangement—the only bond broken is between C4 and C5. Instead we have a new one between C5 and C6: C4 appears to have migrated from C5 to C6. Now for the mechanism. The first step will, of course, be loss of water to generate a secondary cation at C6. The cation is next to a quaternary centre, and migration of any of three bonds could generate a more stable tertiary carbocation. But we know that the new bond in the product is between C4 and C6, so let's migrate carbon 4. Manipulating the diagrams a bit turns up a structure remarkably similar to our product, and all we need to do is lose a proton from C8.

migrate C4 from C5 to C6 to create tertiary cation

> If you are observant, you may ask why the alkyl group migrated in this example and not the methyl group, or the other alkyl group— all three possibilities give similar tertiary carbocations. The reason involves the *alignment* of the orbitals involved, which we will discuss at the end of the chapter.

Although migration of an alkyl group that forms part of a ring leads to much more significant changes in structure than simple migration of a methyl group, the reason why it happens is still just the same.

● **Alkyl migrations occur in order to make a carbocation more stable.**

Ring expansion means rearrangement

'More stable' usually means 'more substituted', but cations can also be made *more stable* if they become *less strained*. So, for example, four-membered rings adjacent to cations readily rearrange to five-membered rings in order to relieve ring strain.

four-membered ring   HCl →   five-membered ring

This time the cation is formed by protonation of an alkene, not departure of a leaving group, but writing a mechanism should now be a straightforward matter to you.

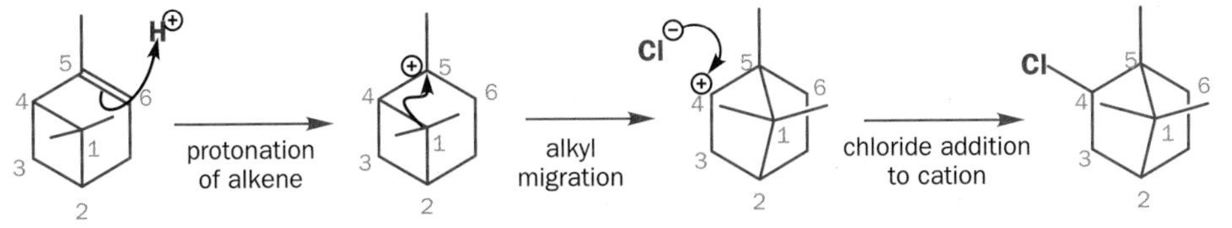

Though the rearrangement step transforms a stable tertiary cation into a less stable secondary cation, relief of strain in expansion from a four- to a five-membered ring makes the alkyl migration favourable. In 1964, E.J. Corey published a synthesis of the natural product α-caryophyllene alcohol that made use of a similar ring expansion. Notice the photochemical [2+2] cycloaddition (Chapter 35) in the synthesis of the starting material.

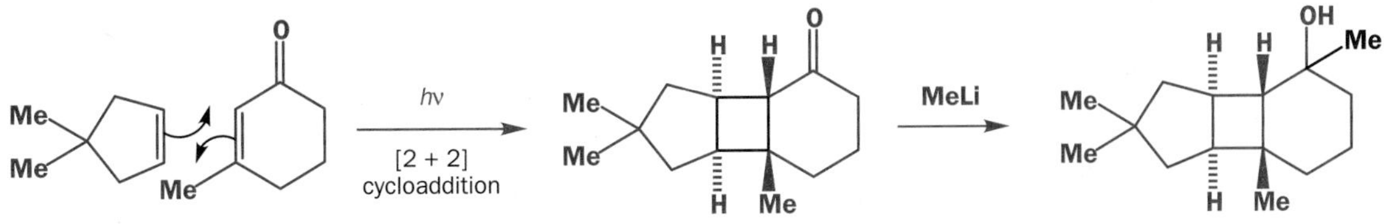

Rearrangement of this tertiary alcohol in acid gives the target natural product. The four-membered ring has certainly disappeared but it may not be obvious at first what has taken its place.

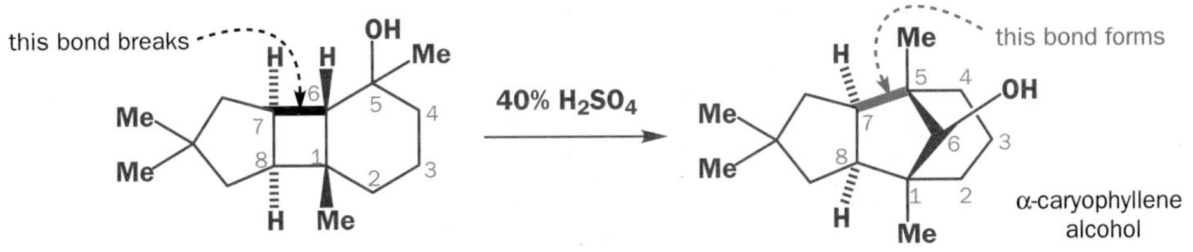

As usual, numbering the atoms makes clear what has happened: carbon 7 has migrated from carbon 6 to carbon 5. Loss of water gives a tertiary carbocation that undergoes rearrangement to a secondary carbocation with expansion of a four- to a five-membered ring.

rearrangement relieves strain in this 4-membered ring

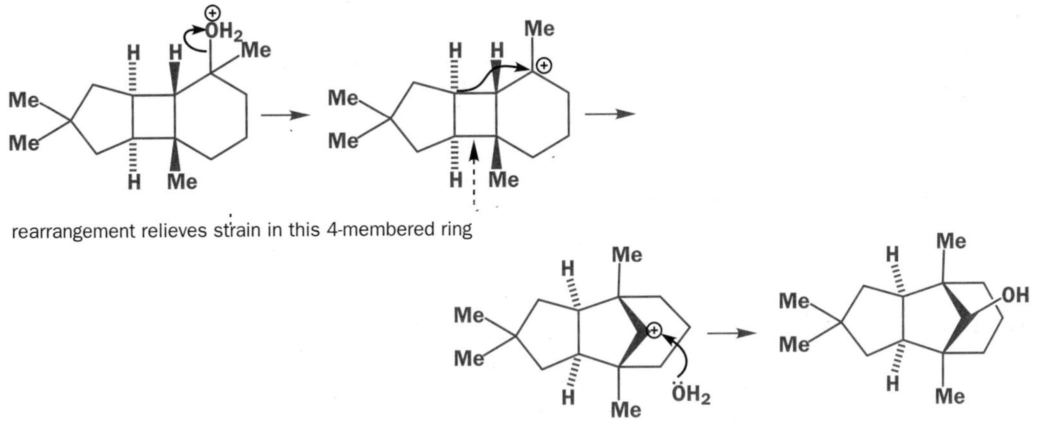

Most compounds are *kinetically* stable precisely because spontaneous rearrangements to more *thermodynamically* stable compounds do not occur—the kinetic barrier to rearrangement is too high. You did meet a few exceptions in the last chapter—cyclopentadienes, for example, undergo rapid [1,5]-sigmatropic shifts of hydrogen, and are unstable with respect to the position of the double bonds. Carbocations are probably the most important class of species that *habitually* undergo rearrangement reactions, even at low temperature.

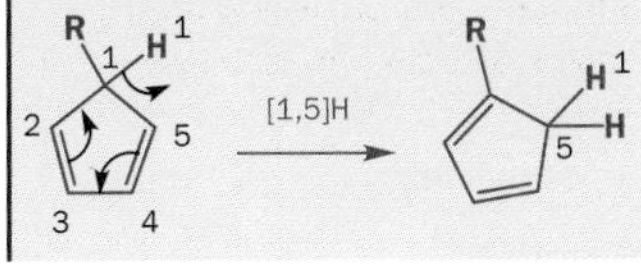

Carbocation rearrangements: blessing or curse?

Well, that depends. You have now seen a few useful carbocation rearrangements that give single products in high yield. But you have also met at least one reaction that *cannot* be done because of carbocation rearrangements: Friedel–Crafts alkylation using primary alkyl halides.

The Friedel–Crafts alkylation illustrates the problems of trying to use carbocation rearrangements to make single products in high yield. We can give three guidelines to spotting this type of reaction.

1   The rearrangement must be fast so that other reactions do not compete

2   The product cation must be sufficiently more stable than the starting one so that the rearrangement happens in high yield

3   Subsequent trapping of the product cation must be reliable: cations are high-energy intermediates, and are therefore unselective about how they react

A reaction is no good if the cation reacts in more than one way—it may react with a nucleophile, eliminate, or undergo further rearrangement—but it must do only one of these! For the rest of the chapter, we will address only reactions that, unlike this Friedel–Crafts reaction, follow these guidelines. The reactions we will talk about all happen in good yield.

The pinacol rearrangement

When the 1,2-diol 'pinacol' is treated with acid, a rearrangement takes place.

> Pinacol, the trivial name for the starting material, which is made from acetone by a reaction you will meet in Chapter 39, gives its name to this class of rearrangements, and to the product, 'pinacolone'.

Whenever you see a rearrangement, you should now think 'carbocation'. Here, protonation of one of the hydroxyl groups allows it to leave as water, giving the carbocation.

You now know that carbocations rearrange by alkyl shifts to get as stable as they can be—but this carbocation is already tertiary, and there is no ring strain, so why should it rearrange? Well, here we have another source of electrons to stabilize the carbocation: lone pairs on an oxygen atom. We pointed out early in the chapter that oxygen is very good at stabilizing a positive charge on an adjacent atom, and somewhat less good at stabilizing a positive charge two atoms away. By rearranging, the first-formed carbocation gets the positive charge into a position where the oxygen can stabilize it, and loss of a proton from oxygen then gives a stable ketone.

> Unlike sulfur, which stabilizes a charge 2 atoms away better than it stabilizes a charge on an adjacent atom.

You can view the pinacol as a rearrangement with a 'push' and a 'pull'. The carbocation left by the departure of water 'pulls' the migrating group across at the same time as the oxygen's lone pair 'pushes' it. A particularly valuable type of pinacol rearrangement forms spirocyclic ring systems. You may find this one harder to follow, though the mechanism is identical with that of the last example. Our 'top tip' of numbering the atoms should help you to see what has happened: atom 2 has migrated from atom 1 to atom 6.

> ■ **Spirocycles** are pairs of rings joined at a single carbon atom (Chapter 33).

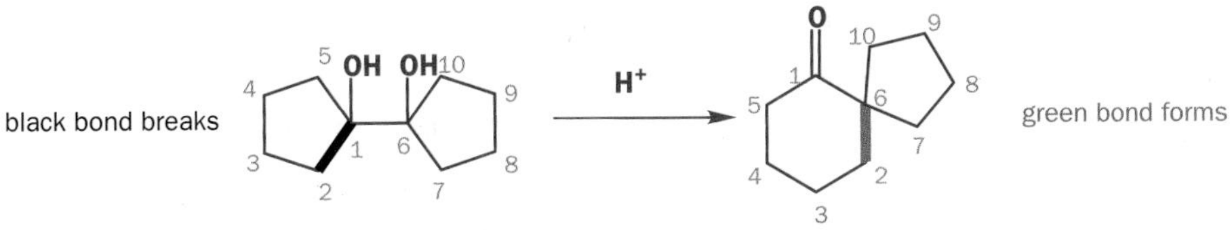

black bond breaks

green bond forms

When drawing the mechanism it doesn't matter which hydroxyl group you protonate or which adjacent C–C bond migrates—they are all the same. One five-membered ring expands to a six-membered ring but the reason this reaction happens is the formation of a carbonyl group, as in all pinacol rearrangements.

> Of course, it doesn't matter how you number the atoms, but the numbering must be consistent. Usually, your initial impression of a greatly changed molecule will come down to just one or two atoms changing their substitution pattern, and numbering will help you to work out which ones they are.

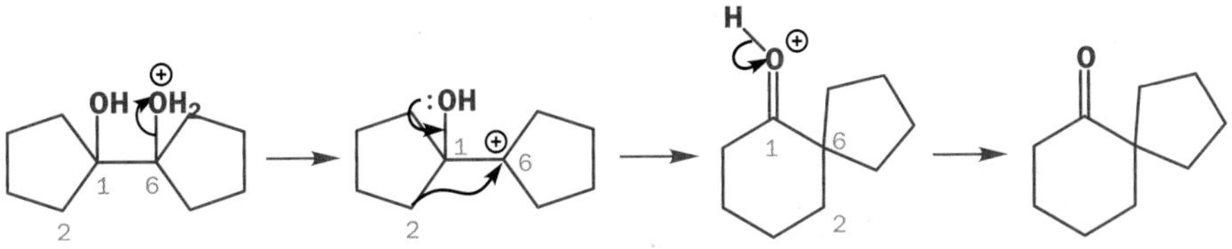

The pinacol reaction in synthesis

A nice synthesis of the bicyclic alkene on the right starts with a pinacol reaction.

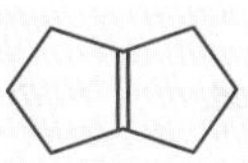

The first step is straightforward—just like the one you have just met. The 'pinacol' dimer from cyclobutanone rearranges with the expansion of one of the rings to give a cyclopentanone fused *spiro* to the remaining four-membered ring. Reduction of the ketone then gives an

alcohol that rearranges to the alkene in acid. Try working out a mechanism for this transformation—start by protonating of the alcohol and allowing water to leave to give a cation. You might also like to think about why the rearrangement happens—for a clue go back to p. 983.

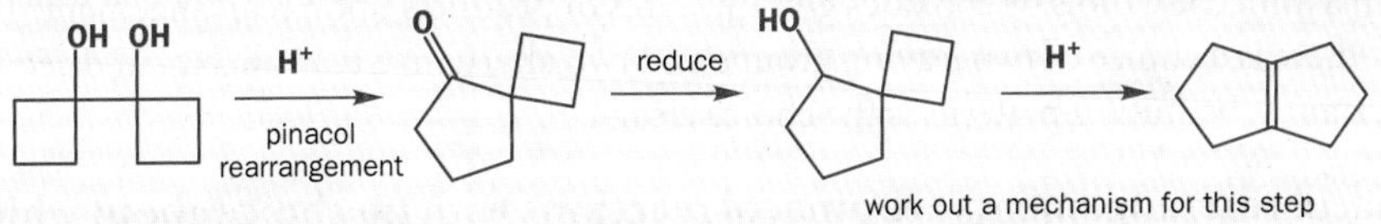

Epoxides rearrange with Lewis acids in a pinacol fashion

The intermediate cation in a pinacol rearrangement can equally well be formed from an epoxide, and treating epoxides with acid, including Lewis acids such as $MgBr_2$, promotes the same type of reaction.

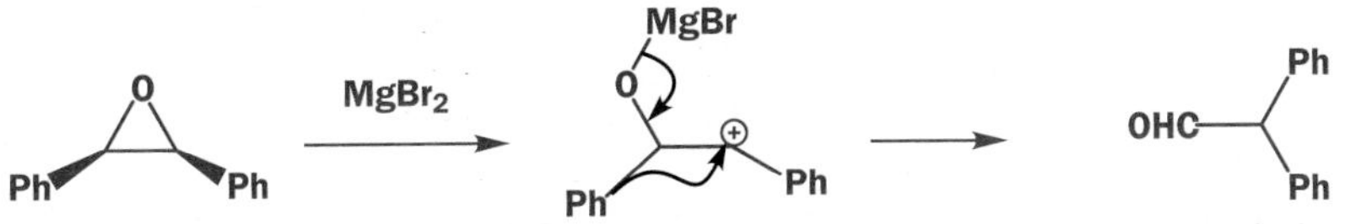

Rearrangement of epoxides with magnesium salts means that opening epoxides with Grignard reagents can give surprising results.

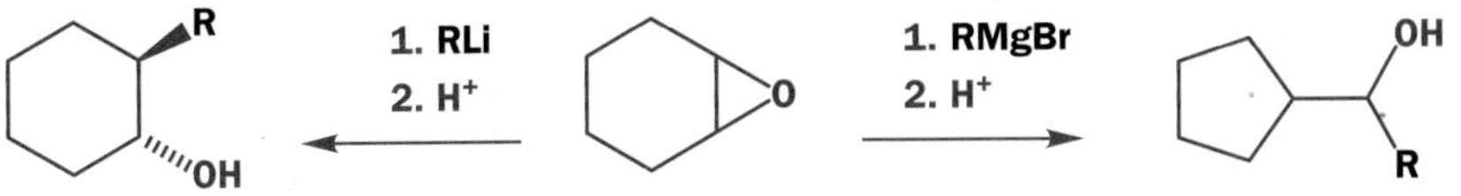

The alkyllithium reaction is quite straightforward as long as the alkyllithium is free of lithium salts. A clue to what has happened with the Grignard reagents comes from the fact that treating this epoxide with just $MgBr_2$ (no RMgBr) gives an aldehyde.

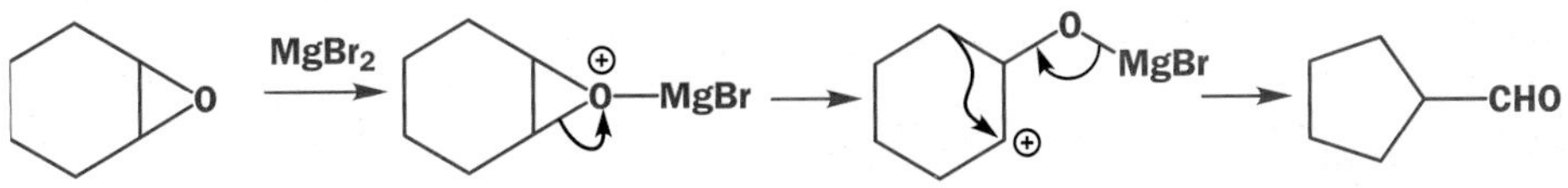

With a Grignard reagent, rearrangement occurs faster than addition to the epoxide, and then the Grignard reagent adds to the aldehyde.

Some pinacol rearrangements have a choice of migrating group

With these symmetrical diols and epoxides, it does not matter which hydroxyl group is protonated and leaves, nor which end the epoxide opens, nor which group migrates. When an unsymmetrical diol or epoxide rearranges, it *is* important which way the reaction goes. Usually, the reaction leaves behind the more stable cation. So, for example, this unsymmetrical diol gives the ring-expanded ketone, a starting material for the synthesis of analogues of the drug methadone.

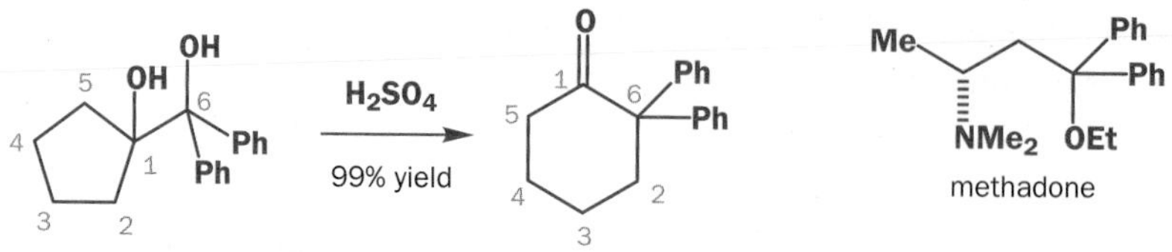

This product is formed because the green OH group leaves more readily than the black because the carbocation stabilized by two phenyl groups forms more readily than the carbocation stabilized by two alkyl groups. The migration step follows without selectivity as both alkyl groups on the black alcohol are the same.

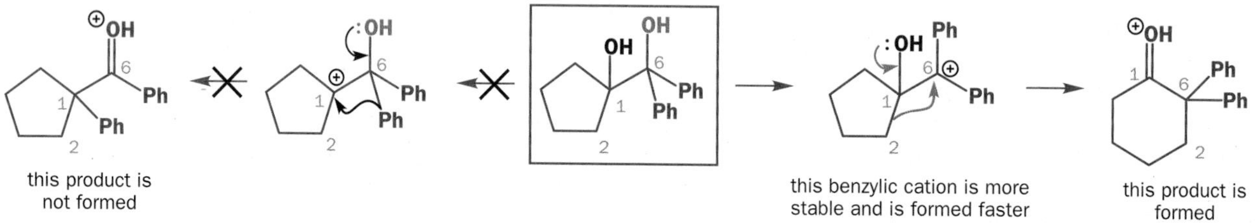

Most unsymmetrical diols or epoxides give mixtures of products upon rearrangement. The problem is that there is a choice of two leaving groups and two alternative rearrangement directions, and only for certain substitution patterns is the choice clear-cut.

Semipinacol rearrangements are pinacol reactions with no choice about which way to go

In 1971, French chemists needed this seven-membered cyclic ketone. A reasonable starting material to use is this diol, because it can be made in two steps from the natural product isonopinone.

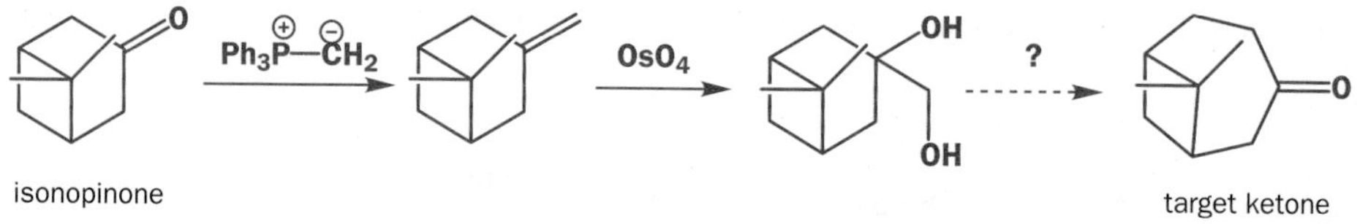

The reaction they needed for the last stage is a pinacol rearrangement—the *primary* hydroxyl group needs persuading to leave as the ring expands. The problem is, of course, that the tertiary hydroxyl group is much more likely to leave since it leaves behind a more stable carbocation.

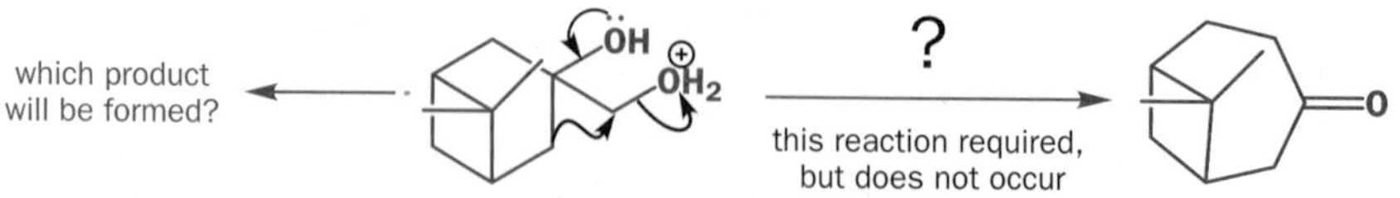

The solution to this problem is to force the primary hydroxyl group to be the leaving group by making it into a tosylate. The primary hydroxyl group reacts more rapidly with TsCl than the tertiary one because it is less hindered. A weak base is now all that is needed to make the compound rearrange in what is known as a **semipinacol rearrangement**.

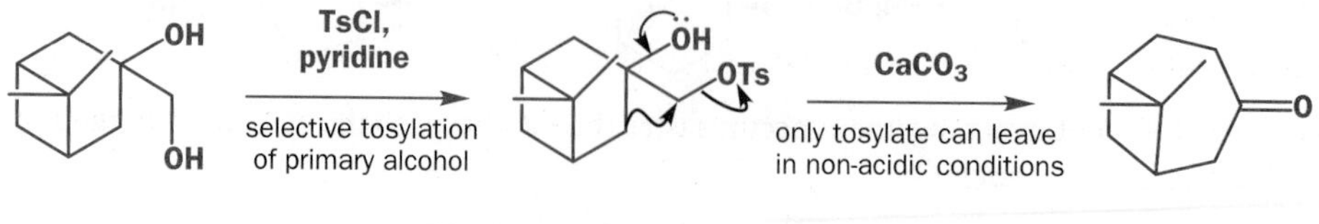

Semipinacol rearrangements are rearrangements in which a hydroxyl group provides the electrons to 'push' the migrating group across, but the 'pull' comes from the departure of leaving groups other than water—tosylate in this example, but typically also halide or nitrogen (N_2). Since tosylation occurs at the *less* hindered hydroxyl group of a diol, not only can semipinacol rearrangements be more regioselective than pinacol rearrangements, but their regioselectivity may be in the opposite direction.

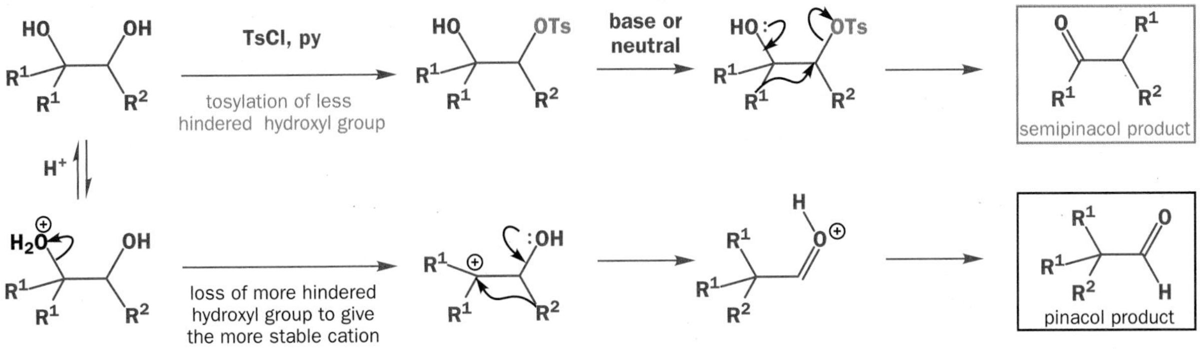

Corey exploited this in a synthesis of the natural product longifolene. He needed to persuade an easily made 6,6-fused ring system to undergo rearrangement to a ring-expanded ketone. Again, a normal acid-catalysed pinacol rearrangement is no good—the tertiary, allylic hydroxyl group is much more likely to ionize, and the acid-sensitive protecting group would be hydrolysed too. Tosylation of the secondary alcohol in the presence of the tertiary is possible, and semipinacol rearrangement gives the required ketone.

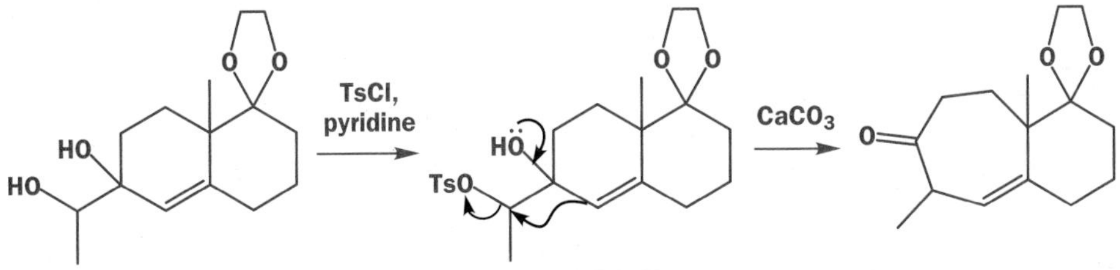

The leaving group need not be tosylate: in the following example, part of a synthesis of bergamotene (a component of valerian root oil and the aroma of Earl Grey tea), a 2-iodo alcohol rearranges.

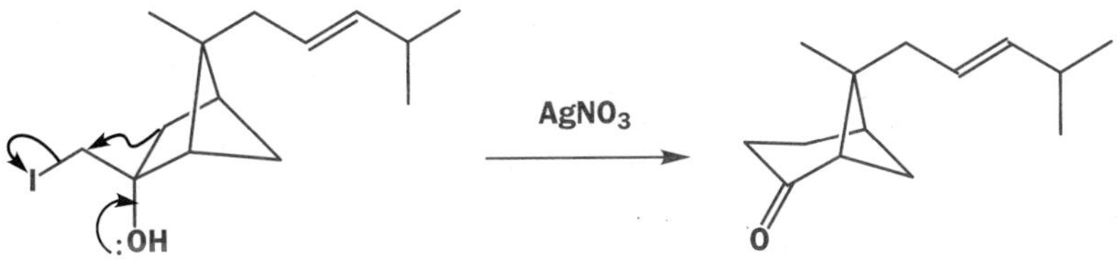

Treating 2-halo alcohols with base is, of course, a good way to make epoxides. Using $AgNO_3$ to improve iodide's leaving ability without increasing the nucleophilicity of the hydroxyl group favours rearrangement at the expense of epoxide formation. There would certainly be a danger of epoxide formation in strong base.

The structure of bergamotene

The structure of bergamotene was, for some years during the 1960s, a matter of debate. The difficult question was the configuration of the chiral centre ringed in black. With modern spectroscopic techniques, we can now solve this type of problem simply, but the only solution then was to synthesise the two isomers and compare them with the natural material. There is more about bergamotene in Chapter 46.

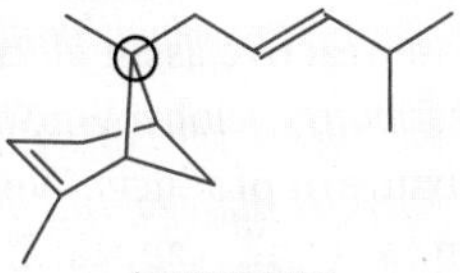

bergamotene

Semipinacol rearrangements of diazonium salts

You saw in Chapter 22 how aromatic amines can be converted to diazonium salts by treatment with acidic sodium nitrite.

It might be an idea to review pp. 597–598 of Chapter 23 to be sure you understand the mechanism of this reaction.

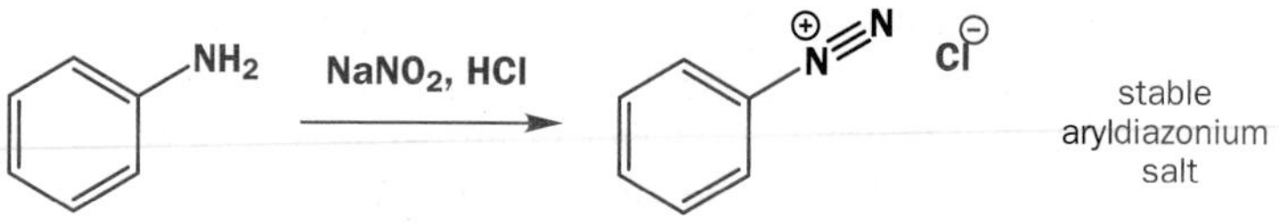

stable
aryldiazonium
salt

Aryldiazonium salts are stable but *alkyl*diazonium salts are not: nitrogen gas is the world's best leaving group, and, when it goes, it leaves behind a carbocation.

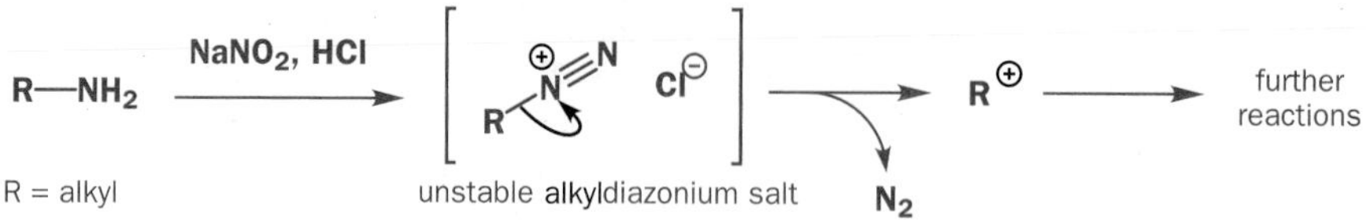

R = alkyl   unstable **alkyl**diazonium salt

Semipinacol rearrangements of diazonium salts derived from 2-amino alcohols are sometimes called **Tiffeneau–Demjanov rearrangements**.

One of the 'further reactions' this carbocation can undergo is rearrangement. If the starting amine is a 2-amino alcohol, the cation can be stabilized by a semipinacol rearrangement.

61% yield

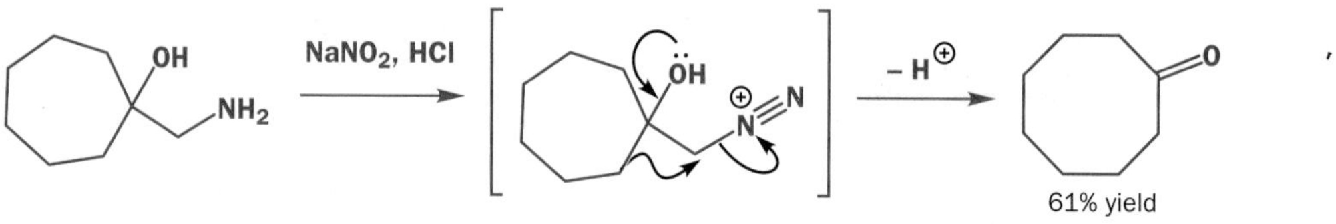

While alkyldiazonium salts are unstable, their conjugate bases, diazoalkanes, are stable enough to be prepared and are nucleophilic towards carbonyl compounds. **Diazoalkanes** are neutral compounds having one fewer proton than diazonium salts and are delocalized structures with a central sp nitrogen atom.

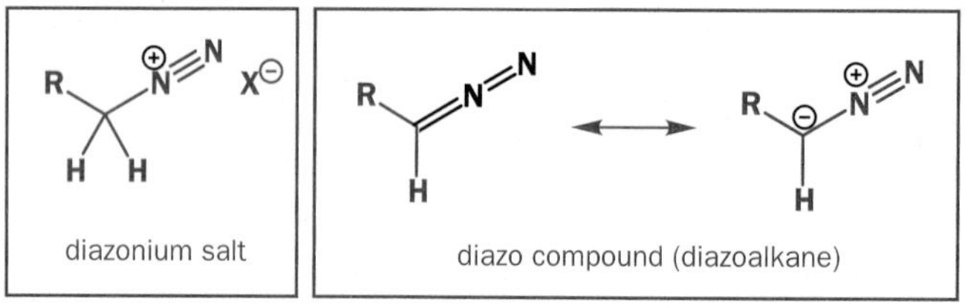

diazonium salt   diazo compound (diazoalkane)

When diazomethane (a compound we will investigate in more detail in Chapter 40) adds to a ketone, the product undergoes a ring expansion by rearrangement of the same type of intermediate.

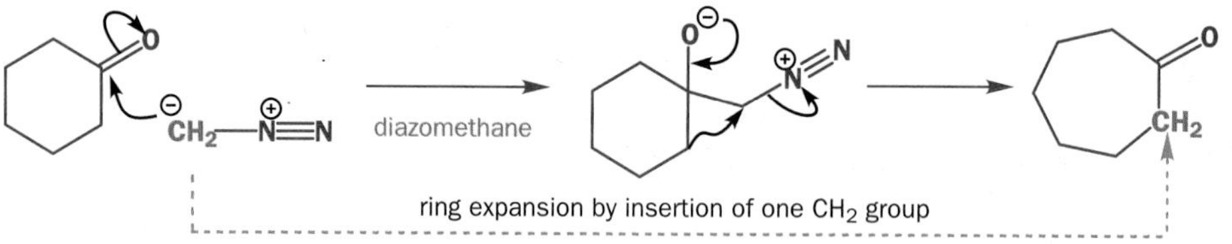

diazomethane

ring expansion by insertion of one CH_2 group

The problem with reactions like this is that both the starting material and product are ketones, so they work cleanly only if the starting material is more reactive than the product. Cyclohexanone is more reactive as an electrophile than either cyclopentanone or cycloheptanone, so it ring expands cleanly to cycloheptanone. But expansion of cyclopentanone to cyclohexanone is messy and gives a mixture of products. We shall come back to diazo compounds in more detail in Chapter 40; diazonium salts will reappear in Chapter 38 where their decomposition will provide the driving force for fragmentation reactions.

The dienone–phenol rearrangement

The female sex hormone oestrone is the metabolic product of another hormone, progesterone, itself made in the body from cholesterol.

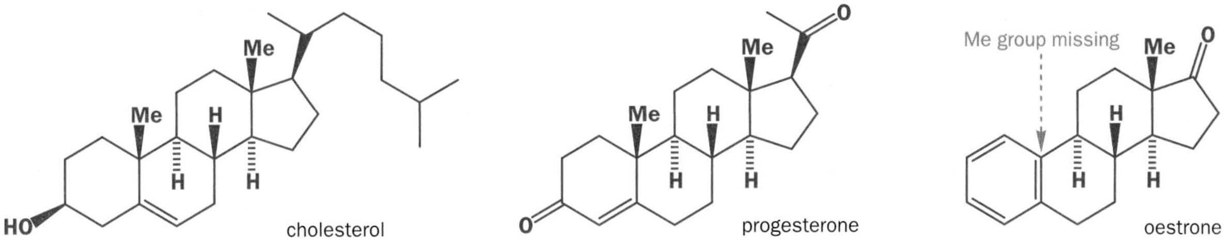

cholesterol   progesterone   oestrone

Oestrone lacks one of progesterone's methyl groups, probably removed in the body as CO_2 after oxidation. In 1946, Carl Djerassi, a man whose work led directly to the invention of the contraceptive pill, showed that another derivative of cholesterol could be rearranged to the oestrone analogue 1-methyloestradiol—notice how the methyl group has this time migrated to an adjacent carbon atom. At the same time, the dienone has become a phenol.

<aside>
Carl Djerassi, an American born in Vienna in 1923, worked chiefly at CIBA, Syntex in Mexico, and at Stanford. He developed syntheses of human steroids from compounds in plants, was a pioneer of mass spectrometry, and is a colourful campaigner for peace and disarmament.
</aside>

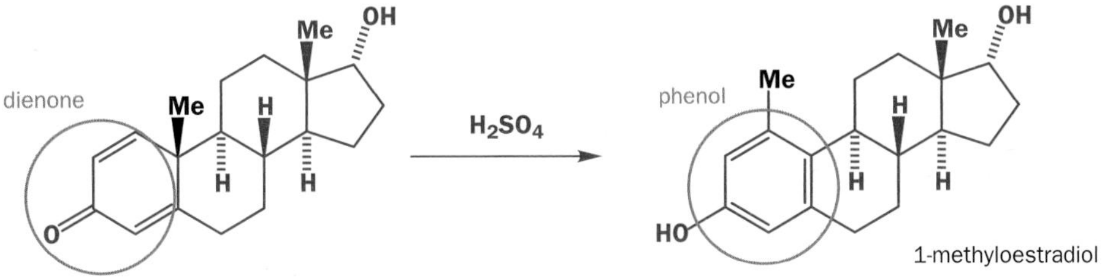

1-methyloestradiol

This type of rearrangement is known helpfully as a **dienone–phenol rearrangement**, and we can consider it quite simply as a type of *reverse* pinacol rearrangement. Pinacol and semipinacol rearrangements are driven by the formation of a carbonyl group. The rearranged cation is stabilized by being next to oxygen, and it can rapidly lose H^+ to give a carbonyl compound. In the key step of a dienone–phenol rearrangement, a protonated carbonyl compound rearranges to a tertiary carbocation.

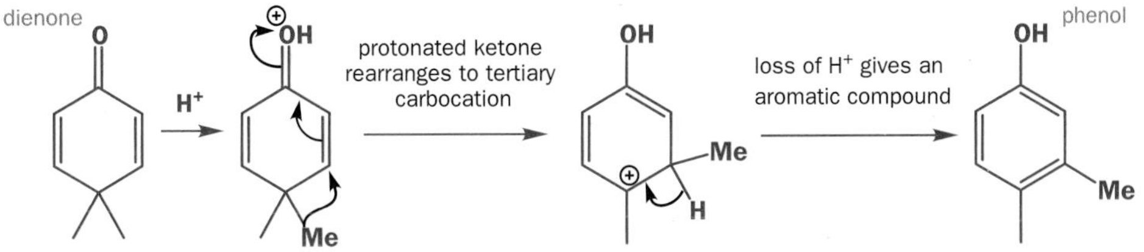

The reaction is driven from dienone to phenol because the product cation can rapidly undergo elimination of H^+ to become aromatic.

The benzilic acid rearrangement

You have seen rearrangements in which carbonyl groups form at the migration origin: the migrating group in the pinacol and semipinacol rearrangements is 'pushed' by the oxygen's lone pair as it forms the new carbonyl group. You have also seen carbonyl groups being destroyed at the migration terminus: the migrating group in the dienone–phenol rearrangement is 'pulled' towards the protonated carbonyl group. The first rearrangement reaction ever to be described has both of these at once.

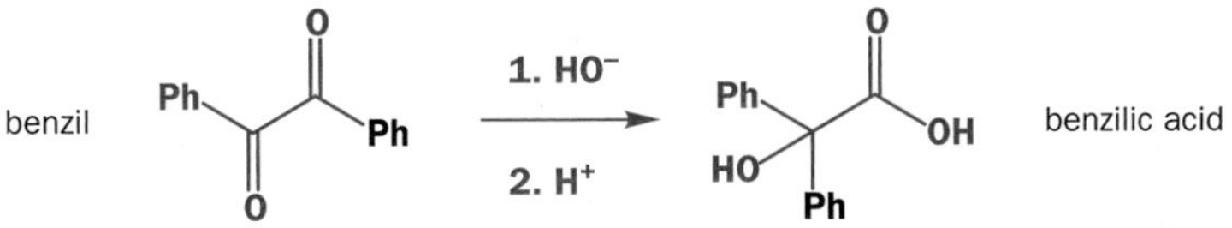

benzil   benzilic acid

In 1838, Justus von Liebig found that treating 'benzil' (1,2-diphenylethan-1,2-dione) with hydroxide gave, after acid quench, 2-hydroxy-2,2-diphenylacetic acid, which he called 'benzilic acid'.

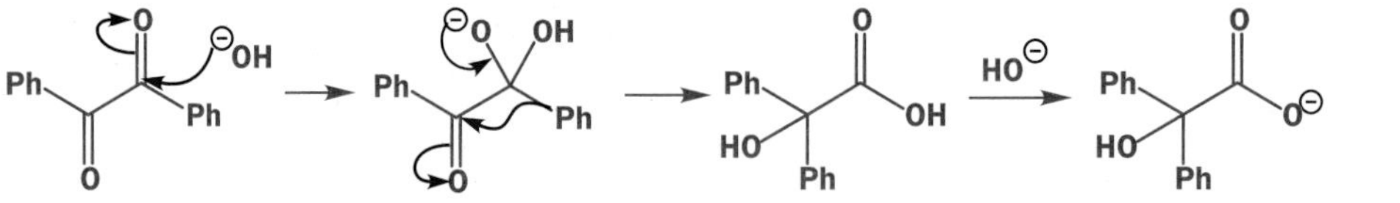

> You may find it helpful to think of the benzilic acid rearrangement as a semipinacol rearrangement in which we have a breaking C=O π bond instead of a leaving group.

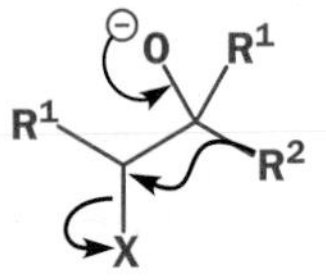

compare the migration step with this semipinacol rearrangement

The mechanism of this **benzilic acid rearrangement** starts with attack of hydroxide on one of the carbonyl groups. The tetrahedral intermediate can collapse in a reaction reminiscent of a semipinacol rearrangement.

carbonyl group is formed here

C=O π bond is broken here   deprotonation of acid makes the reaction irreversible

With alkoxides, the benzilic acid rearrangement can lead directly to esters by the same sort of mechanism.

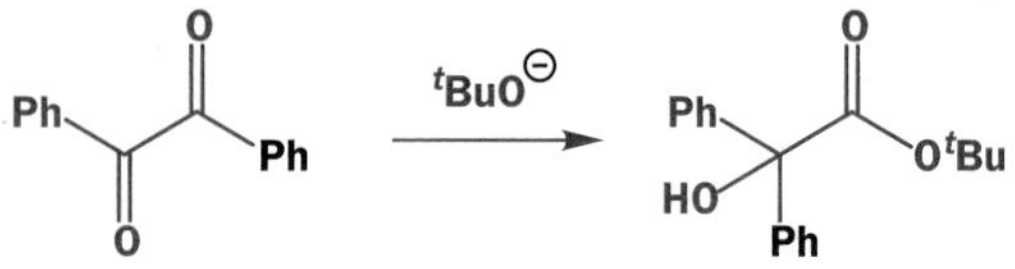

The Favorskii rearrangement

We hope you have appreciated the smooth mechanistic progression so far in this chapter, from Wagner–Meerwein to pinacol and semipinacol through dienone–phenol to benzilic acid. Our aim is to help you gain an overall view of the types of rearrangements that take place (and why) and not to present you with lots of disconnected facts. It is at this point, however, that our mechanistic journey takes a hairpin bend. A surprising one, too, because, when we show you the Favorskii rearrangement, you would be forgiven for wondering what the fuss is about: surely it's rather like a variant of the benzilic acid rearrangement?

the Favorskii rearrangement
rearrangement of an α-halo ketone to an ester

the benzilic acid rearrangement
rearrangement of a diketone to an ester

rearrangement involves breakage of C–X bond

superficially similar; mechanistically quite different!

rearrangement involves breakage of C=O bond

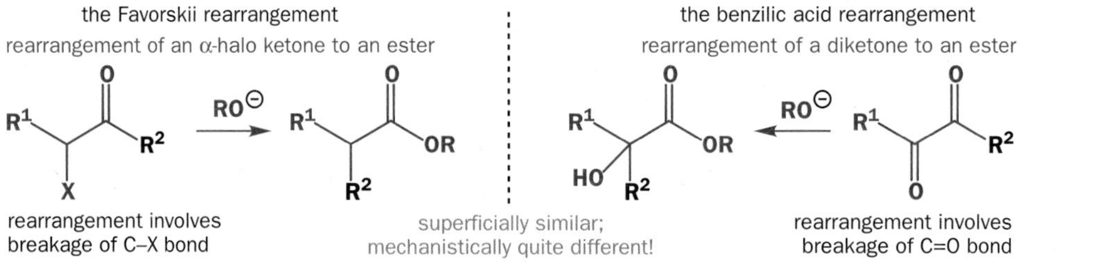

Well, this is what chemists thought until 1944, when some Americans found that two isomeric α-chloro ketones gave exactly the same product on treatment with methoxide. They suggested that both reactions went through the same intermediate.

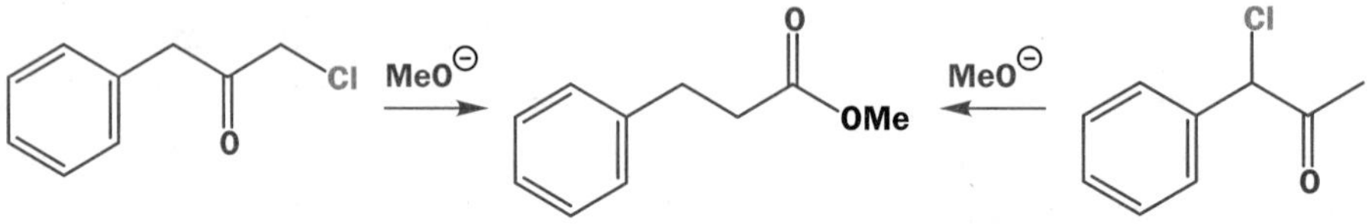

> A full discussion of this point requires Baldwin's rules, which appear in Chapter 41.

That intermediate is a three-membered cyclic ketone, a cyclopropanone: the alkoxide acts not as a nucleophile (its role in the benzilic acid rearrangement) but as a base, enolizing the ketone. The enolate can alkylate itself intramolecularly in a reaction that looks bizarre but that many chemists think is not unreasonable. The product is the same cyclopropanone in each case.

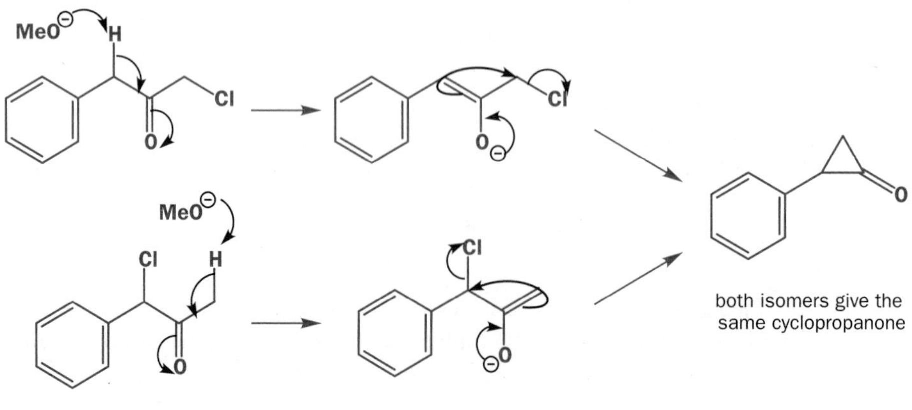

both isomers give the same cyclopropanone

Other chemists prefer a pericyclic description of the ring-closure step. The same enolate simply loses chloride to give an 'oxyallyl cation'—a dipolar species with an oxyanion and a delocalized allylic cation. This species can cyclize in a two-electron disrotatory electrocyclic reaction (Chapter 36) to give the same cyclopropanone. We shall return to this discussion in the next chapter but, whatever the mechanism, there is no doubt that a cyclopropanone is an intermediate.

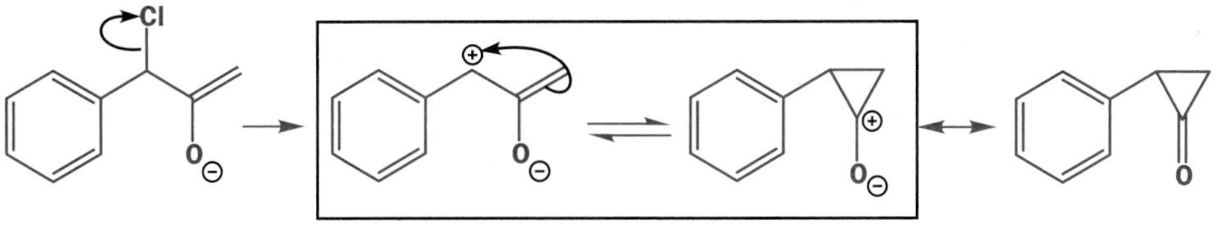

two-electron disrotatory electrocyclic reaction

Cyclopropanones are very reactive towards nucleophiles, and the tetrahedral intermediate arising from the attack of methoxide springs open to give the ester product. The more stable carbanion leaves: though the carbanion is not actually formed as a free species, there must be considerable negative charge at the carbon atom as the three-membered ring opens. Here the benzyl group is the better leaving group.

> Cyclopropanones and cyclobutanones are very reactive, rather like epoxides, because, while the 60° or 90° angle in the ring is nowhere near the tetrahedral angle (108°), it is nearer 108° than the 120° preferred by the sp^2 C of the C=O group. Conversely, the small ring ketones are resistant to enolization, because that would place *two* sp^2 carbon atoms in the ring.

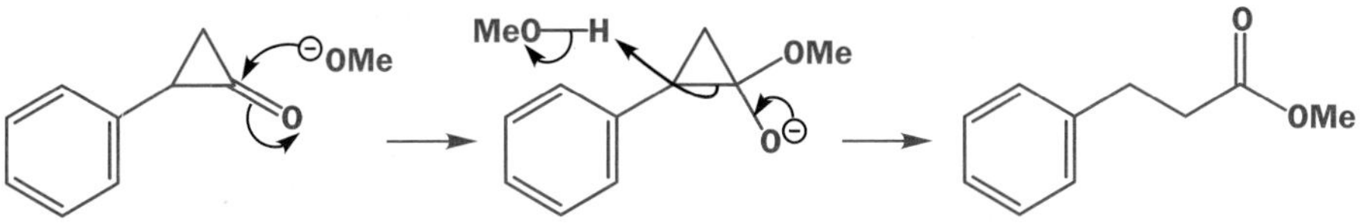

Favorskii rearrangement of cyclic 2-bromoketones leads to ring contraction and this has become one of the most fruitful uses of the rearrangement in synthesis. Bromination of cyclohexanone is a simple reaction (Chapter 21) and treatment with methoxide gives the methyl ester of cyclopentane carboxylic acid in good yield.

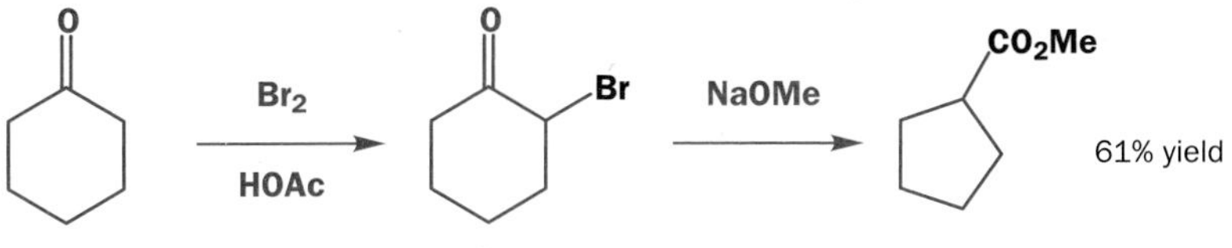

61% yield

Enolization occurs on the side of the ketone away from the bromine atom and the enolate cyclizes as before but the cyclopropanone intermediate is symmetrical so that the product is the same whichever C–C bond breaks after nucleophilic attack by the methoxide ion.

Cubane synthesis

In 1964, two American chemists synthesized for the first time a remarkable molecule, cubane. Two of the key steps were Favorskii rearrangements, which allowed the chemists to contract five-membered rings to four-membered rings. Here is one of them. Two more steps decarboxylate the product to give cubane itself.

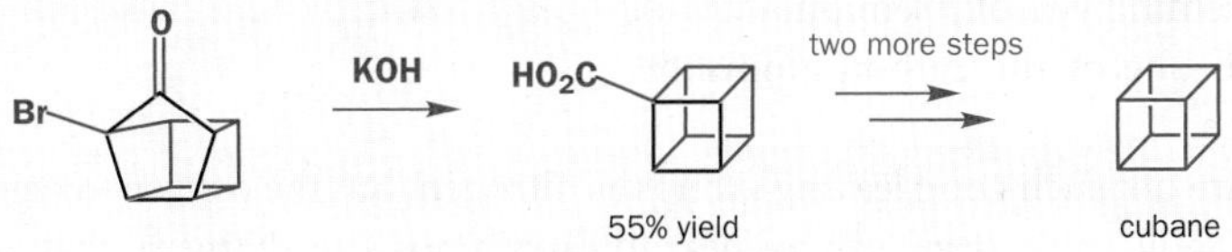

55% yield   cubane

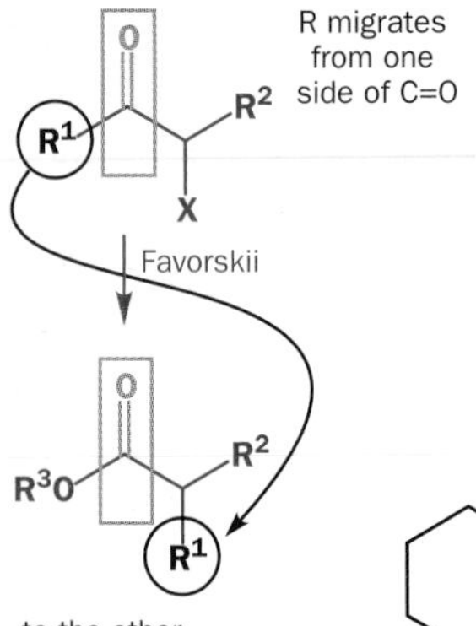

R migrates from one side of C=O

R¹ R²

X

Favorskii

R³O R²

R¹

to the other

The overall consequence of the Favorskii rearrangement is that an alkyl group is transferred from one side of a carbonyl group to the other.

This means that it can be used to build up heavily branched esters and carboxylic acids—the sort that are hard to make by alkylation because of the problems of hindered enolates and unreactive secondary alkyl halides. Heavily substituted acids, where CO_2H is attached to a tertiary carbon atom, would be hard to make by any other method. And the Favorskii rearrangement is a key step in this synthesis of the powerful painkiller Pethidine.

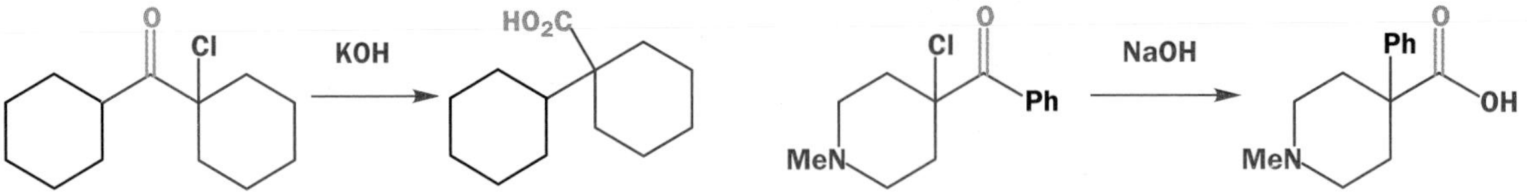

KOH

NaOH

Pethidine

The Favorskii mechanism will help you understand the Ramberg–Bäcklund reaction in Chapter 46—the two reactions have quite similar mechanisms.

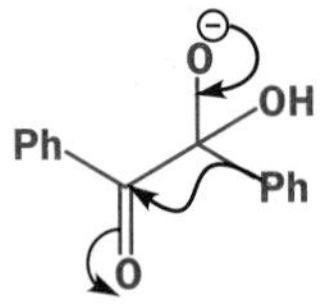

Ph OH Ph

O

compare the migration step with this benzylic acid rearrangement

Try writing a mechanism for this last reaction and you run into a problem—there are no acidic protons so the ketone cannot be enolized! Yet the Favorskii rearrangement still works. Despite our warnings against confusing the mechanisms of the Favorskii and benzilic acid rearrangements, the Favorskii rearrangement may, in fact, follow a benzilic (or 'semibenzilic', by analogy with the semipinacol) rearrangement mechanism, *if* there are no acidic hydrogens available.

'semibenzilic' Favorskii rearrangement of nonenolisable ketones

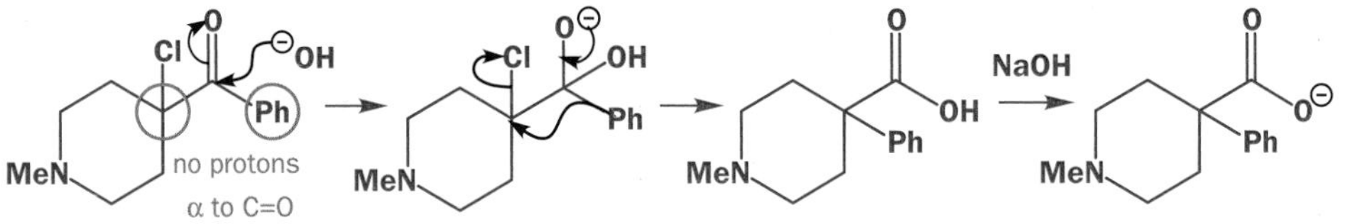

no protons α to C=O

NaOH

Migration to oxygen: the Baeyer–Villiger reaction

In 1899, the Germans, A. Baeyer and V. Villiger, found that treating a ketone with a peroxy-acid (RCO_3H) can produce an ester. An oxygen atom is 'inserted' next to the carbonyl group.

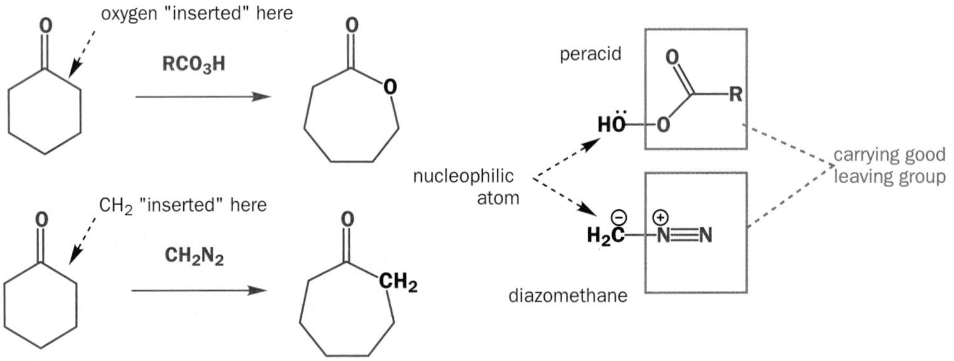

oxygen "inserted" here

RCO_3H

peracid

nucleophilic atom

carrying good leaving group

diazomethane

CH_2 "inserted" here

CH_2N_2

Now, you saw a similar 'insertion' reaction earlier in the chapter, and the mechanism here is not dissimilar. Both peracids and diazomethane contain a nucleophilic centre that carries a good leaving group, and addition of peracid to the carbonyl group gives a structure that should remind you of a semipinacol intermediate with one of the carbon atoms replaced by oxygen.

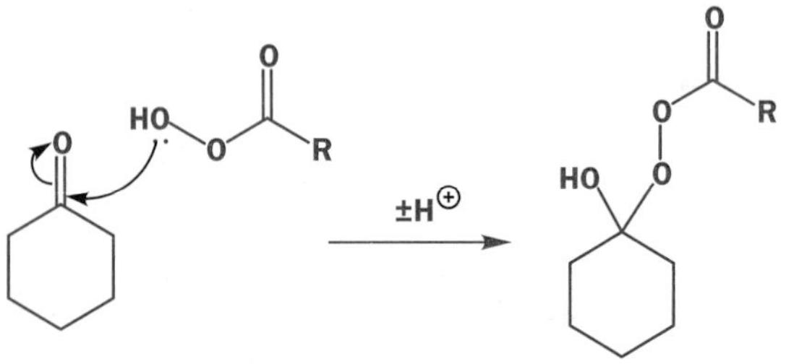

$\pm H^{\oplus}$

Carboxylates are not such good leaving groups as nitrogen, but the oxygen–oxygen single bond is very weak and monovalent oxygen cannot bear to carry a positive charge so that, once the peracid has added, loss of carboxylate is concerted with a rearrangement driven, as in the case of the pinacol and semipinacol, by formation of a carbonyl group.

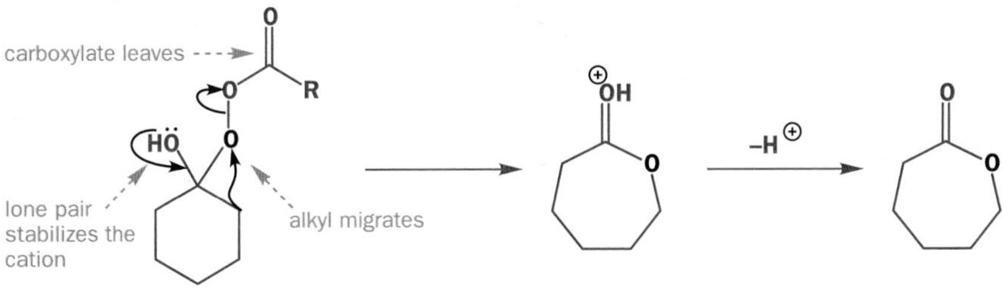

Baeyer–Villiger reactions are among the most useful of all rearrangement reactions, and the most common reagent is *m*-CPBA (*meta*-chloroperbenzoic acid) because it is commercially available.

Which group migrates? (I)—the facts

A question we have deliberately avoided up to this point is this: when there is a competition between two migrating groups, *which group migrates?* This question arises in pinacol, semipinacol, and dienone–phenol rearrangements and in Baeyer–Villiger reactions (in the benzilic acid and Favorskii rearrangements, there is no choice) and the awkward fact is that the answer is different in each case! However, let's start with the Baeyer–Villiger reaction, because here the question is always valid except when the ketone being oxidized is symmetrical. Here are some examples; and you can probably begin to draw up guidelines for yourself.

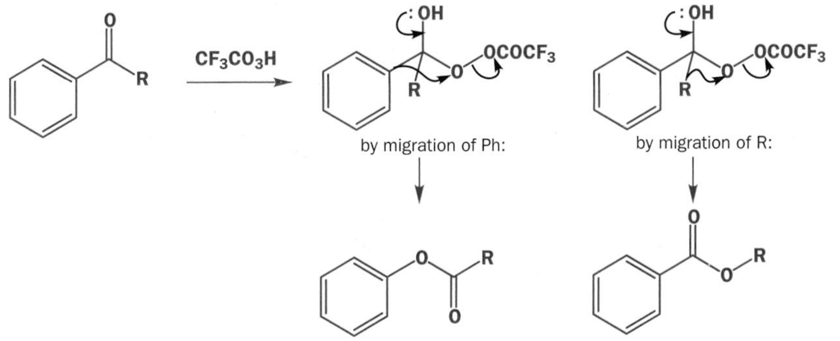

R =	Yield for R (%)	Yield for R (%)
Me	90	0
Et	87	6
i-Pr	33	63
t-Bu	2	77

The order, with *t*-alkyl the best at migrating, then *s*-alkyl closely followed by Ph, then Et, then Me, *very roughly* follows the order in which the groups are able to stabilize a positive charge. Primary groups are much more reluctant to undergo migration than secondary ones or aryl groups, and this makes regioselective Baeyer–Villiger reactions possible.

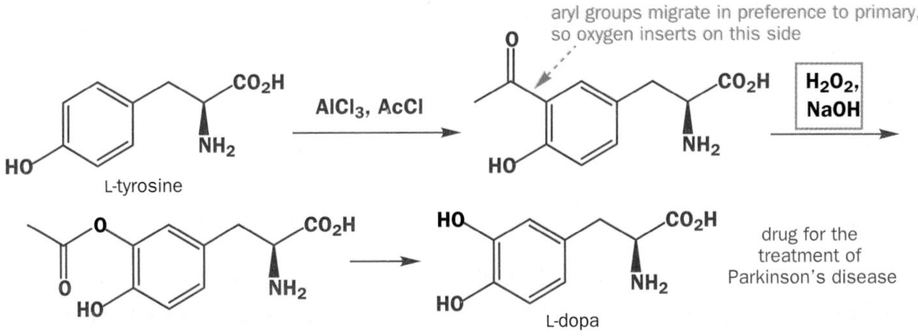

The Baeyer–Villiger reaction has solved a regioselectivity problem here. L-tyrosine, a relatively cheap amino acid, can be converted to the important drug L-dopa provided it can be hydroxylated *ortho* to the OH group. This is where electrophilic substitutions of the phenol take place, but electrophilic substitutions with 'HO$^+$' are not possible. However, after a Friedel–Crafts acylation, the acyl group can be converted to hydroxyl by the Baeyer–Villiger reaction and hydrolysis. The Baeyer–Villiger reaction means that MeCO$^+$ can be used as a synthetic equivalent for 'HO$^+$'. Note the unusual use of the less reactive H$_2$O$_2$ as oxidizing agent in this reaction. This is possible only when the migrating group is an electron-rich aromatic ring; these reactions are sometimes called **Dakin reactions.**

Unsaturated ketones may epoxidize or undergo Baeyer–Villiger rearrangement

Peracids may epoxidize alkenes faster than they take part in Baeyer–Villiger reactions, so unsaturated ketones are not often good substrates for Baeyer–Villiger reactions. The balance is rather delicate. The two factors that matter are: how *electro*philic is the ketone and how *nucleo*philic is the alkene? You might like to consider why this reaction *does* work, and why the C=C double bond here is particularly unreactive.

key intermediate in Corey's prostaglandin syntheses

secondary groups migrate in preference to primary, so oxygen inserts on this side

Small-ring ketones can relieve ring strain by undergoing Baeyer–Villiger reactions—this cyclobutanone (an intermediate in a synthesis of the perfumery compound *cis*-jasmone) is made by a ketene [2+2] cycloaddition, and is so reactive that it needs only H$_2$O$_2$ to rearrange. Unlike CF$_3$CO$_3$H or *m*-CPBA, H$_2$O$_2$ will not epoxidize double bonds (unless they are electron-deficient— see Chapter 23).

One point to note about both of the last two reactions is that the insertion of oxygen goes with retention of stereochemistry. You may think this is unsurprising in a cyclic system like this and, indeed, the first of the two cannot possibly go with inversion. However, this is a general feature of Baeyer–Villiger reactions, even when inversion would give a more stable product.

63% yield of one diastereoisomer

Even when you might imagine that racemization would occur, as in this benzylic ketone, retention is the rule.

87% yield; 98.5% retention of configuration

13% yield of Me-migrated product

By looking at the orbitals involved, you can see why this must be so. The sp^3 orbital of the migrating carbon just slips from one orbital to the next with the minimum amount of structural reorganization. The large lobe of the sp^3 orbital is used so the new bond forms to the same face of the migrating group as the old one, and stereochemistry is retained.

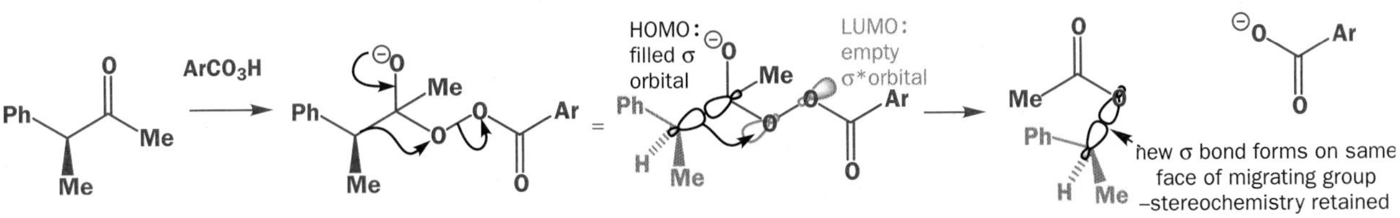

The orbital interactions in all 1,2-migrations are similar, and the migrating group retains its stereochemistry in these too. In the more familiar S_N2 reaction, inversion occurs because the anti-bonding σ* orbital rather than the bonding σ orbital is used. In the S_N2 reaction, carbon undergoes *nucleophilic* attack with *inversion*; in rearrangements the migrating carbon atom undergoes *electrophilic* attack with *retention* of configuration.

> ● In 1,2-migrations, the migrating group retains its stereochemistry.

Which group migrates? (II)—the reasons

Why does the more substituted group migrate in the Baeyer–Villiger reaction? The transition state has a positive charge spread out over the molecule as the carboxylate leaves as an anion. If the migrating group can take some responsibility for the positive charge the transition state will be more stable. The more stable the charge, the faster the rearrangement.

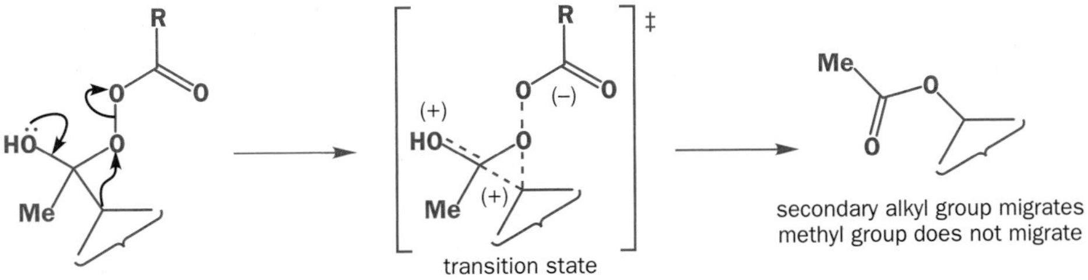

When a benzene ring migrates, π participation is involved as the benzene ring acts as a nucleophile and the positive charge can be spread out even further. Note that the Ph is stabilizing the charge here in the way that it stabilizes the intermediate in an electrophilic aromatic substitution reaction—like a pentadienyl cation rather than like a benzylic cation. What was a transition state in alkyl migration becomes an intermediate in phenyl migration.

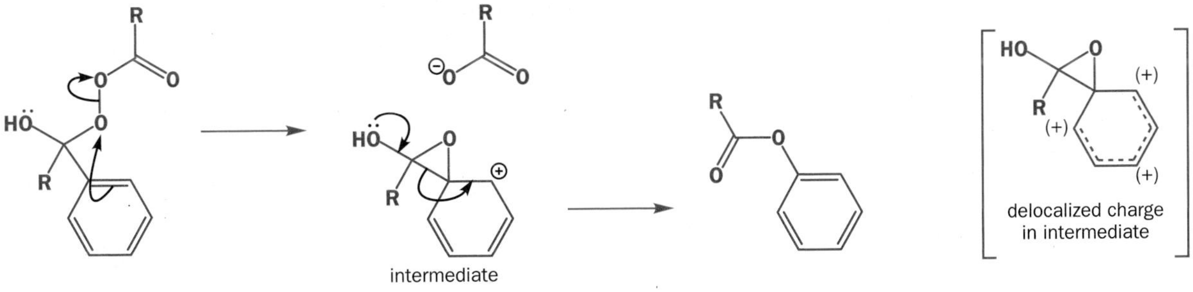

The situation in other rearrangements is much more complicated—and indeed more complicated than many textbooks would have you believe. We shall look just briefly at the dienone–phenol rearrangement again, this time considering reactions in which there is a competition between two different migrating groups. As in the Baeyer–Villiger reaction, the transition state is cationic, so you would expect cation-stabilizing groups to migrate more readily. This appears to be true for Ph versus

Me, but is most definitely not true for Ph versus CO₂Et. The cation-*destabilizing* group CO$_2$Et migrates even though Ph is much better at stabilizing a positive charge!

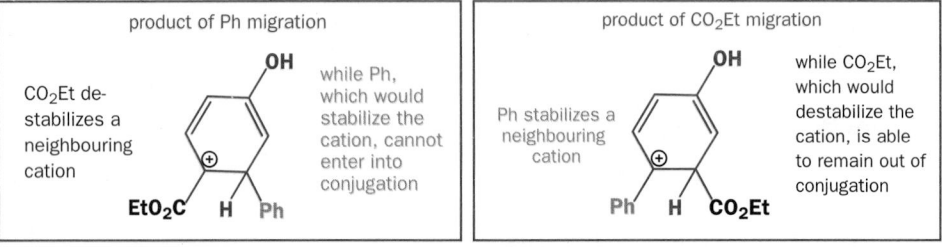

The reason is that CO$_2$Et is so cation-*destabilizing* that it prefers to migrate rather than be left behind next door to a cation. In this case, then, it is the cation-stabilizing ability of the group that *does not* migrate that matters most.

product of Ph migration		product of CO₂Et migration

CO$_2$Et de-stabilizes a neighbouring cation

while Ph, which would stabilize the cation, cannot enter into conjugation

Ph stabilizes a neighbouring cation

while CO$_2$Et, which would destabilize the cation, is able to remain out of conjugation

Which group migrates? (III)—stereochemistry matters too

Selectivity in rearrangement reactions is affected by the electronic nature of *both* the group that migrates *and* the group that is left behind. But there is more! *Stereochemistry* is important too. The outcome of diazotization and semipinacol rearrangement (Tiffeneau–Demjanov rearrangement) of this amino-alcohol depends entirely on the diastereoisomer you start with. There are four diastereoisomers, and we have drawn each one in the only conformation it can reasonably adopt, with the *t*-butyl group equatorial.

In all of these reactions, the OH group provides the electronic 'push'. In the first two reactions, the ring contracts by an alkyl migration from the secondary alcohol, while in the third it is H that migrates from the same position.

The only difference between the compounds is stereochemistry and, if we look at the orbitals involved in the reactions, we can see why this is so important. As the N$_2$ leaving group departs, electrons in the bond to the migrating group have to flow into the C–N σ* orbital—we discussed this on

p. 988. But what we didn't talk about then was the fact that best overlap between these two orbitals (σ and σ*) occurs if they are anti-periplanar to one another—just as in an E2 elimination reaction.

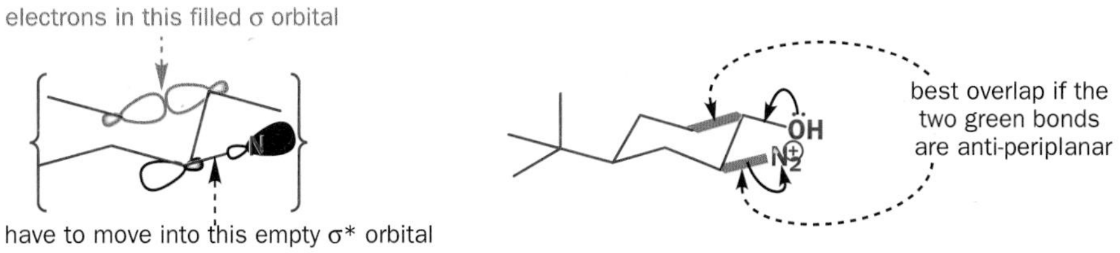

electrons in this filled σ orbital

have to move into this empty σ* orbital

best overlap if the
two green bonds
are anti-periplanar

For the first two compounds, with the $-N_2^+$ group equatorial, the group best placed to migrate is the alkyl group that forms the ring; for the third reaction, there is a hydrogen atom anti-periplanar to the leaving group, so H migrates.

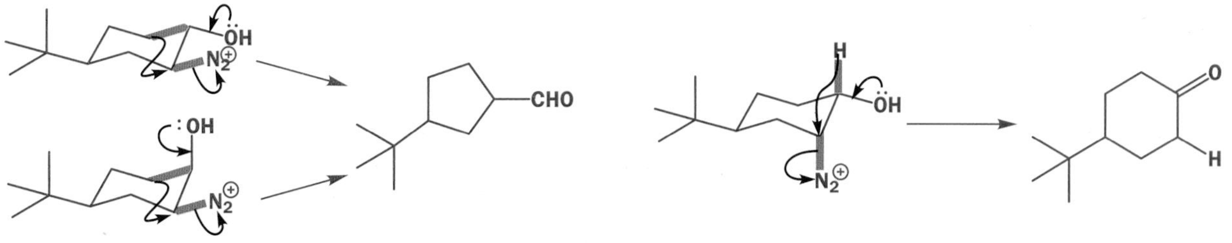

The fourth reaction has, rather than a group that might migrate, the hydroxyl group ideally placed to displace N_2 and form an epoxide—another example of participation.

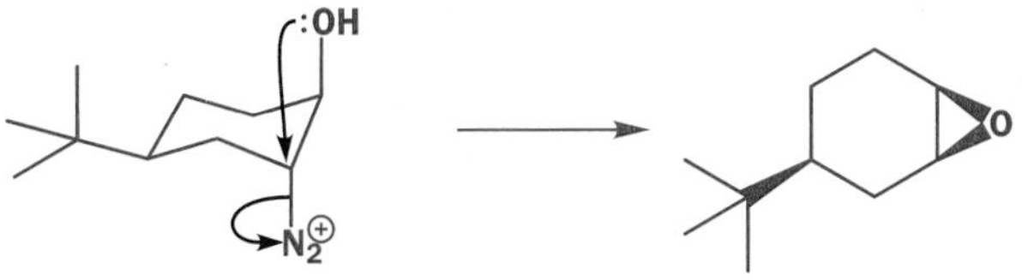

The requirement for the migrating group to be anti-periplanar to the leaving group is quite general in rearrangement reactions. The reason we haven't noticed its effect before is that most of the compounds we have considered have not been conformationally constrained in the way that these are. Free rotation means that the right geometry for rearrangement is always obtainable—stereochemistry is not a factor in the Baeyer–Villiger reaction, for example. We will come back to some more aspects of stereochemical control in the next chapter, on fragmentation reactions. Before then, we will consider one last rearrangement reaction, in which stereochemistry again plays an important controlling role.

The Beckmann rearrangement

The industrial manufacture of nylon relies upon the alkaline polymerization of a cyclic amide known trivially as caprolactam. Caprolactam can be produced by the action of sulfuric acid on the oxime of cyclohexanone in a rearrangement known as the **Beckmann rearrangement**.

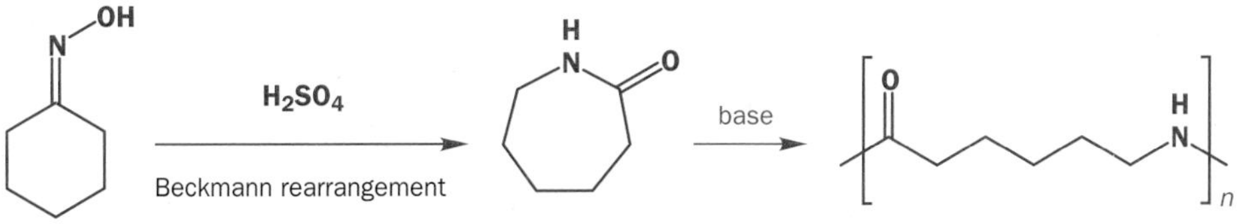

oxime   caprolactam   nylon

The mechanism of the Beckman rearrangement follows the same pattern as a pinacol or Baeyer–Villiger reaction—acid converts the oxime OH into a leaving group, and an alkyl group migrates on to nitrogen as water departs. The product cation is then trapped by water to give an amide.

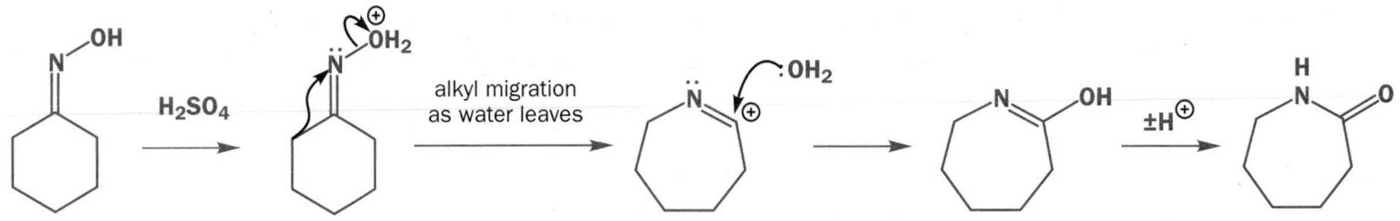

A linear system like this was impossible in the seven-membered ring of the last example.

This rearrangement is not confined to cyclic oximes, and other ways of converting OH to a leaving group also work, such as PCl_5, $SOCl_2$, and other acyl or sulfonyl chlorides. In an acyclic Beckmann rearrangement, the product cation is better represented as this nitrilium ion. When we write the mechanism we can then involve the nitrogen's lone pair to 'push' the migrating group back on to N.

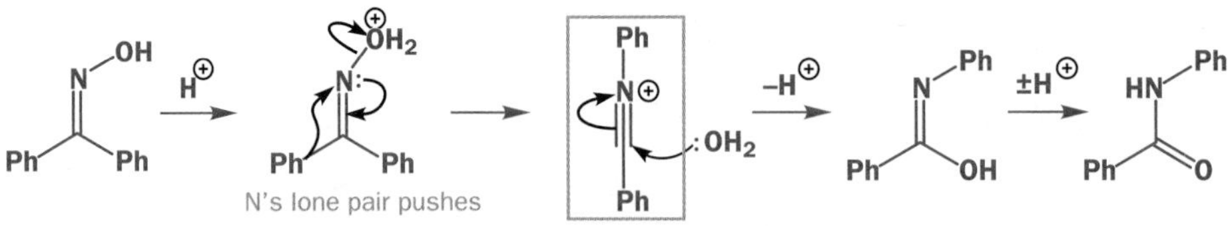

Which group migrates in the Beckmann rearrangement?

In the Beckmann rearrangement of unsymmetrical ketones there are two groups that could migrate. There are also two possible geometrical isomers of an unsymmetrical oxime: C=N double bonds can exhibit *cis/trans* isomerism just as C=C double bonds can. When mixtures of geometrical isomers of oximes are rearranged, mixtures of products result, but the ratio of products mirrors exactly the ratio of geometrical isomers in the starting materials—the group that has migrated is in each case the group *trans* to the OH in the starting material.

75:25 ratio of geometrical isomers   73:27 ratio of products

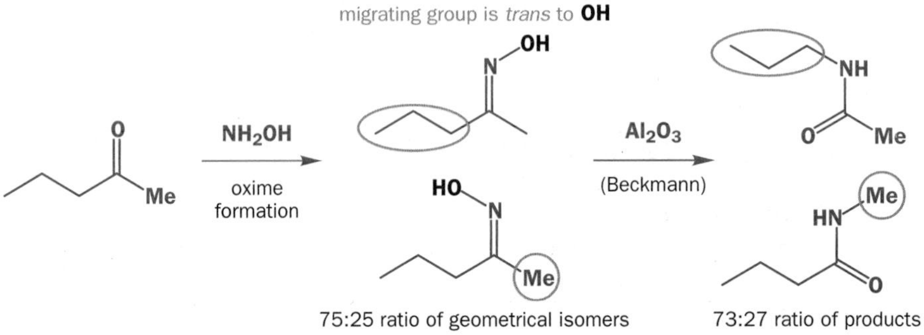

We have already touched on the idea that, for migration to occur, a migrating group has to be able to interact with the σ* of the bond to the leaving group, and this is the reason for the specificity here. In the example a couple of pages back the stereospecificity of the reaction was due to the starting material being constrained in a conformationally rigid ring. Here it is the C=N double bond that provides the constraint. If one of the alkyl chains is branched, more of the oxime with the OH group *anti* to that chain will be formed and correspondingly more of the branched group will migrate.

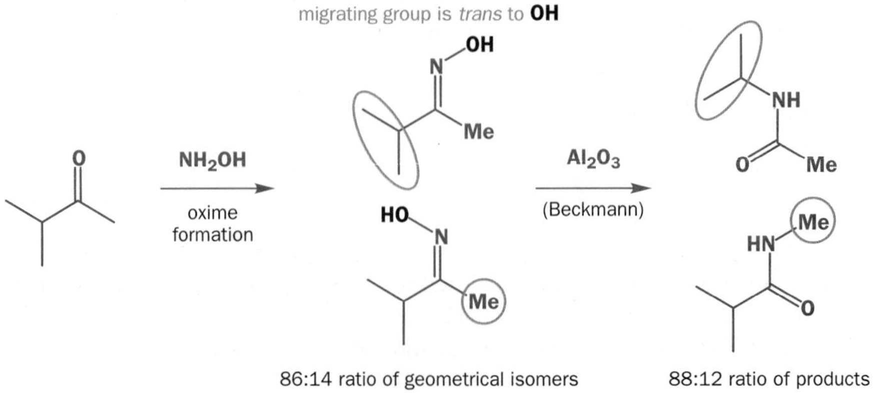

86:14 ratio of geometrical isomers   88:12 ratio of products

Conditions that allow those double isomers to interconvert can allow either group to migrate—which does so will then be decided, as in the Baeyer–Villiger reaction, by electronic factors. Most protic acids allow the oxime isomers to equilibrate—so, for example, this tosylated oxime rearranges with full stereospecificity in Al$_2$O$_3$ (the *anti* methyl group migrates), but with TsOH, equilibration of the oxime geometrical isomers means that either group could migrate—in the event, the propyl group (which is more able to support a positive charge) migrates faster.

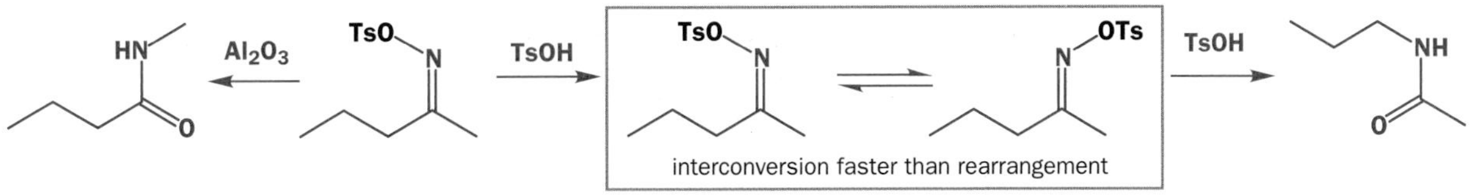

interconversion faster than rearrangement

Notice that the effect of the Beckmann rearrangement is to insert a *nitrogen* atom next to the carbonyl group. It forms a useful trio with the Baeyer–Villiger *oxygen* insertion and the diazoalkane *carbon* insertion.

The diosgenin story: steroids from vegetables

Many of the human steroid hormones are available by 'semisynthesis'—in other words synthesis starting from a natural product similar in structure to the target molecule. One very important starting material for semisynthesis routes to these hormones is diosgenin, a plant steroid which makes up 5% of the dry mass of the roots of Mexican yams. Most of the chemical manipulation necessary to turn diosgenin into human steroids concerns the top right five-membered ring (the 'D' ring). A few steps convert the acetal group of the natural product into a simpler methyl ketone, present in cortisone and progesterone

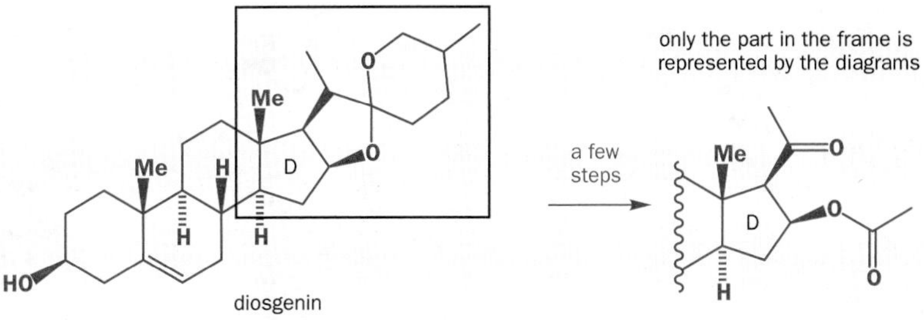

diosgenin

only the part in the frame is represented by the diagrams

a few steps

But for hormones such as oestrone and testerone two carbon atoms need removing to make a cyclopentanone. This is accomplished using a Beckmann rearrangement. The oxime forms with the OH group trans to the more bulky cyclic substituent. Tosylation and Beckmann rearrangement gives an acetylated enamine which hydrolyses to the required cyclopentanone

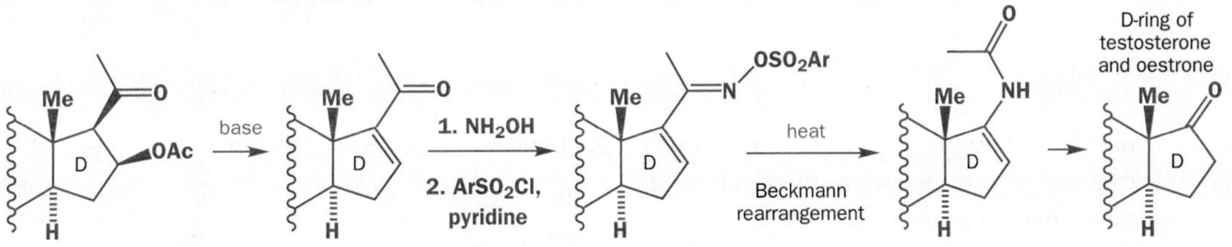

The Beckmann fragmentation

To finish this chapter, a Beckmann rearrangement that is not all that it seems. *t*-Butyl groups migrate well in the Baeyer–Villiger reaction and, indeed, Beckmann rearrangement of this compound appears to be quite normal too.

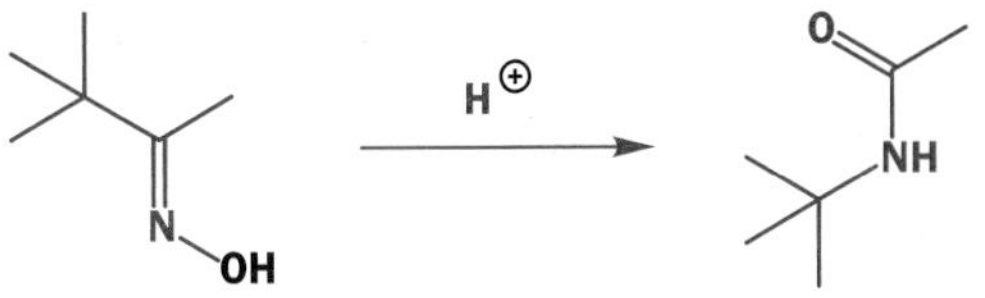

But, when this compound and another compound with a tertiary centre next to the oxime are mixed together and treated with acid, it becomes apparent that what is happening is not an intramolecular reaction.

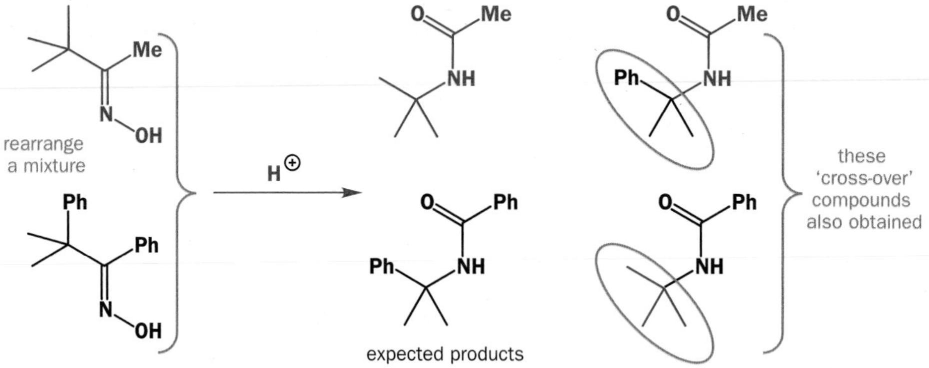

expected products

> The recombination step of this reaction is really just a Ritter reaction: reaction of a nitrile with a carbocation. You came across the Ritter reaction on p. 437.

Each migrating tertiary group must have lost contact with the amide fragment it started out with. Each molecule falls to bits to give a *t*-alkyl cation and a nitrile: the Beckmann rearrangement now goes via a **fragmentation** mechanism.

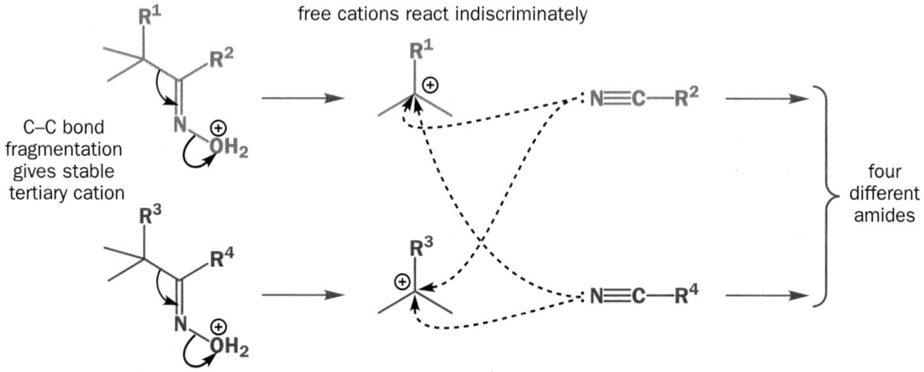

Migrating groups have to provide some degree of cation stabilization. But if they stabilize a cation too well there is a good chance that fragmentation will occur and the 'migrating group' will be lost as a carbocation. It is with this idea that we begin the next chapter.

Problems

1. Rearrangements by numbers. This problem is just to help you acquire the skill of tracking down rearrangements by numbering. There are no complicated new reactions here. Just draw a mechanism.

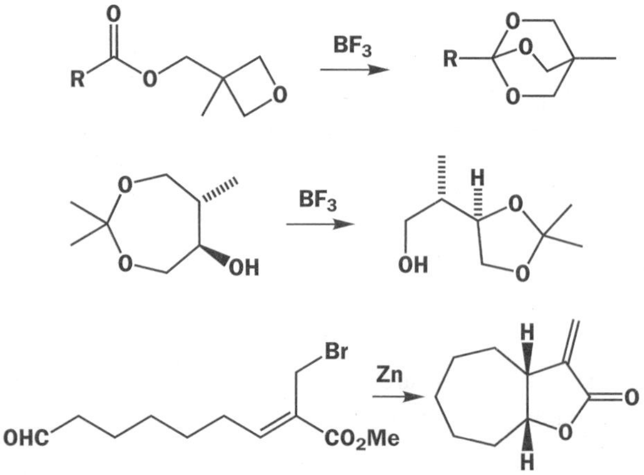

2. Explain this series of reactions.

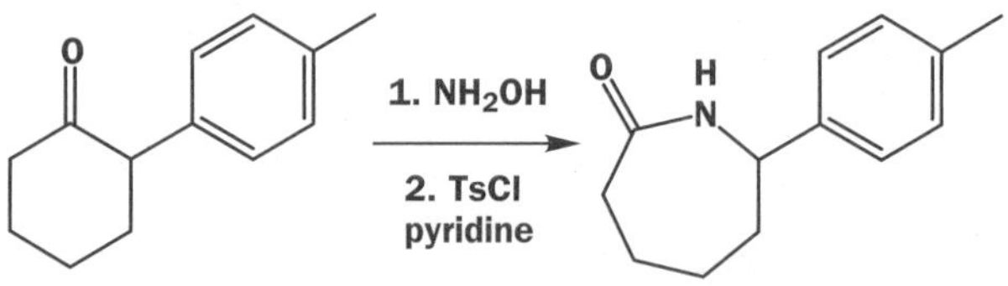

3. Draw mechanisms for the reactions and structures for the intermediates. Explain the stereochemistry, especially of the reactions involving boron. Why was 9-BBN chosen as the hydroborating agent?

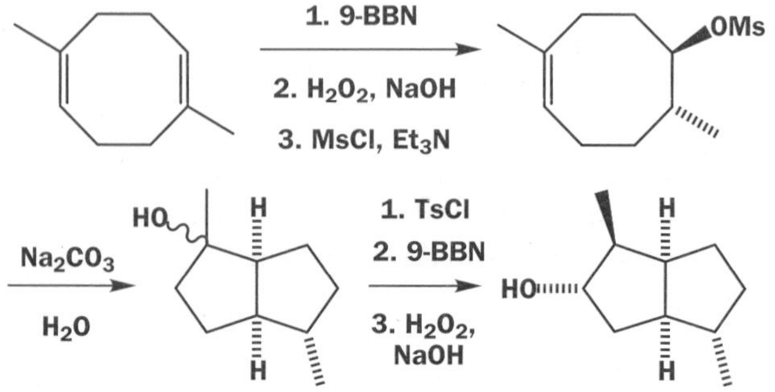

4. It is very difficult to prepare three-membered ring lactones. One attempted preparation, by the epoxidation of di-*t*-butyl ketene, gave an unstable compound with an IR stretch at 1900 cm^{-1} that decomposed rapidly to the four-membered lactone shown. Do you think they made the three-membered ring?

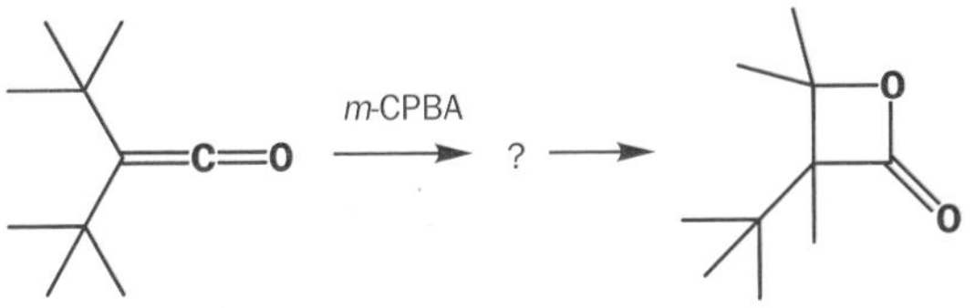

5. Suggest a mechanism for this rearrangement.

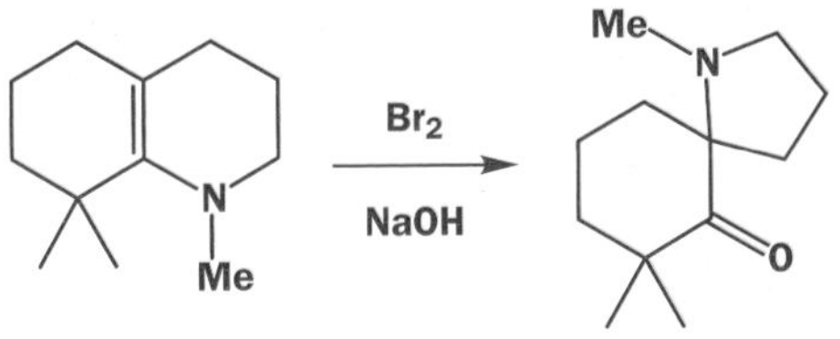

6. A single enantiomer of the epoxide below rearranges with Lewis acid catalysis to give a single enantiomer of product. Suggest a mechanism and comment on the stereochemistry.

7. The 'pinacol' dimer from cyclobutanone rearranges with the expansion of one of the rings to give a cyclopentanone fused *spiro* to the remaining four-membered ring. Draw a mechanism for this reaction. Reduction of the ketone then gives an alcohol that rearranges to the alkene in acid. Try working out a mechanism for this transformation. You might also like to think about why the rearrangement happens.

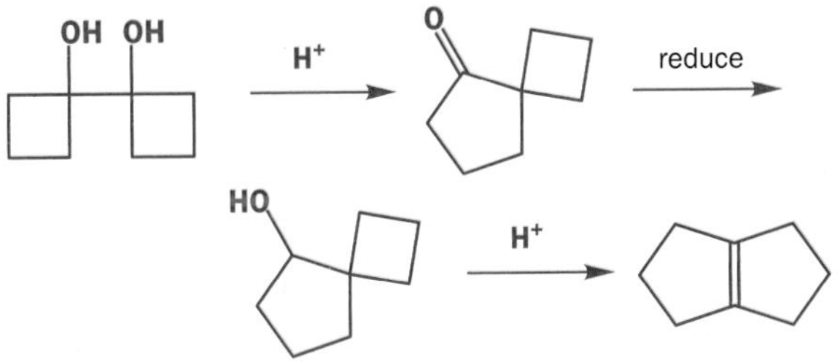

8. Give the products of Baeyer–Villiger rearrangement on these carbonyl compounds with your reasons.

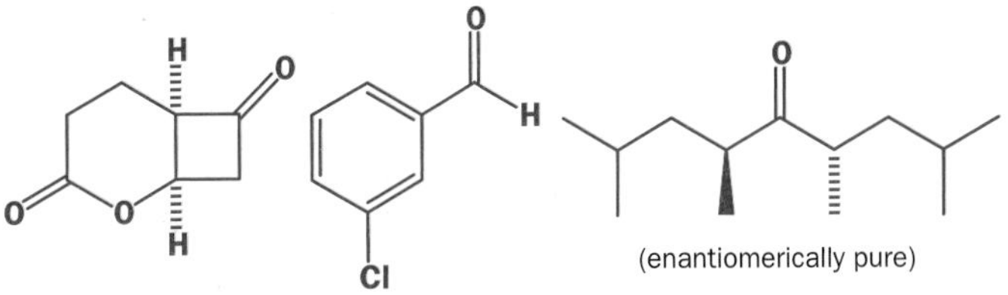

(enantiomerically pure)

9. Suggest mechanisms for these rearrangements explaining the stereochemistry in the second example.

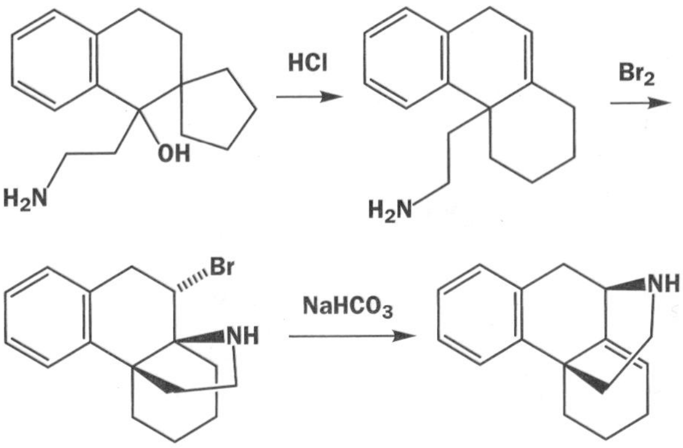

10. Give mechanisms for these reactions, commenting on any regio- and stereoselectivity. What controls the rearrangement?

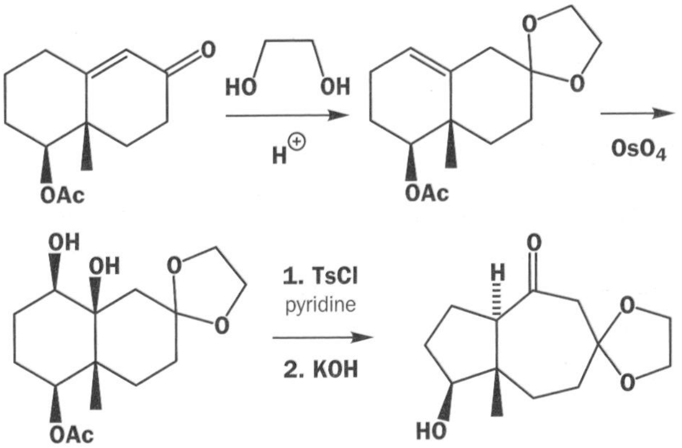

11. Suggest mechanisms for these reactions that explain any selectivity in the migration.

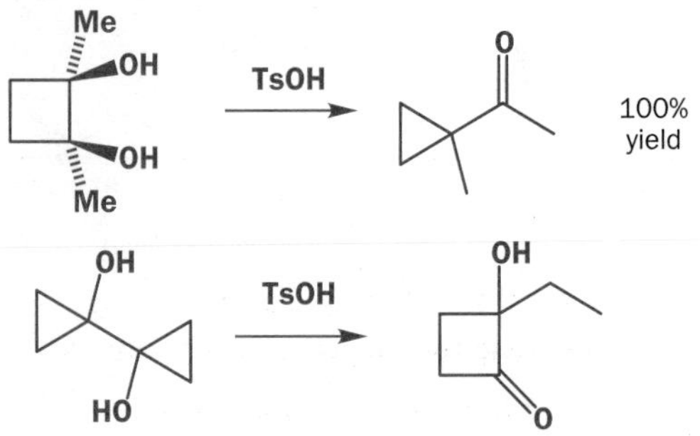

100% yield

12. Attempts to produce the acid chloride from this unusual amino acid by treatment with SOCl$_2$ gave instead a β-lactam. What has happened?

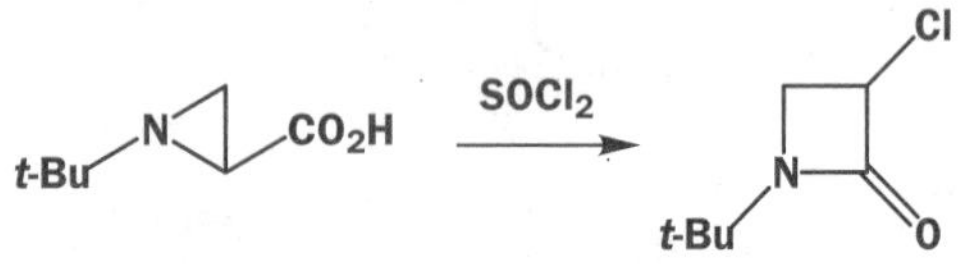

13. Revision content. Suggest mechanisms for these reactions, commenting in detail on the rearrangement step.

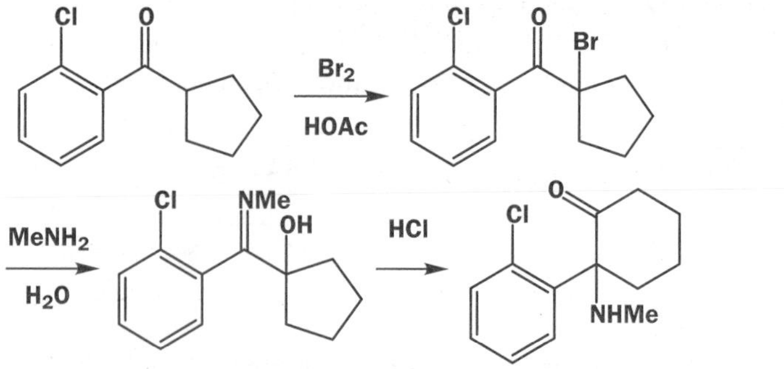

14. Suggest a mechanism for this rearrangement, comparing it with a reaction discussed in the chapter. What controls the stereochemistry?

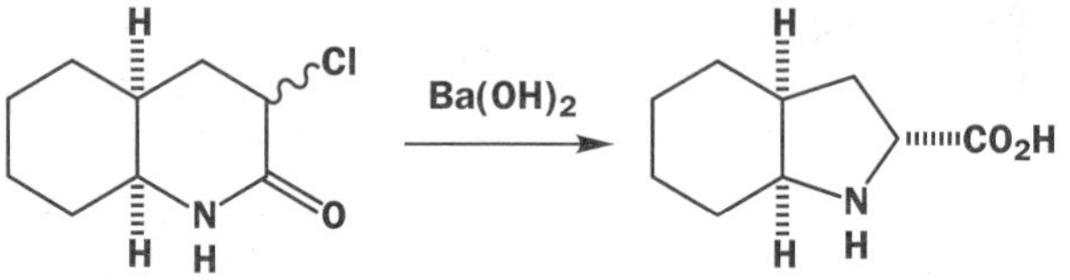

Fragmentation

38

Connections

Building on:

- **Nucleophilic substitution at saturated carbon** ch17
- **Conformational analysis** ch18
- **Elimination reactions** ch19
- **Controlling stereochemistry** ch16, ch33, & ch34
- **Rearrangements** ch37

Arriving at:

- **Electron donation and electron withdrawal combine to create molecules that fragment**
- **Fragmentation literally means the breaking of a molecule into three by the cleavage of a C–C bond**
- **Reactive groups should have a 1,4 relationship**
- **Anti-periplanar conformation is essential**
- **Small rings are easy to fragment**
- **Medium and large rings can be made in this way**
- **Double bond geometry can be controlled**
- **Using fragmentations in synthesis**

Looking forward to:

- **Carbene chemistry** ch40
- **Determination of mechanism** ch41
- **Stereoelectronics** ch42
- **Main group chemistry** ch46–ch47

Polarization of C–C bonds helps fragmentation

We finished the last chapter with an attempted migration that went wrong because the migrating group stabilized a cation too well. Here is a more convincing example of the same reaction: again, the conditions for, but not the result of, a Beckmann rearrangement.

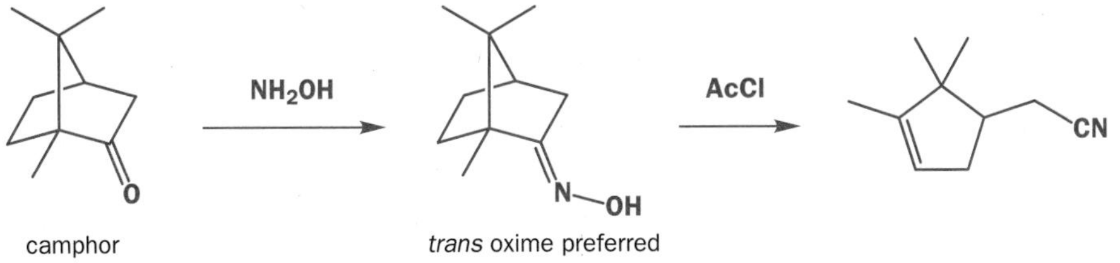

NH2OH
AcCl
O
N—OH
CN
camphor
trans oxime preferred

The starting material is bicyclic, the product monocyclic, so we have broken a C–C bond: the reaction is a **fragmentation**. The mechanism is straightforward once you know what happens to Beckmann rearrangements when the migrating group is tertiary—but hard to follow unless you number the atoms!

Beckmann rearrangements that go with fragmentation are sometimes called 'anomalous' or 'second-order' Beckmann rearrangements. You should not use the second of these names and, in any case, **Beckmann fragmentation** is much better than either.

C–C bond breaks...

...to form stable tertiary carbocation

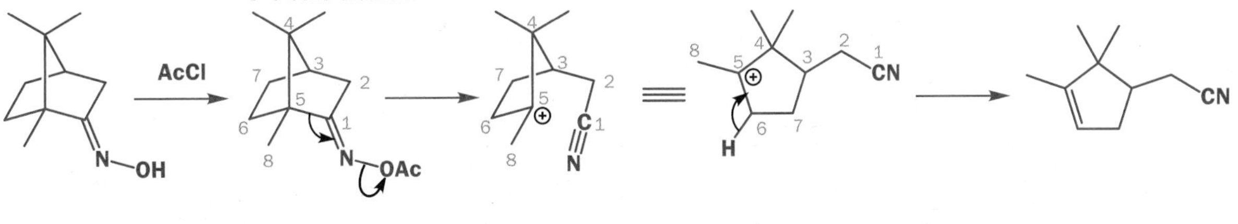

AcCl
N—OH
OAc
CN
H

You have met few fragmentation reactions—reactions in which C–C bonds are broken—largely

Bond	Typical bond energy, kJ mol⁻¹
C–C	339
C–O	351
C–H	418
O–H	460

▶

The bond energies listed in the table are the energies required to break the bonds **homolytically** to give two radicals, not **heterolytically** to give two ions, which is what has happened in most of the reactions we have talked about. We will look at this in much more detail in the next chapter.

because the C–C bond is so strong. Why then does this reaction work? Well, the reason C–C bonds are hard to break is not just because of their strength, as the table of bond energies indicates.

For both carbon and hydrogen, a bond to oxygen is *stronger* than a bond to carbon. Yet we have no hesitation in breaking O–H bonds (of, say, carboxylic acids) with even the weakest of bases and we have spent much of the last chapter showing C–O bonds of protonated alcohols rupturing spontaneously! What is going on?

The answer is **polarization**. Oxygen's electronegativity means that C–O and O–H bonds are polarized and are easy to break with hard nucleophiles and bases; C–C and C–H bonds are (usually) not polarized and, though weaker, are harder to break. It follows that to break a C–C bond it helps a lot if it is polarized—there needs to be a source of electrons at one end and an electron 'sink' (into which they can flow) at the other.

Fragmentations require electron push and electron pull

Fragmentations are reactions in which the molecule breaks into three pieces by the cleavage of a C–C single bond. Now for some examples and comparisons. The first example shows a fragmentation giving only two, not three, molecules. This is because two of the fragments were joined together in a ring. Both diastereoisomers of this cyclic diol fragment in acid to give an aldehyde. Numbering the atoms shows which bond fragments—now we need to provide a source and a sink for the electrons to polarize the bond.

Protonation of a hydroxyl group provides the sink—it can now leave as water. And the lone pair of the other oxygen provides the source. You can think of the electrons in the C–C bond being 'pushed' by the oxygen's lone pair and 'pulled' by the departing water—until the bond breaks. A bit of extra impetus comes from release of ring strain: C–C bonds in three- and four-membered rings are weaker than usual (by about 120 kJ mol⁻¹).

We talked about 'pushing' and 'pulling' electrons when we introduced the pinacol rearrangement, and a very similar thing is happening here *but* the electron source and sink are *separated by one atom* instead of being *adjacent*.

▶

Note the numbering in these diagrams: 1, 2 ,3, 4 from electron source to electron sink. We shall make use of it in many more fragmentation mechanisms.

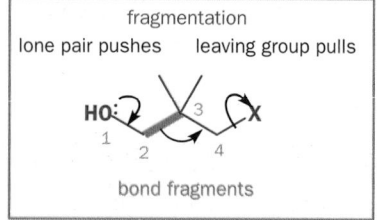

Protonated carbonyl compounds can be electron sinks too (remember the dienone–phenol rearrangement from Chapter 37?), and this bicyclic methoxy ketone fragments to a seven-membered ring in acid. Note the same 1, 2, 3, 4 arrangement, with the bond between carbon atoms 2–3 fragmenting.

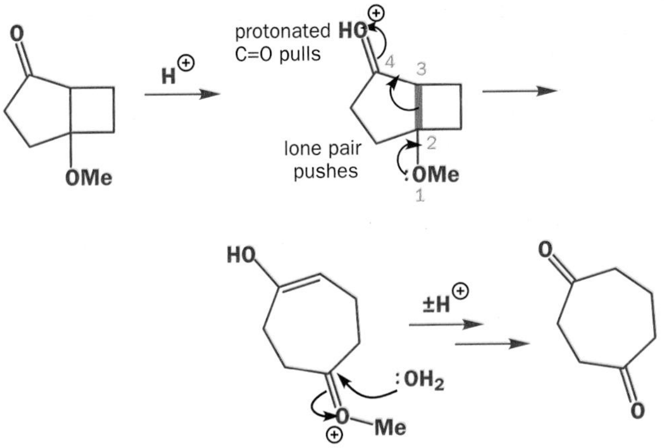

> We should perhaps remind you here of the reversibility of the aldol reaction (Chapter 27): a retro-aldol is a fragmentation reaction with a carbonyl group as electron sink and OH as electron source. The aldol reaction usually goes in the other direction of course, but where steric or ring-strain factors are involved, this may not be the case.

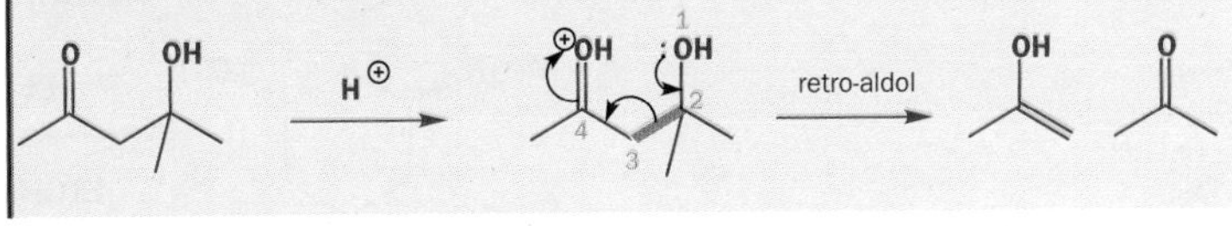

Yet a similar compound to our last example rearranges, and does not fragment because there is an alternative electron sink placed in the right place for migration.

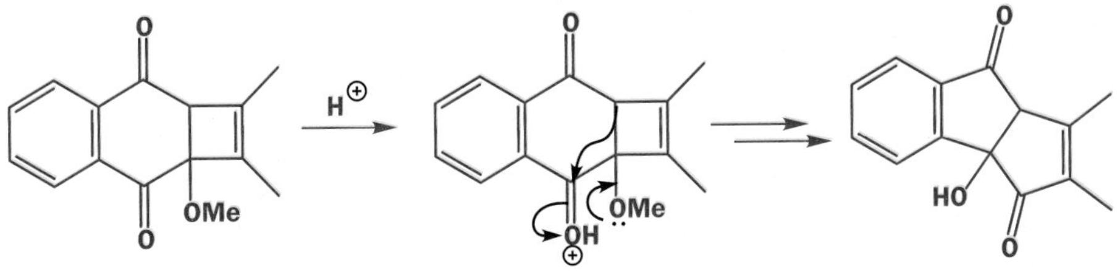

If the MeO group is replaced by a leaving group such as MsO, it can exercise the pull and the carbonyl can provide the push after it has been attacked by a nucleophile. This next five-membered cyclic ketone fragments on treatment with base—can you detect hints of the benzylic acid rearrangement?

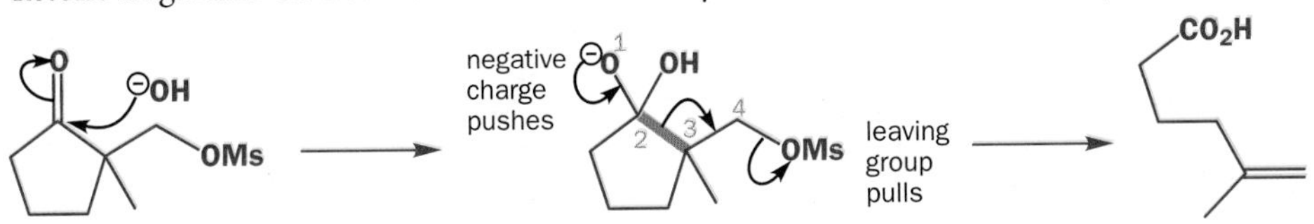

Analysing our Beckmann fragmentation (or anomalous Beckmann rearrangement) in the same way, we can identify the electron sink (the departing acetate group), though the source in this case is a little more obscure. Saying that the tertiary cation is stable is really saying that the neighbouring C–C and C–H bonds provide electrons (through σ conjugation) to stabilize it, so these are the electron sources. A good alternative is to write loss of a proton concerted with fragmentation, which gives one particular C–H bond as the source.

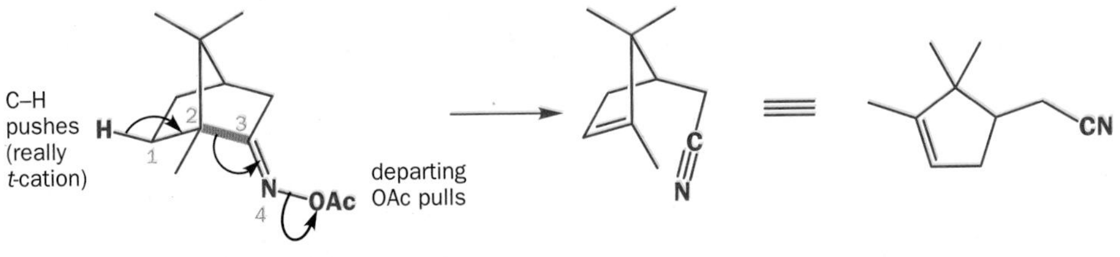

Fragmentations are controlled by stereochemistry

In the last chapter we introduced you to the idea that the control of rearrangements can be stereoelectronic in origin—if a molecule is to rearrange, orbitals have to be able to overlap. This means that, for a Beckmann *rearrangement*, the migrating group has to be *trans* to the leaving group. Not surprisingly, the same is true for Beckmann fragmentations like the one at the end of the last section, where the green fragmenting bond is *trans* to the leaving group.

Before we extend these ideas any further, consider these two quite different reactions of quite similar compounds.

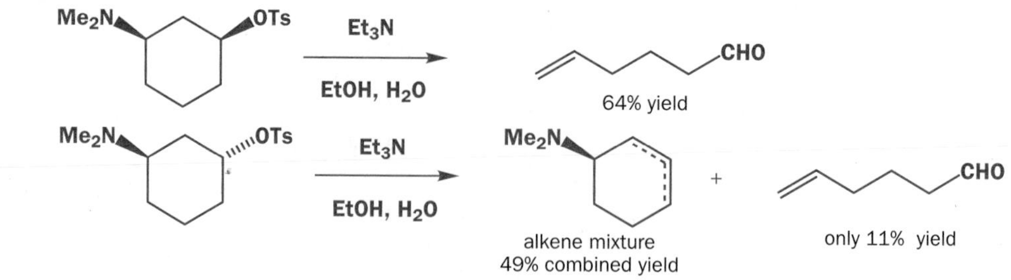

Just as with the rearrangements we looked at on p. 966, we need to draw these compounds in reasonable chair conformations in order to understand what is going on. In the *cis* isomer, both substituents can be equatorial; in the *trans* isomer one has to be axial, and this will be mainly the OTs group, since the two methyl groups of NMe$_2$ suffer greater 1,3-diaxial interactions.

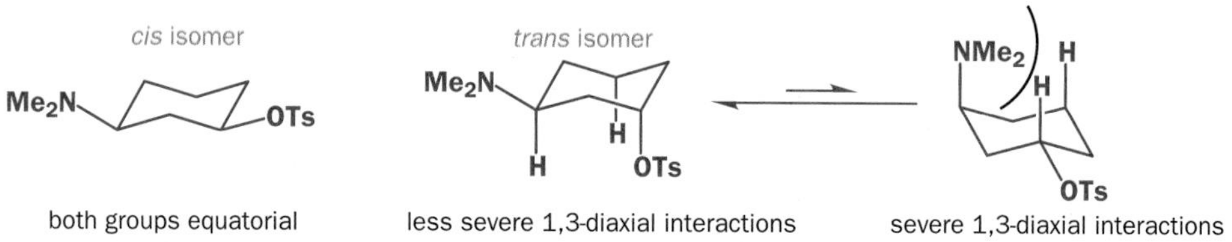

Now, the *cis* isomer has clearly undergone a fragmentation reaction and, as usual, numbering the atoms can help to identify the bond that breaks. The nitrogen lone pair pushes, the departing tosylate pulls, and the resulting iminium ion hydrolyses to the product aldehyde.

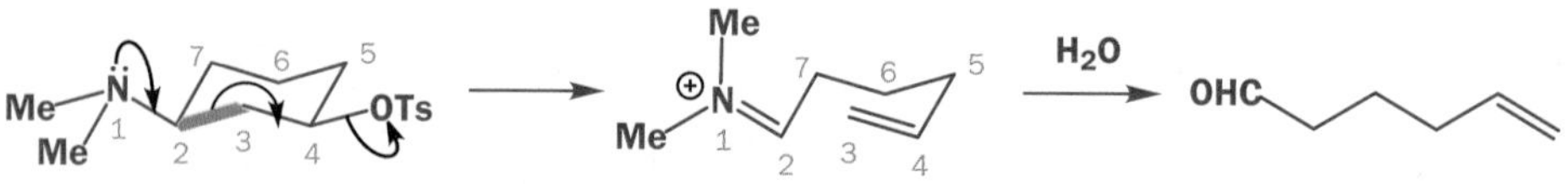

Yet the *trans* isomer only does this in very low yield. Mostly it eliminates TsOH to give a mixture of alkenes. Why? Well, notice that, in the *cis* isomer, the fragmenting bond is *trans* to the leaving group—indeed, it is both parallel and *trans*: in other words **anti-periplanar** to the leaving group. Electrons can flow smoothly from the breaking σ bond into the σ* of the C–OTs bond, forming as they do so, a new π bond.

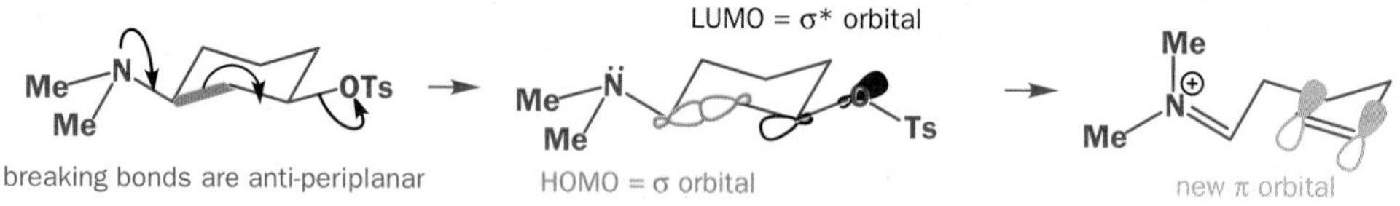

For the *trans* isomer, fragmentation of the most populated conformation is impossible because the leaving group is not anti-periplanar to any C–C bond. The only bonds anti-periplanar to OTs are C–H bonds, making this compound ideally set up for another reaction whose requirement for anti-periplanarity you have already met—E2 elimination.

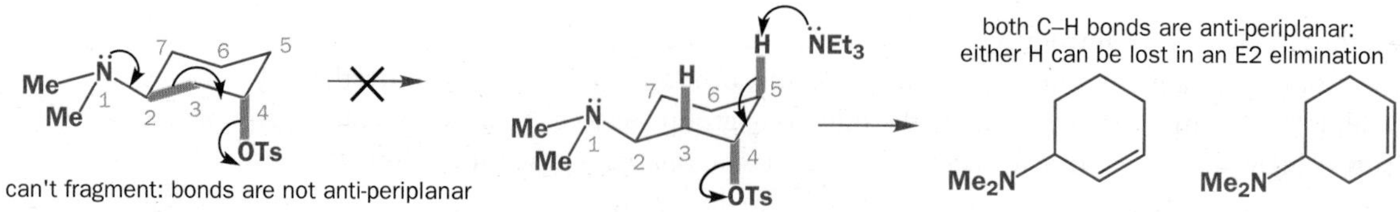

The other conformation can fragment because now the OTs is anti-periplanar to the right C–C bond, and this is probably where the 11% fragmentation product comes from.

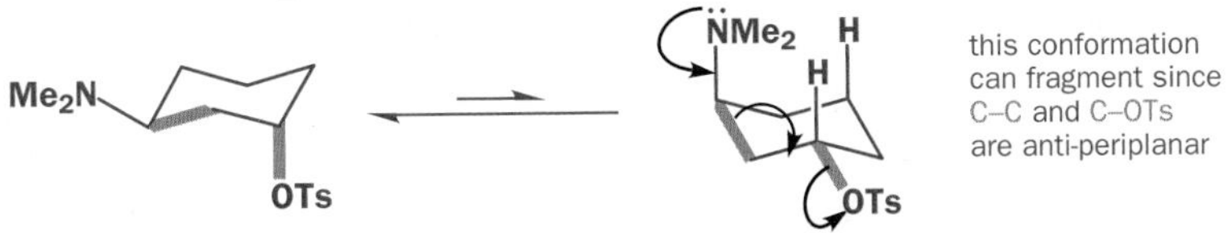

Me2N
OTs
NMe2
H
H
OTs
this conformation can fragment since C–C and C–OTs are anti-periplanar

When McMurry was making longifolene in the early 1970s, a fragmentation reaction saved the day when a conjugate addition reaction using a cuprate gave an unexpected cyclization product through an intramolecular aldol reaction.

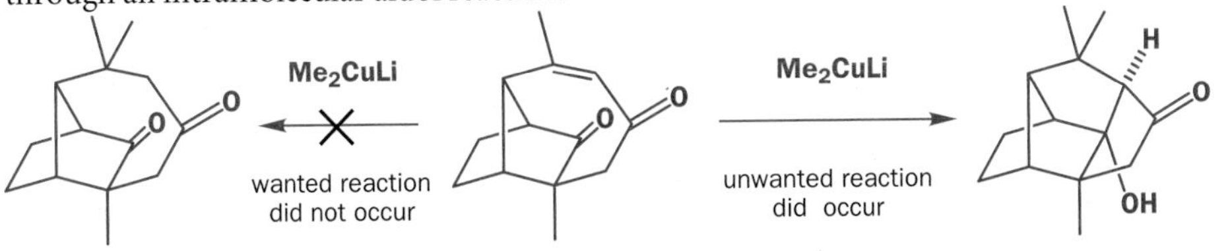

Me2CuLi
wanted reaction did not occur
Me2CuLi
unwanted reaction did occur
O
O
O
H
OH

The actual compound McMurry wanted had the framework of the molecule on the left, but was to be transformed into the alkene below, so he needed to fragment the unexpected product at the green bond.

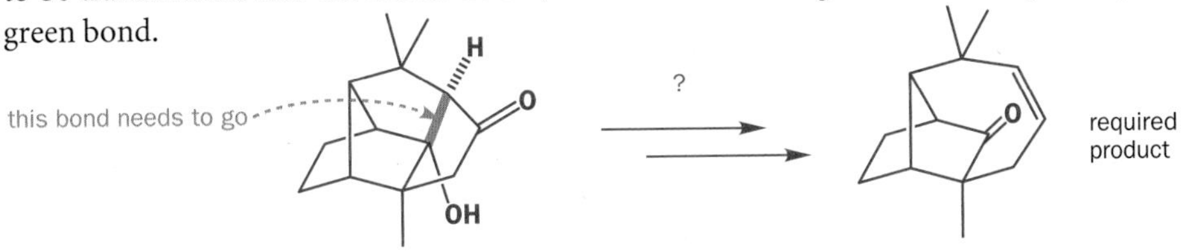

this bond needs to go
H
O
OH
?
O
required product

■ Another synthesis of longifolene is summarized later in this chapter.

Fortunately, reducing the carbonyl group gave a hydroxyl group anti-periplanar to the green bond and therefore set up for fragmentation. Making the hydroxyl a leaving group and treating with base gave the required compound by a fragmentation reaction.

H
O
OH
borohydride approaches from this less hindered face
1. NaBH4
2. MsCl, Et3N
H
OMs
H
OH
KOtBu
H
OMs
H
O
fragmentation
O

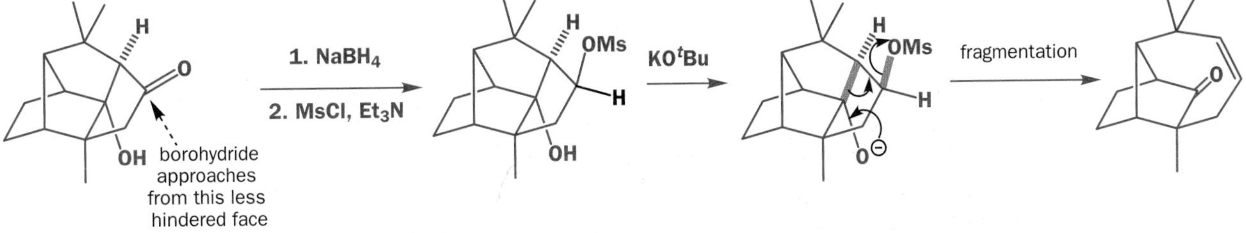

Ring expansion by fragmentation

Ring sizes greater than eight are hard to make. Yet five- and six-membered rings are easy to make. Once you realize that a fused pair of six-membered rings is really a ten-membered ring with a bond across the middle, the potential for making medium rings by fragmentation becomes apparent.

■ This point was discussed in Chapter 18.

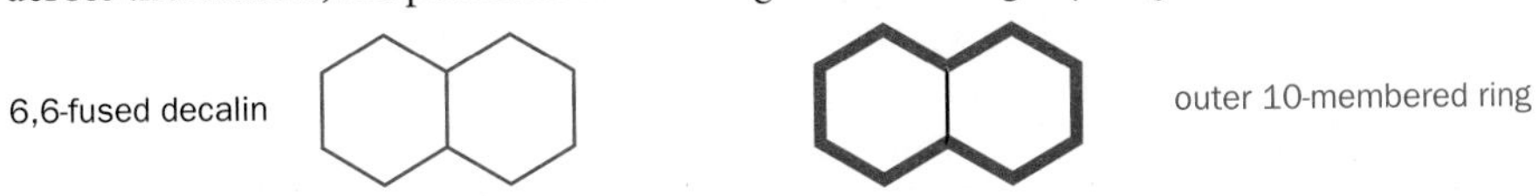

6,6-fused decalin
outer 10-membered ring

All you need to do is to make the bond to be broken the 2–3 bond in a 1, 2, 3, 4 electron source–sink arrangement and the ten-membered ring should appear out of the wreckage of the fragmentation. Here is an example—a decalin that fragments to a ten-membered ring.

OTs
OH
base
OTs
4
3
2
O
1
O
10-membered ring

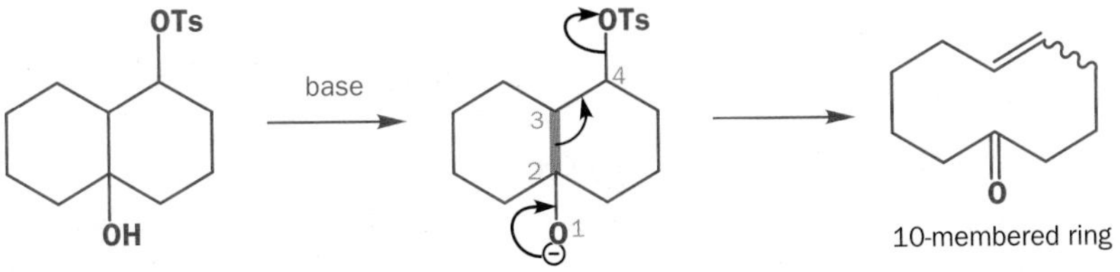

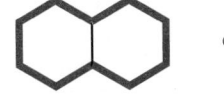

Muscone and exaltone are important perfumery compounds with hard-to-make 15-membered ring structures. Cyclododecanone is commercially available: addition of a fused five-membered ring and fragmentation of the 12,5-ring system is a useful route to these 15-membered ring compounds.

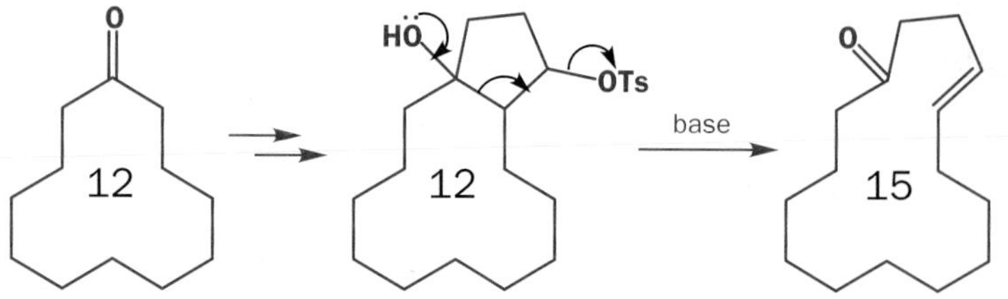

cyclododecanone

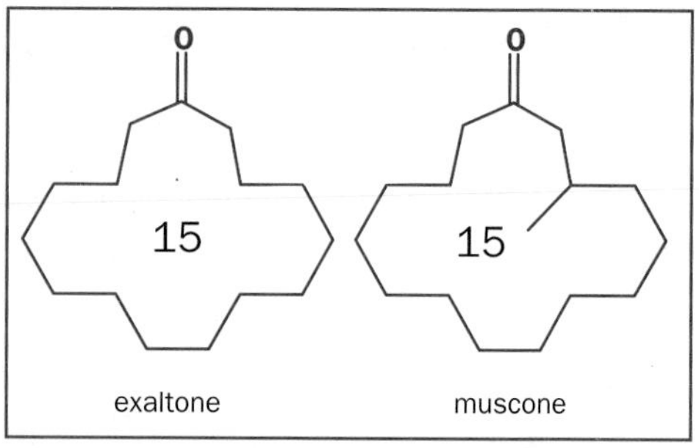

exaltone   muscone

Albert Eschenmoser (1925–), working at the ETH in Zurich, synthesized vitamin B$_{12}$, a cobalt complex and at the time the most complicated molecule yet made, in an unusual international collaboration with Woodward at Harvard.

Electron-poor double bonds can be epoxidized with basic hydrogen peroxide. See Chapter 23.

In the late 1960s, the Swiss chemist Albert Eschenmoser discovered an important reaction that can be used to achieve similar ring expansions and that now bears his name, the **Eschenmoser fragmentation**. The starting material for an Eschenmoser fragmentation is the epoxide of an α,β-unsaturated ketone. The fragmentation happens when this epoxy-ketone is treated with tosyl-hydrazine, and one of the remarkable things about the product is that it is an alkyne. The fragmentation happens across the epoxide (shown in black), and the product contains both a ketone (in a different place to the ketone in the starting material) and an alkyne. You can see how in this case hydrogenation of the triple bond can again give muscone (R = Me) or exaltone (R = H).

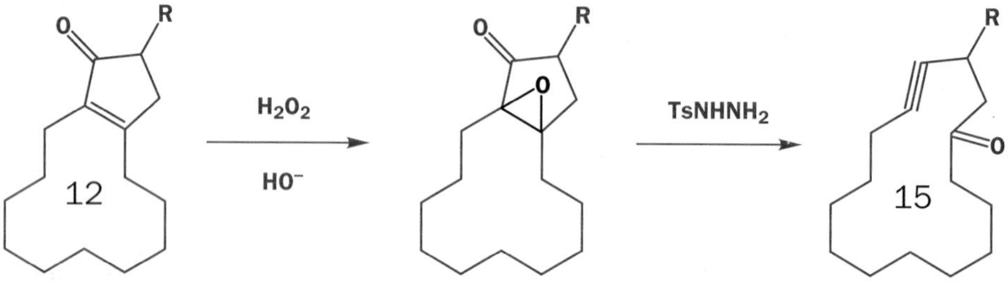

The Eschenmoser fragmentation does not have to be a ring expansion, and it is a useful synthetic method for making keto-alkynes. The following reaction, which we will use to discuss the fragmentation's mechanism, was used to make an intermediate in the synthesis of an insect pheromone, *exo*-brevicomin.

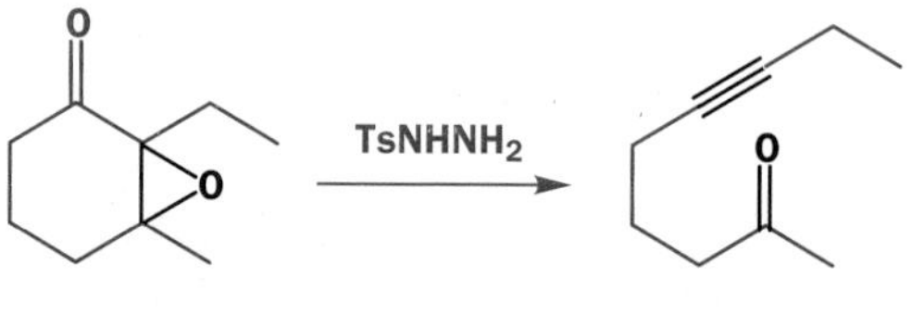

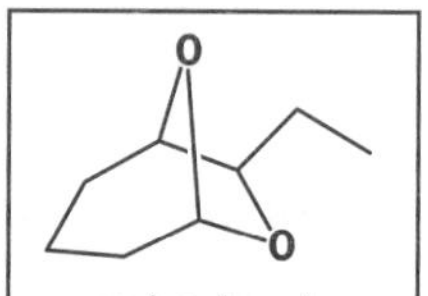

exo-brevicomin

The reaction starts with formation of the tosylhydrazone from the epoxy-ketone. The tosylhydra-zone is unstable with respect to opening of the epoxide in an elimination reaction, and it is this elim-ination that sets up the familiar 1, 2, 3, 4 system ready for fragmentation. The 'push' comes from the newly created hydroxyl group, and the 'pull' from the irresistible concerted loss of a good leaving group (Ts$^-$) and an even better one (N$_2$). Notice how all the (green) bonds that break are parallel to one another, held anti-periplanar by two double bonds. Perfect!

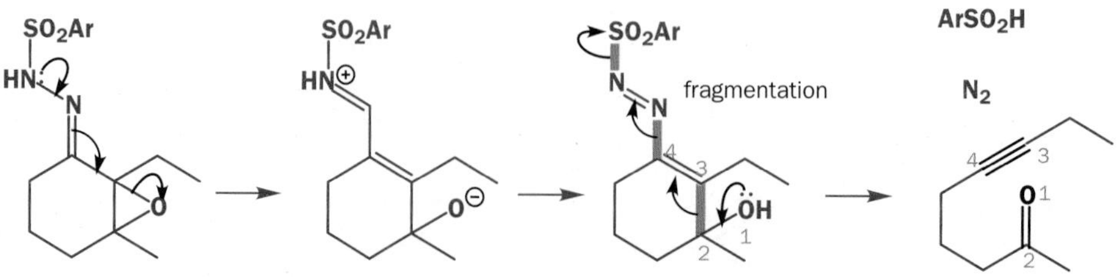

The sulfur-containing leaving group here is not toluenesulfonate (tosylate, ArSO$_3^-$ or TsO$^-$) but toluenesulfinate (ArSO$_2^-$ or Ts$^-$), giving toluenesulfinic acid (TsH or ArSO$_2$H), not toluenesulfonic acid (TsOH or ArSO$_3$H) as a by-product.

More on stereochemistry and fragmentations

You saw, at the beginning of the last section, a ring expansion reaction of a decalin.

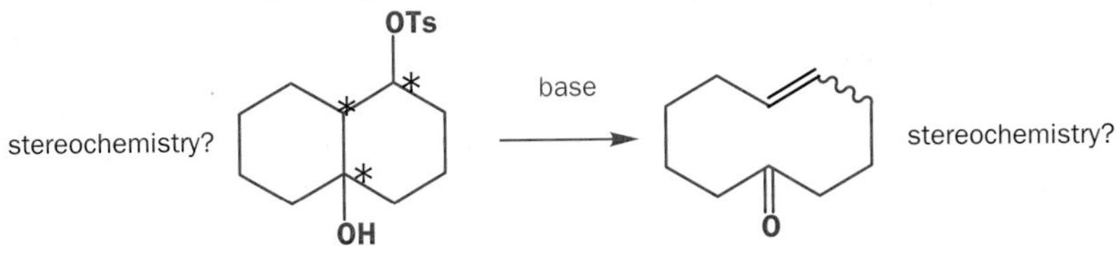

Now, the story of this ring expansion is a little more complex than we led you to believe, because the starting material has three stereogenic centres (*) and hence can exist as four diastereoisomers: two *trans*-decalins and two *cis*-decalins. What is more, the product has a double bond in a ten-membered ring: will it be *cis* or *trans*? (Both are possible—see Chapter 31.)

One of the four diastereoisomers of starting material cannot place the tosylate anti-periplanar to the ring-fusion bond, so it can't fragment.

The other three diastereoisomers all can, but two of them give a *trans* double bond while the third gives *cis*.

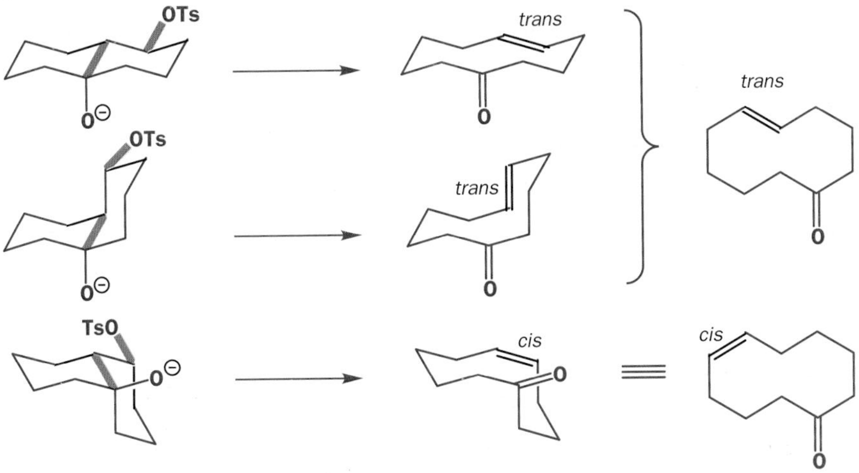

Looking at the alignment of the bonds that end up flanking the double bond in the product shows you where the geometrical isomers come from: these are the black bonds in the starting material, and are *trans* across the forming π system in the first two isomers and *cis* in the third. Fragmentations are stereospecific with regard to double bond geometry, much as E2 elimination reactions are.

Corey applied this stereospecificity in conjunction with a ring expansion reaction to make the natural product caryophyllene. Caryophyllene is a bicyclic molecule with a nine-membered ring containing an *E* trisubstituted double bond. The right relative stereochemistry in the starting material leads both to fragmentation of the right bond and to formation of the alkene with the right stereochemistry.

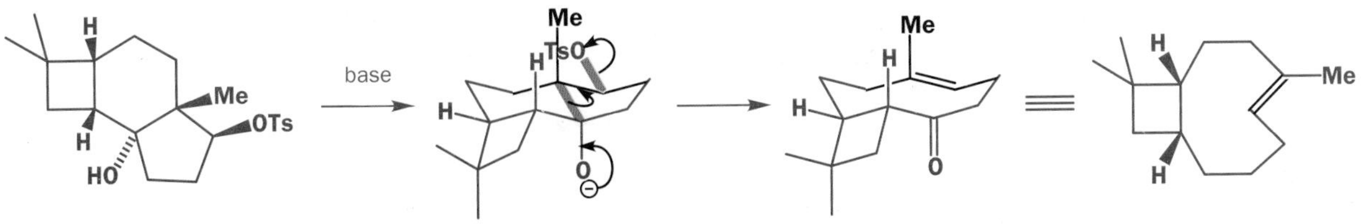

One of the most spectacular demonstrations of the use of fragmentation was the 1968 synthesis of juvenile hormone (a compound you met in Chapter 31) by chemists at Syntex, an American pharmaceutical company.

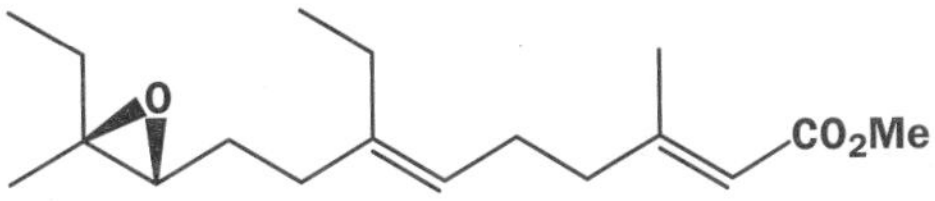

juvenile hormone

> ■ We discussed the conformations of decalins in Chapter 18.
>
>
> green bonds not anti-periplanar: no fragmentation possible

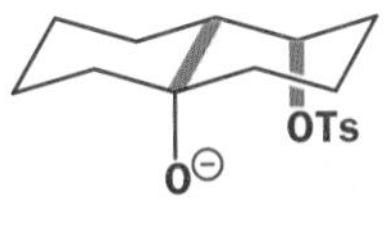

The major challenge in making juvenile hormone is the three trisubstituted double bonds (one of which ends up as an epoxide), and the initial target was to make the related aldehyde, which contains two of them.

The Syntex chemists reasoned that, if this methyl ketone could be made stereospecifically by fragmenting a cyclic starting material, the (hard-to-control) double bond stereochemistry would derive directly from the (easier-to-control) relative stereochemistry of the cyclic compound. The starting material they chose was a 5/6-fused system, which fragments to give one of the double bonds.

The product of this reaction is prepared for another fragmentation by addition of methyllithium (you might like to consider why you get this diastereoisomer) and tosylation of the less hindered secondary alcohol. Base promotes the second fragmentation.

In the next chapter you will meet, among many other reactions, more fragmentations, but they will be radical fragmentations rather than ionic fragmentations, and involve homolytic cleavage of C–C bonds.

A second synthesis of longifolene

In Chapters 28 and 35 we introduced parts of Oppolzer's synthesis of longifolene. We now revise those reactions and bring the synthesis a stage further forward with a fragmentation reaction different from the one used earlier in the chapter for the same molecule. McMurry used a fragmentation to escape from a disaster. Oppolzer had planned to use one right from the start. The first stage in the synthesis involves the building of two five-membered rings into a 1,3-diketone.

82% yield

Next the enol ester of the 1,3-diketone forms a new four-membered ring by a [2 + 2] photo-cycloaddition. This reaction appears in Chapter 35 but you are invited to work out for yourself what is happening here before you refer back to that chapter.

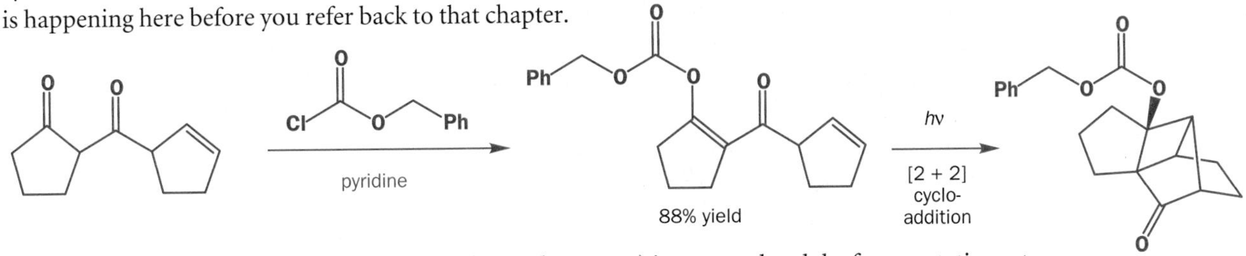

88% yield

Finally the protecting group (a Cbz group from Chapter 24) is removed and the fragmentation set in motion. The four-membered ring is cleaved and the ring system of longifolene revealed. You might like to compare this route with McMurry's route described earlier in this chapter.

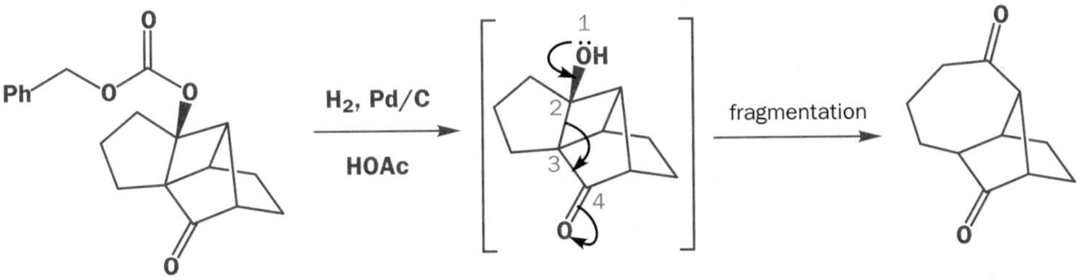

The synthesis of nootkatone

In the 1970s it was supposed that the characteristic sharp fruity scent and flavour of grapefruit came mainly if not entirely from a simple bicyclic enone called nootkatone. There was quite a rush to synthesize this compound in various laboratories and a remarkable feature of many successful syntheses was the use of fragmentation reactions. We shall describe parts of three syntheses involving the fragmentation of a six-, a four-, and a three-membered ring.

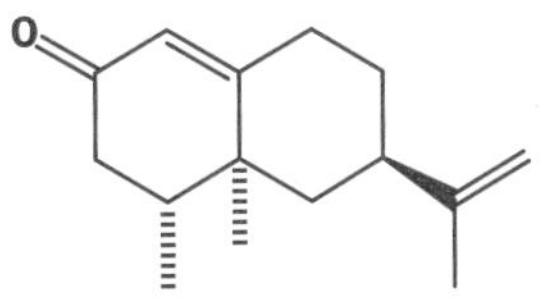

Most syntheses make the side-chain alkene by an elimination reaction so the first 'disconnection' is an FGI adding HX back into the alkene. The last C–C bond-forming operation in most syntheses is an intramolecular aldol reaction to make the enone so that can be disconnected next. It is the starting material for the aldol, a simple monocyclic diketone, which is usually made by a fragmentation reaction because this is a good way to set up the stereochemistry.

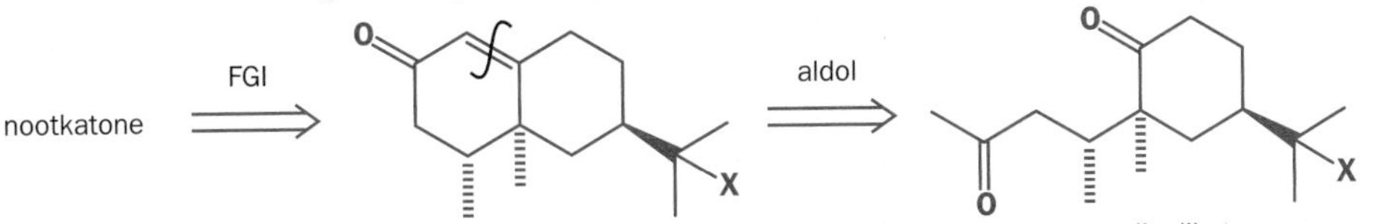

nootkatone
supposed flavour
principle of grapefruit

monocyclic diketone

Fragmentation of a three-membered ring

This synthesis does not look as though it will lead to nootkatone because the fragmentation product requires a great deal of development. It has the advantage that the stereochemistry is correct at one centre at least. The sequence starts from natural (–)-carone: conjugate addition of the enolate to butenone without control leads to a bicyclic diketone with one extra stereogenic centre. The enone adds to the bottom face of the enolate opposite the dimethylcyclopropane ring so the methyl group is forced upwards.

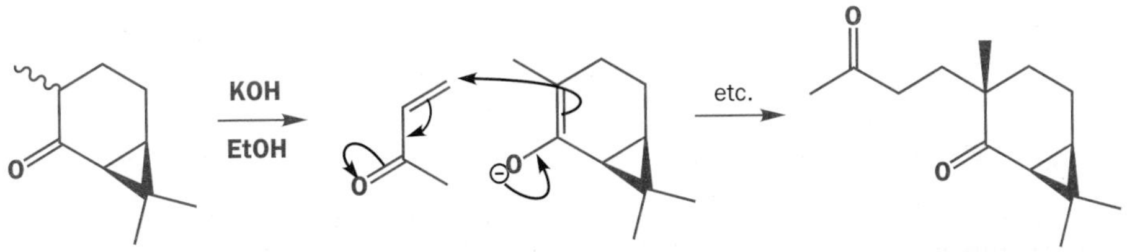

(–)-carone   enolate of (–)-carone   bicyclic diketone

Now the diketone is cyclized in HCl to give a bicyclic enone. A new six-membered ring has been formed but the old three-membered ring has disappeared. First, an intramolecular aldol reaction closes the new six-membered ring to form an enone and then the stage is set for a fragmentation.

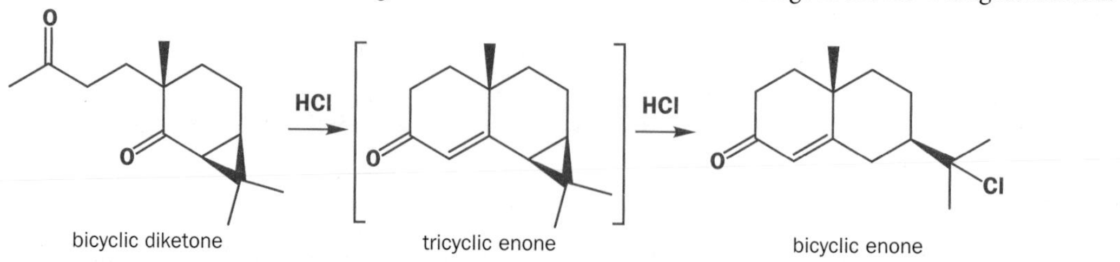

bicyclic diketone   tricyclic enone   bicyclic enone

The fragmentation is pulled by the enone (with some help from the acid) and pushed by the stability of the tertiary carbocation as well as the release of strain as the single bond that is fragmented is in a three-membered ring. The fragmentation product is an enol on the left and a carbocation on the right. Addition of a proton to the end of the enol and a chloride ion to the cation gives the bicyclic enone. The chloroalkyl side chain must be on the top of the molecule because only one of the C–C bonds in the three-membered ring has been broken and the remaining bond cannot change its stereochemistry. The further development of this compound into nootkatone is beyond the scope of this book.

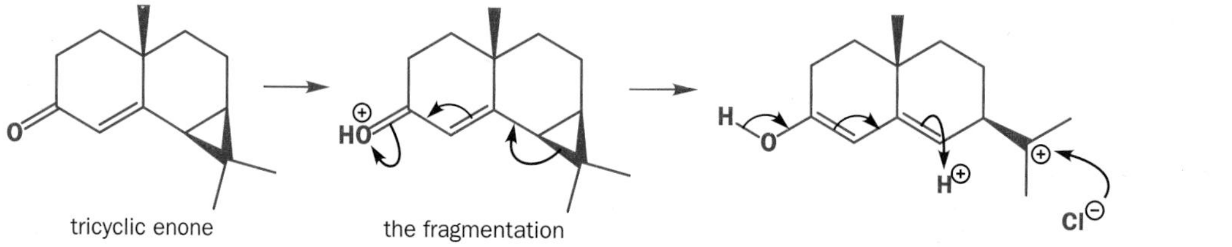

tricyclic enone   the fragmentation

Fragmentation of a four-membered ring

This approach leads directly to the enone needed for nootkatone. A diketone prepared from a natural terpene (Chapter 51) is also treated with HCl and much the same reactions ensue except that the fragmentation now breaks open a four-membered ring. First, the intramolecular aldol reaction to make the second six-membered ring.

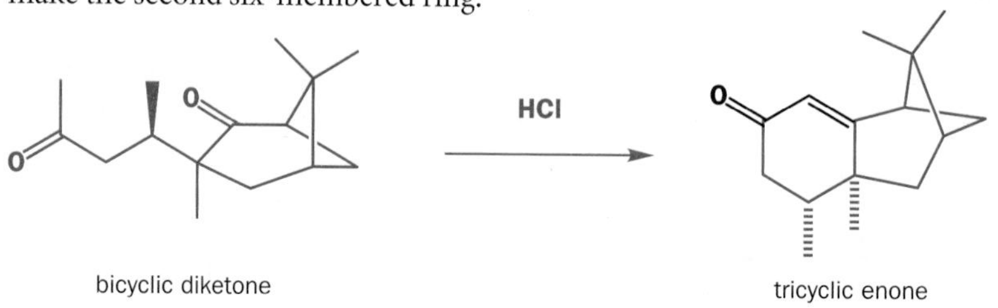

bicyclic diketone   tricyclic enone

Now the fragmentation, which follows much the same course as the last one: the enone again provides the electron pull while the cleavage of a strained C–C single bond in a four-membered ring to give a tertiary carbocation provides the electron push. A simple elimination is all that is needed to make nootkatone from this bicyclic chloroenone.

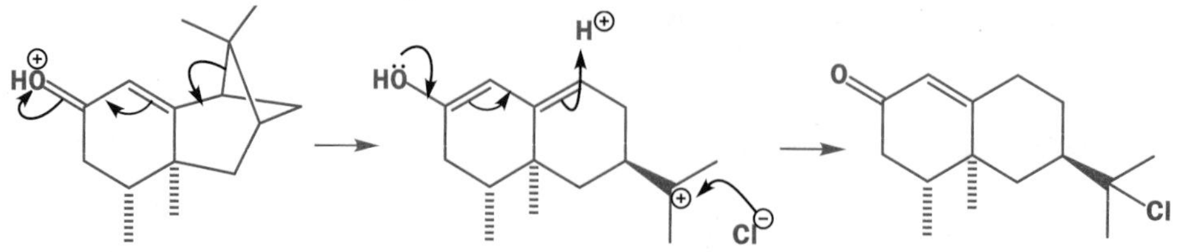

Fragmentation of a six-membered ring

This chemistry is quite different from the examples we have just seen. The starting material has a bridged bicyclic structure and was made by a Diels–Alder reaction (Chapter 35). Fragmentation is initiated by formic acid (HCO$_2$H), which protonates the tertiary alcohol and creates a tertiary carbocation. The ether provides the push. More serious electronic interactions are needed in this fragmentation as the C–C bond being broken is not in a strained ring.

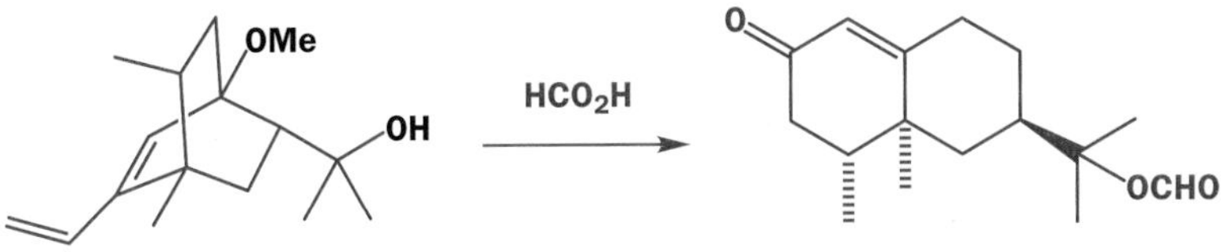

50% yield

The yield of 50% is not wonderful but there is obviously a lot of chemistry going on here so it is acceptable when so much is being achieved. The first stage is the fragmentation itself. Drawing the product first of all in the same shape as the starting material and then redrawing, to ensure that we don't make a mistake, we discover that we are well on the way to nootkatone. Note that the stereochemistry of the two methyl groups comes directly from the stereochemistry of the starting materials and no new stereogenic centres are created in the fragmentation. Though one six-membered ring is fragmented, another remains.

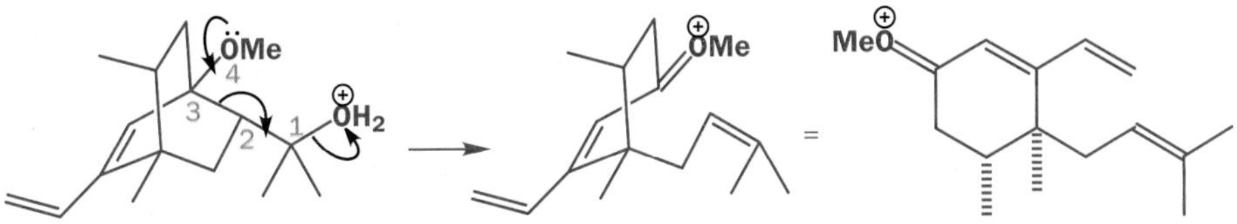

The first formed product now cyclizes to form the second six-membered ring. This recreates a carbocation at the tertiary centre like the one that set off the fragmentation as the more nucleophilic end of the isolated alkene attacks the end of the conjugate electrophile. This is a thermodynamically controlled reaction with the new stereogenic centre choosing an equatorial substituent.

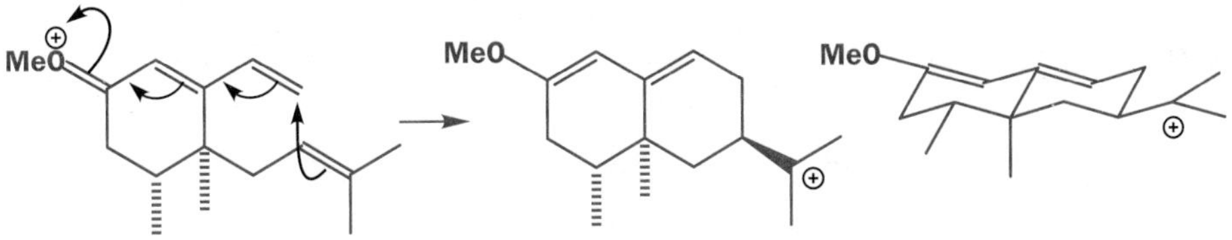

The cation picks up the only nucleophile available—the very weak formic acid. This gives the product of the fragmentation, which contains two unstable functional groups—a tertiary formate ester and an enol ether—and this product is not isolated from the reaction mixture.

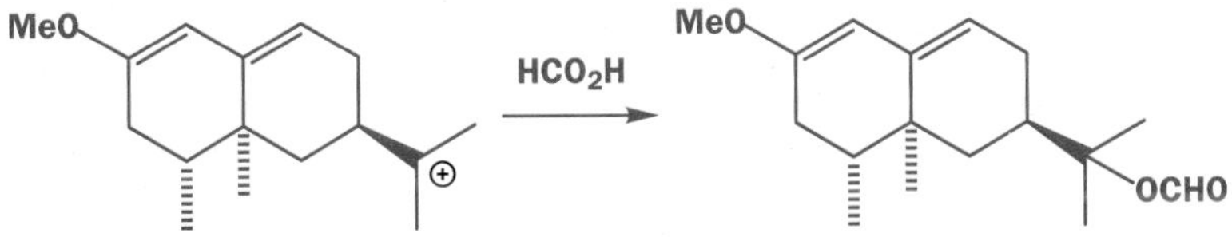

Protonation and hydrolysis of the extended enol ether to release the enone may occur during work-up and the stable enone is the first compound that can be isolated. The 50% yield of this compound represents a much better yield in four steps: fragmentation, olefin cyclization, addition of formic acid, and enol ether hydrolysis.

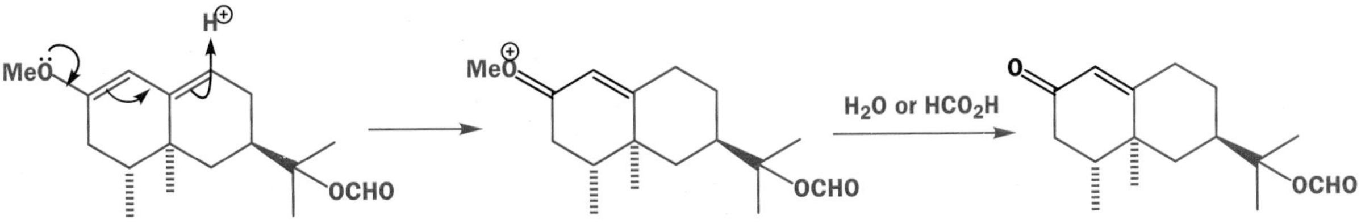

Completion of the synthesis of nootkatone simply requires pyrolysis of the formate ester in refluxing 2,4,6-trimethyl pyridine (b.p. 172 °C). The reaction is a *syn* elimination by a pericyclic mechanism and it gives nootkatone in 79% yield.

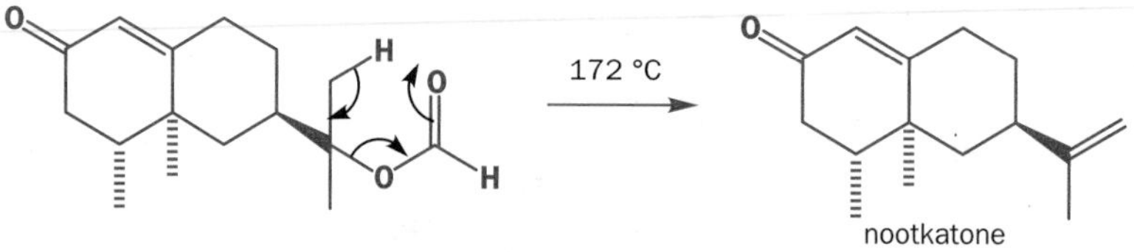

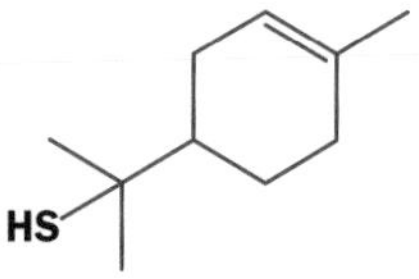

nootkatone

The synthesis of nootkatone occupied many chemists for some years and has given us some excellent examples of fragmentation reactions. However, the synthetic samples of nootkatone failed to deliver the intense grapefruit taste and smell of the material from grapefruits. The reason is simply that nootkatone is not the flavour principle of grapefruit! The samples of nootkatone isolated from grapefruit contained minute traces of the true flavour principle—a simple thiol. Humans can detect 2×10^{-5} p.p.b. (yes, parts per *billion*) of this compound, so even the tiniest trace is very powerful. At least the syntheses allowed chemists to correct an error.

true flavouring principle of grapefruit

■ We first introduced this intense taste in Chapter 1 and we will discuss sulfur compounds in Chapter 46.

A revision example: rearrangements and fragmentation

We shall end this chapter with an example that involves many of the reactions we have been discussing in recent chapters. It culminates in a fragmentation but takes in two different rearrangements (Chapter 37) on the way as well as a cycloaddition (Chapter 35) and an electrocyclic reaction (Chapter 36). Here is the whole scheme with the main changes in each step highlighted in black. You might cast your eye over the scheme and see in general terms what sort of reaction happens at each step (substitution, rearrangement, etc.).

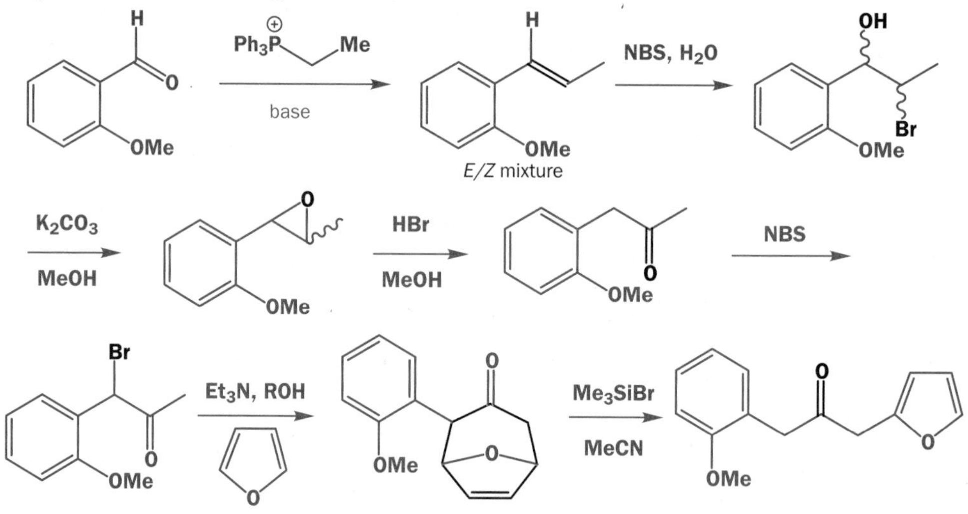

The first step is a simple Wittig reaction with an unstabilized ylid (Chapter 31), which we expect to favour the *Z*-alkene. It does but, as is common with Wittig reactions, an *E/Z* mixture is formed but not separated as both isomers eventually give the same compound. The reaction is kinetically controlled and the decomposition of the oxaphosphetane intermediate is in some ways like a fragmentation.

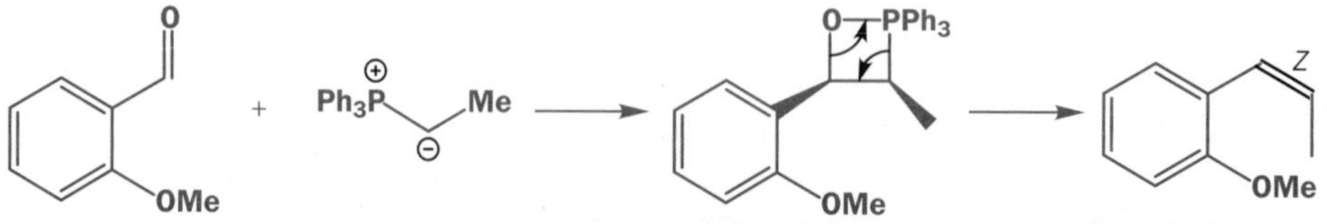

Now the alkene is converted into an epoxide by a slightly unusual sequence. Bromination with NBS (*N*-bromosuccinimide) in water gives a mixture of bromohydrins by electrophilic addition to the double bond. The reaction occurs through a bromonium ion and is stereospecifically *anti* on each iso-mer of the alkene.

■
NBS is a radical generator in nonpolar solvents as we shall see in Chapter 40, but in polar solvents, especially water, it supplies electrophilic bromine, Chapter 20.

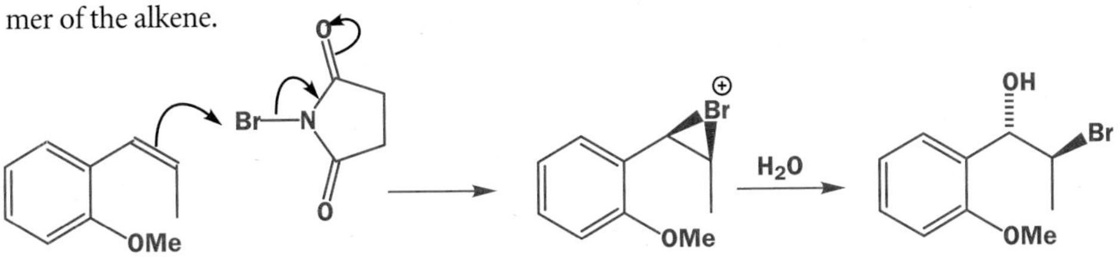

Next, the bromohydrin is treated with base and an intramolecular S_N2 reaction (Chapter 17) closes the epoxide ring. This too is stereospecific and the major isomer only is shown. The mixture of epoxides is a result of the *E/Z*-alkene mixture. Potassium carbonate is too weak a base to generate much of the alkoxide anion but the cyclization may still go this way in methanol. In Chapter 41 you will learn of an alternative type of catalysis by weak bases.

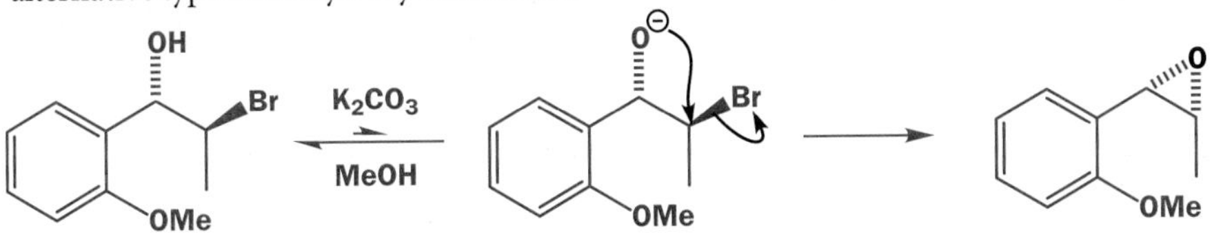

We saw some epoxide rearrangements in Chapter 37 but this reaction seems rather tame by compar-ison. The epoxide opens in acid to give the more stable (secondary and benzylic) of the two possible car-bocations and then a hydrogen atom migrates with the pair of electrons from the C–H bond ('hydride shift') to give a ketone. The rearrangement is useful because it allows the synthesis of aryl ketones, which cannot easily be made by a Friedel–Crafts reaction since the carbonyl group is in the wrong position on the side chain (Chapter 22).

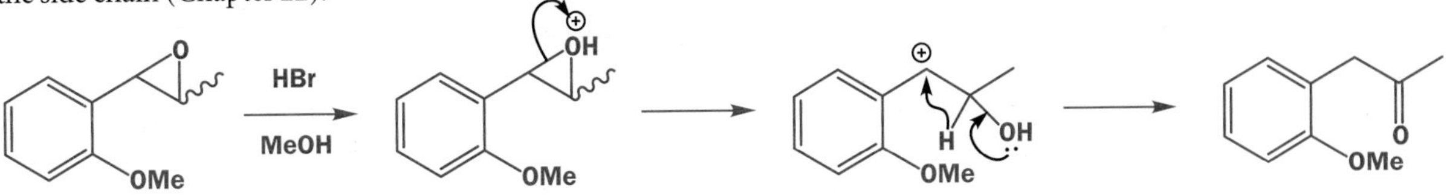

The ketone is then brominated, also with NBS, in a regioselective manner. The more conjugated enol is formed between the carbonyl group and the aromatic ring and this is attacked electrophilically by the bromine atom of the NBS (Chapter 20).

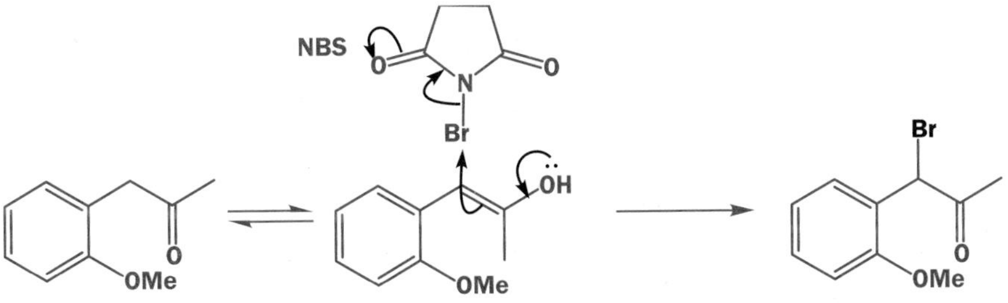

Cycloaddition and rearrangement

Now comes the most interesting step in the whole process—a step that unites a cycloaddition and a rearrangement and sets the scene for a fragmentation. The idea was to treat the bromoketone with base to make an oxyallyl cation as an unstable intermediate.

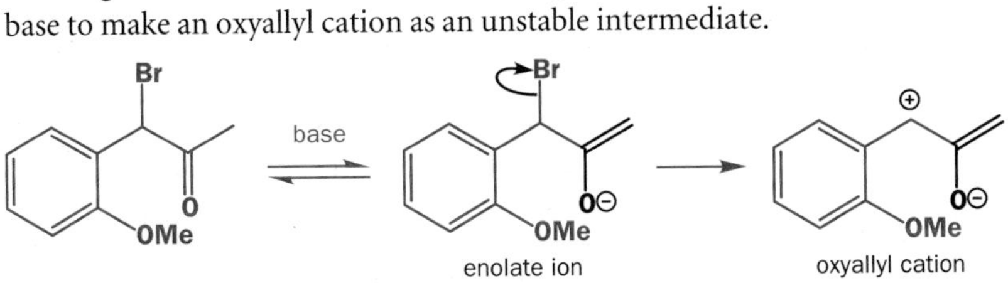

enolate ion   oxyallyl cation

The oxyallyl cation with its two electrons delocalized over the allylic system would add to furan in a [2 + 4] cycloaddition to give a new cation stabilized by the oxyanion or, in more familiar guise, a ketone. The reaction was supposed to go like this.

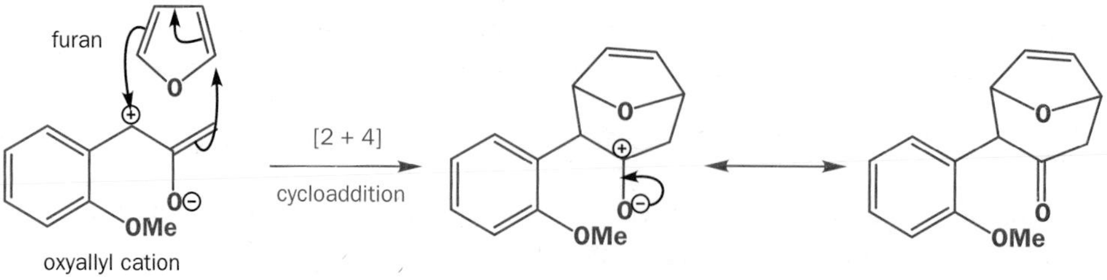

The best base turned out to be the tertiary amine Et₃N and the reaction had to be performed in alcoholic solution as alcohols were the only solvents able to keep the organic and ionic materials in solution. However, a substantial amount of a by-product was formed in ethanol—evidently the product of a Favorskii rearrangement.

cycloaddition product   Favorskii product

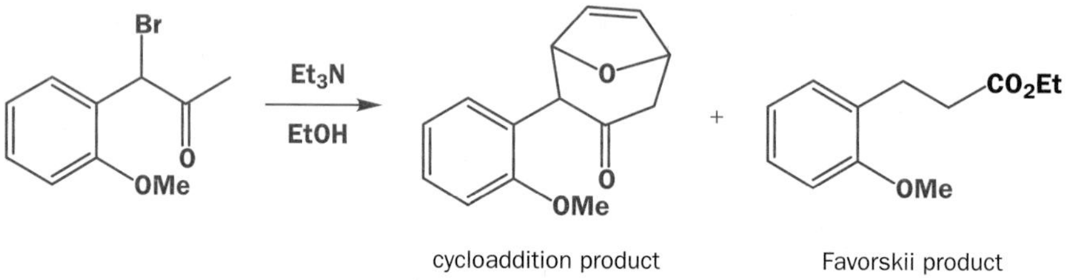

What is happening here is that the oxyallyl cation is in equilibrium with the cyclopropanone by an electrocyclic reaction (Chapter 36) and the alcohol is capturing this unstable ketone by nucleophilic addition. Hemiacetals of cyclopropanones form spontaneously in alcoholic solution (Chapter 6) because of the strain in the ketone. The anion of the hemiacetal decomposes by cleavage of a C–C bond to release what would be the more stable of the two carbanions, that is, the benzylic carbanion. This carbanion is not actually formed as it is protonated by the alcohol as it leaves.

This is an example of GAC (general acid catalysis) as explained in Chapter 41.

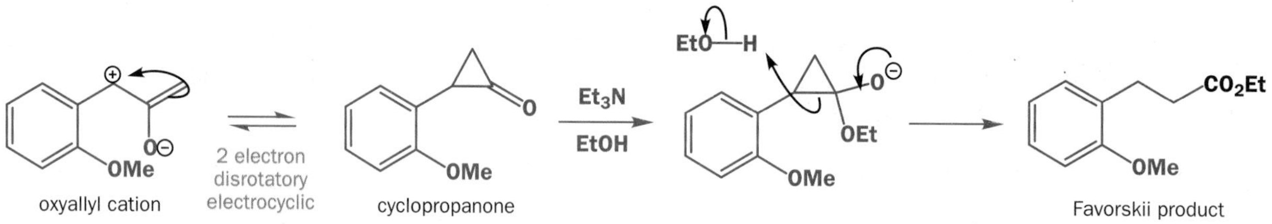

oxyallyl cation   2 electron   cyclopropanone   Favorskii product
                   disrotatory
                   electrocyclic

So how can the cycloaddition be promoted at the expense of the Favorskii rearrangement? Nothing can be done about the equilibrium between the oxyallyl anion and the cyclopropanone—that's a fact of life. The answer is to reduce the nucleophilicity of the alcohol by using trifluoroethanol instead of ethanol. Under these conditions the major product is the cycloadduct, which can be isolated in 73% yield.

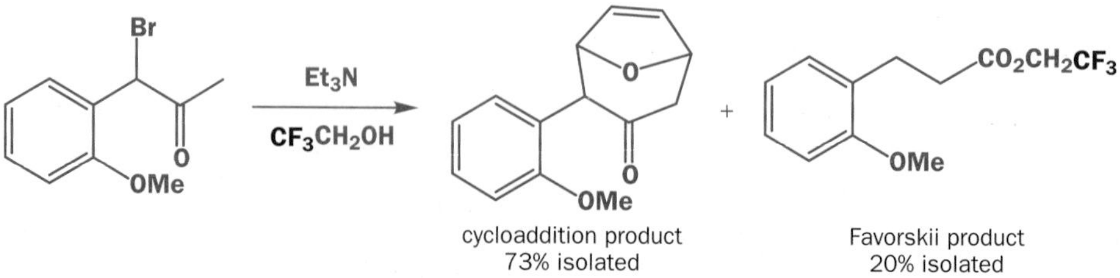

cycloaddition product   Favorskii product
73% isolated   20% isolated

The two compounds can easily be separated as they have completely different structures and are not stereoisomers or indeed isomers of any kind. Now it is time for the fragmentation reaction on the cycloadduct.

The fragmentation reaction

The cycloadduct is fragmented with Me$_3$SiBr in acetonitrile. The electrophilic silicon atom attacks the ketone and the furan oxygen atom provides the electronic push. These two groups have the 1,4 relationship necessary for a fragmentation. First of all, we shall draw the product in the same way as the starting material—this is a good tip in a complicated mechanism. The product may look odd but we can redraw it more realistically in a moment.

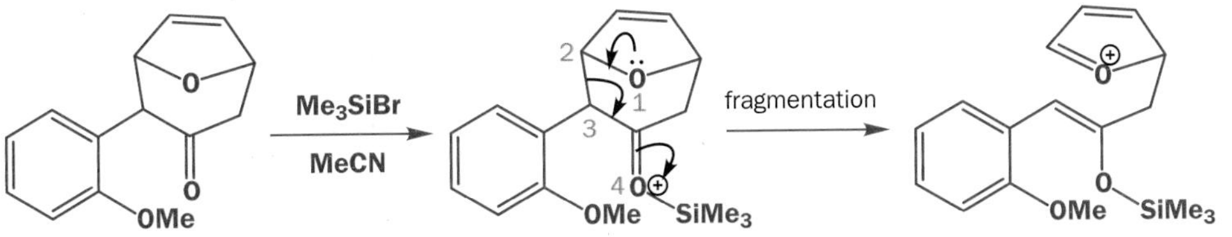

The redrawn product is a silyl enol ether (Chapter 21) at one end and an oxonium ion at the other. Simple proton removal and hydrolysis of the silyl enol ether in the work-up reveals a furan that can be isolated in 81% yield as the true product.

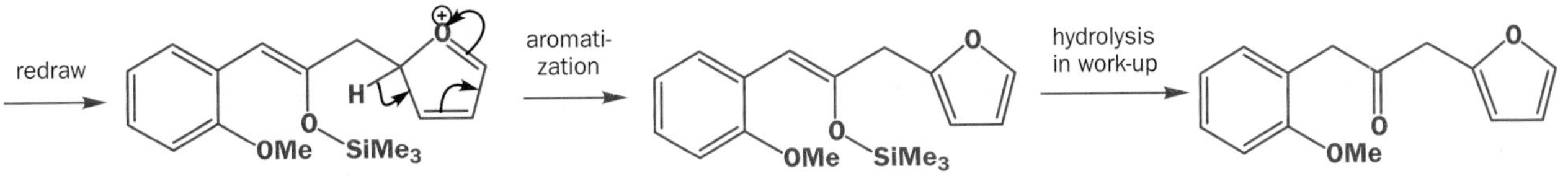

This product is worth a close look. The three-atom chain joining the two aromatic rings has the ketone on the middle carbon atom and it is therefore on C2 (β) with respect to *both* rings. This is the difficult position for a carbonyl group and so this product cannot be made by a Friedel–Crafts reaction on either ring.

Fragmentation reactions cleave C–C single bonds by a combination of electron push and electron pull so that both electrons in the bond move in the same direction as the bond breaks. In the next chapter we shall see reactions that break C–C bonds in a quite different way. No electron push or pull is required because one electron goes one way and one the other. These are radical reactions.

Problems

1. Just to check your skill at finding fragmentations by numbers, draw a mechanism for each of these one-step fragmentations in basic solution (with an acidic work-up).

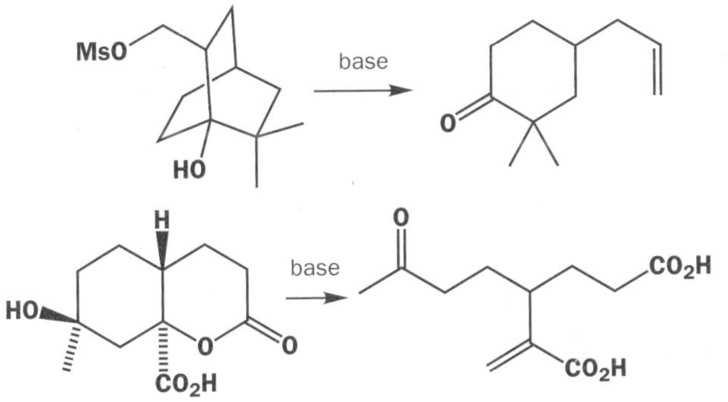

2. Treatment of this hydroxy-ketone with base followed by acid gives the enone shown. What is the structure of the intermediate A, how is it formed, and what is the mechanism of the formation of the final product?

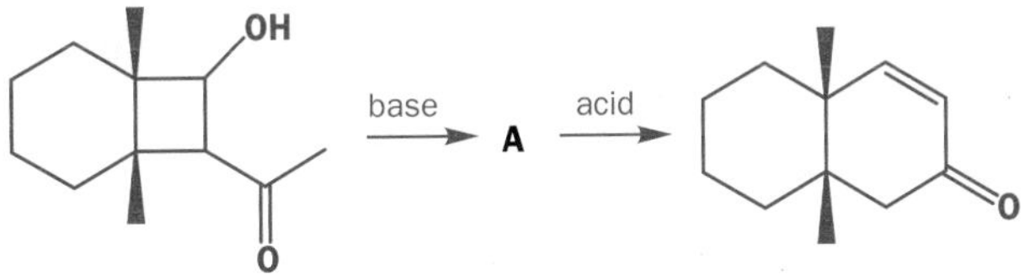

3. Suggest a mechanism for this reaction that involves a fragmentation as a key step.

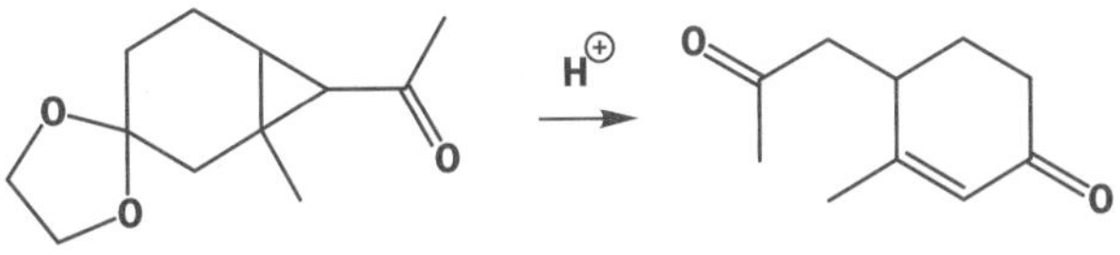

4. Explain why both of these tricyclic ketones fragment to the same diastereoisomer of the same cyclo-octadione.

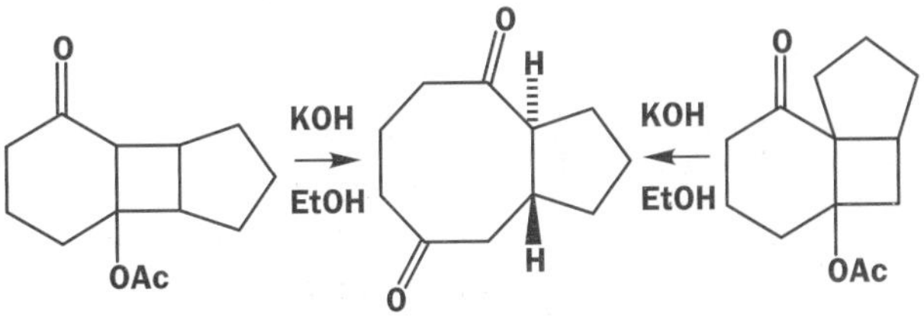

5. Suggest a mechanism for this ring expansion in which fragmentation is one step.

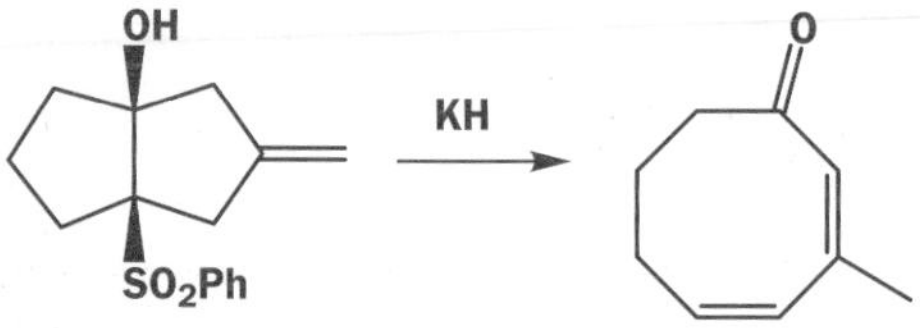

6. Suggest a mechanism for this fragmentation and explain the stereochemistry of the double bonds in the product. This is a tricky problem but find the mechanism and the stereochemistry will follow.

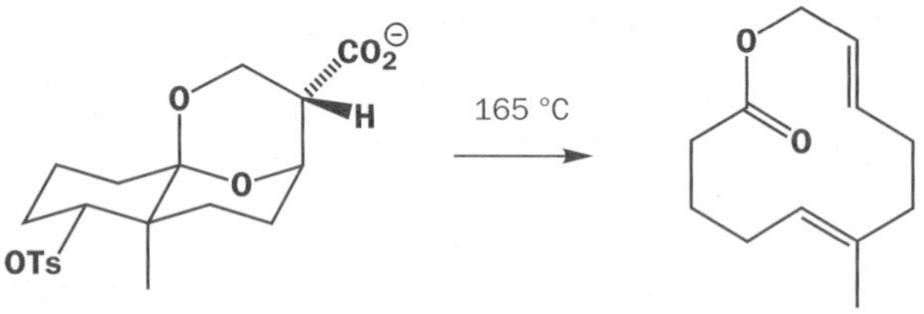

7. Suggest a mechanism for this reaction and explain why the molecule is prepared to abandon a stable six-membered ring for a larger ring.

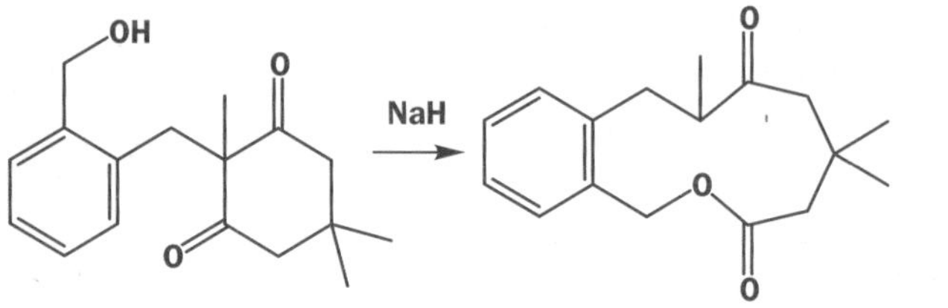

8. Give mechanisms for these reactions, commenting on the fragmentation.

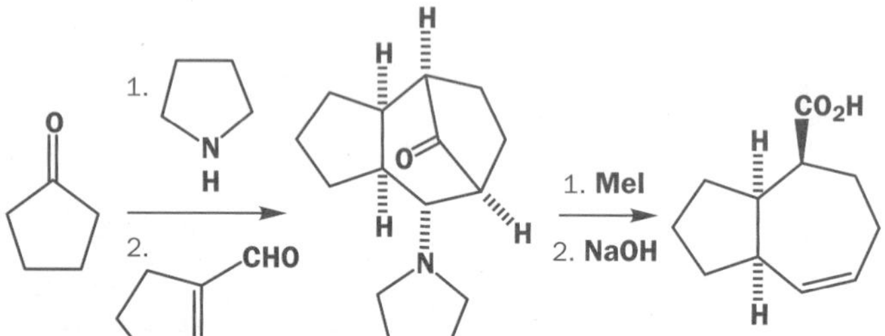

9. Propose mechanisms for the synthesis of the bicyclic intermediate and explain why only one diastereoisomer fragments (which one?).

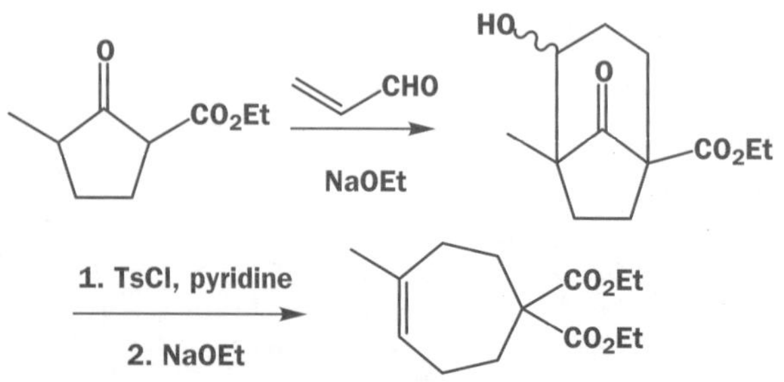

10. Suggest mechanisms for these reactions, explaining the alkene geometry in the first case. Do you consider that they are fragmentations?

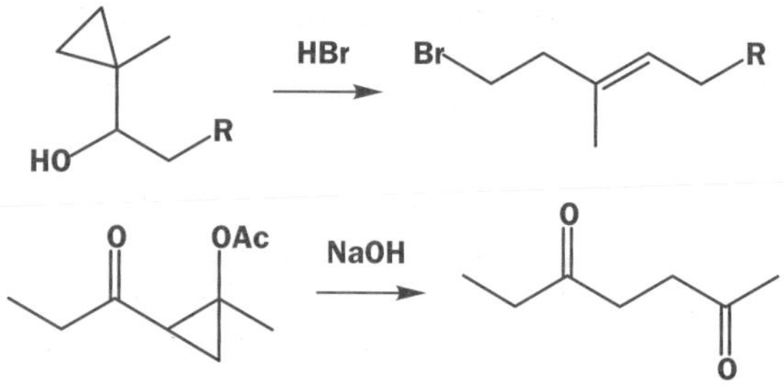

11. What steps would be necessary to carry out an Eschenmoser fragmentation on this ketone and what products would be formed?

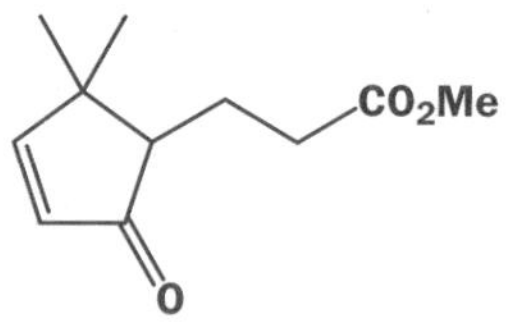

12. These related spirocyclic compounds give different naphthalenes when treated with sodium borohydride or with 5M HCl. Each reaction starts with a different fragmentation. Give mechanisms for the reactions and explain why the fragmentations are different. Treatment of the starting ketone with LiAlH$_4$ instead of NaBH$_4$ gives the alcohol below without fragmentation. Comment on the difference between the two reagents and the stereochemistry of the alcohol.

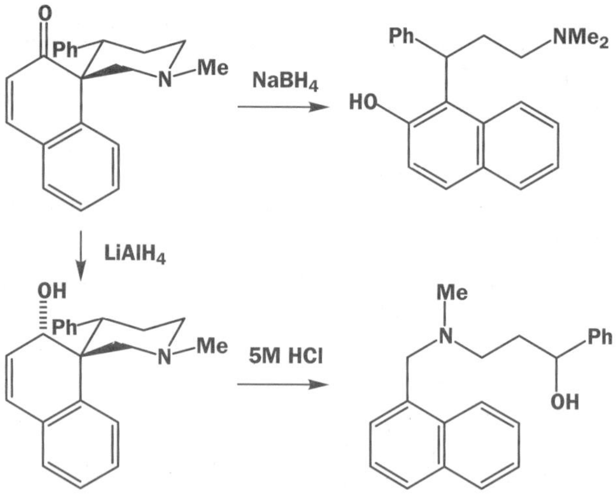

13. Revision content. Suggest mechanisms for these reactions explaining the stereochemistry.

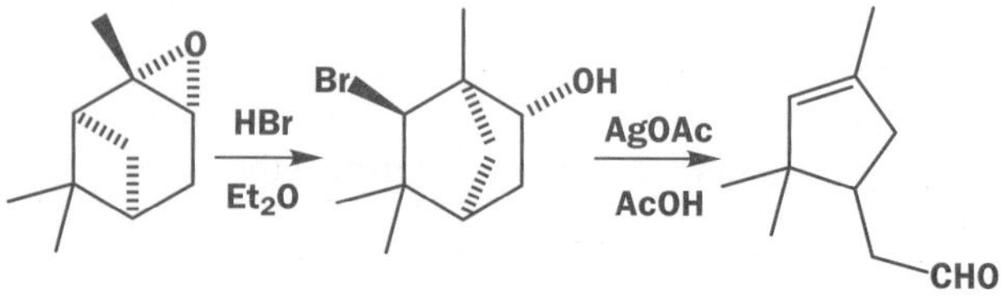

Radical reactions

39

Connections

Building on:

- **Conjugate addition** ch10 & ch23
- **Energy profile diagrams** ch13
- **Nucleophilic substitution** ch17
- **Conformational analysis** ch18
- **Elimination reactions** ch19
- **Controlling stereochemistry** ch16, ch33, & ch34
- **Retrosynthetic analysis** ch30
- **Diastereoselectivity** ch33–ch34

Arriving at:

- Radicals are species with unpaired electrons
- Radical reactions follow different rules to those of ionic reactions
- Bond strength is very important
- Radicals can be formed with Br, Cl, Sn, and Hg
- Efficient radical reactions are chain reactions
- There are electrophilic and nucleophilic radicals
- Radicals favour conjugate addition
- Cyclization is easy with radical reactions

Looking forward to:

- **Carbene chemistry** ch40
- **Determination of mechanism** ch41
- **Stereoelectronics** ch42
- **Main group chemistry** ch46–ch47
- **Natural products** ch51
- **Polymerization** ch52

Radicals contain unpaired electrons

You may remember that at the beginning of Chapter 8 we said that the cleavage of H–Cl into H^+ and Cl^- is possible in solution only because the ions that are formed are solvated: in the gas phase, the reaction is endothermic with $\Delta G = +1347\ \text{kJ mol}^{-1}$, a value so vast that even if the whole universe were made of gaseous HCl at 273 K, not a single molecule would be dissociated into H^+ and Cl^- ions.

$$\text{HCl} \longrightarrow \text{H}^{\oplus} + \text{Cl}^{\ominus}$$

8 electrons in outer shell

At temperatures above about 200 °C, however, HCl does begin to dissociate, but not into ions. Instead of the chlorine atom taking both bonding electrons with it, leaving a naked proton, the electron pair forming the H–Cl bond is shared out between the two atoms. ΔG for this reaction is a much more reasonable $+431\ \text{kJ mol}^{-1}$ and, at high temperatures (above about 200 °C, that is), HCl gas can be dissociated into H and Cl atoms.

$$\text{HCl} \xrightarrow{>200\ °\text{C}} \text{H}^{\bullet} + \text{Cl}^{\bullet}$$

one electron — 7 electrons in outer shell

▶ The single, unpaired electron possessed by each atom is represented by a dot. The Cl atom, of course, has another three pairs of electrons that are not shown.

● **Heterolysis and homolysis**

- When bonds break and one atom gets both bonding electrons, the process is called **heterolysis**
  The products of heterolysis are, of course, **ions.**
- When bonds break and the atoms get one bonding electron each, the process is called **homolysis**
  The products of homolysis are **radicals**, which may be atoms or molecules, and contain an unpaired electron.

It was, in fact, a reaction of a closely related molecule, hydrogen bromide, that was among the first to alert chemists to the possibility that radicals can be formed in chemical reactions even at ambient temperatures, and that they have a distinct pattern of reactivity. In the 1930s, Morris Kharasch found that the regioselectivity of addition of H–Br to isobutene was dependent on whether or not oxygen and peroxides were present in the reaction mixture.

major product in the **absence**
of oxygen and peroxide

major product in the **presence**
of oxygen and peroxide

You should be familiar with this mechanism from Chapter 20.

It turns out that in the *absence* of peroxides the addition takes place by the type of (ionic) mechanism that you have already met. The tertiary bromide is formed because the intermediate, a tertiary cation, is more stable than the alternative primary cation.

tertiary cation

This scheme represents the process schematically: it is not the full mechanism of the reaction!

In the *presence* of peroxides, the mechanism is quite different. Homolysis of the H–Br takes place, and bromine radicals that attack the C=C double bond at its less hindered end are formed. Mostly isobutyl bromide is formed.

radical prefers to attack
this end of double bond

91% yield
iso-butyl bromide

6% yield
tert-butyl bromide

Try to get a feel for bond strengths: we shall refer to them a lot in this chapter as they're very important to radical reactions. Compare this with the situation for ionic reactions, in which the strengths of the bonds involved are often much less important than polar effects (see the example on p. 288, for example).

What does the peroxide do? Why does its presence change the mechanism? The peroxide undergoes homolysis of the weak O–O bond extremely easily, and because of this it initiates a **radical chain reaction**. We said that H–Cl in the gas phase undergoes homolysis in preference to heterolysis: other types of bond are even more susceptible to homolysis. You can see this for yourself by looking at this table of bond dissociation energies (ΔG for X–Y $\rightarrow$ X$^{\bullet}$ + Y$^{\bullet}$).

Dialkyl peroxides (dimethyl peroxide is shown in the table) contain the very weak O–O bond. The radicals formed by homolytic cleavage of these bonds, stimulated by a little heat or

Bond X–Y	ΔG for X–Y $\rightarrow$ X$^{\bullet}$ + Y$^{\bullet}$, kJ mol^{-1}	Bond X–Y	ΔG for X–Y $\rightarrow$ X$^{\bullet}$ + Y$^{\bullet}$, kJ mol^{-1}
H–OH	498	CH$_3$–Br	293
H$_3$C–H	435	CH$_3$–I	234
H$_3$C–OH	383	Cl–Cl	243
H$_3$C–CH$_3$	368	Br–Br	192
H–Cl	431	I–I	151
H–Br	366	HO–OH	213
H–I	298	MeO–OMe	151
CH$_3$–Cl	349		

light, initiate what we call a 'radical chain reaction', which results in the formation of the Br$^{\bullet}$ radicals, which add to the alkene's C=C double bond. We shall return to radical chain reactions and their mechanisms in detail later in this chapter.

Radicals form by homolysis of weak bonds

You've just met the most important way of making radicals: unpairing a pair of electrons by homolysis, making two new radicals. Temperatures of over 200 °C will homolyse most bonds; on the other hand, some weak bonds will undergo homolysis at temperatures little above room temperature. Light is a possible energy source for the homolysis of bonds too. Red light has associated with it 167 kJ mol^{-1}; blue light has about 293 kJ mol^{-1}. Ultraviolet (200 nm), with an associated energy of 586 kJ mol^{-1}, will decompose many organic compounds (including the DNA in skin cells: sunbathers beware!).

It is not sufficient for light to be energetic enough to promote homolysis; the molecule must have a mechanism for absorbing that energy, and the energy must end up concentrated in the vibrational mode that leads to bond breakage. We shall not consider these points further: if you are interested, you will find detailed explanations in specialized books on photochemistry.

There are a number of compounds whose homolysis is particularly important to chemists, and the most important ones are discussed in turn below. They all have weak σ bonds, and generate radicals that can be put to some chemical use. The halogens are quite readily homolysed by light. These process are important in radical halogenation reactions that we shall discuss later.

▶
$\Delta G^{\ddagger}$ is the activation energy for the reaction.

$$\text{Cl—Cl} \xrightarrow{\text{light } (h\nu)} 2 \times \text{Cl}^{\bullet} \qquad \Delta G^{\ddagger} = 243 \text{ kJ mol}^{-1}$$

$$\text{Br—Br} \xrightarrow{\text{light } (h\nu)} 2 \times \text{Br}^{\bullet} \qquad \Delta G^{\ddagger} = 192 \text{ kJ mol}^{-1}$$

$$\text{I—I} \xrightarrow{\text{light } (h\nu)} 2 \times \text{I}^{\bullet} \qquad \Delta G^{\ddagger} = 151 \text{ kJ mol}^{-1}$$

Dibenzoyl peroxide is an important compound because it can act as another initiator of radical reactions; we'll see why later. It undergoes homolysis simply on heating.

dibenzoyl peroxide

$$\xrightarrow[\Delta G^{\ddagger} = 139 \text{ kJ mol}^{-1}]{60\text{–}80 \text{ °C}}$$

Another compound that is often used in synthetic reactions for the same reason (though it reacts with a different set of compounds) is AIBN (*azoisobutyronitrile*).

AIBN

$$\xrightarrow[\Delta G^{\ddagger} = 131 \text{ kJ mol}^{-1}]{66\text{–}72 \text{ °C}}$$

N≡N

Some organometallic compounds, for example organomercuries or organocobalts, have very weak carbon–metal bonds, and are easily homolysed to give carbon-centred radicals. Alkyl mercury hydrides are formed by reducing alkyl mercury halides, but they are unstable at room temperature because the Hg–H bond is very weak. Bonds to hydrogen never break to give radicals *spontaneously* because H$^{\bullet}$ is too unstable to exist, but interaction with almost any radical removes the H atom and breaks the Hg–H bond. This is the process of hydrogen abstraction, which forms the next section of the chapter.

$$\text{R—Hg—R} \longrightarrow \text{R}^{\bullet} + {}^{\bullet}\text{Hg—R} \longrightarrow \text{Hg} + \text{R}^{\bullet}$$

weak C–metal bonds

$$\text{R—Hg—Cl} \xrightarrow{\text{NaBH}_4} \text{R—Hg—H} + {}^{\bullet}\text{R} \xrightarrow{20 \text{ °C}} \text{R—Hg}^{\bullet} + \text{H—R}$$

weak C–metal and metal–H bonds

Radicals in cars

Radicals generated from another organometallic compound, tetraethyllead Et$_4$Pb, were the reason for adding this compound to petrol. These radicals react with other radical species involved in the pre-ignition of petrol vapour in internal combustion engines, and prevent the phenomenon known as 'knocking'. Nowadays simple organic compounds such as MeOBu^t are used instead in 'green' petrol.

Radicals form by abstraction

Notice that we didn't put HBr on the list of molecules that form radicals by homolysis: relative to the weak bonds we have been talking about, the H–Br bond is quite strong (just about as strong as a C–C bond). Yet we said that Br$^{\bullet}$ radicals were involved in the addition reaction we talked about on p. 1020. These radicals are formed by the action of the alkoxy radicals (generated by homolysis of the peroxide) on HBr—a process known as **radical abstraction**. Here is the mechanism.

$$\text{R—O}^{\bullet} \quad \text{H—Br} \longrightarrow \text{ROH} + \text{Br}^{\bullet}$$

The peroxy radical RO$^{\bullet}$ 'abstracts' H$^{\bullet}$ from the HBr to give ROH, leaving behind a new radical Br$^{\bullet}$. We have described this process using arrows with 'half-heads' (also known as 'fish-hook arrows').

They indicate the movement of single electrons among orbitals, by analogy with our normal curly arrows, which indicate the movement of electron pairs.

movement of a pair of electrons     movement of a single electron

Writing radical mechanisms

There is often more than one correct way of drawing a radical mechanism using half-headed arrows. For

The full story shows that the odd electron on RO$^\bullet$ pairs with one of the electrons in the H–Br bond while the other moves on to the bromine atom.

example, we could have represented the abstraction reaction shown above in either of these alternative ways.

Because radical reactions always involve the reorganization of electron pairs, we can choose whether to show what happens to either or both of the members of

each pair. In most examples in this book, we will draw arrows only in one direction.

We use 'spin-paired molecule' to mean a 'normal' molecule, in which all the electrons are paired, in contrast with a radical, which has an unpaired electron.

The ability of radicals to propagate by abstraction is a key feature of radical chain reactions, which we shall come to later. There is an important difference between homolysis and abstraction as a way of making radicals: homolysis is a reaction of a spin-paired molecule that produces *two* radicals; abstraction is a reaction of a radical with a spin-paired molecule that produces *one* new radical and a new spin-paired molecule. Radical abstractions like this are therefore examples of your first radical reaction mechanism: they are in fact substitution reactions at H and can be compared with proton removal or even with an S$_N$2 reaction.

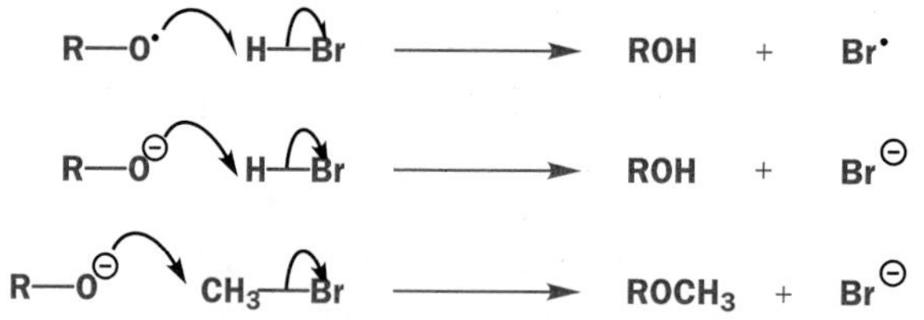

R—O$^\bullet$ H—Br ⟶ ROH + Br$^\bullet$ hydrogen abstraction

R—O$^{\ominus}$ H—Br ⟶ ROH + Br$^{\ominus}$ proton removal

R—O$^{\ominus}$ CH$_3$—Br ⟶ ROCH$_3$ + Br$^{\ominus}$ S$_N$2 reaction

Radical substitutions differ considerably from S$_N$1 or S$_N$2 reactions: importantly, *radical substitutions almost never occur at carbon atoms*. We shall come back to radical substitutions, or abstractions (depending on whether you take the point of view of the H atom or the Br atom), later in the chapter.

First radical detected

The very first radical to be detected, the triphenylmethyl radical, was made in 1900 by abstraction of Cl$^\bullet$ from Ph$_3$CCl by Ag metal.

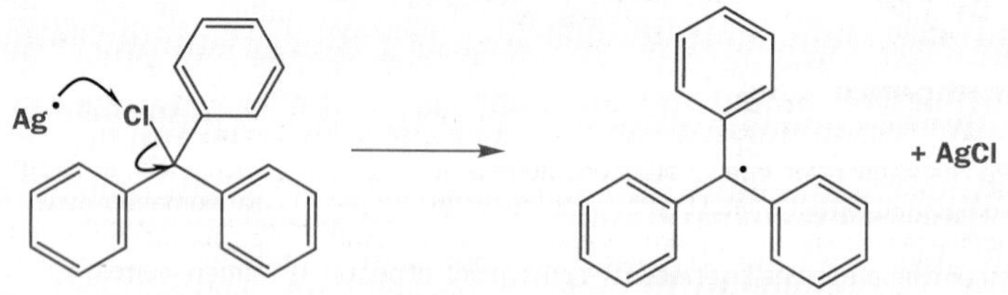

+ AgCl

relatively stable triphenylmethyl radical

This radical is relatively stable (we shall see why shortly), but reacts with itself reversibly in solution. The product of the dimerization of triphenylmethyl was for 70 years believed to be tetraphenyl ethane but, in 1970, NMR showed that it was, in fact, an unsymmetrical dimer.

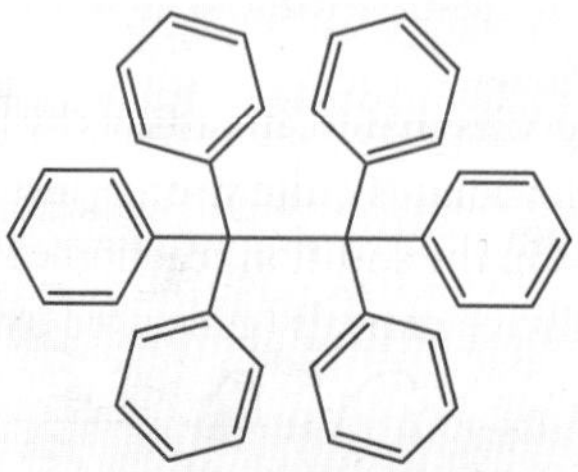

original suggested structure of triphenylmethyl dimer (1900)

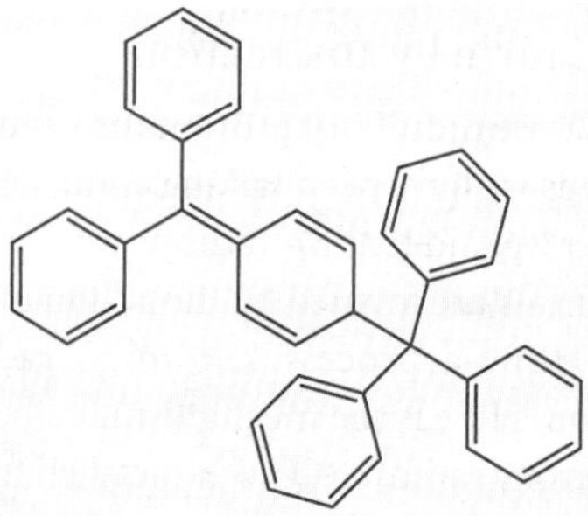

correct structure of dimer (1970)

Radicals form by addition

The key step in the radical reaction with which we started the chapter is the formation of a radical by **radical addition**. The Br$^\bullet$ radical (which, you will remember, was formed by abstraction of H$^\bullet$ from HBr by RO$^\bullet$) adds to the alkene to give a new, carbon-centred radical. This is the mechanism: again, notice that half-headed arrows are used to indicate the movement of single electrons.

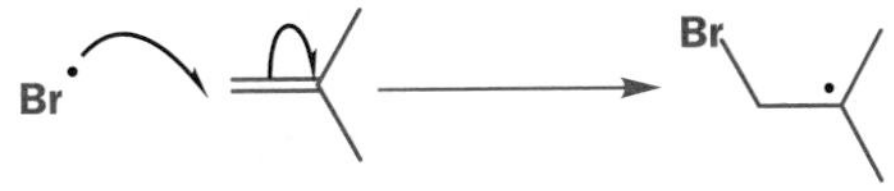

Just as charge must be conserved through a chemical reaction, so must be the spin of the electrons involved. If a reactant carries an unpaired electron, then so must a product. Addition of a radical to a spin-paired molecule always generates a new radical. Radical addition is therefore a second type of radical-forming reaction.

The simplest radical addition reactions occur when a single electron is added to a spin-paired molecule. This process is a reduction. You have already met some examples of single-electron reductions: Birch reductions (Chapter 24) use the single electron formed when a group I metal (sodium, usually) is dissolved in liquid ammonia to reduce organic compounds. Group I metals are common sources of single electrons: by giving up their odd s electron they form a stable M$^+$ ion. They will donate this electron to several classes of molecules; for example, ketones can react with sodium to form ketyl radicals.

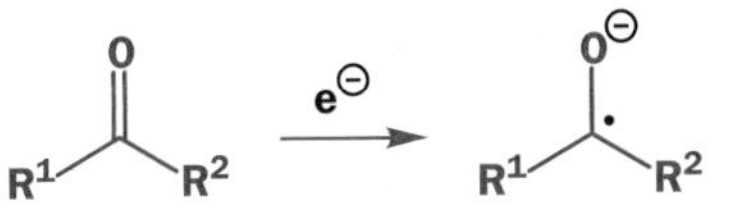

> ■ We shall discuss ketyl radicals and their reactions later in the chapter.

Radicals form by homolytic cleavage of weak bonds

A fourth class of radical-forming reaction is **homolytic cleavage**. For an example, we can go back to dibenzoyl peroxide, the unstable compound we considered earlier in the chapter because it readily undergoes homolysis.

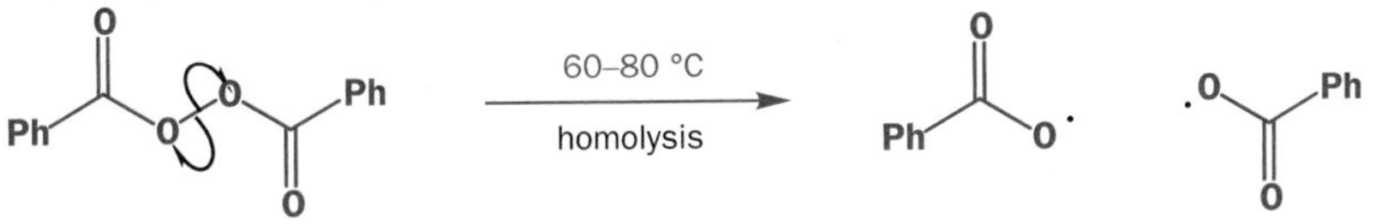

The radicals formed from this homolysis are unstable and each breaks down by cleavage of a C–C bond, generating CO_2 and a phenyl radical. These homolytic bond cleavages are elimination reactions and are the reverse of radical addition reactions.

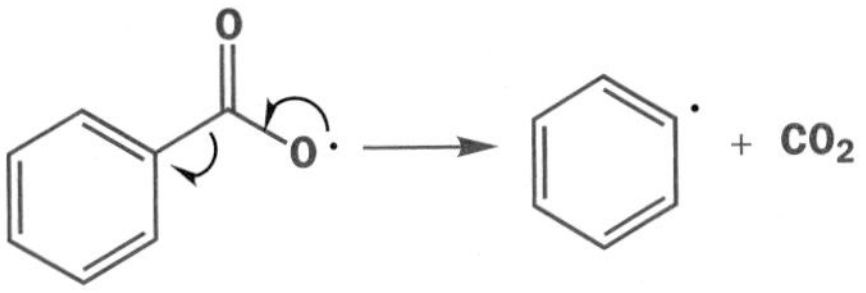

● **To summarize methods of radical formation**

Radicals form from spin-paired molecules by:
- homolysis of weak σ bonds, e.g.   RO—OR ⟶ RO$^\bullet$ (× 2)
- electron transfer, that is, reduction (addition of an electron), e.g.

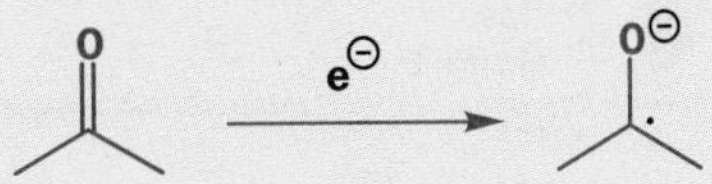

Radicals form from other radicals by:
- substitution (abstraction)   X$^\bullet$ + Y—Z ⟶ X—Y + Z$^\bullet$
- addition   X$^\bullet$ + Y=Z ⟶ X—Y—Z$^\bullet$
- elimination (homolysis) $^\bullet$X—Y—Z ⟶ X=Y + Z$^\bullet$

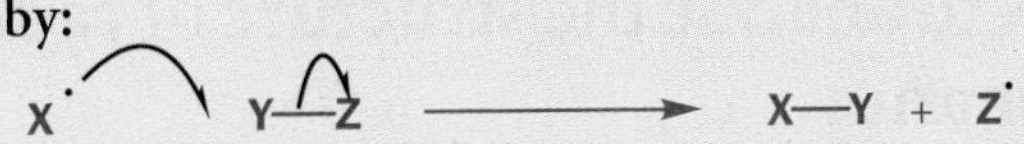

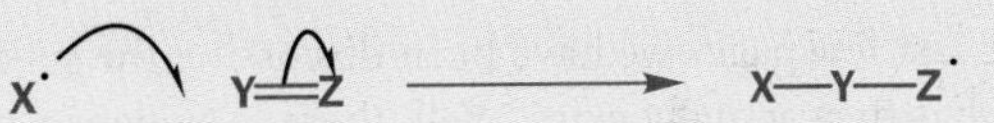

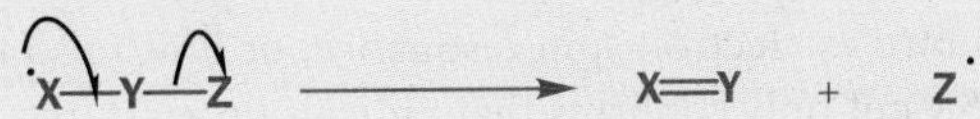

Electron célibataire is the French term for these bachelor electrons searching earnestly for a partner.

triphenylmethyl radical –
stable in solution
in equilibrium with its dimer

Most radicals are extremely reactive . . .

Unpaired electrons are desperate to be paired up again. This means that radicals usually have a very short lifetime; they don't survive long before undergoing a chemical reaction.

Chemists are more interested in radicals that are reactive, because they can be persuaded to do interesting and useful things. However, before we look at their reactions, we shall consider some radicals that are unreactive so that we can analyse the factors that contribute to radical reactivity.

. . . but a few radicals are very unreactive

Whilst simple alkyl radicals are extremely short-lived, some other radicals survive almost indefinitely. Such radicals are known as **persistent radicals**. We mentioned the triphenylmethyl radical on p. 1022: this yellow substance exists in solution in equilibrium with its dimer, but it is persistent enough to account for 2–10% of the equilibrium mixture.

Persistent radicals with the single electron carried by an oxygen or a nitrogen atom are also known: these three radicals can all be handled as stable compounds. The first, known as TEMPO, is a commercial product and can even be sublimed.

TEMPO
TEtraMethylPiperidine *N*-Oxide
m.p. 36–38°C

dark blue solid
m.p. 97°C

violet crystals

There are two reasons why some radicals are more persistent than others: (1) steric hindrance and (2) electronic stabilization. In the four extreme cases above, their exceptional stability is conferred by a mixture of these two effects. Before we can analyse the stability of other radicals, however, we need to look at what is known about the shape and electronic structure of radicals.

Vitamin E tames radicals

Many of the molecules that make up the structure of human tissue are susceptible to homolysis in intense light, and the body makes use of sophisticated chemistry to protect itself from the action of the reactive radical products. Vitamin E plays an important role in the 'taming' of these radicals: abstraction of H from the phenolic hydroxyl group produces a relatively stable radical that does no further damage.

vitamin E

R—O•
dangerous
and reactive
radical

vitamin E

R—OH   +
reactive radical
"tamed" as ROH

more stable delocalized radical

How to analyse the structure of radicals: electron spin resonance

For the last few pages we have been discussing the species we call radicals without offering any evidence that they actually exist. Well, there is evidence, and it comes from a spectroscopic technique known as **electron spin resonance**, or **ESR** (also known as EPR, electron paramagnetic resonance). ESR not only confirms that radicals do exist, but it can also tell us quite a lot about their structure.

Unpaired electrons, like the nuclei of certain atoms, have a magnetic moment associated with them. Proton NMR probes the environment of hydrogen atoms by examining the energy difference between the two possible orientations of their magnetic moments in a magnetic field; ESR works in a similar way for unpaired electrons. The magnetic moment of an electron is much bigger than that of a proton, so the difference in energy between the possible quantum states in an electron field is also much bigger. This means that the magnets used in ESR spectrometers can be weaker than those in NMR spectrometers: usually about 0.3 tesla; even at this low field strength, the resonant frequency of an electron is about 9000 MHz (for comparison, the resonant frequency of a proton at 9.5 tesla is 400 MHz; in other words, a 400 MHz NMR machine has a magnetic field strength of 9.5 tesla).

But there are strong similarities between the techniques. ESR shows us, for example, that unpaired electrons couple with protons in the radical. The spectrum below is that of the methyl radical, $CH_3^{\cdot}$. The 1:3:3:1 quartet pattern is just what you would expect for coupling to three equivalent protons; coupling in ESR is measured in millitesla (or gauss; 1 gauss = 0.1 mT), and for the methyl radical the coupling constant (called a_H) is 2.3 mT.

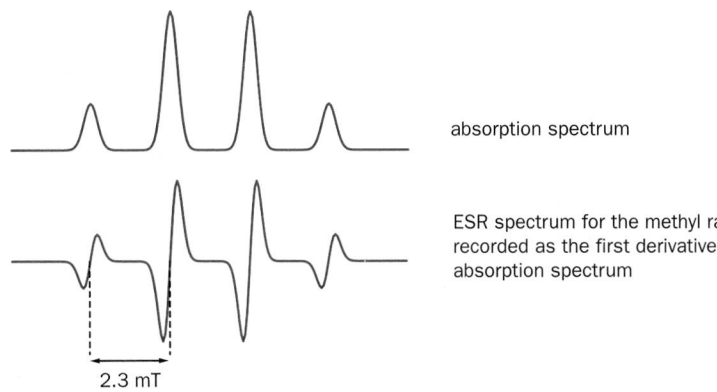

absorption spectrum

ESR spectrum for the methyl radical recorded as the first derivative of the absorption spectrum

2.3 mT

> Notice that, for historical reasons, ESR spectra are recorded in a different way from NMR spectra: the diagram shows the first derivative of the absorption spectrum (the sort of spectrum you would get from a proton NMR machine).

ESR hyperfine splittings (as the coupling patterns are known) can give quite a lot of information about a radical. For example, here is the hyperfine splitting pattern of the cycloheptatrienyl radical. The electron evidently sees all seven protons around the ring as equivalent, and must therefore be fully delocalized. A localized radical would see several different types of proton, resulting in a much more complex splitting pattern.

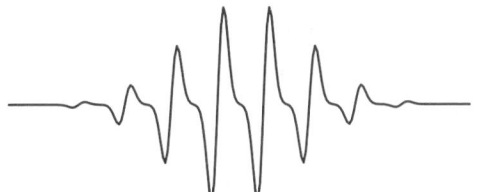

ESR spectrum of cycloheptatrienyl radical

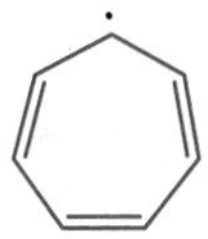

cycloheptatrienyl radical

Even the relatively simple spectrum of the methyl radical tells us quite a lot about the radical. For example, the size of the coupling constant a_H indicates that the methyl radical is planar; the trifluoromethyl radical is, on the other hand, pyramidal. The oxygenated radicals $^{\cdot}CH_2OH$ and $^{\cdot}CMe_2OH$ lie somewhere in between.

> The calculations that show this lie outside the scope of this book!

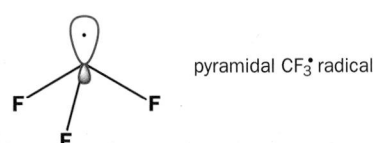

planar $CH_3^{\cdot}$ radical

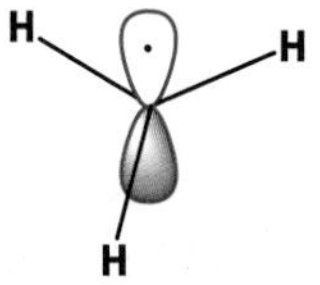

pyramidal $CF_3^{\cdot}$ radical

Radicals have singly occupied molecular orbitals

ESR tells us that the methyl radical is planar: the carbon atom must therefore be sp^2 hybridized, with the unpaired electron in a p orbital. We can represent this in an energy level diagram.

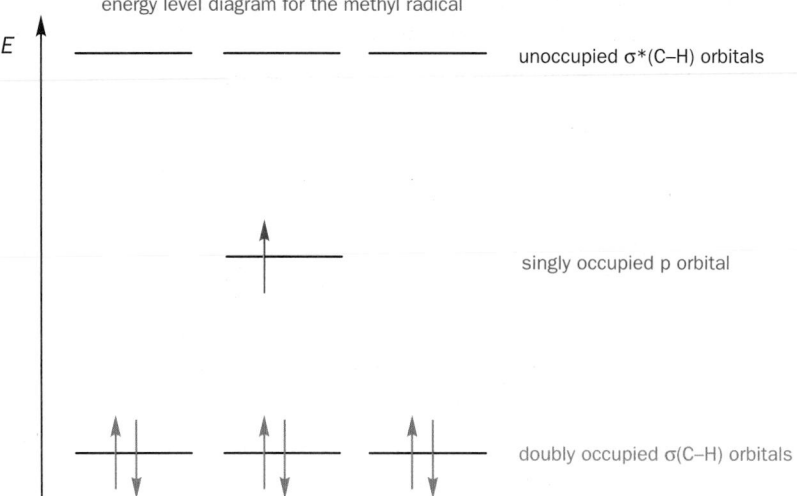

energy level diagram for the methyl radical

unoccupied σ*(C–H) orbitals

singly occupied p orbital

doubly occupied σ(C–H) orbitals

In Chapter 4 we talked about the HOMO (highest occupied molecular orbital) and LUMO (lowest unoccupied molecular orbital) of organic molecules. CH₃ (like all radicals) has an orbital containing one electron, which we call a **Singly Occupied Molecular Orbital (SOMO)**.

As with all molecules, it is the energy of the electrons in the molecular orbitals of the radical that dictate its stability. Any interaction that can decrease the energy levels of the filled molecular orbitals increases the stability of the radical (in other words, decreases its reactivity). Before we use this energy level diagram of the methyl radical to explain the stability of radicals, we need to look at some experimental data that allow us to judge just how stable different radicals are.

Radical stability

On p. 1020 we used bond strength as a guide to the likelihood that bonds will be homolysed by heat or light. Since bond energies give us an idea of the ease with which radicals can form, they can also give us an idea of the stability of those radicals once they have formed.

greater value means stronger bond

ΔG = energy required to homolyse bond

X——Y   ⟶   X˙   Y˙

ΔG = energy released in combining radicals

greater value means higher energy (more unstable) radicals

This is particularly true if we compare the strengths of bonds between the same atoms, for example, carbon and hydrogen, in different molecules; the table does this.

A few simple trends are apparent. For example, C–H bonds decrease in strength in R–H when R goes from primary to secondary to tertiary. Tertiary alkyl radicals are therefore the most stable; methyl radicals the least stable.

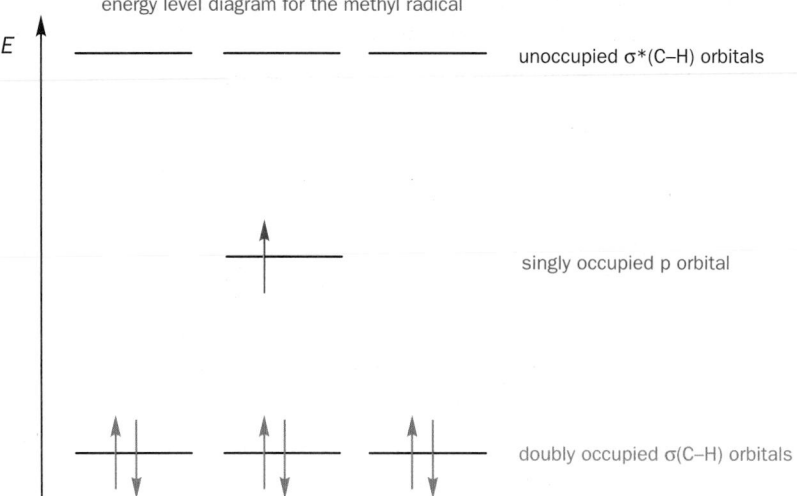

tertiary   is more stable than   secondary   is more stable than   primary   is more stable than   methyl

C–H bonds next to conjugating groups such as allyl or benzyl are particularly weak, so allyl and benzyl radicals are more stable. But C–H bonds to alkynyl, alkenyl, or aryl groups are strong.

allyl   benzyl   vinyl   alkynyl   phenyl

more stable than alkyl radicals   less stable than alkyl radicals

Bond	Dissociation energy, kJ mol⁻¹
CH₃–H	439
MeCH₂–H	423
Me₂CH–H	410
Me₃C–H	397
HC≡C–H	544
H₂C=CH–H	431
Ph–H	464
H₂C=CH₂CH₂–H	364
PhCH₂–H	372
RCO–H	364
EtOCHMe–H	385
N≡CCH₂–H	360
MeCOCH₂–H	385

Adjacent functional groups appear to weaken C–H bonds: radicals next to carbonyl, nitrile, or ether functional groups, or centred on a carbonyl carbon atom, are more stable than even tertiary alkyl radicals.

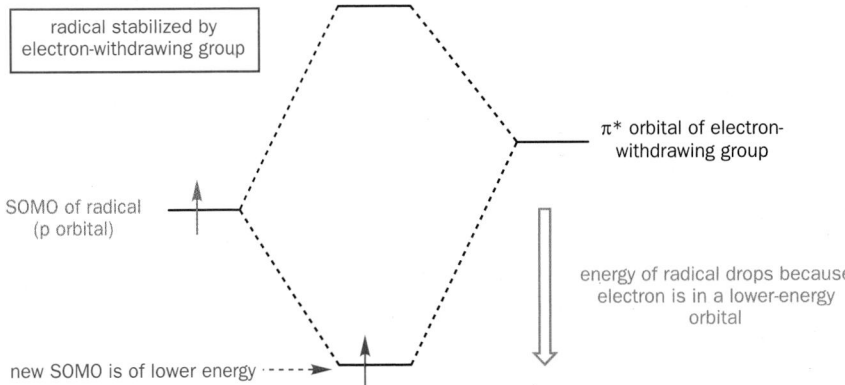

radicals stabilized by functional groups

Whether the functional group is electron-withdrawing or electron-donating is clearly irrelevant here: both types seem to stabilize radicals. We can explain all of this if we look at how the different groups next to the radical centre interact electronically with the radical.

Radicals are stabilized by conjugating, electron-withdrawing, and electron-donating groups

Let's consider first what happens when a radical centre finds itself next to an electron-withdrawing group. Groups like C=O and C≡N are electron-withdrawing because they have a low-lying empty π^* orbital. By overlapping with the (usually p) orbital containing the radical (the SOMO), two new molecular orbitals are generated. One electron (the one in the old SOMO) is available to fill the two new orbitals. It enters the new SOMO, which is of lower energy than the old one, and the radical experiences stabilization because this electron drops in energy.

We can analyse what happens with electron-rich groups, such as RO groups, in a similar way. Ether oxygen atoms have relatively high-energy filled n orbitals, their lone pairs. Interacting this with the SOMO again gives two new molecular orbitals. Three electrons are available to fill them. The SOMO is now higher in energy than it was to start with, but the lone pair is lower. Because two electrons have dropped in energy and only one has risen, there is an overall stabilization of the system, even though the new SOMO is of higher energy than the old one. We shall see later what effect the energy of the SOMO, rather than the overall energy of the radical, has on its reactivity.

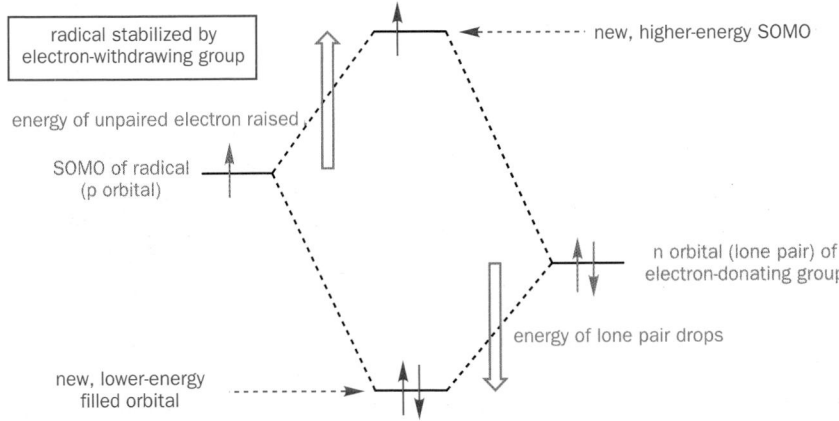

In Chapter 17, you saw how the electrons in C–H σ bonds stabilize cations: they stabilize radicals in the same way, which is why tertiary radicals are more stable than primary ones.

Conjugation, too, is effective at stabilizing radicals. We know that radicals next to double bonds are delocalized from their ESR spectra (p. 1025); that they are more stable is evident from the bond dissociation energies of allylic and benzylic C–H bonds.

● **Anything that would stabilize an anion *or* a cation will stabilize a radical:**

- **electron-withdrawing groups**
- **electron-donating groups (including alkyl groups with C–H σ bonds)**
- **conjugating groups**

Steric hindrance makes radicals less reactive

On p. 1024 we showed you some radicals that are remarkably stable (persistent): some can even be isolated and purified. You should now be able to see at least part of the reason for their exceptional stability: two of them have adjacent powerful electron-donating groups and one has a powerful electron-withdrawing group as well, and three of the four are conjugated.

some persistent (unreactive) radicals

electron-donating groups in green
electron-withdrawing groups in black
conjugating groups in orange

► Both electron-
donating and
electron-withdrawing
effects stabilize
radicals, as you
have just seen.
Some radicals are
stabilized by an
electron-withdrawing
group and an
electron-donating
group at the same
time. These radicals
are known as
**captodative
radicals**.

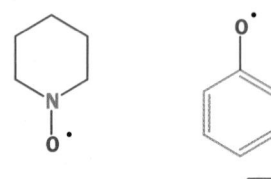

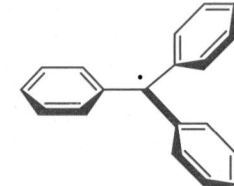

But electronic factors alone are not sufficient to explain the exceptional stability of all four radicals, since the next two radicals (in the margin) receive just about the same electronic stabilization as the first two above, but are much more reactive.

In fact, the stability of the triphenylmethyl radical we know to be due mainly to steric, rather than electronic, factors. X-ray crystallography shows that the three phenyl rings in this compound are not coplanar but are twisted out of a plane by about 30°, like a propeller. This means that the delocalization in this radical is less than ideal (we know that there is some delocalization from the ESR spectrum) and, in fact, it is little more delocalized than the diphenylmethyl or even the benzyl radical.

Yet it is much more stable than either. This must be because the central carbon, which bears most of the radical character, is sterically shielded by the twisted phenyl groups, making it very hard for the molecule to react. And when it does dimerize, we know that it does so through one of its least hindered carbon atoms.

Further evidence for the role of steric effects in helping to stabilize radicals comes from triphenylmethyl derivatives with *ortho* substituents: these force the phenyl rings to twist even more (at 50° or more), decreasing still further the extent of electronic stabilization through delocalization. Yet these *ortho*-substituted radicals are more stable than triphenylmethyl: this must be a steric effect. The rest of this chapter is devoted to the reactions of radicals, and you will see that the two effects we have talked about—electronic stabilization and steric hindrance—are key factors that control these reactions.

How do radicals react?

A reactive radical has a choice: it can either find another radical and combine to form a spin-paired molecule (or more than one spin-paired molecule), or it can react with a spin-paired molecule to form a new radical. Both are possible, and we shall see examples of each. A third alternative is for a radical to decompose in a unimolecular reaction, giving rise to a new radical and a spin-paired molecule.

> ● **Three possibilities**
>
> - Radical + radical → spin-paired molecule
> - Radical + spin-paired molecules → new radical + new spin-paired molecule
> - Radical → new radical + spin-paired molecule

Radical–radical reactions

In view of the energy released when unpaired electrons pair up, you might expect this type of radical reaction to be more common than reaction with a spin-paired molecule, in which no net pairing of electrons takes place. Radical–radical reactions certainly do take place, but they are not the most important type of reaction involving radicals. We shall see why they are not as common as you might expect shortly, but first we can look at some examples.

The pinacol reaction is a radical dimerization

We outlined on p. 1023 a way of making radicals by single electron transfer: effectively, the addition reaction of a single electron to a spin-paired molecule. The types of molecules that undergo this reaction are those with low-lying antibonding orbitals for the electron to go into, in particular, aromatic systems and carbonyl compounds. The radical anion formed by addition of an electron to a ketone is known as a **ketyl**. The single electron is in the π* orbital, so we can represent a ketyl with the radical on oxygen or on carbon and the anion on the other atom.

Ketyls behave in a manner that depends on the solvent that they are in. In protic solvents (ethanol, for example), the ketyl becomes protonated and then accepts a second electron from the metal (sodium is usually used in these cases). An alkoxide anion results, which, on addition of acid at the end of the reaction, gives an alcohol.

> ▶
> This reaction, known as the **Bouveault–Blanc reduction**, used to be used to reduce carbonyl compounds to alcohols, but now aluminium hydrides and borohydrides are usually more convenient. You met an example of the Bouveault–Blanc reduction in Chapter 33 (conformational analysis–reduction of cyclohexanones).
>
>
> Notice that this is a reaction using sodium metal in ethanol, and not sodium ethoxide, which is the basic product that forms once sodium has dissolved in ethanol. It is important that the sodium is *dissolving* as the reaction takes place, since only then are the free electrons available.

reaction of the ketyl radical anion in protic solvents

In aprotic solvents, such as benzene or ether, no protons are available so the concentration of ketyl radical builds up significantly and the ketyl radical anions start to dimerize. As well as being a radical–radical process, this dimerization process is an anion–anion reaction, so why doesn't electrostatic repulsion between the anions prevent them from approaching one another? The key to success is to use a metal such as magnesium or aluminium that forms strong, covalent metal–oxygen bonds and that can coordinate to more than one ketyl at once. Once two ketyls are coordinated to the same metal atom, they react rapidly.

pinacol dimerization of acetone (ketyl radical reaction in hydrocarbon solvent)

diol product known as "pinacol"

The example shows the dimerization of acetone to give a diol (2,3-dimethylbutane-2,3-diol) whose trivial name, pinacol, is used as a name for this type of reaction using any ketone. Sometimes pinacol reactions create new chiral centres: in this example, the two diastereoisomeric diols are formed in a 60:40 mixture. If you want to make a single diastereoisomer of a diol, a pinacol reaction is not a good choice!

> ■ You would be better off using one of the methods described in Chapter 34 on diastereoselectivity.

Benzophenone as an indicator in THF stills

As you should have gathered by now, THF is an important organic solvent in which many low-temperature, inert-atmosphere reactions are conducted. It has a drawback, however: it is quite hygroscopic, and often the reactions for which it is used as a solvent must be kept absolutely free of water. It is therefore always distilled immediately before use from sodium metal, which reacts with any traces of water in the THF. However, it is necessary to have an indicator to show that the THF *is* dry and that the sodium has done its job. The indicator used is a ketone, benzophenone.

highly delocalized, hindered, purple ketyl radical anion

When the THF is dry, the distilling liquid containing the benzophenone becomes bright purple. This colour is due to the ketyl of benzophenone, the formation of which under these conditions should not surprise you. It should also come as no surprise that this ketyl, being stabilized by conjugation and quite hindered, is persistent (long-lived)—it does not undergo pinacol dimerization (as we explained above, you would not normally choose sodium to promote pinacols anyway). However, if water is present, the ketyl is rapidly quenched in the manner of the reduction described above to give the (colourless) alkoxide anion: only when all the water is consumed does the colour return.

Pinacol reactions can be carried out intramolecularly, from compounds containing two carbonyl groups. In fact, the key step of one of the very first syntheses of Taxol® (the important anticancer compound) was an intramolecular pinacol reaction using titanium as the source of electrons.

the pinacol reaction in a synthesis of Taxol

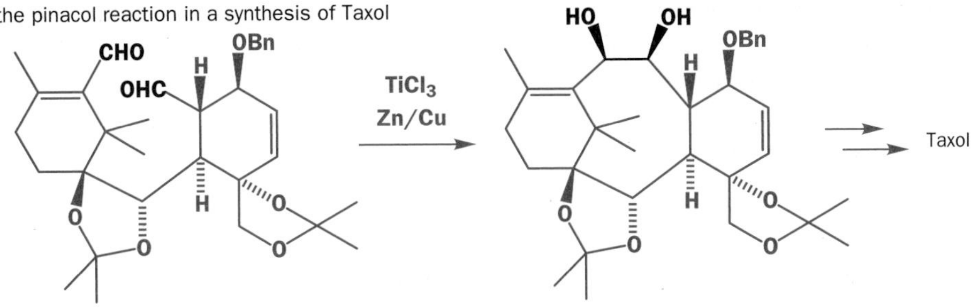

The titanium metal that is the source of electrons is produced during the reaction by reduction of $TiCl_3$ using a zinc–copper mixture. This reaction is, in fact, unusual because, as we shall see below, pinacol reactions using titanium do not normally stop at the diol, but give alkenes.

Titanium promotes the pinacol coupling and then deoxygenates the products: the McMurry reaction

Titanium can be used as the metal source of electrons in the pinacol reaction and, provided the reaction is kept cold and not left for too long, diols can be isolated from the reaction (see the example at the end of the previous section). However, unlike magnesium or aluminium, titanium reacts further with these diol products to give alkenes in a reaction known as the **McMurry reaction**, after its inventor.

McMurry reaction of cyclohexanone

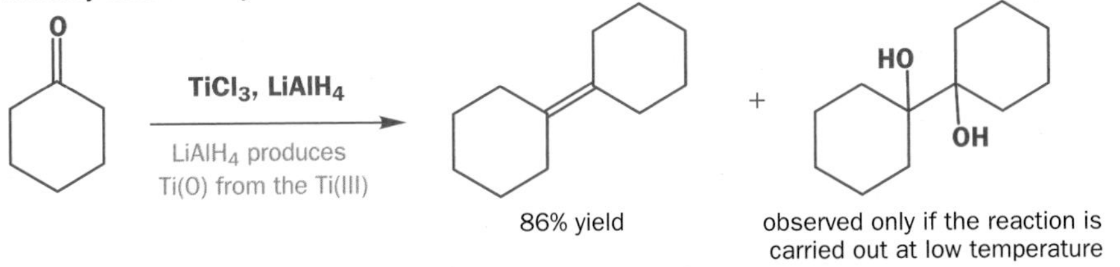

Notice that the titanium(0), which is the source of electrons in the reaction, is produced during the reaction by reacting a Ti(III) salt, usually $TiCl_3$, with a reducing agent such as $LiAlH_4$ or Zn/Cu. The reaction does not work with, say, powdered titanium metal. The McMurry reaction is believed to be a two-stage process involving firstly a pinacol radical–radical coupling. Evidence for this is that the pinacol products (diols) can be isolated from the reaction under certain conditions (you've just seen how this was done during the synthesis of Taxol).

first step of the McMurry reaction

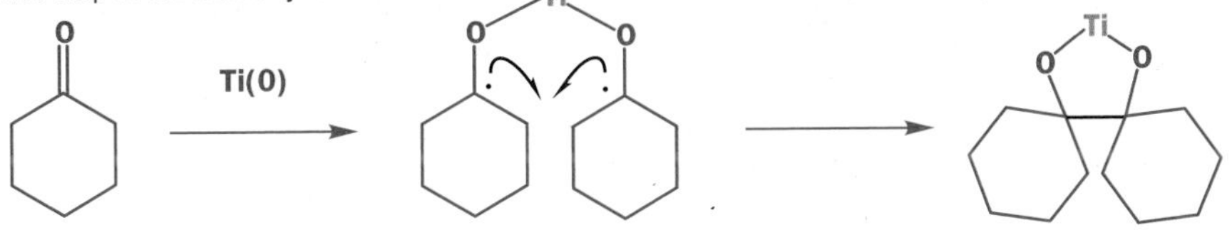

The Ti(0) then proceeds to deoxygenate the diol by a mechanism not fully understood, but thought to involve binding of the diol to the surface of the Ti(0) particles produced in the reduction of $TiCl_3$.

second step of the McMurry reaction: deoxygenation on the surface of a Ti(0) particle

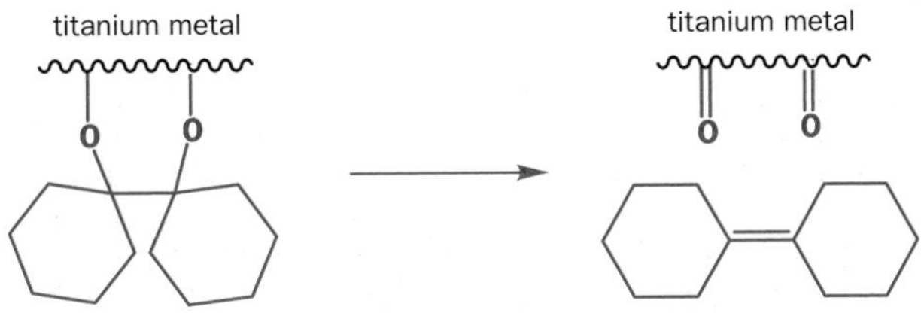

The evidence that it happens on a metal surface is quite nice, though, and if you're interested you can read McMurry's own account of it in *Accounts of Chemical Research*, 1983, p. 405.

We expect you to be mildly horrified by the inadequacy of the mechanism above. But, unfortunately, we can't do much better because no-one really knows quite what is happening. The McMurry reaction is very useful for making tetrasubstituted double bonds—there are few other really effective ways of doing this. However, the double bonds really need to be symmetrical (in other words, have the same substituents at each end) because McMurry reactions between two different ketones are rarely successful.

TiCl₃, Zn/Cu

96% yield

You saw on p. 1031 how the same reagent was used to carry out an intramolecular pinacol reaction, producing an eight-membered ring. Remember that medium rings, in other words ring sizes 7 to 13, are uncommonly difficult to form by cyclization reactions—the reasons for this were discussed in Chapters 18 and 42.

McMurry reactions also work very well intramolecularly, and turn out to be quite a good way of making cyclic alkenes, especially when the ring involved is medium or large (over about eight members). For example, the natural product flexibilene, with a 15-membered ring, can be made by cyclizing a 15-keto-aldehyde.

flexibilene

Esters undergo pinacol-type coupling: the acyloin reaction

You've seen examples of pinacol and McMurry reactions of ketones and aldehydes. What about esters? You would expect the ketyl radical anion to form from an ester in the same way, and then to undergo radical dimerization, and this is indeed what happens.

The product of the dimerization looks very much like a tetrahedral intermediate in a carbonyl addition–elimination reaction, and it collapses to give a 1,2-diketone.

The diketone is however still reducible—in fact, 1,2-diketones are more reactive towards electrophiles and reducing agents than ketones because their π* is lower in energy and straight away two electron transfers take place to form a molecule, which we could term an **enediolate**.

On quenching the reaction with acid, this dianion is protonated twice to give the enol of an α-hydroxy-ketone, and it is this α-hydroxy-ketone that is the final product of the acyloin reaction. The yield in this example is a quite respectable 70%. However, in many other cases, this usefulness of the acyloin reaction is hampered by the formation of by-products that arise because of the reactivity of the enediolate dianion. It is, of course, quite nucleophilic, and is likely to be formed in the presence

of the highly electrophilic diketone. It is also basic, and often catalyses a competing Claisen condensation of the esters being reduced.

first electron transfer to the diketone:

delocalized diketone
ketyl radical anion

second electron transfer:

enediolate

70% yield

The solution to these problems is to add trimethylsilyl chloride to the reaction mixture. The silyl
chloride silylates the enediolate as it is formed, and the product of the acyloin reaction becomes a
bis-silyl ether.

an improved version of the acyloin reaction

95% yield

The silyl ethers are rarely desired as final products, and they can easily be hydrolysed to
α-hydroxy-ketones with aqueous acid. This improved version makes four-membered rings
efficiently.

It's not by accident that these two examples of the acyloin reaction show the formation
of cyclic compounds. It is a particularly powerful method of making carbocyclic rings of from
four members upwards: the energy to be gained by pairing up the two electrons in the radical–
radical reaction step more than compensates for the strain that may be generated in forming the
ring.

The pinacol, McMurry, and acyloin reactions are exceptional

We've already said that this type of reaction, in which two radicals dimerize, is relatively uncommon.
Most radicals are simply too reactive to react with one another! This may sound nonsensical, but the
reason is simply that highly reactive species are unselective about what they react with. Although it
might be energetically favourable for them to find another radical and dimerize, they are much more
likely to collide with a solvent molecule, or a molecule of some other compound present in the mixture, than another radical. Reactive radicals are only ever present in solution in very low concentrations, so the chances of a radical–radical collision are very low. Radical attack on spin-paired
molecules is much more common and, because the product of such reaction is also a radical, they
give rise to the possibility of radical chain reactions.

Radical chain reactions

In looking at how radicals form, you've already seen examples of how radicals react. In fact, we've
already dealt (if only very briefly) with every step of the sequence of reactions that makes up the
mechanism of the radical reaction you met at the beginning of the chapter.

▶
In the absence of the Me₃SiCl,
the main product from this
reaction becomes the cyclic keto-
ester below, which arises from
base-catalysed Dieckmann
cyclization (see Chapter 28) of
the diester.

■
In Chapter 8 we discussed ring strain—
remember that it's not only small
(three- and four-membered) rings that
are strained, but medium (8 to 13
members) ones too.

■
This is known as the **reactivity–
selectivity principle**—see Chapter 41.

▶
Think of radicals as smash-and-
grab raiders. They pick the first
shop that catches their eye,
smash the window, and run off
with a handful of jewellery from
the front of the display. Ions in
solution are stealthy burglars.
They scan all the houses on the
street, choose the most
vulnerable, and then carefully gain
entry to the room that they know
contains the priceless oil painting.

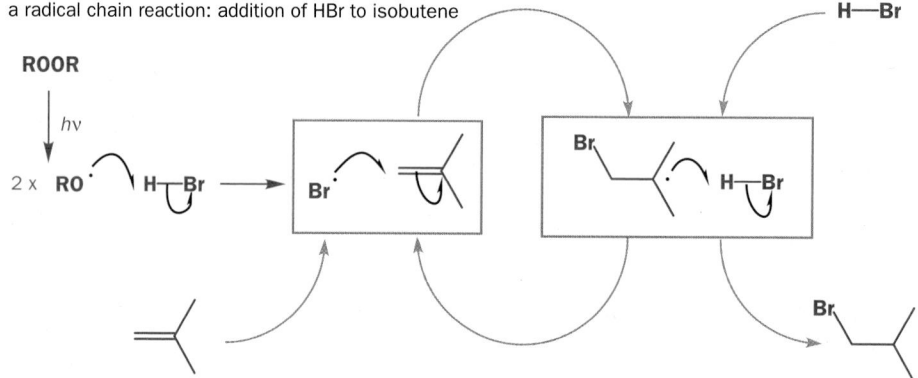

Let's now consider each step in turn and in more detail.

1 The dialkyl peroxide is homolysed (by heat or light) to give two alkoxy radicals

2 RO˙ abstracts H from HBr (radical substitution) to give Br˙

3 Br˙ adds to isobutene to give a carbon-centred radical

4 The carbon-centred radical abstracts a hydrogen atom from H–Br to form the final addition product and regenerate Br˙, which can react with another molecule of alkene

The whole process can conveniently be represented cyclically.

a radical chain reaction: addition of HBr to isobutene

In each step in the cycle a radical is consumed and a new radical is formed. This type of reaction is therefore known as a **radical chain reaction**, and the two steps that form the cyclic process that keeps the chain running are known as the **chain propagation steps**. Only one molecule of peroxide **initiator** is necessary for a large number of product molecules to be formed and, indeed, the peroxide needs to be added in only catalytic quantities (about 10 mol%) for this reaction to proceed in good yield.

Any less than 10 mol%, however, and the yield drops. The problem is that the chain reaction is not 100% efficient. Because the concentration of radicals in the reaction mixture is low, radical–radical reactions are rare, but nonetheless they happen often enough that more peroxide keeps being needed to start the chain off again

possible radical–radical chain termination steps

Reactions like this are known as **termination steps** and are actually an important part of any chain reaction; without termination steps the reaction would be uncontrollable.

● **Radical chain reactions consist of**

   ● **Initiation** steps

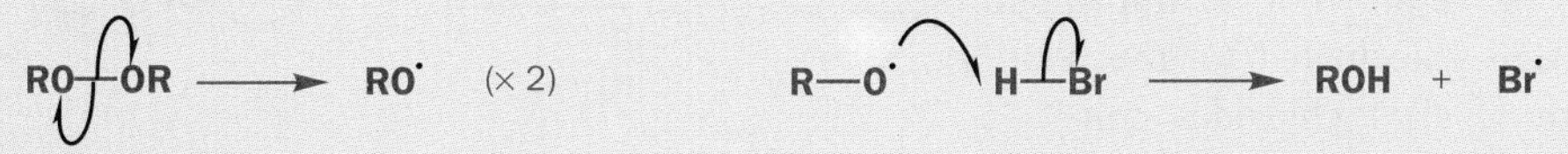

   ● **Propagation** steps

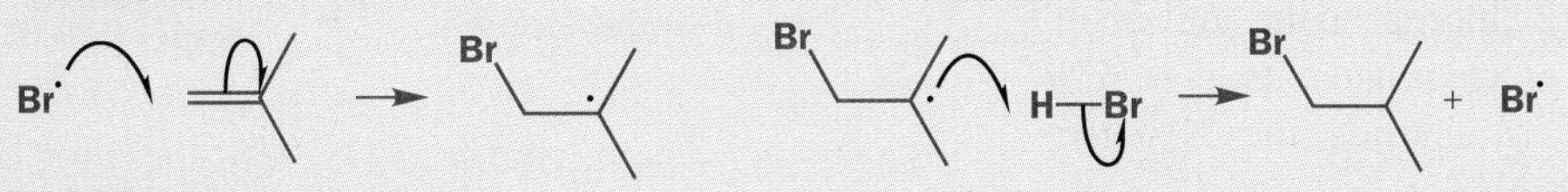

   ● **Termination** steps

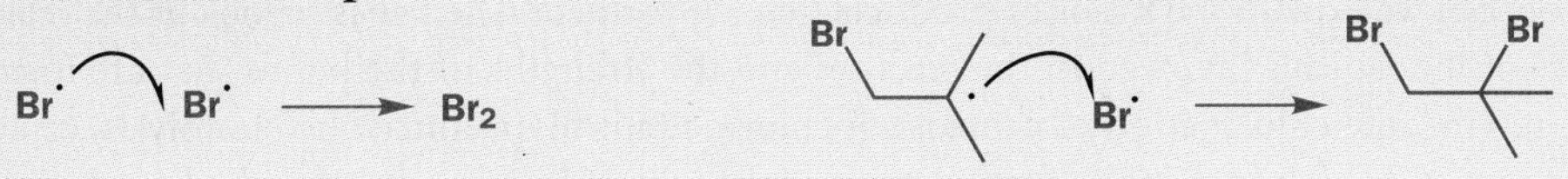

Selectivity in radical chain reactions

In the radical–radical reactions we looked at earlier, there was never any question of what would react with what: only one type of radical was formed and the radicals dimerized in identical pairs. Look at this chain reaction though—there are three types of radical present, Br$^\bullet$, BrCH$_2$Me$_2$CH$^\bullet$, and RO$^\bullet$, and they all react specifically with a chosen spin-paired partner: Br$^\bullet$ with the alkene, and BrCH$_2$Me$_2$CH$^\bullet$ and RO$^\bullet$ with HBr. We need to understand the factors that govern this chemoselectivity. In order to do so we shall look at another radical reaction with chemoselectivity and regioselectivity that is *measurable*.

Chlorination of alkanes

Alkanes will react with chlorine to give alkyl chlorides. For example, cyclohexane plus chlorine gas, in the presence of light, gives cyclohexyl chloride and hydrogen chloride.

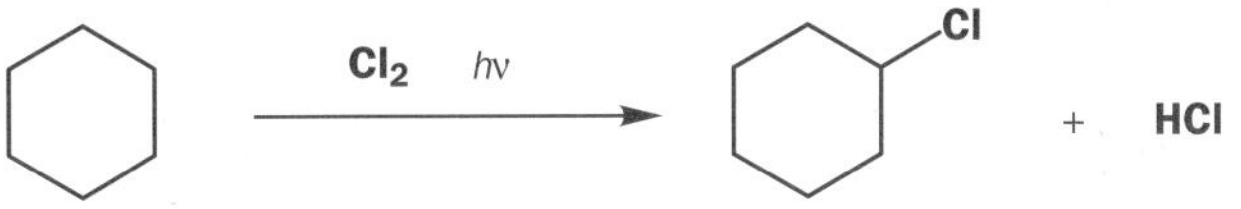

This type of reaction is important industrially since it is one of the few that allows compounds containing functional groups to be made from alkanes. As you might guess, since it needs light for initiation, the process is another example of a radical chain reaction. As with the radical addition of HBr to alkenes, we can identify initiation, propagation, and termination steps in the mechanism.

initiation

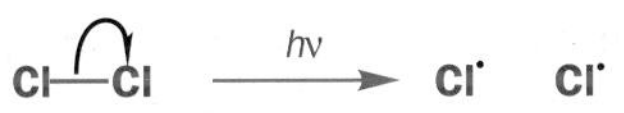

propagation

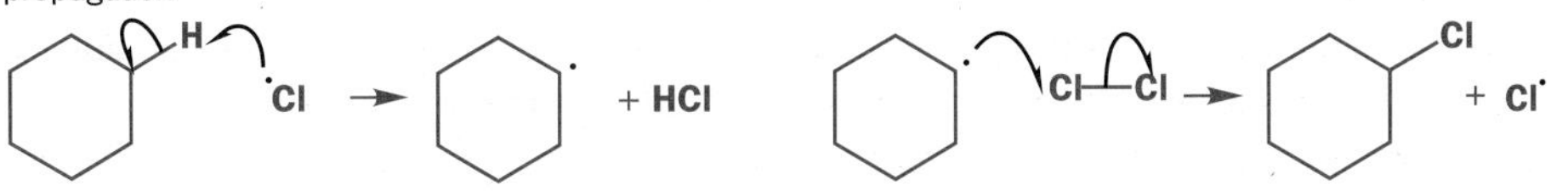

termination

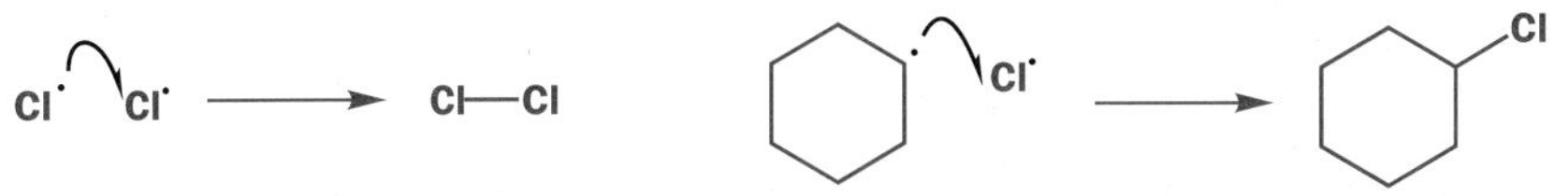

In this case, the termination steps are much less important than in the last case we looked at, and typically the chain reaction can continue for 10^6 steps for each initiation event (photolysis of chlorine). Be warned: reactions like this can be explosive in sunlight.

We have already suggested two reasons why the Br$^\bullet$ radical adds to the alkene with this characteristic regioselectivity, giving a primary alkyl bromide when the polar addition of HBr to an alkene would give a tertiary alkyl bromide: (1) attack at the unsubstituted end of the alkene is less sterically hindered; and (2) the tertiary radical thus formed is more stable than a primary radical. In fact, of all the hydrogen halides, only HBr will add to alkenes in this fashion: HCl and HI will undergo only polar addition to give the tertiary alkyl halide. Why? We need to be able to answer this type of question too.

When the chlorine radical abstracts a hydrogen atom from the cyclohexane, only one product can be formed because all 12 hydrogen atoms are equivalent. For other alkanes, this may not be the case, and mixtures of alkyl chlorides can result. For example, propane is chlorinated to give a mixture of alkyl chlorides containing 45% 1-chloropropane and 55% 2-chloropropane, and *iso*-butane is chlorinated to give 63% *iso*-butyl chloride and 37% *tert*-butyl chloride.

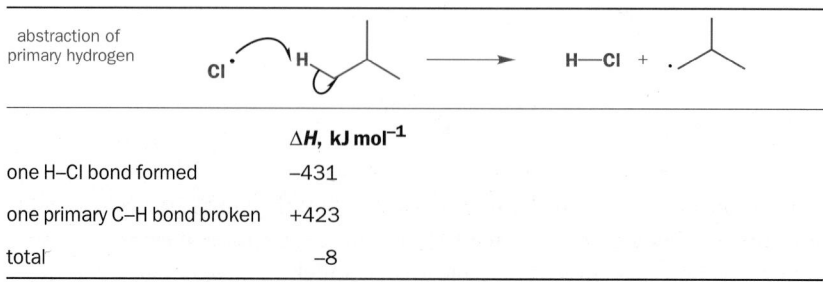

■ These bond energies were given in the tables on pp. 1020 and 1026.

How can we explain the ratios of products that are formed? The key is to look at the relative stabilities of the radicals involved in the reaction and the strengths of the bonds that are formed and broken. First, the chlorination of propane. A chlorine radical, produced by photolysis, can abstract either a primary hydrogen atom, from the end of the molecule, or a secondary hydrogen atom, from the middle. For the first process, we have these energy gains and losses.

First process:

	ΔH, kJ mol^{-1}
one H–Cl bond formed	–431
one primary C–H bond broken	+423
total	–8

For the second process, the energies are given in the table.

Second process:

	ΔH, kJ mol^{-1}
one H–Cl bond formed	–431
one secondary C–H bond broken	+410
total	–21

Abstraction of the secondary hydrogen atom is more exothermic than abstraction of the primary hydrogen atom, for the related reasons that: (1) secondary C–H bonds are weaker than primary ones; and (2) secondary radicals are more stable than primary ones. So, we get more 2-chloropropane than 1-chloropropane. But in this case, that isn't the only factor involved: remember that there are six primary hydrogen atoms and only two secondary ones, so the relative reactivity of the primary and secondary positions is even more different than the simple ratio of products from the reaction suggests. This statistical factor is more evident in the second example we gave above, the chlorination of isobutane. Now the choice is between formation of a tertiary radical and formation of a primary one.

abstraction of primary hydrogen		

	ΔH, kJ mol^{-1}
one H–Cl bond formed	–431
one primary C–H bond broken	+423
total	–8

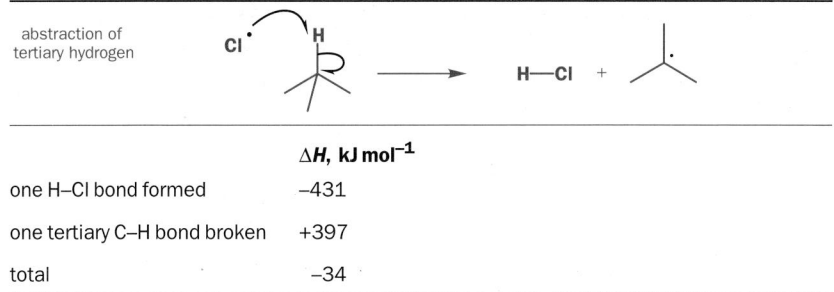

	ΔH, kJ mol^{-1}
one H–Cl bond formed	−431
one tertiary C–H bond broken	+397
total	−34

Tertiary radical formation is more exothermic, yet more primary alkyl chloride is formed than tertiary alkyl chloride. However, once the 9:1 ratio of primary to tertiary hydrogen atoms is taken into account, the relative reactivities, as determined experimentally, turn out to be as shown in the table.

ratio of products formed (tertiary:primary)	37:63
number of hydrogen atoms (tertiary:primary)	1:9
relative reactivity of each C–H bond (tertiary:primary)	37/1:63/9 = 37:7 = ca. 5:1

● Bond strength is all-important in radical reactions

These reactions illustrate a key point about radical reactions—a very important factor affecting selectivity is the strength of the bonds being formed and broken.

The rate of attack by Cl$^{\bullet}$ on a tertiary C–H bond, then, is about five times the rate of attack by Cl$^{\bullet}$ on a primary C–H bond. We said that this is because the formation of the tertiary radical is more exothermic than the formation of the primary radical. But the rate of a reaction depends not on ΔH for that reaction but on the **activation energy** of the reaction; in other words, the energy needed to reach the transition state for the reaction. But we can still use the stability of the product radicals as a guide to the stability of the transition state, because the transition state must have significant radical character.

> Always bear this in mind: bond strength is only a *guide* to selectivity in radical reactions. As we shall see shortly, it's not the only factor involved. Indeed, you've already seen *steric effects* in action when the Br$^{\bullet}$ radical added to the less hindered end of the alkene in the first radical reaction of this chapter, and you will later see how *frontier orbital effects* can operate too.

> We use the symbol (•) to mean a partial radical; a radical that is partially centred on this atom. The symbols (−) and (+) are used to mean a similar thing when a charge is shared by more than one atom.

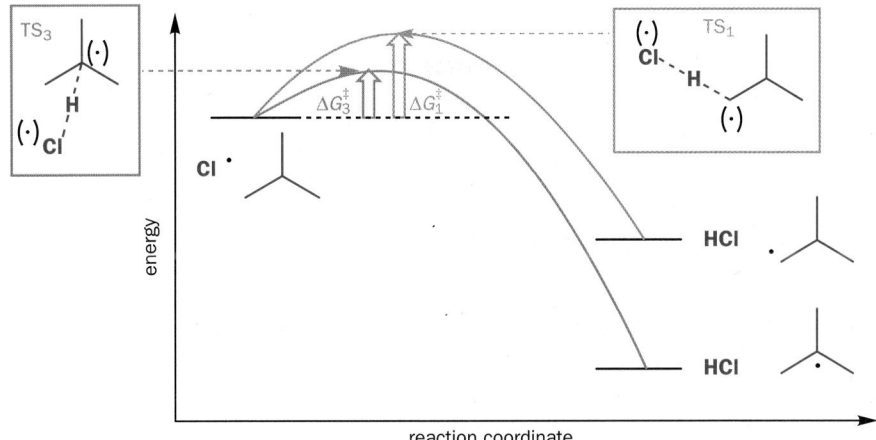

The energy diagram above illustrates this point. As the reactants (Cl$^{\bullet}$ plus isobutane) move towards the products, they pass through a transition state (TS$_1$ for formation of the primary radical, TS$_3$ for formation of the tertiary) in which the radical character of the Cl$^{\bullet}$ starting material is spread over both the Cl and the C centres. The greater stability of a tertiary radical compared with a primary one must be reflected to a lesser degree in these transition states: a radical shared between Cl and a tertiary centre will be more stable than a radical shared between Cl and a primary centre. The

> Of course our calculations involving bond energies only gave us values for ΔH, not ΔG which is what this diagram represents. However, we can assume that the $T\Delta S$ term in the relationship $\Delta G = \Delta H - T\Delta S$ is relatively insignificant.

transition state TS$_3$ for the reaction at the tertiary C–H bond is therefore of lower energy than the transition state TS$_1$ for reaction at the primary C–H bond. In other words, the activation energy $\Delta G_3^\ddagger$ is smaller than $\Delta G_1^\ddagger$, so reaction at the tertiary C–H bond is faster.

The Toray process

A variant of this reaction, known as the Toray process, is used on an industrial scale in Japan to produce caprolactam, a precursor to nylon. Instead of chlorine, nitrosyl chloride is used to form a nitroso compound that rapidly tautomerizes to an oxime. As you saw in Chapter 37, this oxime undergoes a Beckmann rearrangement under acid conditions to form caprolactam.

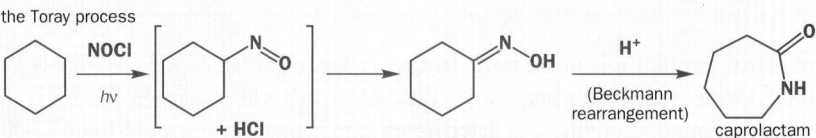

the Toray process

Bromine will also halogenate alkanes, and it does so much more selectively than chlorine. For example, the following reaction yields *tert*-butyl bromide with less than 1% of the primary isomer.

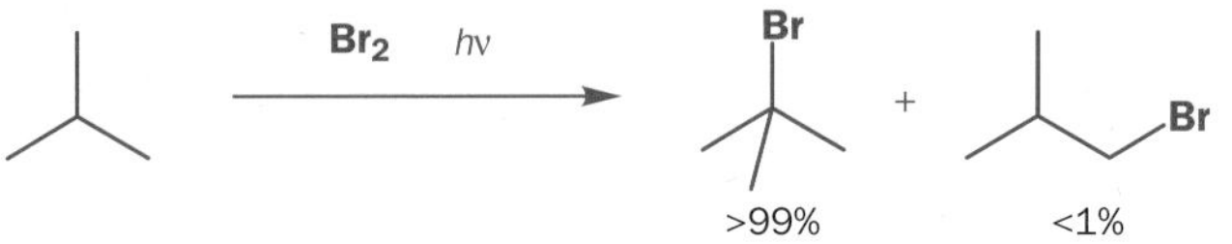

>99%   <1%

In this case, the first step of the radical chain reaction, the abstraction of H by Br•, is endothermic for both the primary and tertiary hydrogen atoms.

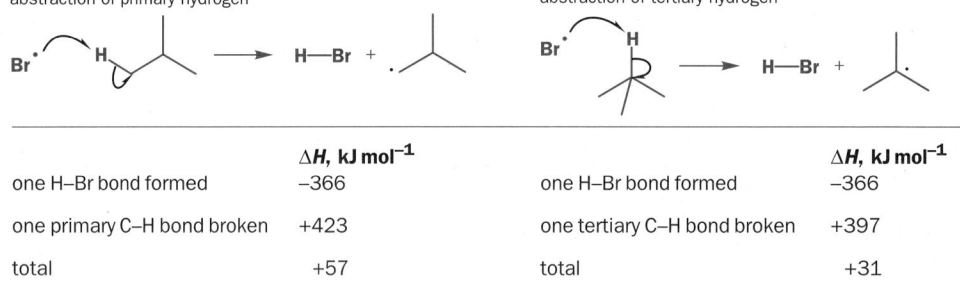

abstraction of primary hydrogen	ΔH, kJ mol^{-1}	abstraction of tertiary hydrogen	ΔH, kJ mol^{-1}
one H–Br bond formed	−366	one H–Br bond formed	−366
one primary C–H bond broken	+423	one tertiary C–H bond broken	+397
total	+57	total	+31

The second step, trapping of the alkyl radical by Br$_2$, is, however, sufficiently exothermic for the reaction to be exothermic overall.

Why is bromination so much more selective than the chlorination of alkanes? This is a good example of how the Hammond postulate applies to real chemistry. Because the products of the first step of the bromination (R• plus HBr) are higher in energy than the starting materials, the transition state must be similar in structure and energy to that product radical; the difference in energies of the primary and tertiary product radicals should therefore be markedly reflected in the different energies of the transition states TS$_1$ and TS$_3$, and $\Delta G_1^\ddagger$ will be significantly larger than $\Delta G_3^\ddagger$. For the chlorination reaction, the products were just slightly lower in energy than the starting materials, so the transition states for the two possible reactions both resembled the starting materials rather more and the products rather less. These are the same for both tertiary and primary hydrogen abstractions, of course, so the difference in

second step of the bromination reaction

R—Br + Br•

	ΔH, kJ mol^{-1}
one C–Br bond formed	−293
one Br–Br bond broken	+192
total	−101

> Of course, the overall ΔH for the reaction of an alkane with chlorine must also take into account the ΔH of this second step, which is −349 + 243 = −106 kJ mol^{-1}, making chlorination much more exothermic overall than bromination. Fluorination continues the trend, and methane–fluorine mixtures are explosive. For iodine, on the other hand, the first step becomes so endothermic, even for formation of a tertiary radical, that the second step (ΔH = −234 + 151 = −83 kJ mol^{-1}) is not exothermic enough to make reaction favourable overall. Radical iodinations therefore do not take place.

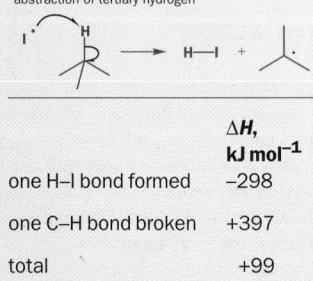

abstraction of tertiary hydrogen

	ΔH, kJ mol^{-1}
one H–I bond formed	−298
one C–H bond broken	+397
total	+99

> The **Hammond postulate** gives information about the structure of transition states. It says that two states that interconvert directly (are directly linked in a reaction profile diagram) and that are close in energy are also similar in structure. So a transition state will be most like the starting material, the intermediate, or the product if it is close in energy to one of these observable structures.

energy of the product radicals exerts a less pronounced effect on the difference in energy of the transition states.

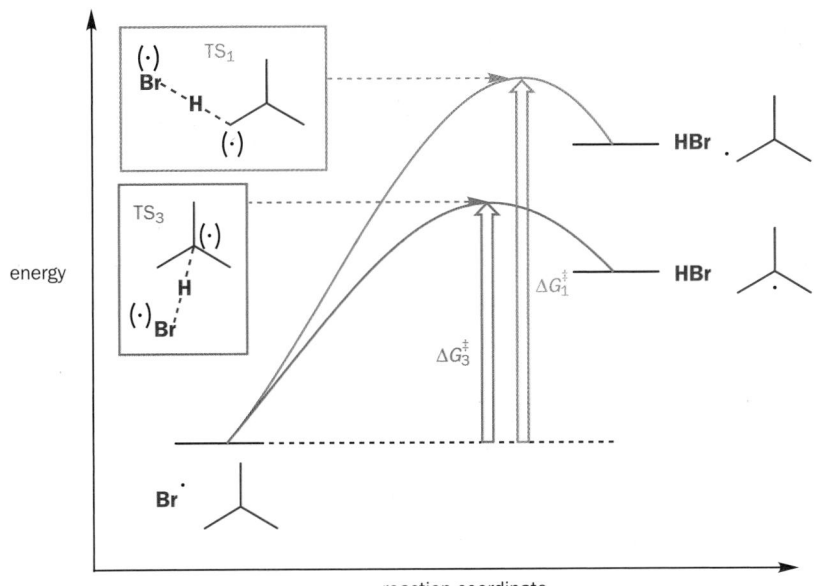

energy

reaction coordinate

Selective radical bromination: allylic substitution of H by Br

Because radical brominations are so selective, they can be used successfully in the lab to make alkyl bromides. There are relatively few ways of functionalizing an unfunctionalized centre, but radical allylic bromination is one of these. Just as tertiary radicals are more stable than primary ones, so allylic radicals are even more stable than tertiary ones (see the table on p. 1026). In the presence of a suitable initiator, bromine will therefore selectively abstract an allylic hydrogen atom to give an allylic radical that can then be trapped by a molecule of bromine to regenerate a bromine radical (chain propagation) and produce the allylic bromide.

initiation Br_2 ⟶ 2 x Br•

propagation

However, there is a problem with this reaction if bromine itself is used, because an alternative radical addition reaction can compete with radical abstraction.

competing addition reaction

The first step of this competing addition reaction is, in fact, reversible; the reaction is driven forward by the participation of a second molecule of bromine that traps the product alkyl radical. This side-reaction can be prevented if the concentration of Br_2 in the reaction is kept very low. One possibility is to add Br_2 very slowly to the reaction mixture, but it is better not to use bromine itself, but a compound that releases molecular bromine slowly during the reaction. That compound is *N*-bromosuccinimide, or NBS.

NBS, CCl₄   *hv*   85% yield

> Bond energy for tertiary C–H: 364 kJ mol⁻¹. Bond energy for allylic C–H: 397 kJ mol⁻¹. Remember though that these figures were determined in the gas phase, and here our reactions are in solution. Nonetheless, because solvation effects are more or less the same for all radicals, we expect the order of the bond strengths to remain the same in both phases.

> This competing reaction is a radical addition across a double bond. You have also met an analogous polar addition across an alkene in Chapter 20: that reaction is suppressed here by using a nonpolar solvent, usually CCl₄.

> NBS (*N*-Bromo Succinimide) is known to be a source of bromine because the ratios of products obtained from its reactions are identical with those obtained from reactions using small amounts of bromine.
>
> *N*-bromosuccinimide (NBS)

In this technique, the NBS acts like a turnstile, allowing only one molecule of bromine to form for every molecule of HBr produced by the reaction. The slow generation of Bu$_3$SnH from Bu$_3$SnCl and NaBH$_4$ is a very similar example and is discussed below.

The HBr produced in the substitution reaction reacts with the NBS to maintain the low concentration of bromine.

While radical halogenation of alkanes is used only rarely in the laboratory, radical allylic bromination of alkenes is a versatile and commonly used way of making allylic bromides. Nucleophilic substitution reactions can then be used to convert the bromide to other functional groups. For example, some chemists in Manchester needed to make the two diastereoisomers of 5-*tert*-butyl-cyclohex-2-en-1-ol to study their reactions with osmium tetroxide. *tert*-Butyl cyclohexene is readily available, so they used a radical allylic bromination to introduce the functional group in the allylic position, which they converted to a hydroxyl group using aqueous base. Steric effects play a role here in the regioselectivity of the reaction: only the less hindered allylic hydrogen atoms further from the *t*-butyl group are removed.

5-*t*-butylcyclohex-2-en-1-ol

Reversing the selectivity: radical substitution of Br by H

We discussed the removal of functional groups, and why you might want to do it, in Chapter 25.

Radical substitution reactions can also be used to *remove* functional groups from molecules. A useful reagent for this (and, as you will see, for other radical reactions too) is tributyltin hydride, Bu$_3$SnH. The Sn–H bond is weak and Bu$_3$SnH will react with alkyl halides to replace the halogen atom with H, producing Bu$_3$SnHal as a by-product.

81% yield

Clearly, for this reaction to be energetically favourable, new bonds formed (Sn–Br and C–H) must be stronger than the old bonds broken (Sn–H and C–halogen). Look at this table of average bond energies and you will see that this is indeed so.

The use of a tin hydride is crucial to this reaction: Sn–H bonds are weaker than Sn–Br bonds, while, for carbon, C–H bonds are stronger. Bu$_3$SnH is therefore an effective source of Bu$_3$Sn• radicals, and the Bu$_3$Sn• radical will abstract halogens, particularly I or Br, but also Cl, from organic halides, breaking a weak C–halogen (C–Hal) bond and forming a strong Sn–Hal bond. The complete mechanism of the reaction reveals a chain reaction.

Bond	Representative bond energy, kJ mol^{-1}
C–Br	280
Sn–H	308
C–H	418
Sn–Br	552

Homolysis of Bu₃SnH is promoted by the initiator AIBN

As you would imagine, the weakest C–Hal bonds are the easiest to cleave, so alkyl bromides are reduced more rapidly than alkyl chlorides, and alkyl fluorides are unreactive. With alkyl iodides and bromides, daylight can be sufficient to initiate the reaction, but with alkyl chlorides, and often with alkyl bromides as well, it is generally necessary to produce a higher concentration of Bu₃Sn• radicals by adding an initiator to the reaction. The best choice is usually AIBN, which you met on p. 1021. This compound undergoes thermal homolysis at 60 °C to give nitrile-stabilized radicals that abstract the hydrogen atom from Bu₃SnH.

Why use AIBN; why not a peroxide? (You came across peroxides as initiators of the addition of H–Br to alkenes.) Since we want to cleave only a weak Sn–H bond, we can get away with using a relatively unreactive, nitrile-stabilized radical. Peroxides, on the other hand, generate RO• radicals. These are highly reactive and will abstract hydrogen from almost any organic molecule, not just the weakly bonded hydrogen atom of Bu₃SnH, and this would lead to side-reactions and lack of selectivity. AIBN is needed only in sufficient quantities to be an initiator of the reaction; it is the Bu₃SnH that provides the hydrogen atoms that end up in the product, so usually you need only 0.02 to 0.05 equivalents of AIBN and a slight excess (1.2 equivalents) of Bu₃SnH.

> The bond energy of H–CH₂CN is only 360 kJ mol⁻¹; a tertiary C–H bond next to a CN group should be even weaker.

> Bond energy of O–H = 460 kJ mol⁻¹; few C–H bonds are stronger than 440 kJ mol⁻¹.

97% yield

Controlling radical chains

You have now met two examples of radical chain reactions:

1 radical addition of halogens to double bonds

2 radical substitution of hydrogen by halogens, or of halogens by hydrogen

You have seen how the selectivity of these reactions depends upon the bond strengths of the bond being formed or broken. Until about 1975, these reactions, with a few exceptions, were all that were expected of radicals. Since that date, however, the use of radicals in synthetic chemistry has increased tremendously, to the point where highly complex ring structures such as the natural product hirsutene and steroids can be made from simple acyclic precursors in one radical-promoted step.

hirsutene

steroid

What has made this all possible is that chemists have learned how to understand the selectivity of radical reactions to such a degree that they can design starting materials and reagents to define precisely the bonds that will break and form during the reactions. We shall now go on to look at the most important consequence of this ability to control radical reactions: they can be used to make carbon–carbon bonds.

Carbon–carbon bond formation using radicals

The following radical reaction forms a new carbon–carbon bond. The mechanism is quite similar to that of the very first radical reaction we showed you, right at the beginning of the chapter. Now, with your additional appreciation of the role of bond strength in the selectivity of radical reactions, you should be able to understand why each step proceeds in the way that it does.

Firstly, the weakest bond, C–Br, is broken by the light being shone on to the reaction. Two radicals form, $CCl_3^{\cdot}$ and $Br^{\cdot}$, and it is the $CCl_3^{\cdot}$ that adds to the (less hindered) unsubstituted end of the alkene to produce a (more stable) secondary benzylic radical.

This radical abstracts a Br, atom from the $BrCCl_3$, breaking the (weakest) C–Br bond, forming the product and regenerating $^{\cdot}CCl_3$, which adds to another molecule of alkene. Notice that the carbon-centred radical abstracts $Br^{\cdot}$ and not $^{\cdot}CCl_3$ from $BrCCl_3$—to abstract $^{\cdot}CCl_3$ would require a radical substitution at carbon—remember, radicals want the easy pickings from the front of the display; they don't go nosing round the back to see if there's anything better to be had.

> ▶
> It is mainly this step that produces the $^{\cdot}CCl_3$ that undergoes addition to the alkene—the initial photolysis, of course, produces both $Br^{\cdot}$ and $^{\cdot}CCl_3$, either of which could add, but, once the radical chain has been initiated, only $^{\cdot}CCl_3$ is reproduced.

This reaction works quite well, giving 78% of the product, but it relies on the fact that the starting material, $BrCCl_3$, has an unusually weak C–Br bond (the $^{\cdot}CCl_3$ radical is highly stabilized by those three chlorine atoms). You can't use most other alkyl bromides for a number of reasons, not least of them being that the product is also an alkyl bromide and, without the selectivity provided by the CCl_3 group, the result would be an awful mixture of polymers. The problem is that we want the product radical to abstract Br from the starting alkyl bromide to make a new alkyl bromide and a new starting radical, and there is no energetic driving force behind this transformation.

For a way of overcoming this problem, let's go back to the reaction we looked at a few pages ago, the dehalogenation of alkyl halides by Bu_3SnH. The mechanism involves formation of an alkyl (carbon-centred) radical by abstraction of Br by $Bu_3Sn^{\cdot}$. This alkyl radical then just abstracted $H^{\cdot}$ from Bu_3SnH.

Is it not possible to use this alkyl radical more constructively, and encourage it to react with another molecule (an alkene, say, like $^{\cdot}CCl_3$ did)? The answer is a qualified yes: look at this reaction.

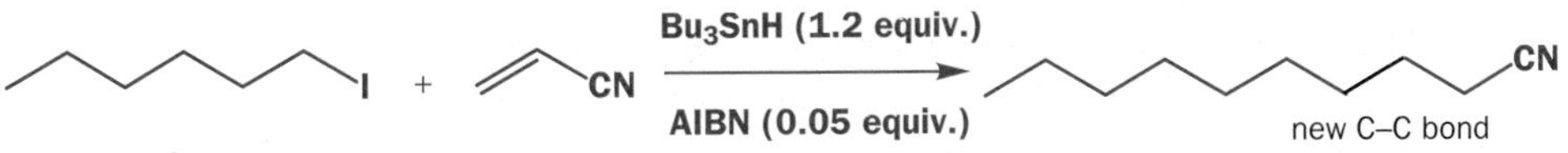

We have added a carbon-centred radical to an alkene in a radical chain reaction! Here is the mechanism.

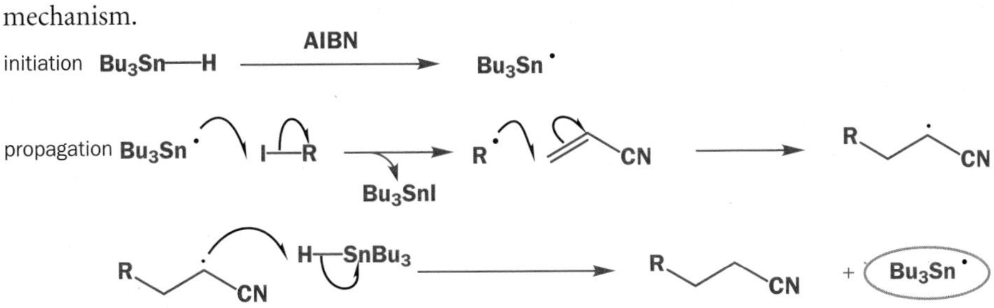

We can alternatively represent the mechanism of the reaction cyclically.

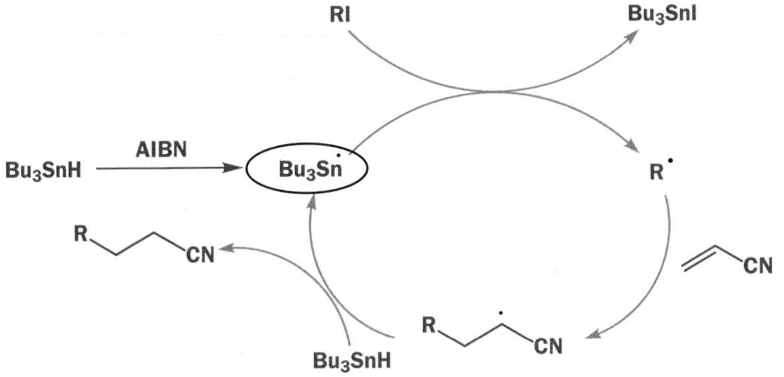

The key point is that the product radical does not have to abstract the halogen from the starting material, but H from Bu$_3$SnH; it is the Bu$_3$Sn$^\bullet$ thus formed that then regenerates the starting radical. The driving force is provided by formation of C–H at the expense of Sn–H and then Sn–Br at the expense of C–Br.

The use of tin hydrides has increased the power of radical reaction in organic synthesis tremendously, and all of the steps in these radical chain processes have been studied in great detail because of the importance of the reactions. We won't dwell excessively on these details, but we need to go back and re-examine some points about this reaction because there are some further subtleties that you need to understand.

Bear in mind that we have four radicals all in the reaction mixture at the same time. Yet each reacts with its chosen partner, forsaking all others.

Let's take each radical in turn, and look at its selectivity. Clearly bond strength has something to do with it, but how do you explain the opposing selectivities of R$^\bullet$ and the nitrile-stabilized radicals? We will see that the origins of the selectivities impose some restrictions on the type of starting material that can be used for these C–C bond-forming reactions.

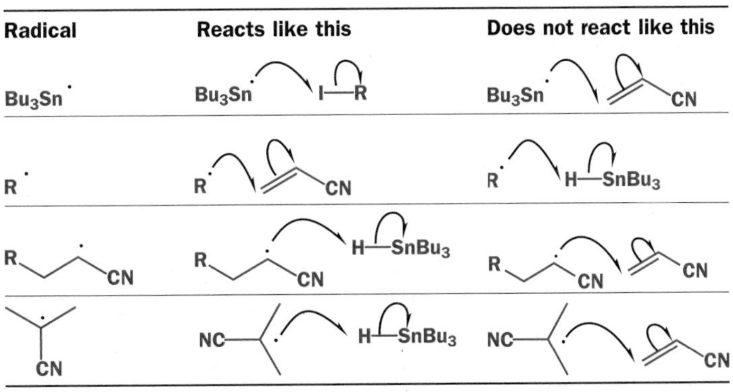

■ We explained on p. 1040 how these same favourable thermodynamics drove the Bu$_3$SnH-promoted reduction of alkyl halides.

Radical	Reacts like this	Does not react like this
Bu$_3$Sn$^\bullet$		
R$^\bullet$		
R⎯CN radical		

For the addition of an alkyl radical
to an alkene:

Reacts like this

R

CN

Does not react like this

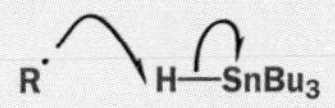

R   H—SnBu₃

Yet for the radical dehalogenation:

Reacts like this

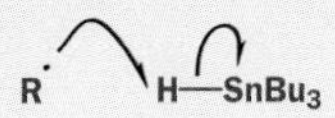

R   H—SnBu₃

1 **Bu₃Sn˙.** Unlike the case of the simple dehalogenation, the tin hydride radical here has a choice of reaction partners: it can either abstract the halide from the starting material or it can add to the alkene. The Sn–C bond is relatively weak, so addition to the alkene becomes a significant reaction only if:

- there is a large excess of alkene present, and
- the starting alkyl halide is relatively unreactive. This means that only alkyl bromides and iodides can be used effectively to form carbon–carbon bonds; alkyl chlorides are just too unreactive

2 **R˙.** On comparing the mechanism of this reaction with that of radical dehalogenation, you may rightly be concerned by the fact that in the dehalogenation the alkyl radical produced from the alkyl bromide was intended to abstract H˙ from the Bu₃SnH, whereas now, the alkyl radical is intended to react with an alkene, despite the fact that Bu₃SnH is still a component of the reaction mixture.

Concentration effects

In fact, the rate *constant* for reaction of R˙ with Bu₃SnH is about the same as that for reaction with acrylonitrile (CH₂=CHCN), so the only way in which good yields can be obtained is by ensuring that the concentration of acrylonitrile is always at least 10 times that of the tin hydride. The difference in rates will then be sufficient to give 10 times as much addition to the alkene as reduction by the tin hydride. Too much acrylonitrile in the reaction mixture causes problems with side-reactions, so a good way of achieving this is to add the tin hydride very slowly during the reaction—often a device known as a syringe pump is used for this. Of course, for complete reaction, a whole equivalent of hydride is necessary, but this can be added over a period of hours.

A useful alternative to NaBH₄ as
a reducing agent, particularly
when there are reactive carbonyl
groups in the molecule, is
NaCNBH₃, which still reduces
Bu₃SnHal but will not touch
aldehydes or ketones.

An elegant alternative is to use a technique conceptually similar to the use of NBS to provide a low concentration of Br₂ for radical allylic substitution. Instead of adding one equivalent of Bu₃SnH, a catalytic amount (usually 0.1–0.2 equivalents) of Bu₃SnCl is added at the beginning of the reaction, with one equivalent of NaBH₄. NaBH₄ will reduce Bu₃SnHal to Bu₃SnH, so about 0.1 equivalent of Bu₃SnH is formed immediately. With each cycle of the chain reaction, a molecule of this Bu₃SnH is converted to Bu₃SnBr, which NaBH₄ can reduce back to Bu₃SnH. Only as much Bu₃SnH is produced as is needed, because the rate of production is limited by the rate of reaction.

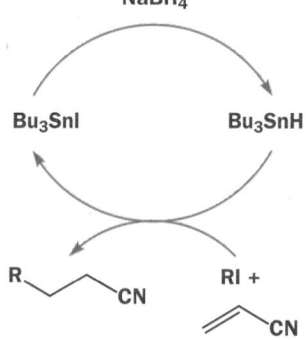

This method was used in the following example, in which an enantiomerically pure lactone, a useful synthetic building block, was made from naturally occurring glyceraldehyde.

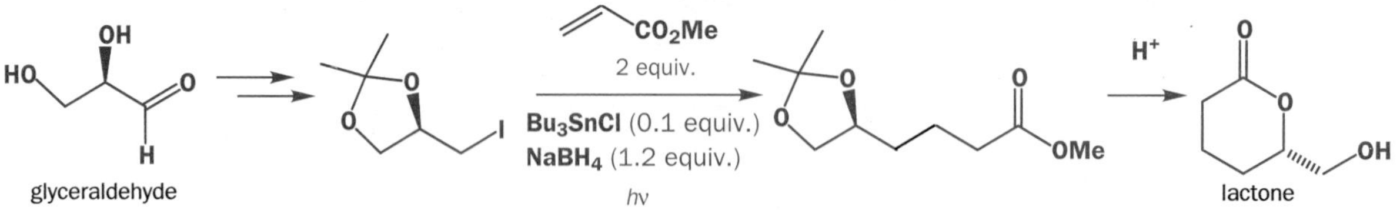

Frontier orbital effects

The second key to success in making sure that the alkyl radical behaves well is to use a reactive radical trap. In fact, this is a major limitation of intermolecular radical carbon–carbon bond-forming reactions: for the trapping of alkyl radicals only electrophilic alkenes (attached to electron-withdrawing groups such as –CN, –CO₂Me, –COMe) will do. This is a limitation, but nonetheless, cyclohexyl iodide adds to all these alkenes with the yields shown and the rate of addition to most of these alkenes is 10^3 to 10^4 times that of addition to 1-hexene.

To explain why, we have to go back to our analysis (on p. 1027) of the electronic structure of radicals and the energy of SOMOs. We said there that, while both electron-withdrawing groups and electron-donating groups will stabilize radicals, electron-withdrawing groups tend to lower the energy of the SOMO, while electron-donating groups tend to raise the energy of the SOMO.

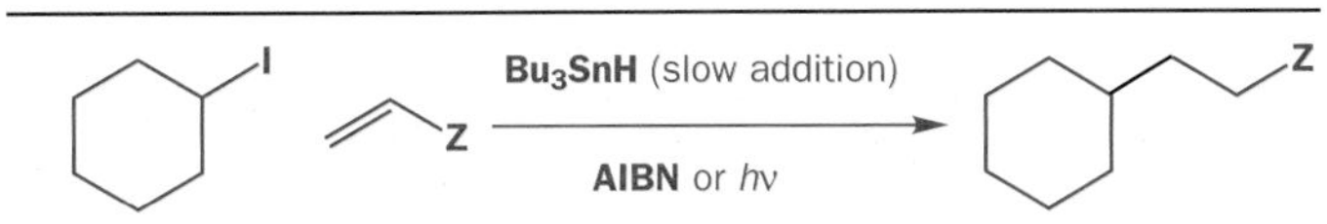

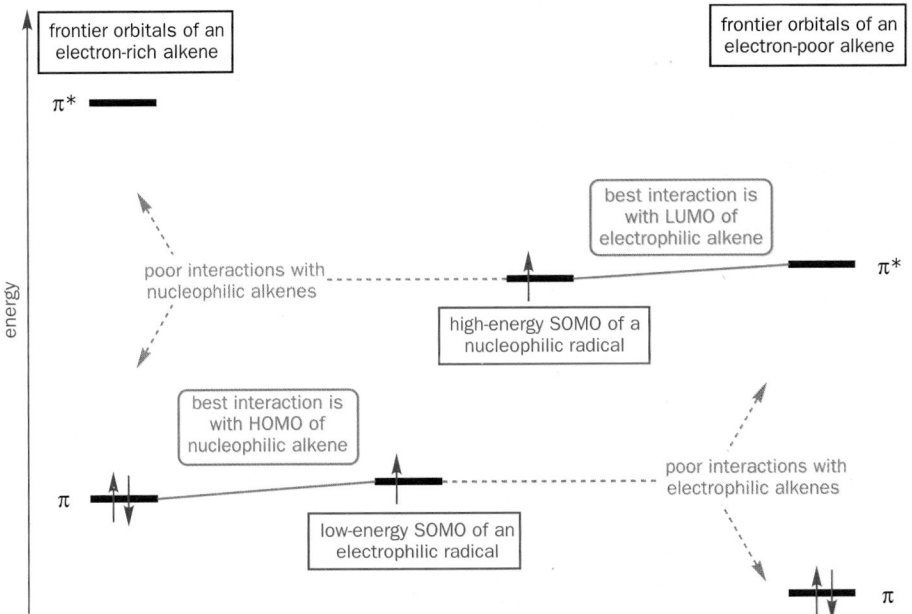

Alkene	% Yield	Alkene	% Yield
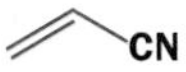 ⟍⟍CN	95	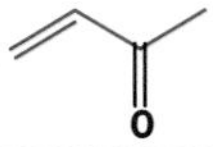 ⟍⟍⟍(=O)	85
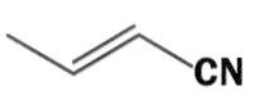 ⟍⟍CN	86	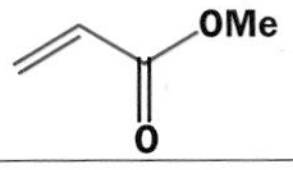 ⟍⟍(=O)OMe	85
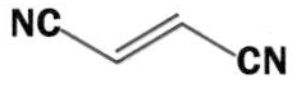 NC⟍⟍CN	72	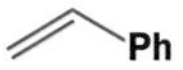 ⟍⟍Ph	83
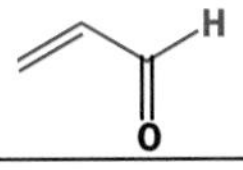 ⟍⟍(=O)H	90	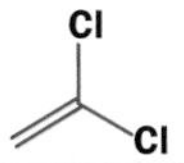 Cl⟍⟍Cl	87

● **Electrophilic and nucleophilic radicals**

- *Low-energy* SOMOs are more willing to accept an electron than to give one up; radicals adjacent to electron-withdrawing groups are therefore *electrophilic*
- *High-energy* SOMOs are more willing to give up an electron than to accept an electron; radicals adjacent to electron-donating groups are therefore *nucleophilic*

Hence the preferred reactivity of these alkyl radicals: they are relatively nucleophilic and therefore prefer to react with electrophilic alkenes. Reaction between a nucleophilic alkyl radical and an unfunctionalized (and therefore nucleophilic) alkene is much slower. Similarly, radicals adjacent to electron-withdrawing groups do not react well with electrophilic alkenes. We can represent all this on an energy level diagram.

frontier orbitals of an electron-rich alkene

frontier orbitals of an electron-poor alkene

π*

best interaction is with LUMO of electrophilic alkene

poor interactions with nucleophilic alkenes

high-energy SOMO of a nucleophilic radical

π*

best interaction is with HOMO of nucleophilic alkene

poor interactions with electrophilic alkenes

π

low-energy SOMO of an electrophilic radical

π

energy

We will now consider a third type of radical—**cyanide-stabilized alkyl radicals**.

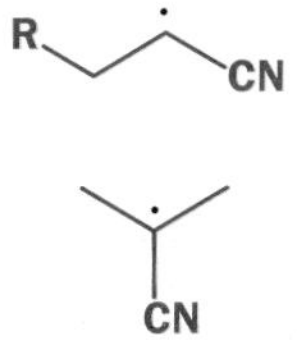

The diagram above explains the third aspect of radical chemoselectivity in this reaction: why both the product radical and the radicals produced by AIBN choose to react with Bu₃SnH and not with acrylonitrile. These radicals are electrophilic—they have an electron-withdrawing nitrile group attached to the radical centre so reaction with an electron-poor alkene is slow.

■ Radical types (1) and (2) were discussed on p. 1027.

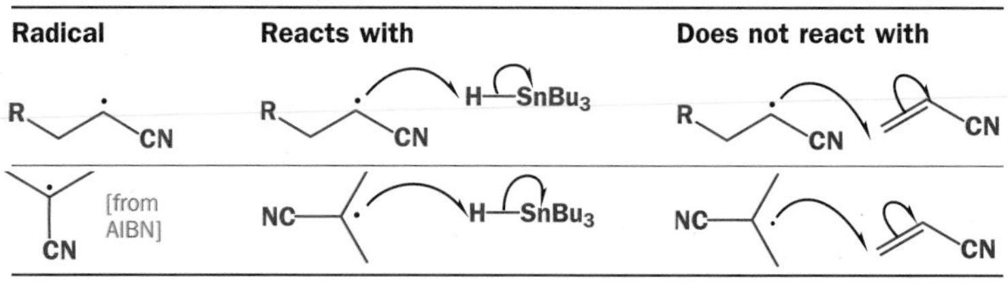

▶

Notice that this reaction works
even though a C–Cl bond needs to
be broken to generate the radical.
Usually only C–I and C–Br bonds
can be used. However, this is a
very weak C–Cl bond because the
radical produced is so stable.

Electrophilic radicals

Having seen the energy diagram above, you will not be surprised to learn that the malonate radical
adds readily not to electrophilic alkenes, but to nucleophilic alkenes, such as this vinyl ether, which
carries an electron-donating oxygen substituent. This electrophilic radical can also be formed by H-
abstraction and by oxidation.

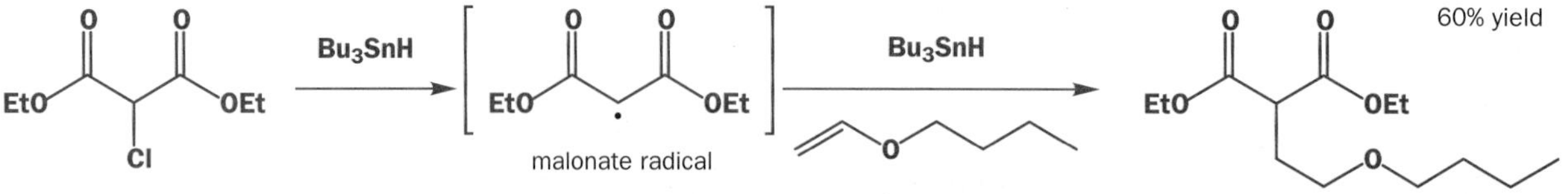

This difference in reactivity applies to non-carbon-centred radicals too. For example, the methyl radical $CH_3^{\bullet}$ and the chlorine radical $Cl^{\bullet}$ will both abstract a hydrogen atom from propionic acid. As you
would expect, the methyl radical abstracts the hydrogen atom from next to the carbonyl group to form a
carbonyl-stabilized radical. Perhaps surprisingly (in view of what we said earlier about the selectivity of
radical chlorinations), the chlorine radical abstracts a hydrogen atom from the terminal methyl group of
the acid, despite the fact that this C–H bond is stronger. The reason has to be to do with HOMO–LUMO
interactions. The methyl radical is nucleophilic, with a high-energy SOMO. It therefore attacks the C–H
bond with the lowest LUMO, in other words, α to the carbonyl group. The chlorine atom, on the other
hand, is electrophilic: it has a low-energy SOMO (because it is an electronegative element) and attacks
the C–H bonds of the terminal methyl group because they have the highest-energy HOMO.
Chlorination of functionalized compounds is not as simple as we implied earlier!

> ● **Summary of requirements for the successful use of the tin method**
>
> - Bu_3SnH must be added or generated slowly
> - R–X starting material   must contain a weak C–X bond (C–I or C–Br)
> - Radical trap   must be an electrophilic alkene
>
>   must be present in a concentration at least 10 times
>   that of Bu_3SnH

Copolymerization

Radical chain reactions are particularly suited to the synthesis of polymers, and
we will look at this rather special type of radical reaction in Chapter 52. But
there is one example of a polymerization that is worth including here since it
demonstrates very nicely the effect of electron-withdrawing or -donating
substituents on radical reactivity. When a mixture of vinyl acetate and methyl
acrylate is treated with a radical initiator, a rather remarkable polymerization
takes place. The polymer produced contains *alternating* vinyl acetate and
methyl acrylate monomers along the length of its chain.

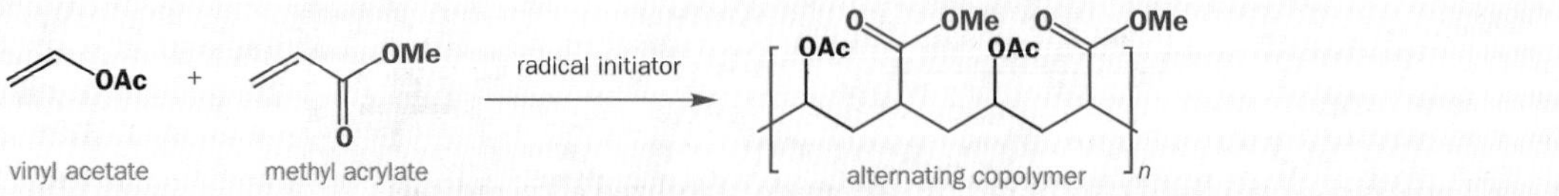

The mechanism of the reaction shows you why. The nucleophilic radical from
vinyl acetate (adjacent to filled n orbital of OAc; high-energy SOMO) prefers to
add to the electrophilic alkene (the acrylate). The new radical (adjacent to the
empty π^* orbital of CO_2Me; low-energy SOMO) is electrophilic and prefers to
add to nucleophilic alkene (the vinyl acetate). This produces a new nucleophilic
radical, which again prefers to add to the electrophilic alkene, and the whole
cycle repeats endlessly.

Copolymerization—continued

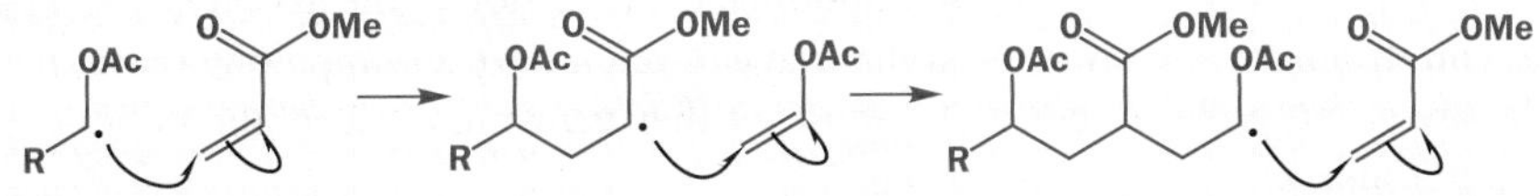

The radical produced by addition to vinyl acetate is nucleophilic, so it adds to methyl acrylate; the radical produced by addition to methyl acrylate is electrophilic, so it adds to vinyl acetate. This reaction is a clear demonstration of the power of frontier orbital theory to explain the reactivity of organic molecules—it would be hard to come up with any other convincing explanation.

The reactivity pattern of radicals is quite different from that of polar reagents

The first reaction that you met in this book, in Chapter 2, was the nucleophilic addition to a carbonyl group. Yet we have shown you no examples of radicals adding to carbonyl groups. This typical reaction of polar reagents is really quite rare with radicals.

In Chapter 8 we introduced the concept of pK_a in which we saw acids and bases exchanging protons. Among the strongest organic acids are those containing O–H bonds. Yet you have seen no radical reactions in which an O–H bond is broken—in fact the reaction on p. 1030 used ethanol as a solvent! Carbon acids tend to be much weaker—yet you've seen plenty of examples of C–H bonds being broken by radical attack.

In Chapter 17 we introduced nucleophilic substitution at saturated carbon, using as an example some alkyl bromides. Now, radicals do react with alkyl halides—but not at carbon! You've seen how alkyl halides undergo substitution at bromine with tin radicals. The difference in reactivity between, say, organolithiums and radicals, both of them highly reactive, is nicely illustrated by the way in which they react with enones.

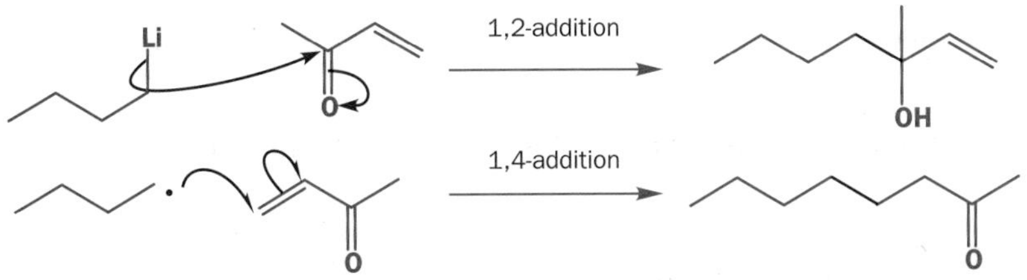

We introduced the terms *hard* and *soft* in Chapters 10 and 17. From all these reactions it's evident that radicals are very soft species: their reactions are driven not by the charge density on an atom but by the coefficient and energy of the frontier orbitals at that atom.

● **Summary of typical reactivity patterns**

With	Polar nucleophiles typically react like this	Radicals typically react like this
unsaturated C=O compounds	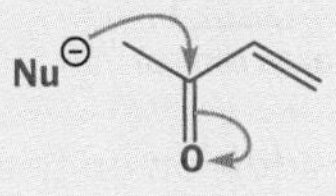	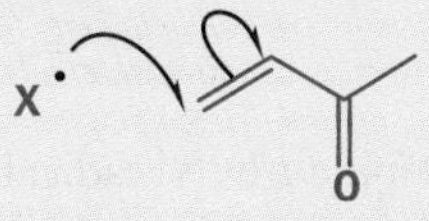
X–H bonds	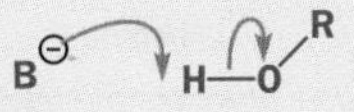	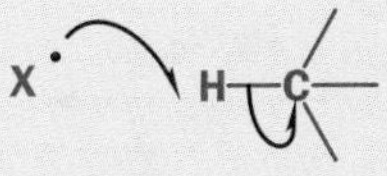
alkyl halides		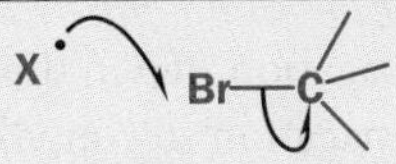

Umpolung

In Chapter 30, you came across the idea of **umpolung**, the inversion of the usual reactivity pattern of a molecule. You may have already noticed that radicals often have an umpolung reactivity pattern. Alkyl halides are electrophiles in polar reactions; yet they generate nucleophilic radicals that react with electrophilic alkenes.

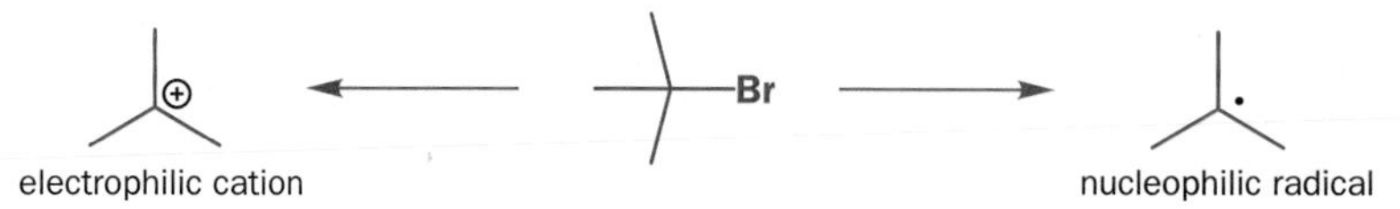

electrophilic cation   ⟵   ─Br   ⟶   nucleophilic radical

Similarly, we consider the carbon atoms α to carbonyl groups to be nucleophilic, because enolization creates a partial negative charge there (in other words, ketones are a^1 reagents). Yet carbonyl-stabilized radicals are electrophilic.

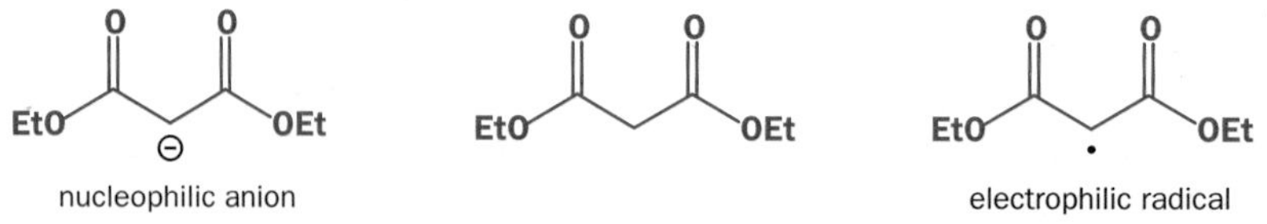

EtO ⟍ ⟋ OEt   EtO ⟍ ⟋ OEt   EtO ⟍ ⟋ OEt
nucleophilic anion   electrophilic radical

An alternative way of making alkyl radicals: the mercury method

Although the tin hydride + alkyl halide method is probably the most important way of making alkyl radicals, we should mention some other methods that are useful. We said at the beginning of the chapter that carbon–metal bonds, particularly carbon–transition metal bonds, are weak and can homolyse to form radicals. Alkyl mercuries are useful sources of alkyl radicals for this reason. They can be made by a number of routes, for example, from Grignard reagents by transmetallation.

▶

This transmetallation works because mercury is softer than magnesium and therefore prefers the softer alkyl ligand.

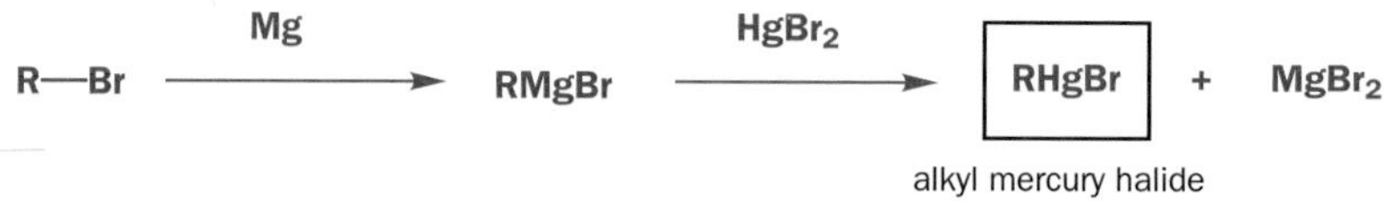

$$R\text{—Br} \xrightarrow{Mg} RMgBr \xrightarrow{HgBr_2} \boxed{RHgBr} + MgBr_2$$

alkyl mercury halide

■

This is an **oxymercuration reaction**. You met it in Chapter 20.

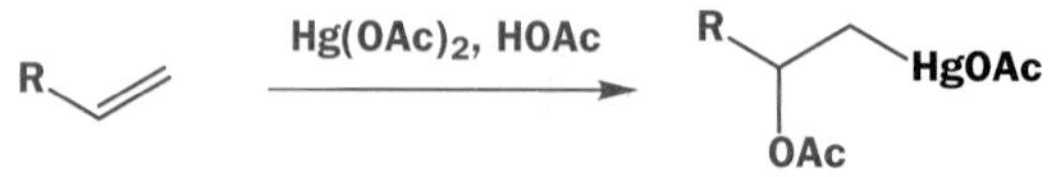

Addition of mercury acetate to a double bond gives an alkyl mercury bearing a functional group.

$$R\diagup\diagdown \xrightarrow{Hg(OAc)_2,\ HOAc} \underset{OAc}{\overset{R}{\diagdown}} HgOAc$$

Alkyl mercury halides and alkyl mercury acetates are quite stable, but reduction with sodium borohydride leads to highly unstable alkyl mercury hydrides, which collapse at room temperature or in the presence of light to yield alkyl radicals. One other product is mercury metal and you might think you would get H• as well but this is too unstable to be formed and is captured by something else (X)—you will see what X is in a moment. This initial decomposition of RHgH initiates the chain but its propagation is by the different mechanism shown below.

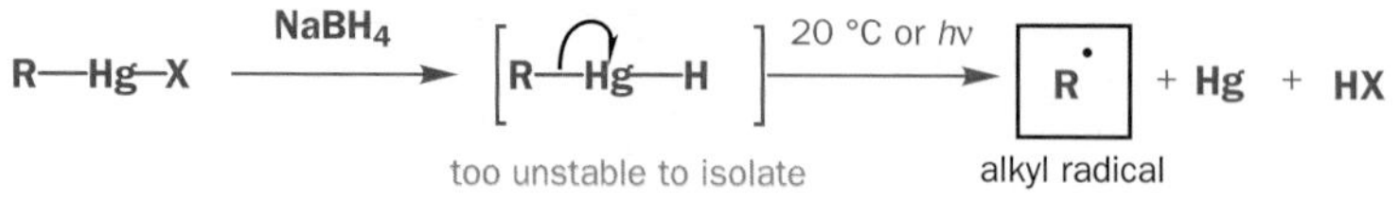

$$R\text{—Hg—X} \xrightarrow{NaBH_4} \left[R\text{—Hg—H} \right] \xrightarrow{20\ °C\ or\ h\nu} \boxed{R^\bullet} + Hg + HX$$

too unstable to isolate   alkyl radical

In this example a *t*-butyl radical does conjugate addition on to acrylonitrile.

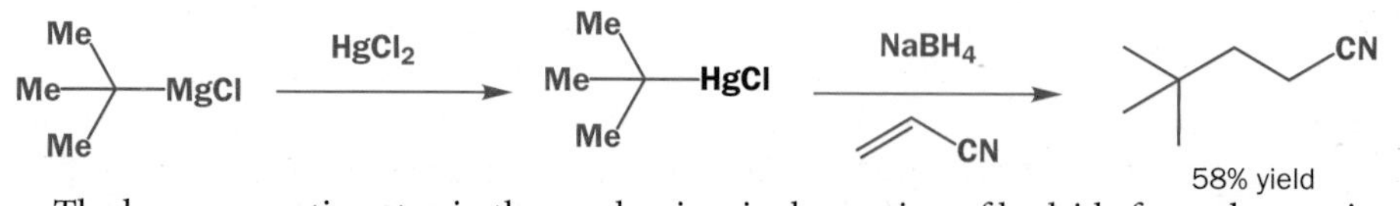

$$\underset{Me}{\overset{Me}{Me}}\text{—MgCl} \xrightarrow{HgCl_2} \underset{Me}{\overset{Me}{Me}}\text{—HgCl} \xrightarrow[\diagup\diagdown CN]{NaBH_4} \diagup\diagdown\diagup\text{CN}$$

58% yield

The key propagation step in the mechanism is abstraction of hydride from the starting alkyl mercury. In the propagation step anything will do to cleave the weak Hg–H bond but once the chain is running it is an alkyl radical that does this job, just as in tin hydride chemistry.

> Notice that *tert*-butyl radicals add readily to electrophilic alkenes: using *tert*-butyl Grignards or *tert*-butyllithium as nucleophiles is much more problematic.

Unfortunately, radicals derived from alkylmercuries are even more limited in what they will react with than radicals made from alkyl halides by the tin hydride method. Styrene, for example, cannot be used to trap alkylmercury-derived radicals efficiently because the radicals react more rapidly with the mercury hydride (which has an even weaker metal–H bond than Bu$_3$SnH) than with the styrene.

■ We discussed this selectivity problem as it applied to the tin hydride method on p. 1043.

Intramolecular radical reactions are more efficient than intermolecular ones

All of the reactions you have met so far involve radical attack between two molecules. We've pointed out some of the drawbacks when C–C bonds are made in this way: the radical trap has to be activated (that is, electrophilic to capture nucleophilic radicals) and must often be present in excess; and the radical starting material must contain very weak C–X bonds (such as C–Br, C–I, C–Hg). The requirements are much less stringent, however, if the radical reaction is carried out intramolecularly. For example, this reaction works.

Notice that the double bond is not activated: in fact, it is nucleophilic, and the reaction still works even though the radical is also substituted with an electron-donating group. The C–S bond that is broken is also relatively strong, yet nonetheless a high yield of product is obtained. Why should this be so? What difference does it make that the reactions are intramolecular?

The key is that the intramolecular cyclization of the radical is now enormously favoured over other possible courses of action for the radical. Remember that when we were carrying out radical reactions *inter*molecularly, addition to the radical trap was encouraged by increasing the concentration of radical trap and decreasing the concentration of Bu$_3$SnH to avoid radical reduction. For *intra*molecular reactions, the double bond that acts as the radical trap is always held close to the radical, and cyclization takes place extremely rapidly, even on to unactivated double bonds. The hydride donor (Bu$_3$SnH) doesn't get a look in, and can be present in higher concentrations than would otherwise be possible. Moreover, as there is only one equivalent of radical trap, and the trap need not be highly reactive, there is little danger of high concentrations of Bu$_3$Sn$^•$ reacting with it, so

Some useful bond strengths

Bond	Typical bond energy, kJ mol^{-1}
C–I	238
C–Br	280
C–Cl	331
C–S	320

the concentration of $Bu_3Sn^\bullet$ can build up to levels where the rate of abstraction of groups like Cl, SPh, and SePh is acceptable, despite their stronger C–X bonds.

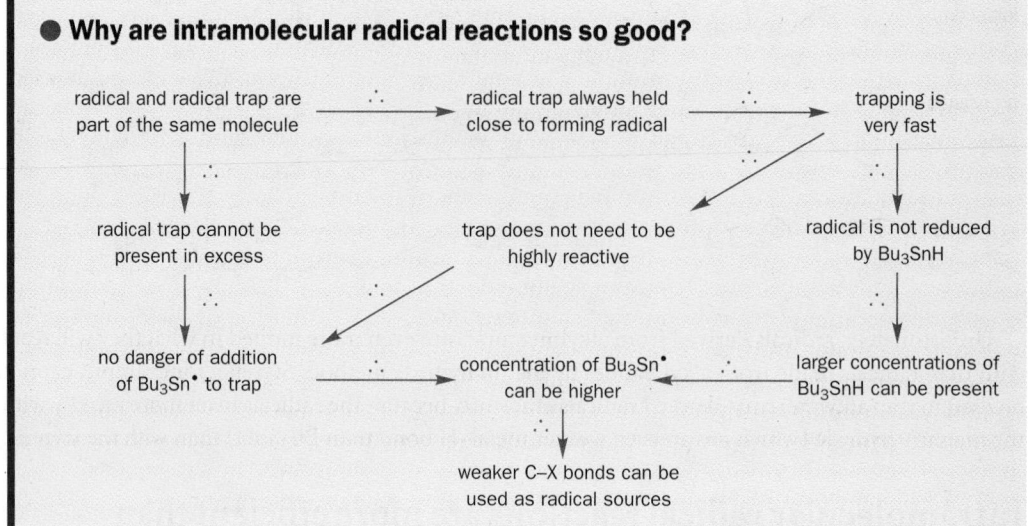

● Why are intramolecular radical reactions so good?

radical and radical trap are part of the same molecule
∴
radical trap always held close to forming radical
∴
trapping is very fast
∴
radical trap cannot be present in excess
trap does not need to be highly reactive
radical is not reduced by Bu3SnH
no danger of addition of Bu3Sn• to trap
concentration of Bu3Sn• can be higher
larger concentrations of Bu3SnH can be present
weaker C–X bonds can be used as radical sources

Overall then, intramolecular radical reactions are very powerful, and are often used to make five-membered rings.

Br
N
SO2Ph
Bu3SnH
AIBN
N
SO2Ph
87% yield

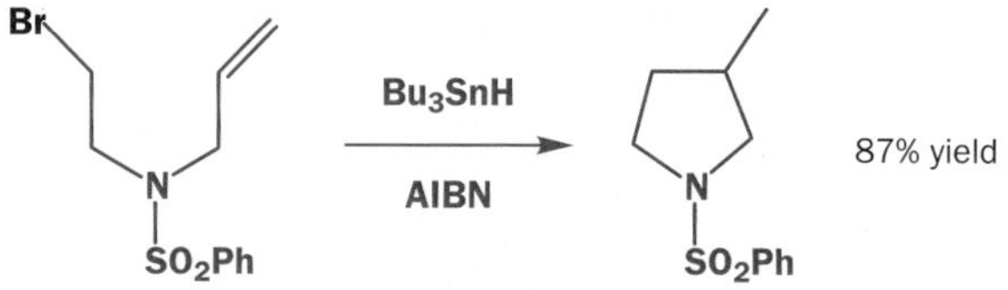

It is possible to make other ring sizes also, but the range is rather limited. Because of ring strain, three- and four-membered rings cannot be formed by radical reactions. Otherwise, smaller rings form faster than larger ones: look at these selectivities.

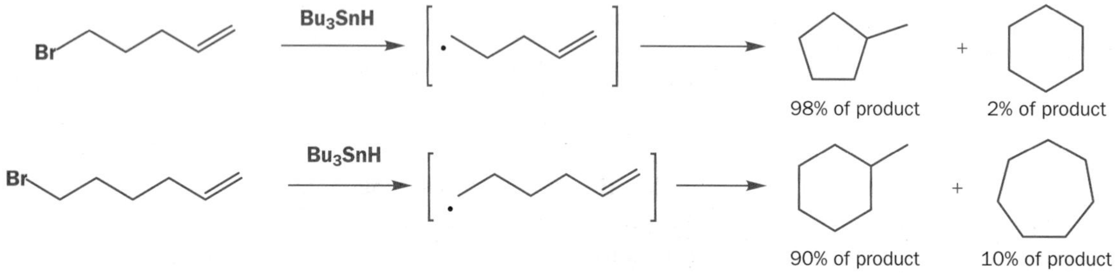

Br
Bu3SnH
98% of product
2% of product
Br
Bu3SnH
90% of product
10% of product

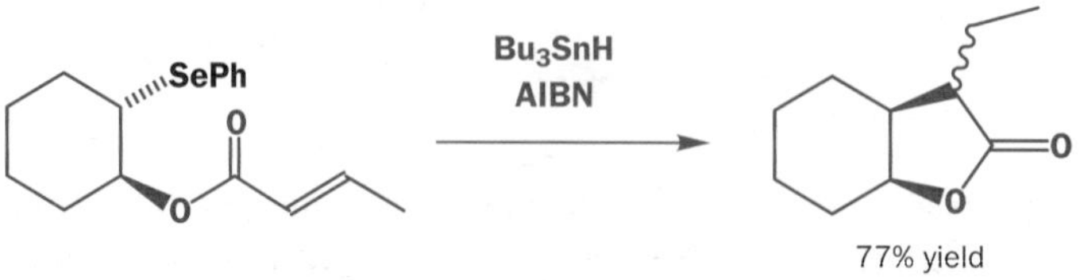

In Chapter 41 we will learn to analyse these situations using Baldwin's rules.

The preference for formation of a smaller ring is a very powerful one: in this reaction, the five-membered ring forms and not the six-membered one, even though cyclization to give a six-membered ring would also give a stabilized radical.

SePh
O
O
Bu3SnH
AIBN
O
O
77% yield

Radicals are important because they react in ways difficult to achieve with anions and cations and with different selectivity. Though radical reactions are less important than ionic reactions you need to understand their mechanisms because they are widespread in an atmosphere of the oxygen diradical. In the next chapter we will move on from carbon atoms carrying seven valence electrons to carbon atoms carrying only six valence electrons called *carbenes*.

Problems

1. In Chapter 33, Problem 13, we used a silylated ene-diol that was actually made in this way. Give a mechanism for the reaction and explain why the Me$_3$SiCl is necessary.

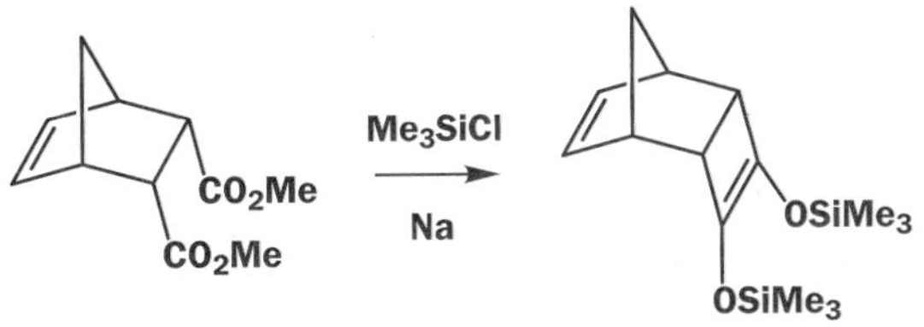

2. Heating the diazonium salt below in the presence of methyl acrylate gives a reasonable yield of a chloroacid. Why is this unlikely to be nucleophilic aromatic substitution by the S$_N$1 mechanism (Chapter 23)? Suggest an alternative mechanism that explains the regioselectivity.

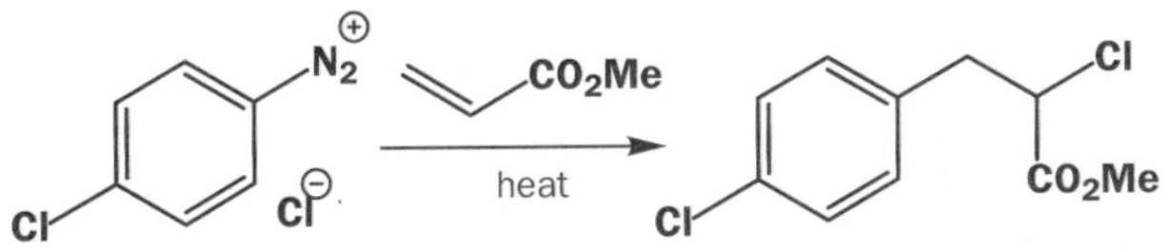

3. Suggest a mechanism for this reaction and comment on the ring size formed. What is the minor product likely to be?

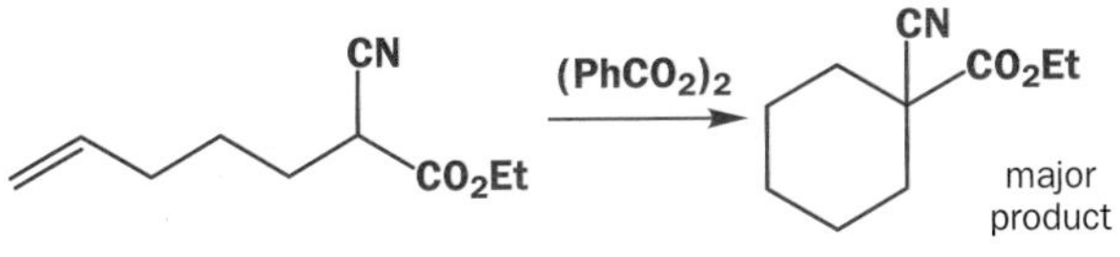

4. Treatment of this aromatic heterocycle with NBS (*N*-bromo-succinimide) and AIBN gives mainly one product but this is difficult to purify from minor impurities containing one or three bromine atoms. Further treatment with 10% aqueous NaOH gives one easily separable product in modest yield (50%). What are the mechanisms for the reactions? What might the minor products be?

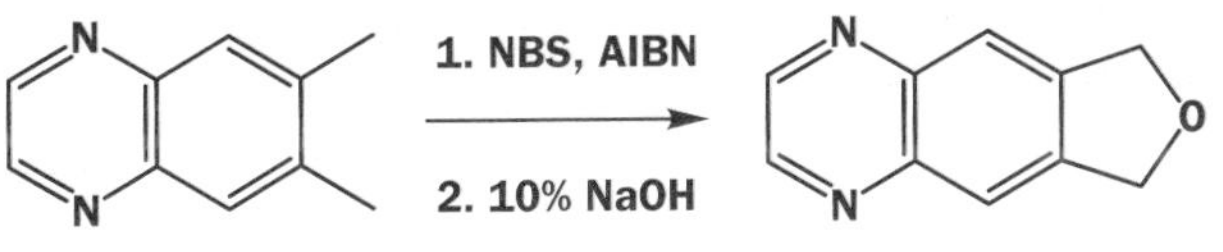

5. Propose a mechanism for this reaction accounting for the selectivity. Include a conformational drawing of the product.

6. An ICI (now AstraZeneca) process for the manufacture of the diene used to make pyrethroid insecticides involves heating these compounds to 500 °C in a flow system. Propose a radical chain mechanism for the reaction.

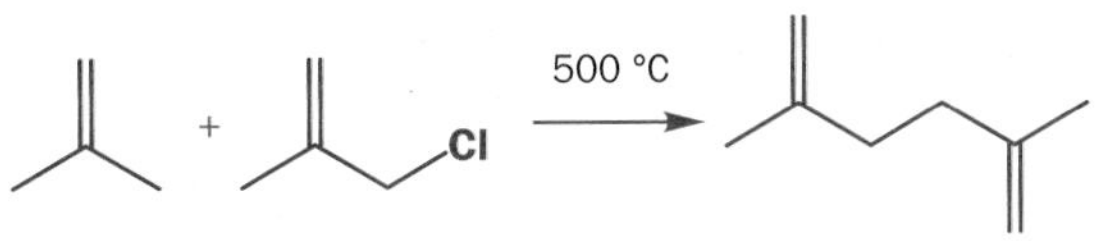

7. Heating this compound at 560 °C gives two products with the spectroscopic data shown below. What are these products and how are they formed?

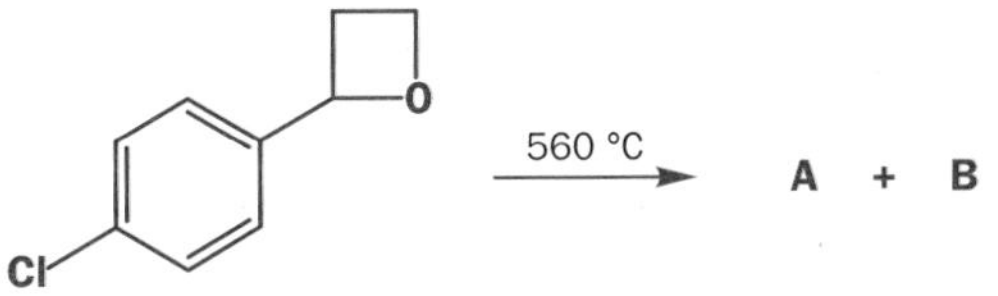

A has IR 1640 cm^{-1}; *m/z* 138 (100%), 140 (33%); δ_H (p.p.m.) 7.1 (4H, s), 6.5 (1H, dd, *J* 17, 11 Hz), 5.5 (1H, dd, *J* 17, 2 Hz), and 5.1 (1H, dd, *J* 11, 2 Hz).

B has IR 1700 cm^{-1}; *m/z* 111 (45%), 113 (15%), 139 (60%), 140 (100%), 141 (20%), and 142 (33%); δ_H (p.p.m.) 9.9 (1H, s), 7.75 (2H, d, *J* 9 Hz), and 7.43 (2H, d, *J* 9 Hz).

8. Treatment of methylcyclopropane with peroxides at very low temperature (−150 °C) gives an unstable species whose ESR spectrum consists of a triplet with coupling 20.7 gauss and fine splitting showing dtt coupling of 2.0, 2.6, and 3.0 gauss. Warming to a mere −90 °C gives a new species whose ESR spectrum consists of a triplet of triplets with coupling 22.2 and 28.5 gauss and fine splitting showing small ddd coupling of less than 1 gauss.

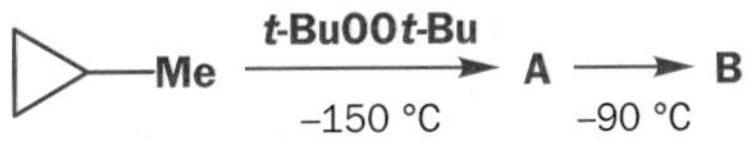

If methylcyclopropane is treated with *t*-BuOCl, various products are obtained, but the two major products are C and D. At lower temperatures more of C is formed and at higher temperatures more of D.

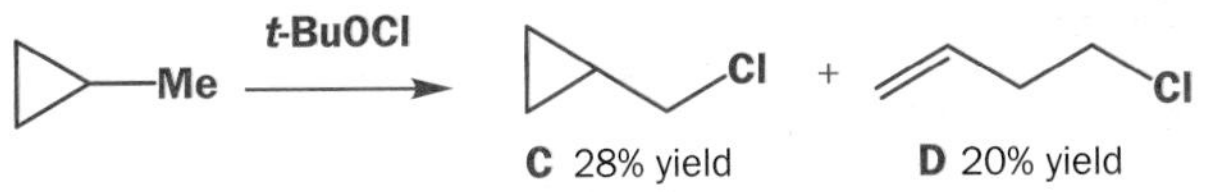

Treatment of the more highly substituted cyclopropane with PhSH and AIBN gives a single product in quantitative yield. Account for all of these reactions, identifying A and B and explaining the differences between the various experiments.

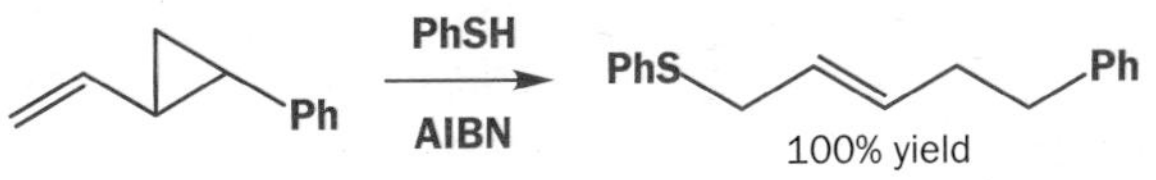

9. The last few stages of Corey's epibatidine synthesis are shown here. Give mechanisms for the first two reactions and suggest a reagent for the last step.

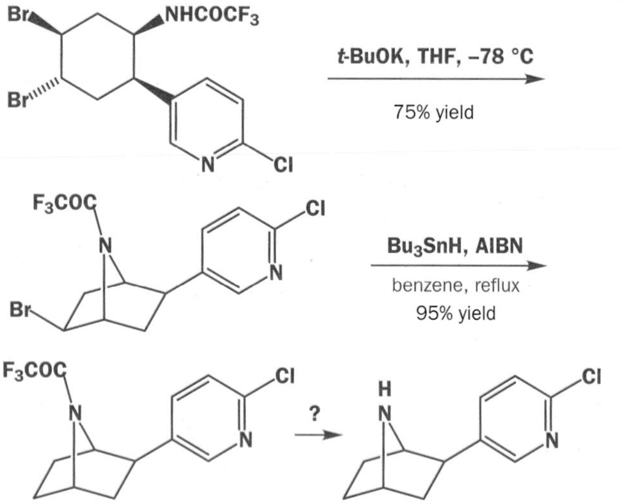

10. How would you make the starting material for this sequence of reactions? Give a mechanism for the first reaction that explains its regio- and stereoselectivity. Your answer should include a conformational drawing of the product. What is the mechanism of the last step? Attempts to carry out this last step by iodine–lithium exchange and reaction with allyl bromide fail. Why? Why is the reaction sequence here successful?

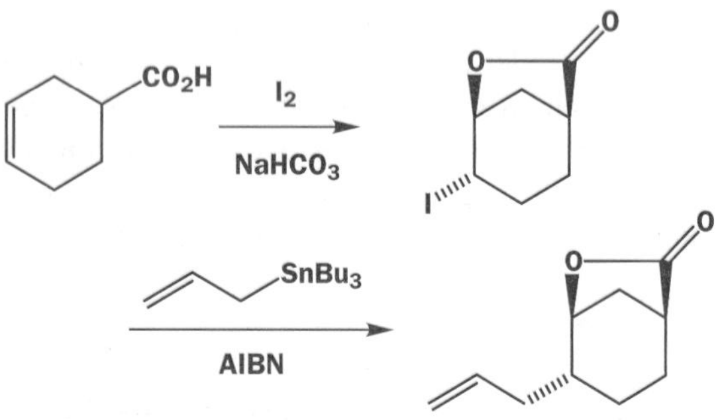

11. Suggest a mechanism for this reaction explaining why a mixture of diastereoisomers of the starting material gives a single diastereoisomer of the product. Is there any other form of selectivity?

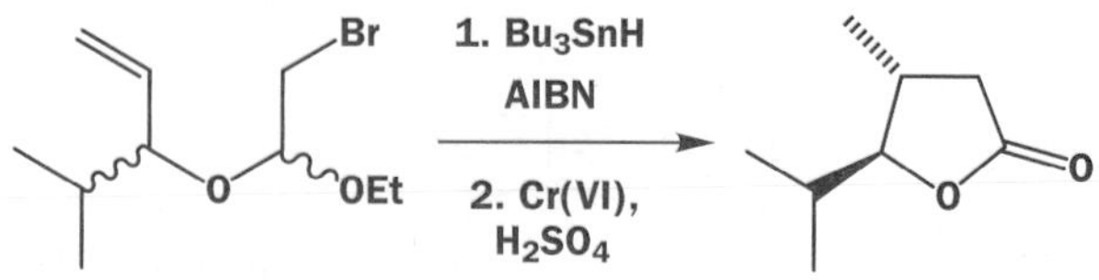

12. On the other hand, why does a single diastereoisomer of this organomercury compound give a mixture of diastereoisomers (68:32) on reduction with borohydride in the presence of acrylonitrile?

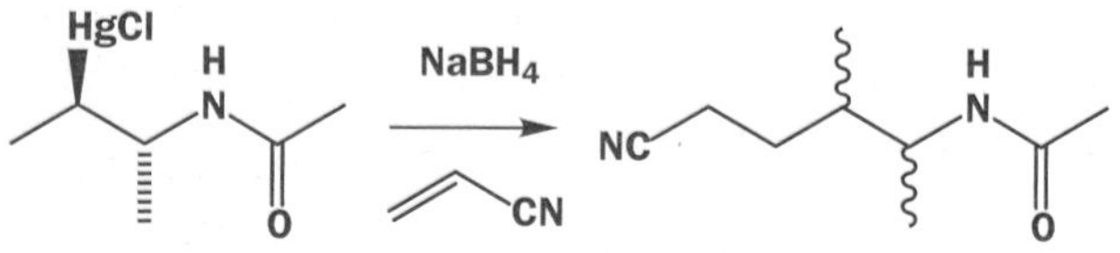

13. Reaction of this carboxylic acid ($C_5H_8O_2$) with bromine in the presence of dibenzoyl peroxide gives an unstable compound ($C_5H_6Br_2O_2$) that gives a stable compound ($C_5H_5BrO_2$) on treatment with base. The stable compound has IR 1735 and 1645 cm^{-1} and NMR δ_H (p.p.m.) 6.18 (1H, s), 5.00 (2H, s), and 4.18 (2H, s). What is the structure of the stable product? Deduce the structure of the unstable compound and mechanisms for the reactions.

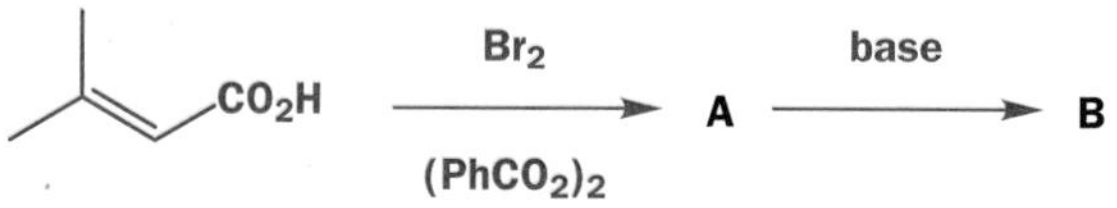

14. The product formed in Problem 9 of Chapter 20 was actually used to make this cyclic ether. What is the mechanism?

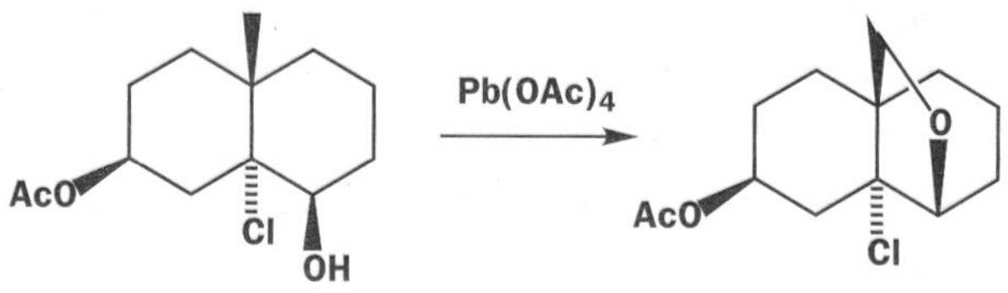

40

Connections

Building on:
- Conjugate addition ch10 & ch23
- Energy profile diagrams ch13
- Elimination reactions ch19
- Controlling stereochemistry ch16 & ch33–ch34
- Retrosynthetic analysis ch30
- Diastereoselectivity ch33–ch34
- Rearrangements ch37
- Radicals ch39

Arriving at:
- Carbenes are neutral species with only six electrons
- Carbenes can have paired or unpaired electrons
- Carbenes are normally electrophilic
- Typical reactions include insertion into C=C bonds
- Insertion into C–H and O–H bonds is possible
- Intramolecular insertion is stereospecific
- Carbenes rearrange easily
- Carbenes are useful in synthesis

Looking forward to:
- Determination of mechanism ch41
- Heterocycles ch42–ch44
- Main group chemistry ch46–ch47
- Organometallic chemistry ch48

Diazomethane makes methyl esters from carboxylic acids

In 1981, some chemists in Pennsylvania needed to convert this carboxylic acid into its methyl ester as part of the synthesis of an antibiotic compound. What reagent did they choose to do the reaction?

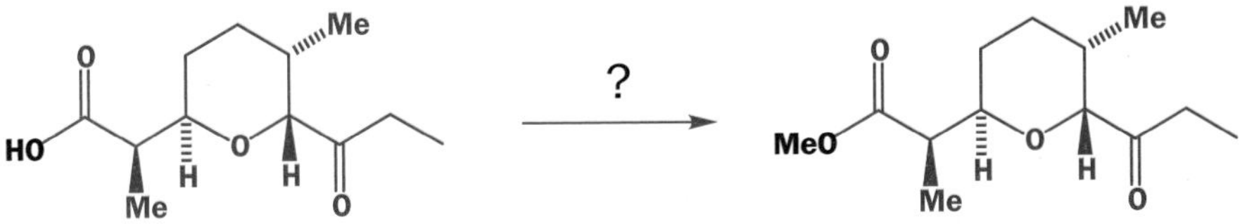

You remember, of course, that esters can be made from carboxylic acids and alcohols under acid catalysis, so you might expect them to use this type of method. On a small scale, it's usually better to convert the acid to an acyl chloride before coupling with an alcohol, using pyridine (or DMAP + Et_3N) as a base; this type of reaction might have been a reasonable choice too.

$$RCO_2H \xrightarrow[\text{or (COCl)}_2]{\text{SOCl}_2} RCOCl \xrightarrow[\text{pyridine}]{\text{MeOH}} RCO_2Me$$

But, in fact, they chose neither of these methods. Instead, they simply treated the carboxylic acid with a compound called diazomethane, CH_2N_2, and isolated the methyl ester.

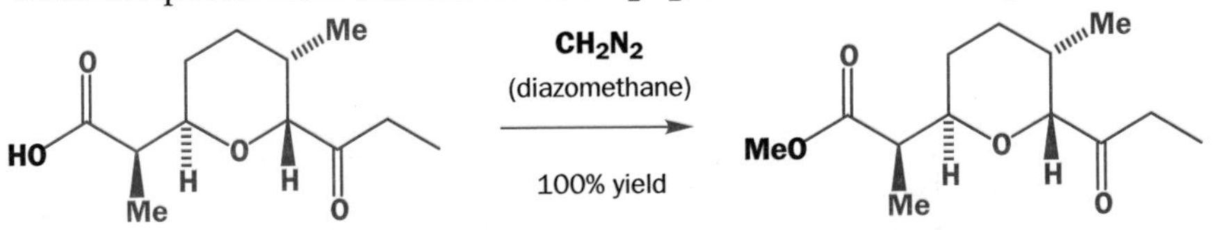

Diazomethane, CH_2N_2, is a rather curious compound that has to be drawn as a dipole. There are several different ways of expressing its structure.

■ Look back at Chapter 12 if you need reminding of any of these reactions.

▶ You might like to think about why the alternatives would not be so suitable in this case.

■ You've met other molecules like this—carbon monoxide is one, and so are nitro compounds and the 1,3-dipoles you met in Chapter 35.

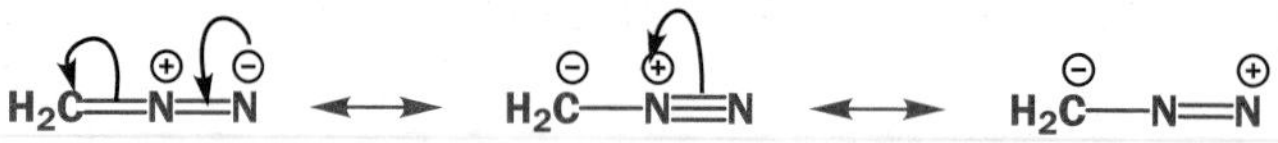

Diazomethane methylates carboxylic acids because carboxylic acids readily protonate it, giving an extremely unstable diazonium cation. This compound is desperate to lose N_2, the world's best leaving group, and so it does, with the N_2 being substituted by the carboxylate anion. The carboxylate anion is in exactly the right position to carry out an S_N2 reaction and that is what we have drawn.

> ▶
> The exact mechanism of this step remains unclear—with such a good leaving group and a bad nucleophile you might expect S_N1, but that would require a methyl carbocation.

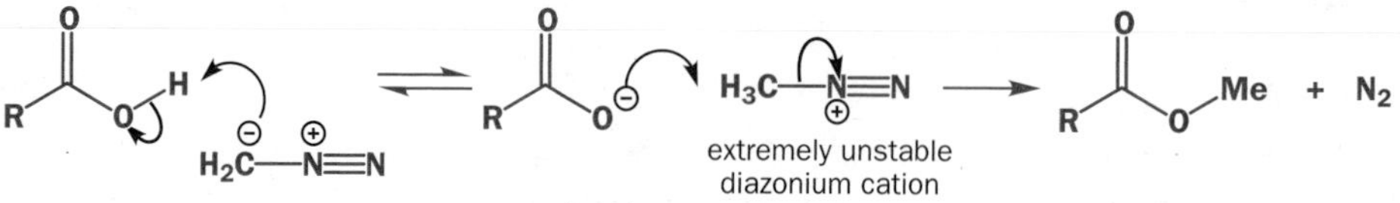

extremely unstable
diazonium cation

Diazomethane methylation is a good way of making methyl esters from carboxylic acids on a small scale because yields are excellent and the only by-product is nitrogen. However, there is a drawback: diazomethane has a boiling point of −24 °C, and it is a toxic and highly explosive gas. It therefore has to be used in solution, usually in ether; the solution must be dilute, because concentrated solutions of diazomethane are also explosive. It is usually produced by reaction of *N*-methyl-*N*-nitrosourea or *N*-methyl-*N*-nitrosotoluenesulfonamide with base, and distilled out of that reaction mixture as an azeotrope with ether, straight into a solution of the carboxylic acid.

> ▶
> Conveniently, solutions containing diazomethane are yellow, so the reaction is **self-titrating**—as the carboxylic acid reacts, the yellow diazomethane is removed, but as long as excess diazomethane remains the yellow colour persists.

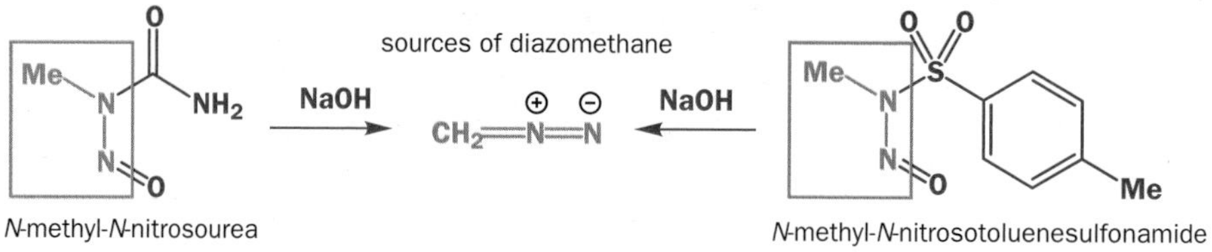

sources of diazomethane

N-methyl-*N*-nitrosourea

N-methyl-*N*-nitrosotoluenesulfonamide

> ■
> There is an alternative mechanism that starts by deprotonation at carbon.

The mechanism of the reaction that forms diazomethane is shown below. The key step is base-catalysed elimination, though the curly arrows we have to draw to represent this are rather tortuous!

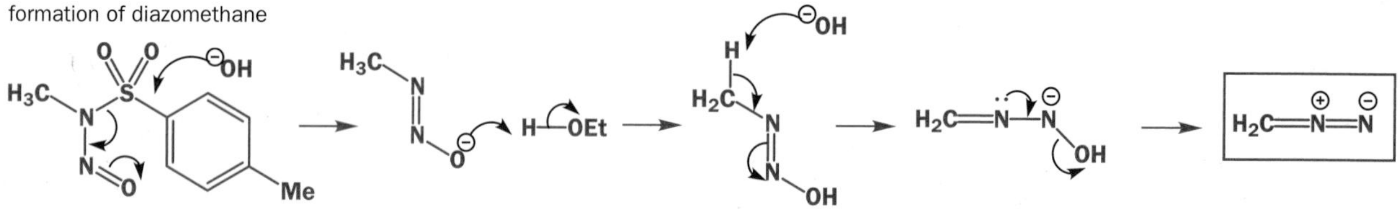

formation of diazomethane

Diazomethane will also methylate phenols, because they too are acidic enough to protonate it. Ordinary alcohols, though, are not methylated because they are not strong enough acids to protonate diazomethane.

| pK_a 10 | | pK_a 16 |

Selective methylation

Chemists studying the hormone degradation products present in the urine of pregnant women needed to methylate the phenolic hydroxyl group of the steroid oestriol. By using diazomethane, they avoided reaction at the two other hydroxylic groups. When, subsequently, they did want to methylate the other two hydroxyl groups, they had to add acid to the reaction to protonate the diazomethane.

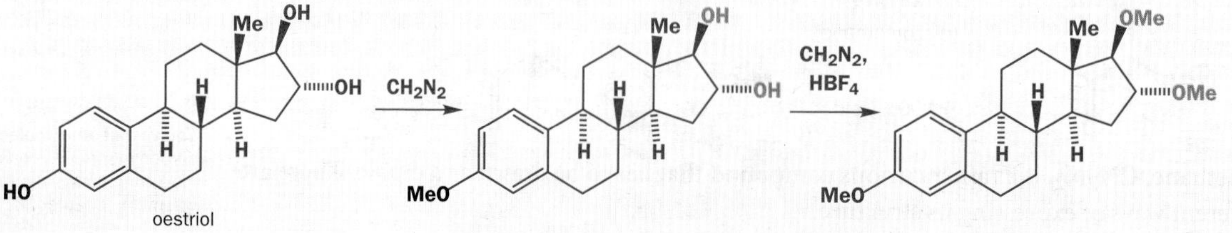

oestriol

Photolysis of diazomethane produces a carbene

Alcohols *can* be methylated by diazomethane if the mixture is irradiated with light.

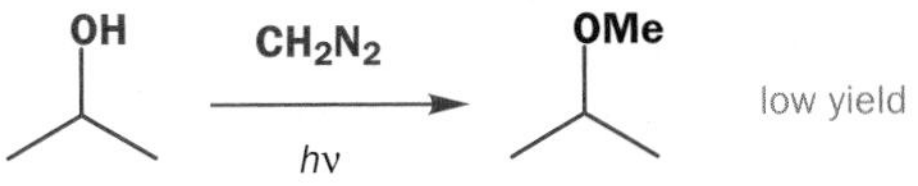

low yield

> ▶
> Although this reaction illustrates an important point, the yield is too low, there are too many by-products, and the potential for serious explosions is too great for it ever to be useful as a way of making methyl ethers.

The mechanism is now totally different, because the light energy promotes loss of nitrogen (N_2) from the molecule *without protonation*. This means that what is left behind is a carbon atom carrying just two hydrogen atoms (CH_2), and having only six electrons. Species like this are called **carbenes**, and they are the subject of this chapter.

> ● **Carbenes are neutral species containing a carbon atom with only six valence electrons.**

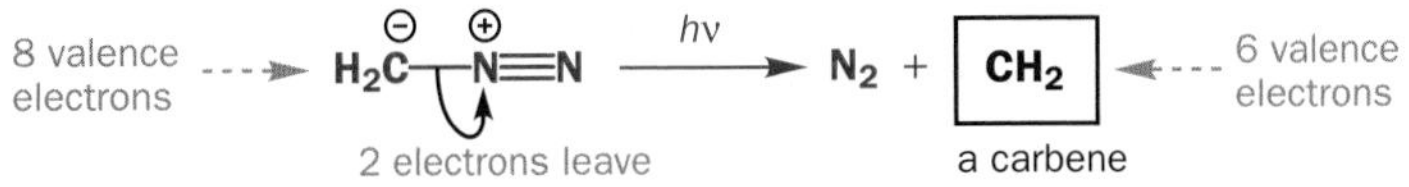

Carbenes have six electrons: two in each bond and two nonbonding electrons, which are often represented as $:CR_2$ (as though they were a lone pair). As you will see later, this can be misleading, but $:CR_2$ is a widely used symbol for a carbene. This carbene is trapped by the alcohol to make an ether.

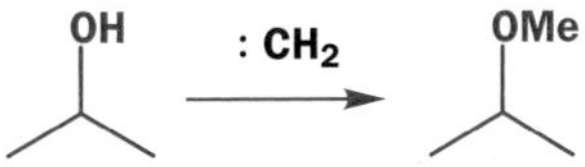

> ▶
> Carbenium ions too have only six valence electrons, but, of course, unlike carbenes they are charged.

Like the radicals in Chapter 39, carbenes are extremely reactive species. As you have just seen, they are trapped by alcohols to make ethers, but more importantly they will react with alkenes to make cyclopropanes, and they will also insert into C–H bonds.

> ● **Typical carbene reactions**
>
> The carbene inserts itself into a σ bond or a π bond. 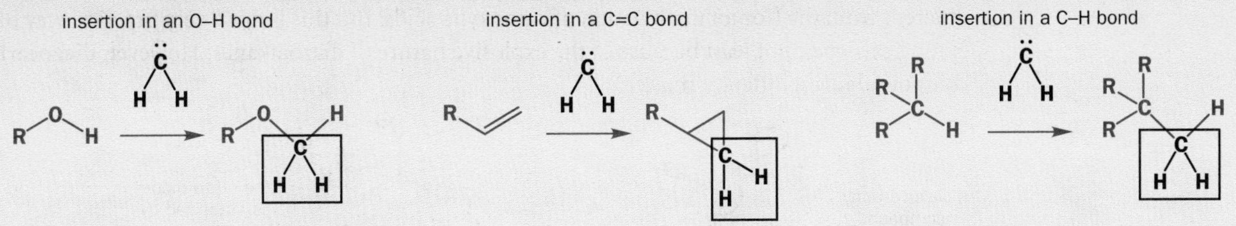
>
>
> insertion in an O–H bond   insertion in a C=C bond   insertion in a C–H bond
>
> We will discuss the mechanisms of these three important reactions shortly, but we have introduced them to you now because they demonstrate that the reactions of carbenes are dominated by *insertion reactions* (here, insertion into O–H, C=C, and C–H) driven by their extreme *electrophilicity*. A carbon atom with only six electrons will do almost anything to get another two!

How do we know that carbenes exist?

The best evidence for the existence of carbenes comes from some very few examples that are stable compounds. An X-ray crystal structure of the second example shows the bond angle at the carbene carbon to be 102°—we will come back to the significance of this later.

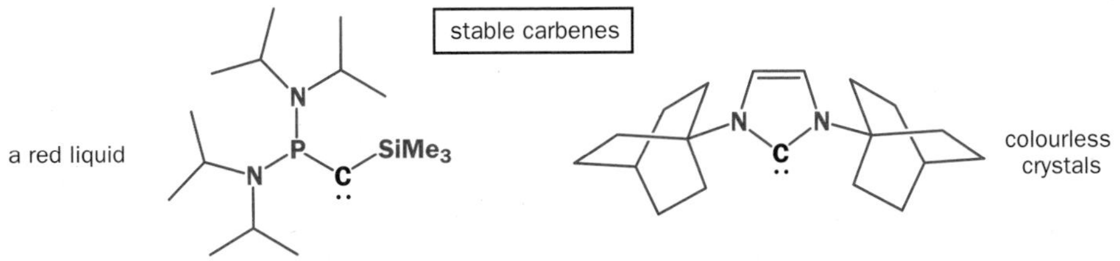

stable carbenes

a red liquid

colourless crystals

But these stable carbenes are very much the exception: most carbenes are too reactive to be observed directly. Electronic and, more importantly, steric effects make these two compounds so stable.

Even reactive carbenes can be observed, however, if they are formed by irradiating precursors (often diazo compounds like diazomethane, which we have just been discussing) trapped in frozen argon at very low temperatures (less than 77 K). IR and ESR spectroscopy can then be used to determine their structure.

How are carbenes formed?

Carbenes are usually formed from precursors by the loss of small, stable molecules. We will discuss some of the most important methods in turn, but you have already seen one in action: the loss of nitrogen from a diazo compound.

> ▶
> You may be somewhat surprised that the structure of carbenes can be investigated by ESR—after all, we explained in Chapter 39 that ESR observes unpaired electrons, and you might expect the six valence electrons of a carbene all to be paired. Indeed, in some carbenes they are, but in many they are not. This is an important point, and we will discuss it at length later in the chapter.

Naming azo compounds

Don't confuse *diazo* compounds with *azo* compounds. Diazomethane has twice as many nitrogen atoms per methyl group as azomethane.

diazomethane   azomethane   benzenediazonium salt   alkyl azide

You met *diazonium* salts in Chapter 23. Arene diazonium salts are stable compounds, but alkyl diazonium salts, which are formed by protonation of diazo compounds, are not. They decompose rapidly to carbocations—this was how the carboxylic acid got methylated at the beginning of the chapter. Other relatives of the azo and diazo compounds are alkyl azides. Alkyl azides have three nitrogen atoms and are usually stable but may explode on impact or heating.

Carbenes from diazo compounds

We showed you the formation of a carbene from diazomethane to illustrate how this reaction was different from the (ionic) methylation of carboxylic acids. But this is not a very practical way of generating carbenes, not least because of the explosive nature of diazoalkanes. However, **diazocarbonyl compounds** are a different matter.

a diazocarbonyl compound   the diazo dipole is stabilized by the carbonyl group

They are much more stable, because the electron-withdrawing carbonyl group stabilizes the diazo dipole, and are very useful sources of carbenes carrying a carbonyl substituent. There are two main ways of making diazocarbonyl compounds:

1   by reacting an acyl chloride with diazomethane

diazocarbonyl compound isolated in 100% yield

2   by reacting the parent carbonyl compound with tosyl azide, TsN_3, in the presence of base.

diazocarbonyl compound isolated in 95% yield

The reaction of diazomethane with acyl chlorides starts as a simple acylation to give a diazonium compound. If there is an excess of diazomethane, a second molecule acts as a base to remove a rather acidic proton between the carbonyl and the diazonium groups to give the diazocarbonyl compound.

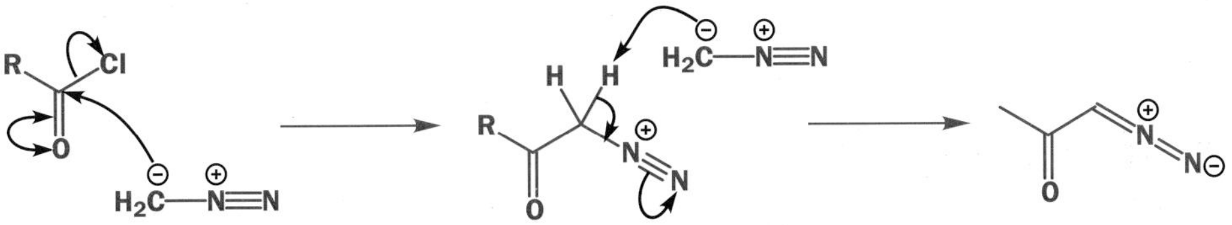

What happens to that second molecule of diazomethane? By collecting a proton it turns into the very reactive diazonium salt, which collects a chloride ion, and MeCl is given off as a gas. The second method uses tosyl azide, which is known as a **diazo transfer reagent**—it's just N_2 attached to a good leaving group.

Diazocarbonyl compounds can be decomposed to carbenes by heat or light. The formation of very stable gaseous nitrogen compensates for the formation of the unstable carbene.

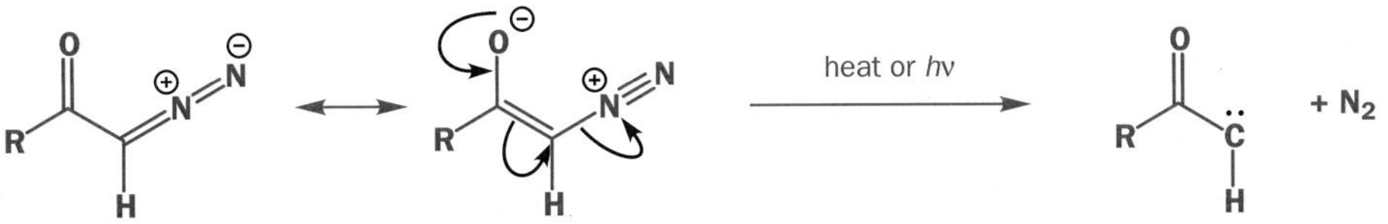

But it is much more common in modern chemistry to use a transition metal such as copper or rhodium, to promote formation of the carbene.

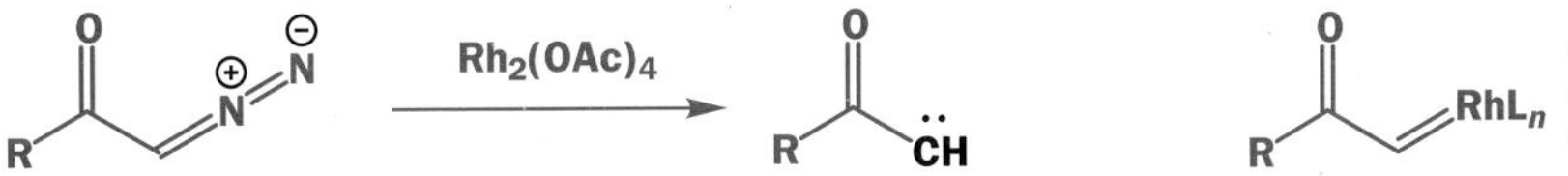

Carbenes formed in this way are, in fact, not true carbenes because it appears that they remain complexed with the metal used to form them. They are known as **carbenoids**, and their reactions are discussed later in the chapter.

> RhL_n means rhodium with an unspecified number of unspecified ligands. This notation is common in organometallic chemistry when the nature of the carbon–metal bonding is important, but the precise structure of the metal complex is not.

> While these rhodium and copper carbenoids are unstable, some transition metals such as tungsten and chromium form stable, isolable carbenoids, called **metallocarbenes** or **Fischer carbenes**.

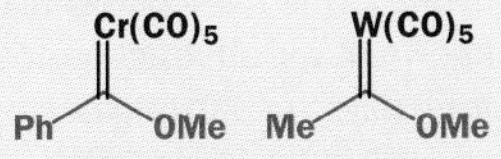

stable 'Fischer carbenes'

Carbenes from tosylhydrazones

Many more carbenes can be made safely from diazoalkanes if the diazoalkane is just an intermediate in the reaction and not the starting material. Good starting materials for these reactions are tosylhydrazones, which produce transient diazo compounds by base-catalysed elimination of toluenesulfinate. The diazo compound is not normally isolated, and decomposes to the carbene on heating.

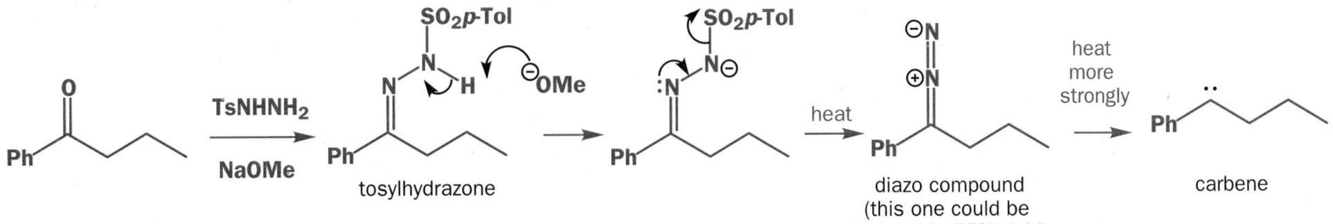

▶
This reaction is sometimes called the **Bamford–Stevens reaction**.

Notice that the leaving group from nitrogen is not the familiar tosylate (toluene-*p*-sulfo*nate* TsO⁻) but the less familiar toluene-*p*-sulf*inate* (Ts⁻).

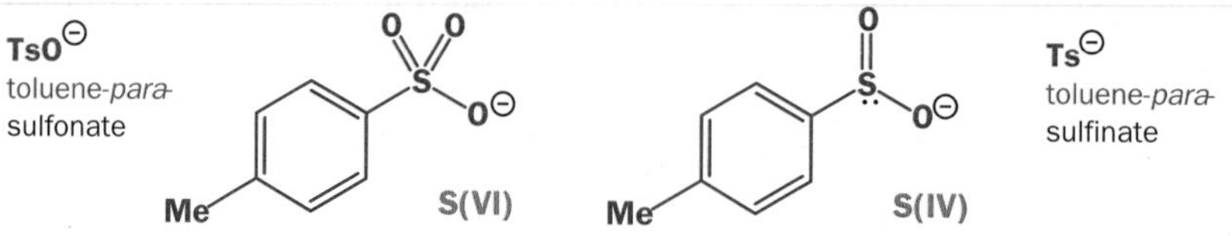

Carbenes are formed in a number of other similar reactions—for example, loss of carbon monoxide from ketenes or elimination of nitrogen from azirines—but these are rarely used as a way of deliberately making carbenes.

Carbene formation by α elimination

In Chapter 19 we discussed **β elimination** in detail, reactions in which a hydrogen atom is removed from the carbon atom β to the leaving group.

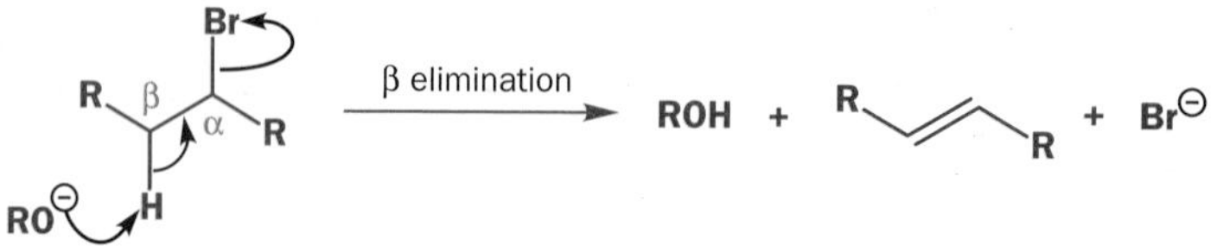

α Eliminations (eliminations in which both the proton and the leaving group are located on the same atom) are also possible—in fact, the reaction we've just been talking about (elimination of toluenesulfinate from tosylhydrazones) is an α elimination. α Eliminations follow a mechanism akin to an E1cB β elimination—a strong base removes an acidic proton adjacent to an electron withdrawing group to give a carbanion. Loss of a leaving group from the carbanion creates a carbene.

One of the best known α elimination reactions occurs when chloroform is treated with base. This is the most important way of making dichlorocarbene, $:CCl_2$, and other dihalocarbenes too, although it must be said that the widespread use of dichlorocarbene in chemistry is due mainly to the ease with which it can be made using this method!

base-catalysed α elimination of HCl from chloroform

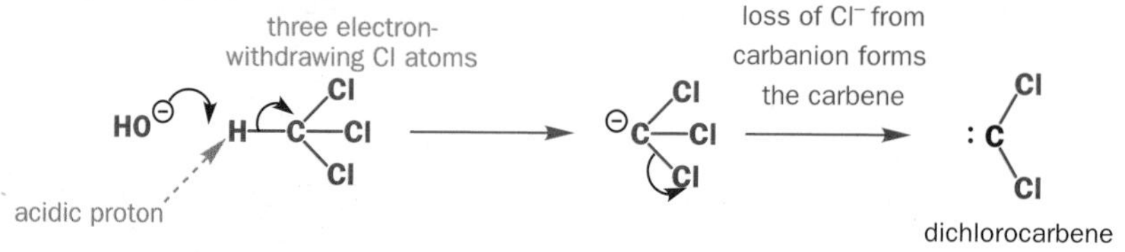

▶
The mixture *t*-BuLi/*t*-BuOK is known as **Schlosser's base**, and is one of the most powerful bases known. It will abstract protons from allylic or benzylic positions, and will even deprotonate benzene.

Hydroxide and alkoxide anions are strong enough bases to promote α elimination from chloroform, and from other trihalomethanes. Carbenes can be formed from dihaloalkanes by deprotonation with stronger bases such as LDA, and even from primary alkyl chlorides using the extremely powerful bases phenylsodium or *t*-BuLi/*t*-BuOK (weaker bases just cause β elimination).

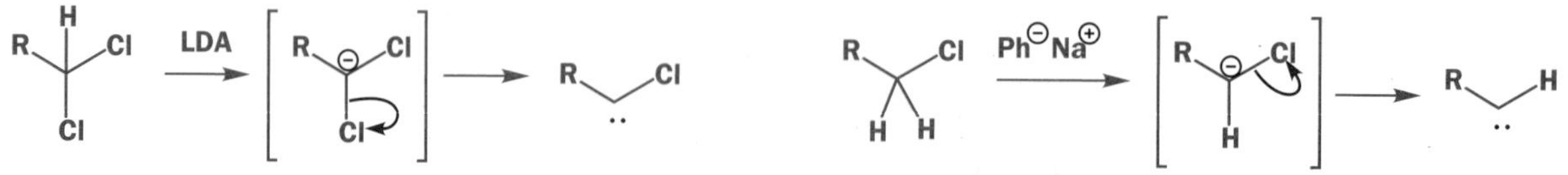

When geminal dibromoalkanes are treated with BuLi, a halogen–metal exchange reaction produces a lithium carbenoid, with a metal atom and a halogen attached to the same carbon atom. Lithium carbenoids are stable at very low temperatures—they can be observed by NMR, but they decompose to carbenes at about −100 °C.

lithium carbenoid   carbene

▶

It is unfortunate that the term carbenoid is used for two distinct classes of molecule—usually it refers to the transition-metal bound carbene formed by metal-catalysed decomposition of diazo compounds (see p. 1057)—and for this reason the carbenoids that we are discussing here are best referred to as 'lithium carbenoids', with the metal specified.

While lithium carbenoids have limited applicability in chemistry, an analogous zinc carbenoid, which can be formed by insertion of zinc into diiodomethane, is a reagent in one of the most widely used carbenoid reactions in chemistry—the Simmons–Smith reaction.

zinc carbenoid

■

The Simmons–Smith reaction, one of the best ways of making cyclopropanes, is discussed later in the chapter.

The essence of this type of carbenoid is that it should have a leaving group, such as a halogen, that can remove a pair of electrons and another, usually a metal, that can donate a pair of electrons. If the metal leaves first, a carbanion is created that can lose the halogen to make a carbene. They might also leave together. Both are α eliminations.

zinc carbenoid   carbene   zinc carbenoid

The problem with many of these reactions is that they require strong bases—either the organometallic compound itself is basic or a base must be used to create the carbanion. Carbenes are so unstable that they must be formed in the presence of the compound they are intended to react with, and this can be a problem if that compound is base-sensitive. For dichlorocarbene, a way round the problem is to make the carbanion by losing CO_2 instead of a metal or a proton. Decarboxylation of sodium trichloroacetate is ideal as it happens at about 80 °C in solution.

carbanion   carbene

● Summary: the most important ways of making carbenes

Carbenes are neutral species containing a carbon atom with only six valence electrons.

Type of carbene	Method of formation
	metal (rhodium or copper)-catalysed decomposition of diazocarbonyl compound
	thermal decomposition of diazo compound, often derived from tosylhydrazone
	α elimination of chloroform with base

This is a good point to remind you of other 'double losses' from molecules. Just as α elimination gives a carbene while β elimination gives an alkene, loss of nitrogen from a diazo compound gives a carbene but loss of nitrogen from an azo compound such as AIBN (*azo*bisisobutyronitrile) gives two radicals (Chapter 39).

diazo compound   carbene   azo compound (AIBN)   radical   radical

bond angle 102°

t-Bu

Carbenes can be divided into two types

We made two important observations earlier regarding the structure of carbenes that we will now return to and seek an explanation for: firstly, we said that the X-ray crystal structure of this stable, crystalline carbene shows that the bond angle at the carbene C is 102° and, secondly, we said that many carbenes can be observed by ESR—in other words, they have unpaired electrons.

Spectroscopic investigations of a number of carbenes of differing structures have shown that they fall broadly into two groups: (1) those (which you will learn to call 'triplets') that ESR spectroscopy demonstrates have unpaired electrons and whose bond angles are 130–150°; and (2) those (like the stable crystalline carbene above which you will learn to call a 'singlet') that have bond angles of 100–110° but cannot be observed by ESR. Many carbenes, like CH_2 itself, can be found in either style, though one may be more common.

Type 1: triplet carbenes	Type 2: singlet carbenes
bond angle 130–150°	bond angle 100–110°
observable by ESR	all electrons paired
:CH_2	:CCl_2
:CHPh	:CHCl
:CHR	:$C(OMe)_2$
:CPh_2	

All these observations can be accounted for by considering the electronic structure of a carbene. Carbenes have 2-coordinate carbon atoms: you might therefore expect them to have a linear (diagonal) structure—like that of an alkyne—with an sp hybridized carbon atom.

> This diagram is for illustration only and is *not* the structure of a carbene.

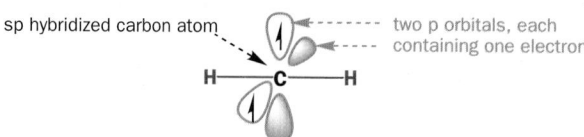

sp hybridized carbon atom — two p orbitals, each containing one electron

Such a linear carbene would have six electrons to distribute amongst two σ orbitals and two (higher-energy) p orbitals. The two electrons in the degenerate p orbitals would remain unpaired because of electron repulsion in the same way as in molecular oxygen •O–O•.

> We usually represent electrons, paired or unpaired, as dots. But here we are using another convention—one little arrow per electron. This allows us to define the electron's spin, up or down.

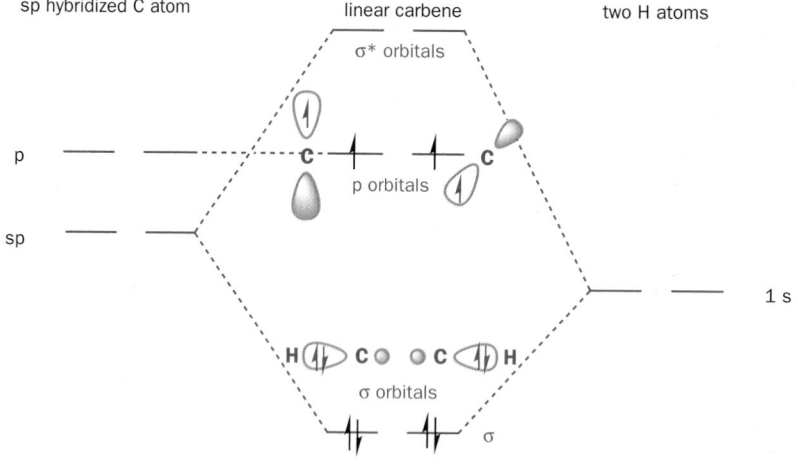

sp hybridized C atom — linear carbene — two H atoms

σ* orbitals

p

p orbitals

sp

σ orbitals

1 s

σ

Yet few carbenes are linear: most are bent, with bond angles between 100° and 150°, suggesting a trigonal (sp^2) hybridization state. An sp^2 hybridized carbene would have three (lower-energy) sp^2 orbitals and one (high-energy) p orbital in which to distribute its six electrons. There are two ways of doing this. Either all of the electrons can be paired, with each pair occupying one of the sp^2 orbitals, or two of the electrons can remain unpaired, with one electron in each of the p orbitals and one of the sp^2 orbitals.

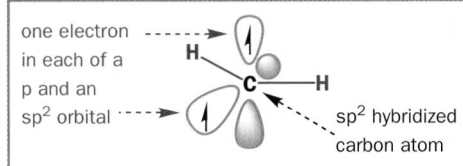

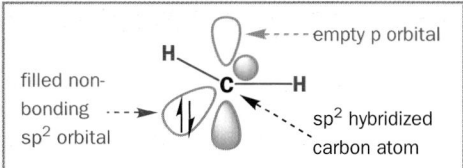

These two possibilities explain our two observed classes of carbene, and the two possible arrangements of electrons (spin states) are termed triplet and singlet. The orbitals are the same in both cases but in **triplet carbenes** we have one electron in each of two molecular orbitals and in **singlet carbenes** both electrons go into the sp^2 orbital.

electronic structure of a bent (sp^2) carbene with 2 unpaired electrons

electronic structure of a bent (sp^2) carbene with 2 paired electrons

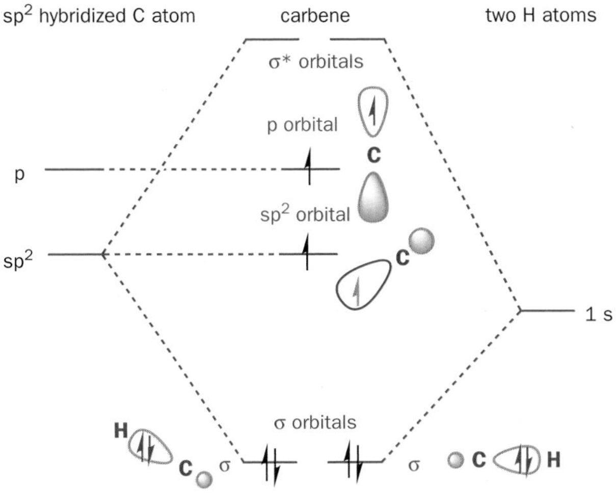

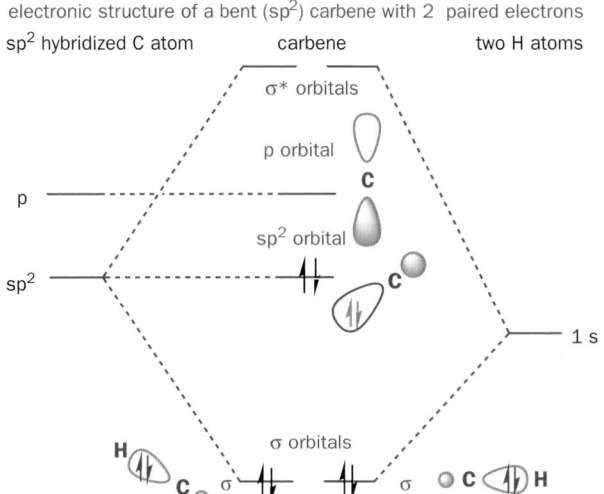

● Singlet and triplet carbenes

Triplet carbenes have two unpaired electrons, one in each of an sp^2 and a p orbital, while **singlet carbenes** have a pair of electrons in a nonbonding sp^2 orbital and have an empty p orbital.

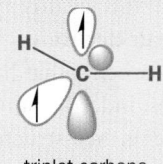

triplet carbene

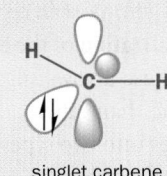

singlet carbene

The existence of the two spin states explains the different behaviour of triplet and singlet carbenes towards ESR spectroscopy; the orbital occupancy also explains the smaller bond angle in singlet carbenes, which have an electron-repelling lone pair in an sp^2 orbital.

Triplet carbenes

two unpaired electrons: observable by ESR

Singlet carbenes

no unpaired electrons: not observable by ESR

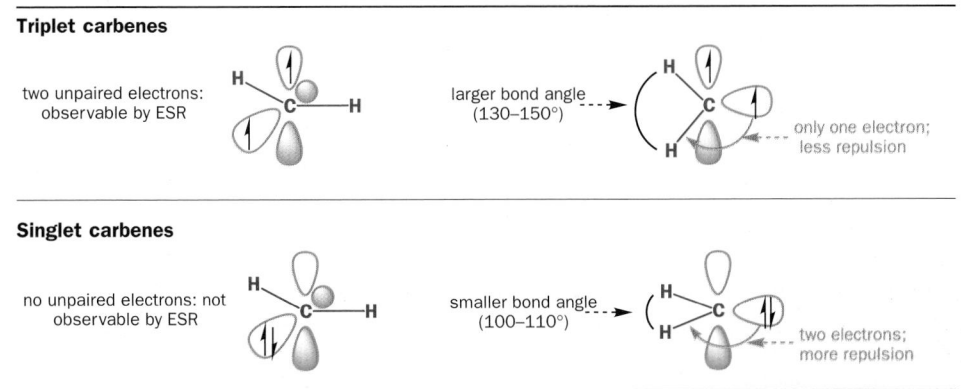

■

A manifestation of Hund's rule—see Chapter 4.

In the table on p. 1060 we saw that the substituents on the carbene affect which of the two classes (which we now call singlet and triplet) it falls into. Why? Most type of carbenes are more stable as triplets because the energy to be gained by bringing the electron in the p orbital down into the sp^2 orbital is insufficient to overcome the repulsion that exists between two electrons in a single orbital.

All carbenes have the potential to exist in either the singlet or the triplet state, so what we mean when we say that a carbene such as :CH$_2$ is a 'triplet carbene' is that the triplet state for this carbene is lower in energy than the singlet state, and vice versa for :CCl$_2$. For most triplet carbenes the singlet spin state that would arise by pairing up the two electrons lies only about 40 kJ mol^{-1} above the ground (triplet) state: in other words, 40 kJ mol^{-1} is required to pair up the two electrons. When a carbene is actually formed in a chemical reaction, it may not be formed in its most stable state, as we shall see.

Carbenes that have singlet ground states (such as :CCl$_2$) all have electron-rich substituents carrying lone pairs adjacent to the carbene centre. These lone pairs can interact with the p orbital of the carbene to produce a new, lower-energy orbital which the two electrons occupy. This stabilization of the lone pair provides the incentive that the electron in the p orbital needs to pair up in the sp^2 orbital.

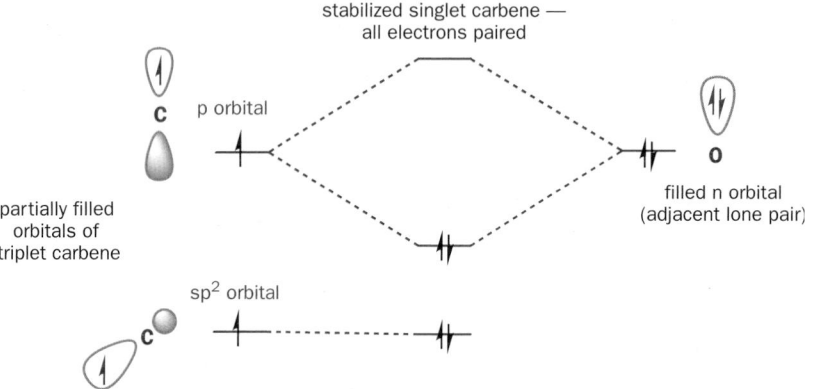

This molecular orbital formation moves electrons localized on oxygen into orbitals shared between carbon and oxygen. We can represent this in curly arrow terms as a delocalization of the lone pair electrons.

■

See p. 1068 for a demonstration that :CHCO$_2$Et is more electrophilic than :CCl$_2$.

As these arrows suggest, carbenes that have heavily electron-donating substituents are less electrophilic than other carbenes: indeed, diamino carbenes can be quite nucleophilic. The division of carbenes into two types explains their structure. It also helps to explain some of their reactions, especially those that have a stereochemical implication. We will spend the rest of this chapter discussing how carbenes react.

The structure of carbenes depends on how they are made

So far we have considered only the most stable possible structure, singlet or triplet, of a given carbene. In real life, a carbene will be formed in a chemical reaction and may well be formed as the less stable of the alternatives. If a reaction occurs by an ionic mechanism on a molecule with all electrons paired (as most molecules are!) then it must be formed as a singlet. Follow the α elimination mechanism, for example.

The starting material, a normal molecule of chloroform CHCl$_3$, has all paired electrons. The C–H σ bond breaks and the two paired electrons from it form the lone pair of the carbanion. The carbanion also has all paired electrons. The two paired electrons of one of the C–Cl bonds leaves the carbanion and the carbene is formed. It has two paired electrons in each of the two remaining C–Cl bonds and the lone pair, also paired. It is formed as a singlet. As it happens, the singlet version of CCl_2 is also the more stable. If the carbene were instead CH_2 and if it reacted rapidly, it might not have a chance to change into the more stable triplet state. And carbenes are very reactive. In explaining their reactions in the next section we shall need to consider:

- how the carbene was formed
- how rapidly it reacts
- whether it can change into the other state (singlet or triplet)

How do carbenes react?

Carbenes are desperate to find another pair of electrons with which to complete their valence shell of electrons. In this respect they are like carbocations. Like carbocations, they are electrophilic but, unlike carbocations, they are uncharged. This has consequences for the type of nucleophiles carbenes choose to react with. Carbocations attack nucleophiles with high charge density—those carrying a negative or partial negative charge (think of the type of nucleophiles that will take part in S_N1 or Friedel–Crafts reactions). Carbenes, on the other hand, attack compounds we'd normally never consider as nucleophiles—even simple alkanes—by taking electrons from their HOMO. Of course, a carbocation will usually react with the HOMO of a molecule, but it will be much more selective about which HOMOs will do—usually these have to be lone pairs or electron-rich alkenes. For carbenes, any HOMO will do—a lone pair, a C=C double bond (electron-rich or -poor), or even a C–H bond.

> In this respect, a carbene is like an electrophilic radical—very reactive and very soft.

As you will see (and as we generalized at the beginning of the chapter), many of these reactions can be considered as **insertion reactions**—overall the carbene appears to have found a bond and inserted itself in the middle of it. It's important to remember that the term 'insertion reaction' describes the outcome of the reaction, though it isn't always an accurate description of the reaction's mechanism.

Carbenes react with alkenes to give cyclopropanes

This reaction is the most important way of making cyclopropanes, and is probably the most important reaction of carbenes.

The mechanism of this type of reaction depends on whether the carbene is a singlet or a triplet, and the outcome of the reaction can provide our first chemical test of the conclusions we came to in the previous section. Singlet carbenes, like this one here (remember that electron-rich substituents stabilize the singlet spin state), can add to alkenes in an entirely concerted manner: the curly arrows for the process can be written to show this.

Because the process is concerted, we expect that the geometry of the alkene should be preserved in the product—the reaction ought to be *stereospecific*. The two examples below show that this is indeed the case. It is more impressive that the Z-alkene gives the *cis* cyclopropane as this is less stable than the *trans* cyclopropane and would change if it could.

The alkene insertion reaction is stereospecific only for singlet carbenes. For triplet carbenes, the reaction is nonstereospecific. Though carbenes formed thermally from diazoalkenes must initially be singlets, photochemistry is one way to provide the energy needed for their transformation to the more stable triplet.

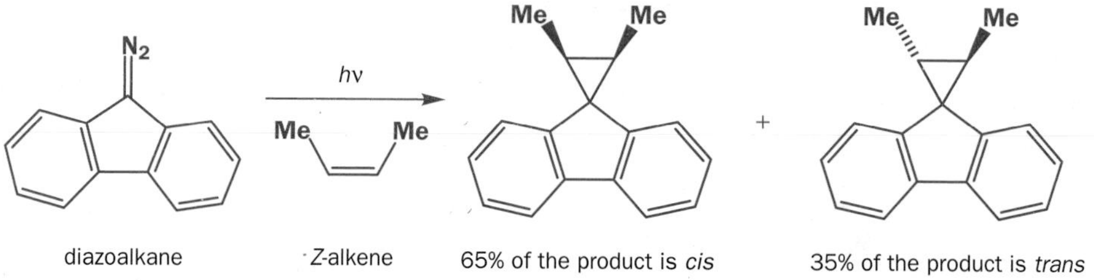

| diazoalkane | Z-alkene | 65% of the product is *cis* | 35% of the product is *trans* |

The mechanism of this nonspecific reaction must be different. In fact, a concerted reaction is impossible for triplet carbenes because of the spins of the electrons involved. After the carbene adds to the alkene in a radical reaction, the diradical (triplet) intermediate must wait until one of the spins inverts so that the second C–C bond can be formed with paired electrons. This intermediate also lives long enough for C–C bond rotation and loss of stereochemistry.

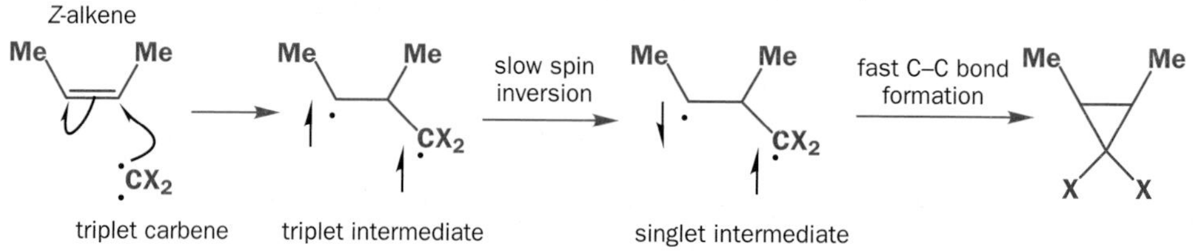

| triplet carbene | triplet intermediate | singlet intermediate | |

A cyclopropane has three σ bonds—in other words, six electrons, all spin-paired (three up, three down). One of these was the σ bond in the starting material; the other two electron pairs come from the π bond and from the carbene. The electrons in the π bond must have been paired, and thus they can form one of the new σ bonds. A singlet carbene (whose electrons are also paired) can then provide the second electron pair.

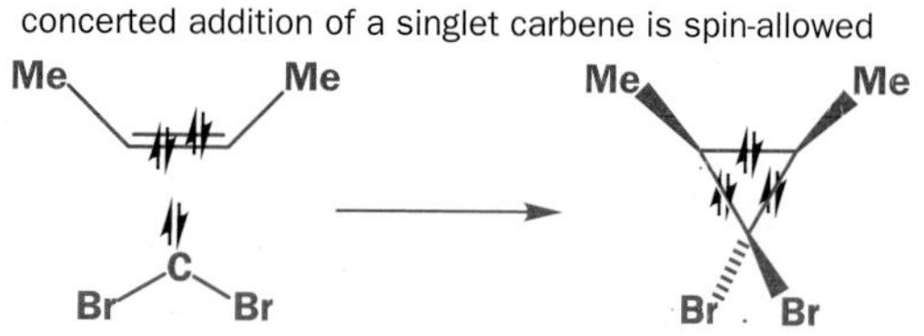

But a triplet carbene cannot, because its electrons are not paired. The second bond can only form once one of the two electrons has flipped its spin. Spin-flipping, which can only occur through collision with another molecule (of solvent, say), is relatively slow on the time-scale of molecular rotations and, by the time the electrons are in a fit state to pair up, the stereochemistry of the starting material has been scrambled by free rotation in the intermediate.

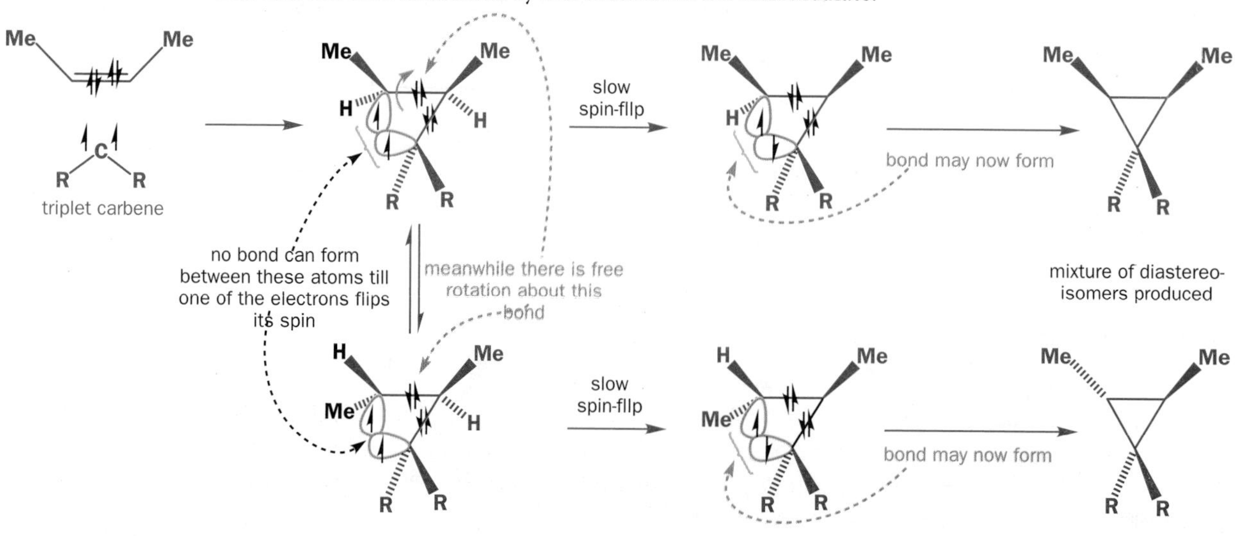

concerted addition of a singlet carbene is spin-allowed

A reminder. The same constraints arising from the need for conservation of electron spin apply to the formation as well as to the reaction of carbenes. When a carbene forms by α elimination, say, from a molecule with all electrons paired, it must be formed as the singlet, whether or not the triplet state is lower in energy. Only later may the carbene undergo spin-flipping to the triplet state. Since most carbene reactions are very rapid, this means that carbenes that are known to have triplet ground states may, in fact, react in their first-formed singlet state because they don't have time to spin-flip to the triplet. This is true for :CH$_2$ produced from CH$_2$N$_2$, which adds stereospecifically to double bonds because it is formed as a singlet and because the singlet state is more reactive than the triplet.

Some evidence for triplet carbenes in cyclopropane formation

If the reaction is diluted with a large amount of an inert solvent such as C$_3$F$_8$ (perfluoropropane) then :CH$_2$ undergoes more collisions before it reacts and so the chances of spin-flipping of singlet :CH$_2$ to triplet :CH$_2$ is increased. Addition to alkenes is then less stereospecific.

Stereospecificity (or lack of it) in the addition of a carbene to an alkene *can* be a good test of whether the carbene reacts as a singlet or triplet: lack of stereospecificity in a carbene addition almost certainly indicates that a triplet carbene is involved, but the fact that an addition *is* stereospecific doesn't mean that the carbene *must* be a singlet. In some cases, bond rotation may be quite slow, and spin-flipping rapid, leading to stereospecific addition. Notice that in this example the less stable *cis* (*Z*-) alkene was used: the reaction will give *trans*-cyclopropane if it can.

> Cycloadditions in which one of the components is a single atom (in other words, [1 + *n*] cycloadditions) are sometimes called **cheletropic reactions**.

The addition of a triplet carbene to an alkene can be considered to be rather like a radical addition to a double bond. The concerted addition of a singlet carbene, on the other hand, is a pericyclic reaction, and from Chapter 35 you should be able to classify it as a [1 + 2] cycloaddition.

addition to alkenes of triplet carbenes is a radical reaction

addition of singlet carbenes is a [1+2] cycloaddition

As a cycloaddition, singlet carbene addition to an alkene must obey the rules of orbital symmetry discussed in Chapters 35 and 36. We might consider the empty p orbital of the carbene (LUMO) interacting with the π bond (HOMO) of the alkene or the lone pair of the carbene in its filled sp^2 orbital (HOMO) interacting with the π* antibonding orbital of the alkene (LUMO).

You can immediately see that there is a problem when we try to interact these orbitals constructively to build two new bonds—direct approach of the carbene to the alkene is impossible because there is always an antibonding interaction. Two new bonds can be formed, however, if the carbene approaches the alkene in a 'sideways-on' manner.

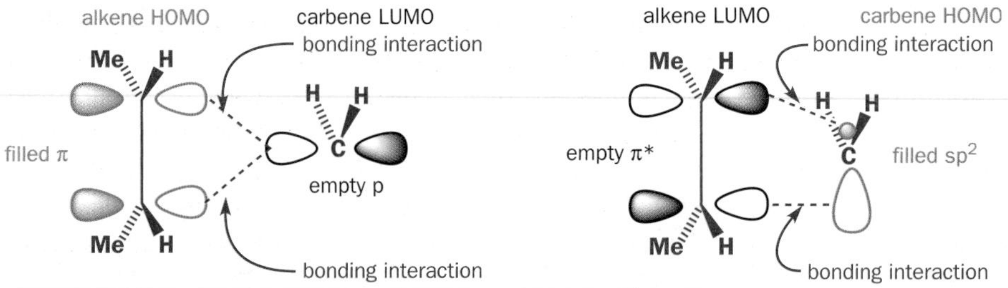

The cyclopropane product must, of course, have a more or less tetrahedral arrangement about the carbon atom that was the carbene so that, even if the carbene approaches in a sideways-on manner, it must then swing round through 90° as the bonds form.

'docking' of the carbene on to the alkene

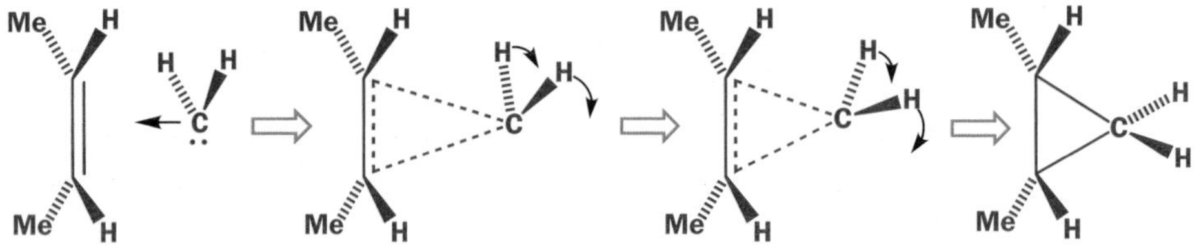

Making cyclopropanes

Many natural products and biologically active compounds contain cyclopropane rings: we shall feature just a few. First, a most important natural insecticide, a pyrethrin from the East African pyrethrum daisy, and its synthetic analogue decamethrin, now the most important insecticide in agriculture (see Chapter 1). Very low doses of this highly active and nonpersistent insecticide are needed.

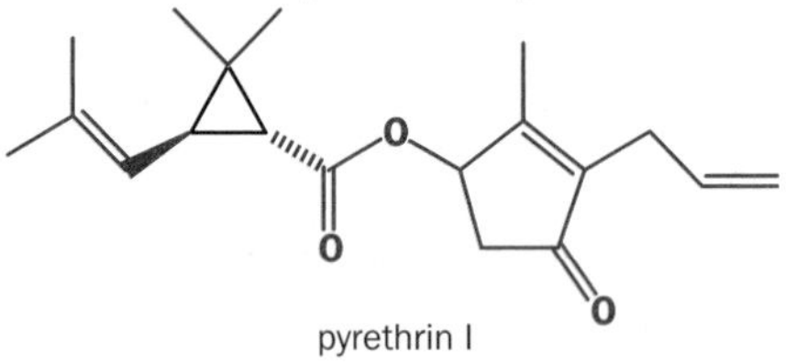

pyrethrin I

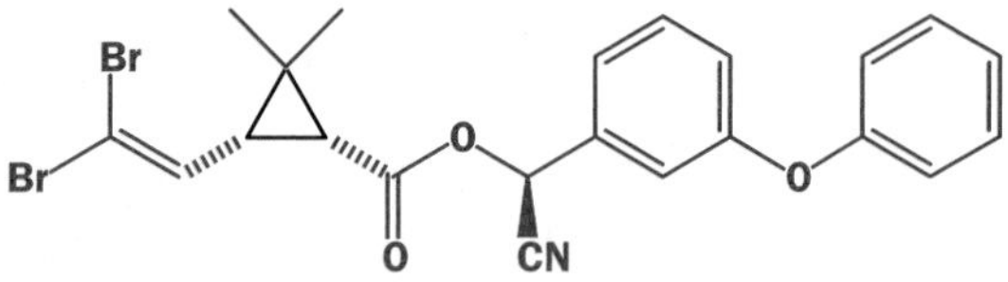

decamethrin

▶ They told John Betjeman: 'Breathe in the ozone, John. It's iodine.' *Summoned by bells*, Murray, 1960, p. 39.

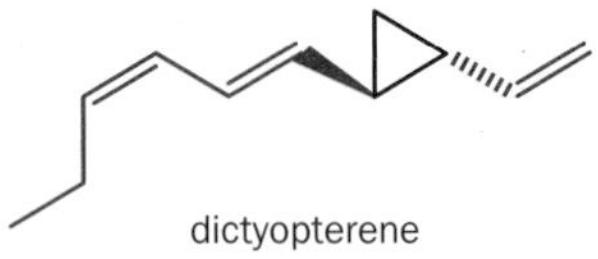

dictyopterene

Ever heard of the 'ozone' or 'iodine' smell of the sea? Well, the smell of the sea is characteristic but has nothing to do with O_3 or I_2. It's more likely to be a dictyopterene, a family of volatile cyclopropanes used by female brown algae to attract male gametes. There is an example in the margin.

Now for two natural but highly unusual amino acids. Hypoglycidin is a blood sugar level lowering agent from the unripe fruit of the ackee tree; the causative agent of Jamaican vomiting sickness. Don't eat the green ackee. Nature makes not only strained cyclopropanes but this even more strained methylene cyclopropane with an sp^2 atom in the ring. The second and simpler amino acid is found in apples, pears, and grapefruit and encourages fruit ripening by degradation to ethylene.

Our last and most extraordinary example is an antifungal antibiotic first synthesized in 1996 and containing no less than five cyclopropanes. It has the prosaic name FR-900848 but is known unofficially in the chemical world as 'jawsamycin'.

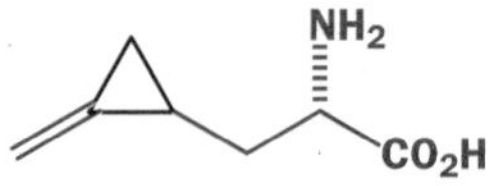

hypoglycin

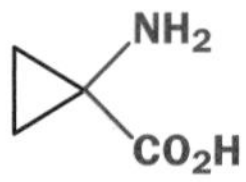

FR-900848 or 'jawsamycin'

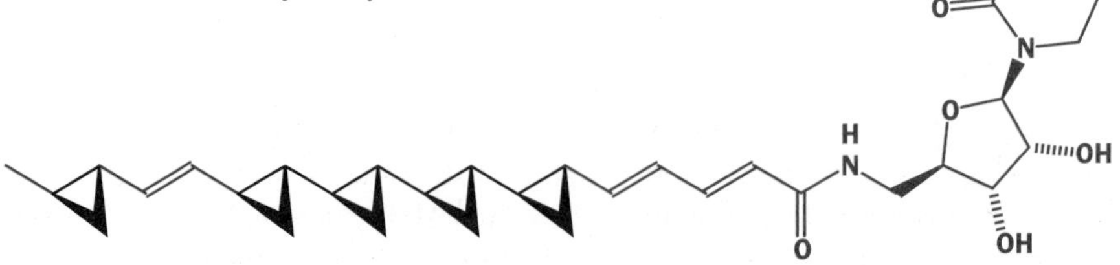

Because of these and other useful molecules containing three-membered rings, methods to make them are important as well as interesting. Most chemical syntheses of compounds containing cyclopropyl groups make use of the addition of a carbene, or carbene equivalent, to an alkene. What do we mean by **carbene equivalent**? Usually, this is a molecule that has the potential to form a carbene, though it may not actually react via a carbene intermediate. One such example is a zinc carbenoid formed when diiodomethane is reacted with zinc metal: it reacts with alkenes just as a carbene would—it undergoes addition to the π bond and produces a cyclopropane.

the Simmons–Smith reagent   the Simmons–Smith reaction

The reaction is known as the **Simmons–Smith reaction**, after the two chemists at the DuPont chemical factory who discovered it in 1958. Even after several decades, it is the most important way of making cyclopropane compounds, though nowadays a variant that uses more easily handled starting materials is often used. Diethyl zinc replaces the Zn/Cu couple of the traditional Simmons–Smith reaction. In this example, a double cyclopropanation on a C_2 symmetric diene derived from tartaric acid gives very good stereoselectivity for reasons we will soon discuss.

diethyl tartrate

The reaction does not involve a free carbene: the zinc is still associated with the carbon atom at the time of the reaction, and the reacting species is a probably a complex of zinc that we can represent as an equilibrium between two zinc carbenoids.

The mechanism of the Simmons–Smith reaction appears to be a carbene transfer from the metal to the alkene without any free carbene being released. It may look something like this.

(X = I or CH₂I)

Some of the evidence for this comes from a reaction that not only throws light on to the mechanism of Simmons–Smith cyclopropanations, but makes them of even greater value in synthesis. When an allylic alcohol is cyclopropanated, the new methylene group adds stereoselectively to the same face of the double bond as the alcohol group.

63% yield
>99% this diastereoisomer

Allylic alcohols also cyclopropanate over 100 times faster than their unfunctionalized alkene equivalents. Coordination between the zinc atom and the hydroxyl group in the transition state explains both the stereoselectivity and the rate increase. Unfortunately, while the Simmons–Smith

■ You could compare this reaction with reduction by sodium borohydride (Chapter 6). Hydride is transferred from a boron atom to a carbonyl group but no free hydride is formed.

■ You might notice the similarity to the epoxidation of allylic alcohols with *m*-CPBA mentioned in Chapter 33.

▶ On the subject of stereochemistry, note that the Simmons–Smith zinc carbenoid behaves like a singlet carbene—its additions to alkenes are stereo*specific* (the product cyclopropane retains the geometry of the alkene) as well as stereo*selective* (the carbenoid adds to the same face as the hydroxyl group).

reaction works well when a methylene (CH_2) group is being transferred, it is less good with substituted methylenes (RCH: or R_2C:).

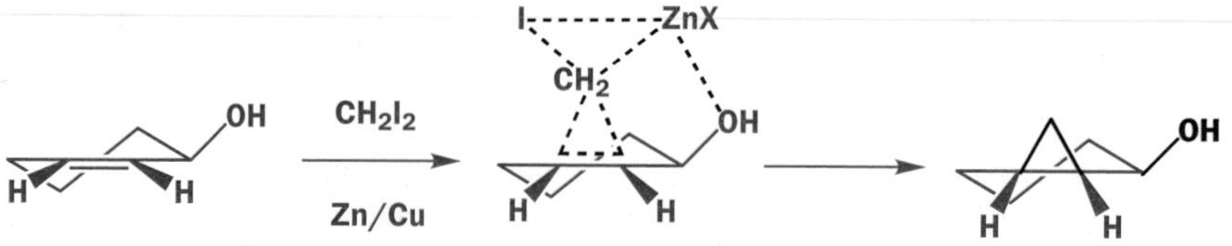

When Ireland wanted to introduce a cyclopropane ring stereoselectively into a pentacyclic system containing an enone, he first reduced the ketone to an alcohol (DIBAL gave only the equatorial alcohol) that controlled the stereochemistry of the Simmons–Smith reaction. Oxidation with Cr(VI) put back the ketone.

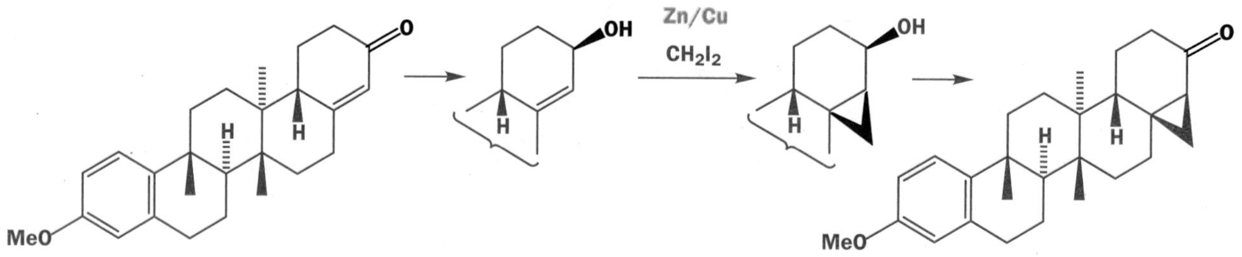

The carbene derived by metal-catalysed decomposition of ethyl diazoacetate attacks alkenes to introduce a two-carbon fragment into a cyclopropane—an industrial synthesis of ethyl chrysanthemate, a precursor to the pyrethrin insecticides (see p. 10), uses this reaction. The diene in the starting material is more nucleophilic (higher-energy HOMO; see Chapter 20) than the single alkene in the product, so the reaction can be stopped after one carbene addition.

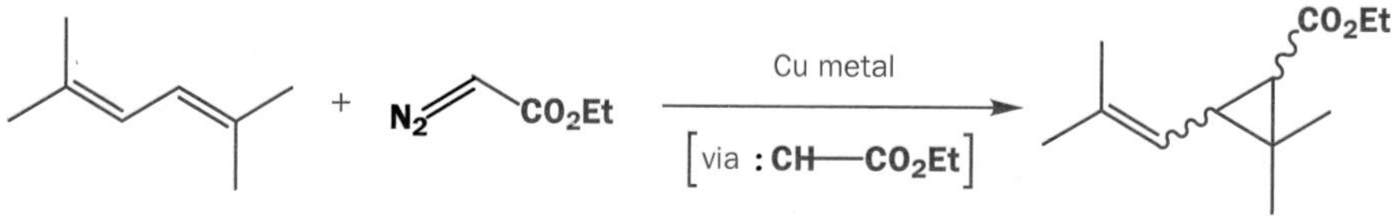

ethyl chrysanthemate

The intramolecular version of this reaction is more reliable, and has often been used to make compounds containing multiply substituted cyclopropanes. Corey made use of it in a synthesis of sirenin, the sperm-attractant of a female water mould.

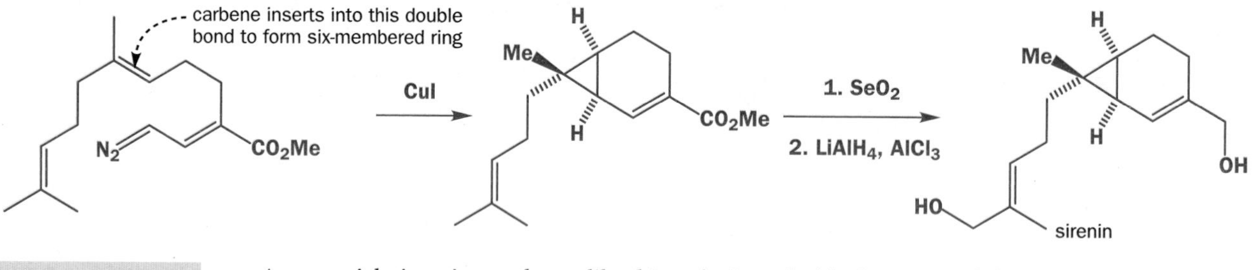

The selenium dioxide oxidation is discussed in Chapter 46.

sirenin

You met reactions like this in Chapter 36.

As you might imagine, carbenes like this, substituted with electron-withdrawing carbonyl groups, are even more powerful electrophiles than carbenes like $:CCl_2$, and will even add to the double bonds of benzene. The product is not stable, but immediately undergoes electrocyclic ring opening.

super-electrophilic carbonyl-substituted carbene

initial product undergoes electrocyclic ring opening

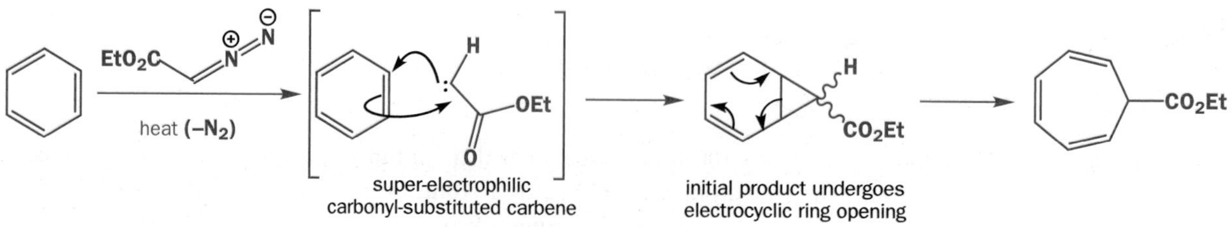

Dichlorocarbene :CCl$_2$ will not add to benzene, but does attack the electron-rich aromatic ring of phenol: the product is not a cyclopropane, but an aldehyde.

the Reimer–Tiemann reaction

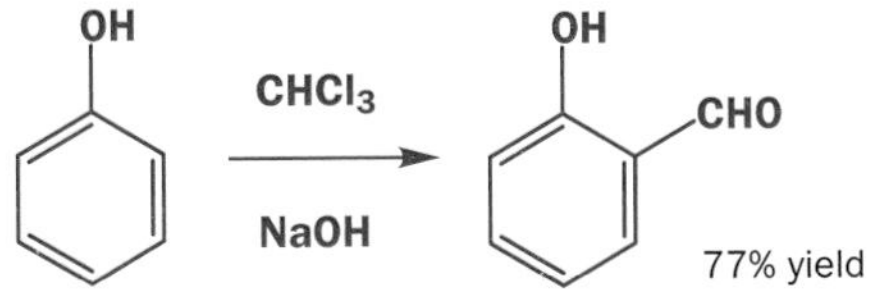

No reaction with benzene

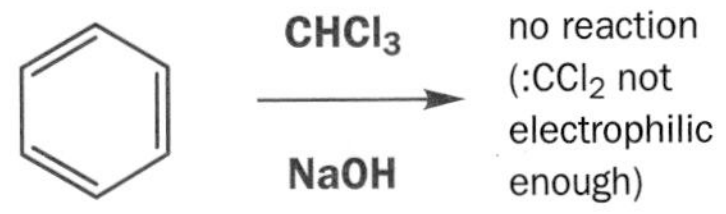

The **Reimer–Tiemann reaction** used to be an important way of making *ortho*-substituted phenols, but the yields are often poor, and modern industry is wary of using large quantities of chlorinated solvents. On a small, laboratory scale it has largely been superseded by ortholithiation (Chapter 9) and by modern methods outside the scope of this book. The mechanism probably goes something like this.

mechanism of the Reimer–Tiemann reaction

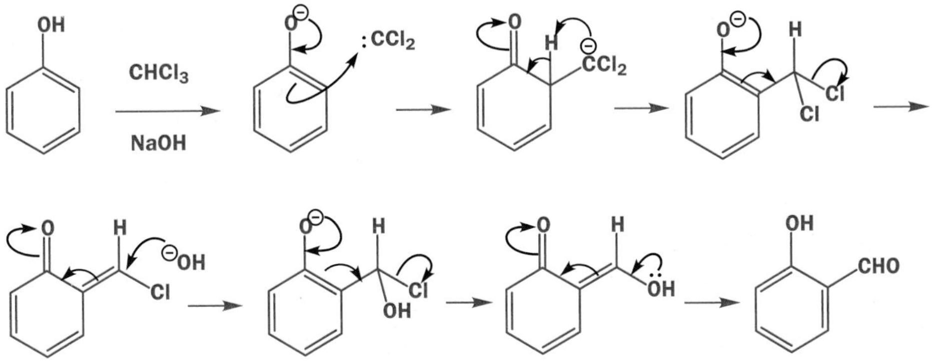

Comparison of '-enoid' reagents

Before we leave this section on cyclopropanes, we want you to take a step back from simply thinking about carbenes, and consider the types of reagents that form three-membered rings generally. They all have something in common, which we could call '-enoid' character. Cyclopropanes form when a carbene (which, in the singlet state, has an empty, electrophilic p orbital and a full, nominally nucleophilic sp^2 orbital) attacks alkenes. The Simmons–Smith carbenoid is not a carbene, but nonetheless has a carbon atom with joint nucleophilic (alkyl zinc) and electrophilic (alkyl iodide) character. When

you think about it, the same is true for peracid epoxidation, which forms the oxygen analogue of a cyclopropane by attacking an alkene with an oxygen atom bearing both a lone pair (nucleophilic) and a carboxylate leaving group (electrophilic). It's an 'oxenoid'. In Chapter 46 you will meet more reagents that form cyclopropanes and epoxides by transferring CH$_2$—sulfonium ylids. These yet again have a schizophrenic carbon atom—carrying a negative charge and a leaving group—and, when you meet them, you can consider them to be particularly stable carbenoids.

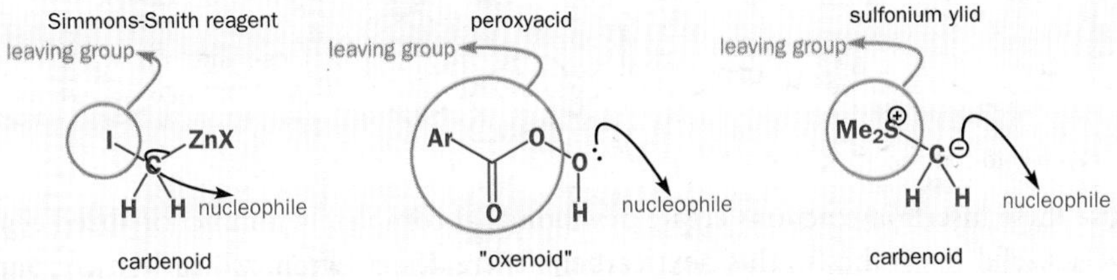

Insertion into C–H bonds

We said that the formation of cyclopropanes by addition of substituted carbenes to alkenes was rare—in fact, alkyl-substituted carbenes undergo very few intermolecular reactions at all because they decompose very rapidly. When primary alkyl halides are treated with base, alkenes are formed by elimination. Having read Chapter 19, you should expect the mechanism of this elimination to be E2 and, if you started with a deuterated compound like this, the alkene product would be labelled with two deuterium atoms at its terminus.

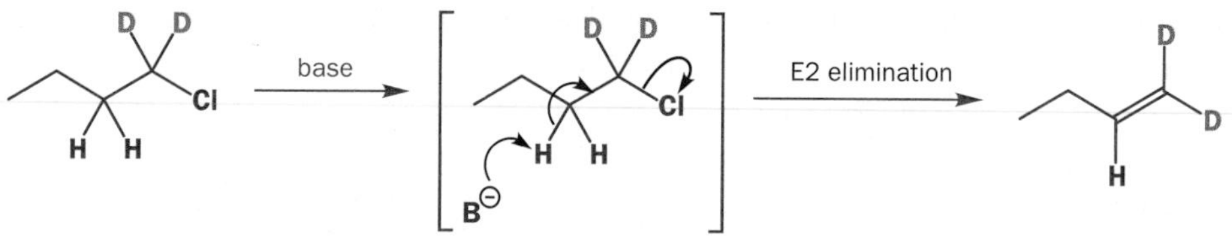

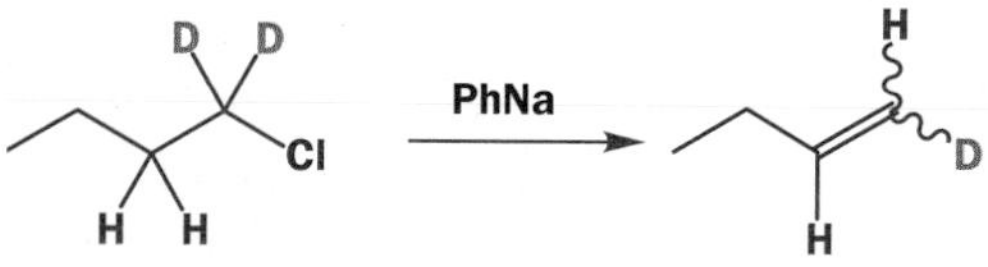

This is indeed what happens if the base is sodium methoxide (pK_a 16). If, however, it is phenylsodium (pK_a about 50), only 6% of the product is labelled in this way while 94% of the product has only one deuterium atom.

A hydrogen atom has 'migrated' from the 2-position to the 1-position. The overall mechanism of the elimination with very strong bases like phenylsodium is believed to be: (1) formation of a carbene by α elimination and then (2) 1,2-migration of a hydrogen atom on to the carbene centre. Carbenes with β hydrogens undergo extremely rapid 1,2-migration of hydrogen to the carbene centre, giving alkenes.

> Migrations were covered in detail in Chapter 37. You will meet examples there of migrations on to electrophilic *carbocationic* centres, but the reactions are in essence very similar to these migrations to carbenes.

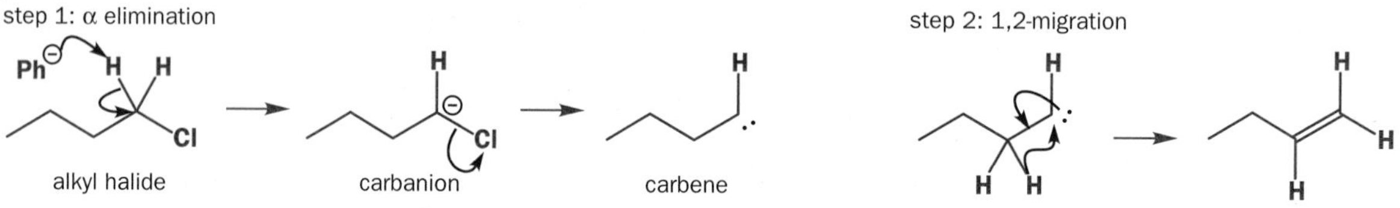

The reason for the rapid migration is that the electrophilic carbene has found a nearby source of electrons—the HOMO of the C–H bond—and it has grabbed the electrons for itself, 'inserting' into the C–H bond.

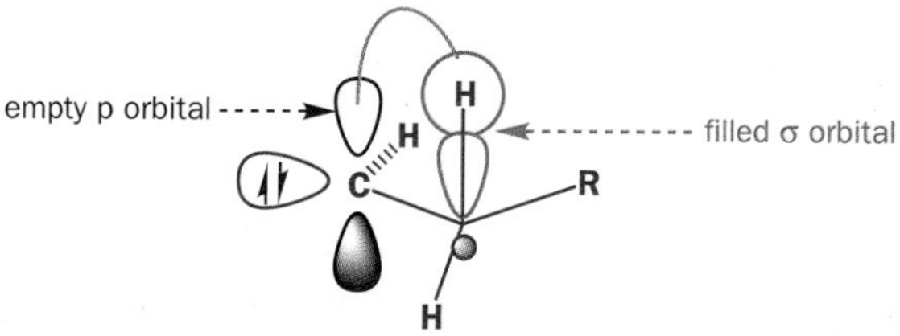

This type of reaction is better demonstrated by two examples in which the 'insertion reaction' is a bit more obvious: when there are no β hydrogens, the carbene inserts into C–H bonds a little further away in the same molecule or even in the solvent (cyclohexane in the second example). In the first case, the carbene is formed by α elimination and, in the second case, by photolysis of a diazoketone.

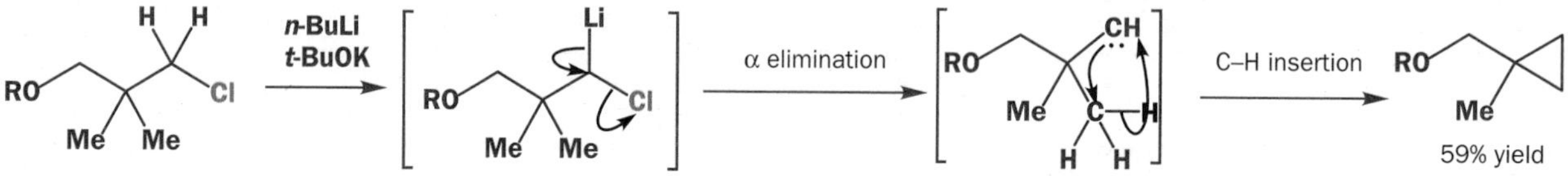

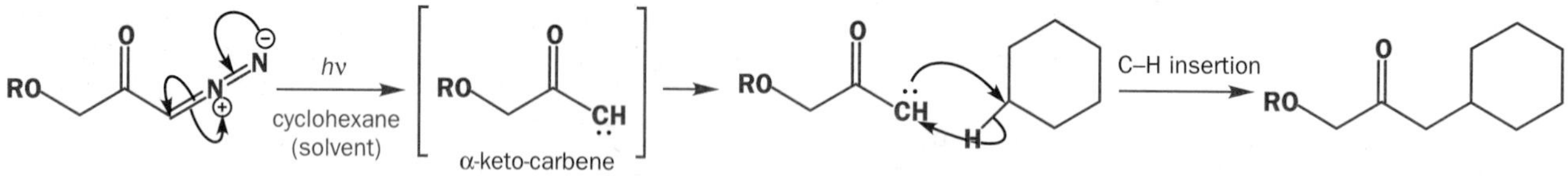

Because these insertion reactions create new bonds at completely unfunctionalized centres, they can be very useful in synthesis. This next carbene is created between two carbonyl groups from a diazocompound with rhodium catalysis and selectively inserts into a C–H bond five atoms away to form a substituted cyclopentanone.

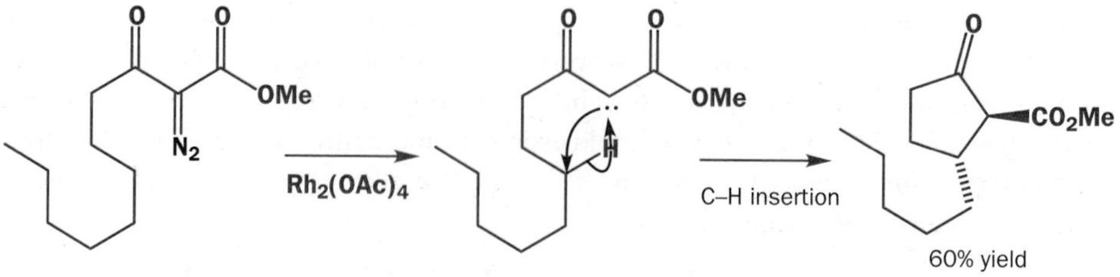

Pentalenolactone synthesis using carbenes

Pentalenolactone is the name given to an antibiotic extracted from *Streptomyces* fungi with an interesting tricyclic structure.

Two groups of chemists, within one year of each other, published syntheses of this compound using rhodium-promoted carbene insertions into C–H bonds. Cane's

insertion reaction (route 1) proceeds stereospecifically with *retention* of stereochemistry. This is excellent evidence for a concerted singlet carbene reaction.

pentalenolactone

route 1: Cane's synthesis of pentalenolactone

route 2: Taber's synthesis of pentalenolactone

In these C–H insertion reactions, the similarity with cyclopropane formation by intramolecular cycloadditions to alkenes is clear, and the mechanisms mirror one another quite closely. As with the cyclopropanation reactions, the path of the reaction differs according to whether the carbene is a singlet or triplet. Singlet carbenes can insert in a concerted manner, with the orbitals overlapping constructively provided the carbene approaches side-on.

orbital interactions during the insertion of a singlet carbene into a C–H bond

This mechanism implies that, if the C–H bond is at a stereogenic centre, the stereochemistry at that centre will be retained through the reaction, as in Cane's synthesis of pentalenolactone. A nice example of this result is the ingenious synthesis of α-cuparenone using a stereospecific carbene insertion.

α-cuparenone

Rearrangement reactions

We talked just at the beginning of this section about migration reactions of hydrogen on to carbenes to give alkenes, and said that these reactions can be viewed as insertion reactions of carbenes into adjacent C–H bonds. Carbenes with no β hydrogens often insert into other C–H bonds in the molecule. However, carbenes with no β-hydrogen atoms can also undergo rearrangement reactions with alkyl or aryl groups migrating.

> In principle, triplet carbene insertions should follow a two-step radical pathway analogous to their insertion into alkenes. However, very few triplet carbene insertions into C–H bonds have been observed, and the stereochemical consequence of the two-step mechanism (which should result in mixtures of stereoisomers on insertion into a C–H bond at a stereogenic centre) has never been verified.

> The migration of alkyl groups to carbene centres has much in common with the migration of alkyl groups to cationic centres discussed in Chapter 37—after all, both carbenes and carbocations are electron-deficient species with a carbon atom carrying only six electrons in its outer shell.

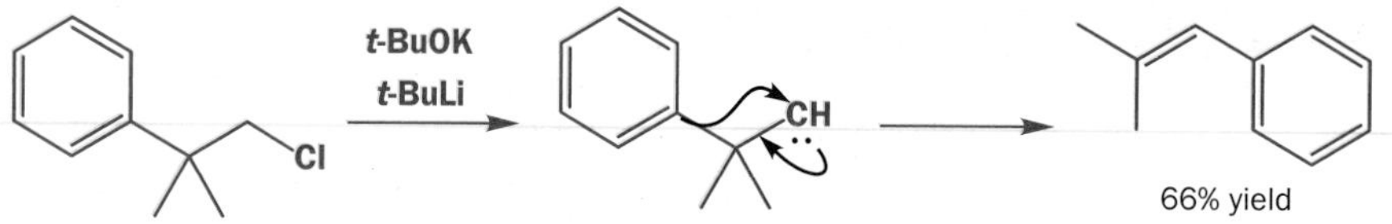

66% yield

You met ketenes in Chapter 35.

The most common example of this type of migration is that in which the carbene is adjacent to a carbonyl group. The initial product of what is known as the **Wolff rearrangement** is a ketene, which cannot be isolated but is hydrolysed to the ester in the work-up. Wolff rearrangement is a typical reaction of diazoketones on heating, though these species do also undergo intramolecular C–H insertion reactions.

the Wolff rearrangement

α-keto-carbene   ketene

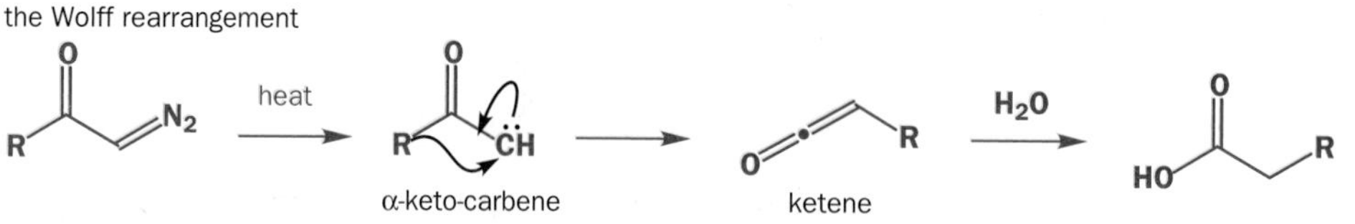

One important application of this reaction is the chain extension of acyl chlorides to their homologous esters, known as the **Arndt–Eistert reaction**. Notice that the starting material for the Wolff rearrangement is easily made from RCO_2H by reaction of the acyl chloride with diazomethane; the product is RCH_2CO_2H—the carboxylic acid with one more carbon atom in the chain. A CH_2 group, marked in black, comes from diazomethane and is inserted into the C–C bond between R and the carbonyl group.

the Arndt–Eistert homologation

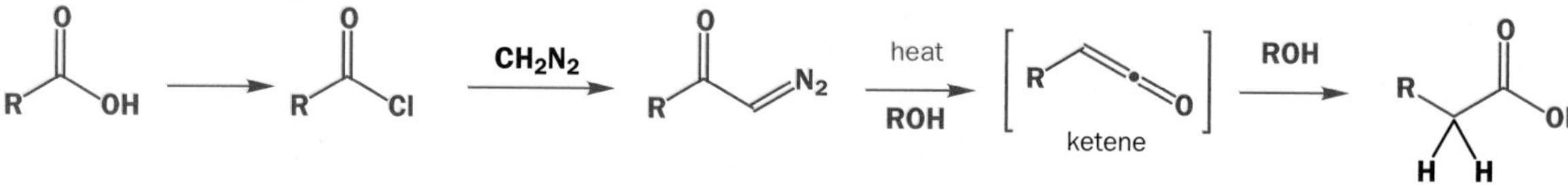

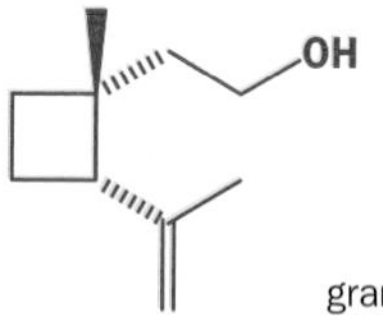

ketene

A synthesis of grandisol using Arndt–Eistert chain extension

The boll weevil is a serious pest of cotton bushes, and it produces a sex pheromone known as grandisol.

Chemists soon showed that it was an easy matter to synthesize a related ester by a conjugate addition of an organocopper derivative (Chapter 10) and then the alkylation of an ester enolate (Chapter 26). The enolate reacts with MeI on the face opposite the propenyl side chain—a good example of stereochemical control with cyclic compounds (Chapter 33).

grandisol

▶
You have already met one synthesis of grandisol—in Chapter 25.

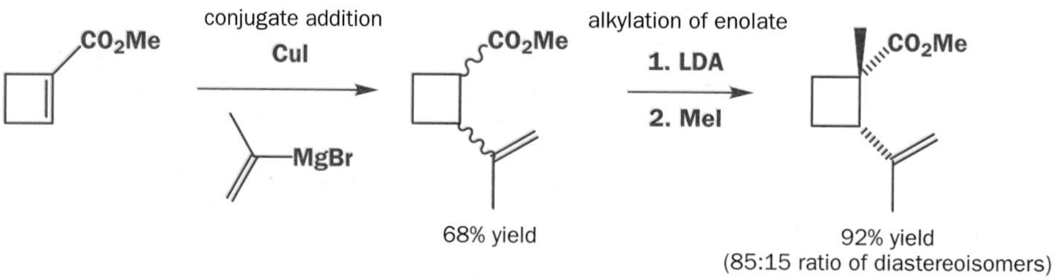

conjugate addition   alkylation of enolate

68% yield   92% yield
(85:15 ratio of diastereoisomers)

This ester is one carbon atom short of the full side chain of grandisol, so an Arndt–Eistert reaction was used to lengthen the chain by one atom. First, the ester was converted into the diazoketone with diazomethane and, then, the Wolff rearrangement was initiated by formation of the carbene with a silver compound at the Ag(II) oxidation state.

Arndt–Eistert chain extension of ester

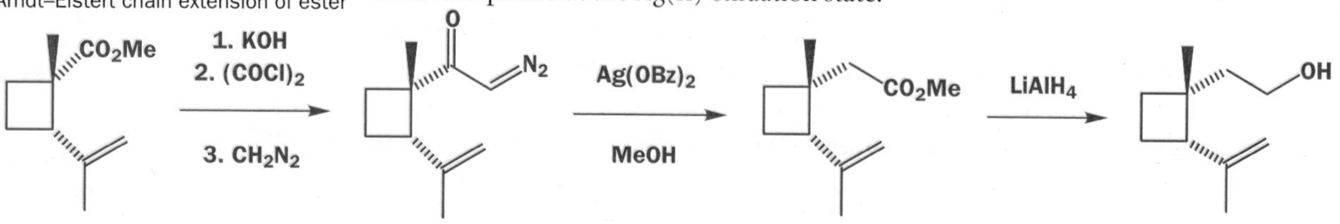

grandisol

Nitrenes are the nitrogen analogues of carbenes

The Wolff rearrangement has some important cousins that we must now introduce to you—they deserve a mention because they bear a family likeness even though they do not, in fact, involve carbenes. They are a group of reactions that proceed through an intermediate **nitrene**—the nitrogen analogue of a carbene. The simplest to understand, because it is the direct nitrogen analogue of the Wolff rearrangement, is the **Curtius rearrangement**. It starts with an acyl azide—which can be made by nucleophilic substitution on an acyl chloride by sodium azide. The acyl azide is what you would get if you just replaced the $-CH=N_2$ of a diazoketone with $-N=N_2$. And, if you heat it, it is not surprising that it decomposes to release nitrogen (N_2), forming the nitrene. The nitrene has two bonds fewer (1) than a normal amine and has two lone pairs making six electrons in all.

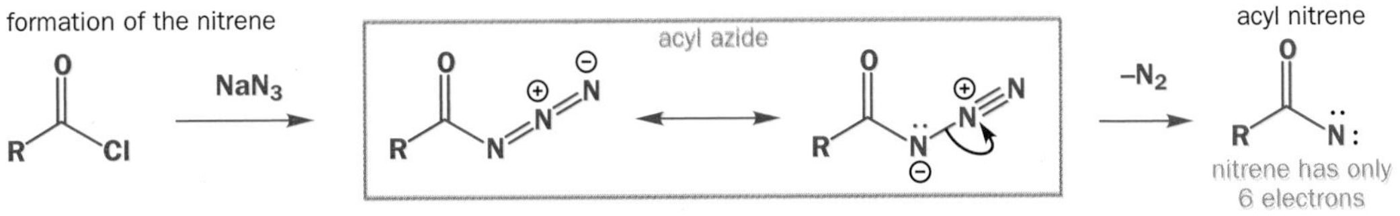

Nitrenes, like carbenes, are immensely reactive and electrophilic, and the same Wolff-style migration takes place to give an isocyanate. The substituent R migrates from carbon to the electron-deficient nitrogen atom of the nitrene. Isocyanates are unstable to hydrolysis: attack by water on the carbonyl group gives a carbamic acid which decomposes to an amine.

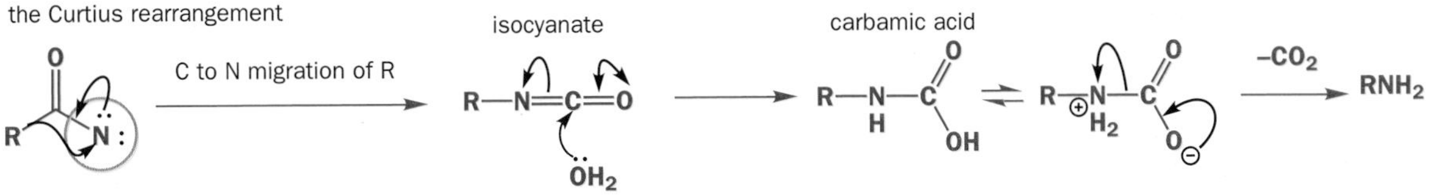

Overall, then, the Curtius rearrangement converts an acid chloride to an amine with loss of a carbon atom—very useful. Also useful is the related **Hofmann rearrangement**, which turns an amide into an amine with loss of a carbon atom. This time we start with a primary amide and make a nitrene by treatment with base and bromine. Notice how close this nitrene-forming reaction is to the carbene-forming reactions we talked about on p. 1072. The nitrene rearranges just as in the Curtius reaction, giving an isocyanate that can be hydrolysed to the amine.

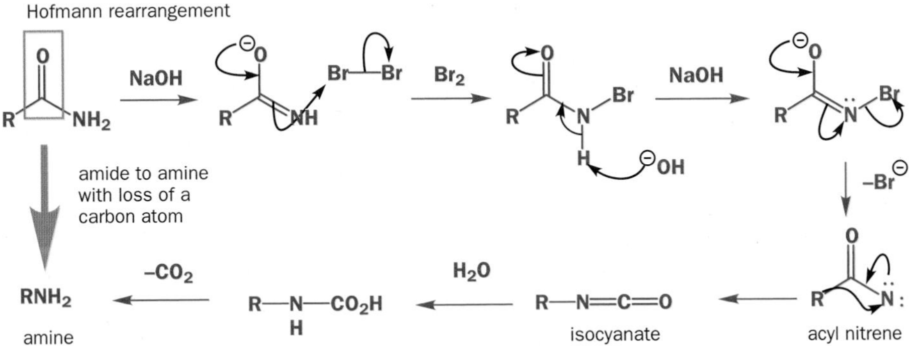

Attack of carbenes on lone pairs

Wolff rearrangements, involving shifts of alkyl groups, are effectively intramolecular insertions into C–C bonds. Carbenes will also insert into other bonds, especially O–H and N–H bonds, though the mechanism in these cases involves initial attack on the lone pair of the heteroatom.

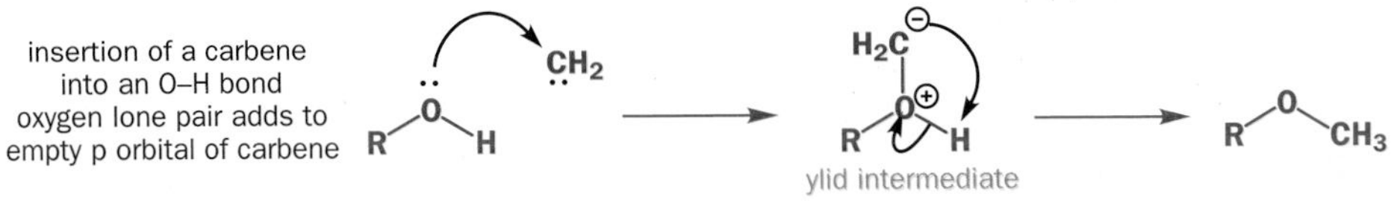

Ylids (or ylides) are zwitterions in which the charges are on adjacent atoms—we mentioned phosphorus ylids in Chapters 14 and 31. A whole chapter, Chapter 46, will be devoted to sulfur ylids and ylid-like species, because they have a special type of chemistry.

We will come back to this in Chapter 46.

Carbene attack is followed by proton transfer to generate a neutral molecule from the first formed zwitterion (or 'ylid'). However, if the heteroatom does not carry a hydrogen, attack on its lone pair generates an ylid that cannot rearrange in this way. Reaction of a carbene with a neutral nucleophile forms an ylid. This type of reaction is, in fact, a very useful way of making reactive ylids that are inaccessible by other means.

stable ylid

As carbonyl-substituted carbenes (like carbonyl-substituted radicals) are electrophilic, their insertion into O–H and N–H bonds can be a useful way of making bonds in an umpolung sense. Because of the difficulties in forming β-lactams (the four-membered rings found in the penicillin classes of antibiotics), Merck decided to design a synthesis of the class of compounds known as carbapenems around a rhodium-catalysed carbene insertion into an N–H bond, building the five-membered ring on to the side of the four-membered ring.

a diazo transfer agent like tosyl azide   75% yield

Alkene (olefin) metathesis

Carbenes can be stabilized as transition metal complexes: decomposition of phenyldiazomethane in the presence of a ruthenium(II) complex gives a carbene complex stable enough to be isolated and stored for months. These complexes are among the most important of carbene-derived reagents because of a remarkable reaction known as **alkene** (or more commonly **olefin**) **metathesis**.

phenyldiazomethane   carbene   'Grubb's' complex

The reaction is most easily understood when a simple diene reacts with a very small amount (in this case 2 mole per cent) of the catalyst. A cyclization reaction occurs and the product is also an alkene. It contains no atoms from the catalyst: indeed, it has lost two carbon atoms, which are given off as ethylene.

The stable 'Fischer carbene' complexes mentioned at the start of this chapter (p. 1057) also catalyse the metathesis reaction but rather less well than these ruthenium complexes.

2 mole% catalyst   98% yield

Any reaction that makes new bonds so efficiently and with so little reagent and so little waste is obviously very important. The yield is also rather good! What happens is a **metathesis**—an exchange of groups between the two arms of the molecule. First, the carbene complex adds to one of the alkenes in what can be drawn as a [2 + 2] cycloaddition (Chapter 35) to give a four-membered ring with the metal atom in the ring.

[2 + 2] cycloaddition

metalla cyclobutane

Now the same reaction happens in reverse (all cycloadditions are, in principle, reversible), either to give the starting materials or, by cleavage of the other two bonds, a new carbene complex and styrene.

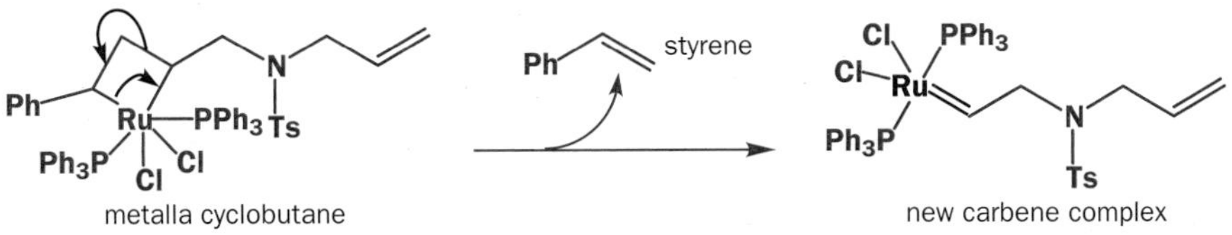

metalla cyclobutane   new carbene complex

Next, an intramolecular [2 + 2] cycloaddition joins up the five-membered ring and produces a second metalla cyclobutane, which decomposes in the same way as the first one to give a third carbene complex and the product.

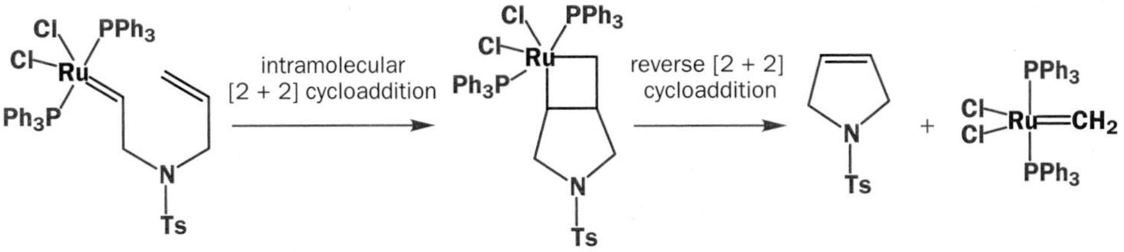

This new carbene complex then attacks another molecule of starting material and the cycle is repeated except that ethylene (ethene) is now lost instead of styrene in all the remaining cycles.

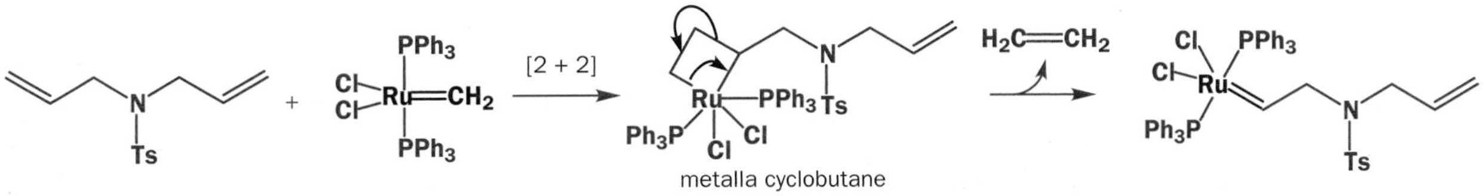

metalla cyclobutane

You will have noticed that the carbene complex appears to exhibit a remarkable selectivity: the ruthenium atom adds to the more substituted end of the first alkene but to the less substituted end of the second. In fact, there is no particular need for selectivity: if the second cycloaddition occurs with the opposite selectivity the metalla cyclobutane has symmetry and can decompose only to the starting materials.

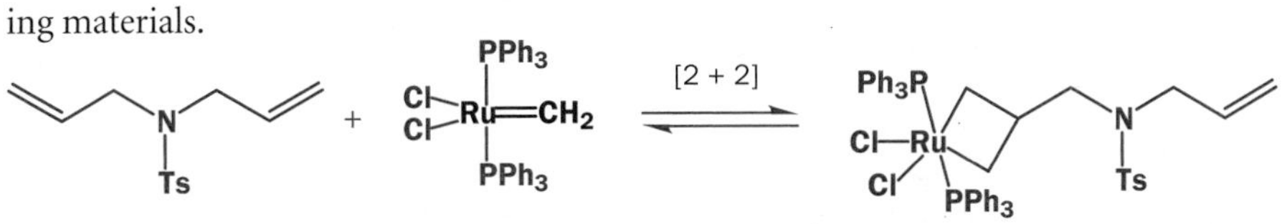

One example that makes a number of points about olefin metathesis is the cyclization of this ester.

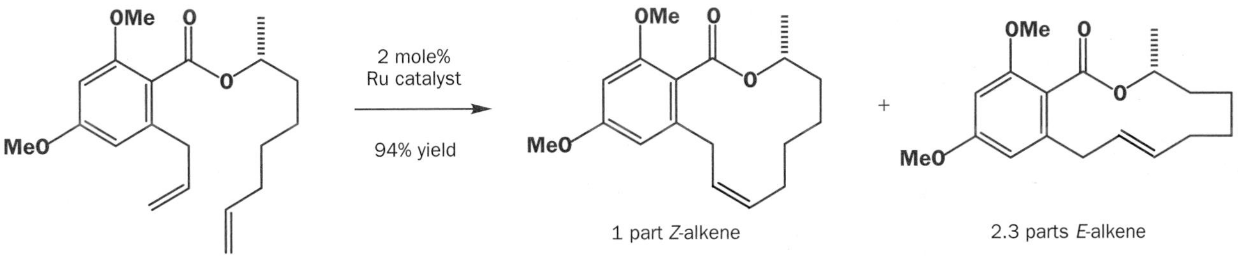

2 mole%
Ru catalyst

94% yield

1 part Z-alkene   2.3 parts E-alkene

The main points are:

- Olefin metathesis is an excellent way to make difficult ring sizes—here a 12-membered ring
- It is compatible with many functional groups—here just an ester and an ether but amines, alcohols, epoxides, and many other carbonyl groups are all right
- The reaction is E-selective. In the previous example only a Z-alkene could be formed but an E-alkene is possible in a 12-membered ring and is the major product
- Stereogenic centres are not racemized

Alkene metathesis is one of the more important of the many new useful reactions that use transition metal complexes as catalysts. You will see more in Chapters 45 and 48.

Summary

We have seen in this chapter how carbenes can be formed from many other reactive intermediates such as carbanions and diazoalkanes and how they can react to give yet more reactive intermediates such as ylids. Here is a summary of the main relationships between carbenes and these other compounds. Note that not all the reactions are reversible. Diazoalkanes lose nitrogen to give carbenes but the addition of nitrogen to carbenes is not a serious reaction.

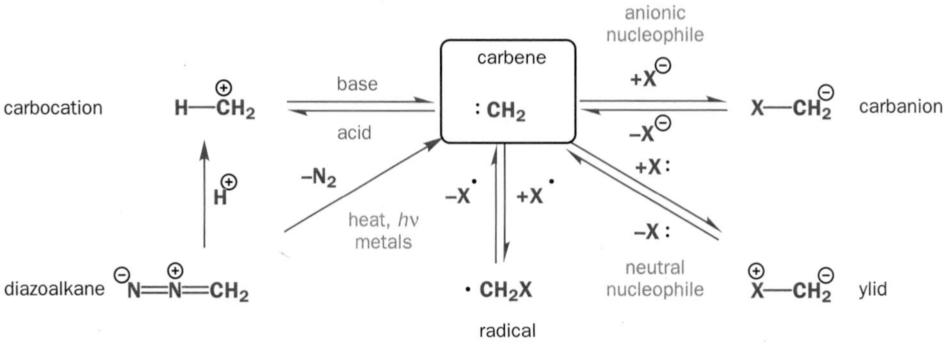

In the last few chapters we have concentrated a lot on what we call reactive intermediates, species like radicals, carbenes, or carbocations that are hard to observe but that definitely exist. Much of the evidence for their existence derives from the study of the mechanisms of reactions—we have discussed some aspects of this as we have met the species concerned, but in the next chapter we will look in detail at how mechanisms are elucidated and the methods used to determine more precisely the structure of reactive intermediates.

Problems

1. Suggest mechanisms for these reactions.

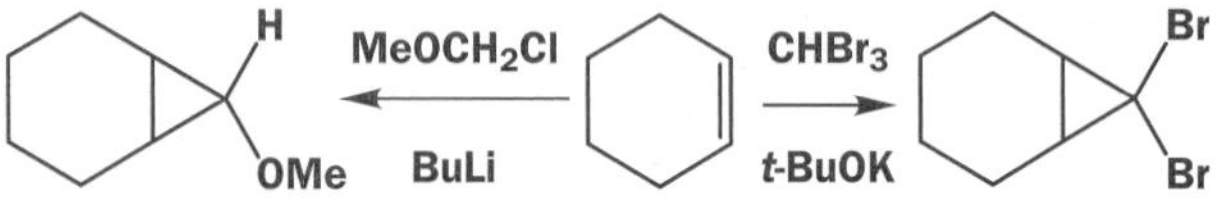

2. Suggest a mechanism and explain the stereochemistry of this reaction.

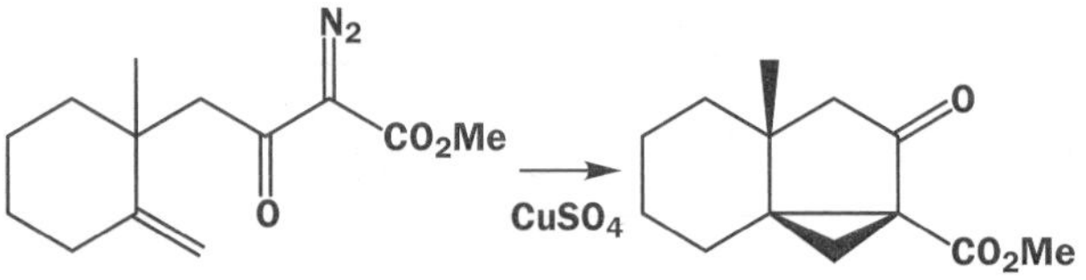

3. Comment on the selectivity shown in these two reactions.

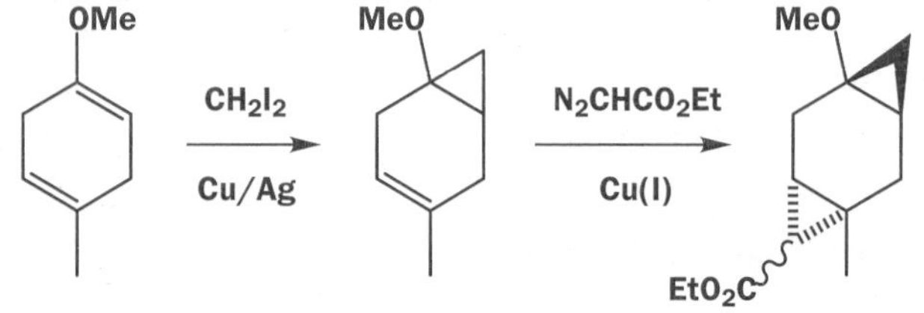

4. Suggest a mechanism for this ring contraction.

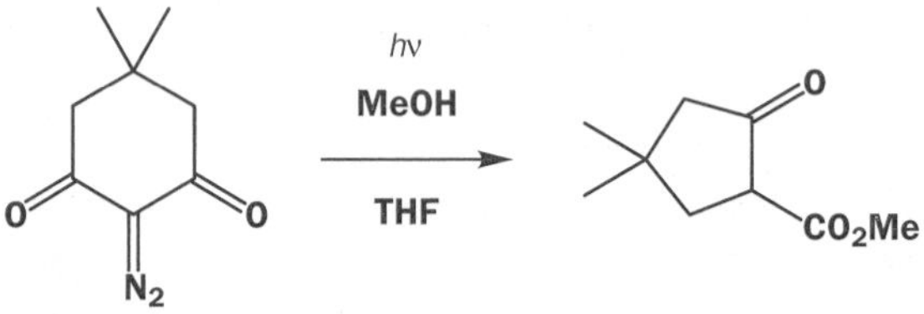

5. Suggest a mechanism for the formation of this cyclopropane.

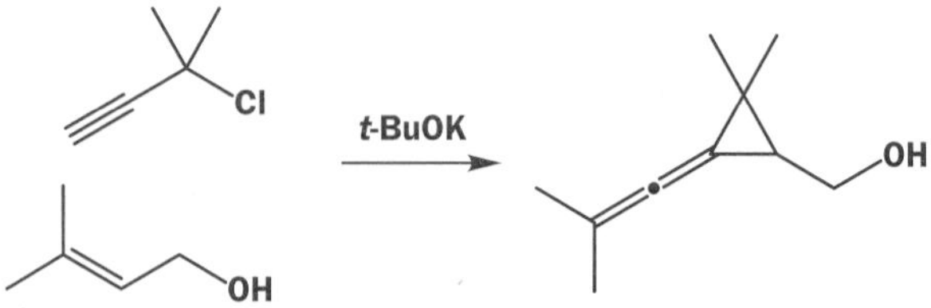

6. Problem 4 in Chapter 31 asked: 'Decomposition of this diazo compound in methanol gives an alkene A ($C_8H_{14}O$) whose NMR spectrum contains two signals in the alkene region: δ_H (p.p.m.) 5.80 (1H, ddd, J 17.9, 9.2, 4.3 Hz), 5.50 (1H, dd, J 17.9, 7.9 Hz), 4.20 (1H, m), 3.50 (3H, s), and 1.3–2.7 (8H, m). What is its structure and geometry?'

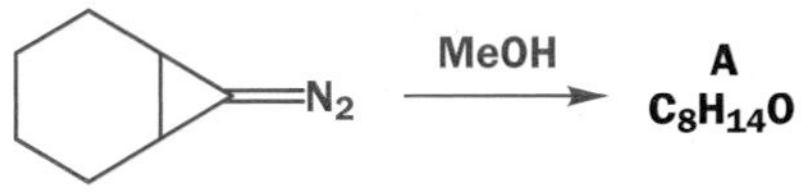

In order to work out the mechanism of the reaction you might like to take these additional facts into account. Compound A is unstable and even at 20 °C isomerizes to B. If the diazo compound is decomposed in methanol containing a diene, compound A is trapped as an adduct. Account for all of these reactions.

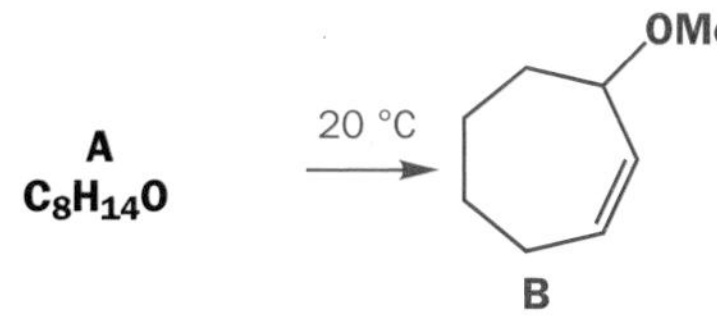

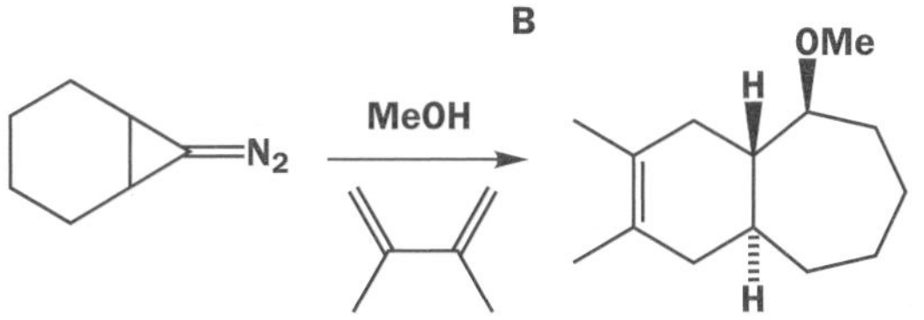

7. Give a mechanism for the formation of the three-membered ring in the first of these reactions and suggest how the ester might be converted into the amine with retention of configuration.

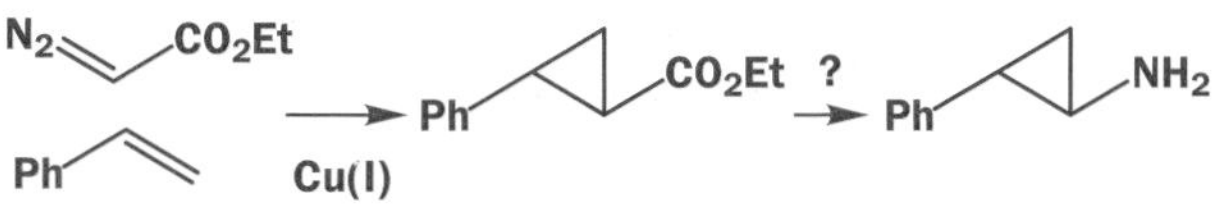

8. Explain how this highly strained ketone is produced, albeit in very low yield, by these reactions. How would you attempt to make the starting material?

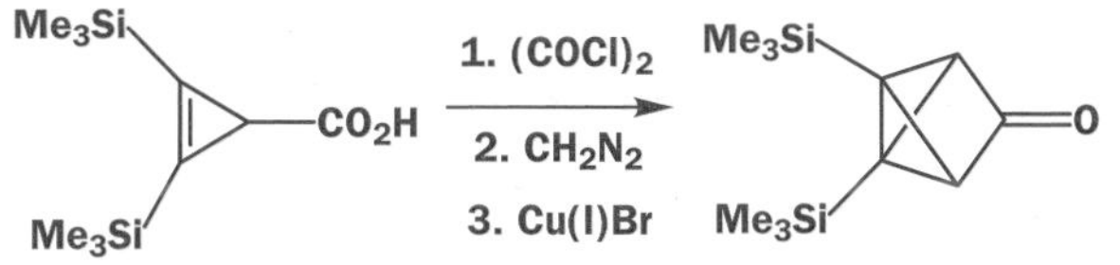

9. Attempts to prepare compound A by a phase-transfer-catalysed cyclization required a solvent immiscible with water. When chloroform ($CHCl_3$) was used, compound B was formed instead and it was necessary to use the more toxic CCl_4 for success. What went wrong?

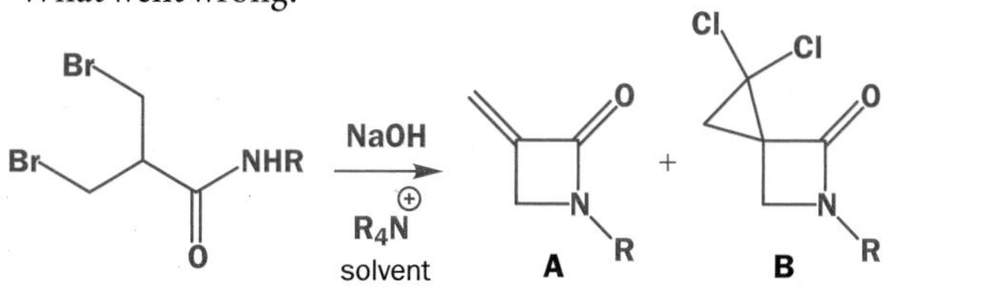

10. Revision content. How would you carry out the first step in this sequence? Propose mechanisms for the remaining steps, explaining any selectivity.

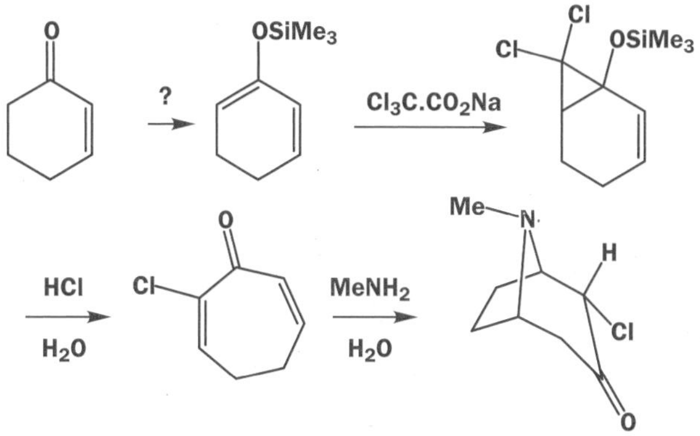

11. How would you attempt to make these alkenes by metathesis?

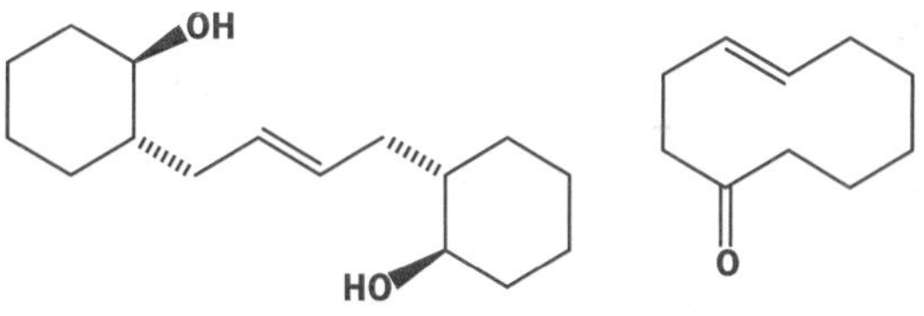

12. Heating this acyl azide in dry toluene under reflux for 3 hours gives a 90% yield of a heterocyclic product. Suggest a mechanism, emphasizing the involvement of any reactive intermediates.

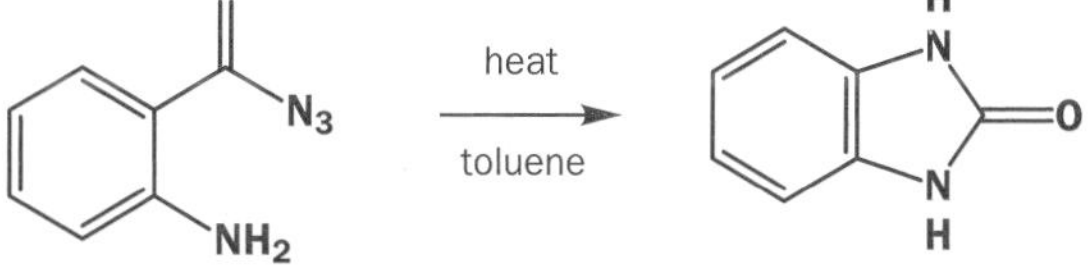

13. Give mechanisms for the steps in this conversion of a five- to a six-membered aromatic heterocycle.

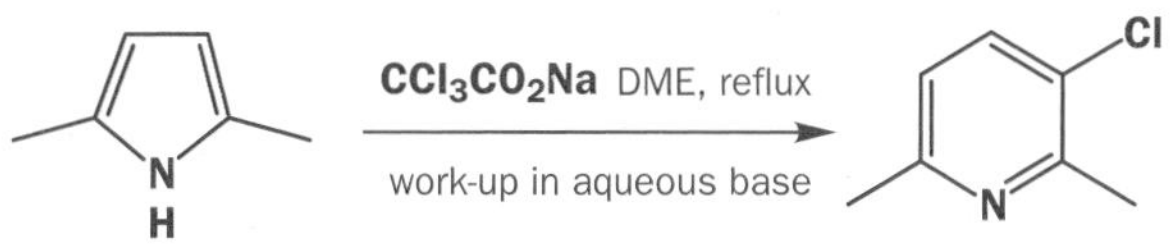

Determining reaction mechanisms

41

Connections

Building on:	Arriving at:	Looking forward to:
• Mainly builds on ch13	• Classes and types of mechanisms	• Saturated heterocycles and stereoelectronics ch42
• Acidity and basicity ch8	• Importance of proposing a mechanism	• Heterocycles ch43–ch44
• Carbonyl reactions ch6, ch12, & ch14	• Structure of the product is all-important	• Asymmetric synthesis ch45
• Nucleophilic substitution at saturated carbon ch17	• Labelling and double labelling	• Chemistry of S, B, Si, and Sn ch46–ch47
• Controlling stereochemistry ch16, ch33, & ch34	• Systematic structure variation and electronic demand	• The chemistry of life ch49–ch51
• Eliminations ch19	• The Hammett correlation explained	
• Electrophilic and nucleophilic aromatic substitution ch22–ch23	• Nonlinear correlations	
• Cycloadditions ch35	• Deuterium isotope effect (kinetic and solvent)	
• Rearrangements ch36–ch37	• Specific acid and specific base catalysis	
• Fragmentations ch38	• General acid and general base catalysis	
	• Detecting and trapping intermediates	
	• A network of related mechanisms	
	• Why stereochemistry matters	

There are mechanisms and there are mechanisms

If you were asked to draw the mechanism of an ester hydrolysis in basic solution you should have no trouble in giving a good answer. It wouldn't matter if you had never seen this particular ester before or even if you knew that it had never actually been made, because you would recognize that the reaction belonged to a class of well known reactions (carbonyl substitution reactions, Chapter 12) and you would assume that the mechanism was the same as that for other ester hydrolyses. And you would be right—nucleophilic attack on the carbonyl group to form a tetrahedral intermediate is followed by loss of the alkoxide leaving group and the formation of the anion of the carboxylic acid.

But someone at some time had to determine this mechanism in full detail. That work was done in the 1940s to 1960s and it was done so well that nobody seriously challenges it. You might also recall from Chapter 13 that, if we change the carbonyl compound to an acid chloride, the mechanism may change to an S_N1 type of reaction with an acylium ion intermediate because the leaving group is now much better: Cl^- is more stable (less basic) than RO^-. It would not be worth using hydroxide for this reaction: as the first step is the slow step, water will do just as well. Again someone had to determine this mechanism, had to show which was the slow step, and had to show that leaving group ability depended on pK_{aH}.

■ The link between leaving group ability and pK_{aH} was discussed in Chapter 12.

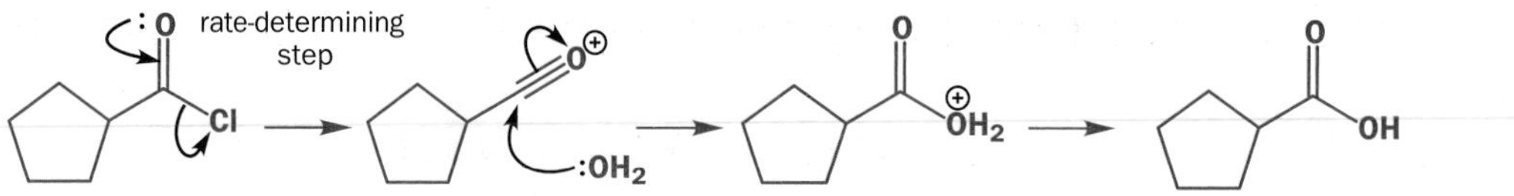

If the reaction were the hydrolysis of an amide, you might remember from Chapter 13 that third-order kinetics are often observed for the expulsion of such bad leaving groups and that this extra catalysis makes it worthwhile using concentrated base. Again, someone had to find out that: (1) the slow step is now the decomposition of the tetrahedral intermediate; (2) there are third-order kinetics involving two molecules of hydroxide; and (3) the first molecule acts as a nucleophile and the second as a base.

> This chemistry was discussed in Chapter 13.

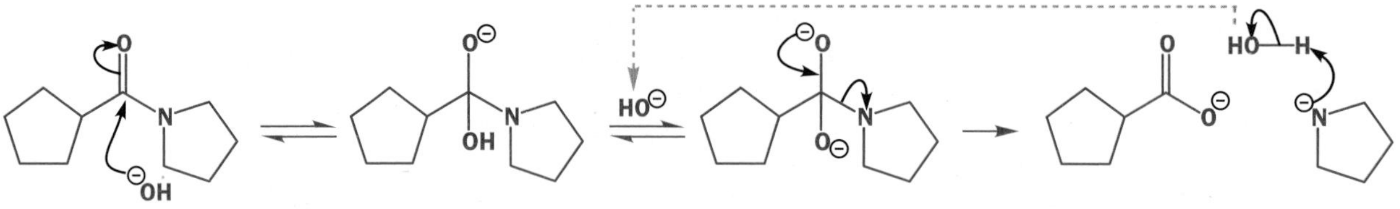

These reactions are versions of the same reaction. For you, writing these mechanisms chiefly means recognizing the type of reaction (nucleophilic substitution at the carbonyl group) and evaluating how good the leaving group is. For the original chemists, determining these reaction mechanisms meant: (1) determining exactly what the product is (that may sound silly, but it is a serious point); (2) discovering how many steps there are and the structures of the intermediates; (3) finding out which is the slow (rate-determining) step; and (4) finding any catalysis. This chapter describes the methods used in this kind of work.

Supposing you were asked what the mechanisms of the next two reactions might be. This is a rather different sort of problem as you probably don't recognize any of these reagents and you probably cannot fit any of the reactions into one of the classes you have seen so far. You probably don't even see at once which of the three main classes of mechanism you should use: ionic; pericyclic; or radical.

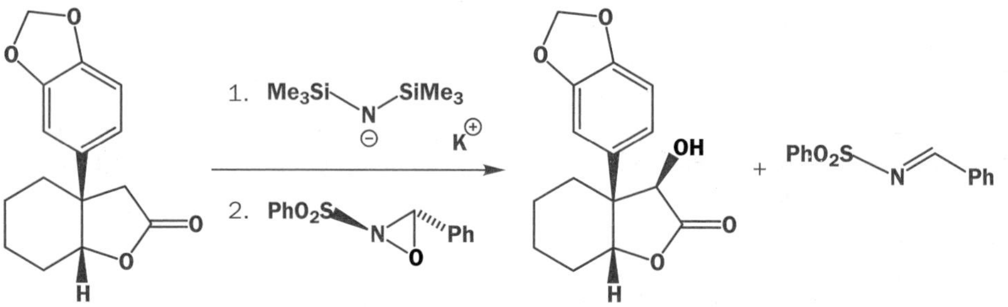

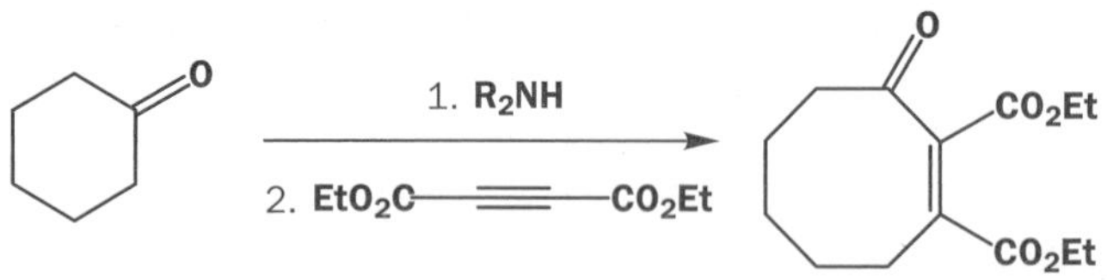

There are two types of answer to the question: 'What is the mechanism of this reaction?' You may do your best to write a mechanism based on your understanding of organic chemistry, moving the electrons from nucleophiles to electrophiles, choosing sensible intermediates, and arriving at the right products. You would not claim any authority for the result, but you would hope, as an organic chemist, to produce one or more reasonable mechanisms. This process is actually an essential preliminary to answering the question in the second way—'What is the real, experimentally verified, mechanism for the reaction?' This chapter is about the second kind of answer.

Determining reaction mechanisms—the Cannizzaro reaction

So how do we know the mechanism of a reaction? The simple answer is that we don't for certain. Organic chemists have to face situations where the structure of a compound is initially thought to be one thing but later corrected to be something different. The same is true of mechanisms. It is the nature of science that all we can do is try to account for observations by proposing theories. We then test the theory by experiment and, when the experiment does not fit the theory, we must start again with a new theory. This is exactly the case with mechanisms. When a new reaction is discovered, one or more mechanisms are proposed; evidence is then sought for and against these mechanisms until one emerges as the best choice and that remains the accepted mechanism for the reaction until fresh evidence comes along that does not fit the mechanism.

We are going to look at one reaction, the Cannizzaro reaction, and use this to introduce the different techniques used in elucidating mechanisms so that you will be able to appreciate the different information each experiment brings to light and how all the pieces fit together to leave us with a probable mechanism. Under strongly basic conditions, an aldehyde with no α hydrogens undergoes disproportionation to give half alcohol and half carboxylate. Disproportionation means one half of the sample is oxidized by the other half, which is itself reduced. In this case, half the aldehyde reduces the other half to the primary alcohol and in the process is oxidized to the carboxylic acid. Before the discovery of LiAlH$_4$ in 1946, this was one of the few reliable ways to reduce aldehydes and so was of some use in synthesis.

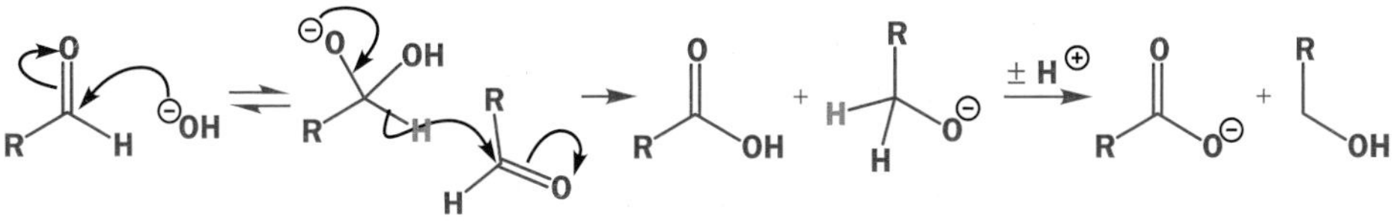

The mechanism we have drawn here is slightly different from that in Chapter 27 where we showed the dianion as an intermediate. The two reactions are related by base catalysis as we shall see. Now for some of the evidence and some of the alternative mechanisms that have been proposed for the Cannizzaro reaction. Most of these have been eliminated, leaving just the ones you have already met. Finally, we will see that even these mechanisms do not explain everything absolutely.

Proposed mechanism A—a radical mechanism

Early on it was thought that the hydrogen transfer might be taking place via a radical chain reaction. If this were the case, then the reaction should go faster if radical initiators are added and it should slow down when radical inhibitors are added. When this was tried, there was no change in the rate, so this proposed mechanism was ruled out.

Kinetic evidence for an ionic mechanism

The first piece of evidence that must be accounted for is the rate law. For the reaction of benzaldehyde with hydroxide, the reaction is first-order with respect to hydroxide ions and second-order with respect to benzaldehyde (third-order overall).

$$\text{rate} = k_3[\text{PhCHO}]^2[\text{HO}^-]$$

For some aldehydes, such as formaldehyde and furfural, the order with respect to the concentration of hydroxide varies between one and two depending on the exact conditions. In high concentrations of base it is fourth-order.

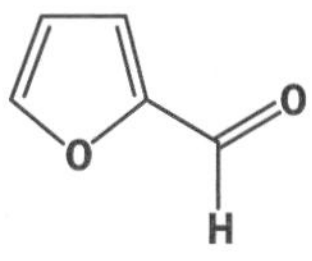

$$\text{rate} = k_4[\text{HCHO}]^2[\text{HO}^-]^2$$

At lower concentrations of base it is a mixture of both third- and fourth-order reactions.

$$\text{rate} = k_3[\text{HCHO}]^2[\text{HO}^-] + k_4[\text{HCHO}]^2[\text{HO}^-]^2$$

Just because the overall order of reaction is third- or fourth-order, it does not mean that all the species must simultaneously collide in the rate-determining step. You saw in Chapter 13 that the rate law actually reveals all the species that are involved *up to and including* the rate-determining step.

The Cannizzaro reaction first appeared in Chapter 27.

For some examples of radical initiators, see Chapter 39. Radical inhibitors are usually stable radicals such as those on p. 1028.

2-furaldehyde (furfural)

Isotopic labelling

When the reaction is carried out in D_2O instead of in H_2O it is found that there is are no C–D bonds in the products. This tells us that the hydrogen must come from the aldehyde and not from the solvent.

Proposed mechanism B—formation of an intermediate dimeric adduct

A possible mechanism that fits all the experimental evidence so far involves nucleophilic attack of the usual tetrahedral intermediate on another aldehyde to give an intermediate adduct. This adduct could then form the products directly by hydride transfer. You may not like the look of this last step, but the mechanism was proposed and evidence is needed to disprove it.

Which step would be rate-determining for this mechanism? It could not be step 1 since, if this were the case, then the rate law would be first-order with respect to the aldehyde rather than the observed second-order relationship. Also, if the reaction is carried out in water labelled with oxygen-18, the oxygen in the benzaldehyde exchanges with the ^{18}O from the solvent much faster than the Cannizzaro reaction takes place. This can only be because of a *rapid* equilibrium in step 1 and so step 1 cannot be rate-determining.

So, for mechanism B, either step 2 or step 3 could be rate-determining—either case would fit the observed rate law. Step 2 is similar to step 1; in both cases an oxyanion nucleophile attacks the aldehyde. Since the equilibrium in step 1 is very rapid, it is reasonable to suggest that the equilibrium in step 2 should also be rapid and thus that the hydride transfer in step 3 must be rate-determining. So mechanism B can fit the rate equation.

How can mechanism B be ruled out? One way is to change the attacking nucleophile. The Cannizzaro reaction works equally well if methoxide is used in a mixture of methanol and water. If mechanism B were correct, the reaction with methoxide would be as follows.

We shall discuss this kind of technique as well as other evidence used to evaluate an intermediate towards the end of this chapter.

One of the products would be different by this mechanism: benzyl methyl ether would be formed instead of benzyl alcohol. None is observed experimentally. Under the conditions of the experiment, benzyl methyl ether does not react to form benzyl alcohol, so it cannot be the case that the ether is formed but then reacts to form the products. Mechanism B can therefore be ruled out.

Proposed mechanism C—formation of an ester intermediate

This mechanism is like mechanism B but the hydride transfer in the adduct formed in step 2 displaces OH^- to form an ester (benzyl benzoate) that is then hydrolysed to the products. This was at one time held to be the correct mechanism for the Cannizzaro reaction. One piece of evidence for this, and at first glance a very good one, is that by cooling the reaction mixture and avoiding excess alkali, some benzyl benzoate could be isolated during the reaction. An important point is that this does not mean that the ester *must* be an intermediate in the reaction—it might be formed at the end of the reaction, for example. However, it does mean that any mechanism we propose must be able to account for its formation. For now though we want to try and establish whether the ester is an *intermediate* rather than a by-product in the Cannizzaro reaction.

An early objection to mechanism C was that the ester would not be hydrolysed fast enough. When someone actually tried it under the conditions of the experiment, they found that benzyl benzoate is very rapidly hydrolysed (the moral here is 'don't just think about it, try it!'). However, just because the ester *could* be hydrolysed, it still did not show that it actually was an intermediate in the reaction. How this was eventually shown was rather clever. The argument goes like this. We can measure the rate constant for step 4 by seeing how quickly pure benzyl benzoate is hydrolysed to benzyl alcohol and benzoate under the same conditions as those of the Cannizzaro reaction. We also know how quickly these products are formed during the Cannizzaro reaction itself. Since, if this mechanism is correct, the only way the products are formed is from this intermediate, it is possible to work out how much of the intermediate ester must be present at any time to give the observed rate of formation of the products. If we can measure the amount of ester that is actually present and it is significantly less than that which we predict, then this cannot be the correct mechanism. It turned out that there was never enough ester present to account for the formation of the products in the Cannizzaro reaction and mechanism C could be ruled out.

The correct mechanism for the Cannizzaro reaction

The only mechanism that has not been ruled out and that appears to fit all the evidence is the one we have already given (p. 1081). The fact that the rate law for this mechanism is overall third- and sometimes fourth-order depending on the aldehyde and the conditions can be explained by the involvement of a second hydroxide ion deprotonating the tetrahedral intermediate to give a dianion. When methoxide is used in a methanol/water mix, some methyl ester is formed. This does not stay around for long—under the conditions of the experiment it is quickly hydrolysed to the carboxylate.

Even this mechanism does not quite fit all the evidence

We said earlier that we can never prove a mechanism—only disprove it. Unfortunately, just as the 'correct' mechanism seems to be found, there are some observations that make us doubt this mechanism. In Chapter 39 you saw how a technique called **electron spin resonance** (ESR) detects radicals and gives some information about their structure. When the Cannizzaro reaction was carried out with benzaldehyde and a number of substituted benzaldehydes in an ESR spectrometer, a radical was detected. For each aldehyde used, the ESR spectrum proved to be identical to that formed when the aldehyde was reduced using sodium metal. The radical formed was the radical anion of the aldehyde.

Our mechanism does not explain this result but small amounts of radicals are formed in many reactions in which the products are actually formed by simple ionic processes. Detection of a species in a reaction mixture does not prove that it is an intermediate. Only a few chemists believe that radicals are involved in the Cannizzaro reaction. Most believe the mechanism we have given.

Variation in the structure of the aldehyde

Before leaving the Cannizzaro reaction, look at these rates of reactions for aromatic aldehydes with different substituents in the *para* position. These aldehydes may be divided into two classes: those that react faster than unsubstituted benzaldehyde and those that react more slowly. Those that go slower all have something in common—they all have substituents on the ring that donate electrons.

We have already seen how substituents on a benzene ring affect the rate of electrophilic substitution (Chapter 22).

Aldehyde	Rate relative to benzaldehyde at 25 °C	Rate relative to benzaldehyde at 100 °C
benzaldehyde	1	1
p-methylbenzaldehyde	0.2	0.2
p-methoxybenzaldehyde	0.05	0.1
p-dimethylaminobenzaldehyde	very slow	0.0004
p-nitrobenzaldehyde	210	2200

Electron-donating groups such as MeO– and Me$_2$N– dramatically speed up the rate at which an aromatic ring is attacked by an electrophile, whereas electron-withdrawing groups, particularly nitro groups, slow the reaction down. The Cannizzaro reaction is not taking place on the benzene ring itself, but substituents on the ring still make their presence known. The fact that the Cannizzaro reaction goes much *slower* with electron-donating groups and faster with electron-withdrawing groups tells us that, for this reaction, rather than a positive charge developing as in the case of electrophilic substitution on an aromatic ring, there must be negative charge accumulating somewhere near the ring. Our mechanism has mono- and dianion intermediates that are stabilized by electron-withdrawing groups. Later in the chapter you will see a more quantitative treatment of this variation of structure.

The rest of the chapter is devoted to discussions of the methods we have briefly surveyed for the Cannizzaro reaction with examples of the use of each method. We give examples of many different types of reaction but we cannot give every type. You may rest assured that all of the mechanisms we have so far discussed in this book have been verified (not, of course, proved) by these sorts of methods.

Be sure of the structure of the product

This seems a rather obvious point. However, there is a lot to be learned from the detailed structure of the product and we will discuss checking which atom goes where as well as the stereochemistry of the product. You will discover that it may be necessary to alter the structure of the starting material in subtle ways to make sure that we know exactly what happens to all its atoms by the time it reaches the product.

Suppose you are studying the addition of HCl to this alkene. You find that you get a good yield of a single adduct and you might be a bit surprised that you do not get a mixture of the two obvious adducts and wonder if there is some participation of the ether oxygen or whether perhaps the ketone enolizes during the reaction and controls the outcome.

If you are cautious you might check on the structure of the product before you start a mechanistic investigation. The NMR spectrum tells you at once that the product is neither of these suggestions. It contains a (CH$_2$)$_3$Cl unit and can no longer have an eight-membered ring. A ring contraction has given a five-membered ring and a mechanistic investigation is hardly needed. Simply knowing what the product is allows us to propose a mechanism. A rearrangement has occurred and we could use the method suggested in Chapter 37, of numbering the atoms in the starting material and finding them in the product. This is quite easy as only one numbering system makes any sense.

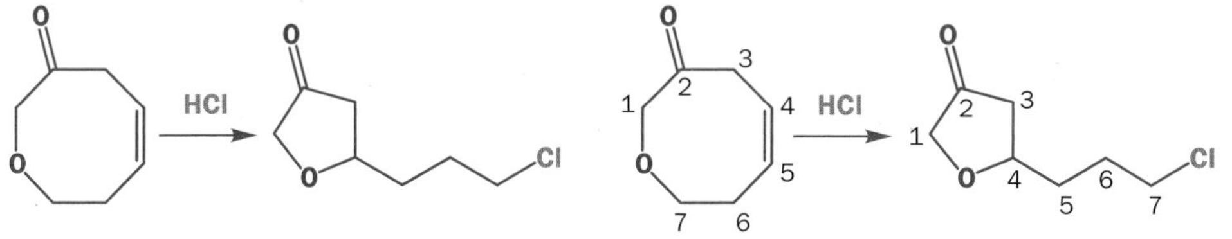

This numbering suggests that the carbon skeleton is unaffected by the reaction, that protonation has occurred at C5, that the ether oxygen has acted as an internal nucleophile across the ring at C4, and that the chloride ion has attacked C7. The mechanism is straightforward.

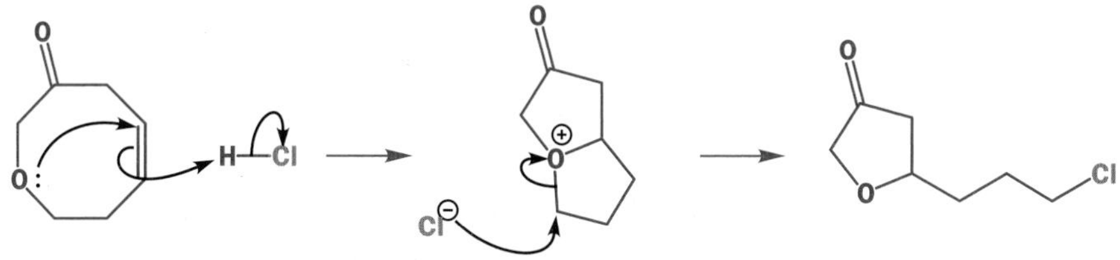

It may be disappointing to find that every step in this mechanism is well known and that the reaction is exactly what we ought to have expected with an eight-membered ring as these rings are famous for their transannular (across-ring) reactions to form 5/5 fused systems. However, it is good that a prolonged investigation is not necessary.

> ● **Find out for sure what the structure of the product is before you start a mechanistic investigation.**

A more subtle distinction occurred in a study of the bromination of alkynes. Bromination of benzyl alkynes in acetic acid gave the products of addition of one molecule of bromine—the 1,2-dibromoalkenes. The reaction was successful with a variety of *para* substituents and there seems at first to be no special interest in the structure of the products.

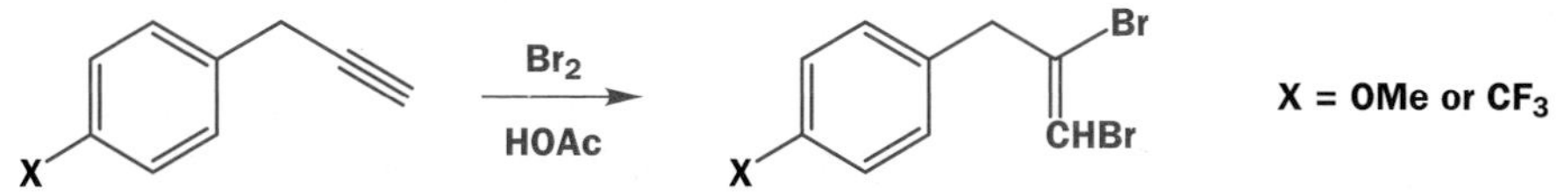

Closer investigation revealed an extraordinary difference between them, not at all obvious from their NMR spectra: the compound from X = OMe was the *Z*-dibromoalkene from *cis* addition of bromine while the product from X = CF$_3$ was the *E*-alkene from *trans* addition. What mechanism could explain this difference?

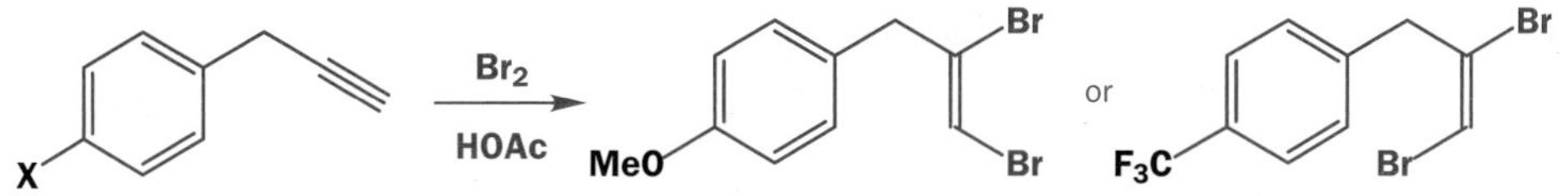

The *anti* addition is more easily explained: it is the result of formation of a bromonium ion, similar, in fact, to the normal mechanism for the bromination of alkenes. Bromine adds from one side of the alkene and the bromide ion must necessarily form the *E*-dibromo product regardless of which atom it attacks.

A similar aryl participation in saturated compounds to give a 'phenonium ion' intermediate appears in Chapter 37, p. 974.

So why does the *p*-MeO– compound behave differently? It cannot react by the same mechanism and a reasonable explanation is that the much more electron-donating ring participates in the reaction to give a carbocyclic three-membered ring intermediate that is attacked in an *anti* fashion to give the *Z*-alkene. Both intermediates are three-membered ring cations and both are attacked with inversion but the *p*-MeO– compound undergoes double inversion by participation of the ring.

Labelling experiments reveal the fate of individual atoms

It often happens that the atoms in starting material and product cannot be correlated without some extra distinction being made by isotopic labelling. The isomerization of *Z*-1-phenylbutadiene to the *E*-diene in acid looks like a simple reaction. Protonation of the *Z*-alkene would give a stabilized secondary benzylic cation that should last long enough to rotate. Loss of the proton would then give the more stable *E*-diene.

However, reaction with D^+ in D_2O reveals that this mechanism is incorrect. The product contains substantial amounts of deuterium at C4, not at C2 as predicted by the proposed mechanism. Protonation must occur at the end of the conjugated system to produce the more stable conjugated cation, which rotates about the same bond and loses H or D from C4 to give the product. More H than D will be lost, partly because there are two Hs and only one D, but also because of the kinetic isotope effect, of which more later.

►

Tritium and ^{14}C are β emitters—they give off electrons—having half-lives of 12 and over 5000 years, respectively. Tritium is made on a large scale by neutron irradiation of ^6Li in a nuclear reactor.

The easiest labels to use for this job are D for H, ^{13}C, and ^{18}O. None of these is radioactive; all can be found by mass spectrometry, while D and ^{13}C can be found by NMR. Old work on mechanisms used radioactive tracers such as T (tritium) for H and ^{14}C. These are isotopes of hydrogen and carbon having extra neutrons. They are, of course, more dangerous to use but they can at least always be found. The real disadvantage is that, to discover exactly where they are in the product, the molecule must be degraded in a known fashion. These radioactive isotopes are not much used nowadays except in determining biological mechanisms as you will see in Chapters 49–51. The first evidence for benzyne as the intermediate in the reaction of chlorobenzene with NH_2^- came from radioactive labelling.

■

Benzyne is discussed in Chapter 23 as an intermediate in nucleophilic aromatic substitution.

● = ^{14}C

benzyne

If benzyne is an intermediate, the product should have 50% label at C1 and 50% at the two identical *ortho* carbons. The labelled aniline was degraded by the reactions shown here, which you must agree was a lot of work for the chemists concerned. Each potentially labelled carbon atom had to be isolated from any other labelled atom and the radioactivity measured. We shall follow the fate of the two labelled atoms with black and green spots. Since the two *ortho* positions are identical, we must put a black spot on both of them.

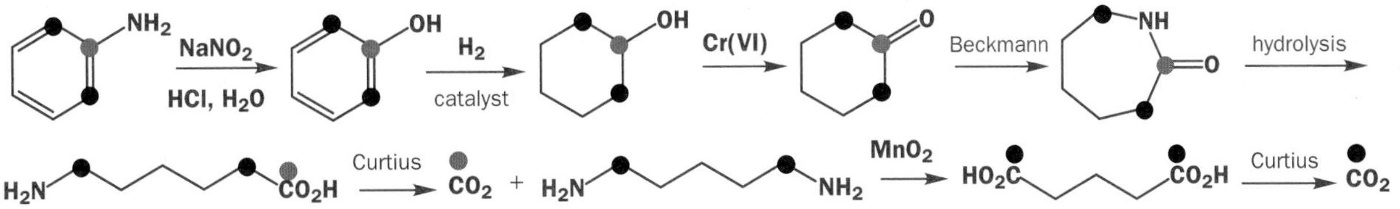

Most of these reactions are well known—the Beckmann rearrangement is described in Chapter 37 and the Curtius reaction in Chapter 40—but the oxidation of the diamine to the dicarboxylic acid is not a standard procedure and is not recommended. All the label came out in the CO_2 and almost exactly half of it was from the black and half from the green labelled carbons. This was the original evidence that convinced organic chemists in 1953 that benzyne was involved in the reaction. The evidence presented in Chapter 23 is more modern.

■
Other symmetrical intermediates originally identified by radioactive labelling include the cyclopropanone in the Favorskii rearrangement in Chapter 37, p. 990, and a spirocyclic intermediate in electrophilic substitution on an indole in Chapter 43, p. 1170.

The value of double labelling experiments

An altogether more modern approach to a labelling study was used in the surprising rearrangement of a hydroxy-acid in acidic solution. The structure of the product suggests a CO_2H migration as the most likely mechanism. This mechanism resembles closely the cationic rearrangements of Chapter 37.

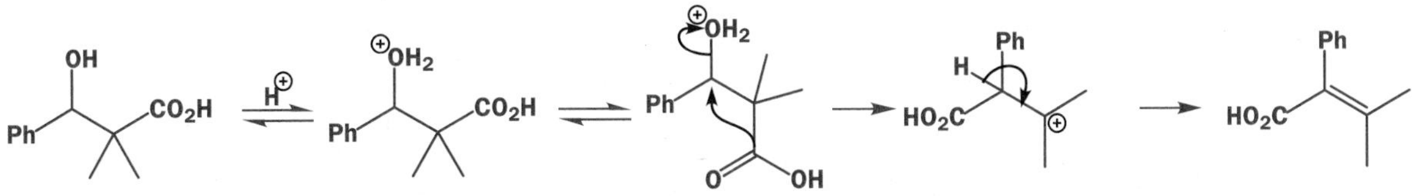

Received wisdom (Chapter 37) objects that the best migrating group in cationic rearrangements is the one best able to bear a positive charge, so that the more familiar Ph and Me migrations ought to be preferred and that a more elaborate mechanism should be sought. Such a mechanism can be written: it involves two methyl migrations and one phenyl migration and is acceptable.

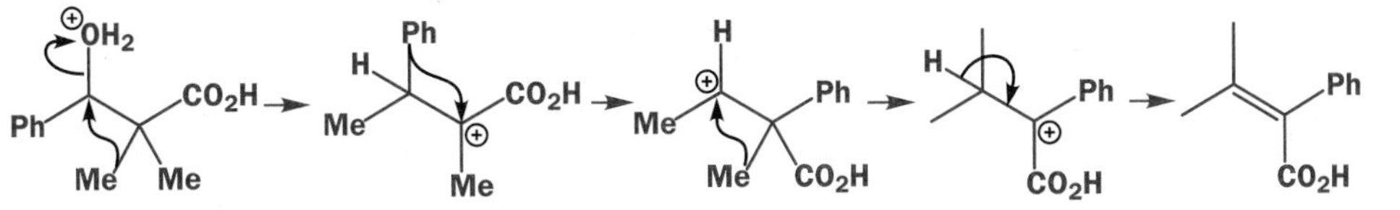

These mechanisms can be tested by finding out whether the CO_2H group remains attached to its original position or becomes attached to the other carbon in the skeleton of the molecule. This can be done by double labelling. If a compound is prepared with two ^{13}C labels, one on the CO_2H group itself and one on the benzylic carbon, the NMR spectrum of the product will show what has happened. In fact, the two ^{13}C labels end up next to each other with a coupling constant $^1J_{CC} = 71$ Hz. It is the CO_2H group that has migrated.

■
This style of double labelling with NMR active isotopes will be seen again in Chapters 49–51.

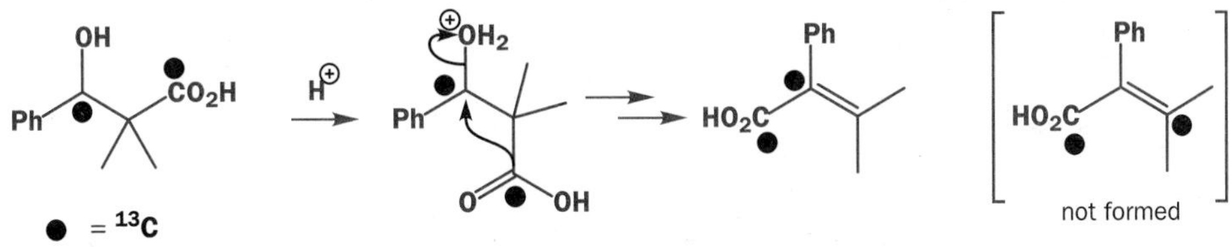

So why does the CO_2H group migrate? It does so not because it is a good migrating group but because it cannot bear to be left behind. The rearranged cation from CO_2H migration is a stable tertiary alkyl cation. The cation from Me migration is a very unstable cation with the positive charge next to the CO_2H group. Such cations are unknown as the carbonyl group is very electron-withdrawing. Received wisdom needs to be amended.

'Crossover' experiments

There is still one tiny doubt. Supposing the reaction is not intramolecular at all, but intermolecular. The CO_2H group might be lost from one molecule as protonated CO_2 and be picked up by another molecule of alkene. No migration would be involved at all.

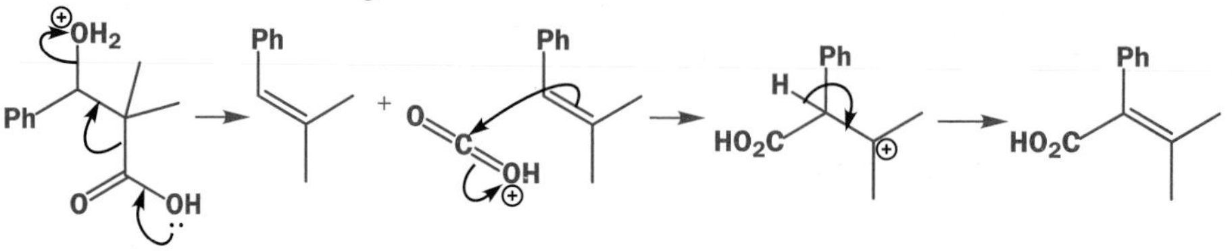

This mechanism can be checked by using a 50:50 mixture of doubly labelled and unlabelled starting material. The molecule of alkene that captures the roving protonated labelled CO_2 might happen to be labelled too but equally well it might be unlabelled. If this last mechanism is correct, we should get a mixture of unlabelled, singly labelled, and doubly labelled product in the ratio 1:2:1 as there are two types of singly labelled product. The two singly labelled compounds are called the **crossover products** and the experiment is called a **crossover experiment** as it discovers whether any parts of one molecule cross over to another.

There is an example of a crossover experiment proving that an S_N2 reaction is intermolecular in Chapter 42, p. 1141.

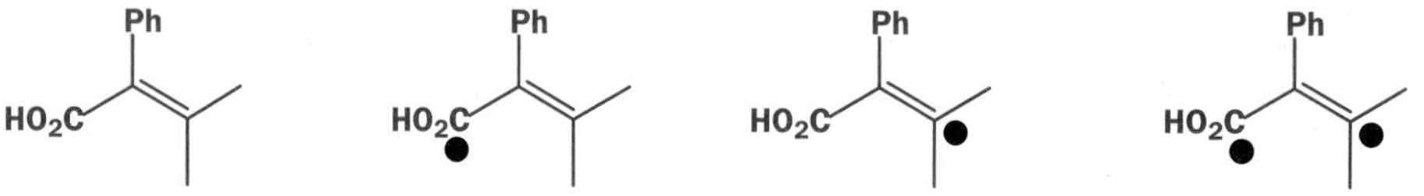

| unlabelled | singly labelled (two types) | | doubly labelled |

In fact, no singly labelled compounds were found: NMR analysis showed that the product consisted entirely of unlabelled or doubly labelled molecules. The CO_2H group remains attached to the same molecule (though not to the same atom) and the first mechanism is correct. Crossover experiments demand some sort of double labelling, which does not have to be isotopic. An example where crossover products are observed is the light-initiated isomerization of allylic sulfides.

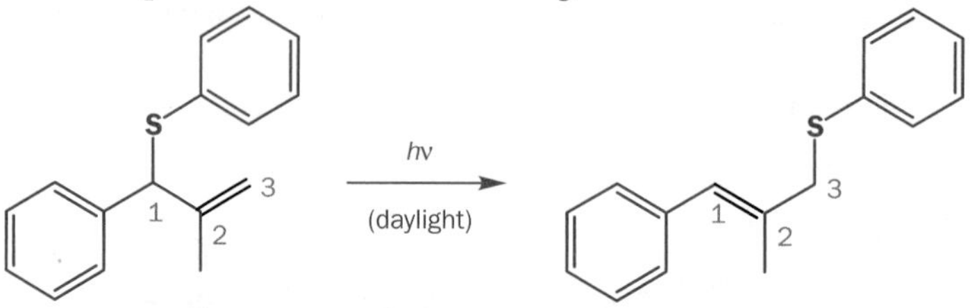

This is formally a [1,3] sigmatropic shift of sulfur (Chapter 36) but that is an unlikely mechanism and a crossover experiment was carried out in which the two molecules had either two phenyl groups or two *para*-tolyl groups.

The mixture was allowed to rearrange in daylight and the products were examined by mass spectroscopy. There was a roughly 1:2:1 mixture of products having two phenyl groups, one phenyl and one *para*-tolyl group, and two *para*-tolyl groups. The diagram shows the starting materials and the two crossover products only.

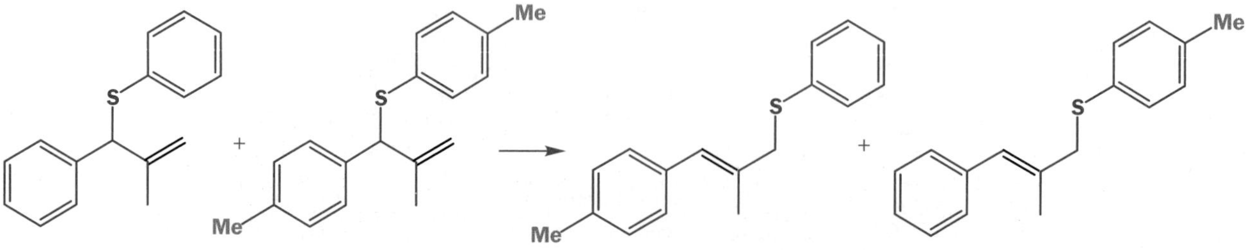

the two crossover products

Clearly, the ArS group had become separated from the rest of the molecule and the most likely explanation was a radical chain reaction (Chapter 39) with the light producing a small amount of ArS• to initiate the chain. The *para*-methyl group acts as a label. The whole system is in equilibrium and the more highly substituted alkene is the product.

Systematic structural variation

In this last example, the hope is that the *para*-methyl group will have too weak an electronic or steric effect and in any case will be too far away to affect the outcome. It is intended to make nearly as slight a change in the structure as an isotopic label. Many structural investigations have exactly the opposite hope. Some systematic change is made in the structure of the molecule in the expectation of a predictable change in rate. A faster or slower reaction will lead to some definite conclusion about the charge distribution in the transition state.

Allylic compounds can react efficiently with nucleophiles by either the S_N1 or S_N2 mechanisms (Chapter 17) as in these two examples.

The carbon skeleton is the same in both reactions but the leaving groups and the nucleophiles are different. These reaction might both go by S_N1 or S_N2 or one might go by S_N1 and the other by S_N2. One way to find out is to make a large change in the electronic nature of the carbon skeleton and see what happens to the rate of each reaction. In these experiments one of the methyl groups was changed for a CF_3 group—exchanging a weakly electron-donating group for a strongly electron-withdrawing group. If a cation is an intermediate, as in the S_N1 reaction, the fluorinated compound will react much more slowly. Here is the result in the first case.

relative rate = 1.0

relative rate $= 1.8 \times 10^{-6}$

The fluorinated compound reacts half a million times more slowly so this looks very much like an S_N1 mechanism. The slow step in an S_N1 mechanism is the formation of a carbocation so any group that destabilizes the positive charge would have (and evidently does have) a large effect on the rate. Rate ratios of several powers of ten are worth noticing and a rate ratio of nearly 10^{-6} is considerable. In the second case the rate difference is much less.

relative rate = 1.0

relative rate = 11.0

A rate ratio of 11 is not worth noticing. The point is not that the fluorinated compound reacts faster but that the two compounds react at about the same rate. This strongly suggests that no charge is generated in the transition state and an S_N1 mechanism is not possible. The S_N2 mechanism makes good sense with its concerted bond formation and bond breaking requiring no charge on the carbon skeleton.

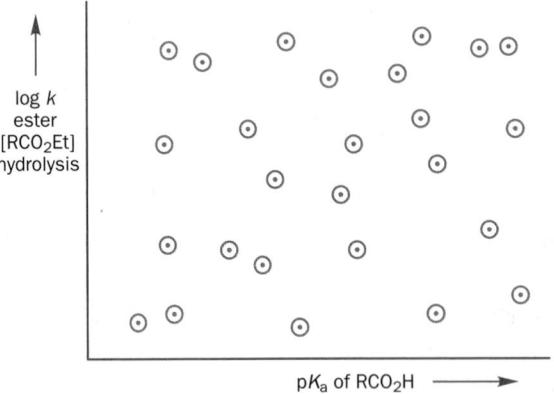

S_N2 transition state

The CF₃ group works well here as a mechanistic probe because it is held well out of the way of the reaction site by a rigid π system but is connected electronically by that same allylic system. Steric effects should be minimized and electronic effects clearly seen. This approach is clearly limited by the small number of groups having properties like those of the CF₃ group and the small number of reactions having such favourable carbon skeletons. We will now present the most important serious correlation between structure and reactivity.

The Hammett relationship

Louis P. Hammett (1894–1987) invented 'physical organic chemistry' and at Columbia University in 1935 derived the Hammett σ/ρ relationship. The impact was enormous and in the 1960s chemists were still working out more such correlations.

What we would ideally like to do is find a way to quantify the effects that electron-donating or -withdrawing groups have on the transition state or intermediate during the course of a reaction. This will then give us an idea of what the transition state is really like. The first question is: can we define exactly how efficient a given group is at donating or withdrawing electrons? Hammett took the arbitrary decision to use the pK_a of an acid as a guide. For example, the rate of hydrolysis of esters might well correlate with the pK_a of the corresponding acid.

substituent on R is mechanistic probe   reaction to be investigated

When Hammett plotted the rates of ethyl ester hydrolyses (as log k since pK_a has a log scale) against the pK_as of the corresponding acids, the initial results were not very encouraging as there was a random scatter of points over the whole graph.

Hammett had used some aliphatic acids (substituted acetic acids) and some aromatic acids (substituted benzoic acids) and he noticed that many of the points towards the top of the graph belonged to the substituted acetic acids. Removing them (brown points) made the graph a lot better. He then noticed that the remaining aromatic compounds were in two classes: the *ortho*-substituted esters reacted more slowly than their *meta*- and *para*-isomers and came towards the bottom of the graph (orange points). Removing them made the graph quite good (remaining green points).

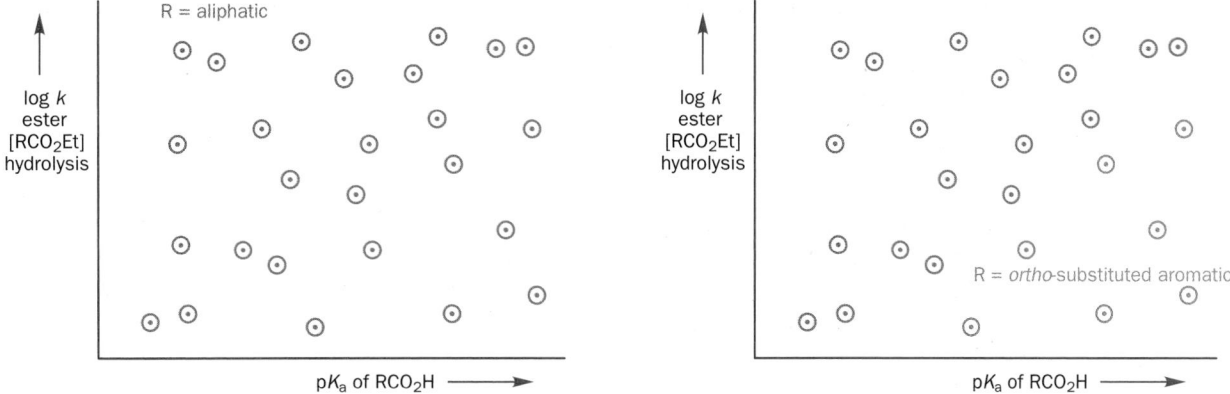

It was not a perfect correlation but Hammett had removed the examples where steric hindrance was important. Aliphatic compounds can adopt a variety of conformations (Chapter 18) and the substituent in some of them will interfere with the reaction. Similarly, in *ortho*-substituted aromatic compounds the nearby substituent might exert steric hindrance on the reaction. Only with *meta*- and *para*-substituted compounds was the substituent held out of the way, on a rigid framework, and in electronic communication with the reaction site through the flat but conjugated benzene ring. The diagrams show the *para* substituent.

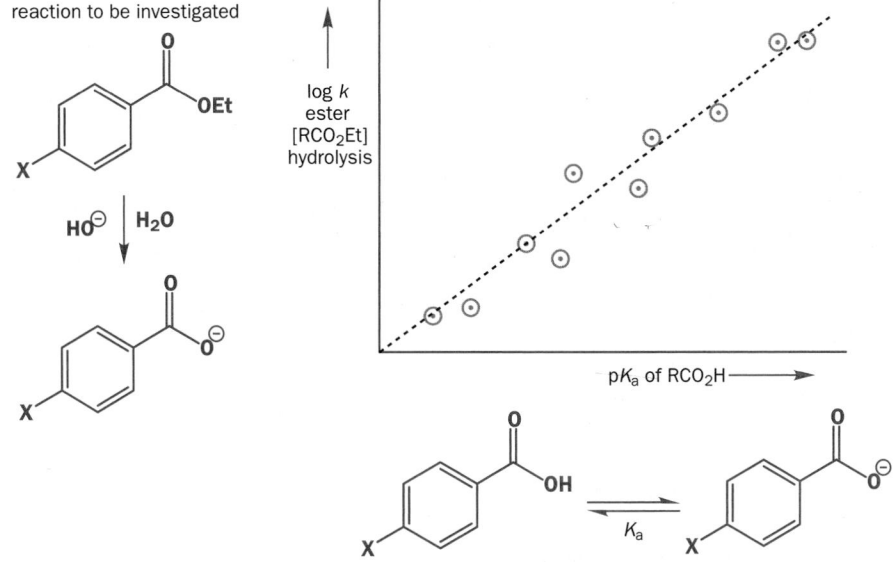

Notice that the straight line is not perfect. This graph is an invention of the human mind. It is a correlation between things that are not directly related. If you determine a rate constant by plotting the right function of concentration against time and get an imperfect straight line, that is your fault because you haven't done your measurements carefully enough. If you make a Hammett plot and the points are not on a straight line (and they won't be) then that is *not* your fault. The points really don't fit on a perfectly straight line. As you will see soon, this does not matter. We need to look at the Hammett correlation in more detail.

> If you plot a graph to correlate the number of miles travelled by jumbo jet against the percentage of births outside of marriage over the twentieth century you will get a sort of straight line. This does not imply a direct causative link!

The Hammett substituent constant σ

A quick glance at the pK_as of some substituted benzoic acids will show how well they correlate electron donation with pK_a. The substituents at the top of the table are electron-donating and the anions of the benzoic acids are correspondingly less stable so these are the weakest acids. At the bottom of the table we have the electron-withdrawing groups, which stabilize the anion and

▶

You cannot push arrows from the negative charge on the carboxylate into the ring. Try it.

make the acid stronger. The whole range is not that great, only one pH unit or so, because the carboxylate anion is not conjugated with the ring.

Hammett decided not to use the pK_as themselves for his correlation but defined a new parameter, which he called σ. This σ shows how electron-donating or -withdrawing a group is relative to H as a ratio of the $\log K_a$s or the difference of the pK_as between the substituent and benzoic acid itself. If the acid required to determine σ for a new substituent was not available, σ could be determined by correlation with other reactions. Here are the equations and the table of σ values for the most important substituents. A different value of σ for any given substituent was needed for the *meta* and the *para* positions and these are called $σ_m$ and $σ_p$, respectively.

Substituent, X	pKa of p-XC6H4COOH	pKa of m-XC6H4COOH
NH$_2$	4.82	4.20
OCH$_3$	4.49	4.09
CH$_3$	4.37	4.26
H	4.20	4.20
F	4.15	3.86
I	3.97	3.85
Cl	3.98	3.83
Br	3.97	3.80
CO$_2$CH$_3$	3.75	3.87
COCH$_3$	3.71	3.83
CN	3.53	3.58
NO$_2$	3.43	3.47

$$\sigma_X = \log\left(\frac{K_a(\text{X–C}_6\text{H}_4\text{COOH})}{K_a(\text{C}_6\text{H}_5\text{COOH})}\right) = pK_a(\text{C}_6\text{H}_5\text{COOH}) - pK_a(\text{X–C}_6\text{H}_4\text{COOH})$$

You need a general idea as to what a σ value means. If σ = 0 the substituent has no effect: it is electronically the same as H. If σ is positive, the substituent is electron-withdrawing. This is unfortunate perhaps, but just remember that the comparison is with acid strength. Positive σ means a stronger acid so the substituent is electron-withdrawing. The more positive the charge induced on the ring by a substituent, the larger its σ value. Negative σ means weaker acid and electron donation. Inductive effects from polarization of σ bonds are greater for $σ_m$ than for $σ_p$ because the substituent is nearer.

Conjugation is generally more effective in the *para* position (see Chapter 22) so $σ_p > σ_m$ for conjugating substituents. Indeed, the NH$_2$ group has a large negative $σ_p$ and a zero $σ_m$. The NH$_2$ group donates electrons strongly to the carbonyl group of benzoic acid from the *para* position but does not conjugate in the *meta* position where its donation happens just to balance the effect of electronegative nitrogen.

The OMe group has a negative $σ_p$ but a positive $σ_m$ because a weaker electron donation from the lone pairs is more important in the *para* position but the effect of very elec-

Substituent, X	$σ_p$	$σ_m$	Comments
NH$_2$	–0.62	0.00	groups that donate electrons have negative σ
OCH$_3$	–0.29	0.11	
CH$_3$	–0.17	–0.06	
H	0.00	0.00	there are no values for *ortho* substituents
F	0.05	0.34	
I	0.23	0.35	
Cl	0.22	0.37	$σ_p < σ_m$ for inductive withdrawal
Br	0.23	0.40	
CO$_2$CH$_3$	0.45	0.33	
COCH$_3$	0.49	0.37	$σ_p > σ_m$ for conjugating substituents
CN	0.67	0.62	
NO$_2$	0.77	0.73	groups that withdraw electrons have positive σ

strong conjugation into carbonyl group: large negative $σ_p$

conjugation into ring not carbonyl group balances weak effect of electronegative N: zero $σ_m$

tronegative oxygen on the σ framework of the ring in the *meta* position is more important than lone pair donation that doesn't reach the carbonyl group. You do not need to learn any σ values but you should be able to work out the sign of σ for well known substituents and estimate a rough value.

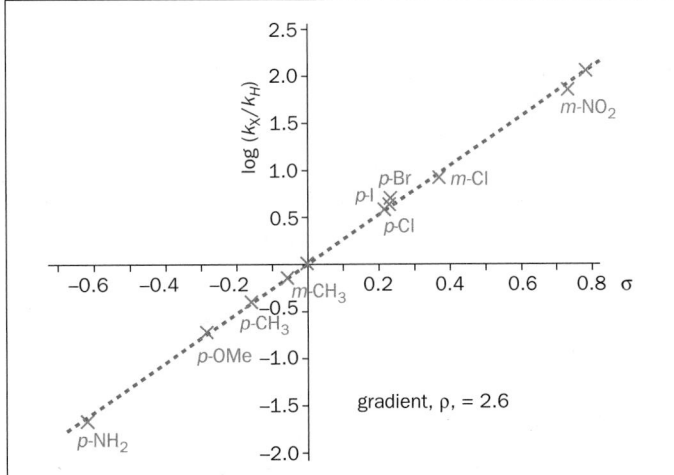

reaction to be investigated

meta- and *para*-X only

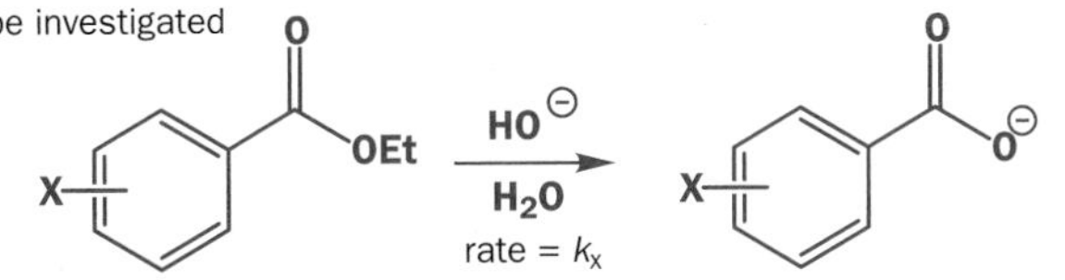

rate = k_x

The Hammett reaction constant ρ

Now we can return to our reaction: the alkaline hydrolysis of various *meta*- and *para*-substituted ethyl benzoates. The rate constants for this second-order reaction have been measured and shown here is a graph of log (k_X/k_H) versus σ, where k_X is the rate constant for the reaction with the substituted benzoate and k_H is that for the unsubstituted reaction (X = H).

We can see straight away that there is a good correlation between how fast the reaction goes and the value of σ; in other words, the points lie more or less on a straight line. The gradient of this best fit line, given the symbol ρ (rho), tells us how sensitive the reaction is to substituent effects in comparison with the ionization of benzoic acids. The gradient is ρ = +2.6. This tells us that the reaction responds to substituent effects in the same way (because it is +) as the ionization of benzoic acids but by much more ($10^{1.6}$ times more) because it is 2.6 instead of 1.0. We already know what the mechanism of this reaction is.

▶ **Getting to grips with logs**

A difference between two values of *x* log units means the values actually differ by a factor of 10^x. From the graph for the hydrolysis of ethyl benzoates we can see that the *p*-NO$_2$ benzoate hydrolyses some 10^2 times faster than the unsubstituted benzoate, while the *p*-NH$_2$ benzoate hydrolyses some 10^2 times slower.

▶
Hammett chose σ (Greek s) for *substituent* and ρ (Greek r) for *reaction*.

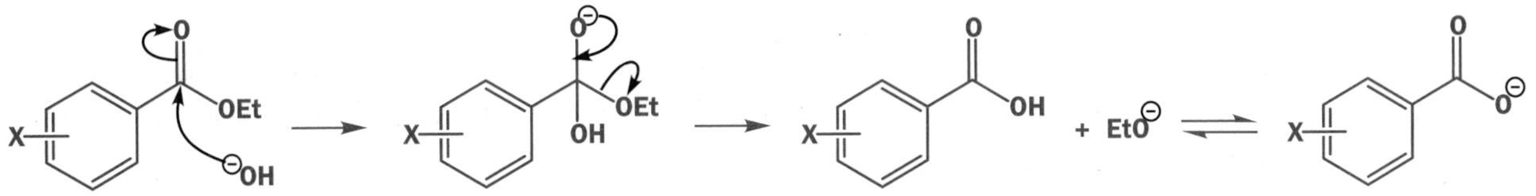

The first step is quite like the ionization of benzoic acid. A negative charge is appearing on the carbonyl oxygen atom and that negative charge will be stabilized by electron-withdrawing X groups. Provided that the first step is rate-determining, a positive ρ is fine. We cannot say much as yet about the value as we are comparing a reaction rate (for the hydrolysis) with an equilibrium position (for the ionization). It will help you a great deal if you think of *positive* ρ values as meaning an *increase* in electron density near to or on the benzene ring. They may mean the appearance of a negative charge but they may not. We need now to look at some other reactions to get a grasp of the meaning of the value of the Hammett ρ.

● **The Hammett reaction constant ρ measures the *sensitivity* of the reaction to electronic effects.**

- A *positive* ρ value means *more* electrons in the transition state than in the starting material
- A *negative* ρ value means *fewer* electrons in the transition state than in the starting material

typical Hammett plots

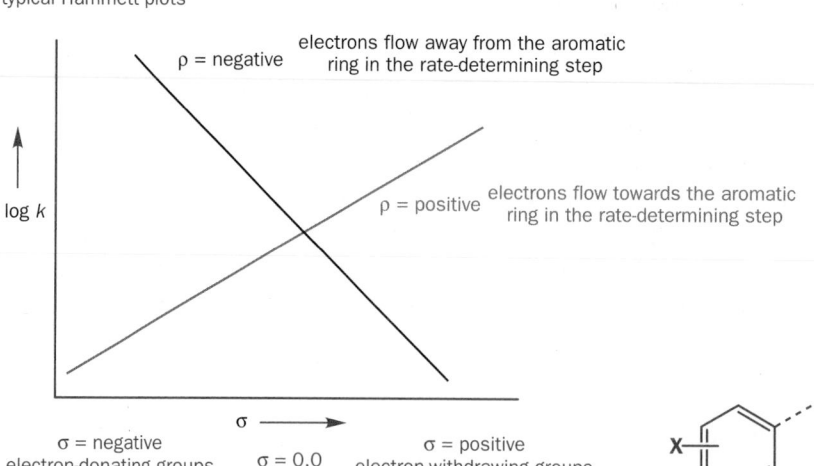

ρ = negative   electrons flow away from the aromatic ring in the rate-determining step

log k

ρ = positive   electrons flow towards the aromatic ring in the rate-determining step

σ

σ = negative σ = 0.0 σ = positive
electron-donating groups   X = H   electron-withdrawing groups
X = MeO, Me, NH$_2$, etc.   X = Cl, CO$_2$Et, CN, NO$_2$, etc.

Equilibria with positive Hammett ρ values

We can compare these directly with the ionization of benzoic acids. If we simply move the carboxylic acid away from the ring, the ρ value for ionization gets less. This is just the effect of a more distant substituent. When there are two saturated carbons between the benzene ring and the carboxylic acid, there is almost no effect. When we are using the aromatic ring as a probe for a reaction mechanism, it must be placed not too far away from the reaction centre. However, if we restore electronic communications with a double bond, ρ goes back up again to a useful value.

ρ = 1.0 (by definition) ρ = 0.5 ρ = 0.2 ρ = 0.5

If the negative charge on the anion can actually be delocalized round the ring, as with substituted phenols, we should expect the size of ρ to increase. Both the phenol and the anion are delocalized but it is more important for the anion. The effect is larger for the ionization of anilinium salts as the acid (ArNH$_3^+$) does not have a delocalized lone pair but the conjugate base (ArNH$_2$) does.

ρ = + 2.3 ρ = + 3.2

Reactions with positive Hammett ρ values

Any reaction that involves nucleophilic attack on a carbonyl group as the rate-determining step is going to have a ρ value of about 2–3, the same as for the hydrolysis of esters that we have already seen. Examples include the Wittig reaction of stabilized ylids (Chapters 14 and 31). Though there is some dispute over the exact mechanism of the Wittig reaction, the ρ value of 2.7 strongly suggests that nucleophilic attack on the aldehyde by the ylid is involved with stabilized ylids and aromatic aldehydes at least. In addition, there is a small variation of rate with the aryl group on phosphorus: if Ar = p-MeOC$_6$H$_4$ the reaction goes about six times faster than if Ar = p-ClC$_6$H$_4$. These groups are a long way from the reaction site but electron donation would be expected to accelerate the donation of electrons from the ylid.

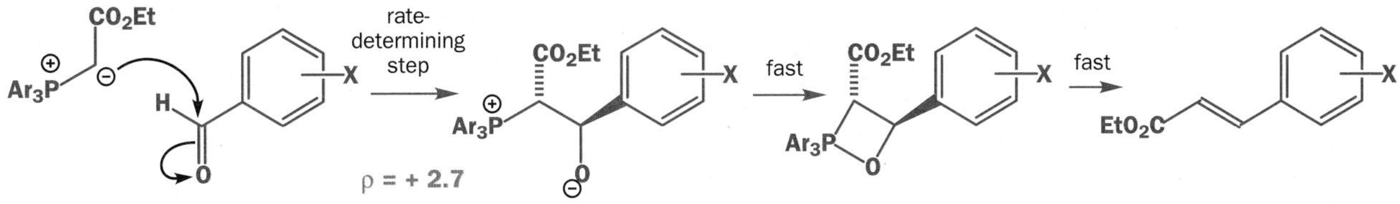

$$\rho = +2.7$$

Large positive ρ values usually indicate extra electrons in the transition state delocalized into the ring itself. A classic example is nucleophilic aromatic substitution by the addition–elimination mechanism (Chapter 23). The ρ value is +4.9, but even this large value does not mean a complete anion on the benzene ring as the nitro group, present in all cases, takes most of the negative charge. The substituent X merely helps.

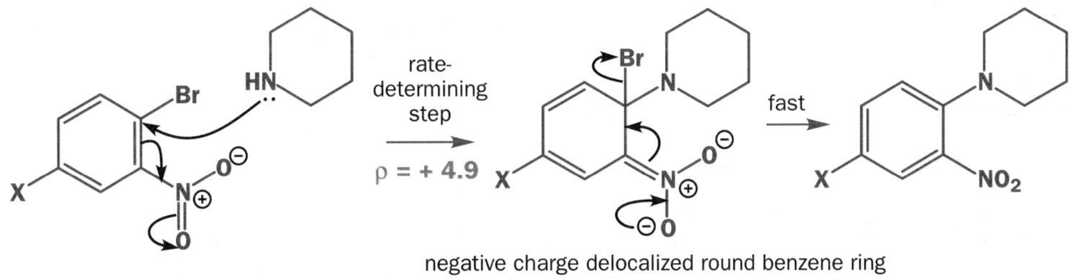

negative charge delocalized round benzene ring

We get the full value when there are no nitro groups to take the brunt of the negative charge. This vinylic substitution (an unusual reaction!) has a ρ value of +9.0. It cannot be an S_N2 reaction or it would have a small ρ value and it cannot be an S_N1 reaction or it would have a negative ρ value (fewer electrons in the transition state). It must be an addition–elimination mechanism through a benzylic anion delocalized round both benzene rings.

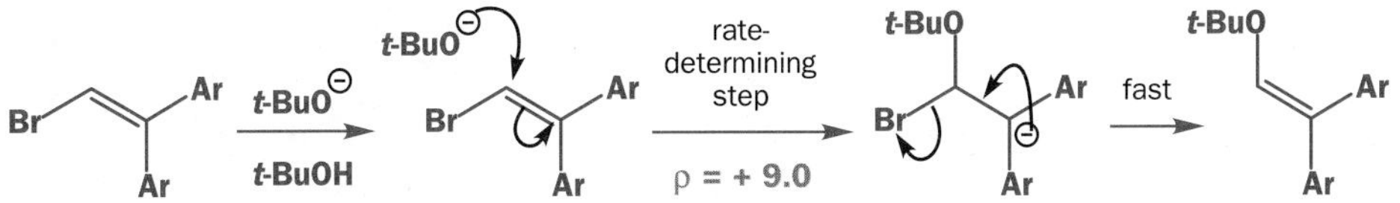

Reactions with negative Hammett ρ values

Negative ρ values mean electrons flowing away from the ring. A useful example is the S_N2 displacement of iodide from EtI by phenoxide anions. This has a ρ value of exactly –1.0. Though the transition state has a negative charge, that charge is decreasing on the aromatic ring as the starting material approaches the transition state.

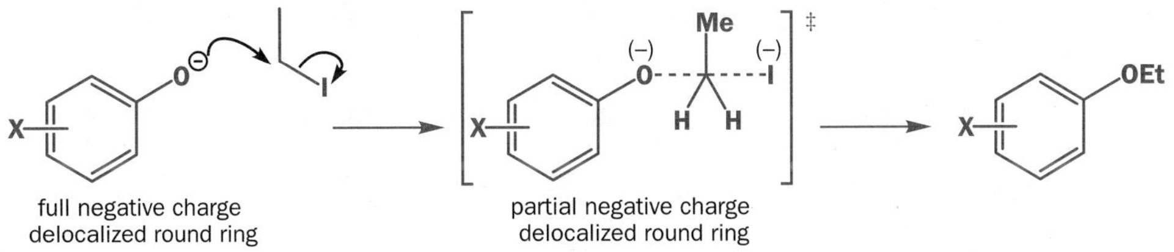

full negative charge delocalized round ring

partial negative charge delocalized round ring

An S_N1 reaction on the carbon atom next to the ring has a large negative ρ value. In this example, a tertiary benzylic cation is the intermediate and the rate-determining step is, of course, the formation of the cation. The cation is next to the ring but delocalized round it and the ρ value is –4.5, about the same value, though negative, as that for the nucleophilic substitution on nitrobenzenes by the addition–elimination mechanism that we saw in the last section.

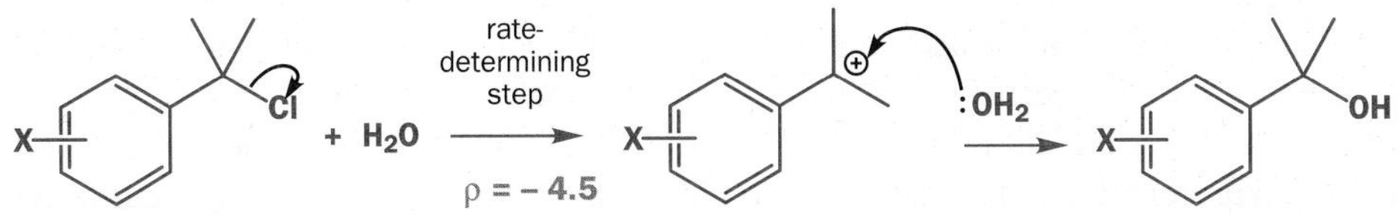

$$\rho = -4.5$$

The largest negative ρ values come from electrophilic aromatic substitution (Chapter 22) where the electrons of the ring are used in the reaction leaving a positive charge on the ring itself in the intermediate. Some of this charge is already there in the transition state. Negative ρ values mean electrons flowing out of the ring. This simple nitration has ρ = − 6.4 and ρ values for electrophilic aromatic substitution are usually in the range −5 to −9.

Reactions with small Hammett ρ values

Small Hammett ρ values arise in three ways. The aromatic ring being used as a probe for the mechanism may simply be too far away for the result to be significant. This trivial case of the alkaline hydrolysis of the 3-aryl propionate ester has a ρ value of +0.5 and it is surprising that it is even that large.

The second case is the informative one where the reaction is not dependent on electrons flowing into or out of the ring. Pericyclic reactions are important examples and the Diels–Alder reaction of arylbutadienes with maleic anhydride shows a small negative ρ value of −0.6. The small value is consistent with a mechanism not involving charge accumulation or dispersal but the sign is interesting. We explained this type of Diels–Alder reaction in Chapter 35 by using the HOMO of the diene and the LUMO of the dienophile. The negative sign of ρ, small though it is, supports this view.

The third case is in many ways the most interesting. We have seen that the alkaline hydrolysis of ethyl esters of benzoic acids ($ArCO_2Et$) has a ρ value of +2.6 and that this is a reasonable value for a reaction involving nucleophilic attack on a carbonyl group conjugated with the aromatic ring. The hydrolysis of the same esters in acid solution, which also involves nucleophilic attack on the same carbonyl group, has a ρ value of +0.1. In other words, all these esters hydrolyse at the same rate in acid solution. Neither of the previous explanations will do. We need to see the full mechanism to explain this remarkable result.

Steps 1, 3, and 5 cannot be slow as they are just proton transfers between oxygen atoms (Chapter 13). That leaves only steps 2 and 4 as possible rate-determining steps. The bimolecular addition of the weak nucleophile water to the low concentration of protonated ester (step 2) is the most attractive candidate, as step 4—the unimolecular loss of ethanol and re-formation of the carbonyl group—should be fast. What ρ value would be expected for the reaction if step 2 were the rate-determining step? It would be made up of two parts. There would be an equilibrium ρ value for the protonation and a reaction ρ value for the addition of water. Step 1 involves electrons flowing out of the molecule and step 2 involves electrons flowing in so the ρ values for these two steps would have opposite charges. We know that the ρ value for step 2 would be about +2.5 and a value of about −2.5 for the equilibrium protonation is reasonable. This is indeed the explanation: step 2 is the rate-determining step and the ρ values for steps 1 and 2 almost cancel each other out. All steps before the rate-determining step are present in the rate equation and also affect the Hammett ρ value.

● **The meaning of Hammett ρ values**

This then is the full picture. You should not, of course, learn these numbers but you need an idea of roughly what each group of values means. You should see now why it is unimportant whether the Hammett correlation gives a good straight line or not. We just want to know whether ρ is + or – and whether it is, say, 3 or 6. It is meaningless to debate the significance of a r value of 3.4 as distinct from one of 3.8.

–6 –5	–4 –3 –2	–1 0 +1	+2 +3 +4	+5 +6
large negative ρ-values	moderate negative ρ-values	small ρ-values	moderate positive ρ-values	large positive ρ-values
positive charge on ring or delocalized round benzene ring	electrons flow out of TS positive charge near ring loss of conjugation	1. Ar too far away 2. No electron change 3. Two ρ-values cancel each other out	electrons flow into TS negative charge near ring loss of conjugation	negative charge on ring or delocalized round benzene ring

Using the Hammett ρ values to discover mechanisms

Electrophilic attack on alkenes by bromine often goes through three-membered ring cyclic bromonium ions and we can sometimes tell that this is so by studying the stereochemistry. Here are two reactions of styrenes that look very similar—a reaction with bromine and one with PhSCl. With no further information, we might be tempted to assume that they both go by the same mechanism. However, the Hammett ρ values for the two reactions are rather different.

Chapter 20 gives a full description of these mechanisms.

There is more about these sulfenyl chlorides in Chapter 46.

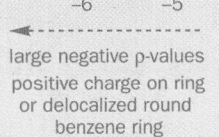 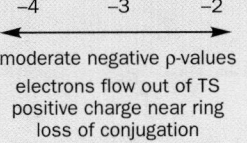 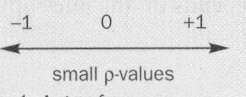

The ρ value for bromination is definitely in the 'large' range and can only mean that a positive charge is formed that is delocalized round the benzene ring. Bromine evidently does not form a bromonium ion with these alkenes but prefers to form a secondary benzylic cation instead.

cation delocalized round ring

The sulfenylation, on the other hand, has a moderate negative ρ value. No cation is formed that is delocalized round the ring, but electrons flow out of the ring and we suspect some loss of conjugation. All this fits well with the formation of a three-membered ring intermediate. From experiments like this we learn that PhSCl is much more likely than bromine to react stereospecifically with alkenes through cyclic cation intermediates.

A complete picture of the transition state from Hammett plots

More information can be gained on the mechanism of the reaction if two separate experiments can be carried out with the mechanistic probe inserted at two different sites on the reagents. If we are studying a reaction between a nucleophile and an electrophile, it may be possible to make Hammett plots from the variation of substituents on both reagents. The acylation of amines with acid chlorides is an example.

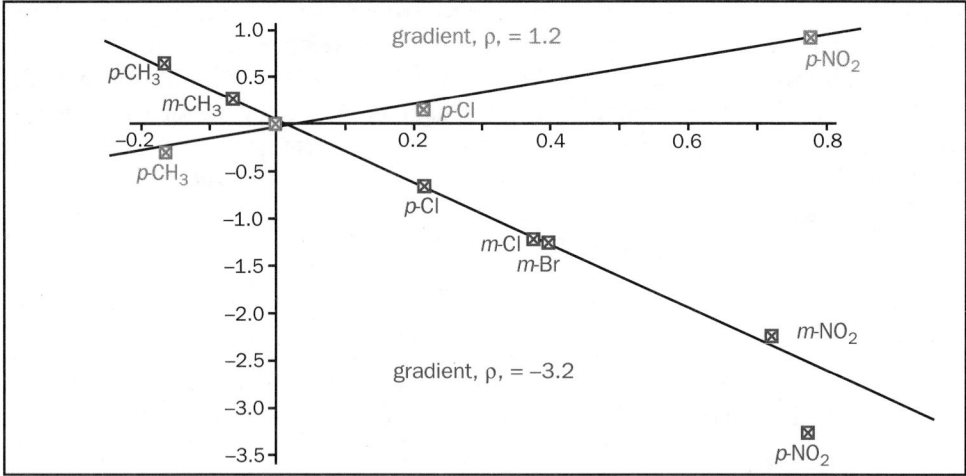

If we vary the structure of the acid chloride we get a ρ value of +1.2, suitable for nucleophilic attack on the carbonyl group. If we vary the amine we get a ρ value of –3.2, again suitable for electrons that were conjugated round the ring moving away to form a new bond. The simple answer is correct but the rate depends on the nucleophilicity of the amine 100 times more than on the electrophilicity of the acid chloride.

Nonlinear Hammett plots

If we look at the hydrolysis of the acid chlorides of benzoic acids in aqueous acetone, we see a very odd Hammett plot indeed. You know that Hammett plots need not be perfectly linear but this one is clearly made up of two intersecting straight lines. This might look like disaster at first but, in fact, it gives us extra information. The right-hand part of the curve, for the more electron-withdrawing substituents, has a slope of +2.5: just what we should expect for rate-determining attack of water on the carbonyl group. As we go to less electron-withdrawing substituents, the rate of the reaction suddenly starts to increase as we pass the *para*-chloro compound and the left-hand part of the curve has a slope of –4.4.

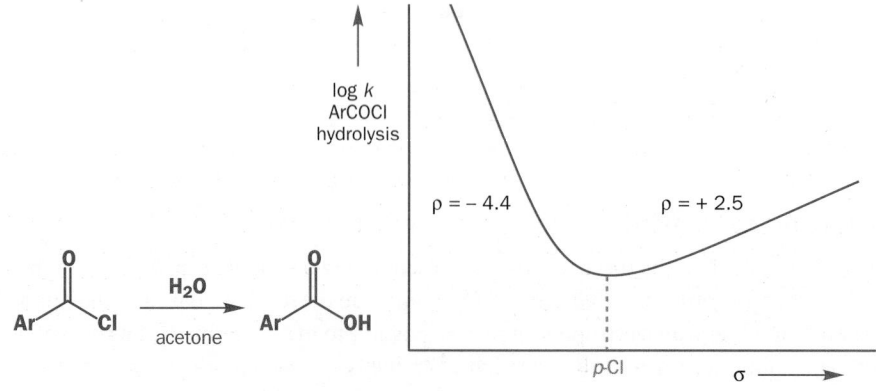

What can this mean? If the reaction becomes faster as we pass the discontinuity in the curve—and it gets faster whether we go from right to left or left to right—there must be a change in mechanism. If there is a choice between two mechanisms, the faster of the two will operate. Mechanism 1 is the rate-determining nucleophilic attack by water on the carbonyl group.

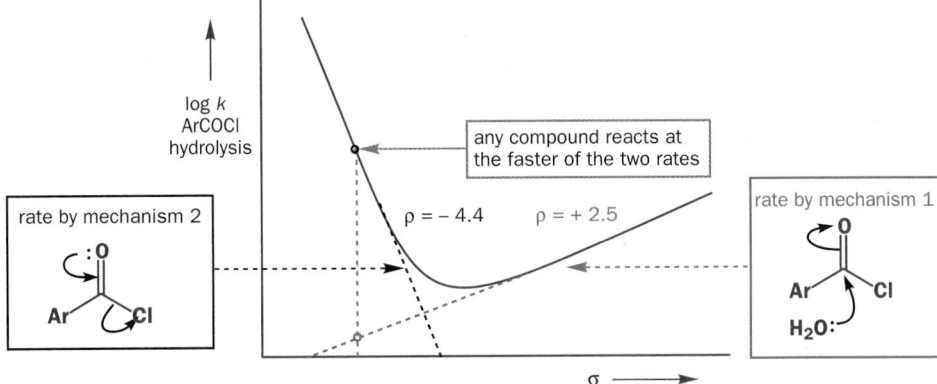

The new mechanism goes faster for more electron-donating substituents and has quite a large negative ρ value suggesting the formation of a cation in the rate-determining step. This mechanism (mechanism 2) must surely be the S_N1-like process of preliminary formation of an acylium ion by loss of chloride ion.

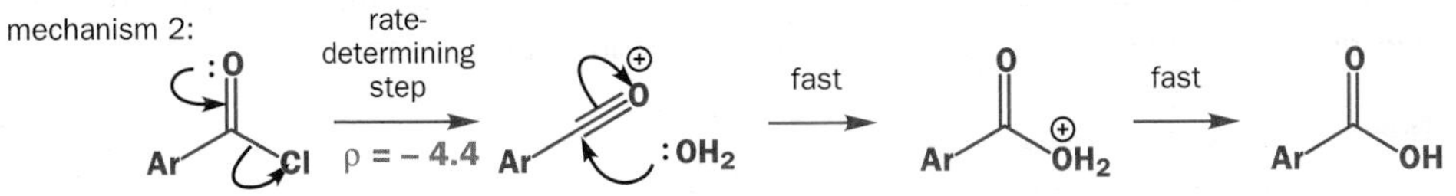

mechanism 2:

When the Hammett plot bends the other way, so that the rate of the reaction decreases as it passes the discontinuity, we have a single mechanism with a change in rate-determining step. A reaction goes by the fastest possible mechanism but its rate is limited by the slowest of the steps in that mechanism. An example is the intramolecular Friedel–Crafts alkylation of a diphenyl derivative where the alkylating agent is a diarylmethanol attached to one of the benzene rings in the *ortho* position.

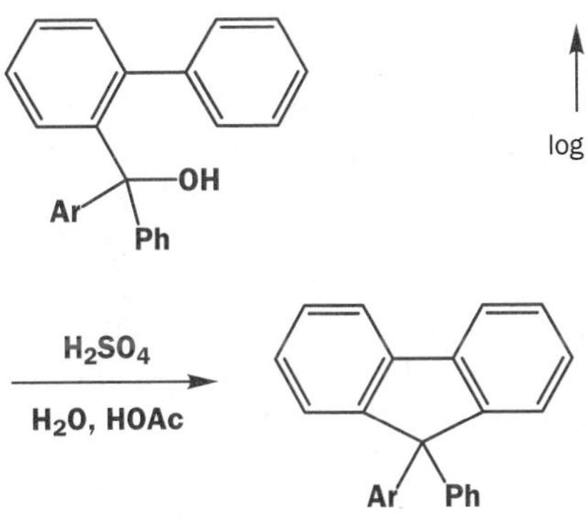

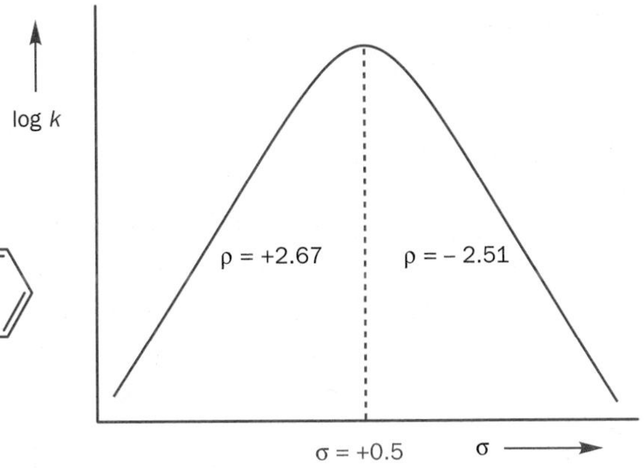

The carbocation intermediate in the Friedel–Crafts reaction (Chapter 22) is rather stable, being tertiary and benzylic, and the formation of the cation, normally the rate-determining step, with inevitably a negative ρ value, goes faster and faster as the electron-donating power of the substituents increases until it is faster than the cyclization which becomes the rate-determining step. The cyclization puts electrons back into the carbocation and has a positive ρ value. As the two steps have more or less the reverse electron flow to and from the same carbon atom, it is reasonable for the size of ρ to be about the same but of opposite sign.

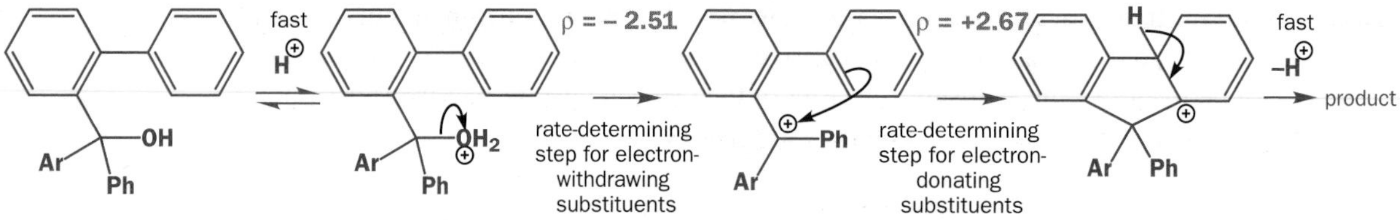

● **A reaction occurs by the faster of two possible mechanisms but by the slower of two possible rate-determining steps.**

We shall see more examples of Hammett ρ values used in conjunction with other evidence as the chapter develops but now it is time to look at what other evidence is available.

Other kinetic evidence

The kinetic deuterium isotope effect

> ►
> Other kinetic isotope effects are known but they are very small: D is twice as heavy as H but ^{13}C only slightly heavier than ^{12}C.

The kinetic isotope effect was introduced in Chapter 19. If a bond to deuterium is formed or broken in the rate-determining step of a reaction, the deuterated compound will react more slowly, usually by a factor of about 2–7. This effect is particularly valuable when C–H bonds are being formed or broken. In Chapter 22 we told you that the rate-determining step in the nitration of benzene was the attack of the electrophile on the benzene ring. This is easily verified by replacing the hydrogen atoms round the benzene ring with deuteriums. The rate of the reaction stays the same.

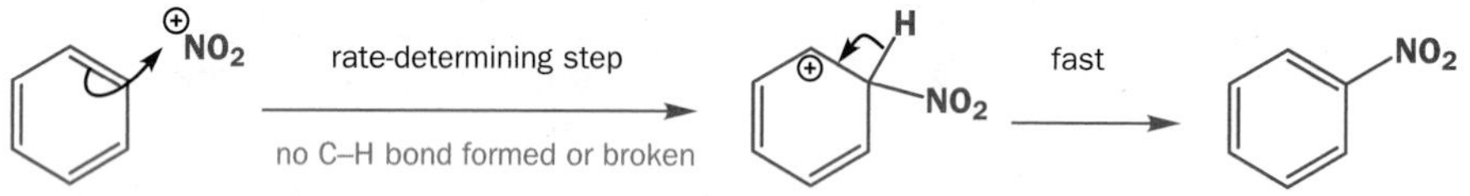

If the second step, which does involve the breaking of a C–H bond, were the rate-determining step it would go more slowly if the H were replaced by D. In this case the deuterium isotope effect is $k_H/k_D = 1.0$. If the reaction is the iodination of phenol in basic solution, there is a deuterium isotope effect of $k_H/k_D = 4.1$. Clearly, the other step must now be the rate-determining step—the phenolate ion reacts so rapidly that the first step is faster than the second.

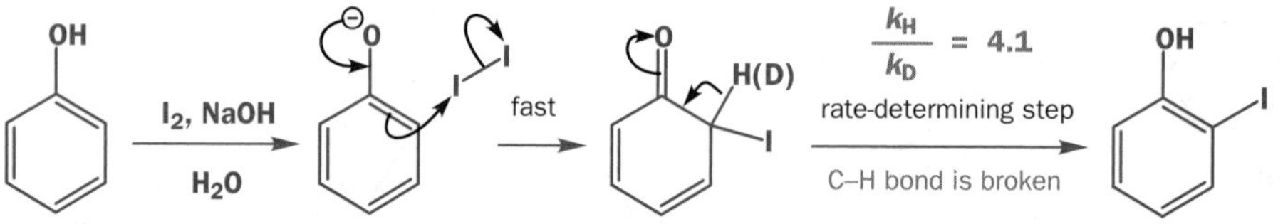

The deuterium isotope effect can add to the information from Hammett plots in building up a picture of a transition state. Three separate Hammett ρ values can be measured for this elimination reaction and this information is very valuable. But it would be sadly incomplete without the information that a large deuterium isotope effect $k_H/k_D = 7.1$ is observed for the hydrogen atom under attack.

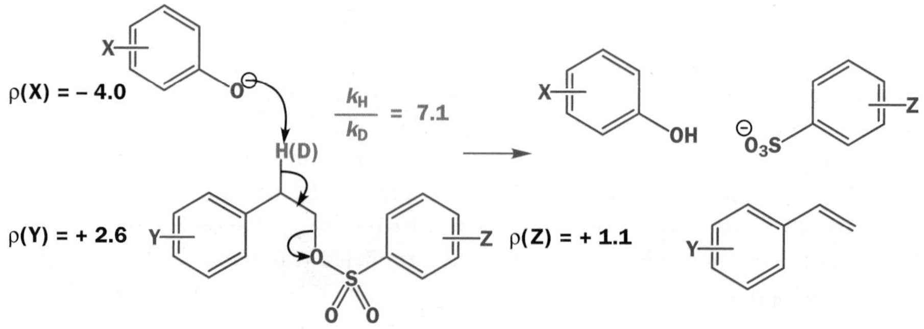

In this E2 reaction, it is no surprise that the base (ArO⁻) donates electrons and the leaving group (ArSO₃⁻) accepts them. But the large deuterium isotope effect and moderate positive ρ(Y) value for an aromatic ring that might have done nothing suggest some build-up of negative charge in the transition state on that carbon atom as well as on the two oxygen atoms.

Entropy of activation

Of all the enthalpies and entropies that we introduced in Chapter 13, the entropy of activation, $\Delta S^{\ddagger}$, is by far the most useful. It tells us about the increase or decrease in order in a reaction as the starting material goes to the transition state. A positive $\Delta S^{\ddagger}$ means an increase in entropy or a decrease in order and a negative $\Delta S^{\ddagger}$ means an increase in order. Normally, unimolecular reactions in which one molecule gives two products have a positive $\Delta S^{\ddagger}$ and bimolecular reactions have a negative $\Delta S^{\ddagger}$. Fragmentations (Chapter 38) such as this decarboxylation in which one molecule fragments to three have positive $\Delta S^{\ddagger}$s. It has $\Delta S^{\ddagger} = +36.8\ \mathrm{J\ mol^{-1}\ K^{-1}}$.

At the other extreme are cycloadditions (Chapter 35) such as the Diels–Alder reaction we examined a few pages back. Not only do two reagents become one product but a very precise orientation is required in the transition state usually meaning a large negative $\Delta S^{\ddagger}$. Diels–Alder reactions usually have $\Delta S^{\ddagger}$ of about -120 to $-160\ \mathrm{J\ mol^{-1}\ K^{-1}}$. The classic cyclopentadiene addition to maleic anhydride has $\Delta S^{\ddagger} = -144\ \mathrm{J\ mol^{-1}\ K^{-1}}$.

> Entropies of activation are measured in units of $\mathrm{J\ mol^{-1}\ K^{-1}}$. All the values in this book are in $\mathrm{J\ mol^{-1}\ K^{-1}}$ but in older books you will see 'entropy units' (e.u.), which are $\mathrm{cal\ mol^{-1}\ K^{-1}}$. Values in e.u. should be multiplied by about 4 to get values in $\mathrm{J\ mol^{-1}\ K^{-1}}$.

These numbers give you the range of entropies of activation you may expect to find. Large negative numbers are common but only small positive numbers are found. The largest negative numbers apply to bimolecular reactions where neither reagent is in great excess. Smaller negative numbers may mean a bimolecular reaction with solvent or some other reagent in large excess. The acid-catalysed opening of styrene oxides in methanol is a good example.

The Hammett ρ value of −4.1 suggests a carbocation intermediate as does the regioselectivity of the reaction (MeOH attacks the benzylic position) but the stereochemistry (the reaction occurs with inversion) and a modest negative entropy of activation ($\Delta S^{\ddagger} = -48\ \text{J mol}^{-1}\text{K}^{-1}$) suggest rather an S_N2 reaction with a loose transition state having substantial positive charge at the benzylic carbon. Neither piece of evidence alone would be enough to define the mechanism.

This example with its acid catalyst brings us to the subject of catalysis. We must now analyse the different sorts of acid and base catalysis and see how the mechanisms can be distinguished using the methods we have discussed.

Acid and base catalysis

Acids and bases provide the best known ways of speeding up reactions. If you want to make an ester—add some acid. If you want to hydrolyse an ester—add some base. It may all seem rather simple. However, there are actually two kinds of acid catalysis and two kinds of base catalysis and this section is intended to explain the difference in concept and how to discover which operates. When we talk about acid catalysis we normally mean **specific acid catalysis**. This is the kind we have just seen—epoxides don't react with methanol but, if we protonate the epoxide first, then it reacts. Specific acid catalysis protonates electrophiles and makes them more electrophilic.

specific acid catalysis

We could, on the other hand, have argued that methanol is not a good enough nucleophile but if deprotonated with a base it becomes the much more nucleophilic methoxide. This is **specific base catalysis**.

specific base catalysis

We shall discuss these two types first because they are straightforward. You need to recognize their characteristics, their strengths, and their weaknesses. We hope you will get into the habit of recognizing these types of catalysis so that you hardly have to think about it—it should become second nature.

Specific acid catalysis

Specific acid catalysis (SAC) involves a rapid protonation of the compound followed by the slow step, which is accelerated in comparison with the uncatalysed reaction because of the greater reactivity of the protonated compound. You have just seen an example with an epoxide. Ester hydrolysis (or formation) is another. Water attacks esters very slowly: it attacks protonated esters much more quickly. This is just the ordinary mechanism for acid-catalysed ester hydrolysis (or formation) given in Chapter 12.

▶
SAC is the usual method by which acids make reactions go faster and, if you think about the acid-catalysed reactions you already know, you will see that you have been using it all along without realizing it.

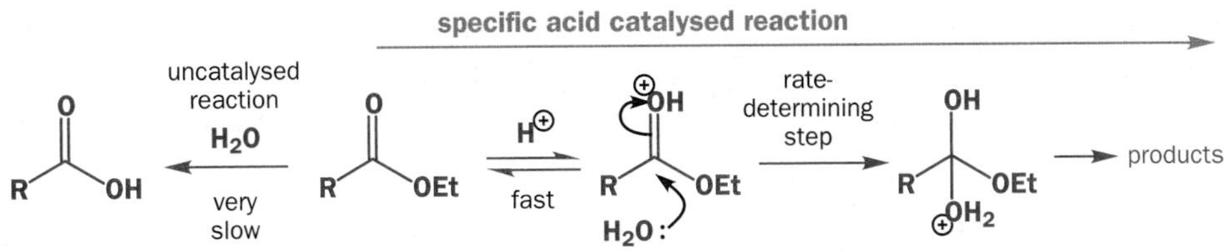

A more interesting reaction is the dienone–phenol rearrangement (Chapter 37). Rearrangement in the absence of acid is very slow but, once the ketone oxygen is protonated, it occurs very rapidly. Again we have fast equilibrium protonation followed by a rate-determining step involving a reaction of the protonated species and again this is the ordinary mechanism that you now know to call SAC.

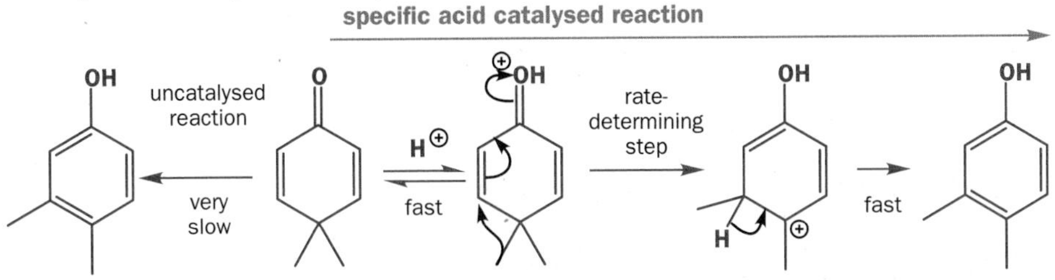

This catalysis depends only on the protonating power of the solution. The compound must be protonated to react so the catalyst must be a strong enough acid to do the job. It is not necessary that every molecule is protonated, just enough to set the reaction going as the acid is regenerated at the end. So the (log of the) rate of the reaction is inversely proportional to the pH of the solution and significant only in the region of, and of course below, the pK_{aH} of the substrate.

There is one special experimental indication of this mechanism. If the reaction is carried out in a deuterated solvent (D_2O instead of H_2O) the rate of the reaction increases. This is a solvent isotope effect rather than a kinetic isotope effect and needs some explanation. If you examine the three examples of SAC in the previous pages you will see that they share these characteristics: a fast proton exchange is followed by a rate-determining step that does *not* involve the making or breaking of any bonds to hydrogen. In general terms:

The rate of the reaction is the rate of the rate-determining step: rate = $k[XH^+]$. The concentration of the intermediate $[XH^+]$ is related to the pH and to the concentration of the substrate by the equilibrium constant, K, of the protonation. So we have: rate = $kK[H^+][X]$. We know that k does not change when hydrogen is replaced by deuterium so K must increase in D_2O.

You will sometimes see in books the statement that D_3O^+ is a stronger acid than H_3O^+. This is partly true. The full truth is that D_3O^+ *in D_2O* is a stronger acid than H_3O^+ *in H_2O*. Water (H_2O) is a better solvating agent for H_3O^+ than D_2O is for D_3O^+, simply because it forms stronger hydrogen bonds due to the greater O–H vibration frequency. So D_3O^+ in D_2O is less well solvated than H_3O^+ in H_2O and is a stronger acid. You need an example.

The Z-allylic alcohol below dehydrates in acid solution to the E-diene. We have lots of data on this mechanism, all summarized in the diagrams. You may like to note as well that the product contains no deuterium after dehydration in D_2O.

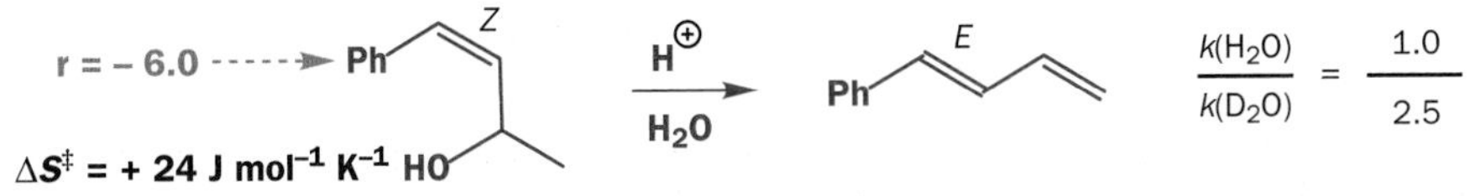

The Hammett ρ value of –6.0 suggests a carbocation intermediate and the positive entropy of activation suggests a rate-determining step in which disorder increases, perhaps one molecule breaking into two. The inverse solvent deuterium isotope effect (faster reaction in D_2O than in H_2O) strongly suggests SAC. Putting all this together we have a mechanism—a simple example of SAC with no protonation at carbon.

▶
A normal kinetic isotope effect has $k_H/k_D > 1$. Deuterium is often put into compounds by exchange with the cheapest source, D_2O, so reactions in D_2O often go slower than reactions in H_2O. Reactions with $k_H/k_D < 1$ have inverse deuterium isotope effects so a reaction that goes faster in D_2O than in H_2O (even when that is the expected pattern) has an *inverse* solvent deuterium isotope effect.

▶
It is not, of course, possible to use D_3O^+ in H_2O as H and D exchange very quickly. The solvent determines which acid is present.

■
You might like to compare this mechanism with the isomerization of the same diene described earlier in this chapter.

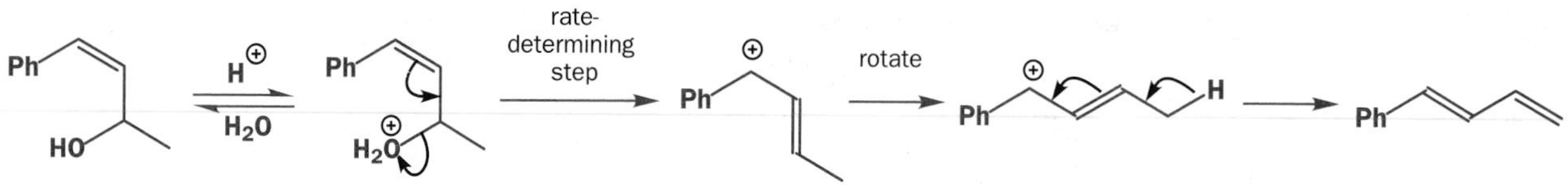

One more thing about this example. The rate-determining step is the second step so the other data, the Hammett ρ value and the entropy of activation, also refer to the combination of K and k. The equilibrium ρ value for the protonation will be fairly small and negative as a positive charge is being created some way from the benzene ring. The kinetic ρ value for the loss of water will be large and negative because a positive charge is being created that is delocalized into the ring. A combined value of –6 looks fine. The equilibrium entropy ΔS^o for the protonation will probably be small and negative as $ROH + H_3O^+ \rightleftharpoons ROH_2^+ + H_2O$ represents little change in order (two molecules going to two) and the $\Delta S^\ddagger$ for the loss of water will be large and positive (one molecule going to two) so a small positive value is about right. It doesn't do to interpret these numbers too closely.

> ● **Summary of features of specific acid catalysis**
>
> 1. Only H_3O^+ is an effective catalyst; pH alone matters
> 2. Usually means rate-determining reaction of protonated species
> 3. Effective only at pHs near or below the pK_{aH} of the substrate
> 4. Proton transfer is not involved in the rate-determining step
> 5. Only simple unimolecular and bimolecular steps—moderate + or $-\Delta S^\ddagger$
> 6. Inverse solvent isotope effect $k(H_2O) < k(D_2O)$

Specific base catalysis

The other side of the coin is specific base catalysis (SBC) which usually involves the removal of a proton from the substrate in a fast pre-equilibrium step followed by a rate-determining reaction of the anion. Most of the base-catalysed reactions you are familiar with work by SBC. Examples include opening of epoxides with thiols.

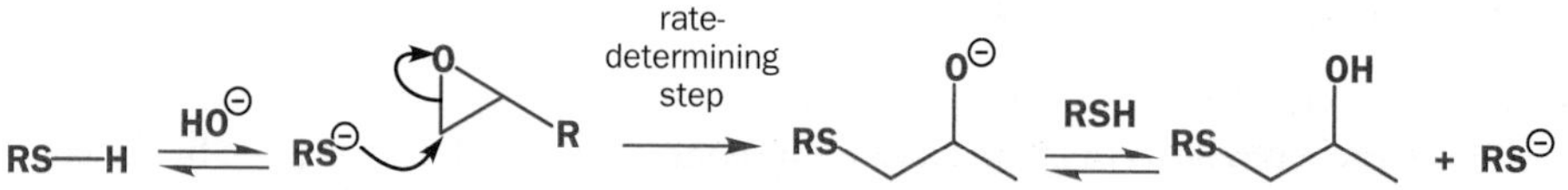

The rate of the reaction depends on the pH of the solution. If it is around or higher than the pK_a of the thiol, thiolate anion will be formed and this opens the epoxide much faster than does the unionized thiol. The nucleophile is regenerated by the oxyanion produced in the rate-determining step. A more familiar example is the base-catalysed hydrolysis of esters we have mentioned several times in this chapter. The full pH–rate profile (Chapter 13) for the hydrolysis of a simple ester such as ethyl acetate shows just two straight lines meeting each other (and zero rate) at about neutrality. Ethyl acetate hydrolysis occurs by SAC or SBC only.

pH–rate profile for ester hydrolysis

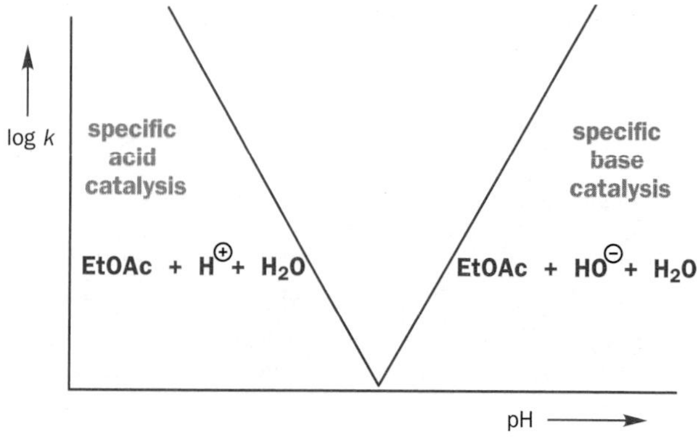

Removal of a proton from heteroatoms by heteroatom bases is always a fast step but removal of a proton from carbon can be the rate-determining step. A remarkably large inverse solvent deuterium isotope effect was found with this elimination of a tertiary amine in basic solution.

$$\frac{k(H_2O)}{k(D_2O)} = \frac{1.0}{7.7}$$

The detailed mechanism cannot, of course, be E2 or the isotope effect, if any, would be the other way round. If it is SBC, the mechanism then becomes the well-known E1cB (Chapter 19) having a carbanion as intermediate.

But 1/7.7 is too large to be a solvent isotope effect and looks much more like a normal kinetic isotope effect. And so it is. The tertiary amine is not a very good leaving group in spite of its positive charge (pK_{aH} about 10) so the carbanion mostly reverts to starting materials. The isotope effect is a kinetic isotope effect on this reverse step—the protonation of the carbanion. This reaction involves a proton transfer from H_2O or D_2O and will be much faster (could be 7.7 times) in H_2O by the ordinary kinetic isotope effect. The *elimination* reaction goes faster in D_2O because the back reaction goes more slowly and more of the carbanion goes on to product.

7.7 times slower in D_2O

> **▶ Microscopic reversibility**
>
> There is only one least-energy pathway between two interconverting compounds such as the starting material and the intermediate here. Every microscopic detail of the back reaction is exactly the same as that for the forward reaction. This is the principle of microscopic reversibility. Here we use evidence from the back reaction (slow proton transfer from water to the carbanion) to tell us about the forward reaction. This principle will be useful in Chapter 42.

● Summary of features of specific base catalysis

1. Only HO^- is an effective catalyst; pH alone matters
2. Usually means rate-determining reaction of deprotonated species
3. Effective only at pHs near or above the pK_a of the substrate
4. Proton transfer is not involved in the rate-determining step, unless C–H bonds are involved
5. Only simple unimolecular and bimolecular steps—moderate + or $-\Delta S^{\ddagger}$
6. Inverse solvent isotope effect $k(H_2O) < k(D_2O)$

General acid/base catalysis

The other kind of acid/base catalysis is called 'general' rather than 'specific' and abbreviated GAC or GBC. As the name implies this kind of catalysis depends not only on pH but also on the concentration of undissociated acids and bases other than hydroxide ion. It is a milder kind of catalysis and is used in living things. The proton transfer is not complete before the rate-determining step but occurs during it. A simple example is the catalysis by acetate ion of the formation of esters from alcohols and acetic anhydride.

> ■
> There was some discussion of this reaction in Chapter 13. Chapter 12 refers to the difficulty of pinpointing proton transfers in mechanisms involving the carbonyl group.

ROH + [electrophile] + [catalyst] → [product] + [catalyst regenerated]

nucleophile   electrophile   catalyst   product   catalyst regenerated

How can this catalysis work? At first sight there seems to be no mechanism available. Acetate cannot act as a specific base—it is far too weak (pK_{aH} 4.7) to remove a proton from an alcohol (pK_a about 15). If it acted as a nucleophile (Chapters 12 and 13) there would be no catalysis as nucleophilic attack on acetic anhydride would be a nonreaction simply regenerating starting materials. The only thing it can do is to remove the proton from the alcohol *as the reaction occurs*.

general base   nucleophile   electrophile
catalyst

You will see at once that there is a great disadvantage in this mechanism: the rate-determining step is termolecular and this is really termolecular—three molecules colliding—and not just some mathematical kinetic trick. This comes out most clearly in the entropy of activation which is an enormous negative value, around $\Delta S^{\ddagger} = -168\ \mathrm{J\,mol^{-1}\,K^{-1}}$ for this reaction. There will also be a normal kinetic isotope effect for ROD against ROH as a bond to hydrogen is being formed and broken in the rate-determining step: it is $k_H/k_D = 2.4$ here. These GBC or GAC reactions are normally effective only if one of the three molecules is present in large excess—this reaction might be done in ROH as a solvent, for example, so that ROH is always present. In understanding how this GBC works it is helpful to look at the mechanism without catalysis.

nucleophile   electrophile   intermediate

The acetate catalyst cannot remove a proton from the starting material but it can easily remove a proton from the intermediate, which has a complete positive charge on the alcohol oxygen atom. The starting material has a pK_a above the pK_{aH} of acetate but the product has a pK_a well below it. Somewhere in the middle of the rate-determining step, the pK_a of the ROH proton passes through the pK_{aH} of acetate and then acetate is a strong enough base to remove it. The GBC is effectively deprotonating the transition state.

general base   nucleophile   electrophile   transition state
catalyst

So how do we find GAC or GBC? Normally, general species catalysis is a weak addition to specific catalysis. We must remove that more powerful style of catalysis by working at a specific pH because SAC or SBC depends on pH alone. If we find that the rate of the reaction changes with the concentration of a weak base at constant pH, we have GBC. Note that, if the proton transfer is between heteroatoms, as in this example, some other bond-making or bond-breaking steps must be happening too as proton transfer between heteroatoms is always a fast process. Proton transfer to or from carbon can be slow.

The formation of three- and five-membered cyclic ethers shows the contrast between GBC and SBC. The formation of epoxides is straightforward SBC with a simple linear dependence on pH between pH 8 and 12 and no acceleration at constant pH by carbonate (CO_3^{2-}) ions. There is an inverse solvent isotope effect and an aryl substituent at the electrophilic carbon atom gives the small positive ρ value expected for S_N2 with an anion.

Formation of tetrahydrofuran (THF) is also faster at higher pH but, by contrast, is also accelerated by various bases at constant pH. If anions of phenols (ArO⁻) are used as catalysts, a Hammett ρ value of +0.8 shows that electrons are flowing away from the aromatic ring. There is a small normal kinetic isotope effect $k_H/k_D = 1.4$. There is SBC and GBC in this reaction. Here is the mechanism with ArO⁻ as GBC.

Why are the two different? The THF is easy to form, the transition state is unstrained, and only a little help is needed to make the reaction go. The epoxide is very strained indeed and the starting material needs to be raised in energy before cyclization will occur. Only the most powerful catalysis is good enough.

> ● **Summary of features of general base catalysis**
>
> 1. Any base is an effective catalyst; pH also matters
> 2. Proton transfer is involved in the rate-determining step
> 3. Effective at neutral pHs even if below the pK_a of the substrate
> 4. Catalyst often much too weak a base to deprotonate reagent
> 5. Catalyst removes proton, which is becoming more acidic in the rate-determining step
> 6. Some other bond-making or bond-breaking also involved unless proton is on carbon
> 7. Often termolecular rate-determining step: large $-\Delta S^{\ddagger}$
> 8. Normal kinetic isotope effect $k(H) > k(D)$

General acid catalysis

We have already discussed this in general terms so a couple of examples will be enough. First, the termolecular problem can be avoided if the reaction is intramolecular. The catalysis is then bimolecular as in the cyclization of this hydroxy-acid. Normally, ester formation and hydrolysis are specific-acid-catalysed only but here there is catalysis by acetic acid; $k(HOAc)/k(DOAc)$ is 2.3 showing that proton transfer occurs in the rate-determining step and there is a large negative $\Delta S^{\ddagger} = -156$ $J\,mol^{-1}\,K^{-1}$. This is general acid catalysis of nucleophilic attack on a carbonyl group, admittedly in a special molecule.

Earlier in the book (Chapter 14) we emphasized the importance of the mechanism for the formation and hydrolysis of acetals. These are SAC reactions: alcohols are bad leaving groups and usually need to be fully protonated by strong acids before they will go, even with the help of a lone pair on another oxygen atom.

specific acid-catalysed acetal hydrolysis

In both these examples the steps after the rate-determining step are omitted and you should look at Chapter 14 for the full details.

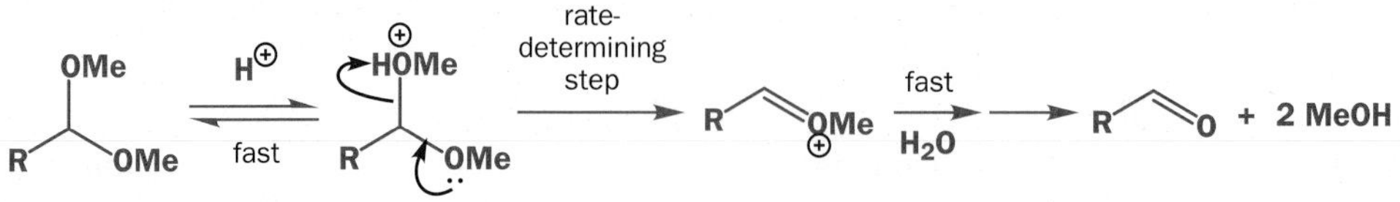

If we speed up the slow step by adding to the molecule some feature that stabilizes the cation intermediate, general acid catalysis may be found. One example is the aromatic cation formed in the hydrolysis of cycloheptatrienone acetals. The normal kinetic isotope effect proclaims GAC.

general acid-catalysed acetal hydrolysis

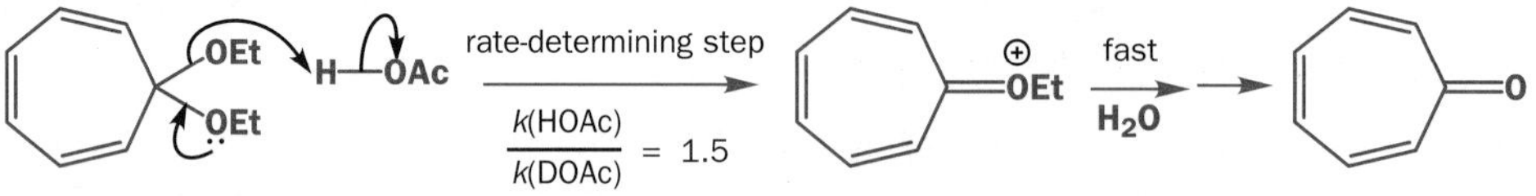

Even adding one extra alkoxy group so that we have an orthoester instead of an acetal is enough. These compounds show catalysis with a variety of weak acids at not very acidic pHs (5–6). As one OMe group is protonated, two others are pushing it out and they both help to stabilize the intermediate cation. Nature prefers these milder methods of catalysis as we will see in Chapter 50.

general acid-catalysed orthoester hydrolysis

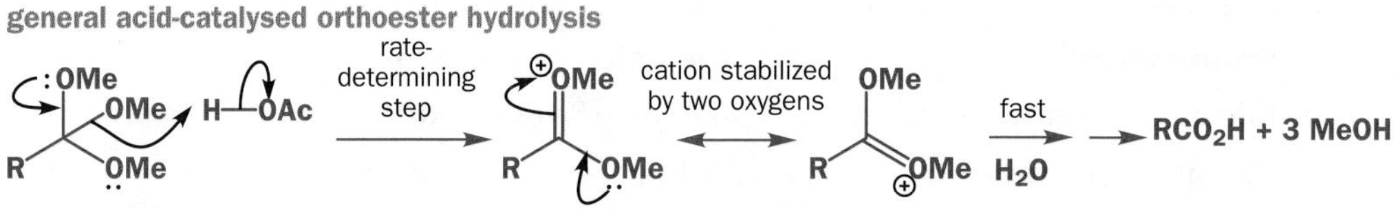

For another contrast between SAC and GAC we need only refer you back to the two *Z*/*E* isomerizations earlier in the chapter. Isomerization of the diene is GAC—protonation at carbon is the slow step—and isomerization of the allylic alcohol is SAC. What we didn't tell you earlier was that the GAC reaction has a normal kinetic isotope effect of $k(\mathrm{H})/k(\mathrm{D}) = 2.5$ and a negative entropy of activation $\Delta S^{\ddagger} = -36\ \mathrm{J\,mol^{-1}\,K^{-1}}$—just what we should expect for a bimolecular reaction involving rate-determining proton transfer from oxygen to carbon. Notice that the intermediate cation is the same whichever the route; only the ways of getting there, including the rate-determining steps, are different.

specific acid catalysis

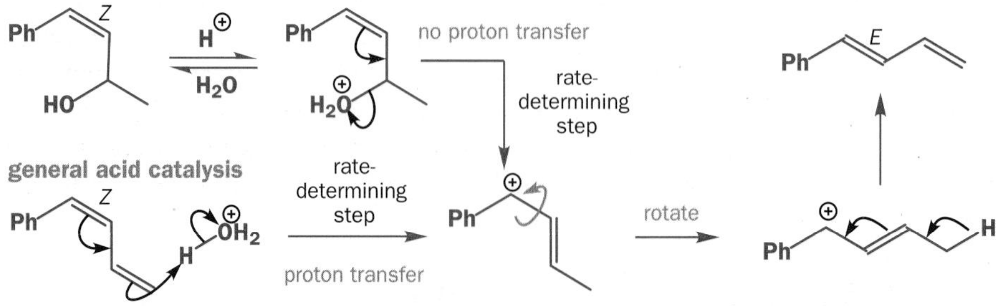

These examples show you that general acid catalysis is possible with strong acids, especially when protonation is at carbon and that, when protonation is at carbon, no other bond-making or -breaking steps need be involved.

● **Summary of features of general acid catalysis**

1. Any acid is an effective catalyst; pH also matters
2. Proton transfer is involved in the rate-determining step
3. Effective at neutral pHs even if above the pK_{aH} of the substrate
4. Catalyst often much too weak an acid to protonate reagent
5. Catalyst adds proton to a site that is becoming more basic in the rate-determining step
6. Some other bond-making or bond-breaking also involved unless proton is on carbon
7. Often termolecular rate-determining step: large $-\Delta S^{\ddagger}$
8. Normal kinetic isotope effect $k(H) > k(D)$

The detection of intermediates

In earlier chapters we revealed how some reactive intermediates can be prepared, usually under special conditions rather different from those of the reaction under study, as a reassurance that some of these unlikely looking species can have real existence. Intermediates of this kind include the carbocation in the S_N1 reaction (Chapter 17), the cations and anions in electrophilic (Chapter 22) and nucleophilic (Chapter 23) aromatic substitutions, and the enols and enolates in various reactions of carbonyl compounds (Chapters 21 and 26–29). We have also used labelling in this chapter to show that symmetrical intermediates are probably involved in, for example, nucleophilic aromatic substitution with a benzyne intermediate (Chapter 23).

intermediate in
S_N1 reactions

intermediates in aromatic substitution reactions
electrophilic   nucleophilic

intermediate in
carbonyl reactions

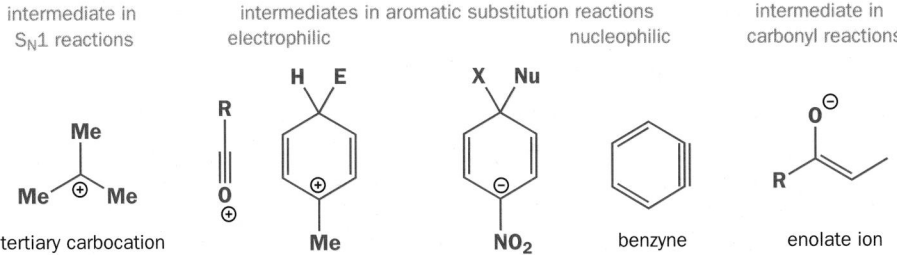

tertiary carbocation   Me   NO$_2$ benzyne   enolate ion

We have hedged this evidence around with caution since the fact that an intermediate can be prepared does not by any means prove that it is involved in a reaction mechanism. In this section we are going to consider other and better evidence for intermediates and at the same time revise some of the earlier material.

Trapping reactions

A more impressive piece of evidence is the design of a molecule that has built into it a functional group that could react with the intermediate in a predictable way but could not reasonably react with other species that might be present. For example, aromatic ethers react with nitrating agents in the *ortho* or *para* positions (Chapter 22). The intermediate has a positive charge delocalized over three of the carbon atoms in the benzene ring. If a nucleophilic group is built into the structure in the right way, it might trap this intermediate and stop it reacting further.

The trapping group is the amide and it has trapped a cation formed by addition of NO_2^+ to the aromatic ring. We are faced with the problem of drawing a mechanism for the formation of this remarkable compound and, when we discover that a necessary intermediate is also an intermediate in our preferred mechanism for aromatic nitration, we feel more confident about that mechanism.

This mechanism explains everything including the stereochemistry. The NO_2^+ attacks the aromatic ring *para* to the OMe group and on the opposite side to the amide. The amide is now in the perfect position to capture the cation at the *meta* position and, because the tether is short, it must form a *cis* bridge.

π complexes in electrophilic aromatic substitution

You will meet the related π complexes of metals in Chapter 48.

The weakness in the experiment is that nitration does not occur in that position without the trap but occurs in the *ortho* position. Nevertheless, many chemists believe that aromatic electrophilic substitution actually starts with a loose association of the electrophile with all of the p orbitals of the benzene ring so that here the NO_2^+ group would initially sit at right angles to the plane of the ring in a 'π complex' and would move afterwards to form a σ bond with one particular carbon atom.

To be convincing, evidence for an intermediate should include:

- detection of the intermediate in the reaction mixture, perhaps by a trapping reaction
- a demonstration that the intermediate gives the product when added to the reaction mixture (this also means that it must be prepared as an at least reasonably stable compound)
- kinetic evidence that the rate of formation and rate of disappearance are adequate
- other suitable evidence of the kind that we have been discussing in this chapter

What is the cyclic acetal for? It is there to make the cyclization more efficient by the Thorpe–Ingold effect; see Chapter 42.

A neat intramolecular trap for benzyne works in this way. A standard benzyne-generating reaction—the diazotization of an *ortho*-amino benzoic acid (Chapter 23) gives a zwitterion that loses nitrogen and CO_2 to release the benzyne. A furan tethered to the next *ortho* position traps the benzyne in an intramolecular Diels–Alder reaction. The yield is impressive and the trap is very efficient.

The argument is that this reaction cannot really be explained without a benzyne intermediate. This same method of making benzyne is used on other *o*-amino benzoic acids and so they presumably create benzynes too.

A collection of reactions linked by a common intermediate

Particularly convincing evidence can develop when a number of chemists suggest the same intermediate for a number of different reactions and show that it is possible to trap the intermediate from one reaction, put it into the others, and get the normal products. We are going to describe one set of such related reactions. In Chapter 37 we suggested a mechanism for the Favorskii rearrangement involving a series of remarkable intermediates. Here is an example.

enolate anion   oxyallyl cation   2-electron electrocyclic

cyclopropanone

A quick summary of the evidence on this particular example. If the reaction is run in MeOD instead of MeOH, the starting material becomes deuterated at the site of enolate formation suggesting that this is a fast and reversible step. The entropy of activation for the reaction is $\Delta S^{\ddagger} = +64 \, J \, mol^{-1} \, K^{-1}$, suggesting that the slow step is one molecule breaking into two. There is only one such step—the second, ionization step. If various substituted phenyl groups are used, the Hammett ρ value is -5. This large negative value also suggests that the ionization is the slow step as the cation is delocalized into the benzene ring.

enolate anion   oxyallyl cation   oxyallyl cation

So there is some evidence for the first intermediate—the exchange of deuterium from the solvent. The formation of the enolate can even become the rate-determining step! If we merely add an extra methyl group to the chloroketone the reaction becomes 220 times faster and the rate-determining step changes. There is no longer any exchange of deuterium from the solvent and the Hammett ρ value changes from -5 to $+1.4$. This small positive value, showing some modest increase in electron density near the ring, matches typical known ρ values for enolate formation.

rate-determining enolate anion formation
$\rho = +1.4$

enolate anion   ionization fast   oxyallyl cation

$\rho = +1.7$
for enolate anion formation

However, we are not surprised that an enolate ion is formed from a ketone in basic solution. The oxyallyl cation is much more surprising. How can we be convinced that it really is an intermediate? There are several alternative ways to make the same intermediate. If basic nucleophiles such as the methoxide ion are avoided and reaction of zinc with an α,α'-dibromoketone in a nonnucleophilic solvent like diglyme is used instead, the oxyallyl cation can be trapped in a Diels–Alder reaction. This is the basis for a good synthesis of seven-membered rings.

diglyme
forms a solid with $ZnBr_2$

Zn/Cu diglyme

enolate   oxyallyl cation

But does the oxyallyl cation go on to give cyclopropanones? In fact, there is good evidence that the two are in equilibrium. If the same method is used to create the diphenyl oxyallyl cation in methanol instead of in diglyme, the normal Favorskii product is produced. Evidently, methoxide is needed only to produce the enolate—methanol is enough to decompose the cyclopropanone.

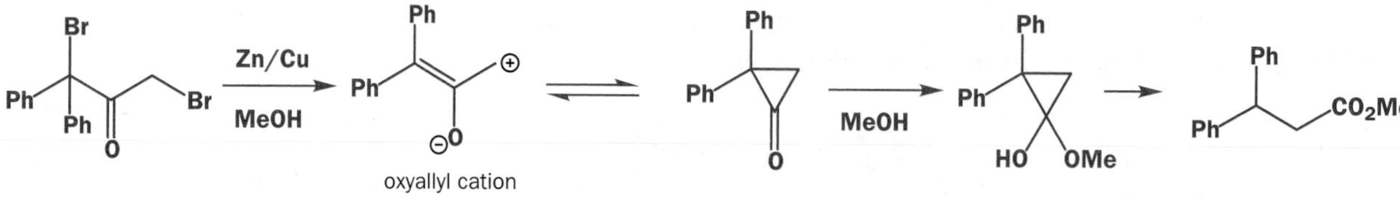

If a suitable (1,3-di-*t*-butyl) allene is epoxidized with *m*-CPBA the unstable allene oxide can actually be isolated. On heating, this epoxide gives a stable *trans*-di-*t*-butylcyclopropanone. It is very difficult to see how this reaction could happen except via the oxyallyl cation intermediate.

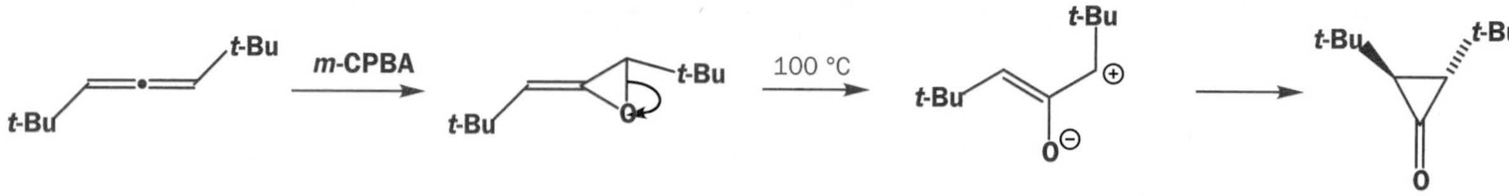

Why draw the oxyallyl cation with this stereochemistry?

If the closure to the cyclopropanone is electrocyclic then it will be disrotatory (Chapter 36). The *E,Z*-isomer we have drawn gives the *anti* cyclopropanone while either the *E,E*- or the *Z,Z*-oxyallyl cation gives the *syn*-di-*t*-butylcyclopropanone.

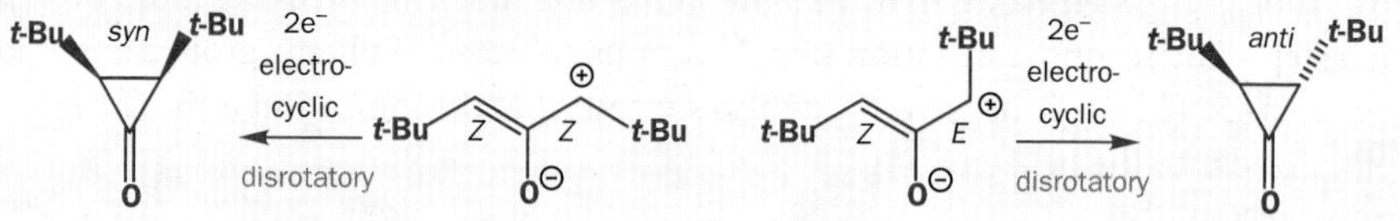

But is the same cyclopropanone an intermediate in the Favorskii reaction? If the bromoketone is treated with methoxide in methanol, it gives the Favorskii product but, if it is treated with a much more hindered base, such as the potassium phenoxide shown, it gives the same cyclopropanone with the same stereochemistry.

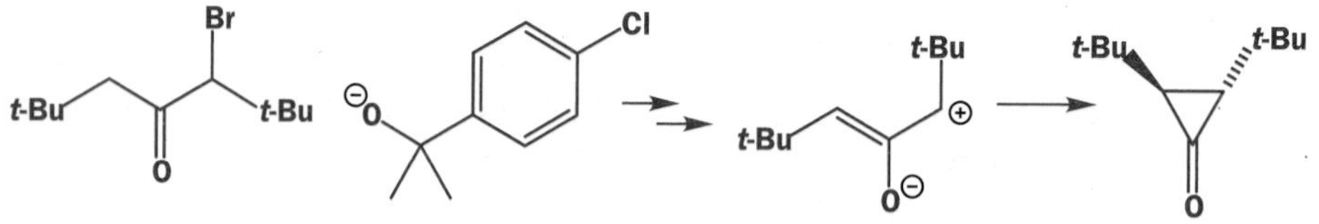

Other, less stable cyclopropanones, such as the 2,2-dimethyl compound, can be made by carbene addition (Chapter 40) to ketenes. This compound did the Favorskii reaction with methoxide in methanol: the only product came from the expected loss of the less unstable carbanion. This will, of course, be general-acid-catalysed by methanol as no free carbanion can be released into an alcoholic solvent.

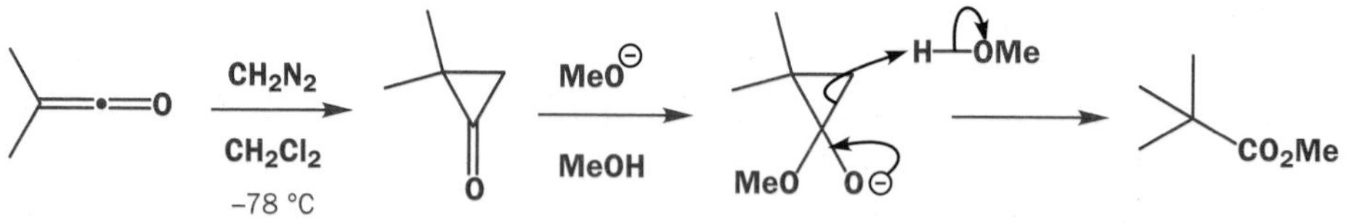

The same cyclopropanone gives a cycloadduct with furans—this must surely be a reaction of the oxyallyl cation and we can conclude that the three isomeric reactive intermediates (allene oxide, cyclopropanone, and oxyallyl cation) are all in equilibrium and give whichever product is appropriate for the conditions.

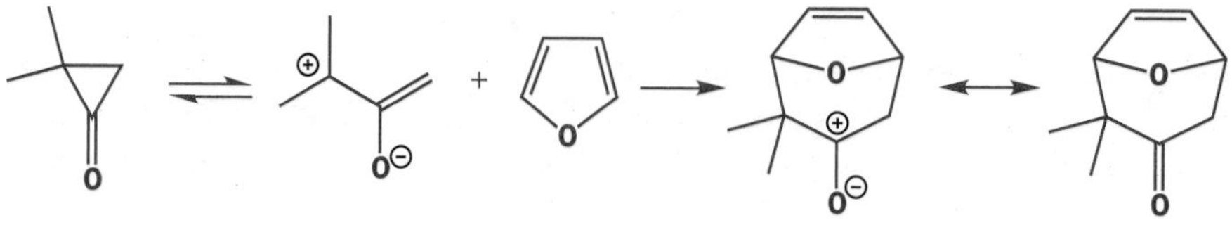

Though it is never possible to prove a mechanism, this interlocking network of intermediates, all known to be formed under the reaction conditions, all being trapped in various ways, and all known to give the products, is very convincing. If any part of the mechanism were not correct, that would throw doubt on all the other reactions as well. Nevertheless, this mechanism is not accepted by all chemists.

Stereochemistry and mechanism

This chapter ends with a survey of the role of stereochemistry in the determination of mechanism. Though we have left stereochemistry to the last, it is one of the most important tools in unravelling complex mechanisms. You have already seen how inversion of configuration is a vital piece of evidence for an S_N2 mechanism (Chapter 17) while retention of configuration is the best evidence for participation (Chapter 37). You have seen the array of stereochemical evidence for pericyclic mechanisms (Chapters 35 and 36). The chapters devoted to diastereoselectivity (33 and 34) give many examples where the mechanism follows from the stereochemistry. We shall not go over that material again, but summarize the types of evidence with new examples. The first example looks too trivial to mention.

Though this reaction looks like a simple S_N2 displacement by the naphthyloxide anion on the primary alkyl chloride, there is, in fact, a reasonable alternative—the opening of the epoxide at the less hindered primary centre followed by closure of the epoxide the other way round. The electrophile is called 'epichlorohydrin' and has two reasonable sites for nucleophilic attack.

It looks difficult to tell these mechanisms apart since both involve the same kind of reaction. Stereochemistry is the answer. If enantiomerically pure epichlorohydrin is used, the two mechanisms give different enantiomers of the product. Though each S_N2 reaction takes place at a primary centre and the stereogenic centre remains the same, from the diagrams the two products are obviously enantiomers.

Finding out the mechanism of this process is not idle curiosity as a group of drugs used to combat high blood pressure and heart disease, such as propranolol, are made from epichlorohydrin and it is essential to know which enantiomer to use to get the right enantiomer of the drug. In fact, the more extended mechanism shown in black is correct. This is an example of determination of mechanism by using enantiomers.

■
The full synthesis of propranolol is given in Chapter 30.

propranolol

A more complicated example arises from the strange reactions used to make malic acid from chloral and ketene. An initial [2 + 2] cycloaddition (Chapter 35) is followed by acid treatment and then treatment with an excess of aqueous NaOH. Neutralization gives malic acid, an acid found naturally in apples (*Malus* spp.).

The mechanism of this reaction also looks straightforward: normal ester hydrolysis followed by hydrolysis of the CCl_3 group to CO_2H. Caution suggests investigation, particularly as four-membered lactones sometimes hydrolyse by S_N2 displacement at the saturated ester carbon rather than by attack on the carbonyl group, like the three-membered lactones discussed in Chapter 37 (p. 972). The solution was urgently needed when it was found that enantiomerically pure lactone could be prepared by asymmetric synthesis (Chapter 45). The sequence was repeated with enantiomerically pure lactone: lactone hydrolysis occurred with retention of configuration and must be normal ester hydrolysis by attack of water at the carbonyl group. But the hydrolysis of the CCl_3 group occurred with inversion of configuration.

You will see in Chapter 42 that this reaction is governed by 'Baldwin's rules' and why attack on even a CCl_2 group is unfavourable.

The answer must be a mechanism related to the one we have just seen for epichlorohydrin. Attack by hydroxide on CCl_3 is almost unknown and it is much more likely that intramolecular attack by alkoxide to give an epoxide should occur. The carboxylate anion can then invert the stereogenic centre by intramolecular S_N2 displacement at the central carbon atom. Notice that the tether ensures attack at the central atom. The second four-membered lactone also hydrolyses by attack at the carbonyl group.

The Ritter reaction was introduced in Chapter 17 and the Beckmann fragmentation in Chapter 38.

The Ritter reaction and the Beckmann fragmentation

Another collection of related intermediates occurs in the Ritter reaction and the Beckmann fragmentation. The **Ritter reaction** involves the combination of a tertiary alcohol and a nitrile in acid solution and the proposed mechanism involves a series of intermediates.

The Beckmann fragmentation also occurs in acid solution upon the fragmentation of an oxime with a *tertiary* alkyl group *anti* to the OH of the oxime. The fragmentation step gives the same cation and the same nitrile together with a molecule of water and these three combine in the same way to give the same amide. We need evidence that the carbocation and the nitrilium ion are genuine intermediates and that the same sequence is found in both reactions.

Evidence that the two reactions are intimately related comes from the formation of the same amide from two different starting materials: a tertiary alcohol and an oxime, both based on the decalin skeleton. The oxime has its OH group *anti* to the ring junction to minimize steric hindrance as oxime formation is under thermodynamic control (Chapter 14).

Decalins are widely used in conformational experiments; see Chapter 18.

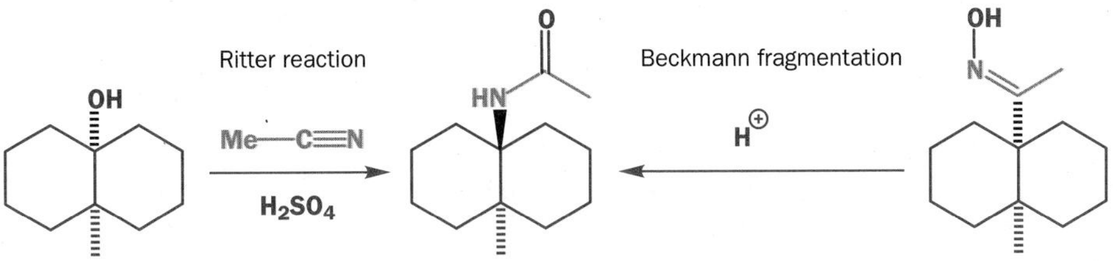

The experiments also provide stereochemical evidence that a carbocation is an intermediate in both reactions. Both starting materials are *cis*-decalins but the product is a *trans*-decalin. The carbocation intermediate has no stereochemistry and can react with the nitrile from either face. Axial attack is preferred and it gives the stable *trans*-decalin. The formation of the carbocation is shown only by the Beckmann fragmentation: formation from the alcohol by the S_N1 mechanism is obvious.

None of these compounds is chiral as there is a plane of symmetry running vertically through each molecule. We are discussing diastereoisomers only.

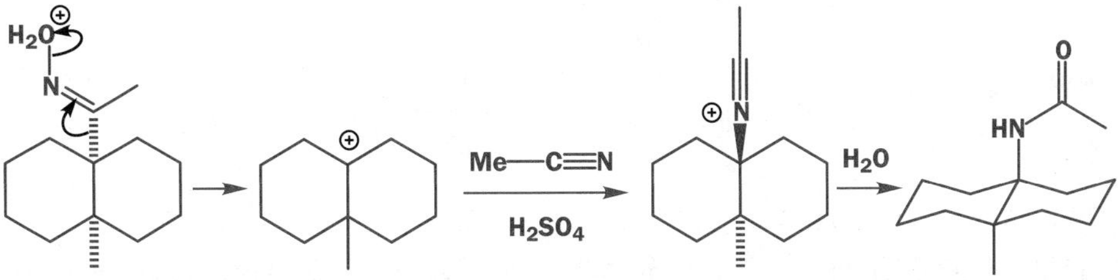

Trapping the carbocation is also possible. The Beckmann fragmentation on this oxime of an aryl seven-membered ring ketone gives a tertiary carbocation that might be expected to cyclize to give an amide. However, this reaction would give an unfavourable eight-membered ring (see Chapter 42) and does not happen. Instead, the chain twists round the other way and forms a much more stable six-membered ring by intramolecular Friedel–Crafts alkylation. Note that the regioselectivity is *meta* to CN and *ortho* to alkyl. These are both favourable but the main factor is the C_4 tether making any other product impossible.

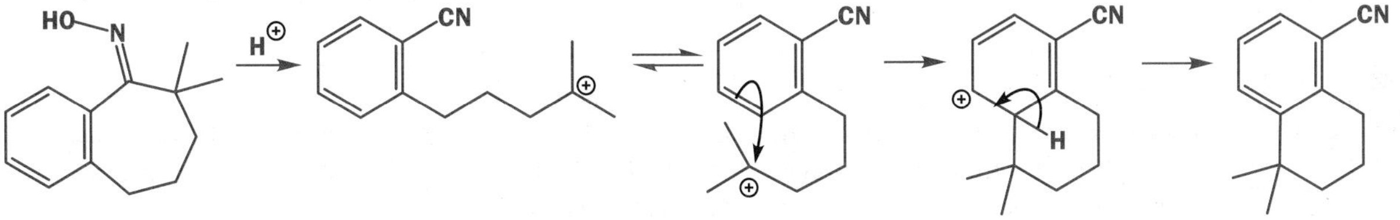

In the Ritter reaction a rather different kind of evidence for the cation is the fact that families of isomeric alcohols all give the same product. In all these cases, rearrangements of the first formed carbocation (Chapter 37) can easily account for the products. Another example in the decalin series is this Ritter reaction with KCN as the nitrile in acidic solution so that HCN is the reagent. The starting material is a spirocyclic tertiary alcohol but the product is a *trans*-decalin formed by rearrangement.

This would be a dangerous experiment to carry out and is not recommended.

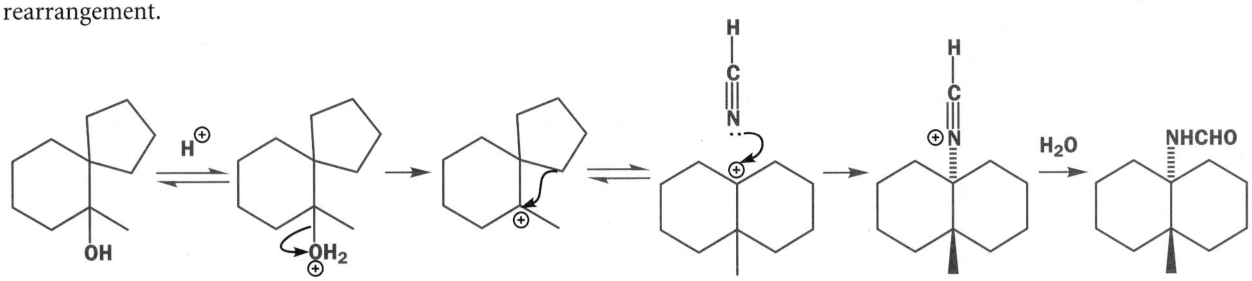

Trapping the nitrilium cation is also possible. The most famous example is probably the heterocycle (an oxazine, Chapter 42) produced by intramolecular capture of the nitrilium ion with a hydroxyl group. Note that the tertiary alcohol reacts to give the cation while the secondary alcohol acts as the nucleophilic trap.

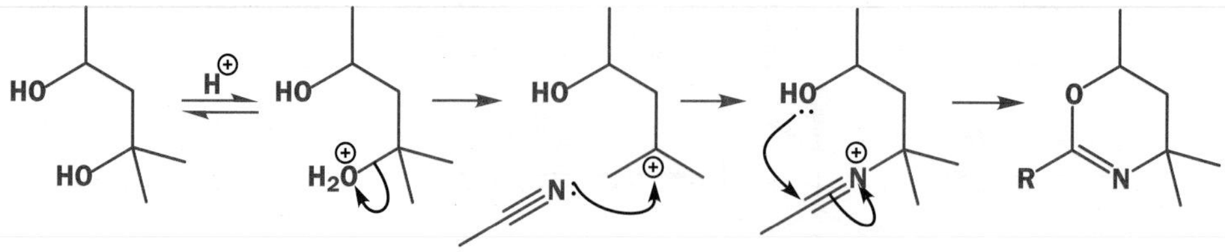

An important example in which the diastereoisomer produced was critical in determining the mechanism is the synthesis of *cis*-aminoindanol, a part of Merck's anti-HIV drug Crixivan (indinavir). The reaction involves treatment of indene epoxide with acetonitrile (MeCN) in acidic solution. The product is a *cis* fused heterocycle. It is easy to see which atoms have come from the nitrile (green) but the substitution of nitrogen for oxygen at one end of the epoxide has occurred with retention of configuration as the *cis*-epoxide has given the *cis* product. Clearly, we have some sort of Ritter reaction and the nitrilium ion has been trapped with an OH group.

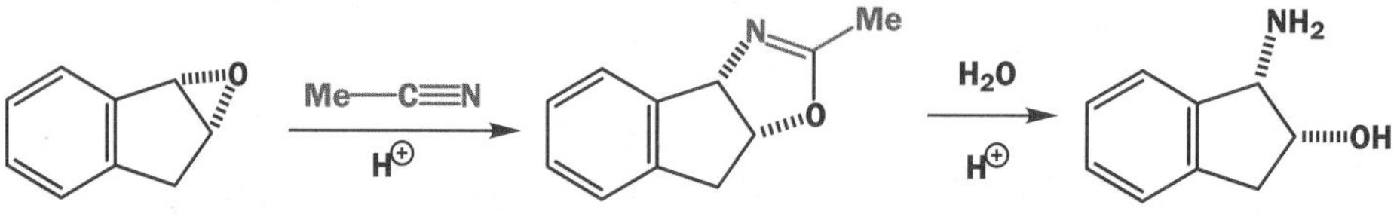

indene epoxide   *cis*-aminoindanol

What about the regioselectivity? The obvious explanation is that a cation is formed from the epoxide in a specific acid-catalysed ring opening. But why should the nitrile attack the bottom face of the cation? We should expect it to attack the top face preferentially as the hydroxyl group partly blocks the bottom face.

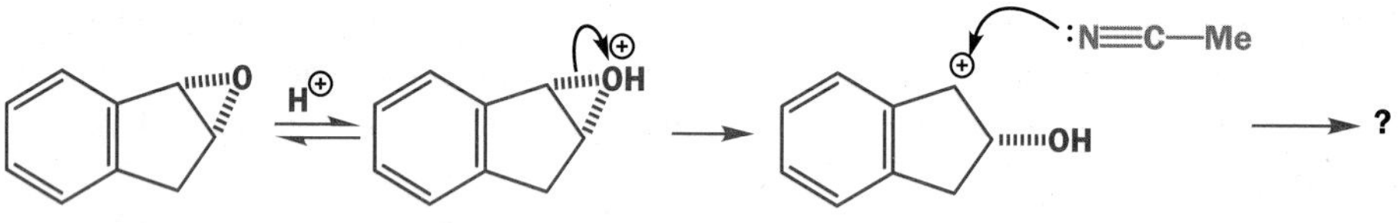

■
This step will be described in Chapter 42 as a favourable '5-*endo-dig*' process (p. 1143).

A reasonable mechanism is that in which the nitrile adds reversibly to the cation. Every time it adds to the top face, it drops off again as the OH group cannot reach it to form the heterocycle. Every time it adds to the bottom face, it is quickly captured by the OH group because 5/5 fused rings are favourable when the ring junction is *cis*. Eventually, all the compound is converted to the heterocycle.

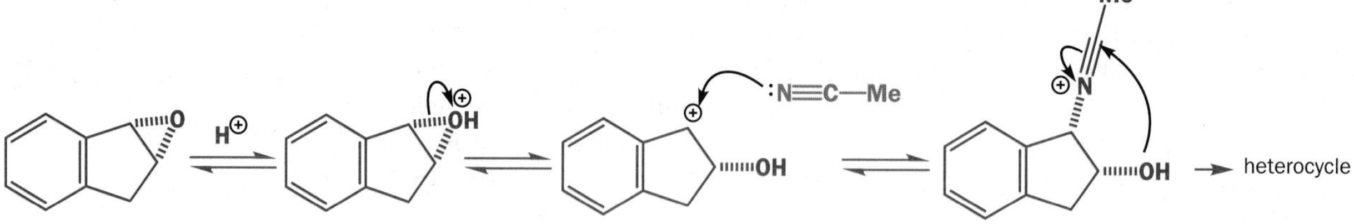

Again, the mechanism of this reaction is of great importance because it is the foundation stone of the synthesis of Crixivan—a drug that is saving thousands of lives. These last examples are of reactions that you would find difficult to classify into any of the familiar types we have met so far in the book. Nevertheless, the organic chemist needs to be able to propose mechanisms for new reactions and to have a general idea of the methods available to test these proposals.

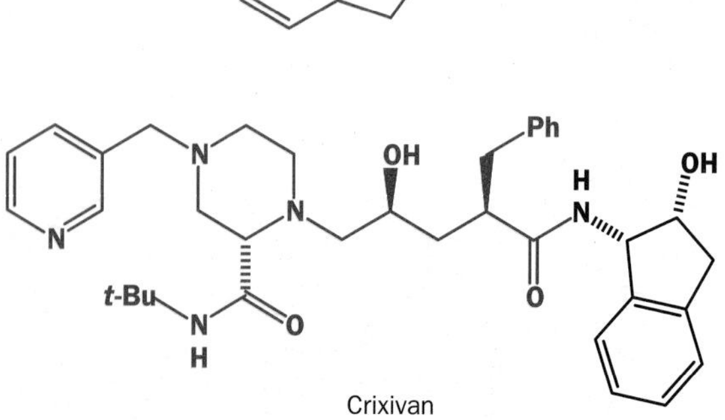

Crixivan

Summary of methods for the investigation of mechanism

This brief summary is for guidance only and the figures quoted are approximate ranges only. The full text above should be used for detail. All methods would not be used in one investigation.

1. Make sure of the structure of the product

- Basic structure (Chapters 4 and 11) and stereochemistry (Chapter 32) by spectroscopic methods
- Detail of fate of individual atoms by labelling with D, ^{13}C, and ^{18}O. Double labelling may help
- Stereochemical course of the reaction (enantio- or diastereoselectivity) may be critical

2. Kinetic methods

- Rate equation gives composition of main transition state
- Deuterium isotope effect: $k_H > k_D$ shows bond to H formed and/or broken in transition state. Values k_H/k_D 2–7 typical
- Entropy of activation shows increase ($\Delta S^{\ddagger}$ positive) or decrease ($\Delta S^{\ddagger}$ negative) in disorder. Typical values and deductions:
    - $\Delta S^{\ddagger}$ positive (rarely larger than $+50 \, J \, mol^{-1} \, K^{-1}$): one molecule breaks into two or three
    - Moderate negative values: no change in number of molecules (one goes to one etc.) or bimolecular reaction with solvent
    - Large negative values: two molecules go to one or unimolecular reaction with ordered $TS^{\ddagger}$ (cycloaddition, etc.)

3. Correlation of structure and reactivity

- Replace one group by another of similar size but different electronic demand (CF_3 for CH_3 or OMe for CH_3)
- Systematic Hammett σ/ρ correlation with m- and p-substituted benzenes:
    - Sign of ρ: $+\rho$ indicates electrons flowing into and $-\rho$ electrons flowing out of ring in transition state
    - Magnitude of ρ shows effect on the benzene ring:
        large (around 5), charge on ring ($+\rho$, anion; $-\rho$, cation)
        moderate (around 2–4), charge on atom next to ring—may be gain or loss of conjugation
        small (<1), ring may be distant from scene of action or ρ may be balance of two ρs of opposite sign

4. Catalysis

- pH–rate profile reveals specific acid or base catalysis
- Rate variation with [HA] or [B] at constant pH reveals GAC or GBC
- Deuterium isotope effect: normal ($k_H > k_D$) shows GA/BC, inverse solvent $k(D_2O) > k(H_2O)$ shows SA/BC
- GA/BC is termolecular and has large negative entropy of activation

5. Intermediates

- Independent preparation or, better, isolation from or detection in reaction mixture helps
- Must show that intermediate gives product under reaction conditions
- Designed trapping experiments often most convincing

Problems

1. Propose three fundamentally different mechanisms (other than variations of the same mechanism with different kinds of catalysis) for this reaction. How would (a) D labelling and (b) ^{18}O labelling help to distinguish the mechanisms? What other experiments would you carry out to eliminate some of these mechanisms?

2. Explain the stereochemistry and labelling pattern in this reaction.

3. The Hammett ρ value for migrating aryl groups in the acid-catalysed Beckmann rearrangement is −2.0. What does this tell us about the rate-determining step?

4. Between pH 2 and 7, the rate of hydrolysis of this thiol ester is independent of pH. At pH 5 the rate is proportional to the concentration of acetate ion $[AcO^-]$ in the solution and the reaction goes twice as fast in D_2O as in H_2O. Suggest a mechanism for the pH-independent hydrolysis. Above pH 7, the rate increases with pH. What kind of change is this?

5. In acid solution, the hydrolysis of this carbodiimide has a Hammett ρ value of −0.8. What mechanism might account for this?

6. Explain the difference between these Hammett ρ values by mechanisms for the two reactions. In both cases the ring marked with the substituent X is varied. When R = H, ρ = −0.3 but, when R = Ph, ρ = −5.1.

7. Explain how chloride ion catalyses this reaction.

8. The hydrolysis of this oxaziridine in 0.1 M sulfuric acid has $k(H_2O)/k(D_2O) = 0.7$ and an entropy of activation of $\Delta S^{\ddagger} = -76$ $J\,mol^{-1}\,K^{-1}$. Suggest a mechanism for the reaction.

9. Explain how both methyl groups in the product of this reaction come to be labelled. If the starting material is re-isolated at 50% reaction, its methyl group is also labelled.

10. The pK_{aH} values of some substituted pyridines are as follows.

X	H	3-Cl	3-Me	4-Me	3-MeO	4-MeO	3-NO₂
pK_{aH}	5.2	2.84	5.68	6.02	4.88	6.62	0.81

Can the Hammett correlation be applied to pyridines using the σ values for benzenes? What equilibrium ρ value does it give and how do you interpret it? Why are no 2-substituted pyridines included in the list?

11. These two reactions of diazo compounds with carboxylic acids give gaseous nitrogen and esters as products. In both cases the rate of the reaction is proportional to [diazo compound]·[RCO_2H]. Use the data for each reaction to suggest mechanisms and comment on the difference between them.

ρ = − 1.6 $k(RCO_2H)/k(RCO_2D)$ = 3.5

$k(RCO_2D)/k(RCO_2H)$ = 2.9

12. Suggest mechanisms for these reactions and comment on their relevance to the Favorskii family of mechanisms.

(a)

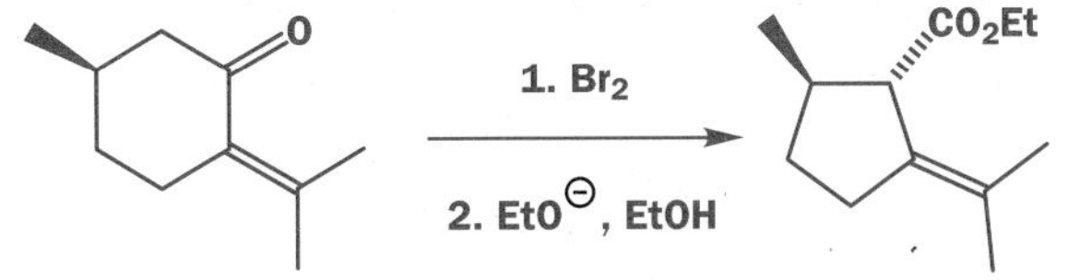

1. Br_2
2. $EtO^{\ominus}$, EtOH
CO_2Et

(b)

$MeO^{\ominus}$ MeOH
bromoketone added to base
Br
Ph
Ph
Me
O
$MeO^{\ominus}$ MeOH
base added to bromoketone
Ph
O
Ph
Ph
CO_2Me

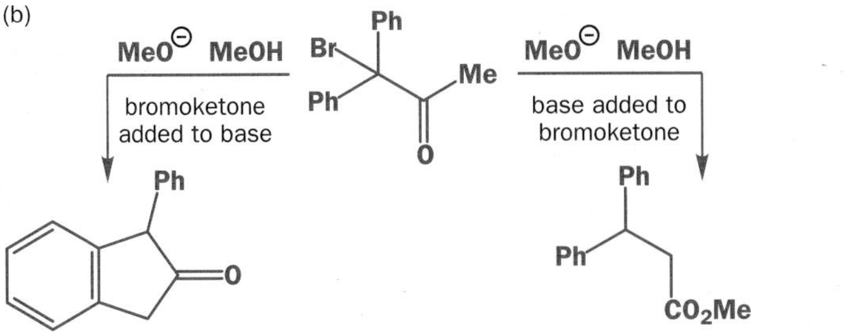

13. Propose mechanisms for the two reactions at the start of the chapter. The other product in the first reaction is the imine $PhCH{=}NSO_2Ph$.

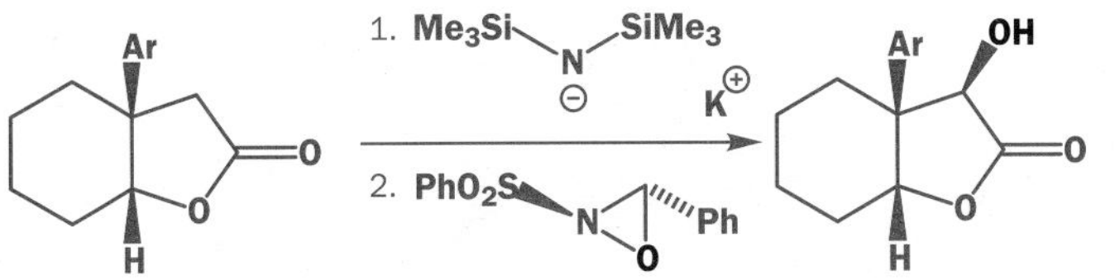

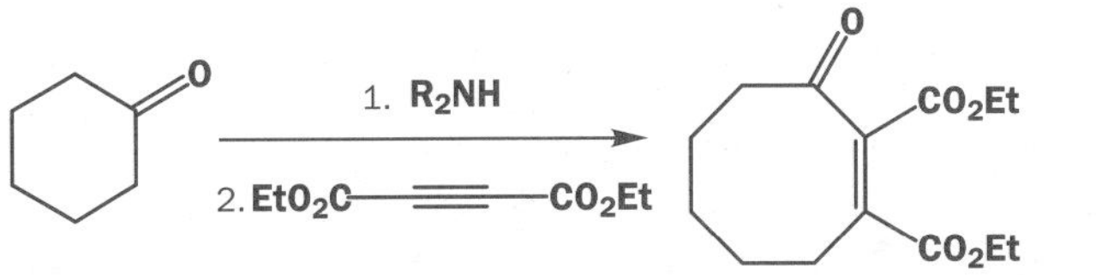

Ar
O
H
1. Me_3Si N $SiMe_3$ $K^{\oplus}$
2. PhO_2S N O Ph
Ar
OH
O
H

O
1. R_2NH
2. EtO_2C CO_2Et
O
CO_2Et
CO_2Et

14. A typical Darzens reaction involves the base-catalysed formation of an epoxide from an α-haloketone and an aldehyde. Suggest a mechanism for the Darzens reaction consistent with the results shown below.

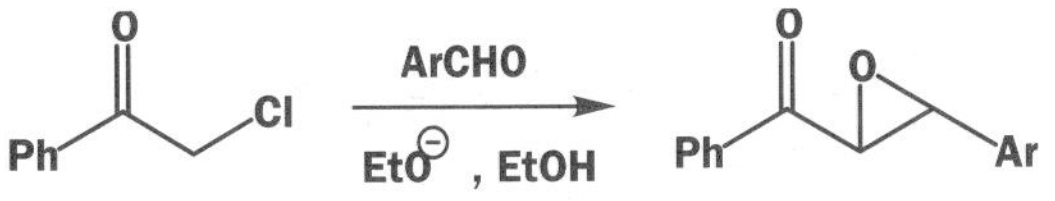

O
Ph
Cl
ArCHO
$EtO^{\ominus}$, EtOH
O
Ph
O
Ar

(**a**) The rate expression is:
rate $= k_3[\text{PhCO·CH}_2\text{Cl}][\text{ArCHO}][\text{EtO}^-]$

(**b**) When Ar is varied, the Hammett ρ value is +2.5.

(**c**) The following attempted Darzens reactions produced unexpected products.

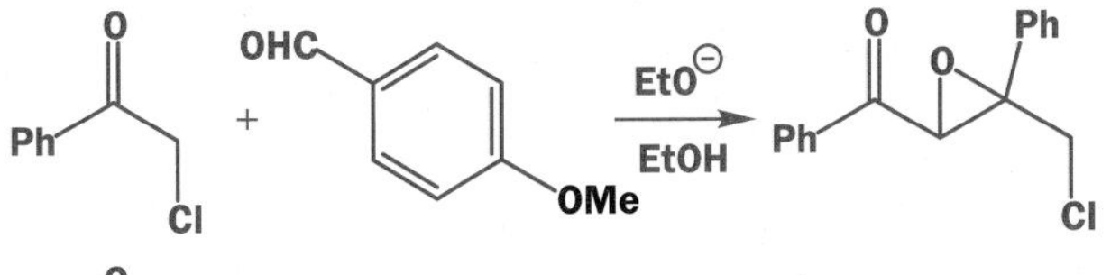

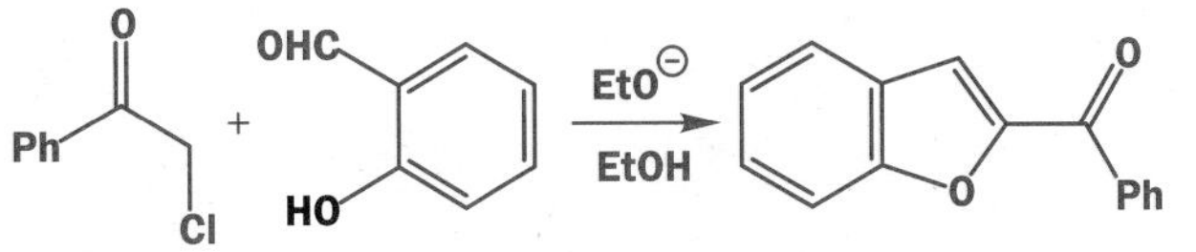

O
Ph
Cl
OHC
OMe
$EtO^{\ominus}$
EtOH
O
Ph
O
Ph
Cl

O
Ph
Cl
OHC
HO
$EtO^{\ominus}$
EtOH
O
O
Ph

15. If you believed that this reaction went by elimination followed by conjugate addition, what experiment would you carry out to try and prove that the enone is an intermediate?

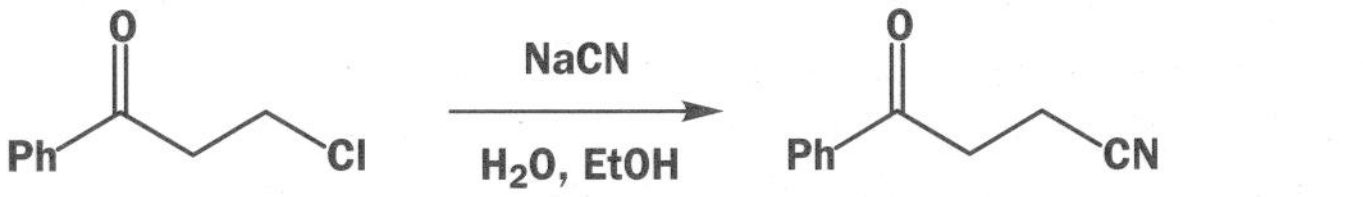

O
Ph
Cl
NaCN
H_2O, EtOH
O
Ph
CN

16. This question is about three related acid-catalysed reactions: (**a**) the isomerization of *Z*-cinnamic acids to *E*-cinnamic acids; (**b**) the dehydration of the related hydroxy-acids; (**c**) the racemization of the same hydroxy-acids. You should be able to use the information provided to build up a complete picture of the interaction of the various compounds and the intermediates in the reactions.

(**a**) Data determined for the acid-catalysed isomerization of *Z*-cinnamic acids in water include the following.

(**i**) The rate is faster in H_2O than in D_2O: $k(H_2O)/k(D_2O) = 2.5$.

(**ii**) The product contains about 80% D at C-2.

(**iii**) The Hammett ρ value is –5.

Suggest a mechanism for the reaction that explains the data.

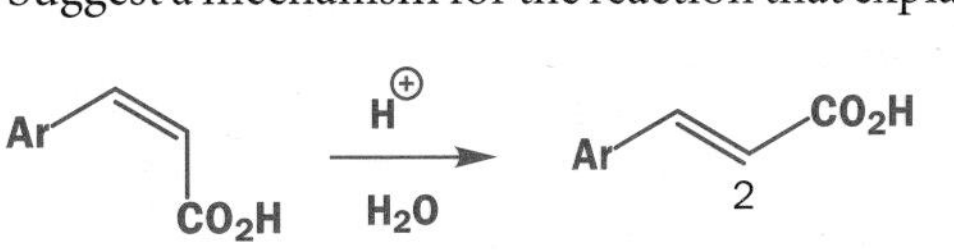

Ar
CO_2H
$H^{\oplus}$
H_2O
Ar
CO_2H
2

(**b**) The dehydration of the related hydroxy-acids also gives *E*-cinnamic acids at a greater rate under the same conditions but the data for the reaction are rather different.

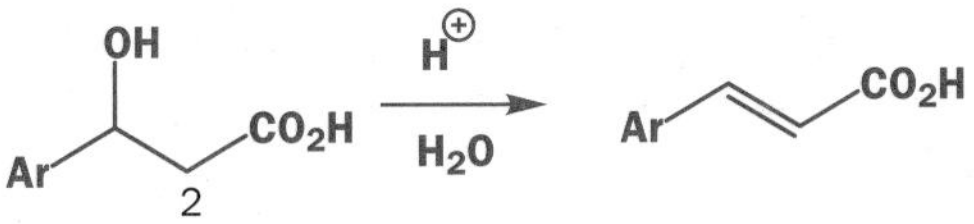

OH
Ar
CO_2H
2
$H^{\oplus}$
H_2O
Ar
CO_2H

Hydroxy-acid deuterated at C2 shows a kinetic isotope effect: $k_H/k_D = 2.5$.

(**c**) If the dehydration reaction is stopped after about 10% conversion to products, the remaining starting material is completely racemized. Data for the *racemization* reaction include:

(**i**) The rate is slower in H_2O than in D_2O.

(**ii**) Hydroxy-acid deuterated at C2 shows practically no kinetic isotope effect.

(**iii**) The Hammett ρ value is –4.5.

What conclusions can you draw about the dehydration?

Recalling that the dehydration goes faster than the isomerization, what would be present in the reaction mixture if the isomerization were stopped at 50% completion?

Saturated heterocycles and stereoelectronics

42

Connections

Building on:

- Acetals and hemiacetals ch14
- Stereochemistry ch16
- The conformation of cyclic molecules ch18
- Stereospecific elimination reactions ch19
- Protecting groups ch24
- NMR and stereochemistry—how orbital overlap affects coupling (the Karplus relationship) ch32
- How rings affect stereoselective reactions ch33
- Ring closing and opening by cycloadditions ch35
- Electrocyclic ring closing and opening ch36
- How alignment of orbitals affects reactivity ch37–ch38
- Determining organic mechanisms ch41

Arriving at:

- Putting a heteroatom in a ring changes the reactivity of the heteroatom
- Ring-opening reactions: the effect of ring strain
- Lone pairs in heterocycles have precise orientations
- Some substituents prefer to be axial on some six-membered saturated heterocycles
- Interactions of lone pairs with empty orbitals can control conformation
- Ring-closing reactions: why five-membered rings form quickly and four-membered rings form slowly
- Baldwin's rules: why some ring closures work well while others don't work at all

Looking forward to:

- Structure and reactions of aromatic heterocycles ch43
- Synthesis of aromatic heterocycles ch44
- Asymmetric synthesis ch45
- Chemistry of life ch49
- Mechanisms in biological chemistry ch50
- Natural products ch51

Introduction

Rings in molecules make a difference, and we have already devoted the whole of one chapter (33) and most of another (18) just to the structure and reactions of rings. In those chapters, the message was that rings have well-defined conformations, and that well-defined conformations allow reactions to be stereoselective.

This chapter and the next two will revisit the ring theme, but the rings will all be **heterocycles**: rings containing not just carbon atoms, but oxygen, nitrogen, or sulfur as well. It may seem strange that this rather narrowly defined class of compounds deserves three whole chapters, but you will soon see that this is justified both by the sheer number and variety of heterocycles that exist and by their special chemical features. Chapters 43 and 44 cover heterocycles that are aromatic, and in this chapter we look at heterocycles that are saturated and flexible. Some examples, a few of which may be familiar to you, are shown below and overleaf.

■ The saturated heterocyclic rings are shown in black, and names for the most important ring types are given: some (like piperidine, morpholine) you will need to remember; others (tetrahydrofuran, pyrrolidine) are more obviously derived from the names for aromatic heterocycles that we will discuss in the next chapter. Some of these compounds (nicotine, coniine, cocaine) are plant products falling into the class called alkaloids. Alkaloids are discussed in Chapter 51. Another important class of saturated heterocycles, sugars, will reappear in Chapter 49.

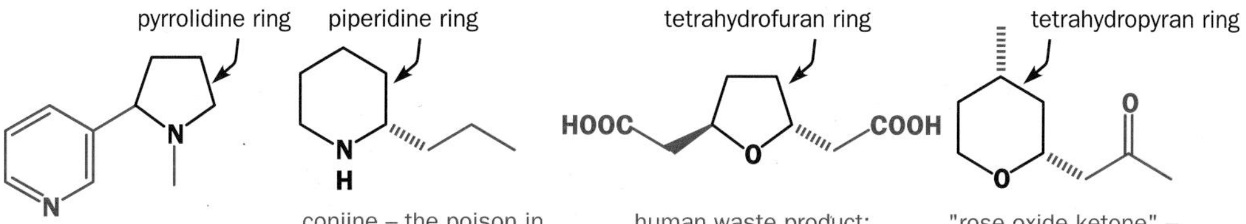

pyrrolidine ring   piperidine ring

tetrahydrofuran ring

tetrahydropyran ring

nicotine

coniine – the poison in hemlock that killed Socrates

human waste product: 3–5 mg per day excreted in urine

"rose oxide ketone" – isolated from geranium oil and used in the perfume industry tetrodotoxin – lethal poison in wrongly prepared and cooked Japanese puffer fish

clavulanic acid – an antibiotic

musty taste of "corked" wine

epoxide or oxirane ring

sex pheromone of the Grey Duiker antelope

isolated from the green alga *Chara globularis*

cocaine

dioxane: a common solvent

morpholine: an important base

But what are the 'special chemical features' of saturated heterocycles? Putting a heteroatom into a ring does two important things, and these lead to the most important new topics in this chapter. Firstly, the heteroatom makes the ring easy to make by a ring-closing reaction, or (in some cases) easy to break by a ring-opening reaction. Closing and opening reactions of rings are subject to constraints that you will need to know about, and the principles that govern these reactions are discussed in the second half of the chapter.

Secondly, the ring fixes the orientation of the heteroatom—and, in particular, the orientation of its lone pairs—relative to the atoms around it. This has consequences for the reactivity and conformation of the heterocycle which can be explained using the concept of **stereoelectronics**.

> ● **Stereoelectronic effects are chemical consequences of the arrangement of orbitals in space.**

Although this is the only chapter in which stereoelectronics appears in the title, you will soon recognize the similarity between the ideas we cover here and concepts like the stereospecificity of E2 elimination reactions (Chapter 19), the Karplus relationship (Chapter 32), the Felkin–Anh transition state (Chapter 33), and the conformational requirements for rearrangement (Chapter 37) and fragmentation (Chapter 38) reactions.

> In Chapters 35 and 36 we discussed pericyclic ring closing and opening (cycloadditions and electrocyclic reactions) that are subject to the Woodward–Hoffmann rules of orbital symmetry. This chapter will be looking at ring closing and opening by the simple addition, substitution, and elimination reactions: orbital symmetry is not an issue in saturated systems, but orbital *shape* and *orientation* is.

Reactions of heterocycles

Nitrogen heterocycles: amines, but more nucleophilic

In many reactions the simple saturated nitrogen heterocycles—piperidine, pyrrolidine, piperazine, and morpholine—behave simply as secondary amines that happen to be cyclic. They do the sorts of things that other amines do, acting as nucleophiles in addition and substitution reactions. Morpholine, for example, is acylated by 3,4,5-trimethoxybenzoyl chloride to form the tranquillizer and muscle relaxant trimetozine, and *N*-methylpiperazine can be alkylated in an S_N1 reaction with diphenylmethyl chloride to give the travel-sickness drug cyclizine.

pyrrolidine   piperidine

piperazine   morpholine

morpholine

base

trimetozine

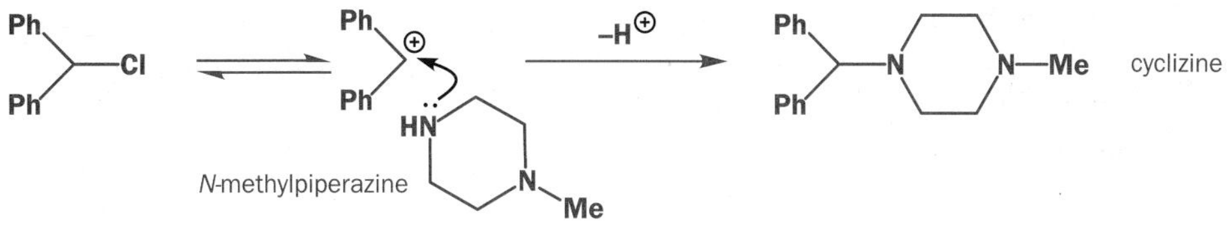

The addition of pyrrolidine to aldehydes and ketones is a particularly important reaction because it leads to enamines, the valuable enol equivalents discussed in Chapter 26.

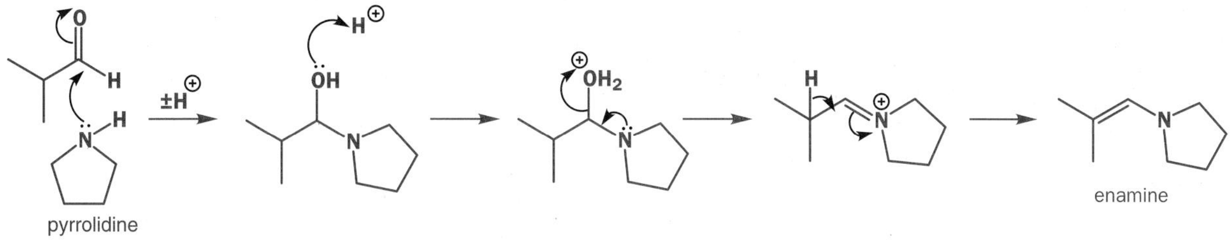

Enamines formed from pyrrolidine and piperidine are particularly stable, because pyrrolidine and piperidine are rather more nucleophilic than comparable acylic amines such as diethylamine. This is a general feature of cyclic amines (and cyclic ethers, too, as you will see shortly), and is a steric effect. The alkyl substituents, being tied back into a ring, are held clear of the nucleophilic lone pair, allowing it to approach an electrophile without hindrance. This effect is well illustrated by comparing the rates of reaction of methyl iodide with three amines—tertiary this time. The two cyclic compounds are bridged—quinuclidine is a bridged piperidine while the diamine known as 'DABCO' (1,4-DiAzaBiCyclo[2.2.2]Octane) is a bridged piperazine. Table 42.1 shows the relative rates, along with pK_{aH} values, for triethylamine, quinuclidine, and DABCO.

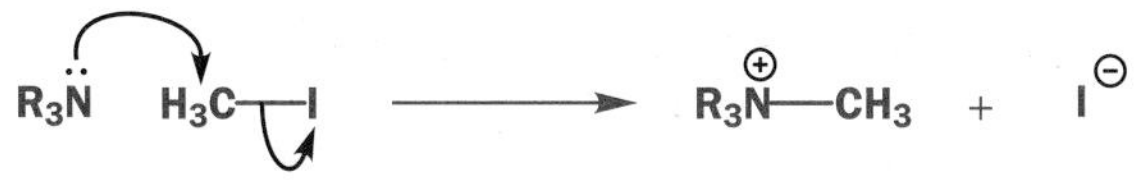

Table 42.1 Rates of reaction of amines with methyl iodide

	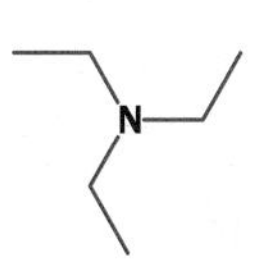 triethylamine	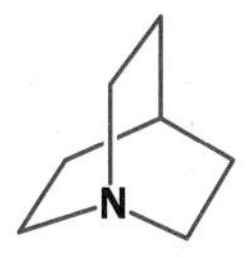 quinuclidine	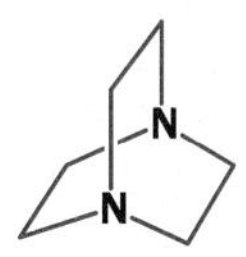 DABCO
relative rate of reaction[a]	1	63	40
pK_{aH}	10.7	11.0	8.8 (and 3.0)

[a] Relative rate of reaction with MeI in MeCN at 20 °C.

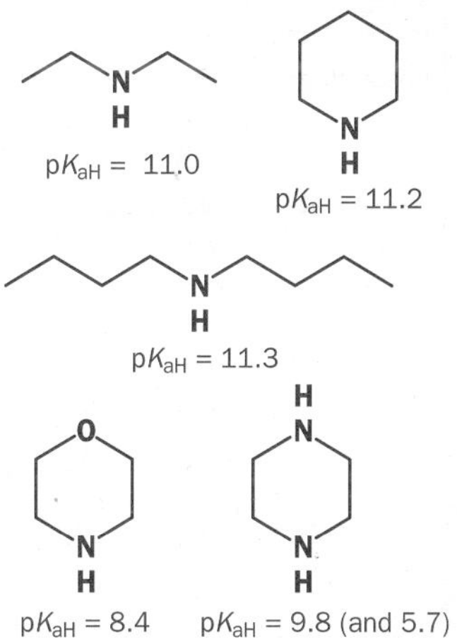

Quinuclidine and DABCO are 40–60 times more reactive than triethylamine. This is again due to the way the ring structures keep the nitrogen's substituents away from interfering with the lone pair as it attacks the electrophile. You should contrast the effect that the cyclic structure has on the pK_{aH} of the amines: none! Triethylamine and quinuclidine are equally basic and, as you can see in the margin, so (more or less) are diethylamine, dibutylamine, and piperidine. A proton is so small that it cares very little whether the alkyl groups are tied back or not.

Much more important in determining pK_{aH} is how electron-rich the nitrogen is, and this is the cause of the glaring discrepancy between the basicity of quinuclidine and that of DABCO, or between the basicities of piperidine (pK_{aH} 11.2) and morpholine (pK_{aH} 9.8) or piperazine (pK_{aH} 8.4). The extra heteroatom, through an inductive effect, withdraws electron density from the nitrogen atom, making it less nucleophilic and less basic. In this

pK_{aH} = 11.0

pK_{aH} = 11.2

pK_{aH} = 11.3

pK_{aH} = 8.4

pK_{aH} = 9.8 (and 5.7)

sense, morpholine can be a very useful base, less basic than triethylamine but somewhat more so than pyridine (pK_{aH} 5.2). Notice how much lower is the second pK_{aH} (that is, the pK_{aH} for protonation of the second nitrogen) of the diamines DABCO and piperazine: the protonated nitrogen of the monoprotonated amine withdraws electrons very effectively from the unprotonated one.

The Baylis–Hillman reaction

One of the most important uses of DABCO is in the **Baylis–Hillman reaction**, discovered in 1972 by two chemists at the Celanese Corporation in New York. Their reaction is a modification of the aldol reaction (Chapter 27), except that, instead of the enolate being formed by deprotonation, it is formed by conjugate addition. You have seen the enolate products of conjugate addition being trapped by alkylating agents in Chapter 26, but in the Baylis–Hillman reaction, the electrophile is an aldehyde and is present right from the start of the reaction, which is done just by stirring the components at room temperature. Here is a typical example.

The reaction starts with the (relatively nucleophilic) DABCO undergoing conjugate addition to ethyl acrylate. This will form an enolate that can then attack the acetaldehyde in an aldol reaction.

E1cB eliminations often follow aldol reactions and lead to α,β-unsaturated products. In this case, though, DABCO is a much better leaving group than the hydroxyl group, so enolization leads to loss of DABCO in an E1cB elimination, giving the product of the reaction. DABCO is recovered unchanged, and is a catalyst.

A disadvantage of the Baylis–Hillman reaction is its rate: typically, several days' reaction time are required. Pressure helps speed the reaction up, but as a catalyst DABCO is about the best. It is nucleophilic, because of the 'tied back' alkyl groups, but importantly it is a good leaving group because it has a relatively low pK_a, meaning that it leaves easily in the last step. As you have seen before, good nucleophiles are usually bad leaving groups, though there are many exceptions. DABCO's combination of nucleophilicity and leaving group ability is perfect here.

The exposed nature of the nitrogen atom in cyclic amines means that nitrogen heterocycles are very frequently encountered in drug molecules, particularly those operating on the central nervous system (cocaine, heroin, and morphine all contain nitrogen heterocycles, as do codeine and many tranquillizers such as Valium). But the ring can also be used as a support for adding substituents that hinder the nitrogen's lone pair. Just as the nitrogen atom of piperidine is permanently exposed, the nitrogen atom of 2,2,6,6-tetramethylpiperidine (TMP) nestles deep in a bed of methyl groups. The lithium salt of TMP (LiTMP) is an analogue of LDA—a base that experiences enormous steric hindrance that can be used in situations where the selectivity even of LDA fails.

> With LDA, one or other of the isopropyl groups always has the option of rotating to place only a C–H group close to the N–Li bond. In LiTMP, there are always four Me groups close to Li.

Aziridine: ring strain promotes ring opening

Aziridine and azetidine are stable, if volatile, members of the saturated nitrogen heterocycle family, and aziridine has some interesting chemistry of its own. Like pyrrolidine and piperidine, aziridine can be acylated by treatment with an acyl chloride, but the product is not stable. The ring opens with attack of chloride, a relatively poor nucleophile, and an open-chain secondary amide results.

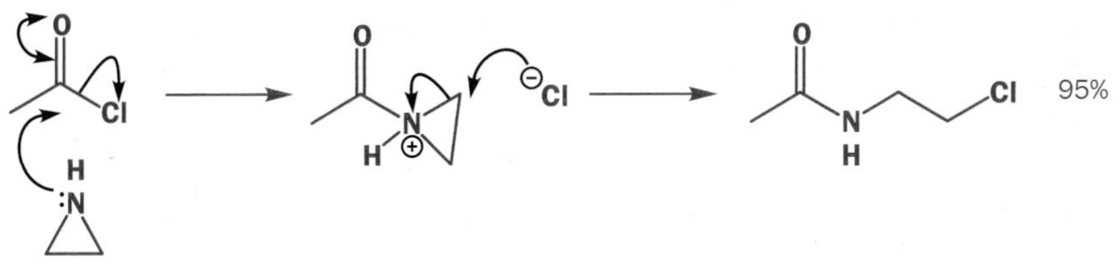

95%

Saturated heterocycles and systematic nomenclature

The names aziridine and azetidine are derived from a reasonably logical system of nomenclature, which assigns three-part heterocycle names according to: (a) the heteroatom ('az-' = nitrogen, 'ox-' = oxygen, 'thi-' = sulfur); (b) the ring size ('-ir-' = 3, from **tri**; '-et-' = 4, from **tetra**; '-ol-' = 5; nothing for 6; '-ep-' = 7, from h**ep**ta; '-oc-' = 8, from **oc**ta; etc.); and (c) the degree of saturation ('-ene' or '-ine' for unsaturated, '-idine' or '-ane' for saturated). Hence az-ir-idine, az-et-idine, di-ox-ol-ane, and ox-ir-ane.

You can view this ring opening as very similar to the ring opening of an epoxide (Chapter 20)—in particular, a *protonated* epoxide, in which the oxygen bears a positive charge. The positive charge is very important for aziridine opening because, when the reaction is done in the presence of a base, removal of the proton leads immediately to the neutral acyl aziridine, which *is* stable.

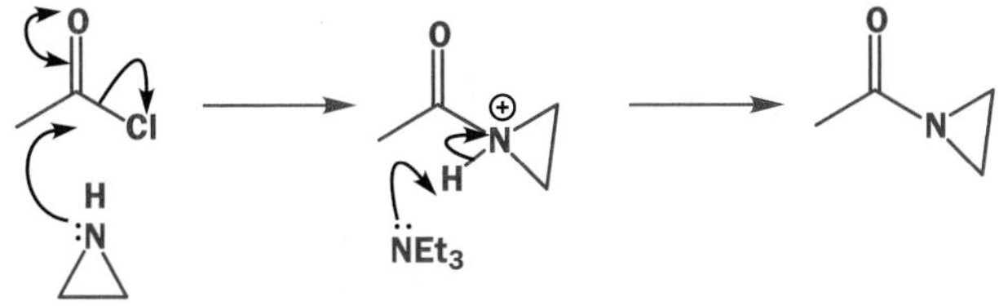

N-acetyl aziridine

The ring opening of aziridine is a useful way of making larger heterocycles: anything that puts a positive charge on nitrogen encourages the opening by making N a better leaving group, whether it's protonation, as shown below, or alkylation.

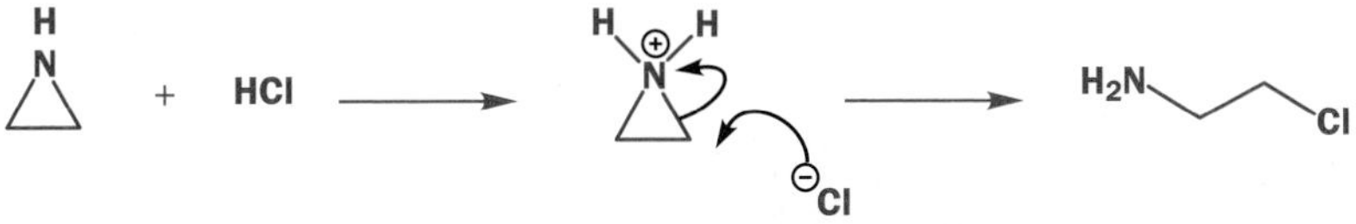

Alkylation of aziridine in base gives the *N*-substituted aziridine as you might expect, but a second alkylation leads to a positively charged aziridinium salt that opens immediately to the useful bromoamine. In this case, the product is an intermediate in the synthesis of two natural products, sandaverine and corgoine.

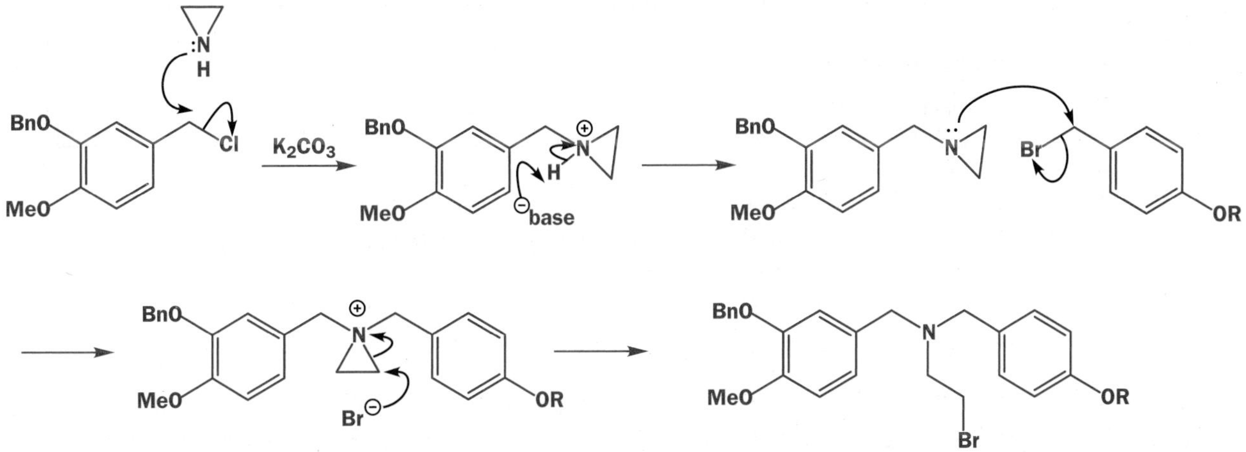

We have just mentioned the protonation of aziridine, and you might imagine from what we said earlier about the comparative nucleophilicity and basicity of nitrogen heterocycles and their acyclic counterparts that aziridine will be even more nucleophilic than pyrrolidine, and about as basic. Well,

For a reminder of the terms associated with ring size, look at Chapter 18, p. 454.

▶

In Chapter 15 we summarized the effect by saying 'Small rings introduce more p character inside the ring and more s character outside it'. Put simply, we can say that, as bond angles decrease, as they must in small rings, the bonds within the ring take up more p character (p orbitals are at an angle of 90° to one another), leaving the bonds (or lone pairs) outside the ring with more s character.

it isn't. The idea that 'tying back' the alkyl groups increases nucleophilicity is only valid for 'normal-sized' (five- or six-membered) rings: with small rings another effect takes over.

Aziridine is, in fact, much less basic than pyrrolidine and piperidine: its pK_{aH} is only 8.0. This is much closer to the pK_{aH} of a compound containing an sp^2 hybridized nitrogen atom—the imine in the margin, for example. This is because the nitrogen's lone pair is in an orbital with much more s character than is typical for an amine, due to the three-membered ring. This is an effect we have discussed before, in Chapter 15, and you should re-read pp. 365–366 if you need to refresh your memory. There we compared three-membered rings with alkynes, explaining that both could be deprotonated relatively easily. The anion carries a negative charge in a low-energy orbital with much s character: the same type of orbital carries aziridine's lone pair.

The s character of the aziridine nitrogen's lone pair has other effects too. The lone pair interacts very poorly with an adjacent carbonyl group, so N-acyl aziridines such as the one you saw on p. 1125 behave not at all like amides. The nitrogen atom is pyramidal and not planar, and the stretching frequency of the C=O bond (1706 cm^{-1}) is much closer to that of a ketone (1710 cm^{-1}) than that of an amide (1650 cm^{-1}).

Lack of conjugation leads to increased reactivity, and N-acyl aziridines are useful in synthesis because they react with organolithium reagents only once to give ketones. No further reactions of the product ketone occur because the N-acyl aziridine is reactive enough to compete with it for the organolithium reagent.

$pK_{aH} = 8.0$ $pK_{aH} = 7.2$

$\nu(C=O) = 1652$ cm^{-1}

planar N

$\nu(C=O) = 1706$ cm^{-1}

pyramidal N

N-acyl aziridine

The s character of the lone pair means that the nitrogen atom inverts very slowly, rather like a phosphine (which also carries its lone pair in an s orbital: see Chapter 4, p. 108). Usually it is not possible for nitrogen to be a stereogenic centre because inversion is too rapid—the transition state for nitrogen inversions (in which the lone pair is in a p orbital) is low in energy. But with an aziridine, getting the lone pair into a p orbital would require an awful lot of energy, so nitrogen can be stereogenic and, for example, these two stereoisomers of an N-substituted aziridine can be separated and isolated.

N-acyl aziridines therefore behave like Weinreb amides (see p. 300), and stabilization of the tetrahedral intermediate by chelation may also play a role. Esters, of course, typically react twice with organolithiums to give tertiary alcohols.

Oxygen heterocycles

Ring-opening chemistry is characteristic of oxygen heterocycles too, and there is no need for us to revisit epoxide opening here. Epoxides are particularly reactive because ring opening releases ring strain, driving the reaction forward. However, we can tell you about some chemistry of the most important simple oxygen heterocycle, THF. You may be surprised that THF does any real chemistry: after all, the very reason it is used as a solvent is precisely because it is so unreactive. Oxygen heterocycles are cyclic ethers, and ethers are the least reactive of all the common functional groups.

Epoxide opening under acidic and basic conditions was covered in Chapter 20.

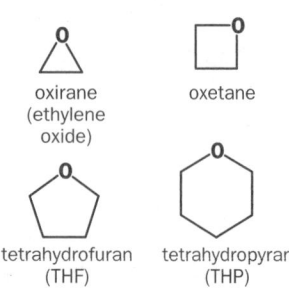

oxirane (ethylene oxide)   oxetane

tetrahydrofuran (THF)   tetrahydropyran (THP)

You have seen (Chapters 17 and 23) the use of HBr, BBr$_3$, and Me$_3$SiCl to deprotect methyl and benzyl ethers of phenols.

To make ethers more reactive, they must be complexed with strong Lewis acids. BF$_3$ is commonly used with cyclic ethers, and even with epoxides it increases the rate and yield of the reaction when organometallic reagents are used as nucleophiles. BF$_3$ is most easily handled as its complex with diethyl ether, written BF$_3$:OEt. BuLi does not react with oxetane, for example, unless a Lewis acid, such as BF$_3$, is added, when it opens the four-membered ring to give a quantitative yield of n-heptanol.

The same reaction happens with THF, but only in much lower yield. Nonetheless, just as cyclic amines are more nucleophilic than acyclic ones, so cyclic ethers are more nucleophilic than acyclic ones. This is one of the reasons why THF is such a good solvent for organolithiums—the nucleophilic lone pair of the oxygen atom stabilizes the electron-deficient lithium atom of the organolithium.

Ways of writing BF$_3$ complex with Et$_2$O. The oxygen lone pair is donated into the boron's empty p orbital.

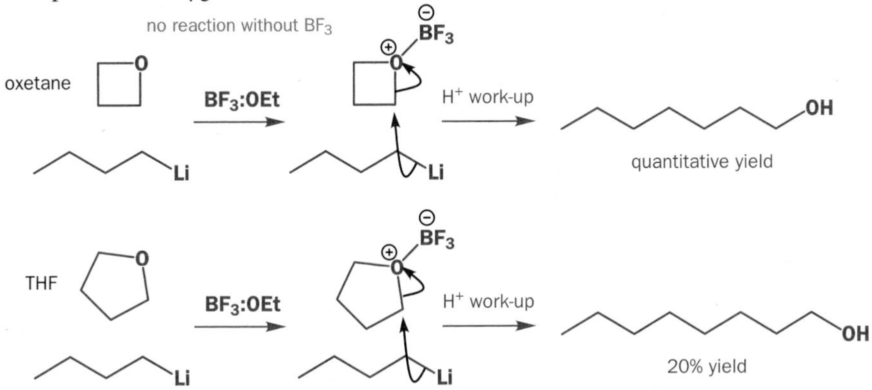

A more important reaction between BuLi and THF is not nucleophilic attack, but deprotonation. You will have noticed that reactions involving BuLi in THF are invariably carried out at temperatures of 0 °C or below—usually –78 °C. This is because, at temperatures above 0 °C, deprotonation of THF begins to take place. You might think that this would not be a problem, if BuLi were being used as a base, because the deprotonated THF could still itself act as a base. The trouble is that deprotonated THF is unstable, and it undergoes a reverse [2+3] cycloaddition. Here is the mechanism (we have represented the organolithium as an anion to help with the arrows). The products are: (1) the (much less basic) enolate of acetaldehyde and (2) ethylene. The first tends to polymerize, and the second usually evaporates from the reaction mixture.

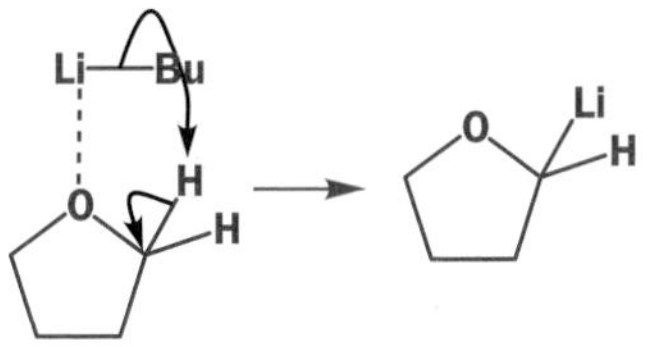

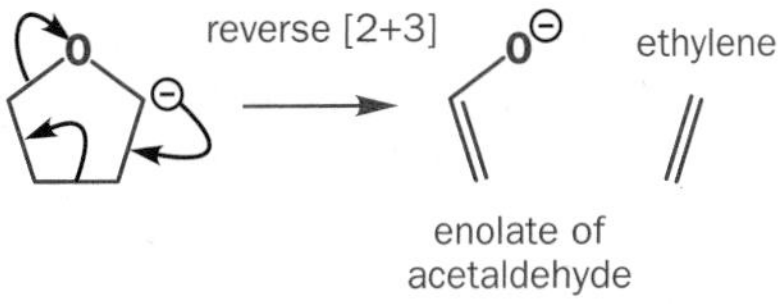

reverse [2+3]   ethylene
enolate of acetaldehyde

The half-life of *n*-BuLi in THF (in the presence of TMEDA) is 40 min at 20 °C, 5.5 h at 0 °C, and 2 days at –20 °C. Diethyl ether is much less readily deprotonated: at 20 °C in ether *n*-BuLi has a half-life of 10 h. With more basic organolithiums, the rate of decomposition of THF is even faster, and *t*-BuLi can be used in THF only at –78 °C. At –20 °C *t*-BuLi has a half-life in THF of only 45 min; in ether its half-life at this temperature is 7.5 h.

The case of the extra ethyl group

Some chemists in Belgium were studying the reactions of the organolithium shown here to find out whether the anionic centre would attack the double bond to form a five-membered ring (like a radical would: see Chapter 39). The reaction was slow, and they stirred the organolithium in THF for 6 hours at 0 °C. When they worked the reaction up they found no five-membered ring products: instead they got a compound with an extra ethyl group attached! They showed that this ethyl group, in fact, comes from THF: the organolithium did not add to the double bond in the same molecule, but it did add slowly and in low yield to the double bond of the ethylene that is formed by decomposition of THF.

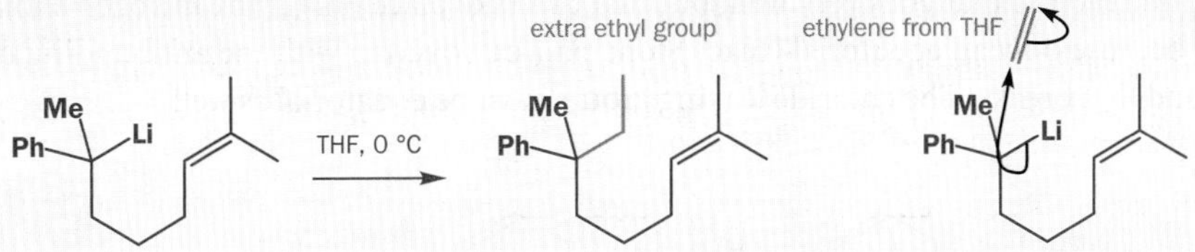

The most common use of tetrahydropyran derivatives is as protecting groups: you met this in Chapter 24 and you can see an example later in the chapter, on p. 1132.

Sulfur heterocycles

The ability of sulfur to stabilize an adjacent anion will be discussed in Chapter 46, and it means that sulfur heterocycles are much easier to deprotonate than THF. The most important of these contains two sulfur atoms: dithiane. Deprotonation of dithiane occurs in between the two heteroatoms, and you can see some chemistry that arises from this on p. 1254. For the moment, we will just show you series of reactions that illustrate nicely both dithiane chemistry and the ring opening of oxygen heterocycles in the presence of BF$_3$. This substituted derivative of dithiane is deprotonated by BuLi in the same way to give a nucleophilic organolithium that will

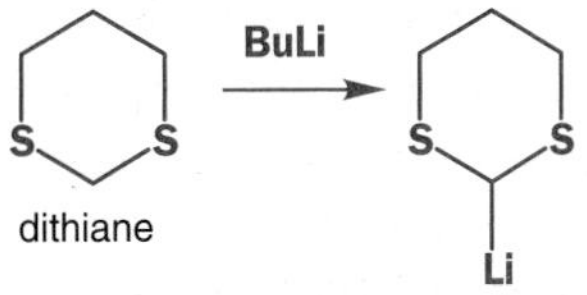

dithiane
BuLi

Dithiolane, the five-membered version of dithiane, cannot be used in this reaction because, although it is easy to deprotonate, once deprotonated it decomposes by the same mechanism as that used by lithiated THF.

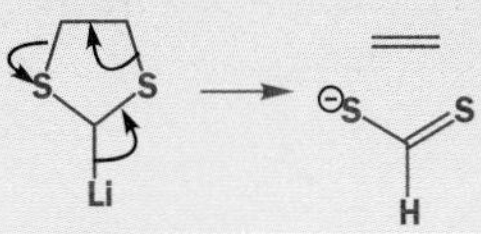

attack electrophiles—even oxygen heterocycles—provided BF_3 is present. The products are formed in excellent yield, even when the electrophile is THP, with no ring strain to drive the reaction. After the addition reaction the dithiane ring can be hydrolysed with mercury(II) (see Chapters 46 and 50 for an explanation) to give a ketone carrying other useful functional groups.

■ The dithiane functions here as an **acyl anion equivalent**. For more on this read Chapter 46.

dithiane derivative

BuLi →

oxygen heterocycle:
$n = 0, 1, 2, \text{ or } 3$

BF$_3$ →

$n = 0$ (from oxirane) 98%
$n = 1$ (from oxetane) 93%
$n = 2$ (from THF) 90%
$n = 3$ (from THP) 78%

HgCl$_2$
MeOH

Conformation of saturated heterocycles: the anomeric effect

Heteroatoms in rings have axial and equatorial lone pairs

equatorial lone pairs in black

dithiane

dioxane

axial lone pairs in green

piperidine

To a first approximation, the conformation of five- and six-membered saturated heterocycles follows very much the same principles as the conformation of carbocyclic compounds that we detailed in Chapter 18. If you feel you need to re-read the parts of that chapter dealing with rings—chairs and boats, or axial and equatorial substituents—now would be a good time to do it. Sticking with dithiane for the moment, then, this is the conformation. Since the sulfur atoms have lone pairs, they too occupy axial and equatorial positions. The same is true of dioxane or of piperidine.

We have coloured the lone pairs green or black according to whether they are axial or equatorial, but you can also consider the colour coding in a different way: black lone pairs are parallel with C–C or C–heteroatom bonds in the ring; green lone pairs are parallel with axial C–H bonds outside the ring, or, if the ring has substituents, with the bonds to those substituents. This substituted tetrahydropyran illustrates all this. Notice that the equatorial substituents next to the heteroatom are parallel with neither the green nor the black lone pair.

green lone pair parallel with bonds to axial substituents
black lone pair parallel with C–C bonds within ring

Why is this important? Well, if you cast your mind back to Chapter 38, you will remember that the overlap of parallel orbitals was very important in fragmentation reactions. Here, for example, is a fragmentation reaction that goes very well, but that can take place only if the nitrogen's lone pair is equatorial, because only an equatorial (black) lone pair can overlap with the antibonding orbital of the C–C bond that breaks. The chloride leaving group must be equatorial as well.

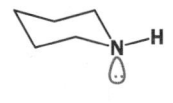

elimination can take place only when lone pair is equatorial and parallel with black bonds

base
$-H^\oplus$ →

H_2O →

+

This is not a problem in this example, because flipping of the ring and inversion of the nitrogen are fast, and enough of the starting material is in this conformation at any one time for the reaction to take place. But compare this bicyclic acetal whose 'fragmentation' (actually just an acetal hydrolysis) looks possible by this mechanism.

→

→

Yet when we try and draw the conformation of the lone pairs we run into a problem: neither overlaps with the C–O bond that is breaking and so neither can donate its electron density into the C–O σ*. (Another way of looking at this is to say that the intermediate oxonium ion—with a C=O double bond formed by one of the oxygen's lone pairs—would be extremely strained.) Not surprisingly, the rate of hydrolysis of this acetal is extremely slow compared with similar ones in which overlap between the oxygen lone pair and the C–O σ* is possible. The acetal in the margin hydrolyses about 10^{10} times faster.

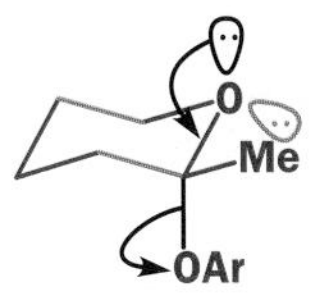

Other situations you have met where overlap between parallel orbitals is important are:

- E2 elimination reactions (Chapter 19)
- NMR coupling constants (Chapter 32)
- reactions of cyclic molecules (Chapter 33)
- the Felkin–Anh transition state conformation (Chapter 34)

Together, these effects are called **stereoelectronic effects**, because they depend on the shape and orientation of orbitals. Most of the examples we have presented you with have been stereoelectronic effects on reactivity, but the next section will deal with how stereoelectronic effects affect **conformation**.

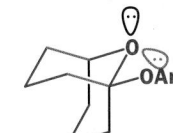

very fast hydrolysis

neither lone pair overlaps with bond to leaving group

Some substituents of saturated heterocycles prefer to be axial: the anomeric effect

Some of the most important saturated oxygen heterocycles are the sugars. Glucose is a cyclic hemiacetal—a pentasubstituted tetrahydropyran if you like—whose major conformation in solution is shown on the right. About two-thirds of glucose in solution exists as this stereoisomer, but hemiacetal formation and cleavage is rapid, and this is in equilibrium with a further one-third that carries the hemiacetal hydroxyl group axial (<1% is in the open-chain form).

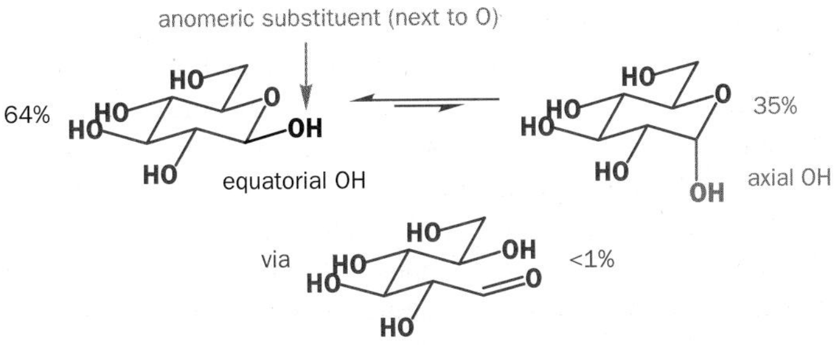

We introduced the hemiacetal structure of glucose in Chapter 6, p. 138.

Having read Chapter 18 you will not be surprised that glucose prefers all its substituents to be equatorial. For four of them, of course, there is no choice: they are either all-equatorial or all-axial, and the only way they can get from one to the other is by ring-flipping. But for the fifth substituent, the hydroxyl group next to the ring oxygen (known as the **anomeric** hydroxyl), the choice of axial or equatorial is made available by hemiacetal cleavage and re-formation—it can invert its **configuration**. What is perhaps surprising is that the equatorial preference of this hydroxyl group is so small—only 2:1. Even more surprising is that, for most derivatives of glucose, the anomeric substituents *prefer* to be axial rather than equatorial.

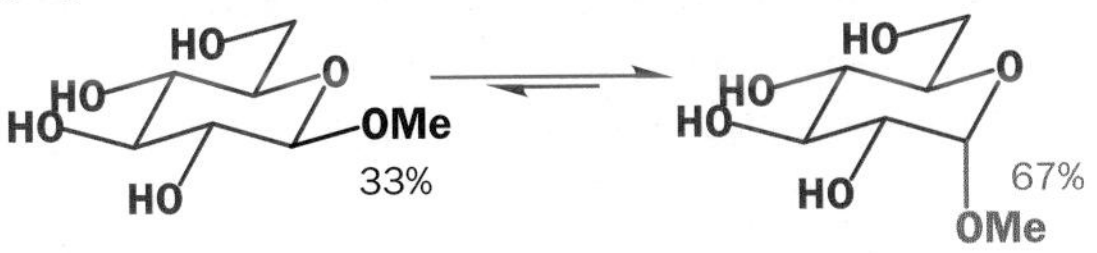

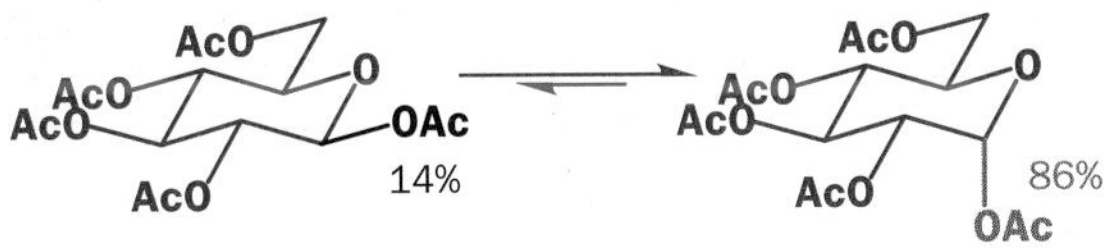

Move away from glucose, and the effect is still there. Here, for example, is the NMR spectrum of this chloro compound. There are now only two possible conformations (no configurational changes are possible because this is not a hemiacetal)—both shown—and from the NMR spectrum you should be able to work out which one this compound has.

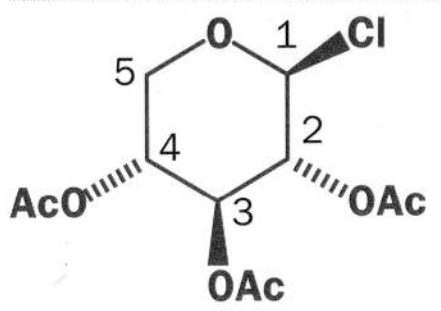

The key point is that axial–axial couplings are large (>8 Hz, say), even with adjacent electronegative atoms (which do tend to lower coupling constants). So if H1

NMR in rings was discussed in Chapter 32: this example should come as revision of that material.

δ			J, Hz	
5.78	1H	t	2.0	H1
5.03	2H	m		H2, H3
4.86	1H	m		H4
4.37	1H	dd	12.9, 3.0	H5a
3.75	1H	ddd	12.9, 3.7, 0.6	H5b
2.10	9H	s		OAc × 3

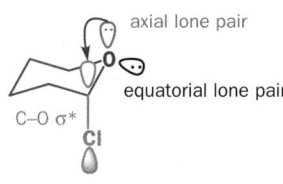

were an axial proton, you would expect it to have a large coupling to H2. But it doesn't—it couples to H2 with *J* of only 2.0 Hz. (The other coupling is a W-coupling to H3, also of 2.0 Hz: see p. 270.) Similarly, we know that the 12.9 Hz coupling shared by the two H5 protons must be a geminal (^2J) coupling. One of H5a or H5b must be axial; yet both couple to H4 with *J* < 4 Hz. So H4 cannot be axial. With this evidence, we have to conclude that H1 and H4 (and therefore H2 and H3) are equatorial, so the compound must exist mainly in the all-axial conformation. (The 0.6 Hz coupling to H5b is another W-coupling, and shows that H5b is the equatorial proton, and H5a therefore the axial one.)

● The anomeric effect

In general, any tetrahydropyran bearing an electronegative substituent in the 2-position will prefer that substituent to be axial. This is is known as the **anomeric effect.**

X axial   is more stable than   X equatorial

But why? This goes against all of what we said in Chapter 18 about axial substituents being more hindered, making conformations carrying axial substituents disfavoured. The key again is stereoelectronics, and we can now link up with the message we left you with at the end of the last section: eliminations and fragmentations can work only when the orbitals involved are parallel.

An amide is more stable (less reactive) than a ketone because the p orbital of the N and the low-lying C=O π* of the carbonyl can lie parallel—they can overlap and electron density can move from nitrogen into the C=O bond, weakening C=O. (Evidence for this comes from the lower IR stretching frequency of an amide C=O, among other things.) But C–X bonds also have low-lying antibonding orbitals—the C–X σ*—so we would expect a molecule to be stabilized if an adjacent heteroatom could donate electrons into this orbital in the same way. Take the generalized tetrahydropyran in the box above, for example, with X = Cl, say. This molecule is most stable if an oxygen lone pair can overlap with C–Cl σ*, like this.

But it can do this only if the chlorine is axial! Remember what we pointed out earlier: the oxygen's equatorial lone pairs are parallel with nothing but bonds in the ring, so the oxygen's axial lone pair is the only one that can help stabilize the molecule, and it can only do this when the Cl is axial. Only the axial conformation benefits from the stabilization, and this is the origin of the anomeric effect.

How shall we represent the stabilization? Comparing again with the amide stabilization, you might think about how to represent it with curly arrows: this is straightforward with the amide and you have seen it many times. But it looks odd with our heterocycle: electron density moves from O to Cl, and the C–Cl bond is weakened. If the process carried right on, Cl⁻ would leave. This is exactly what did happen in the acetal we presented you with as an example on p. 1129: only the axial OAr could leave, however, because of the same requirement for overlap with an oxygen lone pair. In the real structure that we are now looking at, the Cl is still there: the C–Cl bond is weaker, and some of the oxygen's electron density is delocalized on to Cl. This can be seen in crystal structures: compounds exhibiting an anomeric effect have a longer (and therefore weakened) bond outside the ring and a shorter, stronger C–O bond within the ring.

axial lone pair

equatorial lone pair

C–O σ*

equatorial lone pair

axial lone pair

incorrectly aligned
for overlap with
either lone pair

bond weakened
and lengthened

bond weakened
and lengthened

The anomeric effect in some other compounds

Now that you know about the anomeric effect, you should add it to your mental array of ways of explaining 'unexpected' results. Here is an example. Many fruit flies have pheromones based around a 'spiroketal' structure, which we could represent without stereochemistry as shown below. You can imagine the spiroketal (that is, an acetal of a ketone made of two rings joined at a single atom) being made from a dihydroxyketone—and, indeed, this is very often how they are made synthetically. But this is a bad representation because these compounds do have stereochemistry, and the stereochemistry is very interesting.

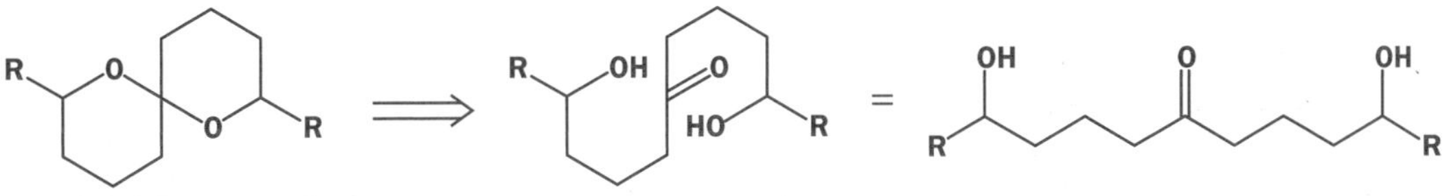

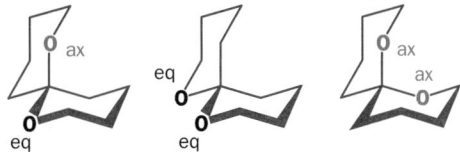

spiroketal structure of
some insect pheromones

Let's start with the simplest example, with R = H (a pheromone of the olive fly). Once you have drawn one ring in its chair conformation, there are three ways of attaching the other ring, shown here. If you think they all look the same, consider the orientation of each C–O bond with respect to the ring that it is not part of: you can have each C–O axial or equatorial, and there are three possible arrangements (three conformations).

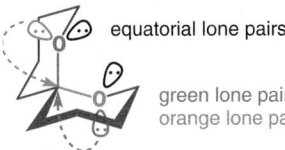

Without knowing about the anomeric effect, you would find it hard to predict which conformation is favoured, and, indeed, you might expect to get a mixture of all three. But NMR tells us that this compound exists entirely in one conformation: the last one here, in which each oxygen is axial on the other ring. Only in this conformation can both C–O bonds benefit from the anomeric effect—this is often known as the **double anomeric effect**.

equatorial lone pairs cannot interact

green lone pair donates into green σ*
orange lone pair donates into orange σ*

Things become even more interesting when the spiroketal carries substituents. The pheromone of *Epeolus crucifer*, for example, carries one additional methyl group at a centre with (*S*) configuration. The spiroketal centre is now a chiral centre, and also exists in a single configuration. Only one possible conformation allows the methyl substituent to be equatorial and the two oxygens to be axial, and that conformation defines the configuration at the spiroketal. Only one diastereoisomer is formed, in which the methyl group controls the *spiro* centre.

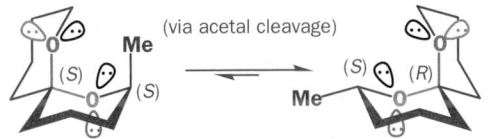

(via acetal cleavage)

Me equatorial: only ketal
configuration and only
conformation observed

The fact that the substituents on the side chains can control the conformation of the spiroketal centre means that it is not necessary to worry about that centre in a synthesis, provided you are trying to make the spiroketal that has the double anomeric stabilization (both oxygens axial) and that has any substituents equatorial on the rings. A recent (1997) synthesis of a single enantiomer of some fruit-fly pheromones from an aspartic acid-derived bromodiol is shown overleaf. It involves three different-sized oxygen heterocycles.

The diol is made into an epoxide by an intramolecular substitution reaction that is S_N2 and so goes with inversion. There are two possible rings that could form, depending on which hydroxyl group attacks, but (as you will shortly see) three-membered rings form faster than four-membered ones, and the reaction gives none of the oxetane. The other hydroxyl group can now be protected as a benzyl ether.

▶

This is a chiral compound, even though the acetal centre is not a chiral centre: no conformation has a plane of symmetry.

▶

If you try to draw these spirocyclic acetals you will soon find there is a trick to getting them to look right: the *spiro* carbon has to be one of the four that aren't at the 'point' of either ring; otherwise one ring ends up looking flat.

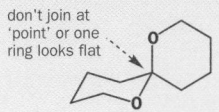

don't join at
'point' or one
ring looks flat

UNHELPFUL REPRESENTATION

■

There is more on asymmetric synthesis, including some pheromone examples, in Chapter 45. The protecting groups used in this synthesis were covered in Chapter 24, and aza-enolate alkylations in Chapter 26.

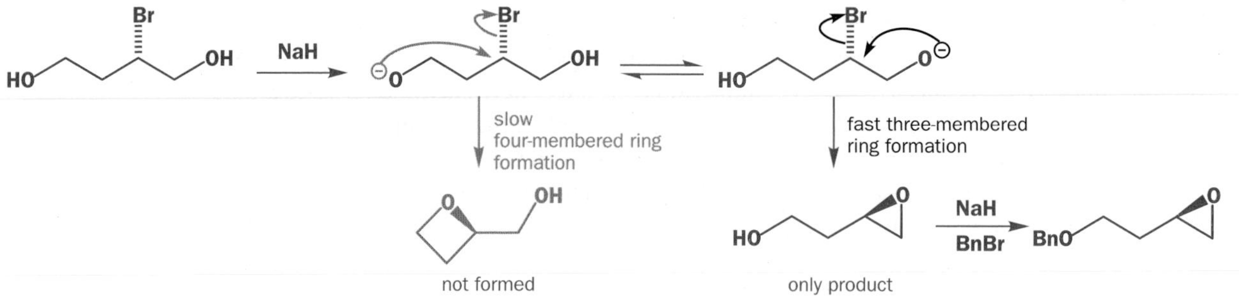

not formed   only product

The epoxide opens well with either a copper derivative (RMgBr + CuI) or simply NaBH$_4$, and the resulting alcohol needs to be protected. A good, and in this instance topical, choice is a THP group, added using dihydropyran in the presence of acid. The disadvantage of THP protecting groups is that they introduce an unwanted chiral centre: this will not be controlled and we expect a mixture of both (*R*) and (*S*) configurations at this centre. However, you should now have no problem predicting the *conformation* of the THP rings, even if it is irrelevant to the synthesis.

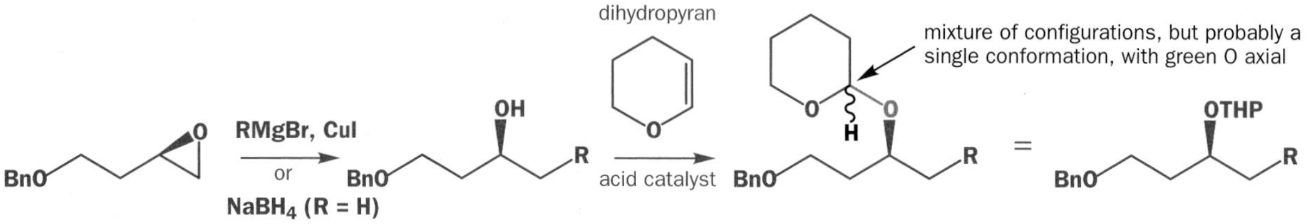

Now the benzyl ether can be deprotected, and the hydroxyl group substituted for iodide via its tosylate. This iodide is an alkylating agent, and is used for two successive alkylations of a hydrazone's aza-enolate.

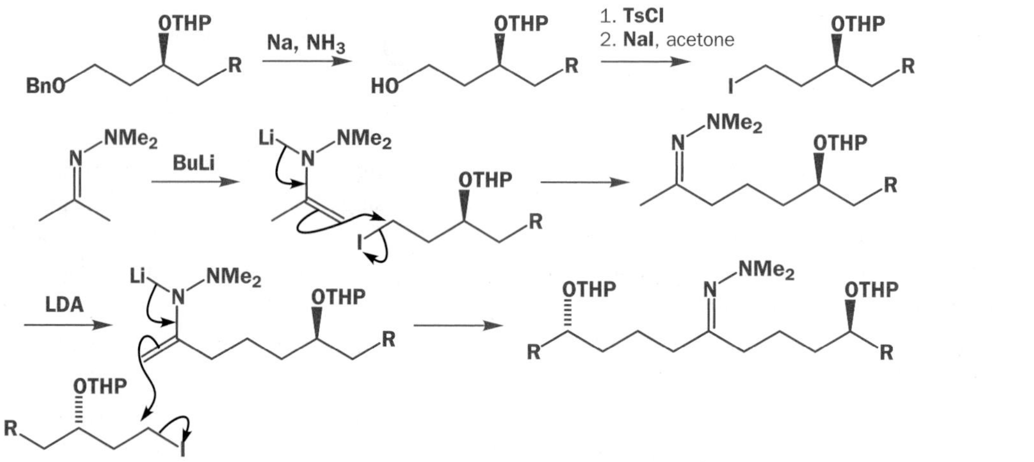

The product is still a hydrazone, and it needs hydrolysing to the ketone with 1 M HCl. These conditions cause immediate hydrolysis of the THP protecting groups and then cyclization to the spiroacetal, which forms with complete control over stereochemistry— a single diastereoisomer is formed in which both alkyl groups go equatorial and both oxygens axial.

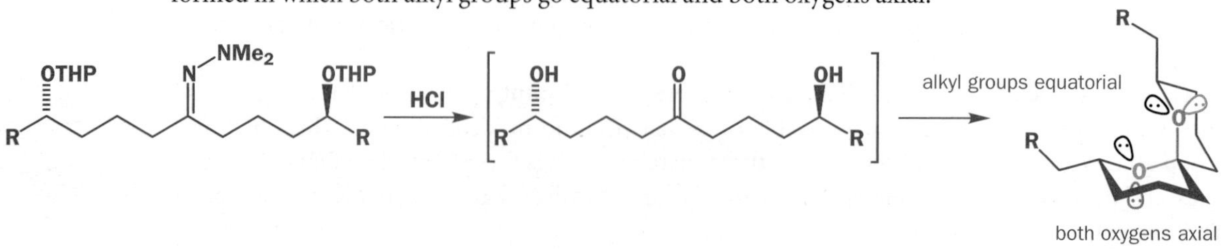

Remember that the key requirement for the anomeric effect is that there is a heteroatom with a lone pair (O, N, S usually) adjacent to (that is, in a position to interact with) a low-lying antibonding orbital—usually a C–X σ* (where X = halogen or O). The C–X bond doesn't have to be within the ring—for example, this nitrogen heterocycle prefers to have the R group axial so that the nitrogen gets an equatorial lone pair. Equatorial lone pairs are parallel with bonds within the ring, one of which is C–O, and this conformation is therefore stabilized by an N lone pair/ C–O σ* interaction.

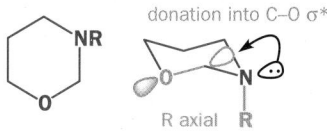
donation into C–O σ*
R axial R

It would be a bit much for this 1,3,5-triazine to have all three *t*-butyl groups axial (too much steric hindrance), but it can get away with having one of them axial, benefiting from the resulting equatorial lone pair, which can overlap with two C–N σ*s in the ring.

t-Bu axial

donation into C–N σ*

Related effects in other types of compounds

> ● **Any conformation in which a lone pair is anti-periplanar to a low-energy antibonding orbital will be stabilized by a stereoelectronic interaction.**

As you will probably realize, it's not only in six-membered rings that stereoelectronic interactions between filled and unfilled orbitals stabilize some conformations more than others. Stereoelectronic effects control the conformations of many types of molecules. We shall look at three common compounds that are stabilized by stereoelectronic effects: in two cases, the stabilization is specific to one conformation, and we can use stereoelectronics to explain what would otherwise be an unexpected result.

But we start with a compound that is so simple that it has only one conformation because it has no rotatable bonds: dichloromethane. You may have wondered why it is that, while methyl chloride (chloromethane) is a reactive electrophile that takes part readily in substitution reactions, dichloromethane is so unreactive that it can be used as a solvent in which substitution reactions of other alkyl halides take place. You may think that this is a steric effect: indeed, Cl is bigger than H. But CH_2Cl_2 is much less reactive as an electrophile than ethyl chloride or propyl chloride: there must be more to its unreactivity. And there is: dichloromethane benefits from a sort of 'permanent anomeric effect'. One lone pair of each chlorine is always anti-periplanar to the other C–Cl bond so that there is always stabilization from this effect.

Among the most widespread classes of acyclic compounds to exhibit stereoelectronic control over conformation are acetals. Take the simple acetal of formaldehyde and methanol, for example: what is its conformation? An obvious suggestion is to draw it fully extended so that every group is fully antiperiplanar to every other—this would be the lowest-energy conformation of pentane, which you get if you just replace the Os with CH_2s.

The trouble is, in this conformation none of the oxygen lone pairs get the chance to donate into the C–O σ* orbitals. Although putting the bonds anti-periplanar to one another makes steric sense, electronically, the molecule much prefers to put the lone pairs anti-periplanar to the C–O bonds, so the bonds themselves end up gauche (synclinal) to one another. This is known as the **gauche effect**, but is really just another way in which the stereoelectronic effects that give rise to the anomeric effect turn up in acyclic systems.

Finally, an even more familiar example that you may never have thought about. You are well aware now that amides are planar, with partially double C–N bonds, and that tertiary amides have one alkyl group *cis* to oxygen and one *trans*. But what about esters? Esters are less reactive than acyl chlorides because of donation from the oxygen p orbital into the carbonyl π*, so we expect them to be planar too, and they are. But there are two possible planar conformations for an ester: one with R *cis* to oxygen and one with R *trans*. Which is preferred?

all bonds shown lie in a plane

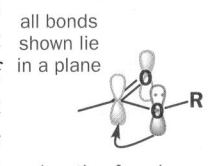
donation from lone pair of O into π* keeps ester planar

> ►
> Dichloromethane will react as an electrophile, but it needs a very powerful nucleophile and long reaction times.
> CH_2Cl_2
> PhSNa $\xrightarrow[\text{several days}]{\text{as solvent}}$ PhS⌒SPh

dichloromethane

permanent donation into C–Cl σ*

Me⌒O⌒O⌒Me
extended conformation of simple acetal

Me⌒⌒⌒Me
lowest-energy conformation of pentane

Me Me

gauche conformation allows donation into σ*

> ■
> We discussed this in relation to DMF in Chapter 7, p. 165.

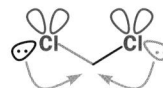
cis   trans
Me   Me

► Some esters—lactones, for example—cannot lie *cis* for steric reasons, and this is one of the reasons why lactones are distinctly more reactive than esters and in many reactions behave more like ketones: lactones are quite easy to reduce with NaBH$_4$, for example.

trans arrangement enforced by ring

Here are the two conformations drawn out for ethyl acetate. When the ethyl group (= R) and O are *cis*, not only can one oxygen lone pair interact with the C=O π*, but the other lone pair can also donate into the σ* of the C=O bond. This is not possible when Et and O are *trans*: they are no longer anti-periplanar. The *cis* conformation of esters is generally the preferred one, even in formate esters, where the alkyl group ends up in what is clearly a more sterically hindered orientation.

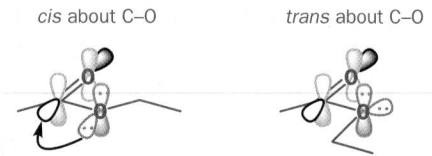

cis about C–O — in this conformation, additional stabilization is possible as second lone pair of O donates into C–O σ*

trans about C–O — in this conformation, no additional stabilization is possible. Second lone pair of O cannot donate into C–O σ*

Making heterocycles: ring-closing reactions

■ *m*-CPBA epoxidation is discussed in Chapters 19 and 33. In Chapter 33 we explained how ring opening of epoxides joined to six-membered rings was controlled by conformational factors: that discussion relates closely to what we will say here.

We have talked about the structure of saturated heterocycles, particularly with regard to stereoelectronic control over conformation, and before that we looked at some of their reactions. In this last section of the chapter we will look at how to make saturated heterocycles. By far the most important way of making them is by **ring-closing reactions**, because we can usually use the heteroatom as the nucleophile in an intramolecular substitution or addition reaction. Ring-closing reactions are, of course, just the opposite of the ring-opening reactions we talked about earlier in the chapter, and we can start with a reaction that works well in both directions: ring closure to form an epoxide. You know well that epoxides can be formed using *m*-CPBA and an alkene, but you have already seen examples (including one earlier in the chapter) where they form by an intramolecular substitution reaction such as this.

The same method can also be used to generate larger cyclic ethers. Oxetane, for example, is conveniently made by adding 3-chloropropyl acetate to hot potassium hydroxide.

The first step in this reaction is the hydrolysis of the ester. The alkoxide produced then undergoes an intramolecular substitution reaction to yield oxetane.

Tetrahydropyran was prepared as early as 1890 by a ring closure that occurs when a mixture of 1,5-pentanediol with sulfuric acid is heated.

These are all S$_N$2 reactions, so you will not be surprised that nitrogen heterocycles can be prepared in the same way. Aziridine itself, for example, was first prepared in 1888 from 2-chloroethylamine.

This method works well to form three-, five-, and six-membered nitrogen heterocycles, but does not work well to form four-membered rings. In fact, four-membered rings are generally among the hardest of all to form. To illustrate this, the first two columns of Table 42.2 show the rates (relative to six-membered ring formation = 1) at which bromoamines of various chain lengths cyclize to saturated nitrogen heterocycles of three to seven members.

Table 42.2 Rates of ring-closing reactions

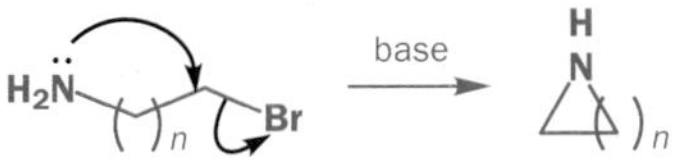

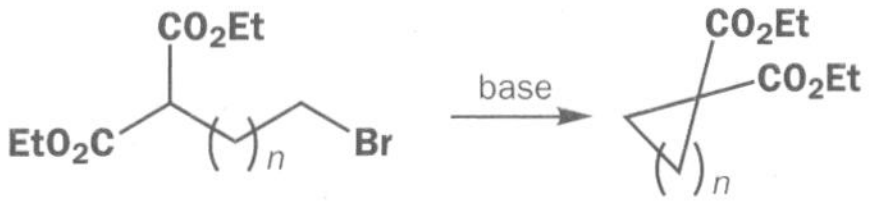

Ring size	Product	Relative rate[a]	Product[b]	Relative rate[a]	Assessment of rate
3		0.07			moderate
4		0.001	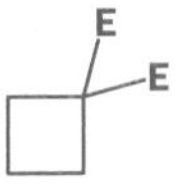	0.58	slow
5		100	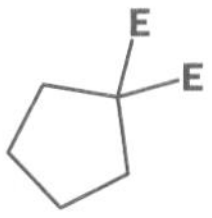	833	very fast
6	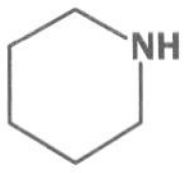	1	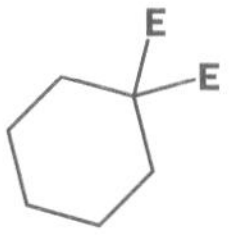	1	fast
7	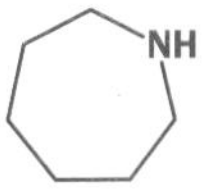	0.002	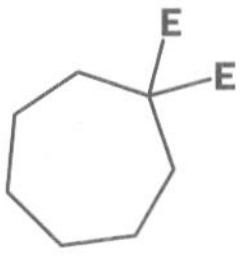	0.0087	slow
8			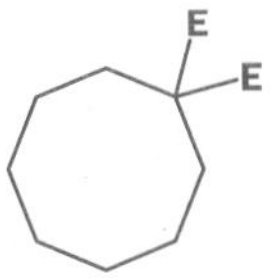	0.00015	very slow

[a] Relative to the six-membered ring formation (= 1).
[b] E = CO_2Et.

The first thing that strikes you perhaps is that the figures in the third column have been produced by a random number generator! There seems to be no rhyme or reason to them, and no consistent trend. To convince you that these numbers mean something, Table 42.2 also shows, in its next two columns, the relative rates for a quite different ring-closing reaction, this time forming four- to seven-membered rings that are not even heterocycles by intramolecular alkylation of a substituted malonate. Though the numbers are quite different in the two cases, the ups and downs are the same, and the final column summarizes the relative rates. Put another way, a rough guide (only rough!—it doesn't work in all cases) to the rate of ring formation is this.

● **Rough guide to the rate of formation of saturated rings**

5 > 6 > 3 > 7 > 4 > 8–10

■ Remind yourself of our definition of small, normal, medium, and large rings, and what ring strain means, by re-reading p. 454. We will deal with what happens in large rings a little later.

We show the numbers in colour to highlight the fact that this seemingly illogical ordering of numbers actually conceals two superimposed trends. Once you get to five-membered rings, the rate of formation drops consistently as the ring size moves from 'normal' to 'medium'. 'Small' (three- and four-membered) rings insert into the sequence below six.

The reason for the two superimposed trends is two opposing factors. Firstly, small rings form slowly because forming them introduces ring strain. This ring strain is there even at the transition state, raising its energy and slowing down the reaction. $\Delta G^{\ddagger}$ is very large for a three-membered ring (due to strain) but decreases as the ring gets larger. This explains why three- and four-membered rings don't fit straightforwardly into the sequence.

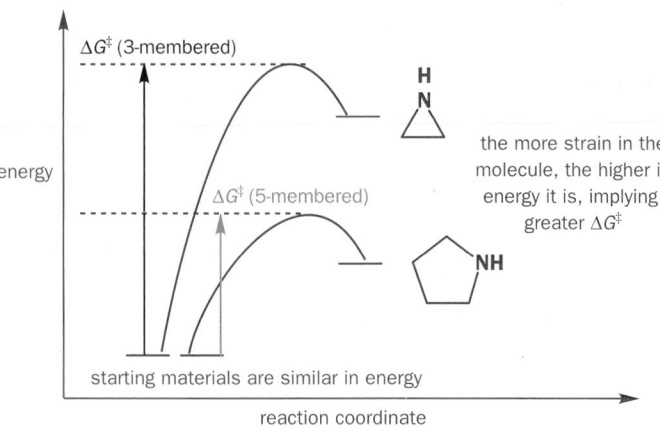

But, if the reaction rate simply depended on the strain of the product, the slowest reaction would be the formation of the three-membered ring, and six-membered rings (which are essentially strain-free) would form fastest. But as it is, four-membered rings form more slowly than three-membered ones, and five-membered ones faster than six-membered ones. To explain this, we need to remind you of an equation we presented in Chapter 13.

$$\Delta G^{\ddagger} = \Delta H^{\ddagger} - T\Delta S^{\ddagger}$$

The activation energy barriers $\Delta G^{\ddagger}$ of our reactions are made up of two parts: an enthalpy of activation $\Delta H^{\ddagger}$, which tells us about the energy required to bring atoms together against the strain and repulsive forces they usually have, and an entropy of activation $\Delta S^{\ddagger}$, which tells us about how easy it is to form an ordered transition state from a wriggling and randomly rotating molecule.

$\Delta G^{\ddagger}$ for three- and four-membered ring formation is large because $\Delta H^{\ddagger}$ is large: energy is needed to bend the molecule into the strained small-ring conformation. $\Delta H^{\ddagger}$ for five-, six-, and seven-membered rings is smaller: this is the quantifiable representation of the 'ring strain' factor we have just introduced. The second factor is one that depends on $\Delta S^{\ddagger}$: how much order must be imposed on the molecule to get it to react. Think of it this way: a long chain has a lot of disorder, and to get its ends to meet up and react means it has to give up a lot of freedom. So, for the formation of medium and large rings, $\Delta S^{\ddagger}$ is large and negative, contributing to a large $\Delta G^{\ddagger}$ and slow reactions. For three-membered rings, on the other hand, the reacting atoms are already very close together and almost no order needs to be imposed on the molecule to get it to cyclize: rotation about just one bond is all that is needed to ensure that the amine group is in the perfect position to attack the σ^* of the C–Br bond in our example above. $\Delta S^{\ddagger}$ is very small for three-membered rings so, while $\Delta H^{\ddagger}$ is large, there is little additional contribution from the $T\Delta S^{\ddagger}$ term and cyclization is relatively fast. Four-membered rings suffer the worst of both worlds: forming a four-membered ring introduces ring strain ($\Delta H^{\ddagger}$) *and* requires order ($\Delta S^{\ddagger}$) to be imposed on the molecule. They form very slowly as a result.

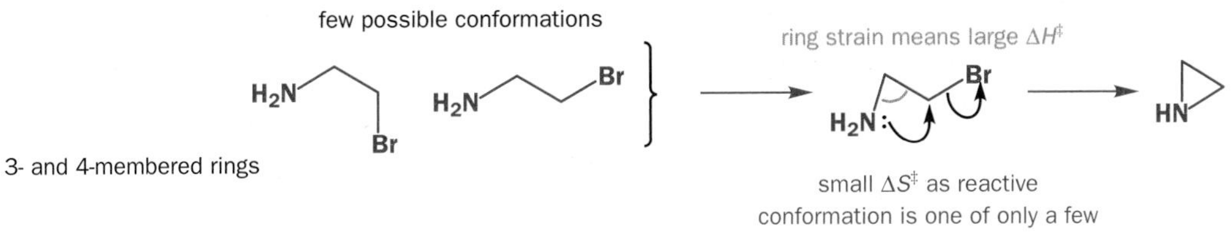

few possible conformations

3- and 4-membered rings

ring strain means large $\Delta H^{\ddagger}$

small $\Delta S^{\ddagger}$ as reactive
conformation is one of only a few

These results are summarized in the following box.

● Ring formation

- Three-membered ring formation is fast—the product is strained so $\Delta H^{\ddagger}$ is large but this is offset by the reacting atoms being as close as they can get in a freely rotating chain
- Four-membered rings form slowly—the product is still significantly strained but the reacting atoms are now not right next to each other to offset this
- Five-membered ring formation is often fastest of all. Significantly less strain and the ends are still not too far apart
- Six-membered ring formation experiences no strain but neither does it have the advantage of the ends being close
- Seven-membered rings and beyond form more slowly as $\Delta S^{\ddagger}$ increases

---

Medium and large rings

Beyond seven-membered rings, the rates stay low, but begin to level off, and may start to rise again when the rings have 10 or 11 members. These are the 'medium rings', of about 8–13 members, and they suffer from a different sort of strain, evident in the graph on p. 455 (Chapter 18), due to interactions between C–H bonds across the ring (**transannular interactions**). These are worst for rings of 8 and 9 members, and begin to be relieved once there are 10

or 11 atoms in the ring. For 14-membered rings and above, there is no transannular strain, and the rates of ring closure remain essentially constant at about the 7-membered ring mark. Rates of reactions in ring sizes of 14 and above are essentially little different from those in acyclic compounds. To get large rings to form, it is often necessary to carry out the cyclization reaction in very dilute solution to discourage competing intermolecular reactions.

transannular interactions
hinder medium-ring formation

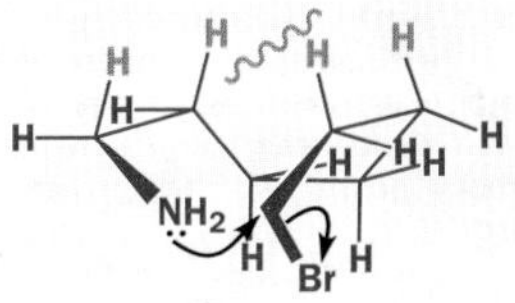

Thermodynamic control

In this section we have discussed the rate at which rings form: in other words the kinetics of ring formation. However, there are many ring-forming reactions that are under thermodynamic and not kinetic control. For example, you have already seen that glucose exists predominantly as a six-membered ring in solution. It could also exist as a five-membered ring: it doesn't because, although five-membered rings form faster than six-membered ones, they are usually less stable (remember, a six-membered ring is essentially strain-free). For similar thermodynamic reasons, it doesn't exist as a seven-membered ring, even though you can draw a reasonable structure for it.

neither of these is formed

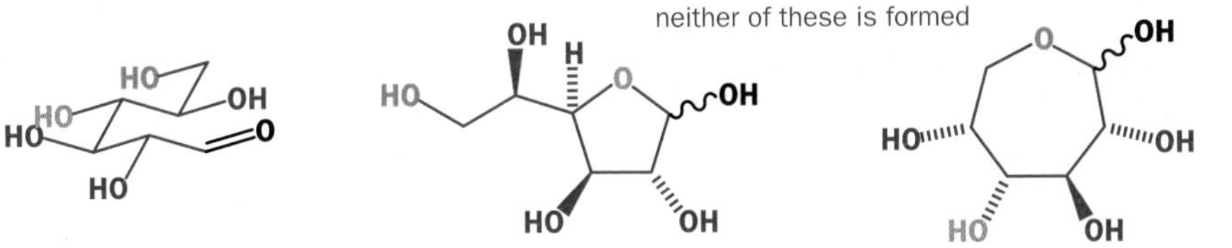

open-chain form of glucose

cyclization of orange
hydroxyl gives
five-membered ring

cyclization of green
hydroxyl gives
seven-membered ring

Thermodynamic control is important in other ways in carbohydrate chemistry, because control over ring size allows selective protection of the hydroxyl groups of sugars. Compare these two reactions. Both of them give acetals from the same starting material, mannitol.

Don't be put off by the way in which we have had to twist half the molecule round to draw the left-hand structure: the stereochemistry hasn't changed. The important thing is that acetone reacts with mannitol to form three five-membered acetals (dioxolanes) while benzaldehyde forms only two six-membered acetals. This is quite a common result: when there is a choice, acetone prefers to react across a 1,2-diol to give a five-membered ring, while aldehydes prefer to react across a 1,3-diol to form a six-membered ring. Drawing a conformational diagram of the product on the right helps to explain why. All of the substituents are equatorial, making this a particularly stable structure. Now imagine what would happen if acetone formed this type of six-membered ring acetal. There would always be an axial methyl group, and the six-membered rings would be less stable.

Aminals are another class of saturated heterocycles that form very readily under thermodynamic control: aminals are nitrogen analogues of acetals. They are usually made by refluxing a 1,2-diamine with an aldehyde in toluene (no acid catalyst is needed because the nitrogens are very nucleophilic), and this makes a very useful way of forming a chiral derivative of an achiral aldehyde. Here is an example: the diamine is made from the amino acid proline. The product has a new chiral centre, and it forms as a single diastereoisomer because the phenyl ring prefers to be on the *exo* face of the bicyclic system (see Chapter 33).

all substituents equatorial

with acetone, axial methyl groups would be inevitable

There is another example of selective protection using thermodynamic control in Chapter 49, p. 1361.

You should be able to write a mechanism for this reaction—it is much the same as the formation of an acetal, but no acid catalyst is present so you will not need to protonate the carbonyl groups before the amines attack.

Refluxing in toluene removes the water as an azeotrope (see p. 347), but, in fact, the aminal forms so readily that, if you do this reaction in cold dichloromethane (in which water is insoluble), the solution becomes cloudy as droplets of water are produced!

Combatting $\Delta S^{\ddagger}$—the Thorpe–Ingold effect

Compare the following relative rates for epoxide-forming cyclization reactions. The second looks as though it suffers more steric hindrance but it is tens of thousands of times faster!

Adding substituents to other ring-forming reactions makes them go faster too: in the next two examples the products are oxetanes and pyrrolidines.

This effect is quite general, and is known as the **Thorpe–Ingold effect** after the first chemists to note its existence, in 1915.

> ● **The Thorpe–Ingold effect**
>
> The Thorpe–Ingold effect is the way in which substituents on the ring increase the rate, or equilibrium constant, for ring-forming reactions.

As the box says, it's not only rate that can be affected by additional substitution. Here are the relative equilibrium constants for the formation of an anhydride from a 1,4-dicarboxylic acid (the unsubstituted acid is called succinic acid, and the values are scaled so that K_{rel} for the formation of succinic anhydride is 1). More substituents mean more cyclized product at equilibrium. The Thorpe–Ingold effect is both a kinetic and a thermodynamic phenomenon.

Now we need to explain why this is. The explanation comes in two parts, one of which may be more important than the other, depending on the ring being formed. The first part is more applicable to the formation of small rings, such as the first example we gave you.

If you measure the bond angles of chains of carbon atoms, you expect them to be close to the tetrahedral angle, 109.5°. The crystal structure of the 1,3-dicarboxylic acid in the margin, for example, shows a C–C–C bond angle of 110°. Now, imagine adding substituents to the chain. They will repel the carbon atoms already there, and force them a little closer than they were, making the bond angle slightly less. X-ray crystallography tells us that adding two methyl groups to our 1,3-dicarboxylic acid decreases the bond angle by about 4°.

We can assume that the same is true in the alcohol starting materials for the epoxide-forming reactions (we can't measure the angle directly because the compounds aren't crystalline). Now consider what happens when both of these alcohols form an epoxide. The bond angle has to become about 60°, which involves about 50° of strain for the first diacid, but only 46° for the second. By distorting the starting material, the methyl groups have made it slightly easier to form a ring.

This part of the argument works only for small rings. For larger rings, we need another explanation, and it involves entropy. We'll use the pyrrolidine-forming reaction as an example. We have explained the effect of $\Delta S^{\ddagger}$ (entropy of activation) on the rate of ring formation: as larger rings form they have to lose more entropy at the transition state, and this contributes to a less favourable $\Delta G^{\ddagger}$.

But, when the starting material has more substituents, it starts off with less entropy anyway. More substituents mean that some conformations are no longer accessible to the starting material—the green arcs below show how the methyl groups hinder rotation of the N and Br substituents into that region of space. Of those fewer conformations, many approximate to the conformation in the transition state, and moving from starting material to transition state involves a small loss of entropy: $\Delta S^{\ddagger}$ is less negative so $\Delta G^{\ddagger} (= \Delta H^{\ddagger} - T\Delta S^{\ddagger})$ is more negative and the ring forms faster.

Because the same arguments apply to ΔS^o for the reaction as a whole (the difference in entropy between starting material and products), increased substitution favours ring closure even under thermodynamic control.

Baldwin's rules

Nearly all of the cyclization reactions that we have discussed have been intramolecular S_N2 reactions where one end of the molecule acted as the nucleophile displacing the leaving group on the other end. We kept to this sort of reaction in order to make valid comparisons between different ring sizes. But you can imagine making saturated heterocycles in plenty of other ways—intramolecular substitution at a carbonyl group, for example, such as happens in this lactonization reaction, or intramolecular addition on to an alkyne.

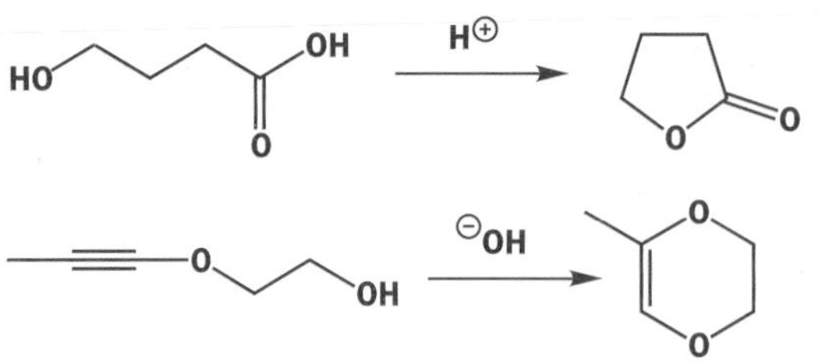

■ Oxyanions add readily to alkynes: see Chapter 31, p. 819.

Cyclization reactions can be classified by a simple system involving: (1) the ring size being formed; (2) whether the bond that breaks as the ring forms is inside (*endo*) or outside (*exo*) the new ring; and (3) whether the electrophile is an sp (digonal), sp^2 (trigonal), or sp^3 (tetrahedral) atom. This system places three of the cyclizations just shown in the following classes.

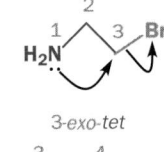

3-exo-tet

5-exo-trig

6-endo-dig

1. The ring being formed has three members; the breaking C–Br bond is outside the new ring (*exo*); the C carrying Br is a tetrahedral (sp^3) atom (*tet*)

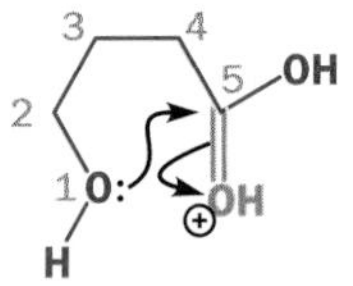

2. The ring being formed has five members; the breaking C=O bond is outside the new ring (*exo*); the C being attacked is a trigonal (sp^2) atom (*trig*)

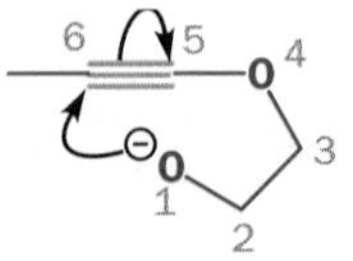

3. The ring being formed has six members; the breaking C≡C bond is inside the new ring (*endo*); the C being attacked is a digonal (sp) atom (*dig*)

■ Professor Sir Jack Baldwin is at Oxford and published his Rules in 1976 while at the Massachusetts Institute of Technology. He has studied biosynthesis (the way living things make molecules) extensively, especially in relation to the penicillins, and has applied many biosynthetic ideas to laboratory synthetic problems.

The classes of cyclization reactions are important, not because we have a compulsive Victorian desire to classify everything, but because which class a reaction falls into determines whether or not it is likely to work. Not all cyclizations are successful, even though they may look fine on paper! The guidelines that describe which reactions will work are known as **Baldwin's rules**: they are not really rules in the Woodward–Hoffmann sense of the term, but more empirical observations backed up by some sound stereoelectronic reasoning. To emphasize this, the rules are couched in terms of 'favoured' and 'disfavoured', rather than 'allowed' and 'forbidden'. We will deal with the rules step by step and then summarize them in a table at the end.

Firstly, and not surprisingly (because we have been talking about them for much of this chapter):

▶ This is a key difference. The Woodward–Hoffmann rules (Chapters 35 and 36) were deduced from theory, and examples were gradually discovered that fitted them. They cannot be violated: a reaction that disobeys the Woodward–Hoffmann rules is getting around them by following a different mechanism. Baldwin's rules were formulated by making observations of reactions that do, or do not, work.

● **All *exo-tet* cyclizations are favoured.**

and, similarly (again you can find many examples in this book):

● **All *exo-trig* cyclizations are favoured.**

Despite the variation in rate we have described for this type of reaction, *exo-tet* cyclizations have no stereoelectronic problems: the lone pair and the C–X σ* (X is the leaving group) can overlap successfully irrespective of ring size. The ring closures in Table 42.2 all fall into this category.

The same is true for *exo-trig* reactions: it is easy for the nucleophilic lone pair to overlap with the C=X π* to form a new bond. Examples include lactone formation such as the one on p. 1140.

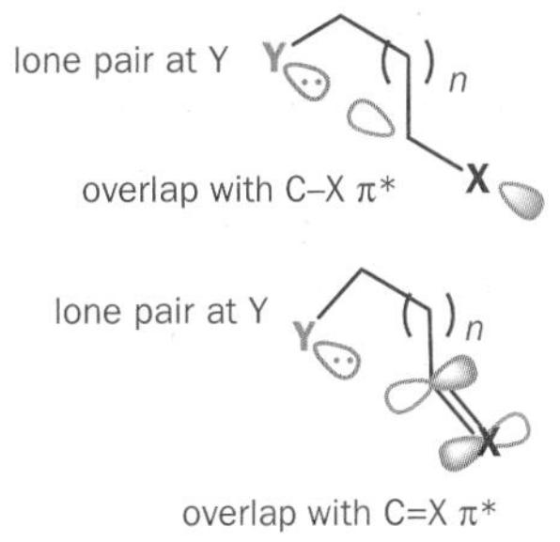

Endo-tet reactions are rather different. For a start:

● **5- and 6-*endo-tet* are disfavoured.**

Endo-tet reactions would not actually make a ring, but they fall conveniently into the system and we will look at them here. Here is a reaction that looks as though it contradicts what we have just said. The arrows in the reasonable-looking mechanism on the right describe a *6-endo-tet* process, because the breaking Me–O bond is within the six-membered ring transition state (even if no ring is formed).

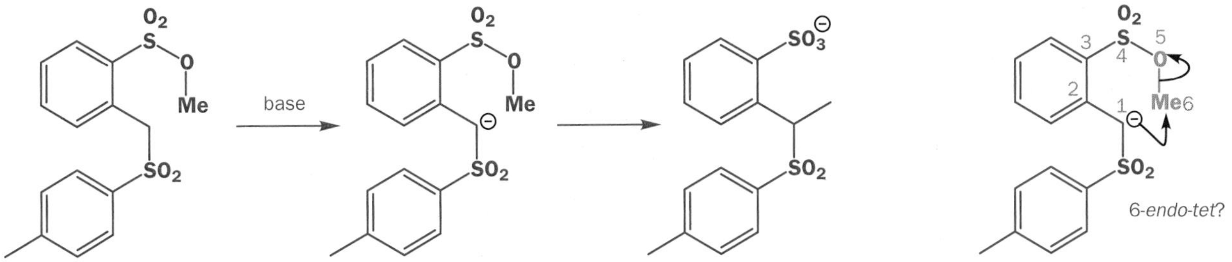

But Eschenmoser showed that, for all its appeal (intramolecular reactions usually outpace all alternatives), this mechanism is wrong. He mixed together the starting material for the reaction above with the hexadeuterated compound shown below, and re-ran the reaction. If the reaction had been intramolecular, the products would have contained either no deuterium, or six deuteriums. In the event, the product mixture contained about 25% of each of these compounds, with a further 50% containing three deuteriums. The products cannot have been formed intramolecularly, and this distribution is exactly what would be expected from an *inter*molecular reaction.

■ This is a **crossover experiment**. See Chapter 41, p. 1088.

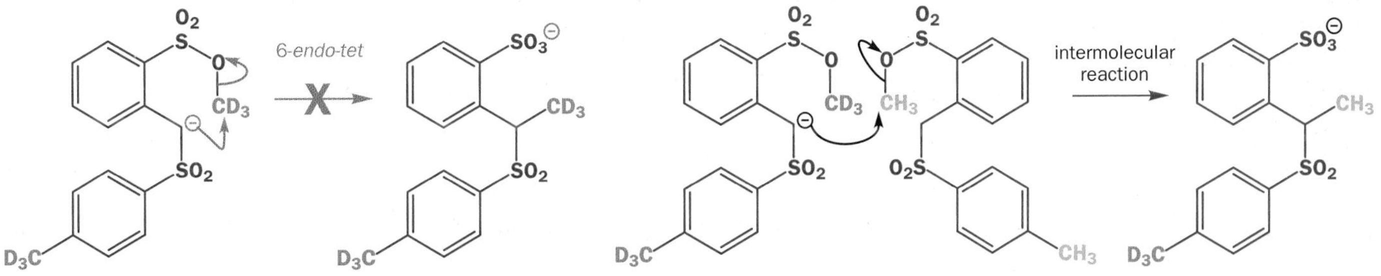

With *endo-trig* reactions, whether they work or not depends on the ring size.

● **3-, 4-, and 5-*endo-trig* are disfavoured; 6- and 7-*endo-trig* are favoured.**

The most important reaction of the *endo-trig* class is the disfavoured 5-*endo-trig* reaction and, if there is one message you take away from this section, it should be that 5-*endo-trig* reactions are disfavoured. The reason we say this is that 5-*endo-trig* cyclizations are reactions that look perfectly fine on paper, and at first sight it seems quite surprising that they won't work. This intramolecular conjugate addition, for example, appears to be a reasonable way of making a substituted pyrrolidine.

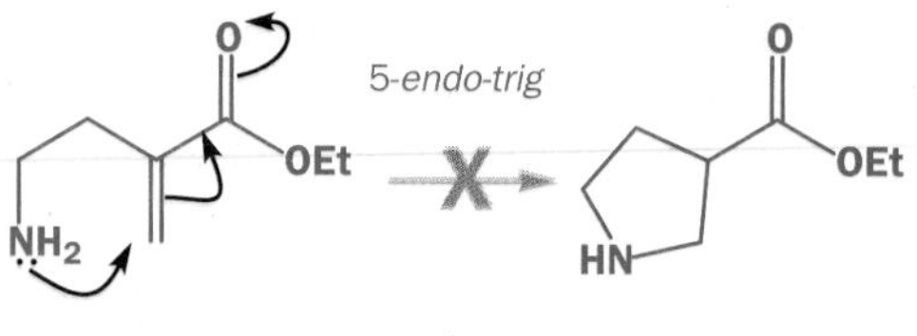

But this reaction doesn't happen: instead, the amine attacks the carbonyl group in a (favoured) 5-*exo-trig* cyclization.

> Amines usually undergo conjugate addition to unsaturated esters: see Chapter 10.

> It's easier to see this with a model, and if you have a set of molecular models you should make one to see for yourself.

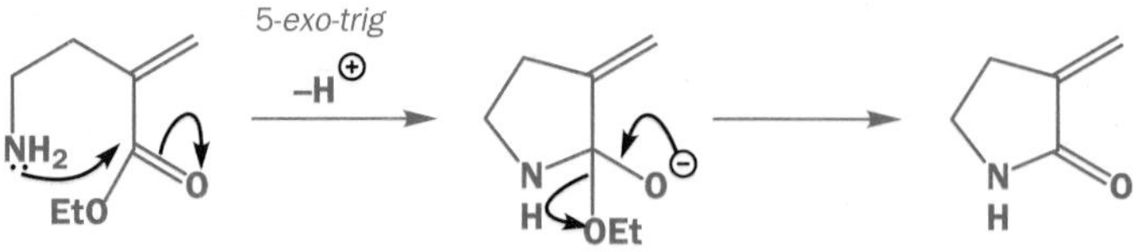

Why is 5-*endo-trig* so bad? The problem is that the nitrogen's lone pair has problems reaching round to the π* orbital of the Michael acceptor. There is no problem reaching as far as the electrophilic carbon in the plane of the substituents but, if it bends out of this plane, which it must if it is to overlap with the π* orbitals, it moves too far away from the methylene carbon to react. It's like a dog chained just out of reach of a bone.

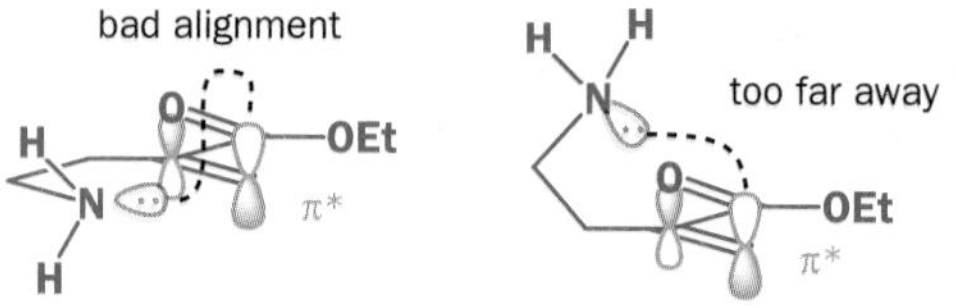

Lengthen the chain, though, and the dog gets his dinner. Here's a perfectly straightforward 6-*endo-trig*, for which orbital overlap presents no problem.

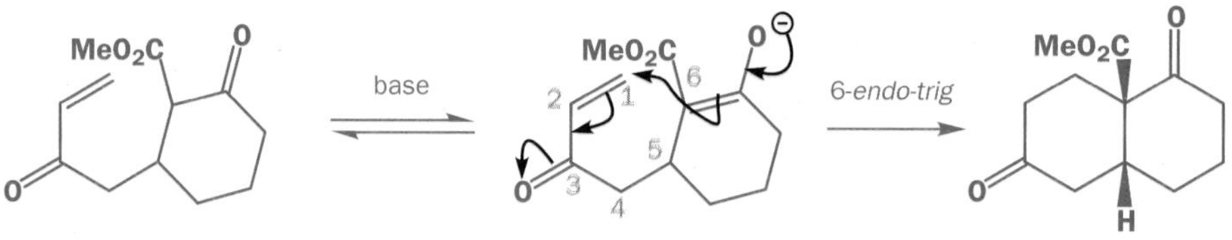

Exceptions to Baldwin's rules

Baldwin's rules are only guidelines and, when a reaction is thermodynamically very favourable (Baldwin's rules, of course, describe the *kinetic* favourability of a reaction) and there is no other possible pathway, 5-*endo-trig* reactions *can* take place. The most striking example is one that you met quite early on in this book (Chapter 14): the formation of a cyclic acetal (dioxolane) from a carbonyl compound and ethylene glycol.

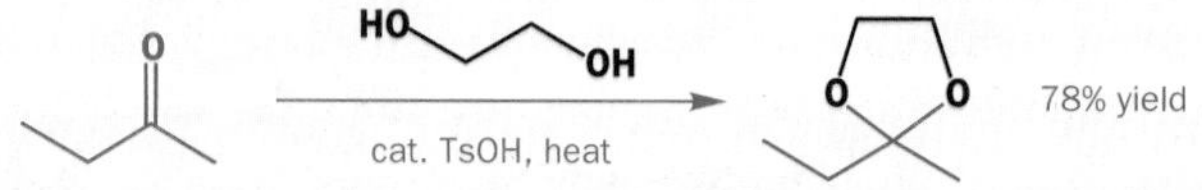

We don't need to give again the full mechanism here, but you should check that you can still write it. The key step with regard to Baldwin's rules is shown with a green arrow. It's a 5-*endo-trig* reaction but it works!

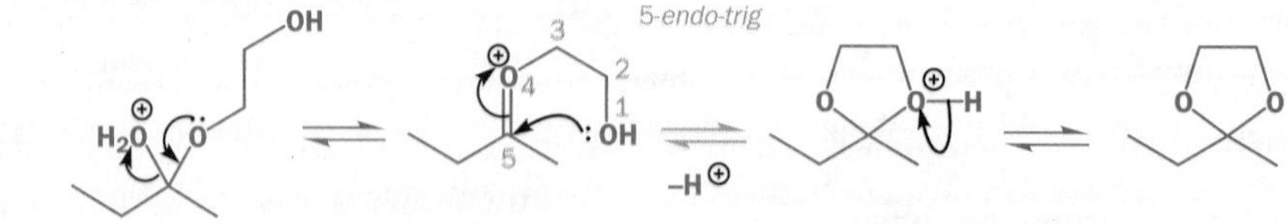

In fact, cations frequently disobey Baldwin's rules. Other well-defined exceptions to Baldwin's rules include pericyclic reactions and reactions in which second-row atoms such as sulfur are included in the ring. This 5-*endo-trig* reaction, the sulfur analogue of the amine cyclization that didn't work, is fine. C–S bonds are long, and the empty 3d orbitals of sulfur may play a role by providing an initial interaction with the C–C π orbital.

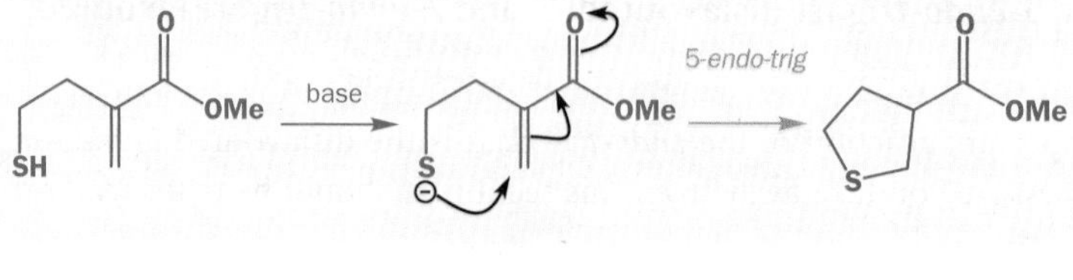

With *tet* and *trig* cyclizations, *exo* is better than *endo*; with *dig* cyclizations, the reverse is true.

> ● All ***endo-dig* cyclizations are favoured.**

Move from 5-*endo-trig* to 5-*endo-dig*, and the reactions become much easier: even 4-*endo-dig* reactions work. Here is an example of 5-*endo-dig*.

We warned you to look out for 5-*endo-trig* reactions because they are disfavoured even though on paper they look fine. Now the alert is the other way round! We expect you'd agree that these *endo-dig* reactions look awful on paper: the linear alkyne seems to put the electrophilic carbon well out of reach of the nucleophile, even further away than in the 5-*endo-trig* reaction. The important thing with *endo-dig* cyclizations, though, is that the alkyne has two π* orbitals, one of which must always lie in the plane of the new ring, making it much easier for the nucleophile to get at.
Conversely:

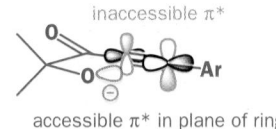

inaccessible π*

accessible π* in plane of ring

> ● 3- and 4-*exo-dig* are disfavoured; 5- to 7-*exo-dig* are favoured.

These reactions are less important and we will not discuss them in detail.

Baldwin's rules and ring opening

Baldwin's rules work because they are based on whether or not orbital overlap can be readily achieved in the conformation required at the transition state. You met in the last chapter the **principle of microscopic reversibility**, which says that, if a reaction goes via a certain mechanism, the reverse reaction must follow exactly the same path in the opposite direction. So Baldwin's rules also work for ring-opening reactions. This is where the unfavourability of 5-*endo-trig* really is important: this tetrahydrofuranyl ester, for example, looks set up to do an E1cB elimination in base. Indeed , when it is treated with methoxide in deuterated methanol it exchanges the proton α to the ester for deuterium, proving that the enolate forms. But is does not eliminate: elimination would be a reverse 5-*endo-trig* process and is disfavoured.

π and σ* orbitals
are orthogonal:
can't interact

Whenever you think about a ring-opening reaction, consider its reverse, and think whether it is favoured according to Baldwin's rules.

To summarize

We shall end by summarizing Baldwin's rules in a chart. You should note the general outline of this chart: commit to memory that, broadly speaking, *endo-tet* and *endo-trig* are disfavoured; *exo-tet* and *exo-trig* are favoured, and the reverse for *dig*. Then you just need to learn the cut-off points that indicate the exceptions to this broad-brush view: 6-*endo-trig* falls into the favoured category while 4-*exo-dig* falls into the disfavoured one. And, if you really can remember only one thing, it should be that 5-*endo-trig* is disfavoured!

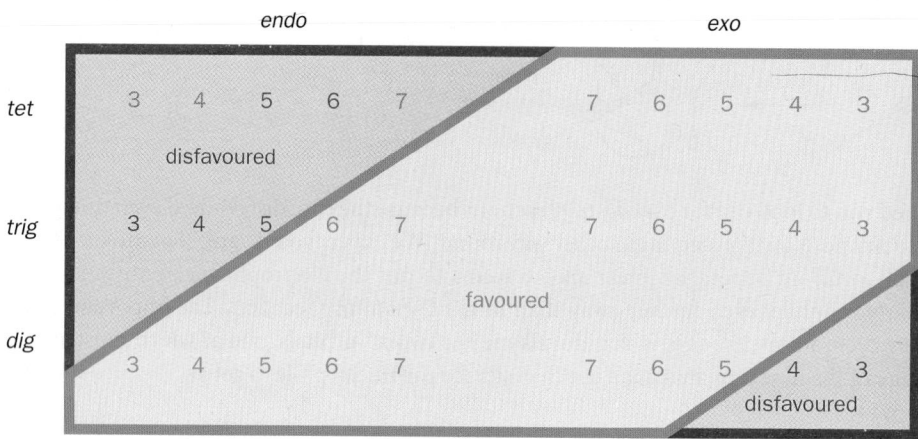

In the next two chapters, we continue with heterocycles, but move from saturated ones to flat, aromatic ones. Conformation and stereoelectronics are no longer issues, but molecular orbitals certainly are. In Chapter 44 you will meet many cyclization reactions: you will find that not a single one is Baldwin-disfavoured.

Problems

1. Predict the most favourable conformations of these insect pheromones.

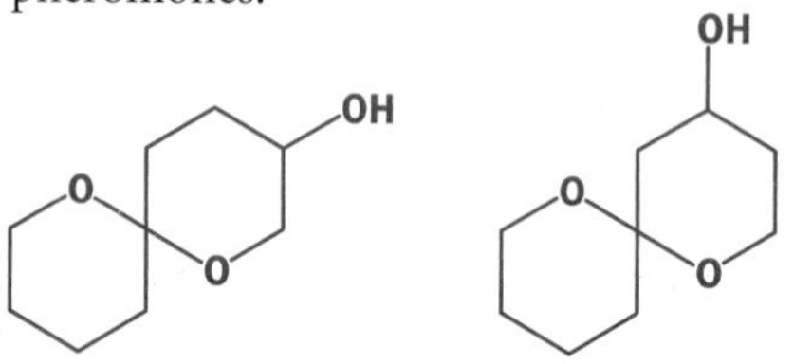

2. Refluxing cyclohexanone with ethanolamine in toluene with a Dean Stark separator to remove the water gives an excellent yield of this spirocycle. What is the mechanism, and why is acid catalysis (or any other kind) unnecessary?

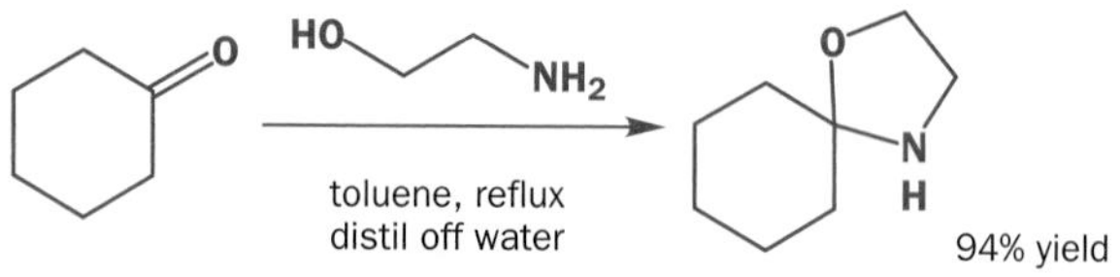

3. What is A in the following reaction scheme and how does it react to give the final product?

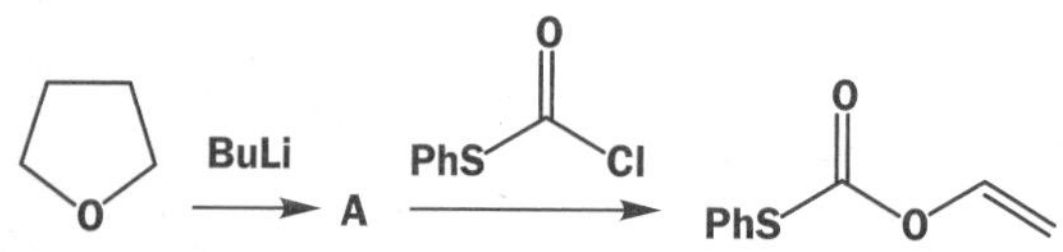

4. Give mechanisms for the formation of this *spiro* heterocycle. Why is the product not formed simply on reacting the starting materials in acid solution without Me₃Al?

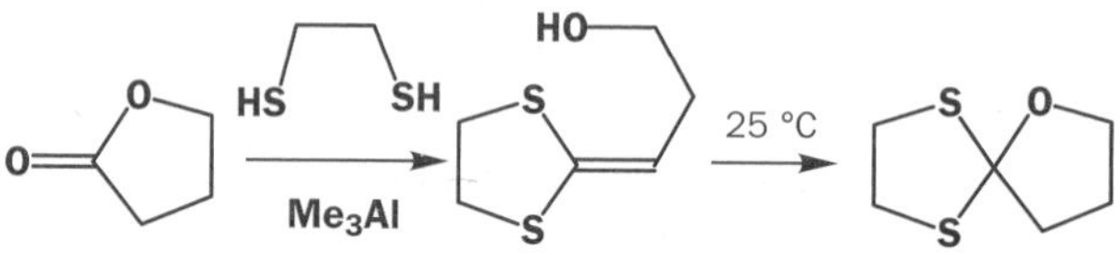

5. The *Lolium* alkaloids have a striking skeleton of saturated heterocycles. One way to make this skeleton is shown below. Explain both the mechanism and the stereochemistry.

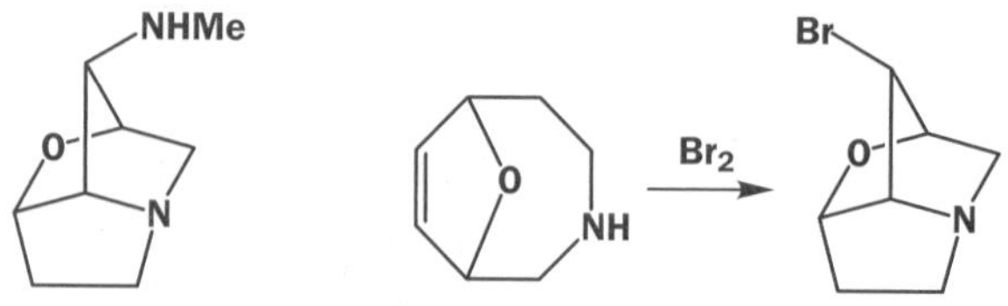

a *Lolium* alkaloid

6. Explain the stereochemical control in this synthesis of a fused bicyclic saturated heterocycle—the trail pheromone of an ant.

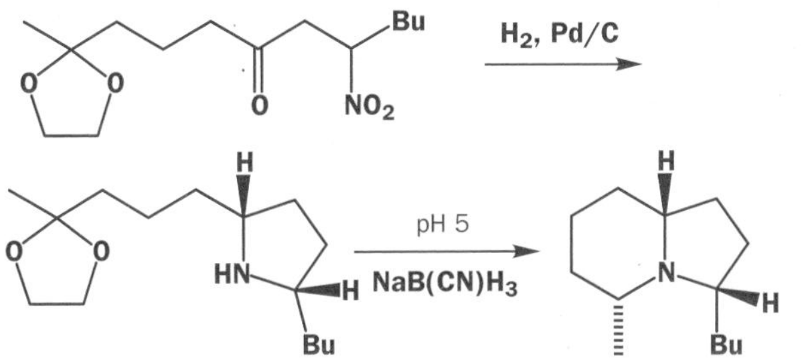

7. In Chapter 31, one of the problems asked you to comment on the difference between these two reactions. Now would you like to comment again and add comments on the way we drew the starting materials.

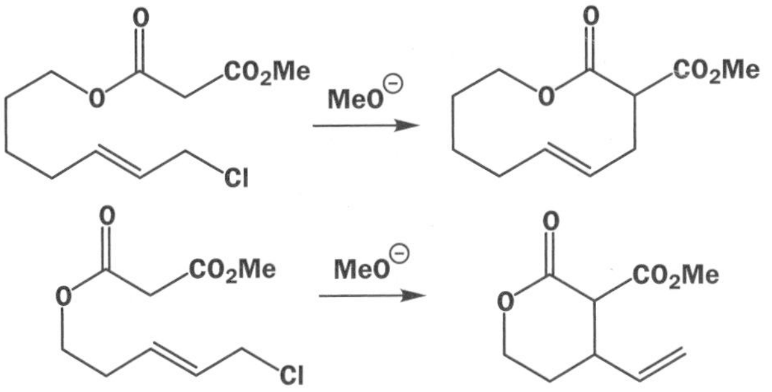

8. In Chapter 32, Problem 3, we asked you to work out the stereochemistry of a sugar. One of the sugar components in the antibiotic kijanimycin has the gross structure and NMR spectrum shown below. What is its stereochemistry?

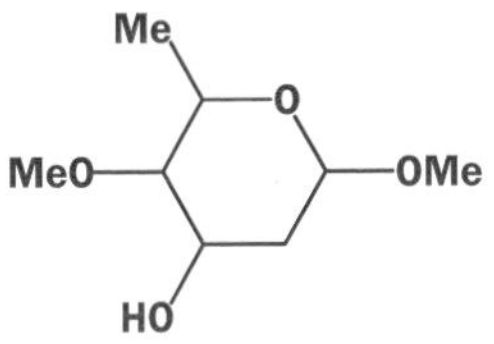

δ_H (p.p.m.) 1.33 (3H, d, J 6 Hz), 1.61* (1H, broad s), 1.87 (1H, ddd, J 14, 3, 3.5 Hz), 2.21 (1H, ddd, J 14, 3, 1.5 Hz), 2.87 (1H, dd, J 10, 3 Hz), 3.40 (3H, s), 3.47 (3H, s), 3.99 (1H, dq, J 10, 6 Hz), 1.33 (3H, d, J 6 Hz), 4.24 (1H, ddd, J 3, 3, 3.5 Hz), and 4.79 (1H, dd, J 3.5, 1.5 Hz). The signal marked * exchanges with D_2O.

When you did this problem, you probably thought about the conformation but now draw it and say why you think the molecule prefers that conformation.

9. Revision of Chapters 35 and 37. Give mechanisms for these reactions, commenting on the formation of that particular saturated heterocycle in the first reaction. What is the alternative product from the migration and why is it not formed?

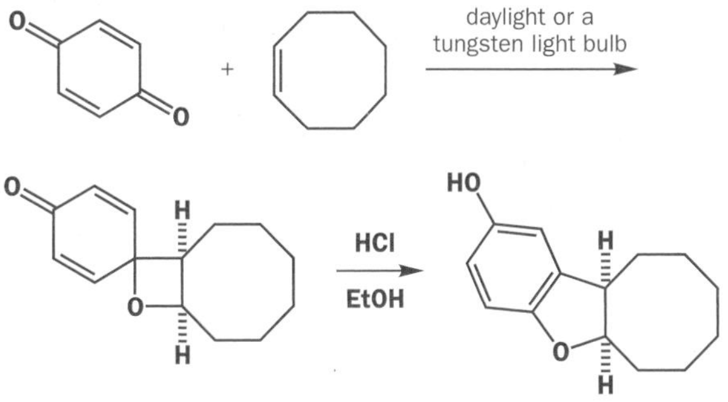

10. Though the anion of dithiolane decomposes as described in the chapter and cannot be used as a d^1 reagent, the example shown here works well without any decomposition. Explain and comment on the regioselectivity of the reaction. Anions of dithianes are notorious for preferring direct to conjugate addition.

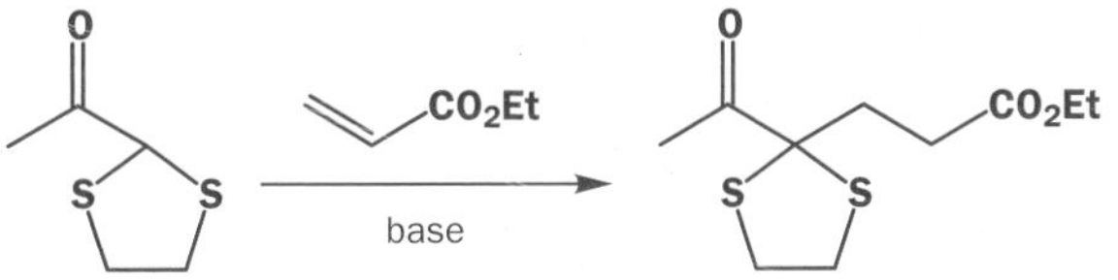

11. Propose a mechanism for this reaction. It does not occur in the absence of an *ortho*- or a *para*-OH group.

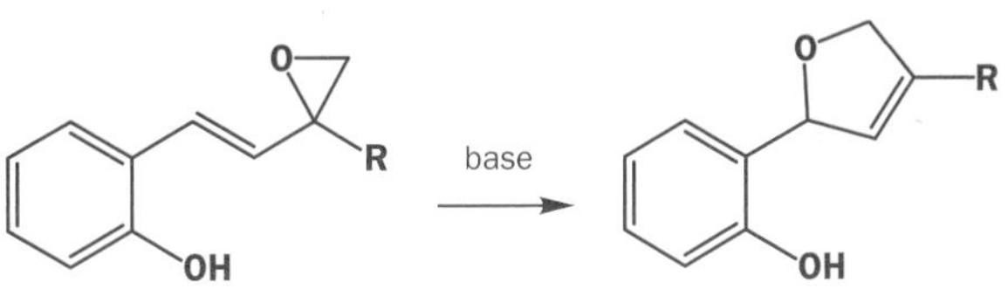

12. Explain why this cyclization gives a preponderance (3:1) of the oxetane though the tetrahydrofuran is much more stable.

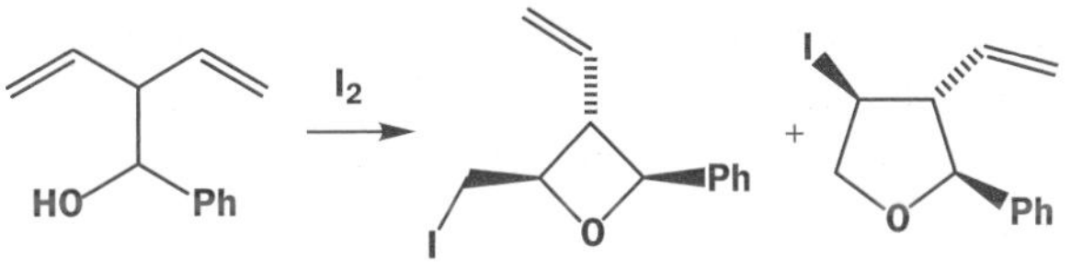

13. Reduction of this keto-ester with LiAlH₄ gives a mixture of diastereoisomers of the diol. Treatment with TsCl and pyridine at −25 °C gives a monotosylate from each. Treatment of these with base leads to the two very different products shown. Explain.

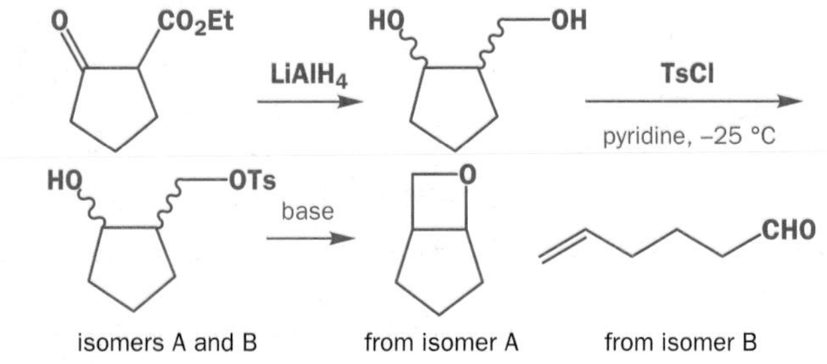

isomers A and B   from isomer A   from isomer B

14. Draw a mechanism for the following multistep reaction. Do the cyclization steps follow Baldwin's rules? What other stereoelectronic effects are involved?

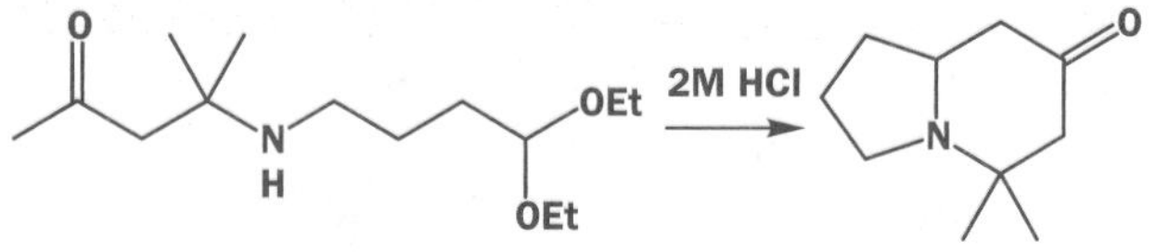

15. Consider the question of Baldwin's rules for each of these reactions. Why do you think they are successful?

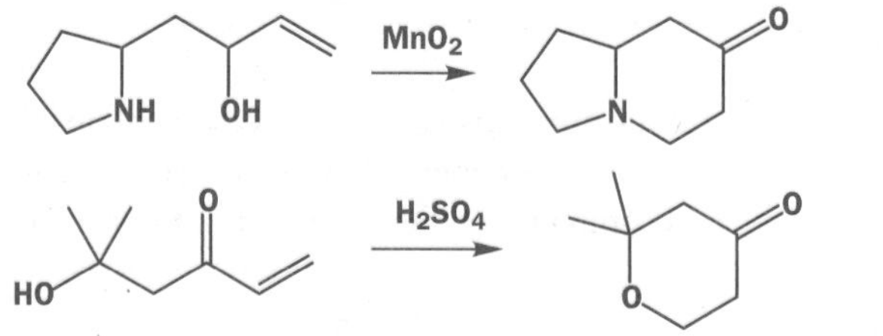

Aromatic heterocycles 1: structures and reactions

43

Connections

Building on:
- Aromaticity ch7
- Electrophilic aromatic substitution ch22
- Nucleophilic attack on aromatic rings ch23
- Saturated heterocycles ch42

Arriving at:
- Aromatic systems conceptually derived from benzene: replacing CH with N to get pyridine
- Replacing CH=CH with N to get pyrrole
- How pyridine reacts
- How pyridine derivatives can be used to extend pyridine's reactivity
- How pyrrole reacts
- How furan and thiophene compare with pyrrole
- Putting more nitrogens in five- and six-membered rings
- Fused rings: indole, quinoline, isoquinoline, and indolizine
- Rings with nitrogen and another heteroatom: oxygen or sulfur
- More complex heterocycles: porphyrins and phthalocyanines

Looking forward to:
- Synthesis of aromatic heterocycles ch44
- Biological chemistry ch49–ch51

Introduction

Benzene is aromatic because it has six electrons in a cyclic conjugated system. We know it is aromatic because it is exceptionally stable and it has a ring current and hence large chemical shifts in the proton NMR spectrum as well as a special chemistry involving substitution rather than addition with electrophiles. This chapter and the next are about the very large number of other aromatic systems in which one or more atoms in the benzene ring are replaced by heteroatoms such as N, O, and S. There are thousands of these systems with five- and six-membered rings, and we will examine just a few.

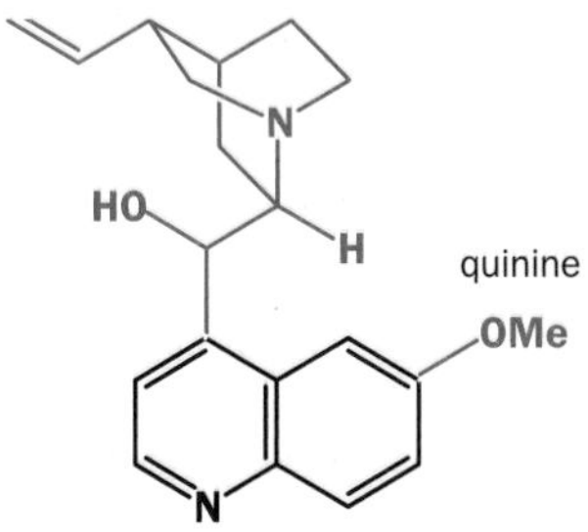

Our subject is **aromatic heterocycles** and it is important that we treat it seriously because most—probably about two-thirds of—organic compounds belong to this class, and they number among them some of the most significant compounds for human beings. If we think only of drugs we can define the history of medicine by heterocycles. Even in the sixteenth century quinine was used to prevent and treat malaria, though the structure of the drug was not known. The first synthetic drug was antipyrine (1887) for the reduction of fevers. The first effective antibiotic was sulfapyridine (1938). The first multi-million pound drug (1970s) was Tagamet, the anti-ulcer drug, and among the most topical of current drugs is Viagra (1997) for treatment of male impotence.

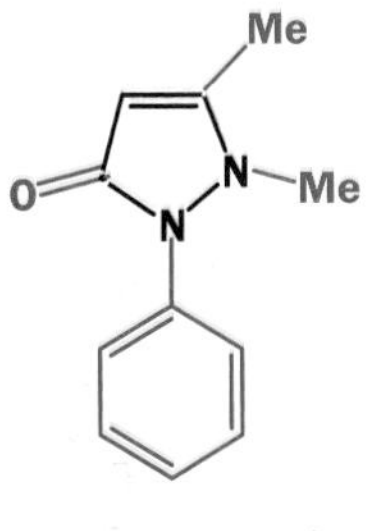

quinine

antipyrine

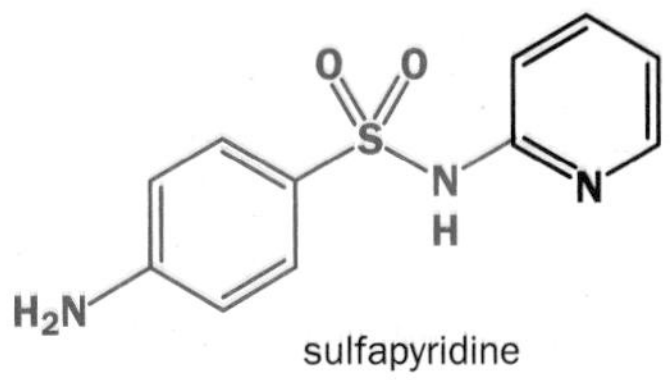

sulfapyridine

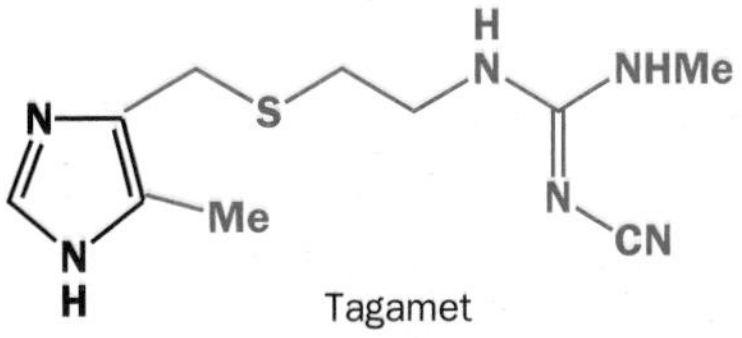

Tagamet

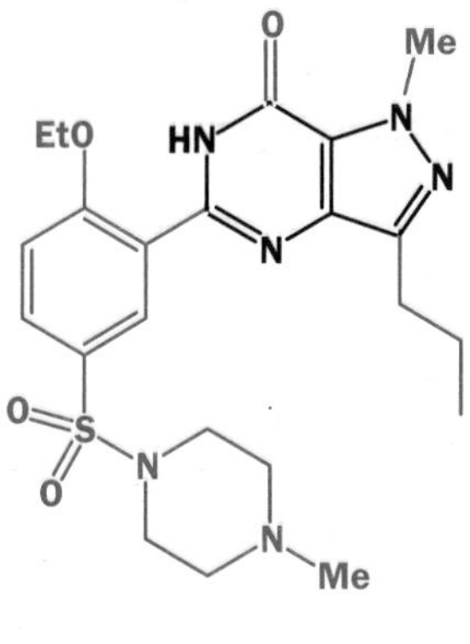

Viagra

■ Sir James Black, discoverer of Tagamet, worked on medicinal chemistry at Smith Kline French (later SmithKline Beecham) and was awarded the Nobel Prize for his analysis of drug receptors in 1988.

All these compounds have heterocyclic aromatic rings shown in black. Three have single rings, five- or six-membered, two have five- or six-membered rings fused together. The number of nitrogens in the rings varies from one to four. We will start by looking at the simple six-membered ring with one nitrogen atom. This is pyridine and the drug sulfapyridine is an example.

Aromaticity survives when parts of benzene's ring are replaced by nitrogen atoms

There is no doubt that benzene is aromatic. Now we must ask: how can we insert a heteroatom into the ring and retain aromaticity? What kind of atom is needed? If we want to replace one of the carbon atoms of benzene with a heteroatom, we need an atom that can be trigonal to keep the flat hexagonal ring and that has a p orbital to keep the six delocalized electrons. Nitrogen is ideal so we can imagine replacing a CH group in benzene with a nitrogen atom.

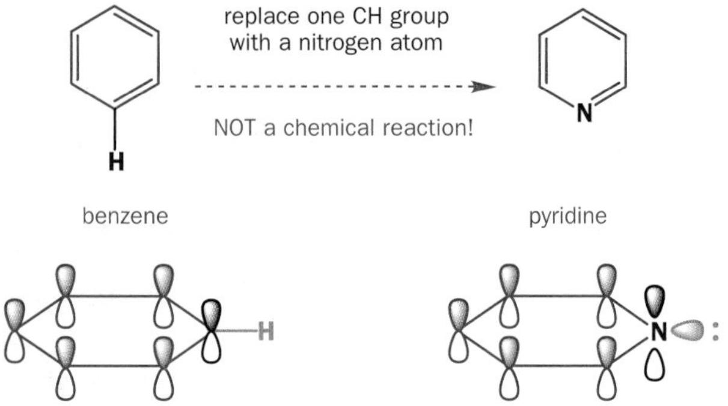

The orbitals in the ring have not changed in position or shape and we still have the six electrons from the three double bonds. One obvious difference is that nitrogen is trivalent and thus there is no NH bond. Instead, a lone pair of electrons occupies the space of the C–H bond in benzene.

^{1}H NMR spectrum of pyridine

In theory then, pyridine is aromatic. But is it in real life? The most important evidence comes from the proton NMR spectrum. The six protons of benzene resonate at δ_H 7.27 p.p.m., some 2 p.p.m. downfield from the alkene region, clear evidence for a ring current (Chapter 11). Pyridine is not as symmetrical as benzene but the three types of proton all resonate in the same region.

As we will see, pyridine is also very stable and, by any reasonable assessment, pyridine is aromatic. We could continue the process of replacing, on paper, more CH groups with nitrogen atoms, and would find three new aromatic heterocycles—pyridazine, pyrimidine, and pyrazine:

▶ One of the most annoying things about heterocyclic chemistry is the mass of what appear to be illogical names. You should not, of course, attempt to learn them all, but a basic idea of how they are designed will help you. We will give you a guide on which names to learn shortly. For the moment accept that 'amine' ends in '-ine' and any heterocyclic compound whose name ends in '-ine' is a nitrogen heterocycle. The syllable 'azo-' also implies nitrogen and 'pyr-' (usually) implies a six-membered ring. (except in pyrrole!)

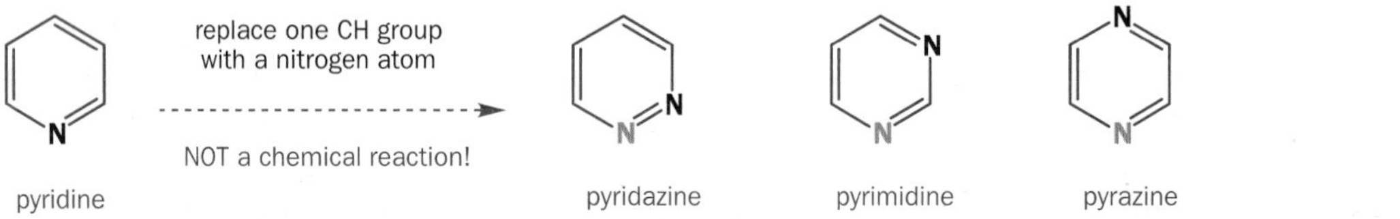

There is another way in which we might transform benzene into a heterocycle. Nitrogen has a lone pair of electrons so we could replace a CH=CH unit in benzene by a nitrogen atom providing that we can use the lone pair in the delocalized system. This means putting it into a p orbital.

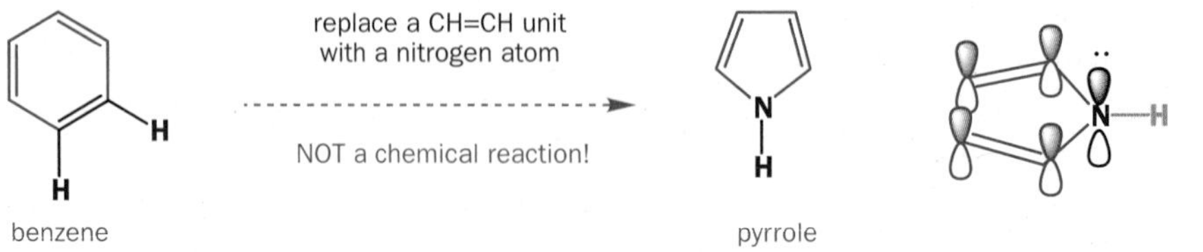

We still have the four electrons from the remaining double bonds and, with the two electrons of the lone pair on nitrogen, that makes six in all. The nitrogen atom must still be trigonal with the lone pair in a p orbital so the N–H bond is in the plane of the five-membered ring.

The NMR of pyrrole is slightly less convincing as the two types of proton on the ring resonate at higher field (6.5 and 6.2 p.p.m.) than those of benzene or pyridine but they still fall in the aromatic rather than the alkene region. Pyrrole is also more reactive towards electrophiles than benzene or

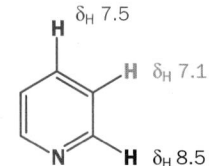

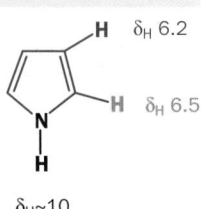

pyridine, but it does the usual aromatic substitution reactions (Friedel–Crafts, nitration, halogenation) rather than addition reactions: pyrrole is also aromatic.

Inventing heterocycles by further replacement of CH groups by nitrogen in pyrrole leads to two compounds, pyrazole and imidazole, after one replacement and to two triazoles after two replacements.

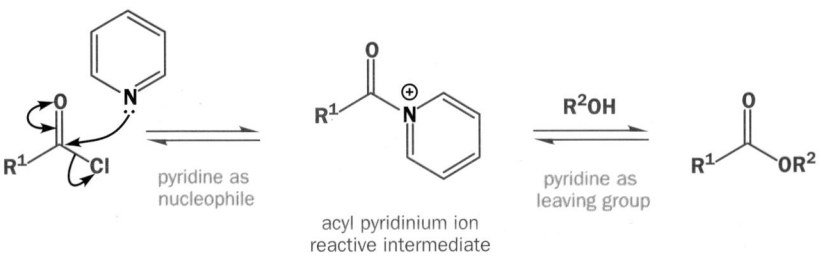

All of these compounds are generally accepted as aromatic too as they broadly have the NMR spectra and reactivities expected for aromatic compounds. As you may expect, introducing heteroatoms into the aromatic ring and, even more, changing the ring size actually affect the chemistry a great deal. We must now return to pyridine and work our way more slowly through the chemistry of these important heterocycles to establish the principles that govern their behaviour.

> The ending '-ole' is systematic and refers to a five-membered heterocyclic ring. All the five-membered aromatic heterocycles with nitrogen in the ring are sometimes called 'the azoles'. Strictly speaking, pyrrole is 'azole', pyrazole is '1,2-diazole', and imidazole is '1,3-diazole'. These names are not used but oxazole and thiazole are used for the oxygen and sulfur analogues of imidazole.

oxazole   thiazole

Pyridine is a very unreactive aromatic imine

The nitrogen atom in the pyridine ring is planar and trigonal with the lone pair in the plane of the ring. This makes it an imine. Most of the imines you have met before (in Chapter 14, for example), have been unstable intermediates in carbonyl group reactions, but in pyridine we have a stable imine—stable because of its aromaticity. All imines are more weakly basic than saturated amines and pyridine is a weak base with a pK_a of 5.5. This means that the pyridinium ion as about as strong an acid as a carboxylic acid.

Pyridine is a reasonable nucleophile for carbonyl groups and is often used as a nucleophilic catalyst in acylation reactions. Esters are often made in pyridine solution from alcohols and acid chlorides (the full mechanism is on p. 281 of Chapter 12).

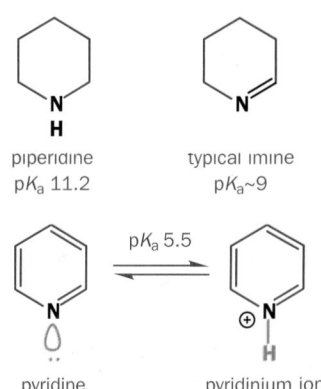

piperidine   typical imine
pK_a 11.2   pK_a~9

pK_a 5.5

pyridine   pyridinium ion

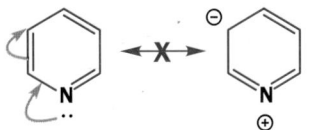

pyridine as nucleophile

acyl pyridinium ion
reactive intermediate

pyridine as leaving group

> Pyridine is also toxic and has a foul smell—so there are disadvantages in using pyridine as a solvent. But it is cheap and remains a popular solvent in spite of the problems.

attempts to delocalize lone pair lead to ridiculous results

Pyridine is nucleophilic at the nitrogen atom because *the lone pair of electrons on nitrogen cannot be delocalized around the ring*. They are in an sp^2 orbital orthogonal to the p orbitals in the ring and there is no interaction between orthogonal orbitals. Try it for yourself, drawing arrows. All attempts to delocalize the electrons lead to impossible results!

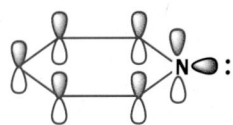

lone pair in sp^2 orbital at right angles to p orbitals in ring: no interaction between orthogonal orbitals

● **The lone pair of pyridine's nitrogen atom is not delocalized.**

Our main interest must be this: what does the nitrogen atom do to the rest of the ring? The important orbitals—the p orbitals of the aromatic system—are superficially the same as in benzene, but the more electronegative nitrogen atom will lower the energy of all the orbitals. Lower-energy filled orbitals mean a *less* reactive nucleophile but a lower-energy LUMO means a *more* reactive electrophile. This is a good guide to the chemistry of pyridine. It is less reactive than benzene in electrophilic aromatic substitution reactions but nucleophilic substitution, which is difficult for benzene, comes easily to pyridine.

Pyridine is bad at electrophilic aromatic substitution

The lower energy of the orbitals of pyridine's π system means that electrophilic attack on the ring is difficult. Another way to look at this is to see that the nitrogen atom destabilizes the cationic would-be intermediate, especially at the 2- and 4-positions.

unstable electron-deficient cation

An equally serious problem is that the nitrogen lone pair is basic and a reasonably good nucleophile—this is the basis for its role as a nucleophilic catalyst in acylations. The normal reagents for electrophilic substitution reactions, such as nitration, are acidic. Treatment of pyridine with the usual mixture of HNO_3 and H_2SO_4 merely protonates the nitrogen atom. Pyridine itself is not very reactive towards electrophiles: the pyridinium ion is totally unreactive.

Other reactions, such as Friedel–Crafts acylations, require Lewis acids and these too react at nitrogen. Pyridine is a good ligand for metals such as Al(III) or Sn(IV) and, once again, the complex with its cationic nitrogen is completely unreactive towards electrophiles.

> ● **Pyridine does not undergo electrophilic substitution**
>
> Aromatic electrophilic substitution on pyridine is not a useful reaction. The ring is unreactive and the electrophilic reagents attack nitrogen making the ring even less reactive. Avoid nitration, sulfonation, halogenation, and Friedel–Crafts reactions on simple pyridines.

Nucleophilic substitution is easy with pyridines

By contrast, the nitrogen atom makes pyridines *more* reactive towards nucleophilic substitution, particularly at the 2- and 4-positions, by lowering the LUMO energy of the π system of pyridine. You can see this effect in action in the ease of replacement of halogens in these positions by nucleophiles.

▶ Contrast the unstable electron-deficient cationic intermediate with the stable pyridinium ion. The nitrogen lone pair is used to make the pyridinium ion but is not involved in the unstable intermediate. Note that reaction at the 3-position is the *best* option but still doesn't occur. Reaction at the 2- and 4-positions is worse.

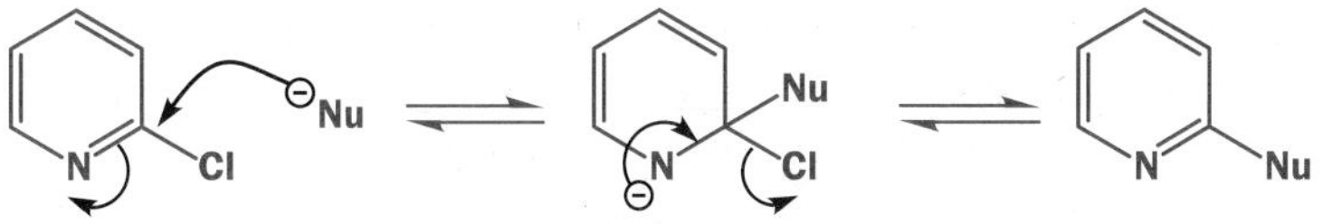

The intermediate anion is stabilized by electronegative nitrogen and by delocalization round the ring. These reactions have some similarity to nucleophilic aromatic substitution (Chapter 23) but are more similar to carbonyl reactions. The intermediate anion is a tetrahedral intermediate that loses the best leaving group to regenerate the stable aromatic system. Nucleophiles such as amines or thiolate anions work well in these reactions.

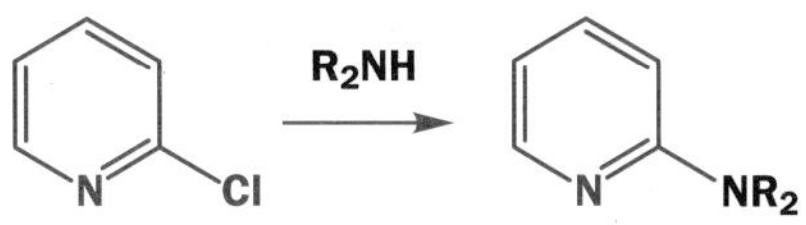

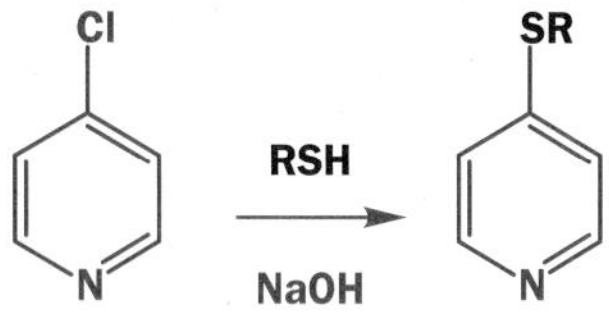

The leaving group does not have to be as good as chloride in these reactions. Continuing the analogy with carbonyl reactions, 2- and 4-chloropyridines are rather like acid chlorides but we need only use less reactive pyridyl ethers, which react like esters, to make amides. The 2- and 4-methoxypyridines allow the completion of the synthesis of flupirtine.

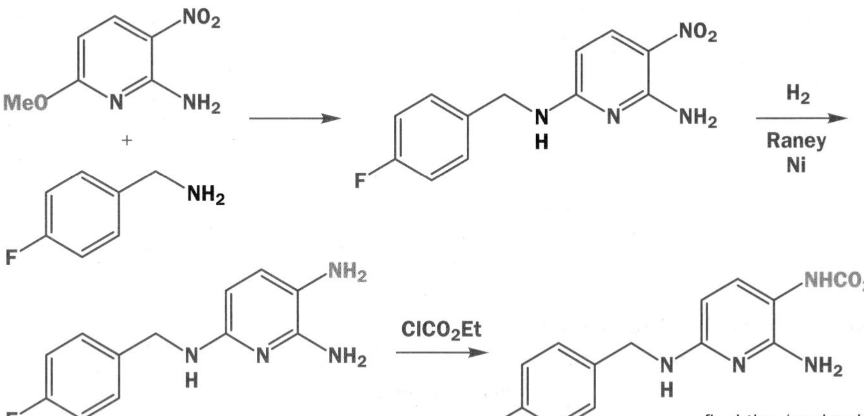

■
Two of the problems at the end of the chapter concern this synthesis: you might like to turn to them now.

flupirtine (analgesic)

The first step is a nucleophilic aromatic substitution. In the second step the nitro group is reduced to an amino group without any effect on the pyridine ring—another piece of evidence for its aromaticity. Finally, one amino group is acylated in the presence of three others.

Pyridones are good substrates for nucleophilic substitution

The starting materials for these nucleophilic substitutions (2- and 4- chloro or methoxypyridines) are themselves made by nucleophilic substitution on pyrid*ones* and we need now to discuss these interesting molecules. If you were asked to propose how 2-methoxypyridine might be made, you would probably suggest, by analogy with the corresponding benzene compound, alkylation of a phenol. Let us look at this in detail.

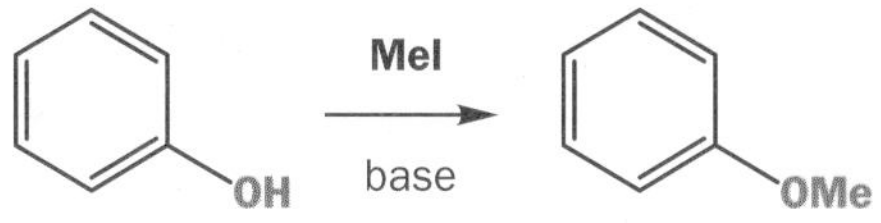

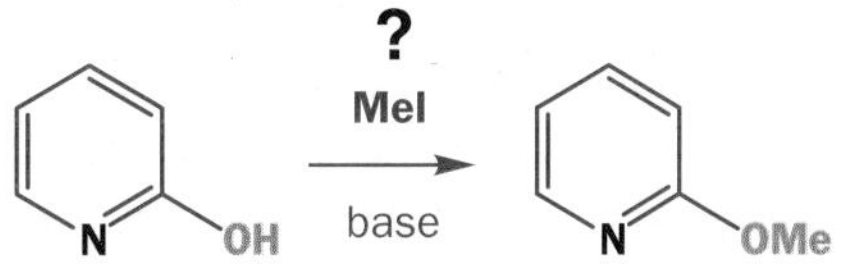

The starting material for this reaction is a 2-hydroxypyridine that can tautomerize to an amide-like structure by the shift of the acidic proton from oxygen to nitrogen. In the phenol series there is no doubt about which structure will be stable as the ketone is not aromatic; for the pyridine both structures are aromatic.

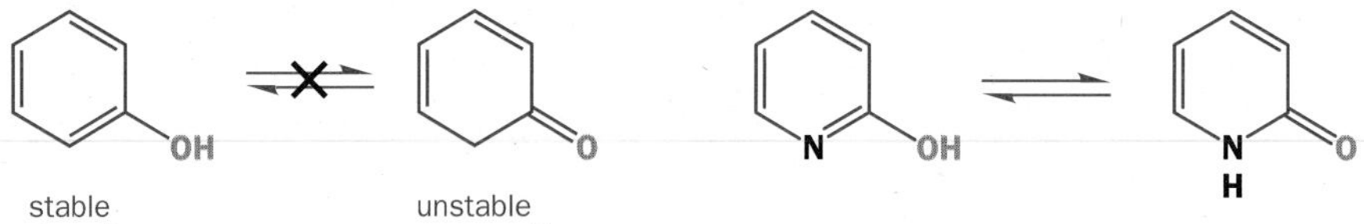

stable
phenol

unstable
non-aromatic

In fact, 2-hydroxypyridine prefers to exist as the 'amide' because that has the advantage of a strong C=O bond and is still aromatic. There are two electrons in each of the C=C double bonds and two also in the lone pair of electrons on the trigonal nitrogen atom of the amide. Delocalization of the lone pair in typical amide style makes the point clearer.

aromatic 2-pyridone

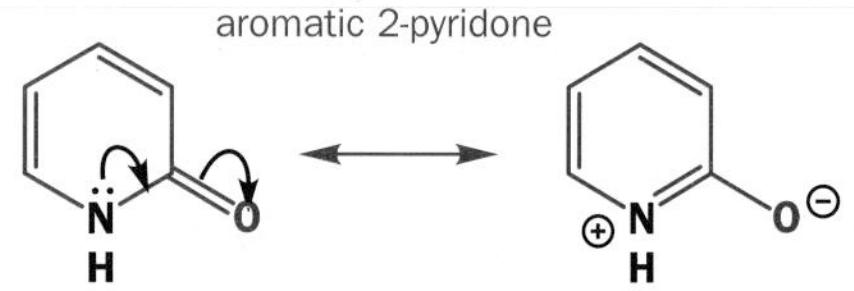

Pyridones are easy to prepare (see Chapter 44) and can be alkylated on oxygen as predicted by their structure. A more important reaction is the direct conversion to chloropyridines with $POCl_3$. The reaction starts by attack of the oxygen atom at phosphorus to create a leaving group, followed by aromatic nucleophilic substitution. The overall effect is very similar to acyl chloride formation from a carboxylic acid.

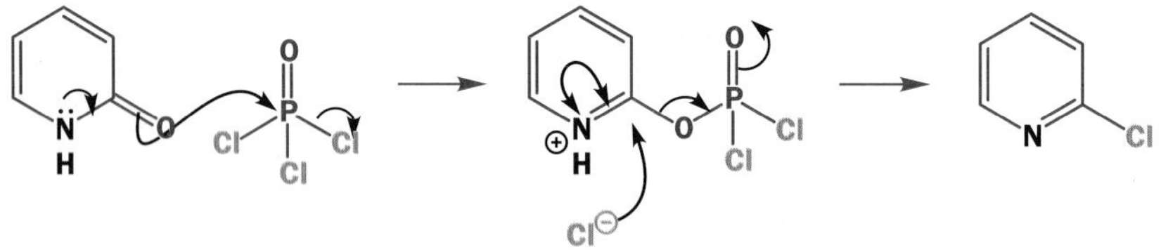

The same reaction occurs with 4-pyridone, which is also delocalized in the same way and exists in the 'amide' form; but not with 3-hydroxypyridine, which exists in the 'phenol' form.

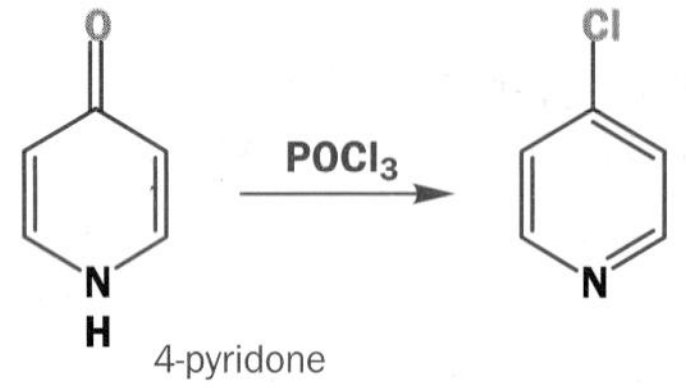

4-pyridone

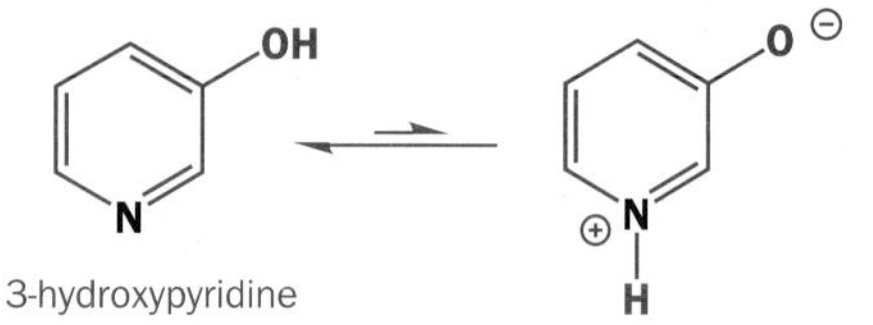

3-hydroxypyridine

● Pyridines undergo nucleophilic substitution

Pyridines can undergo *electrophilic* substitution only if they are activated by electron-donating substituents (see next section) but they readily undergo *nucleophilic* substitution without any activation other than the ring nitrogen atom.

Activated pyridines will do electrophilic aromatic substitution

Useful electrophilic substitutions occur only on pyridines having electron-donating substituents such as NH_2 or OMe. These activate benzene rings too (Chapter 22) but here their help is vital. They supply a nonbonding pair of electrons that becomes the HOMO and carries out the reaction. Simple amino- or methoxypyridines react reasonably well *ortho* and *para* to the activating group. These reactions happen in spite of the molecule being a pyridine, not because of it.

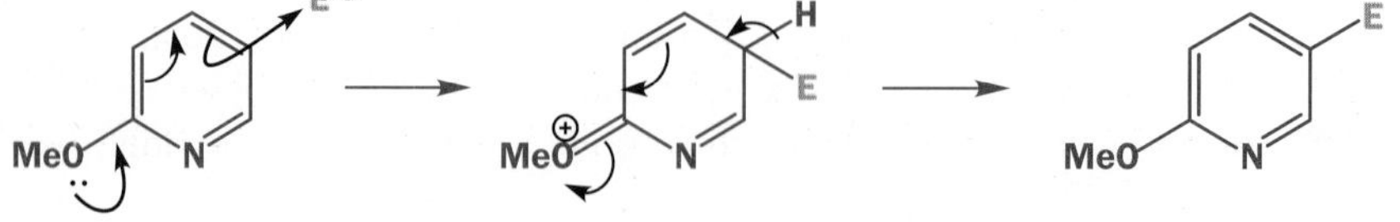

A practical example occurs in the manufacture of the analgesic flupirtine where a doubly activated pyridine having both MeO and NH$_2$ groups is nitrated just as if it were a benzene ring. The nitro group goes in *ortho* to the amino group and *para* to the methoxy group. This sequence is completed in the next section. The activation is evidently enough to compensate for the molecule being almost entirely protonated under the conditions of the reaction.

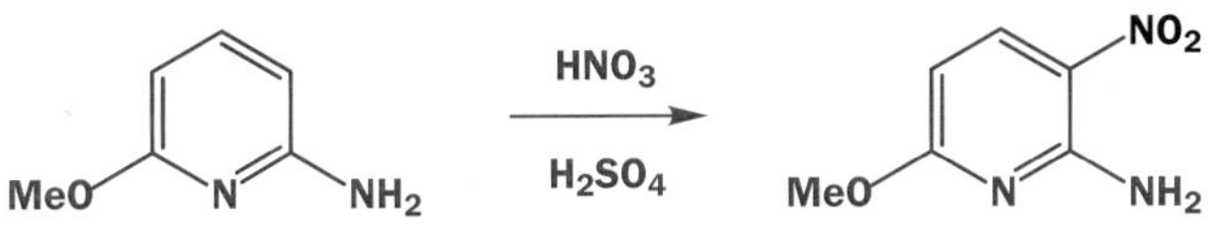

DMAP

One particular amino-pyridine has a special role as a more effective acylation catalyst than pyridine itself. This is DMAP (DiMethylAminoPyridine) in which the amino group is placed to reinforce the nucleophilic nature of the

nitrogen atom. Whereas acylations 'catalysed' by pyridine are normally carried out in solution in pyridine, only small amounts of DMAP in other solvents are needed to do the same job.

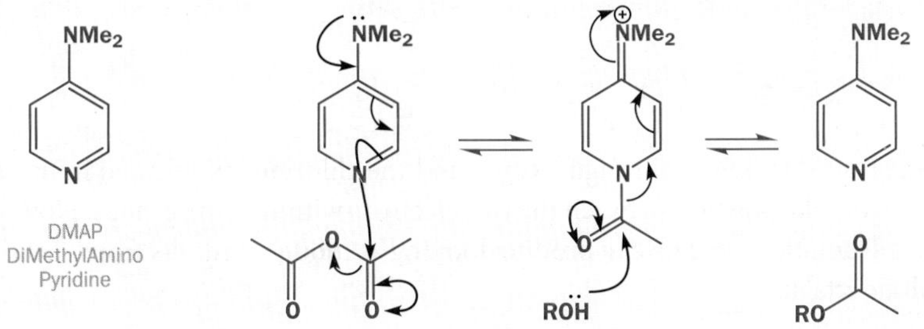

Pyridine *N*-oxides are reactive towards both electrophilic and nucleophilic substitution

This is all very well if the molecule has such activating groups, but supposing it doesn't? How are we to nitrate pyridine itself? The answer involves an ingenious trick. We need to activate the ring with an electron-rich substituent that can later be removed and we also need to stop the nitrogen atom reacting with the electrophile. All of this can be done with a single atom!

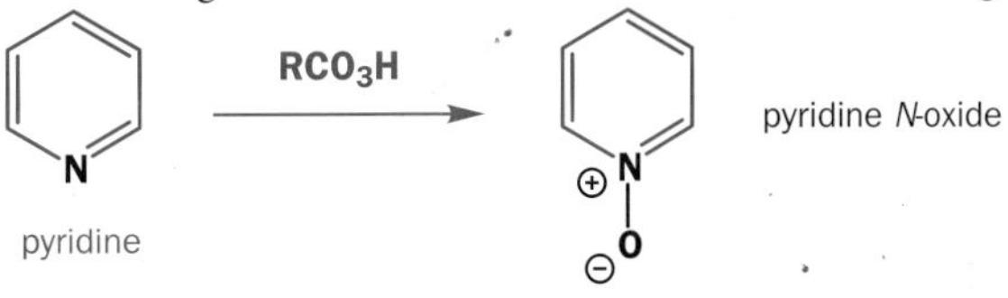

Because the nitrogen atom is nucleophilic, pyridine can be oxidized to pyridine *N*-oxide with reagents such as *m*-CPBA or just H$_2$O$_2$ in acetic acid. These *N*-oxides are stable dipolar species with the electrons on oxygen delocalized round the pyridine ring, raising the HOMO of the molecule. Reaction with electrophiles occurs at the 2- ('*ortho*') and 4- ('*para*') positions, chiefly at the 4-position to keep away from positively charged nitrogen.

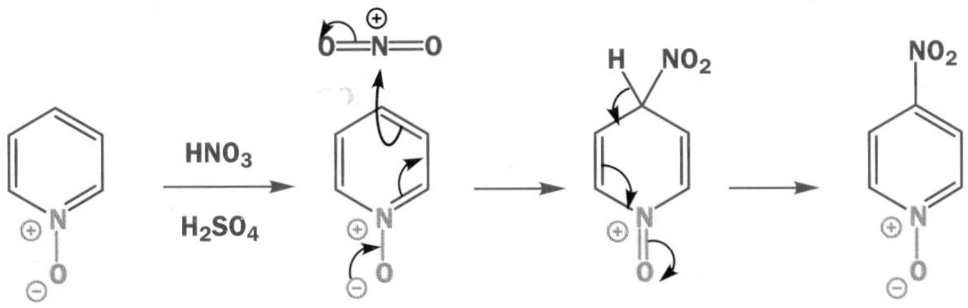

Now the oxide must be removed and this is best done with trivalent phosphorus compounds such as (MeO)$_3$P or PCl$_3$. The phosphorus atom detaches the oxygen atom in a single step to form the very stable P=O double bond. In this reaction the phosphorus atom is acting as both a nucleophile and an electrophile, but mainly as an electrophile since PCl$_3$ is more reactive here than (MeO)$_3$P.

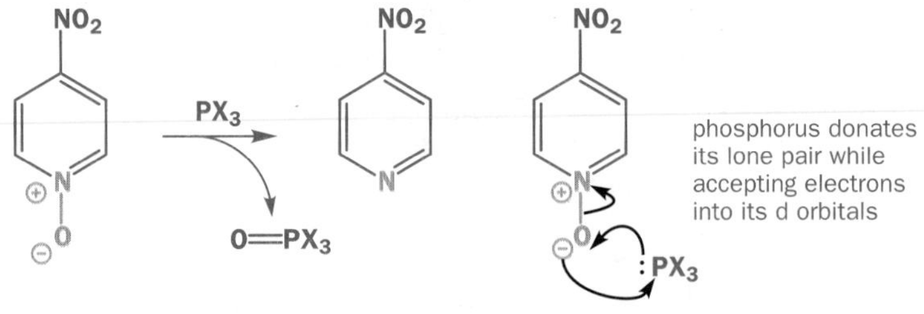
NO2
PX3
O=PX3
phosphorus donates
its lone pair while
accepting electrons
into its d orbitals
:PX3

The same activation that allowed simple electrophilic substitution—oxidation to the *N*-oxide—can also allow a useful nucleophilic substitution. The positive nitrogen atom encourages nucleophilic attack and the oxygen atom can be turned into a leaving group with PCl_3. Our example is nicotinic acid whose biological importance we will discuss in Chapter 50.

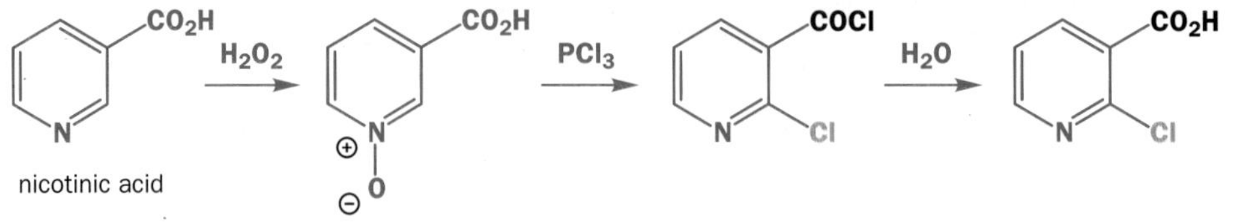
CO2H
H2O2
PCl3
COCl
H2O
Cl
nicotinic acid

The *N*-oxide reacts with PCl_3 through oxygen and the chloride ion released in this reaction adds to the most electrophilic position between the two electron-withdrawing groups. Now a simple elimination restores aromaticity and gives a product looking as though it results from chlorination rather than nucleophilic attack.

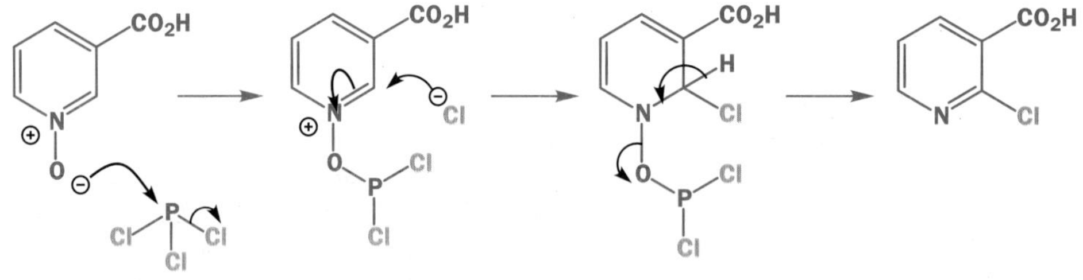

The reagent PCl_3 also converts the carboxylic acid to the acyl chloride, which is hydrolysed back again in the last step. This is a useful sequence because the chlorine atom has been introduced into the 2-position from which it may in turn be displaced by, for example, amines.

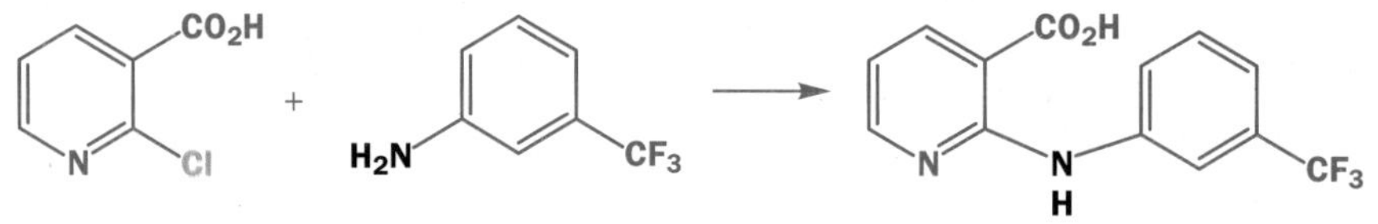
CO2H
H2N
CF3
nifluminic acid - analgesic

● **Pyridine-*N*-oxides**

Pyridine *N*-oxides are useful for both electrophilic and nucleophilic substitutions on the same carbon atoms (2-, 4-, and 6-) in the ring.

Nucleophilic addition at an even more distant site is possible on reaction with acid anhydrides if there is an alkyl group in the 2-position. Acylation occurs on oxygen as in the last reaction but then a proton is lost from the side-chain to give an uncharged intermediate.

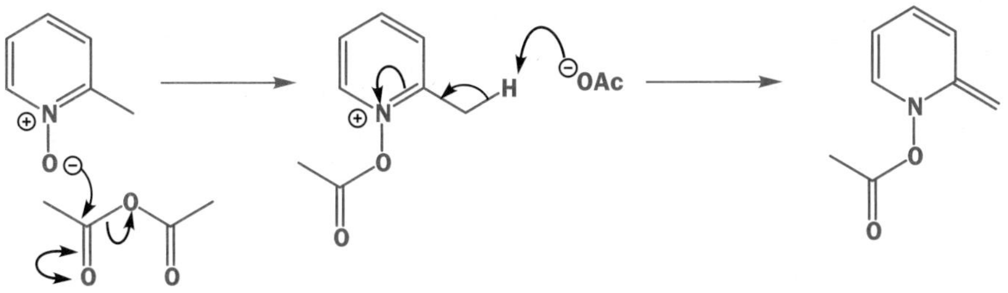

This compound rearranges with migration of the acetate group to the side chain and the restoration of aromaticity. This may be an ionic reaction or a [3,3]-sigmatropic rearrangement.

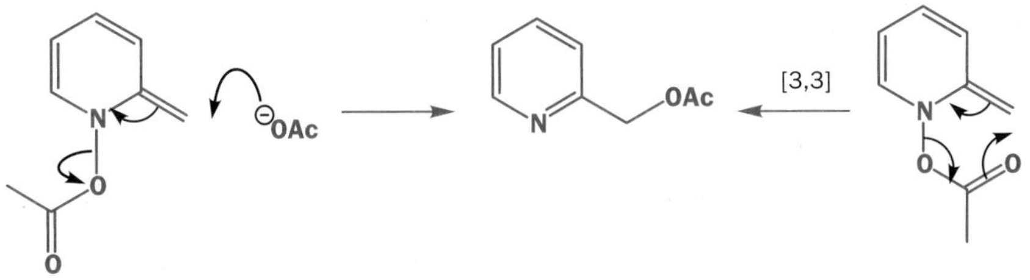

Since pyridine is abundant and cheap and has an extremely rich chemistry, it is not surprising that it has many applications.

Some applications of pyridine chemistry

One of the simplest ways to brominate benzenes is not to bother with the Lewis acid catalysts recommended in Chapter 22 but just to add liquid bromine to the aromatic compound in the presence of a small amount of pyridine. Only about one mole per cent is needed and even then the reaction has to be cooled to stop it getting out of hand.

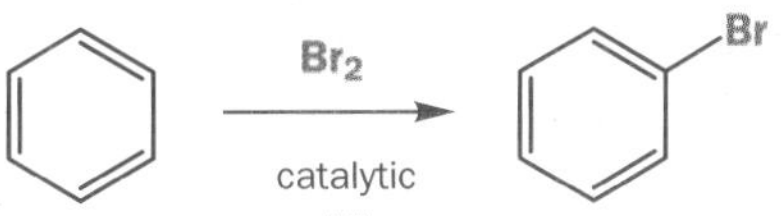

As we have seen, pyridine attacks electrophiles through its nitrogen atom. This produces the reactive species, the N-bromo-pyridinium ion, which is attacked by the benzene. Pyridine is a better nucleophile than benzene and a better leaving group than bromide. This is another example of **nucleophilic catalysis**.

■ Nucleophilic catalysis is discussed on p. 282.

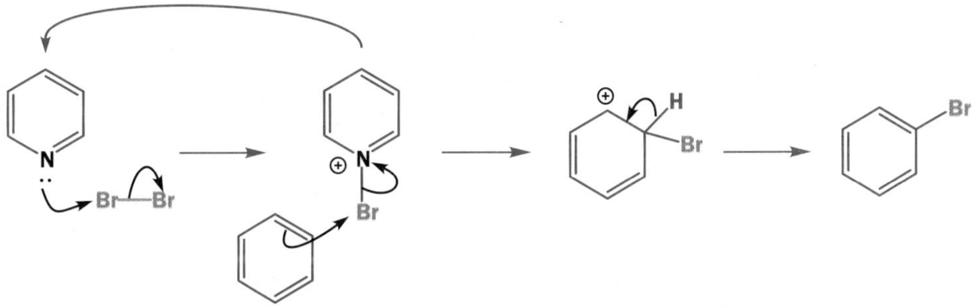

Another way to use pyridine in brominations is to make a stable crystalline compound to replace the dangerous liquid bromine. This compound, known by names such as pyridinium tribromide, is simply a salt of pyridine with the anion Br_3^-. It can be used to brominate reactive compounds such as alkenes (Chapter 20).

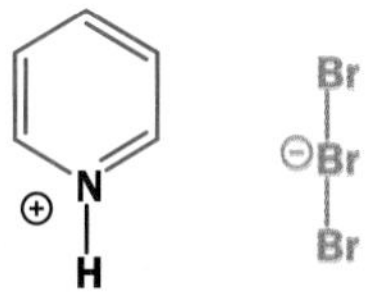

Both of these methods depend on the lack of reactivity of pyridine's π system towards electrophiles such as bromine. Notice that, in the first case, both benzene and pyridine are

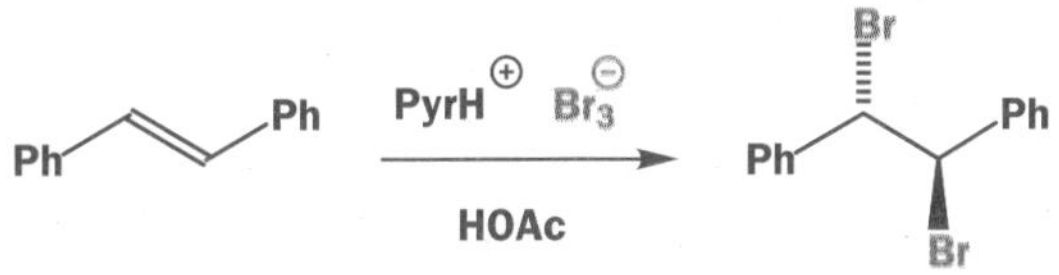

pyridinium tribromide

present together. The pyridine attacks bromine only through nitrogen (and reversibly at that) and never through carbon.

Oxidation of alcohols is normally carried out with Cr(VI) reagents (Chapter 24) but these, like the Jones' reagent ($Na_2Cr_2O_7$ in sulfuric acid), are usually acidic. Some pyridine complexes of Cr(VI) compounds solve this problem by having the pyridinium ion (pK_a 5) as the only acid. The two most famous are 'PDC' (Pyridinium DiChromate) and 'PCC' (Pyridinium Chloro-Chromate). Pyridine forms a complex with CrO_3 but this is liable to burst into flames. Treatment with HCl gives PCC, which is much less dangerous. PCC is particularly useful in the oxidation of primary alcohols to aldehydes as overoxidation is avoided in the only slightly acidic conditions (Chapter 24).

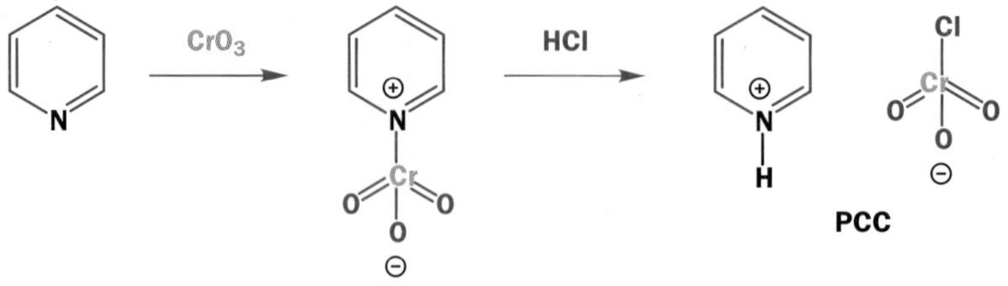

PCC

The ability of pyridine to form metal complexes is greatly enhanced in a dimer—the famous ligand 'bipy' or 2,2′-bipyridyl. It is bidentate and because of its 'bite' it is a good ligand for many transition metals but shows a partiality for Fe(II).

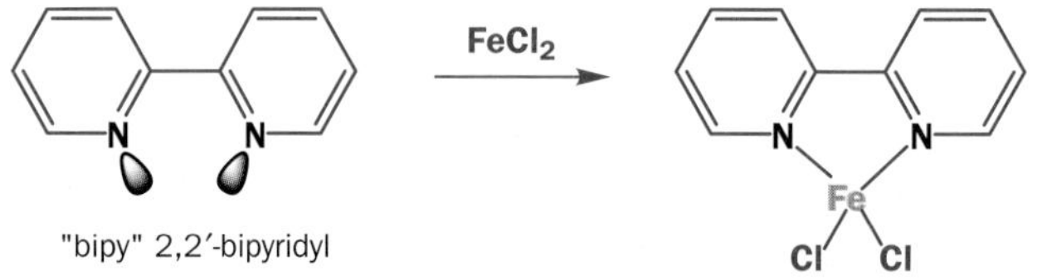

"bipy" 2,2′-bipyridyl

It looks like a rather difficult job to persuade two pyridine rings to join together in this way to form bipy. It is indeed very difficult unless you make things easier by using a reagent that favours the product. And what better than Fe(II) to do the job? ICI manufacture bipy by treating pyridine with $FeCl_2 \cdot 4H_2O$ at high temperatures and high pressures. Only a small proportion of the pyridine is converted to the Fe(II) complex of bipy (about 5%) but the remaining pyridine goes back in the next reaction. This is probably a radical process (Chapter 39) in the coordination sphere of Fe(II).

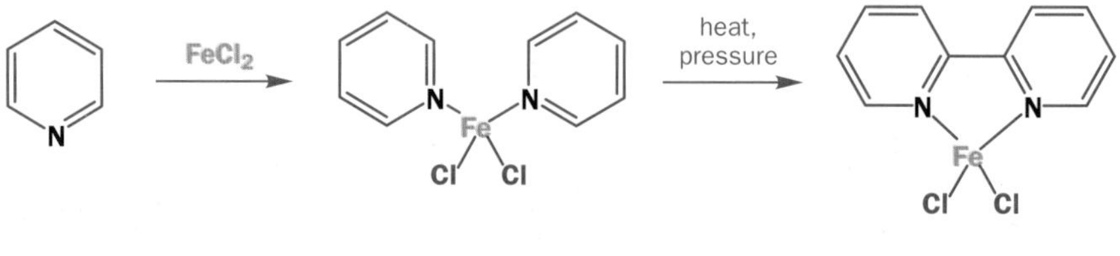

Six-membered aromatic heterocycles can have oxygen in the ring

Though pyridine is overwhelmingly the most important of the six-membered aromatic heterocycles, there are oxygen heterocycles, **pyrones**, that resemble the pyridones. The pyrones are aromatic, though α-pyrone is rather unstable.

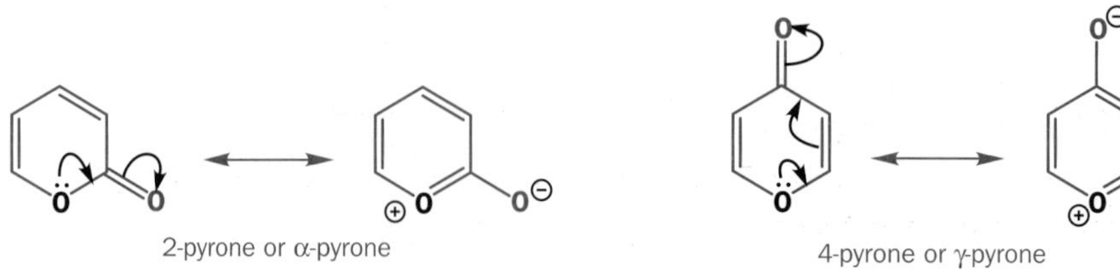

2-pyrone or α-pyrone

4-pyrone or γ-pyrone

The pyrylium salts are stable aromatic cations and are responsible as metal complexes for some flower colours.

Heterocycles with six-membered rings based on other elements (for example, P) do exist but they are outside the scope of this book.

a red pyrylium flower pigment

the pyrylium cation

Five-membered heterocycles are good nucleophiles

Just about everything is the other way round with pyrrole. Electrophilic substitution is much easier than it is with benzene—almost too easy in fact—while nucleophilic substitution is more difficult. Pyrrole is not a base nor can it be converted to an *N*-oxide. We need to find out why this is.

The big difference is that the nitrogen lone pair is delocalized round the ring. The NMR spectrum suggests that all the positions in the ring are about equally electron-rich with chemical shifts about 1 p.p.m. smaller than those of benzene. The ring is flat and the bond lengths are very similar, though the bond opposite the nitrogen atom is a bit longer than the others.

The delocalization of the lone pair can be drawn equally well to any ring atom because of the five-membered ring and we shall soon see the consequences of this. All the delocalization pushes electrons from the nitrogen atom into the ring and we expect the ring to be electron-rich at the expense of the nitrogen atom. The HOMO should go up in energy and the ring become more nucleophilic.

1.43 Å   H δ_H 6.2

1.37 Å

H δ_H 6.5

1.38 Å   H $\delta_H{\sim}10$

An obvious consequence of this delocalization is the decreased basicity of the nitrogen atom and the increased acidity of the NH group as a whole. In fact, the pK_a of pyrrole acting as a base is about −4 and protonation occurs at carbon. The NH proton can be removed by much weaker bases than those that can remove protons on normal secondary amines.

The nucleophilic nature of the ring means that pyrrole is attacked readily by electrophiles. Reaction with bromine requires no Lewis acid and leads to substitution (confirming the aromaticity of pyrrole) at all four free positions.

Br_2

EtOH, 0 °C

This is a fine reaction in its way, but we don't usually want four bromine atoms in a molecule so one problem with pyrrole is to control the reaction to give only monosubstitution. Another problem is that strong acids cannot be used. Though protonation does not occur at nitrogen, it does occur at carbon and the protonated pyrrole then adds another molecule like this.

$H^{\oplus}$

etc.

reaction continues to give polymer

● Pyrrole polymerizes!

Strong acids, those such as H_2SO_4 with a pK_a of less than −4, cannot be used without polymerization of pyrrole.

Some reactions can be controlled to give good yields of monosubstituted products. One is the **Vilsmeier reaction** in which a combination of an N,N-dimethylamide and POCl$_3$ is used to make a carbon electrophile in the absence of strong acid or Lewis acid. It is a substitute for the Friedel–Crafts acylation, and works with aromatic compounds at the more reactive end of the scale (where pyrrole is).

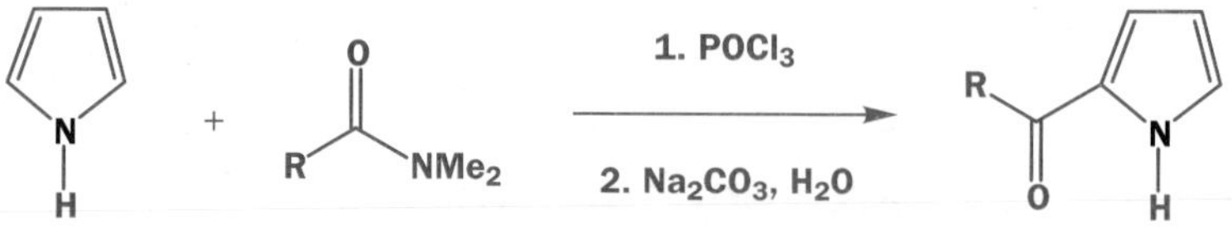

In the first step, the amide reacts with POCl$_3$ which makes off with the amide oxygen atom and replaces it with chlorine. This process would be very unfavourable but for the formation of the strong P–O bond, and is the direct analogy of the chloropyridine-forming reaction you have just seen.

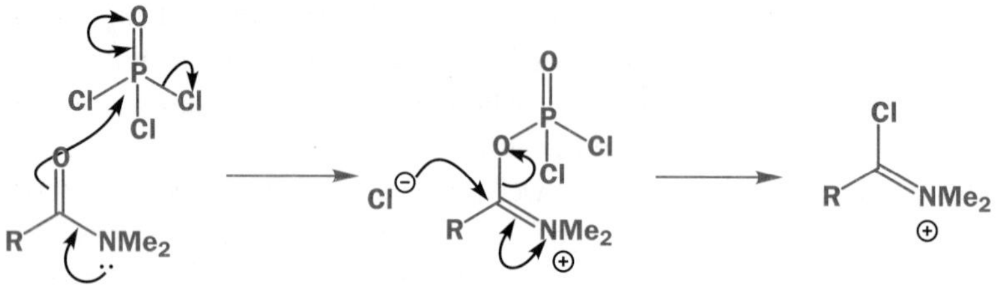

The product from this first step is an iminium cation that reacts with pyrrole to give a more stable iminium salt. The extra stability comes from the conjugation between the pyrrole nitrogen and the iminium group.

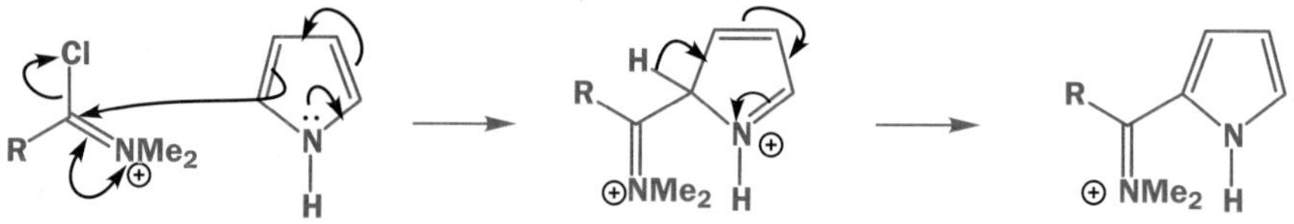

The work-up with aqueous Na$_2$CO$_3$ hydrolyses the imine salt and removes any acid formed. This method is particularly useful because it works well with Me$_2$NCHO (DMF) to add a formyl (CHO) group. This is difficult to do with a conventional Friedel–Crafts reaction.

You may have noticed that the reaction occurred only at the 2-position on pyrrole. Though all positions react with reagents like bromine, more selective reagents usually go for the 2- (or 5-) position and attack the 3- (or 4-) position only if the 2- and 5-positions are blocked. A good example is the Mannich reaction (Chapter 27). In these two examples N-methylpyrrole reacts cleanly at the 2-position while the other pyrrole with both 2- and 5-positions blocked by methyl groups reacts cleanly at the 3-position. These reactions are used in the manufacture of the nonsteroidal anti-inflammatory compounds, tolmetin and clopirac.

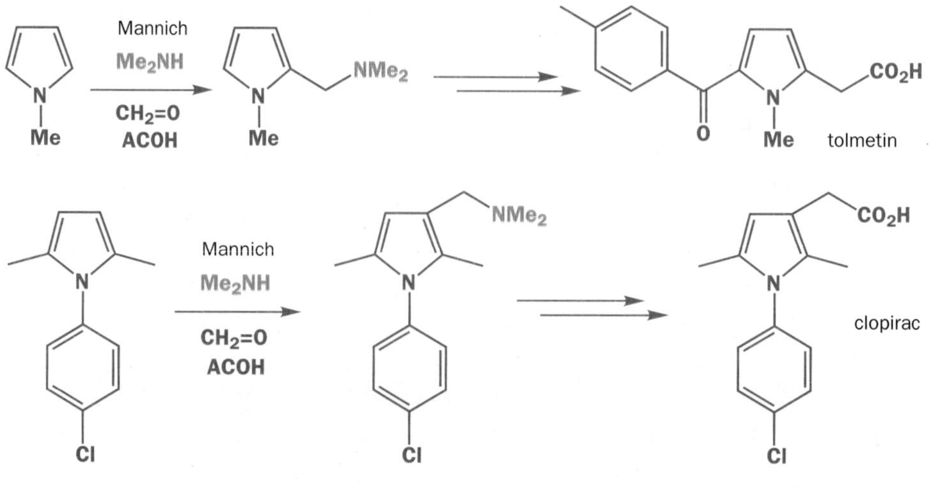

Now we need an explanation. The mechanisms for both 2- and 3-substitutions look good and we will draw both, using a generalized E⁺ as the electrophile.

reaction with electrophiles in the 2-position   reaction with electrophiles in the 3-position

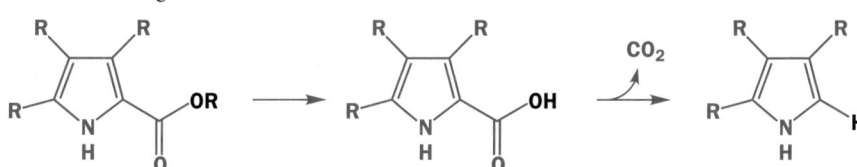

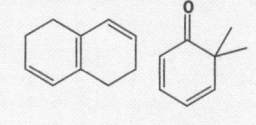

more stable   less stable

Both mechanisms can occur very readily. Reaction in the 2-position is somewhat better than in the 3-position but the difference is small. Substitution is favoured at *all* positions. Calculations show that the HOMO of pyrrole does indeed have a larger coefficient in the 2-position but that is very much a theoretical chemist's answer, which organic chemists cannot reproduce easily. One way to understand the result is to look at the structure of the intermediates. The intermediate from attack at the 2-position has a linear conjugated system. In both intermediates the two double bonds are, of course, conjugated with each other, but only in the first intermediate are both double bonds conjugated with N^+. The second intermediate is 'cross-conjugated', while the first has a more stable linear conjugated system.

Since electrophilic substitution on pyrroles occurs so easily, it can be useful to block substitution with a removable substituent. This is usually done with an ester group. Hydrolysis of the ester (this is particularly easy with *t*-butyl esters—see Chapter 24) releases the carboxylic acid, which decarboxylates on heating.

▶

Cross-conjugation explains other differences in stability too. Here are some examples. The linear conjugated systems are more stable than the cross-conjugated ones.

More stable—linear conjugation

Less stable—cross conjugation

The decarboxylation is a kind of reverse Friedel–Crafts reaction in which the electrophile is a proton (provided by the carboxylic acid itself) and the leaving group is carbon dioxide. The protonation may occur anywhere but it leads to reaction only if it occurs where there is a CO_2H group.

Furan and thiophene are oxygen and sulfur analogues of pyrrole

The other simple five-membered heterocycles are furan, with an oxygen atom instead of nitrogen, and thiophene with a sulfur atom. They also undergo electrophilic aromatic substitution very readily, though not so readily as pyrrole. Nitrogen is the most powerful electron donor of the three, oxygen the next, and sulfur the least. Thiophene is very similar to benzene in reactivity.

You may be surprised that thiophene is the least reactive of the three but this is because the p orbital of the lone pair of electrons on sulfur that conjugates with the ring is a 3p orbital rather than the 2p orbital of N or O, so overlap with the 2p orbitals on carbon is less good. Both furan and thiophene undergo more or less normal Friedel–Crafts reactions though the less reactive anhydrides are used instead of acid chlorides, and weaker Lewis acids than $AlCl_3$ are preferred.

pyrrole   furan   thiophene

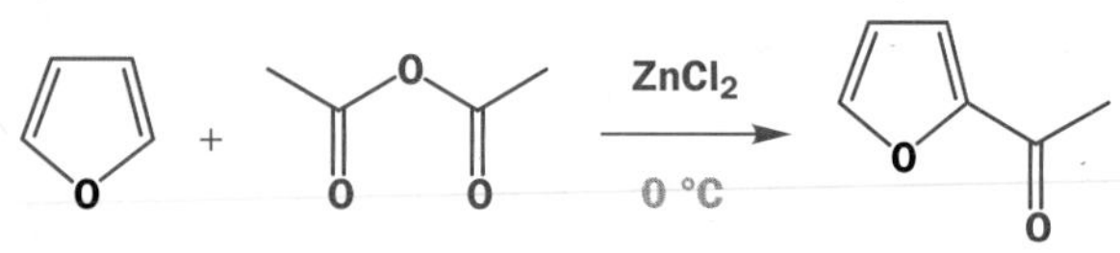

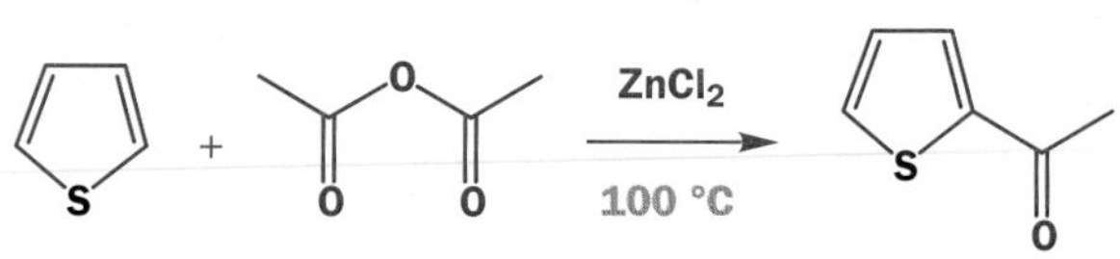

Notice that the regioselectivity is the same as it was with pyrrole—the 2-position is more reactive than the 3-position in both cases. The product ketones are less reactive towards electrophiles than the starting heterocycles and deactivated furans can even be nitrated with the usual reagents used for benzene derivatives. Notice that reaction has occurred at the 5-position in spite of the presence of the ketone. The preference for 2- and 5-substitution is quite marked.

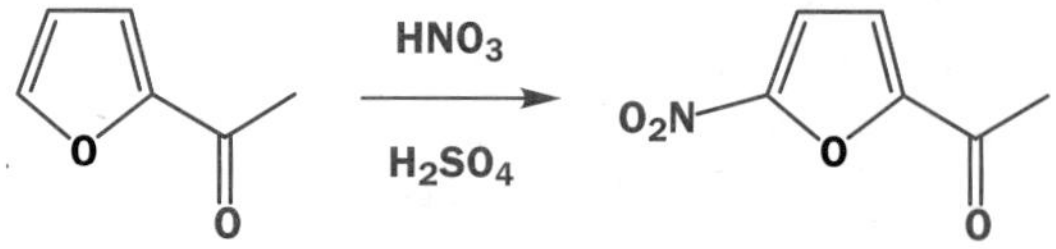

So far, thiophenes and furans look much the same as pyrrole but there are other reactions in which they behave quite differently and we shall now concentrate on those.

Electrophilic addition may be preferred to substitution with furan

Furan is not very aromatic and if there is the prospect of forming stable bonds such as C–O single bonds by addition, this may be preferred to substitution. A famous example is the reaction of furan with bromine in methanol. In nonhydroxylic solvents, polybromination occurs as expected, but in MeOH no bromine is added at all!

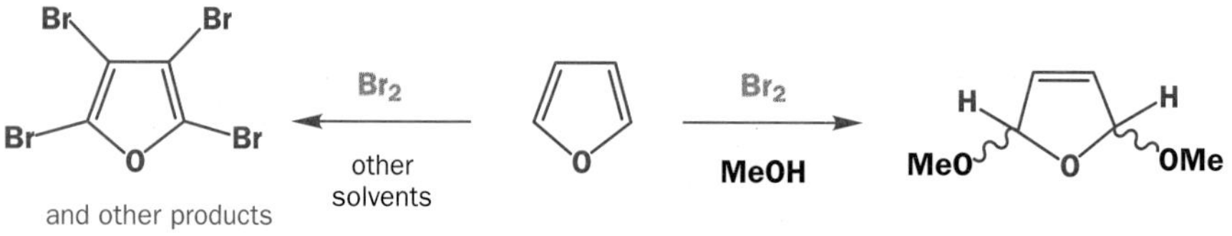

Bromination must start in the usual way, but a molecule of methanol captures the first formed cation in a 1,4-addition to furan.

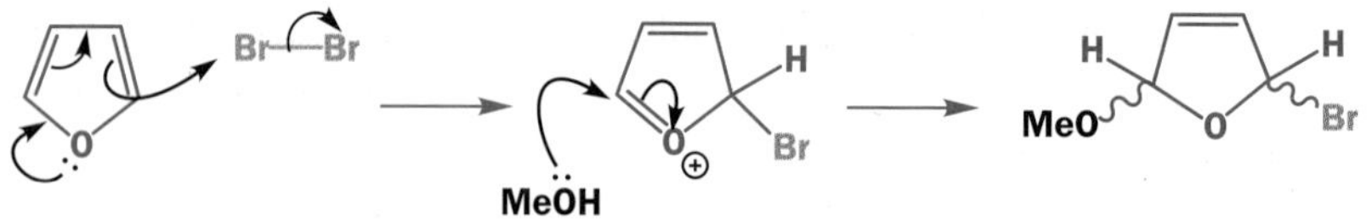

The bromine atom that was originally added is now pushed out by the furan oxygen atom to make a relatively stable conjugated oxonium ion, which adds a second molecule of methanol.

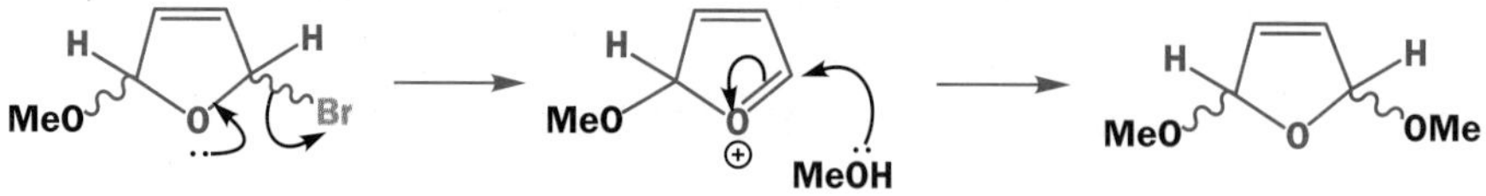

This product conceals an interesting molecule. At each side of the ring we have an acetal, and if we were to hydrolyse the acetals, we would have 'maleic dialdehyde' (*cis*-butenedial)—a molecule that is too unstable to be isolated. The furan derivative may be used in its place.

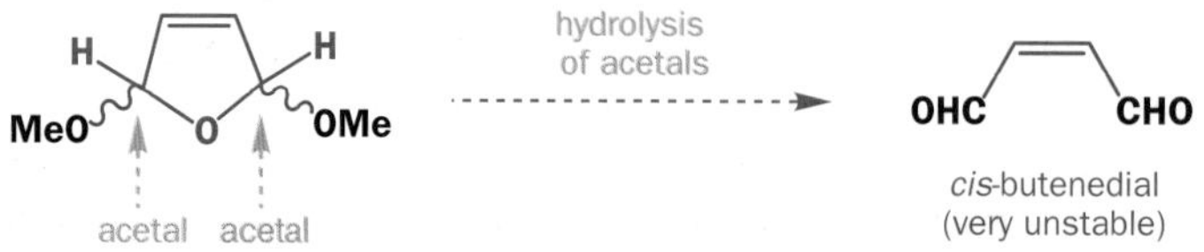

The same 1,4-dialdehyde can be made by oxidizing furan with the mild oxidizing agent dimethyl-dioxirane, which you met on p. 506. In this sequence, it is trapped in a Wittig reaction to give an *E,Z*-diene, which is easily isomerized to *E,E*.

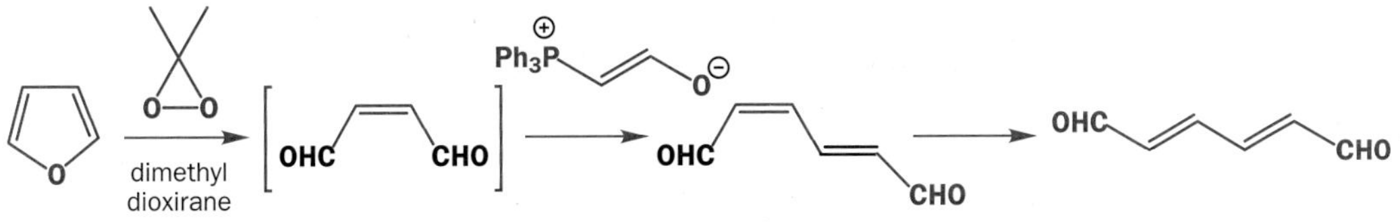

We can extend this idea of furan being the origin of 1,4-dicarbonyl compounds if we consider that furan is, in fact, an enol ether on both sides of the ring. If these enol ethers were hydrolysed we would get a 1,4-diketone.

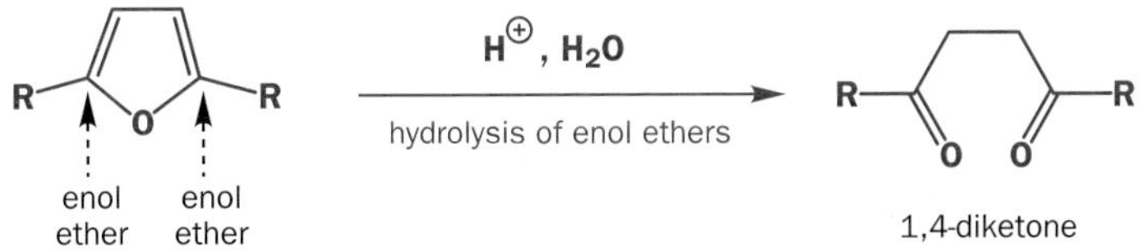

This time the arrow is solid, not dotted, because this reaction really happens. You will discover in the next chapter that furans can also be made from 1,4-diketones so this whole process is reversible. The example we are choosing has other features worth noting. The cheapest starting material containing a furan is furan-2-aldehyde or 'furfural', a by-product of breakfast cereal manufacture. Here it reacts in a typical Wittig process with a stabilized ylid.

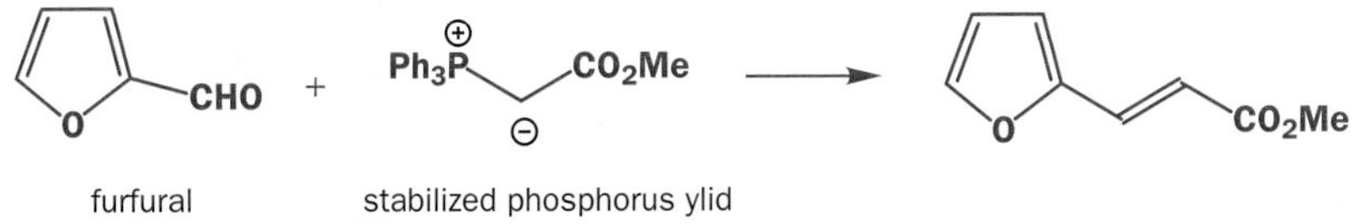

Now comes the interesting step: treatment of this furan with acidic methanol gives a white crystalline compound having two 1,4-dicarbonyl relationships.

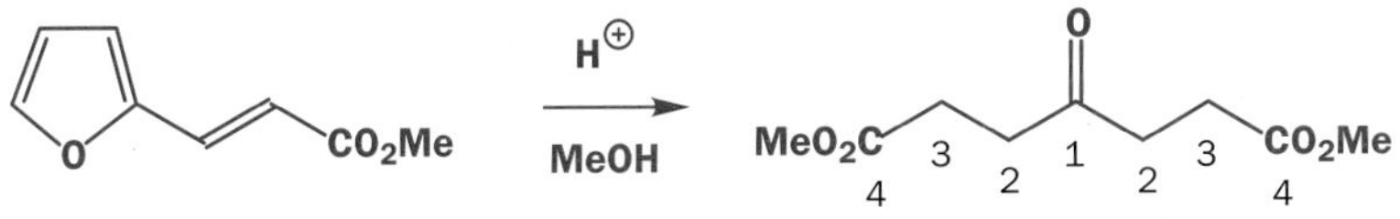

> You should try to draw a mechanism for this reaction.

The thiophene ring can also be opened up, but in a very different way. Reductive removal of the sulfur atom with Raney nickel (Chapter 24) reduces not only the C–S bonds but also the double bonds in the ring and we are left with a saturated alkyl chain.

If the reduction follows two Friedel–Crafts reactions on thiophene the product is a 1,6-diketone instead of the 1,4-diketones from furan. Thiophene is well behaved in Friedel–Crafts acylations, and reaction occurs at the 2- and 5-positions unless these are blocked.

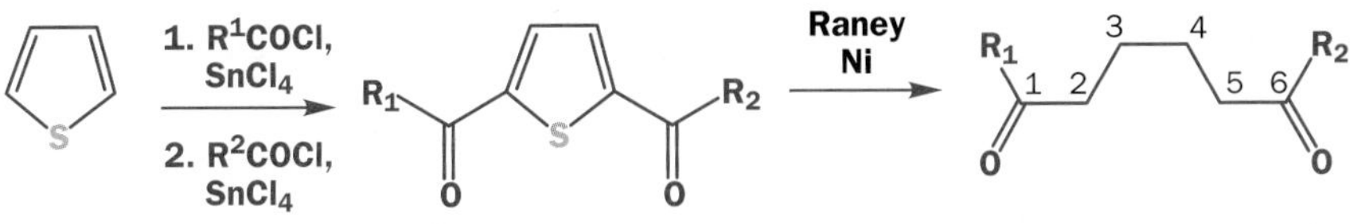

Lithiation of thiophenes and furans

A reaction that furans and thiophenes do particularly well and that fits well with these last two reactions is metallation, particularly lithiation, of a C–H group next to the heteroatom and we will discuss this next. Lithiation of benzene rings (Chapter 9) is carried out by lithium–halogen (Br or I) exchange—a method that works well for heterocycles too as we will see later with pyridine—or by directed ('*ortho*') lithiation of a C–H group next to an activating group such as OMe. With thiophene and furan, the heteroatom in the ring provides the necessary activation.

Activation is by coordination of O or S to Li followed by proton removal by the butyl group so that the by-product is gaseous butane. These lithium compounds have a carbon–lithium σ bond and are soluble in organic solvents with the coordination sphere of Li completed by THF molecules.

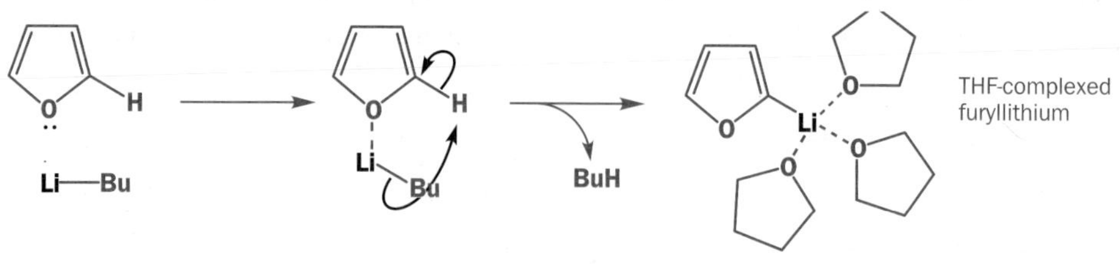

These lithium compounds are very reactive and will combine with most electrophiles—in this example the organolithium is alkylated by a benzylic halide. Treatment with aqueous acid gives the 1,4-diketone by hydrolysis of the two enol ethers.

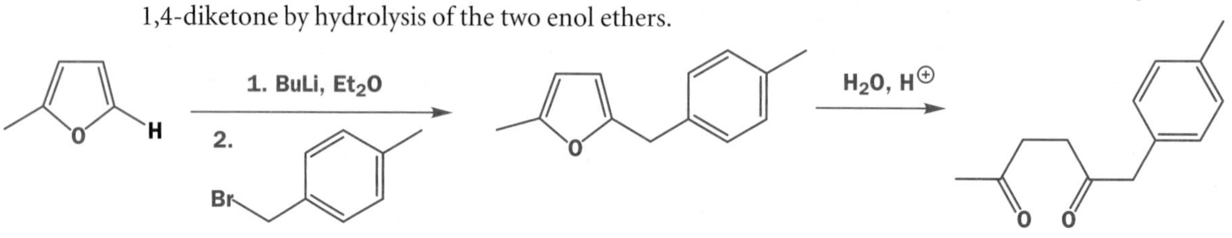

Treatment of this diketone with *anhydrous* acid would cause recyclization to the same furan (see Chapter 44) but it can alternatively be cyclized in base by an intramolecular aldol reaction (Chapter 27) to give a cyclopentenone.

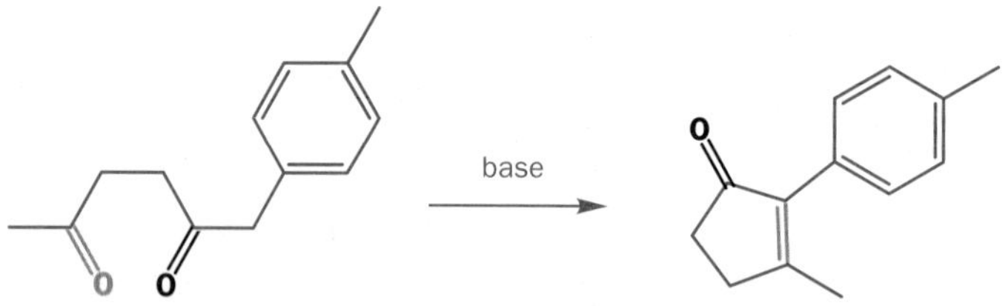

This completes our exploration of chemistry special to thiophene and furan and we now return to all three heterocycles (pyrrole in particular) and look at nucleophilic substitution.

More reactions of five-membered heterocycles

Nucleophilic substitution requires an activating group

Nucleophilic substitution is a relatively rare reaction with pyrrole, thiophene, or furan and requires an activating group such as nitro, carbonyl, or sulfonyl, just as it does with benzene (Chapter 23). Here is an intramolecular example used to make the painkiller ketorolac.

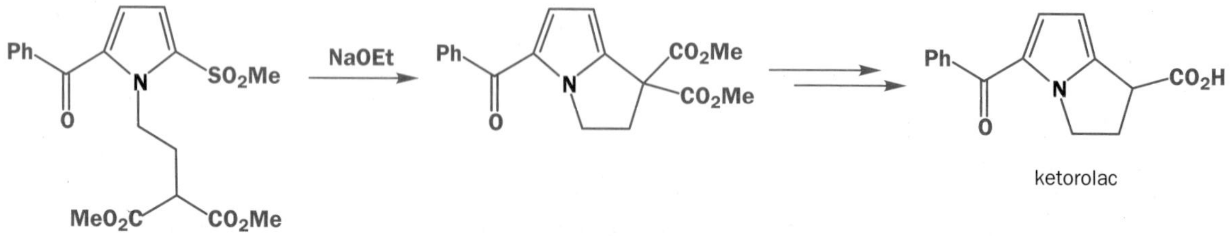

The nucleophile is a stable enolate and the leaving group is a sulfinate anion. An intermediate must be formed in which the negative charge is delocalized on to the carbonyl group on the ring, just as you saw in the benzene ring examples in Chapter 23. Attack occurs at the 2-position because the leaving group is there and because the negative charge can be delocalized on to the ketone from that position—there is no inherent preference for attack at the 2- or 5-position.

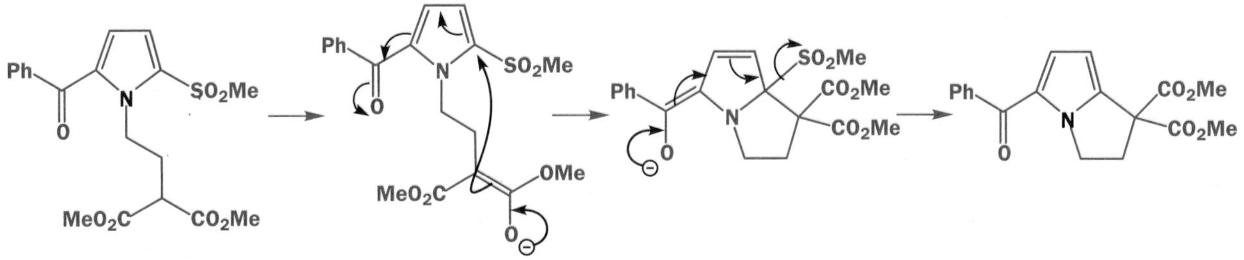

So far, all of the reactions we have discussed have been variations on reactions of benzene. These heterocycles also do reactions totally unlike those of benzene and we are now going to explore two of them.

Five-membered heterocycles act as dienes in Diels–Alder reactions

Furan is particularly good at Diels–Alder reactions but it gives the thermodynamic product, the *exo* adduct, because with this aromatic diene the reaction is reversible (Chapter 35).

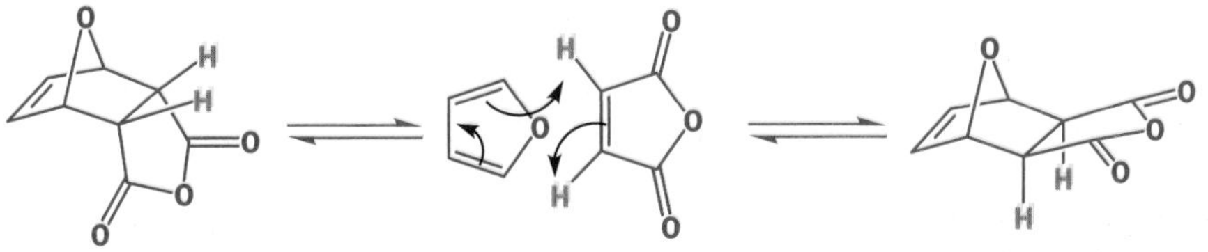

kinetically preferred
endo adduct

thermodynamically preferred
exo adduct

If pyrrole would do a similar thermodynamically controlled *exo* Diels–Alder reaction with a vinyl pyridine, a short route to the interesting analgesic epibatidine could be imagined, with just a simple reduction of the remaining alkene left to do. The reaction looks promising as the pyridine makes the dienophile electron-deficient and pyrrole is an electron-rich 'diene'.

epibatidine

Epibatidine was discovered in the skin of Ecuadorian frogs in 1992. It is an exceptionally powerful analgesic and works by a different mechanism from that of morphine so there is hope that it will not be addictive. The compound can now be synthesized so there is no need to kill the frogs to get it— indeed, they are a protected species.

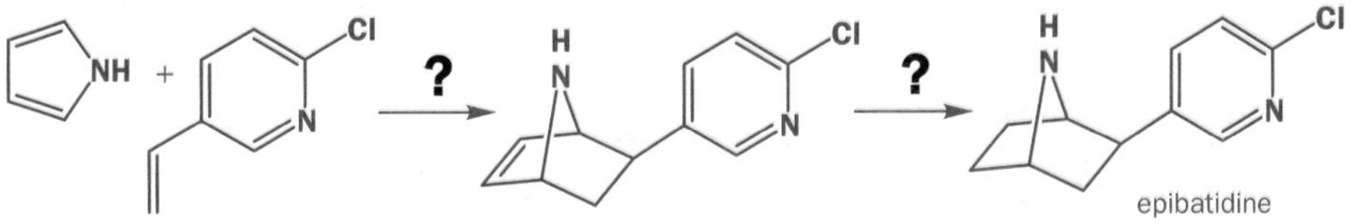

The trouble is that pyrrole will not do this reaction as it is so good at electrophilic substitution. What happens instead is that pyrrole acts as a nucleophile and attacks the electron-deficient alkene. The answer is to make pyrrole less nucleophilic by acylating the nitrogen atom with the famous 'Boc' protecting group (Chapter 24). We will see in the next section how this may be done. A good Diels–Alder reaction then occurs with a alkynyl sulfone.

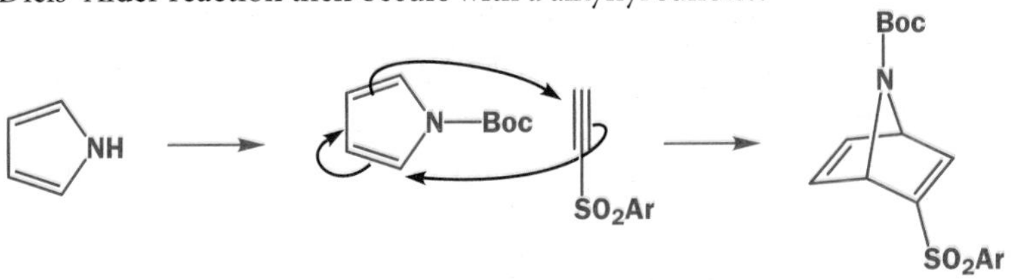

It is then possible to reduce the nonconjugated double bond chemoselectively and add a pyridine nucleophile to the vinyl sulfone. Notice in this step that a lithium derivative can be prepared from a bromopyridine. In general, heterocycles form lithium derivatives rather easily. The skeleton of epibatidine is now complete and you will find some further reactions from the rest of the synthesis in the problems at the end of this chapter.

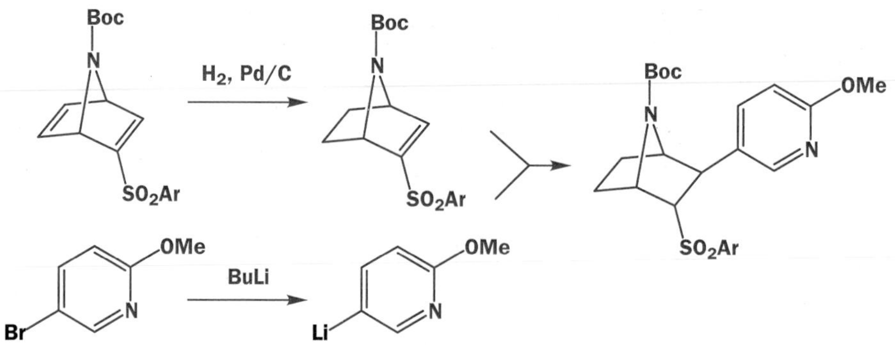

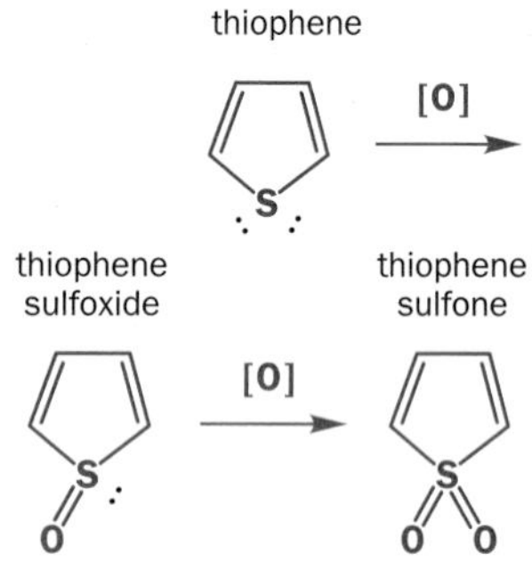

Aromaticity prevents thiophene taking part in Diels–Alder reactions, but oxidation to the sulfone destroys the aromaticity because both lone pairs become involved in bonds to oxygen. The sulfone is unstable and reacts with itself but will also do Diels–Alder reactions with dienophiles. If the dienophile is an alkyne, loss of SO_2 gives a substituted benzene derivative.

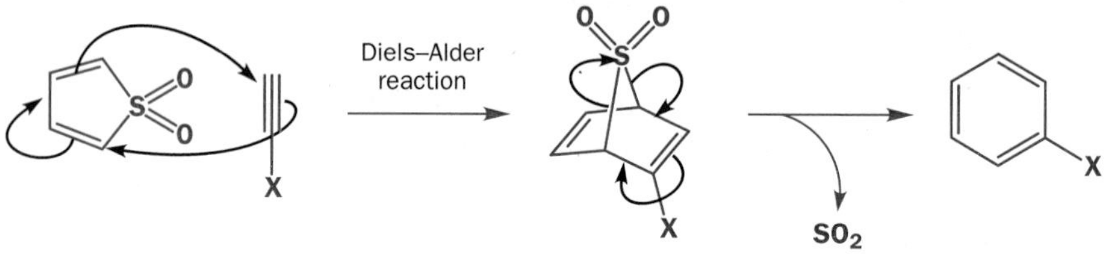

Similar reactions occur with α-pyrones. These are also rather unstable and barely aromatic and they react with alkynes by Diels–Alder reactions followed by reverse Diels–Alder reaction to give benzene derivatives with the loss of CO_2 rather than SO_2.

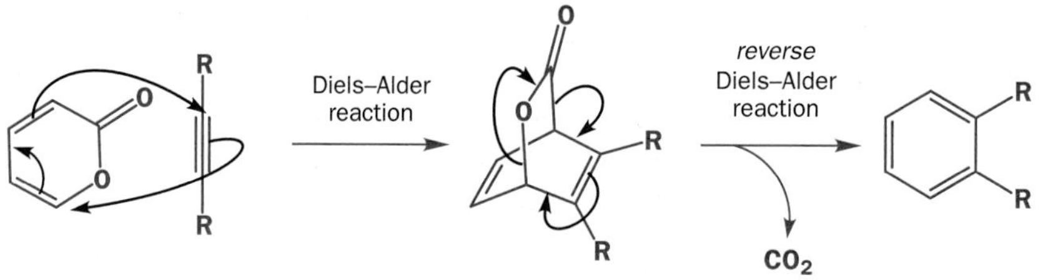

Nitrogen anions can be easily made from pyrrole

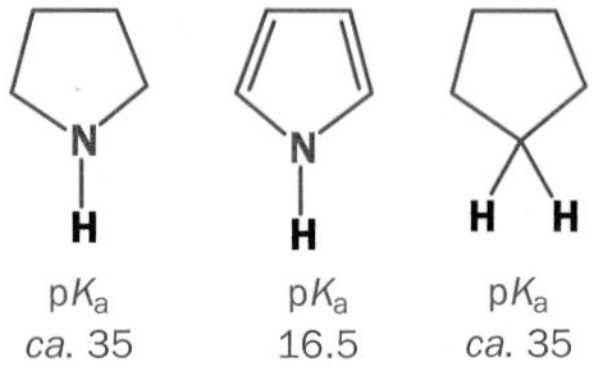

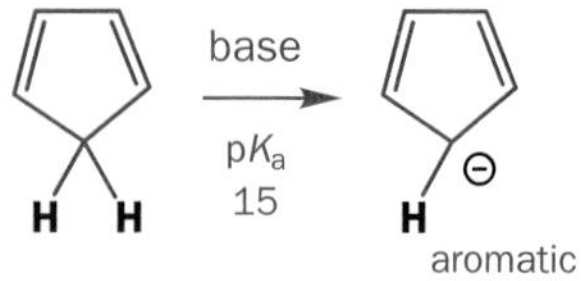

Pyrrole is much more acidic than comparable saturated amines. The pK_a of pyrrolidine is about 35, but pyrrole has a pK_a of 16.5 making it some 10^{23} times more acidic! Pyrrole is about as acidic as a typical alcohol so bases stronger than alkoxides will convert it to its anion. We should not be too surprised at this as the corresponding hydrocarbon, cyclopentadiene, is also extremely acidic with a pK_a of 15. The reason is that the anions are aromatic with six delocalized π electrons. The effect is much greater for cyclopentadiene because the hydrocarbon is not aromatic and much less for pyrrole because it is already aromatic and has less to gain.

In all of the reactions of pyrrole that we have so far seen, new groups have added to the carbon atoms of the ring. The anion of pyrrole is useful because it reacts at nitrogen. The nitrogen atom has two lone pairs of electrons in the anion: one is delocalized around the ring but the other is localized in an sp^2 orbital on nitrogen. This high-energy pair is the new HOMO and this is where the molecule reacts.

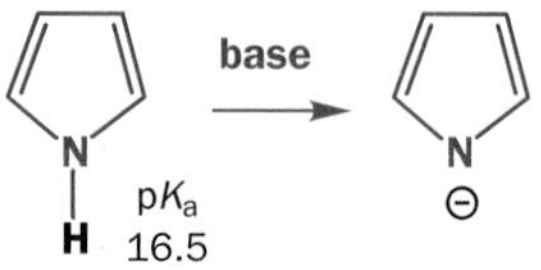

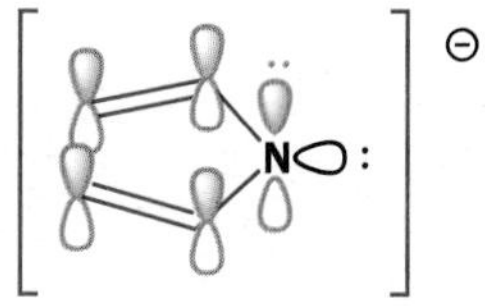

N-acylated derivatives in general can be made in this way. A commonly used base is sodium hydride (NaH) but weaker bases produce enough anion for reaction to occur.

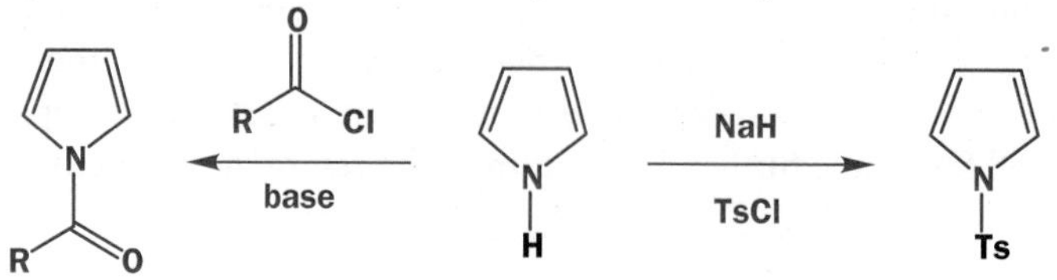

● *Anions* of pyrroles react with electrophiles at the *nitrogen* atom.

This is how the *N*-Boc pyrrole was made for use in the synthesis of epibatidine. The base used was the pyridine derivative DMAP, which you met earlier in the chapter. It has a pK_{aH} of 9.7 and so produces small, equilibrating amounts of the anion as well as acting as a nucleophilic catalyst. 'Boc anhydride' is used as the acylating agent.

► DMAP's pK_{aH} of 9.7 is between those of pyridine (5.5) and tertiary alkyl amines (*ca.* 10) but much closer to the latter.

Anion formation is important in the next main section of this chapter, which is about what happens when we insert more nitrogen atoms into the pyrrole ring.

Five-membered rings with two or more nitrogen atoms

Imidazole

At the beginning of this chapter we imagined adding more nitrogen atoms to the pyrrole ring and noticed then that there were two compounds with two nitrogen atoms: pyrazole and imidazole.

The pyrazole ring is present in Viagra (see the beginning of this chapter for the structure) and we will discuss the synthesis of this compound in the next chapter. In this chapter we will concentrate on imidazole.

Only one nitrogen atom in a five-membered ring can contribute two electrons to the aromatic sextet. The other replaces a CH group, has no hydrogen, and is like the nitrogen atom in pyridine. The black nitrogens are the pyrrole-like nitrogens; the green ones are pyridine-like. The lone pairs on the black nitrogens are delocalized round the ring; those on the green nitrogens are localized in sp^2 orbitals on nitrogen. We can expect these compounds to have properties intermediate between those of pyrrole and pyridine.

Imidazole is a stronger base than either pyrrole or pyridine—it has a pK_{aH} of almost exactly 7, meaning that it is 50% protonated in neutral water. It is also more acidic than pyrrole, with a pK_a of 14.5.

These curious results are a consequence of the 1,3 relationship between the two nitrogen atoms. Both the (protonated) cation and the (deprotonated) anion share the charge equally between the two nitrogen atoms—they are perfectly symmetrical and unusually stable.

Another way to look at the basicity of imidazole would be to say that both nitrogen atoms can act at once on the proton being attacked. It has to be the pyridine-like nitrogen that actually captures the proton but the pyrrole nitrogen can help by using its delocalized electrons like this.

A similar effect accounts for the basicity of DBU and DBN: see p. 202.

Nature makes use of this property by having imidazole groups attached to proteins in the form of the amino acid histidine and using them as nucleophilic, basic and acidic catalytic groups in enzyme reactions (this will be discussed in Chapters 49 and 50). We use this property in the same way when we add a silyl group to an alcohol. Imidazole is a popular catalyst for these reactions.

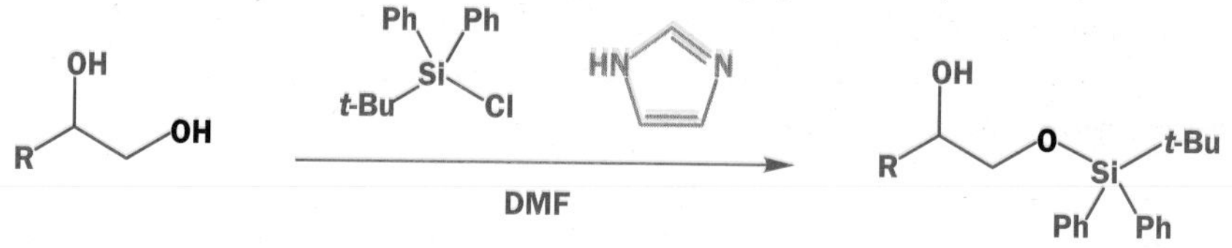

A weakly basic catalyst is needed here because we want to discriminate between the primary and secondary alcohols in the diol. Imidazole is too weak (pK_{aH} 7) to remove protons from an alcohol ($pK_a \sim 16$) but it can remove a proton after the OH group has attacked the silicon atom.

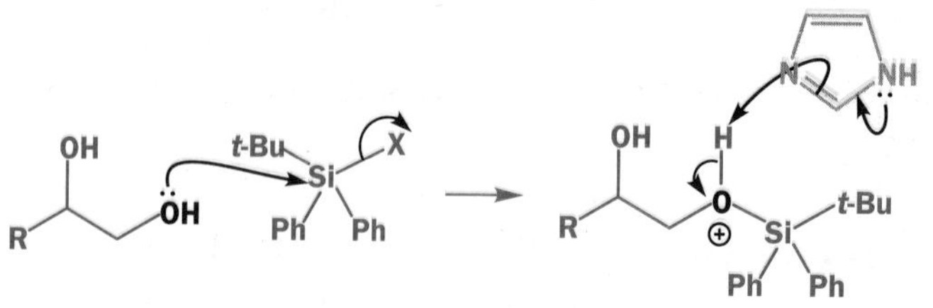

In fact, the imidazole is also a nucleophilic catalyst of this reaction, and the first step is substitution of Cl by imidazole—that is why the leaving group in the last scheme was shown as 'X'. The reaction starts off like this.

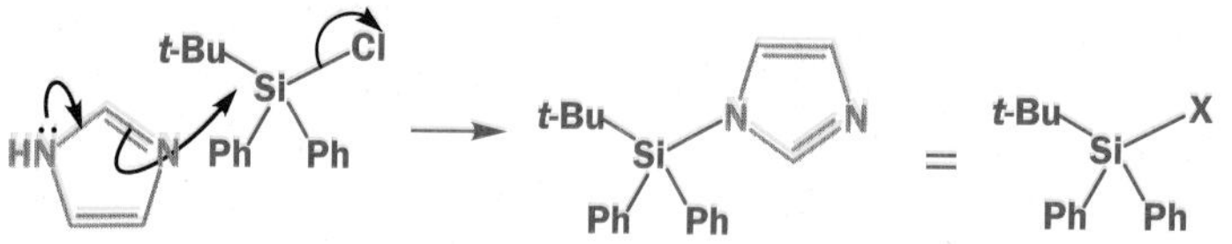

The same idea leads to the use of Carbonyl DiImidazole (CDI) as a double electrophile when we want to link two nucleophiles together by a carbonyl group. Phosgene ($COCl_2$) has been used for this but it is appallingly toxic (it was used in the First World War as a poison gas with dreadful effects). CDI is safer and more controlled. In these reactions imidazole acts (twice) as a leaving group.

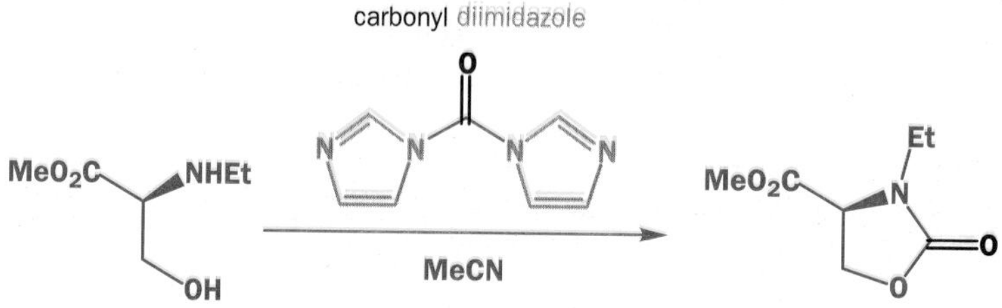

carbonyl diimidazole

The amino group probably attacks first to displace one imidazole anion, which returns to deprotonate the ammonium salt. The alcohol can then attack intramolecularly displacing the second imidazole anion, which deprotonates the OH group in its turn. The other product is just two molecules of imidazole.

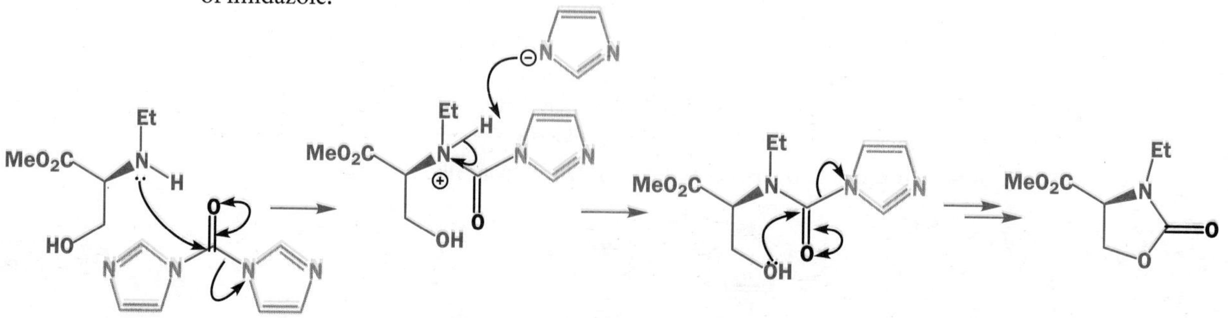

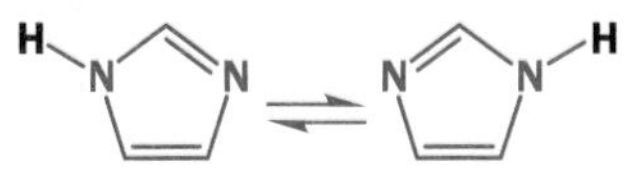

The relationship between the delocalized imidazole anion and imidazole itself is rather like that between an enolate anion and an enol. It will come as no surprise that imidazole tautomerizes rapidly at room temperature in solution. For the parent compound the two tautomers are the same, but with unsymmetrical imidazoles the tautomerism is more interesting. We will explore this question alongside electrophilic aromatic substitution of imidazoles.

two identical tautomers of imidazole

Imidazoles with a substituent between the two nitrogen atoms (position 2) can be nitrated with the usual reagents and the product consists of a mixture of tautomers.

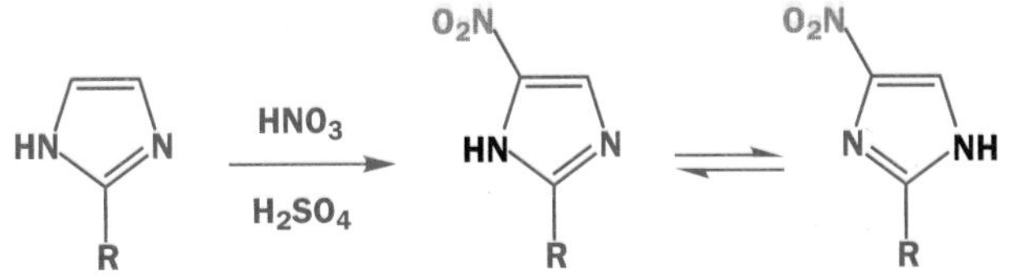

The initial nitration may occur at either of the remaining sites on the ring with the electrons coming from the pyrrole-like nitrogen atom. Tautomerism after nitration gives the mixture.

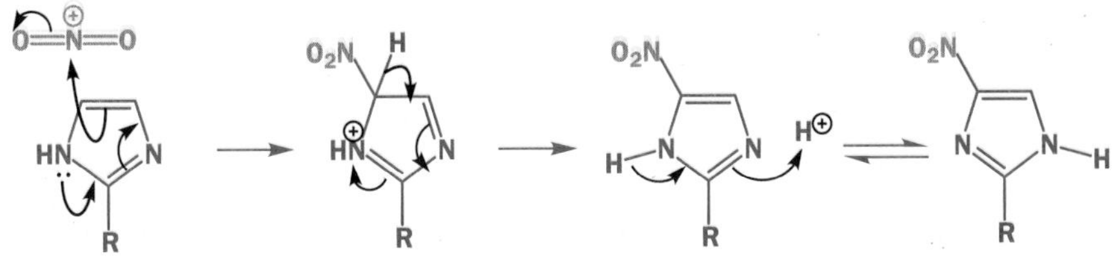

The tautomerism can be stopped by alkylation at one of the nitrogen atoms. If this is done in basic solution, the anion is an intermediate and the alkyl group adds to the nitrogen atom next to the nitro group. Again, it does not matter from which tautomer the anion is derived—there is only one anion delocalized over both nitrogen atoms and the nitro group. One reason for the formation of this isomer is that it has the linear conjugated system between the pyrrole-like nitrogen and the nitro group (see p. 1159).

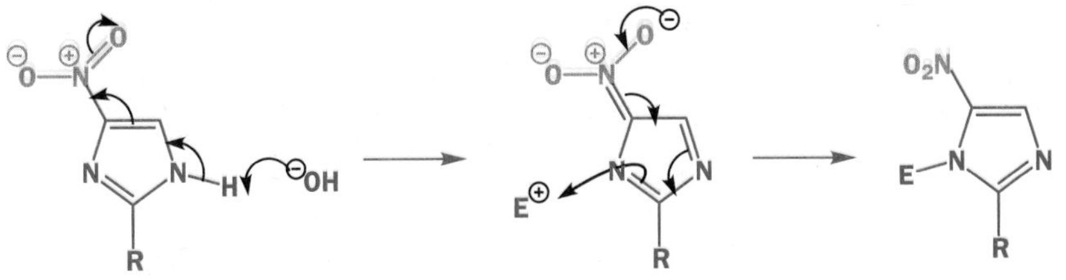

Important medicinal compounds are made in this way. The antiparasitic metronidazole comes from 2-methyl imidazole by nitration and alkylation with an epoxide in base.

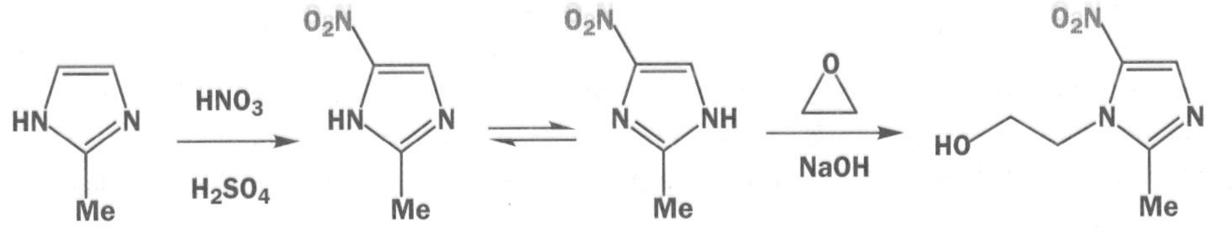

The triazoles

There are two triazoles, and each has one pyrrole-like nitrogen and two pyridine-like nitrogens. Both triazoles have the possibility of tautomerism (in 1,2,3-triazole the tautomers are identical) and both give rise to a single anion.

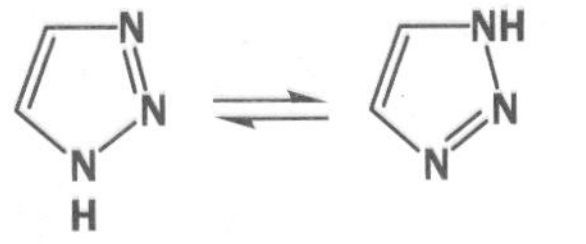

1,2,3-triazole

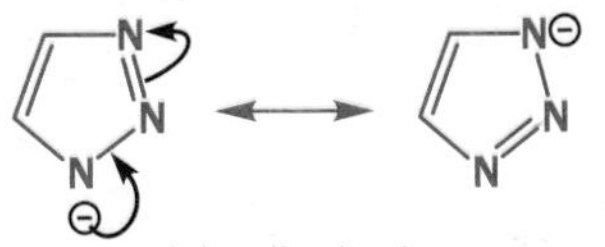

delocalized anion

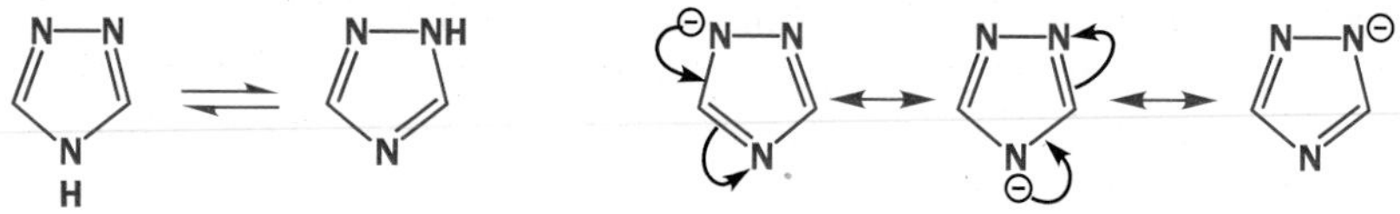

1,2,4-triazole   delocalized anion

The 1,2,4-triazole is more important because it is the basis of the best modern agricultural fungicides as well as drugs for fungal diseases in humans. The extra nitrogen atom makes it more like pyridine and so more weakly basic, but it increases its acidity so that the anion is now easy to make.

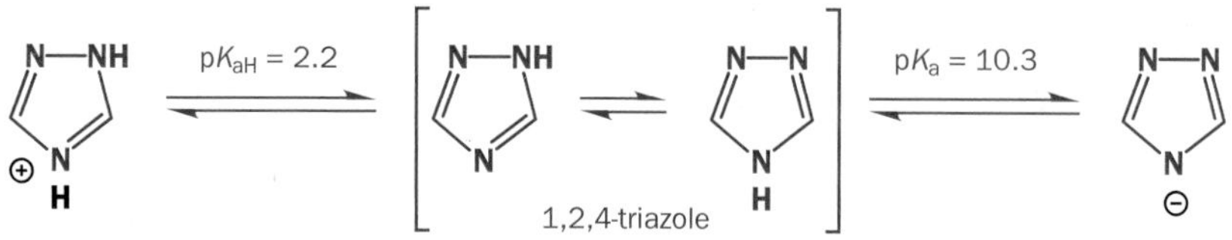

1,2,4-triazole

This may remind you of the α effect—NH₂NH₂ is more nucleophilic than ammonia because of the two linked nitrogen atoms (see p. 588).

The fungicides are usually made by the addition of the triazole anion to an epoxide or other carbon electrophile. The anion normally reacts at one of the two linked nitrogen atoms (it does not matter which—the product is the same).

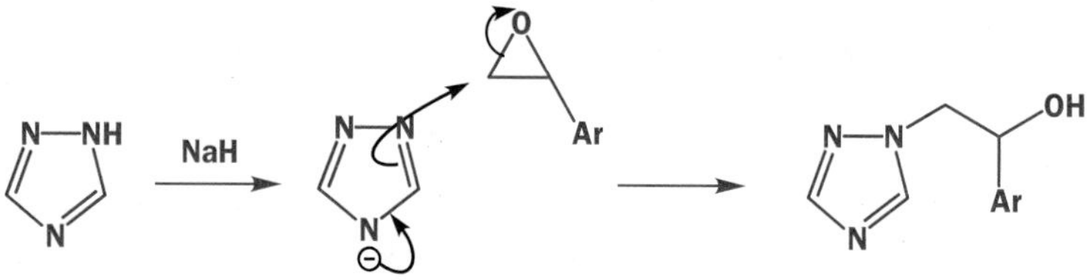

A modern example of an agent used against human fungal infections is Pfizer's fluconazole, which actually contains two triazoles. The first is added as the anion to an α-chloroketone and the second is added to an epoxide made with sulfur ylid chemistry (you will meet this in Chapter 46). Note that weak bases were used to catalyse both of these reactions. Triazole is acidic enough for even $NaHCO_3$ to produce a small amount of the anion.

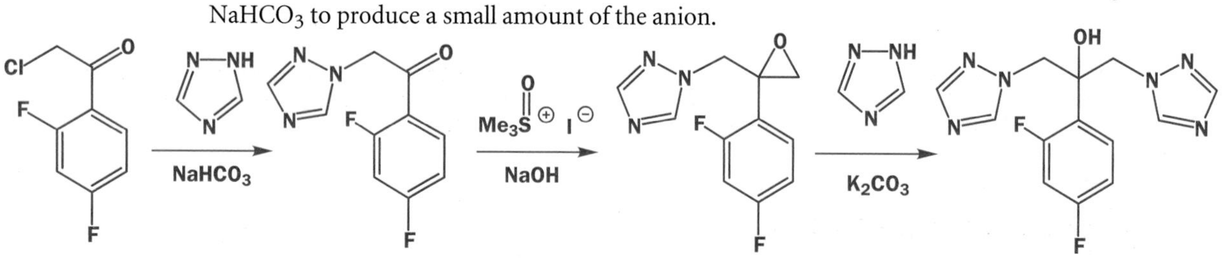

fluconazole

Tetrazole

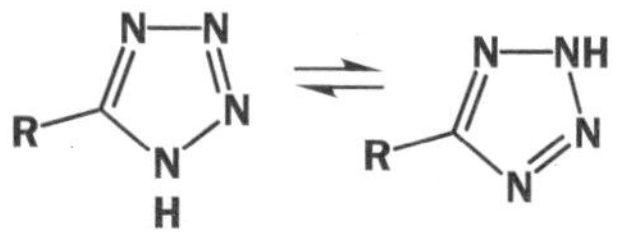

two tautomers of tetrazole

There is only one isomer of tetrazole or of substituted tetrazoles, as there is only one carbon atom in the ring, though there are two tautomers. The main interest in tetrazoles is that they are rather acidic: the pK_a for the loss of the NH proton to form an anion is about 5, essentially the same as that of a carboxylic acid. The anion is delocalized over all four nitrogen atoms (as well as the one carbon atom), and four nitrogen atoms do the work of two oxygen atoms.

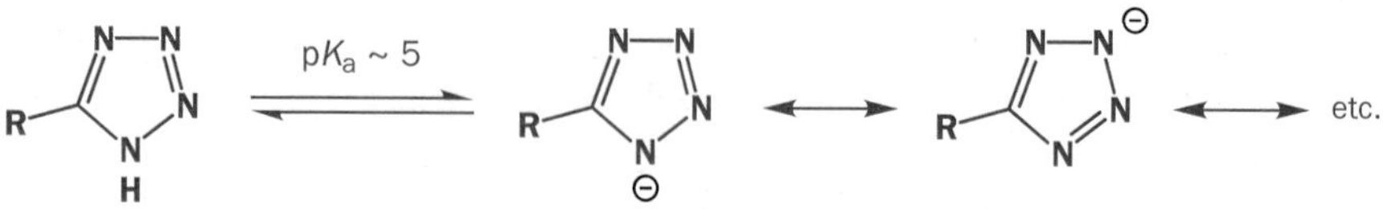

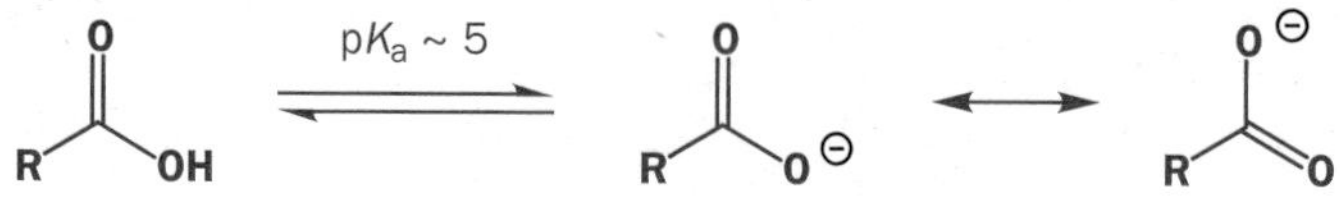

Because tetrazoles have similar acidities to those of carboxylic acids, they have been used in drugs as replacements for the CO_2H unit when the carboxylic acid has unsatisfactory properties for human medicine. A simple example is the anti-arthritis drug indomethacin whose carboxylic acid group may be replaced by a tetrazole with no loss of activity.

indomethacin

tetrazole substitute for indomethacin

Nitrogen atoms and explosions

Compounds with even two or three nitrogen atoms joined together, such as diazomethane (CH_2N_2) or azides (RN_3), are potentially explosive because they can suddenly give off stable gaseous nitrogen. Compounds with more nitrogen atoms, such as tetrazoles, are likely to be more dangerous and few people have attempted to prepare pentazoles. The limit is reached with diazotetrazole, with the amazing formula CN_6! It is made by diazotization of 5-aminotetrazole, which first gives a diazonium salt.

a pentazole
highly explosive!

5-amino-(1H)-tetrazole

the diazonium salt
highly explosive!

The diazonium salt is extremely dangerous: 'It should be emphasised that [the diazonium salt] is extremely explosive and should be handled with great care. We recommend that no more than 0.75 mmol be isolated at one time. Ethereal solutions are somewhat more stable but explosions have occurred after standing at −70 °C for 1 hr.' So much for that, but what about the diazo compound? It is extremely unstable and decomposes to a carbene with loss of one molecule of nitrogen and then loses two more to give...

the diazonium salt
highly explosive!

the diazo compound
highly explosive!

the carbene

All that is left is a carbon atom and this is one of very few ways to make carbon atoms chemically. The carbon atoms have remarkable reactions and these have been briefly studied, but the hazardous preparation of the starting materials discourages too much research. However, you will see in the next chapter that 1-amino tetrazole is a useful starting material for making an anti-allergic drug.

Benzo-fused heterocycles

Indoles are benzo-fused pyrroles

Indomethacin and its tetrazole analogue contain pyrrole rings with benzene rings fused to the side. Such bicyclic heterocyclic structures are called **indoles** and are our next topic. Indole itself has a benzene ring and a pyrrole ring sharing one double bond, or, if you prefer to look at it this way, it is an aromatic system with 10 electrons—eight from four double bonds and the lone pair from the nitrogen atom.

indole

Indole is an important heterocyclic system because it is built into proteins in the form of the amino acid tryptophan (Chapter 49), because it is the basis of important drugs such as indomethacin, and because it provides the skeleton of the **indole alkaloids**—biologically active compounds from plants including strychnine and LSD (alkaloids are discussed in Chapter 51).

tryptophan

LSD
(Lysergic Acid Diethylamide)

▶ Though the first representation is more accurate, you will often see the second used in books and papers.

In many ways the chemistry of indole is that of a reactive pyrrole ring with a relatively unreactive benzene ring standing on one side—electrophilic substitution almost always occurs on the pyrrole ring, for example. But indole and pyrrole differ in one important respect. In indole, electrophilic substitution is preferred in the 3-position with almost all reagents. Halogenation, nitration, sulfonation, Friedel–Crafts acylation, and alkylation all occur cleanly at that position.

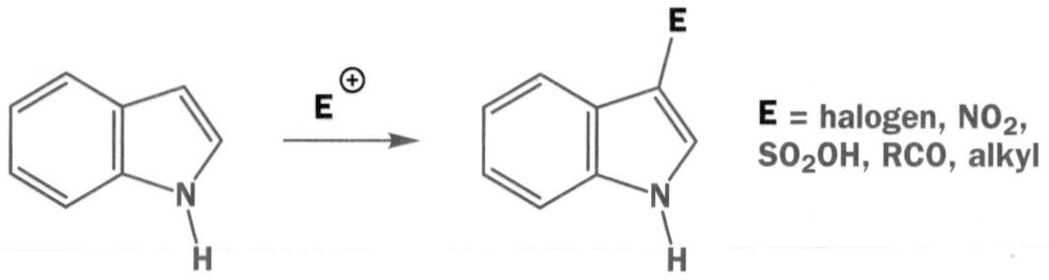

E = halogen, NO_2, SO_2OH, RCO, alkyl

This is, of course, the reverse of what happens with pyrrole. Why should this be? A simple explanation is that reaction at the 3-position simply involves the rather isolated enamine system in the five-membered ring and does not disturb the aromaticity of the benzene ring.

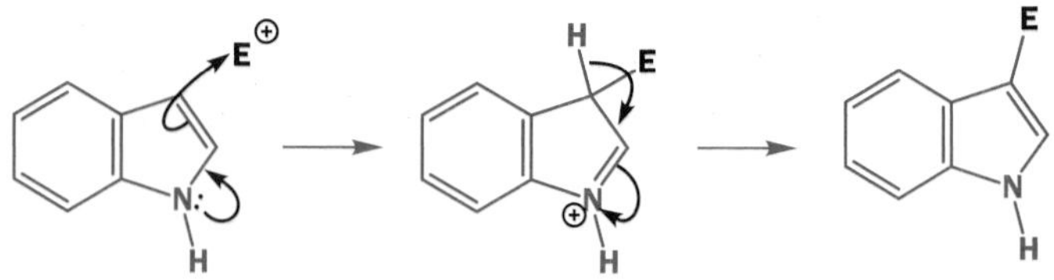

The positive charge in the intermediate is, of course, delocalized round the benzene ring, but it gets its main stabilization from the nitrogen atom. It is not possible to get reaction in the 2-position without seriously disturbing the aromaticity of the benzene ring.

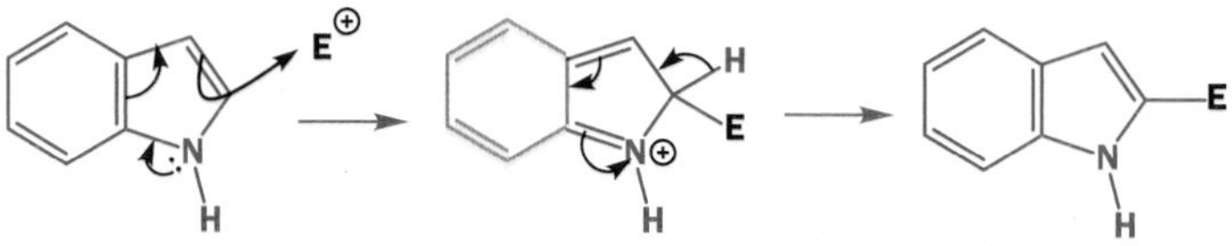

● Electrophilic substitution on pyrrole and indole

Pyrrole reacts with electrophiles at all positions but prefers the 2- and 5-positions, while indole much prefers the 3-position.

A simple example is the Vilsmeier formylation with DMF and $POCl_3$, showing that indole has similar reactivity, if different regioselectivity, to pyrrole.

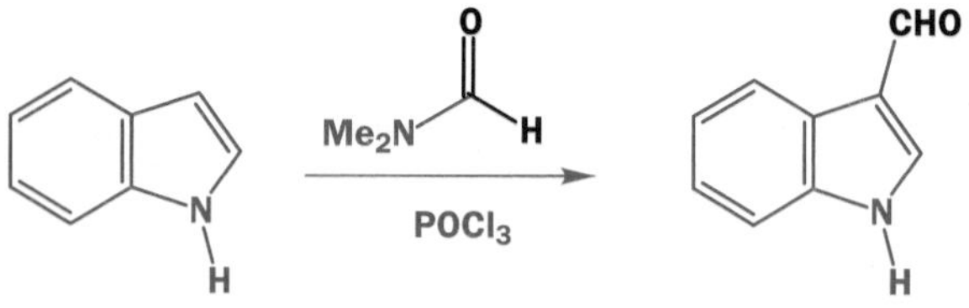

If the 3-position is blocked, reaction occurs at the 2-position and this at first seems to suggest that it is all right after all to take the electrons the 'wrong way' round the five-membered ring. This intramolecular Friedel–Crafts alkylation is an example.

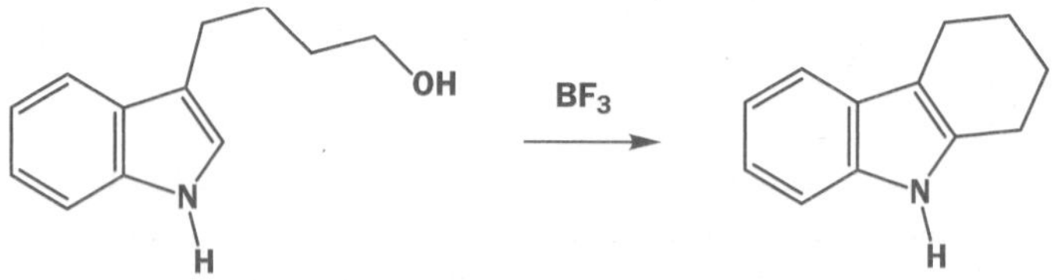

An ingenious experiment showed that this cyclization is not as simple as it seems. If the starting material is labelled with tritium (radioactive ^3H) next to the ring, the product shows exactly 50% of the label where it is expected and 50% where it is not.

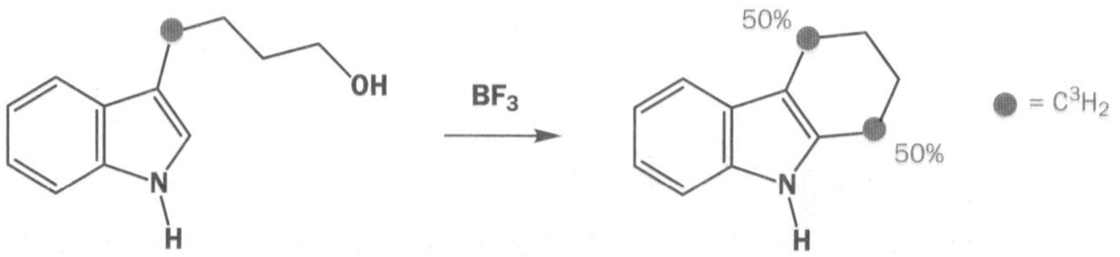

To give this result, the reaction must have a symmetrical intermediate and the obvious candidate arises from attack at the 3-position. The product is formed from the intermediate *spiro* compound, which has the five-membered ring at right angles to the indole ring—each CH_2 group has an exactly equal chance of migrating.

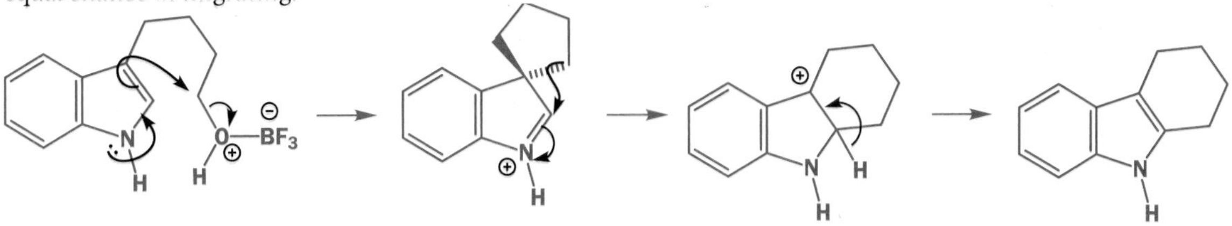

The migration is a pinacol-like rearrangement similar to those in Chapter 37. It is now thought that most substitutions in the 2-position go by this migration route but that some go by direct attack with disruption of the benzene ring.

A good example of indole's 3-position preference is the Mannich reaction, which works as well with indole as it does with pyrrole or furan.

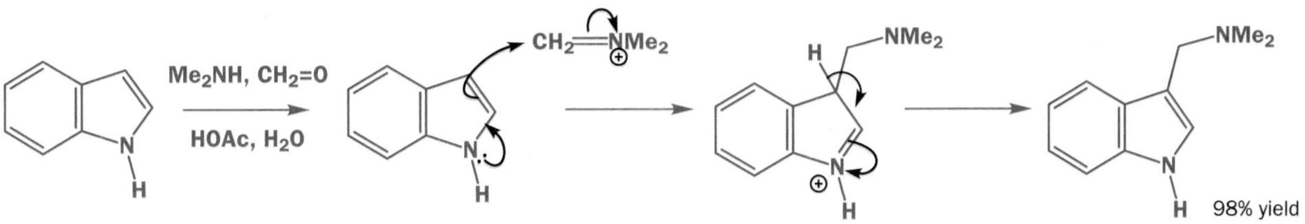

The electron-donating power of the indole and pyrrole nitrogens is never better demonstrated than in the use to which these Mannich bases (the products of the reaction) are put. You may remember that normal Mannich bases can be converted to other compounds by alkylation and substitution (see p. 758). No alkylation is needed here as the indole nitrogen can even expel the Me_2N group when NaCN is around as a base and nucleophile. The reaction is slow and the yield not wonderful but it is amazing that it happens at all. The reaction is even easier with pyrrole derivatives.

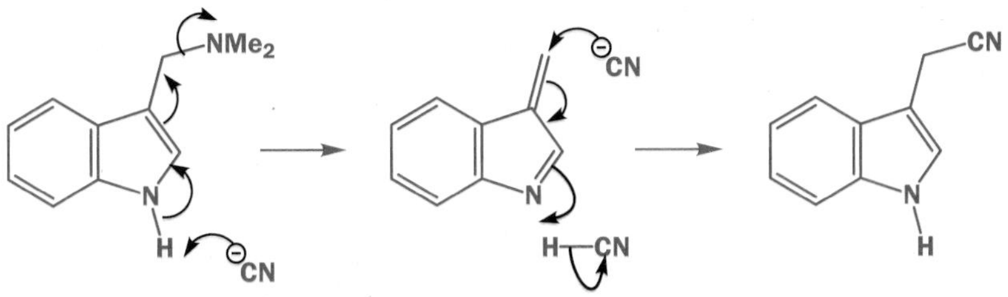

All of the five-membered rings we have looked at have their benzo-derivatives but we will concentrate on just one, 1-hydroxybenzotriazole, both because it is an important compound and because we have said little about simple 1,2,3-triazoles.

HOBt is an important reagent in peptide synthesis

1-Hydroxybenzotriazole (HOBt) is a friend in need in the lives of biochemists. It is added to many reactions where an activated ester of one amino acid is combined with the free amino group of another (see Chapter 25 for some examples). It was first made in the nineteenth century by a remarkably simple reaction.

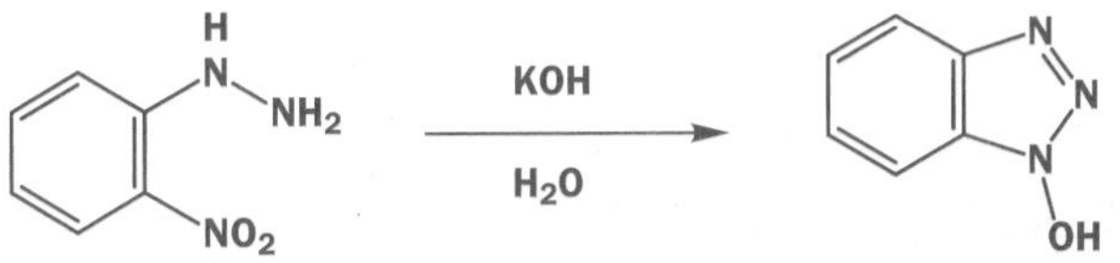

HOBt, 1-hydroxybenzotriazole

■ The mechanism of this reaction forms one of the problems at the end of the chapter.

The structure of HOBt appears quite straightforward, except for the unstable N–O single bond, but we can easily draw some other tautomers in which the proton on oxygen—the only one in the

■
You met some nitrone chemistry in Chapter 35.

heterocyclic ring—can be placed on some of the nitrogen atoms. These structures are all aromatic, the second and third are nitrones, and the third structure looks less good than the other two.

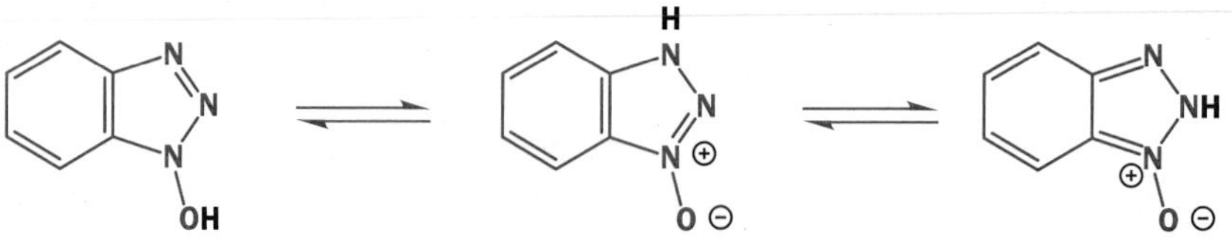

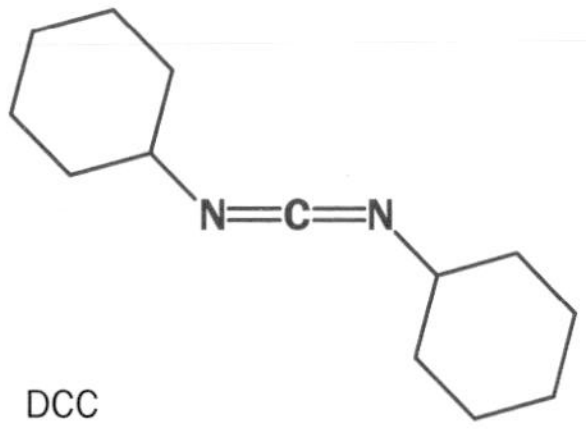

DCC
DiCyclohexylCarbodiimide

HOBt comes into play when amino acids are being coupled together in the lab. The reaction is an amide formation, but in Chapter 25 we mentioned that amino-acyl chlorides cannot be used to make polypeptides—they are too reactive and they lead to side-reactions. Instead, activated amino-esters (with good RO^- leaving groups) are used, such as the phenyl esters of Chapter 25. It even more common to form the activated ester in the coupling reaction, using a **coupling reagent**, the most common being 'DCC', dicyclohexylcarbodiimide. DCC reacts with carboxylic acids like this.

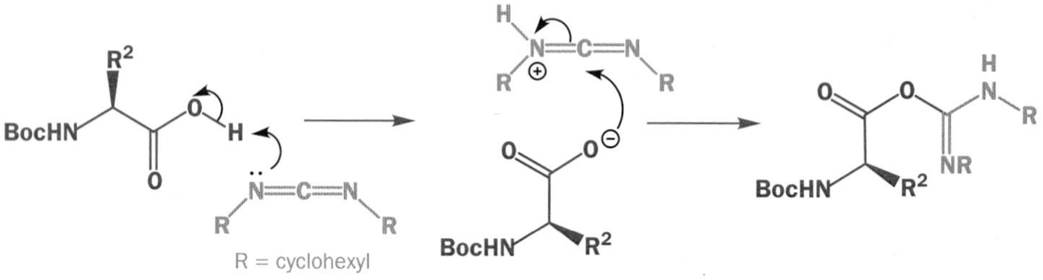

The product ester is activated because substitution with any nucleophile expels this very stable urea as a leaving group.

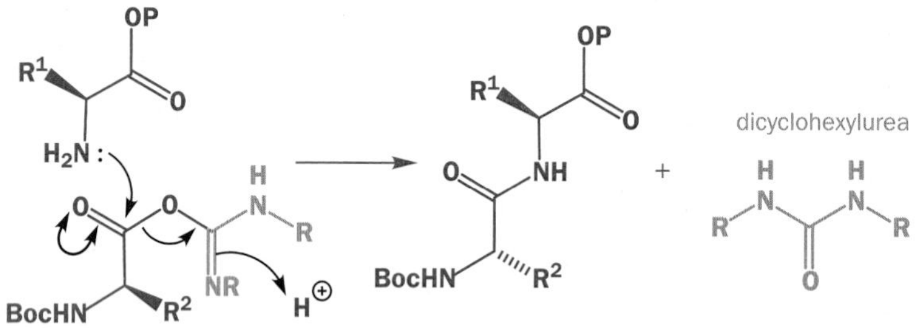

▶
You saw in Chapter 28 that the most electrophilic carboxylic acid derivatives are also the most enolizable.

The problem with attacking this ester directly with the amino group of the second amino acid is that some racemization of the active ester is often found. A better method is to have plenty of HOBt around. It intercepts the activated ester first and the new intermediate does not racemize, mostly because the reaction is highly accelerated by the addition of HOBt. The second amino acid, protected on the carboxyl group, attacks the HOBt ester and gives the dipeptide in a very fast reaction without racemization.

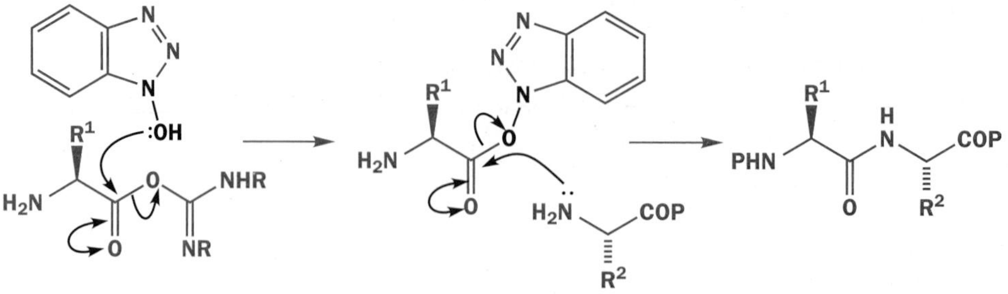

Putting more nitrogen atoms in a six-membered ring

At the beginning of the chapter we mentioned the three six-membered aromatic heterocycles with two nitrogen atoms—pyridazine, pyrimidine, and pyrazine. In these compounds both nitrogen atoms must be of the pyridine sort, with lone pair electrons not delocalized round the ring.

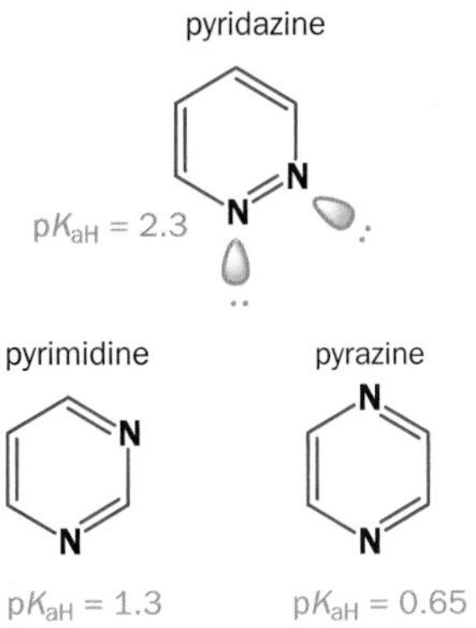

We are going to look at these compounds briefly here. Pyrimidine is more important than either of the others because of its involvement in DNA and RNA—you will find this in Chapter 49. All three compounds are very weak bases—hardly basic at all in fact. Pyridazine is slightly more basic than the other two because the two adjacent lone pairs repel each other and make the molecule more nucleophilic (the α effect again: see p. 588 of Chapter 23).

The chemistry of these very electron-deficient rings mostly concerns nucleophilic attack and displacement of leaving groups such as Cl by nucleophiles such as alcohols and amines. To introduce this subject we need to take one heterocyclic synthesis at this point, though these are properly the subject of the next chapter. The compound 'maleic hydrazide' has been known for some time because it is easily formed when hydrazine is acylated twice by maleic anhydride.

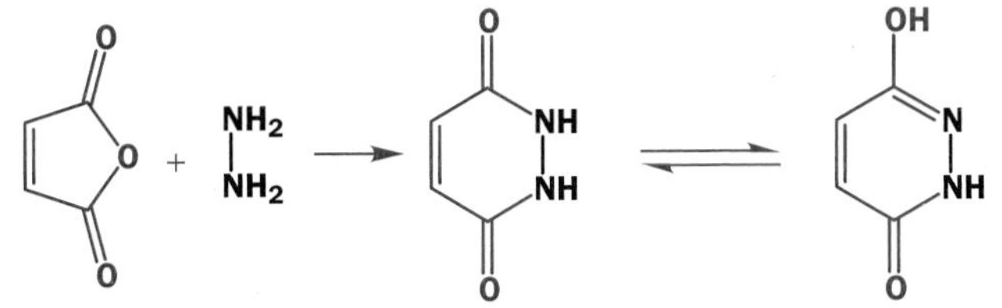

The compound actually prefers to exist as the second tautomer, which is 'more aromatic'. Reaction with POCl$_3$ in the way we have seen for pyridine gives the undoubtedly aromatic pyridazine dichloride.

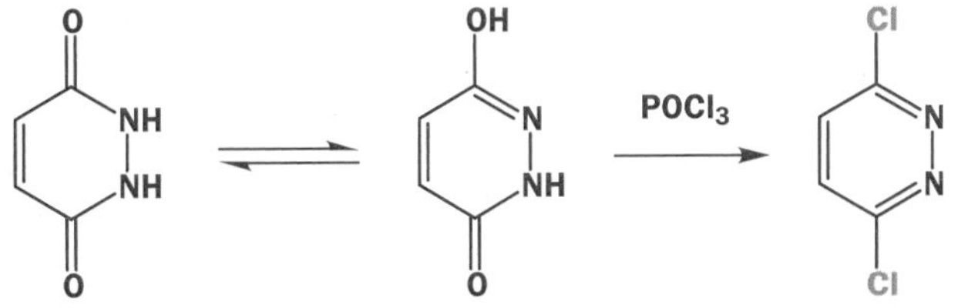

Now we come to the point. Each of these chlorides can be displaced in turn with an oxygen or nitrogen nucleophile. Only one chloride is displaced in the first reaction, if that is required, and then the second can be displaced with a different nucleophile (see reaction on the right).

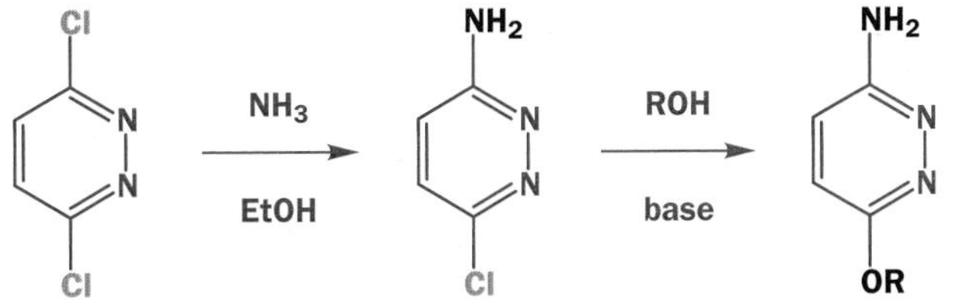

How is this possible? The mechanism of the reactions is addition to the pyridazine ring followed by loss of the leaving group, so the first reaction must go like this.

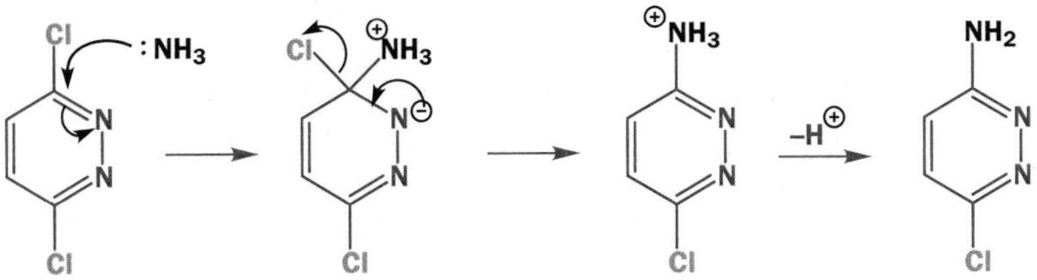

When the second nucleophile attacks it is forced to attack a less electrophilic ring. An electron-withdrawing group (Cl) has been replaced by a strongly electron-donating group (NH$_2$) so the rate-determining step, the addition of the nucleophile, is slower.

The same principle applies to other easily made symmetrical dichloro derivatives of these rings and their benzo-analogues. The nitrogen atoms can be related 1,2, 1,3, or 1,4 as in the examples alongside. The first two are used to link the quinine-derived ligands required for the Sharpless asymmetric dihydroxylation, which will be described in Chapter 45.

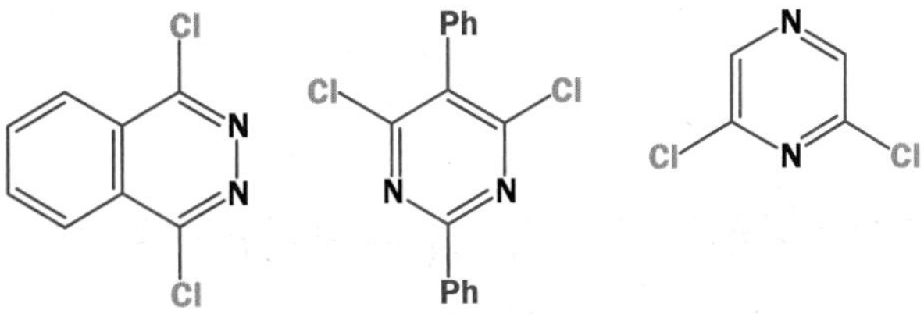

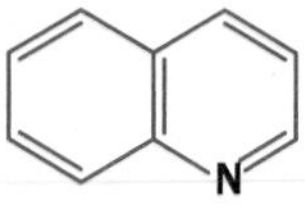

quinoline

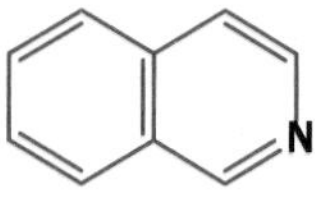

isoquinoline

Fusing rings to pyridines : quinolines and isoquinolines

A benzene ring can be fused on to the pyridine ring in two ways giving the important heterocycles quinoline, with the nitrogen atom next to the benzene ring, and isoquinoline, with the nitrogen atom in the other possible position.

Quinoline forms part of quinine (structure at the head of this chapter) and isoquinoline forms the central skeleton of the isoquinoline alkaloids, which we will discuss at some length in Chapter 51. In this chapter we need not say much about quinoline because it behaves rather as you would expect—its chemistry is a mixture of that of benzene and pyridine. Electrophilic substitution favours the benzene ring and nucleophilic substitution favours the pyridine ring. So nitration of quinoline gives two products—the 5-nitroquinolines and the 8-nitroquinolines—in about equal quantities (though you will realize that the reaction really occurs on protonated quinoline.

> Quinoline numbering, for nomenclature purposes, is shown on this structure.

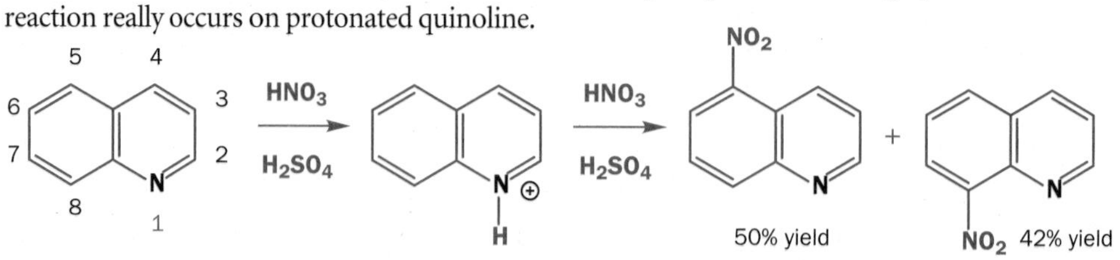

This is obviously rather unsatisfactory but nitration is actually one of the better behaved reactions. Chlorination gives ten products (at least!), of which no fewer than five are chlorinated quinolines of various structures. The nitration of isoquinoline is rather better behaved, giving 72% of one isomer (5-nitroisoquinoline) at 0 °C.

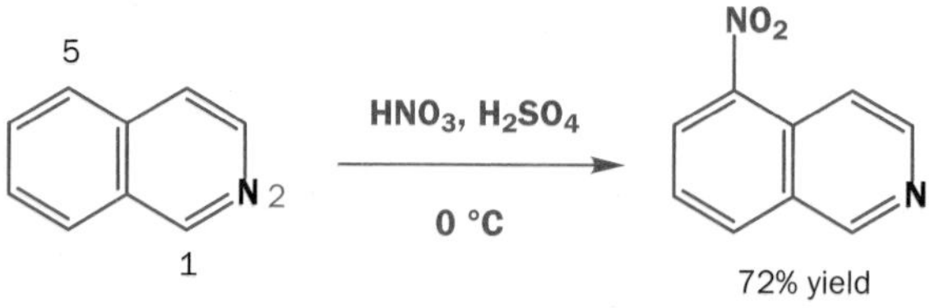

To get reaction on the pyridine ring, the *N*-oxide can be used as with pyridine itself. A good example is acridine, with two benzene rings, which gives four nitration products, all on the benzene rings. Its *N*-oxide, on the other hand, gives just one product in good yield—nitration takes place at the only remaining position on the pyridine ring.

acridine

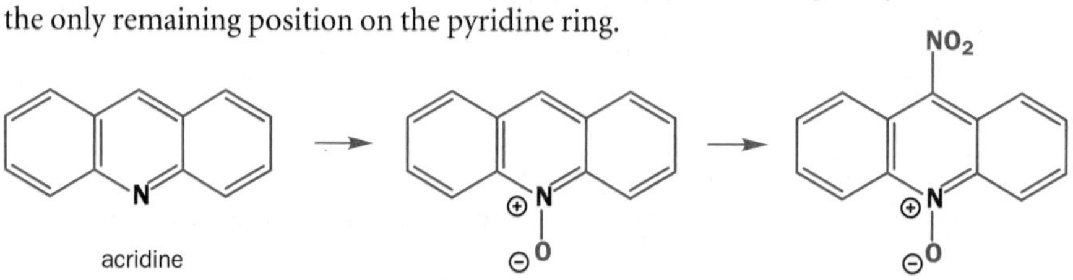

In general, these reactions are not of much use and most substituents are put into quinolines during ring synthesis from simple precursors as we will explain in the next chapter. There are a couple of quinoline reactions that are unusual and interesting. Vigorous oxidation goes for the more electron-rich ring, the benzene ring, and destroys it leaving pyridine rings with carbonyl groups in the 2- and 3-positions.

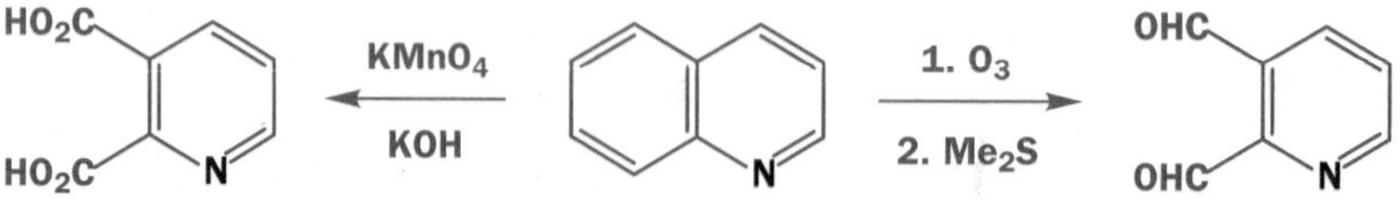

A particularly interesting nucleophilic substitution occurs when quinoline *N*-oxide is treated with acylating agents in the presence of nucleophiles. These two examples show that nucleophilic substitution occurs in the 2-position and you may compare these reactions with those of pyridine *N*-oxide. The mechanism is similar.

In considering quinolines and indoles with their fused rings we kept the benzene and heterocyclic rings separate. Yet there is a way in which they can be combined more intimately, and that is to have a nitrogen atom at a ring junction.

A nitrogen atom can be at a ring junction

It has to be a pyrrole-type nitrogen as it must have three σ bonds, so the lone pair must be in a p orbital. This means that one of the rings must be five-membered and the simplest member of this interesting class is called **indolizine**—it has pyridine and pyrrole rings fused together along a C–N bond.

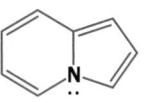

indolizine

If you examine this structure you will see that there is definitely a pyrrole ring but that the pyridine ring is not all there. Of course, the lone pair and the π electrons are all delocalized but this system, unlike indole and quinoline, is much better regarded as a ten-electron outer ring than as two six-electron rings joined together.

Indolizine reacts with electrophiles on the five-membered rings by substitution reactions as expected but it has one special reaction that leads dramatically to a more complex aromatic system. It does a cycloaddition with diethyl acetylenedicarboxylate to give a tricyclic molecule.

The dienophile is the usual sort of unsaturated carbonyl compound—but count the electrons used from the indolizine. The nitrogen lone pair is not used but all the other eight are, so this is a most unusual [2 + 8] cycloaddition. The first formed product is not aromatic (it is not fully conjugated) but it can be dehydrogenated with palladium to make a **cyclazine**.

a cyclazine

> Notice that this is the reverse of a hydrogenation: the catalyst is the same but H_2 is lost, not gained.

Now count the electrons in the cyclazine—there are ten electrons round the outer edge and the nitrogen lone pair is not part of the aromatic system. Cyclazines have NMR spectra and reactions that suggest they are aromatic.

Fused rings with more than one nitrogen

It is easily possible to continue to insert nitrogen atoms into fused ring systems and some important compounds belong to these groups. The **purines** are part of DNA and RNA and are treated in Chapter 49, but simple purines play an important part in our lives. Coffee and tea owe their stimulant properties to caffeine, a simple trimethyl purine derivative. It has an imidazole ring fused to a pyrimidine ring and is aromatic in spite of the two carbonyl groups.

purine   caffeine   uric acid

Uric acid, gout, and allopurinol

allopurinol

Another purine, uric acid, occurs widely in nature—it is used by birds, and to some extent by humans, as a way to excrete excess nitrogen—but it causes much distress in humans when crystalline uric acid is deposited in joints. We call the pain 'gout' and it isn't funny. The solution is a specific inhibitor of the enzyme producing uric acid and it is no surprise that a compound closely resembling uric acid, allopurinol, is the best.

Two of the carbonyl groups have gone and the imidazole ring has been replaced by a pyrazole ring. Purines from DNA are degraded in the body to xanthine, which is oxidized to uric acid. Allopurinol binds to the enzyme xanthine oxidase but inactivates it by not reacting. In fact it imitates not uric acid but the true substrate xanthine in a competitive fashion. This enzyme plays a minor part in human metabolism so inhibiting it is not serious—it just prevents overproduction of uric acid.

guanine → xanthine → (xanthine oxidase, inhibited by allopurinol) → uric acid

Other fused heterocycles have very attractive flavour and odour properties. Pyrazines, in general, are important in many strong food flavours: a fused pyrazine with a ring junction nitrogen atom is one of the most important components in the smell of roast meat. You can read about the simple pyrazine that provides green peppers with their flavour in the Box on the next page.

smell of roast meat

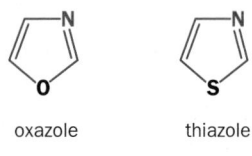

useful medicinal compounds

Finally, the compounds in the margin form a medicinally important group of molecules, which includes antitumour compounds for humans and anthelmintics (compounds that get rid of parasitic worms) for animals. They are derived from a 6/5 fused aromatic ring system that resembles the ten-electron system of the indolizine ring system but has three nitrogen atoms.

All this multiple heteroatom insertion is possible only with nitrogen and we need to look briefly at what happens when we combine nitrogen with oxygen or in heterocycles.

Heterocycles can have many nitrogens but only one sulfur or oxygen in any ring

A neutral oxygen or sulfur atom can have only two bonds and so it can never be like the nitrogen atom in pyridine—it can only be like the nitrogen atom in pyrrole. We can put as many pyridine-like nitrogens as we like in an aromatic ring, but never more than one pyrrole-like nitrogen. Similarly, we can put only one oxygen or sulfur atom in an aromatic ring. The simplest examples are oxazoles and thiazoles and their less stable isomers.

oxazole   thiazole

isoxazole   isothiazole

The instability of the 'iso-' compounds comes from the weak O–N or S–N bond. These bonds can be cleaved by reducing agents, which then usually reduce the remaining functional groups further. The first product from reduction of the N–O bond is an unstable imino-enol. The enol tautomerizes to the ketone and the imine may be reduced further to the amine. We used this sort of chemistry on the products of 1,3-dipolar cycloadditions in Chapter 35 and **isoxazoles** are usually formed by such reactions.

3,5-disubstituted isoxazole —H₂, Pd, C or LiAlH₄→ [imino-enol] → product

Such heterocycles with even more nitrogen atoms exist but are relatively unimportant and we shall mention just one, the 1,2,5-thiadiazole, because it is part of a useful drug, timolol.

1,2,5-thiadiazole

Timolol – a β-blocker

> Timolol is a β-blocker that blocks one action of adrenaline (epinephrine) and keeps heart disease at bay by counteracting high blood pressure.

The flavour of green peppers

As a review of spectroscopy we shall describe the discovery of the compound responsible for the flavour of green peppers. This powerful compound was isolated from the oil of the green pepper (*Capsicum annuum* var. *grossum*). The oil makes up about 0.0001% of the mass of the peppers and the main pepper flavour comes from one compound which is 30% of the oil. It had an even molecular ion at 166 and looks like a compound without nitrogen, perhaps $C_{11}H_{18}O$. But a high-resolution mass spectrum revealed that M^+ was actually 166.1102 which corresponds almost exactly to $C_9H_{14}N_2O$ (166.1106).

The IR had no OH, NH, or C=O peaks, and the proton NMR looked like this.

δ_H, p.p.m.	Integral	Shape	J, Hz	Comments
0.91	6H	d	6.7	Me_2CH-
1.1–2.4	1H	m	?	
2.61	2H	d	7.0	CH_2CH-
3.91	3H	s	—	-O**Me**?
7.80	1H	d	2.4	aromatic
7.93	1H	d	2.4	aromatic

The 'CH' feature in the Me_2CH and CH_2CH signals must be the same CH and it must be the signal at 1.1–2.4 p.p.m. described as a 'multiplet' as it is the only one showing enough coupling. It will be a septuplet of triplets, that is, 21 lines. We can easily reconstruct the aliphatic part of the molecule because it has two methyl groups and a CH_2 group joined to the same CH group.

side chain of green pepper flavour compound

We also have an OMe group (only oxygen is electronegative enough to take a methyl group to nearly 4 p.p.m.). This adds up to $C_5H_{12}O$. What is left? Only $C_4H_2N_2$—and no clue yet as to the nitrogen functionality. We also have an aromatic ring that must have nitrogen in it (because there are only five carbon atoms—not enough for a benzene ring!) and the coupling constant between the two aromatic hydrogens is 2.4 Hz. So could we perhaps have a pyrrole ring? Well, no, and for two reasons. If we try and construct such a molecule, we can't fit in the last nitrogen! If we put it on the end of the dotted line, it would have to be an NH_2 group, and there isn't one.

A better reason is that the chemical shifts are all wrong. The protons on an electron-rich pyrrole ring come at around 6–6.5 p.p.m., upfield from benzene (7.27 p.p.m.). But these protons are at 7.8–8.0 p.p.m., downfield from benzene. We have a deshielded (electron-poor) ring, not a shielded (electron-rich) ring. From what you now know of heterocyclic chemistry, the ring must be a six-membered one, and we must put both nitrogen atoms in the ring. There are three ways to do this.

pyridazine (1,2-nitrogens)

pyrimidine (1,3-nitrogens)

pyrazine (1,4-nitrogens)

The small coupling constant really fits the pyrazine alone and the chemical shifts are about right for that molecule too,

correct structure of the main flavour compound from green peppers

though not as far downfield. But we have a MeO group on the ring feeding electrons into the aromatic system and that will increase the shielding slightly and move the protons upfield. This gives us a unique structure.

There is only one way to be sure and that is to make this compound and see if it is the same as the natural product in all respects including biological activity. The investigators did this but then wished that they hadn't! The structure was indeed correct but the biological activity—the smell of green peppers—was so intense that they had to seal up the laboratory where the work was done as no one would work there. Human beings can detect 2 parts in 10^{12} of this compound in water.

There are thousands more heterocycles out there

But we're not going to discuss them and we hope you're grateful. In fact, it's about time to stop, and we shall leave you with a hint of the complexity that is possible. If pyrrole is combined with benzaldehyde a good yield of a highly coloured crystalline compound is formed. This is a **porphyrin**.

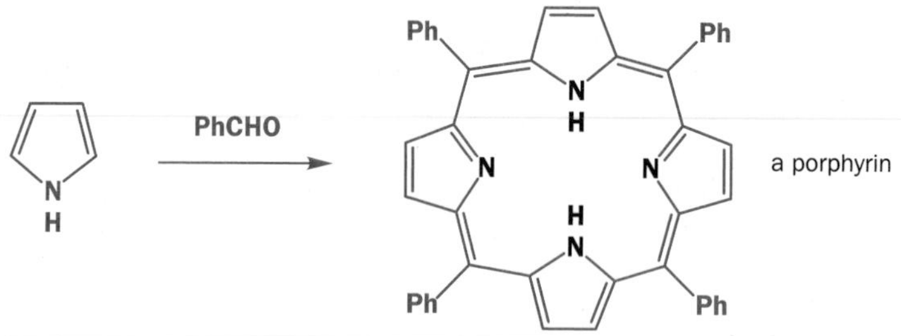

a porphyrin

Now, what about this ring system—is it aromatic? It's certainly highly delocalized and your answer to the question clearly depends on whether you include the nitrogen electrons or not. In fact, if you ignore the pyrrole-like nitrogen atoms but include the pyridine-like nitrogens and weave round the periphery, you have nine double bonds and hence 18 electrons—a $4n + 2$ number. Most people agree that these compounds are aromatic.

They are also more than curiosities. The space in the middle with the four inward-pointing nitrogen atoms is just right for complex formation with divalent metals such as Fe(II). With more varied substituents, this structure forms the reactive part of haemoglobin, and the iron atom in the middle transports the oxygen in blood.

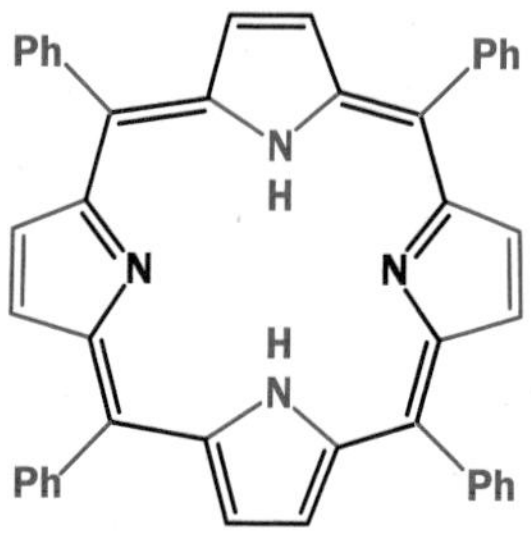

the 18 π-electron system of a porphyrin

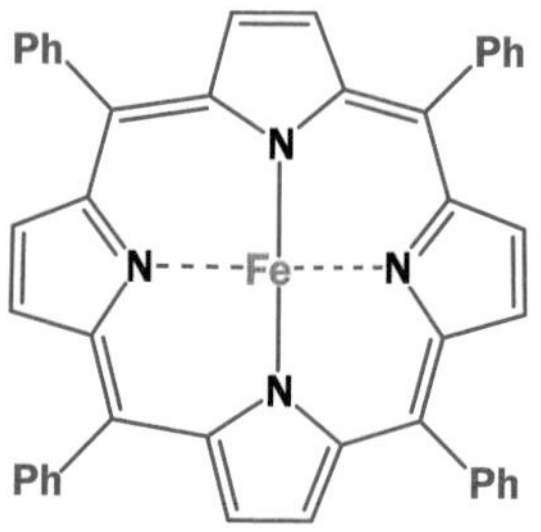

an iron–porphyrin complex

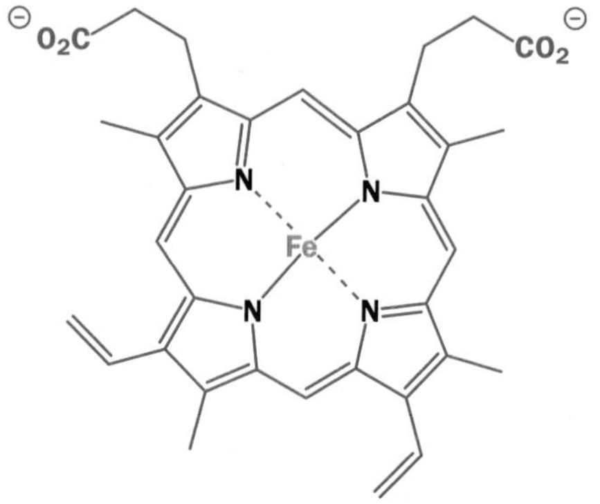

the iron porphyrin in haemoglobin

Iron prefers to be octahedral with six bonds around it and in one of these spare places in haemoglobin that is occupied by oxygen. If you try and make an oxygen complex of the simple porphyrin with four phenyl groups around the edge you get a sandwich dimer that oxidizes itself.

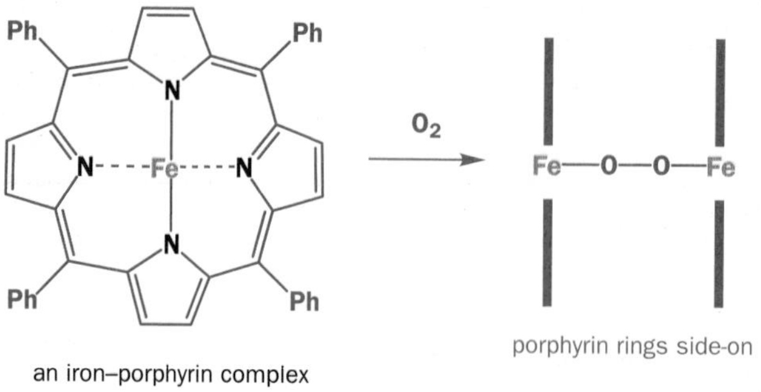

an iron–porphyrin complex

porphyrin rings side-on

The porphyrin in blood avoids this problem by having another heterocycle to hand. Haemoglobin consists of the flat porphyrin bound to a protein by coordination between an imidazole in the protein (a histidine residue: see Chapter 49) and the iron atom. This leaves one face free to bind oxygen and makes the molecule far too big to dimerize.

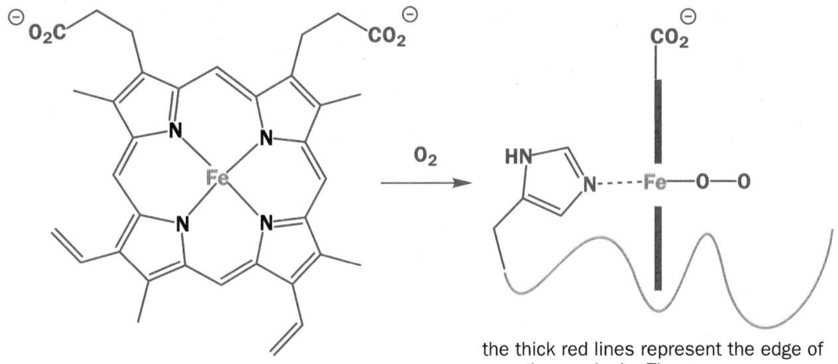

the iron porphyrin in haemoglobin

the thick red lines represent the edge of
the porphyrin. The green curve
represents the protein

Haem–metal complexes are strongly coloured—the iron complex is literally blood red. Some related compounds provide the familiar blue and green pigments used to colour plastic shopping bags. These are the phthalocyanine–metal complexes, which provide intense pigments in these ranges. The basic ring system resembles a porphyrin.

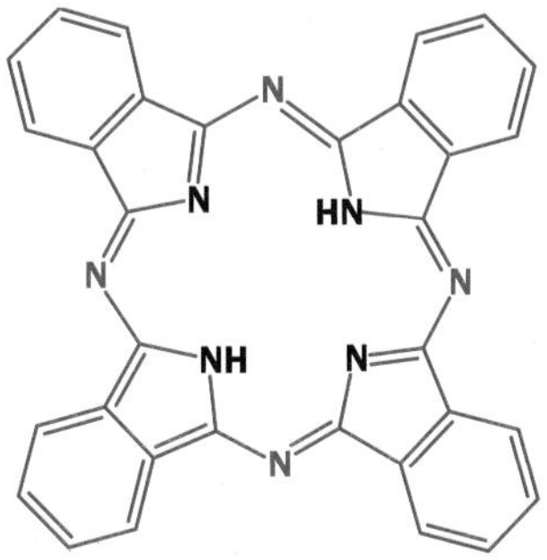

a simple phthalocyanine

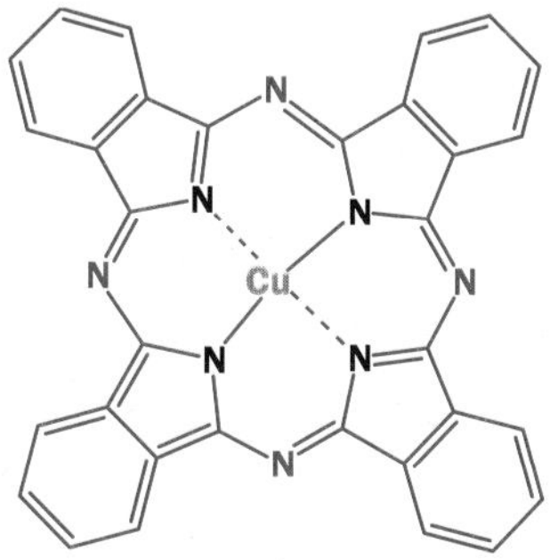

a phthalocyanine–copper complex

The differences are the four extra nitrogen atoms between the rings and the fused benzene rings. These compounds are derivatives of phthalimide, an isoindole derivative that has a nonaromatic five-membered ring. The metal most commonly used with phthalocyanines is Cu(II), and the range of colours is achieved by halogenating the benzene rings. The biggest producer is ICI at Grangemouth in Scotland where they do the halogenation and the phthalocyanine formation to make their range of Procyon™ dyes.

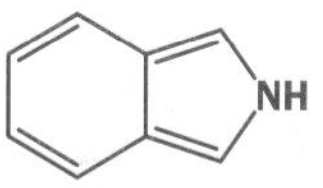

Some heterocycles are simple, some very complex, but we cannot live without them. We shall end this chapter with a wonderful story of heterocyclic chemistry at work. Folic acid is much in the news today as a vitamin that is particularly important for pregnant mothers, but that is involved in the metabolism of all living things. Folic acid is built up in nature from three pieces: a heterocyclic starting material (red), p-aminobenzoic acid (black) and the amino acid glutamic acid (green). Here you see the precursor, dihydrofolic acid.

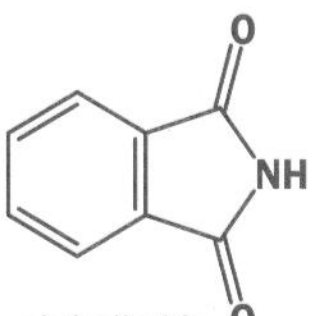

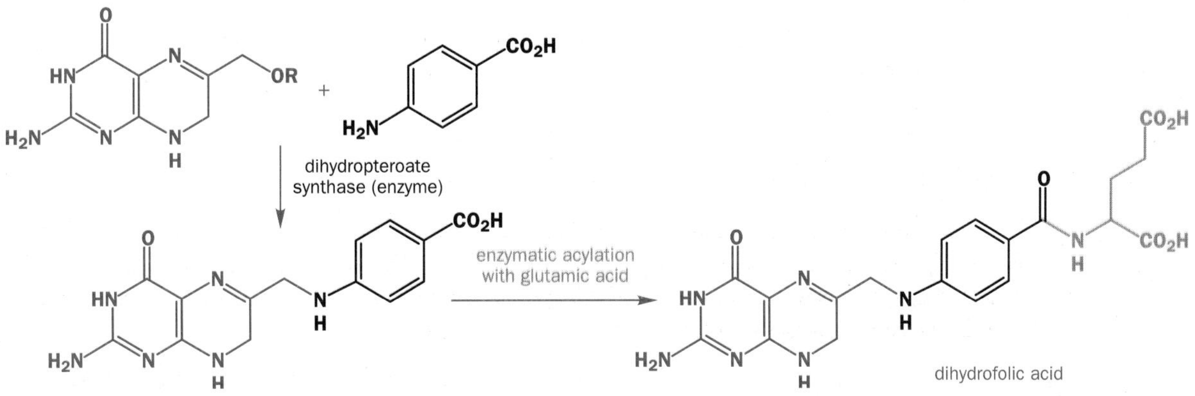

unstable isoindole

phthalimide

dihydropteroate
synthase (enzyme)

enzymatic acylation
with glutamic acid

dihydrofolic acid

Although folic acid is vital for human health, we don't have the enzymes to make it: it's a vitamin, which means we must take it in our diet or we die. Bacteria, on the other hand, do make folic acid. This is very useful, because it means that if we inhibit the enzymes of folic acid synthesis we can kill bacteria but we cannot possibly harm ourselves as we don't have those enzymes. The sulfa drugs, such as sulfamethoxypyridazine or sulfamethoxazole, imitate *p*-aminobenzoic acid and inhibit the enzyme dihydropteroate synthase. Each has a new heterocyclic system added to the sulfonamide part of the drug.

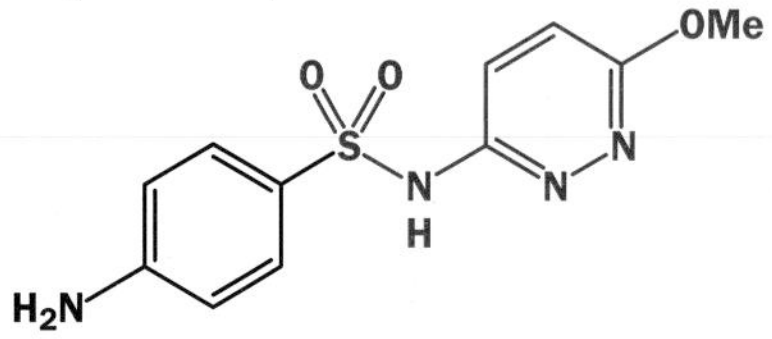

sulfamethoxypyridazine

sulfamethoxazole

The next step in folic acid synthesis is the reduction of dihydrofolate to tetrahydrofolate. This can be done by both humans and bacteria and, although it looks like a rather trivial reaction (see black portion of molecules), it can only be done by the very important enzyme **dihydrofolate reductase**.

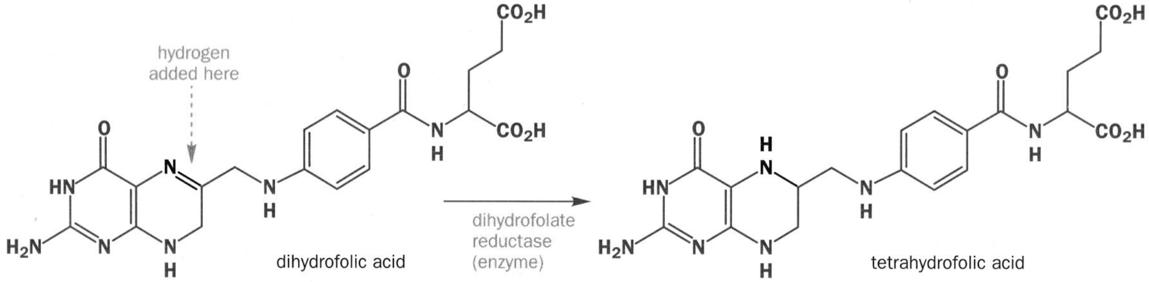

dihydrofolic acid

dihydrofolate reductase (enzyme)

tetrahydrofolic acid

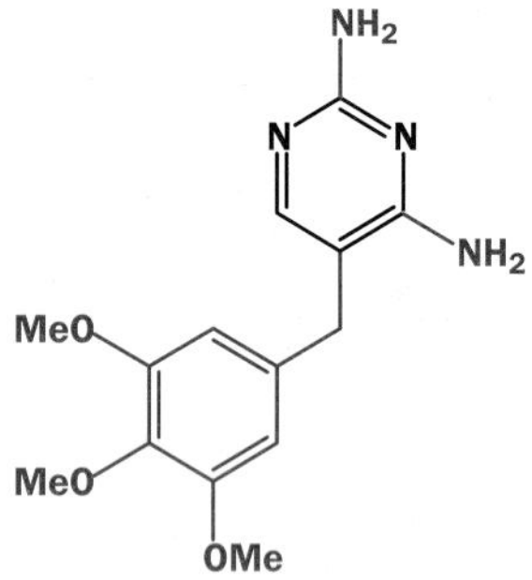

Though both bacteria and humans have this enzyme, the bacterial version is different enough for us to attack it with specific drugs. An example is trimethoprim—yet another heterocyclic compound with a pyrimidine core (black on diagram). These two types of drugs that attack the folic acid metabolism of bacteria are often used together.

We will see in the next chapter how to make these heterocyclic systems and, in Chapters 49–51, other examples of how important they are in living things.

trimethoprim

Which heterocyclic structures should you learn?

This is, of course, nearly a matter of personal choice. Every chemist really must know the names of the simplest heterocycles and we give those below along with a menu of suggestions.

First of all, those every chemist must know:

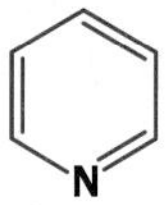

pyridine

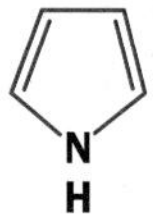

pyrrole

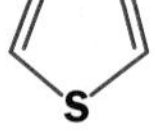

thiophene

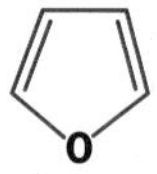

furan

Now the table gives a suggested list of five ring systems that have important roles in the chemistry of life and in human medicine—many drugs are based on these five structures.

1 Imidazole

the most important five-membered ring with two nitrogen atoms

imidazole

part of the amino acid histidine, occurs in proteins and is important in enzyme mechanisms

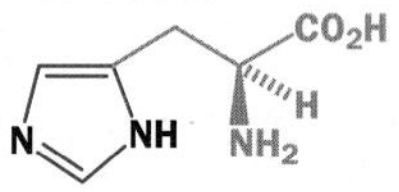

the amino acid histidine

a substituted imidazole is an essential part of the anti-ulcer drug cimetidine

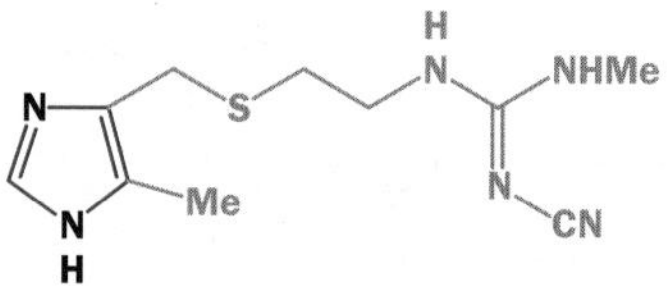

the anti-ulcer drug cimetidine (Tagamet) a selective histidine mimic

2 Pyrimidine

the most important six-membered ring with two nitrogen atoms

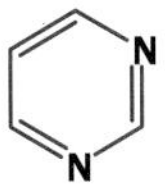

pyrimidine

three functionalized pyrimidines are part of DNA and RNA structure, e.g. uracil

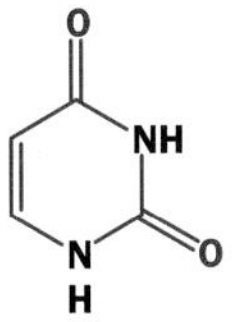

uracil

many antiviral drugs, particularly anti-HIV drugs, are modified pieces of DNA and contain pyrimidines

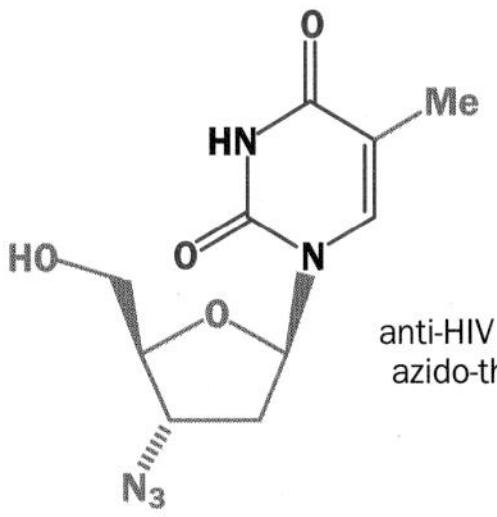

anti-HIV drug AZT azido-thymidine

3 Quinoline

one of two benzo-pyridines with many applications

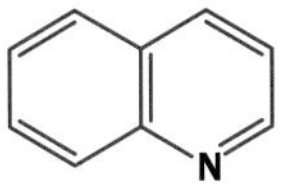

quinoline

occurs naturally in the important antimalarial drug quinine

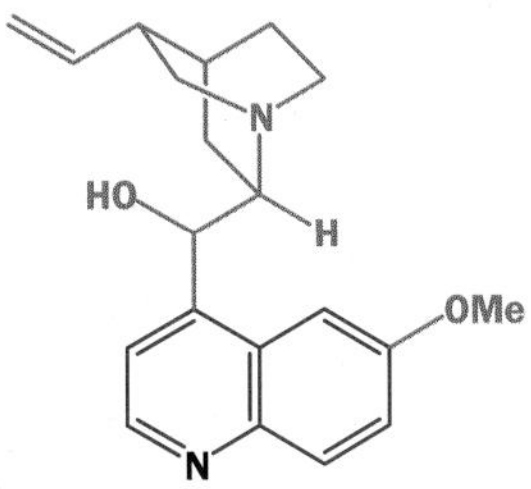

quinine

'cyanine' dyestuffs used as sensitizers for particular light wavelengths in colour photography

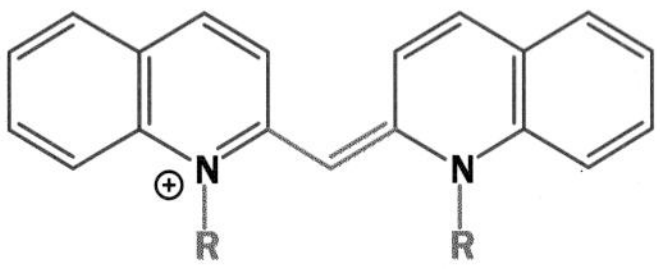

a "cyanine" dyestuff

4 Isoquinoline

the other benzo-pyridine with many applications

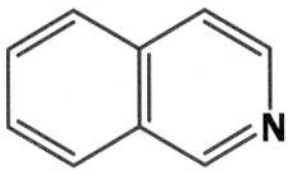

isoquinoline

occurs naturally in the benzyl isoquinoline alkaloids like papaverine

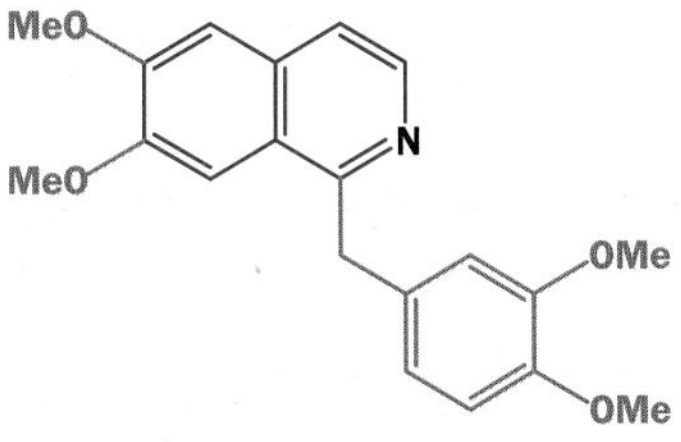

papaverine—a benzyl isoquinoline alkaloid

5 Indole

the more important benzo-pyrrole

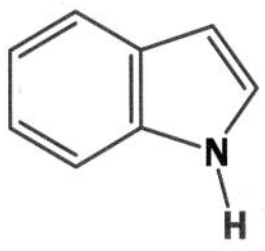

indole

occurs in proteins as tryptophan and in the brain as the neurotransmitter serotonin (5-hydroxy-tryptamine)

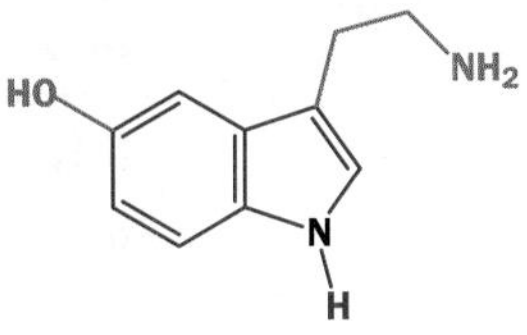

serotonin—a neurotransmitter

important modern drugs are based on serotonin including sumatriptan for migraine and ondansetron, an anti-emetic for cancer chemotherapy

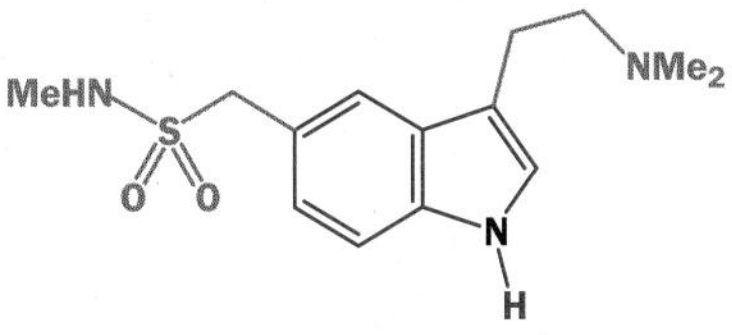

sumatriptan: for treatment of migraine

Problems

1. For each of the following reactions: (a) state what kind of substitution is suggested; (b) suggest what product might be formed if monosubstitution occurs.

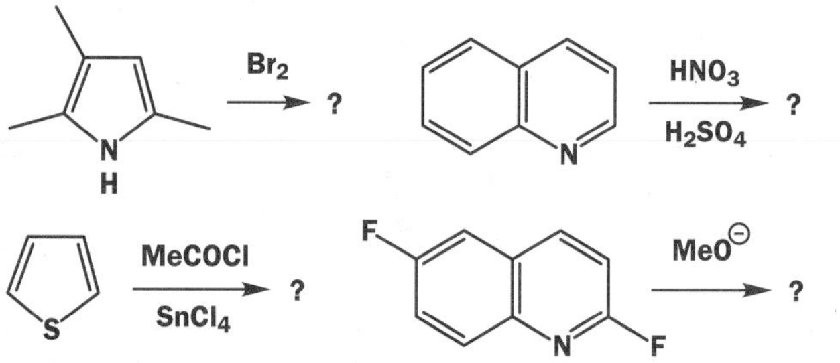

2. Give a mechanism for this side-chain extension of a pyridine.

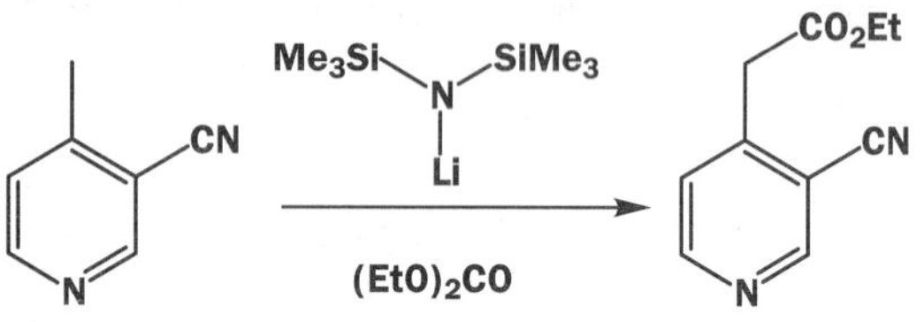

3. Give a mechanism for this reaction, commenting on the position on the furan ring that reacts.

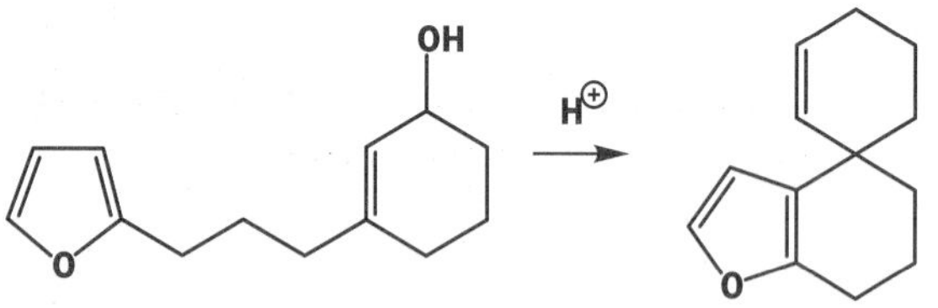

4. Suggest which product might be formed in each of these reactions and justify your choices.

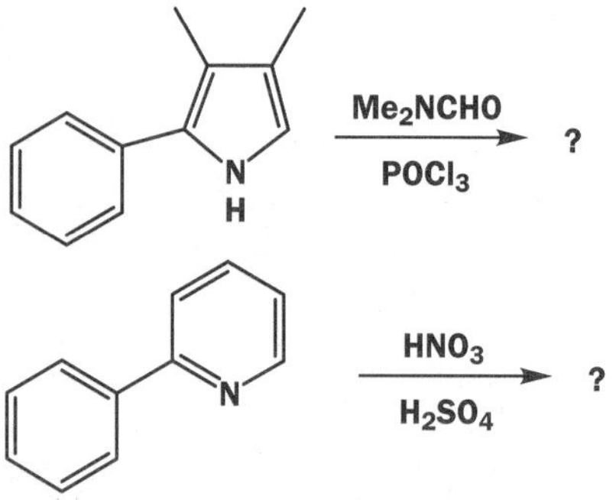

5. Comment on the mechanism and selectivity of this reaction of a pyrrole.

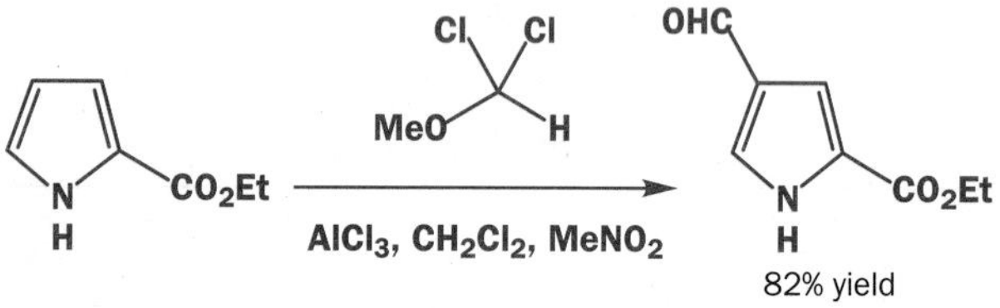

82% yield

6. Explain the formation of the product in this Friedel–Crafts alkylation of an indole.

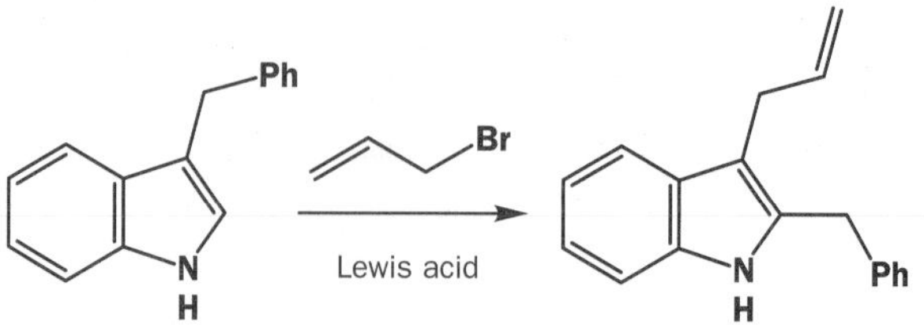

7. Explain the order of events and choice of bases in this sequence.

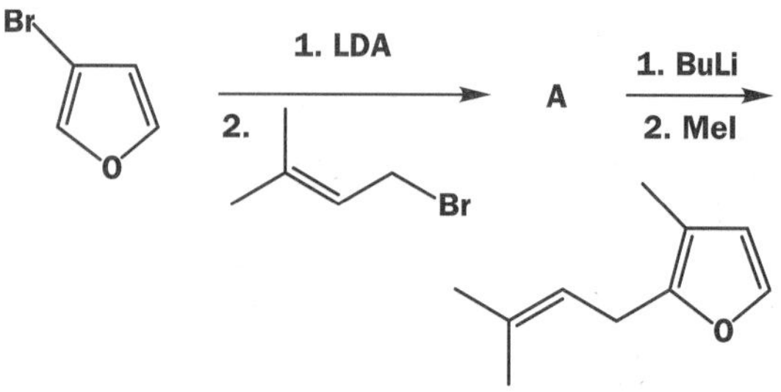

8. Explain the difference between these two pyridine reductions.

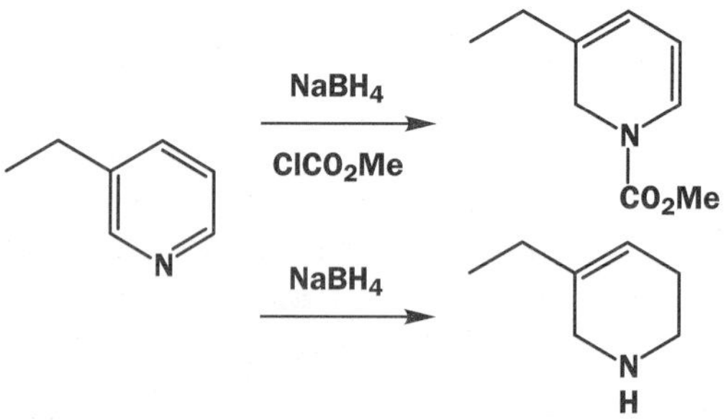

9. Why can this furan not be made by the direct route from available 2-benzylfuran?

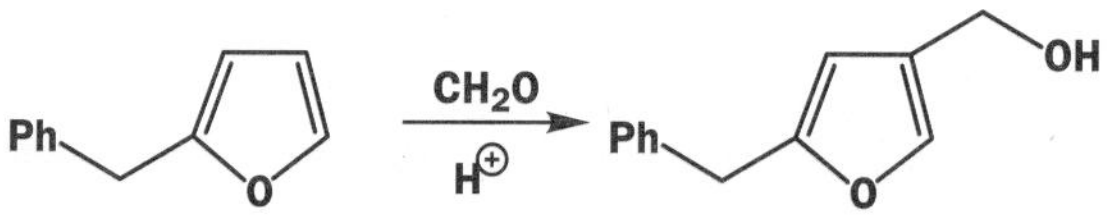

The same furan can be made by the route described below. Suggest mechanisms for the first and the last step. What is the other product of the last step?

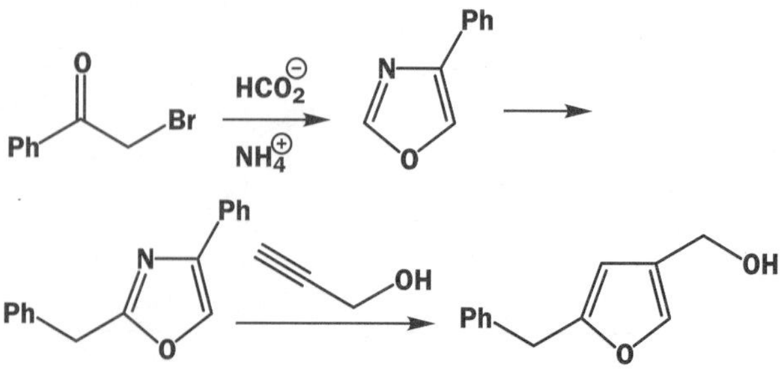

10. What aromatic system might be based on the skeleton given below? What sort of reactivity might it display?

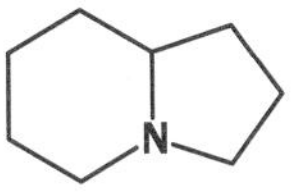

11. The reactions outlined in the chart below describe the early steps in a synthesis of an antiviral drug by the Parke–Davis company.

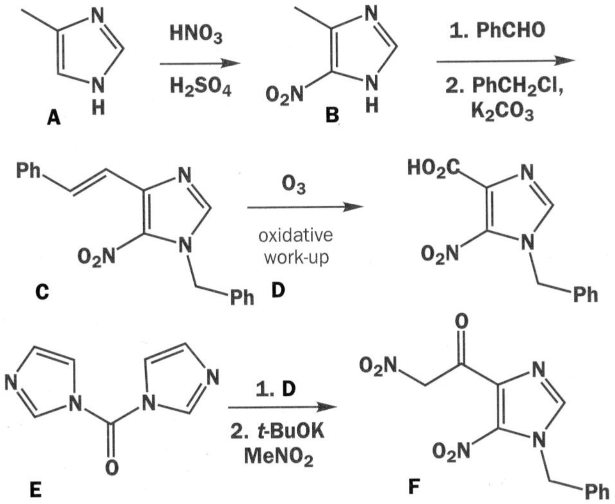

Consider how the reactivity of imidazoles is illustrated in these reactions, which involve not only the skeleton of the molecule but also the reagent E. You will need to draw mechanisms for the reactions and explain how they are influenced by the heterocycles.

12. The synthesis of DMAP, the useful acylation catalyst mentioned in Chapters 8 and 12, is carried out by initial attack of thionyl chloride (SOCl$_2$) on pyridine. Suggest how the reactions might proceed.

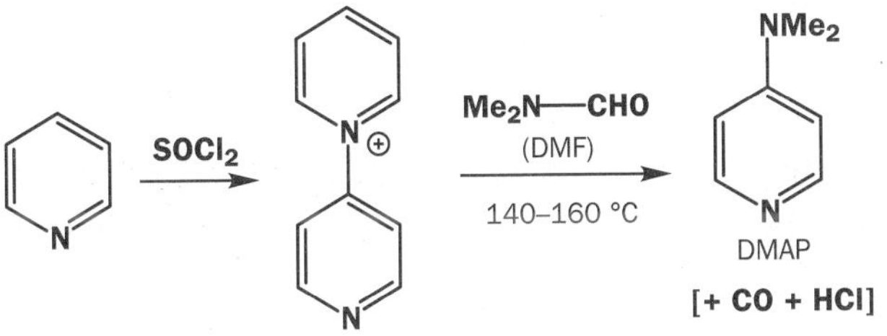

13. Suggest what the products of these nucleophilic substitutions might be.

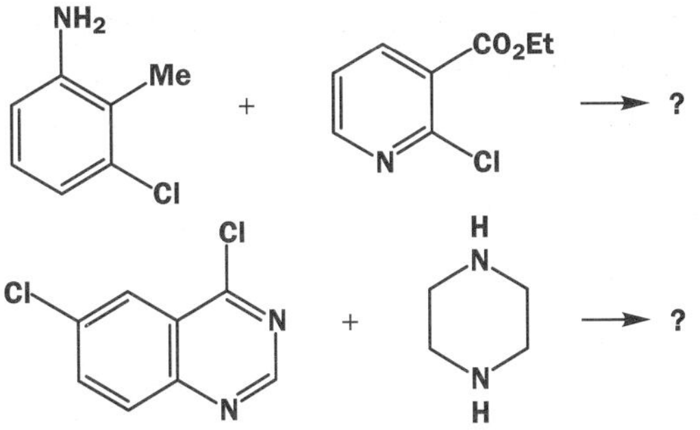

14. Suggest how 2-pyridone might be converted into the amine shown. This amine undergoes mononitration to give compound A with the NMR spectrum given. What is the structure of A? Why is this isomer formed?

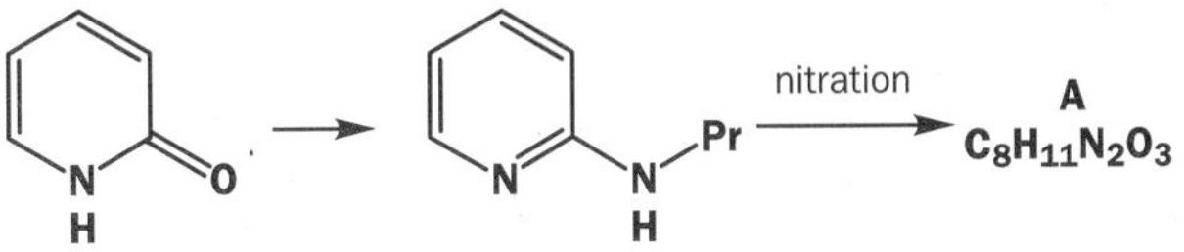

δ_H (p.p.m.) 1.0 (3H, t, J 7 Hz), 1.7 (2H, sextet, J 7 Hz), 3.3 (2H, t, J 7 Hz), 5.9 (1H, broad s), 6.4 (1H, d, J 8 Hz), 8.1 (1H, dd, J 8, 2 Hz), and 8.9 (1H, d, J 2 Hz).

Compound A was needed for conversion into the potential enzyme inhibitor below. How might this be achieved?

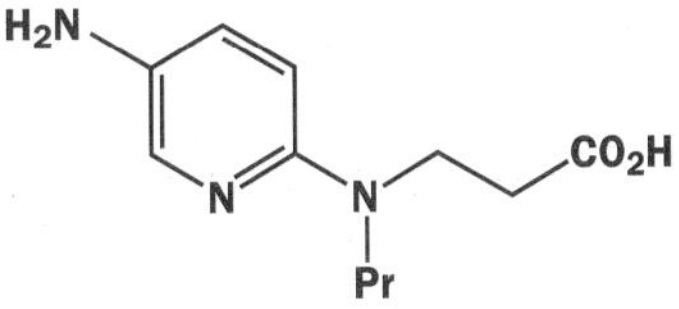

Aromatic heterocycles 2: synthesis

44

Connections

Building on:

- Aromaticity ch7
- Enols and enolates ch21
- The aldol reaction ch27
- Acylation of enolates ch28
- Michael additions of enolates ch29
- Retrosynthetic analysis ch30
- Cycloadditions ch35
- Reactions of heterocycles ch43

Arriving at:

- Thermodynamics is on our side
- Disconnecting the carbon–heteroatom bonds first
- How to make pyrroles, thiophenes, and furans from 1,4-dicarbonyl compounds
- How to make pyridines and pyridones
- How to make pyridazines and pyrazoles
- How to make pyrimidines from 1,3-dicarbonyl compounds and amidines
- How to make thiazoles
- How to make isoxazoles and tetrazoles by 1,3-dipolar cycloadditions
- The Fischer indole synthesis
- Making drugs: Viagra, sumatriptan, ondansetron, indomethacin
- How to make quinolines and isoquinolines

Looking forward to:

- Biological chemistry ch49–ch51

In this chapter you will revisit the heterocyclic systems you have just met and find out how to make them. You'll also meet some new heterocyclic systems and find out how to make those. With so many heterocycles to consider, you'd be forgiven for feeling rather daunted by this prospect, but do not be alarmed. Making heterocycles is easy—that's precisely why there are so many of them. Just reflect . . .

- Making C–O, C–N, and C–S bonds is easy
- Intramolecular reactions are preferred to bimolecular reactions
- Forming five- and six-membered rings is easy
- We are talking about aromatic, that is, very stable molecules

If we are to use those bullet points to our advantage we must think strategy before we start. When we were making benzene compounds we usually started with a preformed simple benzene derivative—toluene, phenol, aniline—and added side chains by electrophilic substitution. In this chapter our strategy will usually be to build the heterocyclic ring with most of its substituents already in place and add just a few others, perhaps by electrophilic substitution, but mostly by nucleophilic substitution.

We will usually make the rings by cyclization reactions with the heteroatom (O, N, S) as a nucleophile and a suitably functionalized carbon atom as the electrophile. This electrophile will almost always be a carbonyl compound of some sort and this chapter will help you revise your carbonyl chemistry from Chapters 6, 12, 14, 21, 23, and 26–29 as well as the approach to synthesis described in Chapter 30.

Thermodynamics is on our side

Some of the syntheses we will meet will be quite surprisingly simple! It sometimes seems that we can just mix a few things together with about the right number of atoms and let thermodynamics do the rest. A commercial synthesis of pyridines combines acetaldehyde and

Allopurinol was discussed in Chapter 43, p. 1176.

ammonia under pressure to give a simple pyridine.

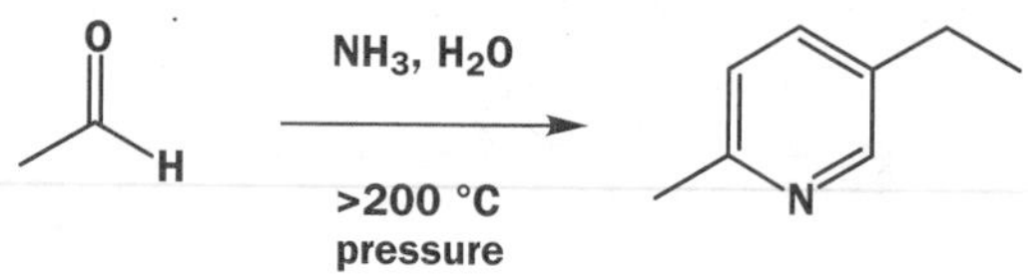

The yield is only about 50%, but what does that matter in such a simple process? By counting atoms we can guess that four molecules of aldehyde and one of ammonia react, but exactly how is a triumph of thermodynamics over mechanism. Much more complex molecules can sometimes be made very easily too. Take allopurinol, for example. One synthesis of this gout remedy goes like this.

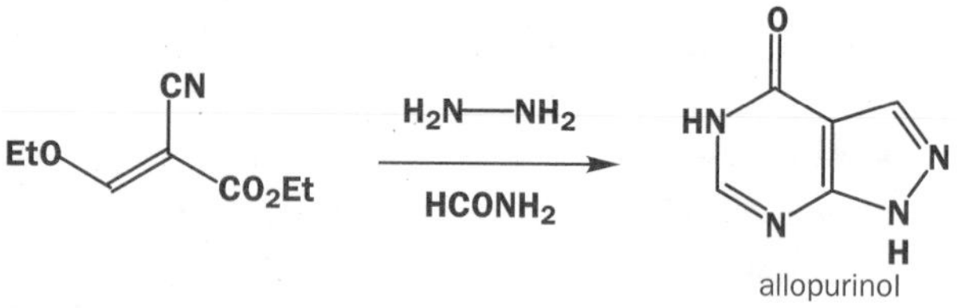

It is not too difficult to work out where the atoms go—the hydrazine obviously gives rise to the pair of adjacent nitrogen atoms in the pyrazole ring and the ester group must be the origin of the carbonyl group (see colours and numbers on the right)—but would you have planned this synthesis?

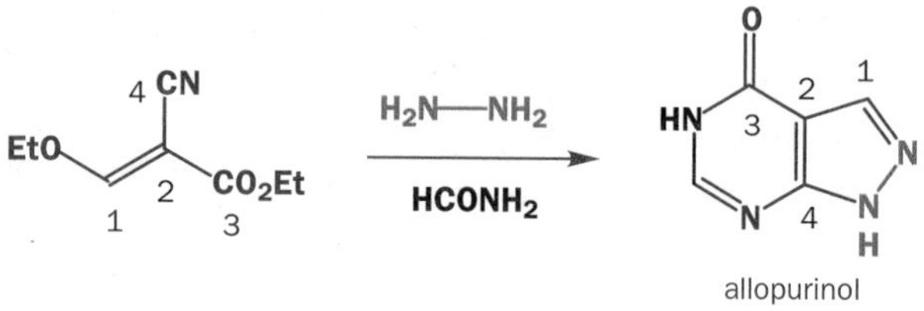

We will see that this sort of 'witch's brew' approach to synthesis is restricted to a few basic ring systems and that, in general, careful planning is just as important here as elsewhere. The difference here is that heterocyclic synthesis is very forgiving—it often 'goes right' instead of going wrong. We'll now look seriously at planning the synthesis of aromatic heterocycles.

Disconnect the carbon–heteroatom bonds first

The simplest synthesis for a heterocycle emerges when we remove the heteroatom and see what electrophile we need. We shall use pyrroles as examples. The nitrogen forms an enamine on each side of the ring and we know that enamines are made from carbonyl compounds and amines.

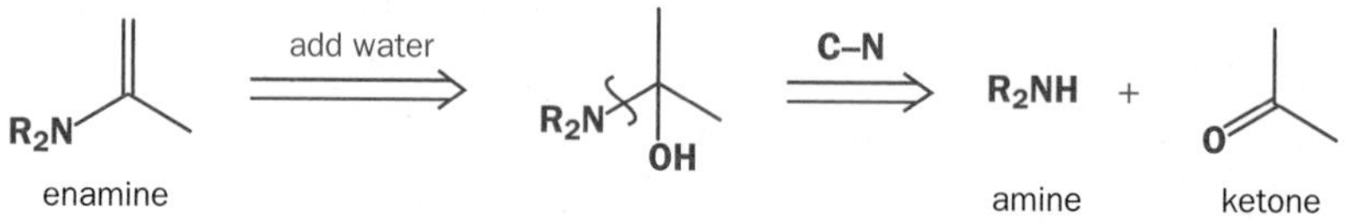

If we do the same disconnection with a pyrrole, omitting the intermediate stage, we can repeat the C–N disconnection on the other side too:

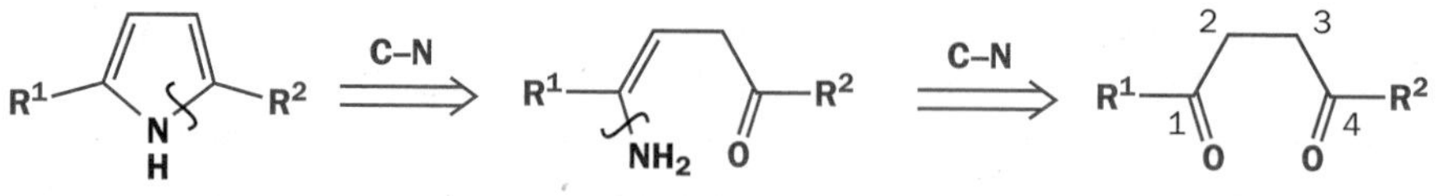

What we need is an amine—ammonia in this case—and a diketone. If the two carbonyl groups have a 1,4 relationship we will get a pyrrole out of this reaction. So hexane-2,5-dione reacts with ammonia to give a high yield of 2,5-dimethyl pyrrole.

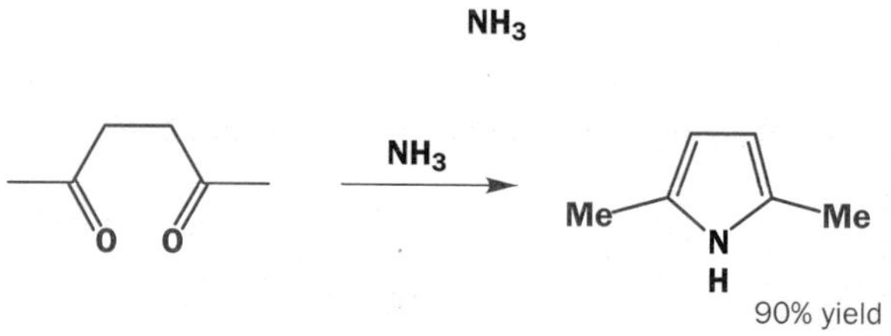

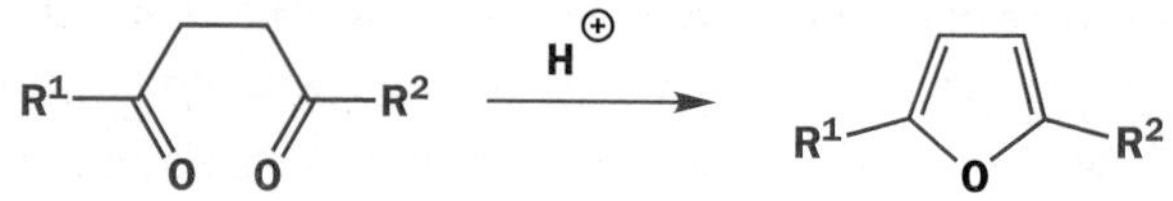

Making furans is even easier because the heteroatom (oxygen) is already there. All we have to do is to dehydrate the 1,4-diketone instead of making enamines from it. Heating with acid is enough.

Avoiding the aldol product

1,4-Diketones also self-condense rather easily in an intramolecular aldol reaction to give a cyclopentenone with an all-carbon five-membered ring. This too is a useful reaction but we need to know how to control it. The usual rule is:

- Base gives the cyclopentenone
- Acid gives the furan

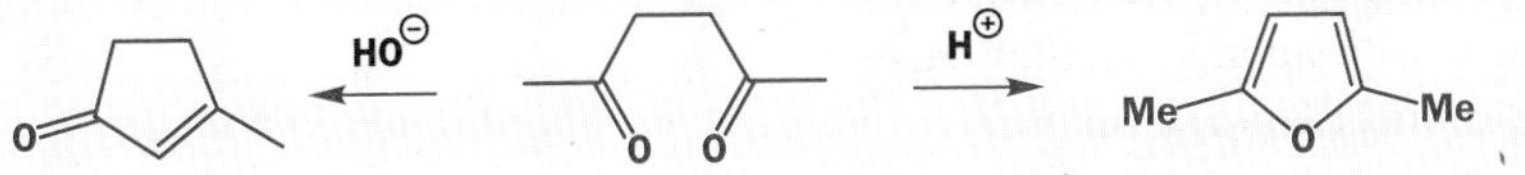

cyclopentenone   1,4-diketone   furan

For thiophenes we could in theory use H_2S or some other sulfur nucleophile but, in practice, an electrophilic reagent is usually used to convert the two C=O bonds to C=S bonds. Thioketones are much less stable than ketones and cyclization is swift. Reagents such as P_2S_5 or Lawesson's reagent are the usual choice here.

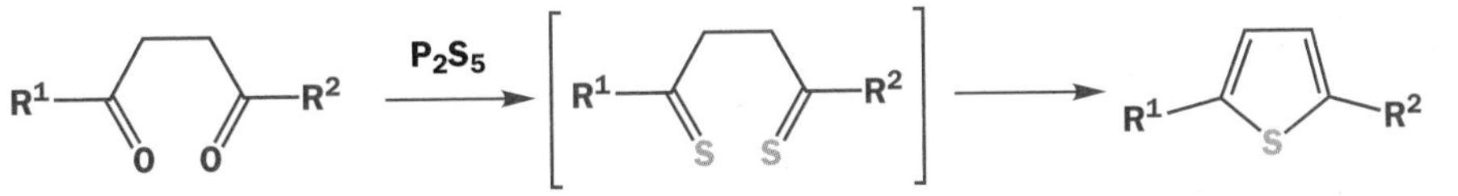

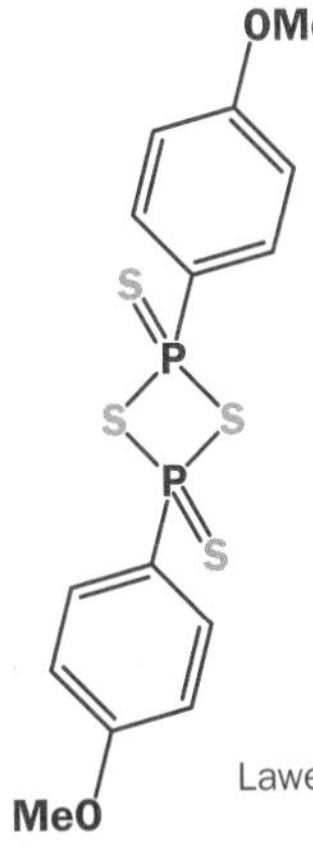

Lawesson's reagent

> Sulfur chemistry is discussed in Chapter 46; all we will say here about the mechanisms of these reactions is that phosphorus is commonly used to remove oxygen and replace it by another element: remember the Mitsunobu reaction?

● Making five-membered heterocycles

Cyclization of 1,4-dicarbonyl compounds with nitrogen, sulfur, or oxygen nucleophiles gives the five-membered aromatic heterocycles pyrrole, thiophene, and furan.

It seems a logical extension to use a 1,5-diketone to make substituted pyridines but there is a slight problem here as we will introduce only two of the required three double bonds when the two enamines are formed.

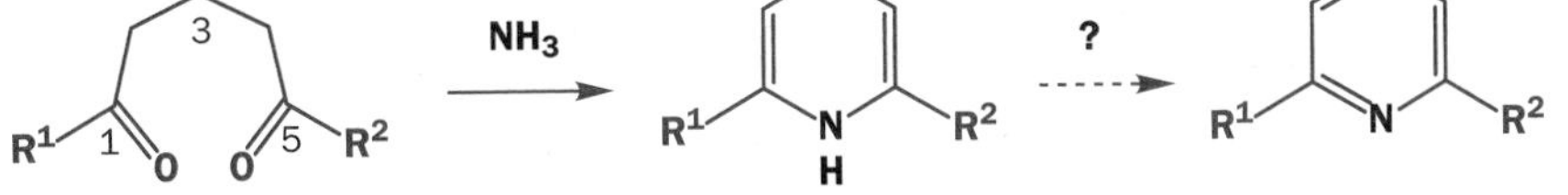

To get the pyridine by enamine formation we should need a double bond somewhere in the chain between the two carbonyl groups. But here another difficulty arises—it will have to be a *cis* (Z) double bond or cyclization would be impossible.

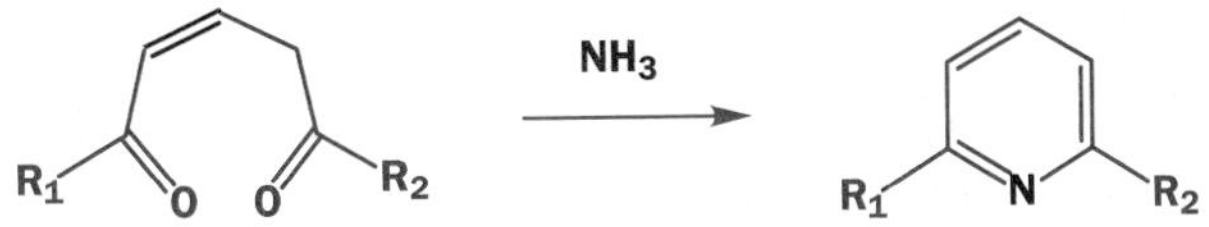

On the whole it is easier to use the saturated 1,5-diketone and oxidize the product to the pyridine. As we are going from a nonaromatic to an aromatic compound, oxidation is easy and we can replace the question mark above with almost any simple oxidizing agent, as we shall soon see.

● Making six-membered heterocycles

Cyclization of 1,5-dicarbonyl compounds with nitrogen nucleophiles leads to the six-membered aromatic heterocycle pyridine.

Heterocycles with two nitrogen atoms come from the same strategy

Reacting a 1,4-diketone with hydrazine (NH_2NH_2) makes a double enamine again and this is only an oxidation step away from a pyridazine. This is again a good synthesis.

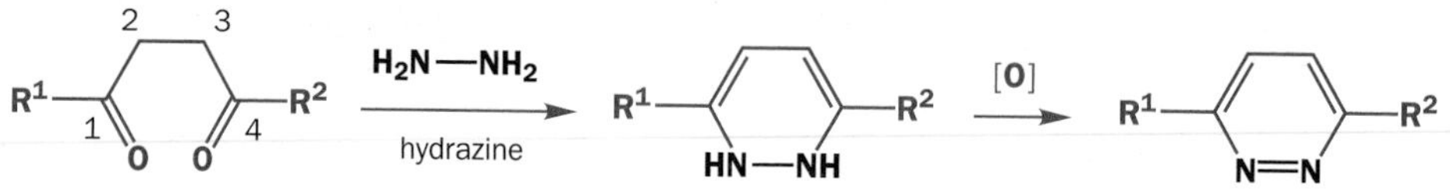

a pyridazine

If we use a 1,3-diketone instead we will get a five-membered heterocycle and the imine and enamine formed are enough to give aromaticity without any need for oxidation. The product is a pyrazole.

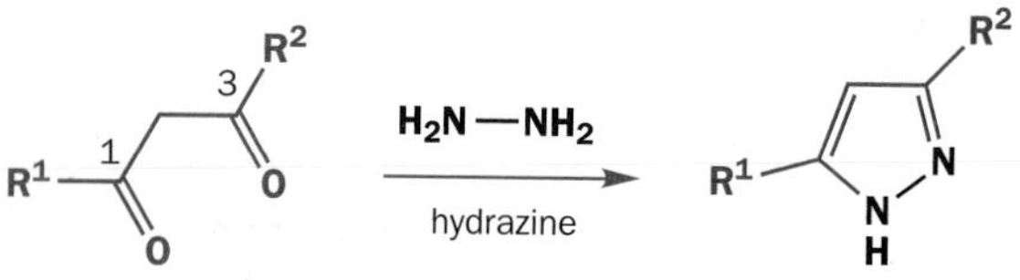

a pyrazole

The two heteroatoms do not, of course, need to be joined together for this strategy to work. If an amidine is combined with the same 1,3-diketone we get a six-membered heterocycle. As the nucleophile contains one double bond already, an aromatic pyrimidine is formed directly.

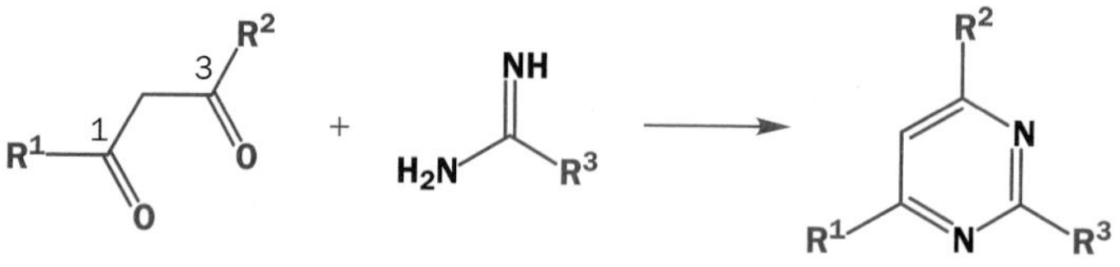

an amidine   a pyrimidine

Since diketones and other dicarbonyl compounds are easily made by enolate chemistry (Chapters 26–30) this strategy has been very popular and we will look at some detailed examples before moving on to more specialized reactions for the different classes of aromatic heterocycles.

Pyrroles, thiophenes, and furans from 1,4-dicarbonyl compounds

We need to make the point that pyrrole synthesis can be done with primary amines as well as with ammonia and a good example is the pyrrole needed for clopirac, a drug we discussed in Chapter 43. The synthesis is very easy.

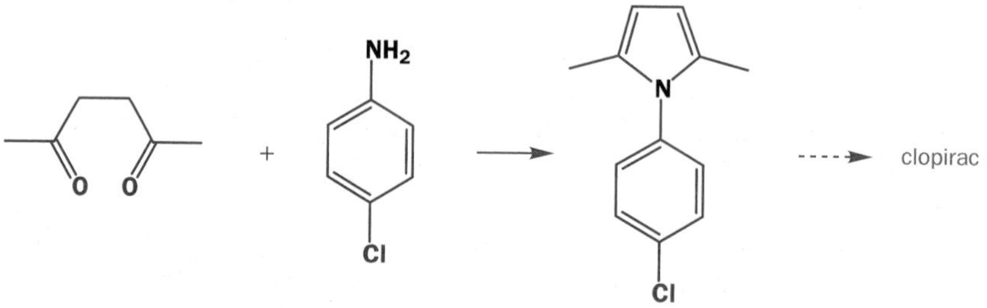

clopirac

For an example of furan synthesis we choose menthofuran, which contributes to the flavour of mint. It has a second ring, but that is no problem if we simply disconnect the enol ethers as we have been doing so far.

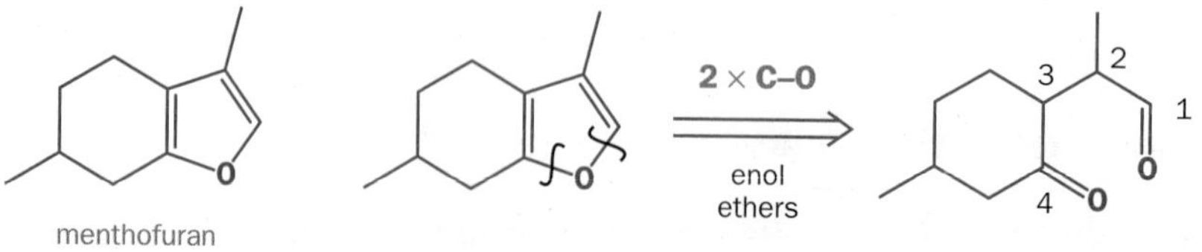

menthofuran

The starting material is again a 1,4-dicarbonyl compound but as there was no substituent at C1 of the furan, that atom is an aldehyde rather than a ketone. This might lead to problems in the synthesis so a few changes (using the notation you met in Chapter 30) are made to the intermediate before further disconnection.

▶

α-Halo aldehydes are unstable and should be avoided.

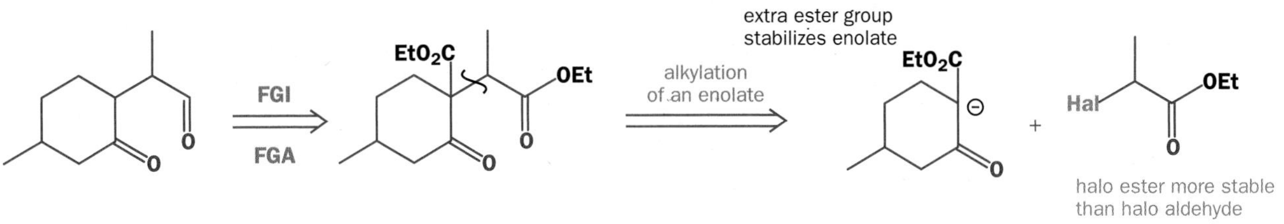

Notice in particular that we have 'oxidized' the aldehyde to an ester to make it more stable—in the synthesis reduction will be needed. Here is the alkylation step of the synthesis, which does indeed go very well with the α-iodo-ester.

menthofuran: synthesis

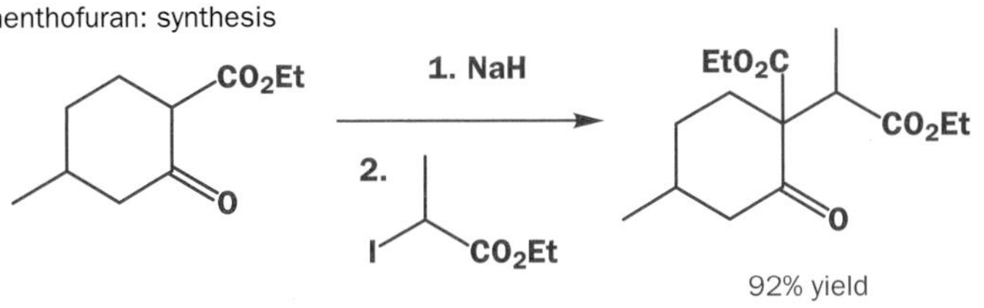

Cyclization with acid now causes a lot to happen. The 1,4-dicarbonyl compound cyclizes to a lactone, not to a furan, and the redundant ester group is lost by hydrolysis and decarboxylation. Notice that the double bond moves into conjugation with the lactone carbonyl group. Finally, the reduction gives the furan. No special precautions are necessary—as soon as the ester is partly reduced, it loses water to give the furan whose aromaticity prevents further reduction even with LiAlH$_4$.

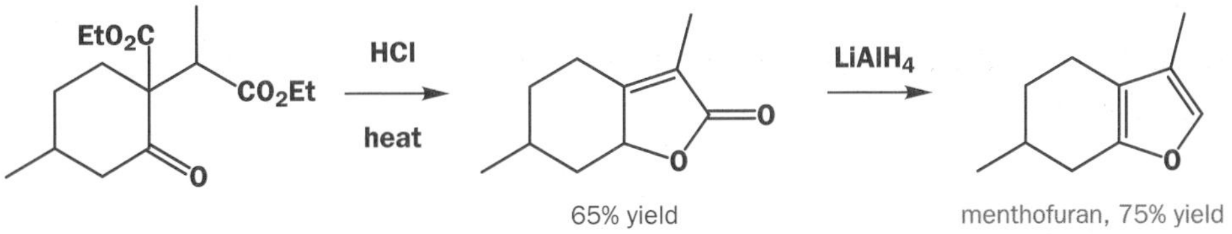

● **A reminder**

Cyclization of 1,4-dicarbonyl compounds with nitrogen, sulfur, or oxygen nucleophiles gives the five-membered aromatic heterocycles pyrrole, thiophene, and furan.

Now we need to take these ideas further and discuss an important pyrrole synthesis that follows this strategy but includes a cunning twist. It all starts with the porphyrin found in blood. In Chapter 43 we gave the structure of that very important compound and showed that it contains four pyrrole rings joined in a macrocycle. We are going to look at one of those pyrroles.

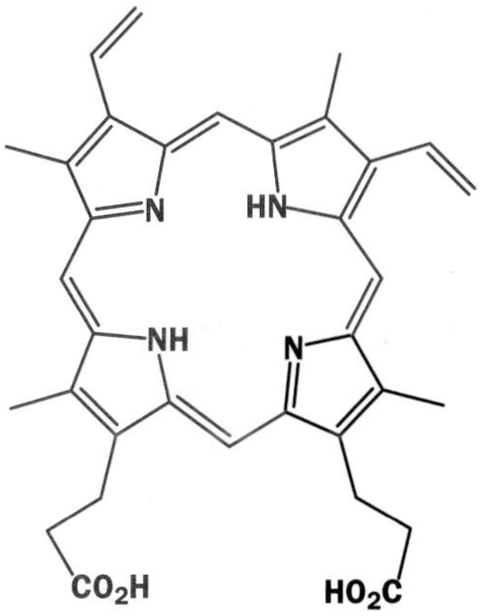

Porphyrins can be made by joining together the various pyrroles in the right order and what is needed for this one (and also, in fact, for another—the one in the north-east corner of the porphyrin) is a pyrrole with the correct substituents in positions 3 and 4, a methyl group in position 5, and a hydrogen atom at position 2. Position 2 must be free. Here is the molecule drawn somewhat more conveniently together with the disconnection we have been using so far.

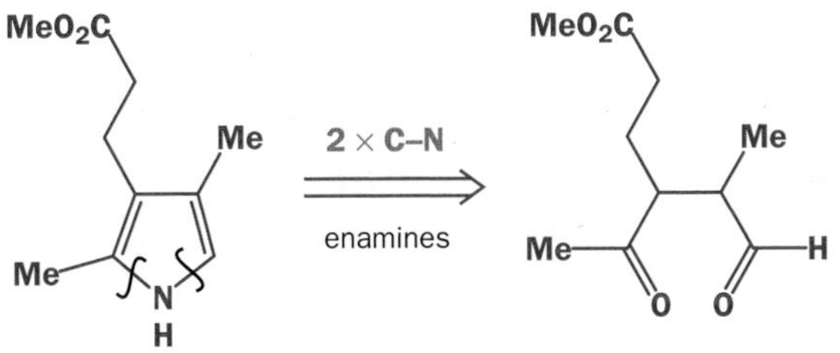

the porphyrin in haemoglobin

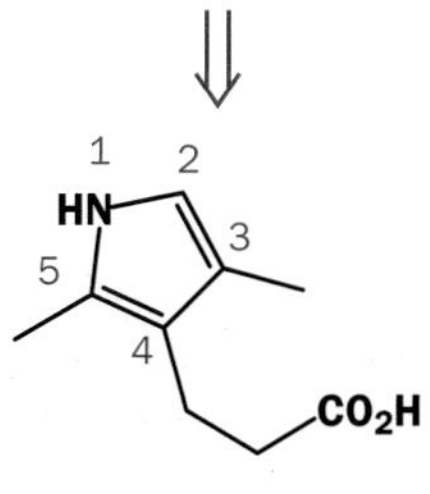

component pyrrole

See p. 1157 for a discussion of pyrrole reactivity.

No doubt such a synthesis could be carried out but it is worth looking for alternatives for a number of reasons. We would prefer not to make a pyrrole with a free position at C2 as that would be very reactive and we know from Chapter 43 that we can block such a position with a *t*-butyl ester group. This gives us a very difficult starting material with four different carbonyl groups.

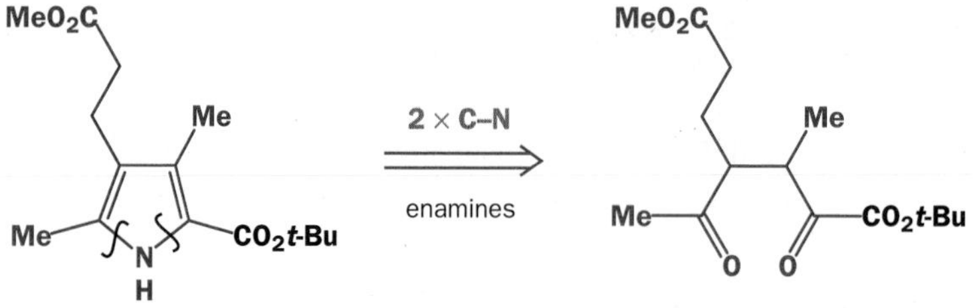

We have made a problem for ourselves by having two carbonyl groups next to each other. Could we escape from that by replacing one of them with an amine? We should then have an ester of an α amino acid, an attractive starting material, and this corresponds to disconnecting just one of the C–N bonds.

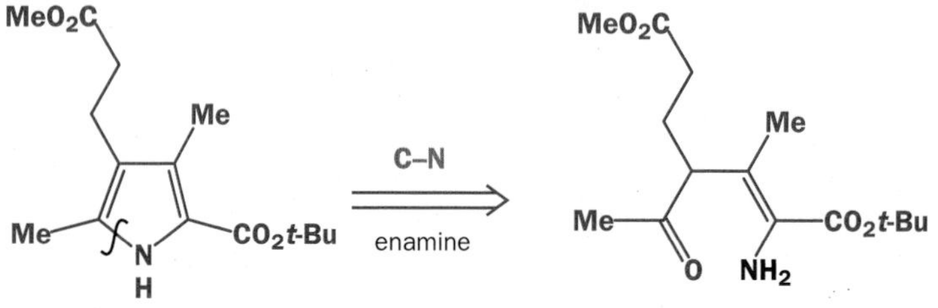

At first we seem to have made no progress but just see what happens when we move the double bond round the ring into conjugation with the ketone. After all, it doesn't matter where the double bond starts out—we will always get the aromatic product.

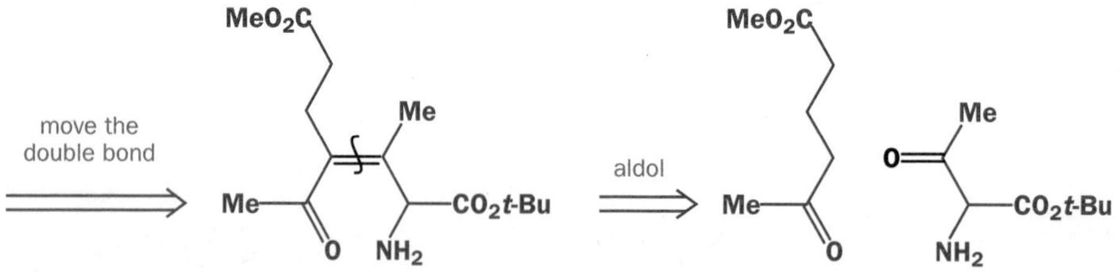

Each of our two much simpler starting materials needs to be made. The keto-ester is a 1,5-dicarbonyl compound so it can be made by a conjugate addition of an enolate, a process greatly assisted by the addition of a second ester group (Chapter 29).

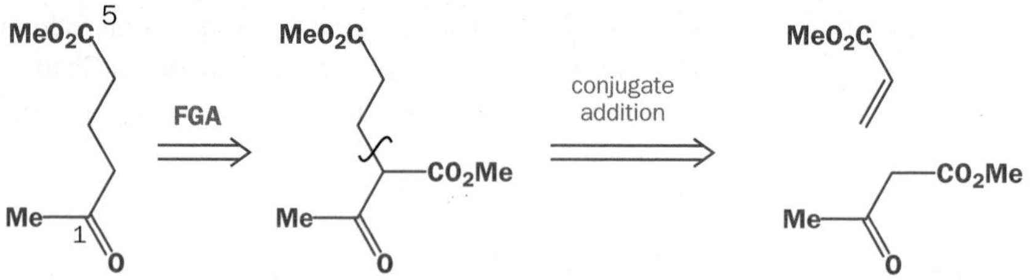

The other compound is an amino-keto-ester and will certainly react with itself if we try to prepare it as a pure compound. The answer is to release it into the reaction mixture and this can be done by nitrosation and reduction (Chapter 21) of another stable enolate.

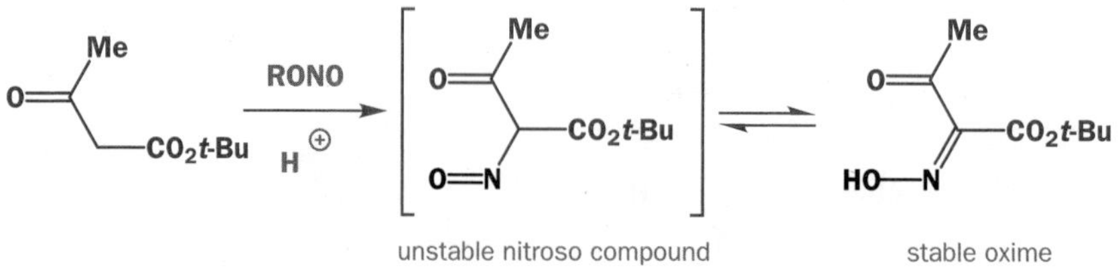

unstable nitroso compound   stable oxime

Zinc in acetic acid (Chapter 24) reduces the oxime to the amine and we can start the synthesis by doing the conjugate addition and then reducing the oxime in the presence of the keto-diester.

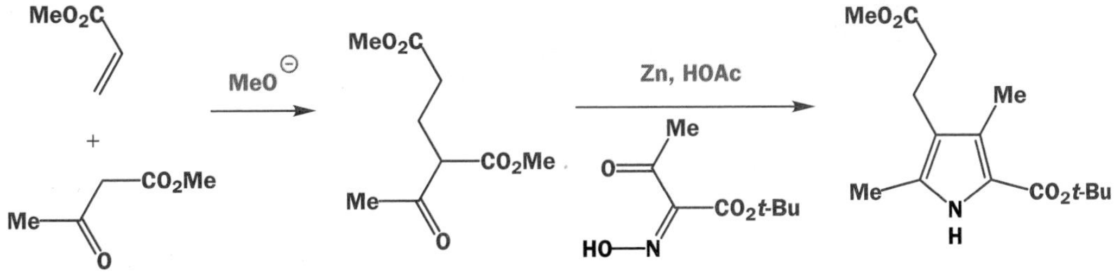

This reaction forms the required pyrrole in one step! First, the oxime is reduced to an amine; then the amino group forms an imine with the most reactive carbonyl group (the ketone) in the keto-diester. Finally, the very easily formed enamine cyclizes on to the other ketone.

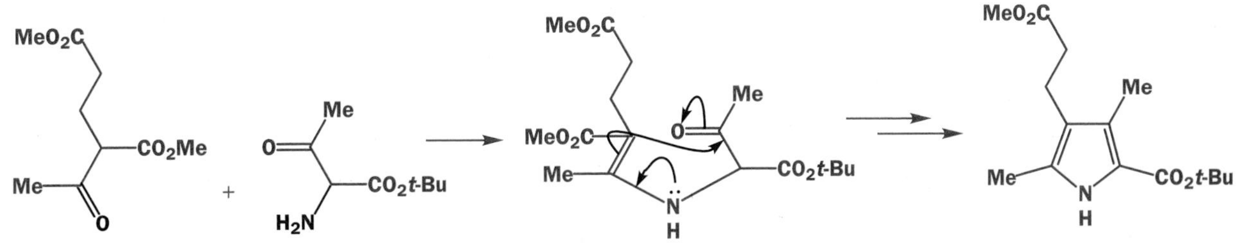

This pyrrole synthesis is important enough to be given the name of its inventor—it is the **Knorr pyrrole synthesis**. Knorr himself made a rather simpler pyrrole in a remarkably efficient reaction. See if you can work out what is happening here.

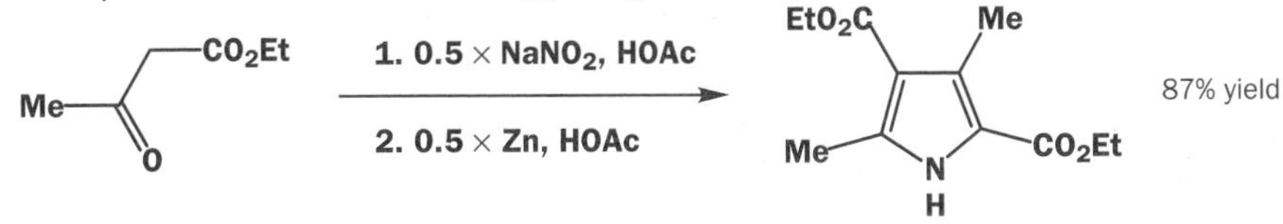

87% yield

How to make pyridines: the Hantzsch pyridine synthesis

The idea of coupling two keto-esters together with a nitrogen atom also works for pyridines except that an extra carbon atom is needed. This is provided as an aldehyde and another important difference is that the nitrogen atom is added as a nucleophile rather than an electrophile. These are features of the **Hantzsch pyridine synthesis**. This is a four-component reaction that goes like this.

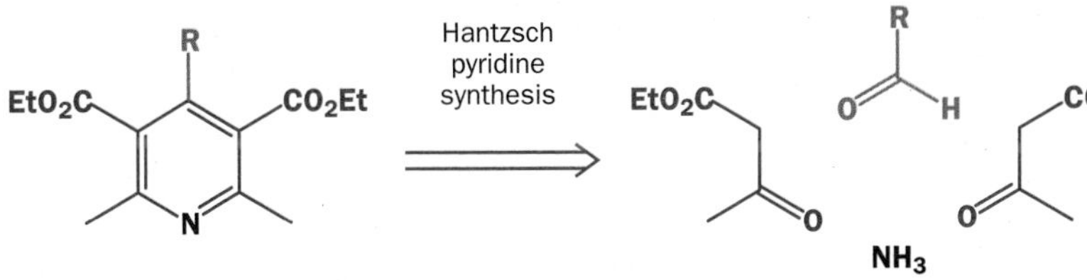

You are hardly likely to understand the rationale behind this reaction from that diagram so let's explore the details. The product of the reaction is actually the dihydropyridine, which has to be oxidized to the pyridine by a reagent such as HNO_3, Ce(IV), or a quinone.

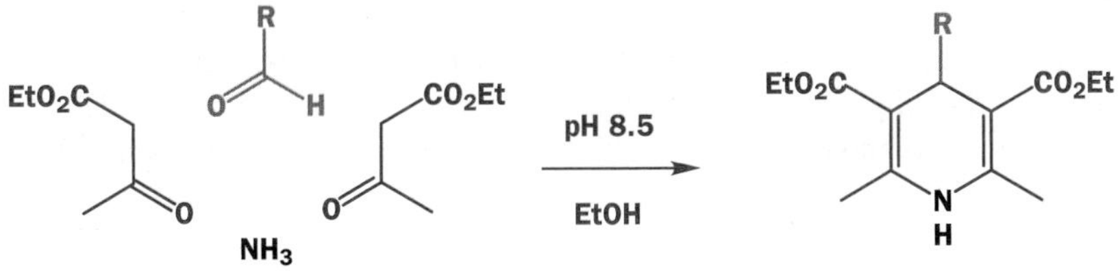

the dihydropyridine

Standard heterocyclic syntheses tend to have a name associated with them and it is simply not worth while learning these names. Few chemists use any but the most famous of them: we will mention the Knorr pyrrole synthesis, the Hantzsch pyridine synthesis, and the Fischer and Reissert indole syntheses. We did not mention that the synthesis of furans from 1,4-dicarbonyl compounds is known as the Feist–Benary synthesis, and there are many more like this. If you are really interested in these other names we suggest you consult a specialist book on heterocyclic chemistry.

Arthur Hantzsch, 1857–1935, the 'fiery stereochemist' of Leipzig, is most famous for the work he did with Werner at the ETH in Zurich where in 1890 he suggested that oximes could exist in *cis* and *trans* forms.

The reaction is very simply carried out by mixing the components in the right proportions in ethanol. The presence of water does not spoil the reaction and the ammonia, or some added amine, ensures the slightly alkaline pH necessary. Any aldehyde can be used, even formaldehyde, and yields of the crystalline dihydropyridine are usually very good.

This reaction is an impressive piece of molecular recognition by small molecules and writing a detailed mechanism is a bold venture. We can see that certain events have to happen. The ammonia has to attack the ketone groups, but it would prefer to attack the more electrophilic aldehyde so this is probably not the first step. The enol or enolate of the keto-ester has to attack the aldehyde (twice!) so let us start there.

▶ **The mechanism of the Hantzsch pyridine synthesis**

Several of these steps could be done in different orders but the essentials are:

• aldol reaction between the aldehyde and the keto-ester
• Michael (conjugate) addition to the enone
• addition of ammonia to one ketone
• cyclization of the imine or enamine on to the other ketone

This adduct is in equilibrium with the stable enolate from the keto-ester and elimination now gives an unsaturated carbonyl compound. Such chemistry is associated with the aldol reactions we discussed in Chapter 27. The new enone has two carbonyl groups at one end of the double bond and is therefore a very good Michael acceptor (Chapter 29). A second molecule of enolate does a conjugate addition to complete the carbon skeleton of the molecule. Now the ammonia attacks either of the ketones and cyclizes on to the other. As ketones are more electrophilic than esters it is to be expected that ammonia will prefer to react there.

the dihydropyridine

The necessary oxidation is easy both because the product is aromatic and because the nitrogen atom can help to expel the hydrogen atom and its pair of electrons from the 4-position. If we use a quinone as oxidizing agent, both compounds become aromatic in the same step. We will show in Chapter 50 that Nature uses related dihydropyridines as reducing agents in living things.

aromatic benzene ring formed

the dihydropyridine

aromatic pyridine ring formed

The Hantzsch pyridine synthesis is an old discovery (1882) which sprang into prominence in the 1980s with the discovery that the dihydropyridine intermediates prepared from aromatic aldehydes are calcium channel blocking agents and therefore valuable drugs for heart disease with useful effects on angina and hypertension.

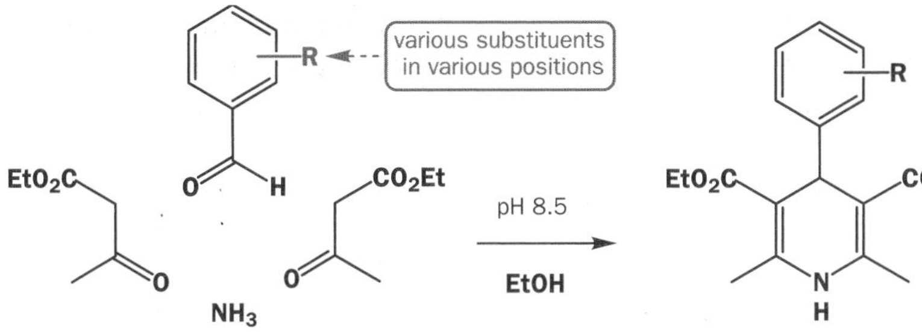

calcium channel blocker
drug for heart disease

These drugs inhibit Ca^{2+} ion transport across cell membranes and relax muscle tissues selectively without affecting the working of the heart. Hence high blood pressure can be reduced. Pfizer's amlodipine (Istin™ or Norvasc™) is a very important drug—it had sales of 1.6 billion dollars in 1996.

So far, so good. But it also became clear that the best drugs were unsymmetrical—some in a trivial way such as felodipine but some more seriously such as Pfizer's amlodipine. At first sight it looks as though the very simple and convenient Hantzsch synthesis cannot be used for these compounds.

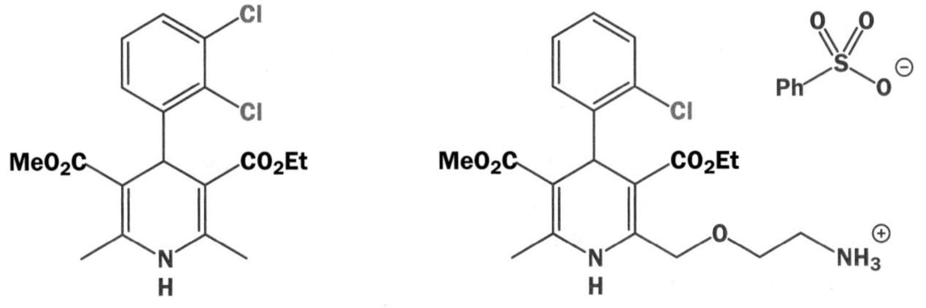

felodipine

amlodipine

Clearly, a modification is needed in which half of the molecule is assembled first. The solution lies in early work by Robinson who made the very first enamines from keto-esters and amines. One half of the molecule is made from an enamine and the other half from a separately synthesized enone. We can use felodipine as a simple example.

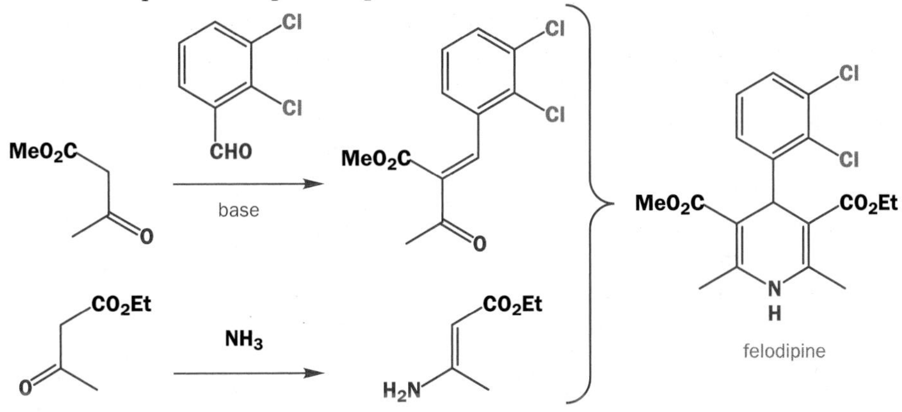

felodipine

Other syntheses of pyridines

The Hantzsch synthesis produces a reduced pyridine but there are many syntheses that go directly to pyridines. One of the simplest is to use hydroxylamine (NH$_2$OH) instead of ammonia as the nucleophile. Reaction with a 1,5-diketone gives a dihydropyridine but then water is lost and no oxidation is needed.

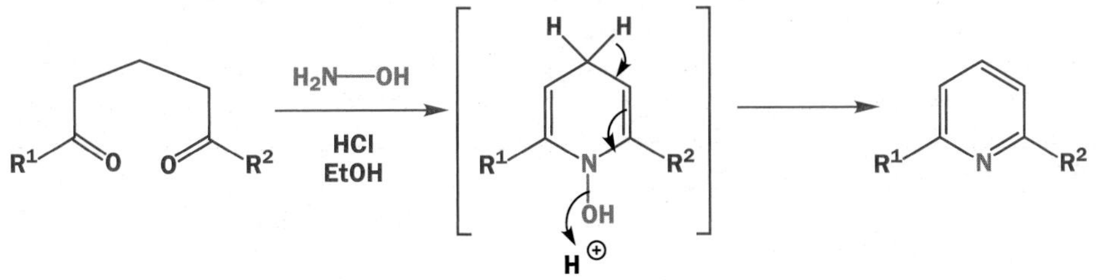

The example below shows how these 1,5-diketones may be quickly made by the Mannich (Chapter 27) and Michael (Chapter 29) reactions. Our pyridine has a phenyl substituent and a fused saturated ring. First we must disconnect to the 1,5-diketone.

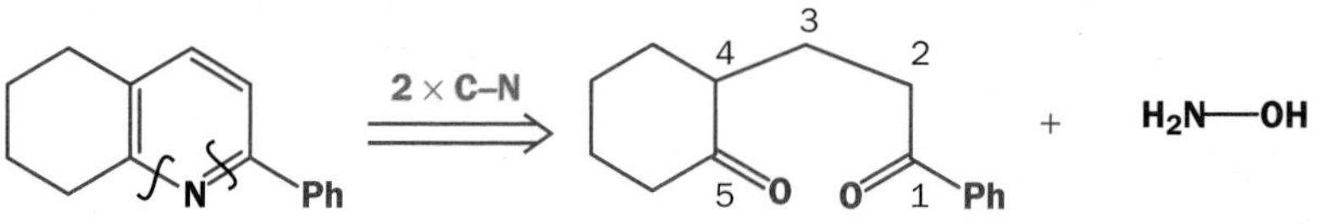

Further disconnection reveals a ketone and an enone. There is a choice here and both alternatives would work well.

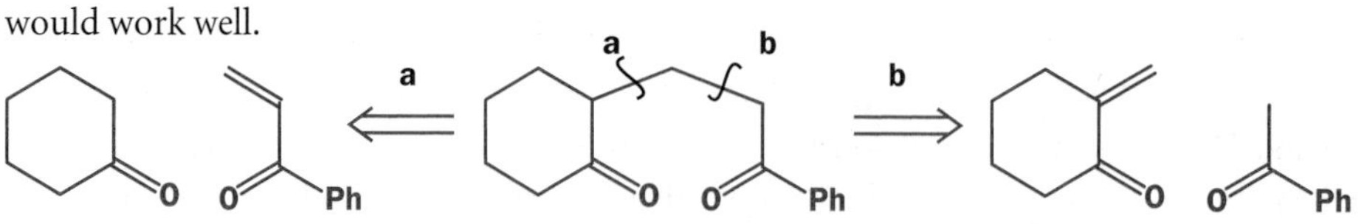

It is convenient to use Mannich bases instead of the very reactive unsaturated ketones and we will continue with disconnection 'a'.

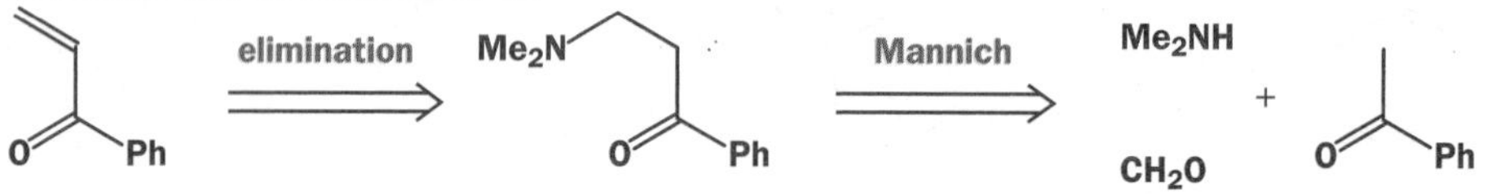

The synthesis is extraordinarily easy. The stable Mannich base is simply heated with the other ketone to give a high yield of the 1,5-diketone. Treatment of that with the HCl salt of NH₂OH in EtOH gives the pyridine directly, also in good yield.

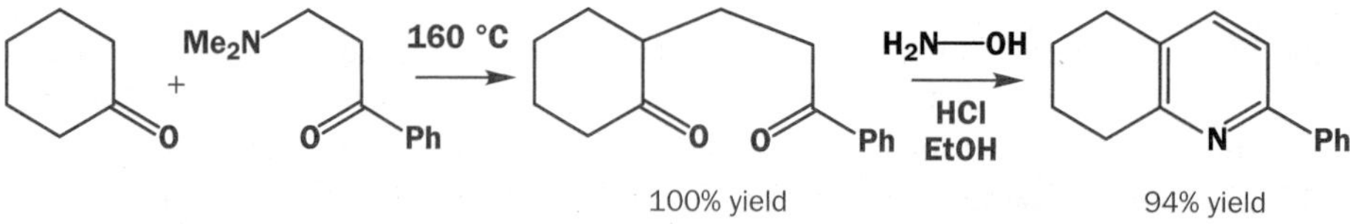

100% yield   94% yield

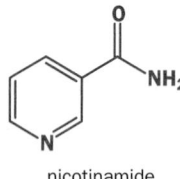

nicotinamide

Another direct route leads, as we shall now demonstrate, to pyridones. These useful compounds are the basis for nucleophilic substitutions on the ring (Chapter 43). We choose an example that puts a nitrile in the 3-position. This is significant because the role of nicotinamide in living things (Chapter 50) makes such products interesting to make. Aldol disconnection of a 3-cyano pyridone starts us on the right path.

If we now disconnect the C–N bond forming the enamine on the other side of the ring we will expose the true starting materials. This approach is unusual in that the nitrogen atom that is to be the pyridine nitrogen is not added as ammonia but is already present in a molecule of cyanoacetamide.

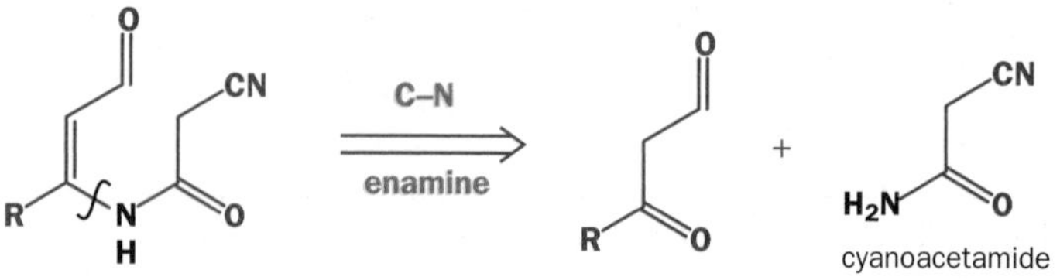

cyanoacetamide

The keto-aldehyde can be made by a simple Claisen ester condensation (Chapter 28) using the enolate of the methyl ketone with ethyl formate (HCO₂Et) as the electrophile. It actually exists as a stable enol, like so many 1,3-dicarbonyl compounds (Chapter 21).

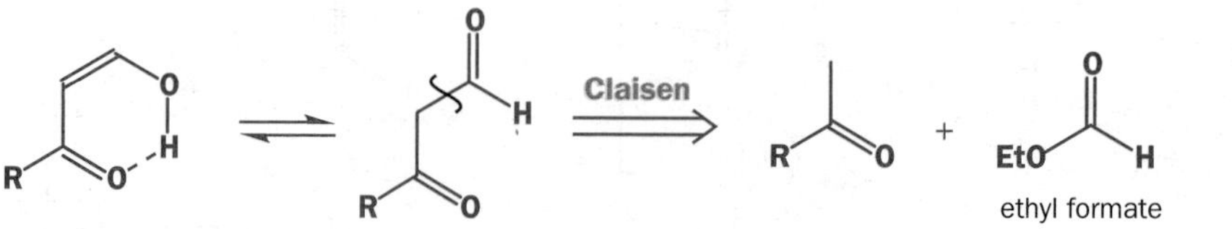

ethyl formate

In the synthesis, the product of the Claisen ester condensation is actually the enolate anion of the keto-aldehyde and this can be combined directly without isolation with cyanoacetamide to give the pyridone in the same flask.

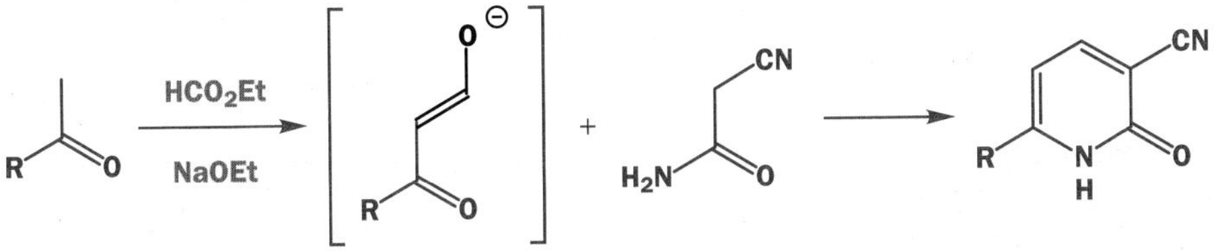

What must happen here is that the two compounds must exchange protons (or switch enolates if you prefer) before the aldol reaction occurs. Cyclization probably occurs next through C–N bond formation and, finally, dehydration is forced to give the *Z*-alkene.

> If dehydration occurred first, only the *Z*-alkene could cyclize and the major product, the *E*-alkene, would be wasted.

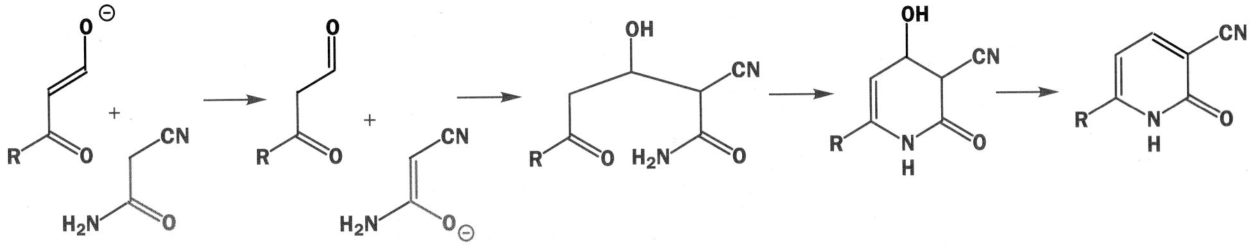

In planning the synthesis of a pyrrole or a pyridine from a dicarbonyl compound, considerable variation in oxidation state is possible. The oxidation state is chosen to make further disconnection of the carbon skeleton as easy as possible. We can now see how these same principles can be applied to pyrazoles and pyridazines.

Pyrazoles and pyridazines from hydrazine and dicarbonyl compounds

Disconnection of pyridazines reveals a molecule of hydrazine and a 1,4-diketone with the proviso that, just as with pyridines, the product will be a dihydropyrazine and oxidation will be needed to give the aromatic compound. As with pyridines, we prefer to avoid the *cis* double bond problem.

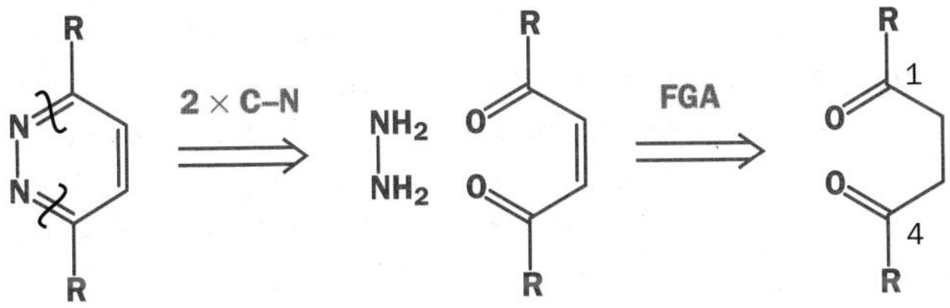

As an example we can take the cotton herbicide made by Cyanamid. Direct removal of hydrazine would require a *cis* double bond in the starting material.

> The herbicide kills weeds in cotton crops rather than the cotton plant itself!

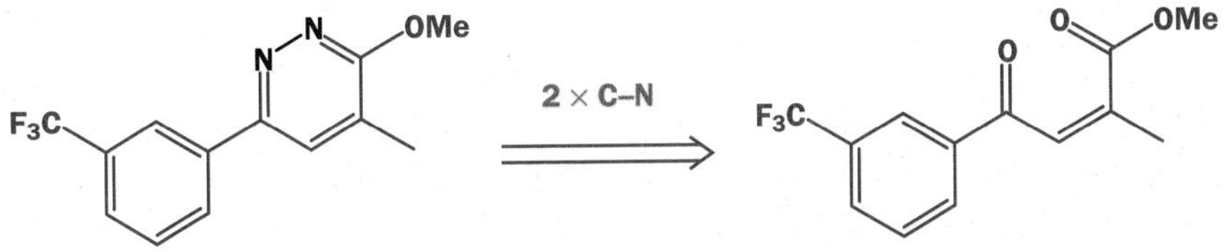

Cyanamid cotton herbicide   reject *cis* alkene as starting material

If we remove the double bond first, a much simpler compound emerges. Note that this is a keto-ester rather than a diketone.

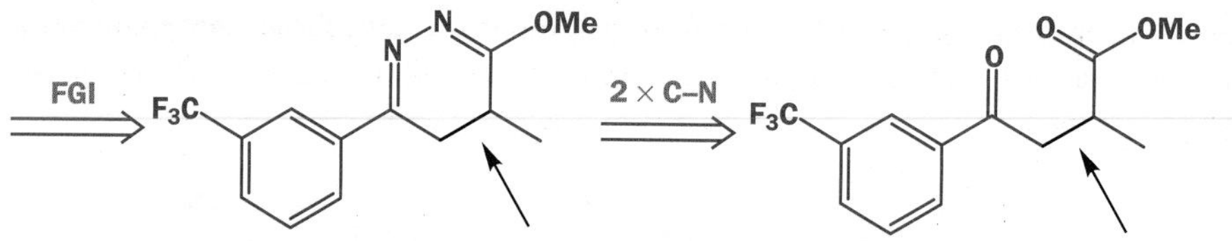

When hydrazine is added to the keto-ester an imine is formed with the ketone but acylation occurs at the ester end to give an amide rather than the imino-ester we had designed. The product is a dihydropyridazolone.

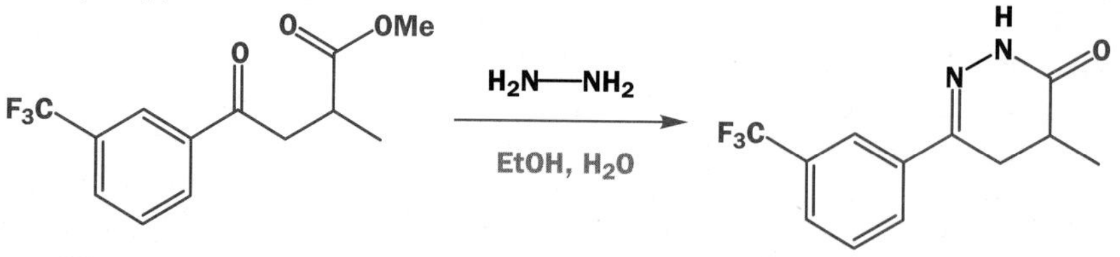

Aromatization with bromine gives the aromatic pyridazolone by bromination and dehydrobromination and now we invoke the nucleophilic substitution reactions introduced in Chapter 43. First we make the chloride with POCl₃ and then displace with methanol.

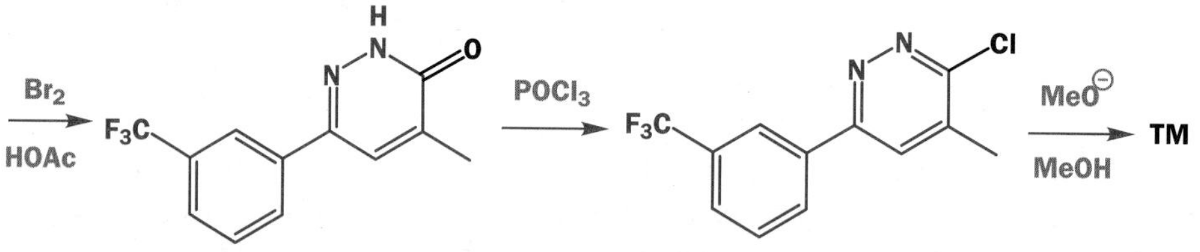

The five-membered ring pyrazoles are even simpler as the starting material is a 1,3-dicarbonyl compound available from the aldol or Claisen ester condensations.

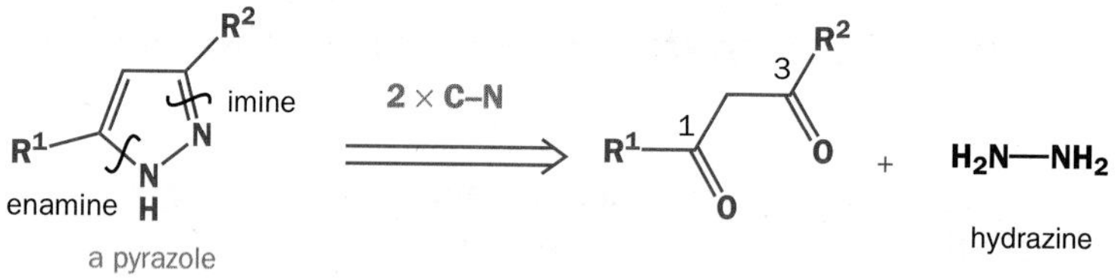

Chemistry hits the headlines—Viagra

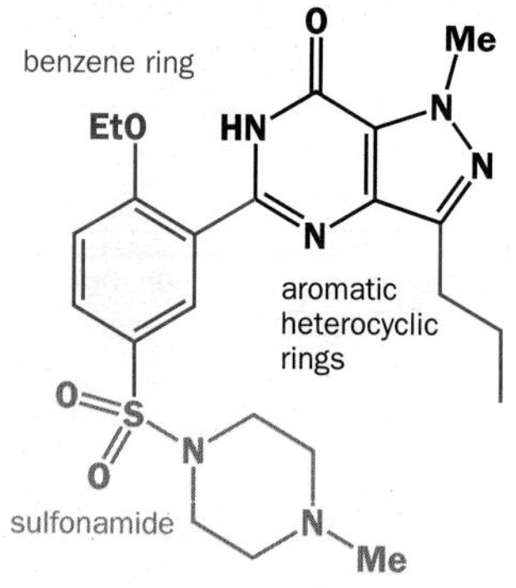

In 1998 chemistry suddenly appeared in the media in an exceptional way. Normally not a favourite of TV or the newspapers, chemistry produced a story with all the right ingredients—sex, romance, human ingenuity—and all because of a pyrazole. In the search for a heart drug, Pfizer uncovered a compound that allowed impotent men to have active sex lives. They called it Viagra.

The molecule contains a sulfonamide and a benzene ring as well as the part that interests us most—a bicyclic aromatic heterocyclic system of a pyrazole fused to a pyrimidine. We shall discuss in detail how Pfizer made this part of the molecule and just sketch in the rest. The sulfonamide can be made from the sulfonic acid that can be added to the benzene ring by electrophilic aromatic sulfonation (Chapter 22).

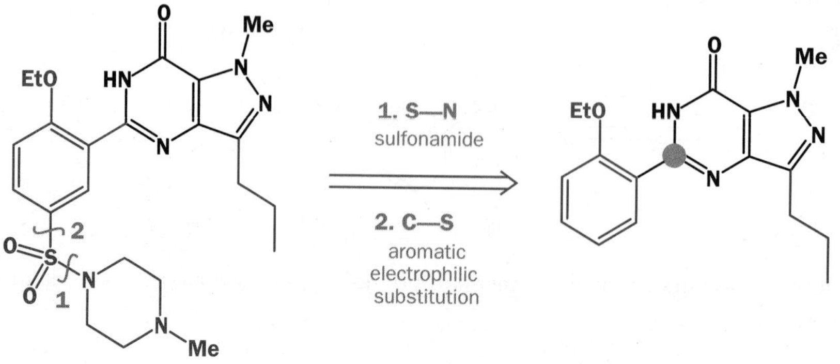

Viagra:
Pfizer's treatment for male
erectile dysfunction

benzene ring

aromatic
heterocyclic
rings

sulfonamide

Inspection of what remains reveals that the carbon atom atom in the heterocycles next to the benzene ring (marked with an orange blob) is at the oxidation level of a carboxylic acid. If, therefore, we disconnect both C–N bonds to this atom we will have two much simpler starting materials.

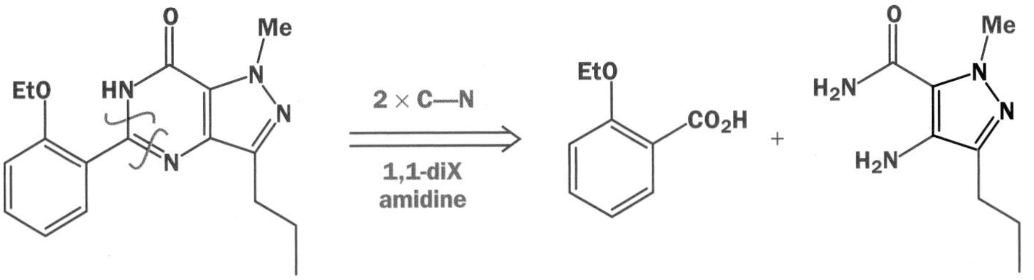

The aromatic acid is available and we need consider only the pyrazole (core pyrazole ring in black in the diagram). The aromatic amino group can be put in by nitration and reduction and the amide can be made from the corresponding ester. This leaves a carbon skeleton, which must be made by ring synthesis.

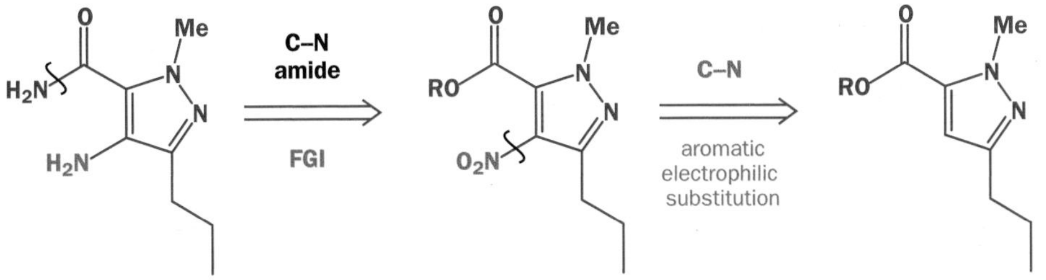

Following the methods we have established so far in this chapter, we can remove the hydrazine portion to reveal a 1,3-dicarbonyl compound. In fact, this is a tricarbonyl compound, a diketo-ester, because of the ester already present and it contains 1,2- 1,3-, and 1,4-dicarbonyl relationships. The simplest synthesis is by a Claisen ester condensation and we choose the disconnection so that the electrophile is a reactive (oxalate) diester that cannot enolize. The only control needed will then be in the enolization of the ketone.

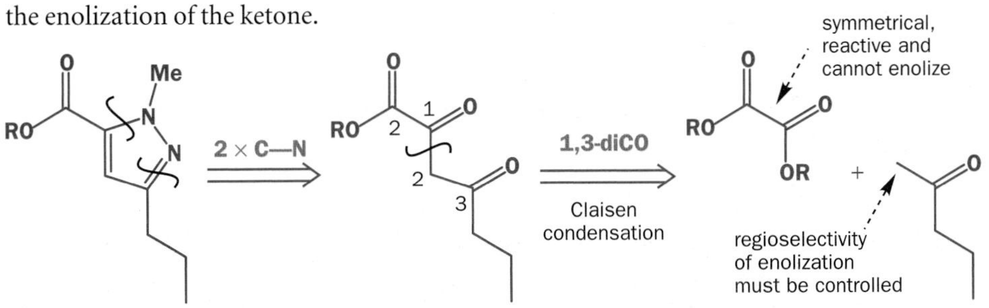

The Claisen ester condensation gives the right product just by treatment with base. The reasons for this are discussed in Chapter 28. We had then planned to react the keto-diester with methylhydrazine but there is a doubt about the regioselectivity of this reaction—the ketones are more electrophilic than the ester all right, but which ketone will be attacked by which nitrogen atom?

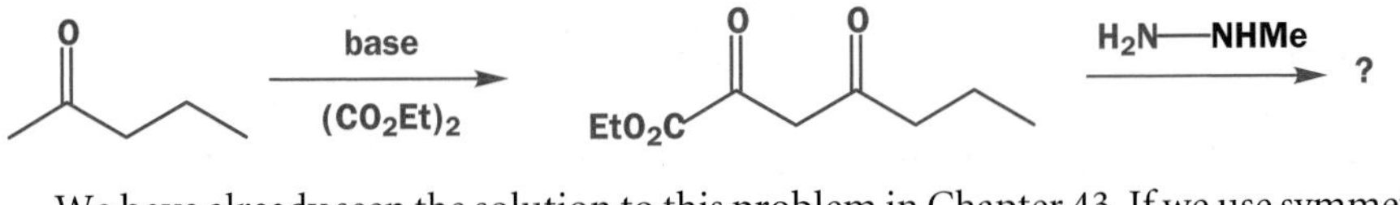

We have already seen the solution to this problem in Chapter 43. If we use symmetrical hydrazine, we can deal with the selectivity problem by alkylation. Dimethyl sulfate turns out to be the best reagent.

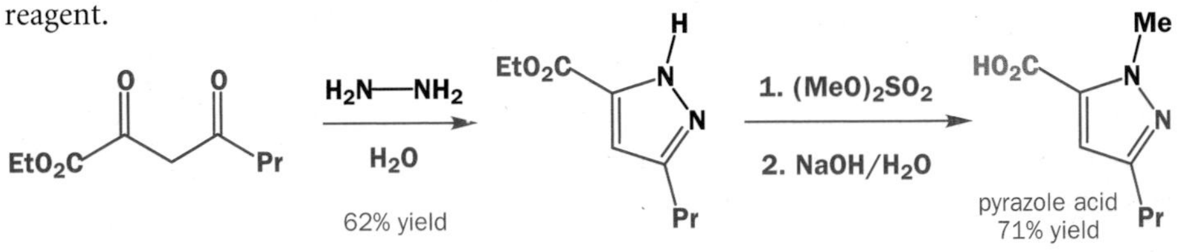

62% yield

pyrazole acid
71% yield

> The alkylation is regioselective because the methylated nitrogen must become the pyrrole-like nitrogen atom and the molecule prefers the longest conjugated system involving that nitrogen and the ester.

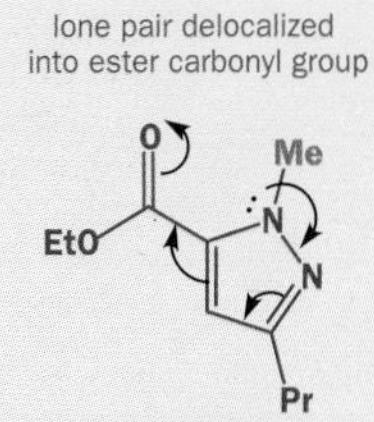

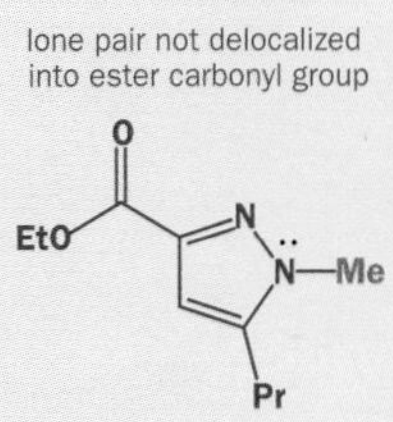

lone pair delocalized into ester carbonyl group

lone pair not delocalized into ester carbonyl group

The stable pyrazole acid from the hydrolysis of this ester is a key intermediate in Viagra production. Nitration can occur only at the one remaining free position and then amide formation and reduction complete the synthesis of the amino pyrazole amide ready for assembly into Viagra.

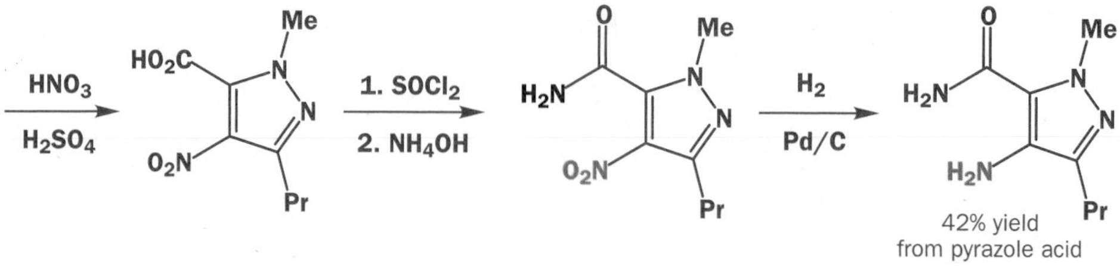

The rest of the synthesis can be summarized very briefly as it mostly concerns material outside the scope of this chapter. You might like to notice how easy the construction of the second heterocyclic ring is—the nucleophilic attack of the nitrogen atom of one amide on to the carbonyl of another would surely not occur unless the product were an aromatic heterocycle.

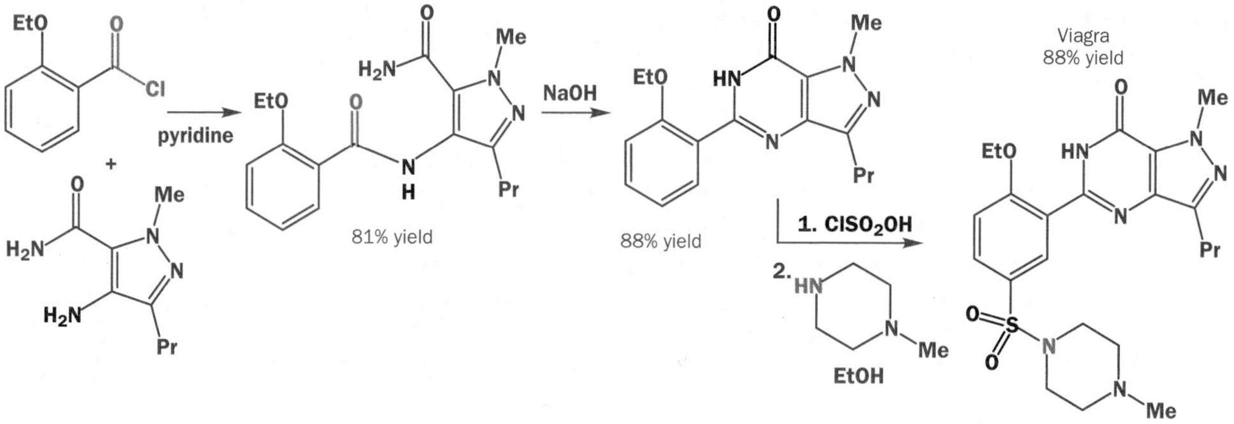

Pyrimidines can be made from 1,3-dicarbonyl compounds and amidines

In Chapter 43 we met some compounds that interfere in folic acid metabolism and are used as antibacterial agents. One of them was trimethoprim and it contains a pyrimidine ring (black on the diagram). We are going to look at its synthesis briefly because the strategy used is the opposite of that used with the pyrimidine ring in Viagra. Here we disconnect a molecule of guanidine from a 1,3-dicarbonyl compound.

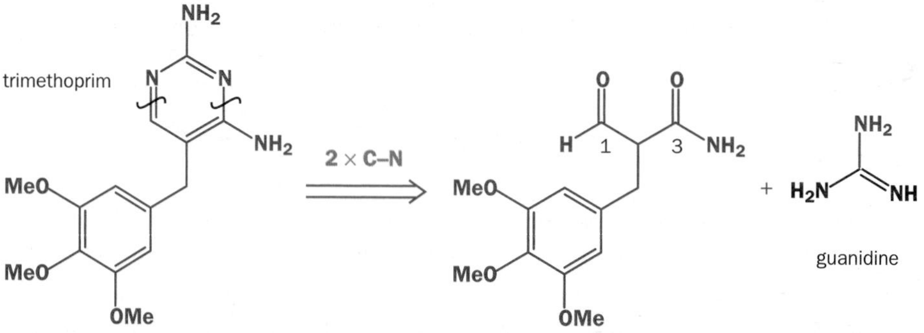

The 1,3-dicarbonyl compound is a combination of an aldehyde and an amide but is very similar to a malonic ester so we might think of making this compound by alkylation of that stable enolate (Chapter 26) with the convenient benzylic bromide.

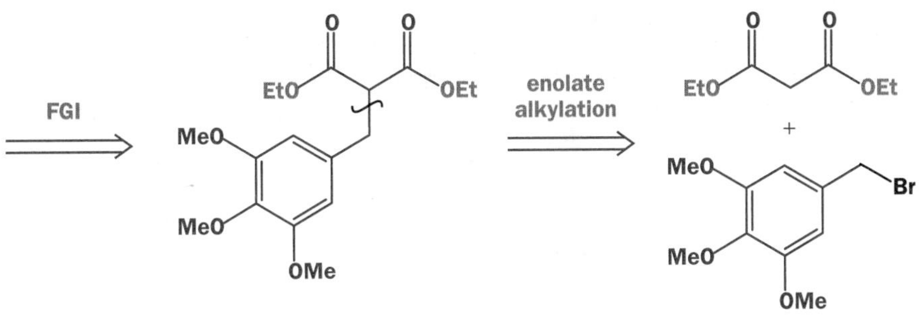

The alkylation works fine but it turns out to be better to add the aldehyde as an electrophile (cf. the pyridone synthesis on p. 1195) rather than try to reduce an ester to an aldehyde. The other ester is already at the right oxidation level. Notice the use of the NaCl method of decarboxylation (Chapter 26).

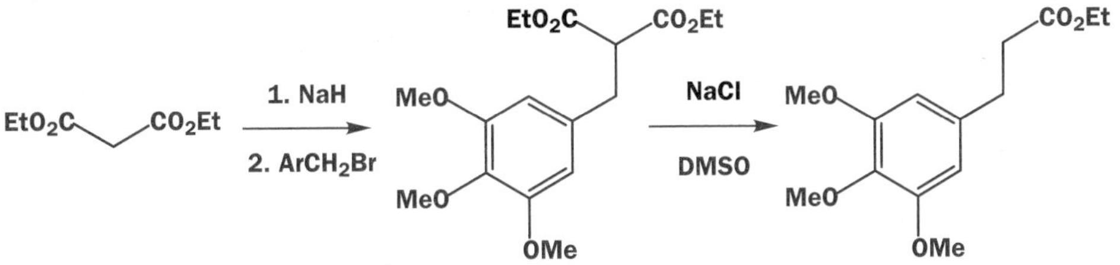

Condensation with ethyl formate (HCO_2Et) and cyclization with guanidine gives the pyrimidine ring system but with an OH instead of the required amino group. Aromatic nucleophilic substitution in the pyrimidone style from Chapter 43 gives trimethoprim.

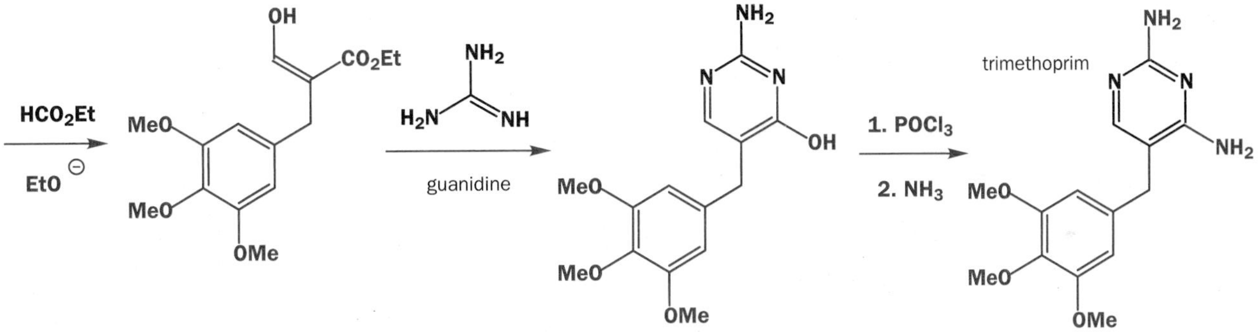

Unsymmetrical nucleophiles lead to selectivity questions

The synthesis of thiazoles is particularly interesting because of a regioselectivity problem. If we try out the two strategies we have just used for pyrimidines, the first requires the reaction of a carboxylic acid derivative with a most peculiar enamine that is also a thioenol. This does not look like a stable compound.

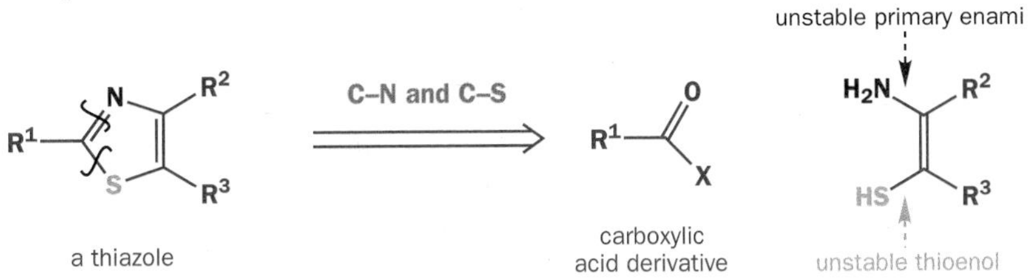

The alternative is to disconnect the C–N and C–S bonds on the other side of the heteroatoms. Here we must be careful what we are about or we will get the oxidation state wrong. We shall do it step by step to make sure. We can rehydrate the double bond in two ways. We can first try putting the OH group next to nitrogen.

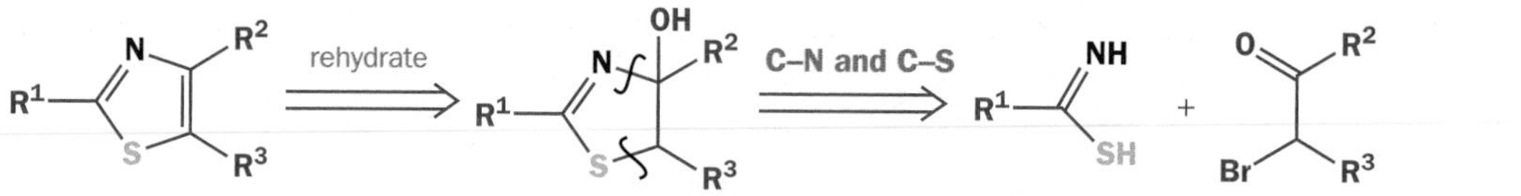

Or we can rehydrate it the other way round, putting the OH group next to the sulfur atom, and disconnect in the same way. In both cases we require an electrophilic carbon atom at the alcohol oxidation level and one at the aldehyde or ketone oxidation level. In other words we need an α-haloketone.

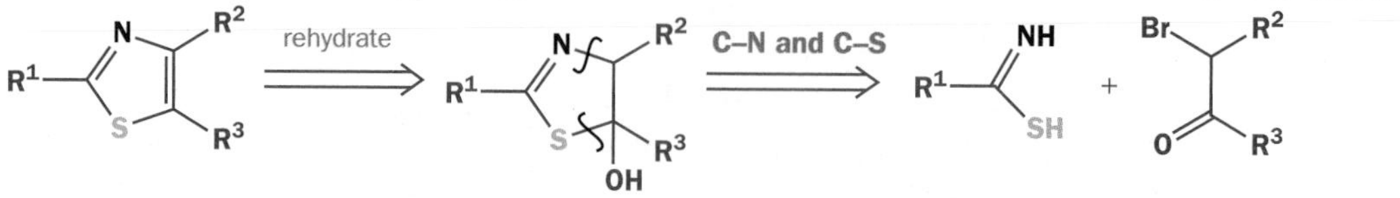

■
This is discussed in Chapter 46 and the structure of Lawesson's reagent is on p. 1266 of that chapter.

The nucleophile is the same in both cases and it is an odd-looking molecule. That is, until we realize that it is just a tautomer of a thioamide. Far from being odd, thioamides are among the few stable thiocarbonyl derivatives and can be easily made from ordinary amides with P_2S_5 or Lawesson's reagent.

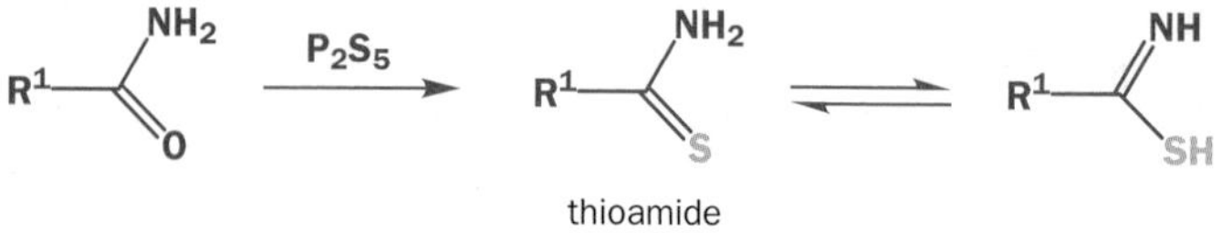

So the only remaining question is: when thioamides combine with α-haloketones, which atom (N or S) attacks the ketone, and which atom (N or S) attacks the alkyl halide? Carbonyl groups are 'hard' electrophiles—their reactions are mainly under charge control and so they react best with basic nucleophiles (Chapter 12). Alkyl halides are 'soft' electrophiles—their reactions are mainly under frontier orbital control and they react best with large uncharged nucleophiles from the lower rows of the periodic table. The ketone reacts with nitrogen and the alkyl halide with sulfur.

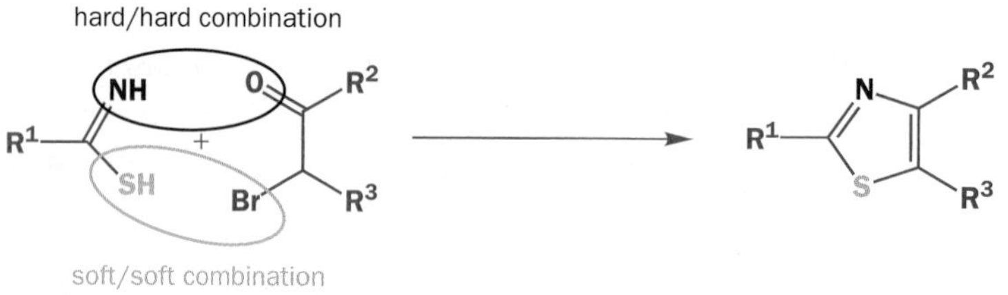

Fentiazac, a nonsteroidal anti-inflammatory drug, is a simple example. Disconnection shows that we need thiobenzamide and an easily made α-haloketone (easily made because the ketone can enolize on this side only—see Chapter 21).

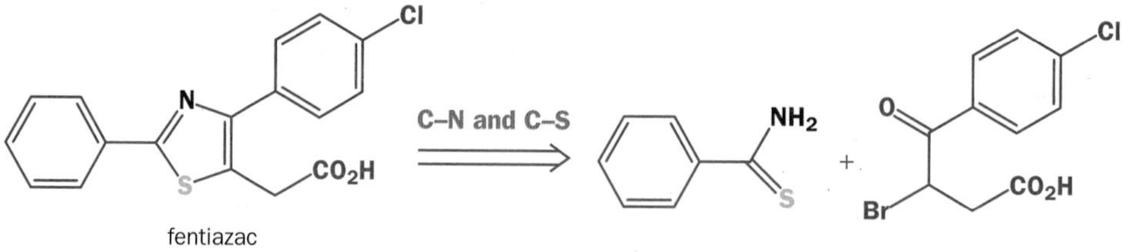

fentiazac

The synthesis involves heating these two compounds together and the correct thiazole forms easily with the double bonds finding their right positions in the product—the only positions for a stable aromatic heterocycle.

Isoxazoles are made from hydroxylamine or by 1,3-dipolar cycloadditions

The two main routes for the synthesis of isoxazoles are the attack of hydroxylamine (NH_2OH) on diketones and 1,3-dipolar cycloadditions of nitrile oxides. They thus form a link between the strategy we have been discussing (cyclization of a nucleophile with two heteroatoms and a compound with two electrophilic carbon atoms) and the next strategy—cycloaddition reactions.

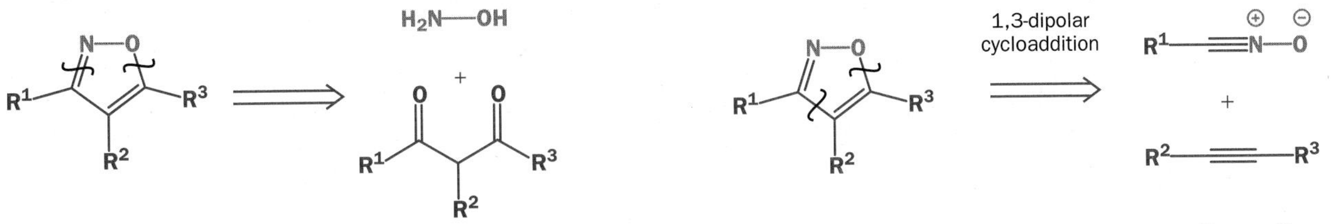

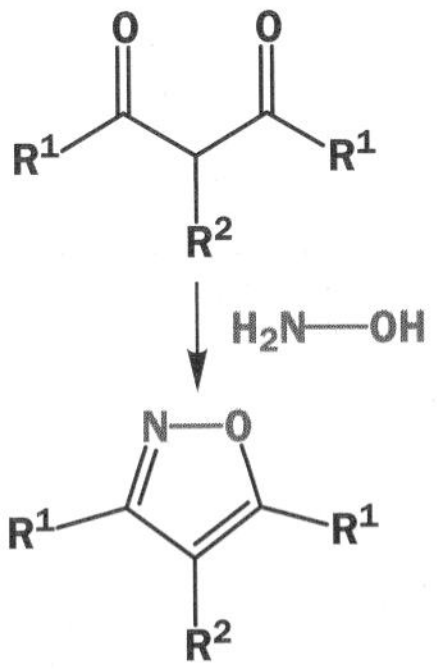

Simple symmetrical isoxazoles are easily made by the hydroxylamine route. If $R^1 = R^3$, we have a symmetrical and easily prepared 1,3-diketone as starting material. The central R^2 group can be inserted by alkylation of the stable enolate of the diketone (Chapter 26).

When $R^1 \neq R^3$, we have an unsymmetrical dicarbonyl compound and we must be sure that we know which way round the reaction will proceed. The more nucleophilic end of NH_2OH will attack the more electrophilic carbonyl group. It seems obvious that the more nucleophilic end of NH_2OH will be the nitrogen atom but that depends on the pH of the solution. Normally, hydroxylamine is supplied as the crystalline hydrochloride salt and a base of some kind added to give the nucleophile. The relevant pK_as are shown in the margin. Bases such as pyridine or sodium acetate produce some of the reactive neutral NH_2OH in the presence of the less reactive cation, but bases such as NaOEt produce the anion. Reactions of keto-aldehydes with acetate-buffered hydroxylamine usually give the isoxazole from nitrogen attack on the aldehyde as expected.

state of hydroxylamine changes with pH: the more nucleophilic atom is marked in black

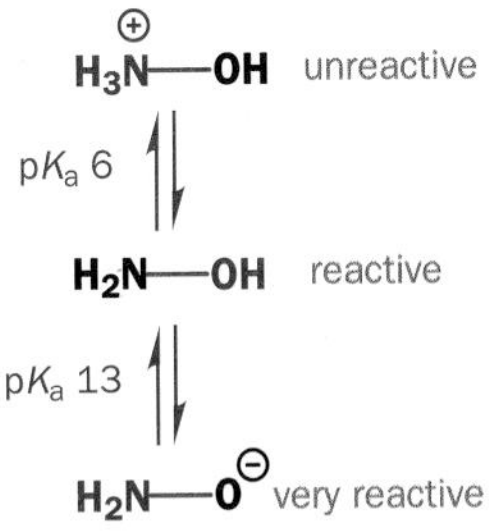

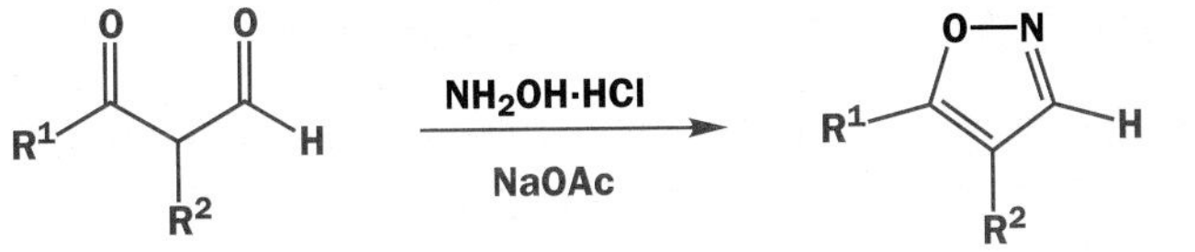

Modification of the electrophile may also be successful. Reaction of hydroxylamine with 1,2,4-diketo-esters usually gives the isoxazole from attack of nitrogen at the more reactive keto group next to the ester.

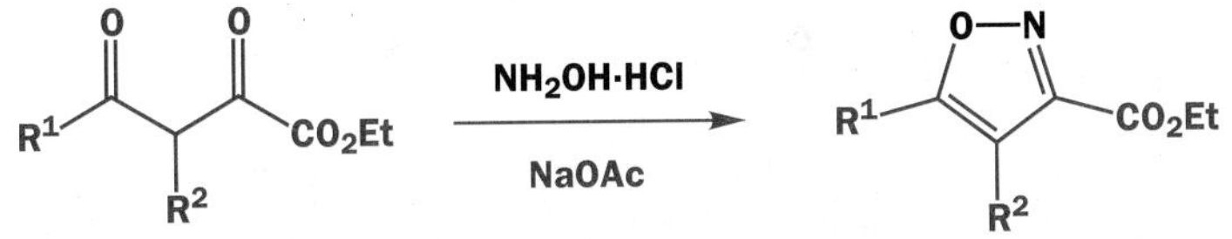

A clear demonstration of selectivity comes from the reactions of bromoenones. It is not immediately clear which end of the electrophile is more reactive but the reactions tell us the answer.

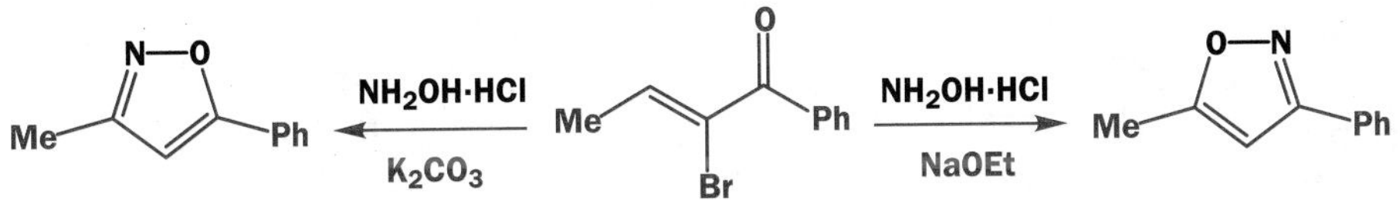

The alternative approach to isoxazoles relies on cycloadditions of nitrile oxides with alkynes. We saw in Chapter 35 that there are two good routes to these reactive compounds, the γ-elimination of chlorooximes or the dehydration of nitroalkanes.

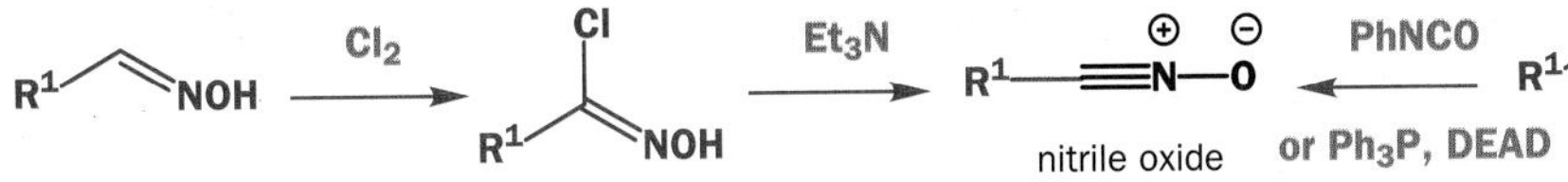

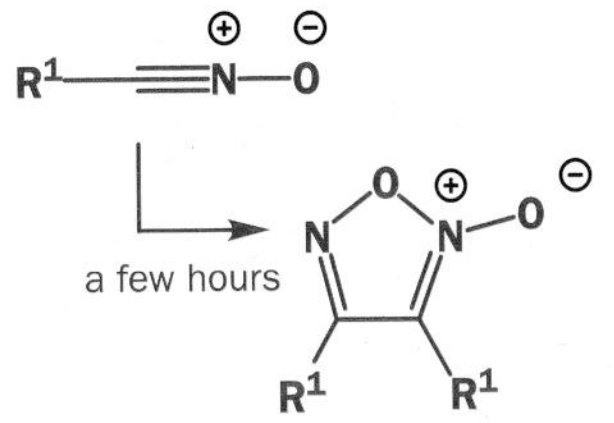

a few hours

A few nitrile oxides are stable enough to be isolated (those with electron-withdrawing or highly conjugating substituents, for example) but most are prepared in the presence of the alkyne by one of these methods because otherwise they dimerize rapidly. Both methods of forming nitrile oxides are compatible with their rapid reactions with alkynes. Reaction with aryl alkynes is usually clean and regioselective.

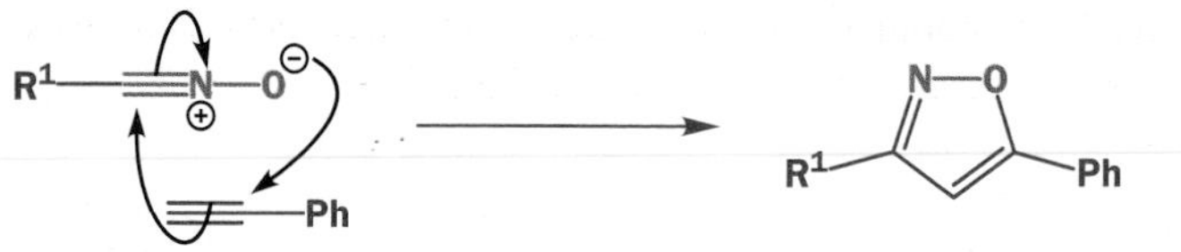

The alkyne is using its HOMO to attack the LUMO of the nitrile oxide (see Chapter 35 for an explanation). If the alkyne has an electron-withdrawing group, mixtures of isomers are usually formed as the HOMO of the nitrile oxide also attacks the LUMO of the alkyne.

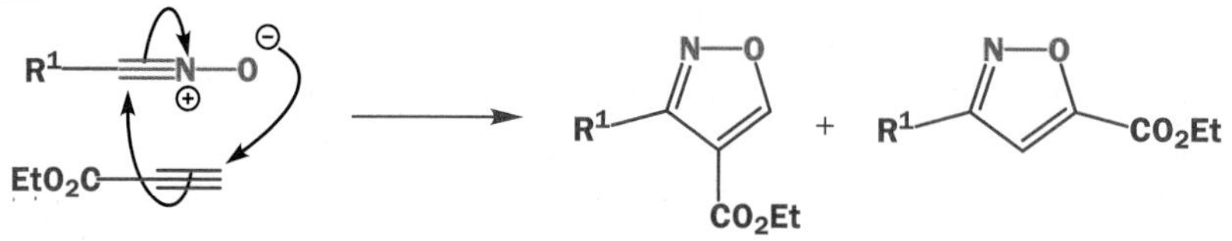

Intramolecular reactions are usually clean regardless of the preferred electronic orientation if the tether is too short to allow any cyclization except one. In this example, even the more favourable orientation looks very bad because of the linear nature of the reacting species, but only one isomer is formed.

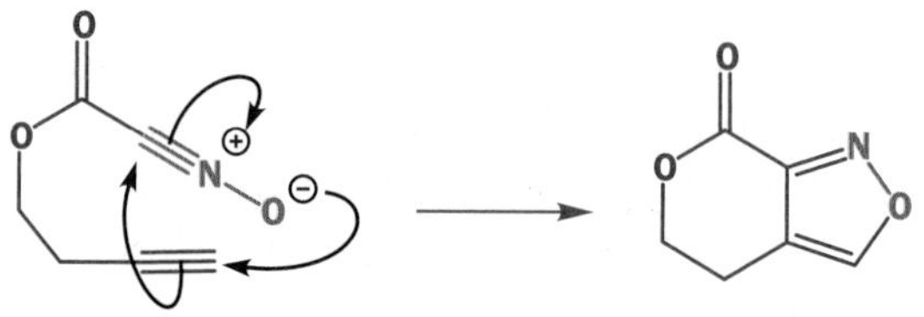

Tetrazoles are also made by 1,3-dipolar cycloadditions

Disconnection of tetrazoles with a 1,3-dipolar cycloaddition in mind is easy to see once we realize that a nitrile (RCN) is going to be one of the components. It can be done in two ways: disconnection of the neutral compound would require hydrazoic acid (HN_3) as the dipole but the anion disconnects directly to azide ion.

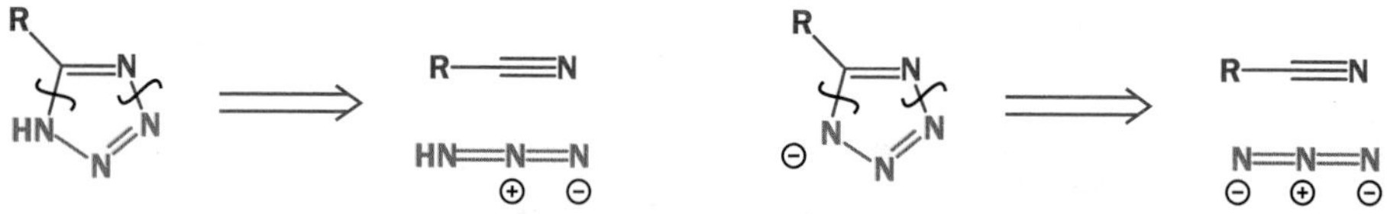

> You saw in Chapter 43 that tetrazoles are about as acidic as carboxylic acids.

Unpromising though this reaction may look, it actually works well if an ammonium-chloride-buffered mixture of sodium azide and the nitrile is heated in DMF. The reagent is really ammonium azide and the reaction occurs faster with electron-withdrawing substituents in R. In the reaction mixture, the anion of the tetrazole is formed but neutralization with acid gives the free tetrazole.

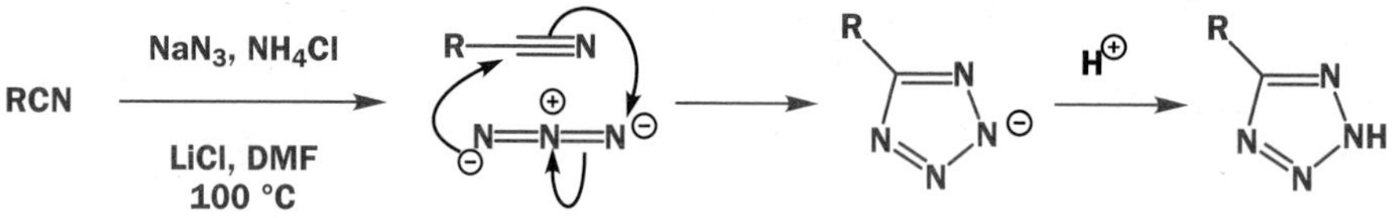

> The synthesis of the indole starting material is in the next section.

As nitriles are generally readily available this is the main route to simple tetrazoles. More complicated ones are made by alkylation of the product of a cycloaddition. The tetrazole substitute for indomethacin that we mentioned in Chapter 43 is made by this approach. First, the nitrile is prepared from the indole. The 1,3-dipolar cycloaddition works well by the azide route we have just discussed, even though this nitrile will form an 'enol' rather easily.

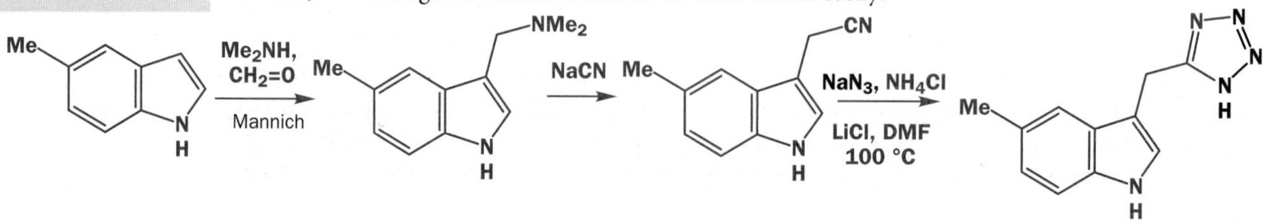

Finally, the indole nitrogen atom must be acylated. The tetrazole is more acidic so it is necessary to form a dianion to get reaction at the right place. The usual rule is followed (see Chapter 24)—the second anion formed is less stable and so it reacts first.

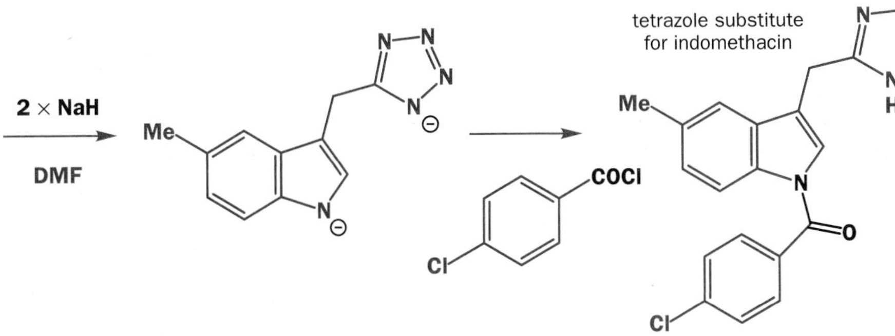

tetrazole substitute for indomethacin

The synthesis of the anti-inflammatory drug broperamole illustrates modification of a tetrazole using its anion. The tetrazole is again constructed from the nitrile—it's an aromatic nitrile with an electron-withdrawing substituent so this will be a good reaction.

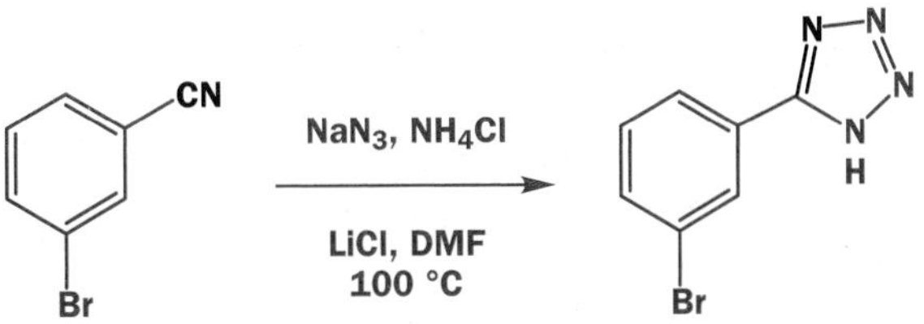

Conjugate addition to acrylic acid (Chapters 10 and 23) occurs to give the other tautomer to the one we have drawn. The anion intermediate is, of course, delocalized and can react at any of the nitrogen atoms. Amide formation completes the synthesis of broperamole.

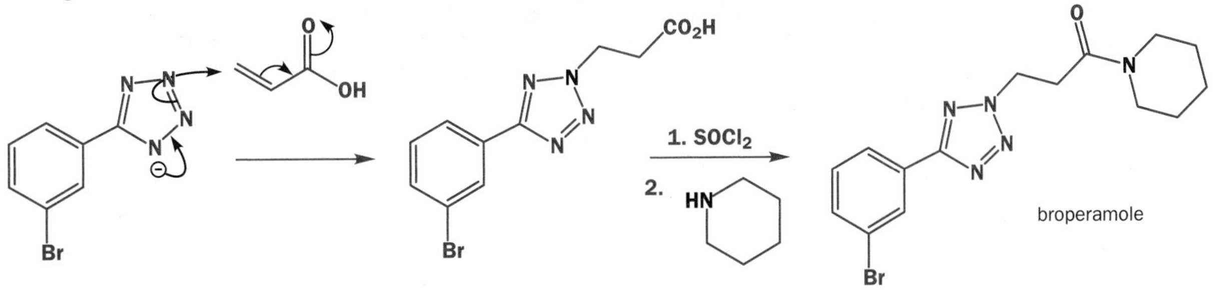

broperamole

The difficulty in trying to forecast which way round a 1,3-dipolar cycloaddition will go is well illustrated when a substituted azide adds to an alkyne in the synthesis of 1,2,3-triazoles. Reaction of an alkyl azide with an unsymmetrical alkyne, having an electron-withdrawing group at one end and an alkyl group at the other, gives mostly a single triazole.

■ 1,2,4-Triazoles are usually made from the reaction of the unsubstituted 1,2,4-triazole anion with electrophiles as described in Chapter 43.

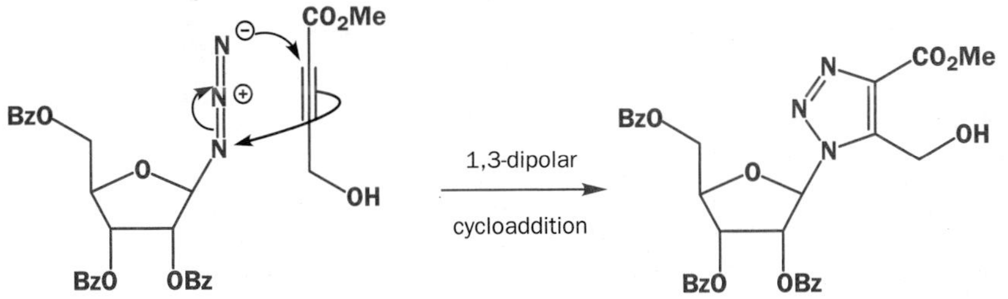

1,3-dipolar cycloaddition

It looks as if the more nucleophilic end of the azide has attacked the wrong end of the alkyne but we must remember that (1) it is very difficult to predict which is the more nucleophilic end of a 1,3-dipole and (2) it may be either HOMO (dipole) and LUMO (alkyne) or LUMO (dipole) and HOMO (alkyne) that dominate the reaction. The reason for doing the reaction was to make analogues of natural nucleosides (the natural compounds are discussed in Chapter 49). In this case the OH group was replaced by a cyanide so that a second aromatic ring, a pyridine, can be fused on to the triazole.

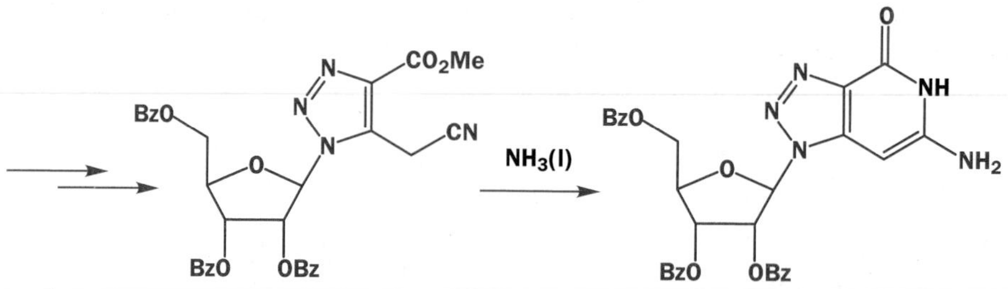

The next section deals with the synthesis of heterocycles where a heterocyclic ring is fused to a benzene ring, the 6/5 system, indole, and the 6/6 systems, quinoline and isoquinoline.

The Fischer indole synthesis

You are about to see one of the great inventions of organic chemistry. It is a remarkable reaction, amazing in its mechanism, and it was discovered in 1883 by one of the greatest organic chemists of all, Emil Fischer. Fischer had earlier discovered phenylhydrazine (PhNHNH$_2$) and, in its simplest form, the **Fischer indole synthesis** occurs when phenylhydrazine is heated in acidic solution with an aldehyde or ketone.

Emil Fischer (1852–1919), discovered phenylhydrazine as a PhD student in 1875, succeeded Hofmann at Berlin in 1900 where he built the then largest chemical institute in the world, and was awarded the Nobel prize in 1902. As well as his work on indoles, he laid the foundations of carbohydrate chemistry by completing the structure and synthesis of the main sugars. If only he hadn't also invented Fischer projections!

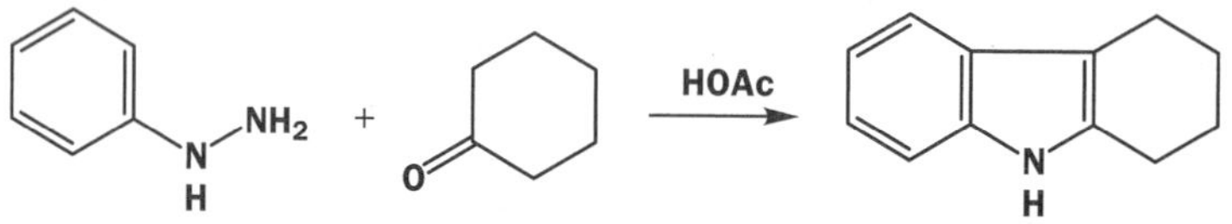

The first step in the mechanism is formation of the phenylhydrazone (the imine) of the ketone. This can be isolated as a stable compound (Chapter 14).

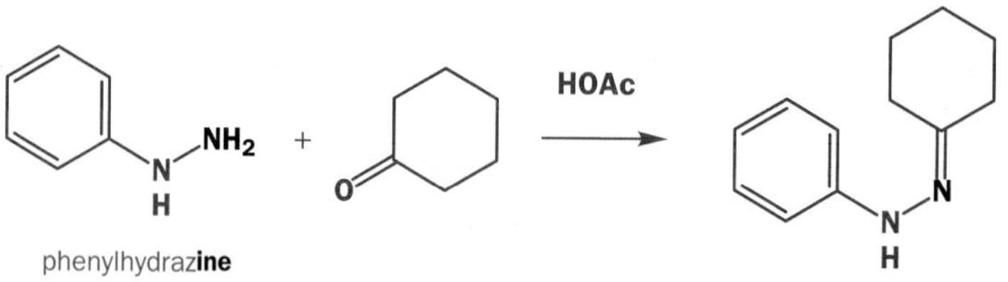

phenylhydraz**ine**

cyclohexanone phenylhydraz**one**

The hydrazone then needs to tautomerize to the enamine, and now comes the key step in the reaction. The enamine can rearrange with formation of a strong C–C bond and cleavage of the weak N–N single bond by moving electrons round a six-membered ring.

This step is a [3,3]-sigmatropic rearrangement (Chapter 36): the new single bond (C–C) bears a 3,3 relationship to the old single bond (N–N).

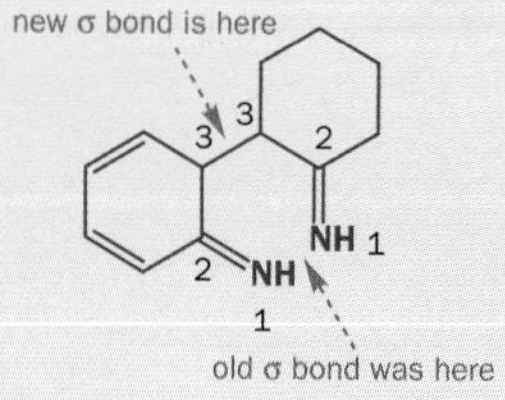

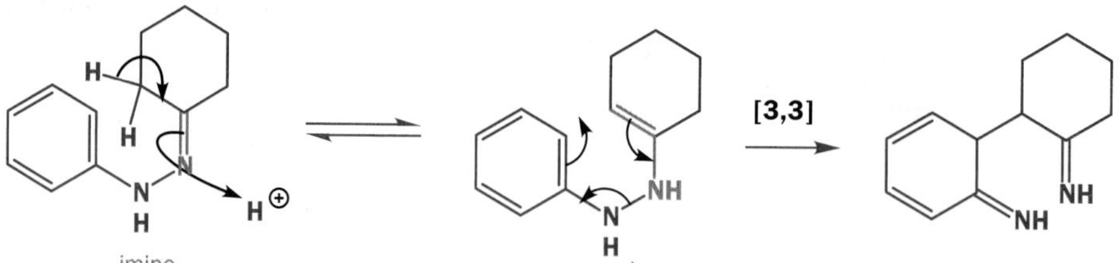

new σ bond is here

old σ bond was here

imine

enamine

Next, re-aromatization of the benzene ring (by proton transfer from carbon to nitrogen) creates an aromatic amine that immediately attacks the other imine. This gives an **aminal**, the nitrogen equivalent of an acetal.

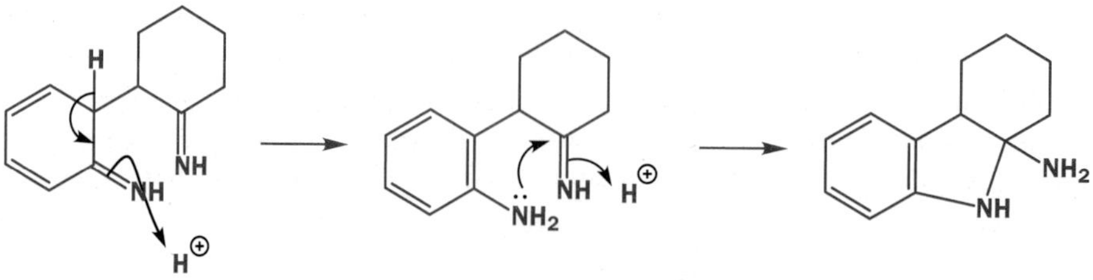

Finally, acid-catalysed decomposition of the aminal in acetal fashion with expulsion of ammonia allows the loss of a proton and the formation of the aromatic indole.

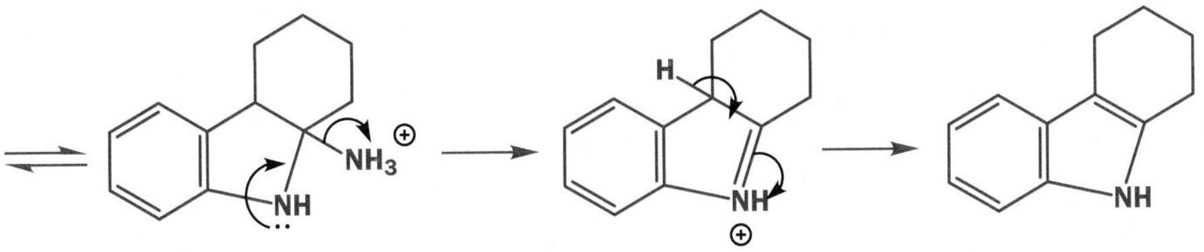

This is admittedly a complicated mechanism but if you remember the central step—the [3,3]-sigmatropic rearrangement—the rest should fall into place. The key point is that the C–C bond is established at the expense of a weak N–N bond. Naturally, Fischer had no idea about [3,3] or any other steps in the mechanism. He was sharp enough to see that something remarkable had happened and skilful enough to find out what it was.

The Fischer method is the main way of making indoles, but it is not suitable for them all. We need now to study its applicability to various substitution patterns. If the carbonyl compound can enolize on one side only, as is the case with an aldehyde, then the obvious product is formed.

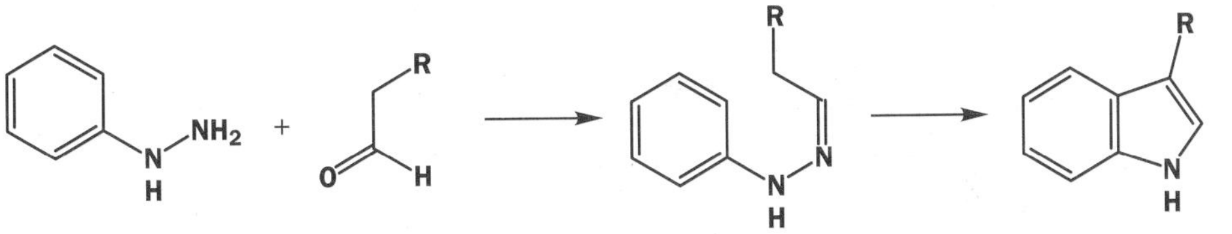

If the benzene ring has only one *ortho* position, then again cyclization must occur to that position. Other substituents on the ring are irrelevant. At this point we shall stop drawing the intermediate phenylhydrazone.

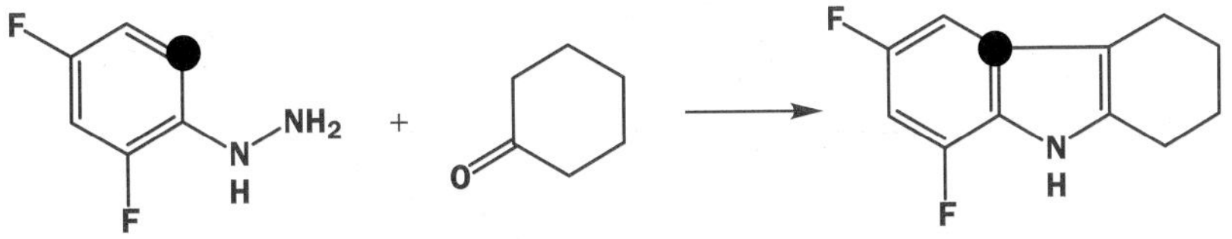

only one free *ortho* position

Another way to secure a single indole as product from the Fischer indole synthesis is to make sure the reagents are symmetrical. These two examples should make plain the types of indole available from symmetrical starting materials.

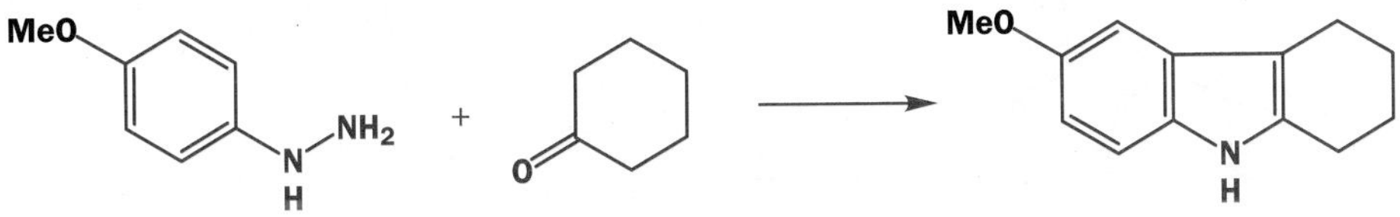

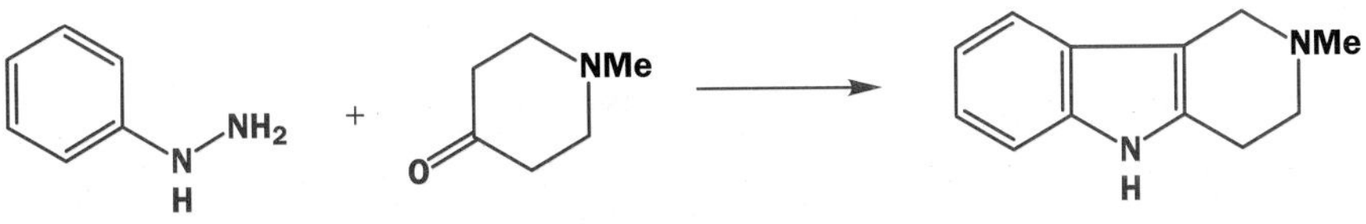

The substitution pattern of the first example is particularly important as the neurotransmitter serotonin is an indole with a hydroxyl group in the 5-position, and many important drugs follow that pattern. Sumatriptan (marketed as Imigran), is an example that we can also use to show that substituted phenylhydrazines are made by reduction of diazonium salts (Chapter 23). The first stage of the synthesis is nitrosation of the aniline and reduction with $SnCl_2$ and HCl to give the salt of the phenylhydrazine.

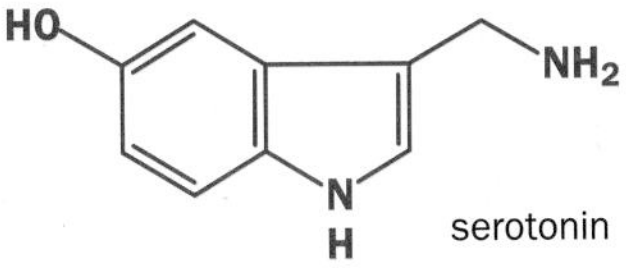

serotonin

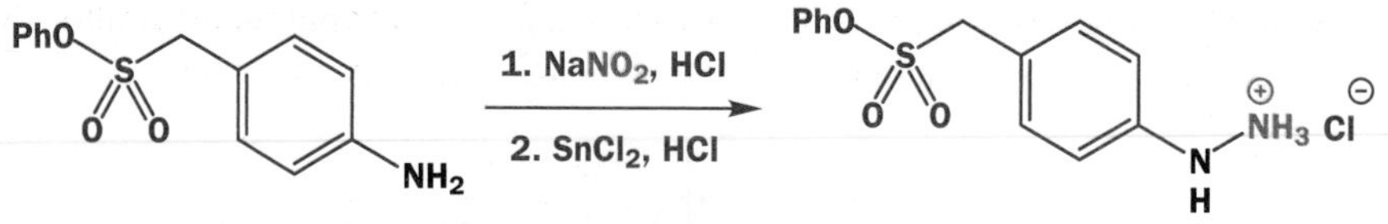

The required aldehyde (3-cyanopropanal) is added as an acetal to prevent self-condensation. The acidic conditions release the aldehyde, which forms the phenylhydrazone ready for the next step.

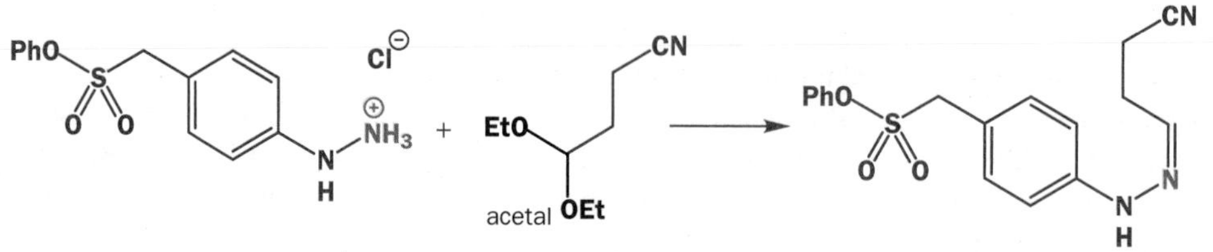

The Fischer indole synthesis itself is catalysed in this case by polyphosphoric acid (PPA), a sticky gum based on phosphoric acid (H_3PO_4) but dehydrated so that it contains some oligomers. It is often used as a catalyst in organic reactions and residues are easily removed in water.

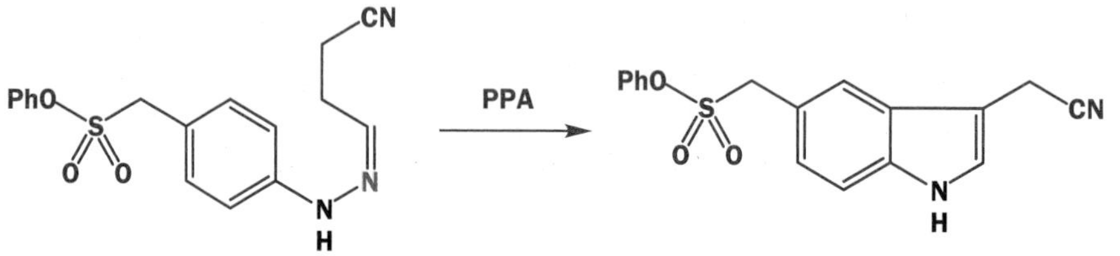

All that remains is to introduce the methylamino and dimethylamino groups. The sulfonate ester is more reactive than the nitrile so the methylamino group must go in first.

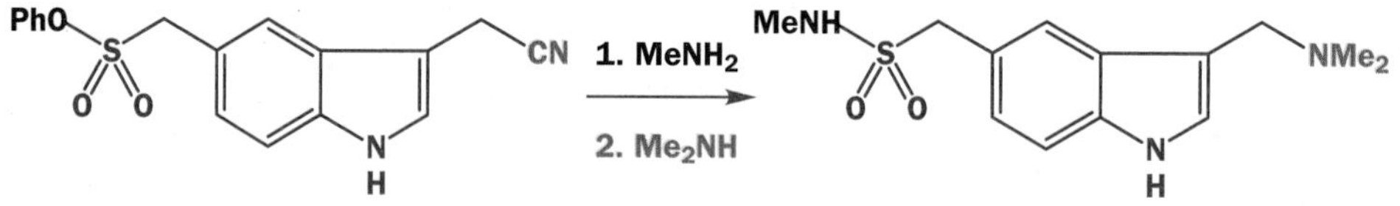

sumatriptan

For some indoles it is necessary to control regioselectivity with unsymmetrical carbonyl compounds. Ondansetron, the anti-nausea compound that is used to help cancer patients take larger doses of antitumour compounds than was previously possible, is an example. It contains an indole and an imidazole ring.

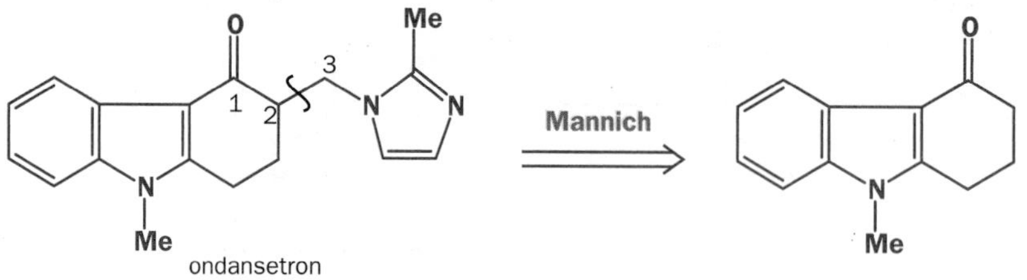

ondansetron

The 1,3 relationship between C–N and C–O suggests a Mannich reaction to add the imidazole ring (Chapter 27), and that disconnection reveals an indole with an unsymmetrical right-hand side, having an extra ketone group. Fischer disconnection will reveal a diketone as partner for phenylhydrazine. We shall leave aside for the moment when to add the methyl group to the indole nitrogen.

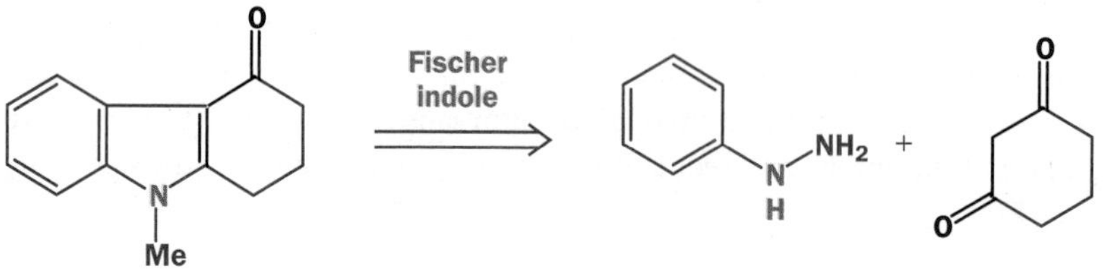

The diketone has two identical carbonyl groups and will enolize (or form an enamine) exclusively towards the other ketone. The phenylhydrazone therefore forms only the enamine we want.

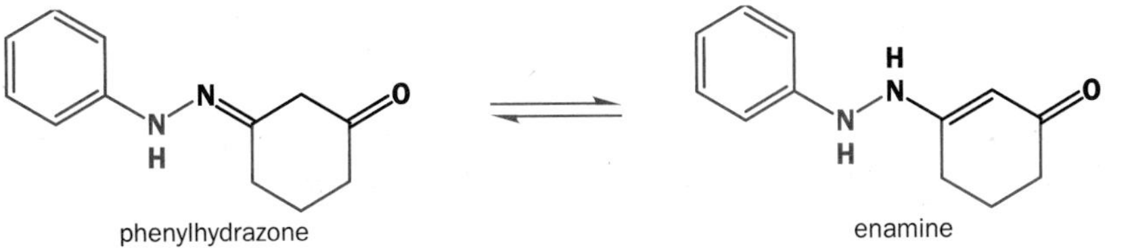

phenylhydrazone   enamine

► Notice that we have drawn a different geometrical isomer of the imine from the one we have previously drawn and you might at first think that this one cannot cyclize. But the geometry of the phenylhydrazone is unimportant—you might like to think why.

In this case, the Fischer indole reaction was catalysed by a Lewis acid, $ZnCl_2$, and base-catalysed methylation followed. The final stages are summarized below.

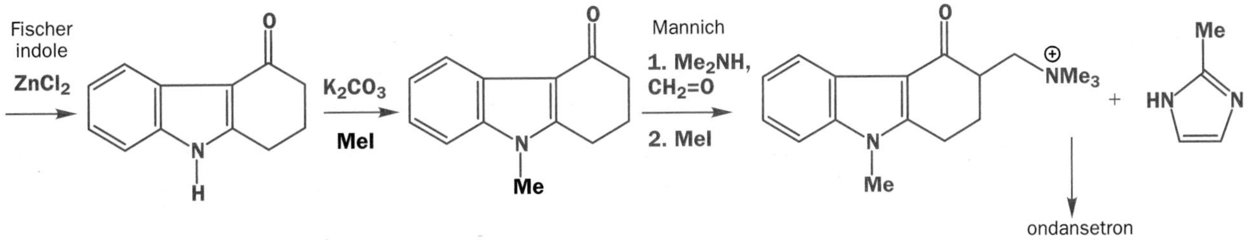

In the worst case, there is no such simple distinction between the two sites for enamine formation and we must rely on other methods of control. The nonsteroidal anti-inflammatory drug indomethacin is a good example. Removing the *N*-acyl group reveals an indole with substituents in both halves of the molecule.

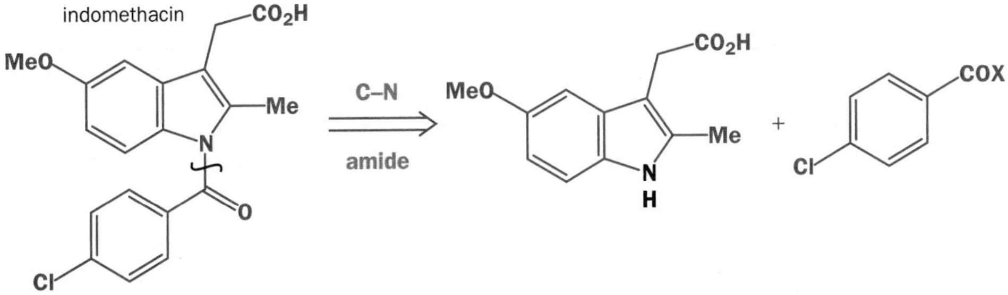

The benzene ring portion is symmetrical and is ideal for the Fischer synthesis but the right-hand half must come from an unsymmetrical open-chain keto-acid. Is it possible to control such a synthesis?

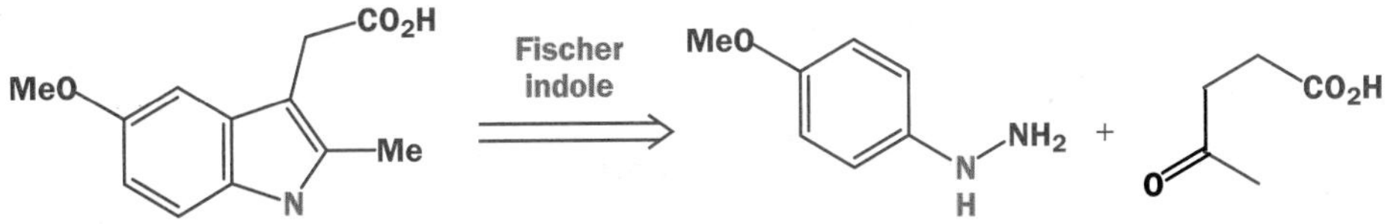

The Fischer indole is acid-catalysed so we must ask: on what side of the ketone is enolization (and therefore enamine formation) expected in acid solution? The answer is away from the methyl group and into the alkyl chain (Chapter 21). This is what we want and the reaction does indeed go this way. In fact, the *t*-butyl ester is used instead of the free acid.

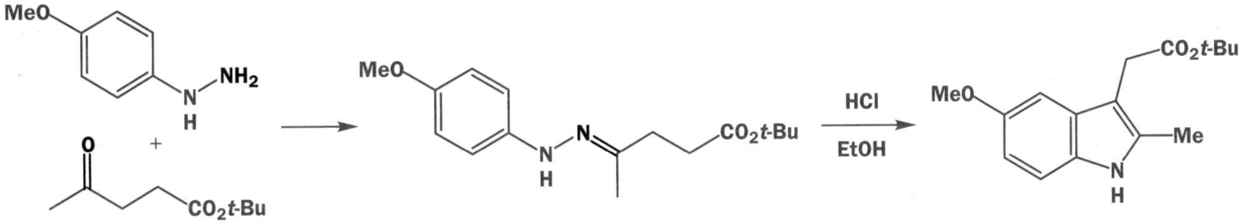

Acylation at the indole nitrogen atom is achieved with acid chloride in base and removal of the *t*-butyl ester gives free indomethacin.

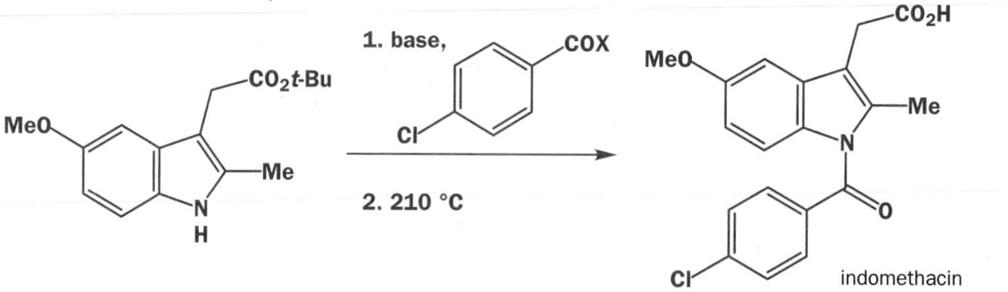

There are many other indole syntheses but we will give a brief mention to only one other and that is because it allows the synthesis of indoles with a different substitution pattern in the benzene ring. If you like names, you may call it the Reissert synthesis, and this is the basic reaction.

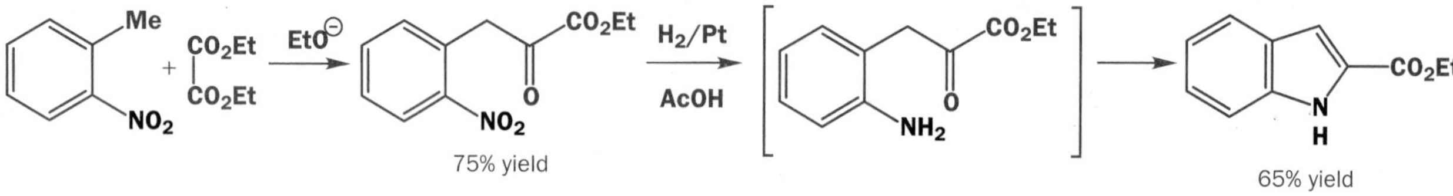

Ethoxide is a strong enough base to remove a proton from the methyl group, delocalizing the negative charge into the nitro group. The anion then attacks the reactive diester (diethyl oxalate) and is acylated by it.

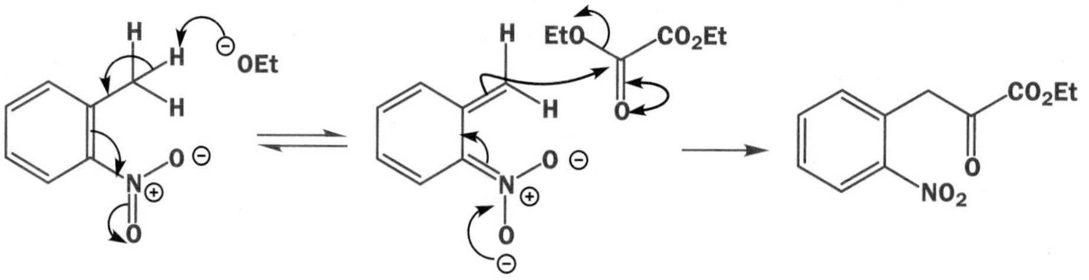

The rest of the synthesis is more straightforward: the nitro group can be reduced to an amine, which immediately forms an enamine by intramolecular attack on the more reactive carbonyl group (the ketone) to give the aromatic indole.

Since the nitro compound is made by nitration of a benzene ring, the preferred symmetry is very different from that needed for the Fischer synthesis. Nitration of *para*-xylene (1,4-dimethylbenzene) is a good example.

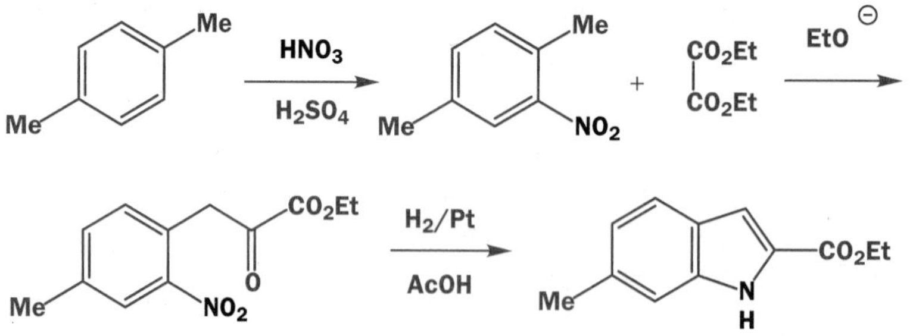

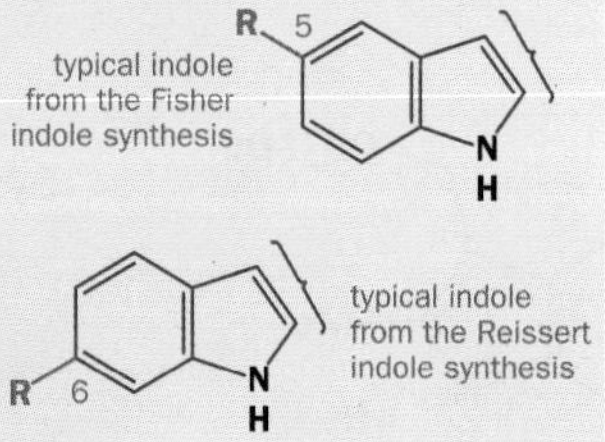

The ester products we have been using so far can be hydrolysed and decarboxylated by the mechanism described in the last chapter if a free indole is required. In any case, it is not necessary to use diethyl oxalate as the electrophilic carbonyl compound. The strange antibiotic chuangxinmycin (which you met in Chapter 32) was made by a Reissert synthesis using the acetal of DMF as the electrophile. Here is part of the synthesis.

► We can contrast the types of indole made by the Fischer and by the Reissert syntheses by the different ideal positions for substituents. These are, of course, not the only possible substitution patterns.

typical indole from the Fisher indole synthesis

typical indole from the Reissert indole synthesis

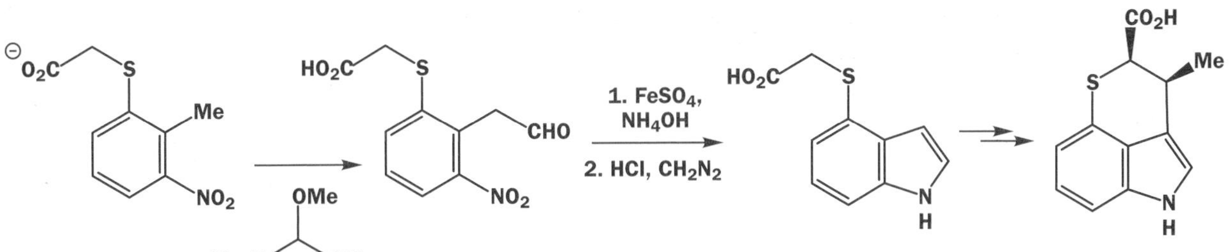

chuangxinmycin

Quinolines and isoquinolines

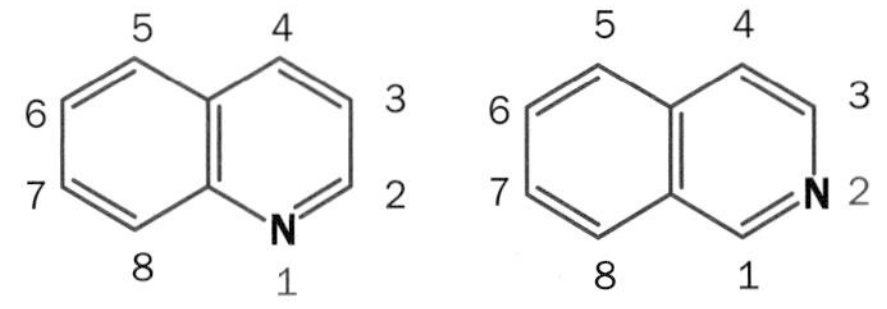

We move from benzo-fused pyrroles to benzo-fused pyridines and meet quinoline and isoquinoline. Isoquinolines will feature as benzylisoquinoline alkaloids in Chapter 51 and their synthesis will mostly be discussed there. In this section we shall concentrate on the quinolines.

Quinoline forms part of the structure of quinine, the malaria remedy found in cinchona bark and known since the time of the Incas. The quinoline in quinine has a 6-MeO substituent and a side chain attached to C4. In discussing the synthesis of quinolines, we will be particularly interested in this pattern. This is because the search for anti-malarial compounds continues and other quinolines with similar structures are among the available anti-malarial drugs.

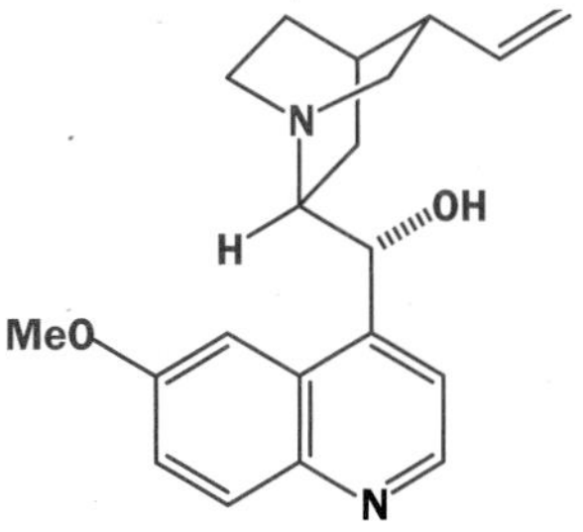

We shall also be very interested in quinolones, analogous to pyridones, with carbonyl groups at positions 2 and 4 as these are useful antibiotics. A simple example is pefloxacin which has a typical 6-F and 7-piperazine substituents.

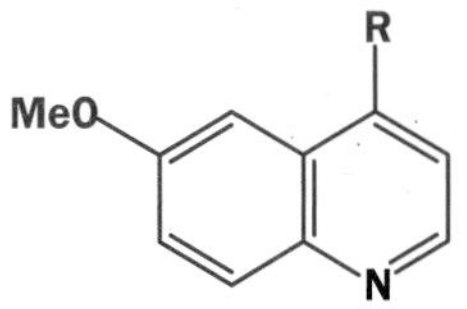

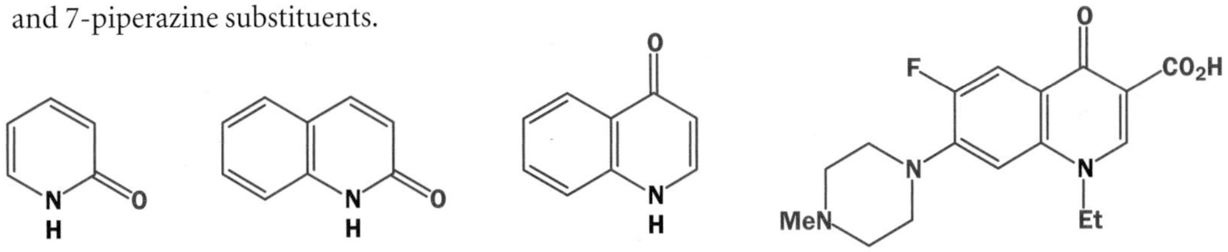

quinoline   isoquinoline

quinine   the quinoline in quinine

2-pyridone   2-quinolone   4-quinolone   a quinolone antibiotic, pefloxacin

When we consider the synthesis of a quinoline, the obvious disconnections are, first, the C–N bond in the pyridine ring and, then, the C–C bond that joins the side chain to the benzene ring. We will need a three-carbon (C$_3$) synthon, electrophilic at both ends, which will yield two double bonds after incorporation. The obvious choice is a 1,3-dicarbonyl compound.

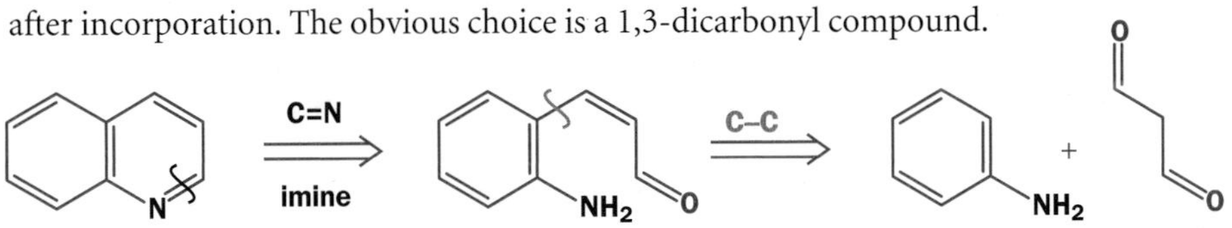

The choice of an aromatic amine is a good one as the NH$_2$ group reacts well with carbonyl compounds and it activates the *ortho* position to electrophilic attack. However, the dialdehyde is malonic dialdehyde, a compound that does not exist, so some alternative must be found. If the quinoline is substituted in the 2- and 4-positions this approach looks better.

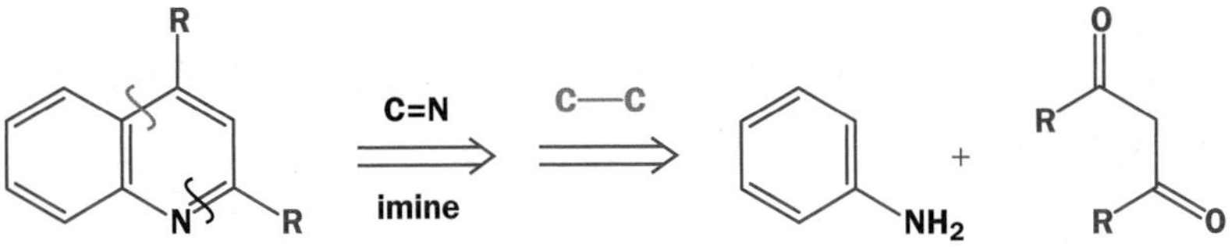

The initially formed imine will tautomerize to a conjugated enamine and cyclization now occurs by electrophilic aromatic substitution.

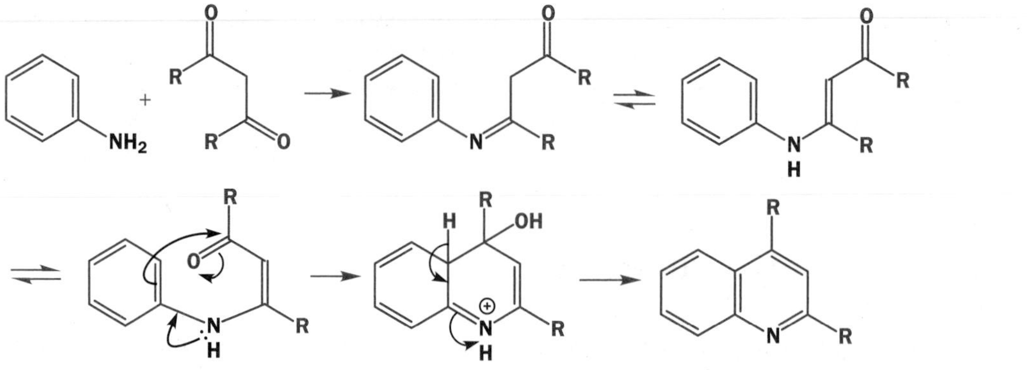

The enamine will normally prefer to adopt the first configuration shown in which cyclization is not possible, and (perhaps for this reason or perhaps because it is difficult to predict which quinoline will be formed from an unsymmetrical 1,3-dicarbonyl compound) this has not proved a very important quinoline synthesis. We shall describe two more important variants on the same theme, one for quinolines and one for quinolones.

In the synthesis of pyridines it proved advantageous to make a dihydropyridine and oxidize it to a pyridine afterwards. The same idea works well in probably the most famous quinoline synthesis, the **Skraup reaction**. The diketone is replaced by an unsaturated carbonyl compound so that the quinoline is formed regiospecifically.

The first step is conjugate addition of the amine. Under acid catalysis the ketone now cyclizes in the way we have just described to give a dihydroquinoline after dehydration. Oxidation to the aromatic quinoline is an easy step accomplished by many possible oxidants.

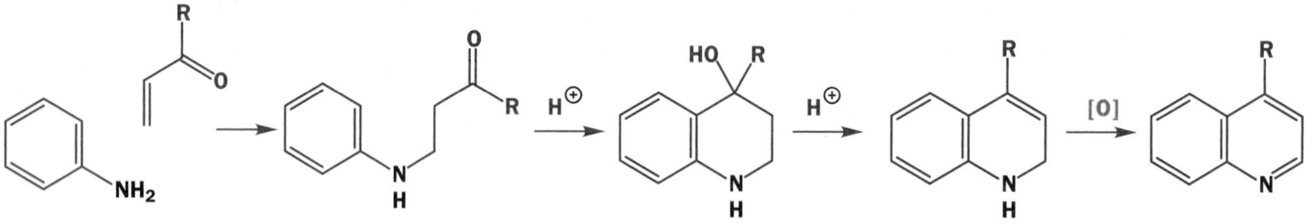

The ugly name of the Skraup reaction appropriately applies to the worst 'witch's brew' of all the heterocyclic syntheses. Some workers have added strange oxidizing agents such as arsenic acid, iron (III) salts, tin (IV) salts, nitrobenzenes of various substitution patterns, or iodine to make it 'go better'.

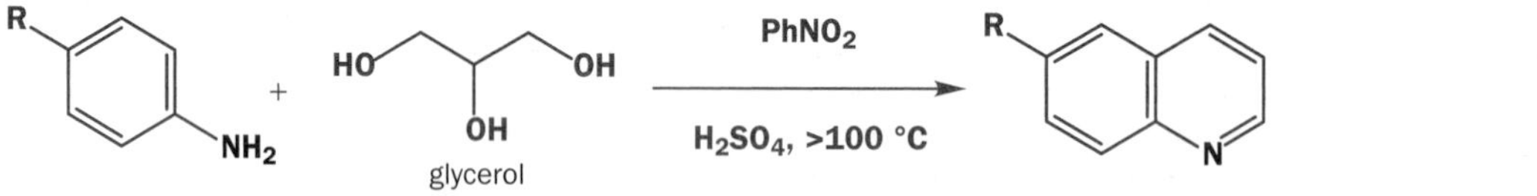

Traditionally, the Skraup reaction was carried out by mixing everything together and letting it rip. A typical mixture to make a quinoline without substituents on the pyridine ring would be the aromatic amine, concentrated sulfuric acid, glycerol, and nitrobenzene all heated up in a large flask at over 100 °C with a wide condenser.

The glycerol was to provide acrolein ($CH_2=CH \cdot CHO$) by dehydration, the nitrobenzene was to act as oxidant, and the wide condenser...? All too often Skraup reactions did let rip—with destructive results. A safer approach is to prepare the conjugate adduct first, cyclize it in acid solution, and then oxidize it with one of the reagents we described for pyridine synthesis, particularly quinones such as DDQ.

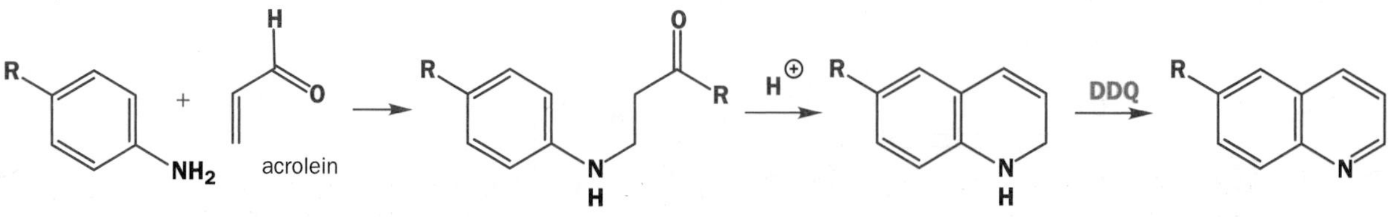

An important use of the traditional Skraup synthesis is to make 6-methoxy-8-nitroquinoline from an aromatic amine with only one free *ortho* position, glycerol, the usual concentrated sulfuric acid, and the oxidant arsenic pentoxide. Though the reported procedure uses 588 grams of As$_2$O$_5$, which might disconcert many chemists, it works well and the product can be turned into other quinolines by reduction of the nitro group, diazotization, and nucleophilic substitution (Chapter 23).

70% yield

Arsenic has a bad reputation because it is traditionally used by poisoners. If arsenic gets into living things it is indeed very poisonous—about 6 mg per kg is needed to kill an animal. However, many other compounds are equally toxic, but you just have to avoid eating them.

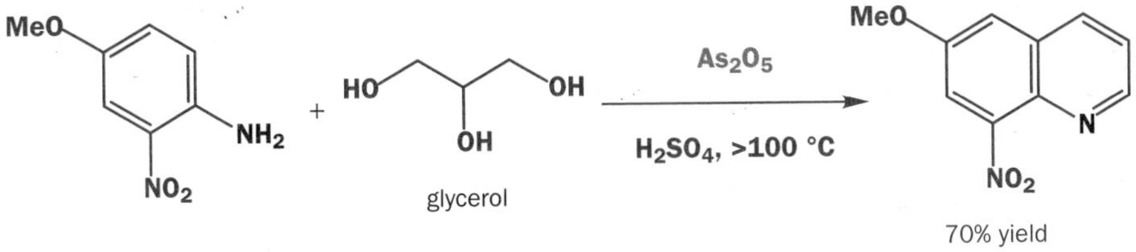

The more modern style of Skraup synthesis is used to make 8-quinolinol or 'oxine'. *ortho*-Aminophenol has only one free position *ortho* to the amino group and is very nucleophilic, so acrolein can be used in weak acid with only a trace of strong acid. Iron(III) is the oxidant with a bit of boric acid for luck, and the yield is excellent.

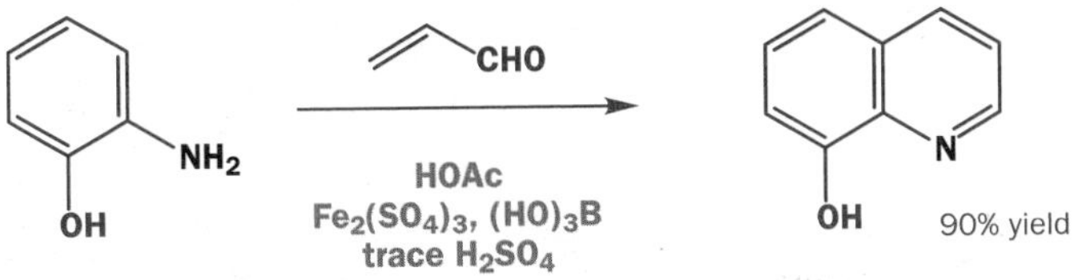

90% yield

oxine complex of copper

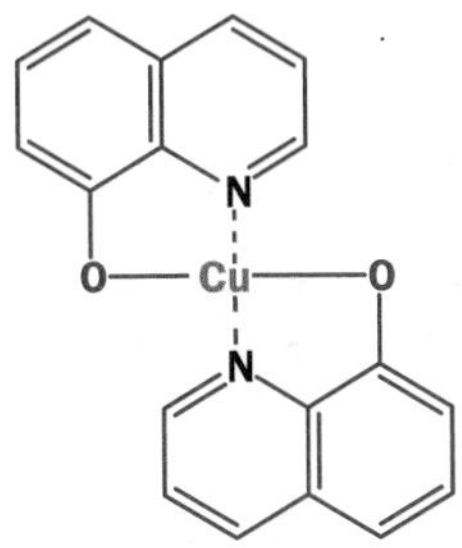

This compound is important because it forms unusually stable metal complexes with metal ions such as Mg(II) or Al(III). It is also used as a corrosion inhibitor on copper because it forms a stable layer of Cu(II) complex that prevents oxidation of the interior.

Quinolones also come from anilines by cyclization to an *ortho* position

The usual method for making quinolone antibiotics is possible because they all have a carboxylic acid in the 3-position. Disconnection suggests a rather unstable malonic ester derivative as starting material.

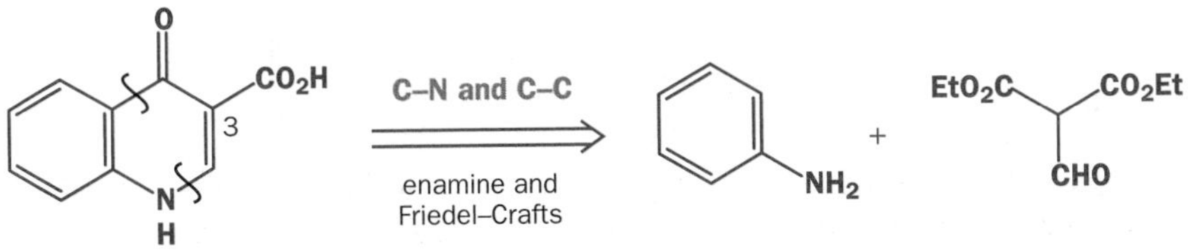

In fact, the enol ether of this compound is easily made from diethyl malonate and ethyl orthoformate [HC(OEt)$_3$]. The aromatic amine reacts with this compound by an addition–elimination sequence giving an enamine that cyclizes on heating. This time there is no worry about the geometry of the enamine.

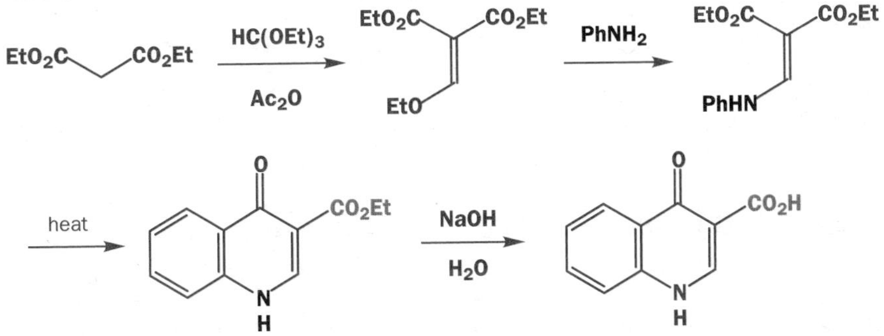

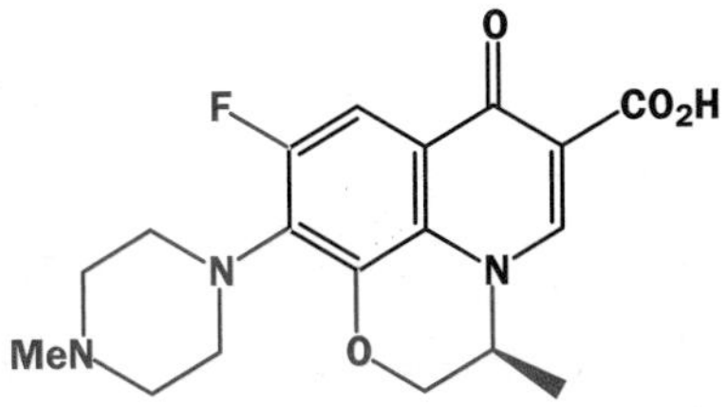

ofloxacin

For examples of quinolone antibiotics we can choose ofloxacin, whose synthesis is discussed in detail in Chapter 23, and rosoxacin whose synthesis is discussed overleaf. Both molecules contain the same quinolone carboxylic acid framework, outlined in black, with another heterocyclic system at position 7 and various other substituents here and there.

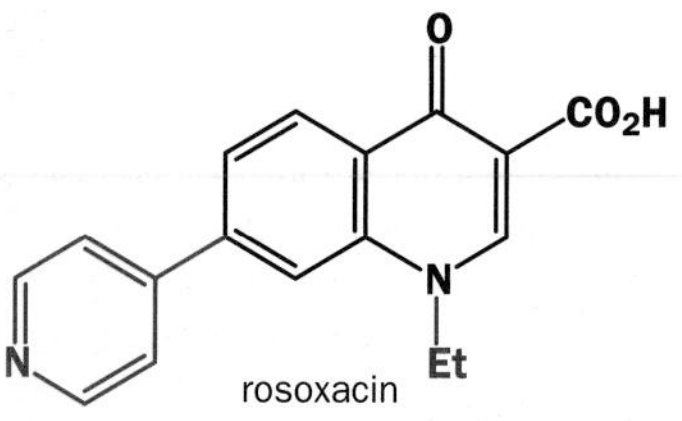

To make rosoxacin two heterocyclic systems must be constructed. Workers at the pharmaceutical company Sterling decided to build the pyridine in an ingenious version of the Hantzsch synthesis using acetylenic esters on 3-nitrobenzaldehyde. The ammonia was added as ammonium acetate. Oxidation with nitric acid made the pyridine, hydrolysis of the esters and decarboxylation removed the acid groups, and reduction with Fe(II) and HCl converted the nitro group into the amino group required for the quinolone synthesis.

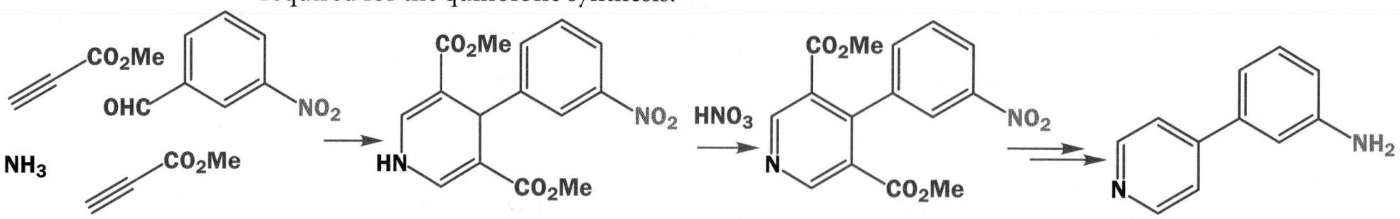

Now the quinolone synthesis can be executed with the same reagents we used before and all that remains is ester hydrolysis and alkylation at nitrogen. Notice that the quinolone cyclization could in theory have occurred in two ways as the two positions *ortho* to the amino group are different. In practice cyclization occurs away from the pyridine ring as the alternative quinolone would be impossibly crowded.

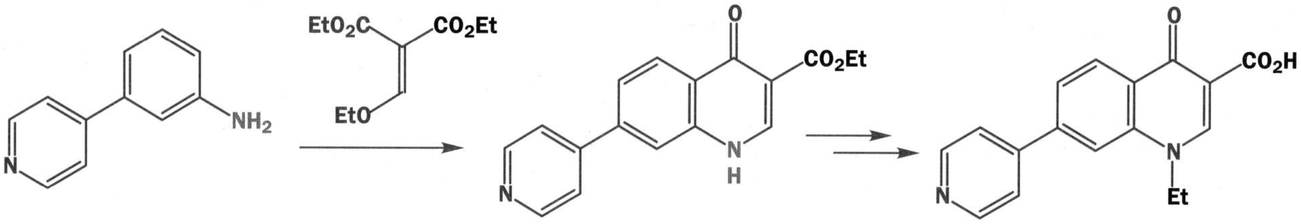

Since quinolones, like pyridones, can be converted into chloro-compounds with $POCl_3$, they can be used in nucleophilic substitution reactions to build up more complex quinolines.

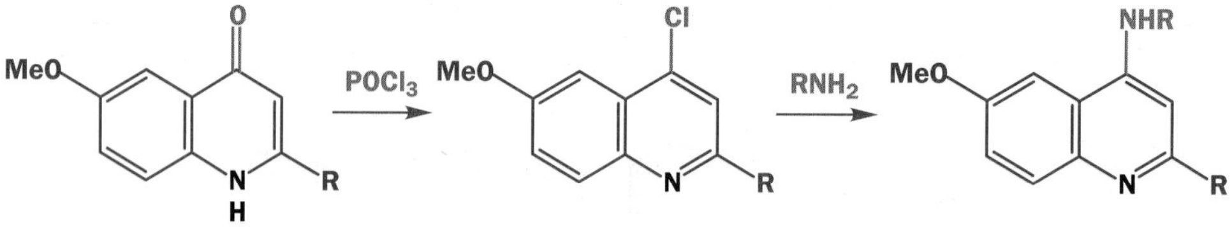

Because isoquinolines are dealt with in more detail in Chapter 51, we will give just one important synthesis here. It is a synthesis of a dihydroisoquinoline by what amounts to an intramolecular Vilsmeier reaction in which the electrophile is made from an amide and $POCl_3$. Since, to make the isoquinoline, two hydrogen atoms must be removed from carbon atoms it makes more sense to use a noble metal such as Pd(0) as the oxidizing agent rather than the reagents we used for pyridine synthesis.

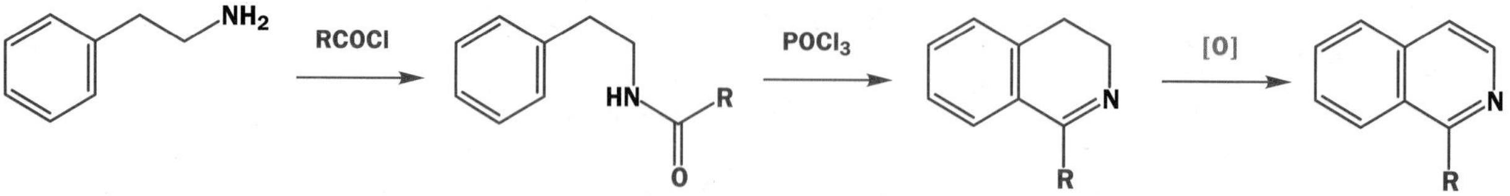

> This dehydrogenation is the reverse of palladium-catalysed hydrogenation.

More heteroatoms in fused rings mean more choice in synthesis

The imidazo-pyridazine ring system forms the basis for a number of drugs in human and animal medicine. The synthesis of this system uses chemistry discussed in Chapter 43 to build the pyridazine ring. There we established that it was easy to make dichloropyridazines and to displace the chlorine atoms one by one with different nucleophiles. Now we will move on from these intermediates to the bicyclic system.

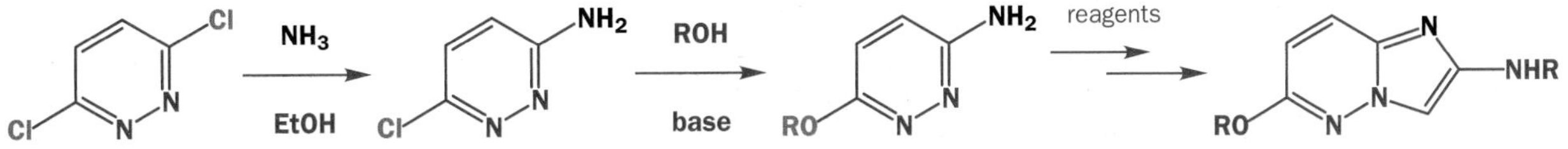

Imidazo[1,2-*b*]pyridazine

A 2-bromo-acid derivative is the vital reagent. It reacts at the amino nitrogen atom with the carbonyl group and at the pyridazine ring nitrogen atom with the alkyl halide. This is the only way the molecule can organize itself into a ten-electron aromatic system.

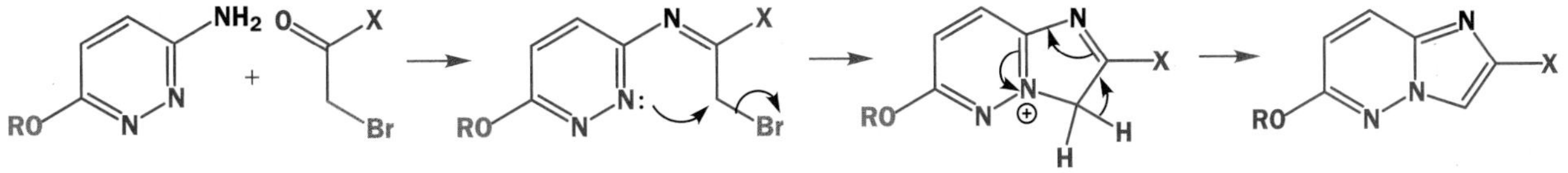

In Chapter 43 we also gave the structure of timolol, a thiadiazole-based β-blocker drug for reduction of high blood pressure. This compound has an aromatic 1,2,5-thiadiazole ring system and a saturated morpholine as well as an aliphatic side chain. Its synthesis relies on ring formation by rather a curious method followed by selective nucleophilic substitution, rather in the style of the last synthesis. The aromatic ring is made by the action of S_2Cl_2 on 'cyanamide'.

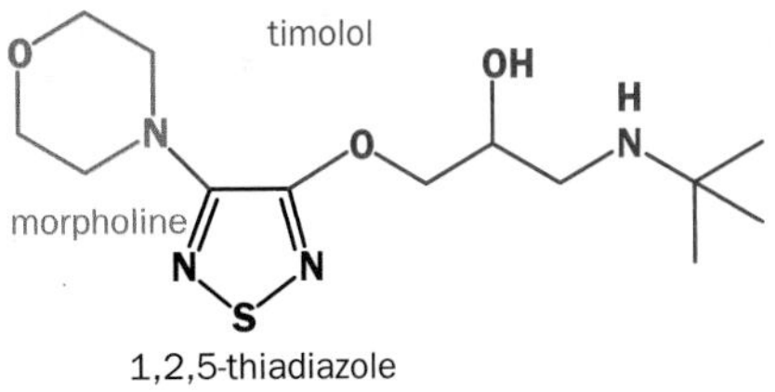

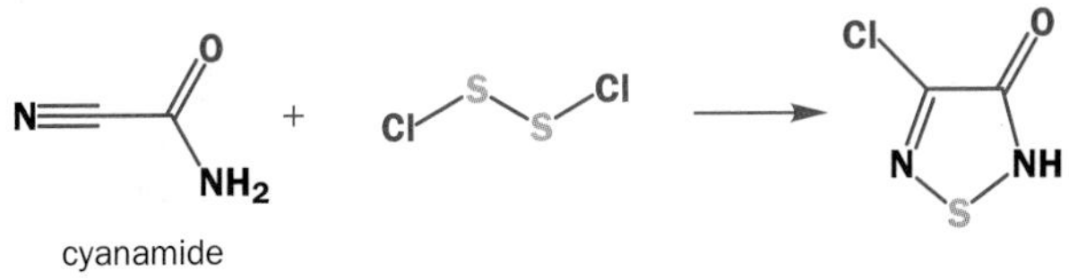

cyanamide

This reaction must start by attack of the amide nitrogen on the electrophilic sulfur atom. Cyclization cannot occur while the linear nitrile is in place so chloride ion must first attack CN. Thereafter cyclization is easy. The chloride ion probably comes from disproportionation of ClS⁻.

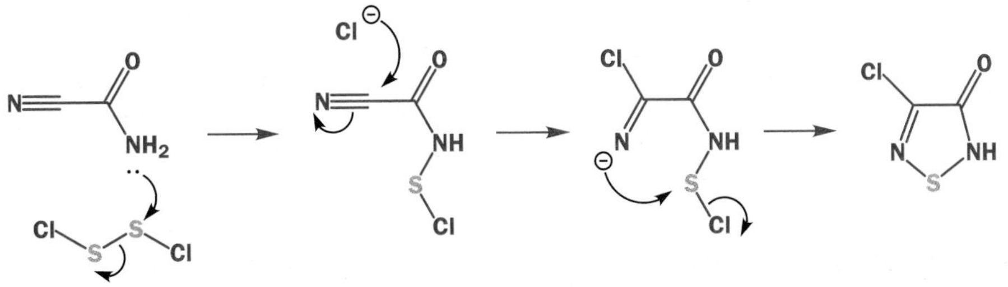

Reaction with epichlorohydrin (the chloroepoxide shown below) followed by amine displacement puts in one of the side chains and nucleophilic substitution with morpholine on the ring completes the synthesis.

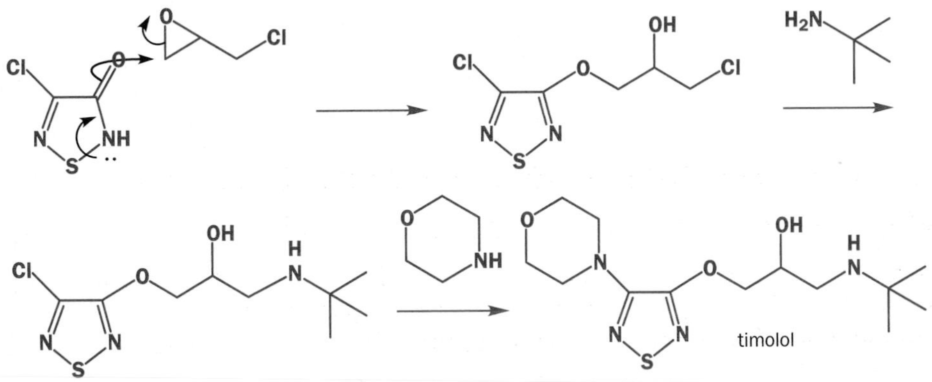

timolol

Summary: the three major approaches to the synthesis of aromatic heterocycles

We end this chapter with summaries of the three major strategies in the synthesis of heterocycles:

- ring construction by ionic reactions
- ring construction by pericyclic reactions
- modification of existing rings by electrophilic or nucleophilic aromatic substitution or by lithiation and reaction with electrophiles

We will summarize the different applications of these strategies, and also suggest cases for which each strategy is not suitable. This section revises material from Chapter 43 as well since most of the ring modifications appear there.

▶ This is only a summary. There are more details in the relevant sections of Chapters 43 and 44. There are also many, many more ways of making all these heterocycles. These methods are just where we suggest you *start*.

Ring construction by ionic cyclization

The first strategy you should try out when faced with the synthesis of an aromatic heterocyclic ring is the disconnection of bonds between the heteroatom or atoms and carbon, with the idea of using the heteroatoms as nucleophiles and the carbon fragment as a double electrophile.

Heterocycles with one heteroatom

five-membered rings

pyrroles, thiophenes, and furans

ideally made by this strategy from 1,4-dicarbonyl compounds

pyrroles

thiophenes

furans

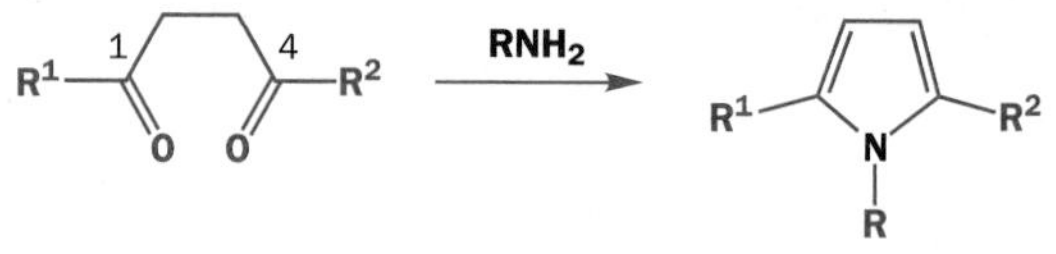

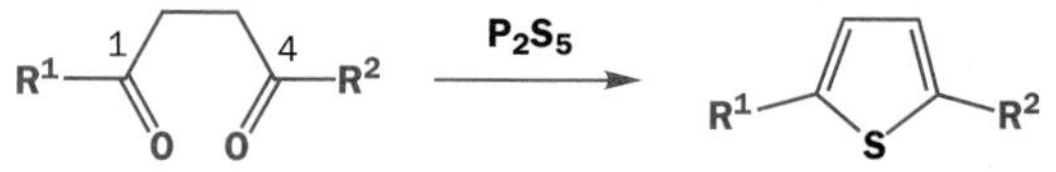

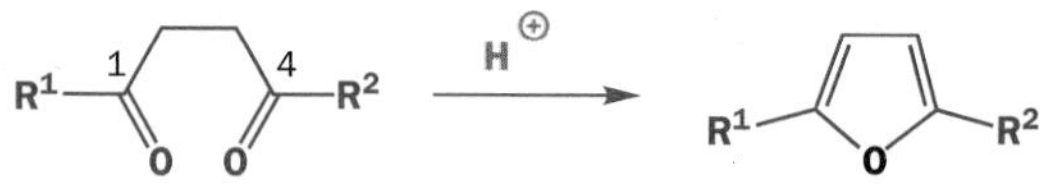

six-membered rings

pyridines

made by this strategy from 1,5-dicarbonyl compounds with oxidation

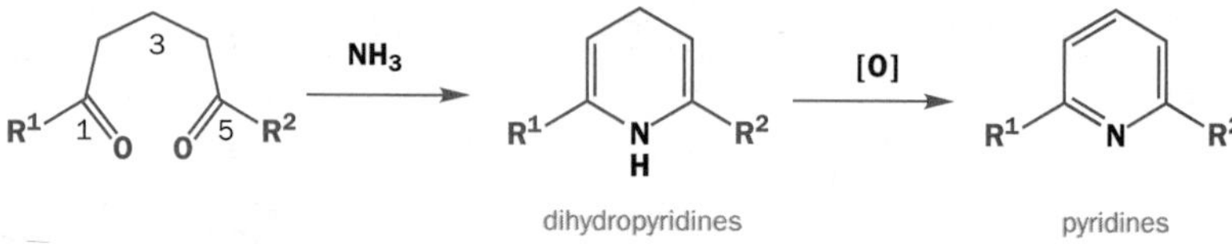

Heterocycles with two adjacent heteroatoms

five-membered rings

pyrazoles and isoxazoles

ideally made by this strategy from 1,3-dicarbonyl compounds

pyrazoles

isoxazoles

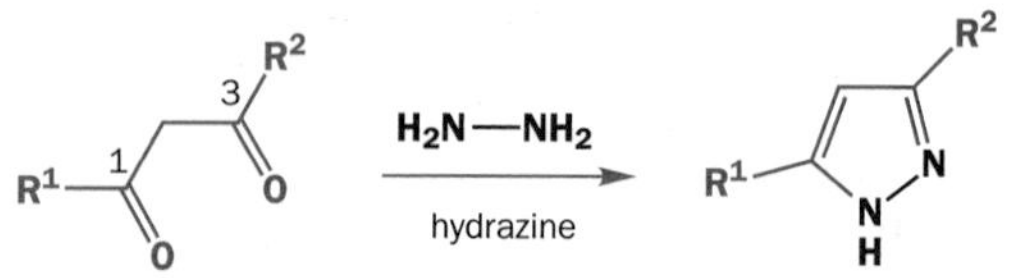

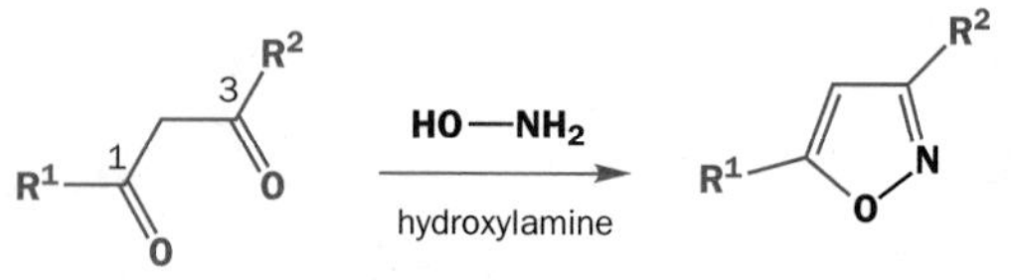

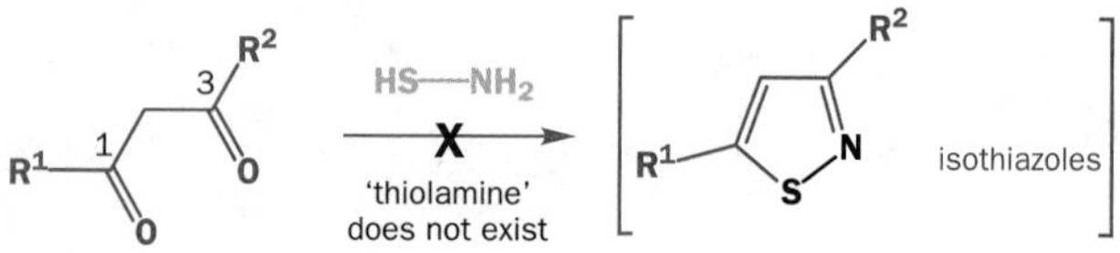

Note. This strategy is *not* suitable for isothiazoles as 'thiolamine' does not exist

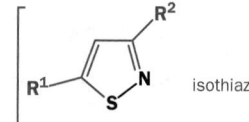

six-membered rings

pyridazines

ideally made by this strategy from
1,4-dicarbonyl compounds with oxidation

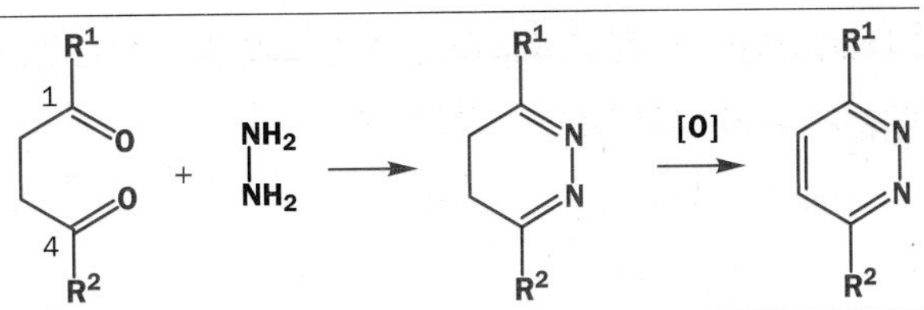

Heterocycles with two non-adjacent heteroatoms

five-membered rings

imidazoles and thiazoles

ideally made by this strategy
from α-halocarbonyl compound

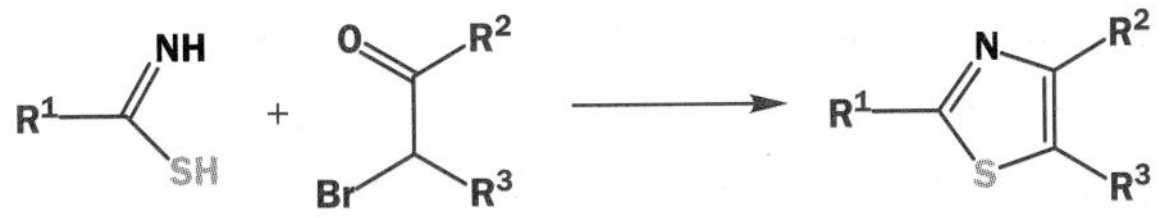

a thiazole

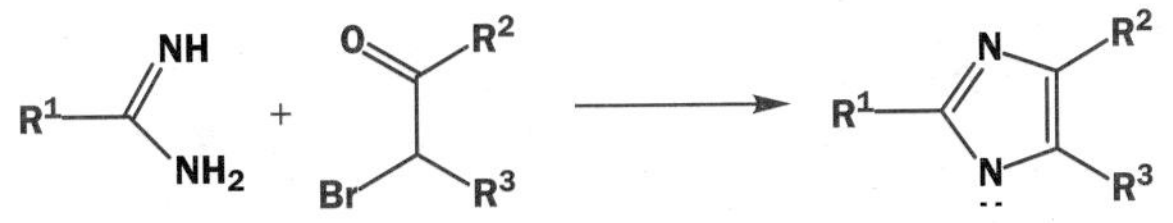

an imidazole

Note. This strategy is *not*
suitable for oxazoles as amides
are not usually reactive enough:
cyclization of acylated carbonyl
compounds is usually preferred

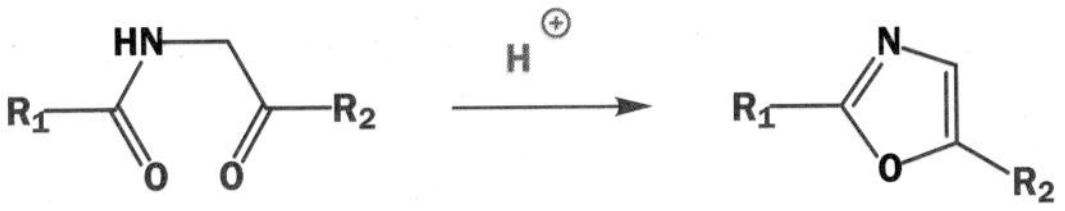

an oxazole

six-membered rings

pyrimidines

ideally made by this strategy from
1,3-dicarbonyl compounds

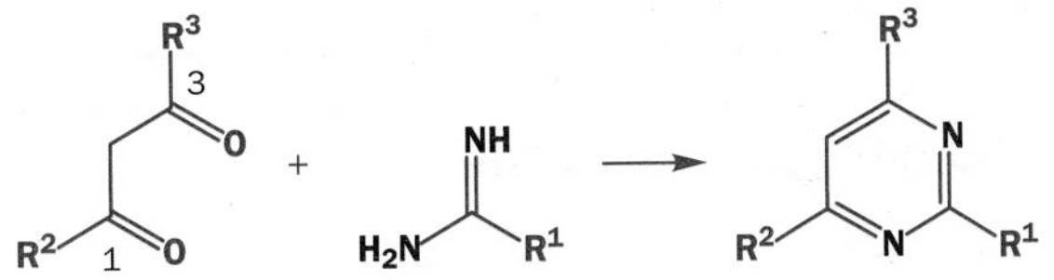

an amidine   a pyrimidine

Ring construction by pericyclic reactions

Cycloaddition reactions

1,3-dipolar cycloaddition
is ideal for the
construction of
isoxazoles,
1,2,3-triazoles, and
tetrazoles

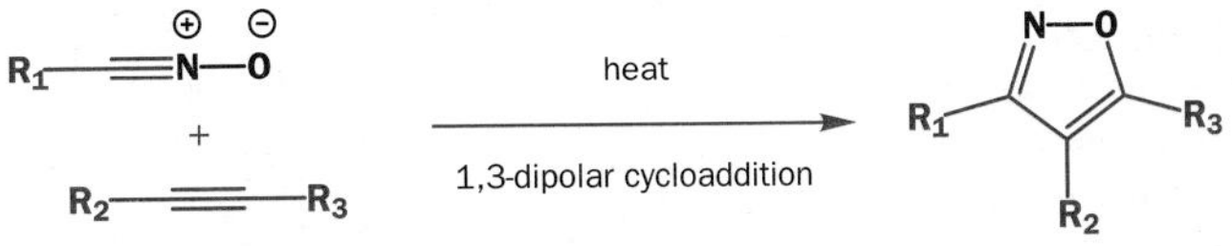

an isoxazole

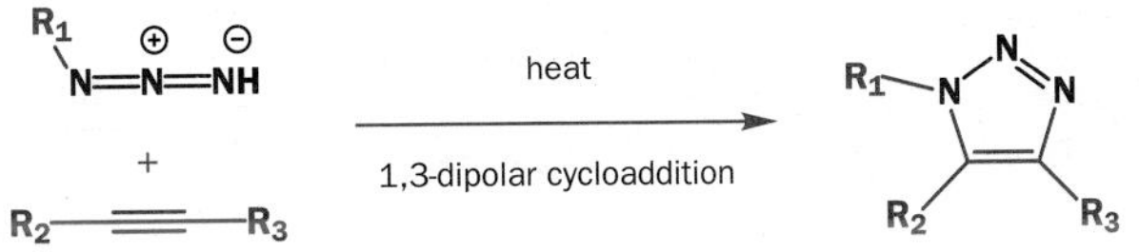

a 1,2,3-triazole

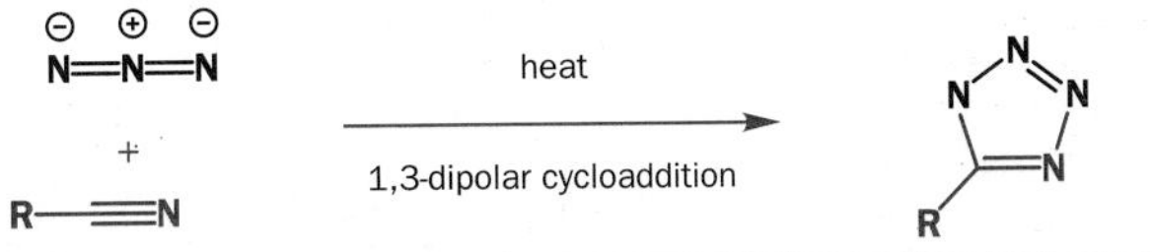

a tetrazole

Sigmatropic rearrangements

a special reaction
that is the vital step
of the Fischer indole
synthesis

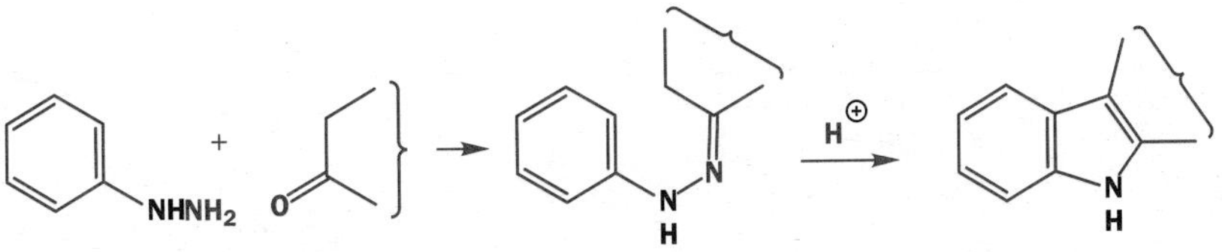

phenylhydrazine   a phenylhydrazone   an indole

Ring modification

Electrophilic aromatic substitution

works very well on pyrroles,
thiophenes and furans where it occurs
best in the 2- and 5-positions and
nearly as well in the 3- and 4-positions.
Often best to block positions where
substitution not wanted

 pyrrole   thiophene   furan   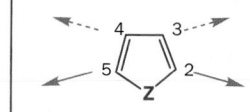

works well for indole—occurs only in
the 3-position but the electrophile
may migrate to the 2-position

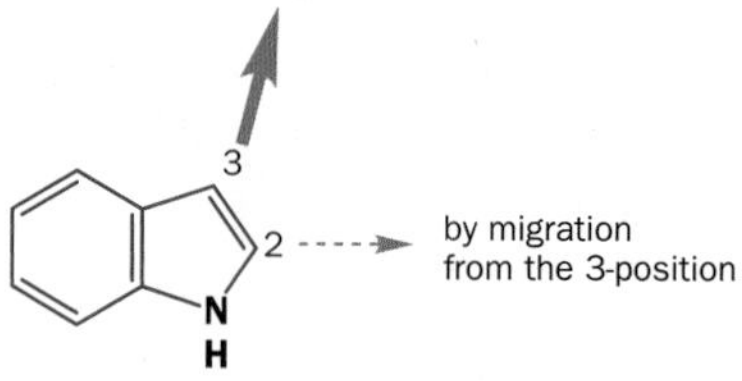

by migration
from the 3-position

indole

works well for five-membered rings with a sulfur, oxygen, or pyrrole-like nitrogen atom and occurs anywhere that is not
blocked (see earlier sections)

Note. Not recommended for pyridine, quinoline, or isoquinoline

Nucleophilic aromatic substitution

works particularly well for pyridine and
quinoline where the charge in the
intermediate can rest on nitrogen

especially important for pyridones and
quinolones with conversion to the
chloro-compound and displacement of
chlorine by nucleophiles and, for
quinolines, displacement of fluorine
atoms on the benzene ring

$POCl_3$ $Nu^{\ominus}$

works well for the six-membered rings
with two nitrogens (pyridazines,
pyrimidines, and pyrazines) in all
positions

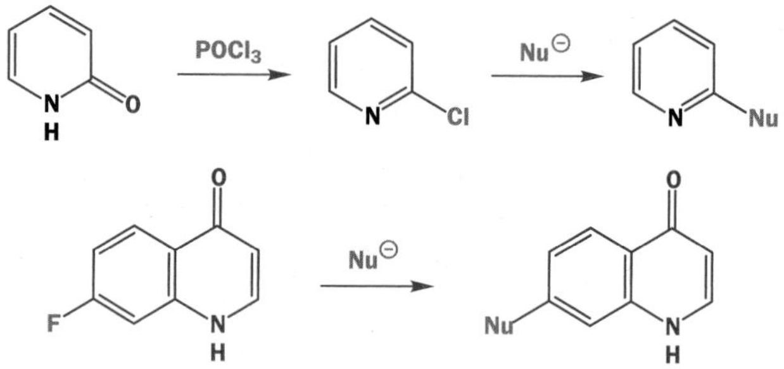

$Nu^{\ominus}$

Lithiation and reaction with electrophiles

works well for pyrrole (if NH blocked),
thiophene, or furan next to the
heteroatom. Exchange of Br or I for Li
works well for most electrophiles
providing any acidic hydrogens
(including the NH in the ring) are blocked

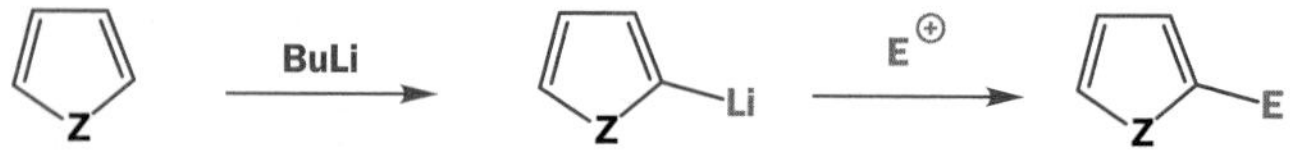

BuLi   Z = NR, S, or O $E^{\oplus}$

Problems

1. In this pyridine synthesis, give a structure for A and mechanisms for the reactions. Why is hydroxylamine used instead of ammonia in the last step?

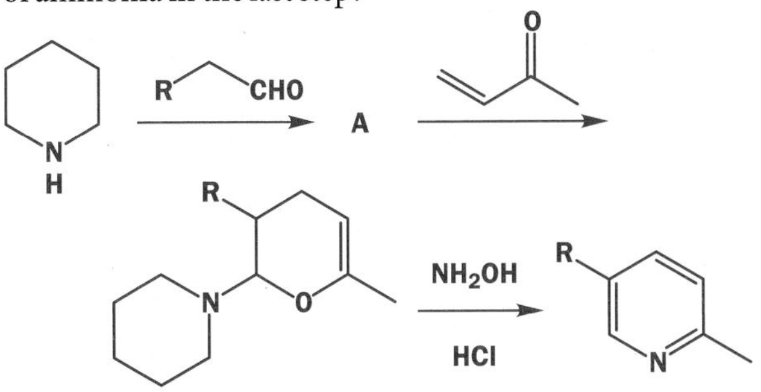

2. Suggest a mechanism for this synthesis of a tricyclic aromatic heterocycle.

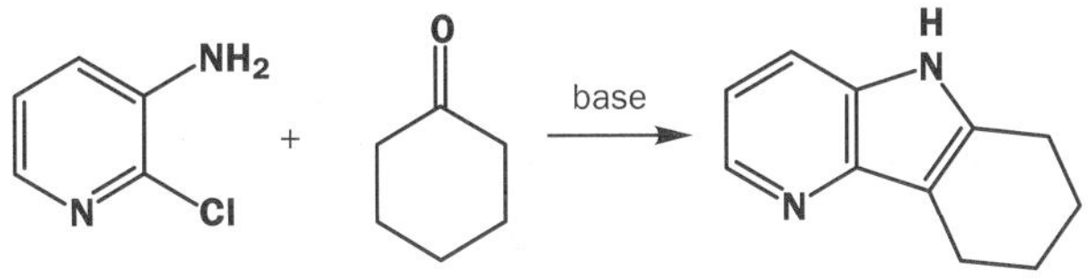

3. How would you synthesize these aromatic heterocycles?

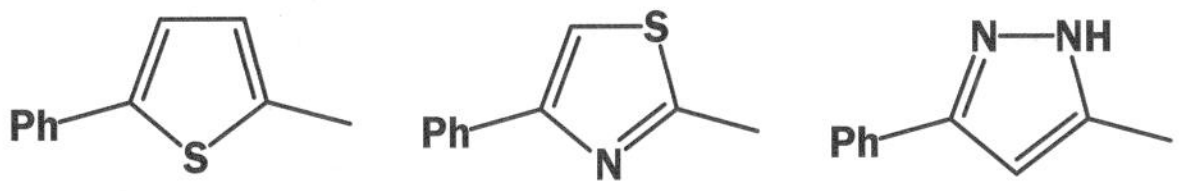

4. Is the heterocyclic ring created in this reaction aromatic? How does the reaction proceed? Comment on the selectivity of the cyclization.

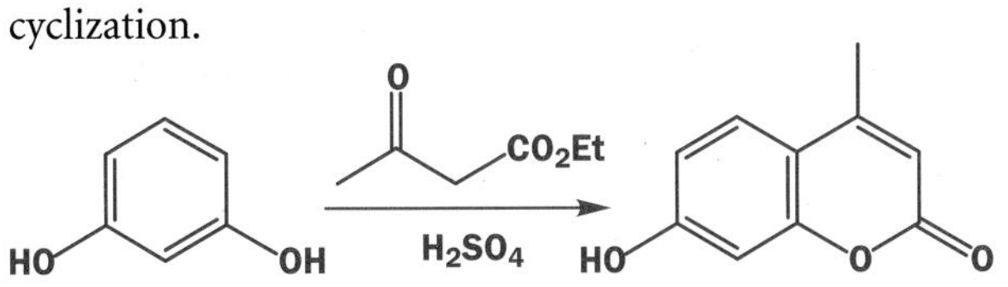

5. Suggest mechanisms for this unusual indole synthesis. How does the second reaction relate to electrophilic substitution at indoles as discussed in Chapter 33?

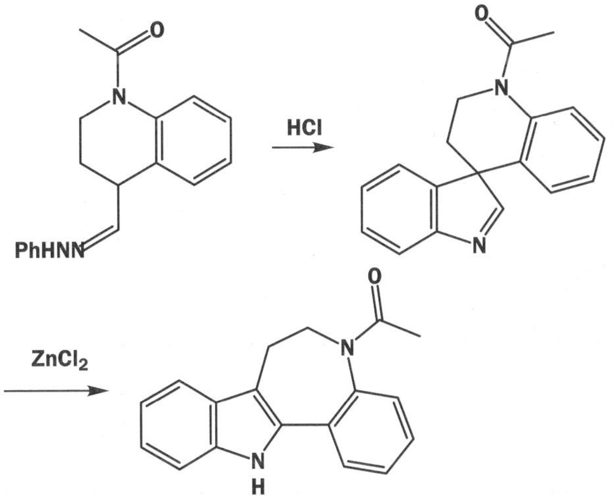

6. Explain the reactions in this partial synthesis of methoxatin, the coenzyme of bacteria living on methanol.

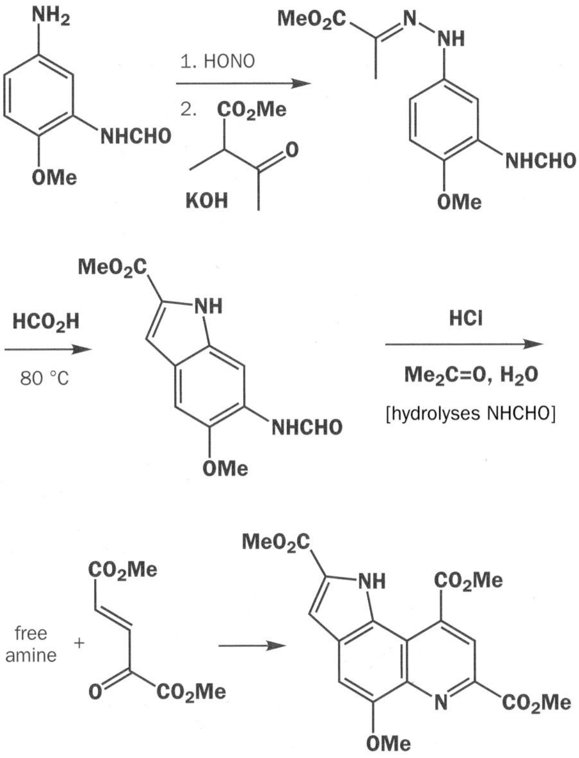

7. Suggest a synthesis of fentiazac, a nonsteroidal anti-inflammatory drug. The analysis is in the chapter but you need to explain why you need these particular starting materials as well as how you would make them.

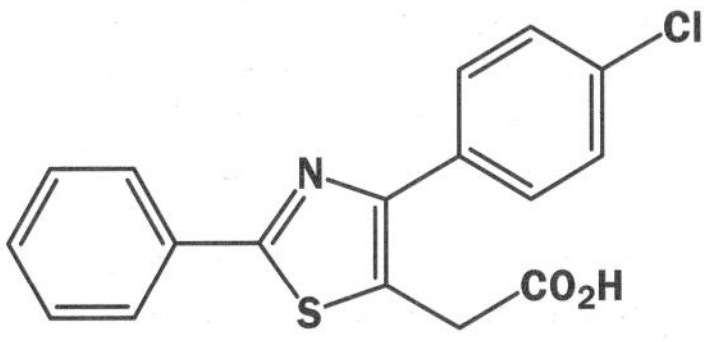

8. Explain why these two quinoline syntheses from the same starting materials give (mainly) different products.

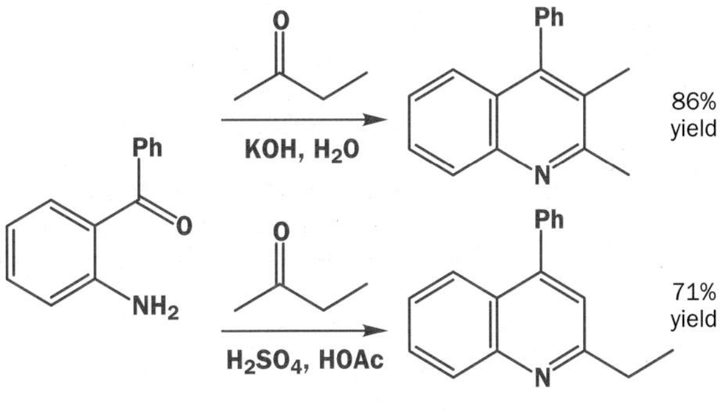

9. Give mechanisms for these reactions used to prepare a fused pyridine. Why is it necessary to use a protecting group?

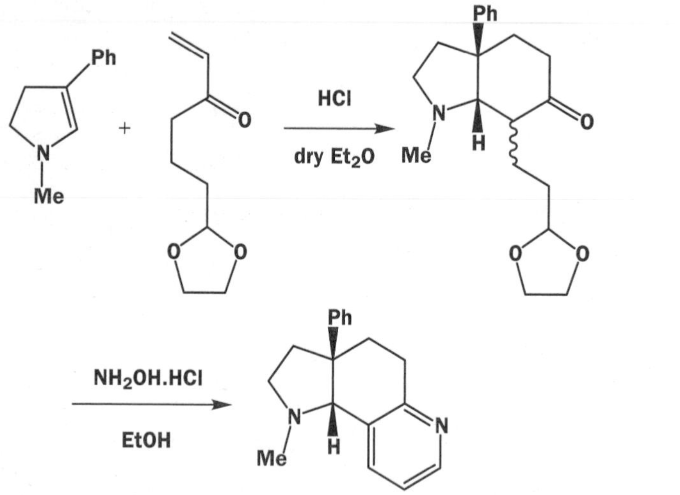

10. Identify the intermediates and give mechanisms for the steps in this synthesis of a triazole.

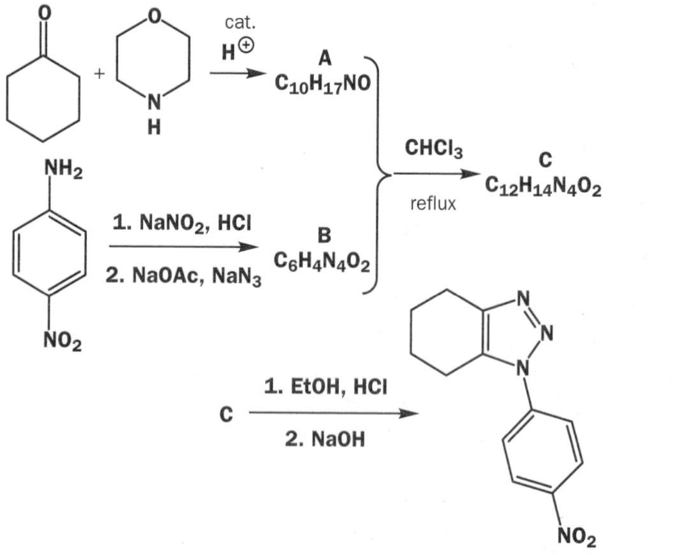

11. Give detailed mechanisms for this pyridine synthesis. The first part revises Chapters 27 and 29.

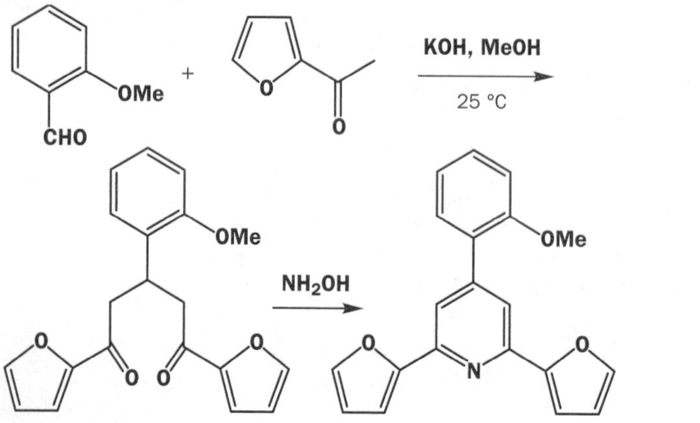

12. This question revises a number of previous chapters, especially 24–26, and 39. Give mechanisms for the reactions in this synthesis of a furan and comment on the choice of reagents for the various steps.

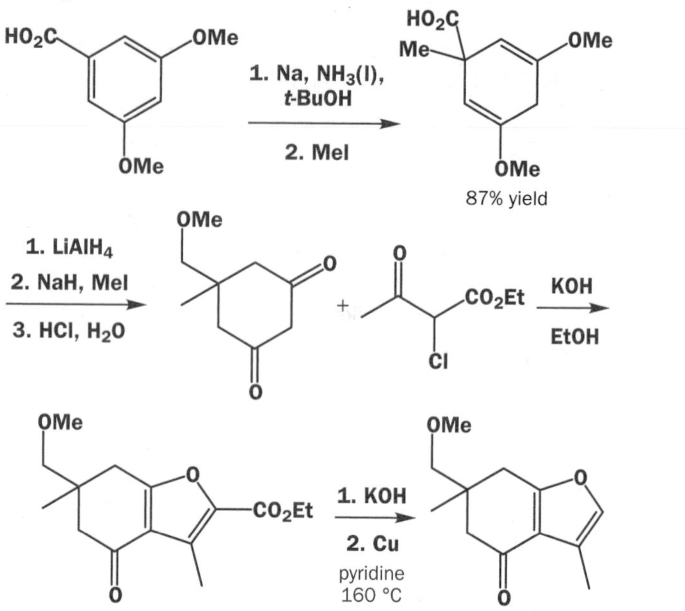

13. Suggest syntheses for this compound, explaining why you choose this particular approach.

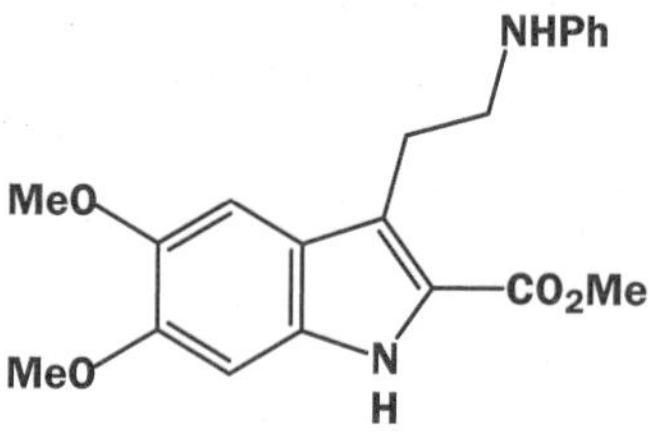

Asymmetric synthesis

45

Connections

Building on:

- **Carbonyl group reactions** ch6, ch9, ch10, ch12, & ch14
- **Controlling stereochemistry** ch16, ch33, & ch34
- **Electrophilic addition to alkenes** ch20
- **Aldol reactions** ch27
- **Diastereoselectivity** ch33–ch34
- **Cycloadditions** ch35

Arriving at:

- **Why making pure enantiomers matters**
- **Chirality derives from nature**
- **Resolution is the last resort**
- **The chiral pool provides starting materials**
- **Chiral auxiliaries are widely used with success**
- **Chiral reagents and catalysts may be even better**
- **Industrial asymmetric synthesis**
- **Two famous methods invented by Sharpless**

Looking forward to:

- **Main group chemistry** ch46–ch47
- **Organometallic chemistry** ch48

Nature is asymmetrical—nature in the looking-glass

'How would you like to live in Looking-glass House, Kitty? I wonder if they'd give you milk in there? Perhaps looking glass milk isn't good to drink...' Lewis Carroll, *Through the looking-glass and what Alice found there*, Macmillan, 1872.

You are chiral, and so are Alice, Kitty, and all living organisms. You may think you look fairly symmetrical in a looking-glass, but as you read this book you are probably turning the pages with your right hand and processing the information with the left side of your brain. Some organisms are rather more obviously chiral: snails, for example, carry shells that could spiral to the left or to the right. Not only is nature chiral, but by and large it exists as just one enantiomer—though some snail shells spiral to the left, the vast majority of marine snail shells spiral to the right; all humans have their stomach on their left and their liver on their right; all honeysuckle climbs by spiralling to the left and all bindweed spirals to the right.

● 'L'univers est dissymmétrique', Louis Pasteur, *ca.* 1860

Nature has a left and a right, and it can tell the difference between them. You may think that human beings are sadly lacking in this respect, since as children we all had to learn, rather laboriously, which is which. Yet at an even earlier age, you could no doubt distinguish the smell of oranges from the smell of lemons, even though this is an achievement at least as remarkable as getting the right shoe on the right foot. The smells of orange and lemon differ in being the left- and right-handed versions of the same molecule, limonene. (*R*)-(+)-Limonene smells rounded and orangey; (*S*)-(−)-limonene is sharp and lemony. Similarly, spearmint and caraway seeds smell quite different, though again this pair of aromas differs only in being the enantiomeric forms of the ketone carvone.

enantiomeric
smells

R-(+)-limonene
smells of oranges

S-(−)-limonene
smells of lemons

S-(+)-carvone
smells of spearmint

R-(−)-carvone
smells of caraway seeds

Even bacteria know their right from their left: *Pseudomonas putida* is a bacterium that can use aromatic hydrocarbons as a foodstuff, degrading them to diols. The diol produced from bromobenzene is formed as one enantiomer only.

Pseudomonas putida

How can this be? We said in Chapter 16 that enantiomers are chemically identical, so how is it that we can distinguish them with our noses and bacteria can produce them selectively? Well, the answer lies in a proviso to our assumption about the identity of enantiomers: they are identical *until they are placed in a chiral environment*. This concept will underlie all we say in this chapter about how to make single enantiomers in the laboratory. We take our lead from Nature: all life is chiral, so all living systems are chiral environments. Nature has chosen to make all its living structures from chiral molecules (amino acids, sugars), and has selected a single enantiomeric form of each. Every amino acid in your body has the *S* and not the *R* configuration, and from this fact, along with the uniform chirality of natural sugars, derives the larger scale chirality of all living structures from the DNA double helix to a blue whale's internal architecture. The answer to Alice's question is most certainly *no*—her kitten will be able to degrade the achiral fats in the milk quite easily, but the proteins (which will be made of *S*-amino acids) and L-lactose will be quite indigestible.

> Some bacteria make their cell walls from 'unnatural' *R*-amino acids to make them unassailable by the (*S*-amino-acid-derived) enzymes used by higher organisms to hydrolyse peptides.

> You might, of course, retort that, in going through the looking-glass, perhaps Alice's kitten has undergone a universal inversion of configuration so that her proteins are all made of *R*-amino acids. Who can tell?

(*S*)-α-amino acid   (*R*)-α-amino acid

■ That is, dopa is not one of the 20 odd amino acids found in proteins; see Chapter 49.

For a perfumer or flavour and fragrance manufacturer, the distinction between enantiomers of the same molecule is clearly of great importance. Nonetheless, we could all get by with caraway-flavoured toothpaste. Yet when it comes to drug molecules, making the right enantiomer can be a matter of life and death. Parkinson's disease sufferers are treated with the non-proteinogenic amino acid dopa (3-(3,4-dihydroxyphenyl)alanine; mentioned in Chapter 51). Dopa is chiral, and only (*S*)-dopa (known as L-dopa) is effective in restoring nerve function. (*R*)-dopa is not only ineffective; it is, in fact, quite toxic, so the drug must be marketed as a single enantiomer. We will look at how L-dopa is made industrially later in the chapter.

> There is no clear relationship between molecular chirality and the chirality of life forms. Right- and left-handed people are made from amino acids and sugars of the same handedness and the rare left-hand-spiralling snails have the same molecular chirality as their more common right-hand-spiralling relatives.

The amphetamine analogue fenfluramine, whose synthesis you designed while you were reading Chapter 31, used to be marketed as an anorectic (appetite-suppressant)—it stimulates the production of the hormone serotonin and makes the body feel satisfied—until it became clear that some undesirable side-effects could be avoided by administering it solely as the (*S*)-enantiomer. Fenfluramine 'relaunched' as the enantiomerically pure dexfenfluramine, and was reputedly 'a turning point for your overweight patients'—was available in the USA as a component of the 'slimming pill' Redux.

L-dopa
marketed as a single enantiomer

D-dopa
toxic

dexfenfluramine is an
appetite suppressant

racemic fenfluramine has
undesirable side-effects

It is not only drugs that have to be manufactured enantiomerically pure. This simple lactone is the pheromone released by Japanese beetles (*Popilia japonica*) as a means of communication. The beetles, whose larvae are serious crop pests, are attracted by the pheromone, and synthetic pheromone is marketed as 'Japonilure' to bait beetle traps. Provided the synthetic pheromone is the stereoisomer shown, with the *Z* double bond and the *R* configuration at the stereogenic centre, only 25 μg per trap catches thousands of beetles. You first met this compound in Chapter 32, where we pointed out that double bond stereocontrol was important since the *E*-isomer of the pheromone is virtually useless as a bait (it retains only about 10% of the activity). Even more important is control over the configuration at the chiral centre, because the *S*-enantiomer of the pheromone is not only inactive in attracting the beetles, but acts as a powerful inhibitor of the *R*-enantiomer—even 1% *S*-enantiomer in a sample of pheromone destroys the activity.

Japanese beetle pheromone

You can see why chemists need to be able to make compounds as single enantiomers. In Chapters 31–34 you looked at relative stereochemistry and how to control it; this chapter is about how to control absolute stereochemistry. In the last 20 years or so, this subject has occupied more organic chemists than probably any other, and we are now at a point where it is not only possible (and in fact essential, because of strict regulatory rules) to make many drug molecules as single enantiomers, but it is also even possible to make some molecules that are indigenous to nature more cheaply in the lab. At least 30% of the world's supply of menthol, for example, is not extracted from plants but is made in Japan using chemical techniques (which you will meet later in this chapter) that produce only a single enantiomer.

Resolution can be used to separate enantiomers

When we first introduced the concept of enantiomers and chirality in Chapter 16, we stressed that any imbalance in enantiomers always derives ultimately from nature. A laboratory synthesis, unless it involves an enantiomerically pure starting material or reagent, will always give a mixture of enantiomers. Here is just such a synthesis of the Japanese beetle pheromone you have just met. You can see the *Z*-selective Lindlar reduction in use—only one geometrical isomer of the double bond is formed—but, of course, the product is necessarily racemic and therefore useless as beetle bait, because in the original addition of the lithiated alkyne to the aldehyde there can be no control over stereochemistry. If all the starting materials and reagents are achiral, the product must be racemic.

racemic pheromone

In Chapter 16 we introduced you to resolution as a means of separating enantiomers, so if we want just the (*R*) compound, we could try that. Resolving the pheromone itself is not straightforward as there are no convenient functional groups to attach a resolving agent to. But the precursor alcohol can be resolved—William Pirkle did this by reacting the racemic alcohol with an enantiomerically pure isocyanate to make a mixture of the two diastereoisomeric amides which he then separated by chromatography. The resolving agent was removed from one of the diastereoisomers to give a single enantiomer of the alcohol, which could be cyclized to the natural (*R*)-pheromone using base and then acid.

This is not, however, the method used to make Japanese beetle pheromone industrially. Resolution, as you have probably realized, is highly wasteful—if you want just one enantiomer, the other ends up being thrown away. In industrial synthesis, this is not an option unless recycling is possible, since chemical plants cannot afford the expense of disposing of such quantities of high-quality waste. So we need alternative methods of making single enantiomers.

> Later in this chapter, you will see an example of resolution of a compound (BINAP) for which there is a demand for *both* enantiomers as components of chiral catalysts. Resolution is the best option there.

The chiral pool—Nature's 'ready-made' chiral centres

A more economical way of making compounds as single enantiomers is to manufacture them using an enantiomerically pure natural product as a *starting material*, rather than just using one as a resolving agent. This method is known as the **chiral pool strategy**, and relies on finding a suitable enantiomerically pure natural product—a member of the chiral pool—that can easily be transformed into the target molecule. The **chiral pool** is that collection of cheap, readily available pure natural products, usually amino acids or sugars, from which pieces containing the required chiral centres can be taken and incorporated into the product.

Sometimes the natural products that are needed are immediately obvious from the structure of the target molecule. An apparently trivial example is the artificial sweetener aspartame (marketed as Nutrasweet), which is a dipeptide. Clearly, an asymmetric synthesis of this compound will start with the two members of the chiral pool, the constituent (natural) (S)-amino acids, aspartic acid and phenylalanine. In fact, because phenylalanine is relatively expensive for an amino acid, significant quantities of aspartame derive from synthetic (S)-phenylalanine made by one of the methods discussed later in the chapter.

Most asymmetric syntheses require rather more than one or two steps from chiral pool constituents. Male bark beetles of the genus *Ips* produce a pheromone that is a mixture of several enantiomerically pure compounds. One is a simple diene alcohol (S)-(−)-ipsenol. Japanese chemists in the 1970s noted the similarity of part of the structure of ipsenol (in black) to the widely available amino acid (S)-leucine and decided to exploit this in a chiral pool synthesis, using the stereogenic centre (green ring) of leucine to provide the stereogenic centre of ipsenol.

> Participation was discussed in Chapters 37 and 41. You will see another example of conversion of NH₂ to OH with retention shortly. This is a useful reaction for converting amino acids to more versatile hydroxy-acids.

The amino group needs to be converted to a hydroxyl group with retention of configuration: diazotization followed by hydrolysis does just this because of neighbouring group participation from the carboxylic acid.

diazotization–hydrolysis of amino acids to give hydroxy-acids proceeds with overall retention

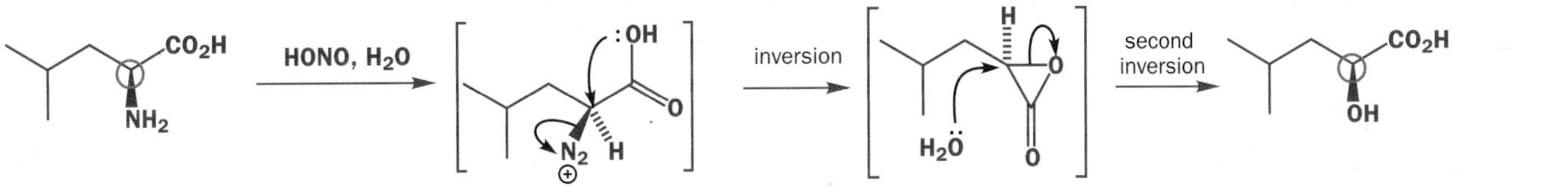

The alcohol was protected as the THP derivative (Chapter 24). Reduction of the acid, via the ester, then allowed introduction of the tosyl leaving group, which was displaced to make an epoxide. The epoxide reacted with a Grignard reagent carrying the diene portion of the target molecule.

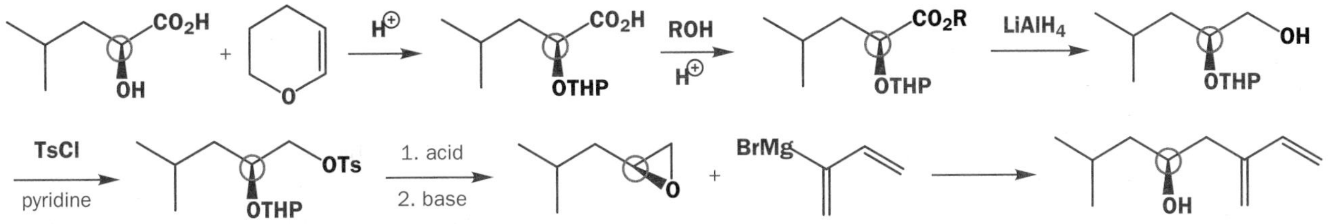

(S)-(–)-ipsenol

Another insect pheromone synthesis illustrates one of the drawbacks of chiral pool approaches. The ambrosia beetle aggregation pheromone is called sulcatol and is a simple secondary alcohol. This pheromone poses a rather unusual synthetic problem: the beetles produce it as a 65:35 mixture of enantiomers so, in order to mimic the pheromone's effect, the chemist has to synthesize both enantiomers separately and mix them together in the right proportion.

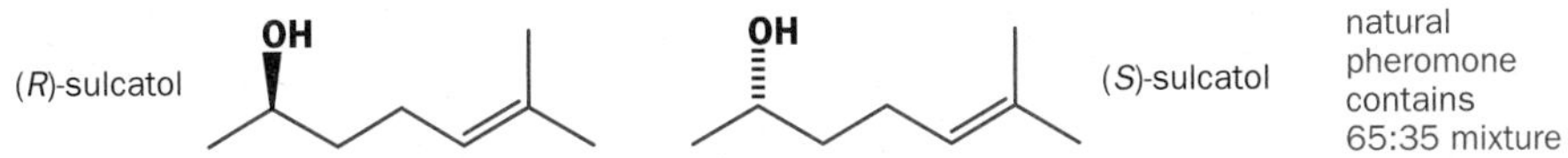

(R)-sulcatol   (S)-sulcatol   natural pheromone contains 65:35 mixture

One approach to the (R)-enantiomer employs the sugar found in DNA, 2-deoxy-D-ribose, as a source of chirality.

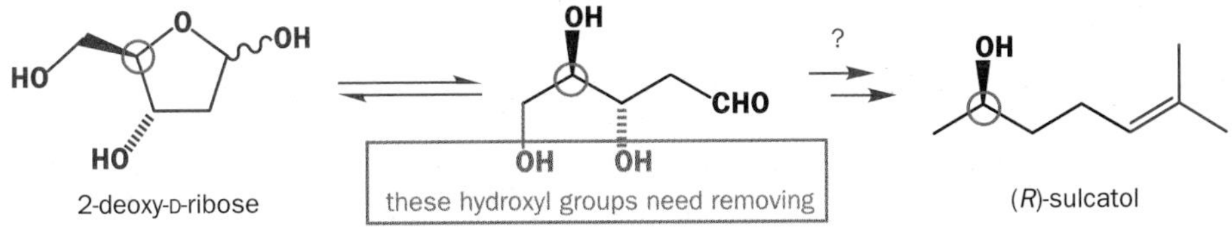

2-deoxy-D-ribose   these hydroxyl groups need removing   (R)-sulcatol

Only one (ringed with green again) of the two defined chiral centres in the sugar appears in the product so, after protecting the hemiacetal, the two free hydroxyl groups were removed by mesylation, substitution by iodide, and reduction. A simple olefination gave (R)-sulcatol. Sugars often need simplifying in this way, because only rarely are all their chiral centres (most have more than two!) needed in the final product.

> The stereochemistry of the iodide formed from the secondary alcohol doesn't matter as it disappears in the next step.

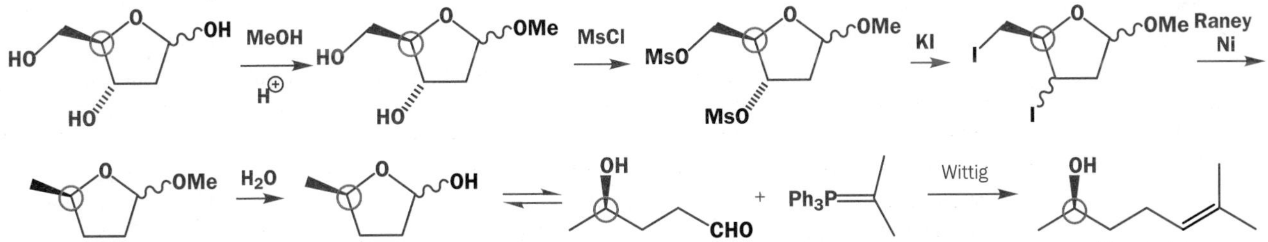

(R)-sulcatol

> Of course, here a resolution strategy would have been ideal!

(S)-Sulcatol cannot be made by this route, because the L-sugar is unavailable (even D-deoxyribose is quite expensive), so an alternative synthesis was needed that could be adapted to give either isomer. The solution is to go back to another hydroxy-acid, ethyl lactate, which is more widely available as its (S)-enantiomer, but which can be converted simply to either enantiomer of a key epoxide intermediate. From (S)-ethyl lactate, protection of the alcohol, reduction of the ester, and tosylation allows ring closure to one enantiomer of the epoxide; tosylation of the secondary hydroxyl group followed by reduction and ring closure gives the other enantiomer.

both enantiomers of propylene oxide can be made from (S)-ethyl lactate

For this reason, the two enantiomers of propylene oxide are commonly used as 'chiral pool' starting materials. These epoxides react with the appropriate Grignard reagent to give either enantiomer of the sulcatol.

(R)- or (S)-propylene oxide   (R)- or (S)-sulcatol

For targets with more than one stereogenic centre, only one need be borrowed from the chiral pool, provided diastereoselective reactions can be used to introduce the others with control over relative stereochemistry. Because the first chiral centre has defined absolute configuration, any diastereoselective reaction that controls the relative stereochemistry of a new chiral centre also defines its absolute configuration. In this synthesis of the rare amino sugar methyl mycaminoside, only one chiral centre comes directly from the chiral pool—the rest are introduced diastereoselectively.

(S)-lactic acid   methyl mycaminoside

The ring was built up from acetylated (S)-lactic acid, and a cyclization step introduced the second chiral centre—the methyl group goes pseudoequatorial while the pseudoaxial position is preferred by the methoxy group because of the anomeric effect (Chapter 42).

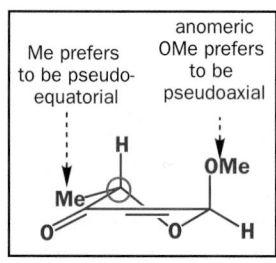

Me prefers to be pseudo-equatorial   anomeric OMe prefers to be pseudoaxial

The third stereogenic centre was controlled by axial reduction of the ketone to give the equatorial alcohol, which then directed introduction of the fourth and fifth stereogenic centres by epoxidation.

In Chapter 18 the conformational factors governing reduction of cyclohexanones are discussed and the directing effects of OH groups in epoxidation are discussed in Chapters 33 and 34.

axial attack leads to more stable equatorial alcohol

hydroxyl group directs epoxidation to top face by hydrogen bonding

Finally, the simple nucleophilic amine Me_2NH attacks the epoxide with inversion of configuration to give methyl mycaminoside. The conformational drawing shows that all substituents are equatorial except the MeO group, which prefers to be axial because of the anomeric effect.

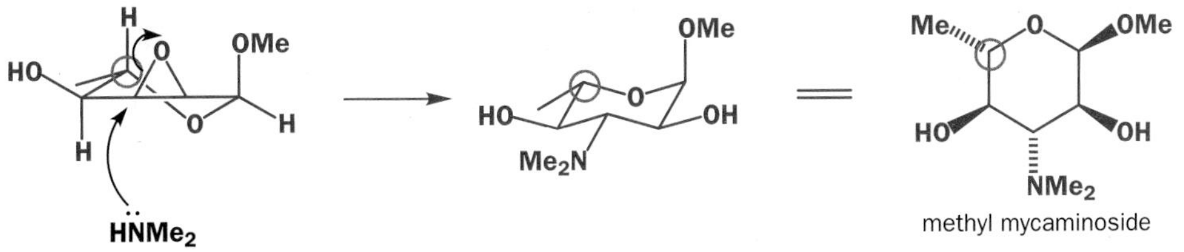

methyl mycaminoside

▶

Normally, axial attack occurs on cyclohexane epoxides as explained in Chapter 18 but the rule is not rigid as you can see here. Equatorial attack occurs here because the transition state already has much of the stability of the product. You should continue to assume axial attack unless told otherwise.

The trouble with chiral pool approaches is that the compound you make has to be pretty close in structure to one of the natural products that are readily available or the synthetic route becomes so tortuous that it's even more wasteful than resolution. The second major drawback is the lack of availability of both enantiomers of most natural products, especially useful starting materials like amino acids and sugars—we have just met this problem with the synthesis of sulcatol from deoxyribose. As a further example, we can return again to our Japanese beetles. Their pheromone can be made from glutamic acid by a short route. Unfortunately, when widely available (*S*)-(+)-glutamic acid is used, the product is the *enantiomer* of the active pheromone, which you will remember is a powerful inhibitor of the natural pheromone. Making the right enantiomer is not economical, because (*R*)-(−)-glutamic acid is about 40 times more expensive than (*S*)-(+)-glutamic acid.

attempted chiral pool synthesis of Japanese beetle pheromone

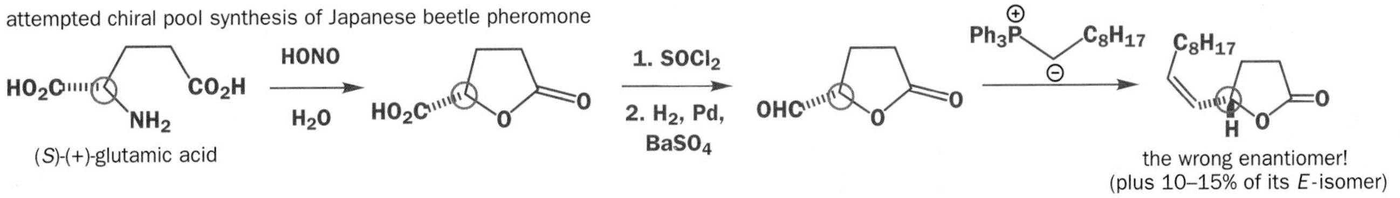

(*S*)-(+)-glutamic acid

the wrong enantiomer!
(plus 10–15% of its *E*-isomer)

Asymmetric synthesis

When we create a new stereogenic centre in a previously achiral molecule using achiral reagents (addition of CN$^-$ to aldehydes was the example you met in Chapter 16), we get a racemic mixture because the transition states leading to the two enantiomers are themselves enantiomeric and therefore equal in energy.

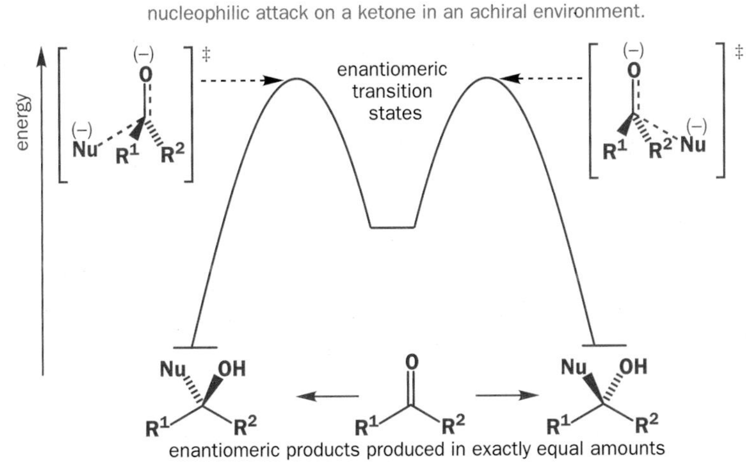

nucleophilic attack on a ketone in an achiral environment.

enantiomeric transition states

enantiomeric products produced in exactly equal amounts

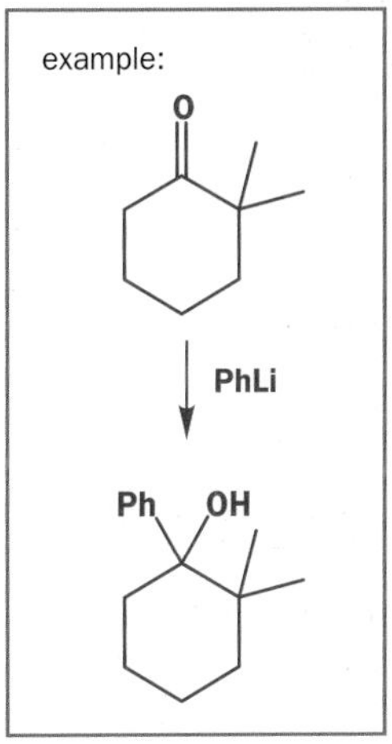

example:

Diastereoselective synthesis, on the other hand, relies on making the transition states for reactions leading to different diastereoisomers as different in energy as possible and therefore favouring the formation of one diastereoisomer over another. You met this type of stereoselectivity in Chapter 33. Here is a simple example: PhLi adds to this ketone to give one diastereoisomer of the tertiary alcohol and not the other. Attack on one or other face of the ketone leads to diastereomeric transition states: this is perhaps most obvious when you realize that one is axial and one equatorial attack. An energy diagram for this type of reduction appears on the next page.

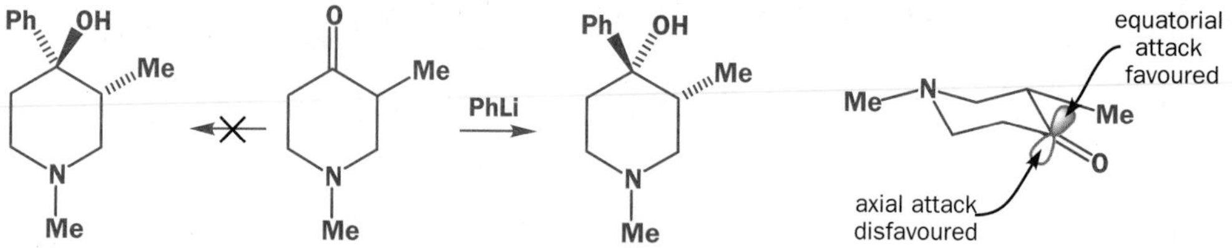

Now, let's go back to the principle of resolution and see how we can devise a way of improving upon it that doesn't require us to throw away 50% of our product. Resolution works because attaching an enantiomerically pure resolving agent to the racemic substrate distinguishes the substrate's two enantiomers as diastereoisomers (diastereoisomers are chemically different; enantiomers are not). Can we use this same idea to make two enantiomeric (and therefore equal in energy) *transition states* into diastereoisomeric ones (which will therefore be unequal in energy)? If we can, the lower-energy transition state will be favoured and we will get more of one enantiomer than the other.

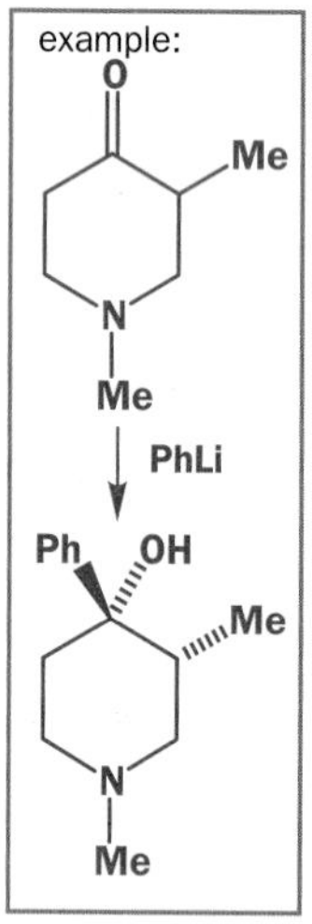

nucleophilic attack on a ketone in a chiral environment.

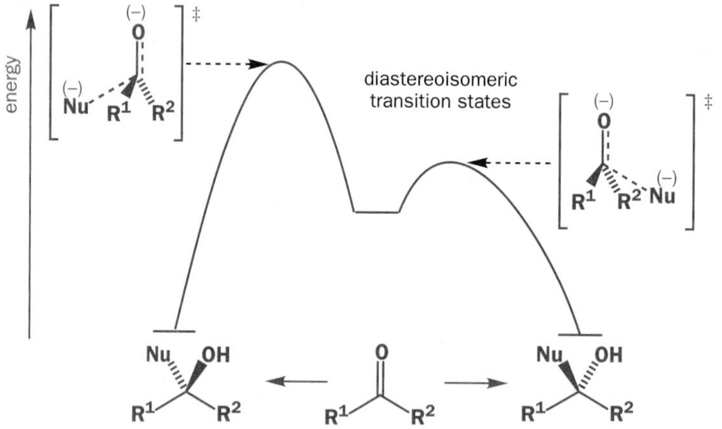

diastereoisomeric transition states

enantiomeric products produced in unequal amounts

The answer is most definitely yes—what is needed is an enantiomerically pure molecule or part of a molecule that will be present during the reaction and will interact with the transition state of the reaction in such a way that it controls the formation of the new stereogenic centre. This molecule might be a reagent or a catalyst, or it might be covalently attached to the starting material. We will consider all of these possibilities, the last first, and you will see that they really are the most powerful and versatile ways of making enantiomerically pure compounds.

Chiral auxiliaries

▶
'Auxiliary' has but one 'l'.

The product of a Diels–Alder reaction between cyclopentadiene and benzyl acrylate must necessarily be racemic as both reagents are achiral. Though only one *diastereoisomer*—the *endo* product—is formed, it must be formed as an exactly 50:50 mixture of *enantiomers*.

Diels–Alder reaction gives a racemic product

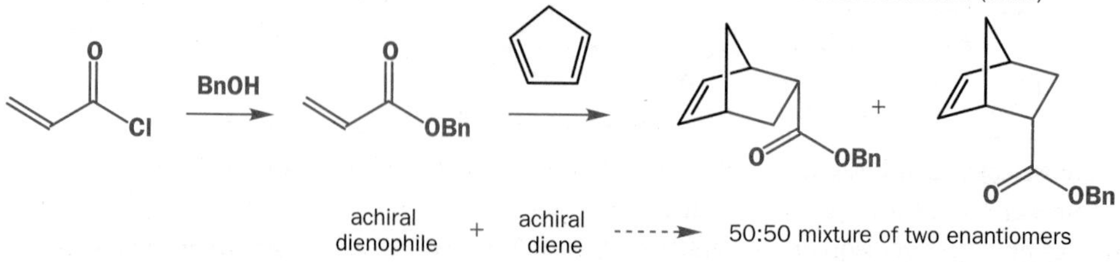

one diastereoisomer (*endo*)

achiral dienophile + achiral diene -----→ 50:50 mixture of two enantiomers

Now see what happens if we replace the achiral benzyl ester group with an amide derived from the natural amino acid valine (Chapter 49). The diastereoselectivity remains the same but the chiral environment created by the single enantiomer covalently bonded to the dienophile has a remarkable effect: only one enantiomer of the product is formed.

chiral auxiliary-controlled Diels–Alder reaction gives a single enantiomer of the product

single enantiomer
derived from (*S*)-valine

single enantiomer
of dienophile

+   achiral
diene

this enantiomer only

As far as stereoselectivity is concerned, the key step is the Diels–Alder reaction—in each case the diene (cyclopentadiene, shown in black) adds across the dienophile, an acrylic acid derivative. As you would expect from what we said in Chapter 35, both reactions are diastereoselective in that they generate mainly the *endo* product. In the first example, that is all there is to say: the product that is formed is necessarily racemic because all the starting materials in the reaction were achiral.

But, in the second example, a green **chiral auxiliary** has been attached to one of the starting materials. It contains another stereogenic centre and is enantiomerically pure—it was, in fact, made by a chiral pool strategy from the amino acid (*S*)–valine (see below). You can see that it has quite an effect on the reaction—the extra stereogenic centre means that there are now *two* possible diastereoisomeric *endo* products, but only one is formed.

this adduct is formed as
a single diastereoisomer

removal of the
chiral auxiliary
reveals this
compound as
a single enantiomer

none of
this enantiomer
is formed

The chiral auxiliary was enantiomerically pure—every molecule had the same configuration at its stereogenic centre. That centre was not involved in the Diels–Alder reaction, so all the products will similarly have the same configuration at the stereogenic centre in the green part of the molecule. So, if one diastereoisomer of the product is formed, all the stereogenic centres in that product must be of a single configuration; in other words the product is diastereoisomerically *and* enantiomerically pure. And when we do the final step of the sequence, to remove the chiral auxiliary, that enantiomeric purity remains, despite the fact that we have removed its source. Overall, by sequential attachment and removal of the auxiliary we have made the same product but as a single enantiomer.

▶
You may note the inclusion of the Et$_2$AlCl Lewis acid catalyst in the second reaction. As we discussed in Chapter 35, the presence of a Lewis acid increases the rate of Diels–Alder reactions, and in this case is also vital for high stereoselectivity.

● **This is what we mean by a chiral auxiliary strategy**

1. An enantiomerically pure compound (usually derived from a simple natural product like an amino acid), called a chiral auxiliary, is attached to the starting material.

2. A diastereoselective reaction is carried out, which, because of the enantiomeric purity of the chiral auxiliary, gives only one enantiomer of the product.

3. The chiral auxiliary is removed by, for example, hydrolysis, leaving the product of the reaction as a single enantiomer. The best chiral auxiliaries (of which the example above is one) can be recycled, so although stoichiometric quantities are needed, there is no waste.

We have introduced you to this chiral auxiliary before any other because it is more commonly used than any other. It is a member of the oxazolidinone (the name of the heterocyclic ring) family of auxiliaries developed by David Evans at Harvard University, and is easily and cheaply made from the amino acid (S)-valine. Not only is it cheaply made: it can also be recycled. The last step of the route above, transesterification with benzyl alcohol, regenerates the auxiliary ready for re-use.

synthesis of Evans's oxazolidinone chiral auxiliary from (S)-valine

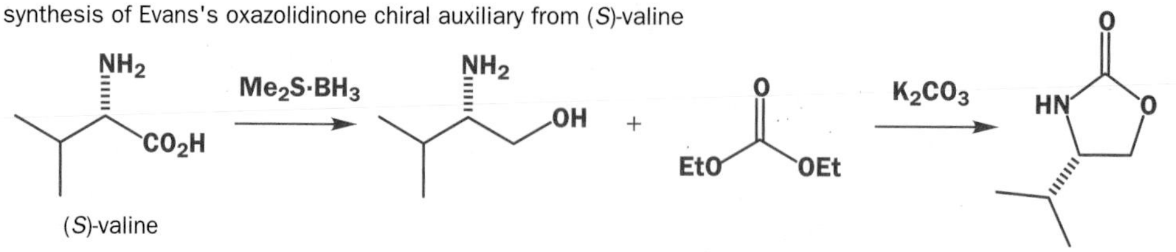

The most versatile chiral auxiliaries should also be available as both enantiomers. Now, for the valine-derived one here, this is not the case—(R)-valine is quite expensive since it is not found in nature. However, by starting with the naturally occurring (and cheap) compound norephedrine, we can make an auxiliary that, although not enantiomeric with the one derived from (S)-valine, acts as though it were. Here is the synthesis of the auxiliary.

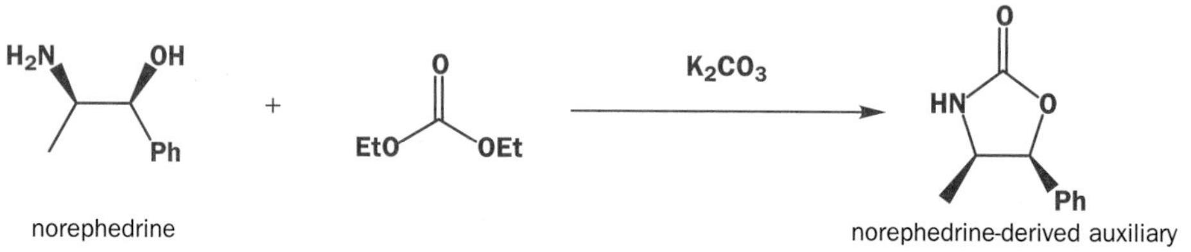

And here it is promoting the same asymmetric Diels–Alder reaction, but giving the enantiomeric product.

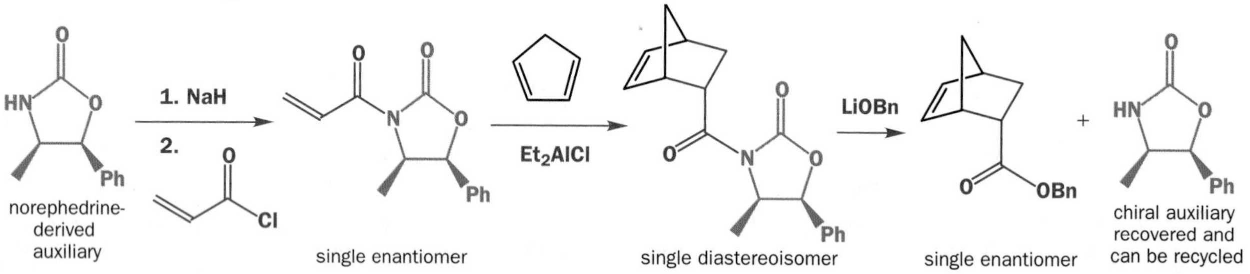

How do these auxiliaries fulfil their role? If we go back to the valine-derived auxiliary and draw the auxiliary-bearing dienophile coordinated with the Lewis acid you can clearly see that the isopropyl group shields the back face of the alkene from attack: when the cyclopentadiene moves in, it must approach from the front face (and remember it will align itself to gain maximum secondary orbital stabilization and therefore give the *endo* product).

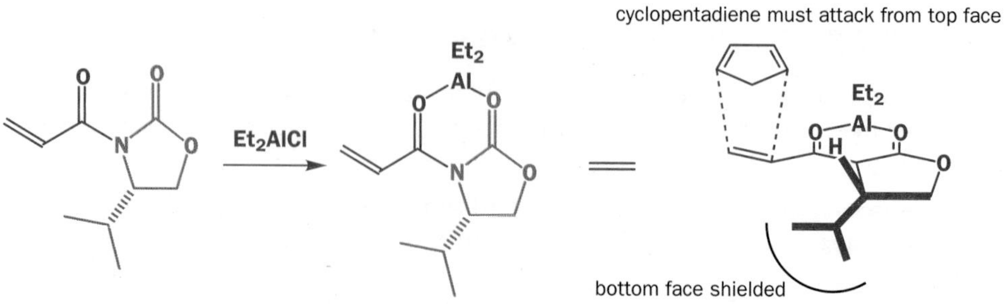

Note that the auxiliary also has the effect of fixing the conformation of the black single bond as s-*cis* (we introduced this nomenclature on p. 907). Attack on the top face of the s-*trans* compound would give the enantiomeric product.

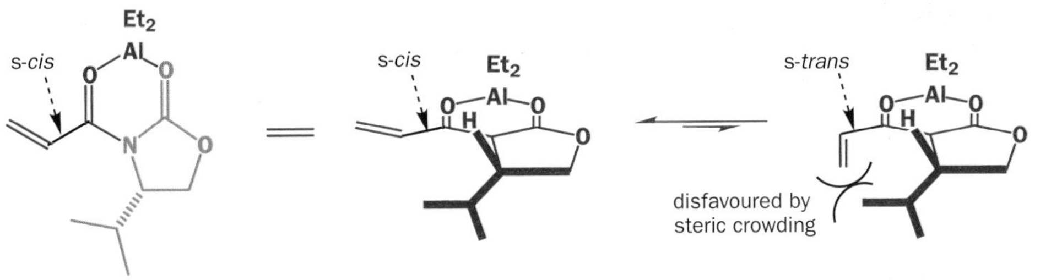

disfavoured by
steric crowding

The auxiliary has succeeded in doing what we set out to do (p. 1227)—it has made diastereo-isomeric the transition states leading to enantiomeric products, the difference in energy arising because of steric crowding of one face of the alkene.

Lest you should imagine that all effective auxiliaries are oxazolidinones, here is a different one—8-phenylmenthol—used by Corey in enantioselective prostaglandin synthesis. 8-Phenylmenthol is made from the natural product pulegone (Chapter 51). Even in the starting material the role of the phenyl group is clearly to crowd one face of the dienophile.

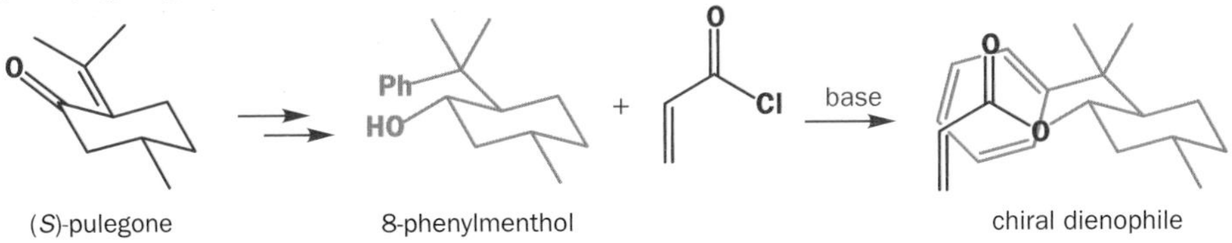

(S)-pulegone   8-phenylmenthol   chiral dienophile

A Lewis acid (AlCl$_3$)-catalysed Diels–Alder reaction with a substituted, but still achiral, cyclopen-tadiene gives a single enantiomer of the adduct. The sense of asymmetry induced in the reaction is seen more clearly if we redraw the product with 'R*' to represent the chiral auxiliary. The phenyl group on the auxiliary shields the back of the dienophile (as drawn) so that the diene has to add from the front to give one of the possible *endo* enantiomers.

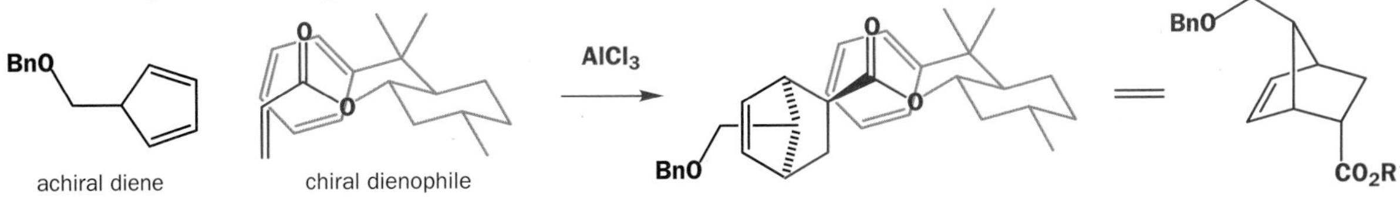

achiral diene   chiral dienophile

Corey used the four chiral centres created in the reaction to provide the chiral centres around the cyclopentanone ring of the prostaglandins (a family of compounds implicated in inflammation; see Chapter 51). After hydroxylation of the ester's enolate, the auxiliary was removed, this time by reduction. Diol cleavage with periodate (mentioned at the end of Chapter 35) gave a ketone that underwent Baeyer–Villiger oxidation on the more substituted side to give a hydrolysable lactone. Iodolactonization gave a substituted cyclopentanone that Corey used as a starting material for sever-al important prostaglandin syntheses.

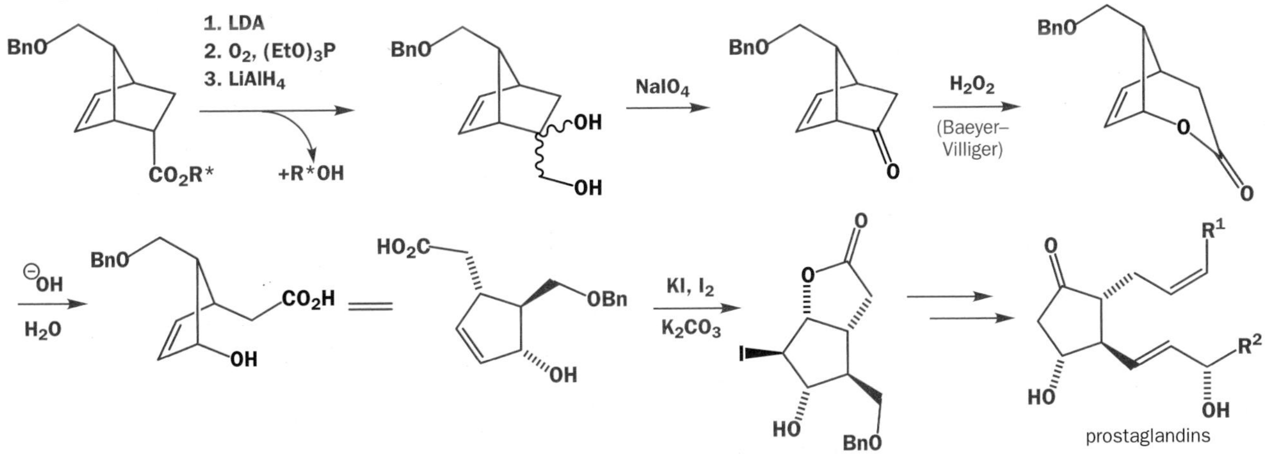

prostaglandins

Alkylation of chiral enolates

Chiral auxiliaries can be used in plenty of other reactions, and one of the most common types is the alkylation of enolates. Evans's oxazolidinone auxiliaries are particularly appropriate here because they are readily turned into enolizable carboxylic acid derivatives.

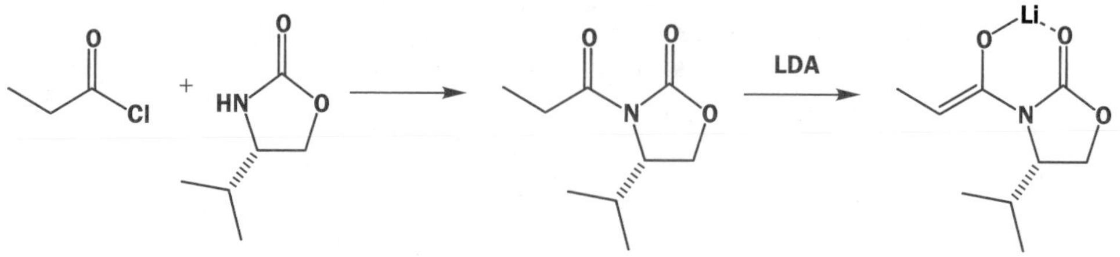

Treatment with base (usually LDA) at low temperature produces an enolate, and you can clearly see that the auxiliary has been designed to favour attack by electrophiles on only one face of that enolate. Notice too that the bulky auxiliary means that only the *Z*-enolate forms: alkylation of the *E*-enolate on the top face would give the diastereoisomeric product. Coordination of the lithium ion to the other carbonyl oxygen makes the whole structure rigid, fixing the isopropyl group where it can provide maximum hindrance to attack on the 'wrong' face.

Electrophile	Ratio of diastereoisomers
PhCH$_2$I	>99:1
allyl bromide	98:2
EtI	94:6

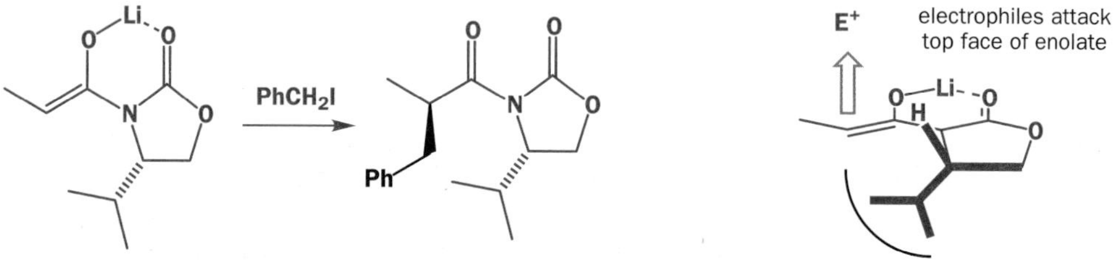

bottom face shielded by isopropyl group

The table in the margin shows the ratio of diastereoisomers produced by this reaction for a few alkylating agents. As you can see, none of these reactions is truly 100% diastereoselective and, indeed, only the best chiral auxiliaries (of which this is certainly one) give >98% of a single diastereoisomer. The problem with less than perfect diastereoselectivity is that, when the chiral auxiliary is removed, the final product is contaminated with some of the other enantiomer. A 98:2 ratio of diastereoisomers will result in a 98:2 ratio of enantiomers.

Enantiomeric excess

When talking about compounds that are neither racemic nor enantiomerically pure (usually called **enantiomerically enriched** or, occasionally, **scalemic**) chemists talk not about ratios of enantiomers but about **enantiomeric excess**. Enantiomeric excess (or **ee**) is defined as the excess of one enantiomer over the other, expressed as a percentage of the whole. So a 98:2 mixture of enantiomers consists of one enantiomer in 96% excess over the other, and we call it an enantiomerically enriched mixture with 96% ee. Why not just say that we have 98% of one enantiomer? Enantiomers are not like other isomers because they are simply mirror images. The 2% of the wrong enantiomer makes a racemate of 2% of the right isomer so the mixture contains 4% racemate and 96% of one enantiomer. 96% ee.

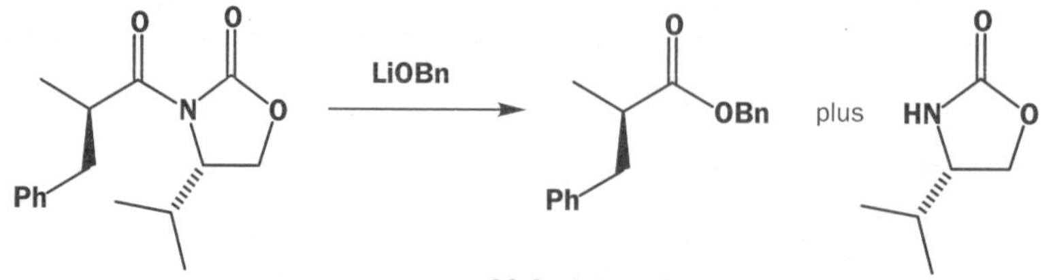

98:2 mixture of diastereoisomers

98:2 mixture of enantiomers
96% enantiomeric excess

We will see shortly how we can make further use of the chiral auxiliary to increase the ee of the reaction products. But, first, we should consider how to measure ee. One way is simply to measure the angle through which the sample rotates plane-polarized light. The angle of rotation is proportional to the enantiomeric excess of the sample (see the Box). The problem with this method is that to measure an actual value for ee you need to know what rotation a sample of 100% ee gives, and that is not always possible. Also, polarimeter measurements are notoriously unreliable—they depend on temperature, solvent, and concentration, and are subject to massive error due to small amounts of highly optically active impurities.

Optical rotation should be proportional to enantiomeric excess

Imagine you have a sample, A, of an enantiomerically pure compound—a natural product perhaps—and, using a polarimeter, you find that it has an $[\alpha]_D$ of +10.0. Another sample, B, of the same compound, which you know to be *chemically* pure (perhaps it is a synthetic sample), shows an $[\alpha]_D$ of +8.0. What is its enantiomeric excess? Well, you would have got the same value of 8.0 for the $[\alpha]_D$ of B if you had mixed 80% of your enantiomerically pure sample A with 20% of a racemic (or achiral) compound with no optical rotation. Since you know that sample B is chemically pure, and is the same compound as A, it must therefore indeed consist of 80% enantiomerically pure material plus 20% racemic material, or 80% of one

enantiomer plus 20% of a 1:1 mixture of the two enantiomers—which is the same as 90% of one enantiomer and 10% of the other, or 80% enantiomeric excess. Optical rotations can give a guide to enantiomeric excess—sometimes called **optical purity** in this context—but slight impurities of compounds with large rotations can distort the result and there are some examples where the linear relationship between ee and optical rotation fails because of what is known as the Horeau effect. You can read more about this in Eliel and Wilen, *Stereochemistry of organic compounds*, Wiley, 1994.

Modern chemists usually use either chromatography or spectroscopy to tell the difference between enantiomers. You may protest that we have told you that this is impossible—enantiomers are chemically identical and have identical NMR spectra, so how can chromatography or spectroscopy tell them apart? Well, again, they are identical unless they are in a chiral environment (the principle on which resolution relies). We introduced HPLC on a chiral stationary phase as a way of separating enantiomers preparatively in Chapter 16. The same method can be used analytically—less than a milligram of chiral compound can be passed down a narrow column containing chirally modified silica. The two enantiomers are separated and the quantity of each can be measured (usually by UV absorption or by refractive index changes) and an ee derived. Gas chromatography can be used in the same way—the columns are packed with a chiral stationary phase such as this isoleucine derivative.

Separating enantiomers spectroscopically relies again on putting them into a chiral environment. One way of doing this, if the compound is, say, an alcohol or an amine, is to make a derivative (an ester or an amide) with an enantiomerically pure acyl chloride. The one most commonly used is known as Mosher's acyl chloride, after its inventor Harry Mosher, though there are many others. The two enantiomers of the alcohol or amine now become diastereoisomers, and give different peaks in the NMR spectrum—the integrals can be used to determine ee and, although the ^1H NMR of such a mixture of diastereoisomers may become quite cluttered because it is a mixture, the presence of the CF_3 group means that the ratio can alternatively be measured by integrating the two singlets in the very simple ^{19}F NMR spectrum.

mixture of enantiomers

diastereoisomeric mixture of Mosher's esters

ratio of diastereoisomers measured by integrating ^1H or ^{19}F NMR spectrum

Another powerful method of discriminating between enantiomers is to add an enantiomerically pure compound to the NMR sample that does not react with the compound under investigation but simply forms a complex with it. The complexes formed from enantiomers are diastereoisomeric and therefore have different chemical shifts and, by integrating the NMR signals, the ratio of enantiomers can be determined. In the past, lanthanide salts of enantiomerically pure weak acids (called **chiral shift reagents**), which formed Lewis acid–base complexes with oxygen atoms in the compound under investigation, were used. More common nowadays is this alcohol, 2,2,2-trifluoro-1-(9-anthryl)ethanol, or TFAE, which can both hydrogen-bond to and form π-stacking complexes with many compounds, and often splits enantiomeric resonances very cleanly. Again the ^{19}F or ^1H NMR spectrum can be used.

Let's go back to chiral auxiliaries. We said that, although we want to get maximum levels of stereo-selectivity in our chiral-auxiliary-controlled reaction, we may still have 1 or 2% of a minor diastereoisomer, which, once we have removed our chiral auxiliary, will compromise the ee of our final product. It is at this point that we can use a trick that essentially employs the chiral auxiliary in a secondary role as a resolving agent. Provided the products are crystalline, it will usually be possible to recrystallize our 98:2 mixture of diastereoisomers to give essentially a single diastereoisomer, rather like carrying out a resolution with an enormous head start. Once this has been done, the chiral auxiliary can be removed and the product may be very close to 100% ee. Of course, the recrystallization sacrifices a few percentage points of yield, but these are invariably much less valuable than the few percentage points of ee gained! Here is an example from the work of Evans himself. During his synthesis of the complex antibiotic X-206 he needed large quantities of the small molecule below. He decided to make it by a chiral-auxiliary-controlled alkylation, followed by reduction to give the alcohol. The auxiliary needed is the one derived from norephedrine, and the alkylation with allyl iodide gives a 98:2 mixture of diastereoisomers. However, recrystallization converted this into an 83% yield of a single diastereoisomer in >99% purity, giving material of essentially 100% ee after removal of the auxiliary.

(S)-(+)-TFAE

At this point we should come clean about the asymmetric Diels–Alder reaction we introduced earlier. In fact, the diastereoisomer in the brown frame is formed in a 7% yield, with the major isomer accounting for 93%. But just one recrystallization gave >99% diastereoisomerically pure material in 81% yield.

This is one big bonus of using a chiral auxiliary—it's much easier to purify diastereoisomers than enantiomers and a chiral auxiliary reaction necessarily produces diastereoisomeric products.

But there are, of course, disadvantages. Chiral auxiliaries must be attached to the compound under construction, and after they have done their job they must be removed. The best auxiliaries can be recycled, but even then there are still at least two 'unproductive' steps in the synthesis. We may have given the impression that successful asymmetric synthesis is made possible by joining any chiral compound to the substrate. This is very far from the truth. Discovering successful chiral auxiliaries requires painstaking research and most potential chiral auxiliaries give low ees in practice. More efficient may be chiral reagents, or, best of all, chiral catalysts, and it is to these that we turn next.

Chiral reagents and chiral catalysts

If we want to create a new chiral centre in a molecule, our starting material must have **prochirality** —the ability to become chiral in one simple transformation. The most common prochiral units that give rise to new chiral centres are the trigonal carbon atoms of alkenes and carbonyl groups, which become tetrahedral by addition reactions. In all of the examples you saw in the last section, a prochiral alkene (we can count enolates as alkenes for this purpose) reacted selectively on one face because of the influence of the chiral auxiliary, which made the faces of the alkene diastereotopic.

One of the simplest transformations you could imagine of a prochiral unit into a chiral one is the reduction of a ketone. Although chiral auxiliary strategies have been used to make this type of reaction asymmetric, you will appreciate that, conceptually, the simplest way of getting the product as a single enantiomer would be to use a chiral reducing agent—in other words, to attach the chiral influence not to the substrate (as we did with chiral auxiliaries) but to the reagent.

■ Go back to Chapters 32–34 if you need reminding about the terms prochiral, enantiotopic, and diastereotopic.

One of the earliest attempts to do this used $LiAlH_4$ as the reducing agent and made it chiral by attaching 'Darvon alcohol' to it. Unfortunately, this reagent is not very effective—successful substrates are confined to acetylenic alcohols, and even then the products are formed with a maximum of about 80% ee.

■ Esters of 'Darvon alcohol' and its enantiomer are the drugs Darvon and Novrad (see p. 403)—hence the ready availability of this compound.

More effective is the chiral borohydride analogue developed by Corey, Bakshi, and Shibita. It is based upon a stable boron heterocycle made from an amino alcohol derived from proline, and is known as the **CBS reagent** after its developers.

The active reducing agent is made by complexing the heterocycle with borane. Only catalytic amounts (usually about 10%) of the boron heterocycle are needed because borane is sufficiently reactive to reduce ketones only when complexed with the nitrogen atom. The rest of the borane just waits until a molecule of catalyst becomes free.

▶ **Catalysts not reagents**

The fact that the reactions are catalytic in the heterocycle means that relatively little is needed and it can be recovered at the end of the reaction. Later in the chapter you will see catalytic reactions that use 1000 times less catalyst than this one and, indeed, none of the reactions we will mention in the rest of this chapter will use chiral reagents—only chiral catalysts. Note the distinction from chiral auxiliaries here: although auxiliaries are recoverable, they always have to be used in stoichiometric quantities, and recovery is usually a separate step.

CBS reductions are best when the ketone's two substituents are well-differentiated sterically—just as Ph and Me are in the example above. Only when the ketone is complexed with the 'other' boron atom (in the ring) is it electrophilic enough to be reduced by the weak hydride source. The hydride is delivered via a six-membered cyclic transition state, with the enantioselectivity arising from the preference of the larger of the ketone's two substituents (R_L) for the pseudoequatorial position on this ring.

larger substituent   smaller substituent

(turn reagent over)

hydride delivered via 6-membered ring

large group pseudoequatorial

Reductions with Nature's CBS reagent—NADH—are discussed in Chapter 51.

The CBS reagent is one of the best asymmetric reducing agents invented by chemists. Yet Nature does asymmetric reductions all the time—and gets 100% ee every time too. Nature uses enzymes as chiral catalysts, and chemists have not been slow to subvert these natural systems to their own ends. The problem with using enzymes is that they are designed to fit into a single biochemical pathway and are often quite substrate-specific, and so are not useful as a general chemical method. However, this can be overcome by using conveniently packaged multienzyme systems, living cells. Yeast is particularly good at reducing ketones, and the best enantioselectivies are obtained when the ketone carries a β-ester group. The reaction is done by stirring the ketone with an aqueous suspension of live yeast, which must be fed with plenty of sugar.

In fact, the enantiomer of the CBS reagent can be made by a resolution strategy.

baker's yeast

glucose

up to 97% ee
55% yield

These reactions are quite messy, and are best done on a large scale! Notice how the selectivity of baker's yeast is the reverse of that of the CBS reagent with respect to the large and small ketone substituents. This is most useful, since (R)-proline is expensive, and an enantiomeric yeast cell would be a rarity indeed.

An important application of this baker's yeast reduction is in the synthesis of citronellol. After reduction and protection of the ester, S_N2 substitution of the secondary tosylate group could be achieved with inversion using a copper nucleophile. The 88% ee obtained here is better than that of many natural samples of citronellol: in common with many other terpenes, citronellol extracted from plants varies greatly in enantiomeric purity. It is quite a compliment to the humble yeast that, with a bit of help from Professor Mori's research group, it can outdo most of the more sophisticated members of the plant kingdom.

1. TsCl
2. LiAlH$_4$
3. NaH, BnBr

1.

CuL$_n$

2. Na, NH$_3$

substitution with inversion

citronellol, 88% ee

Asymmetric hydrogenation

Probably the best-studied way of carrying out enantioselective reduction is to hydrogenate in the presence of a chiral catalyst. You would not normally choose catalytic hydrogenation for reducing a carbonyl group to an alcohol and, indeed, carbonyl reductions using hydrogenation with a chiral catalyst are not usually very enantioselective. Much better are hydrogenations of double bonds, particularly those with nearby heteroatoms (OH, NHR) that can coordinate to the metal.

Here is a simple example: it is, in fact, an asymmetric synthesis of the analgesic drug naproxen. First, look at the reaction—we'll consider the catalyst in a moment.

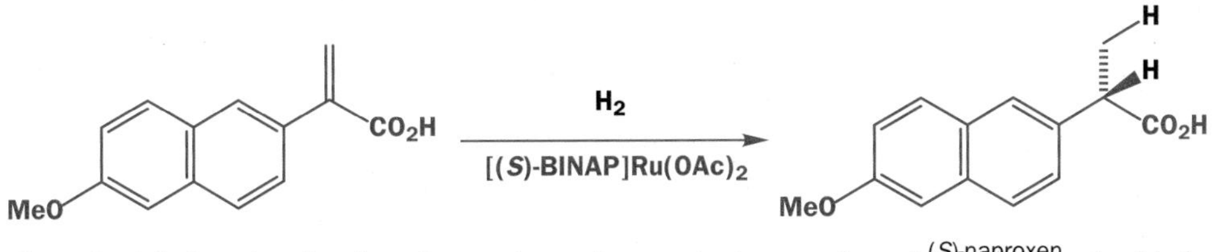

(S)-naproxen

The principle is quite simple—the catalyst selects a single enantiotopic face of the double bond and adds hydrogen across it. Exactly how it does this need not concern you, but we do need to go into more detail about the structure of the catalyst, which consists of a metal atom (Ru) and a ligand, called BINAP.

(R)-BINAP

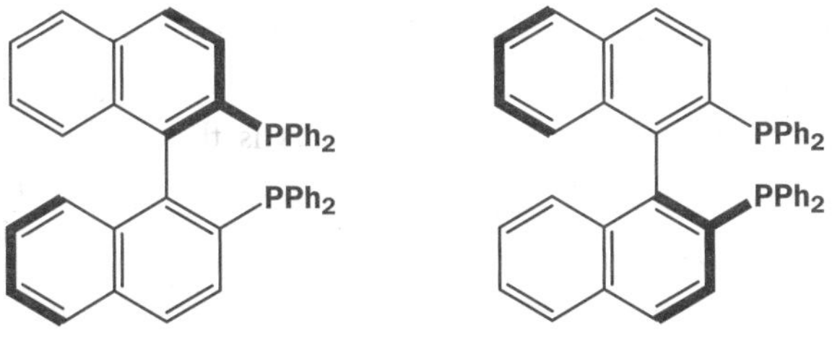

(S)-BINAP

In common with many other ligands for asymmetric hydrogenation, BINAP is a chelating diphosphine: the metal sits between the two phosphorus atoms firmly anchored in a chiral environment. The chirality here is of an unusual sort, since BINAP has no chiral centres. Instead it has **axial chirality** by virtue of restricted rotation about the bond joining the two naphthalene ring systems. In order for the two enantiomers of BINAP to interconvert, the PPh₂ group would have to force its way either past the other PPh₂ group or round the black hydrogen (see next page). Both pathways are too strained for racemization to occur.

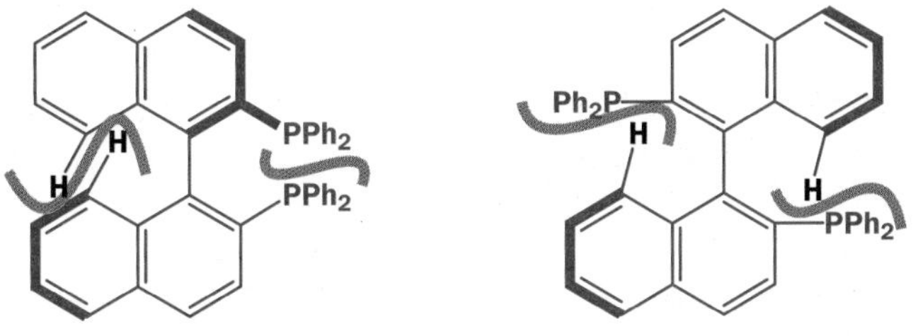

BINAP is not derived from a natural product, and has to be synthesized in the laboratory and resolved.

Resolution of BINAP

The scheme shows one method by which BINAP may be made—the resolution step is unusual because it relies on formation of a molecular complex, not a salt. It is the phosphine oxide that is resolved, and then reduced to the phosphine with trichlorosilane.

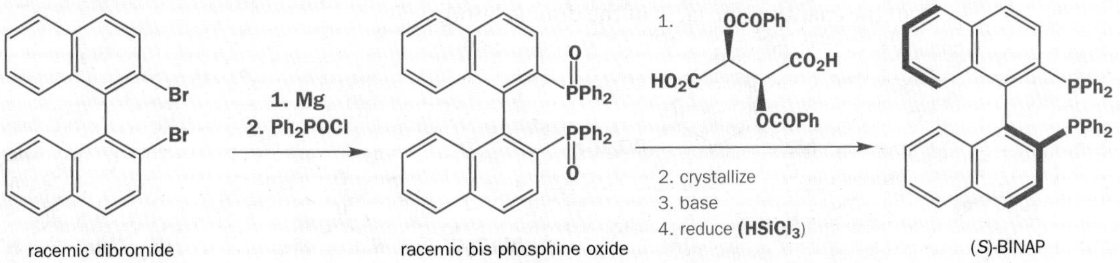

racemic dibromide

racemic bis phosphine oxide

(S)-BINAP

This makes it relatively expensive, but the expense is offset by the economy of catalyst required in such reactions. Whereas about 10 mol% catalyst is needed for CBS reductions, many hydrogenations of this type give high enantiomeric excesses with only 0.0002 mol% BINAP–ruthenium(II) catalyst! Because such minuscule quantities of catalyst are needed, enantioselective hydrogenations are more widely used by industry than any other asymmetric method. The other advantage of the resolution is, of course, that either enantiomer is equally available.

BINAP–ruthenium(II) is particularly good at catalysing the hydrogenation of allylic alcohols, and of α,β-unsaturated carboxylic acids to give acids bearing α stereogenic centres (like naproxen above).

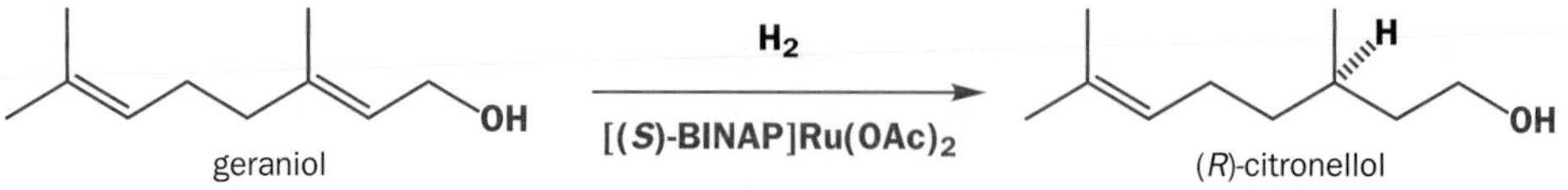

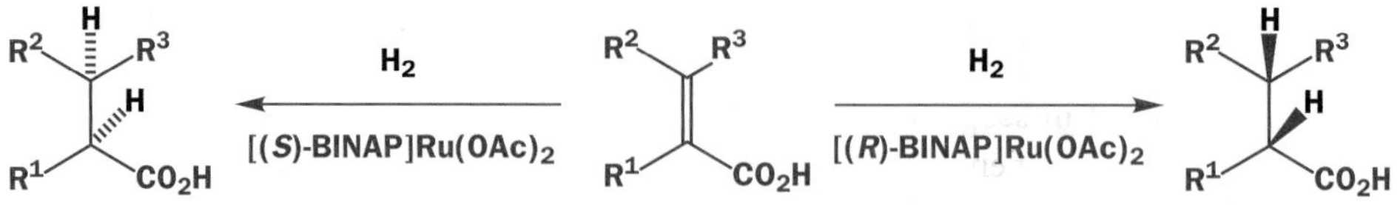

geraniol → (R)-citronellol

If the double bond also bears an amino group, the products of these reactions are α amino acids, and in these cases there is another alternative that works even better, a catalyst based on rhodium. Here is one very important synthesis of an unnatural amino acid using a rhodium catalyst. Again, look first at the reaction and then we will discuss the catalyst.

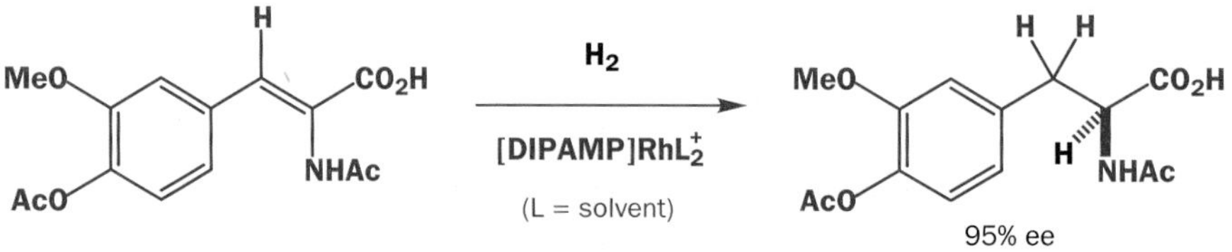

95% ee

The product can be converted into L-dopa, a drug used to treat Parkinson's disease, and it is this reaction and this catalyst, both developed by Monsanto, that convinced many chemical companies that enantioselective synthesis was possible on a large scale.

L-dopa   (R,R)-DIPAMP

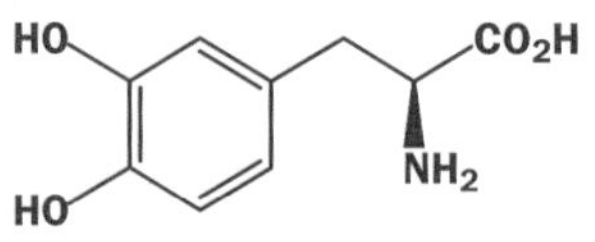

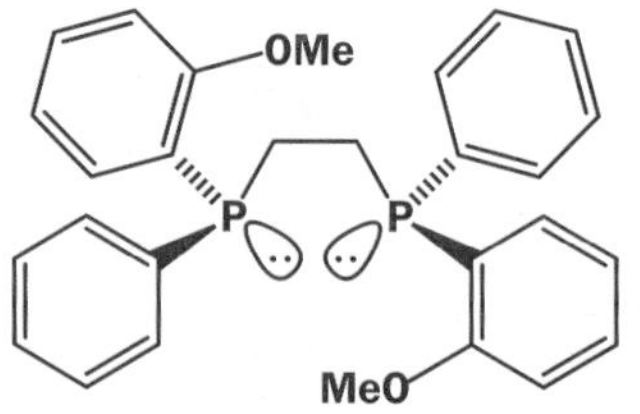

The catalyst is a cationic complex of rhodium with another diphosphine, DIPAMP. DIPAMP's chirality resides in the two stereogenic phosphorus atoms: unlike amines, phosphines are configurationally stable, rather like sulfoxides (which we will discuss in the next chapter). The catalyst imposes chirality on the hydrogenation by coordinating to both the amide group and the double bond of the substrate. Two diastereoisomeric complexes result, since the chiral catalyst can coordinate to either of the enantiotopic faces of the double bond.

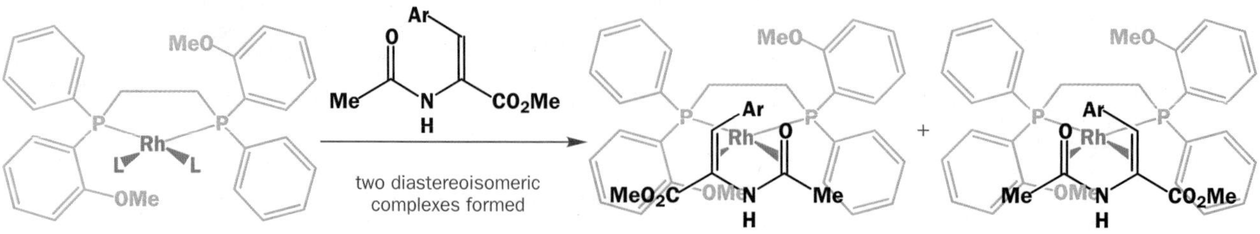

two diastereoisomeric
complexes formed

It turns out that the enantioselectivity in the reaction arises because one of these diastereoisomeric complexes reacts much more rapidly with hydrogen than the other, ultimately transferring both hydrogen atoms to the same face of the double bond.

(An = o-anisyl)

major enantiomer

minor enantiomer

Although more limited in scope than the BINAP–Ru(II)-catalysed hydrogenations, rhodium-catalysed hydrogenations are of enormous commercial importance because of the demand for both natural and unnatural amino acids on a vast scale. It is even economical for the more expensive of the natural amino acids to be made synthetically rather than isolated from natural sources—phenylalanine, for example, of industrial importance as a component of the artificial sweetener aspartame, is manufactured by enantioselective hydrogenation.

N-acetyl L-phenylalanine
83% ee, rising to
97% ee on recrystallization

DNNP

Although DIPAMP is a suitable ligand for this reaction as well, the industrial process uses the diphosphine DNNP. Unfortunately, the product is initially obtained in rather modest enantiomeric excess (83%), but recrystallization improves this to 97%. In the manufacture of aspartame, coupling with natural (and therefore 100% ee) aspartic acid turns the 1.5% of the minor enantiomer into a diastereoisomeric impurity that can be removed by crystallization (essentially a resolution).

Improving ee by recrystallization

This technique is quite frequently used to improve the ee of almost enantiomerically pure samples, since, in general, crystals are most stable if they consist either of a single enantiomer or of a racemic mixture. Recrystallization of samples with ees greater than about 85% has a good chance of improving the ee of the sample (the minor enantiomer remaining in the mother liquors). Samples with ees much less than this tend to decrease in ee on recrystallization. Much depends on the crystal structure—this is quite a complex science and you can read more about it in Eliel and Wilen, *Stereochemistry of organic compounds*, Wiley, 1994. The difficulty of increasing low ees by recrystallization is one disadvantage of chiral reagent techniques as opposed to chiral auxiliary techniques.

Before leaving asymmetric hydrogenation reactions, we should mention one other related process that has acquired immense importance, again because of its industrial application. You have come across citronellol a couple of times in this chapter already: the corresponding aldehyde citronellal is even more important because it is an intermediate in the a synthesis of L-menthol by the Japanese chemical company Takasago. Takasago manufacture about 30% of the 3500 ton annual worldwide demand for L-menthol from citronellal by using an intramolecular ene reaction (a cycloaddition you met in Chapter 35).

(R)-citronellal

L-menthol

The green methyl group prefers to be equatorial in the transition state and directs the formation of the two new chiral centres. The transition state (in the frame) is like a *trans*-decalin with two fused six-membered chair rings. Both new substituents go equatorial in the product while the Lewis acid binds to the oxygen and accelerates the reaction, as it would for a Diels–Alder reaction.

But it is not this step that makes the synthesis remarkable, but rather Takasago's route *to* citronellal. Pinene is another terpene that is produced in only low enantiomeric excess by pine trees (and, indeed, which is the major enantiomer depends on whether it is a European or a North American pine tree). But in the menthol process none of this matters, and cheap, enantiomerically impure pinene can be used, because the first step is to convert it to an achiral terpene, myrcene. Lithium diethylamide adds to this diene to give an allylic amine.

Now for the key step: [(*S*)-BINAP]$_2$Rh$^+$ catalyses the rearrangement of this allylic amine to the enamine, creating a new chiral centre with 98% ee. This reaction is rather like a hydrogenation in which the hydrogen comes from within the same molecule, or you could see it as a [1,3]-sigmatropic shift (usually disallowed) made possible by participation of the metal's orbitals. Whichever way you look at it, the catalyst selects one of two enantiotopic hydrogen atoms (shown in black and green) and allows only the green one to migrate. This reaction can be run on a seven ton scale, needs only 0.01 mol% catalyst, and is a testimony to the power of asymmetric synthesis.

Exactly how this reaction works and exactly what features of [(*S*)-BINAP]$_2$Rh$^+$ make for successful asymmetric induction are not clear. Though we can work out a mechanism for the reaction, we cannot say precisely how the chirality of the ligand directs the formation of the new stereogenic centre. Here, as elsewhere in modern organic chemistry, the experiments get ahead of human understanding.

Rhodium or ruthenium, and which ligands?

The range of diphosphine ligands used in catalytic enantioselective hydrogenation is enormous (though DIPAMP and BINAP are probably the most important), and many of them can be used with Rh or Ru. We can nonetheless give some guidelines to choice of catalyst. In general, Rh demands more of its substrates and less of its diphosphine ligands. Which ligand to choose is a matter of thorough literature searching followed by some experimentation. However, Rh will really give good ees only when hydrogenating electron-poor or conjugated double bonds that carry a β-carbonyl group (necessary for chelation), and the enamides we have been discussing are among the best of these.

Rhodium or ruthenium, and which ligands? – (contd)

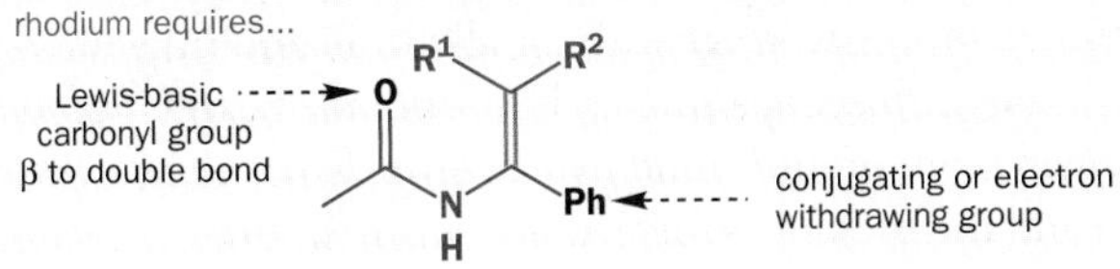

rhodium requires...

Lewis-basic carbonyl group β to double bond

conjugating or electron withdrawing group

Ru is more fussy about ligands (BINAP is the one usually used) but will hydrogenate both electron-rich and electron-poor double bonds. Ru[BINAP] (OAc)$_2$ works best if the double bond carries an α-hydroxyl group—in other words,

if it is an allylic alcohol or an α,β-unsaturated carboxylic acid. The enantioselective hydrogenation of geraniol (p. 1236) is also regioselective, because isolated double bonds are not hydrogenated.

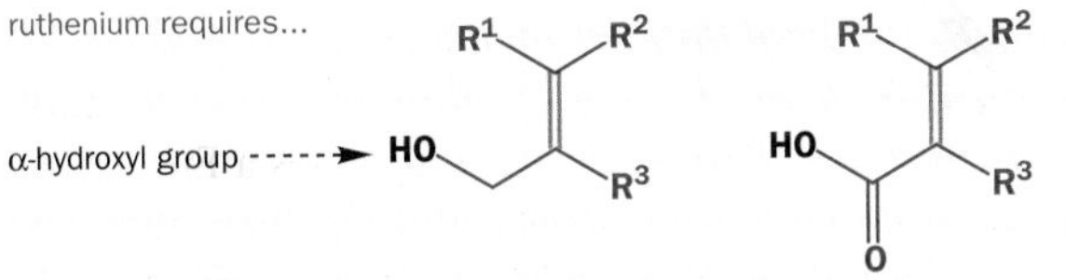

ruthenium requires...

α-hydroxyl group

We now leave asymmetric reductions and move on to two asymmetric oxidations, which are probably the two most important asymmetric reactions known. They are both products of the laboratories of Professor Barry Sharpless.

Asymmetric epoxidation

The first of Sharpless's reactions is an oxidation of alkenes by asymmetric epoxidation. You met vanadium as a transition-metal catalyst for epoxidation with *t*-butyl hydroperoxide in Chapter 33, and this new reaction makes use of titanium, as titanium tetraisopropoxide, Ti(O*i*Pr)$_4$, to do the same thing. Sharpless surmised that, by adding a chiral ligand to the titanium catalyst, he might be able to make the reaction asymmetric. The ligand that works best is diethyl tartrate, and the reaction shown below is just one of many that demonstrate that this is a remarkably good reaction.

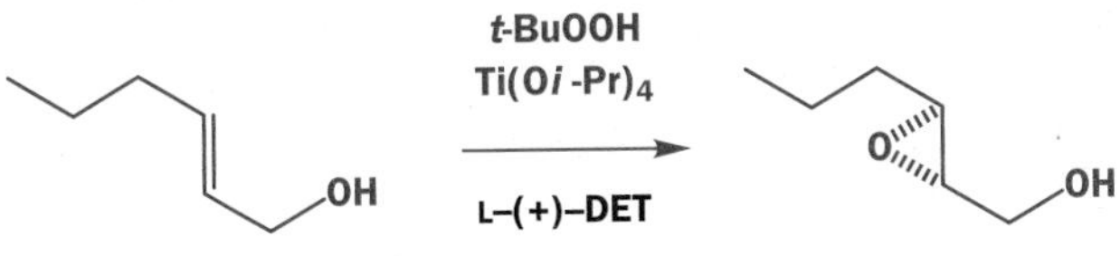

t-BuOOH
Ti(O*i*-Pr)$_4$
L–(+)–DET

85% yield, 94% ee

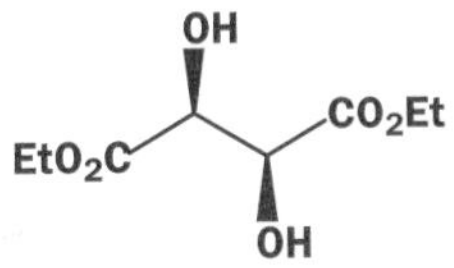

L-(+)-DET = L-(+)-diethyl tartrate

> K.B. Sharpless (1941–) studied at Stanford and was first appointed at MIT but is now at the Scripps Institute in California. His undoubted claim to fame rests on the invention of no fewer than three reactions of immense significance: AE (asymmetric epoxidation) and AD (asymmetric dihydroxylation) are discussed in this chapter. The third reaction, AA (asymmetric aminohydroxylation) has still to reach the perfection of the first two.

Transition-metal-catalysed epoxidations work only on allylic alcohols, so there is one limitation to the method, but otherwise there are few restrictions on what can be epoxidized enantioselectively. When this reaction was discovered in 1981 it was by far the best asymmetric reaction known. Because of its importance, a lot of work went into discovering exactly how the reaction worked, and the scheme below shows what is believed to be the active complex, formed from two titanium atoms bridged by two tartrate ligands (shown in gold). Each titanium atom retains two of its isopropoxide ligands, and is coordinated to one of the carbonyl groups of the tartrate ligand. The reaction works best if the titanium and tartrate are left to stir for a while so that these dimers can form cleanly.

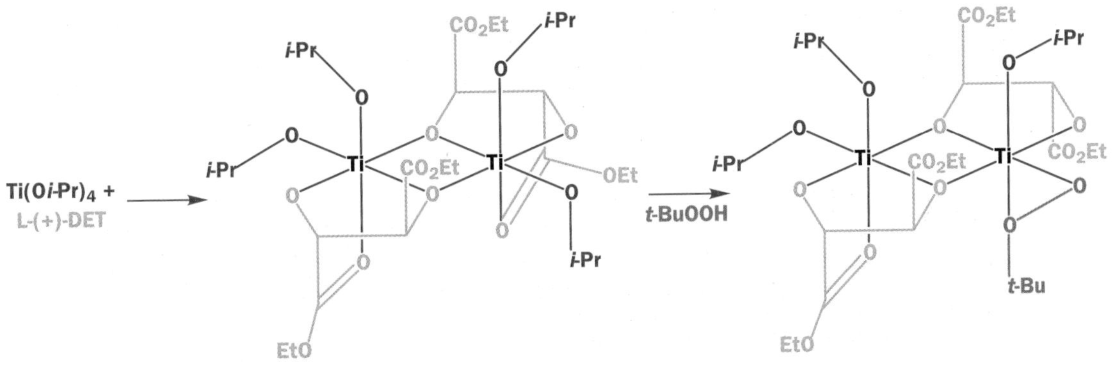

Ti(O*i*-Pr)$_4$ + L-(+)-DET → *t*-BuOOH →

When the oxidizing agent (*t*-BuOOH, shown in green) is added to the mixture, it displaces one of the remaining isopropoxide ligands and one of the tartrate carbonyl groups.

Now, for this oxidizing complex to react with an allylic alcohol, the alcohol must become co-ordinated to the titanium too, displacing a further isopropoxide ligand. Because of the shape of the complex the reactive oxygen atom of the bound hydroperoxide has to be delivered to the lower face of the alkene (as drawn), and the epoxide is formed in high enantiomeric excess.

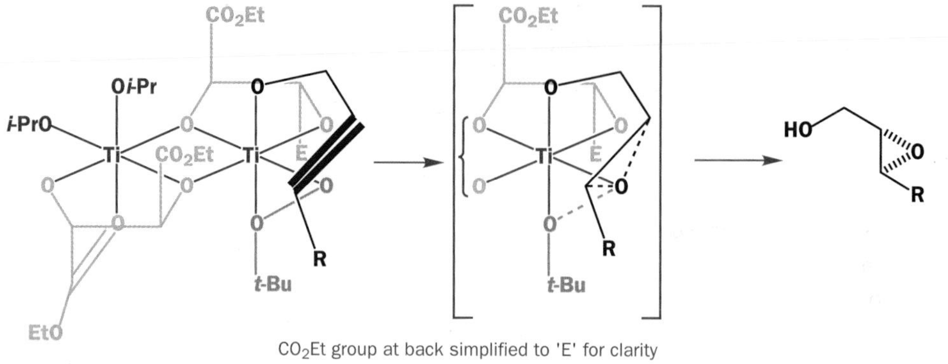

CO₂Et group at back simplified to 'E' for clarity

Different allylic alcohols coordinate in the same way to the titanium and reliably present the same enantiotopic face to the bound oxidizing agent, and the preference for oxidation with L-(+)-DET is shown in the schematic diagram below. Tartrate is ideal as a chiral ligand because it is available relatively cheaply as either enantiomer. L-tartrate is extracted from grapes; D-(−)-tartrate is rarer and more expensive—it is sometimes called unnatural tartrate, but, in fact, it too is natural. By using D-(−)-tartrate it is, of course, possible to produce the other enantiomer of the epoxide equally selectively.

● **Enantioselectivity in the Sharpless asymmetric epoxidation**

enantioselectivity in the Sharpless asymmetric expoxidation

D-(−)-diethyl tartrate delivers oxygen to top face of alkene

arrange allylic alcohol with hydroxyl group top left

L-(+)-diethyl tartrate delivers oxygen to bottom face of alkene

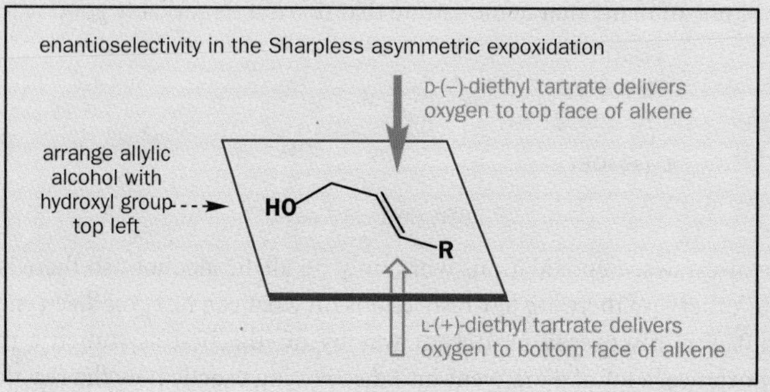

Sharpless also found that this reaction works with only a catalytic amount of titanium–tartrate complex, because the reaction products can be displaced from the metal centre by more of the two reagents. The catalytic version of the asymmetric epoxidation is well suited to industrial exploitation, and the American Company J. T. Baker employs it to make synthetic disparlure, the pheromone of the gypsy moth.

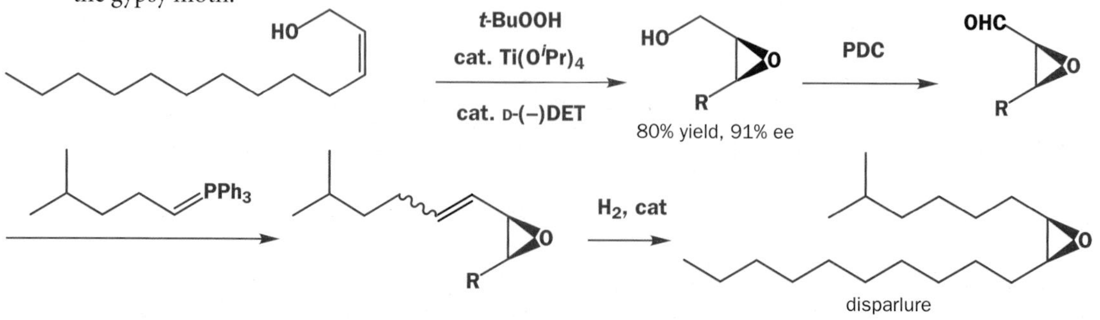

80% yield, 91% ee

disparlure

Not many target molecules are themselves epoxides, but the great thing about the epoxide products is that they are highly versatile—they react with many types of nucleophiles to give 1,2-disubstituted products. You met the chiral β-blocker drug propranolol in Chapter 30, and its 1,2,3-substitution pattern makes it a good candidate for synthesis using asymmetric epoxidation.

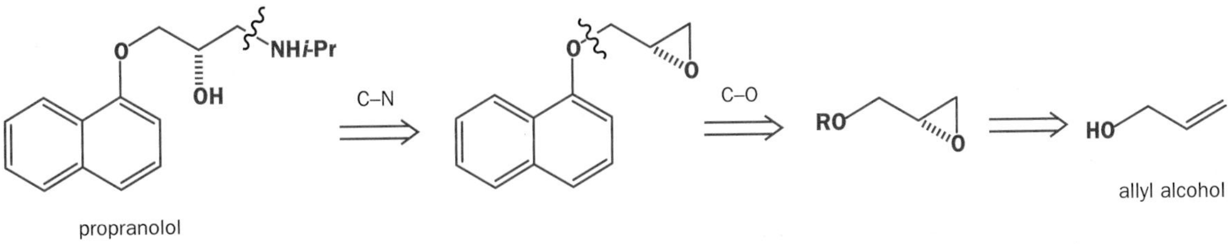

propranolol

allyl alcohol

Unfortunately, the obvious starting material, allyl alcohol itself, gives an epoxide which is hard to handle, so Sharpless, who carried out this synthesis of propranolol, used this silicon-substituted allylic alcohol instead.

The hydroxyl group was mesylated and displaced with 1-naphthoxide and, after treatment with fluoride to remove the silicon, the epoxide was opened with isopropylamine.

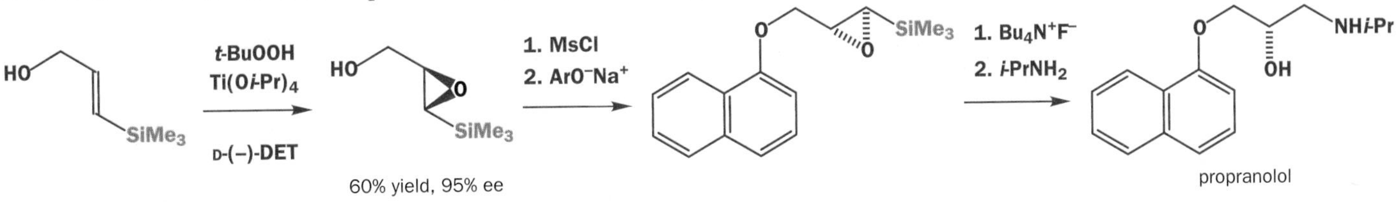

60% yield, 95% ee

propranolol

Asymmetric dihydroxylation

The last asymmetric oxidation we will mention really is probably the best asymmetric reaction of all. It is a chiral version of the *syn* dihydroxylation of alkenes by osmium tetroxide. Here is an example— though the concept is quite simple, the recipe for the reactions is quite complicated so we need to approach it step by step.

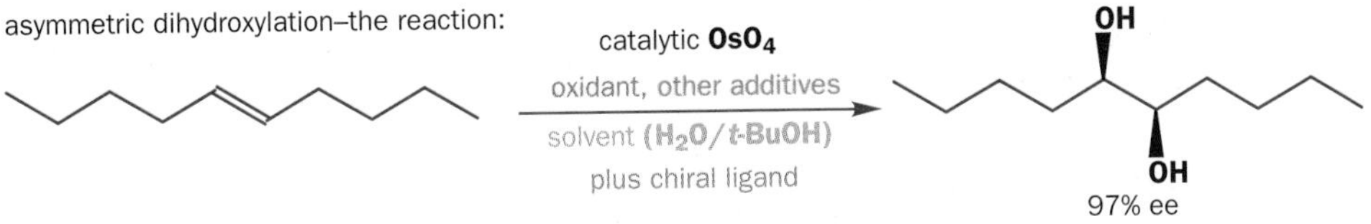

asymmetric dihydroxylation–the reaction:

97% ee

The active reagent is based on osmium(VIII) and is used in just catalytic amounts. This means that there has to be a stoichiometric quantity of another oxidant to reoxidize the osmium after each catalytic cycle—$K_3Fe(CN)_6$ is most commonly used. Because OsO_4 is volatile and toxic, the osmium is usually added as $K_2OsO_2(OH)_4$, which forms OsO_4 in the reaction mixture. The 'other additives' include K_2CO_3 and methanesulfonamide ($MeSO_2NH_2$), which increases the rate of the reaction. Now for the chiral ligand. The best ones are based on the alkaloids dihydroquinidine and dihydroquinine, whose structures are shown below. They coordinate to the osmium through the yellow nitrogen.

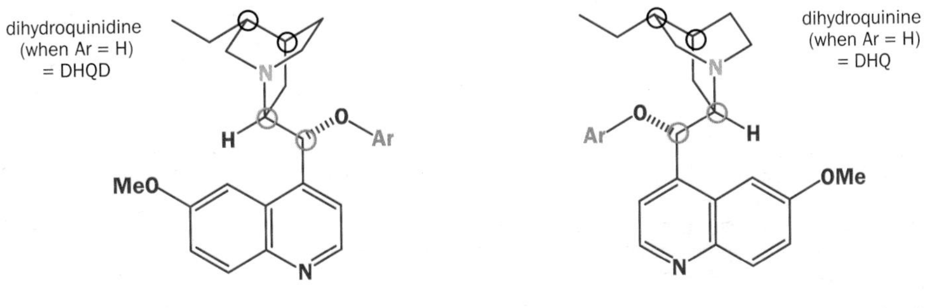

dihydroquinidine
(when Ar = H)
= DHQD

dihydroquinine
(when Ar = H)
= DHQ

The alkaloids (usually abbreviated to DHQD and DHQ, respectively) must be attached to an aromatic group Ar, the choice of which (like the choice of ligand for enantioselective hydrogenation with Rh) varies according to the substrate. The most generally applicable ligands are these two phthalazines in which each aromatic group Ar carries two alkaloid ligands.

"DHQD$_2$PHAL" **DHQD** — DHQD

DHQ — **DHQ** "DHQ$_2$PHAL"

phthalazine-based ligands

Dihydroquinine and dihydroquinidine are not enantiomeric (although the green centres are inverted in dihydroquinidine, the black ones remains the same), but they act on the dihydroxylation as though they were—here, after all that introduction, is a real example, and probably the most remarkable of any in this chapter.

$K_2OsO_2(OH)_4$, $K_3Fe(CN)_6$, K_2CO_3, $MeSO_2NH_2$

^tBuOH, H_2O, 0 °C

DHQD$_2$PHAL

99.8% ee

trans stilbene

$K_2OsO_2(OH)_4$, $K_3Fe(CN)_6$, K_2CO_3, $MeSO_2NH_2$

^tBuOH, H_2O, 0 °C

DHQ$_2$PHAL

>99.5% ee

trans-(*E*)-Stilbene dihydroxylates more selectively than any other alkene, and we would probably not be exaggerating if we said that this particular example is the most enantioselective catalytic reaction ever invented. The dihydroxylation is also much less fussy about the alkenes it will oxidize than the asymmetric epoxidation. Osmium tetroxide itself is a remarkable reagent, since it oxidizes more or less any sort of alkene, electron-rich or electron-poor, and the same is true of the asymmetric dihydroxylation (often abbreviated to AD) reagent. The following example illustrates both this and a synthetic use for the diol product.

OsO_4, $K_3Fe(CN)_6$, K_2CO_3,

DHQD–containing ligand

89% yield, 96% ee

The diol is produced from a double bond that is more electron-poor than most, and can be converted to the antibiotic chloramphenicol in a few more steps.

TsCl, Et$_3$N

K_2CO_3

selective tosylation of more acidic alcohol

NaBH$_4$

THF

NaN$_3$

SiO$_2$

1. Ph$_3$P
2. Cl$_2$CHCO$_2$Me

chloramphenicol

Regioselectivity in this synthesis

This sequence is not only remarkable for the AD reaction—the regioselectivites involved in the formation and reaction of the epoxide need commenting on too. The tosylation is selective because the hydroxyl group near the electron-withdrawing ester is more acidic than the other one—high selectivity here is crucial because tosylation of the other hydroxyl group would lead to the other enantiomer of the epoxide. The regioselectivity of attack of azide on the epoxide must be because of the electron-withdrawing *p*-nitro group—acidic silica encourages the reaction to proceed through an S$_N$1-like (or 'loose S$_N$2') transition state, with cationic character on the reaction centre. Substitution next to the ring is disfavoured, and the 1,3-diol is formed selectively.

disfavoured

favoured

We can sum up the usual selectivity of the AD reaction in another diagram, shown below. With the substrate arranged as shown, with the largest (R_L) and next largest groups (R_M) bottom left and top right, respectively, DHQD-based ligands will direct OsO_4 to dihydroxylate from the top face of the double bond and DHQ-based ligands will direct it to dihydroxylate the bottom.

● Enantioselectivity in the Sharpless asymmetric dihydroxylation

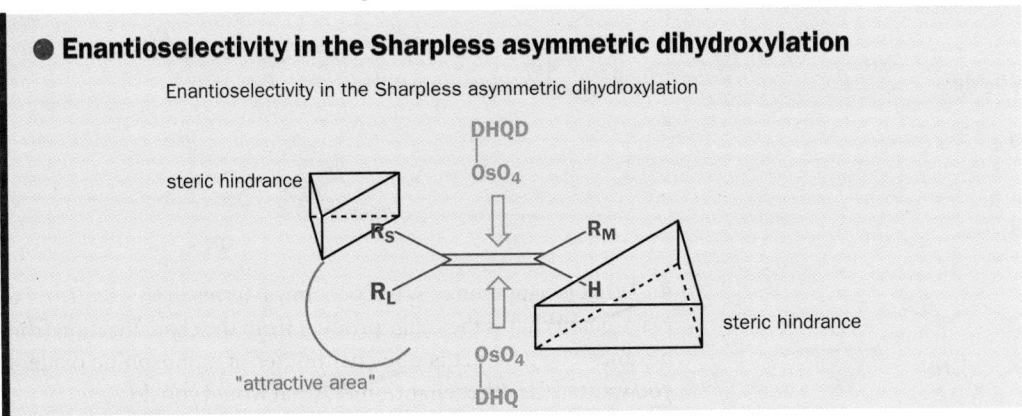

Enantioselectivity in the Sharpless asymmetric dihydroxylation

The reason for this must, of course, lie in the way in which the substrate interacts with the osmium–ligand complex. However, even as we write this book, the detailed mechanism of the asymmetric dihydroxylation is still under discussion. What is known is that the ligand forms some sort of 'chiral pocket', like an enzyme active site, with the osmium sitting at the bottom of it. Alkenes can only approach the osmium if they are correctly aligned in the chiral pocket, and steric hindrance forces the alignment shown in the scheme above. The analogy with an enzyme active site goes even further, since it appears that part of the pocket is 'attractive' to aromatic or strongly hydrophobic groups. This part appears to accommodate R_L, part of the reason why the selectivity in the dihydroxylation of *trans*-stilbene is so high.

This chapter, more than most, deals with topics under active investigation. New and more powerful methods are appearing all the time and it is quite certain that the decade 2000–10 will see many important advances in asymmetric synthesis.

● Summary of methods for asymmetric synthesis

Method	Advantages	Disadvantages	Examples
resolution	both enantiomers available	maximum 50% yield	synthesis of BINAP
chiral pool	100% ee guaranteed	often only 1 enantiomer available	amino acid- and sugar-derived syntheses
chiral auxiliary	often excellent ees; can recrystallize to purify to high ee	extra steps to introduce and remove auxiliary	oxazolidinones
chiral reagent	often excellent ees; can recrystallize to purify to high ee	only a few reagents are successful and often for few substrates	enzymes, CBS reducing agent
chiral catalyst	economical: only small amounts of recyclable material used	only a few reactions are really successful; recrystallization can improve only already high ees	asymmetric hydrogenation, epoxidation, dihydroxylation

Problems

1. Explain how this asymmetric synthesis of amino acids, starting with natural proline, works. Explain the stereoselectivity of each reaction.

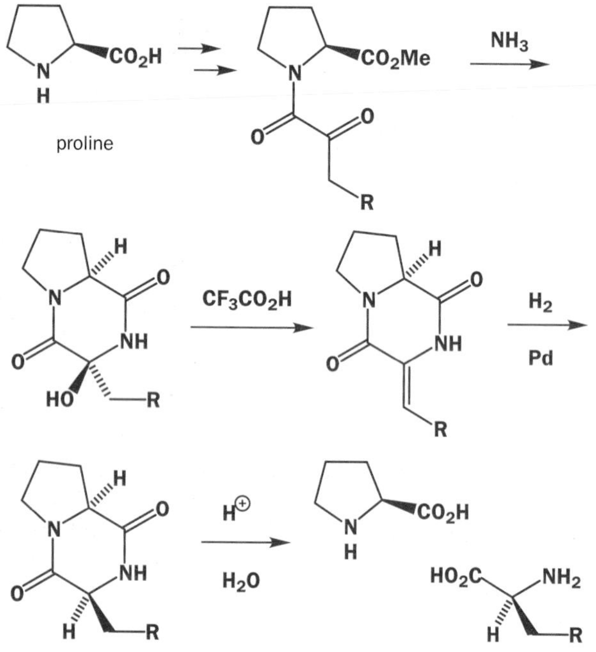

proline

2. This is a synthesis of the racemic drug tazadolene. If the enantiomers of the drug are to be evaluated for biological activity, they must be separated. At which stage would you advocate separating the enantiomers, and how would you do it?

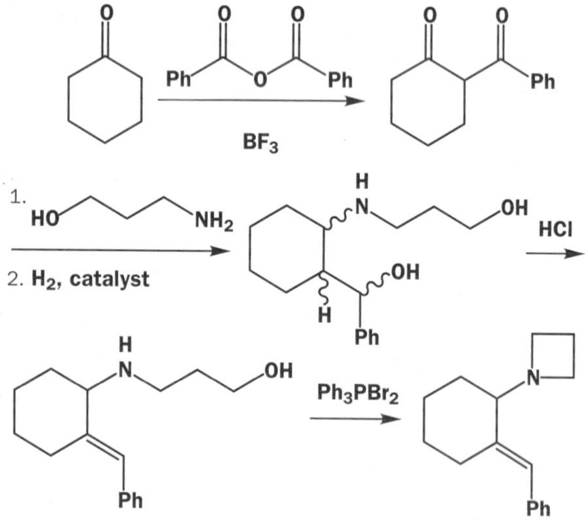

3. How would you make enantiomerically enriched samples of these compounds (either enantiomer)?

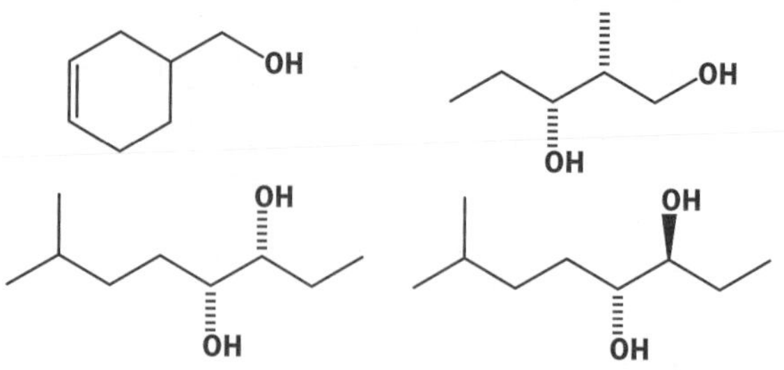

4. What is happening in stereochemical terms in this sequence of reactions? What is the other product from the crystallization from hexane? The product is one enantiomer of a phosphine oxide. If you wanted the other enantiomer, what would you do?

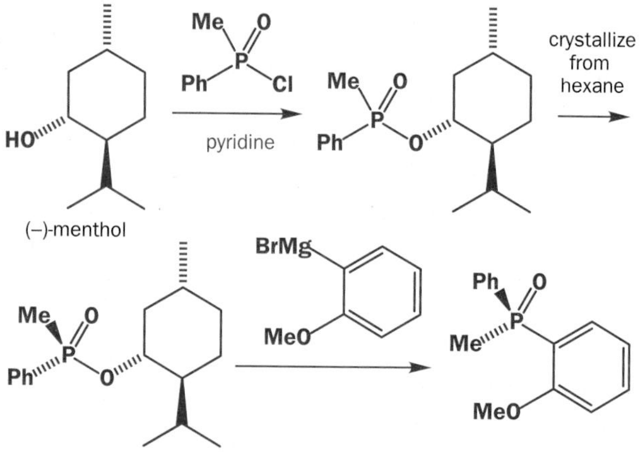

(−)-menthol

Revision. This phosphine oxide is used in the synthesis of DIPAMP, the chiral ligand for asymmetric catalytic hydrogenation mentioned in the chapter. What are the various reagents doing in the conversion into DIPAMP?

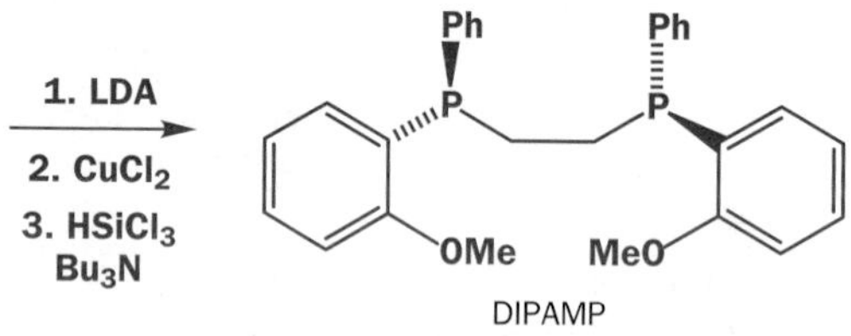

1. LDA
2. CuCl$_2$
3. HSiCl$_3$
   Bu$_3$N

DIPAMP

5. An alternative to the Evans chiral auxiliary described in the chapter is this oxazolidinone, made from natural (S)-(−)-phenylalanine. What strategy is used for this synthesis and why are the conditions and mechanism of the reactions important?

synthesis of Evans's chiral auxiliary from (S)-phenylalanine

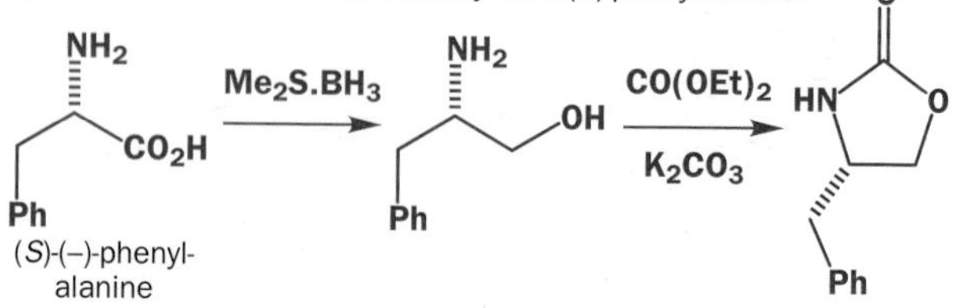

(S)-(−)-phenyl-
alanine

6. In the following reaction sequence, the chirality of mandelic acid is transmitted to a new hydroxy-acid by a sequence of stereochemically controlled reactions. Give mechanisms for the reactions and state whether each is stereospecific or stereoselective. Offer some rationalization for the creation of new stereogenic centres in the first and second reactions.

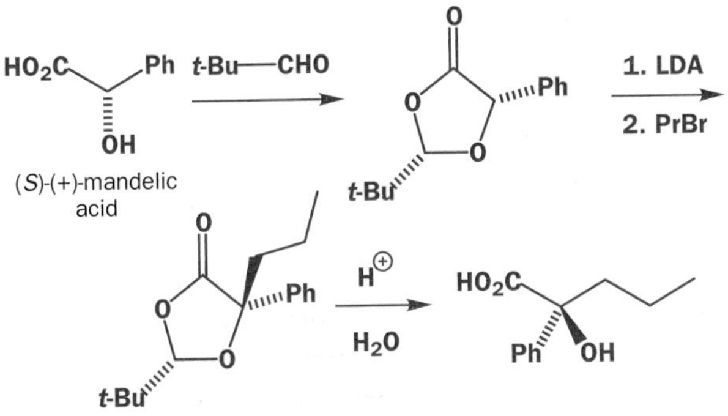

7. This reaction sequence can be used to make enantiomerically enriched amino acids. Which compound is the origin of the chirality and how is it made? Suggest why this particular enantiomer of the amino acid might be made. Suggest reagents for the last stages of the process. Would the enantiomerically enriched starting material be recovered?

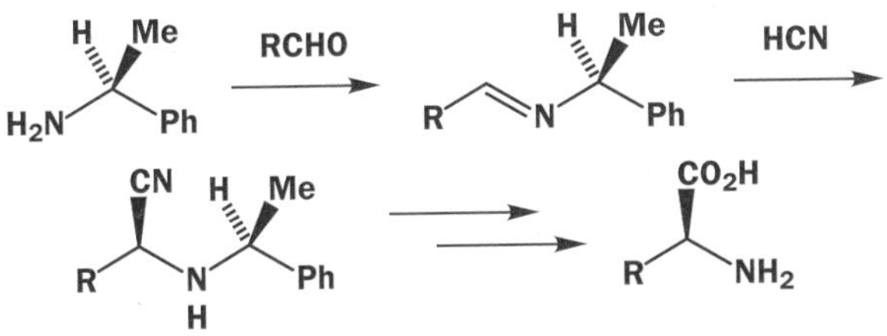

8. Submitting this racemic ester to hydrolysis by an enzyme found in pig pancreas leaves enantiomerically enriched ester with the absolute stereochemistry shown. What are the advantages and disadvantages of this method? Why is the ee not 100%?

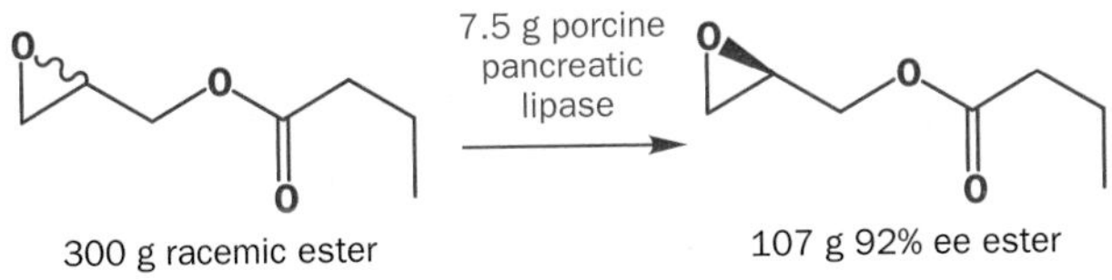

300 g racemic ester   107 g 92% ee ester

How could the same enantiomerically enriched compound be formed by chemical means? What are the advantages and disadvantages of this method?

9. The BINAP-catalysed hydrogenations described in the chapter can also be applied to the reduction of ketones—the same ketones indeed as can be reduced by baker's yeast. Compare these results and comment on the differences between them.

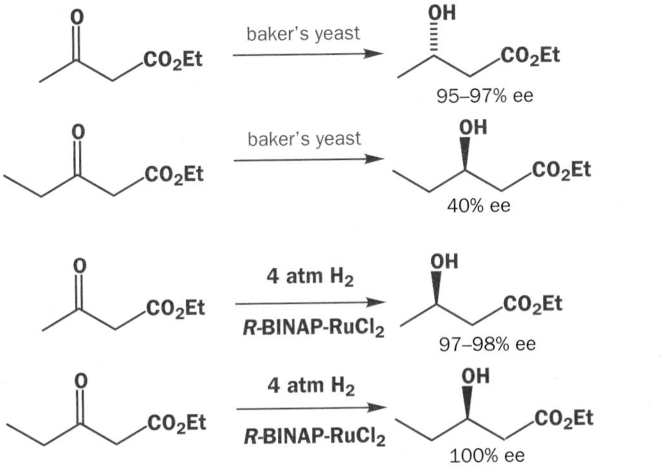

10. Both of these bicyclic compounds readily undergo hydrogenation of the alkene to give the *syn* product. Explain why asymmetric hydrogenation of only one of the compounds would be of much value in synthesis.

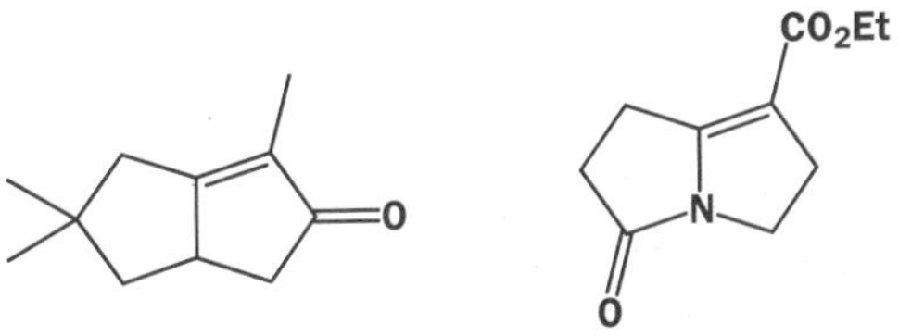

11. Explain the stereochemistry and mechanism in the synthesis of the chiral auxiliary 8-phenylmenthol from (+)-pulegone. After the reduction with Na in *i*-PrOH, what is the minor (13%) component of the mixture?

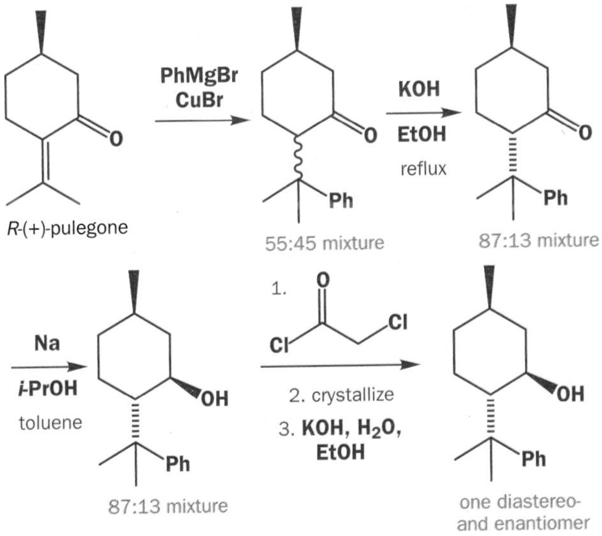

12. Describe the stereochemical happenings in these processes. You should use terms like diastereoselective and diastereotopic where needed. If you wanted to make single enantiomers of the products by these routes, at what stage would you introduce the asymmetry? (You are *not* expected to say *how* you would induce asymmetry!)

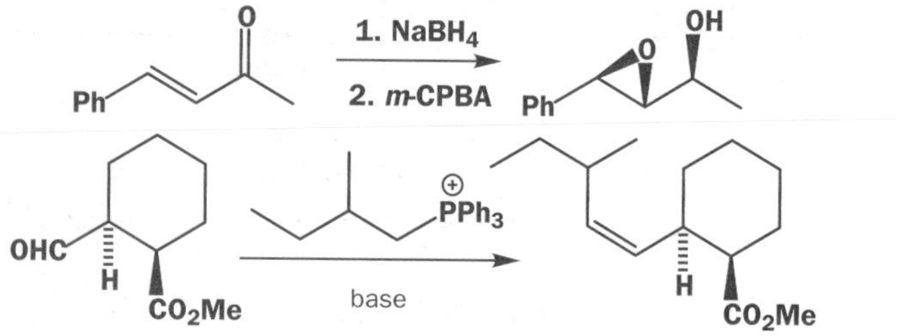

13. The unsaturated amine A, a useful intermediate in the synthesis of the *amaryllidaceae* (daffodil) alkaloids, can be made from the three starting materials shown below. What kind of chemistry is required in each case? Which is best adapted for asymmetric synthesis? Outline your chosen synthesis.

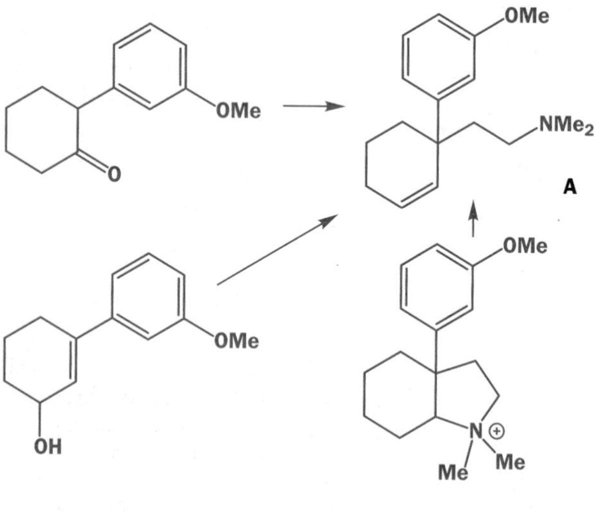

14. Suggest syntheses for single enantiomers of these compounds.

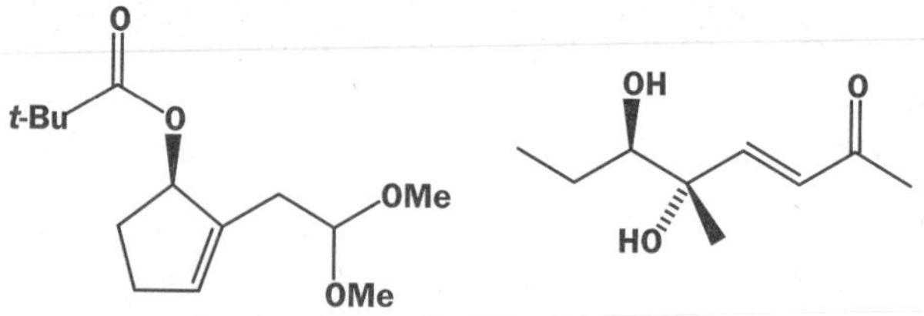

15. Suggest a synthesis of any stereoisomer (for example, *R,Z*) of this compound.

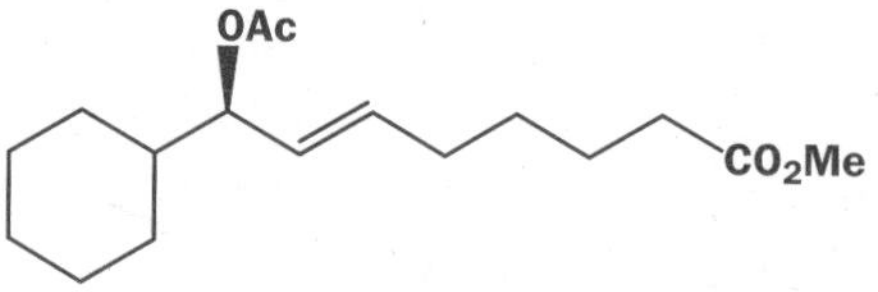

16. Revision. Give mechanisms for the steps in the synthesis of tazadolene in Problem 2.

Connections

Building on:

- Conjugate addition ch10 & ch23
- Nucleophilic substitution at saturated carbon ch17
- Controlling stereochemistry ch16, ch33, & ch34
- Oxidation ch24
- Aldol reactions ch27
- Controlling double bond geometry ch31
- Rearrangements ch36–ch37
- Radicals and carbenes ch39–ch40

Arriving at:

- Sulfur compounds have many oxidation states
- Sulfur is nucleophilic and electrophilic
- Sulfur stabilizes anions and cations
- Sulfur can be removed by reduction or oxidation
- Sulfoxides can be chiral
- Thioacetals provide d^1 reagents
- Allylic sulfides are useful in synthesis
- Epoxides can be made from sulfonium ylids
- Sulfur compounds are good at cationic and [2,3]-sigmatropic rearrangements
- Selenium compounds resemble sulfur compounds

Looking forward to:

- Main group chemistry II: B, Si, and Sn ch47
- Organometallic chemistry ch48
- Biological chemistry ch49–ch51
- Polymerization ch52

Sulfur: an element of contradictions

The first organosulfur compounds in this book were the dreadful smell of the skunk and the wonderful smell of the truffle, which pigs can detect through a metre of soil and which is so delightful that truffles cost more than their weight in gold.

the dreadful smell of the skunk

the delightful smell of the black truffle

▶

If you look in the Oxford English dictionary you will see 'sulphur'. This is a peculiarly British spelling—neither the French nor the Americans for example have the 'ph'. It has recently been decided that chemists the world over should use a uniform spelling 'sulfur'.

More useful sulfur compounds have included the leprosy drug dapsone (Chapter 6), the arthritis drug Feldene (Chapter 21), glutathione (Chapter 23), a scavenger of oxidizing agents that protects most living things against oxidation and contains the natural amino acid cysteine (Chapter 49), and, of course, the famous antibiotics, the penicillins, mentioned in several chapters.

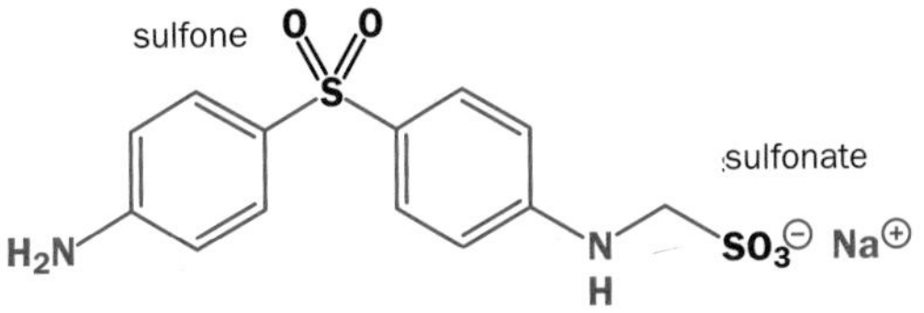

dapsone—water-soluble 'pro-drug' for leprosy

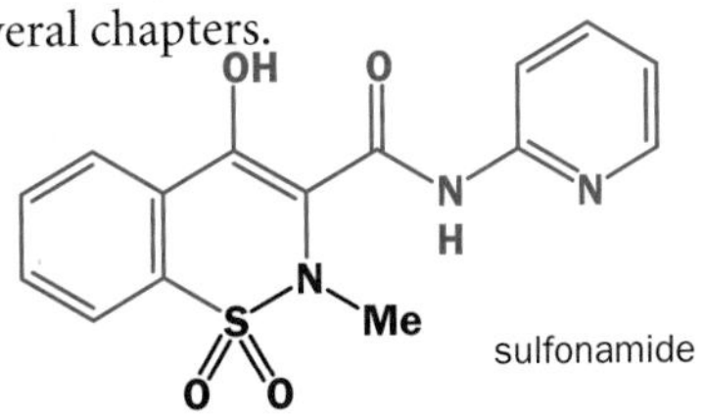

Pfizer's piroxicam or feldene

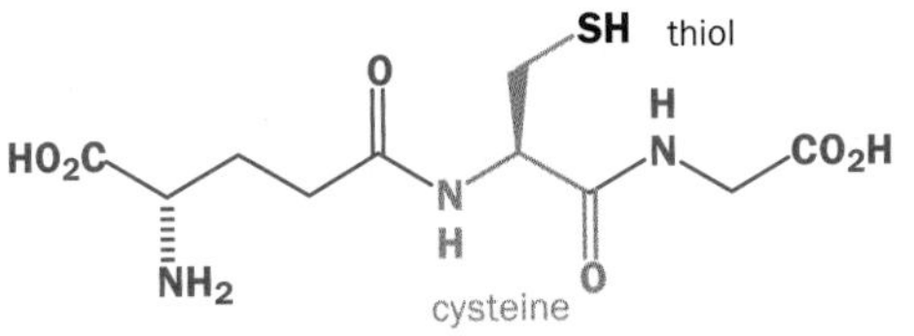

glutathione: scavenger of toxic oxidants

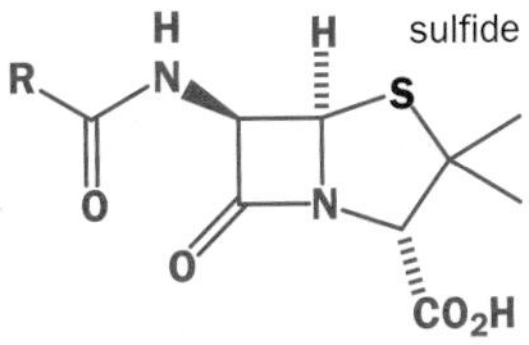

penicillin family of antibiotics

Important reactions have included sulfur as nucleophile and leaving group in the S_N2 reaction (illustrated here; see also Chapter 17), sulfonation of aromatic rings (Chapter 22), formation and reduction of thioacetals (Chapter 24) and Lawesson's reagent for converting carbonyl groups to thiocarbonyl groups (Chapter 44).

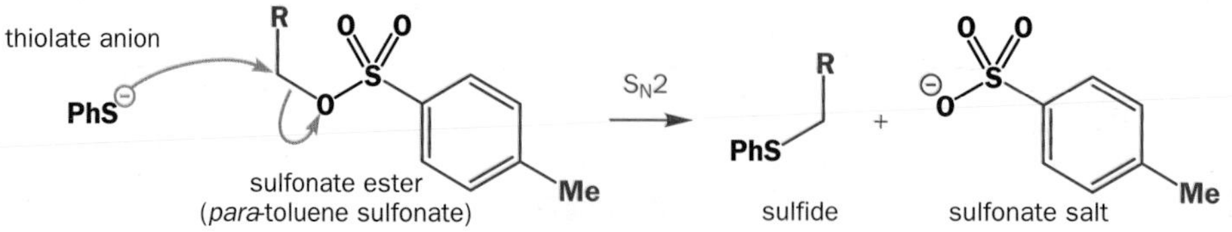

This chapter gathers together the principles behind these examples together with a discussion of what makes organosulfur chemistry special and also introduces new reactions. We have a lot to explain! In Chapter 31 we introduced you to the Julia olefination, a reaction whose first step is the deprotonation of a sulfone.

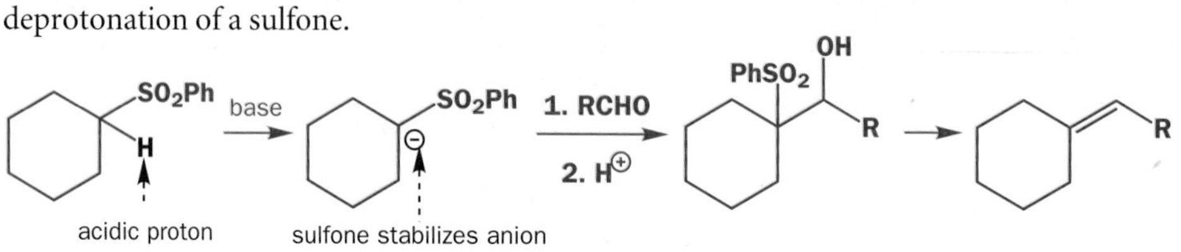

Why is this proton easy to remove? This ability to stabilize an adjacent anion is a property shared by all of the most important sulfur-based functional groups. The anions (or better, lithium derivatives) will react with a variety of electrophiles and here is a selection: a sulfone reacting with a lactone, a sulfoxide with a ketone, and a sulfide with a silyl chloride.

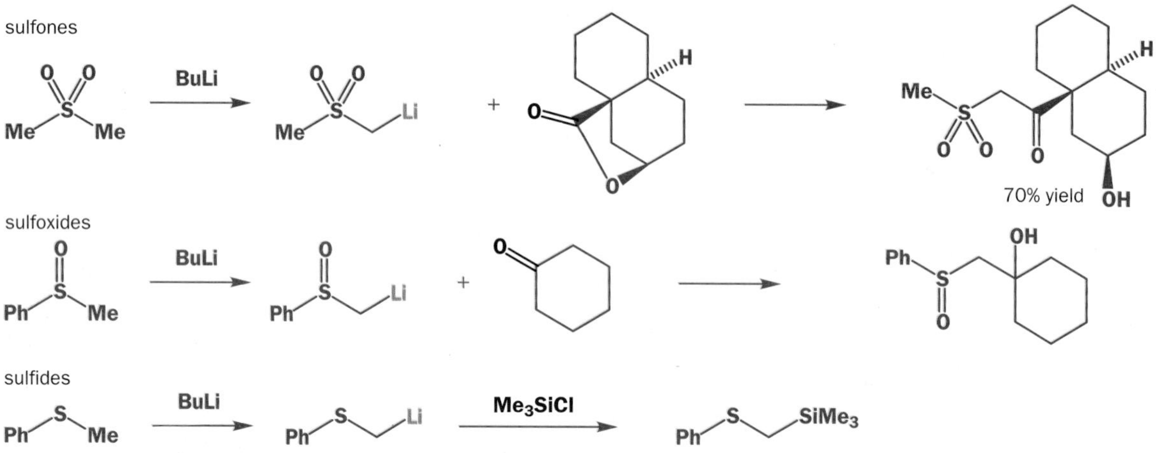

You notice immediately the three main oxidation states of sulfur: S(VI), S(IV), and S(II). You might have expected the S(VI) sulfone and perhaps the S(IV) sulfoxide to stabilize an adjacent anion, but the S(II) sulfide? We will discuss this along with many other unusual features of sulfur chemistry. The interesting aspects are what make sulfur different.

The basic facts about sulfur

Sulfur is a p-block element in group VI (or 16 if you prefer) immediately below oxygen and between phosphorus and chlorine. It is natural for us to compare sulfur with oxygen but we will, strangely, compare it with carbon as well.

Sulfur is much less electronegative than oxygen; in fact, it has the same electronegativity as carbon, so it is no good trying to use the polarization of the C–S bond to explain anything! It forms reasonably strong bonds to carbon—strong enough for the compounds to be stable but weak enough for

Sulfur in the periodic table (electronegativity)

C	N	O	F
(2.5)	(3.0)	(3.5)	(4.0)
Si	P	S	Cl
(1.8)	(2.1)	(2.5)	(3.0)

Bond strengths, kJ mol^{-1}

	X = C	X = H	X = F	X = S
C–X	376	418	452	362
S–X	362	349	384	301

selective cleavage in the presence of the much stronger C–O bonds. It also forms strong bonds to itself. Elemental crystalline yellow sulfur consists of S_8 molecules—eight-membered rings of sulfur atoms!

crystalline sulfur

Because sulfur is in the second row of the periodic table it forms many types of compounds not available to oxygen. Compounds with S–S and S–halogen bonds are quite stable and can be isolated, unlike the unstable and often explosive O–halogen and O–O compounds. Sulfur has d orbitals so it can have oxidation states of 2, 4, or 6 and coordination numbers from 0 to 7. Here is a selection of compounds.

Compounds of sulfur

Oxidation state	S(II)			S(IV)		S(VI)		
coordination number	0	1	2	3	4	4	6	7
example	S^{2-}	RS^-	R_2S	$R_2S=O$	SF_4	R_2SO_2	SF_6	SF_7^-

Sulfur is a very versatile element

As well as this variety of oxidation states, sulfur shows a sometimes surprising versatility in function. Simple S(II) compounds are good nucleophiles as you would expect from the high-energy nonbonding lone pairs ($3sp^3$ rather than the $2sp^3$ of oxygen). A mixture of a thiol (RSH, the sulfur equivalent of an alcohol) and NaOH reacts with an alkyl halide to give the sulfide alone by nucleophilic attack of RS^-.

Thiols (RSH) are more acidic than alcohols so the first step is a rapid proton exchange between the thiol and hydroxide ion. The thiolate anion then carries out a very efficient S_N2 displacement on the alkyl bromide to give the sulfide.

Notice that the thiolate anion does not attack the carbonyl group. Small basic oxyanions have high charge density and low-energy filled orbitals—they are hard nucleophiles that prefer to attack protons and carbonyl groups. Large, less basic thiolate anions have high-energy filled orbitals and are soft nucleophiles. They prefer to attack saturated carbon atoms. Thiols and thiolates are good soft nucleophiles.

● Thiols (RSH) are more acidic than alcohols (ROH) but sulfur compounds are better nucleophiles than oxygen compounds towards saturated carbon atoms (S_N2).

They are also good soft electrophiles. Sulfenyl chlorides (RSCl) are easily made from disulfides (RS–SR) and sulfuryl chloride (SO_2Cl_2). This S(VI) chloride has electrophilic chlorine atoms and is attacked by the nucleophilic disulfide to give two molecules of RSCl and gaseous SO_2. There's a lot of sulfur chemistry here! We start with a nucleophilic attack by one sulfur atom of the disulfide.

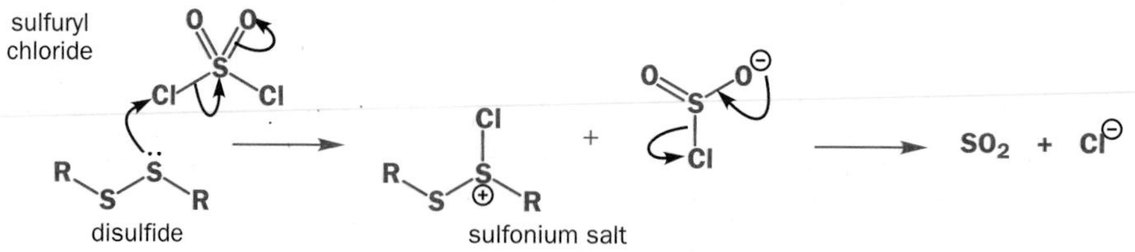

sulfuryl chloride

disulfide   sulfonium salt

The intermediate contains a tricoordinate sulfur cation or sulfonium salt. The chloride ion now attacks the other sulfur atom of this intermediate and two molecules of RSCl result. Each atom of the original disulfide has formed an S–Cl bond. One sulfur atom was a nucleophile towards chlorine and the other an electrophile.

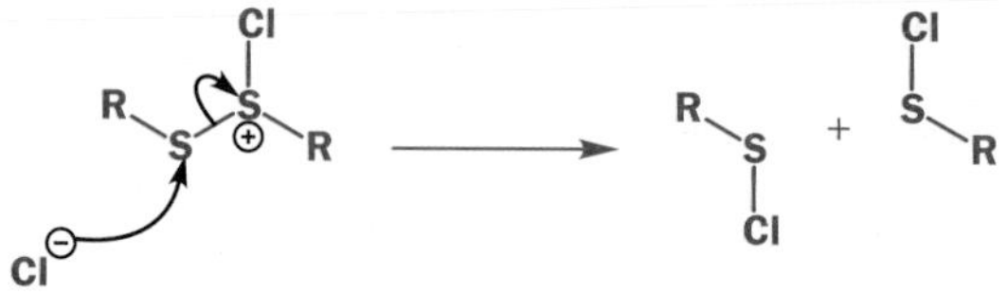

The product of this reaction, the sulfenyl chloride, is also a good soft electrophile towards carbon atoms, particularly towards alkenes. The reaction is very like bromination with a three-membered cyclic sulfonium ion intermediate replacing the bromonium ion of Chapter 20. The reaction is stereo-specific and *anti*.

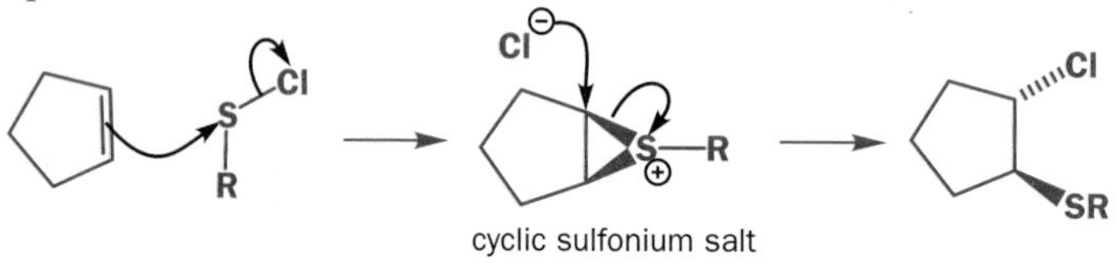

cyclic sulfonium salt

Sulfur at the S(II) oxidation state is both a good nucleophile and a good electrophile. This is also true at higher oxidation states though the compounds become harder electrophiles as the positive charge on sulfur increases. We have already mentioned tosyl (toluene-*para*-sulfonyl) chloride as an electrophile for alkoxide ions in this chapter and in earlier chapters.

At this higher oxidation state it might seem unlikely that sulfur could also be a good nucleophile, but consider the result of reacting TsCl with zinc metal. Zinc provides two electrons and turns the compound into an anion. This anion can also be drawn in two ways.

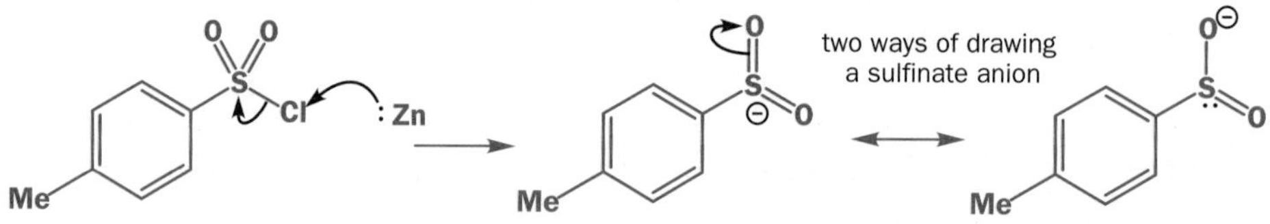

two ways of drawing a sulfinate anion

Surprisingly, this anion is also a good soft nucleophile and attacks saturated carbon atoms through the sulfur atom. In this case attack occurs at the less substituted end of an allylic bromide to give an allylic sulfone, which we will use later on.

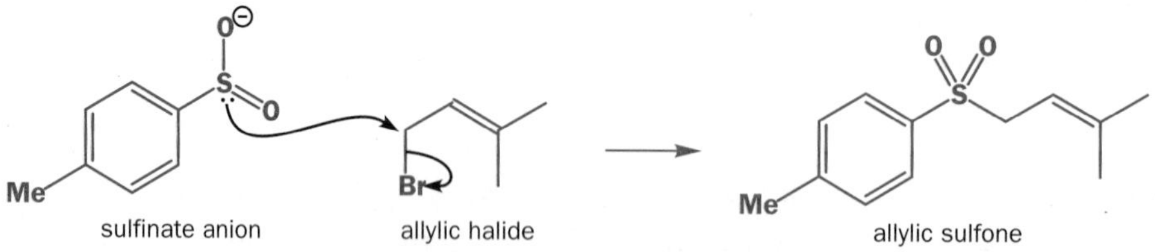

sulfinate anion   allylic halide   allylic sulfone

● **Sulfur compounds are good nucleophiles and good electrophiles.**

As this chapter develops you will see other examples of the versatility of sulfur. You will see how it takes part readily in rearrangements from the simple cationic to the sigmatropic. You will see that it can be removed from organic compounds in either an oxidative or a reductive fashion. You will see that it can stabilize anions or cations on adjacent carbon atoms, and the stabilization of anions is the first main section of the chapter.

Sulfur-based functional groups

We have already met a number of sulfur-containing functional groups and it might be useful to list them for reference.

Name	Structure	Importance	Example	Example details
thiol (or mercaptan)	RSH	strong smell, usually bad, but sometimes heavenly		smell and taste of coffee
thiolate anion	RS⁻	good soft nucleophiles		
disulfide	RS–SR	cross-links proteins		
sulfenyl chloride	RS–Cl	good soft electrophiles		
sulfide (or thioether)	R–S–R	molecular link		smell and taste of pineapple
sulfonium salt	R_3S^+	important reagents		ylid used in epoxidations
sulfoxide	$R_2S=O$ or $R_2S^+–O^-$	many reactions; can be chiral		chiral Michael acceptors
sulfone	R_2SO_2	anion-stabilizing group		
sulfonic acid	RSO_2OH	strong acids		
sulfonyl chloride	RSO_2Cl	turns alcohols into leaving groups		

Sulfur-stabilized anions

In this chapter we shall discuss some of the rich and varied chemistry of these, and other, organosulfur compounds. The stabilization of anions by sulfur is where we begin, and this theme runs right through the chapter. We will start with sulfides, sulfoxides, and sulfones. Sulfur has six electrons in its outer shell. As a sulfide, therefore, the sulfur atom carries two lone pairs. In a sulfoxide, one of these lone pairs is used in a bond to an oxygen atom—sulfoxides can be represented by at least two valence bond structures. The sulfur atom in a sulfone uses both of its lone pairs in bonding to oxygen, and is usually represented with two S=O double bonds.

methyl phenyl sulfide   methyl phenyl sulfoxide   methyl phenyl sulfone

Treatment of any of these compounds with strong base produces an anion (or a lithium derivative if BuLi is used) on what was the methyl group. How does the sulfur stabilize the anion? This question has been the subject of many debates and we have not got space to go into the details of all of them. There are at least two factors involved, and the first is evident from this chart of pK_a values for protons next to sulfone, sulfoxide and sulfide functional groups.

▶

Sulfoxides have the potential for chirality—the tetrahedral sulfur atom is surrounded by four different groups (here Ph, Me, O, and the lone pair) and (unlike, say, the tetrahedral nitrogen atom of an amide) has a stable tetrahedral configuration. We will revisit chirality in sulfoxides later in the chapter.

enantiomers of a chiral sulfoxide

$$\begin{bmatrix}O\\||\\S\end{bmatrix}_n \quad \xrightarrow[\,(n=0,1,2)\,]{\text{base}} \quad \begin{bmatrix}O\\||\\S\end{bmatrix}_n \quad \text{or} \quad \begin{bmatrix}O\\||\\S\end{bmatrix}_n$$
Ph   CH₃   Ph   CH₂⁻   Ph   CH₂Li

increasing acidity →

pKₐ (measured in DMSO)

			going from sulfoxide to sulfone increases the acidity by 4 pKₐ units		
CH_4	Ph–S–CH₃	H₃C–S(O)–CH₃		H₃C–S(O)₂–CH₃	Ph–S(O)₂–CH₃
[65]	48	35		31	29

a PhS group acidifies adjacent protons by *ca.* 17 pKₐ units

adding 2 oxygens increases the acidity by *ca.* 19 pKₐ units

■

If you want to read more about the elegant experiments that have been used to probe the structure of sulfonyl anions, see E. Block, *Reactions of organosulfur compounds*, Academic Press, New York, 1978.

Clearly, the oxygen atoms are important—the best anion-stabilizer is the sulfone, followed by the sulfoxide and then the sulfide. You could compare deprotonation of a sulfone with deprotonation of a ketone to give an enolate (Chapter 21). Enolates have a planar carbon atom and the anion is mainly on the oxygen atom. Sulfone-stabilized carbanions have two oxygen atoms and the anionic centre is probably planar, with the negative charge in a p orbital midway between them. Carbanions next to sulfones are planar, while anions next to sulfoxides and sulfides are believed to be pyramidal (sp³ hybridized).

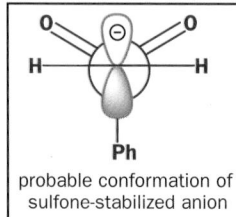

probable conformation of sulfone-stabilized anion

planar enolate   sulfone-stabilized anion

Yet the attached oxygen atoms cannot be the sole reason for the stability of anions next to sulfur because the sulfide functional group also acidifies an adjacent proton quite significantly. There is some controversy over exactly why this should be, but the usual explanation is that polarization of the sulfur's 3s and 3p electrons (which are more diffuse, and therefore more polarizable, than the 2s and 2p electrons of oxygen) contributes to the stabilization.

Anion stabilization by adjacent sulfur

It was long thought that delocalization into sulfur's empty 3d orbitals provided the anion stabilization required, but theoretical work in the last 20 years or so suggests this may not be the case. For example, *ab initio* calculations suggest that the C–S bond in –CH₂SH is longer than in CH₃SH. The converse would be true if delocalization into the sulfur's d orbitals were important. Delocalization would shorten the bond because it would have partial double bond character. More likely as an additional factor is delocalization into the σ* orbital of the C–S bond on the other side of the sulfur atom—the equatorial proton of dithiane (see p. 1254 for more on dithiane) is more acidic than the axial one, and the equatorial anion is more stable because it is delocalized into the C–S bond's σ* orbital.

3d 2p

$HS–CH_2^⊖ \leftrightarrow\ ^⊖HS=CH_2$
          ?

dithiane   →(base)→   σ*

Sulfone-stabilized anions in synthesis

The terpene sesquifenchene is a constituent of Indian valerian root oil. When it was first discovered in 1963, it was assumed to have structure A, related to bergamotene, a constituent of oil of bergamot (the fragrance of Earl Grey tea).

proposed structures of the natural product sesquifenchene

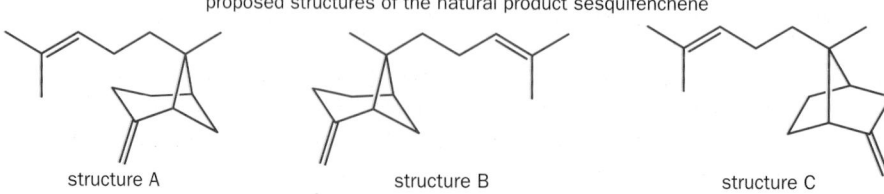

structure A   structure B   structure C

Compound A was synthesized in 1969, but was found not to be identical with sesquifenchene. A new structure was proposed, B, which was synthesized in 1971—but this compound too had different properties from those of natural sesquifenchene! A third structure was proposed, C, and it was made from a bicyclic sulfone.

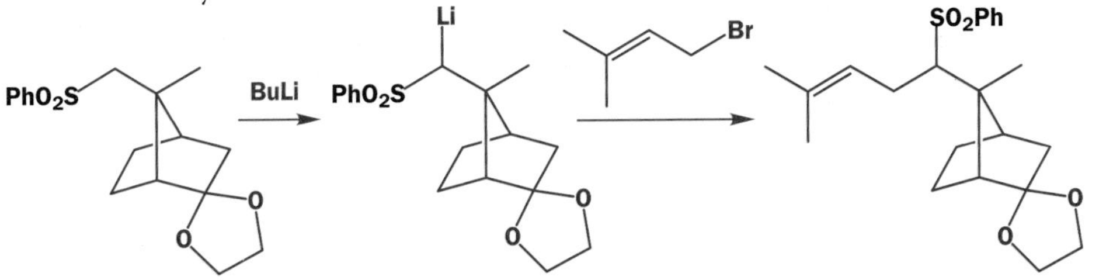

The bicyclic part of the structure was available in a few steps from norbornadiene. Deprotonation of the sulfone made a nucleophile that could be alkylated with prenyl bromide—a convenient way of joining on the extra five carbon atoms needed in the target structure. Next, the sulfone group had to be got rid of—there are a number of ways of doing this, and these chemists chose a Birch reduction with EtNH$_2$ instead of liquid ammonia. They might equally have tried hydrogenation with Raney nickel (see p. 626) or a sodium–amalgam-type reduction as is used in the Julia olefination (p. 810; you will see aluminium amalgam used in this way on p. 1266).

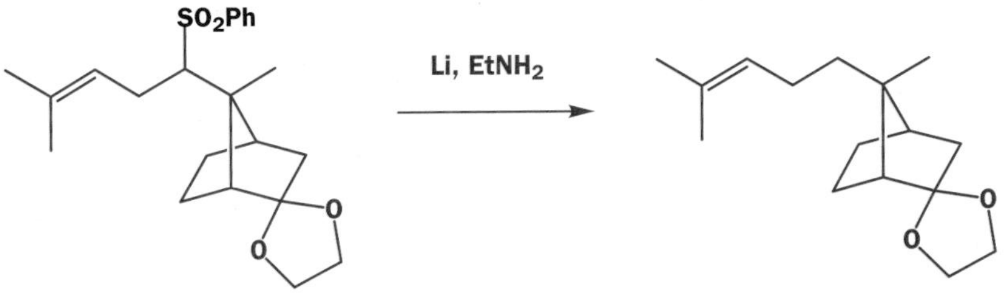

The exocyclic double bond was made by Wittig reaction on the deprotected ketone (aqueous acetic acid removed the dioxolane protecting group). This product had all the characteristics of natural sesquifenchene, confirming its true structure.

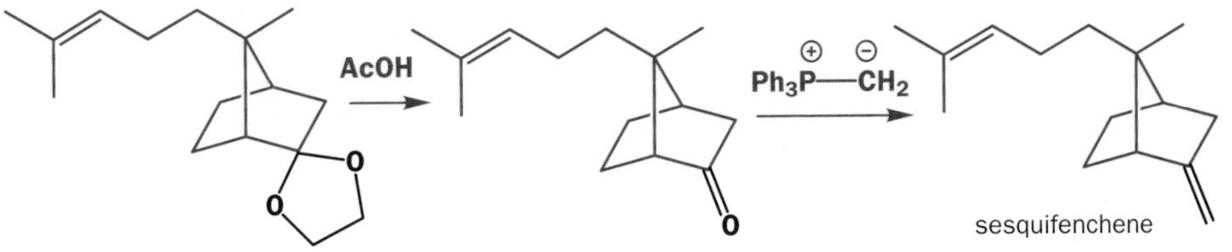

►
Of course, with today's spectroscopic techniques it is rarely necessary to synthesize a compound to confirm its structure, but misinterpretation still takes place and it is only when the compound is synthesized that the error comes to light.

A sulfoxide-stabilized anion in a synthesis

A sulfoxide alkylation formed the key step of a synthesis of the important vitamin biotin. Biotin contains a five-membered heterocyclic sulfide fused to a second five-membered ring, and the bicyclic skeleton was easy to make from a simple symmetrical ester. The vital step is a double S$_N$2 reaction on primary carbon atoms.

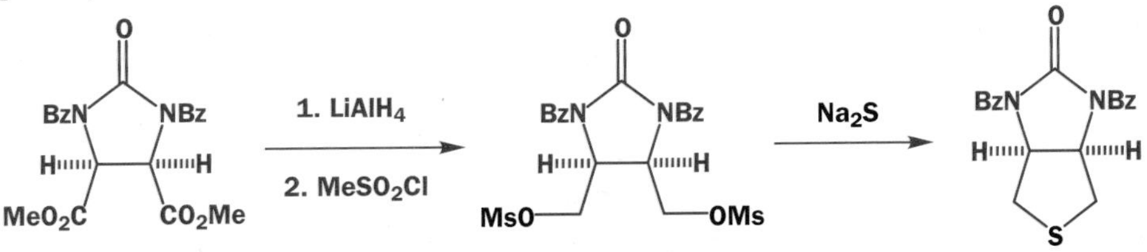

The next step was to introduce the alkyl chain—this was best done by first oxidizing the sulfide to a sulfoxide, using sodium periodate. The sulfoxide was then deprotonated with *n*-BuLi and alkylated with an alkyl iodide containing a carboxylic acid protected as its *t*-butyl ester. Reduction of the sulfoxide and hydrolysis back to the free acid gave biotin.

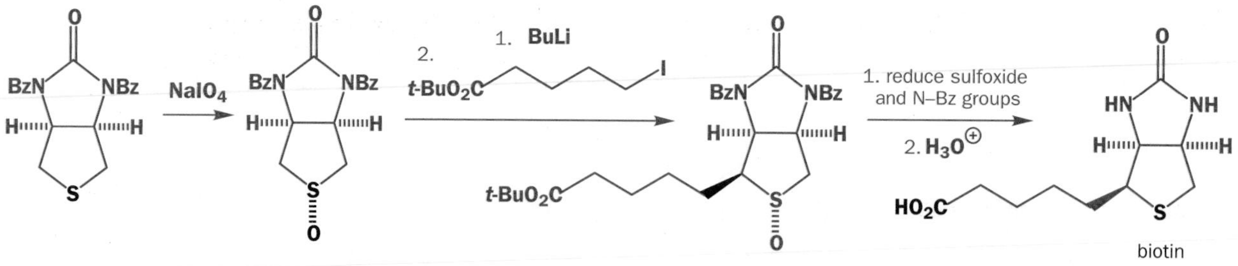

This synthesis involves some stereochemistry. Biotin carries the alkyl chain next to sulfur on the more hindered *endo* face of the molecule, and any successful synthesis has to address this particular problem. Here, the chemists decided to use the fact that alkylations of cyclic sulfoxides result in *trans* stereochemistry between the new alkyl group and the sulfoxide oxygen atom. As expected, oxidation of the sulfide proceeded faster from the *exo* face, giving an 8:1 ratio of *exo:endo* sulfoxides. Alkylation *trans* to the *exo* oxygen gave the desired (*endo*) product.

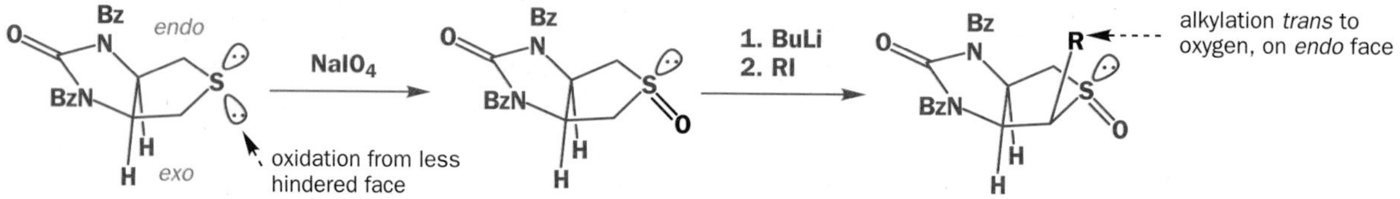

In Chapter 33 we talked about the ways in which cyclic compounds react stereoselectively—the stereochemistry of this sulfide oxidation is what you would expect from the examples we gave there.

The synthesis is diastereoselective—but not enantioselective since there is no way of distinguishing the left and right sides of the symmetrical sulfoxide.

Thioacetals

Although sulfide deprotonations are possible, the protons adjacent to *two* sulfide sulfur atoms are rather more acidic and alkylation of thioacetals is straightforward.

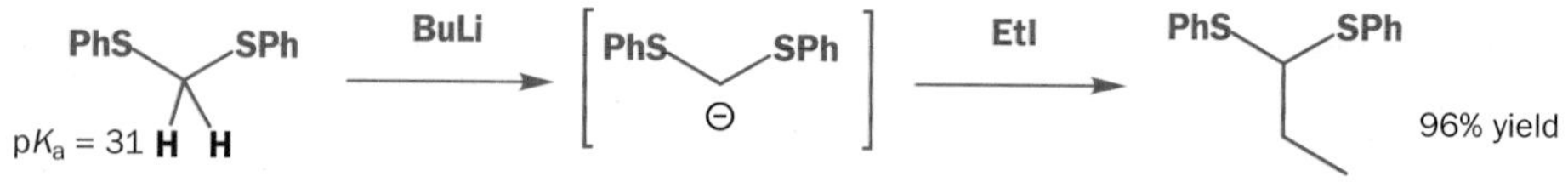

In general, thioacetals can be made in a similar way to 'normal' (oxygen-based) acetals–by treatment of an aldehyde or a ketone with a thiol and an acid catalyst—though a Lewis acid such as BF_3 is usually needed rather than a protic acid. The most easily made, most stable toward hydrolysis, and most reactive towards alkylation are cyclic thioacetals derived from 1,3-propanedithiol, known as **dithianes**.

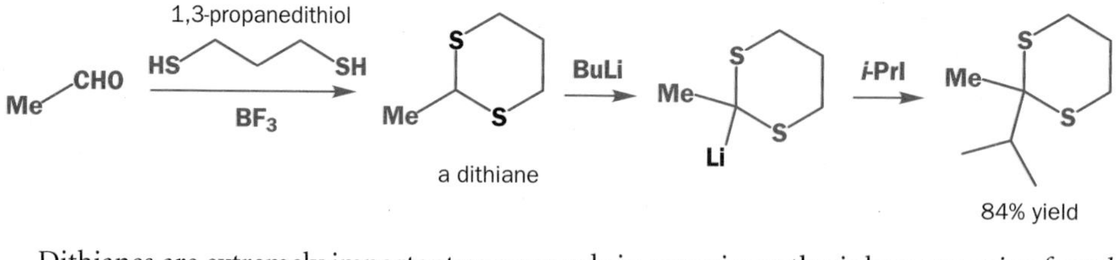

Dithianes are extremely important compounds in organic synthesis because *going from ketone to thioacetal inverts the polarity at the functionalized carbon atom*. Aldehydes, as you are well aware, are electrophiles at the C=O carbon atom, but dithioacetals, through deprotonation to an anion, are nucleophilic at this same atom.

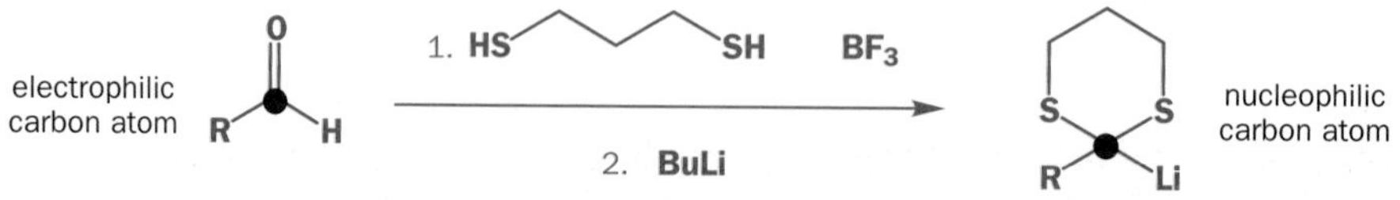

This is a case of umpolung, the concept you met in Chapter 30, and dithianes are among the most important of the umpolung reagents. An example: chemists wanted to make this compound (a 'metacyclophane') because they wanted to study the independent rotation of the two benzene rings, which is hindered in such a small ring. An ideal way would be to join electrophilic benzylic bromides to nucleophilic carbonyl groups, if that were possible.

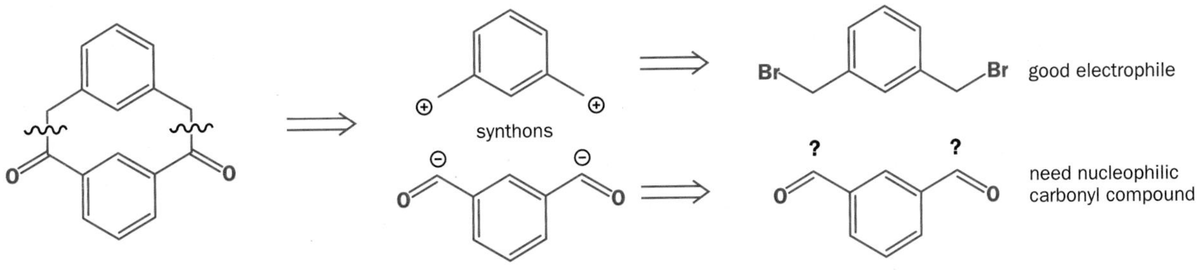

The dibromide and dialdehyde were both available—what they really wanted was a nucleophilic equivalent of the dialdehyde to react with the dibromide. So they made the dithioacetal.

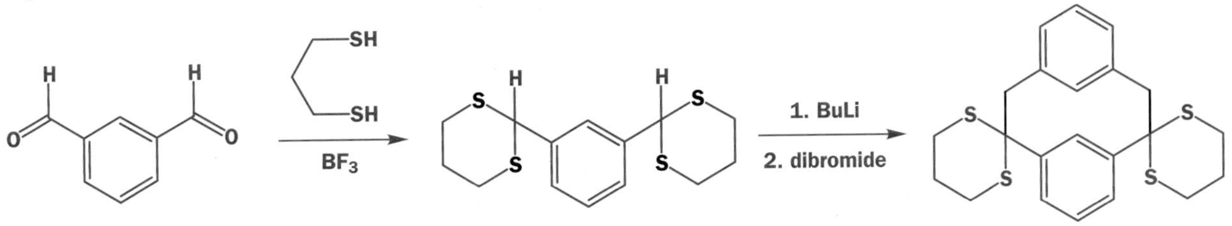

After the dithianes have been alkylated, they can be hydrolysed to give back the carbonyl groups. Alternatively, hydrogenation using Raney nickel replaces the thioacetal with a CH_2 group and gives the unsubstituted cyclophane.

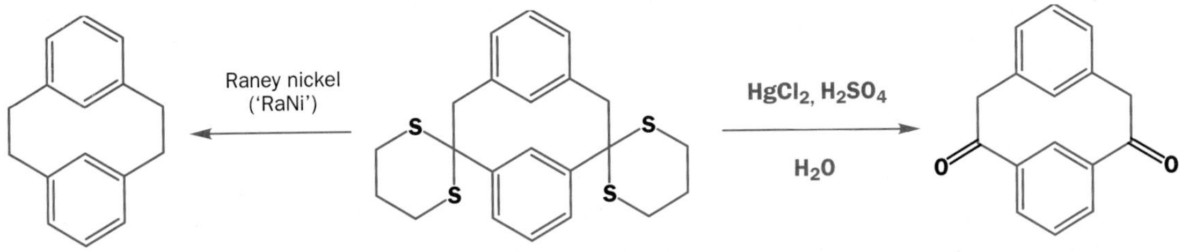

Both of these transformations deserve comment. Dithianes are rather more stable than acetals, and a mercury reagent has to be used to assist their hydrolysis. Mercury(II) and sulfides form strong coordination complexes, and the mercury catalyses the reaction by acting as a sulfur-selective Lewis acid.

■ Thiols are also known as **mercaptans** because of their propensity for 'mercury capture'.

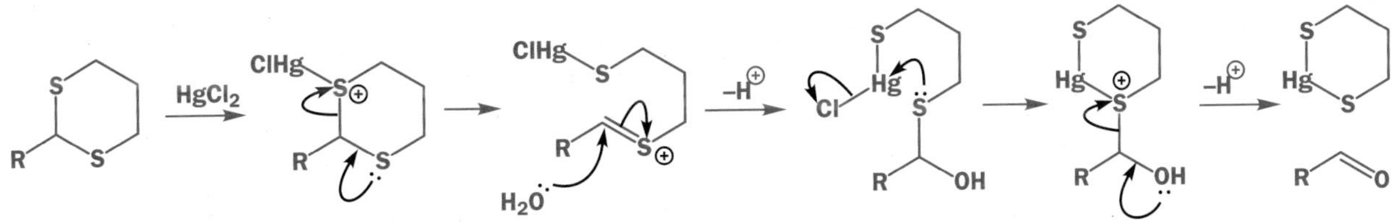

There are two reasons why the normal acid-catalysed hydrolysis of acetals usually fails with thioacetals. Sulfur is less basic than oxygen, so the protonated species is lower in concentration at a given pH, and the sulfur 3p lone pairs are less able to form a stable π bond to carbon than are the oxygen 2p lone pairs.

● Sulfur compounds are less basic than oxygen compounds and C=S compounds are less stable than C=O compounds.

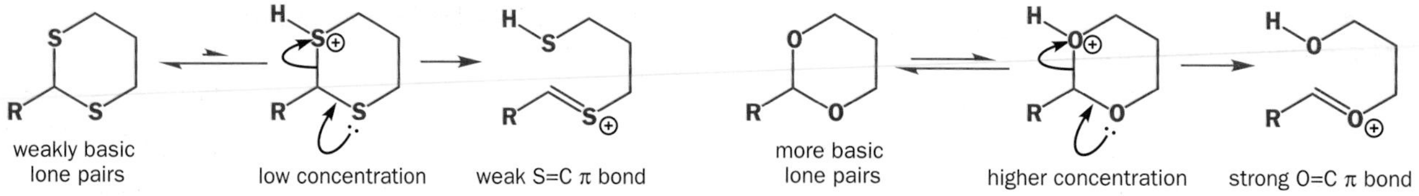

weakly basic lone pairs — low concentration — weak S=C π bond — more basic lone pairs — higher concentration — strong O=C π bond

The most obvious solution to this problem is to provide a better electrophile than the proton for sulfur. Mercury, Hg(II), is one solution. Another is oxidation of one sulfur to the sulfoxide, a process that would be impossible with the oxygen atoms of an ordinary acetal. Protonation can now occur on the more basic oxygen atom of the sulfoxide and the concentration of the vital intermediate is increased.

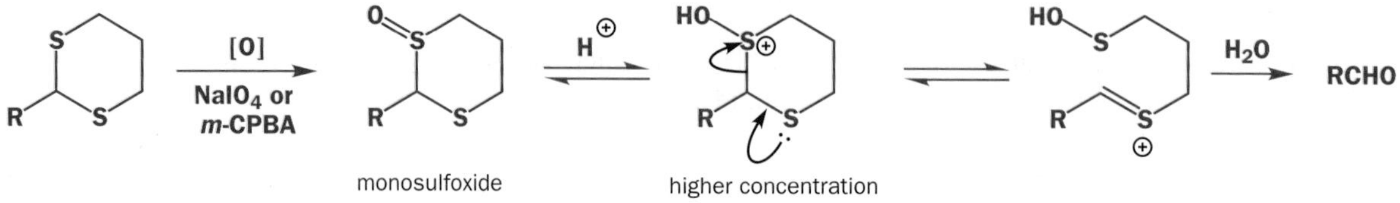

monosulfoxide — higher concentration

A third solution is methylation since sulfur is a better nucleophile for saturated carbon than is oxygen. The sulfonium salt can decompose in the same way to give the free aldehyde. There are many more methods for hydrolysing dithioacetals and their multiplicity should make you suspicious that none is very good. The best is probably the Hg(II) method but not everyone likes to use stoichiometric toxic mercury!

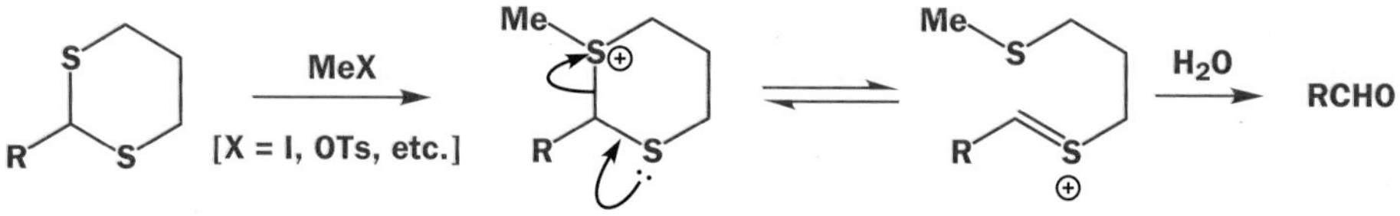

Hydrogenation of C–S bonds in both sulfides and thioacetals is often achieved with **Raney nickel**. This is a finely divided form of nickel made by dissolving away the aluminium from a powdered nickel–aluminium alloy using alkali. It can be used either as a catalyst for hydrogenation with gaseous hydrogen or as a reagent since it often contains sufficient adsorbed hydrogen (from the reaction of aluminium with alkali) to effect reductions alone. Thioacetalization followed by Raney nickel reduction is a useful way of replacing a C=O group with CH_2.

● Dithianes are d^1 reagents (acyl anion equivalents)

A sequence in which a carbonyl group has been masked as a sulfur derivative, alkylated with an electrophile, and then revealed again is a **nucleophilic acylation**. These nucleophilic equivalents of carbonyl compounds are known as **acyl anion equivalents**. In the retrosynthetic terms of Chapter 30 they are d^1 reagents corresponding to the acyl anion synthon.

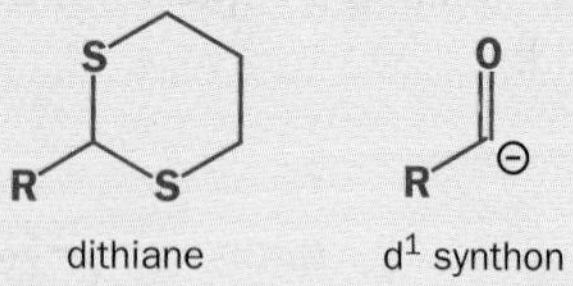

dithiane — d^1 synthon

Allyl sulfides

Apart from thioacetals, allyl sulfides are among the easiest sulfides to deprotonate and alkylate because of the conjugating ability of the allyl group. However, the very delocalization that assists anion formation means that the anions often react unregioselectively: lithiated phenyl allyl sulfide, for instance, reacts with hexyl iodide to give a 3:1 ratio of regioisomers.

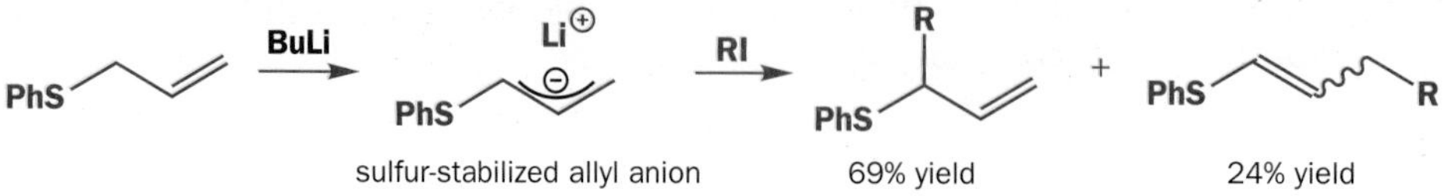

sulfur-stabilized allyl anion — 69% yield — 24% yield

2-Pyridyl allyl sulfide, on the other hand, gives only one regioisomer in its alkylation reactions. It is sensible here to show the 'allyl anion' as a compound with a C–Li bond.

>99:1 this regioisomer

The 'sulfur-stabilized allyl anion' in the previous reaction is probably a mixture of organolithium compounds in unknown proportions and the depiction as an anion avoids this problem.

The same is true for a number of other allylic sulfur compounds in which the sulfur carries a lithium-coordinating heteroatom. Coordination encourages reaction next to sulfur (you might say it makes the lithium more at home there) and means that allyl sulfide alkylations can be made quite regioselective. The importance of this is probably not evident to you, but on p. 1268 you will meet a synthesis of the natural product nuciferal in which this principle is used—the key step will be the alkylation of this allylic sulfide to give an 86% yield of the product with the alkyl group next to sulfur.

86% yield

If the sulfur-based anion-stabilizing group is at a higher oxidation level, it is not usually necessary to provide chelating groups to ensure reaction next to sulfur. The allylic sulfone we made earlier in the chapter (p. 1250) reacts in this way with an unsaturated ester to give a cyclopropane. Notice how much weaker a base (MeO⁻) is needed here, as the anion (and it is an anion if the counterion is Na⁺ or K⁺) is stabilized by sulfone and alkene.

allylic sulfone   anion of allylic sulfone   methyl *trans*-chrysanthemate

The first step is conjugate addition of the highly stabilized anion. The intermediate enolate then closes the three-membered ring by favourable nucleophilic attack on the allylic carbon. The leaving group is the sulfinate anion and the stereochemistry comes from the most favourable arrangement in the transition state for this ring closure. The product is the methyl ester of the important chrysanthemic acid found in the natural pyrethrum insecticides.

In Chapters 10 and 23 we established that more stable nucleophiles, and hence more reversible reactions, are likely to favour conjugate addition.

toluene-*para*-sulfinate   methyl *trans*-chrysanthemate

We shall see more reactions of this sort in which sulfur has a dual role as anion-stabilizing and leaving group in the next section.

Sulfonium salts

Sulfides are nucleophiles even when not deprotonated—the sulfur atom will attack alkyl halides to form sulfonium salts. This may look strange in comparison with ethers, but it is, of course, a familiar pattern of reactivity for amines, and you have seen phosphonium salts formed in a similar way (Chapters 14 and 31).

sulfide → sulfonium salt

This same principle was used in the isolation of stable carbocations described in Chapter 17.

This reaction is an equilibrium and it may be necessary in making sulfonium salts from less reactive sulfides (sterically hindered ones for example) to use more powerful alkylating agents with non-nucleophilic counterions, for example, Me_3O^+ BF_4^-, trimethyloxonium fluoroborate (also known as Meerwein's salt). The sulfur atom captures a methyl group from O^+, but the reverse does not happen and the BF_4^- anion is not a nucleophile.

sulfide   oxonium salt   sulfonium salt   nonnucleophilic counterion   ether

Not only is dimethyl ether a poor nucleophile, it is also a gas and is lost from the reaction mixture. The same principle is used to make sulfides from other sulfides. With that clue, and the position of this reaction in the 'sulfonium salt' section, you should be able to work out the mechanism and say why the reaction works.

The most important chemistry of sulfonium salts is based on one or both of two attributes:

1. Sulfonium salts are electrophiles: nucleophilic substitution displaces a neutral sulfide leaving group

2. Sulfonium salts can be deprotonated to give sulfonium ylids

Sulfonium salts as electrophiles

mustard gas

During the First World War, mustard gas was developed as a chemical weapon—it causes the skin to blister and is an intense irritant of the respiratory tract. Its reactivity towards human tissue is related to the following observation and is gruesome testimony to the powerful electrophilic properties of sulfonium ions.

this reaction goes 600 times more rapidly than . . .   . . . this simple S_N2 reaction

In both cases, intramolecular displacement of the chloride leaving group by the sulfur atom—or, as we should call it, **participation by sulfur** (see Chapter 37)—gives a three-membered cyclic sulfonium ion intermediate (an **episulfonium** or **thiiranium ion**). Nucleophilic attack on this electrophilic sulfonium ion, either by water or by the structural proteins of the skin, is very fast. Of course, mustard gas can react twice in this way. You will see several more examples of reactions in which a sulfonium ion intermediate acts as an electrophile in the next section.

participation by sulfur   sulfonium ion intermediate

Sulfonium ylids

A reminder. An **ylid** is a species with positive and negative charges on adjacent atoms.

The positive charge carried by the sulfur atom means that the protons next to the sulfur atom in a sulfonium salt are significantly more acidic than those in a sulfide, and sulfonium salts can be deprotonated to give **sulfonium ylids**.

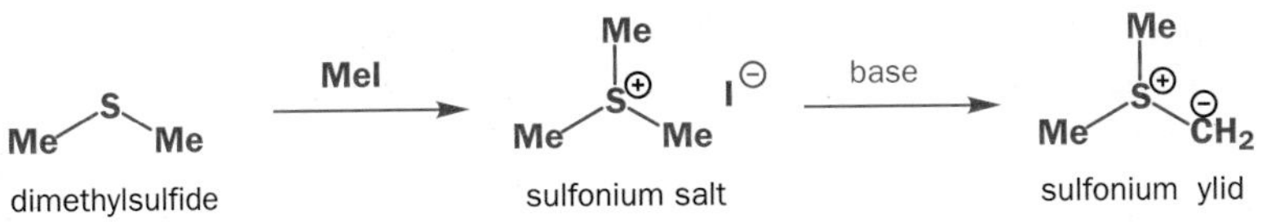

In Chapter 31 we discussed the Wittig reaction of phosphonium ylids with carbonyl compounds. Sulfonium ylids react with carbonyl compounds too, but in quite a different way—compare these two reactions.

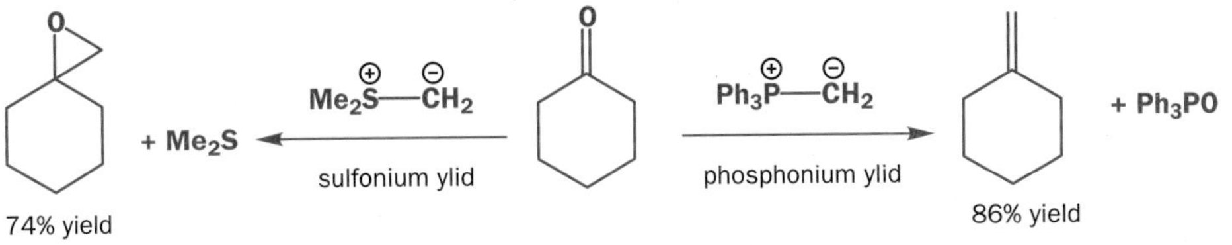

74% yield

86% yield

Phosphonium ylids give alkenes while sulfonium ylids give epoxides. Why should this be the case? The driving force in the Wittig reaction is formation of the strong P=O bond—that force is much less in the sulfur analogues (the P=O bond energy in Ph$_3$PO is 529 kJ mol^{-1}; in Ph$_2$SO the S=O bond energy is 367 kJ mol^{-1}). The first step is the same in both reactions: the carbanion of the ylid attacks the carbonyl group in a nucleophilic addition reaction. The intermediate in the Wittig reaction cyclizes to give a four-membered ring but this does not happen with the sulfur ylids. Instead, the intermediate decomposes by intramolecular nucleophilic substitution of Me$_2$S by the oxyanion.

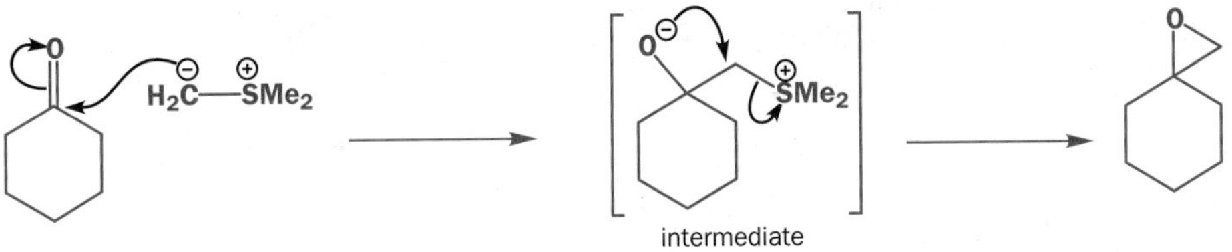

intermediate

We could compare sulfonium ylids with the carbenoids we discussed in Chapter 40—both are nucleophilic carbon atoms carrying a leaving group, and both form three-membered rings by insertion into π bonds. Sulfonium ylids are therefore useful for making epoxides from aldehydes or ketones; other ways you have met of making epoxides (Chapters 20 and 45) started with alkenes that might be made with phosphorus ylids.

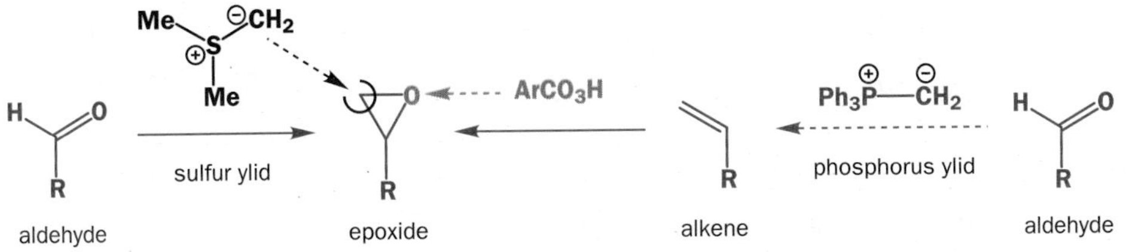

aldehyde   epoxide   alkene   aldehyde

The simplest route to certain potential β-blocker drugs is from an epoxide, and the chemists working on their synthesis decided that, since 4-cyclopropylbenzaldehyde was more readily available than 4-cyclopropyl styrene, they would use the aldehyde as the starting material and make the epoxide in one step using a sulfonium ylid.

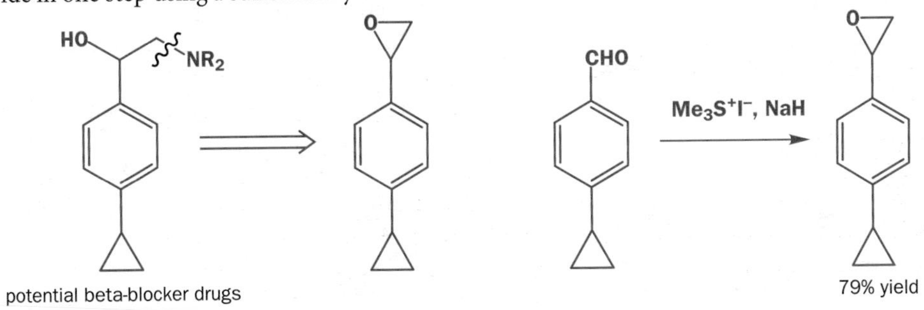

potential beta-blocker drugs

79% yield

You will recall from Chapter 31 that we divided phosphorus ylids into two categories, 'stabilized' and 'unstabilized', in order to explain the stereochemistry of their alkene-forming reactions. Again, there is a similarity with sulfonium ylids: the same sort of division is needed—this time to explain the different regioselectivities displayed by different sulfonium ylids. Firstly, an example.

91% yield, mixture of diastereoisomers

stabilized sulfur ylid

unstabilized sulfur ylid

79% yield

'Stabilized' sulfonium ylids

Changing from the simple sulfonium ylid to one bearing an anion-stabilizing substituent changes the regioselectivity of the reaction. 'Unstabilized' sulfonium ylids give epoxides from α,β-unsaturated carbonyl compounds while 'stabilized' ylids give cyclopropanes. In the absence of the double bond, both types of ylid give epoxides—the ester-stabilized ylid, for example, reacts with benzil to give an epoxide but with methyl vinyl ketone (but-3-en-2-one) to give a cyclopropane.

92% yield

benzil

methyl vinyl ketone

87% yield

Why does the stabilized ylid prefer to react with the double bond? In order to understand this, let's consider first the reaction of a simple, unstabilized ylid with an unsaturated ketone. The enone has two electrophilic sites, but from Chapters 10 and 23, in which we discussed the regioselectivity of attack of nucleophiles on Michael acceptors like this, you would expect that direct 1,2-attack on the ketone is the faster reaction. This step is irreversible, and subsequent displacement of the sulfide leaving group by the alkoxide produces an epoxide. It's unimportant whether a cyclopropane product would have been more stable: the epoxide forms faster and is therefore the kinetic product.

With a stabilized ylid, direct addition to the carbonyl group is, in fact, probably still the faster reaction. But, in this case, the starting materials are sufficiently stable that the reaction is reversible, and the sulfonium ylid is re-expelled before the epoxide has a chance to form. Meanwhile, some ylid adds to the ketone in a 1,4 (Michael or conjugate) fashion. 1,4-Addition, although slower, is energetically more favourable because the new C–C bond is gained at the expense of a (relatively) weak C=C π bond rather than a (relatively) strong C=O π bond, and is therefore irreversible. Eventually, all the ylid ends up adding in a 1,4-fashion, generating an enolate as it does so, which cyclizes to give the cyclopropane, which is the thermodynamic product. This is another classic example of kinetic versus thermodynamic control, and you can add it to the mental list of examples you started when you first read Chapter 13.

continued opposite

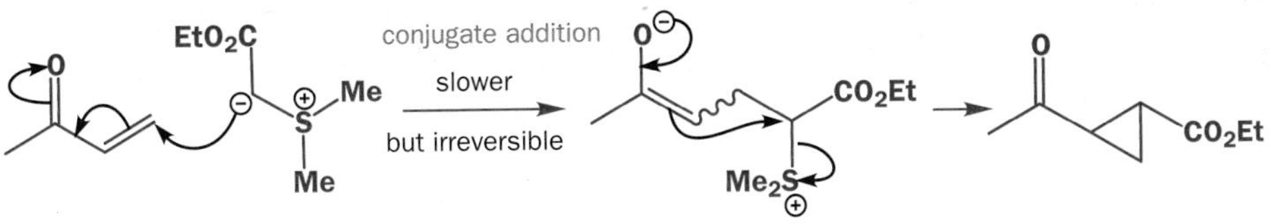

Sulfoxonium ylids

There is another, very important class of stabilized sulfur ylids that owe their stability not to an additional anion-stabilizing substituent but to a more anion-stabilizing sulfur group. These are the **sulfoxonium ylids**, made from dimethylsulfoxide by S_N2 substitution with an alkyl halide. Note that the sulfur atom is the nucleophile rather than the oxygen atom in spite of the charge distribution. The high-energy sulfur lone pair is better at S_N2 substitution at saturated carbon—a reaction that depends very little on charge attraction (Chapter 17).

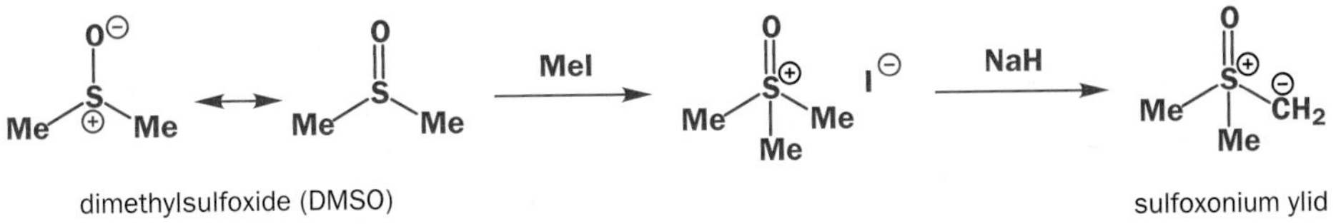

dimethylsulfoxide (DMSO)   sulfoxonium ylid

Sulfoxonium ylids react with unsaturated carbonyl compounds in the same way as the stabilized ylids that you have met already do—they form cyclopropanes rather than epoxides. The example below shows one consequence of this reactivity pattern—by changing from a sulfonium to a sulfoxonium ylid, high yields of either epoxide or cyclopropane can be formed from an unsaturated carbonyl compound (this one is the terpene carvone).

81% yield   sulfoxonium ylid   carvone   sulfonium ylid   89% yield

The table on p. 1253 introduced the idea that anion-stabilization is related to the number of oxygen atoms carried by the sulfur atom.

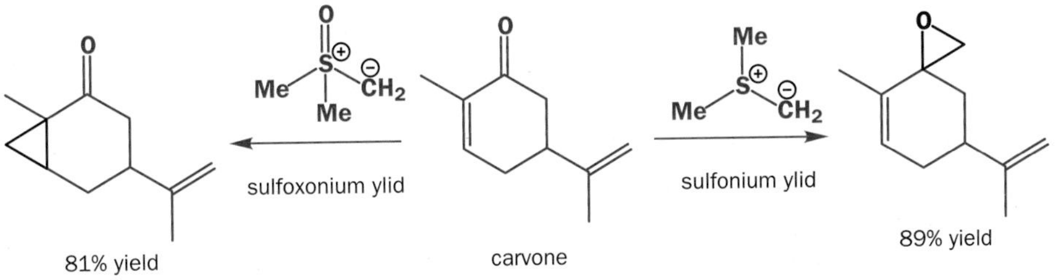

Sulfur-stabilized cations

We have mentioned cations in this chapter several times and now we will gather the various ideas together. Cations are stable on the sulfur atom itself, as you have just seen in sulfonium and sulfoxonium salts. They are stable on adjacent carbon atoms since the sulfur atom contributes a lone pair to form a $C=S^+$ π bond, and they are stable on the next carbon atom along the chain since sulfur contributes a lone pair to form a $C–S^+$ σ bond in a three-membered ring.

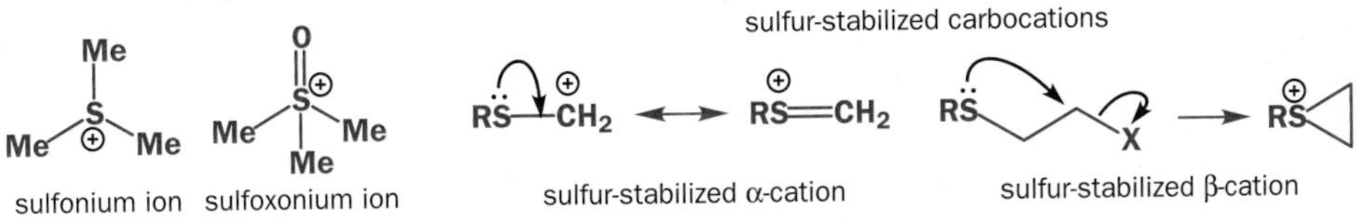

sulfur-stabilized carbocations

sulfonium ion   sulfoxonium ion   sulfur-stabilized α-cation   sulfur-stabilized β-cation

You may protest that these last two species are not *carbo*-cations at all but rather sulfonium ions, and you would be right. However, they can be used in place of carbocations as they are electrophilic at carbon so it is useful to think of them as modified carbocations as well as sulfonium ions. Sulfur-stabilized α-cations are easily made from α-chlorosulfides and are useful in alkylation of silyl enol ethers.

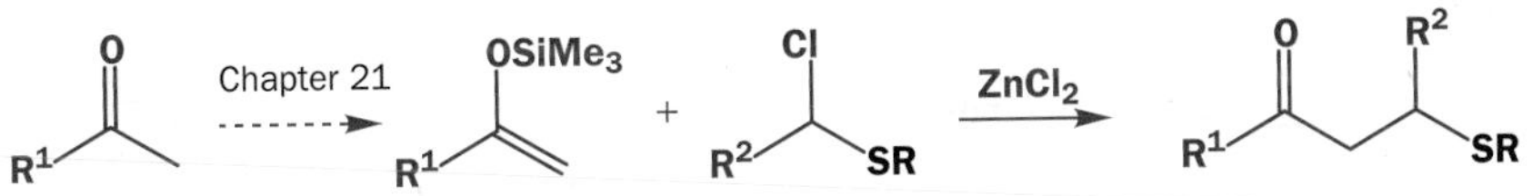

What is the point of this? Silyl enol ethers can be alkylated only by compounds that give carbocations in the presence of Lewis acids. The mechanism for the alkylation therefore involves the formation of a sulfur-stabilized cation.

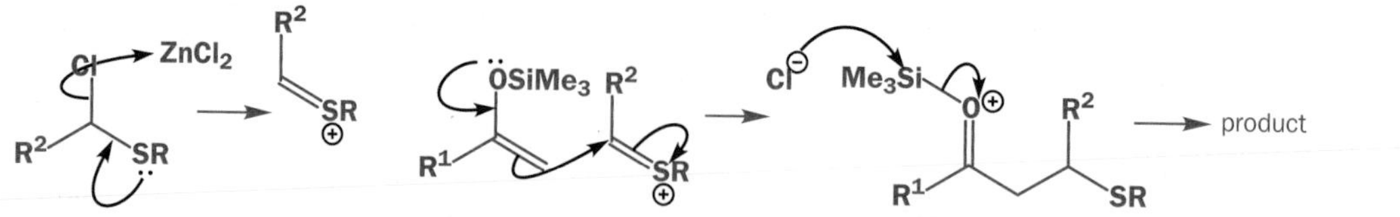

The sulfide (SR) can be removed from the product with Raney nickel to give a simple ketone. This ketone has apparently been made by the alkylation of a silyl enol ether with a primary alkyl group (R^2CH_2). This would be impossible without stabilization of the cation by the sulfur atom.

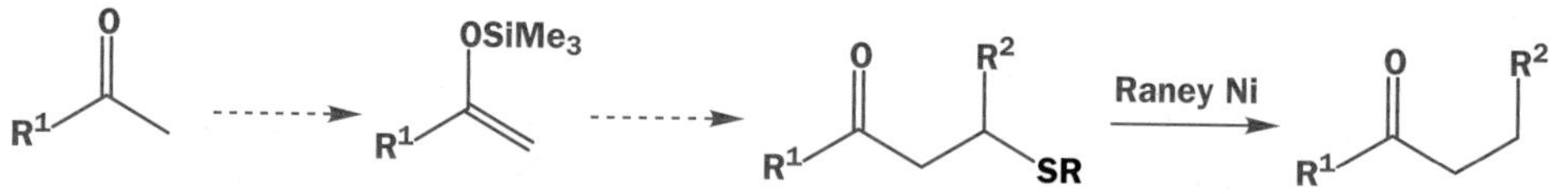

The Pummerer rearrangement

Though the stabilization of the cation by a sulfide is not as good as the stabilization by an ether (the $C=S^+$ bond is weaker than the $C=O^+$ bond), it is still good enough to make the reaction work and, of course, C–O bonds cannot be reduced by any simple reagent. One thing remains—how is the chlorosulfide made in the first place? Remarkably, it is made from the alkyl halide (R^2CH_2Cl) you would use for the (impossible) direct alkylation without sulfur.

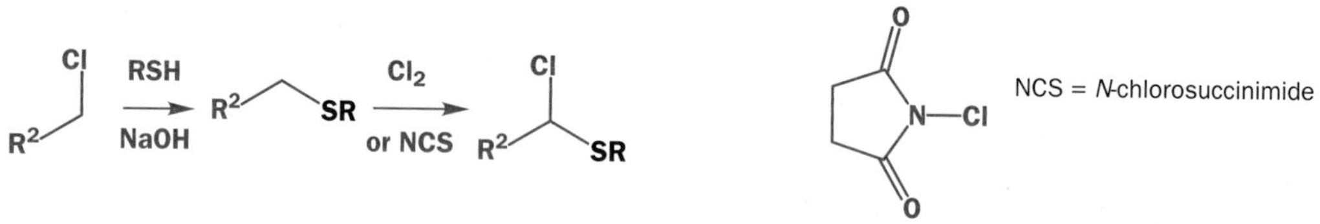

NCS = N-chlorosuccinimide

The first step is just the S_N2 displacement of Cl^- by RS^- that you have already seen. The second step actually involves chlorination at sulfur (you have also seen that sulfides are good soft nucleophiles for halogens) to form a sulfonium salt. Now a remarkable thing happens. The chlorine atom is transferred from the sulfur atom to the adjacent carbon atom by the **Pummerer rearrangement**.

chloro-sulfonium ion

An ylid is first formed by loss of a proton—again, you have seen this—and then chloride is lost to form the same cation that we used in the alkylation reaction. In this step there is no nucleophile available except chloride ion so that adds to the carbon atom.

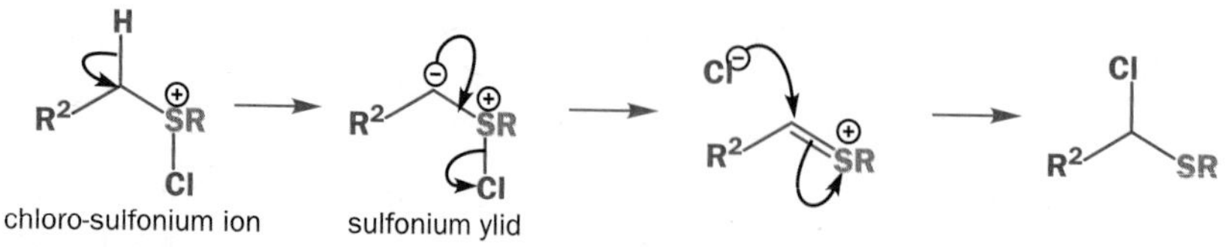

chloro-sulfonium ion   sulfonium ylid

There are many variations on the Pummerer rearrangement but they all involve the same steps: a leaving group is lost from the sulfur atom of a sulfonium ylid to create a cationic intermediate that captures a nucleophile at the α carbon atom. Often the starting material is a sulfoxide.

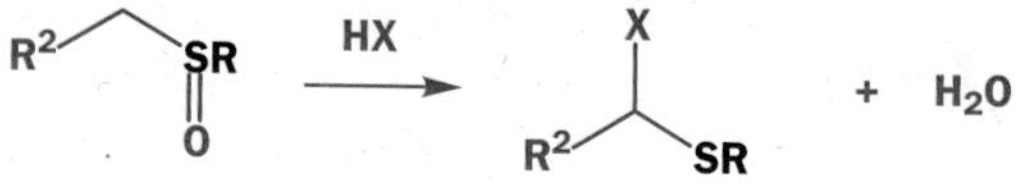

Treatment of a sulfoxide, particularly one with an anion-stabilizing substituent to help ylid formation, produces cations reactive enough to combine with nucleophiles of all sorts, even aromatic rings. The product is the result of electrophilic aromatic substitution (Chapter 22) and, after the sulfur has been removed with Raney nickel, is revealed as a ketone that could not be made without sulfur as the cation required would be too unstable.

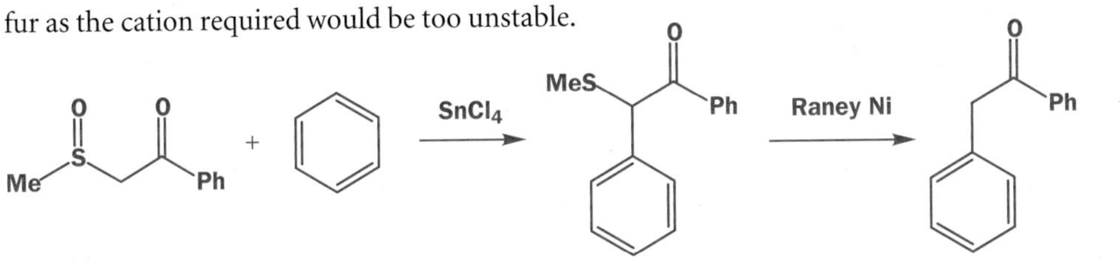

100% yield

A Lewis acid (SnCl₄) is used to remove the oxygen from the sulfoxide and the ketone assists ylid formation. The sulfur atom stabilizes the cation enough to counteract the destabilization by the ketone. The Lewis acid is necessary to make sure that no nucleophile competes with benzene.

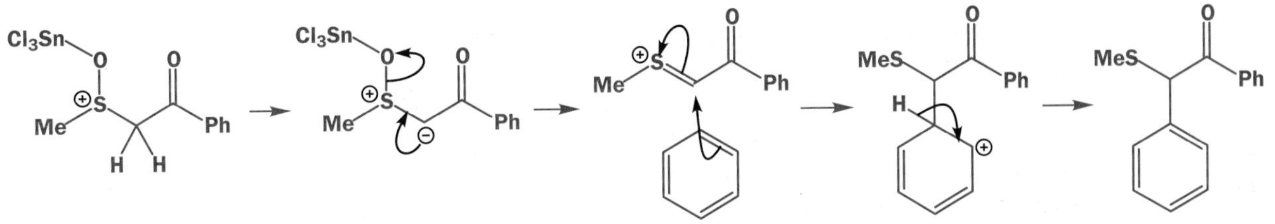

Most commonly of all, a sulfoxide is treated with acetic anhydride and the cation is captured by an internal nucleophile to form a new ring. Here the nitrogen atom of an amide is the nucleophile. The mechanism is very like that of the last example.

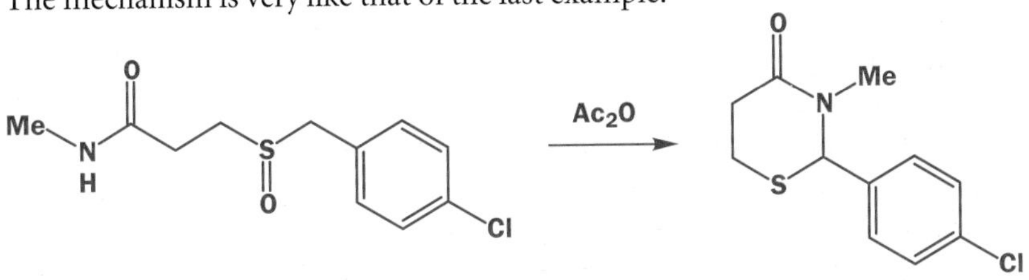

71% yield

Sulfur-stabilized β-carbocations (three-membered rings)

Three-membered cyclic sulfonium ions, representing β carbocations, are often encountered in participation reactions. We have seen this already in the way mustard gas works, but almost any arrangement of a sulfide with a leaving group on the β carbon atom leads to participation and the formation of a three-membered ring. The product is formed by migration of the PhS group from one carbon atom to another (Chapter 37).

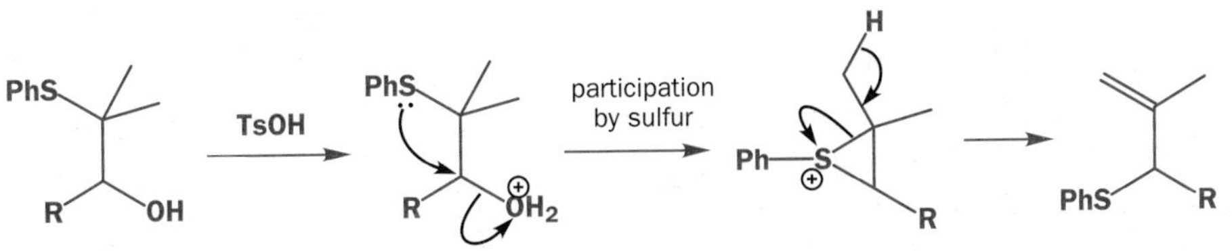

In this case, elimination of a proton from one of the methyl groups leads to an allylic sulfide—you have seen earlier in the chapter how these compounds, and the sulfoxides derived from them, can be used in synthesis. If we make a small change in the structure of the starting material—just joining up the two methyl groups into a cyclopropane—things change quite a bit. It becomes possible to make the starting material by a lithiation reaction because cyclopropyllithiums are significantly stabilized by the three-membered ring (Chapter 8) and the rearrangement goes with carbon rather than sulfur migration.

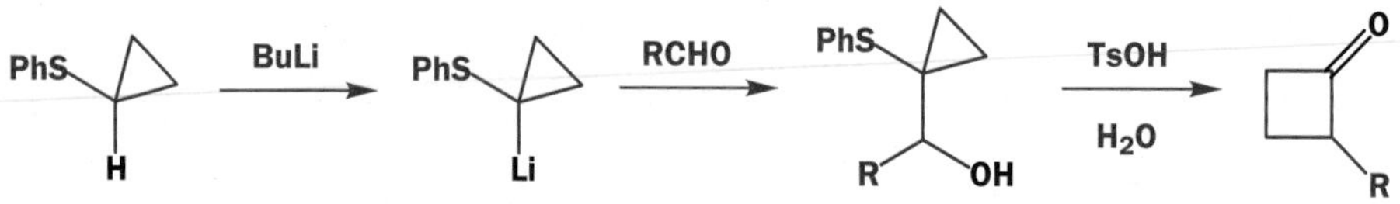

In the rearrangement, the alcohol is protonated as before but no sulfur participation occurs. Instead, a ring expansion, also assisted by sulfur, produces a four-membered ring and hydrolysis of the α cation (an intermediate you have seen several times) gives a cyclobutanone. The difference between participation through space and C=S⁺ bond formation is not that great.

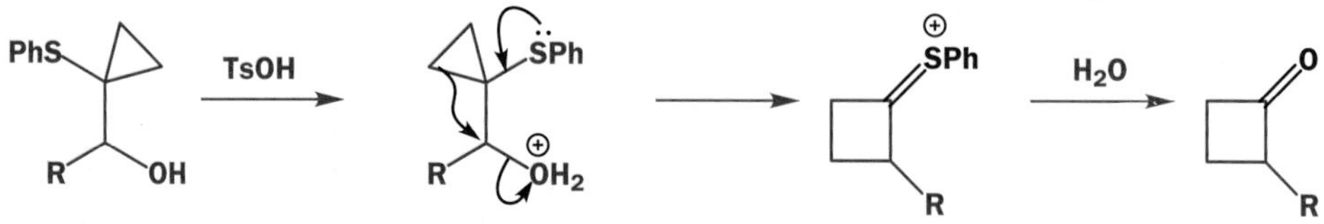

Thiocarbonyl compounds

Simple thioaldehydes and thioketones are too unstable to exist and attempts at their preparation lead to appalling smells (Chapter 1). The problem is the poor overlap between the $2sp^2$ orbital on carbon and the $3sp^2$ orbital on sulfur as well as the more or less equal electronegativities of the two elements. Stable thiocarbonyl compounds include dithioesters and thioamides where the extra conjugation of the oxygen or nitrogen atom helps to stabilize the weak C=S bond.

Dithioesters can be made by a method that would seem odd if you thought only of ordinary esters. Organolithium or Grignard reagents combine well with carbon disulfide (CS_2—the sulfur analogue of CO_2) to give the anion of a dithioacid. This is a much more nucleophilic species than an ordinary carboxylate anion and combines with alkyl halides to give dithioesters.

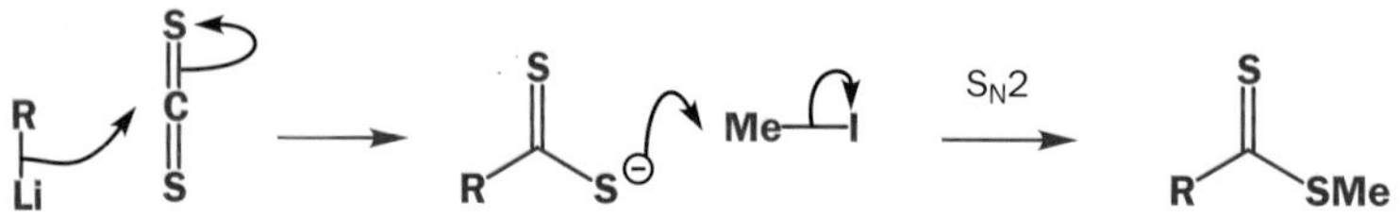

The reaction of dithioesters with Grignard reagents is even more remarkable. Because sulfur and carbon have about the same electronegativity, the Grignard reagent may add to either end of the π bond. If it adds to sulfur, the resulting anion is stabilized by two sulfur atoms, rather like the dithiane anions we have seen earlier in this chapter, and can be used as a d^1 reagent.

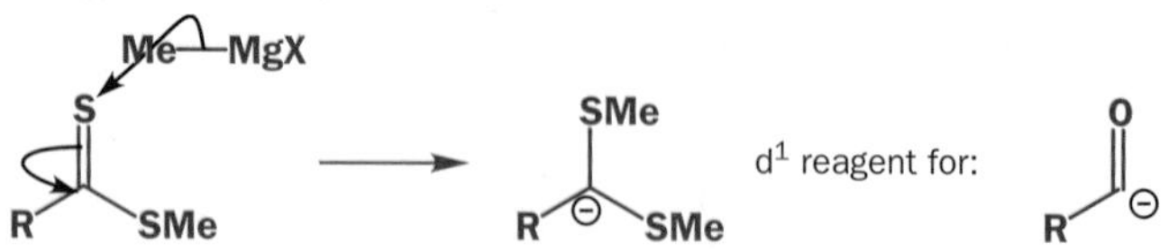

Thioamides are usually made by reaction of ordinary amides with P_2S_5 or **Lawesson's reagent**. Since C=S is so much less stable than C=O, there is a clear case to call in phosphorus to remove the oxygen. The situation is rather like that in the Wittig reaction: C=C is less stable than C=O, so phosphorus is called in to remove the oxygen because of the even greater stability of the P=O bond. Lawesson's reagent has P=S bonds and a slightly surprising structure.

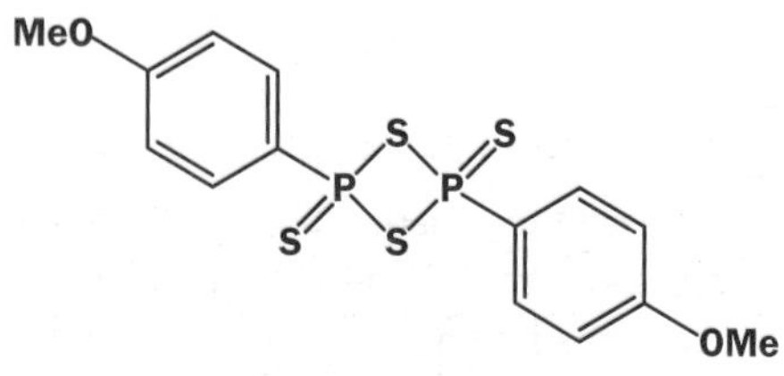

Lawesson's reagent, or, if you prefer...
2,4-bis(4-methoxyphenyl)-1,3-dithia-2,4-phosphetane-2,4-disulfide

We can learn from this compound that sulfur has much less objection to four-membered rings than do oxygen or carbon. We have seen from the structure of sulfur itself (S_8) that it likes eight-membered rings too. Rings of almost any size are acceptable to sulfur as bond angles matter less to second-row elements that are not generally hybridized. Lawesson's reagent converts amides into thioamides and we have seen (Chapter 44) how these are used to make thiazoles.

Sulfoxides

The formation and reactions of sulfoxonium ylids demonstrate how sulfoxides occupy a useful and interesting part of the middle ground between sulfides and sulfones—they are weakly nucleophilic, like sulfides (and can be alkylated with methyl iodide to give sulfoxonium salts as we have just seen), but at the same time they stabilize anions almost as well as sulfones. However, sulfoxides are perhaps the most versatile of the three derivatives because of a good deal of chemistry that is unique to them. There are two reasons why this should be so.

1.   Sulfoxides have the potential to be chiral at sulfur

2.   Sulfoxides undergo some interesting pericyclic reactions

We shall deal with each of these in turn.

Representing S=O compounds

Sulfoxides are sometimes drawn as S=O and sometimes as S^+–O^-. The second representation might remind you of the phosphorus ylids used in the Wittig reaction (Chapters 14 and 31), which can be drawn with a $P=CH_2$ double bond or as P^+–CH_2^-. All of these representations are correct—it is a matter of personal choice which you prefer.

The double bonds are between 2p orbitals of O or C and 3d orbitals of S or P. But when we drew the structure of TsCl we always drew two S=O double bonds. You might think that an alternative structure with two S–O single bonds is not so good and almost nobody draws TsCl that way. Illogical but not unreasonable.

two ways of drawing the sulfonyl chloride TsCl

Sulfoxides are chiral

Providing the two groups attached to sulfur are different, a sulfoxide is chiral at the sulfur atom. There are two important ways of making sulfoxides as single enantiomers, both asymmetric versions of reactions otherwise used to make racemic sulfoxides: oxidation and nucleophilic substitution at sulfur.

Sulfides are easy to oxidize and, depending on the type and quantity of oxidizing agent used, they can be cleanly oxidized either to sulfoxides or sulfones.

The oxidation of sulfides to sulfoxides can be made asymmetric by using one of the important reactions we introduced in the last chapter—the **Sharpless asymmetric epoxidation**. The French chemist Henri Kagan discovered in 1984 that, by treating a sulfide with the oxidant *t*-butyl hydroperoxide in the presence of Sharpless's chiral catalyst ($Ti(O^iPr)_4$ plus one enantiomer of diethyl tartrate), the oxygen atom could be directed to one of the sulfide's two enantiotopic lone pairs to give a sulfoxide in quite reasonable enantiomeric excess (ee).

■ For a definition of 'enantiomeric excess', see Chapter 45.

► Here is an example where drawing a sulfoxide as S⁺–O⁻ is better.

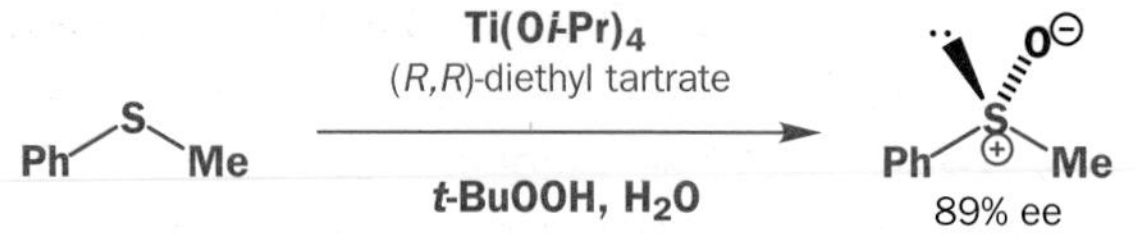

As yet, this asymmetric oxidation is successful only with simple aryl alkyl sulfoxides like this one, and the nucleophilic displacement method is much more widely used since it is more general and gives products of essentially 100% ee.

Sulfoxides can alternatively be made by displacement of RO⁻ from a **sulfinate ester** with a Grignard reagent.

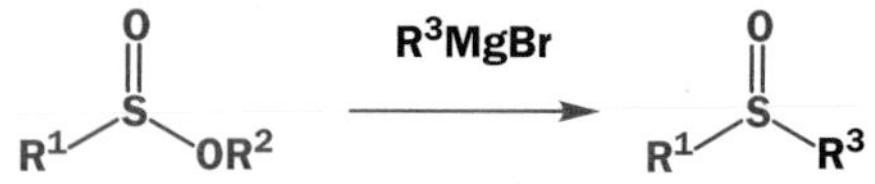

Sulfinate esters, like sulfoxides, are chiral at sulfur and, if the ester is formed from a chiral alcohol (menthol is best), they can be separated into two diastereoisomers by crystallization—this is really a resolution of the type you first met in Chapter 16. Attack by the Grignard reagent takes place with inversion of configuration at sulfur, giving a single enantiomer of the sulfoxide.

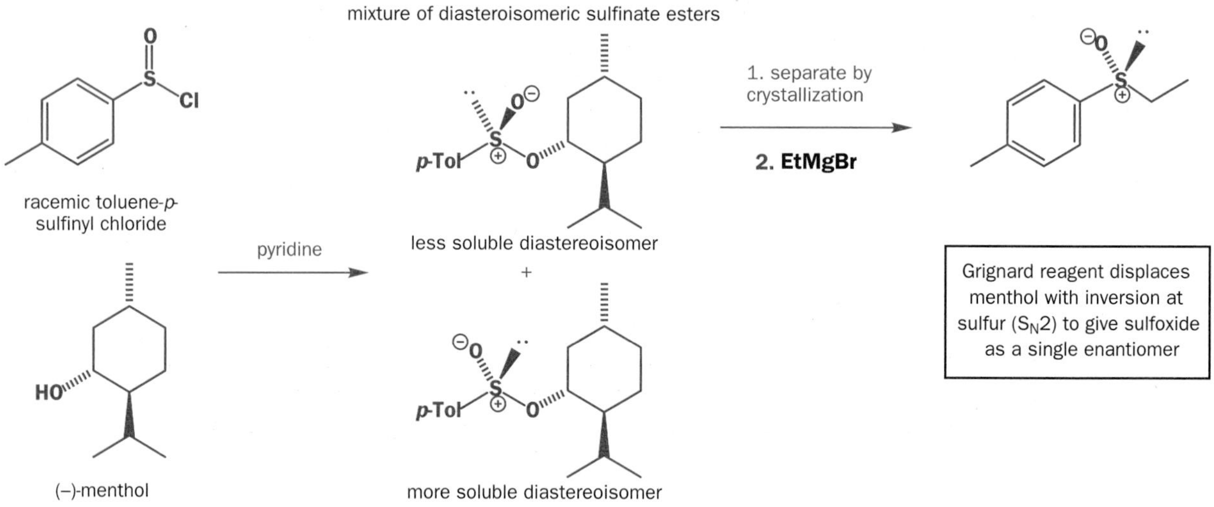

Grignard reagent displaces menthol with inversion at sulfur (S$_N$2) to give sulfoxide as a single enantiomer

Chiral sulfoxides in synthesis

How can the chirality of sulfoxides be made useful? This area of research has received a lot of attention in the last 10–15 years, with many attempts to design reactions in which the chirality at sulfur is transferred to chirality at carbon. Unfortunately, one of the simplest reactions of sulfoxides, the addition of their anions to aldehydes, usually proceeds with no useful stereoselectivity at all.

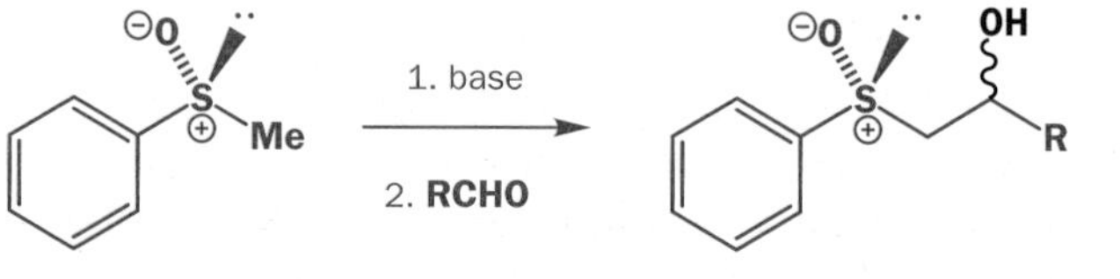

1:1 mixture of diastereoisomers

Some more successful uses of sulfoxides to control new chiral centres at carbon have been developed in Strasbourg by Guy Solladié, and they involve stereoselective reduction of carbonyl groups directed by the sulfoxide's oxygen atom. For example, the synthesis below shows how chirality at sulfur can be transferred to chirality at carbon by using a reduction directed by the S–O bond. If this ketone is treated with the bulky reducing agent DIBAL (*i*-Bu₂AlH), one alcohol is formed, with less than 5% of its diastereoisomer. Remarkably, if ZnCl₂ is added to the mixture, the opposite diastereoisomer is obtained! Reduction of the products with aluminium amalgam removes the sulfoxide (we discussed this process earlier in the chapter) leaving behind enantiomerically enriched samples of the alcohol.

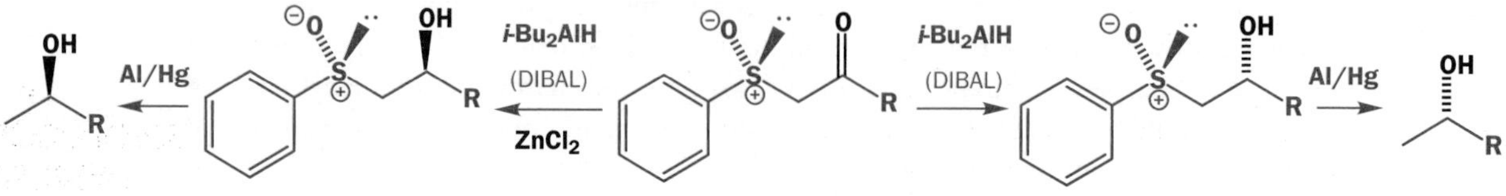

Solladié explained these results by suggesting that, in the absence of $ZnCl_2$, the sulfoxide adopts the conformation that places the two electronegative oxygen atoms as far apart as possible. DIBAL then attacks the less hindered face of the ketone, *syn* to the sulfoxide lone pair. With $ZnCl_2$, on the other hand, the sulfoxide's conformation is fixed by chelation to zinc: attack on the less hindered face of the ketone now gives the other diastereoisomer. Both compounds can be reduced with Al/Hg, which removes the sulfur group, to give opposite enantiomers of a chiral alcohol.

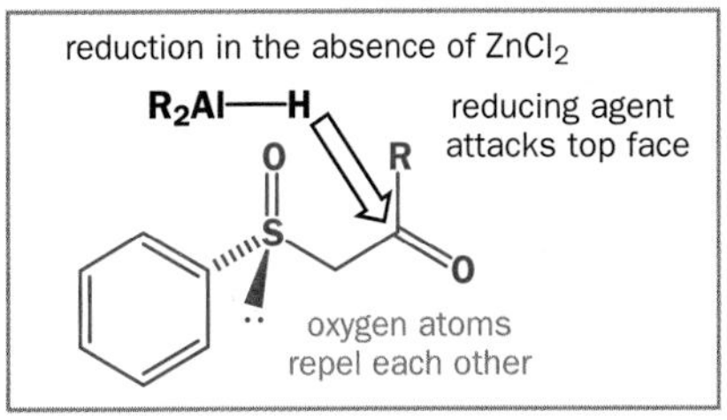

reduction in the absence of $ZnCl_2$ — reducing agent attacks top face — oxygen atoms repel each other

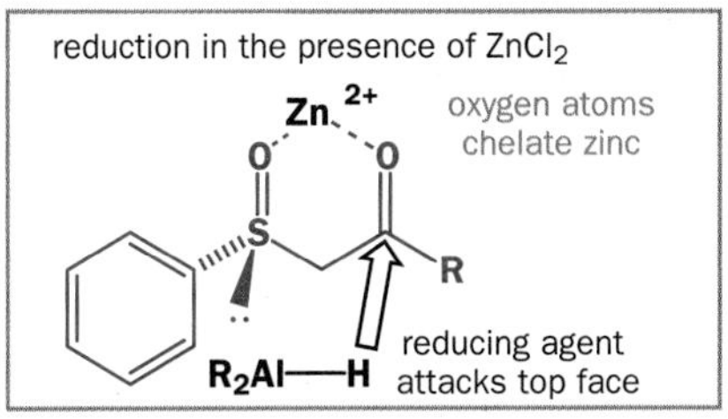

reduction in the presence of $ZnCl_2$ — oxygen atoms chelate zinc — reducing agent attacks top face

Allylic sulfoxides are not configurationally stable

Most sulfoxides will retain their configuration at sulfur up to temperatures of about 200 °C—indeed, it is estimated that the half-life for racemization of an enantiomerically pure sulfoxide is about 5000 years at room temperature. However, sulfoxides carrying allyl groups are much less stable—they racemize rapidly at about 50–70 °C. A clue to why this should be is provided by the reaction of an allylic sulfoxide with trimethyl phosphite, $P(OMe)_3$.

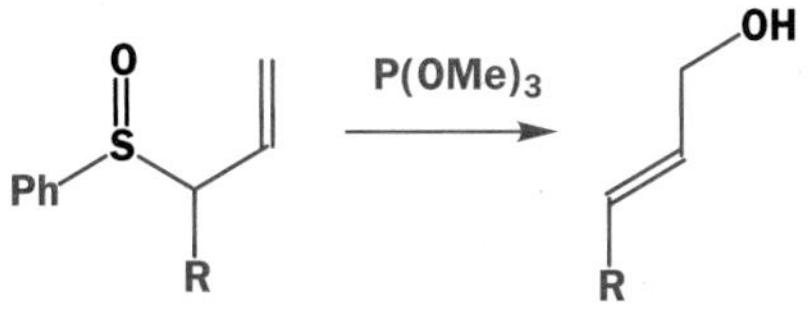

The product obtained is an allylic alcohol with the hydroxyl group at the other end of the allyl system from where the sulfur started—a rearrangement has taken place. We have observed the rearrangement in this case because the $P(OMe)_3$ has trapped the rearrangement product but, even without this reagent, allylic sulfoxides are continually and reversibly rearranging into sulfenate esters by the mechanism shown below.

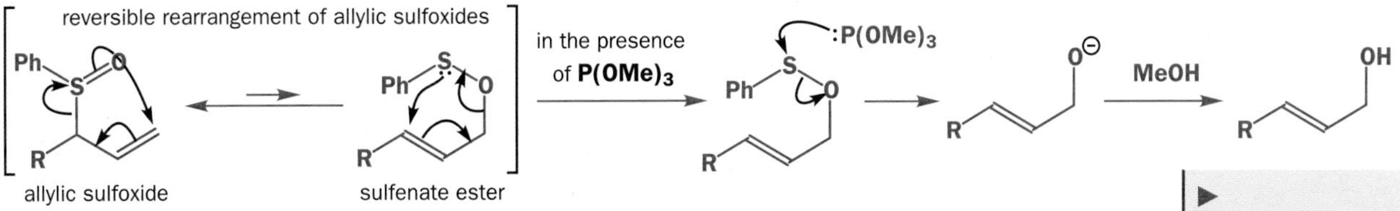

reversible rearrangement of allylic sulfoxides — allylic sulfoxide — sulfenate ester — in the presence of $P(OMe)_3$ — MeOH

The rearrangement product, which is less stable than the sulfoxide and is therefore never observed directly, is a **sulfenate ester**. It has no chirality at sulfur so, when it rearranges back to the sulfoxide, it has no 'memory' of the configuration of the starting sulfoxide, and the sulfoxide becomes racemized.

Having read Chapter 36, you should be able to classify the pericyclic rearrangement reaction: it is a [2,3]-sigmatropic rearrangement (make sure you can see why before you read further) and as such is the first of the pericyclic rearrangements of sulfoxides that we shall talk about.

If our proposal that allylic sulfoxides rearrange reversibly to sulfenate esters is correct, then, if we make the sulfenate ester by another route, it too should rearrange to an allylic sulfoxide—and indeed it does. The sulfenate ester arising from reaction of allylic alcohols with PhSCl (phenylsulfenyl chloride) cannot be isolated: instead, the allylic sulfoxide is obtained, usually in very good yield, and this method is often used to make allylic sulfoxides.

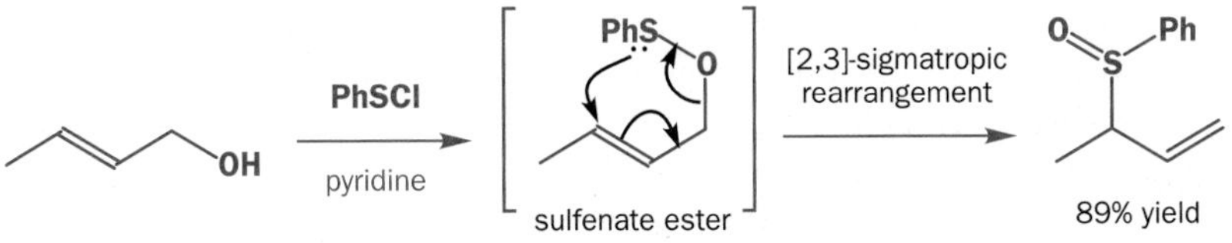

PhSCl / pyridine — sulfenate ester — [2,3]-sigmatropic rearrangement — 89% yield

You shouldn't at this stage try to learn all the names for every type of organosulfur compound—what matters is the structures. Here the names are all very similar and easily confused so, just for reference, here are the structures of a sulfonate ester (such as a tosylate or mesylate), a sulfinate ester, and a sulfenate ester.

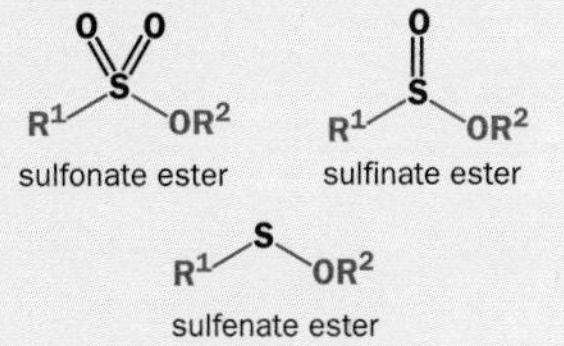

sulfonate ester — sulfinate ester — sulfenate ester

Uses for [2,3]-sigmatropic rearrangements of sulfoxides

Allylic sulfoxides exist in equilibrium with allyl sulfenate esters. The two interconvert by [2,3]-sigmatropic rearrangement, and the equilibrium lies over to the side of the sulfoxide. Allyl sulfenate esters are therefore impossible to isolate, but they can be trapped by adding a compound known as a **thiophile**—P(OMe)$_3$ was the example you just saw, but secondary amines like Et$_2$NH also work—which attacks the sulfur atom to give an allylic alcohol. This can be a very useful way of making allylic alcohols, particularly as the starting sulfoxides can be constructed by using sulfur's anion-stabilizing ability. What is more, the starting allylic sulfoxides can themselves be made from allylic alcohols using PhSCl—overall then we can use allylic sulfoxide to alkylate allylic alcohols! This scheme should make all this clearer.

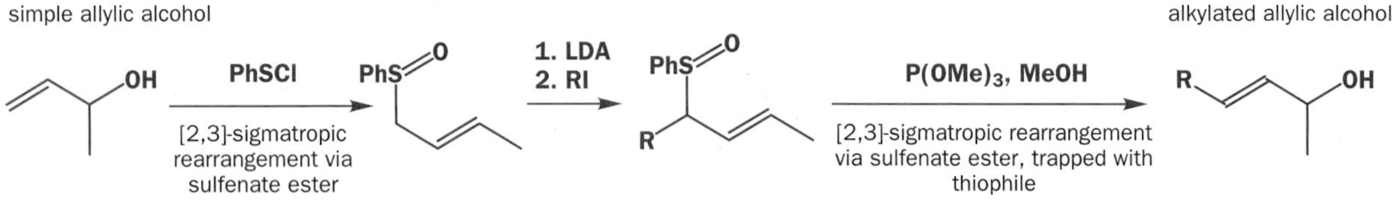

We can illustrate the synthesis of allylic alcohols from allylic sulfoxides with this synthesis of the natural product nuciferal. We mentioned this route on p. 1257 because it makes use of a heterocyclic allyl sulfide to introduce an alkyl substituent regioselectively. The allyl sulfide is oxidized to the sulfoxide, which is converted to the rearranged allylic alcohol with diethylamine as the thiophile. Nuciferal is obtained by oxidizing the allylic alcohol to an aldehyde with manganese dioxide.

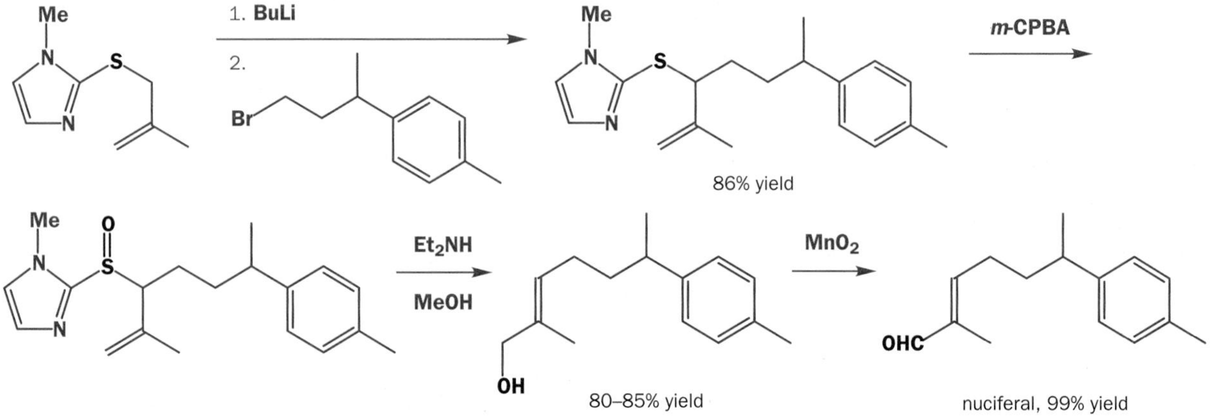

The next example makes more involved use of these [2,3]-sigmatropic allylic sulfoxide–allylic alcohol rearrangements. It comes from the work of Evans (he of the chiral auxiliary) who, in the early 1970s, first demonstrated the synthetic utility of allylic sulfoxides. Here he is using this chemistry to make precursors of the prostaglandins, a family of compounds that modulate hormone activity within the body.

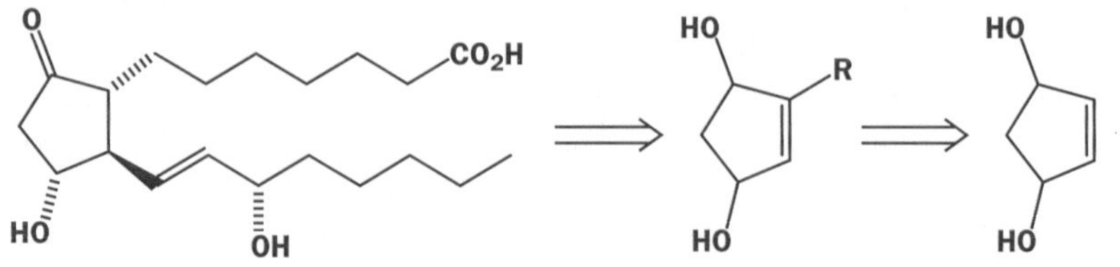

■ Prostaglandins are discussed more thoroughly in Chapter 51.

PGE$_1$, a prostaglandin

Prostaglandins are trisubstituted cyclopentanones, and the aim was to synthesize them from available cyclopentenediol using allylic sulfoxide chemistry to introduce the long alkyl chain R group. Treating *syn*-cyclopentenediol with PhSCl gave the allylic sulfoxide (either hydroxyl can react but the product is the same). The sulfoxide was deprotonated and reacted with an alkyl halide, and then rearranged back to an allylic alcohol using P(OMe)$_3$ as the thiophile.

55% yield · 85% yield · 65% yield

Stereochemistry of sulfoxide reactions

This sequence of reactions contains some interesting stereochemistry. The first rearrangement, from the cyclopentenediol to the allylic sulfoxide, is stereospecific—the *syn*-diol gives the *syn*-hydroxysulfoxide. This is typical of [2,3]-sigmatropic rearrangements—they are suprafacial with respect to the allylic component (see Chapter 36). In the next step, the R group is introduced *trans* to the hydroxyl group. This is a stereoselective reaction, not a stereospecific one,

because the other diastereoisomer of the starting material, with the hydroxyl group and the sulfoxide *trans*, also gives the product with the R group *trans* to the hydroxyl group. Finally, there is another stereospecific (suprafacial) [2,3]-sigmatropic rearrangement, maintaining the *syn* relative stereochemistry of the hydroxysulfoxide in the stereochemistry of the diol product.

Sulfoxide elimination—oxidation to enones

Sulfoxides next to an electron-withdrawing or conjugating group are also unstable on heating, not because they racemize but because they decompose by an elimination process.

81% yield

The rather unstable phenylsulfenic acid (PhS–OH) is eliminated and the reaction occurs partly because of the creation of conjugation and partly because PhSOH decomposes to volatile products. The elimination is a pericyclic reaction—it may not immediately be obvious what sort, but it is, in fact, a reverse cycloaddition. This is clearest if we draw the mechanism of the reverse reaction.

reaction in this direction would be a [3+2] cycloaddition

This reaction provides a useful way of introducing a double bond next to a carbonyl group. Here it is in a synthesis by Barry Trost of the Queen Bee Substance (the compound fed by the workers to those bee larvae destined to become queens). The compound is also a pheromone of the termite and is used to trap these destructive pests. Trost started with the monoester of a dicarboxylic acid, which he converted to a methyl ketone by reacting the acyl chloride with a cuprate. The ketone was then protected as a dioxolane derivative to prevent it enolizing, and the sulfur was introduced by reacting the enolate of the ester with the sulfur electrophile MeSSMe.

In Chapter 9 we discussed ways of making ketones by nucleophilic attack on carboxylic acid derivatives.

Next, the protecting group was removed with acid, and the sulfide was oxidized to the sulfoxide with sodium periodate (NaIO₄) ready for elimination. Heating to 110 °C then gave the Queen Bee Substance in 86% yield.

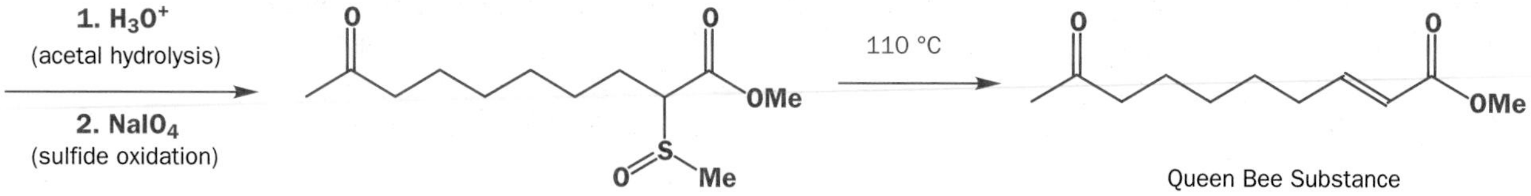

1. H₃O⁺
(acetal hydrolysis)

2. NaIO₄
(sulfide oxidation)

110 °C

Queen Bee Substance

Presumably, the methyl sulfoxide was chosen here because it worked better—it is more usual to use a phenyl sulfoxide, and PhS groups can be introduced in the same way (by reacting enolates with PhSSPh or PhSCl). The cycloheptanone derivative used in our first elimination example was made from cycloheptanone in this way.

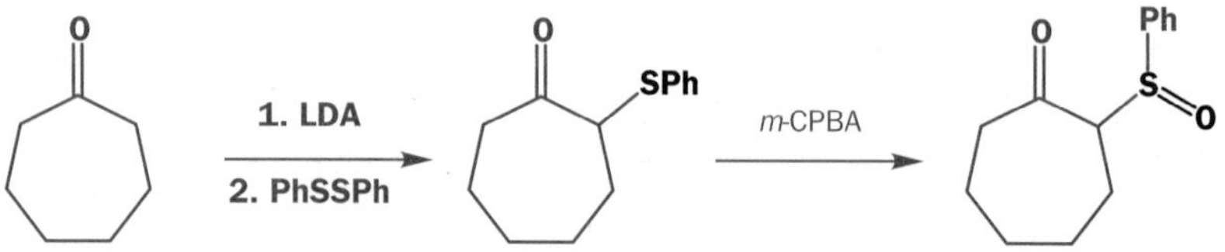

1. LDA

2. PhSSPh

m-CPBA

87% yield   100% yield

This elimination takes place more easily still when sulfur is replaced by a selenium—PhSe groups can be introduced by the same method, and oxidized to selenoxides with *m*-CPBA at low temperature. The selenoxides are rarely isolated, because the elimination takes place rapidly at room temperature.

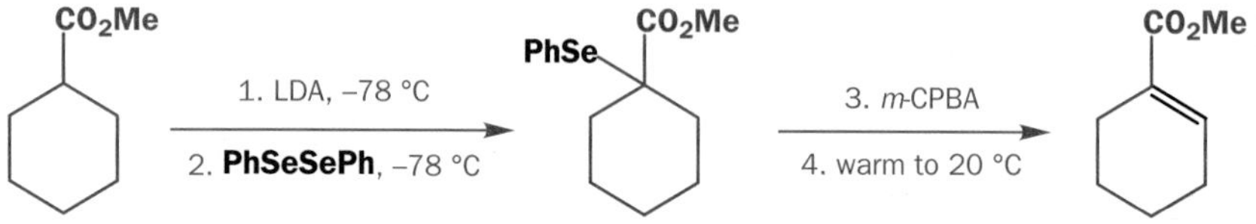

1. LDA, –78 °C

2. PhSeSePh, –78 °C

3. *m*-CPBA

4. warm to 20 °C

96% yield

> Sulfur and selenium have many properties in common, and much sulfur chemistry is mirrored by selenium chemistry. In general, organoselenium compounds tend to be less stable and more reactive than organosulfur ones because the C–Se bond is even weaker than a C–S bond. They also have even fouler odours.

Other oxidations with sulfur and selenium

Selenium dioxide and allylic oxidation

Having introduced selenium, we should at this point mention an important reaction that is peculiar to selenium but that is closely related to these pericyclic reactions. Selenium dioxide will react with alkenes in a [4 + 2] cycloaddition reminiscent of the ene reaction.

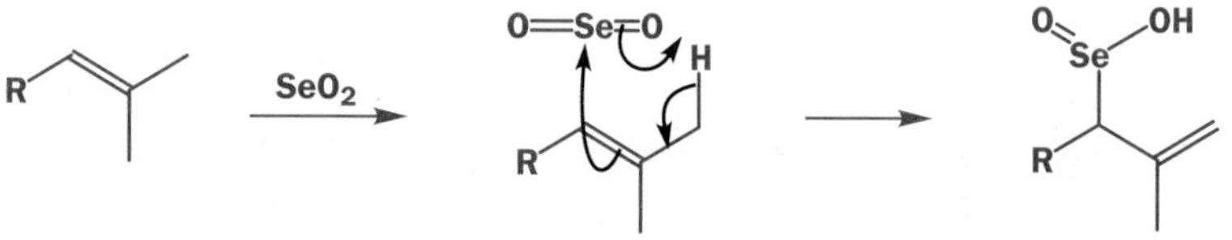

SeO₂

allylic seleninic acid

> In a very few special cases, the seleninic acid intermediate has been isolated.

The initial product is an allylic seleninic acid—and just like an allylic sulfoxide (but more so because the C–Se bond is even weaker) it undergoes allylic rearrangement to give an unstable compound that rapidly decomposes to give an allylic alcohol. In some cases, particularly this most useful oxidation of methyl groups, the oxidation continues to give an aldehyde or ketone.

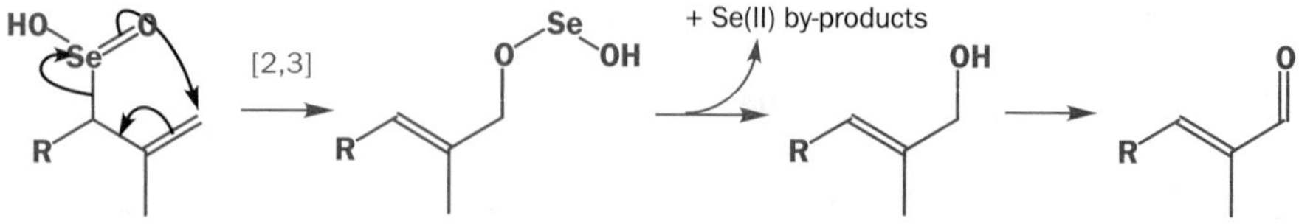

[2,3]   + Se(II) by-products

Overall, CH₃ has been replaced by CH₂OH or CH=O in an allylic position, a transformation similar to the NBS allylic bromination reaction that you met in Chapter 39, but with a very different mechanism. The by-product of the oxidation is a selenium(II) compound, and it can be more practical to carry out the reaction with only a catalytic amount of SeO₂, with a further oxidizing agent, *t*-butyl hydroperoxide, to reoxidize the Se(II) after each cycle of the reaction. This eliminates the need to get rid of large amounts of selenium-containing products, which are toxic and usually smelly.

In Chapter 40 we left the synthesis of sirenin at a tantalizing stage. A carbene insertion into a double bond had formed a three-membered ring and the final stage was the oxidation of a terminal methyl group. This is how it was done.

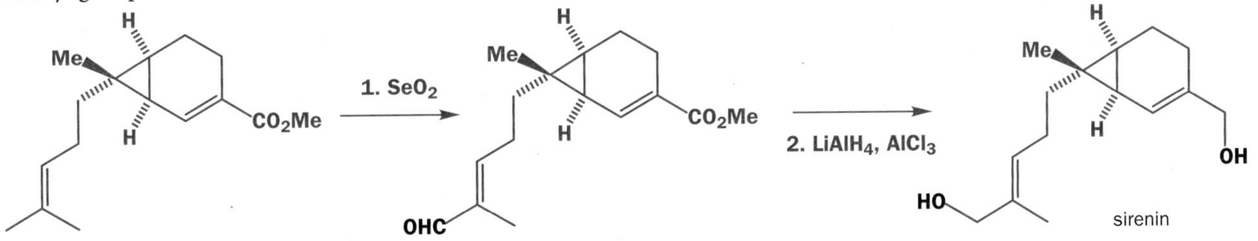

There is some interesting selectivity in this sequence. Only one of the three groups next to the alkene is oxidized and only one (*E*-) isomer of the enal is formed. No position next to the unsaturated ester is oxidized. All these decisions are taken in the initial cycloaddition step. The most nucleophilic double bond uses its more nucleophilic end to attack SeO_2 at selenium. The cycloaddition uses the HOMO (π) of the alkene to attack the LUMO (π^* of Se=O). Meanwhile the HOMO (π) of Se=O attacks the LUMO (C–H σ^*) of the allylic system.

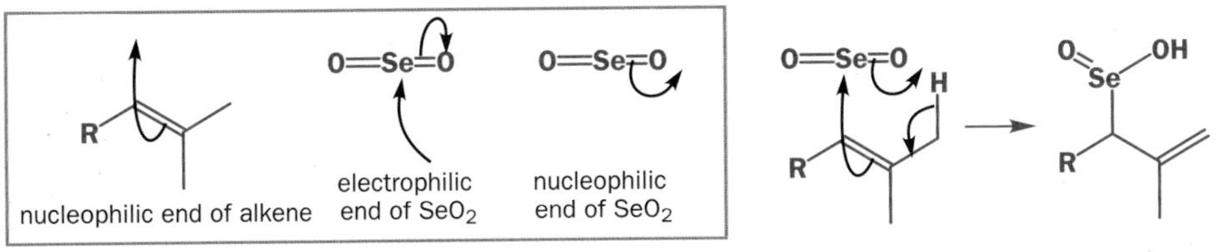

The stereoselectivity also appears to be determined in this step and it is reasonable to assume that the methyl group *trans* to the main chain will react rather than the other for simple steric reasons. Though this is true, the stereochemistry actually disappears in the intermediate and is finally fixed only in the [2,3]-sigmatropic rearrangement step. Both [2,3]- and [3,3]-sigmatropic rearrangements are usually *E*-selective for reasons discussed in Chapter 36.

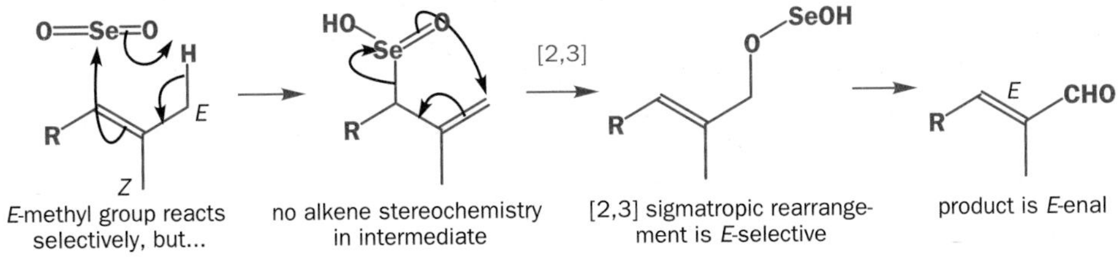

| *E*-methyl group reacts selectively, but... | no alkene stereochemistry in intermediate | [2,3] sigmatropic rearrangement is *E*-selective | product is *E*-enal |

The Swern oxidation

In Chapter 24 we mentioned the Swern oxidation briefly as an excellent method of converting alcohols to aldehydes. We said there that we would discuss this interesting reaction later and now is the time. The mechanism is related to the reactions that we have been discussing and it is relevant that the Swern oxidation is particularly effective at forming enals from allylic alcohols.

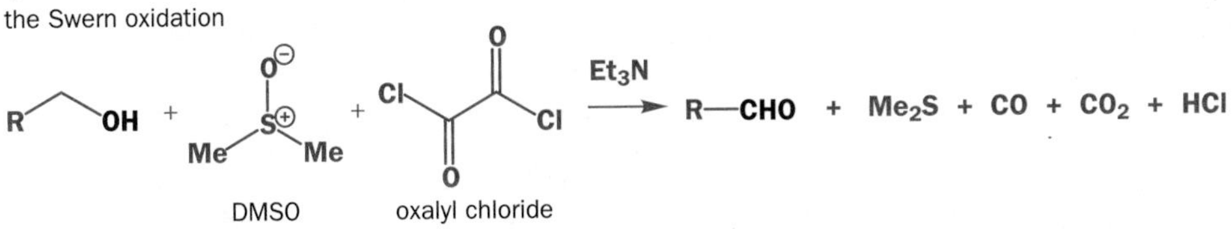

the Swern oxidation

In the first step, DMSO reacts with oxalyl chloride to give an electrophilic sulfur compound. You should not be surprised that it is the charged oxygen atom that attacks the carbonyl group rather than the soft sulfur atom. Chloride is released in this acylation and it attacks the positively charge sulfur atom expelling a remarkable leaving group, which fragments into three pieces: CO_2, CO, and a chloride ion. Entropy favours this reaction.

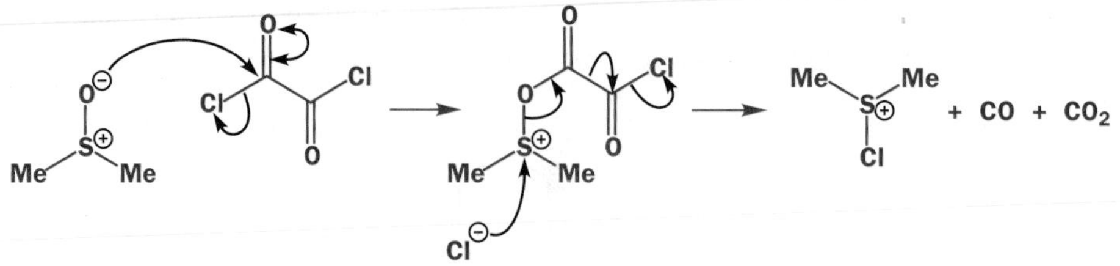

The alcohol has been a spectator of these events so far but the chlorosulfonium ion now formed can react with it to give a new sulfonium salt. This is the sole purpose of all the reactions as this new sulfonium salt is stable enough to survive and to be deprotonated by the base (Et_3N). You will recognize the final step both as the redox step and as a close relative to events in the preceding sections.

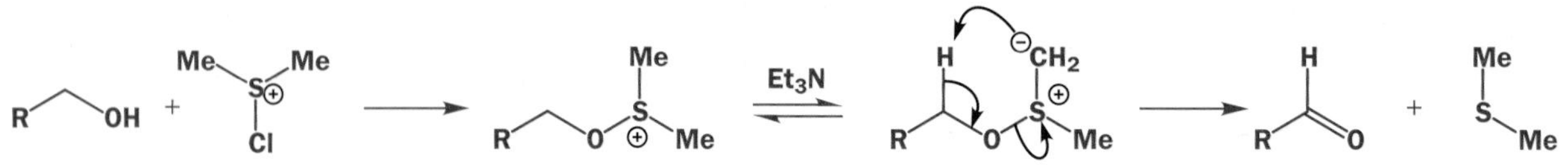

To conclude: the sulfur chemistry of onions and garlic

Traditional medicine suggests that onions and garlic are 'good for you' and modern chemistry has revealed some of the reasons. These bulbs of the genus *Allium* exhibit some remarkable sulfur chemistry and we will end this chapter with a few examples. Both onions and garlic are almost odourless when whole but develop powerful smells and, in the case of onions, tear gas properties when they are cut. These all result from the action of alliinase enzymes released by cell damage on unsaturated sulfoxides in the bulb.

In garlic, a simple sulfoxide elimination creates an unstable sulfenic acid. When we looked at sulfoxide eliminations before, we ignored the fate of the unstable sulfenic acid, but here it is important. It dimerizes with the formation of an S–S bond and the breaking of a weaker S–O bond.

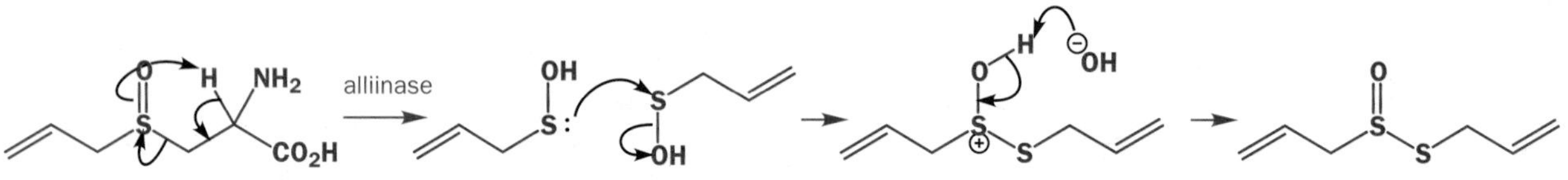

allylic sulfoxide in raw garlic   unstable sulfenic acid   thiosulfinate ester

Another simple elimination reaction on the thiosulfinate ester makes another molecule of the sulfenic acid and a highly unstable unsaturated thioaldehyde, which promptly dimerizes to give a thioacetal found in garlic as a potent platelet aggregation inhibitor.

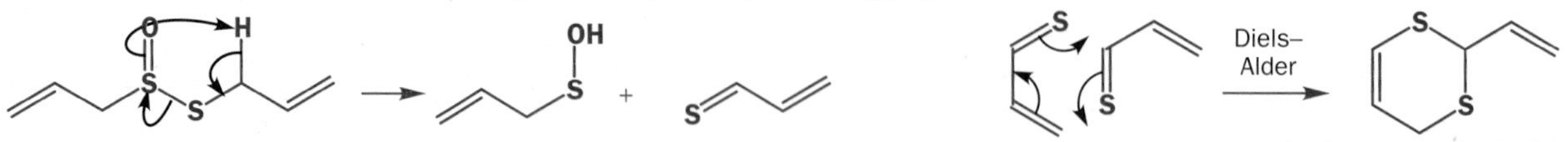

thiosulfinate ester   sulfenic acid   unstable thioaldehyde   platelet aggregation inhibitor

In onions, things start much the same way but the initial amino acid is not quite the same. The skeleton is the same as that of the garlic compound but the double bond is conjugated with the sulfoxide. Elimination and dimerization of the sulfenic acid produce an isomeric thiosulfinate.

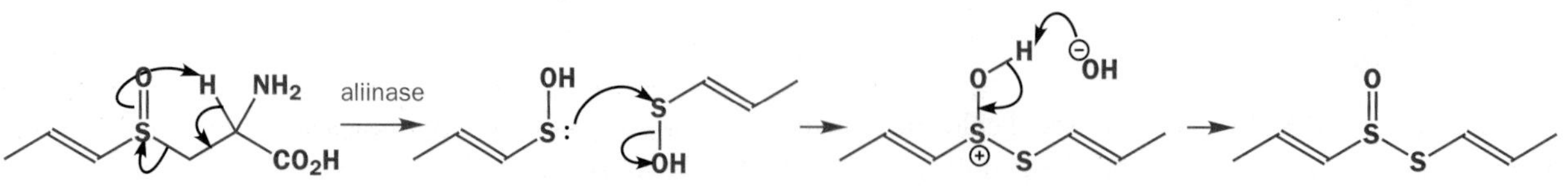

vinylic sulfoxide in raw onions   unstable sulfenic acid   thiosulfinate ester

Oxidation of the thiosulfinate ester up to the sulfonate level gives the compound responsible for the smell of raw onions, while a hydrogen shift on the conjugated sulfenic acid (not possible with the garlic compound) gives a sulfine, the sulfur analogue of a ketene. The compound has the *Z* configuration expected from the mechanism and is the lachrymator that makes you cry when you cut into a raw onion.

> There is still one lone pair on the sulfur atom of a sulfine so the sulfur is trigonal and not linear, rather like the nitrogen in an oxime.

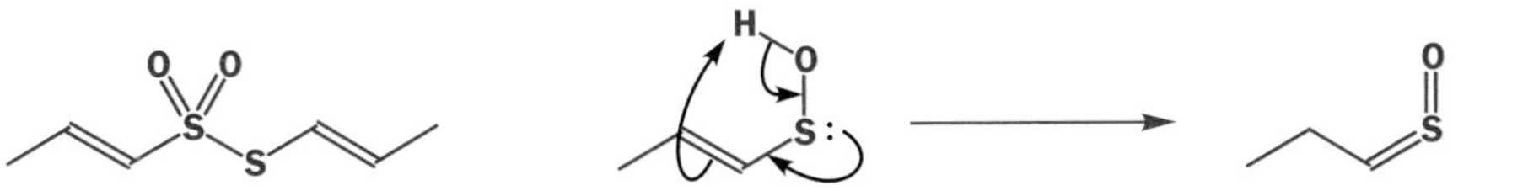

smell of raw onion   lachrymator in raw onion

Even more remarkable is the formation of the 'zwiebelanes', other compounds with potential as drugs for heart disease. They are formed in onions from the conjugated thiosulfinate ester by a [3,3]-sigmatropic rearrangement that gives a compound containing a sulfine and a thioaldehyde. We said that sulfines are the sulfur equivalents of ketenes, so you might expect them to do [2 + 2] cycloadditions (Chapter 35) but you might not expect the thioaldehyde to be the other partner. It is, and the result is a cage compound with one sulfide and one sulfoxide joined in a four-membered ring.

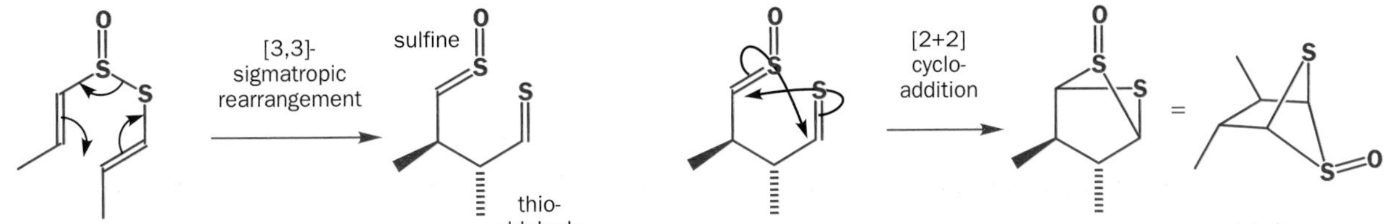

thiosulfinate ester   thio-aldehyde   zwiebelane

Look at onions with respect! They are not only the cornerstone of tasty cooking but are able to do amazing pericyclic reactions as soon as you cut them open. You can read more about the *Allium* family in Eric Block's review in *Angewandte Chemie* (International Edition in English), 1992, Volume 31, p. 1135.

Though you have only seen a couple of examples of the latter, it is clear that organosulfur and organoselenium chemistry are closely related. In the next chapter we will look at the quite different type of chemistry exhibited by organic compounds containing three other heteroatoms—silicon, tin, and boron.

Problems

1. Suggest structures for intermediates A and B and mechanisms for the reactions.

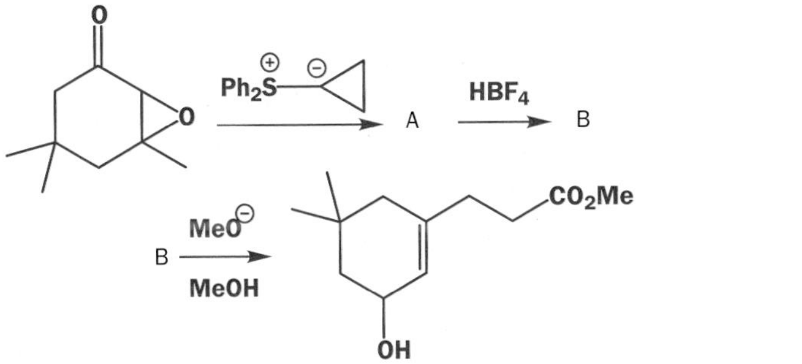

2. Suggest a mechanism for this reaction, commenting on the selectivity and the stereochemistry.

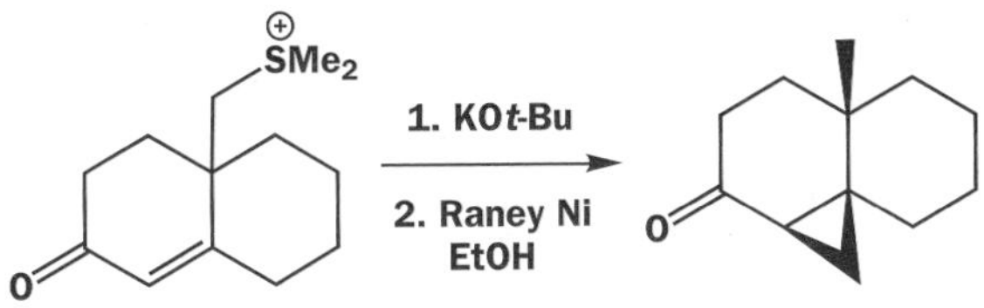

3. The product X of the following reaction has δ_H (p.p.m.) 1.28 (6H, s), 1.63 (3H, d, *J* 4.5 Hz), 2.45 (6H, s), 4.22 (1H, s), 5.41 (1H, d, *J* 15 Hz), and 5.63 (1H, dq, *J* 15, 4.5 Hz). Suggest a structure for X and a mechanism for its formation.

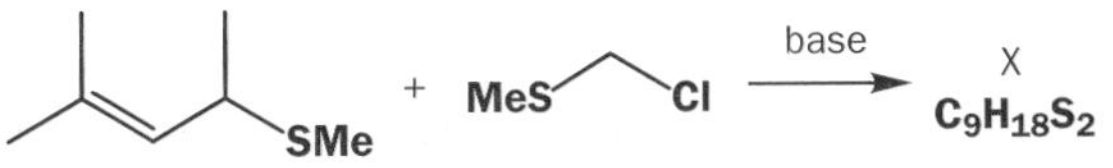

4. The thermal elimination of sulfoxides (example below) is a first-order reaction with almost no rate dependence on substituent at sulfur (Ar) and a modest negative entropy of activation. It is accelerated if R is a carbonyl group (that is, R = COR'). The reaction is (slightly) faster in less polar solvents. Explain.

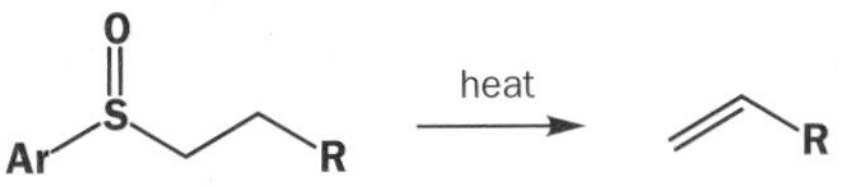

Explain the stereochemistry of the first reaction in the following scheme and the position of the double bond in the final product.

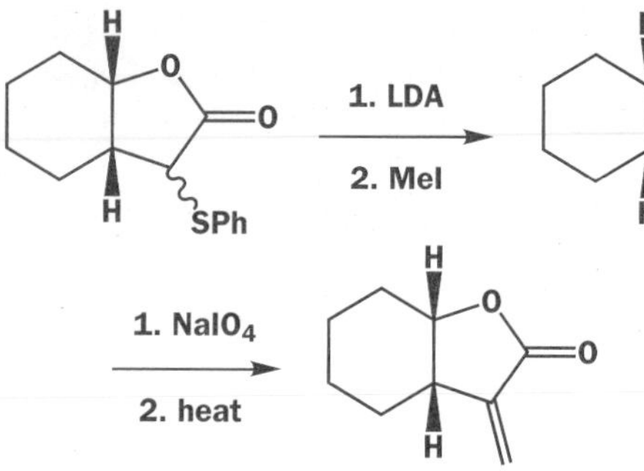

5. Revision content. Explain the reactions and the stereo-chemistry in these first steps in a synthesis of the B vitamin biotin.

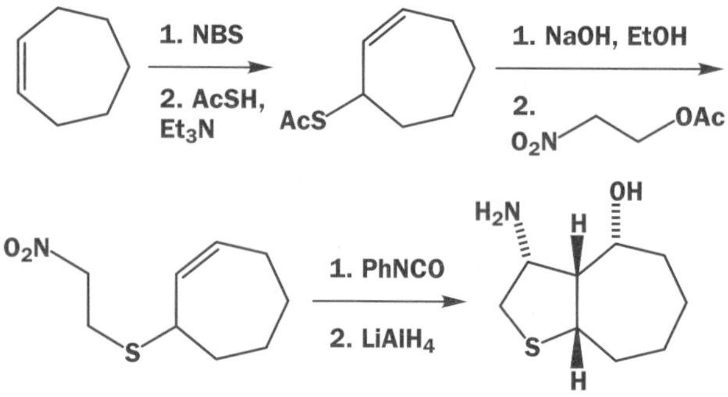

6. Explain the regio- and stereoselectivity of this reaction.

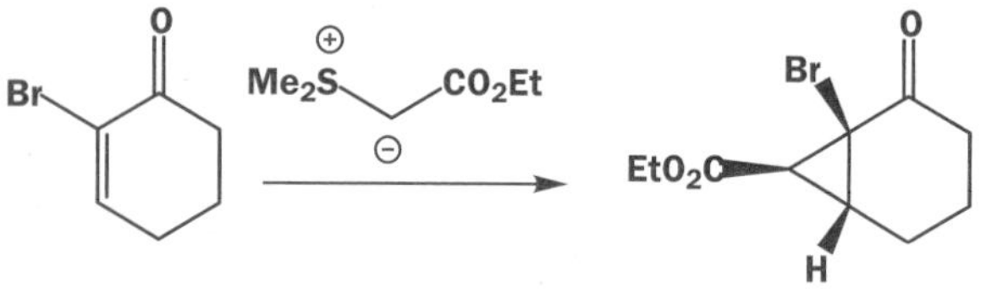

7. Draw mechanisms for these reactions of a sulfonium ylid and the rearrangement of the first product. Why is BF$_4^-$ chosen as the counterion?

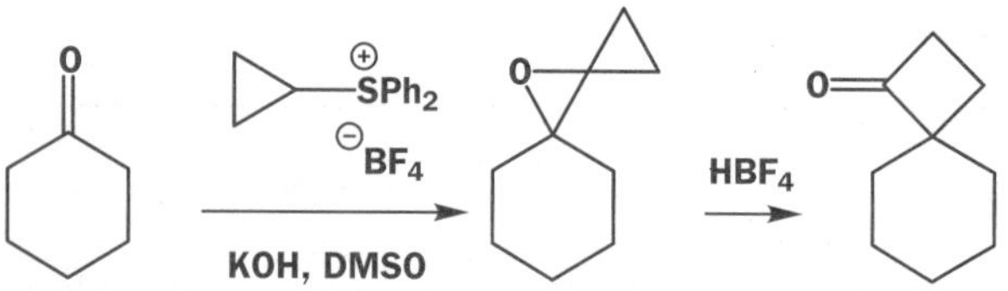

The intermediate may alternatively be reacted with a selenium compound in this sequence of reactions. Explain what is happening, commenting on the regioselectivity. Why is the intermediate in square brackets not usually isolated?

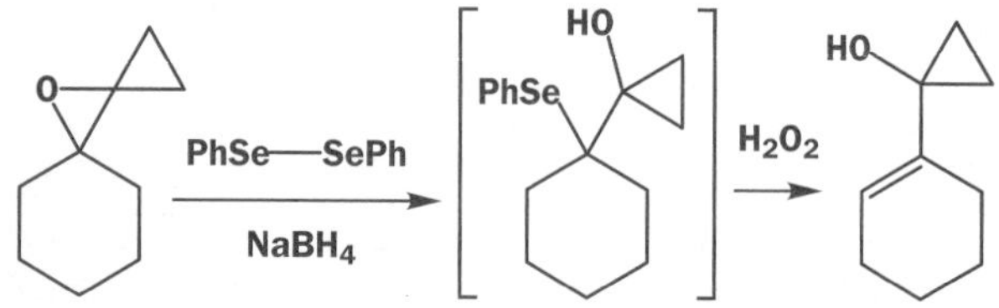

8. Give mechanisms for these reactions, explaining the role of sulfur.

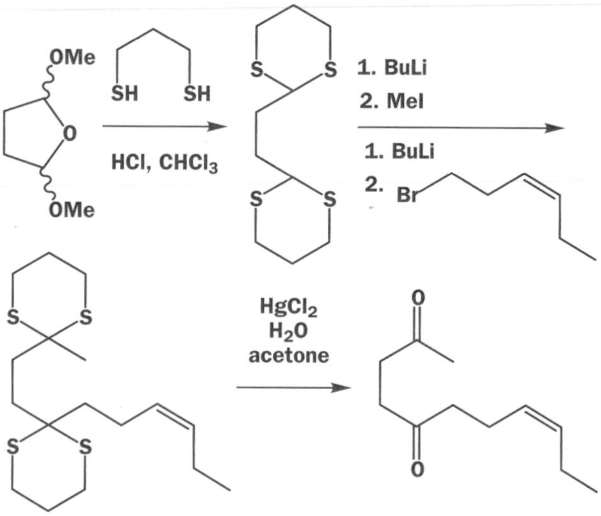

9. Suggest a mechanism for this formation of a nine-membered ring. Warning! The weak hindered base is not strong enough to form an enolate from the lactone.

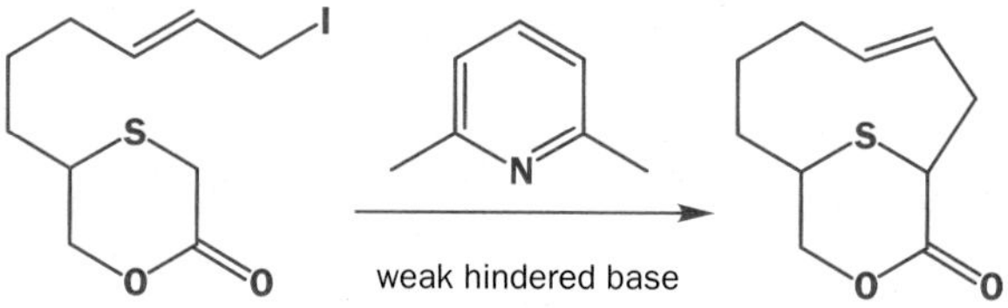

10. Comment on the role of sulfur in the steps in this synthesis of the turmeric flavour compound *ar*-turmerone.

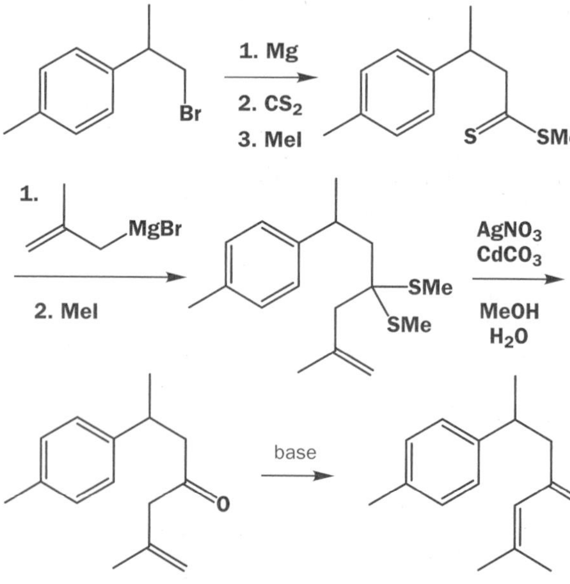

11. Explain how the presence of the sulfur-containing group allows this cyclization to occur regio- and stereoselectively.

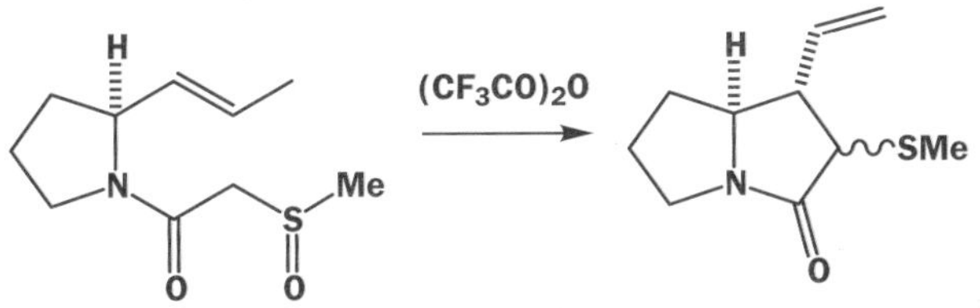

12. Problem 9 in Chapter 32 asked you to interpret the NMR spectrum of a cyclopropane (A). This compound was formed using a sulfur ylid. What is the mechanism of the reaction?

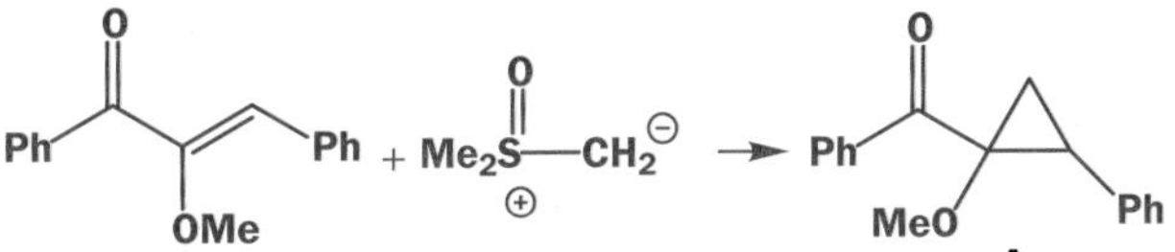

Attempts to repeat this synthesis on the bromo compound below led to a different product. What is different this time?

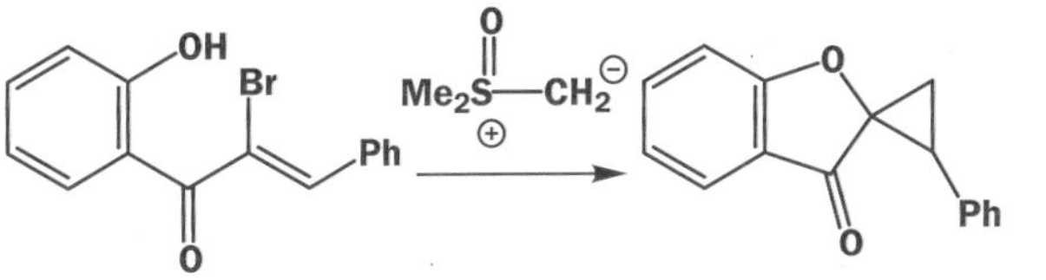

13. Epoxides may be transformed into allylic alcohols by the sequence shown here. Give mechanisms for the reactions and explain why the elimination of the selenium gives an allylic alcohol rather than an enol.

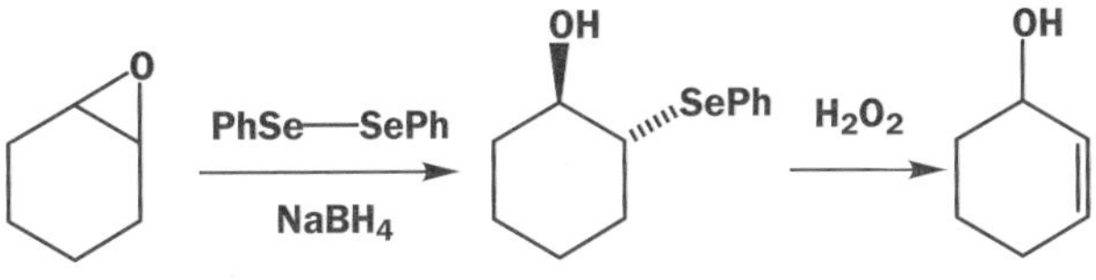

14. In a process resembling the Mitsunobu reaction (Chapter 17), alcohols and acids can be coupled to give esters, even macrocyclic lactones as shown below. In contrast to the Mitsunobu reaction, the reaction leads to retention of stereochemistry at the alcohol. Propose a mechanism that explains the stereochemistry. Why is sulfur necessary here?

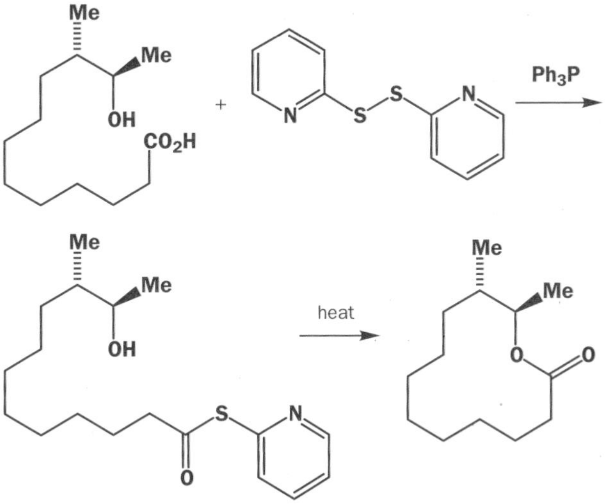

15. Suggest mechanisms for these reactions, explaining any selectivity.

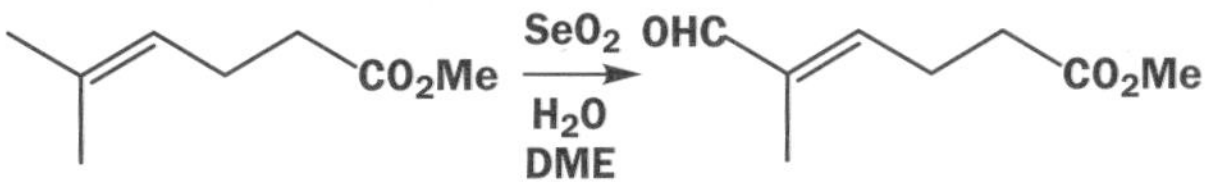

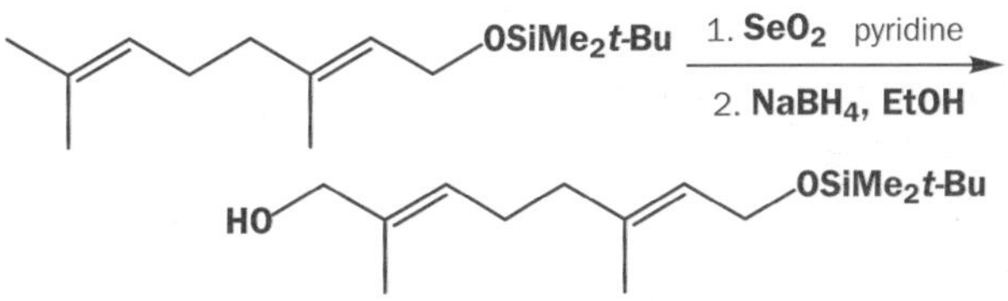

Organo-main-group chemistry 2: boron, silicon and tin

47

Connections

Building on:

- Conjugate addition ch10 & ch23
- Nucleophilic substitution at saturated carbon ch17
- Controlling stereochemistry ch16, ch33, & ch34
- Oxidation, reduction, and protection ch24
- Aldol reactions ch27
- Controlling double bond geometry ch31
- Rearrangements ch36–ch37
- Radicals ch39
- Asymmetric synthesis ch45
- Sulfur chemistry ch46

Arriving at:

- Main group elements in organic chemistry
- Boron is electrophilic because of a vacant orbital
- Hydroboration adds boron selectively
- Oxidation removes boron selectively
- Boron chemistry uses rearrangements
- Allyl B, Si, and Sn compounds are useful in synthesis
- Organo-B, -Si, and -Sn compounds can be used in asymmetric synthesis
- Silicon is more electrophilic than carbon
- Silicon stabilizes β carbocations
- Organo-tin compounds are like Si compounds but more reactive
- Tin is easily exchanged for lithium

Looking forward to:

- Organometallic chemistry ch48
- Polymerization ch52

Organic chemists make extensive use of the periodic table

Although typical organic molecules, such as those of which all living things are composed, are constructed from only a few elements (usually C, H, O, N, S, and P and, on occasion, Cl, Br, I, and a few more), there are very many other elements that can be used as the basis for reagents, catalysts, and as components of synthetic intermediates. The metals will be discussed in the next chapter (48) but many main group (p block) elements are also important. These nonmetals bond covalently to carbon and some of their compounds are important in their own right.

More commonly, elements such as Si, P, and S are used in reagents to carry out some transformation but are not required in the final molecule and so must be removed at a later stage in the synthesis. The fact that organic chemists are prepared to tolerate this additional step demonstrates the importance of these reactions. The Julia olefination is an obvious example. The difficult conversion of aldehydes and ketones into alkenes is important enough to make it worthwhile adding a sulfur atom to the starting material and then removing it at the end of the reaction. So many elements are used like this that the list of nonmetals that are *not* used frequently in organic synthesis would be much shorter than the list of those that *are* useful.

In the previous chapter we described the special chemistry of sulfur, and you have previously met that of phosphorus. These two elements may be thought of as analogues of oxygen and nitrogen but many reactions are possible with S and P that are quite impossible with O and N. This chapter will concentrate on the organic chemistry of three other main group elements: boron, which is unusual in this context because it is a first row element, and silicon and tin, which are in the same group as

■ At the time of writing only Be, Ga, As, Sb, and Bi among the non-radioactive p-block elements are not used extensively and some would argue about As.

■ The organic chemistry of phosphorus is scattered about the book with important reactions in Chapters 14 and 31 (the Wittig reaction), 17 (various nucleophilic substitutions), 23 (conjugate addition of phosphines), and 41 (mechanisms involving phosphorus). In Chapter 48 you will see how important phosphorus compounds are as ligands for transition metals.

carbon in the periodic table but in the second and fourth rows. Here they are surrounded by other familiar elements.

Li	Be	**B**	C	N	O	F	Ne
Na	Mg	Al	**Si**	**P**	**S**	Cl	Ar
			Ge		**Se**		
			Sn				

Boron

Borane has a vacant p orbital

You have already met boron in useful reagents such as sodium borohydride $NaBH_4$ and borane BH_3 (more correctly, B_2H_6). Both display the crucial feature of boron chemistry, which results directly from its position in group IIIB or 13 of the periodic table. Boron has only three electrons in the valence shell and so typically forms three conventional two-centre two-electron bonds with other atoms in a planar structure leaving a vacant 2p orbital. Borane exists as a mixture of B_2H_6—a dimer with hydrogen bridges—and the monomer BH_3. Since most reactions occur with BH_3 and the equilibrium is fast we will not refer to this again.

The vacant orbital is able to accept a lone pair of electrons from a Lewis base to give a neutral species or can combine with a nucleophile to form a negatively charged tetrahedral anion. The reducing agent borane–dimethyl sulfide is an example of the Lewis acid behaviour while the borohydride anion would be the result of the imaginary reaction of borane with a nucleophile hydride. The vacant orbital makes borane a target for nucleophiles.

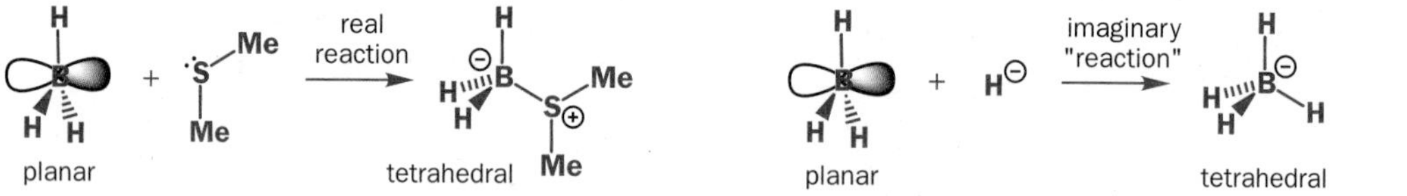

Hydroboration—the addition of boron hydrides to alkenes and alkynes

One of the simplest classes of nucleophiles that attacks borane is that of alkenes. The result, described as **hydroboration**, is an overall addition of borane across the double bond. Unlike most electrophilic additions to alkenes that occur in a stepwise manner via charged intermediates (Chapter 20), this addition is concerted so that both new bonds are formed more or less at the same time. The result is a new borane in which one of the hydrogen atoms has been replaced by an alkane. This monoalkyl borane (RBH_2) is now able to undergo addition with another molecule of the alkene to produce a dialkyl borane (R_2BH) which in turn undergoes further reaction to produce a trialkyl borane (R_3B). All these boranes have a vacant p orbital and are flat so that repeated attack to produce the trialkyl borane is easy and normal if an excess of alkene is present.

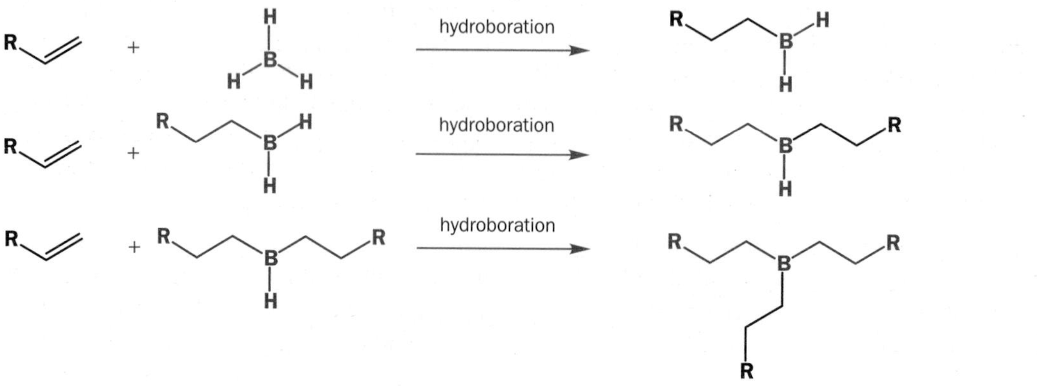

Hydroboration is regioselective

You will notice that the boron atom always adds to the *end* of the alkene. This is just as well; otherwise, three sequential additions would give rise to a complex mixture of products. The boron always becomes attached to the carbon of the double bond that is less substituted. This is what we should expect if the filled π orbital of the alkene adds to the empty orbital of the borane to give the more stable cationic intermediate.

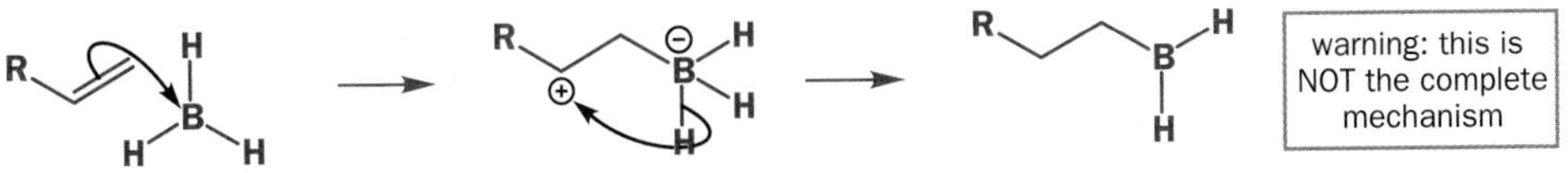

warning: this is NOT the complete mechanism

> The hydrogen must bond to the carbon of the alkene that had fewer hydrogens attached to it. This is a formal violation of Markovnikoff's rule, which you met in Chapter 20, and is a warning to understand the mechanisms of the reactions rather than follow a rule. Boron is less electronegative than hydrogen and so the regioselectivity is normal with the more electronegative atom becoming attached to the more substituted centre.

We know that this is not the whole story because of the stereochemistry. Hydroboration is a *syn* addition across the alkene. As the addition of the empty p orbital to the less substituted end of the alkene gets under way, a hydrogen atom from the boron adds, with its pair of electrons, to the carbon atom, which is becoming positively charged. The two steps shown above are concerted, but formation of the C–B bond goes ahead of formation of the C–H bond so that boron and carbon are partially charged in the four-centred transition state.

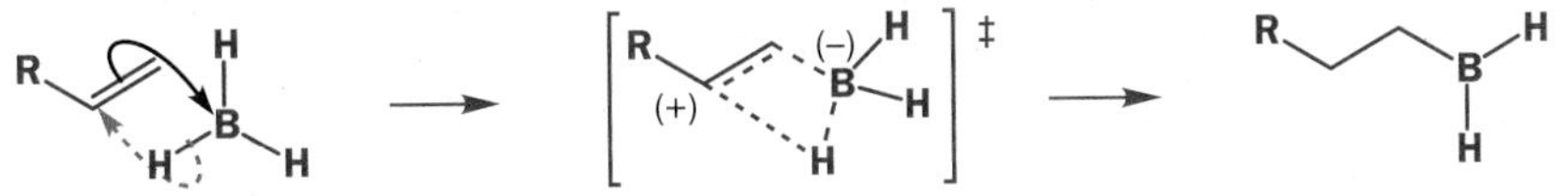

It is, of course, impossible to tell in this case whether the addition is *syn* or *anti* and in any case the alkyl borane products are rather unstable. Although organoboranes can be stored, and some are available commercially, air must be rigorously excluded as they burst into a spectacular green flame in air. A more controlled oxidation is required to remove the boron and reveal the useful organic fragment. The simplest is alkaline hydrogen peroxide, which replaces the carbon–boron bond with a carbon–oxygen bond to give an alcohol.

> More modern alternative reagents, which are stable, inexpensive, safe and easy to handle but achieve the same transformation under mild conditions and often in higher yield, are sodium perborate ($NaBO_3 \cdot 4H_2O$) and sodium percarbonate ($Na_2CO_3 \cdot 1.5H_2O_2$).

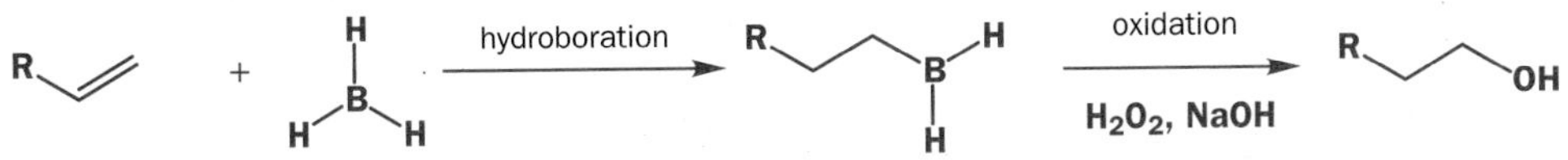

The oxidation occurs by nucleophilic attack of the hydroperoxide ion on the empty orbital of the boron atom followed by a migration of the alkyl chain from boron to oxygen. Do not be alarmed by hydroxide ion as leaving group. It is, of course, a bad leaving group but a very weak bond—the O–O σ bond—is being broken. Finally, hydroxide attacks the now neutral boron to cleave the B–O–alkyl bond and release the alcohol.

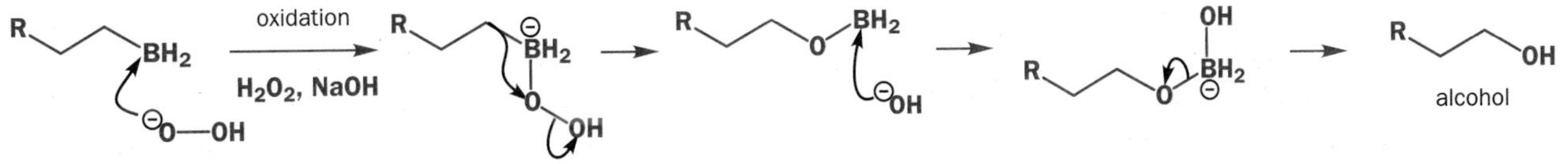

In this sequence boron goes backwards and forwards between planar neutral structures and anionic tetrahedral structures. This is typical of the organic chemistry of boron. The planar structure is neutral but boron has only six valency electrons. The tetrahedral structure gives boron eight valency electrons but it is negatively charged. Boron flits restlessly between these two types of structure, becoming content only when it has three oxygen atoms around it. Returning to the oxidation but concentrating on the boron product, we find that $B(OH)_3$ is the stable product as it is neutral and has three oxygen atoms donating electrons into the empty p orbital on boron.

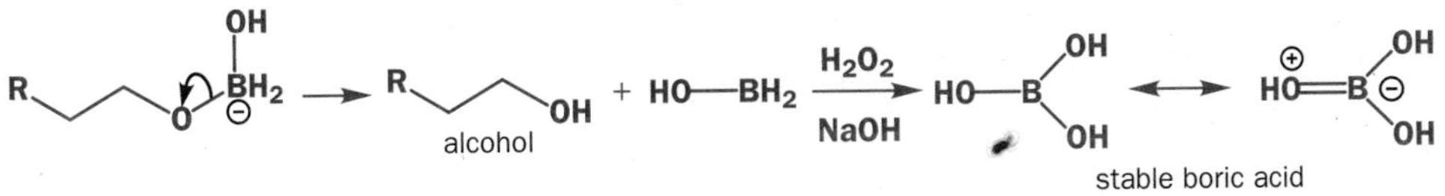

Hydroboration is mostly used for the conversion of alkenes to alcohols by the *cis* addition of water with the OH group going to the less substituted end of the alkene. This is clearest with a cyclic trisubstituted alkene.

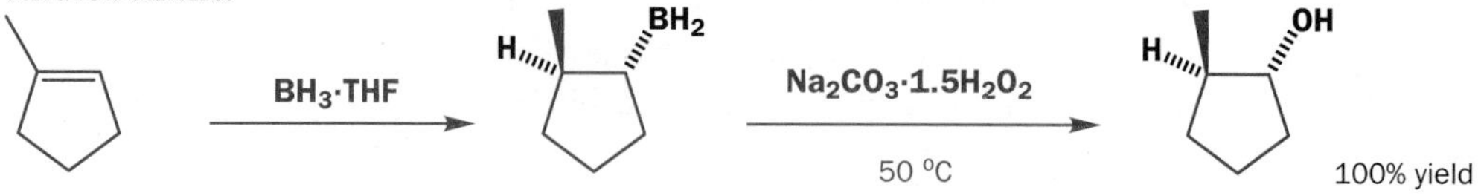

Now we can prove that *cis* addition really does occur in the hydroboration step. The migration of carbon from boron to oxygen might remind you of the Baeyer–Villiger rearrangement (Chapter 37). Both these rearrangements occur with retention of configuration at the migrating group as the bonding (C–C or C–B σ) orbital reacts. Here is the exact analogy.

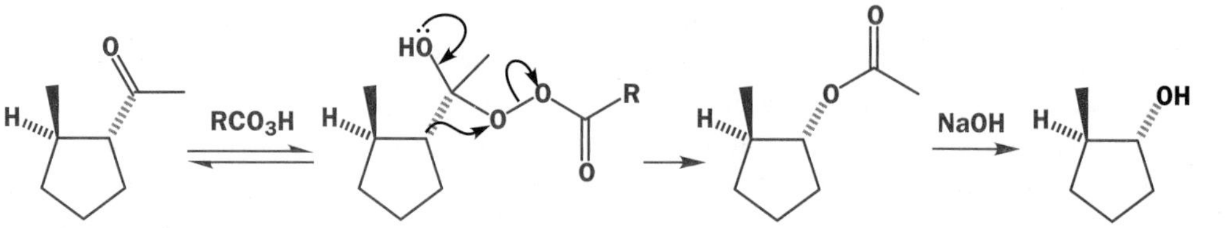

The same alcohol could be made by the Baeyer–Villiger rearrangement but the stereochemistry would have to be set up before the Baeyer–Villiger step. Hydroboration has the advantage that stereochemistry is created in the hydroboration step. We have discussed the details of this step. In drawing the mechanism it is usually best to draw it as a simple concerted four-centre mechanism providing you remember that the regioselectivity is controlled by the initial interaction between the nucleophilic end of the alkene and the empty p orbital on boron.

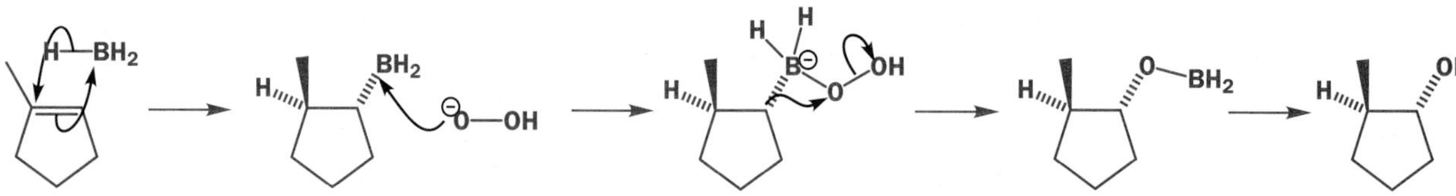

The overall result of the hydroboration–oxidation sequence is addition of water to an alkene with the opposite regiochemistry to that expected for a conventional acid-catalysed hydration. The usual way to do such a hydration is by oxymercuration–reduction.

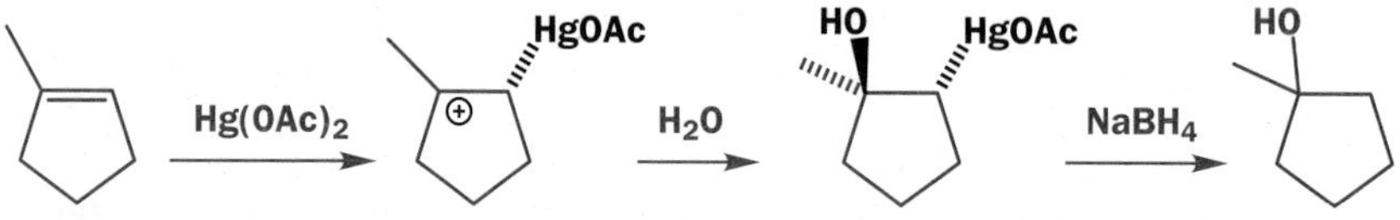

■
Chapter 20 describes the mechanisms for the addition of Hg(II) and suggests that a cyclic mercurinium ion might be involved.

The stereochemical outcome would also be different as the hydroboration adds *syn* to the alkene, whereas oxymercuration gives the *anti* product though in this case the stereochemistry is lost in the reduction step.

Hydroboration–oxidation is normally done via the trialkyl borane

So far we have shown all reactions taking place on the monoalkyl borane. In fact, these compounds are unstable and most hydroborations actually occur via the trialkyl borane. Three molecules of alkene add to the boron atom; three oxidations and three migrations transfer three alkyl groups (R = 2-methylcyclopentyl) from boron to oxygen to give the relatively stable trialkyl borate B(OR)$_3$, which is hydrolysed to give the products.

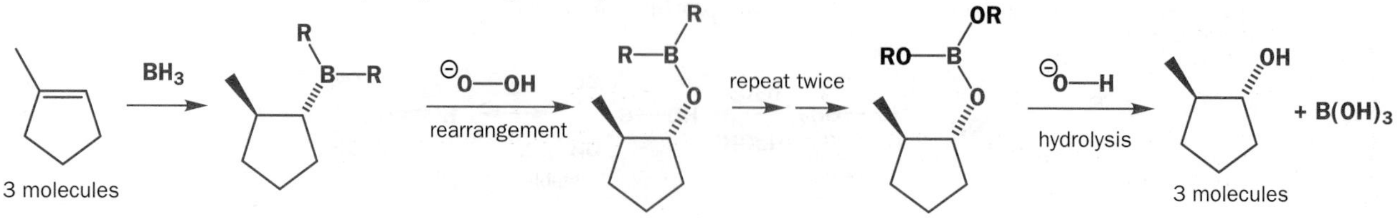

If we have a mixed trialkyl borane, you may be concerned about which of the alkyl groups migrates—the usual answer is that they all do! Oxidation proceeds until the borane is fully oxidized to the corresponding borate, which then breaks down to give the alcohols.

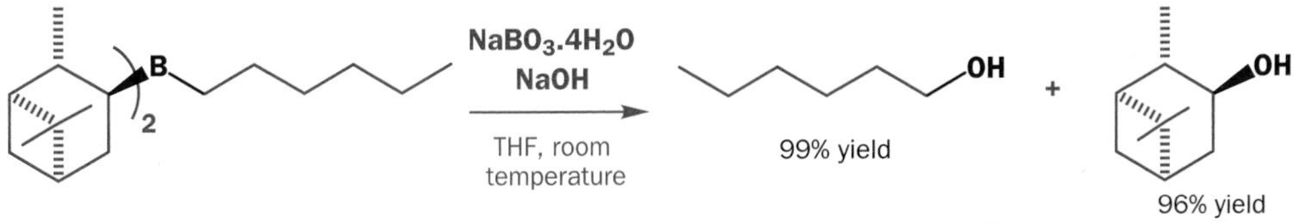

Bulky substituents improve the selectivity of hydroboration

Borane can react one, two, or even three times and this is a disadvantage in many situations so a range of hydroborating reagents has been designed to hydroborate once or twice. Dialkyl boranes R_2BH can hydroborate once only and alkyl boranes RBH_2 twice. In each case the 'dummy' group R must be designed either to migrate badly in the oxidation step or to provide an alcohol that is easily separated from other alcohols. The regioselectivity of hydroboration, good though it is with simple borane, is also improved by very bulky boranes, which explains the choice of dummy groups. Thexyl borane, so-called because the alkyl group is a 'tertiary hexyl' group (*t*-hexyl), is used when two hydroborations are required and it is easily made by hydroboration with borane since the second hydroboration with the tetrasubstituted alkene is very slow.

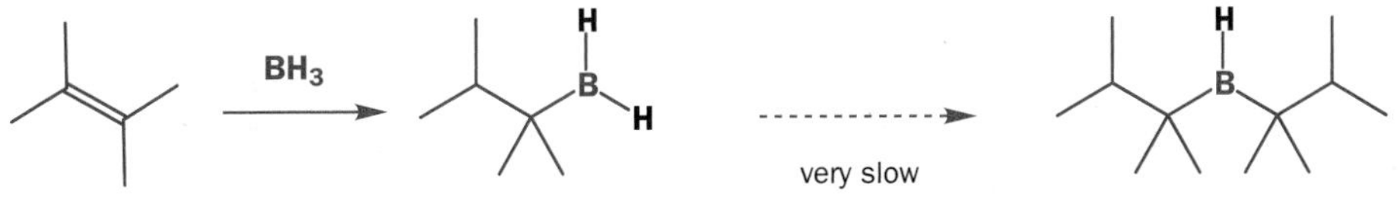

thexyl borane is often written as $ThBH_2$ and drawn as:

Two dialkyl boranes are in common use. The bicyclic 9-borabicyclo[3.3.1]nonane (9-BBN), introduced in Chapter 34 as a reagent for diastereoselective aldol reactions, is a stable crystalline solid. This is very unusual for an alkyl borane and makes it a popular reagent. It is made by hydroboration of cyclo-octa-1,5-diene. The second hydroboration is fast because it is intramolecular but the third would be very slow. The regioselectivity of the second hydroboration is under thermodynamic control.

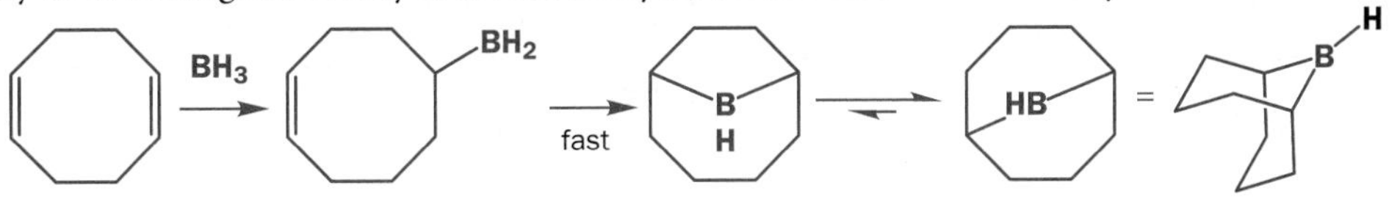

9-BBN is often drawn as:

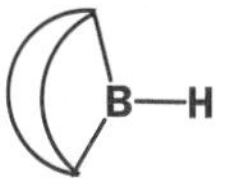

Disiamylborane (an abbreviation for di-*s*-isoamyl borane—not a name we should use now, but the abbreviation has stuck) is also easily made by hydroboration of a simple trialkyl alkene with borane. Two hydroborations occur easily, in contrast to the tetrasubstituted alkene above, but the third is very slow. Disiamylborane is exceptionally regioselective because of its very hindered structure. The structures of these reagents are cumbersome to draw in full and they are often abbreviated.

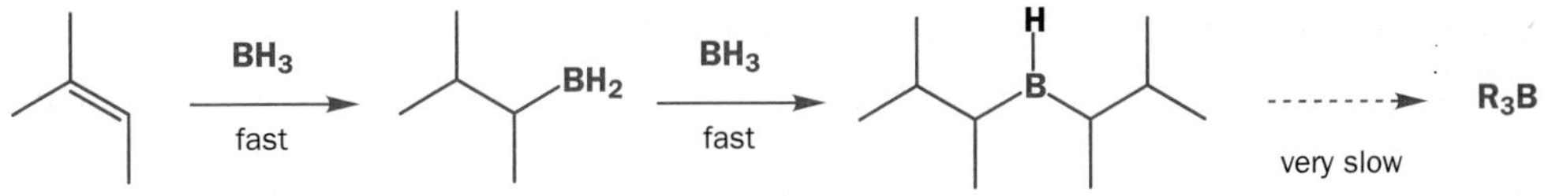

disiamylborane or **Sia₂BH**

● **Hydroboration**

- Hydroboration is a *syn* addition of a borane to an alkene
- Regioselectivity is high: the boron adds to the carbon less able to support a positive charge

● **Hydroboration—contd**

- Oxidation occurs with retention of stereochemistry
- The net result of hydroboration–oxidation is addition of water across the double bond

These bulkier boranes enhance the regioselectivity of hydroboration of trisubstituted alkenes in particular and may also lead to high diastereoselectivity when there is a stereogenic centre next to the alkene. In this next example, an allylic alcohol is hydroborated with thexyl borane. Oxidation reveals complete regioselectivity and a 9:1 stereoselectivity in favour of hydroboration on the same side as the OH group.

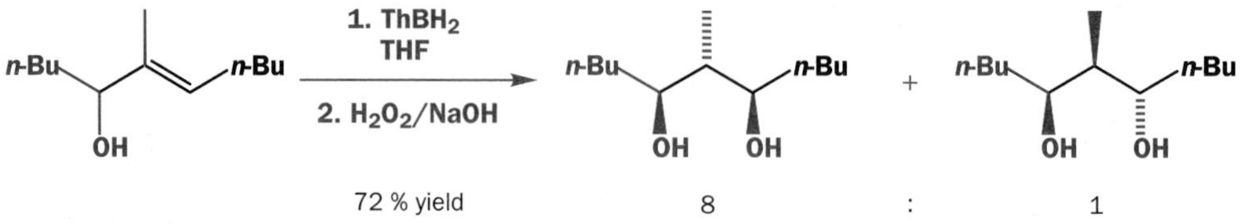

The reactive conformation of the alkene is probably the 'Houk' conformation (Chapter 34) with the hydrogen atom on the stereogenic centre eclipsing the alkene. Attack occurs *syn* to the OH group and *anti* to the larger butyl group.

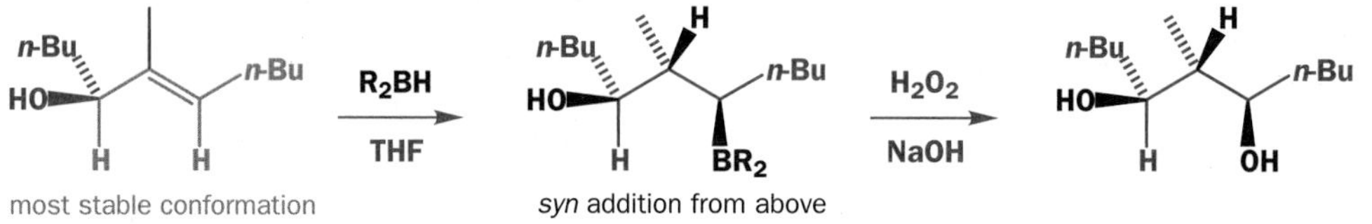

Hydroboration is not restricted to alkenes: alkynes also react well to give vinyl boranes. These may be used directly in synthesis or oxidized to the corresponding enol, which immediately tautomerizes to the aldehyde. An example of this transformation is the conversion of 1-octyne into octanal by hydroboration with disiamylborane and oxidation with sodium perborate under very mild conditions.

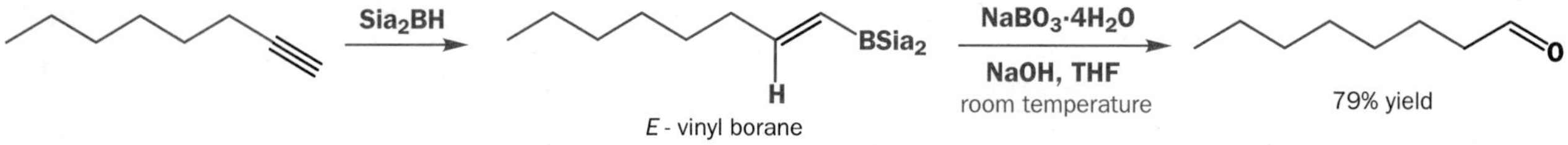

Vinyl boranes, boronates, and boronic acids are important reagents in transition-metal-catalysed processes, as you will see in Chapter 48.

Carbon–boron bonds can be transformed stereospecifically into C–O, C–N, or C–C bonds

Although oxidation to the alcohol is the most common reaction of organoboranes in organic synthesis, the reaction with ⁻O–OH is just one example of a general reaction with a nucleophile of the type ⁻X–Y where the nucleophilic atom X can be O, N, or even C, and Y is a leaving group. We will illustrate the formation of carbon–nitrogen and carbon–carbon bonds by this reaction. The underlying principle is to use the vacant orbital on boron to attack the nucleophile and then rely on the loss of the leaving group to initiate a rearrangement of R groups from B to X similar to that observed from B to O in the hydrogen peroxide oxidation. The overall result is insertion of X into the carbon–boron bond with retention.

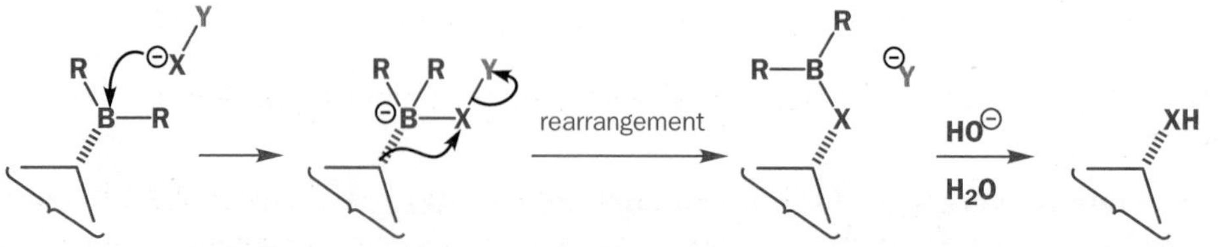

If X is nitrogen then a direct method of amination results. The required reagent is a chloramine or the rather safer O-hydroxylaminesulfonic acid: the leaving group is chloride or sulfonate. The overall process of hydroboration–amination corresponds to a regioselective *syn* addition of ammonia across the alkene. In the case of pinene the two faces of the alkene are very different—one is shielded by the bridge with the geminal dimethyl group. Addition takes place exclusively from the less hindered side to give one diastereoisomer of one regioisomer of the amine.

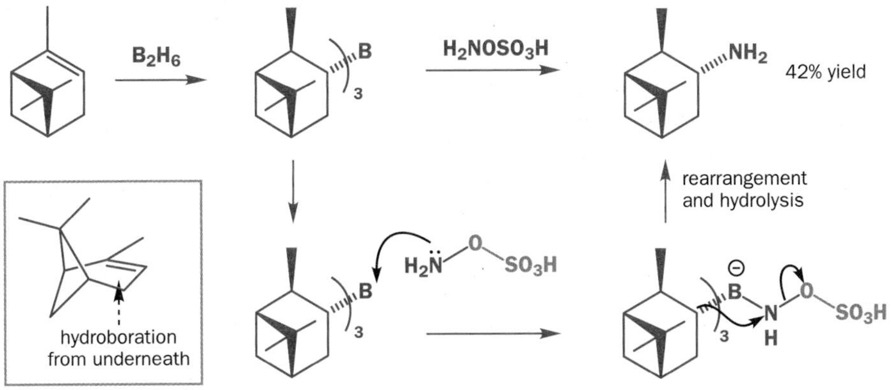

Carbon–carbon bonds can also be made with alkyl boranes. The requirement for a carbon nucleophile that bears a suitable leaving group is met by α-halo carbonyl compounds. The halogen makes enolization of the carbonyl compound easier and then departs in the rearrangement step. The product is a boron enolate with the boron bound to carbon. Under the basic conditions of the reaction, hydrolysis to the corresponding carbonyl compound is rapid.

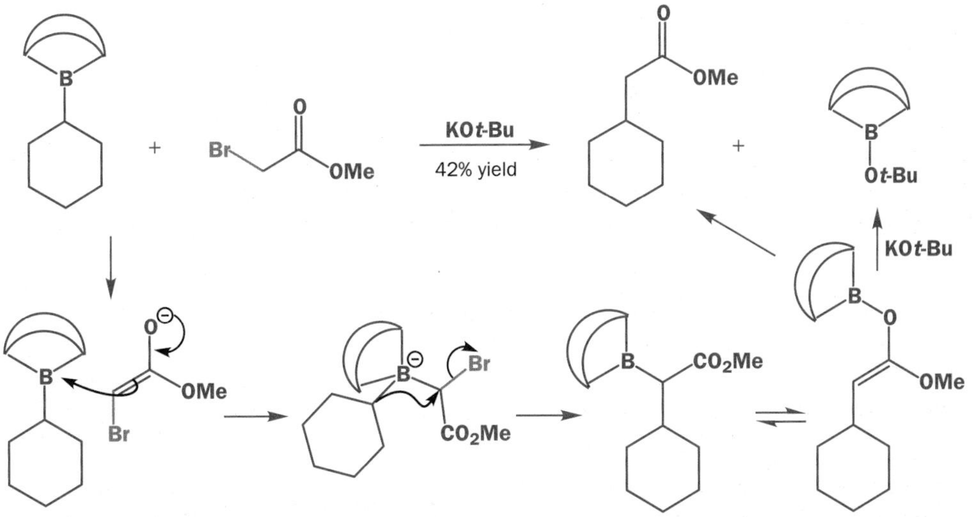

In this example it is important which group migrates from boron to carbon as that is the group that forms the new C–C bond in the product. We previously compared the oxidation of alkyl boranes with the Baeyer–Villiger reaction (Chapter 37) but the order of migrating groups is the opposite in the two reactions. In the Baeyer–Villiger reaction (migration from carbon to oxygen) the more highly substituted carbon atom migrates best so the order is *t*-alkyl > *s*-alkyl > *n*-alkyl > methyl. In organoborane rearrangements it is the reverse order: *n*-alkyl > *s*-alkyl > *t*-alkyl. Methyl does not feature as you cannot make a B–Me bond by hydroboration.

Why the difference between the Baeyer–Villiger rearrangement and boron chemistry?

The transition state for the Baeyer–Villiger rearrangement has a positive charge in the important area. Anything that can help to stabilize the positive charge, such as a tertiary migrating group (R^1), stabilizes the transition state and makes the reaction go better.

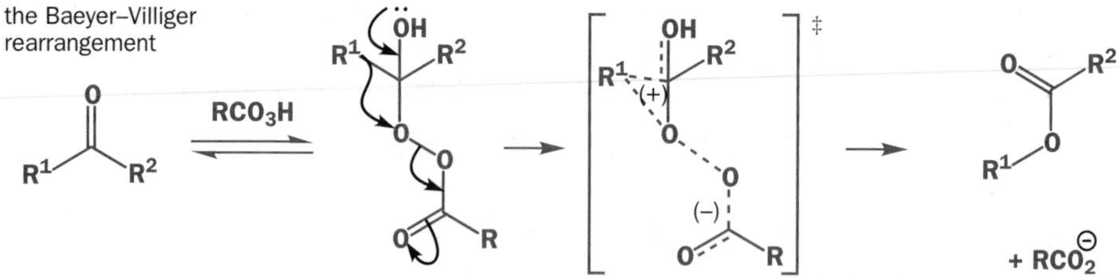

the Baeyer–Villiger
rearrangement

In the boron rearrangements, by contrast, the whole transition state has a negative charge. Alkyl groups destabilize rather than stabilize negative charges, but primary alkyl groups destabilize them less than secondary ones do, and so on. This is another reason for choosing tertiary alkyl 'dummy' groups such as *t*-hexyl—they are less likely to migrate.

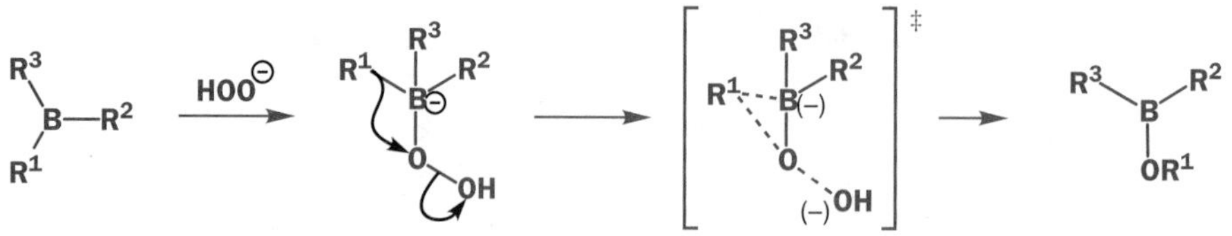

But what about the case we were considering? The migrating group is secondary and the groups that are left behind on the 9-BBN framework are also secondary. What is the distinction? Again we can use the Baeyer–Villiger reaction to help us. The treatment of bridged bicyclic ketones with per-oxy-acids often leads to more migration of the primary alkyl group than of the secondary one.

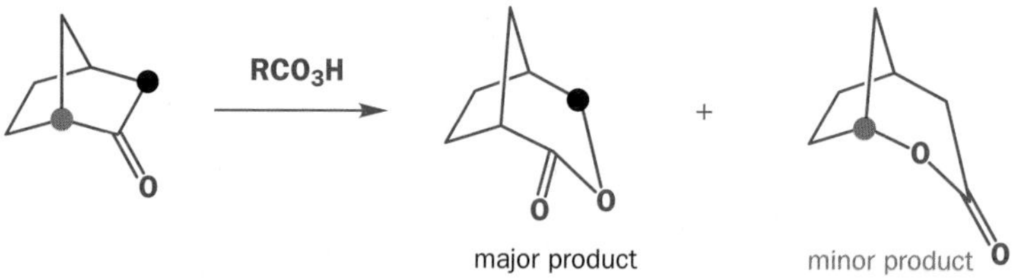

major product   +   minor product

Bridgehead atoms are bad migrating groups. When the green spot carbon migrates, it drags the whole cage structure with it and distorts the molecule a great deal. When the black spot carbon migrates, it simply slides along the O–O bond and disturbs the cage much less. It is the same with 9-BBN. Migration of the bicyclic group is also unfavourable.

> ● **Migration preferences**
>
> * For the Baeyer–Villiger reaction, cation-stabilizing groups migrate best: *t*-alkyl > *s*-alkyl > *n*-alkyl > methyl
> * For boron rearrangements, cation-stabilizing groups migrate worst: *n*-alkyl > *s*-alkyl > *t*-alkyl
> * For both, bridgehead groups migrate badly

Allyl and crotyl boranes react using the double bond

Allylic boron compounds react with aldehydes in a slightly different way. The first step is, as always, coordination of the basic carbonyl oxygen to the Lewis acid boron. This has two important effects: first, the carbonyl is made more electrophilic and, second, the carbon–boron bond in the allylic fragment is weakened so that migration is easier. The difference is that the reaction that follows is not the now familiar 1,2-rearrangement but one involving the allylic double bond as well, rather like a [3,3]-sigmatropic rearrangement (Chapter 36). The negatively charged boron increases the nucleophilicity of the double bond so that it attacks the carbonyl carbon. The result is a six-membered transition state in which transfer of boron from carbon to oxygen occurs with simultaneous carbon–carbon bond formation. Hydrolytic cleavage of the boron–oxygen bond is often accelerated by hydrogen peroxide as in hydroboration. The precise nature of the ligands on boron is not important as this process is successful both for boranes (L = R) and boronates (L = OR).

Other allylic organometallic reagents frequently react with 1,3-rearrangement

It is necessary to have a label of some sort to tell whether an allyl metal has reacted directly or by the mechanism we have just seen, a mechanism common to many metals. An isotopic label such as deuterium or ^{13}C might be used but by far the simplest is a substituent such as a methyl group. The resulting methyl allyl group is known as **crotyl**. Reaction with an aldehyde can follow two pathways; direct addition leads to one product without rearrangement while addition with rearrangement gives an isomeric product.

This sort of rearrangement is often known as **allylic rearrangement**, and even simple Grignard reagents react with aldehydes in this way via a cyclic six-membered transition state (Chapter 9).

Enantioselective allylation is possible with optically pure ligands on boron

You may not think that allylating an aldehyde is much of an achievement—after all, allyl Grignard reagents would do just the same job. The interest in allyl boranes arises because enantiomerically pure ligands derived from naturally occurring chiral terpenes can easily be incorporated into the allyl borane. H.C. Brown, has investigated a range of terpenes as chiral ligands. The reagent below, B-allylbis(2-isocaranyl)borane, has two ligands resulting from hydroboration of carene and delivers the allyl group under such exquisite control that the resulting homoallylic alcohol is virtually a single enantiomer. This reaction is one of the fastest in organic chemistry even at the very low temperature of −100 °C and the product is a useful building block. This makes the process more practical as the cooling is required for only a short time.

Herbert C. Brown (1912–) won the Nobel Prize in 1979 for his invention and development of hydroboration, mostly carried out at Purdue University, which made this new chemistry familiar to the practising organic chemist. The prize was shared with Georg Wittig, showing how important organic chemistry with main group elements had become. He vigorously opposed 'non-classical' carbocation theory.

Asymmetric synthesis is discussed in Chapter 45.

Allyl and crotyl boranes react stereospecifically

The six-membered transition state for the reaction of an allylic borane or boronate is very reminiscent of the cyclic transition state for the aldol reaction you met in Chapter 34. In this case the only change is to replace the oxygen of the enolate with a carbon to make the allyl nucleophile. The transition state for the aldol reaction was a chair and the reaction was stereospecific so that the geometry of the enolate determined the stereochemistry of the product aldol. The same is true in these reactions. *E*-Crotyl boranes (or boronates) give *anti* homoallylic alcohols and *Z*-crotyl boranes (or boronates) give *syn* alcohols via chair transition states in which the aldehyde R group adopts a pseudoequatorial position to minimize steric repulsion. As with the aldol reaction the short bonds to boron create a very tight transition state, which converts the two-dimensional stereochemistry of the reagent into the three-dimensional structure of the product.

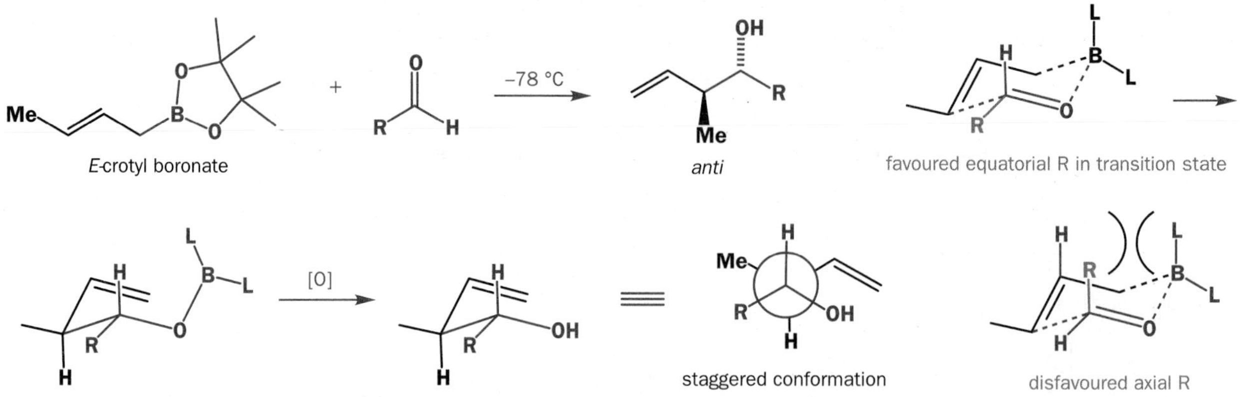

E-crotyl boronate   *anti*   favoured equatorial R in transition state

staggered conformation   disfavoured axial R

The low temperature is a testament to the reactivity of the crotyl boronates and also helps minimize any isomerization of the reagents while maximizing the effect of the energy differences between the favoured and disfavoured transition states.

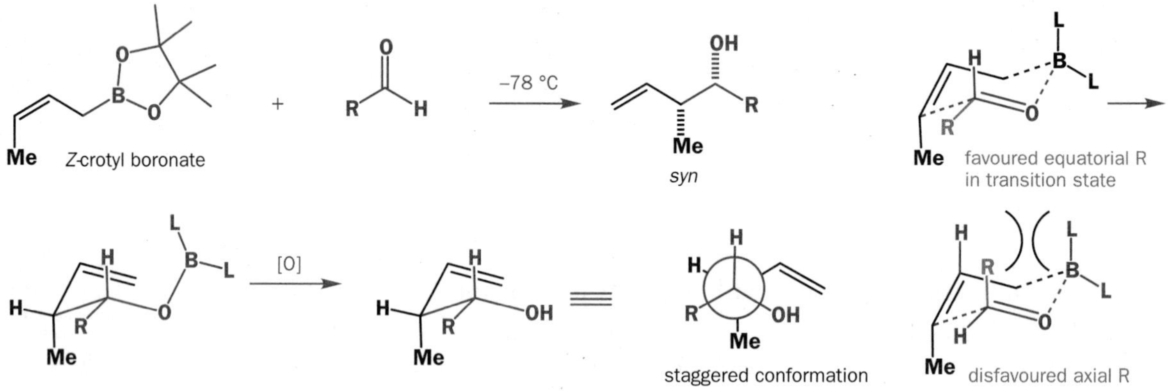

Z-crotyl boronate   *syn*   favoured equatorial R in transition state

staggered conformation   disfavoured axial R

The dramatic diastereoselectivity of this process is noteworthy but, of course, the products are racemic—two *anti* isomers from the *E*-crotyl reagent and two *syn* isomers from the *Z* counterpart. This is inevitable as both starting materials are achiral and there is no external source of chirality. You may be wondering if the use of a chiral ligand on boron would allow the production of a single enantiomer of a single diastereoisomer. The simple answer is that it does, very nicely. In fact, there are a number of solutions to this problem using boranes and boronates but the one illustrated uses the same ligand as that used earlier for allylation derived from carene.

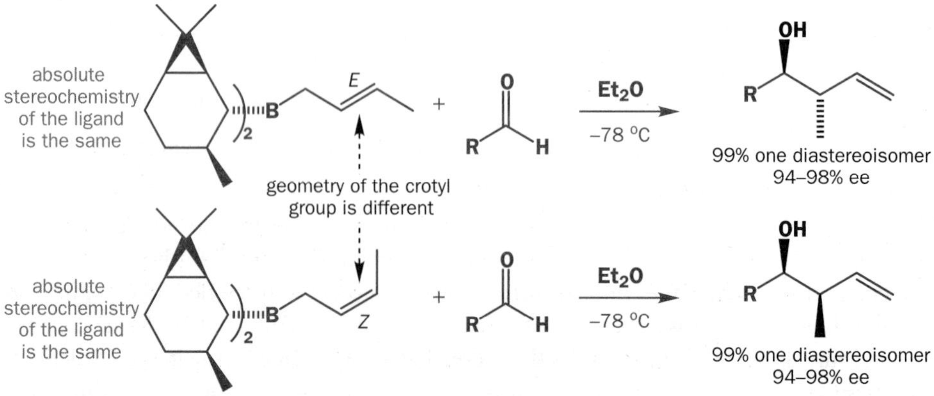

absolute stereochemistry of the ligand is the same

geometry of the crotyl group is different

99% one diastereoisomer
94–98% ee

absolute stereochemistry of the ligand is the same

99% one diastereoisomer
94–98% ee

● **Special features of organoboron chemistry**

- Boron is electrophilic because of its empty p orbital
- Boron forms strong B–O bonds and weak B–C bonds
- Migration of alkyl groups from boron to O, N, or C is stereospecific

Though boron and aluminium form similar reducing agents, such as $NaBH_4$ and $LiAlH_4$, the reactions described so far in this chapter do not occur with aluminium compounds, and compounds with C–Al bonds, other than DIBAL and Me_3Al, are hardly used in organic chemistry. We move on to the other two elements in this chapter, Si and Sn, both members of group IVB (or 14 if you prefer)—the same group as carbon.

Silicon and carbon compared

Silicon is immediately below carbon in the periodic table and the most obvious similarity is that both elements normally have a valency of four and both form tetrahedral compounds. There are important differences in the chemistry of carbon and silicon—silicon is less important and many books are devoted solely to carbon chemistry but relatively few to silicon chemistry. Carbon forms many stable trigonal and linear compounds containing π bonds; silicon forms few. The most important difference is the strength of the silicon–oxygen σ bond (368 kJ mol^{-1}) and the relative weakness of the silicon–silicon (230 kJ mol^{-1}) bond. Together these values account for the absence, in the oxygen-rich atmosphere of earth, of silicon analogues of the plethora of structures possible with a carbon skeleton.

■ Instead, silicon forms compounds containing the very stable O–Si–O linkage giving a variety of structures such as rocks and plastics.

Several of the values in the table are worthy of comment as they give insight into the reactivity differences between carbon and silicon. Bonds to electronegative elements are generally stronger with silicon than with carbon; in

Average bond energies, kJ mol^{-1}

X	H–X	C–X	O–X	F–X	Cl–X	Br–X	I–X	Si–X
C	416	356	336	485	327	285	213	290
Si	323	290	368	582	391	310	234	230
ratio	1.29	1.23	0.91	0.83	0.84	0.92	0.91	1.26

particular, the silicon–fluorine bond is one of the strongest single bonds known, while bonds to electropositive elements are weaker. Silicon–hydrogen bonds are much weaker than their carbon counterparts and can be cleaved easily. This section of Chapter 47 is about organic silicon chemistry. We will mostly discuss compounds with four Si–C bonds. Three of these bonds will usually be the same so we will often have a Me_3Si– group attached to an organic molecule. We shall discuss reactions in which something interesting happens to the organic molecule as one of the Si–C bonds reacts to give a new Si–F or Si–O bond. We shall also discuss organosilicon compounds as reagents, such as triethylsilane (Et_3SiH), which is a reducing agent whereas $Et_3C–H$ is not. Here are a few organosilicon compounds.

a useful electrophile   a protected alcohol   an allyl silane   a silyl benzene

The carbon–silicon bond is strong enough for the trialkyl silyl group to survive synthetic transformations on the rest of the molecule but weak enough for it to be cleaved specifically when we want. In particular, fluoride ion is a poor nucleophile for carbon compounds but attacks silicon very readily. Another important factor is the length of the C–Si bond (1.89 Å)—it is significantly longer than a typical C–C bond (1.54 Å). Silicon has a lower electronegativity (1.8) than carbon (2.5) and therefore C–Si bonds are polarized towards the carbon. This makes the silicon susceptible to attack by nucleophiles.

▶ The strength of the C–Si bond means that alkyl silanes are stable but useful chemistry arises from carbon substituents that are not simple alkyl groups.

Silicon has an affinity for electronegative atoms

The most effective nucleophiles for silicon are the electronegative ones that will form strong bonds to silicon, such as those based on oxygen or halide ions with fluoride being pre-eminent. You saw this in the choice of reagent for the selective cleavage of silyl ethers in Chapter 24. Tetrabutylammonium fluoride is often used as this is an organic-soluble ionic fluoride and forms a silyl fluoride as the by-product. The mechanism is not a simple S_N2 process and has no direct analogue in carbon chemistry. It looks like a substitution at a hindered tertiary centre, which ought to be virtually impossible. Two characteristics of silicon facilitate the process: first, the long silicon–carbon bonds relieve the steric interactions and, second, the d orbitals of silicon provide a target for the nucleophile that does not have the same geometric constraints as a C–O σ^* orbital. Attack of the fluoride on the d orbital leads to a negatively charged pentacoordinate intermediate that breaks down with loss of the alkoxide. There is a discrete intermediate in contrast to the pentacoordinate transition state of a carbon-based S_N2 reaction.

a protected alcohol   pentacovalent Si   alcohol

This process is sometimes abbreviated to S_N2 at silicon to save space. The intermediate is a trigonal bipyramid with negatively charged pentacovalent silicon. It is often omitted in drawings because it is formed slowly and decomposes quickly. This mechanism is similar to nucleophilic substitution at boron except that the intermediate is pentacovalent (Si) rather than tetrahedral (B). The hydrolysis of a boron ester at the end of a hydroboration–oxidation sequence would be an example of an analogous boron reaction.

hydroboration   oxidation   hydrolysis   tetrahedral B   hydrolysis

The silicon Baeyer–Villiger rearrangement

Evidence that the 'S_N2' reaction at silicon does indeed go through a pentacovalent intermediate comes from the silicon analogue of the migration step in hydroboration–oxidation. Treatment of reactive organosilanes (that is, those with at least one heteroatom—F, OR, NR$_2$—attached to silicon to encourage nucleophilic attack of hydroperoxide at silicon)

with the same reagent (alkaline hydrogen peroxide) also gives alkyl migration from Si to O with retention of configuration. It would be difficult to draw a mechanism for this reaction without the intermediate. This is a precise copy of the oxidative cleavage of organoboranes that works on silanes.

oxidation   pentacovalent Si   rearrangement   hydrolysis

● **Silicon forms strong bonds with oxygen and very strong bonds with fluorine.**

Nucleophilic substitution at silicon

You may wonder why trimethylsilyl chloride does not use the S_N1 mechanism familiar from the analogous carbon compound t-butyl chloride. There is, in fact, nothing wrong with the Me_3Si^+

cation—it is often observed in mass spectra, for example. The reason is that the 'S$_N$2' reaction at silicon is too good.

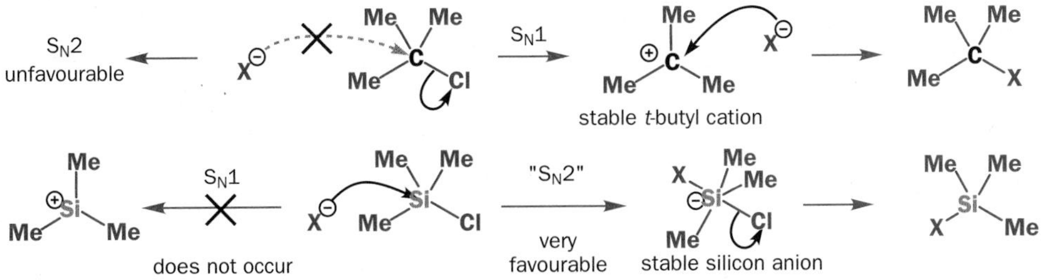

We should compare the 'S$_N$2' reaction at silicon with the S$_N$2 reaction at carbon. There are some important differences. Alkyl halides are soft electrophiles but silyl halides are hard electrophiles. Alkyl halides react only very slowly with fluoride ion but silyl halides react more rapidly with fluoride than with any other nucleophile. The best nucleophiles for saturated carbon are neutral and/or based on elements down the periodic table (S, Se, I). The best nucleophiles for silicon are charged and based on highly electronegative atoms (chiefly F, Cl, and O). A familiar example is the reaction of enolates at carbon with alkyl halides but at oxygen with silyl chlorides (Chapter 21).

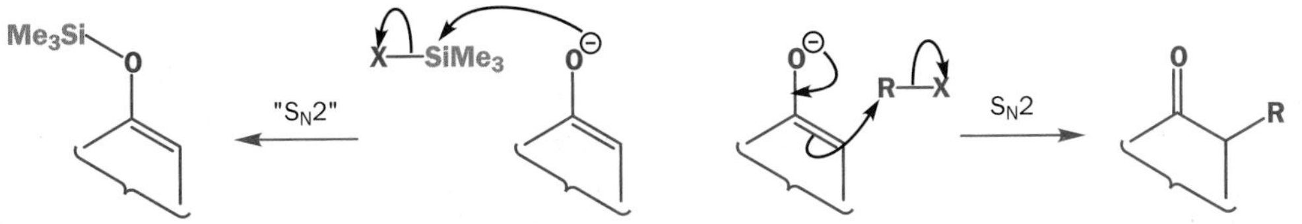

When a Me$_3$Si group is removed from an organic molecule with hydroxide ion, the product is not the silanol as you might expect but the silyl ether 'hexamethyldisiloxane'. The carbon analogue di-*t*-butyl ether could not be formed under these conditions nor by this mechanism, but only by the S$_N$1 mechanism in acid solution.

> You will see in Chapter 52 that Me$_2$SiCl$_2$ polymerizes by repeating this mechanism many many times.

The other side of the coin is that the S$_N$2 reaction at carbon is *not* much affected by partial positive charge ($\delta+$) on the carbon atom. The 'S$_N$2' reaction at silicon *is* affected by the charge on silicon. The most electrophilic silicon compounds are the silyl triflates and it is estimated that they react some 10^8–10^9 times faster with oxygen nucleophiles than do silyl chlorides. Trimethylsilyl triflate is, in fact, an excellent Lewis acid and can be used to form acetals or silyl enol ethers from carbonyl compounds, and to react these two together in aldol-style reactions. In all three reactions the triflate attacks an oxygen atom.

In the acetal formation, silylation occurs twice at the carbonyl oxygen atom and the final leaving group is hexamethyldisiloxane. You should compare this with the normal acid-catalysed mechanism described in Chapter 14 where the carbonyl group is twice protonated and the leaving group is water.

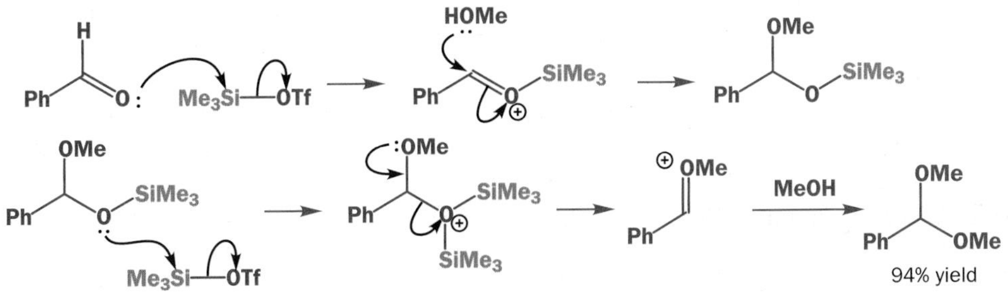

Silyl enol ether formation again results from silylation of carbonyl oxygen but this time no alcohol is added and a weak base, usually a tertiary amine, helps to remove the proton after silylation.

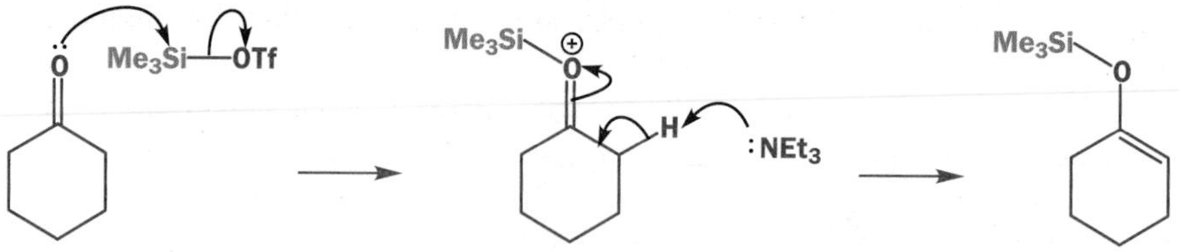

85% yield

When the acetal and the silyl enol ether are mixed with the same Lewis acid catalyst, Noyori found that an efficient aldol-style condensation takes place with the acetal providing the electrophile. The reaction is successful at low temperatures and only a catalytic amount of the Lewis acid is needed. Under these conditions, with no acid or base, few side-reactions occur. Notice that the final desilylation is carried out by the triflate anion to regenerate the Lewis acid Me_3Si–OTf. Triflate would be a very poor nucleophile for saturated carbon but is reasonable for silicon because oxygen is the nucleophilic atom.

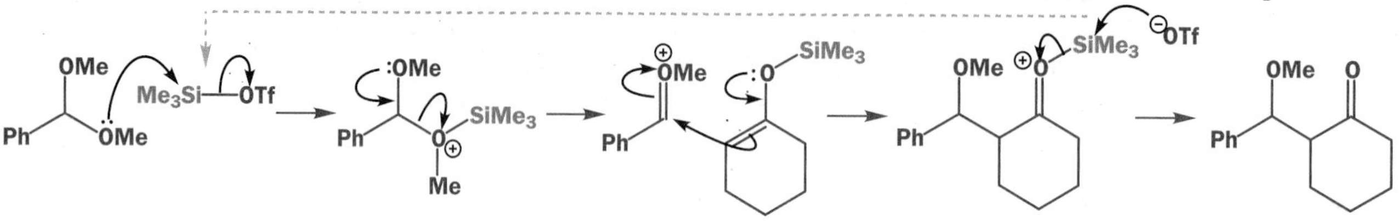

89% yield

Silyl ethers are versatile protecting groups for alcohols

Silicon-based protecting groups for alcohols are the best because they are the most versatile. They are removed by nucleophilic displacement with fluoride or oxygen nucleophiles and the rate of removal depends mostly on the steric bulk of the silyl group. The simplest is trimethylsilyl (Me_3Si or often just TMS) which is also the most easily removed as it is the least hindered. In fact, it is removed so easily by water with a trace of base or acid that special handling is required to keep this labile group in place.

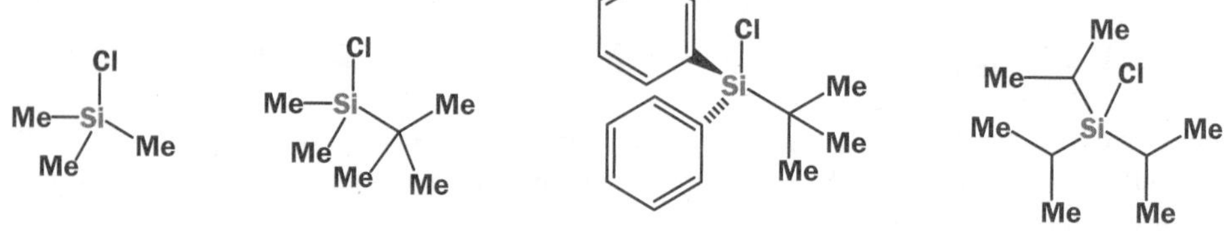

Me₃SiCl
TMSCl

t-BuMe₂SiCl
TBDMSCl

t-BuPh₂SiCl
TBDPSCl

i-Pr₃SiCl
TIPSCl

Replacement of the one of the methyl groups with a much more sterically demanding tertiary butyl group gives the *t*-butyldimethylsilyl (TBDMS) group, which is stable to normal handling and survives aqueous work-up or column chromatography on silica gel. The stability to these isolation and purification conditions has made TBDMS (sometimes over-abbreviated to TBS) a very popular choice for organic synthesis. TBDMS is introduced by a substitution reaction on the corresponding silyl chloride with imidazole in DMF. Yields are usually virtually quantitative and the conditions are mild. Primary alcohols are protected in the presence of secondary alcohols. Removal relies on the strong affinity of fluoride for silicon and is usually very efficient and selective.

However, a protecting group is useful only if it can be introduced and removed in high yield without affecting the rest of the molecule and if it can survive a wide range of conditions in the course of the synthesis. The extreme steric bulk of the *t*-butyldiphenylsilyl (TBDPS) group makes it useful for selective protection of unhindered primary alcohols in the presence of secondary alcohols.

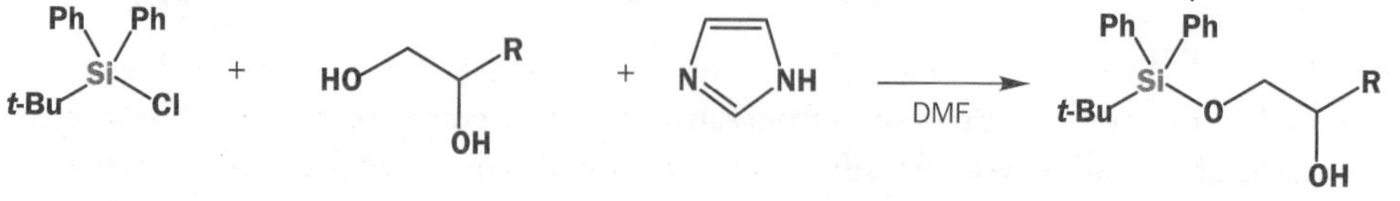

TBDPSCl

diol

imidazole

TBDPS ether

The most stable common silyl protecting group (triisopropylsilyl or TIPS) has three branched alkyl substituents to protect the central silicon from attack by nucleophiles which would lead to cleavage. All three hindered silyl groups (TBDMS, TBDPS, and TIPS) have excellent stability but can still be removed with fluoride.

Alkynyl silanes are used for protection and activation

Alkynyl lithiums and Grignards were made in this way in Chapter 9.

Terminal alkynes have an acidic proton (pK_a *ca.* 25) that can be removed by very strong bases such as organometallic reagents (Grignards, RLi, etc.). While this is often what is intended, in other circumstances it may be an unwanted side-reaction that would consume an organometallic reagent or interfere with the chosen reaction. Exchange of the terminal proton of an alkyne for a trimethylsilyl group exploits the relative acidity of the proton and provides a neat solution to these problems. The $SiMe_3$ group protects the terminus of the alkyne during the reaction but can then be removed with fluoride or sodium hydroxide. A classic case is the removal of a proton next door to a terminal alkyne.

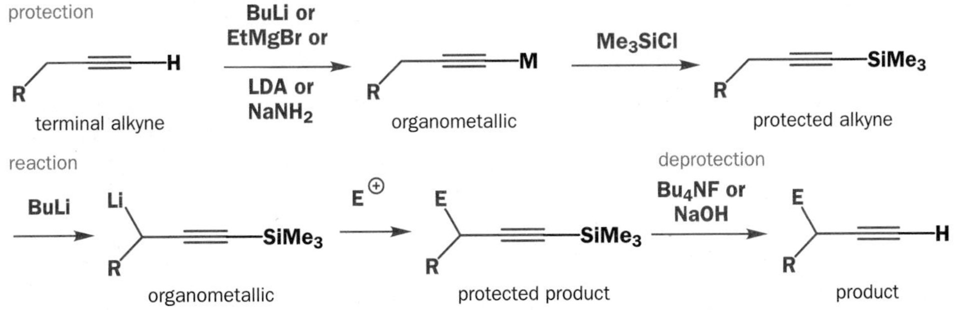

Additionally, acetylene itself is a useful two-carbon building block but is not very convenient to handle as it is an explosive gas. Trimethylsilylacetylene is a distillable liquid that is a convenient substitute for acetylene in reactions involving the lithium derivative as it has only one acidic proton. The synthesis of this alkynyl ketone is an example. Deprotonation with butyl lithium provides the alkynyl lithium that reacted with the alkyl chloride in the presence of iodide as nucleophilic catalyst (see Chapter 17). Removal of the trimethylsilyl group with potassium carbonate in methanol allowed further reaction on the other end of the alkyne.

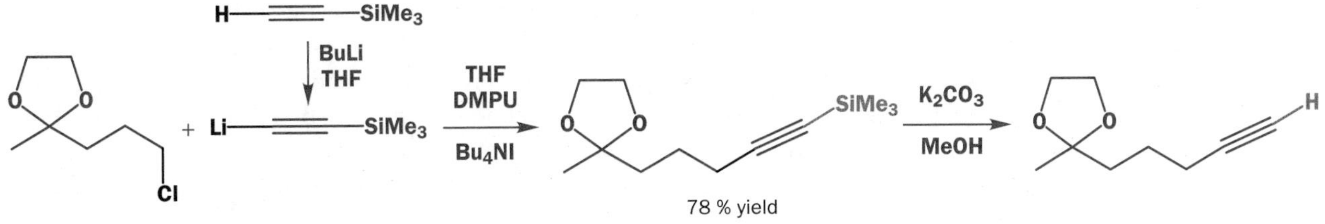

Silicon stabilizes a positive charge on the β carbon

In common with ordinary alkynes, silylated alkynes are nucleophilic towards electrophiles. The presence of the silicon has a dramatic effect on the regioselectivity of this reaction: attack occurs only at the atom directly bonded to the silicon. This must be because the intermediate cation is stabilized.

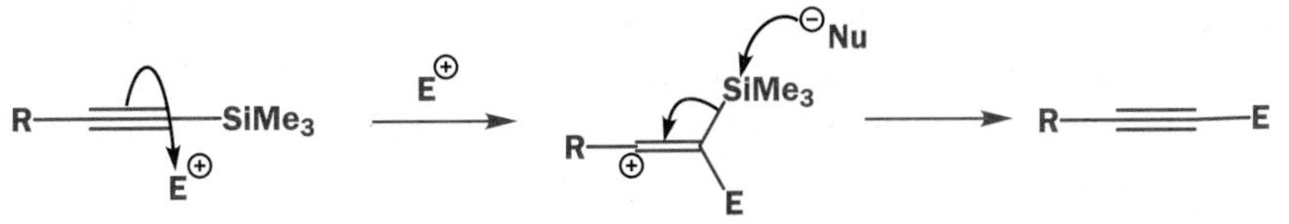

The familiar hierarchy of carbocation stability—tertiary > secondary > primary—is due to the stabilization of the positive charge by donation of electron density from adjacent C–H or C–C bonds (their filled σ orbitals to be precise) that are aligned correctly with the vacant orbital (Chapter 17). The electropositive nature of silicon makes C–Si bonds even more effective donors so that a β-silyl group stabilizes a positive charge so effectively that the course of a reaction involving cationic intermediates is often completely controlled. This is stabilization by σ donation.

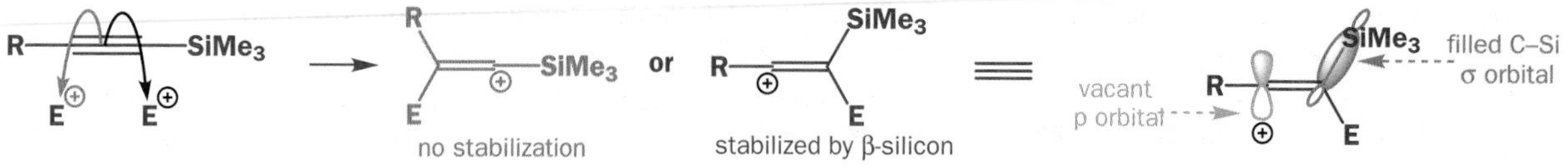

The stabilization of the cation weakens the C–Si bond by the delocalization of electron density so that the bond is more easily broken. Attack of a nucleophile, particularly a halogen or oxygen nucleophile, on silicon removes it from the organic fragment and the net result is electrophilic substitution in which the silicon has been replaced by the electrophile.

> The nucleophile does not need to be very powerful as the C–Si bond is weakened. Many neutral molecules with a lone pair and almost any anion will do, even triflate ($CF_3SO_2O^-$).

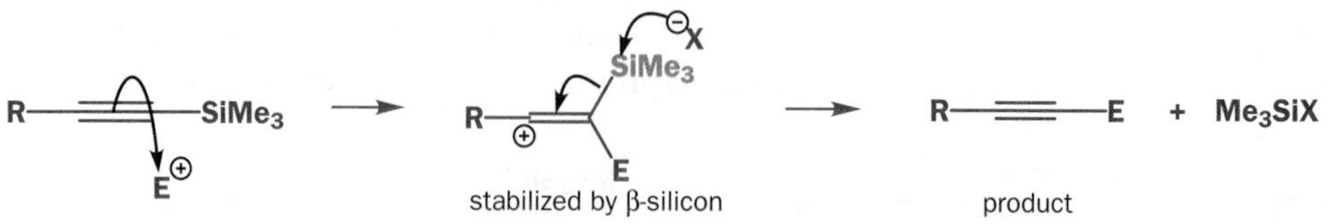

This is useful for the synthesis of alkynyl ketones, which are difficult to make directly with conventional organometallic reagents such as alkynyl–Li or –MgBr because they add to the ketone product. Alkynyl silanes react with acid chlorides in the presence of Lewis acids, such as aluminium chloride, to give the ketones.

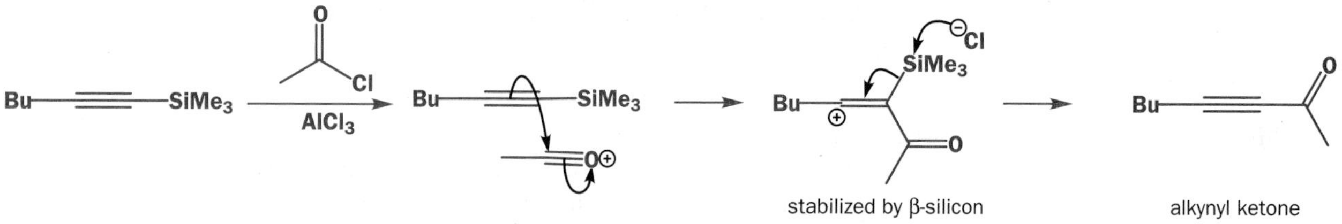

Aryl silanes undergo *ipso* substitution with electrophiles

Exactly the same sort of mechanism accounts for the reactions of aryl silanes with electrophiles under Friedel–Crafts conditions. Instead of the usual rules governing *ortho*, *meta*, and *para* substitution using the directing effects of the substituents, there is just one rule: the silyl group is replaced by the electrophile at the same atom on the ring—this is known as *ipso* substitution. Actually, this selectivity comes from the same principles as those used for ordinary aromatic substitution (Chapter 22): the electrophile reacts to produce the most stable cation—in this case β to silicon. Cleavage of the weakened C–Si bond by any nucleophile leads directly to the *ipso* product.

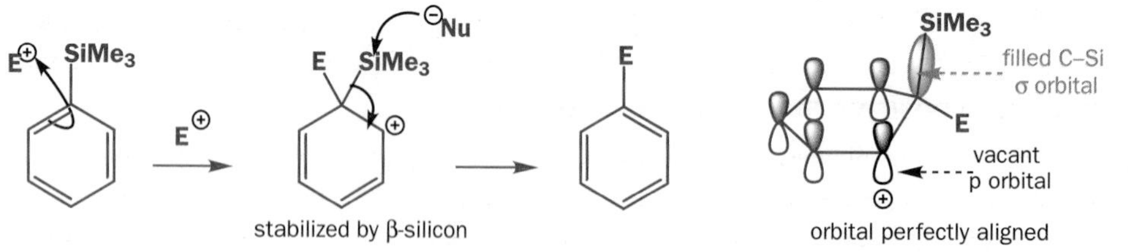

> The Latin word *ipso* means 'the same'—the *same* site as that occupied by the SiR_3 group.

There is an alternative site of attack that would lead to a cation β to silicon, that is, *meta* to silicon. This cation is not particularly stable because the vacant p orbital is orthogonal to the C–Si bond and so cannot interact with it. This illustrates that it is more important to understand the origin of the effect based on molecular orbitals rather than simply to remember the result.

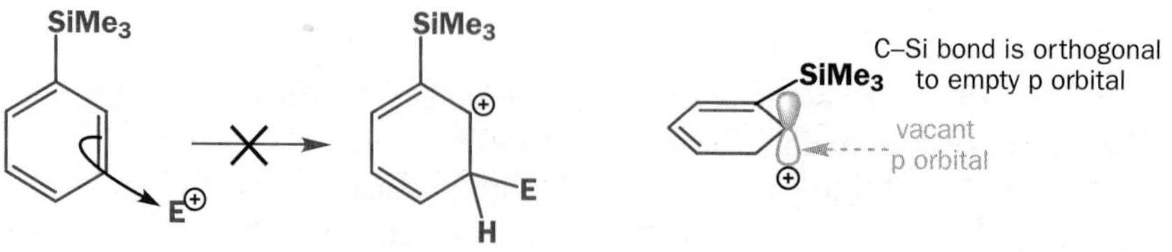

This reactivity of aryl silanes is used to convert the stable phenyl dimethylsilyl group into a more reactive form for conversion into an alcohol by the 'silyl Baeyer–Villiger' reaction described above. Overall this makes the phenyl dimethylsilyl group a bulky masked equivalent for a hydroxyl group. This is useful because the silane will survive reaction conditions that the alcohol might not and the steric bulk allows stereoselective reactions. Ian Fleming at Cambridge has made extensive use of this group and the conversion into an alcohol by several reagents all of which depend on the *ipso* substitution of the phenyl silane. The reaction with bromine is typical. Bromobenzene is produced together with a silyl bromide that is activated towards subsequent oxidation.

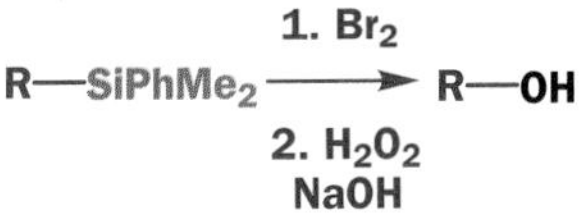

The mechanism of electrophilic desilylation is the same as that for electrophilic aromatic substitution except that the proton is replaced by trimethylsilyl. The important difference is that the silicon stabilizes the intermediate cation, and hence the transition state leading to it, to a dramatic extent so that the rate is much faster. This is the first step with bromine.

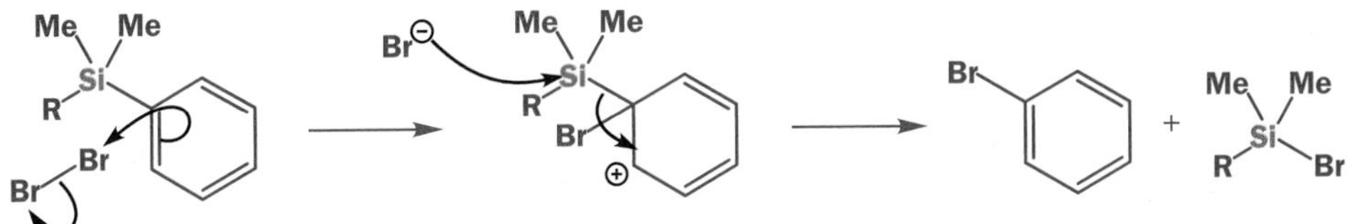

$$R\text{—}SiPhMe_2 \xrightarrow[\text{2. } H_2O_2 \atop \text{NaOH}]{\text{1. } Br_2} R\text{—}OH$$

stabilized by β-silicon

The rest of the reaction sequence involves displacement of Br– by HOO–, addition of hydroxide, rearrangement, and hydrolysis. All these steps involve the silicon atom and the details are given a few pages back.

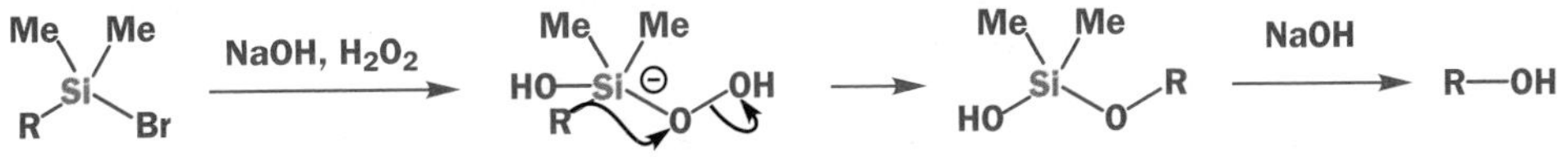

rearrangement   hydrolysis

> ● Trimethylsilyl and other silyl groups stabilize a positive charge on a β carbon and are lost very easily. They can be thought of as very reactive protons or 'super protons'.

Vinyl silanes can be prepared stereospecifically

Controlled reduction of alkynyl silanes produces the corresponding vinyl silanes and the method of reduction dictates the stereochemistry. Lindlar hydrogenation adds a molecule of hydrogen across the alkyne in a *cis* fashion to produce the *Z*-vinyl silane. Red Al reduction of a propargylic alcohol leads instead to the *E*-isomer.

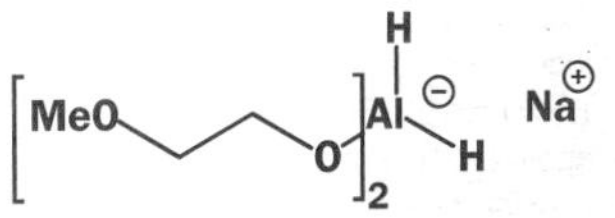

Red Al®

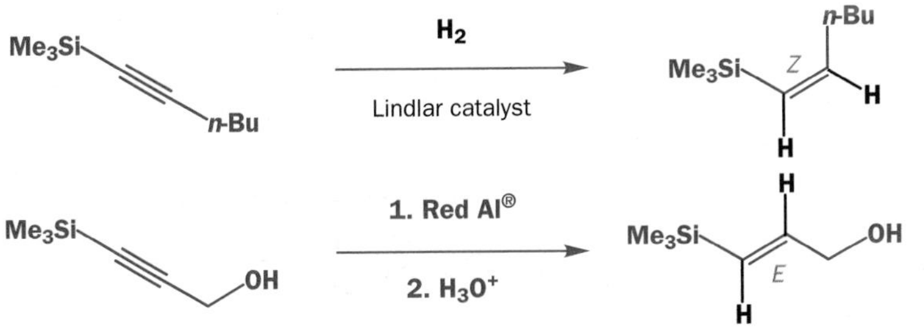

> ■ Another example of this type of reduction appears in Chapter 31.

The mechanism of the second reaction is a *trans* hydroalumination helped by coordination of the alane to the triple bond and external nucleophilic attack. The regioselectivity of the hydroalumination is again determined by silicon: the electrophilic alane attacks the alkyne on the carbon bearing the silyl group (the *ipso* carbon).

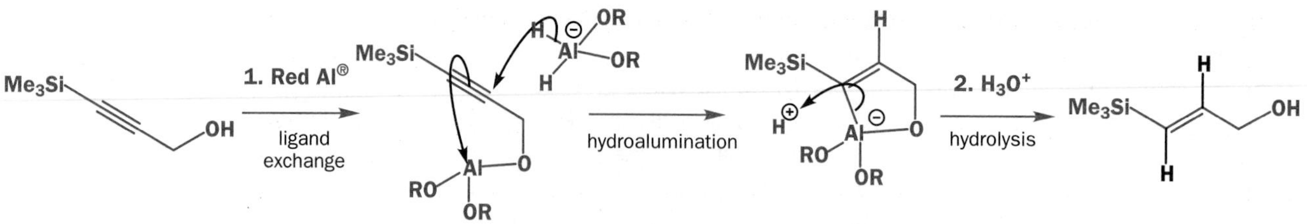

See Chapter 31 for the effect of light on alkenes.

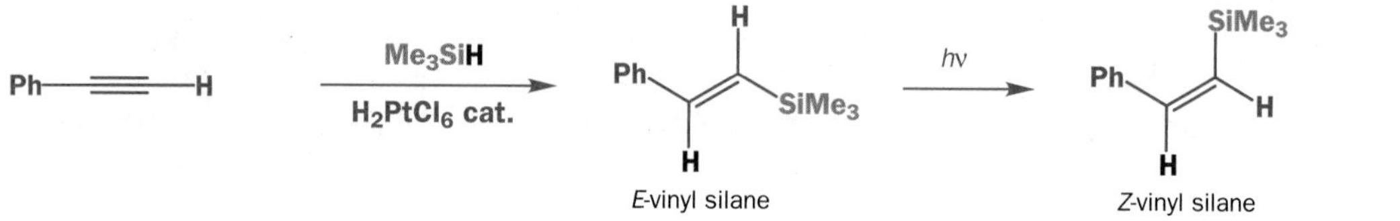

Instead of adding two hydrogen atoms to an alkynyl silane we could add H and SiMe$_3$ to a simple alkyne by hydrosilylation (addition of hydrogen and silicon). This is a *cis* addition process catalysed by transition metals and leads to a *trans* (*E*-) vinyl silane. One of the best catalysts is chloroplatinic acid (H$_2$PtCl$_6$) as in this formation of the *E*-vinyl silane from phenylacetylene. In this case photo-chemical isomerization to the *Z*-isomer makes both available. Other than the need for catalysis, this reaction should remind you of the hydroboration reactions earlier in the chapter. The silicon atom is the electrophilic end of the Si–H bond and is transferred to the less substituted end of the alkyne.

Vinyl silanes can also be prepared from vinyl halides by metal–halogen exchange to form the corresponding vinylic organometallic and coupling with a silyl chloride. Notice that both of these reactions happen with retention of configuration. This route is successful for acyclic and cyclic compounds and even vinyl chlorides, which are much less reactive, can be used with the lithium containing some of the more powerfully reducing sodium as the metal.

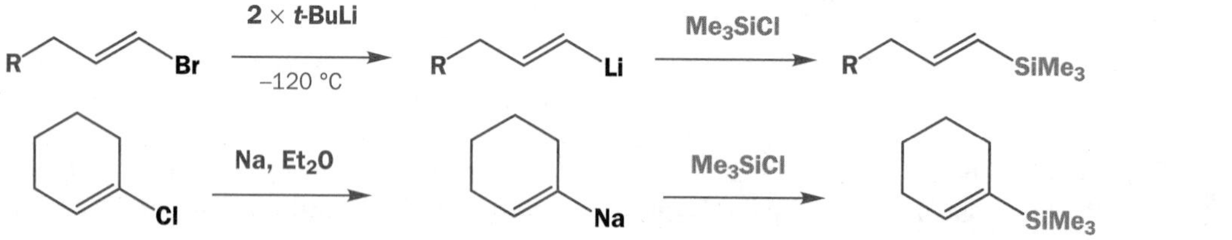

Vinyl silanes can be prepared directly from ketones using the Shapiro reaction

Conversion of ketones into arylsulfonylhydrazones allows preparation of the corresponding vinyl lithiums by base-promoted decomposition of the hydrazone. This is known as the **Shapiro reaction**. Trapping with trimethylsilylchloride gives vinyl silanes, which can be difficult to prepare by other methods.

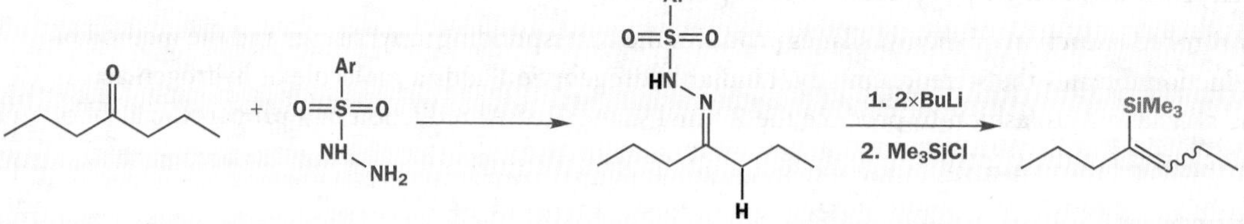

The mechanism involves deprotonation with a very strong base, usually butyllithium or LDA, to form the hydrazone aza-enolate, which then eliminates arylsulfinate to give an unstable anion. Loss of nitrogen, which is extremely favourable, leads to the vinyl lithium.

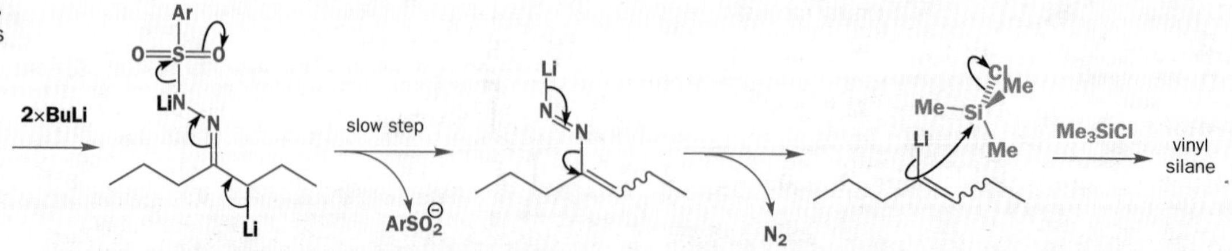

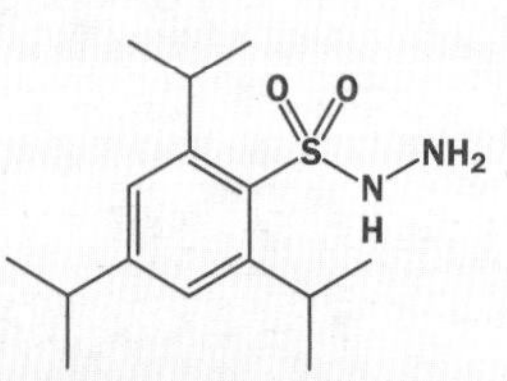

a bulky sulfonyl hydrazine for the Shapiro reaction

The key step is the elimination of the aryl sulfinate and this has been improved by using aryl hydrazones with bulky isopropyl groups on the 2-, 4-, and 6-positions of the aromatic ring to accelerate the elimination. The weakness of this approach to vinyl silanes is that the position of the double bond is governed by the initial site of deprotonation and so the usual problems of regioselective ketone enolate formation arise. However, in symmetrical cases or those where one side is favoured as a result of the structure of the ketone, the Shapiro reaction works well.

Vinyl silanes offer a regio- and stereoselective route to alkenes

Vinyl silanes react with electrophiles in a highly regioselective process in which the silicon is replaced by the electrophile at the *ipso* carbon atom. The stereochemistry of the vinyl silane is important because this exchange usually occurs with retention of geometry as well. Consider the reaction of the two vinyl silanes derived from phenyl acetylene with the simple electrophile D$^+$. Deuterons are chemically very similar to protons but are, of course, distinguishable by NMR.

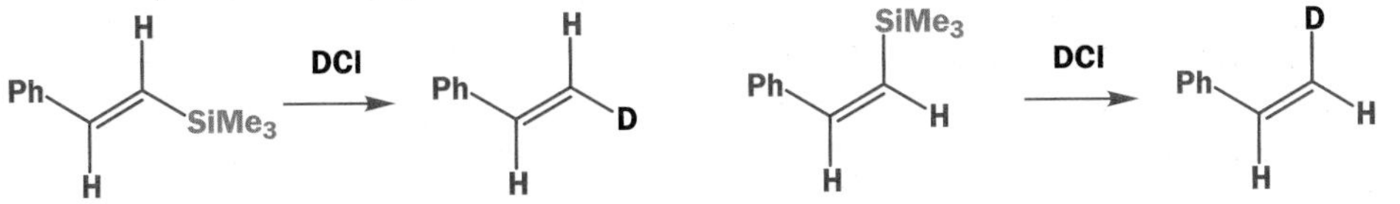

In principle, the alkenes could be protonated at either end but protonation next to silicon leads to the more stable cation β to silicon. In the vinyl silane the C–Si bond is orthogonal to the p orbitals of the π bond, but as the electrophile (D$^+$ here) attacks the π bond, say from underneath, the Me$_3$Si group starts to move upwards. As it rotates, the angle between the C–Si bond and the remaining p orbital decreases from 90°. As the angle decreases, the interaction between the C–Si bond and the empty p orbital of the cation increases. There is every reason for the rotation to continue in the same direction and no reason for it to reverse. The diagram shows that, in the resulting cation, the deuterium atom is in the position formerly occupied by the Me$_3$Si group, *trans* to Ph. Loss of the Me$_3$Si group now gives retention of stereochemistry.

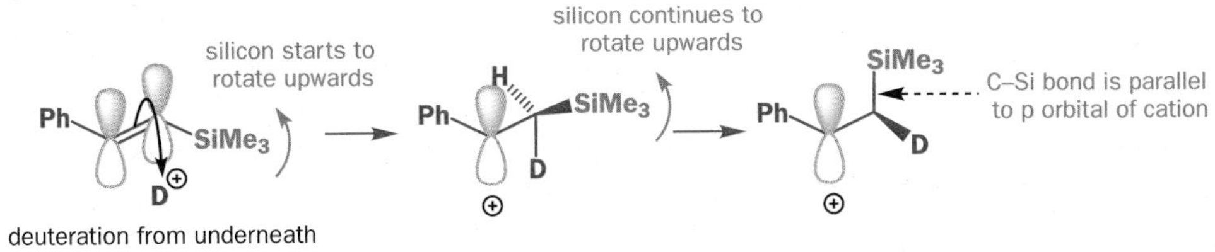

The intermediate cation has only a single bond and so rotation might be expected to lead to a mixture of geometrical isomers of the product but this is not observed. The bonding interaction between the C–Si bond and the empty p orbital means that rotation is restricted. This stabilization weakens the C–Si bond and the silyl group is quickly removed before any further rotation can occur. The stabilization is effective only if the C–Si bond is correctly aligned with the vacant orbital, which means it must be in the same plane—rather like a π bond. Here is the result for both *E*- and *Z*-isomers of the vinyl silane.

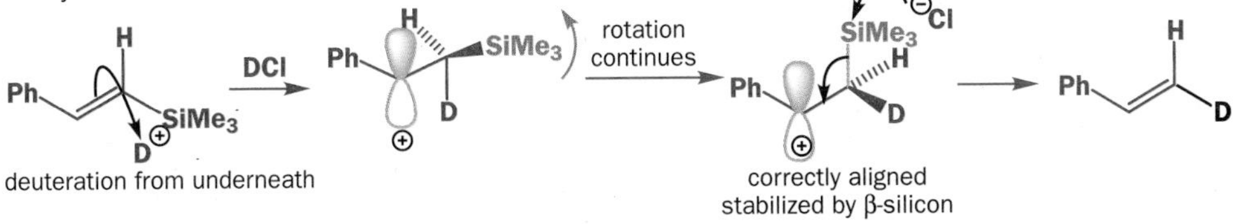

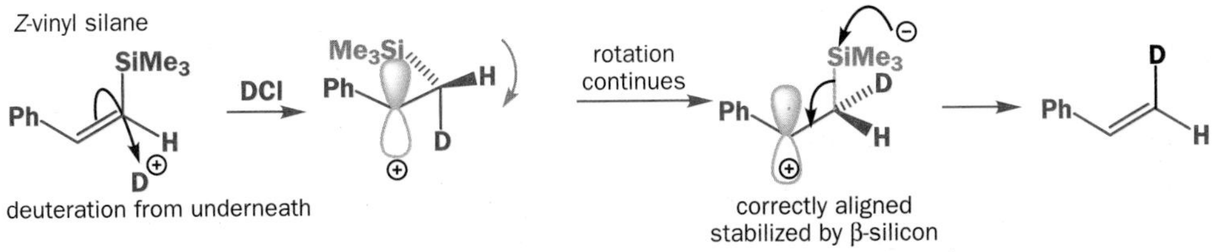

We can illustrate the two alternative rotations with an energy diagram: one rotation leads directly to a stable conformation with the C–Si bonding orbital parallel to the vacant p orbital, while the other passes through a very-high-energy conformation that has the two orbitals orthogonal and so derives no stabilization from the presence of silicon. It is this energy barrier that effectively prevents rotation and leads to electrophilic substitution with retention of double bond geometry. The favoured rotation simply continues the rotation from starting material to cation.

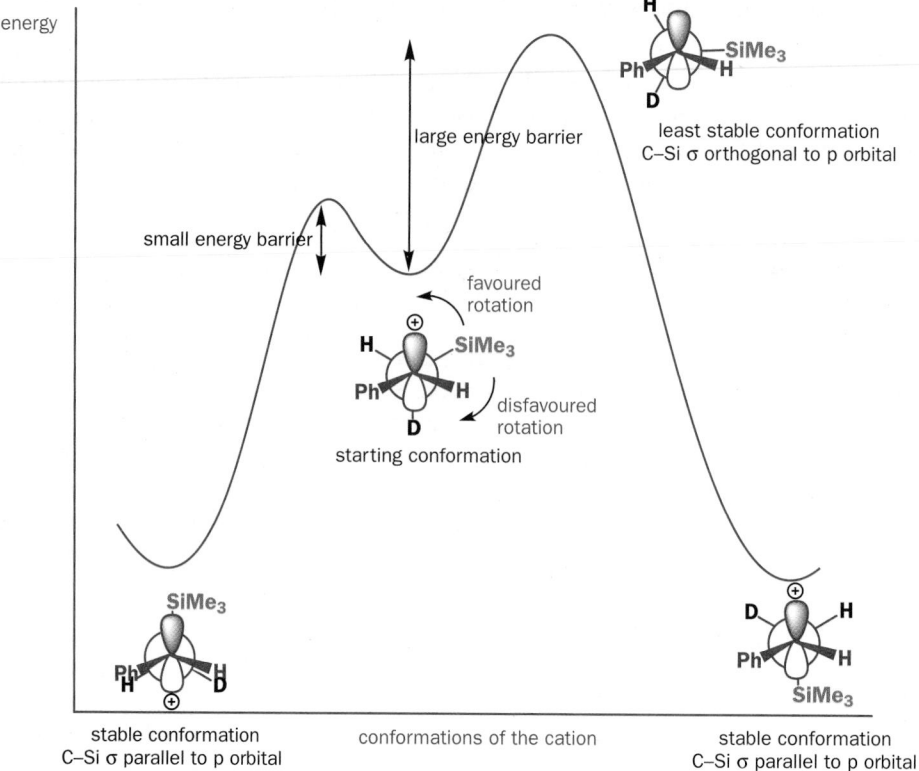

stable conformation
C–Si σ parallel to p orbital
       conformations of the cation
stable conformation
C–Si σ parallel to p orbital

It is unusual for silicon to be required in the final product of a synthetic sequence and the stereospecific removal of silicon from vinyl silanes makes them useful reagents that can be regarded as rather stable vinylic organometallic reagents that will react with powerful electrophiles preserving the double bond location and geometry. **Protodesilylation**, as the process of replacing silicon with a proton is known, is one such important reaction. The halogens are also useful electrophiles while organic halides, particularly acid chlorides, in the presence of Lewis acids, form vinyl halides and unsaturated ketones of defined geometry.

> Geometrically pure vinyl halides are important starting materials for transition-metal-catalysed alkene synthesis (Chapter 48).

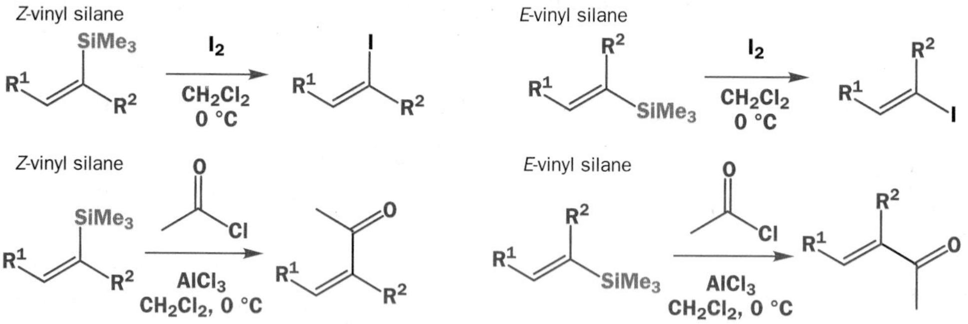

Allyl silanes are readily available

If the silyl group is moved along the carbon chain by just one atom, an allyl silane results. Allyl silanes can be produced from allyl organometallic reagents but there is often a problem over which regioisomer is produced and mixtures often result. Better methods control the position of the double bond using one of the methods introduced in Chapter 31. Two useful examples take advantage of the Wittig reaction and the Peterson olefination to construct the alkene linkage. The reagents are prepared from trimethylsilyl halides either by formation of the corresponding Grignard reagent or alkylation with a methylene Wittig reagent and deprotonation to form a new ylid. The Grignard reagent, with added cerium trichloride, adds twice to esters to give the corresponding tertiary alcohol which loses one of its Me$_3$Si groups in a Peterson elimination to reveal the remaining Me$_3$Si group as part of an allyl silane.

■
The Peterson elimination is another application of organosilicon chemistry. It was discussed in detail in Chapter 31.

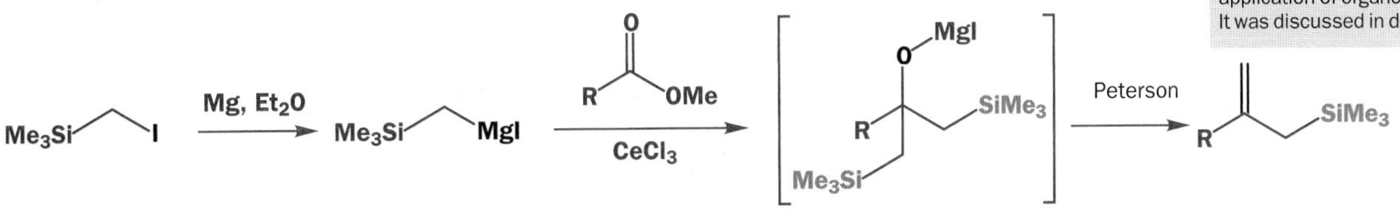

The Wittig reagent is made by alkylation of the simplest ylid with the same silicon reagent. Notice that the leaving group (iodide) is on the carbon next to silicon, not on the silicon itself. Anion formation occurs next to phosphorus, because Ph$_3$P$^+$ is much more anion-stabilizing than Me$_3$Si. The ylid reacts with carbonyl compounds such as cyclohexanone in the usual way to produce the allyl silane with no ambiguity over which end of the allyl system is silylated.

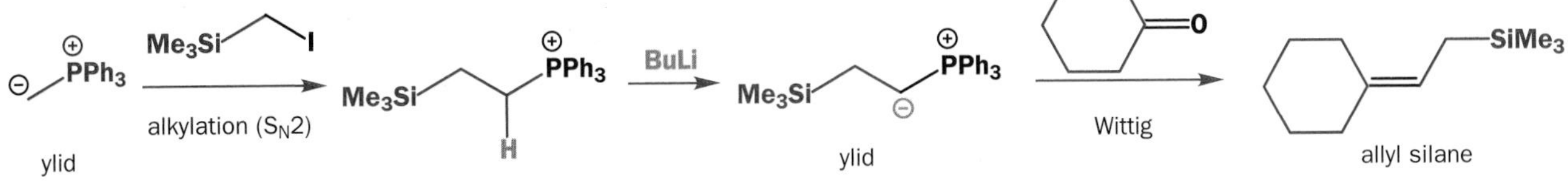

Silicon exerts a surprisingly small steric effect

The Me$_3$Si group is, of course, large. But the C–Si bond is long and the Me$_3$Si group has a smaller steric effect than the Me$_3$C(*t*-butyl) group. For evidence, look at this last sequence: nucleophilic displacement at a carbon atom next to an Me$_3$Si group occurs normally whereas the infamous 'neopentyl' equivalent (see Chapter 17) reacts very slowly if at all. The Me$_3$Si group can get out of the way of the incoming nucleophile.

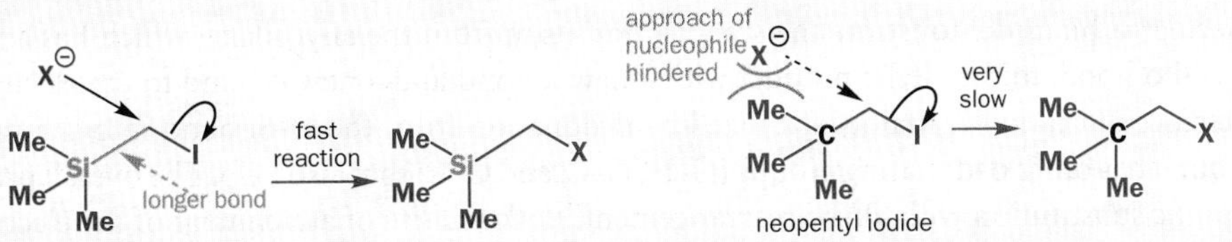

The carbon–silicon bond has two important effects on the adjacent alkene. The presence of a high-energy filled σ orbital of the correct symmetry to interact with the π system produces an alkene that is more reactive with electrophiles, due to the higher-energy HOMO, and the same σ orbital stabilizes the carbocation if attack occurs at the remote end of the alkene. This lowers the transition state for electrophilic addition and makes allyl silanes much more reactive than isolated alkenes.

Allyl silanes are more reactive than vinyl silanes but also react through β-silyl cations

Vinyl silanes have C–Si bonds orthogonal to the p orbitals of the alkene—the C–Si bond is in the nodal plane of the π bond—so there can be no interaction between the C–Si bond and the π bond. Allyl silanes, by contrast, have C–Si bonds that can be, and normally are, parallel to the p orbitals of the π bond so that interaction is possible.

The evidence that such interaction does occur is that allyl silanes are more reactive than vinyl silanes as a result of the increased energy of the HOMO due to the interaction of the π bond with the C–Si bond. Conversely, vinyl silanes are thermodynamically more stable than the allyl isomers by about 8 kJ mol^{-1}. This is evident from the acetylation of a compound having both vinyl silane and allyl silane functional groups. It reacts exclusively as an allyl silane, shown in black, with double bond migration to produce two double bond isomers (*cis* and *trans* cyclononenes) of the vinyl silane product. The vinylic silicon is not involved as the C–Si bond is orthogonal to the π system throughout.

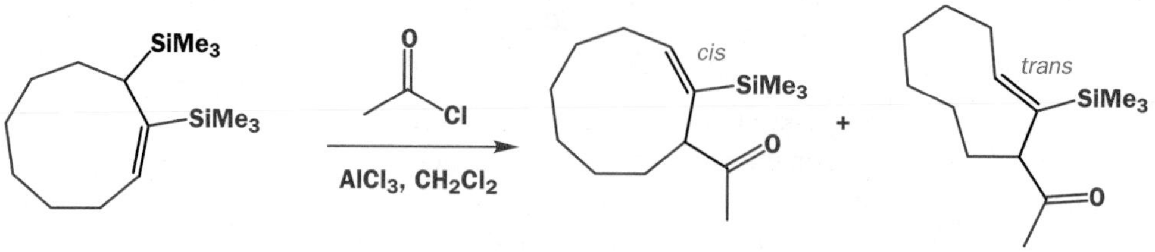

Allyl silanes react with electrophiles with even greater regioselectivity than that of vinyl silanes. The cation β to the silyl group is again formed but there are two important differences. Most obviously, the electrophile attacks at the other end of the allylic system and there is no rotation necessary as the C–Si bond is already in a position to overlap efficiently with the intermediate cation. Electrophilic attack occurs on the face of the alkene *anti* to the silyl group. The process is terminated by loss of silicon in the usual way to regenerate an alkene.

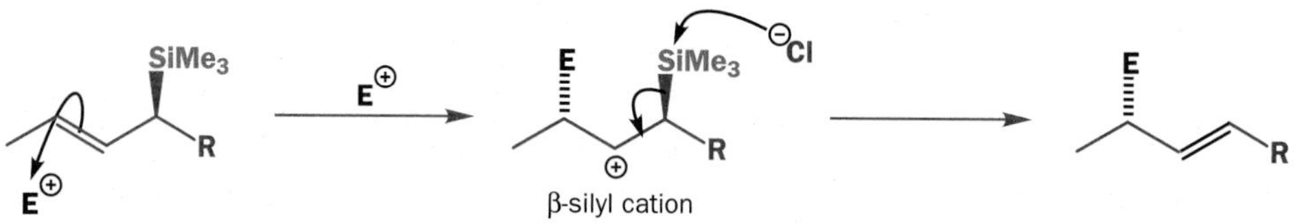

Molecular orbitals demonstrate the smooth transition from the allyl silane, which has a π bond and a C–Si σ bond, to the allylic product with a new π bond and a new σ bond to the electrophile. The intermediate cation is mainly stabilized by σ donation from the C–Si bond into the vacant p orbital but it has other σ-donating groups (C–H, C–C, and C–E) that also help. The overall process is electrophilic substitution with allylic rearrangement. Both the site of attachment of the electrophile and the position of the new double bond are dictated by the silicon.

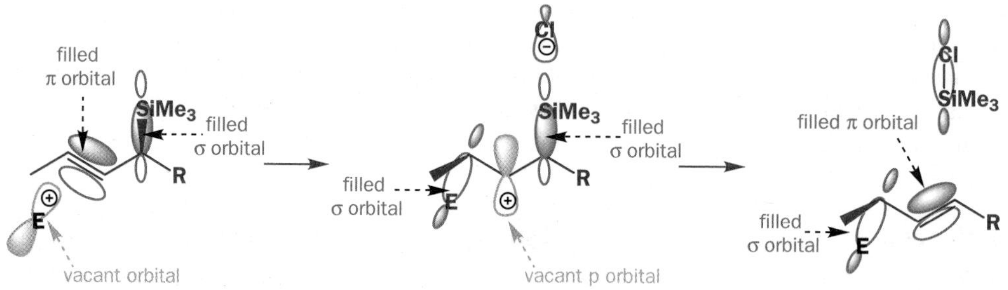

Allyl silanes react with a wide variety of electrophiles, rather like the ones that react with silyl enol ethers, provided they are activated, usually by a Lewis acid. Titanium tetrachloride is widely used but other successful Lewis acids include boron trifluoride, aluminium chloride, and trimethylsilyl triflate. Electrophiles include the humble proton generated from acetic acid. The regiocontrol is complete. No reaction is observed at the other end of the allylic system. All our examples are on the allyl silane we prepared earlier in the chapter.

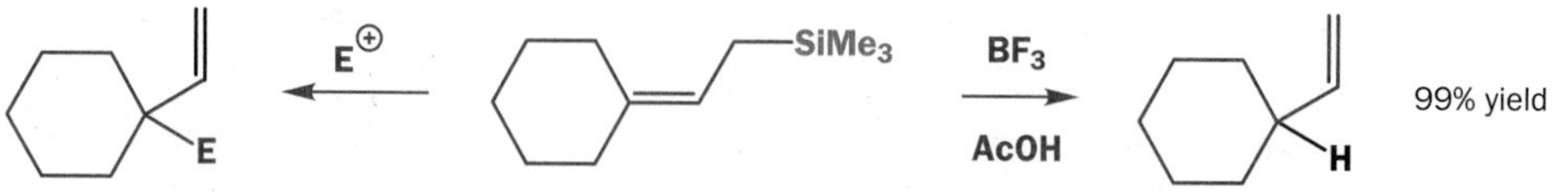

The first reaction is the general reaction with electrophiles and the second shows that even reaction with a proton occurs at the other end of the allyl system with movement of the double bond.

Other electrophiles include acylium ions produced from acid chlorides, carbocations from tertiary halides or secondary benzylic halides, activated enones, and epoxides all in the presence of Lewis acid. In each case the new bond is highlighted in black.

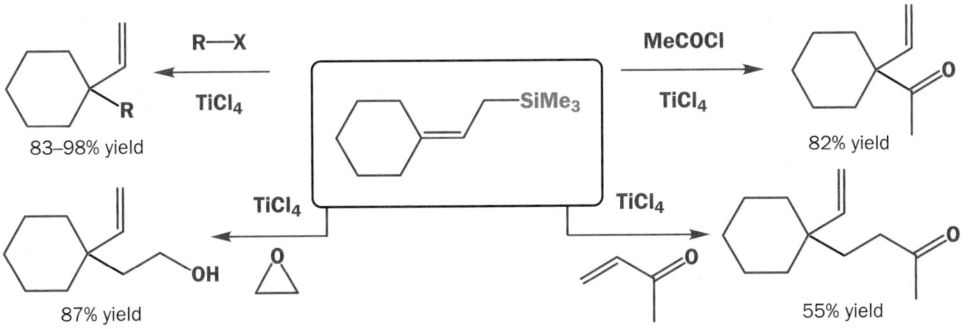

● β-Silyl cations are important intermediates

Vinyl and aryl silanes react with electrophiles at the same (*ipso* or α) atom occupied by silicon. Allyl silanes react at the end of the alkene furthest from silicon (γ). In both cases a β-silyl cation is an intermediate.

In enantiomerically pure systems one enantiomer of the allyl silane gives one enantiomer of the product. The stereogenic centre next to silicon disappears and a new one appears at the other end of the alkene. This is a consequence of the molecule reacting in a well defined conformation by a well defined mechanism. The conformation is controlled by allylic strain (Chapter 34) which compels the proton on the silyl-bearing stereogenic centre to eclipse the alkene in the reactive conformation and the electrophile attacks *anti* to silicon for both steric and stereoelectronic reasons. In these examples of Lewis-acid-promoted alkylation with a *t*-butyl group, *E*- and *Z*-isomers both react highly stereoselectively to give enantiomeric products. The reactions are completely stereospecific.

Lewis acids promote couplings via oxonium ions

Allyl silanes will also attack carbonyl compounds when they are activated by coordination of the carbonyl oxygen atom to a Lewis acid. The Lewis acid, usually a metal halide such as $TiCl_4$ or $ZnCl_2$, activates the carbonyl compound by forming an oxonium ion with a metal–oxygen bond. The allyl silane attacks in the usual way and the β-silyl cation is desilylated with the halide ion. Hydrolysis of the metal alkoxide gives a homoallylic alcohol.

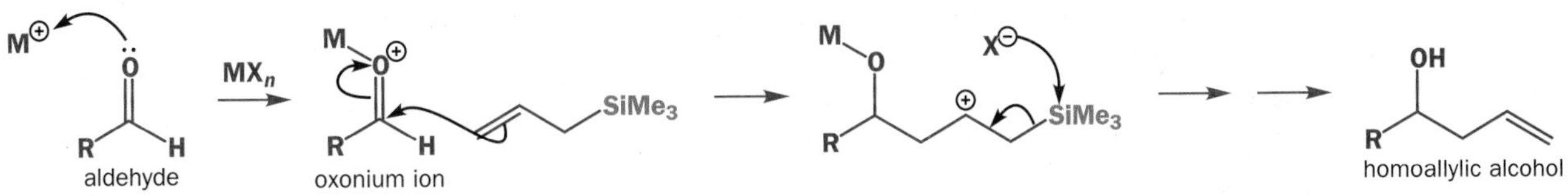

A closely related reactive oxonium ion can be prepared by Lewis-acid-catalysed breakdown of the corresponding acetal. Alternatively, especially if the acetal is at least partly a silyl acetal, the same oxonium ion can be produced *in situ* using yet more silicon in the form of TMSOTf as the Lewis acid catalyst. All these intermediate oxonium ions act as powerful electrophiles towards allyl silanes producing homoallylic alcohols or ethers.

▶ Note how the Me_3Si group mimics the behaviour of a proton even to the extent of producing $(Me_3Si)_2O$—the silicon analogue of water.

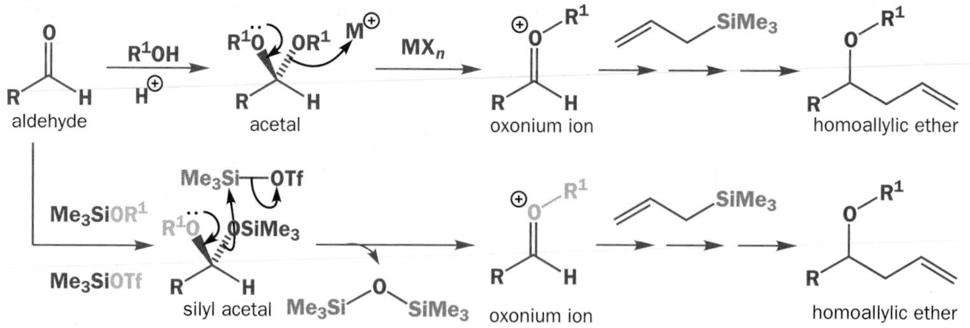

The regiocontrol that results from using an allyl silane to direct the final elimination is illustrated by this example of an intramolecular reaction on to an acetal promoted by tin tetrachloride. The same reaction can be run in the absence of silicon but the intermediate cation can then lose a range of protons to produce five different products!

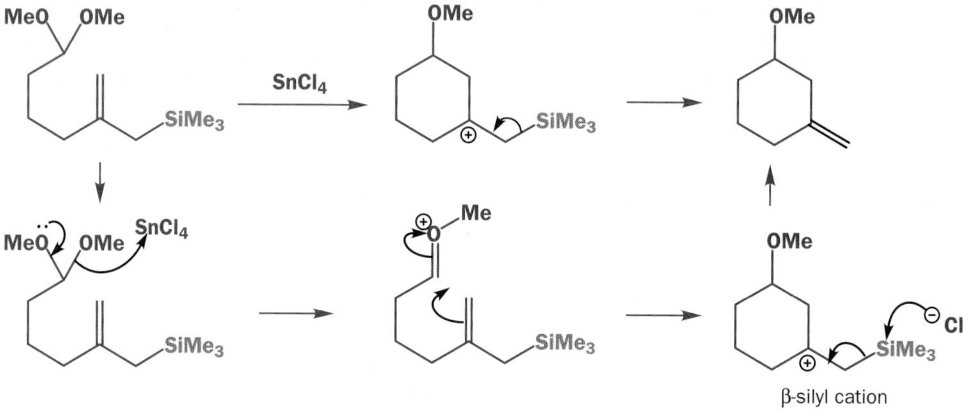

Crotyl silanes are powerful reagents in stereoselective synthesis

Crotyl silanes offer the possibility of diastereoselectivity in reactions with aldehydes in the same way as the corresponding boranes. The mechanism is completely different because crotyl trialkylsilanes react via an open transition state as the silicon is not Lewis acidic enough to bind the carbonyl oxygen of the electrophile. Instead, the aldehyde has to be activated by an additional Lewis acid or by conversion into a reactive oxonium ion by one of the methods described above. The stereoelectronic demands of the allylic silane system contribute to the success of this transformation. Addition takes place in an SE2¢ sense so that the electrophile is attached to the remote carbon on the opposite side of the p system to that originally occupied by silicon and the newly formed double bond is trans to minimize allylic strain.

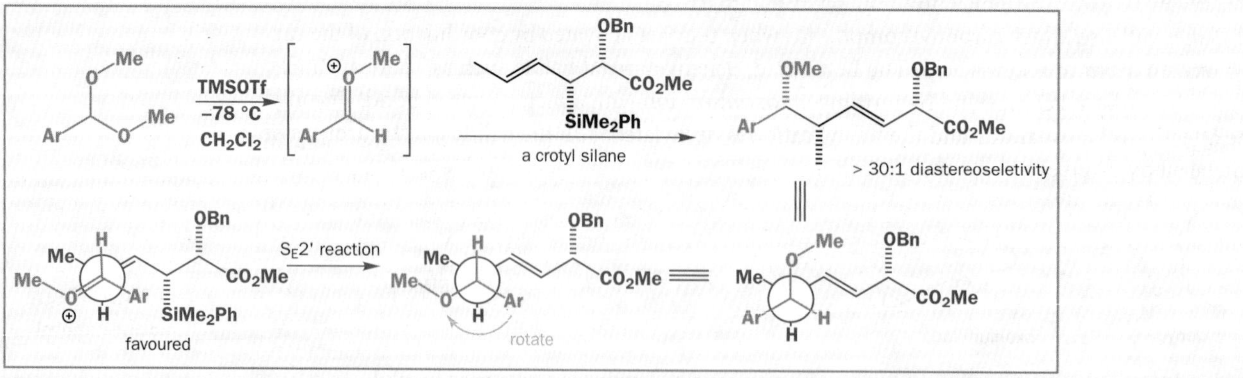

Radicals, anions, and S_N2 transition states stabilized by silicon

In Chapter 31 we discussed the Peterson reaction, which uses carbanions next to silicon, and the reagent Me_3SiCH_2Cl was used to make a Grignard reagent for this reaction. In fact, the chloride can be made directly from Me$_4$Si (tetramethylsilane used as a zero point in NMR spectra) by photo-chemical chlorination. A chlorine atom removes a hydrogen atom from one of the methyl groups to leave a primary radical next to silicon, which reacts in turn with a chlorine molecule, and the radical chain continues.

We might suspect that silicon stabilizes the intermediate carbon-centred radical as primary radicals are not usually stable, but we can prove nothing as there is no alternative. This chloride is a very useful reagent. It readily reacts by the S$_N$2 mechanism, in spite of the large Me$_3$Si group, which makes us suspect that silicon encourages the S$_N$2 reaction at neighbouring carbon. It also readily forms organometallic reagents such as Grignard reagents and lithium derivatives and these were used in the Peterson reaction. This makes us suspect that the Me$_3$Si group stabilizes anions. Can all this really be true?

■
Radical reactions, radical chains, and the stability of radicals are discussed in Chapter 39.

It is all true. Evidence that a silyl group stabilizes the S$_N$2 transition state comes from the reactions of the epoxides of vinyl silanes. These compounds can be made stereospecifically with one equivalent of a buffered peroxy-acid such as *m*-CPBA. Epoxidation is as easy as the epoxidation of simple alkenes. You will see in a moment why acid must be avoided.

These epoxides react stereospecifically with nucleophiles to give single diastereoisomers of adducts. If a carbon nucleophile is used (cuprates are best), it is obvious from the structure of the products that nucleophilic attack has occurred at the end of the epoxide next to silicon. This is obviously an S$_N$2 reaction because it is stereospecific: in any case an S$_N$1 reaction would have occurred at the other end of the epoxide through the β-silyl cation.

When we discussed the Peterson reaction in Chapter 31, we explained that each diastereoisomer of a β-silyl alcohol can eliminate, depending on the reaction conditions, to give either geometrical isomer of the alkene but we did not explain how these diastereoisomers could be made. This is how they are made. Elimination in base is a Wittig-style *syn* process but an *anti* elimination occurs in acid. Here are the reactions on one of the diastereoisomers we have just made.

If the nucleophile is water—as it might be in the work-up of the original epoxidation in acid solution—the product is a diol, which eliminates by the *anti* mechanism in acid solution to give initially an enol and then, under the same conditions, a carbonyl compound. All these steps are often carried out in the one reaction to convert the epoxide to the carbonyl compound in one operation. Stereochemistry does not matter in this reaction.

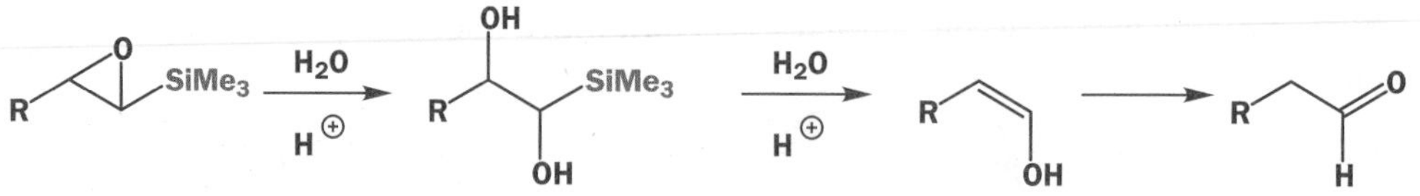

Silicon-stabilized carbanions

We are going to concentrate on the most important of these properties: silyl groups stabilize carbanions. We can show that this is true rather easily. Here are two reactions of carbanions with aldehydes.

The first reagent has a choice: it can do either the Wittig or the Peterson reaction; it prefers the Peterson reaction. This merely tells us that nucleophilic attack at silicon is faster than nucleophilic attack at phosphorus. The carbanion part of the ylid is next to silicon but it could be nowhere else.

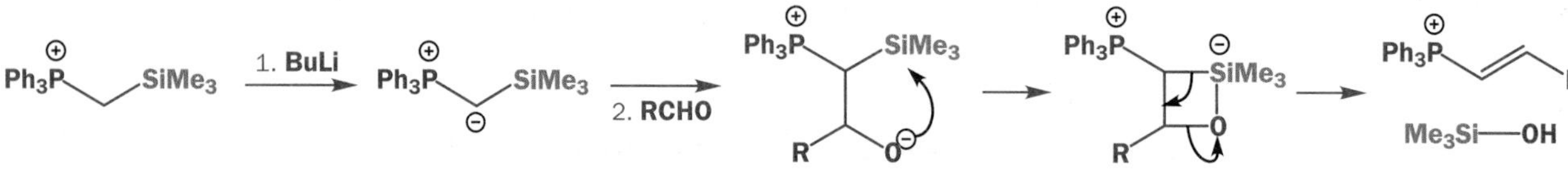

There is, however, a choice in the second reaction. There are six methyl groups on the two Me_3Si groups and one CH_2 between them. That makes eighteen methyl hydrogens and only two on the CH_2 group. Yet the base removes one of the two. It is better to have an anion stabilized by two silicon atoms. Silicon does stabilize a carbanion. There is, of course, no choice in the elimination step—O^- must attack one of the Me_3Si groups and the Peterson reaction must occur.

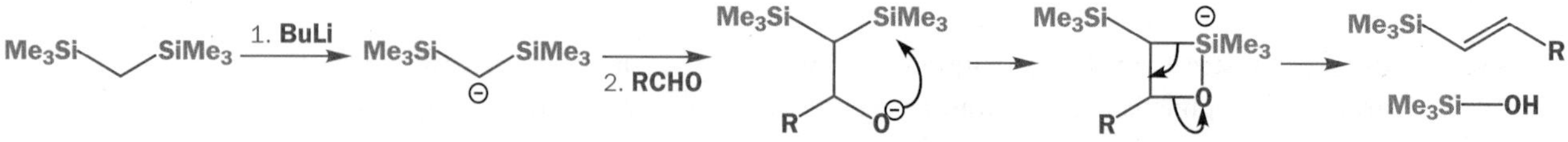

These reactions are also useful syntheses of vinyl phosphine oxides and of vinyl silanes. The stabilization of anions is weak—weaker than from phosphorus or sulfur—but still useful. The Wittig reagent used to make allyl silanes earlier in this chapter illustrates this point.

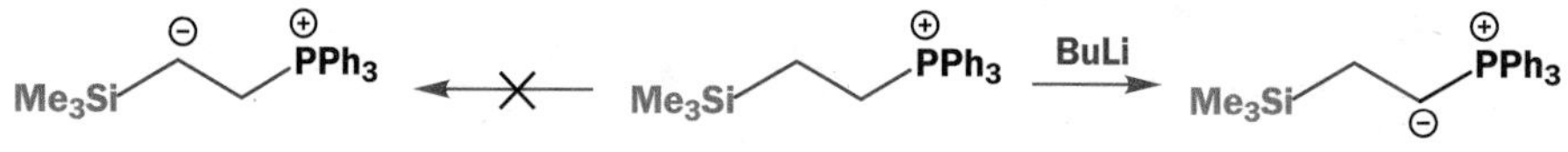

Si-stabilized carbanion

phosphorus ylid

If you want to make an 'anion' stabilized by one Me_3Si group it is better to use an organolithium or organomagnesium compound made from a halide, the most important being the simplest as we have seen. But given just a little extra help—even an alkene—anions can be made with bases. So an allyl silane can give a lithium derivative (using *s*-BuLi as the very strong base) that reacts with electrophiles in the same position as do the allyl silanes themselves—the γ-position relative to the Me_3Si group. In this example the electrophile is a ketone and no Lewis acid is needed.

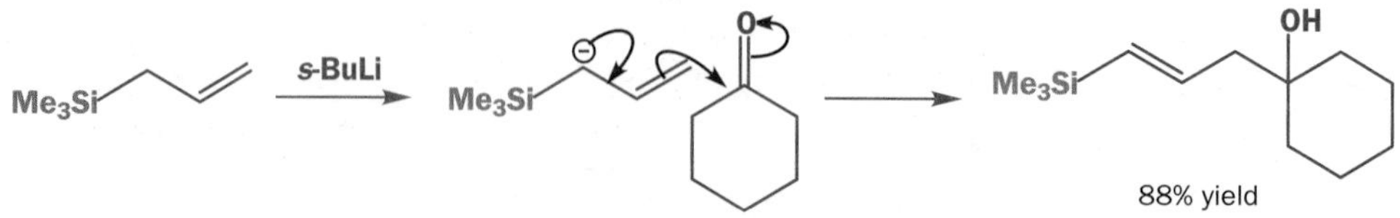

88% yield

The product is a vinyl silane as the Me_3Si group is retained in this reaction of the anion. The reaction is stereoselective in favour of the *E*-alkene as might be expected. The alkene can be epoxidized and the epoxide opened in the reaction we discussed earlier in the chapter. If methanol is used as the nucleophile with BF$_3$ as the Lewis acid, cyclic acetals are formed.

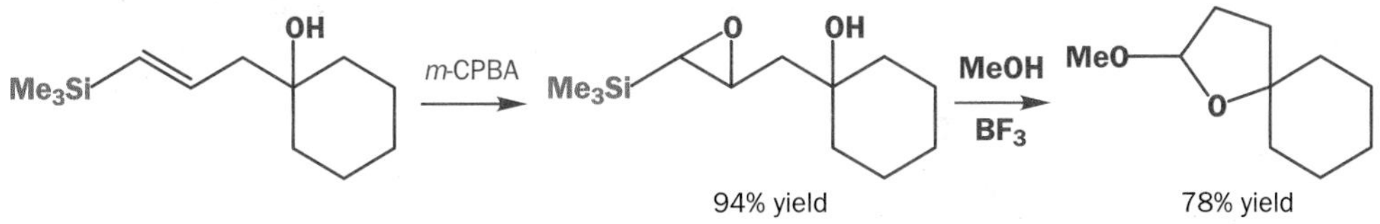

Nucleophilic attack occurs next to silicon and Peterson elimination gives an enol ether that cyclizes to the acetal under the acidic conditions.

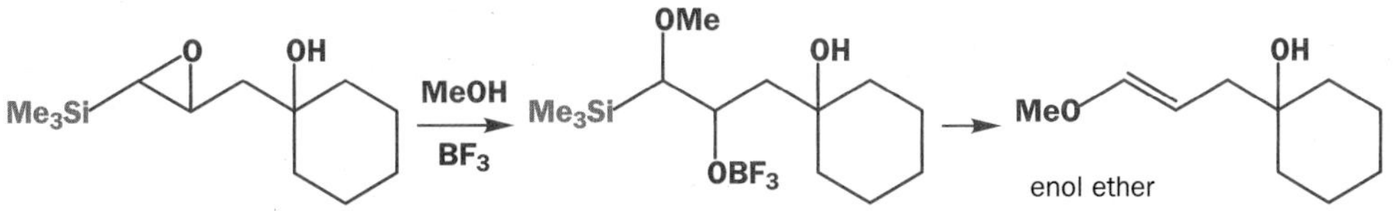

The cyclic acetal is a protected form of the hydroxy-aldehyde and oxidation under acidic conditions (CrO$_3$ in H$_2$SO$_4$) gives a good yield of the spirocyclic lactone. In the whole process from allyl silane to lactone, the allyl silane is behaving as a d^3 synthon or **homoenolate**.

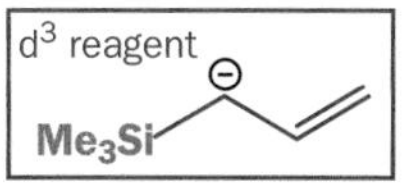

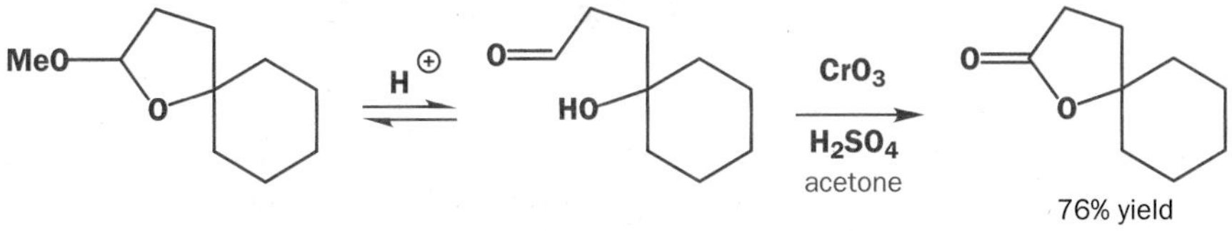

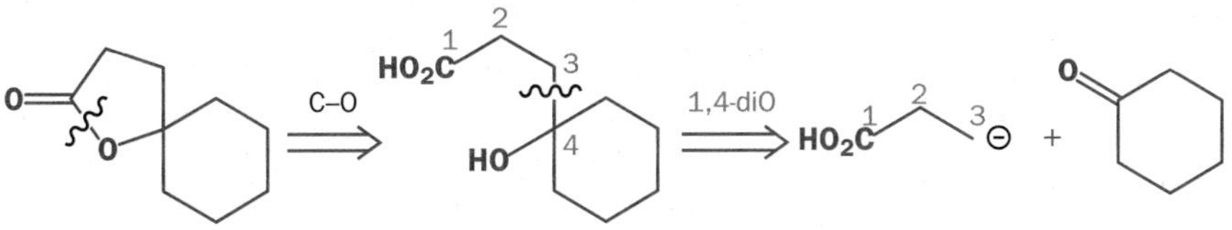

Migration of silicon from carbon to oxygen

Much of silicon chemistry is dominated by the strong Si–O bond and this leads to some surprising reactions. When compounds with an OH and a silyl group on the same carbon atom are treated with a catalytic amount of base, the silyl group migrates from carbon to oxygen. That all sounds reasonable until you realize that it must go through a three-membered ring. It is, in effect, a nucleophilic substitution at silicon. The reaction is known as the **Brook rearrangement**.

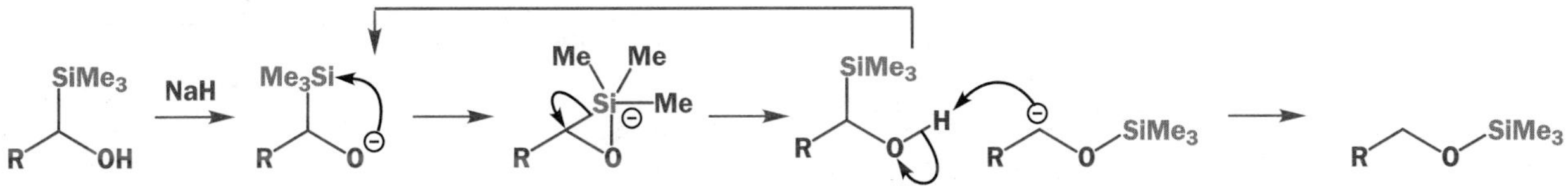

No such reaction could occur at a carbon centre (it would be impossible by Baldwin's rules; see Chapter 42), and the difference is that nucleophilic substitution at silicon goes through a pentacovalent intermediate so that a linear arrangement of nucleophile and leaving group is not required. The product anion is less stable than the oxyanion formed at the start of the reaction but removal of a proton from another molecule of starting material makes the product, with its Si–O bond, more stable than the starting material. The central reaction should really be shown as an equilibrium going to the right with catalytic base and to the left with a full equivalent of base.

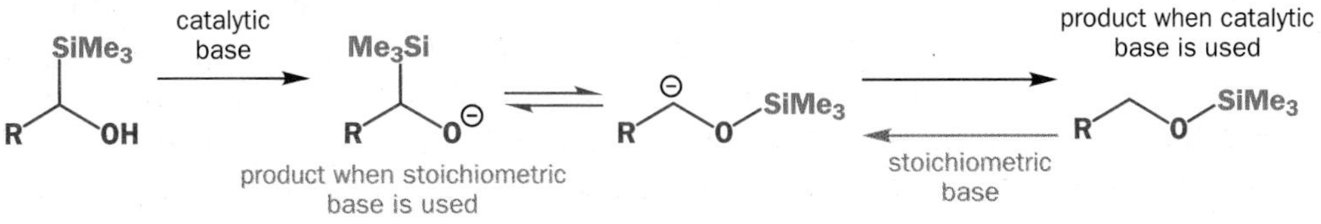

By itself, the Brook rearrangement is not very useful but, if the carbanion can do something else other than just get protonated, something useful may happen. We have seen what happens to the epoxides of vinyl silanes. Dihydroxylation of the same alkenes also gives interesting chemistry when the diols are treated with base.

The overall reaction is the insertion of an oxygen atom between the silicon and the alkene and the product is a useful silyl enol ether (Chapter 21). The Brook rearrangement takes place first but the carbanion has a leaving group (OH) on the neighbouring carbon atom so an E1cB reaction (Chapter 19) occurs next.

It is remarkable that the other OH group does not lose a proton because a Peterson reaction could then follow. Perhaps the three-membered cyclic intermediate is formed more easily than the four-membered ring. This would be the case if carbon were the electrophilic atom. Rearrangements from carbon to oxygen through four-membered rings do occur: examples are the 'sila-Pummerer' rearrangement and the rather annoying tendency of α-silyl carbonyl compounds to rearrange to silyl enol ethers. The **sila-Pummerer rearrangement** is like the normal Pummerer rearrangement (discussed in Chapter 46) except that a silyl group rather than a proton migrates to oxygen.

We could no doubt find uses for α-silyl carbonyl compounds if they did not rearrange with C to O silyl migration simply on heating. The mechanism is similar to that of the sila-Pummerer rearrangement except that the nucleophile that attacks the silicon atom via a four-membered ring intermediate is carbonyl oxygen rather than sulfoxide oxygen. The intermediate might remind you of the intermediate in the Wittig reaction: a C–Si or C–P bond is sacrificed in both cases in favour of an Si–O or a P–O bond.

These last examples show that there is some similarity between silicon and sulfur or phosphorus. Now we shall see similarities with an element further down group IV—tin.

Organotin compounds

Tin is quite correctly regarded as a metal but in the +4 oxidation state it forms perfectly stable organic compounds, known as stannanes, many of which are available commercially. The tin atom is rather large, which means that it forms long covalent bonds that are easily polarized. The table of important bond lengths of the group IV (14) elements C, Si, and Sn shows that all bonds to carbon are shorter than the corresponding ones to silicon, which are in turn shorter and, as a result, stronger than those to tin.

▶
The symbol Sn for tin warns us that there are two sets of names for tin compounds. Stannanes and stannyl are often used but so are tin and, for example, tributyltin hydride. You will meet both and there is no particular significance as to which is chosen.

common organotin compounds

Bu₃SnCl   Bu₃SnH   Me₄Sn

Organotin chemistry exploits the weakness of C–Sn bonds to deliver whatever is attached to the tin to another reagent. You have already seen (Chapter 39) tributyltin hydride used as a radical reducing agent because of the ease with which the Sn–H bond can be broken. Carbon substituents can be transferred by a radical mechanism too but organotins transfer the organic

Bonds to carbon, silicon, and tin compared

X	Bond length, nm					
	C–X	H–X	Cl–X	O–X	S–X	Sn–X
C	0.153	0.109	0.178	0.141	0.180	0.22
Si	0.189	0.148	0.205	0.163	0.214	
Sn	0.22	0.17	0.24	0.21	0.24	0.28

group intact by polar mechanisms as well. This reactivity is closest to that of a conventional organometallic reagent but the organotins are stable distillable liquids that can be stored unlike Grignard reagents. You may be concerned about the fact that there are four substituents on the central tin atom and, in principle, all of them could be transferred. In practice, alkyl groups transfer only very slowly indeed so that the tributylstannyl group (Bu$_3$Sn–), the most popular tin-based functional group, is generally transferred intact during reactions. The exception to this is tetramethyltin which has only methyl groups and therefore must transfer one of them. Methyl ketones may be made from tetramethyltin and acid chlorides. Contrast this with the inert NMR reference tetramethyl silane!

> Tin compounds are often volatile and are usually toxic, so beware! They were very effective in 'antifouling' paints for boats but they killed too many marine creatures and are now banned.

Organotin compounds are like reactive organosilicons

Organotin chemistry is useful because the familiar patterns of organosilicon chemistry are followed but the reactions proceed more easily because the bonds to tin are weaker and tin is more electropositive than silicon. Thus vinyl, allyl, and aryl stannanes react with electrophiles in exactly the same manner as their silicon counterparts but at a faster rate.

● **Organostannanes are more reactive than organosilanes and use the same mechanisms.**

The preparation of organostannanes is also similar to that of organosilanes. Organometallic reagents react with organotin electrophiles such as the trialkyl halides or bis(tributyltin) oxide. This is one method for the preparation of alkyl tributyltin using allyl Grignard and bis(tributyltin) oxide. Alternatively, the polarity can be reversed and a stannyl lithium, generated by deprotonation of the hydride or reductive cleavage of Me$_3$Sn–SnMe$_3$ with lithium metal, will add to organic electrophiles such as alkyl halides and conjugate acceptors. The first reaction is S$_N$2 at tin (probably with a 5-valent tin anion as intermediate) and the second is S$_N$2 at carbon.

Direct hydrostannylation of an alkyne with a tin hydride can be radical-initiated in the way we saw in Chapter 39. The product of kinetic control is the *Z*-isomer but, if there is excess tin hydride or enough radicals are present, isomerization into the more stable *E*-isomer occurs. The regiocontrol of this process is good with terminal alkynes.

Addition of a tributyltin radical to the alkyne gives the more substituted linear (sp) vinyl radical (see Chapter 39). Addition of a hydrogen atom from another molecule of Bu$_3$SnH occurs preferentially from the less hindered side (the Bu$_3$Sn group already in the molecule is in the plane of the p orbital containing the unpaired electron) to give the Z-vinyl stannane. If there is more Bu$_3$SnH around, reversible addition of Bu$_3$Sn• radicals to either end of the vinyl stannane equilibrates it to the more stable E-isomer.

Tin–lithium exchange is rapid

Organotin compounds are usually simply not reactive enough to be useful nucleophiles. Conversion into the corresponding organolithiums provides a much more reactive reagent. This is achieved in the same way as lithium–halogen exchange described in Chapter 9 and has essentially the same mechanism. The principle is simple. A very reactive nucleophile such as butyl lithium reacts at the tin and expels an organolithium species. The process is thermodynamically controlled, so the more stable the organolithium, the more likely it is to form. By having three of the groups on tin as butyl and adding another butyl from the organolithium, the choice is between the re-formation of butyl lithium or creation of an organolithium from the fourth substituent. If this is a vinyl, allyl, aryl, or alkynyl group this emerges as the most stable organolithium and is produced without any lithium halide present. The by-product is tetrabutyltin which is nonpolar and unreactive and can usually be separated by chromatography from the product of the reaction.

■
See Chapters 10 and 23 for a discussion of direct versus conjugate addition.

Such a tin–lithium exchange was the key to the preparation of a functionalized vinyl organolithium that was coupled to an enone in a synthesis of a natural product. Direct addition of the cyclobutenyllithium to the less hindered face of the carbonyl group gave one diastereoisomer of the product.

Crotyl stannanes react with good stereochemical control

Crotyl stannanes are important reagents in organic synthesis because they can be prepared with control over the double bond geometry and will tolerate the presence of additional functional groups. This allows stereoselective synthesis of functionalized acyclic molecules. The control arises from the well-defined transition states for the crotylation reaction. Tin is more electropositive than silicon and can accept a lone pair of electrons in a purely thermal reaction with no added Lewis acid. The carbonyl group of the aldehyde can coordinate to the tin and lead, through a cyclic transition state, to give *anti* products from E-crotyl tin reagents and *syn* products from the Z-crotyl isomer.

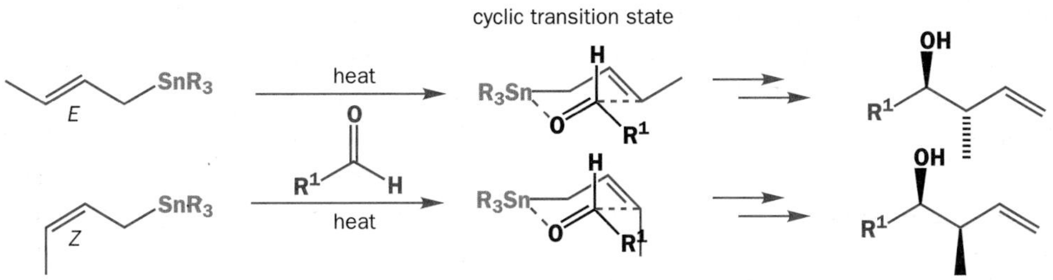

cyclic transition state

Tin–lithium exchange in action

Many organolithium compounds are useful reagents and no doubt many more would be if only they could be made. Tin chemistry allows us to make organolithium compounds that cannot be made by direct lithiation.

An excellent example is a lithium derivative with an oxygen atom on the same carbon. The hydrogen atom is not particularly acidic and cannot be removed by BuLi, while the bromide is unstable and will not survive treatment with BuLi.

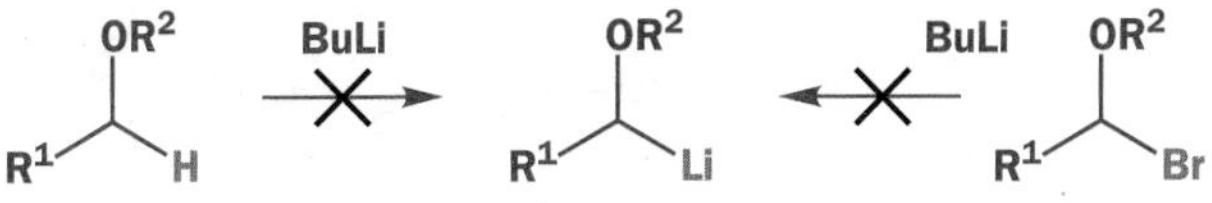

> ● **Organolithium preparations**
>
> - Tin/lithium exchange occurs rapidly and stereospecifically with BuLi
> - Other elements that can be replaced by Li: RX + BuLi gives RLi when $X = SnR_3$, Br, I, SeR

However, the problem should be easily solved with tin chemistry. The idea is to add a tributyltin lithium reagent to the aldehyde, mask the alkoxide formed, and then exchange the tributyl tin group for lithium.

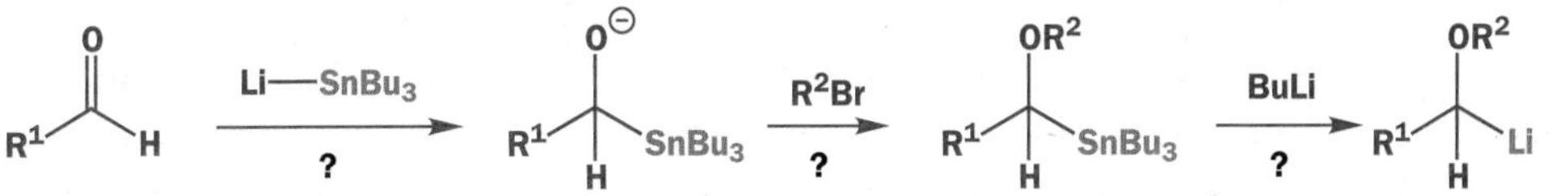

First, the Bu_3Sn–Li reagent has to be made. This can be done in two ways. Treatment of any tin compound with BuLi results in nucleophilic attack at tin but LDA is much less nucleophilic and can be used to remove a proton from tributyltin hydride. Otherwise, we can accept that BuLi will always attack tin and provide two tin atoms so that nucleophilic attack on one expels the other as the lithium derivative.

These THF solutions of Bu_3Sn–Li are stable only at low temperatures so the aldehyde must be added immediately. The lithium alkoxide adduct can be neutralized and the alcohol isolated but it is also unstable and must be quenched immediately with an alkyl halide. The preferred one is ethoxyethyl chloride, which reacts with base catalysis.

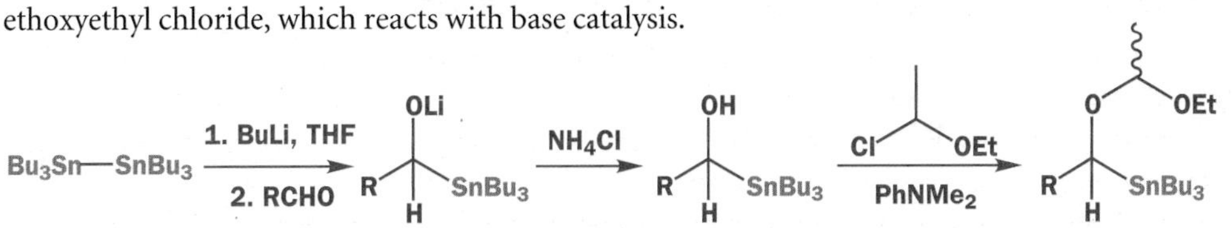

These protected hydroxystannanes are stable compounds and can even be distilled. Treatment with BuLi and an electrophile such as an aldehyde or ketone gives the product from addition of the organolithium derivative to the carbonyl group. Tin–lithium exchange is rapid even at low temperature and no products from addition of BuLi to the carbonyl group are seen.

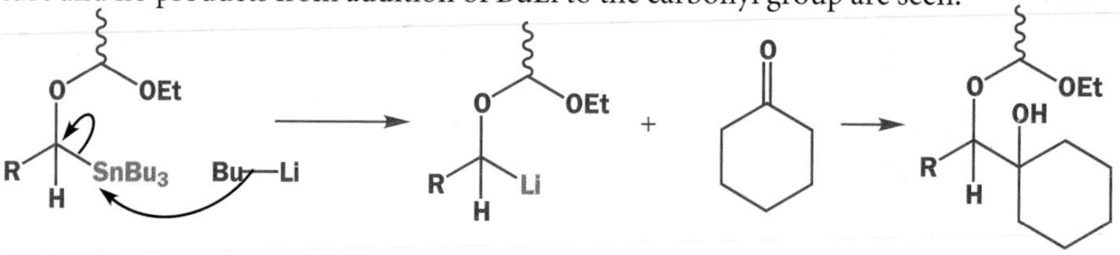

The most surprising thing about these reagents, invented and exploited by W. Clark Still at Columbia University, is that they can be prepared in stable enantiomerically pure forms and that the stereochemistry is preserved through exchange with lithium and reaction with electrophiles. It is very unusual for organolithium compounds to be configurationally stable. Still first quenched the Bu₃SnLi adducts with one enantiomer of an acid chloride and resolved by separating the diastereoisomers.

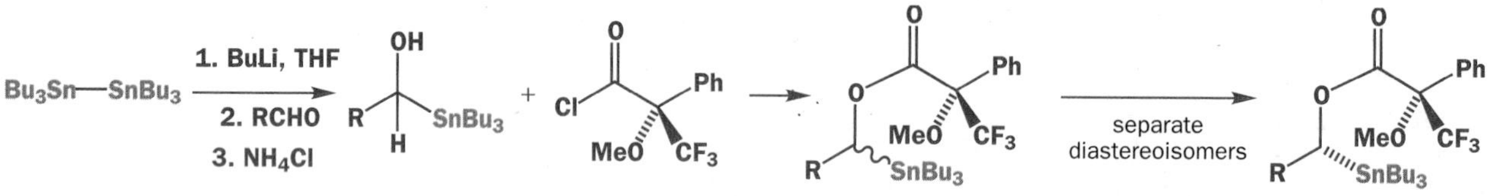

You may recognize this acid as 'Mosher's acid' which we introduced in Chapter 32 as a way of determining enantiomeric excess by NMR.

The ester was cleaved by reduction with DIBAL (*i*-Bu₂AlH) and an achiral version of the normal protecting group put in place. It would obviously be silly to create unnecessary diastereomeric mixtures in these reactions. Then the tin could be exchanged first with lithium and then with an electrophile, even an alkyl halide, with retention of configuration and without loss of enantiomeric purity. The intermediate organolithium compound must have had a stable configuration.

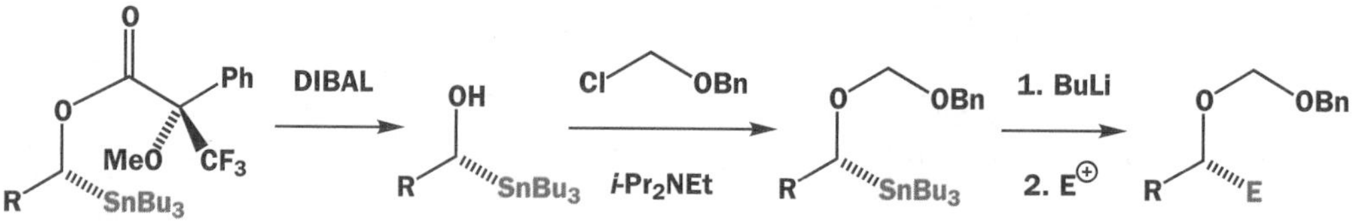

The exchange of tin for lithium or other metals is probably the most valuable job it does. Reagents such as BuLi attack tin or boron directly rather than removing a proton. Silicon is not usually attacked in this way and proton removal is more common. In the next chapter we shall see how transition metals open up a treasure chest of more exotic reactions for which the reactions in this chapter are a preparation.

Problems

1. The Hammett ρ value for the following reaction is –4.8. Explain this in terms of a mechanism. If the reaction were carried out in deuterated solvent, would the rate change and would there be any deuterium incorporation into the product? What is the silicon-containing product?

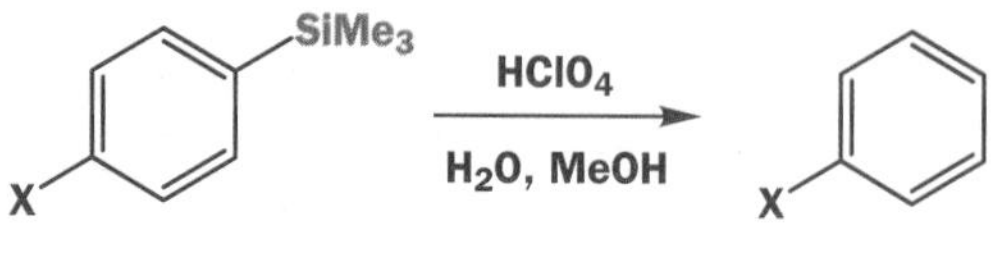

2. Identify the intermediates in this reaction sequence and draw mechanisms for the reactions, explaining the special role of the Me₃Si group.

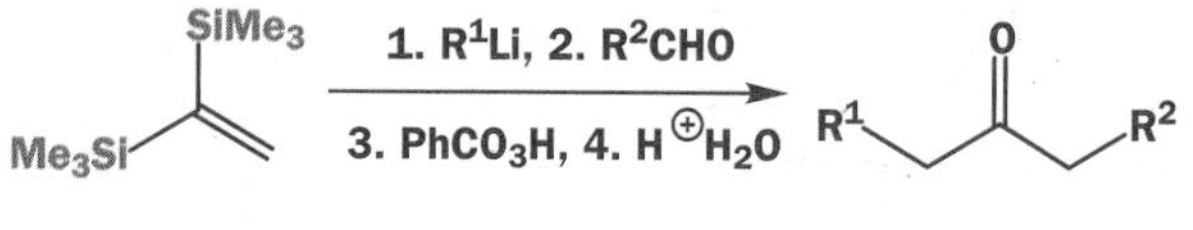

3. The synthesis of a compound used in a problem in Chapter 38 (fragmentation) is given below. Give mechanisms for the reactions explaining the role of silicon.

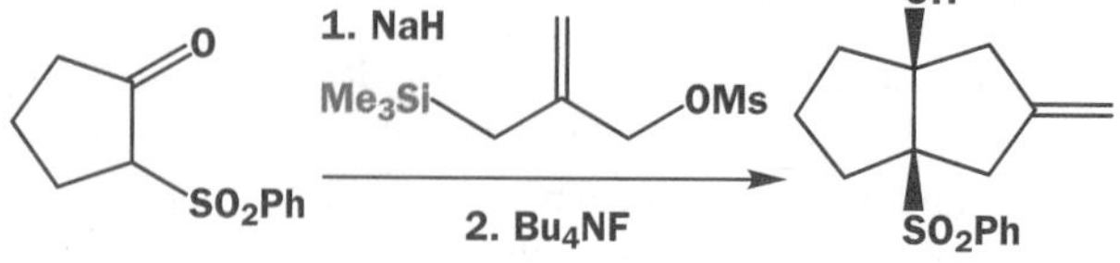

4. Give mechanisms for the following reactions, drawing structures for all the intermediates including stereochemistry. How would the reaction with Bu₃SnH have to be done?

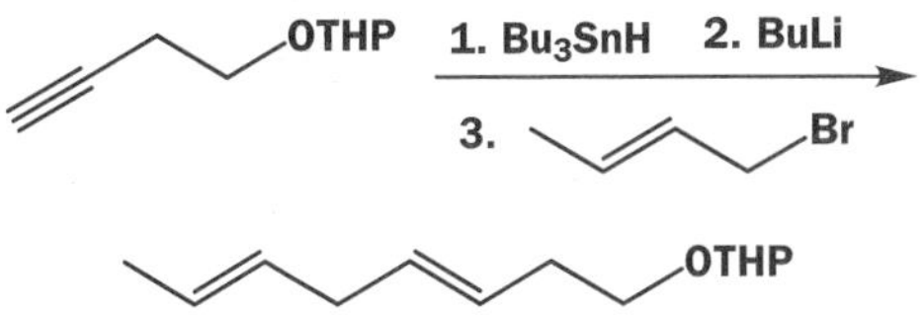

5. Explain the following reactions. In particular, explain the role of tin and why it is necessary and discuss the stereochemistry.

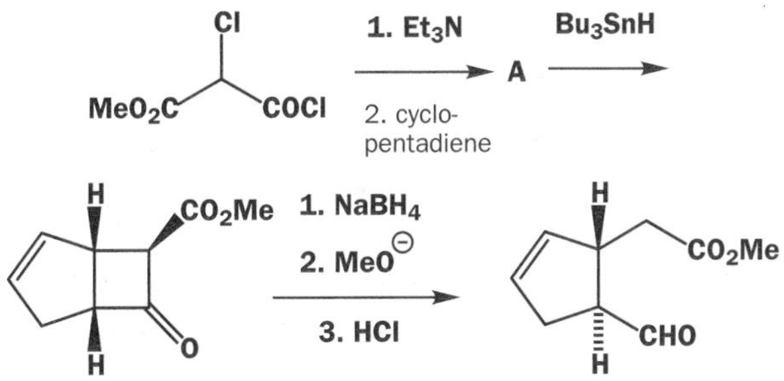

6. Explain the stereochemistry and mechanism of this hydroboration–carbonylation sequence.

7. Revision content. Give mechanisms for these reactions, commenting on the role of silicon and the stereochemistry of the cyclization. The LiAlH₄ simply reduces the ketone to the corresponding alcohol. If you have trouble with the Hg(II)-catalysed step, there is help in Chapter 36.

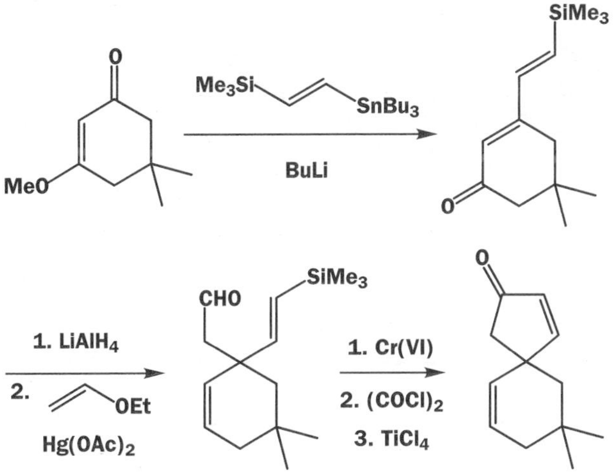

8. Give mechanisms for these reactions explaining: (a) the regio- and stereoselectivity of the hydroboration; (b) why such an odd method was used to close the lactone ring.

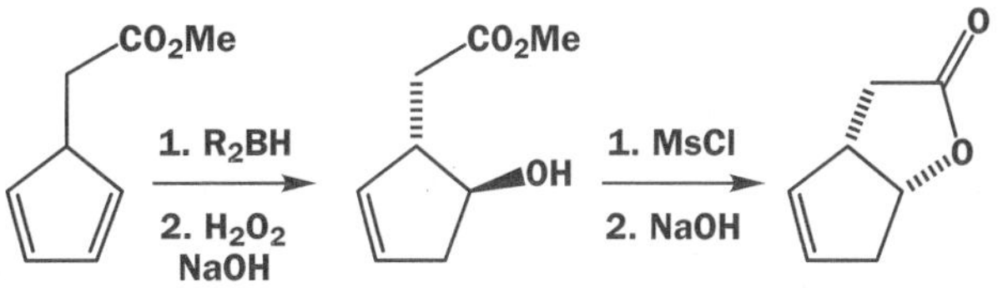

9. Give mechanisms for these reactions, explaining the role of silicon. Why is this type of lactone difficult to make by ordinary acid- or base-catalysed reactions?

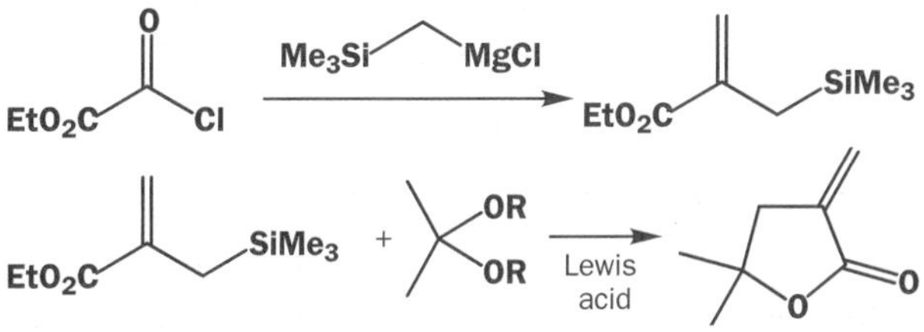

10. Revision of Chapters 38 and 46. How would you prepare the starting material for these reactions? Give mechanisms for the various steps. Why are these sequences useful?

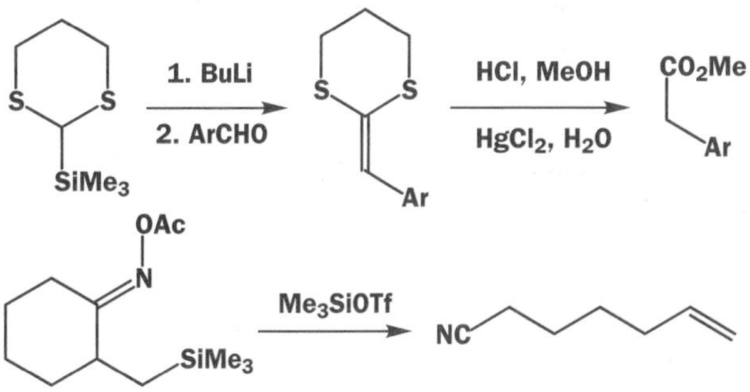

11. How would you carry out the first step in this sequence? Give a mechanism for the second step and suggest an explanation for the stereochemistry. You may find that a Newman projection (Chapters 32 and 33) helps.

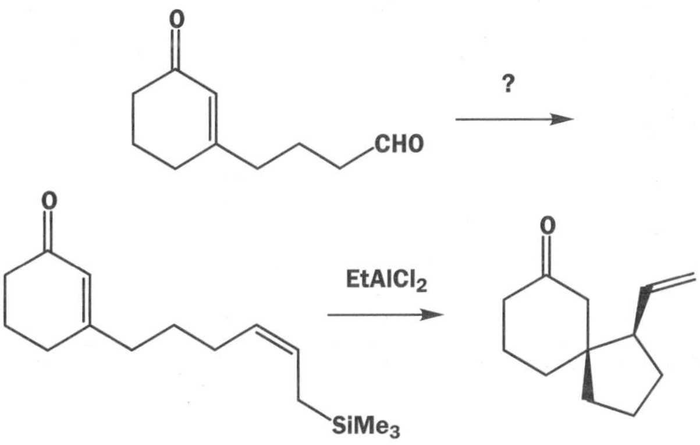

12. Revision of Chapter 36. Give a mechanism for this reaction and explain why it goes in this direction.

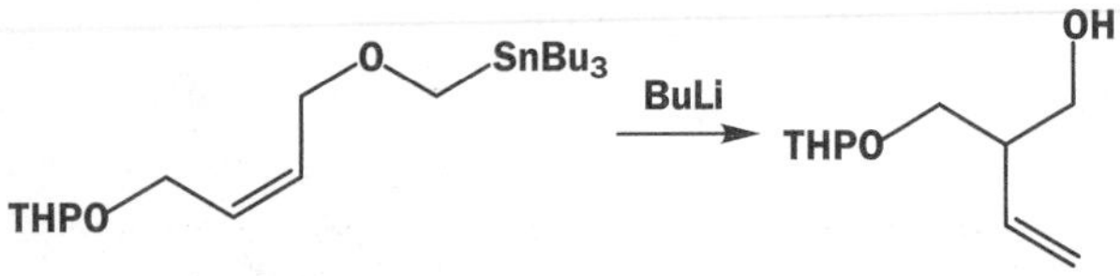

13. The Nazarov cyclization (Chapter 36) normally gives a cyclo-pentenone with the alkene in the more substituted position. That can be altered by the following sequence. Give a mechanism for the reaction and explain why the silicon makes all the difference.

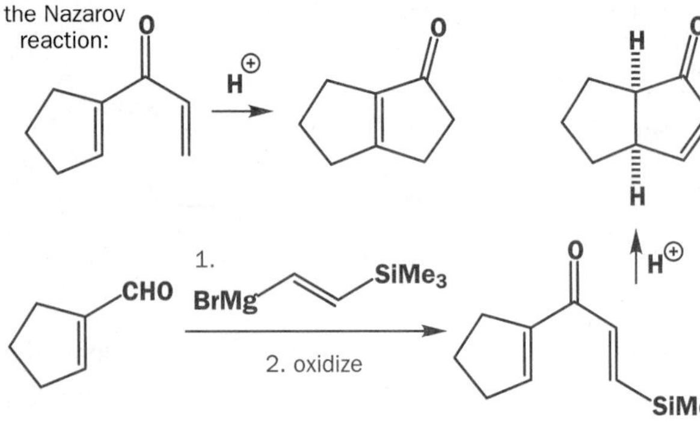

the Nazarov reaction:

14. This is rather a long problem but it gives you the chance to see an advanced piece of chemistry involving several elements—P, Si, Sn, Mg, B, Ni, Cr, Os, and Li—and it revises material from Chapters 23, 33, and 45 at least. It starts with the synthesis of this phosphorus compound: what is the mechanism and selectivity?

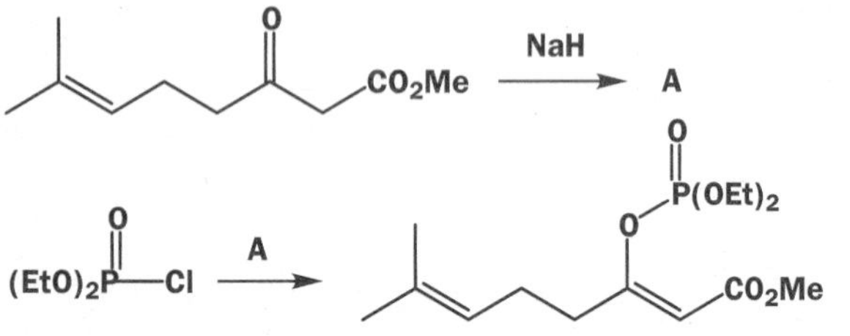

Next, reaction with a silicon-substituted Grignard reagent in the presence of Ni(II) gives an allyl silane. What kind of reaction is this, what was the role of phosphorus, and why was a metal other than sodium added? (You know nothing specific about Ni as yet but you should see the comparison with another metal. Consult Chapter 23 if you need help.)

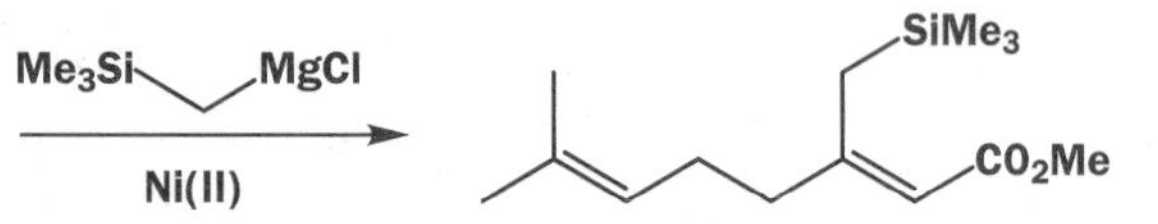

Asymmetric dihydroxylation (Chapter 45) is straightforward though you might like to comment on the chemoselectivity. The diol is converted into the epoxide and you should explain the regio- and chemoselectivity of this step. The next step is perhaps the most interesting: what is the mechanism of the cyclization, what is the role of silicon, and how is the stereochemistry controlled?

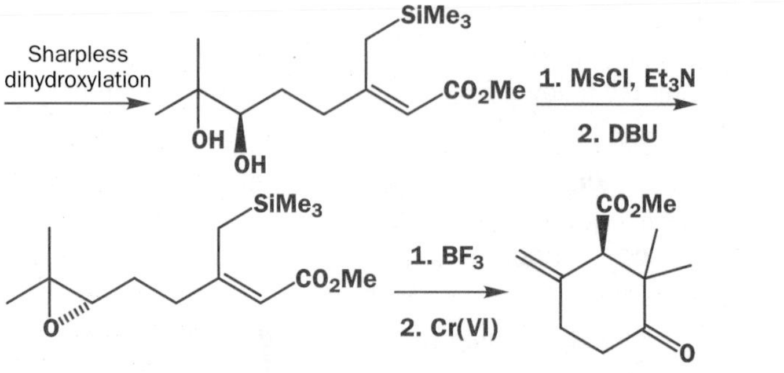

Reaction of this ketone with a stannyl-lithium reagent gives one diastereoisomer of a bridged lactone. Again, give a mechanism for this step and explain the stereochemistry. Make a good conformational drawing of the lactone.

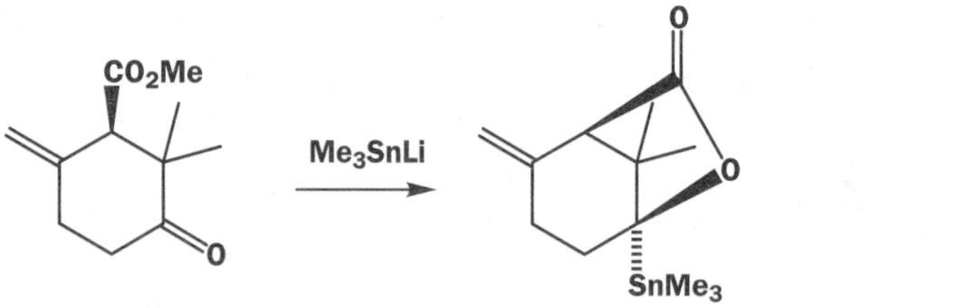

Treatment of the tin compound with MeLi and a complex aldehyde represented as RCHO gave an adduct that was used in the synthesis of some compounds related to Taxol™. What is the mechanism of the reaction, and why is tin necessary?

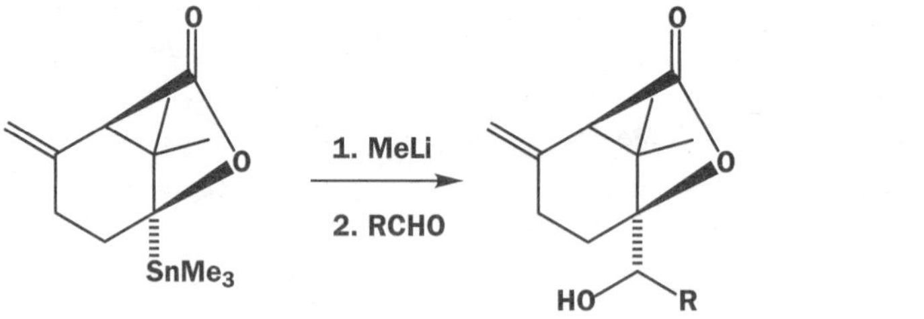

Organometallic chemistry

48

Connections

Building on:

- Conjugate addition ch10 & ch23
- Nucleophilic substitution at saturated carbon ch17
- Controlling stereochemistry ch16, ch33, & ch34
- S$_N$2 and S$_N$2' ch23
- Oxidation and reduction ch24
- Cycloadditions ch35
- Rearrangements ch36–ch37
- Radicals and carbenes ch39–ch40
- Aromatic heterocycles ch43–ch44
- Asymmetric synthesis ch45
- Chemistry of B, Si, and Sn ch47

Arriving at:

- Transition metals form organic compounds
- There are σ- and π-complexes given 'η' numbers
- The bonding is described with the usual orbitals
- Most stable complexes have 18 valency electrons
- Metals catalyse 'impossible' reactions
- Oxidative insertion, reductive elimination, and ligand migration from metal to carbon are key steps
- Carbon monoxide inserts into metal–carbon bonds
- Palladium is the most important metal
- C–C, C–O, and C–N bonds can be made with Pd catalysis
- Cross-coupling of two ligands is common
- Allyl cation complexes are useful electrophiles

Looking forward to:

- The chemistry of life, especially nucleic acids ch49
- Steroids ch51
- Polymerization ch52

Transition metals extend the range of organic reactions

Some of the most exciting reactions in organic chemistry are based on transition metals. How about these two for example? The first is the **Heck reaction**, which allows nucleophilic addition to an unactivated alkene. Catalytic palladium (Pd) is needed to make the reaction go. The second, the **Pauson–Khand reaction**, is a special method of making five-membered rings from three components: an alkene, an alkyne, and carbon monoxide (CO). It requires cobalt (Co). Neither of these reactions is possible without the metal.

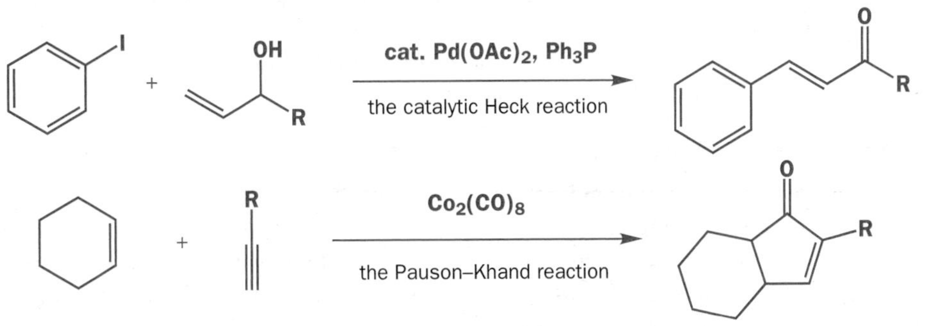

the catalytic Heck reaction

the Pauson–Khand reaction

Reagents and complexes containing transition metals are important in modern organic synthesis because they allow apparently impossible reactions to occur easily. This chemistry complements traditional functional-group-based chemistry and significantly broadens the scope of organic chemistry. This chapter introduces the concepts of metal–ligand interaction, describes the most important reactions that can occur while ligands are bound to the metal, and demonstrates the power of organometallic chemistry in synthesis. Many industries now use transition-metal-catalysed reactions routinely so it is important that you have a basic grounding in what they do.

There is a contradiction in what is required of a metal complex for useful synthetic behaviour. Initially, it is useful to have a stable complex that will have a significant lifetime enabling study and, ideally, storage but, once in the reaction vessel, stability is actually a disadvantage as it implies slow reactivity. An ideal catalyst is a complex that is stable in the resting state, for storage, but quickly becomes activated in solution, perhaps by loss of a ligand, allowing interaction with the substrate. Fortunately, there is a simple guide to the stability of transition metal complexes. If a complex satisfies the 18-electron rule for a stable metal complex it means that the metal at the centre of the complex has the noble gas configuration of 18 electrons in the valence shells. The total of 18 is achieved by combining the electrons that the metal already possesses with those donated by the coordinating ligands. The requirement for 18 electrons comes from the need to fill one 's' orbital, five 'd' orbitals, and three 'p' orbitals with two electrons in each. This table gives you the number of valence electrons each metal starts with before it has acquired any ligands. Notice that the 'new' group numbers 1–18 give you the answer without any calculation. The most important are highlighted.

Group	IVB (4)	VB (5)	VIB (6)	VIIB (7)	VIIIB (8, 9, and 10)			1A (11)
Number of valence electrons	4	5	6	7	8	9	10	11
3d	**Ti**	V	**Cr**	**Mn**	**Fe**	**Co**	**Ni**	**Cu**
4d	**Zr**	Nb	**Mo**	Tc	**Ru**	**Rh**	**Pd**	Ag
5d	Hf	Ta	**W**	Re	**Os**	Ir	**Pt**	Au

Metals to the left-hand side of this list obviously need many more electrons to make up the magic 18. Chromium, for example, forms stable complexes with a benzene ring, giving it six electrons, and three molecules of carbon monoxide, giving it two each: $6 + 6 + 2 + 2 + 2 = 18$. Palladium is happy with just four triphenylphosphines ($Ph_3P:$) giving it two each: $10 + 2 + 2 + 2 + 2 = 18$.

You may already know from your inorganic studies that there are exceptions to the 18-electron rule including complexes of Ti, Zr, Ni, Pd, and Pt, which all form stable 16-electron complexes. An important 16-electron Pd(II) complex with two chlorides and two acetonitriles (MeCN) as ligands appears in the margin. The so-called platinum metals Ni, Pd, and Pt are extremely important in catalytic processes, as you will see later on. The stable 16-electron configuration results from a high-energy vacant orbital caused by the complex adopting a square planar geometry. The benefit of this vacant orbital is that it is a site for other ligands in catalytic reactions.

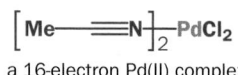

a 16-electron Pd(II) complex

Ligands can be attached in many different ways

Transition metals can have a number of ligands attached to them and each ligand can be attached in more than one place. This affects the reactivity of the ligand and the metal because each additional point of attachment means the donation of more electrons. We usually show the number of atoms involved in bonding to the metal by the **hapto number** η. A simple Grignard reagent is η^1 (pronounced 'eta-one') as the magnesium is attached only to one carbon atom. A metal–alkene complex is η^2 because both carbon atoms of the alkene are equally involved in bonding to the metal. In these cases the η designation is not very useful as there are no alternatives and it is usually omitted.

M━R

η^1

M━║

η^2

Representing bonds in transition metal complexes

It is difficult to know exactly how to draw the bonding in metal complexes and there are often several different acceptable representations. There is no problem when the metal forms a σ bond to atoms such as Cl or C as the simple line we normally use for covalent bonds means exactly what it says. The problems arise with ligands that form σ bonds by donating both their electrons and with π complexes. Everyone writes phosphine–boron compounds with two charges but we normally draw the same sort of bond between a phosphine and, say, Pd as a simple line with no charges.

You will sometimes see π complexes drawn with simpler dotted lines going to the middle of the π bond, sometimes with dotted π bonds, and sometimes with bonds (simple or dotted) going to the ends of the old π bond. These are all acceptable as the bonding is complex as you will see. We might almost say that the ambiguity is helpful: we often don't know either the exact nature of the bonding or the number of other ligands in the complex. In the diagrams in this section we have shown the main bond from metal to ligand as a heavy line in the simplest representation but we also offer alternatives with simple and dotted bonds. Don't worry about this—things should become clearer as the chapter develops. When you have to draw the structure of a complex but you don't know the exact bonding, just draw a line from metal to ligand.

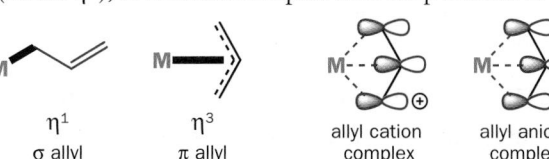

different acceptable ways to draw π complexes

The bonding in these two complexes is very different. In the first there is a simple σ bond between the metal and the alkyl group as in a Grignard reagent R–MgBr and this type of complex is called a σ **complex**. In the alkene complex, bonding is to the p orbitals only. There are no σ bonds to the metal, which sits in the middle of the π bond in between the two p orbitals. This type of complex is called a π **complex**.

These labels are useful where there is a choice of type of bonding as with allylic ligands. The metal can either form a σ bond to a single carbon (hence η^1), or form a π complex with the p orbitals of all three carbons of the allyl system and this would be η^3. If the π complex is made from an allyl cation, the ligand has two electrons, but it has four if it is made from an allyl anion.

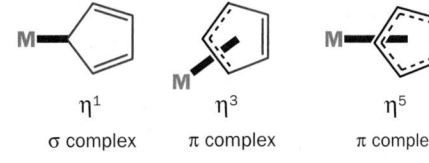

η^1
σ allyl

η^3
π allyl

allyl cation complex

allyl anion complex

Similarly, cyclopentadienyl anion can act as a σ ligand (η^1), an allyl ligand (η^3), or, most usually, as a cyclopentadienyl ligand (η^5). The distinction is very important for electron counting as these three different situations contribute 2, 4, or 6 electrons, respectively, to the complex.

η^1
σ complex

η^3
π complex

η^5
π complex

Neutral ligands can also bond in a variety of ways. Cyclooctatetraene can act as an alkene (η^2), a diene (η^4), a triene (η^6), or a tetraene (η^8), and the reactivity of the ligand changes accordingly. These are all π complexes with the metal above or below the black portion of the ring and with the thick bond to the metal at right angles to the alkene plane.

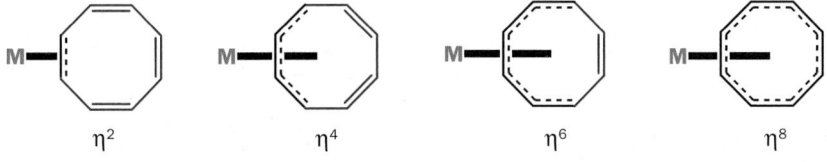

η^2 η^4 η^6 η^8

To determine the number of electrons around the transition metal in a complex the valence electrons from the metal ion are added to those contributed by all the ligands. The numbers of electrons donated by various classes of ligands are summarized in the table. Anions such as halides, cyanide, alkoxide, hydride, and alkyl donate two electrons, as do neutral ligands with a lone pair such as phosphines, amines, ethers, sulfides, carbon monoxide, nitriles, and isonitriles. Unsaturated ligands can contribute as many as eight electrons and can be neutral or negatively charged. If the overall total is eighteen, then the complex is likely to be stable.

Ligand characteristics

anionic ligands						Formal charge	Electrons donated
Cl⁻	Br⁻	I⁻	⁻CN	⁻OR	⁻H ⁻alkyl	–1	2

neutral σ-donor ligands						Formal charge	Electrons donated
R₃P	R₃N	R₂O	R₂S	C=O	C≡N–R C≡N–R	0	2

unsaturated σ- or π-donor ligands	Hapto number	Formal charge	Electrons donated
aryl, σ-allyl	η^1	–1	2
olefins	η^2	0	2
π-allyl cation	η^3	+1	2
π-allyl anion	η^3	–1	4
diene—conjugated	η^4	0	4
dienyls, cyclopentadienyls (anions)	η^5	–1	6
arenes, trienes	η^6	0	6
trienyls, cycloheptatrienyls (anions)	η^7	–1	8
cyclooctatetraene	η^8	0	8
carbene, nitrene, oxo	η^1	0	2

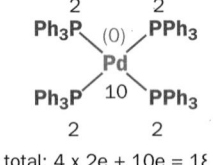

total: 4 x 2e + 10e = 18

►

Note that $(Ph_3P)_4Pd$ is a stable complex and is not a useful catalyst until at least one of the ligands is lost.

$\eta^5 = 6$ (–1)

Fe ◄ - - - 6   Fe(II)

$\eta^5 = 6$ (–1)

total: 3 x 6e = 18e

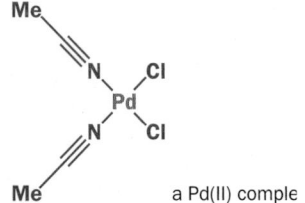

a Pd(II) complex

Electron counting helps to explain the stability of metal complexes

Counting electrons in most complexes is simple if you use the table of ligand characteristics above and the table on p. 1312. Tetrakistriphenylphosphine palladium(0) is an important catalyst as you will see later in the chapter. Each neutral phosphine donates two electrons making a total of eight and palladium still has its full complement of ten valence electrons as it is in the zero oxidation state. Overall, the complex has a total of eighteen electrons and is a stable complex. In the diagrams that follow, the formal charges are highlighted in green and the numbers of electrons contributed shown in black.

All of the different classes of ligands listed in the table can be treated in this way. The cyclopentadienyl ligands contribute six electrons each and have a formal negative charge, shown in green, which means that the iron in ferrocene is in the +2 oxidation state and will have six valence electrons left. The total for the complex is again eighteen and ferrocene is an extremely stable complex.

The oxidation state of metals in complexes

As well as the problem of bond drawing, there is a potential problem over oxidation states too. You can either say that ferrocene is a complex of Fe(II), having two fewer electrons than the normal eight, with two cyclopentadienyl anions contributing six electrons each, or you can say that it is a complex of Fe(0), having eight electrons, with two cyclopentadienyl ligands each contributing five electrons. The simplest approach is to say that a metal is in the (0) oxidation state unless it has σ bonds to ligands such as Cl, AcO, or Me that form bonds with shared electrons. You do not count neutral ligands such as Ph_3P that provide two of their own electrons. Grignard reagents RMgBr have two ligands that share electrons (R and Br) and a number of others, probably two ethers, that donate both their electrons. Magnesium is in the +2 oxidation state.

The useful complex $(MeCN)_2PdCl_2$ has palladium in the +2 oxidation state because of its two chlorine atoms and the number of electrons is 8 for the Pd(II) oxidation state and another two each from the four ligands making 16 in all. This complex does not fulfil the 18-electron rule and is reactive. You would have got the same answer if you had counted ten for the palladium, two each for the nitriles, and one each for the chlorines, but this is not so realistic.

Transition metal complexes exhibit special bonding

The majority of ligands have a lone pair of electrons in a filled sp^n type orbital that can overlap with a vacant metal 'dsp' orbital, derived from the vacant d, p, and s orbitals of the metal, to form a conventional two-electron two-centre σ bond. Ligands of this type increase the electron density on the central metal atom. This is the sort of bond that used to be called 'dative covalent' and represented by an arrow. Nowadays it is more common to represent all bonding to metals of whatever kind by simple lines.

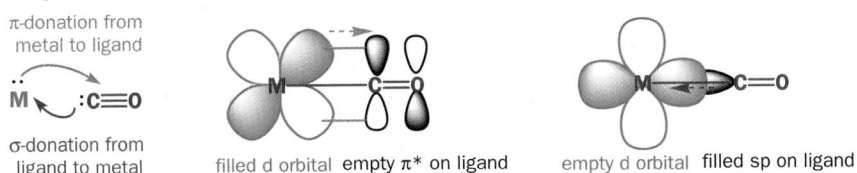

When a bonding interaction is also possible between any filled d orbitals on the metal and vacant ligand orbitals of appropriate symmetry such as π* orbitals. This leads to a reduction of electron density on the metal and is known as **back-bonding**. An example would be a complex with carbon monoxide. Many metals form these complexes and they are known as **metal carbonyls**. The ligand (CO) donates the lone pair on carbon into an empty orbital on the metal while the metal donates electrons into the low-energy π* orbital of CO. Direct evidence for this back-bonding is an increase in the C–O bond length and a lowering of the infrared stretching frequency from the population of the π* orbital of the carbonyl.

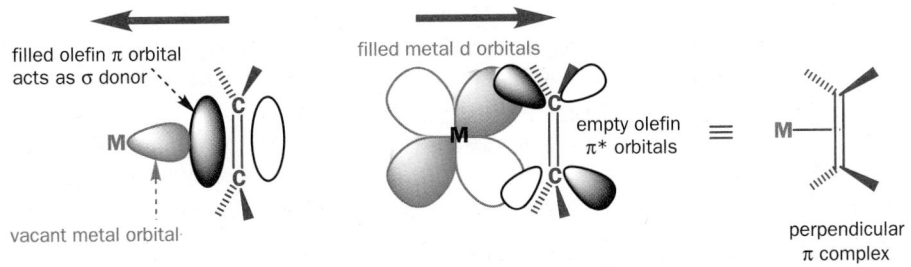

When an unsaturated ligand such as an alkene approaches the metal sideways to form a π complex, similar interactions lead to bonding. The filled π orbitals of the ligand bond to empty d orbitals of the metal, while filled d orbitals on the metal bond to the empty π* orbitals of the ligand. The result is a π complex with the metal–alkene bond perpendicular to the plane of the alkene. The bond has both σ and π character.

Coordination to a metal by any of these bonding methods changes the reactivity of the ligands dramatically and this is exploited in the organometallic chemistry we will be discussing in the rest of the chapter. You do not need to understand all the bonding properties of metal complexes but you need to be able to count electrons, to recognize both σ and π complexes, and to realize that complexes show a balance between electron donation and electron withdrawal by the metal.

Oxidative addition inserts metal atoms into single bonds

Potential ligands that do not have a lone pair or filled π type orbital are still able to interact with transition metal complexes but only by breaking a σ bond. This is the first step in a wide variety of processes and is described as **oxidative addition** because the formal oxidation state of the transition metal is raised by two, for example, M(0) to M(II), in the process. This is the result of having two extra ligands bearing a formal negative charge. You have seen this process in the formation of Grignard reagents (Chapter 9)

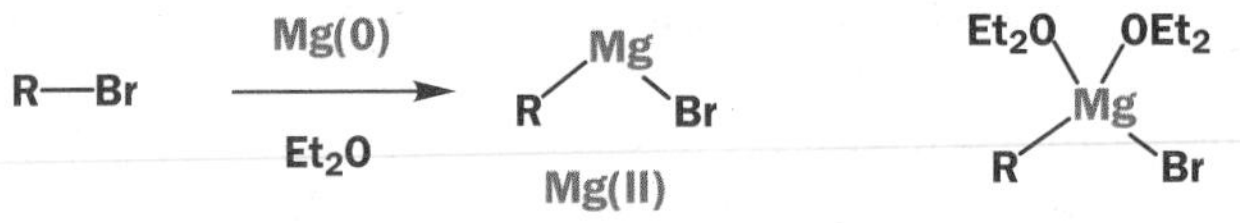

The number of coordinated ligands also increases by two so the starting complex is usually in low oxidation state (0 or 1; the diagram shows 0) and **coordinatively unsaturated**, that is, it has an empty site for a ligand and, say, only 16 electrons, like $(MeCN)_2PdCl_2$, whereas the product is usually **coordinatively saturated**, that is, it cannot accept another ligand unless it loses one first.

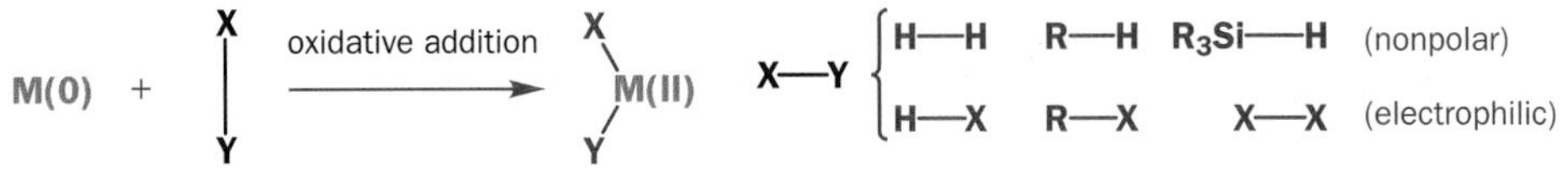

introduces new organic ligands on to metal

Oxidative addition occurs for a number of useful neutral species including hydrogen, carbon–hydrogen bonds, and silanes as well as polarized bonds containing at least one electronegative atom. The resulting species with metal–ligand bonds allow useful chemical transformations to occur. Important examples include the oxidative addition of Pd(0) to aryl iodides and the activation of Wilkinson's catalyst for hydrogenation in solution by oxidative addition to a hydrogen molecule.

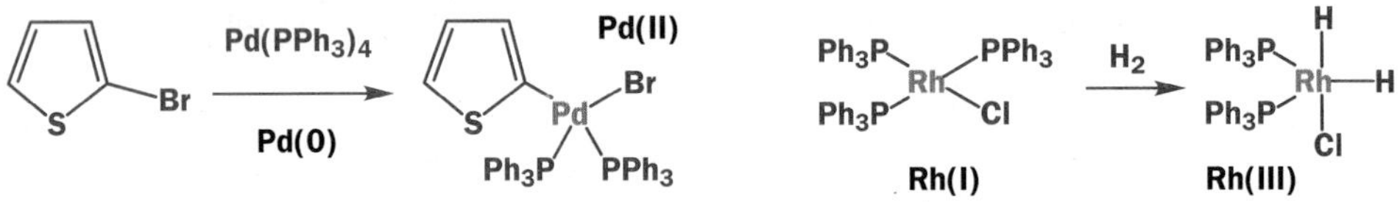

Vaska's complex

There are a number of possible mechanisms for oxidative addition and the precise one followed depends on the nature of the reacting partners. Vaska's complex $[Ir(PPh_3)_2COCl]$ has been extensively studied and it reacts differently with hydrogen and methyl iodide. Hydrogen is added in a *cis* fashion, consistent with concerted formation of the two new iridium–hydrogen bonds. The

16e, d^8, Ir(I) complex becomes a new 18e, d^6, Ir(III) species. With methyl iodide the kinetic product is that of *trans* addition, which is geometrically impossible from a concerted process. Instead, an S_N2-like mechanism is followed involving nucleophilic displacement of iodide followed by ionic recombination.

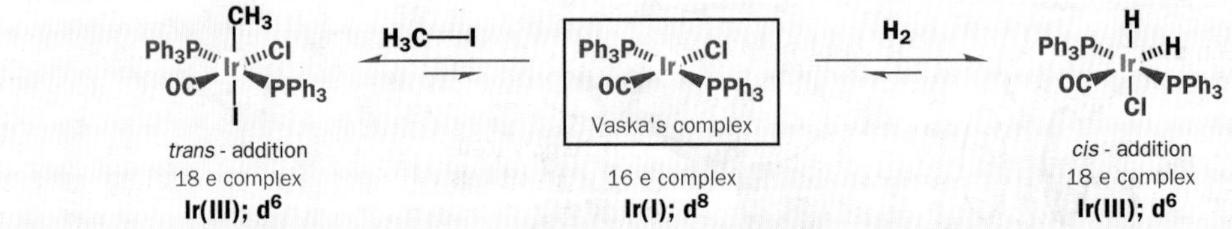

trans - addition	Vaska's complex	*cis* - addition
18 e complex	16 e complex	18 e complex
Ir(III); d^6	**Ir(I); d^8**	**Ir(III); d^6**

Reductive elimination removes metal atoms and forms new single bonds

If we want to use organometallic chemistry to make organic compounds other than those containing metals, we must be able to remove the ligands from the coordination sphere of the metal at the end of the reaction. Neutral organic species such as alkenes, phosphines, and carbon monoxide can simply dissociate in the presence of other suitable ligands but those that are bound to the metal with shared electrons require a more active process. Fortunately, most reactions that occur around a transition metal are reversible and so the reverse of oxidative addition, known as **reductive elimination**, provides a simple route for the release of neutral organic products from a complex. Our general reaction shows M(II) going to M(0) releasing X–Y. These two ligands were separate in the complex but are bound together in the product. A new X–Y σ bond has been formed.

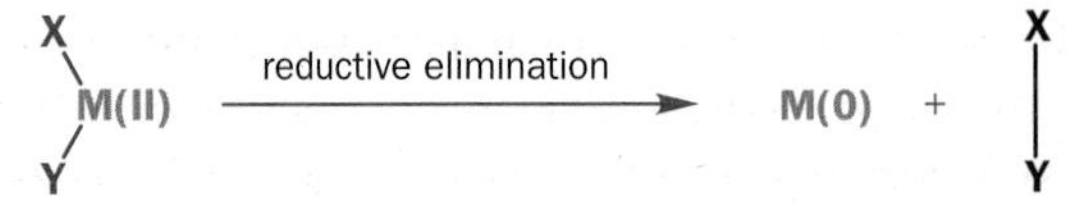

removes organic ligands from metal producing new organic product

The ligands to be eliminated must be *cis* to one another for reductive elimination to occur. This is because the process is concerted. Two examples from palladium chemistry make this point clear. Warming in DMSO causes ethane production from the first palladium complex because the two methyl groups are *cis* in the square planar complex. The more elaborate second bisphosphine forces the two methyl groups to be *trans* and reductive elimination does not occur under the same conditions. Reductive elimination is one of the most important methods for the removal of a transition metal from a reaction sequence leaving a neutral organic product.

In fact, no one wants to make ethane that way (if at all) but many other pairs of ligands can be coupled by reductive elimination. We will see many examples as the chapter develops but here is an indole synthesis that depends on a reductive elimination at palladium as a last step. In the starting material, palladium has two normal σ bonds and is Pd(II). The two substituents bond together to form the indole ring and a Pd(0) species is eliminated. Notice the use of 'L' to mean an undefined ligand of the phosphine sort.

Migratory insertion builds ligand structure

Two ligands can also react together to produce a new complex that still has the composite ligand attached to the metal ready for further modification. This process involves migration of one of the ligands from the metal to the other ligand and insertion of one of the ligands into the other metal–ligand bond and is known as **migratory insertion**. The insertion process is reversible and, as the metal effectively loses a ligand in the process, the overall insertion may be driven by the addition of extra external ligands (L) to produce a coordinatively saturated complex. As with reductive elimination, a *cis* arrangement of the ligands is required and the migrating group (X) retains its stereochemistry (if any) during the migration.

■ Migration normally occurs with retention; see Chapter 37.

Migratory insertion is the principal way of building up the chain of a ligand before elimination. The group to be inserted must be unsaturated in order to accommodate the additional bonds and common examples include carbon monoxide, alkenes, and alkynes producing metal–acyl, metal–alkyl, and metal–alkenyl complexes, respectively. In each case the insertion is driven by additional external ligands, which may be an increased pressure of carbon monoxide in the case of carbonylation or simply excess phosphine for alkene and alkyne insertions. In principle, the chain extension process can be repeated indefinitely to produce polymers by Ziegler–Natta polymerization, which is described in Chapter 52.

carbonylation

$$L_nM \overset{+L}{\underset{-L}{\rightleftharpoons}} L_{n+1}M \overset{O}{\underset{R}{\big|}}$$

carbometallation or hydrometallation, R = H

$$L_nM \overset{+L}{\underset{-L}{\rightleftharpoons}} L_{n+1}M \diagup\diagup R$$

alkyne insertion

$$L_nM \overset{+L}{\underset{-L}{\rightleftharpoons}} L_{n+1}M \diagup R$$

A good example of the carbonylation process is the reaction of the tetracarbonyl ferrate dianion $[Fe(CO)_4^{2-}]$ with alkyl halides. This reagent is made by dissolving metal reduction of the 18-electron Fe(0) compound $Fe(CO)_5$. Addition of two electrons would give an unstable 20-electron species but the loss of one of the ligands with its two electrons restores the stable 18-electron structure.

$$Fe(CO)_5 \xrightarrow[+2\ electrons]{Na/Hg} \left[Fe(CO)_5\right]^{2\ominus} \xrightarrow[-2\ electrons]{-CO} \left[Fe(CO)_4\right]^{2\ominus} \quad 2Na^{\oplus}$$

18 electrons   20 electrons   18 electrons

This iron anion is a good soft nucleophile for alkyl halides and can be used twice over to produce first a monoanion with one alkyl group and then a neutral complex with two alkyl groups and four CO ligands. Each of these complexes has 18 electrons as the electrons represented by the negative charges are retained by the iron to form the new Fe–C bonds. If extra CO is added by increasing the pressure, CO inserts into one Fe–C bond to form an iron acyl complex. Finally, reductive elimination couples the acyl group to the other alkyl group in a conceptually simple ketone synthesis. It does not matter which Fe–C bond accepts the CO molecule: the same unsymmetrical ketone is produced at the end.

tetracarbonyl ferrate anion → (S_N2) → 18 electrons → (S_N2) → 18 electrons → iron acyl complex

Any good two-electron ligand will cause the CO insertion: Ph_3P is often used instead of an increased CO pressure. The phosphine adds to the iron and pushes out the poorest ligand (one of the alkyl groups) on to a CO ligand in a process of **ligand migration**. In simple form it looks like this though the phosphine addition and alkyl migration may be concerted to avoid the formation of a 20-electron complex as intermediate.

18 electrons → (Ph_3P) → 20 electrons → 18 electrons → (reductive elimination) → ketone

Carbon monoxide incorporation extends the carbon chain

Carbonylation (the addition of carbon monoxide to organic molecules) is an important industrial process as carbon monoxide is a convenient one-carbon feedstock and the resulting metal–acyl complexes can be converted into aldehydes, acids, and their derivatives. The **OXO process** is the hydroformylation of alkenes such as propene and uses two migratory insertions to make higher value aldehydes. Though a mixture is formed this is acceptable from very cheap starting materials.

propene $+$ C≡O $\xrightarrow[Rh\ or\ Co\ cat.]{H-H}$ aldehyde $+$ branched aldehyde

A catalytic cycle (going clockwise from the top) shows the various stages of alkene coordination, hydrometallation to produce an alkyl metal species, coordination of carbon monoxide followed by insertion, and finally reductive cleavage with hydrogen to produce the metal–hydride intermediate, which is then ready for another cycle. The steps leading to the other regioisomeric aldehyde and the ligands on the metal are omitted for clarity.

The mechanisms of the two key steps are worth discussion. **Hydrometallation** occurs by initial π-complex formation followed by addition of the metal to one end of the alkene and hydrogen to the other. Both of these regioisomers are formed. The **carbonyl insertion reaction** is another migration from the metal to the carbon atom of a CO ligand.

Insertion reactions are reversible

The reverse process, **decarbonylation**, is also fast but can be arrested by maintaining a pressure of carbon monoxide above the reaction mixture. The reverse of hydrometallation involves the elimination of a hydride from the adjacent carbon of a metal alkyl to form an alkene complex. This process is known as **β-hydride elimination** or simply **β elimination**. It requires a vacant site on the metal as the number of ligands increases in the process and so is favoured by a shortage of ligands as in 16-electron complexes. The metal and the hydride must be *syn* to each other on the carbon chain for the elimination to be possible. The product is an alkene complex that can lose the neutral alkene simply by ligand exchange. So β elimination is an important final step in a number of transition-metal-catalysed processes but can be a nuisance because, say, Pd–Et complexes cannot be used as β elimination is too fast.

Palladium(0) is most widely used in homogeneous catalysis

These elementary steps form the basis for organo-transition-metal chemistry and are the same regardless of which metal is present and the detailed structure of the ligands. This is an enormous and rapidly expanding field that could not be discussed here without doubling the size of the book! Instead, we will concentrate on the chemistry of the most important transition metal, palladium,

> Hydrogenation with homogeneous catalysis involves a soluble catalyst rather than the more common heterogeneous catalysis with, say, Pd metal dispersed on an insoluble charcoal support as in Chapter 24. In general terms **homogeneous catalysts** are those that are soluble in the reaction mixture.

which is the most widely used both in industrial and academic laboratories on both a minute and very large scale. The variety of reactions that can be catalysed together with the range of functional groups tolerated, and usually excellent chemo- and regioselectivity, has meant that an ever increasing amount of research has gone into this area of chemistry. Most syntheses of big organic molecules now involve palladium chemistry in one or more key steps.

Choice of palladium complex

It is supposed that dba complexes the palladium atom through its alkenes.

There are many available complexes of palladium(0) and palladium(II). Tetrakis(triphenylphosphine)palladium(0), Pd(PPh$_3$)$_4$, and tris(dibenzylidene-acetone)dipalladium(0), Pd$_2$(dba)$_3$, or the chloroform complex, Pd$_2$(dba)$_3$·CHCl$_3$, which is air-stable, are the most common sources of palladium(0). The detailed structures of some palladium complexes, particularly the dimers, are beyond the scope of this book but we will discuss the reactions in detail.

a stable **Pd(0)** complex   a stable **Pd(II)** complex

dba
dibenzylidene acetone

Palladium(II) complexes are generally more stable than their palladium(0) counterparts. The dichloride PdCl$_2$ exists as a polymer and is relatively insoluble in most organic solvents. However, (PhCN)$_2$PdCl$_2$ and (MeCN)$_2$PdCl$_2$ (both easily prepared from PdCl$_2$) are soluble forms of PdCl$_2$, as the nitrile ligands are readily displaced in solution. Bis(phosphine)palladium(II) chloride complexes are air-stable and readily prepared from PdCl$_2$. Palladium is, of course, an expensive metal—these complexes cost about £50–100 per gram—but very little is needed for a catalytic reaction.

We should review the basic chemistry of palladium, as you will be seeing many more examples of these steps in specialized situations. Palladium chemistry is dominated by two oxidation states. The lower, palladium(0), present in tetrakis(triphenylphosphine)palladium, for example, is nominally electron-rich, and will undergo oxidative addition with suitable substrates such as halides and triflates (TfO$^-$ = CF$_3$SO$_2$O$^-$), resulting in a palladium(II) complex. Oxidative addition is thought to occur on the coordinatively unsaturated 14-electron species, formed by ligand dissociation in solution.

The resulting σ alkyl bond in such complexes is very reactive, especially towards carbon–carbon π bonds. Thus an alkene in the reacting system will lead to coordination followed by migratory insertion into the palladium–carbon σ bond. This process is like hydrometallation and is called **carbopalladation** as carbon and palladium are attached to the ends of the alkene system. There is no change in oxidation state during this process, although the ligands (often phosphines) must dissociate to allow coordination of the alkene and associate to provide a stable final 16-electron product.

Theoretically, it is possible for the process of olefin coordination and insertion to continue as in Ziegler–Natta polymerization (Chapter 52) but with palladium the metal is expelled from the molecule by a β-hydride elimination reaction and the product is an alkene. For the whole process to be catalytic, a palladium(0) complex must be regenerated from the palladium(II) product of β-hydride elimination. This occurs in the presence of base which removes HX from the palladium(II) species.

This is another example of reductive elimination: one that forms a hydrogen halide rather than a carbon–carbon or carbon–hydrogen bond as described earlier.

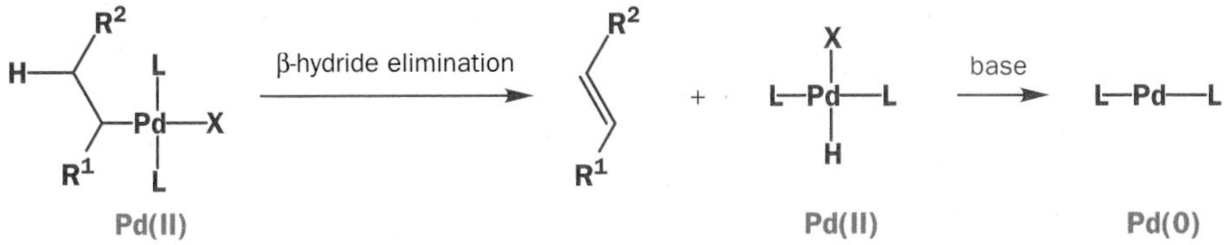

The speed of the intramolecular β-hydride elimination means that the original substrate for the oxidative addition reaction must be chosen with care—the presence of hydrogen at an sp^3 carbon in the β position must be avoided. Thus, substrates for oxidative addition reactions in palladium chemistry are frequently vinylic, allylic, or aromatic and never ethyl or *n*-propyl.

The Heck reaction couples together a halide or triflate and an alkene

All the individual steps outlined above combine to make up the catalytic pathway in the **Heck reaction**, which couples an alkene with a halide or triflate to form a new alkene. The R^1 group in R^1X can be aryl, vinyl, or any alkyl group without β Hs on an sp^3 carbon atom. The group X can be halide (Br or I) or triflate (OSO_2CF_3). The alkene can be mono- or disubstituted and can be electron-rich, -poor, or -neutral. The base need not be at all strong and can be Et_3N, NaOAc, or aqueous Na_2CO_3. The reaction is very accommodating.

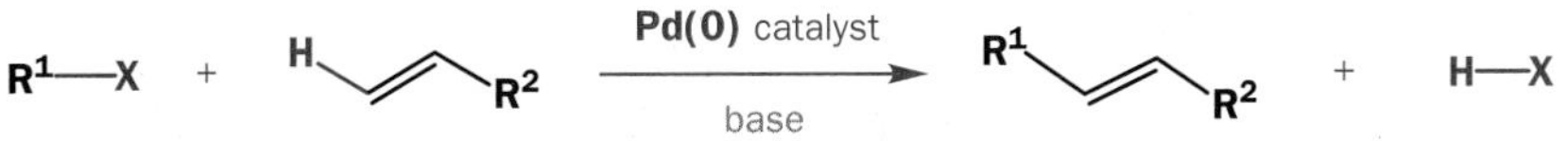

The palladium-catalysed addition of aryl, vinyl, or substituted vinyl groups to organic halides or triflates, the Heck reaction, is one of the most synthetically useful palladium-catalysed reactions. The method is very efficient, and carries out a transformation that is difficult by more traditional techniques. The mechanism involves the oxidative addition of the halide, insertion of the olefin, and elimination of the product by a β-hydride elimination process. A base then regenerates the palladium(0) catalyst. The whole process is a catalytic cycle.

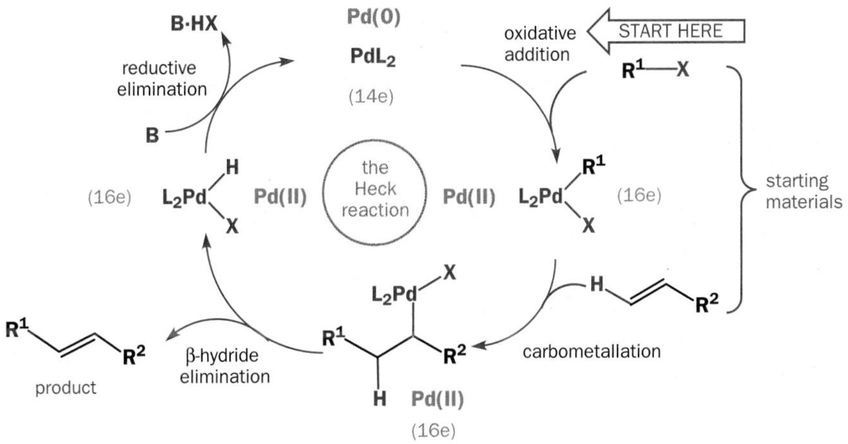

The choice of substrates is limited to aryl, heteroaryl, vinylic, and benzylic halides and triflates, as the presence of an sp^3 carbon in the β position carrying a hydrogen rapidly results in β-hydride elimination. The reaction tolerates a variety of functional groups, and works well with both electron-withdrawing and electron-donating groups on either substrate. Here is an example using a heterocyclic compound we featured earlier reacting with another heterocycle.

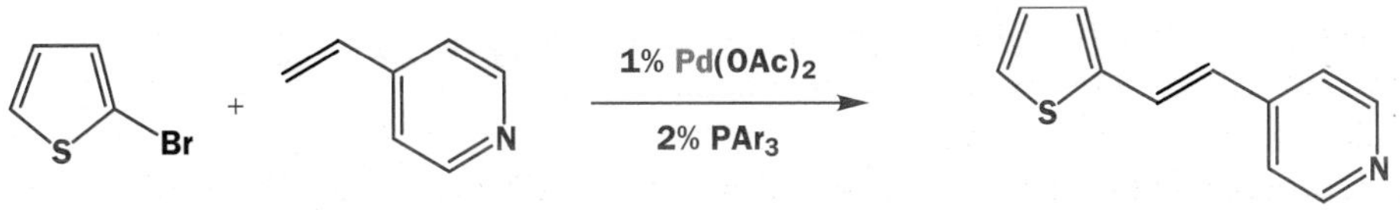

Protected amino acids can be made without any racemization and electron-withdrawing groups such as esters promote excellent regioselectivity in favour of terminal attack. These three examples rely on *in situ* reduction of the palladium(II) acetate by tri(*o*-tolyl)phosphine, a popular more sterically demanding aromatic phosphine.

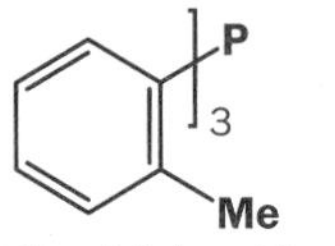

tri(*o*-tolyl)phosphine

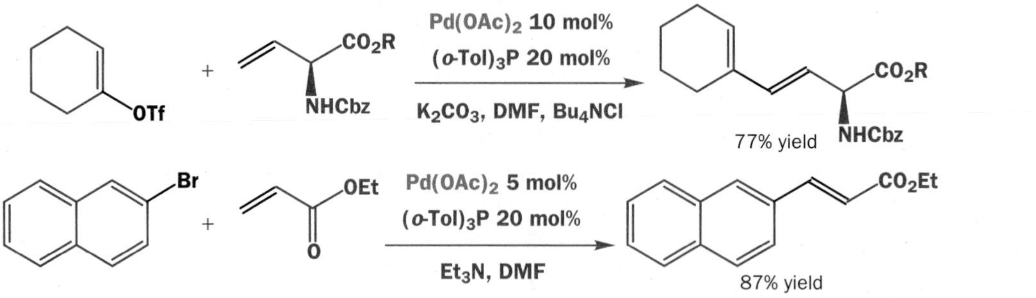

In situ formation of palladium(0) by reduction of Pd(II)

In reactions requiring palladium(0), formation of the active complex may be achieved more conveniently by reduction of a palladium(II) complex, for example, Pd(OAc)$_2$. Any phosphine may then be used in the reaction, without the need to synthesize and isolate the corresponding palladium(0)-phosphine complex. Only 2–3 equivalents of phosphine may be needed, making the palladium(0) complex coordinatively unsaturated and therefore very reactive. The reduction of palladium(II) to palladium(0) can be achieved with amines, phosphines, alkenes, and organometallics such as DIBAL-H, butyl lithium, or trialkyl aluminium. The mechanisms are worth giving as they illustrate the basic steps of organometallic chemistry.

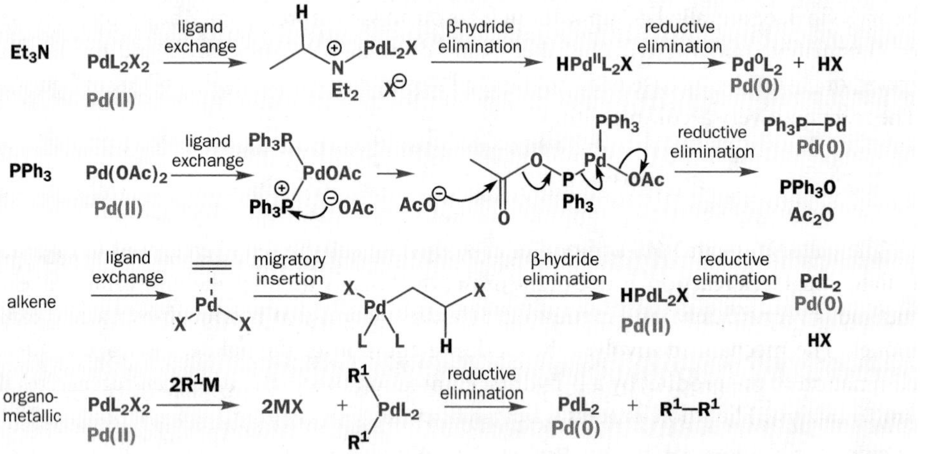

In contrast, electron-donating groups such as ethers lead to attack at the end of the alkene substituted by oxygen to produce in this case the 1,1-disubstituted product. These reactions must be dominated by the interaction of the filled p orbital of the alkene with an empty d orbital on Pd. This is an example of a Heck reaction working in the absence of a phosphine ligand.

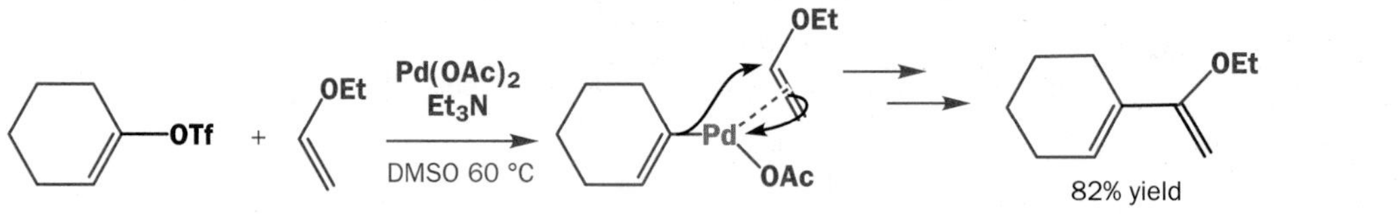

In the β-hydride elimination step, the palladium and hydride must be coplanar for reaction to take place, as this is a *syn* elimination process. For steric reasons, the R group will tend to eclipse the smallest group on the adjacent carbon as elimination occurs, leading predominantly to a *trans* double bond in the product.

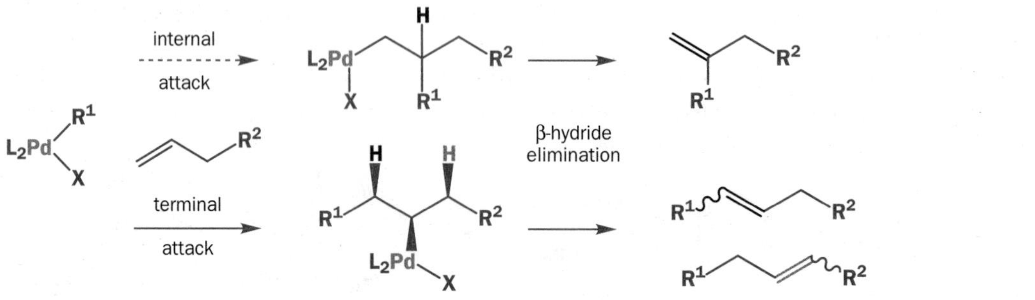

Where there is a choice as to which hydride can be lost to form the alkene, the stability of the possible product alkenes often governs the outcome as the β-hydride elimination is reversible. The reaction of allylic alcohols is particularly important as the more stable of the two alkenes is the enol and a carbonyl compound is formed.

Hydropalladation–dehydropalladation can lead to alkene isomerization

As β-hydride elimination is reversible, hydropalladation with the opposite regiochemistry provides a mechanism for forming regioisomers of the alkene. This allows the most stable alkene that is accessible by the hydropalladation–dehydropalladation sequence to dominate. The only restriction is that all of these processes are *syn*. The migration can be prevented by the addition of bases like silver carbonate, which effectively removes the hydrogen halide from the palladium complex as soon as it is formed. This synthesis of a complex *trans* dihydrofuran involves the Heck reaction followed by alkene isomerization and then a Heck reaction without migration to preserve the stereochemistry.

Oxidative addition of the aryl iodide (Ar^1 = 3,4-dimethoxyphenyl) to a palladium(0) complex, formed from Pd(OAc)$_2$ by reduction (with the phosphine?) gives the active palladium(II) complex ArPdOAcL$_2$. Carbopalladation occurs as expected on an electron-rich alkene to give the product of aryl addition to the oxygen end of the alkene in a *syn* fashion. β-Hydride elimination must occur away from the aryl group to give a new alkene complex as there is no *syn* H on the other side. The alkene has moved one position round the ring. Hydropalladation in the reverse sense gives a new σ complex, which could eliminate either the black or the green hydrogens. Elimination of the green H gives the enol ether, which is the most stable alkene possible due to conjugation.

syn (green) H on one side only
anti (black) Hs on both sides

The second Heck reaction involves a naphthyl iodide (Ar^2 = 2-naphthyl) but the initial mechanism is much the same. However, the enol ether has two diastereotopic faces: *syn* or *anti* to the aromatic substituent (Ar^1) introduced in the first step. Palladium is very sensitive to steric effects and generally forms less hindered complexes where possible. Thus coordination of the palladium(II) intermediate occurs on the face of the enol ether *anti* to Ar^1. This in turn controls all the subsequent steps, which must be *syn*, leading to the *trans* product. The requirement for *syn* β-hydride elimination also explains the regiochemical preference of the elimination. In this cyclic structure there is only one hydrogen (green) that is *syn*; the one on the carbon bearing the naphthyl substituent is *anti* to the palladium and cannot be eliminated.

Heck reactions can be enantioselective

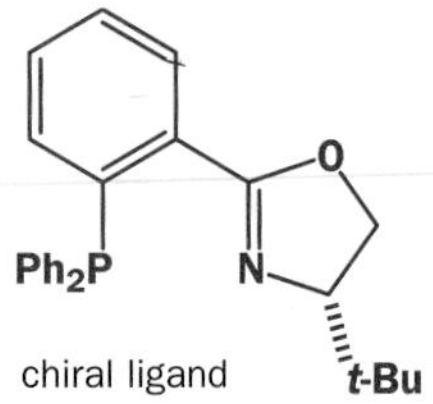

chiral ligand

With chiral ligands the Heck reaction can be enantioselective. The amino-acid-derived phosphine ligand in the margin controls the Heck reaction of phenyl triflate with dihydrofuran. The ligand selects one enantiotopic face of the alkene (see Chapter 45 if you have forgotten this term) and the usual double bond migration and β elimination complete the reaction.

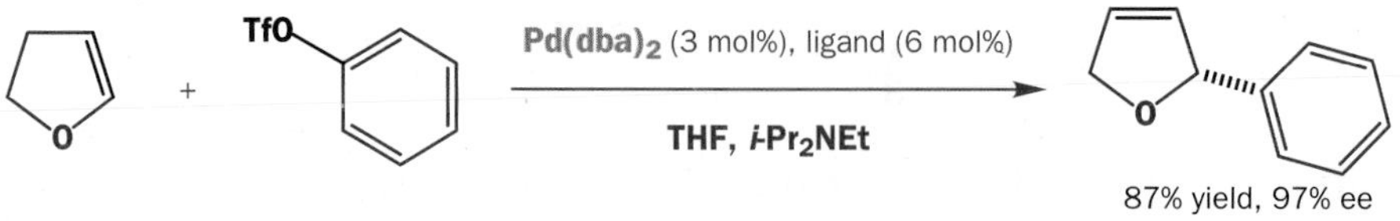

■ BINAP was introduced in Chapter 45.

The famous ligand BINAP controls an intramolecular Heck reaction to give decalin derivatives with good enantiomeric excess. BINAP is the optically pure phosphine built into the palladium catalyst. The presence of silver ions accelerates the reaction as well as preventing double bond isomerization in the original substrate. This time the chiral ligand selects which double bond is to take part in the reaction. The vinyl palladium species is tethered to the alkene and can reach only the same face. The faces of the alkenes are diastereotopic but the two alkenes are enantiotopic and you must know your right from your left to choose one rather than the other.

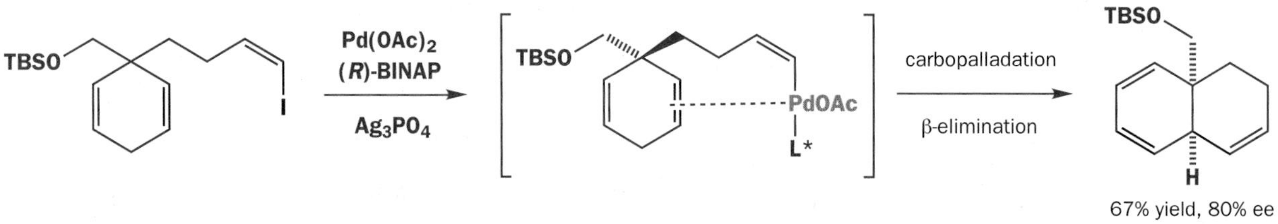

Cross-coupling of organometallics and halides

Other than β-hydride elimination, another important pathway by which palladium(II) intermediates can lead to neutral organic fragments is reductive elimination. This forms the basis of the mechanism for **cross-coupling reactions** between an organometallic reagent and an organic halide or triflate.

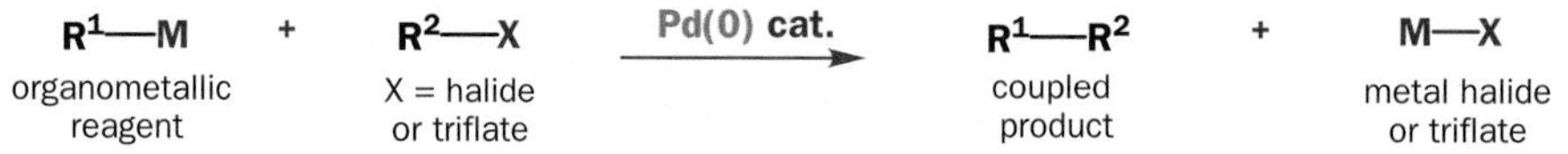

This is a reaction that seems very attractive for synthesis but, in the absence of a transition metal catalyst, the yields are very low. We showed in the last chapter how vinyl silanes can be made with control over stereochemistry and converted into lithium derivatives with retention. Neither of these vinyl metals couple with vinyl halides alone. But in the presence of a transition metal—Cu(I) for Li and Pd(0) for Sn—coupling occurs stereospecifically and in good yield.

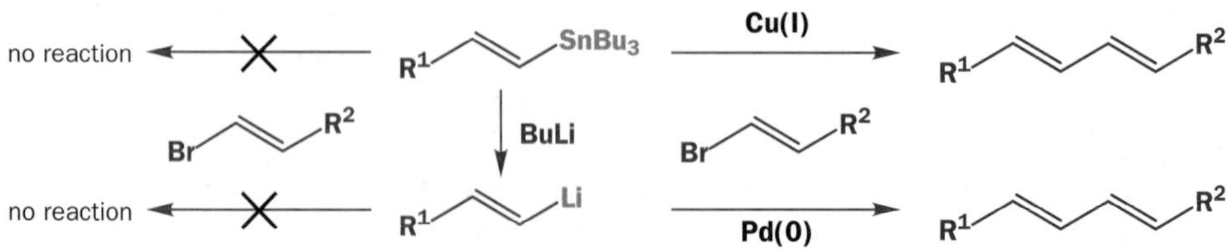

The mechanism involves oxidative addition of the halide or triflate to the initial palladium(0) phosphine complex to form a palladium(II) species. The key slow step is a **transmetallation**, so called because the nucleophile (R^1) is transferred from the metal in the organometallic reagent to the palladium and the counterion (X = halide or triflate) moves in the opposite direction. The new palladium(II) complex with two organic ligands undergoes reductive elimination to give the coupled product and the palladium(0) catalyst ready for another cycle.

R²—X →(Pd(0), PdL₂) R²—PdL₂(X) →(R¹—M) M—X + R²—PdL₂(R¹) → R¹—R² + PdL₂

oxidative addition — Pd(II) — transmetallation (SLOW) — reductive elimination — regenerated catalyst

The reaction is important because it allows the coupling of two different components (R^1 and R^2). If this is to happen, the substituents, M (metal) on R^1 and X (halide or triflate) on R^2, must be different electronically. Both components form σ complexes with Pd but the halide partner (R^2X) bonds first by oxidative addition and the R^2–Pd must survive while the metal partner (R^1M) bonds to the Pd by transmetallation. Once the two components are joined to the palladium atom, only the cross-coupled product can be formed. The essential feature is that X and M are different so that R^2X combines with Pd(0) and R^1M with Pd(II). There can then be no confusion.

PdL₂ →(R^2X reacts with Pd(0), oxidative addition) R²—PdL₂(X) [Pd(II)] →(R^1M reacts with Pd(II), transmetallation (SLOW)) M—X + R²—PdL₂(R¹)

> There is a problem in naming the two partners. The halide partner (R^2X) is sometimes called the electrophile and the organometallic partner (R^1M) the nucleophile. These names describe the nature of the reagents rather than the mechanism of the reaction and we will not use these names.

The halide partner (R^2X) must be chosen with care, as β-hydride elimination would decompose the first intermediate during the slow transmetallation step. The choice for R^2 is restricted to substituents without β-hydrogen atoms: vinyl, allyl, benzyl, and polyfluoroalkyl halides, triflates, and phosphates have all been coupled successfully. The organometallic reagent (R^1M) can be based on magnesium, zinc, copper, tin, silicon, zirconium, aluminium, or boron and the organic fragment can have a wide variety of structures as coupling is faster than β-hydride elimination.

R^1—M R^1 = almost anything including examples with β H

   M = MgX, ZnX, Cu, SnR₃, SiR₃/TASF, ZrCp₂Cl, AlMe₂, B(OR)₂

R^2—X R^2 must not have β Hs that can eliminate   X = I, Br, (Cl), OTf, OPO(OR)₂

Ar—X   (vinyl)—X   (allyl)—X   Ar—CH₂—X   RO—C(O)—CH₂—X R_F—CH₂—X R_F = perfluoroalkyl

The difference in relative reactivity of aromatic iodides and triflates was exploited in this sequential synthesis of substituted terphenyls by repeated coupling with organozinc reagents. The more reactive iodide coupled at room temperature with palladium(0) and tri-*o*-furylphosphine but warming to 65 °C was required for the triflate to participate in the second coupling.

(scheme) Aryl iodide/triflate (TfO, R¹) + ArZnBr (R) →(Pd(dba)₂, THF, RT, tri(2-furyl)phosphine) biaryl (TfO, R, R¹) + ArZnBr (R²) →(Pd(dba)₂, dppf, THF, 65 °C) terphenyl (R², R, R¹)

72–88% yield   74–89% yield

In spite of the wide range of organometallic reagents that can be used there are two classes that have proved particularly popular because they are stable intermediates in their own right and can be prepared separately before the coupling reaction. These cross-couplings are known by the names of the two chemists whose work made the reactions so valuable. The Stille coupling employs a stannane as the organometallic component (R^1M) while the Suzuki coupling relies on a boronic acid.

The Stille coupling uses stannanes as the organometallic component

Since the first reported use in the late 1970s, the Stille coupling has been widely used for the coupling of both aromatic and vinylic systems.

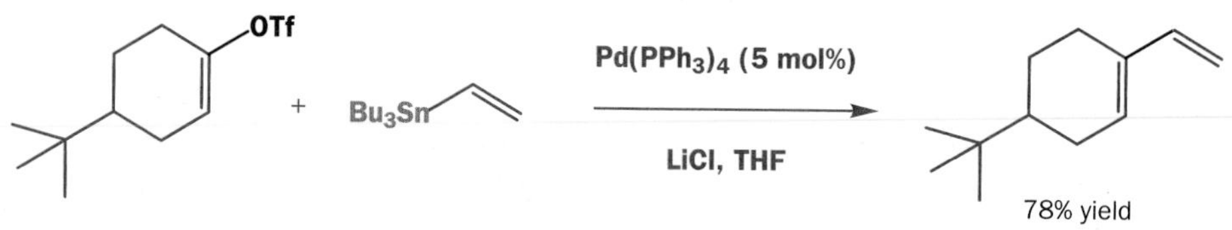

78% yield

The mechanism involves the oxidative addition of the vinyl or aromatic triflate or halide to give a palladium intermediate. This then undergoes a transmetallation reaction with the organostannane, giving an organopalladium intermediate in which both components are σ-bound. This complex then undergoes a reductive elimination step, releasing the product and thereby regenerating the palladium(0) catalyst.

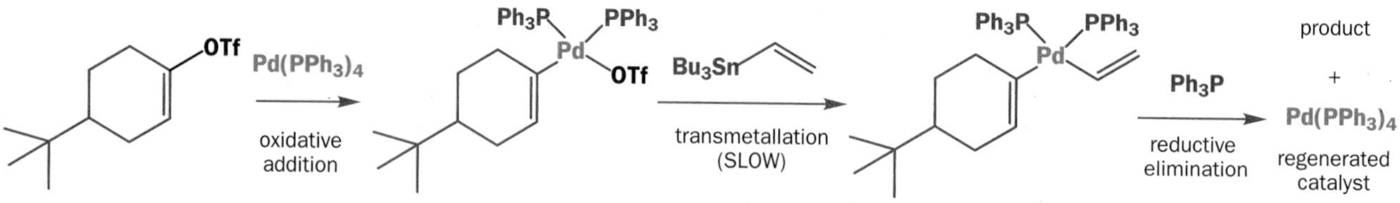

The reaction will also occur if the vinyl or aryl halide is used in place of the triflate. However, the triflates have been more widely used as they are readily prepared from phenols or enolizable aldehydes or ketones. In these reactions, the presence of a source of halide (typically LiCl) is generally required. This may be because the triflate is a counterion and is not bound to the metal as a ligand. If transmetallation is to occur some other ligand must be added to give the necessary square coplanar geometry.

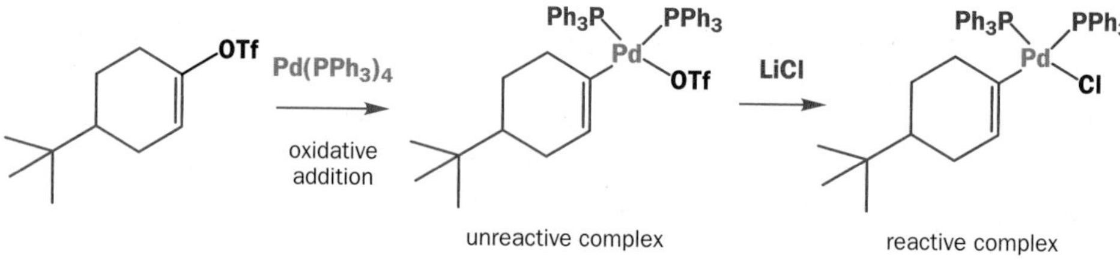

unreactive complex   reactive complex

The Stille reaction, which represents over half of all current cross-coupling reactions, has been used in total synthesis with excellent results. The reaction may also be carried out intramolecularly and with alkynyl stannanes instead of the more usual aryl or vinyl stannanes, even to form medium-sized rings. This example forms a ten-membered ring containing two alkynes.

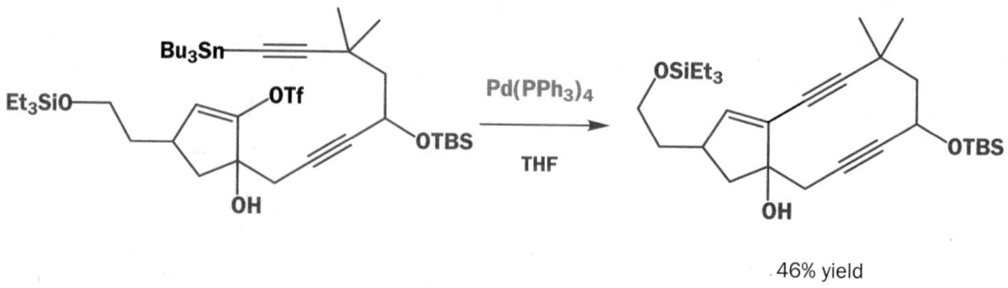

46% yield

Nicolaou's synthesis of rapamycin uses the reaction twice in the **macrocyclization** (cyclization reaction to form a large ring) step. This illustrates an important feature of palladium-catalysed cross-couplings—the geometry of both double bonds involved in the coupling is preserved in the product. This seems a very complex example and the molecule *is* complex. But just inspect the black region and you will see two simple Stille couplings. These reactions work with complex molecules having many functional groups, even if the yield isn't great (26%!).

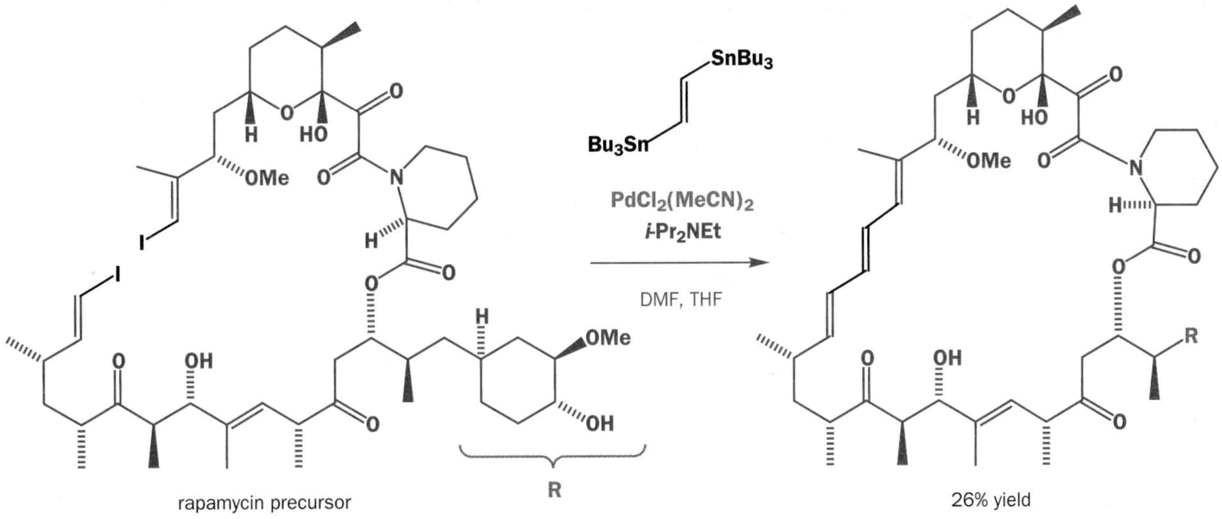

rapamycin precursor

26% yield

The Stille coupling may be combined with carbonylation in two ways. Acid chlorides may be used as substrates for the reaction with vinyl or aryl stannanes. However, an atmosphere of carbon monoxide is frequently required to prevent decarbonylation after the oxidative addition step.

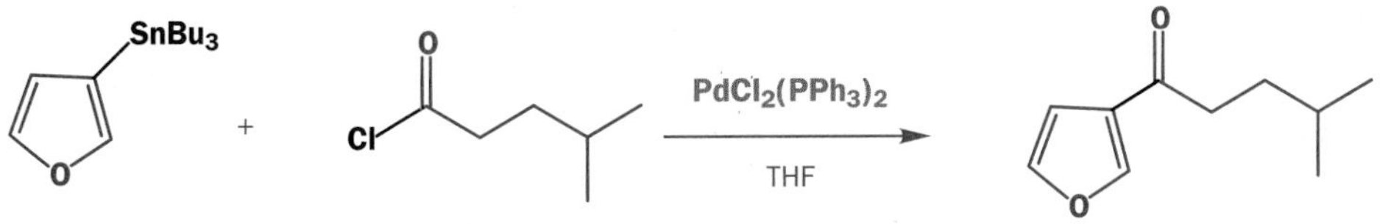

More recently, it has been shown that performing the normal Stille reaction in the presence of carbon monoxide may also lead to carbonylated products. These reactions can take place in a CO saturated solution, under one atmosphere of pressure. Using these conditions, excellent yields of the carbonylated product can be obtained, without any of the normal coupling product being present.

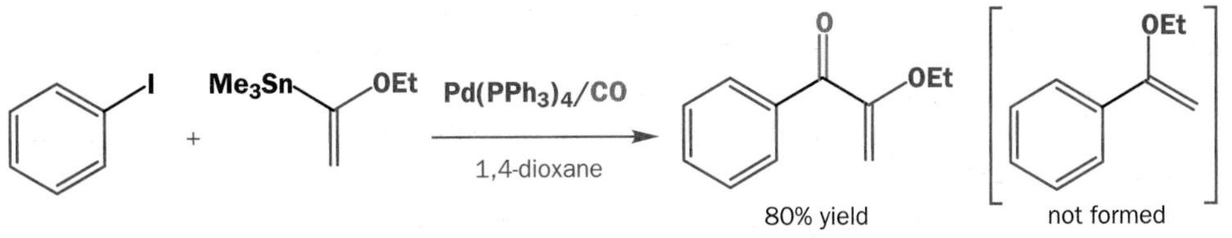

80% yield

not formed

The mechanism is like that of a normal Stille coupling except that the carbon monoxide first exchanges for one of the phosphine ligands and then very rapidly inserts to produce an acyl palladium(II) complex. This then undergoes transmetallation with the vinyl stannane in the usual way forming trimethylstannyl iodide and the palladium complex with two carbon ligands. Reductive elimination gives the masked diketone and regenerates the palladium(0) catalyst. Transmetallation is the slow step in these coupling reactions so that there is time for the carbon monoxide insertion first. The final step—reductive elimination—releases the Pd(0) catalyst for the next cycle.

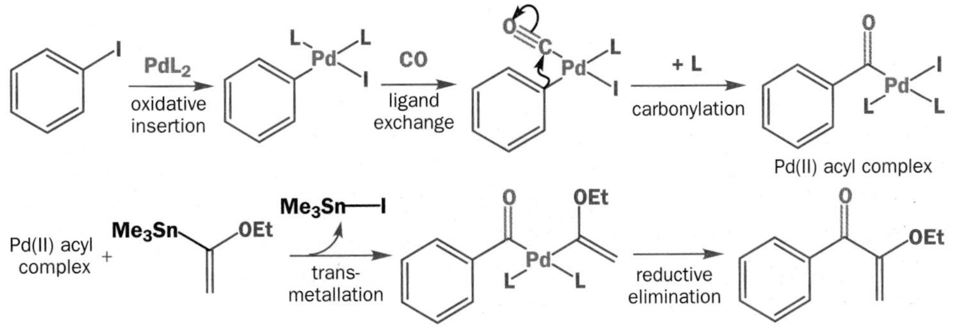

Acyl palladium species react like activated acid derivatives

Carbonylation of a halide or triflate provides a direct route to a range of chain-extended acyl derivatives. A carbonyl group substituted with PdX (X = halide or triflate) is a reactive acylating agent, rather like an acid anhydride, as PdX is a good leaving group. Reaction with alcohols and amines gives esters and amides, while reduction with tributyltin hydride gives the aldehyde. Intramolecular attack by alcohols leads to lactones as demonstrated in the conversion of a vinyl iodide into a 2H-furanone (butenolide). We will see more of these reactions later.

The Suzuki coupling couples boronic acids to halides

Since first being published in 1979, the Suzuki coupling of a boronic acid with a halide or triflate has developed into one of the most important cross-coupling reactions, totalling about a quarter of all current palladium-catalysed cross-coupling reactions. The original version consisted of hydroboration of an alkyne with catecholborane, followed by palladium(0)-catalysed coupling of the resulting vinyl boronate with an aromatic iodide or bromide. The hydroboration is generally regioselective for the less hindered position and addition of boron and hydrogen occurs *cis* stereospecifically.

As in the Stille coupling, the geometry of both unsaturated components is preserved during the coupling so this is an excellent method for stereospecific diene synthesis. Hydroboration of octyne followed by hydrolysis of the boronate gave exclusively the *E*-vinyl boronic acid. Coupling with the *Z*-vinyl bromide in toluene with palladium(0) catalysis with potassium hydroxide as the base gave the *E,Z*-diene in good yield. These dienes are very useful in the Diels–Alder reaction (Chapter 35).

The mechanism is very similar to that of the Stille coupling. Oxidative addition of the vinylic or aromatic halide to the palladium(0) complex generates a palladium(II) intermediate. This then undergoes a transmetallation with the alkenyl boronate, from which the product is expelled by reductive elimination, regenerating the palladium(0) catalyst. The important difference is the transmetallation step, which explains the need for an additional base, usually sodium or potassium ethoxide or hydroxide, in the Suzuki coupling. The base accelerates the transmetallation step leading to the borate directly presumably via a more nucleophilic 'ate' complex.

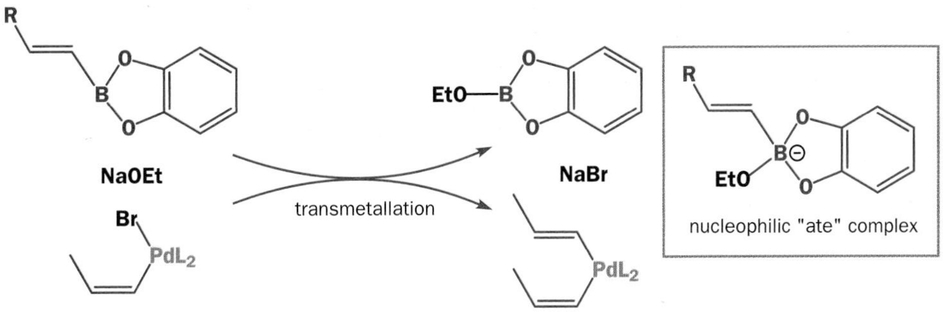

Sterically demanding substrates are tolerated well and Suzuki coupling has been used in a wide range of aryl–aryl cross-couplings. This example has three *ortho* substituents around the newly formed bond (marked in black) and still goes in excellent yield. It also shows that borate esters can be used instead of boronic acids.

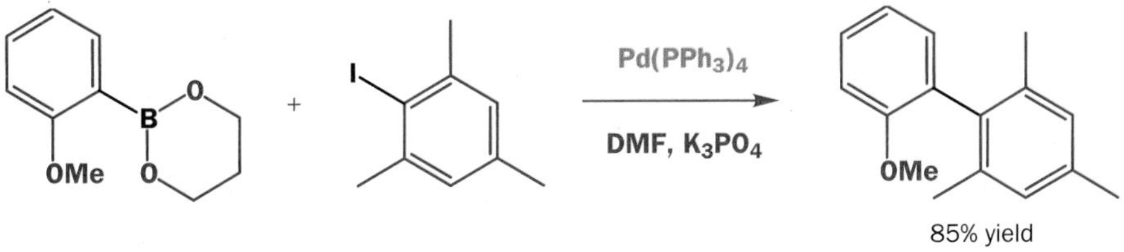

Coupling of aromatic heterocycles goes well. The 2-position of a pyridine is very electrophilic and not at all nucleophilic (Chapter 43) but couplings at this position are fine with either the halide or the boronic acid in that position. Clearly, it is a mistake to see either of these substituents as contributing a 'nucleophilic carbon'. It is better to see the reaction as a coupling of two equal partners and the two substituents (halide and boronic acid) as a control element to ensure cross-coupling and prevent dimerization. In the second example potassium *tert*-butoxide was crucial as weaker and less hindered bases gave poor yields.

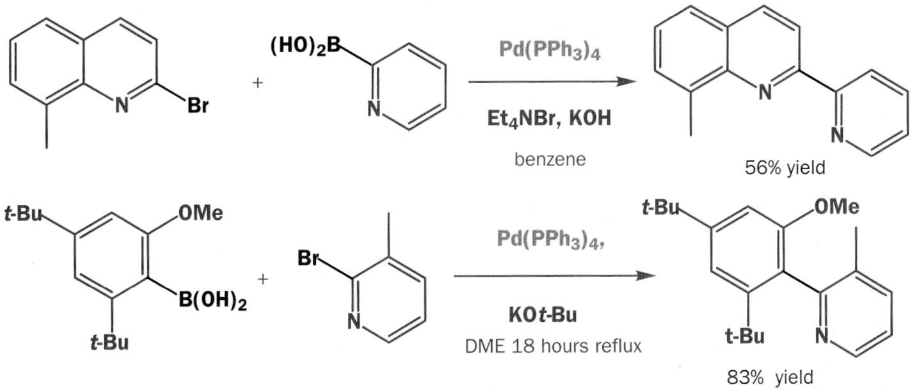

Due to the excellent stereoselectivity of the Suzuki coupling, the reaction has been used in the synthesis of the unsaturated units of a range of natural products including trisporol B. The key step is the stereocontrolled synthesis of an *E,Z*-diene. The geometry of both double bonds comes stereospecifically with retention of configuration from single geometrical isomers of the starting materials.

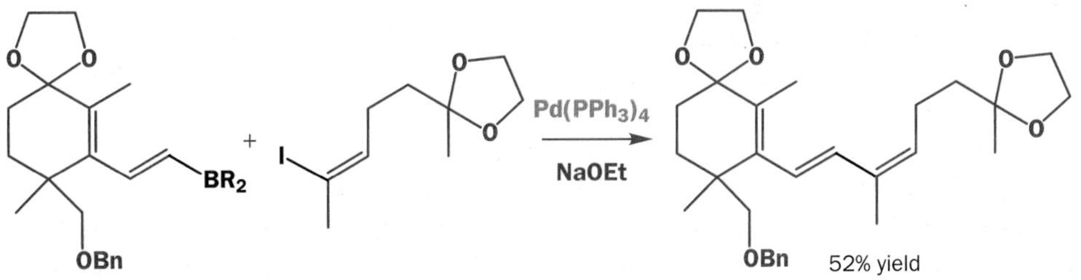

The Sonogashira coupling uses alkynes directly

The coupling of terminal alkynes with aryl or vinyl halides under palladium catalysis is known as the Sonogashira reaction. This catalytic process requires the use of a palladium(0) complex, is performed in the presence of base, and generally uses copper iodide as a co-catalyst. One partner, the aryl or vinyl halide, is the same as in the Stille and Suzuki couplings but the other has hydrogen instead of tin or boron as the 'metal' to be exchanged for palladium.

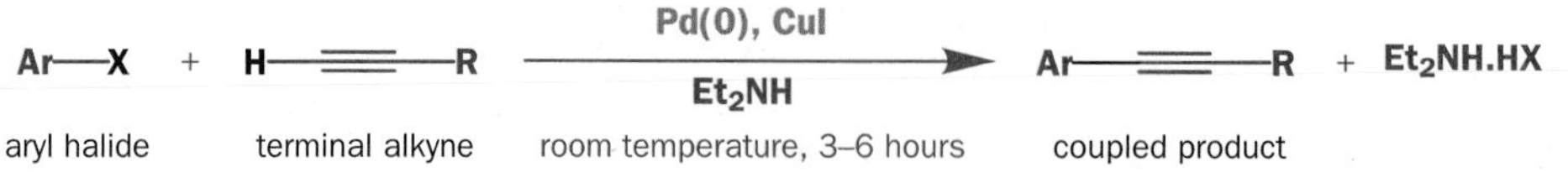

The mild conditions usually employed, frequently room temperature, mean that the reaction can be used with thermally sensitive substrates. The mechanism of the reaction is similar to that of the Stille and Suzuki couplings. Oxidative addition of the organic halide gives a palladium(II) intermediate that undergoes transmetallation with the alkynyl copper (generated from the terminal alkyne, base, and copper iodide). Reductive elimination with coupling of the two organic ligands gives the product and regenerates the palladium(0) catalyst.

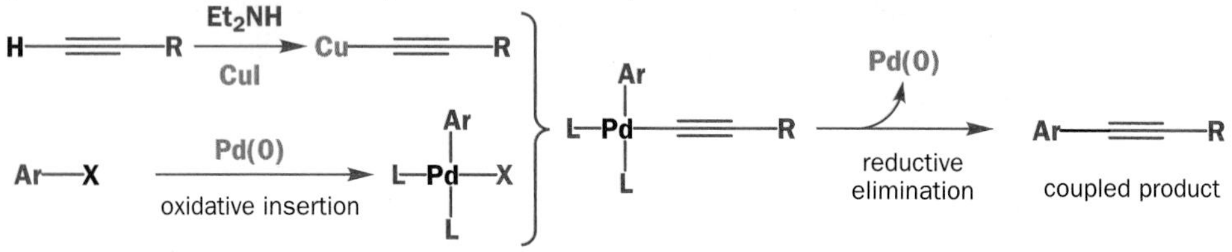

It is often more convenient, as in the Heck reaction, to use a stable and soluble Pd(II) derivative such as bis(triphenylphosphine)palladium(II) chloride instead of Pd(0). This is rapidly reduced *in situ* to give a coordinatively unsaturated, catalytically active, palladium(0) species. The geometry of the alkene is generally preserved so that *cis* (*Z*) and *trans* (*E*) dichloroethylene give the two different geometrical isomers of the enyne below in >99% stereochemical purity as well as excellent yield.

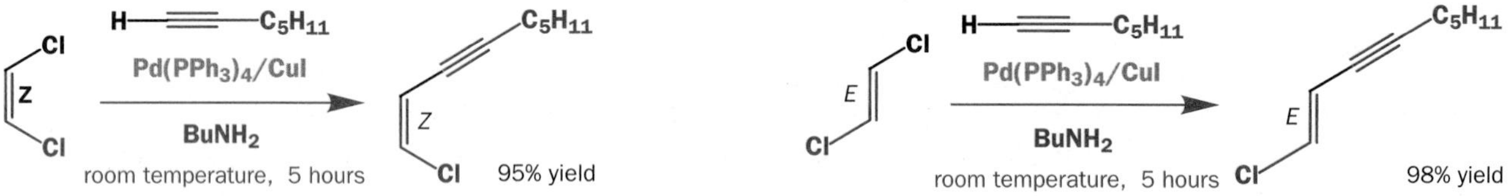

Ene-diynes and the Bergmann cyclization

The Sonogashira reaction is used a lot because of the great potential of **ene-diyne antibiotics**. Symmetrical ene-diynes may be synthesized in one step from two molecules of a terminal alkyne and *Z*-dihaloethene. The ene-diyne part of the molecule does the remarkable Bergmann cyclization to give a benzene diradical: the ene-diyne is able to penetrate DNA and the diradical is able to react with it. These compounds are anticancer drugs of some promise.

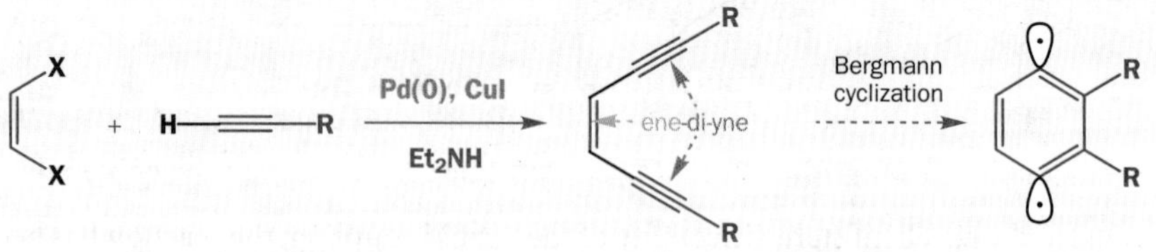

To make useful biologically active compounds, however, the reaction is performed sequentially, allowing different functionality on each of the alkyne units.

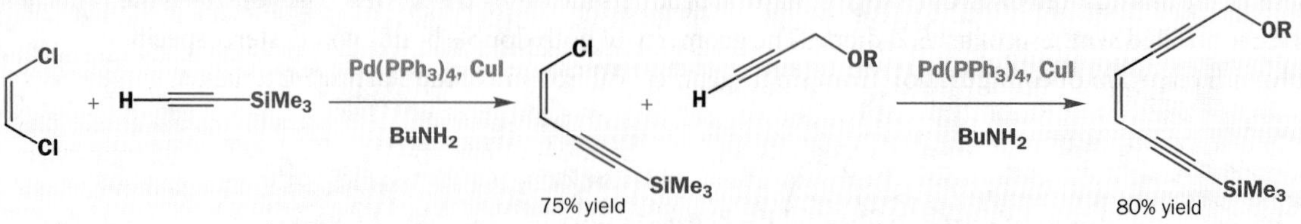

Allylic electrophiles are specifically activated by palladium(0)

Allylic compounds with good leaving groups, such as bromide and iodide, are excellent allylating agents but they suffer from loss of regiochemistry due to competition between the direct S_N2 and

S_N2' reaction. This problem together with the associated stereochemical ambiguity was described in Chapter 23. In contrast, π-allyl cation complexes of palladium allow both the stereochemistry and regiochemistry of nucleophilic displacement reactions to be controlled.

allyl cation complex

In addition, leaving groups (X) that are usually regarded as rather unreactive can be used, which means that the electrophilic partner is more stable in the absence of palladium making handling easier. Acetate (X = OAc) is the most commonly used leaving group, but a wide range of other functional groups (X = OCO_2R, $OPO(OR)_2$, Cl, Br, OPh) will perform a similar role. The full catalytic cycle is shown with the intermediate π-allyl complex in equilibrium between the neutral version, which has the leaving group coordinated to palladium, and the cationic π-allyl, in which one of the phosphine ligands has displaced the anion.

> **The Pd π-allyl cation complex**
>
> You can represent the palladium π-allyl cation complex in two ways. Either you draw a neutral allyl group complexed to Pd⁺ or you draw an allyl cation complexed to neutral Pd. Though the counting is different (Pd⁺ has only 9 electrons: the neutral allyl has 3 but the allyl cation only 2), both come out as η^3 16-electron species, which is just as well as they are different ways of drawing the same thing.
>
> Pd π-allyl cation complex
>
>
> 9 + 3 + 2 + 2 = 16e   10 + 2 + 2 + 2 = 16e

Soft nucleophiles (Nu) generally give the best results so, for carbon–carbon bond formation, stabilized enolates such as malonates are best, but for C–X (X = O, N, S) bond formation the reaction is successful with alkoxides, amines, cyanide, and thioalkoxides. This example shows an amine attacking outside the ring probably because the alkene prefers to be inside the ring.

The intramolecular reaction works well to give heterocyclic rings—the regioselectivity is usually determined by the length of the chain and how far it can reach. Here a 6/5 fused product is preferred to a bridged product containing two seven-membered rings.

The reaction usually proceeds with *retention* of configuration at the reacting centre. As in S_N2 reactions going with retention (Chapter 37), this can mean only a double inversion. Coordination of Pd to the double bond of the allylic acetate occurs on the less hindered face opposite the leaving group and the nucleophile adds to the face of the π-allyl Pd cation complex opposite the Pd. The net result is displacement of the leaving group by the nucleophile with retention. Thereafter, the nucleophile attacks from the less hindered face of the resulting π-allyl complex (that is, away from the metal) leading to overall retention of configuration.

The rather vague arrows on the middle two diagrams are the best we can do to show how Pd(0) uses its electrons to get rid of the leaving group and how it accepts them back again when the nucleophile adds. They are not perfect but it is often difficult to draw precise arrows for organometallic mechanisms. The double inversion process is perhaps more apparent in a perspective view.

Meldrum's acid

The reaction of this allylic acetate with the sodium salt of Meldrum's acid (structure in margin) demonstrates the retention of configuration in the palladium(0)-catalysed process. The tetraacetate and the intermediate π-allyl complex are symmetrical, thus removing any ambiguity in the formation or reaction of the π-allyl complex and hence in the regiochemistry of the overall reaction.

enolate of Meldrum's acid   73% yield

Vinyl epoxides provide their own alkoxide base

Vinyl epoxides and allylic carbonates are especially useful electrophiles because under the influence of palladium(0) they produce a catalytic amount of base since X^- is an alkoxide anion. This is sufficiently basic to deprotonate most nucleophiles that participate in allylic alkylations and thus no added base is required with these substrates. The overall reaction proceeds under almost neutral conditions, which is ideal for complex substrates. The relief of strain in the three-membered ring is responsible for the epoxide reacting with the palladium(0) to produce the zwitterionic intermediate. Attack of the negatively charged nucleophile at the less hindered end of the π-allyl palladium intermediate preferentially leads to overall 1,4-addition of the neutral nucleophile to vinyl epoxides.

Retention of stereochemistry is demonstrated by the reaction of a substituted malonate with epoxycyclopentadiene. Palladium adds to the side opposite the epoxide so the nucleophile is forced to add from the same side as the OH group. This, no doubt, helps 1,4-regioselectivity. The required palladium(0) phosphine complex was formed from a palladium(II) complex as in the Heck reaction.

55% yield

Allylic carbonates produce the required alkoxide by decarboxylation of the carbonate anion that is displaced in the formation of the π-allyl palladium intermediate. Deprotonation creates the active nucleophile, which rapidly traps the π-allyl palladium complex to give the allylated product and regenerates the palladium(0) catalyst.

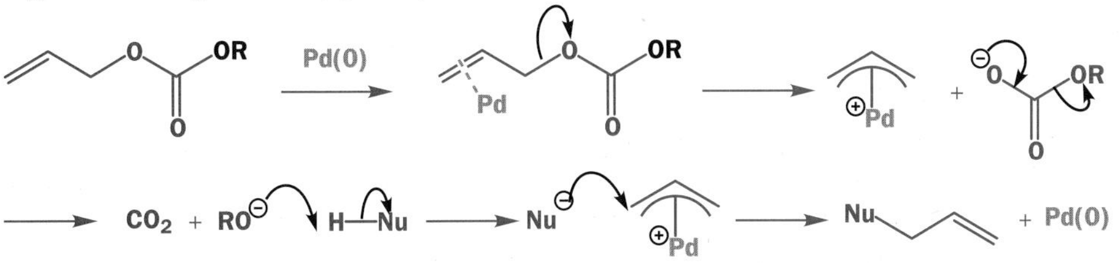

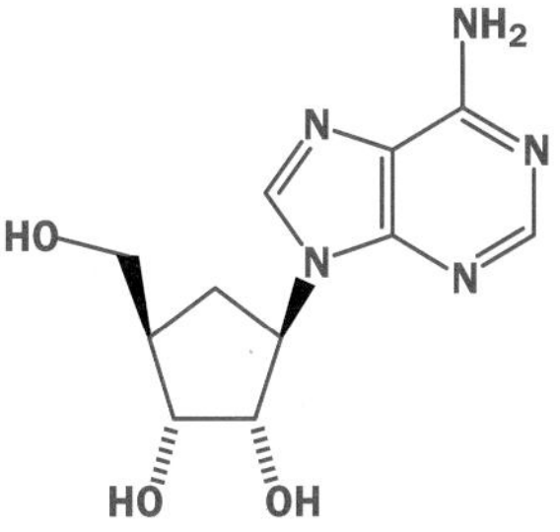

Trost and his group have used both of these palladium-catalysed alkylations in a synthesis of aristeromycin from epoxycyclopentadiene. The *cis* stereochemistry of this carbocyclic nucleotide analogue is of paramount importance and was completely controlled by retention of configuration in both substitutions.

The first reaction is between epoxycyclopentadiene and adenine, one of the heterocyclic building blocks of nucleic acids, and follows the course we have just described to give a *cis*-1,4-disubstituted cyclopentene. The alcohol is then activated by conversion into the carbonate, which reacts with phenylsulfonylnitromethane, which could later be converted into an alcohol. Once again, retention of stereochemistry during the palladium-catalysed substitution gives the *cis* product.

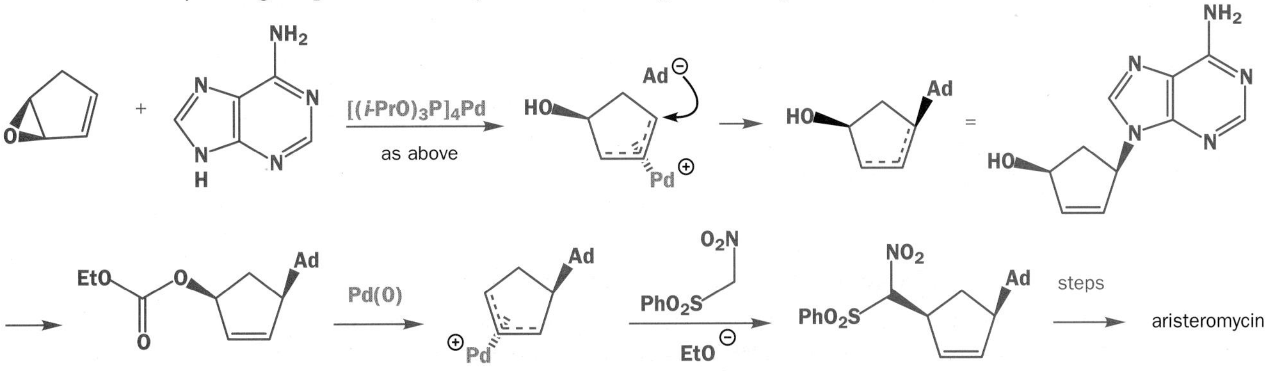

aristeromycin

■ Chapter 49 describes the importance and chemistry of nucleic acids in detail.

Intramolecular alkylations lead to ring synthesis

π-Allyl intermediates may also be used in cyclization reactions including the synthesis of small and medium-sized rings using an intramolecular nucleophilic displacement. Three-membered rings form surprisingly easily taking advantage of the fact that the leaving group can be remote from the nucleophile. The precursors can also be prepared by allylic alkylation. The sodium salts of malonate esters react with the monoacetate under palladium catalysis to give the allylic alcohol. Acetylation activates the second alcohol to displacement so that the combination of sodium hydride as base and palladium(0) catalyst leads to cyclization to the cyclopropane. The regioselectivity of the cyclization is presumably governed by steric hindrance as is usual for allylic alkylations with palladium(0).

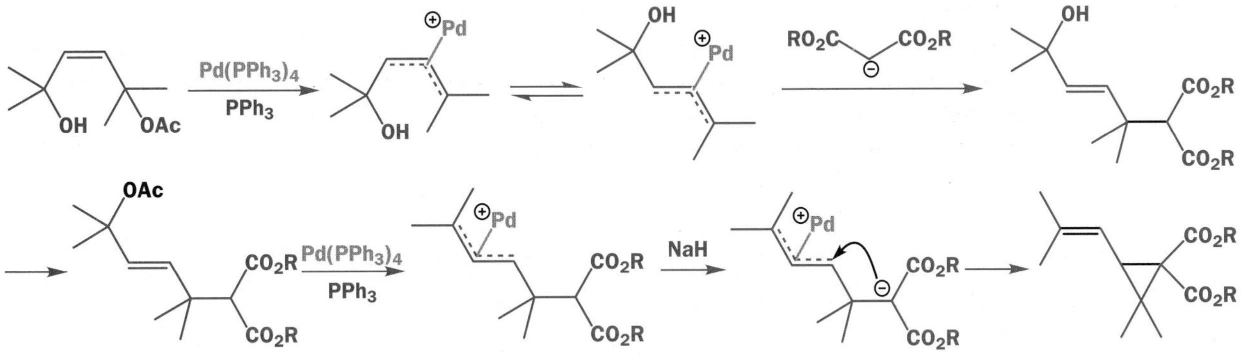

Optically pure ligands on Pd in allylic alkylation can give good enantiomeric excess. You have already seen the first chiral amino-phosphine as the ligand in a chiral Heck reaction and it also gives excellent results in this example. It has to be said, however, that this is a very well behaved example and the next one is more impressive.

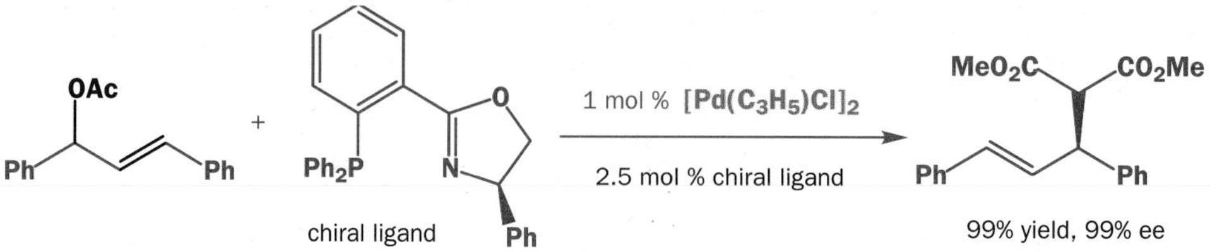

A C_2 symmetric bis(amidophosphine) ligand was used by Trost to prepare the natural nucleoside adenosine (see Chapter 49 for nucleosides) in similar fashion to the carbocyclic analogue described above. The key enantioselective step was the first allylic alkylation that selected between two enantiotopic benzoates in the *meso* dihydrofuran derivative to give one enantiomer the expected *cis* product.

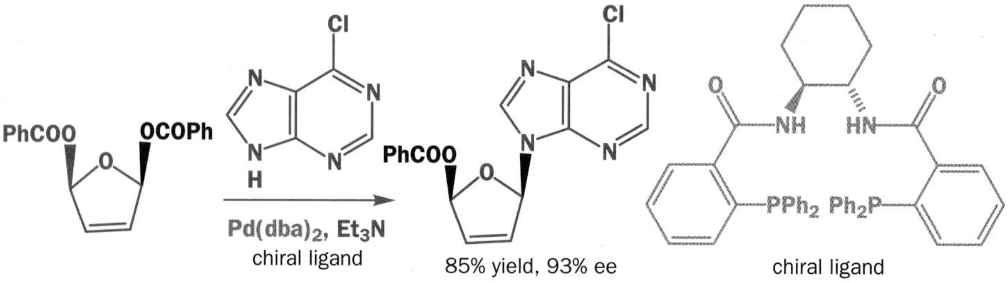

The second benzoate is displaced by a malonate anion, which allows the CH_2OH group to be added at the other side of the dihydrofuran. No enantioselectivity is needed in this step—it is enough to ensure *cis* addition in a 1,4-sense.

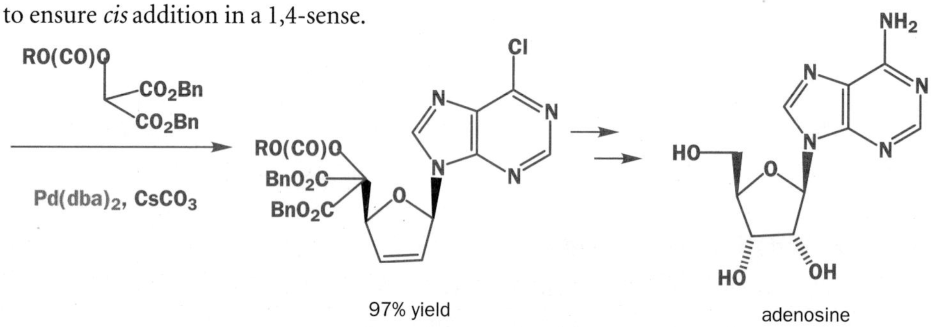

Palladium can catalyse cycloaddition reactions

■
Cycloadditions were described in Chapter 35.

The presence of five-membered rings such as cyclopentanes, cyclopentenes, and dihydrofurans in a wide range of target molecules has led to a variety of methods for their preparation. One of the most successful of these is the use of trimethylenemethane [3 + 2] cycloaddition, catalysed by palladium(0) complexes. The trimethylenemethane unit in these reactions is derived from 2-[(trimethylsilyl)methyl]-2-propen-1-yl acetate which is at the same time an allyl silane and an allylic acetate. This makes it a weak nucleophile and an electrophile in the presence of palladium(0). Formation of the palladium π-allyl complex is followed by removal of the trimethylsilyl group by nucleophilic attack of the resulting acetate ion, thus producing a zwitterionic palladium complex that can undergo cycloaddition reactions.

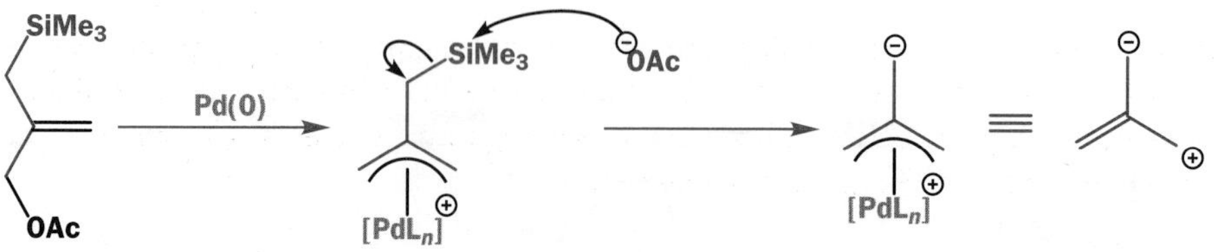

Trimethylene methane

The symmetrical molecule with three CH_2 groups arranged trigonally about a carbon atom is interesting theoretically. It could have a singlet structure with two charges, both of which can be delocalized, but no neutral form can be drawn. Alternatively, it could be a triplet with the two unpaired electrons equally delocalized over the three CH_2

groups. This form is probably preferred and the singlet form is definitely known only as the palladium complex we are now describing. You might compare the singlet and triplet structures of trimethylene methane with those of carbenes in Chapter 40.

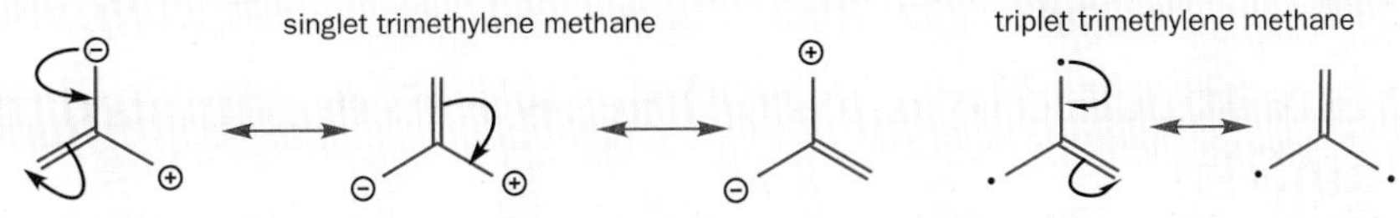

singlet trimethylene methane   triplet trimethylene methane

The normal course of the reaction is to react with an alkene with electron-withdrawing substituents present, which make the substrate prone to Michael-type conjugate addition. The resulting cyclization product has an *exo* methylene group. Cyclopentenones illustrate this overall 'cycloaddition' nicely. The mechanism is thought to be stepwise with conjugate addition of the carbanion followed by attack of the resulting enolate on the π-allyl palladium unit to form a five-membered ring—not a real cycloaddition at all.

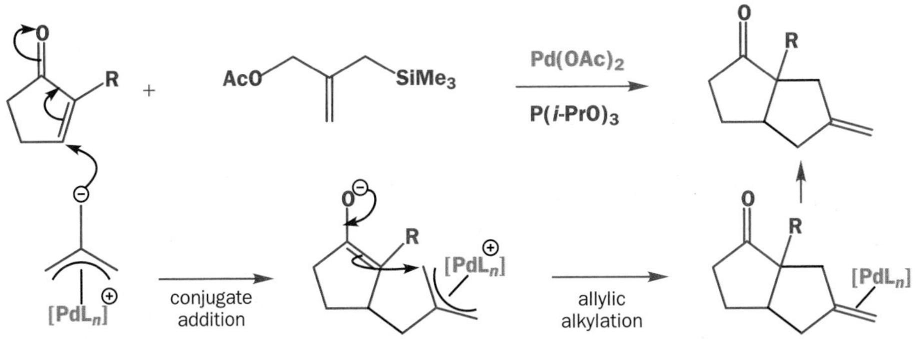

Heteroatom couplings produce aryl– or vinyl– N, –S, or –P bonds

While the major use for palladium catalysis is to make carbon–carbon bonds, which are difficult to make using conventional reactions, the success of this approach has recently led to its application to forming carbon–heteroatom bonds as well. The overall result is a nucleophilic substitution at a vinylic or aromatic centre, which would not normally be possible. A range of aromatic amines can be prepared directly from the corresponding bromides, iodides, or triflates and the required amine in the presence of palladium(0) and a strong alkoxide base. Similarly, lithium thiolates couple with vinylic triflates to give vinyl sulfides provided lithium chloride is present.

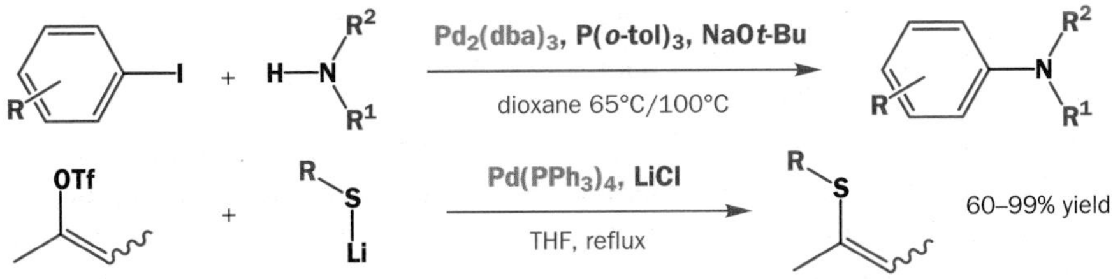

The mechanisms and choice of catalyst, usually a palladium(0) phosphine complex, are the same as those of coupling reactions involving oxidative addition, transmetallation, and reductive elimination. Phosphines do not require additional base for the coupling with aromatic triflates and the reaction has no difficulty in distinguishing the two phosphines present.

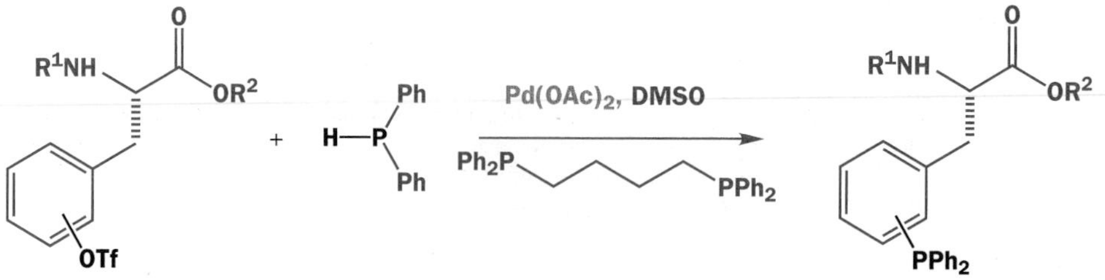

Alkenes are attacked by nucleophiles when coordinated to palladium(II)

The importance of transition-metal-catalysed reactions lies in their ability to facilitate reactions that would not occur under normal conditions. One such reaction is nucleophilic attack on an isolated double bond. While the presence of a conjugating group promotes the attack of nucleophiles, in its absence no such reaction occurs. Coordination of an alkene to a transition metal ion such as palladium(II) changes its reactivity dramatically as electron density is drawn towards the metal and away from the π orbitals of the alkene. This leads to activation towards attack by nucleophiles just as for conjugate addition and unusual chemistry follows. Unusual, that is, for the alkene; the palladium centre behaves exactly as expected.

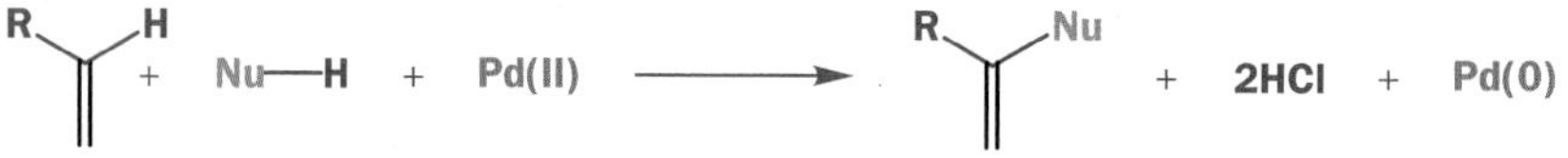

Many nucleophiles, such as water, alcohols, and carboxylates, are compatible with the Pd(II) complex and can attack the complexed alkene from the side opposite the palladium. The attack of the nucleophile is regioselective for the more substituted position. This parallels attack on bromonium ions but is probably governed by the need for the bulky palladium to be in the less hindered position. The resulting Pd(II) σ-alkyl species decomposes by β-hydride elimination to reveal the substituted alkene. Reductive elimination of a proton and the leaving group, usually chloride, leads to palladium(0). The weakness of this reaction is that the catalytic cycle is not complete: Pd(II) not Pd(0) is needed to complex the next alkene.

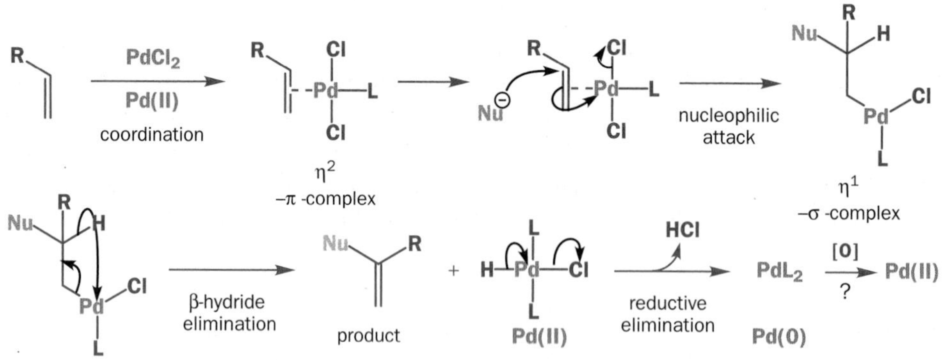

> Unfortunately, this regioselectivity is not the same as in the Heck reaction where attack mostly occurs at the end of the alkene. Internal nucleophiles transferred from the palladium to the alkene usually prefer the end of the alkene but external nucleophiles usually prefer the other end.

A Pd(II) salt such as Pd(OAc)$_2$ adds to an alkene to give, via the π complex, a product with Pd at one end of the alkene and OAc at the other. This is oxypalladation but this product is not usually isolated as it decomposes to the substituted alkene. This reaction is occasionally used with various nucleophiles but it needs a lot of palladium.

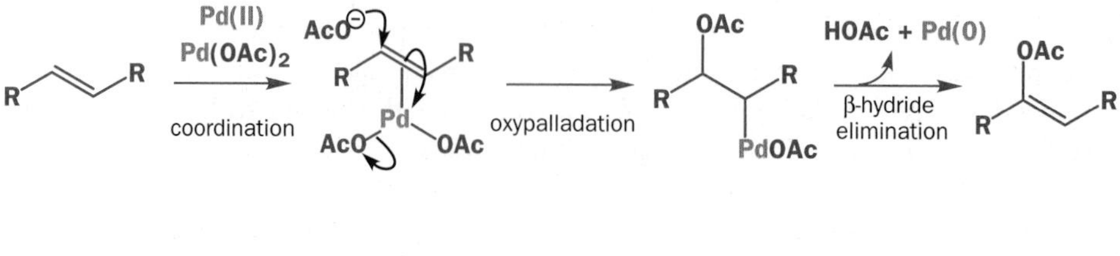

> Please note again that our mechanisms for organometallic steps such as oxypalladation are intended to help organic chemists' understanding and may well be disputed by experts.

Allylic rearrangement by reversible oxypalladation

An example of catalytic oxypalladation is the rearrangement of allylic acetates with Pd(II). The reaction starts with oxypalladation of the alkene and it is the acetate already present in the molecule that provides the nucleophile to attack

the alkene. The intermediate can reverse the oxypalladation in either direction and the product is whichever allylic acetate has the more substituted alkene. In this case, trisubstituted beats monosubstituted easily.

The reaction is *E*-selective, which means that a simple synthesis of an *E*,*Z*-diene is possible from the symmetrical acetate with two *Z*-allylic alkenes. The one that rearranges goes *E* and the one that stays behind remains *Z*. It does not matter which way the acetate goes. The driving force for this rearrangement, from one disubstituted alkene to another, is establishment of conjugation.

There are two solutions to this problem. We could use stoichiometric Pd(II) but this is acceptable only if the product is very valuable or the reaction is performed on a small scale. It is better to use an external oxidant to return the palladium to the Pd(II) oxidation state so that the cycle can continue. Air alone does not react fast enough (even though Pd(0) must be protected from air to avoid oxidation) but, in combination with Cu(II) chloride, oxygen completes the catalytic cycle. The Cu(II) chloride oxidizes Pd(0) to Pd(II) and is itself oxidized back to Cu(II) by oxygen, ready to oxidize more palladium.

This combination of reagents has been used to oxidize terminal vinyl groups to methyl ketones and is known as the **Wacker oxidation**. The nucleophile is simply water, which attacks the activated alkene at the more substituted end in an oxypalladation step. β-Hydride elimination from the resulting σ-alkyl palladium complex releases the enol, which is rapidly converted into the more stable keto form. Overall, the reaction is a hydration of a terminal alkene that can tolerate a range of functional groups.

A related reaction is the oxidation of silyl enol ethers to enones. This requires stoichiometric palladium(II), though reoxidation of Pd(0) with benzoquinone can cut that down to about half an equivalent, but does ensure that the alkene is on the right side of the ketone. The first step is again oxypalladation and β elimination puts the alkene in conjugation with the ketone chiefly because there are no β hydrogens on the other side.

Alcohols and amines are excellent intramolecular nucleophiles

Cyclic ethers and amines can be formed if the nucleophile is an intramolecular alcohol or amine. Stoichiometric palladium can be avoided by using benzoquinone as the stoichiometric oxidant with a catalytic amount of palladium. In this example intramolecular oxypalladation of a diene is followed by attack of an external nucleophile on a π-allyl complex.

1,4-benzoquinone   73% yield; 99:1 *syn*:*anti*

Palladium coordinates to one face of the diene promoting intramolecular attack by the alcohol on the opposite face. The resulting σ-alkyl palladium can form a π-allyl complex with the palladium on the lower face simply by sliding along to interact with the double bond. Nucleophilic attack of chloride from the lithium salt then proceeds in the usual way on the face opposite palladium. The overall addition to the diene is therefore *cis*.

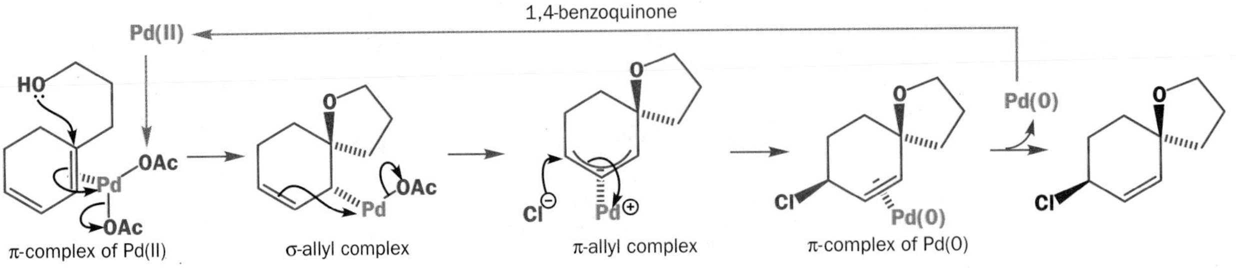

Nitrogen nucleophiles also attack alkenes activated by Pd(II) and benzoquinone can again act as a reoxidant allowing the use of catalytic quantities of palladium. The mechanism follows the same pattern as for oxygen nucleophiles including the final isomerization to produce the most stable regioisomer of product. In this example the product is an aromatic indole (Chapter 43) so the double bond migrates into the five-membered ring.

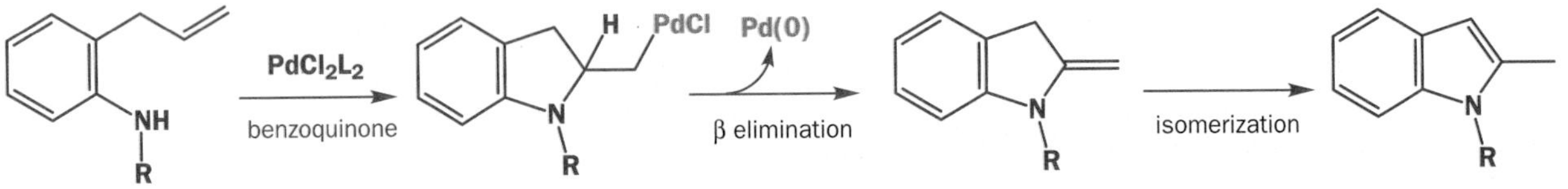

If the substrate lacks a hydrogen suitable for β elimination and there is another alkene present in the molecule, the σ-alkyl palladium intermediate can follow a Heck pathway to form a bicyclic structure in a tandem reaction sequence. Once again, the final step is a palladium-hydride-mediated isomerization to give the endocyclic alkene.

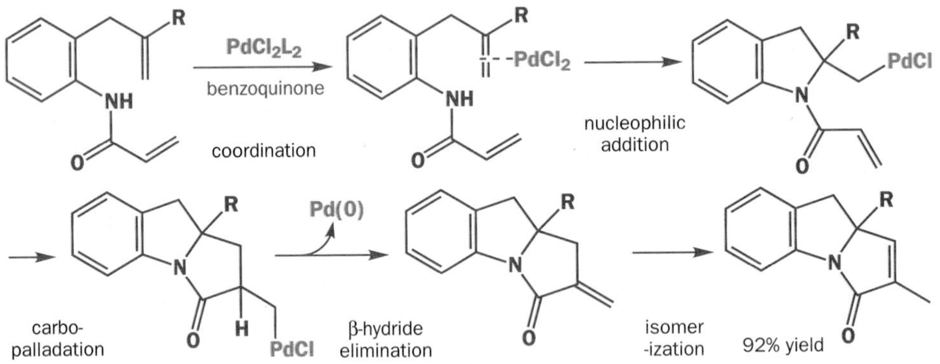

Palladium catalysis in the total synthesis of a natural alkaloid

We end this section with a synthesis of *N*-acetyl clavicipitic acid methyl ester, an ergot alkaloid, by Hegedus. The power of organo-transition-metal chemistry is illustrated in five steps of this seven-step process. Each of the organometallic steps catalysed by Pd(0) or Pd(II) has been described in this chapter. The overall yield is 18%, a good result for a molecule of such complexity.

The first step is to make an indole by Pd(II)-catalysed cyclization in the presence of benzoquinone as reoxidant. The nucleophilic nature of the 3-position of the indole (Chapter 43) was exploited to introduce the required iodine functionality. Rather than direct iodination, a high yielding two-step procedure involving mercuration followed by iodination was employed. The more reactive iodide was then involved in a Heck coupling with an unsaturated side chain in the absence of phosphine ligands. The remaining aromatic bromide then underwent a second Heck reaction with an allylic alcohol to introduce the second side chain. Cyclization of the amide on to the allylic alcohol was achieved with palladium catalysis, not as might have been expected with palladium(0) but instead with palladium(II), to produce the seven-membered ring. Finally, the conjugated double bond was reduced and the sulfonamide removed with sodium borohydride with photolysis.

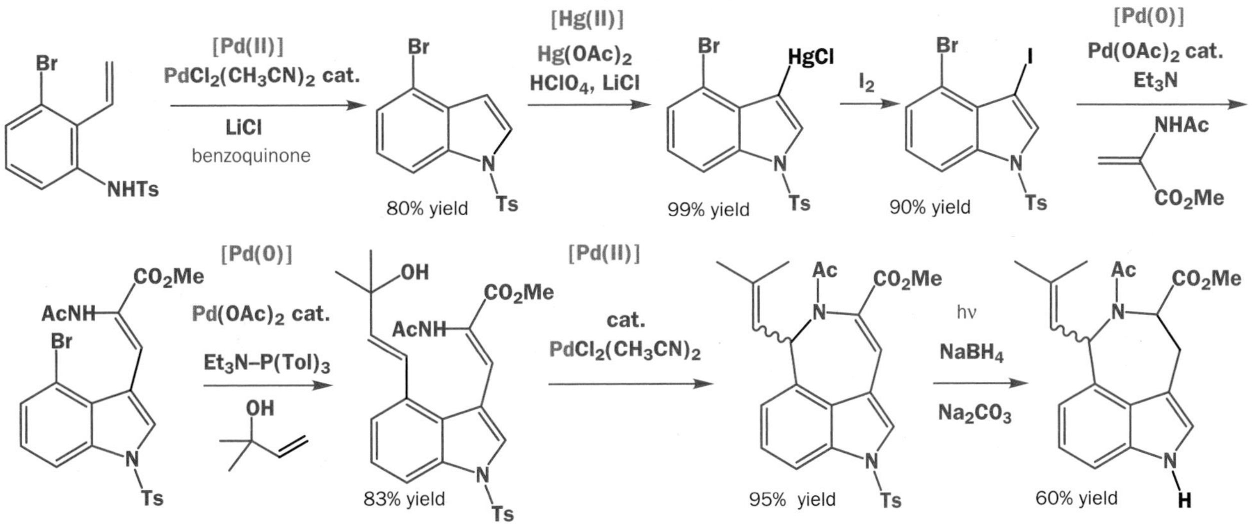

Other transition metals: cobalt

We have concentrated on palladium because it is the most important of the transition metals but we must not leave you with the idea that it is the only one. We shall end with two reactions unique to cobalt—the **Pauson–Khand reaction** that we mentioned right at the start of the chapter and the **Vollhardt co-trimerization**. You will see at once that cobalt has a special affinity with alkynes and with carbon monoxide.

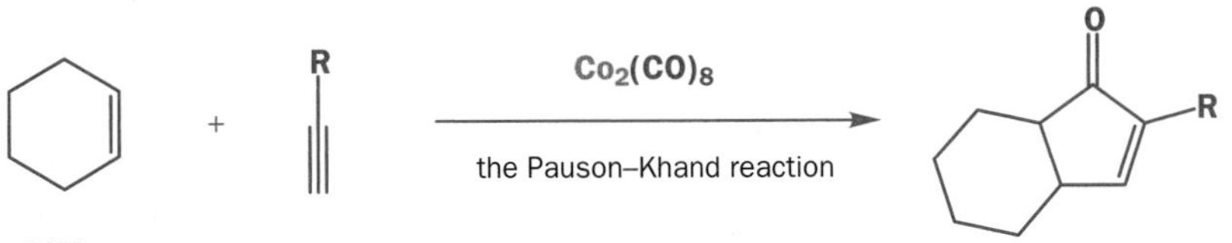

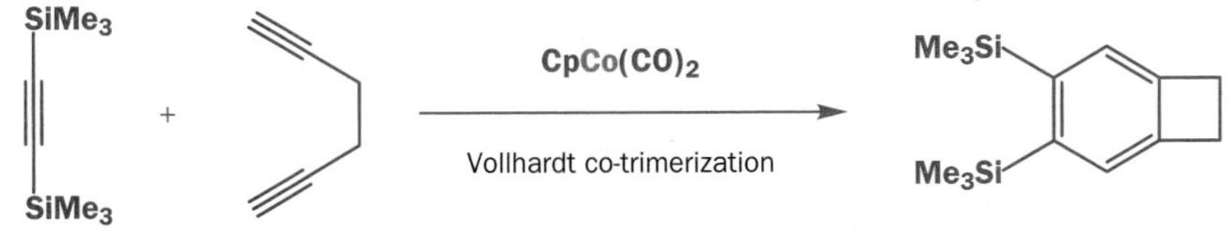

▶
Take care to distinguish between Co and CO in these reactions.

The structure of the cobalt reagents is worth a mention. Cobalt has nine electrons so the second reagent is easy: nine from Co, five from the cyclopentadienyl, and two each from the two COs giving 18 in all. But why is the first reagent a dimer? The monomer $Co(CO)_4$ would have $9 + 8 = 17$ electrons.

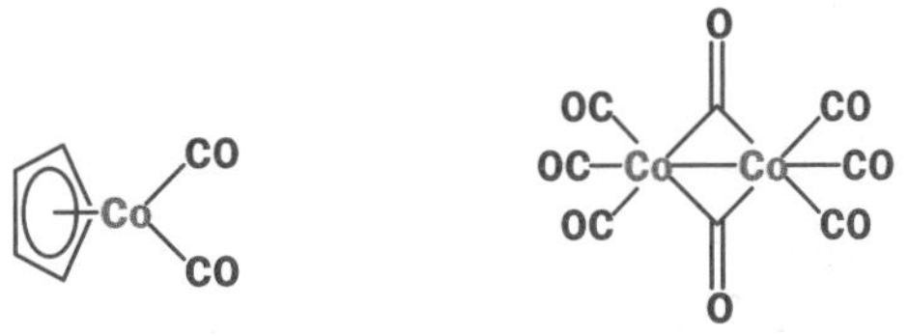

18-electron complex of Co(0)

The Pauson–Khand reaction starts with the replacement of two CO molecules, one from each Co atom, with the alkyne to form a double σ complex with two C–Co σ bonds, again one to each Co atom. One CO molecule is then replaced by the alkene and this π complex in its turn gives a σ complex with one C–Co σ bond and one new C–C σ bond, and a C–Co bond is sacrificed in a ligand coupling reaction. Then a carbonyl insertion follows and reductive elimination gives the product, initially as a cobalt complex.

In the middle few structures, showing the vital steps, we omit all CO molecules except the one that reacts.

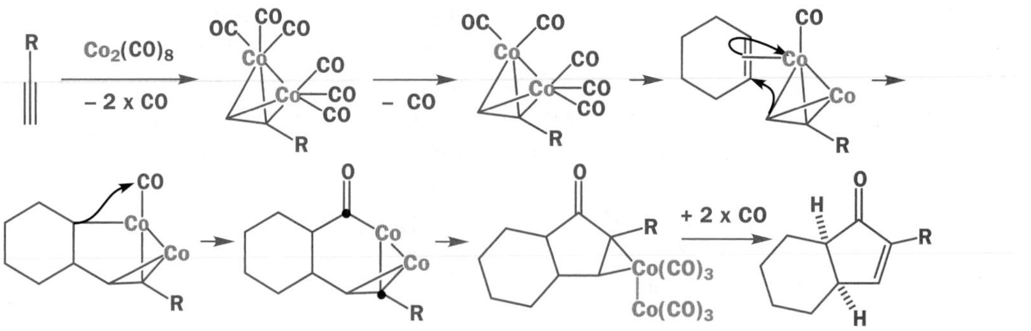

This is an extraordinary reaction because so much seems to happen with no control except the presence of the two cobalt atoms. The alkene reacts so that the more substituted end bonds to the carbonyl group. This is because the ligand coupling occurs to the less substituted end, as in other coupling reactions. The stereochemistry of the alkene is preserved because the coupling step puts the C–C and C–Co bonds in at the same time in a *syn* fashion and the migration to the CO ligand is stereospecific with retention. This is one of the most complicated mechanisms you are likely to meet and few organic chemists can draw it out without looking it up.

The Vollhardt co-trimerization is so-called because it uses cobalt to bring three alkynes into a ring and it is one of the rare ways of making a benzene ring in one step. First, the dialkyne complexes with the cobalt—each alkyne replaces one CO molecule. Then the double π complex rearranges to a double σ complex by a cycloaddition forming a new C–C σ bond. This new five-membered ring cobalt heterocycle has only 16 electrons so it can accept the remaining alkyne to give an 18-electron complex.

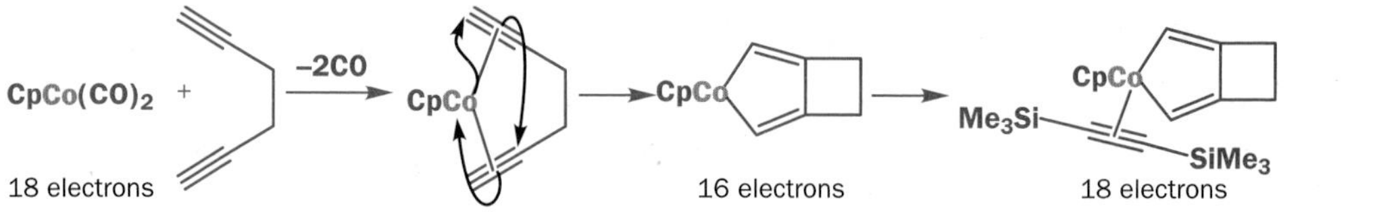

There are now two possible routes to the final product. Reductive elimination would insert the new alkyne into one of the old C–Co bonds and form a seven-membered ring heterocycle. This could close in an electrocyclic reaction to give the new six-membered ring with the cobalt fused on one side and hence the cobalt complex of the new benzene.

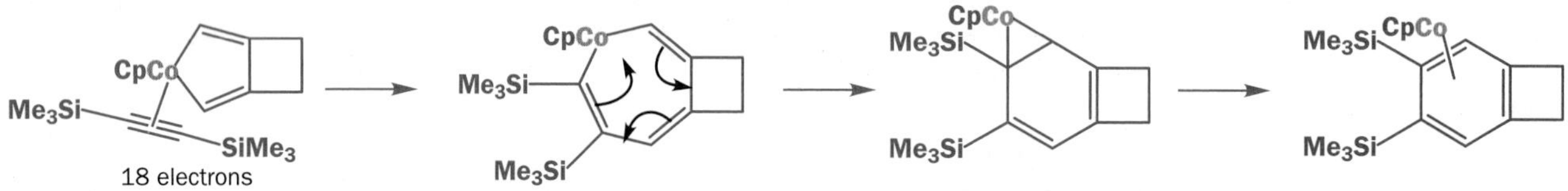

Alternatively, the new alkyne could do a Diels–Alder reaction on the five-membered cobalt heterocycle to give a bridged six-membered ring that could extrude cobalt to give the same benzene complex. The CpCo group can form a stable complex with only four of the benzene electrons and these can be profitably exchanged for two molecules of carbon monoxide to re-form the original catalyst.

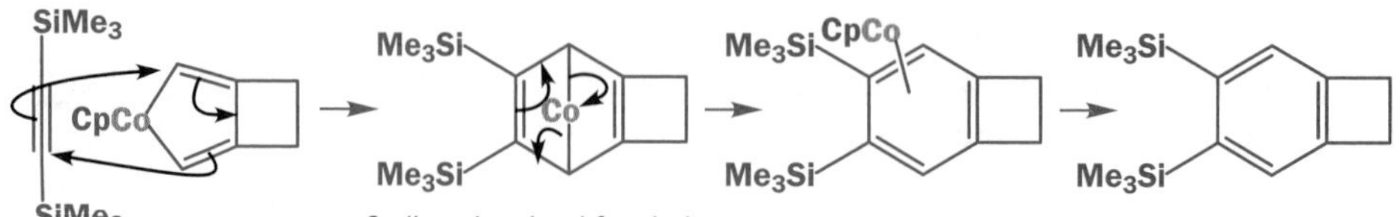

We have selected a few reactions of Co, Fe, and Cu with honourable mentions for Pt, Ir, and Cr. We could have focused on other elements—Ni, W, Ti, Zr, Mn, Ru, and Rh all have special reactions. Transition metal chemistry, particularly involving palladium catalysis, occupies a central role in modern organic synthesis because complex structures can be assembled in few steps with impressive regio- and stereochemical control. There are many books devoted entirely to this subject if you wish to take it further.

Steroid synthesis by the Vollhardt co-trimerization

This product is interesting for two further reactions that revise chemistry from Chapters 36 and 47. If the original acetylene has a special substituent this emerges from the co-trimerization on the four-membered ring.

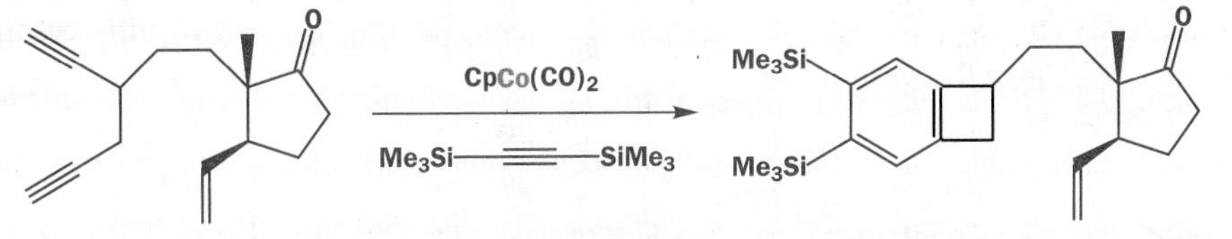

Heating the benzocyclobutene causes an electrocyclic opening (Chapter 36) of the four-membered ring to give a diene that does an intramolecular Diels–Alder rearrangement on the alkene attached to the five-membered ring. The product has the skeleton of the steroids (Chapter 51). The product is not a steroid because steroids do not have Me₃Si groups, but these can be removed (Chapter 47) by

4-electron conrotatory electrocyclic reaction

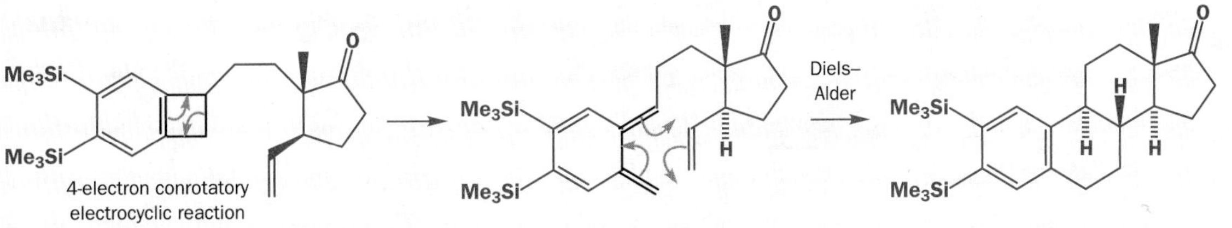

This compound is not a steroid because steroids do not have Me₃Si groups, but these can be removed (Chapter 47) by

protodesilylation and this sequence is a very short synthesis of an important compound.

Problems

1. Suggest mechanisms for these reactions, explaining the role of palladium in the first step.

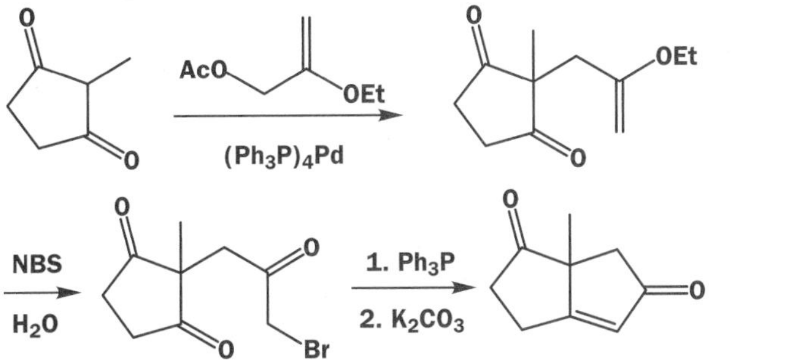

2. This Heck style reaction does not lead to regeneration of the alkene. Why not? What is the purpose of the formic acid (HCO_2H) in the reaction mixture?

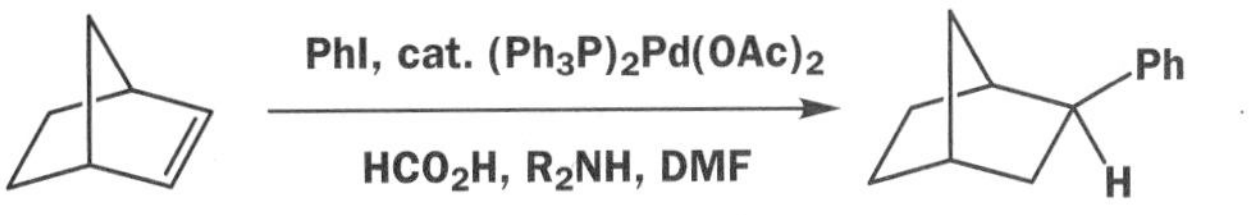

3. Cyclization of this unsaturated amine with catalytic Pd(II) under an atmosphere of oxygen gives a cyclic unsaturated amine in 95% yield. How does the reaction work? Why is the atmosphere of oxygen necessary? Explain the stereo- and regiochemistry of the reaction. How would you remove the CO_2Bn group from the product?

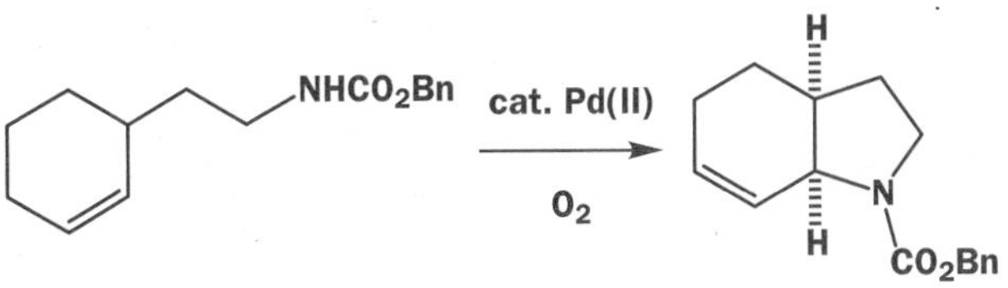

4. Suggest a mechanism for this lactone synthesis.

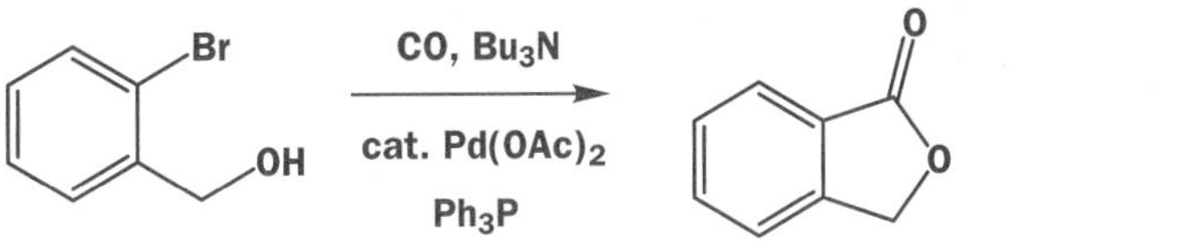

5. Explain why enantiomerically pure lactone gives all *syn* but racemic product in this palladium-catalysed reaction.

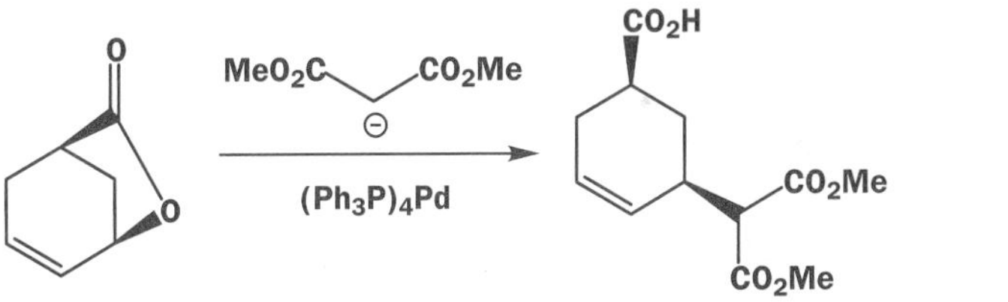

(–)-lactone   all *syn* but racemic

6. Revision of Chapter 47. The synthesis of a bridged tricyclic amine shown below starts with an enantiomerically pure allyl silane. Give mechanisms for the reactions, explaining how the stereochemistry is controlled in each step.

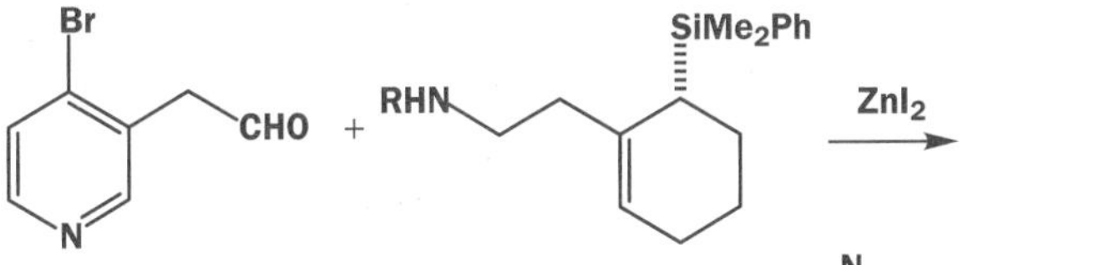

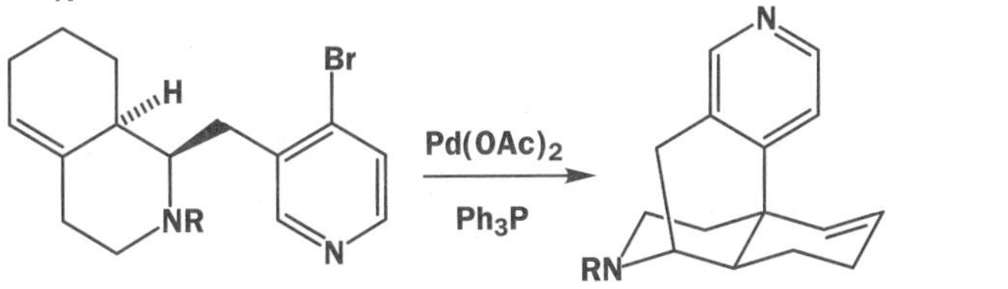

7. Revision of Chapter 44. Explain the reactions in this sequence commenting on the regioselectivity of the organometallic steps.

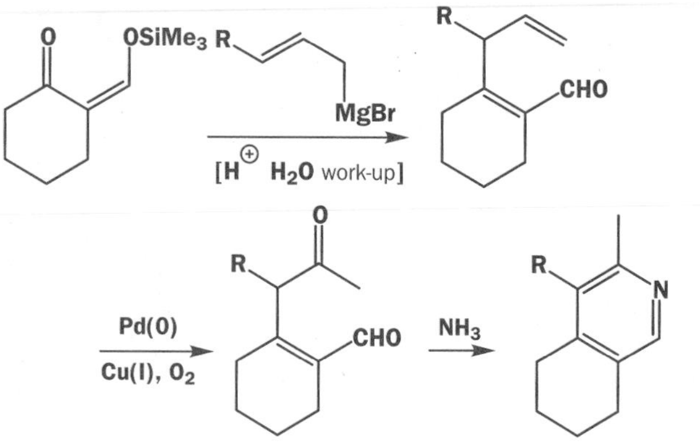

8. Give a mechanism for this carbonylation reaction. Comment on the stereochemistry and explain why the yield is higher if the reaction is carried out under a carbon monoxide atmosphere.

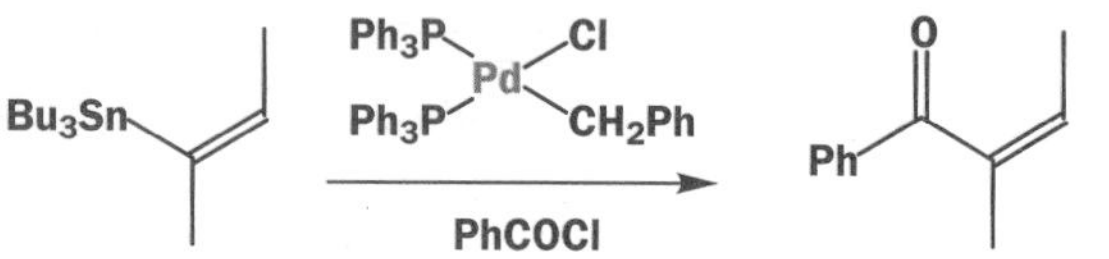

Hence explain this synthesis of part of the antifungal compound pyrenophorin.

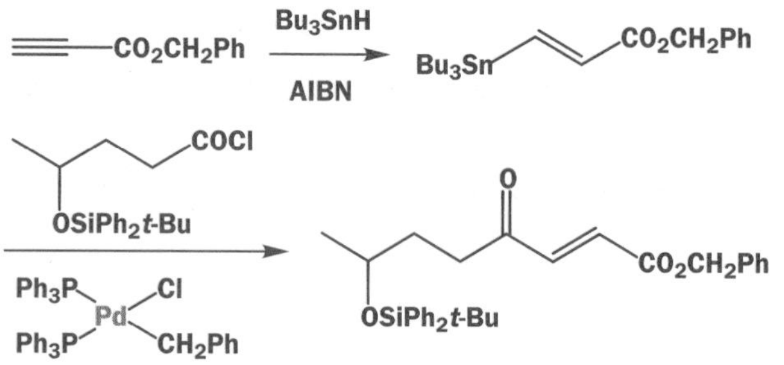

9. Explain the mechanism and stereochemistry of these reactions. The first is revision and the second is rather easy!

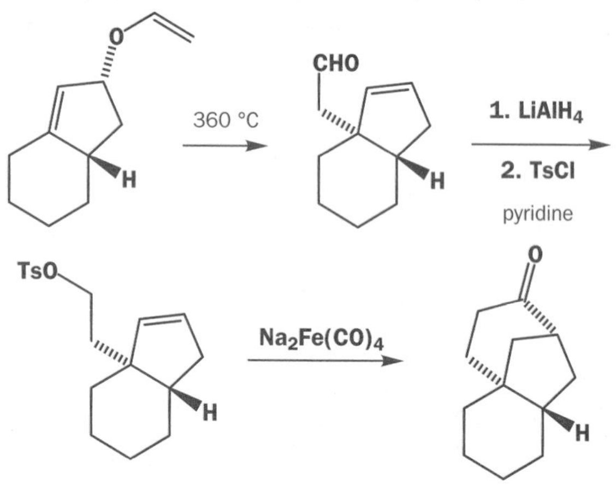

10. The synthesis of an antifungal drug was completed by this palladium-catalysed reaction. Give a mechanism and explain the regio- and stereoselectivity.

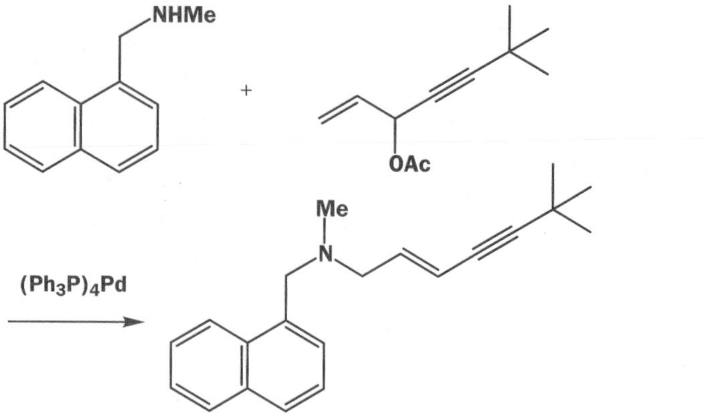

11. Some revision content. Work out the structures of the compounds in this sequence and suggest mechanisms for the reactions, explaining any selectivity.

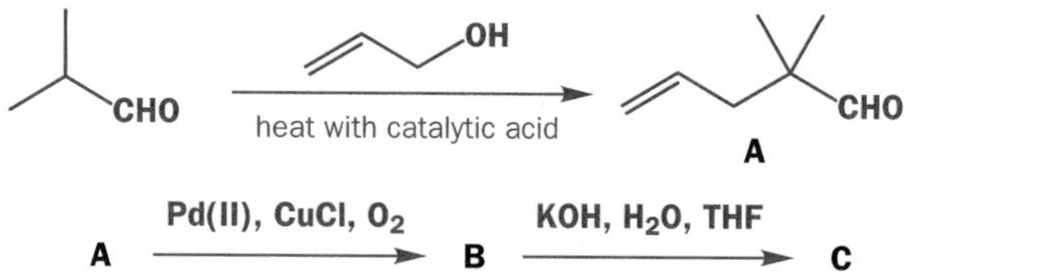

B has IR: 1730, 1710 cm^{-1}; δ_H (p.p.m.) 9.4 (1H, s), 2.6 (2H, s), 2.0 (3H, s), and 1.0 (6H, s).

C has IR: 1710 cm^{-1}; δ_H (p.p.m.) 7.3 (1H, d, J 5.5 Hz), 6.8 (1H, d, J 5.5 Hz), 2.1 (2H, s), and 1.15 (6H, s).

12. Revision of Chapter 36. What would be the starting materials for the synthesis of these cyclopentenones by the Nazarov reaction and by the Pauson–Khand reaction? Which do you prefer in each case?

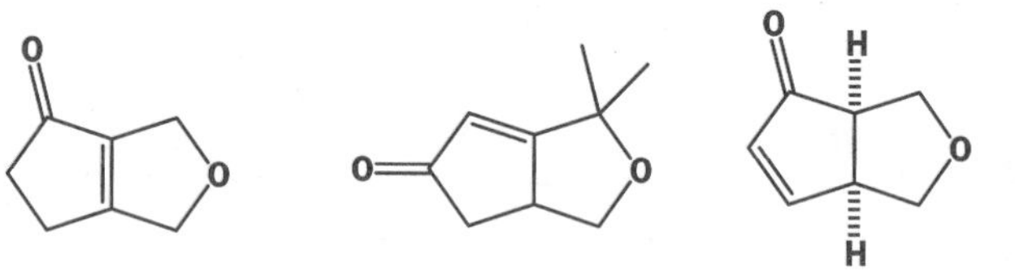

13. A variation on the Vollhardt co-trimerization allows the synthesis of substituted pyridines. Draw the structures of the intermediates in this sequence. In the presence of an excess of the cyanoacetate a second product is formed. Account for this too.

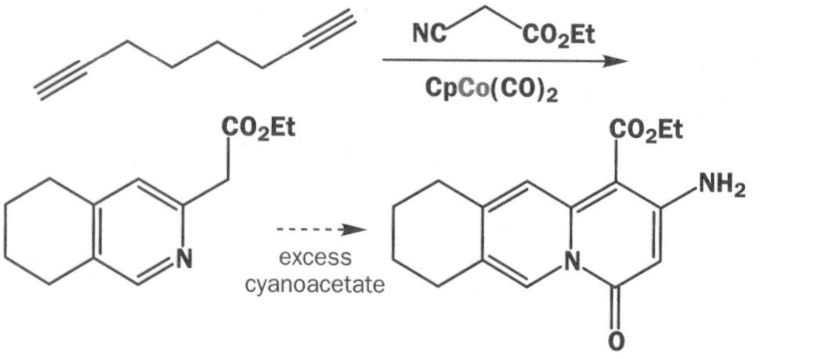

14. The synthesis of the Bristol–Myers Squibb anti-migraine drug Avitriptan (a 5-HT1D receptor antagonist) involves this palladium-catalysed indole synthesis. Suggest a mechanism and comment on the regioselectivity of the alkyne attachment.

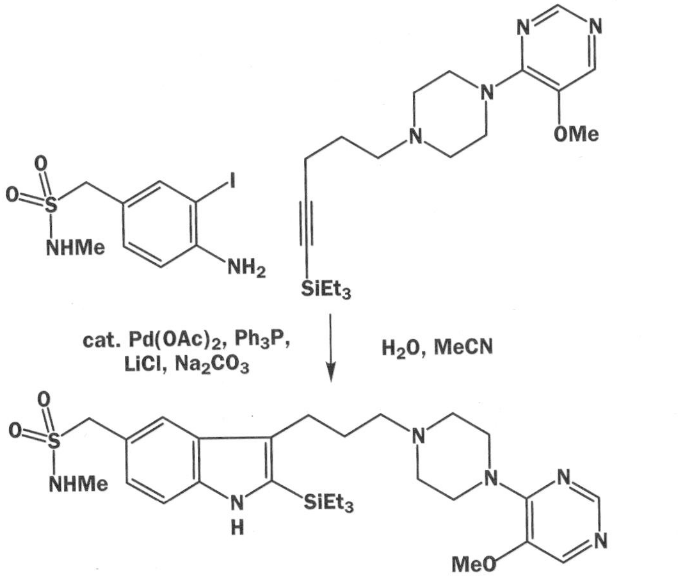

15. A synthesis of the natural product γ-lycorane starts with a palladium-catalysed reaction. What sort of a reaction is this, and how does it work?

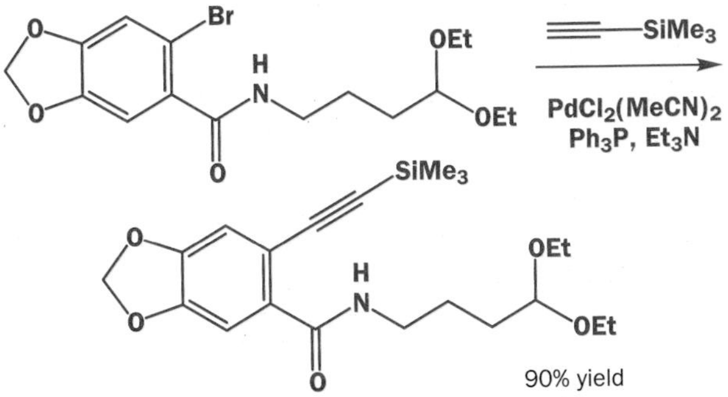

90% yield

The next two steps are a bit of revision: draw mechanisms for them and comment on the survival of the Me₃Si group.

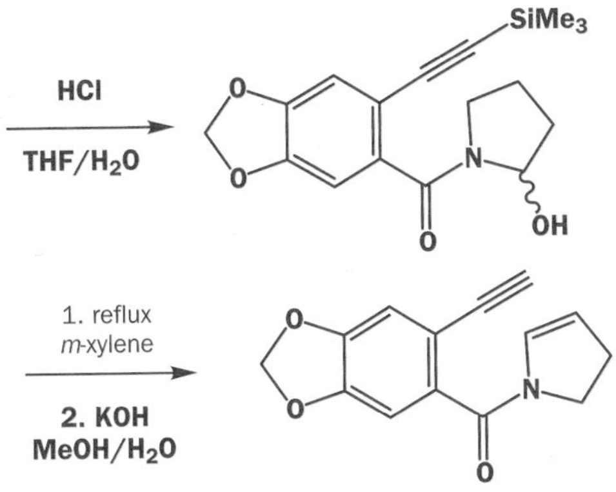

Now the key step—and you should recognize this easily. What is happening here? Though the product is a mixture of isomers, this does not matter. Why not?

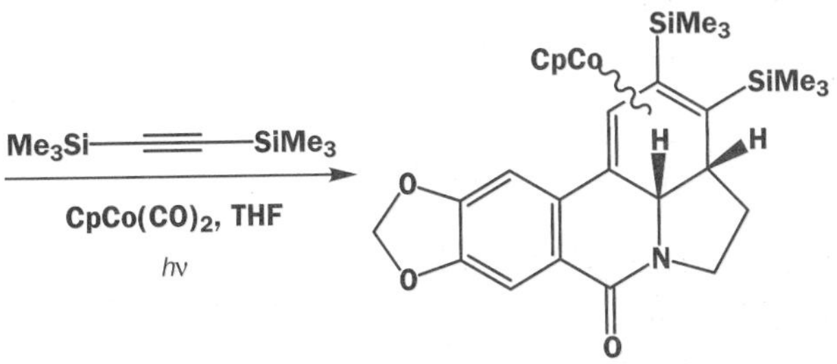

65% yield, 3:2 Co up : Co down

Finally, this mixture must be converted into γ-lycorane: suggest how this might be done.

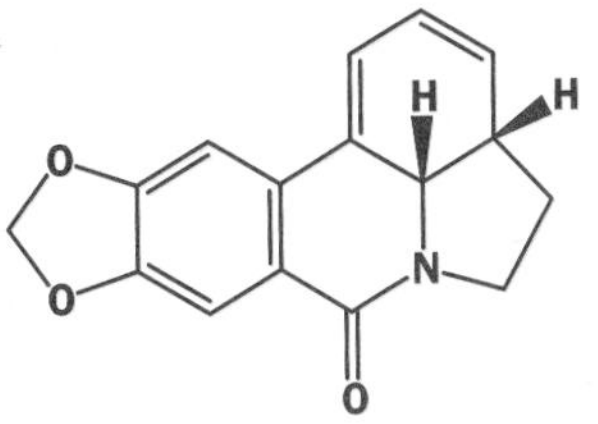

γ-lycorane

The chemistry of life

49

Connections

Building on:
- Acidity and basicity ch8
- Carbonyl chemistry ch12 & ch14
- Stereochemistry ch16
- Conformational analysis ch18
- Enolate chemistry and synthesis ch24–ch30
- Heterocycles ch42–ch44
- Asymmetric synthesis ch45
- Sulfur chemistry ch46

Arriving at:
- Nucleic acids store information for the synthesis of proteins
- Modified nucleosides can be used as antiviral drugs
- Nucleotides have a role in energy storage
- Proteins catalyse reactions and provide structure
- Other amino acid derivatives act as methylating and reducing agents
- Sugars store energy, enable recognition, and protect sensitive functional groups
- How to make and manipulate sugars and their derivatives
- Lipids form the basis of membrane structures

Looking forward to:
- Mechanisms in biological chemistry ch50
- Natural products ch51
- Polymers ch52

Life runs on chemistry, and the chemical side of biology is fascinating for that reason alone. But from the point of view of a textbook, biological chemistry's combination of structures, mechanisms, new reactions, and synthesis is also an ideal revision aid. We shall treat this chemistry of living things in three chapters.

- Chapter 49 introduces the basic molecules of life and explains their roles along with some of their chemistry
- Chapter 50 discusses the mechanisms of biological reactions
- Chapter 51 develops the chemistry of compounds produced by life: natural products

We start with the most fundamental molecules and reactions in what is called **primary metabolism**.

Primary metabolism

It is humbling to realize that the same molecules are present in all living things from the simplest single-cell creatures to ourselves. Nucleic acids contain the genetic information of every organism, and they control the synthesis of proteins. Proteins are partly structural—as in connective tissue—and partly functional—as in enzymes, the catalysts for biological reactions. Sugars and lipids used to be the poor relations of the other two but we now realize that, as well as having a structural role in membranes, they are closely associated with proteins and have a vital part to play in recognition and transport.

The chart overleaf shows the molecules of primary metabolism and the connections between them, and needs some explanation. It shows a simplified relationship between the key structures (emphasized in large black type). It shows their origins—from CO_2 in the first instance—and picks out some important intermediates. Glucose, pyruvic acid, citric acid, acetyl coenzyme A (Acetyl CoA), and ribose are players on the centre stage of our metabolism and are built into many important molecules.

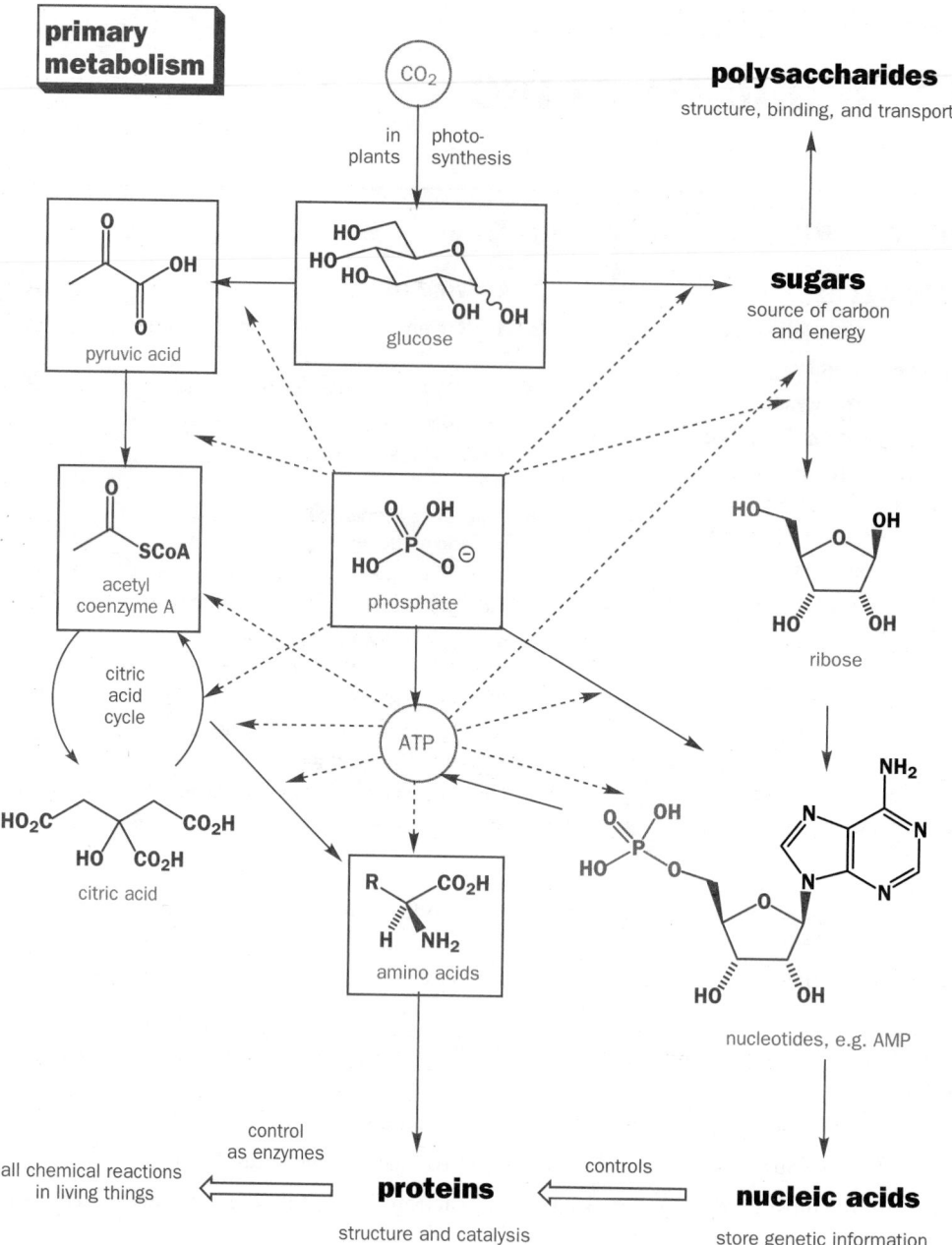

primary metabolism

polysaccharides
structure, binding, and transport

CO_2

in plants │ photo-synthesis

pyruvic acid

glucose

sugars
source of carbon and energy

acetyl coenzyme A

phosphate

ribose

citric acid cycle

ATP

citric acid

amino acids

nucleotides, e.g. AMP

control as enzymes

all chemical reactions in living things

controls

proteins
structure and catalysis

nucleic acids
store genetic information

The arrows used in the chart have three functions.

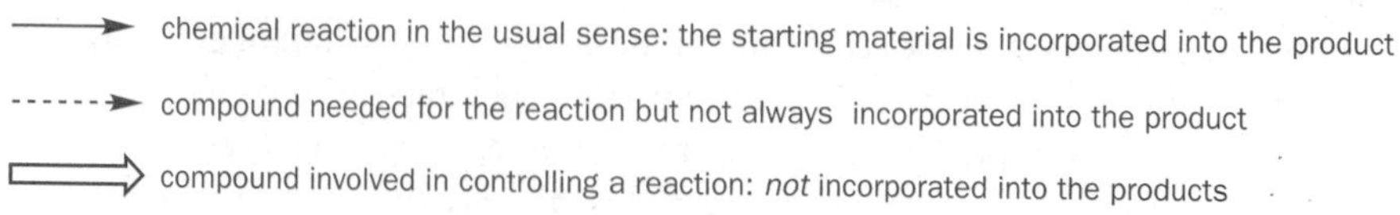

⟶ chemical reaction in the usual sense: the starting material is incorporated into the product

- - - ▸ compound needed for the reaction but not always incorporated into the product

⟹ compound involved in controlling a reaction: *not* incorporated into the products

We hope that this chart will allow you to keep track of the relationships between the molecules of metabolism as you develop a more detailed understanding of them. We will now look briefly at each type of molecule.

Life begins with nucleic acids

Nucleic acids are unquestionably top level molecules because they store our genetic information. They are polymers whose building blocks (monomers) are the **nucleotides**, themselves made of three parts—a heterocyclic base, a sugar, and a phosphate ester. A **nucleo*side*** lacks the phosphate. In the example alongside, adenine is the base (black), adeno*sine* is the nucleo*side* (base and sugar), and the nucleotide is the whole molecule (base + sugar + phosphate).

the phosphate ester group
the pyrimidine base (adenine)
the sugar (ribose)
a nucleotide (AMP)

This nucleotide is called **AMP**—Adenosine MonoPhosphate. Phosphates are key compounds in nature because they form useful stable linkages between molecules and can also be built up into reactive molecules by simply multiplying the number of phosphate residues. The most important of these nucleotides is also one of the most important molecules in nature—Adenosine TriPhosphate or **ATP**.

hard (O, N)   soft (S)

Adenosine TriPhosphate—ATP

ATP is a highly reactive molecule because phosphates are stable anions and good leaving groups. It can be attacked by hard nucleophiles at a phosphate group (usually the end one) or by soft nucleophiles at the CH_2 group on the sugar. We shall see examples of both reactions soon. When a new reaction is initiated in nature, very often the first step is a reaction with ATP to make the compound more reactive. This is rather like our use of TsCl to make alcohols more reactive or converting acids to acid chlorides to make them more reactive.

There are five heterocyclic bases in DNA and RNA

In nucleic acids there are only five bases, two sugars, and one phosphate group possible. The bases are monocyclic pyrimidines or bicyclic purines and are all aromatic.

- There are only two purine bases found in nucleic acids, adenine (A), which we have already met, and guanine (G)

- The three pyrimidine bases are the simpler and they are uracil (U), thymine (T), and cytosine (C). Cytosine is found in DNA and RNA, uracil in RNA only, and thymine in DNA only.

■ You met pyrimidines on p. 1148 and learned how to make them on p. 1198, but the purine ring system may be new to you. It isn't always easy to find the six (or ten!) electrons in these compounds. Check for yourself that you can do this. You may need to draw delocalized structures especially for U, T, and G.

purine bases in nucleic acids

adenine   guanine

pyrimidine bases in nucleic acids

uracil   thymine   cytosine

The stimulants in tea and coffee are methylated nucleic acid purines

An important natural product for most of us is a fully methylated purine present in tea and coffee—caffeine. Theobromine, the partly methylated version, is present in chocolate, and both caffeine and theobromine act as stimulants. Caffeine is a crystalline substance easily extracted from coffee or tea with organic solvents. It is extracted industrially with liquid CO_2 (or if you prefer 'Nature's natural effervescence') to make decaffeinated tea and coffee.

stimulant purines

caffeine

theobromine

If we, as chemists, were to add those methyl groups we should use something like MeI, but Nature uses a much more complicated reagent. There is a great deal of methylating going on in living things—and the methyl groups are usually added by *S*-adenosyl methionine (or **SAM**), formed by reaction of methionine with ATP.

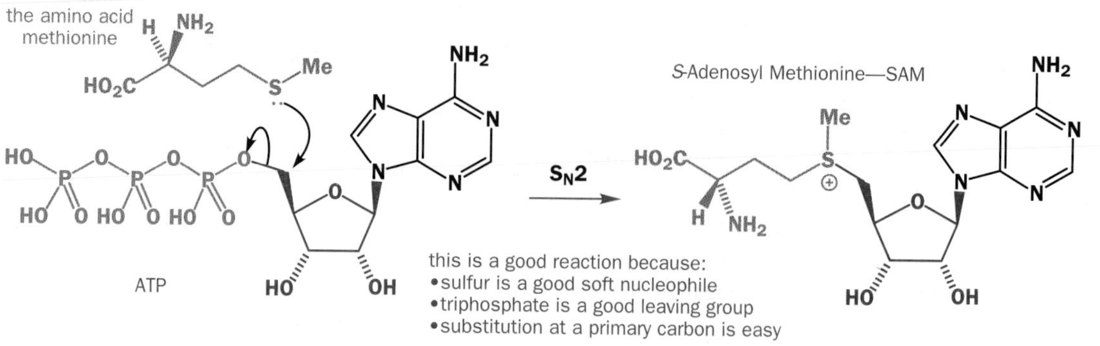

The product (SAM) is a sulfonium salt and could be attacked by nucleophiles at three different carbon atoms. Two are primary centres—good for S_N2 reactions—but the third is the methyl group, which is even better. Many nucleophiles attack SAM in this way.

nucleophilic attack on SAM, Nature's methylating agent

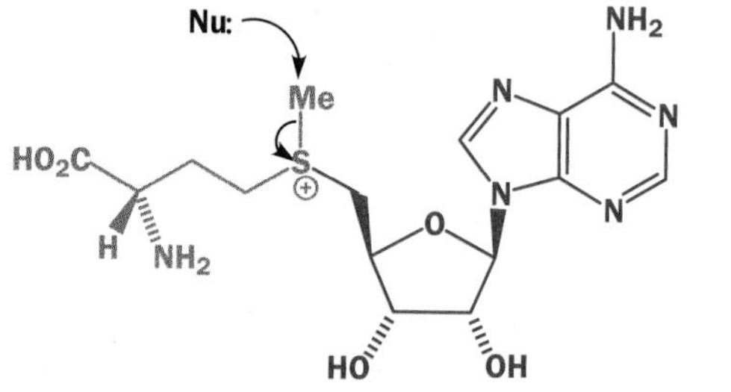

In the coffee plant, theobromine is converted into caffeine with a molecule of SAM. The methylation occurs on nitrogen partly because this preserves both the aromatic ring and the amide functionality and also because the enzyme involved brings the two molecules together in the right orientation for *N*-methylation.

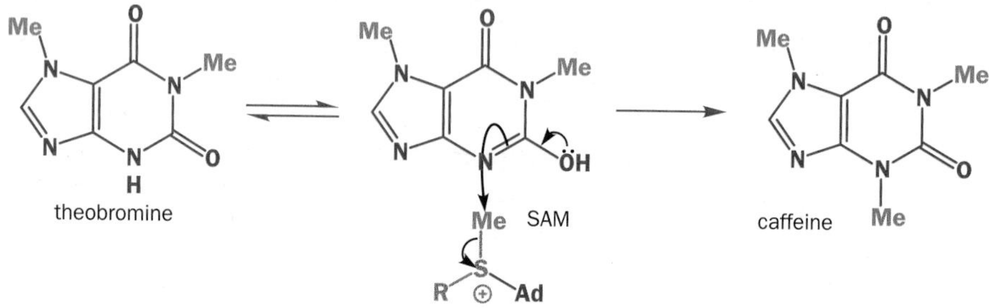

At this point we should just point out something that it's easy to forget: there is *only one chemistry*. There is no magic in biological chemistry, and Nature uses the same chemical principles as we do in the chemical laboratory. All the mechanisms that you have studied so far will help you to draw mechanisms for biological reactions and most reactions that you have met have their counterparts in nature. The difference is that Nature is very very good at chemistry, and all of us are only just learning. We still do much more sophisticated reactions *inside* our bodies without thinking about them than we can do *outside* our bodies with all the most powerful ideas available to us at the beginning of the twenty-first century.

Nucleic acids exist in a double helix

One of the most important discoveries of modern science was the elucidation of the structures of DNA and RNA as the famous double helix by Watson and Crick in 1953. They realized that the basic structure of base–sugar–phosphate was ideal for a three-dimensional coil. The structure of a small part of DNA is shown opposite.

Notice that the 2′ (pronounced 'two prime') position on the ribose ring is vacant. There is no OH group there and that is why it is called *Deoxy*ribo-*N*ucleic *A*cid (DNA). The nucleotides link the two remaining OH groups on the ribose ring and these are called the 3′- and 5′-positions. This piece of DNA has three nucleotides (adenine, adenine, and thymine) and so would be called –AAT– for short.

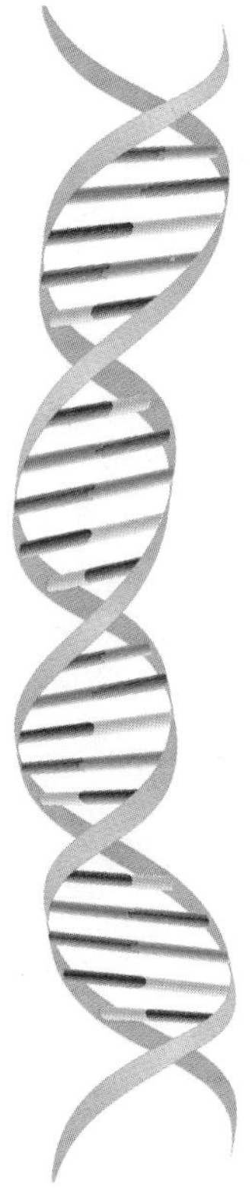

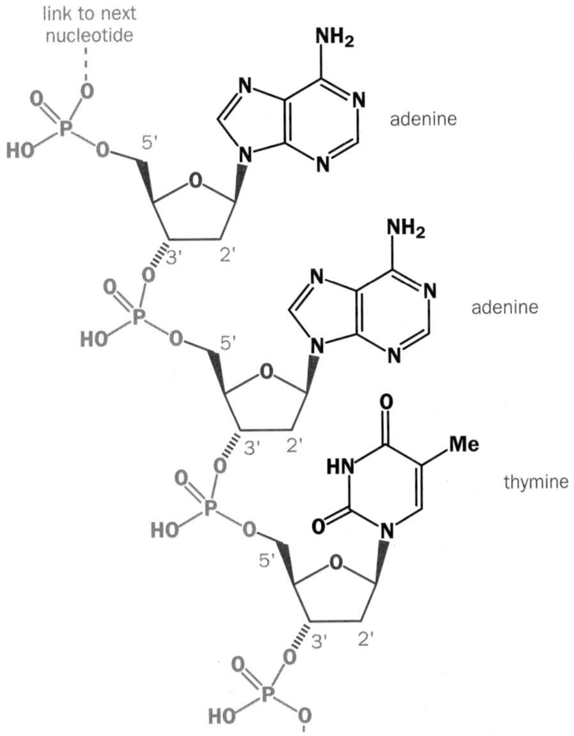

Each polymeric strand of DNA coils up into a helix and is bonded to another strand by hydrogen bonds between the bases. Each base pairs up specifically with another base —adenine with thymine (A–T) and guanine with cytosine (G–C)—like this.

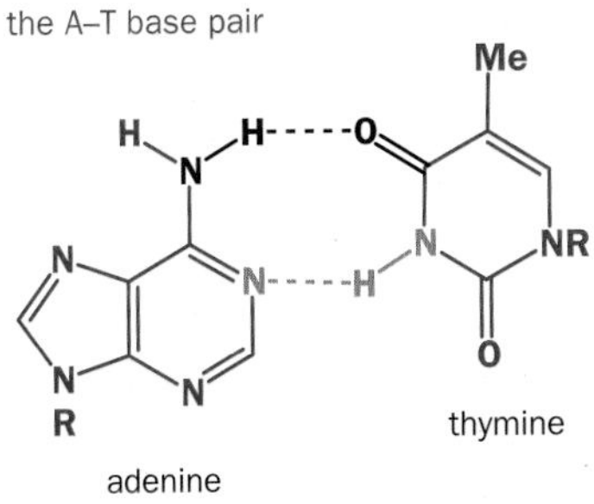

the A–T base pair

adenine

thymine

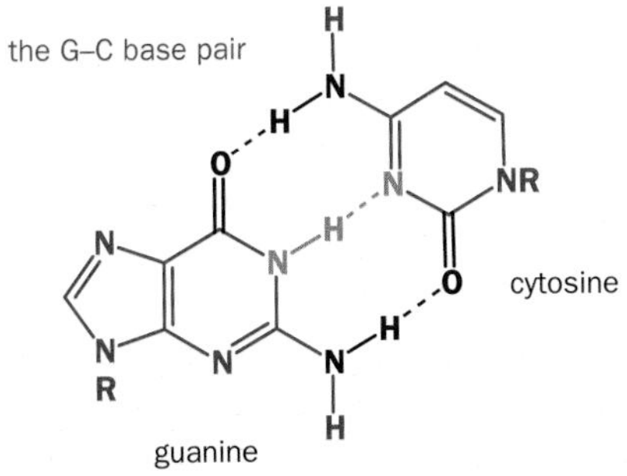

the G–C base pair

cytosine

guanine

There is quite a lot to notice about these structures. Each purine (A or G) is bonded specifically to one pyrimidine (T or C) by two or by three hydrogen bonds. The hydrogen bonds are of two kinds: one links an amine to a carbonyl group (black in the diagram) and one links an amine to an imine (green in the diagram). In this way, each nucleotide reliably recognizes another and reliably pairs with its partner. The short strand of DNA above (–AAT–) would pair reliably with –TTA–.

How the genetic information in DNA is passed to proteins

In the normal structure of DNA each strand is paired with another strand called the **complementary strand** because it has each base paired with its complementary base. When DNA replicates, the strands separate and a new strand with complementary structure grows alongside each. In this way the original double helix now becomes two identical double helices and so on.

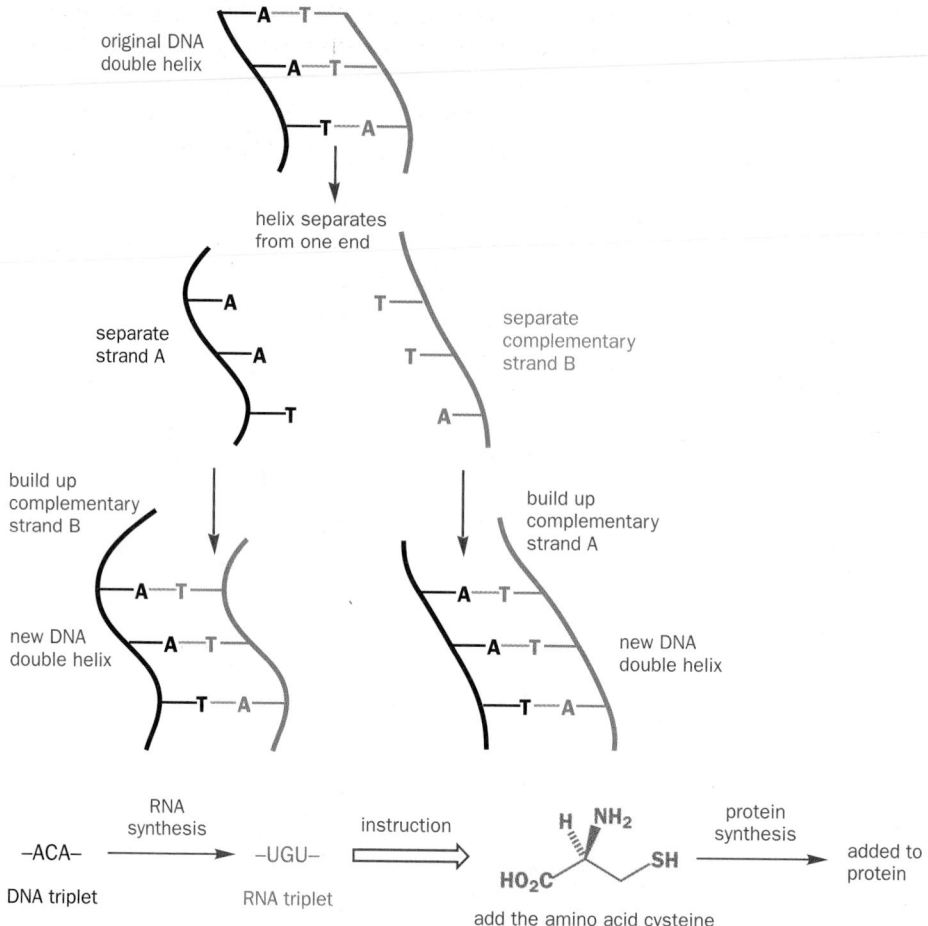

This is a crude simplification of a beautiful process and you should turn to a biochemistry text-book for more details. The actual building up of a strand of DNA obviously involves a complex series of chemical reactions. The DNA is then used to build up a complementary strand of RNA, which *does* have the 2′ hydroxyl group, and the RNA then instructs the cell on protein synthesis using three-nucleotide codes to indicate different amino acids. Again, the details of this process are beyond the scope of this book, but the code is not.

Each set of three nucleotides (called a **triplet** or **codon**) in a DNA molecule tells the cell to do something. Some triplets tell it to start work or stop work but most represent a specific amino acid. The code UGU in RNA tells the cell 'add a molecule of cysteine to the protein you are building'. The code UGA tells the cell 'stop the protein at this point'. So a bit of RNA reading UGUUGA would produce a protein with a molecule of cysteine at the end.

There are four bases available for DNA and so there are $4^3 = 64$ different triplet codons using three bases in each codon. There are only 20 amino acids used in proteins so that gives plenty of spare codons. In fact 61 of the 64 are used as codons for amino acids and the remaining three are 'stop' signals. Thus the code ATT in DNA would produce the complementary UAA and this is another 'stop' signal.

Base	Complementary base in DNA	Complementary base in RNA
A	T	U
C	G	G
G	C	C
U	*	*
T	A	a

* T occurs in DNA only and is replaced by U in RNA.

But that doesn't leave a 'start' signal! This signal is the same (TAC in DNA = AUG in RNA) as that for the amino acid methionine, which you met as a component of SAM, the biological methylating agent. In other words, all proteins start with methionine. At least, they are all made that way, though the methionine is sometimes removed by enzymes before the protein is released. These code letters are the same for all living things except for some minor variations in some microorganisms.

AIDS is being treated with modified nucleosides

Modified nucleosides have proved to be among the best antiviral compounds. The most famous anti-AIDS drug, AZT (zidovudine from GlaxoWellcome), is a slightly modified DNA nucleoside (3′-azidothymidine). It has an azide at C3′ instead of the hydroxyl group in the natural nucleoside.

deoxythymidine
a nucleoside of DNA

AZT
azidothymidine
anti-AIDS drug

deoxycytidine
a nucleoside of DNA

3-TC
lamivudine
anti-AIDS drug

Doctors are having some spectacular success at the moment (1999) against HIV and AIDS by using a combination of AZT and a much more modified nucleoside 3-TC (lamivudine) which is active against AZT-resistant viruses. This drug is based on cytosine but the sugar has been replaced by a different heterocycle though it is recognizably similar especially in the stereochemistry.

The last drug to mention is acyclovir (Zovirax), the cold sore (herpes) treatment. Here is a modified guanosine in which only a ghost of the sugar remains. There is no ring at all and no stereochemistry.

The bottom edge of the sugar ring has been done away with so that a simple alkyl chain remains. This compound has proved amazingly successful as an antiviral agent and it is highly likely that more modified nucleosides will appear in the future as important drugs.

deoxguanosine
a nucleoside of DNA

acyclovir
(Zovirax)
anti-herpes drug

Cyclic nucleosides and stereochemistry

We know the relative stereochemistry around the ribose ring of the nucleosides in DNA and RNA because the bases can be persuaded to cyclize on to the ring in certain reactions. Treatment of deoxythymidine with reagents that make oxygen atoms into leaving groups leads to cyclization by intramolecular S$_N$2 reaction. The amide oxygen of the base attacks the 3′-position in the sugar ring.

deoxythymidine

cyclization with inversion

cyclic nucleoside

This S$_N$2 reaction has to happen with inversion, proving that the base and the 3′-OH group are on opposite sides of the ribose ring. The cyclized product is useful too. If it is reacted with azide ion the ring reopens with inversion in another S$_N$2 reaction and AZT is formed.

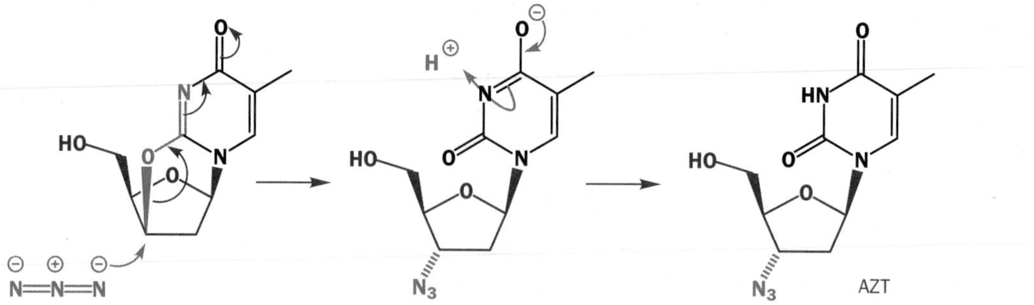

We can show that the primary alcohol is on the same side of the ring as the base by another cyclization reaction. Treatment of the related iodide with a silver(I) salt gives a new seven-membered ring. This reaction can happen only with this stereochemistry of starting material.

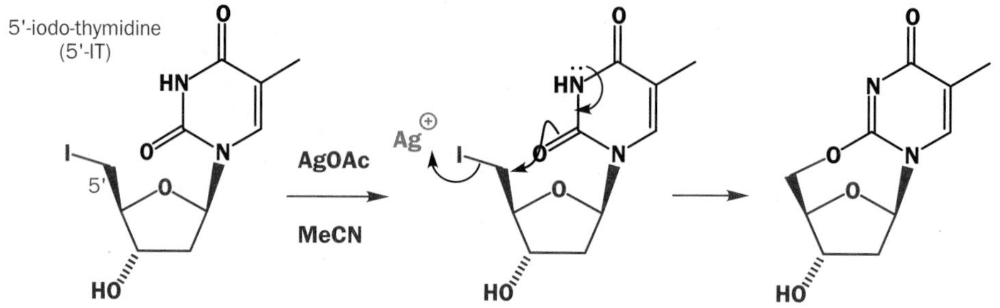

In ribonucleic acids, the fact that the 2′- and 3′-OH groups are on the same side of the ring makes alkaline hydrolysis of such dinucleotides exceptionally rapid by intramolecular nucleophilic catalysis.

> The substituents B¹ and B² represent any purine or pyrimidine base.

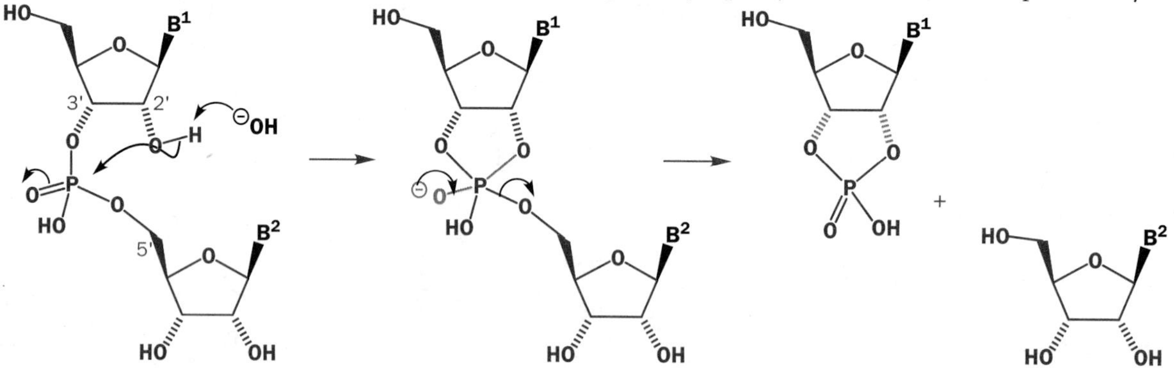

The alkali removes a proton from the 2′-OH group, which cyclizes on to the phosphate link—possible only if the ring fusion is *cis*. The next reaction involves breakdown of the pentacovalent phosphorus intermediate to give a cyclic phosphate. One nucleoside is released by this reaction and the second follows when the cyclic phosphate is itself cleaved by alkali.

The simplest cyclic phosphate that can be formed from a nucleotide is also important biologically as it is a messenger that helps to control such processes as blood clotting and acid secretion in the stomach. It is **cyclic AMP** (**cAMP**), formed enzymatically from ATP by nucleophilic displacement of pyrophosphate by the 3′-OH group.

> Note that cAMP has a trans 6,5-fused ring junction.

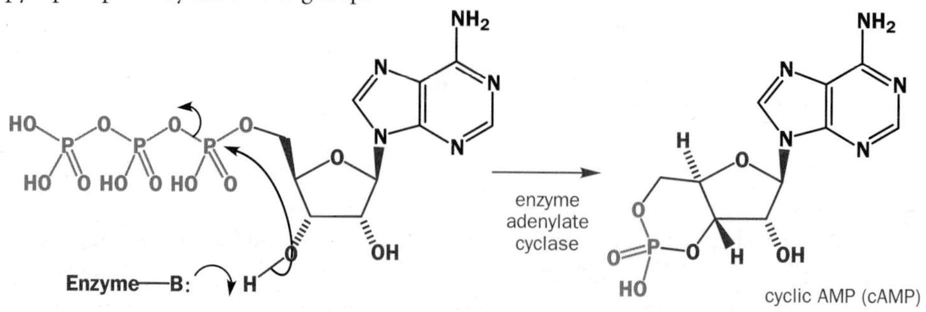

cyclic AMP (cAMP)

Proteins are made of amino acids

The molecule of methionine, which we met as a component of SAM, is a typical amino acid of the kind present in proteins. It is the starter unit in all proteins and is joined to the next amino acid by an amide bond. In general, we could write:

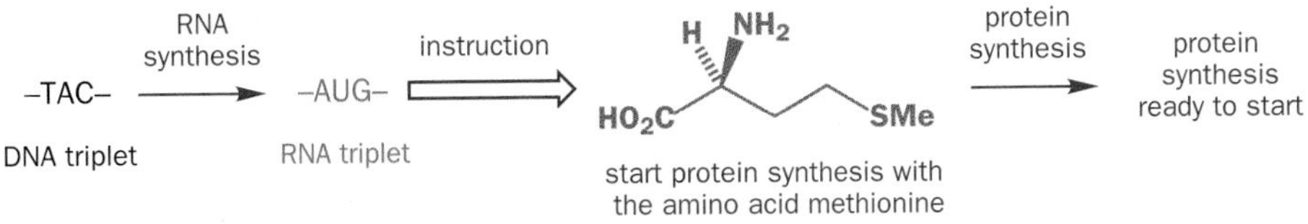

start protein synthesis with the amino acid methionine

Now we can add the next amino acid using its correct codon, but we want to show the process in general so we shall use the general structure in the margin. All amino acids have the same basic structure and differ only in the group 'R'. Both structures are the same and have the same (*S*) stereochemistry.

two views of the general amino acid structure

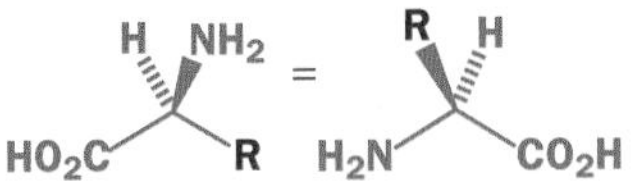

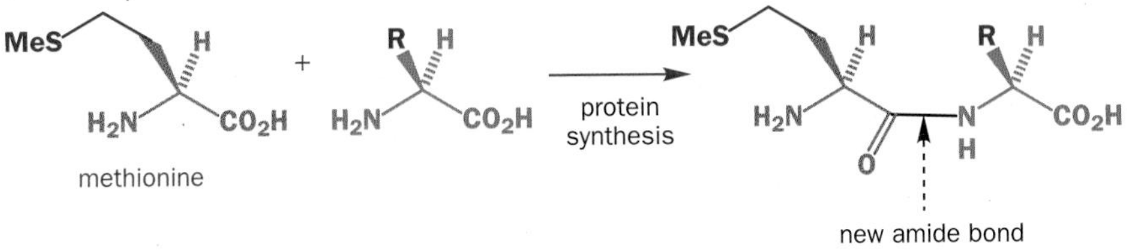

methionine

new amide bond

The process then continues with more amino acids added in turn to the right-hand end of the growing molecule. A section of the final protein drawn in a more realistic conformation might look like this.

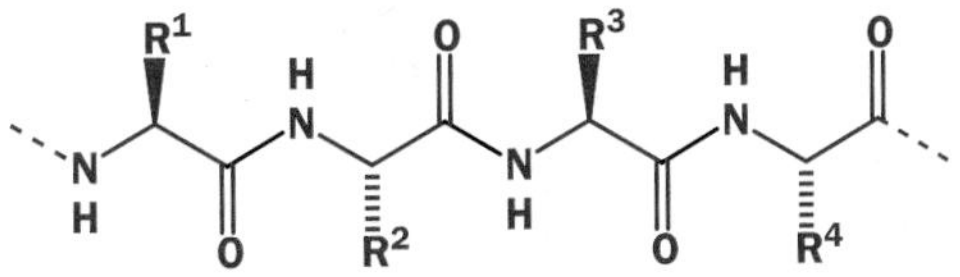

The basic skeleton of the protein zig-zags up and down in the usual way; the amide bonds (shown in black) are rigid because of the amide conjugation and are held in the shape shown. Each amino acid may have a different substituent (R^1, R^2, R^3, etc.) or some may be the same.

A catalogue of the amino acids

So what groups are available when proteins are being made? The simplest amino acid, glycine, has no substituents except hydrogen and is the only amino acid that is not chiral. Four other amino acids have alkyl groups without further functionality. The next table gives their structures together with two abbreviations widely used for them. The three-letter code (which has nothing to do with the codon in DNA!) is almost self-explanatory as are the one-letter codes in this group, but some of the one-letter codes for the other amino acids are not so obvious.

Name	Three-letter code	One-letter code	Structure
glycine	Gly	G	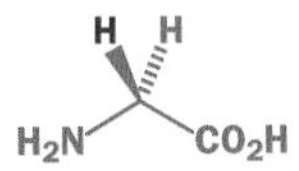
alanine	Ala	A	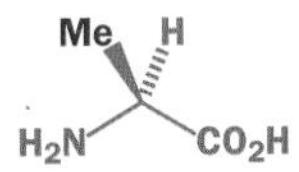
valine	Val	V	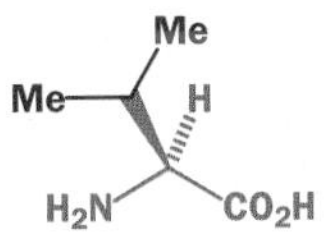
leucine	Leu	L	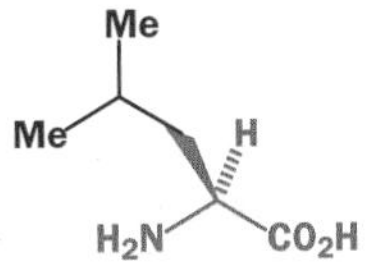
isoleucine	Ile	I	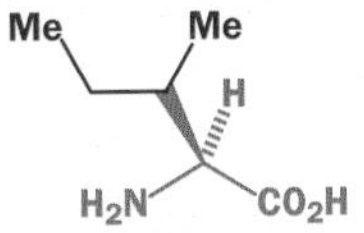

▶

Many of the compounds we discuss in this chapter will be salts under biological conditions. Most carboxylic acids will exist as anions, as will the phosphates you have just seen, and most amines as cations as they would be protonated at pH 7. Amino acids exist in biological systems as zwitterions. For simplicity, we will usually draw functional groups in the simplest and most familiar way, leaving the question of protonation to be addressed separately if required.

These amino acids form hydrophobic (water-repelling) nonpolar regions in proteins. There are three more of this kind with special roles. Phenylalanine and tryptophan have aromatic rings and, though they are still hydrophobic, they can form attractive π-stacking interactions with other aromatic molecules. Enzyme-catalysed hydrolysis of proteins often happens next to one of these residues. Proline is very special. It has its amino group inside a ring and has a different shape from all the other amino acids. It appears in proteins where a bend or a twist in the structure is needed.

Name	Three-letter code	One-letter code	Structure
phenylalanine	Phe	F	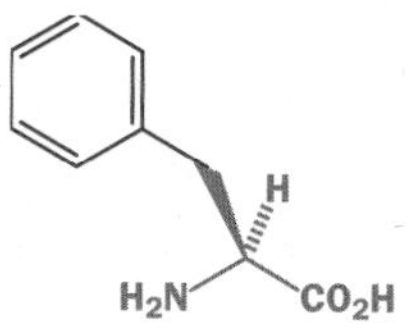
tryptophan	Trp	W	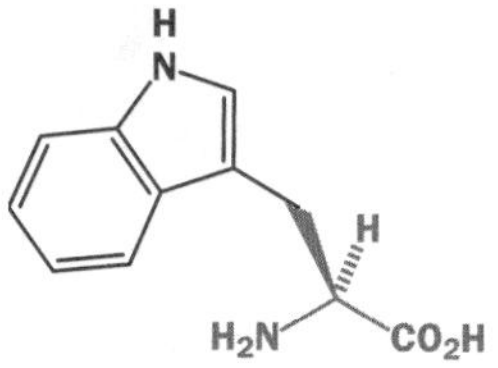
proline	Pro	P	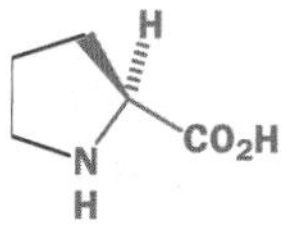

The rest of the amino acids have functional groups of various kinds and we shall deal with them by function. The simplest have hydroxyl groups and there are three of them—two alcohols and a phenol. Serine in particular is important as a reactive group in enzymatic reactions. It is a good nucleophile for carbonyl groups.

Name	Three-letter code	One-letter code	Structure
serine	Ser	S	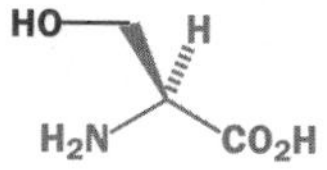
threonine	Thr	T	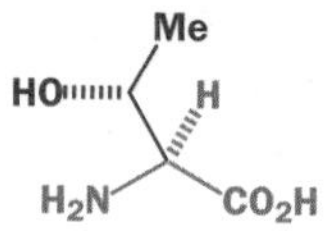
tyrosine	Tyr	Y	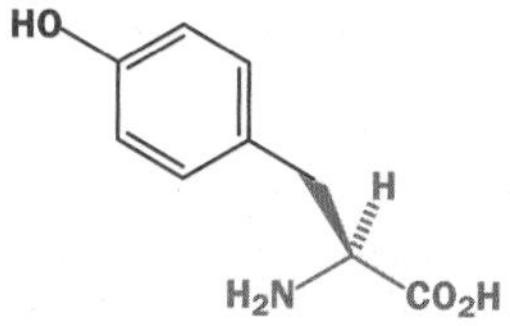

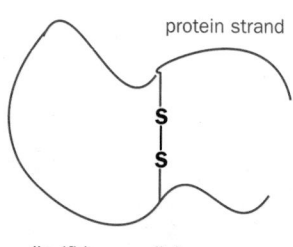

protein strand

disulfide cross-link

Next come the two compounds we have already met, the sulfur-containing cysteine and methionine. Cysteine has a thiol group and methionine a sulfide. These are very important in protein structure—methionine starts off the synthesis of every new protein as its *N*-terminal amino acid, while cysteine forms S–S bridges linking two parts of a protein together. These disulfide links may be important in holding the three-dimensional shape of the molecule.

Name	Three-letter code	One-letter code	Structure
cysteine	Cys	C	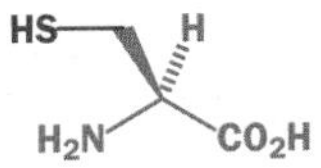
methionine	Met	M	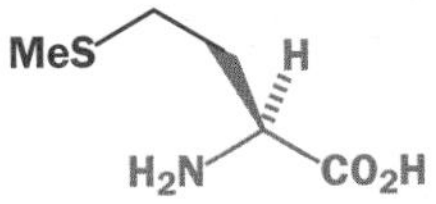

Cysteine and hairdressing

Thiols (RSH) are easily oxidized, by air, for example, to disulfides (RS–SR). This chemistry of cysteine is used by hairdressers to give 'perms' or permanent waves. The hair proteins are first reduced so that any disulfide (cysteine to cysteine) cross-links within each strand are reduced to thiols. Then the hair is styled and the final stage is the 'set' when the hair is oxidized so that disulfide cross-links are established to hold its shape for a good time. The disulfide resulting from cross-links between the thiol groups of *cysteine* is known as *cystine*—beware of confusing the names!

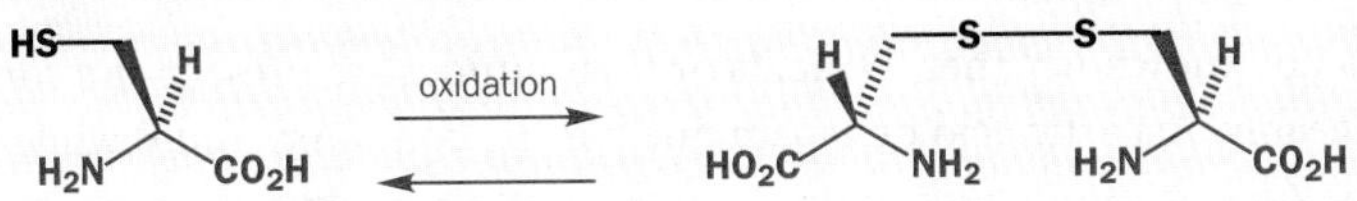

The amino acids with a second amino group are important because of their basicity and they are vital to the catalytic activity of many enzymes. Histidine has a pK_{aH} very close to neutrality (6.5) and can function as an acid or a base. Lysine and arginine are much more basic, but are normally protonated in living things. An extra column in this table gives the pK_{aH} of the extra amino groups.

Name	Three-letter code	One-letter code	pK_{aH}	Structure
histidine	His	H	6.5	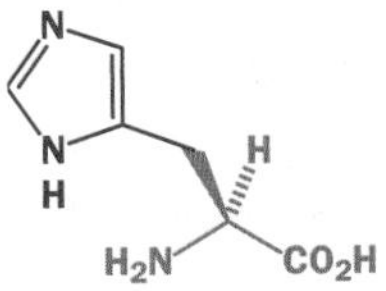
lysine	Lys	K	10.0	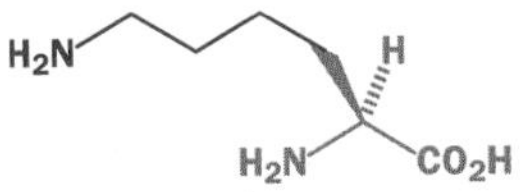
arginine	Arg	R	12.0	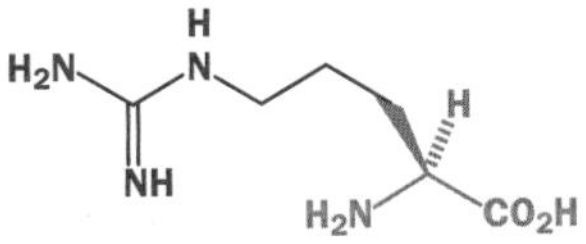

Essential amino acids

If you saw 'Jurassic Park' you may recall that the failsafe device was the 'lysine option'. The dinosaurs were genetically modified so as to need lysine in their diet. The idea was that they would die unless lysine was provided by their keepers. Lysine was a good choice as it is one of the 'essential' amino acids for humans. If we are not given it in our diet, we die. Of course, any normal diet, including the human beings eaten by the escaped dinosaurs, would also contain plenty of lysine. The other essential amino acids (for humans) are His, Ile, Leu, Met, Phe, Thr, Try, and Val.

Finally, we come to the acidic amino acids—those with an extra carboxylic acid group. We are going to include their amides too as they also occur in proteins. This group is again very much involved in the catalytic activity of enzymes. The two acids have pK_as for the extra CO_2H group of about 4.5.

Name	Three-letter code	One-letter code	Structure
aspartic acid	Asp	D	
asparagine	Asn	N	
glutamic acid	Glu	E	
glutamine	Gln	Q	

Sometimes it is not known whether the acids or their amides are present and sometimes they are present interchangeably. Aspartic acid or asparagine has the codes Asx and B while glutamic acid or glutamine is Glx or Z.

Now perhaps you can see that a protein is an assembly of many different kinds of group attached to a polyamide backbone. Some of the groups are purely structural, some control the shape of the protein, some help to bind other molecules, and some are active in chemical reactions.

Most amino acids are readily available to chemists. If proteins are hydrolysed with, say, concentrated HCl, they are broken down into their amino acids. This mixture is tricky to separate, but the acidic ones are easy to extract with base while the aromatic ones crystallize out easily.

Amino acids combine to form peptides and proteins

In nature, the amino acids are combined to give proteins with hundreds or even thousands of amino acids in each one. Small assemblies of amino acids are known as **peptides** and the amide bond that links them is called a **peptide bond**. One important dipeptide is the sweetening agent aspartame, whose synthesis was discussed in Chapter 25. It is composed (and made) of the amino acid aspartic acid (Asp) and the methyl ester of phenylalanine. Only this enantiomer has a sweet taste and it is very sweet indeed—about 160 times as sweet as sucrose. Only a tiny amount is needed to sweeten drinks and so it is much less fattening than sucrose and is 'safe' because it is degraded in the body to Asp and Phe, which are there in larger amounts anyway.

An important tripeptide is **glutathione**. So important is this compound that it is present in almost all tissues of most living things. It is the 'universal thiol' that removes dangerous oxidizing agents by allowing itself to be oxidized to a disulfide.

Glutathione is not quite a simple tripeptide. The left-hand amino acid is normal glutamic acid but it is joined to the next amino acid through its γ-CO_2H group instead of the more normal α-CO_2H group. The middle amino acid is the vital one for the function—cysteine with a free SH group. The C-terminal acid is glycine.

aspartame

aspartic acid   phenylalanine methyl ester

glutathione = RSH

γ-Glu (glutamic acid joined through its γ-CO_2H group)   cysteine   glycine

Thiols are easily oxidized to disulfides, as we have already seen in our discussion on hairdressing (though the redox chemistry of glutathione is a matter of life or death and not merely a bad hair day), and glutathione sacrifices itself if it meets an oxidizing agent. Later, the oxidized form of glutathione is reduced back to the thiol by reagents we shall meet in the next chapter (NADH, etc.).

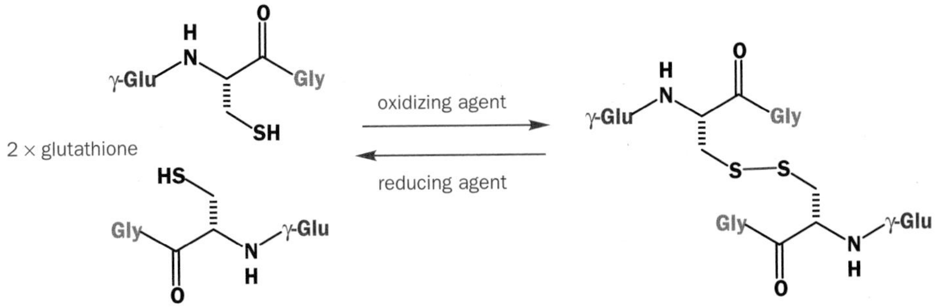

If we imagine that the stray oxidizing agent is a peroxide, say, H_2O_2, we can draw a mechanism to show how this can be reduced to water as glutathione (represented as RSH) is oxidized to a disulfide.

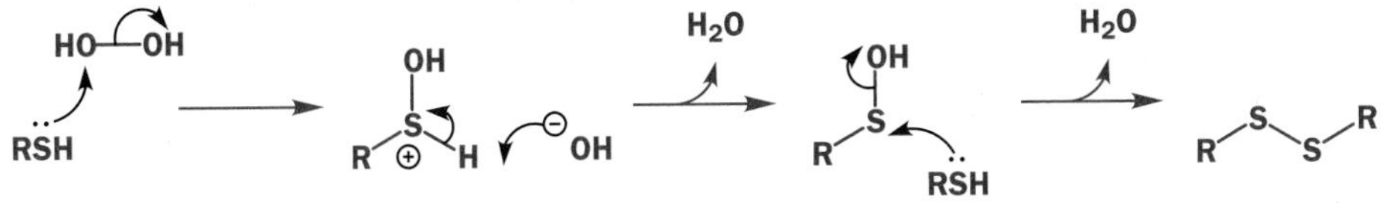

Paracetamol overdoses

Paracetamol is a popular and safe analgesic if used properly but an overdose is insidiously dangerous. The patient often seems to recover only to die later from liver failure. The problem is that paracetamol is metabolized into an oxidized compound that destroys glutathione.

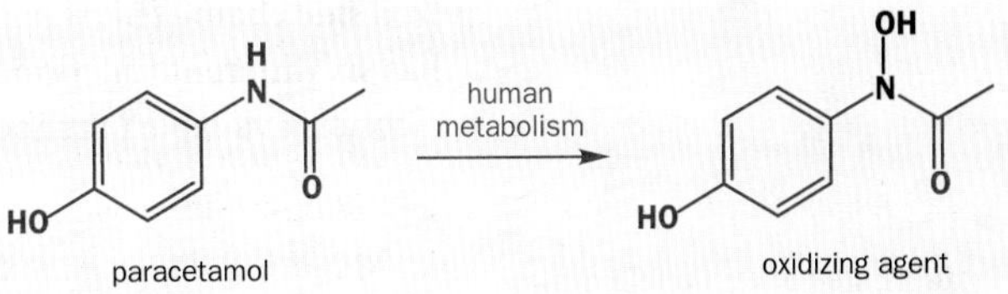

paracetamol

oxidizing agent

Glutathione detoxifies this oxidizing agent by a most unusual mechanism. The unstable hydroxylamine loses water to give a reactive quinone imine that is attacked by glutathione on the aromatic ring. The adduct is stable and safe but, for every molecule of paracetamol, one molecule of glutathione is consumed.

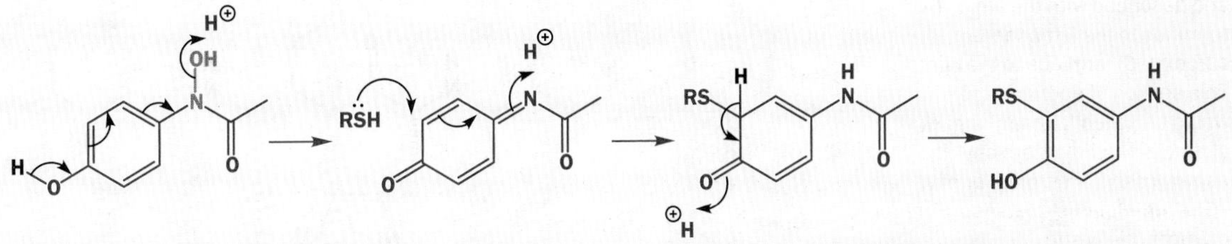

There is no problem if a normal dose is taken—there is plenty of glutathione to deal with that. But if an overdose is taken, all the glutathione may be used up and irreversible liver damage occurs.

Glutathione also detoxifies some of the compounds we have earlier described as very dangerous carcinogens such as Michael acceptors and 2,4-dinitrohalobenzenes. In both cases the thiol acts as a nucleophile for these electrophiles. Most of the time there is enough glutathione present in our cells to attack these poisons before they attack DNA or an enzyme.

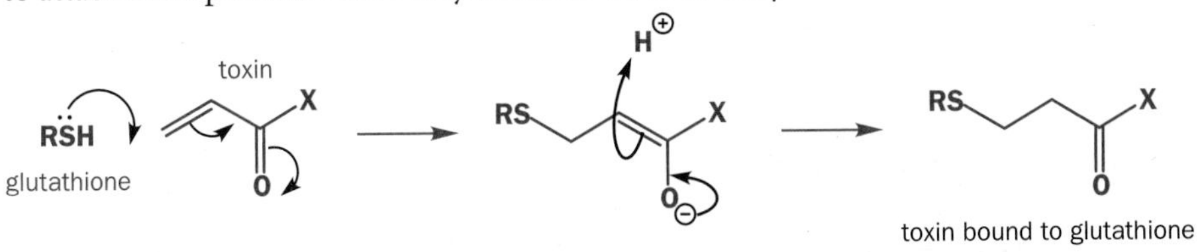

toxin bound to glutathione

The toxin is now covalently bound to glutathione and so is no longer electrophilic. It is harmless and can be excreted. More glutathione will be synthesized from glutamic acid, cysteine, and glycine to replace that which is lost.

Proteins are Nature's chemical laboratories

Longer peptides are called **proteins**, though where exactly the boundary occurs is difficult to say.

The structure of the hormone insulin (many diabetics lack this hormone and must inject themselves with it daily) was deduced in the 1950s by Sanger. It has two peptide chains, one of 21 amino acids and one of 30, linked by three disulfide bridges—just like the links in oxidized glutathione. This is a very small protein.

Enzymes are usually bigger. One of the smaller enzymes—ribonuclease (which hydrolyses RNA) from cows—has a chain of 124 amino acids with four internal disulfide bridges. The abundance of the various amino acids in this enzyme is given in this table.

Type	Amino acid (number)[a]	Total
structural	A (12), F (3), M (4), L (2), P (4), V (9), G (3), I (3)	40
cross-linking	C (8)	8
basic	K (10), R (4), H (4)	18
acids and amides	E (5), Q (7), D (5), N (10)	27
hydroxyl	T (10), S (15), Y (6)	31

[a] See tables earlier in this section for one-letter codes of amino acids.

There are 48 structural and cross-linking amino acids concerned with the shape of the protein but over half of the amino acids have functional groups sticking out of the chain—amino, hydroxy, acid groups, and the like. In fact, the enzyme uses only a few of these functional groups in the reaction it catalyses (the hydrolysis of RNA)—probably only two histidines and one lysine—but it is typical of enzymes that they have a vast array of functional groups available for chemical reactions.

Below is part of the structure of ribonuclease surrounding one of the catalytic amino acids His12. There are seven amino acids in this sequence. Every one is different and every one has a functionalized side chain. This is part of a run of ten amino acids between Phe8 and Ala19. This strip of peptide has six different functional groups (two acids, one each of amide, guanidine, imidazole, sulfide, and alcohol) available for chemical reactions. Only the histidine is actually used.

Proteins are conventionally drawn and described with the amino (N) terminus to the left and the carboxyl (C) terminus to the right. This section of ribonuclease would be called 'glutamyl arginyl glutaminyl histidyl methionyl aspartyl seryl...' or, more briefly, –Glu–Arg–Gln–His–Met–Asp–Ser– or, more briefly still, –ERQHMDS–. The numbers on the diagram such as 'Glu9' tell us that this glutamic acid residue is number 9 from the N-terminus.

One reason for disease is that enzymes may become overactive and it may be necessary to design specific inhibitors for them to treat the disease. Angiotensin-Converting Enzyme (ACE) is a zinc-dependent enzyme that cleaves two amino acids off the end of angiotensin I to give angiotensin II, a protein that causes blood pressure to rise.

It is necessary in some situations for our blood pressure to rise (when we stand up for instance!) but too much too often is a very bad thing leading to heart attacks and strokes. Captopril is a treatment for high blood pressure called an 'ACE inhibitor' because it works by inhibiting the enzyme. It is a dipeptide mimic, having one natural amino acid and something else. The 'something else' is an SH group replacing the NH_2 group in the natural dipeptide. Captopril binds to the enzyme because it is *like* a natural dipeptide but it inhibits the enzyme because it is *not* a natural dipeptide. In particular, the SH group is a good ligand for Zn(II). Many people are alive today because of this simple deception practised on an enzyme.

Structural proteins must be tough and flexible

In contrast with the functional enzymes, there are purely structural proteins such as collagen. Collagen is the tough protein of tendons and is present in skin, bone, and teeth. It contains large amounts of glycine (every third amino acid is glycine), proline, and hydroxyproline (again about a third of the amino acids are either Pro or Hyp).

In the enzyme above there were only three glycines and four prolines and no hydroxyproline at all. Hydroxyproline is a specialized amino acid that appears almost nowhere else and, along with proline, it establishes a very strong triply coiled structure for collagen. The glycine is necessary as there is no room in the inside of the triple coil for any larger amino acid. Functionalized amino acids are rare in collagen.

Hydroxyproline and scurvy

Hydroxyproline is a very unusual amino acid. There is no genetic codon for the insertion of Hyp into a growing protein because collagen is not made that way. The collagen molecule is first assembled with Pro where Hyp ends up. Then some proline residues are oxidized to hydroxyproline. This oxidation requires vitamin C, and without it collagen cannot be formed. This is why vitamin C deficiency causes scurvy—the symptoms of scurvy (teeth falling out, sores, blisters) are caused by the inability to make collagen.

Proteins are enormously diverse in structure and function and we will be looking at a few of their reactions in the next chapter.

Sugars—just energy sources?

Sugars are the building blocks of carbohydrates. They used to be thought of as essential but rather dull molecules whose only functions were the admittedly useful provision of energy and cell wall construction. We have already noted that ribose plays an intimate role in DNA and RNA structure and function. More recently, biochemists have realized that carbohydrates are much more exciting. They are often found in intimate association with proteins and are involved in recognition of one protein by another and in adhesion processes.

That may not sound very exciting, but take two examples. How does a sperm recognize the egg and penetrate its wall? The sperm actually binds to a carbohydrate on the wall of the egg in what was the first event in all of our lives. Then how does a virus get inside a cell? If it fails to do so, it has no life. Viruses depend on host cells to reproduce. Here again, the recognition process involves specific carbohydrates. One of the ways in which AIDS is being tackled with some success is by a combination of the antiviral drugs we met earlier in this chapter with HIV protease inhibitor drugs, which aim to prevent recognition and penetration of cells by HIV.

We now know that many vital activities as diverse as healing, blood clotting, infection, prevention of infection, and fertilization all involve carbohydrates. Mysterious compounds such as 'sialyl Lewis-X', unknown a few years ago, are now known to be vital to our health and happiness. Far from being dull, carbohydrates are exciting molecules and our future depends on them. It is well worthwhile to spend some time exploring their structure and chemistry.

Sugars normally exist in cyclic forms with much stereochemistry

The most important sugar is glucose. It has a saturated six-membered ring containing oxygen and it is best drawn in a chair conformation with nearly all the substituents equatorial. It can also be drawn reasonably as a flat configurational diagram.

We have already met one sugar in this chapter, ribose, because it was part of the structure of nucleic acids. This sugar is a five-membered saturated oxygen heterocycle with many OH groups. Indeed, you can define a sugar as an oxygen heterocycle with every carbon atom bearing an oxygen-based functional group—usually OH, but alternatively C=O.

Both our drawings of glucose and ribose show a number of stereogenic centres and one centre undefined—the OH group is marked with a wavy line. This is because one centre in both sugars is a hemiacetal and therefore the molecule is in equilibrium with an open-chain hydroxy-aldehyde. For glucose, the open-chain form is this.

Sugars have had walk-on roles in a few other chapters, notably Chapter 16 on stereochemistry. They are discussed on pp. 146, 394, and 1220.

two representations of glucose

ribose

a ribonucleotide

furan   pyran

D-glyceraldehyde
(R)–(+)–glyceraldehyde

Glyceraldehyde was the compound used to define the D and L designators before anyone knew what the real configurations of natural compounds were. See the discussion on p. 389 for more on this.

When the ring closes again, any of the OH groups could cyclize on to the aldehyde but there is no real competition—the six-membered ring is more stable than any of the alternatives (which could have three-, four-, five-, or seven-membered rings—check for yourself). However, with ribose there is a reasonable alternative.

ribo-furanoside   ribo-pyranoside

The most important sugars may exist in an open-chain form, as a five-membered oxygen heterocycle (called a furanoside after the aromatic furan) or a six-membered oxygen heterocycle (called a pyranoside after the compound pyran).

From triose to glucose requires doubling the number of carbon atoms

We will return to that in a moment, but let us start from the beginning. The simplest possible sugar is glyceraldehyde, a three-carbon sugar that cannot form a cyclic hemiacetal.

Glyceraldehyde is present in cells as its phosphate which is in equilibrium with dihydroxyacetone phosphate. This looks like a complicated rearrangement but it is actually very simple—the two compounds have a common enol through which they interconvert.

glyceraldehyde-3-phosphate   common enediol   dihydroxyacetone-3-phosphate

Glyceraldehyde is an aldehyde sugar or **aldose** and dihydroxyacetone is a keto-sugar or **ketose**. That ending '-ose' just refers to a sugar. These two molecules combine to form the six-carbon sugar,

fructose, in living things and this reaction is a key step in the synthesis of organic compounds from CO_2 in plants.

When we come to the four-carbon sugars, or **tetroses**, two are important. They are diastereoisomers called erythrose and threose. You can see from this series that each aldose has $n - 2$ stereogenic centres in its carbon chain where n is the total number of carbon atoms in that chain.

this aldo-tetrose has *syn* OH groups

this aldo-tetrose has *anti* OH groups

(2R,3S)-D-threose

(2R,3R)-D-erythrose

These sugars too have given their names to a stereochemical designation—'erythro-' and 'threo-' are used to describe diastereoisomers that resemble these two sugars. We do not use these rather ambiguous terms in this book, preferring more precise or vivid terms such as *R,R* or *anti* as appropriate.

We shall take a longer look at the stereochemistry and reactions of glucose and the important keto-hexose, fructose. These two are often found together in cells and are combined in the same molecule as **sucrose**—ordinary sugar. In this molecule, glucose appears as a pyranose (six-membered ring) and fructose as a furanose (five-membered ring). They are joined through an acetal at what were hemiacetal positions, and sucrose is a single diastereoisomer.

sucrose

acetal

fructose

acetal

glucose

Sugars can be fixed in one shape by acetal formation

This is the simplest way to fix glucose in the pyranose form—any alcohol, methanol, for example, gives an acetal and, remarkably, the acetal has an *axial* OR group.

glucose

MeOH
$\xrightarrow{H^{\oplus}}$

Acetal formation is under thermodynamic control (Chapter 14) so the axial compound must be the more stable. This is because of the **anomeric effect**—so called because this C atom is called the anomeric position and the acetal diastereoisomers are called anomers. The effect is a bonding interaction between the axial lone pair on the oxygen atom in the ring and the σ^* orbital of the OMe group.

The anomeric effect was discussed in Chapter 42, and you should check that you can still write down the mechanism of acetal formation you learned in Chapter 14.

the anomeric effect

axial lone pair

bonding interaction

C–O σ^*

axial anomer

equatorial anomer
no stabilization from anomeric effect

The formation of acetals allows a remarkable degree of control over the chemistry of sugars. Apart from the simple glucoside acetal we have just seen, there are three important acetals worth understanding because of the way in which they illustrate stereoelectronic effects—the interplay of stereochemistry and mechanism. If we make an acetal from methyl glucoside, we get a single compound as a single stereoisomer.

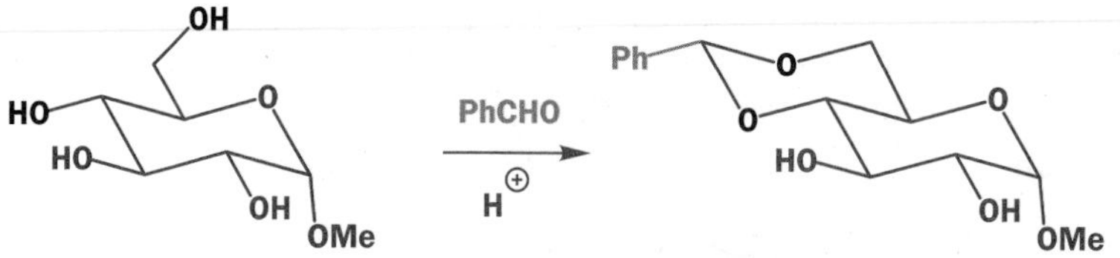

The new acetal could have been formed between any of the adjacent OH groups in the starting material but it chose the only pair (the black OH groups) to give a six-membered ring. The stereochemistry of glucose is such that the new six-membered ring is *trans*-fused on the old so that a beautifully stable all-chair bicyclic structure results, with the phenyl group in an equatorial position in the new chair acetal ring. It does not matter which OH group adds to benzaldehyde first because acetal formation is under thermodynamic control and this product is the most stable possible acetal.

Acetals formed from sugars and acetone have a quite different selectivity. For a start, cyclic acetals of acetone prefer to be five- rather than six-membered rings. In a six-membered ring, one of the acetone's methyl groups would have to be axial, so the five-membered ring is preferred. A 5/5 or 5/6 ring fusion is more stable if it is *cis*, and so acetone acetals ('acetonides') form preferentially from *cis* 1,2-diols. Glucose has no neighbouring *cis* hydroxyls in the pyranose form, but in the furanose form it can have two pairs. Formation of an acetal with acetone fixes glucose in the furanose form. This is all summarized in the scheme below.

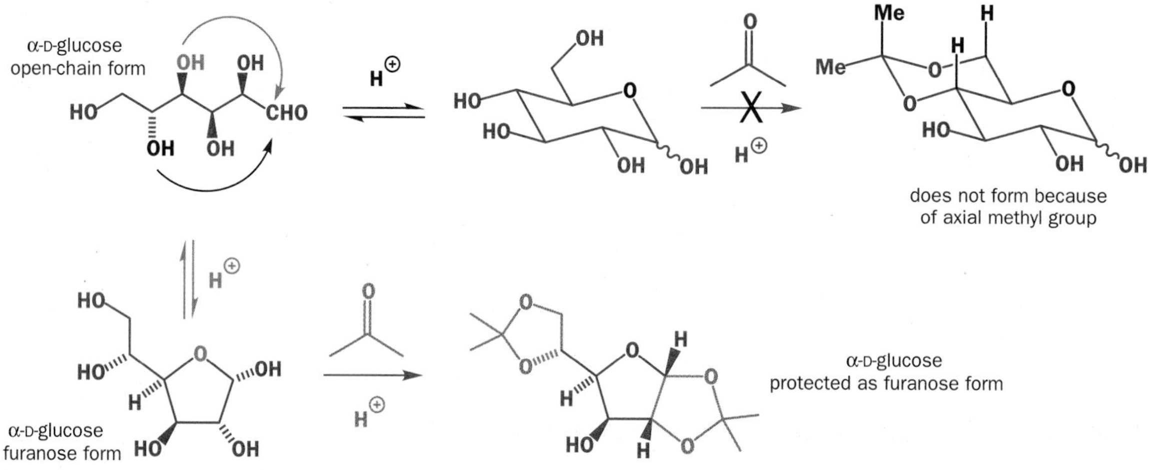

The open-chain form of glucose is in equilibrium with both pyranose and furanose forms by hemiacetal formation with the black and green OH groups, respectively. Normally, the pyranose form is preferred, but the furanose form can form a double acetal with acetone, one acetal having *cis*-fused 5/5 rings and the other being on the side chain. This is the product.

If we want to fix glucose in the open-chain form, we must make an 'acetal' of quite a different kind using a thiol (RSH) instead of an alcohol, an aldehyde, or a ketone.

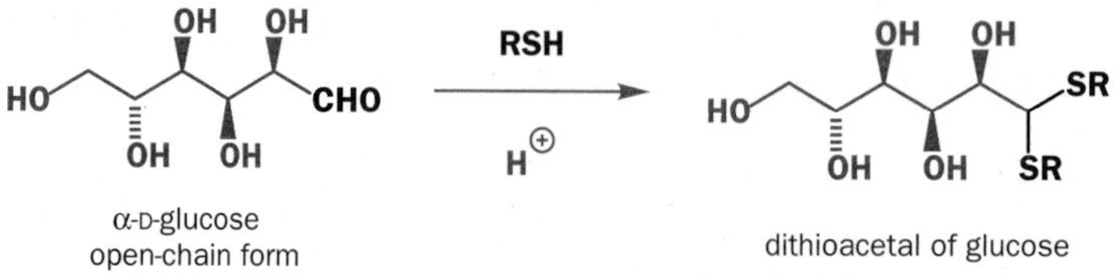

The thiol combines with the aldehyde group of the open-chain form to give a stable dithioacetal. The dithioacetal is evidently more stable than the alternative hemiacetals or monothioacetals that could be formed from the pyranose or furanose forms.

hemithioacetal

dithioacetal

Sugar alcohols are important in food chemistry

Another reaction of the open-chain form of sugars is reduction of the aldehyde group. This leads to a series of polyols having an OH group on each carbon atom. We will use **mannose** as an example. Mannose is a diastereoisomer of glucose having one axial OH group (marked in black) and, like glucose, is in equilibrium with the open-chain form.

If we redraw the open-chain form in a more realistic way, and then reduce it with $NaBH_4$, the product is mannitol whose symmetry is interesting. It has C_2 symmetry with the C_2 axis at right angles to the chain and marked with the orange dot.

mannose:
pyranose structure

mannose:
open-chain structure

axis of C_2 symmetry

$NaBH_4$

mannose:
open-chain structure

mannitol

C_2 means that the axis of symmetry is twofold: rotating 180° gives the same structure.

The simplification of stereochemistry results because the two ends of the sugar both now have CH_2OH groups so that the possibility of C_2 and planar symmetry arises. If we look at the two four-carbon sugars we can establish some important stereochemical correlations. Threose is reduced to threitol which has a C_2 axis like that of mannitol.

(2R,3S)-D-threose
furanose form

(2R,3S)-D-threose
open-chain form

$NaBH_4$

axis of C_2 symmetry

(2R,3R)-threitol

Erythrose on the other hand reduces to erythritol, which is not chiral.

(2R,3R)-D-erythrose
furanose form

(2R,3R)-D-erythrose
open-chain form

$NaBH_4$

(2R,3S)-erythritol

The important correlation is that threose is reduced or oxidized to chiral compounds—the oxidation product is tartaric acid—while erythrose is reduced or oxidized to *meso* compounds. This may help you to remember the labels **erythro-** and **threo-** should you need to.

C_2 axis

(+)-tartaric acid

HNO_3

(2R,3S)-D-threose
open-chain form

$NaBH_4$

C_2 axis

threitol—chiral

This may not be obvious in the normal drawing (which has a centre of symmetry), but rotation around the central C–C bond clearly shows the plane of symmetry. Neither plane nor centre of symmetry may be present in a chiral molecule, but a C_2 axis is allowed (Chapter 16).

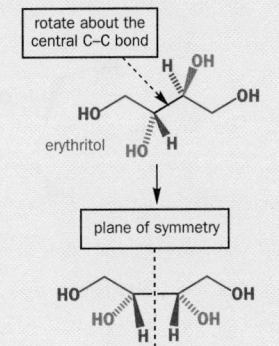

rotate about the central C–C bond

erythritol

plane of symmetry

(continued overleaf)

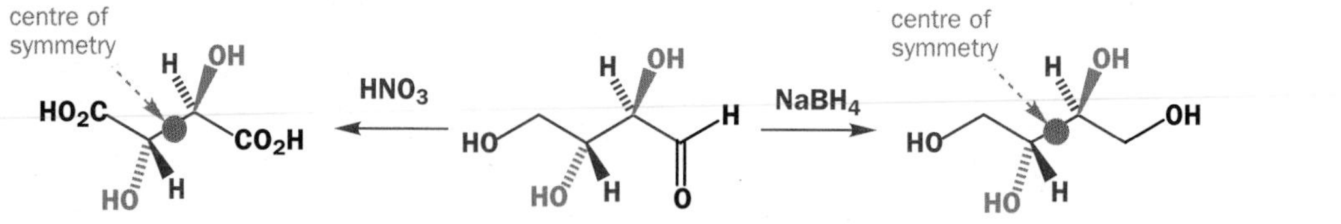

meso-tartaric acid
not chiral

(2*R*,3*R*)-D-erythrose
open-chain form

erythritol—*not* chiral

In the pentoses and hexoses there are again sugars that are reduced to *meso* alcohols and some that are reduced to C_2 symmetric alcohols. The C_5 sugar xylose has the same stereochemistry as glucose from C2 to C4 but lacks the CH_2OH group at C6.

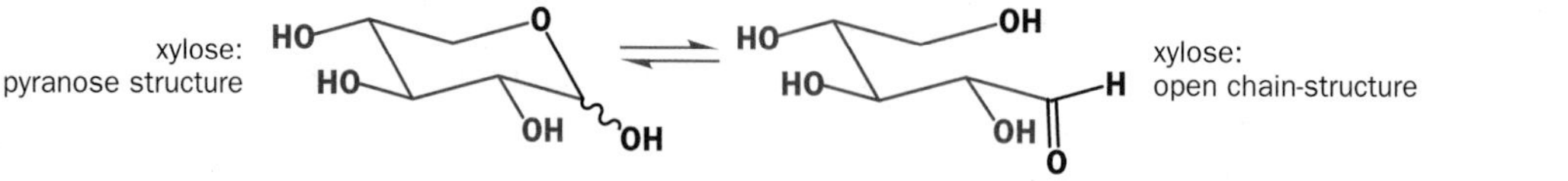

xylose:
pyranose structure

xylose:
open chain-structure

Xylose is reduced to the *meso* alcohol xylitol. This alcohol is more or less as sweet as sugar and, as xylose (which is not sweet) can be extracted in large quantities from waste products such as sawdust or corncobs, xylitol is used as a sweetener in foods. There is an advantage in this. Though we can digest xylitol (so it is fattening), the bacteria on teeth cannot so that xylitol does not cause tooth decay.

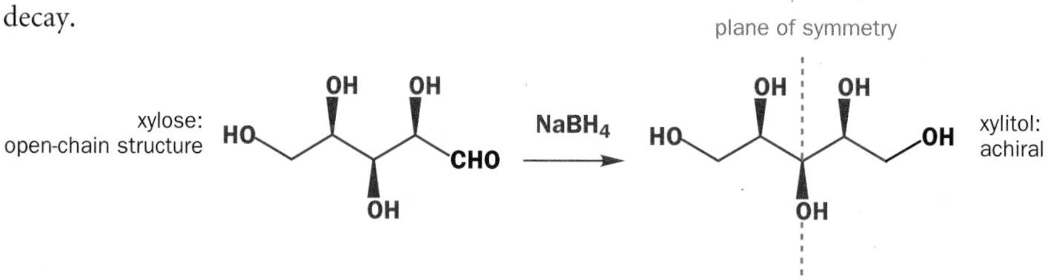

xylose:
open-chain structure

xylitol:
achiral

By careful manipulation of protecting groups such as acetals and reactions such as reduction and oxidation, it is possible to transform sugars into many different organic compounds retaining the natural optical activity of the sugars themselves. As some sugars are also very cheap, they are ideal starting points for the synthesis of other compounds and are widely used in this way (Chapter 45). Sucrose and glucose are very cheap indeed—probably the cheapest optically active compounds available. Here are the relative (to glucose = 1) prices of some other cheap sugars.

Sugar	Price[a]	Sugar	Price[a]
glucose	1	sorbitol	2
mannose	75	mannitol	4
galactose	8	dulcitol[b]	70
xylose	20	xylitol	15
ribose	100	sucrose	1

[a] Prices relative to glucose = 1.
[b] Dulcitol is the reduction product of galactose.

Chemistry of ribose—from sugars to nucleotides

We have said little about selective reactions of pentoses so we shall turn now to the synthesis of nucleotides such as AMP. In nature, ribose is phosphorylated on the primary alcohol to give ribose-5-phosphate. This is, of course, an enzyme-catalysed reaction but it shows straightforward chemoselectivity such as we should expect from a chemical reaction.

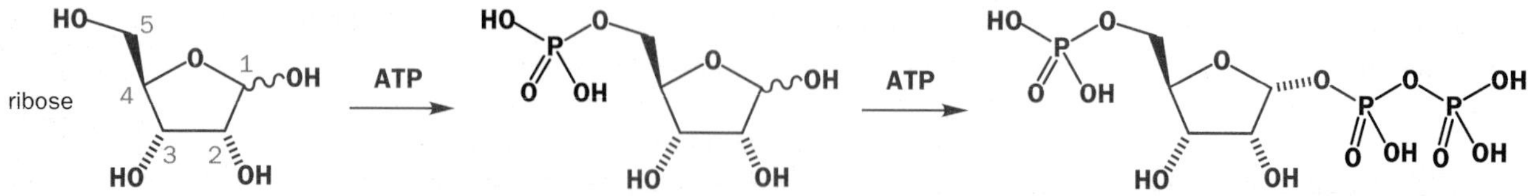

ribose

ribose-5-phosphate

5-phosphoribosyl-1-pyrophosphate (PRPP)

The second step is a pyrophosphorylation at the anomeric position to give PRPP. Only one diastereoisomer is produced so presumably the two anomers interconvert rapidly and only the one isomer reacts under control by the enzyme. This selectivity would be very difficult to achieve chemically.

► You will notice that these two reactions illustrate the flexibility with which ATP can activate biological molecules. In the first reaction, the nucleophilic OH group of ribose attacks the terminal phosphate group, but in the second the OH group must attack the middle phosphate residue. This would be impossible to control chemically.

Now the stage is set for an S_N2 reaction. The nucleophile is actually the amide group of glutamine but the amide is hydrolysed by the same enzyme in the same reaction and the result is as if a molecule of ammonia had done an S_N2 reaction displacing the pyrophosphate from the anomeric position. An NH_2 group is introduced, which is then built into the purine ring-system in a series of reactions involving simple amino acids. These reactions are too complex to describe here.

By contrast, if a pyrimidine is to be made, Nature assembles a general pyrimidine structure first and adds it in one step to the PRPP molecule, again in an S_N2 reaction using a nitrogen nucleophile. This general nucleotide, orotidylic acid, can be converted into the other pyrimidine nucleotides by simple chemistry.

The chemical version—protection all the way

In a chemical synthesis (work that led to Alexander (Lord) Todd's Nobel prize) there are rather different problems. We cannot achieve the remarkable selectivity between the different OH groups achieved in Nature so we have to protect any OH group that is not supposed to react. We also prefer to add pre-formed purines and pyrimidines to a general electrophile derived from ribose. The first step is to form acetate esters from all the OH groups. Since ribose is rather unstable to acetylation conditions, the methyl glycoside (which is formed under very mild conditions) is used. This fixes the sugar in the furanose form. Now the tetraacetate can be made using acetic anhydride in acidic solution. All of the OH groups react by nucleophilic attack on the carbonyl group of the anhydride with retention of configuration except for the anomeric OH, which esterifies by an S_N1 mechanism. This, of course, epimerizes the anomeric centre but the crystalline diastereoisomer shown can be isolated easily.

■ Alexander Todd (1907–97), better known as Lord Todd, was a Scot who pioneered the modern interaction between chemistry and biochemistry in his work at Frankfurt, Oxford, Edinburgh, London, CalTech, Manchester, and Cambridge. He won the Nobel prize in 1957 for his work on the synthesis of the most important coenzymes and nucleotides. This was a remarkable achievement because he had to find out how to do phosphate, ribose, and purine chemistry—none of which was known when he started, and none of which was easy as this brief excursion should show.

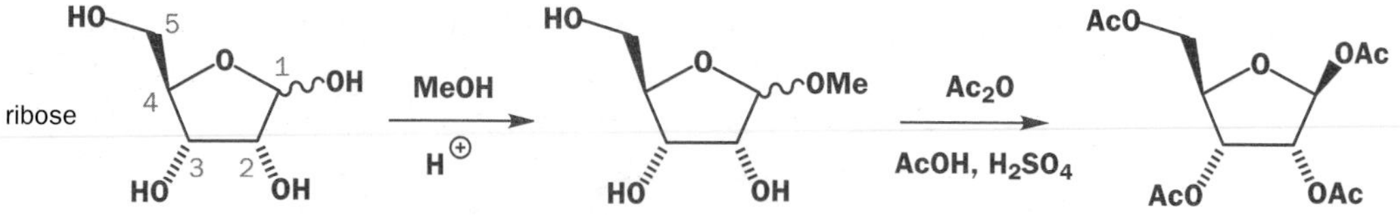

Now the anomeric centre can be activated towards nucleophilic attack by replacement of acetate by chloride. This is again an S_N1 reaction and produces a mixture of chlorides. The other esters are stable to these conditions.

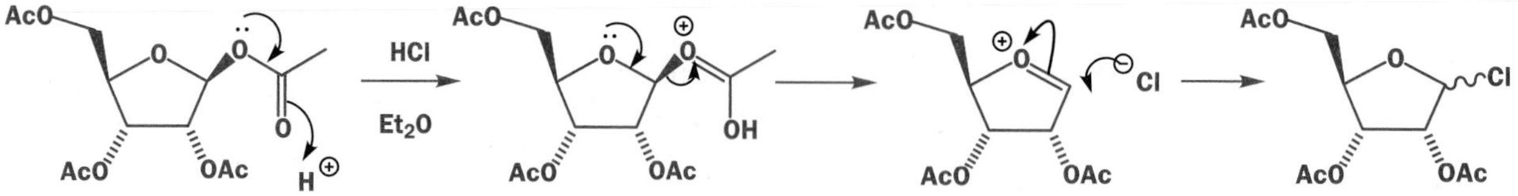

Replacement of the chlorine by the purine or pyrimidine base is sometimes quite tricky and silver or silyl derivatives are often used. Lewis acid catalysis is necessary to help the chloride ion leave in this S_N1 reaction. We shall avoid detailed technical discussion and simply draw the adenosine product from a general reaction.

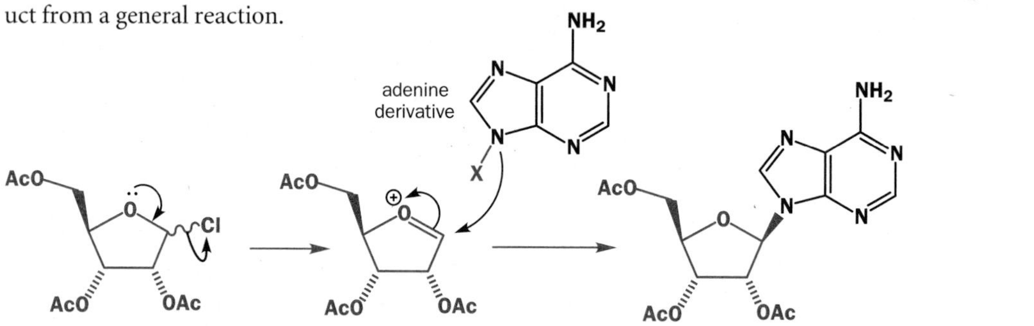

Now we need to remove the acetates and put a phosphate specifically on the 5-position. The acetates can be removed with retention by ester hydrolysis and we already know how to protect the 2-OH and 3-OH groups. They are *cis* to each other so they will form an acetal with acetone leaving the 5-OH group free.

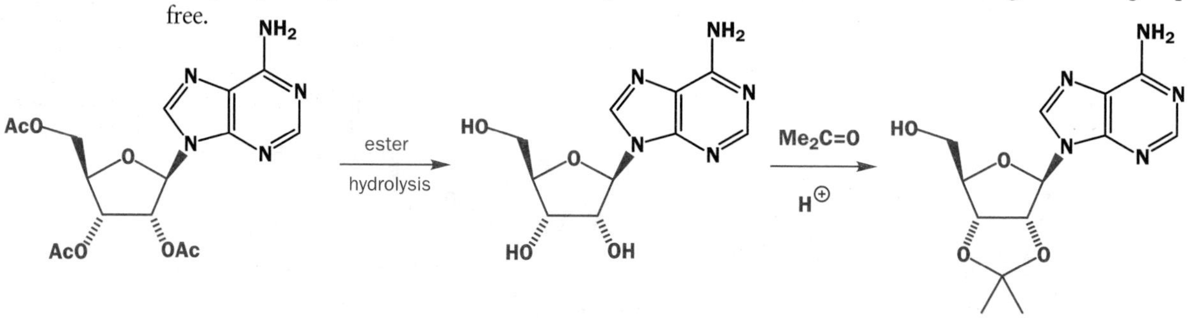

Putting on the phosphate is tricky too and more protection is necessary. This phosphorus compound with one chloride as leaving group and two benzyl esters as protecting groups proved ideal. The benzyl esters can be removed by hydrogenation (Chapter 24) and the acetal by treatment with dilute acid to give AMP.

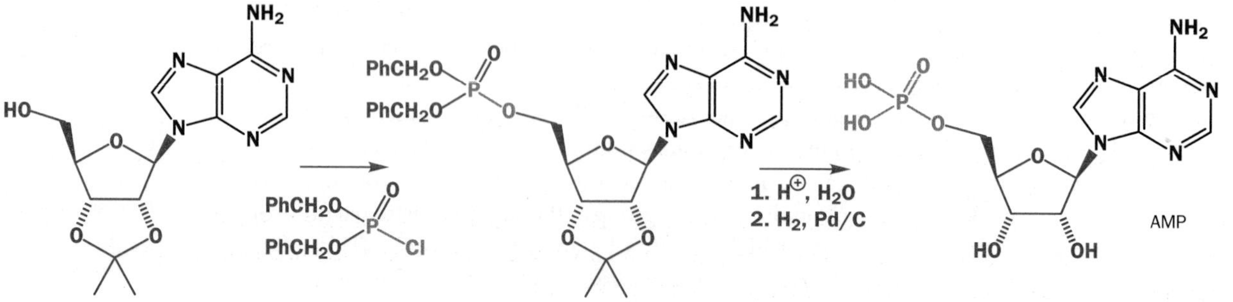

The chemical synthesis involves a lot more selective manipulation of functional groups, particularly by protection, than is necessary in the biological synthesis. However, this synthesis paved the way to the simple syntheses of nucleotides and polynucleotides carried out routinely nowadays. The usual method is to build short runs of nucleotides and then let the enzymes copy them—a real partnership between biology and chemistry.

Glycosides are everywhere in nature

Many alcohols, thiols, and amines occur in nature as **glycosides**, that is as *O*-, *S*-, or *N*-acetals at the anomeric position of glucose. The purpose of attaching these compounds to glucose is often to improve solubility or transport across membranes—to expel a toxin from the cell, for example. Sometimes glucose is attached in order to stabilize the compound so that glucose appears as Nature's protecting group, rather as a chemist would use a THP group (Chapter 24).

> The most important *N*-glycosides are, of course, the nucleotides and we have already described them in some detail.

an *O*-glycoside   an *S*-glycoside   an *N*-glycoside

O-Glycosides occur in immense variety with glucose and other sugars being joined to the OH groups of alcohols and phenols to form acetals. The stereochemistry of these compounds is usually described by the Greek letters α and β. If the OR bond is down, we have an α-glycoside; if up, a β-glycoside.

> We saw an example in Chapter 6 where acetone cyanohydrin is found in the cassava plant as a glucoside and suitable precautions must be taken when eating cassava to avoid poisoning by HCN.

An attractive example is the pigment of red roses, which is an interesting aromatic oxygen heterocycle (an anthocyanidin). Two of the phenolic OH groups are present as β-glycosides.

aromatic pyrilium salt

glycoside linkages

pigment from red roses

a β-glycoside of a phenol

an α-glycoside of a phenol

> It is easy to remember which is which. People who devise nomenclature are maliciously foolish and, just as *E* means *trans* and *Z* means *cis* (each letter has the shape of the *wrong* isomer), so α means *below* and β means *above*—each word begins with the *wrong* letter.

Protect yourself from cancer with green vegetables: *S*-glycosides

We will take an important series of *S*-glycosides for further chemical discussion in this chapter. It is clear that there are special benefits to health in eating broccoli and brussels sprouts because of their potent sulfur-containing anti-cancer compounds. These compounds are unstable isothiocyanates and are not, in fact, present in the plant but are released on damage by, for example, cutting or cooking when a glycosidase (an enzyme which hydrolyses glycosides) releases the sulfur compound from its glucose protection. A simple example is sinigrin.

thioglycoside

sinigrin

When a glycosidase enzyme cleaves an *O*-glycoside, we should expect a simple general acid-catalysed first step followed by fast addition of water to the intermediate oxonium ion, essentially the same mechanism as is shown by the chemical reaction (Chapter 13).

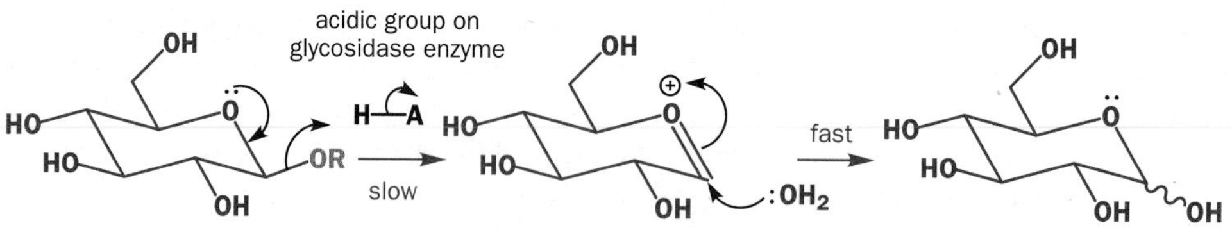

The *S*-glycosides of the sinigrin group start to hydrolyse in the same way. The sulfur atom is the better leaving group when it leaves as an anion (though worse than oxygen when the hydrolysis occurs in acidic conditions—see p. 1255) and these anions are additionally stabilized by conjugation.

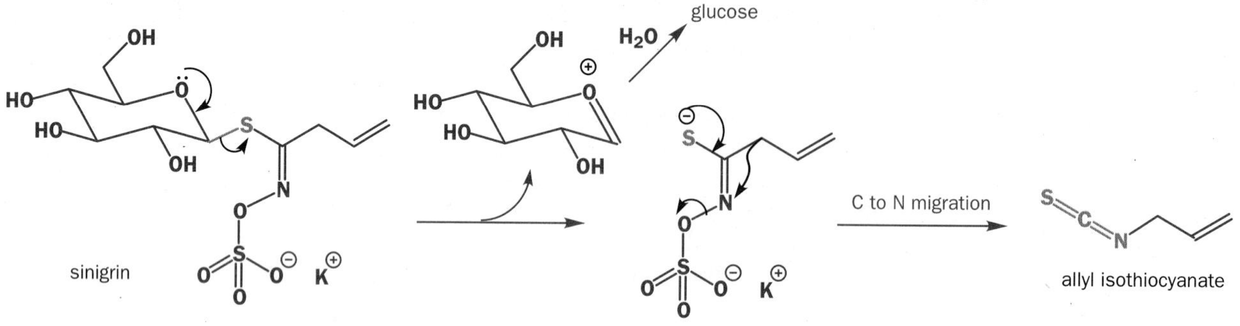

The next step is very surprising. A rearrangement occurs, rather similar to the Beckmann rearrangement (Chapter 37), in which the alkyl group migrates from carbon to nitrogen and an isothiocyanate (R–N=C=S) is formed. Sinigrin occurs in mustard and horseradish and it is the release of the allyl isothiocyanate that gives them their 'hot' taste. When mustard powder is mixed with water, the hot taste develops over some minutes as sinigrin is hydrolysed to the isothiocyanate.

The *S*-glycoside in broccoli and brussels sprouts that protects from cancer is somewhat similar but has one more carbon atom in the chain and contains a sulfoxide group as well. Hydrolysis of the *S*-glycoside is followed by the same rearrangement, producing a molecule called sulforaphane. Sulforaphane protects against cancer-causing oxidants by inducing the formation of a reduction enzyme.

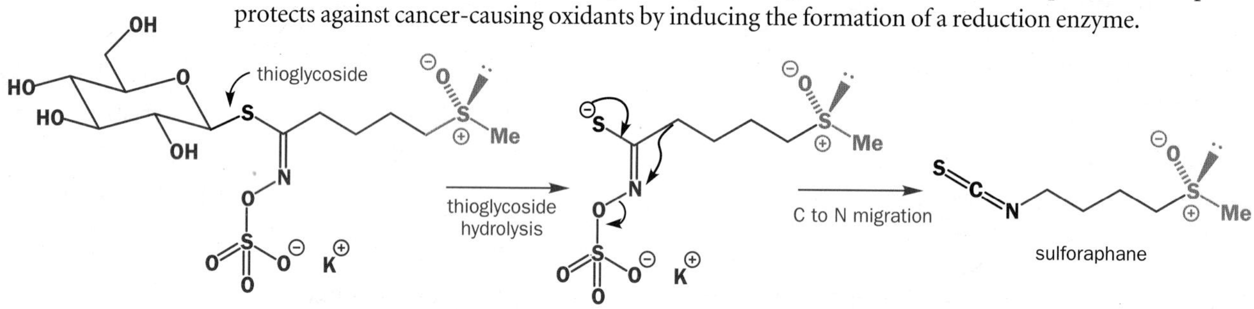

Compounds derived from sugars

Vitamin C

Nature makes some important compounds from simple sugars. Vitamin C—ascorbic acid—is one of these. Like glutathione, it protects us from stray oxidants as well as being involved in primary redox pathways (we mentioned earlier its role in collagen synthesis). Its reduced and oxidized forms are these.

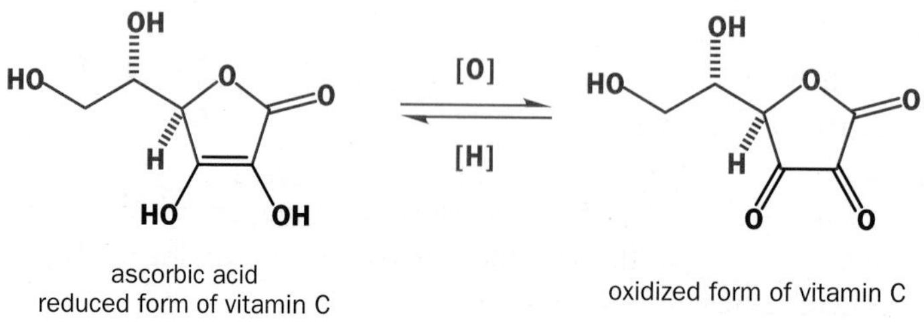

ascorbic acid
reduced form of vitamin C

oxidized form of vitamin C

Vitamin C looks very like a sugar as it has six carbon atoms, each having an oxygen atom as substituent as well as an oxygen heterocycle, and it is no surprise that it is made in nature from glucose. We shall give just an outline of the process, which appropriately involves a lot of oxidation and reduction. The first step takes the primary alcohol of glucose to a carboxylic acid known as glucuronic acid. Next comes a reduction of the masked aldehyde to give 'gulonic acid'. Both reactions are quite reasonable in terms of laboratory chemistry.

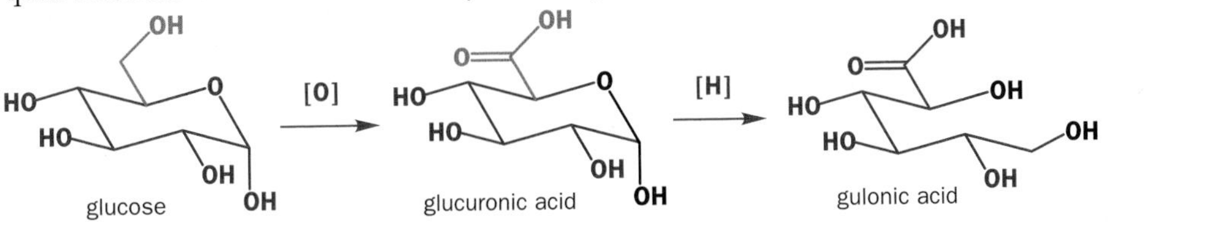

glucose → [O] → glucuronic acid → [H] → gulonic acid

▶ We have given names for these relatively well-known sugar derivatives, but you do not need to learn them.

It is pretty obvious what will happen to this compound as it is an open-chain carboxylic acid with five OH groups. One of the OH groups will cyclize on to the acid to form a lactone. Kinetically, the most favourable cyclization will give a five-membered ring, and that is what happens. Now we are getting quite close to ascorbic acid and it is clear that oxidation must be the next step so that the double bond can be inserted between C2 and C3.

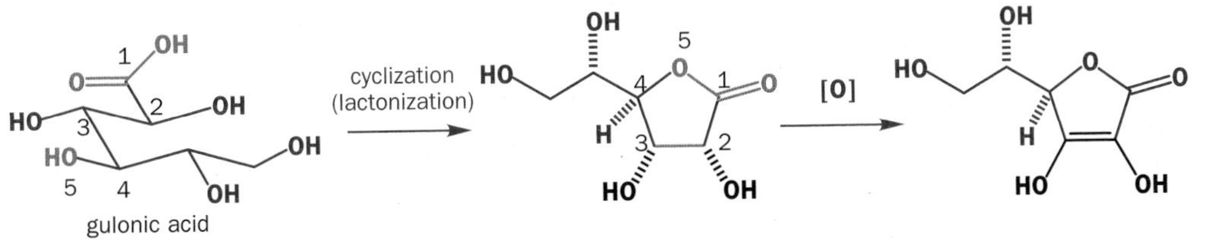

gulonic acid → cyclization (lactonization) → [O]

This looks a strange reaction but it is really quite logical. One of the secondary OH groups must be oxidized to a ketone. This is the 2-OH group and then the resulting ketone can simply enolize to give ascorbic acid.

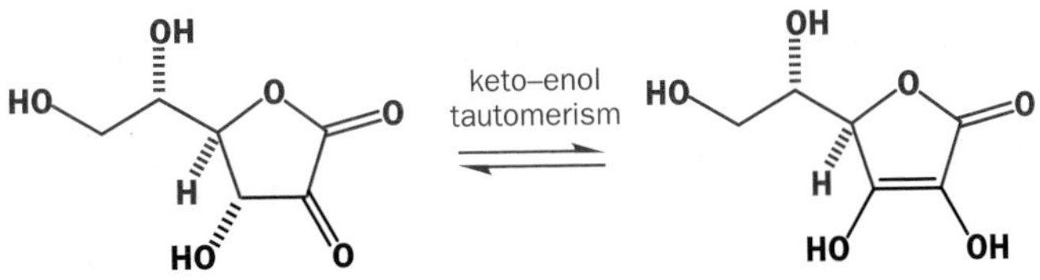

keto–enol tautomerism

Inositols

We have already discussed the widespread sugar alcohols such as mannitol but more important compounds are cyclic sugar alcohols having a carbocyclic ring (**cyclitols**). The most important is **inositol** which controls many aspects of our chemistry that require communication between the inside and the outside of a cell. Inositol-1,4,5-triphosphate (IP_3) can open calcium channels in cell membranes to allow calcium ions to escape from the cell.

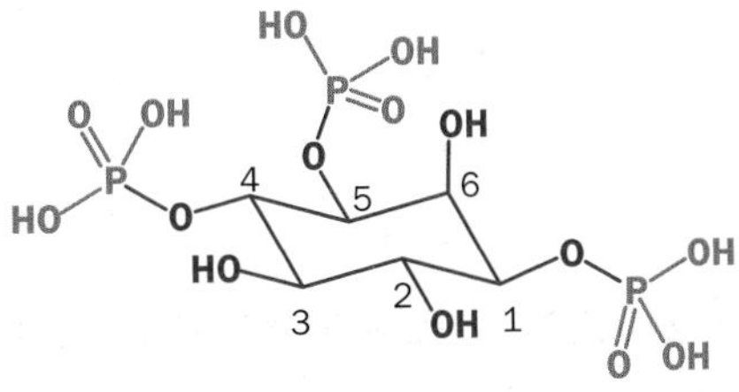

inositol-1,4,5-triphosphate

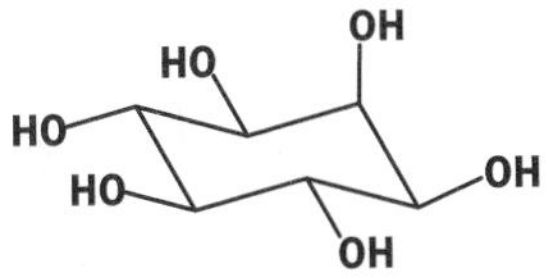

inositol

■ This inositol is known as 'myo-inositol' and is just one of many possible stereoisomers. Inositol was mentioned in Chapter 16.

Inositol is made in nature from glucose-6-phosphate by an aldol reaction that requires preliminary ring opening and selective oxidation (this would be tricky in the lab without protecting groups!).

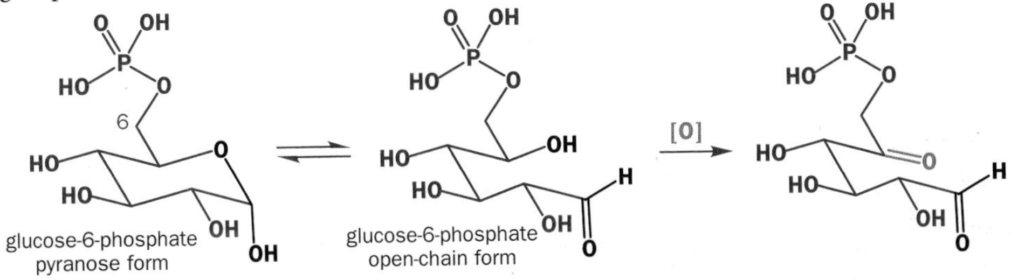

glucose-6-phosphate pyranose form → glucose-6-phosphate open-chain form → [O]

The resulting ketone can be enolized on the phosphate side and added to the free aldehyde group to form the cyclohexane ring. We can draw the mechanism for the aldol reaction easily if we first change the conformation.

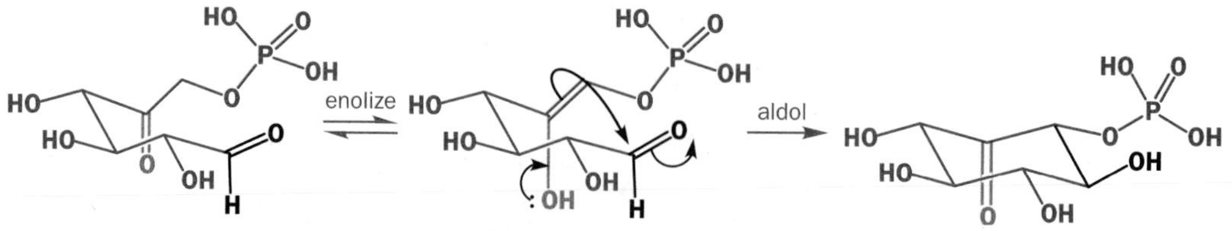

Finally, a stereochemically controlled reduction to give the axial alcohol (this would be the stereo-selectivity expected with NaBH₄ for example: see Chapter 18) gives myo-inositol. The number and position of the phosphate esters can be controlled biochemically. This control is vital in the biological activity and would be difficult in the laboratory.

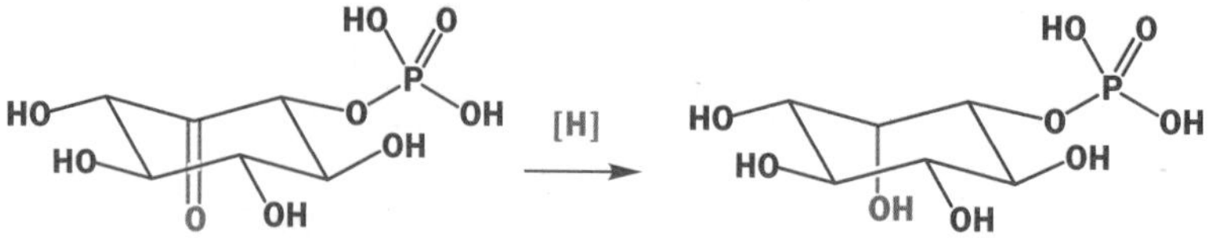

Learning from Nature—the synthesis of inositols

If we wish to devise a chemical version of the biosynthesis of inositols, we need to use cleverly devised protecting groups to make sure that the right OH group is oxidized to a ketone. We can start with glucose trapped in its furanose form by a double acetone acetal as we discussed above. The one remaining OH group is first blocked as a benzyl ether.

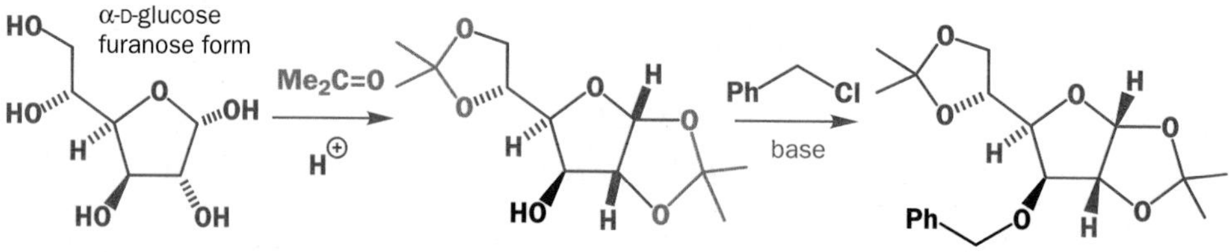

Next, one of the acetals is hydrolysed under very mild conditions, and the primary alcohol is protected as a trityl ether. This is an S$_N$1 reaction with an enormous electrophile—so big that it goes on primary alcohols only.

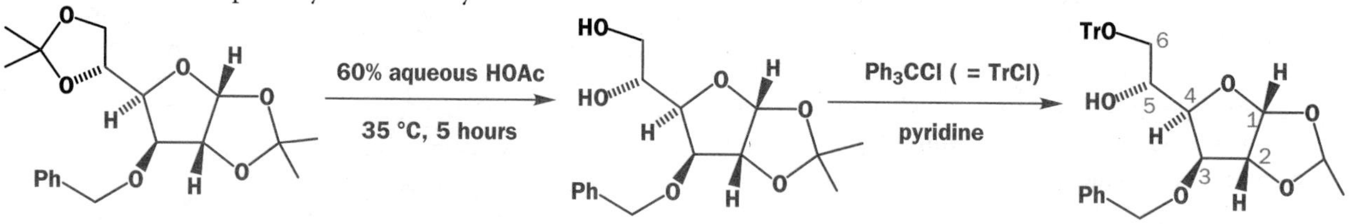

Notice that each oxygen atom in this molecule of protected glucose is now different. Only the OH at C5 is free, and its time has come: it can now be oxidized using a Swern procedure with dimethyl-sulfoxide as the oxidant (Chapter 46).

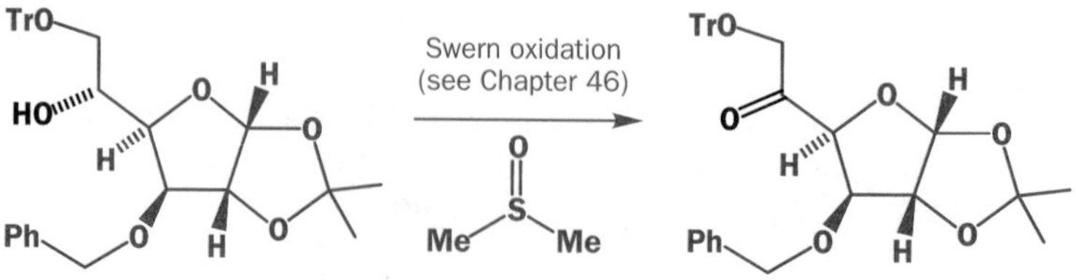

Now we can strip away the protecting groups one by one and it is instructive to see how selective these methods are. The trityl group comes off in aqueous acetic acid by another S$_N$1 reaction in which water captures the triphenylmethyl cation, and the benzyl group is removed by hydrogenolysis—hydrogen gas over a 10% palladium on charcoal catalyst in ethanol.

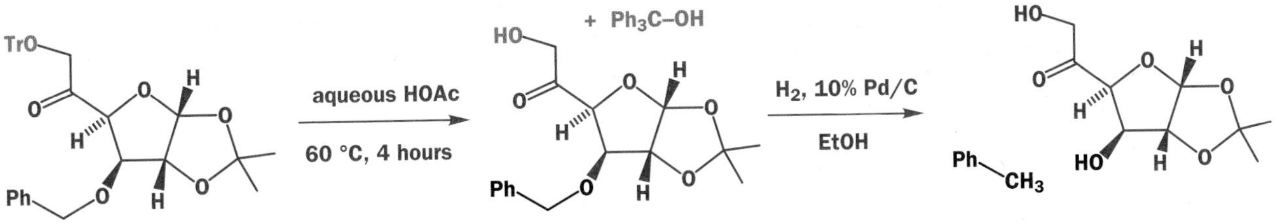

Finally, the acetone acetal is removed by acid hydrolysis. Because free sugars are difficult to isolate it is convenient to use an acidic resin known as 'Dowex'. The resin (whose polymeric structure is discussed in Chapter 52) can simply be filtered off at the end of the reaction and the solid product isolated by **lyophilization**—evaporation of water at low pressure below freezing point. The yield is quantitative.

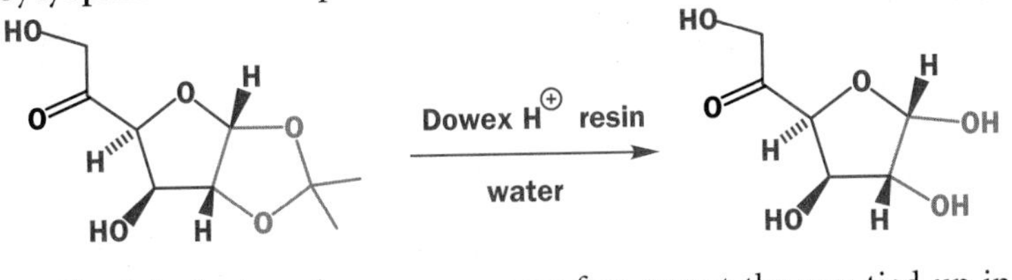

All of the hydroxyl groups are now free except the one tied up in the hemiacetal and that, of course, is in equilibrium with the open-chain hydroxy-aldehyde as we have already seen. Treatment of this free 'glucose ketone' with aqueous NaOH gives the ketone of myo-inositol as the major product together with some of the other diastereoisomers.

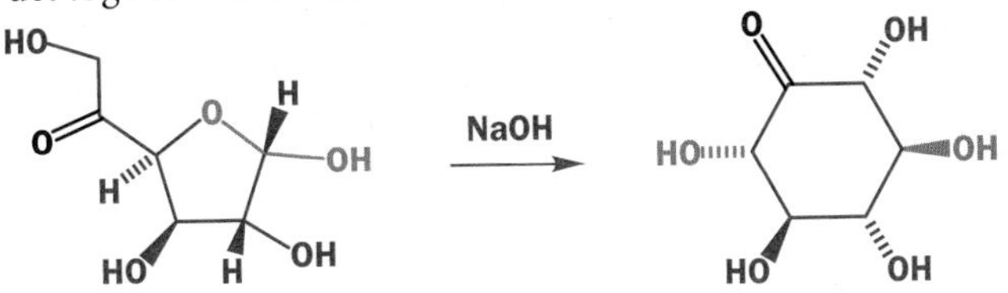

'5-keto-glucose'   'keto-myo-inositol'

The simplest explanation of this result is that the chemical reaction has followed essentially the same course as the biological one. First, the hemiacetal is opened by the base to give the open-chain keto-aldehyde. Rotation about a C–C bond allows a simple aldol condensation between the enolate of the ketone as nucleophile and the aldehyde as electrophile.

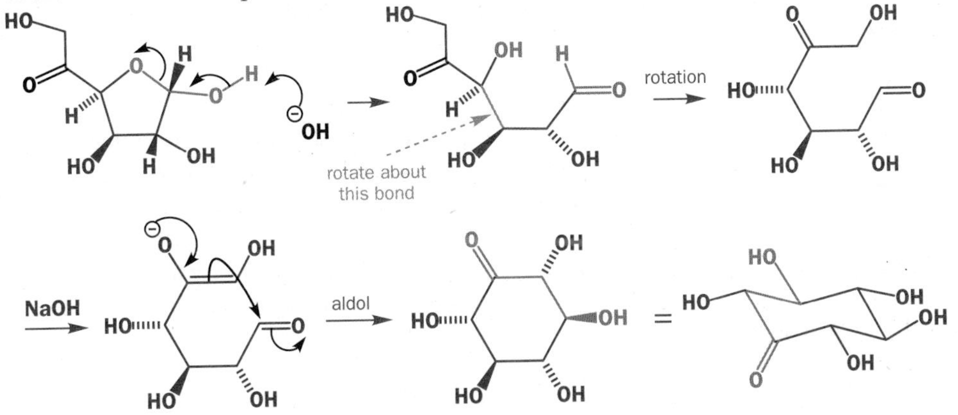

The enolate must prefer to attack the aldehyde in the same way as in the biological reaction to give the all-equatorial product as the conformational drawing shows. The arrangement of the enolate in the aldol reaction itself will be the same as in the cyclization of the phosphate above.

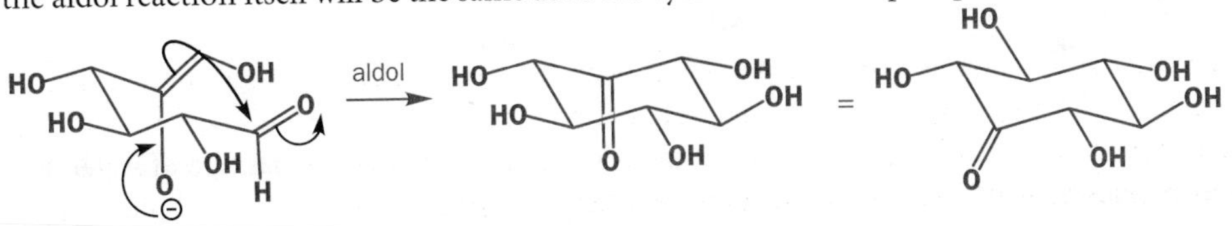

As in many other cases, by improving the rate and perfecting the stereoselectivity, the enzyme makes much better a reaction that already works.

Most sugars are embedded in carbohydrates

Before we leave the sugars we should say a little about the compounds formed when sugars combine together. These are the **saccharides** and they have the same relationship to sugars as peptides and proteins have to amino acids. We have met one simple disaccharide, sucrose, but we need to meet some more important molecules.

One of the most abundant compounds in nature is cellulose, the structural material of plants. It is a glucose polymer and is produced in simply enormous quantities (about 10^{15} kg per year). Each glucose molecule is joined to the next through the anomeric bond (C1) and the other end of the molecule (C4). Here is that basic arrangement.

> ▶
> This is literally an astronomical amount: it's about the mass of one of the moons of Mars, Deimos. Our moon weighs 10^{22} kg.

Notice that the anomeric bonds are all equatorial. This means that the cellulose molecule is linear in general outline. It is made rigid by extra hydrogen bonds between the 3-OH groups and the ring oxygen atoms—like this.

The polymer is also coiled to increase stability still further. All this makes cellulose very difficult to hydrolyse, and humans cannot digest cellulose as we do not have the necessary enzymes. Only ruminants, such as cows, whose many stomachs harbour some helpful bacteria, can manage it.

Amino sugars add versatility to saccharides

To go further in understanding the structural chemistry of life we need to know about **amino sugars**. These molecules allow proteins and sugars to combine and produce structures of remarkable variety and beauty. The most common amino sugars are *N*-acetyl-glucosamine and *N*-acetyl-galactosamine, which differ only in stereochemistry.

The hard outer skeleton of insects and shellfish contains chitin, a polymer very like cellulose but made of acetyl glucosamine instead of glucose itself. It coils up in a similar way and provides the toughness of crab shells and beetle cases.

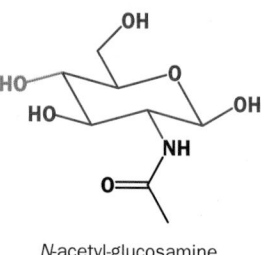

N-acetyl-glucosamine

N-acetyl-galactosamine

Ordinary cell membranes must not be so tough as they need to allow the passage of water and complex molecules through channels that can be opened by molecules such as inositol phosphates.

These membranes contain **glyco-proteins**—proteins with amino-sugar residues attached to asparagine, serine, or threonine in the protein. The attachment is at the anomeric position so that these compounds are *O*- or *N*-glycosides of the amino sugars. Here is *N*-acetyl-galactosamine attached to an asparagine residue as an *N*-glycoside.

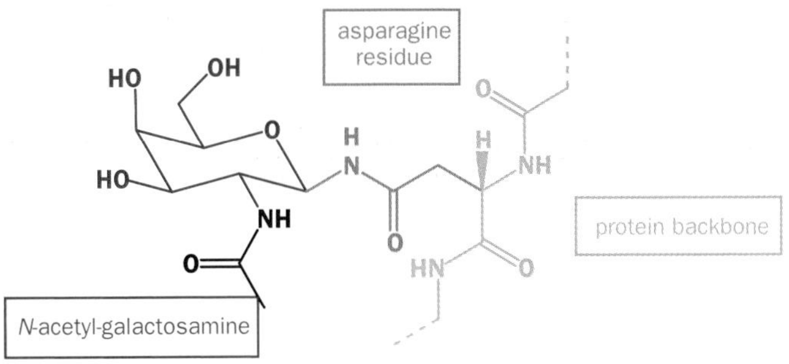

The cell membrane normally contains less than 10% of sugars but these are vital to life. Because the sugars (*N*-acetyl-glucosamine and *N*-acetyl-galactosamine) are covered with very polar groups (OH and amide) they prefer to sit outside the membrane in the aqueous extracellular fluid rather than within the nonpolar membrane itself. When two cells meet, the sugars are the first things they see. We cannot go into the details of the biological processes here, but even the structures of these saccharides dangling from the cell are very interesting. They contain amino sugars, again particularly *N*-acetyl-glucosamine and *N*-acetyl-galactosamine, and they are rich in mannose.

In addition, they are usually branched at one of the mannose residues that is joined to two other mannoses on one side and to one glucosamine on the other. The glucosamine leads back eventually to the protein through a link to asparagine like the one we have just seen. The two mannoses are linked to more sugars at positions marked by the green arrows and provide the recognition site. The structure below is a typical branchpoint.

> Mannose is another glucose diastereoisomer and has one axial OH group at C2.
>
> mannose: axial OH at C2

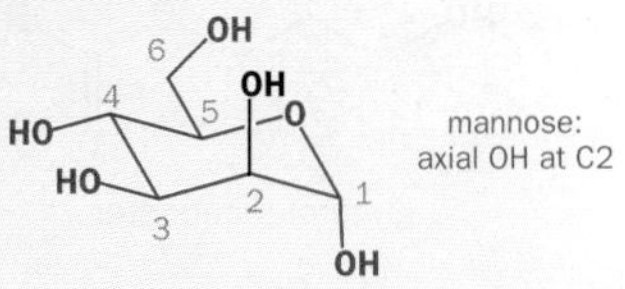

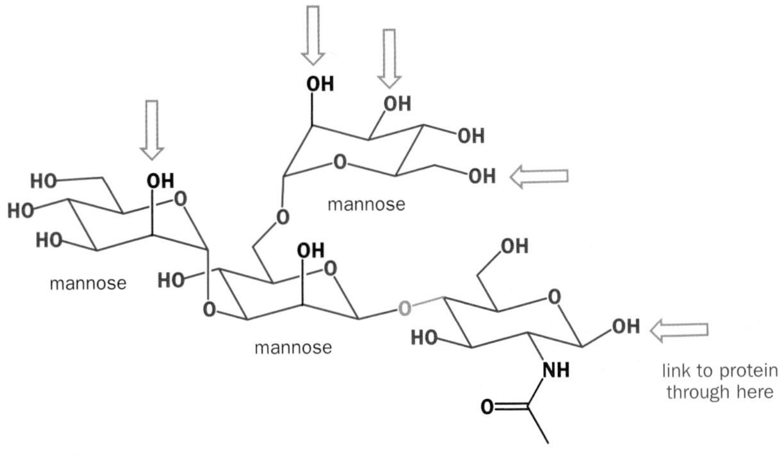

You should begin to see from structures like these just how versatile sugar molecules can be. From just four sugars we have constructed a complex molecule with up to 13 possible link sites. With more sugars added, the possibilities become enormous. It is too early to say what medical discoveries will emerge from these molecules, but one that is likely to be important is **sialyl Lewis X**. This tetrasaccharide is also branched but it contains a different type of molecule—a C_9 sugar with a CO_2H group, called sialic acid.

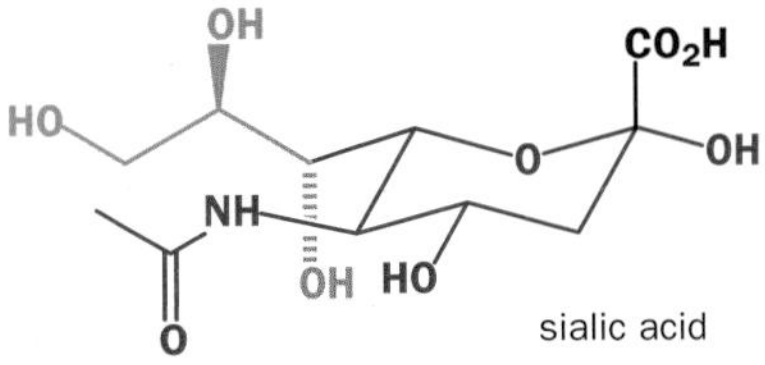

Sialic acid has the CO_2H group at the anomeric position, a typical *N*-acetyl group, and a unique side chain (in green) with three more OH groups. Sialyl Lewis X has sialic acid at the end of a branched sugar chain. The branchpoint is the familiar *N*-acetyl-glucosamine through which the molecule is eventually linked to the glycoprotein. The remaining sugars are galactose, a diastereoisomer of glucose, and a sugar we have not seen before, fucose. Fucose often appears in saccharides of this kind and is a six-carbon sugar without a primary OH group. It is like galactose with Me instead of CH_2OH.

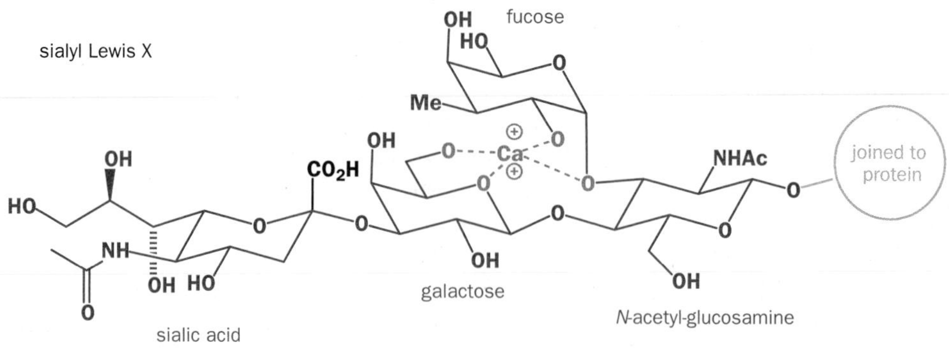
sialyl Lewis X
fucose
OH
HO
O
Me
OH
O
Ca
O
O
NHAc
joined to protein
OH
CO2H
HO
O
O
O
O
O
NH
OH HO
OH
galactose
OH
N-acetyl-glucosamine
sialic acid

Sialyl Lewis X can also form a stable complex with calcium ions as the diagram shows and this may be vital to its activity. It is certainly involved in leukocyte adhesion to cells and is therefore vital in the prevention of infection.

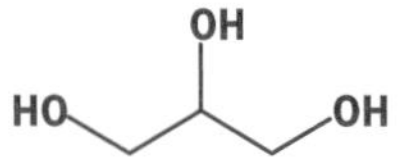
OH
HO
OH
glycerol

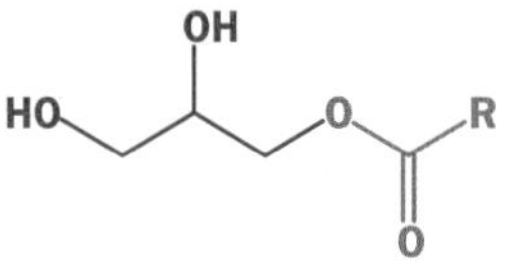
OH
HO
O
R
O
glycerol monoester

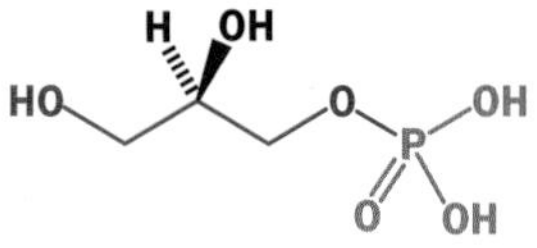
H OH
HO
O
P
OH
O
OH
glycerol 3-phosphate

Lipids

Lipids (fats) are the other important components of cell membranes. Along with cholesterol, also a component of the cell membrane, they have acquired a bad name, but they are nonetheless essential to the function of membranes as selective barriers to the movement of molecules.

The most common types of lipids are esters of glycerol. Glycerol is just propane-1,2,3-triol but it has interesting stereochemistry. It is not chiral as it has a plane of symmetry, but the two primary OH groups are enantiotopic (Chapter 16). If one of them is changed—by esterification, for example—the molecule becomes chiral. Natural glycerol phosphate is such an ester and it is optically active.

A typical lipid in foodstuffs is the triester formed from glycerol and oleic acid, which is the most abundant lipid in olive oil. Oleic acid is a 'mono-unsaturated fatty acid'—it has one *Z* double bond in the middle of the C_{18} chain. This bond gives the molecule a marked kink in the middle. The compound actually present in olive oil is the triester, also kinked.

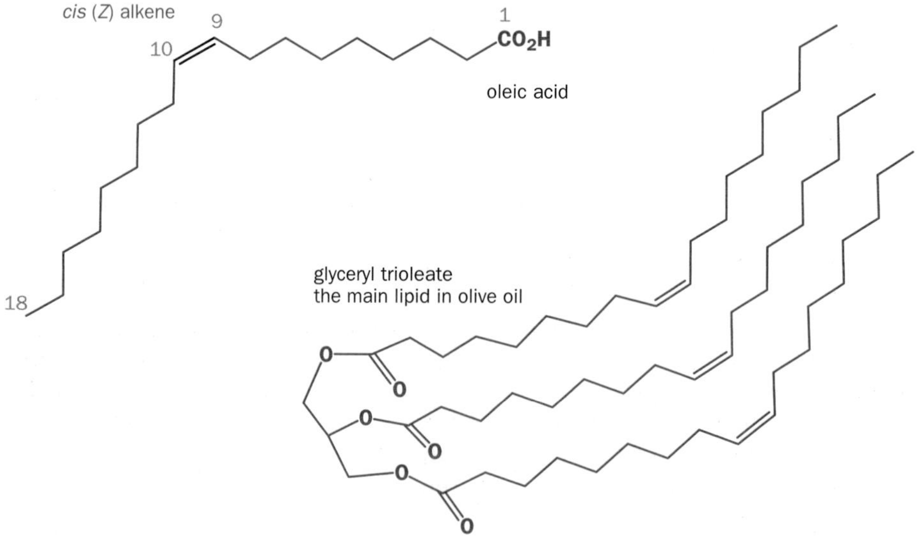
cis (Z) alkene
9
10
1
CO2H
oleic acid
18
glyceryl trioleate
the main lipid in olive oil
O
O
O
O
O
O

Oil and water do not mix

The lipid has, more or less, the conformation shown in the diagram with all the polar ester groups at one end and the hydrocarbon chains bunched together in a nonpolar region. Oil and water do not mix, it is said, but triglyceride lipids associate with water in a special way. A drop of oil spreads out on water in a very thin layer. It does so because the ester groups sit inside the water and the hydrocarbon side chains stick out of the water and associate with each other.

You may have done the 'Langmuir trough' experiment in a physical chemistry practical class. This involves measuring the size of a molecule by allowing an oil to spread on the surface of water in a unimolecular layer.

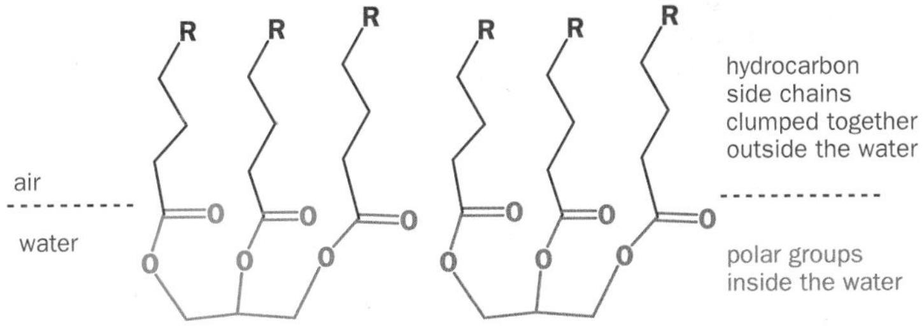

When triglycerides are boiled up with alkali, the esters are hydrolysed and a mixture of carboxylate salts and glycerol is formed. This was how soap was made—hard soap was the sodium salt and soft soap the potassium salt.

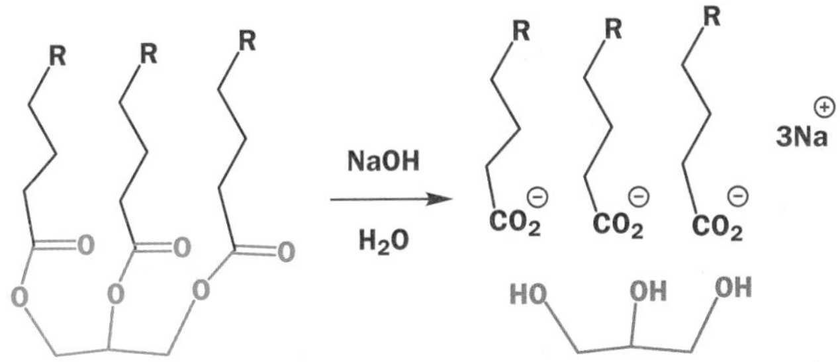

When a soap is suspended in water, the carboxylate groups have a strong affinity for the water and so oily globules or **micelles** are formed with the hydrocarbon side chain inside. It is these globules that remove greasy dirt from you or your clothes.

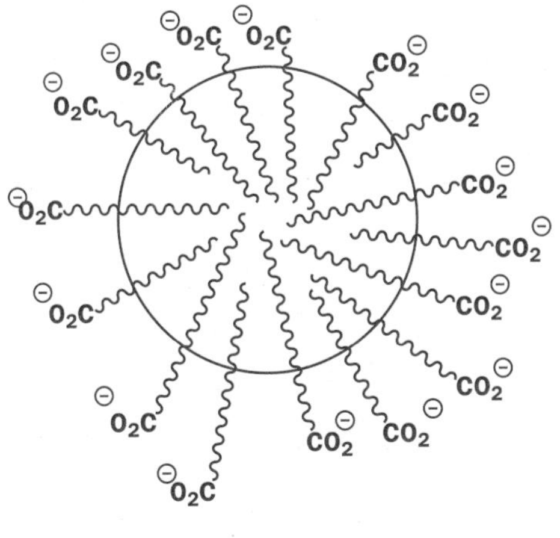

Nature uses thiol esters to make lipids

The repulsion between molecules having oily or aqueous properties is the basis for membrane construction. The lipids found in membranes are mostly based on glyceryl phosphate and normally contain three different side chains—one saturated, one unsaturated, and one very polar.

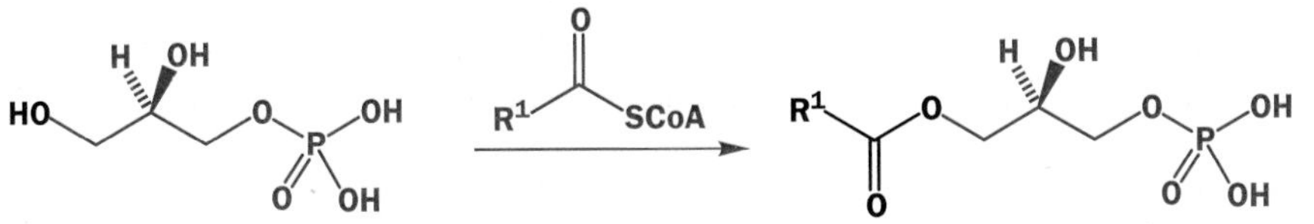

The saturated chain is added first, at C1 of glyceryl phosphate. The reagent is a thiol ester called acyl coenzyme A, whose full structure you will see in the next chapter. This reaction occurs by simple nucleophilic attack on the carbonyl group of the thiol ester followed by loss of the better leaving group, the thiolate anion. Then the process is repeated at the second OH group where an unsaturated fatty acid, perhaps oleic acid, is added by the same mechanism.

■ We discussed acylation by thioesters, the laboratory version of this reaction, in Chapter 27.

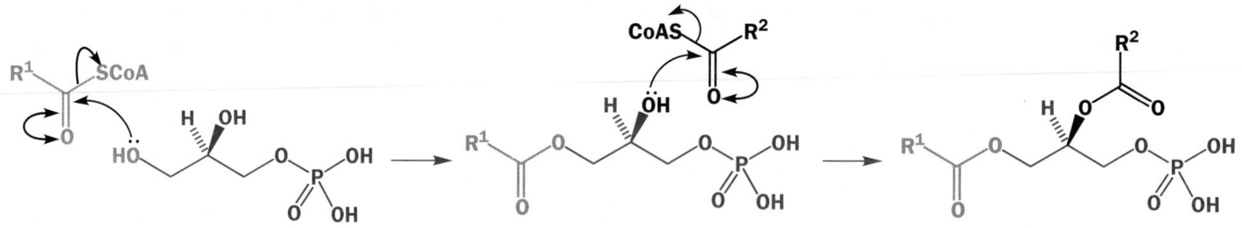

The third acylation requires the phosphate to act as the acylating agent and a polar alcohol to be introduced to form a phosphate ester. This reaction actually occurs by the activation of the phosphate as a pyrophosphate. Pyrophosphates are really acid anhydrides so it is not surprising that they act as acylating agents. The first step is a reaction with cytidine triphosphate (CTP) doing a job we might expect from ATP.

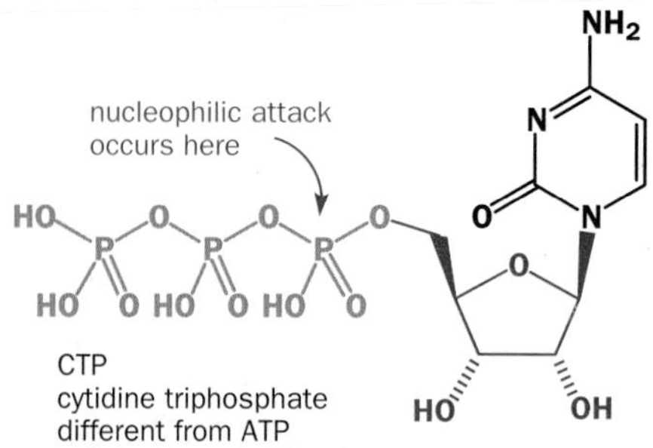

Nucleophilic attack by the phosphate group of the phosphoglyceride at the point indicated on CTP gives the pyrophosphate required for the acylation step.

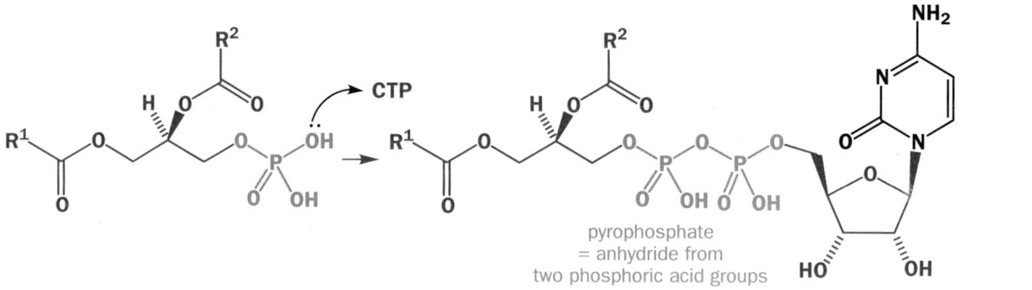

CTP
cytidine triphosphate
different from ATP
only in the pyrimidine base

The anhydride is now attacked by an alcohol acting as a nucleophile. The attack occurs only at the electrophilic phosphorus centre further from the nucleotide. This is an impressive piece of regioselectivity and is presumably controlled by the enzyme.

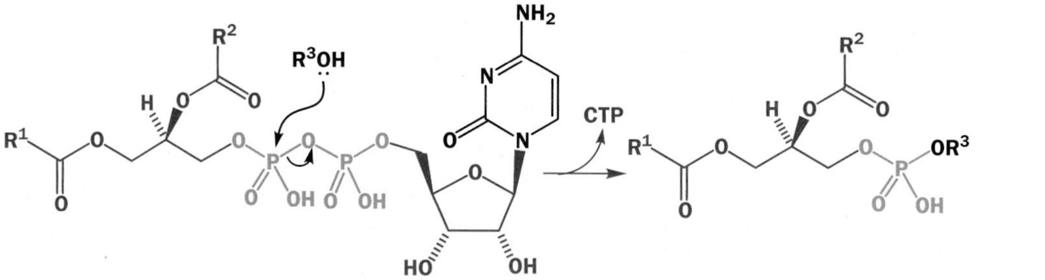

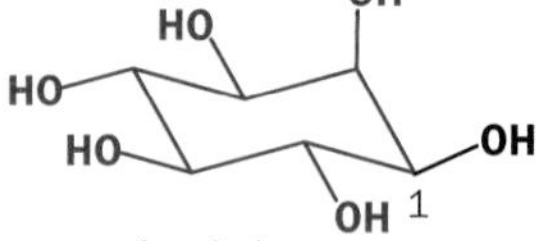

This third chain is rather different from the other two—it's a phosphate diester, and the alcohol portion can be inositol joined through the OH group at C1 or it can be the amino acid serine, joined through its OH group.

The compound formed from serine is particularly important as it can be transformed into the most dramatically contrasted of these phospholipids. A decarboxylation using a coenzyme (we shall look at the mechanism of this reaction in Chapter 51) gives a very simple molecule, phosphatidyl ethanolamine.

inositol

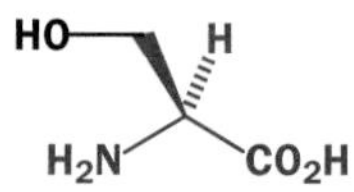

the amino acid
serine (Ser)

phosphatidyl serine

phosphatidyl ethanolamine

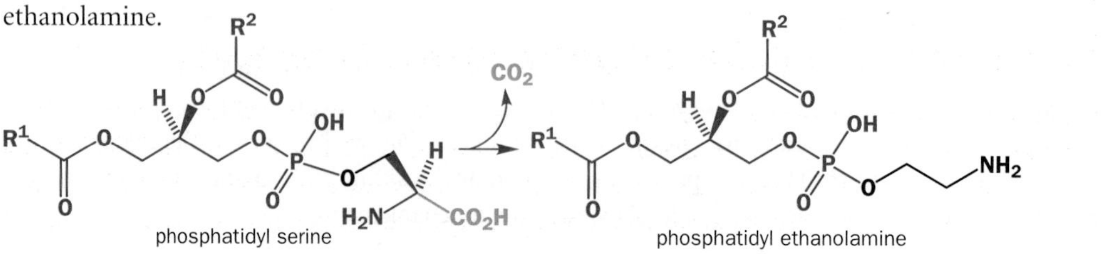

Finally, three methylations on the nitrogen atom by SAM (see p. 1348) gives the zwitterion phosphatidyl choline.

$3 \times$ **SAM**

phosphatidyl choline

► Choline is a tetraalkyl ammonium salt and is important elsewhere in biology.

choline

Phospholipids form membranes spontaneously

The choline terminus of the molecule is very polar indeed. Phosphatidyl choline adopts a shape with the nonpolar chains (R^1 and R^2) close together, and it should be clear that this is an ideal molecule for the construction of membranes.

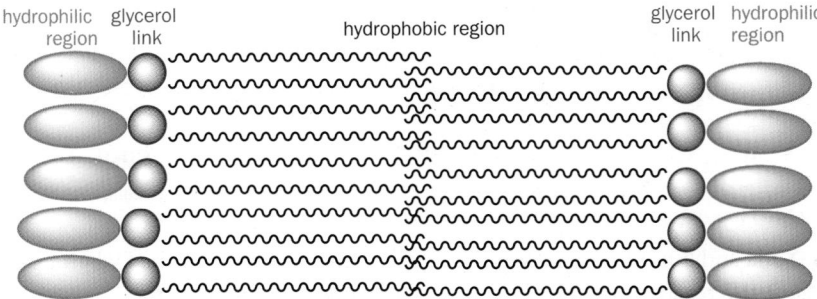

We have already seen how oils such as glyceryl trioleate form thin layers on water while soaps from the alkaline hydrolysis of glycerides form micelles. Phosphatidyl choline forms yet another structure—it spontaneously forms a membrane in water. The hydrophobic hydrocarbon chains line up together on the inside of the membrane with the hydrophilic choline residues on the outside.

This is just a small piece of a cross-section of the membrane. These membranes are called **lipid bilayers** because two rows of molecules line up to form two layers back-to-back. The charged, hydrophilic region on the outside is solvated by the water and the hydrocarbon tails are repelled by the water and attracted to each other by weak forces such as van der Waals attractions.

Full structural analysis of a real cell membrane reveals a chemically diverse thin sheet composed of phospholipid bilayers penetrated by glycoproteins containing the amino sugars we discussed earlier. The amount of each component varies but there is usually about 50:50 phospholipid:protein, with the protein containing about 10% sugar residues. The phospholipids' main role is as a barrier while the glycoproteins have the roles of recognition and transport.

Bacteria and people have slightly different chemistry

We have many times emphasized that all life has very similar chemistry. Indeed, in terms of biochemistry there is little need for the classifications of mammals, plants, and so on. There is only one important division—into **prokaryotes** and **eukaryotes**. Prokaryotes, which include bacteria, evolved first and have simple cells with no nucleus. Eukaryotes, which include plants, mammals, and all

(S)-Ala from normal proteins

(R)-Ala from bacterial cell walls

▶

The reason bacteria use these 'unnatural' D-amino acids in their cell walls is to protect them against the enzymes in animals and plants, which cannot digest proteins containing D-amino acids.

penicillin
β-lactam in black

other multicellular creatures, evolved later and have more complex cells including nuclei. Even so, much of the biochemistry on both sides of the divide is the same.

When medicinal chemists are looking for ways to attack bacteria, one approach is to interfere with chemistry carried out by prokaryotes but not by us. The most famous of these attacks is aimed at the construction of the cell walls of some bacteria that contain 'unnatural' (R)- (or D-) amino acids. Bacterial cell walls are made from glycopeptides of an unusual kind. Polysaccharide chains are cross-linked with short peptides containing (R)-alanine (D-Ala). Before they are linked up, one chain ends with a glycine molecule and the other with D-Ala–D-Ala. In the final step in the cell wall synthesis, the glycine attacks the D-Ala–D-Ala sequence to form a new peptide bond by displacing one D-Ala residue.

The famous molecule that interferes with this step is penicillin, though this was not even suspected when penicillin was discovered. We now know how penicillin works. It inhibits the enzyme that catalyses the D-Ala transfer in a very specific way. It first binds specifically to the enzyme, so it must be a mimic of the natural substrate, and it then reacts with the enzyme and inactivates it by blocking a vital OH group at the active site. If we emphasize the peptide nature of penicillin and compare it with D-Ala–D-Ala, the mimicry may become clearer.

acyl–D-Ala–D-Ala redrawn back to front and upside down

acyl–D-Ala–D-Ala redrawn in the penicilin shape

the peptide part of penicillin

Penicillin imitates D-Ala and binds to the active site of the enzyme, encouraging the OH group of a serine residue to attack the reactive, strained β-lactam. This same OH group of the same serine residue would normally be the catalyst for the D-Ala–D-Ala cleavage used in the building of the bacterial cell wall. The reaction with penicillin 'protects' the serine and irreversibly inhibits the enzyme. The bacterial cell walls cannot be completed, and the bacterial cells literally burst under the pressure of their contents. Penicillin does not kill bacteria whose cell walls are already complete but it does prevent new bacteria being formed.

▶

Our current last line of defence against bacteria resistant to penicillin, and other antibiotics, is vancomycin. Vancomycin works by binding to the D-Ala–D-Ala sequences of the bacterial cell wall.

serine residue
at active site

active site blocked by
covalently bound penicillin molecule

You have seen many instances in this chapter of the importance of a good understanding of both the chemistry and the biochemistry of living things if medicine is to advance: it is at the frontier of chemistry and biology that many of the most important medical advances are being made.

Problems

1. Do you consider that thymine and caffeine are aromatic compounds? Explain.

thymine   caffeine

2. It is important that we draw certain of the purine and pyrimidine bases in their preferred tautomeric forms. The correct pairings are given early in the chapter. What alternative pairings would be possible with these (minor) tautomers of thymine and guanine? Suggest reasons (referring to Chapter 43 if necessary) why the major tautomers are preferred.

thymine tautomer   guanine tautomer

3. Dialkyl phosphates are generally hydrolysed quite slowly at near-neutral pHs but this example hydrolyses much more rapidly. What is the mechanism and what relevance has it to RNA chemistry?

Revision of Chapter 41. This reaction is subject to general base catalysis. Explain.

4. Primary amines are not usually made by displacement reactions on halides with ammonia. Why not? The natural amino acids can be made by this means in quite good yield. Here is an example.

four days
room temperature
67% yield

Why does this example work? Comment on the state of the reagents and products under the reaction conditions. What is the product and how does it differ from the natural amino acid?

5. Human hair is a good source of cystine, the disulfide dimer of cysteine. The hair is boiled with aqueous HCl and HCO_2H for a day, the solution concentrated, and a large amount of sodium acetate added. About 5% of the hair by weight crystallizes out as pure cystine $[\alpha]_D$ –216. How does the process work? Why is such a high proportion of hair cystine? Why is no cysteine isolated by this process? What is the stereochemistry of cystine? Make a good drawing of cystine to show its symmetry. How would you convert the cystine to cysteine?

(S)–cysteine

6. A simple preparation of a dipeptide is given below. Explain the reactions, drawing mechanisms for the interesting steps. Which steps are protection, activation, coupling, and deprotection? Explain the reasons for protection and the nature of the activation. Why is the glycine added to the coupling step as its hydrochloride? What reagent(s) would you use for the final deprotection step?

glycine   66% yield crystalline salt

(S)-(–)-proline   NaOH   Cbz   88% yield Cbz-proline

Cbz-proline → **1.** $ClCO_2Et$, Et_3N **2.** H_3N^+ ...CO_2Et → 99% yield

1M NaOH, H_2O
room temperature
one hour
→ Pro-Gly

7. Suggest how glutathione might detoxify these dangerous chemicals in living things. Why are they still toxic in spite of this protection?

8. Alanine can be resolved by the following method, using a pig kidney acylase. Draw a mechanism for the acylation step. Which isomer of alanine acylates faster? In the enzyme-catalysed reaction, which isomer of the amide hydrolyses faster? In the separation, why is the mixture heated in acid solution, and what is filtered off? How does the separation of the free alanine by dissolution in ethanol work?

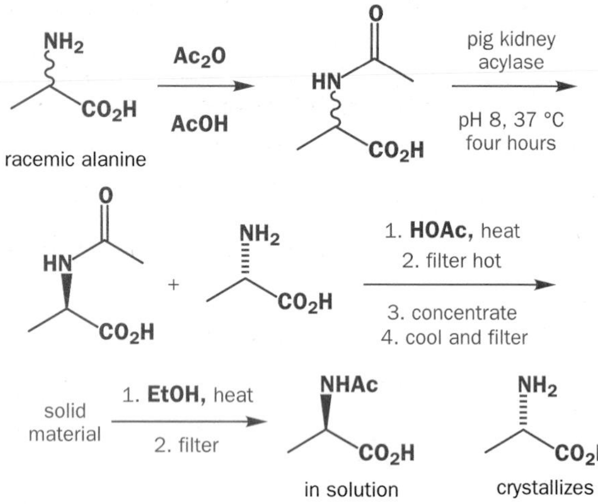

If the acylation is carried out carelessly, particularly if the heating is too long or too strong, a by-product may form that is not hydrolysed by the enzyme. How does this happen?

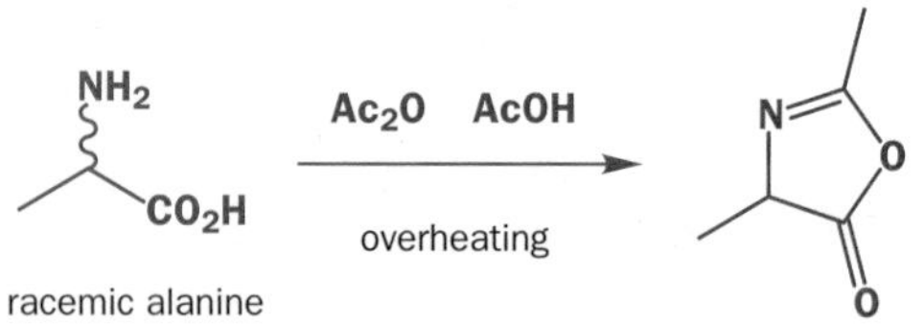

9. A patent discloses this method of making the anti-AIDS drug d4T. The first few stages involve differentiating the three hydroxyl groups of 5-methyluridine as shown below. Explain the reactions, especially the stereochemistry at the position of the bromine atom.

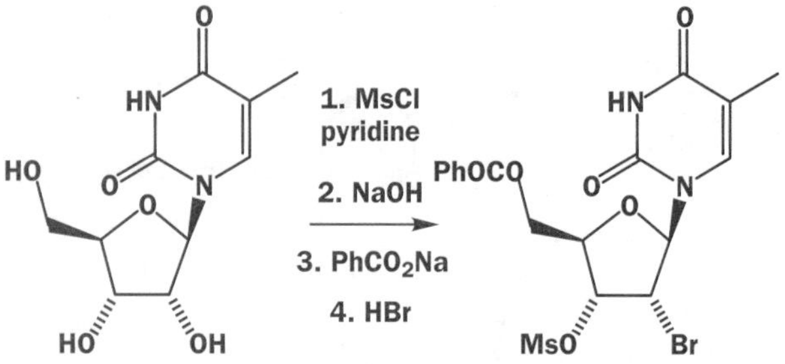

Suggest how the synthesis might be completed.

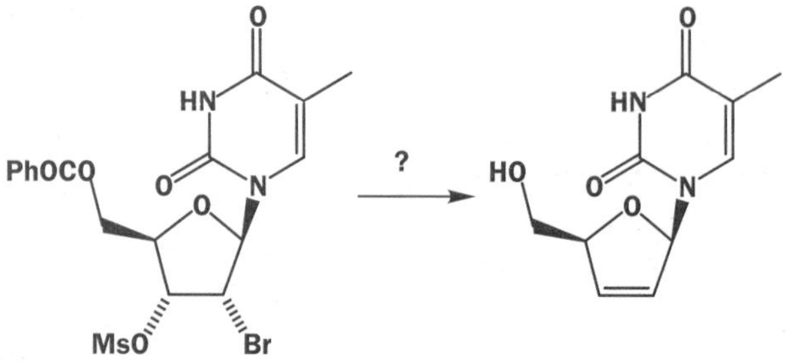

10. Mannose usually exists as the pyranoside shown below. This is in equilibrium with the furanoside. What is the conformation of the pyranoside and what is the stereochemistry of the furanoside? What other stereochemical change will occur more quickly than this isomerization?

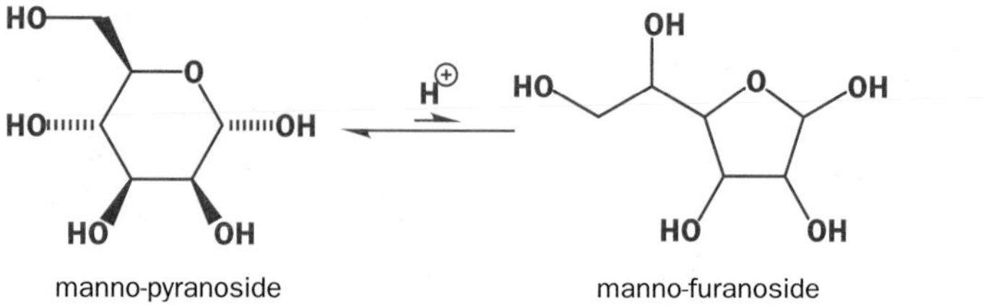

Treatment of mannose with acetone and HCl gives the acetal shown. Explain the selectivity.

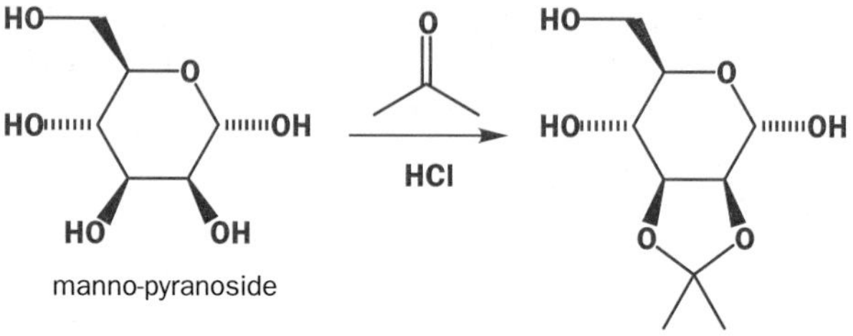

11. How are glycosides formed from phenols (in Nature or in the laboratory)? Why is the stereochemistry of the glycoside not related to that of the original sugar?

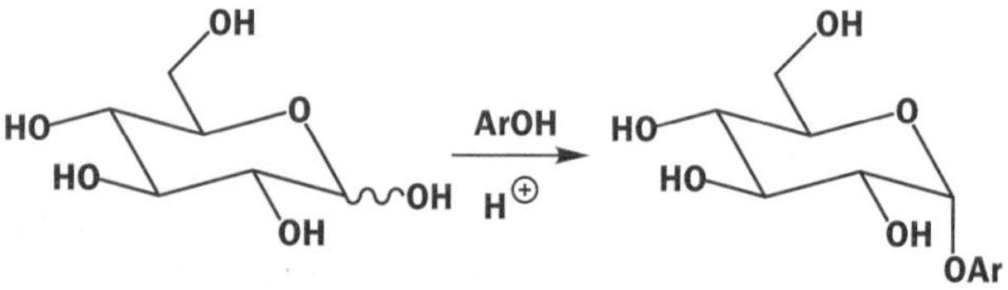

12. Draw all the keto and enol forms of ascorbic acid (vitamin C). Why is the one shown the most stable?

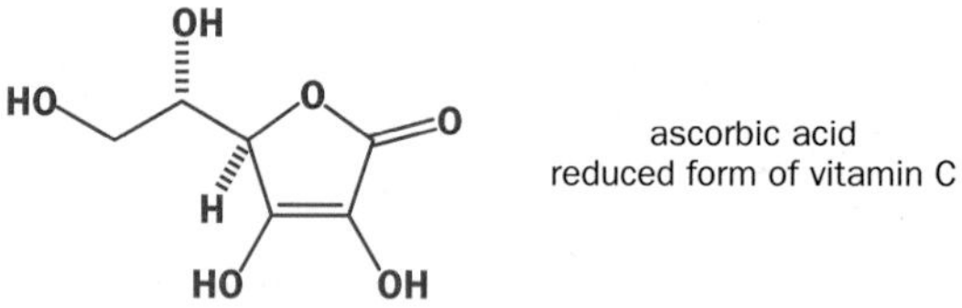

13. 'Caustic soda' (NaOH) was used to clean ovens and clear blocked drains. Many commercial products for these jobs with fancy names still contain NaOH. Even concentrated sodium carbonate (Na_2CO_3) does quite a good job. How do these cleaners work? Why is NaOH so dangerous to humans, particularly if it gets in the eye?

14. Bacterial cell walls contain the unnatural amino acid D-alanine. If you wanted to prepare a sample of D-ala, how would you go about it? (*Hint.* There is not enough in bacteria to make that a worthwhile source, but have you done Problem 8 yet?)

Mechanisms in biological chemistry

50

Connections

Building on:

- Acidity and basicity ch8
- Carbonyl chemistry ch12 & ch14
- Stereochemistry ch16
- Conformational analysis and elimination ch18–ch19
- Enolate chemistry and synthesis ch24–ch30
- Pericyclic reactions ch35–ch36
- Determining mechanisms ch13 & ch41
- Heterocycles ch42–ch44
- Asymmetric synthesis ch45
- Sulfur chemistry ch46
- Chemistry of life ch49

Arriving at:

- How Nature makes small molecules using ordinary organic mechanisms
- Enzymes are Nature's catalysts, speeding up reactions by factors of 10^6 or more
- Coenzymes and vitamins are Nature's versions of common organic reagents
- Reductions with NADH
- Reductive amination, deamination, and decarboxylation with pyridoxal
- Enol chemistry with lysine enamines, with coenzyme A, and with phosphoenolpyruvate
- Umpolung chemistry with thiamine as a d^1 reagent
- Carboxylation with biotin
- Oxidations with FAD
- How Nature makes aromatic amino acids

Looking forward to:

- Natural products ch51

Nature's NaBH$_4$ is a nucleotide: NADH or NADPH

In Chapter 49 we spent some time discussing the structure of nucleotides and their role as codons in protein synthesis. Now we shall see how Nature uses different nucleotides as reagents. Here is the structure of AMP, just to remind you, side by side with a new pyridine nucleotide.

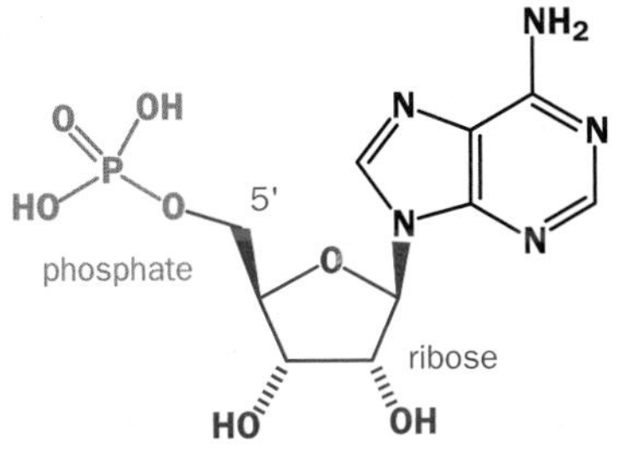

AMP—an adenine nucleotide

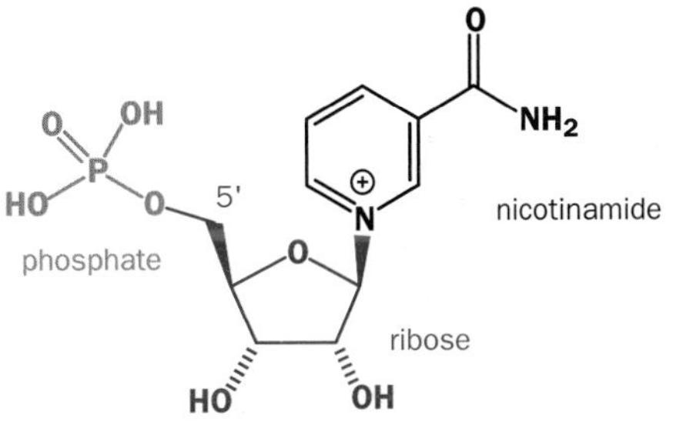

a nicotinamide nucleotide

These two nucleotides can combine together as a pyrophosphate to give a dinucleotide. Notice that the link is not at all the same as in the nucleic acids. The latter are joined by one phosphate that links the 3′–5′ positions. Here we have a *pyro*phosphate link between the two 5′-positions.

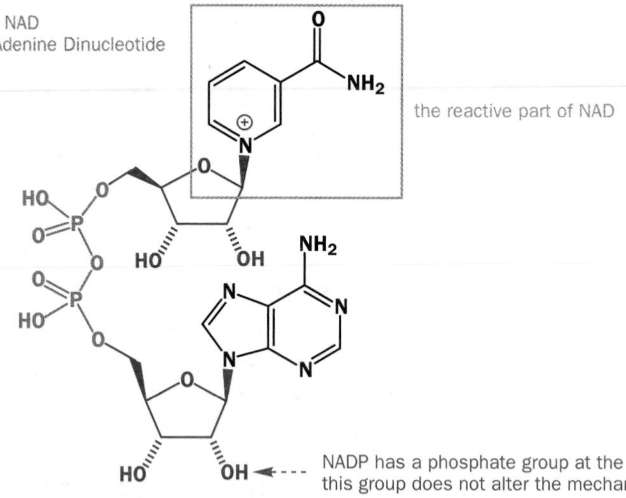

NAD
Nicotinamide Adenine Dinucleotide

the reactive part of NAD

NADP has a phosphate group at the 2' position
this group does not alter the mechanism of action

Notice also the positive charge on the nitrogen atom of the pyridine ring. This part of the molecule does all the work and from now on we will draw only the reactive part for clarity. This is **NAD⁺**, **nicotinamide adenine dinucleotide**, and it is one of Nature's most important oxidizing agents. Some reactions use NADP instead but this differs only in having an extra phosphate group on the adenosine portion so the same part structure will do for both. NAD⁺ and NADP both work by accepting a hydrogen atom and a pair of electrons from another compound. The reduced compounds are called NADH and NADPH.

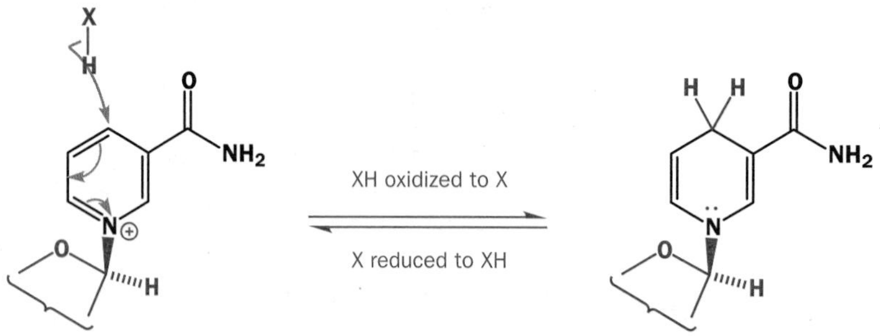

XH oxidized to X

X reduced to XH

NAD⁺—Nature's oxidizing agent

NADH—Nature's reducing agent

The reduction of NAD⁺ (and NADP) is reversible, and NADH is itself a reducing agent. We will first look at one of its reactions: a typical reduction of a ketone. The ketone is pyruvic acid and the reduction product lactic acid, two important metabolites. The reaction is catalysed by the enzyme liver alcohol dehydrogenase.

► The names of enzymes are usually chosen to tell us where they come from and what job they do and the name ends '-ase'. A **dehydrogenase** is clearly a redox enzyme as it removes (or adds) hydrogen.

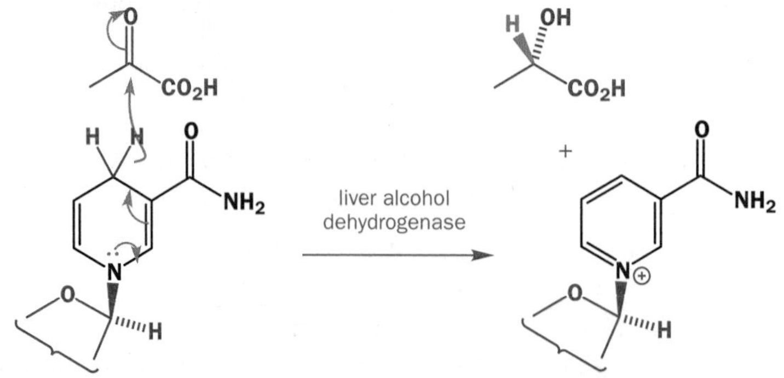

liver alcohol
dehydrogenase

NADH—Nature's reducing agent

NAD⁺—Nature's oxidizing agent

This is a reaction that would also work in the laboratory with NaBH₄ as the reducing agent, but there is a big difference. The product from the NaBH₄ reaction *must* be racemic—no optical activity has been put in from compound, reagent, or solvent.

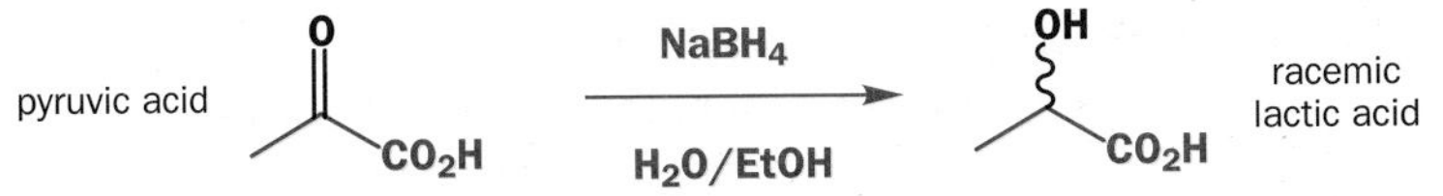

pyruvic acid → (NaBH₄, H₂O/EtOH) → racemic lactic acid

But the product from the enzymatic reaction is optically active. The two faces of pyruvic acid's carbonyl group are enantiotopic and, by controlling the addition so that it occurs from one face only, the reaction gives a single enantiomer of lactic acid.

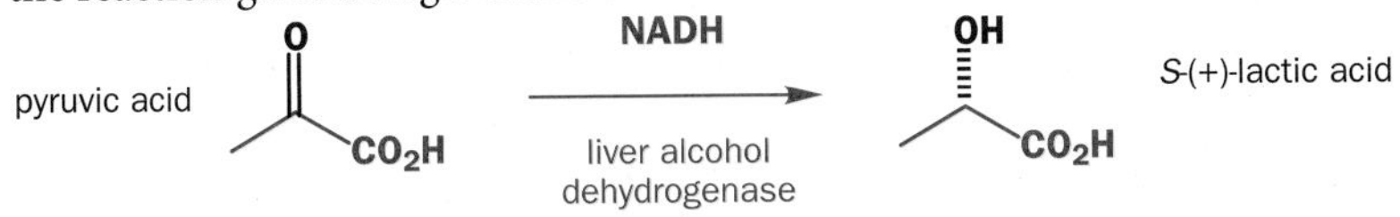

pyruvic acid → (NADH, liver alcohol dehydrogenase) → S-(+)-lactic acid

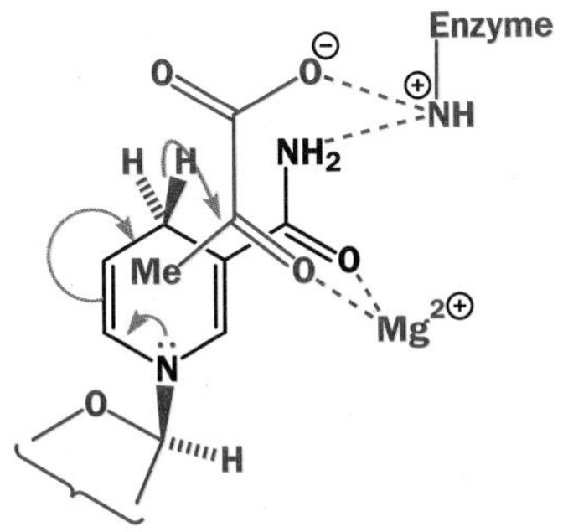

Both the enzyme and the reagent NADH are single enantiomers and they cooperate by binding. The enzyme binds both the substrate (pyruvic acid) and the reagent (NADH) in a specific way so that the hydride is delivered to one enantiotopic face of the ketone. Pyruvic acid under physiological conditions will be the anion, pyruvate, so it is held close to the positively charged amino group of a lysine residue on the enzyme that also binds the amino group of NADH. A magnesium(II) cation, also held by the enzyme, binds the carbonyl group of the amide of NADH and the ketone in pyruvate. If this model is correct, only the top H atom (as drawn) of the diastereotopic CH₂ group in NADH should be transferred to pyruvate. This has been proved by deuterium labelling.

Supporting evidence comes from a model system using a much simpler reducing agent. A dihydropyridine with a primary alcohol replacing the amide group in NADH and a simple benzyl group replacing the nucleotide forms stable esters with keto-acids. As soon as the ester is treated with magnesium(II) ions, intramolecular and stereospecific reduction occurs. The hydride ion is transferred from a stereogenic centre, which replaces the diastereotopic CH₂ group in NADH.

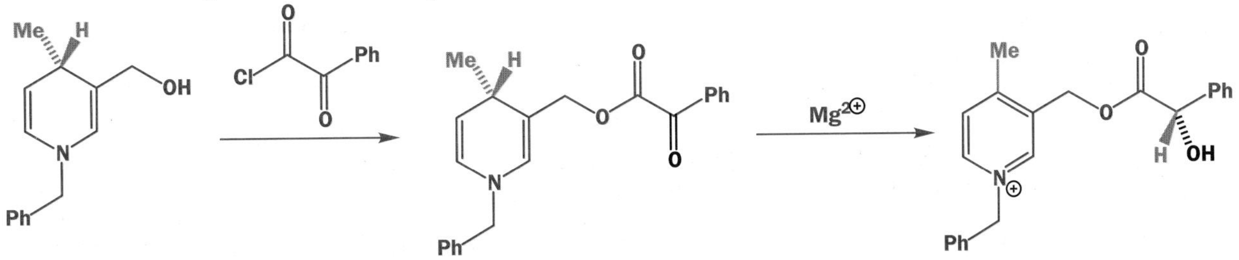

When the ester is cleaved by transesterification with methoxide ion, the newly released hydroxy-ester is optically active.

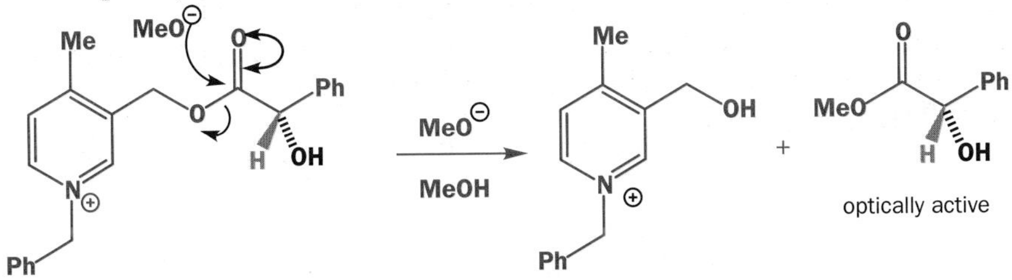

optically active

The details of the reaction are probably a good model for the NADH reaction even down to the activation by magnesium(II) ions. A possible transition state would be very similar to the NADH transition state above.

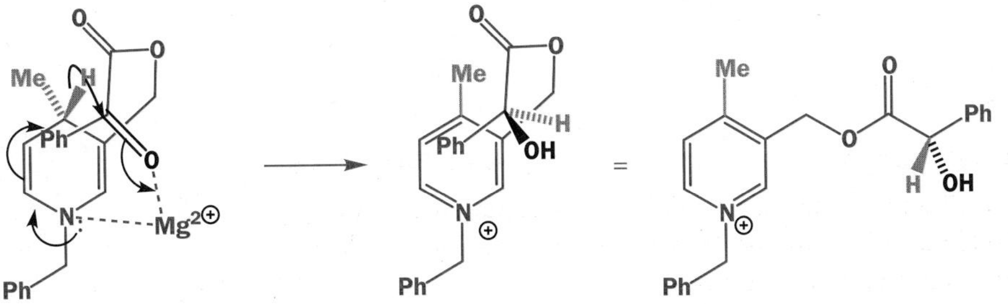

Many other reactions use NADH as a reducing agent or NAD^+ as oxidizing agent. Three molecules of NAD^+ are used in the citric acid cycle (see the chart on p. 1393). One of these oxidations is the simple transformation of a secondary alcohol (malate) to a ketone (oxaloacetate).

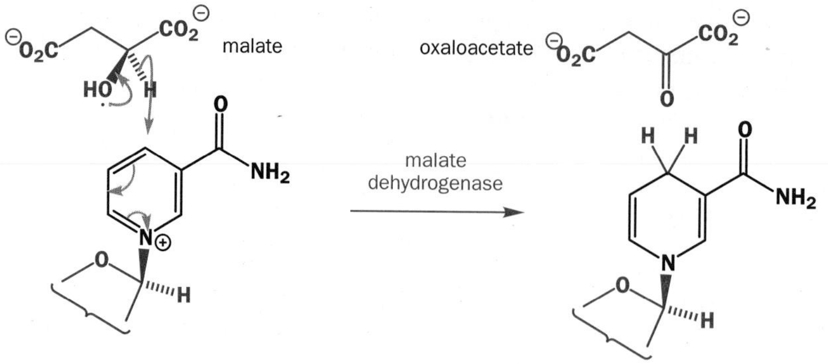

▶

The other two reactions are of a more complex type that we will meet soon when we show how acetyl coenzyme A is a key reagent in the building of carbon–carbon chains.

Other redox reagents include dinucleotides such as FAD (flavine adenine dinucleotide), lipoic acid, which we will meet when we discuss the chemistry of thiamine, and ascorbic acid (vitamin C), which you met in Chapter 49. Ascorbic acid can form a stable enolate anion that can transfer a hydride ion to a suitable oxidant.

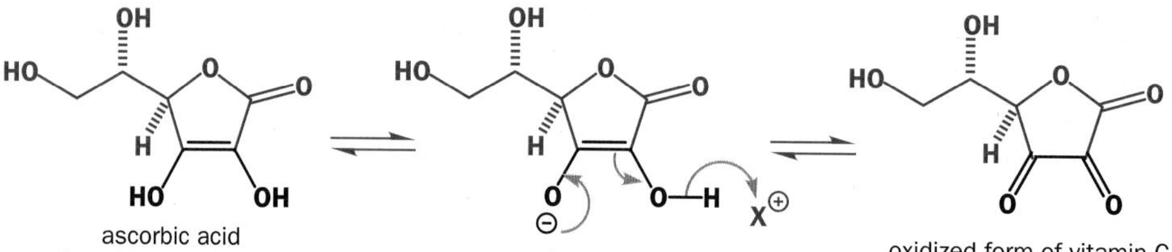

▶

Ascorbic acid is usually described as an antioxidant rather than a reducing agent though mechanistically they are the same.

ascorbic acid
reduced form of vitamin C

oxidized form of vitamin C

In this mechanism 'X^+' represents an oxidant—a dangerously reactive peroxide perhaps, or even Fe(III) which must be reduced to Fe(II) as part of the reaction cycle of many iron-dependent enzymes.

Reductive amination in nature

■

For more on reductive amination, see Chapter 14.

One of the best methods of amine synthesis in the laboratory is **reductive amination**, in which an imine (formed from a carbonyl compound and an amine) is reduced to a saturated amine. Common reducing agents include $NaCNBH_3$ and hydrogen with a catalyst.

reductive amination in the laboratory:

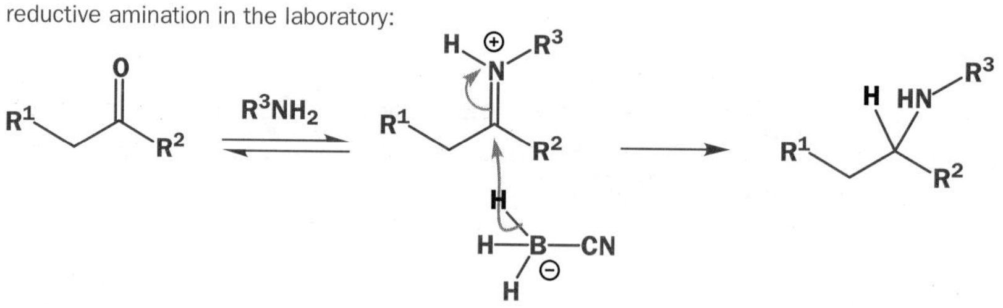

This reaction, of course, produces racemic amines. But Nature transforms this simple reaction into a stereospecific and reversible one that is beautiful in its simplicity and cleverness. The reagents are a pair of substituted pyridines called **pyridoxamine** and **pyridoxal**.

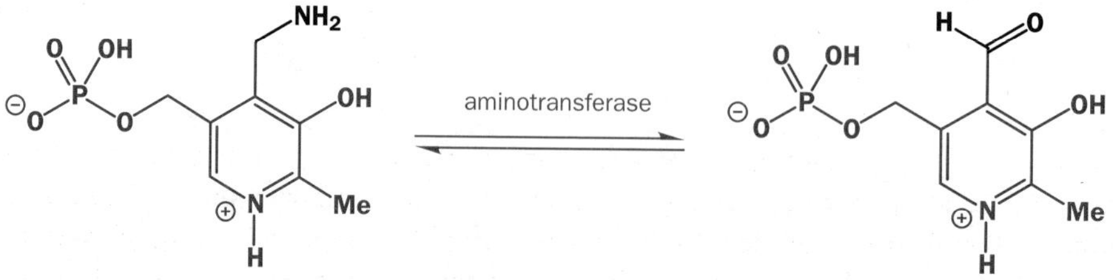

pyridoxamine phosphate

pyridoxal phosphate

You might imagine that pyridoxamine is a product of reductive amination of pyridoxal with ammonia. In practice it doesn't work like that. Nature uses an amine transfer rather than a simple reductive amination, and the family of enzymes that catalyse the process is the family of **aminotransferases**.

Pyridoxal is a coenzyme and it is carried around on the side chain of a lysine residue of the enzyme. Lysine has a long flexible side chain of four CH_2 groups ending with a primary amine (NH_2). This group forms an imine (what biochemists call a 'Schiff base') with pyridoxal. An imine is a good functional group for this purpose as imine formation is easily reversible.

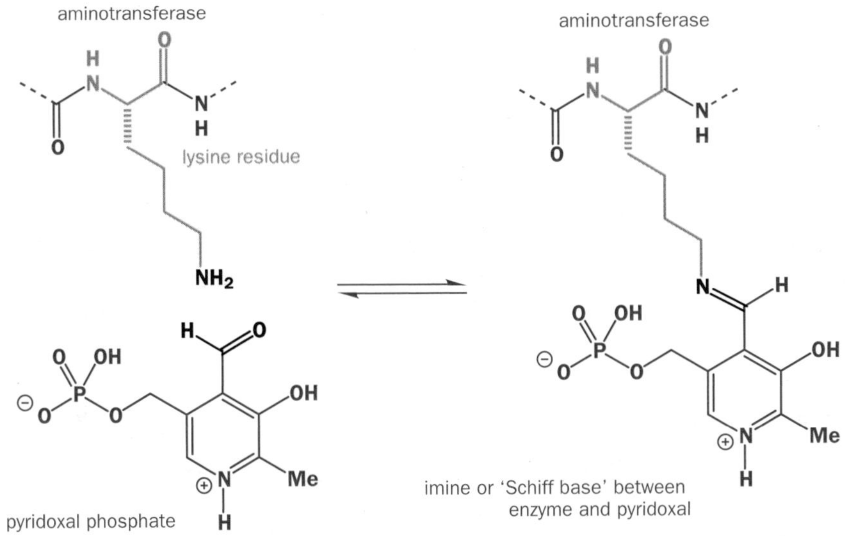

When reductive amination or its reverse is required, the pyridoxal is transferred from the lysine imine to the carbonyl group of the substrate to form a new imine of the same sort. The most important substrates are the amino acids and their equivalent α-keto-acids.

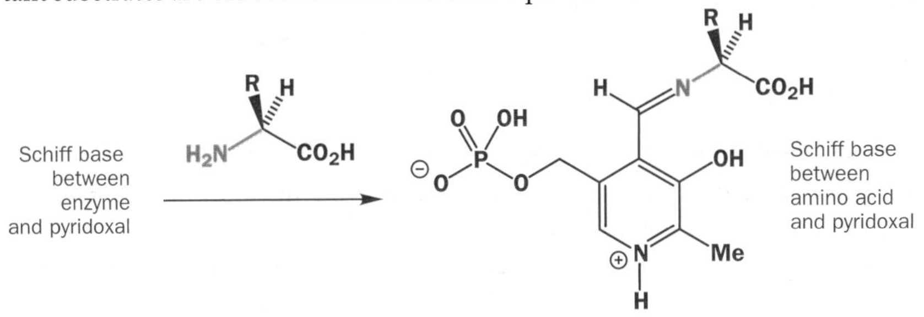

Now the simple but amazing chemistry begins. By using the protonated nitrogen atom of the pyridine as an electron sink, the α proton of the amino acid can be removed to form a new imine at the top of the molecule and an enamine in the pyridine ring.

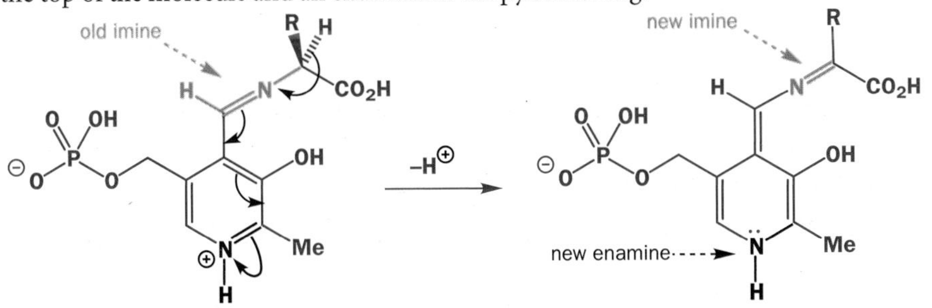

Now the electrons can return through the pyridine ring and pick up a proton at the top of the molecule. The proton can be picked up where it came from, but more fruitfully it can be picked up at the carbon atom on the other side of the nitrogen. Hydrolysis of this imine releases pyridoxamine and the keto-acid. All the natural amino acids are in equilibrium with their equivalent α-keto-acids by this mechanism, catalysed by an aminotransferase.

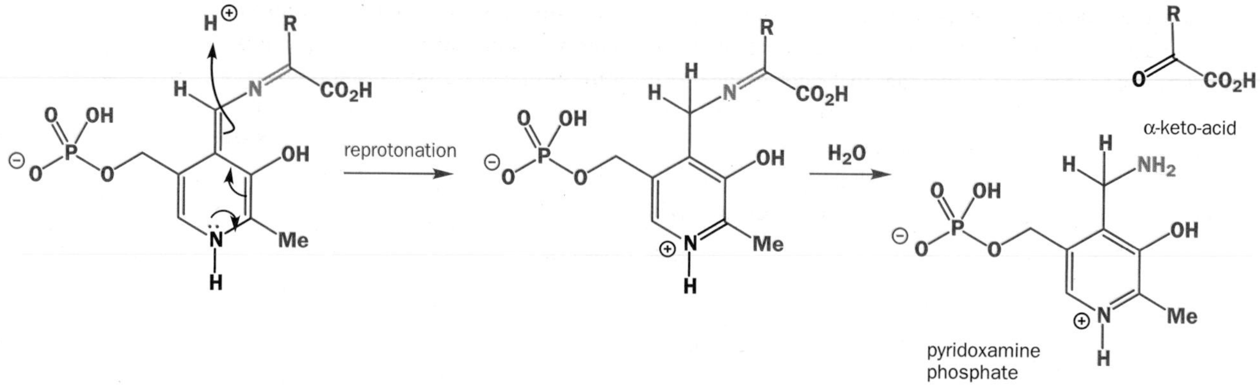

pyridoxamine phosphate

α-keto-acid

Reversing this reaction makes an amino acid stereospecifically out of an α-keto-acid. In fact, a complete cycle is usually set up whereby one amino acid is converted to the equivalent α-keto-acid while another α-keto-acid is converted into its equivalent amino acid. This is true transamination.

Amino acids get used up (making proteins, for example) so, to keep life going, ammonia must be brought in from somewhere. The key amino acid in this link is glutamic acid. A true reductive amina-

tion using NADPH and ammonia builds glutamic acid from α-keto-glutaric acid.

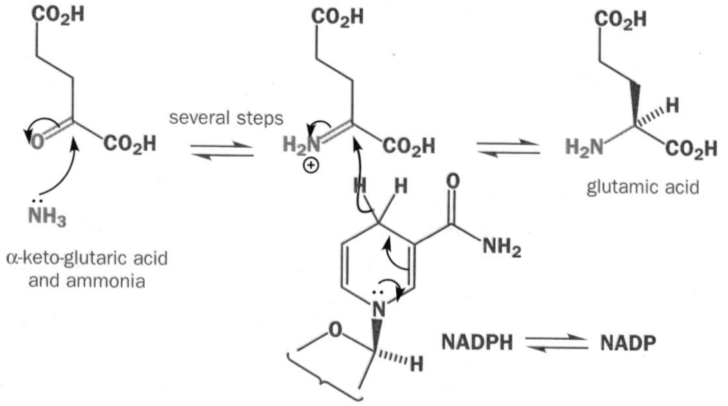

The other amino acids can now be made from glutamic acid by transamination. At the end of their useful life they are transaminated back to glutamic acid which, in mammals at least, gives its nitrogen to urea for excretion.

α-keto-glutaric acid and ammonia

glutamic acid

Pyridoxal is a versatile reagent in the biochemistry of amino acids

Pyridoxal is the reagent in other reactions of amino acids, all involving the imine as intermediate. The simplest is the racemization of amino acids by loss of a proton and its replacement on the other face of the enamine. The enamine, in the middle of the diagram below, can be reprotonated on either face of the prochiral imine (shown in green). Protonation on the bottom face would take us back to the natural amino acid from which the enamine was made in the first place. Protonation on the top face leads to the unnatural amino acid after 'hydrolysis' of the imine (really transfer of pyridoxal to a lysine residue of the enzyme).

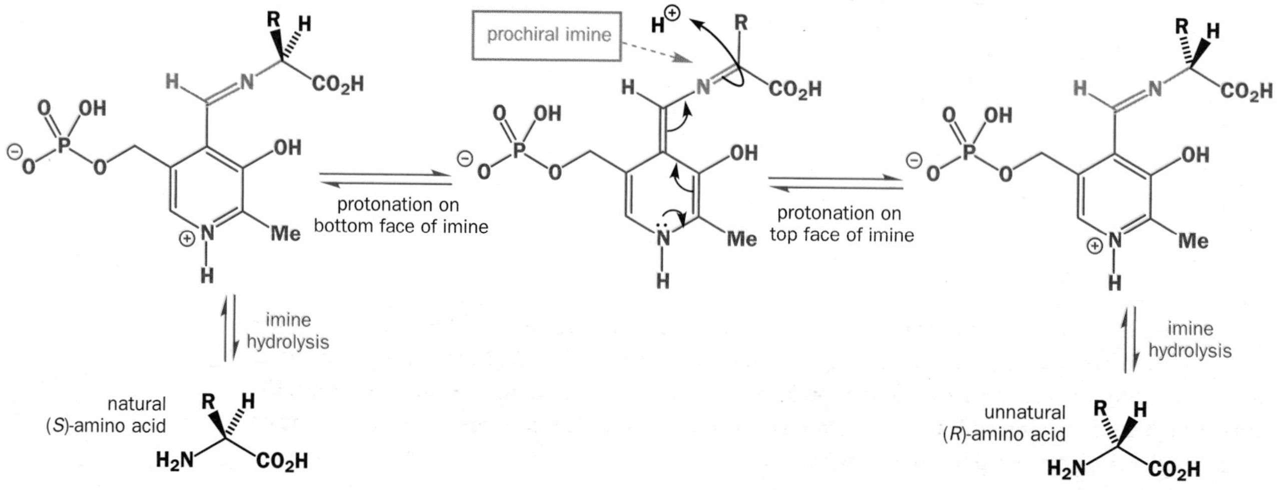

prochiral imine

protonation on bottom face of imine

protonation on top face of imine

imine hydrolysis

imine hydrolysis

natural (S)-amino acid

unnatural (R)-amino acid

A very similar reaction is decarboxylation. Starting from the same imine we could lose carbon dioxide instead of a proton by a very similar mechanism. Reprotonation and imine transfer releases the amine corresponding to the original amino acid. The enzymes catalysing these reactions are called **decarboxylases**.

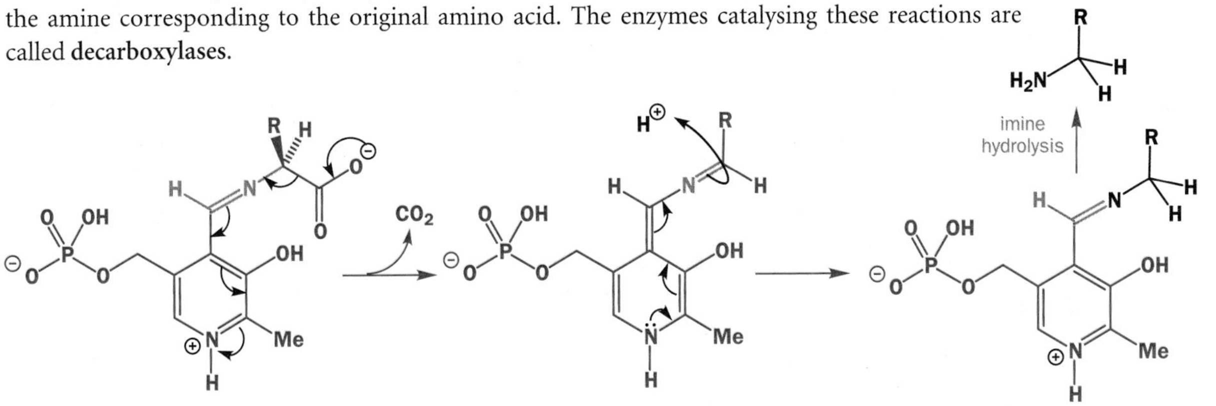

In Chapter 43 we mentioned the role of histamine in promoting acid secretion in the stomach, and its role in causing gastric ulcers. The drug cimetidine was designed to counteract the effect of histamine. Histamine is produced in the body by decarboxylation of histidine using the mechanism you have just seen.

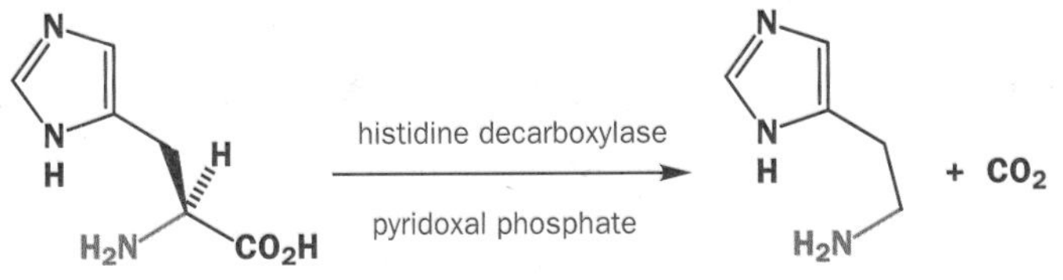

How is it possible for the same reagent operating on the same substrate (an amino acid) to do at will one of two quite different things—removal and/or exchange of a proton and decarboxylation? The answer, of course, lies in the enzymes. These hold pyridoxal exceptionally tightly by using all the available handles: the hydroxy and phosphate groups, the positively charged nitrogen atom, and even the methyl group. The diagram shows the proposed binding of the lysine imine of pyridoxal by an aminotransferase.

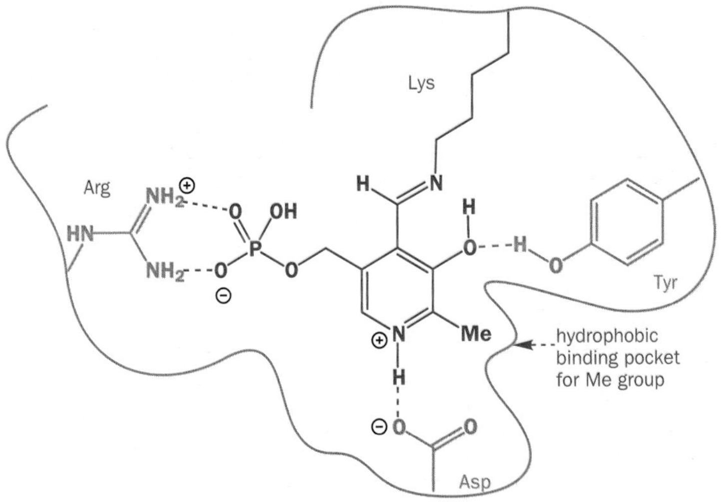

The green line shows an imaginary shape of the enzyme chain into which fit acidic groups and basic groups forming hydrogen bonds to groups on the coenzyme. Around the methyl group are alkyl-substituted amino acids, which form a hydrophobic region. Even when the lysine attachment is exchanged for the substrate, all these interactions remain in place. The substrate is bound by similar interactions with other groups on the enzyme.

Control over the choice of reaction arises because the different enzymes bind the substrate–pyridoxal imine in different ways. Decarboxylases bind so that the C–C bond to be broken is held orthogonal to the pyridine ring and parallel to the p orbitals in the ring. Then the bond can be broken and CO_2 can be lost.

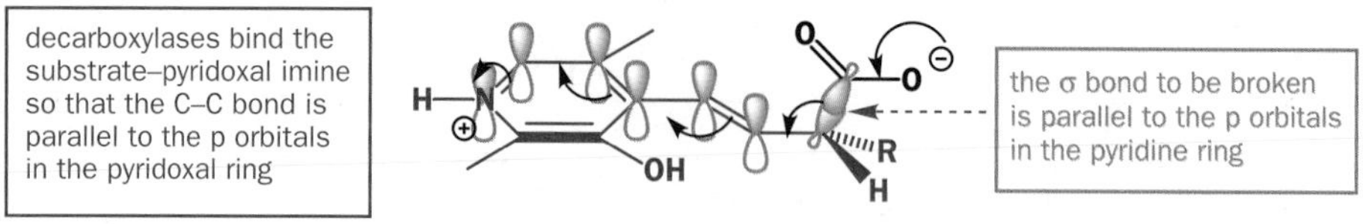

> decarboxylases bind the substrate–pyridoxal imine so that the C–C bond is parallel to the p orbitals in the pyridoxal ring

> the σ bond to be broken is parallel to the p orbitals in the pyridine ring

Racemases and transaminases bind the substrate–pyridoxal imine so that the C–H bond is parallel to the p orbitals in the ring so that proton removal can occur. Enzymes do not speed reactions up indiscriminately—they can selectively accelerate some reactions at the expense of others, even those involving the same reagents.

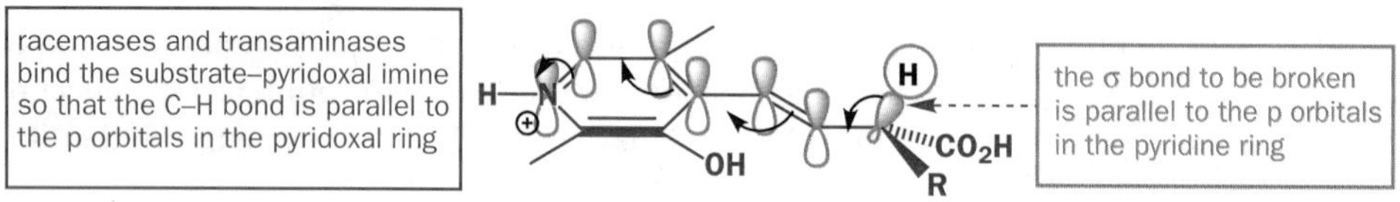

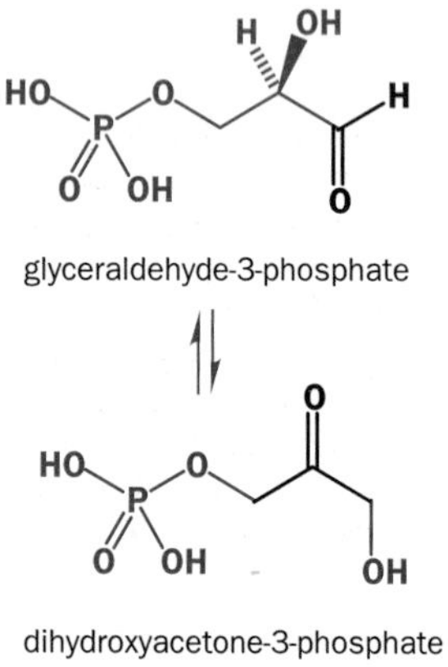

> racemases and transaminases bind the substrate–pyridoxal imine so that the C–H bond is parallel to the p orbitals in the pyridoxal ring

> the σ bond to be broken is parallel to the p orbitals in the pyridine ring

glyceraldehyde-3-phosphate

dihydroxyacetone-3-phosphate

Nature's enols—lysine enamines and coenzyme A

The glycolysis pathway breaks down glucose to produce energy, and in doing so produces smaller molecules for use in the citric acid cycle. In reverse, it allows the synthesis of the six-carbon sugar fructose from two three-carbon fragments. A key reaction is the step in which these two C_3 sugars combine. They are glyceraldehyde and dihydroxyacetone and we met them and their interconversion in the last chapter.

The reaction is effectively an aldol condensation between the enol of the keto-sugar phosphate and the electrophilic aldehyde of glyceraldehyde phosphate and the enzyme is named appropriately **aldolase**. The product is the keto-hexose fructose-1,6-diphosphate.

fructose-1,6-diphosphate

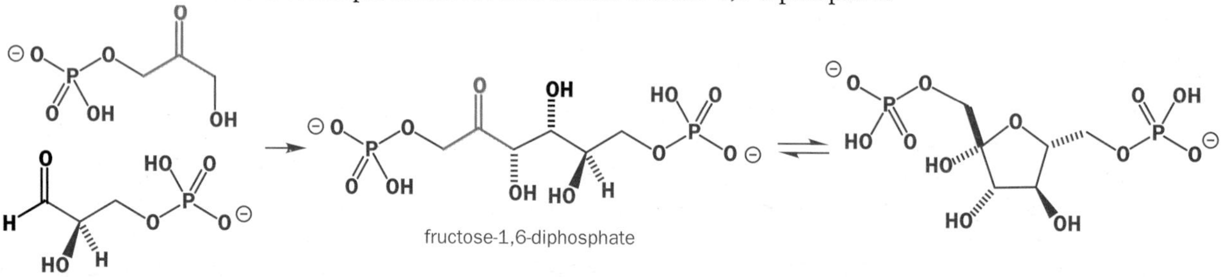

No enolate ion is formed in this aldol. Instead a lysine residue in the enzyme forms an imine with the keto-triose.

> The rest of the aldolase molecule is represented by 'Enz'.

> Lys residue in aldolase

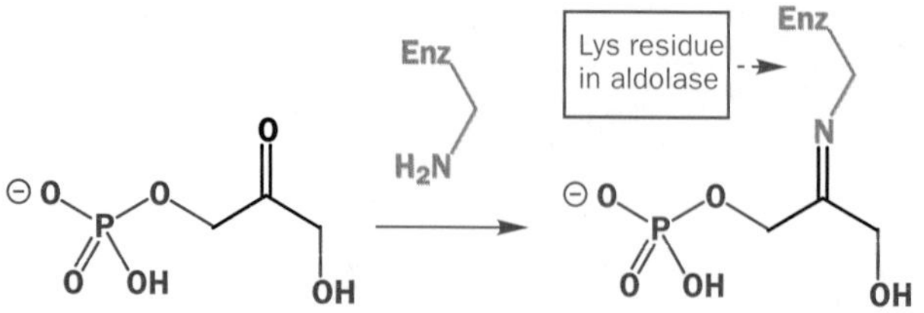

Proton transfers allow this imine to be converted into an enamine, which acts as the nucleophile in the aldol reaction. Stereochemical control (it's a *syn* aldol) comes from the way in which the two molecules are held by the enzyme as they combine. The product is the imine, which is hydrolysed to the open-chain form of fructose-1,6-diphosphate.

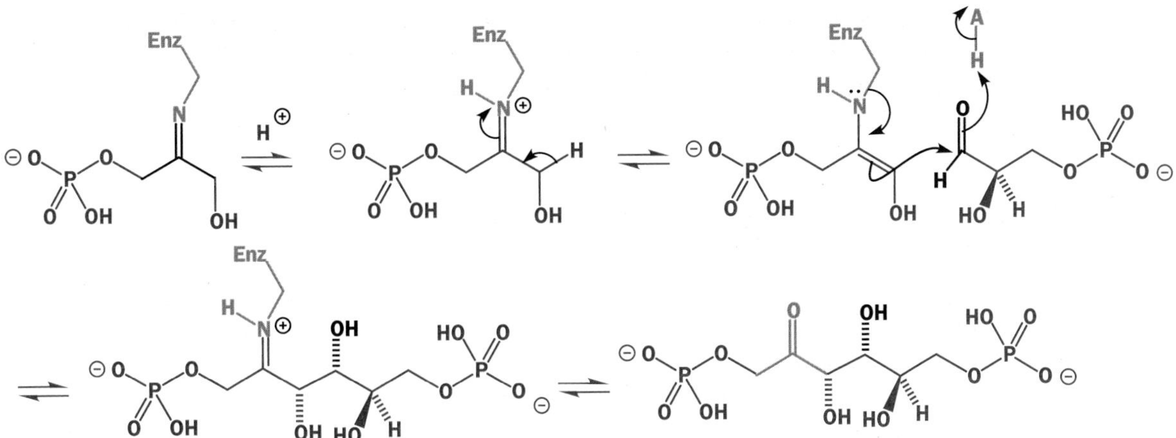

fructose-1,6-diphosphate

Many other reactions in nature use enamines, mostly those of lysine. However, a more common enol equivalent is based on thiol esters derived from coenzyme A.

Coenzyme A and thiol esters

Coenzyme A is an adenine nucleotide at one end, linked by a 5'-pyrophosphate to pantothenic acid, a compound that looks rather like a tripeptide, and then to an amino thiol. Here is the structure broken down into its parts.

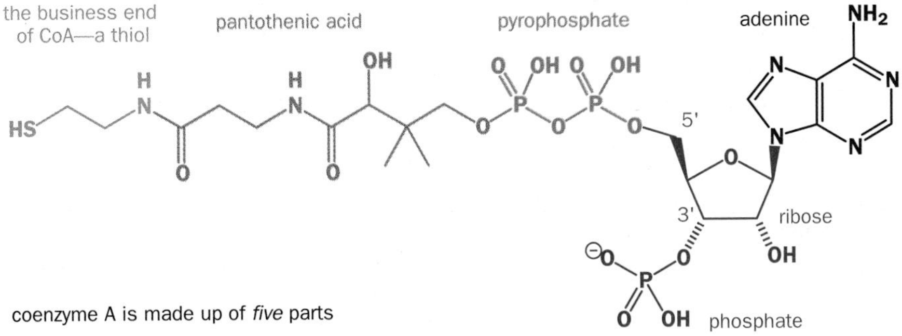

coenzyme A is made up of *five* parts

By now you will realize that most of this molecule is there to allow interaction with the various enzymes that catalyse the reactions of coenzyme A. We will abbreviate it from now on as CoASH where the SH is the vital thiol functional group, and all the reactions we will be interested in are those of esters of CoASH. These are **thiol esters**, as opposed to normal 'alcohol esters', and the difference is worth a few comments.

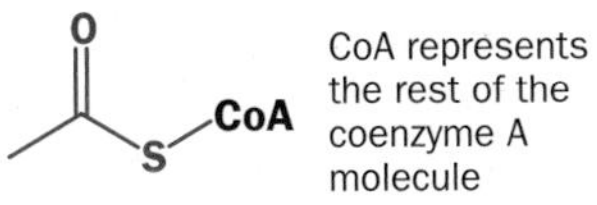

Thiol esters are less conjugated than ordinary esters (see Chapter 28, p. 744), and ester hydrolysis occurs more rapidly with thiol esters than with ordinary esters because in the rate-determining step (nucleophilic attack on the carbonyl group) there is less conjugation to destroy. The thiolate is also a better leaving group.

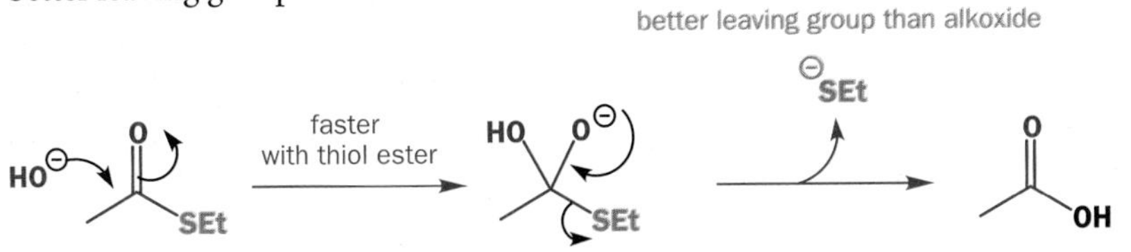

better leaving group than alkoxide

tetrahedral intermediate: conjugation destroyed

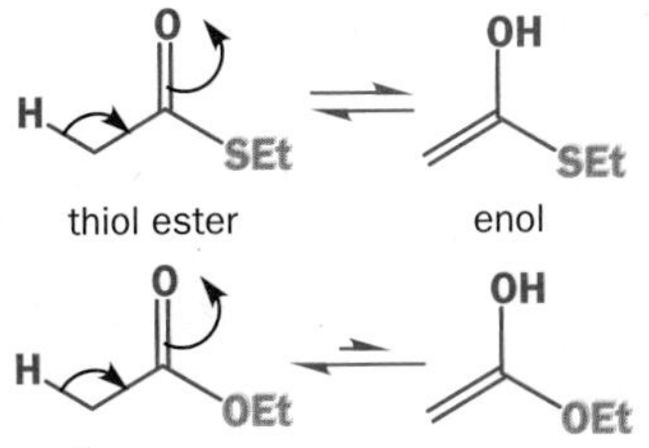

Another reaction that goes better with thiol esters than with ordinary esters is enolization. This is an equilibrium reaction and the enol has lost the conjugation present in the ester. The thiol ester has less to lose so is more enolized. This is the reaction of acetyl CoA that we are now going to discuss. We have mentioned the citric acid cycle several times and it has appeared in two

▶

Compare this structure with that of NAD—the adenine nucleotide is the same, as is the 5'-pyrophosphate link. The difference is at the other end of that link where we find this new tripeptide-like molecule and not another nucleotide. There is also a 3'-phosphate on the ribose ring not present in NAD.

CoA represents the rest of the coenzyme A molecule

acetyl CoA

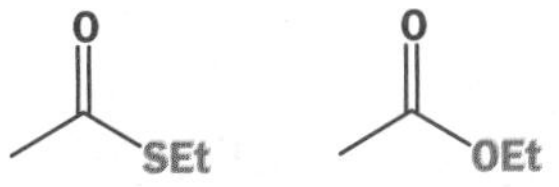

a simple thiol ester

an ordinary ester

thiol ester   enol

normal ester   enol

diagrams but we have not so far discussed the chemistry involved. The key step is the synthesis of citric acid from oxaloacetate and acetyl CoA. The reaction is essentially an aldol reaction between the enol of an acetate ester and an electrophilic ketone and the enzyme is known as **citrate synthase.**

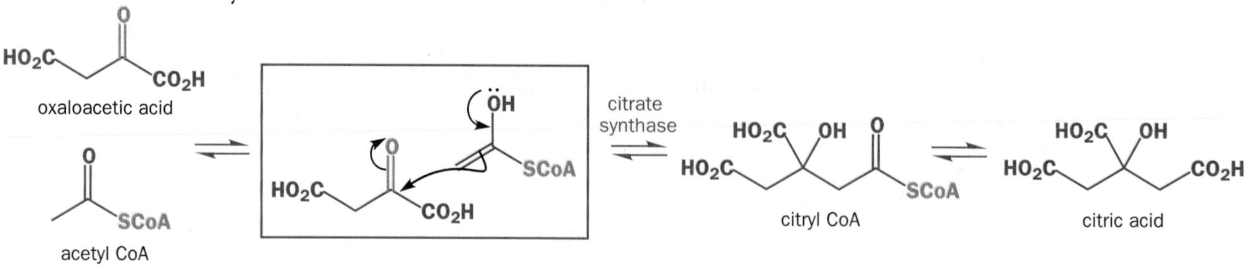

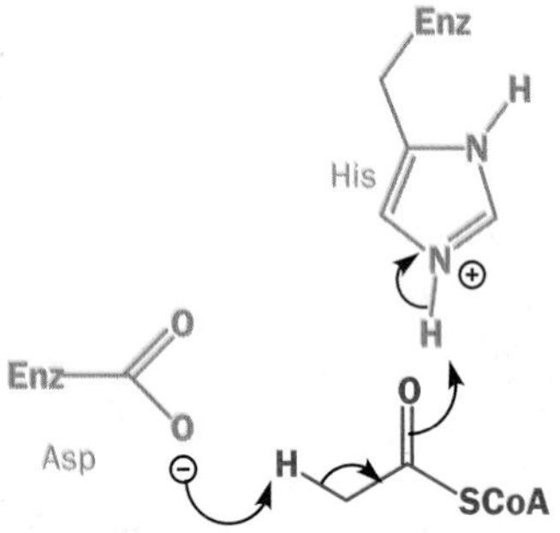

The mechanism in the frame shows the enol of acetyl CoA attacking the reactive ketone. In nature the enolization is catalysed by a basic carboxylate group (Asp) and an acidic histidine, both part of the enzyme, so that even this easy reaction goes faster.

In the C–C bond-forming step, the same histidine is still there to remove the enol proton again and another histidine, in its protonated form, is placed to donate a proton to the oxygen atom of the ketone. You should see now why histidine, with a pK_{aH} of about 7, is so useful to enzymes: it can act either as an acid or as a base.

> This is **general acid catalysis**, as described in Chapter 41.

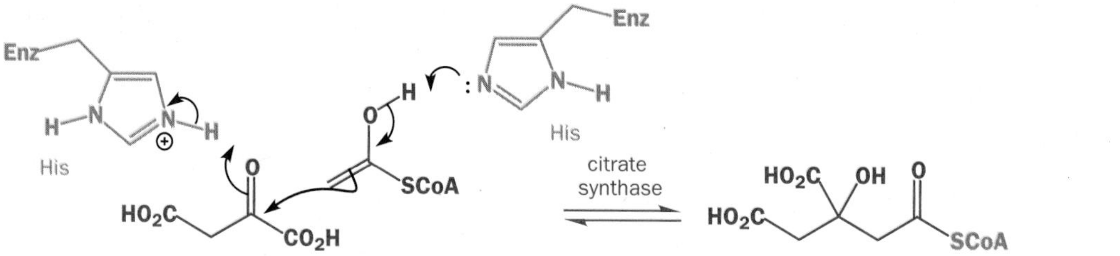

Even the hydrolysis of the reactive thiol ester is catalysed by the enzyme and the original histidine again functions as a proton donor. Acetyl CoA has played its part in all steps. The enolization and the hydrolysis in particular are better with the thiol ester.

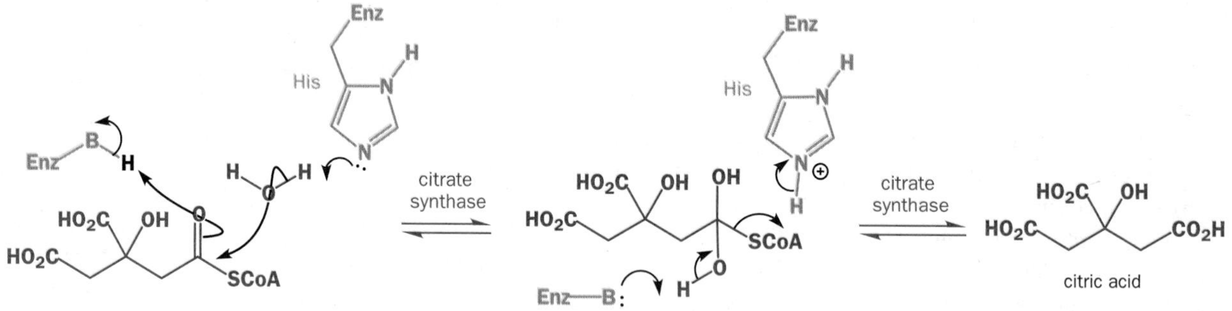

CoA thiol esters are widely used in nature. Mostly they are acetyl CoA, but other thiol esters are also used to make enols. We will see more of this chemistry in the next chapter. The two enol equivalents that we have met so far are quite general: lysine enamines can be used for any aldehyde or ketone and CoA thiol esters for any ester. Another class of enol equivalent—the enol ester—has just one representative but it is a most important one.

Phosphoenolpyruvate

Pyruvic acid is an important metabolite in its own right as we shall see shortly. It is the simplest α-keto-acid (2-oxopropanoic acid). Having the two carbonyl groups adjacent makes them more reactive: the ketone is more electrophilic and enolizes more readily and the acid is stronger. Pyruvate is in equilibrium with the amino acid alanine by an aminotransferase reaction catalysed by pyridoxal (above).

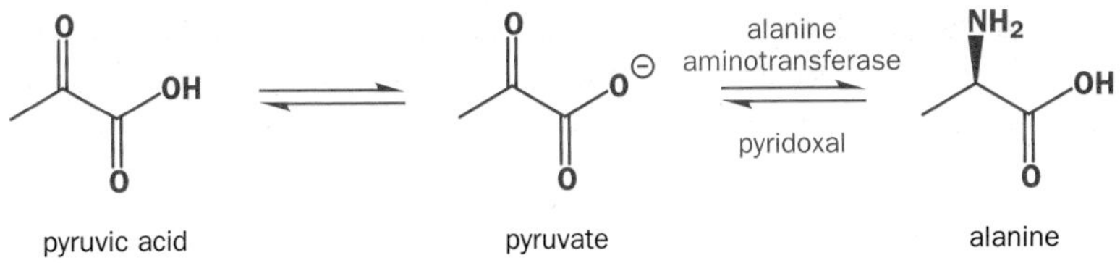

pyruvic acid   pyruvate   alanine

For an explanation of the effect of two adjacent carbonyl groups, see Chapter 28, p. 728.

Nature uses the enol phosphate of pyruvic acid (**phosphoenolpyruvate** or **PEP**) as an important reagent. We might imagine making this compound by first forming the enol and then esterifying on oxygen by some phosphorylating agent such as ATP.

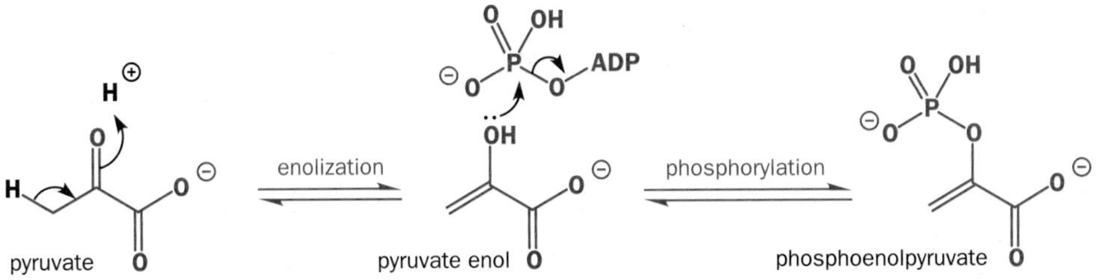

pyruvate   pyruvate enol   phosphoenolpyruvate

Now, in fact, this reaction does occur in nature as part of the glycolysis pathway, but it occurs almost entirely in reverse. PEP is used as a way to make ATP from ADP during the oxidation of energy-storing sugars. An enol is a better leaving group than an ordinary alcohol especially if it can be protonated at carbon. The reverse reaction might look like this.

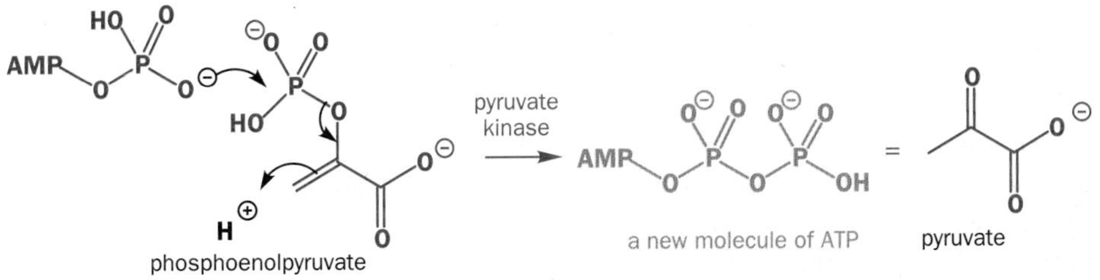

phosphoenolpyruvate   a new molecule of ATP   pyruvate

PEP is also used as an enol in the making of carbon–carbon bonds when the electrophile is a sugar molecule and we will see this reaction in the next chapter. So, if PEP is not made by enolization of pyruvate, how is it made? The answer is by **dehydration**. The phosphate is already in place when the dehydration occurs, catalysed by the enzyme enolase.

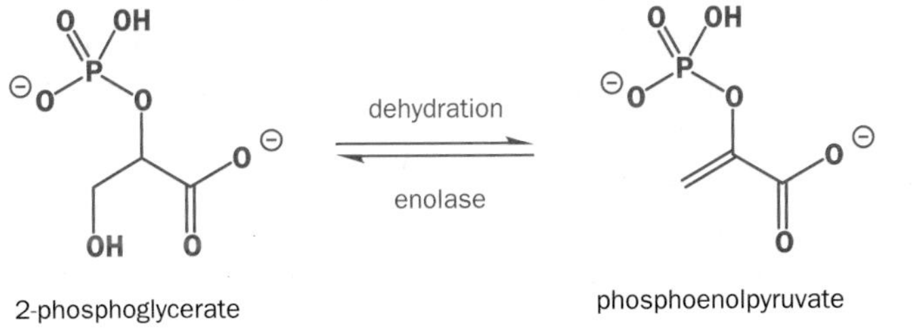

2-phosphoglycerate   phosphoenolpyruvate

You saw in Chapter 19 how simple OH groups could be lost in dehydration reactions. Either the OH group was protonated by strong acid (this is not an option in living things) or an enol or enolate pushed the OH group out in an E1cB-like mechanism. This must be the case here as the better leaving group (phosphate) is ignored and the worse leaving group (OH) expelled.

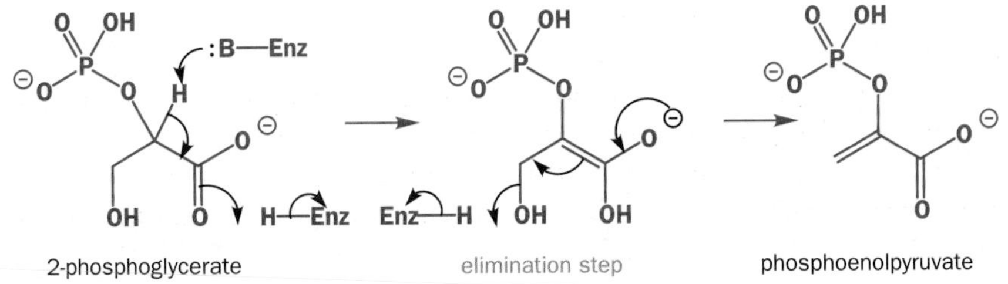

2-phosphoglycerate   elimination step   phosphoenolpyruvate

This would be an unusual way to make an enol in the laboratory but it can be used, usually to make stable enols. An example that takes place under mildly basic conditions is the dehydration of the bicyclic keto-diol in dilute sodium hydroxide—presumably by an E1cB mechanism.

Pyruvic acid and acetyl CoA: the link between glycolysis and the citric acid cycle

We have now examined the mechanism of several steps in glycolysis and one in the citric acid cycle and we have seen enough to look at the outline of these two important processes and the link between them (see opposite).

You have already seen that citric acid is made from acetyl CoA. The acetyl CoA comes in its turn from pyruvic acid. Pyruvic acid comes from many sources but the most important is glycolysis: acetyl CoA is the link between glycolysis and the citric acid cycle. The key reaction involves both CoASH and pyruvate and carbon dioxide is lost. This is an oxidation as well and the oxidant is NAD^+. The overall reaction is easily summarized.

This looks like a simple reaction based on very small molecules. But look again. It is a very strange reaction indeed. The molecule of CO_2 clearly comes from the carboxyl group of pyruvate, but how is the C–C bond cleaved, and how does acetyl CoA join on? If you try to draw a mechanism you will see that there must be more to this reaction than meets the eye. The extra features are two new cofactors, thiamine pyrophosphate and lipoic acid, and the reaction takes place in several stages with some interesting chemistry involved.

Lipoic acid is quite a simple molecule with a cyclic disulfide as its main feature. It is attached to the enzyme as an amide with lysine. Our first concern will be with the much more complex coenzyme thiamine pyrophosphate.

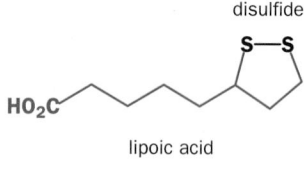

lipoic acid

lipoic acid attached to the enzyme as a lysine amide

Nature's acyl anion equivalent (d^1 reagent) is thiamine pyrophosphate

Thiamine pyrophosphate looks quite like a nucleotide. It has two heterocyclic rings, a pyrimidine similar to those found in DNA and a thiazolium salt. This ring has been alkylated on nitrogen by the pyrimidine part of the molecule. Finally, there is a pyrophosphate attached to the thiazolium salt by an ethyl side chain.

thiamine pyrophosphate

> We will abbreviate pyrophosphate to 'OPP' in structures.

> Do not confuse thiamine with thymine, one of the pyrimidine bases on DNA. The DNA base thymine is just a pyrimidine; thymidine is the corresponding nucleoside. The coenzyme thiamine is a more complicated molecule, that contains a different pyrimidine.

the link between glycolysis and the citric acid cycle

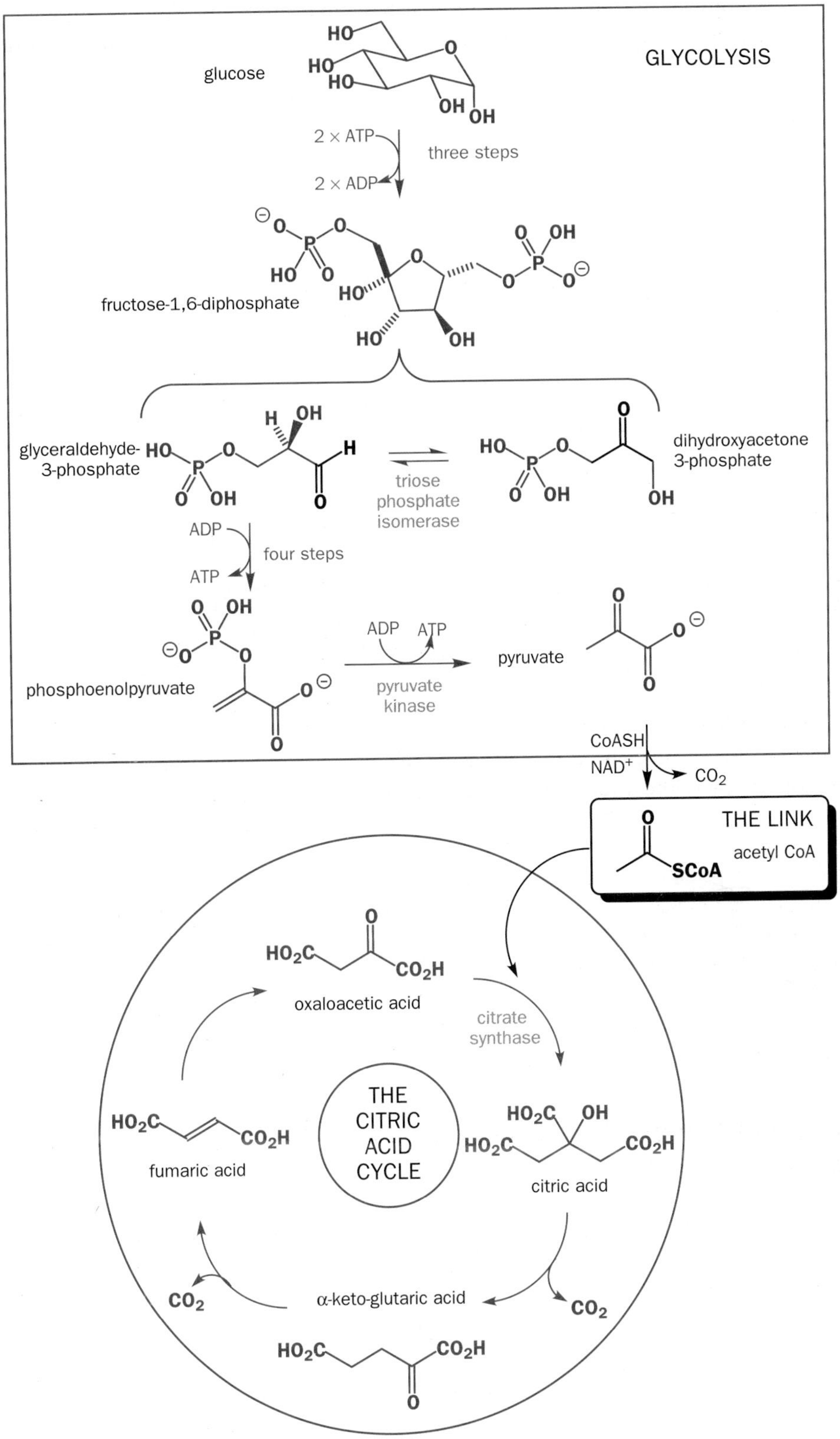

The key part of the molecule for reactivity is the thiazolium salt in the middle. The proton between the N and S atoms can be removed by quite weak bases to form an ylid. You saw sulfonium ylids in Chapter 46, and there is some resemblance here, but this ylid is an ammonium ylid with extra stabilization from the sulfur atom. The anion is in an sp^2 orbital, and it adds to the reactive carbonyl group of pyruvate.

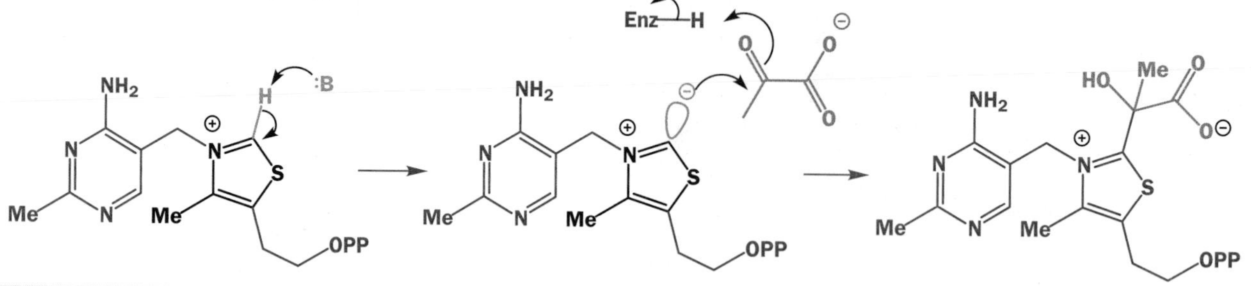

For more on fragmentation reactions see Chapter 38.

Now the carboxylate can be lost from the former pyruvate as the positively charged imine in the thiamine molecule provides a perfect electron sink to take away the electrons from the C–C bond that must be broken.

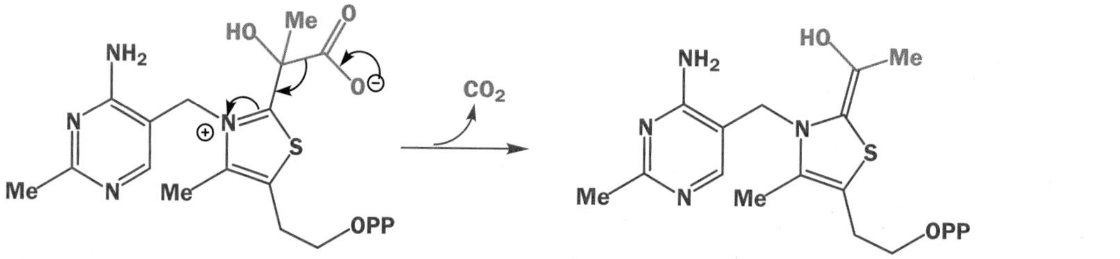

This new intermediate contains a new and strange C=C double bond. It has OH, N, and S substituents making it very electron-rich. As the nitrogen is the most electron-donating you can view it as an enamine, and it attacks the disulfide functional group of lipoic acid, the other cofactor in the reaction.

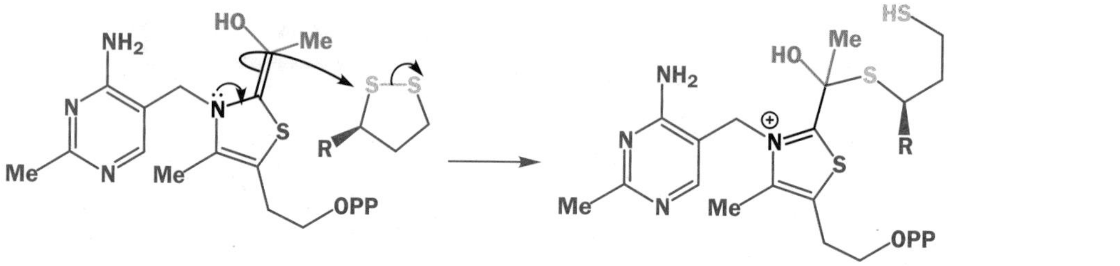

Now the thiamine can be expelled using the green OH group. The leaving group is again the ylid of thiamine, which functions as a catalyst.

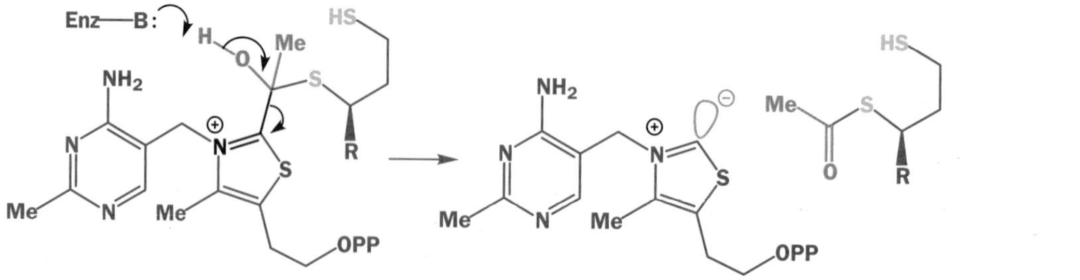

The product is a thiol ester and so can exchange with CoASH in a simple ester exchange reaction. This is a nucleophilic attack on the carbonyl group and will release the reduced form of lipoic acid. All that is necessary to complete the cycle is the oxidation of the dithiol back to the disulfide. This is such an easy reaction to do that it would occur in air anyway but it is carried out in nature by FAD, a close relative of NAD$^+$.

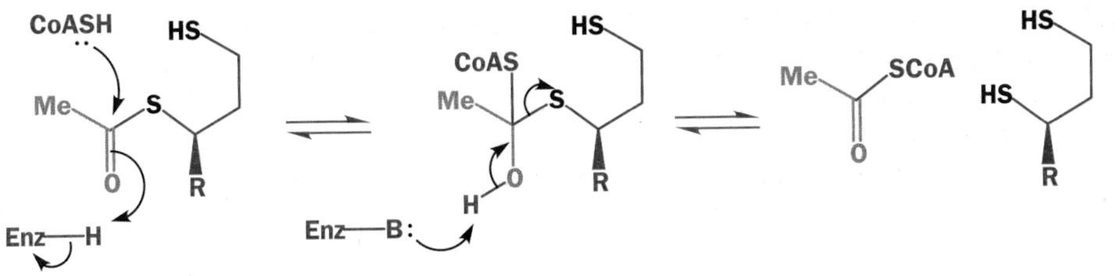

This is one of the most complicated sequences of reactions that we have discussed so far. It is critical to living things because it links glycolysis and the citric acid cycle. Nature has provided not one enzyme but three enzymes to catalyse this process. In the cell they are massed together as a single protein complex.

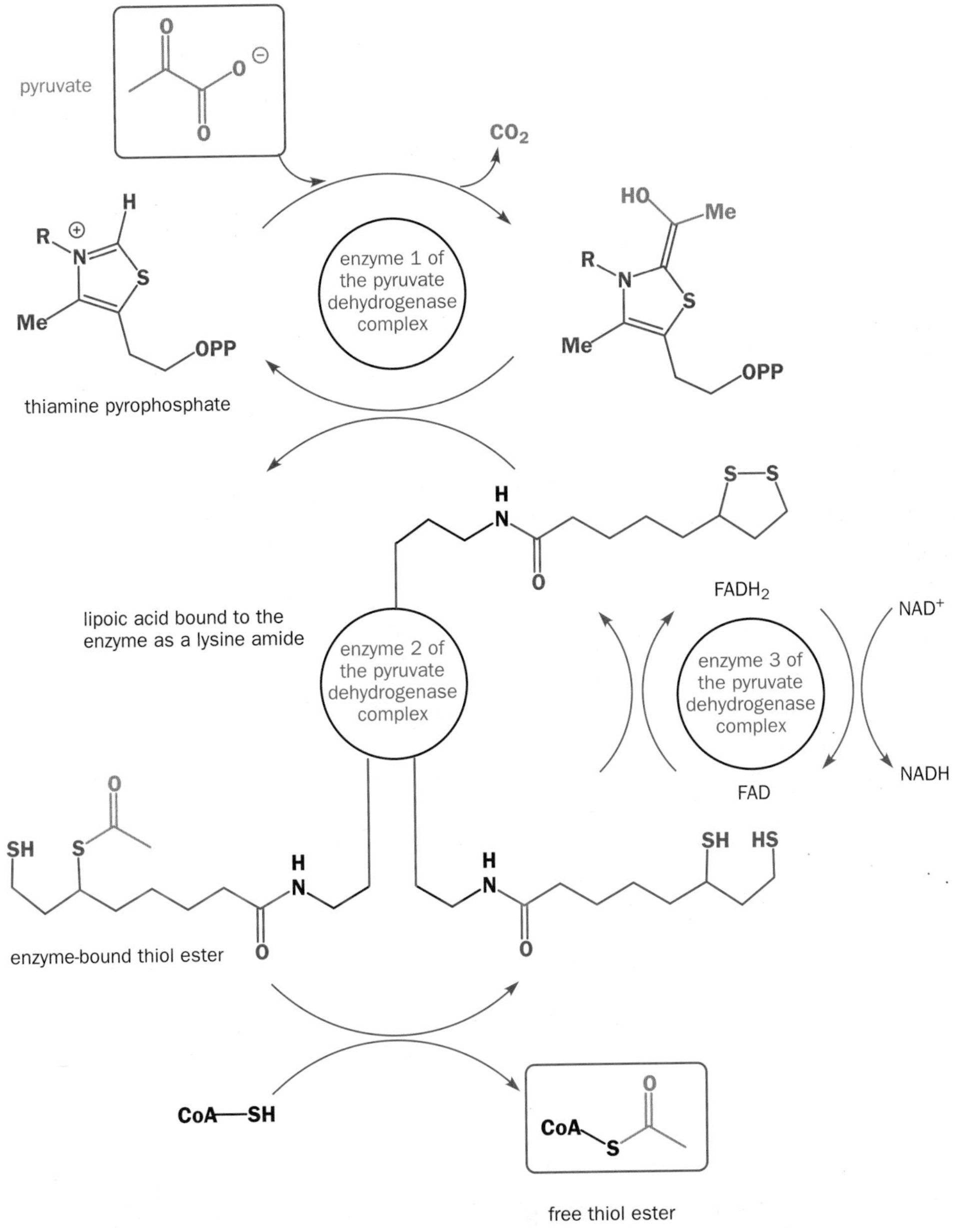

At the centre is 'enzyme 2' which binds the acetyl group through a lipoic acid–lysine amide. On the one side this acetyl group is delivered from pyruvate by the ministrations of thiamine pyrophosphate and 'enzyme 1' and on the other it is delivered to CoA as the free thiol ester. Enzyme 3 recycles the reduced lipoic acid using FAD and then NAD⁺. This remarkable assembly of proteins maintains stocks of acetyl CoA for use in the citric acid cycle and for building complex organic molecules by enol chemistry, as we will see in the next chapter.

One reaction in this sequence is worth detailed analysis. The enzyme-bound lipoic thiol ester is a perfectly normal thiol ester and we would expect it to be formed by acylation of the thiol.

thiol ester formation with normal polarity: nucleophilic S electrophilic carbonyl

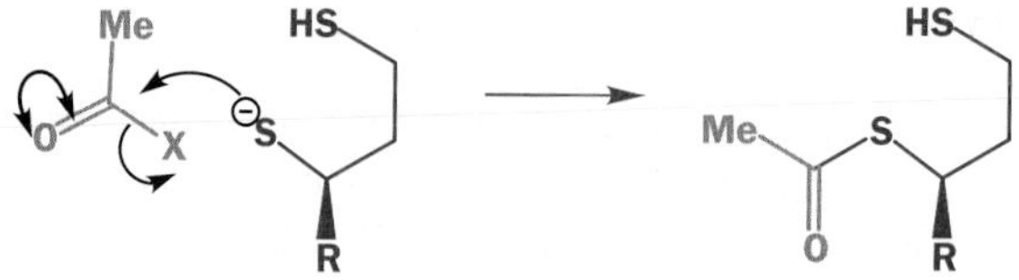

But this thiol ester is not formed by the expected mechanism in the enzymatic reaction. Thiamine delivers a *nucleo*philic acetyl group to an *electro*philic sulfur atom—the reverse polarity to normal ester formation.

thiol ester formation with reverse polarity: electrophilic S nucleophilic carbonyl

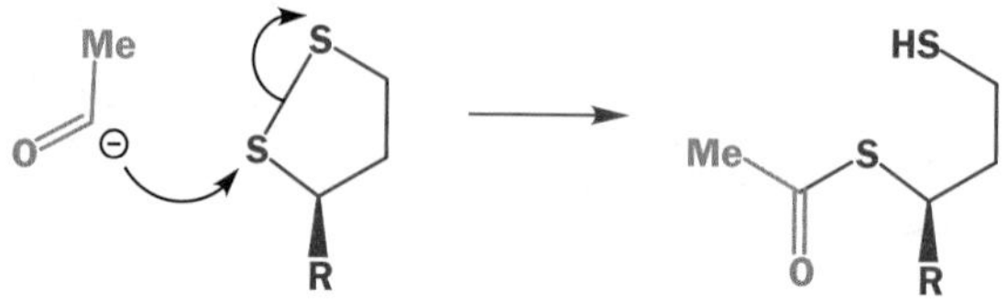

The compound formed from thiamine pyrophosphate and pyruvic acid is Nature's nucleophilic acetyl group. This is a d^1 reagent like the dithiane anion you met in Chapter 46.

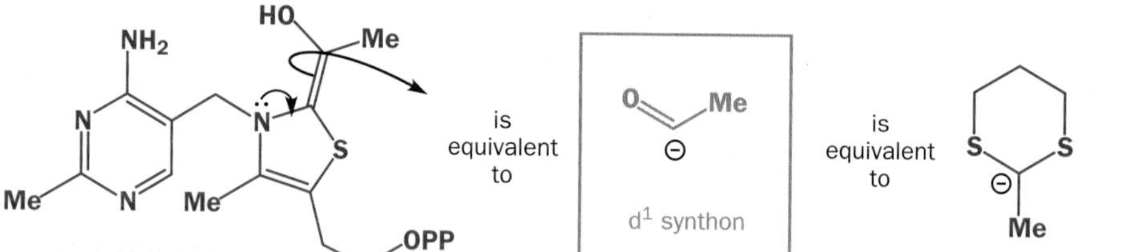

biological reagent   is equivalent to d^1 synthon   is equivalent to   chemical reagent

If this is really true and not just a theoretical analogy, it ought to be possible to learn from Nature and design useful d^1 reagents based on thiamine. This was done by Stetter using simplified thiamines. The pyrimidine is replaced by a benzene ring and the pyrophosphate is removed. This leaves a simple thiazolium salt called a **Stetter reagent**.

thiamine pyrophosphate   Stetter reagent

By analogy with the biological reaction, we need only a weak base (Et₃N) to make the ylid from the thiazolium salt. The ylid adds to aldehydes and creates a d^1 nucleophile equivalent to an acyl anion.

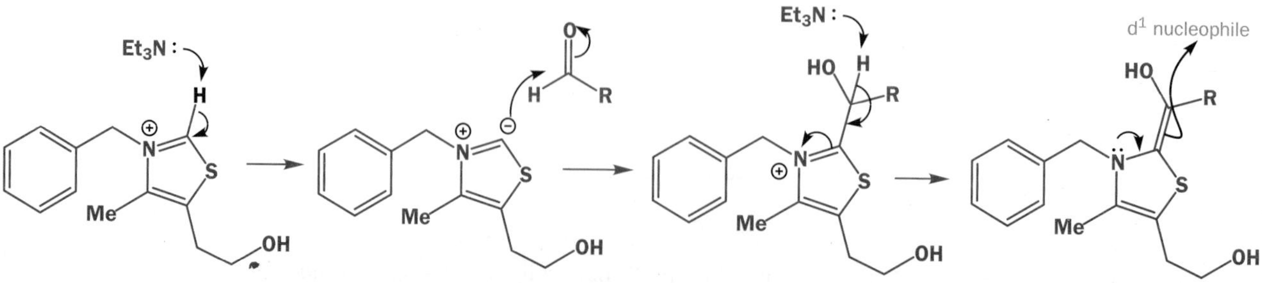

d^1 nucleophile

A useful application of these reagents is in conjugate addition to unsaturated carbonyl compounds. Few d^1 reagents will do this as most are very basic and prefer to add directly to the carbonyl group. Notice that a tertiary amine, pK_{aH} about 10, is strong enough to remove both protons in this sequence.

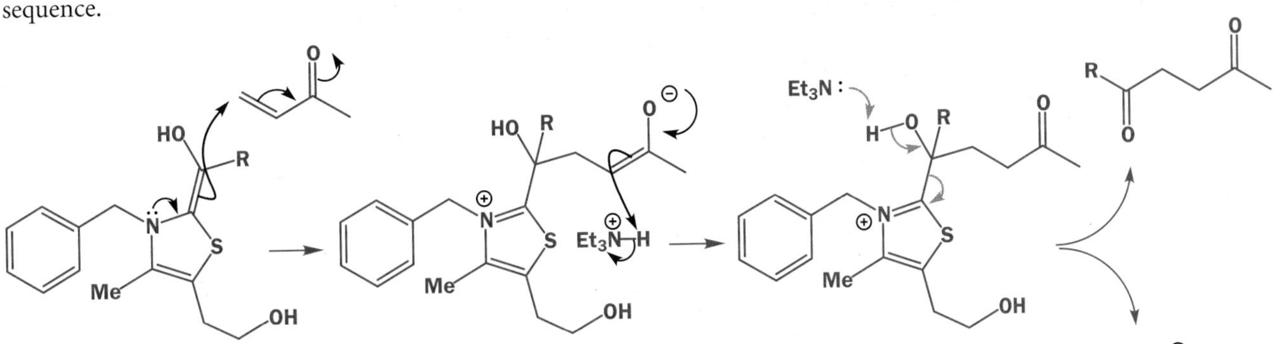

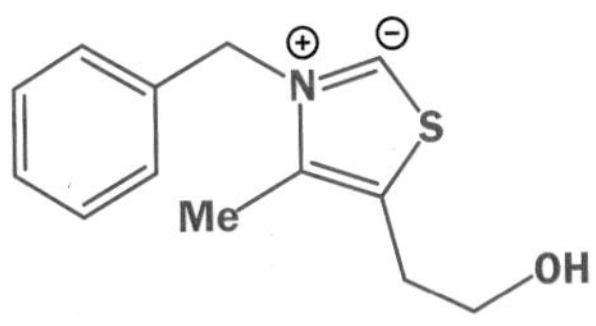

The organic product is a 1,4-diketone and the thiazolium ylid is released to continue with another cycle of the reaction. Like thiamine, the thiazolium ylid is catalyst. Processes like this, which copy nature, are called **biomimetic**.

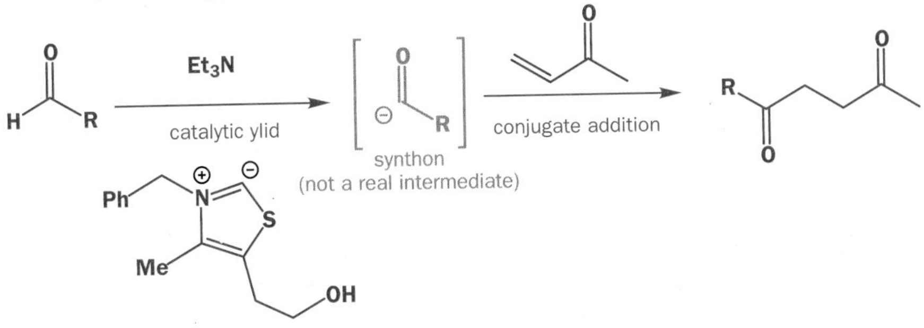

Rearrangements in the biosynthesis of valine and isoleucine

In nature, thiamine pyrophosphate also catalyses reactions of α-keto-acids other than pyruvic acid. One such sequence leads through some remarkable chemistry to the biosynthesis of the branched-chain amino acids valine and isoleucine.

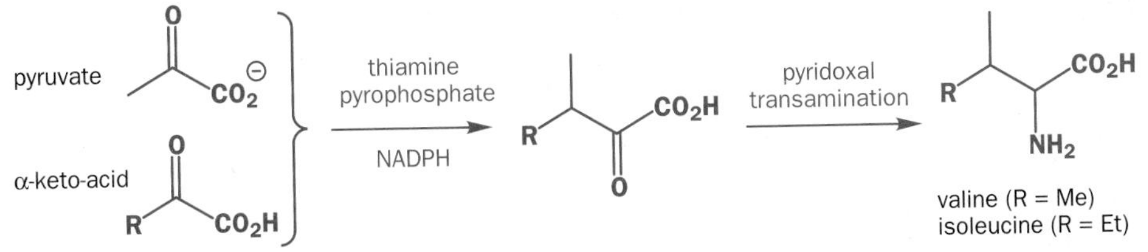

The remarkable aspect of this chemistry is that it involves 1,2-alkyl shifts in pinacol-like rearrangements (Chapter 37). The sequence starts as before and we will pick it up after the addition and decarboxylation of pyruvate and as the resulting d^1 reagent adds to the new α-keto-acid.

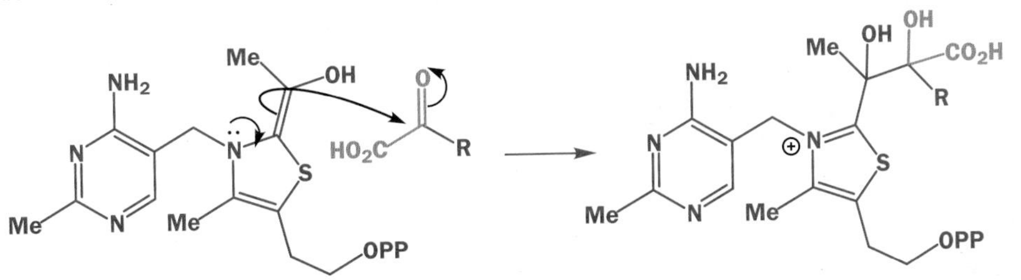

Decomposition of this product with the release of the thiazolium ylid also releases the product of coupling between the two keto-acids: a 1-hydroxy-2-keto-acid (in green). The original keto group of the pyruvate reappears—it's clear that an acetyl anion equivalent (the d^1 reagent) has added to the keto group of the new keto-acid. The thiazolium ylid is free to catalyse the next round of the reaction.

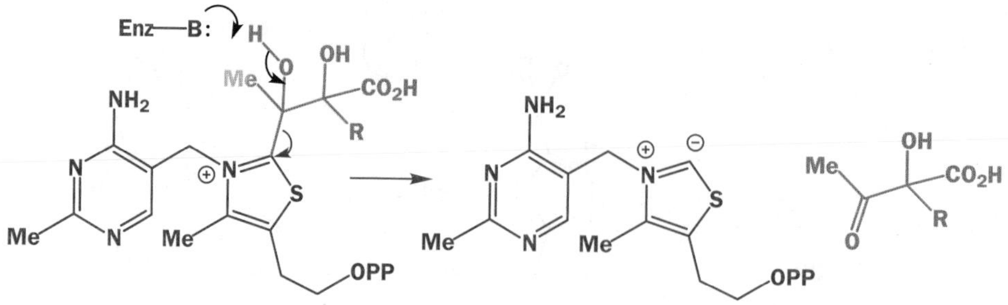

The green hydroxy-keto-acid is now primed for rearrangement. The migration of the group R is pushed by the removal of a proton from the OH group and pulled by the electron-accepting power of the keto group. Notice that the group R (Me or Et) migrates in preference to CO_2H. Usually in rearrangements the group better able to bear a positive charge migrates (Chapter 37).

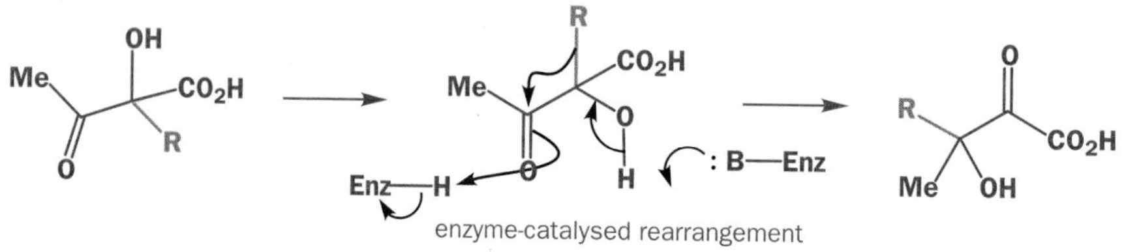

enzyme-catalysed rearrangement

Control in this reaction is likely to be exerted stereoelectronically by the enzyme as it was in the pyridoxal reactions above. Since the C–R bond is held parallel to the p orbitals of the ketone, R migration occurs, but if the CO_2H group were to be held parallel to the p orbitals of the ketone, decarboxylation would occur. Next, a simple reduction with NADPH converts the ketone into an alcohol and prepares the way for a second rearrangement.

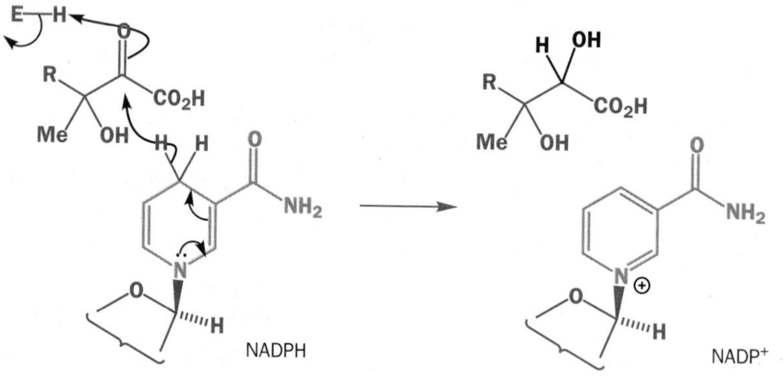

The second rearrangement is even more like a pinacol rearrangement because the starting material is a 1,2-diol. The tertiary alcohol is protonated and leaves, and again the CO_2H group does not migrate even though the alternative is merely hydride.

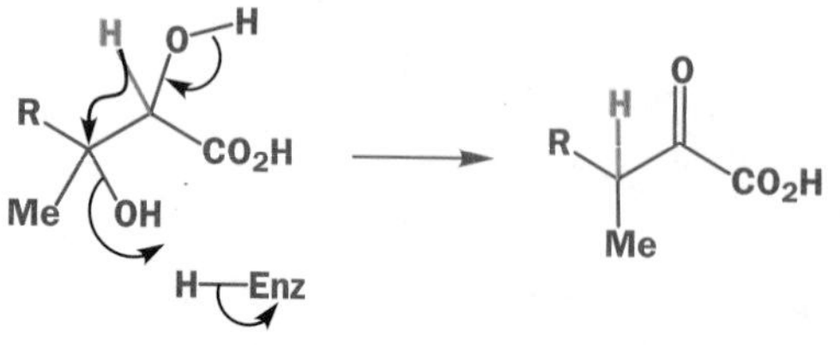

Finally, a pyridoxal transamination converts the two keto-acids stereospecifically to the corresponding amino acids, valine (R = Me) and isoleucine (R = Et). The donor amino acid is probably glutamate—it usually is in amino acid synthesis.

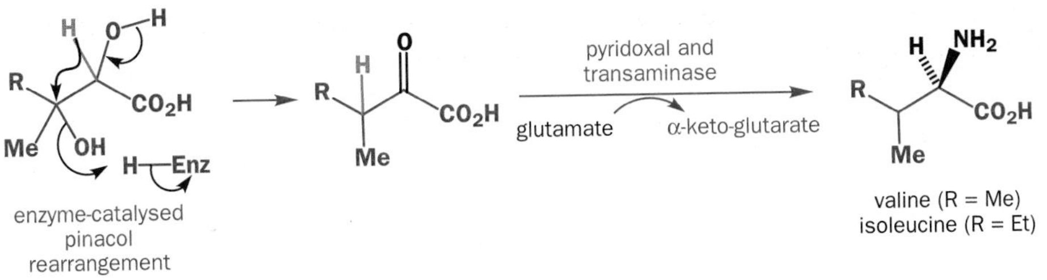

enzyme-catalysed
pinacol
rearrangement

pyridoxal and
transaminase

glutamate α-keto-glutarate

valine (R = Me)
isoleucine (R = Et)

Carbon dioxide is carried by biotin

We have added and removed carbon dioxide on several occasions in this chapter and the last but we have not until now said anything about how this happens. You would not expect gaseous CO_2 to be available inside a cell: instead CO_2 is carried around as a covalent compound with another coenzyme—**biotin**.

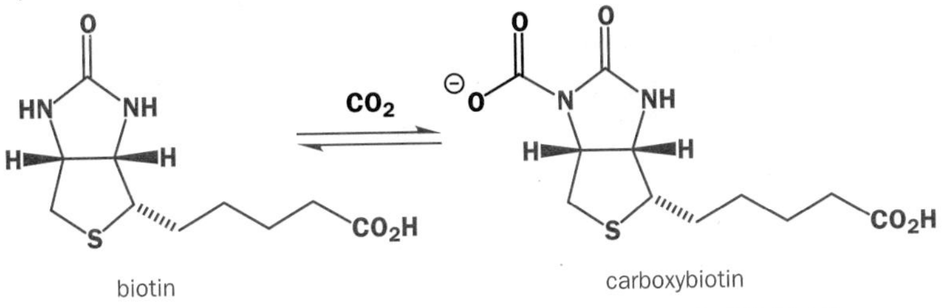

biotin   carboxybiotin

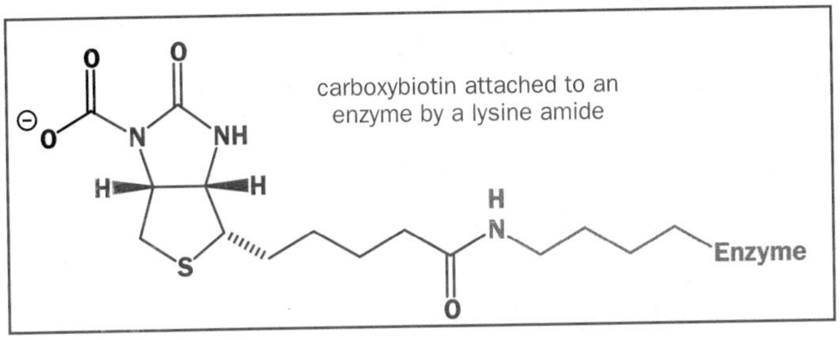

carboxybiotin attached to an
enzyme by a lysine amide

Biotin has two fused five-membered heterocyclic rings. The lower is a cyclic sulfide and has a long side chain ending in a carboxylic acid for attachment—yes, you've guessed it—to a lysine residue of a protein. The upper ring is a urea—it has a carbonyl group flanked by two nitrogen atoms. It is this ring that reversibly captures CO_2, on the nitrogen atom opposite the long side chain. The attachment to the enzyme as a lysine amide gives it an exceptionally long flexible chain and allows it to deliver CO_2 wherever it's needed.

One of the important points at which CO_2 enters as a reagent carried by biotin is in fatty acid biosynthesis where CO_2 is transferred to the enol of acetyl CoA. A magnesium(II) ion is also required and we may imagine the reaction as a nucleophilic attack of the enol on the magnesium salt of carboxybiotin. Most of the CO_2 transfers we have met take place by mechanisms of this sort: nucleophilic attack on a bound molecule of CO_2, usually involving a metal ion.

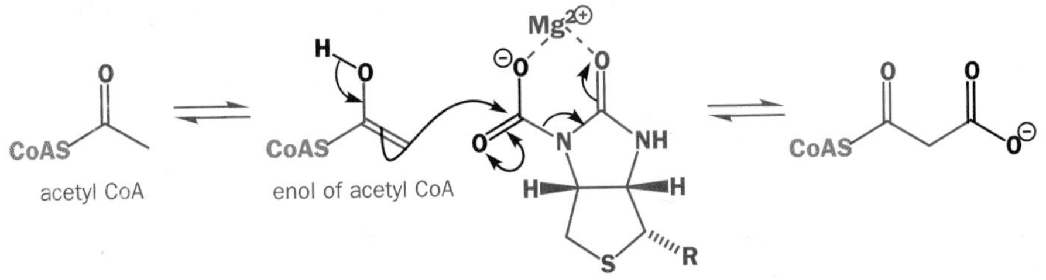

acetyl CoA   enol of acetyl CoA

Very similar reactions can be carried out in the laboratory. This simple cyclic urea reacts twice with the Grignard reagent MeMgBr to give a dimagnesium derivative, probably having the structure shown with one O–Mg and one N–Mg bond.

■ We will see in the next chapter how acetoacetyl CoA is used in the biosynthesis of fatty acids and polyketides.

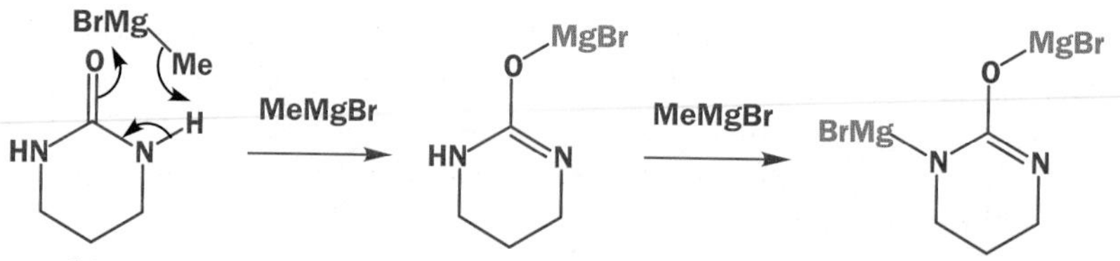

This magnesium derivative reacts with two molecules of CO_2 to give a double adduct with both nitrogens combining with CO_2. The product is stable as the double magnesium salt, which is a white powder.

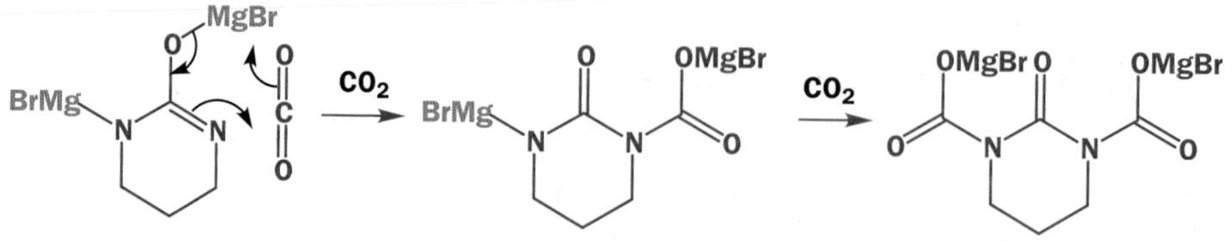

■
Diazomethane esterification appeared
in Chapter 40, p. 1053.

Simply heating this white powder with a ketone leads to efficient carboxylation and the unstable keto-acid may be trapped with diazomethane to form the stable methyl ester. The mechanism is presumably very like that drawn above for the transfer of CO_2 from carboxybiotin to acetyl CoA. Reactions like this *prove* nothing about the biochemical reaction but they at least show us that such reactions are possible and help us to have confidence that we are right about what Nature is doing.

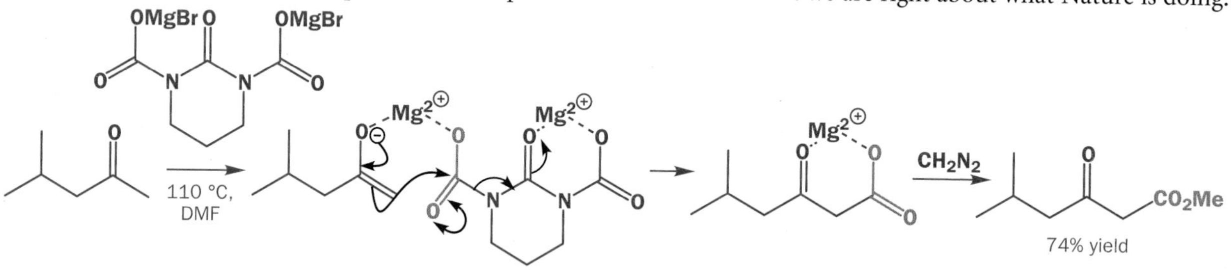

74% yield

The shikimic acid pathway

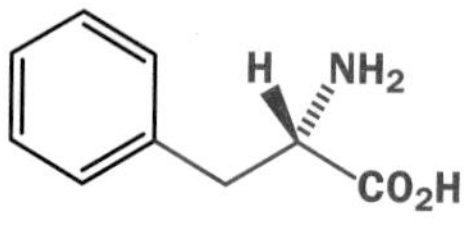

We have described reactions from various different pathways in this chapter so far, but now we are going to look at one complete pathway in detail. It is responsible for the biosynthesis of a large number of compounds, particularly in plants. Most important for us is the biosynthesis of the aromatic amino acids Phe (phenylalanine), Tyr (tyrosine), and Trp (tryptophan). These are 'essential' amino acids for humans—we have to have them in our diet as we cannot make them ourselves. We get them from plants and microorganisms.

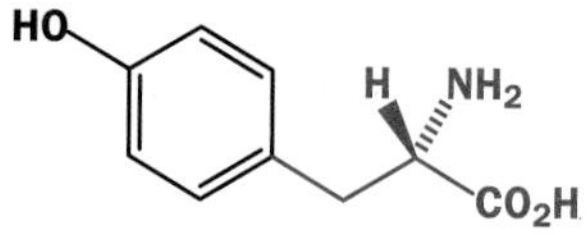

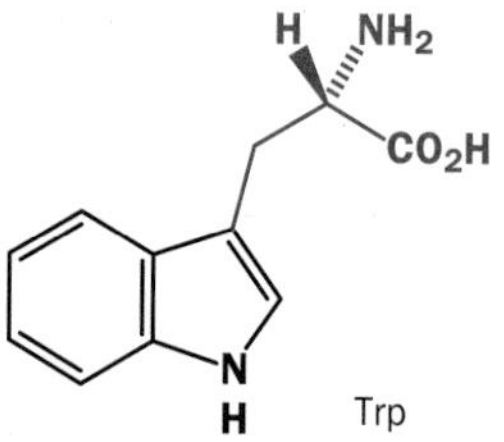

So how do plants make aromatic rings? A clue to the chemistry involved comes from the structure of caffeyl quinic acid, a compound that is present in instant coffee in some quantity. It is usually about 13% of the soluble solids from coffee beans.

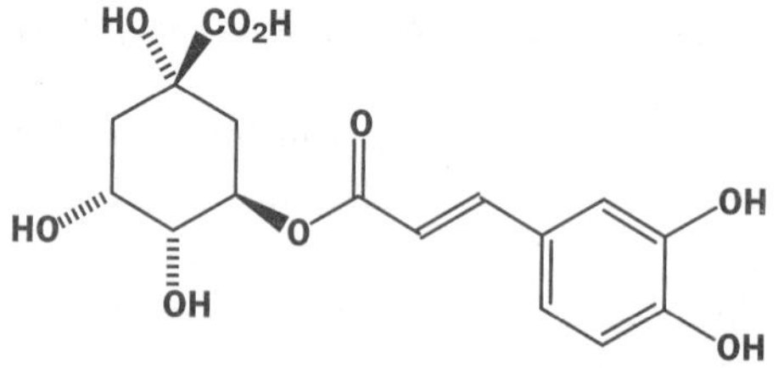

caffeyl quinic acid

This ester has two six-membered rings—one aromatic and one rather like the sugar alcohols we were discussing in the last chapter. You might imagine making an aromatic ring by the dehydration (losing three molecules of water) of a cyclohexane triol and the saturated ring in caffeyl quinic acid looks a good candidate. It is now known that both rings come from the same intermediate, shikimic acid.

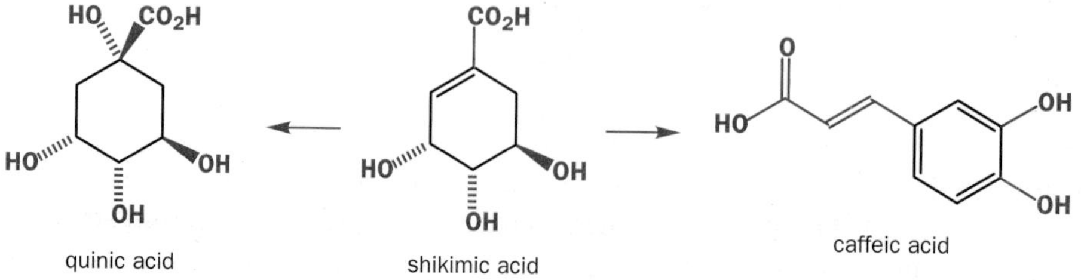

Phe

Tyr

Trp

quinic acid

shikimic acid

caffeic acid

This key intermediate has given its name to Nature's general route to aromatic compounds and many other related six-membered ring compounds: **the shikimic acid pathway**. This pathway contains some of the most interesting reactions (from a chemist's point of view) in biology. It starts with an aldol reaction between phosphoenol pyruvate as the nucleophilic enol component and the C_4 sugar erythrose 4-phosphate as the electrophilic aldehyde.

Wood

Even the structural material of plants, lignin, comes from the shikimic acid pathway. Lignin—from which wood is made—has a variable structure according to the plant and the position in the plant. A typical splinter is shown here.

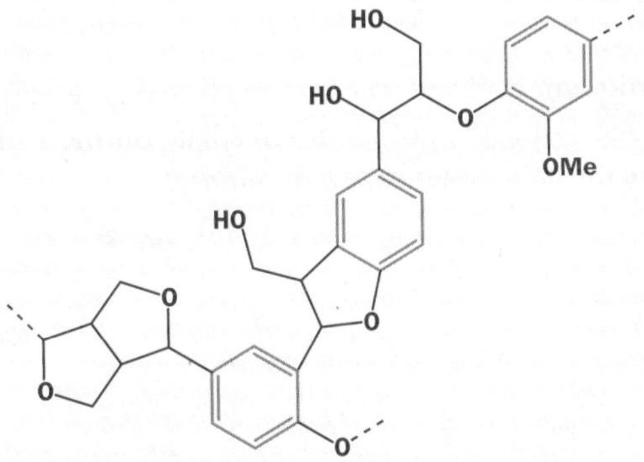

a typical lignin fragment

the aromatic rings are made from shikimic acid

the various structural parts and functional groups might be assembled in many different ways

the dotted bonds show where the structure might continue

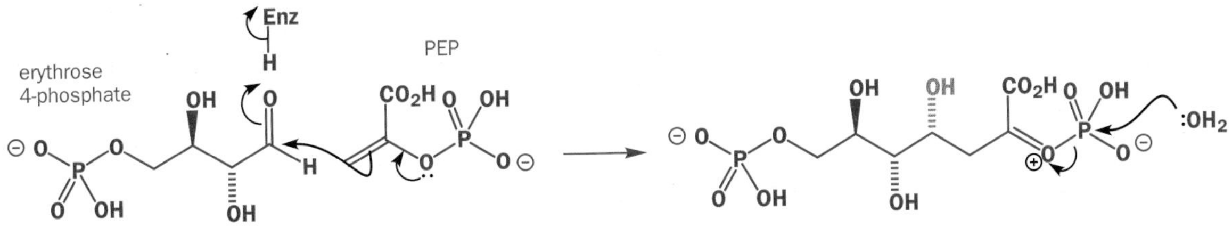

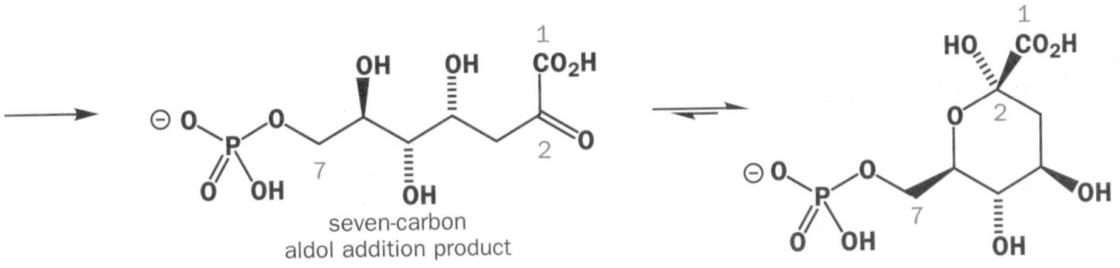

seven-carbon aldol addition product

Hydrolysis of the phosphate releases the aldol product, a C_7 α-keto-acid with one new stereogenic centre, which is in equilibrium with a hemiacetal, just like a sugar. This intermediate has the right number of carbon atoms for shikimic acid and the next stage is a cyclization. If we redraw the C_7 α-keto-acid in the right shape for cyclization we can see what is needed. The green arrow shows only which bond needs to be formed.

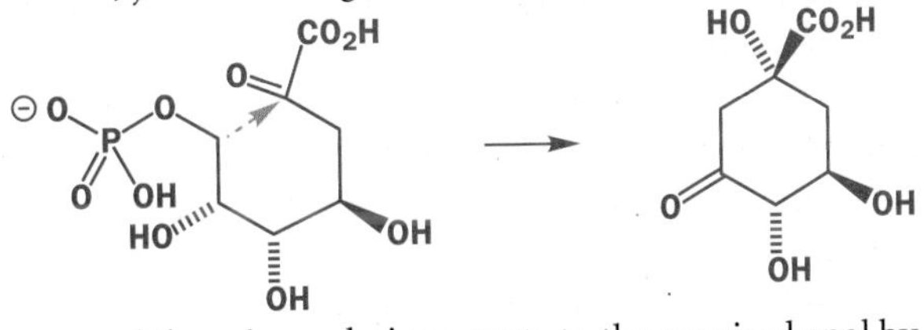

This reaction looks like an aldol reaction too and there is an obvious route to the required enol by elimination of phosphate. This would require the removal of a proton (green in the diagram) that is not at all acidic.

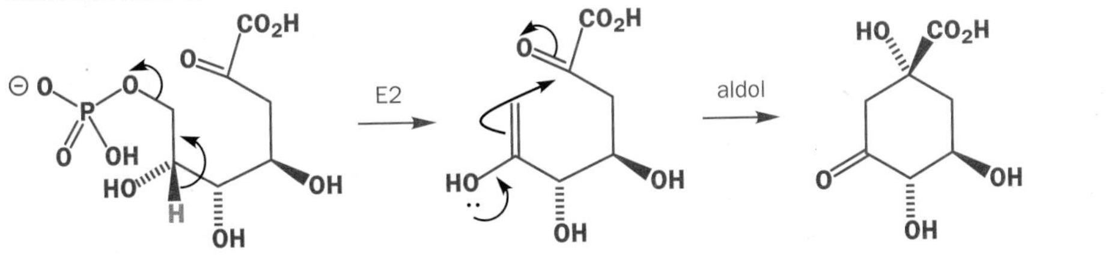

The problem can be avoided if the hydroxyl group at C5 is first oxidized to a ketone (NAD^+ is the oxidant). Then the green proton is much more acidic, and the elimination becomes an E1cB reaction, similar to the one in the synthesis of PEP. True, the ketone must be reduced back to the alcohol afterwards but Nature can deal with that easily.

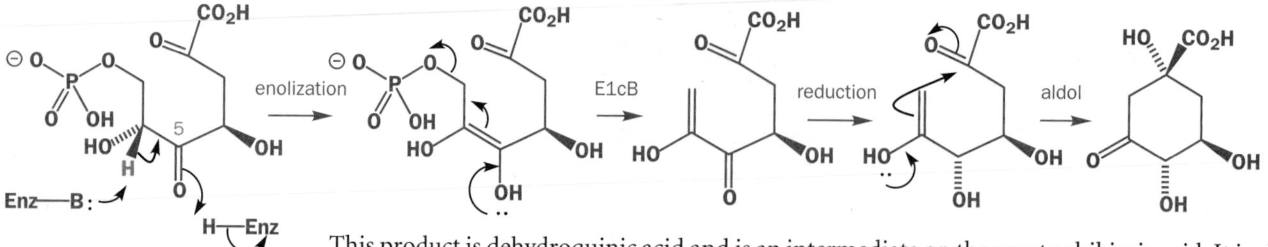

This product is dehydroquinic acid and is an intermediate on the way to shikimic acid. It is also in equilibrium with quinic acid, which is not an intermediate on the pathway but which appears in some natural products like the coffee ester caffeyl quinic acid.

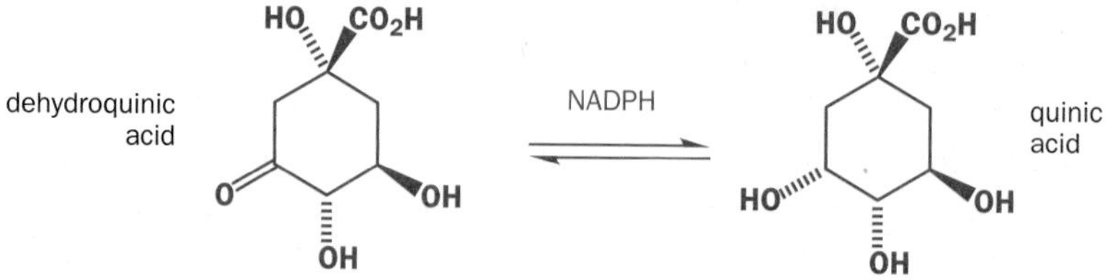

The route to shikimic acid in plants involves, as the final steps, the dehydration of dehydroquinic acid and then reduction of the carbonyl group. Doing the reactions this way round means that the dehydration can be E1cB—much preferred under biological conditions. This is what happens.

dehydroquinic acid — dehydroshikimic acid — shikimic acid

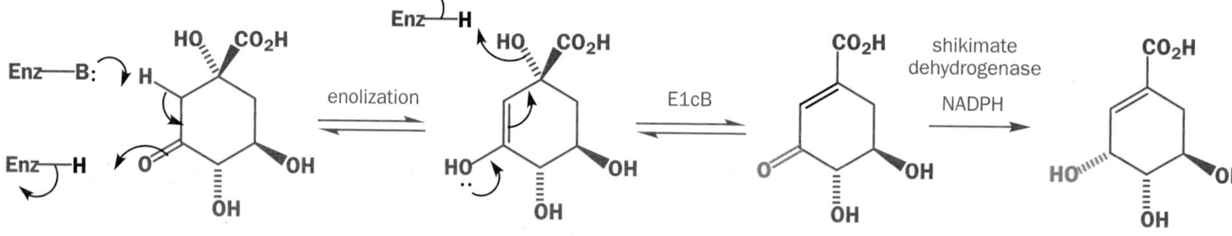

The final reduction uses NADPH as the reagent and is, of course, totally stereoselective with the hydride coming in from the top face of the green ketone as drawn. At last we have arrived at the halfway stage and the key intermediate, shikimic acid.

The most interesting chemistry comes in the second half of the pathway. The first step is a chemoselective phosphorylation of one of the three OH groups by ATP—as it happens, the OH group that has just been formed by reduction of a ketone. This step prepares that OH group for later elimination. Next, a second molecule of PEP appears and adds to the OH group at the other side of the molecule. This is PEP in its enol ether role, forming an acetal under acid catalysis. The reaction occurs with retention of stereochemistry so we know that the OH group acts as a nucleophile and that the ring–OH bond is not broken.

shikimic acid — shikimic acid 3-phosphate — EPSP—5-enolpyruvylshikimate 3-phosphate

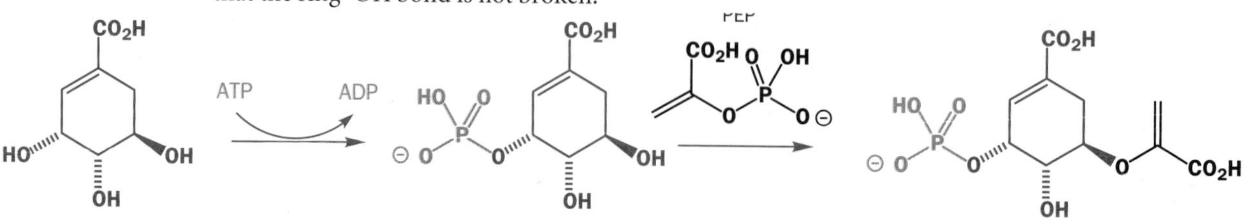

Now a 1,4 elimination occurs. This is known to be a *syn* elimination on the enzyme. When such reactions occur in the laboratory, they can be *syn* or *anti*. The leaving group is the green phosphate added two steps before.

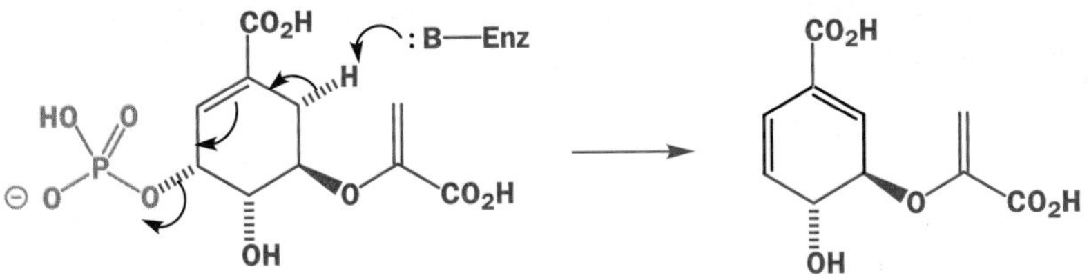

EPSP—5-enolpyruvylshikimate 3-phosphate

chorismic acid

The product is chorismic acid and this undergoes the most interesting step of all—a [3,3]-sigmatropic rearrangement. Notice that the new (black) σ bond forms on the same face of the ring as the old (green) σ bond: this is, as you should expect, a suprafacial rearrangement.

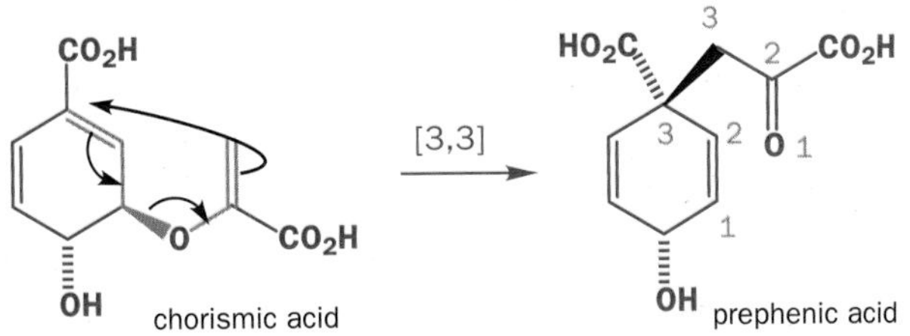

chorismic acid

[3,3]

prephenic acid

For more on sigmatropic rearrangements, see Chapter 36.

The most favourable conformation for chorismic acid has the substituents pseudoequatorial but the [3,3]-sigmatropic rearrangement cannot take place in that conformation. First, the diaxial conformation must be formed and the chair transition state achieved. Then the required orbitals will be correctly aligned.

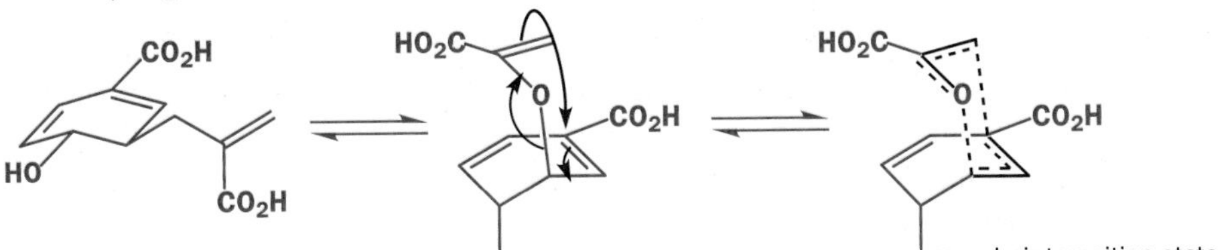

pseudoequatorial conformation of chorismic acid

pseudoaxial conformation of chorismic acid

chair transition state for [3,3]-sigmatropic rearrangement

These reactions occur well without the enzyme (Chapter 36) but the enzyme accelerates this reaction by about a 10^6 increase in rate. There is no acid or base catalysis and we may suppose that the enzyme binds the transition state better than it binds the starting materials. We know this to be the case, because close structural analogues of the six-membered ring transition state also bind to the enzyme and stop it working. An example is shown alongside—a compound that resembles the transition state but can't react.

chair transition state for [3,3]-sigmatropic rearrangement

transition state analogue that inhibits the enzyme

By binding the transition state (not the starting materials) strongly, the enzyme lowers the activation energy for the reaction.

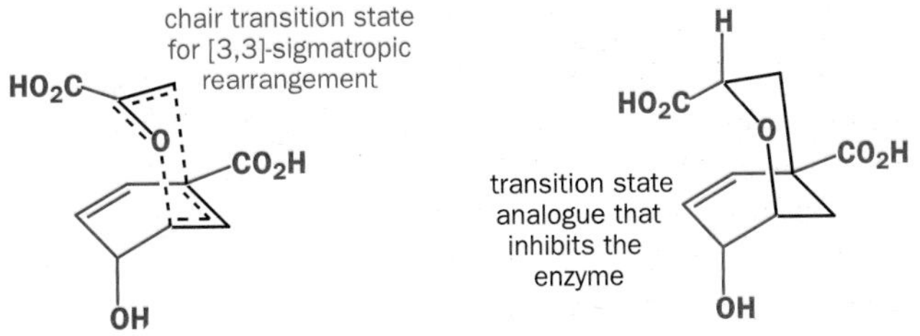

We have arrived at prephenic acid, which as its name suggests is the last compound before aromatic compounds are formed, and we may call this the end of the shikimic acid pathway. The final stages of the formation of phenylalanine and tyrosine start with aromatization. Prephenic acid is unstable and loses water and CO_2 to form phenylpyruvic acid. This α-keto-acid can be converted into the amino acid by the usual transamination with pyridoxal.

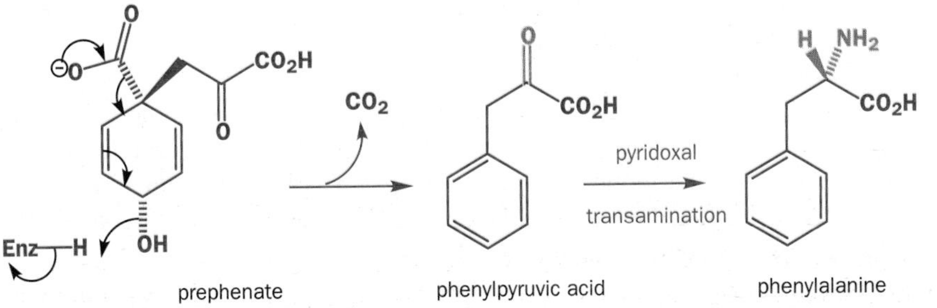

prephenate

CO_2

phenylpyruvic acid

pyridoxal

transamination

phenylalanine

The route to tyrosine requires a preliminary oxidation and then a decarboxylation with the electrons of the breaking C–C bond ending up in a ketone group. Transamination again gives the amino acid.

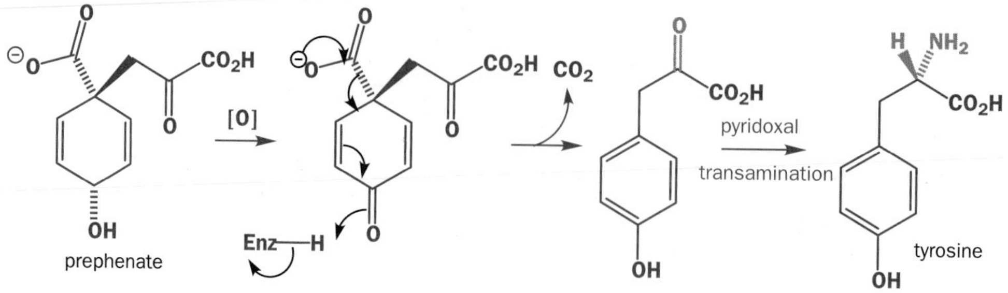

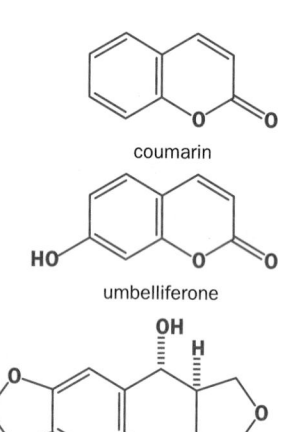

Other shikimate products

Many natural products are formed from the shikimate pathway. Most can be recognized by the aromatic ring joined to a three-carbon atom side chain. Two simple examples are coumarin, responsible for the smell of mown grass and hay, and umbelliferone, which occurs in many plants and is used in suntan oils as it absorbs UV light strongly. These compounds have the same aryl-C_3 structure as Phe and Tyr, but they have an extra oxygen atom attached to the benzene ring and an alkene in the C_3 side chain.

An important shikimate metabolite is podophyllotoxin, an antitumour compound—some podophyllotoxin derivatives are used to combat lung cancer. The compound can be split up notionally into two shikimate-derived fragments (shown in red and green). Both are quite different and there is obviously a lot of chemistry to do after the shikimic acid pathway is finished.

Among the more interesting reactions involved in making all three of these natural products are the loss of ammonia from phenylalanine to give an alkene and the introduction of extra OH groups around the benzene rings. We know how a *para* OH of Tyr is introduced directly by the oxidation of prephenic acid before decarboxylation and it is notable that the extra oxygen functionalities appear next to that point. This is a clue to the mechanism of the oxidation.

Alkenes by elimination of ammonia—phenyl ammonia lyase

Many amino acids can lose ammonia to give an unsaturated acid. The enzymes that catalyse these reactions are known as **amino acid ammonia lyases**. The one that concerns us at the end of the shikimic acid pathway is phenylalanine ammonia lyase, which catalyses the elimination of ammonia from phenylalanine to give the common metabolite cinnamic acid.

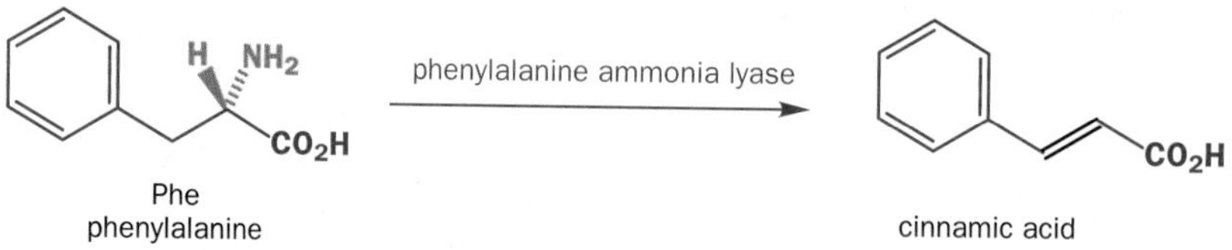

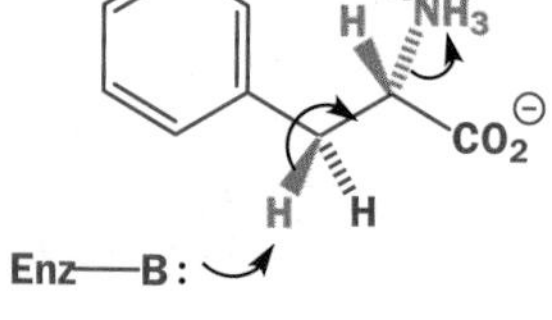

This reaction gives only *E*-cinnamic acid and the proton *anti* to the amino group is lost. This might make us think that we have an E2 reaction with a base on the enzyme removing the required proton. But a closer look at this mechanism makes it very unconvincing. The proton that is removed has no acidity and ammonia is not a good leaving group. It is very unusual for Nature to use an enzyme to make a reaction happen that doesn't happen at all otherwise. It is much more common for Nature to make a good reaction better.

So how does an ammonia lyase work? The enzyme makes the ammonia molecule into a much better leaving group by using a serine residue. This serine is attached to the protein through its carboxyl group by the usual amide bond but its amino group is bound as an imine. This allows it to eliminate water to form a double bond before the phenylalanine gets involved. The elimination converts serine into a dehydroalanine residue. This is an E1cB elimination using only general acid and base catalysis as the proton to be lost is acidic and an enol can be an intermediate.

coumarin

umbelliferone

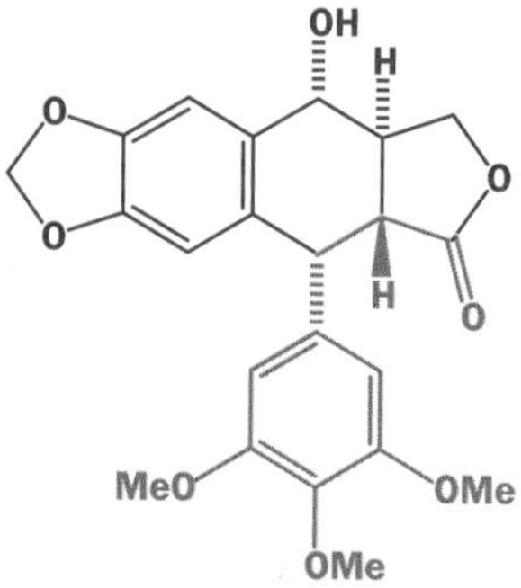

podophyllotoxin

▶ A *lyase* is an enzyme that catalyses *lysis*: it breaks something down.

possible E2 mechanism for phenylalanine ammonia lyase?

▶ Eliminations of ammonium salts (Chapter 19, p. 484) require very strong bases—much stronger than those available to enzymes—and fully alkylated amines. You can't protonate an amine in the presence of strong base.

prephenate

tyrosine

pyridoxal transamination

Phe phenylalanine

phenylalanine ammonia lyase

cinnamic acid

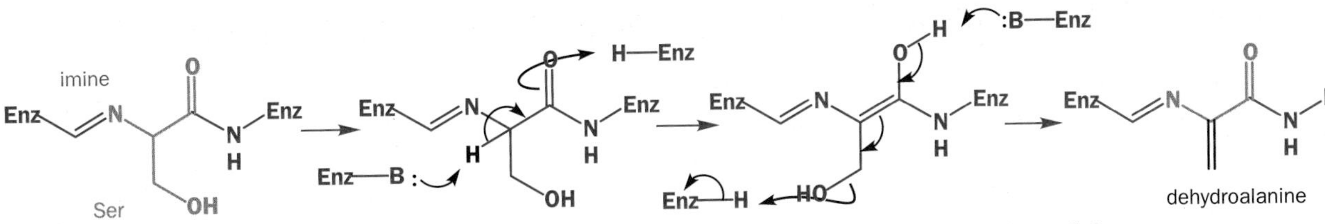

dehydroalanine

The alkene of the dehydroenzyme is conjugated with a carbonyl group—it's electrophilic and the amino group of Phe can add to it in conjugate fashion. When the enol tautomerizes back to a carbonyl compound, it can be protonated on the imine carbon because the imine is conjugated to the enol. This might remind you of pyridoxal's chemistry (p. 1384).

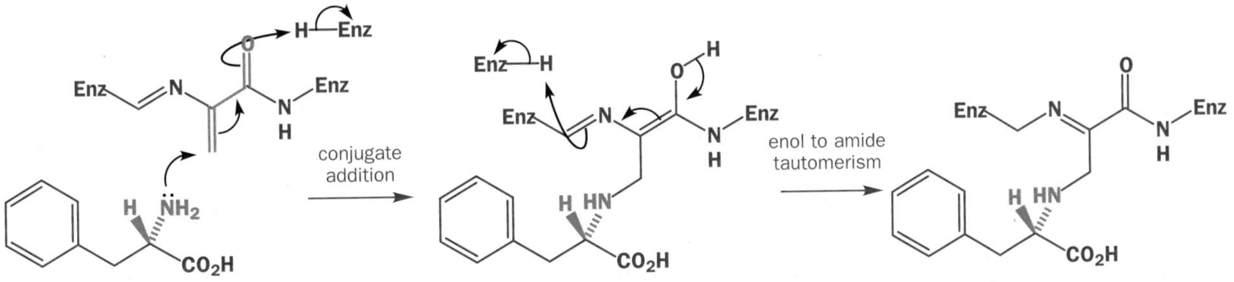

A second tautomerism makes an enamine—again very like the pyridoxal mechanisms you saw earlier.

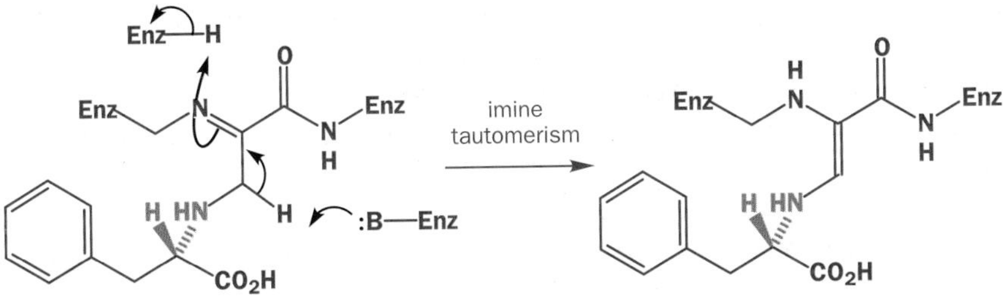

Now at last the secret is revealed. We can break the C–N bond and use the carbonyl group as an electron sink. The acidity of the proton that must be lost is no greater but the nitrogen atom has become a very much better leaving group.

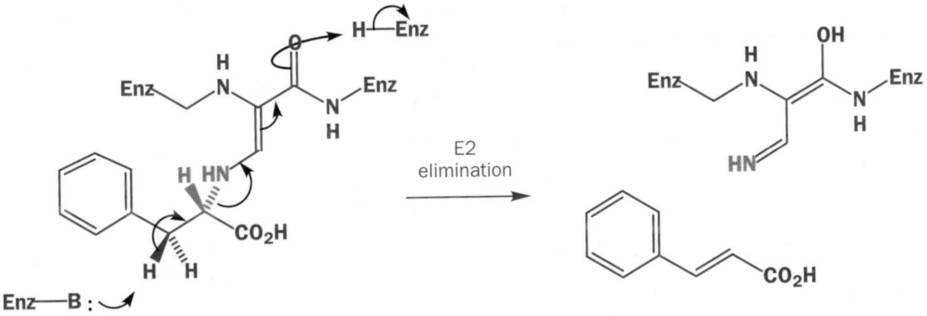

The difficult elimination is accomplished by making it an ammonia transfer reaction rather than an elimination of ammonia. Recycling the enzyme does eventually require elimination of ammonia but in an easy E1cB rather than a difficult E2 reaction. Overall, a difficult reaction—elimination of ammonia—is accomplished in steps that involve no strong acids or strong bases, and most of the steps are simple proton transfers, often tautomerisms between imines, enols, and amides.

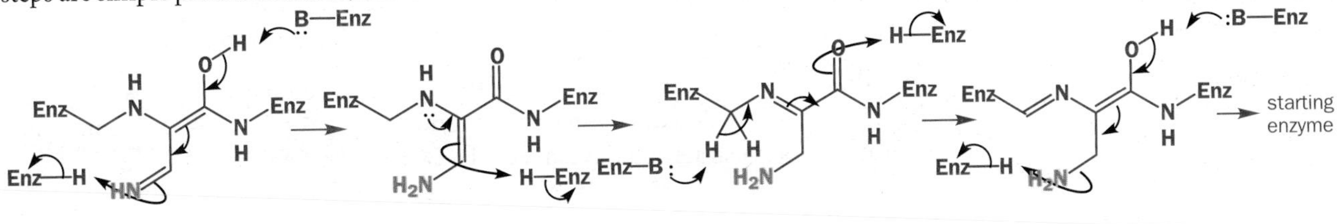

starting enzyme

porphyrin

18 electrons
in a conjugated ring
= 4n + 2 (n = 4)

■ Porphyrins appeared in Chapters 43
and 44, pp. 1178 and 1189.

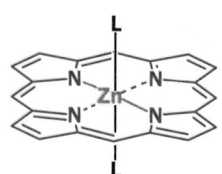

octahedral
zinc(II) porphyrin
with two extra ligands

Haemoglobin carries oxygen as an iron(II) complex

Biological oxidations are very widespread. Human metabolism depends on oxidation, and on getting oxygen, which makes up 20% of the atmosphere, into cells. The oxygen transporter, from atmosphere to cell, is **haemoglobin**.

The reactive part of haemoglobin is a **porphyrin**. These are aromatic molecules with 18 electrons around a conjugated ring formed from four molecules of a five-membered nitrogen heterocycle. Chemically, symmetrical porphyrins are easily made from pyrrole and an aldehyde.

pyrrole   PhCHO   Zn²⁺

The hole in the middle of a porphyrin is just the right size to take a divalent transition metal in the first transition series, and zinc porphyrins, for example, are stable compounds. Once the metal is inside a porphyrin, it is very difficult to get out. Two of the nitrogen atoms form normal covalent bonds (the ones that were NH in the porphyrin) and the other two donate their lone pairs to make four ligands around the metal. The complexed zinc atom is square planar and still has two vacant sites—above and below the (more or less) flat ring. These can be filled with water molecules, ammonia, or other ligands.

The porphyrin part of haemoglobin is called **haem**, and it is an iron(II) complex. It is unsymmetrically substituted with carboxylic acid chains on one side and vinyl groups on the other.

Haem is bound to proteins to make haemoglobin (in blood) and myoglobin (in muscle). The hydrophilic carboxylate groups stick out into the surrounding medium, while the majority of the molecule is embedded in a hydrophobic cleft in the protein, lined with amino acids such a leucine and valine. The octahedral coordination sphere of the iron(II) is completed with a histidine residue from the protein and an oxygen molecule.

hydrophilic region on
surface of protein

haem

The oxygen complex can be drawn like this or, alternatively, as an Fe(III) complex of an oxyanion (below).

protein   protein

oxyhaemoglobin

It is difficult to draw detailed mechanisms for oxidations by iron complexes but it is the oxygen atom further from Fe that reacts. You can see in principle how breakage of the weak O–O bond could deliver an oxygen atom to a substrate and leave an Fe(III)–O⁻ complex behind.

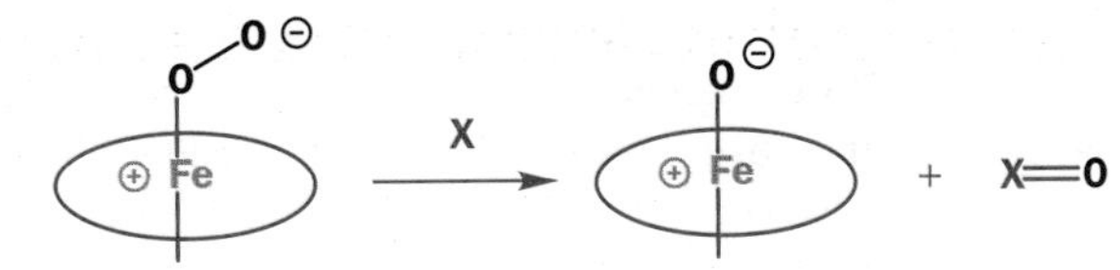

Oxygen molecules are transferred from haemoglobin to other haems, such as the enzyme P450, and to a wide range of oxidizing agents. Almost any molecule we ingest that isn't a nutrient—a drug molecule, for example—is destroyed by oxidation. The details of the mechanisms of these oxidations have proved very difficult to elucidate, but the hydroxylation of benzene is an exception. We do know how it happens, and it's another case of Nature using enzymes to do some really remarkable chemistry.

Aromatic rings are hydroxylated via an epoxide intermediate

The oxidizing agents here are related to FAD. We said little about $FADH_2$ as a reducing agent earlier in this chapter because it is rather similar to NADH which we have discussed in detail. FAD is another dinucleotide and it contains an AMP unit linked through the 5′ position by a pyrophosphate group to another nucleotide. The difference is that the other nucleotide is flavin mononucleotide. Here is the complete structure.

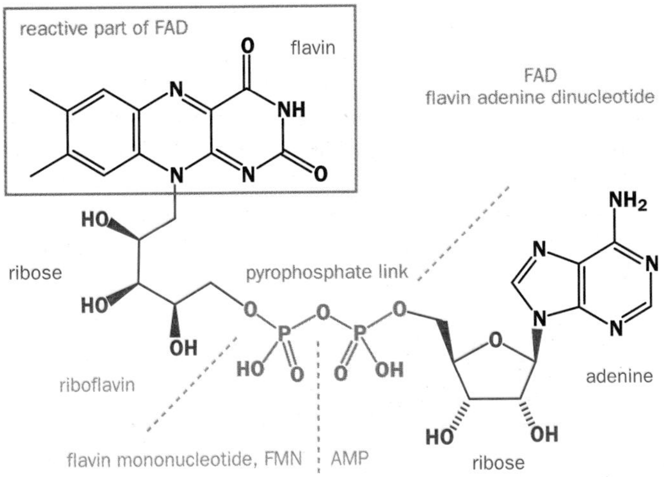

The whole thing is FAD. Cutting FAD in half down the middle of the pyrophosphate link would give us two nucleo*tides*, AMP and FMN (flavin mononucleotide). The sugar in each case is ribose (in its furanose form in AMP but in open-chain form in FMN) so the flavin nucleo*side* is riboflavin. We can abbreviate this complex structure to the reactive part, which is the flavin. The rest we shall just call 'R'.

> ► Riboflavin is also known as vitamin B_2 as you may see on the side of your cornflakes packet.

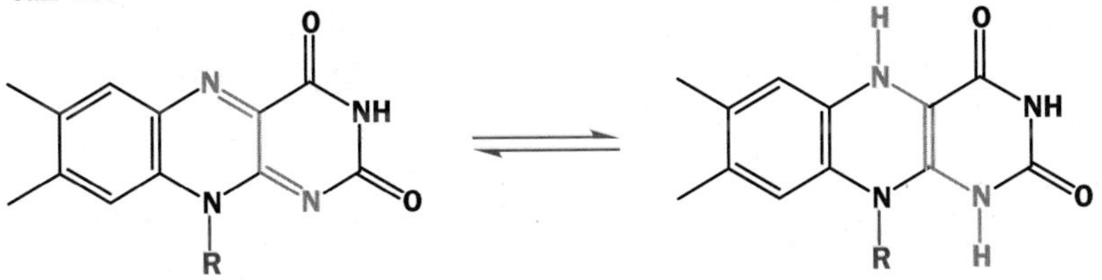

Redox reactions with FAD involve the transfer of two hydrogen atoms to the part of the molecule shown in green. Typical reactions of FAD involve dehydrogenations—as in double bond formation from single bonds. Of course, one of the H atoms can be transferred to FAD as a proton—only one need be a hydride ion H^-, though both could be transferred as radicals ($H^{\bullet}$).

> ► You should contrast this with the redox reactions of NAD where only one hydrogen atom is transferred.

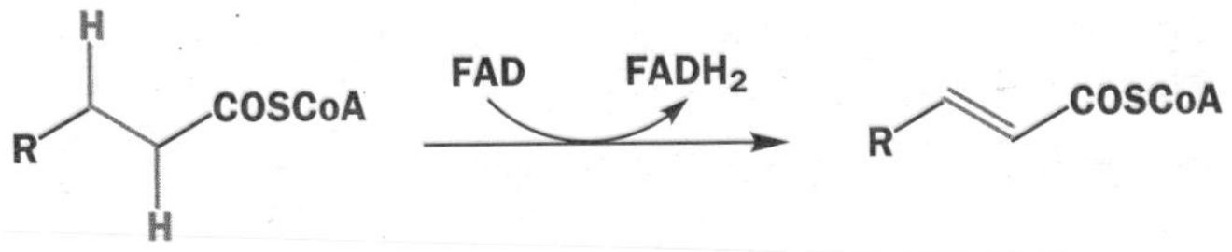

After FAD has been used as an oxidant in this fashion, the $FADH_2$ reacts with molecular oxygen to give a hydroperoxide, which decomposes back to FAD and gives an anion of hydrogen peroxide, which would in turn be reduced by other reagents.

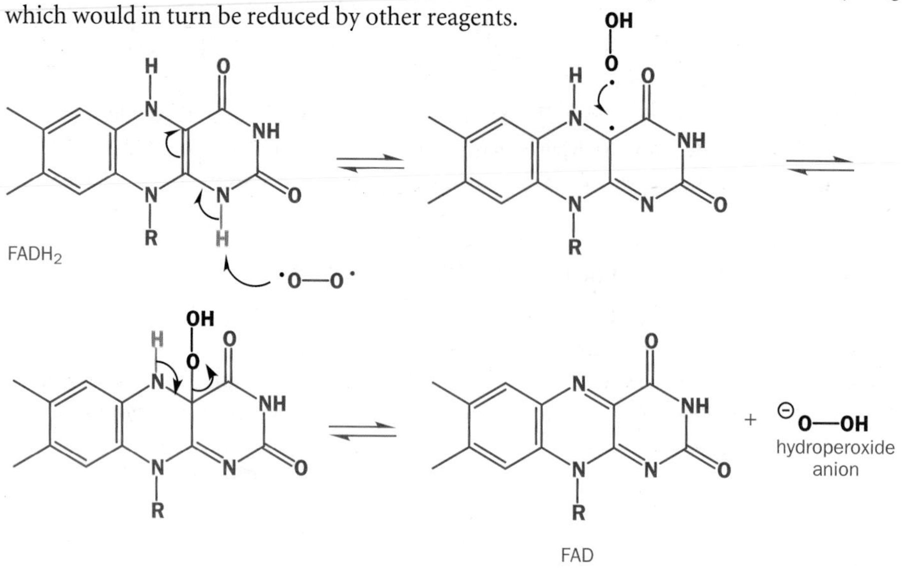

FADH₂

> Note the radical steps in this sequence. The reactions of oxygen, whose ground state is a triplet diradical (see Chapter 4), are typically radical processes.

+ $^{\ominus}O{-}OH$
hydroperoxide anion

FAD

In the reactions we are now concerned with, the hydroperoxide intermediate itself is the important reagent, before it loses hydroperoxide anion. This intermediate is an oxidizing agent—for example, it reacts quite dramatically with benzene to give an epoxide.

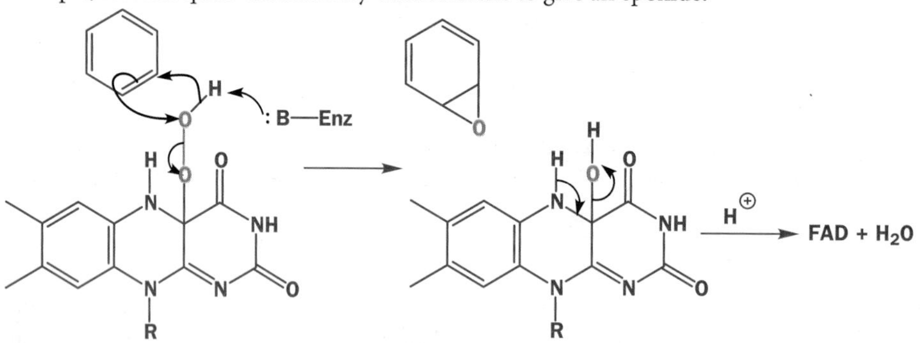

$\xrightarrow{\ \ }$ FAD + H_2O

This benzene oxide may look very dubious and unstable, but benzene oxides can be made in the laboratory by ordinary chemical reactions (though not usually by the direct oxidation of benzene). We can instead start with a Diels–Alder reaction between butadiene and an alkyne. Epoxidation with a nucleophilic reagent ($HO{-}O^-$ from H_2O_2 and NaOH) occurs chemoselectively on the more electrophilic double bond—the one that is conjugated to the electron-withdrawing carbonyl group. Bromination of the remaining alkene gives a dibromo-epoxide.

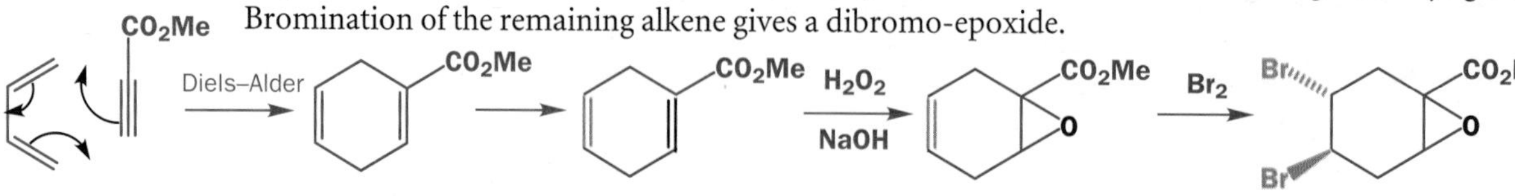

This is an ordinary electrophilic addition to an alkene so the two bromine atoms are *anti* in the product. Elimination under basic conditions with DBN gives the benzene oxide.

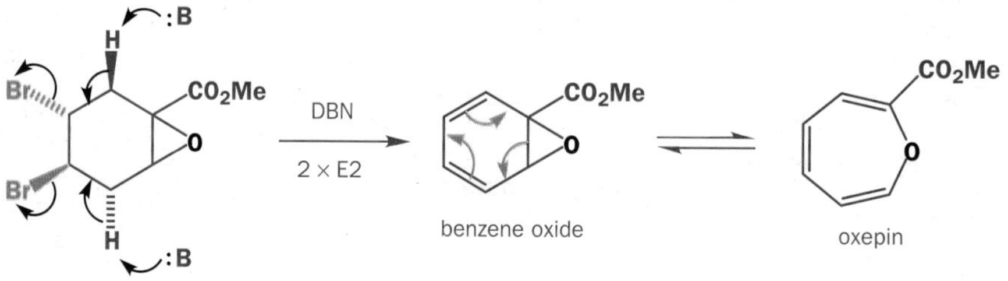

benzene oxide

oxepin

At least, it ought to have given the benzene oxide! The compound turned out to have a fluxional structure—it was a mixture of compounds that equilibrate by a reversible disrotatory electrocyclic reaction.

Treatment with acid turns the benzene oxide/oxepin into an aromatic ring by a very interesting mechanism. The epoxide opens to give the cation, which is *not* conjugated with the electron-withdrawing CO$_2$Me group, and then a migration of that CO$_2$Me group occurs. This has been proved by isotope labelling experiments. The final product is the *ortho*-hydroxy-ester, known as methyl salicylate.

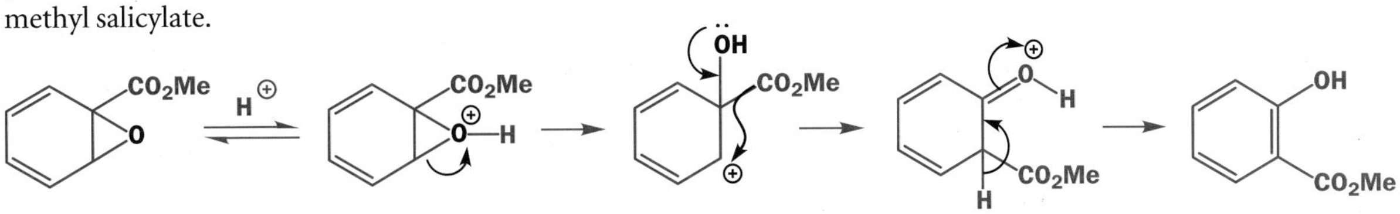

This chemistry seems rather exotic, but in the degradation of phenylalanine two benzene oxide intermediates and two such rearrangements occur one after the other. This is the initial sequence.

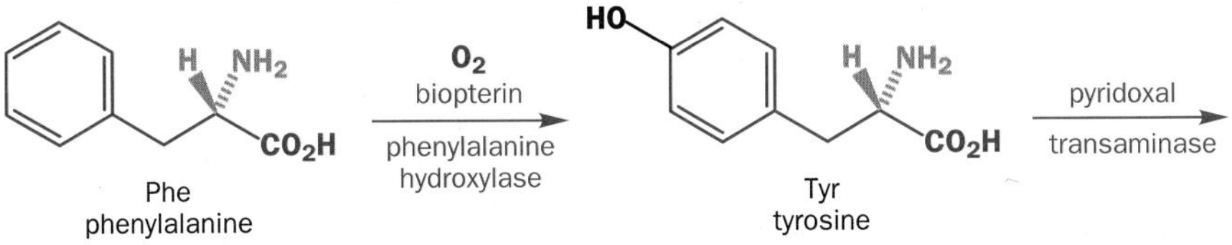

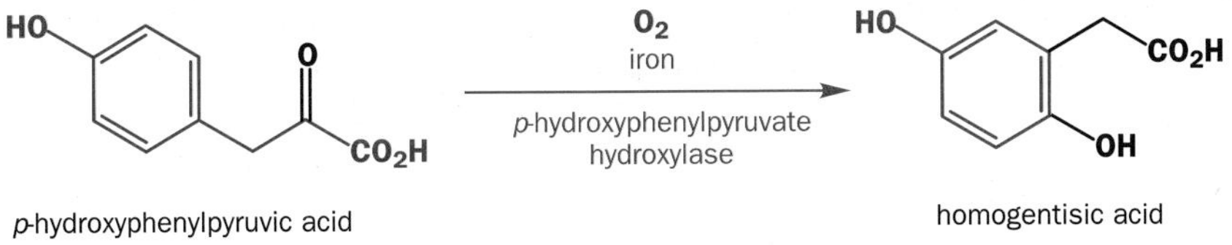

The first reaction involves a hydroperoxide related to the FAD hydroperoxide you have just seen but based on a simpler heterocyclic system, a biopterin. The reaction is essentially the same and a benzene oxide is formed.

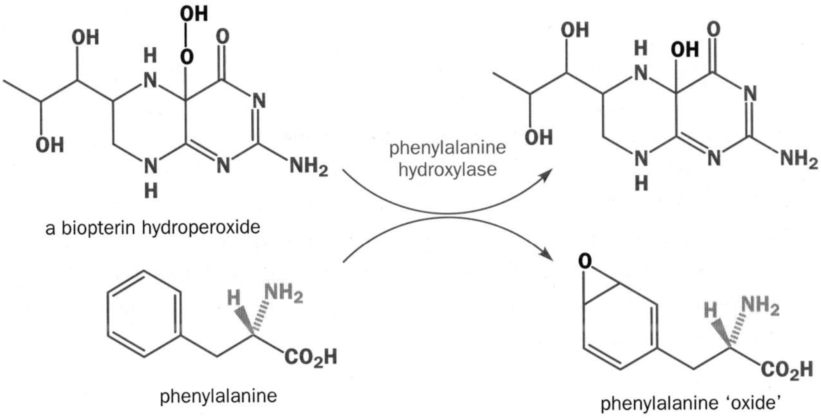

The biopterin product is recycled by elimination of water, reduction using NADPH as the reagent, and reaction with molecular oxygen. The other product, the phenylalanine oxide, rearranges with a hydride shift followed by the loss of a proton to give tyrosine.

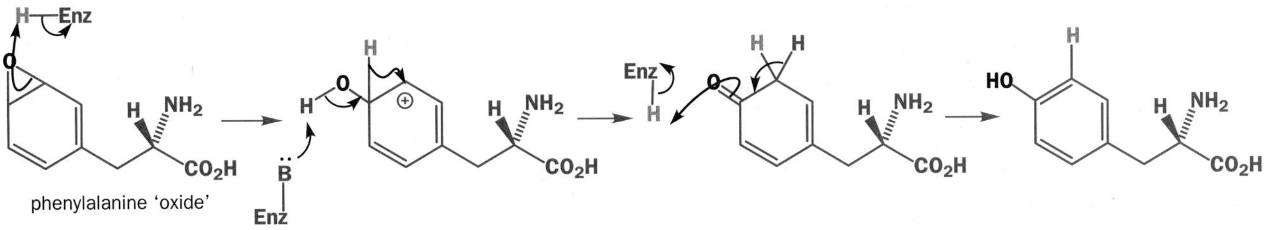

▶

This rearrangement is known as the 'NIH shift', after its discovery at the National Institutes of Health at Bethesda, Maryland.

We know that this is the mechanism because we can make the green H a deuterium atom. We then find that deuterium is present in the tyrosine product *ortho* to the phenolic hydroxyl group. When the migration occurs, the deuterium atom must go as there is no alternative, but in the next step there is a choice and H loss will be preferred to D loss because of the kinetic isotope effect (Chapter 19). Most of the D remains in the product.

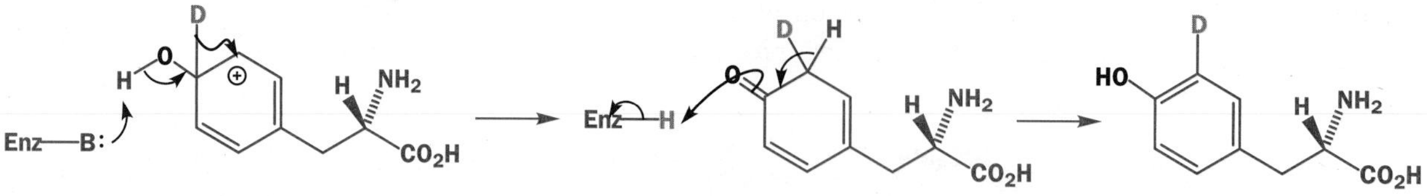

no choice—
deuterium must migrate

choice—H or D could be lost
H preferred because of
kinetic isotope effect

A shift of a larger group comes two steps later in the synthesis of homogentisic acid. Another labelling experiment, this time with $^{18}O_2$, shows that both atoms of oxygen end up in the product.

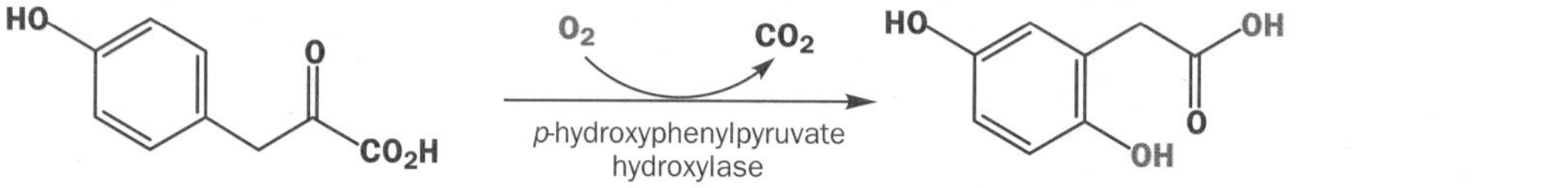

p-hydroxyphenylpyruvic acid

p-hydroxyphenylpyruvate
hydroxylase

homogentisic acid

The key intermediate is a peroxy-acid formed after decarboxylation. The peroxy-acid is perfectly placed for an intramolecular epoxidation of a double bond in the benzene ring next to the side chain.

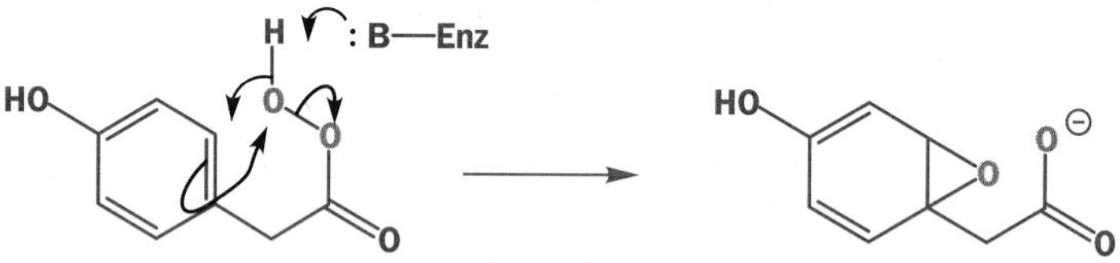

The epoxide can now rearrange with the whole side chain migrating in a reaction very similar to the laboratory rearrangement to give methyl salicylate that you saw on p. 1409.

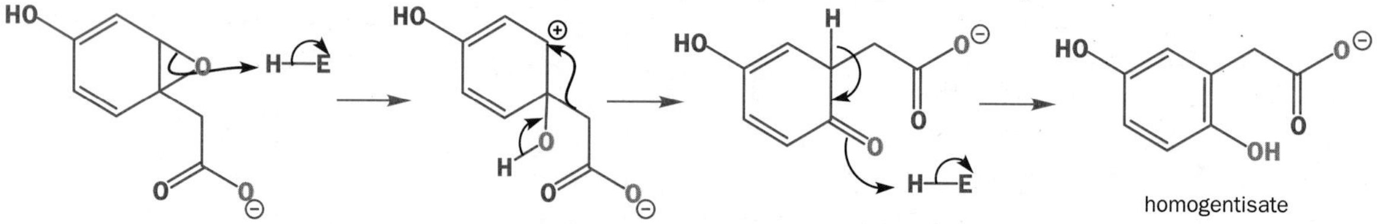

homogentisate

When hydroxylation occurs next to an OH group that is already there, no NIH shift occurs. This is because the epoxide is opened by the push of electrons from the OH group and there is only one H atom to be lost anyway. The cofactor for these enzymes is slightly different, being again the hydroperoxide from FAD, but the principle is the same.

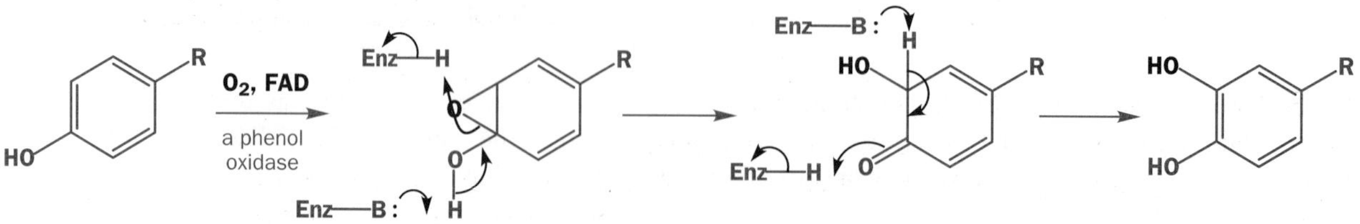

O_2, FAD

a phenol
oxidase

In the next chapter you will see how hydroxylation of benzene rings plays an important part in the biosynthesis of alkaloids and other aromatic natural products.

Problems

1. On standing in alkali in the laboratory, prephenic acid rearranges to 4-hydroxyphenyl-lactic acid with specific incorporation of deuterium label as shown. Suggest a mechanism, being careful to draw realistic conformations.

stand in basic solution

prephenate

4-hydroxyphenyl-lactate
this product is racemic. why?

2. Write a full reaction scheme for the conversion of ammonia and pyruvate to alanine in living things. You will need to refer to the section of the chapter on pyridoxal to be able to give a complete answer.

pyruvate   + NH$_3$ Nature   (S)-alanine

3. Give a mechanism for this reaction. You will find the Stetter catalyst described in the chapter. How is this sequence biomimetic?

Et$_3$N
Stetter catalyst

What starting material would be required for formation of the natural product *cis*-jasmone by an intramolecular aldol reaction (Chapter 27). How would you make this compound using a Stetter reaction?

cis-jasmone

4. The amino acid cyanoalanine is found in leguminous plants (*Lathyrus*) but not in proteins. It is made in the plant from cysteine and cyanide by a two-step process catalysed by pyridoxal phosphate. Suggest a detailed mechanism.

(S)-cysteine   pyridoxal phosphate
Lathyrus spp.   (S)-cyanoalanine

5. This chemical reaction might be said to be similar to a reaction in the shikimic acid pathway. Compare the two mechanisms and suggest how the model might be made closer and more interesting.

EtCO$_2$H

6. Stereospecific deuteration of the substrate for enolase, the enzyme that makes phosphoenol pyruvate, gives the results shown

below. What does this tell us definitely about the reaction and what might it suggest about the mechanism?

enolase

7. This rearrangement was studied as a biomimetic version of the NIH shift. Write a mechanism for the reaction. Do you consider it a good model reaction? If not, how might it be made better?

pH 7

83% yield   17% yield

8. The following experiments relate to the chemical and biological behaviour of NADH. Explain what they tell us.

(**a**) This FAD analogue can be reduced *in vitro* with NADH in D$_2$O with deuterium incorporation in the product as shown.

NADH
D$_2$O

(**b**) NADH does not reduce benzaldehyde *in vitro* but it does reduce this compound.

NADH

9. Oxidation of this simple thiol ester gives a five-membered cyclic disulfide. The reaction is proposed as a model for the behaviour of lipoic acid in living things. Draw a mechanism for the reaction and make the comparison.

I$_2$
MeOH   + RCO$_2$Me

10. This curious compound is chiral—indeed it has been prepared as the (−) enantiomer. Explain the nature of the chirality.

enantiomerically enriched
(−) enantiomer

This compound has been used as a chemical model for pyridoxamine. For example, it transaminates phenylpyruvate under the conditions shown here. Comment on the analogy and the role of Zn(II). In what ways is the model compound worse and in what ways better than pyridoxamine itself?

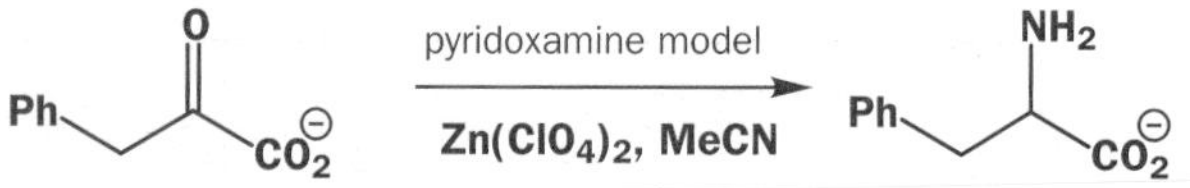

11. Enzymes such as aldolase, thought to operate by the formation of an imine and/or an enamine with a lysine in the enzyme, can be studied by adding $NaBH_4$ to a mixture of enzyme and substrate. For example, treatment of the enzyme with the aldehyde shown below and $NaBH_4$ gives a permanently inhibited enzyme that on hydrolysis reveals a modified amino acid in place of one of the lysines. What is the structure of the modified amino acid, and why is this particular aldehyde chosen?

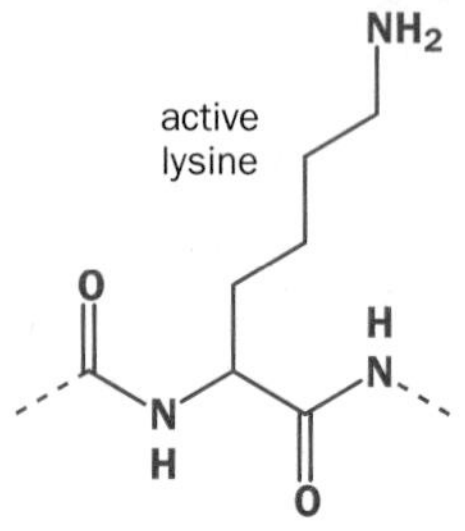

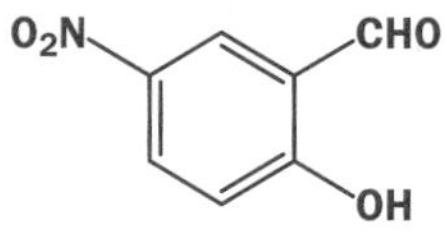

enzyme peptide chain

12. This question is about the hydrolysis of esters by 'serine' enzymes. First, interpret these results: The hydrolysis of this ester is very much faster than that of ethyl benzoate itself. It is catalysed by imidazole and then there is a primary isotope effect (Chapter 41) $k_{(OH)}/k_{(OD)} = 3.5$. What is the mechanism? What is the role of the histidine?

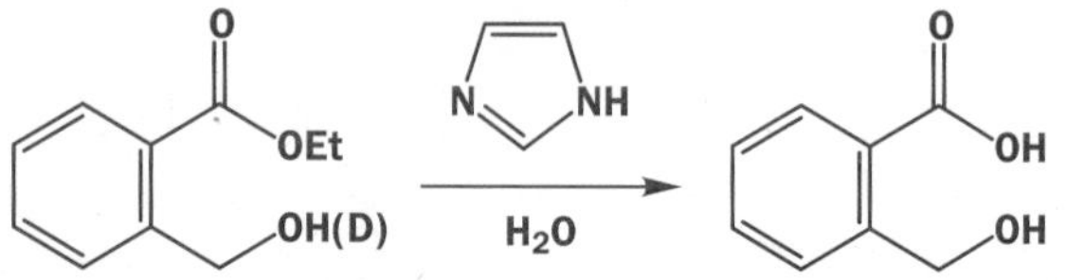

The serine enzymes have a serine residue vital for catalysis. The serine OH group is known to act as a nucleophilic catalyst. Draw out the mechanism for the hydrolysis of *p*-nitrophenyl acetate.

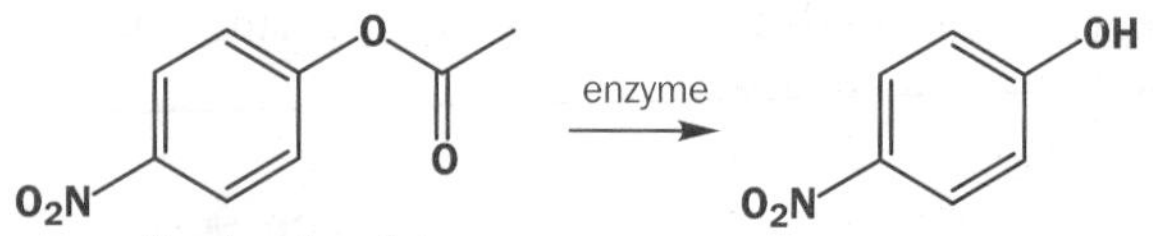

p-nitrophenyl acetate

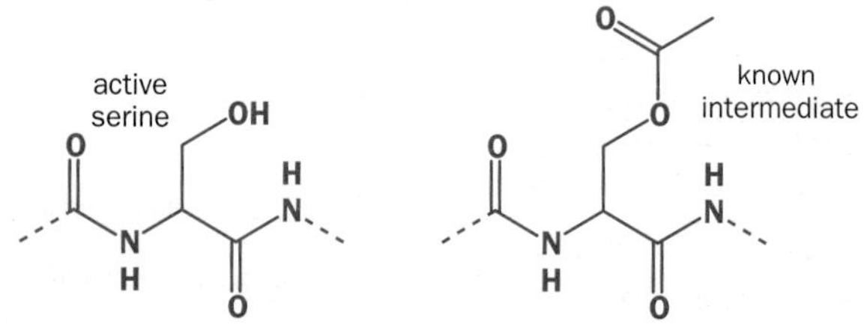

enzyme peptide chain enzyme peptide chain

The enzyme also has a histidine residue vital for catalysis. Use your mechanism from the first part of the question to say how the histidine residue might help. The histidine residue is known to help both the formation and the hydrolysis of the intermediate. The enzyme hydrolyses both *p*-nitrophenyl acetate and *p*-nitrophenyl thiolacetate at the same rate. Which is the rate-determining step?

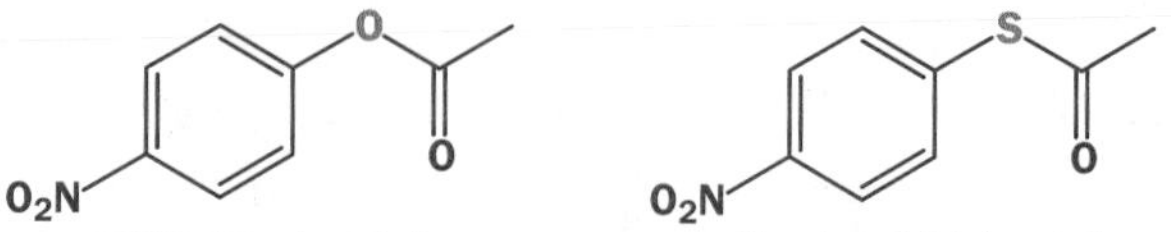

p-nitrophenyl acetate *p*-nitrophenyl thiolacetate

Finally, an aspartic acid residue is necessary for full catalysis and this residue is thought to use its CO_2^- group as a general base. A chemical model shows that the hydrolysis of *p*-nitrophenyl acetate in aqueous acetonitrile containing sodium benzoate and imidazole follows the rate law:

rate = k[*p*-nitrophenyl acetate] [benzoate] [imidazole].

Suggest a mechanism for the chemical reaction.

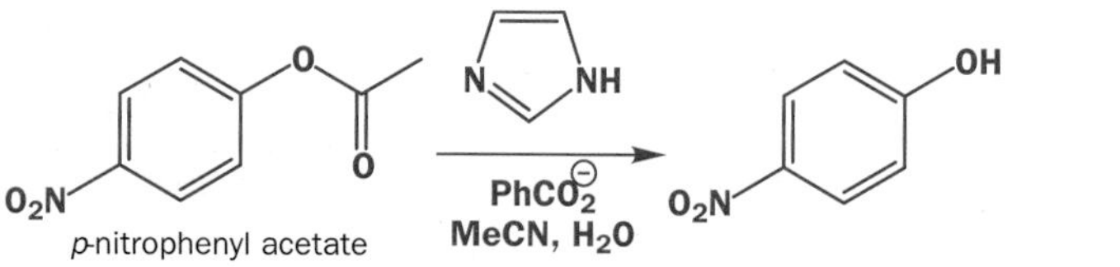

p-nitrophenyl acetate

13. Give mechanisms for the biological formation of biopterin hydroperoxide and its reaction with phenylalanine. The reactions were discussed in the chapter but no details were given.

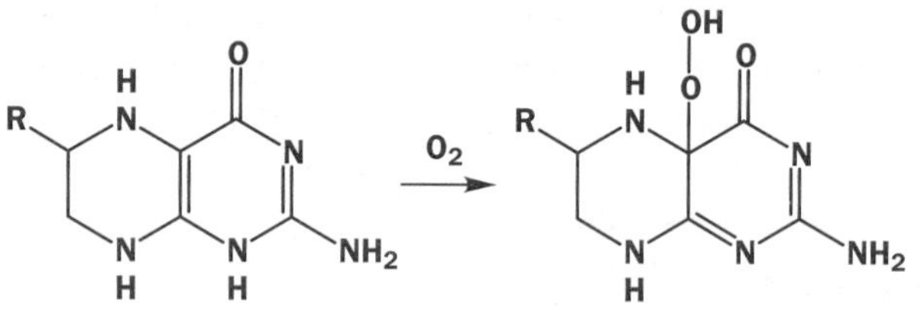

a biopterin a biopterin hydroperoxide

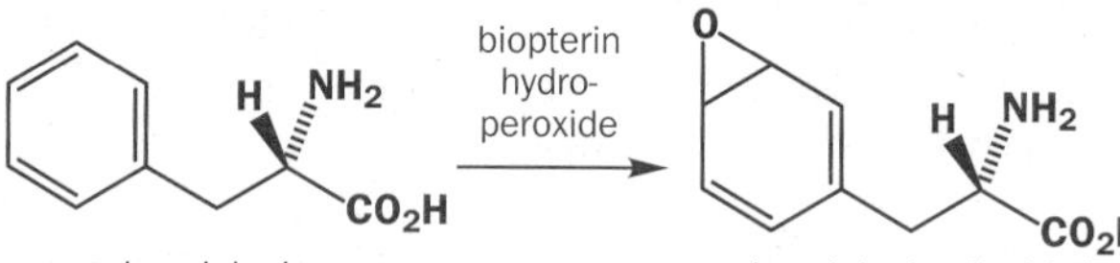

phenylalanine phenylalanine "oxide"

14. Revision of Chapter 48. How many electrons are there on the iron atom in the oxyhaemoglobin structure shown in the chapter? Does it matter if you consider the complex to be of Fe(II) or Fe(III)? Why are zinc porphyrins perfectly stable *without* extra ligands (L in diagram)?

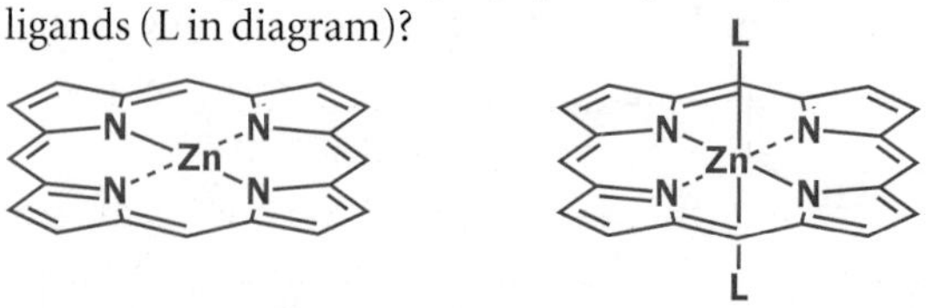

stable zinc (II) porphyrin without extra ligands

octahedral zinc (II) porphyrin with two extra ligands

Connections

Building on:

- Stereochemistry ch16
- Conformational analysis ch18
- Enolate chemistry and synthesis ch24–ch30
- Pericyclic reactions ch35–ch36
- Rearrangement and fragmentation ch37–ch38
- Radicals ch39
- Chemistry of life ch49
- Mechanisms in biological chemistry ch50

Arriving at:

- Natural products are made by secondary metabolism
- Natural products come in enormous variety, but fall mainly into four types: alkaloids, polyketides, terpenes, and steroids
- Alkaloids are amines made from amino acids
- Pyrrolidine alkaloids from ornithine; benzylisoquinoline alkaloids from tyrosine
- Morphine alkaloids are made by radical cyclizations
- Fatty acids are built up from acetyl CoA and malonyl CoA subunits
- Polyketides are unreduced variants of fatty acids
- Terpenes are made from mevalonic acid
- Steroids are tetracyclic terpene derivatives
- Biomimetic synthesis: learning from Nature

Looking forward to:

- Organic synthesis ch53

Introduction

By **natural products**, we mean the molecules of nature. Of course, all life is made of molecules, and we will not be discussing in great detail the major biological molecules, such as proteins and nucleic acids, which we looked at in Chapters 49 and 50. In this chapter we shall talk much more about molecules such as adrenaline (epinephrine). Adrenaline is a human hormone. It is produced in moments of stress and increases our blood pressure and heart rate ready for 'fight or flight'. You've got to sit an exam tomorrow—surge of adrenaline. To an organic chemist adrenaline is intensely interesting because of its remarkable biological activity—but it is also a molecule whose chemical reactions can be studied, whose NMR spectrum can be analysed, which can be synthesized, and which can be imitated in the search for new medicines.

By the end of this chapter we hope you will be able to recognize some basic classes of natural products and know a bit about their chemistry. We will meet **alkaloids** such as coniine, the molecule in hemlock that killed Socrates, and **terpenes** such as thujone, which was probably the toxin in absinthe that killed the nineteenth-century artists in Paris.

Then there are the ambiguous natural products such as the **steroid** cholesterol, which may cause innumerable deaths through heart disease but which is a vital component of cell walls, and the **polyketide** thromboxane, one drop of which would instantaneously clot all the blood in your body but without which you would bleed to death if you cut yourself.

adrenaline

coniine— an alkaloid

thujone— a terpene

cholesterol—a steroid

thromboxane A₂—a polyketide

We will look at the structural variety within these four important classes and beyond, from perhaps the smallest natural product, nitric oxide, NO (which controls penile erections in men), to something approaching the largest—the polyketide brevetoxin, the algal product in 'red tides', which appear in coastal waters from time to time and kill fish and those who eat the fish.

brevetoxin—a toxic polyketide

> Before moving on, just pause to admire brevetoxin, a wonderful and deadly molecule. Look at the alternating oxygen atoms on the top and bottom faces of alternate rings. Look at the rings themselves—six-, seven-, and eight-membered but each with one and no more than one oxygen atom. Trace the continuous carbon chain running from the lactone carbonyl group in the bottom left-hand corner to the aldehyde carbonyl in the top right. There is no break in this chain and, other than the methyl groups, no branch. With 22 stereogenic centres, this is a beautiful piece of molecular architecture. If you want to read more about brevetoxin, read the last chapter in Nicolaou and Sorensen's *Classics in total synthesis*, VCH, 1996.

penicillin
e.g. penicillin G; R = PhCH₂

Many natural products are the source of important life-saving drugs—consider the millions of lives saved by penicillin, a family of **amino acid** metabolites.

Natural products come from secondary metabolism

The chemical reactions common to all living things involve the **primary metabolism** of the 'big four' we met in Chapter 49—nucleic acids, proteins, carbohydrates, and lipids. Now we must look at chemical reactions that are more restricted. They occur perhaps in just one species, though more commonly in several. They are obviously, then, not essential for life, though they usually help survival. These are the products of **secondary metabolism**.

The exploration of the compounds produced by the secondary metabolism of plants, microorganisms, fungi, insects, mammals, and every other type of living thing has hardly begun. Even so, the variety and richness of the structures are overwhelming. Without some kind of classification the task of description would be hopeless. We are going to use a biosynthetic classification, grouping substances not by species but by methods of biological synthesis. Though every species is different, the basic chemical reactions are shared by all. The chart on p. 1415 relates closely to the chart of primary metabolism in the previous chapter.

Alkaloids are basic compounds from amino acid metabolism

Alkaloids were known in ancient times because they are easy to extract from plants and some of them have powerful and deadly effects. Any plant contains millions of chemical compounds, but some plants, like the deadly nightshade, can be mashed up and extracted with aqueous acid to give a few compounds soluble in that medium, which precipitate on neutralization. These compounds were seen to be 'like alkali' and Meissner, the apothecary from Halle, in 1819 named them 'alkaloids'. Lucrezia Borgia already knew all about this and put the deadly nightshade extract atropine in her eyes (to make her look beautiful: atropine dilates the pupils) and in the drinks of her

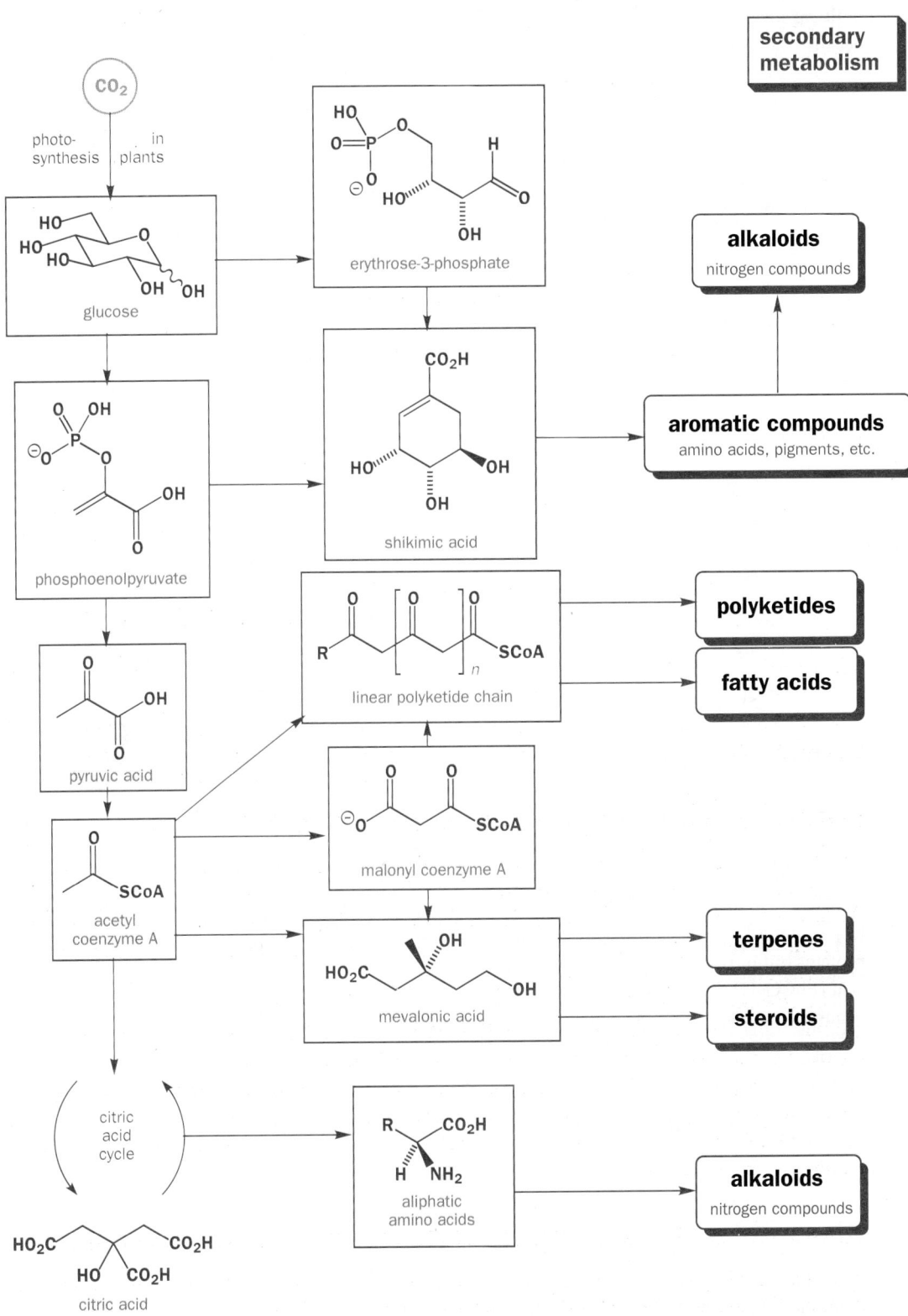

secondary
metabolism

CO₂

photo-   in
synthesis | plants

glucose

erythrose-3-phosphate

alkaloids
nitrogen compounds

phosphoenolpyruvate

shikimic acid

aromatic compounds
amino acids, pigments, etc.

linear polyketide chain

polyketides

fatty acids

pyruvic acid

malonyl coenzyme A

acetyl
coenzyme A

mevalonic acid

terpenes

steroids

citric
acid
cycle

aliphatic
amino acids

alkaloids
nitrogen compounds

citric acid

⟶ chemical reaction in the usual sense: the starting material is incorporated into the product

political adversaries to avoid any trouble in the future. Now, we would simply say that they are basic because they are amines. Here is a selection with the basic amino groups marked in black.

nicotine   morphine   atropine

Natural products are often named by a combination of the name of the organism from which they are isolated and a chemical part name. These compounds are all am*ines* so all their names end in '-ine'. They appear very diverse in structure but all are made in nature from amino acids, and we will look at three types.

Solanaceae alkaloids

The Solanaceae family includes not only deadly nightshade (*Atropa belladonna*—hence atropine) plants but also potatoes and tomatoes. Parts of these plants also contain toxic alkaloids: for example, you should not eat green potatoes because they contain the toxic alkaloid solanine.

the very toxic alkaloid solanine, a mixture of glycosides with R = glucose, mannose, etc.

Atropine is a racemic compound but the (*S*)-enantiomer occurs in henbane (*Hyoscyamus niger*) and was given a different name, hyoscyamine, before the structures were known. In fact, hyoscyamine racemizes very easily just on heating in water or on treatment with weak base. This is probably what happens in the deadly nightshade plant.

Pyrrolidine alkaloids are made from the amino acid ornithine

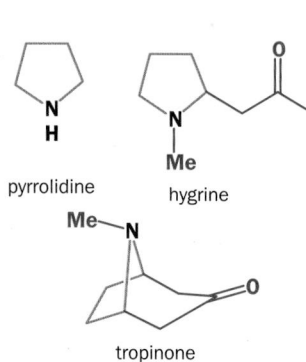

pyrrolidine   hygrine

tropinone

Pyrrolidine is the simple five-membered cyclic amine and pyrrolidine alkaloids contain this ring somewhere in their structure. Both nicotine and atropine contain a pyrrolidine ring as do hygrine and tropinone. All are made in nature from ornithine. Ornithine is an amino acid not usually found in proteins but most organisms use it, often in the excretion of toxic substances. If birds are fed benzoic acid ($PhCO_2H$) they excrete dibenzoyl ornithine. When dead animals decay, the decarboxylation of ornithine leads to putrescine which, as its name suggest, smells revolting. It is the 'smell of death'.

dibenzoyl ornithine   ornithine   putrescine

pyrrolidine alkaloids

Biosynthetic pathways are usually worked out by isotopic labelling of potential precursors and we shall mark the label with a coloured blob. If ornithine is labelled with ^{14}C and fed to the plant, labelled hygrine is isolated.

^{14}C-ornithine   hygrine

If each amino group in ornithine is labelled in turn with ^{15}N, the α amino group is lost but the γ amino group is retained.

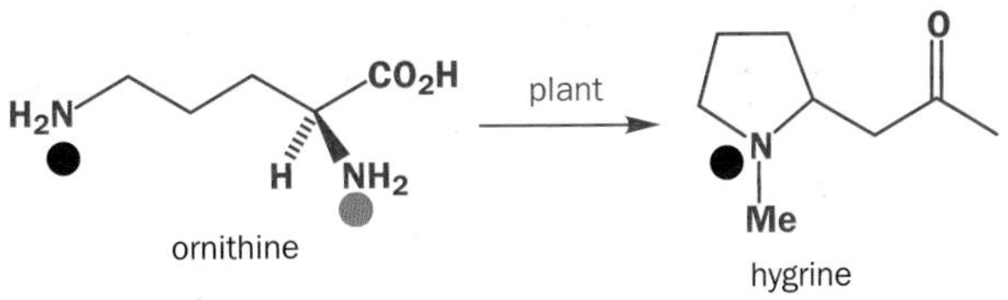

Further labelling experiments along these lines showed that the CO_2H group as well as the α amino group was lost from ornithine and that the rest of the molecule makes the pyrrolidine ring. The three-carbon side-chain in hygrine comes from acetate, or rather from acetyl CoA, and the N-methyl group comes from SAM. We can now work through the biosynthesis.

The first step is a pyridoxal-catalysed decarboxylation of ornithine, which follows the normal sequence up to a point.

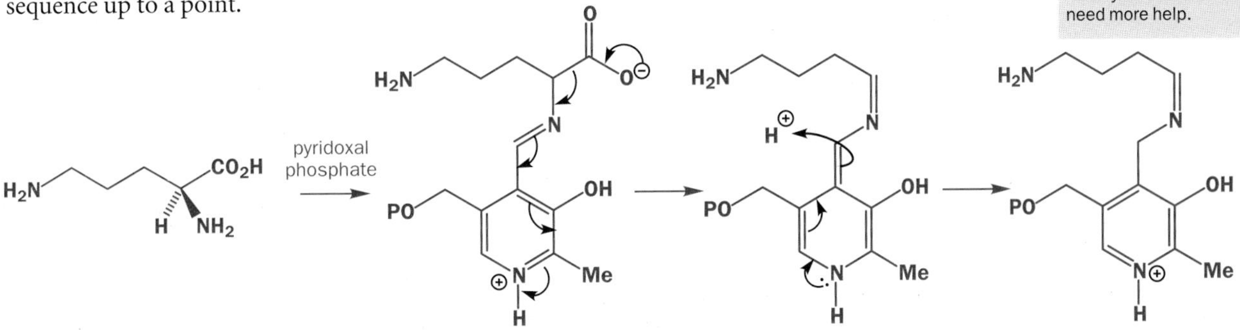

> Both reagents SAM and acetyl CoA were discussed in Chapter 50. We will not be able to repeat at length the details of the chemistry of these and other common biochemical reagents already discussed there. In general, in this chapter we will give only the distinctive or interesting steps and leave you to consult Chapter 50 if you need more help.

Now the terminal amino group is methylated by SAM and the secondary amine cyclizes on to the pyridoxal imine to give an aminal. Decomposition of the aminal the other way round expels pyridoxamine and releases the salt of an electrophilic imine.

> Notice that the methylation step means that the two carbon atoms that eventually become joined to nitrogen in the five-membered ring remain different throughout the sequence. If, say, putrescine had been an intermediate, they would not now be distinguishable.

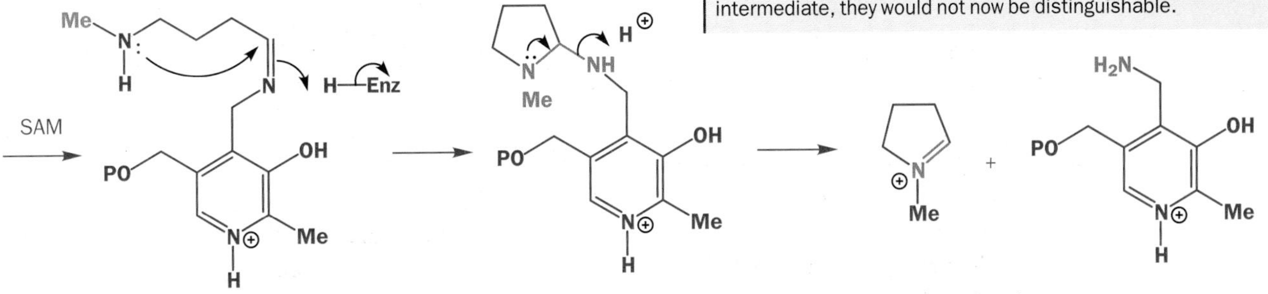

The rest of the biosynthesis does not need pyridoxal, but it does need two molecules of acetyl CoA. In Chapter 50 we noted that this thiol ester is a good electrophile and also enolizes easily. We need both reactivities now in a Claisen ester condensation of acetyl CoA.

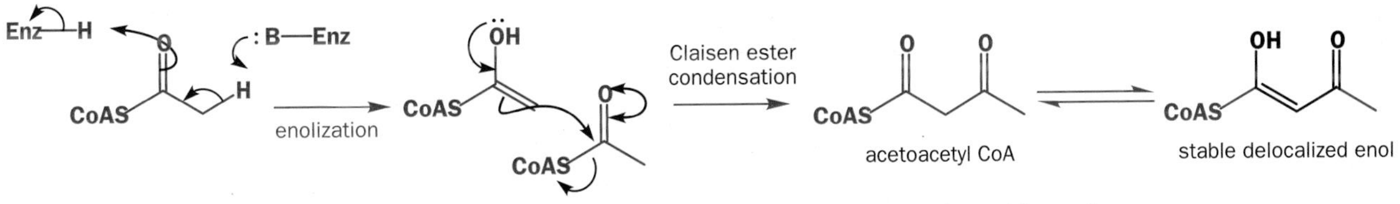

The new keto-ester is very like the acetoacetates we used in Chapter 27 to make stable enolates and the CoA thiol ester will exist mainly as its enol, stabilized by conjugation.

This enol reacts with the imine salt we have previously made and it will be easier to see this reaction if we redraw the enol in a different conformation. The imine salt does not have to wait around for acetoacetyl CoA to be made. The cell has a good stock of acetyl CoA and its condensation product.

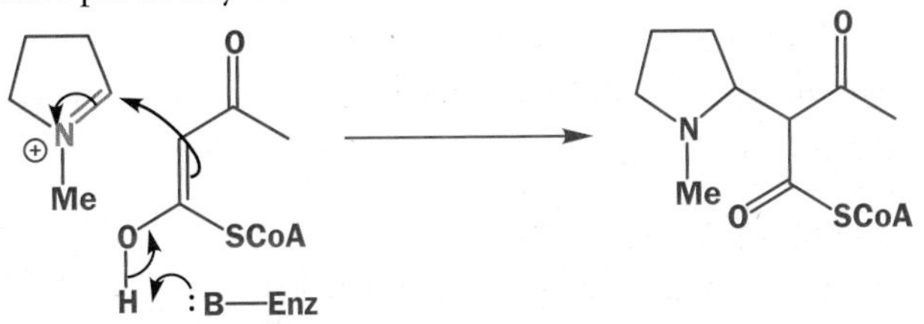

All that remains to form hygrine is the hydrolysis of the CoA thiol ester and decarboxylation of the keto-acid. This is standard chemistry, but you should ensure that you can draw the mechanisms for these steps.

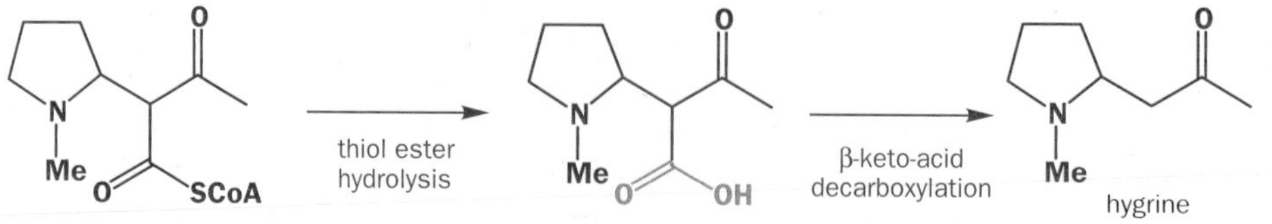

Tropinone is made from hygrine and it is clear what is needed. The methyl ketone must enolize and it must attack another imine salt resembling the first but on the other side of the ring. Such salts can be made chemically by oxidation with Hg(II) and biologically with an oxidizing enzyme and, say, NAD$^+$. The symbol [O] represents an undefined oxidizing agent, chemical or biological.

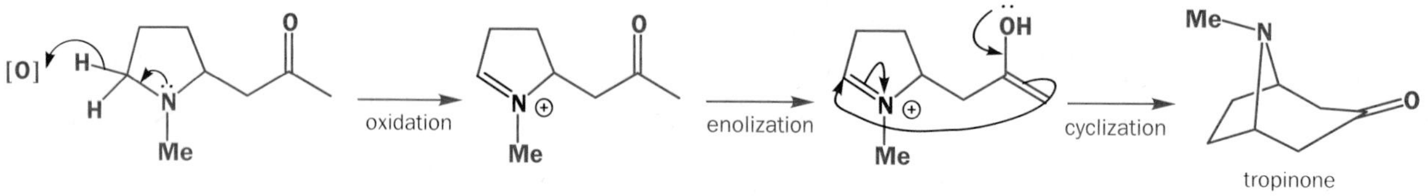

▶

The cyclization step looks dreadful when drawn on a flat molecule, but it looks much better in the conformation of tropinone shown below.

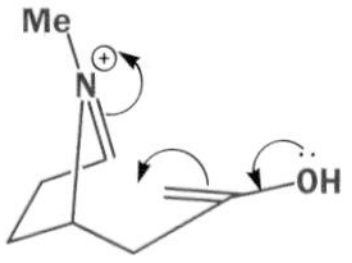

This complex route to tropinone was imitated as long ago as 1917 in one of the most celebrated reactions of all time, Robinson's tropinone synthesis. Robinson argued on purely chemical grounds that the sequence of imine salts and enols, which later (1970) turned out to be Nature's route, could be produced under 'natural' conditions (aqueous solution at pH 7) from a C_4 dialdehyde, MeNH$_2$ and acetone dicarboxylic acid. It worked and the intermediates must be very similar to those in the biosynthesis.

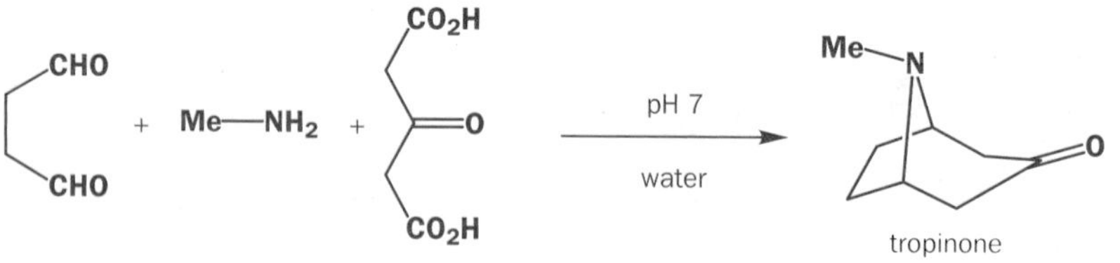

Other pyrrolidine alkaloids

There are many pyrrolidine alkaloids derived from ornithine and another large family of piperidine alkaloids derived from lysine by similar pathways involving decarboxylation and cyclization initiated by pyridoxal. We will not discuss these compounds in detail.

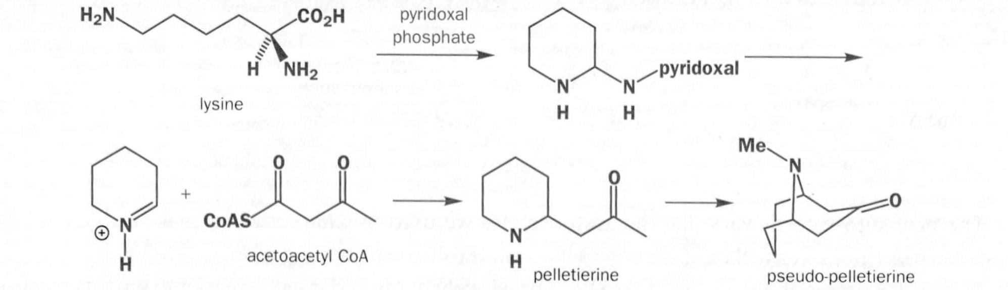

Benzyl isoquinoline alkaloids are made from tyrosine

We switch to a completely different kind of alkaloid made from a different kind of amino acid. The **benzyl isoquinoline alkaloids** have a benzyl group attached to position 2 of an isoquinoline ring. Usually the alkaloids are oxygenated on the benzene ring and many are found in opium poppies (*Papaver somniferum*). For all these reasons papaverine is an ideal example.

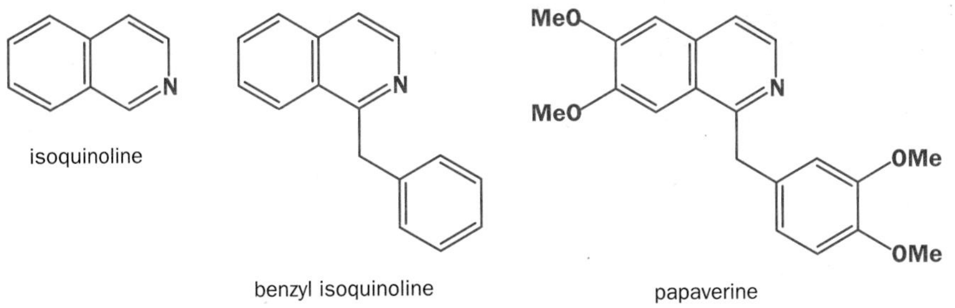

isoquinoline

benzyl isoquinoline

papaverine

Labelling shows that these alkaloids come from two molecules of tyrosine. One must lose CO_2 and the other NH_3. We can easily see how to divide the molecule in half, but the details will have to wait a moment.

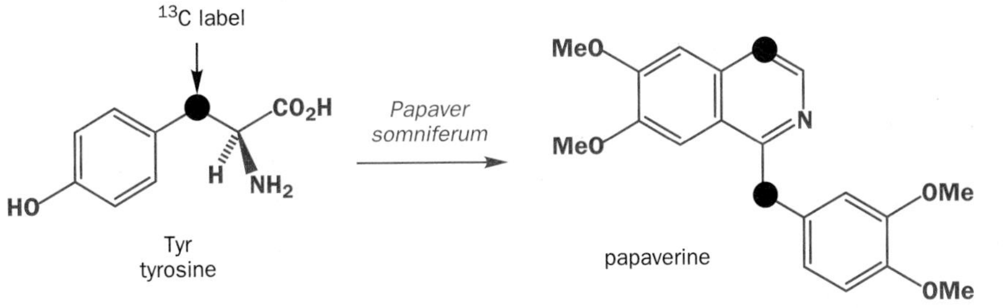

Tyr
tyrosine

papaverine

The question of when the extra OH groups are added was also solved by labelling and it was found that dihydroxyphenyl pyruvate was incorporated into both halves but the dihydroxyphenyl-alanine (an important metabolite usually called 'dopa') was incorporated only into the isoquinoline half.

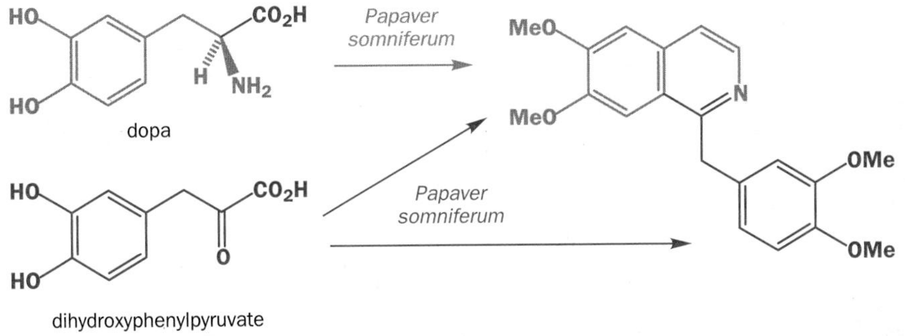

dopa

dihydroxyphenylpyruvate

The amino acid and the keto-acid are, of course, related by a pyridoxal-mediated transaminase and the hydroxylation must occur right at the start. Both of these reactions are discussed in Chapter 50.

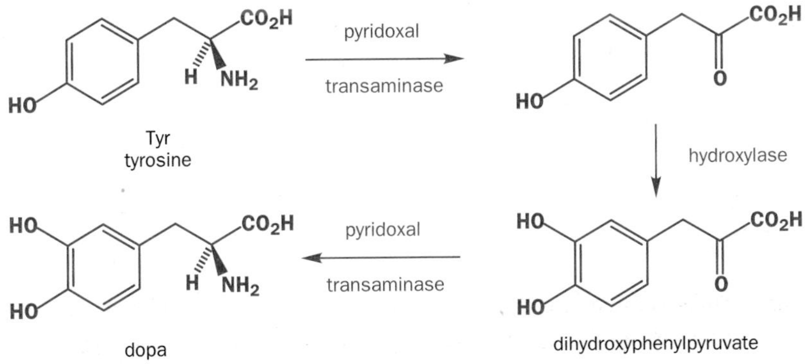

Tyr
tyrosine

dopa

dihydroxyphenylpyruvate

Catecholamines

Dopa and dopamine are important compounds because they are the precursors to adrenaline in humans. Decarboxylation of dopa gives dopamine, which an oxidase (Chapter 50) hydroxylates stereospecifically at the benzylic position to give noradrenaline (norepinephrine).

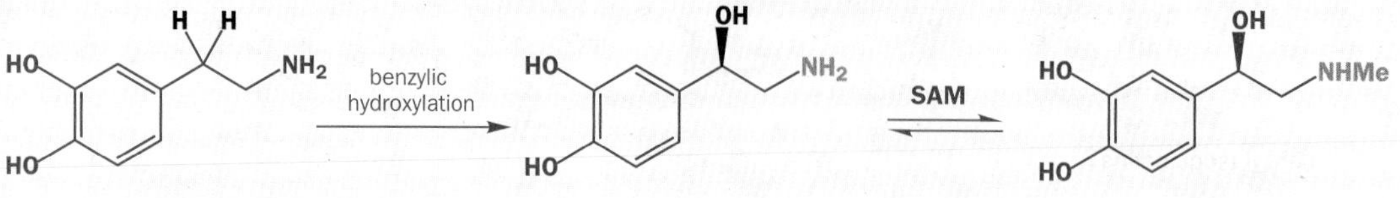

dopamine   noradrenaline (norepinephrine)   adrenaline (epinephrine)

The family of hormones that includes adrenaline and noradrenaline is often called the **catecholamines** (catechol is 1,2-dihydroxybenzene). The hormones are produced in the adrenal gland around the kidneys and regulate several important aspects of metabolism: they help to control the breakdown of stored sugars to release glucose and they have a direct effect on blood pressure, heart rate, and breathing. The relative proportion of noradrenaline and its N-methylated analogue, adrenaline, controls these things.

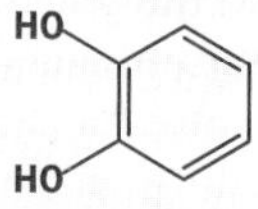

catechol

Pyridoxal-mediated decarboxylation of dopa gives dopamine and this reacts with the keto-acid to form an imine salt. This is an open-chain imine salt unlike the cyclic ones we saw in the pyrrolidine alkaloids, but it will prove to have similar reactivity.

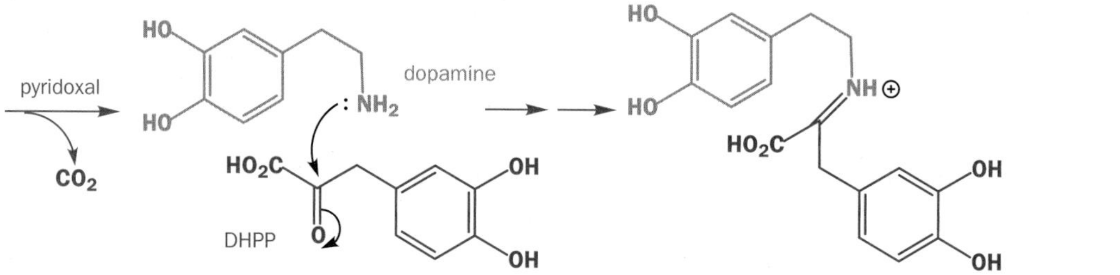

The imine salt is perfectly placed for an intramolecular electrophilic aromatic substitution by the electron-rich dihydroxyphenyl ring. This closes the isoquinoline ring in a Mannich-like process (Chapter 27) with the phenol replacing the enol in the pyrrolidine alkaloid biosynthesis.

Even in biological electrophilic aromatic substitutions, it is still important to remember to write in the hydrogen atom at the place of substitution (Chapter 22)!

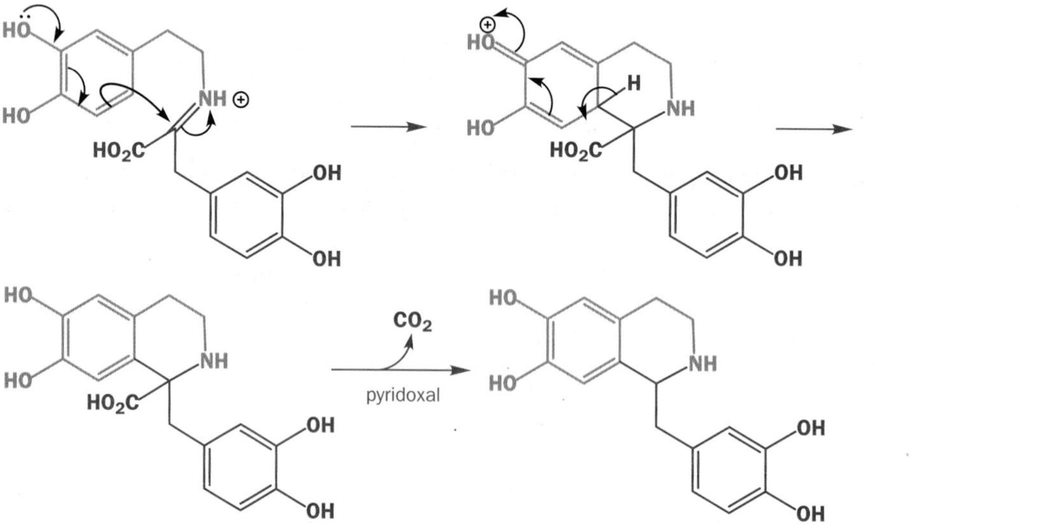

The cyclization product is still an amino acid and it can be decarboxylated by pyridoxal. Now we have something quite like papaverine but it lacks the methyl groups and the aromatic heterocyclic ring. Methylation needs SAM and is done in two stages for a reason we will discover soon. The final oxidation should again remind you of the closing stages of the tropinone route.

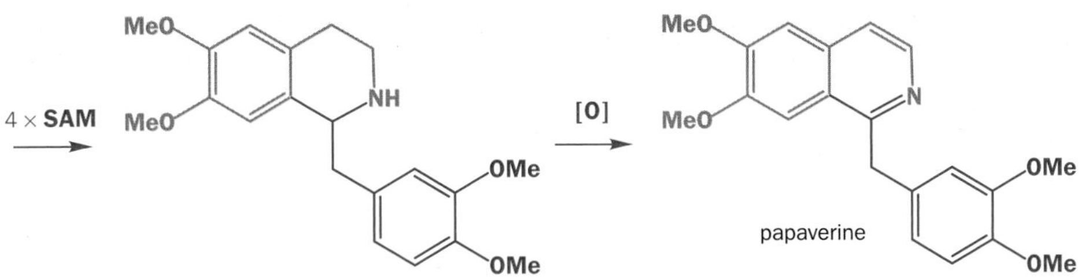

papaverine

The reaction to make the isoquinoline ring can be carried out chemically under very mild conditions providing that we use an aldehyde as the carbonyl component. Then it works very well with rather similar compounds.

The reaction also works with an aryl pyruvic acid, but the decarboxylation is more difficult to organize without pyridoxal.

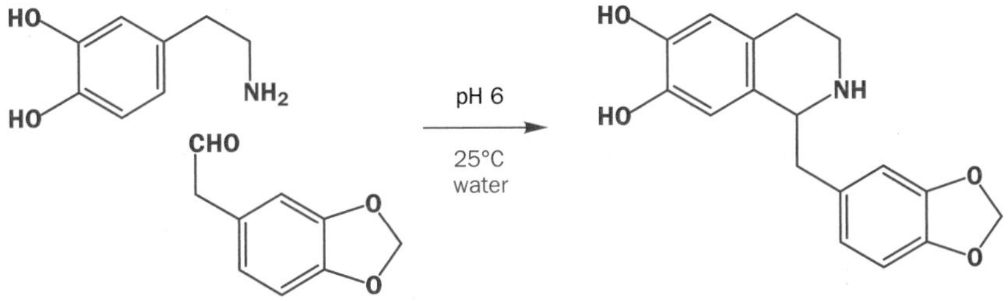

The mechanism is straightforward—the imine is formed and will be protonated at pH 6, ready for the C–C bond formation, which is both a Mannich reaction and an electrophilic aromatic substitution.

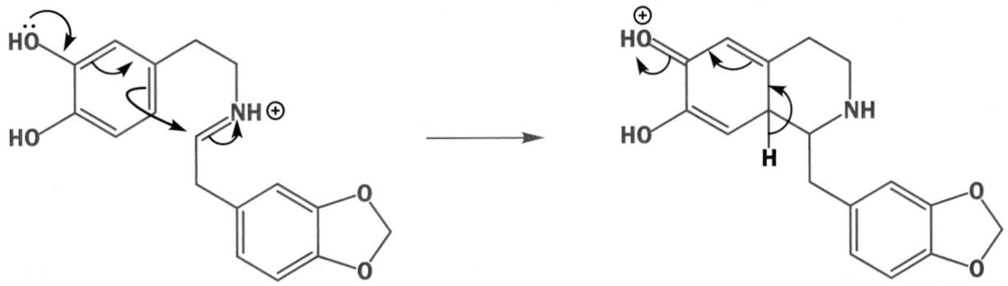

Notice that it was not necessary to protect the OH groups—the acetal on the lower ring is not for protection, and this group (methylenedioxy or dioxolan) is present in many benzyl isoquinoline alkaloids. It is formed in nature by oxidation of an MeO group *ortho* to an OH group on a benzene ring.

Complex benzyl isoquinoline alkaloids are formed by radical coupling

A more interesting series of alkaloids arises when benzyl isoquinoline alkaloids cyclize by radical reactions. Phenols easily form radicals when treated with oxidizing agents such as Fe(III), and benzyl isoquinoline alkaloids with free phenolic hydroxyl groups undergo radical reactions in an intramolecular fashion through a similar mechanism. Here are the details of some methylations of a class of alkaloids closely related to papaverine.

See Chapter 39.

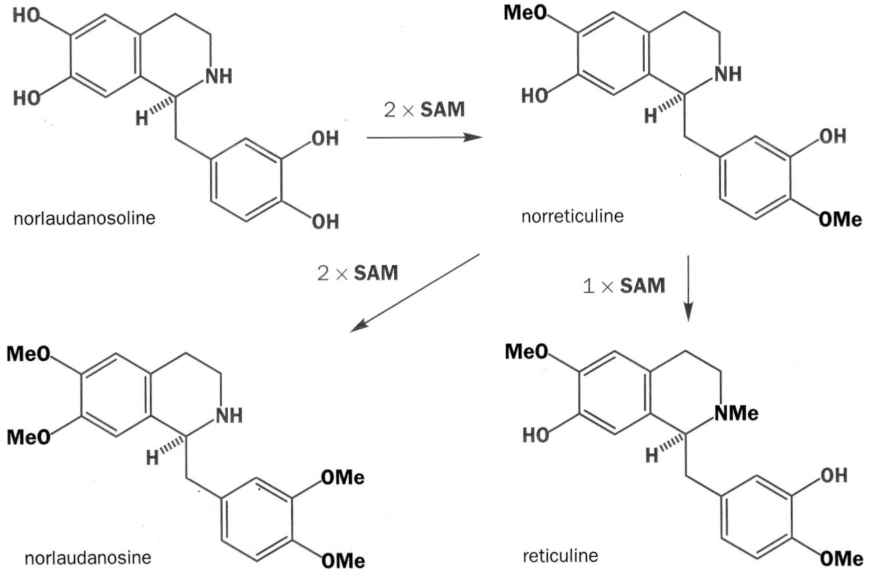

norlaudanosoline

norreticuline

norlaudanosine

reticuline

The names of the alkaloids should not, of course, be learned, but they are a convenient handle for quick reference. The prefix 'nor' means without a methyl group, in this case the *N*-Me group, as you can see with norreticuline and reticuline.

Methylating only one phenol on each ring of norreticuline leaves the other one free for radical coupling. Reticuline is oxidized in the plant to isoboldine by a radical cyclization with the formation of a new C–C bond.

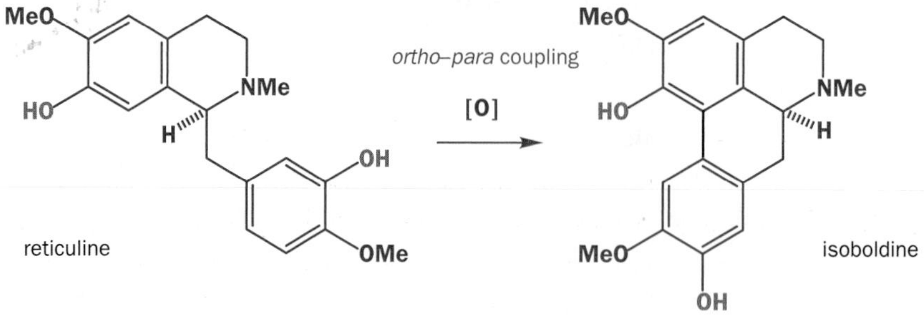

The new C–C bond is marked in black and the free phenolic OHs in green. Notice the relationship between them. The new bond is between a carbon atom *ortho* to one OH group and a carbon atom *para* to the other. We shall see in all these phenolic couplings that the *ortho* and *para* positions are the only activated ones (*ortho/ortho*, *ortho/para*, and *para/para* couplings are all possible). Oxidation occurs at the phenolic hydroxyl groups, and the resulting oxygen radicals couple.

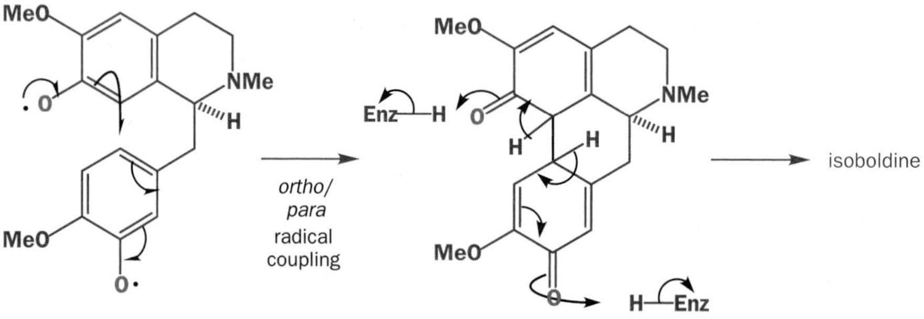

Phenol coupling occurs chemically under oxidation with Fe(III). The most famous example is the coupling of 2-naphthol to give binaphthol—an *ortho/ortho* coupling. The stereochemistry of binaphthyls like this was discussed in Chapter 45.

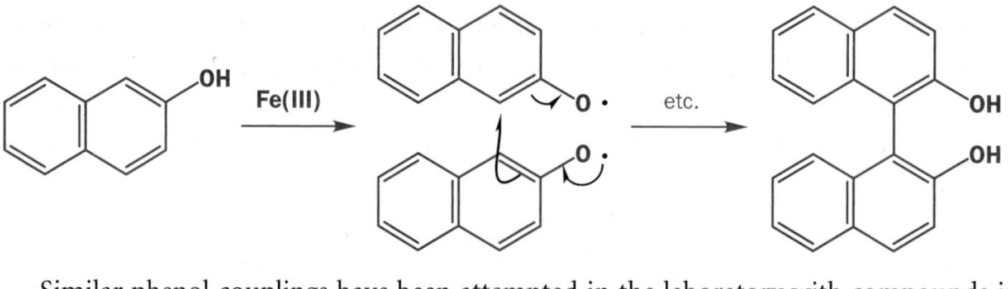

Similar phenol couplings have been attempted in the laboratory with compounds in the benzyl isoquinoline series but the nitrogen atom interferes if it is at all basic. When it has a carbonyl substituent the reactions do work reasonably well, but the yields are poor. Nature is still much better at this reaction than we are.

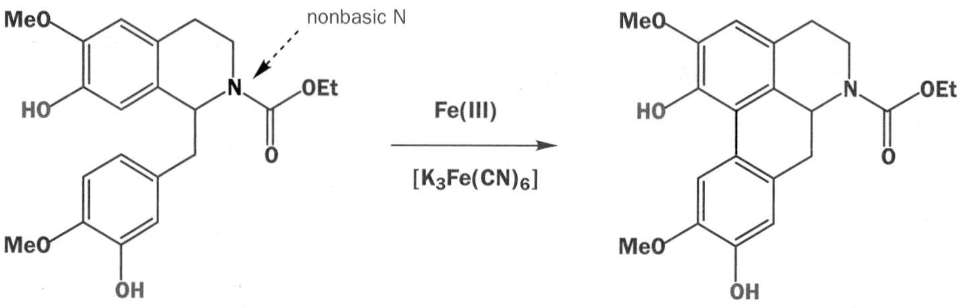

Reticuline is also the source of the morphine alkaloids by *ortho/para* radical coupling. The roles of the two rings are reversed this time and it is quite difficult to see at first how the structures are related.

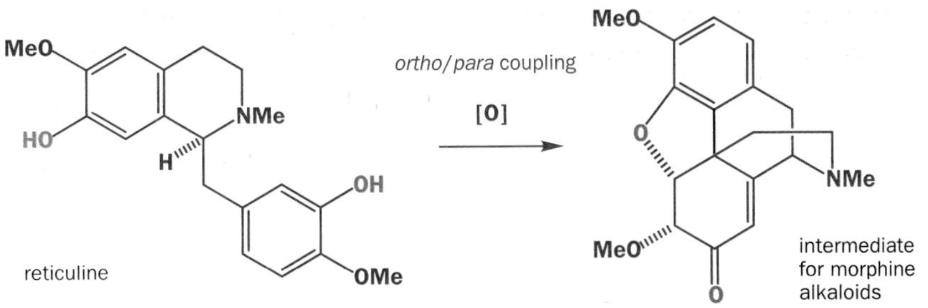

reticuline

ortho/para coupling

[O]

intermediate for morphine alkaloids

A great deal has happened in this reaction, but the new C–C bond (black) is *ortho* to the green oxygen atom in the top ring and *para* to the green oxygen atom in the bottom ring, so *ortho/para* coupling has occurred. To draw the reaction mechanism we need to draw reticuline in the right conformation.

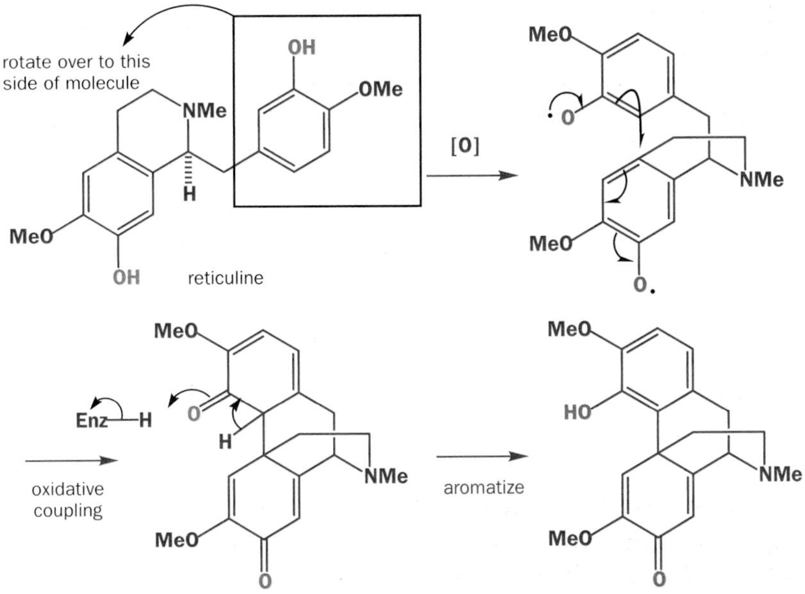

rotate over to this side of molecule

reticuline

[O]

oxidative coupling

aromatize

One of the two rings can re-aromatize but the other has a quaternary carbon atom so no proton can be lost from this site. Instead, the OH group in the top ring adds in conjugate fashion to the enone in the bottom ring.

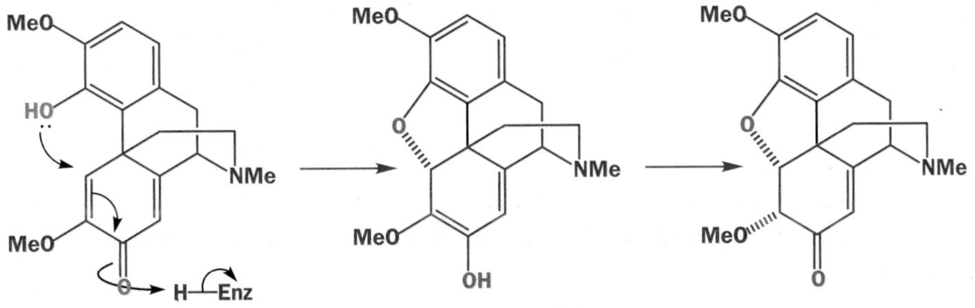

This intermediate gives rise to the important alkaloids codeine and morphine, which differ only by a methyl group. Nature can remove methyl groups as well as add them.

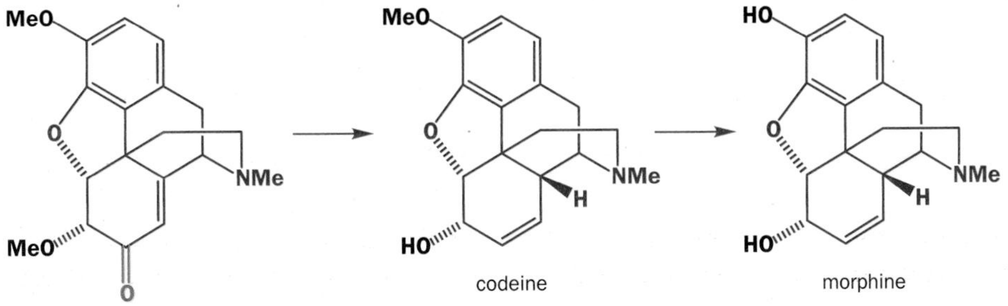

codeine

morphine

These alkaloids have plenty of stereochemistry. Indeed, if we compare the structures of reticuline and morphine, we can see that the one stereogenic centre in reticuline (marked in green) is still there in morphine (it hasn't been inverted—that part of the molecule has just been turned over) and that four new stereogenic centres marked in black have been added. These centres all result from the original twisting of reticuline to allow phenol coupling except for the one bearing an OH group, which comes from a stereoselective reduction.

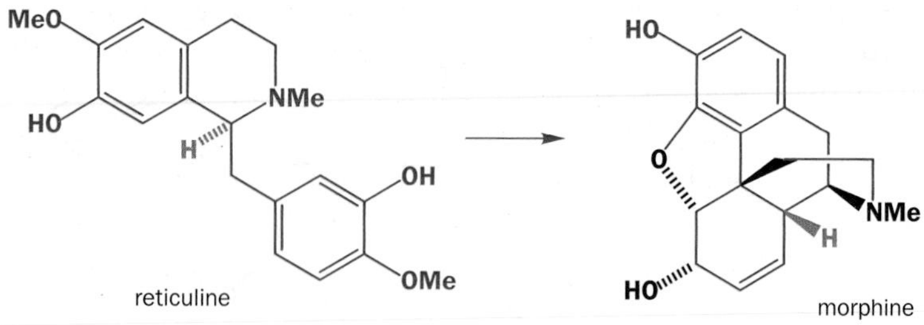

Boldine, an isomer of isoboldine, is formed by rearrangement

We mentioned isoboldine a while back, so there must be a boldine as well. This alkaloid is also formed from norlaudanosoline by a different methylation sequence and oxidative radical coupling. Looking at the structure of boldine you may see what appears to be a mistake on someone's part.

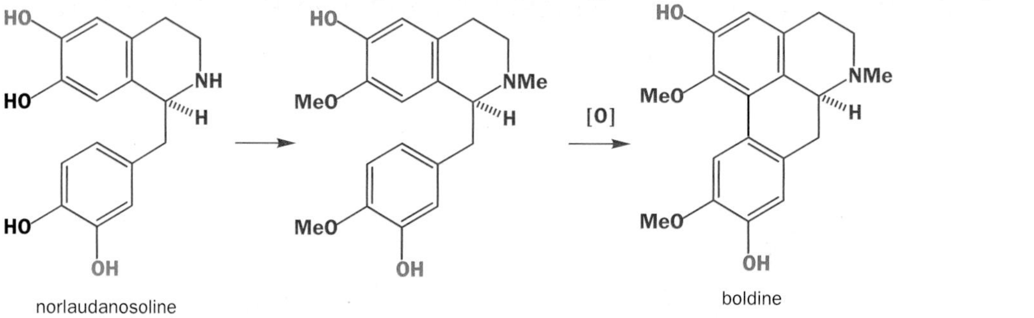

The coupling is correctly *para* in the bottom ring but is *meta* in the top ring. But there is no mistake (neither by the authors nor by Nature!)—this structure is correct and it has been made by *para/para* coupling.

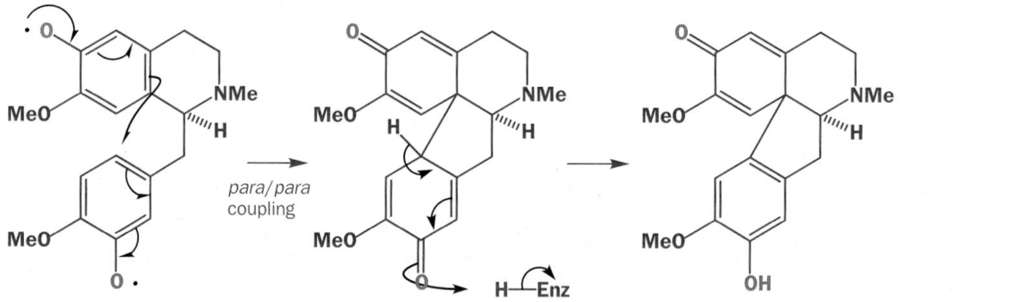

One of the rings has aromatized, but the other cannot—this should remind you of the morphine biosynthesis. However, there is no nucleophilic OH group here capable of conjugate addition to the enone so a rearrangement occurs instead. The new bond to the lower ring migrates across the top ring. You might even say that the lower ring does an intramolecular conjugate addition on the upper ring.

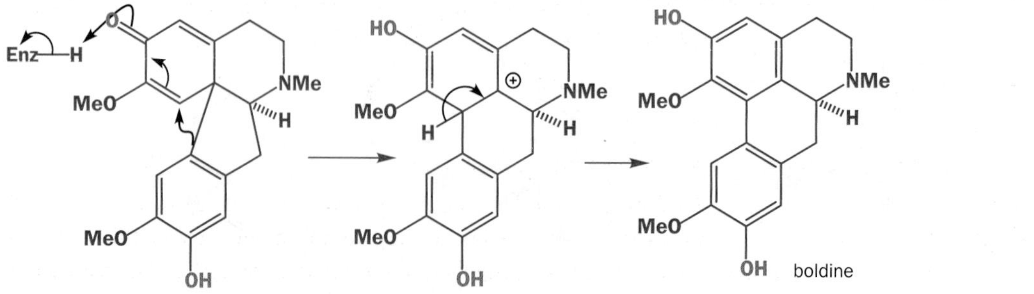

After the rearrangement there is a proton available to be lost and the cation can aromatize. The *para* relationship in the original coupling product has become a *meta* relationship by rearrangement. You should be able to recognize this rearrangement from Chapter 37: it is a dienone–phenol rearrangement.

In rearrangements like these with cationic intermediates, the group that can best support a positive charge usually prefers to migrate. The reasons for this are discussed in Chapter 37. Here is a purely chemical example of the same reaction, giving 82% yield in acidic solution. The bond that migrates is marked in black.

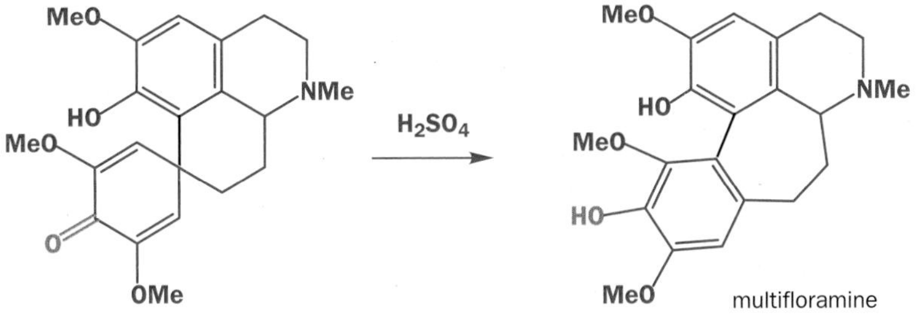

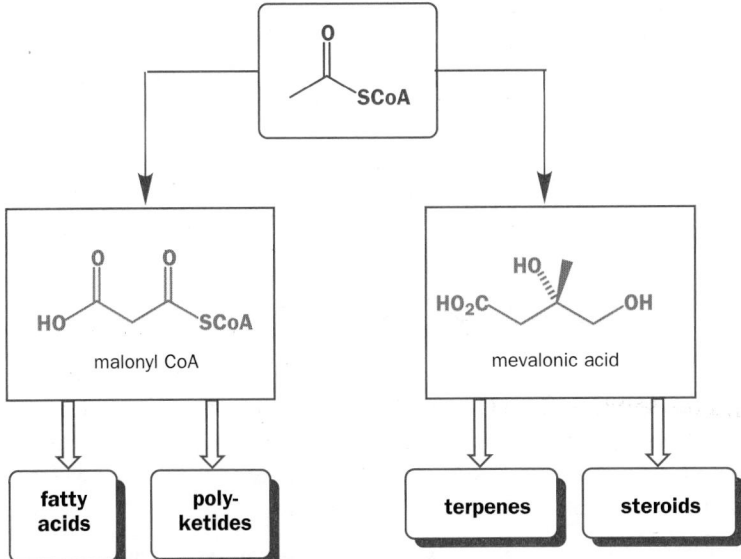

multifloramine

Fatty acids and other polyketides are made from acetyl CoA

The sections that remain in this chapter show how Nature can take a very simple molecule—acetyl CoA—and build it up into an amazing variety of structures. There are two main pathways from acetyl CoA and each gives rise to two important series of natural products.

malonyl CoA

mevalonic acid

fatty acids

poly- ketides

terpenes

steroids

We shall discuss these four types of compounds in the order shown so that we start with the simplest, the fatty acids. You met these compounds in Chapter 49 as their glyceryl esters, but you now need to learn about the acids in more detail and outline their biosynthesis. Compare the structures of the typical fatty acids in the chart overleaf.

These are just a few of the fatty acids that exist, but all are present in our diet and you'll find many referred to on the labels of processed foods. You should notice a number of features.

- They have straight chains with no branching

- They have even numbers of carbon atoms

- They may be saturated with no double bonds in the chain, or

- They may have one or more C=C double bonds in the chain, in which case they are usually *cis* (*Z*) alkenes. If there is more than one C=C double bond, they are not conjugated (either with the CO_2H group or with each other)—there is normally one saturated carbon atom between them.

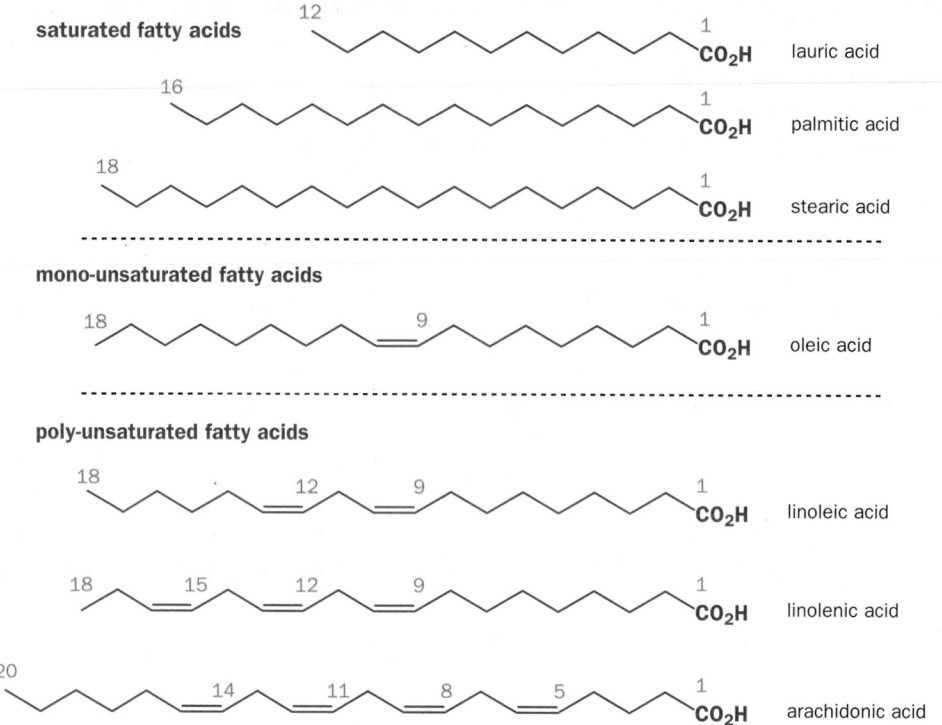

Palmitic acid (C_{16} saturated) is the most common fatty acid in living things. Oleic acid (C_{18} mono-unsaturated) is the major fatty acid in olive oil. Arachidonic acid (C_{20} tetra-unsaturated) is a rare fatty acid, which is the precursor of the very important prostaglandins, thromboxanes, and leukotrienes, of which more later.

The prevalence of fatty acids with even numbers of carbon atoms suggests a two-carbon building block, the most obvious being acetate. If labelled acetate is fed to plants, the fatty acids emerge with labels on alternate carbons like this.

The green blob might represent deuterium (as a CD_3 group) and the black blob ^{13}C. In fact, the reactions are more complex than this suggests as CO_2 is also needed as well as CoA and it turns out that only the first two-carbon unit is put in as acetyl CoA. The remainder are added as malonyl CoA. If labelled malonyl CoA is fed, the starter unit, as it is called, is not labelled.

Malonyl CoA is made from acetyl CoA and CO_2 carried, as usual, on a molecule of biotin (Chapter 50). The first stage in the fatty acid biosynthesis proper is a condensation between acetyl CoA (the starter unit) and malonyl CoA with the loss of CO_2. This reaction could be drawn like this.

Notice that CO_2 is lost as the new C–C bond is formed. When chemists use malonates, we like to make the stable enol using both carbonyl groups, condense, and only afterwards release CO_2 (Chapter 26). Nature does this in making acetoacetyl CoA during alkaloid biosynthesis, but here she works differently.

The next step is reduction of the ketone group.

This NADPH reaction is typically stereo- and chemoselective, though the stereochemistry is rather wasted here as the next step is a dehydration, typical of what is now an aldol product, and occurring by an enzyme-catalysed E1cB mechanism.

The elimination is known to be a *cis* removal of H and OH and the double bond is exclusively *trans* (*E*). Only later in the nonconjugated unsaturated fatty acids do we get *Z*-alkenes. Finally, in this cycle, the double bond is reduced using another molecule of NADPH to give the saturated side-chain.

Now the whole cycle can start again using this newly made C_4 fatty acid as the starter unit and building a C_6 fatty acid and so on. Each time the cycle turns, two carbon atoms are added to the acyl end of the growing chain.

Fatty acid synthesis uses a multienzyme complex

We have not told you the whole truth so far. Did you notice that 'SCoA' in the structures had been replaced by 'SR' and that a mysterious 'ACP' had crept into the enzyme names? That was because these reactions actually happen while the growing molecule is attached as a thiol ester to a long side-chain on an **acyl carrier protein** (ACP). The long side-chain is closely related to CoA and is attached through a phosphate to a serine residue of the ACP.

All of the enzymes needed for one cycle are clumped together to form two large proteins (ACP, the acyl carrier protein, and **CE**, the **condensing enzyme**) which associate in a stable dimer. The long side-chain passes the substrate from enzyme to enzyme so that synthesis can be continuous until the chain is finished and only then is the thiol ester hydrolysed. The chart on p. 1428 illustrates this.

■ You saw a smaller multienzyme complex in Chapter 50 (p. 1395), but this one is much more complex. More are being discovered all the time—Nature invented the production line well before Henry Ford.

There are three ways of making unsaturated fatty acids

Conjugated unsaturated fatty acids are made simply by stopping the acylation cycle at that stage and hydrolysing the thiol ester linkage between the unsaturated acyl chain and ACP. They always have the *E* (*trans*) configuration and are the starting points for other biosynthetic pathways.

fatty acid biosynthesis: schematic diagram of the multienzyme dimer

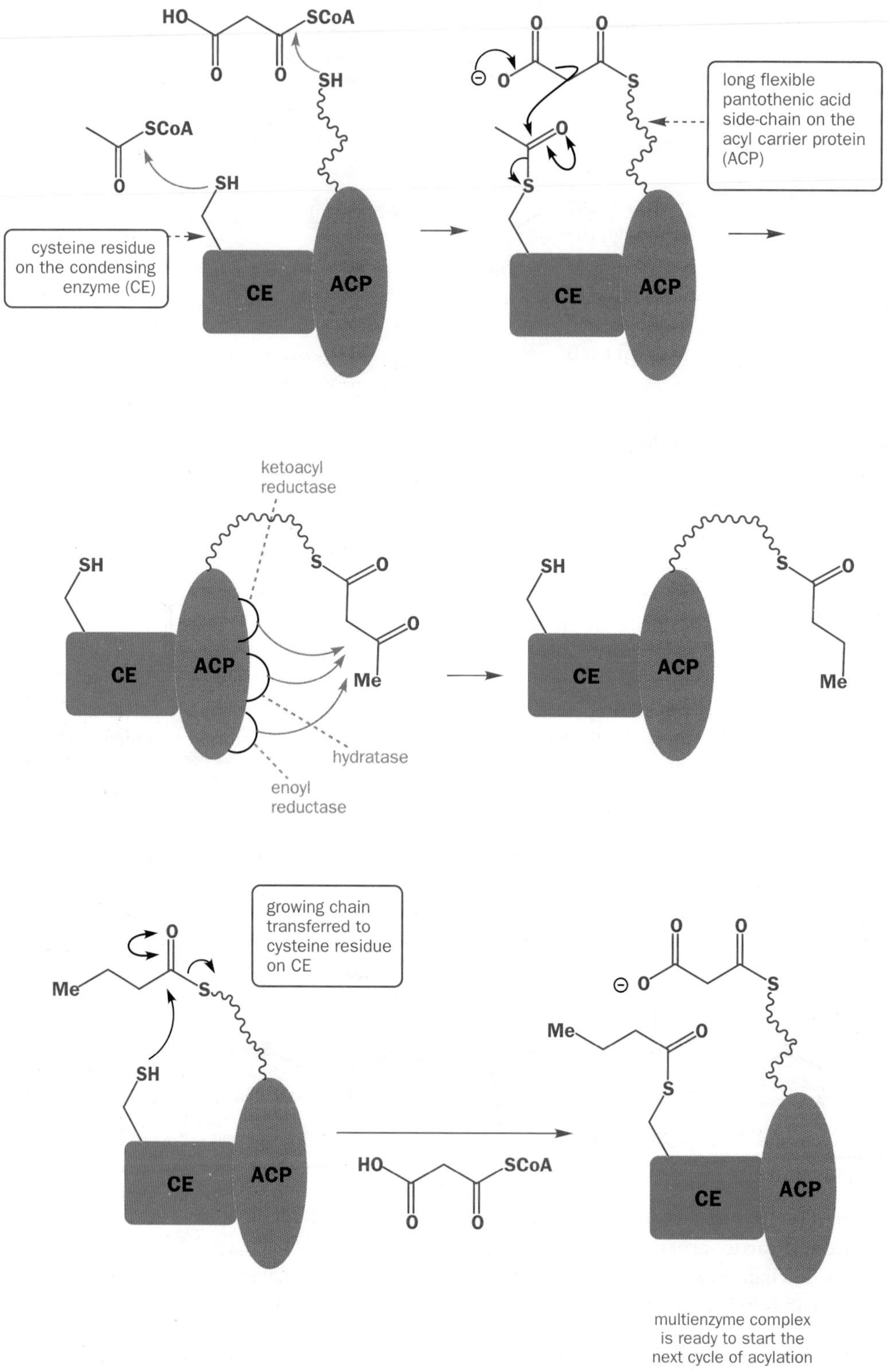
HO
SCoA
SH
SCoA
SH
cysteine residue on the condensing enzyme (CE)
CE
ACP
long flexible pantothenic acid side-chain on the acyl carrier protein (ACP)
ketoacyl reductase
hydratase
enoyl reductase
Me
growing chain transferred to cysteine residue on CE
multienzyme complex is ready to start the next cycle of acylation

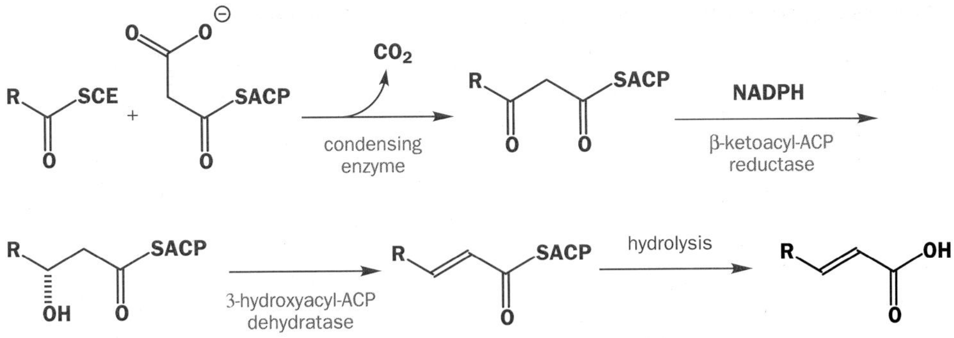

The second method makes *Z*-3,4-unsaturated acids by deconjugation from the *E*-2,3-unsaturated acids catalysed by an isomerase while the acyl chain is still attached to ACP. This is an anaerobic route as no oxidation is required (the double bond is already there—it just has to be moved) and is used by prokaryotes such as bacteria.

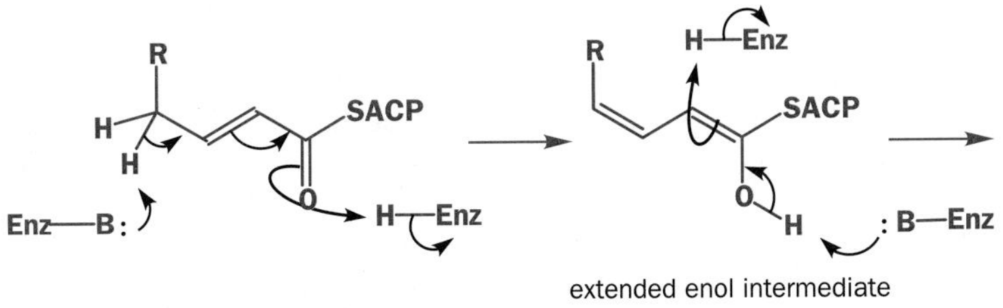

extended enol intermediate

Removal of a proton from C4 forms an extended enol, which can be protonated at C2 or C4. Protonation at C4 is thermodynamically favoured as it leads to the conjugated alkene. But protonation at C2 is kinetically favoured, and this leads to the nonconjugated alkene. The geometry of the new alkene depends on the conformation of the chain when the first (deprotonation) step occurs. It is thought that this is the best conformation for the previous reaction, the dehydration step, and that no rotation of the chain occurs before the isomerase gets to work.

■
For more on this, read a specialized book, such as Ian Fleming's, *Frontier orbitals and organic chemical reactions*, Wiley, Chichester, 1976. Similar regioselectivity is evident in the protonation of the Birch reduction products on p. 628.

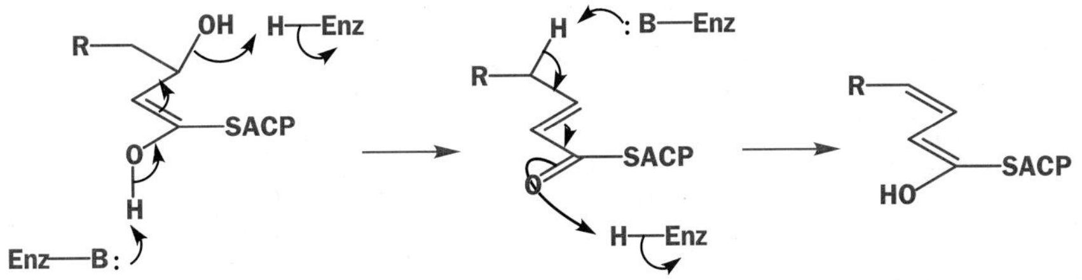

You may think this a rather unlikely reaction, but the same thing can be done in the laboratory. If a simple unsaturated ester is converted into its lithium enolate and then reprotonated with water, the major product is the ester of the *Z*-3,4-enoic acid. Yields and steroselectivities are excellent.

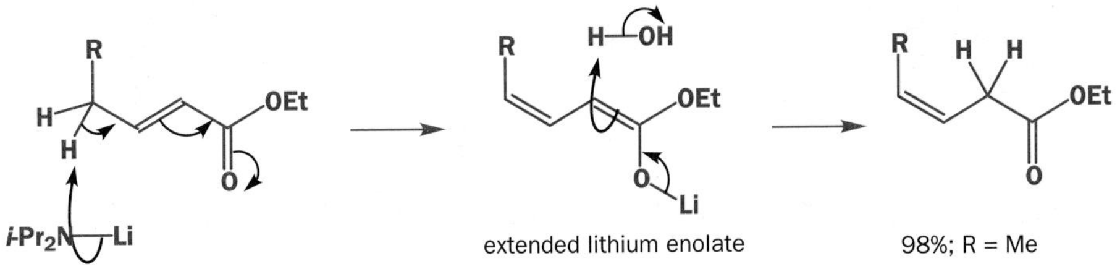

extended lithium enolate   98%; R = Me

One explanation suggests that control is exercised by a favourable conformation in which 1,3-allylic strain is preferred to 1,2-strain. It looks as though Nature has again seized on a natural chemical preference and made it even better.

■
$A^{1,3}$ strain (1,3-allylic strain) was discussed in Chapter 34, p. 896.

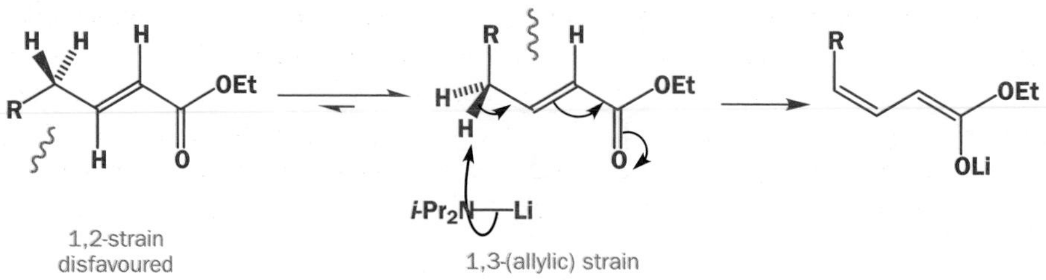

1,2-strain
disfavoured

1,3-(allylic) strain

The third method is a concerted stereospecific removal of two adjacent hydrogen atoms from the chain of a fatty acid after synthesis. This is an aerobic route as oxidation is required and is used by mammals such as ourselves. The stereochemistry of the reaction is known from labelling studies to be *cis* elimination.

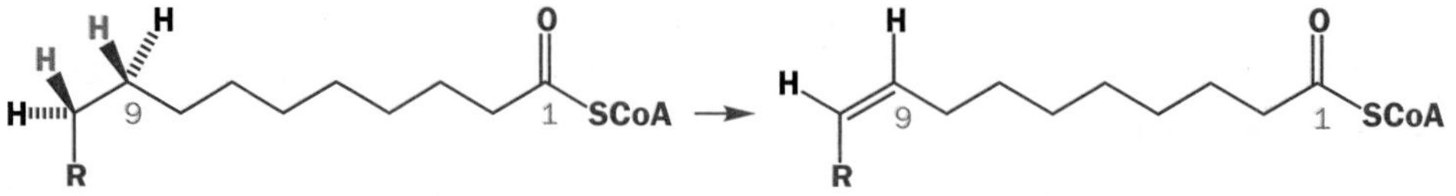

This oxidation involves a chain of reagents including molecular oxygen, Fe(III), FAD, and NAD^+. A hydroxylation followed by a dehydration or a sulfur-promoted dehydrogenation has been suggested for the removal of the hydrogen atoms. The chemical reaction corresponding to the biological reaction has not yet been discovered.

What is so important about unsaturated fatty acids?

Mammals can insert a *cis*-alkene into the chain, providing that it is no further away from the carbonyl group than C9. We cannot synthesize linoleic or linolenic acids (see chart a few pages back) directly as they have alkenes at C12 and C15. These acids must be present in our diet. And why are we so keen to have them? They are needed for the synthesis of arachidonic acid, a C_{20} tetraenoic acid that is the precursor for some very interesting and important compounds. Here is the biosynthesis of arachidonic acid.

synthesis of unsaturated fatty acids

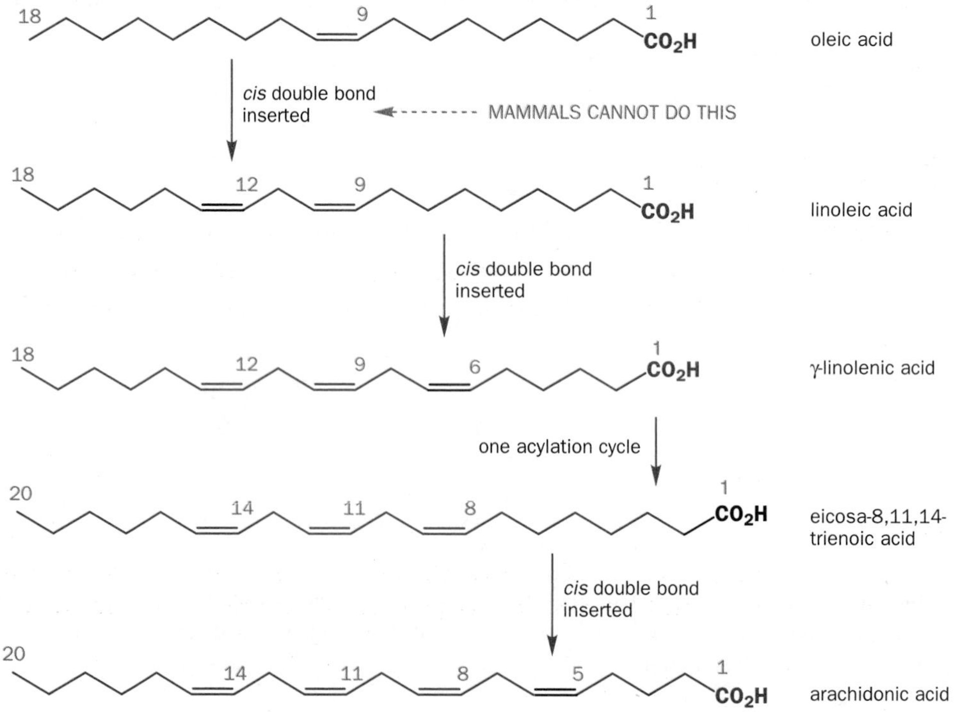

The final product of this chain of events—arachidonic acid—is one of the **eicosanoids**, so called because *eicosa* is Greek for 'twenty', and the systematic names for these compounds contain 'eicosanoic acid' in some form. The leukotrienes resemble arachidonic acid most closely, the prostaglandins have a closed chain forming a five-membered ring, and the thromboxanes resemble the prostaglandins but have a broken chain. All are C_{20} compounds with the sites of the alkenes (C5, C8, C11, and C14) marked by functionality or some other structural feature.

compounds synthesized from arachidonic acid

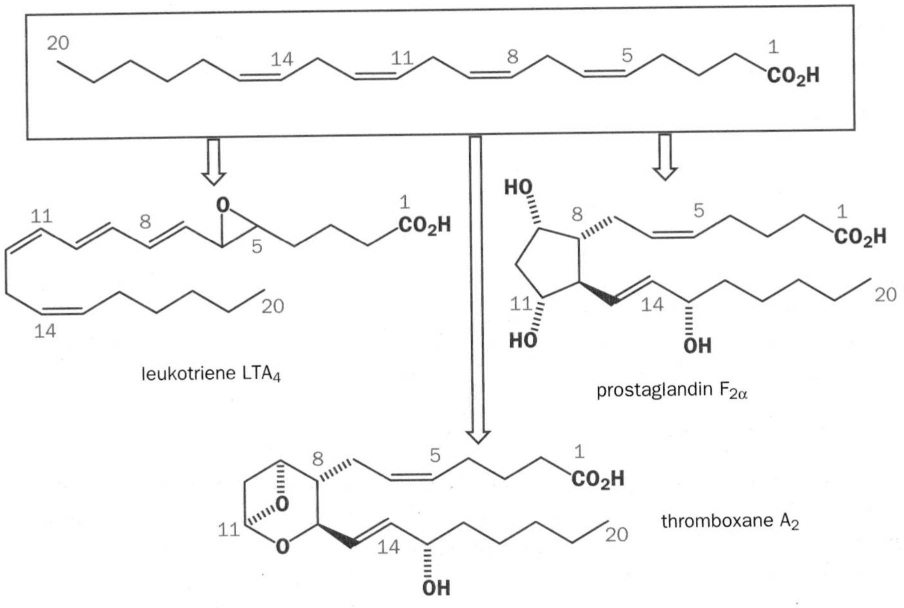

leukotriene LTA$_4$

prostaglandin F$_{2\alpha}$

thromboxane A$_2$

These compounds are all unstable and all are involved in transient events such as inflammation, blood clotting, fertilization, and immune responses. They are produced locally and decay quickly and are implicated in autoimmune diseases like asthma and arthritis. They are made by oxidation of arachidonic acid—you can see this best if you redraw the molecule in a different conformation.

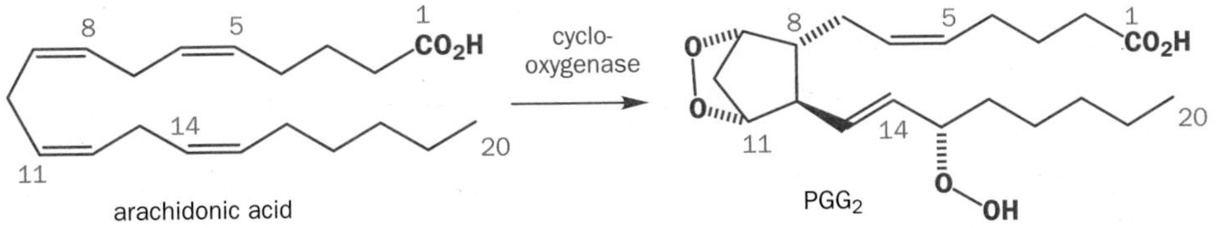

arachidonic acid · cyclo-oxygenase · PGG$_2$

The first step is a radical abstraction of a hydrogen atom from an allylic position by oxygen (perhaps carried on an iron atom in a haem). The atom removed is between two alkenes so that the resulting radical is doubly allylic.

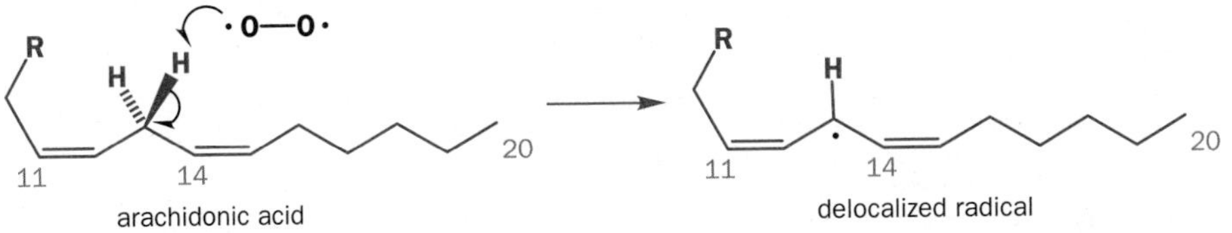

arachidonic acid · delocalized radical

This allylic radical captures a molecule of oxygen at C11 to form a new oxyradical. The reaction occurs at one end of the delocalized radical so that the product is a conjugated diene and the new alkene is *trans* (*E*).

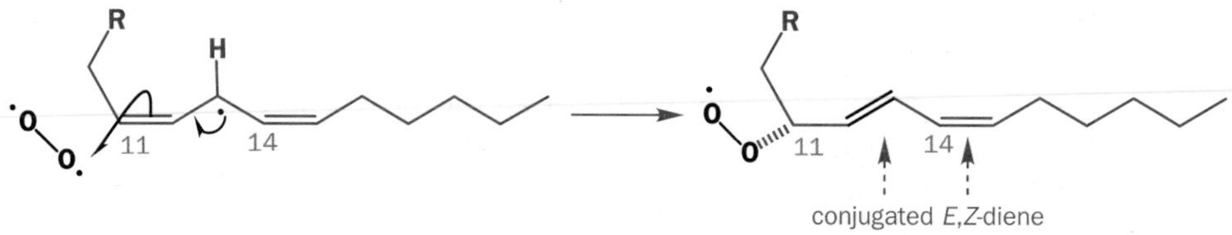

conjugated *E,Z*-diene

Now we need to resume the full structure of the intermediate because the oxyradical does an elaborate addition to the C8 alkene and then to the newly formed diene to form a new stable allylic radical.

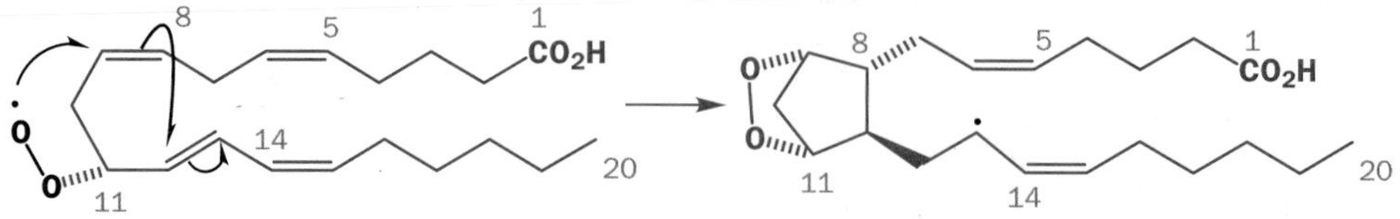

Three new stereogenic centres are created in this cyclization, at C8, C9, and C12, and all are under full control both from the centre already present and from the way in which the molecule folds up under the guidance of the enzyme. Now the allylic radical reacts with oxygen to give the unstable hydroperoxide PGG$_2$.

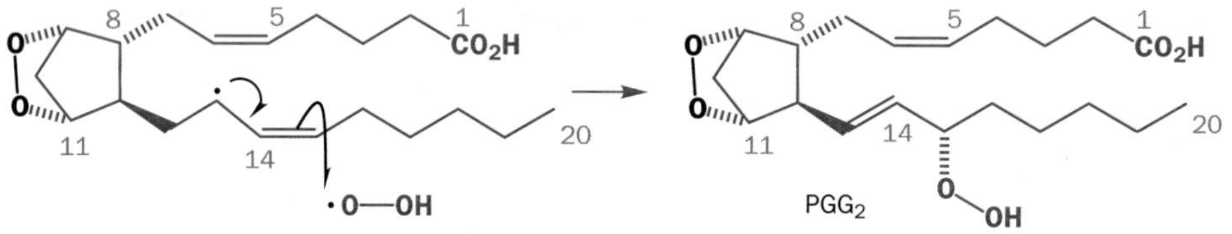

This unstable prostaglandin has been isolated from sheep but, as it has a half-life of only 5 minutes, this is no trivial matter. Both weak O–O bonds are now reduced enzymatically to give the first reasonably stable compound, PGF$_{2\alpha}$ (PG just means prostaglandin).

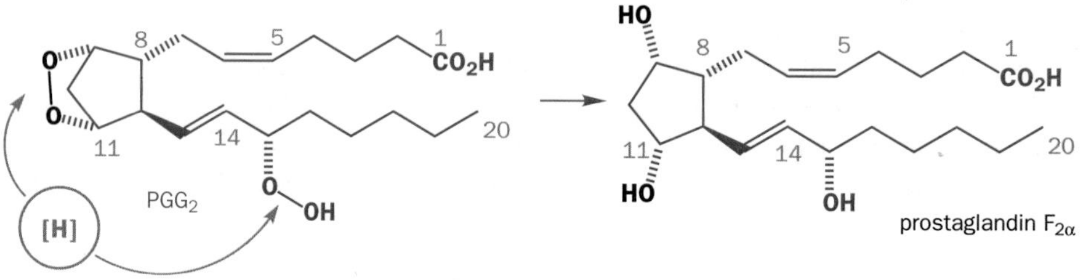

prostaglandin F$_{2\alpha}$

The best evidence for this pathway comes from labelled oxygen molecules. If a mixture of ^{16}O–^{16}O (ordinary oxygen) and ^{18}O–^{18}O is supplied to an organism making PGF$_{2\alpha}$, the product has either both black OHs as ^{16}O or both as ^{18}O but no molecules are formed with one ^{16}O and one ^{18}O. These isotopes are easily measured by mass spectrometry. Both black OHs then come from one and the same molecule of oxygen—not an obvious conclusion when you inspect the molecule of PGF$_{2\alpha}$, and thus good evidence for this pathway.

How aspirin works

The enzyme that catalyses these remarkable reactions, cyclooxygenase, is an important target for medicinal chemists. Inhibiting PG synthesis can bring about a reduction of inflammation and pain. In fact, this is how aspirin works. It was not, of course, *designed* to work that way and its mode of action was discovered decades after its use began. There is a price to pay for such a useful drug. PGs also control acid secretion in the stomach and aspirin inhibits their synthesis there too so stomach ulceration can result.

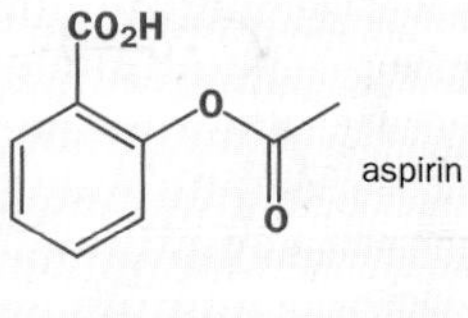

aspirin

Each of the other families of eicosanoids—thromboxanes and leukotrienes—has interesting biosynthetic pathways too, but we will mention only one small detail. A completely different oxidation enzyme, lipoxygenase, initiates a separate pathway leading to the leukotrienes, but the first steps are very similar. They just occur elsewhere in the arachidonic acid molecule.

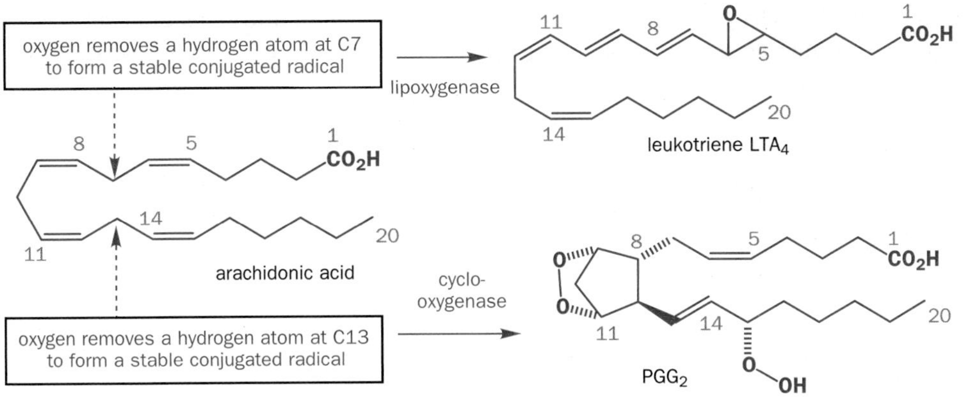

oxygen removes a hydrogen atom at C7
to form a stable conjugated radical

lipoxygenase

leukotriene LTA$_4$

arachidonic acid

cyclo-
oxygenase

oxygen removes a hydrogen atom at C13
to form a stable conjugated radical

PGG$_2$

The initially formed radical is stabilized by two double bonds in the same way as that we have just seen and reacts with oxygen in the same way again to give a *trans*-alkene and a new hydroperoxide.

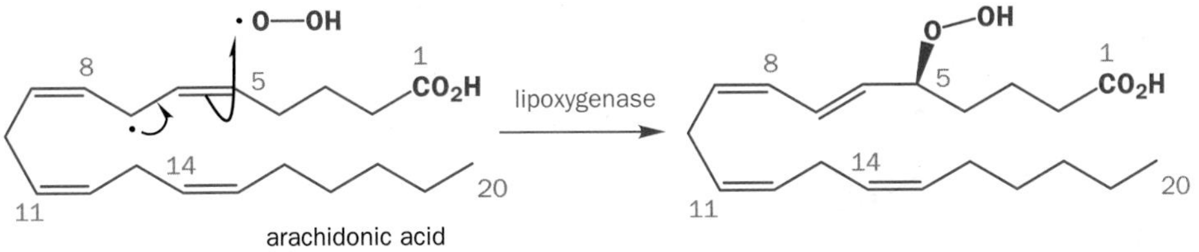

lipoxygenase

arachidonic acid

The next step is something quite new. No new C–C bond is formed: instead, the diene attacks the hydroperoxide to give an epoxide and a fully conjugated triene. The new double bond is *cis* this time, which is what we should expect from the conformation we have been using. This is LTA$_4$ and all the other leukotrienes are made from this compound.

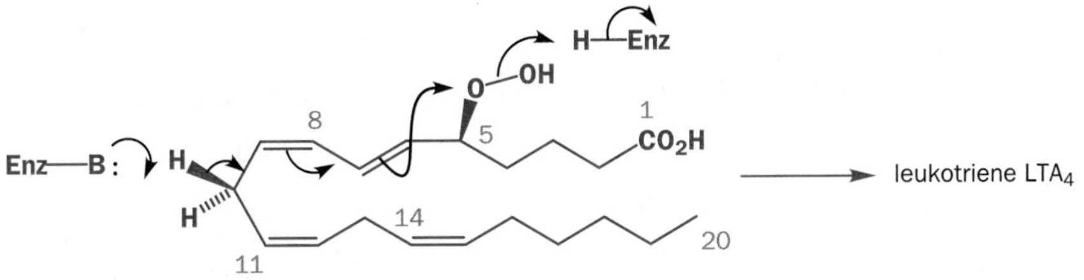

leukotriene LTA$_4$

The relatively recent discovery of these unstable molecules of incredibly powerful biological activity means that we by no means know all about them yet. They are very important to our well-being and important medical advances are bound to follow from a better understanding.

■
Prostaglandins and leukotrienes have appeared several times before in this book, and you can read about aspects of their laboratory synthesis on pp. 686, 1229, 1268, and 883.

Aromatic polyketides come in great variety

The **fatty acid pathway** or, as we should call it now, the **acyl polymalonate pathway**, also gives rise to an inexhaustible variety of aromatic and other compounds belonging to the family of the polyketides. You saw in Chapter 50 how the shikimic acid pathway makes aromatic compounds but the compounds below are from the polyketide route.

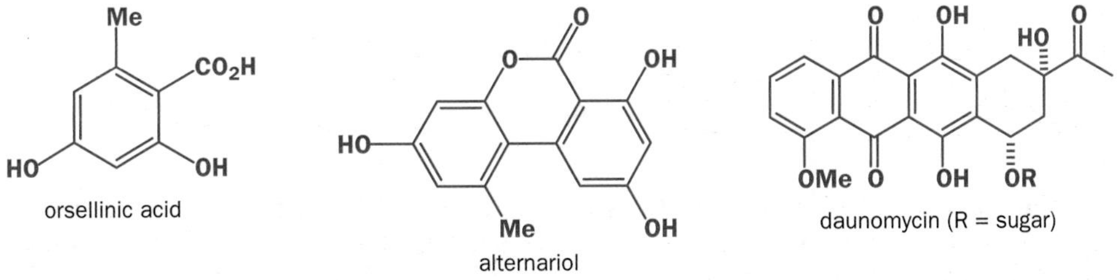

orsellinic acid

alternariol

daunomycin (R = sugar)

You might immediately be struck by the extent of oxygenation in these compounds. The shikimic acid route produced Ar–C$_3$ compounds with at most one OH group in the *para* position and others added *ortho* to that first OH group. Here we have multiple oxygenation with a predominant 1,3 pattern. If we try to arrange an acyl polymalonate product to make orsellinic acid, this is what we shall need.

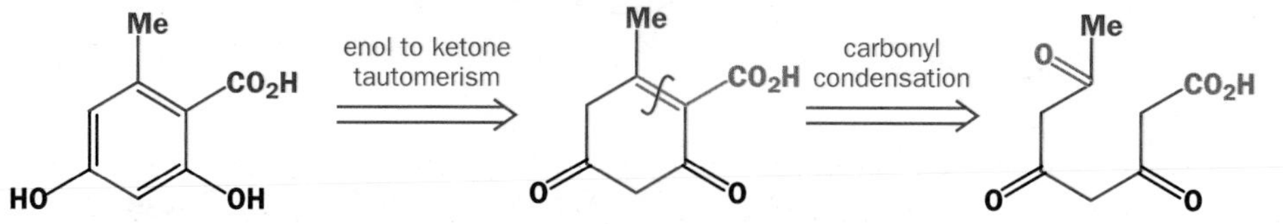

Merely by writing ketones instead of phenols and doing one disconnection corresponding to a simple carbonyl condensation, we have reached a possible starting material which is a typical acyl polymalonate product without any reductions. This is what polyketides are. The fatty acids are assembled with full reduction at each stage. Polyketides are assembled from the same process but without full reduction; indeed, as the name polyketide suggests, many are made without any reduction at all. This is the biosynthesis of orsellinic acid.

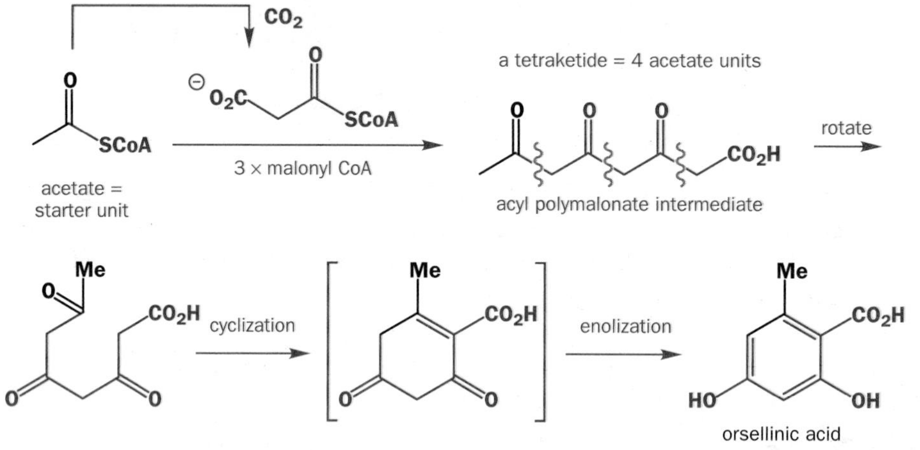

This route has been demonstrated by feeding ^{13}C-labelled malonyl CoA to a microorganism. The orsellinic acid produced has three ^{13}C atoms only, seen by an $M + 3$ peak in the mass spectrum. The location of the labels can be proved by NMR. The starter unit, acetate, is not labelled.

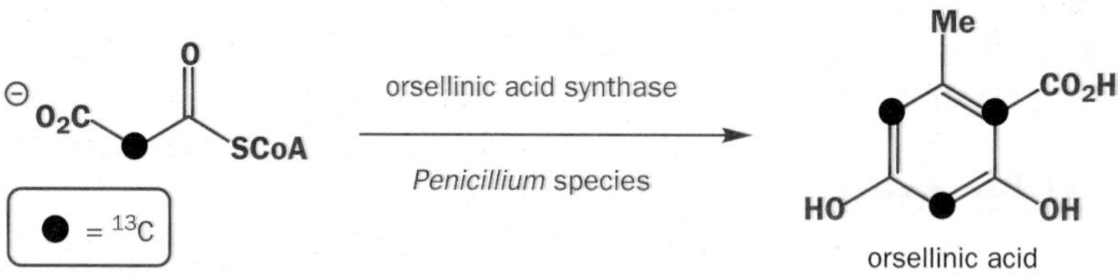

As the polyketide chain is built up, any of the reductions or eliminations from fatty acid biosynthesis can occur at any stage. The simple metabolite 6-methyl salicylic acid (6-MSA) is made in the microorganism *Penicillium patulum*, and it could come from the same intermediate as orsellinic acid with one reduction.

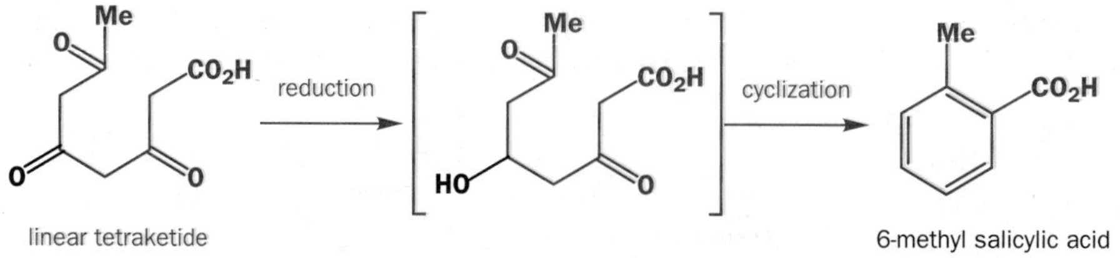

Reduction to the alcohol or to the unsaturated acid or ketone would give the right oxidation level and could occur as the chain is built, after it is completed, or after cyclization. In fact, reduction to the conjugated unsaturated triketide occurs as the third acetate unit is added, just as the fatty acid route would lead us to expect.

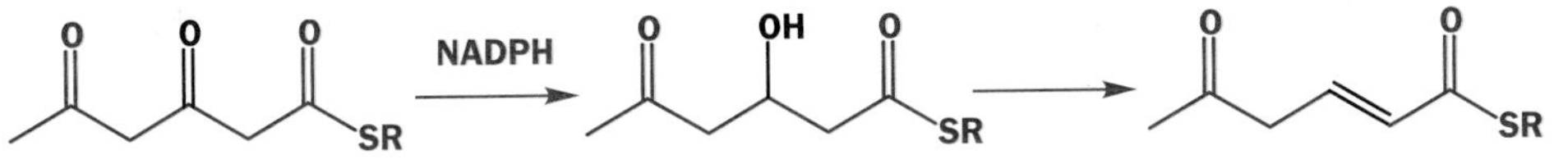

linear triketide

This intermediate cannot cyclize as it has a *trans* double bond and the ends cannot reach each other. First, the double bond is moved out of conjugation with the COSR group, again as in the fatty acids, except that here the new *Z* double bond moves into conjugation with the remaining keto group.

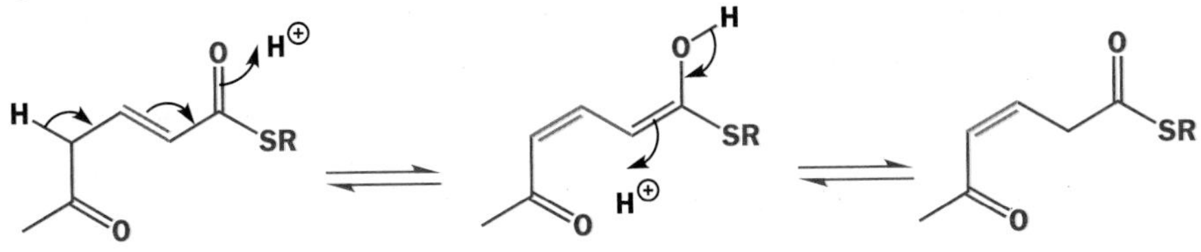

E-alkene conjugated with acid group　　　　　extended enol　　　　　*Z*-alkene conjugated with ketone group

Now the last chain extension occurs and the completed *Z*-tetraketide cyclizes to 6-methyl salicylic acid. Chemically, we would prefer not to carry the unstable *Z*-enone through several steps, but Nature controls these reactions very precisely.

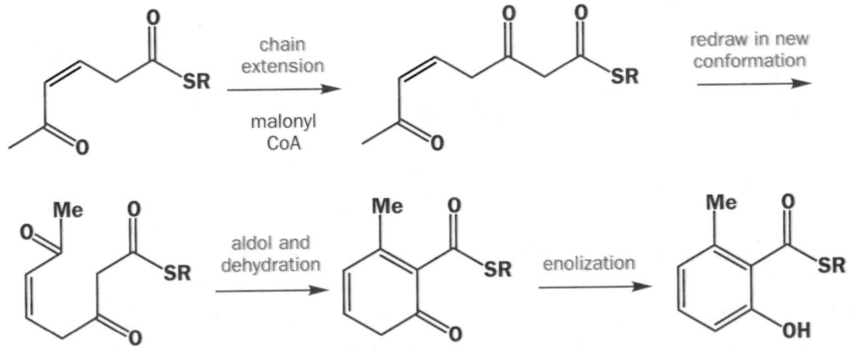

This precise sequence was discovered only through very careful double labelling experiments and after the discovery of specific inhibitors for the enzyme. Since polyketides can be made from the acyl polymalonate pathway with or without reduction and elimination at any step, the number of possible structures is vast. With more reduction, no aromatic ring can be formed: macrolide antibiotics such as brefeldin A come from this route.

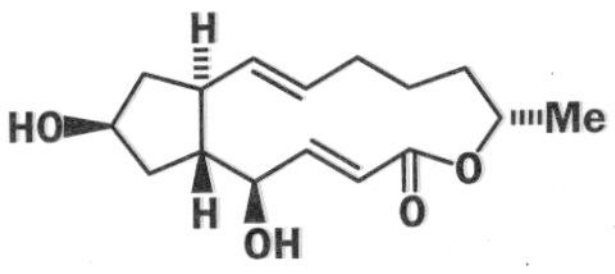

brefeldin A

If you examine this structure, you should be able to find a continuous carbon chain made from an acetate starter unit and seven malonyl CoA units with full or partial reduction occurring after many acylation steps.

Other starter units

So far we have started the chain with acetate, but many other starter units are used. Some important groups of compounds use shikimic acid metabolites such as cinnamic acid (Chapter 50) as starter units. They include the widespread plant flavones and the anthocyanidin flower pigments.

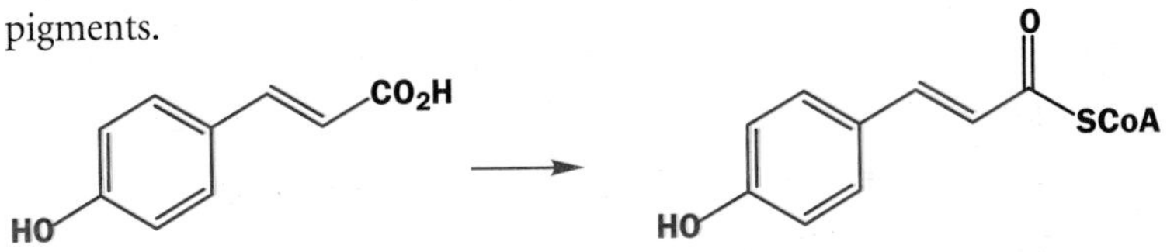

4-hydroxy cinnamic acid from shikimic acid　　　　　activated for polymalonate addition

The most common sequence uses three malonyl CoA acylations followed by cyclization to a new aromatic ring. The simplest type is exemplified by resveratrole, the compound in red wine that helps to prevent heart disease. Each step in this sequence is a simple reaction that you have met before.

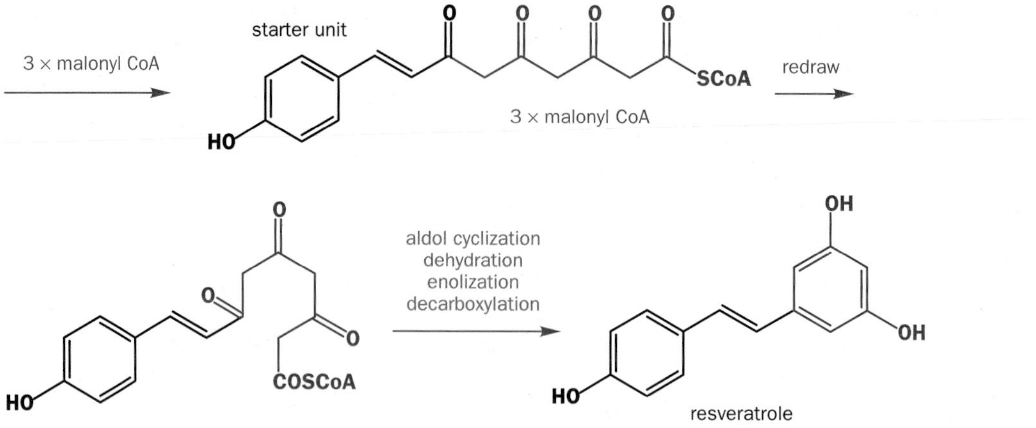

A different cyclization leads to the flavones and anthocyanidins. Reaction of the stable enol from a 1,3-diketone with the thiol ester as electrophile results in acylation at carbon in the manner of the Claisen ester condensation (Chapter 28) with loss of CoASH and the formation of a trihydroxyben-zene ring.

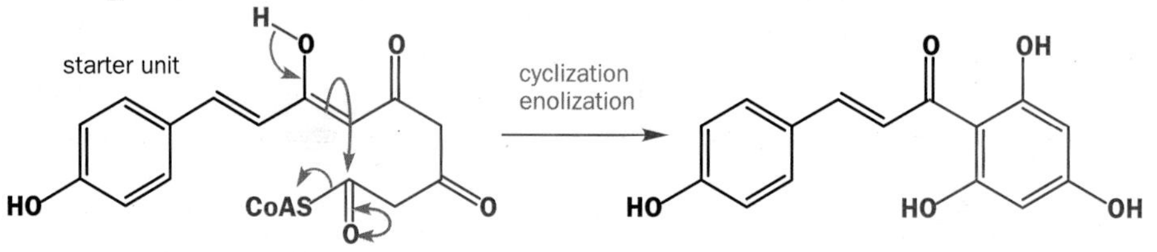

This cyclization is followed by a conjugate addition of an *ortho* phenolic OH group on to the enone system. The product is a flavanone structure, which is always drawn a different way up to the molecules we have just been discussing. Redrawing the last product shows the cyclization.

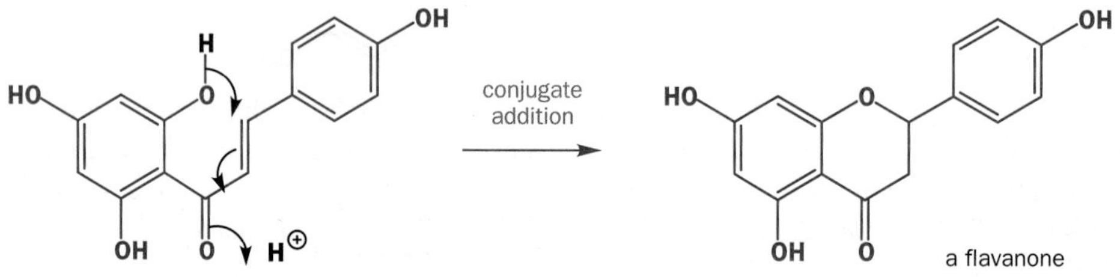

Aromatization of the central oxygen heterocycle by oxidation leads to the flavones, which are yellow or orange depending on their substituents. Dehydration leads to the red or blue anthocyanidins, pigments of flowers and fruit. This important group of molecules also includes plant growth hormones and defence compounds.

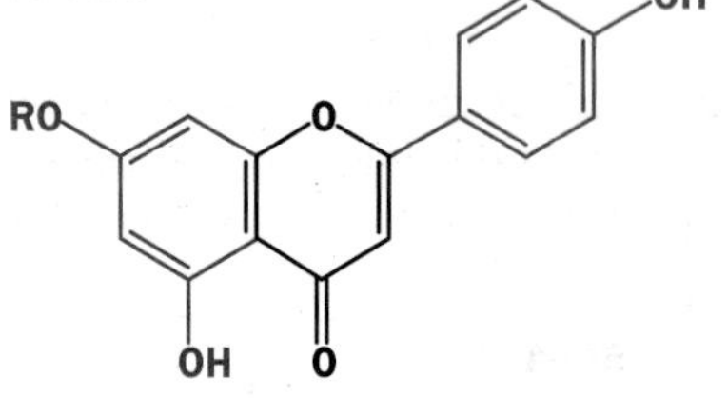

R = H; naringenin, R = glucose; naringin
—a bitter substance from grapefruit peel

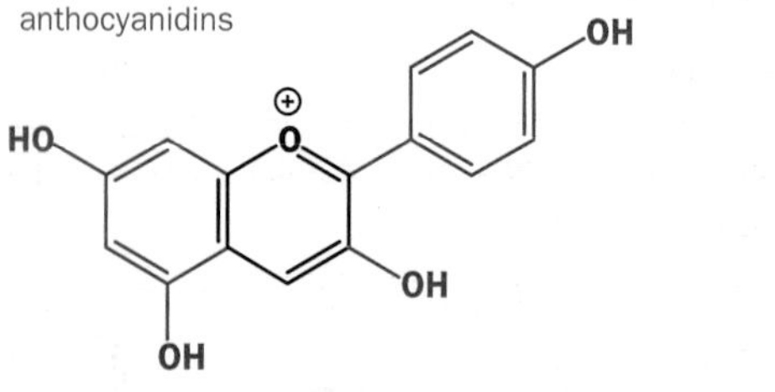

pelargonidin, pigment of raspberries, geraniums, and red grape skins

Terpenes are volatile constituents of plant resins and essential oils

α-pinene

camphor

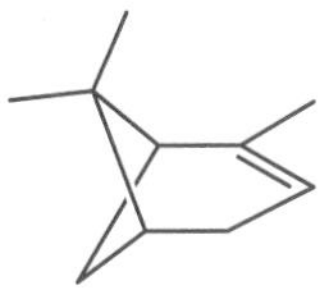

Terpenes were originally named after turpentine, the volatile oil from pine trees used in oil painting, whose major constituent is α-pinene. The term was rather vaguely used for all the volatile oily compounds, insoluble in water and usually with resiny smells from plants. The oils distilled from plants, which often contain perfumery or flavouring materials, are called **essential oils** and these too contain terpenes. Examples include camphor from the camphor tree, used to preserve clothes from moths, humulene from hops, which helps to give beer its flavour, and phytol, found in many plants.

phytol

humulene

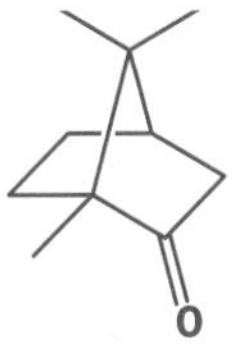

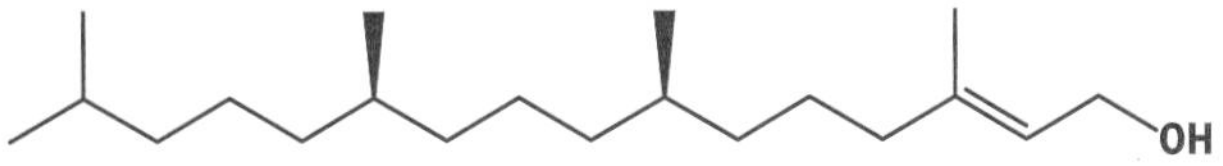

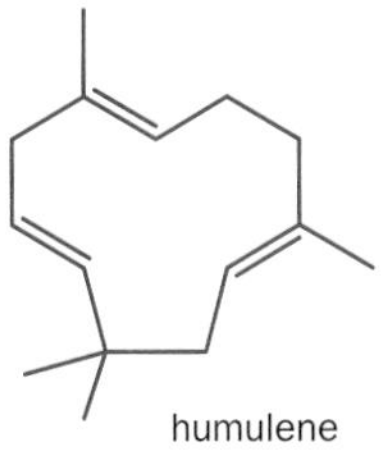

You will notice that they are all aliphatic compounds with a scattering of double bonds and rings, few functional groups, and an abundance of methyl groups. A better definition (that is, a biosynthetically based definition) arose when it was noticed that all these compounds have 5n carbon atoms. Pinene and camphor are C_{10} compounds, humulene is C_{15}, and phytol is C_{20}. It seemed obvious that terpenes were made from a C_5 precursor and the favourite candidate was isoprene (2-methylbuta-1,3-diene) as all these structures can be drawn by joining together 2-, 3-, or 4-isoprene skeletons end to end. Humulene illustrates this idea.

3 isoprene units humulene

isoprene

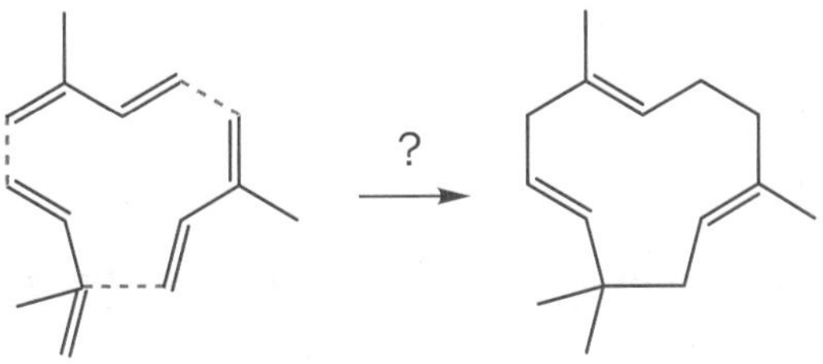

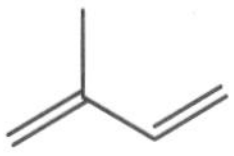

In fact, this is not correct. Isoprene is not an intermediate, and the discovery of the true pathway started when acetate was, rather surprisingly, found to be the original precursor for all terpenes. The key intermediate is mevalonic acid, formed from three acetate units and usually isolated as its lactone.

acetyl CoA mevalonic acid mevalonolactone

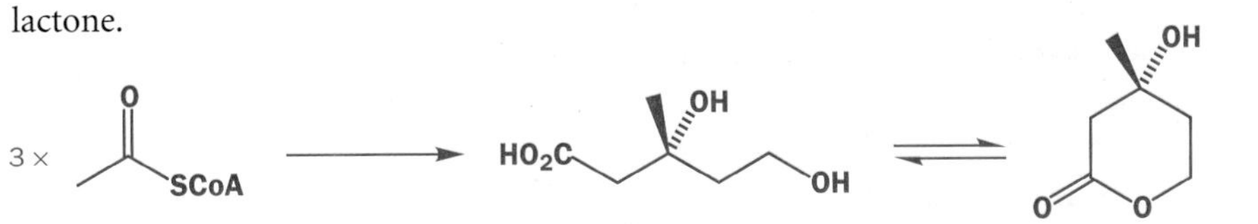

The first step is the Claisen ester condensation of two molecules of acetyl CoA, one acting as an enol and the other as an electrophilic acylating agent to give acetoacetyl CoA. We saw the same reaction in the biosynthesis of the pyrrolidine alkaloids earlier in this chapter.

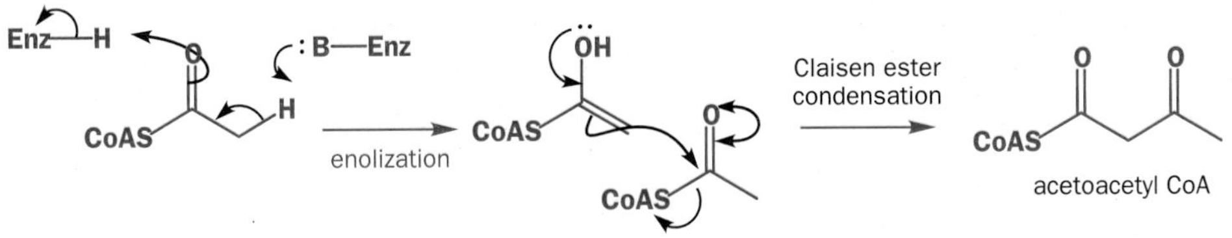

The third molecule of acetyl CoA also functions as a nucleophilic enol and attacks the keto group of acetoacetyl CoA. This is not a Claisen ester condensation—it is an aldol reaction between the enol of a thiol ester and an electrophilic ketone.

acetoacetyl CoA

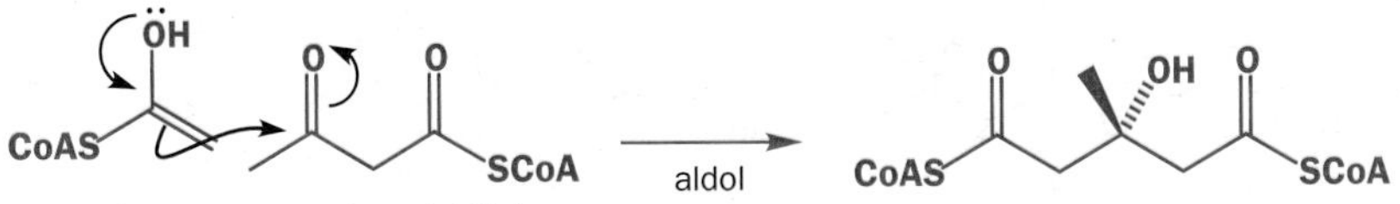

We have drawn the product with stereochemistry even though it is not chiral. This is because one of the two enantiotopic thiol esters is hydrolysed while this intermediate is still bound to the enzyme, so a single enantiomer of the half-acid/half-thiol ester results.

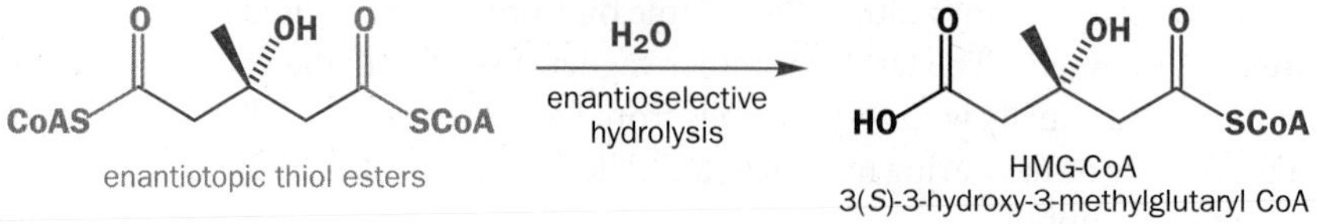

enantiotopic thiol esters   →(H₂O, enantioselective hydrolysis)→   HMG-CoA
3(S)-3-hydroxy-3-methylglutaryl CoA

The remaining thiol ester is more electrophilic than the acid and can be reduced by the nucleophilic hydride from NADPH. Just as in $LiBH_4$ reductions of esters (Chapter 24), the reaction does not stop at the aldehyde level, and two molecules of NADPH are used to make the alcohol. This is mevalonic acid.

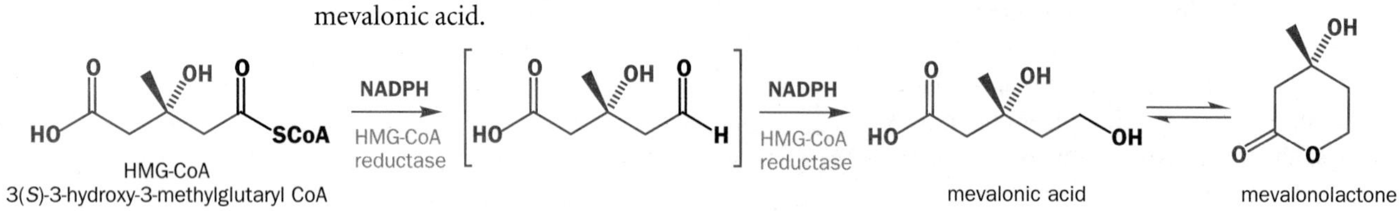

HMG-CoA
3(S)-3-hydroxy-3-methylglutaryl CoA   →(NADPH, HMG-CoA reductase)→   →(NADPH, HMG-CoA reductase)→   mevalonic acid   ⇌   mevalonolactone

Different pathways; different reactivity

Acetyl CoA (as an enol) and malonyl CoA are both acylated by acetyl CoA as an electrophile, but the behaviour of the two nucleophiles is different when they react with acetoacetyl CoA. Malonyl CoA is acylated while acetyl CoA does the aldol reaction. This could be enzymatic control.

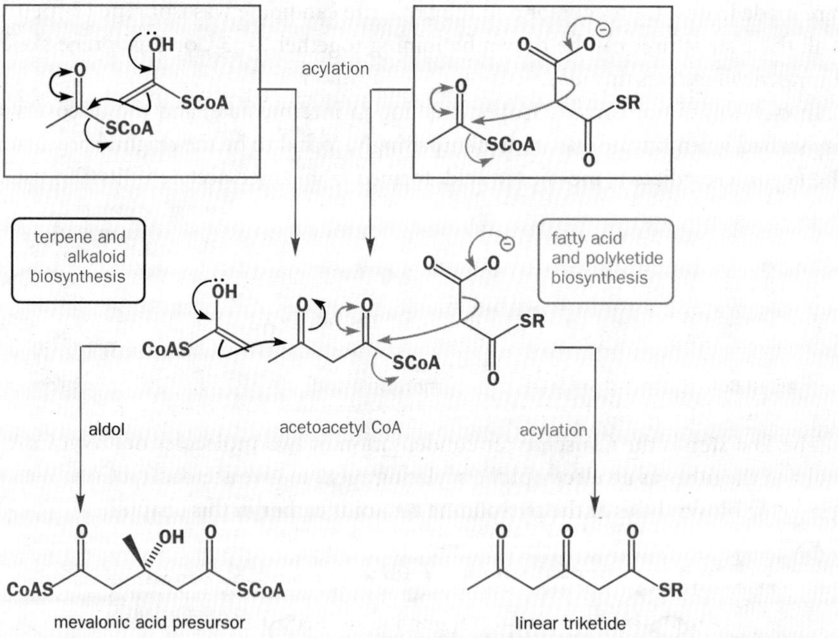

terpene and alkaloid biosynthesis

fatty acid and polyketide biosynthesis

acetoacetyl CoA

aldol   acylation

mevalonic acid presursor   linear triketide

Mevalonic acid is indeed the true precursor of the terpenes but it is a C_6 compound and so it must lose a carbon atom to give the C_5 precursor. The spare carbon atom becomes CO_2 by an elimination reaction. First, the primary alcohol is pyrophosphorylated with ATP (Chapter 49); then the CO_2H group and the tertiary alcohol are lost in a concerted elimination. We know it is concerted because labelling the diastereotopic hydrogen atoms on the CH_2CO_2H group reveals that the elimination is stereospecific.

mevalonic acid   →(ATP)→   →(elimination)→   isopentenyl pyrophosphate

▶

'PP' indicates the pyrophosphate group transferred from ATP.

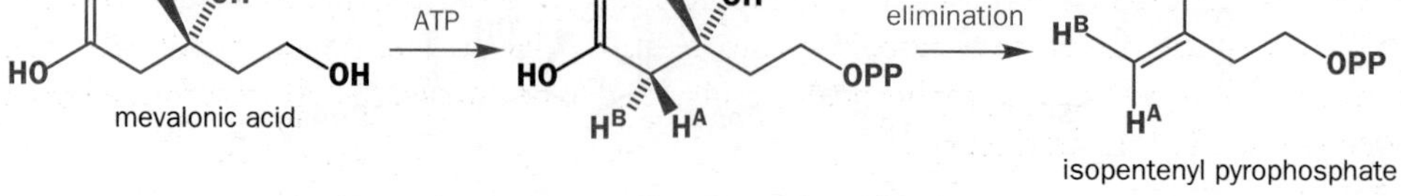

So is isopentenyl pyrophosphate the C_5 intermediate at last? Well, yes and no. There are actually two closely related C_5 intermediates, each of which has a specific and appropriate role in terpene biosynthesis. Isopentenyl pyrophosphate is in equilibrium with dimethylallyl pyrophosphate by a simple allylic proton transfer.

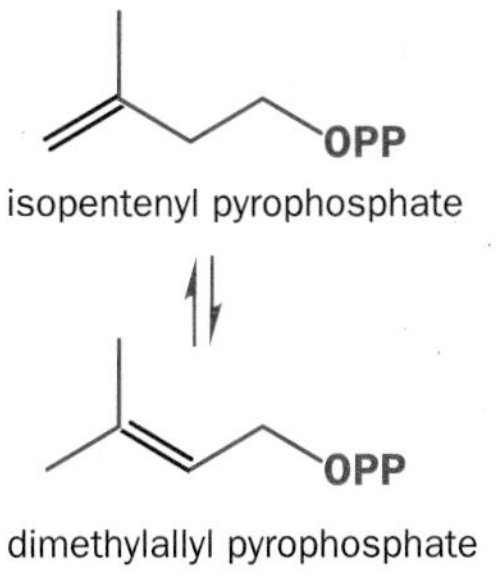

This is again a concerted reaction and again we know that by proton labelling. One of the two enantiotopic protons (H^S in the diagram) is lost from the bottom face of the allylic CH_2 group while the new proton is added to the top face of the alkene. This is an *anti* rearrangement overall.

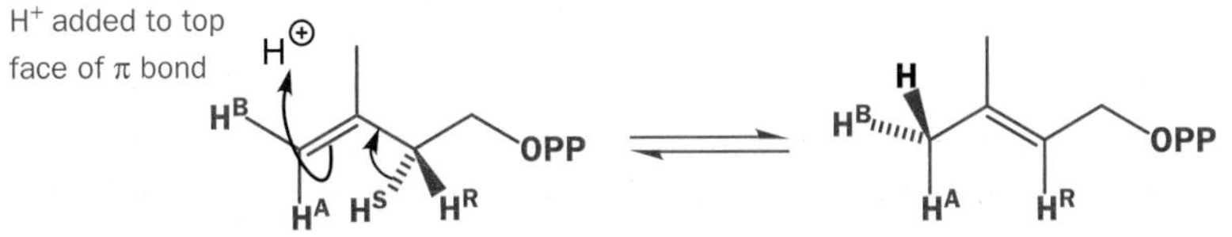

isopentenyl pyrophosphate

dimethylallyl pyrophosphate

H⁺ added to top face of π bond

The stereochemical details are interesting in establishing the mechanism but not important to remember. What is important is that the origin of the two methyl groups in dimethylallyl pyrophosphate is quite distinct and can easily be traced if you always draw the intermediates in the way we have drawn them. We will now switch to ^{13}C labelling to make the point.

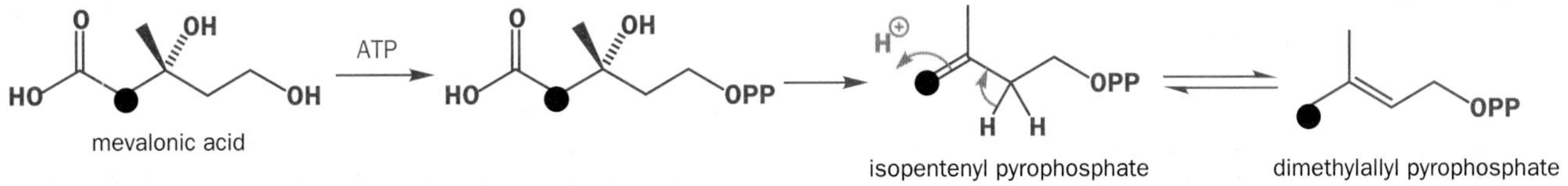

mevalonic acid

ATP

isopentenyl pyrophosphate

dimethylallyl pyrophosphate

The two C_5 intermediates now react with each other. The dimethylallyl pyrophosphate is the better electrophile because it is allylic, and allylic compounds are good at both S_N1 and S_N2 reactions (Chapter 17). Isopentenyl pyrophosphate is the better nucleophile because it can react through an unhindered primary carbon atom to produce a tertiary cation. This is what we have in mind.

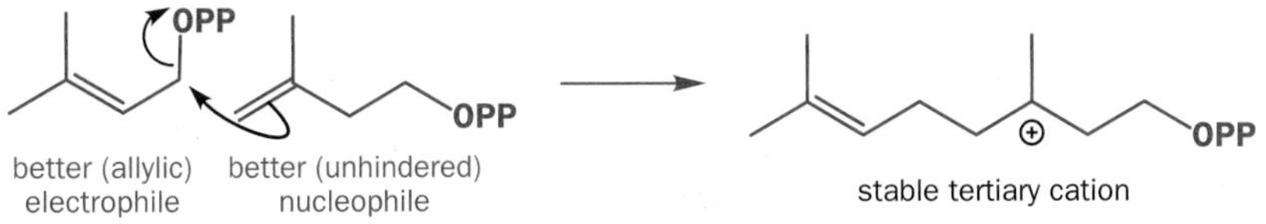

better (allylic) electrophile

better (unhindered) nucleophile

stable tertiary cation

Though this idea reveals the thinking behind the reaction, in fact it does not go quite like this. The product is one particular positional and geometrical isomer of an alkene and the cation is not an intermediate. Indeed, the reaction is also stereospecific (discovered again by proton labelling, but we will not give the rather complex details) and this too suggests a concerted process.

geranyl pyrophosphate

Geranyl pyrophosphate is the starting point for all the monoterpenes. It is still an allylic pyrophosphate and repeating the alkylation with another molecule of isopentenyl pyrophosphate gives farnesyl pyrophosphate, the starting point for the sesquiterpenes, and so on.

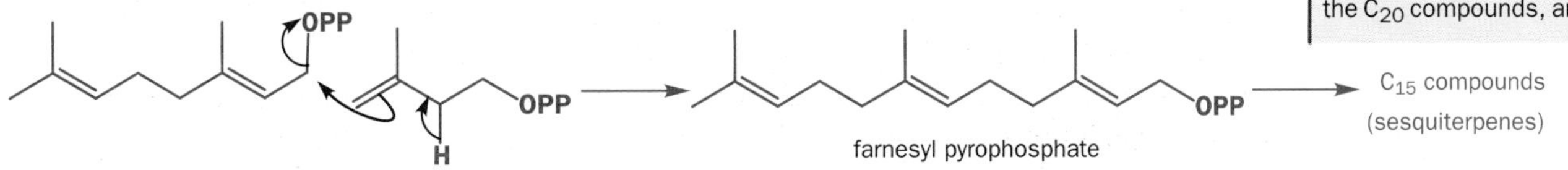

> Though terpenes are made from C_5 units, they are classified in C_{10} units. The monoterpenes are the C_{10} compounds, the sesquiterpenes (*sesqui* is Latin for one-and-a-half) are the C_{15} compounds, the diterpenes are the C_{20} compounds, and so on.

farnesyl pyrophosphate

C_{15} compounds (sesquiterpenes)

As soon as we start to make typical cyclic monoterpenes from geranyl pyrophosphate we run into a snag. We cannot cyclize geranyl pyrophosphate because it has a *trans* double bond! We *could* cyclize the *cis* compound (neryl pyrophosphate), and it used to be thought that this was formed from the *trans* compound as an intermediate.

many of these names are derived from
fragrant plant oils:

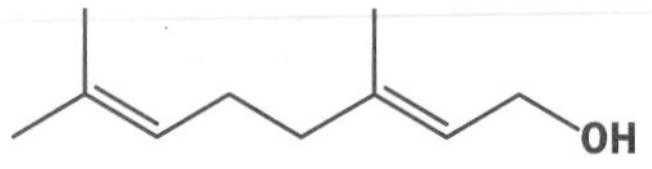

geraniol from geraniums

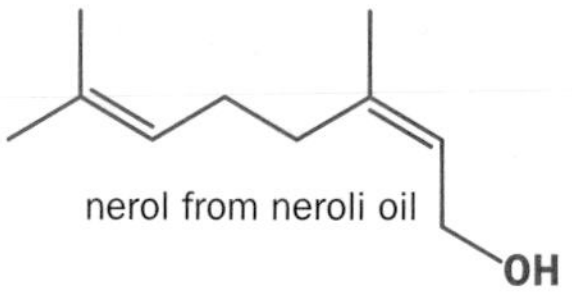

nerol from neroli oil

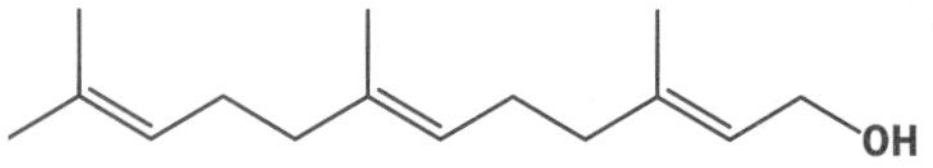

farnesol, also present in neroli oil

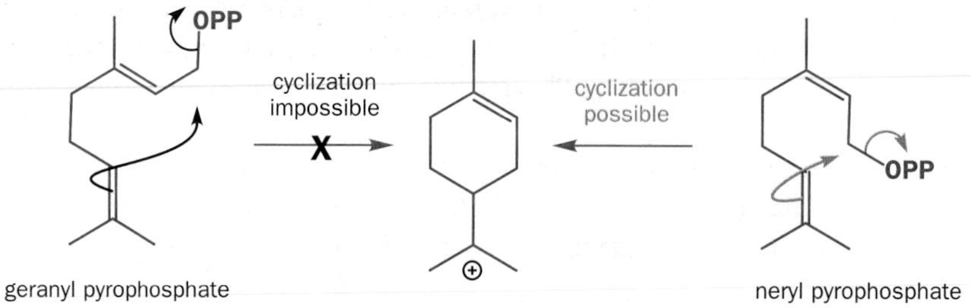

geranyl pyrophosphate

cyclization
impossible

X

cyclization
possible

neryl pyrophosphate

It is now known that Nature gets round this problem without making neryl pyrophosphate. An allylic rearrangement occurs to move the pyrophosphate group to the tertiary centre. This is an unfavourable rearrangement thermodynamically and probably occurs via the allyl cation and catalysed by Mg(II). There is no longer any geometry about the alkene. The molecule can now rotate freely about a single bond and cyclization can occur. Even if only a small amount of the rearranged allylic pyrophosphate is present, that can rearrange and more can isomerize.

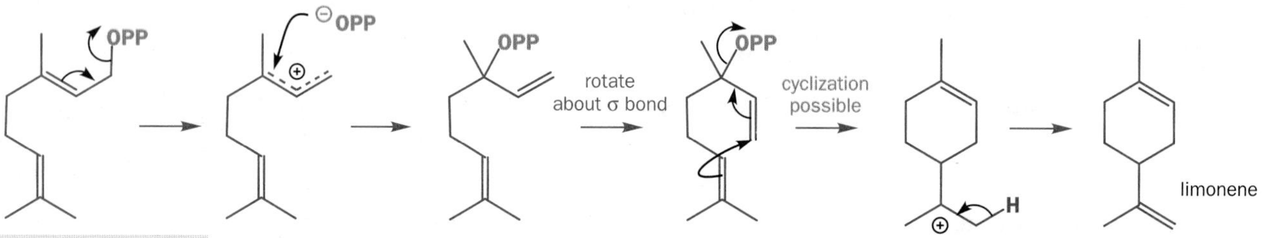

rotate
about σ bond

cyclization
possible

limonene

The product here is limonene—a terpene of the peel of citrus fruits. One enantiomer occurs in lemon peel—the other in orange peel. See Chapter 45.

More interesting compounds come from the cyclization of the first formed cation. The remaining alkene can attack the cation to form what looks at first to be a very unstable compound but which is actually a tertiary carbocation with the pinene skeleton.

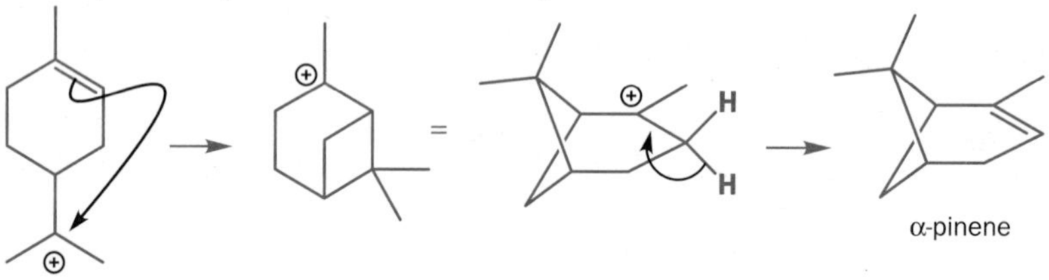

α-pinene

The camphor skeleton looks as though it might be formed by cyclization of the wrong end of the alkene on to the cation. This would certainly give the right skeleton but the intermediate secondary cation is rather unlikely.

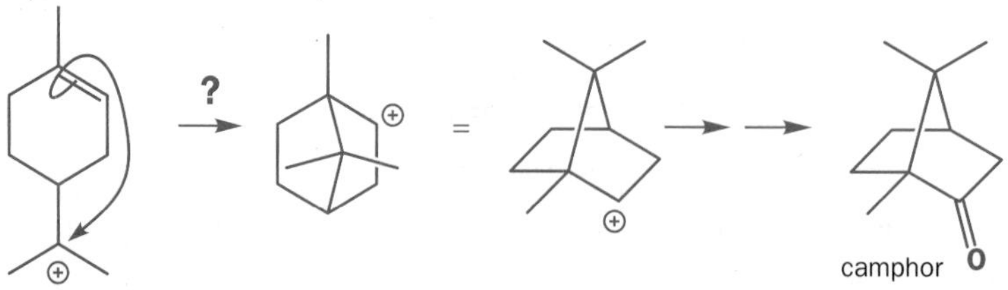

camphor

There is a better route. The more likely cation formed on the way to pinene could rearrange to the camphor cation. This is a known chemical reaction and is a simple 1,2-shift of the kind discussed in Chapter 37. However the new cation is formed, addition of water and oxidation would give camphor.

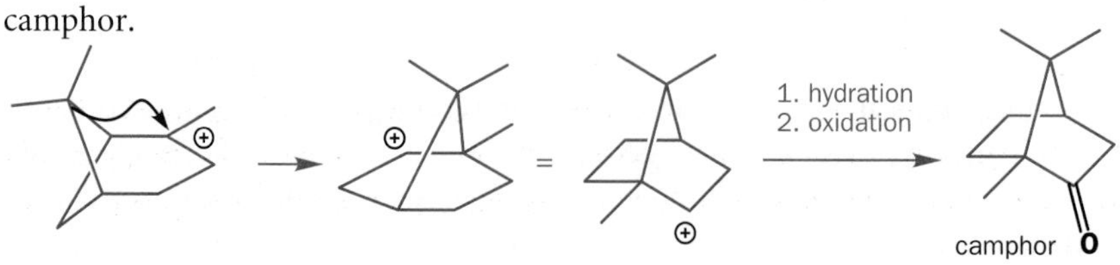

1. hydration
2. oxidation

camphor

In the sesquiterpene series, similar cyclizations lead to an amazing variety of products. After the initial unfavourable allylic rearrangement of the pyrophosphate group, farnesyl pyrophosphate can give a six-membered ring cation known as the bisabolyl cation.

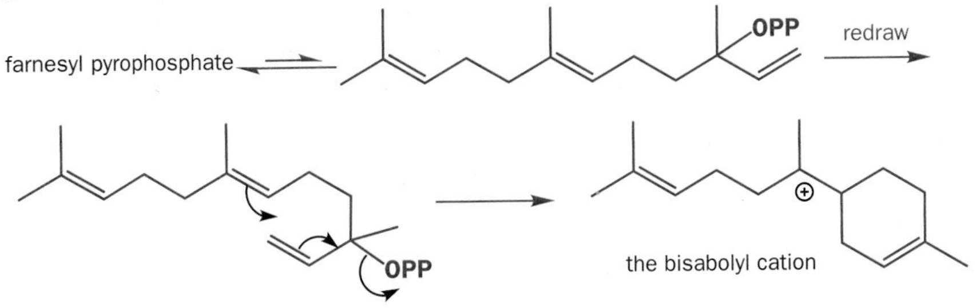

This cation does many things but it takes its name from the three fairly random proton losses that lead to the α-, β-, and γ-bisabolenes.

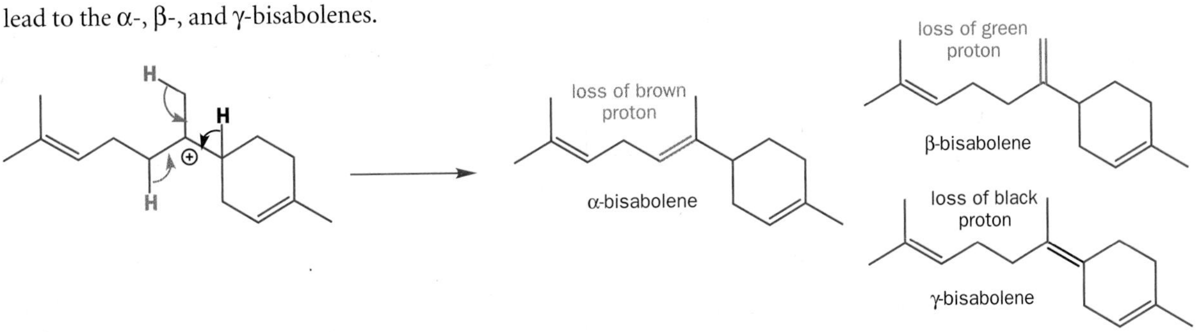

Many other reactions give even larger and more complex terpenes with a variety of functionalization but we will treat only one group in detail. These compounds are so important to us that they are given a different name.

Steroids are metabolites of terpene origin

Two types of human hormone are steroidal—the sex hormones such as oestradiol and testosterone and the adrenal hormones such as cortisone. Cholesterol is a steroid too, as is vitamin D, derived from ergosterol.

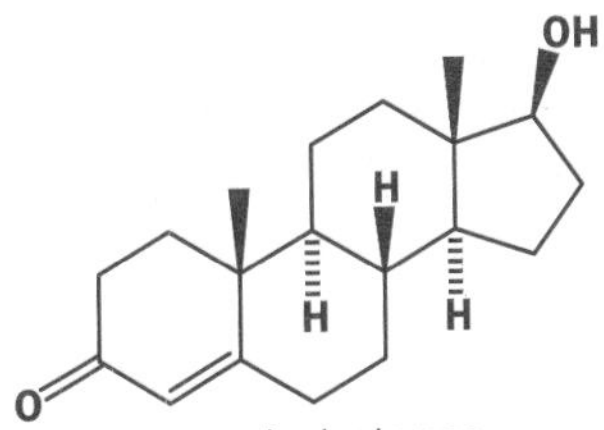

testosterone

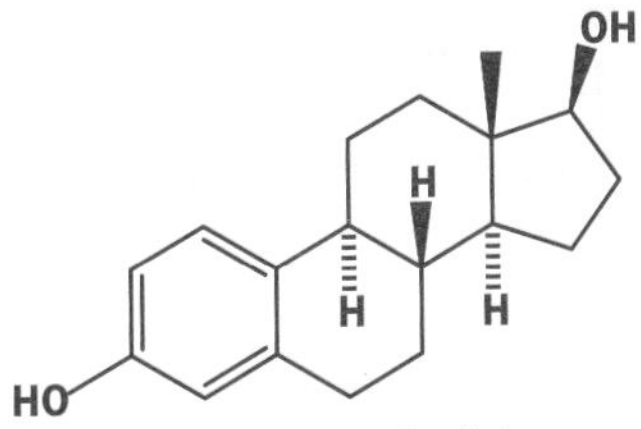

oestradiol

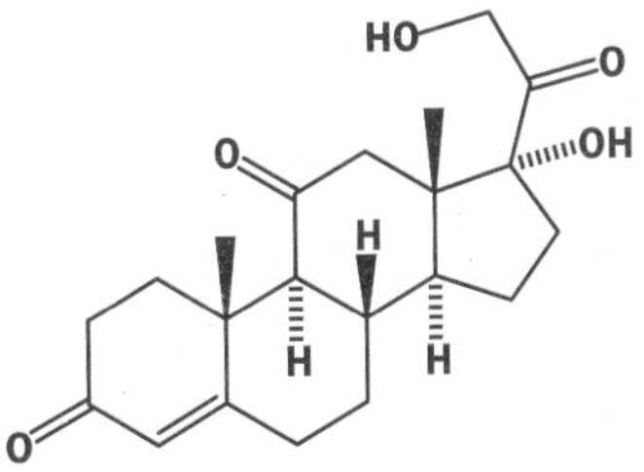

cortisone

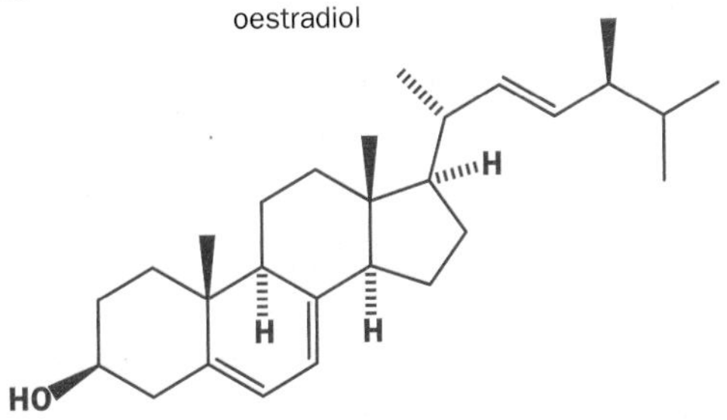

ergosterol

For reference, here is the numbering of the steroid nucleus, not because we want you to learn it, but because it is often used without explanation in books and it is not obvious.

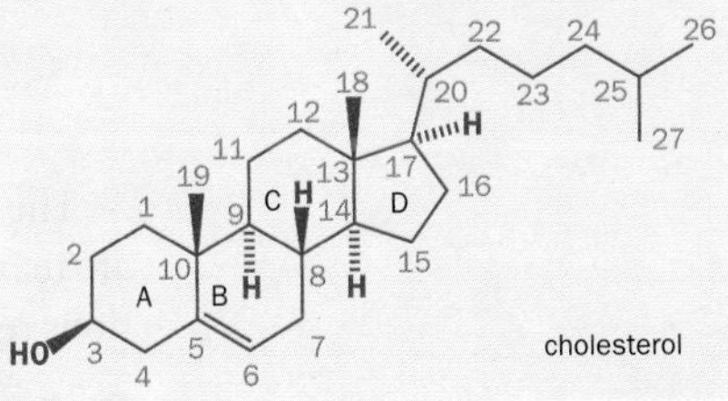

cholesterol

All share the skeleton of four fused rings, three six-membered and one five-membered and conventionally lettered A–D. Beyond the ring stereochemistry and some common oxygenation patterns they share little else. Some (such as the female sex hormones) have an aromatic A ring; some have side-chains on the five-membered ring.

At first glance, it is not at all clear that steroids are terpenoid in origin. The $5n$ numbers are absent—cholesterol is a C_{27} compound while the others variously have 20, 21, or 23 carbon atoms. Studies with labelled mevalonic acid showed that cholesterol is terpenoid, and that it is formed from two molecules of farnesyl pyrophosphate ($2 \times C_{15} = C_{30}$ so three carbon atoms must be lost). Labelling of one or other of the methyl groups (two experiments combined in one diagram!) showed that two of the green carbon atoms and one of the black carbon atoms were lost during the biosynthesis.

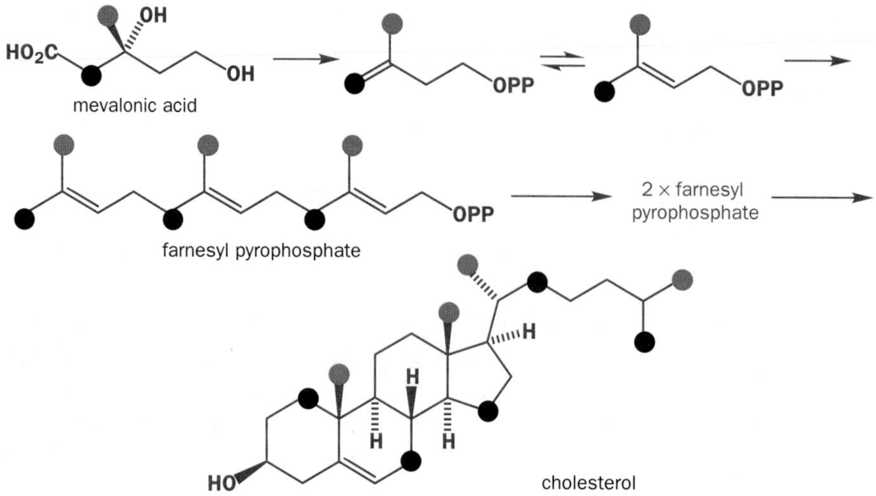

mevalonic acid

farnesyl pyrophosphate

2 × farnesyl pyrophosphate

cholesterol

It is not obvious how the two farnesyl pyrophospate molecules could be combined to make the steroid skeleton, and the chemistry involved is extraordinary and very interesting. The first clues came from the discovery of the intermediates squalene and lanosterol. Squalene is obviously the farnesyl pyrophosphate dimer we have been looking for while lanosterol looks like cholesterol but still has all 30 carbon atoms.

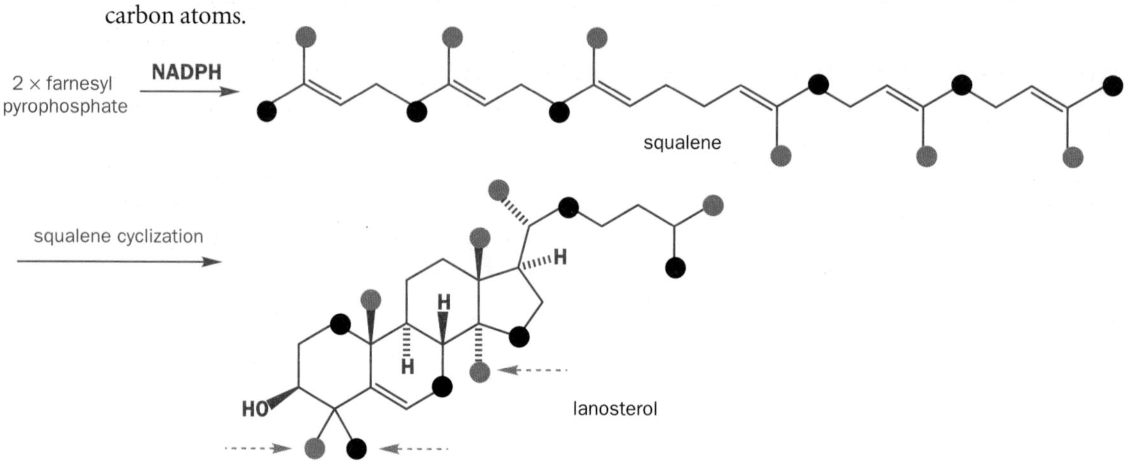

2 × farnesyl pyrophosphate   **NADPH**

squalene

squalene cyclization

lanosterol

The three carbon atoms that are lost from lanosterol (C_{30}) in its conversion to cholesterol (C_{27}) are marked with brown arrows. Now at least we know which carbon atoms are lost. But many questions remain to be answered.

- How does farnesyl pyrophosphate dimerize so that two electrophilic carbon atoms (CH_2OPP) join together?
- Why does the formation of squalene require the reducing agent NADPH?
- How does squalene cyclize to lanosterol so that the very odd labelling pattern can be achieved?
- Where do the three lost carbon atoms go?
- How is the sterochemistry controlled?

Before we tell you the answers, be warned: prepare for some surprises, and be ready to hold back outright disbelief!

The formation of squalene from farnesyl pyrophosphate

If the reducing agent NADPH is omitted from the cell preparation, squalene is not formed. Instead, another farnesyl pyrophosphate dimer accumulates—presqualene pyrophosphate—which has a three-membered ring and in which we can see that the two molecules of farnesyl pyrophosphate are joined in a slightly more rational way.

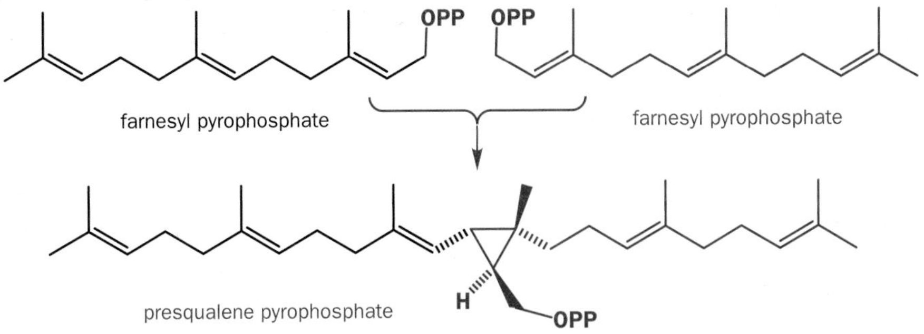

farnesyl pyrophosphate

farnesyl pyrophosphate

presqualene pyrophosphate

Maybe it's not so obvious that this is more rational! The first C–C bond formation is quite straight-forward. The alkene in the red molecule attacks the allylic pyrophosphate in the black molecule in a simple S_N2 reaction. The product is a stable carbocation. Only one C–C bond remains to be formed to close the three-membered ring and this occurs by the loss of a proton from the black molecule.

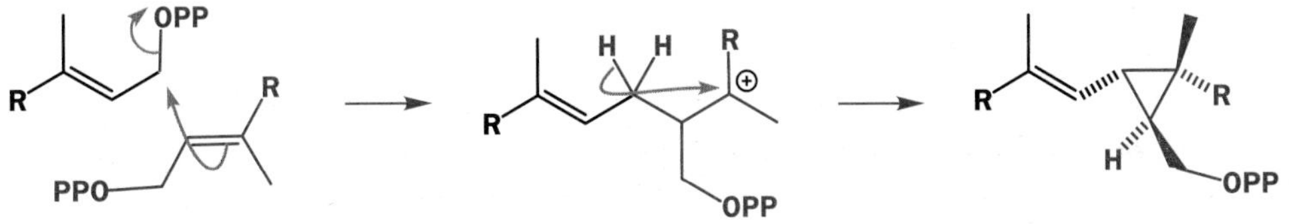

■
We will abbreviate the long terpene side-chain to 'R' from now on.

This is a very remarkable reaction. Such reactions do not occur chemically: this biological one occurs only because the molecule is held in the right shape by the enzyme and because the new ring is three-membered. Three-membered rings are very easily formed but also very easily opened—and that is what happens to this ring. In the presence of NADPH, a series of rearrangements gives a series of carbocations, the last of which is trapped by reduction.

The first step is the migration of one of the bonds (shown in green) of the three-membered ring to displace the pyrophosphate leaving group, expand the ring to four-membered, and release some strain. Now the cyclobutyl cation breaks down to give an open-chain allylic cation stabilized by one of the alkenes. This is the cation that is reduced by NADPH.

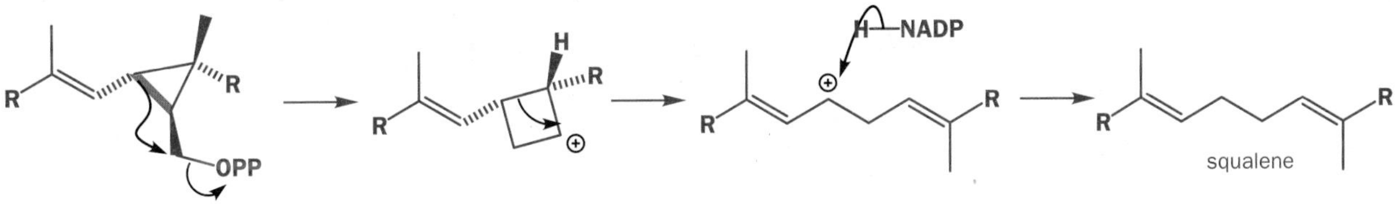

squalene

If you follow this sequence backwards, you will see that the originally formed 'rational' bond (shown in green) is the one that migrated and is retained in squalene, while the second bond is cleaved in the last step.

This may all seem far-fetched, but it happens in laboratory reactions too! Treatment of the simplest cyclopropyl alcohol with HBr gives cyclobutyl bromide by a similar rearrangement.

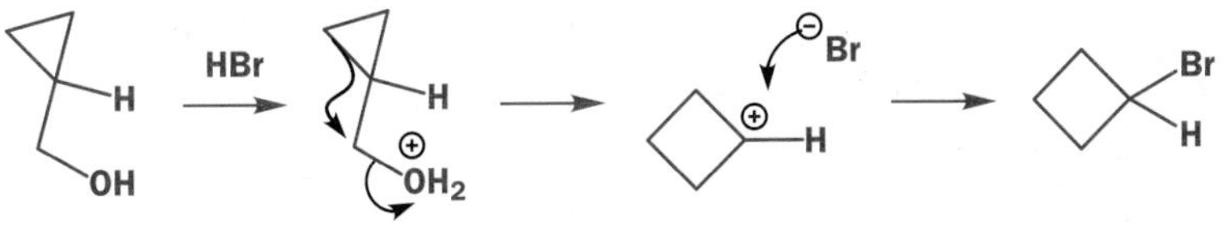

In fact, cyclopropylmethyl compounds, cyclobutyl compounds, and homoallyl compounds are all in equilibrium in acid solution and mixtures of products are often formed. The delocalized cation

shown has been suggested as an intermediate. Make sure that you can draw mechanisms for each
starting material to give the intermediate cation and from the cation to each product.

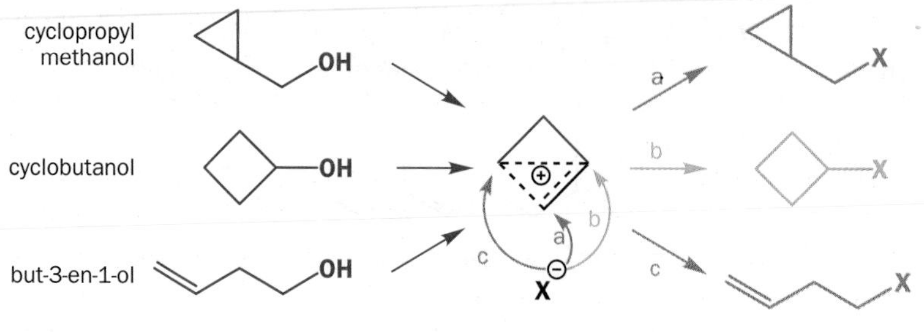

Squalene to lanosterol

The next step is simple—the epoxidation of one of the terminal double bonds—but it leads to two of the
most remarkable reactions in all of biological chemistry. Squalene is not chiral, but enzymatic epoxidation
of one of the enantiotopic alkenes gives a single enantiomer of the epoxide with just one stereogenic centre.

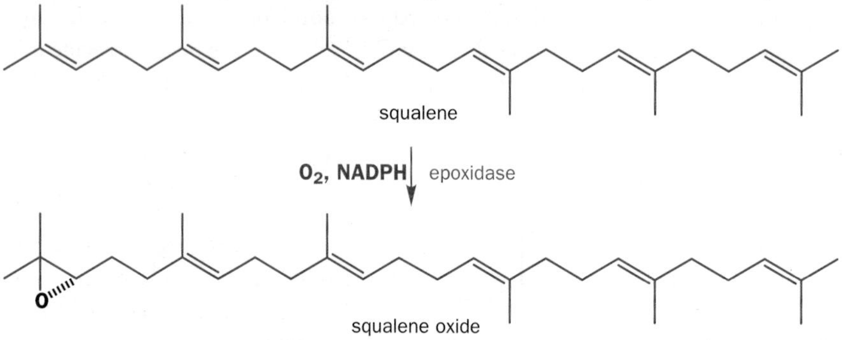

We will start now to draw squalene in a coiled up way as the next step is the polycyclization of the
epoxide. The basic reaction is best seen first in the flat, though we will draw the stereochemistry
immediately. The first alkene cyclizes on to the epoxide and then each remaining alkene cyclizes on
to the next to give a stable tertiary cation.

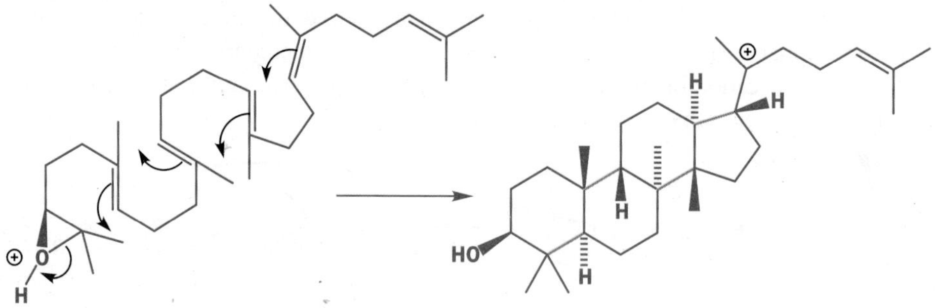

By analogy with what has gone before, you might now expect a tame hydration or reduction of
this cation. Nothing of the sort! A rearrangement occurs in which *five* consecutive 1,2-shifts are followed by an elimination. Since this reaction organizes the backbone of the steroids, it is often called
the **steroid backbone rearrangement**.

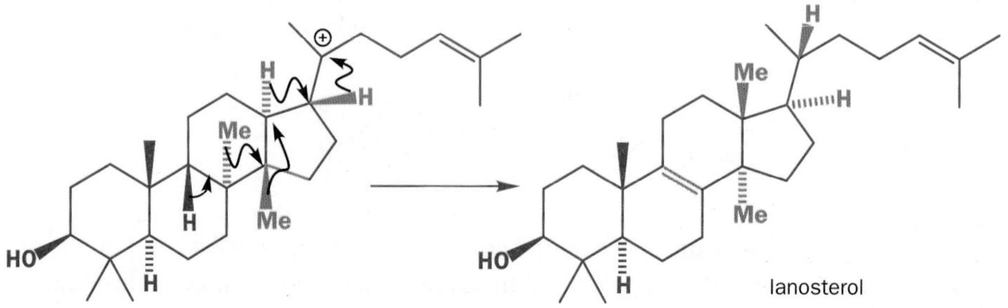

Finally, we have reached lanosterol. Now we will go back over these two steps and discuss them a bit more. Consider first the regiochemistry of the cyclization. The epoxide opens in the way we would expect to give positive charge at the more substituted carbon atom and then all the alkenes attack through their less substituted end (again as we would expect to give positive charge at the more substituted carbon atom)—all except one. The third alkene cyclizes the 'wrong' way—this is presumably a result of the way the molecule is folded.

We learn much more about the folding by examining the stereochemistry of the product cation. First, all of the stereochemistry of each alkene is faithfully reproduced in the product: the cyclization is stereospecific. This is emphasized in colour in the diagram. The green stereochemistry arises because the green Me and H were *trans* in the first alkene of squalene, the black Me and H *trans* in the second, and the brown *trans* in the third. But what about the relationship between the green methyl and the black H? Or between the black and brown methyls? These were determined by the folding and the key observation is that all the relationships are *trans* except that between the green Me and the black H. Now we can draw a conformation for the cyclization.

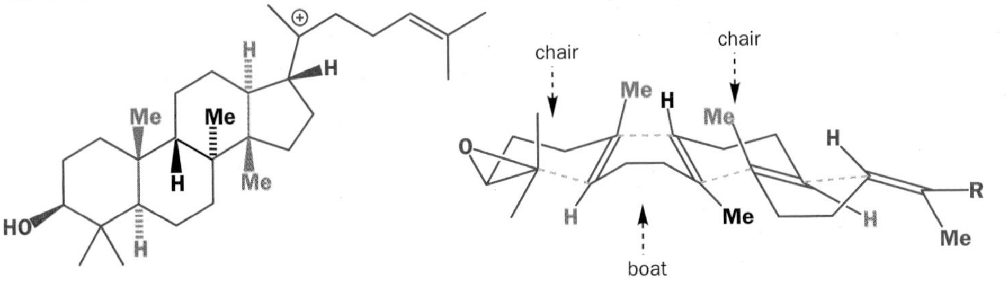

When the transition state for a ring closure forms a chair then a *trans* relationship results. This is the case for the black Me and brown Me. When a boat is formed a *cis* relationship results. This is the case for the green Me and black H. Squalene folds up in a chair–boat–chair conformation and that leads to the observed stereochemistry.

Next, we need to look at the stereochemistry of the rearrangement step. If we draw the product cation as nearly as possible in the conformation of folded squalene, we will see which substituents are axial and which equatorial.

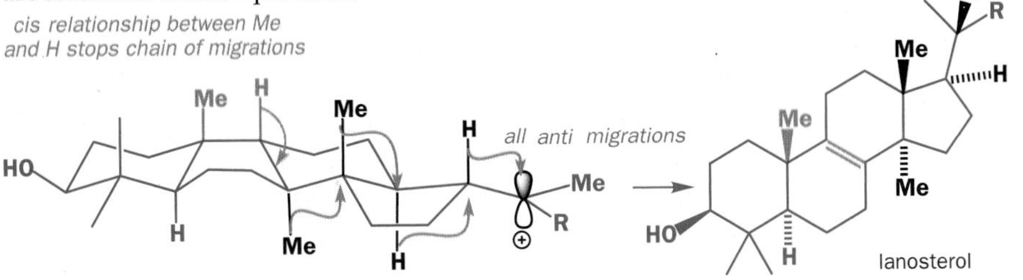

Each group that migrates (black) is axial and is anti-periplanar to the one before so that each migrating group does an S_N2 reaction on the migration terminus with inversion. The chain stops because of the *cis* relationship between the green Me and H in ring B and an elimination of the green H is all that can happen.

The remainder of the biosynthesis of cholesterol requires various redox reactions and is a bit of an anticlimax: the details are summarized in the scheme below.

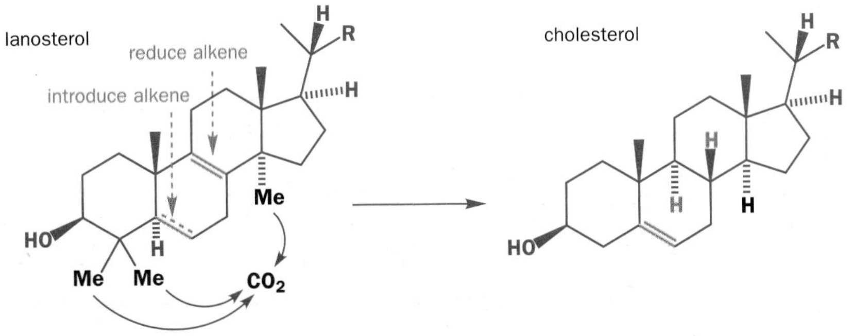

Biomimetic synthesis: learning from Nature

When new and academic-looking reactions are discovered in the laboratory, it often seems only a short time before they are found in nature as well. However, the development of polyolefin cyclization reactions in synthesis occurred by the reverse philosophy—it was inspiration from Nature that led W. S. Johnson to use the reactions in synthesis, including steroid synthesis. This is **biomimetic synthesis**, a strategy that is bound to work provided we can just master the practical details.

There are quite a lot of differences between the chemical and the biochemical versions so far—the chemical ones are less complex and less sophisticated but more versatile. The reactions are just cyclizations without the backbone rearrangements. The most important points of difference are:

- The cyclization is usually begun with a cation from treatment of a cyclic tertiary alcohol rather than an epoxide
- The cyclization sequence is terminated with an alkyne or an allyl silane rather then with simple alkene
- The substituents are placed in the correct positions in the starting material as no rearrangement follows cyclizations
- The cyclizations are all stereospecific as in nature but the rings coil up in an all-chair fashion rather than in a chair–boat–chair fashion as there is no enzyme to shape the molecule
- The product cation is quenched by addition of water rather than loss of a proton

Here is one of Johnson's best examples which leads eventually to a biomimetic synthesis of the human hormone progesterone. The cyclization occurs just on treatment of the tertiary alcohol with acid.

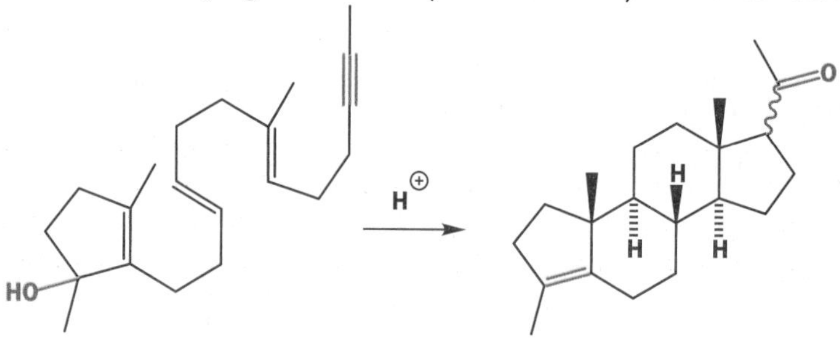

The first step is the formation of a symmetrical allyl cation, which then initiates the cyclization. The next double bond is disubstituted so that it has no built-in regioselectivity but prefers to form a six-membered rather than a five-membered ring B. The next double bond is trisubstituted and directs the formation of a six-membered ring C. The alkyne, being linear, can reach only through its inner end and so a five-membered ring D is formed. The resulting linear vinyl cation picks up a molecule of water to give the ketone via its enol.

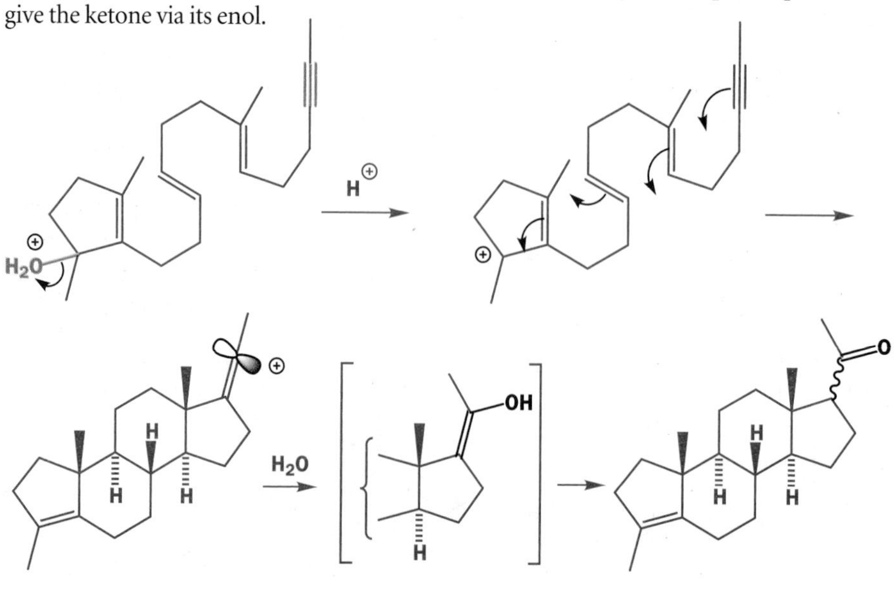

The five-membered ring A is there to ensure efficient initiation of the cyclization by the symmetrical allylic cation. It can easily be opened with ozone and the product cyclized to progesterone.

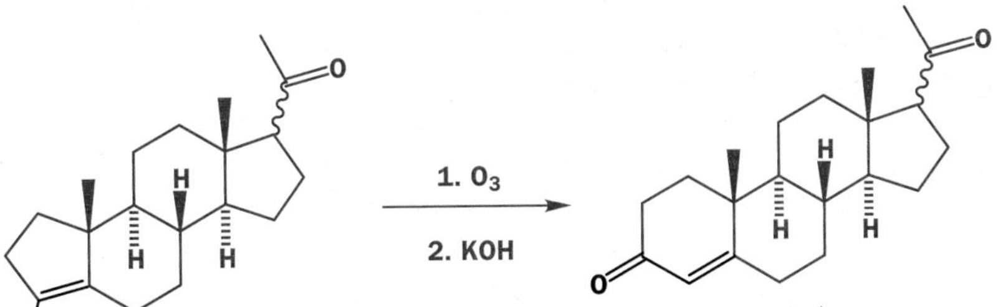

The conformation of the molecule in the moment of cyclization can be seen easily by working backwards from the product. The green dashed lines show new bonds that are being formed. All the six-membered rings in the transition state are chairs and all the ring junctions *trans*. This is an impressive result as there is no enzyme to help the molecule fold up in this way.

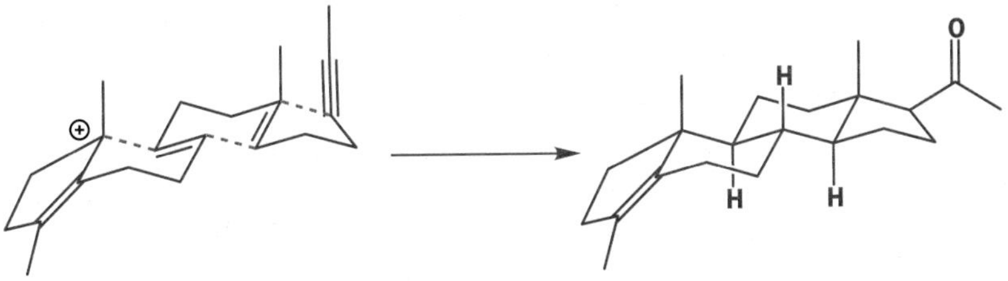

By studying the chemistry that Nature uses in living things we can learn new reactions as well as new ways in which to carry out known reactions. Many of the reactions in this chapter would be laughed at by worldly wise chemists if they appeared in a research proposal, but they have been evolved over millions of years to do precise jobs under mild conditions. Humans have been doing complex organic chemistry for only about a hundred years so that learning from Nature is one of the most important ways in which organic chemistry is advancing at the beginning of the twenty-first century.

Problems

1. Assign each of these natural products to a general class (such as amino acid metabolite, terpene, polyketide) explaining what makes you choose that class. Then assign them to a more specific part of the general class (for example, tetraketide, sesquiterpene).

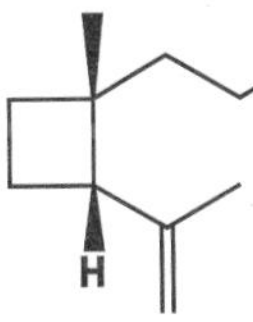

grandisol

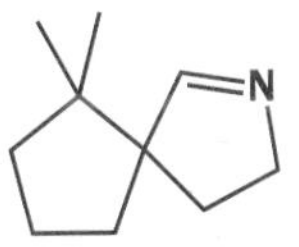

polyzonimine

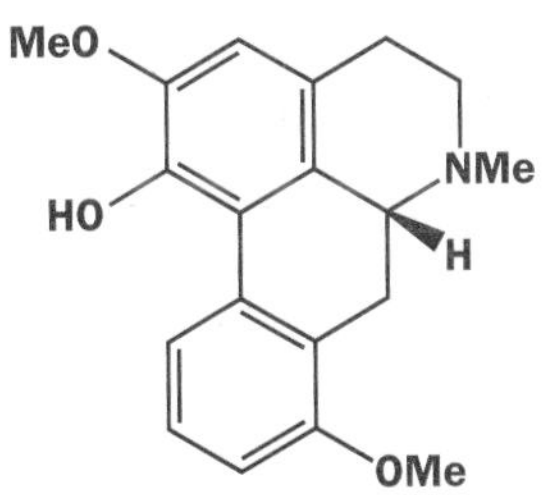

stephanine

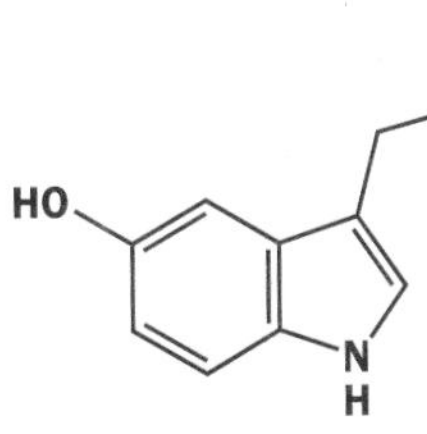

serotonin

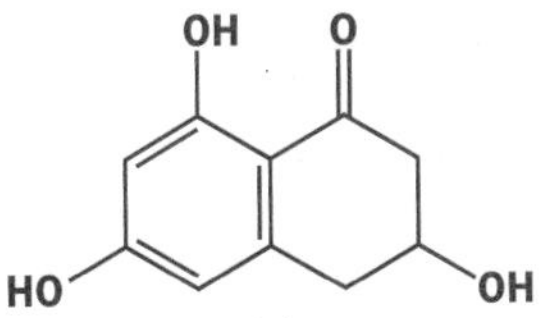

scytalone

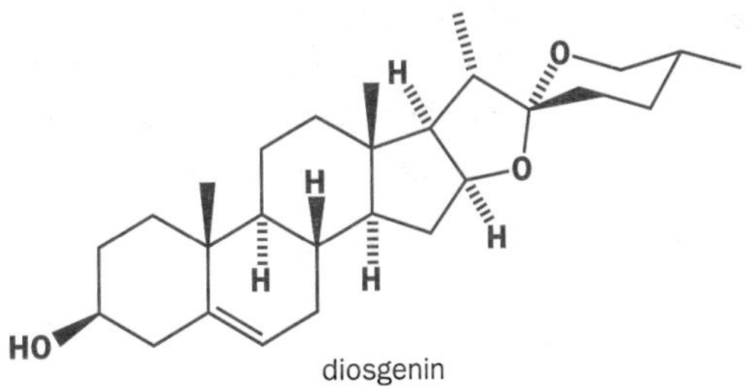

diosgenin

2. Some compounds can arise from different sources in different organisms. 2,5-Dihydroxybenzoic acid comes from shikimic acid (Chapter 50) in *Primula acaulis* but from acetate in *Penicillium* species. Outline details.

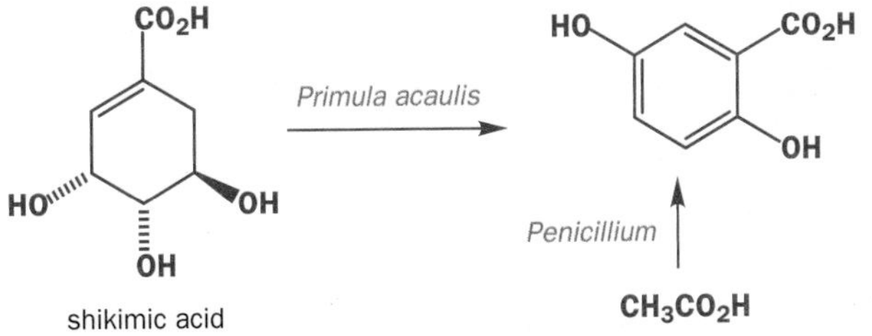

shikimic acid

Primula acaulis

Penicillium

CH_3CO_2H

3. The piperidine alkaloid pelletierine was mentioned in the chapter but full details of its biosynthesis were not given. There follows an outline of the intermediates and reagents used. Fill in the details. Pyridoxal chemistry is discussed in Chapter 50.

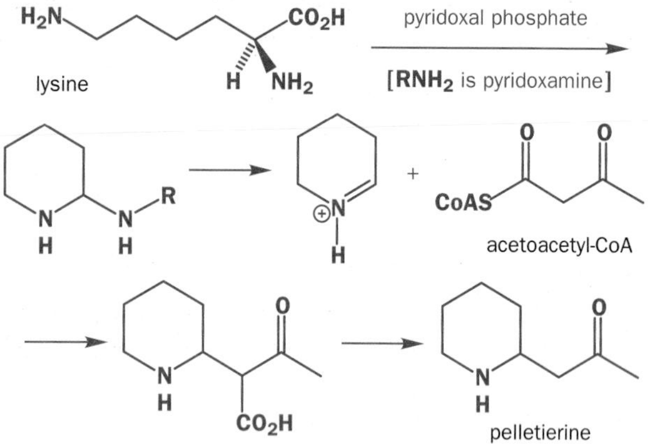

4. The rather similar alkaloids anabasine and anatabine come from different biosynthetic pathways. Labelling experiments outlined below show the origin of one carbon atom from lysine and others from nicotinic acid. Suggest detailed pathways. (*Hint.* Nicotinic acid and the intermediate you have been using in Problem 3 in the biosynthesis of the piperidine alkaloid are both electrophilic at position 2. You also need an intermediate derived from nicotinic acid which is nucleophilic at position 3. The biosynthesis involves reduction.)

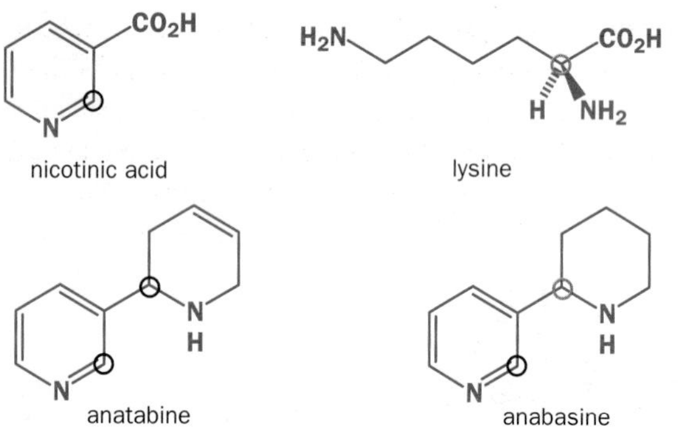

5. The three steps in the biosynthesis of papaverine set out below involve pyridoxal (or pyridoxamine). Write detailed mechanisms.

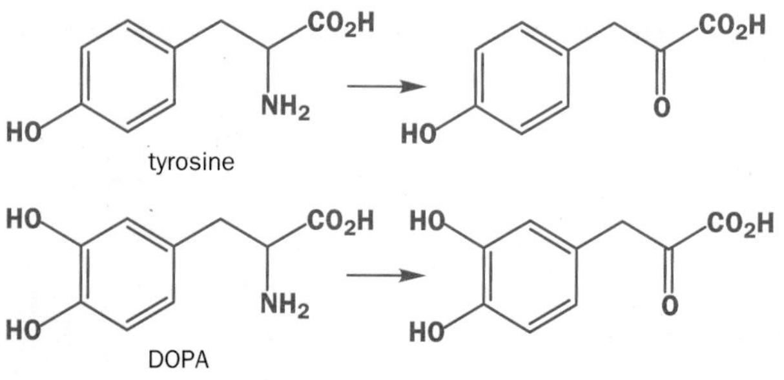

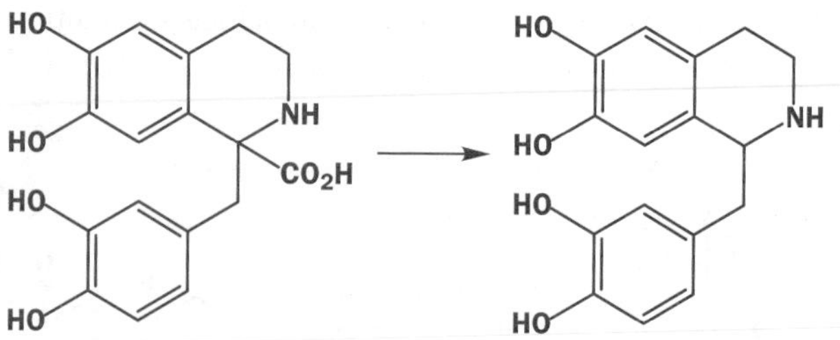

6. Concentrate now on the biosynthesis of scytalone in the first problem. You should have identified it as a pentaketide. Now consider how many different ways the pentaketide chain might be folded to give scytalone.

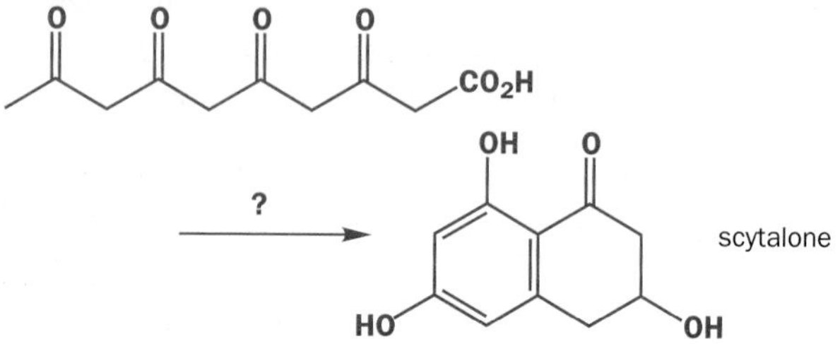

7. This question concerns the biosynthesis of stephanine, another compound mentioned in Problem 1. You should have deduced that it is a benzylisoquinoline alkaloid. Now suggest a biosynthesis from orientaline.

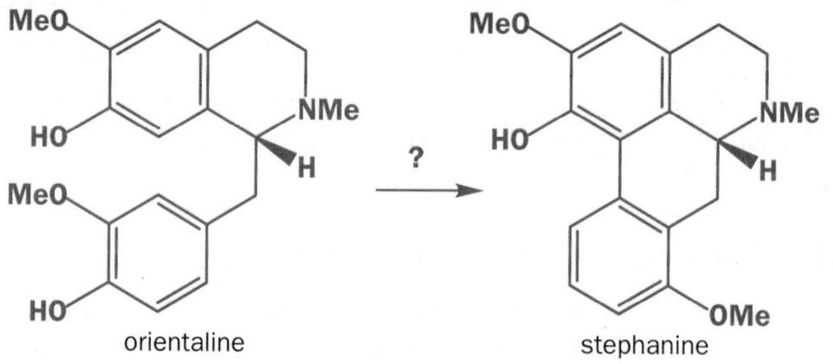

8. Suggest a biosynthesis of olivetol.

olivetol

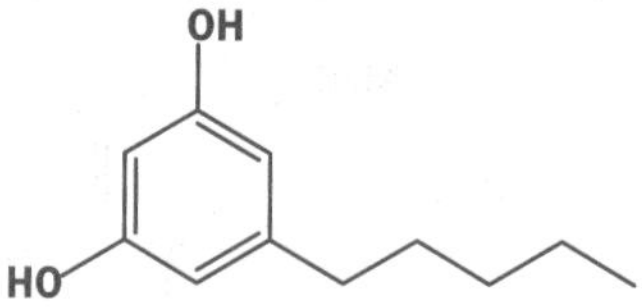

9. Tetrahydrocannabinol, the major psychoactive compound in marijuana, is derived in the *Cannabis* plant from olivetol and geranyl pyrophosphate. Details of the pathway are unknown. Make some suggestions and outline a labelling experiment to establish whether your suggestions are correct.

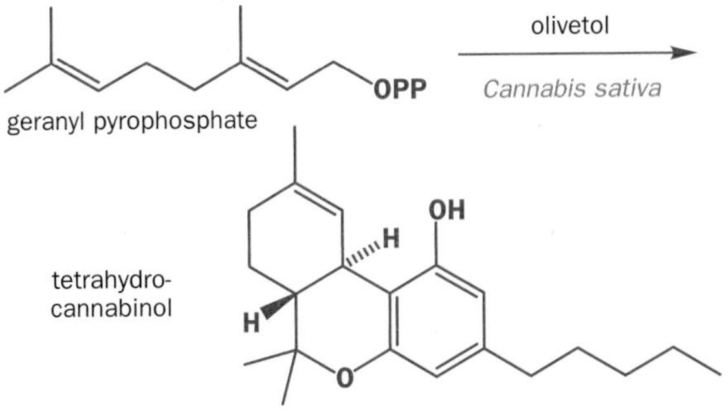

geranyl pyrophosphate

tetrahydro-cannabinol

10. Both humulene, mentioned in the chapter, and caryophyllene are made in nature from farnesyl pyrophosphate in different plants. Suggest detailed pathways. How do the enzymes control which product is formed?

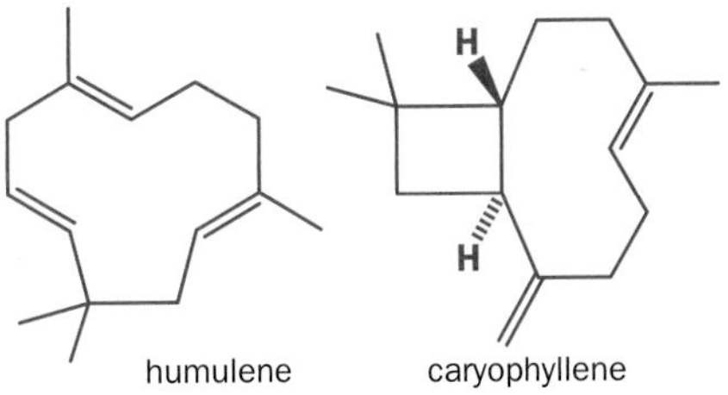

humulene   caryophyllene

11. Abietic acid is formed in nature from mevalonate via the intermediates shown. Give some more details of the cyclization and rearrangement steps and compare this route with the biosynthesis of the steroids.

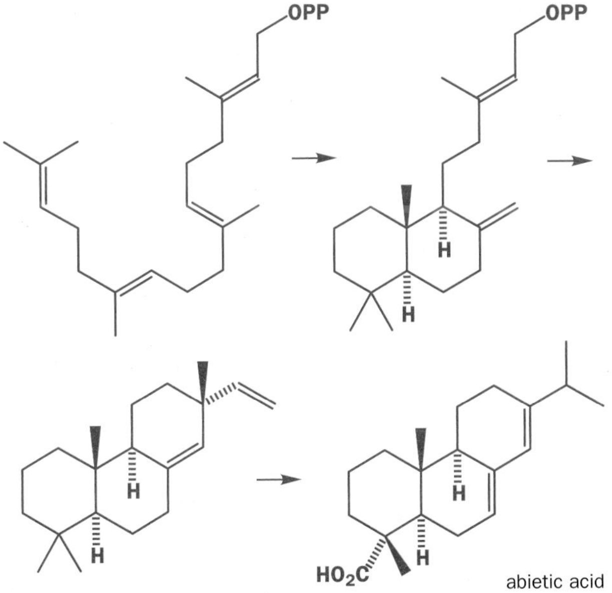

abietic acid

12. Borneol, camphene, and α-pinene are made in nature from geranyl pyrophosphate. The biosynthesis of α-pinene and the related camphor is described in the chapter. In the laboratory bornyl chloride and camphene can be made from α-pinene by the reactions described below. Give mechanisms for these reactions and say whether you consider them to be biomimetic.

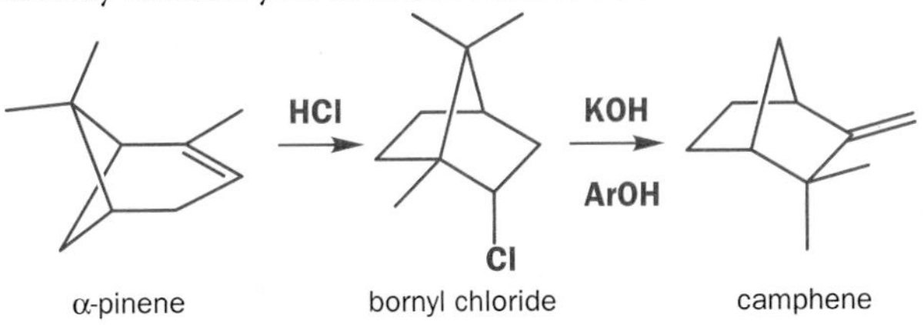

α-pinene   bornyl chloride   camphene

13. Suggest a biosynthetic route to the monoterpene chrysanthemic acid that uses a reaction similar to the formation of squalene in steroid biosynthesis.

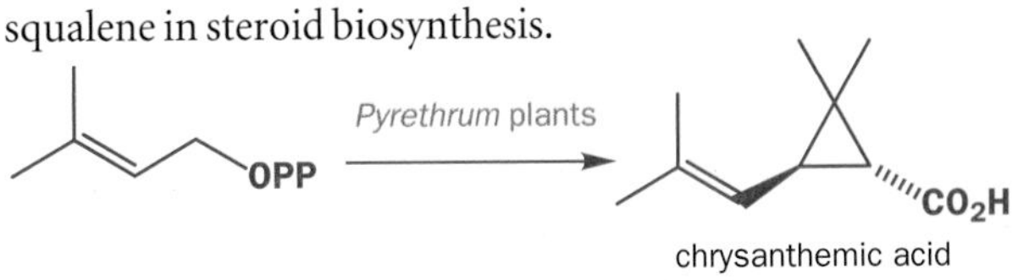

chrysanthemic acid

How could the same route also lead to the natural products yomogi alcohol and artemisia ketone?

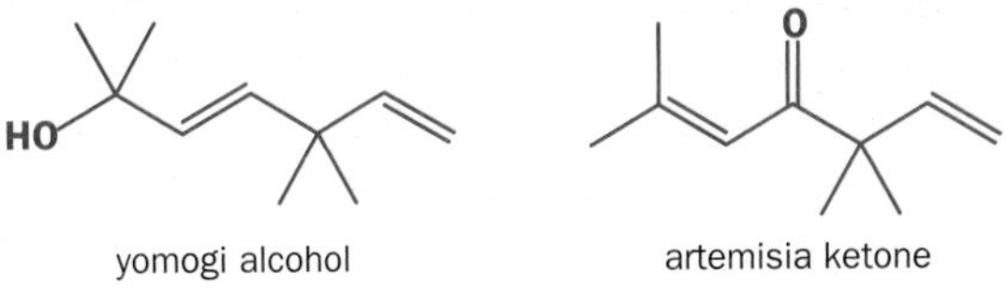

yomogi alcohol   artemisia ketone

14. In the chapter we suggested that you could detect an acetate starter unit and seven malonate additional units in the skeleton of brefeldin. Give the mechanism of the addition of the first malonyl CoA unit to acetate. Draw out the structure of the complete acyl polymalonate chain and state clearly what must happen to each section of it (reduction, elimination, etc.) to get brefeldin A.

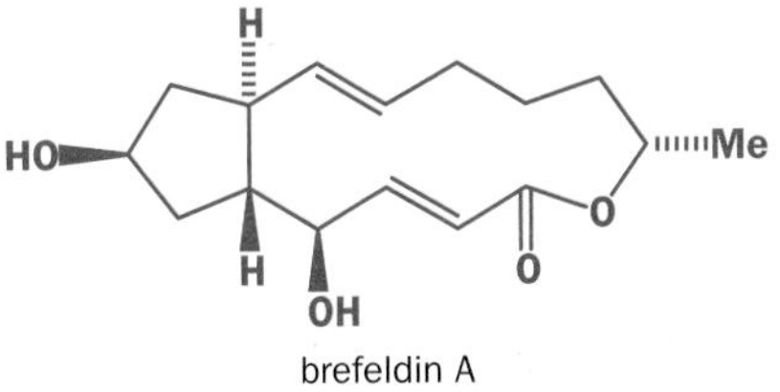

brefeldin A

15. This chemical experiment aims to imitate the biosynthesis of terpenes. A mixture of products results. Draw a mechanism for the reaction. To what extent is it biomimetic, and what can the natural system do better?

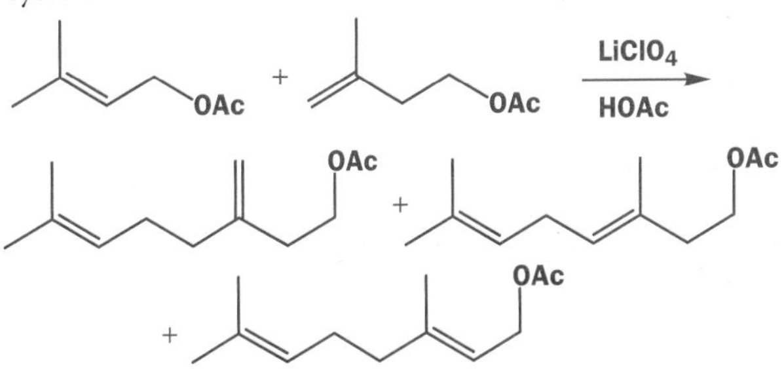

Polymerization

Connections

Building on:
- Carbonyl chemistry ch12 & ch14
- Substitution reactions ch17
- Radical reactions ch39
- Protecting groups and synthesis ch24–ch25
- The aldol reaction ch27
- Making double bonds ch31
- Cycloadditions ch35
- Heterocycles ch43–ch44
- Organometallics ch48
- The chemistry of life ch49
- Natural products ch51

Arriving at:
- Some molecules react together to form oligomers
- Some molecules spontaneously polymerize
- Polyamides, polyesters, and polycarbonates are formed by substitution reactions at carbonyl groups
- Polyurethane foams are formed by nucleophilic attack on isocyanates
- Epoxy adhesives work by polymerization via substitution reactions at saturated carbon
- The most important polymers are derived from alkene monomers
- Alkenes can be polymerized by radical, cationic, anionic, or organometallic methods
- Cross-linking or co-polymerization changes the physical properties of polymers
- Reactions on polymers are involved in paint drying, rubber strengthening, and the chemical synthesis of peptides

Most of the things you can see about you at this moment are made of organic polymers. Skin, clothes, paper, hair, wood, plastic, and paint are among them. Teeth, muscle, glue, cling film, starch, crab shells, and marmalade are all polymer-based too. In this chapter we will explore the world of polymers. We will ask questions like these:

- What makes a molecule prefer to react with others of its kind to form a polymer?
- What mechanisms are available for polymerization reactions?
- How can polymerization reactions be controlled?
- How are the properties of polymers related to their molecular structure?

Monomers, dimers, and oligomers

Cyclopentadiene featured in Chapter 35 as an important diene in the Diels–Alder reaction. If you try to buy 'cyclopentadiene' you will find that the catalogues list only 'dicyclopentadiene' or 'cyclopentadiene dimer'. The dimerization of cyclopentadiene is reversible: the monomer dimerizes by a Diels–Alder reaction at room temperature to give the dimer and the reaction is reversed on heating. So the dimer is a good source of the monomer.

cyclopentadiene dimer

heat (boil) / stand at room temperature

2 cyclopentadiene monomers

Other familiar cases of stable dimers are neutral boron and aluminium hydrides. DIBAL, for example, exists as two molecules linked by Al–H–Al bonds in a four-membered ring. Again, the dimer is a practical source of monomer for chemical reactions.

DIBAL dimer

DIBAL monomer

Simple aldehydes easily form trimers. When cyclopentanecarbaldehyde is prepared, it is a colourless liquid. On standing, particularly with traces of acid, it forms the crystalline trimer. The trimer is a stable six-membered heterocycle with all substituents equatorial

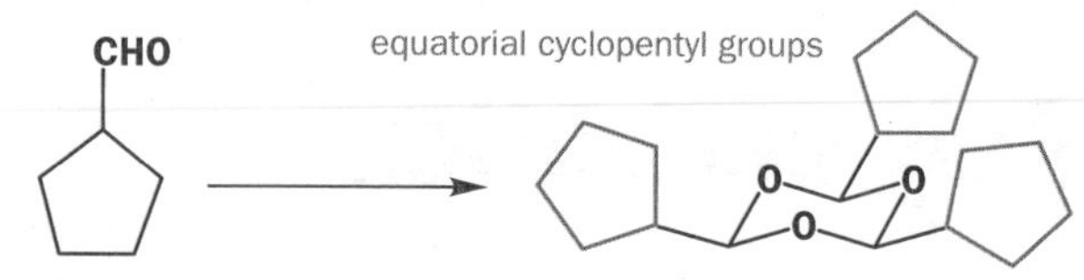

cyclopentane carbaldehyde

cyclopentane carbaldehyde trimer 2,4,6-tricyclopentyl-1,3,5-trioxane

Acetaldehyde (ethanal) forms a liquid trimer called 'paraldehyde', which reverts to the monomer on distillation with catalytic acid. More interesting is 'metaldehyde', the common slug poison,

'metaldehyde'

which is an all-*cis* tetramer (2,4,6,8-tetramethyl-1,3,5,7-tetroxocane) formed from acetaldehyde with dry HCl at below 0 °C. Metaldehyde is a white crystalline solid that has all the methyl groups pseudoequatorial, and it reverts to acetaldehyde on heating.

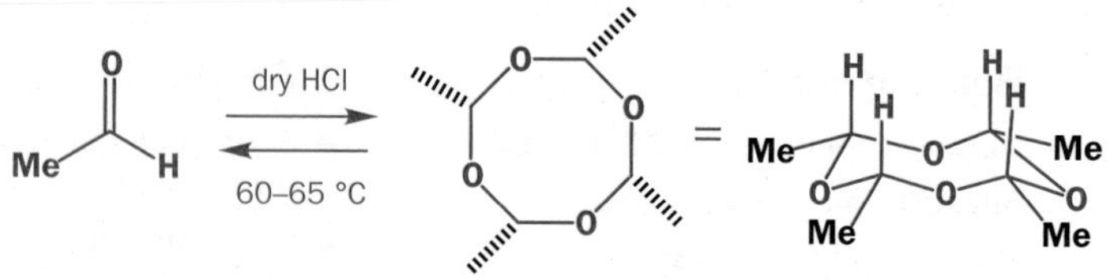

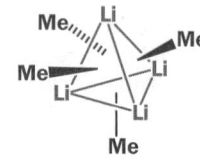

Another tetramer is methyl lithium. MeLi is a very reactive compound in the monomeric state, and it crystallizes as a tetramer: a tetrahedron of lithium atoms with a methyl group 'plugged in' to the centre of each face.

Whereas oxygen gas consists of diatomic molecules O_2, crystalline sulfur is S_8, a cyclic octomer. Such multiples are usually called **oligomers** (*oligo* = a few). The monomer in this case would be the sulfur atom. The shape of the S_8 ring is very similar to that of the eight-membered ring of metaldehyde.

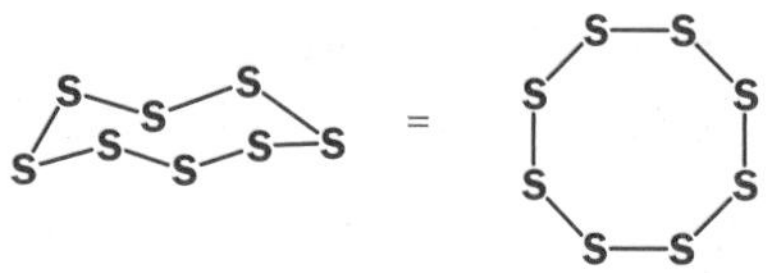

If you buy formaldehyde (methanal), which is in fact a gas, b.p. −19 °C, you have four choices. You can buy a 37% aqueous solution 'formalin' which is mostly hydrate in equilibrium with a small amount of formaldehyde, or the crystalline trimer (1,3,5-trioxane), or a white solid called (misleadingly) 'paraformaldehyde', or another white solid called polyoxymethylene.

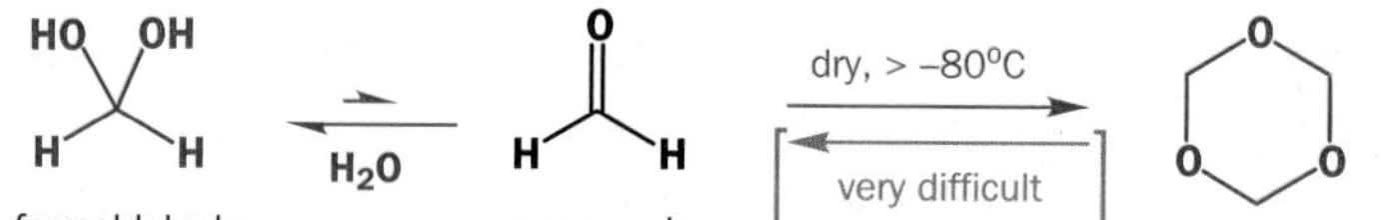

formaldehyde hydrate

monomeric formaldehyde

1,3,5-trioxane

Trioxane is not a good source of formaldehyde as it is very stable but the two other solids are good sources. Both paraformaldehyde and, more obviously, polyoxymethylene are **polymers**. Each molecule of either polymer consists of a large number of formaldehyde molecules reacted together.

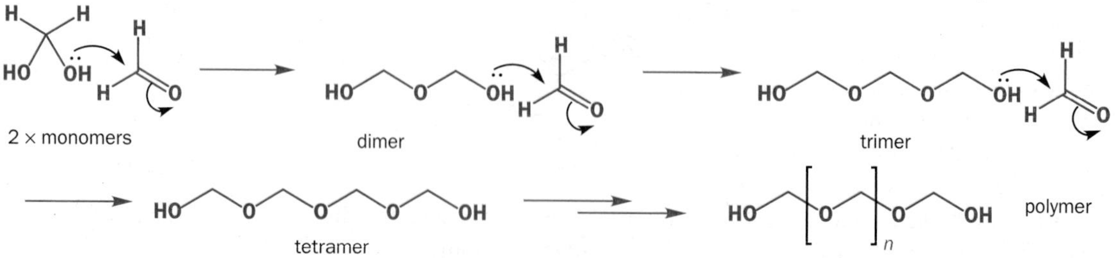

2 × monomers   dimer   trimer

tetramer   polymer

Paraformaldehyde is made by evaporation of aqueous formaldehyde to dryness and is a water-soluble polymer. Polyoxymethylene is made by heating formaldehyde with catalytic sulfuric acid and is *not* soluble in water. They are both linear polymers of formaldehyde, so how can they be so different? The answer is in the polymer chain length—the n in the diagram. Paraformaldehyde is water-soluble because it has short chain lengths, about $n = 8$ on average, and so it has many hydrophilic OH groups. Polyoxymethylene has much longer chain lengths, $n > 100$ on average, and so has very few OH groups per monomer of formaldehyde.

Trioxane is formed when the trimer cyclizes instead of continuing to polymerize. All the oligomers and polymers of formaldehyde have this potential as there is a hemiacetal group at each end of the chain.

● Summary of what we know so far

Not much, you might think. Actually we have mentioned some important things about polymerization, which we will discuss further in the pages that follow.

- Polymerization tends to occur at low temperature
- Depolymerization tends to occur at high temperature
- Polymerization competes with cyclic oligomer formation
- Different polymers of the same monomer can have different chain lengths
- The chain length varies about a mean value in a given polymer
- The properties of polymers depend on chain length (among other things!)

Check back over these last few pages to make sure you see which pieces of evidence establish each of these points.

> There is no exact limit to the terms oligomer and polymer. You have just seen us refer to paraformaldehyde—on average an octomer—as a polymer. The terms monomer, dimer, trimer, tetramer, etc. do have exact meanings. Oligomer usually means > 3 and < 25 but different authors will use the term in different ways.

Polymerization by carbonyl substitution reactions

In general, carbonyl compounds do not polymerize by themselves. It is only the exceptional reactivity of formaldehyde as an electrophile that allows repeated nucleophilic addition of hemiacetal intermediates. A more common way to polymerize carbonyl compounds is to use two different functional groups that react together by carbonyl substitution to form a stable functional group such as an amide or an ester. Nylon is just such a polymer.

Polyamides

You may have carried out the nylon rope trick in a practical class. The diacid chloride of adipic acid is dissolved in a layer of a heavy organic solvent such as CCl_4 and a layer of aqueous hexane-1,6-diamine is carefully placed on top. With a pair of tweezers you can pick up the film of polymer that forms at the interface and draw it out to form a fibre. The reaction is a simple amide formation.

After the first amide is formed, one end of the new molecule is nucleophilic and the other electrophilic so that it can grow at both ends. The polymer is made up of alternating $-NH(CH_2)_6NH-$ and $-(CH_2)_4CO-$ units, each having six carbon atoms, and is called 'nylon 6.6'. Another and much simpler way to make nylon is to polymerize caprolactam. This monomer is a cyclic amide and the polymer does not have alternating units—instead, each unit is the same.

Caprolactam

Caprolactam can be made by the Beckmann rearrangement of the oxime of cyclohexanone. (Check that you can draw the mechanisms, of both these reactions and look at Chapters 14 and 37 if you find you can't.) Cyclohexanone used to be made by the oxidation of cyclohexane with molecular oxygen until the explosion at Flixborough in Lincolnshire on 1 June 1974 that killed 28 people. Now cyclohexanone is made from phenol.

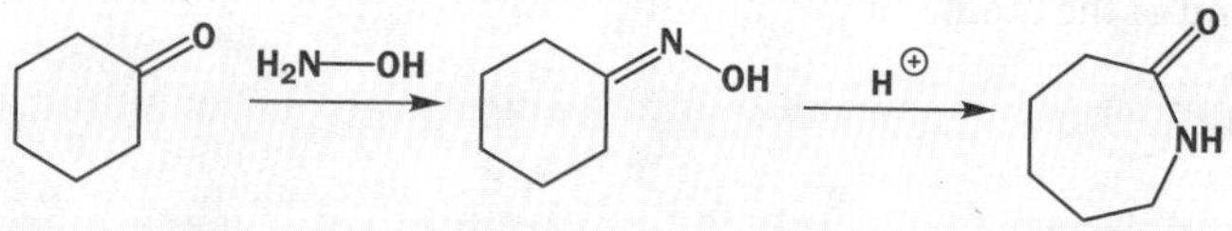

So how is this polymerization initiated? A small amount of water is added to hydrolyse some of the caprolactam to 6-aminohexanoic acid. The amino group can then attack another molecule of caprolactam and so on. The amount of water added influences the average chain length of the polymer.

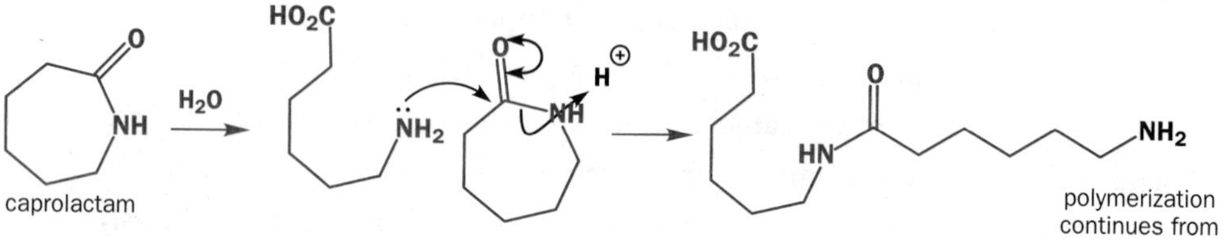

These synthetic polyamides are made up of the same repeating unit but will inevitably have a range of molecular weights as the polymer length will vary. This is a different story from that of the natural polyamides—peptides and proteins—that you met in Chapter 49. Those polymers were made of twenty or so different monomers (the amino acids) combined in a precise order with a precise stereochemistry and all molecules of the same protein have the same length. Nonetheless, some of their uses are almost identical: both nylon and wool are polyamides, for example.

Polyesters

Much the same act can be carried out with dicarboxylic acids and diols. The most famous example is the polymer of ethylene glycol (ethane-1,2-diol) and terephthalic acid, which can be made simply by melting the two components together so that water is lost in the esterification reaction. The mechanism is obvious.

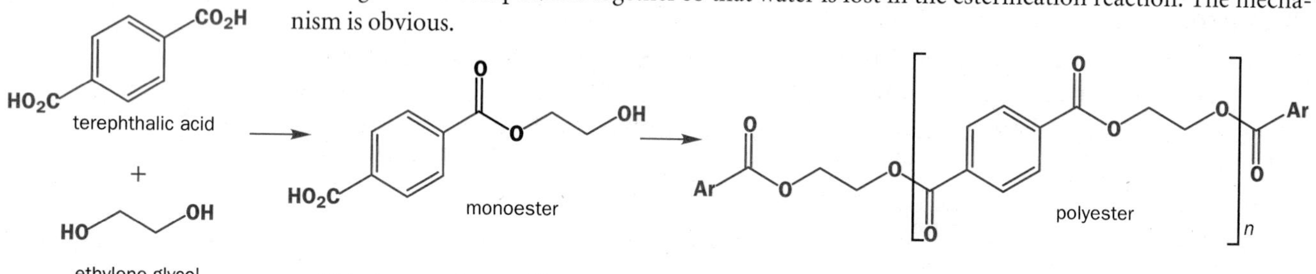

This linear polymer, like nylon, is well shaped for making long fibres and is now so important for making clothes that it is usually just called 'polyester' rather than by the older names such as 'Terylene'.

Polycarbonates

These too are made by carbonyl substitution reactions, but this time the nucleophile is aromatic and the electrophile is an aliphatic derivative of carbonic acid such as phosgene ($COCl_2$) or a carbonate diester [$CO(OR)_2$]. The aromatic nucleophile is a diphenol but the two OH groups are on separate rings joined together by an electrophilic aromatic substitution. This compound is called bisphenol A and has many other applications.

Make sure that you can draw the mechanism for this reaction—two electrophilic aromatic substitutions are involved (Chapter 22). If you need a hint, look at the synthesis of Bakelite on the next page.

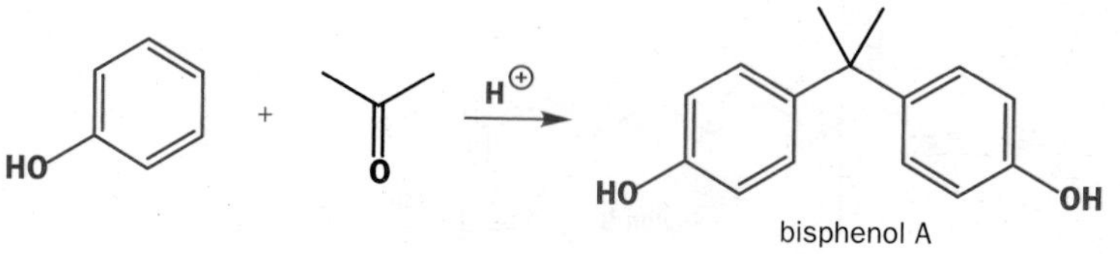

The diphenol reacts with the carbonic acid derivative, which is doubly electrophilic at the same carbon atom.

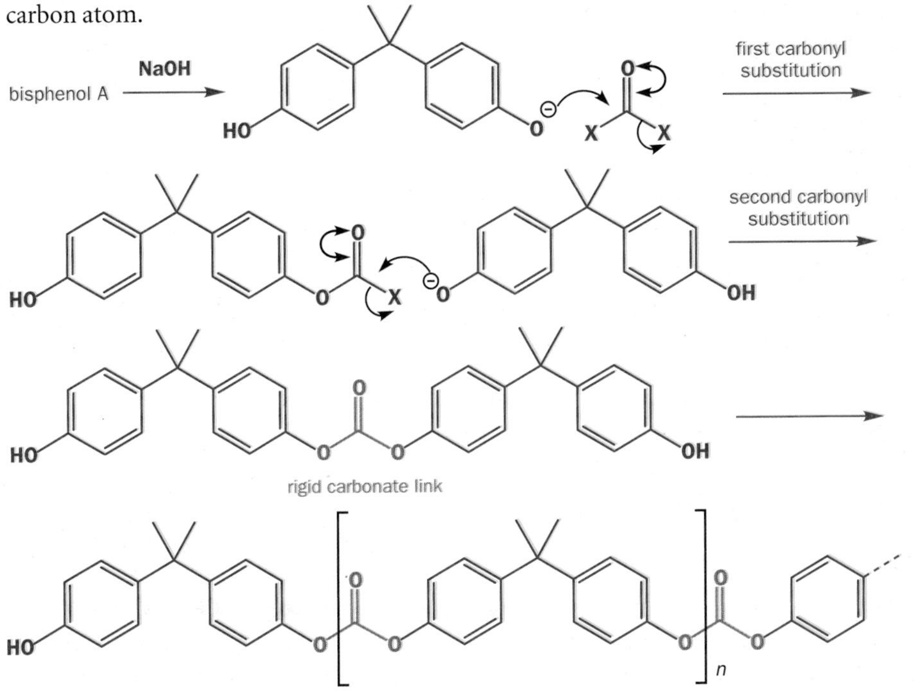

After two carbonyl substitutions the rigid carbonate ester group is formed. This polymer is neither as flexible nor as linear as the previous examples. The carbonate portion is conjugated to the benzene rings and held rigidly in the conformation shown by the anomeric effect (Chapter 42). The only flexibility is where the CMe$_2$ group links the two benzene rings. This is a polymer that combines transparency, lightness, and strength with just enough flexibility not to be brittle. Your safety glasses are probably made of polycarbonate.

Polymerization by electrophilic aromatic substitution

The first synthetic polymers to be of any use were the 'phenol formaldehyde resins' of which the most famous, Bakelite, was discovered by Bäkeland at the turn of the century. He combined phenol and formaldehyde in acid solution and got a reaction that starts like the bisphenol A synthesis.

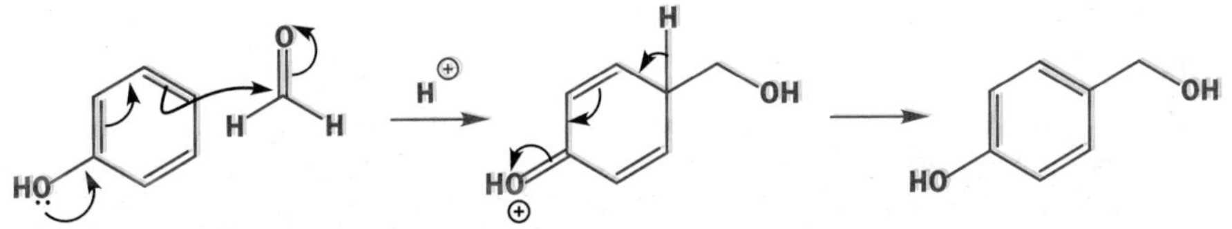

A second acid-catalysed electrophilic aromatic substitution now occurs to link a second phenol to the first. The rather stable benzylic cation makes a good intermediate.

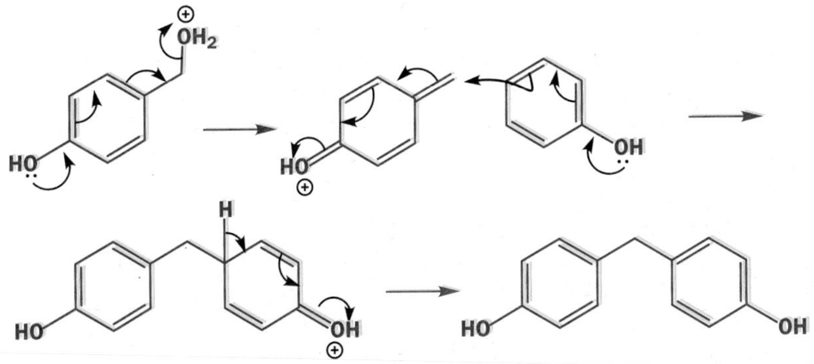

■
If you tried a moment ago, as we suggested, to write the mechanism for the formation of bisphenol A, this is what you should have done (but with acetone, of course, instead of formaldehyde).

Formaldehyde is reactive enough to continue and put another substituent ortho to the OH group in one of the rings. The mechanisms are the same as those we have just written.

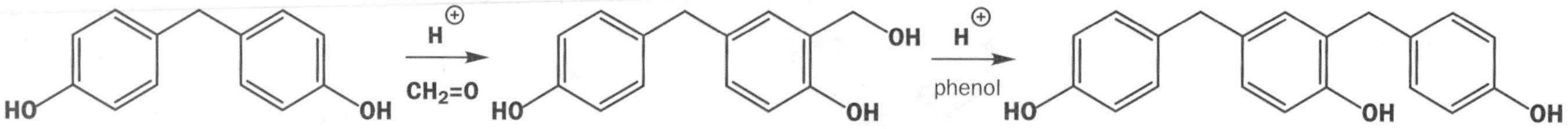

growing points for the Bakelite resin

The carbon chains are *meta* related on the central ring so for the first time we have a **branched polymer**. Complexity can rapidly increase as more phenols linked through more formaldehydes can be joined on to this core structure at several points. Each benzene ring could, in theory, form three new C–C bonds.

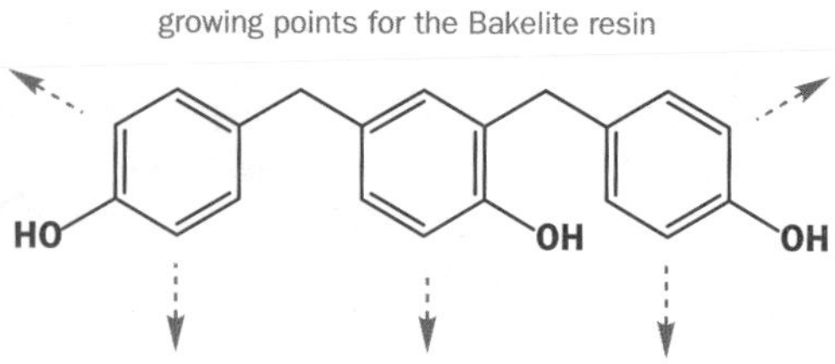

These polymers have the useful property of being **thermosetting**—they are made from liquid mixtures that polymerize on heating to form a solid polymer, and can therefore be moulded easily.

Polymerization by the S$_N$2 reaction

In principle, co-polymerization of a 1,2-diol and a 1,2-dihalide might lead to a polyether.

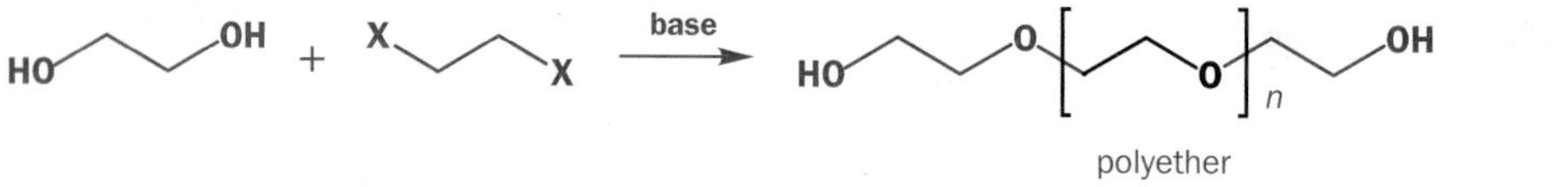

polyether

We mentioned the inventor of crown ethers, Charles Pedersen, on p. 974.

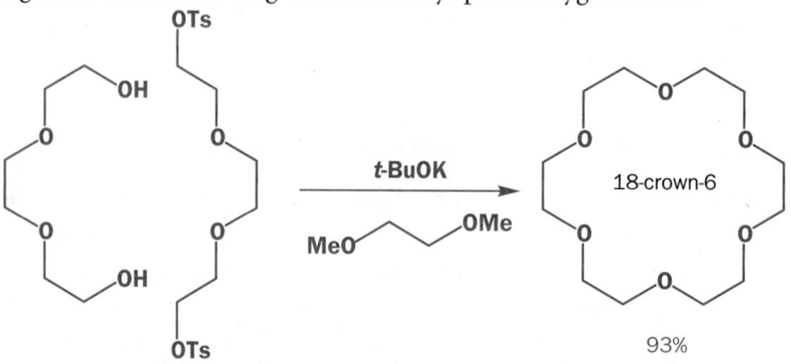

This route is not used because of the large amounts of base needed. One molecule of base is consumed for each new C–O bond made, and these reactions terminate quickly before long chains are made. It is more useful for making the cyclic oligomers called 'crown ethers'. 18-Crown-6 has an eighteen-membered ring with six evenly spaced oxygen atoms.

18-crown-6

93%

These crown ethers have cavities ideal for complex formation with metal ions. They can even carry metal ions into solution in organic solvents. This one, 18-crown-6, is the right size for potassium ions, and a solution of KMnO$_4$ and 18-crown-6 in benzene, so-called 'purple benzene', is a useful oxidizing agent. The high-yielding oligomerization is a template reaction with a potassium ion holding the two reagents together. If a base such as Bu$_4$N$^+$OH$^-$ (which cannot form complexes) is used with the same reagents, linear polymers result.

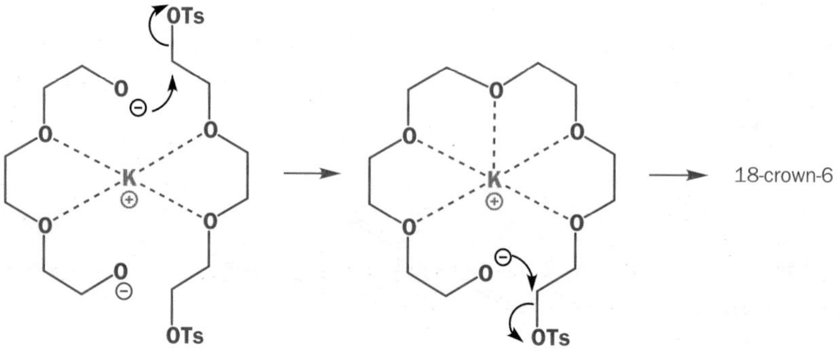

18-crown-6

A more practical way to make linear polyethers is by polymerization of epoxides. Each time an epoxide is opened by a nucleophile, it releases a nucleophilic oxyanion that can attack another epoxide, and so on. The whole process can be initiated by just a catalytic amount of a nucleophile such as an alkoxide or an amine.

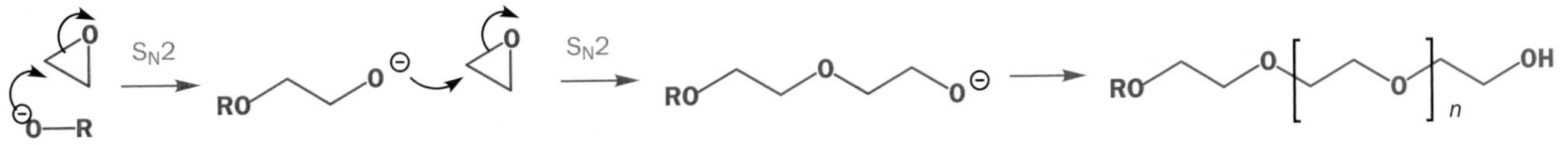

This reaction cannot be controlled—once it is initiated, it runs to completion. Treatment of ethylene oxide with controlled amounts of water does lead to the important coolant ethylene glycol (excess water) and the oligomers di-, tri-, and tetraethylene glycol. These are important solvents for polar compounds. Triethylene glycol is also the starting material for the synthesis of 18-crown-6 above.

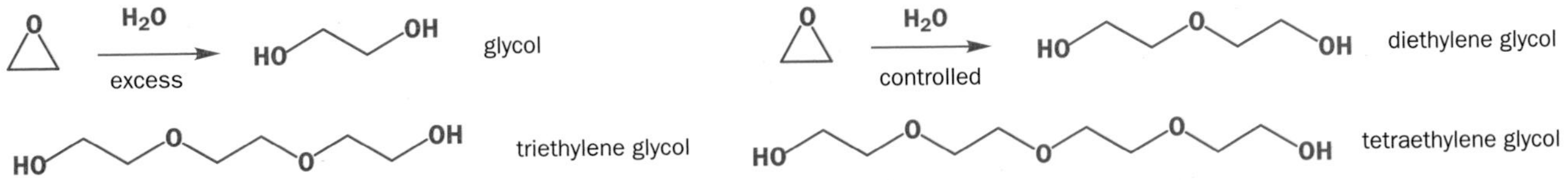

A subtle method of controlling the reaction so that it can be made to run at will is to use bisphenol A as the diol and epichlorohydrin as the epoxide. Epichlorohydrin reacts with nucleophiles at the epoxide end, but the released alkoxide ion immediately closes down at the other end to give a new epoxide.

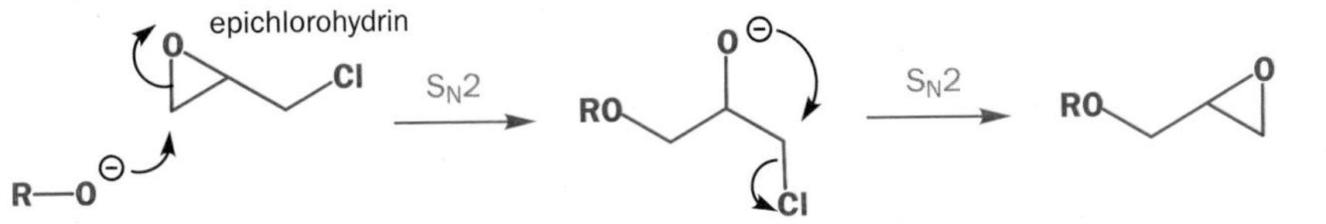

With bisphenol A in alkaline solution, this reaction happens twice and a bis adduct is formed. Further reaction with more bisphenol A creates oligomers with about 8–10 bisphenol A molecules and an epoxide at each end. This is a reasonably stable neutral compound with two terminal epoxides, just waiting for initiation for polymerization to start.

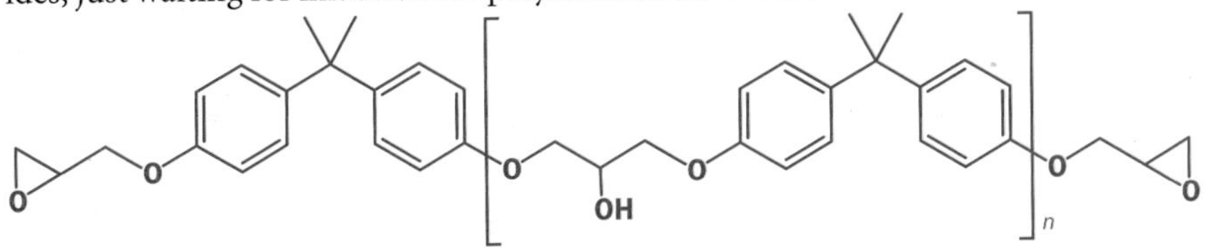

In the CIBA–Geigy glue Araldite, strong enough to glue aeroplane wings on to the fuselage, a solution of this oligomer is mixed with a solution of a polyfunctional amine such as diethylenetriamine. Since each NH$_2$ group can react twice and the NH group once with epoxides, the final polymer has a densely cross-linked structure and is very strong. The reaction is again a simple S$_N$2 process.

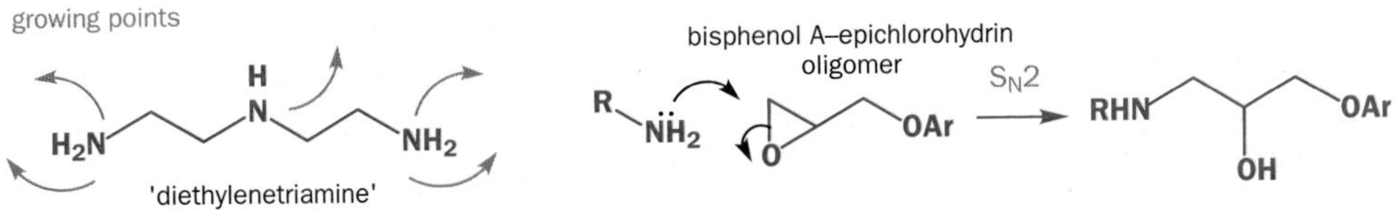

A totally different kind of polymer is a poly-silylether. Dimethylsilyl dichloride polymerizes easily on treatment with hydroxide. Silicon is more susceptible to the S$_N$2 reaction than is carbon and long chains grow quickly.

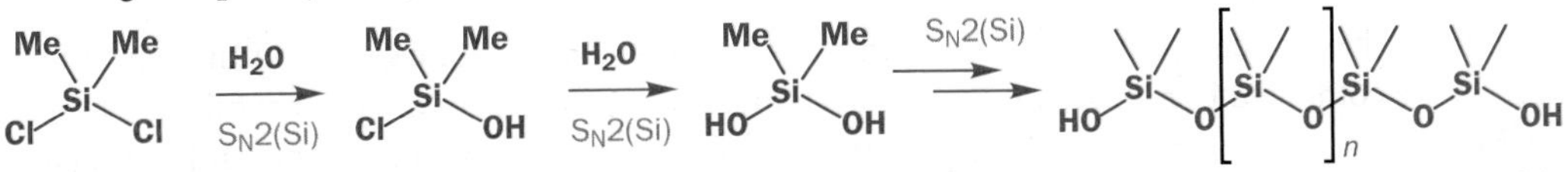

This linear poly(dimethylsiloxane) is an oil and is used in the lab in oil baths as it is more stable and less smelly than conventional paraffin baths at high temperatures.

Polymerization by nucleophilic attack on isocyanates

Isocyanates react with alcohol nucleophiles to give **urethanes**—hybrids between carbonates and ureas—half-esters and half-amides of carbonic acid. Nucleophilic attack occurs at the very reactive linear (sp) carbon in the centre of the isocyanate.

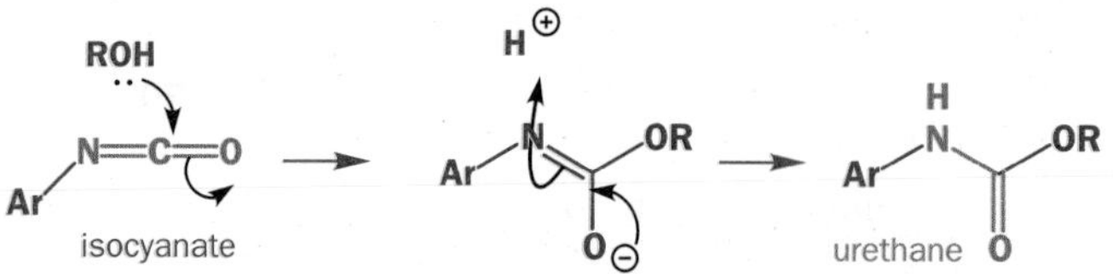

To make a polymer it is necessary to react aryl diisocyanates with diols. Some important polymers of the type, called **elastanes**, are made by using long-chain aliphatic diols from partly polymerized epoxides, rather like those discussed in the last section, and reacting them with diaryl diisocyanates to give a 'pre-polymer'.

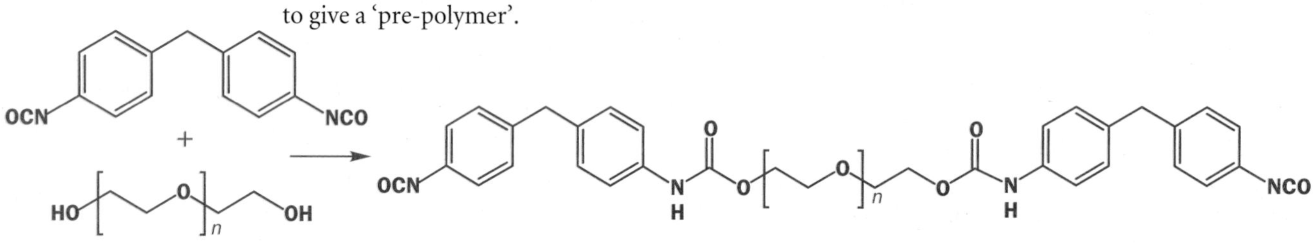

The next stage is to initiate an exothermic linking of the residual terminal isocyanates with simple diamines. The reaction is again nucleophilic attack on the isocyanate, but the new functional group is now a urea rather than a urethane. Showing just one end of the growing polymer:

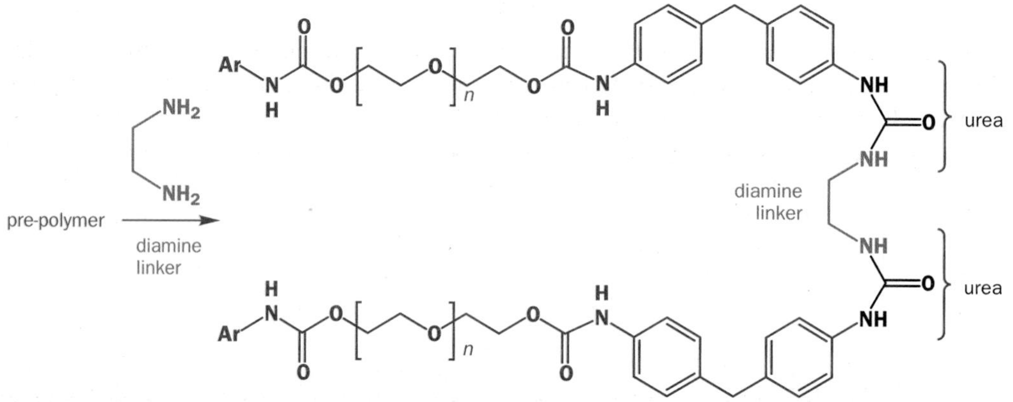

These polymers have short rigid portions (the aromatic rings and the ureas) joined by short flexible 'hinges' (the diamine linker and the CH_2 group between the aromatic ring) and long very flexible portions (the polyether) whose length can be adjusted. The polymer is easily stretched and regains its shape on relaxation—it is an **elastomer**.

Why should it matter that the second polymerization is exothermic? If the diamine linker is added as a solution in a volatile hydrocarbon such as heptane, the heat of the polymerization causes the heptane to boil and the polymer becomes a foam. What is more, the length of the polyether chain determines what kind of foam results. Shorter (~500 $-OCH_2CH_2O-$ units) chains give rigid foams but longer chains (>1000 $-OCH_2CH_2O-$ units) give soft foams. This is only a bare outline of one of the many skills polymer chemists now have in the design of materials. The results are all around us.

So far we have discussed polymerization that has been essentially of one kind—bifunctional molecules have combined in normal ionic reactions familiar from the rest of organic chemistry where a nucleophilic functional group attacks an electrophilic functional group. The new bonds have generally been C–O or C–N. We need now to look at the polymerization of alkenes. In these reactions, C–C bonds will be formed and many of the reactions may be new to you.

Polymerization of alkenes

Formaldehyde polymerizes because the two resulting C–O σ bonds are very slightly more stable than its C=O π bond, but the balance is quite fine. Alkenes are different: two C–C σ bonds are always considerably more stable than an alkene, so thermodynamics is very much on the side of alkene polymerization. However, there is a kinetic problem. Formaldehyde polymerizes without our intervention, but alkenes do not. We will discuss four quite distinct mechanisms by which alkene polymerization can be initiated—two ionic, one organometallic, and one radical.

Radical polymerization of alkenes: the most important polymerization of all

We will start with the radical mechanism simply because it is the most important. A bigger tonnage of polymers is made by this method than by any other, including the three most familiar ones—polythene (polyethylene), PVC (poly(vinyl chloride)), and polystyrene.

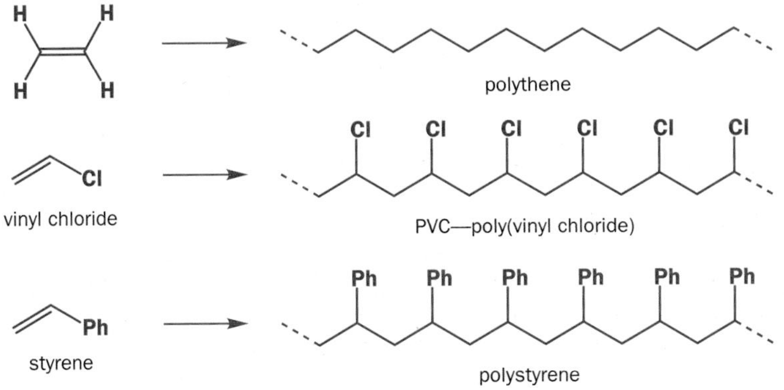

Polythene is difficult to make and was discovered only when chemists at ICI were attempting to react ethylene with other compounds under high pressure. Even with the correct reagents, radical initiators like AIBN or peroxides (Chapter 39), high pressures and temperatures are still needed. At 75 °C and 1700 atmospheres pressure ethylene polymerization, initiated by dibenzoyl peroxide, is a radical chain reaction. The peroxide is first cleaved homolytically to give two benzoate radicals.

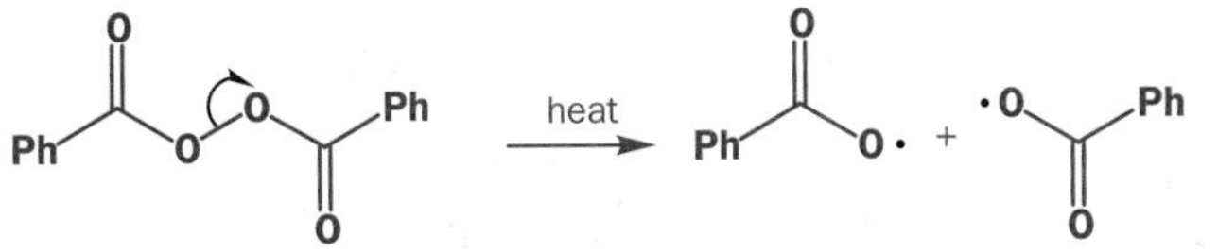

These oxyradicals add to the alkene to give an unstable primary carbon radical that adds to another molecule of alkene, and so on.

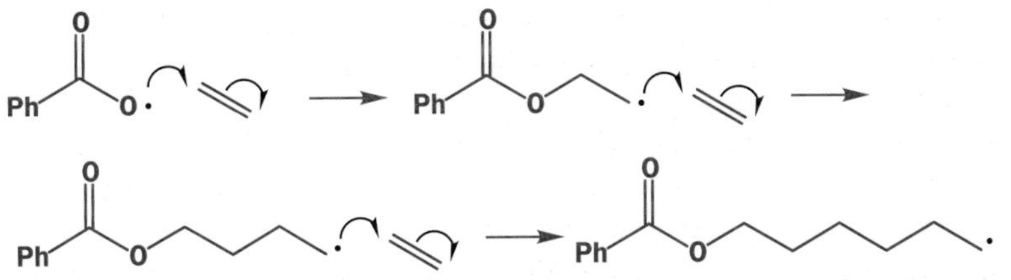

Eventually, the chain is terminated by combination with another radical (unlikely) or by hydrogen abstraction from another polymer molecule. This approach to polythene synthesis, using ethylene liquefied by pressure and small amounts (<0.005% by weight) of peroxide, produces relatively low molecular weight polymer as a white solid.

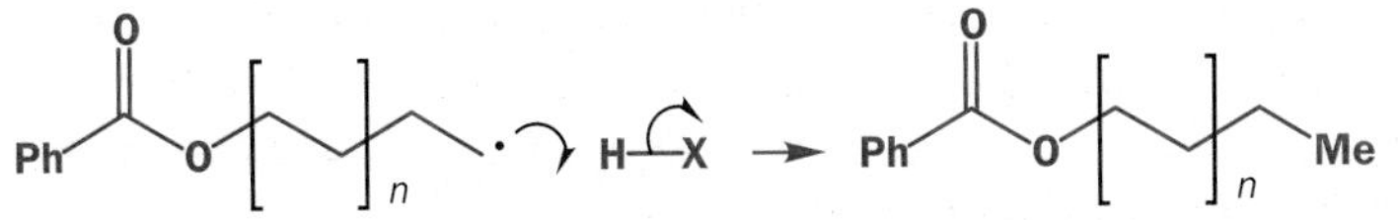

Radical polymerization can lead to branched polymers by intramolecular hydrogen atom transfer, a process sometimes called **backbiting**. Removal of H through a six-membered transition state moves the growing radical atom five atoms back down the chain, and leads to butyl side-chains. A more stable secondary radical is produced and chain growth then occurs from that point.

Radical polymerization of vinyl chloride and styrene is much easier than that of ethylene because the intermediate radicals are more stable. You saw in Chapter 39 that any substituent stabilizes a radical, but Cl and Ph are particularly good because of conjugation of the unpaired electron with a lone pair on chlorine or the π bonds in the benzene ring.

Neither PVC nor polystyrene is very crystalline and polystyrene often has poor mechanical strength. Both of these may be results of the stereorandom nature of the polymerization process. The substituents (Cl or Ph) are randomly to one side or other of the polymer chain and so the polymer is a mixture of many diastereoisomers as well as having a range of chain lengths. Such polymers are called **atactic**. In some polymerizations, it is possible to control stereochemistry, giving (instead of atactic polymers) **isotactic** (where all substituents are on the same side of the zig-zag chain) or **syndiotactic** (where they alternate) polymers.

A unique polymer is formed by the radical polymerization of tetrafluoroethylene and is called PTFE or Teflon. The outside of the polymer consists of a layer of fluorine atoms which repel all other molecules. It is used as the coating in nonstick pans and as a bearing that needs no lubrication. Two pieces of Teflon slide across one another almost without friction.

Something else is special about this polymerization—it is done in solution. Normally, no solvent is used because it would be difficult to separate from the polymer product. However, PTFE interacts with no other molecules. It precipitates from all known solvents and can be isolated easily by filtration.

Acrylics—easily made polymers of acrylate esters

Alkenes conjugated with carbonyl groups, such as acrylates (derivatives of acrylic acid), are easily polymerized by a variety of mechanisms. Indeed, these compounds are often difficult to store because they polymerize spontaneously when traces of weak nucleophiles (even water) or radicals (even oxygen) are present. Radical polymerization occurs very easily because the intermediate carbon radical is stabilized by conjugation with the carbonyl group.

Polymerization follows the mechanism that we have seen several times already, and each radical has the same additional stabilization from the carbonyl group.

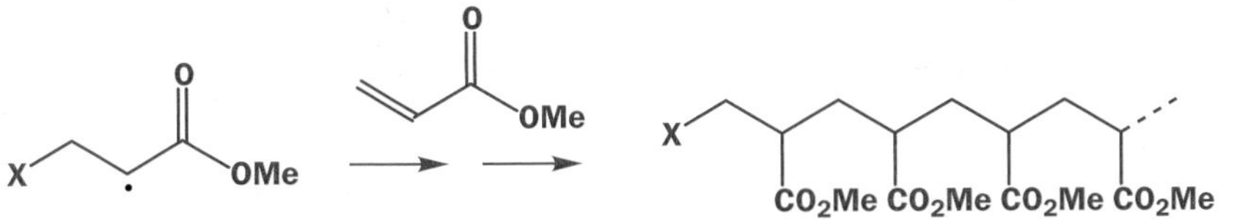

With two stabilizing groups on the carbon radical, polymerization becomes even easier. A famous example is 'SuperGlue', which is methyl 2-cyanoacrylate. The monomer in the tube polymerizes on to any surface (wood, metal, plastic, fingers, eyelids, lips, ...) catalysed by traces of moisture or air, and the bonds, once formed, are very difficult to break. The intermediate radical in this polymerization is stabilized by both CN and CO_2Me groups.

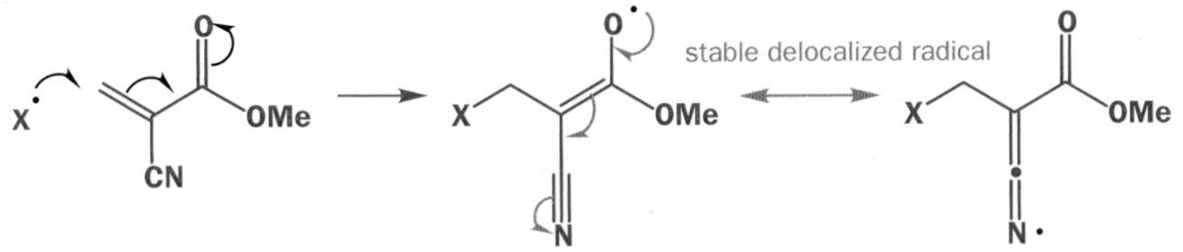

Though there are many other polymers made by radical pathways, we need now to look at the two main ionic routes—anionic and cationic polymerization.

Anionic polymerization is multiple conjugate addition

We have seen in Chapter 23 how alkenes conjugated with electron-withdrawing groups undergo conjugate addition to give an enolate anion as an intermediate. This enolate anion is itself nucleophilic and could attack another molecule of the conjugate alkene. Acrylonitrile is polymerized in liquid ammonia at low temperature by this method. Small amounts of alkali metal are added to generate NH_2^-, initiating polymerization.

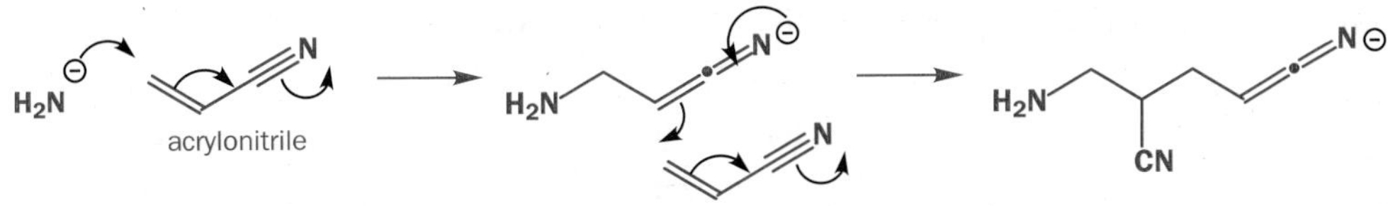

The chain grows by repetition of the last step: each new C–C bond-forming step produces a new anion stabilized by the nitrile group. Termination probably occurs most frequently by proton capture from the solvent. The result is poly(acrylonitrile).

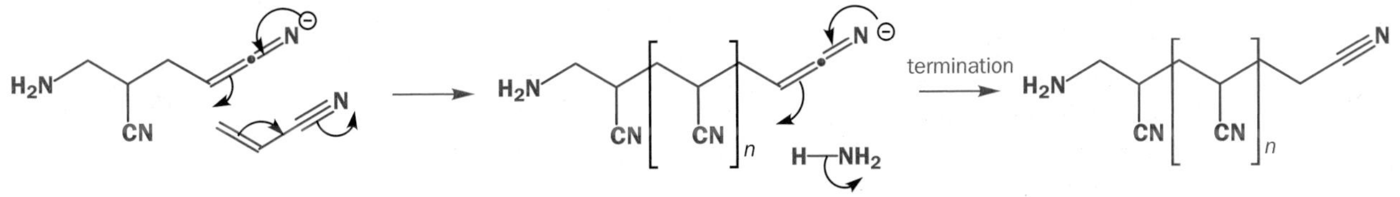

'Living polymers' by the anionic polymerization of styrene

Nucleophilic addition to styrene is possible only because the intermediate carbanion is stabilized by conjugation into the benzene ring. It needs a more reactive carbanion than the benzyl anion to initiate the polymerization, and an unstabilized nonconjugated organolithium compound like butyl lithium is the answer.

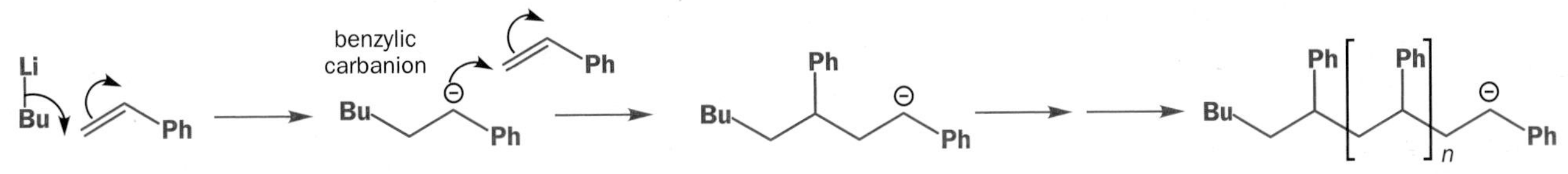

It is clear enough how the chain is propagated, but how is it terminated? You might expect protonation to bring things to a close, but there cannot be any acid (even a weak one) present—if there were, it would have already been destroyed by the butyl lithium. To terminate the polymerization, a weak acid must be added in a separate step—water will do.

When this polystyrene sample is analysed, it is found to consist of a remarkably narrow range of chain lengths—almost all the chains are the same. Such polymers are known as **monodisperse**. This result must mean that all the BuLi molecules must add immediately to a styrene molecule and that chain growth then occurs at the same rate for each chain until the styrene is used up.

There is a useful expansion of this idea. Under the conditions of the polymerization (before the water is added), these almost identical chain lengths all end with a carbanion. If, instead of adding water, we add another monomer (say, 4-chlorostyrene) it too will add to the end of the chain and polymerize until it is used up, producing new chains again of about the same length. This will be the situation after the second polymerization.

And still the polymer is active towards further polymerization. Indeed, these polymers are called 'living polymers' because they can go on growing when a new monomer is added. The final result, after as many monomers have been added as is required and the living polymer has been quenched, is a polymer with blocks of one monomer followed by blocks of another. These polymers are called **block co-polymers** for obvious reasons.

Cationic polymerization requires stabilized carbocations

Cationic polymerization is used only for alkenes that can give a tertiary carbocation on protonation or for vinyl ethers that can give an oxonium ion. In other words, the cation intermediate must be quite stable. If it isn't, the chain is terminated too quickly by loss of a proton.

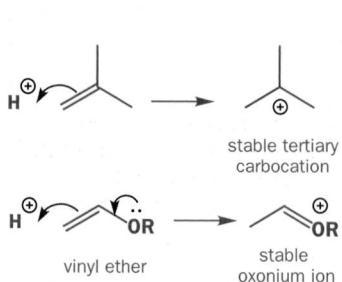
stable tertiary carbocation

vinyl ether   stable oxonium ion

The initiator for isobutene (2-methylpropene) polymerization is usually a Lewis acid with a proton source. We shall illustrate isobutene polymerization with BF₃ as the Lewis acid and water as the proton source.

The tertiary carbocation can now act as an electrophile and attack the alkene to form another tertiary carbocation of similar stability and reactivity to the first. So the polymerization continues.

Chain propagation (polymerization)

The termination will be the loss of a proton to form an alkene (an E1 reaction). Providing that the tertiary carbocation is reasonably stable, this will be a slower process than chain elongation, especially as there are no good bases around, and long polymer chains may result.

termination

The polymerization of vinyl ethers follows much the same mechanism, using the oxonium ion as an intermediate instead of the tertiary carbocation. Termination might again be by loss of a proton or by picking up a nucleophile at the oxonium ion centre.

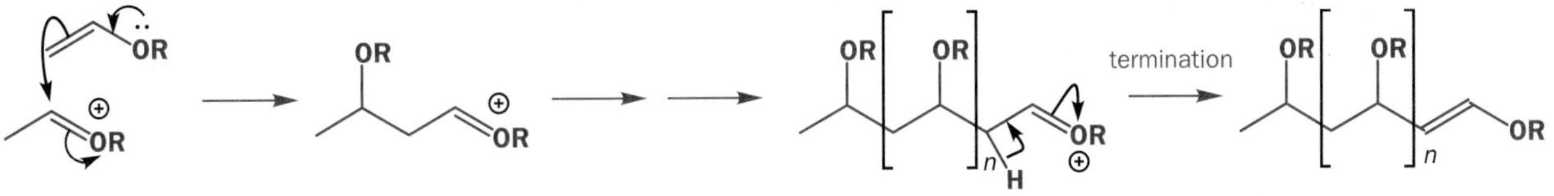

One of the best polymers for building strong rigid heat-resistant objects is polypropylene but this can be made by none of the methods we have examined so far. We need now to look at the polymerization of alkenes in the coordination sphere of a transition metal.

Ziegler–Natta polymerization gives isotactic polypropylene

Propylene can be polymerized by a titanium/aluminium catalyst developed by Ziegler and Natta. The mere fact that polymerization is possible is remarkable, but this polymer also has stereo-regularity and can be isotactic. The overall process is shown on the right.

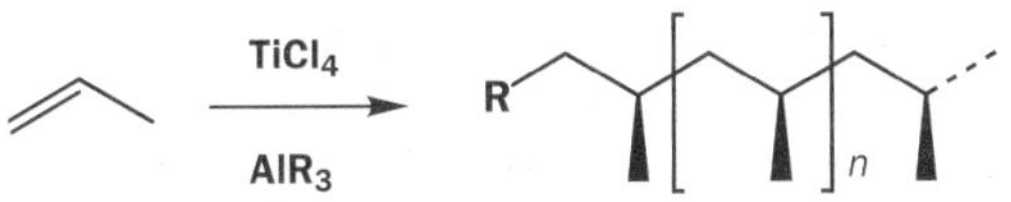

> ■
> The organometallic principles relating to this section can be found in Chapter 48.

The mixed metal compounds react to form a titanium σ complex that is the true catalyst for the polymerization. An alkyl group is transferred from aluminium to titanium in exchange for a chloride.

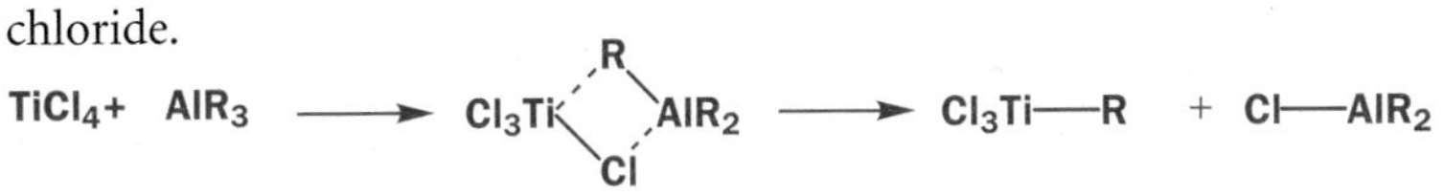

The alkyl-Ti σ complex can form a π complex with the first molecule of propene and then carry out a carbo-titanation of the π bond. This establishes the first C–C bond.

> ▶
> This is a simplification as the catalyst is a solid and the active Ti atom almost certainly Ti(III) rather than Ti(IV) as we have shown here. The third Cl ligand is in fact shared with other Ti atoms in the crystal. Coordination of the active Ti(III) atom must be such that each σ complex is a 16-electron species while the π complexes are 18-electron species.

Insertion of the next propene by a repeat of the previous step now starts the polymerization. Each new C-C bond is formed on the coordination sphere of the Ti atom by transformation of a π complex into a σ complex. Repetition of this process leads to polymerization. We have shown the polymer with isotactic stereochemistry, and this control over the stereochemistry reflects the close proximity of the new propene molecule and the growing polymer.

> ▶
> In fact, the reaction can lead either to isotactic or syndiotactic polymer depending on the detailed structure of the catalyst.

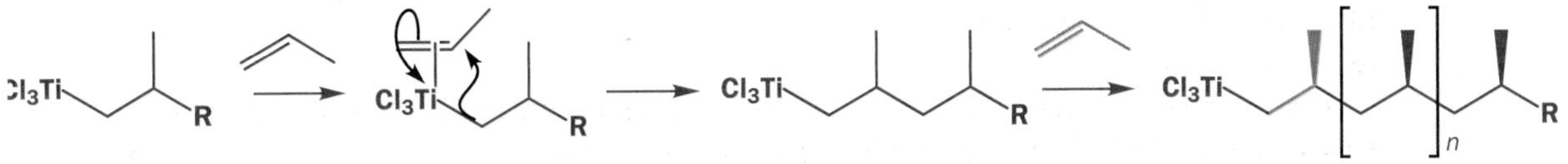

One important elastane polymer that can be made by polymerization in a Ziegler–Natta fashion is rubber. Natural rubber is a polymeric terpene (Chapter 51) made from mevalonic acid and has a branched structure with regular trisubstituted alkenes, which are all in the Z-configuration.

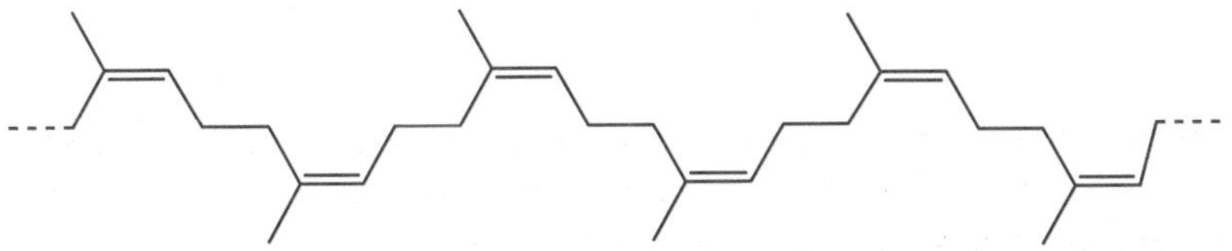

Looked at as a polymer, rubber is made up of C_5 units joined together by C–C bonds. We should naturally expect to make a hydrocarbon polymer from alkenes, so if we separate these C_5 units we find that they are dienes rather than simple alkenes. If you have read Chapter 51, they might be familiar to you as the isoprene units from which terpenes were originally supposed to be made.

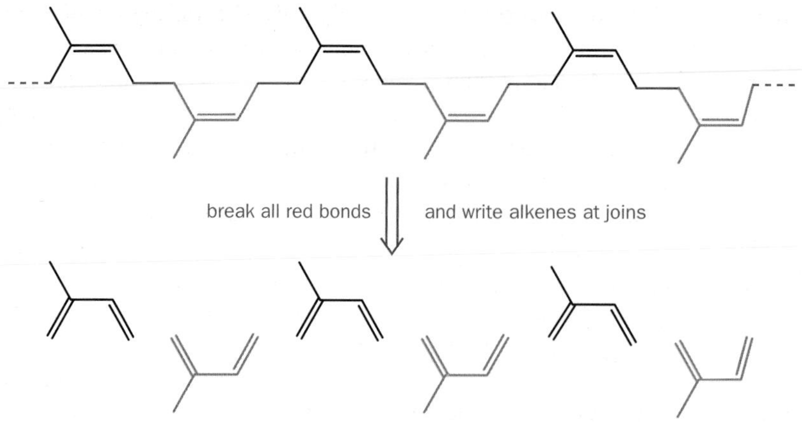

break all red bonds ⇓ and write alkenes at joins

The all-*cis* structure of natural rubber is vital to its elasticity. The all-*trans* compound is known and it is hard and brittle. Though dienes such as isoprene can easily be polymerized by cationic methods, the resulting 'rubber' is not all-*cis* and has poor elasticity and durability. However, polymerization of isoprene in the Ziegler–Natta way gives an all-*cis* (90–95% at least) polyisoprene very similar to natural rubber.

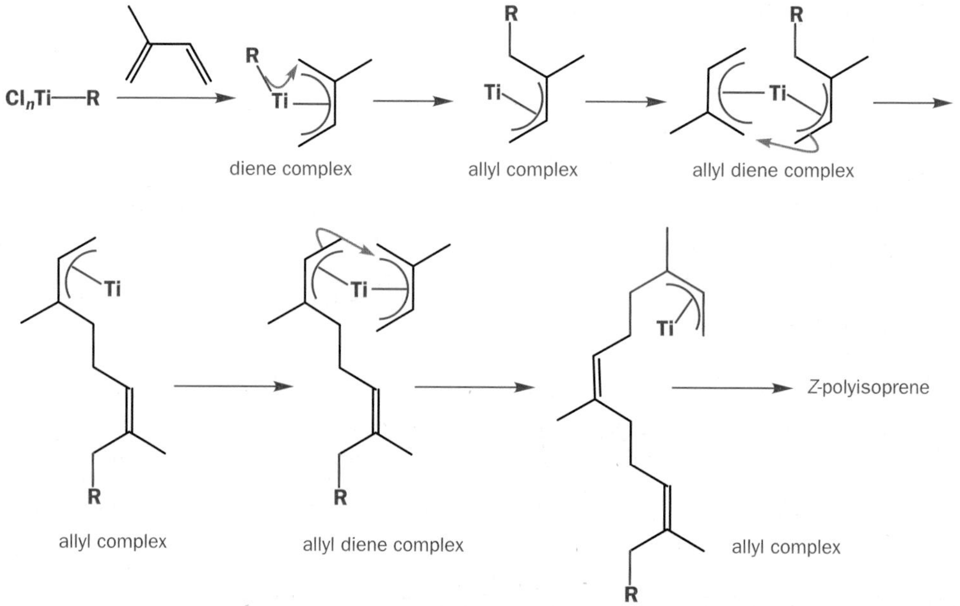

diene complex   allyl complex   allyl diene complex

allyl complex   allyl diene complex   allyl complex

One possible explanation is that each isoprene unit adds to the titanium (and we will drop the pretence at this point that we have any idea which other ligands are on the Ti atom) to form an η^4 diene complex. This must necessarily have the s-*cis* conformation. Addition of R to one end of this complex gives an η^3 allyl complex still maintaining the *cis* configuration. The next diene then adds to form a new η^4 diene complex, couples to the allyl complex, and so on. As the chain grows, each diene is added as an η^4 complex and an all-*cis* polymer results.

Co-polymerization

If two or more monomers polymerize to give a single polymer containing different subunits, the product is a **co-polymer** and the process is called **co-polymerization**. Protein synthesis is an example from nature: amino acids are polymerized stepwise to give proteins of precise sequence and precise length. We can do the same thing chemically providing that we do it in a stepwise fashion—we shall discuss this later. In most cases, chemical co-polymerization cannot be precisely ordered, but still gives useful results.

It may have surprised you, when you read the fine print on packaging, that some quite different materials are made out of the same polymer. PVC, for example, is widely used in clothing, 'vinyl' floor and seat coverings, pipework, taps, and lab stopcocks. Some of these applications require strength and rigidity; others flexibility. How is this possible with the same polymer? Some variation can be achieved by the addition of **plasticizers**—additives that are blended into the polymer mixture but are not chemically bonded to it. Another approach is to use a co-polymer with a smaller amount of a different (but often similar) monomer built randomly into the growing polymer chain. This is quite different from the alternating co-polymers that we saw under carbonyl substitution polymerization, such as nylon 6.6 or the block co-polymers we met a page or two back.

We will choose the example of elastane films for food wrapping—'ClingFilm'. These can be made from poly(vinylidene dichloride) (this is poly(1,1-dichloroethene)) into which a small amount of vinyl chloride is co-polymerized. The method is radical polymerization and the initiator usually a peroxide in aqueous suspension.

> A polymer is a chemical compound while a plastic is a mixture of a polymer and other substances (plasticizers, pigments, fibres, etc.), which allow it to be used in a certain way.

vinylidene dichloride (1,1-dichloroethene) vinyl chloride

radical formation / radical stabilized by two chlorine atoms

Polymerization continues adding vinyl chloride or vinylidene dichloride more or less at random. At first, several dichloroalkene molecules will add, simply because there are more of them.

Every now and then a vinyl chloride adds in, followed again by a number of dichloroalkenes to give the co-polymer.

Eventually, polymerization will be terminated by the usual methods and the final co-polymer will have a random mixture of dichloroalkene (mostly) and monochloroalkene, roughly in proportion to their availabilities in the polymerization mixture. The precise properties of the resulting polymer will depend on the ratio of the two monomers.

Synthetic rubbers can be made by co-polymerization of alkenes and dienes

Radical co-polymerization of styrene and butadiene produces a material that is very like natural rubber. The initiator is a one-electron oxidizing agent, and a thiol (RSH) is used to start the polymerization process. The mixture is about 3:1 butadiene:styrene so there are no long runs of one monomer in the product. We will use butadiene as the starter unit.

stabilized allylic radical

The first radical is an allylic radical, stabilized by conjugation with the remaining alkene in the old butadiene molecule. Addition could now occur to another butadiene or to styrene.

stabilized benzylic radical

The product is the stabilized benzylic radical with the more stable *trans* double bond. Stabilization of radicals in allylic and benzylic groups is about the same, so the two monomers will react roughly in proportion to their concentration. The final product will be a random co-polymer of about 3:1 butadiene to styrene with mostly *E*-alkenes. It is an elastomer used for tyres and other applications where a tough and flexible 'rubber' is needed.

Cross-linked polymers

Many linear polymers are too flexible to be of use in making everyday objects because they lack the strength, the rigidity, or the elasticity for the job. Linear polymers can be stiffened and strengthened by bonds between the chains. This process is known as **cross-linking** and we will look now at some ways in which this can be achieved.

All that is really needed is a co-polymer with a small amount of a compound similar to the main monomer but with at least one more functional group than is strictly necessary to form a linear polymer. For example, a small amount of 1,4-divinylbenzene co-polymerized with styrene leads to a linear polymer in which some of the phenyl rings carry a 4-vinyl group.

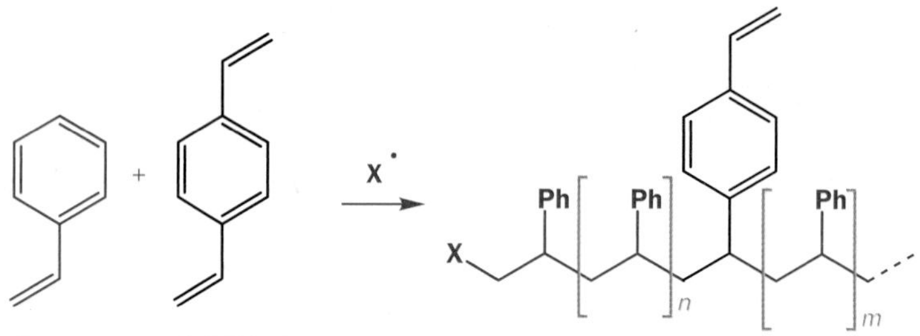

styrene   1,4-divinyl
               benzene

When another chain polymerizes nearby, the spare vinyl group in the first chain may be incorporated into the new chain of polystyrene.

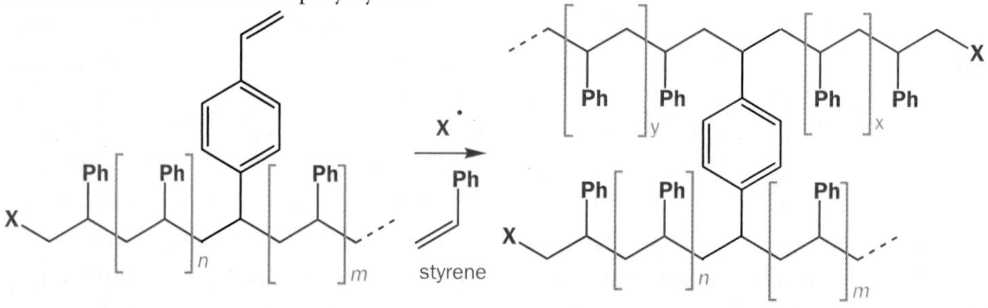

cross-linked polystyrene

Not all of the spare vinyl groups will be caught up in a new chain of polymerizing styrene, but that need not matter if there are enough of them. It is simply a question of adding enough 1,4-divinyl benzene to get the required degree of cross-linking. These cross-linked styrenes are often made into small beads for polymer-supported reagents, as described below.

Divinyl benzene has two identical 'arms', which become growing points in polymerization. In the polymerization of Me_2SiCl_2 we had two growing points (the two chlorine atoms) on each monomer. To get cross-linking we need a third, provided by (a small amount of) $MeSiCl_3$.

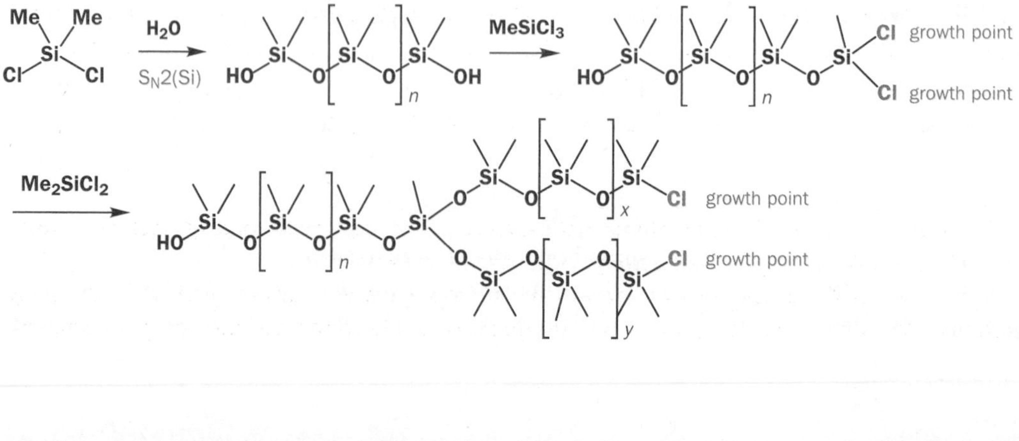

The four-armed cross-linking agent known as pentaerythritol is made from acetaldehyde and formaldehyde in aqueous base. The four arms are arranged in a tetrahedron around a quaternary carbon atom.

Co-polymerization of pentaerythritol and two other monomers—an unsaturated acid and benzene 1,3-dicarboxylic acid—gives a network of polymer chains branching out from the quaternary carbon atom at the centre of pentaerythritol. The reaction is simply ester formation by a carbonyl substitution reaction at high temperature ($> 200\,°C$). Ester formation between acids and alcohols is an equilibrium reaction but at high temperatures water is lost as steam and the equilibrium is driven over to the right.

pentaerythritol

Pentaerythritol is made by a Cannizzaro reaction: see Chapter 27, p. 713.

The black pentaerythritol at the centre of the polymer is shown with two each of the ester side chains, though this need not be the case, of course. The green pentaerythritol molecules are the growing points of the network of polymer chains. It is obvious why the benzene dicarboxylic acid is helpful in linking growing points together, but what is the point of the long-chain unsaturated acid? These are naturally occurring acids as described in Chapter 51 and the alkenes are used for further cross-linking under oxidative conditions as described in the next section. Such polymers are called 'alkyd resins' and are used in paints. They form emulsions in water ('emulsion paints') and the ester groups do not hydrolyse under these conditions as water cannot penetrate the polymer network. As the paint 'dries' it is cross-linked by oxygen in the air.

It is not necessary to have quite such a highly branched cross-linking agent to make a network of polymer chains. A triply branched compound is the basis for one of the strongest polymers known—one that we take for granted every time we use the kitchen. It is made by a very simple reaction.

Melamine

You saw a carbonyl addition reaction forming a polymer right at the beginning of the chapter—the polymerization of formaldehyde. If an amine is added to formaldehyde, condensation to form imines and imine salts occurs readily. These intermediates are themselves electrophilic so we have the basis for **ionic polymerization**—electrophilic and nucleophilic molecules present in the same mixture. Reaction with a second molecule of amine gives an **aminal**, the nitrogen equivalent of an acetal.

This is also the first step of the Mannich reaction: Chapter 27, p. 712.

imine salt   aminal

There are now two nucleophilic atoms in the molecule. Each can react with formaldehyde to form more C–N bonds and so on, making two growth points for the polymer.

We do better if we have two or even three nucleophilic amino groups present in the same molecule. With three amino groups we will produce a branching polymer of great strength

growth points for amine–formaldehyde polymer

and the most important of the triamines is melamine. This compound is itself produced by the trimerization of a simple compound, cyanamide $H_2N–CN$, and has given its name to a group of plastics.

You should be able to write the full mechanism for the formation of melamine: the first step is given.

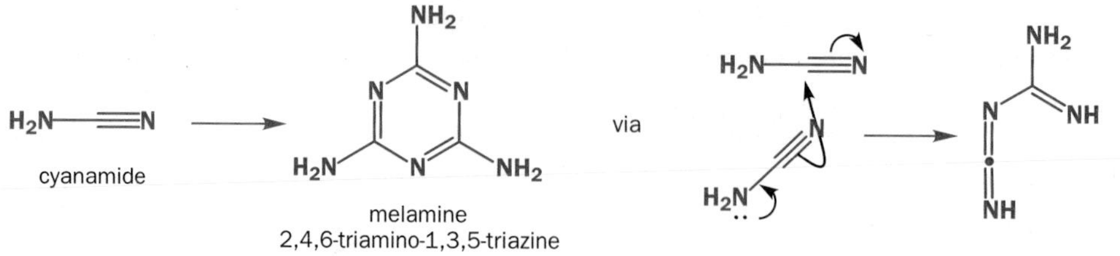

When the triamine reacts with formaldehyde, branched polymerization can occur by the same mechanism as the one we drew above for simple amines. Further condensations with formaldehyde allow amines to be attached in many places, and each new amine itself adds many new growing points. An exceptionally strong polymer results.

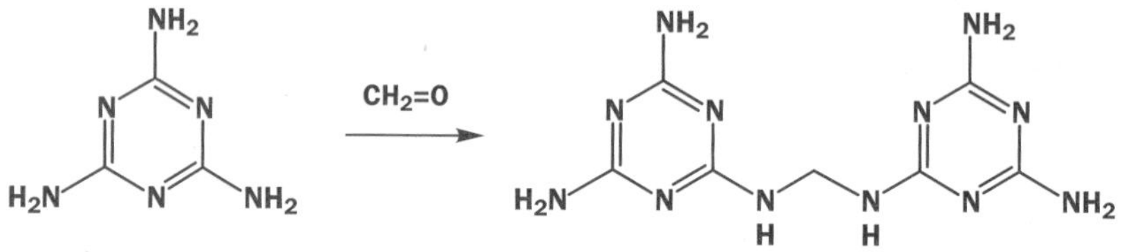

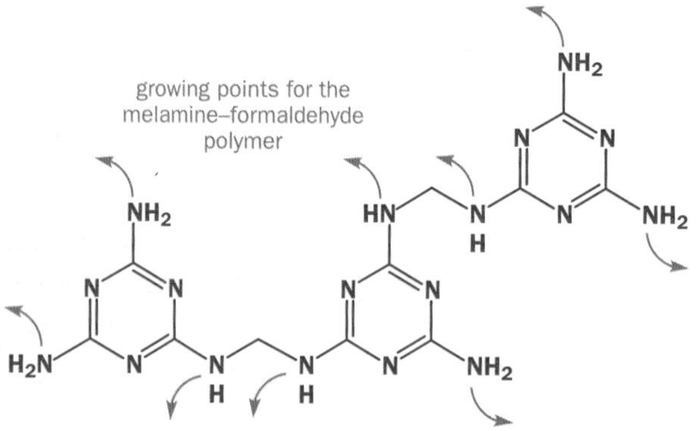

growing points for the melamine–formaldehyde polymer

These resins are used to make 'unbreakable' plastic plates and for the famous kitchen surface 'Formica'. Partly polymerized melamine–formaldehyde mixtures are layered with other polymers such as cellulose (Chapter 49) and phenol–formaldehyde resins and the polymerization is completed under pressure with heat. The result is the familiar, tough, heat-resistant surface.

Reactions of polymers

We have so far given the impression that all polymers are formed fully armed, as it were, from monomers already having the correct functionality. This is, indeed, often the case because it can be very difficult to persuade polymers to carry out any reactions—reagents cannot penetrate their interiors. Polyester fabrics can be washed without any of the ester linkages being hydrolysed in the washing machine because the water cannot penetrate the fibres. However, some useful reactions, including ester hydrolysis, can be carried out on complete polymers.

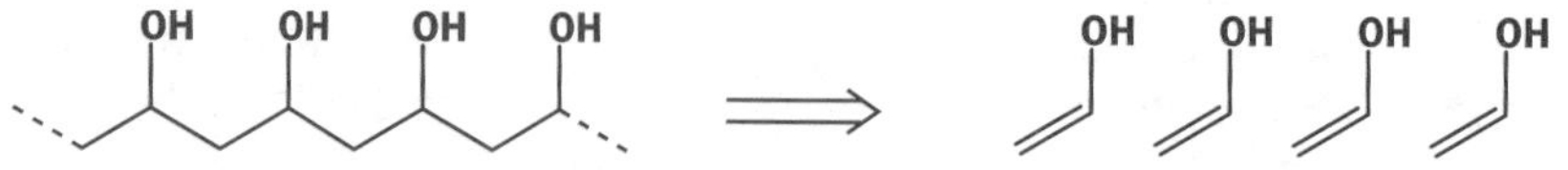

PVA–poly(vinyl alcohol)

Poly(vinyl alcohol) is an important example. Inspection of the structure reveals that this is a typical alkene polymer but the monomer would have to be vinyl alcohol—the unstable enol of acetaldehyde. The way to make the polymer is to start with something else and only later convert the polymer product into poly(vinyl alcohol). The most common method of doing this is to use radical polymerization of vinyl acetate, the enol ester of acetaldehyde, and hydrolyse the ester afterwards.

Vinyl acetate

Vinyl acetate is manufactured on a large scale by two routes.

Satisfy yourself that you can at least see what is happening here—if you are stuck on the Pd(II)-catalysed reaction, refer to Chapter 48 and look at oxypalladation and the Wacker reaction for clues.

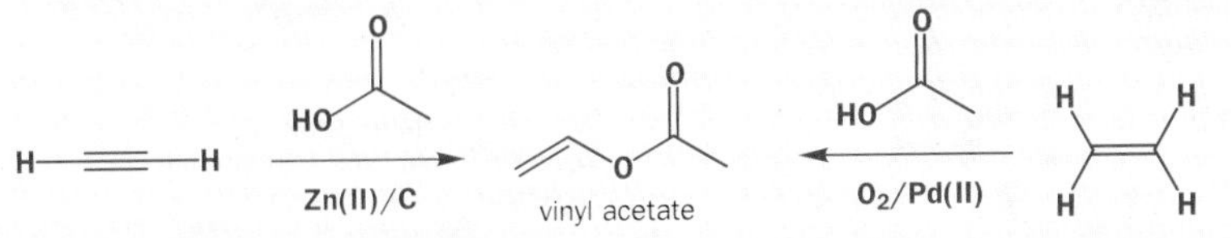

The polymerization of the enol acetate goes in the usual way.

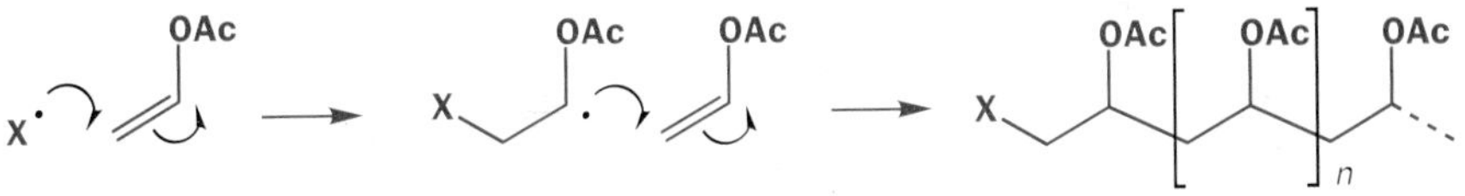

The complete polymer may now be attacked by reagents that cleave the ester groups. Water is a possibility, but methanol penetrates the polymer better and ester exchange in alkaline solution gives poly(vinyl alcohol).

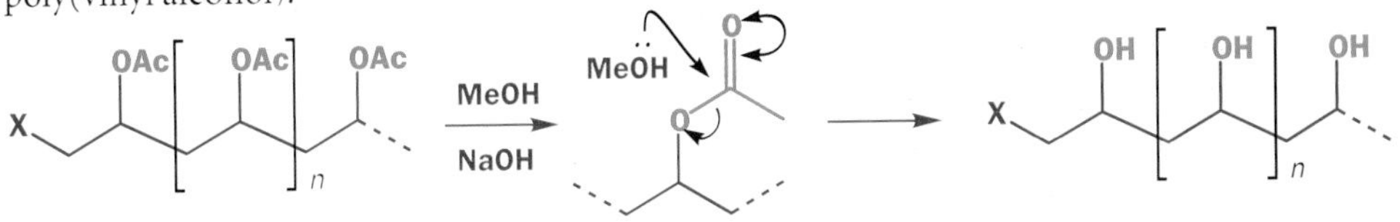

Poly(vinyl alcohol) is soluble in water, unlike almost all other polymers, and that gives it many uses in glues and even as a solublizing agent in chemical reactions to make other polymers. Poly(vinyl acetate) is used in paints.

Cross-linking of pre-formed polymers

We have already discussed cross-linking during polymerization but cross-linking is often carried out after the initial polymer is made. You saw earlier how poly(dimethylsiloxane) can be cross-linked by co-polymerization with MeSiCl₃. An alternative way of cross-linking the linear polymer uses radical reactions to convert silicone oil into silicone putty. Peroxides are used in this process.

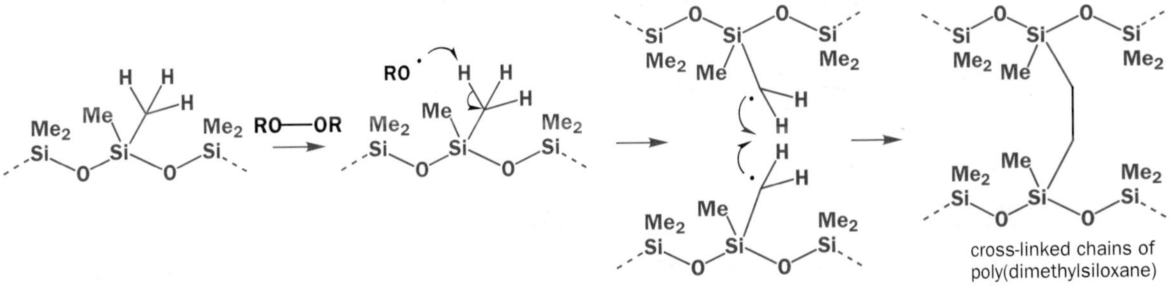

cross-linked chains of poly(dimethylsiloxane)

A similar sort of reaction occurs during the cross-linking of alkyd resins for paint manufacture. You may recall that the alkenes are incorporated in these resins for a reason not yet made clear. Now these alkene units come into their own. Oxygen is the reagent and it works by radical dimerization of the chains (see overleaf).

The most important of all of these types of reactions is the vulcanization of rubber. Originally, the raw rubber was just heated with sulfur (S₈) and cross-linking of the polyisoprene chains with short chains of sulfur atoms gave it extra strength without destroying the elasticity. Nowadays, a vulcanizing initiator, usually a thiol or a simple disulfide, is added as well. Some examples are

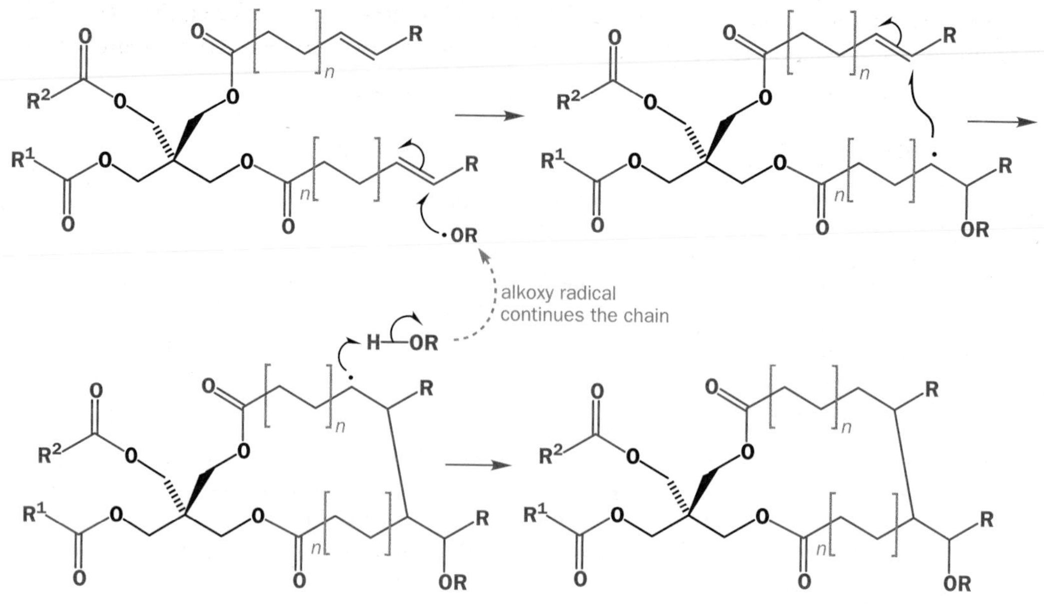

cross-linked alkyd resin

alkoxy radical
continues the chain

vulcanization initiators

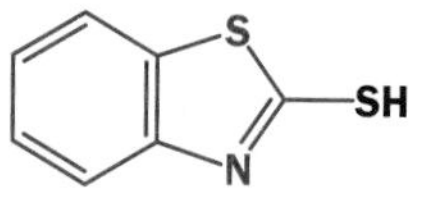

benzothiazole-2-thiol

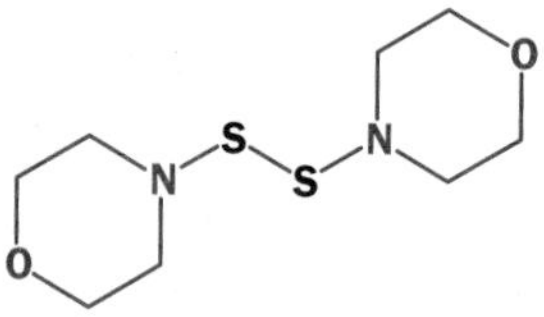

dithio-*bis*-morpholine

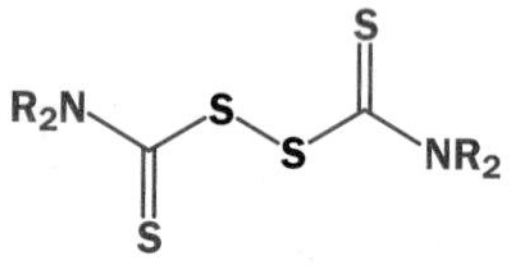

tetraalkyl thiuram disulfide

shown in the margin. The thiols give sulfur radicals with oxygen and the disulfides cleave easily as the S–S bond is weak (about 140 kJ mol^{-1} in S_8). We will write all these as RS$^{\bullet}$. The initiators either attack the rubber directly or attack sulfur to open the S_8 ring.

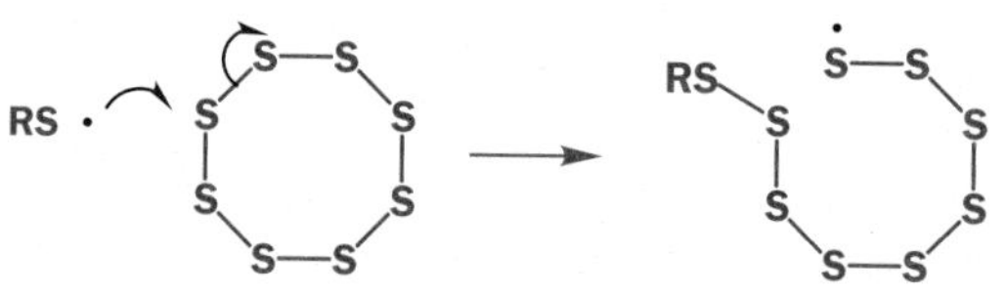

The newly released sulfur radical can bite back on to the sulfur chain and close a ring of 5–7 sulfur atoms, releasing a short chain of sulfur atoms attached to the initiator and terminating in a sulfur radical.

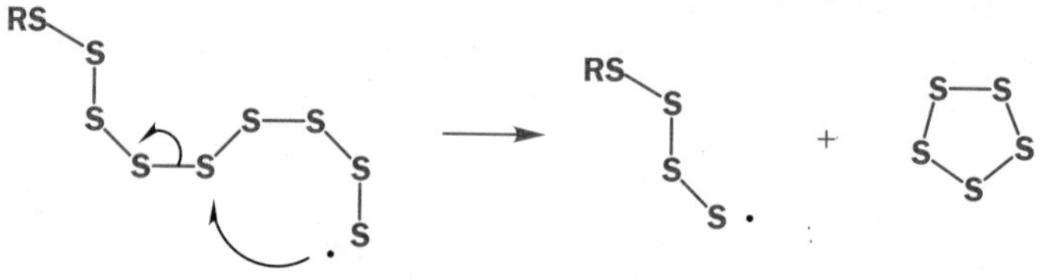

Now the attack on rubber can start. We know that vul-canized rubber has many *E*-alkenes, whereas unvulcanized rubber is all *Z*-alkenes. This suggests that

the sulfur radicals do not add to the alkenes but rather abstract allylic hydrogen atoms. Writing only a small section of rubber, we have:

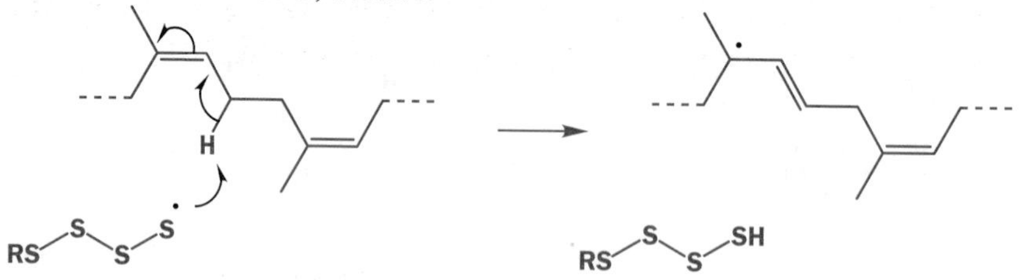

The new allylic radical can do many things, but it might, for example, capture one of the sulfur rings present (S_5 to S_8). We will use the S_5 ring we have just made.

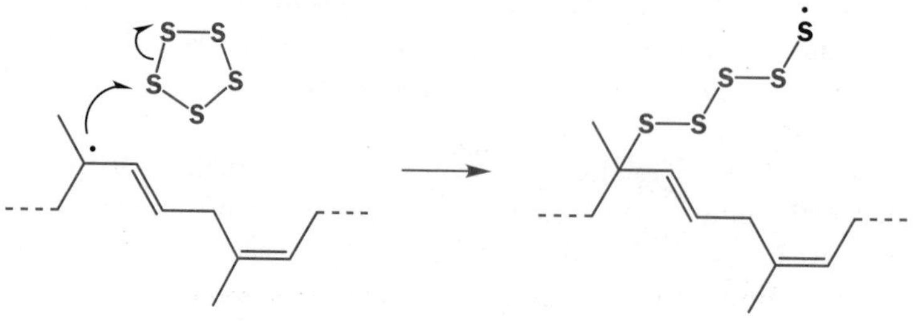

This sulfur radical can attack another chain to give a cross-link or bite back to give a link within the same chain. Many different sulfur links are formed and the next diagram summarizes a part of the vulcanized rubber struc-ture. There is some license here: in reality the links would not be as dense as this, and more than two chains would be involved. Notice the two chains joined by one cross-link, the inter-nal cross-link in the black chain, the attachment of the initiator (RS) to the green chain, and the (*E*,*E*)-dienes in both chains.

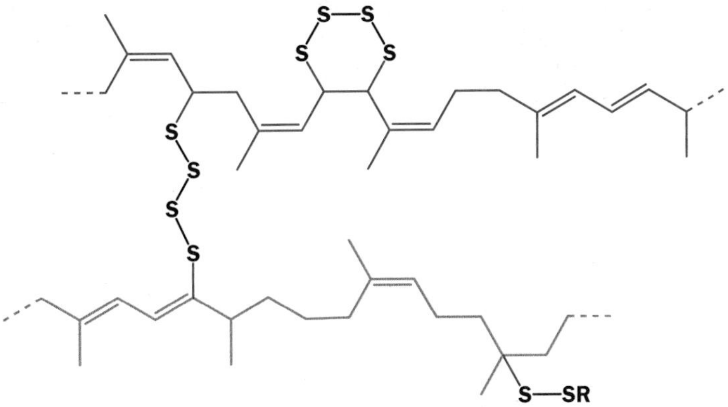

We have not given compositions of complete plastics in general, but you might like to know the typical composition of a motor tyre. Notice that the ratio of sulfur to rubber is about 1:40—that gives an idea of how many cross-links there are. Notice also that the rubber contains a great deal of carbon to improve the wear of the rubber. The roles of the other materials are explained in the table.

Typical composition of rubber motor tyre

Component	Parts by weight, %	Function
rubber	61	basic structure
carbon black	27	reinforcement
oils and waxes	4.9	processing aid
sulfur	1.5	vulcanizing agent
organic disulfide	0.4	accelerates vulcanization
zinc oxide (ZnO)	3	activates vulcanization
stearic acid	0.6	activates vulcanization

This makes only 98.4% in total and there are small amounts of other materials such as antioxidants to prolong the life of the rubber.

Though synthetic diene polymers have now replaced natural rubber in many applications, they too need to be cross-linked by vulcanization using essentially the same reactions, though the details vary from product to product and from company to company.

Chemical reactions of cellulose

We met cellulose, the bulk polysaccharide of woody plants, in Chapter 49. It is a strong and flexible polymer but no use for making fabrics or films as it cannot be processed. One solution to this prob-lem is to carry out chemical reactions that transform its properties. Acid-catalysed acetylation with acetic anhydride gives a triacetate with most of the free OH groups converted into esters.

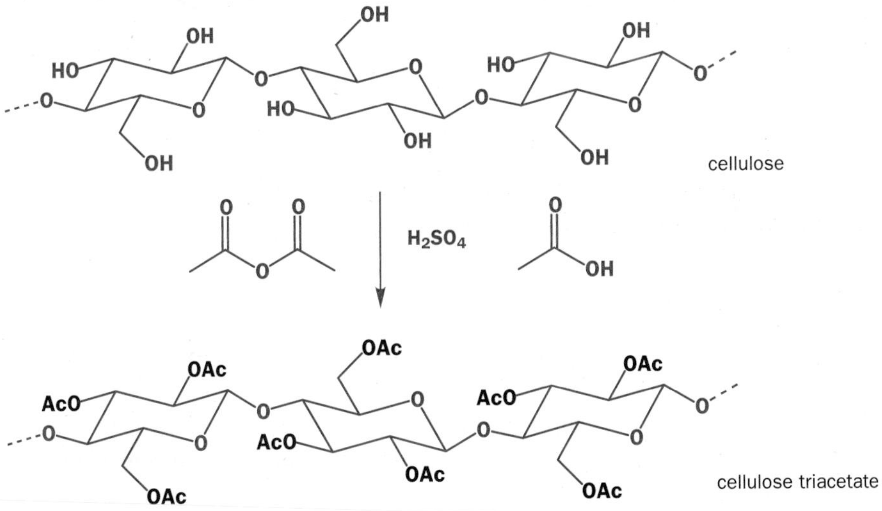

The starting material for this process is wood pulp, cloth, or paper waste and the acetic acid is added first to 'swell' the material and allow it to take up the reagents better. Organic solvents often do this to polymers. The anhydride now carries out the acid-catalysed acetylation and the cellulose triacetate, unlike the cellulose, dissolves in the reaction mixture. The new polymer is often known simply as 'acetate'.

Another cellulose product is rayon. This is really cellulose itself, temporarily modified so that it can be dissolved and processed to give films or fibres. The starting material (from wood, cloth, or paper) is impregnated with concentrated NaOH solution. Addition of CS_2 allows some of the OH groups to react to give a 'xanthate' salt that is soluble in water.

Injection of the viscous solution of cellulose xanthate into an acidic (H_2SO_4) bath regenerates the cellulose by the reverse of this reaction, as a film or a fibre depending on the process. The result is known as 'cellophane' if it is a film or 'viscose rayon' if it is a fibre.

Biodegradable polymers and plastics

It is necessary to take only a short walk in most cities to see that plastics are not very easily degraded biologically, and it is becoming more important to design plastics, for packaging at least, that have built-in susceptibility to bacteria or fungi. Natural polymers based on proteins and polysaccharides do have that advantage, and one approach is to use a near-natural polymer, poly(hydroxybutyrate) or P(3-HB). This compound is found in some microorganisms as massive (by microorganism standards!) whitish granules occupying substantial parts of the cell—up to 80% of its dry weight of the cell. It seems that it is used as a storage compound (like starch or fat in our case) for excess carbohydrates in the diet.

3-hydroxybutyric acid

polyhydroxybutyrate [P(3-HB)]

A co-polymer of P(3-HB) and poly(hydroxyvalerate) P(3-HV) is also found in microorganisms and performs the same function. This polyester forms the basis for a good strong but flexible plastic for containers such as toiletries, and is produced by ICI under the name 'BIOPOL'. Microorganisms must be able to degrade both P(3-HB) and BIOPOL since they themselves use them to store energy.

3-hydroxyvaleric acid
(3-hydroxypentanoic acid)

BIOPOL
P(3-HB/3-HV)

BIOPOL and the two simple polymers P(3-HB) and P(3-HV) are manufactured by fermentation. They can also be produced chemically by the polymerization of a four-membered lactone (β-butyrolactone). The polymerization is initiated by a water molecule that opens the first lactone ring. The reaction is catalysed by Et_3Al and continues by repeated esterification of the released OH group.

β-butyrolactone

Biological degradation requires that fungi or microorganisms can attack the polymer with their enzymes. This happens efficiently with very few polymers (because these enzymes do not exist) and is, of course, the reason that they are used: people tolerate ugly plastic window frames because they don't rot.

One way in which most polymers do decay is by the action of oxygen in the air and of light. You will be familiar with the way that some polymers go yellow after a time and some become brittle. Coloured plastics, in particular, absorb light and oxygen-induced radical reactions follow. The polymer becomes too cross-linked and loses flexibility. One ingenious application of this natural process helps to degrade the polythene rings that hold cans of beer in packs. These are often discarded and decay quite quickly because some carbon monoxide has been incorporated into the polyethylene to make it more sensitive to photolysis.

Chemical reagents can be bonded to polymers

We have left this subject to the end of the chapter because it uses all of the principles we have established earlier on. It requires an understanding of radical polymerization, co-polymerization, cross-linking, functionalization of polymers after they have been made, and so on. This is a rapidly growing subject and we can only outline the basics.

If you are already wondering why anyone would bother to attach reagents to polymers, just think of the problems you have had in the lab in separating the product you want from the other products of the reaction, often the spent reagent and inorganic by-products. If the reagent is attached to a polymer, the work-up becomes easier as the spent reagent will still be attached and can just be filtered off. Polymer-supported reagents can often be reused and their reactions can even be automated.

You may already be familiar with ion-exchange resins and we will start with them. They are commonly based on the co-polymer of styrene and 1,4-divinyl benzene we discussed earlier. The polymerization is carried out in an emulsion in water so that the organic molecules are in tiny droplets. The resulting polymer forms as more or less spherical beads of less than a millimetre in diameter. They can be put through a series of sieves to ensure even sizes if required. The surface of each bead bristles with benzene rings (attached to the polymer backbone) that can be sulfonated in the *para* position just like toluene.

A good proportion of the rings become sulfonated, and the outside of each bead is now coated with strongly acidic sulfonic acid groups. The polymer is an acidic reagent that is not soluble in any normal solvent. It can be packed into a column or simply used as a heterogeneous reagent. In any case, whatever reaction we are doing, there is no difficulty in separating the organic product from the acid.

A useful *basic* polymer is made by co-polymerization of 4-vinyl pyridine and styrene.

These polymers are reagents in themselves, but a new style of chemistry is being developed around the idea of attaching reagents to the polymer. Poly 4-bromostyrene (or a co-polymer with styrene itself) allows a number of different groups to be attached in the place of the bromine atom. One example is a polymer-bound Wittig reagent. The phosphine can be introduced by nucleophilic displacement with Ph$_2$P–Li, an excellent nucleophile, by the addition–elimination mechanism (Chapter 23).

Though we have shown only one bromine atom and hence only one Ph₂P group on the polymer, almost all of the benzene rings in polystyrene can be functionalized if the bromopolymer is made by bromination of polystyrene in the presence of a Lewis acid. Now the phosphine can be alkylated with an alkyl halide of your choice to form a phosphonium salt, still on the polymer.

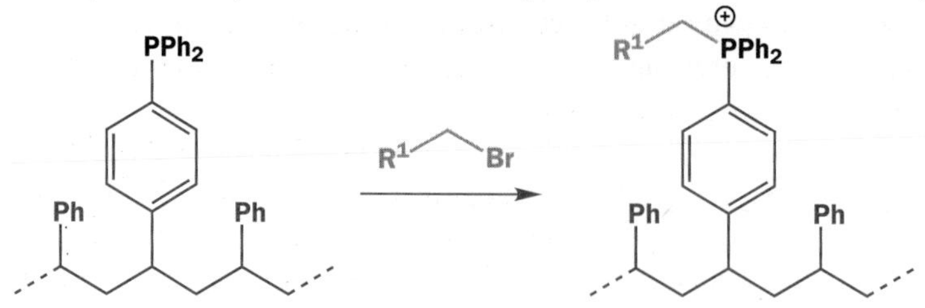

Treatment of the polymer with BuLi and then the aldehyde gives a Wittig reaction (Chapter 31) that releases the alkene product but leaves the phosphine oxide bound to the polymer.

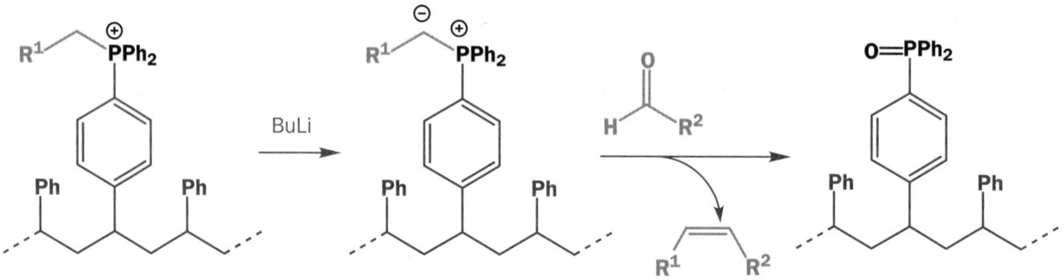

The phosphine oxide can be reduced back to the phosphine (for example, with Cl₃SiH) while still bound to the polymer and the polymer-bound reagent can be used again. Separation of Ph₃P=O from alkene products after a Wittig reaction can be quite a nuisance so the ease of work-up alone makes this an attractive procedure.

It is not necessary to attach the functional group directly to the benzene ring. There are some advantages in separating the reaction from the polymer by a 'spacer', normally a chain of aliphatic carbon atoms. It may allow reagents to approach more easily and it may allow a higher 'loading' of functional groups per bead. Even a spacer of one CH₂ group makes S_N2 reactions not only possible but favourable at the benzylic position and the most important of these spacers is introduced by chloromethylation. Reaction of the cross-linked polystyrene with MeOCH₂Cl and a Lewis acid gives the benzylic chloride via the ether.

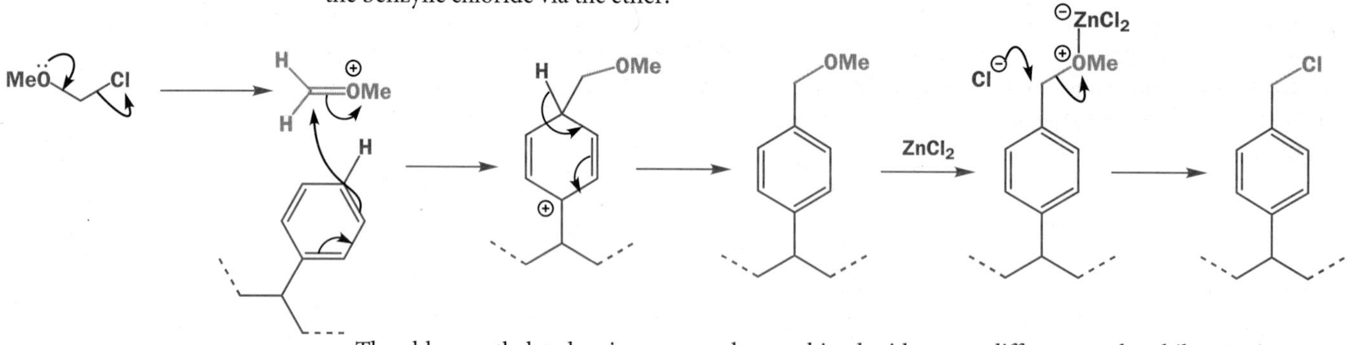

The chloromethylated resin can now be combined with many different nucleophiles. Amines give basic ion-exchange resins while Ph₂P–Li gives a phosphine suitable for complexation to transition metals.

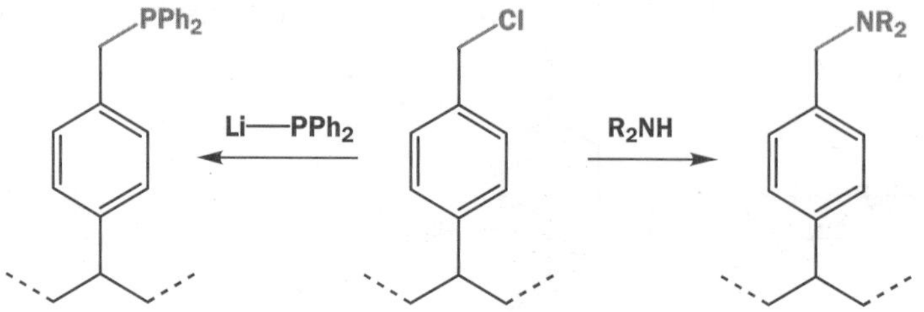

Automated peptide synthesis uses polymer-bound reagents

Automated polymer-based synthesis comes into its own when a stepwise polymerization is required with precise control over the addition of particular monomers in a specific sequence. This is almost a definition of peptide synthesis. Nature attaches each amino acid to a different 'polymer' (transfer RNA) and uses a 'computer program' (the genetic code) to assemble the polymers in the right order so that the amino acids can be joined together while bound to another polymer (a ribosome). No protection of any functional groups is necessary in this process.

Chemical synthesis of peptides uses a similar approach but our more primitive chemistry has not yet escaped from the need for full protection of all functional groups not involved in the coupling step. The idea is that the first amino acid is attached to a polymer bead through its carboxyl group (and a spacer) and then each *N*-protected amino acid is added in turn. After each addition, the *N*-protection must be removed before the next amino acid is added. The growing peptide chain is attached to the polymer so that all waste products, removed protecting groups, excess reagents, and inorganic rubbish can be washed out after each operation.

> ■ This subject was introduced in Chapter 25 and we will not repeat here all the details of how protecting groups are added and removed. Please refer to that chapter if you need more explanations of the reactions. We will concentrate here on the role of the polymer.

stage 1: attachment of the first (C-terminal) amino acid

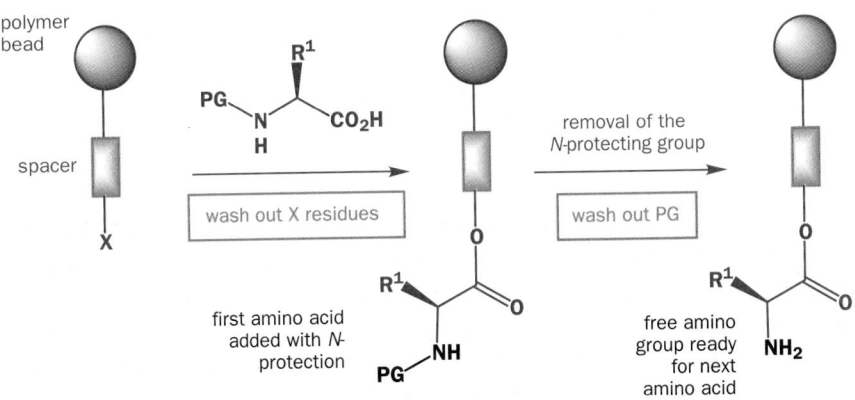

polymer bead

spacer

PG–N(H)–CHR^1–CO$_2$H

wash out X residues

first amino acid added with *N*-protection

removal of the *N*-protecting group

wash out PG

free amino group ready for next amino acid

Stage 1 involves two chemical reactions—linking the first amino acid to the polymer and removing the *N*-protecting group—and two washing operations. These four steps would take time if everything were in solution but, with the compounds attached to polystyrene beads, they can be carried out simply by packing the beads into a column chromatography-style and passing reagents and solvents through.

Stage 2 involves the addition of the second *N*-protected amino acid with a reagent to couple it to the free amino group of the amino acid already in place. Removal of the protecting group from the new amino acid is needed, followed by washes, as in stage 1.

stage 2: formation of the first peptide bond

coupling agent

wash out X residues

second amino acid added with *N*-protection

removal of the *N*-protecting group

wash out PG

free amino group ready for next amino acid

This process must now be repeated until all of the amino acids have been added. Finally, all the side-chain protecting groups must be removed and the bond joining the peptide chain to the polymer must be broken to give the free peptide. That is the process in outline, but we need now to look at some of the chemistry involved.

It is obviously important that all reactions are very efficient. Suppose that the coupling step joining the second amino acid on to the first goes in 80% yield. This may not seem bad for a chemical reaction, but it would mean that 20% of the chains consisted of only the first amino acid while 80% contained correctly both first and second. Now what happens when the third amino acid is added?

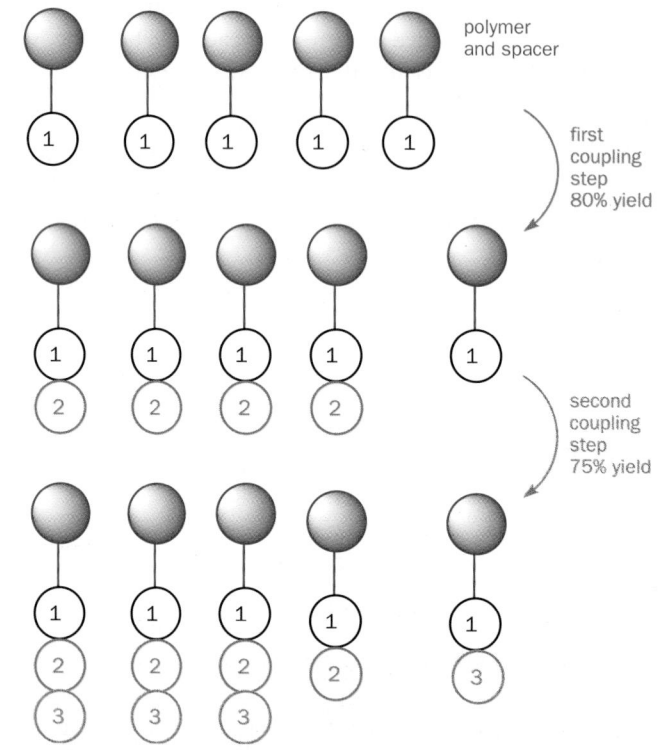

The diagram shows that four out of five growing chains will be right (1–2) after the first coupling step, but after the second (we have put this one at 75% yield for convenience) only three of the five are correct (1–2–3). One of the others has the sequence 1–2 and the other 1–3. This situation will rapidly deteriorate and the final peptide will be a mixture of thousands of different peptides. So, for a start, each reaction must occur in essentially 100% yield. This can be achieved with efficient reactions and an excess of reagents (which are not a problem in polymer-supported reactions as the excess is washed away).

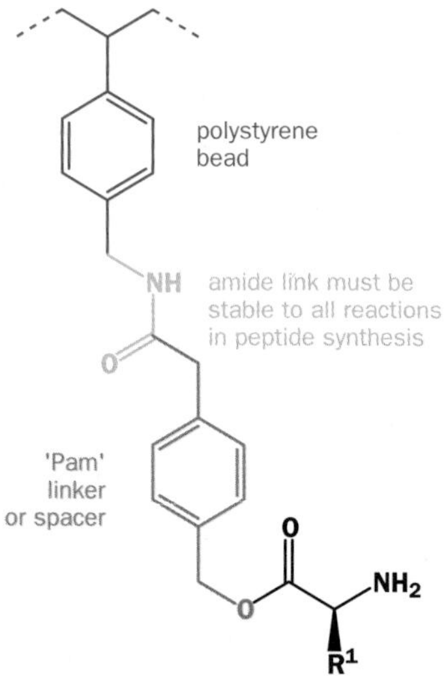

Now some detail—and we will discuss the Merrifield version of peptide synthesis. Spherical cross-linked polystyrene beads of about 50 μm in diameter are used and attached to various spacers of which the simplest is just a CH_2 group from the chloromethylated polystyrene we have just discussed. The caesium (Cs) salt of the amino acid is used to displace the chloride as it is a better nucleophile than the Na or K salts. A better alternative is 'Pam' (shown in the margin). It can be used as the nucleophile to displace the chloride first. The amino acid is then added after purification. No chloromethyl groups can remain on the polymer with this spacer.

The next stage is to link the carboxyl group of the second amino acid on to the amino group of the first. The Boc group (Chapter 24) is usually used for amino group protection in the Merrifield method and DCC (dicyclohexylcarbodiimide) is used to activate the new amino acid. Here is a summary of this step, using symbols again for polymer and spacer.

polystyrene bead

spacer

first amino acid

'Pam' linker or spacer

polystyrene bead

amide link must be stable to all reactions in peptide synthesis

first amino acid

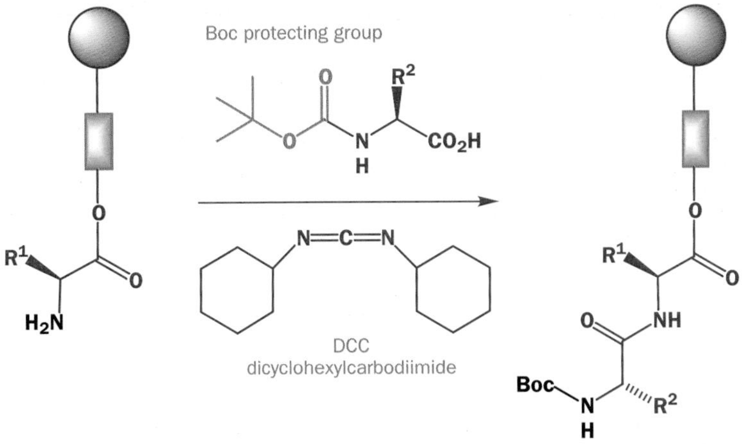

The details of the reaction mechanism with DCC were given in Chapter 43, p. 1172, and can be shown more easily if we mark the polymer and spacer as 'P' and the cyclohexyl groups as 'R'. The DCC is protonated by the free carboxylic acid and is then attacked by the carboxylate anion. The intermediate is rather like an anhydride with a C=NR group replacing one of the carbonyl groups. It is attacked by the amino group of the polymer-bound amino acid. The by-product is dicyclohexyl-lurea, which is washed off the column of resin.

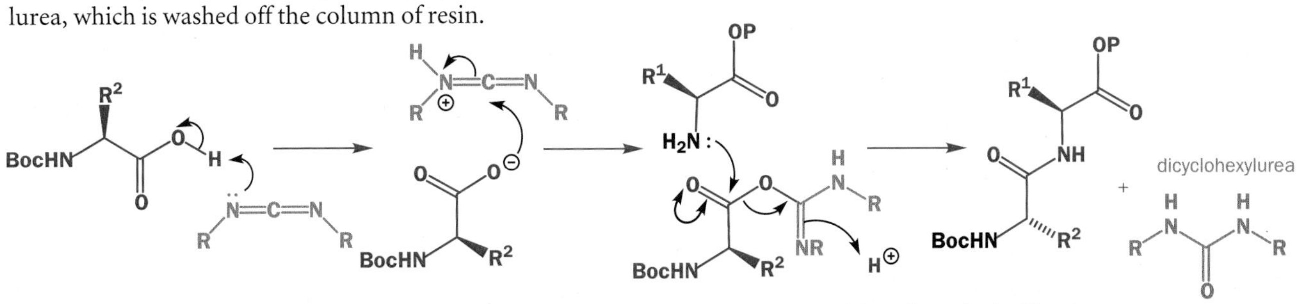

Now the Boc group must be removed with acid (such as CF_3CO_2H in CH_2Cl_2) and washed off the column leaving the free NH_2 group of amino acid number two ready for the next step.

> ■
> The mechanism of this reaction is discussed in Chapter 25.

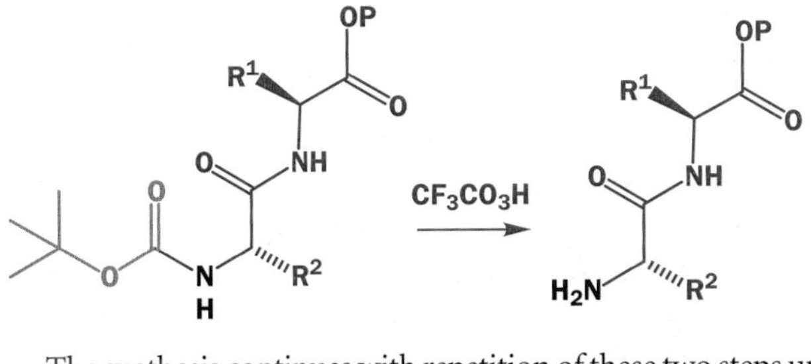

The synthesis continues with repetition of these two steps until the peptide chain is complete. The peptide is cleaved from the resin, usually with HF in pyridine or CF_3SO_2OH in CF_3CO_2H and given a final purification from small amounts of peptides of the wrong sequence by chromatography, usually HPLC.

This process is routinely automated in commercially available machines. Solutions of all of the protected amino acids required are stored in separate containers and a programmed sequence of coupling and deprotection leads rapidly to the complete peptide in days rather than the years needed for solution chemistry. The most dramatic illustration of this came with the publication of a heroic traditional synthesis of bovine pancreatic ribonuclease A (an enzyme with 124 amino acids) by Hirschmann, side-by-side with one by Merrifield using functionalized polystyrene as we have described. The traditional method required 22 co-workers, while the Merrifield method needed only one.

Peptide synthesis on polyacrylamide gel

Another method of polymer-supported peptide synthesis has been developed by Sheppard. Most things are different in this approach, which is better adapted for polar solvents and automated operation. The polymer is a polyacrylamide cross-linked with bis-acrylamides joined by –NCH₂CH₂N– groups.

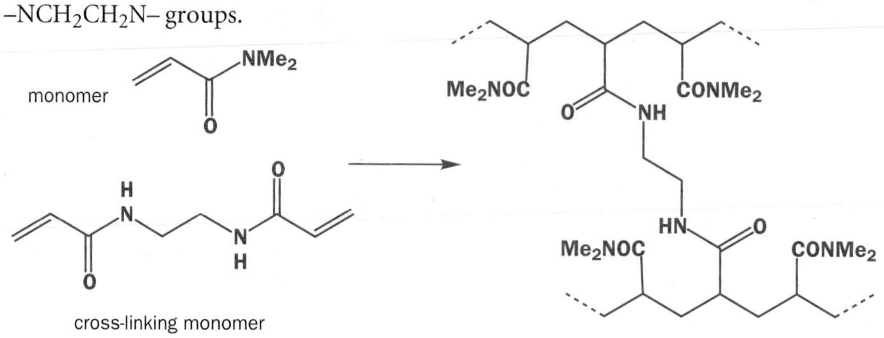

monomer

cross-linking monomer

Polar solvents such as water or DMF penetrate the beads, making them swell much more than do the polystyrene resins. This exposes more reactive groups and increases the loading of peptide chains on each bead. The first amino acid is attached through its carboxyl group to an amino group on the polymer, added during or after polymerization by incorporating more 1,2-diaminoethane. The favoured amino protecting group is now Fmoc (see Chapter 24), which has the advantage that it can be removed under basic conditions (piperidine) which do not affect acid-labile side-chain protecting groups.

Methods like these have made polymer-supported synthesis so valuable a method that it is now being developed for many reactions old and new. A recent (1999) issue of the journal *Perkin Transactions 1* reported two syntheses of natural products in which every step was carried out using a polymer-supported reagent. Polymers are vital to us in everyday life in a multitude of ways and new polymers are being invented all the time. We have done no more than scratch the surface of this subject and you should turn to more specialized books if you want to go further.

Problems

1. The monomer bisphenol A is made by the following reaction. Suggest a detailed mechanism.

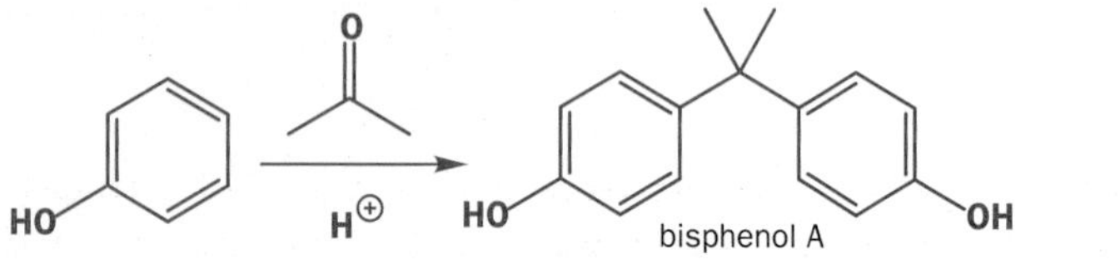

bisphenol A

2. An alternative synthesis of 18-crown-6 to the one given in the chapter is outlined below. How would you describe the product in polymer terms? What is the monomer? How would you make 15-crown-5?

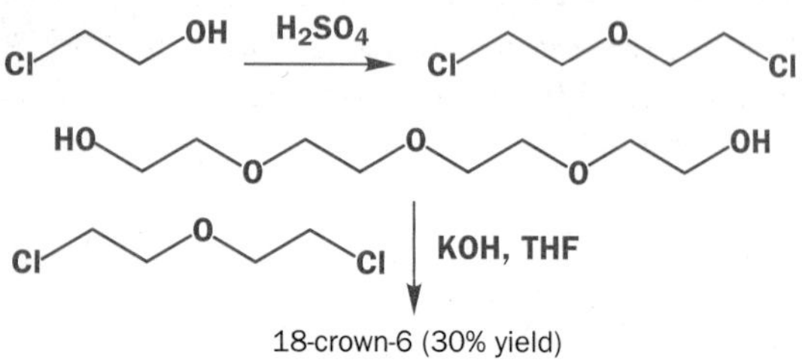

18-crown-6 (30% yield)

3. Melamine is formed by the trimerization of cyanamide and a hint was given in the chapter as to the mechanism of this process. Expand that hint into a full mechanism.

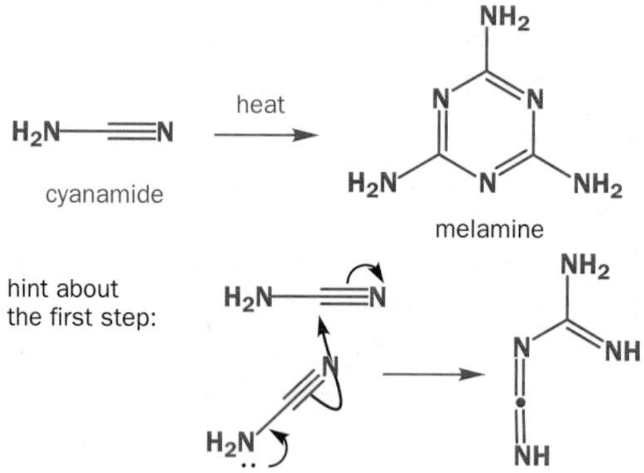

cyanamide

melamine

hint about the first step:

Melamine is polymerized with formaldehyde to make formica. Draw a mechanism for the first step in this process.

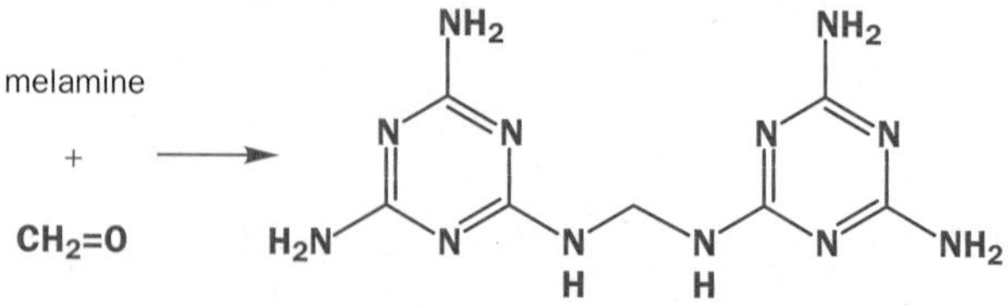

melamine

4. An acidic resin can be made by the polymerization of 4-vinylpyridine initiated by AIBN and heat followed by treatment of the polymer with bromoacetate. Explain what is happening and give a representative part structure of the acidic resin.

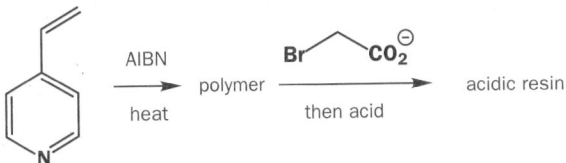

5. An artificial rubber may be made by cationic polymerization of isobutene using acid initiation with BF_3 and water. What is the mechanism of the polymerization, and what is the structure of the polymer?

This rubber is too weak to be used commercially and 5–10% isoprene is incorporated into the polymerizing mixture to give a different polymer that can be cross-linked by heating with sulfur (or other radical generators). Draw representative structures for sections of the new polymer and show how it can be cross-linked with sulfur.

6. When sodium metal is dissolved in a solution of naphthalene in THF, a green solution of a radical anion is produced. What is its structure?

This green solution initiates the polymerization of butadiene to give a 'living polymer'. What is the structure of this polymer and why is it called 'living'?

7. We introduced the idea of a spacer between a benzene ring (in a polystyrene resin) and a functional group in the chapter. If a polymer is being designed to do Wittig reactions, why would it be better to have a Ph_2P group joined directly to the benzene ring than to have a CH_2 spacer between them?

polymer — useful for Wittig reactions

polymer — useless for Wittig reactions

If you need a hint, draw out the reagents that you would add to the polymer to do a Wittig reaction and work out what you would get in each case.

8. A useful reagent for the oxidation of alcohols is 'PCC' (pyridinium chlorochromate). Design a polymeric (or at least polymer-bound) reagent that should show similar reactivity.

What would be the advantage of the polymer-bound reagent over normal PCC?

PCC pyridinium chlorochromate

9. A polymer that might bind specifically to metal ions and be able to extract them from solution would be based on a crown ether. How would you make a polymer such as this?

polymer chain

10. What is a 'block co-polymer'? What polymer would be produced by this sequence of reactions? What special physical properties would it have?

$$\triangle \xrightarrow[\text{controlled amounts}]{H_2O} \text{polymer A}$$

OCN — in excess — NCO → polymer B

H_2N — NH_2 → polymer C

11. Why does polymerization occur only at relatively low temperatures often below 200 °C? What occurs at higher temperatures? Formaldehyde polymerizes only below about 100 °C but ethylene still polymerizes up to about 500 °C. Why the difference?

12. Poly(vinyl chloride) (PVC) is used for rigid structures like window frames and gutters with only small amounts of additives such as pigments. If PVC is used for flexible things like plastic bags, about 20–30% of dialkyl phthalates such as the compound below are incorporated during polymerization. Why is this?

dialkyl phthalate

Organic chemistry today

Connections

Building on:
- The rest of the book ch1–ch52

Arriving at:
- How organic chemistry produced an AIDS treatment in collaboration with biologists

Looking forward to:
- Life as a chemist

Modern science is based on interaction between disciplines

Organic chemistry has transformed the materials of everyday life, as we have seen in Chapter 52, but this is merely a glimpse of the future of organic materials where light-emitting polymers, polymers that conduct electricity, self-reproducing organic compounds, molecules that work (nano-engineering), and even molecules that think may transform our world in ways not yet imagined. These developments are the result of cooperation between organic chemists and physicists, engineers, material scientists, computer experts, and many others.

The most dramatic developments at the beginning of the twenty-first century are new methods in medicine from collaborations between organic chemists and biologists. (The biochemical background is sketched out in Chapters 49–51.) The media's favourite 'a cure for cancer' is already not just 'a cure' but hundreds of successful cures for the hundreds of diseases collectively called 'cancer'. A newspaper headline in 1999 revealed that there was *some* chance of survival for all known types of childhood cancer. We are going to discuss just one equally dramatic medical development, the treatment of AIDS. Like the treatment of cancer, this is a story that is only just starting, but enough is known to make it a gripping story full of hope.

When AIDS (Acquired Immune Deficiency Syndrome) first came into the news in the 1980s it was a horror story of mysterious deaths from normally harmless diseases after the patient's immune system had been weakened and eventually destroyed. The cause was identified by biologists as a new virus: HIV (Human Immunodeficiency Virus) and antiviral drugs, notably AZT (Chapter 49), were used with some success. These drugs imitate natural nucleosides (AZT imitates deoxythymidine) and inhibit the virus from copying its RNA into DNA inside human cells by inhibiting the enzyme 'reverse transcriptase'.

These drugs also inhibit our own enzymes and are very toxic. Biologists then discovered an alternative point of attack. An enzyme unique to the virus cuts up long proteins into small pieces essential for the formation of new HIV particles. If this enzyme could be inhibited, no new viruses would be formed, and the inhibitor should not damage human chemistry. Several companies invented HIV protease inhibitors, which looked more like small pieces of proteins with the weak link of the amide bond replaced by a more stable C–C bond.

Real peptides are usually poor drugs because we have our own peptidases which quickly cut up ingested proteins into their constituent amino acids by hydrolysis of the amide link. Drugs that imitate peptides may avoid this ignominious fate by replacing the amide bond with another bond less susceptible to hydrolysis. This part structure of one HIV protease inhibitor makes the point.

deoxythymidine—
a nucleoside of DNA

AZT
azidothymidine
anti-AIDS drug

section of protein   intermediate in amide hydrolysis   section of inhibitor

On the left is a section of normal protein with glycine and phenylalanine residues (Chapter 49). In the middle is the intermediate formed when a molecule of water attacks the amide carbonyl group. On the right is a piece of the HIV protease inhibitor. The amide nitrogen atom has been replaced by a CH_2 group (ringed in black) so that no 'hydrolysis' of the C–C bond can occur. The inhibitor may bind but it cannot react.

Enzymes ideally bind their substrates strongly and the product of the reaction much more weakly. If they are to accelerate the reaction they need to lower the energy of the transition state (Chapters 13 and 41) and they can do this by binding the transition state of the reaction strongest of all. We cannot literally synthesize a transition state analogue because transition states are by definition unstable, but intermediate analogues can be synthesized. The inhibitor above has one OH group instead of the two in the genuine intermediate but this turns out to be the vital one. This knowledge was acquired from an X-ray crystal structure showing how the enzyme binds the substrate. The inhibitor binds well to the enzyme but cannot react so it blocks the active site.

These compounds are a good deal more sophisticated than this simple analysis suggests. For example, HIV protease is a dimeric enzyme and experience with this class of protease suggested correctly that more or less symmetrically placed heterocyclic rings (Chapters 42–44) would greatly improve binding. Here are two of the inhibitors with the active site binding portion framed in black and the heterocyclic binding portions framed in green.

Crixivan (Merck)

enzyme binding groups

Norvir (Abbott)

hydrolysis intermediate mimics

deoxycytidine
a nucleoside of DNA

3-TC
Lamivudine
anti-AIDS drug

These developments looked so promising that Merck even set up a completely new research station at West Point, Pennsylvania, dedicated to this work. The biochemist in charge, Dr Irving Segal, was one of the victims of the Lockerbie bombing in 1988 but his work lives on as Crixivan (indinavir) is now one of the cocktail of three drugs (AZT and 3TC, shown with the nucleoside it imitates, are the others) that has revolutionized the treatment of HIV. Before this treatment most HIV victims were dead within 2 years. Now no one knows how long they will survive as the combination of the three drugs reduces the amount of virus below detectable levels.

Crixivan was not the first compound that Merck discovered. Many others fell by the wayside because they were not active enough, were too toxic, didn't last long enough in the body, or for other reasons. Crixivan was developed from cooperation between biochemists, virologists, X-ray crystallographers, and molecular modellers as well as organic chemists. When the choice of Crixivan from the various drug candidates had been made and the chemists were trying to make enough of it for trials and use, theirs was an exceptionally urgent task. They knew that a kilo of compound was needed to keep each patient alive and well for a year. Merck built a dedicated plant for the manufacture of Crixivan at Elkton, Virginia, in 1995. Within 1 year, production was running at full blast and there are thousands of people alive today as a result.

The AIDS crisis led to cooperation between the pharmaceutical companies unparalleled since the development of penicillin during the Second World War. Fifteen companies set up an AIDS drug development collaboration programme and government agencies and universities have all joined in.

The battle is not yet won, of course, but the HIV protease inhibitors are being followed by ~~~~
eration of nonnucleoside reverse transcriptase inhibitors, which promise to be less toxic to hu~~~
An example is the DuPont–Merck compound DMP-266, made as a single enantiomer and now
under clinical trials. This compound, though it contains a most unusual cyclopropane and alkyne
combination, is nevertheless a much simpler compound than Crixivan. We shall devote most of this
final chapter to the synthesis of the established and chemically more interesting drug Crixivan.

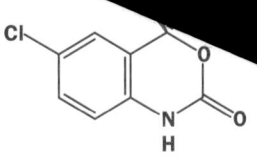

DMP-266

The synthesis of Crixivan

Crixivan is a formidable synthetic target. It is probably the most complex compound ever made in
quantity by organic synthesis and very large amounts must be made because one kilo is needed per
patient per year. The complexity largely aris-
es from the stereochemistry. There are five
stereogenic centres, marked with coloured
circles on this diagram, and their disposition
means that three separate pieces of asymmetric synthesis must be devised. There are,
of course, also many functional groups and
four different rings.

Crixivan (Merck)
stereogenic centres marked with circles

The two black centres are 1,2-related and we have already discussed them in part at the end of
Chapter 41. The green centres are 1,3-related and we saw in Chapter 45 that this type of control is
possible though difficult. The orange centre is 1,4-related to the nearer green centre and must be
considered separately.

two amine alkylations
(disconnection next to
heteroatom)

amide formation
(disconnection next to
heteroatom)

enolate alkylation
(creates ringed
stereogenic centre)

—use

the
piperazine
fragment

—use

the
central
epoxide

—use

cis
amino
indanol

The challenge with Crixivan, as with any drug, is to make it efficiently—high yields; few steps. It
has five stereogenic centres, so the chemists developing the synthesis needed to address the issue of
diastereoselectivity. And it is a single enantiomer, so an asymmetric synthesis was required. We can
start by looking at some likely disconnections, summarized in the scheme above. They are all disconnections of the sorts you met in Chapter 30, and they all correspond to reliable reactions.

These disconnections split the molecule into five manageable chunks (synthons), three of which
contain stereogenic centres and will have to be made as single enantiomers. The final stereogenic

...am) would have to be made in the enolate alkylation step,
...reoselectively.

...thons in turn. First, the simplest one: the central epoxide. The
...a leaving group, such as a tosylate, and it can easily be made from the
...very good way of making this compound as a single enantiomer—a
...xidation of allyl alcohol.

Next, the piperazine fragment. This has two nucleophilic nitrogen atoms and they will both need protecting with different protecting groups to allow them to be revealed one at a time. It will also need to be made as a single enantiomer. In an early route to Crixivan, this was done by resolution, but enantioselective hydrogenation provides a better alternative. Starting from a pyrazine derivative, a normal hydrogenation over palladium on charcoal could be stopped at the tetrahydropyrazine stage. The two nitrogens in this compound are quite different because one is conjugated with the amide while one is not (the curly arrows in the margin show this). The more nucleophilic nitrogen— the one *not* conjugated with the amide—was protected with benzyl chloroformate to give the Cbz derivative. Now the less reactive nitrogen can be protected with a Boc group, using DMAP as a nucleophilic catalyst.

You met asymmetric hydrogenation using BINAP–metal complexes in Chapter 45 as a method for the synthesis of amino acids. The substrate and catalyst are slightly different here, but the principle is the same: the chiral ligand, BINAP, directs addition of hydrogen across the double bond with almost perfect enantioselectivity and in very high yield. In Chapter 45 we described this as addition to one enantiotopic face of the alkene. A further hydrogenation step allowed selective removal of the Cbz group, preparing one of the two nitrogen atoms for alkylation.

COD = cyclooctadiene.

96% yield; 99% ee

H_2O_2 and MeCN react to give a 'peroxyimidic acid'—the C=N analogue of a peroxy-acid—as the true epoxidizing agent.

peroxyimidic acid

The remaining chiral fragment is a compound whose synthesis was discussed in Chapter 41, and you should turn to p. 1116 for more details of the mechanisms in the reaction sequence. It can be made on a reasonably large scale (600 kg) in one reaction vessel, starting from indene. First, the double bond is epoxidized, not with a peroxy-acid but with the cheaper hydrogen peroxide in an acetonitrile–methanol mixture. Acid-catalysed opening of the epoxide leads to a cation, which takes part in a reversible Ritter reaction with the acetonitrile solvent, leading to a single diastereoisomer of a heterocyclic intermediate which is hydrolysed to the amino-alcohol.

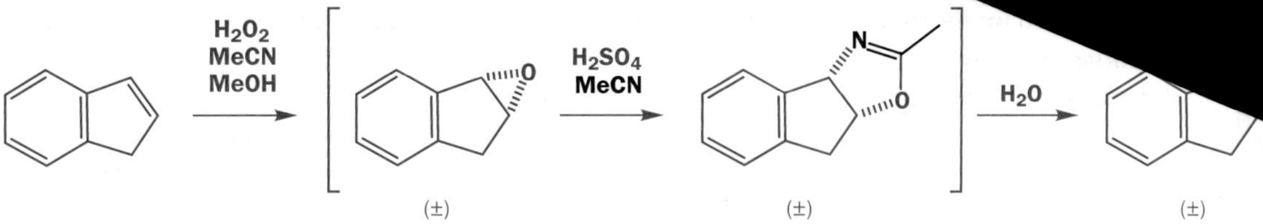

The product is, of course, racemic but, as it is an amine, resolution with an acid should be straightforward. Crystallization of its tartrate salt, for example, leads to the required single enantiomer in 99.9% ee. With such cheap starting materials, resolution is just about acceptable, even though it wastes half the material. It would be better to oxidize the indene enantioselectively, and retain the enantiomeric purity through the sequence: it is indeed possible to carry out a very selective Sharpless asymmetric dihydroxylation (Chapter 45) of indene, and the diol serves as an equally good starting material for the Ritter reaction. The stereogenic centre carrying the green hydroxyl group remains firmly in place throughout the route, and controls the absolute configuration of the final product.

> ■ The Ritter reaction was described in Chapter 17, p. 435. The reason for the formation of the *cis* diastereoisomer in this example is discussed in Chapter 41, p. 1116.

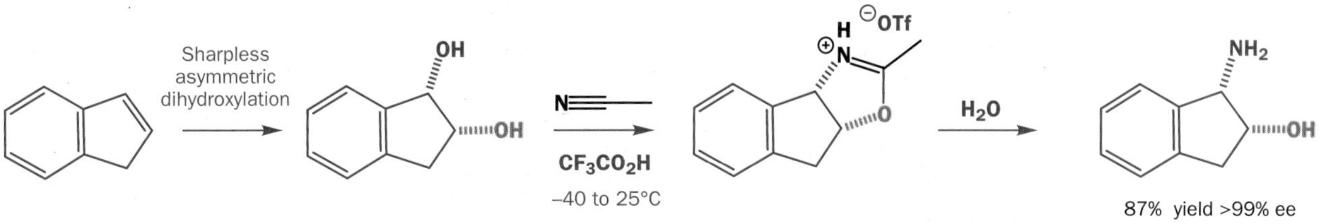

87% yield >99% ee

Both resolution and Sharpless asymmetric dihydroxylation were successful in the synthesis of Crixivan but the best method is one we shall keep till later. Only one stereogenic centre remains, and its stereoselective formation turns out to be the most remarkable reaction of the whole synthesis. The centre is the one created in the planned enolate alkylation step.

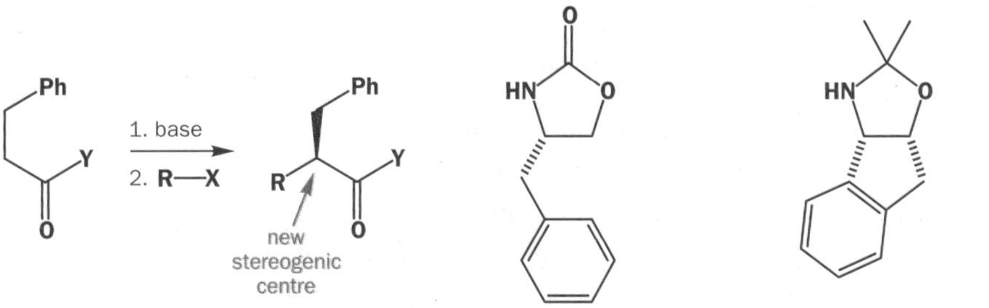

Evans' phenylalanine-derived oxazolidinone auxiliary

protected amino-alcohol

The obvious way to make this centre is to make Y a chiral auxiliary; the required acyl chloride could be used to acylate the auxiliary, which would direct a diastereoselective alkylation, before being removed and replaced with the amino-alcohol portion. But the amino-alcohol itself, certainly once protected, has a remarkable similarity to Evans' oxazolidinone auxiliaries (Chapter 45), and it turns out that this amino-alcohol will function very successfully as a chiral auxiliary, which does not need to be removed, avoiding waste and saving steps! The amino-alcohol was acylated with the acyl chloride, and the amide was protected as the nitrogen analogue of an acetonide by treating with 2-methoxypropene (the methyl enol ether of acetone) and an acid catalyst. The enolate of this amide reacts highly diastereoselectively with alkylating agents, including, for example, allyl bromide.

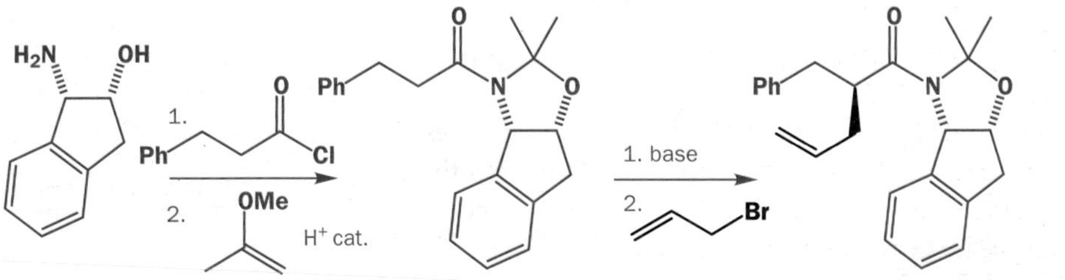

96:4 ratio of diastereoisomers

altogether clear, but we would expect the bulky nitrogen
... the *cis* enolate. With the amino-alcohol portion arranged as
... attack by electrophiles.

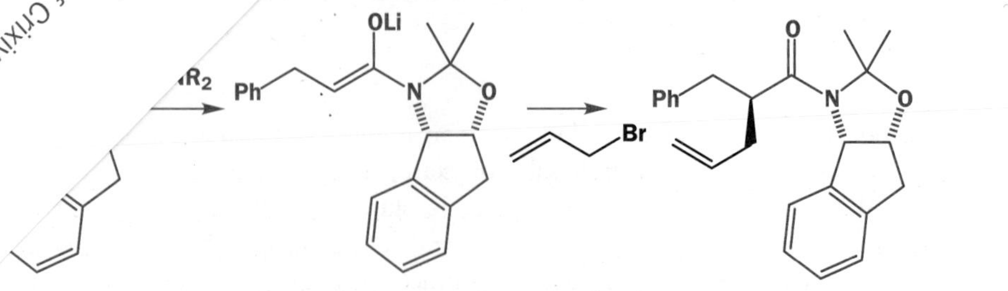

enolate also reacted diastereoselectively with the epoxy-tosylate prepared earlier. The epox-
... being more electrophilic than the tosylate, is opened first, giving an alkoxide, which closes again
... to give a new epoxide.

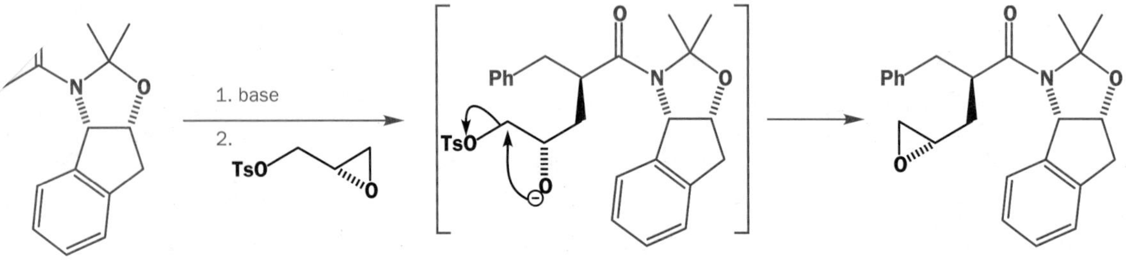

The absolute configuration at the stereogenic centre in the epoxide was, of course, already
fixed (by the earlier *enantioselective* Sharpless epoxidation). However, it also turned out to be possible to make this compound by a different route involving a *diastereoselective* reaction of the alkylation product from allyl bromide, again directed by the amino-alcohol-derived auxiliary. The
reagents make the reaction look like an iodolactonization—and, indeed, there are many similarities
with the diastereoselective iodolactonizations of Chapter 33. NIS (*N*-iodosuccinimide, the iodine
analogue of NBS) provides an 'I$^+$' source, reacting reversible and non-stereoselectively with the
alkene. Of the two diastereoisomeric iodonium ions, one may cyclize rapidly by intramolecular
attack of the amide carbonyl group. Cyclization of the other diastereoisomer is prevented by steric
hindrance between the parts of the molecule coloured green. Opening of the five-membered ring
gives a single diastereoisomer of the iodoalcohol, which was closed to the epoxide by treatment with
base.

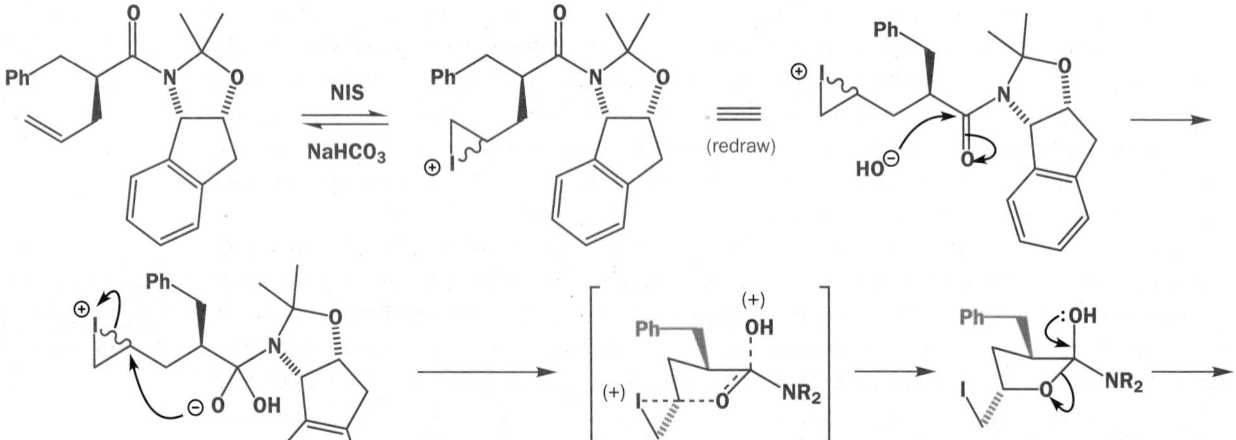

green groups on opposite sides of ring

continued opposite

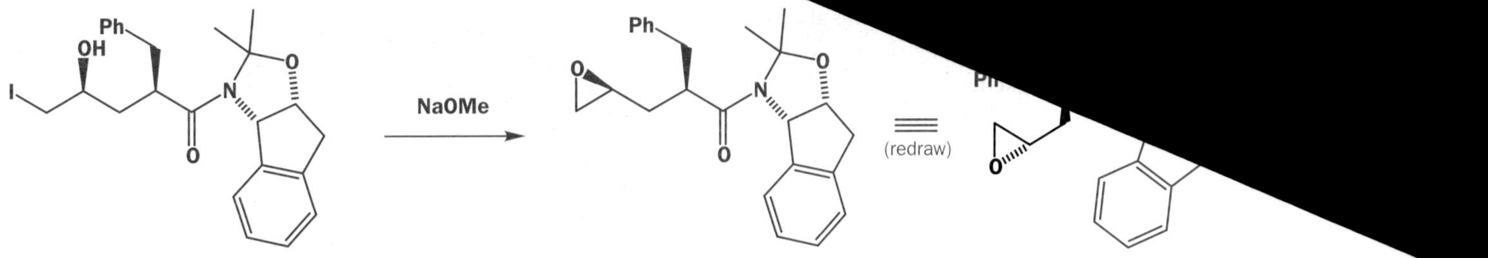

Three of the five fragments have now been assembled, and only the two amine alkylations remain. The first alkylation makes use of the epoxide to introduce the required 1,2-amino-alcohol functionality. The protected enantiomerically pure piperazine reacted with the epoxide, and the product was treated with acid to deprotect both the second piperazine nitrogen and the 'acetonide' group left over from the earlier chiral auxiliary step. The newly liberated secondary amine was alkylated with the reactive electrophile 3-chloromethyl pyridine, and the final product was crystallized as its sulfate salt.

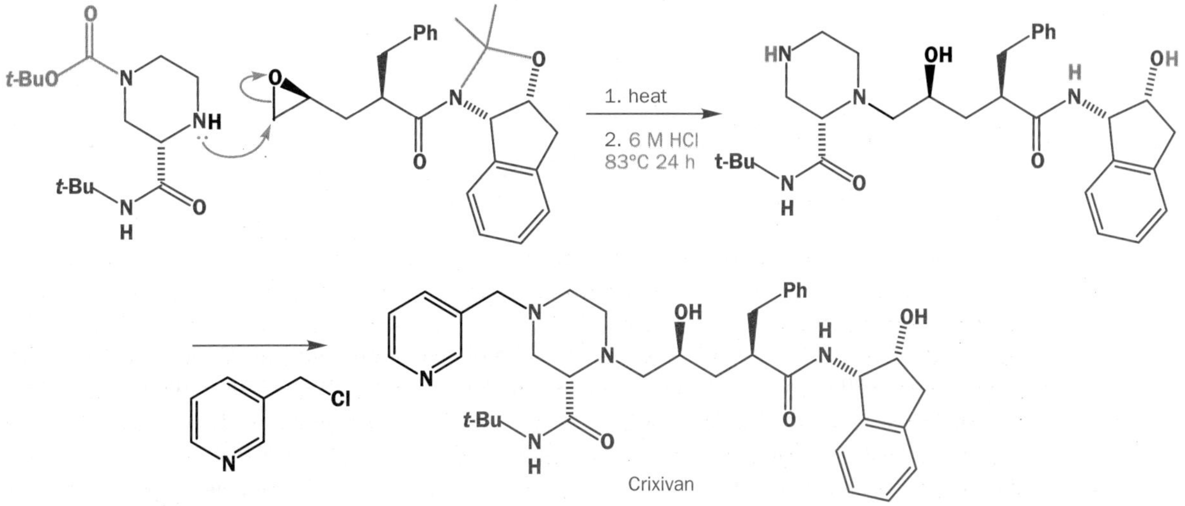

Crixivan

The future of organic chemistry

Not all organic chemists can be involved in such exciting projects as the launching of a new anti-AIDS drug. But the chemistry used in this project was invented by chemists in other institutions who had no idea that it would eventually be used to make Crixivan. The Sharpless asymmetric epoxidation, the catalytic asymmetric reduction, the stereoselective enolate alkylation, and the various methods tried out for the enantiomerically pure amino indanol (resolution, enzymatic kinetic resolution) were developed by organic chemists in research laboratories. Some of these famous chemists like Sharpless invented new methods, some made new compounds, some studied new types of molecules, but all built on the work of other chemists.

In 1980 Giovanni Casiraghi, a rather less famous chemist from the University of Parma, published a paper in the *Journal of the Chemical Society* about selective reactions between phenols and formaldehyde. He and his colleagues made the modest discovery that controlled reactions to give salicylaldehydes could be achieved in toluene with $SnCl_4$ as catalyst. The reaction is regioselective for the *ortho* isomer and the paper described the rather precise conditions needed to get a good yield.

■ In Chapter 52 you met Bakelite, the first synthetic polymer, which results from unselective reactions between these two compounds.

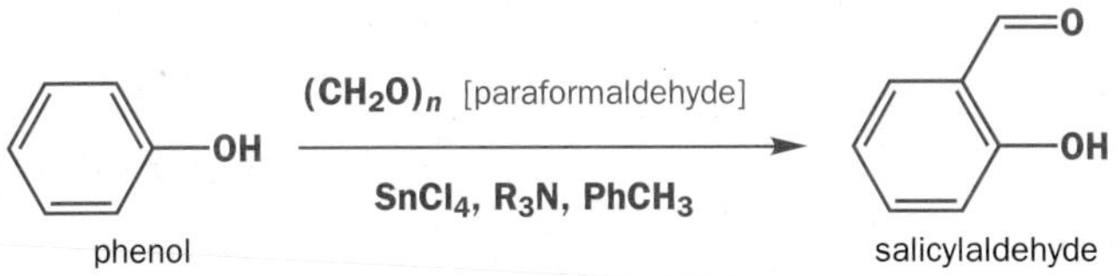

...uted salicylaldehydes. When Jacobsen came to develop ... the Sharpless asymmetric epoxidation, works for simple ..., he chose 'salens' as his catalysts, partly because they could be ... For example:

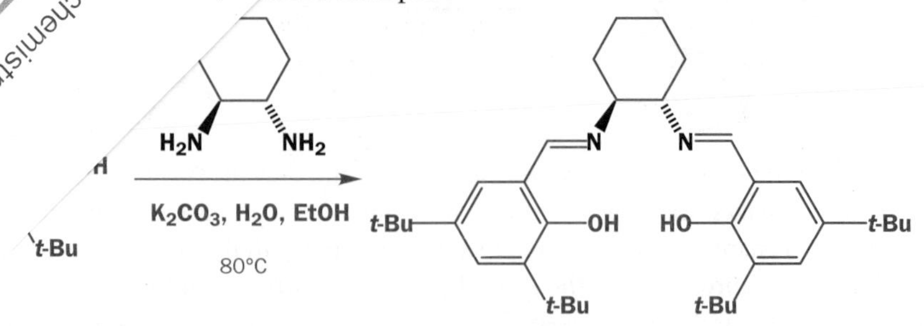

This 'salen' is the ligand for manganese in the asymmetric epoxidation. The stable brown Mn(III) complex can be made from it with $Mn(OAc)_3$ in excellent yield and this can be oxidized to the active complex used above with domestic bleach (NaOCl).

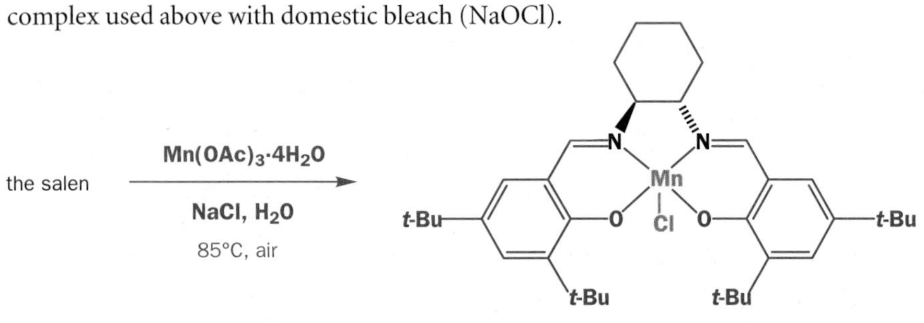

stable Mn(III) complex

Jacobsen epoxidation turned out to be the best large-scale method for preparing the *cis*-amino-indanol for the synthesis of Crixivan. This process is very much the cornerstone of the whole synthesis. During the development of the first laboratory route into a route usable on a very large scale, many methods were tried and the final choice fell on this relatively new type of asymmetric epoxidation. The Sharpless asymmetric epoxidation works only for allylic alcohols (Chapter 45) and so is no good here. The Sharpless asymmetric dihydroxylation works less well on *cis*-alkenes than on *trans*-alkenes. The Jacobsen epoxidation works best on *cis*-alkenes. The catalyst is the Mn(III) complex easily made from a chiral diamine and an aromatic salicylaldehyde (a 2-hydroxybenzaldehyde).

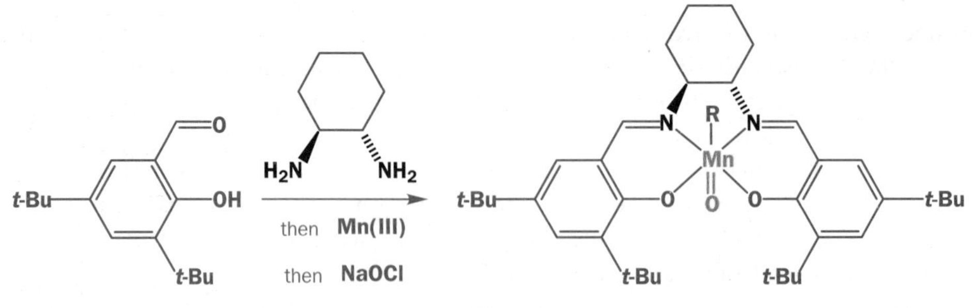

The chirality comes from the diamine and the oxidation from ordinary domestic bleach (NaOCl), which continually recreates the Mn=O bond as it is used in the epoxidation. Only 0.7% catalyst is needed to keep the cycle going efficiently. The epoxide is as good as the diol in the Ritter reaction and the whole process gives a 50% yield of enantiomerically pure *cis*-amino-indanol on a very large scale.

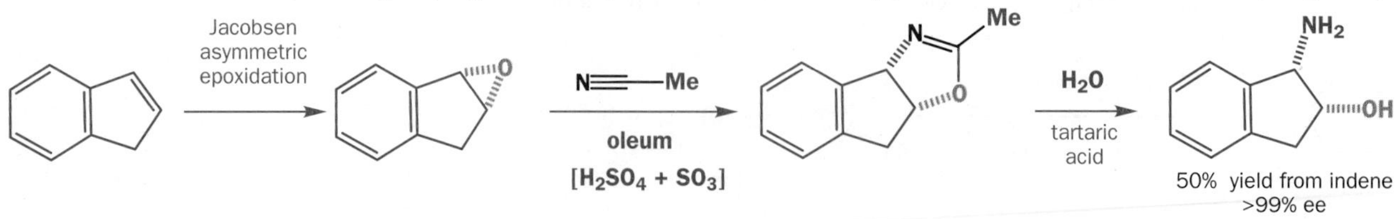

In the same year (1990) that Jacobsen reported his asymmetric epoxidation Tsutomu Katsuki at the University of Kyushu in Japan reported a closely related asymmetric epoxidation. The chiral catalyst is also a salen and the metal manganese. The oxidant is iodosobenzene (PhI=O) but this method works best for E-alkenes. It is no coincidence that Katsuki and Jacobsen both worked for Sharpless. It is not unusual for similar discoveries to be made independently in different parts of the world.

the Katsuki manganese salen complex

It did not enter Casiraghi's wildest dreams that his work might some day be useful in a matter of life and death. Nor did his four co-workers nor Jacobsen's more numerous co-workers see clearly the future applications of their work. By its very nature it is impossible to predict the outcome or the applications of research. But be quite sure of one thing. Good research and exciting discoveries come from a thorough understanding of the fundamentals of organic chemistry and require chemists to work as a team. The Italian work is a model of careful experimentation and a thorough study of reaction conditions together with sensible explanations of their discoveries using the same curly arrows we have been using. The Harvard team probably had a clearer idea that they were into something significant and worked with equal care and precision. Jacobsen's name is famous but both teams at Parma and Harvard Universities were needed to make the work available to Merck.

Hexamethylenetetramine

Hexamethylenetetramine is a co-polymer (oligomer really such as those we met in Chapter 52) of formaldehyde and ammonia containing six formaldehyde and four ammonia molecules. It has a beautifully symmetrical cage structure belonging to the adamantane series.

Hexamethylenetetramine is a crystalline compound used as a convenient source of formaldehyde for, among other things, polymerization reactions. It has a tetrahedral symmetry, as does adamantane, which might be regarded as the basic structural unit (not the same as the monomer!) of diamond. Diamond is of course a polymer of carbon atoms.

When Jacobsen's epoxidation was fully described in 1998–99, the Casiraghi method was abandoned in favour of an even older method discovered in the 1930s by Duff. The remarkable Duff reaction uses hexamethylenetetramine, the oligomer of formaldehyde and ammonia, to provide the extra carbon atom. The otherwise unknown Duff worked at Birmingham Technical College. Later in 1972, a William E. Smith, working in the GEC chemical laboratories at Schenectady, New York, found how to make the Duff reaction more general and better yielding by using CF_3CO_2H as catalyst. Even so, this method gives a lower yield than the Casiraghi method but it uses no dangerous reagents (particularly no stoichiometric tin) and is more suitable for large-scale work. When Duff was inventing his reaction or Smith was modifying the conditions, asymmetric synthesis was not even a gleam in anyone's eyes. It is impossible even for the inventor to predict whether a discovery is important or not.

the Duff reaction

works best for *trans* disubstituted alkenes, while the
ostituted alkenes. Even in this small area, there is a need
ganic chemistry has a long way to go.

ganic chemistry beyond the scope of this book, you will want to
cialized areas. Your university library should have a selection of
s and chemical reactions; NMR spectroscopy; enzyme mechanisms;
osynthesis; asymmetric synthesis; combinatorial chemistry; and molec-
k should equip you with enough fundamental organic chemistry to explore
erstanding and enjoyment and, perhaps, to discover what you want to do for
e. All of the chemists mentioned in this chapter and throughout the book began
students of chemistry at universities somewhere in the world. You have the good for-
dy chemistry at a time when more is understood about the subject than ever before, when
tion is easier to retrieve than ever before, and when organic chemistry is more interrelated
other disciplines than ever before. Duff, Smith, and Casiraghi felt themselves part of an international community of organic chemists in industry and universities but never has that community
been so well founded as it is nowadays. Travel to laboratories in other countries is commonplace for
students of organic chemistry now and even at home you can travel on the internet to other countries and see what is going on in chemistry there. You might try the web pages of our institutions for
a start: Cambridge is http://www.ch.cam.ac.uk/; Liverpool is http://www.liv.ac.uk/Chemistry/; and
Manchester is http://www.ch.man.ac.uk/. There is a general index to chemistry all over the world on
http://www.ch.cam.ac.uk/ChemSitesIndex.html.

Illustrations acknowledgements

Oxford University Press is grateful to the following for permission to reproduce the pictures indicated.

Page 3	Skunk, Tom Ulrich/Oxford Scientific Films
Page 22	Mona Lisa, Bridgeman Art Library. Cartoon by Jeremy Dennis
Page 28	Geodesic dome, Martin Jones/Arcaid
Page 57	Brain scanner, Science Photo Library
Page 383	Study of feet (pastel on paper) by Evelyn de Morgan (1855–1919). The De Morgan Foundation, London, UK/Bridgeman Art Library
Page 384	Nefertari, Egyptian Queen. Making an Offering, wall painting, 14th century BC, Valley of the Queens, Thebes, Egypt/Giraudon/Bridgeman Art Library
Page 447	Conformation and configuration cartoons by David Beeby
Page 1416	Atropa Belladonna, Garden Matters

Index

Abbreviations

Ac	Acetyl		HPLC	High performance liquid chromatography
acac	Acetylacetonate		HIV	Human immunodeficiency virus
AD	Asymmetric dihydroxylation		IR	Infrared
ADP	Adenosine 5'-diphosphate		KHMDS	Potassium Hexamethyldisilazide
AE	Asymmetric epoxidation		LCAO	Linear combination of atomic orbitals
AIBN	2,2'-Azobisisobutyronitrile		LDA	Lithium diisopropylamide
AO	Atomic orbital		LHMDS	Lithium hexamethyldisilazide
Ar	Aryl		LICA	Lithium isopropylcyclohexylamide
ATP	Adenosine 5'triphosphate		LTMP, LiTMP	Lithium 2,2,6,6-tetramethylpiperidide
9-BBN	9-Borabicyclo[3.3.1]nonane		LUMO	Lowest unoccupied molecular orbital
BHT	Butylated hydroxy toluene (2,6-di- t-butyl-4-methylphenol)		m-CPBA	*meta*-Chloroperoxybenzoic acid
			Me	Methyl
BINAP	2,2'-Bis(diphenylphosphino)-1,1'-binaphthyl		MO	Molecular orbital
Bn	Benzyl		MOM	Methoxymethyl
Boc, BOC	tert-Butyloxycarbonyl		Ms	Methanesulfonyl (mesyl)
Bu	Butyl		NAD	Nicotinamide adenine dinucleotide
s-Bu	*sec*-Butyl		NADH	Reduced NAD
t-Bu	*tert*-Butyl		NBS	*N*-Bromosuccinimide
Bz	Benzoyl		NIS	*N*-Iodosuccinimide
Cbz	Carboxybenzyl		NMO	*N*-Methylmorpholine-*N*-oxide
CDI	1,1'-Carbonyldiimidazole		NMR	Nuclear magnetic resonance
CI	Chemical ionization		NOE	Nuclear Overhauser effect
CoA	Coenzyme A		PCC	Pyridinium chlorochromate
COT	Cyclooctatetraene		PDC	Pyridinium dichromate
Cp	Cyclopentadienyl		Ph	Phenyl
DABCO	1,4-Diazabicyclo[2.2.2]octane		PPA	Polyphosphoric acid
DBE	Double bond equivalent		Pr	Propyl
DBN	1,5-Diazabicyclo[4.3.0]non-5-ene		*i*-Pr	*iso*-Propyl
DBU	1,8-Diazabicyclo[5.4.0]undec-7-ene		PTC	Phase transfer catalysis
DCC	*N,N*-dicyclohexylcarbodiimide		PTSA	p-Toluenesulfonic acid
DDQ	2,3-Dichloro-5,6-dicyano-1,4-benzoquinone		py	pyridine
			Red Al	Sodium bis(2-methoxyethoxy)aluminum hydride
DEAD	Diethyl azodicarboxylate		RNA	Ribonucleic acid
DIBAL	Diisobutylaluminum hydride		SAC	Specific acid catalysis
DMAP	4-Dimethylaminopyridine		SAM	*S*-Adenosyl methionine
DME	1,2-Dimethoxyethane		SBC	Specific base catalysis
DMF	*N,N*-Dimethylformamide		S_N1	Unimolecular nucleophilic substitution
DMPU	1,3-Dimethyl-3,4,5,6-tetrahydro-2(1*H*)-pyrimidinone		S_N2	Bimolecular nucleophilic substitution
			SOMO	Singly occupied molecular orbital
DMS	Dimethyl sulfide		STM	Scanning tunnelling microscopy
DMSO	Dimethyl sulfoxide		TBDMS	Tert-butyldimethylsilyl
DNA	Deoxyribonucleic acid		TBDPS	Tert-butyldiphenylsilyl
E1	Unimolecular elimination		Tf	Trifluoromethanesulfonyl (triflyl)
E2	Bimolecular elimination		THF	Tetrahydrofuran
E_a	Activation energy		THP	Tetrahydropyran
EDTA	Ethylenediaminetetraacetic acid		TIPS	Triisopropylsilyl
EPR	Electron paramagnetic resonance		TMEDA	*N,N,N',N'*-tetramethyl-1,2-ethylenediamine
ESR	Electron spin resonance		TMP	2,2,6,6-Tetramethylpiperidine
Et	Ethyl		TMS	Trimethylsilyl, tetramethylsilane
FGI	Functional group interconversion		TMSOTf	Trimethylsilyl triflate
Fmoc	Fluorenylmethyloxycarbonyl		TPAP	Tetra-*N*-propylammonium perruthenate
GAC	General acid catalysis		Tr	Triphenylmethyl (trityl)
GBC	General base catalysis		TS	Transition state
HMPA	Hexamethylphosphoramide		Ts	p-Toluenesulfonyl. Tosyl
HMPT	Hexamethylphosphorous triamide		UV	Ultraviolet
HOBt	1-Hydroxybenzotriazole		VSEPRT	Valence shell electron pair repulsion
HOMO	Highest occupied molecular orbital			

Periodic table

	1 I	**2** II	**3** III	**4** IV	**5** V	**6** VI	**7** VII	**8**	**9** VIII

s

	3 **Li** RAM: 6.941 P: 0.98 Lithium	4 **Be** RAM: 9.012182 P: 1.57 Beryllium
2		

	11 **Na** RAM: 22.98977 P: 0.93 Sodium	12 **Mg** RAM: 24.305 P: 1.31 Magnesium
3		

d

4	19 **K** RAM: 39.0983 P: 0.82 Potassium	20 **Ca** RAM: 40.078 P: 1 Calcium	21 **Sc** RAM: 44.95591 P: 1.36 Scandium	22 **Ti** RAM: 47.88 P: 1.54 Titanium	23 **V** RAM: 50.9415 P: 1.63 Vanadium	24 **Cr** RAM: 51.9961 P: 1.66 Chromium	25 **Mn** RAM: 54.93805 P: 1.55 Manganese	26 **Fe** RAM: 55.847 P: 1.83 Iron	27 **Co** RAM: 58.9332 P: 1.88 Cobalt
5	37 **Rb** RAM: 85.4678 P: 0.82 Rubidium	38 **Sr** RAM: 87.62 P: 0.95 Strontium	39 **Y** RAM: 88.90585 P: 1.22 Yttrium	40 **Zr** RAM: 91.224 P: 1.33 Zirconium	41 **Nb** RAM: 92.90638 P: 1.6 Niobium	42 **Mo** RAM: 95.94 P: 2.16 Molybdenum	43 **Tc** RAM: 98 P: 1.9 Technetium	44 **Ru** RAM: 101.07 P: 2.2 Ruthenium	45 **Rh** RAM: 102.905 P: 2.28 Rhodium
6	55 **Cs** RAM: 132.9054 P: 0.79 Cesium	56 **Ba** RAM: 137.327 P: 0.89 Barium	71 **Lu** RAM: 174.967 P: 1.27 Lutetium	72 **Hf** RAM: 178.49 P: 1.3 Hafnium	73 **Ta** RAM: 180.9479 P: 1.5 Tantalum	74 **W** RAM: 183.85 P: 2.36 Tungsten	75 **Re** RAM: 186.207 P: 1.9 Rhenium	76 **Os** RAM: 190.2 P: 2.2 Osmium	77 **Ir** RAM: 192.22 P: 2.2 Iridium
7	87 **Fr** RAM: 223 P: 0.7 Francium	88 **Ra** RAM: 226.0254 P: 0.9 Radium	103 **Lr** RAM: 260 P: Lawrencium	104 **Rf** RAM: 261 P: Rutherfordium	105 **Db** RAM: 262 P: Dubnium	106 **Sg** RAM: 263 P: Seaborgium	107 **Bh** RAM: 262 P: Bohrium	108 **Hs** RAM: 265 P: Hassium	109 **Mt** RAM: 266 P: Meitnerium

f

	57 **La** RAM: 138.9055 P: 1.1 Lanthanum	58 **Ce** RAM: 140.115 P: 1.12 Cerium	59 **Pr** RAM: 140.9077 P: 1.13 Praseodymium	60 **Nd** RAM: 144.24 P: 1.14 Neodymium	61 **Pm** RAM: 145 P: 1.13 Promethium	62 **Sm** RAM: 150.36 P: 1.17 Samarium	63 **Eu** RAM: 151.965 P: 1.2 Europium
	89 **Ac** RAM: 227 P: 1.1 Actinium	90 **Th** RAM: 232.0381 P: 1.3 Thorium	91 **Pa** RAM: 213.0359 P: 1.5 Protactinium	92 **U** RAM: 238.0289 P: 1.38 Uranium	93 **Np** RAM: 237.0482 P: 1.36 Neptunium	94 **Pu** RAM: 244 P: 1.28 Plutonium	95 **Am** RAM: 243 P: 1.3 Americium

Key

	Symbol
Atomic number	00 **Xx**
Relative Atomic Mass	RAM: 0.000
Electronegativity (Pauling)	P: 0.0
Element	Name

III